Michael [illegible] 040380

SELECTED CRC HANDBOOK SERIES

CRC HANDBOOK OF BIOCHEMISTRY AND MOLECULAR BIOLOGY
Gerald D. Fasman
Brandeis University

CRC HANDBOOK SERIES IN CLINICAL LABORATORY SCIENCE
David Seligson
Yale University

CRC HANDBOOK OF ELECTROPHORESIS
Lena A. Lewis
Cleveland Clinic
J. J. Opplt
Metropolitan General Hospital

CRC HANDBOOK SERIES IN ENGINEERING IN MEDICINE AND BIOLOGY
David G. Fleming
Case Western Reserve University
Barry N. Feinberg
Purdue University

CRC HANDBOOK OF ENVIRONMENTAL CONTROL
Richard G. Bond
Univeristy of Minnesota
Conrad P. Straub
University of Minnesota

CRC FENAROLI'S HANDBOOK OF FLAVOR INGREDIENTS
Nicolo Bellanca
CIBA-GEIGY Corp.
Giovanni Fenaroli
University of Milano, Italy
Thomas E. Furia
Dynapol

CRC HANDBOOK OF MARINE SCIENCE
F. G. Walton Smith
International Oceanographic Foundation
Frederick A. Kalber
Hydrobiological Services
Joseph T. Baker
Vreni Murphy
Roche Institute of Marine Pharmacology, Australia

CRC HANDBOOK OF MATERIALS SCIENCE
C. T. Lynch
Wright-Patterson Air Force Base

CRC STANDARD MATH TABLES
William H. Beyer
University of Akron

CRC HANDBOOK OF MICROBIOLOGY
Allen I. Laskin
Esso Research and Engineering Co.
Hubert Lechevalier
Rutgers University

CRC HANDBOOK SERIES IN NUTRITION AND FOOD
Miloslav Rechcigl, Jr.
Agency for International Development

CRC ATLAS OF SCINTIMAGING FOR CLINICAL NUCLEAR MEDICINE
Henry N. Wellman
Indiana University School of Medicine

CRC ATLAS OF SPECTRAL DATA AND PHYSICAL CONSTANTS FOR ORGANIC COMPOUNDS
Jeanette Grasselli
Standard Oil Company (Ohio)
William M. Ritchey
Case Western Reserve University

CRC HANDBOOK SERIES IN ZOONOSES
James H. Steele
University of Texas

CRC Handbook
OF
Chemistry and Physics

A Ready-Reference Book of Chemical and Physical Data

HANDBOOK OF CHEMISTRY AND PHYSICS
1978-1979
59th
EDITION
CRC PRESS, INC.

EDITOR

ROBERT C. WEAST, Ph.D.

Vice President, Research, Consolidated Natural Gas Service Company, Inc.
Formerly Professor of Chemistry at Case Institute of Technology

ASSOCIATE EDITOR

MELVIN J. ASTLE, Ph.D.

Formerly Professor of Organic Chemistry at Case Institute of Technology
and
Manager of Research at Glidden-Durkee Division of SCM Corporation

In collaboration with a large number of professional chemists and physicists whose assistance is acknowledged in the list of general collaborators and in connection with the particular tables or sections involved.

CRC

CRC PRESS, Inc.
Boca Raton, Florida 33431

Second Printing, 1979

Library of Congress Card No. 13-11056

PRINTED IN U.S.A.

ISBN-0-8493-0459-8

Preface to the 59th Edition

There continues to be an increasing abundance of technical data and information of variable reliability in a variety of publications throughout the world. Some people inquire as to the value, need, and advisability of this growth of such data and information. One international agency, CODATA, an international council of scientific unions, speaks in favor of the need for these data and for the proper handling of them. This agency recently stated,* "The world depends on scientific and technological progress to provide a viable economy, a sound physical environment, a high personal standard of living, adequate health care, collective security, and effective transportation and communication networks. These efforts require a continual growth of our understanding of the physical world and the laws of nature. Such understanding is the goal of scientific research. Its application in the solution of practical problems, in industrial development, and in the achievement of other international objectives constitutes technology.

"One key to the effectiveness of science and technology is the ability to apply yesterday's discoveries to today's problems. The reservoir of knowledge already gained — the scientific and technical literature — is a major global asset. Since this reservoir is now growing at a rate of more than two million pages per year, finding the precise information relevant to a specific question and assuring its reliability become major problems. Great effort has gone into the development of abstracting and indexing services, and preparation of specialized bibliographies. Such work partially alleviates the situation but does not address the central problem of reliability. The problem-solving scientist or engineer has neither the time nor the all-emcompassing expertise to read through a mass of reports of original research, select and appraise portions relevant to his needs, and resolve the many conflicts and inconsistencies reported by different workers. If the number he needs is presented in three different papers with three different values, he faces a dilemma he cannot resolve without expert guidance."

The foregoing plus other portions of the material in the CODATA Bulletin indicate concerns of this important international agency for ensuring the adequacy of published data for critical evaluation, for compression of the literature to ensure the ready availability of data, for enhanced presentation in the literature, and an appreciation of the role of statistical analysis both in design of experiments and in indicating the reliability of the data. The Editors and Publishers of the Handbook of Chemistry have for the past 64 years put forth their best effort to place into the Handbook only data and information which could be considered best and most reliable. We shall continue to do so in the future. However, with the advent of the activities sponsored by CODATA, UNESCO, and other international agencies for more critical evaluation of experimental results and for standarization of advanced techniques for treatment of experimental data a still greater reservoir of critically evaluated data should be available for compilers of handbooks and other reference literature. As in previous editions of the Handbook of Chemistry and Physics we shall in the future editions replace as rapidly as feasible old data as new and better data become available.

In keeping with this tradition of deleting certain information, revising certain other information, and adding timely new data we have for this 59th edition made certain changes. Among these are:

1. Deletion of certain mathematical tables. The ready availability of low cost handheld calculators makes inclusion of some mathematical tables in the Handbook unnecessary.

* CODATA Bulletin, No. 26, January 1978.

2. Rearrangement of the format of several tables to provide for up to 20% expansion of the content of the Handbook in future editions over the next several years.
3. Revision of the section, "The Elements". This 55 page section has again been updated and slightly expanded.
4. A supplementary table for Key Values for Thermodynamics has been added. This information lists the international recommended values for some ions, elements, and compounds not included in the table presently in the Handbook.
5. Addition of a table of Lattice Energies. This extensive table contains calculated values of the lattice energies of crystalline salts based on the best thermodynamic data currently available. In addition, the lattice energy calculated from the Born-Fajans-Haber cycle is given.
6. The Strengths of Chemical Bonds table has been revised, updated, and expanded. Data for this table include those available through June 1977. Part of the expanded information includes data for 80 diatomic molecules not included in the previous table. Literature references are provided for each entry in the table.
7. Tables of Line Spectra of the Elements and Atomic Tranisition Probabilities have been added. Information is presented for 96 elements. There are more than 40,000 entries in this table which was compiled by many experts throughout the U.S.
8. Information regarding the weight of water has been revised by including a table listing the weight of one gallon of water, expressed in grams and pounds, and with the water being exposed to air and also being in vacuo.
9. In keeping with the recognition of the importance of knowing better the possible interaction of humans and their physical environment we have revised the information listing the Limits of Human Exposure to Toxic and Hazardous Substances.
10. Table of Emmisivity of Total Radiation for Various Materials has been revised.
11. A list of publications available from the National Standard Reference Data System has been expanded and updated. This U.S. agency has extensive and specialized data and information useful to many investigators in the area of the physical sciences. The information is frequently too specialized and too extensive for inclusion in the Handbook of Chemistry and Physics, but users of the Handbook should find it useful to be aware of these NSRDS publications.

We are appreciative of our Contributors, Collaborators, and Advisory Board Members. We are also appreciative of users of the Handbook who write to us making helpful suggestions and pointing out typographical errors. We hope that such cooperation continues for it can only result in still further improvement of the book.

Robert C. Weast
April 1, 1978

Physics Editorial Board

Collaborators and Contributors

L. ALBERTS, Ph.D.
Director-General
National Institute for Metallurgy
Randburg, South Africa

C. J. ALLEN
Illuminating Engineer
General Electric Company
Nela Park
Cleveland, Ohio

W. A. ANDERSON, Ph.D.
Analytical Instrument Research
Varian Associates
611 Hansen Way
Palo Alto, California

W. J. ARMENTO
Oak Ridge National Laboratory
P.O. Box X
Oak Ridge, Tennessee

J. ASKILL, Ph.D.
Chairman, Physics Department
Millikin University
Decatur, Illinois

J. C. BAILAR, JR., Ph.D.
Professor of Inorganic Chemistry
University of Illinois
Urbana, Illinois

R. A. BAXTER, M.S.
Professor of Chemistry
Colorado School of Mines
Golden, Colorado

J. A. BEARDEN, Ph.D.
Department of Physics
Johns Hopkins University
Baltimore, Maryland

A. H. BENADE, Ph.D.
Department of Physics
Case Western Reserve University
Cleveland, Ohio

F. F. BENTLEY
Air Force Materials Laboratory
Wright-Patterson Air Force Base, Ohio

J. H. BILLMAN, Ph.D.
Professor of Chemistry
Indiana University
Bloomington, Indiana

A. L. BLOOM, Ph.D.
Spectra-Physics, Inc.
1255 Terra Bella Avenue
Mountain View, California

J. A. BRADLEY, Ph.D.
Dean, Newark College of Engineering
323 High Street
Newark, New Jersey

B. H. BROWN, Ph.D.
Dartmouth College
Hanover, New Hampshire

J. E. BROWN
Eastman Kodak Company
Rochester, New York

A. B. BURG, Ph.D.
Department of Chemistry
University of Southern California
Los Angeles, California

G. P. BURNS, Ph.D.
Department of Physics
Mary Washington College
Fredericksburg, Virginia

A. F. BURR, Ph.D.
Department of Physics
New Mexico State University
Las Cruces, New Mexico

J. P. CATCHPOLE, Ph.D.
Admiralty Materials Laboratory
Holton Heath
Dorset, England

E. RICHARD COHEN, Ph.D.
Associate Director
Science Center/Aerospace and Systems Group
North American Rockwell Corporation
Thousand Oaks, California

CHARLES H. CORLISS
Spectroscopy Section
Optical Physics Division
National Bureau of Standards
Washington, D.C.

M. DAVIES, Ph.D.
Edward Davies Chemical Laboratories
University of Wales
Aberystwyth, Wales

COLLABORATORS AND CONTRIBUTORS

J. DEMENT, D. SC.
Dement Laboratories
4847 S.E. Division St.
Portland, Oregon

H. G. DEMING, Ph.D.
2316 Tuttle Terrace
Sarasota, Florida

E. DICYAN, Ph.D.
Consulting Chemist
420 Lexington Ave
New York, New York

HANS DOLEZALEK
Department of the Navy
Office of Naval Research
Arlington, Virginia

A. P. DUNLOP, Ph.D.
Director of Chemical Research and Development
John Stuart Research Laboratories
The Quaker Oats Company
Barrington, Illinois

L. M. FOSTER, Ph.D.
Thomas J. Watson Research Center
International Business Machines Corp.
Yorktown Heights, New York

J. L. FRANKLIN, Ph.D.
Welch Professor of Chemistry
William Marsh Rice University
Houston, Texas

G. FULFORD, Ph.D.
Assistant Professor of Chemical Engineering
University of Waterloo
Waterloo, Ontario,
Canada

GLADYS H. FULLER
NBS Institute of Science and Technology
National Bureau of Standards
Washington, D.C.

E. F. FURTSCH, Ph.D.
Department of Chemistry
Virginia Polytechnic Institute
Blacksburg, Virginia

R. J. GETTENS, M. A.
Head Curator
Freer Gallery Laboratory
Smithsonian Institution
Washington, D.C.

L. A. GILLETTE, Ph.D.
Manager, Product Development
Pennsalt Chemicals Corporation
Philadelphia, Pennsylvania

H. GILMAN, Ph.D.
Department of Chemistry
Iowa State University
Ames, Iowa

B. GIRLING, M.Sc., F.I.M.A.
Department of Mathematics
The City University
London E.C.1, England

E. C. GREGG, Ph.D.
School of Medicine
Case Western Reserve University
Cleveland, Ohio

R. R. GUPTA, M.Sc., Ph.D.
Department of Chemistry
University of Rajasthan
Jaipur-4, India

C. R. HAMMOND
Emhart Industries, Inc.
P.O. Box 700
Hartford, Connecticut

W. E. HARRIS
Professor of Chemistry
University of Alberta
Edmonton, Alberta, Canada

H. J. HARWOOD, Ph.D.
Head, Durkee Famous Foods
Organic Chemistry Division
Glidden Company
Chicago, Illinois

R. L. HEATH
Atomic Energy Division
Phillips Petroleum Co.
Idaho Falls, Idaho

R. W. HOFFMAN, Ph.D.
Department of Physics
Case Western Reserve University
Cleveland, Ohio

COLLABORATORS AND CONTRIBUTORS

JESSE F. HUNSBERGER
RD #1
East Cedarville Road
Pottstown, Pennsylvania

C. D. HURD, Ph.D.
Chemical Laboratory
Northwestern University
Evanston, Illinois

H. D. B. JENKINS, Ph.D.
Department of Molecular Sciences
University of Warwick
Coventry CV4 7AL
England

J. L. KASSNER, Ph.D.
Department of Physics
The University of Missouri-Rolla
Rolla, Missouri

OLGA KENNARD, Ph.D.
University Chemical Laboratory
Cambridge, England

T. G. KENNARD, Ph.D.
20747 E. Palm Drive
Glendora, California

J. A. KERR, Ph.D.
Chemistry Department
The University of Birmingham
Haworth Building
Birmingham 15, England

A. L. KING, Ph.D.
Department of Physics
Dartmouth College
Hanover, New Hampshire

C. R. KINNEY, Ph.D.
1318 27th Street
Des Moines, Iowa

R. KRETZ, Ph.D.
Department of Science
and Engineering
University of Ottawa
Ottawa, Ontario, Canada K1N 6N5

G. LANG, DIPL. ING.
A–4963
St. Peter am Hart
Austria

D. F. LAWDEN, Sc.D.
Department of Mathematics
University of Aston
Birmingham, England

K. LEE, Ph.D.
IBM Research Laboratory
San Jose, California

A. P. LEVITT
75 Lovett Road
Newton Center, Massachusetts

M. M. MACMASTERS, Ph.D.
Department of Grain Science and
Industry
Kansas State University
Manhattan, Kansas

W. MAHLIG
Assistant Sales Manager
Laboratory Equipment Division
The W. S. Tyler Company
Cleveland, Ohio

C. J. MAJOR, Ph.D.
Department of Chemical Engineering
University of Akron
Akron, Ohio

G. A. MARTIN
National Bureau of Standards
Washington, D.C.

J. C. McGOWAN, Ph.D., D.Sc.
"Quantock"
13 Moreton Avenue
Harpenden, Herts AL5 2EU, England

L. MEITES, Ph.D.
Head, Department of Chemistry
Clarkson College of Technology
Potsdam, New York

M. G. MELLON, Ph.D.
Professor Emeritus, Analytical Chemistry
Purdue University
West Lafayette, Indiana

H. B. MICHAELSON
IBM Journal of Research and Development
Armonk, New York

KARL Z. MORGAN, Ph.D.
Director, Health Physics Division
Oak Ridge National Laboratory
Oak Ridge, Tennessee

W. M. MORGAN, Ph.D.
Professor Emeritus
Mount Union College
Alliance, Ohio

COLLABORATORS AND CONTRIBUTORS

R. R. NIMMO, Ph.D.
Professor of Physics
University of Otago
Dunedin, New Zealand

F. J. NORTON, Ph.D.
General Electric Company
1133 Eastern Avenue
Schenectady, New York

F. M. PAGE
Department of Chemistry
The University of Aston
Birmingham B4 7ET, England

B. R. PAMPLIN, Ph.D.
Scientific Advisers
15 Park Lane
Bath A1 2XH, England

W. PARKER, Ph.D.
Department of Physics
University of California
Irvine, California

M. J. PARSONAGE
Chemistry Department
The University of Birmingham
Birmingham 15, England

SAUL PATAI, Ph.D.
Department of Organic Chemistry
Hebrew University of Jerusalem
Jerusalem, Israel

I. A. PEARL, Ph.D.
The Institute of Paper Chemistry
Appleton, Wisconsin

R. F. PEART, Ph.D.
The Plessey Co. Ltd.
Allen Clark Research Centre
Northampton, England

A. C. PEED, JR.
Eastman Kodak Company
243 State Street
Rochester, New York

R. PEPINSKY, Ph.D.
Department of Physics and Astronomy
University of Florida
Gainesville, Florida

H. A. POUL, Ph.D.
Professor of Physics
Oklahoma State University
Stillwater, Oklahoma

RICHARD L. PRATT
Staff Analyst
Data Corporation
Dayton, Ohio

H. J. PREBLUDA, Ph.D.
Consulting Biochemist
3 Belmont Circle
Trenton, New Jersey

I. B. PRETTYMAN, M.S.
The Firestone Tire and Rubber Co.
1200 Firestone Parkway
Akron, Ohio

ZVI RAPPOPORT, Ph.D.
Department of Organic Chemistry
Hebrew University of Jerusalem
Jerusalem, Israel

E. H. RATCLIFFE
Water Research Centre
Stevenage Laboratory
Elder Way
Stevenage, Hertfordshire SG1 1TH, England

JOSEPH READER, Ph.D.
Spectroscopy Section
Optical Physics Division
National Bureau of Standards
Washington, D.C.

M. C. REED, Ph.D.
1368 Wood Valley Road
Mountainside, New Jersey

B. W. ROBERTS, Ph.D.
General Electric Research Laboratory
Schenectady, New York

R. C. ROBERTS, Ph.D.
Professor Emeritus of Chemistry
Colgate University
Hamilton, New York

R. A. ROBINSON
School of Chemistry
The University
Newcastle-upon-Tyne NE1 7RU, England

R. J. ROSEN
Consulting Chemist
9301 Parkhill Drive
Los Angeles, California

A. H. ROSENFELD, Ph.D.
Lawrence Radiation Laboratory
University of California
Berkeley, California

COLLABORATORS AND CONTRIBUTORS

GORDON D. ROWE
Specialist Lighting of GE Properties
General Electric Company
Cleveland, Ohio

A. L. ROZEK
Velsicol Chemical Corporation
Chicago, Illinois

S. I. SALEM, Ph.D.
Professor, Department of Physics and Astronomy
California State College
Long Beach, California

G. T. SEABORG, Ph.D.
Lawrence Berkeley Laboratory
University of California
Berkeley, California

R. S. SHANKLAND, Ph.D.
Department of Physics
Case Western Reserve University
Cleveland, Ohio

R. SHAW
Chemical Physicist
Physical Organic Program
Stanford Research Institute
Menlo Park, California

J. R. SHELTON, Ph.D.
Department of Chemistry
Case Western Reserve University
Cleveland, Ohio

G. W. SMITH, Ph.D.
General Motors Corporation
Research Laboratories
Warren, Michigan

J. M. SMITH, B.S.E.E.
Product Planning Large Lamp Department
General Electric Company
Cleveland, Ohio

L. D. SMITHSON
Air Force Materials Laboratory
Wright-Patterson Air Force Base, Ohio

F. H. SPEDDING
Director, Ames Laboratory
Iowa State University
Ames, Iowa

R. H. STOKES, Ph.D.
Department of Chemistry
The University of New England
Armidale, N.S.W., Australia

DONALD F. SWINEHART, Ph.D.
Department of Chemistry
University of Oregon
Eugene, Oregon

A. TARPINIAN
Army Materials and Mechanics Research Center
Arsenal Street
Watertown, Massachusetts

B. N. TAYLOR, Ph.D.
A-247-Metrology
National Bureau of Standards
Washington, D.C.

D. H. TOMLIN, Ph.D.
Department of Physics
University of Reading
Reading, Berkshire, England

A. F. TROTMAN-DICKENSON, Ph.D.
Institute of Technology
Cardiff, England

H. B. VICKERY, Ph.D.
Connecticut Agricultural Experimental Station
112 Huntington Street
New Haven, Connecticut

W. W. WENDLANDT, Ph.D.
Professor of Chemistry
University of Houston
Houston, Texas

N. R. WHETTEN, Ph.D.
Research and Development Center
General Electric Company
P.O. Box 8
Schenectady, New York

WOLFGANG L. WIESE, Ph.D.
Plasma Spectroscopy Section
National Bureau of Standards
Washington, D.C.

COLLABORATORS AND CONTRIBUTORS

J. H. YOE, Ph.D.
Professor Emeritus of Chemistry
University of Virginia
Charlottesville, Virginia

G. R. YOHE, Ph.D.
Illinois State Geological Survey
University of Illinois Campus
Urbana, Illinois

T. F. YOUNG, Ph.D.
Division of Chemical Engineering
Argonne National Laboratory
Argonne, Illinois

J. ZABICKY, Ph.D.
Department of Biophysics
The Weizmann Institute of Science
Rehovoth, Israel

S. ZUFFANTI, A.M.
Professor of Chemistry
Northeastern University
Boston, Massachusetts

The Publishers and Editors will be grateful to readers of this Handbook who will call their attention to errors that may be discovered. Suggestions for improvement are also welcome.

TABLE OF CONTENTS

FOUR-PLACE MANTISSAS FOR COMMON LOGARITHMS

N	0	1	2	3	4	5	6	7	8	9	1	2	3	4	5	6	7	8	9
											Proportional Parts								
10	0000	0043	0086	0128	0170	0212	0253	0294	0334	0374	*4	8	12	17	21	25	29	33	37
11	0414	0453	0492	0531	0569	0607	0645	0682	0719	0755	4	8	11	15	19	23	26	30	34
12	0792	0828	0864	0899	0934	0969	1004	1038	1072	1106	3	7	10	14	17	21	24	28	31
13	1139	1173	1206	1239	1271	1303	1335	1367	1399	1430	3	6	10	13	16	19	23	26	29
14	1461	1492	1523	1553	1584	1614	1644	1673	1703	1732	3	6	9	12	15	18	21	24	27
15	1761	1790	1818	1847	1875	1903	1931	1959	1987	2014	*3	6	8	11	14	17	20	22	25
16	2041	2068	2095	2122	2148	2175	2201	2227	2253	2279	3	5	8	11	13	16	18	21	24
17	2304	2330	2355	2380	2405	2430	2455	2480	2504	2529	2	5	7	10	12	15	17	20	22
18	2553	2577	2601	2625	2648	2672	2695	2718	2742	2765	2	5	7	9	12	14	16	19	21
19	2788	2810	2833	2856	2878	2900	2923	2945	2967	2989	2	4	7	9	11	13	16	18	20
20	3010	3032	3054	3075	3096	3118	3139	3160	3181	3201	2	4	6	8	11	13	15	17	19
21	3222	3243	3263	3284	3304	3324	3345	3365	3385	3404	2	4	6	8	10	12	14	16	18
22	3424	3444	3464	3483	3502	3522	3541	3560	3579	3598	2	4	6	8	10	12	14	15	17
23	3617	3636	3655	3674	3692	3711	3729	3747	3766	3784	2	4	6	7	9	11	13	15	17
24	3802	3820	3838	3856	3874	3892	3909	3927	3945	3962	2	4	5	7	9	11	12	14	16
25	3979	3997	4014	4031	4048	4065	4082	4099	4116	4133	2	3	5	7	9	10	12	14	15
26	4150	4166	4183	4200	4216	4232	4249	4265	4281	4298	2	3	5	7	8	10	11	13	15
27	4314	4330	4346	4362	4378	4393	4409	4425	4440	4456	2	3	5	6	8	9	11	13	14
28	4472	4487	4502	4518	4533	4548	4564	4579	4594	4609	2	3	5	6	8	9	11	12	14
29	4624	4639	4654	4669	4683	4698	4713	4728	4742	4757	1	3	4	6	7	9	10	12	13
30	4771	4786	4800	4814	4829	4843	4857	4871	4886	4900	1	3	4	6	7	9	10	11	13
31	4914	4928	4942	4955	4969	4983	4997	5011	5024	5038	1	3	4	6	7	8	10	11	12
32	5051	5065	5079	5092	5105	5119	5132	5145	5159	5172	1	3	4	5	7	8	9	11	12
33	5185	5198	5211	5224	5237	5250	5263	5276	5289	5302	1	3	4	5	6	8	9	10	12
34	5315	5328	5340	5353	5366	5378	5391	5403	5416	5428	1	3	4	5	6	8	9	10	11
35	5441	5453	5465	5478	5490	5502	5514	5527	5539	5551	1	2	4	5	6	7	9	10	11
36	5563	5575	5587	5599	5611	5623	5635	5647	5658	5670	1	2	4	5	6	7	8	10	11
37	5682	5694	5705	5717	5729	5740	5752	5763	5775	5786	1	2	3	5	6	7	8	9	10
38	5798	5809	5821	5832	5843	5855	5866	5877	5888	5899	1	2	3	5	6	7	8	9	10
39	5911	5922	5933	5944	5955	5966	5977	5988	5999	6010	1	2	3	4	5	7	8	9	10
40	6021	6031	6042	6053	6064	6075	6085	6096	6107	6117	1	2	3	4	5	6	8	9	10
41	6128	6138	6149	6160	6170	6180	6191	6201	6212	6222	1	2	3	4	5	6	7	8	9
42	6232	6243	6253	6263	6274	6284	6294	6304	6314	6325	1	2	3	4	5	6	7	8	9
43	6335	6345	6355	6365	6375	6385	6395	6405	6415	6425	1	2	3	4	5	6	7	8	9
44	6435	6444	6454	6464	6474	6484	6493	6503	6513	6522	1	2	3	4	5	6	7	8	9
45	6532	6542	6551	6561	6571	6580	6590	6599	6609	6618	1	2	3	4	5	6	7	8	9
46	6628	6637	6646	6656	6665	6675	6684	6693	6702	6712	1	2	3	4	5	6	7	7	8
47	6721	6730	6739	6749	6758	6767	6776	6785	6794	6803	1	2	3	4	5	5	6	7	8
48	6812	6821	6830	6839	6848	6857	6866	6875	6884	6893	1	2	3	4	4	5	6	7	8
49	6902	6911	6920	6928	6937	6946	6955	6964	6972	6981	1	2	3	4	4	5	6	7	8
50	6990	6998	7007	7016	7024	7033	7042	7050	7059	7067	1	2	3	3	4	5	6	7	8
51	7076	7084	7093	7101	7110	7118	7126	7135	7143	7152	1	2	3	3	4	5	6	7	8
52	7160	7168	7177	7185	7193	7202	7210	7218	7226	7235	1	2	2	3	4	5	6	7	7
53	7243	7251	7259	7267	7275	7284	7292	7300	7308	7316	1	2	2	3	4	5	6	6	7
54	7324	7332	7340	7348	7356	7364	7372	7380	7388	7396	1	2	2	3	4	5	6	6	7
N	0	1	2	3	4	5	6	7	8	9	1	2	3	4	5	6	7	8	9

* Interpolation in this section of the table is inaccurate.

FOUR-PLACE MANTISSAS FOR COMMON LOGARITHMS (Continued)

N	0	1	2	3	4	5	6	7	8	9	Proportional Parts								
											1	2	3	4	5	6	7	8	9
55	7404	7412	7419	7427	7435	7443	7451	7459	7466	7474	1	2	2	3	4	5	5	6	7
56	7482	7490	7497	7505	7513	7520	7528	7536	7543	7551	1	2	2	3	4	5	5	6	7
57	7559	7566	7574	7582	7589	7597	7604	7612	7619	7627	1	2	2	3	4	5	5	6	7
58	7634	7642	7649	7657	7664	7672	7679	7686	7694	7701	1	1	2	3	4	4	5	6	7
59	7709	7716	7723	7731	7738	7745	7752	7760	7767	7774	1	1	2	3	4	4	5	6	7
60	7782	7789	7796	7803	7810	7818	7825	7832	7839	7846	1	1	2	3	4	4	5	6	6
61	7853	7860	7868	7875	7882	7889	7896	7903	7910	7917	1	1	2	3	4	4	5	6	6
62	7924	7931	7938	7945	7952	7959	7966	7973	7980	7987	1	1	2	3	3	4	5	6	6
63	7993	8000	8007	8014	8021	8028	8035	8041	8048	8055	1	1	2	3	3	4	5	5	6
64	8062	8069	8075	8082	8089	8096	8102	8109	8116	8122	1	1	2	3	3	4	5	5	6
65	8129	8136	8142	8149	8156	8162	8169	8176	8182	8189	1	1	2	3	3	4	5	5	6
66	8195	8202	8209	8215	8222	8228	8235	8241	8248	8254	1	1	2	3	3	4	5	5	6
67	8261	8267	8274	8280	8287	8293	8299	8306	8312	8319	1	1	2	3	3	4	5	5	6
68	8325	8331	8338	8344	8351	8357	8363	8370	8376	8382	1	1	2	3	3	4	4	5	6
69	8388	8395	8401	8407	8414	8420	8426	8432	8439	8445	1	1	2	2	3	4	4	5	6
70	8451	8457	8463	8470	8476	8482	8488	8494	8500	8506	1	1	2	2	3	4	4	5	6
71	8513	8519	8525	8531	8537	8543	8549	8555	8561	8567	1	1	2	2	3	4	4	5	5
72	8573	8579	8585	8591	8597	8603	8609	8615	8621	8627	1	1	2	2	3	4	4	5	5
73	8633	8639	8645	8651	8657	8663	8669	8675	8681	8686	1	1	2	2	3	4	4	5	5
74	8692	8698	8704	8710	8716	8722	8727	8733	8739	8745	1	1	2	2	3	4	4	5	5
75	8751	8756	8762	8768	8774	8779	8785	8791	8797	8802	1	1	2	2	3	3	4	5	5
76	8808	8814	8820	8825	8831	8837	8842	8848	8854	8859	1	1	2	2	3	3	4	5	5
77	8865	8871	8876	8882	8887	8893	8899	8904	8910	8915	1	1	2	2	3	3	4	4	5
78	8921	8927	8932	8938	8943	8949	8954	8960	8965	8971	1	1	2	2	3	3	4	4	5
79	8976	8982	8987	8993	8998	9004	9009	9015	9020	9025	1	1	2	2	3	3	4	4	5
80	9031	9036	9042	9047	9053	9058	9063	9069	9074	9079	1	1	2	2	3	3	4	4	5
81	9085	9090	9096	9101	9106	9112	9117	9122	9128	9133	1	1	2	2	3	3	4	4	5
82	9138	9143	9149	9154	9159	9165	9170	9175	9180	9186	1	1	2	2	3	3	4	4	5
83	9191	9196	9201	9206	9212	9217	9222	9227	9232	9238	1	1	2	2	3	3	4	4	5
84	9243	9248	9253	9258	9263	9269	9274	9279	9284	9289	1	1	2	2	3	3	4	4	5
85	9294	9299	9304	9309	9315	9320	9325	9330	9335	9340	1	1	2	2	3	3	4	4	5
86	9345	9350	9355	9360	9365	9370	9375	9380	9385	9390	1	1	2	2	3	3	4	4	5
87	9395	9400	9405	9410	9415	9420	9425	9430	9435	9440	0	1	1	2	2	3	3	4	4
88	9445	9450	9455	9460	9465	9469	9474	9479	9484	9489	0	1	1	2	2	3	3	4	4
89	9494	9499	9504	9509	9513	9518	9523	9528	9533	9538	0	1	1	2	2	3	3	4	4
90	9542	9547	9552	9557	9562	9566	9571	9576	9581	9586	0	1	1	2	2	3	3	4	4
91	9590	9595	9600	9605	9609	9614	9619	9624	9628	9633	0	1	1	2	2	3	3	4	4
92	9638	9643	9647	9652	9657	9661	9666	9671	9675	9680	0	1	1	2	2	3	3	4	4
93	9685	9689	9694	9699	9703	9708	9713	9717	9722	9727	0	1	1	2	2	3	3	4	4
94	9731	9736	9741	9745	9750	9754	9759	9763	9768	9773	0	1	1	2	2	3	3	4	4
95	9777	9782	9786	9791	9795	9800	9805	9809	9814	9818	0	1	1	2	2	3	3	4	4
96	9823	9827	9832	9836	9841	9845	9850	9854	9859	9863	0	1	1	2	2	3	3	4	4
97	9868	9872	9877	9881	9886	9890	9894	9899	9903	9908	0	1	1	2	2	3	3	4	4
98	9912	9917	9921	9926	9930	9934	9939	9943	9948	9952	0	1	1	2	2	3	3	4	4
99	9956	9961	9965	9969	9974	9978	9983	9987	9991	9996	0	1	1	2	2	3	3	3	4
N	0	1	2	3	4	5	6	7	8	9	1	2	3	4	5	6	7	8	9

*FOUR-PLACE MANTISSAS FOR COMMON LOGARITHMS OF DECIMAL FRACTIONS

N	0	1	2	3	4	5	6	7	8	9
.10	−1.000	−.9957	−.9914	−.9872	−.9830	−.9788	−.9747	−.9706	−.9666	−.9626
.11	−.9586	−.9547	−.9508	−.9469	−.9431	−.9393	−.9355	−.9318	−.9281	−.9245
.12	−.9208	−.9172	−.9136	−.9101	−.9066	−.9031	−.8996	−.8962	−.8928	−.8894
.13	−.8861	−.8827	−.8794	−.8761	−.8729	−.8697	−.8665	−.8633	−.8601	−.8570
.14	−.8539	−.8508	−.8477	−.8447	−.8416	−.8386	−.8356	−.8327	−.8297	−.8268
.15	−.8239	−.8210	−.8182	−.8153	−.8125	−.8097	−.8069	−.8041	−.8013	−.7986
.16	−.7959	−.7932	−.7905	−.7878	−.7852	−.7825	−.7799	−.7773	−.7747	−.7721
.17	−.7696	−.7670	−.7645	−.7620	−.7595	−.7570	−.7545	−.7520	−.7496	−.7471
.18	−.7447	−.7423	−.7399	−.7375	−.7352	−.7328	−.7305	−.7282	−.7258	−.7235
.19	−.7212	−.7190	−.7167	−.7144	−.7122	−.7100	−.7077	−.7055	−.7033	−.7011
.20	−.6990	−.6968	−.6946	−.6925	−.6904	−.6882	−.6861	−.6840	−.6819	−.6799
.21	−.6778	−.6757	−.6737	−.6716	−.6696	−.6676	−.6655	−.6635	−.6615	−.6596
.22	−.6576	−.6556	−.6536	−.6517	−.6498	−.6478	−.6459	−.6440	−.6421	−.6402
.23	−.6383	−.6364	−.6345	−.6326	−.6308	−.6289	−.6271	−.6253	−.6234	−.6216
.24	−.6198	−.6180	−.6162	−.6144	−.6126	−.6108	−.6091	−.6073	−.6055	−.6038
.25	−.6021	−.6003	−.5986	−.5969	−.5952	−.5935	−.5918	−.5901	−.5884	−.5867
.26	−.5850	−.5834	−.5817	−.5800	−.5784	−.5768	−.5751	−.5735	−.5719	−.5702
.27	−.5686	−.5670	−.5654	−.5638	−.5622	−.5607	−.5591	−.5575	−.5560	−.5544
.28	−.5528	−.5513	−.5498	−.5482	−.5467	−.5452	−.5436	−.5421	−.5406	−.5391
.29	−.5376	−.5361	−.5346	−.5331	−.5317	−.5302	−.5287	−.5272	−.5258	−.5243
.30	−.5229	−.5214	−.5200	−.5186	−.5171	−.5157	−.5143	−.5129	−.5114	−.5100
.31	−.5086	−.5072	−.5058	−.5045	−.5031	−.5017	−.5003	−.4989	−.4976	−.4962
.32	−.4949	−.4935	−.4921	−.4908	−.4895	−.4881	−.4868	−.4855	−.4841	−.4828
.33	−.4815	−.4802	−.4789	−.4776	−.4763	−.4750	−.4737	−.4724	−.4711	−.4698
.34	−.4685	−.4672	−.4660	−.4647	−.4634	−.4622	−.4609	−.4597	−.4584	−.4572
.35	−.4559	−.4547	−.4535	−.4522	−.4510	−.4498	−.4486	−.4473	−.4461	−.4449
.36	−.4437	−.4425	−.4413	−.4401	−.4389	−.4377	−.4365	−.4353	−.4342	−.4330
.37	−.4318	−.4306	−.4295	−.4283	−.4271	−.4260	−.4248	−.4237	−.4225	−.4214
.38	−.4202	−.4191	−.4179	−.4168	−.4157	−.4145	−.4134	−.4123	−.4112	−.4101
.39	−.4089	−.4078	−.4067	−.4056	−.4045	−.4034	−.4023	−.4012	−.4001	−.3990
.40	−.3979	−.3969	−.3958	−.3947	−.3936	−.3925	−.3915	−.3904	−.3893	−.3883
.41	−.3872	−.3862	−.3851	−.3840	−.3830	−.3820	−.3809	−.3799	−.3788	−.3778
.42	−.3768	−.3757	−.3747	−.3737	−.3726	−.3716	−.3706	−.3696	−.3686	−.3675
.43	−.3665	−.3655	−.3645	−.3635	−.3625	−.3615	−.3605	−.3595	−.3585	−.3575
.44	−.3565	−.3556	−.3546	−.3536	−.3526	−.3516	−.3507	−.3497	−.3487	−.3478
.45	−.3468	−.3458	−.3449	−.3439	−.3429	−.3420	−.3410	−.3401	−.3391	−.3382
.46	−.3372	−.3363	−.3354	−.3344	−.3335	−.3325	−.3316	−.3307	−.3298	−.3288
.47	−.3279	−.3270	−.3261	−.3251	−.3242	−.3233	−.3224	−.3215	−.3206	−.3197
.48	−.3188	−.3179	−.3170	−.3161	−.3152	−.3143	−.3134	−.3125	−.3116	−.3107
.49	−.3098	−.3089	−.3080	−.3072	−.3063	−.3054	−.3045	−.3036	−.3028	−.3019
.50	−.3010	−.3002	−.2993	−.2984	−.2976	−.2967	−.2958	−.2950	−.2941	−.2933
.51	−.2924	−.2916	−.2907	−.2899	−.2890	−.2882	−.2874	−.2865	−.2857	−.2848
.52	−.2840	−.2832	−.2823	−.2815	−.2807	−.2798	−.2790	−.2782	−.2774	−.2765
.53	−.2757	−.2749	−.2741	−.2733	−.2725	−.2716	−.2708	−.2700	−.2692	−.2684
.54	−.2676	−.2668	−.2660	−.2652	−.2644	−.2636	−.2628	−.2620	−.2612	−.2604

* This table gives the logarithms of the decimal fractions which are negative numbers.

For example log 0.61 = −0.2147 = 9.7853 − 10. It should be noted that the entries as given can be used conveniently to find cologarithms of positive numbers. Every positive number $N = P \cdot (10)^k$, where $0 < P \leq 1$. Since colog N = − log N, it follows that colog N = − log P − k.

For example colog 0.61 = − log 0.61 = 0.2147; colog 61 = 0.2147 − 2; and colog 0.00061 = 3.2147.

*FOUR-PLACE MANTISSAS FOR COMMON LOGARITHMS OF DECIMAL FRACTIONS (Continued)

N	0	1	2	3	4	5	6	7	8	9
.55	−.2596	−.2588	−.2581	−.2573	−.2565	−.2557	−.2549	−.2541	−.2534	−.2526
.56	−.2518	−.2510	−.2503	−.2495	−.2487	−.2480	−.2472	−.2464	−.2457	−.2449
.57	−.2441	−.2434	−.2426	−.2418	−.2411	−.2403	−.2396	−.2388	−.2381	−.2373
.58	−.2366	−.2358	−.2351	−.2343	−.2336	−.2328	−.2321	−.2314	−.2306	−.2299
.59	−.2291	−.2284	−.2277	−.2269	−.2262	−.2255	−.2248	−.2240	−.2233	−.2226
.60	−.2218	−.2211	−.2204	−.2197	−.2190	−.2182	−.2175	−.2168	−.2161	−.2154
.61	−.2147	−.2140	−.2132	−.2125	−.2118	−.2111	−.2104	−.2097	−.2090	−.2083
.62	−.2076	−.2069	−.2062	−.2055	−.2048	−.2041	−.2034	−.2027	−.2020	−.2013
.63	−.2007	−.2000	−.1993	−.1986	−.1979	−.1972	−.1965	−.1959	−.1952	−.1945
.64	−.1938	−.1931	−.1925	−.1918	−.1911	−.1904	−.1898	−.1891	−.1884	−.1878
.65	−.1871	−.1864	−.1858	−.1851	−.1844	−.1838	−.1831	−.1824	−.1818	−.1811
.66	−.1805	−.1798	−.1791	−.1785	−.1778	−.1772	−.1765	−.1759	−.1752	−.1746
.67	−.1739	−.1733	−.1726	−.1720	−.1713	−.1707	−.1701	−.1694	−.1688	−.1681
.68	−.1675	−.1669	−.1662	−.1656	−.1649	−.1643	−.1637	−.1630	−.1624	−.1618
.69	−.1612	−.1605	−.1599	−.1593	−.1586	−.1580	−.1574	−.1568	−.1561	−.1555
.70	−.1549	−.1543	−.1537	−.1530	−.1524	−.1518	−.1512	−.1506	−.1500	−.1494
.71	−.1487	−.1481	−.1475	−.1469	−.1463	−.1457	−.1451	−.1445	−.1439	−.1433
.72	−.1427	−.1421	−.1415	−.1409	−.1403	−.1397	−.1391	−.1385	−.1379	−.1373
.73	−.1367	−.1361	−.1355	−.1349	−.1343	−.1337	−.1331	−.1325	−.1319	−.1314
.74	−.1308	−.1302	−.1296	−.1290	−.1284	−.1278	−.1273	−.1267	−.1261	−.1255
.75	−.1249	−.1244	−.1238	−.1232	−.1226	−.1221	−.1215	−.1209	−.1203	−.1198
.76	−.1192	−.1186	−.1180	−.1175	−.1169	−.1163	−.1158	−.1152	−.1146	−.1141
.77	−.1135	−.1129	−.1124	−.1118	−.1113	−.1107	−.1101	−.1096	−.1090	−.1085
.78	−.1079	−.1073	−.1068	−.1062	−.1057	−.1051	−.1046	−.1040	−.1035	−.1029
.79	−.1024	−.1018	−.1013	−.1007	−.1002	−.0996	−.0991	−.0985	−.0980	−.0975
.80	−.0969	−.0964	−.0958	−.0953	−.0947	−.0942	−.0937	−.0931	−.0926	−.0921
.81	−.0915	−.0910	−.0904	−.0899	−.0894	−.0888	−.0883	−.0878	−.0872	−.0867
.82	−.0862	−.0857	−.0851	−.0846	−.0841	−.0835	−.0830	−.0825	−.0820	−.0814
.83	−.0809	−.0804	−.0799	−.0794	−.0788	−.0783	−.0778	−.0773	−.0768	−.0762
.84	−.0757	−.0752	−.0747	−.0742	−.0737	−.0731	−.0726	−.0721	−.0716	−.0711
.85	−.0706	−.0701	−.0696	−.0691	−.0685	−.0680	−.0675	−.0670	−.0665	−.0660
.86	−.0655	−.0650	−.0645	−.0640	−.0635	−.0630	−.0625	−.0620	−.0615	−.0610
.87	−.0605	−.0600	−.0595	−.0590	−.0585	−.0580	−.0575	−.0570	−.0565	−.0560
.88	−.0555	−.0550	−.0545	−.0540	−.0535	−.0531	−.0526	−.0521	−.0516	−.0511
.89	−.0506	−.0501	−.0496	−.0491	−.0487	−.0482	−.0477	−.0472	−.0467	−.0462
.90	−.0458	−.0453	−.0448	−.0443	−.0438	−.0434	−.0429	−.0424	−.0419	−.0414
.91	−.0410	−.0405	−.0400	−.0395	−.0391	−.0386	−.0381	−.0376	−.0372	−.0367
.92	−.0362	−.0357	−.0353	−.0348	−.0343	−.0339	−.0334	−.0329	−.0325	−.0320
.93	−.0315	−.0311	−.0306	−.0301	−.0297	−.0292	−.0287	−.0283	−.0278	−.0273
.94	−.0269	−.0264	−.0259	−.0255	−.0250	−.0246	−.0241	−.0237	−.0232	−.0227
.95	−.0223	−.0218	−.0214	−.0209	−.0205	−.0200	−.0195	−.0191	−.0186	−.0182
.96	−.0177	−.0173	−.0168	−.0164	−.0159	−.0155	−.0150	−.0146	−.0141	−.0137
.97	−.0132	−.0128	−.0123	−.0119	−.0114	−.0110	−.0106	−.0101	−.0097	−.0092
.98	−.0088	−.0083	−.0079	−.0074	−.0070	−.0066	−.0061	−.0057	−.0052	−.0048
.99	−.0044	−.0039	−.0035	−.0031	−.0026	−.0022	−.0017	−.0013	−.0009	−.0004

*See footnote page A-3.

NATURAL OR NAPERIAN LOGARITHMS

0.000–0.499

N	0	1	2	3	4	5	6	7	8	9
0.00	−∞	−6‡	−6	−5	−5	−5	−5	−4	−4	−4
		.90776	.21461	.80914	.52146	.29832	.11600	.96185	.82831	.71053
.01	−4.60517	.50986	.42285	.34281	.26870	.19971	.13517	.07454	.01738	*.96332
.02	−3.91202	.86323	.81671	.77226	.72970	.68888	.64966	.61192	.57555	.54046
.03	.50656	.47377	.44202	.41125	.38139	.35241	.32424	.29684	.27017	.24419
.04	.21888	.19418	.17009	.14656	.12357	.10109	.07911	.05761	.03655	.01593
.05	−2.99573	.97593	.95651	.93746	.91877	.90042	.88240	.86470	.84731	.83022
.06	.81341	.79688	.78062	.76462	.74887	.73337	.71810	.70306	.68825	.67365
.07	.65926	.64508	.63109	.61730	.60369	.59027	.57702	.56395	.55105	.53831
.08	.52573	.51331	.50104	.48891	.47694	.46510	.45341	.44185	.43042	.41912
.09	.40795	.39690	.38597	.37516	.36446	.35388	.34341	.33304	.32279	.31264
0.10	−2.30259	.29263	.28278	.27303	.26336	.25379	.24432	.23493	.22562	.21641
.11	.20727	.19823	.18926	.18037	.17156	.16282	.15417	.14558	.13707	.12863
.12	.12026	.11196	.10373	.09557	.08747	.07944	.07147	.06357	.05573	.04794
.13	.04022	.03256	.02495	.01741	.00992	.00248	*.99510	*.98777	*.98050	*.97328
.14	−1.96611	.95900	.95193	.94491	.93794	.93102	.92415	.91732	.91054	.90381
.15	.89712	.89048	.88387	.87732	.87080	.86433	.85790	.85151	.84516	.83885
.16	.83258	.82635	.82016	.81401	.80789	.80181	.79577	.78976	.78379	.77786
.17	.77196	.76609	.76026	.75446	.74870	.74297	.73727	.73161	.72597	.72037
.18	.71480	.70926	.70375	.69827	.69282	.68740	.68201	.67665	.67131	.66601
.19	.66073	.65548	.65026	.64507	.63990	.63476	.62964	.62455	.61949	.61445
0.20	−1.60944	.60445	.59949	.59455	.58964	.58475	.57988	.57504	.57022	.56542
.21	.56065	.55590	.55117	.54646	.54178	.53712	.53248	.52786	.52326	.51868
.22	.51413	.50959	.50508	.50058	.49611	.49165	.48722	.48281	.47841	.47403
.23	.46968	.46534	.46102	.45672	.45243	.44817	.44392	.43970	.43548	.43129
.24	.42712	.42296	.41882	.41469	.41059	.40650	.40242	.39837	.39433	.39030
.25	.38629	.38230	.37833	.37437	.37042	.36649	.36258	.35868	.35480	.35093
.26	.34707	.34323	.33941	.33560	.33181	.32803	.32426	.32051	.31677	.31304
.27	.30933	.30564	.30195	.29828	.29463	.29098	.28735	.28374	.28013	.27654
.28	.27297	.26940	.26585	.26231	.25878	.25527	.25176	.24827	.24479	.24133
.29	.23787	.23443	.23100	.22758	.22418	.22078	.21740	.21402	.21066	.20731
0.30	−1.20397	.20065	.19733	.19402	.19073	.18744	.18417	.18091	.17766	.17441
.31	.17118	.16796	.16475	.16155	.15836	.15518	.15201	.14885	.14570	.14256
.32	.13943	.13631	.13320	.13010	.12701	.12393	.12086	.11780	.11474	.11170
.33	.10866	.10564	.10262	.09961	.09661	.09362	.09064	.08767	.08471	.08176
.34	.07881	.07587	.07294	.07002	.06711	.06421	.06132	.05843	.05555	.05268
.35	−1.04982	.04697	.04412	.04129	.03846	.03564	.03282	.03002	.02722	.02443
.36	.02165	.01888	.01611	.01335	.01060	.00786	.00512	.00239	*.99967	*.99696
.37	−0.99425	.99155	.98886	.98618	.98350	.98083	.97817	.97551	.97286	.97022
.38	.96758	.96496	.96233	.95972	.95711	.95451	.95192	.94933	.94675	.94418
.39	.94161	.93905	.93649	.93395	.93140	.92887	.92634	.92382	.92130	.91879
0.40	−0.91629	.91379	.91130	.90882	.90634	.90387	.90140	.89894	.89649	.89404
.41	.89160	.88916	.88673	.88431	.88189	.87948	.87707	.87467	.87227	.86988
.42	.86750	.86512	.86275	.86038	.85802	.85567	.85332	.85097	.84863	.84630
.43	.84397	.84165	.83933	.83702	.83471	.83241	.83011	.82782	.82554	.82326
.44	.82098	.81871	.81645	.81419	.81193	.80968	.80744	.80520	.80296	.80073
.45	.79851	.79629	.79407	.79186	.78966	.78746	.78526	.78307	.78089	.77871
.46	.77653	.77436	.77219	.77003	.76787	.76572	.76357	.76143	.75929	.75715
.47	.75502	.75290	.75078	.74866	.74655	.74444	.74234	.74024	.73814	.73605
.48	.73397	.73189	.72981	.72774	.72567	.72361	.72155	.71949	.71744	.71539
.49	.71335	.71131	.70928	.70725	.70522	.70320	.70118	.69917	.69716	.69515

‡ Note that the whole number values are **given above** the decimal values for the first line. In the second and following lines they are given at **the left.**

NATURAL OR NAPERIAN LOGARITHMS (Continued)
0.500–0.999

N	0	1	2	3	4	5	6	7	8	9
0.50	−0.69315	.69115	.68916	.68717	.68518	.68320	.68122	.67924	.67727	.67531
.51	.67334	.67139	.66934	.66748	.66553	.66359	.66165	.65971	.65778	.65585
.52	.65393	.65201	.65009	.64817	.64626	.64436	.64245	.64055	.63866	.63677
.53	.63488	.63299	.63111	.62923	.62736	.62549	.62362	.62176	.61990	.61804
.54	.61619	.61434	.61249	.61065	.60881	.60697	.60514	.60331	.60148	.59966
.55	.59784	.59602	.59421	.59240	.59059	.58879	.58699	.58519	.58340	.58161
.56	.57982	.57803	.57625	.57448	.57270	.57093	.56916	.56740	.56563	.56387
.57	.56212	.56037	.55862	.55687	.55513	.55339	.55165	.54991	.54818	.54645
.58	.54473	.54300	.54128	.53957	.53785	.53614	.53444	.53273	.53103	.52933
.59	.52763	.52594	.52425	.52256	.52088	.51919	.51751	.51584	.51416	.51249
0.60	−0.51083	.50916	.50750	.50584	.50418	.50253	.50088	.49923	.49758	.49594
.61	.49430	.49266	.49102	.48939	.48776	.48613	.48451	.48289	.48127	.47965
.62	.47804	.47642	.47482	.47321	.47160	.47000	.46840	.46681	.46522	.46362
.63	.46204	.46045	.45887	.45728	.45571	.45413	.45256	.45099	.44942	.44785
.64	.44629	.44473	.44317	.44161	.44006	.43850	.43696	.43541	.43386	.43232
.65	.43078	.42925	.42771	.42618	.42465	.42312	.42159	.42007	.41855	.41703
.66	.41552	.41400	.41249	.41098	.40947	.40797	.40647	.40497	.40347	.40197
.67	.40048	.39899	.39750	.39601	.39453	.39304	.39156	.39008	.38861	.38713
.68	.38566	.38419	.38273	.38126	.37980	.37834	.37688	.37542	.37397	.37251
.69	.37106	.36962	.36817	.36673	.36528	.36384	.36241	.36097	.35954	.35810
0.70	−0.35667	.35525	.35382	.35240	.35098	.34956	.34814	.34672	.34531	.34390
.71	.34249	.34108	.33968	.33827	.33687	.33547	.33408	.33268	.33129	.32989
.72	.32850	.32712	.32573	.32435	.32296	.32158	.32021	.31883	.31745	.31608
.73	.31471	.31334	.31197	.31061	.30925	.30788	.30653	.30517	.30381	.30246
.74	.30111	.29975	.29841	.29706	.29571	.29437	.29303	.29169	.29035	.28902
.75	.28768	.28635	.28502	.28369	.28236	.28104	.27971	.27839	.27707	.27575
.76	.27444	.27312	.27181	.27050	.26919	.26788	.26657	.26527	.26397	.26266
.77	.26136	.26007	.25877	.25748	.25618	.25489	.25360	.25231	.25103	.24974
.78	.24846	.24718	.24590	.24462	.24335	.24207	.24080	.23953	.23826	.23699
.79	.23572	.23446	.23319	.23193	.23067	.22941	.22816	.22690	.22565	.22439
0.80	−0.22314	.22189	.22065	.21940	.21816	.21691	.21567	.21433	.21319	.21196
.81	.21072	.20949	.20825	.20702	.20579	.20457	.20334	.20212	.20089	.19967
.82	.19845	.19723	.19601	.19480	.19358	.19237	.19116	.18995	.18874	.18754
.83	.18633	.18513	.18392	.18272	.18152	.18032	.17913	.17793	.17674	.17554
.84	.17435	.17316	.17198	.17079	.16960	.16842	.16724	.16605	.16487	.16370
.85	−0.16252	.16134	.16017	.15900	.15782	.15665	.15548	.15432	.15315	.15199
.86	.15082	.14966	.14850	.14734	.14618	.14503	.14387	.14272	.14156	.14041
.87	.13926	.13811	.13697	.13582	.13467	.13353	.13239	.13125	.13011	.12897
.88	.12783	.12670	.12556	.12443	.12330	.12217	.12104	.11991	.11878	.11766
.89	.11653	.11541	.11429	.11317	.11205	.11093	.10981	.10870	.10759	.10647
0.90	−0.10536	.10425	.10314	.10203	.10093	.09982	.09872	.09761	.09651	.09541
.91	.09431	.09321	.09212	.09102	.08992	.08883	.08744	.08665	.08556	.08447
.92	.08338	.08230	.08121	.08013	.07904	.07796	.07688	.07580	.07472	.07365
.93	.07257	.07150	.07042	.06935	.06828	.06721	.06614	.06507	.06401	.06294
.94	.06188	.06081	.05975	.05869	.05763	.05657	.05551	.05446	.05340	.05235
.95	.05129	.05024	.04919	.04814	.04709	.04604	.04500	.04395	.04291	.04186
.96	.04082	.03978	.03874	.03770	.03666	.03563	.03459	.03356	.03252	.03149
.97	.03046	.02943	.02840	.02737	.02634	.02532	.02429	.02327	.02225	.02122
.98	.02020	.01918	.01816	.01715	.01613	.01511	.01410	.01309	.01207	.01106
.99	.01005	.00904	.00803	.00702	.00602	.00501	.00401	.00300	.00200	.00100

To find the natural logarithm of a number which is $\frac{1}{10}$, $\frac{1}{100}$, $\frac{1}{1000}$, etc. of a number whose logarithm is given, subtract from the given logarithm $\log_e 10$, $2 \log_e 10$, $3 \log_e 10$, etc.

To find the natural logarithm of a number which is 10, 100, 1000, etc. times a number whose logarithm is given, add to the given logarithm $\log_e 10$, $2 \log_e 10$, $3 \log_e 10$, etc.

$\log_e 10$ = 2.30258 50930	$6 \log_e 10$ = 13.81551 05580
$2 \log_e 10$ = 4.60517 01860	$7 \log_e 10$ = 16.11809 56510
$3 \log_e 10$ = 6.90775 52790	$8 \log_e 10$ = 18.42068 07440
$4 \log_e 10$ = 9.21034 03720	$9 \log_e 10$ = 20.72326 58369
$5 \log_e 10$ = 11.51292 54650	$10 \log_e 10$ = 23.02585 09299

See preceding table for logarithms for numbers between 0.000 and 0.999.

1.00–4.99

N	0	1	2	3	4	5	6	7	8	9
1.0	0.00000	.00995	.01980	.02956	.03922	.04879	.05827	.06766	.07696	.08618
.1	.09531	.10436	.11333	.12222	.13103	.13976	.14842	.15700	.16551	.17395
.2	.18232	.19062	.19885	.20701	.21511	.22314	.23111	.23902	.24686	.25464
.3	.26236	.27003	.27763	.28518	.29267	.30010	.30748	.31481	.32208	.32930
.4	.33647	.34359	.35066	.35767	.36464	.37156	.37844	.38526	.39204	.39878
.5	.40547	.41211	.41871	.42527	.43178	.43825	.44469	.45108	.45742	.46373
.6	.47000	.47623	.48243	.48858	.49470	.50078	.50682	.51282	.51879	.52473
.7	.53063	.53649	.54232	.54812	.55389	.55962	.56531	.57098	.57661	.58222
.8	.58779	.59333	.59884	.60432	.60977	.61519	.62058	.62594	.63127	.63658
.9	.64185	.64710	.65233	.65752	.66269	.66783	.67294	.67803	.68310	.68813
2.0	0.69315	.69813	.70310	.70804	.71295	.71784	.72271	.72755	.73237	.73716
.1	.74194	.74669	.75142	.75612	.76081	.76547	.77011	.77473	.77932	.78390
.2	.78846	.79299	.79751	.80200	.80648	.81093	.81536	.81978	.82418	.82855
.3	.83291	.83725	.84157	.84587	.85015	.85442	.85866	.86289	.86710	.87129
.4	.87547	.87963	.88377	.88789	.89200	.89609	.90016	.90422	.90826	.91228
.5	.91629	.92028	.92426	.92822	.93216	.93609	.94001	.94391	.94779	.95166
.6	.95551	.95935	.96317	.96698	.97078	.97456	.97833	.98208	.98582	.98954
.7	.99325	.99695	*.00063	*.00430	*.00796	*.01160	*.01523	*.01885	*.02245	*.02604
.8	1.02962	.03318	.03674	.04028	.04380	.04732	.05082	.05431	.05779	.06126
.9	.06471	.06815	.07158	.07500	.07841	.08181	.08519	.08856	.09192	.09527
3.0	1.09861	.10194	.10526	.10856	.11186	.11514	.11841	.12168	.12493	.12817
.1	.13140	.13462	.13783	.14103	.14422	.14740	.15057	.15373	.15688	.16002
.2	.16315	.16627	.16938	.17248	.17557	.17865	.18173	.18479	.18784	.19089
.3	.19392	.19695	.19996	.20297	.20597	.20896	.21194	.21491	.21788	.22083
.4	.22378	.22671	.22964	.23256	.23547	.23837	.24127	.24415	.24703	.24990
.5	.25276	.25562	.25846	.26130	.26413	.26695	.26976	.27257	.27536	.27815
.6	.28093	.28371	.28647	.28923	.29198	.29473	.29746	.30019	.30291	.30563
.7	.30833	.31103	.31372	.31641	.31909	.32176	.32442	.32708	.32972	.33237
.8	.33500	.33763	.34025	.34286	.34547	.34807	.35067	.35325	.35584	.35841
.9	.36098	.36354	.36609	.36864	.37118	.37372	.37624	.37877	.38128	.38379
4.0	1.38629	.38879	.39128	.39377	.39624	.39872	.40118	.40364	.40610	.40854
.1	.41099	.41342	.41585	.41828	.42070	.42311	.42552	.42792	.43031	.43270
.2	.43508	.43746	.43984	.44220	.44456	.44692	.44927	.45161	.45395	.45629
.3	.45862	.46094	.46326	.46557	.46787	.47018	.47247	.47476	.47705	.47933
.4	.48160	.48387	.48614	.48840	.49065	.49290	.49515	.49739	.49962	.50185
.5	.50408	.50630	.50851	.51072	.51293	.51513	.51732	.51951	.52170	.52388
.6	.52606	.52823	.53039	.53256	.53471	.53687	.53902	.54116	.54330	.54543
.7	.54756	.54969	.55181	.55393	.55604	.55814	.56025	.56235	.56444	.56653
.8	.56862	.57070	.57277	.57485	.57691	.57898	.58104	.58309	.58515	.58719
.9	.58924	.59127	.59331	.59534	.59737	.59939	.60141	.60342	.60543	.60744

NATURAL OR NAPERIAN LOGARITHMS (Continued)

5.00–9.99

N	0	1	2	3	4	5	6	7	8	9
5.0	1.60944	.61144	.61343	.61542	.61741	.61939	.62137	.62334	.62531	.62728
.1	.62924	.63120	.63315	.63511	.63705	.63900	.64094	.64287	.64481	.64673
.2	.64866	.65058	.65250	.65441	.65632	.65823	.66013	.66203	.66393	.66582
.3	.66771	.66959	.67147	.67335	.67523	.67710	.67896	.68083	.68269	.68455
.4	.68640	.68825	.69010	.69194	.69378	.69562	.69745	.69928	.70111	.70293
.5	.70475	.70656	.70838	.71019	.71199	.71380	.71560	.71740	.71919	.72098
.6	.72277	.72455	.72633	.72811	.72988	.73166	.73342	.73519	.73695	.73871
.7	.74047	.74222	.74397	.74572	.74746	.74920	.75094	.75267	.75440	.75613
.8	.75786	.75958	.76130	.76302	.76473	.76644	.76815	.76985	.77156	.77326
.9	.77495	.77665	.77834	.78002	.78171	.78339	.78507	.78675	.78842	.79009
6.0	1.79176	.79342	.79509	.79675	.79840	.80006	.80171	.80336	.80500	.80665
.1	.80829	.80993	.81156	.81319	.81482	.81645	.81808	.81970	.82132	.82294
.2	.82455	.82616	.82777	.82938	.83098	.83258	.83418	.83578	.83737	.83896
.3	.84055	.84214	.84372	.84530	.84688	.84845	.85003	.85160	.85317	.85473
.4	.85630	.85786	.85942	.86097	.86253	.86408	.86563	.86718	.86872	.87026
.5	.87180	.87334	.87487	.87641	.87794	.87947	.88099	.88251	.88403	.88555
.6	.88707	.88858	.89010	.89160	.89311	.89462	.89612	.89762	.89912	.90061
.7	.90211	.90360	.90509	.90658	.90806	.90954	.91102	.91250	.91398	.91545
.8	.91692	.91839	.91986	.92132	.92279	.92425	.92571	.92716	.92862	.93007
.9	.93152	.93297	.93442	.93586	.93730	.93874	.94018	.94162	.94305	.94448
7.0	1.94591	.94734	.94876	.95019	.95161	.95303	.95445	.95586	.95727	.95869
.1	.96009	.96150	.96291	.96431	.96571	.96711	.96851	.96991	.97130	.97269
.2	.97408	.97547	.97685	.97824	.97962	.98100	.98238	.98376	.98513	.98650
.3	.98787	.98924	.99061	.99198	.99334	.99470	.99606	.99742	.99877	*.00013
.4	2.00148	.00283	.00418	.00553	.00687	.00821	.00956	.01089	.01223	.01357
.5	.01490	.01624	.01757	.01890	.02022	.02155	.02287	.02419	.02551	.02683
.6	.02815	.02946	.03078	.03209	.03340	.03471	.03601	.03732	.03862	.03992
.7	.04122	.04252	.04381	.04511	.04640	.04769	.04898	.05027	.05156	.05284
.8	.05412	.05540	.05668	.05796	.05924	.06051	.06179	.06306	.06433	.06560
.9	.06686	.06813	.06939	.07065	.07191	.07317	.07443	.07568	.07694	.07819
8.0	2.07944	.08069	.08194	.08318	.08443	.08567	.08691	.08815	.08939	.09063
.1	.09186	.09310	.09433	.09556	.09679	.09802	.09924	.10047	.10169	.10291
.2	.10413	.10535	.10657	.10779	.10900	.11021	.11142	.11263	.11384	.11505
.3	.11626	.11746	.11866	.11986	.12106	.12226	.12346	.12465	.12585	.12704
.4	.12823	.12942	.13061	.13180	.13298	.13417	.13535	.13653	.13771	.13889
.5	.14007	.14124	.14242	.14359	.14476	.14593	.14710	.14827	.14943	.15060
.6	.15176	.15292	.15409	.15524	.15640	.15756	.15871	.15987	.16102	.16217
.7	.16332	.16447	.16562	.16677	.16791	.16905	.17020	.17134	.17248	.17361
.8	.17475	.17589	.17702	.17816	.17929	.18042	.18155	.18267	.18380	.18493
.9	.18605	.18717	.18830	.18942	.19054	.19165	.19277	.19389	.19500	.19611
9.0	2.19722	.19834	.19944	.20055	.20166	.20276	.20387	.20497	.20607	.20717
.1	.20827	.20937	.21047	.21157	.21266	.21375	.21485	.21594	.21703	.21812
.2	.21920	.22029	.22138	.22246	.22354	.22462	.22570	.22678	.22786	.22894
.3	.23001	.23109	.23216	.23324	.23431	.23538	.23645	.23751	.23858	.23965
.4	.24071	.24177	.24284	.24390	.24496	.24601	.24707	.24813	.24918	.25024
.5	.25129	.25234	.25339	.25444	.25549	.25654	.25759	.25863	.25968	.26072
.6	.26176	.26280	.26384	.26488	.26592	.26696	.26799	.26903	.27006	.27109
.7	.27213	.27316	.27419	.27521	.27624	.27727	.27829	.27932	.28034	.28136
.8	.28238	.28340	.28442	.28544	.28646	.28747	.28849	.28950	.29051	.29152
.9	.29253	.29354	.29455	.29556	.29657	.29757	.29858	.29958	.30058	.30158

EXPONENTIAL FUNCTIONS

x	e^x	$\text{Log}_{10}(e^x)$	e^{-x}
0.00	1.0000	0.00000	1.000000
0.01	1.0101	.00434	0.990050
0.02	1.0202	.00869	.980199
0.03	1.0305	.01303	.970446
0.04	1.0408	.01737	.960789
0.05	1.0513	0.02171	0.951229
0.06	1.0618	.02606	.941765
0.07	1.0725	.03040	.932394
0.08	1.0833	.03474	.923116
0.09	1.0942	.03909	.913931
0.10	1.1052	0.04343	0.904837
0.11	1.1163	.04777	.895834
0.12	1.1275	.05212	.886920
0.13	1.1388	.05646	.878095
0.14	1.1503	.06080	.869358
0.15	1.1618	0.06514	0.860708
0.16	1.1735	.06949	.852144
0.17	1.1853	.07383	.843665
0.18	1.1972	.07817	.835270
0.19	1.2092	.08252	.826959
0.20	1.2214	0.08686	0.818731
0.21	1.2337	.09120	.810584
0.22	1.2461	.09554	.802519
0.23	1.2586	.09989	.794534
0.24	1.2712	.10423	.786628
0.25	1.2840	0.10857	0.778801
0.26	1.2969	.11292	.771052
0.27	1.3100	.11726	.763379
0.28	1.3231	.12160	.755784
0.29	1.3364	.12595	.748264
0.30	1.3499	0.13029	0.740818
0.31	1.3634	.13463	.733447
0.32	1.3771	.13897	.726149
0.33	1.3910	.14332	.718924
0.34	1.4049	.14766	.711770
0.35	1.4191	0.15200	0.704688
0.36	1.4333	.15635	.697676
0.37	1.4477	.16069	.690734
0.38	1.4623	.16503	.683861
0.39	1.4770	.16937	.677057
0.40	1.4918	0.17372	0.670320
0.41	1.5068	.17806	.663650
0.42	1.5220	.18240	.657047
0.43	1.5373	.18675	.650509
0.44	1.5527	.19109	.644036
0.45	1.5683	0.19543	0.637628
0.46	1.5841	.19978	.631284
0.47	1.6000	.20412	.625002
0.48	1.6161	.20846	.618783
0.49	1.6323	.21280	.612626
0.50	1.6487	0.21715	0.606531

x	e^x	$\text{Log}_{10}(e^x)$	e^{-x}
0.50	1.6487	0.21715	0.606531
0.51	1.6653	.22149	.600496
0.52	1.6820	.22583	.594521
0.53	1.6989	.23018	.588605
0.54	1.7160	.23452	.582748
0.55	1.7333	0.23886	0.576950
0.56	1.7507	.24320	.571209
0.57	1.7683	.24755	.565525
0.58	1.7860	.25189	.559898
0.59	1.8040	.25623	.554327
0.60	1.8221	0.26058	0.548812
0.61	1.8404	.26492	.543351
0.62	1.8589	.26926	.537944
0.63	1.8776	.27361	.532592
0.64	1.8965	.27795	.527292
0.65	1.9155	0.28229	0.522046
0.66	1.9348	.28663	.516851
0.67	1.9542	.29098	.511709
0.68	1.9739	.29532	.506617
0.69	1.9937	.29966	.501576
0.70	2.0138	0.30401	0.496585
0.71	2.0340	.30835	.491644
0.72	2.0544	.31269	.486752
0.73	2.0751	.31703	.481909
0.74	2.0959	.32138	.477114
0.75	2.1170	0.32572	0.472367
0.76	2.1383	.33006	.467666
0.77	2.1598	.33441	.463013
0.78	2.1815	.33875	.458406
0.79	2.2034	.34309	.453845
0.80	2.2255	0.34744	0.449329
0.81	2.2479	.35178	.444858
0.82	2.2705	.35612	.440432
0.83	2.2933	.36046	.436049
0.84	2.3164	.36481	.431711
0.85	2.3396	0.36915	0.427415
0.86	2.3632	.37349	.423162
0.87	2.3869	.37784	.418952
0.88	2.4109	.38218	.414783
0.89	2.4351	.38652	.410656
0.90	2.4596	0.39087	0.406570
0.91	2.4843	.39521	.402524
0.92	2.5093	.39955	.398519
0.93	2.5345	.40389	.394554
0.94	2.5600	.40824	.390628
0.95	2.5857	0.41258	0.386741
0.96	2.6117	.41692	.382893
0.97	2.6379	.42127	.379083
0.98	2.6645	.42561	.375311
0.99	2.6912	.42995	.371577
1.00	2.7183	0.43429	0.367879

EXPONENTIAL FUNCTIONS (Continued)

x	e^x	$\text{Log}_{10}(e^x)$	e^{-x}	x	e^x	$\text{Log}_{10}(e^x)$	e^{-x}
1.00	2.7183	0.43429	0.367879	**1.50**	4.4817	0.65144	0.223130
1.01	2.7456	.43864	.364219	1.51	4.5267	.65578	.220910
1.02	2.7732	.44298	.360595	1.52	4.5722	.66013	.218712
1.03	2.8011	.44732	.357007	1.53	4.6182	.66447	.216536
1.04	2.8292	.45167	.353455	1.54	4.6646	.66881	.214381
1.05	2.8577	0.45601	0.349938	**1.55**	4.7115	0.67316	0.212248
1.06	2.8864	.46035	.346456	1.56	4.7588	.67750	.210136
1.07	2.9154	.46470	.343009	1.57	4.8066	.68184	.208045
1.08	2.9447	.46904	.339596	1.58	4.8550	.68619	.205975
1.09	2.9743	.47338	.336216	1.59	4.9037	.69053	.203926
1.10	3.0042	0.47772	0.332871	**1.60**	4.9530	0.69487	0.201897
1.11	3.0344	.48207	.329559	1.61	5.0028	.69921	.199888
1.12	3.0649	.48641	.326280	1.62	5.0531	.70356	.197899
1.13	3.0957	.49075	.323033	1.63	5.1039	.70790	.195930
1.14	3.1268	.49510	.319819	1.64	5.1552	.71224	.193980
1.15	3.1582	0.49944	0.316637	**1.65**	5.2070	0.71659	0.192050
1.16	3.1899	.50378	.313486	1.66	5.2593	.72093	.190139
1.17	3.2220	.50812	.310367	1.67	5.3122	.72527	.188247
1.18	3.2544	.51247	.307279	1.68	5.3656	.72961	.186374
1.19	3.2871	.51681	.304221	1.69	5.4195	.73396	.184520
1.20	3.3201	0.52115	0.301194	**1.70**	5.4739	0.73830	0.182684
1.21	3.3535	.52550	.298197	1.71	5.5290	.74264	.180866
1.22	3.3872	.52984	.295230	1.72	5.5845	.74699	.179066
1.23	3.4212	.53418	.292293	1.73	5.6407	.75133	.177284
1.24	3.4556	.53853	.289384	1.74	5.6973	.75567	.175520
1.25	3.4903	0.54287	0.286505	**1.75**	5.7546	0.76002	0.173774
1.26	3.5254	.54721	.283654	1.76	5.8124	.76436	.172045
1.27	3.5609	.55155	.280832	1.77	5.8709	.76870	.170333
1.28	3.5966	.55590	.278037	1.78	5.9299	.77304	.168638
1.29	3.6328	.56024	.275271	1.79	5.9895	.77739	.166960
1.30	3.6693	0.56458	0.272532	**1.80**	6.0496	0.78173	0.165299
1.31	3.7062	.56893	.269820	1.81	6.1104	.78607	.163654
1.32	3.7434	.57327	.267135	1.82	6.1719	.79042	.162026
1.33	3.7810	.57761	.264477	1.83	6.2339	.79476	.160414
1.34	3.8190	.58195	.261846	1.84	6.2965	.79910	.158817
1.35	3.8574	0.58630	0.259240	**1.85**	6.3598	0.80344	0.157237
1.36	3.8962	.59064	.256661	1.86	6.4237	.80779	.155673
1.37	3.9354	.59498	.254107	1.87	6.4883	.81213	.154124
1.38	3.9749	.59933	.251579	1.88	6.5535	.81647	.152590
1.39	4.0149	.60367	.249075	1.89	6.6194	.82082	.151072
1.40	4.0552	0.60801	0.246597	**1.90**	6.6859	0.82516	0.149569
1.41	4.0960	.61236	.244143	1.91	6.7531	.82950	.148080
1.42	4.1371	.61670	.241714	1.92	6.8210	.83385	.146607
1.43	4.1787	.62104	.239309	1.93	6.8895	.83819	.145148
1.44	4.2207	.62538	.236928	1.94	6.9588	.84253	.143704
1.45	4.2631	0.62973	0.234570	**1.95**	7.0287	0.84687	0.142274
1.46	4.3060	.63407	.232236	1.96	7.0993	.85122	.140858
1.47	4.3492	.63841	.229925	1.97	7.1707	.85556	.139457
1.48	4.3929	.64276	.227638	1.98	7.2427	.85990	.138069
1.49	4.4371	.64710	.225373	1.99	7.3155	.86425	.136695
1.50	4.4817	0.65144	0.223130	**2.00**	7.3891	0.86859	0.135335

EXPONENTIAL FUNCTIONS (Continued)

x	e^x	$\text{Log}_{10}(e^x)$	e^{-x}
2.00	7.3891	0.86859	0.135335
2.01	7.4633	.87293	.133989
2.02	7.5383	.87727	.132655
2.03	7.6141	.88162	.131336
2.04	7.6906	.88596	.130029
2.05	7.7679	0.89030	0.128735
2.06	7.8460	.89465	.127454
2.07	7.9248	.89899	.126186
2.08	8.0045	.90333	.124930
2.09	8.0849	.90768	.123687
2.10	8.1662	0.91202	0.122456
2.11	8.2482	.91636	.121238
2.12	8.3311	.92070	.120032
2.13	8.4149	.92505	.118837
2.14	8.4994	.92939	.117655
2.15	8.5849	0.93373	0.116484
2.16	8.6711	.93808	.115325
2.17	8.7583	.94242	.114178
2.18	8.8463	.94676	.113042
2.19	8.9352	.95110	.111917
2.20	9.0250	0.95545	0.110803
2.21	9.1157	.95979	.109701
2.22	9.2073	.96413	.108609
2.23	9.2999	.96848	.107528
2.24	9.3933	.97282	.106459
2.25	9.4877	0.97716	0.105399
2.26	9.5831	.98151	.104350
2.27	9.6794	.98585	.103312
2.28	9.7767	.99019	.102284
2.29	9.8749	.99453	.101266
2.30	9.9742	0.99888	0.100259
2.31	10.074	1.00322	.099261
2.32	10.176	1.00756	.098274
2.33	10.278	1.01191	.097296
2.34	10.381	1.01625	.096328
2.35	10.486	1.02059	0.095369
2.36	10.591	1.02493	.094420
2.37	10.697	1.02928	.093481
2.38	10.805	1.03362	.092551
2.39	10.913	1.03796	.091630
2.40	11.023	1.04231	0.090718
2.41	11.134	1.04665	.089815
2.42	11.246	1.05099	.088922
2.43	11.359	1.05534	.088037
2.44	11.473	1.05968	.087161
2.45	11.588	1.06402	0.086294
2.46	11.705	1.06836	.085435
2.47	11.822	1.07271	.084585
2.48	11.941	1.07705	.083743
2.49	12.061	1.08139	.082910
2.50	12.182	1.08574	0.082085

x	e^x	$\text{Log}_{10}(e^x)$	e^{-x}
2.50	12.182	1.08574	0.082085
2.51	12.305	1.09008	.081268
2.52	12.429	1.09442	.080460
2.53	12.554	1.09877	.079659
2.54	12.680	1.10311	.078866
2.55	12.807	1.10745	0.078082
2.56	12.936	1.11179	.077305
2.57	13.066	1.11614	.076536
2.58	13.197	1.12048	.075774
2.59	13.330	1.12482	.075020
2.60	13.464	1.12917	0.074274
2.61	13.599	1.13351	.073535
2.62	13.736	1.13785	.072803
2.63	13.874	1.14219	.072078
2.64	14.013	1.14654	.071361
2.65	14.154	1.15088	0.070651
2.66	14.296	1.15522	.069948
2.67	14.440	1.15957	.069252
2.68	14.585	1.16391	.068563
2.69	14.732	1.16825	.067881
2.70	14.880	1.17260	0.067206
2.71	15.029	1.17694	.066537
2.72	15.180	1.18128	.065875
2.73	15.333	1.18562	.065219
2.74	15.487	1.18997	.064570
2.75	15.643	1.19431	0.063928
2.76	15.800	1.19865	.063292
2.77	15.959	1.20300	.062662
2.78	16.119	1.20734	.062039
2.79	16.281	1.21168	.061421
2.80	16.445	1.21602	0.060810
2.81	16.610	1.22037	.060205
2.82	16.777	1.22471	.059606
2.83	16.945	1.22905	.059013
2.84	17.116	1.23340	.058426
2.85	17.288	1.23774	0.057844
2.86	17.462	1.24208	.057269
2.87	17.637	1.24643	.056699
2.88	17.814	1.25077	.056135
2.89	17.993	1.25511	.055576
2.90	18.174	1.25945	0.055023
2.91	18.357	1.26380	.054476
2.92	18.541	1.26814	.053934
2.93	18.728	1.27248	.053397
2.94	18.916	1.27683	.052866
2.95	19.106	1.28117	0.052340
2.96	19.298	1.28551	.051819
2.97	19.492	1.28985	.051303
2.98	19.688	1.29420	.050793
2.99	19.886	1.29854	.050287
3.00	20.086	1.30288	0.049787

EXPONENTIAL FUNCTIONS (Continued)

x	e^x	$\text{Log}_{10}(e^x)$	e^{-x}
3.00	20.086	1.30288	0.049787
3.01	20.287	1.30723	.049292
3.02	20.491	1.31157	.048801
3.03	20.697	1.31591	.048316
3.04	20.905	1.32026	.047835
3.05	21.115	1.32460	0.047359
3.06	21.328	1.32894	.046888
3.07	21.542	1.33328	.046421
3.08	21.758	1.33763	.045959
3.09	21.977	1.34197	.045502
3.10	22.198	1.34631	0.045049
3.11	22.421	1.35066	.044601
3.12	22.646	1.35500	.044157
3.13	22.874	1.35934	.043718
3.14	23.104	1.36368	.043283
3.15	23.336	1.36803	0.042852
3.16	23.571	1.37237	.042426
3.17	23.807	1.37671	.042004
3.18	24.047	1.38106	.041586
3.19	24.288	1.38540	.041172
3.20	24.533	1.38974	0.040762
3.21	24.779	1.39409	.040357
3.22	25.028	1.39843	.039955
3.23	25.280	1.40277	.039557
3.24	25.534	1.40711	.039164
3.25	25.790	1.41146	0.038774
3.26	26.050	1.41580	.038388
3.27	26.311	1.42014	.038006
3.28	26.576	1.42449	.037628
3.29	26.843	1.42883	.037254
3.30	27.113	1.43317	0.036883
3.31	27.385	1.43751	.036516
3.32	27.660	1.44186	.036153
3.33	27.938	1.44620	.035793
3.34	28.219	1.45054	.035437
3.35	28.503	1.45489	0.035084
3.36	28.789	1.45923	.034735
3.37	29.079	1.46357	.034390
3.38	29.371	1.46792	.034047
3.39	29.666	1.47226	.033709
3.40	29.964	1.47660	0.033373
3.41	30.265	1.48094	.033041
3.42	30.569	1.48529	.032712
3.43	30.877	1.48963	.032387
3.44	31.187	1.49397	.032065
3.45	31.500	1.49832	0.031746
3.46	31.817	1.50266	.031430
3.47	32.137	1.50700	.031117
3.48	32.460	1.51134	.030807
3.49	32.786	1.51569	.030501
3.50	33.115	1.52003	0.030197
3.50	33.115	1.52003	0.030197
3.51	33.448	1.52437	.029897
3.52	33.784	1.52872	.029599
3.53	34.124	1.53306	.029305
3.54	34.467	1.53740	.029013
3.55	34.813	1.54175	0.028725
3.56	35.163	1.54609	.028439
3.57	35.517	1.55043	.028156
3.58	35.874	1.55477	.027876
3.59	36.234	1.55912	.027598
3.60	36.598	1.56346	0.027324
3.61	36.966	1.56780	.027052
3.62	37.338	1.57215	.026783
3.63	37.713	1.57649	.026516
3.64	38.092	1.58083	.026252
3.65	38.475	1.58517	0.025991
3.66	38.861	1.58952	.025733
3.67	39.252	1.59386	.025476
3.68	39.646	1.59820	.025223
3.69	40.045	1.60255	.024972
3.70	40.447	1.60689	0.024724
3.71	40.854	1.61123	.024478
3.72	41.264	1.61558	.024234
3.73	41.679	1.61992	.023993
3.74	42.098	1.62426	.023754
3.75	42.521	1.62860	0.023518
3.76	42.948	1.63295	.023284
3.77	43.380	1.63729	.023052
3.78	43.816	1.64163	.022823
3.79	44.256	1.64598	.022596
3.80	44.701	1.65032	0.022371
3.81	45.150	1.65466	.022148
3.82	45.604	1.65900	.021928
3.83	46.063	1.66335	.021710
3.84	46.525	1.66769	.021494
3.85	46.993	1.67203	0.021280
3.86	47.465	1.67638	.021068
3.87	47.942	1.68072	.020858
3.88	48.424	1.68506	.020651
3.89	48.911	1.68941	.020445
3.90	49.402	1.69375	0.020242
3.91	49.899	1.69809	.020041
3.92	50.400	1.70243	.019841
3.93	50.907	1.70678	.019644
3.94	51.419	1.71112	.019448
3.95	51.935	1.71546	0.019255
3.96	52.457	1.71981	.019063
3.97	52.985	1.72415	.018873
3.98	53.517	1.72849	.018686
3.99	54.055	1.73283	.018500
4.00	54.598	1.73718	0.018316

EXPONENTIAL FUNCTIONS (Continued)

x	e^x	$\text{Log}_{10}(e^x)$	e^{-x}	x	e^x	$\text{Log}_{10}(e^x)$	e^{-x}
4.00	54.598	1.73718	0.018316	**4.50**	90.017	1.95433	0.011109
4.01	55.147	1.74152	.018133	4.51	90.922	1.95867	.010998
4.02	55.701	1.74586	.017953	4.52	91.836	1.96301	.010889
4.03	56.261	1.75021	.017774	4.53	92.759	1.96735	.010781
4.04	56.826	1.75455	.017597	4.54	93.691	1.97170	.010673
4.05	57.397	1.75889	0.017422	**4.55**	94.632	1.97604	0.010567
4.06	57.974	1.76324	.017249	4.56	95.583	1.98038	.010462
4.07	58.557	1.76758	.017077	4.57	96.544	1.98473	.010358
4.08	59.145	1.77192	.016907	4.58	97.514	1.98907	.010255
4.09	59.740	1.77626	.016739	4.59	98.494	1.99341	.010153
4.10	60.340	1.78061	0.016573	**4.60**	99.484	1.99775	0.010052
4.11	60.947	1.78495	.016408	4.61	100.48	2.00210	.009952
4.12	61.559	1.78929	.016245	4.62	101.49	2.00644	.009853
4.13	62.178	1.79364	.016083	4.63	102.51	2.01078	.009755
4.14	62.803	1.79798	.015923	4.64	103.54	2.01513	.009658
4.15	63.434	1.80232	0.015764	**4.65**	104.58	2.01947	0.009562
4.16	64.072	1.80667	.015608	4.66	105.64	2.02381	.009466
4.17	64.715	1.81101	.015452	4.67	106.70	2.02816	.009372
4.18	65.366	1.81535	.015299	4.68	107.77	2.03250	.009279
4.19	66.023	1.81969	.015146	4.69	108.85	2.03684	.009187
4.20	66.686	1.82404	0.014996	**4.70**	109.95	2.04118	0.009095
4.21	67.357	1.82838	.014846	4.71	111.05	2.04553	.009005
4.22	68.033	1.83272	.014699	4.72	112.17	2.04987	.008915
4.23	68.717	1.83707	.014552	4.73	113.30	2.05421	.008826
4.24	69.408	1.84141	.014408	4.74	114.43	2.05856	.008739
4.25	70.105	1.84575	0.014264	**4.75**	115.58	2.06290	0.008652
4.26	70.810	1.85009	.014122	4.76	116.75	2.06724	.008566
4.27	71.522	1.85444	.013982	4.77	117.92	2.07158	.008480
4.28	72.240	1.85878	.013843	4.78	119.10	2.07593	.008396
4.29	72.966	1.86312	.013705	4.79	120.30	2.08027	.008312
4.30	73.700	1.86747	0.013569	**4.80**	121.51	2.08461	0.008230
4.31	74.440	1.87181	.013434	4.81	122.73	2.08896	.008148
4.32	75.189	1.87615	.013300	4.82	123.97	2.09330	.008067
4.33	75.944	1.88050	.013168	4.83	125.21	2.09764	.007987
4.34	76.708	1.88484	.013037	4.84	126.47	2.10199	.007907
4.35	77.478	1.88918	0.012907	**4.85**	127.74	2.10633	0.007828
4.36	78.257	1.89352	.012778	4.86	129.02	2.11067	.007750
4.37	79.044	1.89787	.012651	4.87	130.32	2.11501	.007673
4.38	79.838	1.90221	.012525	4.88	131.63	2.11936	.007597
4.39	80.640	1.90655	.012401	4.89	132.95	2.12370	.007521
4.40	81.451	1.91090	0.012277	**4.90**	134.29	2.12804	0.007447
4.41	82.269	1.91524	.012155	4.91	135.64	2.13239	.007372
4.42	83.096	1.91958	.012034	4.92	137.00	2.13673	.007299
4.43	83.931	1.92392	.011914	4.93	138.38	2.14107	.007227
4.44	84.775	1.92827	.011796	4.94	139.77	2.14541	.007155
4.45	85.627	1.93261	0.011679	**4.95**	141.17	2.14976	0.007083
4.46	86.488	1.93695	.011562	4.96	142.59	2.15410	.007013
4.47	87.357	1.94130	.011447	4.97	144.03	2.15844	.006943
4.48	88.235	1.94564	.011333	4.98	145.47	2.16279	.006874
4.49	89.121	1.94998	.011221	4.99	146.94	2.16713	.006806
4.50	90.017	1.95433	0.011109	**5.00**	148.41	2.17147	0.006738

EXPONENTIAL FUNCTIONS (Continued)

x	e^x	$\text{Log}_{10}(e^x)$	e^{-x}
5.00	148.41	2.17147	0.006738
5.01	149.90	2.17582	.006671
5.02	151.41	2.18016	.006605
5.03	152.93	2.18450	.006539
5.04	154.47	2.18884	.006474
5.05	156.02	2.19319	0.006409
5.06	157.59	2.19753	.006346
5.07	159.17	2.20187	.006282
5.08	160.77	2.20622	.006220
5.09	162.39	2.21056	.006158
5.10	164.02	2.21490	0.006097
5.11	165.67	2.21924	.006036
5.12	167.34	2.22359	.005976
5.13	169.02	2.22793	.005917
5.14	170.72	2.23227	.005858
5.15	172.43	2.23662	0.005799
5.16	174.16	2.24096	.005742
5.17	175.91	2.24530	.005685
5.18	177.68	2.24965	.005628
5.19	179.47	2.25399	.005572
5.20	181.27	2.25833	0.005517
5.21	183.09	2.26267	.005462
5.22	184.93	2.26702	.005407
5.23	186.79	2.27136	.005354
5.24	188.67	2.27570	.005300
5.25	190.57	2.28005	0.005248
5.26	192.48	2.28439	.005195
5.27	194.42	2.28873	.005144
5.28	196.37	2.29307	.005092
5.29	198.34	2.29742	.005042
5.30	200.34	2.30176	0.004992
5.31	202.35	2.30610	.004942
5.32	204.38	2.31045	.004893
5.33	206.44	2.31479	.004844
5.34	208.51	2.31913	.004796
5.35	210.61	2.32348	0.004748
5.36	212.72	2.32782	.004701
5.37	214.86	2.33216	.004654
5.38	217.02	2.33650	.004608
5.39	219.20	2.34085	.004562
5.40	221.41	2.34519	0.004517
5.41	223.63	2.34953	.004472
5.42	225.88	2.35388	.004427
5.43	228.15	2.35822	.004383
5.44	230.44	2.36256	.004339
5.45	232.76	2.36690	0.004296
5.46	235.10	2.37125	.004254
5.47	237.46	2.37559	.004211
5.48	239.85	2.37993	.004169
5.49	242.26	2.38428	.004128
5.50	244.69	2.38862	0.004087

x	e^x	$\text{Log}_{10}(e^x)$	e^{-x}
5.50	244.69	2.38862	0.0040868
5.55	257.24	2.41033	.0038875
5.60	270.43	2.43205	.0036979
5.65	284.29	2.45376	.0035175
5.70	298.87	2.47548	.0033460
5.75	314.19	2.49719	0.0031828
5.80	330.30	2.51891	.0030276
5.85	347.23	2.54062	.0028799
5.90	365.04	2.56234	.0027394
5.95	383.75	2.58405	.0026058
6.00	403.43	2.60577	0.0024788
6.05	424.11	2.62748	.0023579
6.10	445.86	2.64920	.0022429
6.15	468.72	2.67091	.0021335
6.20	492.75	2.69263	.0020294
6.25	518.01	2.71434	0.0019305
6.30	544.57	2.73606	.0018363
6.35	572.49	2.75777	.0017467
6.40	601.85	2.77948	.0016616
6.45	632.70	2.80120	.0015805
6.50	665.14	2.82291	0.0015034
6.55	699.24	2.84463	.0014301
6.60	735.10	2.86634	.0013604
6.65	772.78	2.88806	.0012940
6.70	812.41	2.90977	.0012309
6.75	854.06	2.93149	0.0011709
6.80	897.85	2.95320	.0011138
6.85	943.88	2.97492	.0010595
6.90	992.27	2.99663	.0010078
6.95	1043.1	3.01835	.0009586
7.00	1096.6	3.04006	0.0009119
7.05	1152.9	3.06178	.0008674
7.10	1212.0	3.08349	.0008251
7.15	1274.1	3.10521	.0007849
7.20	1339.4	3.12692	.0007466
7.25	1408.1	3.14863	0.0007102
7.30	1480.3	3.17035	.0006755
7.35	1556.2	3.19206	.0006426
7.40	1636.0	3.21378	.0006113
7.45	1719.9	3.23549	.0005814
7.50	1808.0	3.25721	0.0005531
7.55	1900.7	3.27892	.0005261
7.60	1998.2	3.30064	.0005005
7.65	2100.6	3.32235	.0004760
7.70	2208.3	3.34407	.0004528
7.75	2321.6	3.36578	0.0004307
7.80	2440.6	3.38750	.0004097
7.85	2565.7	3.40921	.0003898
7.90	2697.3	3.43093	.0003707
7.95	2835.6	3.45264	.0003527
8.00	2981.0	3.47436	0.0003355

EXPONENTIAL FUNCTIONS (Continued)

x	e^x	$\text{Log}_{10}(e^x)$	e^{-x}
8.00	2981.0	3.47436	0.0003355
8.05	3133.8	3.49607	.0003191
8.10	3294.5	3.51779	.0003035
8.15	3463.4	3.53950	.0002887
8.20	3641.0	3.56121	.0002747
8.25	3827.6	3.58293	0.0002613
8.30	4023.9	3.60464	.0002485
8.35	4230.2	3.62636	.0002364
8.40	4447.1	3.64807	.0002249
8.45	4675.1	3.66979	.0002139
8.50	4914.8	3.69150	0.0002035
8.55	5166.8	3.71322	.0001935
8.60	5431.7	3.73493	.0001841
8.65	5710.1	3.75665	.0001751
8.70	6002.9	3.77836	.0001666
8.75	6310.7	3.80008	0.0001585
8.80	6634.2	3.82179	.0001507
8.85	6974.4	3.84351	.0001434
8.90	7332.0	3.86522	.0001364
8.95	7707.9	3.88694	.0001297
9.00	8103.1	3.90865	0.0001234
9.05	8518.5	3.93037	.0001174
9.10	8955.3	3.95208	.0001117
9.15	9414.4	3.97379	.0001062
9.20	9897.1	3.99551	.0001010
9.25	10405	4.01722	0.0000961
9.30	10938	4.03894	.0000914
9.35	11499	4.06065	.0000870
9.40	12088	4.08237	.0000827
9.45	12708	4.10408	.0000787
9.50	13360	4.12580	0.0000749
9.55	14045	4.14751	.0000712
9.60	14765	4.16923	.0000677
9.65	15522	4.19094	.0000644
9.70	16318	4.21266	.0000613
9.75	17154	4.23437	0.0000583
9.80	18034	4.25609	.0000555
9.85	18958	4.27780	.0000527
9.90	19930	4.29952	.0000502
9.95	20952	4.32123	0.0000477
10.00	22026	4.34294	0.0000454

HYPERBOLIC FUNCTIONS AND THEIR COMMON LOGARITHMS

The logarithms given below show the mantissa only. The proper characteristic must be added.

x	Sinh x		Cosh x		Tanh x		Coth x	
	Value	$\log_{10}$	Value	$\log_{10}$	Value	$\log_{10}$	Value	$\log_{10}$
0.00	0.00000	$-\infty$	1.00000	.00000	0.00000	$-\infty$	∞	∞
0.01	.01000	.00001	1.00005	.00002	.01000	.99999	100.003	.00001
0.02	.02000	.30106	1.00020	.00009	.02000	.30097	50.007	.69903
0.03	.03000	.47719	1.00045	.00020	.02999	.47699	33.343	.52301
0.04	.04001	.60218	1.00080	.00035	.03998	.60183	25.013	.39817
0.05	0.05002	.69915	1.00125	.00054	0.04996	.69861	20.017	.30139
0.06	.06004	.77841	1.00180	.00078	.05993	.77763	16.687	.22237
0.07	.07006	.84545	1.00245	.00106	.06989	.84439	14.309	.15561
0.08	.08009	.90355	1.00320	.00139	.07983	.90216	12.527	.09784
0.09	.09012	.95483	1.00405	.00176	.08976	.95307	11.141	.04693
0.10	0.10017	.00072	1.00500	.00217	0.09967	.99856	10.0333	.00144
0.11	.11022	.04227	1.00606	.00262	.10956	.03965	9.1275	.96035
0.12	.12029	.08022	1.00721	.00312	.11943	.07710	8.3733	.92290
0.13	.13037	.11517	1.00846	.00366	.12927	.11151	7.7356	.88849
0.14	.14046	.14755	1.00982	.00424	.13909	.14330	7.1895	.85670
0.15	0.15056	.17772	1.01127	.00487	0.14889	.17285	6.7166	.82715
0.16	.16068	.20597	1.01283	.00554	.15865	.20044	6.3032	.79956
0.17	.17082	.23254	1.01448	.00625	.16838	.22629	5.9389	.77371
0.18	.18097	.25762	1.01624	.00700	.17808	.25062	5.6154	.74938
0.19	.19115	.28136	1.01810	.00779	.18775	.27357	5.3263	.72643
0.20	0.20134	.30392	1.02007	.00863	0.19738	.29529	5.0665	.70471
0.21	.21155	.32541	1.02213	.00951	.20697	.31590	4.8317	.68410
0.22	.22178	.34592	1.02430	.01043	.21652	.33549	4.6186	.66451
0.23	.23203	.36555	1.02657	.01139	.22603	.35416	4.4242	.64584
0.24	.24231	.38437	1.02894	.01239	.23550	.37198	4.2464	.62802
0.25	0.25261	.40245	1.03141	.01343	0.24492	.38902	4.0830	.61098
0.26	.26294	.41986	1.03399	.01452	.25430	.40534	3.9324	.59466
0.27	.27329	.43663	1.03667	.01564	.26362	.42099	3.7933	.57901
0.28	.28367	.45282	1.03946	.01681	.27291	.43601	3.6643	.56399
0.29	.29408	.46847	1.04235	.01801	.28213	.45046	3.5444	.54954
0.30	0.30452	.48362	1.04534	.01926	0.29131	.46436	3.4327	.53564
0.31	.31499	.49830	1.04844	.02054	.30044	.47775	3.3285	.52225
0.32	.32549	.51254	1.05164	.02187	.30951	.49067	3.2309	.50933
0.33	.33602	.52637	1.05495	.02323	.31852	.50314	3.1395	.49686
0.34	.34659	.53981	1.05836	.02463	.32748	.51518	3.0536	.48482
0.35	0.35719	.55290	1.06188	.02607	0.33638	.52682	2.9729	.47318
0.36	.36783	.56564	1.06550	.02755	.34521	.53809	2.8968	.46191
0.37	.37850	.57807	1.06923	.02907	.35399	.54899	2.8249	.45101
0.38	.38921	.59019	1.07307	.03063	.36271	.55956	2.7570	.44044
0.39	.39996	.60202	1.07702	.03222	.37136	.56980	2.6928	.43020
0.40	0.41075	.61358	1.08107	.03385	0.37995	.57973	2.6319	.42027
0.41	.42158	.62488	1.08523	.03552	.38847	.58936	2.5742	.41064
0.42	.43246	.63594	1.08950	.03723	.39693	.59871	2.5193	.40129
0.43	.44337	.64677	1.09388	.03897	.40532	.60780	2.4672	.39220
0.44	.45434	.65738	1.09837	.04075	.41364	.61663	2.4175	.38337
0.45	0.46534	.66777	1.10297	.04256	0.42190	.62521	2.3702	.37479
0.46	.47640	.67797	1.10768	.04441	.43008	.63355	2.3251	.36645
0.47	.48750	.68797	1.11250	.04630	.43820	.64167	2.2821	.35833
0.48	.49865	.69779	1.11743	.04822	.44624	.64957	2.2409	.35043
0.49	.50984	.70744	1.12247	.05018	.45422	.65726	2.2016	.34274

HYPERBOLIC FUNCTIONS AND THEIR COMMON LOGARITHMS (Continued)

x	Sinh x Value	Sinh x $\log_{10}$	Cosh x Value	Cosh x $\log_{10}$	Tanh x Value	Tanh x $\log_{10}$	Coth x Value	Coth x $\log_{10}$
0.50	0.52110	.71692	1.12763	.05217	0.46212	.66475	2.1640	.33525
0.51	.53240	.72624	1.13289	.05419	.46995	.67205	2.1279	.32795
0.52	.54375	.73540	1.13827	.05625	.47770	.67916	2.0934	.32084
0.53	.55516	.74442	1.14377	.05834	.48538	.68608	2.0602	.31392
0.54	.56663	.75330	1.14938	.06046	.49299	.69284	2.0284	.30716
0.55	0.57815	.76204	1.15510	.06262	0.50052	.69942	1.9979	.30058
0.56	.58973	.77065	1.16094	.06481	.50798	.70584	1.9686	.29416
0.57	.60137	.77914	1.16690	.06703	.51536	.71211	1.9404	.28789
0.58	.61307	.78751	1.17297	.06929	.52267	.71822	1.9133	.28178
0.59	.62483	.79576	1.17916	.07157	.52990	.72419	1.8872	.27581
0.60	0.63665	.80390	1.18547	.07389	0.53705	.73001	1.8620	.26999
0.61	.64854	.81194	1.19189	.07624	.54413	.73570	1.8378	.26430
0.62	.66049	.81987	1.19844	.07861	.55113	.74125	1.8145	.25875
0.63	.67251	.82770	1.20510	.08102	.55805	.74667	1.7919	.25333
0.64	.68459	.83543	1.21189	.08346	.56490	.75197	1.7702	.24803
0.65	0.69675	.84308	1.21879	.08593	0.57167	.75715	1.7493	.24285
0.66	.70897	.85063	1.22582	.08843	.57836	.76220	1.7290	.23780
0.67	.72126	.85809	1.23297	.09095	.58498	.76714	1.7095	.23286
0.68	.73363	.86548	1.24025	.09351	.59152	.77197	1.6906	.22803
0.69	.74607	.87278	1.24765	.09609	.59798	.77669	1.6723	.22331
0.70	0.75858	.88000	1.25517	.09870	0.60437	.78130	1.6546	.21870
0.71	.77117	.88715	1.26282	.10134	.61068	.78581	1.6375	.21419
0.72	.78384	.89423	1.27059	.10401	.61691	.79022	1.6210	.20978
0.73	.79659	.90123	1.27849	.10670	.62307	.79453	1.6050	.20547
0.74	.80941	.90817	1.28652	.10942	.62915	.79875	1.5895	.20125
0.75	0.82232	.91504	1.29468	.11216	0.63515	.80288	1.5744	.19712
0.76	.83530	.92185	1.30297	.11493	.64108	.80691	1.5599	.19309
0.77	.84838	.92859	1.31139	.11773	.64693	.81086	1.5458	.18914
0.78	.86153	.93527	1.31994	.12055	.65271	.81472	1.5321	.18528
0.79	.87478	.94190	1.32862	.12340	.65841	.81850	1.5188	.18150
0.80	0.88811	.94846	1.33743	.12627	0.66404	.82219	1.5059	.17781
0.81	.90152	.95498	1.34638	.12917	.66959	.82581	1.4935	.17419
0.82	.91503	.96144	1.35547	.13209	.67507	.82935	1.4813	.17065
0.83	.92863	.96784	1.36468	.13503	.68048	.83281	1.4696	.16719
0.84	.94233	.97420	1.37404	.13800	.68581	.83620	1.4581	.16380
0.85	0.95612	.98051	1.38353	.14099	0.69107	.83952	1.4470	.16048
0.86	.97000	.98677	1.39316	.14400	.69626	.84277	1.4362	.15723
0.87	.98398	.99299	1.40293	.14704	.70137	.84595	1.4258	.15405
0.88	.99806	.99916	1.41284	.15009	.70642	.84906	1.4156	.15094
0.89	1.01224	.00528	1.42289	.15317	.71139	.85211	1.4057	.14789
0.90	1.02652	.01137	1.43309	.15627	0.71630	.85509	1.3961	.14491
0.91	1.04090	.01741	1.44342	.15939	.72113	.85801	1.3867	.14199
0.92	1.05539	.02341	1.45390	.16254	.72590	.86088	1.3776	.13912
0.93	1.06998	.02937	1.46453	.16570	.73059	.86368	1.3687	.13632
0.94	1.08468	.03530	1.47530	.16888	.73522	.86642	1.3601	.13358
0.95	1.09948	.04119	1.48623	.17208	0.73978	.86910	1.3517	.13090
0.96	1.11440	.04704	1.49729	.17531	.74428	.87173	1.3436	.12827
0.97	1.12943	.05286	1.50851	.17855	.74870	.87431	1.3356	.12569
0.98	1.14457	.05864	1.51988	.18181	.75307	.87683	1.3279	.12317
0.99	1.15983	.06439	1.53141	.18509	.75736	.87930	1.3204	.12070

HYPERBOLIC FUNCTIONS AND THEIR COMMON LOGARITHMS (Continued)

x	Sinh x Value	Sinh x $\log_{10}$	Cosh x Value	Cosh x $\log_{10}$	Tanh x Value	Tanh x $\log_{10}$	Coth x Value	Coth x $\log_{10}$
1.00	1.17520	.07011	1.54308	.18839	0.76159	.88172	1.3130	.11828
1.01	1.19069	.07580	1.55491	.19171	.76576	.88409	1.3059	.11591
1.02	1.20630	.08146	1.56689	.19504	.76987	.88642	1.2989	.11358
1.03	1.22203	.08708	1.57904	.19839	.77391	.88869	1.2921	.11131
1.04	1.23788	.09268	1.59134	.20176	.77789	.89092	1.2855	.10908
1.05	1.25386	.09825	1.60379	.20515	0.78181	.89310	1.2791	.10690
1.06	1.26996	.10379	1.61641	.20855	.78566	.89524	1.2728	.10476
1.07	1.28619	.10930	1.62919	.21197	.78946	.89733	1.2667	.10267
1.08	1.30254	.11479	1.64214	.21541	.79320	.89938	1.2607	.10062
1.09	1.31903	.12025	1.65525	.21886	.79688	.90139	1.2549	.09861
1.10	1.33565	.12569	1.66852	.22233	0.80050	.90336	1.2492	.09664
1.11	1.35240	.13111	1.68196	.22582	.80406	.90529	1.2437	.09471
1.12	1.36929	.13649	1.69557	.22931	.80757	.90718	1.2383	.09282
1.13	1.38631	.14186	1.70934	.23283	.81102	.90903	1.2330	.09097
1.14	1.40347	.14720	1.72329	.23636	.81441	.91085	1.2279	.08915
1.15	1.42078	.15253	1.73741	.23990	0.81775	.91262	1.2229	.08738
1.16	1.43822	.15783	1.75171	.24346	.82104	.91436	1.2180	.08564
1.17	1.45581	.16311	1.76618	.24703	.82427	.91607	1.2132	.08393
1.18	1.47355	.16836	1.78083	.25062	.82745	.91774	1.2085	.08226
1.19	1.49143	.17360	1.79565	.25422	.83058	.91938	1.2040	.08062
1.20	1.50946	.17882	1.81066	.25784	0.83365	.92099	1.1995	.07901
1.21	1.52764	.18402	1.82584	.26146	.83668	.92256	1.1952	.07744
1.22	1.54598	.18920	1.84121	.26510	.83965	.92410	1.1910	.07590
1.23	1.56447	.19437	1.85676	.26876	.84258	.92561	1.1868	.07439
1.24	1.58311	.19951	1.87250	.27242	.84546	.92709	1.1828	.07291
1.25	1.60192	.20464	1.88842	.27610	0.84828	.92854	1.1789	.07146
1.26	1.62088	.20975	1.90454	.27979	.85106	.92996	1.1750	.07004
1.27	1.64001	.21485	1.92084	.28349	.85380	.93135	1.1712	.06865
1.28	1.65930	.21993	1.93734	.28721	.85648	.93272	1.1676	.06728
1.29	1.67876	.22499	1.95403	.29093	.85913	.93406	1.1640	.06594
1.30	1.69838	.23004	1.97091	.29467	0.86172	.93537	1.1605	.06463
1.31	1.71818	.23507	1.98800	.29842	.86428	.93665	1.1570	.06335
1.32	1.73814	.24009	2.00528	.30217	.86678	.93791	1.1537	.06209
1.33	1.75828	.24509	2.02276	.30594	.86925	.93914	1.1504	.06086
1.34	1.77860	.25008	2.04044	.30972	.87167	.94035	1.1472	.05965
1.35	1.79909	.25505	2.05833	.31352	0.87405	.94154	1.1441	.05846
1.36	1.81977	.26002	2.07643	.31732	.87639	.94270	1.1410	.05730
1.37	1.84062	.26496	2.09473	.32113	.87869	.94384	1.1381	.05616
1.38	1.86166	.26990	2.11324	.32495	.88095	.94495	1.1351	.05505
1.39	1.88289	.27482	2.13196	.32878	.88317	.94604	1.1323	.05396
1.40	1.90430	.27974	2.15090	.33262	0.88535	.94712	1.1295	.05288
1.41	1.92591	.28464	2.17005	.33647	.88749	.94817	1.1268	.05183
1.42	1.94770	.28952	2.18942	.34033	.88960	.94919	1.1241	.05081
1.43	1.96970	.29440	2.20900	.34420	.89167	.95020	1.1215	.04980
1.44	1.99188	.29926	2.22881	.34807	.89370	.95119	1.1189	.04881
1.45	2.01427	.30412	2.24884	.35196	0.89569	.95216	1.1165	.04784
1.46	2.03686	.30896	2.26910	.35585	.89765	.95311	1.1140	.04689
1.47	2.05965	.31379	2.28958	.35976	.89958	.95404	1.1116	.04596
1.48	2.08265	.31862	2.31029	.36367	.90147	.95495	1.1093	.04505
1.49	2.10586	.32343	2.33123	.36759	.90332	.95584	1.1070	.04416

HYPERBOLIC FUNCTIONS AND THEIR COMMON LOGARITHMS (Continued)

x	Sinh x Value	Sinh x $\log_{10}$	Cosh x Value	Cosh x $\log_{10}$	Tanh x Value	Tanh x $\log_{10}$	Coth x Value	Coth x $\log_{10}$
1.50	2.12928	.32823	2.35241	.37151	0.90515	.95672	1.1048	.04328
1.51	2.15291	.33303	2.37382	.37545	.90694	.95758	1.1026	.04242
1.52	2.17676	.33781	2.39547	.37939	.90870	.95842	1.1005	.04158
1.53	2.20082	.34258	2.41736	.38334	.91042	.95924	1.0984	.04076
1.54	2.22510	.34735	2.43949	.38730	.91212	.96005	1.0963	.03995
1.55	2.24961	.35211	2.46186	.39126	0.91379	.96084	1.0943	.03916
1.56	2.27434	.35686	2.48448	.39524	.91542	.96162	1.0924	.03838
1.57	2.29930	.36160	2.50735	.39921	.91703	.96238	1.0905	.03762
1.58	2.32449	.36633	2.53047	.40320	.91860	.96313	1.0886	.03687
1.59	2.34991	.37105	2.55384	.40719	.92015	.96386	1.0868	.03614
1.60	2.37557	.37577	2.57746	.41119	0.92167	.96457	1.0850	.03543
1.61	2.40146	.38048	2.60135	.41520	.92316	.96528	1.0832	.03472
1.62	2.42760	.38518	2.62549	.41921	.92462	.96597	1.0815	.03403
1.63	2.45397	.38987	2.64990	.42323	.92606	.96664	1.0798	.03336
1.64	2.48059	.39456	2.67457	.42725	.92747	.96730	1.0782	.03270
1.65	2.50746	.39923	2.69951	.43129	0.92886	.96795	1.0766	.03205
1.66	2.53459	.40391	2.72472	.43532	.93022	.96858	1.0750	.03142
1.67	2.56196	.40857	2.75021	.43937	.93155	.96921	1.0735	.03079
1.68	2.58959	.41323	2.77596	.44341	.93286	.96982	1.0720	.03018
1.69	2.61748	.41788	2.80200	.44747	.93415	.97042	1.0705	.02958
1.70	2.64563	.42253	2.82832	.45153	.93541	.97100	1.0691	.02900
1.71	2.67405	.42717	2.85491	.45559	.93665	.97158	1.0676	.02842
1.72	2.70273	.43180	2.88180	.45966	.93786	.97214	1.0663	.02786
1.73	2.73168	.43643	2.90897	.46374	.93906	.97269	1.0649	.02731
1.74	2.76091	.44105	2.93643	.46782	.94023	.97323	1.0636	.02677
1.75	2.79041	.44567	2.96419	.47191	0.94138	.97376	1.0623	.02624
1.76	2.82020	.45028	2.99224	.47600	.94250	.97428	1.0610	.02572
1.77	2.85026	.45488	3.02059	.48009	.94361	.97479	1.0598	.02521
1.78	2.88061	.45948	3.04925	.48419	.94470	.97529	1.0585	.02471
1.79	2.91125	.46408	3.07821	.48830	.94576	.97578	1.0574	.02422
1.80	2.94217	.46867	3.10747	.49241	0.94681	.97626	1.0562	.02374
1.81	2.97340	.47325	3.13705	.49652	.94783	.97673	1.0550	.02327
1.82	3.00492	.47783	3.16694	.50064	.94884	.97719	1.0539	.02281
1.83	3.03674	.48241	3.19715	.50476	.94983	.97764	1.0528	.02236
1.84	3.06886	.48698	3.22768	.50889	.95080	.97809	1.0518	.02191
1.85	3.10129	.49154	3.25853	.51302	0.95175	.97852	1.0507	.02148
1.86	3.13403	.49610	3.28970	.51716	.95268	.97895	1.0497	.02105
1.87	3.16709	.50066	3.32121	.52130	.95359	.97936	1.0487	.02064
1.88	3.20046	.50521	3.35305	.52544	.95449	.97977	1.0477	.02023
1.89	3.23415	.50976	3.38522	.52959	.95537	.98017	1.0467	.01983
1.90	3.26816	.51430	3.41773	.53374	0.95624	.98057	1.0458	.01943
1.91	3.30250	.51884	3.45058	.53789	.95709	.98095	1.0448	.01905
1.92	3.33718	.52338	3.48378	.54205	.95792	.98133	1.0439	.01867
1.93	3.37218	.52791	3.51733	.54621	.95873	.98170	1.0430	.01830
1.94	3.40752	.53244	3.55123	.55038	.95953	.98206	1.0422	.01794
1.95	3.44321	.53696	3.58548	.55455	0.96032	.98242	1.0413	.01758
1.96	3.47923	.54148	3.62009	.55872	.96109	.98276	1.0405	.01724
1.97	3.51561	.54600	3.65507	.56290	.96185	.98311	1.0397	.01689
1.98	3.55234	.55051	3.69041	.56707	.96259	.98344	1.0389	.01656
1.99	3.58942	.55502	3.72611	.57126	.96331	.98377	1.0381	.01623

HYPERBOLIC FUNCTIONS AND THEIR COMMON LOGARITHMS (Continued)

x	Sinh x		Cosh x		Tanh x		Coth x	
	Value	$\log_{10}$	Value	$\log_{10}$	Value	$\log_{10}$	Value	$\log_{10}$
2.00	3.62686	.55953	3.76220	.57544	0.96403	.98409	1.0373	.01591
2.01	3.66466	.56403	3.79865	.57963	.96473	.98440	1.0366	.01560
2.02	3.70283	.56853	3.83549	.58382	.96541	.98471	1.0358	.01529
2.03	3.74138	.57303	3.87271	.58802	.96609	.98502	1.0351	.01498
2.04	3.78029	.57753	3.91032	.59221	.96675	.98531	1.0344	.01469
2.05	3.81958	.58202	3.94832	.59641	0.96740	.98560	1.0337	.01440
2.06	3.85926	.58650	3.98671	.60061	.96803	.98589	1.0330	.01411
2.07	3.89932	.59099	4.02550	.60482	.96865	.98617	1.0324	.01383
2.08	3.93977	.59547	4.06470	.60903	.96926	.98644	1.0317	.01356
2.09	3.98061	.59995	4.10430	.61324	.96986	.98671	1.0311	.01329
2.10	4.02186	.60443	4.14431	.61745	0.97045	.98697	1.0304	.01303
2.11	4.06350	.60890	4.18474	.62167	.97103	.98723	1.0298	.01277
2.12	4.10555	.61337	4.22558	.62589	.97159	.98748	1.0292	.01252
2.13	4.14801	.61784	4.26685	.63011	.97215	.98773	1.0286	.01227
2.14	4.19089	.62231	4.30855	.63433	.97269	.98798	1.0281	.01202
2.15	4.23419	.62677	4.35067	.63856	0.97323	.98821	1.0275	.01179
2.16	4.27791	.63123	4.39323	.64278	.97375	.98845	1.0270	.01155
2.17	4.32205	.63569	4.43623	.64701	.97426	.98868	1.0264	.01132
2.18	4.36663	.64015	4.47967	.65125	.97477	.98890	1.0259	.01110
2.19	4.41165	.64460	4.52356	.65548	.97526	.98912	1.0254	.01088
2.20	4.45711	.64905	4.56791	.65972	0.97574	.98934	1.0249	.01066
2.21	4.50301	.65350	4.61271	.66396	.97622	.98955	1.0244	.01045
2.22	4.54936	.65795	4.65797	.66820	.97668	.98975	1.0239	.01025
2.23	4.59617	.66240	4.70370	.67244	.97714	.98996	1.0234	.01004
2.24	4.64344	.66684	4.74989	.67668	.97759	.99016	1.0229	.00984
2.25	4.69117	.67128	4.79657	.68093	0.97803	.99035	1.0225	.00965
2.26	4.73937	.67572	4.84372	.68518	.97846	.99054	1.0220	.00946
2.27	4.78804	.68016	4.89136	.68943	.97888	.99073	1.0216	.00927
2.28	4.83720	.68459	4.93948	.69368	.97929	.99091	1.0211	.00909
2.29	4.88684	.68903	4.98810	.69794	.97970	.99109	1.0207	.00891
2.30	4.93696	.69346	5.03722	.70219	0.98010	.99127	1.0203	.00873
2.31	4.98758	.69789	5.08684	.70645	.98049	.99144	1.0199	.00856
2.32	5.03870	.70232	5.13697	.71071	.98087	.99161	1.0195	00839
2.33	5.09032	.70675	5.18762	.71497	.98124	.99178	1.0191	.00822
2.34	5.14245	.71117	5.23878	.71923	.98161	.99194	1.0187	.00806
2.35	5.19510	.71559	5.29047	.72349	0.98197	.99210	1.0184	.00790
2.36	5.24827	.72002	5.34269	.72776	.98233	.99226	1.0180	.00774
2.37	5.30196	.72444	5.39544	.73203	.98267	.99241	1.0176	.00759
2.38	5.35618	.72885	5.44873	.73630	.98301	.99256	1.0173	.00744
2.39	5.41093	.73327	5.50256	.74056	.98335	.99271	1.0169	.00729
2.40	5.46623	.73769	5.55695	.74484	0.98367	.99285	1.0166	.00715
2.41	5.52207	.74210	5.61189	.74911	.98400	.99299	1.0163	.00701
2.42	5.57847	.74652	5.66739	.75338	.98431	.99313	1.0159	.00687
2.43	5.63542	.75093	5.72346	.75766	.98462	.99327	1.0156	.00673
2.44	5.69294	.75534	5.78010	.76194	.98492	.99340	1.0153	.00660
2.45	5.75103	.75975	5.83732	.76621	0.98522	.99353	1.0150	.00647
2.46	5.80969	.75415	5.89512	.77049	.98551	.99366	1.0147	.00634
2.47	5.86893	.76856	5.95352	.77477	.98579	.99379	1.0144	.00621
2.48	5.92876	.77296	6.01250	.77906	.98607	.99391	1.0141	.00609
2.49	5.98918	.77737	6.07209	.78334	.98635	.99403	1.0138	.00597

HYPERBOLIC FUNCTIONS AND THEIR COMMON LOGARITHMS (Continued)

x	Sinh x		Cosh x		Tanh x		Coth x	
	Value	$\log_{10}$	Value	$\log_{10}$	Value	$\log_{10}$	Value	$\log_{10}$
2.50	6.05020	.78177	6.13229	.78762	0.98661	.99415	1.0136	.00585
2.51	6.11183	.78617	6.19310	.79191	.98688	.99426	1.0133	.00574
2.52	6.17407	.79057	6.25453	.79619	.98714	.99438	1.0130	.00562
2.53	6.23692	.79497	6.31658	.80048	.98739	.99449	1.0128	.00551
2.54	6.30040	.79937	6.37927	.80477	.98764	.99460	1.0125	.00540
2.55	6.36451	.80377	6.44259	.80906	0.98788	.99470	1.0123	.00530
2.56	6.42926	.80816	6.50656	.81335	.98812	.99481	1.0120	.00519
2.57	6.49464	.81256	6.57118	.81764	.98835	.99491	1.0118	.00509
2.58	6.56068	.81695	6.63646	.82194	.98858	.99501	1.0115	.00499
2.59	6.62738	.82134	6.70240	.82623	.98881	.99511	1.0113	.00489
2.60	6.69473	.82573	6.76901	.83052	0.98903	.99521	1.0111	.00479
2.61	6.76276	.83012	6.83629	.83482	.98924	.99530	1.0109	.00470
2.62	6.83146	.83451	6.90426	.83912	.98946	.99540	1.0107	.00460
2.63	6.90085	.83890	6.97292	.84341	.98966	.99549	1.0104	.00451
2.64	6.97092	.84329	7.04228	.84771	.98987	.99558	1.0102	.00442
2.65	7.04169	.84768	7.11234	.85201	0.99007	.99566	1.0100	.00434
2.66	7.11317	.85206	7.18312	.85631	.99026	.99575	1.0098	.00425
2.67	7.18536	.85645	7.25461	.86061	.99045	.99583	1.0096	.00417
2.68	7.25827	.86083	7.32683	.86492	.99064	.99592	1.0094	.00408
2.69	7.33190	.86522	7.39978	.86922	.99083	.99600	1.0093	.00400
2.70	7.40626	.86960	7.47347	.87352	0.99101	.99608	1.0091	.00392
2.71	7.48137	.87398	7.54791	.87783	.99118	.99615	1.0089	.00385
2.72	7.55722	.87836	7.62310	.88213	.99136	.99623	1.0087	.00377
2.73	7.63383	.88274	7.69905	.88644	.99153	.99631	1.0085	.00369
2.74	7.71121	.88712	7.77578	.89074	.99170	.99638	1.0084	.00362
2.75	7.78935	.89150	7.85328	.89505	0.99186	.99645	1.0082	.00355
2.76	7.86828	.89588	7.93157	.89936	.99202	.99652	1.0080	.00348
2.77	7.94799	.90026	8.01065	.90367	.99218	.99659	1.0079	.00341
2.78	8.02849	.90463	8.09053	.90798	.99233	.99666	1.0077	.00334
2.79	8.10980	.90901	8.17122	.91229	.99248	.99672	1.0076	.00328
2.80	8.19192	.91339	8.25273	.91660	0.99263	.99679	1.0074	.00321
2.81	8.27486	.91776	8.33506	.92091	.99278	.99685	1.0073	.00315
2.82	8.35862	.92213	8.41823	.92522	.99292	.99691	1.0071	.00309
2.83	8.44322	.92651	8.50224	.92953	.99306	.99698	1.0070	.00302
2.84	8.52867	.93088	8.58710	.93385	.99320	.99704	1.0069	.00296
2.85	8.61497	.93525	8.67281	.93816	0.99333	.99709	1.0067	.00291
2.86	8.70213	.93963	8.75940	.94247	.99346	.99715	1.0066	.00285
2.87	8.79016	.94400	8.84686	.94679	.99359	.99721	1.0065	.00279
2.88	8.87907	.94837	8.93520	.95110	.99372	.99726	1.0063	.00274
2.89	8.96887	.95274	9.02444	.95542	.99384	.99732	1.0062	.00268
2.90	9.05956	.95711	9.11458	.95974	0.99396	.99737	1.0061	.00263
2.91	9.15116	.96148	9.20564	.96405	.99408	.99742	1.0060	.00258
2.92	9.24368	.96584	9.29761	.96837	.99420	.99747	1.0058	.00253
2.93	9.33712	.97021	9.39051	.97269	.99431	.99752	1.0057	.00248
2.94	9.43149	.97458	9.48436	.97701	.99443	.99757	1.0056	.00243
2.95	9.52681	.97895	9.57915	.98133	0.99454	.99762	1.0055	.00238
2.96	9.62308	.98331	9.67490	.98565	.99464	.99767	1.0054	.00233
2.97	9.72031	.98768	9.77161	.98997	.99475	.99771	1.0053	.00229
2.98	9.81851	.99205	9.86930	.99429	.99485	.99776	1.0052	.00224
2.99	9.91770	.99641	9.96798	.99861	.99496	.99780	1.0051	.00220

HYPERBOLIC FUNCTIONS AND THEIR COMMON LOGARITHMS (Continued)

x	Sinh x		Cosh x		Tanh x		Coth x	
	Value	$\log_{10}$	Value	$\log_{10}$	Value	$\log_{10}$	Value	$\log_{10}$
3.0	10.0179	.00078	10.0677	.00293	0.99505	.99785	1.0050	.00215
3.1	11.0765	.04440	11.1215	.04616	.99595	.99824	1.0041	.00176
3.2	12.2459	.08799	12.2866	.08943	.99668	.99856	1.0033	.00144
3.3	13.5379	.13155	13.5748	.13273	.99728	.99882	1.0027	.00118
3.4	14.9654	.17509	14.9987	.17605	.99777	.99903	1.0022	.00097
3.5	16.5426	.21860	16.5728	.21940	0.99818	.99921	1.0018	.00079
3.6	18.2855	.26211	18.3128	.26275	.99851	.99935	1.0015	.00065
3.7	20.2113	.30559	20.2360	.30612	.99878	.99947	1.0012	.00053
3.8	22.3394	.34907	22.3618	.34951	.99900	.99957	1.0010	.00043
3.9	24.6911	.39254	24.7113	.39290	.99918	.99964	1.0008	.00036
4.0	27.2899	.43600	27.3082	.43629	0.99933	.99971	1.0007	.00029
4.1	30.1619	.47946	30.1784	.47970	.99945	.99976	1.0005	.00024
4.2	33.3357	.52291	33.3507	.52310	.99955	.99980	1.0004	.00020
4.3	36.8431	.56636	36.8567	.56652	.99963	.99984	1.0004	.00016
4.4	40.7193	.60980	40.7316	.60993	.99970	.99987	1.0003	.00013
4.5	45.0030	.65324	45.0141	.65335	0.99975	.99989	1.0002	.00011
4.6	49.7371	.69668	49.7472	.69677	.99980	.99991	1.0002	.00009
4.7	54.9690	.74012	54.9781	.74019	.99983	.99993	1.0002	.00007
4.8	60.7511	.78355	60.7593	.78361	.99986	.99994	1.0001	.00006
4.9	67.1412	.82699	67.1486	.82704	.99989	.99995	1.0001	.00005
5.0	74.2032	.87042	74.2099	.87046	0.99991	.99996	1.0001	.00004
5.1	82.008	.91386	82.014	.91389	.99993	.99997	1.0001	.00003
5.2	90.633	.95729	90.639	.95731	.99994	.99997	1.0001	.00003
5.3	100.17	.00072	100.17	.00074	.99995	.99998	1.0000	.00002
5.4	110.70	.04415	110.71	.04417	.99996	.99998	1.0000	.00002
5.5	122.34	.08758	122.35	.08760	0.99997	.99999	1.0000	.00001
5.6	135.21	.13101	135.22	.13103	.99997	.99999	1.0000	.00001
5.7	149.43	.17444	149.44	.17445	.99998	.99999	1.0000	.00001
5.8	165.15	.21787	165.15	.21788	.99998	.99999	1.0000	.00001
5.9	182.52	.26130	182.52	.26131	.99998	.99999	1.0000	.00001
6.0	201.71	.30473	201.72	.30474	0.99999	.00000	1.0000	.00000
6.1	222.93	.34817	222.93	.34817	.99999	.00000	1.0000	.00000
6.2	246.37	.39159	246.38	.39161	.99999	.00000	1.0000	.00000
6.3	272.29	.43503	272.29	.43503	.99999	.00000	1.0000	.00000
6.4	300.92	.47845	300.92	.47845	.99999	.00000	1.0000	.00000
6.5	332.57	.52188	332.57	.52188	1.0000	.00000	1.0000	.00000
6.6	367.55	.56532	367.55	.56532	1.0000	.00000	1.0000	.00000
6.7	406.20	.60874	406.20	.60874	1.0000	.00000	1.0000	.00000
6.8	448.92	.65217	448.92	.65217	1.0000	.00000	1.0000	.00000
6.9	496.14	.69560	496.14	.69560	1.0000	.00000	1.0000	.00000
7.0	548.32	.73903	548.32	.73903	1.0000	.00000	1.0000	.00000
7.1	605.98	.78246	605.98	.78246	1.0000	.00000	1.0000	.00000
7.2	669.72	.82589	669.72	.82589	1.0000	.00000	1.0000	.00000
7.3	740.15	.86932	740.15	.86932	1.0000	.00000	1.0000	.00000
7.4	817.99	.91275	817.99	.91275	1.0000	.00000	1.0000	.00000
7.5	904.02	.95618	904.02	.95618	1.0000	.00000	1.0000	.00000
7.6	999.10	.99961	999.10	.99961	1.0000	.00000	1.0000	.00000
7.7	1104.2	.04305	1104.2	.04305	1.0000	.00000	1.0000	.00000
7.8	1220.3	.08647	1220.3	.08647	1.0000	.00000	1.0000	.00000
7.9	1348.6	.12988	1348.6	.12988	1.0000	.00000	1.0000	.00000

HYPERBOLIC FUNCTIONS AND THEIR COMMON LOGARITHMS (Continued)

x	Sinh x		Cosh x		Tanh x		Coth x	
	Value	$\log_{10}$	Value	$\log_{10}$	Value	$\log_{10}$	Value	$\log_{10}$
8.0	1490.5	.17333	1490.5	.17333	1.0000	.00000	1.0000	.00000
8.1	1647.2	.21675	1647.2	.21675	1.0000	.00000	1.0000	.00000
8.2	1820.5	.26019	1820.5	.26019	1.0000	.00000	1.0000	.00000
8.3	2011.9	.30360	2011.9	.30360	1.0000	.00000	1.0000	.00000
8.4	2223.5	.34704	2223.5	.34704	1.0000	.00000	1.0000	.00000
8.5	2457.4	.39048	2457.4	.39048	1.0000	.00000	1.0000	.00000
8.6	2715.8	.43390	2715.8	.43390	1.0000	.00000	1.0000	.00000
8.7	3001.5	.47734	3001.5	.47734	1.0000	.00000	1.0000	.00000
8.8	3317.1	.52076	3317.1	.52076	1.0000	.00000	1.0000	.00000
8.9	3666.0	.56419	3666.0	.56419	1.0000	.00000	1.0000	.00000
9.0	4051.5	.60762	4051.5	.60762	1.0000	.00000	1.0000	.00000
9.1	4477.6	.65105	4477.6	.65105	1.0000	.00000	1.0000	.00000
9.2	4948.6	.69448	4948.6	.69448	1.0000	.00000	1.0000	.00000
9.3	5469.0	.73791	5469.0	.73791	1.0000	.00000	1.0000	.00000
9.4	6044.2	.78134	6044.2	.78134	1.0000	.00000	1.0000	.00000
9.5	6679.9	.82477	6679.9	.82477	1.0000	.00000	1.0000	.00000
9.6	7382.4	.86820	7382.4	.86820	1.0000	.00000	1.0000	.00000
9.7	8158.8	.91163	8158.8	.91163	1.0000	.00000	1.0000	.00000
9.8	9016.9	.95506	9016.9	.95506	1.0000	.00000	1.0000	.00000
9.9	9965.2	.99849	9965.2	.99849	1.0000	.00000	1.0000	.00000
10.0	11013.2	.04191	11013.2	.04191	1.0000	.00000	1.0000	.00000

NATURAL TRIGONOMETRIC FUNCTIONS FOR ANGLES IN x RADIANS

x	Sin	Tan	Cot	Cos	x	Sin	Tan	Cot	Cos
.00	.00000	.00000	∞	1.00000	**.50**	.47943	.54630	1.8305	.87758
.01	.01000	.01000	99.997	0.99995	.51	.48818	.55936	1.7878	.87274
.02	.02000	.02000	49.993	.99980	.52	.49688	.57256	1.7465	.86782
.03	.03000	.03001	33.323	.99955	.53	.50553	.58592	1.7067	.86281
.04	.03999	.04002	24.987	.99920	.54	.51414	.59943	1.6683	.85771
.05	.04998	.05004	19.983	.99875	.55	.52269	.61311	1.6310	.85252
.06	.05996	.06007	16.647	.99820	.56	.53119	.62695	1.5950	.84726
.07	.06994	.07011	14.262	.99755	.57	.53963	.64097	1.5601	.84190
.08	.07991	.08017	12.473	.99680	.58	.54802	.65517	1.5263	.83646
.09	.08988	.09024	11.081	.99595	.59	.55636	.66956	1.4935	.83094
.10	.09983	.10033	9.9666	.99500	**.60**	.56464	.68414	1.4617	.82534
.11	.10978	.11045	9.0542	.99396	.61	.57287	.69892	1.4308	.81965
.12	.11971	.12058	8.2933	.99281	.62	.58104	.71391	1.4007	.81388
.13	.12963	.13074	7.6489	.99156	.63	.58914	.72911	1.3715	.80803
.14	.13954	.14092	7.0961	.99022	.64	.59720	.74454	1.3431	.80210
.15	.14944	.15114	6.6166	.98877	.65	.60519	.76020	1.3154	.79608
.16	.15932	.16138	6.1966	.98723	.66	.61312	.77610	1.2885	.78999
.17	.16918	.17166	5.8256	.98558	.67	.62099	.79225	1.2622	.78382
.18	.17903	.18197	5.4954	.98384	.68	.62879	.80866	1.2366	.77757
.19	.18886	.19232	5.1997	.98200	.69	.63654	.82534	1.2116	.77125
.20	.19867	.20271	4.9332	.98007	**.70**	.64422	.84229	1.1872	.76484
.21	.20846	.21314	4.6917	.97803	.71	.65183	.85953	1.1634	.75836
.22	.21823	.22362	4.4719	.97590	.72	.65938	.87707	1.1402	.75181
.23	.22798	.23414	4.2709	.97367	.73	.66687	.89492	1.1174	.74517
.24	.23770	.24472	4.0864	.97134	.74	.67429	.91309	1.0952	.73847
.25	.24740	.25534	3.9163	.96891	.75	.68164	.93160	1.0734	.73169
.26	.25708	.26602	3.7591	.96639	.76	.68892	.95045	1.0521	.72484
.27	.26673	.27676	3.6133	.96377	.77	.69614	.96967	1.0313	.71791
.28	.27636	.28755	3.4776	.96106	.78	.70328	.98926	1.0109	.71091
.29	.28595	.29841	3.3511	.95824	.79	.71035	1.0092	.99084	.70385
.30	.29552	.30934	3.2327	.95534	**.80**	.71736	1.0296	.97121	.69671
.31	.30506	.32033	3.1218	.95233	.81	.72429	1.0505	.95197	.68950
.32	.31457	.33139	3.0176	.94924	.82	.73115	1.0717	.93309	.68222
.33	.32404	.34252	2.9195	.94604	.83	.73793	1.0934	.91455	.67488
.34	.33349	.35374	2.8270	.94275	.84	.74464	1.1156	.89635	.66746
.35	.34290	.36503	2.7395	.93937	.85	.75128	1.1383	.87848	.65998
.36	.35227	.37640	2.6567	.93590	.86	.75784	1.1616	.86091	.65244
.37	.36162	.38786	2.5782	.93233	.87	.76433	1.1853	.84365	.64483
.38	.37092	.39941	2.5037	.92866	.88	.77074	1.2097	.82668	.63715
.39	.38019	.41105	2.4328	.92491	.89	.77707	1.2346	.80998	.62941
.40	.38942	.42279	2.3652	.92106	**.90**	.78333	1.2602	.79355	.62161
.41	.39861	.43463	2.3008	.91712	.91	.78950	1.2864	.77738	.61375
.42	.40776	.44657	2.2393	.91309	.92	.79560	1.3133	.76146	.60582
.43	.41687	.45862	2.1804	.90897	.93	.80162	1.3409	.74578	.59783
.44	.42594	.47078	2.1241	.90475	.94	.80756	1.3692	.73034	.58979
.45	.43497	.48306	2.0702	.90045	.95	.81342	1.3984	.71511	.58168
.46	.44395	.49545	2.0184	.89605	.96	.81919	1.4284	.70010	.57352
.47	.45289	.50797	1.9686	.89157	.97	.82489	1.4592	.68531	.56530
.48	.46178	.52061	1.9208	.88699	.98	.83050	1.4910	.67071	.55702
.49	.47063	.53339	1.8748	.88233	.99	.83603	1.5237	.65631	.54869
.50	.47943	.54630	1.8305	.87758	**1.00**	.84147	1.5574	.64209	.54030
Rad.	Sin	Tan	Cot	Cos	Rad.	Sin	Tan	Cot	Cos

NATURAL TRIGONOMETRIC FUNCTIONS
FOR ANGLES IN x RADIANS (Continued)

x	Sin	Tan	Cot	Cos	x	Sin	Tan	Cot	Cos
1.00	.84147	1.5574	.64209	.54030	**1.50**	.99749	14.101	.07091	.07074
1.01	.84683	1.5922	.62806	.53186	1.51	.99815	16.428	.06087	.06076
1.02	.85211	1.6281	.61420	.52337	1.52	.99871	19.670	.05084	.05077
1.03	.85730	1.6652	.60051	.51482	1.53	.99917	24.498	.04082	.04079
1.04	.86240	1.7036	.58699	.50622	1.54	.99953	32.461	.03081	.03079
1.05	.86742	1.7433	.57362	.49757	1.55	.99978	48.078	.02080	.02079
1.06	.87236	1.7844	.56040	.48887	1.56	.99994	92.621	.01080	.01080
1.07	.87720	1.8270	.54734	.48012	1.57	1.00000	1255.8	.00080	.00080
1.08	.88196	1.8712	.53441	.47133	1.58	.99996	−108.65	−.00920	−.00920
1.09	.88663	1.9171	.52162	.46249	1.59	.99982	−52.067	−.01921	−.01920
1.10	.89121	1.9648	.50897	.45360	**1.60**	.99957	−34.233	−.02921	−.02920
1.11	.89570	2.0143	.49644	.44466	1.61	.99923	−25.495	−.03922	−.03919
1.12	.90010	2.0660	.48404	.43568	1.62	.99879	−20.307	−.04924	−.04918
1.13	.90441	2.1198	.47175	.42666	1.63	.99825	−16.871	−.05927	−.05917
1.14	.90863	2.1759	.45959	.41759	1.64	.99761	−14.427	−.06931	−.06915
1.15	.91276	2.2345	.44753	.40849	1.65	.99687	−12.599	−.07937	−.07912
1.16	.91680	2.2958	.43558	.39934	1.66	.99602	−11.181	−.08944	−.08909
1.17	.92075	2.3600	.42373	.39015	1.67	.99508	−10.047	−.09953	−.09904
1.18	.92461	2.4273	.41199	.38092	1.68	.99404	−9.1208	−.10964	−.10899
1.19	.92837	2.4979	.40034	.37166	1.69	.99290	−8.3492	−.11977	−.11892
1.20	.93204	2.5722	.38878	.36236	**1.70**	.99166	−7.6966	−.12993	−.12884
1.21	.93562	2.6503	.37731	.35302	1.71	.99033	−7.1373	−.14011	−.13875
1.22	.93910	2.7328	.36593	.34365	1.72	.98889	−6.6524	−.15032	−.14865
1.23	.94249	2.8198	.35463	.33424	1.73	.98735	−6.2281	−.16056	−.15853
1.24	.94578	2.9119	.34341	.32480	1.74	.98572	−5.8535	−.17084	−.16840
1.25	.94898	3.0096	.33227	.31532	1.75	.98399	−5.5204	−.18115	−.17825
1.26	.95209	3.1133	.32121	.30582	1.76	.98215	−5.2221	−.19149	−.18808
1.27	.95510	3.2236	.31021	.29628	1.77	.98022	−4.9534	−.20188	−.19789
1.28	.95802	3.3413	.29928	.28672	1.78	.97820	−4.7101	−.21231	−.20768
1.29	.96084	3.4672	.28842	.27712	1.79	.97607	−4.4887	−.22278	−.21745
1.30	.96356	3.6021	.27762	.26750	**1.80**	.97385	−4.2863	−.23330	−.22720
1.31	.96618	3.7471	.26687	.25785	1.81	.97153	−4.1005	−.24387	−.23693
1.32	.96872	3.9033	.25619	.24818	1.82	.96911	−3.9294	−.25449	−.24663
1.33	.97115	4.0723	.24556	.23848	1.83	.96659	−3.7712	−.26517	−.25631
1.34	.97348	4.2556	.23498	.22875	1.84	.96398	−3.6245	−.27590	−.26596
1.35	.97572	4.4552	.22446	.21901	1.85	.96128	−3.4881	−.28669	−.27559
1.36	.97786	4.6734	.21398	.20924	1.86	.95847	−3.3608	−.29755	−.28519
1.37	.97991	4.9131	.20354	.19945	1.87	.95557	−3.2419	−.30846	−.29476
1.38	.98185	5.1774	.19315	.18964	1.88	.95258	−3.1304	−.31945	−.30430
1.39	.98370	5.4707	.18279	.17981	1.89	.94949	−3.0257	−.33051	−.31381
1.40	.98545	5.7979	.17248	.16997	**1.90**	.94630	−2.9271	−.34164	−.32329
1.41	.98710	6.1654	.16220	.16010	1.91	.94302	−2.8341	−.35284	−.33274
1.42	.98865	6.5811	.15195	.15023	1.92	.93965	−2.7463	−.36413	−.34215
1.43	.99010	7.0555	.14173	.14033	1.93	.93618	−2.6632	−.37549	−.35153
1.44	.99146	7.6018	.13155	.13042	1.94	.93262	−2.5843	−.38695	−.36087
1.45	.99271	8.2381	.12139	.12050	1.95	.92896	−2.5095	−.39849	−.37018
1.46	.99387	8.9886	.11125	.11057	1.96	.92521	−2.4383	−.41012	−.37945
1.47	.99492	9.8874	.10114	.10063	1.97	.92137	−2.3705	−.42185	−.38868
1.48	.99588	10.983	.09105	.09067	1.98	.91744	−2.3058	−.43368	−.39788
1.49	.99674	12.350	.08097	.08071	1.99	.91341	−2.2441	−.44562	−.40703
1.50	.99749	14.101	.07091	.07074	**2.00**	.90930	−2.1850	−.45766	−.41615
Rad.	**Sin**	**Tan**	**Cot**	**Cos**	**Rad.**	**Sin**	**Tan**	**Cot**	**Cos**

NATURAL TRIGONOMETRIC FUNCTIONS
SECANTS AND COSECANTS
FOR ANGLES IN RADIANS

x	sec x	csc x	x	sec x	csc x	x	sec x	csc x
0.00	1.00000	∞	0.55	1.17299	1.91319	1.10	2.20460	1.12207
0.01	1.00005	100.00167	0.56	1.18028	1.88258	1.11	2.24890	1.11645
0.02	1.00020	50.00333	0.57	1.18779	1.85311	1.12	2.29525	1.11099
0.03	1.00045	33.33833	0.58	1.19551	1.82474	1.13	2.34379	1.10569
0.04	1.00080	25.00667	0.59	1.20346	1.79739	1.14	2.39467	1.10055
0.05	1.00125	20.00834	0.60	1.21163	1.77103	1.15	2.44806	1.09557
0.06	1.00180	16.67667	0.61	1.22004	1.74560	1.16	2.50413	1.09075
0.07	1.00246	14.29739	0.62	1.22868	1.72107	1.17	2.56311	1.08607
0.08	1.00321	12.51334	0.63	1.23758	1.69738	1.18	2.62519	1.08154
0.09	1.00406	11.12613	0.64	1.24673	1.67449	1.19	2.69063	1.07716
0.10	1.00502	10.01669	0.65	1.25615	1.65238	1.20	2.75970	1.07292
0.11	1.00608	9.10927	0.66	1.26583	1.63101	1.21	2.83271	1.06881
0.12	1.00724	8.35337	0.67	1.27580	1.61034	1.22	2.90997	1.06485
0.13	1.00851	7.71402	0.68	1.28605	1.59035	1.23	2.99188	1.06102
0.14	1.00988	7.16624	0.69	1.29660	1.57100	1.24	3.07885	1.05732
0.15	1.01136	6.69173	0.70	1.30746	1.55227	1.24	3.17136	1.05376
0.16	1.01294	6.27675	0.71	1.31863	1.53413	1.26	3.26993	1.05032
0.17	1.01463	5.91078	0.72	1.33013	1.51657	1.27	3.37518	1.04701
0.18	1.01642	5.58567	0.73	1.34197	1.49954	1.28	3.48778	1.04382
0.19	1.01833	5.29496	0.74	1.35415	1.48305	1.29	3.60853	1.04076
0.20	1.02034	5.03349	0.75	1.36670	1.46705	1.30	3.73833	1.03782
0.21	1.02246	4.79709	0.76	1.37962	1.45154	1.31	3.87822	1.03500
0.22	1.02470	4.58233	0.77	1.39293	1.43650	1.32	4.02941	1.03230
0.23	1.02705	4.38640	0.78	1.40664	1.42191	1.33	4.19329	1.02971
0.24	1.02951	4.20694	0.79	1.42077	1.40775	1.34	4.37153	1.02724
0.25	1.03209	4.04197	0.80	1.43532	1.39401	1.35	4.56607	1.02488
0.26	1.03478	3.88983	0.81	1.45033	1.38067	1.36	4.77923	1.02264
0.27	1.03759	3.74909	0.82	1.46580	1.36772	1.37	5.01379	1.02050
0.28	1.04052	3.61853	0.83	1.48175	1.35514	1.38	5.27313	1.01848
0.29	1.04358	3.49709	0.84	1.49821	1.34293	1.39	5.56133	1.01657
0.30	1.04675	3.38386	0.85	1.51519	1.33106	1.40	5.88349	1.01477
0.31	1.05005	3.27806	0.86	1.53271	1.31954	1.41	6.24593	1.01307
0.32	1.05348	3.17898	0.87	1.55080	1.30834	1.42	6.65666	1.01148
0.33	1.05704	3.08601	0.88	1.56949	1.29746	1.43	7.12598	1.00999
0.34	1.06072	2.99862	0.89	1.58878	1.28688	1.44	7.66732	1.00862
0.35	1.06454	2.91632	0.90	1.60873	1.27661	1.45	8.29856	1.00734
0.36	1.06849	2.83870	0.91	1.62934	1.26662	1.46	9.04406	1.00617
0.37	1.07258	2.76537	0.92	1.65065	1.25691	1.47	9.93782	1.00510
0.38	1.07682	2.69600	0.93	1.67271	1.24747	1.48	11.02881	1.00414
0.39	1.08119	2.63027	0.94	1.69552	1.23830	1.49	12.39028	1.00327
0.40	1.08570	2.56793	0.95	1.71915	1.22938	1.50	14.13683	1.00251
0.41	1.09037	2.50872	0.96	1.74362	1.22072	1.51	16.45850	1.00185
0.42	1.09518	2.45242	0.97	1.76897	1.21229	1.52	19.69493	1.00129
0.43	1.10015	2.39882	0.98	1.79526	1.20410	1.53	24.51881	1.00083
0.44	1.10528	2.34775	0.99	1.82252	1.19614	1.54	32.47654	1.00047
0.45	1.11056	2.29903	1.00	1.85082	1.18840	1.55	48.08888	1.00022
0.46	1.11601	2.25252	1.01	1.88019	1.18087	1.56	92.62589	1.00006
0.47	1.12162	2.20806	1.02	1.91071	1.17356	1.57	+1255.76599	1.00000
0.48	1.12740	2.16554	1.03	1.94243	1.16645	1.58	− 108.65381	1.00004
0.49	1.13336	2.12483	1.04	1.97542	1.15955	1.59	− 52.07657	1.00018
0.50	1.13949	2.08583	1.05	2.00976	1.15284	1.60	− 34.24714	1.00043
0.51	1.14581	2.04844	1.06	2.04552	1.14632			
0.52	1.15231	2.01256	1.07	2.08279	1.13999			
0.53	1.15901	1.97811	1.08	2.12166	1.13384			
0.54	1.16590	1.94501	1.09	2.16223	1.12787			
0.55	1.17299	1.91319	1.10	2.20460	1.12207			

NATURAL TRIGONOMETRIC FUNCTIONS
SINE, TANGENT, COTANGENT, COSINE
FOR ANGLES IN π RADIANS

x		Sin (πx)	Tan (πx)	Cot (πx)	Cos (πx)
.00 or	1.00	.00000	.00000	inf	1.00000
.01	.99	.03141	.03143	31.821	.99951
.02	.98	.06279	.06291	15.895	.99803
.03	.97	.09411	.09453	10.579	.99556
.04	.96	.12533	.12633	7.9158	.99211
.05	.95	.15643	.15838	6.3138	.98769
.06	.94	.18738	.19076	5.2422	.98229
.07	.93	.21814	.22353	4.4737	.97592
.08	.92	.24869	.25676	3.8947	.96858
.09	.91	.27899	.29053	3.4420	.96029
.10	.90	.30902	.32492	3.0777	.95106
.11	.89	.33874	.36002	2.7776	.94088
.12	.88	.36812	.39593	2.5257	.92978
.13	.87	.39715	.43274	2.3109	.91775
.14	.86	.42578	.47056	2.1251	.90483
.15	.85	.45399	.50953	1.9626	.89101
.16	.84	.48175	.54975	1.8190	.87631
.17	.83	.50904	.59140	1.6909	.86074
.18	.82	.53583	.63462	1.5757	.84433
.19	.81	.56208	.67960	1.4715	.82708
.20	.80	.58779	.72654	1.3764	.80902
.21	.79	.61291	.77568	1.2892	.79016
.22	.78	.63742	.82727	1.2088	.77051
.23	.77	.66131	.88162	1.1343	.75011
.24	.76	.68455	.93906	1.0649	.72897
.25	.75	.70711	1.0000	1.0000	.70711
.26	.74	.72897	1.0649	.93906	.68455
.27	.73	.75011	1.1343	.88162	.66131
.28	.72	.77051	1.2088	.82727	.63742
.29	.71	.79016	1.2892	.77568	.61291
.30	.70	.80902	1.3764	.72654	.58779
.31	.69	.82708	1.4715	.67960	.56208
.32	.68	.84433	1.5757	.63462	.53583
.33	.67	.86074	1.6909	.59140	.50904
.34	.66	.87631	1.8190	.54975	.48175
.35	.65	.89101	1.9626	.50953	.45399
.36	.64	.90483	2.1251	.47056	.42578
.37	.63	.91775	2.3109	.43274	.39715
.38	.62	.92978	2.5257	.39593	.36812
.39	.61	.94088	2.7776	.36002	.33874
.40	.60	.95106	3.0777	.32492	.30902
.41	.59	.96029	3.4420	.29053	.27899
.42	.58	.96858	3.8947	.25676	.24869
.43	.57	.97592	4.4737	.22353	.21814
.44	.56	.98229	5.2422	.19076	.18738
.45	.55	.98769	6.3138	.15838	.15643
.46	.54	.99211	7.9158	.12633	.12533
.47	.53	.99556	10.579	.09453	.09411
.48	.52	.99803	15.895	.06291	.06279
.49	.51	.99951	31.821	.03143	.03141
.50	.50	1.0000	inf	.00000	.00000

These functions are useful in the solution of wave equations such as the displacement equation of a sound wave in the form:

$$y = A \sin 2\pi nx$$

without the necessity of reducing the angular rotation either to radians or to degrees in order to find the value of the function.

The algebraic sign of the function follows the familiar Quadrant Law for the particular function desired. Thus a numerical value of 9.13π radians becomes (by the subtraction of the greatest multiple of 2π radians) 1.13π radians. This is the same as $.13\pi$ radians in the 3rd Quadrant, which would give, from the tables above, a value of the sine function of $-.39715$.

Submitted by J. A. Blythe Jr.

SINE AND COSINE FUNCTIONS FOR SPECIAL MULTIPLES OF π RADIANS

k	$\sin \frac{2\pi k}{m}$	$\cos \frac{2\pi k}{m}$	$\sin \frac{2\pi k}{m}$	$\cos \frac{2\pi k}{m}$	$\sin \frac{2\pi k}{m}$	$\cos \frac{2\pi k}{m}$
	$m = 3$		$m = 4$		$m = 5$	
1	0.86603	−0.50000	1.00000	+0.00000	0.95106	+0.30902
2			0.00000	−1.00000	0.58779	−0.80902
	$m = 6$		$m = 7$		$m = 8$	
1	0.86603	+0.50000	0.78183	+0.62349	0.70711	0.70711
2	0.86603	−0.50000	0.97493	−0.22252	1.00000	+0.00000
3	0.00000	−1.00000	0.43388	−0.90097	0.70711	+0.70711
4					0.00000	−1.00000
	$m = 9$		$m = 10$		$m = 11$	
1	0.64279	0.76604	0.58779	0.80902	0.54064	0.84125
2	0.98481	+0.17365	0.95106	+0.30902	0.90963	+0.41542
3	0.86603	−0.50000	0.95106	−0.30902	0.98982	−0.14231
4	0.34202	−0.93969	0.58779	−1.80902	0.75575	−0.65486
5			0.00000	−1.00000	0.28173	−0.95949
	$m = 12$		$m = 13$		$m = 14$	
1	0.50000	0.86603	0.46472	0.88546	0.43388	0.90097
2	0.86603	0.50000	0.82289	0.56806	0.78183	0.62349
3	1.00000	+0.00000	0.99271	+0.12054	0.97493	+0.22252
4	0.86603	−0.50000	0.93502	−0.35460	0.97493	−0.22252
5	0.50000	−0.86603	0.66312	−0.74851	0.78183	−0.62349
6	0.00000	−1.00000	0.23932	−0.97904	0.43388	−0.90097
7					0.00000	−1.00000
	$m = 15$		$m = 16$		$m = 17$	
1	0.40674	0.91355	0.38268	0.92388	0.36124	0.93247
2	0.74314	0.66913	0.70711	0.70711	0.67370	0.73901
3	0.95106	+0.30902	0.92388	0.38268	0.89516	0.44574
4	0.99452	−0.10453	1.00000	+0.00000	0.99573	+0.09227
5	0.86603	−0.50000	0.92388	−0.38268	0.96183	−0.27366
6	0.58779	−0.80902	0.70711	−0.70711	0.79802	−0.60263
7	0.20791	−0.97815	0.38268	−0.92388	0.52643	−0.85022
8			0.00000	−1.00000	0.18375	−0.98297
	$m = 18$		$m = 19$		$m = 20$	
1	0.34202	0.93969	0.32470	0.94582	0.30902	0.95106
2	0.64279	0.76604	0.61421	0.78914	0.58779	0.80902
3	0.86603	0.50000	0.83717	0.54695	0.80902	0.58779
4	0.98481	+0.17365	0.96940	+0.24549	0.95106	0.30902
5	0.98481	−0.17365	0.99658	−0.08258	1.00000	+0.00000
6	0.86603	−0.50000	0.91577	−0.40170	0.95106	−0.30902
7	0.64279	−0.76604	0.73572	−0.67728	0.80902	−0.58779
8	0.34202	−0.93969	0.47595	−0.87947	0.58779	−0.80902
9	0.00000	−1.00000	0.16459	−0.98636	0.30902	−0.95106
10					0.00000	−1.00000
	$m = 21$		$m = 22$		$m = 23$	
1	0.29476	0.95557	0.28173	0.95949	0.26980	0.96292
2	0.56332	0.82624	0.54064	0.84125	0.51958	0.85442
3	0.78183	0.62349	0.75575	0.65486	0.73084	0.68255
4	0.93087	0.36534	0.90963	0.41542	0.88789	0.46007
5	0.99720	+0.07473	0.98982	+0.14231	0.97908	+0.20346
6	0.97493	−0.22252	0.98982	−0.14231	0.99767	−0.06824
7	0.86603	−0.50000	0.90963	−0.41542	0.94226	−0.33488
8	0.68017	−0.73305	0.75575	−0.65486	0.81697	−0.57668
9	0.43388	−0.90097	0.54064	−0.84125	0.63109	−0.77571
10	0.14904	−0.98883	0.28173	−0.95949	0.39840	−0.91721
11			0.00000	−1.00000	0.13617	−0.99069
	$m = 24$		$m = 25$			
1	0.25882	0.96593	0.24869	0.96858		
2	0.50000	0.86603	0.48175	0.87631		
3	0.70711	0.70711	0.68455	0.72897		
4	0.86603	0.50000	0.84433	0.53583		
5	0.96593	0.25882	0.95106	0.30902		
6	1.00000	+0.00000	0.99803	+0.06279		
7	0.96593	−0.25882	0.98229	−0.18738		
8	0.86603	−0.50000	0.90483	−0.42578		
9	0.70711	−0.70711	0.77051	−0.63742		
10	0.50000	−0.86603	0.58779	−0.80902		
11	0.25882	−0.96593	0.36812	−0.92978		
12	0.00000	−1.00000	0.12533	−0.99211		

DEGREES, MINUTES, AND SECONDS TO RADIANS

Units in degrees, minutes *or* seconds	Degrees to Radians	Minutes to Radians	Seconds to Radians
10	0.174 5329	0.002 9089	0.000 0485
20	0.349 0659	0.005 8178	0.000 0970
30	0.523 5988	0.008 7266	0.000 1454
40	0.698 1317	0.011 6355	0.000 1939
50	0.872 6646	0.014 5444	0.000 2424
60	1.047 1976	0.017 4533	0.000 2909
70	1.221 7305	(0.020 3622)	(0.000 3394)
80	1.396 2634	(0.023 2711)	(0.000 3879)
90	1.570 7963	(0.026 1800)	(0.000 4364)
100	1.745 3293	....	....
200	3.490 6585	....	....
300	5.235 9878	....	....

where n = 1, 2, 3, 4, etc. n (100°) = n (1.745 3293)

*RADIANS TO DEGREES, MINUTES, AND SECONDS

Radians	1.0	0.1	0.01	0.001	0.0001
1	57° 17′ 44.8″	5° 43′ 46.5″	0° 34′ 22.6″	0° 03′ 26.3″	0° 00′ 20.6″
2	114° 35′ 29.6″	11° 27′ 33.0″	1° 08′ 45.3″	0° 06′ 52.5″	0° 00′ 41.3″
3	171° 53′ 14.4″	17° 11′ 19.4″	1° 43′ 07.9′	0° 10′ 18.8″	0° 01′ 01.9″
4	229° 10′ 59.2″	22° 55′ 05.9″	2° 17′ 30.6″	0° 13′ 45.1′	0° 01′ 22.5″
5	286° 28′ 44.0″	28° 38′ 52.4″	2° 51′ 53.2″	0° 17′ 11.3″	0° 01′ 43.1″
6	343° 46′ 28.8″	34° 22′ 38.9″	3° 26′ 15.9″	0° 20′ 37.6″	0° 02′ 03.8″
7	401° 04′ 13.6″	40° 06′ 25.4″	4° 00′ 38.5″	0° 24′ 03.9″	0° 02′ 24.4″
8	458° 21′ 58.4″	45° 50′ 11.8″	4° 35′ 01.2″	0° 27′ 30.1″	0° 02′ 45.0″
9	515° 39′ 43.3″	51° 33′ 58.3″	5° 09′ 23.8″	0° 30′ 56.4″	0° 03′ 05.6″

*If 3.214 is desired in degrees, minutes and seconds it is obtained as follows:

3 = 171° 53′ 14.4″
.2 = 11° 27′ 33.0″
.01 = 0° 34′ 22.6″
.004 = 0° 13′ 45.1″
3.214 = 184° 8′ 55.1″

MILS—RADIANS—DEGREES

1 mil = 0.00098175 radians = 0.05625° = 3.375′ = 202.5″
1000 mils = 0.98175 radians = 56.25°
6400 mils = 360° = 2π radians
1 radian = 1018.6 mils
1° = 17.777778 mils
1′ = 0.296296 mils
1″ = 0.0049382 mils

DEGREES—RADIANS

1 radian = 57° 17′ 44.80625″

		log
1 radian =	57.29577 95131 degrees	1.75812 26324
1 radian =	3437.74677 07849 minutes	3.53627 38828
1 radian =	206264.80625 seconds	5.31442 51332
1 degree =	0.01745 32925 19943 radians	8.24187 73676–10
1 minute =	0.00029 08882 08666 radians	6.46372 61172–10
1 second =	0.00000 48481 36811 radians	4.68557 48668–10

MINUTES AND SECONDS TO DECIMAL PARTS OF A DEGREE

MINUTES AND SECONDS TO DECIMAL PARTS OF A DEG.

Min.	Degrees	Sec.	Degrees
0	0.00000	**0**	0.00000
1	.01667	1	.00028
2	.03333	2	.00056
3	.05	3	.00083
4	.06667	4	.00111
5	.08333	5	.00139
6	.10	6	.00167
7	.11667	7	.00194
8	.13333	8	.00222
9	.15	9	.0025
10	0.16667	**10**	0.00278
11	.18333	11	.00306
12	.20	12	.00333
13	.21667	13	.00361
14	.23333	14	.00389
15	.25	15	.00417
16	.26667	16	.00444
17	.28333	17	.00472
18	.30	18	.005
19	.31667	19	.00528
20	0.33333	**20**	0.00556
21	.35	21	.00583
22	.36667	22	.00611
23	.38333	23	.00639
24	.40	24	.00667
25	.41667	25	.00694
26	.43333	26	.00722
27	.45	27	.0075
28	.46667	28	.00778
29	.48333	29	.00806
30	0.50	**30**	0.00833
31	.51667	31	.00861
32	.53333	32	.00889
33	.55	33	.00917
34	.56667	34	.00944
35	.58333	35	.00972
36	.60	36	.01
37	.61667	37	.01028
38	.63333	38	.01056
39	.65	39	.01083
40	0.66667	**40**	0.01111
41	.68333	41	.01139
42	.70	42	.01167
43	.71667	43	.01194
44	.73333	44	.01222
45	.75	45	.0125
46	.76667	46	.01278
47	.78333	47	.01306
48	.80	48	.01333
49	.81667	49	.01361
50	0.83333	**50**	0.01389
51	.85	51	.01417
52	.86667	52	.01444
53	.88333	53	.01472
54	.90	54	.015
55	.91667	55	.01528
56	.93333	56	.01556
57	.95	57	.01583
58	.96667	58	.01611
59	.98333	59	.01639
60	1.00	**60**	0.01667

DECIMAL PARTS OF A DEGREE TO MINUTES AND SECONDS

Deg.	′	″	Deg.	′	″
0.00	0	00	**0.60**	36	
.01	0	36	.61	36	36
.02	1	12	.62	37	12
.03	1	48	.63	37	48
.04	2	24	.64	38	24
.05	3		.65	39	
.06	3	36	.66	39	36
.07	4	12	.67	40	12
.08	4	48	.68	40	48
.09	5	24	.69	41	24
0.10	6		**0.70**	42	
.11	6	36	.71	42	36
.12	7	12	.72	43	12
.13	7	48	.73	43	48
.14	8	24	.74	44	24
.15	9		.75	45	
.16	9	36	.76	45	36
.17	10	12	.77	46	12
.18	10	48	.78	46	48
.19	11	24	.79	47	24
0.20	12		**0.80**	48	
.21	12	36	.81	48	36
.22	13	12	.82	49	12
.23	13	48	.83	49	48
.24	14	24	.84	50	24
.25	15		.85	51	
.26	15	36	.86	51	36
.27	16	12	.87	52	12
.28	16	48	.88	52	48
.29	17	24	.89	53	24
0.30	18		**0.90**	54	
.31	18	36	.91	54	36
.32	19	12	.92	55	12
.33	19	48	.93	55	48
.34	20	24	.94	56	24
.35	21		.95	57	
.36	21	36	.96	57	36
.37	22	12	.97	58	12
.38	22	48	.98	58	48
.39	23	24	.99	59	24
0.40	24		**1.00**	60	
.41	24	36			
.42	25	12			
.43	25	48			
.44	26	24			
.45	27				
.46	27	36			
.47	28	12			
.48	28	48			
.49	29	24			
0.50	30				
.51	30	36			
.52	31	12			
.53	31	48			
.54	32	24			
.55	33				
.56	33	36			
.57	34	12			
.58	34	48			
.59	35	24			
0.60	36				

Deg.	Sec.
0.000	0.0
.001	3.6
.002	7.2
.003	10.8
.004	14.4
.005	18.
.006	21.6
.007	25.2
.008	28.8
.009	32.4
0.010	36.

*Derivatives

In the following formulas u, v, w represent functions of x, while a, c, n represent fixed real numbers. All arguments in the trigonometric functions are measured in radians, and all inverse trigonometric and hyperbolic functions represent principal values.

1. $\frac{d}{dx}(a) = 0$

2. $\frac{d}{dx}(x) = 1$

3. $\frac{d}{dx}(au) = a\frac{du}{dx}$

4. $\frac{d}{dx}(u + v - w) = \frac{du}{dx} + \frac{dv}{dx} - \frac{dw}{dx}$

5. $\frac{d}{dx}(uv) = u\frac{dv}{dx} + v\frac{du}{dx}$

6. $\frac{d}{dx}(uvw) = uv\frac{dw}{dx} + vw\frac{du}{dx} + uw\frac{dv}{dx}$

7. $\frac{d}{dx}\left(\frac{u}{v}\right) = \frac{v\frac{du}{dx} - u\frac{dv}{dx}}{v^2} = \frac{1}{v}\frac{du}{dx} - \frac{u}{v^2}\frac{dv}{dx}$

8. $\frac{d}{dx}(u^n) = nu^{n-1}\frac{du}{dx}$

9. $\frac{d}{dx}(\sqrt{u}) = \frac{1}{2\sqrt{u}}\frac{du}{dx}$

10. $\frac{d}{dx}\left(\frac{1}{u}\right) = -\frac{1}{u^2}\frac{du}{dx}$

11. $\frac{d}{dx}\left(\frac{1}{u^n}\right) = -\frac{n}{u^{n+1}}\frac{du}{dx}$

12. $\frac{d}{dx}\left(\frac{u^n}{v^m}\right) = \frac{u^{n-1}}{v^{m+1}}\left(nv\frac{du}{dx} - mu\frac{dv}{dx}\right)$

13. $\frac{d}{dx}(u^n v^m) = u^{n-1}v^{m-1}\left(nv\frac{du}{dx} + mu\frac{dv}{dx}\right)$

14. $\frac{d}{dx}[f(u)] = \frac{d}{du}[f(u)] \cdot \frac{du}{dx}$

* Let $y = f(x)$ and $\frac{dy}{dx} = \frac{d[f(x)]}{dx} = f'(x)$ define respectively a function and its derivative for any value x in their common domain. The differential for the function at such a value x is accordingly defined as

$$dy = d[f(x)] = \frac{dy}{dx}dx = \frac{d[f(x)]}{dx}dx = f'(x)\,dx$$

Each derivative formula has an associated differential formula. For example, formula 6 above has the differential formula

$$d(uvw) = uv\,dw + vw\,du + uw\,dv$$

15. $\frac{d^2}{dx^2}[f(u)] = \frac{df(u)}{du} \cdot \frac{d^2u}{dx^2} + \frac{d^2f(u)}{du^2} \cdot \left(\frac{du}{dx}\right)^2$

16. $\frac{d^n}{dx^n}[uv] = \binom{n}{0} v\frac{d^nu}{dx^n} + \binom{n}{1}\frac{dv}{dx}\frac{d^{n-1}u}{dx^{n-1}} + \binom{n}{2}\frac{d^2v}{dx^2}\frac{d^{n-2}u}{dx^{n-2}}$

$$+ \cdots + \binom{n}{k}\frac{d^kv}{dx^k}\frac{d^{n-k}u}{dx^{n-k}} + \cdots + \binom{n}{n} u\frac{d^nv}{dx^n}$$

where $\binom{n}{r} = \frac{n!}{r'(n-r)'}$ the binomial coefficient, n non-negative integer and $\binom{n}{0} = 1$.

17. $\frac{du}{dx} = \frac{1}{\frac{dx}{du}}$ if $\frac{dx}{du} \neq 0$

18. $\frac{d}{dx}(\log_a u) = (\log_a e)\frac{1}{u}\frac{du}{dx}$

19. $\frac{d}{dx}(\log_e u) = \frac{1}{u}\frac{du}{dx}$

20. $\frac{d}{dx}(a^u) = a^u(\log_e a)\frac{du}{dx}$

21. $\frac{d}{dx}(e^u) = e^u\frac{du}{dx}$

22. $\frac{d}{dx}(u^v) = vu^{v-1}\frac{du}{dx} + (\log_e u)u^v\frac{dv}{dx}$

23. $\frac{d}{dx}(\sin u) = \frac{du}{dx}(\cos u)$

24. $\frac{d}{dx}(\cos u) = -\frac{du}{dx}(\sin u)$

25. $\frac{d}{dx}(\tan u) = \frac{du}{dx}(\sec^2 u)$

26. $\frac{d}{dx}(\cot u) = -\frac{du}{dx}(\csc^2 u)$

27. $\frac{d}{dx}(\sec u) = \frac{du}{dx}\sec u \cdot \tan u$

28. $\frac{d}{dx}(\csc u) = -\frac{du}{dx}\csc u \cdot \cot u$

29. $\frac{d}{dx}(\text{vers } u) = \frac{du}{dx}\sin u$

30. $\frac{d}{dx}(\text{arc sin } u) = \frac{1}{\sqrt{1-u^2}}\frac{du}{dx}$, $\left(-\frac{\pi}{2} \leq \text{arc sin } u \leq \frac{\pi}{2}\right)$

31. $\dfrac{d}{dx}(\text{arc}\cos u) = -\dfrac{1}{\sqrt{1-u^2}}\dfrac{du}{dx}, \qquad (0 \leq \text{arc}\cos u \leq \pi)$

32. $\dfrac{d}{dx}(\text{arc}\tan u) = \dfrac{1}{1+u^2}\dfrac{du}{dx}, \qquad \left(-\dfrac{\pi}{2} < \text{arc}\tan u < \dfrac{\pi}{2}\right)$

33. $\dfrac{d}{dx}(\text{arc}\cot u) = -\dfrac{1}{1+u^2}\dfrac{du}{dx}, \qquad (0 \leq \text{arc}\cot u \leq \pi)$

34. $\dfrac{d}{dx}(\text{arc}\sec u) = \dfrac{1}{u\sqrt{u^2-1}}\dfrac{du}{dx}, \qquad \left(0 \leq \text{arc}\sec u < \dfrac{\pi}{2}, -\pi \leq \text{arc}\sec u < -\dfrac{\pi}{2}\right)$

35. $\dfrac{d}{dx}(\text{arc}\csc u) = -\dfrac{1}{u\sqrt{u^2-1}}\dfrac{du}{dx}, \qquad \left(0 < \text{arc}\csc u \leq \dfrac{\pi}{2}, -\pi < \text{arc}\csc u \leq -\dfrac{\pi}{2}\right)$

36. $\dfrac{d}{dx}(\text{arc vers } u) = \dfrac{1}{\sqrt{2u-u^2}}\dfrac{du}{dx}, \qquad (0 \leq \text{arc vers } u \leq \pi)$

37. $\dfrac{d}{dx}(\sinh u) = \dfrac{du}{dx}(\cosh u)$

38. $\dfrac{d}{dx}(\cosh u) = \dfrac{du}{dx}(\sinh u)$

39. $\dfrac{d}{dx}(\tanh u) = \dfrac{du}{dx}(\text{sech}^2 u)$

40. $\dfrac{d}{dx}(\coth u) = -\dfrac{du}{dx}(\text{csch}^2 u)$

41. $\dfrac{d}{dx}(\text{sech } u) = -\dfrac{du}{dx}(\text{sech } u \cdot \tanh u)$

42. $\dfrac{d}{dx}(\text{csch } u) = -\dfrac{du}{dx}(\text{csch } u \cdot \coth u)$

43. $\dfrac{d}{dx}(\sinh^{-1} u) = \dfrac{d}{dx}[\log(u + \sqrt{u^2+1})] = \dfrac{1}{\sqrt{u^2+1}}\dfrac{du}{dx}$

44. $\dfrac{d}{dx}(\cosh^{-1} u) = \dfrac{d}{dx}[\log(u + \sqrt{u^2-1})] = \dfrac{1}{\sqrt{u^2-1}}\dfrac{du}{dx}, \qquad (u > 1, \cosh^{-1} u > 0)$

45. $\dfrac{d}{dx}(\tanh^{-1} u) = \dfrac{d}{dx}\left[\dfrac{1}{2}\log\dfrac{1+u}{1-u}\right] = \dfrac{1}{1-u^2}\dfrac{du}{dx}, \qquad (u^2 < 1)$

46. $\dfrac{d}{dx}(\coth^{-1} u) = \dfrac{d}{dx}\left[\dfrac{1}{2}\log\dfrac{u+1}{u-1}\right] = \dfrac{1}{1-u^2}\dfrac{du}{dx}, \qquad (u^2 > 1)$

47. $\dfrac{d}{dx}(\text{sech}^{-1} u) = \dfrac{d}{dx}\left[\log\dfrac{1+\sqrt{1-u^2}}{u}\right] = -\dfrac{1}{u\sqrt{1-u^2}}\dfrac{du}{dx}, \qquad (0 < u < 1, \text{sech}^{-1} u > 0)$

48. $\dfrac{d}{dx}(\text{csch}^{-1} u) = \dfrac{d}{dx}\left[\log\dfrac{1+\sqrt{1+u^2}}{u}\right] = -\dfrac{1}{|u|\sqrt{1+u^2}}\dfrac{du}{dx}$

49. $\frac{d}{dq}\int_p^q f(x)\,dx = f(q),\qquad [p \text{ constant}]$

50. $\frac{d}{dp}\int_p^q f(x)\,dx = -f(p),\qquad [q \text{ constant}]$

51. $\frac{d}{da}\int_p^q f(x,a)\,dx = \int_p^q \frac{\partial}{\partial a}[f(x,a)]\,dx + f(q,a)\frac{dq}{da} - f(p,a)\frac{dp}{da}$

INTEGRATION

The following is a brief discussion of some integration techniques. A more complete discussion can be found in a number of good text books. However, the purpose of this introduction is simply to discuss a few of the important techniques which may be used, in conjunction with the integral table which follows, to integrate particular functions.

No matter how extensive the integral table, it is a fairly uncommon occurrence to find in the table the exact integral desired. Usually some form of transformation will have to be made. The simplest type of transformation, and yet the most general, is substitution. Simple forms of substitution, such as $y = ax$, are employed almost unconsciously by experienced users of integral tables. Other substitutions may require more thought. In some sections of the tables, appropriate substitutions are suggested for integrals which are similar to, but not exactly like, integrals in the table. Finding the right substitution is largely a matter of intuition and experience.

Several precautions must be observed when using substitutions:

1. Be sure to make the substitution in the dx term, as well as everywhere else in the integral.
2. Be sure that the function substituted is one-to-one and continuous. If this is not the case, the integral must be restricted in such a way as to make it true. See the example following.
3. With definite integrals, the limits should also be expressed in terms of the new dependent variable. With indefinite integrals, it is necessary to perform the reverse substitution to obtain the answer in terms of the original independent variable. This may also be done for definite integrals, but it is usually easier to change the limits.

Example:

$$\int \frac{x^4}{\sqrt{a^2 - x^2}}\, dx$$

Here we make the substitution $x = |a| \sin\theta$. Then $dx = |a| \cos\theta \, d\theta$, and

$$\sqrt{a^2 - x^2} = \sqrt{a^2 - a^2 \sin^2\theta} = |a|\sqrt{1 - \sin^2\theta} = |a \cos\theta|$$

Notice the absolute value signs. It is very important to keep in mind that a square root radical always denotes the positive square root, and to assure the sign is always kept positive. Thus $\sqrt{x^2} = |x|$. Failure to observe this is a common cause of errors in integration.

Notice also that the indicated substitution is not a one-to-one function, that is, it does not have a unique inverse. Thus we must restrict the range of θ in such a way as to make the function one-to-one. Fortunately, this is easily done by solving for θ

$$\theta = \sin^{-1}\frac{x}{|a|}$$

and restricting the inverse sine to the principal values, $-\frac{\pi}{2} \leq \theta \leq \frac{\pi}{2}$.

Thus the integral becomes

$$\int \frac{a^4 \sin^4\theta |a| \cos\theta \, d\theta}{|a|\,|\cos\theta|}$$

Now, however, in the range of values chosen for θ, $\cos\theta$ is always positive. Thus we may remove the absolute value signs from $\cos\theta$ in the denominator. (This is one of the reasons that the principal values of the inverse trigonometric functions are defined as they are.)

Then the $\cos\theta$ terms cancel, and the integral becomes

$$a^4 \int \sin^4\theta\, d\theta$$

By application of integral formulas 299 and 296, we integrate this to

$$-a^4 \frac{\sin^3\theta\cos\theta}{4} - \frac{3a^4}{8}\cos\theta\sin\theta + \frac{3a^4}{8}\theta + C$$

We now must perform the inverse substitution to get the result in terms of x. We have

$$\theta = \sin^{-1}\frac{x}{|a|}$$

$$\sin\theta = \frac{x}{|a|}$$

Then

$$\cos\theta = \pm\sqrt{1-\sin^2\theta} = \pm\sqrt{1-\frac{x^2}{a^2}} = \pm\frac{\sqrt{a^2-x^2}}{|a|}.$$

Because of the previously mentioned fact that $\cos\theta$ is positive, we may omit the $\pm$ sign. The reverse substitution then produces the final answer

$$\int \frac{x^4}{\sqrt{a^2-x^2}}\,dx = -\tfrac{1}{4}x^3\sqrt{a^2-x^2} - \tfrac{3}{8}a^2x\sqrt{a^2-x^2} + \frac{3a^4}{8}\sin^{-1}\frac{x}{|a|} + C.$$

Any rational function of x may be integrated, if the denominator is factored into linear and irreducible quadratic factors. The function may then be broken into partial fractions, and the individual partial fractions integrated by use of the appropriate formula from the integral table. See the section on partial fractions for further information.

Many integrals may be reduced to rational functions by proper substitutions. For example,

$$z = \tan\frac{x}{2}$$

will reduce any rational function of the six trigonometric functions of x to a rational function of z. (Frequently there are other substitutions which are simpler to use, but this one will always work. See integral formula number 484.)

Any rational function of x and $\sqrt{ax+b}$ may be reduced to a rational function of z by making the substitution

$$z = \sqrt{ax+b}.$$

Other likely substitutions will be suggested by looking at the form of the integrand.

The other main method of transforming integrals is integration by parts. This involves applying formula number 5 or 6 in the accompanying integral table. The critical factor in this method is the choice of the functions u and v. In order for the method to be successful, $v = \int dv$ and $\int v\,du$ must be easier to integrate than the original integral. Again, this choice is largely a matter of intuition and experience.

Example:

$$\int x\sin x\,dx$$

Two obvious choices are $u = x$, $dv = \sin x\,dx$, or $u = \sin x$, $dv = x\,dx$. Since a preliminary mental calculation indicates that $\int v\,du$ in the second choice would be more, rather than less,

complicated than the original integral (it would contain x^2), we use the first choice.

$$u = x \qquad du = dx$$

$$dv = \sin x\,dx \qquad v = -\cos x$$

$$\int x \sin x\,dx = \int u\,dv = uv - \int v\,du = -x\cos x + \int \cos x\,dx$$

$$= \sin x - x\cos x$$

Of course, this result could have been obtained directly from the integral table, but it provides a simple example of the method. In more complicated examples the choice of u and v may not be so obvious, and several different choices may have to be tried. Of course, there is no guarantee that any of them will work.

Integration by parts may be applied more than once, or combined with substitution. A fairly common case is illustrated by the following example.

Example:

$$\int e^x \sin x\,dx$$

Let

$$u = e^x \qquad \text{Then} \quad du = e^x\,dx$$

$$dv = \sin x\,dx \qquad v = -\cos x$$

$$\int e^x \sin x\,dx = \int u\,dv = uv - \int v\,du = -e^x\cos x + \int e^x \cos x\,dx$$

In this latter integral,

$$\text{let } u = e^x \qquad \text{Then} \quad du = e^x\,dx$$

$$dv = \cos x\,dx \qquad v = \sin x$$

$$\int e^x \sin x\,dx = -e^x \cos x + \int e^x \cos x\,dx = -e^x\cos x + \int u\,dv$$

$$= -e^x\cos x + uv - \int v\,du$$

$$= -e^x\cos x + e^x \sin x - \int e^x \sin x\,dx$$

This looks as if a circular transformation has taken place, since we are back at the same integral we started from. However, the above equation can be solved algebraically for the required integral:

$$\int e^x \sin x\,dx = \tfrac{1}{2}(e^x \sin x - e^x \cos x)$$

In the second integration by parts, if the parts had been chosen as $u = \cos x$, $dv = e^x\,dx$, we would indeed have made a circular transformation, and returned to the starting place. In general, when doing repeated integration by parts, one should never choose the function u at any stage to be the same as the function v at the previous stage, or a constant times the previous v.

The following rule is called the extended rule for integration by parts. It is the result of $n + 1$ successive applications of integration by parts.

If

$$g_1(x) = \int g(x)\,dx, \qquad g_2(x) = \int g_1(x)\,dx,$$

$$g_3(x) = \int g_2(x)\,dx, \ldots, g_m(x) = \int g_{m-1}(x)\,dx, \ldots,$$

then

$$\int f(x)\cdot g(x)\,dx = f(x)\cdot g_1(x) - f'(x)\cdot g_2(x) + f''(x)\cdot g_3(x) - + \cdots$$
$$+ (-1)^n f^{(n)}(x) g_{n+1}(x) + (-1)^{n+1}\int f^{(n+1)}(x) g_{n+1}(x)\,dx.$$

A useful special case of the above rule is when $f(x)$ is a polynomial of degree n. Then $f^{(n+1)}(x) = 0$, and

$$\int f(x)\cdot g(x)\,dx = f(x)\cdot g_1(x) - f'(x)\cdot g_2(x) + f''(x)\cdot g_3(x) - + \cdots + (-1)^n f^{(n)}(x) g_{n+1}(x) + C$$

Example:

If $f(x) = x^2$, $g(x) = \sin x$

$$\int x^2 \sin x\,dx = -x^2 \cos x + 2x \sin x + 2 \cos x + C$$

Another application of this formula occurs if

$$f''(x) = af(x) \quad \text{and} \quad g''(x) = bg(x),$$

where a and b are unequal constants. In this case, by a process similar to that used in the above example for $\int e^x \sin x\,dx$, we get the formula

$$\int f(x) g(x)\,dx = \frac{f(x)\cdot g'(x) - f'(x)\cdot g(x)}{b - a} + C$$

This formula could have been used in the example mentioned. Here is another example.

Example:

If $f(x) = e^{2x}$, $g(x) = \sin 3x$, then $a = 4$, $b = -9$, and

$$\int e^{2x} \sin 3x\,dx = \frac{3\,e^{2x} \cos 3x - 2\,e^{2x} \sin 3x}{-9 - 4} + C = \frac{e^{2x}}{13}(2 \sin 3x - 3 \cos 3x) + C$$

The following additional points should be observed when using this table.

1. A constant of integration is to be supplied with the answers for indefinite integrals.
2. Logarithmic expressions are to base $e = 2.71828\ldots$, unless otherwise specified, and are to be evaluated for the absolute value of the arguments involved therein.
3. All angles are measured in radians, and inverse trigonometric and hyperbolic functions represent principal values, unless otherwise indicated.
4. If the application of a formula produces either a zero denominator or the square root of a negative number in the result, there is usually available another form of the answer which avoids this difficulty. In many of the results, the excluded values are specified, but when such are omitted it is presumed that one can tell what these should be, especially when difficulties of the type herein mentioned are obtained.
5. When inverse trigonometric functions occur in the integrals, be sure that any replacements made for them are strictly in accordance with the rules for such functions. This causes

little difficulty when the argument of the inverse trigonometric function is positive, since then all angles involved are in the first quadrant. However, if the argument is negative, special care must be used. Thus if $u > 0$,

$$\sin^{-1} u = \cos^{-1}\sqrt{1-u^2} = \csc^{-1}\frac{1}{u}, \text{ etc.}$$

However, if $u < 0$,

$$\sin^{-1} u = -\cos^{-1}\sqrt{1-u^2} = -\pi - \csc^{-1}\frac{1}{u}, \text{ etc.}$$

See the section on inverse trigonometric functions for a full treatment of the allowable substitutions.

6. In integrals 340–345 and some others, the right side includes expressions of the form

$$A \tan^{-1}[B + C \tan f(x)].$$

In these formulas, the $\tan^{-1}$ does not necessarily represent the principal value. Instead of always employing the principal branch of the inverse tangent function, one must instead use that branch of the inverse tangent function upon which $f(x)$ lies for any particular choice of x.

Example:

$$\int_0^{4\pi} \frac{dx}{2+\sin x} = \frac{2}{\sqrt{3}}\tan^{-1}\frac{2\tan\frac{x}{2}+1}{\sqrt{3}}\Bigg]_0^{4\pi}$$

$$= \frac{2}{\sqrt{3}}\left[\tan^{-1}\frac{2\tan 2\pi + 1}{\sqrt{3}} - \tan^{-1}\frac{2\tan 0 + 1}{\sqrt{3}}\right]$$

$$= \frac{2}{\sqrt{3}}\left[\frac{13\pi}{6} - \frac{\pi}{6}\right] = \frac{4\pi}{\sqrt{3}} = \frac{4\sqrt{3}\pi}{3}$$

Here

$$\tan^{-1}\frac{2\tan 2\pi + 1}{\sqrt{3}} = \tan^{-1}\frac{1}{\sqrt{3}} = \frac{13\pi}{6},$$

since $f(x) = 2\pi$; and

$$\tan^{-1}\frac{2\tan 0 + 1}{\sqrt{3}} = \tan^{-1}\frac{1}{\sqrt{3}} = \frac{\pi}{6},$$

since $f(x) = 0$.

INTEGRALS

ELEMENTARY FORMS

1. $\int a\,dx = ax$

2. $\int a \cdot f(x)\,dx = a\int f(x)\,dx$

3. $\int \phi(y)\,dx = \int \frac{\phi(y)}{y'}\,dy, \qquad \text{where } y' = \frac{dy}{dx}$

4. $\int (u + v)\,dx = \int u\,dx + \int v\,dx, \qquad$ where u and v are any functions of x

5. $\int u\,dv = u\int dv - \int v\,du = uv - \int v\,du$

6. $\int u\frac{dv}{dx}dx = uv - \int v\frac{du}{dx}dx$

7. $\int x^n\,dx = \frac{x^{n+1}}{n+1}, \qquad \text{except } n = -1$

8. $\int \frac{f'(x)\,dx}{f(x)} = \log f(x), \qquad (df(x) = f'(x)\,dx)$

9. $\int \frac{dx}{x} = \log x$

10. $\int \frac{f'(x)\,dx}{2\sqrt{f(x)}} = \sqrt{f(x)}, \qquad (df(x) = f'(x)\,dx)$

11. $\int e^x\,dx = e^x$

12. $\int e^{ax}\,dx = e^{ax}/a$

13. $\int b^{ax}\,dx = \frac{b^{ax}}{a\log b}, \qquad (b > 0)$

14. $\int \log x\,dx = x\log x - x$

15. $\int a^x \log a\,dx = a^x, \qquad (a > 0)$

16. $\int \frac{dx}{a^2 + x^2} = \frac{1}{a}\tan^{-1}\frac{x}{a}$

17. $$\int \frac{dx}{a^2 - x^2} = \begin{cases} \dfrac{1}{a}\tanh^{-1}\dfrac{x}{a} \\ \text{or} \\ \dfrac{1}{2a}\log\dfrac{a+x}{a-x}, \qquad (a^2 > x^2) \end{cases}$$

18. $$\int \frac{dx}{x^2 - a^2} = \begin{cases} -\dfrac{1}{a}\coth^{-1}\dfrac{x}{a} \\ \text{or} \\ \dfrac{1}{2a}\log\dfrac{x-a}{x+a}, \qquad (x^2 > a^2) \end{cases}$$

19. $$\int \frac{dx}{\sqrt{a^2 - x^2}} = \begin{cases} \sin^{-1}\dfrac{x}{|a|} \\ \text{or} \\ -\cos^{-1}\dfrac{x}{|a|}, \qquad (a^2 > x^2) \end{cases}$$

20. $$\int \frac{dx}{\sqrt{x^2 \pm a^2}} = \log\left(x + \sqrt{x^2 \pm a^2}\right)$$

21. $$\int \frac{dx}{x\sqrt{x^2 - a^2}} = \frac{1}{|a|}\sec^{-1}\frac{x}{a}$$

22. $$\int \frac{dx}{x\sqrt{a^2 \pm x^2}} = -\frac{1}{a}\log\left(\frac{a + \sqrt{a^2 \pm x^2}}{x}\right)$$

FORMS CONTAINING $(a + bx)$

For forms containing $a + bx$, but not listed in the table, the substitution $u = \dfrac{a + bx}{x}$ may prove helpful.

23. $$\int (a + bx)^n\,dx = \frac{(a + bx)^{n+1}}{(n + 1)b}, \qquad (n \neq -1)$$

24. $$\int x(a + bx)^n\,dx$$

$$= \frac{1}{b^2(n + 2)}(a + bx)^{n+2} - \frac{a}{b^2(n + 1)}(a + bx)^{n+1}, \qquad (n \neq -1, -2)$$

25. $$\int x^2(a + bx)^n\,dx = \frac{1}{b^3}\left[\frac{(a + bx)^{n+3}}{n + 3} - 2a\frac{(a + bx)^{n+2}}{n + 2} + a^2\frac{(a + bx)^{n+1}}{n + 1}\right]$$

26. $$\int x^m(a+bx)^n\,dx = \begin{cases} \dfrac{x^{m+1}(a+bx)^n}{m+n+1} + \dfrac{an}{m+n+1}\displaystyle\int x^m(a+bx)^{n-1}\,dx \\ \text{or} \\ \dfrac{1}{a(n+1)}\left[-x^{m+1}(a+bx)^{n+1} + (m+n+2)\displaystyle\int x^m(a+bx)^{n+1}\,dx\right] \\ \text{or} \\ \dfrac{1}{b(m+n+1)}\left[x^m(a+bx)^{n+1} - ma\displaystyle\int x^{m-1}(a+bx)^n\,dx\right] \end{cases}$$

27. $$\int \frac{dx}{a+bx} = \frac{1}{b}\log(a+bx)$$

28. $$\int \frac{dx}{(a+bx)^2} = -\frac{1}{b(a+bx)}$$

29. $$\int \frac{dx}{(a+bx)^3} = -\frac{1}{2b(a+bx)^2}$$

30. $$\int \frac{x\,dx}{a+bx} = \begin{cases} \dfrac{1}{b^2}[a+bx-a\log(a+bx)] \\ \text{or} \\ \dfrac{x}{b} - \dfrac{a}{b^2}\log(a+bx) \end{cases}$$

31. $$\int \frac{x\,dx}{(a+bx)^2} = \frac{1}{b^2}\left[\log(a+bx) + \frac{a}{a+bx}\right]$$

32. $$\int \frac{x\,dx}{(a+bx)^n} = \frac{1}{b^2}\left[\frac{-1}{(n-2)(a+bx)^{n-2}} + \frac{a}{(n-1)(a+bx)^{n-1}}\right], \quad n \neq 1, 2$$

33. $$\int \frac{x^2\,dx}{a+bx} = \frac{1}{b^3}\left[\frac{1}{2}(a+bx)^2 - 2a(a+bx) + a^2\log(a+bx)\right]$$

34. $$\int \frac{x^2\,dx}{(a+bx)^2} = \frac{1}{b^3}\left[a+bx-2a\log(a+bx) - \frac{a^2}{a+bx}\right]$$

35. $$\int \frac{x^2\,dx}{(a+bx)^3} = \frac{1}{b^3}\left[\log(a+bx) + \frac{2a}{a+bx} - \frac{a^2}{2(a+bx)^2}\right]$$

36. $$\int \frac{x^2\,dx}{(a+bx)^n} = \frac{1}{b^3}\left[\frac{-1}{(n-3)(a+bx)^{n-3}} + \frac{2a}{(n-2)(a+bx)^{n-2}} - \frac{a^2}{(n-1)(a+bx)^{n-1}}\right], \quad n \neq 1, 2, 3$$

37. $\int \frac{dx}{x(a+bx)} = -\frac{1}{a}\log\frac{a+bx}{x}$

38. $\int \frac{dx}{x(a+bx)^2} = \frac{1}{a(a+bx)} - \frac{1}{a^2}\log\frac{a+bx}{x}$

39. $\int \frac{dx}{x(a+bx)^3} = \frac{1}{a^3}\left[\frac{1}{2}\left(\frac{2a+bx}{a+bx}\right)^2 + \log\frac{x}{a+bx}\right]$

40. $\int \frac{dx}{x^2(a+bx)} = -\frac{1}{ax} + \frac{b}{a^2}\log\frac{a+bx}{x}$

41. $\int \frac{dx}{x^3(a+bx)} = \frac{2bx-a}{2a^2x^2} + \frac{b^2}{a^3}\log\frac{x}{a+bx}$

42. $\int \frac{dx}{x^2(a+bx)^2} = -\frac{a+2bx}{a^2x(a+bx)} + \frac{2b}{a^3}\log\frac{a+bx}{x}$

FORMS CONTAINING $c^2 \pm x^2$, $x^2 - c^2$

43. $\int \frac{dx}{c^2+x^2} = \frac{1}{c}\tan^{-1}\frac{x}{c}$

44. $\int \frac{dx}{c^2-x^2} = \frac{1}{2c}\log\frac{c+x}{c-x}, \qquad (c^2 > x^2)$

45. $\int \frac{dx}{x^2-c^2} = \frac{1}{2c}\log\frac{x-c}{x+c}, \qquad (x^2 > c^2)$

46. $\int \frac{x\,dx}{c^2 \pm x^2} = \pm\frac{1}{2}\log(c^2 \pm x^2)$

47. $\int \frac{x\,dx}{(c^2 \pm x^2)^{n+1}} = \mp\frac{1}{2n(c^2 \pm x^2)^n}$

48. $\int \frac{dx}{(c^2 \pm x^2)^n} = \frac{1}{2c^2(n-1)}\left[\frac{x}{(c^2 \pm x^2)^{n-1}} + (2n-3)\int\frac{dx}{(c^2 \pm x^2)^{n-1}}\right]$

49. $\int \frac{dx}{(x^2 - c^2)^n} = \frac{1}{2c^2(n-1)}\left[-\frac{x}{(x^2 - c^2)^{n-1}} - (2n-3)\int\frac{dx}{(x^2 - c^2)^{n-1}}\right]$

50. $\int \frac{x\,dx}{x^2-c^2} = \frac{1}{2}\log(x^2-c^2)$

51. $\int \frac{x\,dx}{(x^2-c^2)^{n+1}} = -\frac{1}{2n(x^2-c^2)^n}$

FORMS CONTAINING $a + bx$ and $c + dx$

$$u = a + bx, \qquad v = c + dx, \qquad k = ad - bc$$

If $k = 0$, then $v = \frac{c}{a}u$

52. $$\int \frac{dx}{u \cdot v} = \frac{1}{k} \cdot \log\left(\frac{v}{u}\right)$$

53. $$\int \frac{x\,dx}{u \cdot v} = \frac{1}{k}\left[\frac{a}{b}\log(u) - \frac{c}{d}\log(v)\right]$$

54. $$\int \frac{dx}{u^2 \cdot v} = \frac{1}{k}\left(\frac{1}{u} + \frac{d}{k}\log\frac{v}{u}\right)$$

55. $$\int \frac{x\,dx}{u^2 \cdot v} = \frac{-a}{bku} - \frac{c}{k^2}\log\frac{v}{u}$$

56. $$\int \frac{x^2\,dx}{u^2 \cdot v} = \frac{a^2}{b^2ku} + \frac{1}{k^2}\left[\frac{c^2}{d}\log(v) + \frac{a(k - bc)}{b^2}\log(u)\right]$$

57. $$\int \frac{dx}{u^n \cdot v^m} = \frac{1}{k(m-1)}\left[\frac{-1}{u^{n-1} \cdot v^{m-1}} - (m + n - 2)b\int \frac{dx}{u^n \cdot v^{m-1}}\right]$$

58. $$\int \frac{u}{v}\,dx = \frac{bx}{d} + \frac{k}{d^2}\log(v)$$

59. $$\int \frac{u^m\,dx}{v^n} = \begin{cases} \dfrac{-1}{k(n-1)}\left[\dfrac{u^{m+1}}{v^{n-1}} + b(n - m - 2)\displaystyle\int \frac{u^m}{v^{n-1}}\,dx\right] \\ \text{or} \\ \dfrac{-1}{d(n - m - 1)}\left[\dfrac{u^m}{v^{n-1}} + mk\displaystyle\int \frac{u^{m-1}}{v^n}\,dx\right] \\ \text{or} \\ \dfrac{-1}{d(n-1)}\left[\dfrac{u^m}{v^{n-1}} - mb\displaystyle\int \frac{u^{m-1}}{v^{n-1}}\,dx\right] \end{cases}$$

FORMS CONTAINING $(a + bx^n)$

60. $$\int \frac{dx}{a + bx^2} = \frac{1}{\sqrt{ab}}\tan^{-1}\frac{x\sqrt{ab}}{a}, \qquad (ab > 0)$$

61. $$\int \frac{dx}{a + bx^2} = \begin{cases} \dfrac{1}{2\sqrt{-ab}}\log\dfrac{a + x\sqrt{-ab}}{a - x\sqrt{-ab}}, & (ab < 0) \\ \text{or} \\ \dfrac{1}{\sqrt{-ab}}\tanh^{-1}\dfrac{x\sqrt{-ab}}{a}, & (ab < 0) \end{cases}$$

62. $$\int \frac{dx}{a^2 + b^2x^2} = \frac{1}{ab} \tan^{-1} \frac{bx}{a}$$

63. $$\int \frac{x\,dx}{a + bx^2} = \frac{1}{2b} \log (a + bx^2)$$

64. $$\int \frac{x^2\,dx}{a + bx^2} = \frac{x}{b} - \frac{a}{b} \int \frac{dx}{a + bx^2}$$

65. $$\int \frac{dx}{(a + bx^2)^2} = \frac{x}{2a(a + bx^2)} + \frac{1}{2a} \int \frac{dx}{a + bx^2}$$

66. $$\int \frac{dx}{a^2 - b^2x^2} = \frac{1}{2ab} \log \frac{a + bx}{a - bx}$$

67. $$\int \frac{dx}{(a + bx^2)^{m+1}} = \begin{cases} \dfrac{1}{2ma} \dfrac{x}{(a + bx^2)^m} + \dfrac{2m - 1}{2ma} \displaystyle\int \frac{dx}{(a + bx^2)^m} \\ \text{or} \\ \dfrac{(2m)!}{(m!)^2} \left[\dfrac{x}{2a} \displaystyle\sum_{r=1}^{m} \frac{r!(r - 1)!}{(4a)^{m-r}(2r)!(a + bx^2)^r} + \frac{1}{(4a)^m} \int \frac{dx}{a + bx^2} \right] \end{cases}$$

68. $$\int \frac{x\,dx}{(a + bx^2)^{m+1}} = -\frac{1}{2bm(a + bx^2)^m}$$

69. $$\int \frac{x^2\,dx}{(a + bx^2)^{m+1}} = \frac{-x}{2mb(a + bx^2)^m} + \frac{1}{2mb} \int \frac{dx}{(a + bx^2)^m}$$

70. $$\int \frac{dx}{x(a + bx^2)} = \frac{1}{2a} \log \frac{x^2}{a + bx^2}$$

71. $$\int \frac{dx}{x^2(a + bx^2)} = -\frac{1}{ax} - \frac{b}{a} \int \frac{dx}{a + bx^2}$$

72. $$\int \frac{dx}{x(a + bx^2)^{m+1}} = \begin{cases} \dfrac{1}{2am(a + bx^2)^m} + \dfrac{1}{a} \displaystyle\int \frac{dx}{x(a + bx^2)^m} \\ \text{or} \\ \dfrac{1}{2a^{m+1}} \left[\displaystyle\sum_{r=1}^{m} \frac{a^r}{r(a + bx^2)^r} + \log \frac{x^2}{a + bx^2} \right] \end{cases}$$

73. $$\int \frac{dx}{x^2(a + bx^2)^{m+1}} = \frac{1}{a} \int \frac{dx}{x^2(a + bx^2)^m} - \frac{b}{a} \int \frac{dx}{(a + bx^2)^{m+1}}$$

74. $$\int \frac{dx}{a + bx^3} = \frac{k}{3a} \left[\frac{1}{2} \log \frac{(k + x)^3}{a + bx^3} + \sqrt{3} \tan^{-1} \frac{2x - k}{k\sqrt{3}} \right], \qquad \left(k = \sqrt[3]{\frac{a}{b}} \right)$$

75. $$\int \frac{x\,dx}{a + bx^3} = \frac{1}{3bk} \left[\frac{1}{2} \log \frac{a + bx^3}{(k + x)^3} + \sqrt{3} \tan^{-1} \frac{2x - k}{k\sqrt{3}} \right], \qquad \left(k = \sqrt[3]{\frac{a}{b}} \right)$$

76. $$\int \frac{x^2\,dx}{a+bx^3} = \frac{1}{3b}\log(a+bx^3)$$

77. $$\int \frac{dx}{a+bx^4} = \frac{k}{2a}\left[\frac{1}{2}\log\frac{x^2+2kx+2k^2}{x^2-2kx+2k^2} + \tan^{-1}\frac{2kx}{2k^2-x^2}\right], \qquad \left(ab>0,\ k=\sqrt[4]{\frac{a}{4b}}\right)$$

78. $$\int \frac{dx}{a+bx^4} = \frac{k}{2a}\left[\frac{1}{2}\log\frac{x+k}{x-k} + \tan^{-1}\frac{x}{k}\right], \qquad \left(ab<0,\ k=\sqrt[4]{-\frac{a}{b}}\right)$$

79. $$\int \frac{x\,dx}{a+bx^4} = \frac{1}{2bk}\tan^{-1}\frac{x^2}{k}, \qquad \left(ab>0,\ k=\sqrt{\frac{a}{b}}\right)$$

80. $$\int \frac{x\,dx}{a+bx^4} = \frac{1}{4bk}\log\frac{x^2-k}{x^2+k}, \qquad \left(ab<0,\ k=\sqrt{-\frac{a}{b}}\right)$$

81. $$\int \frac{x^2\,dx}{a+bx^4} = \frac{1}{4bk}\left[\frac{1}{2}\log\frac{x^2-2kx+2k^2}{x^2+2kx+2k^2} + \tan^{-1}\frac{2kx}{2k^2-x^2}\right], \qquad \left(ab>0,\ k=\sqrt[4]{\frac{a}{4b}}\right)$$

82. $$\int \frac{x^2\,dx}{a+bx^4} = \frac{1}{4bk}\left[\log\frac{x-k}{x+k} + 2\tan^{-1}\frac{x}{k}\right], \qquad \left(ab<0,\ k=\sqrt[4]{-\frac{a}{b}}\right)$$

83. $$\int \frac{x^3\,dx}{a+bx^4} = \frac{1}{4b}\log(a+bx^4)$$

84. $$\int \frac{dx}{x(a+bx^n)} = \frac{1}{an}\log\frac{x^n}{a+bx^n}$$

85. $$\int \frac{dx}{(a+bx^n)^{m+1}} = \frac{1}{a}\int\frac{dx}{(a+bx^n)^m} - \frac{b}{a}\int\frac{x^n\,dx}{(a+bx^n)^{m+1}}$$

86. $$\int \frac{x^m\,dx}{(a+bx^n)^{p+1}} = \frac{1}{b}\int\frac{x^{m-n}\,dx}{(a+bx^n)^p} - \frac{a}{b}\int\frac{x^{m-n}\,dx}{(a+bx^n)^{p+1}}$$

87. $$\int \frac{dx}{x^m(a+bx^n)^{p+1}} = \frac{1}{a}\int\frac{dx}{x^m(a+bx^n)^p} - \frac{b}{a}\int\frac{dx}{x^{m-n}(a+bx^n)^{p+1}}$$

88. $$\int x^m(a+bx^n)^p\,dx = \begin{cases} \dfrac{1}{b(np+m+1)}\left[x^{m-n+1}(a+bx^n)^{p+1} - a(m-n+1)\displaystyle\int x^{m-n}(a+bx^n)^p\,dx\right] \\ \text{or} \\ \dfrac{1}{np+m+1}\left[x^{m+1}(a+bx^n)^p + anp\displaystyle\int x^m(a+bx^n)^{p-1}\,dx\right] \\ \text{or} \\ \dfrac{1}{a(m+1)}\left[x^{m+1}(a+bx^n)^{p+1} - (m+1+np+n)b\displaystyle\int x^{m+n}(a+bx^n)^p\,dx\right] \\ \text{or} \\ \dfrac{1}{an(p+1)}\left[-x^{m+1}(a+bx^n)^{p+1} + (m+1+np+n)\displaystyle\int x^m(a+bx^n)^{p+1}\,dx\right] \end{cases}$$

FORMS CONTAINING $c^3 \pm x^3$

89. $$\int \frac{dx}{c^3 \pm x^3} = \pm\frac{1}{6c^2}\log\frac{(c\pm x)^3}{c^3\pm x^3} + \frac{1}{c^2\sqrt{3}}\tan^{-1}\frac{2x\mp c}{c\sqrt{3}}$$

90. $$\int \frac{dx}{(c^3\pm x^3)^2} = \frac{x}{3c^3(c^3\pm x^3)} + \frac{2}{3c^3}\int\frac{dx}{c^3\pm x^3}$$

91. $$\int \frac{dx}{(c^3\pm x^3)^{n+1}} = \frac{1}{3nc^3}\left[\frac{x}{(c^3\pm x^3)^n} + (3n-1)\int\frac{dx}{(c^3\pm x^3)^n}\right]$$

92. $$\int \frac{x\,dx}{c^3\pm x^3} = \frac{1}{6c}\log\frac{c^3\pm x^3}{(c\pm x)^3} \pm \frac{1}{c\sqrt{3}}\tan^{-1}\frac{2x\mp c}{c\sqrt{3}}$$

93. $$\int \frac{x\,dx}{(c^3\pm x^3)^2} = \frac{x^2}{3c^3(c^3\pm x^3)} + \frac{1}{3c^3}\int\frac{x\,dx}{c^3\pm x^3}$$

94. $$\int \frac{x\,dx}{(c^3\pm x^3)^{n+1}} = \frac{1}{3nc^3}\left[\frac{x^2}{(c^3\pm x^3)^n} + (3n-2)\int\frac{x\,dx}{(c^3\pm x^3)^n}\right]$$

95. $$\int \frac{x^2\,dx}{c^3\pm x^3} = \pm\frac{1}{3}\log(c^3\pm x^3)$$

96. $\int \frac{x^2\,dx}{(c^3 \pm x^3)^{n+1}} = \mp \frac{1}{3n(c^3 \pm x^3)^n}$

97. $\int \frac{dx}{x(c^3 \pm x^3)} = \frac{1}{3c^3} \log \frac{x^3}{c^3 \pm x^3}$

98. $\int \frac{dx}{x(c^3 \pm x^3)^2} = \frac{1}{3c^3(c^3 \pm x^3)} + \frac{1}{3c^6} \log \frac{x^3}{c^3 \pm x^3}$

99. $\int \frac{dx}{x(c^3 \pm x^3)^{n+1}} = \frac{1}{3nc^3(c^3 \pm x^3)^n} + \frac{1}{c^3} \int \frac{dx}{x(c^3 \pm x^3)^n}$

100. $\int \frac{dx}{x^2(c^3 \pm x^3)} = -\frac{1}{c^3 x} \mp \frac{1}{c^3} \int \frac{x\,dx}{c^3 \pm x^3}$

101. $\int \frac{dx}{x^2(c^3 \pm x^3)^{n+1}} = \frac{1}{c^3} \int \frac{dx}{x^2(c^3 \pm x^3)^n} \mp \frac{1}{c^3} \int \frac{x\,dx}{(c^3 \pm x^3)^{n+1}}$

FORMS CONTAINING $c^4 \pm x^4$

102. $\int \frac{dx}{c^4 + x^4} = \frac{1}{2c^3\sqrt{2}} \left[\frac{1}{2} \log \frac{x^2 + cx\sqrt{2} + c^2}{x^2 - cx\sqrt{2} + c^2} + \tan^{-1} \frac{cx\sqrt{2}}{c^2 - x^2} \right]$

103. $\int \frac{dx}{c^4 - x^4} = \frac{1}{2c^3} \left[\frac{1}{2} \log \frac{c + x}{c - x} + \tan^{-1} \frac{x}{c} \right]$

104. $\int \frac{x\,dx}{c^4 + x^4} = \frac{1}{2c^2} \tan^{-1} \frac{x^2}{c^2}$

105. $\int \frac{x\,dx}{c^4 - x^4} = \frac{1}{4c^2} \log \frac{c^2 + x^2}{c^2 - x^2}$

106. $\int \frac{x^2\,dx}{c^4 + x^4} = \frac{1}{2c\sqrt{2}} \left[\frac{1}{2} \log \frac{x^2 - cx\sqrt{2} + c^2}{x^2 + cx\sqrt{2} + c^2} + \tan^{-1} \frac{cx\sqrt{2}}{c^2 - x^2} \right]$

107. $\int \frac{x^2\,dx}{c^4 - x^4} = \frac{1}{2c} \left[\frac{1}{2} \log \frac{c + x}{c - x} - \tan^{-1} \frac{x}{c} \right]$

108. $\int \frac{x^3\,dx}{c^4 \pm x^4} = \pm \frac{1}{4} \log (c^4 \pm x^4)$

FORMS CONTAINING $(a + bx + cx^2)$

$X = a + bx + cx^2$ and $q = 4ac - b^2$

If $q = 0$, then $X = c\left(x + \frac{b}{2c}\right)^2$, and formulas starting with 23 should be used in place of these.

109. $\int \frac{dx}{X} = \frac{2}{\sqrt{q}} \tan^{-1} \frac{2cx + b}{\sqrt{q}}, \qquad (q > 0)$

110. $$\int \frac{dx}{X} = \begin{cases} \dfrac{-2}{\sqrt{-q}} \tanh^{-1} \dfrac{2cx+b}{\sqrt{-q}} \\ \text{or} \\ \dfrac{1}{\sqrt{-q}} \log \dfrac{2cx+b-\sqrt{-q}}{2cx+b+\sqrt{-q}}, \qquad (q<0) \end{cases}$$

111. $$\int \frac{dx}{X^2} = \frac{2cx+b}{qX} + \frac{2c}{q}\int \frac{dx}{X}$$

112. $$\int \frac{dx}{X^3} = \frac{2cx+b}{q}\left(\frac{1}{2X^2} + \frac{3c}{qX}\right) + \frac{6c^2}{q^2}\int \frac{dx}{X}$$

113. $$\int \frac{dx}{X^{n+1}} = \begin{cases} \dfrac{2cx+b}{nqX^n} + \dfrac{2(2n-1)c}{qn}\displaystyle\int \frac{dx}{X^n} \\ \text{or} \\ \dfrac{(2n)!}{(n!)^2}\left(\dfrac{c}{q}\right)^n \left[\dfrac{2cx+b}{q} \displaystyle\sum_{r=1}^{n} \left(\frac{q}{cX}\right)^r \left(\frac{(r-1)!r!}{(2r)!}\right) + \int \frac{dx}{X}\right] \end{cases}$$

114. $$\int \frac{x\,dx}{X} = \frac{1}{2c}\log X - \frac{b}{2c}\int \frac{dx}{X}$$

115. $$\int \frac{x\,dx}{X^2} = -\frac{bx+2a}{qX} - \frac{b}{q}\int \frac{dx}{X}$$

116. $$\int \frac{x\,dx}{X^{n+1}} = -\frac{2a+bx}{nqX^n} - \frac{b(2n-1)}{nq}\int \frac{dx}{X^n}$$

117. $$\int \frac{x^2}{X}\,dx = \frac{x}{c} - \frac{b}{2c^2}\log X + \frac{b^2-2ac}{2c^2}\int \frac{dx}{X}$$

118. $$\int \frac{x^2}{X^2}\,dx = \frac{(b^2-2ac)x+ab}{cqX} + \frac{2a}{q}\int \frac{dx}{X}$$

119. $$\int \frac{x^m\,dx}{X^{n+1}} = -\frac{x^{m-1}}{(2n-m+1)cX^n} - \frac{n-m+1}{2n-m+1}\cdot\frac{b}{c}\int \frac{x^{m-1}\,dx}{X^{n+1}} + \frac{m-1}{2n-m+1}\cdot\frac{a}{c}\int \frac{x^{m-2}\,dx}{X^{n+1}}$$

120. $$\int \frac{dx}{xX} = \frac{1}{2a}\log\frac{x^2}{X} - \frac{b}{2a}\int \frac{dx}{X}$$

121. $$\int \frac{dx}{x^2X} = \frac{b}{2a^2}\log\frac{X}{x^2} - \frac{1}{ax} + \left(\frac{b^2}{2a^2} - \frac{c}{a}\right)\int \frac{dx}{X}$$

122. $$\int \frac{dx}{xX^n} = \frac{1}{2a(n-1)X^{n-1}} - \frac{b}{2a}\int \frac{dx}{X^n} + \frac{1}{a}\int \frac{dx}{xX^{n-1}}$$

123. $$\int \frac{dx}{x^m X^{n+1}} = -\frac{1}{(m-1)ax^{m-1}X^n} - \frac{n+m-1}{m-1}\cdot\frac{b}{a}\int \frac{dx}{x^{m-1}X^{n+1}} - \frac{2n+m-1}{m-1}\cdot\frac{c}{a}\int \frac{dx}{x^{m-2}X^{n+1}}$$

FORMS CONTAINING $\sqrt{a+bx}$

124. $$\int \sqrt{a+bx}\,dx = \frac{2}{3b}\sqrt{(a+bx)^3}$$

125. $$\int x\sqrt{a+bx}\,dx = -\frac{2(2a-3bx)\sqrt{(a+bx)^3}}{15b^2}$$

126. $$\int x^2\sqrt{a+bx}\,dx = \frac{2(8a^2-12abx+15b^2x^2)\sqrt{(a+bx)^3}}{105b^3}$$

127. $$\int x^m\sqrt{a+bx}\,dx = \begin{cases} \dfrac{2}{b(2m+3)}\left[x^m\sqrt{(a+bx)^3} - ma\displaystyle\int x^{m-1}\sqrt{a+bx}\,dx\right] \\ \quad\text{or} \\ \dfrac{2}{b^{m+1}}\sqrt{a+bx}\displaystyle\sum_{r=0}^{m}\frac{m!(-a)^{m-r}}{r!(m-r)!(2r+3)}(a+bx)^{r+1} \end{cases}$$

128. $$\int \frac{\sqrt{a+bx}}{x}dx = 2\sqrt{a+bx} + a\int \frac{dx}{x\sqrt{a+bx}}$$

129. $$\int \frac{\sqrt{a+bx}}{x^2}dx = -\frac{\sqrt{a+bx}}{x} + \frac{b}{2}\int \frac{dx}{x\sqrt{a+bx}}$$

130. $$\int \frac{\sqrt{a+bx}}{x^m}dx = -\frac{1}{(m-1)a}\left[\frac{\sqrt{(a+bx)^3}}{x^{m-1}} + \frac{(2m-5)b}{2}\int \frac{\sqrt{a+bx}}{x^{m-1}}dx\right]$$

131. $$\int \frac{dx}{\sqrt{a+bx}} = \frac{2\sqrt{a+bx}}{b}$$

132. $$\int \frac{x\,dx}{\sqrt{a+bx}} = -\frac{2(2a-bx)}{3b^2}\sqrt{a+bx}$$

133. $$\int \frac{x^2\,dx}{\sqrt{a+bx}} = \frac{2(8a^2-4abx+3b^2x^2)}{15b^3}\sqrt{a+bx}$$

134. $$\int \frac{x^m\,dx}{\sqrt{a+bx}} = \begin{cases} \dfrac{2}{(2m+1)b}\left[x^m\sqrt{a+bx} - ma\displaystyle\int \frac{x^{m-1}\,dx}{\sqrt{a+bx}}\right] \\ \text{or} \\ \dfrac{2(-a)^m\sqrt{a+bx}}{b^{m+1}} \displaystyle\sum_{r=0}^{m} \frac{(-1)^r m!(a+bx)^r}{(2r+1)r!(m-r)!a^r} \end{cases}$$

135. $$\int \frac{dx}{x\sqrt{a+bx}} = \frac{1}{\sqrt{a}} \log\left(\frac{\sqrt{a+bx}-\sqrt{a}}{\sqrt{a+bx}+\sqrt{a}}\right), \qquad (a>0)$$

136. $$\int \frac{dx}{x\sqrt{a+bx}} = \frac{2}{\sqrt{-a}} \tan^{-1}\sqrt{\frac{a+bx}{-a}}, \qquad (a<0)$$

137. $$\int \frac{dx}{x^2\sqrt{a+bx}} = -\frac{\sqrt{a+bx}}{ax} - \frac{b}{2a}\int \frac{dx}{x\sqrt{a+bx}}$$

138. $$\int \frac{dx}{x^n\sqrt{a+bx}} = \begin{cases} -\dfrac{\sqrt{a+bx}}{(n-1)ax^{n-1}} - \dfrac{(2n-3)b}{(2n-2)a}\displaystyle\int \frac{dx}{x^{n-1}\sqrt{a+bx}} \\ \text{or} \\ \dfrac{(2n-2)!}{[(n-1)!]^2}\left[-\dfrac{\sqrt{a+bx}}{a}\displaystyle\sum_{r=1}^{n-1} \frac{r!(r-1)!}{x^r(2r)!}\left(-\frac{b}{4a}\right)^{n-r-1} + \left(-\frac{b}{4a}\right)^{n-1}\int \frac{dx}{x\sqrt{a+bx}}\right] \end{cases}$$

139. $$\int (a+bx)^{\pm\frac{n}{2}}\,dx = \frac{2(a+bx)^{\frac{2\pm n}{2}}}{b(2\pm n)}$$

140. $$\int x(a+bx)^{\pm\frac{n}{2}}\,dx = \frac{2}{b^2}\left[\frac{(a+bx)^{\frac{4\pm n}{2}}}{4\pm n} - \frac{a(a+bx)^{\frac{2\pm n}{2}}}{2\pm n}\right]$$

141. $$\int \frac{dx}{x(a+bx)^{\frac{m}{2}}} = \frac{1}{a}\int \frac{dx}{x(a+bx)^{\frac{m-2}{2}}} - \frac{b}{a}\int \frac{dx}{(a+bx)^{\frac{m}{2}}}$$

142. $$\int \frac{(a+bx)^{\frac{n}{2}}\,dx}{x} = b\int (a+bx)^{\frac{n-2}{2}}\,dx + a\int \frac{(a+bx)^{\frac{n-2}{2}}}{x}\,dx$$

143. $$\int f(x,\sqrt{a+bx})\,dx = \frac{2}{b}\int f\left(\frac{z^2-a}{b}, z\right) z\,dz, \qquad (z=\sqrt{a+bx})$$

FORMS CONTAINING $\sqrt{a+bx}$ and $\sqrt{c+dx}$

$$u = a + bx \qquad v = c + dx \qquad k = ad - bc$$

If $k = 0$, then $v = \frac{c}{a}u$, and formulas starting with 124 should be used in place of these.

144. $$\int \frac{dx}{\sqrt{uv}} = \begin{cases} \frac{2}{\sqrt{bd}} \tanh^{-1} \frac{\sqrt{bduv}}{bv} \\ \text{or} \\ \frac{1}{\sqrt{bd}} \log \frac{bv + \sqrt{bduv}}{bv - \sqrt{bduv}} \\ \text{or} \\ \frac{1}{\sqrt{bd}} \log \frac{(bv + \sqrt{bduv})^2}{v}, \qquad (bd > 0) \end{cases}$$

145. $$\int \frac{dx}{\sqrt{uv}} = \begin{cases} \frac{2}{\sqrt{-bd}} \tan^{-1} \frac{\sqrt{-bduv}}{bv} \\ \text{or} \\ -\frac{1}{\sqrt{-bd}} \sin^{-1} \left(\frac{2bdx + ad + bc}{|k|} \right), \qquad (bd < 0) \end{cases}$$

146. $$\int \sqrt{uv}\, dx = \frac{k + 2bv}{4bd} \sqrt{uv} - \frac{k^2}{8bd} \int \frac{dx}{\sqrt{uv}}$$

147. $$\int \frac{dx}{v\sqrt{u}} = \begin{cases} \frac{1}{\sqrt{kd}} \log \frac{d\sqrt{u} - \sqrt{kd}}{d\sqrt{u} + \sqrt{kd}} \\ \text{or} \\ \frac{1}{\sqrt{kd}} \log \frac{(d\sqrt{u} - \sqrt{kd})^2}{v}, \qquad (kd > 0) \end{cases}$$

148. $$\int \frac{dx}{v\sqrt{u}} = \frac{2}{\sqrt{-kd}} \tan^{-1} \frac{d\sqrt{u}}{\sqrt{-kd}}, \qquad (kd < 0)$$

149. $$\int \frac{x\, dx}{\sqrt{uv}} = \frac{\sqrt{uv}}{bd} - \frac{ad + bc}{2bd} \int \frac{dx}{\sqrt{uv}}$$

150. $$\int \frac{dx}{v\sqrt{uv}} = \frac{-2\sqrt{uv}}{kv}$$

151. $$\int \frac{v\,dx}{\sqrt{uv}} = \frac{\sqrt{uv}}{b} - \frac{k}{2b}\int \frac{dx}{\sqrt{uv}}$$

152. $$\int \sqrt{\frac{v}{u}}\,dx = \frac{v}{|v|}\int \frac{v\,dx}{\sqrt{uv}}$$

153. $$\int v^m\sqrt{u}\,dx = \frac{1}{(2m+3)d}\left(2v^{m+1}\sqrt{u} + k\int \frac{v^m\,dx}{\sqrt{u}}\right)$$

154. $$\int \frac{dx}{v^m\sqrt{u}} = -\frac{1}{(m-1)k}\left(\frac{\sqrt{u}}{v^{m-1}} + \left(m - \frac{3}{2}\right)b\int \frac{dx}{v^{m-1}\sqrt{u}}\right)$$

155. $$\int \frac{v^m\,dx}{\sqrt{u}} = \begin{cases} \dfrac{2}{b(2m+1)}\left[v^m\sqrt{u} - mk\displaystyle\int \frac{v^{m-1}}{\sqrt{u}}\,dx\right] \\ \text{or} \\ \dfrac{2(m!)^2\sqrt{u}}{b(2m+1)!}\displaystyle\sum_{r=0}^{m}\left(-\frac{4k}{b}\right)^{m-r}\frac{(2r)!}{(r!)^2}v^r \end{cases}$$

FORMS CONTAINING $\sqrt{x^2 \pm a^2}$

156. $$\int \sqrt{x^2 \pm a^2}\,dx = \tfrac{1}{2}[x\sqrt{x^2 \pm a^2} \pm a^2 \log(x + \sqrt{x^2 \pm a^2})]$$

157. $$\int \frac{dx}{\sqrt{x^2 \pm a^2}} = \log(x + \sqrt{x^2 \pm a^2})$$

158. $$\int \frac{dx}{x\sqrt{x^2 - a^2}} = \frac{1}{|a|}\sec^{-1}\frac{x}{a}$$

159. $$\int \frac{dx}{x\sqrt{x^2 + a^2}} = -\frac{1}{a}\log\left(\frac{a + \sqrt{x^2 + a^2}}{x}\right)$$

160. $$\int \frac{\sqrt{x^2 + a^2}}{x}\,dx = \sqrt{x^2 + a^2} - a\log\left(\frac{a + \sqrt{x^2 + a^2}}{x}\right)$$

161. $$\int \frac{\sqrt{x^2 - a^2}}{x}\,dx = \sqrt{x^2 - a^2} - |a|\sec^{-1}\frac{x}{a}$$

162. $$\int \frac{x\,dx}{\sqrt{x^2 \pm a^2}} = \sqrt{x^2 \pm a^2}$$

163. $$\int x\sqrt{x^2 \pm a^2}\,dx = \tfrac{1}{3}\sqrt{(x^2 \pm a^2)^3}$$

164. $\int \sqrt{(x^2 \pm a^2)^3}\,dx = \frac{1}{4}\left[x\sqrt{(x^2 \pm a^2)^3} \pm \frac{3a^2x}{2}\sqrt{x^2 \pm a^2} + \frac{3a^4}{2}\log\left(x + \sqrt{x^2 \pm a^2}\right)\right]$

165. $\int \frac{dx}{\sqrt{(x^2 \pm a^2)^3}} = \frac{\pm x}{a^2\sqrt{x^2 \pm a^2}}$

166. $\int \frac{x\,dx}{\sqrt{(x^2 \pm a^2)^3}} = \frac{-1}{\sqrt{x^2 \pm a^2}}$

167. $\int x\sqrt{(x^2 \pm a^2)^3}\,dx = \tfrac{1}{5}\sqrt{(x^2 \pm a^2)^5}$

168. $\int x^2\sqrt{x^2 \pm a^2}\,dx = \frac{x}{4}\sqrt{(x^2 \pm a^2)^3} \mp \frac{a^2}{8}x\sqrt{x^2 \pm a^2} - \frac{a^4}{8}\log\left(x + \sqrt{x^2 \pm a^2}\right)$

169. $\int x^3\sqrt{x^2 + a^2}\,dx = \left(\tfrac{1}{5}x^2 - \tfrac{2}{15}a^2\right)\sqrt{(a^2 + x^2)^3}$

170. $\int x^3\sqrt{x^2 - a^2}\,dx = \frac{1}{5}\sqrt{(x^2 - a^2)^5} + \frac{a^2}{3}\sqrt{(x^2 - a^2)^3}$

171. $\int \frac{x^2\,dx}{\sqrt{x^2 \pm a^2}} = \frac{x}{2}\sqrt{x^2 \pm a^2} \mp \frac{a^2}{2}\log\left(x + \sqrt{x^2 \pm a^2}\right)$

172. $\int \frac{x^3\,dx}{\sqrt{x^2 \pm a^2}} = \frac{1}{3}\sqrt{(x^2 \pm a^2)^3} \mp a^2\sqrt{x^2 \pm a^2}$

173. $\int \frac{dx}{x^2\sqrt{x^2 \pm a^2}} = \mp\frac{\sqrt{x^2 \pm a^2}}{a^2x}$

174. $\int \frac{dx}{x^3\sqrt{x^2 + a^2}} = -\frac{\sqrt{x^2 + a^2}}{2a^2x^2} + \frac{1}{2a^3}\log\frac{a + \sqrt{x^2 + a^2}}{x}$

175. $\int \frac{dx}{x^3\sqrt{x^2 - a^2}} = \frac{\sqrt{x^2 - a^2}}{2a^2x^2} + \frac{1}{2|a^3|}\sec^{-1}\frac{x}{a}$

176. $\int x^2\sqrt{(x^2 \pm a^2)^3}\,dx = \frac{x}{6}\sqrt{(x^2 \pm a^2)^5} \mp \frac{a^2x}{24}\sqrt{(x^2 \pm a^2)^3} - \frac{a^4x}{16}\sqrt{x^2 \pm a^2} \mp \frac{a^6}{16}\log\left(x + \sqrt{x^2 \pm a^2}\right)$

177. $\int x^3\sqrt{(x^2 \pm a^2)^3}\,dx = \frac{1}{7}\sqrt{(x^2 \pm a^2)^7} \mp \frac{a^2}{5}\sqrt{(x^2 \pm a^2)^5}$

178. $$\int \frac{\sqrt{x^2 \pm a^2}\,dx}{x^2} = -\frac{\sqrt{x^2 \pm a^2}}{x} + \log(x + \sqrt{x^2 \pm a^2})$$

179. $$\int \frac{\sqrt{x^2 + a^2}}{x^3}\,dx = -\frac{\sqrt{x^2 + a^2}}{2x^2} - \frac{1}{2a}\log\frac{a + \sqrt{x^2 + a^2}}{x}$$

180. $$\int \frac{\sqrt{x^2 - a^2}}{x^3}\,dx = -\frac{\sqrt{x^2 - a^2}}{2x^2} + \frac{1}{2|a|}\sec^{-1}\frac{x}{a}$$

181. $$\int \frac{\sqrt{x^2 \pm a^2}}{x^4}\,dx = \mp\frac{\sqrt{(x^2 \pm a^2)^3}}{3a^2x^3}$$

182. $$\int \frac{x^2\,dx}{\sqrt{(x^2 \pm a^2)^3}} = \frac{-x}{\sqrt{x^2 \pm a^2}} + \log(x + \sqrt{x^2 \pm a^2})$$

183. $$\int \frac{x^3\,dx}{\sqrt{(x^2 \pm a^2)^3}} = \sqrt{x^2 \pm a^2} \pm \frac{a^2}{\sqrt{x^2 \pm a^2}}$$

184. $$\int \frac{dx}{x\sqrt{(x^2 + a^2)^3}} = \frac{1}{a^2\sqrt{x^2 + a^2}} - \frac{1}{a^3}\log\frac{a + \sqrt{x^2 + a^2}}{x}$$

185. $$\int \frac{dx}{x\sqrt{(x^2 - a^2)^3}} = -\frac{1}{a^2\sqrt{x^2 - a^2}} - \frac{1}{|a^3|}\sec^{-1}\frac{x}{a}$$

186. $$\int \frac{dx}{x^2\sqrt{(x^2 \pm a^2)^3}} = -\frac{1}{a^4}\left[\frac{\sqrt{x^2 \pm a^2}}{x} + \frac{x}{\sqrt{x^2 \pm a^2}}\right]$$

187. $$\int \frac{dx}{x^3\sqrt{(x^2 + a^2)^3}} = -\frac{1}{2a^2x^2\sqrt{x^2 + a^2}} - \frac{3}{2a^4\sqrt{x^2 + a^2}} + \frac{3}{2a^5}\log\frac{a + \sqrt{x^2 + a^2}}{x}$$

188. $$\int \frac{dx}{x^3\sqrt{(x^2 - a^2)^3}} = \frac{1}{2a^2x^2\sqrt{x^2 - a^2}} - \frac{3}{2a^4\sqrt{x^2 - a^2}} - \frac{3}{2|a^5|}\sec^{-1}\frac{x}{a}$$

189. $$\int \frac{x^m}{\sqrt{x^2 \pm a^2}}\,dx = \frac{1}{m}x^{m-1}\sqrt{x^2 \pm a^2} \mp \frac{m-1}{m}a^2\int \frac{x^{m-2}}{\sqrt{x^2 \pm a^2}}\,dx$$

190. $$\int \frac{x^{2m}}{\sqrt{x^2 \pm a^2}}\,dx = \frac{(2m)!}{2^{2m}(m!)^2}\left[\sqrt{x^2 \pm a^2}\sum_{r=1}^{m}\frac{r!(r-1)!}{(2r)!}(\mp a^2)^{m-r}(2x)^{2r-1} + (\mp a^2)^m\log(x + \sqrt{x^2 \pm a^2})\right]$$

191. $$\int \frac{x^{2m+1}}{\sqrt{x^2 \pm a^2}}\,dx = \sqrt{x^2 \pm a^2}\sum_{r=0}^{m}\frac{(2r)!(m!)^2}{(2m+1)!(r!)^2}(\mp 4a^2)^{m-r}x^{2r}$$

192. $$\int \frac{dx}{x^m\sqrt{x^2 \pm a^2}} = \mp \frac{\sqrt{x^2 \pm a^2}}{(m-1)a^2x^{m-1}} \mp \frac{(m-2)}{(m-1)a^2}\int \frac{dx}{x^{m-2}\sqrt{x^2 \pm a^2}}$$

193. $$\int \frac{dx}{x^{2m}\sqrt{x^2 \pm a^2}} = \sqrt{x^2 \pm a^2}\sum_{r=0}^{m-1} \frac{(m-1)!m!(2r)!2^{2m-2r-1}}{(r!)^2(2m)!(\mp a^2)^{m-r}x^{2r+1}}$$

194. $$\int \frac{dx}{x^{2m+1}\sqrt{x^2 + a^2}} = \frac{(2m)!}{(m!)^2}\left[\frac{\sqrt{x^2 + a^2}}{a^2}\sum_{r=1}^{m} (-1)^{m-r+1}\frac{r!(r-1)!}{2(2r)!(4a^2)^{m-r}x^{2r}} + \frac{(-1)^{m+1}}{2^{2m}a^{2m+1}}\log\frac{\sqrt{x^2 + a^2} + a}{x}\right]$$

195. $$\int \frac{dx}{x^{2m+1}\sqrt{x^2 - a^2}} = \frac{(2m)!}{(m!)^2}\left[\frac{\sqrt{x^2 - a^2}}{a^2}\sum_{r=1}^{m} \frac{r!(r-1)!}{2(2r)!(4a^2)^{m-r}x^{2r}} + \frac{1}{2^{2m}|a|^{2m+1}}\sec^{-1}\frac{x}{a}\right]$$

196. $$\int \frac{dx}{(x-a)\sqrt{x^2 - a^2}} = -\frac{\sqrt{x^2 - a^2}}{a(x-a)}$$

197. $$\int \frac{dx}{(x+a)\sqrt{x^2 - a^2}} = \frac{\sqrt{x^2 - a^2}}{a(x+a)}$$

198. $$\int f(x, \sqrt{x^2 + a^2})\,dx = a\int f(a\tan u, a\sec u)\sec^2 u\,du, \qquad \left(u = \tan^{-1}\frac{x}{a}, a > 0\right)$$

199. $$\int f(x, \sqrt{x^2 - a^2})\,dx = a\int f(a\sec u, a\tan u)\sec u\tan u\,du, \qquad \left(u = \sec^{-1}\frac{x}{a}, a > 0\right)$$

FORMS CONTAINING $\sqrt{a^2 - x^2}$

200. $$\int \sqrt{a^2 - x^2}\,dx = \frac{1}{2}\left[x\sqrt{a^2 - x^2} + a^2\sin^{-1}\frac{x}{|a|}\right]$$

201. $$\int \frac{dx}{\sqrt{a^2 - x^2}} = \begin{cases} \sin^{-1}\dfrac{x}{|a|} \\ \text{or} \\ -\cos^{-1}\dfrac{x}{|a|} \end{cases}$$

202. $$\int \frac{dx}{x\sqrt{a^2 - x^2}} = -\frac{1}{a}\log\left(\frac{a + \sqrt{a^2 - x^2}}{x}\right)$$

203. $$\int \frac{\sqrt{a^2 - x^2}}{x}\,dx = \sqrt{a^2 - x^2} - a\log\left(\frac{a + \sqrt{a^2 - x^2}}{x}\right)$$

204. $$\int \frac{x\,dx}{\sqrt{a^2 - x^2}} = -\sqrt{a^2 - x^2}$$

205. $$\int x\sqrt{a^2 - x^2}\,dx = -\tfrac{1}{3}\sqrt{(a^2 - x^2)^3}$$

206. $$\int \sqrt{(a^2 - x^2)^3}\,dx = \frac{1}{4}\left[x\sqrt{(a^2 - x^2)^3} + \frac{3a^2x}{2}\sqrt{a^2 - x^2} + \frac{3a^4}{2}\sin^{-1}\frac{x}{|a|}\right]$$

207. $$\int \frac{dx}{\sqrt{(a^2 - x^2)^3}} = \frac{x}{a^2\sqrt{a^2 - x^2}}$$

208. $$\int \frac{x\,dx}{\sqrt{(a^2 - x^2)^3}} = \frac{1}{\sqrt{a^2 - x^2}}$$

209. $$\int x\sqrt{(a^2 - x^2)^3}\,dx = -\tfrac{1}{5}\sqrt{(a^2 - x^2)^5}$$

210. $$\int x^2\sqrt{a^2 - x^2}\,dx = -\frac{x}{4}\sqrt{(a^2 - x^2)^3} + \frac{a^2}{8}\left(x\sqrt{a^2 - x^2} + a^2\sin^{-1}\frac{x}{|a|}\right)$$

211. $$\int x^3\sqrt{a^2 - x^2}\,dx = \left(-\tfrac{1}{5}x^2 - \tfrac{2}{15}a^2\right)\sqrt{(a^2 - x^2)^3}$$

212. $$\int x^2\sqrt{(a^2 - x^2)^3}\,dx = -\frac{1}{6}x\sqrt{(a^2 - x^2)^5} + \frac{a^2x}{24}\sqrt{(a^2 - x^2)^3} + \frac{a^4x}{16}\sqrt{a^2 - x^2} + \frac{a^6}{16}\sin^{-1}\frac{x}{|a|}$$

213. $$\int x^3\sqrt{(a^2 - x^2)^3}\,dx = \frac{1}{7}\sqrt{(a^2 - x^2)^7} - \frac{a^2}{5}\sqrt{(a^2 - x^2)^5}$$

214. $$\int \frac{x^2\,dx}{\sqrt{a^2 - x^2}} = -\frac{x}{2}\sqrt{a^2 - x^2} + \frac{a^2}{2}\sin^{-1}\frac{x}{|a|}$$

215. $$\int \frac{dx}{x^2\sqrt{a^2 - x^2}} = -\frac{\sqrt{a^2 - x^2}}{a^2x}$$

216. $$\int \frac{\sqrt{a^2 - x^2}}{x^2}\,dx = -\frac{\sqrt{a^2 - x^2}}{x} - \sin^{-1}\frac{x}{|a|}$$

217. $$\int \frac{\sqrt{a^2 - x^2}}{x^3}\,dx = -\frac{\sqrt{a^2 - x^2}}{2x^2} + \frac{1}{2a}\log\frac{a + \sqrt{a^2 - x^2}}{x}$$

218. $$\int \frac{\sqrt{a^2 - x^2}}{x^4}\,dx = -\frac{\sqrt{(a^2 - x^2)^3}}{3a^2x^3}$$

219. $$\int \frac{x^2\,dx}{\sqrt{(a^2-x^2)^3}} = \frac{x}{\sqrt{a^2-x^2}} - \sin^{-1}\frac{x}{|a|}$$

220. $$\int \frac{x^3\,dx}{\sqrt{a^2-x^2}} = -\frac{2}{3}(a^2-x^2)^{\frac{3}{2}} - x^2(a^2-x^2)^{\frac{1}{2}} = -\frac{1}{3}\sqrt{a^2-x^2}(x^2+2a^2)$$

221. $$\int \frac{x^3\,dx}{\sqrt{(a^2-x^2)^3}} = 2(a^2-x^2)^{\frac{1}{2}} + \frac{x^2}{(a^2-x^2)^{\frac{1}{2}}} = \frac{a^2}{\sqrt{a^2-x^2}} + \sqrt{a^2-x^2}$$

222. $$\int \frac{dx}{x^3\sqrt{a^2-x^2}} = -\frac{\sqrt{a^2-x^2}}{2a^2x^2} - \frac{1}{2a^3}\log\frac{a+\sqrt{a^2-x^2}}{x}$$

223. $$\int \frac{dx}{x\sqrt{(a^2-x^2)^3}} = \frac{1}{a^2\sqrt{a^2-x^2}} - \frac{1}{a^3}\log\frac{a+\sqrt{a^2-x^2}}{x}$$

224. $$\int \frac{dx}{x^2\sqrt{(a^2-x^2)^3}} = \frac{1}{a^4}\left[-\frac{\sqrt{a^2-x^2}}{x} + \frac{x}{\sqrt{a^2-x^2}}\right]$$

225. $$\int \frac{dx}{x^3\sqrt{(a^2-x^2)^3}} = -\frac{1}{2a^2x^2\sqrt{a^2-x^2}} + \frac{3}{2a^4\sqrt{a^2-x^2}} - \frac{3}{2a^5}\log\frac{a+\sqrt{a^2-x^2}}{x}$$

226. $$\int \frac{x^m}{\sqrt{a^2-x^2}}\,dx = -\frac{x^{m-1}\sqrt{a^2-x^2}}{m} + \frac{(m-1)a^2}{m}\int\frac{x^{m-2}}{\sqrt{a^2-x^2}}\,dx$$

227. $$\int \frac{x^{2m}}{\sqrt{a^2-x^2}}\,dx = \frac{(2m)!}{(m!)^2}\left[-\sqrt{a^2-x^2}\sum_{r=1}^{m}\frac{r!(r-1)!}{2^{2m-2r+1}(2r)!}a^{2m-2r}x^{2r-1} + \frac{a^{2m}}{2^{2m}}\sin^{-1}\frac{x}{|a|}\right]$$

228. $$\int \frac{x^{2m+1}}{\sqrt{a^2-x^2}}\,dx = -\sqrt{a^2-x^2}\sum_{r=0}^{m}\frac{(2r)!(m!)^2}{(2m+1)!(r!)^2}(4a^2)^{m-r}x^{2r}$$

229. $$\int \frac{dx}{x^m\sqrt{a^2-x^2}} = -\frac{\sqrt{a^2-x^2}}{(m-1)a^2x^{m-1}} + \frac{m-2}{(m-1)a^2}\int\frac{dx}{x^{m-2}\sqrt{a^2-x^2}}$$

230. $$\int \frac{dx}{x^{2m}\sqrt{a^2-x^2}} = -\sqrt{a^2-x^2}\sum_{r=0}^{m-1}\frac{(m-1)!m!(2r)!2^{2m-2r-1}}{(r!)^2(2m)!a^{2m-2r}x^{2r+1}}$$

231. $$\int \frac{dx}{x^{2m+1}\sqrt{a^2-x^2}} = \frac{(2m)!}{(m!)^2}\left[-\frac{\sqrt{a^2-x^2}}{a^2}\sum_{r=1}^{m}\frac{r!(r-1)!}{2(2r)!(4a^2)^{m-r}x^{2r}} + \frac{1}{2^{2m}a^{2m+1}}\log\frac{a-\sqrt{a^2-x^2}}{x}\right]$$

232. $$\int \frac{dx}{(b^2 - x^2)\sqrt{a^2 - x^2}} = \frac{1}{2b\sqrt{a^2 - b^2}} \log \frac{(b\sqrt{a^2 - x^2} + x\sqrt{a^2 - b^2})^2}{b^2 - x^2}, \qquad (a^2 > b^2)$$

233. $$\int \frac{dx}{(b^2 - x^2)\sqrt{a^2 - x^2}} = \frac{1}{b\sqrt{b^2 - a^2}} \tan^{-1} \frac{x\sqrt{b^2 - a^2}}{b\sqrt{a^2 - x^2}}, \qquad (b^2 > a^2)$$

234. $$\int \frac{dx}{(b^2 + x^2)\sqrt{a^2 - x^2}} = \frac{1}{b\sqrt{a^2 + b^2}} \tan^{-1} \frac{x\sqrt{a^2 + b^2}}{b\sqrt{a^2 - x^2}}$$

235. $$\int \frac{\sqrt{a^2 - x^2}}{b^2 + x^2}\, dx = \frac{\sqrt{a^2 + b^2}}{|b|} \sin^{-1} \frac{x\sqrt{a^2 + b^2}}{|a|\sqrt{x^2 + b^2}} - \sin^{-1} \frac{x}{|a|}$$

236. $$\int f(x, \sqrt{a^2 - x^2})\, dx = a \int f(a \sin u, a \cos u) \cos u\, du, \qquad \left(u = \sin^{-1} \frac{x}{a},\ a > 0\right)$$

FORMS CONTAINING $\sqrt{a + bx + cx^2}$

$$X = a + bx + cx^2,\ q = 4ac - b^2, \text{ and } k = \frac{4c}{q}$$

If $q = 0$, then $\sqrt{X} = \sqrt{c}\left|x + \frac{b}{2c}\right|$

237. $$\int \frac{dx}{\sqrt{X}} = \begin{cases} \dfrac{1}{\sqrt{c}} \log (2\sqrt{cX} + 2cx + b) \\ \quad \text{or} \\ \dfrac{1}{\sqrt{c}} \sinh^{-1} \dfrac{2cx + b}{\sqrt{q}}, \qquad (c > 0) \end{cases}$$

238. $$\int \frac{dx}{\sqrt{X}} = -\frac{1}{\sqrt{-c}} \sin^{-1} \frac{2cx + b}{\sqrt{-q}}, \qquad (c < 0)$$

239. $$\int \frac{dx}{X\sqrt{X}} = \frac{2(2cx + b)}{q\sqrt{X}}$$

240. $$\int \frac{dx}{X^2\sqrt{X}} = \frac{2(2cx + b)}{3q\sqrt{X}}\left(\frac{1}{X} + 2k\right)$$

241. $$\int \frac{dx}{X^n\sqrt{X}} = \begin{cases} \dfrac{2(2cx + b)\sqrt{X}}{(2n - 1)qX^n} + \dfrac{2k(n - 1)}{2n - 1} \displaystyle\int \frac{dx}{X^{n-1}\sqrt{X}} \\ \quad \text{or} \\ \dfrac{(2cx + b)(n!)(n - 1)!4^n k^{n-1}}{q[(2n)!]\sqrt{X}} \displaystyle\sum_{r=0}^{n-1} \frac{(2r)!}{(4kX)^r (r!)^2} \end{cases}$$

242. $$\int \sqrt{X}\,dx = \frac{(2cx+b)\sqrt{X}}{4c} + \frac{1}{2k}\int \frac{dx}{\sqrt{X}}$$

243. $$\int X\sqrt{X}\,dx = \frac{(2cx+b)\sqrt{X}}{8c}\left(X + \frac{3}{2k}\right) + \frac{3}{8k^2}\int \frac{dx}{\sqrt{X}}$$

244. $$\int X^2\sqrt{X}\,dx = \frac{(2cx+b)\sqrt{X}}{12c}\left(X^2 + \frac{5X}{4k} + \frac{15}{8k^2}\right) + \frac{5}{16k^3}\int \frac{dx}{\sqrt{X}}$$

245. $$\int X^n\sqrt{X}\,dx = \begin{cases} \dfrac{(2cx+b)X^n\sqrt{X}}{4(n+1)c} + \dfrac{2n+1}{2(n+1)k}\displaystyle\int X^{n-1}\sqrt{X}\,dx \\ \text{or} \\ \dfrac{(2n+2)!}{[(n+1)!]^2(4k)^{n+1}}\left[\dfrac{k(2cx+b)\sqrt{X}}{c}\displaystyle\sum_{r=0}^{n}\dfrac{r!(r+1)!(4kX)^r}{(2r+2)!} \right. \\ \qquad\qquad\qquad\qquad\qquad\qquad\qquad\qquad \left. + \displaystyle\int \dfrac{dx}{\sqrt{X}}\right] \end{cases}$$

246. $$\int \frac{x\,dx}{\sqrt{X}} = \frac{\sqrt{X}}{c} - \frac{b}{2c}\int \frac{dx}{\sqrt{X}}$$

247. $$\int \frac{x\,dx}{X\sqrt{X}} = -\frac{2(bx+2a)}{q\sqrt{X}}$$

248. $$\int \frac{x\,dx}{X^n\sqrt{X}} = -\frac{\sqrt{X}}{(2n-1)cX^n} - \frac{b}{2c}\int \frac{dx}{X^n\sqrt{X}}$$

249. $$\int \frac{x^2\,dx}{\sqrt{X}} = \left(\frac{x}{2c} - \frac{3b}{4c^2}\right)\sqrt{X} + \frac{3b^2-4ac}{8c^2}\int \frac{dx}{\sqrt{X}}$$

250. $$\int \frac{x^2\,dx}{X\sqrt{X}} = \frac{(2b^2-4ac)x+2ab}{cq\sqrt{X}} + \frac{1}{c}\int \frac{dx}{\sqrt{X}}$$

251. $$\int \frac{x^2\,dx}{X^n\sqrt{X}} = \frac{(2b^2-4ac)x+2ab}{(2n-1)cqX^{n-1}\sqrt{X}} + \frac{4ac+(2n-3)b^2}{(2n-1)cq}\int \frac{dx}{X^{n-1}\sqrt{X}}$$

252. $$\int \frac{x^3\,dx}{\sqrt{X}} = \left(\frac{x^2}{3c} - \frac{5bx}{12c^2} + \frac{5b^2}{8c^3} - \frac{2a}{3c^2}\right)\sqrt{X} + \left(\frac{3ab}{4c^2} - \frac{5b^3}{16c^3}\right)\int \frac{dx}{\sqrt{X}}$$

253. $$\int \frac{x^n\,dx}{\sqrt{X}} = \frac{1}{nc}x^{n-1}\sqrt{X} - \frac{(2n-1)b}{2nc}\int \frac{x^{n-1}\,dx}{\sqrt{X}} - \frac{(n-1)a}{nc}\int \frac{x^{n-2}\,dx}{\sqrt{X}}$$

254. $$\int x\sqrt{X}\,dx = \frac{X\sqrt{X}}{3c} - \frac{b(2cx+b)}{8c^2}\sqrt{X} - \frac{b}{4ck}\int\frac{dx}{\sqrt{X}}$$

255. $$\int xX\sqrt{X}\,dx = \frac{X^2\sqrt{X}}{5c} - \frac{b}{2c}\int X\sqrt{X}\,dx$$

256. $$\int xX^n\sqrt{X}\,dx = \frac{X^{n+1}\sqrt{X}}{(2n+3)c} - \frac{b}{2c}\int X^n\sqrt{X}\,dx$$

257. $$\int x^2\sqrt{X}\,dx = \left(x - \frac{5b}{6c}\right)\frac{X\sqrt{X}}{4c} + \frac{5b^2 - 4ac}{16c^2}\int\sqrt{X}\,dx$$

258. $$\int\frac{dx}{x\sqrt{X}} = -\frac{1}{\sqrt{a}}\log\frac{2\sqrt{aX} + bx + 2a}{x}, \qquad (a > 0)$$

259. $$\int\frac{dx}{x\sqrt{X}} = \frac{1}{\sqrt{-a}}\sin^{-1}\left(\frac{bx+2a}{|x|\sqrt{-a}}\right), \qquad (a < 0)$$

260. $$\int\frac{dx}{x\sqrt{X}} = -\frac{2\sqrt{X}}{bx}, \qquad (a = 0)$$

261. $$\int\frac{dx}{x^2\sqrt{X}} = -\frac{\sqrt{X}}{ax} - \frac{b}{2a}\int\frac{dx}{x\sqrt{X}}$$

262. $$\int\frac{\sqrt{X}\,dx}{x} = \sqrt{X} + \frac{b}{2}\int\frac{dx}{\sqrt{X}} + a\int\frac{dx}{x\sqrt{X}}$$

263. $$\int\frac{\sqrt{X}\,dx}{x^2} = -\frac{\sqrt{X}}{x} + \frac{b}{2}\int\frac{dx}{x\sqrt{X}} + c\int\frac{dx}{\sqrt{X}}$$

FORMS INVOLVING $\sqrt{2ax - x^2}$

264. $$\int\sqrt{2ax - x^2}\,dx = \frac{1}{2}\left[(x-a)\sqrt{2ax - x^2} + a^2\sin^{-1}\frac{x-a}{|a|}\right]$$

265. $$\int\frac{dx}{\sqrt{2ax - x^2}} = \begin{cases}\cos^{-1}\dfrac{a-x}{|a|}\\ \text{or}\\ \sin^{-1}\dfrac{x-a}{|a|}\end{cases}$$

266. $$\int x^n\sqrt{2ax-x^2}\,dx = \begin{cases} -\dfrac{x^{n-1}(2ax-x^2)^{\frac{3}{2}}}{n+2} + \dfrac{(2n+1)a}{n+2}\displaystyle\int x^{n-1}\sqrt{2ax-x^2}\,dx \\ \text{or} \\ \sqrt{2ax-x^2}\left[\dfrac{x^{n+1}}{n+2} - \displaystyle\sum_{r=0}^{n}\dfrac{(2n+1)!(r!)^2a^{n-r+1}}{2^{n-r}(2r+1)!(n+2)!n!}x^r\right] \\ \qquad + \dfrac{(2n+1)!a^{n+2}}{2^n n!(n+2)!}\sin^{-1}\dfrac{x-a}{|a|} \end{cases}$$

267. $$\int\frac{\sqrt{2ax-x^2}}{x^n}dx = \frac{(2ax-x^2)^{\frac{3}{2}}}{(3-2n)ax^n} + \frac{n-3}{(2n-3)a}\int\frac{\sqrt{2ax-x^2}}{x^{n-1}}dx$$

268. $$\int\frac{x^n\,dx}{\sqrt{2ax-x^2}} = \begin{cases} \dfrac{-x^{n-1}\sqrt{2ax-x^2}}{n} + \dfrac{a(2n-1)}{n}\displaystyle\int\frac{x^{n-1}}{\sqrt{2ax-x^2}}dx \\ \text{or} \\ -\sqrt{2ax-x^2}\displaystyle\sum_{r=1}^{n}\frac{(2n)!r!(r-1)!a^{n-r}}{2^{n-r}(2r)!(n!)^2}x^{r-1} \\ \qquad + \dfrac{(2n)!a^n}{2^n(n!)^2}\sin^{-1}\dfrac{x-a}{|a|} \end{cases}$$

269. $$\int\frac{dx}{x^n\sqrt{2ax-x^2}} = \begin{cases} \dfrac{\sqrt{2ax-x^2}}{a(1-2n)x^n} + \dfrac{n-1}{(2n-1)a}\displaystyle\int\frac{dx}{x^{n-1}\sqrt{2ax-x^2}} \\ \text{or} \\ -\sqrt{2ax-x^2}\displaystyle\sum_{r=0}^{n-1}\frac{2^{n-r}(n-1)!n!(2r)!}{(2n)!(r!)^2a^{n-r}x^{r+1}} \end{cases}$$

270. $$\int\frac{dx}{(2ax-x^2)^{\frac{3}{2}}} = \frac{x-a}{a^2\sqrt{2ax-x^2}}$$

271. $$\int\frac{x\,dx}{(2ax-x^2)^{\frac{3}{2}}} = \frac{x}{a\sqrt{2ax-x^2}}$$

MISCELLANEOUS ALGEBRAIC FORMS

272. $$\int\frac{dx}{\sqrt{2ax+x^2}} = \log\left(x+a+\sqrt{2ax+x^2}\right)$$

273. $$\int\sqrt{ax^2+c}\,dx = \frac{x}{2}\sqrt{ax^2+c} + \frac{c}{2\sqrt{a}}\log\left(x\sqrt{a}+\sqrt{ax^2+c}\right), \qquad (a>0)$$

274. $$\int\sqrt{ax^2+c}\,dx = \frac{x}{2}\sqrt{ax^2+c} + \frac{c}{2\sqrt{-a}}\sin^{-1}\left(x\sqrt{-\frac{a}{c}}\right), \qquad (a<0)$$

275. $\int \sqrt{\frac{1+x}{1-x}}\,dx = \sin^{-1} x - \sqrt{1-x^2}$

276. $\int \frac{dx}{x\sqrt{ax^n+c}} = \begin{cases} \frac{1}{n\sqrt{c}} \log \frac{\sqrt{ax^n+c}-\sqrt{c}}{\sqrt{ax^n+c}+\sqrt{c}} \\ \text{or} \\ \frac{2}{n\sqrt{c}} \log \frac{\sqrt{ax^n+c}-\sqrt{c}}{\sqrt{x^n}}, \qquad (c>0) \end{cases}$

277. $\int \frac{dx}{x\sqrt{ax^n+c}} = \frac{2}{n\sqrt{-c}} \sec^{-1} \sqrt{-\frac{ax^n}{c}}, \qquad (c<0)$

278. $\int \frac{dx}{\sqrt{ax^2+c}} = \frac{1}{\sqrt{a}} \log (x\sqrt{a} + \sqrt{ax^2+c}), \qquad (a>0)$

279. $\int \frac{dx}{\sqrt{ax^2+c}} = \frac{1}{\sqrt{-a}} \sin^{-1}\left(x\sqrt{-\frac{a}{c}}\right), \qquad (a<0)$

280. $\int (ax^2+c)^{m+\frac{1}{2}}\,dx = \begin{cases} \frac{x(ax^2+c)^{m+\frac{1}{2}}}{2(m+1)} + \frac{(2m+1)c}{2(m+1)} \int (ax^2+c)^{m-\frac{1}{2}}\,dx \\ \text{or} \\ x\sqrt{ax^2+c} \sum_{r=0}^{m} \frac{(2m+1)!(r!)^2 c^{m-r}}{2^{2m-2r+1} m!(m+1)!(2r+1)!}(ax^2+c)^r \\ \quad + \frac{(2m+1)!c^{m+1}}{2^{2m+1} m!(m+1)!} \int \frac{dx}{\sqrt{ax^2+c}} \end{cases}$

281. $\int x(ax^2+c)^{m+\frac{1}{2}}\,dx = \frac{(ax^2+c)^{m+\frac{3}{2}}}{(2m+3)a}$

282. $\int \frac{(ax^2+c)^{m+\frac{1}{2}}}{x}\,dx = \begin{cases} \frac{(ax^2+c)^{m+\frac{1}{2}}}{2m+1} + c \int \frac{(ax^2+c)^{m-\frac{1}{2}}}{x}\,dx \\ \text{or} \\ \sqrt{ax^2+c} \sum_{r=0}^{m} \frac{c^{m-r}(ax^2+c)^r}{2r+1} + c^{m+1} \int \frac{dx}{x\sqrt{ax^2+c}} \end{cases}$

283. $\int \frac{dx}{(ax^2+c)^{m+\frac{1}{2}}} = \begin{cases} \frac{x}{(2m-1)c(ax^2+c)^{m-\frac{1}{2}}} + \frac{2m-2}{(2m-1)c} \int \frac{dx}{(ax^2+c)^{m-\frac{1}{2}}} \\ \text{or} \\ \frac{x}{\sqrt{ax^2+c}} \sum_{r=0}^{m-1} \frac{2^{2m-2r-1}(m-1)!m!(2r)!}{(2m)!(r!)^2 c^{m-r}(ax^2+c)^r} \end{cases}$

284. $$\int \frac{dx}{x^m\sqrt{ax^2+c}} = -\frac{\sqrt{ax^2+c}}{(m-1)cx^{m-1}} - \frac{(m-2)a}{(m-1)c}\int \frac{dx}{x^{m-2}\sqrt{ax^2+c}}$$

285. $$\int \frac{1+x^2}{(1-x^2)\sqrt{1+x^4}}\,dx = \frac{1}{\sqrt{2}}\log\frac{x\sqrt{2}+\sqrt{1+x^4}}{1-x^2}$$

286. $$\int \frac{1-x^2}{(1+x^2)\sqrt{1+x^4}}\,dx = \frac{1}{\sqrt{2}}\tan^{-1}\frac{x\sqrt{2}}{\sqrt{1+x^4}}$$

287. $$\int \frac{dx}{x\sqrt{x^n+a^2}} = -\frac{2}{na}\log\frac{a+\sqrt{x^n+a^2}}{\sqrt{x^n}}$$

288. $$\int \frac{dx}{x\sqrt{x^n-a^2}} = -\frac{2}{na}\sin^{-1}\frac{a}{\sqrt{x^n}}$$

289. $$\int \sqrt{\frac{x}{a^3-x^3}}\,dx = \frac{2}{3}\sin^{-1}\left(\frac{x}{a}\right)^{\frac{3}{2}}$$

FORMS INVOLVING TRIGONOMETRIC FUNCTIONS

290. $$\int (\sin ax)\,dx = -\frac{1}{a}\cos ax$$

291. $$\int (\cos ax)\,dx = \frac{1}{a}\sin ax$$

292. $$\int (\tan ax)\,dx = -\frac{1}{a}\log\cos ax = \frac{1}{a}\log\sec ax$$

293. $$\int (\cot ax)\,dx = \frac{1}{a}\log\sin ax = -\frac{1}{a}\log\csc ax$$

294. $$\int (\sec ax)\,dx = \frac{1}{a}\log(\sec ax + \tan ax) = \frac{1}{a}\log\tan\left(\frac{\pi}{4}+\frac{ax}{2}\right)$$

295. $$\int (\csc ax)\,dx = \frac{1}{a}\log(\csc ax - \cot ax) = \frac{1}{a}\log\tan\frac{ax}{2}$$

296. $$\int (\sin^2 ax)\,dx = -\frac{1}{2a}\cos ax\sin ax + \frac{1}{2}x = \frac{1}{2}x - \frac{1}{4a}\sin 2ax$$

297. $$\int (\sin^3 ax)\,dx = -\frac{1}{3a}(\cos ax)(\sin^2 ax + 2)$$

298. $$\int (\sin^4 ax)\,dx = \frac{3x}{8} - \frac{\sin 2ax}{4a} + \frac{\sin 4ax}{32a}$$

299. $$\int (\sin^n ax)\,dx = -\frac{\sin^{n-1} ax\cos ax}{na} + \frac{n-1}{n}\int (\sin^{n-2} ax)\,dx$$

300. $$\int (\sin^{2m} ax)\,dx = -\frac{\cos ax}{a}\sum_{r=0}^{m-1}\frac{(2m)!(r!)^2}{2^{2m-2r}(2r+1)!(m!)^2}\sin^{2r+1} ax + \frac{(2m)!}{2^{2m}(m!)^2}x$$

301. $$\int (\sin^{2m+1} ax)\,dx = -\frac{\cos ax}{a}\sum_{r=0}^{m}\frac{2^{2m-2r}(m!)^2(2r)!}{(2m+1)!(r!)^2}\sin^{2r} ax$$

302. $$\int (\cos^2 ax)\,dx = \frac{1}{2a}\sin ax\cos ax + \frac{1}{2}x = \frac{1}{2}x + \frac{1}{4a}\sin 2ax$$

303. $$\int (\cos^3 ax)\,dx = \frac{1}{3a}(\sin ax)(\cos^2 ax + 2)$$

304. $$\int (\cos^4 ax)\,dx = \frac{3x}{8} + \frac{\sin 2ax}{4a} + \frac{\sin 4ax}{32a}$$

305. $$\int (\cos^n ax)\,dx = \frac{1}{na}\cos^{n-1} ax\sin ax + \frac{n-1}{n}\int (\cos^{n-2} ax)\,dx$$

306. $$\int (\cos^{2m} ax)\,dx = \frac{\sin ax}{a}\sum_{r=0}^{m-1}\frac{(2m)!(r!)^2}{2^{2m-2r}(2r+1)!(m!)^2}\cos^{2r+1} ax + \frac{(2m)!}{2^{2m}(m!)^2}x$$

307. $$\int (\cos^{2m+1} ax)\,dx = \frac{\sin ax}{a}\sum_{r=0}^{m}\frac{2^{2m-2r}(m!)^2(2r)!}{(2m+1)!(r!)^2}\cos^{2r} ax$$

308. $$\int \frac{dx}{\sin^2 ax} = \int (\csc^2 ax)\,dx = -\frac{1}{a}\cot ax$$

309. $$\int \frac{dx}{\sin^m ax} = \int (\csc^m ax)\,dx = -\frac{1}{(m-1)a}\cdot\frac{\cos ax}{\sin^{m-1} ax} + \frac{m-2}{m-1}\int \frac{dx}{\sin^{m-2} ax}$$

310. $$\int \frac{dx}{\sin^{2m} ax} = \int (\csc^{2m} ax)\,dx = -\frac{1}{a}\cos ax\sum_{r=0}^{m-1}\frac{2^{2m-2r-1}(m-1)!m!(2r)!}{(2m)!(r!)^2\sin^{2r+1} ax}$$

311. $$\int \frac{dx}{\sin^{2m+1} ax} = \int (\csc^{2m+1} ax)\,dx =$$
$$-\frac{1}{a}\cos ax\sum_{r=0}^{m-1}\frac{(2m)!(r!)^2}{2^{2m-2r}(m!)^2(2r+1)!\sin^{2r+2} ax} + \frac{1}{a}\cdot\frac{(2m)!}{2^{2m}(m!)^2}\log\tan\frac{ax}{2}$$

312. $$\int \frac{dx}{\cos^2 ax} = \int (\sec^2 ax)\,dx = \frac{1}{a}\tan ax$$

313. $$\int \frac{dx}{\cos^n ax} = \int (\sec^n ax)\,dx = \frac{1}{(n-1)a}\cdot\frac{\sin ax}{\cos^{n-1} ax} + \frac{n-2}{n-1}\int \frac{dx}{\cos^{n-2} ax}$$

314. $$\int \frac{dx}{\cos^{2m} ax} = \int (\sec^{2m} ax)\,dx = \frac{1}{a}\sin ax\sum_{r=0}^{m-1}\frac{2^{2m-2r-1}(m-1)!m!(2r)!}{(2m)!(r!)^2\cos^{2r+1} ax}$$

315. $$\int \frac{dx}{\cos^{2m+1} ax} = \int (\sec^{2m+1} ax)\, dx = \frac{1}{a} \sin ax \sum_{r=0}^{m-1} \frac{(2m)!(r!)^2}{2^{2m-2r}(m!)^2(2r+1)! \cos^{2r+2} ax} + \frac{1}{a} \cdot \frac{(2m)!}{2^{2m}(m!)^2} \log(\sec ax + \tan ax)$$

316. $$\int (\sin mx)(\sin nx)\, dx = \frac{\sin(m-n)x}{2(m-n)} - \frac{\sin(m+n)x}{2(m+n)}, \qquad (m^2 \neq n^2)$$

317. $$\int (\cos mx)(\cos nx)\, dx = \frac{\sin(m-n)x}{2(m-n)} + \frac{\sin(m+n)x}{2(m+n)}, \qquad (m^2 \neq n^2)$$

318. $$\int (\sin ax)(\cos ax)\, dx = \frac{1}{2a} \sin^2 ax$$

319. $$\int (\sin mx)(\cos nx)\, dx = -\frac{\cos(m-n)x}{2(m-n)} - \frac{\cos(m+n)x}{2(m+n)}, \qquad (m^2 \neq n^2)$$

320. $$\int (\sin^2 ax)(\cos^2 ax)\, dx = -\frac{1}{32a} \sin 4ax + \frac{x}{8}$$

321. $$\int (\sin ax)(\cos^m ax)\, dx = -\frac{\cos^{m+1} ax}{(m+1)a}$$

322. $$\int (\sin^m ax)(\cos ax)\, dx = \frac{\sin^{m+1} ax}{(m+1)a}$$

323. $$\int (\cos^m ax)(\sin^n ax)\, dx = \begin{cases} \dfrac{\cos^{m-1} ax \sin^{n+1} ax}{(m+n)a} + \dfrac{m-1}{m+n} \displaystyle\int (\cos^{m-2} ax)(\sin^n ax)\, dx \\ \text{or} \\ -\dfrac{\sin^{n-1} ax \cos^{m+1} ax}{(m+n)a} + \dfrac{n-1}{m+n} \displaystyle\int (\cos^m ax)(\sin^{n-2} ax)\, dx \end{cases}$$

324. $$\int \frac{\cos^m ax}{\sin^n ax}\, dx = \begin{cases} -\dfrac{\cos^{m+1} ax}{(n-1)a \sin^{n-1} ax} - \dfrac{m-n+2}{n-1} \displaystyle\int \frac{\cos^m ax}{\sin^{n-2} ax}\, dx \\ \text{or} \\ \dfrac{\cos^{m-1} ax}{a(m-n) \sin^{n-1} ax} + \dfrac{m-1}{m-n} \displaystyle\int \frac{\cos^{m-2} ax}{\sin^n ax}\, dx \end{cases}$$

325. $$\int \frac{\sin^m ax}{\cos^n ax}\,dx = \begin{cases} \dfrac{\sin^{m+1} ax}{a(n-1)\cos^{n-1} ax} + \dfrac{m-n+2}{n-1}\displaystyle\int \frac{\sin^m ax}{\cos^{n-2} ax}\,dx \\ \text{or} \\ -\dfrac{\sin^{m-1} ax}{a(m-n)\cos^{n-1} ax} - \dfrac{m-1}{m-n}\displaystyle\int \frac{\sin^{m-2} ax}{\cos^n ax}\,dx \end{cases}$$

326. $$\int \frac{\sin ax}{\cos^2 ax}\,dx = \frac{1}{a\cos ax} = \frac{\sec ax}{a}$$

327. $$\int \frac{\sin^2 ax}{\cos ax}\,dx = -\frac{1}{a}\sin ax + \frac{1}{a}\log\tan\left(\frac{\pi}{4} + \frac{ax}{2}\right)$$

328. $$\int \frac{\cos ax}{\sin^2 ax}\,dx = -\frac{1}{a\sin ax} = -\frac{\csc ax}{a}$$

329. $$\int \frac{dx}{(\sin ax)(\cos ax)} = \frac{1}{a}\log\tan ax$$

330. $$\int \frac{dx}{(\sin ax)(\cos^2 ax)} = \frac{1}{a}\left(\sec ax + \log\tan\frac{ax}{2}\right)$$

331. $$\int \frac{dx}{(\sin ax)(\cos^n ax)} = \frac{1}{a(n-1)\cos^{n-1} ax} + \int \frac{dx}{(\sin ax)(\cos^{n-2} ax)}$$

332. $$\int \frac{dx}{(\sin^2 ax)(\cos ax)} = -\frac{1}{a}\csc ax + \frac{1}{a}\log\tan\left(\frac{\pi}{4} + \frac{ax}{2}\right)$$

333. $$\int \frac{dx}{(\sin^2 ax)(\cos^2 ax)} = -\frac{2}{a}\cot 2ax$$

334. $$\int \frac{dx}{\sin^m ax\cos^n ax} = \begin{cases} -\dfrac{1}{a(m-1)(\sin^{m-1} ax)(\cos^{n-1} ax)} \\ \qquad + \dfrac{m+n-2}{m-1}\displaystyle\int \frac{dx}{(\sin^{m-2} ax)(\cos^n ax)} \\ \text{or} \\ \dfrac{1}{a(n-1)\sin^{m-1} ax\cos^{n-1} ax} \\ \qquad - \dfrac{m+n-2}{n-1}\displaystyle\int \frac{dx}{\sin^m ax\cos^{n-2} ax} \end{cases}$$

335. $$\int \sin(a+bx)\,dx = -\frac{1}{b}\cos(a+bx)$$

336. $$\int \cos(a+bx)\,dx = \frac{1}{b}\sin(a+bx)$$

337. $$\int \frac{dx}{1 \pm \sin ax} = \mp\frac{1}{a}\tan\left(\frac{\pi}{4} \mp \frac{ax}{2}\right)$$

338. $\displaystyle\int \frac{dx}{1+\cos ax} = \frac{1}{a}\tan\frac{ax}{2}$

339. $\displaystyle\int \frac{dx}{1-\cos ax} = -\frac{1}{a}\cot\frac{ax}{2}$

***340.** $$\int \frac{dx}{a+b\sin x} = \begin{cases} \dfrac{2}{\sqrt{a^2-b^2}}\tan^{-1}\dfrac{a\tan\frac{x}{2}+b}{\sqrt{a^2-b^2}} \\ \quad\text{or} \\ \dfrac{1}{\sqrt{b^2-a^2}}\log\dfrac{a\tan\frac{x}{2}+b-\sqrt{b^2-a^2}}{a\tan\frac{x}{2}+b+\sqrt{b^2-a^2}} \end{cases}$$

***341.** $$\int \frac{dx}{a+b\cos x} = \begin{cases} \dfrac{2}{\sqrt{a^2-b^2}}\tan^{-1}\dfrac{\sqrt{a^2-b^2}\tan\frac{x}{2}}{a+b} \\ \quad\text{or} \\ \dfrac{1}{\sqrt{b^2-a^2}}\log\left(\dfrac{\sqrt{b^2-a^2}\tan\frac{x}{2}+a+b}{\sqrt{b^2-a^2}\tan\frac{x}{2}-a-b}\right) \end{cases}$$

***342.** $$\int \frac{dx}{a+b\sin x+c\cos x}$$
$$= \begin{cases} \dfrac{1}{\sqrt{b^2+c^2-a^2}}\log\dfrac{b-\sqrt{b^2+c^2-a^2}+(a-c)\tan\frac{x}{2}}{b+\sqrt{b^2+c^2-a^2}+(a-c)\tan\frac{x}{2}}, & \text{if } a^2<b^2+c^2,\ a\neq c \\ \quad\text{or} \\ \dfrac{2}{\sqrt{a^2-b^2-c^2}}\tan^{-1}\dfrac{b+(a-c)\tan\frac{x}{2}}{\sqrt{a^2-b^2-c^2}}, & \text{if } a^2>b^2+c^2 \\ \quad\text{or} \\ \dfrac{1}{a}\left[\dfrac{a-(b+c)\cos x-(b-c)\sin x}{a-(b-c)\cos x+(b+c)\sin x}\right], & \text{if } a^2=b^2+c^2,\ a\neq c. \end{cases}$$

***343.** $\displaystyle\int \frac{\sin^2 x\,dx}{a+b\cos^2 x} = \frac{1}{b}\sqrt{\frac{a+b}{a}}\tan^{-1}\left(\sqrt{\frac{a}{a+b}}\tan x\right) - \frac{x}{b}, \qquad (ab>0, \text{ or } |a|>|b|)$

*See p. A-39.

*344. $$\int \frac{dx}{a^2\cos^2 x + b^2\sin^2 x} = \frac{1}{ab}\tan^{-1}\left(\frac{b\tan x}{a}\right)$$

*345. $$\int \frac{\cos^2 cx}{a^2 + b^2\sin^2 cx}\,dx = \frac{\sqrt{a^2+b^2}}{ab^2c}\tan^{-1}\frac{\sqrt{a^2+b^2}\tan cx}{a} - \frac{x}{b^2}$$

346. $$\int \frac{\sin cx\cos cx}{a\cos^2 cx + b\sin^2 cx}\,dx = \frac{1}{2c(b-a)}\log(a\cos^2 cx + b\sin^2 cx)$$

347. $$\int \frac{\cos cx}{a\cos cx + b\sin cx}\,dx = \int \frac{dx}{a + b\tan cx} = \frac{1}{c(a^2+b^2)}[acx + b\log(a\cos cx + b\sin cx)]$$

348. $$\int \frac{\sin cx}{a\sin cx + b\cos cx}\,dx = \int \frac{dx}{a + b\cot cx} = \frac{1}{c(a^2+b^2)}[acx - b\log(a\sin cx + b\cos cx)]$$

*349. $$\int \frac{dx}{a\cos^2 x + 2b\cos x\sin x + c\sin^2 x} = \begin{cases} \dfrac{1}{2\sqrt{b^2-ac}}\log\dfrac{c\tan x + b - \sqrt{b^2-ac}}{c\tan x + b + \sqrt{b^2-ac}}, & (b^2 > ac) \\ \text{or} & \\ \dfrac{1}{\sqrt{ac-b^2}}\tan^{-1}\dfrac{c\tan x + b}{\sqrt{ac-b^2}}, & (b^2 < ac) \\ \text{or} & \\ -\dfrac{1}{c\tan x + b}, & (b^2 = ac) \end{cases}$$

350. $$\int \frac{\sin ax}{1 \pm \sin ax}\,dx = \pm x + \frac{1}{a}\tan\left(\frac{\pi}{4} \mp \frac{ax}{2}\right)$$

351. $$\int \frac{dx}{(\sin ax)(1 \pm \sin ax)} = \frac{1}{a}\tan\left(\frac{\pi}{4} \mp \frac{ax}{2}\right) + \frac{1}{a}\log\tan\frac{ax}{2}$$

352. $$\int \frac{dx}{(1+\sin ax)^2} = -\frac{1}{2a}\tan\left(\frac{\pi}{4} - \frac{ax}{2}\right) - \frac{1}{6a}\tan^3\left(\frac{\pi}{4} - \frac{ax}{2}\right)$$

353. $$\int \frac{dx}{(1-\sin ax)^2} = \frac{1}{2a}\cot\left(\frac{\pi}{4} - \frac{ax}{2}\right) + \frac{1}{6a}\cot^3\left(\frac{\pi}{4} - \frac{ax}{2}\right)$$

354. $$\int \frac{\sin ax}{(1+\sin ax)^2}\,dx = -\frac{1}{2a}\tan\left(\frac{\pi}{4} - \frac{ax}{2}\right) + \frac{1}{6a}\tan^3\left(\frac{\pi}{4} - \frac{ax}{2}\right).$$

*See p. A-39.

355. $\int \frac{\sin ax}{(1-\sin ax)^2}dx = -\frac{1}{2a}\cot\left(\frac{\pi}{4}-\frac{ax}{2}\right) + \frac{1}{6a}\cot^3\left(\frac{\pi}{4}-\frac{ax}{2}\right)$

356. $\int \frac{\sin x\,dx}{a+b\sin x} = \frac{x}{b} - \frac{a}{b}\int \frac{dx}{a+b\sin x}$

357. $\int \frac{dx}{(\sin x)(a+b\sin x)} = \frac{1}{a}\log\tan\frac{x}{2} - \frac{b}{a}\int\frac{dx}{a+b\sin x}$

358. $\int \frac{dx}{(a+b\sin x)^2} = \frac{b\cos x}{(a^2-b^2)(a+b\sin x)} + \frac{a}{a^2-b^2}\int\frac{dx}{a+b\sin x}$

359. $\int \frac{\sin x\,dx}{(a+b\sin x)^2} = \frac{a\cos x}{(b^2-a^2)(a+b\sin x)} + \frac{b}{b^2-a^2}\int\frac{dx}{a+b\sin x}$

*360. $\int \frac{dx}{a^2+b^2\sin^2 cx} = \frac{1}{ac\sqrt{a^2+b^2}}\tan^{-1}\frac{\sqrt{a^2+b^2}\tan cx}{a}$

*361. $\int \frac{dx}{a^2-b^2\sin^2 cx} = \begin{cases} \frac{1}{ac\sqrt{a^2-b^2}}\tan^{-1}\frac{\sqrt{a^2-b^2}\tan cx}{a}, & (a^2>b^2) \\ \text{or} & \\ \frac{1}{2ac\sqrt{b^2-a^2}}\log\frac{\sqrt{b^2-a^2}\tan cx + a}{\sqrt{b^2-a^2}\tan cx - a}, & (a^2<b^2) \end{cases}$

362. $\int \frac{\cos ax}{1+\cos ax}dx = x - \frac{1}{a}\tan\frac{ax}{2}$

363. $\int \frac{\cos ax}{1-\cos ax}dx = -x - \frac{1}{a}\cot\frac{ax}{2}$

364. $\int \frac{dx}{(\cos ax)(1+\cos ax)} = \frac{1}{a}\log\tan\left(\frac{\pi}{4}+\frac{ax}{2}\right) - \frac{1}{a}\tan\frac{ax}{2}$

365. $\int \frac{dx}{(\cos ax)(1-\cos ax)} = \frac{1}{a}\log\tan\left(\frac{\pi}{4}+\frac{ax}{2}\right) - \frac{1}{a}\cot\frac{ax}{2}$

366. $\int \frac{dx}{(1+\cos ax)^2} = \frac{1}{2a}\tan\frac{ax}{2} + \frac{1}{6a}\tan^3\frac{ax}{2}$

367. $\int \frac{dx}{(1-\cos ax)^2} = -\frac{1}{2a}\cot\frac{ax}{2} - \frac{1}{6a}\cot^3\frac{ax}{2}$

368. $\int \frac{\cos ax}{(1+\cos ax)^2}dx = \frac{1}{2a}\tan\frac{ax}{2} - \frac{1}{6a}\tan^3\frac{ax}{2}$

369. $\int \frac{\cos ax}{(1-\cos ax)^2}dx = \frac{1}{2a}\cot\frac{ax}{2} - \frac{1}{6a}\cot^3\frac{ax}{2}$

*See p. A-39.

370. $$\int \frac{\cos x\, dx}{a + b\cos x} = \frac{x}{b} - \frac{a}{b}\int \frac{dx}{a + b\cos x}$$

371. $$\int \frac{dx}{(\cos x)(a + b\cos x)} = \frac{1}{a}\log\tan\left(\frac{x}{2} + \frac{\pi}{4}\right) - \frac{b}{a}\int \frac{dx}{a + b\cos x}$$

372. $$\int \frac{dx}{(a + b\cos x)^2} = \frac{b\sin x}{(b^2 - a^2)(a + b\cos x)} - \frac{a}{b^2 - a^2}\int \frac{dx}{a + b\cos x}$$

373. $$\int \frac{\cos x}{(a + b\cos x)^2}\, dx = \frac{a\sin x}{(a^2 - b^2)(a + b\cos x)} - \frac{b}{a^2 - b^2}\int \frac{dx}{a + b\cos x}$$

***374.** $$\int \frac{dx}{a^2 + b^2 - 2ab\cos cx} = \frac{2}{c(a^2 - b^2)}\tan^{-1}\left(\frac{a + b}{a - b}\tan\frac{cx}{2}\right)$$

***375.** $$\int \frac{dx}{a^2 + b^2\cos^2 cx} = \frac{1}{ac\sqrt{a^2 + b^2}}\tan^{-1}\frac{a\tan cx}{\sqrt{a^2 + b^2}}$$

***376.** $$\int \frac{dx}{a^2 - b^2\cos^2 cx} = \begin{cases} \dfrac{1}{ac\sqrt{a^2 - b^2}}\tan^{-1}\dfrac{a\tan cx}{\sqrt{a^2 - b^2}}, & (a^2 > b^2) \\ \text{or} & \\ \dfrac{1}{2ac\sqrt{b^2 - a^2}}\log\dfrac{a\tan cx - \sqrt{b^2 - a^2}}{a\tan cx + \sqrt{b^2 - a^2}}, & (b^2 > a^2) \end{cases}$$

377. $$\int \frac{\sin ax}{1 \pm \cos ax}\, dx = \mp\frac{1}{a}\log(1 \pm \cos ax)$$

378. $$\int \frac{\cos ax}{1 \pm \sin ax}\, dx = \pm\frac{1}{a}\log(1 \pm \sin ax)$$

379. $$\int \frac{dx}{(\sin ax)(1 \pm \cos ax)} = \pm\frac{1}{2a(1 \pm \cos ax)} + \frac{1}{2a}\log\tan\frac{ax}{2}$$

380. $$\int \frac{dx}{(\cos ax)(1 \pm \sin ax)} = \mp\frac{1}{2a(1 \pm \sin ax)} + \frac{1}{2a}\log\tan\left(\frac{\pi}{4} + \frac{ax}{2}\right)$$

381. $$\int \frac{\sin ax}{(\cos ax)(1 \pm \cos ax)}\, dx = \frac{1}{a}\log(\sec ax \pm 1)$$

382. $$\int \frac{\cos ax}{(\sin ax)(1 \pm \sin ax)}\, dx = -\frac{1}{a}\log(\csc ax \pm 1)$$

383. $$\int \frac{\sin ax}{(\cos ax)(1 \pm \sin ax)}\, dx = \frac{1}{2a(1 \pm \sin ax)} \pm \frac{1}{2a}\log\tan\left(\frac{\pi}{4} + \frac{ax}{2}\right)$$

384. $$\int \frac{\cos ax}{(\sin ax)(1 \pm \cos ax)}\, dx = -\frac{1}{2a(1 \pm \cos ax)} \pm \frac{1}{2a}\log\tan\frac{ax}{2}$$

*See p. A-39.

385. $\int \frac{dx}{\sin ax \pm \cos ax} = \frac{1}{a\sqrt{2}} \log \tan\left(\frac{ax}{2} \pm \frac{\pi}{8}\right)$

386. $\int \frac{dx}{(\sin ax \pm \cos ax)^2} = \frac{1}{2a} \tan\left(ax \mp \frac{\pi}{4}\right)$

387. $\int \frac{dx}{1 + \cos ax \pm \sin ax} = \pm \frac{1}{a} \log\left(1 \pm \tan \frac{ax}{2}\right)$

388. $\int \frac{dx}{a^2 \cos^2 cx - b^2 \sin^2 cx} = \frac{1}{2abc} \log \frac{b \tan cx + a}{b \tan cx - a}$

389. $\int x(\sin ax)\, dx = \frac{1}{a^2} \sin ax - \frac{x}{a} \cos ax$

390. $\int x^2(\sin ax)\, dx = \frac{2x}{a^2} \sin ax - \frac{a^2x^2 - 2}{a^3} \cos ax$

391. $\int x^3(\sin ax)\, dx = \frac{3a^2x^2 - 6}{a^4} \sin ax - \frac{a^2x^3 - 6x}{a^3} \cos ax$

392. $$\int x^m \sin ax\, dx = \begin{cases} -\frac{1}{a} x^m \cos ax + \frac{m}{a} \int x^{m-1} \cos ax\, dx \\ \text{or} \\ \cos ax \sum_{r=0}^{\left[\frac{m}{2}\right]} (-1)^{r+1} \frac{m!}{(m-2r)!} \cdot \frac{x^{m-2r}}{a^{2r+1}} \\ \quad + \sin ax \sum_{r=0}^{\left[\frac{m-1}{2}\right]} (-1)^r \frac{m!}{(m-2r-1)!} \cdot \frac{x^{m-2r-1}}{a^{2r+2}} \end{cases}$$

Note: $[s]$ means greatest integer $\leq s$; $[3\tfrac{1}{2}] = 3$, $[\tfrac{1}{2}] = 0$, etc.

393. $\int x(\cos ax)\, dx = \frac{1}{a^2} \cos ax + \frac{x}{a} \sin ax$

394. $\int x^2(\cos ax)\, dx = \frac{2x \cos ax}{a^2} + \frac{a^2x^2 - 2}{a^3} \sin ax$

395. $\int x^3(\cos ax)\, dx = \frac{3a^2x^2 - 6}{a^4} \cos ax + \frac{a^2x^3 - 6x}{a^3} \sin ax$

396. $$\int x^m(\cos ax)\, dx = \begin{cases} \frac{x^m \sin ax}{a} - \frac{m}{a} \int x^{m-1} \sin ax\, dx \\ \text{or} \\ \sin ax \sum_{r=0}^{\left[\frac{m}{2}\right]} (-1)^r \frac{m!}{(m-2r)!} \cdot \frac{x^{m-2r}}{a^{2r+1}} \\ \quad + \cos ax \sum_{r=0}^{\left[\frac{m-1}{2}\right]} (-1)^r \frac{m!}{(m-2r-1)!} \cdot \frac{x^{m-2r-1}}{a^{2r+2}} \end{cases}$$

See note integral 392.

397. $\int \frac{\sin ax}{x}\, dx = \sum_{n=0}^{\infty} (-1)^n \frac{(ax)^{2n+1}}{(2n+1)(2n+1)!}$

398. $\int \frac{\cos ax}{x}\, dx = \log x + \sum_{n=1}^{\infty} (-1)^n \frac{(ax)^{2n}}{2n(2n)!}$

399. $\int x(\sin^2 ax)\, dx = \frac{x^2}{4} - \frac{x \sin 2ax}{4a} - \frac{\cos 2ax}{8a^2}$

400. $\int x^2(\sin^2 ax)\, dx = \frac{x^3}{6} - \left(\frac{x^2}{4a} - \frac{1}{8a^3}\right) \sin 2ax - \frac{x \cos 2ax}{4a^2}$

401. $\int x(\sin^3 ax)\, dx = \frac{x \cos 3ax}{12a} - \frac{\sin 3ax}{36a^2} - \frac{3x \cos ax}{4a} + \frac{3 \sin ax}{4a^2}$

402. $\int x(\cos^2 ax)\, dx = \frac{x^2}{4} + \frac{x \sin 2ax}{4a} + \frac{\cos 2ax}{8a^2}$

403. $\int x^2(\cos^2 ax)\, dx = \frac{x^3}{6} + \left(\frac{x^2}{4a} - \frac{1}{8a^3}\right) \sin 2ax + \frac{x \cos 2ax}{4a^2}$

404. $\int x(\cos^3 ax)\, dx = \frac{x \sin 3ax}{12a} + \frac{\cos 3ax}{36a^2} + \frac{3x \sin ax}{4a} + \frac{3 \cos ax}{4a^2}$

405. $\int \frac{\sin ax}{x^m}\, dx = -\frac{\sin ax}{(m-1)x^{m-1}} + \frac{a}{m-1} \int \frac{\cos ax}{x^{m-1}}\, dx$

406. $\int \frac{\cos ax}{x^m}\, dx = -\frac{\cos ax}{(m-1)x^{m-1}} - \frac{a}{m-1} \int \frac{\sin ax}{x^{m-1}}\, dx$

407. $\int \frac{x}{1 \pm \sin ax}\, dx = \mp \frac{x \cos ax}{a(1 \pm \sin ax)} + \frac{1}{a^2} \log (1 \pm \sin ax)$

408. $\int \frac{x}{1 + \cos ax}\, dx = \frac{x}{a} \tan \frac{ax}{2} + \frac{2}{a^2} \log \cos \frac{ax}{2}$

409. $\int \frac{x}{1 - \cos ax}\, dx = -\frac{x}{a} \cot \frac{ax}{2} + \frac{2}{a^2} \log \sin \frac{ax}{2}$

410. $\int \frac{x + \sin x}{1 + \cos x}\, dx = x \tan \frac{x}{2}$

411. $\int \frac{x - \sin x}{1 - \cos x}\, dx = -x \cot \frac{x}{2}$

412. $\int \sqrt{1 - \cos ax\, dx} = -\frac{2 \sin ax}{a\sqrt{1 - \cos ax}} = -\frac{2}{a}\sqrt{1 + \cos ax}$

413. $\int \sqrt{1 + \cos ax\, dx} = \frac{2 \sin ax}{a\sqrt{1 + \cos ax}} = \frac{2}{a}\sqrt{1 - \cos ax}$

414. $\int \sqrt{1 + \sin x}\, dx = \pm 2\left(\sin\frac{x}{2} - \cos\frac{x}{2}\right),$

[use + if $(8k-1)\frac{\pi}{2} < x \le (8k+3)\frac{\pi}{2}$, otherwise −; k an integer]

415. $\int \sqrt{1 - \sin x}\, dx = \pm 2\left(\sin\frac{x}{2} + \cos\frac{x}{2}\right),$

[use + if $(8k-3)\frac{\pi}{2} < x \le (8k+1)\frac{\pi}{2}$, otherwise −; k an integer]

416. $\int \frac{dx}{\sqrt{1 - \cos x}} = \pm \sqrt{2} \log \tan\frac{x}{4},$

[use + if $4k\pi < x < (4k+2)\pi$, otherwise −; k an integer]

417. $\int \frac{dx}{\sqrt{1 + \cos x}} = \pm \sqrt{2} \log \tan\left(\frac{x + \pi}{4}\right),$

[use + if $(4k-1)\pi < x < (4k+1)\pi$, otherwise −; k an integer]

418. $\int \frac{dx}{\sqrt{1 - \sin x}} = \pm \sqrt{2} \log \tan\left(\frac{x}{4} - \frac{\pi}{8}\right),$

[use + if $(8k+1)\frac{\pi}{2} < x < (8k+5)\frac{\pi}{2}$, otherwise −; k an integer]

419. $\int \frac{dx}{\sqrt{1 + \sin x}} = \pm \sqrt{2} \log \tan\left(\frac{x}{4} + \frac{\pi}{8}\right),$

[use + if $(8k-1)\frac{\pi}{2} < x < (8k+3)\frac{\pi}{2}$, otherwise −; k an integer]

420. $\int (\tan^2 ax)\, dx = \frac{1}{a} \tan ax - x$

421. $\int (\tan^3 ax)\, dx = \frac{1}{2a} \tan^2 ax + \frac{1}{a} \log \cos ax$

422. $\int (\tan^4 ax)\, dx = \frac{\tan^3 ax}{3a} - \frac{1}{a} \tan x + x$

423. $\int (\tan^n ax)\, dx = \frac{\tan^{n-1} ax}{a(n-1)} - \int (\tan^{n-2} ax)\, dx$

424. $\int (\cot^2 ax)\, dx = -\frac{1}{a} \cot ax - x$

425. $\int (\cot^3 ax)\, dx = -\frac{1}{2a} \cot^2 ax - \frac{1}{a} \log \sin ax$

426. $\int (\cot^4 ax)\, dx = -\frac{1}{3a} \cot^3 ax + \frac{1}{a} \cot ax + x$

427. $$\int (\cot^n ax)\,dx = -\frac{\cot^{n-1} ax}{a(n-1)} - \int (\cot^{n-2} ax)\,dx$$

428. $$\int \frac{x}{\sin^2 ax}\,dx = \int x(\csc^2 ax)\,dx = -\frac{x \cot ax}{a} + \frac{1}{a^2}\log \sin ax$$

429. $$\int \frac{x}{\sin^n ax}\,dx = \int x(\csc^n ax)\,dx = -\frac{x \cos ax}{a(n-1)\sin^{n-1} ax} - \frac{1}{a^2(n-1)(n-2)\sin^{n-2} ax} + \frac{(n-2)}{(n-1)}\int \frac{x}{\sin^{n-2} ax}\,dx$$

430. $$\int \frac{x}{\cos^2 ax}\,dx = \int x(\sec^2 ax)\,dx = \frac{1}{a} x \tan ax + \frac{1}{a^2}\log \cos ax$$

431. $$\int \frac{x}{\cos^n ax}\,dx = \int x(\sec^n ax)\,dx = \frac{x \sin ax}{a(n-1)\cos^{n-1} ax} - \frac{1}{a^2(n-1)(n-2)\cos^{n-2} ax} + \frac{n-2}{n-1}\int \frac{x}{\cos^{n-2} ax}\,dx$$

432. $$\int \frac{\sin ax}{\sqrt{1 + b^2 \sin^2 ax}}\,dx = -\frac{1}{ab}\sin^{-1}\frac{b \cos ax}{\sqrt{1+b^2}}$$

433. $$\int \frac{\sin ax}{\sqrt{1 - b^2 \sin^2 ax}}\,dx = -\frac{1}{ab}\log\left(b \cos ax + \sqrt{1 - b^2 \sin^2 ax}\right)$$

434. $$\int (\sin ax)\sqrt{1 + b^2 \sin^2 ax}\,dx = -\frac{\cos ax}{2a}\sqrt{1 + b^2 \sin^2 ax} - \frac{1+b^2}{2ab}\sin^{-1}\frac{b \cos ax}{\sqrt{1+b^2}}$$

435. $$\int (\sin ax)\sqrt{1 - b^2 \sin^2 ax}\,dx = -\frac{\cos ax}{2a}\sqrt{1 - b^2 \sin^2 ax} - \frac{1-b^2}{2ab}\log\left(b \cos ax + \sqrt{1 - b^2 \sin^2 ax}\right)$$

436. $$\int \frac{\cos ax}{\sqrt{1 + b^2 \sin^2 ax}}\,dx = \frac{1}{ab}\log\left(b \sin ax + \sqrt{1 + b^2 \sin^2 ax}\right)$$

437. $$\int \frac{\cos ax}{\sqrt{1 - b^2 \sin^2 ax}}\,dx = \frac{1}{ab}\sin^{-1}(b \sin ax)$$

438. $$\int (\cos ax)\sqrt{1 + b^2 \sin^2 ax}\,dx = \frac{\sin ax}{2a}\sqrt{1 + b^2 \sin^2 ax} + \frac{1}{2ab}\log\left(b \sin ax + \sqrt{1 + b^2 \sin^2 ax}\right)$$

439. $$\int (\cos ax)\sqrt{1 - b^2 \sin^2 ax}\, dx = \frac{\sin ax}{2a}\sqrt{1 - b^2 \sin^2 ax} + \frac{1}{2ab}\sin^{-1}(b \sin ax)$$

440. $$\int \frac{dx}{\sqrt{a + b \tan^2 cx}} = \frac{\pm 1}{c\sqrt{a - b}} \sin^{-1}\left(\sqrt{\frac{a - b}{a}} \sin cx\right), \qquad (a > |b|)$$

$$\left[\text{use + if } (2k - 1)\frac{\pi}{2} < x \leq (2k + 1)\frac{\pi}{2}, \text{ otherwise } -;\ k \text{ an integer}\right]$$

FORMS INVOLVING INVERSE TRIGONOMETRIC FUNCTIONS

441. $$\int (\sin^{-1} ax)\, dx = x \sin^{-1} ax + \frac{\sqrt{1 - a^2x^2}}{a}$$

442. $$\int (\cos^{-1} ax)\, dx = x \cos^{-1} ax - \frac{\sqrt{1 - a^2x^2}}{a}$$

443. $$\int (\tan^{-1} ax)\, dx = x \tan^{-1} ax - \frac{1}{2a} \log(1 + a^2x^2)$$

444. $$\int (\cot^{-1} ax)\, dx = x \cot^{-1} ax + \frac{1}{2a} \log(1 + a^2x^2)$$

445. $$\int (\sec^{-1} ax)\, dx = x \sec^{-1} ax - \frac{1}{a} \log(ax + \sqrt{a^2x^2 - 1})$$

446. $$\int (\csc^{-1} ax)\, dx = x \csc^{-1} ax + \frac{1}{a} \log(ax + \sqrt{a^2x^2 - 1})$$

447. $$\int \left(\sin^{-1} \frac{x}{a}\right) dx = x \sin^{-1} \frac{x}{a} + \sqrt{a^2 - x^2}, \qquad (a > 0)$$

448. $$\int \left(\cos^{-1} \frac{x}{a}\right) dx = x \cos^{-1} \frac{x}{a} - \sqrt{a^2 - x^2}, \qquad (a > 0)$$

449. $$\int \left(\tan^{-1} \frac{x}{a}\right) dx = x \tan^{-1} \frac{x}{a} - \frac{a}{2} \log(a^2 + x^2)$$

450. $$\int \left(\cot^{-1} \frac{x}{a}\right) dx = x \cot^{-1} \frac{x}{a} + \frac{a}{2} \log(a^2 + x^2)$$

451. $$\int x[\sin^{-1}(ax)]\, dx = \frac{1}{4a^2}[(2a^2x^2 - 1) \sin^{-1}(ax) + ax\sqrt{1 - a^2x^2}]$$

452. $$\int x[\cos^{-1}(ax)]\, dx = \frac{1}{4a^2}[(2a^2x^2 - 1) \cos^{-1}(ax) - ax\sqrt{1 - a^2x^2}]$$

453. $\int x^n[\sin^{-1}(ax)]\,dx = \frac{x^{n+1}}{n+1}\sin^{-1}(ax) - \frac{a}{n+1}\int \frac{x^{n+1}\,dx}{\sqrt{1-a^2x^2}}, \qquad (n \neq -1)$

454. $\int x^n[\cos^{-1}(ax)]\,dx = \frac{x^{n+1}}{n+1}\cos^{-1}(ax) + \frac{a}{n+1}\int \frac{x^{n+1}\,dx}{\sqrt{1-a^2x^2}}, \qquad (n \neq -1)$

455. $\int x(\tan^{-1} ax)\,dx = \frac{1+a^2x^2}{2a^2}\tan^{-1} ax - \frac{x}{2a}$

456. $\int x^n(\tan^{-1} ax)\,dx = \frac{x^{n+1}}{n+1}\tan^{-1} ax - \frac{a}{n+1}\int \frac{x^{n+1}}{1+a^2x^2}\,dx$

457. $\int x(\cot^{-1} ax)\,dx = \frac{1+a^2x^2}{2a^2}\cot^{-1} ax + \frac{x}{2a}$

458. $\int x^n(\cot^{-1} ax)\,dx = \frac{x^{n+1}}{n+1}\cot^{-1} ax + \frac{a}{n+1}\int \frac{x^{n+1}}{1+a^2x^2}\,dx$

459. $\int \frac{\sin^{-1}(ax)}{x^2}\,dx = a\log\left(\frac{1-\sqrt{1-a^2x^2}}{x}\right) - \frac{\sin^{-1}(ax)}{x}$

460. $\int \frac{\cos^{-1}(ax)\,dx}{x^2} = -\frac{1}{x}\cos^{-1}(ax) + a\log\frac{1+\sqrt{1-a^2x^2}}{x}$

461. $\int \frac{\tan^{-1}(ax)\,dx}{x^2} = -\frac{1}{x}\tan^{-1}(ax) - \frac{a}{2}\log\frac{1+a^2x^2}{x^2}$

462. $\int \frac{\cot^{-1} ax}{x^2}\,dx = -\frac{1}{x}\cot^{-1} ax - \frac{a}{2}\log\frac{x^2}{a^2x^2+1}$

463. $\int (\sin^{-1} ax)^2\,dx = x(\sin^{-1} ax)^2 - 2x + \frac{2\sqrt{1-a^2x^2}}{a}\sin^{-1} ax$

464. $\int (\cos^{-1} ax)^2\,dx = x(\cos^{-1} ax)^2 - 2x - \frac{2\sqrt{1-a^2x^2}}{a}\cos^{-1} ax$

465. $$\int (\sin^{-1} ax)^n\,dx = \begin{cases} x(\sin^{-1} ax)^n + \frac{n\sqrt{1-a^2x^2}}{a}(\sin^{-1} ax)^{n-1} \\ \qquad\qquad -n(n-1)\int (\sin^{-1} ax)^{n-2}\,dx \\ \qquad\qquad \text{or} \\ \sum_{r=0}^{\left[\frac{n}{2}\right]} (-1)^r \frac{n!}{(n-2r)!}\,x(\sin^{-1} ax)^{n-2r} \\ \qquad + \sum_{r=0}^{\left[\frac{n-1}{2}\right]} (-1)^r \frac{n!\sqrt{1-a^2x^2}}{(n-2r-1)!\,a}(\sin^{-1} ax)^{n-2r-1} \end{cases}$$

Note: $[s]$ means greatest integer $\leq s$. Thus $[3.5]$ means 3; $[5] = 5$, $[\frac{1}{2}] = 0$.

466. $$\int (\cos^{-1} ax)^n\, dx = \begin{cases} x(\cos^{-1} ax)^n - \dfrac{n\sqrt{1-a^2x^2}}{a}(\cos^{-1} ax)^{n-1} \\ \qquad\qquad - n(n-1)\displaystyle\int (\cos^{-1} ax)^{n-2}\, dx \\ \quad \text{or} \\ \displaystyle\sum_{r=0}^{\left[\frac{n}{2}\right]} (-1)^r \frac{n!}{(n-2r)!} x(\cos^{-1} ax)^{n-2r} \\ \qquad - \displaystyle\sum_{r=0}^{\left[\frac{n-1}{2}\right]} (-1)^r \frac{n!\sqrt{1-a^2x^2}}{(n-2r-1)!\,a}(\cos^{-1} ax)^{n-2r-1} \end{cases}$$

467. $$\int \frac{1}{\sqrt{1-a^2x^2}}(\sin^{-1} ax)\, dx = \frac{1}{2a}(\sin^{-1} ax)^2$$

468. $$\int \frac{x^n}{\sqrt{1-a^2x^2}}(\sin^{-1} ax)\, dx = -\frac{x^{n-1}}{na^2}\sqrt{1-a^2x^2}\sin^{-1} ax + \frac{x^n}{n^2a} + \frac{n-1}{na^2}\int \frac{x^{n-2}}{\sqrt{1-a^2x^2}}\sin^{-1} ax\, dx$$

469. $$\int \frac{1}{\sqrt{1-a^2x^2}}(\cos^{-1} ax)\, dx = -\frac{1}{2a}(\cos^{-1} ax)^2$$

470. $$\int \frac{x^n}{\sqrt{1-a^2x^2}}(\cos^{-1} ax)\, dx = -\frac{x^{n-1}}{na^2}\sqrt{1-a^2x^2}\cos^{-1} ax - \frac{x^n}{n^2a} + \frac{n-1}{na^2}\int \frac{x^{n-2}}{\sqrt{1-a^2x^2}}\cos^{-1} ax\, dx$$

471. $$\int \frac{\tan^{-1} ax}{a^2x^2+1}\, dx = \frac{1}{2a}(\tan^{-1} ax)^2$$

472. $$\int \frac{\cot^{-1} ax}{a^2x^2+1}\, dx = -\frac{1}{2a}(\cot^{-1} ax)^2$$

473. $$\int x\sec^{-1} ax\, dx = \frac{x^2}{2}\sec^{-1} ax - \frac{1}{2a^2}\sqrt{a^2x^2-1}$$

474. $$\int x^n \sec^{-1} ax\, dx = \frac{x^{n+1}}{n+1}\sec^{-1} ax - \frac{1}{n+1}\int \frac{x^n\, dx}{\sqrt{a^2x^2-1}}$$

475. $$\int \frac{\sec^{-1} ax}{x^2}\, dx = -\frac{\sec^{-1} ax}{x} + \frac{\sqrt{a^2x^2-1}}{x}$$

476. $$\int x\csc^{-1} ax\, dx = \frac{x^2}{2}\csc^{-1} ax + \frac{1}{2a^2}\sqrt{a^2x^2-1}$$

477. $$\int x^n \csc^{-1} ax\, dx = \frac{x^{n+1}}{n+1}\csc^{-1} ax + \frac{1}{n+1}\int \frac{x^n\, dx}{\sqrt{a^2x^2-1}}$$

478. $$\int \frac{\csc^{-1} ax}{x^2}\,dx = -\frac{\csc^{-1} ax}{x} - \frac{\sqrt{a^2x^2-1}}{x}$$

FORMS INVOLVING TRIGONOMETRIC SUBSTITUTIONS

479. $$\int f(\sin x)\,dx = 2\int f\left(\frac{2z}{1+z^2}\right)\frac{dz}{1+z^2}, \qquad \left(z = \tan\frac{x}{2}\right)$$

480. $$\int f(\cos x)\,dx = 2\int f\left(\frac{1-z^2}{1+z^2}\right)\frac{dz}{1+z^2}, \qquad \left(z = \tan\frac{x}{2}\right)$$

***481.** $$\int f(\sin x)\,dx = \int f(u)\frac{du}{\sqrt{1-u^2}}, \qquad (u = \sin x)$$

***482.** $$\int f(\cos x)\,dx = -\int f(u)\frac{du}{\sqrt{1-u^2}}, \qquad (u = \cos x)$$

***483.** $$\int f(\sin x, \cos x)\,dx = \int f(u, \sqrt{1-u^2})\frac{du}{\sqrt{1-u^2}}, \qquad (u = \sin x)$$

484. $$\int f(\sin x, \cos x)\,dx = 2\int f\left(\frac{2z}{1+z^2}, \frac{1-z^2}{1+z^2}\right)\frac{dz}{1+z^2}, \qquad \left(z = \tan\frac{x}{2}\right)$$

LOGARITHMIC FORMS

485. $$\int (\log x)\,dx = x\log x - x$$

486. $$\int x(\log x)\,dx = \frac{x^2}{2}\log x - \frac{x^2}{4}$$

487. $$\int x^2(\log x)\,dx = \frac{x^3}{3}\log x - \frac{x^3}{9}$$

488. $$\int x^n(\log ax)\,dx = \frac{x^{n+1}}{n+1}\log ax - \frac{x^{n+1}}{(n+1)^2}$$

489. $$\int (\log x)^2\,dx = x(\log x)^2 - 2x\log x + 2x$$

490. $$\int (\log x)^n\,dx = \begin{cases} x(\log x)^n - n\int (\log x)^{n-1}\,dx, & (n \neq -1) \\ \text{or} \\ (-1)^n n!\,x\sum_{r=0}^{n}\frac{(-\log x)^r}{r!} \end{cases}$$

* The square roots appearing in these formulas may be plus or minus, depending on the quadrant of x. Care must be used to give them the proper sign.

491. $$\int \frac{(\log x)^n}{x}\,dx = \frac{1}{n+1}(\log x)^{n+1}$$

492. $$\int \frac{dx}{\log x} = \log(\log x) + \log x + \frac{(\log x)^2}{2\cdot 2!} + \frac{(\log x)^3}{3\cdot 3!} + \cdots$$

493. $$\int \frac{dx}{x\log x} = \log(\log x)$$

494. $$\int \frac{dx}{x(\log x)^n} = -\frac{1}{(n-1)(\log x)^{n-1}}$$

495. $$\int \frac{x^m\,dx}{(\log x)^n} = -\frac{x^{m+1}}{(n-1)(\log x)^{n-1}} + \frac{m+1}{n-1}\int \frac{x^m\,dx}{(\log x)^{n-1}}$$

496. $$\int x^m(\log x)^n\,dx = \begin{cases} \dfrac{x^{m+1}(\log x)^n}{m+1} - \dfrac{n}{m+1}\displaystyle\int x^m(\log x)^{n-1}\,dx \\ \text{or} \\ (-1)^n\dfrac{n!}{m+1}x^{m+1}\displaystyle\sum_{r=0}^{n}\frac{(-\log x)^r}{r!(m+1)^{n-r}} \end{cases}$$

497. $$\int \sin(\log x)\,dx = \tfrac{1}{2}x\sin(\log x) - \tfrac{1}{2}x\cos(\log x)$$

498. $$\int \cos(\log x)\,dx = \tfrac{1}{2}x\sin(\log x) + \tfrac{1}{2}x\cos(\log x)$$

499. $$\int [\log(ax+b)]\,dx = \frac{ax+b}{a}\log(ax+b) - x$$

500. $$\int \frac{\log(ax+b)}{x^2}\,dx = \frac{a}{b}\log x - \frac{ax+b}{bx}\log(ax+b)$$

501. $$\int x^m[\log(ax+b)]\,dx = \frac{1}{m+1}\left[x^{m+1} - \left(-\frac{b}{a}\right)^{m+1}\right]\log(ax+b) - \frac{1}{m+1}\left(-\frac{b}{a}\right)^{m+1}\sum_{r=1}^{m+1}\frac{1}{r}\left(-\frac{ax}{b}\right)^r$$

502. $$\int \frac{\log(ax+b)}{x^m}\,dx = -\frac{1}{m-1}\frac{\log(ax+b)}{x^{m-1}} + \frac{1}{m-1}\left(-\frac{a}{b}\right)^{m-1}\log\frac{ax+b}{x} + \frac{1}{m-1}\left(-\frac{a}{b}\right)^{m-1}\sum_{r=1}^{m-2}\frac{1}{r}\left(-\frac{b}{ax}\right)^r,\ (m>2)$$

503. $$\int \left[\log\frac{x+a}{x-a}\right]dx = (x+a)\log(x+a) - (x-a)\log(x-a)$$

504. $$\int x^m\left[\log\frac{x+a}{x-a}\right]dx = \frac{x^{m+1}-(-a)^{m+1}}{m+1}\log(x+a) - \frac{x^{m+1}-a^{m+1}}{m+1}\log(x-a) + \frac{2a^{m+1}}{m+1}\sum_{r=1}^{\left[\frac{m+1}{2}\right]}\frac{1}{m-2r+2}\left(\frac{x}{a}\right)^{m-2r+2}$$

505. $\int \frac{1}{x^2}\left[\log\frac{x+a}{x-a}\right]dx = \frac{1}{x}\log\frac{x-a}{x+a} - \frac{1}{a}\log\frac{x^2-a^2}{x^2}$

506. $$\int (\log X)\,dx = \begin{cases} \left(x+\frac{b}{2c}\right)\log X - 2x + \frac{\sqrt{4ac-b^2}}{c}\tan^{-1}\frac{2cx+b}{\sqrt{4ac-b^2}}, & (b^2-4ac<0) \\ \text{or} \\ \left(x+\frac{b}{2c}\right)\log X - 2x + \frac{\sqrt{b^2-4ac}}{c}\tanh^{-1}\frac{2cx+b}{\sqrt{b^2-4ac}}, & (b^2-4ac>0) \\ \text{where} \\ X = a+bx+cx^2 \end{cases}$$

507. $\int x^n(\log X)\,dx = \frac{x^{n+1}}{n+1}\log X - \frac{2c}{n+1}\int\frac{x^{n+2}}{X}dx - \frac{b}{n+1}\int\frac{x^{n+1}}{X}dx$

where $X = a + bx + cx^2$

508. $\int [\log(x^2+a^2)]\,dx = x\log(x^2+a^2) - 2x + 2a\tan^{-1}\frac{x}{a}$

509. $\int [\log(x^2-a^2)]\,dx = x\log(x^2-a^2) - 2x + a\log\frac{x+a}{x-a}$

510. $\int x[\log(x^2\pm a^2)]\,dx = \frac{1}{2}(x^2\pm a^2)\log(x^2\pm a^2) - \frac{1}{2}x^2$

511. $\int [\log(x+\sqrt{x^2\pm a^2})]\,dx = x\log(x+\sqrt{x^2\pm a^2}) - \sqrt{x^2\pm a^2}$

512. $\int x[\log(x+\sqrt{x^2\pm a^2})]\,dx = \left(\frac{x^2}{2}\pm\frac{a^2}{4}\right)\log(x+\sqrt{x^2\pm a^2}) - \frac{x\sqrt{x^2\pm a^2}}{4}$

513. $$\int x^m[\log(x+\sqrt{x^2\pm a^2})]\,dx = \frac{x^{m+1}}{m+1}\log(x+\sqrt{x^2\pm a^2}) - \frac{1}{m+1}\int\frac{x^{m+1}}{\sqrt{x^2\pm a^2}}dx$$

514. $\int \frac{\log(x+\sqrt{x^2+a^2})}{x^2}dx = -\frac{\log(x+\sqrt{x^2+a^2})}{x} - \frac{1}{a}\log\frac{a+\sqrt{x^2+a^2}}{x}$

515. $\int \frac{\log(x+\sqrt{x^2-a^2})}{x^2}dx = -\frac{\log(x+\sqrt{x^2-a^2})}{x} + \frac{1}{|a|}\sec^{-1}\frac{x}{a}$

516. $$\int x^n \log(x^2 - a^2)\,dx = \frac{1}{n+1}\left[x^{n+1}\log(x^2 - a^2) - a^{n+1}\log(x-a) - (-a)^{n+1}\log(x+a) - 2\sum_{r=0}^{\left[\frac{n}{2}\right]} \frac{a^{2r}x^{n-2r+1}}{n-2r+1}\right]$$

EXPONENTIAL FORMS

517. $$\int e^x\,dx = e^x$$

518. $$\int e^{-x}\,dx = -e^{-x}$$

519. $$\int e^{ax}\,dx = \frac{e^{ax}}{a}$$

520. $$\int x\,e^{ax}\,dx = \frac{e^{ax}}{a^2}(ax-1)$$

521. $$\int x^m e^{ax}\,dx = \begin{cases} \dfrac{x^m e^{ax}}{a} - \dfrac{m}{a}\displaystyle\int x^{m-1} e^{ax}\,dx \\ \text{or} \\ e^{ax}\displaystyle\sum_{r=0}^{m} (-1)^r \frac{m!x^{m-r}}{(m-r)!a^{r+1}} \end{cases}$$

522. $$\int \frac{e^{ax}\,dx}{x} = \log x + \frac{ax}{1!} + \frac{a^2x^2}{2\cdot 2!} + \frac{a^3x^3}{3\cdot 3!} + \cdots$$

523. $$\int \frac{e^{ax}}{x^m}\,dx = -\frac{1}{m-1}\frac{e^{ax}}{x^{m-1}} + \frac{a}{m-1}\int \frac{e^{ax}}{x^{m-1}}\,dx$$

524. $$\int e^{ax}\log x\,dx = \frac{e^{ax}\log x}{a} - \frac{1}{a}\int \frac{e^{ax}}{x}\,dx$$

525. $$\int \frac{dx}{1+e^x} = x - \log(1+e^x) = \log\frac{e^x}{1+e^x}$$

526. $$\int \frac{dx}{a+be^{px}} = \frac{x}{a} - \frac{1}{ap}\log(a+be^{px})$$

527. $$\int \frac{dx}{ae^{mx}+be^{-mx}} = \frac{1}{m\sqrt{ab}}\tan^{-1}\left(e^{mx}\sqrt{\frac{a}{b}}\right), \qquad (a>0, b>0)$$

528. $$\int \frac{dx}{ae^{mx}-be^{-mx}} = \begin{cases} \dfrac{1}{2m\sqrt{ab}}\log\dfrac{\sqrt{a}\,e^{mx}-\sqrt{b}}{\sqrt{a}\,e^{mx}+\sqrt{b}} \\ \text{or} \\ \dfrac{-1}{m\sqrt{ab}}\tanh^{-1}\left(\sqrt{\dfrac{a}{b}}\,e^{mx}\right), \qquad (a>0, b>0) \end{cases}$$

529. $\int (a^x - a^{-x})\,dx = \frac{a^x + a^{-x}}{\log a}$

530. $\int \frac{e^{ax}}{b + ce^{ax}}\,dx = \frac{1}{ac}\log(b + ce^{ax})$

531. $\int \frac{x\,e^{ax}}{(1 + ax)^2}\,dx = \frac{e^{ax}}{a^2(1 + ax)}$

532. $\int x\,e^{-x^2}\,dx = -\frac{1}{2}e^{-x^2}$

533. $\int e^{ax}[\sin(bx)]\,dx = \frac{e^{ax}[a\sin(bx) - b\cos(bx)]}{a^2 + b^2}$

534. $\int e^{ax}[\sin(bx)][\sin(cx)]\,dx = \frac{e^{ax}[(b - c)\sin(b - c)x + a\cos(b - c)x]}{2[a^2 + (b - c)^2]} - \frac{e^{ax}[(b + c)\sin(b + c)x + a\cos(b + c)x]}{2[a^2 + (b + c)^2]}.$

535. $\int e^{ax}[\sin(bx)][\cos(cx)]\,dx = \begin{cases} \frac{e^{ax}[a\sin(b - c)x - (b - c)\cos(b - c)x]}{2[a^2 + (b - c)^2]} + \frac{e^{ax}[a\sin(b + c)x - (b + c)\cos(b + c)x]}{2[a^2 + (b + c)^2]} \\ \text{or} \\ \frac{e^{ax}}{\rho}[(a\sin bx - b\cos bx)[\cos(cx - \alpha)] - c(\sin bx)\sin(cx - \alpha)] \\ \text{where} \\ \rho = \sqrt{(a^2 + b^2 - c^2)^2 + 4a^2c^2}, \\ \rho\cos\alpha = a^2 + b^2 - c^2, \qquad \rho\sin\alpha = 2ac \end{cases}$

536. $\int e^{ax}[\sin(bx)][\sin(bx + c)]\,dx$

$$= \frac{e^{ax}\cos c}{2a} - \frac{e^{ax}[a\cos(2bx + c) + 2b\sin(2bx + c)]}{2(a^2 + 4b^2)}$$

537. $\int e^{ax}[\sin(bx)][\cos(bx + c)]\,dx$

$$= \frac{-e^{ax}\sin c}{2a} + \frac{e^{ax}[a\sin(2bx + c) - 2b\cos(2bx + c)]}{2(a^2 + 4b^2)}$$

538. $\int e^{ax}[\cos(bx)]\,dx = \frac{e^{ax}}{a^2 + b^2}[a\cos(bx) + b\sin(bx)]$

539. $$\int e^{ax}[\cos(bx)][\cos(cx)]\,dx = \frac{e^{ax}[(b-c)\sin(b-c)x + a\cos(b-c)x]}{2[a^2+(b-c)^2]} + \frac{e^{ax}[(b+c)\sin(b+c)x + a\cos(b+c)x]}{2[a^2+(b+c)^2]}$$

540. $$\int e^{ax}[\cos(bx)][\cos(bx+c)]\,dx = \frac{e^{ax}\cos c}{2a} + \frac{e^{ax}[a\cos(2bx+c) + 2b\sin(2bx+c)]}{2(a^2+4b^2)}$$

541. $$\int e^{ax}[\cos(bx)][\sin(bx+c)]\,dx = \frac{e^{ax}\sin c}{2a} + \frac{e^{ax}[a\sin(2bx+c) - 2b\cos(2bx+c)]}{2(a^2+4b^2)}$$

542. $$\int e^{ax}[\sin^n bx]\,dx = \frac{1}{a^2+n^2b^2}\left[(a\sin bx - nb\cos bx)\,e^{ax}\sin^{n-1} bx + n(n-1)b^2\int e^{ax}[\sin^{n-2} bx]\,dx\right]$$

543. $$\int e^{ax}[\cos^n bx]\,dx = \frac{1}{a^2+n^2b^2}\left[(a\cos bx + nb\sin bx)\,e^{ax}\cos^{n-1} bx + n(n-1)b^2\int e^{ax}[\cos^{n-2} bx]\,dx\right]$$

544. $$\int x^m e^x \sin x\,dx = \frac{1}{2}x^m e^x(\sin x - \cos x) - \frac{m}{2}\int x^{m-1} e^x \sin x\,dx + \frac{m}{2}\int x^{m-1} e^x \cos x\,dx$$

545. $$\int x^m e^{ax}[\sin bx]\,dx = \begin{cases} x^m e^{ax}\dfrac{a\sin bx - b\cos bx}{a^2+b^2} - \dfrac{m}{a^2+b^2}\displaystyle\int x^{m-1} e^{ax}(a\sin bx - b\cos bx)\,dx \\ \text{or} \\ e^{ax}\displaystyle\sum_{r=0}^{m}\frac{(-1)^r m!\,x^{m-r}}{\rho^{r+1}(m-r)!}\sin[bx - (r+1)\alpha] \\ \text{where} \\ \rho = \sqrt{a^2+b^2}, \qquad \rho\cos\alpha = a, \qquad \rho\sin\alpha = b \end{cases}$$

546. $$\int x^m e^x \cos x\,dx = \frac{1}{2}x^m e^x(\sin x + \cos x) - \frac{m}{2}\int x^{m-1} e^x \sin x\,dx - \frac{m}{2}\int x^{m-1} e^x \cos x\,dx$$

547. $$\int x^m e^{ax}\cos bx\,dx = \begin{cases} x^m e^{ax}\dfrac{a\cos bx + b\sin bx}{a^2+b^2} \\ \qquad -\dfrac{m}{a^2+b^2}\displaystyle\int x^{m-1}e^{ax}(a\cos bx + b\sin bx)\,dx \\ \qquad \text{or} \\ e^{ax}\displaystyle\sum_{r=0}^{m}\frac{(-1)^r m!\,x^{m-r}}{\rho^{r+1}(m-r)!}\cos[bx-(r+1)\alpha] \\ \qquad \text{where} \\ \rho=\sqrt{a^2+b^2},\qquad \rho\cos\alpha = a,\qquad \rho\sin\alpha = b \end{cases}$$

548. $$\int e^{ax}(\cos^m x)(\sin^n x)\,dx = \begin{cases} \dfrac{e^{ax}\cos^{m-1}x\sin^n x[a\cos x + (m+n)\sin x]}{(m+n)^2+a^2} \\ \qquad -\dfrac{na}{(m+n)^2+a^2}\displaystyle\int e^{ax}(\cos^{m-1}x)(\sin^{n-1}x)\,dx \\ \qquad +\dfrac{(m-1)(m+n)}{(m+n)^2+a^2}\displaystyle\int e^{ax}(\cos^{m-2}x)(\sin^n x)\,dx \\ \qquad \text{or} \\ \dfrac{e^{ax}\cos^m x\sin^{n-1}x[a\sin x - (m+n)\cos x]}{(m+n)^2+a^2} \\ \qquad +\dfrac{ma}{(m+n)^2+a^2}\displaystyle\int e^{ax}(\cos^{m-1}x)(\sin^{n-1}x)\,dx \\ \qquad +\dfrac{(n-1)(m+n)}{(m+n)^2+a^2}\displaystyle\int e^{ax}(\cos^m x)(\sin^{n-2}x)\,dx \\ \qquad \text{or} \\ \dfrac{e^{ax}(\cos^{m-1}x)(\sin^{n-1}x)(a\sin x\cos x + m\sin^2 x - n\cos^2 x)}{(m+n)^2+a^2} \\ \qquad +\dfrac{m(m-1)}{(m+n)^2+a^2}\displaystyle\int e^{ax}(\cos^{m-2}x)(\sin^n x)\,dx \\ \qquad +\dfrac{n(n-1)}{(m+n)^2+a^2}\displaystyle\int e^{ax}(\cos^m x)(\sin^{n-2}x)\,dx \\ \qquad \text{or} \\ \dfrac{e^{ax}(\cos^{m-1}x)(\sin^{n-1}x)(a\cos x\sin x + m\sin^2 x - n\cos^2 x)}{(m+n)^2+a^2} \\ \qquad +\dfrac{m(m-1)}{(m+n)^2+a^2}\displaystyle\int e^{ax}(\cos^{m-2}x)(\sin^{n-2}x)\,dx \\ \qquad +\dfrac{(n-m)(n+m-1)}{(m+n)^2+a^2}\displaystyle\int e^{ax}(\cos^m x)(\sin^{n-2}x)\,dx \end{cases}$$

549. $$\int x\,e^{ax}(\sin bx)\,dx = \frac{x\,e^{ax}}{a^2+b^2}(a\sin bx - b\cos bx) - \frac{e^{ax}}{(a^2+b^2)^2}[(a^2-b^2)\sin bx - 2ab\cos bx]$$

550. $$\int x\,e^{ax}(\cos bx)\,dx = \frac{x\,e^{ax}}{a^2+b^2}(a\cos bx + b\sin bx) - \frac{e^{ax}}{(a^2+b^2)^2}[(a^2-b^2)\cos bx + 2ab\sin bx]$$

551. $$\int \frac{e^{ax}}{\sin^n x}\,dx = -\frac{e^{ax}[a\sin x + (n-2)\cos x]}{(n-1)(n-2)\sin^{n-1} x} + \frac{a^2+(n-2)^2}{(n-1)(n-2)}\int \frac{e^{ax}}{\sin^{n-2} x}\,dx$$

552. $$\int \frac{e^{ax}}{\cos^n x}\,dx = -\frac{e^{ax}[a\cos x - (n-2)\sin x]}{(n-1)(n-2)\cos^{n-1} x} + \frac{a^2+(n-2)^2}{(n-1)(n-2)}\int \frac{e^{ax}}{\cos^{n-2} x}\,dx$$

553. $$\int e^{ax}\tan^n x\,dx = e^{ax}\frac{\tan^{n-1} x}{n-1} - \frac{a}{n-1}\int e^{ax}\tan^{n-1} x\,dx - \int e^{ax}\tan^{n-2} x\,dx$$

HYPERBOLIC FORMS

554. $$\int (\sinh x)\,dx = \cosh x$$

555. $$\int (\cosh x)\,dx = \sinh x$$

556. $$\int (\tanh x)\,dx = \log\cosh x$$

557. $$\int (\coth x)\,dx = \log\sinh x$$

558. $$\int (\operatorname{sech} x)\,dx = \tan^{-1}(\sinh x)$$

559. $$\int \operatorname{csch} x\,dx = \log\tanh\left(\frac{x}{2}\right)$$

560. $$\int x(\sinh x)\,dx = x\cosh x - \sinh x$$

561. $$\int x^n(\sinh x)\,dx = x^n\cosh x - n\int x^{n-1}(\cosh x)\,dx$$

562. $$\int x(\cosh x)\,dx = x\sinh x - \cosh x$$

563. $$\int x^n(\cosh x)\,dx = x^n\sinh x - n\int x^{n-1}(\sinh x)\,dx$$

564. $\int (\operatorname{sech} x)(\tanh x)\,dx = -\operatorname{sech} x$

565. $\int (\operatorname{csch} x)(\coth x)\,dx = -\operatorname{csch} x$

566. $\int (\sinh^2 x)\,dx = \frac{\sinh 2x}{4} - \frac{x}{2}$

567. $$\int (\sinh^m x)(\cosh^n x)\,dx = \begin{cases} \frac{1}{m+n}(\sinh^{m+1} x)(\cosh^{n-1} x) \\ \qquad + \frac{n-1}{m+n}\int (\sinh^m x)(\cosh^{n-2} x)\,dx \\ \text{or} \\ \frac{1}{m+n}\sinh^{m-1} x \cosh^{n+1} x \\ \qquad - \frac{m-1}{m+n}\int (\sinh^{m-2} x)(\cosh^n x)\,dx, \quad (m+n \neq 0) \end{cases}$$

568. $$\int \frac{dx}{(\sinh^m x)(\cosh^n x)} = \begin{cases} -\frac{1}{(m-1)(\sinh^{m-1} x)(\cosh^{n-1} x)} \\ \qquad - \frac{m+n-2}{m-1}\int \frac{dx}{(\sinh^{m-2} x)(\cosh^n x)}, \quad (m \neq 1) \\ \text{or} \\ \frac{1}{(n-1)\sinh^{m-1} x \cosh^{n-1} x} \\ \qquad + \frac{m+n-2}{n-1}\int \frac{dx}{(\sinh^m x)(\cosh^{n-2} x)}, \quad (n \neq 1) \end{cases}$$

569. $\int (\tanh^2 x)\,dx = x - \tanh x$

570. $\int (\tanh^n x)\,dx = -\frac{\tanh^{n-1} x}{n-1} + \int (\tanh^{n-2} x)\,dx, \quad (n \neq 1)$

571. $\int (\operatorname{sech}^2 x)\,dx = \tanh x$

572. $\int (\cosh^2 x)\,dx = \frac{\sinh 2x}{4} + \frac{x}{2}$

573. $\int (\coth^2 x)\,dx = x - \coth x$

574. $\int (\coth^n x)\,dx = -\frac{\coth^{n-1} x}{n-1} + \int \coth^{n-2} x\,dx, \quad (n \neq 1)$

575. $\int (\operatorname{csch}^2 x)\,dx = -\operatorname{ctnh} x$

576. $\int (\sinh mx)(\sinh nx)\,dx = \frac{\sinh(m+n)x}{2(m+n)} - \frac{\sinh(m-n)x}{2(m-n)}, \quad (m^2 \neq n^2)$

577. $\int (\cosh mx)(\cosh nx)\,dx = \frac{\sinh(m+n)x}{2(m+n)} + \frac{\sinh(m-n)x}{2(m-n)}, \quad (m^2 \neq n^2)$

578. $\int (\sinh mx)(\cosh nx)\,dx = \frac{\cosh(m+n)x}{2(m+n)} + \frac{\cosh(m-n)x}{2(m-n)}, \quad (m^2 \neq n^2)$

579. $\int \left(\sinh^{-1}\frac{x}{a}\right) dx = x\sinh^{-1}\frac{x}{a} - \sqrt{x^2+a^2}, \quad (a>0)$

580. $\int x\left(\sinh^{-1}\frac{x}{a}\right) dx = \left(\frac{x^2}{2}+\frac{a^2}{4}\right)\sinh^{-1}\frac{x}{a} - \frac{x}{4}\sqrt{x^2+a^2}, \quad (a>0)$

581. $\int x^n(\sinh^{-1} x)\,dx = \frac{x^{n+1}}{n+1}\sinh^{-1} x - \frac{1}{n+1}\int \frac{x^{n+1}}{(1+x^2)^{\frac{1}{2}}}\,dx, \quad (n \neq -1)$

582. $\int \left(\cosh^{-1}\frac{x}{a}\right) dx = \begin{cases} x\cosh^{-1}\frac{x}{a} - \sqrt{x^2-a^2}, & \left(\cosh^{-1}\frac{x}{a} > 0\right) \\ \text{or} & \\ x\cosh^{-1}\frac{x}{a} + \sqrt{x^2-a^2}, & \left(\cosh^{-1}\frac{x}{a} < 0\right), \quad (a>0) \end{cases}$

583. $\int x\left(\cosh^{-1}\frac{x}{a}\right) dx = \frac{2x^2-a^2}{4}\cosh^{-1}\frac{x}{a} - \frac{x}{4}(x^2-a^2)^{\frac{1}{2}}$

584. $\int x^n(\cosh^{-1} x)\,dx = \frac{x^{n+1}}{n+1}\cosh^{-1} x - \frac{1}{n+1}\int \frac{x^{n+1}}{(x^2-1)^{\frac{1}{2}}}\,dx, \quad (n \neq -1)$

585. $\int \left(\tanh^{-1}\frac{x}{a}\right) dx = x\tanh^{-1}\frac{x}{a} + \frac{a}{2}\log(a^2-x^2), \quad \left(\left|\frac{x}{a}\right| < 1\right)$

586. $\int \left(\coth^{-1}\frac{x}{a}\right) dx = x\coth^{-1}\frac{x}{a} + \frac{a}{2}\log(x^2-a^2), \quad \left(\left|\frac{x}{a}\right| > 1\right)$

587. $\int x\left(\tanh^{-1}\frac{x}{a}\right) dx = \frac{x^2-a^2}{2}\tanh^{-1}\frac{x}{a} + \frac{ax}{2}, \quad \left(\left|\frac{x}{a}\right| < 1\right)$

588. $\int x^n\left(\tanh^{-1} x\right) dx = \frac{x^{n+1}}{n+1}\tanh^{-1} x - \frac{1}{n+1}\int \frac{x^{n+1}}{1-x^2}\,dx, \quad (n \neq -1)$

589. $\int x\left(\coth^{-1}\frac{x}{a}\right) dx = \frac{x^2-a^2}{2}\coth^{-1}\frac{x}{a} + \frac{ax}{2}, \quad \left(\left|\frac{x}{a}\right| > 1\right)$

590. $\int x^n(\coth^{-1} x)\,dx = \frac{x^{n+1}}{n+1}\coth^{-1} x + \frac{1}{n+1}\int \frac{x^{n+1}}{x^2-1}\,dx, \quad (n \neq -1)$

591. $\int (\operatorname{sech}^{-1} x)\,dx = x \operatorname{sech}^{-1} x + \sin^{-1} x$

592. $\int x \operatorname{sech}^{-1} x\,dx = \frac{x^2}{2} \operatorname{sech}^{-1} x - \frac{1}{2}(1 - x^2)$

593. $\int x^n \operatorname{sech}^{-1} x\,dx = \frac{x^{n+1}}{n+1} \operatorname{sech}^{-1} x + \frac{1}{n+1} \int \frac{x^n}{(1 - x^2)^{\frac{1}{2}}}\,dx, \qquad (n \neq -1)$

594. $\int \operatorname{csch}^{-1} x\,dx = x \operatorname{csch}^{-1} x + \frac{x}{|x|} \sinh^{-1} x$

595. $\int x \operatorname{csch}^{-1} x\,dx = \frac{x^2}{2} \operatorname{csch}^{-1} x + \frac{1}{2} \frac{x}{|x|} \sqrt{1 + x^2}$

596. $\int x^n \operatorname{csch}^{-1} x\,dx = \frac{x^{n+1}}{n+1} \operatorname{csch}^{-1} x + \frac{1}{n+1} \frac{x}{|x|} \int \frac{x^n}{(x^2 + 1)^{\frac{1}{2}}}\,dx, \qquad (n \neq -1)$

DEFINITE INTEGRALS

597. $\int_0^\infty x^{n-1} e^{-x}\,dx = \int_0^1 \left(\log \frac{1}{x}\right)^{n-1} dx = \frac{1}{n} \prod_{m=1}^{\infty} \frac{\left(1 + \frac{1}{m}\right)^n}{1 + \frac{n}{m}}$

$= \Gamma(n), n \neq 0, -1, -2, -3, \ldots \qquad$ (Gamma Function)

598. $\int_0^\infty t^n p^{-t}\,dt = \frac{n!}{(\log p)^{n+1}}, \qquad (n = 0, 1, 2, 3, \ldots \text{ and } p > 0)$

599. $\int_0^\infty t^{n-1} e^{-(a+1)t}\,dt = \frac{\Gamma(n)}{(a+1)^n}, \qquad (n > 0, a > -1)$

600. $\int_0^1 x^m \left(\log \frac{1}{x}\right)^n dx = \frac{\Gamma(n+1)}{(m+1)^{n+1}}, \qquad (m > -1, n > -1)$

601. $\Gamma(n)$ is finite if $n > 0$, $\Gamma(n+1) = n\Gamma(n)$

602. $\Gamma(n) \cdot \Gamma(1 - n) = \frac{\pi}{\sin n\pi}$

603. $\Gamma(n) = (n - 1)!$ if $n =$ integer > 0

604. $\Gamma(\frac{1}{2}) = 2 \int_0^\infty e^{-t^2}\,dt = \sqrt{\pi} = 1.7724538509 \cdots = (-\frac{1}{2})!$

605. $\Gamma(n + \frac{1}{2}) = \frac{1 \cdot 3 \cdot 5 \ldots (2n-1)}{2^n} \sqrt{\pi} \qquad n = 1, 2, 3, \ldots$

606. $\Gamma(-n + \frac{1}{2}) = \frac{(-1)^n 2^n \sqrt{\pi}}{1 \cdot 3 \cdot 5 \ldots (2n-1)} \qquad n = 1, 2, 3, \ldots$

607. $\displaystyle\int_0^1 x^{m-1}(1-x)^{n-1}\,dx = \int_0^\infty \frac{x^{m-1}}{(1+x)^{m+n}}\,dx = \frac{\Gamma(m)\Gamma(n)}{\Gamma(m+n)} = B(m,n)$

(Beta function)

608. $B(m,n) = B(n,m) = \dfrac{\Gamma(m)\Gamma(n)}{\Gamma(m+n)}$, where m and n are any positive real numbers.

609. $\displaystyle\int_a^b (x-a)^m(b-x)^n\,dx = (b-a)^{m+n+1}\frac{\Gamma(m+1)\cdot\Gamma(n+1)}{\Gamma(m+n+2)},$

$(m > -1, n > -1, b > a)$

610. $\displaystyle\int_1^\infty \frac{dx}{x^m} = \frac{1}{m-1}, \qquad [m > 1]$

611. $\displaystyle\int_0^\infty \frac{dx}{(1+x)x^p} = \pi \csc p\pi, \qquad [p < 1]$

612. $\displaystyle\int_0^\infty \frac{dx}{(1-x)x^p} = -\pi \cot p\pi, \qquad [p < 1]$

613. $\displaystyle\int_0^\infty \frac{x^{p-1}\,dx}{1+x} = \frac{\pi}{\sin p\pi}$

$= \mathrm{B}(p, 1-p) = \Gamma(p)\Gamma(1-p), \qquad [0 < p < 1]$

614. $\displaystyle\int_0^\infty \frac{x^{m-1}\,dx}{1+x^n} = \frac{\pi}{n\sin\dfrac{m\pi}{n}}, \qquad [0 < m < n]$

615. $\displaystyle\int_0^\infty \frac{x^a\,dx}{(m+x^b)^c} = \frac{m^{\frac{a+1}{b}-c}}{b}\left[\frac{\Gamma\left(\frac{a+1}{b}\right)\Gamma\left(c-\frac{a+1}{b}\right)}{\Gamma(c)}\right]$

$\left(a > -1, b > 0, m > 0, c > \dfrac{a+1}{b}\right)$

616. $\displaystyle\int_0^\infty \frac{dx}{(1+x)\sqrt{x}} = \pi$

617. $\displaystyle\int_0^\infty \frac{a\,dx}{a^2+x^2} = \frac{\pi}{2}$, if $a > 0$; 0, if $a = 0$; $-\dfrac{\pi}{2}$, if $a < 0$

618. $\displaystyle\int_0^a (a^2-x^2)^{\frac{n}{2}}\,dx = \frac{1}{2}\int_{-a}^a (a^2-x^2)^{\frac{n}{2}}\,dx = \frac{1\cdot 3\cdot 5\ldots n}{2\cdot 4\cdot 6\ldots(n+1)}\cdot\frac{\pi}{2}\cdot a^{n+1} \qquad (n \text{ odd})$

619. $\displaystyle\int_0^a x^m(a^2-x^2)^{\frac{n}{2}}\,dx = \begin{cases} \dfrac{1}{2}a^{m+n+1}B\left(\dfrac{m+1}{2}, \dfrac{n+2}{2}\right) \\ \text{or} \\ \dfrac{1}{2}a^{m+n+1}\dfrac{\Gamma\left(\frac{m+1}{2}\right)\Gamma\left(\frac{n+2}{2}\right)}{\Gamma\left(\frac{m+n+3}{2}\right)} \end{cases}$

620. $$\int_0^{\pi/2} (\sin^n x)\, dx = \begin{cases} \displaystyle\int_0^{\pi/2} (\cos^n x)\, dx \\ \text{or} \\ \dfrac{1 \cdot 3 \cdot 5 \cdot 7 \ldots (n-1)}{2 \cdot 4 \cdot 6 \cdot 8 \ldots (n)} \dfrac{\pi}{2}, & (n \text{ an even integer, } n \neq 0) \\ \text{or} \\ \dfrac{2 \cdot 4 \cdot 6 \cdot 8 \ldots (n-1)}{1 \cdot 3 \cdot 5 \cdot 7 \ldots (n)}, & (n \text{ an odd integer, } n \neq 1) \\ \text{or} \\ \dfrac{\sqrt{\pi}}{2} \dfrac{\Gamma\left(\frac{n+1}{2}\right)}{\Gamma\left(\frac{n}{2}+1\right)}, & (n > -1) \end{cases}$$

621. $$\int_0^\infty \frac{\sin mx\, dx}{x} = \frac{\pi}{2}, \text{ if } m > 0;\ 0, \text{ if } m = 0;\ -\frac{\pi}{2}, \text{ if } m < 0$$

622. $$\int_0^\infty \frac{\cos x\, dx}{x} = \infty$$

623. $$\int_0^\infty \frac{\tan x\, dx}{x} = \frac{\pi}{2}$$

624. $$\int_0^\pi \sin ax \cdot \sin bx\, dx = \int_0^\pi \cos ax \cdot \cos bx\, dx = 0, \qquad (a \neq b;\, a, b \text{ integers})$$

625. $$\int_0^{\pi/a} [\sin(ax)][\cos(ax)]\, dx = \int_0^\pi [\sin(ax)][\cos(ax)]\, dx = 0$$

626. $$\int_0^\pi [\sin(ax)][\cos(bx)]\, dx = \frac{2a}{a^2 - b^2}, \text{ if } a - b \text{ is odd, or 0 if } a - b \text{ is even}$$

627. $$\int_0^\infty \frac{\sin x \cos mx\, dx}{x}$$

$$= 0, \text{ if } m < -1 \text{ or } m > 1;\ \frac{\pi}{4}, \text{ if } m = \pm 1;\ \frac{\pi}{2}, \text{ if } m^2 < 1$$

628. $$\int_0^\infty \frac{\sin ax \sin bx}{x^2}\, dx = \frac{\pi a}{2}, \qquad (a \leq b)$$

629. $$\int_0^\pi \sin^2 mx\, dx = \int_0^\pi \cos^2 mx\, dx = \frac{\pi}{2}$$

630. $$\int_0^\infty \frac{\sin^2(px)}{x^2}\, dx = \frac{\pi p}{2}$$

631. $\int_0^\infty \frac{\sin x}{x^p}\,dx = \frac{\pi}{2\Gamma(p)\sin(p\pi/2)}, \qquad 0 < p < 1$

632. $\int_0^\infty \frac{\cos x}{x^p}\,dx = \frac{\pi}{2\Gamma(p)\cos(p\pi/2)}, \qquad 0 < p < 1$

633. $\int_0^\infty \frac{1-\cos px}{x^2}\,dx = \frac{\pi p}{2}$

634. $\int_0^\infty \frac{\sin px \cos qx}{x}\,dx = \left\{0, \quad q > p > 0; \quad \frac{\pi}{2}, \quad p > q > 0; \quad \frac{\pi}{4}, \quad p = q > 0\right\}$

635. $\int_0^\infty \frac{\cos(mx)}{x^2+a^2}\,dx = \frac{\pi}{2a}e^{-|m\,a|}$

636. $\int_0^\infty \cos(x^2)\,dx = \int_0^\infty \sin(x^2)\,dx = \frac{1}{2}\sqrt{\frac{\pi}{2}}$

637. $\int_0^\infty \sin ax^n\,dx = \frac{1}{na^{1/n}}\Gamma(1/n)\sin\frac{\pi}{2n}, \qquad n > 1$

638. $\int_0^\infty \cos ax^n\,dx = \frac{1}{na^{1/n}}\Gamma(1/n)\cos\frac{\pi}{2n}, \qquad n > 1$

639. $\int_0^\infty \frac{\sin x}{\sqrt{x}}\,dx = \int_0^\infty \frac{\cos x}{\sqrt{x}}\,dx = \sqrt{\frac{\pi}{2}}$

640. $\int_0^\infty \frac{\sin^3 x}{x^2}\,dx = \frac{3}{4}\log 3$

641. $\int_0^\infty \frac{\sin^3 x}{x^3}\,dx = \frac{3\pi}{8}$

642. $\int_0^\infty \frac{\sin^4 x}{x^4}\,dx = \frac{\pi}{3}$

643. $\int_0^{\pi/2} \frac{dx}{1+a\cos x} = \frac{\cos^{-1} a}{\sqrt{1-a^2}}, \qquad (a < 1)$

644. $\int_0^{\pi} \frac{dx}{a+b\cos x} = \frac{\pi}{\sqrt{a^2-b^2}}, \qquad (a > b \geq 0)$

645. $\int_0^{2\pi} \frac{dx}{1+a\cos x} = \frac{2\pi}{\sqrt{1-a^2}}, \qquad (a^2 < 1)$

646. $\int_0^\infty \frac{\cos ax - \cos bx}{x}\,dx = \log\frac{b}{a}$

647. $\int_0^{\pi/2} \frac{dx}{a^2\sin^2 x + b^2\cos^2 x} = \frac{\pi}{2ab}$

648. $\displaystyle\int_0^{\pi/2} \frac{dx}{(a^2 \sin^2 x + b^2 \cos^2 x)^2} = \frac{\pi(a^2 + b^2)}{4a^3b^3}, \qquad (a, b > 0)$

649. $\displaystyle\int_0^{\pi/2} \sin^{n-1} x \cos^{m-1} x \, dx = \frac{1}{2}\mathrm{B}\left(\frac{n}{2}, \frac{m}{2}\right),$ m and n positive integers

650. $\displaystyle\int_0^{\pi/2} (\sin^{2n+1} \theta)\, d\theta = \frac{2 \cdot 4 \cdot 6 \ldots (2n)}{1 \cdot 3 \cdot 5 \ldots (2n+1)}, \qquad (n = 1, 2, 3 \ldots)$

651. $\displaystyle\int_0^{\pi/2} (\sin^{2n} \theta)\, d\theta = \frac{1 \cdot 3 \cdot 5 \ldots (2n-1)}{2 \cdot 4 \ldots (2n)}\left(\frac{\pi}{2}\right), \qquad (n = 1, 2, 3 \ldots)$

652. $\displaystyle\int_0^{\pi/2} \frac{x}{\sin x} dx = 2\left\{\frac{1}{1^2} - \frac{1}{3^2} + \frac{1}{5^2} - \frac{1}{7^2} + \cdots\right\}$

653. $\displaystyle\int_0^{\pi/2} \frac{dx}{1 + \tan^m x} = \frac{\pi}{4}$

654. $\displaystyle\int_0^{\pi/2} \sqrt{\cos \theta}\, d\theta = \frac{(2\pi)^{\frac{3}{2}}}{[\Gamma(\frac{1}{4})]^2}$

655. $\displaystyle\int_0^{\pi/2} (\tan^h \theta)\, d\theta = \frac{\pi}{2 \cos\left(\dfrac{h\pi}{2}\right)}, \qquad (0 < h < 1)$

656. $\displaystyle\int_0^{\infty} \frac{\tan^{-1}(ax) - \tan^{-1}(bx)}{x} dx = \frac{\pi}{2} \log \frac{a}{b}, \qquad (a, b > 0)$

657. The area enclosed by a curve defined through the equation $x^{\frac{b}{c}} + y^{\frac{b}{c}} = a^{\frac{b}{c}}$ where $a > 0$, c a positive odd integer and b a positive even integer is given by

$$\frac{\left[\Gamma\left(\dfrac{c}{b}\right)\right]^2}{\Gamma\left(\dfrac{2c}{b}\right)}\left(\frac{2ca^2}{b}\right)$$

658. $I = \displaystyle\iiint_R x^{h-1} y^{m-1} z^{n-1}\, dv$, where R denotes the region of space bounded by the co-ordinate planes and that portion of the surface $\left(\dfrac{x}{a}\right)^p + \left(\dfrac{y}{b}\right)^q + \left(\dfrac{z}{c}\right)^k = 1$, which lies in the first octant and where $h, m, n, p, q, k, a, b, c$, denote positive real numbers is given by

$$\int_0^a x^{h-1}\, dx \int_0^{b\left[1-\left(\frac{x}{a}\right)^p\right]^{\frac{1}{q}}} y^m\, dy \int_0^{c\left[1-\left(\frac{x}{a}\right)^p-\left(\frac{y}{b}\right)^q\right]^{\frac{1}{k}}} z^{n-1}\, dz$$

$$= \frac{a^h b^m c^n}{pqk} \frac{\Gamma\left(\dfrac{h}{p}\right)\Gamma\left(\dfrac{m}{q}\right)\Gamma\left(\dfrac{n}{k}\right)}{\Gamma\left(\dfrac{h}{p} + \dfrac{m}{q} + \dfrac{n}{k} + 1\right)}$$

659. $\int_0^\infty e^{-ax}\,dx = \frac{1}{a}, \qquad (a > 0)$

660. $\int_0^\infty \frac{e^{-ax} - e^{-bx}}{x}\,dx = \log\frac{b}{a}, \qquad (a, b > 0)$

661. $\int_0^\infty x^n e^{-ax}\,dx = \begin{cases} \frac{\Gamma(n+1)}{a^{n+1}}, & (n > -1, a > 0) \\ \text{or} & \\ \frac{n!}{a^{n+1}}, & (a > 0, n \text{ positive integer}) \end{cases}$

662. $\int_0^\infty x^n \exp(-ax^p)\,dx = \frac{\Gamma(k)}{pa^k}, \qquad \left(n > -1, p > 0, a > 0, k = \frac{n+1}{p}\right)$

663. $\int_0^\infty e^{-a^2x^2}\,dx = \frac{1}{2a}\sqrt{\pi} = \frac{1}{2a}\Gamma\left(\frac{1}{2}\right), \qquad (a > 0)$

664. $\int_0^\infty x\,e^{-x^2}\,dx = \frac{1}{2}$

665. $\int_0^\infty x^2 e^{-x^2}\,dx = \frac{\sqrt{\pi}}{4}$

666. $\int_0^\infty x^{2n} e^{-ax^2}\,dx = \frac{1 \cdot 3 \cdot 5 \ldots (2n-1)}{2^{n+1}a^n}\sqrt{\frac{\pi}{a}}$

667. $\int_0^\infty x^{2n+1} e^{-ax^2}\,dx = \frac{n!}{2a^{n+1}}, \qquad (a > 0)$

668. $\int_0^1 x^m e^{-ax}\,dx = \frac{m!}{a^{m+1}}\left[1 - e^{-a}\sum_{r=0}^{m}\frac{a^r}{r!}\right]$

669. $\int_0^\infty e^{\left(-x^2 - \frac{a^2}{x^2}\right)}\,dx = \frac{e^{-2a}\sqrt{\pi}}{2}, \qquad (a \geq 0)$

670. $\int_0^\infty e^{-nx}\sqrt{x}\,dx = \frac{1}{2n}\sqrt{\frac{\pi}{n}}$

671. $\int_0^\infty \frac{e^{-nx}}{\sqrt{x}}\,dx = \sqrt{\frac{\pi}{n}}$

672. $\int_0^\infty e^{-ax}(\cos mx)\,dx = \frac{a}{a^2 + m^2}, \qquad (a > 0)$

673. $\int_0^\infty e^{-ax}(\sin mx)\,dx = \frac{m}{a^2 + m^2}, \qquad (a > 0)$

674. $\int_0^\infty x\, e^{-ax}[\sin(bx)]\, dx = \frac{2ab}{(a^2 + b^2)^2}, \qquad (a > 0)$

675. $\int_0^\infty x\, e^{-ax}[\cos(bx)]\, dx = \frac{a^2 - b^2}{(a^2 + b^2)^2}, \qquad (a > 0)$

676. $\int_0^\infty x^n e^{-ax}[\sin(bx)]\, dx = \frac{n![(a + ib)^{n+1} - (a + ib)^{n+1}]}{2\mathrm{i}(a^2 + b^2)^{n+1}}, \qquad (i^2 = -1, a > 0)$

677. $\int_0^\infty x^n e^{-ax}[\cos(bx)]\, dx = \frac{n![(a - ib)^{n+1} + (a + ib)^{n+1}]}{2(a^2 + b^2)^{n+1}}, \qquad (i^2 = -1, a > 0)$

678. $\int_0^\infty \frac{e^{-ax} \sin x}{x}\, dx = \cot^{-1} a, \qquad (a > 0)$

679. $\int_0^\infty e^{-a^2x^2} \cos bx\, dx = \frac{\sqrt{\pi}}{2a} \exp\left(-\frac{b^2}{4a^2}\right), \qquad (ab \neq 0)$

680. $\int_0^\infty e^{-t\cos\phi}\, t^{b-1} \sin(t \sin\phi)\, dt = [\Gamma(b)] \sin(b\phi), \qquad \left(b > 0, -\frac{\pi}{2} < \phi < \frac{\pi}{2}\right)$

681. $\int_0^\infty e^{-t\cos\phi}\, t^{b-1}[\cos(t \sin\phi)]\, dt = [\Gamma(b)] \cos(b\phi), \qquad \left(b > 0, -\frac{\pi}{2} < \phi < \frac{\pi}{2}\right)$

682. $\int_0^\infty t^{b-1} \cos t\, dt = [\Gamma(b)] \cos\left(\frac{b\pi}{2}\right), \qquad (0 < b < 1)$

683. $\int_0^\infty t^{b-1} (\sin t)\, dt = [\Gamma(b)] \sin\left(\frac{b\pi}{2}\right), \qquad (0 < b < 1)$

684. $\int_0^1 (\log x)^n\, dx = (-1)^n \cdot n!$

685. $\int_0^1 \left(\log \frac{1}{x}\right)^{\frac{1}{2}} dx = \frac{\sqrt{\pi}}{2}$

686. $\int_0^1 \left(\log \frac{1}{x}\right)^{-\frac{1}{2}} dx = \sqrt{\pi}$

687. $\int_0^1 \left(\log \frac{1}{x}\right)^{n} dx = n!$

688. $\int_0^1 x \log(1 - x)\, dx = -\frac{3}{4}$

689. $\int_0^1 x \log(1 + x)\, dx = \frac{1}{4}$

690. $\int_0^1 x^m (\log x)^n\, dx = \frac{(-1)^n n!}{(m + 1)^{n+1}}, \qquad m > -1, n = 0, 1, 2, \ldots$

If $n \neq 0, 1, 2, \ldots$ replace $n!$ by $\Gamma(n + 1)$.

691. $\int_0^1 \frac{\log x}{1+x}\,dx = -\frac{\pi^2}{12}$

692. $\int_0^1 \frac{\log x}{1-x}\,dx = -\frac{\pi^2}{6}$

693. $\int_0^1 \frac{\log(1+x)}{x}\,dx = \frac{\pi^2}{12}$

694. $\int_0^1 \frac{\log(1-x)}{x}\,dx = -\frac{\pi^2}{6}$

695. $\int_0^1 (\log x)[\log(1+x)]\,dx = 2 - 2\log 2 - \frac{\pi^2}{12}$

696. $\int_0^1 (\log x)[\log(1-x)]\,dx = 2 - \frac{\pi^2}{6}$

697. $\int_0^1 \frac{\log x}{1-x^2}\,dx = -\frac{\pi^2}{8}$

698. $\int_0^1 \log\left(\frac{1+x}{1-x}\right)\cdot\frac{dx}{x} = \frac{\pi^2}{4}$

699. $\int_0^1 \frac{\log x\,dx}{\sqrt{1-x^2}} = -\frac{\pi}{2}\log 2$

700. $\int_0^1 x^m\left[\log\left(\frac{1}{x}\right)\right]^n dx = \frac{\Gamma(n+1)}{(m+1)^{n+1}}$, if $m+1>0, n+1>0$

701. $\int_0^1 \frac{(x^p - x^q)\,dx}{\log x} = \log\left(\frac{p+1}{q+1}\right)$, $(p+1>0, q+1>0)$

702. $\int_0^1 \frac{dx}{\sqrt{\log\left(\frac{1}{x}\right)}} = \sqrt{\pi}$

703. $\int_0^\infty \log\left(\frac{e^x+1}{e^x-1}\right)dx = \frac{\pi^2}{4}$

704. $\int_0^{\pi/2} (\log\sin x)\,dx = \int_0^{\pi/2} \log\cos x\,dx = -\frac{\pi}{2}\log 2$

705. $\int_0^{\pi/2} (\log\sec x)\,dx = \int_0^{\pi/2} \log\csc x\,dx = \frac{\pi}{2}\log 2$

706. $\int_0^{\pi} x(\log\sin x)\,dx = -\frac{\pi^2}{2}\log 2$

707. $\int_0^{\pi/2} (\sin x)(\log\sin x)\,dx = \log 2 - 1$

708. $\displaystyle\int_0^{\pi/2} (\log \tan x)\,dx = 0$

709. $\displaystyle\int_0^{\pi} \log(a \pm b\cos x)\,dx = \pi \log\left(\frac{a + \sqrt{a^2 - b^2}}{2}\right), \qquad (a \geq b)$

710. $\displaystyle\int_0^{\pi} \log(a^2 - 2ab\cos x + b^2)\,dx = \begin{cases} 2\pi \log a, & a \geq b > 0 \\ 2\pi \log b, & b \geq a > 0 \end{cases}$

711. $\displaystyle\int_0^{\infty} \frac{\sin ax}{\sinh bx}\,dx = \frac{\pi}{2b}\tanh\frac{a\pi}{2b}$

712. $\displaystyle\int_0^{\infty} \frac{\cos ax}{\cosh bx}\,dx = \frac{\pi}{2b}\operatorname{sech}\frac{\alpha\pi}{2b}$

713. $\displaystyle\int_0^{\infty} \frac{dx}{\cosh ax} = \frac{\pi}{2a}$

714. $\displaystyle\int_0^{\infty} \frac{x\,dx}{\sinh ax} = \frac{\pi^2}{4a^2}$

715. $\displaystyle\int_0^{\infty} e^{-ax}(\cosh bx)\,dx = \frac{a}{a^2 - b^2}, \qquad (0 \leq |b| < a)$

716. $\displaystyle\int_0^{\infty} e^{-ax}(\sinh bx)\,dx = \frac{b}{a^2 - b^2}, \qquad (0 \leq |b| < a)$

717. $\displaystyle\int_0^{\infty} \frac{\sinh ax}{e^{bx} + 1}\,dx = \frac{\pi}{2b}\csc\frac{a\pi}{b} - \frac{1}{2a}$

718. $\displaystyle\int_0^{\infty} \frac{\sinh ax}{e^{bx} - 1}\,dx = \frac{1}{2a} - \frac{\pi}{2b}\cot\frac{a\pi}{b}$

719. $$\int_0^{\pi/2} \frac{dx}{\sqrt{1 - k^2\sin^2 x}} = \frac{\pi}{2}\left[1 + \left(\frac{1}{2}\right)^2 k^2 + \left(\frac{1\cdot 3}{2\cdot 4}\right)^2 k^4 + \left(\frac{1\cdot 3\cdot 5}{2\cdot 4\cdot 6}\right)^2 k^6 + \cdots\right], \text{ if } k^2 < 1$$

720. $$\int_0^{\pi/2} \sqrt{1 - k^2\sin^2 x}\,dx = \frac{\pi}{2}\left[1 - \left(\frac{1}{2}\right)^2 k^2 - \left(\frac{1\cdot 3}{2\cdot 4}\right)^2 \frac{k^4}{3} - \left(\frac{1\cdot 3\cdot 5}{2\cdot 4\cdot 6}\right)^2 \frac{k^6}{5} - \cdots\right], \text{ if } k^2 < 1$$

721. $\displaystyle\int_0^{\infty} e^{-x}\log x\,dx = -\gamma = -0.5772157\ldots$

722. $\displaystyle\int_0^{\infty} e^{-x^2}\log x\,dx = -\frac{\sqrt{\pi}}{4}(\gamma + 2\log 2)$

723. $$\int_0^\infty \left(\frac{1}{1-e^{-x}} - \frac{1}{x}\right) e^{-x}\,dx = \gamma = 0.5772157\ldots \qquad \text{[Euler's Constant]}$$

724. $$\int_0^\infty \frac{1}{x}\left(\frac{1}{1+x} - e^{-x}\right) dx = \gamma = 0.5772157\ldots$$

FOURIER SERIES

1. If $f(x)$ is a bounded periodic function of period $2L$ (i.e. $f(x+2L)=f(x)$), and satisfies the *Dirichlet conditions*:

a) In any period $f(x)$ is continuous, except possibly for a finite number of jump discontinuities.

b) In any period $f(x)$ has only a finite number of maxima and minima.

Then $f(x)$ may be represented by the *Fourier series*

$$\frac{a_0}{2} + \sum_{n=1}^{\infty} \left(a_n \cos \frac{n\pi x}{L} + b_n \sin \frac{n\pi x}{L} \right),$$

where a_n and b_n are as determined below. This series will converge to $f(x)$ at every point where $f(x)$ is continuous, and to

$$\frac{f(x^+) + f(x^-)}{2}$$

(i.e. the average of the left-hand and right-hand limits) at every point where $f(x)$ has a jump discontinuity.

$$a_n = \frac{1}{L} \int_{-L}^{L} f(x) \cos \frac{n\pi x}{L}\, dx, \quad n = 0, 1, 2, 3, \ldots;$$

$$b_n = \frac{1}{L} \int_{-L}^{L} f(x) \sin \frac{n\pi x}{L}\, dx, \quad n = 1, 2, 3, \ldots$$

We may also write

$$a_n = \frac{1}{L} \int_{\alpha}^{\alpha+2L} f(x) \cos \frac{n\pi x}{L}\, dx \text{ and } b_n = \frac{1}{L} \int_{\alpha}^{\alpha+2L} f(x) \sin \frac{n\pi x}{L}\, dx,$$

where α is any real number. Thus if $\alpha = 0$,

$$a_n = \frac{1}{L} \int_{0}^{2L} f(x) \cos \frac{n\pi x}{L}\, dx, \quad n = 0, 1, 2, 3, \ldots;$$

$$b_n = \frac{1}{L} \int_{0}^{2L} f(x) \sin \frac{n\pi x}{L}\, dx, \quad n = 1, 2, 3, \ldots$$

2. If in addition to the above restrictions, $f(x)$ is even (i.e. $f(-x) = f(x)$), the Fourier series reduces to

$$\frac{a_0}{2} + \sum_{n=1}^{\infty} a_n \cos \frac{n\pi x}{L}.$$

That is, $b_n = 0$. In this case, a simpler formula for a_n is

$$a_n = \frac{2}{L} \int_{0}^{L} f(x) \cos \frac{n\pi x}{L}\, dx, \quad n = 0, 1, 2, 3, \ldots$$

3. If in addition to the restrictions in (1), $f(x)$ is an odd function (i.e. $f(-x) = -f(x)$), then the Fourier series reduces to

$$\sum_{n=1}^{\infty} b_n \sin \frac{n\pi x}{L}.$$

That is, $a_n = 0$. In this case, a simpler formula for the b_n is

$$b_n = \frac{2}{L} \int_{0}^{L} f(x) \sin \frac{n\pi x}{L}\, dx, \quad n = 1, 2, 3, \ldots$$

4. If in addition to the restrictions in (2) above, $f(x) = -f(L - x)$, then a_n will be 0 for all even values of n, including $n = 0$. Thus in this case, the expansion reduces to

$$\sum_{m=1}^{\infty} a_{2m-1} \cos \frac{(2m - 1)\pi x}{L}.$$

5. If in addition to the restrictions in (3) above, $f(x) = f(L - x)$, then b_n will be 0 for all even values of n. Thus in this case, the expansion reduces to

$$\sum_{m=1}^{\infty} b_{2m-1} \sin \frac{(2m - 1)\pi x}{L}.$$

(The series in (4) and (5) are known as *odd-harmonic series*, since only the odd harmonics appear. Similar rules may be stated for even-harmonic series, but when a series appears in the even-harmonic form, it means that $2L$ has not been taken as the smallest period of $f(x)$. Since any integral multiple of a period is also a period, series obtained in this way will also work, but in general computation is simplified if $2L$ is taken to be the smallest period.)

6. If we write the Euler definitions for $\cos\theta$ and $\sin\theta$, we obtain the complex form of the Fourier Series known either as the "Complex Fourier Series" or the "Exponential Fourier Series" of $f(x)$. It is represented as

$$f(x) = \tfrac{1}{2} \sum_{n=-\infty}^{n=+\infty} c_n e^{i\omega_n x},$$

where

$$c_n = \frac{1}{L} \int_{-L}^{L} f(x) e^{-i\omega_n x} dx, \quad n = 0, \pm 1, \pm 2, \pm 3, \ldots$$

with $\omega_n = \dfrac{n\pi}{L}$, $\quad n = 0, \pm 1, \pm 2, \ldots$

The set of coefficients $\{c_n\}$ is often referred to as the Fourier spectrum.

7. If both sine and cosine terms are present and if $f(x)$ is of period $2L$ and expandable by a Fourier series, it can be represented as

$$f(x) = \frac{a_0}{2} + \sum_{n=1}^{\infty} c_n \sin\left(\frac{n\pi x}{L} + \phi_n\right), \text{ where } a_n = c_n \sin\phi_n,$$

$$b_n = c_n \cos\phi_n, \quad c_n = \sqrt{a_n^2 + b_n^2}, \quad \phi_n = \arctan\left(\frac{a_n}{b_n}\right)$$

It can also be represented as

$$f(x) = \frac{a_0}{2} + \sum_{n=1}^{\infty} c_n \cos\left(\frac{n\pi x}{L} + \phi_n\right), \text{ where } a_n = c_n \cos\phi_n,$$

$$b_n = -c_n \sin\phi_n, \quad c_n = \sqrt{a_n^2 + b_n^2}, \quad \phi_n = \arctan\left(-\frac{b_n}{a_n}\right)$$

where ϕ_n is chosen so as to make a_n, b_n, and c_n hold.

FOURIER SERIES (Continued)

8. The following table of trigonometric identities should be helpful for developing Fourier Series.

	n	n even	n odd	$n/2$ odd	$n/2$ even
$\sin n\pi$	0	0	0	0	0
$\cos n\pi$	$(-1)^n$	+1	−1	+1	+1
*$\sin \frac{n\pi}{2}$		0	$(-1)^{(n-1)/2}$	0	0
*$\cos \frac{n\pi}{2}$		$(-1)^{n/2}$	0	−1	+1
$\sin \frac{n\pi}{4}$			$\frac{\sqrt{2}}{2}(-1)^{(n^2+4n+11)/8}$	$(-1)^{(n-2)/4}$	0

*A useful formula for $\sin \frac{n\pi}{2}$ and $\cos \frac{n\pi}{2}$ is given by

$$\sin \frac{n\pi}{2} = \frac{(i)^{n+1}}{2}[(-1)^n - 1] \text{ and } \cos \frac{n\pi}{2} = \frac{(i)^n}{2}[(-1)^n + 1], \text{ where } i^2 = -1.$$

AUXILIARY FORMULAS FOR FOURIER SERIES

$$1 = \frac{4}{\pi}\left[\sin \frac{\pi x}{k} + \frac{1}{3}\sin \frac{3\pi x}{k} + \frac{1}{5}\sin \frac{5\pi x}{k} + \cdots\right] \qquad [0 < x < k]$$

$$x = \frac{2k}{\pi}\left[\sin \frac{\pi x}{k} - \frac{1}{2}\sin \frac{2\pi x}{k} + \frac{1}{3}\sin \frac{3\pi x}{k} - \cdots\right] \qquad [-k < x < k]$$

$$x = \frac{k}{2} - \frac{4k}{\pi^2}\left[\cos \frac{\pi x}{k} + \frac{1}{3^2}\cos \frac{3\pi x}{k} + \frac{1}{5^2}\cos \frac{5\pi x}{k} + \cdots\right] \qquad [0 < x < k]$$

$$x^2 = \frac{2k^2}{\pi^3}\left[\left(\frac{\pi^2}{1} - \frac{4}{1}\right)\sin \frac{\pi x}{k} - \frac{\pi^2}{2}\sin \frac{2\pi x}{k} + \left(\frac{\pi^2}{3} - \frac{4}{3^3}\right)\sin \frac{3\pi x}{k} - \frac{\pi^2}{4}\sin \frac{4\pi x}{k} + \left(\frac{\pi^2}{5} - \frac{4}{5^3}\right)\sin \frac{5\pi x}{k} + \cdots\right] \qquad [0 < x < k]$$

$$x^2 = \frac{k^2}{3} - \frac{4k^2}{\pi^2}\left[\cos \frac{\pi x}{k} - \frac{1}{2^2}\cos \frac{2\pi x}{k} + \frac{1}{3^2}\cos \frac{3\pi x}{k} - \frac{1}{4^2}\cos \frac{4\pi x}{k} + \cdots\right] \qquad [-k < x < k]$$

$$1 - \frac{1}{3} + \frac{1}{5} - \frac{1}{7} + \cdots = \frac{\pi}{4}$$

$$1 + \frac{1}{2^2} + \frac{1}{3^2} + \frac{1}{4^2} + \cdots = \frac{\pi^2}{6}$$

$$1 - \frac{1}{2^2} + \frac{1}{3^2} - \frac{1}{4^2} + \cdots = \frac{\pi^2}{12}$$

$$1 + \frac{1}{3^2} + \frac{1}{5^2} + \frac{1}{7^2} + \cdots = \frac{\pi^2}{8}$$

$$\frac{1}{2^2} + \frac{1}{4^2} + \frac{1}{6^2} + \frac{1}{8^2} + \cdots = \frac{\pi^2}{24}$$

DIFFERENTIAL EQUATIONS

SPECIAL FORMULAS

Certain types of differential equations occur sufficiently often to justify the use of formulas for the corresponding particular solutions. The following set of tables I to XIV covers all first, second and nth order ordinary linear differential equations with constant coefficients for which the right members are of the form $P(x)e^{rx}\sin sx$ or $P(x)e^{rx}\cos sx$, where r and s are constants and $P(x)$ is a polynomial of degree n.

When the right member of a reducible linear partial differential equation with constant coefficients is not zero, particular solutions for certain types of right members are contained in tables XV to XXI. In these tables both F and P are used to denote polynomials, and it is assumed that no denominator is zero. In any formula the roles of x and y may be reversed throughout, changing a formula in which x dominates to one in which y dominates. Tables XIX, XX, XXI are applicable whether the equations are reducible or not. The symbol $\binom{m}{n}$ stands for $\dfrac{m!}{(m-n)!n!}$ and is the $n+1$ st coefficient in the expansion of $(a+b)^m$. Also $0! = 1$ by definition.

The tables as herewith given are those contained in the text *Differential Equations* by Ginn and Company (1955) and are published with their kind permission and that of the author, Professor Frederick H. Steen.

Solution of Linear Differential Equations with Constant Coefficients

Any linear differential equation with constant coefficients may be written in the form

$$p(D)y = R(x),$$

where D is the differential operation

$$Dy = \frac{dy}{dx},$$

$p(D)$ is a polynomial in D,
y is the dependent variable,
x is the independent variable,
$R(x)$ is an arbitrary function of x.

A power of D represents repeated differentiation, that is

$$D^n y = \frac{d^n y}{dx^n}.$$

For such an equation, the general solution may be written in the form

$$y = y_c + y_p,$$

where y_p is any particular solution, and y_c is called the *complementary function*. This complementary function is defined as the general solution of the *homogeneous equation*, which is the original differential equation with the right side replaced by zero, i.e.

$$p(D)y = 0$$

The complementary function y_c may be determined as follows:

1. Factor the polynomial $p(D)$ into real and complex linear factors, just as if D were a variable instead of an operator.

2. For each non-repeated linear factor of the form $(D - a)$, where a is real, write down a term of the form

$$ce^{ax},$$

where c is an arbitrary constant.

3. For each repeated real linear factor of the form $(D - a)^n$, write down n terms of the form

$$c_1 e^{ax} + c_2 x e^{ax} + c_3 x^2 e^{ax} + \cdots + c_n x^{n-1} e^{ax},$$

where the c_i's are arbitrary constants.

4. For each non-repeated conjugate complex pair of factors of the form $(D - a + ib)(D - a - ib)$, write down 2 terms of the form

$$c_1 e^{ax} \cos bx + c_2 e^{ax} \sin bx$$

5. For each repeated conjugate complex pair of factors of the form $(D - a + ib)^n (D - a - ib)^n$, write down $2n$ terms of the form

$$c_1 e^{ax} \cos bx + c_2 e^{ax} \sin bx + c_3 x e^{ax} \cos bx + c_4 x e^{ax} \sin bx + \cdots + c_{2n-1} x^{n-1} e^{ax} \cos bx + c_{2n} x^{n-1} e^{ax} \sin bx$$

6. The sum of all the terms thus written down is the complementary function y_c.

To find the particular solution y_p, use the following tables, as shown in the examples. For cases not shown in the tables, there are various methods of finding y_p. The most general method is called *variation of parameters*. The following example illustrates the method:

Find y_p for $(D^2 - 4)y = e^x$.

This example can be solved most easily by use of equation 63 in the tables following. However it is given here as an example of the method of variation of parameters.

The complementary function is

$$y_c = c_1 e^{2x} + c_2 e^{-2x}$$

To find y_p, replace the constants in the complementary function with unknown functions,

$$y_p = ue^{2x} + ve^{-2x}.$$

We now prepare to substitute this assumed solution into the original equation. We begin by taking all the necessary derivatives:

$$y_p = ue^{2x} + ve^{-2x}$$
$$y_p' = 2ue^{2x} - 2ve^{-2x} + u'e^{2x} + v'e^{-2x}$$

For each derivative of y_p except the highest, we set the sum of all the terms containing u' and v' to 0. Thus the above equation becomes

$$u'e^{2x} + v'e^{-2x} = 0 \quad \text{and} \quad y_p' = 2ue^{2x} - 2ve^{-2x}$$

Continuing to differentiate, we have

$$y_p'' = 4ue^{2x} + 4ve^{-2x} + 2u'e^{2x} - 2v'e^{-2x}$$

When we substitute into the original equation, all the terms not containing u' or v' cancel out. This is a consequence of the method by which y_p was set up.

Thus all that is necessary is to write down the terms containing u' or v' in the highest order derivative of y_p, multiply by the constant coefficient of the highest power of D in $p(D)$, and set it equal to $R(x)$. Together with the previous terms in u' and v' which were set equal to 0, this gives us as many linear equations in the first derivatives of the unknown functions as there are unknown functions. The first derivatives may then

be solved for by algebra, and the unknown functions found by integration. In the present example, this becomes

$$u'e^{2x} + v'e^{-2x} = 0$$
$$2u'e^{2x} - 2v'e^{-2x} = e^x.$$

We eliminate v' and u' separately, getting

$$4u'e^{2x} = e^x$$
$$4v'e^{-2x} = -e^x.$$

Thus

$$u' = \tfrac{1}{4}e^{-x}$$
$$v' = -\tfrac{1}{4}e^{3x}.$$

Therefore, by integrating

$$u = -\tfrac{1}{4}e^{-x}$$
$$v = -\tfrac{1}{12}e^{3x}.$$

A constant of integration is not needed, since we need only one particular solution. Thus

$$y_p = ue^{2x} + ve^{-2x} = -\tfrac{1}{4}e^{-x}e^{2x} - \tfrac{1}{12}e^{3x}e^{-2x}$$
$$= -\tfrac{1}{4}e^x - \tfrac{1}{12}e^x = -\tfrac{1}{3}e^x,$$

and the general solution is

$$y = y_c + y_p = c_1e^{2x} + c_2e^{-2x} - \tfrac{1}{3}e^x.$$

The following examples illustrate the use of the tables.

Example 1. Solve $(D^2 - 4)y = \sin 3x$.

Substitution of $q = -4$, $s = 3$ in formula 24 gives

$$y_p = \frac{\sin 3x}{-9 - 4},$$

wherefore the general solution is

$$y = c_1e^{2x} + c_2e^{-2x} - \frac{\sin 3x}{13}.$$

Example 2. Obtain a particular solution of $(D^2 - 4D + 5)y = x^2e^{3x}\sin x$.

Applying formula 40 with $a = 2$, $b = 1$, $r = 3$, $s = 1$, $P(x) = x^2$, $s + b = 2$, $s - b = 0$, $a - r = -1$, $(a - r)^2 + (s + b)^2 = 5$, $(a - r)^2 + (s - b)^2 = 1$, we have

$$y_p = \frac{e^{3x}\sin x}{2}\left[\left(\frac{2}{5} - \frac{0}{1}\right)x^2 + \left(\frac{2(-1)2}{25} - \frac{2(-1)0}{1}\right)2x + \left(\frac{3\cdot1\cdot2 - 2^3}{125} - \frac{3\cdot1\cdot0 - 0}{1}\right)2\right]$$

$$- \frac{e^{3x}\cos x}{2}\left[\left(\frac{-1}{5} - \frac{-1}{1}\right)x^2 + \left(\frac{1-4}{25} - \frac{1-0}{1}\right)2x + \left(\frac{-1 - 3(-1)4}{125} - \frac{-1 - 3(-1)0}{1}\right)2\right]$$

$$= (\tfrac{1}{5}x^2 - \tfrac{4}{25}x - \tfrac{2}{125})e^{3x}\sin x + (-\tfrac{2}{5}x^2 + \tfrac{28}{25}x - \tfrac{136}{125})e^{3x}\cos x.$$

The special formulas effect a very considerable saving of time in problems of this type.

Example 3. Obtain a particular solution of $(D^2 - 4D + 5)y = x^2e^{2x}\cos x$. (Compare with Example 2.)

Formula 40 is not applicable here since for this equation $r = a$, $s = b$, wherefore the denominator $(a - r)^2 + (s - b)^2 = 0$. We turn instead to formula 44. Substituting $a = 2$, $b = 1$, $P(x) = x^2$ and replacing sin by cos, cos by $-$sin, we obtain

$$y_p = \frac{e^{2x}\cos x}{4}(x^2 - \tfrac{2}{4}) + \frac{e^{2x}\sin x}{2}\int (x^2 - \tfrac{1}{2})\,dx$$

$$= \left(\frac{x^2}{4} - \frac{1}{8}\right)e^{2x}\cos x + \left(\frac{x^3}{6} - \frac{x}{4}\right)e^{2x}\sin x,$$

which is the required solution.

Example 4. Find z_p for $(D_x - 3D_y)z = \ln(y + 3x)$.

Referring to Table XV we note that formula 69 (not 68) is applicable. This gives

$$z_p = x\ln(y + 3x).$$

It is easily seen that $-\dfrac{y}{3}\ln(y + 3x)$ would serve equally well.

Example 5. Solve $(D_x + 2D_y - 4)z = y\cos(y - 2x)$.

Since R in formula 76 contains a polynomial in x, not y, we rewrite the given equation in the form

$$(D_y + \tfrac{1}{2}D_x - 2)z = \tfrac{1}{2}y\cos(y - 2x).$$

Then

$$z_c = e^{2y}F(x - \tfrac{1}{2}y) = e^{2y}f(2x - y),$$

and by the formula

$$z_p = -\tfrac{1}{2}\cos(y - 2x)\cdot\left(\frac{y}{2} + \frac{\frac{1}{2}}{2}\right)$$

$$= -\tfrac{1}{8}(2y + 1)\cos(y - 2x).$$

Example 6. Find z_p for $(D_x + 4D_y)^3 z = (2x - y)^2$.

Using formula 79, we obtain

$$z_p = \frac{\int\int\int u^2 du^3}{[2 + 4(-1)]^3} = \frac{u^5}{5\cdot 4\cdot 3\cdot(-8)} = -\frac{(2x - y)^5}{480}.$$

Example 7. Find z_p for $(D_x^3 + 5D_x^2D_y - 7D_x + 4)z = e^{2x+3y}$.

By formula 87

$$z_p = \frac{e^{2x+3y}}{2^3 + 5\cdot 2^2\cdot 3 - 7\cdot 2 + 4} = \frac{e^{2x+3y}}{58}.$$

Example 8. Find z_p for

$$(D_x^4 + 6D_x^3D_y + D_xD_y + D_y^2 + 9)z = \sin(3x + 4y).$$

Since every term in the left member is of *even* degree in the two operators D_x and D_y, formula 90 is applicable.

It gives

$$z_p = \frac{\sin(3x + 4y)}{(-9)^2 + 6(-9)(-12) + (-12) + (-16) + 9}$$

$$= \frac{\sin(3x + 4y)}{710}.$$

TABLE I: $(D - a)y = R$

R	y_p
1. e^{rx}	$\dfrac{e^{rx}}{r-a}$
2. $\sin sx$*	$-\dfrac{a \sin sx + s \cos sx}{a^2+s^2} = -\dfrac{1}{\sqrt{a^2+s^2}} \sin\left(sx + \tan^{-1}\dfrac{s}{a}\right)$
3. $P(x)$	$-\dfrac{1}{a}\left[P(x) + \dfrac{P'(x)}{a} + \dfrac{P''(x)}{a^2} + \cdots + \dfrac{P^{(n)}(x)}{a^n}\right]$
4. $e^{rx} \sin sx$*	Replace a by $a - r$ in formula 2 and multiply by e^{rx}.
5. $P(x)e^{rx}$	Replace a by $a - r$ in formula 3 and multiply by e^{rx}.
6. $P(x) \sin sx$*	$-\sin sx\left[\dfrac{a}{a^2+s^2}P(x) + \dfrac{a^2-s^2}{(a^2+s^2)^2}P'(x) + \dfrac{a^3-3as^2}{(a^2+s^2)^3}P''(x) + \cdots + \dfrac{a^k - \binom{k}{2}a^{k-2}s^2 + \binom{k}{4}a^{k-4}s^4 - \cdots}{(a^2+s^2)^k}P^{(k-1)}(x) + \cdots\right]$ $-\cos sx\left[\dfrac{s}{a^2+s^2}P(x) + \dfrac{2as}{(a^2+s^2)^2}P'(x) + \dfrac{3a^2s-s^3}{(a^2+s^2)^3}P''(x) + \cdots + \dfrac{\binom{k}{1}a^{k-1}s - \binom{k}{3}a^{k-3}s^3 + \cdots}{(a^2+s^2)^k}P^{(k-1)}(x) + \cdots\right]$
7. $P(x)e^{rx} \sin sx$*	Replace a by $a - r$ in formula 6 and multiply by e^{rx}.
8. e^{ax}	xe^{ax}
9. $e^{ax} \sin sx$*	$-\dfrac{e^{ax} \cos sx}{s}$
10. $P(x)e^{ax}$	$e^{ax}\int P(x)\,dx$
11. $P(x)e^{ax} \sin sx$*	$\dfrac{e^{ax} \sin sx}{s}\left[\dfrac{P'(x)}{s^3} - \dfrac{P'''(x)}{s^3} + \dfrac{P^{v}(x)}{s^5} - \cdots\right] - \dfrac{e^{ax} \cos sx}{s}\left[P(x) - \dfrac{P''(x)}{s^2} + \dfrac{P^{iv}(x)}{s^4} - \cdots\right]$

* For $\cos sx$ in R replace "sin" by "cos" and "cos" by "– sin" in y_p.

$$D^n = \frac{d^n}{dx^n} \qquad \binom{m}{n} = \frac{m!}{(m-n)!\,n!} \qquad 0! = 1$$

TABLE II: $(D - a)^2 y = R$

R	y_p
12. e^{rx}	$\dfrac{e^{rx}}{(r-a)^2}$
13. $\sin sx$*	$\dfrac{1}{(a^2+s^2)^2}[(a^2-s^2)\sin sx + 2as\cos sx] = \dfrac{1}{a^2+s^2}\sin\left(sx + \tan^{-1}\dfrac{2as}{a^2-s^2}\right)$
14. $P(x)$	$\dfrac{1}{a^2}\left[P(x) + \dfrac{2P'(x)}{a} + \dfrac{3P''(x)}{a^2} + \cdots + \dfrac{(n+1)P^{(n)}(x)}{a^n}\right]$
15. $e^{rx}\sin sx$*	Replace a by $a - r$ in formula 13 and multiply by e^{rx}.
16. $P(x)e^{rx}$	Replace a by $a - r$ in formula 14 and multiply by e^{rx}.
17. $P(x)\sin sx$*	$\sin sx\left[\dfrac{a^2-s^2}{(a^2+s^2)^2}P(x) + 2\dfrac{a^3-3as^2}{(a^2+s^2)^3}P'(x) + 3\dfrac{a^4-6a^2s^2+s^4}{(a^2+s^2)^4}P''(x) + \cdots + (k-1)\dfrac{a^k - \binom{k}{2}a^{k-2}s^2 + \binom{k}{4}a^{k-4}s^4 - \cdots}{(a^2+s^2)^k}P^{(k-2)}(x) + \cdots\right]$ $+ \cos sx\left[\dfrac{2as}{(a^2+s^2)^2}P(x) + 2\dfrac{3a^2s-s^3}{(a^2+s^2)^3}P'(x) + 3\dfrac{4a^3s-4as^3}{(a^2+s^2)^4}P''(x) + \cdots + (k-1)\dfrac{\binom{k}{1}a^{k-1}s - \binom{k}{3}a^{k-3}s^3 + \cdots}{(a^2+s^2)^k}P^{(k-2)}(x) + \cdots\right]$
18. $P(x)e^{rx}\sin sx$*	Replace a by $a - r$ in formula 17 and multiply by e^{rx}.
19. e^{ax}	$\frac{1}{2}x^2e^{ax}$
20. $e^{ax}\sin sx$*	$-\dfrac{e^{ax}\sin sx}{s^2}$
21. $P(x)e^{ax}$	$e^{ax}\displaystyle\iint P(x)\,dx\,dx$
22. $P(x)e^{ax}\sin sx$*	$-\dfrac{e^{ax}\sin sx}{s^2}\left[P(x) - \dfrac{3P''(x)}{s^2} + \dfrac{5P^{iv}(x)}{s^4} - \dfrac{7P^{vi}(x)}{s^6} + \cdots\right] - \dfrac{e^{ax}\cos sx}{s^2}\left[\dfrac{2P'(x)}{s} - \dfrac{4P'''(x)}{s^3} + \dfrac{6P^{v}(x)}{s^5} - \cdots\right]$

* For $\cos sx$ in R replace "sin" by "cos" and "cos" by "− sin" in y_p.

TABLE III: $(D^2 + q)y = R$

R	y_p
23. e^{rx}	$\dfrac{e^{rx}}{r^2+q}$
24. $\sin sx$*	$\dfrac{\sin sx}{-s^2+q}$
25. $P(x)$	$\dfrac{1}{q}\left[P(x) - \dfrac{P''(x)}{q} + \dfrac{P^{\text{iv}}(x)}{q^2} - \cdots + (-1)^k \dfrac{P^{(2k)}(x)}{q^k} \cdots\right]$
26. $e^{rx}\sin sx$*	$\dfrac{(r^2-s^2+q)e^{rx}\sin sx - 2rse^{rx}\cos sx}{(r^2-s^2+q)^2+(2rs)^2} = \dfrac{e^{rx}}{\sqrt{(r^2-s^2+q)^2+(2rs)^2}}\sin\left[sx - \tan^{-1}\dfrac{2rs}{r^2-s^2+q}\right]$
27. $P(x)e^{rx}$	$\dfrac{e^{rx}}{r^2+q}\left[P(x) - \dfrac{2r}{r^2+q}P'(x) + \dfrac{3r^2-q}{(r^2+q)^2}P''(x) - \dfrac{4r^3-4qr}{(r^2+q)^3}P'''(x) + \cdots + (-1)^{k-1}\dfrac{\binom{k}{1}r^{k-1} - \binom{k}{3}r^{k-3}q + \binom{k}{5}r^{k-5}q^2 - \cdots}{(r^2+q)^{k-1}}P^{(k-1)}(x) + \cdots\right]$
28. $P(x)\sin sx$*	$\dfrac{\sin sx}{(-s^2+q)}\left[P(x) - \dfrac{3s^2+q}{(-s^2+q)^2}P''(x) + \dfrac{5s^4+10s^2q+q^2}{(-s^2+q)^4}P^{\text{iv}}(x) + \cdots + (-1)^k \dfrac{\binom{2k+1}{1}s^{2k} + \binom{2k+1}{3}s^{2k-2}q + \binom{2k+1}{5}s^{2k-4}q^2 + \cdots}{(-s^2+q)^{2k}}P^{(2k)}(x) + \cdots\right]$ $- \dfrac{s\cos sx}{(-s^2+q)}\left[\dfrac{2P'(x)}{(-s^2+q)} - \dfrac{4s^2+4q}{(-s^2+q)^3}P'''(x) + \cdots + (-1)^{k+1}\dfrac{\binom{2k}{1}s^{2k-2} + \binom{2k}{3}s^{2k-4}q + \cdots}{(-s^2+q)^{2k-1}}P^{(2k-1)}(x) + \cdots\right]$

TABLE IV: $(D^2 + b^2)y = R$

R	y_p
29. $\sin bx$*	$-\dfrac{x\cos bx}{2b}$
30. $P(x)\sin bx$*	$\dfrac{\sin bx}{(2b)^2}\left[P(x) - \dfrac{P''(x)}{(2b)^2} + \dfrac{P^{\text{iv}}(x)}{(2b)^4} - \cdots\right] - \dfrac{\cos bx}{2b}\int\left[P(x) - \dfrac{P''(x)}{(2b)^2} + \cdots\right]dx$

* For $\cos sx$ in R replace "sin" by "cos" and "cos" by "$-$ sin" in y_p.

TABLE V: $(D^2 + pD + q)y = R$

R	y_p
31. e^{rx}	$\dfrac{e^{rx}}{r^2 + pr + q}$
32. $\sin sx$*	$\dfrac{(q - s^2)\sin sx - ps\cos sx}{(q - s^2)^2 + (ps)^2} = \dfrac{1}{\sqrt{(q - s^2)^2 + (ps)^2}}\sin\left(sx - \tan^{-1}\dfrac{ps}{q - s^2}\right)$
33. $P(x)$	$\dfrac{1}{q}\left[P(x) - \dfrac{p}{q}P'(x) + \dfrac{p^2 - q}{q^2}P''(x) - \dfrac{p^3 - 2pq}{q^3}P'''(x) + \cdots + (-1)^n \dfrac{p^n - \binom{n-1}{1}p^{n-2}q + \binom{n-2}{2}p^{n-4}q^2 - \cdots}{q^n}P^{(n)}(x)\right]$
34. $e^{rx}\sin sx$*	Replace p by $p + 2r$, q by $q + pr + r^2$ in formula 32 and multiply by e^{rx}.
35. $P(x)e^{rx}$	Replace p by $p + 2r$, q by $q + pr + r^2$ in formula 33 and multiply by e^{rx}.

TABLE VI: $(D - b)(D - a)y = R$

36. $P(x)\sin sx$*	$\dfrac{\sin sx}{b - a}\left[\left(\dfrac{a}{a^2 + s^2} - \dfrac{b}{b^2 + s^2}\right)P(x) + \left(\dfrac{a^2 - s^2}{(a^2 + s^2)^2} - \dfrac{b^2 - s^2}{(b^2 + s^2)^2}\right)P'(x) + \left(\dfrac{a^3 - 3as^2}{(a^2 + s^2)^3} - \dfrac{b^3 - 3bs^2}{(b^2 + s^2)^3}\right)P''(x) + \cdots\right] + \dfrac{\cos sx}{b - a}\left[\left(\dfrac{s}{a^2 + s^2} - \dfrac{s}{b^2 + s^2}\right)P(x) + \left(\dfrac{2as}{(a^2 + s^2)^2} - \dfrac{2bs}{(b^2 + s^2)^2}\right)P'(x) + \left(\dfrac{3a^2s - s^3}{(a^2 + s^2)^3} - \dfrac{3b^2s - s^3}{(b^2 + s^2)^3}\right)P''(x) + \cdots\right]$†
37. $P(x)e^{rx}\sin sx$*	Replace a by $a - r$, b by $b - r$ in formula 36 and multiply by e^{rx}.
38. $P(x)e^{ax}$	$\dfrac{e^{ax}}{a - b}\left[\int P(x)dx + \dfrac{P(x)}{(b - a)} + \dfrac{P'(x)}{(b - a)^2} + \dfrac{P''(x)}{(b - a)^3} + \cdots + \dfrac{P^{(n)}(x)}{(b - a)^{n+1}}\right]$

* For $\cos sx$ in R replace "sin" by "cos" and "cos" by "− sin" in y_p.
† For additional terms, compare with formula 6.

TABLE VII: $(D^2 - 2aD + a^2 + b^2)y = R$

R	y_p
39. $P(x)\sin sx$*	$\frac{\sin sx}{2b}\left[\left(\frac{s+b}{a^2+(s+b)^2} - \frac{s-b}{a^2+(s-b)^2}\right)P(x) + \left(\frac{2a(s+b)}{[a^2+(s+b)^2]^2} - \frac{2a(s-b)}{[a^2+(s-b)^2]^2}\right)P'(x) + \left(\frac{3a^2(s+b)-(s+b)^3}{[a^2+(s+b)^2]^3} - \frac{3a^2(s-b)-(s-b)^3}{[a^2+(s-b)^2]^3}\right)P''(x) + \cdots\right]$ $- \frac{\cos sx}{2b}\left[\left(\frac{a}{a^2+(s+b)^2} - \frac{a}{a^2+(s-b)^2}\right)P(x) + \left(\frac{a^2-(s+b)^2}{[a^2+(s+b)^2]^2} - \frac{a^2-(s-b)^2}{[a^2+(s-b)^2]^2}\right)P'(x) + \left(\frac{a^3-3a(s+b)^2}{[a^2+(s+b)^2]^3} - \frac{a^3-3a(s-b)^2}{[a^2+(s-b)^2]^3}\right)P''(x) + \cdots\right]^{\dagger}$
40. $P(x)e^{rx}\sin sx$*	Replace a by $a - r$ in formula 39 and multiply by e^{rx}.
41. $P(x)e^{ax}$	$\frac{e^{ax}}{b^2}\left[P(x) - \frac{P''(x)}{b^2} + \frac{P^{iv}(x)}{b^4} - \cdots\right]$
42. $e^{ax}\sin sx$*	$\frac{e^{ax}\sin sx}{-s^2+b^2}$
43. $e^{ax}\sin bx$*	$-\frac{xe^{ax}\cos bx}{2b}$
44. $P(x)e^{ax}\sin bx$*	$\frac{e^{ax}\sin bx}{(2b)^2}\left[P(x) - \frac{P''(x)}{(2b)^2} + \frac{P^{iv}(x)}{(2b)^4} - \cdots\right] - \frac{e^{ax}\cos bx}{2b}\int\left[P(x) - \frac{P''(x)}{(2b)^2} + \frac{P^{iv}(x)}{(2b)^4} - \cdots\right]dx$

* For $\cos sx$ in R replace "sin" by "cos" and "cos" by "− sin" in y_p.
† For additional terms, compare with formula 6.

TABLE VIII: $f(D)y = [D^n + a_{n-1}D^{n-1} + \cdots + a_1D + a_0]y = R$

R	y_p
45. e^{rx}	$\dfrac{e^{rx}}{f(r)}$
46. $\sin sx$*	$\dfrac{[a_0 - a_2s^2 + a_4s^4 - \cdots]\sin sx - [a_1s - a_3s^3 + a_5s^5 + \cdots]\cos sx}{[a_0 - a_2s^2 + a_4s^4 - \cdots]^2 + [a_1s - a_3s^3 + a_5s^5 - \cdots]^2}$

TABLE IX: $f(D^2)y = R$

47. $\sin sx$*	$\dfrac{\sin sx}{f(-s^2)} = \dfrac{\sin sx}{a_0 - a_2s^2 + \cdots \pm s^{2n}}$

TABLE X: $(D - a)^n y = R$

48. e^{rx}	$\dfrac{e^{rx}}{(r-a)^n}$
49. $\sin sx$*	$\dfrac{(-1)^n}{(a^2+s^2)^n}\{[a^n - \binom{n}{2}a^{n-2}s^2 + \binom{n}{4}a^{n-4}s^4 - \cdots]\sin sx + [\binom{n}{1}a^{n-1}s - \binom{n}{3}a^{n-3}s^3 + \cdots]\cos sx\}$
50. $P(x)$	$\dfrac{(-1)^n}{a^n}\left[P(x) + \binom{n}{1}\dfrac{P'(x)}{a} + \binom{n+1}{2}\dfrac{P''(x)}{a^2} + \binom{n+2}{3}\dfrac{P'''(x)}{a^3} + \cdots\right]$
51. $e^{rx}\sin sx$*	Replace a by $a - r$ in formula 49 and multiply by e^{rx}.
52. $e^{rx}P(x)$	Replace a by $a - r$ in formula 50 and multiply by e^{rx}.

* For $\cos sx$ in R replace "sin" by "cos" and "cos" by "− sin" in y_p.

53. $P(x)\sin sx$*

$$(-1)^n \sin sx\left[A_nP(x) + \binom{n}{1}A_{n+1}P'(x) + \binom{n+1}{2}A_{n+2}P''(x) + \binom{n+2}{3}A_{n+3}P'''(x) + \cdots\right]$$
$$+ (-1)^n \cos sx\left[B_nP(x) + \binom{n}{1}B_{n+1}P'(x) + \binom{n+1}{2}B_{n+2}P''(x) + \binom{n+2}{3}B_{n+3}P'''(x) + \cdots\right]$$

$$A_1 = \frac{a}{a^2+s^2},\ A_2 = \frac{a^2-s^2}{(a^2+s^2)^2},\ \cdots,\ A_k = \frac{a^k - \binom{k}{2}a^{k-2}s^2 + \binom{k}{4}a^{k-4}s^4 - \cdots}{(a^2+s^2)^k}.$$

$$B_1 = \frac{s}{a^2+s^2},\ B_2 = \frac{2as}{(a^2+s^2)^2},\ \cdots,\ B_k = \frac{\binom{k}{1}a^{k-1}s - \binom{k}{3}a^{k-3}s^3 + \cdots}{(a^2+s^2)^k}.$$

54. $P(x)e^{rx}\sin sx$* Replace a by $a - r$ in formula 53 and multiply by e^{rx}.

55. $e^{ax}P(x)$ $\quad e^{ax}\iint \cdots \int P(x)dx^n$

56. $P(x)e^{ax}\sin sx$*

$$\frac{(-1)^{\frac{n-1}{2}}e^{ax}\sin sx}{s^n}\left[\binom{n}{n-1}\frac{P'(x)}{s} - \binom{n+2}{n-1}\frac{P'''(x)}{s^3} + \binom{n+4}{n-1}\frac{P^{\text{v}}(x)}{s^5} - \cdots\right]$$
$$+ \frac{(-1)^{\frac{n+1}{2}}e^{ax}\cos sx}{s^n}\left[\binom{n-1}{n-1}P(x) - \binom{n+1}{n-1}\frac{P''(x)}{s^2} + \binom{n+3}{n-1}\frac{P^{\text{iv}}(x)}{s^4} - \cdots\right] \quad (n \text{ odd})$$

$$\frac{(-1)^{\frac{n}{2}}e^{ax}\sin sx}{s^n}\left[\binom{n-1}{n-1}P(x) - \binom{n+1}{n-1}\frac{P''(x)}{s^2} + \binom{n+3}{n-1}\frac{P^{\text{iv}}(x)}{s^4} - \cdots\right]$$
$$+ \frac{(-1)^{\frac{n}{2}}e^{ax}\cos sx}{s^n}\left[\binom{n}{n-1}\frac{P'(x)}{s} - \binom{n+2}{n-1}\frac{P'''(x)}{s^3} + \binom{n+4}{n-1}\frac{P^{\text{v}}(x)}{s^5} - \cdots\right] \quad (n \text{ even})$$

TABLE XI: $(D - a)^n f(D)y = R$

57. e^{ax} $\quad \dfrac{x^n}{n!}\cdot\dfrac{e^{ax}}{f(a)}$

* For $\cos sx$ in R replace "sin" by "cos" and "cos" by "− sin" in y_p.

TABLE XII: $(D^2 + q)^n y = R$

R	y_p
58. e^{rx}	$e^{rx}/(r^2 + q)^n$
59. $\sin sx$*	$\sin sx/(q - s^2)^n$
60. $P(x)$	$\frac{1}{q^n}\left[P(x) - \binom{n}{1}\frac{P''(x)}{q} + \binom{n+1}{2}\frac{P^{\text{iv}}(x)}{q^2} - \binom{n+2}{3}\frac{P^{\text{vi}}(x)}{q^3} + \cdots\right]$
61. $e^{rx}\sin sx$*	$\frac{e^{rx}}{(A^2+B^2)^n}\{[A^n - \binom{n}{2}A^{n-2}B^2 + \binom{n}{4}A^{n-4}B^4 - \cdots]\sin sx - [\binom{n}{1}A^{n-1}B - \binom{n}{3}A^{n-3}B^3 + \cdots]\cos sx\}$ $A = r^2 - s^2 + q, \quad B = 2rs$

TABLE XIII: $(D^2 + b^2)^n y = R$

R	y_p
62. $\sin bx$*	$(-1)^{\frac{n+1}{2}}\frac{x^n \cos bx}{n!(2b)^n}$ (n odd), $\quad (-1)^{\frac{n}{2}}\frac{x^n \sin bx}{n!(2b)^n}$ (n even)

TABLE XIV: $(D^n - q)y = R$

R	y_p
63. e^{rx}	$e^{rx}/(r^n - q)$
64. $P(x)$	$-\frac{1}{q}\left[P(x) + \frac{P^{(n)}(x)}{q} + \frac{P^{(2n)}(x)}{q^2} + \cdots\right]$
65. $\sin sx$*	$-\frac{q\sin sx + (-1)^{\frac{n-1}{2}} s^n \cos sx}{q^2 + s^{2n}}$ (n odd), $\quad \frac{\sin sx}{(-s^2)^{n/2} - q}$ (n even)
66. $e^{rx}\sin sx$*	$\frac{Ae^{rx}\sin sr - Be^{rx}\cos sx}{A^2 + B^2} = \frac{e^{rx}}{\sqrt{A^2 + B^2}}\sin\left(sx - \tan^{-1}\frac{B}{A}\right)$ $A = [r^n - \binom{n}{2}r^{n-2}s^2 + \binom{n}{4}r^{n-4}s^4 - \cdots] - q, \quad B = [\binom{n}{1}r^{n-1}s - \binom{n}{3}r^{n-3}s^3 + \cdots]$

* For $\cos sx$ in R replace "sin" by "cos" and "cos" by "− sin" in y_p.

TABLE XV: $(D_x + mD_y)z = R$

R	z_p
67. e^{ax+by}	$\dfrac{e^{ax+by}}{a+mb}$
68. $f(ax+by)$	$\dfrac{\int f(u)du}{a+mb}$, $u = ax + by$
69. $f(y-mx)$	$xf(y-mx)$
70. $\phi(x, y)f(y-mx)$	$f(y-mx)\int \phi(x, a+mx)dx$ ($a = y - mx$ after integration)

TABLE XVI: $(D_x + mD_y - k)z = R$

71. e^{ax+by}	$\dfrac{e^{ax+by}}{a+mb-k}$
72. $\sin(ax+by)$*	$-\dfrac{(a+bm)\cos(ax+by)+k\sin(ax+by)}{(a+bm)^2+k^2}$
73. $e^{\alpha x+\beta y}\sin(ax+by)$*	Replace k in 72 by $k-\alpha-m\beta$ and multiply by $e^{\alpha x+\beta y}$
74. $e^{xk}f(ax+by)$	$\dfrac{e^{kx}\int f(u)du}{a+mb}$, $u = ax + by$
75. $f(y-mx)$	$-\dfrac{f(y-mx)}{k}$
76. $P(x)f(y-mx)$	$-\dfrac{1}{k}f(y-mx)\left[P(x)+\dfrac{P'(x)}{k}+\dfrac{P''(x)}{k^2}+\cdots+\dfrac{P^{(n)}(x)}{k^n}\right]$
77. $e^{kx}f(y-mx)$	$xe^{kx}f(y-mx)$

* For cos $(ax + by)$ replace "sin" by "cos," and "cos" by "− sin" in z_p.

$$D_x = \frac{\partial}{\partial x}; \quad D_y = \frac{\partial}{\partial y}; \quad D_x{}^k D_y{}^r = \frac{\partial^{k+r}}{\partial_x{}^k \partial_y{}^r}$$

TABLE XVII: $(D_x + mD_y)^n z = R$

R	z_p
78. e^{ax+by}	$\dfrac{e^{ax+by}}{(a+mb)^n}$
79. $f(ax+by)$	$\dfrac{\int\int \cdots \int f(u)du^n}{(a+mb)^n}$, $u = ax + by$
80. $f(y-mx)$	$\dfrac{x^n}{n!} f(y-mx)$
81. $\phi(x, y)f(y+mx)$	$f(y-mx)\int\int \cdots \int \phi(x, a+mx)dx^n$ $\quad (a = y - mx$ after integration)

TABLE XVIII: $(D_x + mD_y - k)^n z = R$

82. e^{ax+by}	$\dfrac{e^{ax+by}}{(a+mb-k)^n}$
83. $f(y-mx)$	$\dfrac{(-1)^n f(y-mx)}{k^n}$
84. $P(x)f(y-mx)$	$\dfrac{(-1)^n}{k^n} f(y-mx)\left[P(x) + \binom{n}{1}\dfrac{P'(x)}{k} + \binom{n+1}{2}\dfrac{P''(x)}{k^2} + \binom{n+2}{3}\dfrac{P'''(x)}{k^3} + \cdots\right]$
85. $e^{kx}f(ax+by)$	$\dfrac{e^{kx}\int\int \cdots \int f(u)du^n}{(a+mb)^n}$, $u = ax + by$
86. $e^{kx}f(y-mx)$	$\dfrac{x^n}{n!} e^{kx} f(y-mx)$

TABLE XIX: $[D_x^n + a_1 D_x^{n-1} D_y + a_2 D_x^{n-2} D_y^2 + \cdots + a^n D_y^n]z = R$

87. e^{ax+by} $\qquad \dfrac{e^{ax+by}}{a + a_1 a^{n-1} b + a_2 a^{n-2} b^2 + \cdots + a_n b^n}$

88. $f(ax+by)$ $\qquad \dfrac{\int\int \cdots \int f(u)\,du^n}{a^n + a_1 a^{n-1} b + a_2 a^{n-2} b^2 + \cdots + a^n b^n}, \; (u = ax + by)$

TABLE XX: $F(D_x, D_y)z = R$

89. e^{ax+by} $\qquad \dfrac{e^{ax+by}}{F(a, b)}$

TABLE XXI: $F(D_x^2, D_x D_y, D_y^2)z = R$

90. $\sin(ax+by)$* $\qquad \dfrac{\sin(ax+by)}{F(-a^2, -ab, -b^2)}$

* For $\cos(ax + by)$ replace "sin" by "cos", and "cos" by "$-$sin" in z_p.

BESSEL FUNCTIONS $J_0(x)$ AND $J_1(x)$

x	$J_0(x)$	$J_1(x)$	x	$J_0(x)$	$J_1(x)$	x	$J_0(x)$	$J_1(x)$
0.0	1.0000	.0000	**5.0**	−.1776	−.3276	**10.0**	−.2459	.0435
0.1	.9975	.0499	5.1	−.1443	−.3371	10.1	−.2490	.0184
0.2	.9900	.0995	5.2	−.1103	−.3432	10.2	−.2496	−.0066
0.3	.9776	.1483	5.3	−.0758	−.3460	10.3	−.2477	−.0313
0.4	.9604	.1960	5.4	−.0412	−.3453	10.4	−.2434	−.0555
0.5	.9385	.2423	**5.5**	−.0068	−.3414	**10.5**	−.2366	−.0789
0.6	.9120	.2867	5.6	.0270	−.3343	10.6	−.2276	−.1012
0.7	.8812	.3290	5.7	.0599	−.3241	10.7	−.2164	−.1224
0.8	.8463	.3688	5.8	.0917	−.3110	10.8	−.2032	−.1422
0.9	.8075	.4059	5.9	.1220	−.2951	10.9	−.1881	−.1603
1.0	.7652	.4401	**6.0**	.1506	−.2767	**11.0**	−.1712	−.1768
1.1	.7196	.4709	6.1	.1773	−.2559	11.1	−.1528	−.1913
1.2	.6711	.4983	6.2	.2017	−.2329	11.2	−.1330	−.2039
1.3	.6201	.5220	6.3	.2238	−.2081	11.3	−.1121	−.2143
1.4	.5669	.5419	6.4	.2433	−.1816	11.4	−.0902	−.2225
1.5	.5118	.5579	**6.5**	.2601	−.1538	**11.5**	−.0677	−.2284
1.6	.4554	.5699	6.6	.2740	−.1250	11.6	−.0446	−.2320
1.7	.3980	.5778	6.7	.2851	−.0953	11.7	−.0213	−.2333
1.8	.3400	.5815	6.8	.2931	−.0652	11.8	.0020	−.2323
1.9	.2818	.5812	6.9	.2981	−.0349	11.9	.0250	−.2290
2.0	.2239	.5767	**7.0**	.3001	−.0047	**12.0**	.0477	−.2234
2.1	.1666	.5683	7.1	.2991	.0252	12.1	.0697	−.2157
2.2	.1104	.5560	7.2	.2951	.0543	12.2	.0908	−.2060
2.3	.0555	.5399	7.3	.2882	.0826	12.3	.1108	−.1943
2.4	.0025	.5202	7.4	.2786	.1096	12.4	.1296	−.1807
2.5	−.0484	.4971	**7.5**	.2663	.1352	**12.5**	.1469	−.1655
2.6	−.0968	.4708	7.6	.2516	.1592	12.6	.1626	−.1487
2.7	−.1424	.4416	7.7	.2346	.1813	12.7	.1766	−.1307
2.8	−.1850	.4097	7.8	.2154	.2014	12.8	.1887	−.1114
2.9	−.2243	.3754	7.9	.1944	.2192	12.9	.1988	−.0912
3.0	−.2601	.3391	**8.0**	.1717	.2346	**13.0**	.2069	−.0703
3.1	−.2921	.3009	8.1	.1475	.2476	13.1	.2129	−.0489
3 2	−.3202	.2613	8.2	.1222	.2580	13.2	.2167	−.0271
3.3	−.3443	.2207	8.3	.0960	.2657	13.3	.2183	−.0052
3.4	−.3643	.1792	8.4	.0692	.2708	13.4	.2177	.0166
3.5	−.3801	.1374	**8.5**	.0419	.2731	**13.5**	.2150	.0380
3.6	−.3918	.0955	8.6	.0146	.2728	13.6	.2101	.0590
3.7	−.3992	.0538	8.7	−.0125	.2697	13.7	.2032	.0791
3.8	−.4026	.0128	8.8	−.0392	.2641	13.8	.1943	.0984
3.9	−.4018	−.0272	8.9	−.0653	.2559	13.9	.1836	.1165
4.0	−.3971	−.0660	**9.0**	−.0903	.2453	**14.0**	.1711	.1334
4.1	−.3887	−.1033	9.1	−.1142	.2324	14.1	.1570	.1488
4.2	−.3766	−.1386	9.2	−.1367	.2174	14.2	.1414	.1626
4.3	−.3610	−.1719	9.3	−.1577	.2004	14.3	.1245	.1747
4.4	−.3423	−.2028	9.4	−.1768	.1816	14.4	.1065	.1850
4.5	−.3205	−.2311	**9.5**	−.1939	.1613	**14.5**	.0875	.1934
4.6	−.2961	−.2566	9.6	−.2090	.1395	14.6	.0679	.1999
4.7	−.2693	−.2791	9.7	−.2218	.1166	14.7	.0476	.2043
4.8	−.2404	−.2985	9.8	−.2323	.0928	14.8	.0271	.2066
4.9	−.2097	−.3147	9.9	−.2403	.0684	14.9	.0064	.2069

BESSEL FUNCTIONS FOR SPHERICAL COORDINATES

$$j_n(x) = \sqrt{\frac{\pi}{2x}}\, J_{(n+\frac{1}{2})}(x), \quad y_n(x) = \sqrt{\frac{\pi}{2x}}\, Y_{(n+\frac{1}{2})}(x) = (-1)^{n+1}\sqrt{\frac{\pi}{2x}}\, J_{-(n+\frac{1}{2})}(x)$$

x	$j_0(x)$	$y_0(x)$	$j_1(x)$	$y_1(x)$	$j_2(x)$	$y_2(x)$
0.0	1.0000	$-\infty$	0.0000	$-\infty$	0.0000	$-\infty$
0.1	0.9983	−9.9500	0.0333	−100.50	0.0007	−3005.0
0.2	0.9933	−4.9003	0.0664	−25.495	0.0027	−377.52
0.4	0.9735	−2.3027	0.1312	−6.7302	0.0105	−48.174
0.6	0.9411	−1.3756	0.1929	−3.2337	0.0234	−14.793
0.8	0.8967	−0.8709	0.2500	−1.9853	0.0408	−6.5740
1.0	0.8415	−0.5403	0.3012	−1.3818	0.0620	−3.6050
1.2	0.7767	−0.3020	0.3453	−1.0283	0.0865	−2.2689
1.4	0.7039	−0.1214	0.3814	−0.7906	0.1133	−1.5728
1.6	0.6247	+0.0182	0.4087	−0.6133	0.1416	−1.1682
1.8	0.5410	0.1262	0.4268	−0.4709	0.1703	−0.9111
2.0	0.4546	0.2081	0.4354	−0.3506	0.1984	−0.7340
2.2	0.3675	0.2675	0.4345	−0.2459	0.2251	−0.6028
2.4	0.2814	0.3072	0.4245	−0.1534	0.2492	−0.4990
2.6	0.1983	0.3296	0.4058	−0.0715	0.2700	−0.4121
2.8	0.1196	0.3365	0.3792	+0.0005	0.2867	−0.3359
3.0	+0.0470	0.3300	0.3457	0.0630	0.2986	−0.2670
3.2	−0.0182	0.3120	0.3063	0.1157	0.3054	−0.2035
3.4	−0.0752	0.2844	0.2622	0.1588	0.3066	−0.1442
3.6	−0.1229	0.2491	0.2150	0.1921	0.3021	−0.0890
3.8	−0.1610	0.2081	0.1658	0.2158	0.2919	−0.0378
4.0	−0.1892	0.1634	0.1161	0.2301	0.2763	+0.0091
4.2	−0.2075	0.1167	0.0673	0.2353	0.2556	0.0514
4.4	−0.2163	0.0698	+0.0207	0.2321	0.2304	0.0884
4.6	−0.2160	+0.0244	−0.0226	0.2213	0.2013	0.1200
4.8	−0.2075	−0.0182	−0.0615	0.2037	0.1691	0.1456
5.0	−0.1918	−0.0567	−0.0951	0.1804	0.1347	0.1650
5.2	−0.1698	−0.0901	−0.1228	0.1526	0.0991	0.1781
5.4	−0.1431	−0.1175	−0.1440	0.1213	0.0631	0.1850
5.6	−0.1127	−0.1385	−0.1586	0.0880	+0.0277	0.1856
5.8	−0.0801	−0.1527	−0.1665	0.0538	−0.0060	0.1805
6.0	−0.0466	−0.1600	−0.1678	+0.0199	−0.0373	0.1700
6.2	−0.0134	−0.1607	−0.1629	−0.0125	−0.0654	0.1547
6.4	+0.0182	−0.1552	−0.1523	−0.0425	−0.0896	0.1353
6.6	0.0472	−0.1440	−0.1368	−0.0690	−0.1094	0.1126
6.8	0.0727	−0.1278	−0.1172	−0.0915	−0.1243	0.0875
7.0	0.0939	−0.1077	−0.0943	−0.1092	−0.1343	0.0609
7.2	0.1102	−0.0845	−0.0692	−0.1220	−0.1391	0.0337
7.4	0.1215	−0.0593	−0.0429	−0.1295	−0.1388	+0.0068
7.6	0.1274	−0.0331	−0.0163	−0.1317	−0.1338	−0.0189
7.8	0.1280	−0.0069	+0.0095	−0.1289	−0.1244	−0.0427
8.0	0.1237	+0.0182	0.0336	−0.1214	−0.1111	−0.0637

HYPERBOLIC BESSEL FUNCTIONS

$$I_m(x) = i^{-m}J_m(ix)$$

x	$I_0(x)$	$I_1(x)$	$I_2(x)$
0.0	1.0000	0.0000	0.0000
0.1	1.0025	0.0501	0.0012
0.2	1.0100	0.1005	0.0050
0.4	1.0404	0.2040	0.0203
0.6	1.0920	0.3137	0.0464
0.8	1.1665	0.4329	0.0844
1.0	1.2661	0.5652	0.1357
1.2	1.3937	0.7147	0.2026
1.4	1.5534	0.8861	0.2875
1.6	1.7500	1.0848	0.3940
1.8	1.9896	1.3172	0.5260
2.0	2.2796	1.5906	0.6889
2.2	2.6291	1.9141	0.8891
2.4	3.0493	2.2981	1.1342
2.6	3.5533	2.7554	1.4337
2.8	4.1573	3.3011	1.7994
3.0	4.8808	3.9534	2.2452
3.2	5.7472	4.7343	2.7883
3.4	6.7848	5.6701	3.4495
3.6	8.0277	6.7927	4.2540
3.8	9.5169	8.1404	5.2325
4.0	11.302	9.7595	6.4222
4.2	13.442	11.706	7.8684
4.4	16.010	14.046	9.6258
4.6	19.093	16.863	11.761
4.8	22.794	20.253	14.355
5.0	27.240	24.336	17.506
5.2	32.584	29.254	21.332
5.4	39.009	35.182	25.978
5.6	46.738	42.328	31.620
5.8	56.039	50.946	38.470
6.0	67.234	61.342	46.787
6.2	80.718	73.886	56.884
6.4	96.962	89.026	69.141
6.6	116.54	107.30	84.021
6.8	140.14	129.38	102.08
7.0	168.59	156.04	124.01
7.2	202.92	188.25	150.63
7.4	244.34	227.17	182.94
7.6	294.33	274.22	222.17
7.8	354.68	331.10	269.79
8.0	427.56	399.87	327.60

Taken from Vibration and Sound with the permission of Philip Morse, author, and McGraw-Hill Book Company, Inc., publisher.

GAMMA FUNCTION*

Values of $\Gamma(n) = \int_0^\infty e^{-x}x^{n-1}\,dx;\ \Gamma(n+1) = n\Gamma(n)$

n	$\Gamma(n)$	n	$\Gamma(n)$	n	$\Gamma(n)$	n	$\Gamma(n)$
1.00	1.00000	1.25	.90640	1.50	.88623	1.75	.91906
1.01	.99433	1.26	.90440	1.51	.88659	1.76	.92137
1.02	.98884	1.27	.90250	1.52	.88704	1.77	.92376
1.03	.98355	1.28	.90072	1.53	.88757	1.78	.92623
1.04	.97844	1.29	.89904	1.54	.88818	1.79	.92877
1.05	.97350	1.30	.89747	1.55	.88887	1.80	.93138
1.06	.96874	1.31	.89600	1.56	.88964	1.81	.93408
1.07	.96415	1.32	.89464	1.57	.89049	1.82	.93685
1.08	.95973	1.33	.89338	1.58	.89142	1.83	.93969
1.09	.95546	1.34	.89222	1.59	.89243	1.84	.94261
1.10	.95135	1.35	.89115	1.60	.89352	1.85	.94561
1.11	.94739	1.36	.89018	1.61	.89468	1.86	.94869
1.12	.94359	1.37	.88931	1.62	.89592	1.87	.95184
1.13	.93993	1.38	.88854	1.63	.89724	1.88	.95507
1.14	.93642	1.39	.88785	1.64	.89864	1.89	.95838
1.15	.93304	1.40	.88726	1.65	.90012	1.90	.96177
1.16	.92980	1.41	.88676	1.66	.90167	1.91	.96523
1.17	.92670	1.42	.88636	1.67	.90330	1.92	.96878
1.18	.92373	1.43	.88604	1.68	.90500	1.93	.97240
1.19	.92088	1.44	.88580	1.69	.90678	1.94	.97610
1.20	.91817	1.45	.88565	1.70	.90864	1.95	.97988
1.21	.91558	1.46	.88560	1.71	.91057	1.96	.98374
1.22	.91311	1.47	.88563	1.72	.91258	1.97	.98768
1.23	.91075	1.48	.88575	1.73	.91466	1.98	.99171
1.24	.90852	1.49	.88595	1.74	.91683	1.99	.99581
						2.00	1.00000

* For large positive values of x, $\Gamma(x)$ approximates the asymptotic series

$$x^x e^{-x}\sqrt{\frac{2\pi}{x}}\left[1+\frac{1}{12x}+\frac{1}{288x^2}-\frac{139}{51840x^3}-\frac{571}{2488320x^4}+\cdots\right].$$

MISCELLANEOUS MATHEMATICAL CONSTANTS

π CONSTANTS

π = 3.14159 26535 89793 23846 26433 83279 50288 41971 69399 37511

$1/\pi$ = 0.31830 98861 83790 67153 77675 26745 02872 40689 19291 48091

π^2 = 9.86960 44010 89358 61883 44909 99876 15113 53136 99407 24079

$\log_e \pi$ = 1.14472 98858 49400 17414 34273 51353 05871 16472 94812 91531

$\log_{10} \pi$ = 0.49714 98726 94133 85435 12682 88290 89887 36516 78324 38044

$\log_{10} \sqrt{2\pi}$ = 0.39908 99341 79057 52478 25035 91507 69595 02099 34102 92128

CONSTANTS INVOLVING e

e = 2.71828 18284 59045 23536 02874 71352 66249 77572 47093 69996

$1/e$ = 0.36787 94411 71442 32159 55237 70161 46086 74458 11131 03177

e^2 = 7.38905 60989 30650 22723 04274 60575 00781 31803 15570 55185

$M = \log_{10} e$ = 0.43429 44819 03251 82765 11289 18916 60508 22943 97005 80367

$1/M = \log_e 10$ = 2.30258 50929 94045 68401 79914 54684 36420 76011 01488 62877

$\log_{10} M$ = 9.63778 43113 00536 78912 29674 98645 − 10

π^e AND e^π CONSTANTS

π^e = 22.45915 77183 61045 47342 71522

e^π = 23.14069 26327 79269 00572 90864

$e^{-\pi}$ = 0.04321 39182 63772 24977 44177

$e^{\frac{1}{2}\pi}$ = 4.81047 73809 65351 65547 30357

$i^i = e^{-\frac{1}{2}\pi}$ = 0.20787 95763 50761 90854 69556

NUMERICAL CONSTANTS

$\sqrt{2}$ = 1.41421 35623 73095 04880 16887 24209 69807 85696 71875 37695

$\sqrt[3]{2}$ = 1.25992 10498 94873 16476 72106 07278 22835 05702 51464 70151

$\log_e 2$ = 0.69314 71805 59945 30941 72321 21458 17656 80755 00134 36026

$\log_{10} 2$ = 0.30102 99956 63981 19521 37388 94724 49302 67681 89881 46211

$\sqrt{3}$ = 1.73205 08075 68877 29352 74463 41505 87236 69428 05253 81039

$\sqrt[3]{3}$ = 1.44224 95703 07408 38232 16383 10780 10958 83918 69253 49935

$\log_e 3$ = 1.09861 22886 68109 69139 52452 36922 52570 46474 90557 82275

$\log_{10} 3$ = 0.47712 12547 19662 43729 50279 03255 11530 92001 28864 19070

OTHER CONSTANTS

Euler's Constant γ = 0.57721 56649 01532 86061

$\log_e \gamma$ = −0.54953 93129 81644 82234

Golden Ratio ϕ = 1.61803 39887 49894 84820 45868 34365 63811 77203 09180

MELTING AND BOILING POINTS, AND ATOMIC WEIGHTS OF THE ELEMENTS

Based on the Assigned Relative Atomic Mass of $^{12}C = 12$

The following values apply to elements as they exist in materials of terrestrial origin and to certain artificial elements. When used with the footnotes, they are reliable to ±1 in the last digit, or ±3 if that digit is in small type.

Name	Symbol	Atomic number	Atomic weight	Melting point, °C	Boiling point, °C
Actinium[k]	Ac	89	227.028	1,050	3,200 ± 300
Aluminum	Al	13	26.98154[b]	660.37	2,467
Americium	Am	95	(243)	994 ± 4	2,607
Antimony	Sb	51	121.75*	630.74	1,750
Argon[h,i]	Ar	18	39.948*[b,c,d,g]	−189.2	−185.7
Arsenic (gray)	As	33	74.9216[a]	817(28 atm)	613(sub.)
Astatine	At	85	(210)	302	337
Barium[i]	Ba	56	137.33	725	1,640
Berkelium	Bk	97	(247)	–	–
Beryllium	Be	4	9.01218[a]	1,278 ± 5	2,970(5 mm)
Bismuth	Bi	83	208.9804[a]	271.3	1,560 ± 5
Boron[h,j]	B	5	10.81[c,d,e]	2,300	2,550(sub.)
Bromine	Br	35	79.904[c]	−7.2	58.78
Cadmium[i]	Cd	48	112.41	320.9	765
Calcium[i]	Ca	20	40.08	839 ± 2	1,484
Californium	Cf	98	(251)	–	–
Carbon[h]	C	6	12.011[b,d]	~3,550	4,827
Cerium[i]	Ce	58	140.12	798 ± 3	3,257
Cesium	Cs	55	132.9054[c]	28.40 ± 0.01	678.4
Chlorine	Cl	17	35.453[c]	−100.98	−34.6
Chromium	Cr	24	51.996[c]	1,857 ± 20	2,672
Cobalt	Co	27	58.9332[a]	1,495	2,870
Copper[h]	Cu	29	63.546*[c,d]	1,083.4 ± 0.2	2,567
Curium	Cm	96	(247)	1,340 ± 40	
Dysprosium	Dy	66	162.50*	1,409	2,335
Einsteinium	Es	99	(254)	–	–
Erbium	Er	68	167.26*	1,522	2,510
Europium[i]	Eu	63	151.96	822 ± 5	1,597
Fermium	Fm	100	(257)	–	–
Fluorine	F	9	18.998403[a]	−219.62	−188.14
Francium	Fr	87	(223)	(27)	(677)
Gadolinium[i]	Gd	64	157.25*	1,311 ± 1	3,233
Gallium	Ga	31	69.72	29.78	2,403
Germanium	Ge	32	72.59*	937.4	2,830
Gold	Au	79	196.9665[a]	1,064.43	2,807
Hafnium	Hf	72	178.49*	2,227 ± 20	4,602
Helium[i]	He	2	4.00260[b]	$-272.2^{26\ \text{atm}}$	−268.934
Holmium	Ho	67	164.9304[a]	1,470	2,720
Hydrogen[h]	H	1	1.0079[b,d]	−259.14	−252.87
Indium[i]	In	49	114.82	156.61	2,080
Iodine	I	53	126.9045[a]	113.5	184.35
Iridium	Ir	77	192.22*	2,410	4,130
Iron	Fe	26	55.847*	1,535	2,750
Krypton[i,j]	Kr	36	83.80	−156.6	−152.30 ± 0.10
Lanthanum[i]	La	57	138.9055*[b]	920 ± 5	3,454
Lawrencium	Lr	103	(260)	–	–
Lead[h,j]	Pb	82	207.2[d,g]	327.502	1,740
Lithium[h,i,j]	Li	3	6.941*[c,d,e]	180.54	1,347
Lutetium	Lu	71	174.967 ± 0.003	1,656 ± 5	3,315
Magnesium[i]	Mg	12	24.305[c]	648.8 ± 0.5	1,090
Manganese	Mn	25	54.9380[a]	1,244 ± 3	1,962
Mendelevium	Md	101	(258)	–	–
Mercury	Hg	80	200.59*	−38.87	356.58
Molybdenum	Mo	42	95.94	2,617	4,612

Note: See end of table for footnotes.

MELTING AND BOILING POINTS, AND ATOMIC WEIGHTS OF THE ELEMENTS (Continued)

Name	Symbol	Atomic number	Atomic weight	Melting point, °C	Boiling point, °C
Neodymium[i]	Nd	60	144.24*	1,010	3,127
Neon[j]	Ne	10	20.179*[c]	−248.67	−246.048
Neptunium[k]	Np	93	237.0482[b]	640 ± 1	3,902
Nickel	Ni	28	58.70	1,453	2,732
Niobium (Columbium)	Nb	41	92.9064[a]	2,468 ± 10	4,742
Nitrogen	N	7	14.0067[b,c]	−209.86	−195.8
Nobelium	No	102	(259)	—	—
Osmium[i]	Os	76	190.2	3,045 ± 30	5,027 ± 100
Osygen[h]	O	8	15.9994*[b,c,d]	−218.4	−182.962
Palladium[i]	Pd	46	106.4	1,552	3,140
Phosphorus	P	15	30.97376	44.1 (white)	280 (white)
Platinum	Pt	78	195.09*	1,772	3,827 ± 100
Plutonium	Pu	94	(244)	641	3,232
Polonium	Po	84	(209)	254	962
Potassium	K	19	39.0983*	63.65	774
Praeseodymium	Pr	59	140.9077[a]	931 ± 4	3,212
Promethium	Pm	61	(145)	~1,080	2,460(?)
Protactinium[k]	Pa	91	231.0359[a]	<1,600	—
Radium[i,k]	Ra	88	226.0254[a,f,g]	700	1,140
Radon	Rn	86	(222)	−71	−61.8
Rhenium	Re	75	186.2	3,180	5,627(est.)
Rhodium	Rh	45	102.9055[a]	1,966 ± 3	3,727 ± 100
Rubidium[i]	Rb	37	85.4678*[c]	38.89	688
Ruthenium[i]	Ru	44	101.07*	2,310	3,900
Samarium[i]	Sm	62	150.4	1,072 ± 5	1,778
Scandium	Sc	21	44.9559[a]	1,539	2,832
Selenium	Se	34	78.96*	217	684.9 ± 1.0
Silicon	Si	14	28.0855*	1,410	2,355
Silver[i]	Ag	47	107.868[c]	961.93	2,212
Sodium	Na	11	22.98977[a]	97.81 ± 0.03	882.9
Strontium[i]	Sr	38	87.62[g]	769	1,384
Sulfur[h]	S	16	32.06[d]	112.8	444.674
Tantalum	Ta	73	180.9479*[b]	2,996	5,425 ± 100
Technetium	Tc	43	(97)[f]	2,172	4,877
Tellurium[i]	Te	52	127.60*	449.5 ± 0.3	989.8 ± 3.8
Terbium	Tb	65	158.9254[a]	1,360 ± 4	3,041
Thallium	Tl	81	204.37*	303.5	1,457 ± 10
Thorium[i,k]	Th	90	232.0381[a]	1,750	~4,790
Thulium	Tm	69	168.9342[a]	1,545 ± 15	1,727
Tin	Sn	50	118.69*	231.9681	2,270
Titanium	Ti	22	47.90*	1,660 ± 10	3,287
Tungsten	W	74	183.85*	3,410 ± 20	5,660
Uranium[i,j]	U	92	238.029[b,c,e]	1,132.3 ± 0.8	3,818
Vanadium	V	23	50.9415*[b,c]	1,890 ± 10	3,380
Wolfram (see Tungsten)					
Xenon[i,j]	Xe	54	131.30	−111.9	−107.1 ± 3
Ytterbium	Yb	70	173.04*	824 ± 5	1,193
Yttrium	Y	39	88.9059[a]	1,523 ± 8	3,337
Zinc	Zn	30	65.38	419.58	907
Zirconium[i]	Zr	40	91.22	1,852 ± 2	4,377

[a]Mononuclidic element.

[b]Element with one predominant isotope (about 99 to 100% abundance).

[c]Element for which the atomic weight is based on calibrated measurements.

[d]Element for which variation in isotropic abundance in terrestrial samples limits the precision of the atomic weight given.

[e]Element for which users are cautioned against the possibility of large variations in atomic weight due to inadvertent or undisclosed artificial isotropic separation in commercially available materials.

[f]Most commonly available long-lived isotope.

[g]In some geological specimens this element has a highly anomalous isotopic composition, corresponding to an atomic weight significantly different from that given.

[h]Element for which known variations in isotopic composition in normal terrestrial material prevent a more precise atomic weight given; A_r (E) values should be applicable to any "normal" material.
[i]Element for which geological specimens are known in which the element has an anomalous isotopic composition, such that the difference in atomic weight of the element in such specimens from that given in the Table may exceed considerably the implied uncertainty.
[j]Element for which substantial variations in A_r from the value given can occur in commercially available material because of inadvertent or undisclosed change of isotopic composition.
[k]Element for which the value A_r is that of the radioisotope of longest half-life.

ELECTRONIC CONFIGURATION OF THE ELEMENTS

By Laurence S. Foster

References: F. H. Spedding and A. H. Daane, Editors, *The Rare Earths*, John Wiley and Sons, Inc. Publishers, New York, 1961. R. F. Gould, Editor, *Lanthanide–Actinide Chemistry*, Advances in Chemistry Series, No. 71, American Chemical Society, Washington, D.C., 1967: Paper No. 14, Mark Fred, *Electronic Structure of the Actinide Elements.*

Atomic No.	Element	K	L	M	N	O	P	Q
		1	2	3	4	5	6	7
		s	s p	s p d	s p d f	s p d f	s p d f	s p d f
1	H	1						
2	He	2						
3	Li	2	1					
4	Be	2	2					
5	B	2	2 1					
6	C	2	2 2					
7	N	2	2 3					
8	O	2	2 4					
9	F	2	2 5					
10	Ne	2	2 6					
11	Na	2	2 6	1				
12	Mg	2	2 6	2				
13	Al	2	2 6	2 1				
14	Si	2	2 6	2 2				
15	P	2	2 6	2 3				
16	S	2	2 6	2 4				
17	Cl	2	2 6	2 5				
18	Ar	2	2 6	2 6				
19	K	2	2 6	2 6 ..	1			
20	Ca	2	2 6	2 6 ..	2			
21	Sc	2	2 6	2 6 1	2			
22	Ti	2	2 6	2 6 2	2			
23	V	2	2 6	2 6 3	2			
24	Cr	2	2 6	2 6 5*	1			
25	Mn	2	2 6	2 6 5	2			
26	Fe	2	2 6	2 6 6	2			
27	Co	2	2 6	2 6 7	2			
28	Ni	2	2 6	2 6 8	2			
29	Cu	2	2 6	2 6 10*	1			
30	Zn	2	2 6	2 6 10	2			
31	Ga	2	2 6	2 6 10	2 1			
32	Ge	2	2 6	2 6 10	2 2			
33	As	2	2 6	2 6 10	2 3			
34	Se	2	2 6	2 6 10	2 4			
35	Br	2	2 6	2 6 10	2 5			
36	Kr	2	2 6	2 6 10	2 6			
37	Rb	2	2 6	2 6 10	2 6 ..	1		
38	Sr	2	2 6	2 6 10	2 6 ..	2		
39	Y	2	2 6	2 6 10	2 6 1	2		
40	Zr	2	2 6	2 6 10	2 6 2 ..	2		
41	Nb	2	2 6	2 6 10	2 6 4* ..	1		
42	Mo	2	2 6	2 6 10	2 6 5 ..	1		
43	Tc	2	2 6	2 6 10	2 6 6 ..	1		
44	Ru	2	2 6	2 6 10	2 6 7 ..	1		
45	Rh	2	2 6	2 6 10	2 6 8 ..	1		
46	Pd	2	2 6	2 6 10	2 6 10* ..	0		
47	Ag	2	2 6	2 6 10	2 6 10 ..	1		
48	Cd	2	2 6	2 6 10	2 6 10 ..	2		
49	In	2	2 6	2 6 10	2 6 10 ..	2 1		
50	Sn	2	2 6	2 6 10	2 6 10 ..	2 2		
51	Sb	2	2 6	2 6 10	2 6 10 ..	2 3		
52	Te	2	2 6	2 6 10	2 6 10 ..	2 4		
53	I	2	2 6	2 6 10	2 6 10 ..	2 5		
54	Xe	2	2 6	2 6 10	2 6 10 ..	2 6		
55	Cs	2	2 6	2 6 10	2 6 10 ..	2 6	1	
56	Ba	2	2 6	2 6 10	2 6 10 ..	2 6	2	
57	La	2	2 6	2 6 10	2 6 10 ..	2 6 1 ..	2	
58	Ce	2	2 6	2 6 10	2 6 10 2*	2 6	2	
59	Pr	2	2 6	2 6 10	2 6 10 3	2 6	2	
60	Nd	2	2 6	2 6 10	2 6 10 4	2 6	2	
61	Pm	2	2 6	2 6 10	2 6 10 5	2 6	2	
62	Sm	2	2 6	2 6 10	2 6 10 6	2 6	2	
63	Eu	2	2 6	2 6 10	2 6 10 7	2 6	2	
64	Gd	2	2 6	2 6 10	2 6 10 7	2 6 1 ..	2	
65	Tb	2	2 6	2 6 10	2 6 10 9*	2 6	2	
66	Dy	2	2 6	2 6 10	2 6 10 10	2 6	2	
67	Ho	2	2 6	2 6 10	2 6 10 11	2 6	2	
68	Er	2	2 6	2 6 10	2 6 10 12	2 6	2	
69	Tm	2	2 6	2 6 10	2 6 10 13	2 6	2	
70	Yb	2	2 6	2 6 10	2 6 10 14	2 6	2	
71	Lu	2	2 6	2 6 10	2 6 10 14	2 6 1 ..	2	
72	Hf	2	2 6	2 6 10	2 6 10 14	2 6 2 ..	2	
73	Ta	2	2 6	2 6 10	2 6 10 14	2 6 3 ..	2	
74	W	2	2 6	2 6 10	2 6 10 14	2 6 4 ..	2	
75	Re	2	2 6	2 6 10	2 6 10 14	2 6 5 ..	2	
76	Os	2	2 6	2 6 10	2 6 10 14	2 6 6 ..	2	
77	Ir	2	2 6	2 6 10	2 6 10 14	2 6 7 ..	2	
78	Pt	2	2 6	2 6 10	2 6 10 14	2 6 9 ..	1	
79	Au	2	2 6	2 6 10	2 6 10 14	2 6 10 ..	1	
80	Hg	2	2 6	2 6 10	2 6 10 14	2 6 10 ..	2	
81	Ti	2	2 6	2 6 10	2 6 10 14	2 6 10 ..	2 1	
82	Pb	2	2 6	2 6 10	2 6 10 14	2 6 10 ..	2 2	
83	Bi	2	2 6	2 6 10	2 6 10 14	2 6 10 ..	2 3	
84	Po	2	2 6	2 6 10	2 6 10 14	2 6 10 ..	2 4	
85	At	2	2 6	2 6 10	2 6 10 14	2 6 10 ..	2 5	
86	Rn	2	2 6	2 6 10	2 6 10 14	2 6 10 ..	2 6	
87	Fr	2	2 6	2 6 10	2 6 10 14	2 6 10 ..	2 6.. ..	1
88	Ra	2	2 6	2 6 10	2 6 10 14	2 6 10 ..	2 6.. ..	2
89	Ac	2	2 6	2 6 10	2 6 10 14	2 6 10 ..	2 6 1 ..	2
90	Th	2	2 6	2 6 10	2 6 10 14	2 6 10 ..	2 6 2 ..	2
91	Pa	2	2 6	2 6 10	2 6 10 14	2 6 10 2*	2 6 1 ..	2
92	U	2	2 6	2 6 10	2 6 10 14	2 6 10 3	2 6 1 ..	2
93	Np	2	2 6	2 6 10	2 6 10 14	2 6 10 4	2 6 1 ..	2
94	Pu	2	2 6	2 6 10	2 6 10 14	2 6 10 6	2 6.. ..	2
95	Am	2	2 6	2 6 10	2 6 10 14	2 6 10 7	2 6.. ..	2
96	Cm	2	2 6	2 6 10	2 6 10 14	2 6 10 7	2 6 1 ..	2
97	Bk	2	2 6	2 6 10	2 6 10 14	2 6 10 9*	2 6.. ..	2
98	Cf	2	2 6	2 6 10	2 6 10 14	2 6 10 10	2 6.. ..	2
99	Es	2	2 6	2 6 10	2 6 10 14	2 6 10 11	2 6.. ..	2
100	Fm	2	2 6	2 6 10	2 6 10 14	2 6 10 12	2 6.. ..	2
101	Md	2	2 6	2 6 10	2 6 10 14	2 6 10 13	2 6.. ..	2
102	No	2	2 6	2 6 10	2 6 10 14	2 6 10 14	2 6.. ..	2
103	Lr	2	2 6	2 6 10	2 6 10 14	2 6 10 14	2 6 1 ..	2
104	—	2	2 6	2 6 10	2 6 10 14	2 6 10 14	2 6 2 ..	2

* Note irregularity.

THE ELEMENTS

C.R. Hammond

One of the most striking facts about the elements is their unequal distribution and occurrence in nature. Present knowledge of the chemical composition of the universe, obtained from the study of the spectra of stars and nebulae, indicates that hydrogen is by far the most abundant element and may account for more than 90% of the atoms or about 75% of the mass of the universe. Helium atoms make up most of the remainder. All of the other elements together contribute only slightly to the total mass.

The chemical composition of the universe is undergoing continuous change. Hydrogen is being converted into helium, and helium is being changed into heavier elements. As time goes on, the ratio of heavier elements increases relative to hydrogen. Presumably, the process is not reversible.

Burbidge, Burbidge, Fowler, and Hoyle have studied the synthesis of elements in stars. To explain all of the features of the nuclear abundance curve — obtained by studies of the composition of the earth, meteorites, stars, etc. — it is necessary to postulate that the elements were originally formed by at least eight different processes: (1) hydrogen burning, (2) helium burning, (3) *α* process, (4) *e* process, (5) *s* process, (6) *r* process, (7) *p* process, and (8) the X process. The X process is thought to account for the existence of light nuclei such as D, Li, Be, and B. Common metals such as Fe, Cr, Ni, Cu, Ti, Zn, etc. were likely produced early in the history of our galaxy. It is also probable that most of the heavy elements on earth and elsewhere in the universe were originally formed in supernovae, or in the hot interior of stars.

Studies of the solar spectrum have led to the identification of 67 elements in the sun's atmosphere; however, all elements cannot be identified with the same degree of certainty. Other elements may be present in the sun, although they have not yet been detected spectroscopically. The element helium was discovered on the sun before it was found on earth. Some elements such as scandium are relatively more plentiful in the sun and stars than here on earth.

Minerals in lunar rocks brought back from the moon on the Apollo missions consist predominantly of *plagioclase* {$(Ca,Na)(Al,Si)O_4O_8$} and *pyroxene* {$(Ca,Mg,Fe)_2Si_2O_6$} — two minerals common in terrestrial volcanic rock. No new elements have been found on the moon that cannot be accounted for on earth; however, two minerals, *armalcolite* {$(Fe,Mg)Ti_2O_5$} and *pyroxferroite* {$CaFe_6(SiO_3)_7$}, are new. The oldest known terrestrial rocks are about 3.75 billion years old. One rock, known as the "Genesis Rock," brought back from the Apollo 15 Mission, is about 4.15 billion years old. This is only about one-half billion years younger than the supposed age of the moon and solar system. Lunar rocks appear to be relatively enriched in refractory elements such as chromium, titanium, zirconium, and the rare earths, and impoverished in volatile elements such as the alkali metals, in chlorine, and in noble metals such as nickel, platinum, and gold.

Even older than the "Genesis Rock" are *carbonaceous chondrites,* a type of meteorite that has fallen to earth and has been studied. These are some of the most primitive objects of the solar system yet found. The grains making up these objects probably condensed directly out of the gaseous nebula from which the sun and planets were born. Most of the condensation of the grains probably was completed within 50,000 years of the time the disk of the nebula was first formed — about 4.6 billion years ago. The relative abundances of the elements in these meteorites are about the same as the abundances found in the sun.

Early returns from the X-ray fluorescent spectrometer sent with the Viking I spacecraft to Mars show that the Martian soil contains about 12 to 16% iron, 14 to 15% silicon, 3 to 8% calcium, 2 to 7% aluminum, and one half to 2% titanium. The gas chromatograph — mass spectrometer on Viking II found no trace of organic compounds that should be present if life ever existed there.

F. W. Clarke and others have carefully studied the composition of rocks making up the

crust of the earth. Oxygen accounts for about 47% of the crust, by weight, while silicon comprises about 28%, and aluminum about 8%. These elements, plus iron, calcium, sodium, potassium, and magnesium, account for about 99% of the composition of the crust.

Many elements such as tin, copper, zinc, lead, mercury, silver, platinum, antimony, arsenic, and gold, which are so essential to our needs and civilization, are among some of the rarest elements in the earth's crust. These are made available to us only by the processes of concentration in ore bodies. Some of the so-called *rare-earth* elements have been found to be much more plentiful than originally thought and are about as abundant as uranium, mercury, lead, or bismuth. The least abundant rare-earth or *lanthanide* element, thulium, is now believed to be more plentiful on earth than silver, cadmium, gold, or iodine, for example. Rubidium, the 16th most abundant element, is more plentiful than chlorine, while its compounds are little known in chemistry and commerce.

It is now thought that at least 24 elements are essential to living matter. The four most abundant in the human body are hydrogen, oxygen, carbon, and nitrogen. The seven next most common, in order of abundance, are calcium, phosphorous, chlorine, potassium, sulfur, sodium, and magnesium. Iron, copper, zinc, silicon, iodine, cobalt, manganese, molybdenum, fluorine, tin, chromium, selenium, and vanadium are needed and play a role in living matter. Boron is also thought essential for some plants, and it is possible that aluminum, nickel, and germanium may turn out to be necessary.

Ninety-one elements occur naturally on earth. Minute traces of plutonium-244 have been discovered in rocks mined in Southern California. This discovery supports the theory that heavy elements were produced during creation of the solar system. While technetium and promethium have not yet been found naturally on earth, they have been found to be present in stars. Technetium has been identified in the spectra of certain "late" type stars, and promethium lines have been identified in the spectra of a faintly visible star HR465 in Andromeda. Promethium must have been made very recently near the star's surface for no known isotope of this element has a half-life longer than 17.7 years.

It has been suggested that californium is present in certain stellar explosions known as supernovae; however, this has not been proved. At present no elements are found elsewhere in the universe that cannot be accounted for here on earth.

All atomic mass numbers from 1 to 238 are found naturally on earth except for masses 5 and 8. About 280 stable and 67 naturally radioactive isotopes occur on earth totalling 347. In addition, the neutron, technetium, promethium, and the transuranic elements (lying beyond uranium) up to Element 106 have been produced artificially. Soviet scientists have announced synthesis of Element 107; however, the discovery has not been confirmed by scientists of other nations. Laboratory processes have now extended the radioactive mass numbers beyond 238 to about 260. Each element from atomic number 1 to 106 is known to have at least one radioactive isotope. About 1700 different nuclides (the name given to different kinds of nuclei, whether they are of the same or different elements) are now recognized. Many stable and radioactive isotopes are now produced and distributed by the Oak Ridge National Laboratory, Oak Ridge, Tenn., U.S.A., to customers licensed by the U.S. Atomic Energy Commission.

The nucleus of an atom is characterized by the number of protons it contains, usually denoted by Z, and by the number of neutrons, N. Isotopes of an element have the same value of Z, but different values of N. The *mass number* A, is the sum of Z and N. For example, Uranium-238 has a mass number of 238, and would contain 92 protons and 146 neutrons.

In addition to the proton, neutron, and electron, there are considerably more than 100 other fundamental particles which have been discovered or hypothesized. The majority of these fall into one of two classes, *leptons* or *hadrons*. The leptons comprise just four known particles, the electron, the *muon* (μ meson), and two kinds of *neutrinos*. The muon is essentially similar to the electron and has a charge of −1, but it is 200 times heavier. The neutrino is either

of two stable particles of small (probably zero) rest mass, carrying no charge. Also there are four *antileptons,* identical to the corresponding leptons in some respects, such as mass, but they have properties exactly opposite those of the leptons. The *positron,* for example, is an antilepton, with a charge of + 1. Leptons cannot be broken into smaller units and are considered to be elementary. On the other hand, hadrons are complex and thought to have internal structure. Protons and neutrons, which make up atomic nuclei, are hadrons.

Elementary particle physics is not yet clearly understood, but groupings and arrangements of these particles have been made resembling the periodic table of chemical elements. This has led to the speculation that hadrons are composed of three (or possibly more) simpler components called *quarks.* Quarks are presumed to be elementary particles. There is presently no evidence that quarks exist in isolation. Only particles, such as the electron, muon, and all the quarks, that have electric charge, are acted on by the electromagnetic force. This force binds atoms together and is responsible for many properties of matter, including chemical properties.

The available evidence leads to the conclusion that elements 89 (actinum) through 103 (lawrencium) are chemically similar to the rare-earth or lanthanide elements (elements 57 to 71, inclusive). These elements therefore have been named *actinides* after the first member of this series. Those elements beyond uranium that have been produced artificially have the following names and symbols: neptunium, 93 (Np); plutonium, 94 (Pu); americium, 95 (Am); curium, 96 (Cm); berkelium, 97 (Bk); californium, 98 (Cf); einsteinium, 99 (Es); fermium, 100 (Fm); mendelevium, 101 (Md); nobelium, 102 (No); and lawrencium 103 (Lr). It is now claimed that Elements 104, 105, and 106 have been produced and positively identified. Element 107 is claimed to have been produced. Names and chemical symbols have been suggested for Elements 104 and 105, but have not been officially adopted. Names for Elements 106 and 107 have not yet been suggested.

Element 104 is expected to have chemical properties similar to those of hafnium and would not be a member of the actinide series. Element 105 probably would have chemical properties similar to those of tantalum, Element 106 similar to tungsten, and Element 107 similar to rhenium.

There is presently some reason for optimism in producing elements beyond Element 107 either by bombardment of heavy isotopic targets with heavy ions, or by the irradiation of uranium or transuranic elements with the instantaneous high flux of neutrons produced by underground nuclear explosions. The limit will be set by the yields of the nuclear reactions and by the half-lives of radioactive decay. It has been suggested that Elements 102 and 103 have abnormally short lives only because they are in a pocket of instability, and that this region of instability might begin to "heal" around Element 105. If so, it may be possible to produce heavier isotopes with longer half-lives. It has also been suggested that Element 114, with a mass number of 298, and Element 126, with a mass number of 310, may be sufficiently stable to make discovery and identification possible. Calculations indicate that Element 110, a homolog of platinum, may have a half-life of as long as 100 million years. Searches have already been made by workers in a number of laboratories for Element 110 and its neighboring elements in naturally occurring platinum. Recent studies of the xenon component ($Xe^{131-136}$) of certain carbonaceous chronditic meteorites suggest that Elements 113, 114, or 115 may have been its progenitor.

There are many claims in the literature of the existence of various allotropic modifications of the elements, some of which are based on doubtful or incomplete evidence. Also, the physical properties of an element may change drastically by the presence of small amounts of impurities. With new methods of purification, which are now able to produce elements with 99.9999 + % purity, it has been necessary to restudy the properties of the elements. For example, the melting point of thorium changes by several hundred degrees by the presence of a small percentage of ThO_2 as an impurity. Ordinary commercial tungsten is brittle and can

be worked only with difficulty. Pure tungsten, however, can be cut with a hacksaw, forged, spun, drawn, or extruded. In general, the value of physical properties given here applies to the pure element, when it is known.

Many of the chemical elements and their compounds are toxic and should be handled with due respect and care. In recent years there has been a greatly increased knowledge and awareness of the health hazards associated with chemicals, radioactive materials, and other agents. Anyone working with the elements and certain of their compounds should become thoroughly familiar with the proper safeguards to be taken. Reference should be made to publications such as the following:

1. *Code of Federal Regulations,* Title 29, Labor, chapter XVII, section 1910.93 of subpart G, redesignated as 1910.1000 at 40 FR (Federal Register) 23072. May 28, 1975; amended at 41 FR 11505, March 19, 1976; 41 FR 35184, August 20, 1976; FR 46784, October 22, 1976; 42 FR 3304, January 18, 1977 (corrections) and additional amendments and corrections as issued, U.S. Government Printing Office, Supt. of Documents, Washington, D.C.
2. *Code of Federal Regulations,* Title 10, Energy, Chapter 1, Nuclear Regulatory Commission, section 20.103 — 20.108; 20.201 — 207; 20.301 — 305; 20.401 — 409; 20.501 — 2; 20.601; appendices, corrections, and amendments.
3. *Occupational Safety and Health Reporter* (latest edition with amendments and corrections), Bureau of National Affairs, Washington, D.C.
4. *Atomic Energy Law Reporter,* Commerce Clearing House, Chicago, Il.
5. *Nuclear Regulation Reporter,* Commerce Clearing House, Chicago, Il.
6. *Maximum Permissible Body Burdens and Maximum Permissible Concentrations of Radionuclides in Air and in Water for Occupational Exposure,* with addenda, U.S. Department of Commerce, N.B.S, Handbook No. 69, (NCRP Report No. 22), latest edition, National Council on Radiation Protection and Measurements (NCRP), Bethesda, MD.; also refer to Permissible Quarterly Intakes of Radionuclides. Handbook of Chemistry and Physics, 59th Edition.
7. *TLVs® Threshold Limit Values for Chemical Substances and Physical Agents in Workroom Environment with Intended Changes,* latest edition, American Conference of Governmental Industrial Hygienists, Cincinnati, Ohio.

Actinium — (Gr. *aktis, aktinos,* beam or ray), Ac; at. wt. (227); at no. 89; m.p. 1050°C; b.p. 3200± 300°C (est.); sp. gr. 10.07 (calc.). Discovered by Andre Debierne in 1899 and independently by F. Giesel in 1902. Occurs naturally in association with uranium minerals. Actinium-227, a decay product of uranium-235, is a beta emitter with a 21.6-year half-life. Its principal decay products are thorium-227 (18.5-day half-life), radium-223 (11.4-day half-life), and a number of short-lived products including radon, bismuth, polonium, and lead isotopes. In equilibrium with its decay products, it is a powerful source of alpha rays. Actinium metal has been prepared by the reduction of actinium fluoride with lithium vapor at about 1100 to 1300°C. The chemical behavior of actinium is similar to that of the rare earths, particularly lanthanum. Purified actinium comes into equilibrium with its decay products at the end of 185 days, and then decays according to its 21.6-year half-life. It is about 150 times as active as radium, making it of value in the production of neutrons.

Aluminum — (L. *alumen, alum*), Al; at. wt. 26.98154; at. no. 13; m.p. 660.37°C; b.p. 2467°C; sp. gr. 2.6989 (20°C); valence 3. The ancient Greeks and Romans used *alum* in medicine as an astringent, and as a mordant in dyeing. In 1761 de Morveau proposed the name *alumine* for the base in alum, and Lavoisier, in 1787, thought this to be the oxide of a still undiscovered metal. Wohler is generally credited with having isolated the metal in 1827, although an impure form was prepared by Oersted 2 years earlier. In 1807, Davy proposed the name *alumium* for the metal, undiscovered at that time, and later agreed to change it to *aluminum.* Shortly thereafter, the name *aluminium*

was adopted to conform with the "ium" ending of most elements, and this spelling is now in use elsewhere in the world. *Aluminium* was also the accepted spelling in the U.S. until 1925, at which time the American Chemical Society officially decided to use the name *aluminum* thereafter in their publications. The method of obtaining aluminum metal by the electrolysis of alumina dissolved in *cryolite* was discovered in 1886 by Hall in the U.S. and about the same time by Heroult in France. Cryolite, a natural ore found in Greenland, is no longer widely used in commercial production, but has been replaced by an artifical mixture of sodium, aluminum, and calcium fluorides. *Bauxite,* an impure hydrated oxide ore, is found in large deposits in Jamaica, Australia, Surinam, Guyana, Arkansas, and elsewhere. The Bayer process is most commonly used today to refine bauxite so it can be accommodated in the Hall-Heroult refining process, used to produce most aluminum. Aluminum can now be produced from clay, but the process is not economically feasible at present. Aluminum is the most abundant metal to be found in the earth's crust (8.1%), but is never found free in nature. In addition to the minerals mentioned above, it is found in feldspars, granite, and in many other common minerals. Pure aluminum, a silvery-white metal, possesses many desirable characteristics. It is light, nontoxic, has a pleasing appearance, can easily be formed, machined, or cast, has a high thermal conductivity, and has excellent corrosion resistance. It is nonmagnetic and nonsparking, stands second among metals in the scale of malleability, and sixth in ductility. It is extensively used for kitchen utensils, outside building decoration, and in thousands of industrial applications where a strong, light, easily constructed material is needed. Although its electrical conductivity is only about 60% that of copper per area of cross section, it is used in electrical transmission lines because of its light weight. Pure aluminum is soft and lacks strength, but it can be alloyed with small amounts of copper, magnesium, silicon, manganese, and other elements to impart a variety of useful properties. These alloys are of vital importance in the construction of modern aircraft and rockets. Aluminum, evaporated in a vacuum, forms a highly reflective coating for both visible light and radiant heat. These coatings soon form a thin layer of the protective oxide and do not deteriorate as do silver coatings. They have found application in coatings for telescope mirrors, in making decorative paper, packages, toys, and in many other uses. The compounds of greatest importance are aluminum oxide, the sulfate, and the soluble sulfate with potassium (alum). The oxide, alumina, occurs naturally as ruby, sapphire, corundum, and emery, and is used in glassmaking and refractories. Synthetic ruby and sapphire have found application in the construction of lasers for producing coherent light. In 1852, the price of aluminum was about $545/lb, and just before Hall's discovery in 1886, about $11.00. The price rapidly dropped to 30c and has been as low as 15c/lb.

Americium — (the Americas), Am; at. wt. 243; at. no. 95; m.p. 994± 4°C; b.p. 2607°C; sp. gr. 13.67 (20°C); valence 2, 3, 4, 5, or 6. Americium was the fourth transuranium element to be discovered; the isotope Am^{241} was identified by Seaborg, James, Morgan, and Ghiorso late in 1944 at the wartime Metallurgical Laboratory (now the Argonne National Laboratory) of the University of Chicago as the result of successive neutron capture reactions by plutonium isotopes in a nuclear reactor:

$$Pu^{239}(n\gamma Pu^{240}(n\gamma Pu^{241} \xrightarrow{\beta} Am^{241}$$

Since the isotype Am^{241}can be prepared in relatively pure form by extraction as a decay product over a period of years from strongly neutron-bombarded plutonium Pu^{241} this isotope is used for much of the chemical investigation of this element. Better suited is the isotope Am^{243} due to its longer half-life (8.8×10^3 years as compared to 470 years for Am^{241}). A mixture of the isotopes Am^{241}, Am^{242}, and Am^{243} can be prepared by intense neutron irradiation of Am^{241} according to the reactions Am^{241} (n, γ) Am^{242} (n, γ) Am^{243}. Nearly isotopically pure Am^{243} can be prepared by a sequence of neutron bombardments and chemical separations as follows: neutron bombardment of Am^{241} yields Pu^{242} by the reactions Am^{241} (n, γ) Am^{242} EC; Pu^{242}, after chemical separation the

Pu^{242} can be transformed to Am^{243} via the reactions Pu^{242} (n, γ) Pu^{243} B−; Am ^{243}, and the Am^{243} can be chemically sparated. Fairly pure Pu^{242} can be prepared more simply by very intense neutron irradiation of Pu^{239} as the result of successive neutron-capture reactions. Americium metal has been prepared by reducing the trifluoride with barium vapor at 1000 to 1200°C or the dioxide by lanthanum metal. The luster of freshly prepared americium metal is whiter and more silvery than plutonium or neptunium prepared in the same manner. It appears to be more malleable than uranium or neptunium and tarnishes slowly in dry air at room temperature. Americium is thought to exist in two forms: an alpha form which has a double hexagonal close-packed structure and a loose-packed cubic beta form. Americium must be handled with great care to avoid personal contamination. As little as 0.03 μg of Am^{241} is the maximum permissible total body burden. The alpha activity from Am^{241} is about three times that of radium. When gram quantities of Am^{241} are handled, the intense gamma activity makes exposure a serious problem. Americium dioxide, AmO_2, is the most important oxide. AmF_3, AmF_4, $AmCl_3$, $AmBr_3$, AmI_3, and other compounds have been prepared. The isotope Am^{241} has been used as a portable source for gamma radiography. It has also been used as a radioactive glass thickness gage for the flat glass industry, and as a source of ionization for smoke detectors. Americium-241 is available from the A.E.C. at a cost of $150/g and Americium-243 at a cost of $100/mg.

Antimony — (Gr. *anti* plus *monos* — a metal not found alone), sb; at. wt. 121.75; at. no. 51; m.p. 630.74°C; b.p. 1750°C; sp. gr. 6.691 (20°C); valence 0, −3, +3, or +5. Antimony was recognized in compounds by the ancients and was known as a metal at the beginning of the 17th century and possibly much earlier. It is not abundant, but is found in over 100 mineral species. It is sometimes found native, but more frequently as the sulfide, *stibnite* (Sb_2S_3); it is also found as antimonides of the heavy metals, and as oxides. It is extracted from the sulfide by roasting to the oxide, which is reduced by salt and scrap iron; from its oxides it is also prepared by reduction with carbon. Two allotropic forms of antimony exist: the normal stable, metallic form, and the amorphous gray form. The so-called explosive antimony is an ill-defined material always containing an appreciable amount of halogen; therefore, it no longer warrants consideration as a separate allotrope. The yellow form, obtained by oxidation of *stibine*, SbH_3, is probably impure, and is not a distinct form. Metallic antimony is an extremely brittle metal of a flaky, crystalline texture. It is bluish white and has a metallic luster. It is not acted on by air at room temperature, but burns brilliantly when heated with the formation of white fumes of Sb_2O_3. It is a poor conductor of heat and electricity, and has a hardness of 3 to 3.5. Antimony, available commercially with a purity of 99.999 + %, is finding use in semiconductor technology for making infrared detectors, diodes, and Hall-effect devices. Commercial-grade antimony is widely used in alloys with percentages ranging from 1 to 20. It greatly increases the hardness and mechanical strength of lead. Batteries, antifriction alloys, type metal, small arms and tracer bullets, cable sheathing, and minor products use about half the metal produced. Compounds taking up the other half are oxides, sulfides, sodium antimonate, and antimony trichloride. These are used in manufacturing flame- proofing compounds, paints, ceramic enamels, glass, and pottery. Tartar emetic (hydrated potassium antimonyltartate) has been used in medicine. Antimony and many of its compounds are toxic. Exposure to antimony and its compounds should not exceed 0.5 mglM^3 (8-hr time weighted average 40-hr work week).

Argon — (Gr. *argos*, inactive), Ar; at. wt. 39.948; at. no. 18; freezing pt. −189.2°C; b.p.−185.7°C; density 1.7837 g/l. Its presence in air was suspected by Cavendish in 1785, discovered by Lord Rayleigh and Sir William Ramsay in 1894. The gas is prepared by fractionation of liquid air, the atmosphere containing 0.94% argon. The atmosphere of Mars contains 1.6% of Ar^{40} and 5 p.p.m. of Ar^{36}. Argon is two and one half times as soluble in water as nitrogen, having about the same solubility as oxygen. It is recognized by the characteristic lines in the red end of the spectrum. It is used in electric light bulbs and in fluorescent tubes

at a pressure of about 3 mm, and in filling photo tubes, glow tubes, etc. Argon is also used as an inert gas shield for arc welding and cutting, as a blanket for the production of titanium and other reactive elements, and as a protective atmosphere for growing silicon and germanium crystals. Argon is colorless and odorless, both as a gas and liquid. It is available in high-purity form. Commercial argon is available at a cost of about 10c per cubic foot. Argon is considered to be a very inert gas and is not now known to form true chemical compounds, as do krypton, xenon, and radon. However, it does form a hydrate having a dissociation pressure of 105 atm at 0°C. Ion molecules such as $(ArKr)^+$, $(ArXe)^+$, $(NeAr)^+$ have been observed spectroscopically. Argon also forms a clathrate with β hydroquinone. This clathrate is stable and can be stored for a considerable time, but a true chemical bond does not exist. Van der Waals' forces act to hold the argon. Naturally occurring argon is a mixture of three isotopes. Five other radioactive isotopes are now known to exist.

Arsenic — (L. *arsenicum,* Gr. *arsenikon,* yellow orpiment, identified with *arsenikos,* male, from the belief that metals were different sexes; Arab, *az-zernikh,* the orpiment from Persian *zerni-zar,* gold), As; at. wt. 74.9216; at. no. 33; valence −3, 0, +3 or +5. Elemental arsenic occurs in two solid modifications: yellow, and gray or metallic, with specific gravities of 1.97, and 5.73, respectively. Gray arsenic, the ordinary stable form, has a m.p. of 817°C (28 atm) and sublimes at 613°C. Several other allotropic forms of arsenic are reported in the literature. It is believed that Albertus Magnus obtained the element in 1250 A.D. In 1649 Schroeder published two methods of preparing the element. It is found native, in the sulfides *realgar* and *orpiment,* as arsenides and sulfarsenides of heavy metals, as the oxide, and as arsenates. *Mispickel* or arsenopyrite (FeSAs) is the most common mineral, from which on heating the arsenic sublimes leaving ferrous sulfide. The element is a steel gray, very brittle, crystalline, semimetallic solid; it tarnishes in air, and when heated is rapidly oxidized to arsenous oxide (As_2O_3) with the odor of garlic. Arsenic and its compounds are poisonous. Exposure to arsenic and its compounds (as As) should not exceed 0.5 mg/M^3 (8-hr time-weighted average-40 hr work week.) These values, however, are being studied, and may be lowered. Arsenic is also used in bronzing, pyrotechny, and for hardening and improving the sphericity of shot. The most important compounds are white arsenic (As_2O_3), the sulfide, Paris green $3Cu(AsO_2)_2 \cdot Cu(C_2H_3O_2)_2$, calcium arsenate, and lead arsenate; the last three have been used as agricultural insecticides and poisons. Marsh's test makes use of the formation and ready decomposition of arsine (AsH_3). Arsenic is available in high-purity form. It is finding increasing uses as a doping agent in solid-state devices such as transistors. Gallium arsenide is used as a laser material to convert electricity directly into coherent light.

Astatine — (Gr. *astatos,* unstable), At; at. wt. ~210; at. no. 85; m.p.302°C; b.p. 337°C; valence probably 1, 3, 5, or 7. Synthesized in 1940 by D.R. Corson, K. R. MacKenzie, and E. Segre at the University of California by bombarding bismuth with alpha particles. The longest-lived isotope, At^{210}, has a half-life of only 8.3 hr. Twenty isotopes are now known. Minute quantities of At^{215}, At^{218}, and At^{219} exist in equilibrium in nature with naturally occurring uranium and thorium isotopes, and traces of At^{217} are in equilibrium with U^{233} and Np^{239} resulting from interaction of thorium and uranium with naturally produced neutrons. The total amount of astatine present in the earth's crust, however, totals less than 1 oz. Astatine can be produced by bombarding bismuth with energetic alpha particles to obtain the relatively long-lived $At^{209-211}$, which can be distilled from the target by heating it in air. Only about 0.05 μg of astatine have been prepared to date. The "time of flight" mass spectrometer has been used to confirm that this highly radioactive halogen behaves chemically very much like other halogens, particularly iodine. The interhalogen compounds AtI, AtBr, and AtCl are known to form, but it is not yet known if astatine forms diatomic astatine molecules. HAt and CH_3At (methyl astatide) have been detected. Astatine is said to be more metallic than iodine, and, like iodine, it probably accumulates in the thyroid gland. Workers at the Brookhaven National

Laboratory have recently used reactive scattering in crossed molecular beams to identify and measure elementary reactions involving astatine.

Barium — (Gr. *barys,* heavy), Ba; at. wt. 137.33 ±0.01, at. no. 56; m.p. 725°C; b.p.1640°C; sp. gr. 3.5 (20°C); valence 2. Baryta was distinguished from lime by Scheele in 1774; the element was discovered by Sir Humphrey Davy in 1808. It is found only in combination with other elements, chiefly in *barite* or *heavy spar* (sulfate) and *witherite* (carbonate) and is prepared by electrolysis of the chloride. Barium is a metallic element, soft, and when pure is silvery white like lead; it belongs to the alkaline earth group, resembling calcium chemically. The metal oxidizes very easily and should be kept under petroleum or other suitable oxygen-free liquids to exclude air. It is decomposed by water or alcohol. The metal is used as a "getter" in vacuum tubes. The most important compounds are the peroxide (BaO_2), chloride, sulfate, carbonate, nitrate, and chlorate. Lithopone, a pigment containing barium sulfate and zinc sulfide, has good covering power, and does not darken in the presence of sulfides. The sulfate, as permanent white or *blanc fixe,* is also used in paint, in X-ray diagnostic work, and in glassmaking. *Barite* is extensively used as a wetting agent in oilwell drilling fluids, and also in making rubber. The carbonate has been used as a rat poison, while the nitrate and chlorate give colors in pyrotechny. The impure sulfide phosphoresces after exposure to the light. The compounds and the metal are not expensive. Barium metal (99.5 + % pure) costs about $20.00/ lb. All barium compounds that are water or acid soluble are poisonous. Naturally occurring barium is a mixture of seven stable isotopes. Thirteen other radioactive isotopes are known to exist.

Berkelium — (*Berkeley,* home of the University of California), Bk; at. wt. 247; at. no. 97; valence 3 or 4; sp. gr. 14 (est.). Berkelium, the eighth member of the actinide transition series, was discovered in December 1949 by Thompson, Ghiorso, and Seaborg, and was the fifth transuranium element synthesized. It was produced by cyclotron bombardment of milligram amounts of Am^{241} with helium ions at Berkeley, California. The first isotope produced had a mass number of 243 and decayed with a half-life of 4.6 hr. Eight isotopes are now known and have been synthesized. The existence of Bk^{249}, with a half-life of 314 days, makes it feasible to isolate berkelium in weighable amounts so that its properties can be investigated with macroscopic quantities. One of the first visible amounts of a pure berkelium compound, berkelium chloride, was produced in 1962. It weighed 3 billionth of a gram. Berkelium has not yet been prepared in elemental form, but it is expected to be a silvery metal, easily soluble in dilute mineral acids, and readily oxidized by air or oxygen at elevated temperatures to form the oxide. X-ray diffraction methods have been used to identify the following compounds: BkO_2, Bk_2O_3, BkF_3, $BkCl_3$, and BkOCl. As with other actinide elements, berkelium tends to accumulate in the skeletal system. The maximum permissible body burden of Bk^{249} in the human skeleton is about 0.0004 μg. Because of its rarity, berkelium presently has no commercial or technological use.

Beryllium — (Gr. *beryllos, beryl;* also called Glucinium or Glucinum, Gr. *glykys,* sweet), Be; at. wt. 9.01218; at. no. 4; m.p. 1278±5°C; b.p. 2970°C; sp. gr. 1.848 (20° C); valence 2. Discovered as the oxide by Vauquelin in beryl and in emeralds in 1798. The metal was isolated in 1828 by Wohler and by Bussy independently by the action of potassium on beryllium chloride. Beryllium is found in some 30 mineral species, the most important of which are *bertrandite, beryl, chrysoberyl,* and *phenacite. Aquamarine* and *emerald* are precious forms of *beryl. Beryl* ($3BeO \cdot Al_2O_3 \cdot 6SiO_2$) and *bertrandite* ($4BeO \cdot 2SiO_2 \cdot H_2O$) are the most important commercial sources of the element and its compounds. Most of the metal is now prepared by reducing beryllium fluoride with magnesium metal. Beryllium metal did not become readily available to industry until 1957. The metal, steel gray in color, has many desirable properties. It is one of the lightest of all metals, and has one of the highest melting points of the light metals. Its modulus of elasticity is about one third greater than that of steel. It resists attack by concentrated nitric acid, has excellent thermal conductivity, and is nonmagnetic. It

has a high permeability to X-rays, and when bombarded by alpha particles, as from radium or polonium, neutrons are produced in the ratio of about 30 neutrons/million alpha particles. At ordinary temperatures beryllium resists oxidation in air, although its ability to scratch glass is probably due to the formation of a thin layer of the oxide. Beryllium is used as an alloying agent in producing beryllium copper, which is extensively used for springs, electrical contacts, spot-welding electrodes, and nonsparking tools. It has found application as a structural material for high-speed aircraft, missiles, spacecraft, and communication satellites. It is being used in the windshield frame, brake discs, support beams, and other structural components of the space shuttle. Because beryllium is relatively transparent to X-rays, ultra-thin Be-foil is finding use in X-ray lithography for reproduction of micro-miniature integrated circuits.

Beryllium is used in nuclear reactors as a reflector or moderator for it has a low thermal neutron absorption cross section. It is used in gyroscopes, computer parts, and instruments where lightness, stiffness, and dimensional stability are required. The oxide has a very high melting point and is also used in nuclear work and ceramic applications. Beryllium and its salts are toxic and should be handled with the greatest of care. Beryllium and its compounds should not be tasted to verify the sweetish nature of beryllium (as did early experimenters). The metal, its alloys, and its salts can be handled safely if certain work codes are observed, but no attempt should be made to work with beryllium before becoming familiar with proper safeguards. Exposure to beryllium dust in air should be limited to $2 \mu g/M^3$ (8-hr time-weighted average — 40-hr week), with a ceiling concentration of $5 \mu g/M^3$. A maximum peak above the acceptable ceiling concentration for an 8-hr shift is $25 \mu g/M^3$ for a maximum duration of 30 min. These values are being reviewed and studied. Beryllium metal in vacuum cast billet form is priced roughly at $150/lb. Fabricated forms are more expensive.

Bismuth — (Ger. *Weisse Masse,* white mass; later *Wismuth* and *Bisemutum*), Bi; at. wt. 208.9808; at. no. 83; m.p. 271.3°C; b.p. 1560± 5°C; sp. gr. 9.747 (20°C); valence 3 or 5. In early times bismuth was confused with tin and lead. Claude Geoffroy the Younger showed it to be distinct from lead in 1753. It is a white, crystalline, brittle metal with a pinkish tinge. It occurs native. The most important ores are *bismuthinite* or *bismuth glance* (Bi_2S_3) and *bismite* (Bi_2O_3). Peru, Japan, Mexico, Bolivia, and Canada are major bismuth producers. Much of the bismuth produced in the U.S. is obtained as a by-product in refining lead, copper, tin, silver, and gold ores. Bismuth is the most diamagnetic of all metals, and the thermal conductivity is lower than any metal, except mercury. It has a high electrical resistance, and has the highest Hall effect of any metal (i.e., greatest increase in electrical resistance when placed in a magnetic field). "Bismanol" is a permanent magnet of high coercive force, made of MnBi, by the U.S. Naval Ordnance Laboratory. Bismuth expands 3.32% on solidification. This property makes bismuth alloys particularly suited to the making of sharp castings of objects subject to damage by high temperatures. With other metals such as tin, cadmium, etc. bismuth forms low-melting alloys which are extensively used for safety devices used in fire detection and extinguishing systems. Bismuth is used in producing malleable irons and is finding use as a catalyst for making acrylic fibers. When bismuth is heated in air it burns with a blue flame, forming yellow fumes of the oxide. The metal is also used as a thermocouple material (has highest negativity known), and has found application as a carrier for U^{235} or U^{233} fuel in atomic reactors. Its soluble salts are characterized by forming insoluble basic salts on the addition of water, a property sometimes used in detection work. Bismuth oxychloride is used extensively in cosmetics. Bismuth subnitrate and subcarbonate are used in medicine. Bismuth metal costs about $8/lb.

Boron — (Ar. *Buraq,* Pers. *Burah*), B; at. wt. 10.81; at. no. 5; m.p. 2079°C; b.p. sublimes 2550°C; sp. gr. of crystals 2.34, of amorphous variety 2.37; valence 3. Boron compounds have been known for thousands of years, but the element was not discovered until 1808 by Sir Humphry Davy and by Gay-Lussac and Thenard. The element is not found free in nature, but oc-

curs as orthoboric acid usually in certain volcanic spring waters and as borates in *borax* and *colemanite. Ulexite,* another boron mineral, is interesting as it is nature's own version of "fiber optics." By far the most important source of boron is the mineral *rasorite,* also known as *kernite,* found in the Mojave dersert of California. Extensive *borax* deposits are also found in Turkey. Boron exists naturally as 19.78%$_5$ B^{10} isotope and 80.22%$_5$$B^{11}$ isotope. High-purity crystalline boron may be prepared by the vapor phase reduction of boron trichloride or tribromide with hydrogen on electrically heated filaments. The impure, or amorphous, boron, a brownish- black powder, can be obtained by heating the trioxide with magnesium powder. Boron of 99.9999% purity has been produced and is available commercially. Elemental boron has an energy band gap of 1.50 to 1.56 eV, which is higher than that of either silicon or germanium. It has interesting optical characteristics, transmitting portions of the infrared, and is a poor conductor of electricity at room temperature, but a good conductor at high temperature. Amorphous boron is used in pyrotechnic flares to provide a distinctive green color, and in rockets as an igniter. The most important compounds of boron are boric, or boracic, acid, widely used as a mild antiseptic, and borax ($Na_2B_4O_7 \cdot 10H_2O$), which serves as a cleansing flux in welding and as a water softener in washing powders. Boron compounds are used in production of enamels for covering steel of refrigerators, washing machines, and like products. Boron compounds are also extensively used in the manufacture of borosilicate glasses. The isotope boron 10 is used as a control for nuclear reactors, as a shield for nuclear radiation, and in instruments used for detecting neutrons. Boron nitride has remarkable properties and can be used to make a material as hard as diamond. The nitride also behaves like an electrical insulator but conducts heat like a metal. It also has lubricating properties similar to graphite. The hydrides are easily oxidized with considerable energy liberation, and have been studied for use as rocket fuels. Demand is increasing for boron filaments, a high-strength, lightweight material chiefly employed for advanced aerospace structures. Boron is similar to carbon in that it has a capacity to form stable covalently bonded molecular networks. Carboranes, metalloboranes, phosphacarboranes, and other families comprise thousands of compounds.Crystalline boron (99%) costs about $/g. Amorphous boron costs about $/1lb. Elemental boron is not considered to be a poison, but assimilation of its compounds has a cumulative poisonous effect. Care must be taken in handling these.

Bromine — (Gr. *bromos,* stench), Br; at. wt. 79.904; at. no. 35; m.p. −7.2°C; b.p. 58.78°C; density of gas 7.59 g/*l*, liquid 3.12 (20°C); valence 1, 3, 5, or 7. Discovered by Balard in 1826, but not prepared in quantity until 1860. A member of the halogen group of elements, it is obtained from natural brines from wells in Michigan and Arkansas. Little bromine is extracted today from seawater, which contains only about 85 ppm. Bromine is the only liquid nonmetallic element. It is a heavy, mobile, reddish-brown liquid, volatilizing readily at room temperature to a red vapor with a strong disagreeable odor, resembling chlorine, and having a very irritating effect on the eyes and throat; it is readily soluble in water or carbon disulfide, forming a red solution; it is less active than chlorine but more so than iodine; it unites readily with many elements and has a bleaching action; when spilled on the skin it produces painful sores. It presents a serious health hazard, and maximum safety precautions should be taken when handling it. Much of the bromine output in the U.S. is used in the production of ethylene dibromide, a lead scavenger used in making gasoline antiknock compounds. Lead in gasoline, however, is presently being drastically reduced, due to environmental considerations. This will greatly affect future production of bromine. Bromine is also used in making fumigants, flameproofing agents, water purification compounds, dyes, medicinals, sanitizers, inorganic bromides for photography, etc. Organic bromides are also important.

Cadmium — (L. *cadmia;* Gr. *kadmeia* — ancient name for calamine, zinc carbonate), Cd; at. wt. 112.41 ± 0.01; at. no. 48; m.p. 320.9°C; b.p. 765°C; sp. gr. 8.65 (20°C); valence 2. Discovered by Stromeyer in 1817 from an impurity in zinc carbonate. Cadmium most often occurs

in small quantities associated with zinc ores, such as *sphalerite* (ZnS). *Greenockite* (CdS) is the only mineral of any consequence bearing cadmium. Almost all cadmium is obtained as a by-product in the treatment of zinc, copper, and lead ores. It is a soft, bluish-white metal which is easily cut with a knife. It is similar in many respects to zinc. It is a component of some of the lowest melting alloys; it is used in bearing alloys with low coefficients of friction and great resistance to fatigue; it is used extensively in electroplating, which accounts for about 60% of its use. It is also used in many types of solder, for standard E.M.F. cells, for Ni-Cd batteries, and as a barrier to control atomic fission. Cadmium compounds are used in black and white television phosphors and in blue and green phosphors for color TV tubes. It forms a number of salts, of which the sulfate is the most common; the sulfide is used as a yellow pigment. Cadmium and solutions of its compounds are toxic. Failure to appreciate the toxic properties of cadmium may cause workers to be unwittingly exposed to dangerous fumes. Silver solder, for example, which contains cadmium, should be handled with care. Serious toxicity problems have been found from long-term exposure and work with cadmium plating baths. Exposure to cadmium dust should not exceed 0.05 mg/M^3 (8-hr time-weighted average, 40-hr week). The ceiling concentration (maximum), for a period of 15 min, should not exceed 0.15 mg/M^3. Cadmium oxide fume exposure (8-hr, 40-hr week) should not exceed 0.05 mg/M^3, and the maximum concentration should not exceed 0.05 mg/M^3. These values are presently being restudied and recommendations have been made to reduce the exposure.The current price of cadmium is about $3/lb. It is available in high purity form.

Calcium — (L. *calx,* lime), Ca; at. wt. 40.08; at. no. 20; m.p. 839 ± 2°C; b.p. 1484°C; sp. gr. 1.55 (20°C); valence 2. Though lime was prepared by the Romans in the first century under the name calx, the metal was not discovered until 1808. After learning that Berzelius and Pontin prepared calcium amalgam by electrolyzing lime in mercury, Davy was able to isolate the impure metal. Calcium is a metallic element, fifth in abundance in the earth's crust, of which it forms more than 3%. It is an essential constituent of leaves, bones, teeth, and shells. Never found in nature uncombined, it occurs abundantly as *limestone* ($CaCO_3$), *gypsum* ($CaSO_4 \cdot 2H_2O$), and *fluorite* (CaF_2); *apatite* is the fluophosphate or chlorophosphate of calcium. The metal has a silvery color, is rather hard, and is prepared by electrolysis of the fused chloride to which calcium fluoride is added to lower the melting point. Chemically it is one of the alkaline earth elements; it readily forms a white coating of nitride in air, reacts with water, burns with a yellow-red flame, forming largely the nitride. The metal is used as a reducing agent in preparing other metals such as thorium, uranium, zirconium, etc., and is used as a deoxidizer, desulfurizer, or decarburizer for various ferrous and nonferrous alloys. It is also used as an alloying agent for aluminum, beryllium, copper, lead, and magnesium alloys, and serves as a "getter" for residual gases in vacuum tubes, etc. Its natural and prepared compounds are widely used. Quicklime (CaO), made by heating limestone and changed into slaked lime by the careful addition of water, is the great cheap base of chemical industry with countless uses. Mixed with sand it hardens as mortar and plaster by taking up carbon dioxide from the air. Calcium from limestone is an important element in Portland cement. The solubility of the carbonate in water containing carbon dioxide causes the formation of caves with stalactites and stalagmites and hardness in water. Other important compounds are the carbide (CaC_2), chloride ($CaCl_2$), cyanamide ($Ca(CN_2)$), hypochlorite ($Ca(OCl)_2$), nitrate ($Ca(NO_3)_2$), and sulfide (CaS).

Californium — (State and University of California), Cf; at. wt. 251; at. no. 98. Californium, the sixth transuranium element to be discovered, was produced by Thompson, Street, Ghiorso, and Seaborg in 1950 by bombarding microgram quantities of Cm^{242} with 35 MeV helium ions in the Berkeley 60-in. cyclotron. Californium (III) is the only ion stable in aqueous solutions, all attempts to reduce or oxidize californium (III) having failed. The isotope Cf^{249} results from the beta decay of Bk^{249} while the heavier isotopes are produced by intense neutron irradiation by the reactions:

$Bk^{249}(n\gamma)Bk^{250} \xrightarrow{\beta} Cf^{250}$ and $Cf^{249}(n,\gamma)Cf^{250}$

followed by

$Cf^{250}(n,\gamma)Cf^{251}(n,\gamma)Cf^{252}$

The existence of the isotopes Cf^{249}, Cf^{250}, Cf^{251}, and Cf^{252} makes it feasible to isolate californium in weighable amounts so that its properties can be investigated with macroscopic quantities. Californium-252 is a very strong neutron emitter. One microgram releases 170 million neutrons per minute, which presents biological hazards. Proper safeguards should be used in handling californium. In 1960 a few tenths of a microgram of californium trichloride $CfCl_3$, californium oxychloride, CfOCl, and californium oxide, Cf_2O_3, were first prepared. Reduction of californium to its metallic state has not yet been accomplished. Because californium is a very efficient source of neutrons, many new uses are expected for it. It has already found use in neutron moisture gages and in well-logging (the determination of water and oil-bearing layers.) It is also being used as a portable neutron source for discovery of metals such as gold or silver by on-the-spot activation analysis. Cf^{252} is now being offered for sale by the A.E.C. at a cost of \$10/0.1μg. As of May 1975, more than 63 mg have been produced and sold. It has been suggested that californium may be produced in certain stellar explosions, called *supernovae*, for the radioactive decay of Cf^{254} (55-day half-life) agrees with the characteristics of the light curves of such explosions observed through telescopes. This suggestion, however, is questioned.

Carbon — (L. *carbo,* charcoal), C; at. wt. 12 exactly (C^{12}); at. wt. (natural carbon) 12.011; at. no. 6; m.p. ~ 3550°C, graphite sublimes at 3367 ± 25°C; b.p. 4827°C; sp. gr. amorphous 1.8 to 2.1, graphite 1.9 to 2.3, diamond 3.15 to 3.53 (depending on variety); gem diamond 3.513 (25°C); valence 2, 3, or 4. Carbon, an element of prehistoric discovery, is very widely distributed in nature. It is found in abundance in the sun, stars, comets, and atmospheres of most planets. Carbon in the form of microscopic diamonds is found in some meteorites. Natural diamonds are found in *kimberlite* of ancient volcanic "pipes," such as found in South Africa, Arkansas, and elsewhere. Diamonds are now also being recovered from the ocean floor off the Cape of Good Hope. About 30% of all industrial diamonds used in the U.S. are now made synthetically. The energy of the sun and stars can be attributed at least in part to the well-known carbon-nitrogen cycle. Carbon is found free in nature in three allotropic forms: amorphous, graphite, and diamond. A fourth form, known as "white" carbon, is now thought to exist. Graphite is one of the softest known materials while diamond is one of the hardest. Graphite exists in two forms: alpha and beta. These have identical physical properties, except for their crystal structure. Naturally occurring graphites are reported to contain as much as 30% of the rhombohedral (beta) form, whereas synthetic materials contain only the alpha form. The hexagonal alpha type can be converted to the beta by mechanical treatment, and the beta form reverts to the alpha on heating it above 1000°C. In 1969 a new allotropic form of carbon was produced during the sublimation of pyrolytic graphite at low pressures. Under free-vaporization conditions above ~2550 K, "white" carbon forms as small transparent crystals on the edges of the basal planes of graphite. The interplanar spacings of "white" carbon are identical to those of a carbon form noted in the graphitic gneiss from the Ries (meteoritic) Crater of Germany. "White" carbon is a transparent birefringent material. Little information is presently available about this allotrope. In combination, carbon is found as carbon dioxide in the atmosphere of the earth and dissolved in all natural waters. It is a component of great rock masses in the form of carbonates of calcium (limestone), magnesium, and iron. Coal, petroleum, and natural gas are chiefly hydrocarbons. Carbon is unique among the elements in the vast number of variety of compounds it can form. With hydrogen, oxygen, and nitrogen, and other elements, it forms a very large number of compounds, carbon atom often being linked to carbon atom. There are upwards of a million or more known carbon compounds, many thousands of which are vital to organic and life processes. Without carbon, the basis for life would be impossible. While it has been thought that silicon might take the

place of carbon in forming a host of similar compounds, it is now not possible to form stable compounds with very long chains of silicon atoms. The atmosphere of Mars contains 96.2% CO_2. Some of the most important compounds of carbon are carbon dioxide (CO_2), carbon monoxide (CO), carbon disulfide (CS_2), chloroform ($CHCl_3$), carbon tetrachloride (CCl_4), methane (CH_4), ehtylene (C_2H_4), acetylene (C_2H_2), benzene (C_6H_6), ethyl alcohol (C_2H_5OH), acetic acid (CH_3COOH), and their derivatives. Carbon has seven isotopes. In 1961 the International Union of Pure and Applied Chemistry adopted the isotope carbon-12 as the basis for atomic weights. Carbon-14, an isotope with a half-life of 5730 years, has been widely used to date such materials as wood, archeological speciments, etc. Carbon-13 is now commercially available at a cost of $700/g.

Cerium — (named for the asteroid *Ceres,* which was discovered in 1801 only 2 years before the element), Ce; at. wt. 140.12; at. no. 58; m.p. 799°C; b.p. 3426°C; sp. gr. 6.657 (25°C); valence 3 or 4. Discovered in 1803 by Klaproth and by Berzelius and Hisinger; metal prepared by Hillebrand and Norton in 1875. Cerium is the most abundant of the metals of the so-called rare earths; it is found in a number of minerals including *allanite* (also known as *orthite*), *monazite, bastnasite, cerite,* and *samarskite.* Monazite and bastnasite are presently the two most important sources of cerium. Large deposits of monazite found on the beaches of Travancore, India, in river sands in Brazil, and deposits of *allanite* in the western United States, and *bastnasite* in Southern California will supply cerium, thorium, and the other rare-earth metals for many years to come. Metallic cerium is prepared by metallothermic reduction techniques, such as by reducing cerous fluoride with calcium, or by electrolysis of molten cerous chloride or other cerous halides. The metallothermic technique is used to produce high-purity cerium. Cerium is especially interesting because of its variable electronic structure. The energy of the inner 4f level is nearly the same as that of the outer or valence electrons, and only small amounts of energy are required to change the relative occupancy of these electronic levels. This gives rise to dual valency states. For example, a volume change of about 10% occurs when cerium is subjected to high pressures or low temperatures. It appears that the valence changes from about 3 to 4 when it is cooled or compressed. The low temperature behavior of cerium is complex. Four allotropic modifications are thought to exist: cerium at room temperature and at atmospheric pressure is known as γ cerium. Upon cooling to −23°C, γ cerium changes to β cerium. The remaining γ cerium starts to change to α cerium when cooled to −158°C, and the transformation is complete at −196°C. α Cerium has a density of 8.24; δ cerium exists above 726°C. At atmospheric pressure, liquid cerium is more dense than its solid form at the melting point. Cerium is an iron-gray lustrous metal. It is malleable, and oxidizes very readily at room temperature, especially in moist air. Except for europium, cerium is the most reactive of the "rare-earth" metals. It slowly decomposes in cold water, and rapidly in hot water. Alkali solutions and dilute and concentrated acids attack the metal rapidly. The pure metal is likely to ignite if scratched with a knife. Ceric salts are orange red or yellowish; cerous salts are usually white. Cerium is a component of misch metal, which is extensively used in the manufacture of pyrophoric alloys for cigarette lighters, etc. While cerium is not radioactive, the impure commercial grade may contain traces of thorium, which is radioactive. The oxide is an important constituent of incandescent gas mantles and it is emerging as a hydrocarbon catalyst in "self-cleaning" ovens. In this application it can be incorporated into oven walls to prevent the collection of cooking residues. As ceric sulfate it finds extensive use as a volumetric oxidizing agent in quantitative analysis. Cerium compounds are used in the manufacture of glass, both as a component and as a decolorizer. The oxide is finding increased use as a glass polishing agent instead of rouge, for it is much faster than rouge in polishing glass surfaces. Cerium, with other rare earths, is used in carbon-arc lighting, especially in the motion picture industry. It is also finding use as an important catalyst in petroleum refining and in metallurgical and nuclear applications. Commerical cerium metal costs about $75/lb. In small lots, 99.9% cerium costs about 30/g.

Cesium — (L. *caesius,* sky blue), Cs; at. wt.

132.9054; at. no. 55; m.p. 28.40± 0.01°C; b.p. 678.4°C; sp. gr. 1.873 (20°C); valence 1. Cesium was discovered spectroscopically by Bunsen and Kirchhoff in 1860 in mineral water from Durkheim. Cesium, an alkali metal, occurs in *lepidolite, pollucite* (a hydrated silicate of aluminum and cesium), and in other sources. One of the world's richest sources of cesium is located at Bernic Lake, Manitoba. The deposits are estimated to contain 300,000 tons of pollucite, averaging 20% cesium. It can be isolated by electrolysis of the fused cyanide and by a number of other methods. Very pure, gas-free cesium can be prepared by thermal decomposition of cesium azide. The metal is characterized by a spectrum containing two bright lines in the blue along with several others in the red, yellow, and green. It is silvery white, soft, and ductile. It is the most electropositive and most alkaline element. Cesium, gallium, and mercury are the only three metals that are liquid at room temperature. Cesium reacts explosively with cold water, and reacts with ice at temperatures above −116°C. Cesium hydroxide, the strongest base known, attacks glass. Because of its great affinity for oxygen the metal is used as a "getter" in radio tubes. It is also used in photoelectric cells, as well as a catalyst in the hydrogenation of certain organic compounds. The metal has recently found application in ion propulsion systems. Although these are not usable in the earth's atmosphere, 1 lb of cesium in outer space theoretically will propel a vehicle 140 times as far as the burning of the same amount of any known liquid or solid. Cesium is used in atomic clocks, which are accurate to 5 sec in 300 years. Its chief compounds are the chloride and the nitrate. The present price of cesium is about $100 to $375/ lb depending on quantity and purity.

Chlorine — (Gr. *chloros,* greenish yellow), Cl; at. wt. 35.453; at. no. 17; m.p. −100.98°C; b.p.−34.6°C; density 3.214 g/l; sp. gr. 1.56(−33.6°C); valence 1, 3, 5, or 7. Discovered in 1774 by Scheele, who thought it contained oxygen; named in 1810 by Davy, who insisted it was an element. In nature it is found in the combined state only, chiefly with sodium as common salt (NaCl), *carnallite* ($KMgCl_3 \cdot {}_2O$), and *sylvite* (KCl). It is a member of the halogen (salt-forming) group of elements and is obtained from chlorides by the action of oxidizing agents and more often by electrolysis; it is a greenish-yellow gas, combining directly with nearly all elements. At 10°C one volume of water dissolves 3.10 volumes of chlorine, at 30°C only 1.77 volumes. Chlorine is widely used in making many everyday products. It is used for producing safe drinking water the world over. Even the smallest water supplies are now usually chlorinated. It is also extensively used in the production of paper products, dyestuffs, textiles, petroleum products, medicines, antiseptics, insecticides, foodstuffs, solvents, paints, plastics, and many other consumer products. Most of the chlorine produced is used in the manufacture of chlorinated compounds of sanitation, pulp bleaching, disinfectants, and textile processing. Further use is in the manufacture of chlorates, chloroform, carbon tetrachloride, and in the extraction of bromine. Organic chemistry demands much from chlorine, both as an oxidizing agent and in substitution, since it often brings desired properties in an organic compound when substituted for hydrogen, as in one from of synthetic rubber. Chlorine is a respiratory irritant. The gas irritates the mucous membranes and the liquid burns the skin. As little as 3.5 ppm can be detected as an odor, and 1000 ppm is likely to be fatal after a few deep breaths. It was used as a war gas in 1915. Exposure to chlorine should not exceed 1 ppm(8-hr time-weighted average — 40 hr week.)

Chromium — (Gr. *chroma,* color), Cr; at. wt. 51.996; at. no. 24; m.p. 1857 ± 20°C; b.p. 2672°C; sp. gr. 7.18 to 7.20 (20°C); valence chiefly 2, 3, or 6. Discovered in 1797 by Vauquelin, who prepared the metal the next year. Chromium is a steel-gray, lustrous, hard metal that takes a high polish. The principal ore is *chromite* ($FeO \cdot {}_2O_3$), which is found in Southern Rhodesia, U.S.S.R., Transvaal, Turkey, Iran, Albania, Finland, Malagasy, and the Philippines. The metal is usually produced by reducing the oxide with aluminum. Chromium is used to harden steel, to manufacture stainless steel, and to form many useful alloys. Much is used in plating to produce a hard, beautiful surface and to prevent corrosion. Chromium is

used to give glass an emerald green color. It finds wide use as a catalyst. All compounds of chromium are colored; the most important are the chromates of sodium and potassium (K_2CrO_4) and the dichromates ($K_2Cr_2O_7$) and the potassium and ammonium chrome alums, as $KCr(SO_4)_2 \cdot 12H_2O$. The dichromates are used as oxidizing agents in quantitative analysis, also in tanning leather. Other compounds are of industrial value; lead chromate is chrome yellow, a valued pigment. Chromium compounds are used in the textile industry as mordants, and by the aircraft and other industries for anodizing aluminum. The refractory industry has found chromite useful for forming bricks and shapes, as it has a high melting point, moderate thermal expansion, and stability of crystalline structure. Chromium compounds are toxic and should be handled with proper safeguards.

Cobalt — (*Kobald,* from the German, goblin or evil spirit, *cobalos,* Greek, mine), Co; at. wt. 58.9332; at. no. 27; m.p. 1495°C; b.p. 2870°C; sp. gr. 8.9 (20°C); valence 2 or 3. Discovered by Brandt about 1735. Cobalt occurs in the minerals *cobaltite, smaltite,* and *erythrite,* and is often associated with nickel, silver, lead, copper, and iron ores, from which it is most frequently obtained as a by-product. It is also present in meteorites. Important ore deposits are found in Zaire, Morocco, and Canada. Cobalt is a brittle, hard metal, closely resembling iron and nickel in appearance. It has a magnetic permeability of about two thirds that of iron. Cobalt tends to exist as a mixture of two allotropes over a wide temperature range; the β-form predominates below 400°C, and the α above that temperature. The transformation is sluggish and accounts in part for the wide variation in reported data on physical properties of cobalt. It is alloyed with iron, nickel and other metals to make Alnico, an alloy of unusual magnetic strength with many important uses. Stellite ® alloys, containing cobalt, chromium, and tungsten, are used for high-speed, heavy-duty, high-temperature cutting tools, and for dies. Cobalt is also used in other magnet steels and stainless steels, and in alloys used in jet turbines and gas turbine generators. The metal is used in electroplating because of its appearance, hardness, and resistance to oxidation. The salts have been used for centuries for the production of brilliant and permanent blue colors in porcelain, glass, pottery, tiles, and enamels. It is the principal ingredient in Serves and Thenard's blue. A solution of the chloride ($CoCl_2 \cdot 6H_2O$) is used as sympathetic ink. The cobalt ammines are of interest; the oxide and the nitrate are important. Cobalt carefully used in the form of the chloride, sulfate, acetate, or nitrate has been found effective in correcting a certain mineral deficiency disease in animals. Soils should contain 0.13 to 0.30 ppm of cobalt for proper animal nutrition. Cobalt-60, an artifical isotope, is an important gamma ray source, and is extensively used as a tracer and a radiotherapeutic agent. Single compact sources of Cobalt-60 are readily available and have an equivalent gamma ray output equal to thousands of grams of radium. The cost of Cobalt-60 varies from about 40c to $7.00/curie, depending on quantity and specific activity. Exposure to cobalt (metal fumes and dust) should be limited to 0.05 mg/M³ (8-hr time-weighted average, 40 hr week).

Columbium — see Niobium.

Copper — (L. *cuprum,* from the island of Cyprus), Cu; at. wt. 63.546; at. no. 29; m.p. 1083.4 ± 0.2°C; b.p. 2567 °C; sp. gr. 8.96 (20°C); valence 1 or 2. The discovery of copper dates from prehistoric times; it is said to have been mined for more than 5000 years. It is one of man's most important metals. Copper is reddish colored, takes on a bright metallic luster, and is malleable, ductile, and a good conductor of heat and electricity (second only to silver in electrical conductivity). The electrical industry is one of the greatest users of copper. Copper occasionally occurs native, and is found in many minerals such as *cuprite, malachite, azurite, chalcopyrite,* and *bornite.* Large copper ore deposits are found in the U.S., Chile, Zambia, Zaire, Peru, and Canada. The most important copper ores are the sulfides, oxides, and carbonates. From these, copper is obtained by smelting, leaching, and by electrolysis. Its alloys, brass and bronze, long used, are still very important; all American coins are now copper alloys; monel and gun metals also contain copper. The most important compounds are the

oxide and the sulfate, blue vitriol; the latter has wide use as an agricultural poison and as an algicide in water purification. Copper compounds such as Fehling's solution are widely used in analytical chemistry in tests for sugar. High-purity copper (99.999 + %) is available commercially.

Curium — (Pierre and Marie Curie), Cm; at. wt. 247; at. no. 96; m.p. 1340 ± 40°C; sp. gr. 13.51 (calc.); valence 3 and 4. Although curium follows americium in the periodic system, it was actually known before americium and was the third transuranium element to be discovered. It was identified by Seaborg, James, and Ghiorso in 1944 at the wartime Metallurgical Laboratory in Chicago as a result of helium-ion bombardment of Pu^{239} in the Berkeley, California, 60-in. cyclotron. Visible amounts (30 μg) of Cm^{242}, in the form of the hydroxide, were first isolated by Werner and Perlman of the University of California in 1947. In 1950, Crane, Wallmann, and Cunningham found that the magnetic susceptibility of microgram samples of CmF_3 was of the same magnitude as that of GdF_3. This provided direct experimental evidence for assigning an electronic configuration to Cm^{+3}. In 1951, the same workers prepared curium in its elemental form for the first time. Thirteen isotopes of curium are now known. The most stable, Cm^{247}, with a half-life of 16 million years, is so short compared to the earth's age that any primordial curium must have disappeared long ago from the natural scene. Minute amounts of curium probably exist in natural deposits of uranium, as a result of a sequence of neutron captures and β decays sustained by the very low flux of neutrons naturally present in uranium ores. The presence of natural curium, however, has never been detected. Cm^{242} and Cm^{244} are available in multigram quantities. Cm^{248} has been produced only in milligram amounts. Curium is smiliar in some regards to gadolinium, its rare-earth homolog, but it has a more complex crystal structure. Curium is silver in color, is chemically reactive, and is more electropositive than aluminum. CmO_2, Cm_2O_3, CmF_3, CmF_4, $CmCl_3$, $CmBr_3$, and CmI_3 have been prepared. Most compounds of trivalent curium are faintly yellow in color. The A.E.C. is attempting to produce several kilograms of Cm^{244}, an isotope with a 17.6-year half-life, by neutron irradiation of plutonium in a nuclear reactor. Ultimately, it is possible that it may be produced in ton quantities by converting a number of plutonium production reactors to the manufacture of Cm^{244}. Cm^{242} generates about three thermal watts of energy per gram. This compares to one-half thermal watt per gram of Pu^{238}. This suggests use for curium as an isotope power source. Cm^{244} is now offered for sale by the A.E.C. at $100/ mg. Curium absorbed into the body accumulates in the bones, and is therefore very toxic as its radiation destroys the red-cell forming mechanism. The maximum permissible total body burden of Cm^{244}(soluble) in a human being is 0.3 μCi (microcurie).

Deuterium, an isotope of hydrogen — see Hydrogen.

Dysprosium — (Gr. *dysprositos*, hard to get at), Dy; at. wt. 162.50; at. no. 66; m.p. 1412°C; b.p.2562°C; sp. gr. 8.550 (25°C); valence 3. Dysprosium was discovered in 1886 by Lecoq de Boisbaudran, but not isolated. Neither the oxide nor the metal was available in relatively pure form until the development of ion-exchange separation and metallographic reduction techniques by Spedding and associates about 1950. Dysprosium occurs along with other so-called rare-earth or lanthanide elements in a variety of minerals such as *xenotime, fergusonite, gadolinite, euxenite, polycrase,* and *blomstrandine.* The most important sources, however, are from *monazite* and *bastnasite.* Dysprosium can be prepared by reduction of the trifluoride with calcium. The element has a metallic, bright silver luster. It is relatively stable in air at room temperature, and is readily attacked and dissolved, with the evolution of hydrogen, by dilute and concentrated mineral acids. The metal is soft enough to be cut with a knife and can be machined without sparking if overheating is avoided. Small amounts of impurities can greatly affect its physical properties. While dysprosium has not yet found many applications, its thermal neutron absorption cross-section and high melting point suggest metallurgical uses in nuclear control applications for alloying with special stainless steels. A dysprosium oxide-nickel cermet

has found use in cooling nuclear reactor control rods. This cermet absorbs neutrons readily without swelling or contracting under prolonged neutron bombardment. In combination with vanadium and other rare earths, dysprosium has been used in making laser materials. Dysprosium-cadmium calcogenides, as sources of infrared radiation, have been used for studying chemical reactions. The cost of dysprosium metal has dropped in recent years since the development of ion- exchange and solvent extraction techniques, and the discovery of large ore bodies. The metal is still expensive, however, and costs about 70c/g or $190/lb in purities of 99 + %.

Einsteinium — (Albert Einstein), Es; at wt. 254; at no. 99. Einsteinium, the seventh transuranic element of the actinide series to be discovered, was identified by Ghiorso and coworkers at Berkeley in December 1952 in debris from the first large thermonuclear or "hydrogen" bomb explosion, which took place in the Pacific in November 1952. The isotope produced was the 20-day Es^{253} isotope. In 1961, a sufficient amount of einsteinium was produced to permit separation of a macroscopic amount of Es^{253}. This sample weighed about 0.01 μg. A special magnetic-type balance was used in making this determination. Es^{253} so produced was used to produce mendelevium (Element 101). About 3 μg of einsteinium has been produced at Oak Ridge National Laboratores by irradiating for several years kilogram quantities of Pu^{239} in a reactor to produce $Pu.^{242}$ This was then fabricated into pellets of plutonium oxide and aluminum powder, and loaded into target rods for an initial 1-year irradiation at the A.E.C.'s Savannah River Plant, followed by irradiation in a HFIR (High Flux Isotopic Reactor). After 4 months in the HFIR the targets were removed for chemical separation of the einsteinium from californium. Eleven isotopes of einsteinium are now recognized. Es^{254} has the longest half-life (276 days). Tracer studies using Es^{253} show that einsteinium has chemical properties typical of a heavy trivalent, actinide element.

Element 104 — In 1964, workers of the Joint Nuclear Research Institute at Dubna (U.S.S.R.) bombarded plutonium with accelerated 113 to 115 MeV neon ions. By measuring fission tracks in a special glass with a microscope, they detected an isotope that decays by spontaneous fission. They suggested that this isotope, which had a half-life of 0.3 ± 0.1 sec might be 104^{260}, produced by the following reaction:

$$_{94}Pu^{242} + _{10}Ne^{22} \rightarrow 104^{260} + 4n$$

Element 104, the first *transactinide* element, is expected to have chemical properties similar to those of hafnium. It would, for example, form a relatively volatile compound with chlorine (a tetrachloride). The Soviet scientists have performed experiments aimed at chemical identification, and have attempted to show that the 0.3-sec activity is more volatile than that of the relatively nonvolatile actinide trichlorides. This experiment does not fulfill the test of chemically separating the new element from all others, but it provides important evidence for evaluation. New data, reportedly issued by Soviet scientists, have reduced the half-life of the isotope they worked with from 0.3 to 0.15 sec. The Dubna scientists suggest the name *kurchatovium* and symbol *Ku* for Element 104, in honor of Igor Vasilevich Kurchatov (1903-1960), late Head of Soviet Nuclear Research. In 1969, Ghiorso, Nurmia, Harris, K.A.Y. Eskola, and P.L. Eskola of the University of California at Berkeley reported they had positively identified two, and possibly three, isotopes of Element 104. The group also indicated that after repeated attempts so far they have been unable to produce isotope 104^{260} reported by the Dubna group in 1964. The discoveries at Berkeley were made by bombarding a target of Cf^{249} with C^{12} nuclei of 71 MeV, and C^{13} nuclei of 69 MeV. The combination of C^{12} with Cf^{249} followed by instant emission of four neutrons produced Element 104^{257}. This isotope has a half-life of 4 to 5 sec, decaying by emitting an alpha particle into No^{253}, with a half-life of 105 sec. The same reaction, except with the emission of three neutrons, was thought to have produced 104^{258}. with a half-life of about 1/100 sec. Element 104^{259} is formed by the merging of a C^{13} nuclei with Cf^{249}, followed by emission of three neutrons. This isotope has a half-life of 3 to 4 sec, and decays by emitting an alpha particle into

No 255, which has a half-life of 185 sec. Thousands of atoms of 104^{257} and 104^{259} have been detected. The Berkeley group believe their identification of 104^{258} is correct, but they do not attach the same degree of confidence to this work as to their work on 104^{257} and 104^{259}. The Berkeley group proposes for the new element the name *rutherfordium* (symbol R*f*), in honor of Ernest R. Rutherford, New Zealand physicist. The claims for discovery and the naming of Element 104 are still in question.

Element 105 — In 1967 G.N. Flerov reported that a Soviet team working at the Joint Institute for Nuclear Research at Dubna may have produced a few atoms of 105^{260} and 105^{261} by bombarding Am^{243} with Ne^{22}. Their evidence was based on time-coincidence measurements of alpha energies. More recently, it was reported that early in 1970 Dubna scientists synthesized Element 105 and that by the end of April 1970 "had investigated all the types of decay of the new element and had determined its chemical properties." The Soviet groups has not proposed a name for Element 105. In late April 1970, it was announced that Ghiorso, Nurmia, Harris, K.A.Y. Eskola, and P.L.Eskola, working at the University of California at Berkeley, had positively identified Element 105. The discovery was made by bombarding a target of Cf^{249} with a beam of 84 MeV nitrogen nuclei in the Heavy Ion Linear Accelerator (HILAC). When a N^{15} nuclei is absorbed by a Cf^{249} nucleus, four neutrons are emitted and a new atom of 105^{260} with a half-life of 1.6 sec is formed. While the first atoms of Element 105 are said to have been detected conclusively on March 5, 1970, there is evidence that Element 105 had been formed in Berkeley experiments a year earlier by the method described. Ghiorso and his associates have attempted to confirm Soviet findings by more sophisticated methods without success. The Berkeley Group proposes the name *hahnium,* after the late German scientist Otto Hahn (1879-1968), and *Ha* for the chemical symbol.

More recently, in October 1971, it was announced that two new isotopes of Element 105 were synthesized with the heavy ion linear accelerator by A. Ghiorso and co-workers at Berkeley. Element 105^{261} was produced both by bombarding Cl^{250} with N^{15} and by bombarding Bk^{249} with O^{16}. The isotope emits 8.93-MeV α particles and decays to Lr^{257} with a half-life of about 1.8 sec. Element 105^{262} was produced by bombarding Bk^{249} with O^{18}. It emits 8.45 MeV α particles and decays to Lr^{258} with a half-life of about 40 sec.

Element 106 — In June 1974, members of the Joint Institute for Nuclear Research in Dubna, U.S.S.R., reported their discovery of Element 106, which they claim to have synthesized. In September 1974, workers of the Lawrence Berkeley and Livermore Laboratories also reported creation of Element 106 "without any scientific doubt." The LBL and LLL Group used the SuperHILAC to accelerate O^{18} ions onto a Cf^{249} target. Element 106 was created by the reaction Cf^{249} $(O^{18}\ 4n)(0^{18}\ 4n)_{106}$ X^{263}, which decayed by α emission to rutherfordium, and then by α emission to nobelium, which in turn further decayed by α emission. An elaborate detection system not only looked for correlations between the new element and its daughter, but also between daughter and granddaughter. The element so identified had α energies of 9.06 and 9.25 MeV with a half-life of 0.9 ± 0.2 sec. At Dubna, 280-MeV ions of Cr^{54} from the 310-cm cyclotron were used to strike targets of Pb^{206}, Pb^{207}, and Pb^{208}, in separate runs. Foils exposed to a rotating target disc were used to detect spontaneous fission activities, the foils being etched and examined microscopically to detect the number of fission tracts and the half-life of the fission activity. Other experiments were made to aid in confirmation of the discovery. Neither the Dubna team nor the Berkeley-Livermore Group has proposed a name as yet for Element 106.

Element 107 — In 1976 Soviet scientists at Dubna announced they had synthesized Element 107 by bombarding Bi^{204} with heavy nuclei of Cr^{54}. It is reported that earlier experiments in 1975 had allowed scientists "to glimpse" the new element for 2/1000 sec. A rapidly rotating cylinder, coated with a thin layer of the bismuth metal, was used as the target. This was bombarded by a stream of Cr^{54} ions fired tangentially. The existence of Element 107 has not been confirmed by scientists of other nations, nor is it certain this claim constitutes "discovery."

Erbium — (*Ytterby,* a town in Sweden), Er;

at. wt. 167.26; at no. 68; m.p. 159°C; b.p. 2863°C; sp. gr. 9.066 (25°C); valence 3. Erbium, one of the so-called rare-earth elements of the lanthanide series, is found in the minerals mentioned under dysprosium above. In 1842 Mosander separated "yttria," found in the mineral *gadolinite,* into three fractions which he called *yttria, erbia,* and *terbia.* The names *erbia* and *terbia* became confused in this early period. After 1860, Mosander's *terbia* was known as *erbia,* and after 1877, the earlier known *erbia* became *terbia.* The *erbia* of this period was later shown to consist of five oxides, now known as *erbia, scandia, holmia, thulia* and *ytterbia.* By 1905 Urbain and James independently succeeded in isolating fairly pure Er_2O_3. Klemm and Bommer first produced reasonably pure erbium metal in 1934 by reducing the anhydrous chloride with potassium vapor. The pure metal is soft and malleable and has a bright, silvery, metallic luster. As with other rare-earth metals, its properties depend to a certain extent on the impurities present. The metal is fairly stable in air and does not oxidize as rapidly as some of the other rare-earth metals. Naturally occurring erbium is a mixture of six isotopes, all of which are stable. Nine radioactive isotopes of erbium are also recognized. Recent production techniques, using ion-exchange reactions, have resulted in much lower prices of the rare-earth metals and their compounds in recent years. The cost of 99 + % erbium metal is about $1.00/g in small quantities. Erbium is finding nuclear and metallurgical uses. Added to vanadium, for example, erbium lowers the hardness and improves workability. Most of the rare-earth oxides have sharp absorption bands in the visible, ultraviolet, and near infrared. This property, associated with the electronic structure, gives beautiful pastel colors to many of the rare-earth salts. Erbium oxide gives a pink color and has been used as a colorant in glasses and porcelain enamel glazes.

Europium — (Europe), Eu; at. wt. 151.96; at. no. 63; m.p. 822°C; b.p. 1597°C; sp. gr. 5.243 (25°C); valence 2 or 3. In 1890 Boisbaudran obtained basic fractions from samarium-gadolinium concentrates which had spark spectral lines not accounted for by samarium or gadolinium. These lines subsequently have been shown to belong to europium. The discovery of europium is generally credited to Demarcay, who separated the earth in reasonably pure form in 1901. The pure metal was not isolated until recent years. Europium is now prepared by mixture Eu_2O_3 with a 10%-excess of lanthanum metal and heating the mixture in a tantalum crucible under high vacuum. The element is collected as a silvery-white metallic deposit on the walls of the crucible. As with other rare-earth metals, except for lanthanum, europium ignites in air at about 150 to 180°C. Europium is about as hard as lead and is quite ductile. It is the most reactive of the rare-earth metals, quickly oxidizing in air. It resembles calcium in its reaction with water. *Bastnasite* and *monazite* are the principal ores containing europium. Europium has been identified spectroscopically in the sun and certain stars. Seventeen isotopes are now recognized. Europium isotopes are good neutron absorbers and are being studied for use in nuclear control applications. Europium oxide is now widely used as a phosphor activator and europium-activated yttrium vanadate is in commercial use as the red phosphor in color TV tubes. Europium-doped plastic has been used as a laser material. With the development of ion-exchange techniques and special processes, the cost of the metal has been greatly reduced in recent years. Europium is one of the rarest and most costly of the rare-earth metals. It is priced at about $11 to $15/g or $3000/lb.

Fermium — (Enrico Fermi), Fm; at. wt. 257; at. no. 100. Fermium, the eighth transuranium element of the actinide series to be discovered, was identified by Ghiorso and co-workers in 1952 in the debris from a thermonuclear explosion in the Pacific in work involving the University of California Radiation Laboratory, the Argonne National Laboratory, and the Los Alamos Scientific Laboratory. The isotope produced was the 20-hr Fm^{255}. During 1953 and early 1954, while discovery of elements 99 and 100 was withheld from publication for security reasons, a group from the Nobel Institute of Physics in Stockholm bombarded U^{238} with O^{16} ions, and isolated a 30-min α-emitter, which they ascribed to 100^{250}, without claiming discovery of the element. This isotope has since been identified positively, and the 30-min half-

life confirmed. The chemical properties of fermium have been studied solely with tracer amounts, and in normal aqueous media only the (III) oxidation state appears to exist. The isotope Fm^{254} and heavier isotopes can be produced by intense neutron irradiation of lower elements such as plutonium by a process of successive neutron capture interspersed with beta decays until these mass numbers and atomic numbers are reached. Ten isotopes of fermium are known to exist. Fm^{257}, with a half-life of about 80 days, is the longest lived. Fm^{250}, with a half-life of 30 min, has been shown to be a product of decay of Element 102^{254}. It was by chemical identification of Fm^{250} that it was certain that Element 102 (nobelium) had been produced.

Fluorine — (L. and F. *fluere,* flow, or flux), F; at. wt. 18.998403 ± 0.000001; at. no. 9; m.p.−219.62°C (1 atm.); b.p. −188.14°C (1 atm); density 1.696 g/l (O°C, 1 atm.); sp. gr. of liquid 1.108 at b.p.; valence 1. In 1529, Georgius Agricola described the use of fluorspar as a flux, and as early as 1670 Schwandhard found that glass was etched when exposed to fluorspar treated with acid. Scheele and many later investigators, including Davy, Gay-Lussac, Lavoisier, and Thenard, experimented with hydrofluoric acid, some experiments ending in tragedy. The element was finally isolated in 1886 by Moisson after nearly 74 years of continuous effort. Fluorine occurs chiefly in *fluorspar* (CaF_2) and *cryolite* (Na_2AlF_6), but is rather widely distributed in other minerals. It is a member of the halogen family of elements, and obtained by electrolyzing a solution of potassium hydrogen fluoride in anhydrous hydrogen fluoride in a vessel of metal or transparent fluorspar. Modern commercial production methods are essentially variations on the procedures first used by Moisson. Fluorine is the most electronegative and reactive of all elements. It is a pale yellow, corrosive gas, which reacts with practically all organic and inorganic substances. Finely divided metals, glass, ceramics, carbon, and even water burn in fluorine with a bright flame. Until World War II, there was no commercial production of elemental fluorine. The atom bomb project and nuclear energy applications, however, made it necessary to produce large quantities. Safe handling techniques have now been developed and it is possible at present to transport liquid fluorine by the ton. Fluorine and its compounds are used in producing uranium (from the hexafluoride) and more than 100 commercial fluorochemicals, including many well-known high-temperature plastics. Hydrofluoric acid is extensively used for etching the glass of light bulbs, etc. Fluorochloro hydrocarbons are extensively used in air conditioning and refrigeration. It has been suggested that fluorine can be substituted for hydrogen wherever it occurs in organic compounds, which could lead to an astronomical number of new fluorine compounds. The presence of fluorine as a soluble fluoride in drinking water to the extent of 2 ppm may cause mottled enamel in teeth, when used by children acquiring permanent teeth; in smaller amounts, however, fluorides are said to be beneficial and used in water supplies to prevent dental cavities. Elemental fluorine is being studied as a rocket propellant as it has an exceptionally high specific impulse value. Compounds of fluorine with rare gases have now been confirmed. Fluorides of xenon, radon, and krypton are among those reported. Elemental fluorine and the fluoride ion are highly toxic. The free element has a characteristic pungent odor, detectable in concentrations as low as 20 ppb, which is below the safe working level. The recommended maximum allowable concentration for a daily 8-hr time-weighted exposure is 0.1 ppm.

Francium — (France), Fr; at. no. 87; at. wt. 223; m.p. 27°C; b.p. 677°C; valence 1. Discovered in 1939 by Mlle. Marguerite Perey of the Curie Institute, Paris. Francium, the heaviest known member of the alkali metal series, occurs as a result of an alpha disintegration of actinium. It can also be made artificially by bombarding thorium with protons. While it occurs naturally in uranium minerals, there is probably less than an ounce of francium at any time in the total crust of the earth. It has the highest equivalent weight of any element, and is the most unstable of the first 101 elements of the periodic system. Twenty isotopes of francium are recognized. The longest lived, Fr^{223}(AcK), a daughter of Ac^{227}, has a half-life of 22 min.

This is the only isotope of francium occurring in nature. Because all known isotopes of francium are highly unstable, knowledge of the chemical properties of this element comes from radiochemical techniques. No weighable quantity of the element has been prepared or isolated. The chemical properties of francium most closely resemble cesium.

Gadolinium — (*gadolinite,* a minteral named for Gadolin, a Finnish chemist), Gd; at. wt. 157.25; at. no. 64; m.p. 1313±1°C; b.p. 3266°C; sp. gr. 7.9004 (25°C); valence 3. Gadolinia, the oxide of gadolinium, was separated by Marignac in 1880 and Lecoq de Boisbaudran independently isolated the element from Mosanders "yttria" in 1886. The element was named for the mineral *gadolinite* from which this rare earth was originally obtained. Gadolinium is found in several other minerals, including *monazite* and *bastnasite,* which are of commercial importance. The element has been isolated only in recent years. With the development of ion-exchange and solvent extraction techniques, the availability and price of gadolinium and the other rare-earth metals have greatly improved. Seventeen isotopes of gadolinium are now recognized; seven occur naturally. The metal can be prepared by the reduction of the anhydrous fluoride with metallic calcium. As with other related rare-earth metals, it is silvery white, has a metallic luster, and is malleable and ductile. At room temperature, gadolinium crystallizes in the hexagonal, close-packed α form. Upon heating to 1262°C, α gadolinium transforms into the β form, which has a body-centered cubic structure. The metal is relatively stable in dry air, but in moist air it tarnishes with the formation of a loosely adhering oxide film which spalls off and exposes more surface to oxidation. The metal reacts slowly with water and is soluble in dilute acid. Gadolinium has the highest thermal neutron capture cross-section of any known element (49,000 barns). Natural gadolinium is a mixture of seven isotopes. Two of these, Gd^{155} and Gd^{157}, have excellent capture characteristics, but they are present naturally in low concentrations. As a result, gadolinium has a very fast burnout rate and has limited use as a nuclear control rod material. It has been used in making gadolinium yttrium garnets, which have microwave applications. Compounds of gadolinium are used in making phosphors for color TV tubes. The metal has unusual superconductive properties. As little as 1% gadolinium has been found to improve the workability and resistance of iron, chromium, and related alloys to high temperatures and oxidation. Gadolinium ethyl sulfate has extremely low noise characteristics and may find use in duplicating the performance of h.f. amplifiers, such as the maser. The metal is ferromagnetic. Gadolinium is unique for its high magnetic moment and for its special Curie temperature (above which ferromagnetism vanishes) lying just at room temperature. This suggests uses as a magnetic component that senses hot and cold. The price of the metal is $1.25/g or $250/lb.

Gallium — (L. *Gallia,* France), Ga; at. wt. 69.737 ± 0.006; at. no. 31; m.p. 29.78°C; b.p.2403°C; sp. gr. 5.904 (29.6°C) solid; sp. gr. 6.095 (29.8°C) liquid; valence 2 or 3. Predicted and described by Mendeleev as ekaaluminum, and discovered spectroscopically by Lecoq de Boisbaudran in 1875, who in the same year obtained the free metal by electrolysis of a solution of the hydroxide in KOH. Gallium is often found as a trace element in *diaspore, sphalerite, germanite, bauxite,* and *coal.* Some flue dusts from burning coal have been shown to contain as much as 1.5% gallium. It is the only metal, except for mercury, cesium, and rubidium, which can be liquid near room temperatures; this makes possible its use in high-temperature thermometers. It has one of the longest liquid ranges of any metal and has a low vapor pressure even at high temperatures. There is a strong tendency for gallium to supercool below its freezing point. Therefore, seeding may be necessary to initiate solidification. Ultra-pure gallium has a beautiful, silvery appearance, and the solid metal exhibits a conchoidal fracture similar to glass. The metal expands 3.1% on solidifying; therefore, it should not be stored in glass or metal containers, as they may break as the metal solidifies. Gallium wets glass or porcelain, and forms a brilliant mirror when it is painted on glass. It has found recent use in doping semiconductors and producing solid-state devices such as transistors. High-purity gallium

is attacked only slowly by mineral acids. Magnesium gallate containing divalent impurities such as Mn^{+2} is finding use in commercial ultraviolet activated powder phosphors. Gallium arsenide is capable of converting electricity directly into coherent light. Gallium readily alloys with most metals, and has been used as a component in low-melting alloys. Its toxicity appears to be of a low order, but should be handled with care until more data are forthcoming. The metal can be supplied in ultrapure form (99.99999 + %). The cost is about $1/g.

Germanium —(L. *Germania*, Germany), Ge; at. wt. 72.59; at. no. 32; m.p. 937.4°C; b.p. 2,830°C; sp. gr. 5.323 (25°C); valence 2 and 4. Predicted by Mendeleev in 1871 as ekasilicon, and discovered by Winkler in 1886. The metal is found in *argyrodite*, a sulfide of germanium and silver; in *germanite*, which contains 8% of the element; in zinc ores; in coal; and in other minerals. The element is frequently obtained commercially from flue dusts of smelters processing zinc ores, and has been recovered from the by-products of combustion of certain coals. Its presence in coal insures a large reserve of the element in the years to come. Germanium can be separated from other metals by fractional distillation of its volatile tetrachloride. The tetrachloride may then be hydrolyzed to give GeO_2; then the dioxide can be reduced with hydrogen to give the metal. Recently developed zone-refining techniques permit the production of germanium of ultra-high purity. The element is a gray-white metalloid, and in its pure state is crystalline and brittle, retaining its luster in air at room temperature. It is a very important semiconductor material. Zone-refining techniques have led to production of crystalline germanium for semiconductor use with an impurity of only one part in 10^{10}. Doped with arsenic, gallium, or other elements, it is used as a transistor element in thousands of electronic applications. Its application as a semiconductor element now provides the largest use for germanium. Germanium is also finding many other applications including use as an alloying agent, as a phosphor in fluorescent lamps, and as a catalyst. Germanium and germanium oxide are transparent to the infrared and are used in infrared spectroscopes and other optical equipment, including extremely sensitive infrared detectors. Germanium oxide's high index of refraction and dispersion has made it useful as a component of glasses used in wide-angle camera lenses and microscope objectives. The field of organogermanium chemistry is becoming increasingly important. Certain germanium compounds have a low mammalian toxicity, but a marked activity against certain bacteria, which makes them of interest as chemotherapeutic agents. The cost of germanium is about $300/kg.

Gold — (Sanskrit *Jval;* Anglo-Saxon *gold*), Au (L. *aurum*, shining dawn); at. wt. 196.9665; at. no. 79; m.p. 1064.43°C; b.p. 2807°C; sp. gr. 19.32 (20°C); valence 1 or 3. Known and highly valued from earliest times, gold is found in nature as the free metal and in tellurides; it is very widely distributed and is almost always associated with quartz or pyrite. It occurs in veins and alluvial deposits, and is often separated from rocks and other minerals by sluicing or panning operations. About two thirds of the world's gold output now comes from South Africa, and about two thirds of the total U.S. production comes from South Dakota and Nevada. The metal is recovered from its ores by cyaniding, amalgamating, and smelting processes. Refining is also frequently done by electrolysis. Gold occurs in sea water to the extent of 0.1 to 2 mg/ton, depending on the location where the sample is taken. As yet, no method has been found for recovering gold from sea water profitably. It is estimated that all the gold in the world, so far refined, could be placed in a single cube 50 ft. on a side. Of all the elements, gold in its pure state is undoubtedly the most beautiful. It is metallic, having a yellow color when in a mass, but when finely divided it may be black, ruby, or purple. The Purple of Cassius is a delicate test for auric gold. It is the most malleable and ductile metal; 1 oz of gold can be beaten out to 300 ft^2. It is a soft metal and is usually alloyed to give it more strength. It is a good conductor of heat and electricity, and is unaffected by air and most reagents. It is used in coinage and is a standard for monetary systems in many countries. It is also extensively used for jewelry, decoration, dental work, and for plating. It is used for coating cer-

tain space satellites, as it is a good reflector of infrared and is inert. Gold, like other precious metals, is measured in troy weight; when alloyed with other metals, the term *carat* is used to express the amount of gold present, 24 carats being pure gold. For many years the value of gold was set by the U.S. at $20.67 /troy ounce; in 1934 this value was fixed by law at $35.00/ troy ounce, 9/10th fine. On March 17, 1968, because of a gold crisis, a two-tiered pricing system was established whereby gold was still used to settle international accounts at the old $35.00/troy ounce price while the price of gold on the private market would be allowed to fluctuate. Since this time, the price of gold on the free market has fluctuated widely. On March 19, 1968, President Johnson signed into law a bill removing the last statutory requirement for a gold backing against U.S. currency. On August 15, 1971, President Nixon announced an embargo on U.S. gold to settle international accounts, and on May 18, 1972, U.S. monetary gold was revalued at $38/ troy ounce. In February 1973 the U.S. in effect devalued the dollar by another 10%, decreasing the dollar in relation to gold from $38.00 to $42.22 / troy ounce. On September 21, 1973, President Nixon signed a bill ratifying the action taken earlier in February; the bill also restored to private U.S. citizens the right to own gold after December 31, 1973 (private possession, except for gold in the form of jewelry, certain coins, etc., had been prohibited since 1933 when the U.S. went off the gold standard). The final version of the bill gave the President discretion to lift the ban when he determined that private ownership would not impair the monetary position during this period of instability. President Ford signed final legislation on August 14, 1974, lifting the 41-year ban, to be effective after December 31, 1974. By this date the price of gold on the free market had reached an all-time high of about $200/troy ounce. The most common gold compounds are auric chloride ($AuCl_3$) and chlorauric acid ($HAuCl_4$), the latter being used in photography for toning the silver image. Gold has 18 isotopes; Au^{198}, with a half-life of 2.7 days, is used for treating cancer and other diseases. Disodium aurothiomalate is administered intramuscularly as a treatment for arthritis. A mixture of one part nitric acid with three of hydrochloric acid is called *aqua regia* (because it dissolved gold, the King of Metals). Gold is available commercially with a purity of 99.999 + %. For many years the temperature assigned to the freezing point of gold has been 1063.0°C; this has served as a calibration point for the International Temperature Scales (ITS-27 and ITS-48) and the International Practical Temperature Scale (ITPS-48). In 1968, a new International Practical Temperature Scale (ITPS-68) was adopted, which demands that the freezing point of gold be changed to 1064.43°C. Although workers in precision temperature measurement should adopt IPTS-68 immediately, many of the scale changes are of minor significance to the routine user. IPTS-68 has defined several other fixed temperature points, among which are the boiling points of hydrogen, neon, oxygen, and sulfur, and the freezing points of zinc, silver, tin, lead, antimony, and aluminum.

Hafnium — (*Hafnia*, Latin name for Copenhagen), Hf; at. wt. 178.49; at. no. 72; m.p. 2227 ± 20°C; b.p. 4602°C; sp. gr. 13.31 (20°C); valence 4. Hafnium was thought to be present in various minerals and concentrations many years prior to its discovery, in 1923, credited to D. Coster and G. von Hevesey. On the basis of the Bohr theory, the new element was expected to be associated with zirconium. It was finally identified in *zircon* from Norway, by means of X-ray spectroscope analysis. It was named in honor of the city in which the discovery was made. Most zirconium minerals contain 1 to 5% hafnium. It was originally separated from zirconium by repeated recrystallization of the double ammonium or potassium fluorides by von Hevesey and Jantzen. Metallic hafnium was first prepared by van Arkel and deBoer by passing the vapor of the tetraiodide over a heated tungsten filament. Almost all hafnium metal now produced is made by reducing the tetrachloride with magnesium or with sodium (Kroll Process). Hafnium is a ductile metal with abrilliant silver luster. Its properties are considerably influenced by the impurities of zirconium present. Of all the elements, zirconium and hafnium are two of the most difficult to separate. Their chemistry is almost identical, how-

ever, the density of zirconium is about half that of hafnium. Very pure hafnium has been produced, with zirconium being the major impurity. Because hafnium has a good absorption cross section for thermal neutrons (almost 600 times that of zirconium), has excellent mechanical properties, and is extremely corrosion resistant, it is used for reactor control rods. Such rods are used in nuclear submarines. Hafnium has been successfully alloyed with iron, titanium, niobium, tantalum, and other metals. Hafnium carbide is the most refractory binary composition known, and the nitride is the most refractory of all known metal nitrides (m.p. 3310° C). Hafnium is used in gas-filled and incandescent lamps, and is an efficient "getter" for scavenging oxygen and nitrogen. Finely divided hafnium is pyrophoric and can ignite spontaneously in air. Care should be taken when machining the metal or when handling hot sponge hafnium. At 700°C hafnium rapidly absorbs hydrogen to form the composition $HfH_{1,86}$. Hafnium is resistant to concentrated alkalis, but at elevated temperatures reacts with oxygen, nitrogen, carbon, boron, sulfur, and silicon. Halogens react directly to form tetrahalides. Exposure to hafnium should not exceed 0.5 mg/M^3 (8-hr time-weighted average — 40-hr week). The price of the metal is in the broad range of \$50 to \$125/lb, depending on purity and quantity. The yearly demand for hafnium in the U.S. is now in excess of 175,000 lb.

Hahnium — see Element 105.

Helium — (Gr. *helios,* the sun), He; at. wt. 4.00260; at. no. 2; m.p. below −272.2°C (26atm); b.p. − 268.934°C; density 0.1785 g/1 (0°C, 1 atm); liquid density 7.62 lb/ft^3 at. b.p.; valence usually 0. Evidence of the existence of helium was first obtained by Janssen during the solar eclipse of 1868 when he detected a new line in the solar spectrum; Lockyer and Frankland suggested the name *helium* for the new element; in 1895, Ramsay discovered helium in the uranium mineral *clevite,* and it was independently discovered in clevite by the Swedish chemists Cleve and Langlet about the same time. Rutherford and Royds in 1907 demonstrated that α particles are helium nuclei. Except for hydrogen, helium is the most abundant element found throughout the universe. It has been detected spectroscopically in great abundance, especially in the hotter stars, and it is an important component in both the proton-proton reaction and the carbon cycle, which account for the energy of the sun and stars. The fusion of hydrogen into helium provides the energy of the hydrogen bomb. The helium content of the atmosphere is about 1 part in 200,000. While it is present in various radioactive minerals as a decay product, the bulk of the Free World's supply is obtained from wells in Texas, Oklahoma, and Kansas. The only helium plant in the Free World outside the U.S. is near Swift River, Saskatchewan. The cost of helium fell from \$2500/ft^3 in 1915 to 1.5c/ft^3 in 1940. The U.S. Bureau of Mines has set the price of Grade A helium at \$35/1000 ft^3. Helium has the lowest melting point of any element and has found wide use in cryogenic research, as its boiling point is close to absolute zero. Its use in the study of superconductivity is vital. Using liquid helium, Kurti and co-workers, and others, have succeeded in obtaining temperatures of a few microdegrees K by the adiabatic demagnetization of copper nuclei, starting from about 0.01 K. Five isotopes of helium are known. Liquid helium (He^4) exists in two forms: He^4 I and He^4 II, with a sharp transition point at 2.174 K (3.83 cm Hg). He^4I (above this temperature) is a normal liquid, but He^4II (below it) is unlike any other known substance. It expands on cooling; its conductivity for heat is enormous; and neither its heat conduction nor viscosity obeys normal rules. It has other peculiar properties. Helium is the only liquid that cannot be solidified by lowering the temperature. It remains liquid down to absolute zero at ordinary pressures, but it can readily be solidified by increasing the pressure. Solid He^3 and He^4 are unusual in that both can readily be changed in volume by more than 30% by application of pressure. The specific heat of helium gas is unusually high. The density of helium vapor at the normal boiling point is also very high, with the vapor expanding greatly when heated to room temperature. Containers filled with helium gas at 5 to 10° K should be treated as though they contained liquid helium due to the large increase in pressure resulting from warming the gas to room tem-

perature. While helium normally has a 0 valence, it seems to have a weak tendency to combine with certain other elements. Means of preparing helium difluoride have been studied, and species such as HeNe and the molecular ions He^{+} and He^{++} have been investigated. Helium is widely used as an inert gas shield for arc welding; as a protective gas in growing silicon and germanium crystals, and in titanium and zirconium production; as a cooling medium for nuclear reactors, and as a gas for supersonic wind tunnels. A mixture of 80% helium and 20% oxygen is used as an artificial atmosphere for divers and others working under pressure. Helium is extensively used for filling balloons as it is a much safer gas than hydrogen. While its density is almost twice that of hydrogen, it has about 98% of the lifting power of hydrogen. At sea level, 1000 ft^3 of helium lifts 68.5 lb. One of the recent largest uses for helium has been for pressuring liquid fuel rockets. A saturn booster such as used on the Apollo lunar missions requires about 13 million cubic feet of helium for a firing, plus more for checkouts.

Holmium — (L. *Holmia,* for Stockholm), Ho; at. wt. 164.9304; at no. 67; m.p. 1474°C; b.p.2695°C; sp. gr. 8.795 (25°C); valence +3. The spectral absorption bands of holmium were noticed in 1878 by the Swiss chemists Delafontaine and Soret, who announced the existence of an "Element X." Cleve, of Sweden, later independently discovered the element while working on erbia earth. The element is named after Cleve's native city. Pure holmia, the yellow oxide, was prepared by Homberg in 1911. Holmium occurs in *gadolinite, monazite,* and in other rare-earth minerals. It is commercially obtained from *monazite,* occurring in that mineral to the extent of about 0.05%. It has been isolated in pure form only in recent years. It can be separated from other rare earths by ion-exchange and solvent extraction techniques, and isolated by the reduction of its anhydrous chloride or fluoride with calcium metal. Pure holmium has a metallic to bright silver luster. It is relatively soft and malleable, and is stable in dry air at room temperature, but rapidly oxidizes in moist air and at elevated temperatures. The metal has unusual magnetic properties. Few uses have yet been found for the element. The element, as with other rare earths, seems to have a low acute toxic rating. The price of 99+% holmium metal is about $1.50/g or $350 /lb.

Hydrogen — (Gr. *hydro,* water, and *genes,* forming), H; at. wt. (natural) 1.0079; at. wt. (H^1)1.007822; at. no. 1; m.p. −259.14°C; b.p. −252.87°C; density 0.08988 g/l; density (liquid) 70.8 g/l (−253°C); density (solid) 70.6 g/l (−262°C); valence 1. Hydrogen was prepared many years before it was recognized as a distinct substance by Cavendish in 1766. It was named by Lavoisier. Hydrogen is the most abundant of all elements in the universe, and it is thought that the heavier elements were, and still are, being built from hydrogen and helium. It has been estimated that hydrogen makes up more than 90% of all the atoms or three quarters of the mass of the universe. It is found in the sun and most stars, and plays an important part in the proton-proton reaction and carbon-nitrogen cycle, which accounts for the energy of the sun and stars. It is thought that hydrogen is a major component of the planet Jupiter and that at some depth in the planet's interior the pressure is so great that solid molecular hydrogen is converted into solid metallic hydrogen. In 1973, it was reported that a group of Russian experimenters may have produced metallic hydrogen at a pressure of 2.8 Mbar. At the transition the density changed from 1.08 to 1.3 g/cm^3. Earlier, in 1972, a Livermore (California) group also reported on a similar experiment in which they observed a pressure-volume point centered at 2 Mbar. It has been predicted that metallic hydrogen may may be metastable; others have predicted it would be a superconductor at room temperature. On earth, hydrogen occurs chiefly in combination with oxygen in water, but it is also present in organic matter such as living plants, petroleum, coal, etc. It is present as the free element in the atmosphere, but only to the extent of less than 1 ppm, by volume. It is the lightest of all gases, and combines with other elements, sometimes explosively, to form compounds. Great quantities of hydrogen are required commercially for the fixation of nitrogen from the air in the Haber ammonia process and for the hydrogenation of fats and oils. It is also used in large quantities

in methanol production, in hydrodealkylation, hydrocracking, and hydrodesulfurization. It is also used as a rocket fuel, for welding, for production of hydrochloric acid, for the reduction of metallic ores, and for filling balloons. The lifting power of 1 ft^3 of hydrogen gas is about 0.076 lb at 0°C, 760 mm pressure. Production of hydrogen in the U.S. alone now amounts to about 3 billion cubic feet per year. It is prepared by the action of steam on heated carbon, by decomposition of certain hydrocarbons with heat, by the electrolysis of water, or by the displacement from acids by certain metals. It is also produced by the action of sodium or potassium hydroxide on aluminum. Liquid hydrogen is important in cryogenics and in the study of superconductivity as its melting point is only a few degrees above absolute zero. The ordinary isotope of hydrogen, ${}_1H^1$, is known as *protium.* In 1932, Urey announced the preparation of a stable isotope, deuterium (${}_1H^2$ or D) with an atomic weight of 2. Two years later an unstable isotope, tritium (${}_1H^3$), with an atomic weight of 3 was discovered. Tritium has a half-life of about 12.5 years. One atom of deuterium is found mixed in with about 6000 ordinary hydrogen atoms. Tritium atoms are also present but in much smaller proportion. Tritium is readily produced in nuclear reactors and is used in the production of the hydrogen bomb. It is also used as a radioactive agent in making luminous paints, and as a tracer. The current price of tritium, to authorized personnel, is about \$2/cr; deuterium gas is readily available, without permit, at about \$1/l. Heavy water, deuterium oxide (D_2O), which is used as a moderator to slow down neutrons, is available without permit at a cost of 6c to \$1/g, depending on quantity and purity. Quite apart from isotopes, it has been shown that hydrogen gas under ordinary conditions is a mixture of two kinds of molecules, known as *ortho-* and *para-*hydrogen, which differ from one another by the spins of their electrons and nuclei. Normal hydrogen at room temperature contains 25% of the *para* form and 75% of the *ortho* form. The *ortho* form cannot be prepared in the pure state. Since the two forms differ in energy, the physical properties also differ. The melting and boiling points of *para*-hydrogen are about 0.1°C lower than those of normal hydrogen. Consideration is being given to an entire economy based on solar- and nuclear-generated hydrogen. Located in remote regions, power plants would electrolyze sea water; the hydrogen produced would travel to distant cities by pipelines. Pollution-free hydrogen could replace natural gas, gasoline, etc., and could serve as a reducing agent in metallurgy, chemical processing, refining, etc. It could also be used to hydrolyze trash into methane and ethylene. Public acceptance, high capital investment, and the high present cost of hydrogen with respect to present fuels are but a few of the problems facing establishment of such an economy.

Indium — (from the brilliant indigo line in its spectrum), In; at. wt. 114.82; at. no. 49; m.p. 156.61°C; b.p. 2080°C; sp. gr. 7.31 (20°C); valence 1, 2, or 3. Discovered by Reich and Richter, who later isolated the metal. Indium is most frequently associated with zinc minerals, and it is from these that most commercial indium is now obtained; however, it is also found in iron, lead, and copper ores. Until 1924, a gram or so constituted the world's supply of this element in isolated form. It is probably about as abundant as silver. About 4 million troy ounces of indium are now produced annually in the Free World. Canada is presently producing more than 1,000,000 troy ounces annually. The present cost of indium is about \$10.00/troy ounce, depending on quantity and purity. It is available in ultrapure form. Indium is a very soft, silvery-white metal with a brilliant luster. The pure metal gives a high-pitched "cry" when bent. It wets glass, as does gallium. It has found application in making low-melting alloys; an alloy of 24% indium-76% gallium is liquid at room temperature. It is used in making bearing alloys, germanium transistors, rectifiers, thermistors, and photoconductors. It can be plated onto metal and evaporated onto glass, forming a mirror as good as that made with silver but with more resistance to atmospheric corrosion. There is evidence that indium has a low order of toxicity; however, care should be taken until further information is available.

Iodine — (Gr. *iodes,* violet), I; at. wt. 126.9045; at. no. 53; m.p. 113.5°C; b.p.

184.35°C; density of the gas 11.27 g/l; sp. gr. solid 4.93 (20°C); valence 1, 3, 5, or 7. Discovered by Courtois in 1811. Iodine, a halogen, occurs sparingly in the form of iodides in sea water from which it is assimilated by seaweeds, in Chilean saltpeter and nitrate-bearing earth, known as *caliche,* in brines from old sea deposits, and in brackish waters from oil and salt wells. Ultrapure iodine can be obtained from the reaction of potassium iodide with copper sulfate. Several other methods of isolating the element are known. Iodine is a bluish-black, lustrous solid, volatilizing at ordinary temperatures into a blue-violet gas with an irritating odor; it forms compounds with many elements, but is less active than the other halogens, which displace it from iodides. Iodine exhibits some metallic-like properties. It dissolves readily in chloroform, carbon tetrachloride, or carbon disulfide to form beautiful purple solutions. It is only slightly soluble in water. Iodine compounds are important in organic chemistry and very useful in medicine. Twenty-three isotopes are recognized. Only one stable isotope, I^{127}, is found in nature. The artificial radioisotope I^{131}, with a half-life of 8 days, has been used in treating the thyroid gland. The most common compounds are the iodides of sodium and potassium (KI) and the iodates (KIO_3). Lack of iodine is the cause of goiter. The iodide, and thyroxin which contains iodine, are used internally in medicine, and a solution of KI and iodine in alcohol is used for external wounds. Potassium iodide finds use in photography. The deep blue color with starch solution is characteristic of the free element. Care should be taken in handling and using iodine, as contact with the skin can cause lesions; iodine vapor is intensely irritating to the eyes and mucous membranes. The maximum allowable concentration of iodine in air should not exceed 1 mg/M^3 (8-hr time-weighted average — 40-hr).

Iridium — (L. *iris,* rainbow), Ir; at. wt. 192.22; at. no. 77; m.p. 2410°C; b.p. 4130°C; sp. gr. 22.42 (17°C); valence 3 or 4. Discovered in 1803 by Tennant in the residue left when crude platinum is dissolved by aqua regia. The name iridium is appropriate, for its salts are highly colored. Iridium, a metal of the platinum family, is white, similar to platinum, but with a slight yellowish cast. It is very hard and brittle, making it very hard to machine, form, or work. It is the most corrosion-resistant metal known, and was used in making the standard meter bar of Paris, which is a 90% platinum-10% iridium alloy. This meter bar has since been replaced as a fundamental unit of length (see under Krypton). Iridium is not attacked by any of the acids nor by aqua regia, but is attacked by molten salts, such as NaCl and NaCN. Iridium occurs uncombined in nature with platinum and other metals of this family in alluvial deposits. It is recovered as a by-product from the nickel mining industry. Iridium has found use in making crucibles and apparatus for use at high temperatures. It is also used for electrical contacts. Its principal use is as a hardening agent for platinum. With osmium, it forms an alloy which is used for tipping pens and compass bearings. The specific gravity of iridium is only very slightly lower than that of osmium, which has been generally credited as being the heaviest known element. Calculations of the densities of iridium and osmium from the space lattices give values of 22.65 and 22.61 g/cm^3, respectively. These values may be more reliable than actual physical measurements. At present, therefore, we know that either iridium or osmium is the densest known element, but the data do not yet allow selection between the two. Iridium costs about $300/ troy ounce.

Iron — (Anglo-Saxon, *iron*), Fe (L. *ferrum*); at. wt. 55.847; at. no. 26; m.p. 1535°C; b.p. 2750°C; sp. gr. 7.874 (20°C); valence 2, 3, 4, or 6. The use of iron is prehistoric. Genesis mentions that Tubal-Cain, seven generations from Adam, was "an instructor of every artificer in brass and iron." A remarkable iron pillar, dating to about A.D. 400, remains standing today in Delhi, India. This solid shaft of wrought iron is about 7¼ m high by 40 cm in diameter. Corrosion to the pillar has been minimal although it has been exposed to the weather since its erection. Iron is a relatively abundant element in the universe. It is found in the sun and many types of stars in considerable quantity. Its nuclei are very stable. Iron is found native as a principal component of a class of meteorites known as *siderites,* and is a minor constituent of the other two classes. The

core of the earth, 2150 mi in radius, is thought to be largely composed of iron with about 10% occluded hydrogen. The metal is the fourth most abundant element, by weight, making up the crust of the earth. The most common ore is *hematite* (Fe_2O_3), from which the metal is obtained by reduction with carbon. Iron is found in other widely distributed minerals such as *magnetite,* which is frequently seen as *black sands* along beaches and banks of streams. *Taconite* is becoming increasingly important as a commercial ore. Common iron is a mixture of four isotopes. Six other isotopes are known to exist. Iron is a vital constituent of plant and animal life, and appears in hemoglobin. The pure metal is not often encountered in commerce, but is usually alloyed with carbon or other metals. The pure metal is very reactive chemically, and rapidly corrodes, especially in moist air or at elevated temperatures. It has four allotropic forms, or ferrites, known as α, β, γ, and δ, with transition points at 770, 928, and 1530°C. The α form is magnetic, but when transformed into the β form, the magnetism disappears although the lattice remains unchanged. The relations of these forms are peculiar. Pig iron is an alloy containing about 3% carbon with varying amounts of S, Si, Mn, and P. It is hard, brittle, fairly fusible, and is used to produce other alloys, including steel. Wrought iron contains only a few tenths of a percent of carbon, is tough, malleable, less fusible, and has usually a "fibrous" structure. Carbon steel is an alloy of iron with carbon, with small amounts of Mn, S, P, and Si. Alloy steels are carbon steels with other additives such as nickel, chromium, vanadium, etc. Iron is the cheapest and most abundant, useful, and important of all metals.

Krypton — (Gr. *kryptos,* hidden), Kr; at. wt. 83.80; at. no. 36; m.p. −156.6°C; b.p. −152.30± 0.10°C; density 3.733 g/l (O°C); valence usually 0. Discovered in 1898 by Ramsay and Travers in the residue left after liquid air had nearly boiled away. Krypton is present in the air to the extent of about 1 ppm. The atmosphere of Mars has been found to contain 0.3 ppm of Krypton. It is one of the "noble" gases. It is characterized by its brilliant green and orange spectral lines. Naturally occurring krypton contains six stable isotopes. Fifteen other unstable isotopes are now recognized. The spectral lines of krypton are easily produced and some are very sharp. In 1960 it was internationally agreed that the fundamental unit of length, the meter, should be defined in terms of the orange-red spectral line of Kr^{86}, corresponding to the transition $5p[O_{1/2}]_1 - 6d[O_{1/2}]_1$, as follows: *1 m = 1,650,763.73 wavelengths (in vacuo) of the orange-red line of* Kr^{86}. This replaces the standard meter of Paris, which was defined in terms of a bar made of a platinum-iridium alloy. Solid krypton is a white crystalline substance with a face-centered cubic structure which is common to all the "rare gases." While krypton is generally thought of as a rare gas that normally does not combine with other elements to form compounds, it now appears that the existence of some krypton compounds is established. Krypton difluoride has been prepared in gram quantities and can be made by several methods. A higher fluoride of krypton and a salt of an oxyacid of krypton also have been reported. Molecule ions of $ArKr^+$ and KrH^+ have been identified and investigated, and evidence is provided for the formation of KrXe or $KrXe^+$. Krypton clathrates have been prepared with hydroquinone and phenol. Kr^{85} has found recent application in chemical analysis. By imbedding the isotope in various solids, *kryptonates* are formed. The activity of these kryptonates is sensitive to chemical reactions at the surface. Estimates of the concentration of reactants are therefore made possible. Krypton is used commercially with argon as a low-pressure filling gas for fluorescent lights. It is used in certain photographic flash lamps for high-speed photography. Uses thus far have been limited because of its high cost. Krypton gas presently costs about $20/l.

Kurchatovium — see Element 104.

Lanthanum — (Gr. *lanthanein,* to lie hidden), La; at. wt. 138.9055; at. no. 57; m.p. 921°C; b.p. 3457°C; sp. gr. 6.145 (25°C); valence 3. Mosander in 1839 extracted a new earth, *lanthana,* from impure cerium nitrate, and recognized the new element. Lanthanum is found in rare-earth minerals such as *cerite, monazite, allanite,* and *bastnasite.* Monazite and bastnasite are principal ores in which lanthanum occurs in percentages up to 25 and 38%, respectively.

Misch metal, used in making lighter flints, contains about 25% lanthanum. Lanthanum was isolated in relatively pure form in 1923. Ion-exchange and solvent extraction techniques have led to much easier isolation of the so-called "rare-earth" elements. The availability of lanthanum and other rare earths has improved greatly in recent years. The metal can be produced by reducing the anhydrous fluoride with calcium. Lanthanum is silvery white, malleable, ductile, and soft enough to be cut with a knife. It is one of the most reactive of the rare-earth metals. It oxidizes rapidly when exposed to air. Cold water attacks lanthanum slowly, and hot water attacks it much more rapidly. The metal reacts directly with elemental carbon, nitrogen, boron, selenium, silicon, phosphorous, sulfur, and with halogens. At 310°C, lanthanum changes from a hexagonal to a face-centered cubic structure, and at 865°C it again transforms into a body-centered cubic structure. Natural lanthanum is a mixture of two stable isotopes, $La^{138}$and La^{139}. Seventeen other radioactive isotopes are recognized. Rare-earth compounds containing lanthanum are extensively used in carbon lighting applications, especially by the motion picture industry for studio lighting and projection. This application consumes about 25% of the rare-earth compounds produced. La_2O_3 improves the alkali resistance of glass, and is used in making special optical glasses. Small amounts of lanthanum, as an additive, can be used to produce nodular cast iron. There is current interest in hydrogen sponge alloys containing lanthanum. These alloys take up to 400 times their own volume of hydrogen gas, and the process is reversible. Heat energy is released every time they do so; therefore these alloys have possibilities in energy conservation systems.Lanthanum and its compounds have a low to moderate acute toxicity rating; therefore, care should be taken in handling them. The price of 99.9% La_2O_3 is about $8/lb. The metal costs about $30/lb.

Lawrencium — (Ernest O. Lawrence, inventor of the cyclotron), Lr; at. no. 103; at. mass no. 257; valence +3(?). This member of the 5f transition elements (actinide series) was discovered in March 1961 by A. Ghiorso, T. Sikkeland, A.E. Larsh, and R.M. Latimer. A 3-μg californium target, consisting of a mixture of isotopes of mass number 249, 250, 251, and 252, was bombarded with either B^{10} or B^{11}. The electrically charged transmutation nuclei recoiled with an atmosphere of helium and were collected on a thin copper conveyor tape which was then moved to place collected atoms in front of a series of solid-state detectors. The isotope of element 103 produced in this way decayed by emitting an 8.6-MeV alpha particle with a half-life of 8 sec. In 1967, Flerov and associates of the Dubna Laboratory reported their inability to detect an alpha emitter with a half-life of 8 sec which was assigned by the Berkeley group to 103^{257}. This assignment has been changed to Lr^{258} or Lr^{259}. In 1965, the Dubna workers found a longer-lived lawrencium isotope, Lr^{256}, with a half-life of 35 sec. In 1968, Ghiorso and associates at Berkeley were able to use a few atoms of this isotope to study the oxidation behavior of lawrencium. Using solvent extraction techniques and working very rapidly, they extracted lawrencium ions from a buffered aqueous solution into an organic solvent, completing each extraction in about 30 sec. It was found that lawrencium behaves differently from dipositive nobelium and more like the tripositive elements earlier in the actinide series.

Lead — (Anglo-Saxon *lead*), Pb (L.*plumbum*); at. wt. 207.2; at. no. 82; m.p. 327.502°C; b.p. 1740°C; so.gr. 11.35 (20°C); valence 2 or 4. Long known, mentioned in Exodus. The alchemists believed lead to be the oldest metal and associated it with the planet Saturn. Native lead occurs in nature, but it is rare. Lead is obtained chiefly from *galena* (PbS) by a roasting process. *Anglesite* ($PbSO_4$), *cerrusite* ($PbC0_3$), amd *minim* ($Pb;30_4$) are other common lead minerals. Lead is a bluish-white metal of bright luster, is very soft, highly malleable, ductile, and a poor conductor of electricity. It is very resistant to corrosion; lead pipes bearing the insignia of Roman emperors, used as drains from the baths, are still in service. It is used in containers for corrosive liquids (such as in sulfuric acid chambers) and may be toughened by the addition of a small percentage of antimony or other metals. Natural lead is a mixture of four stable isotopes: Pb^{204} (1.48%),

Pb^{206} (23.6%), Pb^{207} (22.6%), and Pb^{208} (52.3%). Lead isotopes are the end products of each of the three series of naturally occurring radioactive elements: Pb^{206} for the uranium series, Pb^{207} for the actinium series, and Pb^{208} for the thorium series. Seventeen other iostopes of lead, all of which are radioactive, are recognized. Its alloys include solder, type metal, and various antifriction metals. Great quantities of lead, both as the metal and as the dioxide, are used in storage batteries. Much metal also goes into cable covering, plumbing, ammunition, and in the manufacture of lead tetraethyl, used as an antiknock compound in gasoline. The metal is very effective as a sound absorber, is used as a radiation shield around X-ray equipment and nuclear reactors, and is used to absorb vibration. White lead, the basic carbonate, sublimed white lead ($PbSO_4$), chrome yellow ($PbCrO_4$), red lead (Pb_3O_4), and other lead compounds are used extensively in paints, although in recent years the use of lead in paints has been drastically curtailed to eliminate or reduce health hazards. Lead oxide is used in producing fine "crystal glass" and "flint glass" of a high index of refraction for achromatic lenses. The nitrate and the acetate are soluble salts. Lead salts such as lead arsenate have been used as insecticides, but their use in recent years has been practically eliminated in favor of less harmful organic compounds. Care must be used in handling lead as it is a cumulative poison. Environmental concern with lead poisoning has resulted in a national program to reduce the concentration of lead in gasoline.

Lithium — (Gr. *lithos,* stone), Li; at. wt. 6.941; at. no. 3; m.p. 180.54°C; b.p. 1347°C; sp. gr. 0.534 (20 °C); valence 1. Discovered by Arfvedson in 1817. Lithium is the lightest of all metals, with a density only about half that of water. It does not occur free in nature; combined it is found in small amounts in nearly all igneous rocks and in the waters of many mineral springs. *Lepidolite, spodumene, petalite,* and *amblygonite* are the more important minerals containing it. Lithium is presently being recovered from brines of Searles Lake, in California, and from those in Nevada. Large deposits of spodumene are found in North Carolina. The metal is produced electrolytically from the fused chloride. Lithium is silvery in appearance, much like Na and K, other members of the alkali metal series. It reacts with water, but not as vigorously as sodium. Lithium imparts a beautiful crimson color to a flame, but when the metal burns strongly the flame is a dazzling white. Since World War II, the production of lithium metal and its compounds has increased greatly. Because the metal has the highest specific heat of any solid element, it has found use in heat transfer applications; however, it is corrosive and requires special handling. The metal has been used as an alloying agent, is of interest in synthesis of organic compounds, and has nuclear applications. It ranks as a leading contender as a battery anode material as it has a high electrochemical potential. Lithium is used in special glasses and ceramics. The glass for the 200-in. telescope at Mt. Palomar contains lithium as a minor ingredient. Lithium chloride is one of the most hygroscopic materials known, and it, as well as lithium bromide, is used in air conditioning and industrial drying systems. Lithium stearate is used as an all-purpose and high-temperature lubricant. Other lithium compounds are used in dry cells and storage batteries. The metal is priced at about $12/lb.

Lutetium — (Lutetia, ancient name for Paris, sometimes called *cassiopeium* by the Germans), Lu; at. wt. 174.967 ± 0.003; at. no. 71; m.p. 1663°C; b.p. 3395°C; sp. gr. 9.840 (25°C); valence 3. In 1907, Urbain described a process by which Marignac's ytterbium (1879) could be separated into the two elements, ytterbium (neoytterbium) and lutetium. These elements were identical with "aldebaranium" and "cassiopeium, " independently discovered by von Welsbach about the same time. Charles James of the University of New Hampshire also independently prepared the very pure oxide, *lutecia,* at this time. The spelling of the element was changed from *lutecium* to *lutetium* in 1949. Lutetium occurs in very small amounts in nearly all minerals containing yttrium, and is present in *monazite* to the extent of about 0.003%, which is a commercial source. The pure metal has been isolated only in recent years and is one of the most difficult to prepare. It can be prepared by the reduction of an-

hydrous $LuCl_3$ or LuF_3 by an alkali or alkaline earth metal. The metal is silvery white and relatively stable in air. While new techniques, including ion-exchange reactions, have been developed to separate the various rare-earth elements, lutetium is still the most costly of all naturally occurring rare earths. It is slightly more abundant than thulium. It is now priced at about \$20/g or \$6000/lb. Lu^{176}occurs naturally (2.6%) with Lu^{175} (97.4%). It is radioactive with a half-life of about 3×10^{10} years. Stable lutetium nuclides, which emit pure beta radiation after thermal neutron activation, can be used as catalysts in cracking, alkylation, hydrogenation, and polymerization. Virtually no other commercial uses have been found yet for lutetium, as it is one of the most costly of the rare earth metals. While lutetium, like other rare-earth metals, is thought to have a low toxicity rating, it should be handled with care until more information is available.

Magnesium — (*Magnesia,* district in Thessaly) Mg; at. wt. 24.305; at. no. 12; m.p. 648.8±0.5°C; b.p. 1090°C; sp. gr. 1.738 (20°C); valence 2. Compounds of magnesium have long been known. Black recognized magnesium as an element in 1755. It was isolated by Davy in 1808, and prepared in coherent form by Bussy in 1831. Magnesium is the eighth most abundant element in the earth's crust. It does not occur uncombined, but is found in large deposits in the form of *magnesite, dolomite,* and other minerals. The metal is now principally obtained in the U.S. by electrolysis of fused magnesium chloride derived from brines, wells, and sea water. Magnesium is a light, silvery-white, and fairly tough metal. It tarnishes slightly in air, and finely divided magnesium readily ignites upon heating in air and burns with a dazzling white flame. It is used in flashlight photography, flares, and pyrotechnics, including incendiary bombs. It is one third lighter than aluminum, and in alloys is essential for airplane and missile construction. The metal improves the mechanical, fabrication, and welding characteristics of aluminum when used as an alloying agent. Magnesium is used in producing nodular graphite in cast iron, and is used as an additive to conventional propellants. It is also used as a reducing agent in the production of pure uranium and other metals from their salts. The hydroxide (*milk of magnesia*), chloride, sulfate (*Epsom salts),* and citrate are used in medicine. Dead-burned magnesite is employed for refractory purposes such as brick and liners in furnaces and converters. Organic magnesium compounds (Grignard's reaction) are important. Magnesium is an important element in both plant and animal life. Chlorophylls are magnesium-centered porphyrins. The adult daily requirement of magnesium is about 300 mg/day, but this is affected by various factors. Great care should be taken in handling magnesium metal, especially in the finely divided state, as serious fires can occur. Water should not be used on burning magnesium or on magnesium fires.

Manganese — (L. *magnes,* magnet, from magnetic properties of pyrolusite; It. *manganese,* corrupt form of *magnesia*), Mn; at. wt. 54.9380; at. no. 25; m.p. 1244±3°C; b.p. 1962°C; sp. gr. 7.21 to 7.44, depending on allotropic form; valence 1, 2, 3, 4, 6, or 7. Recognized by Scheele, Bergman, and others as an element and isolated by Gahn in 1774 by reduction of the dioxide with carbon. Manganese minerals are widely distributed; oxides, silicates, and carbonates are the most common. The discovery of large quantities of manganese nodules on the floor of the oceans holds promise as a source of manganese. These nodules contain about 24% manganese together with many other elements in lesser abundance. Most manganese today is obtained from ores found in the U.S.S.R., Brazil, Australia, Republic of So. Africa, Gabon, and India. *Pyrolusite* (MnO_2) and *rhodochrosite* ($MnCO_3$) are among the most common manganese minerals. The metal is obtained by reduction of the oxide with sodium, magnesium, aluminum, or by electrolysis. It is gray-white, resembling iron, but is harder and very brittle. The metal is reactive chemically, and decomposes cold water slowly. Manganese is used to form many important alloys. In steel, manganese improves the rolling and forging qualities, strength, toughness, stiffness, wear resistance, hardness, and hardenability. With aluminum and antimony, especially with small amounts of copper, it forms highly ferromagnetic alloys. Manganese metal is fer-

romagnetic only after special treatment. The pure metal exists in four allotropic forms. The alpha form is stable at ordinary temperature; gamma manganese, which changes to alpha at ordinary temperatures, is said to be flexible, soft, easily cut, and capable of being bent. The dioxide (pyrolusite) is used as a depolarizer in dry cells, and is used to "decolorize" glass that is colored green by impurities of iron. Manganese by itself colors glass an amethyst color, and is responsible for the color of true amethyst. The dioxide is also used in the preparation of oxygen, chlorine, and in drying black paints. The permanganate is a powerful oxidizing agent and is used in quantitative analysis and in medicine. Manganese is widely distributed throughout the animal kingdom. It is an important trace element and may be essential for utilization of vitamin B_1. Exposure to manganese dusts, fume, and compounds (as Mn) should not exceed the ceiling value of 5 mg/M^3 for even short periods because of the toxicity of the element in larger quantity.

Mendelevium — (Dmitri Mendeleev), Md; at. wt. 256; at. no. 101; valence +2, +3. Mendelevium, the ninth transuranium element of the actinide series to be discovered, was first identified by Ghiorso, Harvey, Choppin, Thompson, and Seaborg early in 1955 as a result of the bombardment of the isotope Es^{253} with helium ions in the Berkeley 60-in. cyclotron. The isotope produced was Md^{256}, which has a half-life of 77 min. This first identification was notable in that Md^{256} was synthesized on a one-atom-at-a-time basis. Four isotopes are now recognized. Md^{258} has a half-life of 2 months. This isotope has been produced by the bombardment of an isotope of einsteinium with ions of helium. It now appears possible that eventually Md^{258} can be made so that some of its physical properties can be determined. Md^{256} has been used to elucidate some of the chemical properties of mendelevium in aqueous solution. Experiments seem to show that the element possesses a moderately stable dipositive (II) oxidation state in addition to the tripositive (III) oxidation state, which is characteristic of actinide elements.

Mercury — (Planet *Mercury*), Hg (*hydrargyrum*, liquid silver); at. wt. 200.59; at. no. 80; m.p. −38.842°C; b.p. 356.58°C; sp. gr. 13.546 (20°C); valence 1 or 2. Known to ancient Chinese and Hindus; found in Egyptian tombs of 1500 B.C. Mercury is the only common metal liquid at ordinary temperatures. It only rarely occurs free in nature. The chief ore is *cinnabar* (HgS). Spain and Italy produce about 50% of the world's supply of the metal. The commercial unit for handling mercury is the "flask," which weighs 76 lb. The metal is obtained by heating cinnabar in a current of air and by condensing the vapor. It is a heavy, silvery-white metal; a rather poor conductor of heat, as compared with other metals, and a fair conductor of electricity. It easily forms alloys with many metals, such as gold, silver, and tin, which are called *amalgams*. Its ease in amalgamating with gold is made use of in the recovery of gold from its ores. The metal is widely used in laboratory work for making thermometers, barometers, diffusion pumps, and many other instruments. It is used in making mercury-vapor lamps and advertising signs, etc. and is used in mercury switches and other electrical apparatus. Other uses are in making pesticides, mercury cells for caustic soda and chlorine production, dental preparations, antifouling paint, batteries, and catalysts. The most important salts are mercuric chloride $HgCl_2$ (corrosive sublimate — a violent poison), mercurous chloride Hg_2Cl_2 (calomel, occasionally still used in medicine), mercury fulminate ($Hg(ONC)_2$, a detonator widely used in explosives), and mercuric sulfide (HgS, vermillion, a high-grade paint pigment). Organic mercury compounds are important. It has been found that an electrical discharge causes mercury vapor to combine with neon, argon, krypton, and xenon. These products, held together with van der Waals' forces, correspond to HgNe, HgAr, HgKr, and HgXe. Mercury is a virulent poison and is readily absorbed through the respiratory tract, the gastrointestinal tract, or through unbroken skin. It acts as a cumulative poison since only small amounts of the element can be eliminated at a time by the human organism. The acceptable ceiling concentration of mercury in air has been set at 1mg/10M^3. Since Mercury is a very volatile element, dangerous levels are readily attained in air. Air saturated

with mercury vapor at 20°C contains a concentration that exceeds the toxic limit many times. The danger increases at higher temperatures. *It is therefore important that mercury be handled with care.* Containers of mercury should be securely covered and spillage should be avoided. If it is necessary to heat mercury or mercury compounds, it should be done in a well-ventilated hood. Methyl mercury is a dangerous pollutant and is now widely found in water and streams. The U.S. National Bureau of Standards has recently redetermined the triple point of mercury and found it to be −38.84168°C.

Molybdenum — (Gr. *molybdos,* lead)), Mo; at. wt. 95.94 ± 0.01; at. no. 42; m.p. 2617°C; b.p. 4612°C; sp. gr. 10.22 (20°C); valence 2, 3, 4?, 5?, or 6. Before Scheele recognized molybdenite as a distinct ore of a new element in 1778, it was confused with graphite and lead ore. The metal was prepared in an impure form in 1782 by Hjelm. Molybdenum does not occur native, but is obtained principally from *molybdenite* (MoS_2). *Wulfenite* ($PbMoO_4$), and *powellite* ($Ca(MoW)O_4$) are also minor commercial ores. Molybdenum is also recovered as a byproduct of copper and tungsten mining operations. The metal is prepared from the powder made by the hydrogen reduction of purified molybdic trioxide or ammonium molybdate. The metal is silvery white, very hard, but is softer and more ductile than tungsten. It has a high elastic modulus, and only tungsten and tantalum, of the more readily available metals, have higher melting points. It is a valuable alloying agent, as it contributes to the hardenability and toughness of quenched and tempered steels. It also improves the strength of steel at high temperatures. It is used in certain nickel-based alloys, such as the "Hastelloys,®" which are heat-resistant and corrosion-resistant to chemical solutions. Molybdenum oxidizes at elevated temperatures. The metal has found recent application as electrodes for electrically heated glass furnaces and orehearths. The metal is also used in nuclear energy applications and for missile and aircraft parts. Molybdenum is valuable as a catalyst in the refining of petroleum. It has found application as a filament material in electronic and electrical applications. Molybdenum is an essential trace element in plant nutrition. Some lands are barren for lack of this element in the soil. Molybdenum sulfide is useful as a lubricant, especially at high temperatures where oils would decompose. Almost all ultra-high strength steels with minimum yield points up to 300,000 psi(lb/in.^{2}) contain molybdenum in amounts from 0.25 to 8%. Molybdenum powder is priced at about $4/lb, and bars rolled from arc-cast ingots cost about $15/lb.

Neodumium — (Gr. *neos,* new, and *didymos,* twin), Nd; at. wt. 144.24; at. no. 60; m.p. 1021°C; b.p.3068°C; sp. gr. 6.80 and 7.007, depending on allotropic form; valence 3. In 1841, Mosander, extracted from *cerite* a new rose-colored oxide, which he believed contained a new element. He named the element *didymium,* as it was "an inseparable twin brother of lanthanum." In 1885 von Welsbach separated didymium into two new elemental components, *neodymia* and *praseodymia,* by repeated fractionation of ammonium didymium nitrate. While the free metal is in *misch metal,* long known and used as a pyrophoric alloy for lighter flints, the element was not isolated in relatively pure form until 1925. Neodymium is present in misch metal to the extent of about 18%. It is present in the minerals *monazite* and *bastnasite,* which are principal sources of rare-earth metals. The element may be obtained by separating neodymium salts from other rare earths by ion-exchange or solvent extraction techniques, and by reducing anhydrous halides such as NdF_3 with calcium metal. Other separation techniques are possible. The metal has a bright silvery metallic luster. Neodymium is one of the more reactive rare-earth metals and quickly tarnishes in air, forming an oxide that spalls off and exposes metal to oxidation. The metal, therefore, should be kept under light mineral oil or sealed in a plastic material. Neodymium exists in two allotropic forms, with a transformation from a double hexagonal to a body-centered cubic structure taking place at 860°C. Natural neodymium is a mixture of seven stable isotopes. Seven other radioactive isotopes are recognized. Didymium, of which neodymium is a component, is used for coloring glass to make welder's goggles. By itself, neodymium colors glass delicate shades ranging

from pure violet through wine-red and warm gray. Light transmitted through such glass shows unusually sharp absorption bands. The glass has been used in astronomical work to produce sharp bands by which spectral lines may be calibrated. Glass containing neodymium can be used as a laser material in place of ruby to produce coherent light. Neodymium salts are also used as a colorant for enamels. The price of the metal is about 50c/g or $120/lb. Neodymium has a low-to-moderate acute toxic rating. As with other rare earths, neodymium should be handled with care.

Neon — (Gr. *neos,* new), Ne; at. wt. 20.179; at. no. 10; m.p. −248.67°C; b.p. −246.048°C (1 atm); density of gas 0.89990 g/l (1 atm, 0°C); density of liquid at b.p. 1.207 g/cm³; valence O. Discovered by Ramsay and Travers in 1898. Neon is a rare gaseous element present in the atmosphere to the extent of 1 part in 65,000 of air. It is obtained by liquefaction of air and separated from the other gases by fractional distillation. Natural neon is a mixture of three isotopes. Five other unstable isotopes are known. It is a very inert element; however, it is said to form a compound with fluorine. It is still questionable if true compounds of neon exist, but evidence is mounting in favor of their existence. The following ions are known from optical and mass spectrometric studies: Ne^+, $(NeAr)^+$, $(NeH)^+$, and $(HeNe)^+$. Neon also forms an unstable hydrate. In a vacuum discharge tube, neon glows reddish orange. Of all the rare gases, the discharge of neon is the most intense at ordinary voltages and currents. Neon is used in making the common neon advertising signs, which accounts for its largest use. It is also used to make high-voltage indicators, lightning arrestors, wave meter tubes, and TV tubes. Neon and helium are used in making gas lasers. Liquid neon is now commercially available and is finding important application as an economical cryogenic refrigerant. It has over 40 times more refrigerating capacity per unit volume than liquid helium and more than three times that of liquid hydrogen. It is compact, inert, and is less expensive than helium when it meets refrigeration requirements. Neon has costs about $1.50/l.

Neptunium — (Planet *Neptune*), Np; at. wt. 237.0482; at. no. 93; m.p. 640± 1°C; b.p. 3902°C (est.); sp. gr. 20.25 (20°C); valence 3, 4, 5, and 6. Neptunium was the first synthetic transuranium element of the actinide series discovered; the isotope Np^{239} was produced by McMillan and Abelson in 1940 at Berkeley, California, as the result of bombarding uranium with cyclotron-produced neutrons. The isotope Np^{237} (half-life of 2.14 × 10^6 years) is currently obtained in gram quantities as a by-product from nuclear reactors in the production of plutonium. Trace quantities of the element are actually found in nature due to transmutation reactions in uranium ores produced by the neutrons which are present. Neptunium is prepared by the reduction of NpF_3 with barium or lithium vapor at about 1200°C. Neptunium metal has a silvery appearance, is chemically reactive, and exists in at least three structural modifications: α-neptunium, orthorhombic, density 20.25 gl/cm³; β-neptunium (above 280°C), tetragonal, density (313° C) 19.36 g/cm³; γ-neptunium (above 577°), cubic, density (600°C) 18.0 g/cm³. Neptunium has four ionic oxidation states in solution: Np^{+3} (pale purple), analogous to the rare earth ion Pm^{+3}, Np^{+4} (yellow green); NpO^+ (green blue); and NpO^{++} (pale pink). These latter oxygenated species are in contrast to the rare earths which exhibit only simple ions of the (II), (III), and (IV) oxidation states in aqueous solution. The element forms tri- and tetrahalides such as NpF_3, NpF_4, $NpCl_4$, $NpBr_3$, NpI_3, and oxides of various compositions such as are found in the uranium-oxygen system, including Np_3O_8 and NpO_2. Thirteen isotopes of neptunium are now recognized. The A.E.C. has Np^{237} available for sale to its licensees and for export. This isotope can be used as a component in neutron detection instruments. It is offered at a price of $225/g or $0.25/mg for quantities less than 1g.

Nickel — (Ger, *Nickel,* Satan or "Old Nick's and from *kupfernickel,* Old Nicks copper), Ni; at. wt. 58.71; at. no. 28; m.p. 1453°C; b.p. 2732°C; sp. gr. 8.902 (25° C); valence 0,1,2,3. Discovered by Cronstedt in 1751 in kupfernickel(*niccolite*). Nickel is found as a constituent in most meteorites and often serves as one of the criteria for distinguishing a meteorite from other minerals. Iron meteorites, or *sider-*

ites, may contain iron alloyed with from 5 to nearly 20% nickel. Nickel is obtained commercially from *pentlandite* and *pyrrhotite* of the Sudbury region of Ontario, a district that produces about 50% of the nickel for the Free World. Other deposits are found in New Caledonia, Australia, Cuba, Indonesia, and elsewhere. Nickel is silvery white and takes on a high polish. It is hard, malleable, ductile, somewhat ferromagnetic, and a fair conductor of heat and electricity. It belongs to the iron-cobalt group of metals and is chiefly valuable for the alloys it forms. It is extensively used for making stainless steel and other corrosion-resistant alloys such as Invar®, Monel®, Inconel®, and the Hastelloys®. Tubing made of a copper-nickel alloy is extensively used in making desalination plants for converting sea water into fresh water. Nickel is also now used extensively in coinage and in making nickel steel for armor plate and burglar-proof vaults, and is a component in Nichrome®, Permalloy®, and constantan. Nickel added to glass gives a green color. Nickel plating is often used to provide a protective coating for other metals, and finely divided nickel is a catalyst for hydrogenating vegetable oils. It is also used in ceramics, in the manufacture of Alnico magnets, and in the Edison® storage battery. The sulfate and the oxides are important compounds. Natural nickel is a mixture of five stable isotopes; seven other unstable isotopes are known.Exposure to nickel metal and soluble compounds (as Ni) should not exceed 1mg/M^3 (8-hr time-weighted average — 40-hr week). Nickel carbonyl exposure, however, should not exceed 0.007mg/M^3, and is considered to be a very toxic material. Nickel sulphide fume and dust is recognized as having carcinogenic potential.

Niobium —*(Niobe,* daughter of Tantalus), Nb; or **Columbium** (*Columbia,* name for America), Cb; at. wt. 92.9064; at. no. 41; m.p. 2468± 10°C; b.p. 4742°C; sp. gr. 8.57 (20°C); valence 2, 3, 4?, 5. Discovered in 1801 by Hatchett in an ore sent to England more than a century before by John Winthrop the Younger, first goveror of Connecticut. The metal was first prepared in 1864 by Blomstrand, who reduced the chloride by heating it in a hydrogen atmosphere. The name "niobium" was adopted by the International Union of Pure and Applied Chemistry in 1950 after 100 years of controversy. Many leading chemical societies and government organizations refer to it by this name. Most metallurgists, leading metal societies, and all but one of the leading U.S. commercial producers, however, still refer to the metal as "columbium." The element is found in *niobite* (or *columbite*), *niobite-tantalite, pyrochlore,* and *euxenite.* Large deposits of niobium have been found associated with *carbonatites* (carbon-silicate rocks), as a constituent of *pyrochlore.* Extensive ore reserves are found in Canada, Brazil, Nigeria, Zaire, and in the U.S.S.R. The metal can be isolated from tantalum, and prepared in several ways. It is a shiny, white, soft, and ductile metal, and takes on a bluish cast when exposed to air at room temperatures for a long time. The metal starts to oxidize in air at 200°C, and when processed at even moderate temperatures must be placed in a protective atmosphere. It is used as an alloying agent in carbon and alloy steels and in nonferrous metals. These alloys have improved strength and other desirable properties, and have been used in pipeline construction. The metal has a low capture cross-section for thermal neutrons. It is used in arc-welding rods for stabilized grades of stainless steel. Thousands of pounds of niobium have been used in advanced air frame systems such as were used in the Gemini space program. The element has superconductive properties; superconductive magnets have been made with Nb-Zr wire, which retains its superconductivity in strong magnetic fields. This type of application offers hope of direct large-scale generation of electric power. Sixteen isotopes of niobium are known. Niobium metal (99.5% pure) is priced at about $15/lb.

Nitrogen — (L. *nitrum,* Gr. *nitron,* native soda; genes, *forming*), N; at. wt. 14.0067; at. no. 7; m.p. −209.86 °C; b.p. −195.8°C; density 1.2506 g/l; sp. gr. liquid 0.808 (−195.8°C), solid 1.026 (−252°C); valence 3 or 5. Discovered by Daniel Rutherford in 1772, but Scheele, Cavendish, Priestley, and others about the same time studied "burnt or dephlogisticated air," as air without oxygen was then called. Nitrogen makes up 78% of the air, by volume. The at-

mosphere of Mars, by comparison, is 2.6% nitrogen. The estimated amount of this element in our atmosphere is more than 4000 billion tons. From this inexhaustible source it can be obtained by liquefaction and fractional distillation. Nitrogen molecules give the orange-red, blue-green, blue-violet, and deep violet shades to the aurora. The element is so inert that Lavoisier named it *azote,* meaning without life, yet its compounds are so active as to be most important in foods, poisons, fertilizers, and explosives. Nitrogen can be also easily prepared by heating a water solution of ammonium nitrite. Nitrogen, as a gas, is colorless, odorless, and a generally inert element. As a liquid it is also colorless and odorless, and is similar in appearance to water. Two allotropic forms of solid nitrogen exist, with the transition from the α to the β form taking place at −237 °C. When nitrogen is heated, it combines directly with magnesium, lithium, or calcium; when mixed with oxygen and subjected to electric sparks, it forms first nitric oxide (NO) and then the dioxide (NO_2); when heated under pressure with a catalyst with hydrogen, ammonia is formed (Haber process). The ammonia thus formed is of the utmost importance as it is used in fertilizers, and it can be oxidized to nitric acid (Ostwald process). The ammonia industry is the largest consumer of nitrogen. Large amounts of the gas are also used by the electronics industry, which uses the gas as a blanketing medium during production of such components as transistors, diodes, etc. Large quantities of nitrogen are used in annealing stainless steel and other steel mill products. The drug industry also uses large quantities. Nitrogen is used as a refrigerant both for the immersion freezing of food products and for transportation of foods. Liquid nitrogen is also used in missile work as a purge for components, insulators for space chambers, etc., and by the oil industry to build up great pressures in wells to force crude oil upward. Sodium and potassium nitrates are formed by the decomposition of organic matter with compounds of the metals present. In certain dry areas of the world these saltpeters are found in quantity. Ammonia, nitric acid, the nitrates, the five oxides (N_2O, NO, N_2O_3, NO_2, and N_2O_5), TNT, the cyanides, etc. are but a few of the important compounds. Nitrogen gas prices vary from 2c to \$2.75 per 100 ft^3, depending on purity, etc. Production of elemental nitrogen in the U.S. is more than 9 million short tons per year.

Nobelium — (Alfred Nobel, discoverer of dynamite), No; at. no. 102; valence +2, +3. Nobelium was unambiguously discovered and identified in April 1958 at Berkeley by A. Ghiorso, T. Sikkeland, J. R. Walton, and G.T. Seaborg, who used a new double-recoil technique. A heavy-ion linear accelerator (HILAC) was used to bombard a thin target of curium (95% Cm^{244} and 4.5% Cm^{246}) with C^{12} ions to produce 102^{254} according to the Cm^{246} (C^{12} 4*n*) reaction. Earlier in 1957 workers of the U.S., Britain, and Sweden announced the discovery of an isotope of element 102 with a 10-min half-life at 8.5 MeV, as a result of bombarding $Cm^{244}$with C^{13} nuclei. On the basis of this experiment the name *nobelium* was assigned and accepted by the Commission on Atomic Weights of the International Union of Pure and Applied Chemistry. The acceptance of the name was premature, for both Russian and American efforts now completely rule out the possibility of any isotope of element 102 having a half-life of 10 min in the vicinity of 8.5 MeV. Early work in 1957 on the search for this element, in Russia at the Kurchatov Institute, was marred by the assignment of 8.9± 0.4 MeV alpha radiation with a half-life of 2 to 40 sec, which was too indefinite to support claim to discovery. Confirmatory experiments at Berkeley in 1966 have shown the existence of 102^{254} with a 55-sec half-life, 102^{252} with a 2.3-sec half-life, and 102^{257} with a 23-sec half-life. Four other isotopes are now recognized, one of which — 102^{255} — has a half-life of 3 min. In view of the discoverer's traditional right to name an element, the Berkeley group, in 1967, suggested that the hastily given name *nobelium,* along with the symbol No, be retained.

Osmium — (Gr. *osme,* a smell), Os; at. wt. 190.2; at. no. 76; m.p. 3045± 30°C; b.p. 5027 ± 100°C; sp. gr. 22.57; valence 0 to +8, more usually +3, +4, +6, and +8. Discovered in 1803 by Tennant in the residue left when crude platinum is dissolved by *aqua regia.* Osmium occurs in *iridosmine* and in platinum-bearing

river sands of the Urals, North America, and South America. It is also found in the nickel-bearing ores of the Sudbury, Ontario region along with other platinum metals. While the quantity of platinum metals in these ores is very small, the large tonnages of nickel ores processed make commercial recovery possible. The metal is lustrous, bluish white, extremely hard, and brittle even at high temperatures. It has the highest melting point and lowest vapor pressure of the platinum group. The metal is very difficult to fabricate, but the powder can be sintered in a hydrogen atmostphere at a temperature of 2000°C. The solid metal is not affected by air at room temperature, but the powdered or spongy metal slowly gives off osmium tetroxide, which is a powerful oxidizing agent and has a strong smell. The tetroxide is highly toxic, and boils at 130° C (760 mm). Concentrations in air as low as $10^{-7}g/m^3$ can cause lung congestion, skin damage, or eye damage. Exposure to osmium tetroxide should not exceed 0.002 mg/M^3 (8-hr time-weighted average — 40-hr work week). The tetroxide has been used to detect fingerprints and to stain fatty tissue for microscope slides. The metal is almost entirely used to produce very hard alloys, with other metals of the platinum group, for fountain pen tips, instrument pivots, phonograph needles, and electrical contacts. The price of 99% pure osmium powder — the form usually supplied commercially — is about $5/g or $300 to $450/troy ounce, depending on quantity and supplier. The measured densities of iridium and osmium seem to indicate that osmium is slightly more dense than iridium, so osmium has generally been credited with being the heaviest known element. Calculations of the density from the space lattice, which may be more reliable for these elements than actual measurements, however, give a density of 22.65 for iridium compared to 22.61 for osmium. At present, therefore, we know either iridium or osmium is the heaviest element, but the data do not allow selection between the two.

Oxygen — (GR. *oxys,* sharp, acid, and *genes,* forming; acid former), O; at. wt. (natural) 15.99994; at. no. 8; m.p. −218.4°C; b.p. −182.962°C; density 1.429 g/l (O °C); sp. gr. liquid 1.14 (−182.96°C); valence 2. For many centuries, workers occasionally realized air was composed of more than one component. The behavior of oxygen and nitrogen as components of air led to the advancement of the phlogiston theory of combustion, which captured the minds of chemists for a century. Oxygen was prepared by several workers, including Bayen and Borch, but they did not know how to collect it, did not study its properties, and did not recognize it as an elementary substance. Priestley is generally credited with its discovery, although Scheele also discovered it independently. Oxygen is the third most abundant element found in the sun, and it plays a part in the carbon-nitrogen cycle, one process thought to give the sun and stars their energy. Oxygen under excited conditions is responsible for the bright red and yellow-green colors of the aurora. Oxygen, as a gaseous element, forms 21% of the atmosphere by volume from which it can be obtained by liquefaction and fractional distillation. The atmosphere of Mars contains about 0.15% oxygen. The element and its compounds make up 49.2 %, by weight, of the earth's crust. About two thirds of the human body and nine tenths of water is oxygen. In the laboratory it can be prepared by the electrolysis of water or by heating potassium chlorate with manganese dioxide as a catalyst. The gas is colorless, odorless, and tasteless. The liquid and solid forms are a pale blue color and are strongly paramagnetic. Ozone (O_3), a highly active allotropic form of oxygen, is formed by the action of an electrical discharge or ultraviolet light on oxygen. Ozone's presence in the atmosphere (amounting to the equivalent of a layer 3 mm thick at ordinary pressures and temperatures) is of vital importance in preventing harmful ultraviolet rays of the sun from reaching the earth's surface. There has been recent concern that aerosols in the atomosphere may have a detrimental effect on this ozone layer. Ozone is toxic and exposure should not exceed 0.2mg/M^3 (8-hr time-weighted average — 40-hr work week.) Undiluted ozone has a bluish color. Liquid ozone is bluish black, and solid ozone is violet-black. Oxygen is very reactive and capable of combining with most elements. It is a component of hundreds of thousands of organic compounds. It is essential for respiration of all

plants and animals and for practically all combustion. In hospitals it is frequently used to aid respiration of patients. Its atomic weight was used as a standard of comparison for each of the other elements until 1961 when the International Union of Pure and Applied Chemistry adopted carbon 12 as the new basis. Oxygen has eight isotopes. Natural oxygen is a mixture of three isotopes. Oxygen 18 occurs naturally, is stable, and is available commercially. Water (H_2O with 1.5% O^{18}) is also available. Commercial oxygen consumption in the U.S. is estimated to be 20 million short tons per year and the demand is expected to increase substantially in the next few years. Oxygen enrichment of steel blast furnaces accounts for the greatest use of the gas. Large quantities are also used in making synthesis gas for ammonia and methanol, ethylene oxide, and for oxy-acetylene welding. Air separation plants produce about 99% of the gas, electrolysis plants about 1%. The gas costs 5c/ ft^3 in small quantities, and about $15/ton.

Palladium — (named after the asteroid *Pallas,* discovered about the same time; Gr. *Pallas,* goddess of wisdom), Pd; at. wt. 106.4; at. no. 46; m.p. 1552°C; b.p. 3140°C; sp. gr. 12.02 (20°C); valence 2, 3, or 4. Discovered in 1803 by Wollaston. Palladium is found along with platinum and other metals of the platinum group in placer deposits of the U.S.S.R., South and North America, Ethiopia, and Australia. It is also found associated with the nickel-copper deposits of South Africa and Ontario. Its separation from the platinum metals depends upon the type of ore in which it is found. It is a steel-white metal, does not tarnish in air, and is the least dense and lowest melting of the platinum group of metals. When annealed, it is soft and ductile; cold working greatly increases its strength and hardness. Palladium is attacked by nitric and sulfuric acid. At room temperatures the metal has the unusual property of absorbing up to 900 times its own volume of hydrogen, possibly forming Pd_2H. It is not yet clear if this is a true compound. Hydrogen readily diffuses through heated palladium and this provides a means of purifying the gas. Finely divided palladium is a good catalyst and is used for hydrogenation and dehydrogenation reactions. It is alloyed and used in jewelry trades. White gold is an alloy of gold decolorized by the addition of palladium. Like gold, palladium can be beaten into leaf as thin as 1/250,000 in. The metal is used in dentistry, watchmaking, and in making surgical instruments and electrical contacts. The metal sells for about $60/troy ounce.

Phosphorus — (Gr. *phosphoros,* light bearing; ancient name for the planet Venus when appearing before sunrise), P; at. wt. 30.97376; at. no. 15; m.p. (white) 44.1°C; b.p. (white) 280°C; sp. gr. (white) 1.82, (red) 2.20, (black) 2.25 to 2.69; valence 3 or 5. Discovered in 1669 by Brand, who prepared it from urine. Phosphorus exists in four or more allotropic forms: white (or yellow), red, and black (or violet). White phosphorus has two modifications: α and β with a transition temperature at −3.8°C. Never found free in nature, it is widely distributed in combination with minerals. *Phosphate* rock, which contains the mineral *apatite,* an impure tri-calcium phosphate, is an important source of the element. Large deposits are found in the U.S.S.R., in Morocco, and in Florida, Tennessee, Utah, Idaho, and elsewhere. Phosphorus is an essential ingredient of all cell protoplasm, nervous tissue, and bones. Ordinary phosphorus is a waxy white solid; when pure it is colorless and transparent. It is insoluble in water, but soluble in carbon disulfide. It takes fire spontaneously in air, burning to the pentoxide. It is very poisonous, 50mg constituting an approximate fatal dose. Exposure to white phosphorus should not exceed 0.1mg/M^3 (8-hr time-weighted average — 40-hr work week). White phosphorus should be kept under water, as it is dangerously reactive in air, and it should be handled with forceps, as contact with the skin may cause severe burns. When exposed to sunlight or when heated in its own vapor to 250°C, it is converted to the red variety, which does not phosphoresce in air as does the white variety. This form does not ignite spontaneously and it is not as dangerous as white phosphorus. It should, however, be handled with care as it does convert to the white form at some temperatures and it emits highly toxic fumes of the oxides of phosphorus when heated. The red modification is fairly stable, sublimes with a vapor pressure of 1 atm at

417°C, and is used in the manufacture of safety matches, pyrotechnics, pesticides, incendiary shells, smoke bombs, tracer bullets, etc. White phosphorus may be made by several methods. By one process, tri-calcium phosphate, the essential ingredient of phosphate rock, is heated in the presence of carbon and silica in an electric furnace or fuel-fired blast furnace. Elementary phosphorus is liberated as vapor and may be collected under water. If desired, the phosphorus vapor and carbon monoxide produced by the reaction can be oxidized at once in the presence of moisture or water to produce phosphoric acid, an important compound in making super-phosphate fertilizers. In recent years, concentrated phosphoric acids, which may contain as much as 70 to 75% P_2O_5 content, have become of great importance to agriculture and farm production. World-wide demand for fertilizers has caused record phosphate production in recent years. Phosphates are used in the production of special glasses, such as those used for sodium lamps. Bone-ash, calcium phosphate, is also used to produce fine chinaware and to produce mono-calcium phosphate used in baking powder. Phosphorus is also important in the production of steels, phosphor bronze, and many other products. Trisodium phosphate is important as a cleaning agent, as a water softener, and for preventing boiler scale and corrosion of pipes and boiler tubes. Organic compounds of phosphorus are important.

Platinum — (Sp. *platina*, silver), Pt; at. wt. 195.09; at. no. 78; m.p. 1772°C; b.p. 3827 ± 100°C; sp. gr. 21.45 (20°C); valence 1?, 2, 3, or 4. Discovered in South America by Ulloa in 1735 and by Wood in 1741. The metal was used by pre-Columbian Indians. Platinum occurs native, accompanied by small quantities of iridium, osmium, palladium, ruthenium, and rhodium, all belonging to the same group of metals. These are found in the alluvial deposits of the Ural mountains, of Columbia, and of certain western American states. *Sperrylite* ($PtAs_2$), occurring with the nickel-bearing deposits of Sudbury, Ontario, is the source of a considerable amount of the metal. The large production of nickel offsets there being only one part of the platinum metals in two million parts of ore. Platinum is a beautiful silvery-white metal, when pure, and is malleable and ductile. It has a coefficient of expansion almost equal to that of soda-lime-silica glass, and is therefore used to make sealed electrodes in glass systems. The metal does not oxidize in air at any temperature, but is corroded by halogens, cyanides, sulfur, and caustic alkalis. It is insoluble in hydrochloric and nitric acid, but dissolves when they are mixed as *aqua regia*, forming chloroplatinic acid (H_2PtCl_6), an important compound. The metal is extensively used in jewelry, wire, and vessels for laboratory use, and in many valuable instruments including thermocouple elements. It is also used for electrical contacts, corrosion-resistant apparatus, and in dentistry. Platinum-cobalt alloys have magnetic properties. One such alloy made of 76.7% Pt and 23.3% Co, by weight, is an extremely powerful magnet that offers a B-H (max) almost twice that of Alnico V. Platinum resistance wires are used for constructing high-temperature electric furnaces. The metal is used for coating missile nose cones, jet engine fuel nozzles, etc., which must perform reliably for long periods of time at high temperatures. The metal, like palladium, absorbs large volumes of hydrogen, retaining it at ordinary temperatures but giving it up at red heat. In the finely divided state platinum is an excellent catalyst, having long been used in the contact process for producing sulfuric acid. It is also used as a catalyst in cracking petroleum products. There is also much current interest in the use of platinum as a catalyst in fuel cells and in antipollution devices for automobiles. Platinum anodes are extensively used in cathodic protection systems for large ships and ocean-going vessels, pipelines, steel piers, etc. Fine platinum wire will glow red hot when placed in the vapor of methyl alcohol. It acts here as a catalyst, converting the alcohol to formaldehyde. This phenomenon has been used commercially to produce cigarette lighters and hand warmers. Hydrogen and oxygen explode in the presence of platinum. The price of platinum has varied widely; more than a century ago it was used to adulterate gold. It was nearly eight times as valuable as gold in 1920; the present price is about $180/troy ounce.

Plutonium — (Planet *pluto*), Pu; at. no. 94;

isotopic mass Pu^{239} 239.13 (physical scale); sp. gr. (α modification) 19.84 (25°C); m.p. 641°C; b.p. 3232°C; valence 3, 4, 5, or 6. Plutonium was the second transuranium element of the actinide series to be discovered. The isotope Pu^{238} was produced in 1940 by Seaborg, McMillan, Kennedy, and Wahl by deuteron bombardment of uranium in the 60-in. cyclotron at Berkeley, California. Plutonium also exists in trace quantities in naturally occurring uranium ores. It is formed in much the same manner as neptunium, by irradiation of natural uranium with the neutrons which are present. By far of greatest importance is the isotope Pu^{239}, with a half-life of 24,360 years, produced in extensive quantities in nuclear reactors from natural uranium:

$$U^{238}(n,\gamma)U^{239} \xrightarrow{\beta} Np^{239} \xrightarrow{\beta} Pu^{239}$$

Fifteen isotopes of plutonium are known. Plutonium has assumed the position of dominant importance among the transuranium elements because of its successful use as an explosive ingredient in nuclear weapons and the place which it holds as a key material in the development of industrial use of nuclear power. One kilogram is equivalent to about 22 million kilowatt hours of heat energy. The complete detonation of a kilogram of plutonium produces an explosion equal to about 20,000 tons of chemical explosive. Its importance depends on the nuclear property of being readily fissionable with neutrons and its availability in quantity. The world's nuclear-power reactors are now producing about 20,000 kg of plutonium/yr. By 1982 it is estimated that about 300,000 kg will have been accumulated. The various nuclear applications of plutonium are well known. Pu^{238} has been used in the Apollo lunar missions to power seismic and other equipment on the lunar surface. As with neptunium and uranium, plutonium metal can be prepared by reduction of the trifluoride with alkaline-earth metals. The metal has a silvery appearance and takes on a yellow tarnish when slightly oxidized. It is chemically reactive. A relatively large piece of plutonium is warm to the touch because of the energy given off in alpha decay. Larger pieces will produce enough heat to boil water. The metal readily dissolves in concentrated hydrochloric acid, hydroiodic acid, or perchloric acid with formation of the Pu^{+3} ion. The metal exhibits six allotropic modifications having various crystalline structures. The densities of these vary from 16.00 to 19.86 g/cm³. Plutonium also exhibits four ionic valence states in aqueous solutions: Pu^{+3} (blue lavender), Pu^{+4} (yellow brown), PuO^{+} (pink?), and PuO^{+2} (pink orange). The ion PuO^{+} is unstable in aqueous solutions, disproportionating into Pu^{+4} and PuO^{+2} the Pu^{+4} thus formed, however, oxidizes the PuO^{+} into PuO^{+2}, itself being reduced to Pu^{+3}, giving finally Pu^{+3} and PuO^{+2}. Plutonium forms binary compounds with oxygen: PuO, PuO_2, and intermediate oxides of variable composition; with the halides: PuF_3, PuF_4, $PuCl_3$, $PuBr_3$, PuI_3; with carbon, nitrogen, and silicon: PuC, PuN, $PuSi_2$. Oxyhalides are also well known: PuOCl, PuOBr, PuOI. Because of the high rate of emission of alpha particles and the element being specifically absorbed by bone marrow, plutonium, as well as all of the other transuranium elements except neptunium, are radiological poisons and must be handled with very special equipment and precautions. Plutonium is a very dangerous radiological hazard. Precautions must also be taken to prevent the unintentional formation of a critical mass. Plutonium in liquid solution is more likely to become critical than solid plutonium. The shape of the mass must also be considered where criticality is concerned. Plutonium-238 is available from the A.E.C. at a cost of about $700/g (80 to 89% enriched.)

Polonium — (Poland, native country of Mme. Curie), Po; at. mass. (~210); at. no. 84; m.p. 254°C; b.p. 962°C; sp. gr. (alpha modification) 9.32; valence −2, 0, +2, +3 (?), +4, and +6. Polonium was the first element discovered by Mme. Curie in 1898, while seeking the cause of radioactivity of pitchblende from Joachimsthal, Bohemia. The electroscope showed it separating with bismuth. Polonium is also called Radium F. Polonium is a very rare natural element. Uranium ores contain only about 100μg of the element per ton. Its abundance is only about 0.2% of that of radium. In 1934 it was found that when natural bismuth (Bi^{209}) was bombarded by neutrons, Bi^{210}, the parent of polonium, was obtained. Milligram

amounts of polonium may now be prepared this way, by using the high neutron fluxes of nuclear reactors. Polonium-210 is a low-melting, fairly volatile metal, 50% of which is vaporized in air in 45 hr at 55°C. It is an alpha emitter with a half-life of 138.39 days. A milligram emits as many alpha particles as 5 g of radium. The energy released by its decay is so large (27.5 cal /Ci/day or 140 W/g) that a capsule containing about half a gram reaches a temperature above 500°C. The capsule also presents a contact gamma-ray dose rate of 1.2R/hr. A few curies of polonium exhibit a blue glow, caused by excitation of the surrounding gas. Because almost all alpha radiation is stopped within the solid source and its container, giving up its energy, polonium has attracted attention for uses as a lightweight heat source for thermoelectric power in space satellites. Polonium has more isotopes than any other element. Twenty-seven isotopes of polonium are known, with atomic masses ranging from 192 to 218. Polonium-210 is the most readily available. Isotopes of mass 209 (half-life 103 years) and mass 208 (half-life 2.9 years) can be prepared by alpha, proton, or deuteron bombardment of lead or bismuth in a cyclotron, but these are expensive to produce. Metallic polonium has been prepared from polonium hydroxide and some other polonium compounds in the presence of concentrated aqueous or anhydrous liquid ammonia. Two allotropic modifications are known to exist. Polonium is readily dissolved in dilute acids, but is only slightly soluble in alkalis. Polonium salts of organic acids char rapidly; halide ammines are reduced to the metal. Polonium can be mixed or alloyed with beryllium to provide a source of neutrons. It has been used in devices for eliminating static charges in textile mills, etc.; however, beta sources are more commonly used and are less dangerous. It is also used on brushes for removing dust from photographic films. The polonium for these is carefully sealed and controlled, minimizing hazards to the user. Polonium-210 is very dangerous to handle in even milligram or microgram amounts, and special equipment and strict control is necessary. Damage arises from the complete absorption of the energy of the alpha particle into tissue. The maximum permissible body burden for ingested polonium is only 0.03μCi, which represents a particle weighing only 6.8×10^{-12}g. Weight for weight it is about 2.5×10^{11} times as toxic as hydrocyanic acid. The maximum allowable concentration for soluble polonium compounds in air is about 2×10^{-11} μCi/cm^3. Polonium is available commercially on special order with an A.E.C. permit from the Oak Ridge National Laboratory.

Potassium — (English, *potash* — pot ashes; L. *kalium,* Arab. *qali,* alkali), K; at. wt. 39.0983±0.0003; at. no. 19; m.p. 63.65°C; b.p.774°C; sp. gr. 0.862 (20°C); valence 1. Discovered in 1807 by Davy, who obtained it from caustic potash (KOH); this was the first metal isolated by electrolysis. The metal is the seventh most abundant and makes up about 2.4% by weight of the earth's crust. Most potassium minerals are insoluble and the metal is obtained from them only with great difficulty. Certain minerals, however, such as *sylvite, carnallite, langbeinite,* and *polyhalite* are found in ancient lake and sea beds and form rather extensive deposits from which potassium and its salts can readily be obtained. Potash is mined in Germany, New Mexico, California, Utah, and elsewhere. Large deposits of potash, found at a depth of some 3000 ft in Saskatchewan, promise to be important in coming years. Potassium is also found in the ocean, but is present only in relatively small amounts, compared to sodium. The greatest demand for potash has been in its use for fertilizers. Potassium is an essential constituent for plant growth and it is found in most soils. Potassium is never found free in nature, but is obtained by electrolysis of the hydroxide, much in the same manner as prepared by Davy. Thermal methods also are commonly used to produce potassium (such as by reduction of potassium compounds with CaC_2, C, Si, or Na). It is one of the most reactive and electropositive of metals; except for lithium, it is the lightest known metal. It is soft, easily cut with a knife, and is silvery in appearance immediately after a fresh surface is exposed. It rapidly oxidizes in air and must be preserved in a mineral oil such as kerosene. As with other metals of the alkali group, it decomposes in water with the evolution of hydrogen. It catches

fire spontaneously on water. Potassium and its salts impart a violet color to flames. Nine isotopes of potassium are known. Ordinary potassium is composed of three isotopes, one of which is K^{40} (.00118%), a radioactive isotope with a half-life of 1.28×10^9 years. The radioactivity presents no appreciable hazard. An alloy of sodium and potassium (NaK) is used as a heat-transfer medium. Many potassium salts are of utmost importance, including the hydroxide, nitrate, carbonate, chloride, chlorate, bromide, iodide, cyanide, sulfate, chromate, and dichromate. Metallic potassium is available commercially for about $15/oz in small quantities.

Praseodymium — (Gr. *prasios,* green, and *didymos,* twin), Pr; at. wt. 140.9077; at. no. 59; m.p. 931°C; b.p. 3512 °C; sp. gr. (α) 6.773 (β) 6.64; valence 3 or 4. In 1841 Mosander extracted the rare earth *didymia* from *lanthana;* in 1879, Lecoq de Boisbaudran isolated a new earth, *samaria,* from didymia obtained from the mineral *samarskite.* Six years later, in 1885, von Welsbach separated didymia into two other earths, *praseodymia* and *neodymia,* which gave salts of different colors. As with other rare earths, compounds of these elements in solution have distinctive sharp spectral absorption bands or lines, some of which are only a few Angstroms wide. The element occurs along with other rare-earth elements in a variety of minerals. *Monazite* and *bastnasite* are the two principal commercial sources of the rare-earth metals. Ion-exchange and solvent extraction techniques have led to much easier isolation of the rare earths and the cost has dropped greatly in the past few years. Praseodymium can be prepared by several methods, such as by calcium reduction of the anhydrous chloride of fluoride. Misch metal, used in making cigarette lighters, contains about 5% praseodymium metal. Praseodymium is soft, silvery, malleable, and ductile. It was prepared in relatively pure form in 1931. It is somewhat more resistant to corrosion in air than europium, lanthanum, cerium, or neodymium, but it does develop a green oxide coating that spalls off when exposed to air. As with other rare-earth metals it should be kept under a light mineral oil or sealed in plastic. The rare-earth oxides, including Pr_2O_3, are among the most refractory substances known. Along with other rare earths, it is widely used as a core material for carbon arcs used by the motion picture industry for studio lighting and projection. Salts of praseodymium are used to color glasses and enamels; when mixed with certain other materials, praseodymium produces an intense and unusually clean yellow color in glass. Didymium glass, of which praseodymium is a component, is a colorant for welder's goggles. The metal (99 + % pure) is priced at about $1.00/g.

Promethium — (*Prometheus,* who, according to mythology, stole fire from heaven), Pm; at. no. 61; m.p. 1168± 6°C; b.p. 2460°C; sp. gr. 7.22±0.02 (25°C); valence 3. In 1902 Branner predicted the existence of an element between neodymium and samarium, and this was confirmed by Moseley in 1914. In 1941, workers at Ohio State University irradiated neodymium and praseodymium with neutrons, deuterons, and alpha particles, resp., and produced several new radioactivities, which most likely were those of element 61. Wu and Segre, and Bethe, in 1942, confirmed the formation; however, chemical proof of the production of element 61 was lacking because of the difficulty in separating the rare earths from each other at that time. In 1945, Marinsky, Glendenin, and Coryell made the first chemical identification by use of ion-exchange chromatography. Their work was done by fission of uranium and by neutron bombardment of neodymium. Searches for the element on earth have been fruitless, and it now appears that promethium is completely missing from the earth's crust. Promethium, however, has been identified in the spectrum of the star HR^{465} in Andromeda. This element is being formed recently near the star's surface, for no known isotope of promethium has a half-life longer than 17.7 years. Thirteen isotopes of promethium, with atomic masses from 141 to 154, are now known. Promethium — 147, with a half- life of 2.5 years, is the most generally useful. Promethium — 145 is the longest lived, and has a specific activity of 940 Ci /g. It is a soft beta emitter; although no gamma rays are emitted, X-radiation can be generated when beta particles impinge on elements of a high atomic number, and great

care must be taken in handling it. Promethium salts luminesce in the dark with a pale blue or greenish glow, due to their high radioactivity. Ion-exchange methods led to the preparation of about 10 g of promethium from atomic reactor fuel processing wastes in early 1963. Little is yet generally known about the properties of metallic promethium. Two or more allotropic modifications are thought to exist. The element has applications as a beta source for thickness gages, and it can be absorbed by a phosphor to produce light. Light produced in this manner can be used for signs or signals that require dependable operation; it can be used as a nuclear-powered battery by capturing light in photocells which convert it into electric current. Such a battery, using Pm^{147}, would have a useful life of about 5 years. Promethium shows promise as a portable X-ray unit, and it may become useful as a heat source to provide auxiliary power for space probes and satellites. More than 30 promethium compounds have been prepared. Most are colored. Promethium—147 is available to A.E.C. licensees at a cost of about 50c/Ci.

Protactinium — (Gr. *protos,* first), Pa; at. wt. 231.0359; at. no. 91; m.p.< 1600°C; b.p. . . . ; sp. gr. 15.37 (calc.); valence 4 or 5. The first isotope of element 91 to be discovered was Pa^{234m}, also known as UX_2, a short-lived member of the naturally occurring U^{238} decay series. It was identified by K. Fajans and O. H. Gohring in 1913 and they named the new element *brevium.* When the longer-lived isotope Pa^{231} was identified by Hahn and Meitner in 1918, the name protoactinium was adopted as being more consistent with the characteristics of the most abundant isotope. Soddy, Cranston, and Fleck were also active in this work. The name *protoactinium* was shortened to *protactinium* in 1949. In 1927, Grosse prepared 2 mg of a white powder, which was shown to be Pa_2O_5. Later, in 1934, from 0.1 g of pure Pa_2O_5 he isolated the element by two methods, one of which was by converting the oxide to an iodide and "cracking" it in a high vacuum by an electrically heated filament by the reaction

$$2PaI_5 \rightarrow 2Pa + 5I_2$$

Protactinium has a bright metallic luster which it retains for some time in air. The element occurs in *pitchblende* to the extent of about 1 part Pa^{231} to 10 million of ore. Ores from Zaire have about 3 ppm. Protactinium has 13 isotopes, the most common of which is Pr^{231} with a half-life of 32,500 years. A number of protactinium compounds are known, some of which are colored. The element is superconductive below 1.4 K. An indirect measurement indicates that protactinium has a vapor pressure of 5.1×10^{-5} at 1927°C. The element is a dangerous toxic material and requires precautions similar to those used when handling plutonium. In 1959 and 1961, it was announced that the Great Britain Atomic Energy Authority extracted by a 12-stage process 125 g of 99.9 % protactinium, the world's only stock of the metal for many years to come. The extraction was made from 60 tons of waste material at a cost of about $500,000. Protactinium is one of the rarest and most expensive naturally occurring elements. It was reported that this stock was being distributed to laboratories around the world at a cost of about $2800/g. The element is an alpha emitter (5.0 MeV), and is a radiological hazard similar to polonium.

Radium — (L. *radius,* ray), Ra; at. wt. 226.0254; at. no. 88; m.p. 700°C; b.p. 1140°C; sp. gr. 5?; valence 2. Radium was discovered in 1898 by M. and Mme. Curie in the *pitchblende* or *uraninite* of North Bohemia, in which it occurs. There is about 1 g of radium in 7 tons of pitchblende. The element was isolated in 1911 by Mme. Curie and Debierne by the electrolysis of a solution of pure radium chloride, employing a mercury cathode; on distillation in an atmosphere of hydrogen this amalgam yielded the pure metal. Originally, radium was obtained from the rich pitchblende ore found at Joachimsthal, Bohemia. The *carnotite* sands of Colorado furnish some radium, but richer ores are found in the Democractic Republic of Zaire and the Great Bear Lake region of Canada. Radium is present in all uranium minerals, and could be extracted, if desired, from the extensive wastes of uranium processing. Large uranium deposits are located in Ontario, New Mexico, Utah, Australia, and elsewhere. Rad-

ium is obtained commercially as the bromide or chloride; it is doubtful if any appreciable stock of the isolated element now exists. The pure metal is brilliant white when freshly prepared, but blackens on exposure to air, probably due to formation of the nitride. It exhibits luminescence, as do its salts; it decomposes in water and is somewhat more volatile than barium. It is a member of the alkaline-earth group of metals. Radium imparts a carmine red color to a flame. Radium emits alpha, beta, and gamma rays and when mixed with beryllium produces neutrons. One gram of Ra^{226} undergoes 3.7×10^{10} disintegrations per sec. The *curie (Ci)* is defined as that amount of radioactivity which has the same disintegration rate as 1 g of Ra^{226}. Sixteen isotopes are now known; radium 226, the common isotope, has a half-life of 1620 years. One gram of radium produces about 0.0001 ml(stp) of emanation, or radon gas, per day. This is pumped from the radium and sealed in minute tubes, which are used in the treatment of cancer and other diseases. One gram of radium yields about 1000 cal per year. Radium is used in producing self-luminous paints, neutron sources, and in medicine for the treatment of disease. Some of the more recently discovered radioisotopes, such as Co^{60}, are now being used in place of radium. Some of these sources are much more powerful, and others are safer to use. Radium loses about 1% of its activity in 25 years, being transformed into elements of lower atomic weight. Lead is a final product of disintegration. The study of radium has greatly altered our ideas of the structure of the atom. Radium is a radiological hazard. (Stored radium should be ventilated to prevent build-up of radon.) Inhalation, injection, or body exposure to radium can cause cancer and other body disorders. The maximum permissible burden in the total body for Ra^{226} is 0.2μCi (microcuries).

Radon — (from *radium;* called *niton* at first, L. *nitens,* shining), Rn; at. wt. (~ 222); at. no. 86; m.p. − 71°C; b.p. −61.8°C; density of gas 9.73 g/l; sp. gr. liquid 4.4 at −62°C, solid 4; valence usually 0. The element was discovered in 1900 by Dorn, who called it *radium emanation.* In 1908 Ramsay and Gray, who named it *niton,* isolated the element and determined its density, finding it to be the heaviest known gas. It is essentially inert and occupies the last place in the zero group of gases in the Periodic Table. Since 1923, it has been called radon. Twenty isotopes are known. Radon—222, coming from radium, has a half-life of 3.823 days and is an alpha emitter; radon—220, emanating naturally from thorium and called *thoron,* has a half-life of 54.5 sec and is also an alpha emitter. Radon—219 emanates from actinium and is called *actinon.* It has a half-life of 3.92 sec and is also an alpha emitter. It is estimated that every square mile of soil to a depth of 6 in. contains about 1 g of radium, which releases radon in tiny amounts to the atmosphere. Radon is present in some spring waters, such as those at Hot Springs, Arkansas. On the average, one part of radon is present to 1 sextillion parts of air. At ordinary temperatures radon is a colorless gas; when cooled below the freezing point, radon exhibits a brilliant phosphorescence which becomes yellow as the temperature is lowered and orange-red at the temperature of liquid air. It has been reported that fluorine reacts with radon, forming radon fluoride. Radon clathrates have also been reported. Radon is still produced for therapeutic use by a few hospitals by pumping it from a radium source and sealing it in minute tubes, called seeds or needles, for application to patients. This practice has now been largely discontinued as hospitals can order the seeds directly from suppliers, who make up the seeds with the desired activity for the day of use. Radon is available at a cost of about $4/mCi. Care must be taken in handling radon, as with other radioactive materials. The main hazard is from inhalation of the element and its solid daughters, which are collected on dust in the air. The maximum permissible concentration of Rn^{222} in air has been set at 3×10^{-8} μ Ci/cc (lung) for an 8-hr day, 40-hr work week. Good ventilation should be provided where radium, thorium, or actinium is stored to prevent build-up of this element. Radon build-up is also a health consideration in uranium mines.

Rhenium — (L. *Rhenus,* Rhine), Re; at. wt. 186.2; at. no. 75; m.p. 3180°C; b.p. 5627°C (est.); sp. gr. 21.02 (20°C); valence −1, +1, 2, 3, 4, 5, 6, 7. Discovery of rhenium is generally attributed to Noddack, Tacke, and Berg, who

announced in 1925 they had detected the element in platinum ores and *columbite*. They also found the element in *gadolinite* and *molybdenite*. By working up 660 kg of molybdenite they were able in 1928 to extract 1 g of rhenium. The price in 1928 was $ 10,000/g. Rhenium does not occur free in nature or as a compound in a distinct mineral species. It is, however, widely spread throughout the earth's crust to the extent of about 0.001 ppm. Commercial rhenium in the U.S. today is obtained from molybdenite roaster-flue dusts obtained from copper-sulfide ores mined in the vicinity of Miami, Arizona, and elsewhere in Arizona and Utah. Some molybdenites contain from 0.002 to 0.2% rhenium. More than 120,000 troy ounces of rhenium are now being produced yearly in the United States. The total estimated Free World reserve of rhenium metal is 100 tons. Natural rhenium is a mixture of two stable isotopes. Sixteen other unstable iostopes are recognized. Rhenium metal is prepared by reducing ammonium perrhenate with hydrogen at elevated temperatures. The element is silvery white with a metallic luster; its density is exceeded only by that of platinum, iridium, and osmium, and its melting point is exceeded only by that of tungsten and carbon. It has other useful properties. The usual commercial form of the element is a powder, but it can be consolidated by pressing and resistance-sintering in a vacuum or hydrogen atmosphere. This produces a compact shape in excess of 90% of the density of the metal. Annealed rhenium is very ductile, and can be bent, coiled, or rolled. Rhenium is used as an additive to tungsten and molybdenum-based alloys to impart useful properties. It is widely used for filaments for mass spectrographs and ion gages. Rhenium-molybdenum alloys are superconductive at 10 K. Rhenium is also used as an electrical contact material as it has good wear resistance and withstands arc corrosion. Thermocouples made of Re-W are used for measuring temperatures up to 2200°C, and rhenium wire is used in photoflash lamps for photography. Rhenium catalysts are exceptionally resistant to poisoning from nitrogen, sulfur, and phosphorus, and are used for hydrogenation of fine chemicals, hydrocracking, reforming, and the disproportionation of olefins. Rhenium costs about $50/troy ounce. Little is known of its toxicity; therefore, it should be handled with care until more data are available.

Rhodium — (Gr. *rhodon*, rose), Rh; at. wt. 102.9055; at. no. 45; m.p. 1966 ± 3°C; b.p. 3727 ±100°C; sp. gr. 12.41 (20°C); valence 2, 3, 4, 5, and 6. Wollaston discovered rhodium in 1803-4 in crude platinum ore he presumably obtained from South America. Rhodium occurs native with other platinum metals in river sands of the Urals and in North and South America. It is also found with other platinum metals in the copper-nickel sulfide ores of the Sudbury, Ontario region. Although the quantity occurring here is very small, the large tonnages of nickel processed make the recovery commercially feasible. The annual world production of rhodium is only 2 or 3 tons. The metal is silvery white and at red heat slowly changes in air to the sesquioxide. At higher temperatures it converts back to the element. Rhodium has a higher melting point and lower density than platinum. Its major use is as an alloying agent to harden platinum and palladium. Such alloys are used for furnace windings, thermocouple elements, bushings for glass fiber production, electrodes for aircraft spark plugs, and laboratory crucibles. It is useful as an electrical contact material as it has a low electrical resistance, a low and stable contact resistance, and is highly resistant to corrosion. Plated rhodium, produced by electroplating or evaporation, is exceptionally hard and is used for optical instruments. It has a high reflectance and is hard and durable. Rhodium is also used for jewelry, for decoration, and as a catalyst. Exposure to rhodium (metal fume and dust, as Rh) should not exceed 0.1mg/M^3 (8-hr. time-weighted average, 40-hr wk.). Soluble salts should not exceed 0.001mg/M^3. Rhodium costs about $450/troy ounce.

Rubidium — (L. *rubidius*, deepest red), Rb; at. wt. 85.4678; at. no. 37; m.p. 38.89°C; b.p. 688°C; sp. gr. (solid) 1.532(20°C), (liquid) 1.475 (39°C); valence 1, 2, 3, 4. Discovered in 1861 by Bunsen and Kirchoff in the mineral *lepidolite* by use of the spectroscope. The element is much more abundant than was thought several years ago. It is now considered to be the 16th most abundant element in the earth's

crust. Rubidium occurs in *pollucite, carnallite, leucite,* and *zinnwaldite,* which contains traces up to 1%, in the form of the oxide. It is found in lepidolite to the extent of about 1.5%, and is recovered commercially from this source. Potassium minerals, such as those found at Searles Lake, California, and potassium chloride recovered from brines in Michigan also contain the element and are commercial sources. It is also found along with cesium in the extensive deposits of *pollucite* at Bernic Lake, Manitoba. Rubidium can be liquid at room temperature. It is a soft, silvery-white metallic element of the alkali group and is the second most electropositive and alkaline element. It ignites spontaneously in air and reacts violently in water, setting fire to the liberated hydrogen. As with other alkali metals, it forms amalgams with mercury and it alloys with gold, cesium, sodium, and potassium. It colors a flame yellowish violet. Rubidium metal can be prepared by reducing rubidium chloride with calcium, and by a number of other methods. It must be kept under a dry mineral oil or in a vacuum or inert atmosphere. Seventeen isotopes of rubidium are known. Naturally occurring rubidium is made of two isotopes, Rb^{85} and Rb^{87}. Rubidium—87 is present to the extent of 27.85% in natural rubidium and is a beta emitter with a half-life of 5×10^{11} years. Ordinary rubidium is sufficiently radioactive to expose a photographic film in about 30 to 60 days. Rubidium forms four oxides: Rb_2O, Rb_2O_2, Rb_2O_3, Rb_2O_4. Because rubidium can be easily ionized, it is being considered for use in "ion engines" for space vehicles; however, cesium is somethwat more efficient for this purpose. It is also proposed for use as a working fluid for vapor turbines and for use in a thermoelectric generator using the magnetohydrodynamic principle where rubidium ions are formed by heat at high temperature and passed through a magnetic field. These conduct electricity and act like an armature of a generator and cause electricity to be generated. Rubidium is used as a getter in vacuum tubes and as a photocell component. It has been used in making special glasses. $RbAg_4I_5$ is important, as it has the highest room conductivity of any known ionic crystal. At 20°C its conductivity is about the same as dilute sulfuric acid. This suggests use in thin film batteries and other applications. The present cost in small quanitites is about $10/g (99.9%), or $300/lb.

Ruthenium — (L. *Ruthenia,* Russia), Ru; at. wt. 101.07; at. no. 44, m.p. 2310°C; b.p. 3900°C; sp. gr. 12.41 (20°C); valence 0,1,2,3,4,5,6,7,8. Berzelius and Osann in 1827 examined the residues left after dissolving crude platinum from the Ural mountains in *aqua regia.* While Berzelius found no unusual metals, Osann thought he found three new metals, one of which he named ruthenium. In 1844 Klaus, generally recognized as the discoverer, showed that Osann's ruthenium oxide was very impure and that it contained a new metal. Klaus obtained 6 g of ruthenium from the portion of crude platinum that is insoluble in *aqua regia.* A member of the platinum group, ruthenium occurs native with other members of the group in ores found in the Ural mountains and in North and South America. It is also found along with other platinum metals in small but commercial quantities in *pentlandite* of the Sudbury, Ontario, nickel-mining region, and in *pyroxinite* deposits of South Africa. The metal is isolated commercially by a complex chemical process, the final stage of which is the hydrogen reduction of ammonium ruthenium chloride, which yields a powder. The powder is consolidated by powder metallurgy techniques or by argon-arc welding. Ruthenium is a hard, white metal and has four crystal modifications. It does not tarnish at room temperatures, but oxidizes in air at about 800°C. The metal is not attacked by hot or cold acids or *aqua regia,* but when potassium chlorate is added to the solution, it oxidizes explosively. It is attacked by halogens, hydroxides, etc. Ruthenium can be plated by electrodeposition or by thermal decomposition methods. The metal is one of the most effective hardeners for platinum and palladium, and is alloyed with these metals to make electrical contacts for severe wear resistance. A ruthenium-molybdenum alloy is said to be superconductive at 10.6 K. The corrosion resistance of titanium is improved a hundredfold by addition of 0.1% ruthenium. It is a versatile catalyst. Compounds in at least eight oxidation states have been found, but of these, the +2,

+3, and +4 states are the most common. Ruthenium tetroxide, like osmium tetroxide, is highly toxic. In addition, it may explode. Ruthenium compounds show a marked resemblance to those of osmium. The metal is priced at about $4/g or $60/troy ounce.

Rutherfordium — See Element 104.

Samarium — (*Samarskite,* a mineral), Sm; at. wt. 150.4; at. no. 62; m.p. 1077 ± 5°C; b.p. 1791°C; sp. gr. (α) 7.520 (β) 7.40; valence 2 or 3. Discovered spectroscopically by its sharp absorption lines in 1879 by Lecoq de Boisbaudran in the mineral *samarskite,* named in honor of a Russian mine official, Col. Samarski. Samarium is found along with other members of the rare-earth elements in many minerals, including *monazite* and *bastnasite,* which are commercial sources. It occurs in monazite to the extent of 2.8%. While *misch metal* containing about 1% of samarium metal, has long been used, samarium has not been isolated in relatively pure form until recent years. Ion-exchange and solvent extraction techniques have recently simplified separation of the rare earths from one another; more recently, electrochemical deposition, using an electrolytic solution of lithium citrate and a mercury electrode, is said to be a simple, fast, and highly specific way to separate the rare earths. Samarium metal can be produced by reducing the oxide with barium or lanthanum. Samarium has a bright silver luster and is reasonably stable in air. Two crystal modifications of the metal exist, with a transformation point at 917°C. The metal ignites in air at about 150°C. Sixteen isotopes of samarium exist. Natural samarium is a mixture of seven isotopes, three of which are unstable with long half-lives. Samarium, along with other rare earths, is used for carbon-arc lighting for the motion picture industry. The sulfide has excellent high-temperature stability and good thermoelectric efficiencies up to 1100°C. $SmCo_5$ has been used in making a new permanent magnet material with the highest resistance to demagnetization of any known material. It is said to have an intrinsic coercive force as high as 28,000 oersteds. Samarium oxide has been used in optical glass to absorb the infrared. Samarium is used to dope calcium fluoride crystals for use in optical masers or lasers. Compounds of the metal act as sensitizers for phosphors excited in the infrared; the oxide exhibits catalytic properties in the dehydration and dehydrogenation of ethyl alcohol. It is used in infrared absorbing glass and as a neutron absorber in nuclear reactors. The metal is priced at about 75c/g or $200/ lb. Little is known of the toxicity of samarium; therefore, it should be handled carefully.

Scandium — (L. *Scandia,* Scandinavia), Sc; at. wt. 44.9559; at. no. 21; m.p. 1541°C; b.p. 2831°C; sp. gr. 2.989 (25°C); valence 3. On the basis of the Periodic System, Mendeleev predicted the existence of *ekaboron,* which would have an atomic weight between 40 of calcium and 48 of titanium. The element was discovered by Nilson in 1876 in the minerals *euxenite* and *gadolinite,* which had not yet been found anywhere except in Scandinavia. By processing 10 kg of euxenite and other residues of rare-earth minerals, Nilson was able to prepare about 2 g of scandium oxide of high purity. Cleve later pointed out that Nilson's scandium was identical with Mendeleev's ekaboron. Scandium is apparently a much more abundant element in the sun and certain stars than here on earth. It is about the 23rd most abundant element in the sun, compared to the 50th most abundant on earth. It is widely distributed on earth, occurring in very minute quantities in over 800 mineral species. The blue color of beryl (aquamarine variety) is said to be due to scandium. It occurs as a principal component in the rare mineral *thortveitite,* found in Scandinavia and Malagasy. It is also found in the residues remaining after the extraction of tungsten from Zinnwald *wolframite,* and in *wiikite* and *bazzite.* Most scandium is presently being recovered from *thortveitite* or is extracted as a by-product from uranium mill tailings. Metallic scandium was first prepared in 1937 by Fischer, Brunger, and Grieneisen, who electrolyzed a eutectic melt of potassium, lithium, and scandium chlorides at 700 to 800°C. Tungsten wire and a pool of molten zinc served as the electrodes in a graphic crucible. Pure scandium is now produced by reducing scandium fluoride with calcium metal. The production of the first pound of 99% pure scandium metal was announced in 1960 as having been made under a

U.S. Air Force contract. Scandium is a silvery-white metal which develops a slightly yellowish or pinkish cast upon exposure to air. It is relatively soft, and resembles yttrium and the rare-earth metals more than it resembles aluminum or titanium. It is a very light metal and has a higher melting point than aluminum, making it of interest to designers of space missiles. Scandium is not attacked by a 1:1 mixture of conc. HNO_3 and 48% HF. Scandium reacts rapidly with many acids. Eleven isotopes of scandium are recognized. The metal is still expensive, costing about $10/g with a purity of about 99.9%.Scandium oxide costs about :5/g. About 12 Kg. of scandium)as Sc;2O_3) are now being used yearly in the U.S. to produce high-intensity lights and the radioactive isotope Sc^{46} used as a tracing agent in refinery crackers for crude oil, etc. Scandium iodide added to mercury vapor lamps produces a highly efficient light source resembling sunlight, which is important for indoor or night-time color TV transmission. Little is yet known about the toxicity of scandium; therefore, it should be handled with care.

Selenium — (Gr. *Selene,* moon), Se; at. wt. 78.96; at. no. 34; m.p. (gray) 217°C; b.p. (gray) 684.9 ± 1.0°C; sp. gr. (gray) 4.79, (vitreous) 4.28; valence −2, +4, or + 6. Discovered by Berzelius in 1817, who found it associated with tellurium, named for the earth. Selenium is found in a few rare minerals, such as *crooksite* and *clausthalite.* In years past it has been obtained from flue dusts remaining from processing copper sulfide ores, but the anode muds from electrolytic copper refineries now provide the source of most of the world's selenium. Selenium is recovered by roasting the muds with soda or sulfuric acid, or by smelting them with soda and niter. Selenium exists in several allotropic forms. Three are generally recognized, but as many as six have been claimed. Selenium can be prepared with either an amorphous or crystalline structure. The color of amorphous selenium is either red, in powder form, or black, in vitreous form. Crystalline monoclinic selenium is a deep red; crystalline hexagonal selenium, the most stable variety, is a metallic gray. Natural selenium contains six stable isotopes. Fourteen other nuclides and isomers have been characterized. The element is a member of the sulfur family and resembles sulfur both in its various forms and in its compounds. Selenium exhibits both photovoltaic action, where light is converted directly into electricity, and photoconductive action, where the electrical resistance decreases with increased illumination. These properties make selenium useful in the production of photocells and exposure meters for photographic use, as well as solar cells. Selenium is also able to convert a.c. electricity to d.c., and is extensively used in rectifiers. Below its melting point selenium is a p-type semiconductor and is finding many uses in electronic and solid-state applications. It is used in Xerography for reproducing and copying documents, letters, etc. It is used by the glass industry to decolorize glass and to make ruby-colored glasses and enamels. It is also used as a photographic toner, and as an additive to stainless steel. Elemental selenium has been said to be practically nontoxic and is considered to be an essential trace element; however, hydrogen selenide and other selenium compounds are extremely toxic, and resemble arsenic in their physiological reactions. Hydrogen selenide in a concentration of 1.5 ppm is intolerable to man. Selenium occurs in some soils in amounts sufficient to produce serious effects on animals feeding on plants, such as locoweed, grown in such soils. Exposure to ,selenium compounds (as Se) in air should not exceed 0.2 mg/M^3 (8-hr time-weighted average — 40-hr week). Selenium is priced at about $20/lb. It is also available in high-purity form at a somewhat higher cost.

Silicon — (L. *silex, silicis,* flint), Si; at. wt. 28.0855 ± 0.0003; at. no. 14; m.p. 1410°C; b.p. 2355°C; sp. gr. 2.33 (25°C); valence 4. Davy in 1800 thought silica to be a compound and not an element; later in 1811, Gay Lussac and Thenard probably prepared impure amorphous silicon by heating potassium with silicon tetrafluoride. Berzelius, generally credited with the discovery, in 1824 succeeded in preparing amorphous silicon by the same general method as used earlier, but he purified the product by removing the fluosilicates by repeated washings. Deville in 1854 first prepared crystalline silicon, the second allotropic form of the ele-

ment. Silicon is present in the sun and stars and is a principal component of a class of meteorites known as *aerolites.* It is also a component of *tektites,* a natural glass of uncertain origin. Silicon makes up 25.7% of the earth's crust, by weight, and is the second most abundant element, being exceeded only by oxygen. Silicon is not found free in nature, but occurs chiefly as the oxide, and as silicates. *Sand, quartz, rock crystal, amethyst, agate, flint, jasper,* and *opal* are some of the forms in whch the oxide appears. *Granite, hornblende, asbestos, feldspar, clay, mica,* etc. are but a few of the numerous silicate minerals. Silicon is prepared commercially by heating silica and carbon in an electric furnace, using carbon electrodes. Several other methods can be used for preparing the element. Amorphous silicon can be prepared as a brown powder, which can be easily metled or vaporized. Crystalline silicon has a metallic luster and grayish color. The Czochralski process is commonly used to produce single crystals of silicon used for solid-state or semiconductor devices. Hyperpure silicon can be prepared by the thermal decomposition of ultra-pure trichlorosilane in a hydrogen atmosphere, and by a vacuum float zone process. This product can be doped with boron, gallium, phosphorus, or arsenic, etc. to produce silicon for use in transistors, solar cells, rectifiers, and other solid-state devices which are used extensively in the electronics and space-age industries. Hydrogenated amorphous silicon has shown promise in producing economical cells for converting solar energy into electricity. Silicon is a relatively inert element, but it is attacked by halogens and dilute alkali. Most acids, except hydrofluoric, do not affect it. Silicones are important products of silicon. They may be prepared by hydrolyzing a silicon organic chloride, such as dimethyl silicon chloride. Hydrolysis and condensation of various substituted chlorosilanes can be used to produce a very great number of polymeric products, or silicones, ranging from liquids to hard, glasslike solids with many useful properties. Elemental silicon transmits more than 95% of all wavelengths of infrared, from 1.3 to 6.7μm. Silicon is one of man's most useful elements. In the form of sand and clay it is used to make concrete and brick; it is a useful refractory material for high-temperature work, and in the form of silicates it is used in making enamels, pottery, etc. Silica, as sand, is a principal ingredient of glass, one of the most inexpensive of materials with excellent mechanical, optical, thermal, and electrical properties. Glass can be made in a very great variety of shapes, and is used as containers, window glass, insulators, and thousands of other uses. Silicon tetrachloride can be used to iridize glass. Silicon is important in plant and animal life. Diatoms in both fresh and salt water extract silica from the water to build up their cell walls. Silica is present in ashes of plants and in the human skeleton. Silicon is an important ingredient in steel; silicon carbide is one of the most important abrasives and has been used in lasers to produce coherent light of 4560 A. Regular grade silicon (97%) costs about 50c. Silicon 99.7% pure costs about $7/lb; hyperpure silicon may cost as much as $100/lb. Miners, stonecutters, and others engaged in work where siliceous dust is breathed in large quantities often develop a serious lung disease known as *silicosis.*

Silver — (Anglo-Saxon, *Seolfor siolfur*), Ag (L. argentum), at. wt. 107.868; at. no. 47; m.p. 961.93°C; b.p. 2212°C; sp. gr. 10.50 (20°C); valence 1, 2, Silver has been known since ancient times. It is mentioned in Genesis. Slag dumps in Asia Minor and on islands in the Aegean Sea indicate that man learned to separate silver from lead as early as 3000 B.C. Silver occurs native and in ores such as *argentite* (Ag_2S) and *horn silver* (AgCl); lead, lead-zinc, copper, gold, and copper-nickel ores are principal sources. Mexico, Canada, Peru, and the U.S. are the principal silver producers in the western hemisphere. Silver is also recovered during electrolytic refining of copper. Commercial fine silver contains at least 99.9% silver. Purities of 99.999+% are available commercially. Pure silver has a brilliant white metallic luster. It is a little harder than gold and is very ductile and malleable, being exceeded only by gold and perhaps palladium. Pure silver has the highest electrical and thermal conductivity of all metals, and possesses the lowest contact resistance. It is stable in pure air and water, but tarnishes when exposed to ozone, hydrogen sulfide, or air containing sulfur. The alloys of silver are

important. Sterling silver is used for jewelry, silverware, etc. where appearance is paramount. This alloy contains 92.5% silver, the remainder being copper or some other metal. Silver is of utmost importance in photography, about 30% of the U.S. industrial consumption going into this application. It is used for dental alloys. Silver is used in making solder and brazing alloys, electrical contacts, and high capacity silver-zinc and silver-cadmium batteries. Silver paints are used for making printed circuits. It is used in mirror production and may be deposited on glass or metals by chemical deposition, electrodeposition, or by evaporation. When freshly deposited, it is the best reflector of visible light known, but it rapidly tarnishes and loses much of its reflectance. It is a poor reflector of ultraviolet. Silver fulminate ($Ag_2C_2N_2O_2$), a powerful explosive, is sometimes formed during the silvering process. Silver iodide is used in seeding clouds to produce rain. Silver chloride has interesting optical properties as it can be made transparent; it also is a cement for glass. Silver nitrate, or *lunar caustic,* the most important silver compound, is used extensively in photography. While silver itself is not considered to be toxic, most of its salts are poisonous due to the anions present. Exposure to silver (metal and soluble compounds, as Ag) in air should not exceed 0.01mg/ M^3, (8-hr time-weighted average — 40-hr week). Silver compounds can be absorbed in the circulatory system and reduced silver deposited in the various tissues of the body. A condition, known as *argyria,* results, with a greyish pigmentation of the skin and mucous membranes. Silver has germicidal effects and kills many lower organisms effectively without harm to higher animals. Silver for centuries has been used traditionally for coinage by many countries of the world. In recent times, however, consumption of silver has greatly exceeded the output. In 1939, the price of silver was fixed by the U.S. Treasury at 71c/troy ounce, and at 90.5c/ troy ounce in 1946. In November 1961 the U.S. Treasury suspended sales of nonmonetized silver, and the price stabilized for a time at about $1.29, the melt-down value of silver U.S. coins. The Coinage Act of 1965 authorized a change in the metallic composition of the three U.S. subsidiary denominations to clad or composite type coins. This was the first change in U.S. coinage since the monetary system was established in 1792. Clad dimes and quarters are made of an outer layer of 75% Cu and 25% Ni bonded to a central core of pure Cu. The composition of the one- and five-cent pieces remains unchanged. One-cent coins are 95% Cu and 5% Zn. Five-cent coins are 75% Cu and 25% Ni. Old silver dollars are 90% Ag and 10% Cu. Earlier subsidiary coins of 90% Ag and 10% Cu officially are to circulate alongside the clad coins; however, in practice they have largely disappeared (Gresham's Law), as the value of the silver is now greater than their exchange value. Silver coins of other countries have largely been replaced with coins made of other metals. On June 24, 1968, the U.S. Government ceased to redeem U.S. Silver Certificates with silver. Since that time, the price of silver has fluctuated widely. The U.S. Government discontinued selling silver to domestic users and foreign buyers on November 10, 1970. As of December 31, 1977, the price of silver was about $5/troy ounce.

Sodium — (English, *soda;* Medieval Latin, *sodanum,* headache remedy), Na (L. *natrium*); at. wt. 22.9898; at. no. 11; m.p. 97.81± 0.03°C; b.p. 882.9°C; sp. gr. 0.971 (20°C); valence 1. Long recognized in compounds, sodium was first isolated by Davy in 1807 by electrolysis of caustic soda. Sodium is present in fair abundance in the sun and stars. The D lines of sodium are among the most prominent in the solar spectrum. Sodium is the sixth most abundant element on earth, comprising about 2.6% of the earth's crust; it is the most abundant of the alkali group of metals of which it is a member. The most common compound is sodium chloride, but it occurs in many other minerals, such as *soda niter, cryolite, amphibole, zeolite, sodalite,* etc. It is a very reactive element and is never found free in nature. It is now obtained commercially by the electrolysis of absolutely dry fused sodium chloride. This method is much cheaper than that of electrolyzing sodium hydroxide, as was used several years ago. Sodium is a soft, bright, silvery metal which floats on water, decomposing it with the evolution of hydrogen and the formation of the hydroxide. It may or may not ignite spontaneously on water, depending on the amount of

oxide and metal exposed to the water. It normally does not ignite in air at temperatures below 115°C. Sodium should be handled with respect, as it can be dangerous when improperly handled. Metallic sodium is vital in the manufacture of sodamide and sodium cyanide, sodium peroxide, and sodium hydride. It is used in preparing tetraethyl lead, in the reduction of organic esters, and in the preparation of organic compounds. The metal may be used to improve the structure of certain alloys, to descale metal, to purity molten metals, and as a heat transfer agent. An alloy of sodium with potassium, NaK, is also an important heat transfer agent. Sodium compounds are important to the paper, glass, soap, textile, petroleum, chemical, and metal industries. Soap is generally a sodium salt of certain fatty acids. The importance of common salt to animal nutrition has been recognized since prehistoric times. Among the many compounds that are of the greatest industrial importance are common salt (NaCl), soda ash (Na_2CO_3), baking soda ($NaHCO_3$), caustic soda (NaOH), Chile saltpeter ($NaNO_3$), di- and tri-sodium phosphates, sodium thiosulfate (hypo, $Na_2S_2O_3 \cdot 5H_2O$), and borax ($Na_2B_4O_7 \cdot 10H_2O$). Seven isotopes of sodium are recognized. Metallic sodium is priced at about 15 to 20c/lb in quantity. On a per cubic inch basis, it is the cheapest of all metals. Sodium metal should be handled with great care. It should be maintained in an inert atmosphere and contact with water and other substances with which sodium reacts should be avoided.

Strontium — (*Strontian,* town in Scotland), Sr; at. wt. 87.62; at. no. 38; m.p. 769°C; b.p. 1384°C; sp. gr. 2.54; valence 2. Isolated by Davy by electrolysis in 1808; however, Adair Crawford in 1790 recognized a new mineral (strontianite) as differing from other barium minerals (baryta). Strontium is found chiefly as *celestite* ($SrSO_4$) and *strontianite* ($SrCO_3$). The metal can be prepared by electrolysis of the fused chloride mixed with potassium chloride, or is made by reducing strontium oxide with aluminum in a vacuum at a temperature at which strontium distills off. Three allotropic forms of the metal exist, with transition points at 235 and 540°C. Strontium is softer than calcium and decomposes water more vigorously. It does not absorb nitrogen below 380°C. It should be kept under kerosene to prevent oxidation. Freshly cut strontium has a silvery appearance, but rapidly turns a yellowish color with the formation of the oxide. The finely divided metal ignites spontaneously in air. Volatile strontium salts impart a beautiful crimson color to flames, and these salts are used in pyrotechnics. Natural strontium is a mixture of four stable isotopes. Twelve other unstable isotopes are known to exist. Of greatest importance is Sr^{90} with a half-life of 28 years. It is a product of nuclear fallout and presents a health problem. This isotope is one of the best long-lived high-energy beta emitters known, and is used in SNAP (Systems for Nuclear Auxiliary Power) devices. These devices hold promise for use in space vehicles, remote weather stations, navigational buoys, etc., where a lightweight, long-lived, nuclear-electric power source is needed. Strontium hydroxide has been used in sugar refining; however, lime is replacing its use as it is cheaper. Strontium titanate is an interesting optical material as it has an extremely high refractive index and an optical dispersion greater than that of diamond. It has been used as a gemstone, but it is very soft. It does not occur naturally. The applications of strontium are similar to those of barium and calcium, but there are few advantages and the cost is much higher. Strontium metal costs about \$6 to \$8/lb.

Sulfur — (Sanskrit, *sulvere;* L. *sulphurium*), S; at. wt. 32.06; at. no. 16; m.p. (rhombic) 112.8°C, (monoclinic) 119.0°C; b.p. 444.674°C; sp. gr. (rhombic) 2.07, (monoclinic) 1.957 (20°C); valence 2, 4, or 6. Known to the ancients; referred to in Genesis as *brimstone.* Sulfur is found in meteorites. A dark area near the crater Aristarchus on the moon has been studied by R. W. Wood with ultraviolet light. This study suggests strongly that it is a sulfur deposit. Sulfur occurs native in the vicinity of volcanoes and hot springs. It is widely distributed in nature as *iron pyrites, galena, sphalerite, cinnabar, stibnite, gypsum, Epsom salts, celestite, barite,* etc. Sulfur is commercially recovered from wells sunk into the salt domes along the Gulf Coast of the U.S. It is obtained from these wells by the Frasch process, which forces heated water into the wells to melt the sulfur that is then brought to the surface. Sulfur also occurs in natural gas and petroleum

crudes and must be removed from these products. Formerly this was done chemically, which wasted the sulfur. New processes now permit recovery, and these sources promise to be very important. Large amounts of sulfur are being recovered from Alberta gas fields. Sulfur is a pale yellow, odorless, brittle solid, which is insoluble in water but soluble in carbon disulfide. In every state, whether gas, liquid or solid, elemental sulfur occurs in more than one allotropic form or modification; these present a confusing multitude of forms whose relations are not yet fully understood. Amorphous or "plastic" sulfur is obtained by fast cooling of the crystalline form. X-ray studies indicate that amorphous sulfur may have a helical structure with eight atoms per spiral. Crystalline sulfur seems to be made of rings, each containing eight sulfur atoms, which fit together to give a normal X-ray pattern. Ten isotopes of sulfur exist. Four occur in natural sulfur, none of which is radioactive. A finely divided form of sulfur, known as *flowers of sulfur,* is obtained by sublimation. Sulfur readily forms sulfides with many elements. sulfur is a component of black gunpowder, and is used in the vulcanization of natural rubber and as a fungicide. It is also used extensively in making phosphatic fertilizers. A tremendous tonnage is used to produce sulfuric acid, the most important manufactured chemical. It is used in making sulfite paper and other papers, as a fumigant, and in the bleaching of dried fruits. The element is a good electrical insulator. Organic compounds containing sulfur are very important. Calcium sulfate, ammonium sulfate, carbon disulfide, sulfur dioxide, and hydrogen sulfide are but a few of the other many important compounds of sulfur. Sulfur is essential to life. It is a minor constituent of fats, body fluids, and skeletal minerals. Carbon disulfide, hydrogen sulfide, and sulfur dioxide should be handled carefully. Hydrogen sulfide in small concentrations can be metabolized, but in higher concentrations it quickly can cause death by respiratory paralysis. It is insidious in that it quickly deadens the sense of smell. Sulfur dioxide is a dangerous component in atmospheric air pollution. In 1975, University of Pennsylvania scientists reported synthesis of polymeric sulfur nitride, which has the properties of a metal, although it contains no metal atoms. The material has unusual optical and electrical properties. High-purity sulfur is commercially available in purities of 99.999 + %.

Tantalum — (Gr. *Tanalos,* mythological character, father of *Niobe*), Ta; at. wt. 180.9479; at. no. 73; m.p. 2996 °C; b.p. 5425 ± 100°C; sp. gr. 16.654; valence 2?,3,4?, or 5. Discovered in 1802 by Ekeberg, but many chemists thought niobium and tantalum were identical elements until Rose, in 1844, and Marignac, in 1866, indicated and showed that niobic and tantalic acids were two different acids. The early investigators only isolated the impure metal. The first relatively pure ductile tantalum was produced by von Bolton in 1903. Tantalum occurs principally in the mineral *columbite-tantalite* $(Fe, Mn)(Nb,Ta)_2O_6$. Tantalum ores are found in the Republic of Zaire, Brazil, Mozambique, Thailand, Portugal, Nigeria, and Canada. Mines at Bernic Lake, Manitoba, have reserves of 900,000 tons of ore averaging about 0.15% tantalum oxide. Separation of tantalum from niobium requires several complicated steps. Several methods are used to commercially produce the element, including electrolysis of molten potassium fluotantalate, reduction of potassium fluotantalate with sodium, or reacting tantalum carbide with tantalum oxide. Sixteen isotopes of tantalum are known to exist. Natural tantalum contains two isotopes; one of these, Ta^{180}, is present in very small quantity (0.0123%) and is unstable with a very long half-life of $> 10^{13}$ years. Tantalum is a gray, heavy, and very hard metal. When pure, it is ductile and can be drawn into fine wire, which is used as a filament for evaporating metals such as aluminum. Tantalum is almost completely immune to chemical attack at temperatures below 150°C, and is attacked only by hydrofluoric acid, acidic solutions containing the fluoride ion, and free sulfur trioxide. Alkalis attack it only slowly. At high termperatures, tantalum becomes much more reactive. The element has a melting point exceeded only by tungsten and rhenium. Tantalum is used to make a variety of alloys with desirable properties such as high melting point, high strength, good ductility, etc. Scientists at Los Alamos have produced a tantalum carbide graphite composite material, which is said to be one of the hardest materials ever made. The compound has a melting point of 6760°F. Tantalum

has good "gettering" ability at high temperatures, and tantalum oxide films are stable and have good rectifying and dielectric properties. Tantalum is used to make electrolytic capacitors and vacuum furnace parts, which account for about 60% of its use. The metal is also widely used to fabricate chemical process equipment, nuclear reactors, and aircraft and missile parts. Tantalum is completely immune to body liquids and is a nonirritating metal. It has, therefore, found wide use in making surgical appliances. Tantalum oxide is used to make special glass with a high index of refraction for camera lenses. The metal has many other uses. In powdered form it costs about $40/lb. Sheet tantalum and fabricated forms are more expensive.

Technetium — (Gr. *technetos,* artifical), Tc; at. wt. 98.9062; at. no. 43; m.p. 2172°C; b.p. 4877°C; sp. gr. 11.50 (calc.); valence 0, +2, +4, +5, +6, and +7. Element 43 was predicted on the basis of the periodic table, and was erroneously reported as having been discovered in 1925, at which time it was named *masurium.* The element was actually discovered by Perrier and Segre in Italy in 1937. It was found in a sample of molybdenum, which was bombarded by deuterons in the Berkeley cyclotron, and which E. Lawrence sent to these investigators. Technetium was the first element to be produced artificially. Since its discovery, searches for the element in terrestrial materials have been made without success. If it does exist, the concentration must be very small. Surprisingly, it has been found in the spectrum of S-, M-, and N-type stars, and its presence in stellar matter is leading to new theories of the production of heavy elements in the stars. Sixteen isotopes of technetium, with atomic masses ranging from 92 to 107, are known. Tc^{97} has a half-life of 2.6×10^6 years. Tc^{98} has a half-life of 1.5×10^6 years. The isomeric isotope Tc^{95m}, with a half-life of 61 days, is useful for tracer work, as it produces energetic gamma rays. Technetium metal has been produced in kilogram quantities. The metal was first prepared by passing hydrogen gas at 1100°C over Tc_2S_7. It is now conveniently prepared by the reduction of ammonium pertechnetate with hydrogen. Technetium is a silvery-gray metal that tarnishes slowly in moist air. Until 1960, technetium was available only in small amounts and the price was as high as $2800/g. It is now offered commercially to holders of A.E.C. permits at a price of $90 to $100/g. The chemistry of technetium is said to be similar to that of rhenium. Technetium dissolves in nitric acid, aqua regia, and conc. sulfuric acid, but is not soluble in hydrochloric acid of any strength. The element is a remarkable corrosion inhibitor for steel. It is reported that mild carbon steels may be effectively protected by as little as 5 ppm of $KTcO_4$ in aerated distilled water at temperatures up to 250°C. This corrosion protection is limited to closed systems, since technetium is radioactive and must be confined. Tc^{99} has a specific activity of 6.2×10^8 disintegrations per sec/g. Activity of this level must not be allowed to spread. Tc^{99} is a contamination hazard and should be handled in a glove box. The metal is an excellent superconductor at 11 K and below.

Tellurium — (L. *tellus,* earth), Te; at. wt. 127.60; at. no. 52; m.p. 449.5 ± 0.3 °C; b.p. 989.8 ± 3.8°C; sp. gr. 6.24 (20°C); valence 2, 4, or 6. Discovered by Muller von Reichenstein in 1782; named by Klaproth, who isolated it in 1798. Tellurium is occasionally found native, but is more often found as the telluride of gold (*calaverite*), and combined with other metals. It is recovered commercially from the anode muds produced during the electrolytic refining of blister copper. The U.S., Canada, Peru, and Japan are the largest Free World producers of the element. Crystalline tellurium has a silvery-white appearance, and when pure exhibits a metallic luster. It is brittle and easily pulverized. Amorphous tellurium is formed by precipitating tellurium from a solution of telluric or tellurous acid. Whether this form is truly amorphous, or made of minute crystals, is open to question. Tellurium is a p-type semiconductor, and shows greater conductivity in certain directions, depending on alignment of the atoms. Its conductivity increases slightly with exposure to light. It can be doped with silver, copper, gold, tin, or other elements. In air, tellurium burns with a greenish-blue flame, forming the dioxide. Molten tellurium corrodes iron, copper, and stainless steel. Tellurium and its compounds are probably toxic and should be handled with care. Workmen exposed to as little as 0.0l mg/m³ of air, or less, develop "tellurium breath," which has a garlic-like odor. Twenty-

one isotopes of tellurium are known, with atomic masses ranging from 115 to 135. Natural tellurium consists of eight isotopes, one of which, Te^{127},is unstable. It is present to the extent of 0.87% and has a half-life of 1.2×10^{13} years. Tellurium improves the machinability of copper and stainless steel, and its addition to lead decreases the corrosive action of sulfuric acid to lead and improves its strength and hardness. Tellurium is used as a basic ingredient in blasting caps, and is added to cast iron for chill control. Tellurium is used in ceramics. Bismuth telluride has been used in thermoelectric devices. Tellurium costs about $20/lb, with a purity of about 99.5%.

Terbium — (*Ytterby,* village in Sweden), Tb; at. wt. 158.9254; at. no. 65; m.p. 1356°C; b.p. 3123°C; sp. gr. 8.229; valence 3, 4. Discovered by Mosander in 1843. Terbium is a member of the lanthanide or "rare earth" group of elements. It is found in *cerite, gadolinite,* and other minerals along with other rare earths. It is recovered commercially from *monazite* in which it is present to the extent of 0.03%, from *xenotime,* and from *euxenite,* a complex oxide containing 1% or more of terbia. Terbium has been isolated only in recent years with the development of ion-exchange techniques for separating the rare-earth elements. As with other rate earths, it can be produced by reducing the anhydrous chloride or fluoride with calcium metal in a tantalum crucible. Calcium and tantalum impurities can be removed by vacuum remelting. Other methods of isolation are possible. Terbium is reasonably stable in air. It is a silvery-gray metal, and is malleable, ductile, and soft enough to be cut with a knife. Two crystal modifications exist, with a transformation temperature of 1315°C. Nineteen isotopes with atomic masses ranging from 147 to 164 are recognized. The oxide is a chocolate or dark maroon color. Sodium terbium borate is used as a laser material and emits coherent light at 5460A. Terbium is used to dope calcium fluoride, calcium tungstate, and strontium molybdate, used in solid-state devices. The oxide has potential application as an activator for green phosphors used in color TV tubes. It can be used with ZrO_2 as a crystal stabilizer of fuel cells which operate at elevated temperature. Few other uses have been found. The element is priced at about $3/g or $900/lb. Little is known of the toxicity of terbium. It should be handled with care as with other lanthanide elements.

Thallium — (Gr. *thallos,* a green shoot or twig), Tl; at. wt. 204.37; at. no. 81; m.p. 303.5°C; b.p. 1457 ± 10°C; sp. gr. 11.85 (20°C); valence 1, or 3. Thallium was discovered spectroscopically in 1861 by Crookes. The element was named after the beautiful green spectral line, which identified the element. The metal was isolated both by Crookes and Lamy in 1862 about the same time. Thallium occurs in *crooksite, lorandite,* and *hutchinsonite.* It is also present in *pyrites* and is recovered from the roasting of this ore in connection with the production of sulfuric acid. It is also obtained from the smelting of lead and zinc ores. Extraction is somewhat complex and depends on the source of the thallium. Manganese nodules, found on the ocean floor, contain thallium. When freshly exposed to air, thallium exhibits a metallic luster, but soon develops a bluish-gray tinge, resembling lead in appearance. A heavy oxide builds up on thallium if left in air, and in the presence of water the hydroxide is formed. The metal is very soft and malleable. It can be cut with a knife. Twenty isotopic forms of thallium, with atomic masses ranging from 191 to 210, are recognized. Natural thallium is a mixture of two isotopes. The element and its compounds are toxic and should be handled carefully. Contact of the metal with the skin is dangerous, and when melting the metal adequate ventilation should be provided. Exposure to Thallium (soluble compounds) — skin, as TI, should not exceed $0.1 mg/M^3$ (8-hr time-weighted average — 40-hr week). Thallium is suspect of carcinogenic potential for man. Thallium sulfate has been widely employed as a rodenticide and ant killer. It is odorless and tasteless, giving no warning of its presence. Its use, however, has been prohibited in the U.S. since 1975 as a household insecticide and rodenticide. Continued sales subjects dealers to civil and criminal penalties. The electrical conductivity of thallium sulfide changes with exposure to infrared light, and this compound is used in photocells. Thallium bromide-iodide crystals have been used as infrared detectors. Thallium has been used, with sulfur or selenium and arsenic, to produce low melting glasses which become fluid between 125 and 150°C.

These glasses have properties at room temperatures similar to ordinary glasses and are said to be durable and insoluble in water. Thallium oxide has been used to produce glasses with a high index of refraction. Thallium has been used in treating ringworm and other skin infections; however, its use has been limited because of the narrow margin between toxicity and therapeutic benefits. A mercury-thallium alloy, which forms a eutectic at 8.5% thallium, is reported to freeze at −60°C, some 20° below the freezing point of mercury, Commercial thallium metal costs about $8/lb.

Thorium — (*Thor,* Scandinavian god of war), Th; at. wt. 232.0381; at. no. 90; m.p. 1750°C; b.p. ~4790°C; sp. gr. 11.72; valence +2(?), +3(?), +4. Discovered by Berzelius in 1828. Thorium occurs in *thorite* ($ThSiO_4$) and in *thorianite* ($ThO_2 + UO_2$). Large deposits of thorium minerals have been reported in New England and elsewhere, but these have not yet been exploited. Thorium is now thought to be about three times as abundant as uranium and about as abundant as lead or molybdenum. The metal is a source of nuclear power. There is probably more energy available for use from thorium in the minerals of the earth's crust than from both uranium and fossil fuels. Any sizable demand for thorium as a nuclear fuel is still several years in the future. Work has been done in developing thorium cycle converter-reactor systems. Several prototypes, including the HTGR (high-temperature gas-cooled reactor) and MSRE (molten salt converter reactor experiment), have operated. While the HTGR reactors are efficient, they are not expected to become important commercially for 10 or more years because of certain operating difficulties. Thorium is recovered commercially from the mineral *monazite,* which contains from 3 to 9% ThO_2 along with most rare-earth minerals. Much of the internal heat the earth has been attributed to thorium and uranium. Several methods are available for producing thorium metal: it can be obtained by reducing thorium oxide with calcium, by electrolysis of anhydrous thorium chloride in a fused mixture of sodium and potassium chlorides, by calcium reduction of thorium tetrachloride mixed with anhydrous zinc chloride, and by reduction of thorium tetrachloride with an alkali metal. Thorium was originally assigned a position in Group IV of the periodic table. Because of its atomic weight, valence, etc., it is now considered to be the second member of the *actinide* series of elements. When pure, thorium is a silvery-white metal which is air-stable and retains its luster for several months. When contaminated with the oxide, thorium slowly tarnishes in air, becoming gray and finally black. The physical properties of thorium are greatly influenced by the degree of contamination with the oxide. The purest speciments often contain several tenths of a percent of the oxide. High-purity thorium has been made. Pure thorium is soft, very ductile, and can be cold-rolled, swaged, and drawn. Thorium is dimorphic, changing at 1400°C from a cubic to a body-centered cubic structure. Thorium oxide has a melting point of 3300°C, which is the highest of all oxides. Only a few elements, such as tungsten, and a few compounds, such as tantalum carbide, have higher melting points. Thorium is slowly attacked by water, but does not dissolve readily in most common acids, except hydrochloric. Powdered thorium metal is often pyrophoric and should be carefully handled. When heated in air, thorium turnings ignite and burn brilliantly with a white light. The principal use of thorium has been in the preparation of the Welsbach mantle, used for portable gas lights. These mantles, consisting of thorium oxide with about 1% cerium oxide and other ingredients, glow with a dazzling light when heated in a gas flame. Thorium is an important alloying element in magnesium, imparting high strength and creep resistance at elevated temperatures. Because thorium has a low work-function and high electron emission, it is used to coat tungsten wire used in electronic equipment. The oxide is also used to control the grain size of tungsten used for electric lamps; it is also used for high-temperature laboratory crucibles. Glasses containing thorium oxide have a high refractive index and low dispersion. Consequently, they find application in high quality lenses for cameras and scientific instruments. Thorium oxide has also found use as a catalyst in the conversion of ammonia to nitric acid, in petroleum cracking, and in producing sulfuric acid. Twelve isotopes of thorium are known with atomic masses ranging from 223 to 234. All are unstable. Th^{232} occurs naturally and has a half-life of 1.41×10^{10} years. It is an alpha emitter.

Th^{232} goes through six alpha and four beta decay steps before becoming the stable isotope Pb^{208}. Th^{232} is sufficiently radioactive to expose a photographic plate in a few hours. Thorium disintegrates with the production of thoron (radon^{220}), which is an alpha emitter and presents a radiation hazard. Good ventilation of areas where thorium is stored or handled is therefore essential. Thorium and is compounds are subject to licensing and control by the U.S. Atomic Energy Commission. Thorium metal (99.9%) costs about $10/lb.

Thulium — (*Thule,* the earliest name for Scandinavia), Tm; at. wt. 168.9342; at. no. 69; m.p. 1,545 ± 15° C; b.p. 1947°C; sp. gr. 9.321(25°); valence 2, 3. Discovered in 1879 by Cleve. Thulium occurs in small quantities along with other rare earths in a number of minerals. It is obtained commercially from *monazite,* which contains about 0.007% of the element. Thulium is the least abundant of the earth elements, but with new sources recently discovered, it is now considered to be about as rare as silver, gold, or cadmium. Ion-exchange and solvent extraction techniques have recently permitted much easier separation of the rare earths, with much lower costs. Only a few years ago, thulium metal was not obtainable at any cost; in 1950 the oxide sold for $450/g. Thulium metal now costs from $3 to $20/g, depending on the purity, quantity, and supplier. Thulium can be isolated by reduction of the oxide with lanthanum metal or by calcium reduction of the anhydrous fluoride. The pure metal has a bright, silvery luster. It is reasonably stable in air, but the metal should be protected from moisture in a closed container. The element is silver-gray, soft, malleable, and ductile, and can be cut with a knife. Sixteen isotopes are known, with atomic masses ranging from 161 to 176. Natural thulium, Tm^{169} is stable. Because of the relatively high price of the metal, thulium has not yet found many practical applications. Tm^{169} bombarded in a nuclear reactor can be used as a radiation source in portable X-ray equipment. Tm^{171} is potentially useful as an energy source. Natural thulium also has possible use in *ferrites* (ceramic magnetic materials) used in microwave equipment. As with other lanthanides, thulium has a low-to-moderate acute toxic rating. It should be handled with care.

Tin — (Anglo-Saxon, *tin*), Sn (L. *stannum*); at. wt. 118.69; at. no. 50; m.p. 231.9681°C; b.p. 2270°C; sp. gr. (gray) 5.75, (white) 7.31; valence 2, 4. Known to the ancients. Tin is found chiefly in *cassiterite* (SnO_2). Most of the world's supply comes from Malaya, Boliva, Indonesia, Zaire, Thailand, and Nigeria. The U.S. produces almost none, although occurrences have been found in Alaska and California. Tin is obtained by reducing the ore with coal in a reverberatory furnace. Ordinary tin is composed of nine stable isotopes; 13 unstable isotopes are also known. Ordinary tin is a silvery-white metal, is malleable, somewhat ductile, and has a highly crystalline structure. Due to the breaking of these crystals, a "tin cry" is heard when a bar is bent. The element has two or perhaps three allotropic forms. On warming, gray, or α tin, with a cubic structure, changes at 13.2°C into white, or β tin, the ordinary form of the metal. White tin has a tetragonal structure. Some authorities believe a γ form exists between 161°C and the melting point; however, other authorities discount its existence. When tin is cooled below 13.2°C, it changes slowly from white to gray. This change is affected by impurities such as aluminum and zinc, and can be prevented by small additions of antimony or bismuth. This change from the α to β form is called the tin pest. There are few if any uses for gray tin. Tin takes a high polish and is used to coat other metals to prevent corrosion or other chemical action. Such tin plate over steel is used in the so-called tin can for preserving food. Alloys of tin are very important. Soft solder, type metal, fusible metal, pewter, bronze, bell metal, Babbitt metal, White metal, die casting alloy, and phosphor bronze are some of the important alloys using tin. Tin resists distilled, sea, and soft tap water, but is attacked by strong acids, alkalis, and acid salts. Oxygen in solution accelerates the attack. When heated in air, tin forms SnO_2, which is feebly acid, forming stannate salts with basic oxides. The most important salt is the chloride ($SnCl_2 \cdot H_2O$), which is used as a reducing agent and as a mordant in calico printing. Tin salts sprayed onto glass are used to produce electrically conductive coatings. These have been used for panel lighting and for frost-free windshields. Most window glass is now made by floating molten glass on molten tin (float glass)

to produce a flat surface (Pilkington process). Of recent interest is a crystalline tin-niobium alloy that is superconductive at very low temperatures. This promises to be important in the construction of superconductive magnets that generate enormous field strengths but use practically no power. Such magnets, made of tin-niobium wire, weigh but a few pounds and produce magnetic fields that, when started with a small battery, are comparable to that of a 100 ton electromagnet operated continuously with a large power supply. The small amount of tin used in canned foods is quite harmless. The agreed limit of tin content in U.S. foods is 300 mg/kg. The trialkyl and triaryl tin compounds are used as biocides and must be handled carefully. Over the past 25 years the price of tin has varied from 50c/lb to its present price of $6/lb.

Titanium — (L. *Titans,* the first sons of the Earth, myth.), Ti; at. wt. 47.90; at. no. 22; m.p. 1660 ± 10°C; b.p. 3287°C; sp. gr. 4.54; valence 2, 3, or 4. Discovered by Gregor in 1791; named by Klaproth in 1795. Impure titanium was prepared by Nilson and Pettersson in 1887; however, the pure metal (99.9%) was not made until 1910 by Hunter by heating $TiCl_4$ with sodium in a steel bomb. Titanium is present in meteorites and in the sun. Rocks obtained during the Apollo 17 lunar mission showed presence of 12.1% TiO_2. Analyses of rocks obtained during earlier Apollo missions show lower percentages. Titanium oxide bands are prominent in the spectra of M-type stars. The element is the ninth most abundant in the crust of the earth. Titanium is almost always present in igneous rocks and in the sediments derived from them. It occurs in the minerals *rutile, ilmenite,* and *sphene,* and is present in titanates and in many iron ores. Titanium is present in the ash of coal, in plants, and in the human body. The metal was a laboratory curiosity until Kroll, in 1946, showed that titanium could be produced commercially by reducing titanium tetrachloride with magnesium. This method is largely used for producing the metal today. The metal can be purified by decomposing the iodide. Titanium, when pure, is a lustrous, white metal. It has a low density, good strength, is easily fabricated, and has excellent corrosion resistance. It is ductile only when it is free of oxygen. The metal burns in air and is the only element that burns in nitrogen. Titanium is resistant to dilute sulfuric and hydrochloric acid, most organic acids, moist chlorine gas, and chloride solutions. Natural titanium consists of five isotopes with atomic masses from 46 to50. All are stable. Four other unstable isotopes are known. Natural titanium is reported to become very radioactive after bombardment with deuterons. The emitted radiations are mostly positrons and hard gamma rays. The metal is dimorphic. The hexagonal α form changes to the cubic β form very slowly at about 880°C. The metal combines with oxygen at red heat, and with chlorine at 550°C. Titanium is important as an alloying agent with aluminum, molybdenum, manganese, iron, and other metals. Alloys of titanium are principally used for aircraft and missiles where lightweight strength and ability to withstand extremes of temperature are important. Titanium is a strong as steel, but 45% lighter. It is 60% heavier than aluminum, but twice as strong. Titanium has potential use in desalination plants for converting sea water into fresh water. The metal has excellent resistance to sea water and is used for propeller shafts, rigging, and other parts of ships exposed to salt water. A titanium anode coated with platinum has been used to provide cathodic protection from corrosion by salt water. Titanium metal is considered to be physiologically inert. When pure, titanium dioxide is relatively clear and has an extremely high index of refraction with an optical dispersion higher than diamond. It is produced artificially for use as a gemstone, but it is relatively soft. Star sapphires and rubies exhibit their asterism as a result of the presence of TiO_2. Titanium dioxide is extensively used for both house paint and artist's paint, as it is permanent and has good covering power. Titanium oxide pigment accounts for the largest use of the element. Titanium paint is an excellent reflector of infrared, and is extensively used in solar observatories where heat causes poor seeing conditions. Titanium tetrachloride is used to iridize glass. This compound fumes strongly in air and has been used to produce smoke screens. The price of titanium metal powder (99.7%) is about $25/lb.

Tungsten — (Swedish, *tung sten,* heavy stone); also known as *wolfram*(from *wolframite,* said to be named from *wolf rahm* or *spumi lupi,* because the ore interfered with the smelt-

ing of tin and was supposed to devour the tin), W; at. wt. 183.85; at. no. 74; m.p. 3410 ± 20°C; b.p. 5660°C; sp. gr. 19.3 (20°C); valence 2, 3, 4, 5, or 6. In 1779 Peter Woulfe examined the mineral now known as *wolframite* and concluded it must contain a new substance. Scheele, in 1781, found that a new acid could be made from *tung sten* (a name first applied about 1758 to a mineral now known as *scheelite*). Scheele and Bergman suggested the possibility of obtaining a new metal by reducing this acid. The de Elhuyar brothers found an acid in *wolframite* in 1783 that was identical to the acid of *tung sten* (tungstic acid) of Scheele, and in that year they succeeded in obtaining the element by reduction of this acid with charcoal. Tungsten occurs in *wolframite,* (Fe, Mn)WO_4; *scheelite,* $CaWO_4$; *huebnerite,* $MnWO_4$; and *ferberite,* $FeWO_4$. Important deposits of tungsten occur in California, Colorado, South Korea, Bolivia, U.S.S.R., and Portugal. China is reported to have about 75% of the world's tungsten resources. Natural tungsten contains five stable isotopes. Twelve other unstable isotopes are recognized. The metal is obtained commercially by reducing tungsten oxide with hydrogen or carbon. Pure tungsten is a steel-gray to tin-white metal. Very pure tungsten can be cut with a hacksaw, and can be forged, spun, drawn, and extruded. The impure metal is brittle and can be worked only with difficulty. Tungsten has the highest melting point and lowest vapor pressure of all metals, and at temperatures over 1650°C has the highest tensile strength. The metal oxidizes in air and must be protected at elevated temperatures. It has excellent corrosion resistance and is attacked only slightly by most mineral acids. The thermal expansion is about the same as borosilicate glass, which makes the metal useful for glass-to-metal seals. Tungsten and its alloys are used extensively for filaments for electric lamps, electron and television tubes, and for metal evaporation work; for electrical contact points for automobile distributors; X-ray targets; windings and heating elements for electrical furnaces; and for numerous space missile and high-temperature applications. High-speed tool steels, Hastelloy ®, Stellite ®, and many other alloys contain tungsten. Tungsten carbide is of great importance to the metal-working, mining, and petroleum industries. Calcium and magnesium tungstates are widely used in fluorescent lighting; other salts of tungsten are used in the chemical and tanning industries. Tungsten disulfide is a dry, high-temperature lubricant, stable to 500°C. Tungsten bronzes and other tungsten compounds are used in paints. Tungsten powder costs about $15/lb.

Uranium — (Planet *Uranus*), U; at. wt. 238.029; at. no. 92; m.p. 1132.3 ± 0.8°C; b.p. 3818°C; sp. gr. ~ 18.95; valence 2, 3, 4, 5, or 6. Yellow-colored glass, containing more than 1% uranium oxide and dating back to 79 A.D., has been found near Naples, Italy. Klaproth recognized an unknown element in *pitchblende* and attempted to isolate the metal in 1789. The metal apparently was first isolated in 1841 by Peligot, who reduced the anhydrous chloride with potassium. Uranium is not as rare as it was once thought. It is now considered to be more plentiful than mercury, antimony, silver, or cadmium, and is about as abundant as molybdenum or arsenic. It occurs in numerous minerals such as *pitchblende, uraninite, carnotite, autunite, uranophane, davidite,* and *tobernite.* It is also found in *phosphate rock, lignite, monazite sands,* and can be recovered commercially from these sources. The A.E.C. purchases uranium in the form of acceptable U_3O_8 concentrates. This incentive program has greatly increased the known uranium reserves. Uranium can be prepared by reducing uranium halides with alkali or alkaline earth metals or by reducing uranium oxides by calcium, aluminum, or carbon at high temperatures. The metal can also be produced by electrolysis of KUF_5 or UF_4, dissolved in a molten mixture of $CaCl_2$ NaCl. High-purity uranium can be prepared by the thermal decomposition of uranium halides on a hot filament. Uranium exhibits three crystallographic modifications as follows:

$$\alpha \xrightarrow{667°C} \beta \xrightarrow{772°C} \gamma$$

Uranium is a heavy, silvery-white metal which is pyrophoric when finely divided. It is a little softer than steel, and is attacked by cold water in a finely divided state. It is malleable, ductile, and slightly paramagnetic. In air, the metal becomes coated with a layer of oxide. Acids dissolve the metal, but it is unaffected by alkalis, Uranium has fourteen isotopes, all of which are radioactive. Naturally occurring uranium nom-

inally contains 99.2830% by weight U^{238}, 0.7110% U^{235}, and 0.0054% U^{234}. Studies show that the percentage weight of U^{235} in natural uranium varies by as much as 0.1%, depending on the source. The A.E.C. has adopted the value of 0.711 as being their "official" percentage of U^{235} in natural uranium. Natural uranium is sufficiently radioactive to expose a photographic plate in an hour or so. Much of the internal heat of the earth is thought to be attributable to the presence of uranium and thorium. U^{238} with a half-life of 4.51×10^9 years, has been used to estimate the age of igneous rocks. The origin of uranium, the highest member of the naturally occurring elements — except perhaps for traces of neptunium or plutonium — is not clearly understood, although it may be presumed that uranium is a decay prouct of elements of higher atomic weight, which may have once been present on earth or elsewhere in the universe. These original elements may have been created as a result of a primordial "creation," known as "the big bang," in a supernovae, or in some other stellar processes. Uranium is of great importance as a nuclear fuel. U^{238} can be converted into fissionable plutonium by the following reactions:

$$U^{238}(n,\gamma)U^{239} \xrightarrow{\beta-} Np^{239} \xrightarrow{\beta-} Pu^{239}$$

This nuclear conversion can be brought about in "breeder" reactors where it is possible to produce more new fissionable material than the fissionable material used in maintaining the chain reaction. U^{235} is of even greater importance, for it is the key to the utilization of uranium. U^{235}, while occurring in natural uranium to the extent of only 0.71%, is so fissionable with slow neutrons that a self-sustaining fission chain reaction can be made to occur in a reactor constructed from natural uranium and a suitable moderator, such as heavy water or graphite, alone. U^{235} can be concentrated by gaseous diffusion and other physical processes, if desired, and used directly as a nuclear fuel, instead of natural uranium, or used as an explosive. Natural uranium, slightly enriched with U^{235} by a small percentage, is used to fuel nuclear power reactors for the generation of electricity. Natural thorium can be irradiated with neutrons as follows to produce the important isotope U^{233}.

$$Th^{232}(n,\gamma)Th^{233} \xrightarrow{\beta-} Pa^{233} \xrightarrow{\beta-} U^{233}$$

While thorium itself is not fissionable, U^{233} is, and in this way may be used as a nuclear fuel. One pound of completely fissioned uranium has the fuel value of over 1500 tons of coal. The uses of nuclear fuels to generate electrical power, to make isotopes for peaceful purposes, and to make explosives are well known. The estimated world-wide capacity of the 153 nuclear power reactors in operation in 1977 amounted to about 83 million kilowatts. Uranium in the U.S.A. is controlled by the Atomic Energy Commission. New uses are being found for "depleted uranium, i.e., uranium with the percentage of U^{235} lowered to about 0.2%. It has found use in inertial guidance devices, gyro compasses, counterweights for aircraft control surfaces, as ballast for missile reentry vehicles, and as a shielding material. Uranium metal is used for X-ray targets for production of high-energy X-rays; the nitrate has been used as photographic toner, and the acetate is used in analytical chemistry. Crystals of uranium nitrate are triboluminescent. Uranium salts have also been used for producing yellow "vaseline" glass and glazes. Uranium and its compounds are highly toxic, both from a chemical and radiological standpoint. Finely divided uranium metal, being pyrophoric, presents a fire hazard. The maximum recommended allowable concentration of soluble uranium compound in air (based on chemical toxicity) is 0.05 mg/M^3 (8-hr time-weighted average — 40-hr week); for insoluble compounds the concentration is set at 0.25 mg/M^3 of air. The maximum permissible total body burden of natural uranium (based on radiotoxicity) is 0.2μCi for soluble compounds.

Vanadium — (Scandinavian goddess, *Vanadis*), V; at. wt. 50.9415; at. no. 23; m.p. 1890 ± 10°C; b.p. 3380°C; sp. gr. 6.ll (18.7°C); valence 2, 3, 4, or 5. Vanadium was first discovered by del Rio in 1801. Unfortunately, a French chemist incorrectly declared del Rio's new element was only impure chromium; del Rio thought himself to be mistaken and accepted the French chemist's statement. The element was rediscovered in 1830 by Sefstrom, who named the element in honor of the Scandinavian goddess *Vanadis* because of its beautiful multicolored compounds. It was isolated

in nearly pure form by Roscoe, in 1867, who reduced the chloride with hydrogen. Vanadium of 99.3 to 99.8% purity was not produced until 1927. Vanadium is found in about 65 different minerals among which are *carnotite, roscoelite, vanadinite,* and *patronite* — important sources of the metal. Vanadium is also found in phosphate rock and certain iron ores, and is present in some crude oils in the form of organic complexes. It is also found in small percentages in meteorites. Commercial production from petroleum ash holds promise as an important source of the element. High-purity ductile vanadium can be obtained by reduction of vanadium trichloride with magnesium or with magnesium-sodium mixtures. Much of the vanadium metal being produced is now made by calcium reduction of V_2O_5 in a pressure vessel, an adaption of a process developed by McKechnie and Seybolt. Natural vanadium is a mixture of two isotopes, V^{50} (0.24%) and V^{51} (99.76%). V^{50} is slightly radioactive, having a half-life of 6×10^{15} years. Seven other unstable isotopes are recognized. Pure vanadium is a bright white metal, and is soft and ductile. It has good corrosion resistance to alkalis, sulfuric and hydrochloric acid, and salt waters, but the metal oxidizes readily above 660°C. The metal has good structural strength and a low-fission neutron cross section, making it useful in nuclear applications. Vanadium is used in producing rust-resistant, spring, and high-speed tool steels. It is an important carbide stabilizer in making steels. About 80% of the vanadium now produced is used as ferrovanadium or as a steel additive. Vanadium foil is used as a bonding agent in cladding titanium to steel. Vanadium pentoxide is used in ceramics and as a catalyst. It is also used as a mordant in dyeing and printing fabrics and in the manufacture of aniline black. Vanadium-gallium tape has been used in producing a superconductive magnet with a field of 175,000 gauss. Vanadium and its compounds are toxic and should be handled with care. Exposure to V_2O_2 dust (as V) should not exceed the ceiling value of 0.05mg/M^3, and exposure to V_2O_2 fume (as V) should not exceed 0.1mg/M^3 (8-hr time-weighted average — 40-hr week). Ductile vanadium is commercially available at a cost of about $40/lb. Commercial vanadium metal, of about 95% purity, costs about $5/lb.

Wolfram — see Tungsten.

Xenon — (Gr. *xenon,* stranger), Xe; at. wt. 131.30; at. no. 54; m.p. −111.9°C; b.p. −107.1 ± 3°C; density (gas) 5.887 ± 0.009 g/l, sp. gr. (liquid) 3.52 (−109°C); valence usually O. Discovered by Ramsay and Travers in 1898 in the residue left after evaporating liquid air components. Xenon is a member of the so-called noble or "inert" gases. It is present in the atmosphere to the extent of about one part in twenty million. Xenon is present in the Martian atmosphere to the extent of 0.08 ppm. The element is found in the gases evolved from certain mineral springs, and is commercially obtained by extraction from liquid air. Natural xenon is composed of nine stable isotopes. In addition to these, 22 unstable nuclides and isomers have been characterized. Before 1962, it had generally been assumed that xenon and other noble gases were unable to form compounds. Evidence has been mounting in the past few years that xenon, as well as other members of the zero valence elements, do form compounds. Among the "compounds" of xenon now reported are xenon hydrate, sodium perxenate, xenon deuterate, difluoride, tetrafluoride, hexafluoride, and $XePtF_6$ and $XeRhF_6$. Xenon trioxide, which is highly explosive, has been prepared. More recently, compounds such as $FXeN(So_2F)_2$ with a xenon-nitrogen bond have been made. Xenon in a vacuum tube produces a beautiful blue glow when excited by an electrical discharge. The gas is used in making electron tubes, stroboscopic lamps, bactericidal lamps, and lamps used to excite ruby lasers for generating coherent light. Xenon is used in the atomic energy field in bubble chambers, probes, and other applications where its high molecular weight is of value. It is also potentially useful as a gas for ion engines. The perxenates are used in analytical chemistry as oxidizing agents. $Xe^{133}$and Xe^{135} are produced by neutron irradiation in aircooled nuclear reactors. Xe^{133} has useful applications as a radioisotope. The element is available in sealed glass containers for about $20/l of gas at standard pressure. Xenon is not toxic, but its compounds are highly toxic because of their strong oxidizing characteristics.

Ytterbium — (Ytterby, village in Sweden), Yb; at. wt. 173.04; at. no. 70; m.p. 819°C; b.p. 1194°C; sp. gr. (α) 6.965 (β) 6.54; valence 2, 3.

Marignac in 1878 discovered a new component, which he called *ytterbia,* in the earth then known as *erbia.* In 1907, Urbain separated ytterbia into two components, which he called *neoytterbia* and *lutecia.* The elements in these earths are now known as *ytterbium* and *lutetium,* respectively. These elements are identical with *aldebaranium* and *cassiopeium,* discovered independently and at about the same time by von Welsbach. Ytterbium occurs along with other rare earths in a number of rare minerals. It is commercially recovered principally from *monazite sand,* which contains about 0.03%. Ion-exchange and solvent extraction techniques developed in recent years have greatly simplified the separation of the rare earths from one another. The element was first prepared by Klemm and Bonner in 1937 by reducing ytterbium trichloride with potassium. Their metal was mixed, however, with KCI. Daane, Dennison, and Spedding prepared a much purer form in 1953 from which the chemical and physical properties of the element could be determined. Ytterbium has a bright silvery luster, is soft, malleable, and quite ductile. While the element is fairly stable, it should be kept in closed containers to protect it from air and moisture. Ytterbium is readily attacked and dissolved by dilute and concentrated mineral acids and reacts slowly with water. Ytterbium normally has two allotropic forms with a transformation point at 798°C. The alpha form is a room-temperature, face-centered, cubic modification, while the high-temperature beta form is a body-centered cubic form. Another body-centered cubic phase has recently been found to be stable at high pressures at room temperatures. The alpha form ordinarily has metallic-type conductivity, but becomes a semiconductor when the pressure is increased above 16,000 atm. The electrical resistance increases tenfold as the pressure is increased to 39,000 atm and drops to about 80% of its standard temperature-pressure resistivity at a pressure of 40,000 atm. Natural ytterbium is a mixture of seven stable isotopes. Seven other unstable isotopes are known. Ytterbium metal has possible use in improving the grain refinement, strength, and other mechanical properties of stainless steel. One isotope is reported to have been used as a radiation source as a substitute for a portable X-ray machine where electricity is unavailable. Few other uses have been found. Ytterbium metal is commercially available with a purity of about 99+% for about $1.00/g or $300/lb. Ytterbium has a low acute toxic rating.

Yttrium — (*Ytterby,* village in Sweden near Vauxholm), Y; at. wt. 88.9059; at. no. 39; m.p. 1522 ± 8°C; b.p. 3338°C; sp. gr. 4.469 (25°C); valence 3. *Yttria,* which is an earth containing yttrium, was discovered by Gadolin in 1794. *Ytterby* is the site of a quarry which yeilded many unusual minerals containing rare earths and other elements. This small town, near Stockholm, bears the honor of giving names to *erbium, terbium,* and *ytterbium* as well as *yttrium.* In 1843 Mosander showed that yttria could be resolved into the earths of three elements. The name yttria was reserved for the most basic one; the others were named *erbia* and *terbia.* Yttrium occurs in nearly all of the rare-earth minerals. Analysis of lunar rock samples obtained during the Apollo missions show a relatively high yttrium content. It is recovered commercially from *monazite sand,* which contains about 3%, and from *bastnasite,* which contains about 0.2%. Wohler obtained the impure element in 1828 by reduction of the anhydrous chloride with potassium. The metal is now produced commercially by reduction of the fluoride with calcium metals. It can also be prepared by other techniques. Yttrium has a silvery-metallic luster and is relatively stable in air if their temperature exceeds 400°C, and finely divided yttrium is very unstable in air. Turnings of the metal, however, ignite in air. Yttrium oxide is one of the most important compounds of yttrium and accounts for the largest use. It is widely used in making YVO_4europium, and Y_2O_3europium phosphors to give the red color in color television tubes. Many hundreds of thousands of pounds are now used in this application. Yttrium oxide also is used to produce yttrium-iron-garnets, which are very effective microwave filters. Yttrium iron, aluminum, and gadolinium garnets, with formulas such as $Y_3Fe_5O_{12}$ and $Y_3Al_5O_{12}$, have interesting magnetic properties. Yttrium iron garnet is also exceptionally efficient as both a transmitter and transducer of acoustic energy. Yttrium aluminum garnet, with a hardness of 8.5, is also finding use as a gemstone (simulated diamond). Small amounts of yttrium (0.1 to 0.2%) can be used to reduce the grain size in chromium, mo-

lybdenum, zirconium, and titanium, and to increase strength of aluminum and magnesium alloys. Alloys with other useful properties can be obtained by using yttrium as an additive. The metal can be used as a deoxidizer for vanadium and other nonferrous metals. The metal has a low cross section for nuclear capture. Y^{90}, one of the isotopes of yttrium, exists in equilibrium with its parent Sr^{90}, a product of atomic explosions. Yttrium has been considered for use as a nodulizer for producing nodular cast iron, in which the graphite forms compact nodules instead of the usual flakes. Such iron has increased ductility. Yttrium is also finding application in laser systems and as a catalyst for ethylene polymerization. It has also potential use in ceramic and glass formulas as the oxide has a high melting point and imparts shock resistance and low expansion characteristics to glass. Natural yttrium contains but one isotope, Y^{89}. Twenty other unstable nuclides and isomers have been characterized. Yttrium metal of 99+% purity is commercially available at a cost of about 60c/g or $150/lb.

Zinc — (Ger. *Zink,* of obscure origin), Zn; at. wt. 65.38; at. no. 30; m.p. 419.58°C; b.p. 907°C; sp. gr. 7.133 (25°C); valence 2. Centuries before zinc was recognized as a distinct element, zinc ores were used for making brass. Tubal-Cain, seven generations from Adam, is mentioned as being an "instructor in every artificer in brass and iron." An alloy containing 87% zinc has been found in prehistoric ruins in Transylvania. Metallic zinc was produced in the 13th century A.D. in India by reducing calamine with organic substances such as wool. The metal was rediscovered in Europe by Marggraf in 1746, who showed that it could be obtained by reducing *calamine* with charcoal. The principal ores of zinc are *sphalerite* or *blende* (sulfide), *smithsonite* (carbonate), *calamine* (silicate), and *franklinite* (zinc, manganese, iron oxide). Zinc can be obtained by roasting its ores to form the oxide and by reduction of the oxide with coal or carbon, with subsequent distillation of the metal. Other methods of extraction are possible. Naturally occurring zinc contains five stable isotopes. Ten other unstable nuclides and isomers are recognized. Zinc is a bluish-white, lustrous metal. It is brittle at ordinary temperatures but malleable at 100 to 150°C. It is a fair conductor of electricity, and burns in air at high red heat with evolution of white clouds of the oxide. The metal is employed to form numerous alloys with other metals. Brass, nickel silver, typewriter metal, commercial bronze, spring brass, German silver, soft solder, and aluminum solder are some of the more important alloys. Large quantities of zinc are used to produce die castings, used extensively by the automotive, electrical, and hardware industries. An alloy called *Prestal,* consisting of 78% zinc and 22% aluminum is reported to be almost as strong as steel but as easy to mold as plastic. It is said to be so plastic that it can be molded into form by relatively inexpensive die casts made of ceramics and cement. It exhibits superplasticity. Zinc is also extensively used to galvanize other metals such as iron to prevent corrosion. Neither zinc nor zirconium is ferromagnetic, but $ZrZn_2$ exhibits ferromagnetism at temperatures below 35 K. Zinc oxide is a unique and very useful material to modern civilization. It is widely used in the manufacture of paints, rubber products, cosmetics, pharmaceuticals, floor coverings, plastics, printing inks, soap, storage batteries, textiles, electrical equipment, and other products. It has unusual electrical, thermal, optical, and solid-state properties that have not yet been fully investigated. Lithopone, a mixture of zinc sulfide and barium sulfate, is an important pigment. Zinc sulfide is used in making luminous dials, X-ray and TV screens, and fluorescent lights. The chloride and chromate are also important compounds. Zinc is an essential element in the growth of human beings and animals. Tests show that zinc-deficient animals require 50% more food to gain the same weight of an animal supplied with sufficient zinc. Zinc is not considered to be toxic, but when freshly formed ZnO is inhaled a disorder known as the *oxide shakes* or *zinc chills* sometimes occurs. It is recommended that where zinc oxide is encountered good ventilation be provided to avoid concentrations exceeding 5 mg/M^3, (time-weighted over an 8-hr exposure, 40-hr work week). Zinc costs roughly 40c/lb

Zirconium — (Arabic *zargun,* gold color), Zr; at. wt. 91.22; at. no. 40; m.p. 1852 ± 2°C; b.p. 4377°C; sp. gr. 6.506 (20°C); valence +2, +3, and +4. The name *zircon* probably originated from the arabic word *zargun,* which describes the color of the gemstone now known

as *zircon, jargon, hyacinth, jacinth,* or *ligure.* This mineral, or its variations, is mentioned in biblical writings. The mineral was not known to contain a new element until Klaproth, in 1789, analyzed a jargon from Ceylon and found a new earth, which Werner named zircon (*silex circonius*), and Klaproth called *Zirkonerde* (*zirconia*). The impure metal was first isolated by Berzelius in 1824 by heating a mixture of potassium and potassium zirconium fluoride in a small iron tube. Pure zirconium was first prepared in 1914. Very pure zirconium was first produced in 1925 by van Arkel and de Boer by an iodide decomposition process they developed. Zirconium is found in abundance in S-type stars, and has been identified in the sun and meteorites. Analyses of lunar rock samples obtained during the various Apollo missions to the moon show a surprisingly high zirconium oxide content, compared with terrestial rocks. Naturally occurring zirconium contains five isotopes, one of which, Zr^{96} (abundant to the extent of 2.80%,) is unstable with a very long half-life of $> 3.6 \times 10^{17}$ years. Fifteen other unstable nuclides and isomers of zirconium have been characterized. Zircon, $ZrSiO_4$, the principal ore, is found in deposits in Florida, South Carolina, Australia, and Brazil. *Baddeleyite,* found in Brazil, is an important zirconium mineral. It is principally pure ZrO_2 in crystalline form having a hafnium content of about 1%. Zirconium also occurs in some 30 other recognized mineral species. Zirconium is produced commercially by reduction of the chloride with magnesium (the Kroll Process), and by other methods. It is a grayish-white lustrous metal. When finely divided, the metal may ignite spontaneously in air, especially at elevated temperatures. The solid metal is much more difficult to ignite. The inherent toxicity of zirconium compounds is low. Hafnium is invariably found in zirconium ores, and the separation is difficult. Commercial-grade zirconium contains from 1 to 3% hafnium. Zirconium has a low absorption cross section for neutrons, and is therefore used for nuclear energy applications, such as for cladding fuel elements. Zirconium has been found to be extremely resistant to the corrosive environment inside atomic reactors, and it allows neutrons to pass through the internal zirconium construction material without appreciable absorption of energy. Commercial nuclear power generation now takes more than 90% of zirconium metal production. Reactors of the size now being made may use as much as a half-million lineal feet of zirconium alloy tubing. Reactor-grade zirconium is essentially free of hafnium. *Zircaloy* is an important alloy developed specifically for nuclear applications. Zirconium is exceptionally resistant to corrosion by many common acids and alkalis, by sea water, and by other agents. It is used extensively by the chemical industry where corrosive agents are employed. Zirconium is used as a getter in vacuum tubes, as an alloying agent in steel, in making surgical appliances, photoflash bulbs, explosive primers, rayon spinnerets, lamp filaments, etc. It is used in poison ivy lotions in the form of the carbonate as it combines with *urushiol.* With niobium, zirconium is superconductive at low temperatures and is used to make superconductive magnets, which offer hope of direct large-scale generation of electric power. Alloyed with zinc, zirconium becomes magnetic at temperatures below 35 K. Zirconium oxide (zircon) has a high index of refraction and is used as a gem material. The impure oxide, zirconia, is used for laboratory crucibles that will withstand heat shock, for linings of metallurgical furnaces, and by the glass and ceramic industries as a refractory material. Its use as a refractory material accounts for a large share of all zirconium consumed. Commercial zirconium metal sponge is priced at about $7/lb. Fabricated zirconium parts are higher in cost.

NOMENCLATURE OF INORGANIC CHEMISTRY

AMERICAN VERSION*

By permission of the Committee on Publications of the International Union of Pure and Applied Chemistry.

INDEX TO IUPAC INORGANIC RULES

*This is the United States presentation of these IUPAC Inorganic Rules as published in J. Am. Chem. Soc., 82, 5525 (1960)

1. ELEMENTS

1.1. Names and Symbols of the Elements

1.11.—The elements should have the symbols given in the following table (Table I). It is desirable that the names should differ as little as possible among the different languages, but as complete uniformity is hard to achieve, separate lists have been drawn up in English and in French. The English list only is reproduced here.

TABLE I

ELEMENTS

Name	Symbol	Atomic number	Name	Symbol	Atomic number	Name	Symbol	Atomic number
Actinium	Ac	89	Gold (Aurum)	Au	79	Praseodymium	Pr	59
Aluminum	Al	13	Hafnium	Hf	72	Promethium	Pm	61
Americium	Am	95	Helium	He	2	Protactinium	Pa	91
Antimony	Sb	51	Holmium	Ho	67	Radium	Ra	88
Argon	Ar	18	Hydrogen	H	1	Radon	Rn	86
Arsenic	As	33	Indium	In	49	Rhenium	Re	75
Astatine	At	85	Iodine	I	53	Rhodium	Rh	45
Barium	Ba	56	Iridium	Ir	77	Rubidium	Rb	37
Berkelium	Bk	97	Iron (Ferrum)	Fe	26	Ruthenium	Ru	44
Beryllium	Be	4	Krypton	Kr	36	Samarium	Sm	62
Bismuth	Bi	83	Lanthanum	La	57	Scandium	Sc	21
Boron	B	5	Lead (Plumbum)	Pb	82	Selenium	Se	34
Bromine	Br	35	Lithium	Li	3	Silicon	Si	14
Cadmium	Cd	48	Lutetium	Lu	71	Silver (Argentum)	Ag	47
Calcium	Ca	20	Magnesium	Mg	12	Sodium	Na	11
Californium	Cf	98	Manganese	Mn	25	Strontium	Sr	38
Carbon	C	6	Mendelevium	Md	101	Sulfur	S	16
Cerium	Ce	58	Mercury	Hg	80	Tantalum	Ta	73
Cesium	Cs	55	Molybdenum	Mo	42	Technetium	Tc	43
Chlorine	Cl	17	Neodymium	Nd	60	Tellurium	Te	52
Chromium	Cr	24	Neon	Ne	10	Terbium	Tb	65
Cobalt	Co	27	Neptunium	Np	93	Thallium	Tl	81
Copper (Cuprum)	Cu	29	Nickel	Ni	28	Thorium	Th	90
Curium	Cm	96	Niobium	Nb	41	Thulium	Tm	69
Dysprosium	Dy	66	Nitrogen	N	7	Tin (Stannum)	Sn	50
Einsteinium	Es	99	Nobelium	No	102	Titanium	Ti	22
Erbium	Er	68	Osmium	Os	76	Tungsten (Wolfram)	W	74
Europium	Eu	63	Oxygen	O	8	Uranium	U	92
Fermium	Fm	100	Palladium	Pd	46	Vanadium	V	23
Fluorine	F	9	Phosphorus	P	15	Xenon	Xe	54
Francium	Fr	87	Platinum	Pt	78	Ytterbium	Yb	70
Gadolinium	Gd	64	Plutonium	Pu	94	Yttrium	Y	39
Gallium	Ga	31	Polonium	Po	84	Zinc	Zn	30
Germanium	Ge	32	Potassium	K	19	Zirconium	Zr	40

◆**The committees reaffirm the name niobium for element 41 in spite of the fact that many in the United States, particularly outside of chemical circles, still retain the name columbium.**

1.12.—The names placed in parentheses (after the trivial names) in the list in Table I shall always be used when forming names derived from those of the elements, *e.g.*, aurate, ferrate, wolframate and not goldate, ironate, tungstate.

For some compounds of sulfur, nitrogen, and antimony, derivatives of the Greek name *θεῖον*, the French name *azote*, and the Latin name *stibium*, respectively, are used.

Although the name nickel agrees with the chemical symbol, it is essentially a trivial name and is spelled so differently in various languages (niquel, nikkel, *etc.*) that it is recommended that names of derivatives be formed from the Latin name *niccolum, e.g.*, niccolate instead of nickelate. The name mercury should be used as the root name also in languages where the element has another name (mercurate, *not* hydrargyrate).

In the cases in which different names have been used, the Commission has selected one based upon considerations of prevailing usage and practicability. It should be emphasized that their selection carries no implication regarding priority of discovery.

◆**Tungstate and nickelate are both so well established in American practice that the committees object to changing them to wolframate and niccolate.**

1.13.—Any new metallic elements should be given names ending in -ium. Molybdenum and a few other elements have long been spelled without an "i" in most languages, and the Commission hesitates to insert it.

1.14.—All new elements shall have two-letter symbols.

1.15.—All isotopes of an element should have the same name. For hydrogen the isotope names protium, deuterium, and tritium may be retained, but it is undesirable to assign isotopic names instead of numbers to other elements. They should be designated by mass numbers as, for example, "oxygen-18."

◆**The list in 3.21 implies that D is an acceptable symbol for deuterium, whereas ^{2}H is used in 1.32. It is recommended that D and T be allowed for deuterium and tritium, respectively. *Cf.* comments made at 1.31, 1.32.**

1.2. Names for Groups of Elements and their Subdivisions

1.21.—The use of the collective names: halogens (F, Cl, Br, I, and At), chalcogens (O, S, Se, Te, and Po), and halides and chalcogenides for their compounds, alkali metals (Li to Fr), alkaline earth metals (Ca to Ra), and inert gases may be continued. The name rare earth metals may be used for the elements Sc, Y, and La to Lu inclusive; the name lanthanum series for the elements no. 57–71 (La to Lu inclusive), and the name lanthanides for the elements 58–71 (Ce to Lu inclusive) are recommended. Elements no. 89 (Ac) to 103 form the actinium series, and the name actinides is reserved for the elements in which the 5*f* shell is being filled. The name transuranium elements is also approved for the elements following uranium.

◆The collective term halogenides used in the Rules has been replaced in this version by halides, which is almost universally used in English and is unambiguous.

The inclusion of Sc with the rare earths is questioned by some. No need is seen for the terms lanthanum series and actinium series, particularly since the latter term is used for a radioactive series. The use of a collective term for elements 58–71 is approved, although it is suggested that lanthan*oid* is preferable to lanthan*ide* because of the use of -ide for binary compounds; similarly, actin*oid* is preferable to actin*ide*. Definition by means of atomic numbers is recommended in both cases rather than on the basis of interpretation (*e.g.*, filling of 5*f* shells).

1.22.—The word metalloid should not be used to denote nonmetals.

1.3. Indication of Mass, Charge, etc., on Atomic Symbols

1.31.—The mass number, atomic number, number of atoms, and ionic charge of an element may be indicated by means of four indices placed around the symbol. The positions are to be occupied thus

left upper index	right lower index	mass number	number of atoms
left lower index	right upper index	atomic number	ionic charge

Ionic charge should be indicated by A^{n+} rather than by A^{+n}.

Example: ${}^{32}_{16}S_2^{2+}$ represents a doubly ionized molecule containing two atoms of sulfur, each of which has the atomic number 16 and mass number 32.

The following is an example of an equation for a nuclear reaction

$$ {}^{26}_{12}Mg + {}^{4}_{2}He = {}^{29}_{13}Al + {}^{1}_{1}H $$

◆Although the practice of American chemists and physicists in general has been to put the mass number at the upper right of the symbol, the committees recognize the advantage of putting it at the upper left so that the upper right is available for the ionic charge as needed.

1.32.—Isotopically labeled compounds may be described by adding to the name of the compounds the symbol of the isotope in parentheses.

Examples:

${}^{32}PCl_3$ phosphorus(${}^{32}P$) trichloride (spoken: phosphorus-32 trichloride)

$H^{36}Cl$ hydrogen chloride^{36}(Cl) (spoken: hydrogen chloride-36)

${}^{15}NH_3$ ammonia(${}^{15}N$) (spoken: ammonia nitrogen-15)

The position of the labeled atom may be indicated by placing the isotope symbol immediately after the locant (name of the group concerned).

Example: ${}^{2}H_2{}^{35}SO_4$ sulfuric(${}^{35}S$) acid(${}^{2}H$)

If this method gives names which are ambiguous or difficult to pronounce, the whole group containing the labeled atom may be indicated.

Examples:

$HOSO_2{}^{35}SH$	thiosulfuric(${}^{35}SH$) acid
${}^{15}NO_2NH_2$	nitramide(${}^{15}NO_2$), not nitr(${}^{15}N$)amide
$NO_2{}^{15}NH_2$	nitramide(${}^{15}NH_2$)
$HO_3S^{18}O{-}^{18}OSO_3H$	peroxo(${}^{18}O_2$)disulfuric acid

1.4. Allotropes

If systematic names for gaseous and liquid modifications are required, they should be based on the size of the molecule, which can be indicated by Greek numerical prefixes (listed in 2.251). If the number of atoms is large and unknown, the prefix poly may be used. To indicate ring and chain structures the prefixes cyclo and catena may be used.

Examples:

Symbol	Trivial Name	Systematic Name
H	atomic hydrogen	monohydrogen
O_2	(common) oxygen	dioxygen
O_3	ozone	trioxygen
P_4	white phosphorus (yellow phosphorus)	tetraphosphorus
S_8	λ-sulfur	cycloöctasulfur or octasulfur
S_n	μ-sulfur	catenapolysulfur or polysulfur

For the nomenclature of solid allotropic forms the rules in Section 8 may be applied.

◆The use of prefixes to indicate ring and chain structures is favored by the committees, but it should be pointed out that ino (not catena) has been used by mineralogists for indicating chain structures in silicates, along with cyclo and other prefixes (neso, phyllo, tecto, soro) denoting structure, all used without italics or hyphens (*e.g.*, inosilicates). *Cf.* use of catena in 7.42 for chains of alternating, not self-linking, atoms.

2. FORMULAS AND NAMES OF COMPOUNDS IN GENERAL

Many chemical compounds are essentially binary in nature and can be regarded as combinations of ions or radicals; others may be treated as such for the purpose of nomenclature.

Some chemists have expressed the opinion that the name of a compound should indicate whether it is ionic or covalent. Such a distinction is made in some languages (*e.g.*, in German: Natriumchlorid *but* Chlorwasserstoff), but it has not been made consistently, and indeed it seems impossible to introduce this distinction into a consistent system of nomenclature, because the line of demarcation between these two categories is not sharp. In these rules a system of nomenclature has been built on the basis of the endings -ide and -ate, and it should be emphasized that these are intended to be applied both to ionic and covalent compounds. If it is desired to avoid such endings for neutral molecules, names can be given as coördination compounds in accordance with **2.24** and Section **7.**

2.1. Formulas

2.11.—Formulas provide the simplest and clearest method of designating inorganic compounds. They are of particular importance in chemical equations and in descriptions of chemical procedure. However, their general use in text is not recommended, although in some cases a formula, on account of its compactness, may be preferable to a cumbersome and awkward name.

2.12.—The *empirical formula* is formed by juxtaposition of the atomic symbols to give the *simplest possible* formula expressing the stoichiometric composition of the compound in question. The empirical formula may be supplemented by indication of the crystal structure—see Section **8.**

2.13.—For compounds consisting of discrete molecules the *molecular formula, i.e.*, a formula corresponding with the correct molecular weight of the compound, should be used, *e.g.*, S_2Cl_2 and $H_4P_2O_6$ and not SCl and H_2PO_3. When the molecular weight varies with temperature, *etc.*, the simplest possible formula generally may be chosen, *e.g.*, S, P, and NO_2 instead of S_8, P_4, and N_2O_4, unless it is desirable to indicate the molecular complexity.

2.14.—In the *structural formula* the sequence and spatial arrangement of the atoms in a molecule are indicated.

2.15.—In formulas the *electropositive constituent* (cation) should always be placed first, *e.g.*, KCl, $CaSO_4$. This also applies in Romance languages even though the electropositive constituent is placed last in the *name, e.g.*, KCl, chlorure de potassium.

If the compound contains more than one electropositive or more than one electronegative constituent, their sequence is determined by Rules **6.32** and **6.33.**

2.16.—In the case of binary compounds between nonmetals that constituent should be placed first which appears earlier in the sequence: B, Si, C, Sb, As, P, N, H, Te, Se, S, At, I, Br, Cl, O, F.

Examples: NH_3, H_2S, N_4S_4, S_2Cl_2, Cl_2O, OF_2

◆Because N_4S_4 is definitely a nitride, not a sulfide, the committees prefer to write the formula S_4N_4, name it sulfur nitride, and cite it as an exception rather than as an example.

2.161.—For compounds containing three or more elements, however, the sequence should in general follow the order in which the atoms are actually bound in the molecule or ion, *e.g.*, NCS^-, not CNS^-, HOCN (cyanic acid), and HONC (fulminic acid).

Although formulas such as HNO_3, $HClO_4$, H_2SO_4, do not agree with this rule and HNO_3 does not even follow the main rule in **2.16**, the Commission does not at this time wish to break the old custom of putting the central atom immediately after the hydrogen atom in such cases (*cf* Section **5**). The formula for hypochlorous acid may be written HOCl or HClO.

2.17.—In intermetallic compounds the constituents should be placed in the order

Fr, Cs, Rb, K, Na, Li	Pt, Ir, Os, Pd, Rh, Ru, Ni, Co, Fe
Ra, Ba, Sr, Ca, Mg, Be	Au, Ag, Cu
103, No, Md, Fm, Es, Cf, Bk, Cm, Am, Pu,	Hg, Cd, Zn
Np, U, Pa, Th, Ac, Lu–La, Y, Sc	Tl, In, Ga, Al
Hf, Zr, Ti	Pb, Sn, Ge
Ta, Nb, V	Bi, Sb
W, Mo, Cr	Po
Re, Tc, Mn	

Nonmetals (except Sb) in the order given in **2.16.**

Deviations from this order may be allowed, *e.g.*, when compounds with analogous structures are compared (AgZn and AgMg).

2.18.—The number of identical atoms or atomic groups in a formula is indicated by means of Arabic numerals, placed below and to the right of the symbol or symbols in parentheses ( ) or brackets [ ] to which they refer. Water of crystallization and similar loosely bound molecules, however, are designated by means of Arabic numerals before their formulas.

Examples: $CaCl_2$ not $CaCl^2$
$[Co(NH_3)_6]Cl_3$ not $[Co6NH_3]Cl_3$
$[Co(NH_3)_6]_2(SO_4)_3$
$Na_2SO_4.10H_2O$

2.19.—The prefixes *cis, trans, sym, asym* may be used in their usual senses. The prefixes may be connected with the formula by a hyphen and it is recommended that they be italicized.

Example: *cis*-$[PtCl_2(NH_3)_2]$

2.2. Systematic Names

Systematic names of compounds are formed by indicating the constituents and their proportions according to the following rules. (For the order of the constituents see also the later sections.)

2.21.—The name of the *electropositive constituent* (or that treated as such according to **2.16**) will not be modified (see, however, **2.2531**).

In Germanic languages the electropositive constituent is placed first, but in Romance languages it is customary to place the electronegative constituent first.

2.22.—If the *electronegative constituent* is monatomic its name is modified to end in -ide. For binary compounds of the nonmetals the name of the element standing later in the sequence in **2.16** is modified to end in -ide.

Examples: Sodium chloride, calcium sulfide, lithium nitride, arsenic selenide, calcium phosphides, nickel arsenide, aluminum borides, iron carbides, boron hydrides, phosphorus hydrides, hydrogen chloride, hydrogen sulfide, silicon carbide, carbon disulfide, sulfur hexafluoride, chlorine dioxide, oxygen difluoride.

Certain polyatomic groups are also given the ending -ide—see **3.22**.

In the Romance languages the endings -ure, -uro, and -eto are used instead of -ide. In some languages the word *oxyde* is used, whereas the ending -ide is used in the names of other binary compounds; it is recommended that the ending -ide be universally adopted in these languages.

◆**Nitrogen sulfide has been taken out of the examples.** ***Cf.*** **comment at 2.16.**

2.23.—If the electronegative constituent is polyatomic it should be designated by the termination -ate. In certain exceptional cases the terminations -ide and -ite are used—see **3.22**.

2.24.—In inorganic compounds it is generally possible in a polyatomic group to indicate a *characteristic atom* (as in ClO^-) or a *central atom* (as in ICl_4^-). Such a polyatomic group is designated a *complex*, and the atoms, radicals, or molecules bound to the characteristic or central atom are termed *ligands*.

In this case the name of a negatively charged complex should be formed from the name of the characteristic or central element (as indicated in **1.12**) modified to end in -ate.

Anionic ligands are indicated by the termination -o. Further details concerning the designation of ligands, the definition of "central atom," *etc.*, appear in Section **7**.

Although the terms sulfate, phosphate, *etc.*, were originally the names of the anions of particular oxo acids, the names sulfate, phosphate, *etc.*, should now designate quite generally a negative group containing sulfur or phosphorus, respectively, as the central atom, irrespective of its oxidation state (the designation of the oxidation state is discussed in later rules) and the number and nature of the ligands. The complex is indicated by brackets [ ], but this is not always necessary.

Examples:

$Na_2[SO_4]$	sodium tetraoxosulfate	$Na_3[PS_4]$	sodium tetrathiophosphate
$Na_2[SO_3]$	sodium trioxosulfate	$Na[PCl_6]$	sodium hexachlorophosphate
$Na_2[S_2O_3]$	sodium trioxothiosulfate	$K[PO_2F_2]$	potassium dioxodifluorophosphate
$Na[SO_3F]$	sodium trioxofluorosulfate	$K[POCl_2(NH)]$	potassium oxodichloroimidophosphate
$Na_3[PO_4]$	sodium tetraoxophosphate		

In many cases these names may be abbreviated, *e.g.*, sodium sulfate, sodium thiosulfate (see **2.26**), and in other cases trivial names may be used (*cf.* **2.3**, **3.224**, and Section **5**). It should be pointed out, however, that the principle is quite generally applicable, to compounds containing organic ligands also, and its use is recommended in all cases where trivial names do not exist.

The coördination principle applied in this rule may also be applied to complexes which are positive or neutral (*cf.* **3.1** and Section **7**). However, neutral complexes which are as a rule considered as binary compounds are given names according to **2.22**, **2.16**. Thus, SO_3, sulfur trioxide, not trioxosulfur.

◆**In the examples it would seem that full cöordination-type names should be given,** ***e.g.*****, either sodium tetraoxosulfate (VI) or disodium tetraoxosulfate.** ***Cf.*** **7.32 and comment at 7.312.**

2.25.—Indication of the Proportions of the Constituents.

2.251.—The *stoichiometric proportions* may be denoted by means of Greek numerical prefixes (mono, di, tri, tetra, penta, hexa, hepta, octa, ennea, deca, hendeca, and dodeca) preceding without hyphen the names of the elements to which they refer. It may be necessary in some languages to supplement these numerals with hemi ($^1/_2$) and the Latin sesqui ($^3/_2$).

The prefix mono may generally be omitted. Beyond 12, Greek prefixes are replaced by Arabic numerals (with or without hyphen according to the custom of the language), because they are more readily understood.

This sytem is applicable to all types of compounds and is especially suitable for binary compounds of the nonmetals.

When it is required to indicate the number of entire groups of atoms, particularly when the name includes a numerical prefix with a different significance, the multiplicative numerals (Latin bis, Greek tris, tetrakis, *etc.*) are used and the whole group to which they refer may be placed in parentheses if necessary.

Examples:

N_2O	dinitrogen oxide	Fe_3O_4	triiron tetraoxide
NO_2	nitrogen dioxide	U_3O_8	triuranium octaoxide
N_2O_4	dinitrogen tetraoxide	MnO_2	manganese dioxide
N_2S_5	dinitrogen pentasulfide	$Ca_3[PO_4]_2$	tricalcium diorthophosphate
S_2Cl_2	disulfur dichloride	$Ca[PCl_6]_2$	calcium bis(hexachlorophosphate)

In indexes it may be convenient to italicize a numerical prefix at the beginning of the name and connect it to the rest of the name with a hyphen, but this is not desirable in text, *e.g.*, *tri*-Uranium octaoxide.

Since the degree of polymerization of many substances varies with temperature, state of aggregation, *etc.*, the name to be used should normally be based upon the simplest possible formula of the substance except when it is required specifically to draw attention to the degree of polymerization.

Example: The name nitrogen dioxide may be used for the equilibrium mixture of NO_2 and N_2O_4. Dinitrogen tetraoxide means specifically N_2O_4.

◆In accordance with the organic nomenclature rules and well-established practice, it is recommended that "or Latin" be inserted between "Greek" and "numerical prefixes" in the first sentence and that "nona" replace "ennea," and "undeca" replace "hendeca."

Extreme caution is advised in the omission of numerical prefixes, including mono (*cf.* second set of examples in 5.23), because of the frequent use of names such as chloroplatinate (*cf.* 2.26 and the last sentence in 5.24).

2.252.—The proportions of the constituents also may be indicated indirectly by *Stock's system*, that is, by Roman numerals representing the oxidation number or stoichiometric valence of the element, placed in parentheses immediately following the name. For zero the Arabic 0 will be used. When used in conjunction with symbols the Roman numeral may be placed above and to the right.

The Stock notation can be applied to both cations and anions, but preferably should *not* be applied to compounds between nonmetals.

In employing the Stock notation, use of the Latin names of the elements (or Latin roots) is considered advantageous.

Examples:

$FeCl_2$	iron(II) chloride or ferrum(II) chloride
$FeCl_3$	iron(III) chloride or ferrum(III) chloride
MnO_2	manganese(IV) oxide
BaO_2	barium(II) peroxide
$Pb^{II}_2Pb^{IV}O_4$	dilead(II) lead(IV) oxide or trilead tetraoxide
$K_4[Ni(CN)_4]$	potassium tetracyanonicollate(0)
$K_4[Fe(CN)_6]$	potassium hexacyanoferrate(II)
$Na_2[Fe(CO)_4]$	sodium tetracarbonylferrate(—II)

◆While the committees favor the extended use of the Stock notation, they suggest that in some cases the system of Ewens and Bassett (designation of the aggregate charge of a complex ion by an Arabic numeral in parentheses following the name, similar to the use as superior notations with formulas) is advantageous and should be allowed as an alternate (*cf.* 3.17 and comment at 7.323). Mixed use of the two systems, while not desirable in any one context, does not affect indexing and should not lead to confusion. See comment at 1.12.

2.253.—The following systems are in use but are not recommended:

2.2531.—The system of indicating valence by means of the suffixes *-ous* and *-ic* added to the root of the name of the cation may be retained for elements exhibiting not more than two valences.

2.2532.—*"Functional"* nomenclature (such as "nitric anhydride" for N_2O_5) is not recommended apart from the name *acid* to designate the acid function (Section **5**).

◆Apparently there is no objection to acid anhydride as a class name (*cf.* 5.32). Other functional derivatives of acids are named as such in the Rules (5.3).

2.26.—In systematic names it is not always necessary to indicate stoichiometric proportions. In many instances it is permissible to omit the numbers of atoms, oxidation numbers, *etc.*, when they are not required in the particular circumstances. For instance, these indications are not generally necessary with elements of essentially constant valence.

Examples:

sodium sulfate instead of sodium tetraoxosulfate
aluminum sulfate instead of aluminum(III) sulfate
potassium chloroplatinate(IV) instead of potassium hexachloroplatinate(IV)
potassium cyanoferrate(III) instead of potassium hexacyanoferrate(III)
phosphorus pentaoxide instead of diphosphorus pentaoxide

2.3. Trivial Names

Certain well-established trivial names for oxo acids (Section **5**) and for hydrogen compounds (water, ammonia, hydrazine) are still acceptable. For some other hydrogen compounds these names are approved:

B_2H_6	diboran	PH_3	phosphine	SbH_3	stibine	P_2H_4	diphosphine
SiH_4	silane	AsH_3	arsine	Si_2H_6	disilane, *etc.*	As_2H_4	diarsine

In some languages names of the type "Chlorwasserstoff" are in use and may be retained if national nomenclature committees so wish.

Purely trivial names, free from false scientific implications, such as soda, Chile saltpeter, quicklime, are harmless in industrial and popular literature; but old incorrect scientific names such as sulfate of magnesia, Natronhydrat, sodium muriate, carbonate of lime, should be avoided under all circumstances, and they should be eliminated from technical and patent literature.

◆For BH_3 (omitted in the Rules) borane rather than the previously used borine has been recommended by the Advisory Committee on the Nomenclature of Organic Boron Compounds of the ACS in a report not yet published.

Because soda is an ambiguous term, it is suggested that it be replaced by soda ash.

3. NAMES FOR IONS AND RADICALS

3.1. Cations

3.11.—Monatomic cations should be named like the corresponding element, without change or suffix, except as provided by **2.2531**.

Examples:

Cu^+ the copper(I) ion Cu^{2+} the copper(II) ion I^+ the iodine cation

◆For I^+, iodide(I) cation is more consistent with recommended practice. *Cf.* 3.21.

3.12.—The preceding principle should apply also to polyatomic cations corresponding to radicals for which special names are given in **3.32**, *i.e.*, these names should be used without change or suffix.

Examples: NO^+ the nitrosyl cation NO_2^+ the nitryl cation

◆Polyatomic here and in 3.13, 3.14, 3.223, 3.32, and 5.2 seems to be limited to more than one *kind* of atom, and hence heteratomic would be a more precise term. It is agreed that nitryl and not nitronium should be used in all cases (*cf.* 3.151).

3.13.—Polyatomic cations formed from monatomic cations by the addition of other ions or neutral atoms or molecules (ligands) will be regarded as complex and will be named according to the rules given in Section 7.

Examples:

$[Al(H_2O)_6]^{3+}$ the hexaaquoaluminum ion $[CoCl(NH_3)_5]^{2+}$ the chloropentamminecobalt ion

For some important polyatomic cations which fall in this section, radical names given in **3.32** may be used alternatively, *e.g.*, for UO_2^{2+} the name uranyl(VI) ion in place of dioxouranium(VI) ion.

3.14.—Names for polyatomic cations derived by addition of protons to monatomic anions are formed by adding the ending -onium to the root of the name of the anion element.

Examples: phosphonium, arsonium, stibonium, oxonium, sulfonium, selenonium, telluronium, and iodonium ions.

Organic ions derived by substitution in these parent cations should be named as such, whether the parent itself is a known compound or not: for example $(CH_3)_4Sb^+$, the tetramethylstibonium ion.

The ion H_3O^+, which is in fact the monohydrated proton, is to be known as the oxonium ion when it is believed to have this constitution, as for example in $H_3O^+ClO_4^-$, oxonium perchlorate. The widely used term hydronium should be kept for the cases where it is wished to denote an indefinite degree of hydration of the proton, as, for example, in aqueous solution. If, however, the hydration is of no particular importance to the matter under consideration, the simpler term hydrogen ion may be used. The latter also may be used for the indefinitely solvated proton in nonaqueous solvents; but definite ions such as $CH_3OH_2^+$ and $(CH_3)_2OH^+$ should be named as derivatives of the oxonium ion, *i.e.*, as methyl- and dimethyloxonium ions, respectively.

◆The committees concur in oxonium for the ion H_3O^+, but see little reason for encouraging retention of the term hydronium ion because hydrogen ion adequately designates an indeterminate degree of hydration.

3.15.—Ions from Nitrogen Bases.

3.151.—The name ammonium for the ion NH_4^+ does not conform to **3.14**, but should be retained. This decision does *not* release the word nitronium for other uses: this would lead to inconsistencies when the rules are applied to other elements.

3.152.—Substituted ammonium ions derived from nitrogen bases with names ending in -amine will receive names formed by changing -amine to -ammonium. For example, $HONH_3^+$, the hydroxylammonium ion.

3.153.—When the nitrogen base is known by a name ending otherwise than in -amine, the cation name is to be formed by adding the ending -ium to the name of the base (if necessary omitting a final -e or other vowel).

Examples: hydrazinium, anilinium, glycinium, pyridinium, guanidinium, imidazolium.

The names uronium and thiouronium, though inconsistent with this rule, may be retained.

3.16.—Cations formed by adding protons to nonnitrogenous bases may also be given names formed by adding -ium to the name of the compound to which the proton is added.

Examples: dioxanium, acetonium.

In the case of cations formed by adding protons to acids, however, their names are to be formed by adding the word acidium to the name of the corresponding anion, and not that of the acid itself. For example, $H_2NO_3^+$, the nitrate acidium ion; $H_2NO_2^+$, the nitrite acidium ion; and $CH_3COOH_2^+$, the acetate acidium ion. Note, however, that when the anion of the acid is monatomic **3.14** will apply; for example, FH_2^+ is the fluoronium ion.

◆In accord with present practice, nitric acidium ion, *etc.*, are preferred to nitrate acidium ion, *etc.* $CH_3COOH_2^+$ is organic.

3.17.—Where more than one ion is derived from one base, as, for example, $N_2H_5^+$ and $N_2H_6^{2+}$, their charges may be indicated in their names as the hydrazinium(1+) and the hydrazinium(2+) ion, respectively.

◆*Cf.* comment at 2.252 and the use of Stock notation with ions or radicals in 3.13 and 3.32.

3.2. Anions

3.21.—The names for monatomic anions shall consist of the name (sometimes abbreviated) of the element, with the termination -ide. Thus

H^-	hydride ion	Br^-	bromide ion	Se^{2-}	selenide ion	As^{3-}	arsenide ion
D^-	deuteride ion	I^-	iodide ion	Te^{2-}	telluride ion	Sb^{3-}	antimonide ion
F^-	fluoride ion	O^{2-}	oxide ion	N^{3-}	nitride ion	C^{4-}	carbide ion
Cl^-	chloride ion	S^{2-}	sulfide ion	P^{3-}	phosphide ion	Si^{4-}	silicide ion
				B^{3-}	boride ion		

Expressions of the type "chlorine ion" are used particularly in connection with crystal structure work and spectroscopy; the Commission recommends that whenever the charge corresponds with that indicated above, the termination -ide should be used.

◆*Cf.* comments at 1.15 and 1.32 regarding ^2H and D.

3.22. Polyatomic Anions.

3.221.—Certain polyatomic anions have names ending in -ide. These are

OH^-	hydroxide ion	I_3^-	triiodide ion	$NHOH^-$	hydroxylamide ion
O_2^{2-}	peroxide ion	HF_2^-	hydrogen difluoride ion	$N_2H_3^-$	hydrazide ion
O_2^-	hyperoxide ion	N_3^-	azide ion	CN^-	cyanide ion
O_3^-	ozonide ion	NH^{2-}	imide ion	C_2^{2-}	acetylide ion
S_2^{2-}	disulfide ion	NH_2^-	amide ion		

Names for other polysulfide, polyhalide, *etc.*, ions may be formed in analogous manner. The OH^- ion should not be called the hydroxyl ion. The name hydroxyl is reserved for the OH group when neutral or positively charged, whether free or as a substituent (*cf.* **3.12** and **3.32**).

◆Superoxide is well established in English for O_2^- and no advantage is seen in changing to hyperoxide.

3.222.—Ions such as SH^- and O_2H^- will be called the hydrogen sulfide ion and the hydrogen peroxide ion, respectively. This agrees with **6.2**, and names such as hydrosulfide are not needed.

◆*Cf.* comment at 6.2. All "fused" names (as hydrogensulfide here and methylisocyanide in 5.33) in the original version have been written as two words in this version. For a rule on the written form of the names of compounds see *J. Chem. Education*, 8, 1336–8 (1931)

3.223.—The names for other polyatomic anions shall consist of the name of the central atom with the termination -ate, which is used quite generally for complex anions. Atoms and groups attached to the central atom shall generally be treated as ligands in a complex (*cf.* **2.24** and Section **7**) as, for example, $[Sb(OH)_6]^-$, the hexahydroxoantimonate(V) ion.

This applies also when the exact composition of the anion is not known; *e.g.*, by solution of aluminum hydroxide or zinc hydroxide in sodium hydroxide, aluminate and zincate ions are formed.

3.224.—It is quite practicable to treat oxygen in the same manner as other ligands (**2.24**), but it has long been customary to ignore the name of this element altogether in anions and to indicate its presence and proportion by means of a series of prefixes (hypo-, per-, *etc.*, see Section **5**) and sometimes also by the suffix -ite in place of -ate.

The termination -ite has been used to denote a lower oxidation state and may be retained in trivial names in these cases

NO_2^-	nitrite	$PH_2O_2^-$	hypophosphite	$S_2O_2^{2-}$	thiosulfite
$N_2O_2^{2-}$	hyponitrite	AsO_3^{3-}	arsenite	SeO_3^{2-}	selenite
NOO_2^-	peroxonitrite	SO_3^{2-}	sulfite	ClO_2^-	chlorite
PHO_3^{2-}	phosphite	$S_2O_5^{2-}$	disulfite (pyrosulfite)	ClO^-	hypochlorite
$P_2H_2O_5^{2-}$	diphosphite (pyrophosphite)	$S_2O_4^{2-}$	dithionite		

(and correspondingly for the other halogens)

The Commission does not recommend the use of any such names other than those listed. A number of other names ending in -ite have been used, *e.g.*, antimonite, tellurite, stannite, plumbite, ferrite, manganite, but in many cases such compounds are known in the solid state to be double oxides and are to be treated as such (*cf.* **6.5**), *e.g.*, $Cu(CrO_2)_2$ copper(II) chromium(III) oxide, not copper chromite. Where there is reason to believe that a definite salt with a discrete anion exists, the name is formed in accordance with **2.24**. By dissolving, for example, Sb_2O_3, SnO, or PbO in sodium hydroxide an antimonate(III), a stannate(II), or a plumbate(II) is formed in the solution.

Concerning the use of prefixes hypo-, per-, *etc.*, see the list of acids (table in **5.214**). For all new compounds and even for the less common ones listed in the table in **3.224** or derived from the acids listed in the table in **5.214**, it is preferable to use the system given in **2.24** and in Sections **5** and **7**.

◆For phosphite names see comment at 6.2.

3.3. Radicals

3.31.—A radical is here regarded as a group of atoms which occurs repeatedly in a number of different compounds. Sometimes the same radical fulfils different functions in different cases, and accordingly different names often have been assigned to the same group. The Commission considers it desirable to reduce this diversity and recommends that formulas or systematic names be used to denote all new radicals, instead of introducing new trivial names. The list of names for ions and radicals on page B-169 gives an extensive selection of radical names at present in use in inorganic chemistry.

◆The list of names for ions and radicals following the Rules is very useful and can be made more so by additions (as of dithio, nitrilo, and azido in the last column) and by greater attempt at uniformity with organic usage. Some of the terms (as nitride and amide) listed as anions are used also of covalent compounds. A single atom may function like a radical as defined above and may be named similarly, as chloro and oxo.

3.32.—Certain radicals containing oxygen or other chalcogens have special names ending in -yl, and the Commission approves the provisional retention of

HO	hydroxyl	S_2O_5	pyrosulfuryl	PO	phosphoryl	ClO_2	chloryl
CO	carbonyl	SeO	seleninyl	VO	vanadyl	ClO_3	perchloryl
NO	nitrosyl	SeO_2	selenonyl	PuO_2	plutonyl		(and similarly
SO	sulfinyl	CrO_2	chromyl		(similarly for		for other
	(thionyl)	UO_2	uranyl		other actinide		halogens)
SO_2	sulfonyl	NpO_2	neptunyl		elements)		
	(sulfuryl)	NO_2	nitryl[1]	ClO	chlorosyl		

[1]The name nitroxyl should not be used for this group since the name nitroxylic acid has been used for H_2NO_2. Although the word nitryl is firmly established in English, nitroyl may be a better model for many other languages.

Names such as the above should be used only to designate compounds containing these discrete groups. The use of thionyl and sulfuryl should be restricted to the halides. Names such as bismuthyl and antimonyl are not approved because the compounds do not contain BiO and SbO groups, respectively; such compounds are to be designated as oxide halides (**6.4**).

Radicals analogous to the above containing other chalcogens in place of oxygen are named by adding the prefixes thio-, seleno-, *etc.*

Examples:

PS	thiophosphoryl	CSe	selenocarbonyl

In cases where radicals may have different valences, the oxidation number of the characteristic element should be indicated by means of the Stock notation. For example, the uranyl group UO_2 may refer either to the ion UO_2^{2+} or to the ion UO_2^{+}; these can be distinguished as uranyl(VI) and uranyl(V), respectively. In like manner, VO may be vanadyl(V), vanadyl(IV), and vanadyl(III).

These polyatomic radicals always are treated as forming the positive part of the compound.

Examples:

$COCl_2$	carbonyl chloride	$NO_2HS_2O_7$	nitryl hydrogen disulfate
NOS	nitrosyl sulfide	S_2O_5ClF	pyrosulfuryl chloride fluoride
PON	phosphoryl nitride	$SO_2(N_3)_2$	sulfonyl azide
$PSCl_3$	thiophosphoryl chloride	SO_2NH	sulfonyl imide
POCl	phosphoryl(III) chloride	IO_2F	iodyl fluoride

By using the same radical names regardless of unknown or controversial polarity relationships, names can be formed without entering into any controversy. Thus, for example, the compounds NOCl and $NOClO_4$ are quite unambiguously denoted by the names nitrosyl chloride and nitrosyl perchlorate, respectively.

◆Caution is urged in the use of some of these radical names: Vanadyl, for example, has been used for VO_2 as well as for VO (*cf.* also the naming of $VOSO_4$ in 6.42). Most of these radical names (except hydroxyl and thionyl) can be regarded as derived from the names of acids which have lost all of their hydroxyls (analogous to -yl or -oyl organic acid radical names) by the use of -yl and -osyl for radicals from -ic and -ous acids, respectively; this is implied in the footnote about nitroxyl. The use of the Stock notation in only one example (phosphoryl(III) chloride) seems confusing. It might be clearer to indicate stoichiometric proportions, *e.g.*, phosphoryl (mono)chloride, thiophosphoryl trichloride (*cf.* phosphoryl triamide in 5.34) or to use phosphorosyl for phosphoryl(III).

The restriction of the use of thionyl and sulfuryl to the halides was agreed upon at a joint meeting of the inorganic and organic nomenclature commissions of the IUPAC in 1951.

3.33.—It should be noted that the same radical may have different names in inorganic and organic chemistry. To draw attention to such differences the prefix names of radicals as substituents in organic compounds have been listed together with the inorganic names in the list of names printed at the end of the Rules. Names of purely organic compounds, of which many are important in the chemistry of coördination compounds (Section **7**), should agree with the nomenclature of organic chemistry.

Organic chemical nomenclature is to a large extent based on the principle of substitution, *i.e.*, replacement of hydrogen atoms by other atoms or groups. Such "substitutive names" are extremely rare in inorganic chemistry; they are used, *e.g.*, in the following cases: NH_2Cl is called chloramine, and $NHCl_2$ dichloramine. These names may be retained in the absence of better terms. Other substitutive names (derived from "sulfonic acid" as a name for HSO_3H) are fluoro- and chlorosulfonic acid, aminosulfonic acid, iminodisulfonic acid, and nitrilotrisulfonic acid. These names should preferably be replaced by the following

FSO_3H	fluorosulfuric acid	$NH(SO_3H)_2$	imidodisulfuric acid
$ClSO_3H$	chlorosulfuric acid	$N(SO_3H)_3$	nitridotrisulfuric acid
NH_2SO_3H	amidosulfuric acid		

Names such as chlorosulfuric acid and amidosulfuric acid might be considered to be substitutive names derived by substitution of *hydroxyl* groups in sulfuric acid. From a more fundamental point of view, however (see **2.24**), such names are formed by adding hydroxyl, amide, imide, *etc.*, groups together with oxygen atoms to a sulfur atom, "sulfuric acid" in this connection standing as an abbreviation for "trioxosulfuric acid."

Another organic-chemical type of nomenclature, the formation of "conjunctive names," is also met in only a few cases in inorganic chemistry, *e.g.*, the hydrazine- and hydroxylaminesulfonic acids. According to the principles of inorganic chemical nomenclature these compounds should be called hydrazido- and hydroxylamidosulfuric acid.

◆These are not true "conjunctive names" since sulfonic acid is not a compound. For the naming of partial amides *cf.* also 5.34.

4. CRYSTALLINE PHASES OF VARIABLE COMPOSITION

Isomorphous replacement, interstitial solutions, intermetallic compounds, and other nonstoichiometric compounds (berthollides)

4.1.—If an intermediate crystalline phase occurs in a two-component (or more complex) system, it may obey the law of constant composition very closely, as in the case of sodium chloride, or it may be capable of varying in composition over an appreciable range, as occurs for example with FeS. A substance showing such a variation is called a *berthollide.*

In connection with the berthollides the concept of a characteristic or ideal composition is frequently used. A unique definition of this concept seems to be lacking. In one case it may be necessary to use a definition based upon lattice geometry and in another to base it on the ratio of valence electrons to atoms. Sometimes one can state several characteristic compositions, and at other times it is impossible to say whether a phase corresponds to a characteristic composition or not.

In spite of these difficulties it seems that the concept of a characteristic composition can be used in its present undefined form for establishing a system of notation for phases of variable composition. It also seems possible to use the concept even if the characteristic composition is not included in the known homogeneity range of the phase.

4.2.—For the present, mainly formulas should be used for berthollides and solid solutions, since strictly logical names tend to become inconveniently cumbersome. The latter should be used only when unavoidable (*e.g.*, for indexing), and may be written in the style of iron(II) sulfide (iron-deficient); molybdenum dicarbide (excess carbon), or the like. Mineralogical names should be used only to designate actual minerals and not to define chemical composition; thus the name calcite refers to a particular mineral (contrasted with other minerals of similar composition) and is not a term for the chemical compound whose composition is properly expressed by the name calcium carbonate. (The mineral name may, however, be used to indicate the structure type—see **6.52.**)

4.3.—A general notation for the berthollides, which can be used even when the mechanism of the variation in composition is unknown, is to put the sign ~ (read as *circa*) before the formula. (In special cases it may also be printed above the formula.)

Examples: ~FeS, $\widetilde{\mathrm{CuZn}}$

The direction of the deviation may be indicated when required:

~FeS (iron-deficient); ~MoC_2 (excess carbon)

4.4.—For a phase where the variable composition is solely or partially caused by replacement, atoms or atomic groups which replace each other are separated by a comma and placed together between parentheses.

If possible the formula ought to be written so that the limits of the homogeneity range are represented when one or other of the two atoms or groups is lacking. For example the symbol (Ni,Cu) denotes the complete range from pure Ni to pure Cu; likewise K(Br,Cl) comprises the range from pure KBr to pure KCl. If only part of the homogeneity range is referred to, the major constituent should be placed first.

Substitution accompanied by the appearance of vacant positions (combination of substitutional and interstitial solution) receives an analogous notation. For example, $(Li_2,Mg)Cl_2$ denotes the homogeneous phase from LiCl to $MgCl_2$ where the anion lattice structure remains the same but one vacant cation position appears for every substitution of $2Li^+$ by Mg^{2+}.

The formula $(Mg_3,Al_2)Al_6O_{12}$ represents the homogeneous phase from the spinel $MgAl_2O_4$ (= $Mg_3Al_6O_{12}$) to the spinel form of Al_2O_3 (= $Al_2Al_6O_{12}$).

The solid solutions between CaF_2 and YF_3, where cation substitution is accompanied by interstitial addition of F^-, would be represented by the formula $(Ca,YF)F_2$. It is important to note that this formula is based purely on considerations of composition, and it does not imply that YF^{2+} takes over the actual physical position of Ca^{2+}. On the same basis a notation for the plagioclases would be $(NaSi,CaAl)Si_2AlO_8$.

4.5.—A still more complete notation, which should always be used in more complex cases, may be constructed by indicating in a formula the variables that define the composition. Thus, a phase involving simple substitution may be written A_xB_{1-x}.

Examples: Ni_xCu_{1-x} and KBr_xCl_{1-x}

This shows immediately that the total number of atoms in the lattice is constant. Combined substitutional and interstitial or subtractive solution can be shown in an analogous way. The commas and parentheses called for in **4.4** are not required in this case.

For example, the homogeneous phase between LiCl and $MgCl_2$ becomes $Li_{2x}Mg_{1-x}Cl_2$ and the phase between $MgAl_2O_4$ and Al_2O_3 can be written $Mg_{3x}Al_{2(1-x)}Al_6O_{12}$, which shows that it cannot contain more Mg than that corresponding to $MgAl_2O_4$ ($x = 1$). The other examples given in **4.4** will be given the formulas $Ca_xY_{1-x}F_{3-x}$ and $Na_xCa_{1-x}Si_{2+x}Al_{2-x}O_8$. In the case of the γ-phase of the Ag–Cd system, which has the characteristic formula Ag_5Cd_8, the Ag and Cd atoms can replace one another to some extent and the notation would be $Ag_{5\pm x}Cd_{8\mp x}$.

Further examples:

$Fe_{1-x}Sb$ $Fe_{1-x}O$ $Fe_{1-x}S$ $Cu_{2-x}O$ $Na_{1-x}WO_3$ (sodium tungsten bronzes)

For $x = 0$ each of these formulas corresponds to a characteristic composition. If it is desired to show that the variable denoted by x can attain only small values, this may be done by substituting ϵ for x.

Likewise a solid solution of hydrogen in palladium can be written as PdH_x, and a phase of the composition M which has dissolved a variable amount of water can be written $M(H_2O)_x$.

When this notation is used, a particular composition can be indicated by stating the actual value of the variable x. Probably the best way of doing this is to put the value in parentheses after the general formula. For example, $Li_{4-x}Fe_{3x}Ti_{2(1-x)}O_6$ ($x = 0.35$). If it is desired to introduce the value of x into the formula itself, the mechanism of solution is more clearly understood if one writes $Li_{4-0.35}Fe_{3\times 0.35}Ti_{2(1-0.35)}O_6$ instead of $Li_{3.65}Fe_{1.05}Ti_{1.30}O_6$.

5. ACIDS

Many of the compounds which now according to some definitions are called acids do not fall into the classical province of acids. In other parts of inorganic chemistry functional names are disappearing and it would have been most satisfactory to abolish them also for those compounds generally called acids. Names for these acids may be derived from the names of the anions as in Section 2, *e.g.*, hydrogen sulfate instead of sulfuric acid. The nomenclature of acids has, however, a long history of established custom, and it appears impossible to systematize acid names without drastic alteration of the accepted names of many important and well-known substances.

The present rules are aimed at preserving the more useful of the older names while attempting to guide further development along directions which should allow new compounds to be named in a more systematic manner.

5.1. Binary and Pseudobinary Acids

Acids giving rise to the -ide anions defined by **3.21** and **3.221** will be named as binary and pseudobinary compounds of hydrogen, *e.g.*, hydrogen chloride, hydrogen sulfide, hydrogen cyanide.

For the compound HN_3 the name hydrogen azide is recommended in preference to hydrazoic acid.

5.2. Acids Derived from Polyatomic Anions

Acids giving rise to anions bearing names ending in -ate or in -ite may also be treated as in **5.1**, but names more in accordance with custom are formed by using the terminations -ic acid and -ous acid corresponding with the anion terminations -ate and -ite, respectively. Thus chloric acid corresponds to chlorate, sulfuric acid to sulfate, and phosphorous acid to phosphite.

This nomenclature may also be used for less common acids, *e.g.*, hexacyanoferric acids correspond to hexacyanoferrate ions. In such cases, however, systematic names of the type hydrogen hexacyanoferrate are preferable.

Most of the common acids are oxo acids, *i.e.*, they contain only oxygen atoms bound to the characteristic atom. It is a long-established custom not to indicate these oxygen atoms. It is mainly for these acids that long-established names will have to be retained. Most other acids may be considered as coördination compounds and be named as such.

◆Polyatomic means heteratomic here, as opposed to pseudobinary. *Cf.* comment at 3.12.

5.21. Oxo Acids.—For the oxo acids the ous–ic notation to distinguish between different oxidation states is applied in many cases. The -ous acid names are restricted to acids corresponding to the -ite anions listed in the table in **3.224**.

Further distinction between different acids with the same characteristic element is in some cases effected by means of prefixes. This notation should not be extended beyond the cases listed below.

5.211.—The prefix hypo- is used to denote a lower oxidation state, and may be retained in these cases

$H_4B_2O_4$	hypoboric acid	HPH_2O_2	hypophosphorous acid
$H_2N_2O_2$	hyponitrous acid	$HOCl$	hypochlorous acid (and similarly for the other halogens)
$H_4P_2O_6$	hypophosphoric acid		

5.212.—The prefix per- is used to designate a higher oxidation state and may be retained for $HClO_4$, perchloric acid, and similarly for the other elements in Group VII.

The prefix per- should not be confused with the prefix peroxo- (see **5.22**).

5.213.—The prefixes ortho- and meta- have been used to distinguish acids differing in "water content." These names are approved

H_3BO_3	orthoboric acid	H_5IO_6	orthoperiodic acid	$(H_2SiO_3)_n$	metasilicic acids
H_4SiO_4	orthosilicic acid	H_6TeO_6	orthotelluric acid	$(HPO_3)_n$	metaphosphoric acids
H_3PO_4	orthophosphoric acid	$(HBO_2)_n$	metaboric acids		

For the acids derived by removing water from orthoperiodic or orthotelluric acid, the systematic names should be used, *e.g.*, HIO_4 tetraoxoiodic(VII) acid.

The prefix pyro- has been used to designate an acid formed from two molecules of an ortho acid minus one molecule of water. Such acids can now generally be regarded as the simplest cases of isopoly acids (*cf.* **7.5**). The prefix pyro- may be retained for pyrosulfurous and pyrosulfuric acids and for pyrophosphorous and pyrophosphoric acids, although in these cases also the prefix di- is preferable.

◆The use of orthoperiodic acid and orthotelluric acid is approved, but the question of names for HIO_4, *etc.*, needs further study because of confusion in the literature.

5.214.—The accompanying Table II contains the accepted names of the oxo acids (whether known in the free state or not) and some of their thio and peroxo derivatives (**5.22** and **5.23**).

For the less common of these acids systematic names would seem preferable, for example

H_2MnO_4	manganic(VI) acid, to distinguish it from H_3MnO_4, manganic(V) acid
$HReO_4$	tetraoxorhenic(VII) acid, to distinguish it from H_3ReO_5, pentaoxorhenic(VII) acid
H_2ReO_4	tetraoxorhenic(VI) acid, to distinguish it from $HReO_3$, trioxorhenic(V) acid; H_3ReO_4, tetraoxorhenic(V) acid; and $H_4Re_2O_7$, heptaoxodirhenic(V) acid
H_2NO_2	dioxonitric(II) acid instead of nitroxylic acid.

Trivial names should not be given to such acids as HNO, $H_2N_2O_3$, $H_2N_2O_4$, of which salts have been described. These salts are to be designated systematically as oxonitrates(I), trioxodinitrates(II), tetraoxodinitrates(III), respectively.

The names germanic acid, stannic acid, antimonic acid, bismuthic acid, vanadic acid, niobic acid, tantalic acid, telluric acid, molybdic acid, wolframic acid, and uranic acid may be used for substances with indefinite "water content" and degree of polymerization.

◆Unless trivial names clash with good nomenclature practices or are ambiguous, retention of well-established ones or the use of formulas is urged (especially for HNO, $H_2N_2O_3$, *etc.*) until structures are known. Systematic coordination-type names in the case of these nitrogen acids, for example, imply a structure that is ruled out by our present state of knowledge.

If hexahydroxoantimonic acid is considered a trivial name, hexahydrox*y*- might be preferable (*cf.* the systematic name hexahydroxoantimonate(V) ion in 3.223). For the analogous use of peroxo and peroxy, see comment at 5.22.

Table II

Names For Oxo Acids

H_3BO_3	orthoboric acid or (mono)boric acid	H_2SO_4	sulfuric acid
$(HBO_2)_n$	metaboric acids	$H_2S_2O_7$	disulfuric or pyrosulfuric acid
$(HBO_2)_3$	trimetaboric acid	H_2SO_5	peroxo(mono)sulfuric acid
$H_4B_2O_4$	hypoboric acid	$H_2S_2O_8$	peroxodisulfuric acid
H_2CO_3	carbonic acid	$H_2S_2O_3$	thiosulfuric acid
HOCN	cyanic acid	$H_2S_2O_6$	dithionic acid
HNCO	isocyanic acid	H_2SO_3	sulfurous acid
HONC	fulminic acid	$H_2S_2O_5$	disulfurous or pyrosulfurous acid
H_4SiO_4	orthosilicic acid	$H_2S_2O_2$	thiosulfurous acid
$(H_2SiO_3)_n$	metasilicic acids	$H_2S_2O_4$	dithionous acid
HNO_3	nitric acid	H_2SO_2	sulfoxylic acid
HNO_4	peroxonitric acid	$H_2S_xO_6$ (x = 3,4...)	polythionic acids
HNO_2	nitrous acid	H_2SeO_4	selenic acid
HOONO	peroxonitrous acid	H_2SeO_3	selenious acid
H_2NO_2	nitroxylic acid	H_6TeO_6	(ortho)telluric acid
$H_2N_2O_2$	hyponitrous acid	H_2CrO_4	chromic acid
H_3PO_4	(ortho)phosphoric acid	$H_2Cr_2O_7$	dichromic acid
$H_4P_2O_7$	diphosphoric or pyrophosphoric acid	$HClO_4$	perchloric acid
$H_5P_3O_{10}$	triphosphoric acid	$HClO_3$	chloric acid
$H_{n+2}P_nO_{3n+1}$	polyphosphoric acids	$HClO_2$	chlorous acid
$(HPO_3)_n$	metaphosphoric acids	HClO	hypochlorous acid
$(HPO_3)_3$	trimetaphosphoric acid	$HBrO_3$	bromic acid
$(HPO_3)_4$	tetrametaphosphoric acid	$HBrO_2$	bromous acid
H_3PO_5	peroxo(mono)phosphoric acid	HBrO	hypobromous acid
$H_4P_2O_8$	peroxodiphosphoric acid	H_5IO_6	(ortho)periodic acid
$(HO)_2OP{-}PO(OH)_2$	hypophosphoric acid	HIO_3	iodic acid
$(HO)_2P{-}O{-}PO(OH)_2$	diphosphoric(III,V) acid	HIO	hypoiodous acid
H_2PHO_3	phosphorous acid	$HMnO_4$	permanganic acid
$H_4P_2O_5$	diphosphorous or pyrophosphorous acid	H_2MnO_4	manganic acid
HPH_2O_2	hypophosphorous acid	$HTcO_4$	pertechnetic acid
H_3AsO_4	arsenic acid	H_2TcO_4	technetic acid
H_3AsO_3	arsenious acid	$HReO_4$	perrhenic acid
$HSb(OH)_6$	hexahydroxoantimonic acid	H_2ReO_4	rhenic acid

5.22. Peroxo Acids.—The prefix peroxo, when used in conjunction with the trivial names of acids, indicates substitution of –O– by –O–O– (*cf.* **7.312**).

Examples:

HNO_4	peroxonitric acid	H_2SO_5	peroxosulfuric acid
H_3PO_5	peroxophosphoric acid	$H_2S_2O_8$	peroxodisulfuric acid
$H_4P_2O_8$	peroxodiphosphoric acid		

◆Peroxy, as recommended in the 1940 Rules (inorganic), is more acceptable than peroxo to organic chemists. It is not necessary that the use with trivial names conform with the use of peroxo denoting a coördinated ligand; *e.g.*, peroxysulfuric acid or trioxoperoxosulfuric(VI) acid.

5.23. Thio Acids.—Acids derived from oxo acids by replacement of oxygen by sulfur are called *thio acids* (*cf.* **7.312**).

Examples: $H_2S_2O_2$ thiosulfurous acid $H_2S_2O_3$ thiosulfuric acid HSCN thiocyanic acid

When more than one oxygen atom can be replaced by sulfur the number of sulfur atoms generally should be indicated

H_3PO_3S	monothiophosphoric acid	H_3AsS_3	trithioarsenious acid
$H_3PO_2S_2$	dithiophosphoric acid	H_3AsS_4	tetrathioarsenic acid
H_2CS_3	trithiocarbonic acid		

The prefixes seleno- and telluro- may be used in a similar manner.

5.24. Chloro Acids, etc.—Acids containing ligands other than oxygen and sulfur are generally designated according to the rules in Section 7.

Examples:

$HAuCl_4$	hydrogen tetrachloraurate(III) or tetrachloroauric(III) acid
H_2PtCl_4	hydrogen tetrachloroplatinate(II) or tetrachloroplatinic(II) acid
H_2PtCl_6	hydrogen hexachloroplatinate(IV) or hexachloroplatinic(IV) acid
$H_4Fe(CN)_6$	hydrogen hexacyanoferrate(II) or hexacyanoferric(II) acid
$H[PHO_2F]$	hydrogen hydridodioxofluorophosphate or hydridodioxofluorophosphoric acid
HPF_6	hydrogen hexafluorophosphate or hexafluorophosphoric acid
H_2SiF_6	hydrogen hexafluorosilicate or hexafluorosilicic acid
H_2SnCl_6	hydrogen hexachlorostannate(IV) or hexachlorostannic(IV) acid
HBF_4	hydrogen tetrafluoroborate or tetrafluoroboric acid
$H[B(OH)_2F_2]$	hydrogen dihydroxodifluoroborate or dihydroxodifluoroboric acid
$H[B(C_6H_5)_4]$	hydrogen tetraphenylborate or tetraphenylboric acid

It is preferable to use names of the type hydrogen tetrachloroaurate(III).

For some of the more important acids of this type abbreviated names may be used, *e.g.*, chloroplatinic acid, fluorosilicic acid.

◆For the use of hydrido in the fifth example see comment at 7.312.

5.3. Functional Derivatives of Acids

Functional derivatives of acids are compounds formed from acids by substitution of OH and sometimes also O by other groups. In this borderline between organic and inorganic chemistry organic-chemical nomenclature principles prevail.

◆The intention of the statement "organic-chemical nomenclature principles prevail" is not clear, since most of the examples given in the sections immediately following are not named according to organic practice. *Cf.* 3.33.

5.31. Acid Halides.—The names of acid halides are formed from the name of the corresponding acid radical if this has a special name, *e.g.*, sulfuryl chloride, phosphoryl chloride.

In other cases these compounds are named as oxide halides according to rule **6.41**, *e.g.*, MoO_2Cl_2, molybdenum dioxide dichloride.

5.32. Acid Anhydrides.—Anhydrides of inorganic acids generally should be given names as oxides, *e.g.*, N_2O_5 dinitrogen pentaoxide, *not* nitric anhydride or nitric acid anhydride.

5.33. Esters.—Esters of inorganic acids are given names in the same way as the salts, *e.g.*, dimethyl sulfate, diethyl hydrogen phosphate.

If, however, it is desired to specify the constitution of the compound, a name based on the nomenclature for cöordination compounds should be used.

Example:

$(CH_3)_4[Fe(CN)_6]$ tetramethyl hexacyanoferrate(II)

or

$[Fe(CN)_2(CH_3NC)_4]$ dicyanotetrakis(methyl isocyanide)iron(II)

◆According to common organic practice for esters (ethers, sulfides, *etc.*) and to the naming of inorganic salts (*e.g.*, sodium sulfate, not disodium sulfate), methyl sulfate would be used instead of dimethyl sulfate. However, no objection is seen to the more specific name. Such names as methyl sulfate are better for alphabetic listing, as in indexes.

5.34. Amides.—The names for amides may be derived from the names of acids by replacing acid by amide, or from the names of the acid radicals.

Examples:

$SO_2(NH_2)_2$	sulfuric diamide or sulfonyl diamide
$PO(NH_2)_3$	phosphoric triamide or phosphoryl triamide

If not all hydroxyl groups of the acid have been replaced by NH_2 groups, names ending in -amidic acid may be used: this is an alternative to naming the compounds as complexes.

Examples:

NH_2SO_3H	amidosulfuric acid or sulfamidic acid
$NH_2PO(OH)_2$	amidophosphoric acid or phosphoramidic acid
$(NH_2)_2PO(OH)$	diamidophosphoric acid or phosphorodiamidic acid

Abbreviated names (sulfamide, phosphamide, sulfamic acid) are often used but are not recommended.

◆The use of adjectives from names of inorganic acids may lead to confusion because an -ic or -ous adjective (as chromic) may refer to a higher- or lower-valent form of the element as well as to the acid (a possibility that does not arise, of course, with adjectives from names of organic acids).

Names of the type phosphoramidic acid are recommended in the report of the Advisory Committee on the Nomenclature of Organic Phosphorus Compounds of the Division of Organic Chemistry of the ACS published in 1952, but are not as acceptable to the inorganic nomenclature committees as a whole as the amido- or cöordination-type names. *Cf.* 3.33.

The retention of sulfamic acid and sulfamide as trivial names is favored by the committees. An acceptable systematic name for NH_2SO_3H would be ammonia-sulfur trioxide, in keeping with its probable structure.

5.35. Nitriles.—The suffix -nitrile has been used in the names of a few inorganic compounds, *e.g.* $(PNCl_2)_3$, trimeric phosphonitrile chloride. According to **2.22** such compounds can be designated as nitrides, *e.g.*, phosphorus nitride dichloride. Accordingly there seems to be no reason for retention of the name nitrile (and nitrilo, *c.f.* **3.33**) in inorganic chemistry.

◆Nitrilo is used in organic chemistry, though not given in the list at the end of the Rules.

6. SALTS AND SALT-LIKE COMPOUNDS

Among salts particularly there persist many old names which are bad and misleading, and the Commission wishes to emphasize that any which do not conform to these Rules should be discarded.

6.1. Simple Salts

Simple salts fall under the broad definition of binary compounds given in Section **2**, and their names are formed from those of the constituent ions (given in Section **3**) in the manner set out in Section **2**.

6.2. Salts Containing Acid Hydrogen ("Acid" Salts[1])

Names are formed by adding the word hydrogen, to denote the replaceable hydrogen present, immediately in front of the name of the anion.

The nonacidic hydrogen present, *e.g.*, in the phosphite ion, is included in the name of the anion and is not explicitly cited (*e.g.*, Na_2PHO_3, sodium phosphite).

Examples:

$NaHCO_3$	sodium hydrogen carbonate
NaH_2PO_4	sodium dihydrogen phosphate
$NaH[PHO_3]$	sodium hydrogen phosphite

◆The use of "fused" hydrogen names (as hydrogencarbonate in the original version) is not acceptable in English; the present practice of running hydrogen as a separate word is preferred and has been followed throughout this version. *Cf.* 3.222, 6.324, 6.333. If necessary for clarity, parentheses can be used, as in naming ligands, *e.g.*, (hydrogen carbonato). The use of hydro (as in hydrocarbonato) is not acceptable because of conflicts with organic usage, where hydro denotes addition of hydrogen to unsaturated compounds.

It seems safer to cite even nonacidic hydrogen present in an anion like PHO_3^{2-} (unless the ion has a specific name), because of current usage. (*Cf.* triethyl phosphite in 7.412, seventh example).

6.3. Double Salts, Triple Salts, etc.

6.31.—In formulas all the cations shall precede the anions; in names the principles embodied in Section 2 shall be applied. In those languages where cation names are placed after anion names the adjectives double, triple, *etc.* (their equivalents in the language concerned) may be added immediately after the anion name. The number so implied concerns the number of *kinds* of cation present and *not* the total number of such ions.

[1]For "basic" salts see 6.4.

6.32.—Cations.

6.321.—Cations shall be arranged in order of increasing valence (except hydrogen, *cf.* **6.2** and **6.324**).

6.322. The cations of each valence group shall be arranged in order of decreasing atomic number, with the polyatomic radical ions (*e.g.*, ammonium) at the end of their appropriate group.

◆Alphabetical order would be simpler here and even for 6.321.

6.323. Hydration of Cations.—Owing to the prevalence of hydrated cations, many of which are in reality complex, it seems unnecessary to disturb the cation order in order to allow for this; but if it is necessary to draw attention specifically to the presence of a particular hydrated cation this may be done by writing, for example, "hexaaquo" or "tetraaquo" before the name of the simple ion. Apart from this exception, however, all complex ions should be placed after simple ones in the appropriate valence group.

◆*Cf.* comment at 7.322.

6.324. Acidic Hydrogen.—When hydrogen is considered to be present as a cation its name shall be cited last among the cations. Actually acidic hydrogen will in most cases be bound to an anion and shall be cited together with this (**6.2**). If the salt contains only one anion, acidic hydrogen shall be cited in the same place whichever view is taken of the function of this hydrogen. Nonacidic hydrogen shall be either not explicitly cited (*cf.* **6.2**) or designated hydrido (*cf.* **5.24** and **7.311**). For salts with more than one anion see **6.333.**

Examples:

$KMgF_3$	potassium magnesium fluoride
$TlNa(NO_3)_2$	thallium(I) sodium nitrate or thallium sodium dinitrate
$KNaCO_3$	potassium sodium carbonate
$NH_4MgPO_4.6H_2O$	ammonium magnesium phosphate hexahydrate
$NaZn(UO_2)_3(C_2H_3O_2)_9.6H_2O$	sodium zinc triuranyl acetate hexahydrate
$Na[Zn(H_2O)_6](UO_2)_3(C_2H_3O_2)_9$	sodium hexaaquozinc triuranyl acetate
$NaNH_4HPO_4.4H_2O$	sodium ammonium hydrogen phosphate tetrahydrate

◆*Cf.* comments at 6.2, 7.312. In the fifth and sixth examples, either triuranyl(VI) or nonaacetate should be specified. *Cf.* 3.32, 6.34.

6.33.—Anions.

6.331.—Anions are to be cited in this group order

1. H^-
2. O_2^- and OH^- (in that order)
3. Simple (*i.e.*, one element only) inorganic anions, other than H^- and O^{2-}
4. Inorganic anions containing two or more elements, other than OH^-
5. Anions of organic acids and organic substances exerting an acid function

◆The committees consider it preferable to cite H^- last in accordance with usage.

6.332.—Within group 3 the ions shall be cited in the order given in **2.16**, the inclusion of O in that list being taken as referring to all oxygen anions apart from O^{2-} (*i.e.*,O_2^{2-}, *etc.*).

Within group 4, anions containing the smallest number of atoms shall be cited first, and in the case of two ions containing the same number of atoms they shall be cited in order of decreasing atomic number of the central atoms. Thus CO_3^{2-} should precede CrO_4^{2-}, and the latter should precede SO_4^{2-}.

Within group 5 the anions shall be cited in alphabetical order.

◆Again alphabetical order would be simpler, within groups 3 and 4 as well as 5. *Cf.* comment at 7.251.

6.333.—Acidic hydrogen should be cited together with the anion to which it is attached. If it is not known to which anion the hydrogen is bound, it should be cited last among the cations.

◆*Cf.* comment at 6.2.

6.34.—The stoichiometric method is the most practicable for indicating the proportions of the constituents. It is not always essential to give the numbers of all the anions, provided the valences of all the cations are either known or indicated.

Examples:

$NaCl.NaF.2Na_2SO_4$ or $Na_6ClF(SO_4)_2$	(hexa)sodium chloride fluoride (bis)sulfate
$Ca_5F(PO_4)_3$	(penta)calcium fluoride (tris)phosphate

The parentheses in these cases mean that numerical prefixes may not be necessary. The multiplicative numerical prefixes bis, tris, *etc.*, should be used in connection with anions, because disulfate, triphosphate, *etc.*, designate isopoly anions.

6.4. Oxide and Hydroxide Salts ("Basic" salts formerly oxy and hydroxy salts)

6.41.—For the purposes of nomenclature, these should be regarded as double salts containing O^{2-} and OH^- anions, and Section **6.3** may be applied in its entirety.

6.42. Use of the Prefixes Oxy and Hydroxy.—In some languages the citation in full of all the separate anion names presents no trouble and is strongly recommended (*e.g.*, copper oxide chloride), to the exclusion of the oxy form wherever possible. In some other languages, however, such names as "oxyde et chlorure double de cuivre" are so far removed from current practice that the present system of using oxy- and hydroxy-, *e.g.*, oxychlorure de cuivre, may be retained in such cases.

Examples:

$Mg(OH)Cl$	magnesium hydroxide chloride
$BiOCl$	bismuth oxide chloride
$LaOF$	lanthanum oxide fluoride
$VOSO_4$	vanadium(IV) oxide sulfate
$CuCl_2.3Cu(OH)_2$ or $Cu_2(OH)_3Cl$	dicopper trihydroxide chloride
$ZrOCl_2.8H_2O$	zirconium oxide (di)chloride octahydrate

6.5. Double Oxides and Hydroxides

The terms "mixed oxides" and "mixed hydroxides" are not recommended. Such substances preferably should be named double, triple, *etc.*, oxides or hydroxides as the case may be.

Many double oxides and hydroxides belong to several distinct groups, each having its own characteristic structure type, which is sometimes named after some well-known mineral of the same group (*e.g.*, perovskite, ilmenite, spinel). Thus, $NaNbO_3$, $CaTiO_3$, $CaCrO_3$, $CuSnO_3$, $YAlO_3$, $LaAlO_3$, and $LaGaO_3$ all have the same structure as perovskite, $CaTiO_3$. Names such as calcium titanate may convey false implications and it is preferable to name such compounds as double oxides and double hydroxides unless there is clear and generally accepted evidence of cations and oxo or hydroxo anions in the structure. This does not mean that names such as titanates or aluminates should always be abandoned, because such substances may exist in solution and in the solid state (*cf.* **3.223**).

◆"Multiple" has been used in English as a class term including double, triple, etc. (oxides or the like).

6.51.—In the double oxides and hydroxides the metals shall be named in the same order as for double salts (**6.32**).

6.52.—When required the structure type may be added in parentheses and in italics after the name, except that when the type name is also the mineral name of the substance itself then the italics should not be used (*cf.* **4.2**).

Examples:

$NaNbO_3$	sodium niobium trioxide (*perovskite* type)
$MgTiO_3$	magnesium titanium trioxide (*ilmenite* type)
$FeTiO_3$	iron(II) titanium trioxide (ilmenite)
$4CaO.AL_2O_3.nH_2O$ or $Ca_2Al(OH)_7.nH_2O$	dicalcium aluminum hydroxide hydrate
but $Ca_3[Al(OH)_6]_2$	(tri)calcium (bis)-[hexahydroxoaluminate]
$LiAl(OH)_4.2MnO_2$ or $LiAlMn^{IV}_2O_4(OH)_4$	lithium aluminum dimanganese(IV) tetraoxide tetrahydroxide

7. COÖRDINATION COMPOUNDS

7.1. Definitions

In its oldest sense the term *coördination compound* is taken as referring to molecules or ions in which there is an atom (A) to which are attached other atoms (B) or groups (C) to a number in excess of that corresponding to the oxidation number of the atom A. However, the system of nomenclature originally evolved for the compounds within this narrow definition has proved useful for a much wider class of compounds, and for the purposes of nomenclature the restriction "in excess . . . oxidation number" is to be omitted. Any compound formed by addition of one or several ions and/or molecules to one or more ions or/and molecules may be named according to the same system as strict coördination compounds.

The effect of this is to bring many simple and well-known compounds under the same nomenclature rules as accepted coöordination compounds; the result is to reduce the diversity of names and avoid many controversial issues, because it should be understood that there is no intention of implying that any structural analogy necessarily exists between different compounds merely on account of a common system of nomenclature. The system extends also to many addition compounds.

In the rules which follow certain terms are used in the senses here indicated: the atom referred to above as (A) is known as the *central* or *nuclear* atom, and all other atoms which are directly attached to A are known as *coördinating atoms.* Atoms (B) and groups (C) are called *ligands.* A group containing more than one *potential* coördinating atom is termed a *multidentate* ligand, the number of potential coöordinating atoms being indicated by the terms *unidentate, bidentate, etc.* A *chelate* ligand is a ligand *attached* to *one* central atom through *two or more* coördinating atoms, while a *bridging group* is attached to *more than one* atom. The whole assembly of one or more central atoms with their attached ligands is referred to as a *complex,* which may be an uncharged molecule or an ion of either polarity. A *polynuclear complex* is a complex which contains *more than one* nuclear atom, their number being designated by the terms *mononuclear, dinuclear, etc.*

◆Some dissatisfaction with the definition of coördination compounds was expressed, though the broad definition (last sentence of first paragraph) was generally approved.

Central atom or *center of coördination* is to be preferred to *nuclear* atom because of other senses of nucleus, especially of an atom. Possible replacements for polynuclear, etc., are polycentric, *etc.,* or bridged complex since bridging group is in common use.

In the United States the Greek-Latin hybrid terms polydentate and monodentate seem to be used more than the all-Latin multidentate and unidentate.

7.2. Formulas and Names for Complex Compounds in General

7.21. Central Atoms.—In *formulas* the symbol for the central atom(s) should be placed *first* (except in formulas which are primarily structural), the anionic and neutral, *etc.,* ligands following as prescribed in **7.25**, and the formula for the whole complex entity (ion or molecule) should be placed in brackets [ ].

In *names* the central atom(s) should be placed *after* the ligands.

◆It is considered preferable to place the whole complex in brackets, especially when more than one central atom is present, but not essential with only one central atom or with complex ions or nonionic complexes, because brackets may be needed sometimes for indicating concentrations.

7.22. Indication of Valence and Proportion of Constituents.—The oxidation number of the central atom is indicated by means of the Stock notation (**2.252**). Alternatively the proportion of constituents may be given by means of stoichiometric prefixes (**2.251**).

7.23.—Formulas and names may be supplemented with the prefixes *cis, trans, etc.* (**2.19**).

7.24. Terminations.—Complex anions shall be given the termination -ate (*cf.* **2.23**, **2.24** and **3.223**). Complex cations and neutral molecules are given no distinguishing termination. For further details concerning the names of ligands see **7.3.**

7.25. Order of Citation of Ligands in Complexes.—

first: anionic ligands
next: neutral and cationic ligands

7.251.—The anionic ligands shall be cited in the order

1. H^-
2. O^{2-}, OH^- (in that sequence)
3. Other simple anions (*i.e.,* one element only)
4. Polyatomic anions
5. Organic anions in alphabetical order

The sequence within categories 3 and 4 should be that given in **6.332.**

◆H^- is preferably named last, not first (*cf.* comment at 6.331). Alphabetical order is strongly recommended for simplicity at least within 3 and 4, where only rare uses are involved. The intention seems to be to include under 3 monatomic ions (*i.e.,* one *atom*—rather than one *element*—only), in other words to exclude N_3, I_3, etc. Under 4 the insertion of "inorganic" between "Polyatomic" and "anions" is recommended.

7.252.—Neutral and cationic ligands shall be cited in the order given

first: water, ammonia (in that sequence)
then: other inorganic ligands in the sequence in which their coördinating elements appear in the list given in **2.16**
last: organic ligands in alphabetical order.

◆For the inorganic ligands alphabetical order again is recommended. The use of parentheses wherever there is any possibility of ambiguity should be stressed, as illustrated in examples under 7.321: potassium trichloro(ethylene)platinate(II), where parentheses are given in the Rules, and tetra(pyridine)platinum(II) tetrachloroplatinate(II), where they have been added in this version. Parentheses might also be helpful with "thiocyanato" preceded by a numerical prefix (see last example in 7.311) and are definitely required with two-word names for ligands recommended instead of the fused names of the original version (6.2), as in the seventh example in 7.412: di-μ-carbonyl-bis{carbonyl(triethyl phosphite)cobalt}. Since brackets denote complexes, braces can be used in formulas or names where needed in order to avoid the use of two sets of parentheses (braces were so used in some but not all such cases in the original version).

7.3. Names for Ligands

7.31.—Anionic Ligands.

7.311.—The names for anionic ligands, whether inorganic or organic, end in -o (see, however, **7.324**). In general, if the anion name ends in -ide, -ite, or -ate, the final -e is replaced by -o, giving -ido, -ito, and -ato, respectively.

Examples:

$Li[AlH_4]$	lithium tetrahydridoaluminate
$Na[BH_4]$	sodium tetrahydridoborate
$K_2[OsNCl_5]$	potassium nitridopentachloroösmate(VI)
$[Co(NH_2)_2(NH_3)_4]OC_2H_5$	diamidotetraamminecobalt(III) ethanolate
$[CoN_3(NH_3)_5]SO_4$	azidopentaamminecobalt(III) sulfate
$Na_3[Ag(S_2O_3)_2]$	sodium bis(thiosulfato)argentate(I)
$[Ru(HSO_3)_2(NH_3)_4]$	bis(hydrogen sulfito)tetraammineruthenium
(II) $NH_4[Cr(SCN)_4(NH_3)_2]$	ammonium tetrathiocyanatodiamminechromate (III)

◆For "ethanolate" in the fourth example "ethoxide" may be preferred.

7.312.—These anions do not follow exactly the above rule; modified forms have become established:

F^-	fluoride	fluoro (*not* fluo)	O_2^{2-}	peroxide	peroxo
Cl^-	chloride	chloro	HS^-	hydrogen sulfide	thiolo
Br^-	bromide	bromo	S^{2-}	sulfide	thio[1] (sulfido)
I^-	iodide	iodo	(but:S_2^{2-}	disulfide	disulfido)
O^{2-}	oxide	oxo	CN^-	cyanide	cyano
OH^-	hydroxide	hydroxo			

[1]The name thio has long been used to denote the ligand S^{2-} when it can be regarded as replacing O^{2-} in an oxo acid or its anion. The general use of this name will prevent confusion between the two interpretations of disulfido as S_2^{2-} or two S^{2-} ligands.

By analogy with hydroxo, CH_3O^-, *etc.*, are called methoxo, *etc.* For CH_3S^-, *etc.*, the systematic names methanethiolato, *etc.*, are used.

Examples:

$K[AgF_4]$	potassium tetrafluoroargentate(III)
$K_2[NiF_6]$	potassium hexafluoroniccolate(IV)
$Ba[BrF_4]_2$	barium tetrafluorobromate(III)
$Na[AlCl_4]$	sodium tetrachloroaluminate
$Cs[ICl_4]$	cesium tetrachloroiodate(III)
$K[Au(OH)_4]$	potassium tetrahydroxoaurate(III)
$K[CrOF_4]$	potassium oxotetrafluorochromate(V)
$K_2[Cr(O)_2O_2(CN)_2(NH_3)]$	potassium dioxoperoxodicyanoamminechromate(VI)
$Na[BH(OCH_3)_3]$	sodium hydridotrimethoxoborate
$K_2[Fe_2S_2(NO)_4]$	dipotassium dithiotetranitrosyldiferrate

◆It is strongly recommended on the basis of past and present usage, analogy with chloro, *etc.*, and euphony that hydro be added to the list of modified forms of names for anionic ligands and that hydrido be abandoned (*cf.* also examples in 7.311).

While the Subcommittee on Coördination Compounds recognizes the usefulness of the invariable -o ending for anionic ligands, some members of the general committees do not see a sharp enough distinction between such ligands and the same groups in organic compounds to justify a departure from organic usage by using hydroxo, methoxo, *etc.*, instead of hydroxy, methoxy, *etc.* These members therefore favor adding hydroxy, methoxy, *etc.*, to the hydrocarbon radicals excepted from the rule of -o endings for anions (7.324). For the use of peroxo, see comment at 5.22.

The -o of a negative ligand should not be elided before another vowel (chloroösmate, chloroiodo). This agrees with organic practice. *Cf.* comment at 7.322.

It should be pointed out here as well as in the list of names for ions and radicals at the end of the Rules that the approved organic name for unsubstituted HS is mercapto (not thiol, as also given in the list) and for the alkyl- and aryl-substituted radicals methylthio, *etc.*

7.313.—Ligands derived from organic compounds not normally called acids, but which function as such in complex formation by loss of a proton, should be treated as anionic and given the ending -ato. If, however, no proton is lost, the ligand must be treated as neutral—see **7.32.**

Examples:

$[Ni(C_4H_7N_2O_2)_2]$ bis(dimethylglyoximato)nickel(II) $[Cu(C_5H_7O_2)_2]$ bis(acetylacetonato)copper(II)

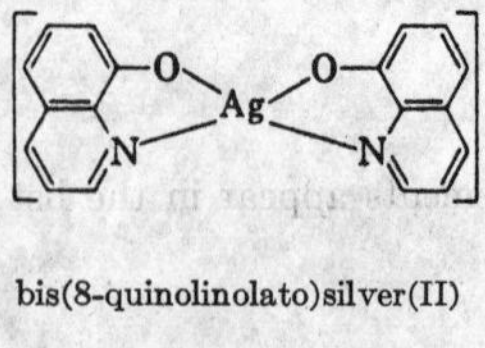

bis(8-quinolinolato)silver(II)

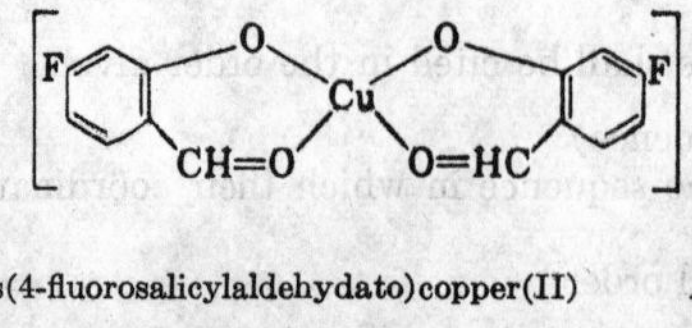

bis(4-fluorosalicylaldehydato)copper(II)

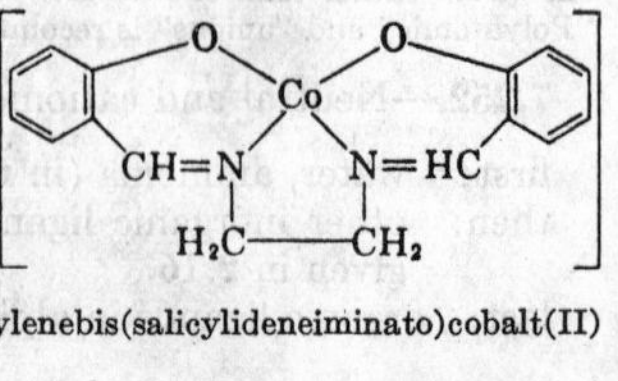

N,N'-ethylenebis(salicylideneiminato)cobalt(II)

◆Although according to 3.33 the name for the ligand in the second example should be derived from the systematic name 2,4-pentanedione instead of from acetylacetone, acetylacetonato perhaps conveys better the idea that it is the enol form that is involved. However, the coördination subcommittee questions the use of the termination -ato rather than plain -o especially in cases where the -ate terms (as dimethylglyoximate) are not accepted organic practice. Such -ato terms are especially misleading in the last two examples, where the -ato belongs with the hydroxyl part of the name, not the aldehyde or imine part to which it is attached.

7.32.—Neutral and Cationic Ligands.

7.321.—The name of the coördinated molecule or cation is to be used without change, except in the special cases provided for in **7.322.**

Examples:

$[CoCl_2(C_4H_8N_2O_2)_2]$	dichlorobis(dimethylglyoxime)cobalt(II) (*cf.* nickel derivative given in **7.313**)
cis-$[PtCl_2(Et_3P)_2]$	*cis*-dichlorobis(triethylphosphine)platinum(II)
$[CuCl_2(CH_3NH_2)_2]$	dichlorobis(methylamine)copper(II)
$[Pt\ py_4]\ [PtCl_4]$	tetra(pyridine)platinum(II) tetrachloroplatinate(II)
$[Fe(dipy)_3]Cl_2$	tris(dipyridyl)iron(II) chloride
$[Co\ en_3]_2(SO_4)_3$	tris(ethylenediamine)cobalt(III) sulfate
$[Zn\{NH_2CH_2CH(NH_2)CH_2NH_2\}_2]I_2$	bis(1,2,3-triaminopropane)zinc iodide
$K[PtCl_3(C_2H_4)]$	potassium trichloro(ethylene)platinate(II) or potassium trichloromonoethyleneplatinate(II)
$[PtCl_2\{H_2NCH_2CH(NH_2)CH_2NH_3\}]Cl$	dichloro(2,3 - diaminopropylammonium)platinum(II) chloride
$[Cr(C_6H_5NC)_6]$	hexakis(phenyl isocyanide)chromium

◆In the fifth example, bipyridine is preferred to dipyridyl in organic practice, and in the seventh example 1,2,3-propanetriamine to 1,2,3-triaminopropane.

7.322.—Water and ammonia as neutral ligands in coördination complexes are called "aquo" and "ammine," respectively.

In the tentative rules it was proposed to change the old-established "aquo" to "aqua," thus keeping the -o termination consistently for anionic ligands alone. However, as the old form is so widely used, many regarded the change as too pedantic, and the Commission has decided to retain "aquo" as an exception.

Examples:

$[Cr(H_2O)_6]Cl_3$	hexaaquochromium(III) chloride or hexaaquochromium trichloride
$[Al(OH)(H_2O)_5]^{++}$	the hydroxopentaaquoaluminum ion
$[Co(NH_3)_6]ClSO_4$	hexaamminecobalt(III) chloride sulfate
$[CoCl(NH_3)_5]Cl_2$	chloropentaamminecobalt(III) chloride
$[CoCl_3(NH_3)_2\{(CH_3)_2NH\}]$	trichlorodiammine(dimethylamine)cobalt(III)

◆Hexaquo, pentaquo, *etc.*, are used in the examples in the original version, but have been changed in this version to hexaaquo, *etc.*, for conformity with the latest approved organic practice. (*Cf.* hexaammonium, given with two a's separated by a hyphen in 7.6, second example, in the original version.)

7.323.—The groups NO, NS, CO, and CS, when linked directly to a metal atom, are to be called nitrosyl, thionitrosyl, carbonyl, and thiocarbonyl, respectively. In computing the oxidation number these radicals are treated as neutral.

Examples:

$Na_2[Fe(CN)_5NO]$	disodium pentacyanonitrosylferrate
$K_3[Fe(CN)_5CO]$	tripotassium pentacyanocarbonylferrate
$K[Co(CN)(CO)_2(NO)]$	potassium cyanodicarbonylnitrosylcobaltate(0)
$HCo(CO)_4$	hydrogen tetracarbonylcobaltate(−I)
$[Ni(CO)_2(Ph_3P)_2]$	dicarbonylbis(triphenylphosphine)nickel(0)
$[Fe\ en_3]\ [Fe(CO)_4]$	tris(ethylenediamine)iron(II) tetracarbonylferrate (−II)
$Mn_2(CO)_{10}$ or $[CO_5)Mn\text{–}Mn(CO)_5]$	decacarbonyldimanganese(0) or bis(pentacarbonylmanganese)

◆The necessity of arbitrarily considering these groups as always neutral can be avoided by not using the oxidation number but instead the stoichiometric proportions (as in some of these examples) or the Ewens-Bassett system (*cf.* comment at 2.252). Thus the anion in the last example in 7.312 (where NO is known to be positive) would by this system be named dithiotetranitrosyldiferrate (2–).

7.324.—Anions derived from hydrocarbons are given radical names without -o, but are counted as negative when computing the oxidation number.

The consistent introduction of the ending -o would in this case lead to unfamiliar names, *e.g.*, phenylato or phenido for $C_6H_5^-$. On the other hand, if the radicals were counted as neutral ligands, the central atom would have to be given an unusual oxidation number, *e.g.*, −I for boron in $K[B(C_6H_5)_4]$, instead of III.

Examples:

$K[B(C_6H_5)_4]$	potassium tetraphenylborate
$K[SbCl_5C_6H_5]$	potassium pentachloro(phenyl)antimonate(V)
$K_2[Cu(C_2H)_3]$	potassium triethynylcuprate(I)
$K_4[Ni(C_2C_6H_5)_4]$	potassium tetrakis(phenylethynyl)niccolate(0)
$[Fe(CO)_4(C_2C_6H_5)_2]$	tetracarbonylbis(phenylethynyl)iron(II)
$Fe(C_5H_5)_2$	bis(cyclopentadienyl)iron(II)
$[Fe(C_5H_5)_2]Cl$	bis(cyclopentadienyl)iron(III) chloride
$[Ni(NO)(C_5H_5)]$	nitrosylcyclopentadienylnickel

◆The use of radical names such as cyclopentadienyl in the examples does not seem consistent with the use of Stock notations; for the sixth example ("ferrocene"), iron(II) cyclopentadienide is preferred by some. This matter is presumably part of the whole organometallic problem to be dealt with by the new IUPAC joint organic-inorganic subcommittee.
For niccolate, see comment at 1.12.

7.33.—Alternative Modes of Linkage of Some Ligands.—Where ligands are capable of attachment by different atoms this may be denoted by adding the symbol for the atom by which attachment occurs at the end of the name of the ligand. Thus the dithioöxalato group may be attached through S or O, and these are distinguished as dithioöxalato-S,S' and dithioöxalato-O,O', respectively.

In some cases different names are already in use for alternative modes of attachment, as, for example, thiocyanato ($-SCN$) and isothiocyanato ($-NCS$), nitro ($-NO_2$), and nitrito ($-ONO$). In these cases existing custom may conveniently be retained.

Examples:

$$K_2\left[Ni\binom{S-CO}{S-CO}_2\right]$$

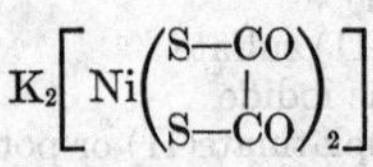

potassium bis(dithioöxalato-S,S')niccolate(II)

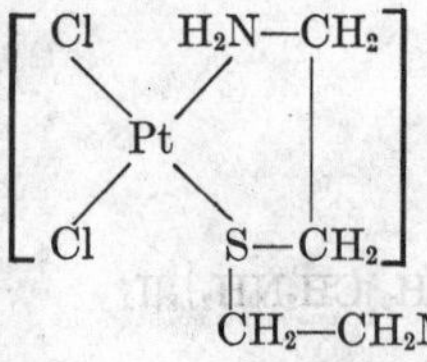

dichloro(2-N,N-dimethylaminoethyl 2-aminoethyl sulfide-N',S)platinum(II)

$K_2[Pt(NO_2)_4]$	potassium tetranitroplatinate(II)
$Na_3[Co(NO_2)_6]$	sodium hexanitrocobaltate(III)
$[Co(NO_2)_3(NH_3)_3]$	trinitrotriamminecobalt(III)
$[Co(ONO)(NH_3)_5]SO_4$	nitritopentaamminecobalt(III) sulfate
$[Co(NCS)(NH_3)_5]Cl_2$	isothiocyanatopentaamminecobalt(III) chloride

◆Thioöxalato in the original version has been changed to dithioöxalato.

7.4.   Di- and Polynuclear Compounds

7.41.—Bridging Groups.

7.411.—A bridging group shall be indicated by adding the Greek letter μ immediately before its name and separating this from the rest of the complex by a hyphen. Two or more bridging groups of the same kind are indicated by di-μ-, etc.

7.412.—If the number of central atoms bound by one bridging group exceeds two, the number shall be indicated by adding a subscript numeral to the μ.

This system of notation allows simply of distinction between, for example, μ-disulfido (one S_2 bridge) and di-μ-sulfido (two S bridges). It is also capable of extension to much more complex and unsymmetrical structures by use of the conventional prefixes cis, trans, asym, and sym where necessary.

Examples:

$[(NH_3)_5Cr-OH-Cr(NH_3)_5]Cl_5$
μ-hydroxo-bis{pentaamminechromium(III)}chloride

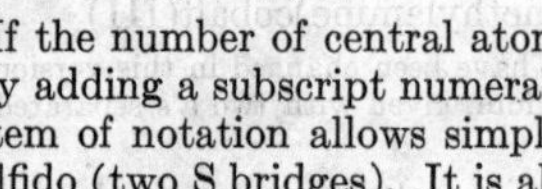

di-μ-chloro-dichlorobis(triethylarsine)diplatinum (II) (three possible isomers: asym, sym-cis, and sym-trans; the last is shown)

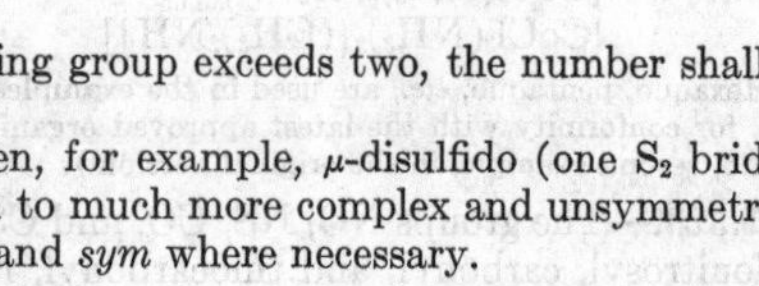

di-μ-thiocyanato-dithiocyanatobis(tripropylphosphine)diplatinum(II)

$[(CO)_3Fe(CO)_3Fe(CO)_3]$	tri-μ-carbonyl-bis(tricarbonyliron)
$[(CO)_3Fe(SEt)_2Fe(CO)_3]$	di-μ-ethanethiolato-bis(tricarbonyliron)
$[(C_5H_5)(CO)Fe(CO)_2Fe(CO)(C_5H_5)]$	di-μ-carbonyl-bis(carbonylcyclopentadienyliron)
$[(CO)\{P(OEt)_3\}Co(CO)_2Co(CO)\{P(OEt)_3\}]$	di-μ-carbonyl-bis{carbonyl(triethyl phosphite)-cobalt}
$[Au(CN)(C_3H_7)_2]_4$	cyclo-tetra-μ-cyano-tetrakis(dipropylgold)
$[CuI(Et_3As)]_4$	tetra-μ_3-iodo-tetrakis{triethylarsinecopper(I)}
$[Be_4O(CH_3COO)_6]$	μ_4-oxo-hexa-μ-acetato-tetraberyllium

7.42. Extended Structures.—Where bridging causes an indefinite extension of the structure it is best to name compounds primarily on the basis of their over-all composition; thus the compound having the composition represented by the formula $CsCuCl_3$ has an anion with the structure:

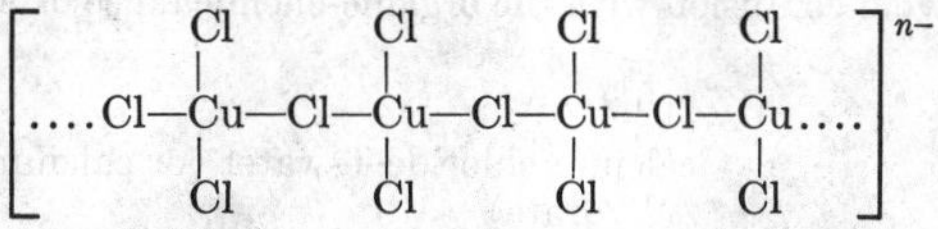

This may be expressed in the formula $(Cs^+)_n$ $[(CuCl_3)_n]^{n-}$ which leads to the simple name cesium catena-μ-chloro-dichlorocuprate(II). If the structure were in doubt, however, the substance would be called cesium copper(II) chloride (as a double salt).

◆*Cf.* comment on *catena* at 1.4.

7.5. Isopoly Anions

The structure of many complicated isopoly anions has now been cleared up by X-ray work and it turns out that the indication of the several μ-oxo and oxo atoms in the name does not convey any clear picture of the structure and is therefore of little value.

For the time being it is sufficient to indicate the number of atoms by Greek prefixes, at least until isomers are found. When all atoms have their "normal" oxidation states (*e.g.*, W^{VI}), it is not necessary to give the numbers of the oxygen atoms, if all the others are indicated.

Examples:

$K_2S_2O_7$	dipotassium disulfate
$K_2S_3O_{10}$	dipotassium trisulfate
$Na_5P_3O_{10}$	pentasodium triphosphate
$K_2Cr_4O_{13}$	dipotassium tetrachromate
$Na_2B_4O_7$	disodium tetraborate
NaB_5O_8	sodium pentaborate
$Ca_3Mo_7O_{24}$	tricalcium heptamolybdate
$Na_7HNb_6O_{19}.15H_2O$	heptasodium monohydrogen hexaniobate-15-water
$K_2Mg_2V_{10}O_{28}.16H_2O$	dipotassium dimagnesium decavanadate-16-water

7.6. Heteropoly Anions

The central atom or atoms should be cited last in the name and first in the formula of the anion (*cf.* **7.21**), *e.g.*, wolframophosphate, *not* phosphowolframate.

If the oxidation number has to be given, it may be necessary to place it immediately after the atom referred to and not after the ending -ate, in order to avoid ambiguity.

The method formerly recommended for naming iso- and heteropoly anions by giving the number of atoms in parentheses is not practicable in more complicated cases.

Examples:

$(NH_4)_3PW_{12}O_{40}$	triammonium dodecawolframophosphate
$(NH_4)_6TeMo_6O_{24}.7H_2O$	hexaammonium hexamolybdotellurate heptahydrate
$Li_3HSiW_{12}O_{40}.24H_2O$	trilithium (mono)hydrogen dodecawolframosilicate-24-water
$K_6Mn^{IV}Mo_9O_{32}$	hexapotassium enneamolybdomanganate(IV)
$Na_6P^V{}_2Mo_{18}O_{62}$	hexasodium 18-molybdodiphosphate(V)
$Na_4P^{III}{}_2Mo_{12}O_{41}$	tetrasodium dodecamolybdodiphosphate(III)
$K_7Co^{II}Co^{III}W_{12}O_{42}.16H_2O$	heptapotassiumdodecawolframocobalt(II)-cobalt(III)ate-16-water
$K_3PV_2Mo_{10}O_{39}$	tripotassium decamolybdodivanadophosphate

◆The coördination subcommittee would prefer not to have sections 7.5 and 7.6 included under 7 and recommends that they should be studied by a special subcommittee. Some cyclic isomers are already known, and isopoly cations also are being investigated.

Cf. comment at 1.12 for stand on wolframate and wolframo, and 2.251 for nona instead of ennea (fourth example).

Isopoly and heteropoly are not separate words in the original version.

7.7. Addition Compounds

The ending -ate is now the accepted ending for *anions* and should generally not be used for addition compounds. Alcoholates are the *salts* of alcohols and this name should not be used to indicate alcohol of crystallization. Analogously addition compounds containing ether, ammonia, *etc.*, should *not* be termed etherates, ammoniates, *etc.*

However, one exception has to be recognized. According to the commonly accepted meaning of the ending -ate, "hydrate" would be, and was formerly regarded as, the name for a *salt* of water, *i.e.*, what is now known as a hydroxide; the name hydrate has now a very firm position as the name of a compound containing water of crystallization and is allowed also in these Rules to designate water bound in an unspecified way; it is considered to be preferable even in this case to avoid the ending -ate by using the name "water" (or its equivalent in other languages) when possible.

The names of addition compounds may be formed by connecting the names of individual compounds by hyphens (short dashes) and indicating the number of molecules by Arabic numerals. When the added molecules are organic, however, it is recommended to use multiplicative numeral prefixes (bis, tris, tetrakis, *etc.*) instead of Arabic figures to avoid confusion with the organic-chemical use of Arabic figures to indicate position of substituents.

Examples:

$CaCl_2.6H_2O$	calcium chloride–6-water (or calcium chloride hexahydrate)
$3CdSO_4.8H_2O$	3-cadmium sulfate–8-water
$Na_2CO_3.10H_2O$	sodium carbonate–10-water (or sodium carbonate decahydrate)
$AlCl_3.4C_2H_5OH$	aluminum chloride–4-ethanol or –tetrakisethanol
$BF_3.(C_2H_5)_2O$	boron trifluoride–diethyl ether
$BF_3.2CH_3OH$	boron trifluoride–bismethanol
$BF_3.H_3PO_4$	boron trifluoride–phosphoric acid
$BiCl_3.3PCl_5$	bismuth trichloride–3-(phosphorus pentachloride)
$TeCl_4.2PCl_5$	tellurium tetrachloride–2-(phosphorus pentachloride)
$(CH_3)_4NAsCl_4.2AsCl_3$	tetramethylammonium tetrachloroarsenate(III)–2-(arsenic trichloride)
$CaCl_2.8NH_3$	calcium chloride–8-ammonia
$8H_2S.46H_2O$	8-(hydrogen sulfide)–46-water
$8Kr.46H_2O$	8-krypton–46-water
$6Br_2.46H_2O$	6-dibromine–46-water
$8CHCl_3.16H_2S.136H_2O$	8-chloroform–16-(hydrogen sulfide)–136-water

These names are not very different from a pure verbal description which may in fact be used, *e.g.*, calcium chloride with 6 water, compound of aluminum chloride with 4 ethanol, *etc.*

If it needs to be shown that added molecules form part of a complex, the names are given according to **7.2** and **7.3.**

Examples:

$FeSO_4.7H_2O$ or $[Fe(H_2O)_6]SO_4.H_2O$	iron(II) sulfate heptahydrate or hexaaquoiron(II) sulfate monohydrate
$PtCl_2.2PCl_3$ or $[PtCl_2(PCl_3)_2]$	platinum(II) chloride–2-(phosphorus trichloride) or dichlorobis(phosphorus trichloride)-platinum(II)
$AlCl_3.NOCl$ or $NO[AlCl_4]$	aluminum chloride–nitrosyl chloride or nitrosyl tetrachloroaluminate
$BF_3.Et_3N$ or $[BF_3(Et_3N)]$	boron trifluoride–triethylamine or trifluoro(triethylamine)boron

◆Only some so-called "addition compounds" are known to be coordination compounds, and the formulas and names can show such structures. Those that are lattice compounds and those of unknown structure do not really belong in 7.

The committees like hydrate and ammoniate and see no particular advantage in dropping them (*cf.* the use of hydrate terms in examples in 6.324; 6-hydrate, *etc.*, as well as hexahydrate, *etc.*, are considered acceptable).

No reason can be seen for deviating here from usual organic practice by using multiplicative prefixes with simple names: tetraethanol is just as clear as tetrakisethanol, and dimethanol as bismethanol. Parentheses can always be used if there is danger of any ambiguity.

The use of short dashes ("en" dashes) instead of hyphens between the names (as in the examples here, but not in the original version) makes for greater ease in reading.

It is considered preferable by some to place the electron donor first in both the formulas and name: $(C_2H_5)_2O.BF_3$, diethyl ether–boron trifluoride (note that organic usage favors ethyl ether over diethyl ether).

8. POLYMORPHISM

Minerals occurring in nature with similar compositions have different names according to their crystal structures; thus, zinc blende, wurtzite; quartz, tridymite, and cristobalite. Chemists and metallographers have designated polymorphic modifications with Greek letters or with Roman numerals (α-iron, ice-I, *etc.*). The method is similar to the use of trivial names, and is likely to continue to be of use in the future in cases where the existence of polymorphism is established, but not the structures underlying it. Regrettably there has been no consistent system, and some investigators have designated as α the form stable at ordinary temperatures, while others have used α for the form stable immediately below the melting point, and some have even changed an already established usage and renamed α-quartz β-quartz, thereby causing confusion. If the α–β nomenclature is used for two substances A and B, difficulties are encountered when the binary system A–B is studied.

LIST OF NAMES FOR IONS AND RADICALS

In inorganic chemistry substitutive names seldom are used, but the organic-chemical names are shown to draw attention to certain differences between organic and inorganic nomenclature.

	Name				
Atom or group	as neutral molecule	as cation or cationic radical[1]	as anion	as ligand	as prefix for substituent in organic compounds
H	monohydrogen	hydrogen	hydride	hydrido	
F	monofluorine		fluoride	fluoro	fluoro
Cl	monochlorine	chlorine	chloride	chloro	chloro
Br	monobromine	bromine	bromide	bromo	bromo
I	monoiodine	iodine	iodide	iodo	iodo
I_3			triiodide		
ClO		chlorosyl	hypochlorite	hypochlorito	
ClO_2	chlorine dioxide	chloryl	chlorite	chlorito	
ClO_3		perchloryl	chlorate	chlorato	
ClO_4			perchlorate		
IO		iodosyl	hypoiodite		iodoso
IO_2		iodyl			iodyl or iodoxy
O	monoöxygen		oxide	oxo	oxo or keto
O_2	dioxygen		O_2^{2-}: peroxide O_2^-: hyperoxide	peroxo	peroxy
HO	hydroxyl		hydroxide	hydroxo	hydroxy
HO_2	(perhydroxyl)		hydrogen peroxide	hydrogen peroxo	hydroperoxy
S	monosulfur		sulfide	thio (sulfido)	thio
HS	(sulfhydryl)		hydrogen sulfide	thiolo	thiol or mercapto
S_2	disulfur		disulfide	disulfido	
SO	sulfur monoxide	sulfinyl (thionyl)			sulfinyl
SO_2	sulfur dioxide	sulfonyl (sulfuryl)	sulfoxylate		sulfonyl
SO_3	sulfur trioxide		sulfite	sulfito	
HSO_3			hydrogen sulfite	hydrogen sulfito	
S_2O_3			thiosulfate	thiosulfato	
SO_4			sulfate	sulfato	
Se	selenium		selenide	seleno	seleno
SeO		seleninyl			seleninyl
SeO_2		selenonyl			selenonyl
SeO_3	selenium trioxide		selenite	selenito	
SeO_4			selenate	selenato	
Te	tellurium		telluride	telluro	telluro
CrO_2		chromyl			
UO_2		uranyl			
NpO_2		neptunyl			
PuO_2		plutonyl			
AmO_2		americyl			
N	mononitrogen		nitride	nitrido	
N_3			azide	azido	
NH			imide	imido	imino
NH_2			amide	amido	amino
NHOH			hydroxylamide	hydroxylamido	hydroxylamino
N_2H_3			hydrazide	hydrazido	hydrazino
NO	nitrogen oxide	nitrosyl		nitrosyl	nitroso
NO_2	nitrogen dioxide	nitryl		nitro	nitro
ONO			nitrite	nitrito	
NS		thionitrosyl			
$(NS)_n$		thiazyl (*e.g.*, trithiazyl)			
NO_3			nitrate	nitrato	
N_2O_2			hyponitrite	hyponitrito	
P	phosphorus		phosphide	phosphido	
PO		phosphoryl			phosphoroso
PS		thiophosphoryl			
PH_2O_2			hypophosphite	hypophosphito	
PHO_3			phosphite	phosphito	
PO_4			phosphate	phosphato	
AsO_4			arsenate	arsenato	
VO		vanadyl			
CO	carbon monoxide	carbonyl		carbonyl	carbonyl
CS		thiocarbonyl			
CH_3O	methoxyl		methanolate	methoxo	methoxy
C_2H_5O	ethoxyl		ethanolate	ethoxo	ethoxy
CH_3S			methanethiolate	methanethiolato	methylthio
C_2H_5S			ethanethiolate	ethanethiolato	ethylthio
CN		cyanogen	cyanide	cyano	cyano
OCN			cyanate	cyanato	cyanato
SCN			thiocyanate	thiocyanato and isothiocyanato	thiocyanato and isothiocyanato
SeCN			selenocyanate	selenocyanato	selenocyanato
TeCN			tellurocyanate	tellurocyanato	
CO_3			carbonate	carbonato	
HCO_3			hydrogen carbonate	hydrogen carbonato	
CH_3CO_2			acetate	acetato	acetoxy
CH_3CO	acetyl	acetyl			acetyl
C_2O_4			oxalate	oxalato	

[1]If necessary, oxidation state is to be given by Stock notation.

◆Although some additions might be made to this list, especially in the last column, and a few changes suggested, no attempt has been made to do so at this time. *Cf.* comments at such rules as 3.31, 3.32, 5.35, 6.2, and 7.312.

A rational system should be based upon crystal structure, and the designations *α*, *β*, *γ*, *etc.*, should be regarded as provisional, or as trivial names. The designations should be as short and understandable as possible, and convey a maximum of information to the reader. The rules suggested here have been framed as a basis for future work, and it is hoped that experience in their use may enable more specific rules to be formulated at a later date.

8.1.—For chemical purposes (*i.e.*, when particular mineral occurrences are not under consideration) polymorphs should be indicated by adding the crystal system after the name or formula. For example, zinc sulfide(cub.) or ZnS(cub.) corresponds to zinc blende or sphalerite, and ZnS(hex.) to wurtzite. The Commission considers that these abbreviations might with advantage be standardized internationally:

cub.	=	cubic	hex.	=	hexagonal
c.	=	body-centered	trig.	=	trigonal
f.	=	face-centered	mon.	=	monoclinic
tetr.	=	tetragonal	tric.	=	triclinic
o-rh.	=	orthorhombic			

Slightly distorted lattices may be indicated by use of the *circa* sign, ∼. Thus, for example, a slightly distorted face-centered cubic lattice would be expressed as ∼f.cub.

8.2.—Crystallographers may find it valuable to add the space-group; it is doubtful whether this system would commend itself to chemists where **8.1** is sufficient.

8.3.—Simple well-known structures may also be designated by giving the type-compound in italics in parentheses; but this system often breaks down because many structures are not referable to a type in this way. Thus, AuCd above 70° may be written as AuCd(cub.) or as AuCd(*CsCl-type*); but at low temperature only as AuCd(o-rh.), as its structure cannot be referred to a type.

◆*Cf.* 6.5 and 6.52.

ABBREVIATIONS USED IN TABLE OF PHYSICAL CONSTANTS OF INORGANIC COMPOUNDS

Abbreviation	Meaning
a	acid
abs	absolute
ac. a	acetic acid
acet	acetone
act	active
al	alcohol
alk	alkali
amm	ammonium
amor	amorphous
anh	anhydrous
appr	approximately
aq	aqua, water
aq. reg	aqua regia
asym	asymmetrical
atm	atmospheres
bipyr	bipyramidal
bl	blue
blk	black
boil	boiling
br., brn	brown
brnsh	brownish
bz	benzene
c	cold
calc	calculated
carb	carbon
caust	caustic
chl	chloroform
choc	chocolate
cit. a	citric acid
col	colorless
coll	colloidal
com'l	commercial
comp	compounds
compl	completely
conc	concentrated
const	constant
cont	contains
corros	corrosive
cr	crystalline
cub	cubic
d., dec	decomposes
deliq	deliquescent
deriv	derivative
dibas	dibasic
di-H	dihydrogen
dil	dilute
dimorph	dimorphous
disg	disagreeable
dk	dark
doubt	doubtful
duct	ductile
effl	efflorescent
em	emerald
eth	ether
ev	evolves
evln	evolution
ex	excess
exist	existence
exp	explodes
extr	extreme(ly)
f., fr	from
feath	feathery
fl	flakes
floc	flocculent
fluo, fluores	fluorescent
form	formic
fum	fuming
fus	fused
fxd	fixed
gel., gelat	gelatinous
gl	glass
glac	glacial
glit	glittering
glob	globular
glyc	glycerin
gran	granular
greas	greasy
grn	green
h	hot
hex	hexagonal
ht	heat
hyd	hydrolyzed
hydx	hydroxides
hyg	hygroscopic
i	insoluble
ign	ignites
ind	indigo
indef	indefinite
infl., inflam	inflammable
infus	infusible
irid	iridescent
leaf	leaflets
lem	lemon
lgr	ligroin
lng	long
lq., liq	liquid
lt	light
lum	luminous
lust	lustrous
me., meth	methyl
met	metal or metallic
micr	microscopic
min	mineral
misc	miscible
mixt	mixture
mod	modifications
monbas	monobasic
mon-H	monohydrogen
monocl	monoclinic
near	nearly
need	needles
nit	nitrate
oct	octahedral
odorl	odorless
offen	offensive
olv	olive
opt	optical or optically
or	orange
ord	ordinary
org	organic
oxal	oxalate or oxalic
pa	pale
pet	petroleum
pl	plates
pois	poisonous
polymorph	polymorphous
powd	powder
ppt	precipitate
pr	prisms
press	pressure
prob	probably
prop	properties
purp	purple
pyr	pyridine
quad	quadrilateral
quest	questioned
rect	rectangular
redsh	reddish
reg	regular
rhbdr	rhombohedral
rhomb	rhombic, orthorhombic
s	soluble
satd	saturated
sld	solid
sensit	sensitive
sc	scales
sec	secondary
silv	silver
sl	slightly
sly	slowly
sm	small
sod	sodium
soln	solution
solv	solvents
spont	spontaneous
st	steel
stab	stable
subl	sublimes
suffoc	suffocating
sulfd	sulfides
sulf	sulfur
sym	symmetrical
tabl	tablets
tart. a	tartaric acid
tetr	tetragonal
tetrah	tetrahedral
tol	toluene
trac	trace, traces
trans	transparent
translu	translucent
tri., trig	trigonal
tribas	tribasic
tricl	triclinic
trim	trimetric
tr	transition point
turp	turpentine
unpleas	unpleasant
unst	unstable
v	very
vac	vacuum
var	various
viol	violent, violence
visc	viscous
vitr	vitreous
vlt	violet
volt., volat	volatizes
wh	white
wh. lt	white light
yel	yellow
yelsh	yellowish
∞	soluble in all proportions
>	above
<	below

PHYSICAL CONSTANTS OF INORGANIC COMPOUNDS

No.	Name	Synonyms and Formulae	Mol. wt.	Crystalline form, properties and index of refraction	Density or spec. gravity	Melting point, °C	Boiling point, °C	Solubility, in grams per 100 cc		
								Cold water	Hot water	Other solvents
	Actinium									
a1	**Actinium**	Ac	227	silv wh met, cub		1050	3200 ± 300	d to $Ac(OH)_3$		
a2	bromide	$AcBr_3$	466.73	wh, hex	5.85	subl 800		s		
a3	chloride, tri-	$AcCl_3$	333.36	wh cr, hex	4.81	subl 960				
a4	fluoride, tri-	AcF_3	284	wh cr, hex	7.88			i	i	
a5	hydroxide	$Ac(OH)_3$	278.02	wh				i		
a6	iodide	AcI_3	607.71	wh		subl 700–800		s		
a7	oxalate	$Ac_2(C_2O_4)_3$	718.06					i		
a8	oxide, sesqui-	Ac_2O_3	502	wh cr, hex	9.19			i		
a9	sulfide, sesqui-	Ac_2S_3	550.19	dk, cub	6.75					
a10	**Aluminum**	Al	26.98154	silv wh duct met, cub	2.702	660.37	2467	i	i	s alk, HCl, H_2SO_4; i conc HNO_3, h ac a
a11	acetate, tri-	$Al(C_2H_3O_2)_3$	204.12	wh solid		d		v sl s	d	
a12	acetylacetonate	$Al(C_5H_7O_2)_3$	324.31	col, monocl	1.27	193, subl vac	314	i	i	v s al; s eth, bz
a13	*ortho*arsenate	$AlAsO_4.8H_2O$	310.02	wh powd	3.001	$-H_2O$		i	i	sl s a
a14	benzoate	$Al(C_7H_5O_2)_3$	390.33	wh cr powd				v sl s		
a15	benzyloxide	Aluminum benzylate. $Al(C_7H_7O)_3$	348.38			59–60	$283–284^{0.5}$			
a16	boride	AlB_{12}	156.71	dk red-blk, monocl	2.55_{4}^{18}			i		s hot HNO_3; i a, alk
a17	boride, di-	AlB_2	48.60	copper red, hex	3.19					
a18	bromate	$Al(BrO_3)_3.9H_2O$	572.84	wh cr, hygr		62.3	d 100	s	s	sl s a
a19	bromide	$AlBr_3$ (or Al_2Br_6)	266.71	col, rhomb pl, deliq	2.64^{10} (fused)	97.5	263.3^{747}	s with viol	d	s al, acet, CS_2
a20	bromide, hexahydrate	$AlBr_3.6H_2O$	374.80	col-yelsh need, deliq	2.54	93	d 135	s	d	s al, amyl al; sl s CS_2
a21	bromide, pentadecylhydrate	$AlBr_3.15H_2O$	536.94	col need		−7.5	d 7	s	s	s al
a22	butoxide, tert-	$Al(C_4H_9O)_3$	246.33	wh cr	1.0251_{0}^{20}	subl 180, m.p >300, sealed tube.				v s org solv
a23	carbide	Al_4C_3	143.96	yel-grn, hex	2.36	stab to 1400	d 2200^{400}	d to CH_4		d dil a; i acet
a24	chlorate	$Al(ClO_3)_3.6H_2O$	385.43	col, rhbdr, deliq		d		vs	vs	s dil HCl
a25	*per*chlorate	$Al(ClO_4)_3.6H_2O$	433.43	col, hygr	2.020	82	$-6H_2O$, 178 d 262	s	s	
a26	chloride	$AlCl_3$ (or Al_2Cl_6)	133.34	wh to col, hex, odor HCl, v deliq	fus 2.44^{25} liq 1.31^{200}	$190^{2.5}$ atm	182.7^{752} subl. 177.8	69.9^{15} with viol	s d	$100^{12.5}$ abs al; 0.072^{25} chl; sCCl_4, eth sl s bz
a27	chloride, hexahydrate	$AlCl_3.6H_2O$	241.43	col, rhomb, deliq, 1.6	2.398	d 100		s	v s ev HCl	50 abs al; s eth; sl s HCl
a28	chloride, hexammine	$AlCl_3.6NH_3$	235.52	col cr, hygr	1.412_{4}^{25}	d		s		
a29	diethylmalonate deriv	$Al(C_7H_{11}O_4)_3$	504.47		1.084^{100}	98		i		s org solv
a30	ethoxide	$Al(C_2H_5O)_3$	162.14	wh cr	1.142_{0}^{20}	134	205^{14}	d	s	v sl s al, eth
a31	α-ethylacetoacetate deriv	$Al(C_6H_9O_3)_3$	414.39	wh cr	1.101^{80}	78–79	$190–200^{11}$	s		s lgr
a32	ferrocyanide	$Al_4[Fe(CN)_6]_3.17H_2O$	1050.05	br powd				sl s	sl s	s dil a
a33	fluoride	AlF_3	83.98	col, tricl	2.882_{4}^{25}	1291 $subl^{760}$		0.559^{25}	s	i a, alk, al, acet
a34	fluoride	$AlF_3.3\frac{1}{2}H_2O$	147.03	wh cr powd	1.914_{4}^{25}	$-2H_2O$ 100	anhydr 250	i	sl s	d ac a
a35	fluoride, monohydrate	Nat. fluellite. $AlF_3.H_2O$.	101.99	col, rhomb, 1.473, 1.490, 1.511	2.17			sl s	sl s	
a36	fluosilicate	Nat topaz. $2AlFO.SiO_2$	184.04	rhomb, 1.619, 1.620, 1.627	3.58			i	i	
a37	hydroxide	Nat. boehmite. AlO(OH)	59.99	wh, orthorhomb microcr	3.01	$-H_2O$, trans to γ-Al_2O_3		i	i	s h a, h alk
a38	hydroxide	Nat. diaspore. AlO(OH)	59.99	col, rhomb cr	3.3–3.5	$-H_2O$, trans to Al_2O_3		i	i	s h a, h alk
a39	hydroxide	$Al(OH)_3$	78.00	wh, monocl	2.42	$-H_2O$, 300		i	i	s a, alk; i al
a40	iodide	AlI_3 (or Al_2I_6)	407.69	br pl, cont free I_2, deliq	3.98^{25}	191	360	s d	s	s al, eth, CS_2, liq NH_3
a41	iodide, hexahydrate	$AlI_3.6H_2O$	515.79	wh-yel cr, hygr	2.63	d 185	d	v s	v s	s al, CS_2
a42	isopropoxide	$Al(C_3H_7O)_3$	204.25	wh cr	1.0346_{0}^{20}	118.5	140.5^{8}	d		s al, bz, chl
a43	lactate	$Al(C_3H_5O_3)_3$	294.20	wh-yelsh powd				v s		

No.	Name	Synonyms and Formulae	Mol. wt.	Crystalline form, properties and index of refraction	Density or spec. gravity	Melting point, °C	Boiling point, °C	Solubility, in grams per 100 cc		
								Cold water	Hot water	Other solvents
	Aluminum									
a44	nitrate	$Al(NO_3)_3.9H_2O$	375.13	col, rhomb, deliq, 1.54		73.5	d 150	63.7^{25}	v s d	100 al; s alk, acet, HNO_3. a
a45	nitride	AlN	40.99	wh cr, hex	3.26	>2200 (in N_2)	subl 2000	d (NH_3)	d	d a, alk
a46	oleate (com'l)	$Al(C_{18}H_{33}O_2)_3(?)$	871.37	wh powd, existence doubted except as basic salt				d	s	i al; v sl s bz
a47	oxalate	$Al_2(C_2O_4)_3.4H_2O$	390.08	wh powd				i	i	i al; s a
a48	oxide	Al_2O_3	101.96	col, hex, 1.768, 1.760	3.965^{25}	2072	2980	i		v sl s a, alk
a49	oxide	α-Alumina, nat. corundum. Al_2O_3	101.96	col, rhomb cr, 1.765	3.97	2015 ± 15	2980 ± 60	0.000098^{29}	i	v sl s a, alk
a50	oxide	γ-Alumina. Al_2O_3	101.96	wh micr cr, 1.7	3.5–3.9	tr to α		i	i	sl s a, alk
a51	oxide, monohydrate	$Al_2O_3.H_2O$	119.98	col, rhomb, 1.624 ± 0.003	3.014			i	i	
a52	oxide, trihydrate	Nat. gibbsite, hydrargilite. $Al_2O_3.3H_2O$	156.01	wh monocl cr, 1.577, 1.577, 1.595	2.42	tr to $Al_2O_3.H_2O$ (Boehmite)		i	i	s h a, alk
a53	oxide, trihydrate	Nat. bayerite. $Al_2O_3.3H_2O$	156.01	wh micr cr, 1.583	2.53	tr to $Al_2O_3H_2O$ (Boehmite)		i	i	s hot a, alk
a54	*meta*phosphate	$Al(PO_3)_3$	263.90	col, tetr	2.779			i	i	i a
a55	palmitate, mono- (com'l)	$Al(OH)_2C_{16}H_{31}O_2$	316.41	wh	1.095	200		i		s alk, hydrocarb
a56	1-phenol-4-sulfonate	$Al(C_6H_5O_4S)_3$	546.49	redsh-wh powd				s		s al, glyc
a57	phenoxide	$Al(C_6H_5O)_3$	306.27	grayish-wh cr mass	1.23	d 265		d		s al, eth, chl
a58	*ortho*phosphate	$AlPO_4$	121.95	wh rhomb pl, 1.546, 1.556, 1.578	2.566	>1500		i	i	s a, alk, al
a59	propoxide	$Al(C_3H_7O)_3$	204.25	wh cr	1.0578^{20}_{0}	106	248^{14}	d	d	s al
a60	salicylate	$Al(C_7H_5O_3)_3$	438.33	redsh-wh powd				i		i al; s alk
a61	selenide	Al_2Se_3	290.84	lt brn powd, unstable in air	3.437^{15}_{4}			d	d	d a
a62	silicate	Nat. sillimanite, andalusite, cyanite. $Al_2O_3.SiO_2$	162.04	wh, rhomb, 1.66	3.247	1545 tr to $Al_2O_3.2SiO_2$	>1545	i	i	d HF; i HCl; s fus alk
a63	silicate	Nat. mullite. $3Al_2O_3.2SiO_2$	426.05	col, rhomb, 1.638, 1.642, 1.653	3.156	1920		i	i	i a, HF
a64	stearate, tri-	$Al(C_{18}H_{35}O_2)_3$	877.42	wh powd	1.010	103		i		s al, bz, turp, alk
a65	sulphate	$Al_2(SO_4)_3$	342.15	wh powd, 1.47	2.71	d 770		31.3^{0}	98.1^{100}	s dil a; sl s al
a66	sulfate, hydrate	Nat. alunogenite. $Al_2(SO_4)_3.18H_2O$	666.43	col, monocl, 1.474, 1.467, 1.483	1.69^{17}	d 86.5		86.9^{0}	1104^{100}	i al
a67	sulfide	Al_2S_3	150.16	yel, hex, odor H_2S, d moist air	2.02^{13}	1100	subl 1500 (N_2)	d		s a; i acet
a68	thallium sulfate	Aluminum thallium alum. $AlTl(SO_4)_2.12H_2O$	639.66	col, oct, 1.50112	2.325^{20}_{4}	91		4.84^{0}	65.19^{60}	
a69	**Americium**	Am	243.13	silvery, hex		994±4	2607 (extrap)			s dil a
a70	bromide	$AmBr_3$	482.86	wh, orthorhomb		subl		s		
a71	chloride	$AmCl_3$	349.49	pink, hex	5.78	subl 850		s		
a72	fluoride	AmF_3	300.12	pink, hex	9.53			i		
a73	iodide	AmI_3	623.84	yel, orthoromb	6.9			s		
a74	oxide	Am_2O_3	534.26	redsh-brn, cub or tan, or hex						s min a
a75	oxide, di-	AmO_2	275.13	blk, cub	11.68					s min a
a76	**Ammonia**	NH_3	17.03	col gas; liq, 0.817^{-79}, $1.325^{16.5}$	0.7710 g/l 760 mm	−77.7	−33.35	89.9	7.4^{100}	13.20^{20} al; s eth, org solv
a77	Ammonia-d_3	Trideuterio ammonia. ND_3	20.05			−74	−30.9			
	Ammonium									
v78	acetate	$NH_4C_2H_3O_2$	77.08	wh cr, hygr	1.17^{20}_{4}	114	d	148^{4}	d	7.89^{15} MeOH; s al; sl s acet
a79	acetate, hydrogen	$(NH_4)H(C_2H_3O_2)_2$	137.14	col need, deliq		66		s		s al
a80	aluminum chloride	$NH_4Cl.AlCl_3$	186.83	wh cr		304		s		
a81	aluminum sulfate	$NH_4Al(SO_4)_2$	237.14	col, hex	2.45^{20}			s		s glyc; i al

No.	Name	Synonyms and Formulae	Mol. wt.	Crystalline form, properties and index of refraction	Density or spec. gravity	Melting point, °C	Boiling point, °C	Solubility, in grams per 100 cc		
								Cold water	Hot water	Other solvents
	Ammonium									
a82	aluminum sulfate, hydrate	Nat. tschernigite. $NH_4Al(SO_4)_2.12H_2O$	453.33	col, cub, 1.459	1.64	93.5	$-10H_2O$, 120	15^{20}	v s	s dil a; i al
a83	*ortho*arsenate	$(NH_4)_3AsO_4.3H_2O$	247.08	rhomb cr		d, $-NH_3$		sl s		
a84	*ortho*arsenate, di-H	$NH_4H_2AsO_4$	158.98	col, tetr, 1.577, 1.522	2.311^{9}	d, $-NH_3^{300}$		33.74^{0}	122.4^{90}	
a85	*ortho*arsenate mono-H	$(NH_4)_2HAsO_4$	176.00	col, monocl, odor NH_3	1.989	d		s	d	i al
a86	*meta*arsenite	NH_4AsO_2	124.96	col, rhomb pr, hygr				v s	d	i al, acet; sl s NH_4OH
a87	azide	NH_4N_3	60.06	col pl	1.346	160	subl 134 expl	20.16^{30}	27.04^{40}	1.06^{20} 80 % al; i eth, bz
a88	benzene sulfonate	$NH_4C_6H_5SO_3$	175.21	rhomb	1.342	d 271–275		98	320	19 c al; i eth, bz
a89	benzoate	$NH_4C_7H_5O_2$	139.16	col, rhomb	1.260	d 198	subl 160	$19.6^{14.5}$	83.3^{100}	1.63^{25} al; s glyc; i eth
a90	*penta*borate	Ammonium decaborate. $NH_4B_5O_8.4H_2O$	272.15					7.03^{18}		
a91	*peroxy*borate	$NH_4BO_3.\frac{1}{2}H_2O$	85.86	wh cr		d		$1.55^{17.5}$		i al
a92	*tetra*borate	Ammonium biborate. $(NH_4)_2B_4O_7.4H_2O$	263.38	col, tetr		d		7.27^{18}	52.68^{90}	sl s acet; i al
a93	bromate	NH_4BrO_3	145.95	col, hex		expl		v s	vs	
a94	bromide	NH_4Br	97.95	cub, coll hygr, 1.712^{25}	2.429	subl 452	235 vac	97^{25}	145.6^{100}	10^{78} al; s acet, eth, NH_3
a95	*di*bromoiodide	NH_4IBr_2	304.75	met-grn pr, hygr		198		v s	s	s eth
a96	bromoplatinate	$(NH_4)_2PtBr_6$	710.62	red-brn cub	4.265^{24}	d 145		0.59^{20}	0.36^{100}	
a97	bromoselenate	$(NH_4)_2SeBr_6$	594.49	red, oct cr	3.326			d	d	sl s eth
a98	bromostannate	$(NH_4)_2SnBr_6$	634.22	col, cub	3.50	d		v s		
a99	cadmium chloride	$4NH_4Cl.CdCl_2$	397.27	col, rhomb, 1.6038	2.01			s		
a100	calcium arsenate	$NH_4CaAsO_4.6H_2O$	305.13	col, monocl	1.905^{15}	d 140		0.02	s	s NH_4Cl; i NH_4OH
a101	calcium phosphate	$NH_4CaPO_4.7H_2O$	279.20	1.561^{15}		d		i	d	s a
a102	carbamate	$NH_4NH_2CO_2$	78.07	col, rhomb		subl 60		v s	d	v s NH_4OH; sl s al; i acet
a103	carbamate acid carbonate	Sal volatile. $NH_4NH_2CO_2.NH_4HCO_3$	157.13	wh cr		subl		25^{15}	67^{65}	d al; s glyc; i acet
a104	carbonate	$(NH_4)_2CO_3.H_2O$	114.10	col, cub		d 58		100^{15}	d	i al, NH_3, CS_2; s dil MeOH
a105	carbonate, hydrogen	Ammonium bicarbonate. NH_4HCO_3	79.06	col, rhomb or monocl, 1.423, 1.536, 1.555	1.58	107.5 (d 36–60)	subl	11.9^{0}	d	i al, acet
a106	cerium nitrate(ic)	$(NH_4)_2Ce(NO_3)_6$	548.23	or, monocl				141^{25}	227^{80}	s HNO_3, al
a107	cerium nitrate(ous)	$2NH_4NO_3.Ce(NO_3)_3.4H_2O$	558.28	monocl		74		318.20	817.4^{65}	
a108	cerium sulfate(ous)	$(NH_4)_2SO_4.Ce_2(SO_4)_3.8H_2O$	844.69	monocl	2.523	$-6H_2O$, 100	$-8H_2O$,150	3.29^{49} (anhydr)		
a109	chlorate	NH_4ClO_3	101.49	col, monocl need.	1.80	102 expl		28.7^{0}	115^{75}	sl s al
a110	*per*chlorate	NH_4ClO_4	117.49	col, rhomb, 1.482.	1.95	d		10.74^{0}	42.45^{85}	s acet; sl s al
a111	chloride	Sal ammoniac. NH_4Cl	53.49	col, cub, 1.642	1.527	subl 340	520	29.7^{0}	75.8^{100}	0.6^{19} al; s liq NH_3; i acet, eth
a112	chloroaurate	NH_4AuCl_4	356.82	yel, monocl or rhomb			520	s		sl s al
a113	chloroaurate, hydrate	$(NH_4AuCl_4)_4.5H_2O$	1517.35	yel, monocl		$-5H_2O$, 100		s		s al
a114	chlorogallate	NH_4GaCl_4	229.57	wh cr		275		v s	v s	s al; i pet eth
a115	chloroiridate	$(NH_4)_2IrCl_6$	441.00	red-blk, cub	2.856	d		0.69^{14}	4.38^{80}	i al; s HCl
a116	chloroiridite	$(NH_4)_3IrCl_6.1\frac{1}{2}H_2O$	486.06					s		
a117	chloroosmate	$(NH_4)_2OsCl_6$	439.00	cub	2.93					
a118	chloropalladate	$(NH_4)_2PdCl_6$	355.20	red-brn, cub	2.418	d		sl s		
a119	chloropalladite	$(NH_4)_2PdCl_4$	284.29	olive grn, tetr	2.17	d		s		i al
a120	*hexa*chloroplatinate	$(NH_4)_2PtCl_6$	443.89	yel, cub, 1.8	3.065	d		0.7^{15}	1.25^{100}	0.005 al; i eth, conc HCl
a121	chloroplatinite	$(NH_4)_2PtCl_4$	372.98	red, rhomb (tetr).	2.936	d 140–150		s	s	i al
a122	chloroplumbate	$(NH_4)_2PbCl_6$	455.98	yel, cub	2.925	d 120		sl s	d	s a
a123	chlorostannate	$(NH_4)_2SnCl_6$	367.49	wh, cub	2.4	d		$33.^{14.5}$	v s	
a124	*tetra*chlorozincate	$ZnCl_2.2NH_4Cl$	243.26	wh pl, rhomb, hygr	1.879	d 150		v s		
a125	chromate	$(NH_4)_2CrO_4$	152.08	yel, monocl	1.91^{12}	d 180		40.5^{30}	d	i al; sl s NH_3, acet
a126	*di*chromate	$(NH_4)_2Cr_2O_7$	252.06	or, monocl	2.15^{25}	d 170		30.8^{15}	89^{30}	s al; i acet
a127	*peroxy*chromate	$(NH_4)_3CrO_8$	234.11	red-brn, cub		d 40	expl 50	sl s	d	i al, eth, sl s NH_3; expl H_2SO_4
a129	chromium sulfate(ic)	$(NH_4)Cr(SO_4)_2.12H_2O$	478.34	grn or vlt, cub; 1.4842	1.72	94, $-9H_2O$, 100		21.2^{25}	32.8^{40}	s al, dil a
a130	citrate, di(*sec.*)	$(NH_4)_2HC_6H_5O_7$	226.19	wh gran or powd	1.48			100		sl s al
a131	citrate, tri-(*tert.*)	$(NH_4)_3C_6H_5O_7$	243.22	wh cr, deliq		d		v s		i al, eth, acet
a132	cobalt *ortho*phosphate(ous)	$(NH_4)CoPO_4.H_2O$	189.96	vlt cr powd				i	d	s a

No.	Name	Synonyms and Formulae	Mol. wt.	Crystalline form, properties and index of refraction	Density or spec. gravity	Melting point, °C	Boiling point, °C	Solubility, in grams per 100 cc Cold water	Hot water	Other solvents
	Ammonium									
a133	cobalt (II)-selenate	$(NH_4)_2SeO_4.CoSO_4.6H_2O$	489.02	ruby-red, monocl, 1.526, 1.532, 1.541	2.228^{20}_{4}					
a134	cobalt sulfate(ous)	$(NH_4)_2SO_4.CoSO_4.6H_2O$	395.23	ruby-red, monocl, 1.490, 1.495, 1.503	1.902			20.5^{20}	45.4^{80}	i al
a135	copper chloride(ic)	$2NH_4Cl.CuCl_2.2H_2O$	277.46	blue, tetrag, 1.744, 1.724	1.993	d 110		33.8^{0}	99.3^{80}	s a, al; sl s NH_3
a136	copper iodide(ous)	$NH_4I.CuI.H_2O$	353.40	rhomb pl				d	d	s NH_4I
a137	cyanate	NH_4OCN	60.06	wh cr	1.342^{20}_{4}	d 60		v s	d	sl s al; i eth
a138	cyanide	NH_4CN	44.06	col, cub	1.02^{100}	d 36	subl 40	v s	d	v s al
a139	cyanaurate	$NH_4Au(CN)_4.H_2O$	337.09	col pl		d 200		v s		v s al; i eth
a140	cyanaurite	$NH_4Au(CN)_2$	267.04	col, cub		d 100		v s	v s	s al: i eth
a141	cyanoplatinite	$(NH_4)_2Pt(CN)_4.H_2O$	353.61	yel cr				s		
a142	ethylsulfate	$NH_4C_2H_5SO_4$	143.16			99		s		
a143	ferricyanide	$(NH_4)_3Fe(CN)_6$	266.06	red cr, rhomb		d		v s		
a144	ferrocyanide	$(NH_4)_4Fe(CN)_6.3H_2O$	338.15	yel, monocl, turns bl in air		d		s	d	i al
a145	fluoantimonite	$(NH_4)_2SbF_5$	252.84	col, rhomb		d, subl		108		
a146	fluoborate	NH_4BF_4	104.84	wh, rhomb	1.871^{15}	subl		25^{16}	97^{100}	s NH_4OH
a147	fluogallate	$(NH_4)_3GaF_6$	237.83	wh oct cr		d >250–GaF_3			sl s	
a148	fluogermanate	$(NH_4)_2GeF_6$	222.66	col hex pr and bipyram, 1.428, 1.425	2.564^{25}_{25}			s		i al, MeOH
a149	fluophosphate, di-	$NH_4PO_2F_2$	119.01	col, rhomb		213		s	s	s al, acet
a150	fluophosphate, hexa.	NH_4PF_6	163.00	col, pl	2.180^{18}	d		s	s	s al, acet
a151	fluoride	NH_4F	37.04	col, hex, deliq	1.009^{25}	subl		100^{0}	d	s al; i NH_3
a152	fluoride, hydrogen	NH_4HF_2	57.04	rhomb or tetr, deliq, 1.390	1.50	125.6		v s	v s	sl s al
a153	fluorosulfonate	NH_4FSO_3	117.10	wh need		d 245		v s	v s	sl s al; s MeOH
a154	fluosilicate	α-Nat. cryptohalite. $(NH_4)_2SiF_6$	178.14	α oct, β hex, col α 1.3696	α 2.011 β 2.152	d		18.6^{17}	55.5^{100}	sl s al; i acet
a155	fluosulfonate	NH_4SO_3F	117.10	col, need		244.7		s	s	sl s al; s MeOH; d NH_4OH
a156	fluotitanate	$(NH_4)_2TiF_6$	197.97	hex pr		d		s	s	i al, eth
a157	fluozirconate	$(NH_4)_2ZrF_6$	241.29	rhomb, hex	1.154					
a158	fluozirconate	$(NH_4)_3ZrF_7$	278.32	col, cub	1.433			sl s		
a159	formate	NH_4CHO_2	63.06	wh, monocl, deliq.	1.280	116	d 180	102^{0}	531^{80}	s al, eth, NH_3
a160	gallium sulfate	Ammonium gallium alum. $Ga(NH_4)(SO_4)_2.12H_2O$	496.07	cub, 1.468	1.777			30.84^{25}		0.00875^{25} 70 % EtOH
a161	*tri*hydrogen para-*per*iodate	$(NH_4)_2H_3IO_6$	262.00	col, rhomb	2.85					
a162	hydroxide	NH_4OH	35.05	at ord temp in sol only		−77		s		
a163	iodate	NH_4IO_3	192.94	col, rhomb or monocl	3.309^{21}	d 150		2.06^{15}	14.5^{101}	
a164	iodide	NH_4I	144.94	col, cub, hygr, 1.7031	2.514^{25}	subl 551	220 vac	154.2^{0}	250.3^{100}	v s al, acet, NH_3; sl s eth
a165	*tri*iodide	NH_4I_3	398.75	dk br, rhomb	3.749	d 175		s d	d	
a166	iodoplatinate	$(NH_4)_2PtI_6$	992.59	blk, cub	4.61					i al
a167	iridium chloride (III)	$(NH_4)_3IrCl_6.H_2O$	477.05	grn-br-bl, cub		d 350		10.5^{19}		
a168	iridium sulfate	$NH_4Ir(SO_4)_2.12H_2O$	618.54	yel-red cr			106	s		
a169	iron (III) chloride	$2NH_4Cl.FeCl_3.H_2O$	287.20	ruby-red, rhomb, hygr, 1.78	1.99	234		v s	v s	
a170	iron (III) fluoride	$3NH_4F.FeF_3$	223.95	col to lt yel, oct	1.96			sl s	sl s	
a171	iron (II) selenate	$(NH_4)_2SeO_4.FeSO_4.6H_2O$	485.93	lt grn, monocl, 1.5226, 1.5260, 1.5334	2.191^{20}_{4}					
a172	iron (III) sulfate	$(NH_4)_2SO_4.Fe_2(SO_4)_3$	532.02	wh, hex	2.49^{22}	d 420		44.15^{25}		
a173	iron sulfate(ic)	$NH_4Fe(SO_4)_2.12H_2O$	482.19	vlt, cub oct, 1.4854	1.71	39–41	$-12H_2O$, 230	124.0^{25}	400^{100}	i al; s dil a
a174	iron sulfate(ous)	$(NH_4)_2SO_4.FeSO_4.6H_2O$	392.14	grn, monocl, 1.487, 1.492, 1.499	1.864^{20}_{4}	d 100–110		26.9^{20}	73.0^{80}	i al
a175	lactate	$NH_4C_3H_5O_3$	107.11	col-yelsh liq	$1.19–1.21^{15}$			∞		∞ al
a176	laurate, acid (mixt.)	$NH_4C_{12}H_{23}O_2.C_{12}H_{24}O_2$	417.68	wh		75	d	s	s	4.8^{7} al; sl s eth, acet
a177	magnesium arsenate	$NH_4MgAsO_4.6H_2O$	289.36	col, tetrg, 1.608	1.932^{15}	d		0.038^{20}	0.024^{80}	s a; i al
a178	magnesium carbonate	$(NH_4)_2CO_3.MgCO_3.4H_2O$	252.47	wh				s	v s	s a; i al

No.	Name	Synonyms and Formulae	Mol. wt.	Crystalline form, properties and index of refraction	Density or spec. gravity	Melting point, °C	Boiling point, °C	Solubility, in grams per 100 cc: Cold water	Hot water	Other solvents
	Ammonium									
a179	magnesium chloride	$NH_4Cl.MgCl_2.6H_2O$	256.80	col, rhomb, deliq	1.456	$-2H_2O$, 100	d	16.7	...	d al
a180	magnesium chromate	$(NH_4)_2CrO_4.MgCrO_4.6H_2O$	400.51	yel, monocl, 1.636, 1.637, 1.653	1.84	d	...	v s	v s	...
a181	magnesium phosphate	Guanite, struvite $NH_4MgPO_4.6H_2O$	245.41	col, rhomb, 1.495, 1.496, 1.504	1.711	d	...	0.0231^{0}	0.0195^{80}	v s dil a; i al
a182	magnesium selenate	$(NH_4)_2SeO_4.MgSeO_4.6H_2O$	454.40	col monocl pr, 1.507, 1.509, 1.517	2.058_{4}^{20}	...	...	sl s	...	...
a183	magnesium sulfate	Nat. boussingaulite. $(NH_4)_2SO_4.MgSO_4.6H_2O$	360.60	col, monocl, 1.472, 1.473, 1.479	1.723	>120	d 250	17.92^{0}	64.7^{100} (anhydr)	...
a184	*l*-malate, hydrogen	$NH_4HC_4H_4O_5$	151.08	col, rhomb	1.5	161	d	$32.2^{15.7}$	...	...
a185	*per*manganate	NH_4MnO_4	136.97	rhomb	2.208^{10}	d 110	...	7.9^{15}	d	...
a186	manganese phosphate(ic)	$NH_4MnPO_4.H_2O$	185.96	wh cr	...	...	...	0.0031	0.05	i al, NH_4 salts
a187	manganese sulfate(ous)	$(NH_4)_2SO_4.MnSO_4.6H_2O$	391.23	pa red, monocl, 1.480, 1.484, 1.491	1.83	...	...	51.3^{25}	v s	...
a188	molybdate	$(NH_4)_2MoO_4$	196.01	col, monocl pr	2.276_{4}^{25}	d	...	s, d	d	s a; i al, NH_3, acet
a189	*para*molybdate	"Molybdic acid" com'l. $(NH_4)_6Mo_7O_{24}.4H_2O$	1235.86	col-yelsh, monocl	2.498	$-H_2O$, 90	d 190	43	d	i al; s a, alk
a190	*per*molybdate	$3(NH_4)_2.O.5MoO_3.2MoO_4.6H_2O$	608.17	lt yel monocl pr	2.975	d 170	...	v s	v s	sl s al
a191	molybdotellurate	$(NH_4)_6TeMo_6O_{24}.7H_2O$	1321.56	col, rhomb	2.78	d 550	d	s	s	...
a192	myristate, acid (mixt.)	$NH_4C_{14}H_{27}O_2.C_{14}H_{28}O_2$	473.78	wh solid	...	75–90	d	sl s	s	s al; i eth
a193	nickel chloride	$NH_4Cl.NiCl_2.6H_2O$	291.20	grn, monocl, deliq	1.654	...	...	v s	v s	...
a194	nickel sulfate	Double nickel salt. $(NH_4)_2SO_4.NiSO_4.6H_2O$	395.00	dk bl-grn, monocl, 1.495, 1.501, 1.508	1.923	...	...	10.4^{20}	30^{80}	i al; s $(NH_4)_2SO_4$
a195	nitrate	NH_4NO_3	80.04	col, rhomb, (monocl >32.1°)	1.725^{25}	169.6	210^{11}	118.3^{0}	871^{100}	3.8^{20} al; 17.1^{20} MeOH s acet, NH_3; i eth
a196	nitratocerate	$(NH_4)_2Ce(NO_3)_6$	548.23	yel-red, monocl	...	...	...	142.6^{25}	232^{90}	s al; sl s HNO_3
a197	nitrite	NH_4NO_2	64.04	wh-yelsh cr	1.69	60–70 expl	30 subl vac	v s	d	s dil al; i eth
a198	oleate, acid (mixt.)	$NH_4C_{18}H_{33}O_2.C_{18}H_{34}O_2$	581.97	wh powd	...	d 78	...	s	s	80^{60} al; 13.3^{15} eth
a199	osmium chloride	$(NH_4)_2OsCl_6$	439.00	blk, oct	2.93^{25}	subl 170	...	d	d	s HCl
a200	oxalate	$(NH_4)_2C_2O_4.H_2O$	142.11	col, rhomb, 1.439, 1.546, 1.594	1.50	d	...	2.54^{0}	11.8^{50}	i NH_3
a201	oxalate, acid	Ammonium binoxalate. $NH_4HC_2O_4.H_2O$	125.08	col, rhomb	1.556	$-H_2O$, 170	...	s	...	i eth, bz
a202	oxalatoferrate (III)	$(NH_4)_3Fe(C_2O_4)_3.3H_2O$	428.07	grn, monocl	1.78	d 165	...	42.7^{0}	345^{100}	...
a203	palladium (II)-chloride	$(NH_4)_2PdCl_4$	284.29	grnsh-yel, tetr	2.17	d	...	v s	v s	s dil al; i abs al
a204	palladium (IV)-chloride	$(NH_4)_2PdCl_6$	355.20	red oct cr	2.418	d −Cl	...	sl s	sl s	...
a205	palmitate, (acid)- (mixt.)	$NH_4C_{16}H_{31}O_2.C_{16}H_{32}O_2$	529.90	yelsh soapy mass or yel powd	...	>100	d	sl s	s	8.8^{50} al; 0.23^{13} eth
a206	*meta*periodate	NH_4IO_4	208.94	col, tetr	3.056^{18}	expl	...	2.7^{16}	...	...
a207	*hypo*phosphate	$(NH_4)_2H_2P_2O_6$	196.04	col cr	...	170	...	7^{25}	25^{100}	...
a208	*ortho*phosphate	$(NH_4)_3PO_4.3H_2O$	203.13	wh pr	...	...	...	26.1^{25}	...	sl s dil NH_4OH; i NH_3, acet
a209	*ortho*phosphate, di-H	$NH_4H_2PO_4$	115.03	col, tetr, 1.525, 1.479	1.803^{19}	190	...	22.7^{0}	173.2^{100}	i acet
a210	*ortho*phosphate-mono-H	$(NH_4)_2HPO_4$	132.05	col, monocl, 1.52	1.619	d 155	d	57.5^{10}	106.0^{70}	i al, acet, NH_3
a211	*hypo*phosphite	$NH_4H_2PO_2$	83.03	rhomb tabl	1.634	200	d 240	s	s	s al, NH_3; i acet
a212	*ortho*phosphite, di-H	$NH_4H_2PO_3$	99.03	col, monocl pr	...	123	d 145	171^{0}	260^{31}	i al
a213	phosphofluoride, hexa-	NH_4PF_6	163.00	col, cub	2.180_{4}^{18}	d	...	74.8^{20}	...	s al, acet; d h a
a214	phosphomolybdate	Ammonium molybdo phosphate. $(NH_4)_3P(Mo_3O_{10})_4$	1876.35	yel powd	...	d	...	sl s	sl s	s alk; i al, HNO_3
a215	phosphotungstate	$(NH_4)_3P(W_3O_{10})_4$	2931.27	wh	...	...	...	sl s	sl s	...
a216	picramate	$NH_4C_6H_4N_3O_5$	216.15	redsh-brn cr powd	...	...	...	s	...	...
a217	picrate	$NH_4C_6H_2N_3O_7$	246.14	red or yel, rhomb	1.719	d	expl 423	1.1^{20}	s	sl s al
a219	praseodymium sulfate	$(NH_4)_2SO_4.Pr_2(SO_4)_3.8H_2O$	846.26	cr	$2.531^{16.5}$	$-8H_2O$, 170	...	sl s	...	...
a220	propionate	$NH_4C_3H_5O_2$	91.11	pr, deliq	1.108^{25}	45	...	v s	...	s al, ac a
a221	*per*rhenate	NH_4ReO_4	268.24	wh, hex pl	3.97	d	d	6.1^{20}	32.34^{80}	...

No.	Name	Synonyms and Formulae	Mol. wt.	Crystalline form, properties and index of refraction	Density or spec. gravity	Melting point, °C	Boiling point, °C	Solubility, in grams per 100 cc: Cold water	Hot water	Other solvents
	Ammonium									
a222	rhodanid	NH_4NCS	76.12	monocl cr, to rhomb at 92, ~1.685	1.305	149.6	d 170	128^{0}	347^{60}	v s al; s MeOH, acet; i $CHCl_3$
a223	rhodium chloride	$(NH_4)_3RhCl_6.H_2O$	387.75	dk red rhomb need		$-H_2O$, 140			s	sl s al; s dil NH_4Cl
a224	rhodium sulfate	Ammonium rhodium alum. $Rh(NH_4)(SO_4)_2.12H_2O$	529.25	orange, 1.5150		102				
a225	*d*-saccharate, hydrogen	$NH_4HC_6H_8O_8$	227.17	wh need or monocl pr				1.22^{15}	24.35^{100}	i c al; s h al
a226	salicylate	$NH_4C_7H_5O_3$	155.15	col, monocl			subl	111^{25}	v s	28.8^{25} al
a227	selenate	$(NH_4)_2SeO_4$	179.03	col, monocl, 1.561, 1.563, 1.585	2.194^{20}_{4}	d		117^{7}	197^{100}	i al, acet, NH_3
a228	selenate, hydrogen	NH_4HSeO_4	162.00	rhomb	2.162	d				
a229	selenide	$(NH_4)_2Se$	115.04	col, or wh cr		d		s		
a230	sodium phosphate, hydrate	Microsmic salt. $NH_4NaHPO_4.4H_2O$	209.07	col, monocl, 1.439, 1.442, 1.469	1.574	d 97				
a231	stearate, acid (mixt.)	$NH_4C_{18}H_{35}O_2.C_{18}H_{36}O_2$	586.00	wh cr		d 110		v s		0.3^{25} al; 0.19^{25} eth; 0.08^{25} acet
a232	succinate	$(NH_4)_2.C_4H_4O_4$	152.15	col cr	1.37			s		sl s al
a233	sulfamate	$NH_4NH_2SO_3$	114.12	large, pl, deliq		125	d 160	166.6^{10}	357^{50}	i al
a234	sulfate	Nat. mascagnite. $(NH_4)_2SO_4$	132.14	col, rhomb, 1.521, 1.523, 1.533	1.769^{50}	d 235		70.6^{0}	103.8^{100}	i al, acet, NH_3
a235	sulfate, hydrogen	Ammonium bisulfate. NH_4HSO_4	115.11	col, rhomb, 1.473	1.78	146.9	d	100	v s	sl s al; i acet
a236	*peroxy*disulfate	$(NH_4)_2S_2O_8$	228.18	col, monocl, 1.498, 1.502, 1.587	1.982	d 120		58.2^{0}	v s	
a237	sulfide, hydro-	NH_4HS	51.11	wh rhomb, 1.74	1.17	118^{150atm}	88.4^{19atm}	128.1^{0}	d	s al; NH_3
a238	sulfide, mono-	$(NH_4)_2S$	68.14	col yel cr (> −18), hygr				v s	d	s al; v s NH_3
a239	sulfide, penta-	$(NH_4)_2S_5$	196.40	yel pr		d 115		v s		s al, i eth, CS_2
a240	sulfite	$(NH_4)_2SO_3.H_2O$	134.15	col, monocl, 1.515	1.41^{25}	d 60–70	subl 150	32.4^{0}	60.4^{100}	sl s al; i acet
a241	sulfite, hydrogen	Ammonium bisulfite. NH_4HSO_3	99.10	rhomb pr, deliq	2.03	subl 150 (in N_2)		71.8^{0}	84.7^{60}	
a242	*dl*-tartrate	$(NH_4)_2C_4H_4O_6$	184.15	col, monocl, *d*, α 1.55, β 1.581	1.601	d		58.01^{15}	81.17^{60}	sl s al
a243	*dl*-tartrate, hydrogen	$NH_4HC_4H_4O_6$	167.12	col, monocl pr, 1.561, 1.591	1.636	d		2.35^{15}	3.24^{25}	i al, s a, alk
a244	tellurate	$(NH_4)_2TeO_4$	227.67	wh powd	$3.024^{24.5}$	d		s	s	i al; s dil a
a245	thallium chloride	$3NH_4Cl.TlCl_3.2H_2O$	507.23	col	2.39			s		
a246	thioantimonate	$(NH_4)_3SbS_4.4H_2O$	376.18	yel pr				71.2^{0}	d	i al, eth
a247	thiocarbamate	$NH_4CS_2NH_2$	110.20	yel cr		d 50		v s		s al; sl s eth
a248	thiocarbonate, tri-	$(NH_4)_2CS_3$	144.28	yel cr, hygr		subl		v s	d	sl s al, eth
a249	thiocyanate	NH_4SCN	76.12	col, monocl, deliq	1.305	149.6	d 170	128^{0}	v s	s al, acet, NH_3
a250	*di*thionate	$(NH_4)_2S_2O_6.\frac{1}{2}H_2O$	205.21	col, monocl	1.704	d 130		135^{0}	v s	i al
a251	thiosulfate	$(NH_4)_2S_2O_3$	148.20	col, monocl, hygr	1.679	d 150		v s	103.3^{100}	i al; sl s acet
a252	titanium oxalate, basic	$(NH_4)_2TiO(C_2O_4)_2.H_2O$	294.03					v s		
a253	uranylcarbonate	$2(NH_4)_2CO_3.UO_2CO_3.2H_2O$	558.24	yel, monocl	2.773	d 100		5.8^{19}	d	$s(NH_4)_2CO_3$, aq.SO_2
a254	uranylfluoride, penta-	$(NH_4)_3UO_2F_5$	419.14	tetr cr, 1.495	3.186	subl		s		
a255	valerate	$NH_4C_5H_9O_2$	119.16	col or wh cr		d		s		s al, eth
a256	*meta*vanadate	NH_4VO_3	116.98	wh-yelsh or col cr	2.326	d 200		0.52^{15}	6.95^{96}, d	
a257	vanadium sulfate	$NH_4V(SO_4)_2.12H_2O$	477.28	red to bl	1.687	49		28.45^{20}		
a258	zinc sulfate	$(NH_4)_2SO_4.ZnSO_4.6H_2O$	401.66	wh monocl, 1.489, 1.493, 1.499	1.931			7^{0} (anhydr)	42^{80} (anhydr)	
a259	**Antimony**	Sb	121.75	silv wh met, hex	6.684^{25}	630.74	1750	i	i	s hot conc H_2SO_4, aq reg
a260	bromide, tri-	$SbBr_3$	361.48	col rhomb, 1.74	4.148^{23}	96.6	280	d	d	s HCl, HBr, CS_2, NH_3, al, acet
a261	chloride, penta-	$SbCl_5$	299.02	wh liq or monocl, 1.601^{14}	liq 2.336^{20}_{4}	2.8	79^{22}	d	d	s HCl, tarta, $CHCl_3$
a262	chloride, tri-	Butter of antimony. $SbCl_3$	228.11	col, rhomb, deliq	3.140^{25}	73.4	283	601.6^{0}	∞^{80}	s abs al, HCl, tart a, $CHCl_3$, bz, acet
a263	fluoride, penta-	SbF_5	216.74	col oily liq	2.99^{23} liq	7	149.5	s		s KF
a264	fluoride, tri-	SbF_3	178.75	col, rhomb	$4.379^{20.9}$	292	subl 319	384.7^{0}	563.6^{30}	i NH_3
a265	hydride	SbH_3	124.77	inflamm gas	gas 4.36^{15}; liq 2.26^{-25}	−88	-17.1^{751}	0.41^{0}		1500 ml A; 2500 ml CS_2
a266	iodide, penta-	SbI_5	756.27	br		79	400.6			
a267	iodide, tri-	SbI_3	502.46	ruby-red, hex, 2.78*Li*, 2.36*Li*	4.917^{17}	170	401	d	d	i al; s CS_2, bz, HI, HCl
a268	iodosulfide	SbSI	280.72	dk red		392	d	i	i	d conc HCl; i CS_2

No.	Name	Synonyms and Formulae	Mol. wt.	Crystalline form, properties and index of refraction	Density or spec. gravity	Melting point, °C	Boiling point, °C	Solubility, in grams per 100 cc: Cold water	Hot water	Other solvents
	Antimony									
a269	mercaptoacetamide	Antimony thioglycolamide. $Sb(C_2H_4NOS)_3$	392.12	wh cr		139		200		
a270	nitrate, basic	$2Sb_2O_3.N_2O_5$(?)	691.01	wh glossy cr		d		d		d a; v sl s conc HNO_3
a271	nitride	SbN	135.76	or powd		d		d		
a272	oxide, penta-	Sb_2O_5 (or Sb_4O_{10})	323.50	yel powd	3.80 (dep on temp)	−O, 380 −O_2, 930		v sl s	v sl s	v sl s KOH, HCl, HI
a273	oxide, tetra-	Nat. cervantite. Sb_2O_4 (or $Sb_2O_3.Sb_2O_5$)	307.50	wh powd, 2.00	5.82	−O, 930		v sl s	v sl s	v sl s KOH, HCl, HI
a274	oxide, tri-	Nat. senarmontite. Sb_2O_3 (or Sb_4O_6)	291.50	wh, cub, 2.087	5.2	656	1550 subl	v sl s	sl s	s KOH HCl(3%), 0.03^{20} tart a, ac a
a275	oxide, tri-	Nat. valentinite. Sb_2O_3(or Sb_4O_6)	291.50	col, rhomb, 2.18, 2.35, 2.35	5.67	656	1550	v sl s	sl s	s KOH, HCl, tart, a, ac a
a276	III oxychloride(ous)	SbOCl	173.20	wh monocl		d 170		i	d	s acet, HCl, CS_2; i al, eth, $CHCl_3$
a277	III oxychloride(ous)	$Sb_4O_5Cl_2$	637.90	monocl	5.01	d 320		i		s HCl; i al, eth
a278	oxyhydrate	$H_4Sb_2O_7$	359.13	wh, amorph		−H_2O, 200		sl s	sl s	s alk
a279	III oxysulfate, di-(ous)	$Sb_2O_2SO_4$	371.56	wh	4.89			d	d	s, H_2SO_4
a280	potassium tartrate	Tartar emetic. $K(SbO)C_4H_4O_6.\frac{1}{2}H_2O$	333.93	col cr	2.6	−$\frac{1}{2}H_2O$, 100		8.3	33.3	i al; 6.7 glyc
a281	selenide	Sb_2Se_3	480.38	gray cr		611		v sl s		s conc HCl
a282	III sulfate	$Sb_2(SO_4)_3$	531.68	wh powd, deliq	3.625^4	d		i	d	s a
a283	sulfide, penta-	Sb_2S_5	403.82	yel powd, prism	4.120	d 75		i	i	i al; s HCl, alk, NH_4HS
a284	sulfide, tri-	Nat. stibnite. Sb_2S_3	339.69	blk, rhomb, 3.194, 4.064, 4.303	4.64	550	*ca* 1150	0.000175^{18}		s al, NH_4HS. K_2S, HCl; i ac a
a285	sulfide, tri-	Sb_2S_3	339.69	yel-red, amorph	4.12	550	*ca* 1150	0.000175^{18}		s al, NH_4SH, K_2S, HCl; i ac a
a286	*d*-tartrate	$Sb_2(C_4H_4O_6)_3.6H_2O$	795.81	wh cr powd				s		
a287	telluride, tri-	Sb_2Te_3	626.30	gray	6.50^{13}	629				s HNO_3, aq reg
a288	**Argon**	Ar	39.948	col inert gas.	1.784^0g/l cr: 1.65^{-233}, liq: 1.40^{-186}	−189.2	−185.7	5.6^0 cm³	3.01^{50} cm³	
a289	**Arsenic**	As	74.9216	gray met, hex-rhomb	5.727^{14}	817 (28 atm.)	613 subl	i	i	s HNO_3
a290	**Arsenic**	As_4	299.69	yel, cub	2.026^{18}	d 358		i	i	s CS_2, bz
a291	**Arsenic acid**, meta-	$HAsO_3$	123.93	wh, hygr		d	forms ortho-arsenic acid	d		
a292	ortho-	$H_3AsO_4.\frac{1}{2}H_2O$	150.95	wh translu hygr cr	2.0–2.5	35.5	−H_2O 160	$302^{12.5}$		s al, alk, glyc
a293	pyro-	$H_4As_2O_7$	265.87	col pr		forms ortho-arsenic acid 206 d				
	Arsenic									
a294	bromide, tri-	$AsBr_3$	314.65	col-yel hygr pr	3.54^{25}	32.8	221	d	d	s HCl, HBr, CS_2
a295	chloride, tri-	$AsCl_3$	181.28	oily liq or need, 1.621^{14}_{F}	liq 2.163^{20}	−8.5	130.2	d	d	s HBr, HCl, PCl_3, al, eth
a296	fluoride, penta-	AsF_5	169.91	col gas	7.71 g/l	−80	−53	s		s alk, al, eth, bz
a297	fluoride, tri-	AsF_3	131.92	oily liq	liq 2.666	−8.5	-63^{752}	d	d	s al, eth, bz, NH_4OH
a298	hydride	Arsine. AsH_3	77.95	col gas	gas 2.695 liq $1.689^{64.9}$	−116.3	−55 (d 300)	20 ml		s $CHCl_3$, bz
a299	iodide, di-	AsI_2	328.73	red pr		d 136		d		s al, eth, $CHCl_3$, bz, CS_2
a300	iodide, penta-	AsI_5	709.44		3.93	76				
a301	iodide, tri-	AsI_3	455.64	red hex, *ca.* 2.59, *ca.* 2.23	4.39^{13}	146	403	6^{25}	30 d	5.2 CS_2; s al, eth, bz, $CHCl_3$
a302	oxide, pent-	As_2O_5	229.84	wh amor, deliq	4.32	d 315		150^{16}	76.7^{100}	s al, a, alk
a303	oxide, tri-	As_2O_3	197.84	amor or vitreous	3.738	312.3		3.7^{20}	10.14^{100}	s alk, alk carb, HCl
a304	oxide, tri-	Nat. arsenolite. As_2O_3 (or As_4O_6)	197.84	col, cub or fibr, 1.755	3.865^{25}	subl 193		1.2^2	11.46^{100}	s al, alk, HCl
a305	oxide, tri-	Nat. claudetite. As_2O_3 (or As_4O_6)	197.84	col, monocl, 1.871, 1.92, 2.01	4.15	193, subl 312.3	457.2	1.2^2	11.46^{100}	s al, alk, HCl
a306	(III)oxychloride(ous)	AsOCl	126.37	brnsh		d		d	d	
a307	phosphide, mono-	AsP	105.90	br-red powd		subl, d		d	d	al s CS_2; s H_2SO_4, HCl; i al, eth, $CHCl_3$
a308	selenide	As_2Se_3	386.72	br cr	4.75	*ca.* 360		i	d	s alk
a309	sulfide, di-	Nat. realgar. As_2S_2	213.97	red-br monocl, 2.46, 2.59, 2.61	α 3.506^{19} β 3.254^{19}	α tr 267 β 307	565	i	i	s K_2S, $NaHCO_3$
a310	sulfide, penta-	As_2S_5	310.16	yel		subl, d 500		0.000136^0	i	s alk, alk sulf, HNO_3

No.	Name	Synonyms and Formulae	Mol. wt.	Crystalline form, properties and index of refraction	Density or spec. gravity	Melting point, °C	Boiling point, °C	Solubility, in grams per 100 cc		
								Cold water	Hot water	Other solvents
	Arsenic									
a311	sulfide, tri-	Nat. orpiment. As_2S_3	246.04	yel or red, monocl, 2.4, 2.81, 3.02 (Li)	3.43	300	707	0.00005^{18}	sl s	s al, alk, alk carb
	Auric or Aurous	*See Gold*								
b1	**Barium**	Ba	137.33	yel-silv met	3.51^{20}	725	1640	d, ev H_2	d	s al; i bz
b2	acetate	$Ba(C_2H_3O_2)_2$	255.43	col cr	2.468			58.8^{0}	75^{100}	
b3	acetate, hydrate	$Ba(C_2H_3O_2)_2.H_2O$	273.45	col, tricl, 1.500, 1.517, 1.525	2.19	$-H_2O$, 150		76.4^{26}	75^{100}	sl s al
b4	amide	$Ba(NH_2)_2$	169.39	gray-wh cr		280		d	d	i liq NH_3
b5	*ortho*arsenate	$Ba_3(AsO_4)_2$	689.86		5.10	1605		0.055		s a, NH_4Cl
b6	*ortho*arsenate, mono-H	$BaHAsO_4.H_2O$	295.28	col, rhomb or monocl, 1.635	1.93^{15}	$-H_2O$, 150		sl s	d	s HCl
b7	arsenide	Ba_3As_2	561.86	br	4.1^{15}			d		d Cl_2, F_2, Br_2
b8	azide	$Ba(N_3)_2$	221.38	monocl pr	2.936	$-N_2$, 120	expl	17.3^{17}		abs al 0.017^{16}; i eth
b9	azide, hydrate	$Ba(N_3)_2.H_2O$	239.40	tricl, 1.7		expl		v s	v s	sl s al; i eth
b10	benzoate	$Ba(C_7H_5O_2)_2.2H_2O$	415.60	col, nacreous leaf		$-2H_2O$, 100		s		sl s al
b11	boride, hexa-	BaB_6	202.21	met-blk, cub	4.36^{16}	2270		i	i	s HNO_3; i HCl
b12	bromate	$Ba(BrO_3)_2.H_2O$	411.16	col, monocl	3.99^{18}	d 260		0.3^{0}	5.67^{100}	i al; s acet
b13	bromide	$BaBr_2$	297.16	col cr, 1.75	4.781^{24}	847	d	104.1^{20}	149^{100}	v s al, MeOH
b14	bromide, dihydrate	$BaBr_2.2H_2O$	333.19	col, monocl, 1.713, 1.727, 1.744	3.58^{24}	880, $(-H_2O$, 75)	$-2H_2O$, 120	151^{20}	204^{100}	v s MeOH, s al
b15	bromofluoride	$BaBr_2.BaF_2$	472.49	col pl	4.96^{18}			d	d	i al; s conc HCl, conc HNO_3
b16	bromoplatinate	$BaPtBr_6.10H_2O$	992.04	monocl	3.71					
b17	butyrate	$Ba(C_4H_7O_2)_2.2H_2O$	347.57	col				37.42^{0}	42.12^{80}	
b18	carbide	BaC_2	161.36	gray, tetr	3.75			d to C_2H_2		d a
b19	carbonate(α)	$BaCO_3$	197.35	wh, hex	4.43	1740^{90atm}	d	0.002^{20}	0.006^{100}	s a, NH_4Cl; i al
b20	carbonate(β)	$BaCO_3$	197.35			tr to α, 982		0.0022^{18}	0.0065^{100}	s a, NH_4Cl; i al
b21	carbonate(γ)	Nat. witherite. $BaCO_3$	197.35	wh rhomb, 1.529, 1.676, 1.677	4.43	to β, 811	d 1450	0.0022^{18}	0.0065^{100}	s a, NH_4Cl; i al
b22	chlorate	$Ba(ClO_3)_2.H_2O$	322.26	col monocl, 1.5622, 1.577, 1.635	3.18	414 $(-H_2O$, 120)	$-O$, 250	27.4^{15}	111.2^{100}	sl s al, acet, HCl
b23	*per*chlorate	$Ba(ClO_4)_2$	336.24	col, hex	3.2	505		198.5^{25}	562.3^{100}	v s al
b24	*per*chlorate, hydrate	$Ba(ClO_4)_2.3H_2O$	390.29	col, hex, 1.533	2.74	d 400		198^{23}	s	al 124^{62}
b25	chloride α	$BaCl_2$	208.25	col, monocl; 1.7303, 1.7367, 1.7420	3.856^{24}	tr to cub 962	1560	37.5^{26}	59^{100}	sl s HCl, HNO_3; v sl s al
b26	chloride β	$BaCl_2$	208.25	col, cub	3.917	963	1560			v sl s al
b27	chloride, hydrate	$BaCl_2.2H_2O$	244.28	col, monocl, 1.629, 1.642, 1.658 n_D^{25}	3.097_4^{24}	$-2H_2O$, 113	35.7^{20}		58.7^{100}	sl s HCl, HNO_3; v sl s al
b28	*hypo*chlorite	$Ba(ClO)_2.2H_2O$	276.28	col cr		d				
b29	chlorofluoride	$BaCl_2.BaF_2$	383.58	col, tetr, 1.640	4.51^{18}			d	d	i al; s conc HCl, HNO_3
b30	chloroplatinate	$BaPtCl_6.6H_2O$	653.24	or-yel, monocl	2.868	$-5H_2O$, 70	d	s		d a, al; i MeOH, eth
b31	chloroplatinite	$BaPtCl_4.3H_2O$	528.35	dk red pr	2.868	$-3H_2O$, 150		v s		s al
b32	chromate	$BaCrO_4$	253.33	yel, rhomb	4.498^{15}			0.00034^{16}	0.00044^{28}	s min a
b33	*di*chromate	$BaCr_2O_7$	353.33	red, monocl				sl s		s h conc H_2SO_4
b34	*di*chromate, hydrate	$BaCr_2O_7.2H_2O$	389.36	bright red-yel need		$-2H_2O$, 120		d		s conc CrO_3 soln
b35	chromite	$BaO.4Cr_2O_3$	761.30	grn-blk, hex	5.4^{15}			i		s a, fus carb
b26	citrate	$Ba_3(C_6H_5O_7)_2.7H_2O$	916.33	wh powd		$-7H_2O$, 150		0.0406^{18}	0.0572^{25}	sl s al; s HCl
b37	cyanide	$Ba(CN)_2$	189.38	wh cr powd				80^{14}		18^{14} 70 % al
b38	cyanoplatinite	$BaPt(CN)_4.4H_2O$	508.56	(a) monocl, yel, α 1.6704 (b) grn, rhomb	a) 2.076 b) 2.085	$-2H_2O$, 100		3^{16}	25^{100}	i al
b39	dithionate	$Ba(SO_3)_2.2H_2O$	333.50	col, rhomb or monocl, 1.586, 1.595, 1.607	$4.536^{13.5}$	d 120		24.75^{18}	90.9^{100}	sl s al
b40	ethylsulfate	$Ba(C_2H_5SO_4)_2.2H_2O$	423.62	wh lust leaf				s		sl s al
b41	ferrocyanide	$Ba_2Fe(CN)_6.6H_2O$	594.73	yel, monocl	2.666	$-H_2O$, 40		0.17^{15}	0.9^{100}	i al
b42	fluogallate	$Ba_3(GaF_6)_2.H_2O$	797.46	wh cr	4.06	$-\frac{1}{2}H_2O$, 110; $-\frac{1}{2}H_2O$, 230		i		s HF
b43	fluoride	BaF_2	175.34	col, cub, 1.4741 n_D^{25}	4.89	1355	2137	0.12^{25}	sl s	s a, NH_4Cl
b44	fluoroiodide	$BaF_2.BaI_2$	566.49	pl	5.21^{18}			d	d	i al; s conc HCl, conc NHO_3

No.	Name	Synonyms and Formulae	Mol. wt.	Crystalline form, properties and index of refraction	Density or spec. gravity	Melting point, °C	Boiling point, °C	Solubility, in grams per 100 cc		
								Cold water	Hot water	Other solvents
	Barium									
b45	fluosilicate	$BaSiF_6$	279.42	rhomb need	4.29^{21}_{4}	d 300		0.026^{17}	0.09^{100}	i al; sl s a, NH_4Cl
b46	formate	$Ba(CHO_2)_2$	227.38	col, rhomb, 1.573, 1.597, 1.636	3.21	d		27.76^{0}	39.71^{80}	i al, eth
b47	gluconate	$Ba(C_6H_{11}O_7)_2.3H_2O$	581.69	pr or rhomb leaf		$-3H_2O$, 100; d 120		$3.3^{15.5}$		i al
b48	hydride	BaH_2	139.36	gray cr	4.21^{0}	d 675	1400(?)	d to $Ba(OH)_2$ $+H_2$		d a
b49	hydroxide	$Ba(OH)_2.8H_2O$	315.48	col, monocl, 1.471, 1.502, 1.50	2.18^{16}	78	$-8H_2O$, 78	5.6^{15}	94.7^{78}	sl s al; i acet
b50	hyponitrite	$BaN_2O_2.4H_2O$	269.41	wh cr powd	2.742^{25}					
b51	iodate	$Ba(IO_3)_2$	487.15	monocl	4.998	d		0.008^{0}	197^{100}	s HNO_3, HCl
b52	iodate, hydrate	$Ba(IO_3)_2.H_2O$	505.17	col, monocl	4.657^{15}	$-H_2O$, 200		v sl s	sl s	s HNO_3, HCl; i al, acet, H_2SO_4
b53	iodide	BaI_2	391.15	col cr	5.15^{25}_{4}	740		170^{0}		al 77^{20}
b54	iodide, hydrate	$BaI_2.2H_2O$	427.18	col rhomb, deliq	5.15	$-H_2O$, 98.9; $-2H_2O$, 539; d 740	200^{15}		269^{100}	1.07^{15} al; s acet
b55	iodide, hydrate	$BaI_2.6H_2O$	499.24	col, hex		25.7		410^{0}	v s	v s al
b56	laurate	$Ba(C_{12}H_{23}O_2)_2$	535.97	wh leaf cr		260		$0.008^{15.5}$	0.011^{50}	0.008^{25} al; 0.006^{25} eth
b57	*l*-malate	$BaC_4H_4O_5$	269.41					0.883^{20}	1.044^{80}	
b58	malonate	$BaC_3H_2O_4.H_2O$	257.40	col				0.143^{0}	0.326^{80}	
b59	manganate	$BaMnO_4$	256.28	gray-grn, hex	4.85			v sl s	s a	
b60	*per*-manganate	$Ba(MnO_4)_2$	375.21	br-vlt cr	3.77	d 200		62.5^{11}	75.4^{25}	d al
b61	methylsulfate	$Ba(CH_3SO_4)_2.2H_2O$	395.56	col effl cr				s		s al
b62	molybdate	$BaMoO_4$	297.28	wh powd	4.65	1480		0.0058^{23}		sl s a
b63	myristate	$Ba(C_{14}H_{27}O_2)_2$	592.08					0.007^{25}	0.010^{50}	0.009^{25} al; 0.003^{25} eth 0.046^{15} MeOH
b64	nitrate	Nitrobarite. $Ba(NO_3)_2$	261.35	col cub, 1.572	3.24^{23}	592	d	8.7^{20}	34.2^{100}	i al; sl s a
b65	nitride	Ba_3N_2	440.03	yel-br	4.783^{4}_{4}		1000 vac	d	d	
b66	nitrite	$Ba(NO_2)_2$	229.35	col, hex	3.23^{23}	d 217		67.5^{20}	300^{100}	sl s al
b67	nitrite, hydrate	$Ba(NO_2)_2.H_2O$	247.37	col-yelsh, hex	3.173^{30}	d 115		63^{30}	109.6^{80}	1.6 al; v s HCl; i acet
b68	oxalate	BaC_2O_4	225.36	cr	2.658	d 400		0.0093^{18}	0.0228^{100}	i al; s NH_4Cl, a
b69	oxide	BaO	153.34	col, cub, wh-yelsh powd, 1.98	5.72	1918	*ca* 2000	3.48^{20}	90.8^{100}	s dil a, al; i acet, NH_3
b70	oxide, per-	BaO_2	169.34	wh-gray powd	4.96		$-O$, 800	v sl s	d	s dil a; i acet
b71	oxide, per-, hydrate	BaO_28H_2O	313.46	col, hex	2.292	$-8H_2O$, 100		0.168	d	s dil a; i al, eth, acet
b72	palmitate	$Ba(C_{16}H_{31}O_2)_2$	648.19	wh cr powd		d		0.004^{15}	0.007^{50}	$0.008^{16.5}$ al; 0.001^{15} eth
b73	*hypo*phosphate	$BaPO_3$	216.31	need				sl s		s al; v sl s ac a
b74	*ortho*phosphate di-	$BaHPO_4$	233.32	wh, rhomb, 1.635, 1.617	4.165^{15}	d 410^{710}		0.01–0.02		s a, NH_4Cl
b75	*ortho*phosphate, mono-	$Ba(H_2PO_4)_2$	331.31	tricl	2.9^{4}			d	d	s a
b76	*ortho*phosphate, tri-	$Ba_3(PO_4)_2$	601.96	wh, cub	4.1^{16}			i	i	s a
b77	*pyro*phosphate	$Ba_2P_2O_7$	448.62	wh, rhomb	3.9^{20}			0.01	sl s	s a, NH_4 salts
b78	*hypo*phosphite	$Ba(H_2PO_2)_2.H_2O$	285.33	wh, monocl	2.90^{17}_{4}	d 100−150		30^{15}	33^{100}	i al
b79	propionate	$Ba(C_3H_5O_2)_2.H_2O$	301.50	rhomb, β 1.518		d 300		48^{0}	67.9^{80}	0.08 al
b80	salycylate	$Ba(C_7H_5O_3)_2.H_2O$	437.65	wh need				s		
b81	selenate	$BaSeO_4$	280.30	wh, rhomb	4.75	d		0.0118	0.138^{100}	s HCl; i HNO_3
b82	selenide	BaSe	216.30	wh cub disc, n_D, 2.268	5.02			d	d	d HCl
b83	*meta*silicate	$BaSiO_3$	213.42	col, rhomb, 1.673, 1.674, 1.678	4.399	1604		i	d	s HCl
b84	*meta*silicate, hydrate	$BaSiO_3.6H_2O$	351.52	rhomb, 1.542, 1.548, 1.548	2.59			0.17	d	
b85	stearate	$Ba(C_{18}H_{35}O_2)_2$	704.13	wh powd				0.004^{15}	0.006^{50}	$0.005^{16.5}$ al, 0.008^{25} al, 0.001^{25} eth
b86	succinate	$BaC_4H_4O_4$	253.37	wh powd				0.421^{0}	0.237^{80}	sl s al
b87	sulfate	Nat. barite, prec. blanc fixe. $BaSO_4$	233.40	wh, rhomb (monocl), 1.637, 1.638, 1.649	4.50^{15}	1580	tr 1149 monocl	0.000222^{18} 0.000246^{25}	0.000336^{50} 0.000413^{100}	0.006 s 3 % HCl; sl s H_2SO_4
b88	*peroxydi*sulfate	$BaS_2O_8.4H_2O$	401.52	wh, monocl		d		52.2^{0}	d	d al
b89	sulfide, hydro-	$Ba(HS)_2.4H_2O$	275.56	yel, rhomb		d 50		s		i al
b90	sulfide, mono-	BaS	169.40	col, cub, n_D 2.155	4.25^{15}	1200		d	d	i al
b91	sulfide, tetra-	$BaS_4.H_2O$	283.61	red or yel, rhomb	2.988	d 300		41^{15}	v s	i al, CS_2
b92	sulfide, tri-	BaS_3	233.53	yel-grn cr		d 554		s	s	

PHYSICAL CONSTANTS OF INORGANIC COMPOUNDS (Continued)

No.	Name	Synonyms and Formulae	Mol. wt.	Crystalline form, properties and index of refraction	Density or spec. gravity	Melting point, °C	Boiling point, °C	Solubility, in grams per 100 cc		
								Cold water	Hot water	Other solvents
	Barium									
b93	sulfite	$BaSO_3$	217.40	col, cub hex		d		0.02^{20}	0.002^{20}	v s HCl
b94	tartrate	$BaC_4H_4O_6.H_2O$	303.52	wh cr	$2.980^{20.8}$			0.026^{18}	0.058^{80}	0.032^{18} al
b95	tellurate	$BaTeO_4.3H_2O$	383.07	volum wh	4.2^{200}	d >200		sl s	sl s	s HCl, NHO_3
b96	*pyro*tellurate, hydrogen	$Ba(HTe_2O_7)_2.H_2O$	891.76	volum ppt; hot: yel; cold: wh				s	s	s a
b97	telluride	BaTe	264.94	yel-wh, cub, disc, n_D 2.440	5.13					d a
b98	thiocarbonate	$BaCS_3$	245.54	yel, hex		d		1.08^{0}	d	i al
b99	thiocyanate	$Ba(SCN)_2.2H_2O$	289.53	need, deliq	2.286^{18}	$-H_2O$, 160		4.3^{20}	s	35^{20} al
b100	thiosulfate	BaS_2O_3	249.47	wh, rhomb		d 220		0.2		
b101	thiosulfate, hydrate	$BaS_2O_3.H_2O$	267.48	wh, rhomb	3.5^{18}	d 100		0.208^{20}		i al
b102	titanate	$BaTiO_3$	233.24	tetr and hex, 2.40	tetr 6.017, hex 5.806					
b103	tungstate	$BaWO_4$	385.19	col, tetr	5.04			sl s	sl s	d a
b104	*pyro*vanadate	$Ba_2V_2O_7$	488.56	wh cr		863				
b105	**Beryllium**	Glucinum. Be	9.0122	grey met, hex	1.85^{20}	1278 ± 5	$2970^{(5mm)}$	i	sl s, d	s dil a, alk; i Hg
b106	acetate	$Be(C_2H_3O_2)_2$	127.10	col pl		d 300		i		i al, eth, CCl_4
b107	acetate, basic	$BeO(C_2H_3O_2)_6$	406.32	oct	1.36^{4}	284	331	sl d	d	s chl, ac a; sl s al, eth
b108	acetate propionate, basic	$Be_4O(C_2H_3O_2)_3.(C_3H_5O_2)_3$	448.40			127	330			
b109	aluminate	Nat. chrysoberyl. $BeAl_2O_4$	126.97	rhomb, 1.747, 1.748, 1.757	3.76	1870				i a
b110	aluminum silicate	Nat. beryl. $Be_3Al_2(SiO_3)_6$	537.51	transp, hex, col, 1.580, 1.547	2.66	1410 ± 100				i a
b111	aluminum silicate	Nat. euclase. $Be_2Al_2(SiO_4)_2.(OH)_2$	290.17	monocl, 1.652, 1.655, 1.671	3.1					
b112	benzenesulfonate	$Be(C_6H_5O_3S)_2$	323.35	monocl				v s	v s	v s al, acet, ac a; i eth, bz, CS_2, CCl_4
b113	*ortho*borate, basic	Nat. hambergite. $Be_2(OH)BO_3$	93.84	rhomb, 1.560, 1.591, 1.631	2.35					
b114	bromide	$BeBr_2$	168.83	wh need, deliq	3.465^{25}	490 ± 10 subl	520	s	v s	s al, eth; i bz; 18.56 pyr
b115	butyrate, basic	$Be_4O(C_4H_7O_2)_6$	574.64				239^{19}			
b116	carbide	Be_2C	30.04	yel, hex	1.90^{15}	>2100 d		d	d	s a; d alk
b117	carbonate, basic	$BeCO_3+Be(OH)_2$	112.05	wh powd				i	d	s a, alk
b118	chloride	$BeCl_2$	79.92	col need, deliq	1.899^{25}	405	520 (488)	v s	v s, d	v s al, eth, pyr; sl s bz, chl, CS_2; i NH_3, acet
b119	fluoride	BeF_2	47.01	col, amorph, <1.33	1.986^{25}	subl 800		∞	∞	sl s al; s H_2SO_4
b120	hydride	BeH_2	11.03	wh cr		d 125		d	d	i eth, tol
b121	iodide	BeI_2	262.82	col need	4.325^{25}	510 ± 10	590	d	d	s al, eth, CS_2
b122	nitrate	$Be(NO_3)_2.3H_2O$	187.07	wh-yel cr, deliq	1.557	60	142	v s	v s	v s al
b123	nitride	Be_3N_2	55.05	col, cub		2200 ± 100	d 2240	d	d	d a, conc alk; i al
b124	oxalate	$BeC_2O_4.3H_2O$	151.08	rhomb, β 1.487		$-H_2O$, 100; $-3H_2O$, 220	d 350	38.22^{25}		
b125	oxide	Nat. bromellite. BeO	25.01	wh hex, 1.719, 1.733	3.01	2530 ± 30	*ca* 3900	0.00002^{20}		s conc H_2SO_4, fus KOH
b126	oxide	$BeO.xH_2O$		wh amorph powd or gel		d		i	i	s a, alk, $(NH_4)_2CO_3$
b127	2,4-pentanedione deriv	Beryllium acetylacetonate $Be(C_5H_7O_2)_2$	207.23	wh, monocl	1.168^{4}	108	270	sl s	d	s al, eth, a
b128	*ortho*phosphate	$Be_3(PO_4)_2.3H_2O$	271.03			$-H_2O$, 100		s	s	s ac a
b129	propionate basic	$Be_4O(C_3H_5O_2)_6$	490.43			120				
b130	selenate	$BeSeO_4.4H_2O$	224.03	col, rhomb, 1.466, 1.501, 1.503	2.03	$-2H_2O$, 100; $-4H_2O$, 300		56.7^{25}	s	
b131	*di*-silicate	Nat. bertrandite. $Be_4Si_2O_7(OH)_2$	238.23	rhomb, 1.591, 1.605, 1.604	2.6					
b132	*ortho*silicate	Nat. phenazite. Be_2SiO_4	110.11	tricl, col, 1.654, 1.670	3.0					i a
b133	stearate (com'l)	$Be(C_{18}H_{35}O_2)_2$	575.97	wh, waxy		45		i	i	i al; s eth, CCl_4
b134	sulfate	$BeSO_4$	105.07		2.443	d 550–600		i	d to $BeSO_4$ $4H_2O$	
b135	sulfate, hydrate	$BeSO_4.4H_2O$	177.14	col, tetr, 1.472, 1.440	$1.713^{10.5}$	$-2H_2O$, 100	$-4H_2O$, 400	42.5^{25}	100^{100}	sl s conc H_2SO_4; i al
b136	sulfide	BeS	41.08	reg	2.36			d	d	
b137	**Bismuth**	Bi	208.9808	rhomb silver-wh or redsh met	9.80	271.3	1560 ± 5^{760}	i	i	s h H_2SO_4, HNO, aq, reg; sl s h HCl
b138	acetate	$Bi(C_2H_3O_2)_3$	386.12	wh cr		d		i	i	s ac a

No.	Name	Synonyms and Formulae	Mol. wt.	Crystalline form, properties and index of refraction	Density or spec. gravity	Melting point, °C	Boiling point, °C	Solubility, in grams per 100 cc		
								Cold water	Hot water	Other solvents
	Bismuth									
b139	*ortho*arsenate	$BiAsO_4$	347.90	wh, monocl, 2.14, 2.15, 2.18	7.14			i	i	sl s h conc HNO_3
b140	benzoate	$Bi(C_7H_5O_2)_3$	572.33	wh powd						s a; i eth
b141	bromide, tri-	$BiBr_3$	448.71	yel cr powd, deliq	5.72^{25}_{4}	218	453	d to BiOBr	d	i al; s HCl, HBr, eth; v s liq NH_3
b142	carbonate, basic	Bismuth oxycarbonate. $Bi_2O_2CO_3$	509.97	wh powd	6.86		d	i	i	s a
b143	chloride, tetra-	$BiCl_4$	350.79	col cr		226				
b144	chloride, tri-	$BiCl_3$	315.34	wh cr, deliq	4.75^{25}_{4}	230–232	447	d to BiOCl	d	s a, al, eth, acet
b145	*di*chromate, basic	$(BiO)_2.Cr_2O_7$	665.94	yel-or-red cr				i	i	s a; i alk
b146	citrate	$BiC_6H_5O_7$	398.08	wh cr	3.458	d		sl s	sl s	sl s al; s NH_4OH
b147	fluoride, tri-	BiF_3	265.98	gray cr, cub, 1.74	5.32^{20}	727		i		s inorg a; i liq NH_3, al
b148	gallate, basic	Com'l dermatol. $Bi(OH_2)C_7H_5O_5$ (appr)	412.11	yel, amorph		d		i		i al, eth
b149	hydroxide	$Bi(OH)_3$ (appr)	260.00	wh amorph powd.	4.36	$-H_2O$, 100; d 415	$-1\frac{1}{2}H_2O$, 400	0.00014	d	s a; i or sl s conc alk
b150	iodate	$Bi(IO_3)_3$	733.69	wh				i		i HNO_3
b151	iodide, di-	BiI_2	462.79	red need, rhomb.		d 400	subl vac	s		s al, MeOH
b152	iodide, tri-	BiI_3	589.69	redsh, hex	5.778^{15}	408	*ca* 500	i	d	3.5^{20} al; i a; sl s NH_3; s HCl, HI
b153	lactate, *dl*	$Bi(C_3H_5O_3).7H_2O$	512.22	pr, need				14.4^{25}		i al
b154	molybdate	$Bi(MoO_4)_3$	897.78	yel-wh tetr need	6.07	643				v s a
b155	nitrate	$Bi(NO_3)_3.5H_2O$	485.07	col tricl, sl hygr	2.83	d 30	$-5H_2O$, 80	d	d	v s HNO_3; s a, glyc; i al, 42^{19} acet
b156	nitrate, basic	$BiONO_3.H_2O$	305.00	hex leaf	4.928^{15}	$-H_2O$, 105	$-HNO_3$, 260	i	i	s a; i al
b157	oxalate	$Bi_2(C_2O_4)_3.7H_2O$	808.73	wh powd		$-6H_2O$, *ca* 130		d		s inorg a; i al, eth
b158	oxide, mono-	BiO	224.97	dk-gray powd	7.15^{19}	d *ca* 180 (Bi_2O_3)		sl d	d	d dil a; s dil KOH
b159	oxide, pent-	Bi_2O_5	497.96	dk red or br	5.10	−O, 150	−2O, 357	i	i	s KOH
b160	oxide, tetr-	$Bi_2O_4.2H_2O$	517.99	br powd	5.6	$-H_2O$, 110	$-2H_2O$, 180	i	i	s a
b161	oxide, tri-	Bi_2O_3	465.96	yel, rhomb	8.9	825 ± 3	1890 (?)	i	i	s a
b162	oxide, tri-	Bi_2O_3	465.96	gray-blk, cub	8.20	tr 704		i	i	s a
b163	oxide, tri-	Bi_2O_3	465.96	wh-lt yel, rhomb, 1.91	8.55	860		i	i	sl s a
b164	oxybromide	BiOBr	304.89	col cr or wh powd	8.082^{15}	d red ht		i	i	i al; s a
b165	oxychloride	BiOCl	260.43	wh cr or powd, 2.15	7.72^{15}	red ht.		i	i	s a; i NH_3, tart a, acet
b166	oxyfluoride	BiOF	243.98	wh cr or powd	7.5^{20}_{20}	d red ht		i		s a
b167	oxyiodide	BiOI	351.88	red cr, tetr	7.922	d red ht		i	i	s a; i al, $CHCl_3$
b168	*ortho*phosphate	$BiPO_4$	303.95	wh, monocl.	6.323^{15}	d		i	i	s HCl; i al, dil HNO_3
b169	propionate, basic	$BiOC_3H_5O_2$	297.97	wh powd				i		v s dil HCl; i al
b170	salicylate, basic	Bismuth subsalicylate. $Bi(C_7H_5O_3)_3.Bi_2O_3$ (appr)	1086.29	wh micr cr (variable comp)				i		s a, alk; i al, eth
b171	selenide, tri-	Nat. guanajuatite. Bi_2Se_3	654.84	blk, rhomb	6.82	710	d	i		i alk
b172	silicate	Nat. eulytite. $2Bi_2O_3.3SiO_2$	1112.17	yel, cub, 2.05	6.11			i	i	s, d HCl, HNO_3
b173	sulfate	$Bi_2(SO_4)_3$	706.14	wh need	5.08^{15}	d 405		d	d	s a
b174	mono-sulfide,	BiS	241.04	dk gray powd	7.6–7.8	680 (in CO_2)	d	v sl s		
b175	sulfide, tri-	Nat. bismuthinite, bismuthglance. Bi_2S_3	514.15	br-blk, rhomb, 1.340, 1.456, 1.459	7.39	d 685		0.000018^{18}		s HNO_3: i dil al
b176	tartrate	$Bi_2(C_4H_4O_6)_3.6H_2O$	970.27	wh powd	2.595^{25}_{5}	$-3H_2O$, 105		i	i	s a, alk; i al
b177	tellurate	Montanite. $Bi_2TeO_6.2H_2O$	677.59	biaxial, β: 2.09	3.79					
b178	telluride, tri-	Nat. tetradymite. Bi_2Te_3	800.76	gray, rhbdr	7.7^{20}_{4}	573				d HNO_3
b179	vandate	Nat. pucherite. $Bi_2O_3.V_2O_5$	647.84	red-grn, rhomb, 2.41, 2.50, 2.51 (Li)	6.25^{25}					
b180	**Bismuthic acid**	$HBiO_3$	257.99	red	5.75	$-H_2O$, 120	−2O, 357	i	i	s a, KOH
b181	**Borazole**	$B_3N_3H_6$	80.50	col liq	1^{-65}, 0.8614^{0}	−58	55	sl d	d	
b182	**Boric Acid**, **meta-**	HBO_2	43.82	wh cr, cub, 1.619	2.486	236 ± 1		v sl s	sl s	

No.	Name	Synonyms and Formulae	Mol. wt.	Crystalline form, properties and index of refraction	Density or spec. gravity	Melting point, °C	Boiling point, °C	Solubility, in grams per 100 cc Cold water	Hot water	Other solvents
	Boric acid									
b183	ortho-	Boracic acid. H_3BO_3	61.83	col, tricl, 1.337, 1.461, 1.462	1.435^{15}	169 ± 1 tr to HBO_2	−1½H_2O, 300	6.35^{20}	27.6^{100}	28^{20} glyc; 0.0078 eth; 5.56 al; 20. 20^{25} MeOH; 1.92^{25} liq NH_3; sl s acet
b184	tetra- (pyro-)	$H_2B_4O_7$	157.26	vitr or wh powd				s	s	s al
b185	fluo-	HBF_4	87.81	col liq			d 130	∞	∞	s al
b186	**Borinoaminoborine**	B_2H_7N	42.68	col liq		−66.5	76.2			s triborine triamine
b187	**Boron**	B	10.811	yel monocl or br amorph powd	2.34, 2.37 amorph	2300	2550 (sub)	i	i	v sl s HNO_3
b188	arsenate	$BAsO_4$	149.73	wh cr tetrag, 1.681, 1.690	3.64	subl *ca* 700		v sl s	1.4^{100}	i al; s inorg a
b189	bromide, tri-	BBr_3	250.54	col fum liq, $n_D^{16.8}$ 1.5312	$2.6431^{18.4}_4$	−46	91.3 ± 0.25	d		s al, CCl_4
b190	bromide, di-, iodide	BBr_2I	297.53	col liq			125	d	d	
b191	bromide, mono-, diiodide	$BBrI_2$	344.53	col liq			180	d	d	
b192	(di-) bromide, mono-, pentahydride	B_2H_5Br	106.67	col gas		−104	*ca* 10	hyd to HBO_2, HBr + H_2		
b193	(tetra-) carbide	B_4C	55.26	blk rhbdr	2.52	2350	>3500	i	i	i a; s fus alk
b194	chloride, tri-	BCl_3	117.17	col fum liq, $1.4195^{8.7}$ α line H_2	1.349^{11}_4	−107.3	12.5		d to HCl + H_3BO_3	d al
b195	fluoride, tri-	BF_3	67.81	col gas	2.99 g/l	−126.7	−99.9	106 (762 mm)	d	d al; s conc H_2SO_4
b196	fluoride dihydrate	$BF_3.2H_2O$	103.84	col liq; n_{HE}^{20} 1.31498	1.6316^{20}_4	6		d	d	s eth, dioxan
b197	hydride	Diborane, boroethane. B_2H_6	27.67	col gas	liq: 0.447^{-11} sol: 0.577^{-183}	−165.5	−92.5	sl s d to H_3BO_3 + H_2		d 1.6^0 al; s NH_4OH, conc H_2SO_4
b198	hydride	Dihydrotetraborane, borobutane. B_4H_{10}	53.32	col gas, pois	0.56^{-35}	−120.8	16	sl s, d		s bz; d al
b199	hydride	Pentaborane. B_5H_9	63.13	col liq	0.66^0	−46.82	58.4			
b200	hydride	Hexaborane. B_6H_{10}	74.95	col liq	0.69^0	−65	0^{73}	d		
b201	hydride	Decaborane. $B_{10}H_{14}$	122.22	wh cr	0.94^{25}	99.5	213	sl s	d	v s CS_2; s al, eth bz
b202	iodide, tri-	BI_3	391.52	col pl, hygr	3.35^{50}	49.9	210	d	d	d al; v s CS_2, bz, CCl_4
b203	nitride	BN	24.82	wh, hex	2.25	subl *ca* 3000		i	sl d	sl s h a
b204	oxide	B_2O_3	69.62	rhomb cr, 1,64, 1.61	2.46 ± 0.01	450 ± 2	*ca* 1860	sl s	s	
b205	oxide glass	B_2O_3	69.62	col, vitr 1.485	1.812^{25}_4	*ca* 450		1.1^0	15.7^{100}	s al, a
b206	phosphide	BP	41.78	maroon powd		ign 200		i	i	i all solv
b207	*tri*selenide	B_2Se_3	258.50	yel-gray powd				d	d	
b208	(hexa-) silicide	B_6Si	92.95	blk cr	2.47			i		s HNO_3; d H_2SO_4; i KOH
b209	(tri-) silicide	B_3Si	60.52	blk rhomb	2.52			i		sl s HNO_3; d H_2SO_4, KOH
b210	sulfide, penta-	B_2S_5	181.94	col, tetrag	1.85	390		d	d	d al
b211	sulfide, tri-	B_2S_3	117.81	wh cr or vitr	1.55	310		d		sl s PCl_3, SCl_2; d al
b212	**Borotungstic acid**	$H_5BW_{12}O_{40}.30H_2O$	3402.49	tetr, cr	3	45–51		s		s al, eth
b213	**Bromic acid**	$HBrO_3$	128.92	known in sol only, col or yelsh		d 100		v s	d	
b214	**Bromine**	Br_2	159.808	dk red liq, 1.661	2.928^{59}, 3.119^{20}	−7.2	58.78	4.17^0, 3.58^{20}	3.52^{50}	v s al, eth, chl, CS_2
b215	azide	Bromoazide. BrN_3	121.93	cr, red liq		*ca* 45	exp			s eth, KI; sl s bz, ligr
b216	chloride	BrCl	115.36	red-col liq or gas		*ca* −66 d 10	*ca* 5	s d		s eth, CS_2
b217	fluoride, mono-	BrF	98.91	red-br gas		d −33	−20			
b218	fluoride, penta-	BrF_5	174.90	col liq	2.466^{25}	−61.3	40.5	d	d	
b219	fluoride, tri-	BrF_3	136.90	col-gray-yel liq	2.49^{135}	(−2) 8.8	135	d viol. to O_2, HOBr, HF, $HBrO_3$		d alk
b220	hydrate	$Br_2.10H_2O$	339.97	red oct	1.49	d 6.8		s		
b221	oxide, di-	BrO_2	111.91	lt yel		d 0				
b222	oxide, mon-	Br_2O	175.82	dk br		−17 to −18				s, d CCl_4
b223	(tri-) oxide, oct-	Br_3O_8(or Br_3O_8)$_n$	367.72	wh		stable at −40				

No.	Name	Synonyms and Formulae	Mol. wt.	Crystalline form, properties and index of refraction	Density or spec. gravity	Melting point, °C	Boiling point, °C	Solubility, in grams per 100 cc Cold water	Hot water	Other solvents
	Bromauric acid									
b224	**Bromauric acid**	$HAuBr_4.5H_2O$	607.69	red-br cr		27		v s		s al
b225	**Bromous acid, hypo-**	HBrO	96.92	known only in soln, col-yel		40 (vac)		s	s d	s al, eth, chl
b226	**Bromoplatinic acid**	$H_2PtBr_6.9H_2O$	838.70	red monocl, deliq		d −100		v s	v s	
c1	**Cadmium**	Cd	112.41	hex silv-wh malleable met	8.642	320.9	765	i	i	s a, NH_4NO_3, h H_2SO_4
c2	acetate	$Cd(C_2H_3O_2)_2$	230.50	col	2.341	256	d	v s		s MeOH
c3	acetate, hydrate	$Cd(C_2H_3O_2)_2.2H_2O$	266.54	col monocl, odor ac a	2.01	$-H_2O$, 130		v s	v s	v s al
c4	amide	$Cd(NH_2)_2$	144.45		3.05^{25}	d 120				
c5	ammonium chloride	$CdCl_2.NH_4Cl$	236.79	col need, rhomb	2.93	289		33.45^{16}	$43.99^{62.8}$	s al, MeOH
c6	ammonium sulfate	$Cd(NH_4)_2(SO_4)_2.6H_2O$	448.69	col monocl pr	2.061^{20}_{4}	100 d $-H_2O$		s	s	
c7	arsenate, hydrogen	$CdHAsO_4.H_2O$	270.34		4.164^{15}_{4}	>120				
c8	arsenide	Cd_3As_2	487.04	dk gray cub	6.21^{15}_{4}	721		i	i	sl s HCl; s HNO_3; i aq reg
c9	benzoate	$Cd(C_7H_5O_2)_2.2H_2O$	390.66					3.34^{20}		sl s al
c10	borate	$Cd(BO_2)_2.H_2O$	248.03	wh rhomb	3.758	d		125^{17}		i al
c11	borotungstate	$Cd_5(BW_{12}O_{40}).18H_2O$	6600.25	yel tricl		75		1250^{19}	v s	
c12	bromide	$CdBr_2$	272.22	yel cr	5.192^{25}	567	863	57^{10}	162^{104}	26.6^{15} al; 0.4^{15} eth; s HCl; 1.6^{18} acet
c13	bromide, tetrahydrate	$CdBr_2.4H_2O$	344.28	sm wh need, effl		tr 36		121^{10}	v s	25 al; s acet; sl s eth
c14	carbonate	$CdCO_3$	172.41	wh, trig	4.258^{4}	d <500		i	i	s a, KCN, NH_4 salts; i NH_3
c15	chlorate	$Cd(ClO_3)_2.2H_2O$	315.33	col pr, deliq	2.28^{18}	80		298^{0}	487^{65}	s al, a, acet
c16	chloride	$CdCl_2$	183.32	col, hex	4.047^{25}	568	960	140^{20}	150^{100}	1.52^{15} al; $1.7^{15.5}$ MeOH; i acet, eth
c17	chloride	$CdCl_2.2\frac{1}{2}H_2O$	228.35	col monocl, 1.6513	3.327	tr 34		168^{20}	180^{100}	2.05^{15} MeOH; sl s al
c18	chloroacetate, di-	$Cd(C_2HCl_2O_2).2H_2O$	386.29	need	2.132^{15}					
c19	chloroacetate mono-	$Cd(C_2H_2ClO_2)_2.6H_2O$	407.47		1.942^{25}					
c20	chloroacetate, tri-	$Cd(C_2Cl_3O_2)_2.1\frac{1}{2}H_2O$	464.18	rhomb	2.093^{25}					
c21	chloroplatinate	$CdPtCl_6.3H_2O$	574.25	yel trig need	2.882	$-H_2O$, 170, d		s	s	
c23	chromite	$CdCr_2O_4$	280.39	grn to blk, cub	5.79^{17}			i	i	i a
c24	cyanide	$Cd(CN)_2$	164.44	cr		dec >200		1.7^{15}	s	s a, KCN, NH_4OH; i al
c25	ferrocyanide	$Cd_2Fe(CN)_6.xH_2O$						i	i	s HCl
c26	fluogallate	$[Cd(H_2O)_6].[GaF_5H_2O]$	403.22	col cr, 1.45	2.79	$-5H_2O$, 110		v s		
c27	fluoride	CdF_2	150.40	wh cub, 1.56	6.64	1100	1758	4.35^{25}		s a, HF; i al, NH_3
c28	fluosilicate	$CdSiF_6.6H_2O$	362.57	col hex				s	s	s 50 % al
c29	formate	$Cd(CHO_2)_2.2H_2O$	238.47	monocl	2.44	d		v s		
c30	fumarate	$Cd(C_4H_2O_4)$	226.48					0.9^{20}		
c31	hydroxide	$Cd(OH)_2$	146.41	wh, trig or amorph	4.79^{15}	d 300		0.00026^{25}		s a, NH_4 salts; i alk
c32	iodate	$Cd(IO_3)_2$	462.21	wh cr	6.43	d		s		s NHO_3; NH_4OH
c33	iodide	CdI_2	366.21	grn-yel powd	5.670^{30}	387	796	86.2^{25}	125^{100}	110.5^{20} al; 41^{25} acet; sl s NH_3; 206.7^{25} MeOH
c34	lactate	$Cd(C_3H_5O_3)_2$	290.54	need				10	12.5	i al
c35	maleate	$Cd(C_4H_2O_4).2H_2O$	262.49					0.66^{20}		
c36	*per*manganate	$Cd(MnO_4)_2.6H_2O$	458.36		2.81	d 95		v s	v s	
c37	molybdate	$CdMoO_4$	272.34	yel pl	5.347			sl s		s a, NH_4OH, KCN
c38	nitrate	$Cd(NO_3)_2$	236.41	col		350		109^{0}	326^{60} 682^{100}	v s a; s et ac
c39	nitrate, tetrahydrate	$Cd(NO_3)_2.4H_2O$	308.47	wh pr, need, hygr	2.455^{17}_{4}	59.4	132	215		s al, NH_3; i HNO_3
c40	nitrocobaltate (III)	Cadium cobaltinitrite. $Cd[Co(NO_2)_6]_2$	1007.13	yel		d 175		sl s	v s	d a, alk, org solv
c41	oxalate	CdC_2O_4	200.42	col cr	3.32^{18}_{4}	d 340				s a; i al
c42	oxalate, trihydrate	$CdC_2O_4.3H_2O$	254.45	col cr		d		0.005^{18}	0.009	
c43	oxide	CdO	128.40	br, amorph	6.95	>1500		i	i	s a, NH_4 salts; i alk
c44	oxide	CdO	128.40	br cub, 2.49 (Li)	8.15	>1500	subl 1559	i	i	s a, NH_4 salts; i alk
c45	*ortho*phosphate	$Cd_3(PO_4)_2$	527.14	col, amorph		1500		i		s a, NH_4 salts

No.	Name	Synonyms and Formulae	Mol. wt.	Crystalline form, properties and index of refraction	Density or spec. gravity	Melting point, °C	Boiling point, °C	Solubility, in grams per 100 cc		
								Cold water	Hot water	Other solvents
	Cadmium									
c46	*pyro*phosphate	$Cd_2P_2O_7$	398.74	wh cr leaf	4.965^{15}	above red heat		sl s	s	s a, NH_3
c47	phosphate, dihydrogen	$Cd(H_2PO_4)_2.2H_2O$	342.41	col, tricl	2.74^{15}_{4}	d 100				i al, eth; s HCl
c48	phosphide	Cd_3P_2	399.15	grn, tetr need	5.60	700				s d HCl; s exp conc HNO_3
c49	potassium cyanide	$Cd(CN)_2.2KCN$	294.68	col, glossy, oct	1.847			33.3	100	i al
c50	potassium sulfate	$CdK_2(SO_4)_2.2H_2O$	322.69	tricl col tab	2.922^{16}			42.89^{16}	47.40^{40}	
c51	salycilate	$Cd(C_7H_5O_3)_2.H_2O$	404.65	wh need				sl s	s	s al, eth, glycerol a, NH_4OH
c52	selenate	$CdSeO_4.2H_2O$	291.39	rhomb	3.63	$-1H_2O$, 100; $-2H_2O$, 170		v s		
c53	selenide	CdSe	191.36	grn-br or red powd, hex	5.81^{15}_{4}	>1350		i		d a
c54	*meta*silicate	$CdSiO_3$	188.48	col, rhomb, 1.739	4.93	1242		v sl s		
c55	sulfate	$CdSO_4$	208.46	wh, rhomb	4.691^{20}_{4}	1000		75.5^{0}	60.8^{100}	i al, acet, NH_3
c56	sulfate, hydrate	$CdSO_4.H_2O$	226.48	monocl	3.79^{20}	tr 108		s	s	i al
c57	sulfate, hydrate	$CdSO_4.7H_2O$	334.57	col, monocl	2.48	tr 4		s		i al
c58	sulfate, hydrate	$3CdSO_4.8H_2O$	769.50	col, monocl, 1.565	3.09	tr 41.5		113^{0}	s	
c59	sulfide	Nat. greenockite. CdS	144.46	yel-or, hex, 2.506, 2.529	4.82	$1750^{100 atm}$	subl in N_2, 980	0.00013^{18}	colloid	s a; v sl s NH_4OH
c60	sulfite	$CdSO_3$	192.46	cr		d		sl s		i al; s a, NH_4OH
c61	tartrate	$CdC_4H_4O_6$	260.47	wh cr powd				sl s		s a, NH_4OH
c62	telluride	CdTe	240.00	blk, cub	6.20^{15}	1041		i		i a; d HNO_3
c63	tungstate	$CdWO_4$	360.25	yel cr				0.05		s NH_4OH
	Cadmium complexes									
c64	tetramminecadmium *per*rhenate	$[Cd(NH_3)_4](ReO_4)_2$	681.16		3.714^{25}_{4}					0.037 conc NH_4OH
c65	tetrapyridine cadmiumfluosilicate	$[Cd(C_5H_5N_4)]SiF_6$	570.89	wh, tricl	2.282					
c66	**Calcium**	Ca	40.08	silv wh soft met, cub	1.54	842-8	1484	d to H_2+ $Ca(OH)_2$	d	s a, liq NH_3; sl s al; i bz
c67	acetate	$Ca(C_2H_3O_2)_2$	158.17	col cr; 1.55, 1.56, 1.57		d		37.4^{0}	29.7^{100}	sl s al
c68	acetate, dihydrate	$Ca(C_2H_3O_2)_2.2H_2O$	194.21	col need		$-1H_2O$, 84		34.7^{20}	33.5^{80}	
c69	acetate, monohydrate	$Ca(C_2H_3O_2)_2.H_2O$	176.19	col need		d		43.6^{0}	34.3^{100}	sl s al
c70	aluminate	$CaAl_2O_4$ (or $CaO.Al_2O_3$)	158.04	wh monocl, tricl or rhomb; 1.643, 1.665, 1.663	2.981^{25}	1600		d		s HCl; i HNO_3, H_2SO_4
c71	(tri-)aluminate	$Ca_3Al_2O_6$ (or $3CaO.Al_2O_3$)	270.20	wh, cub, 1.710	3.038^{25}	d 1535		i		s a
c72	(tri-)aluminate hexahydrate	$3CaO.Al_2O_3.6H_2O$	378.29	col, oct, 1.603	2.52^{20}	d 700–800		d		
c73	aluminosilicate	$2Ca.Al_2O_3.SiO_2$	274.20	col, tetr, 1.669, 1.658	3.048	1590 ± 2				d a
c74	aluminosilicate	Nat. anorthite. $CaAl_2.Si_2O_8$ (or $CaO.Al_2O_3.2SiO_2$)	278.21	wh, tricl, 1.5832	2.765	1551				
c75	*ortho*arsenate	$Ca_3(AsO_4)_2$	398.08	col amorph powd	3.620			0.013^{25}		
c76	arsenate, trihydrate	Nat. haidingerite. $2CaO.As_2O_5.3H_2O$	396.04	col, rhomb, 1.590, 1.602, 1.638	2.967					
c77	arsenide	Ca_3As_2	270.08	red cr	3.031^{25}	d		d	d	d a; s h HNO_3
c78	azide	$Ca(N_3)_2$	124.12	col, rhomb, hyg		$-3H_2O$, 110; exp 144–156		38.1^{0}	45^{15}	0.211^{16} al; i eth
c79	benzoate	$Ca(C_7H_5O_2)_2.3H_2O$	336.36	col, rhomb	1.436	$-3H_2O$, 110		2.7^{0}	8.3^{90}	
c80	*meta*borate	$Ca(BO_2)_2$	125.70	col, flat rhomb pr, 1.550, 1.660, 1.680		1154		sl s		s a, NH_4 salts; sl s ac a
c81	*meta*borate, hexahydrate	$Ca(BO_2)_2.6H_2O$	233.79	col, tetr, 1.520, 1.502	1.88			0.25^{20}		
c82	*tetra*borate	CaB_4O_7	195.32	readily vitrified		986				
c83	boride	CaB_6	104.95	blk, cub	2.3^{20}	2235		i	i	s HNO_3; sl s conc H_2SO_4
c84	bromide	$CaBr_2$	199.90	col, rhomb need, deliq	3.353^{25}	sl d 730	806–812	142^{20}	312^{105}	s al, acet, a; sl s NH_3, MeOH
c85	bromate	$Ca(BrO_3)_2.H_2O$	313.91	monocl cr	3.329	$-H_2O$, 180		v s	v s	
c86	bromide, hexahydrate	$CaBr_2.6H_2O$	307.99	col, hex cr	2.295	38.2	149	594^{0}	1360^{25}	s al, acet, a

No.	Name	Synonyms and Formulae	Mol. wt.	Crystalline form, properties and index of refraction	Density or spec. gravity	Melting point, °C	Boiling point, °C	Solubility, in grams per 100 cc		
								Cold water	Hot water	Other solvents
	Calcium									
c87	butyrate	$Ca(C_4H_7O_2)_2.3H_2O$	268.32	col cr				s	sl s	
c88	carbide	CaC_2	64.10	col, tetr, 1.75	2.22	stab	2300	d	d	
c89	carbonate	Nat. aragonite. $CaCO_3$	100.09	col, rhomb, 1.530, 1.681, 1.685	2.930	25–447 tr to calcite 520	d 825	0.00153^{25}	0.00190^{75}	s a, NH_4Cl
c90	carbonate	Nat. calcite. $CaCO_3$	100.09	col, rhomb or hex, 1.6583, 1.4864	2.710^{18}	1339^{1025}	d 898.6	0.0014^{25}	0.0018^{75}	s a, NH_4Cl
c91	carbonate, hexahydrate	$CaCO_3.6H_2O$	208.18	col, monocl, 1.460, 1.535, 1.545	1.771^{0}					
c92	chlorate	$Ca(ClO_3)_2$	206.99	wh cr, hyg		340 ± 10 (−some O)		s	s	s al, acet
c93	chlorate, dihydrate	$Ca(ClO_3)_2.2H_2O$	243.01	wh-yelsh, rhomb, or monocl, deliq	2.711	$-H_2O$, 100		177.7^{8}	v s	s al, acet
c94	*per*chlorate	$(CaClO_4)_2$	238.98	col cr	2.651	d 270		188.6^{25}	v s	166.2^{25}_{25} al; 237.4 MeOH
c95	chloride	$CaCl_2$	110.99	col, cub, deliq 1.52	2.15^{25}_{4}	782	>1600	74.5^{20}	159^{100}	s al, acet, ace a
c96	chloride aluminate	$3CaO.Al_2O_3.CaCl_2.10H_2O$	561.33	col, monocl or hex, hex, 1.550, 1.535	1.892^{14}	$-H_2O$, 105	$-8H_2O$, 350	sl s	d	s a
c97	chloride, dihydrate	$CaCl_2.2H_2O$	147.02	col cr	0.835			97.7^{0}	326^{60}	50^{80} al
c98	chloride, hexahydrate	$CaCl_2.6H_2O$	219.08	col, trig, deliq, 1.417, 1.393	1.71^{25}	29.92	$-4H_2O$, 30, $-6H_2O$, 200	279^{0}	536^{20}	s al
c99	chloride, monohydrate	$CaCl_2.H_2O$	129.00	col cr, deliq		260		76.8^{0}	249^{100}	s al; i acet
c100	chloride fluoride *ortho*phosphate	$3Ca_3(PO_4)_2.CaClF$	1025.08	col cr, 1.634, 1.631	3.14	1270		v sl s		
c101	chlorite	$Ca(ClO_2)_2$	174.98	wh, cub	2.71			d	d	i al
c102	*hypo*chlorite	$Ca(ClO)_2$	142.98	wh powd or flat pl, 1.545, 1.69	2.35	d 100		s		i al
c103	chlorite, basic	$Ca(ClO)_2.2Ca(OH)_2$	257.16	wh, hex, 1.51, 1.585	2.10			sl s solns with 5–6 % avail Cl	d	d a
c104	*hypo*chlorite, basic	Bleaching powder, chlorinated lime. $Ca(ClO)_2.CaCl_2.xCa(OH)_2.xH_2O$	comp varies	wh powd strong Cl odor		d		d evln Cl		d a
c105	*hypo*chlorite, trihydrate	$Ca(ClO)_2.3H_2O$	197.03	tetr pl, 1.535, 1.63	2.1	$-3H_2O$, 60				
c106	chromate	$CaCrO_4.2H_2O$	192.09	yel, monocl pr.		$-2H_2O$, 200		16.3^{20}	18.2^{45}	s a, al
c107	chromite	$CaCr_2O_4$	208.07	ol grn, cub need	4.8^{18}	2090		i	i	i a; s fus K_2CO_3
c108	cinnamate	$Ca(C_9H_7O_2)_2.3H_2O$	388.44	col cr				0.22^{2}	1.34^{100}	
c109	citrate	$Ca_3(C_6H_5O_7)_2 4H_2O$	570.51	wh need		$-4H_2O$, 120		0.085^{18}	0.096^{23}	0.0065^{18} al
c110	cyanamide	$CaCN_2$	80.10	col, hex, rhbdr		1300 subl >1150		d evl NH_3	d	
c111	cyanide	$Ca(CN)_2$	92.12	wh powd		d >350		d	d	
c112	cyanoplatinite	$CaPt(CN)_4.5H_2O$	429.31	yel-grn fluoresc, rhomb, 1.6226		$-5H_2O$, 100		s		
c113	ferricyanide	$Ca_3[Fe(CN)_6]_2 12H_2O$	760.42	red need, deliq				v s	v s	
c114	ferrite, mono-	$CaO.Fe_2O_3$	215.77	dk redsh r, rhomb, 2.58, 2.43 (Na)	5.08	1250		i	i	v sl s a
c115	ferrocyanide	$Ca_2Fe(CN)_6 11$ or $12H_2O$	490.28	yel tricl, 1.570, 1.582, 1.596	1.68	d		86.8^{25}	115^{65}	i al
c116	fluosilicate	$CaSiF_6$	182.16	col, tetr	2.66^{18}			sl s		s al, HF, HCl
c117	fluoride	Nat. fluorite. CaF_2	78.08	col, cub luminisc w heat, 1.434	3.180	1423	*ca* 2500	0.0016^{18}	0.0017^{26}	s NH_4 salts; sl s a; i acet
c118	fluosilicate, dihydrate	$CaSiF_6.2H_2O$	218.19	col, tetrag	2.254			sl s d		s HCl, HF; i al
c119	formate	$Ca(CHO_2)_2$	130.12	col, rhomb, 1.510, 1.514, 1.578	2.015	d		16.2^{0}	18.4^{100}	i al
c120	fumarate	$CaC_4H_2O_4.3H_2O$	208.18	col, rhomb				2.11^{30}		
c121	*d*-gluconate	$Ca(C_6H_{11}O_7)_2.H_2O$	448.40	wh cr powd, need		$-H_2O$, 120		3.3^{15}		v sl s al
c122	glycerophosphate	$CaC_3H_5(OH)_2PO_4$	210.16	wh cr powd, hyg		d 170		2^{25}	sl s	i al
c123	hydride	CaH_2	42.10	wh, rhomb cr	1.9	816 (in H_2) d *ca* 600		d H_2+ $Ca(OH)_2$		d a
c124	hydroxide	$Ca(OH)_2$	74.09	col, hex, 1.574, 1.545	2.24	$-H_2O$, 580	d	0.185^{0}	0.077^{100}	s NH_4 salts, a; i al
c125	hyponitrite	$CaN_2O_2.4H_2O$	172.15	wh cr	1.834	d 320				d dil a
c126	iodate	Nat. lautarite. $Ca(IO_3)_2$	389.89	col, monocl	4.519^{15}	d 540		0.20^{15}	0.67^{90}	s HNO_3; i al
c127	iodate, hexahydrate	$Ca(IO_3)_2.6H_2O$	497.98	col, rhomb		d 35		0.13^{0}	1.22^{100}	s HNO_3
c128	iodide	CaI_2	293.89	yelsh-wh, hex, deliq	3.956^{25}_{4}	784	*ca* 1100	209^{20}	426^{100}	126^{20} MeOH; s al, acet, a

No.	Name	Synonyms and Formulae	Mol. wt.	Crystalline form, properties and index of refraction	Density or spec. gravity	Melting point, °C	Boiling point, °C	Solubility, in grams per 100 cc		
								Cold water	Hot water	Other solvents
	Calcium									
c129	iodide, hexahydrate	$CaI_2.6H_2O$	401.98	yel, hex need	2.55	d 42	160	757^{0}	1680^{20}	s a, al, acet
c130	iron (III) aluminate	Calcium (tetra-) alumino-ferite, nat. celite. $4CaO.Fe_2O_3.Al_2O_3$	485.97	brn, rhomb, 1.98, 2.05, 2.08 all for λ	3.77	1418		3		
c131	isobutyrate	$Ca(C_4H_7O_2)_2.5H_2O$	304.35	col powd				20	sl s	
c132	lactate	$Ca(C_3H_5O_3)_2.5H_2O$	308.30	wh need, effl		$-3H_2O$, 100		3.1^{0}	7.9^{30}	sl s a; i al, eth
c133	laurate	$Ca(C_{12}H_{23}O_2)_2.H_2O$	456.73	wh need, effl		182–183		0.004^{15}	0.055^{100}	0.059^{15}, 1.72^{78} al
c134	linoleate	$Ca(C_{18}H_{31}O_2)_2$	598.97	wh amorph powd				i		s al, eth
c135	magnesium carbonate	Nat. dolomite. $CaCO_3.MgCO_3$	184.41	col, trig, 1.6817, 1.5026	2.872	d 730–760		0.032^{18}		
c136	magnesium *meta*silicate	Nat. diopside. $CaO.MgO.2SiO_2$	216.56	col, monocl, 1.665, 1.672, 1.695	3.275	1391		i	i	i HCl
c137	magnesium *ortho*silicate	Nat. mervinite. $3CaO.MgO.SiO_2$	328.72	col to pa grn, monocl, 1.708, 1.711, 1.718	3.150					
c138	dl-malate	$CaC_4H_4O_5.3H_2O$	226.20	col, rhomb, 1.545, 1.555, 1.575				0.321^{0}	$0.451^{37.5}$	i al
c139	*l*-malate	$CaC_4H_4O_5.2H_2O$	208.18	col				0.812^{0}	$1.224^{37.5}$	s al
c140	malate, dihydrogen	$Ca(HC_4H_4O_5)_2.6H_2O$	414.33	rhomb, or wh cr powd, 1.493, 1.507, 1.545				sl s		
c141	maleate	$CaC_4H_2O_4.H_2O$	172.15	col rhomb, 1.495, 1.575, 1.640				2.89^{25}	3.21^{40}	
c142	malonate	$CaC_3H_2O_4.4H_2O$	214.19					0.44^{0}	0.72^{100}	
c143	*per*manganate	$Ca(MnO_4)_2.5H_2O$	368.03	purp cr	2.4	d		331^{14}	338^{25}	s NH_4OH
c144	α-methylbutyrate	Calvium ethylmethyl-acetate. $Ca(C_5H_9O_2)_2$	242.34					24.24^{0}	25.65^{70}	
c145	molybdate	Nat. pawellite. $CaMoO_4$	200.01	col, tetr, 1.967, 1.978	4.38–4.53			i	d	s a i al, eth
c146	nitrate	$Ca(NO_3)_2$	164.09	col, cub, hyg	2.504^{18}	561		121.2^{18}	376^{100}	14^{15} al; s MeOH, liq NH_3, acet; i eth
c147	nitrate, tetrahydrate	$Ca(NO_3)_2.4H_2O$	236.15	col, monocl, deliq, 1.465, 1.498, 1.504	α 1.896, β 1.82	α 42.7, β 39.7	d 132	266^{0}	660^{30}	s al, acet
c148	nitrate, trihydrate	$Ca(NO_3)_2.3H_2O$	218.14	col, tricl		51.1				
c149	nitride	Ca_3N_2	148.25	brn cr, hex	2.63^{17}	1195		d	d	s dil a; d abs al
c150	nitrite	$Ca(NO_2)_2.H_2O$	150.11	col-yelsh, hex, deliq	2.23^{34}	$-H_2O$, 100		45.9^{0}	89.6^{91}	sl s al
c151	nitrite, tetrahydrate	$Ca(NO_2)_2.4H_2O$	204.15	col cr, tetr	1.674_{0}^{0}	$-2H_2O$, 44		74.9^{0}	106^{42}	s al
c152	oleate	$Ca(C_{18}H_{33}O_2)_2$	603.01	wh wax-like cr		83–84		0.04^{25}	0.03^{50}	sl s eth
c153	oxalate	CaC_2O_4	128.10	col, cub	2.2^{4}	d		0.00067^{13}	0.0014^{95}	s a; i ac a
c154	oxalate, hydrate	$CaC_2O_4.H_2O$	146.12	col	2.2	$-H_2O$, 200		i	i	s a; i ac a
c155	oxide	Lime, calcia. CaO	56.08	col, cub, 1.838	3.25–3.38	2614	2850	0.131^{10} d	0.07^{80} d	s a
c156	oxide, per-	CaO_2	72.08	wh, tetr, 1.895	2.92_{4}^{25}	d 275		sl s		s a
c157	oxide, per-octahydrate	$CaO_2.8H_2O$	216.20	wh, tetr, pearly	1.70	$-8H_2O$, 200	d 275 expl	sl s	d	s a, NH_4 salts; i al, eth
c158	palmitate	$Ca(C_{16}H_{31}O_2)_2$	550.93	wh or yelsh, wh fatty powd				0.003^{25}		v sl s al; 0.008^{25} eth
c159	1-phenol-4 sulfonate(p-)	$Ca[C_6H_4(OH)SO_3]_2.H_2O$	404.43	wh to pinkish powd				s		s al
c160	phenoxide	$Ca(OC_6H_5)_2$	226.29	redsh powd				sl s		sl s al
c161	*hypo*phosphate	$Ca_2P_2O_6.2H_2O$	274.13	gel				i		s HCl
c162	*meta*phosphate	$Ca(PO_3)_2$	198.02	col, 1.588, 1.595	2.82	975		i	i	i a
c163	*ortho*phosphate, di-(sec)	Nat. brushite. $CaHPO_4.2H_2O$	172.09	wh, tricl, 1.5576, 1.5457, 1.5392	2.306^{16}	$-H_2O$, 109		0.0316^{38}	0.075^{100}	i al, s a
c164	*ortho*phosphate, mono-(prim.)	$Ca(H_2PO_4)_2.H_2O$	252.07	col, tricl, deliq, 1.5292, 1.5176, 1.4392	2.220^{16}	$-H_2O$, 109	d 203	1.8^{30}	d	s a
c165	*ortho*phosphate, tri-(tert.)	Nat. whitlockite. $Ca_3(PO_4)_2$	310.18	wh amorph powd, 1.629, 1.626	3.14	1670		0.002	d	i al; s a
c166	*pyro*phosphate	$Ca_2P_2O_7$	254.10	col, biax, 1.585, 1.604	3.09	1230		i		s a
c167	*pyro*phosphate, pentahydrate	$Ca_2P_2O_7.5H_2O$	344.18	col, monocl, 1.539, 1.545, 1.551	2.25			sl s		s a; i NH_4Cl
c168	phosphide	Ca_3P_2	182.19	gray lumps	2.51	*ca* 1600		d ev PH_3		s a; i al, eth, bs
c169	*hypo*phosphite	$Ca(H_2PO_2)_2$	170.06	wh-gray, monocl		d		15.4^{25}	12.5^{100}	i al
c170	*ortho*phosphite, di-	$2CaHPO_3.3H_2O$	294.17					sl s	d	s NH_4Cl
c171	*ortho*plumbate	Ca_2PbO_4	351.35	red-br cr	5.71	d		i	d	s a
c172	propionate	$Ca(C_3H_5O_2)_2.H_2O$	204.24	col, monocl tabl				49^{0}	55.8^{100}	i al
c173	*l*-quinate	$Ca(C_7H_{11}O_6)_2.10H_2O$	602.56	rhomb leaf		50, $-10H_2O$ 120		16^{18}		i al

No.	Name	Synonyms and Formulae	Mol. wt.	Crystalline form, properties and index of refraction	Density or spec. gravity	Melting point, °C	Boiling point, °C	Solubility, in grams per 100 cc Cold water	Hot water	Other solvents
	Calcium									
c174	salicylate	$Ca(C_7H_5O_3)_2.2H_2O$	350.34	wh, oct		$-2H_2O$, 120		4^{25}	s	s al
c175	selenate	$CaSeO_4$	183.04	col	2.88			7.9^{4}	5.4^{67}	
c176	selenate, dihydrate	$CaSeO_4.2H_2O$	219.07	col, monocl	2.68					
c177	selenide	CaSe	119.04	cub, 2.274	3.57					
c178	*meta*silicate(α)	Nat. pseudowollastonite. $CaSiO_3$	116.16	col, monocl, 1.610, 1.611, 1.664	2.905	1540		0.0095^{17}		s HCl
c179	*meta*silicate(β)	Nat. wollastonite. $CaSiO_3$	116.16	col, monocl, 1.616, 1.629, 1.631	2.5	tr 1200				
c180	di-*ortho*silicate (I)	Ca_2SiO_4	172.24	col, monocl, 1.717, 1.735	3.27	2130				
c181	di-*ortho*silicate (II)	Ca_2SiO_4	172.24	col, rhomb, 1.717, 1.735	3.28	tr to (I) 1420				
c182	di-*ortho*silicate (III)	Ca_2SiO_4	172.24	col, monocl, 1.642, 1.645, 1.654	2.97	tr to 675				
c183	(tri-)silicate	Nat. alite. Ca_3SiO_5 or $(3CaO.SiO_2)$	228.32	col, monocl, α 1.718, β 1.724		1900 (incogr)				
c184	silicide	$CaSi_2$	96.25		2.5			i	d	s a, alk
c185	stearate	$Ca(C_{18}H_{35}O_2)_2$	607.04	cr powd		179–180		0.004^{15}		i al, eth
c186	succinate	$CaC_4H_4O_4.3H_2O$	212.22	col, 1.460, 1.540, 1.610				0.193^{10}	0.89^{80}	
c187	sulfate	Nat. anhydrite. $CaSO_4$	136.14	col, rhomb, or monocl, 1.569, 1.575, 1.613	2.960	monocl 1450	rhomb tr to monocl 1193	0.209^{30}	0.1619^{100}	s a, NH_4 salts, $Na_2S_2O_3$, glyc
c188	sulfate	Soluble anhydrite. $CaSO_4$	136.14	col, hex or tricl, 1.505, 1.548	2.61	tr to rhomb >200				
c189	sulfate half-hydrate	Plaster of Paris. $CaSO_4.\frac{1}{2}H_2O$	145.15	wh powd		$-\frac{1}{2}H_2O$, 163		0.3^{20}	sl s	s a, NH_4 salts, $Na_2S_2O_3$, glyc
c190	sulfate dihydrate	Nat. gypsum. $CaSO_4.2H_2O$	172.17	col, monocl, 1.521, 1.523, 1.530	2.32	$-1\frac{1}{2}H_2O$, 128	$-2H_2O$, 163	0.241	0.222^{100}	s a, NH_4 salts, $Na_2S_2O_3$, glyc
c191	sulfide	Nat. oldhamite. CaS	72.14	col, cub, 2.137	2.5	d		0.021^{15} d	0.048^{60} d	d a
c192	sulfide, hydro-	$Ca(HS)_2.6H_2O$	214.32	col pr		d 15–18		v s		s al,
c193	sulfite	$Ca(SO)_3 \cdot \frac{1}{2}H_2O$	129.15	col, hex		$-\frac{1}{2}H_2O$, >250		0.0043^{18}	0.0011^{100}	s H_2SO_4
c194	sulfite, dihydrogen	$Ca(HSO_3)_2$	202.22	yelsh liq, strong SO_2 odor				s		s a
c195	*d*-tartrate	$CaC_4H_4O_6.4H_2O$	260.21	col, rhomb, 1.525, 1.535, 1.550		d		0.0266^{0}	$0.0689^{37.5}$	sl s al
c196	*dl*-tartrate	$CaC_4H_4O_6.4H_2O$	260.21	tricl, powd or need		$-4H_2O$, 200		0.0032^{0}	$0.0078^{37.5}$	s HCl; i ac, a
c197	*meso*tartrate	$CaC_4H_4O_6.3H_2O$	242.20	wh, monocl or tricl pr		$-3H_2O$ <170		i	0.16^{100}	0.28^{18}, 0.85^{100} ac a
c198	telluride	CaTe	167.68	cub, 2.51, 2.58	4.873					
c199	tellurite	$CaTeO_3$	215.68	wh fl		>960		sl s	s	s a
c200	thiocarbonate, tri-	$CaCS_3$	148.28	yel cr				s		i al
c201	thiocyanate	$Ca(SCN)_2.3H_2O$	210.29	wh cr, deliq				v s	v s	v s al
c202	*di*-thionate	$Ca(SO_3)_2.4H_2O$	272.27	col, trig, 1.5496	2.176			16^{0}	30^{80}	
c203	thiosulfate	$CaS_2O_3.6H_2O$	260.30	tricl	1.872	d		100^{3}	d	s al
c204	*meta*titanate	Nat. perovskite. $CaTiO_3$	135.98	col, cub, rhomb, β 2.34	4.10	1975				
c205	tungstate	$CaWO_4$	287.93	wh, tetr, 1.9263, 1.9107	6.062^{20}			0.00064^{15}	0.00012^{100}	
c206	tungstate	Nat. scheelite. $CaWO_4$	287.93	col or w sc, tetr. 1.918, 1.934	6.06			0.2		i al, a; s NH_4Cl
c207	*meta*tungstate	$Ca_3H_4[H_2(W_2O_7)_6].27H_2O$	3500.96	col, tric		$-7H_2O$, 105	$-10H_2O$, d			d a
c208	valerate	$Ca(C_5H_9O_2)_2$	242.33					8.28^{0}	7.39^{100}	
c209	*meta*zirconate	$CaZrO_3$	179.30	col, monocl	4.78	2550				
c210	**Carbon**	Diamond. C	12.01	col, cub, 2.4173	3.51	>3550	4827	i	i	i a, alk
c211	carbon	Graphite. C	12.01	blk, hex	2.25^{20}	subl 3652–97	4827	i	i	s liq Fe; i a, alk
c212	carbon, amorphous	C	12.01	amorph, blk	1.8–2.1	subl 3652–97	4827	i	i	i a, alk
c213	(di)-bromide, hexa-	Hexabromomethane. C_2Br_6	503.48	rhomb pr, 1.740, 1.847, 1.863	3.823	148–149 d	210	i		s CS_2; v sl s al, eth
c214	bromide, tetra-	Tetrabromomethane. CBr_4	331.65	col, monocl or oct	3.42	tr to oct 48.4; m.p. 90.1	189.5	0.024^{30}		s al, eth, chl
c215	(di)-bromide, tetra-	Tetrabromethylene. C_2Br_4	343.66			57.5	227			
c216	(di)-chloride, hexa-	Hexachloro ethane. C_2Cl_6	236.74	col, rhmb, tricl or cub	2.091	subl 187		i		s al, eth, oils
c217	chloride, tetra-	Tetrachloromethane. CCl_4	153.81	col liq, 1.4601.	1.5867^{20}_{20}	−23	76.8		v sl s	s al, bz, chl, eth
c218	(di)-chloride, tetra,-	Tetrachloroethylene. C_2Cl_4	165.83	col liq, eth odor, 1.5055	1.6311^{15}_{4}	−22.4	120.8			s al, eth

No.	Name	Synonyms and Formulae	Mol. wt.	Crystalline form, properties and index of refraction	Density or spec. gravity	Melting point, °C	Boiling point, °C	Solubility, in grams per 100 cc: Cold water	Hot water	Other solvents
	Carbon									
c219	fluoride, tetra-	Tetrafluoromethane. CF_4	87.99	col gas	1.96^{-184}	−184	−128	sl s		
c220	iodide, tetra-	Tetraiodomethane. CI_4	519.63	dk red, cub	4.34^{20}	d 171		i	d	s al, CS_2, eth, MeOH, bz
c221	oxide, di-	CO_2	44.01	col gas or col liq	1.977^{0}g/l, liq 1.101^{-37}, solid 1.56^{-79}	$-56.6^{5.2atm}$	−78.5 subl	171.3^{0} cm³ 0.348^{0}g 0.145^{25}g	90.1^{20} cm³ 0.097^{40} g 0.058^{60}g	31^{15} cm³ al, s acet
c222	oxide, mon-	CO	28.01	col odorl pois gas	1.250^{0}g/l liq 0.793	−199	−191.5	3.5^{0} cm	2.32^{20} cm³	s al, bz, ac a, Cu_2Cl_2
c223	oxide, sub-	C_3O_2	68.03	col gas or liq, 1.4538	liq 1.114^{0}	−111.3	7	d		
c224	oxysulfide	COS	60.07	col gas, pois	gas 1.073 g/l^{0} liq 1.24^{-87}	−138.2	−50.2	54^{20} ml		s al; v s CS_2
c225	selenide, di-	CSe_2	169.93	golden yel liq, 1.845^{20}	2.6626^{25}_{4}	−45.5	125–126	i		d al; s CS_2, tol
c226	selenide, sulfide	CSeS	123.04	yel oily liq	1.9874	−85	84.5	i	i	sl s al; s CS_2
c227	sulfide, di-	CS_2	76.14	col liq, inflamm, 1.62950^{18}	1.261^{22}_{20}	−110.8	46.3	0.22^{22}	0.14^{50}	s al, eth
c228	sulfide, mono-	CS	44.08	red powd	1.66	d 200		i		i al; s eth, CS_2
c229	sulfide, sub-	C_3S_2	100.16	red liq	1.274	−0.5	d 90			
c230	sulfide telluride	CSTe	171.68	yel-red liq	2.9^{-50}	−54	d > −54			s CS_2, bz
c231	sulfochloride	Thiophosgene. $CSCl_2$	114.98	yel-red liq	1.509^{15}		73.5			
c232	**Carbonic acid**	H_2CO_3	62.03	exists in solution only				s		
c233	**Carbonyl bromide**	Carbon oxybromide. $COBr_2$	187.83	col liq			64.5			
c234	**Carbonyl chloride**	Phosgene, carbon oxychloride, $COCl_2$	98.92	col gas, pois	1.392	−104	8.3	d		d al, a; v s bz, tol s ac a
c235	**Carbonyl fluoride, di-**	COF_2	66.01	col gas, hyg	sol 1.388^{-190} liq 1.139^{-114}	−114	−83.1	d		
c236	**Carbonyl selenide**	COSe	106.97	col gas, very pois	liq $1.812^{4.1}$	−124.4	−21.7	d		s $COCl_2$
c237	**Cerium**	Ce	140.12	gray met, cub or hex	hex 6.657 cub 6.757	799	3426	sl d	d	s dil min a, i alk
c238	(III) acetate	$Ce(C_2H_3O_2)_3$	317.26	col		d 308		20^{15}	12^{75}	
c239	(III) acetate hydrate	$Ce(C_2H_3O_2)_3.1\frac{1}{2}H_2O$	344.28	wh-redsh cr powd		$-1\frac{1}{2}$ H_2O, 115	d	26.5^{15}	16.2^{75}	
c240	boride, hexa-	CeB_6	204.98	blue met, cub		2190	d	i	i	i HCl
c241	boride, tetra-	CeB_4	183.36	tetr	5.74					
c242	III bromate	$Ce(BrO_3)_3.9H_2O$	685.98	redsh-wh, hex		49		s		
c243	bromide	$CeBr_3.H_2O$	397.86	col need, deliq		d		v s	v s	v s al
c244	carbide	CeC_2	164.14	red, hex	5.23			d	d	s a
c245	carbonate	$Ce_2(CO_3)_3.5H_2O$	550.37	wh cr				i		s a; sl s $(NH_4)_2CO_3$
c246	carbonate fluoride	Nat. bastnaesite. $CeFCO_3$	219.13	hex, 1.717, 1.817	5					
c247	chloride	$CeCl_3$	246.48	col cr, deliq	3.92^{0}	848	1727	100	d	30 al; s acet
c248	citrate	$Ce(C_6H_5O_7).3\frac{1}{2}H_2O$	392.28	wh powd				i		s dil min a
c249	(III) cyanoplatinite	$Ce_2[Pt(CN)_4]_3.18H_2O$	1502.00	yel-bl lust, monocl	2.657	$-13\frac{1}{2}H_2O$ 100–110	d	s		
c250	(III) fluoride	CeF_3	197.12	wh, hex	6.16	1460	2300	i		
c251	(IV) fluoride	$CeF_4.H_2O$	234.13	col microcr, 1.614	4.77	*ca* 650	d	i		s a
c252	hydride	CeH_3	143.14	dk bl amorph powd		ign		d		
c253	(III) hydroxide	$Ce(OH)_3$	191.14	wh gelat ppt						s a, $(NH_4)_2CO_3$; i alk
c254	(IV) iodate	$Ce(IO_3)_4$	839.73	yel cr				0.015^{20}		
c255	(III) iodate	$Ce(IO_3)_3.2H_2O$	700.86	cr				0.16^{25}		s HNO_3
c256	(III) iodide	$CeI_3.9H_2O$	682.97	wh or redsh-wh cr		752	1397	v s		v s al
c257	(III) molybdate	$Ce_2(MoO_4)_3$	760.05	yel, tetr, 2.019, 2.007	4.83	973				
c258	(III) nitrate	$Ce(NO_3)_3.6H_2O$	434.23	col or redsh cr (trac La, Di), deliq		$-3H_2O$, 150	d 200	v s	v s	50 al; s acet
c259	(IV) nitrate, basic	$Ce(OH)(NO_3)_3.3H_2O$	397.07	long red need				s		
c260	(III) oxalate	$Ce_2(C_2O_4)_3.9H_2O$	706.44	yel-wh cr		d		v sl s		s H_2SO_4, HCl; i $H_2C_2O_4$, alk, eth, al
c261	(IV) oxide	$CeO_2.xH_2O$		yelsh gelat ppt						s a; sl s alk carb, i alk
c262	(III) oxide	Ce_2O_3	328.24	gray-grn, trig	6.86	1692, ign 200		i	i	s H_2SO_4, i HCl

No.	Name	Synonyms and Formulae	Mol. wt.	Crystalline form, properties and index of refraction	Density or spec. gravity	Melting point, °C	Boiling point, °C	Solubility, in grams per 100 cc: Cold water	Hot water	Other solvents
	Cerium									
c263	(IV) oxide(di-)	Ceria. CeO_2	172.12	brn-wh, cub	7.132^{23}	*ca* 2600		i	i	s H_2SO_4, HNO_3; i dil a
c264	oxychloride	CeOCl	191.57	purp leaf				i		s dil a
c265	(III) 2,4-pentanedione	Cerium acetylacetonate. $Ce(C_5H_7O_2)_3.3H_2O$	491.50	lt yel cr ppt		131–132		d		v s al
c266	(III) *meta*phosphate	$Ce(PO_3)_3$	377.04	micr need	3.272					i a
c267	(III) *ortho*phosphate	Nat. monazite. $CePO_4$	235.09	red, monocl or yel, rhomb, 1.795	5.22			i	i	s a; i al
c268	(III) salicylate	$Ce(C_7H_5O_3)_3$	551.47	wh-redsh wh powd				i		i al
c269	(III) selenate	$Ce_2(SeO_4)_3$	709.11	rhomb	4.456			39.55^{0}	2.513^{100}	
c270	silicide	$CeSi_2$	196.29		5.67^{17}			i		
c271	(IV) sulfate	$Ce(SO_4)_2$	332.24	deep yel cr	3.91^{18}	d 195		sl d, forms basic salts		
c272	(III) sulfate	$Ce_2(SO_4)_3$	568.42	col to grn, monocl or rhomb	3.912	d 920^{746}		10.1^{0}	2.25^{100}	
c273	(IV) sulfate, dihydrate	$Ce(SO_4)_2.4H_2O$	404.31	yel, rhomb				v s d		s dil H_2SO_4
c274	(III) sulfate, monohydrate	$Ce_2(SO_4)_3.9H_2O$	730.56	hex need	2.831			11.87^{15}	0.42^{30}	
c275	(III) sulfate, octahydrate	$Ce_2(SO_4)_3.8H_2O$	712.54	pink cr, tricl	2.886^{17}	$-8H_2O$, 630		12^{20}	6^{50}	
c276	(III) sulfate, pentahydrate	$Ce_2(SO_4)_3.5H_2O$	658.50	monocl	3.17			3.90^{60}	0.514^{100}	
c277	(III) sulfide	Ce_2S_3	376.43	red cr, br-dk powd, purp	5.020^{11}	d 2100 (vac)		d		s dil a
c278	(III) tungstate	$Ce_2(WO_4)_3$	1023.78	yel, tetr	6.77^{17}	1089				
	Cerium complexes									
c280	hexaantipyrinecerium perchlorate	$[Ce(C_{11}H_{12}N_2O)_6].(ClO_4)_3$	1567.86	col, hex cr		d 295–300		1.08^{20}		
c281	hexaantipyrinecerium iodide (III)	$[Ce(C_{11}H_{12}H_2O)_6].I_3$	1650.22	large yel cr		268–270		15.10^{20}		
c282	**Cesium**	Cs	132.9054	silv met cr hex	1.8785^{15}	28.40±0.01	678.4	d		s liq NH_3
c283	acetate	$CsC_2H_3O_2$	191.95	deliq		194		$945.1^{-2.5}$	$1345.5^{88.5}$	
c284	aluminum sulfate	$CsAl(SO_4)_2.12H_2O$	568.19	col, cub, 1.4587	1.97	117		0.34^{0}	42.54^{100}	s dil al
c285	amide	$CsNH_2$	148.93	wh need	3.44^{25}_{4}	262 ± 1		d		s liq NH_3
c286	azide	CsN_3	174.93	col need, deliq		310		224.2^{0}		1.037^{16} al; i eth
c287	benzoate	$CsC_7H_5O_2$	254.02					294.5^{0}	398.5^{100}	
c288	borofluoride	$CsBF_4$	219.71	rhomb, 1.350	3.20	550 d		1.6^{17}	*ca* 30^{100}	s dil NH_3
c289	borohydride	$CsBH_4$	147.75	wh, cub, 1.498	2.404			v s		sl s al; i eth, bz
c290	bromate	$CsBrO_3$	260.81	hex, *ca* 2.15	4.109^{16}	*ca* 420 d		3.66^{25}	5.32^{35}	
c291	bromide, mono-	CsBr	212.81	col, cub, 1.6984	4.44, liq 3.04^{700}	636	1300	124.3^{25}	v s	s a
c292	bromide, tri-	$CsBr_3$	372.63	rhomb		180				
c293	*di*bromochloride	$CsBr_2Cl$	328.18	yel-red, rhomb		191	150, $-Br_2$	s		d al, acet
c294	bromochloride iodide	CsIBrCl	375.17	yel-red, rhomb		235	d 290	s		s al
c295	bromoiodide di-	$CsIBr_2$	419.63	rhomb	4.25	248	d 320	4.61^{20}		s al
c296	carbonate	Cs_2CO_3	325.82	col cr, deliq		d 610		260.5^{15}	v s	11^{19} al; s eth
c297	carbonate, hydrogen	$CsHCO_3$	193.92	rhomb		$175-\frac{1}{2}H_2O$		209.3^{15}	v s	s al
c298	chlorate	$CsClO_3$	216.36	sm cr	3.57			$6.28^{19.8}$	76.5^{90}	s al
c299	*per*chlorate	$CsClO_4$	232.36	rhomb, at 219 cub, 1.4752, 1.4788, 1.4804	3.327^{4}	d 250		2.00^{25}	28.57^{99}	0.093^{25} al; 0.7878^{25} al; 0.150^{25} acet
c300	chlorobromide	$CsBrCl_2$	283.72	glossy-yel, rhomb.		205		s		d al, eth
c301	chloride	CsCl	168.36	col, cub, deliq, 1.6418	3.988	645	1290	$162.22^{0.7}$	$259.56^{99.5}$	33.7^{25} MeOH; v s al; i acetone
c302	chloroiodide	$CsICl_2$	330.72	or, trig	3.86	230	d 290	s		s al
c303	chloroaurate	$CsAuCl_4$	471.68	yel, monocl				0.5^{10}	27.5^{100}	s al
c304	chlorobromide, di-	$CsBrCl_2$	283.72	glossy-yel, rhomb		205		s		
c305	chloro*di*bromide	$CsBr_2Cl$	328.18	yel.		191				
c306	chloroiodide, di-	$CsICl_2$	330.72	or, trig	3.86	230	d 290	s		s al
c307	chloroplatinate	Cs_2PtCl_6	673.62	yel, cub	4.197 ± 0.004	d 570		0.024^{0}	0.377^{100}	i al
c309	chlorostannate	Cs_2SnCl_6	597.22	wh, cub	3.33					
c310	chromate	Cs_2CrO_4	381.80	yel pr, rhomb	4.237			71.4^{13}	95.5^{60}	
c311	chromium sulfate	Cesium chromium alum. $Cs[Cr(H_2O)_6](SO_4)_2.6H_2O$	593.21	vlt cr	2.064	116		9.4^{25}		
c312	cyanide	CsCN	158.92	very sm wh cr	2.93			v s	v s	
c313	fluoride	CsF	151.90	cub, deliq, 1.478 ± 0.005^{18}	4.115	682	1251	367^{18}		191^{15} MeOH; i Diox, Pyr
c314	fluoride	$CsF.1\frac{1}{2}H_2O$	178.93			703		366.5^{18}		
c315	fluorogermanate	Cs_2GeF_6	452.39	isotrop cr, reg oct.	4.10			sl s	v s	sl s a
c316	fluosilicate	Cs_2SiF_6	407.89	wh, cub	3.372^{17}			60^{17}	sl s	i al
c317	fluotellurite	$CsTeF_5$	355.50	col need				d	d	s HF soln
c318	formate	$CsCHO_2$	172.92		1.0169^{21}_{4}				$2012^{95.4}$	

No.	Name	Synonyms and Formulae	Mol. wt.	Crystalline form, properties and index of refraction	Density or spec. gravity	Melting point, °C	Boiling point, °C	Solubility, in grams per 100 cc Cold water	Hot water	Other solvents
	Cesium									
c319	formate	$CsCHO_2.H_2O$	195.94			41, $-H_2O$				
c320	gallium selenate	$CsGa(SeO_4)_2.2H_2O$	704.72	col cr				4.14^{25}		
c321	gallium sulfate	$CsGa(SO_4)_2.12H_2O$	610.93	col cub, 1.46495		2.113		1.21^{25}		0.0035^{25} 75 % al
c322	hydride	CsH	133.91	wh cr, cub	3.41	d		d	d	d a; i org solv
c323	hydrofluoride	$CsF.HF$	171.91	need, deliq		160		v s		v s a; i al
c324	hydrogencarbide	$CsHC_2$	157.94	trsp cr		300		d		
c325	hydroxide	$CsOH$	149.91	lt yel, deliq	3.675	272.3		395.5^{15}		s al
c326	iodate	$CsIO_3$	307.81	wh, monocl	4.85			2.6^{24}		
c327	*meta*periodate	$CsIO_4$	323.81	wh rhomb pl	4.259^{15}_{4}			2.15^{15}	s	
c328	iodide	CsI	259.81	rhomb, deliq, 1.7876	4.510^{25}_{4}	626	1280	44^{0}	160^{61}	s al
c329	iodide, penta-	CsI_5	767.43	bl, tricl		73				
c330	iodide, tri-	CsI_3	513.62	blk, rhomb	4.47	207.5		sl s	sl s	s al
c331	iodotetrachloride	$CsICl_4$	401.62	pale or needles	3.374^{-10}	228	d	sl s	sl s	
c332	iron (II) sulfate	$Cs_2SO_4.FeSO_4.6H_2O$	621.87	lt grn, monocl, 1.500, 1.504, 1.509	2.791^{20}_{4}	*ca* 70		101.1^{25} (anhyd)		
c333	iron (III) sulfate	$CsFe(SO_4)_2.12H_2O$	597.06	pa-vlt cr, 1.484	2.061^{20}	*ca* 90		s	s	
c334	magnesium sulfate	$Cs_2SO_4.MgSO_4.6H_2O$	590.34	col, monocl, 1.486, 1.452	2.676^{20}_{4}					
c335	*per*manganate	$CsMnO_4$	251.84		3.597	d 320		0.097^{1}	1.27^{50}	
c336	mercury bromide(ic)	$CsBr.2HgBr_2$	933.63	rhomb				0.807^{17}		sl s al
c337	mercury chloride(ic)	$CsCl.HgCl_2$	439.85	col, cub or rhomb, 1.792				1.44^{17}		i abs al
c338	nitrate	$CsNO_3$	194.91	col, hex or cub, 1.55, 1.56	3.685 liq 2.71^{100}	414	d	9.16^{0}	196.8^{100}	s acet, sl s al
c339	nitrate, hydrogen	$CsNO_3.NHO_3$	257.92	oct		100				
c340	nitrate, dihydrogen	$CsNO_3.2HNO_3$	320.92	col pl		32–36				
c341	nitrate	$CsNO_2$	178.91	yel cr				v s	v s	
c342	oxalate	$Cs_2C_2O_4$	353.82		3.230^{15}			282.9^{25}		
c343	oxide	Cs_2O	281.81	or need	4.25	d 400; m.p. 490 (in N_2)		v s	d	s a
c344	oxide, per	Cs_2O_2	297.81	pa yel need	4.25	400	650. $-O_2$	s	d	s a
c345	oxide, tri-	Cesium oxide, sesqui- Cs_2O_3	313.81	choc br cr, cub	4.25	400		d		s a
c346	phthalate, hydrogen	$CsHC_8H_4O_4$	298.03	rhomb	2.178					
c347	polonium chloride	Cs_2PoCl_6	688.53	cub, 1.86	3.82					
c348	rhodium sulfate	$CsRh(SO_4)_2.12H_2O$	644.12	yel, oct	2.238	110–111		sl s		
c349	rhodium sulfate	$CsRh(SO_4)_2.12H_2O$	644.12	or cr	2.22^{20}_{4}	111		sl s	s	
c350	salicylate	$CsC_7.H_5O_3$	270.02					196.2^{0}	1522^{100}	
c351	selenate	Cs_2SeO_4	408.77	rhomb, deliq, 1.5950, 1.5060, 1.5964	4.4528^{20}_{4}			244.8^{12}		
c352	sulfate	Cs_2SO_4	361.87	col rhomb, or hex, 1.560, 1.564, 1.566	4.243	1010	tr hex 600	167^{0}	220^{100}	i al, acet
c353	sulfate, hydrogen	$CsHSO_4$	229.97	col rhomb pr	3.352^{16}	d		s		
c354	sulfide	$Cs_2S.4H_2O$	369.94	wh cr, deliq				v s	v s	
c355	sulfide, di-	Cs_2S_2	329.94	dk red, amorph		460	>800	hgr		
c356	sulfide, di-	$Cs_2S_2.H_2O$	347.95	tetr				s		
c357	sulfide, hexa-	Cs_2S_6	458.19	br red		186				
c358	sulfide, penta-	Cs_2S_5	426.13		2.806^{15}	210				
c359	sulfide, tetra-	Cs_2S_4	394.06	yel		d 160				
c360	sulfide, tri-	Cs_2S_3	362.00	yel leaf		217	780			
c361	tartrate, hydrogen	$CsHC_4H_4O_6$	281.99	wh, rhomb cr				9.7^{25}	98^{100}	
c362	*l*-tartrate	$Cs_2C_4H_4O_6$	413.88	col, trig	3.03^{14}			v s d	v s	
c363	vanadium sulfate	Cesium vanadium alum· $VCs(SO_4)_2.12H_2O$	592.15	red, cub, 1.4780	2.033^{20}_{4}	82	$-12H_2O$, 230; d 300	0.464^{10}	sl s	
c364	**Chloramine, mono-**	NH_2Cl	51.48	yel liq		−66		s		s al, eth; v sl s CCl_4, bz
c365	**Chloric acid**	$HClO_3.7H_2O$	210.57	known only as col sol	$1.282^{14.2}$	<−20	d 40	v s		
c366	**Chloric acid, per**	$HClO_4$	100.46	col liq unstable	1.764^{22}	−112	39^{56}	∞		
c367	**Chloric acid, per**	Hydronium perchlorate. $HClO_4.H_2O$ or $(H_3O)^+(ClO_4)^-$	118.47	need, fairly stab	1.88, liq 1.776^{50}	50	exp 110	v s	v s	
c368	**per, dihydrate**	$HClO_4.2HC$	136.49	stab liq	1.65	−17.8	200	v s	v s	s al

No.	Name	Synonyms and Formulae	Mol. wt.	Crystalline form, properties and index of refraction	Density or spec. gravity	Melting point, °C	Boiling point, °C	Solubility, in grams per 100 cc		
								Cold water	Hot water	Other solvents
	Chlorine									
c369	**Chlorine**	Cl_2	70.906	grnsh-yel gas, or liq, or rhomb cr; gas 1.000768, liq 1.367	3.214^{0}	−100.98	−34.6	310^{10} cm³ 1.46^{0} g	177^{30} cm³ 0.57^{30} g	s alk
c370	azide	chlor(o)azide ClN_3	77.48	gas, expl				sl s		d alk
c371	fluoride, mono-	ClF	54.45	col gas	1.62^{-100}	−154 ± 5	−100.8	d	d	
c372	fluoride, tri-	ClF_3	92.45	col gas	1.77^{13}	−83	11.3	d	d	
c373	hydrate	$Cl_2.8H_2O$	215.03	lt yel, rhomb	1.23	d 9.6		i		s alk
c374	oxide, di-	ClO_2	67.45	yel red gas, or red cr, expl	3.09^{11} g/1	−59.5	9.9^{731} exp	2000^{4} cm³	d to $HClO_3$, Cl_2, O_2	s alk, H_2SO_4
c375	oxide, hept-	Cl_2O_7	182.90	col oil		−91.5	82	s d		s bz
c376	oxide, mono-	Cl_2O	86.91	yel-red gas, or red-br liq	3.89^{0} g/1	−20	3.8^{766} exp	200 cm³	d to HOCl	s alk, H_2SO_4
c377	oxide, tetr-	ClO_4 or Cl_2O_8	99.45				d	s d		s bz
c378	chloroauric acid	$HAuCl_4.4H_2O$	411.85	brt yel need, deliq		d		s	v s	s al, eth
c379	chloroplatinic acid	$H_2PtCl_6.6H_2O$	517.92	red br pr, deliq	2.431	60		v s	v s	s al, eth
c380	chlorostannic acid	$H_2SnCl_6.6H_2O$	441.52	col leaf	1.93	9		s		
c381	**Chlorosulfonic acid**	$ClSO_3H$	116.52	col fum liq, 1.437^{14}	1.766^{18}	−80	158	d to H_2SO_4+ HCl		d al, a; i CS_2
c382	**Chlorotetroxy fluoride**	ClO_4F	118.45	col gas, v exp		−167.3	−15.9			
c383	**Chloryl (per-)fluoride**	ClO_3F	102.45	gas	1.392^{25}	−146	−46.8			
c384	**Chromium**	Cr	51.996	steel gray, cub v hard	7.20^{28}	1857±20	2672	i	i	s dil H_2SO_4, HCl; i HNO_3, aq reg
c385	(II) acetate	$Cr(C_2H_3O_2)_2$	170.09	red cr				sl s	s	sl s al
c386	(III) acetate	$Cr(C_2H_3O_2)_3.H_2O$	247.15	gray-grn powd or blsh-grn pasty mass				s		i al
c387	arsenide, mon-	CrAs	126.92	gray, hex	6.35^{16}			i	i	i a
c388	boride, mono-	CrB	62.81	silv cr, orthorhomb	6.17	2760(?)		i	i	s fus Na_2O_2
c389	(II) bromide	$CrBr_2$	211.81	wh cr	4.356	842		s	s	s al
c390	(III) bromide	$CrBr_3$	291.72	olv gr, hex	4.250^{25}_{4}	subl		i	s	v s al; d alk
c391	bromide, hexahydrate	$[CrBr_2(H_2O)_4]Br.2H_2O$	399.81	grn cr, deliq				s	s tr to vlt	s al; i eth
c392	bromide, hexahydrate	$[Cr(H_2O)_6]Br_3$	399.81	blsh gray to vlt	5.4^{17}			v s	v s	i al
c393	(tri-)carbide, di-	Cr_3C_2	180.02	gray, rhomb	6.68	1890	3800	i	i	
c394	carbonyl	$Cr(CO)_6$	220.06	col, orthorhomb	1.77	d 110	210 exp	i	i	i al, eth, ac a; sl s CHI_3, CCl_4
c395	(II) chloride	$CrCl_2$	122.90	wh need, deliq	2.878^{25}	824		v s	v s	i al, eth
c396	(III) chloride	$CrCl_3$	158.35	vlt, trig	2.76^{15}	*ca* 1150	subl 1300	i	sl s	i al, acet, MeOH, eth
c397	chloride, hexahydrate	$[Cr(H_2O)_4Cl_2].2H_2O$	266.45	vlt, monocl	1.76	83		58.5^{25}	s	s al; i eth; sl s acet
c398	(II) fluoride	CrF_2	89.99	grn, cr, monocl	4.11	1100	>1300	sl s		i al; s h HCl
c399	(III) fluoride	CrF_3	108.99	grn, rhomb	3.8	>1000	subl 1100−1200	i		i al, NH_3; sl s a; s HF
c400	(II) hydroxide	$Cr(OH)_2$	86.01	yel-br		v		d		s a
c401	iodate, hydrate	$[Cr(H_2O)_6]I_3.3H_2O$	594.85	dk vlt cr, hygro	4.915^{25}_{4}	41 − HI		v s	v s	s al, acet; i CHI
c402	(II) iodide	CrI_2	305.80	graysh powd	5.196	856	subl vac 800	s		
c403	(III) iodide	CrI_3	432.71	shiny blk cr	4.915^{25}_{4}	>600	$-I_2$, vac 350			
c404	(III) nitrate	$Cr(NO_3)_3.7\frac{1}{2}H_2O$	373.13	br, monocl		100	d	s	s	
c405	(III) nitrate	$Cr(NO_3)_3.9H_2O$	400.15	purple, monocl		60	d 100	s	s	s a, alk, al, acet
c406	nitride, mono-	CrN	66.00	cub or amorph	5.9	d 1700		i	i	sl s aq reg
c407	(II) oxalate	$CrC_2O_4.H_2O$	158.03	yel cr powd	2.468			sl s	s	s dil a
c408	(III) oxalate	$Cr_2(C_2O_4)_3.6H_2O$	476.14	red, amorph, hyg		120, $-H_2O$ tr to grn		s	s	v s (red) al, eth; i (grn) al
c409	oxide, di-	CrO_2	84.00	br-blk powd		300, −O		i		s HNO_3
c410	(II) oxide, mon-	CrO	68.00	blk powd				i	i	i dil HNO_3
c411	(III) oxide, sesqui-	Cr_2O_3	151.99	grn, hex, 2.551	5.21	2266 ± 25	4000	i	i	i a, alk, al
c412	(III) oxide, sesqui-	$Cr_2O_3.xH_2O$	varies	vlt, amorph or bl-gray grn gel				i	i	s a, alk; sl s NH_4OH
c413	oxide, tri-	Chromic anhydride, "chromic acid", CrO_3	99.99	red, rhomb, deliq	2.70	196	d	61.7°	67.45^{100}	d al, eth; s H_2SO_4 HNO_3
c414	oxychloride	CrO_2Cl_2	154.90	dk red liq	1.911	−96.5	117	d	d	d al; s eth, ac a
c415	2,4-pentandione	Chromium acetyl-acetonate. $Cr(C_5H_7O_2)_3$	349.33			216	340	i		s org solv; i lgr

No.	Name	Synonyms and Formulae	Mol. wt.	Crystalline form, properties and index of refraction	Density or spec. gravity	Melting point, °C	Boiling point, °C	Solubility, in grams per 100 cc: Cold water	Hot water	Other solvents
	Chromium									
c416	(III) *ortho*phosphate	$CrPO_4.2H_2O$	183.00	vlt cr	$2.42^{32.5}$			sl s		s a, alk; i ac a
c417	(III) *ortho*phosphate	$CrPO_4.6H_2O$	255.06	vlt, tricl, 1.568, 1.591, 1.699	2.121^{14}	100				
c418	*pyro*phosphate	$Cr_4(P_2O_7)_3$	729.81	pa grn, monocl	3.2			i	i	s alk
c419	phosphide, mono-	CrP	82.97	gray-blk cr	5.7^{15}			i		s HNO_3, HF
c420	silicide	Cr_3Si_2	212.17	gray, tetr pr	5.5^{0}			i	i	s HCl, HF; i H_2SO_4, HNO_3
c421	(II) sulfate	$CrSO_4.7H_2O$	274.17	bl cr				12.35^{0}	d	s ls al; s NH_4OH
c422	(III) sulfate	$Cr_2(SO_4)_3$	392.18	vlt or red powd	3.012			i, s*		sl s al; i a
c423	(III) sulfate	$Cr_2(SO_4)_3.15H_2O$	662.41	vlt, amorph sc	1.867^{17}	100	$-10H_2O$, 100	s	d 67	i al
c424	(III) sulfate	$Cr_2(SO_4)_3.18H_2O$	716.45	bl vlt, cub oct, 1.564	1.7^{22}	$-12H_2O$, 100		120^{20}	s	s al
c425	(II) sulfide, mono-	CrS	84.06	blk powd, hex	4.85	1550		i		v s a
c426	(III) sulfide, sesqui-	Cr_2S_3	200.18	brn-blk powd	3.77^{19}	−S, 1350		i, d	i	s HNO_3; d al
c427	(III) sulfite	$Cr_2(SO_3)_3$	344.18	grnsh-wh	2.2	d				
c428	(II) tartrate	$CrC_4H_4O_6$	200.07	bl powd	2.33			i	i	sl s a; i ac a
c429	**Chromium complexes** *hex*ammine chromium-(III) chloride	$[Cr(NH_3)_6]Cl_3.H_2O$	278.55	yel cr	1.585			s		
c430	*hex*aureachromium-(III) fluosilicate	$[Cr(CON_2H_4)_6]_2.[SiF_6]_3.3H_2O$	1304.94	lt grn leaf				0.522^{20}		i al
c431	*hex*aureachromium-(III) *per*rhenate	$[C_2(CON_2H_4)_6].(ReO_4)_3$	1162.92	grn need	2.652^{25}_{4}			1.786		0.667 al
c432	chloropentammine chromium chloride	$[Cr(NH_3)_5Cl]Cl_2$	243.51	red, oct	1.696			0.65^{16}		i HCl
c433	**Cobalt**	Co	58.933	silv gray met, cub	8.9	1495	2870	i	i	s a
c434	(III) acetate	$Co(C_2H_3O_2)_3$	236.07	grn, oct		d 100		hydr readily		s a, glac ac a
c435	(II) acetate	$Co(C_2H_3O_2)_2.4H_2O$	249.08	red-vlt, monocl, deliq, 1.542	1.705^{19}	$-4H_2O$, 140		s	s	s a, al
c436	aluminate	(approx) Thenard's blue. $CoAl_2O_4$	176.89	bl, cub				i	i	
c437	(II) *ortho*arsenate	$Co_3(AsO_4)_2.8H_2O$	598.75	vlt-red, monocl, 1.626, 1.661, 1.669	3.178^{15}	d		i	i	s dil a, NH_4OH
c438	arsenic sulfide	Nat. cobaltite. CoAsS	156.92	gray-redsh	6.2–6.3	d				
c439	arsenide	Co_2As	192.79	cr powd	8.28	950		i	i	i HCl, H_2SO_4; s HNO_3, aq reg
c440	(II) benzoate	$Co(C_7H_5O_2)_2.4H_2O$	373.23	gray red leaf		$-4H_2O$, 115		v s		
c441	boride, mono-	CoB	69.74	pr	7.25^{18}			d	d	s HNO_3, aq reg
c442	(II) bromate	$Co(BrO_3)_2.6H_2O$	422.84	red, oct				45.5^{17}		s NH_4OH
c443	(II) bromide	$CoBr_2$	218.75	grn, hex, deliq	4.909^{25}_{4}	678 (in N_2)		66.7^{59}	68.1^{97}	77.1^{20} al; 58.6^{30} MeOH; s eth, acet
c444	(II) bromide hexahydrate	$CoBr_2.6H_2O$	326.84	red-vlt pr, deliq	2.46	47–48, $-4H_2O$ 100	$-6H_2O$, 130	s red color	153.2^{97}	s blk color, al, a, eth
c445	bromoplatinate	$CoPtBr_6.12H_2O$	949.66	trig	2.762					
c446	carbonate	Nat. spherocobaltite. $CoCO_3$	118.94	red, trig, 1.855, 1.60	4.13	d		i	i	s a; i NH_3
c447	(II) carbonate, basic	$2CoCO_3.Co(OH)_2.H_2O$	534.74	vlt-red pr				i	d	s a, $(NH_4)_2CO_3$
c448	carbonyl tetra-	Dicobalt octacarbonyl. $[Co(CO)_4]_2$ or $Co_2(CO)_8$	341.95	or cr or dk br, microcr	1.73^{18}	51	d 52	i	i	sl s al; s CS_2, eth
c449	carbonyl, tri-	Tetracobalt dodecacarbonyl. $[Co(CO)_3]_4$ or $Co_4(CO)_{12}$	571.86	blk cr				sl s		s bz; d Br
c450	(II) chlorate	$Co(ClO_3)_2.6H_2O$	333.93	red, cub, deliq, 1.55	1.92	50	d 100	558.3^{0}	v s	s al
c451	(II) *per*chlorate	$Co(ClO_4)_2$	257.83	red need 1.510, 1.490	3.327			100^{0}	115^{45}	s al, acet
c452	(II) *per*chlorate	$Co(ClO_4)_2.5H_2O$	347.88	red, hex		143		100.13^{0}	115.10^{65}	v s al, acet i $CHCl_3$
c453	*per*chlorate	$Co(ClO_4)_2.6H_2O$	365.93	red pr		d 1534	d	259^{18}		s al, acet
c454	(II) *per*chlorate	$Co(ClO_4)_2.6H_2O$	365.93	red, oct, deliq, 1.55		d 182		255^{18}		s al, acet
c455	(II) chloride	$CoCl_2$	129.84	bl, hex, hygr	3.356^{25}_{4}	724 (in HCl gas)	1049	45^{7}	105^{96}	54.4 al; 8.6 acet; 38.5 MeOH; sl s eth
c456	(III) chloride	$CoCl_3$	165.29	red cr or yel cr	2.94	subl		s		

* Several chromic salts exist in two forms, a soluble and insoluble modification.

No.	Name	Synonyms and Formulae	Mol. wt.	Crystalline form, properties and index of refraction	Density or spec. gravity	Melting point, °C	Boiling point, °C	Solubility, in grams per 100 cc		
								Cold water	Hot water	Other solvents
	Cobalt									
c457	(II) chloride, dihydrate	$CoCl_2.2H_2O$	165.87	red-vlt, monocl or tricl, 1.625, 1.671, 1.67	2.477^{25}_{25}			s	s	v sl s eth
c458	(II) chloride, hexahydrate	$CoCl_2.6H_2O$	237.93	red, monocl	1.924^{25}_{25}	86	$-6H_2O$, 110	76.7^{0}	190.7^{100}	v s (bl col) al; s acet; 0.29 eth
c459	chloroplatinate	$CoPtCl_6.6H_2O$	574.83	trig	2.699	d				
c460	chlorostannate	$CoSnCl_6.6H_2O$	498.43	rhomb or trig		d 100				
c461	(II) chromate	$CoCrO_4$	174.93	gray blk cr		d		i	d	s a, NH_4OH
c462	(II) citrate	$Co_3(C_6H_5O_7)_2.2H_2O$	591.04	rose-red		$-2H_2O$, 150		0.8		
c463	(II) cyanide dihydrate	$Co(CN)_2.2H_2O$	147.00	buff anhydr bl-vlt powd	anhydr 1.872^{25}	$-2H_2O$, 280	d 300	0.00418^{18}		s KCN, HCl, NH_4OH
c464	(II) cyanide, trihydrate	$Co(CN)_2.3H_2O$	165.01	red-gray powd, amorph		$-3H_2O$, 250		i		s KCN
c465	(II) ferricyanide	$Co_3[Fe(CN)_6]_2$	600.71	red need				i		s NH_4OH; i HCl
c466	(II) ferrocyanide	$Co_2Fe(CN)_6.xH_2O$		gray-grn				i		s KCN; i HCl
c467	(II) fluogallate	$[Co(H_2O)_6][GaF_5.H_2O]$	349.75	pink cr, monocl (?), 1.45	2.35	$-5H_2O$, 110		sl s		d a
c468	(II) fluoride	CoF_2	96.93	pink monocl	4.46^{25}_{4}	*ca* 1200	1400	1.5^{25}	s	sl s a; i al, eth, bz
c469	(III) fluoride	CoF_3	115.93	br, hex	3.88			d to $Co(OH)_3$		i, al, eth, bz
c470	fluoride	$Co_2F_5.7H_2O$	357.96	grn powd	2.314^{25}			d		s H_2SO_4
c471	(II) fluoride, tetrahydrate	$CoF_2.4H_2O$	168.99	α: red, rhomb oct, β: rose cr powd	2.192^{25}_{4}	d 200		s	s	i al
c472	fluosilicate	$CoSiF_6.6H_2O$	309.10	pink trig, 1.382, 1.387	2.113^{19}			$118.1^{21.5}$		
c473	(II) formate	$Co(CHO_2)_2.2H_2O$	185.00	red cr	2.129^{22}	$-2H_2O$, 140	d 175	5.03^{20}		
c474	(II) hydroxide	$Co(OH)_2$	92.95	rose-red, rhomb	3.597^{15}	d		0.00032		s a, NH_4 salts; i alk
c475	(III) hydroxide	$Co_2O_3.3H_2O$	219.91	blk-brn powd	4.46	d	$-H_2O$, 100	0.00032		s a; i al
c476	(II) iodate	$Co(IO_3)_2$	408.74	bl-vlt need	5.008^{18}	d 200		0.45^{18}	1.33^{100}	s HCl, HNO_3, h H_2SO_4
c477	(II) iodate, hexahydrate	$Co(IO_3)_2.6H_2O$	516.83	red, oct	3.689^{21}	d 61	$-4H_2O$, 135	s		
c478[1]	(II) iodide (α) stable	CoI_2	312.74	blk hex, hyg	5.68	515(vac)	570 (vac)	159^{0}	420^{100}	v s al, acet
c478[2]	(II) iodide (β)	CoI_2	312.74	yel need, unstab	5.45^{25}	d 400		s		
c479	(II) iodide, dihydrate	$CoI_2.2H_2O$	348.77	grn, deliq		d 100		376.2^{45}	s	
c480	(II) iodide, hexahydrate	$CoI_2.6H_2O$	420.83	br-red hex, hygr	2.90	d 27, $-6H_2O$, 130		s	s	s al, eth, chl
c481	iodoplatinate	$CoPtI_6.9H_2O$	1177.59	trig	3.618					
c482	linoleate	$Co(C_{18}H_{31}O_2)_2$	617.83	br, amorph				i		s al, eth, acet
c483	(II) nitrate	$Co(NO_3)_2.6H_2O$	291.04	red, monocl, 1.52	1.87^{25}_{4}	55–56	$-3H_2O$, 55	133.8^{0}	217^{80}	$100.0^{12.5}$ al; s acet; sl s NH_3
c484	nitrosylcarbonyl	$Co(NO)(CO)_3$	172.97	cherryred liq		−1.05	48.6; d 55	i		s al, eth, acet, bz
c485	(II) oleate	$Co(C_{18}H_{33}O_2)_2$	621.86	br, amorph						s al, eth, oils, bz
c486	(II) oxalate	CoC_2O_4	146.95	wh or redsh	3.021^{25}	d 250		i		s a, NH_4OH
c487	oxalate, dihydrate	$CoC_2O_4.2H_2O$	182.98	pink cr		$-H_2O$, *ca* 190		v sl s	sl s	v sl s a; s NH_4OH
c488	(II) oxide	CoO	74.93	grn-brn cub	6.45	1795 ± 20		i	i	s a; i al, NH_4OH
c489	(III) oxide	Co_2O_3	165.86	blk-gray, hex, or rhomb	5.18	d 895		i	i	s a; i al
c490	(II, III) oxide	Co_3O_4	240.80	blk, cub	6.07	tr to CoO 900–950		i	i	v sl s a; i aq reg
c491	palmitate	$Co(C_{16}H_{31}O_2)_2$	569.78			70.5				s pyr, hot CS_2, CCl_4; sl s eth; i MeOH, acet
c492	(II) *ortho*phosphate	$Co_3(PO_4)_2$	366.74	redsh cr	2.587^{25}			i	i	s H_3PO_4, NH_4OH
c493	(II) *ortho*phosphate, dihydrate	$Co_3(PO_4)_2.2H_2O$	402.77	pink powd				i		s H_3PO_4
c494	(II) *ortho*phosphate, octahydrate	$Co_3(PO_4)_2.8H_2O$	510.87	redsh powd	2.769^{25}	$-8H_2O$, 200		sl s		s min a, H_3PO_4; i al
c495	phosphide	Co_2P	148.84	gray need	6.4^{15}	1386		i	i	s HNO_3, aq reg
c496	(II) propionate	$Co(C_3H_5O_2).3H_2O$	259.12	dk-red cr		*ca* 250		anh 33.5^{11}		v s al
c497	(II) *per*rhenate	$Co(ReO_4)_2.5H_2O$	649.41	dk pink		d		d		
c498	(II) selenate, heptahydrate	$CoSeO_4.7H_2O$	328.00	monocl	2.135					
c499	selenate, hexahydrate	$CoSeO_4.6H_2O$	309.98	red, monocl, 1.5225	2.25^{17}			s	s	
c500	(II) selenate, pentahydrate	$CoSeO_4.5H_2O$	291.97	ruby red, tricl-	2.512	d		v s		

No.	Name	Synonyms and Formulae	Mol. wt.	Crystalline form, properties and index of refraction	Density or spec. gravity	Melting point, °C	Boiling point, °C	Solubility, in grams per 100 cc: Cold water	Hot water	Other solvents
	Cobalt									
c501	selenide, mono-	CoSe	137.89	yel, hex	7.65	red heat				s HNO_3, aq reg; i alk
c502	(II) *ortho*silicate	Co_2SiO_4	209.95	vlt cr, rhomb	4.63	1345		i	i	s dil HCl
c503	silicide	CoSi	87.03	rhomb		1395				s HCl; i HNO_3, H_2SO_4
c504	silicide, di-	$CoSi_2$	115.11	rhomb	5.3	1277				
c505	(di-)silicide	Co_2Si	145.95	gray cr	7.28^{0}	1327				d a
c506	(II) *ortho*stannate	Co_2SnO_4	300.55	grnsh-bl, cub	6.30^{18}					i H_2SO_4; s h HCl
c507	(II) sulfate	$CoSO_4$	155.00	dk blsh, cub	3.71^{25}_{25}	d 735		36.2^{20}	83^{100}	1.04^{18} MeOH; i NH_3
c508	(II) sulfate, heptahydrate	Nat. bieberite. $CoSO_4.7H_2O$	281.10	red-pink, monocl, 1.477, 1.483, 1.489	1.948^{25}_{25}	96.8	$-7H_2O$, 420	60.4^{3}	67^{70}	2.5^{3} al; 54.5^{18} MeOH
c509	(II) sulfate, hexahydrate	$CoSO_4.6H_2O$	263.09	red, monocl, 1.531 1.549, 1.552	2.019^{15}_{15}	$-2H_2O$, 95				
c510	(II) sulfate, monohydrate	$CoSO_4.H_2O$	173.01	red cr, 1.603, 1.639, 1.683	3.075^{25}	d		s	s	
c511	(III) sulfate	$Co_2(SO_4)_3.18H_2O$	730.33	bl-grn		d 35		s d		s H_2SO_4; i pyr
c512	sulfide, di-	CoS_2	123.06	blk, cub	4.269			i		s HNO_3, aq reg
c513	sulfide, mono-	Nat. sycoporite. CoS	91.00	redsh, silv-wh, oct	5.45^{18}	>1116		0.00038^{18}		sl s a
c514	(III) sulfide, sesqui	Co_2S_3	214.06	blk cr	4.8					d a, aq reg
c515	(tri-) sulfide	Cobalt sulfide, tetra-(ous, ic) Nat. linneite. Co_3S_4	305.06	dk gray, cub	4.86	d 480				
c516	(II) sulfite	$CoSO_3.5H_2O$	229.07	red				i		s H_2SO_3
c517	tartrate	$CoC_4H_4O_6$	207.01	redsh, monocl				sl s		s dil a
c518	thiocyanate	$Co(SCN)_2.3H_2O$	229.14	vlt, rhomb		$-3H_2O$, 105		s		s al, MeOH, eth
c519	*ortho*titanate	Co_2TiO_4	229.76	grnsh-blk, cub	5.07–5.12					s conc HCl; sl s dil HCl
c520	(II) tungstate	$CoWO_4$	306.78	bl-grn, monocl	8.42			i		s h conc a; sl s c dil a
	Cobalt complexes									
c521	hexammine cobalt (II) bromide	$CoBr_2.6NH_3$	320.93	dk pink cr	1.871^{25}_{4}	d 258		d		
c522	diamminecobalt (II) chloride [α]	$CoCl_2.2NH_3$	163.90	rose cr	2.097	273				
c523	diamminecobalt (II) chloride(β)	$CoCl_2.2NH_3$	163.90	bl-vlt	2.073	tr to α, 210 (in NH_3)				
c524	hexamminecobalt (II) chloride	$[Co(NH_3)_6]Cl_2$	232.02	rose red, oct	1.497	d		d		s NH_4OH; i abs al
c525	hexamminecobalt (III) chloride	$Co(NH_3)_6Cl_3$	267.46	wine-red, monocl	1.710^{25}_{4}	$-1NH_3$, 215		5.9^{10}	$12.74^{46.6}$	s conc HCl; i al, NH_4OH
c526	hexamminecobalt (II) iodide	$CoI_2.6NH_3$	414.93	dk pink, cub	2.096^{25}_{4}	141^{100mm}				
c527	hexamminecobalt (III) nitrate	$Co(NH_3)_6.(NO_3)_3$	347.13	yel, tetr	1.804^{25}_{4}			1.7^{25}	v s	v sl s dil a
c528	hexamminecobalt (III) *per*rhenate	$[Co(NH_3)_6](ReO_4)_3.2H_2O$	947.74	or-yel pr	3.329^{25}			0.0469		
c529	hexamminecobalt (II) sulfate	$CoSO_4.6NH_3$	257.18	pink powd	1.654^{25}_{4}	d 116^{760}		d		v s dil NH_3
c530	hexamminecobalt (III) sulfate	$[Co(NH_3)_6]_2(SO_4)_3.5H_2O$	700.50	dk yel, monocl	1.797^{25} anhydr	$-4H_2O$, 100	$-5H_2O$, 150	$1.4^{17.4}$		
c531	ammonium tetranitrodiammine (III) cobaltate	Erdmann's salt. $NH_4[Co(NH_3)_2(NO_2)_4]$	295.12	redsh-pa brn, rhomb, 1.78, 1.78. 1.74	1.876^{25}					
c532	aquapentammine-cobalt (III) chloride (roseo)	$[Co(NH_3)_5.H_2O]Cl_3$	268.45	brick red cr	1.7^{25}	d 100		24.87^{25}		sl s HCl; i al
c533	aquapentammine-cobalt (III)-sulfate (roseo)	$[Co(NH_3)_5H_2O]_2(SO_4)_3.2H_2O$	638.34	red, tetr	1.854^{20}	$-3H_2O$, 99	d 110	$1^{17.2}$	1.72^{27}	s H_2SO_4
c534	*cis*-chloroaquo-tetramminecobalt (III) chloride	$[Co(NH_3)_4(H_2O)Cl]Cl_2$	251.42	vlt, rhomb	1.847	d		1.4^{0}		s a; i al
c535	chloropentammine cobalt-(III) chloride (purpureo)	$[Co(NH_3)_5Cl]Cl_2$	250.45	dk red-vlt, rhomb	1.819^{25}_{25}	d		0.4^{25}	$1.031^{46.6}$	s conc H_2SO_4; i al

No.	Name	Synonyms and Formulae	Mol. wt.	Crystalline form, properties and index of refraction	Density or spec. gravity	Melting point, °C	Boiling point, °C	Solubility, in grams per 100 cc Cold water	Hot water	Other solvents
	Cobalt complexes									
c536	triethylenediam-minecobalt-(III) chloride	$Co[C_2H_4(NH_2)_2]_3Cl_3.3H_2O$	399.64	br pr	1.542^{17}	256; $-3H_2O$, 100		v s		
c537	trinitrotriammine-cobalt	$Co(NH_3)_3(NO_2)_3$	248.04	yel, rhomb pl or leaf	1.992^{25}_{4}	d 158	exp 164	$0.177^{16.5}$	0.28^{25}	
c538	trinitrotetrammine-cobalt-(III) nitrate	$[Co(NH_3)_4(NO_2)_2]NO_2$	265.07	yel, rhomb	1.922^{17}			3^{20}		
c539	potassium tetra-nitrodiammine-cobaltate (III)	$K[Co(NH_3)_2(NO_2)_4]$	316.12	yel, rhomb	2.076^{15}			$1.758^{16.5}$		
	Colombium	see Niobium.								
c540	**Copper**	Cu	63.546	redsh met, cub	8.92	1083.4±0.2	2567	i	i	s HNO_3, h H_2SO_4; v sl s HCl, NH_4OH
c541	acetate, basic	Blue verdigris. $Cu(C_2H_3O_2)_2.CuO.6H_2O$	369.26	grnsh-bl powd				sl s		s dil a, NH_4OH; sl s al
c542	(II) acetate	Neutral verdigris. $Cu(C_2H_3O_2)_2.H_2O$	199.65	dk grn powd, 1.545, 1.550	1.882, anhydr 1.93	115	d 240	7.2	20	7.14 al; s eth
c543	(II) acetate meta-arsenate	Paris green. $Cu(C_2H_3O_2)_2.3Cu(AsO_2)_2$ (approx)	1013.77	em grn powd				i		s a, NH_4OH; i al
c544	(III) acetylide	Cu_2C_2	151.10	red, amorph, expl.		exp		v sl s		s a, KCN
c545	amine azide	$Cu(NH_3)_2(N_3)_2$	181.64	dk grn cr, exp		d 100–105	exp 202	i	d	d a; i MeOH
c546	(II) *di*ammine-chloride, di-	$Cu(NH_3)_2Cl_2$	168.51	grn cr	2.32^{25}_{4}	260–270	d 300	i		s NH_4OH; i abs a
c547	(II) *hex*ammine-chloride, di-	$Cu(NH_3)_6Cl_2$	236.63	bl, cub	1.48^{25}_{4}			v s		
c548	*tet*rammine dithionate	$[Cu(NH_3)_4]S_2O_6$	291.79	vlt-bl cr		d 160		s	d	
c549	(II) *tet*rammine nitrate	$[Cu(NH_3)_4](NO_3)_2$	255.67	dk-bl, oct	1.91^{25}_{4}	d 210 exp		s		
c550	(II) amine nitrate	$[Cu(NH_3)_4](NO_2)_2$	223.61	vlt-bl, tetr		$-2NH_3$ 97		v s		
c551	*tet*rammine sulfate	Cuprum ammoniacale. $[Cu(NH_3)_4]SO_4.H_2O$	245.74	dk-bl, rhomb, unstab	1.79^{25}_{4}	$-NH_3.H_2O$, 30		$18.05^{21.5}$		
c552	(tri-)antimonide	Cu_3Sb	312.37	gray	8.51	687				
c553	(II) *ortho*arsenate	$Cu_3(AsO_4)_2.4H_2O$	540.52	blsh-grn				i	i	s a, NH_4OH
c554	(II) *ortho*arsenate, di-H	$Cu_5H_2(AsO_4)_4.2H_2O$	911.42	bl				i		s a, NH_4OH
c555	arsenide	Cu_5As_2	467.54	bl, oct	7.56			i	i	s a, NH_4OH
c556	tri-arsenide	Nat. domeykite. Cu_3As	265.54	hex	8.0	830				
c557	(II) *ortho*arsenite, hydrogen(?)	Scheele's green. $CuHAsO_3$(?)	187.47	grn powd		d		i	i	s a, NH_4OH; i al
c558	(I) azide	CuN_3	105.56	col cr, v exp	3.26			0.00075^{20}		d conc H_2SO_4; s NH_4Cl
c559	(II) azide	$Cu(N_3)_2$	147.58	brn-red or brn-yel cr, exp	2.604	exp 215		0.008^{20}		v s dil a
c560	(II) benzoate	$Cu(C_7H_5O_2)_2.2H_2O$	341.80	lt bl cr powd		$-H_2O$, 110		sl s		s dil a; sl s al
c561	(II) *meta*borate	$Cu(BO_2)_2$	149.16	blsh grn cr powd	3.859			s		
c562	boride	Cu_3B_2	212.24	yel	8.116					
c563	(II) bromate	$Cu(BrO_3)_2.6H_2O$	427.45	bl-grn, cub	2.583	d 180	$-6H_2O$, 200	v s		s NH_4OH
c564	(I) bromide	CuBr (or Cu_2Br_2)	143.45	wh, cub, 2.116	4.98	492	1345	v sl s	d	s HBr, HCl, HNO_3, NH_4OH; i acet
c565	(II) bromide	$CuBr_2$	223.31	blk, monocl, deliq	4.77^{25}_{4}	498				s al, acet, NH_3, pyr; i bz
c566	*trioxy*bromide	$CuBr_2.3Cu(OH)_2$	516.02	em grn, rhomb	4.00	$-H_2O$, 210–215	d 240–250	i	d	s dil min a, NH_4OH; v s ac a
c567	(II) butyrate	$Cu(C_4H_7O_2)_2.2H_2O$	273.77	dk grn cr				v sl s		s al, eth, NH_4OH, dil a
c568	(I) carbonate	Cu_2CO_3	187.09	yel	4.40	d		i	i	s a, NH_4OH
c569	(II) carbonate, basic	Nat. malachite. $CuCO_3.Cu(OH)_2$	221.11	dk grn, monocl, 1.655, 1.875, 1.909	4.0	d 200		i	d	0.026 aq CO_2; s a, NH_4OH, KCN; i al
c570	(II) carbonate, basic	Nat. azurite, chessylite. $2CuCO_3.Cu(OH)_2$	344.65	bl, monocl, 1.730, 1.758, 1.838	3.88	d 220		i	d	s NH_4OH, h $NaHCO_3$
c571	(II) chlorate	$Cu(ClO_3)_2.6H_2O$	338.53	grn, cub, deliq		65	d 100	207^{0}	v s	s al, acet
c572	(II) chlorate, basic	$Cu(ClO_3)_2.3Cu(OH)_2$	523.11	grn cr or amorph	3.55	d		i	i	s dil a

No.	Name	Synonyms and Formulae	Mol. wt.	Crystalline form, properties and index of refraction	Density or spec. gravity	Melting point, °C	Boiling point, °C	Solubility, in grams per 100 cc: Cold water	Hot water	Other solvents
	Copper									
c573	*per*chlorate	$Cu(ClO_4)_2$	262.43	monocl, 1.495, 1.505, 1.522	2.225^{23}	82.3		s	s	
c574	*per*chlorate, hexahydrate	$Cu(ClO_4)_2.6H_2O$	370.53	lt bl, tricl, deliq, 1.505	2.225^{25}_{4}	82	d 120	v s		s al, eth
c575	(I) chloride(ous)	Nat. nantokite. CuCl (or Cu_2Cl_2)	98.99	wh, cub, 1.93	4.14	430	1490	0.0062		s HCl, NH_4OH, eth; i al
c576	(II) chloride	$CuCl_2$	134.44	br, yel powd, hygr	3.386^{25}_{4}	620	993 d to CuCl	70.6^{0}	107.9^{100}	53^{15} al; 68^{15} MeOH; s h H_2SO_4, acet
c577	(II) chloride, basic	$CuCl_2.Cu(OH)_2$	232.00	yel-grn, hex	3.78	$-H_2O$, 250	d red heat	d	d	
c578	(II) chloride, dihydrate	Nat. eriochalcite. $CuCl_2.2H_2O$	170.47	bl-grn, rhomb, deliq. 1.644, 1.683, 1.731	2.54	$-2H_2O$, 100	d	110.4^{0}	192.4^{100}	s al, NH_4OH
c579	chloride, thioureate	$CuCl.3[CS(NH_2)_2]$	327.35	col pr, 1.758, 1.17719	1.73	168		v s		
c580	(II) chromate, basic	$CuCrO_4.2CuO.2H_2O$	374.64	yel br		$-2H_2O$, 260		i		s dil a, NH_4OH; i al
c581	(II) *di*chromate	$CuCr_2O_7.2H_2O$	315.56	blk cr, deliq	2.283	$-2H_2O$, 100		v s	d	s a, NH_4OH, al
c582	(I) chromite	$Cu_2Cr_2O_4$	295.07	gray blk cub pl.	5.24^{20}			i	i	s HNO_3
c583	(II) citrate	$2Cu_2C_6H_4O_7.5H_2O$	720.43	blsh grn powd		$-H_2O$, 100		i		s a, NH_4OH
c584	(I) cyanide	CuCN	89.56	wh, monocl pr	2.92	473 (in N_2)	d	i	i	s HCl, KCN, NH_4OH; sl s liq NH_3
c585	(II) cyanide	$Cu(CN)_2$	115.58	yel-grn powd		d		i		s a, alk, KCN, pyr,
c586	ethylacetoacetate	$Cu(C_6H_9O_3)_2$	321.81	grn need		192–193	subl	i		v s al, eth; 10^{80} bz
c587	(I) ferricyanide	$Cu_3Fe(CN)_6$	402.57	br red				i		i HCl; s NH_4OH
c588	(II) ferricyanide	$Cu_3[Fe(CN)_6]_2.14H_2O$	866.74	yel-grn				i		i HCl; s NH_4OH
c589	(II) ferrocyanide	Hatchett's brown. $Cu_2Fe(CN)_6.xH_2O$		red brn				i	i	i a, NH_3; s NH_4OH
c590	(I) fluogallate	$[Cu(H_2O)_6][GaF_5.H_2O]$	354.36	pa bl, monocl(?), 1.45	2.20	$-5H_2O$, 110		sl s		s HF
c591	(I) fluoride	CuF (or Cu_2F_2)	82.54	red cr, (exist?)		908	subl, 1100	i		s HCl, HF; d HNO_3; i al
c592	(II) fluoride	CuF_2	101.54	wh, tricl	4.23	d 950		4.7^{20}	s	s dil min a; i al
c593	(II) fluoride dihydrate	$CuF_2.2H_2O$	137.57	bl, monocl	2.93^{25}_{4}	d		4.7^{20}	d	s HCl, HF, HNO_3, al; i acet, NH_3
c594	(I) fluosilicate	Cu_2SiF_6	269.16	red powd		d to SiF_4			d 100	
c595	(II) fluosilicate	$CuSiF_6.4H_2O$	277.68	monocl pr	2.158			42.8		
c596	(II) fluosilicate hexahydrate	$CuSiF_6.6H_2O$	313.71	bl, rhomb, deliq, 1.409, 1.408	2.207			232^{17}		0.16^{20} 92 % al
c597	(II) formate	$Cu(CHO_2)_2$	153.55	bl, monocl	1.831			12.5	d	0.25 al
c598	(II) formate tetrahydrate	$Cu(CHO_2)_2.4H_2O$	225.61	bl cr	1.81	$-H_2O$, 130		6.2		s alk; sl s al
c599	(II) glycerine deriv	$Cu(C_2H_4NO_2)_2.H_2O$	229.67	bl need		$-H_2O$, 130		0.57^{15}		s alk
c600	hydride	CuH (or Cu_2H_2)	64.55	red-brn, (exist?)	6.38	d sl 55–60		i	d 65	d HCl
c601	(II) hydroxide	$Cu(OH)_2$	97.56	bl gel cr powd	3.368	$-H_2O$, d		i	d	s a, NH_4OH, KCN
c602	(II) *tri*hydroxy-chloride	γ: Paratacamite δ: atacamite $CuCl_2.3Cu(OH)_2$	427.11	γ: grn, hex, δ: grn rhomb; γ: 1.743, 1.849, δ:1.861, 1.861, 1.880, grn lt	(γ)3.75	$-H_2O$, 250		i	i	v s a
c603	(II) *tri*hydroxy-nitrate	$Cu(NO_3)_2.3Cu(OH)_2$	480.27	dk grn, rhomb or moncl	rhomb, 3.41 monocl, 3.378	$-H_2O$ ~400		i	d	v s a
c604	(II) iodate	$Cu(IO_3)_2$	413.35	grn, moncl	5.241^{15}	d		0.1364^{15}	i	s dil HNO_3, dil H_2SO_4
c605	(II) iodate, basic	$Cu(OH)IO_3$	255.45	grn, rhomb	4.873	d 290		i	i	s dil H_2SO_4
c606	(II) iodate, monohydrate	Nat. bellingerite. $Cu(IO_3)_2.H_2O$	431.36	bl, tricl	4.872	$-H_2O$, 248	d 290	0.33^{15}	0.65^{100}	s dil H_2SO_4, NH_4OH; i al, dil HNO_3
c607	*para*periodate	Cu_2HIO_6	350.00	grn cr powd		d 110		i	i	s HNO_3, NH_4OH
c608	(I) iodide	Nat. marshite. CuI (or Cu_2I_2)	190.44	wh or brnsh-wh, cub, 2.346	5.62	605	1290	0.0008^{18}		s dil HCl, KI, KCN, conc H_2SO_4, liq NH_3
c609	(II) lactate	$Cu(C_3H_5O_3)_2.2H_2O$	277.71	dk bl, monocl				16.7	45^{100}	s NH_4OH; sl s al
c610	(II) laurate	$Cu(C_{12}H_{23}O_2)_2$	462.17	lt bl powd		111–113		sl s	sl s	
c611	mercury iodide (α)	Cu_2HgI_4	835.29	red, tetr	6.116	tr *ca* 67		i		
c612	mercury iodide (β)	Cu_2HgI_4	835.29	choc, cub	6.102			i		

No.	Name	Synonyms and Formulae	Mol. wt.	Crystalline form, properties and index of refraction	Density or spec. gravity	Melting point, °C	Boiling point, °C	Solubility, in grams per 100 cc: Cold water	Hot water	Other solvents
	Copper									
c613	(II) nitrate, hexahydrate	$Cu(NO_3)_2.6H_2O$	295.64	bl cr, deliq	2.074	$-3H_2O$, 26.4		243.7^{0}	∞	s al
c614	(II) nitrate, trihydrate	$Cu(NO_3)_2.3H_2O$	241.60	bl cr, deliq	2.32^{25}_{4}	114.5	$-HNO_3$, 170	137.8^{0}	1270^{100}	$100^{12.5}$ al; v s liq NH_3
c615	nitride	Cu_3N	204.63	dk grn powd	5.84^{25}_{4}	d 300		d		d a
c616	(II) nitrite, basic	$Cu(NO_2)_2.3Cu(OH)_2$	448.22	grn powd		d 120		i	d	v s dil a; sl s al; s NH_4OH
c617	(II) *hypo*nitrite, basic	$Cu(NO)_2.Cu(OH)_2$	221.11	pea grn amorph, hygr		d<100		i		s dil a; v s NH_4OH; d NaOH
c618	(II) nitroprusside	$CuFe(CN)_5NO.2H_2O$	315.51	wh-grnsh powd				i		s alk; i al
c619	(II) oleate	$Cu(C_{18}H_{33}O_2)_2$	626.47	br powd or grn-bl mass, pois				i	i	s eth
c620	(II) oxalate	$CuC_2O_4.\frac{1}{2}H_2O$	160.57	bl wh				0.00253^{25}		s NH_4OH; i ac a
c621	(I) oxide	Nat. cuprite. Cu_2O	143.08	red, oct cub, 2.705	6.0	1235	−O, 1800	i	i	s HCl, NH_4Cl, NH_4OH; sl s HNO_3; i al
c622	(II) oxide	Nat. tenorite. CuO	79.54	blk, monocl, β 2.63	6.3–6.49	1326		i	i	s a, NH_4Cl, KCN
c623	oxide, per-	$CuO_2.H_2O$	113.55	br or brnsh-blk cr		d 60		i		i al, s d a
c624	oxide, sub-	Cu_4O	270.16	olv grn, (exist?)		d		i		d a
c625	(II) oxychloride	Nat. atacamite. $Cu_2(OH)_3Cl$ (or $CuCl_2.3Cu(OH)_2$)	213.56	grn, orthorhomb.	3.76–3.78					
c626	(II) oxychloride	Brunswick green. $CuCl_2.3CuO.4H_2O$(?)	445.13	grn powd, or em grn to grnsh-blk, rhomb		$-3H_2O$, 140		d 100		s a, NH_4OH
c627	(II) palmitate	$Cu(C_{16}H_{31}O_2)_2$	574.39	grn-bl powd		120		i		s h bz, CS_2, CCl_4; sl s al, eth; i MeOH, acet
c628	2,4-pentanedione	Copper acetylacetonate. $Cu(C_5H_7O_2)_2$	261.76	bl cr		>230	subl	i		sl s al; s chl
c629	(I) phenyl	C_6H_5Cu	140.65	col powd		d 80		d	d	i al, CS_2; s pyr
c630	(II) *ortho*phosphate	$Cu_3(PO_4)_2.3H_2O$	434.61	bl, rhomb		d		i	sl s	s a, NH_4OH, H_3PO_4; i NH_3
c631	(tri-) phosphide	Cu_3P	221.59	gray-blk	6.4–6.8	d		i		s HNO_3; i HCl
c632	(di-) pyridine chloride(di)	$Cu(C_5H_5N)_2Cl_2$	292.65	grn-bluish, monocl, 1.60, 1.75	1.76	d 263		s	d	sl s c al, chl
c633	(II) salicylate	$Cu(C_7H_5O_3)_2.4H_2O$	409.83	bl-grn need				v s		v s al, NH_4OH
c634	(II) selenate	$CuSeO_4.5H_2O$	296.57	bl, tricl, 1.56	2.559	$-4H_2O$, 50–100	$-5H_2O$, 150	25.7^{15}	d	s a, NH_4OH; v sl s acet; i a
c635	(I) selenide	Cu_2Se	206.04	blk, cub	6.749^{30}_{4}	1113				d HCl
c636	(II) selenide	CuSe	142.50	grn-blk hex pl, unstab	5.99	d red heat		i	i	sl s HCl, NH_4OH; s h HNO_3
c637	selenite	$CuSeO_3.2H_2O$	226.53	bl-grn, rhomb	3.31^{25}_{4}	$-H_2O$, 100		i	i	
c638	silicide	Cu_4Si	282.25	wh met	7.53	850				i HCl; d HNO_3
c639	(II) stearate	$Cu(C_{18}H_{35}O_2)_2$	630.50	lt grn-bl amorph powd		125		i		s eth, h bz, chl, turp; sl s pyr; i MeOH, acet
c640	(I) sulfate	Cu_2SO_4	223.14	gray powd, 1.724, 1.733, 1.739	3.605	+O, 200		d		s conc HCl, NH_3 glac ac a
c641	(II) sulfate	Nat. hydrocyanite. $CuSO_4$	159.60	grn, wh, rhomb, 1.733	3.603	sl d above 200	d 650 to CuO	14.3^{0}	75.4^{100}	1.04^{18} MeOH; i al
c642	(II) sulfate, basic	Nat. brochantite. $CuSO_4.3Cu(OH)_2$	452.27	grn, monocl, 1.728, 1.771, 1.800	3.78	d 300		i	i	s a, NH_4OH
c643	(II) sulfate, pentahydrate	Blue vitriol, nat. chalcanthite. $CuSO_4.5H_2O$	249.68	bl, tricl, 1.514, 1.537, 1.543	2.284	$-4H_2O$, 110	$-5H_2O$, 150	31.6^{0}	203.3^{100}	15.6^{18} MeOH; i al
c644	(I) sulfide	Nat. chalcocite. Cu_2S	159.14	blk, rhomb	5.6	1100		$\times 10^{-14}$		s HNO_3, NH_4OH; i acet
c645	(II) sulfide	Nat. covellite. CuS	95.60	blk, monocl or hex, 1.45	4.6	tr 103	d 220	0.000033^{18}		s HNO_3, KCN, h HCl, H_2SO_4; i al, alk
c646	(I) sulfite, monohydrate	$Cu_2SO_3.H_2O$	225.16	red pr	4.46^{15}			i		d dil a; s NH_4OH
c647	(I) sulfite, monohydrate	$Cu_2SO_3.H_2O$	225.16	wh, hex	3.83^{15}	d		sl s		s HCl, NH_4OH; i al, eth
c648	(I, II) sulfite, dihydrate	Chevreul's salt. $Cu_2SO_3.CuSO_3.2H_2O$	386.78	red cr	3.57	d 200		i	i	s HCl, NH_4OH; sl s HNO_3
c649	(II) tartrate	$CuC_4H_4O_6$	211.61	lt bl powd				v sl s		s a, alk
c650	(II) tartrate, trihydrate	$CuC_4H_4O_6.3H_2O$	265.66	lt gray-bl powd		d		0.02^{15}	0.14^{85}	s a, alk

No.	Name	Synonyms and Formulae	Mol. wt.	Crystalline form, properties and index of refraction	Density or spec. gravity	Melting point, °C	Boiling point, °C	Solubility, in grams per 100 cc: Cold water	Hot water	Other solvents
	Copper									
c651	telluride	Cu_2Te	254.68	bl-blk, oct	7.27					s $Br+H_2O$; i HCl, H_2SO_4
c652	telluride	Nat. rickardite. Cu_4Te_3	636.96	purp, tetr	7.54					
c653	tellurite	$CuTeO_3$	239.10	blk glass				i	i	s conc HCl
c654	(I) thiocyanate	CuSCN	121.62	wh	2.843	1084		0.0005^{18}		s NH_4OH; sl s ac a; i al; d min a
c655	(II) thiocyanate	$Cu(SCN)_2$	179.70	blk		d 100		d	d	s a, NH_4OH
c656	(II) tungstate	$CuWO_4.2H_2O$	347.42	lt-grn, oct		red heat		0.1^{15}		s NH_4OH; sl s ac a; i al; d min a
c657	xanthate	Copper ethylxanthogenate. $Cu(C_3H_5OS_2)_2$	305.94	yel ppt		d		i		s NH_4OH; v sl s al; i CS_2
	Copper complexes									
c658	diamminecopper (II) acetate	$Cu(C_2H_3O_2)_2.2NH_3$	215.69	vlt bl cr		d *ca* 175		s d		s ac a, NH_4OH; i al
c659	tetrammine copper (II) sulfate	$[Cu(NH_3)_4]SO_4.H_2O$	245.74	bl, rhomb	1.81	d 150		$18.5^{21.5}$	d	i al
c660	tetrapyridine copper-(II) fluosilicate	$[Cu(C_5H_5N)_4]SiF_6$	522.03	purp-bl, rhomb	2.108					
c661	tetrapyridine copper-(II) *per*rhenate	$[Cu(C_5H_5N)_4](ReO_4)_2$	880.34	bl cr, monocl	2.338			0.5555		
c662	**Cyanic acid isocyanic acid**	HOCN	43.03	liq	1.14^{20}_{4}			s d		s eth, bz, tol
c663	**Cyanoauric acid**	$HAu(CN)_4.3H_2O$	356.09	tab		50	d	s		s al, eth
c664	**Cobalticyanic acid**	$[H_3Co(CN)_6]_2.H_2O$	454.14	col need, deliq		d 100		s		s al, HCl, dil HNO_3, dil H_2SO_4
c665	**Cyanogen**	$(CN)_2$	52.04	col gas, pungent odor v pois	2.335 g/l, liq: $0.9577^{-21.17}$	−27.9	−20.7	450^{20} cm^3		230 cm^3 al, 500 cm^3 eth
c666	**Cyanogen compounds**	See organic tables								
d1	**Deuterium**	Heavy hydrogen. D_2	4.032	col gas	liq $0.169^{-250.9}$	−254.6	−249.7	sl s		
d2	deuterium chloride	DCl	37.47	gas		−114.8	−81.6	24.1 cm^3 11.9^{25} cm^3	8.4^{50} cm^3 7.12^{80} cm^3	
d3	deuterium oxide	Heavy water. D_2O	20.031	col liq or hex cr, 1.33844^{20}	1.105^{20}_{4}	3.82	101.42^{760}			
d4	**Dysprosium**	Dy	162.50	met, hex	8.5500	1412	2562	i	i	s a
d5	acetate	$Dy(C_2H_3O_2)_3.4H_2O$	411.64	yel need		d 120		s		v sl s al
d6	bromate	$Dy(BrO_3)_3.9H_2O$	708.36	yel hex need		78	$-6H_2O$, 110	v s		sl s al
d7	bromide	$DyBr_3$	402.23	col cr		881	1480	s	s	
d8	carbonate	$Dy_2(CO_3)_3.4H_2O$	577.06			$-3H_2O$, 15		i		
d9	chloride	$DyCl_3$	268.85	shining yel pl	3.67^{0}_{4}	718	1500	s	s	
d10	chromate	$Dy_2(CrO_4)_3.10H_2O$	853.13	yel cr		$-3\frac{1}{2}H_2O$, 150	d		1.002^{25}	
d11	fluoride	DyF_3	219.50	col cr		1360	>2200	i	i	i dil a
d12	iodide	DyI_3	543.21	yelsh grn cr		955	1320	s	s	
d13	nitrate	$Dy(NO_3)_3.5H_2O$	438.58	yel cr		88.6		s		
d14	oxalate	$Dy_2(C_2O_4)_3.10H_2O$	769.21	wh pr		$-H_2O$, 40		i	i	s dil a
d15	oxide	Dysprosia. Dy_2O_3	373.00	wh powd	7.81^{27}	2340 ± 10				grn soln a
d16	*ortho*phosphate	$DyPO_4.5H_2O$	347.55	yelsh-wh powd		$-5H_2O$> 200		i		s dil a, ac a
d17	selenate	$Dy_2(SeO_4)_3.8H_2O$	898.00	yel need		$-8H_2O$, 200		v s		i al
d18	sulfate	$Dy_2(SO_4)_3.8H_2O$	757.31	bril yel cr		stab 110	$-8H_2O$, 360	5.072^{20}	3.34^{40}	
e1	**Erbium**	Er	167.28	dk gray powd	9.006	1529	2863	i	i	s a
e2	acetate	$Er(C_2H_3O_2)_3.4H_2O$	416.48	wh cr, tricl	2.114					
e3	bromide	$ErBr_3.9H_2O$	569.15	vlt-rose cr		950	1460	s	s	
e4	chloride	$ErCl_3.6H_2O$	381.73	pink cr, deliq		774	1500	s	s	s al
e5	fluoride	ErF_3	224.28	rose cr		1350	2200	i	i	i dil a
e6	iodide	ErI_3	547.99	vlt-red cr		1020	1280	s	s	
e7	nitrate	$Er(NO_3)_3.5H_2O$	443.37	redsh cr		$-4H_2O$, 130		s		s al, eth, acet
e8	oxalate	$Er_2(C_2O_4)_3.10H_2O$	778.77	redsh micr powd	2.64(?)	d 575		i	i	i dil a
e9	oxide	Erbia. Er_2O_3	382.56	rose red powd, tr to cub at 1300	8.640	infus		0.00049^{24}		sl s min a
e10	sulfate	$Er_2(SO_4)_3$	622.74	wh powd, hygr	3.678	d 630		43^{0}		
e11	sulfate, octahydrate	$Er_2(SO_4)_3.8H_2O$	766.87	rose red, monocl	3.217	$-8H_2O$, 400		16^{20}	6.53^{40}	
e12	**Europium**	Eu	151.96	steel gray met, cub	5.2434	822	1597	i	i	
e13	(II) bromide	$EuBr_2$	311.78			677	1880	s	s	
e14	(III) bromide	$EuBr_3$	391.69			702	d	s	s	
e15	(II) chloride	$EuCl_2$	222.87	wh, amorph		727	>2000	s	s	s a
e16	(III) chloride	$EuCl_3$	258.32	yel need	4.89^{20}	850				
e17	(II) fluoride	EuF_2	189.96	brt yel	6.495	1380	>2400	i	i	
e18	(III) fluoride	EuF_3	208.96	col		1390	2280	i	i	i dil a

No.	Name	Synonyms and Formulae	Mol. wt.	Crystalline form, properties and index of refraction	Density or spec. gravity	Melting point, °C	Boiling point, °C	Solubility, in grams per 100 cc Cold water	Hot water	Other solvents
	Europium									
e19	(II) iodide	EuI_2	405.77	br to olv grn cr	5.50^{25}_{4}	527	1580	s	s	
e20	(III) iodide	EuI_3	532.68			877	d	s	s	
e21	(III) nitrate	$Eu(NO_3)_3.6H_2O$	446.07	col		85 (sealed tube)		v s	v s	
e22	oxide	Eu_2O_3	351.92	pa rose powd	7.42	2050 ± 30				
e23	(II) sulfate	$EuSO_4$	248.02	col, orthorhomb	4.989^{20}			i	i	i dil a
e24	(III) sulfate	$Eu_2(SO_4)_3.8H_2O$	736.23	pa rose cr	4.95 (anh)	$-8H_2O$, 375		2.563^{20}	1.93^{40}	
—	**Ferric or ferrous**	See *Iron*								
f1	**Ferricyanic acid**	$H_3Fe(CN)_6$	214.98	grn-brn need, deliq		d		s	s	s al
f2	**Ferrocyanic acid**	$H_4Fe(CN)_6$	215.99	wh need, bl in moist air		d		s	s	s al; i eth
f3	**Fluoboric acid**	HBF_4	87.81	col liq		d 130		∞	s	∞ al
f4	**Fluorophosphoric acid, di-**	HPO_2F_2	101.98	col fum liq	1.583^{25}_{4}	−75	116	s		
f5	**Fluorophosphoric acid, hexa-**	HPF_6	145.97	col fum liq	*ca* 1.65 (65%)	31 ($6H_2O$)				
f6	**Fluorophosphonic acid, mono-**	H_2PO_3F	99.99	col visc liq	1.818^{25}_{4}			∞		
f7	**Fluorine**	F	18.998403	grn yel gas, pois, 1.000195	1.69^{15}g/l 1.51^{-188}	−219.62	−188.14	HF+O_3	d	
f8	(di-)oxide	F_2O	54.00	col gas or yel brn liq	liq 1.65^{-190}	−223.8	−144.8	sl s d	i	sl s alk, a
f9	oxide, di-	Dioxygen fluoride. F_2O_2	70.00	brn gas, red liq, orange solid	sol 1.912^{-160} liq 1.45^{-57}	−163.5	−57			
f10	**Fluosilicic acid**	$H_2SiF_6.xH_2O$	hydr (not known)	col fum coros liq, 1.3465^{25} (60.97% soln)	1.4634^{25} (60.97% soln)		d	s	s	sl s alk
f11	dihydrate	$H_2SiF_6.2H_2O$	180.12	wh cr, fum, deliq		d		s	s	s alk
f12	**Fluosulfonic acid**	HSO_3F	100.07	col liq	1.743^{15}	−87.3	165.5	s		
g1	**Gadolinium**	Gd	157.25	col or lt yel met, hex	7.9004	1313	3266	i	i	s a
g2	acetate tetrahydrate	$Gd(C_2H_3O_2)_3.4H_2O$	406.45	col, tricl	1.611			11.6^{25}		
g3	acetylacetonate, trihydrate	$Gd[CH(COCH_3)_2]_3.3H_2O$	508.63			143.5–145		i	i	
g4	bromide, hexahydrate	$GdBr_3.6H_2O$	505.07	rhomb pl	2.844^{16}			s	s	s HBr
g5	chloride	$GdCl_3$	263.61	col monocl, pr	4.52^{0}	609		s	s	
g6	chloride, hexahydrate	$GdCl_3.6H_2O$	371.70	wh pr, deliq	2.424^{0}			s	s	
g7	iodide	GdI_3	537.96	citr yel		926	1340	s	s	
g8	fluoride	GdF_3	214.25					i		sl s h HF
g9	dimethylphosphate	$Gd[(CH_3)_2PO_4]_3$	532.37					23.0^{25}	6.7^{100}	
g10	nitrate, hexahydrate	$Gd(NO_3)_3.6H_2O$	451.36	tricl, deliq	2.332	91		v s	v s	s al
g11	nitrate, pentahydrate	$Gd(NO_3)_3.5H_2O$	433.34	pr	2.406^{15}	92		i	i	v sl s conc HNO_3
g12	oxalate	$Gd_2(C_2O_4)_3.10H_2O$	758.71	monocl		$-6H_2O$, 110		i		s HNO_3; v sl s H_2SO_4
g13	oxide	Gadolinia. Gd_2O_3	362.50	wh amorph powd, hygr	7.407^{15}	2330 ± 20		v sl s		s a
g14	selenate	$Gd_2(SeO_4)_3.8H_2O$	887.50	monocl, pearly	3.309	$-8H_2O$, 130		s	s	
g15	sulfate	$Gd_2(SO_4)_3$	602.68	col	$4.139^{14.5}$	d 500		3.98^{0}	$2.26^{34.4}$	
g16	sulfate, octahydrate	$Gd_2(SO_4)_3.8H_2O$	746.81	col, monocl	$3.010^{14.6}$			3.28^{20}	2.54^{40}	
g17	sulfide	Gd_2S_3	410.69	yel mass, hyg	3.8			d		d a
g18	**Gallium**	Ga	69.72	gray-blk ortho-rhomb, tendency to undercool	sol $5.904^{29.6}$ liq $6.095^{29.8}$	29.78	2403	i	i	s a, i alk
g19	acetate, basic	$4Ga(C_2H_3O_2)_3.2Ga_2O_3.5H_2O$	1452.37	wh micro cr		d 160		s d	d	i ac a
g20	acetylacetonate	2,4-Pentanedione deriv. $Ga(C_5H_7O_2)_3$	367.05	monocl or pl, rhomb or pyram, rhomb pyram	1.42, 1.41	194–195	subl 140^{10}	s	s	s acet
g21	arsenide	GaAs	144.64	dk-gray cub cr		1238				
g22	bromide, tri-	$GaBr_3$	309.45	wh cr	3.69^{25}_{4}	121.5 ± 0.6	278.8	s	s	sl s NH_3
g23	bromide, tri-, hexammine	$GaBr_3.6NH_3$	411.63	wh powd				d	d	sl s NH_3
g24	bromide, tri-, monammine	$GaBr_3.NH_3$	326.48	wh powd	3.112^{25}	124		d	d	sl s NH_3
g25	perchlorate	$Ga(ClO_4)_3.6H_2O$	476.16			d 175		v s		v s al
g26	chloride, di-	$GaCl_2$	140.63	wh cr, deliq		164	535	d	d	s bz
g27	chloride, tri-	$GaCl_3$	176.03	wh need, deliq	2.47^{25}_{4} liq 2.36^{80}_{80}	77.9 ± 02	201.3	v s	v s	s bz, CCl_4, CS_2

No.	Name	Synonyms and Formulae	Mol. wt.	Crystalline form, properties and index of refraction	Density or spec. gravity	Melting point, °C	Boiling point, °C	Solubility, in grams per 100 cc		
								Cold water	Hot water	Other solvents
	Gallium									
g28	chloride, tri-, hexammine	$GaCl_3.6NH_3$	278.26					d	d	s NH_3
g29	chloride, tri-, monoammine	$GaCl_3.NH_3$	193.11	wh powd	2.189[25]	124		d	d	s NH_3
g30	ferrocyanide	$Ga_4[Fe(CN)_6]_3$	914.74			d		d	d	i conc HCl
g31	fluoride, tri-	GaF_3	126.72	wh powd	4.47 ± 0.01	subl 800 (in N_2)	*ca* 1000	0.002	i	v sl s dil a; s HF
g32	fluoride, tri-, trihydrate	$GaF_3.3H_2O$	180.76	wh powd		$-H_2O$(vac) 140		i	sl s	sl s dil H_2F_2; v s dil HCl
g33	fluoride, tri-, triammine	$GaF_3.3NH_3$	177.81	wh powd		$-NH_3$, 100		d	d	
g34	hydride	Digallane, galloethane. Ga_2H_6	145.49	col liq		−21.4	139, d>130	d	d	d a, alk
g35	hydroxide	$Ga(OH)_3$	120.74	wh		d 440		i	i	s dil a
g36	hydroxyquinoline deriv.	$Ga(C_9H_6NO)_3$	502.18	grn-yel cr		>150	subl vac	0.0001	0.0012	s a, alk; sl s al
g37	iodide, tri-	GaI_3	450.43	lt yel cr	4.15^{25}_{4}	212 ± 1	subl 345	d		
g38	iodide, tri-, hexammine	$GaI_3.6HN_3$	552.62	wh powd				d	d	
g39	iodide, tri-, monoammine	$GaI_3.NH_3$	467.46	wh powd	3.635^{25}_{4}			d	d	
g40	nitrate	$Ga(NO_3)_3.xH_2O$		wh cr, deliq		d 110	d to Ga_2O_3 200	v s	v s	s abs al; i eth
g41	nitride	GaN	83.73	dk gray powd	6.1	subl 800		i	i	i dil a; sl s h conc H_2SO_4, h conc NaOH
g42	oxide, sesqui-(α)	Ga_2O_3	187.44	wh, hex, rhomb, 1.92, 1.95	6.44	1900;tr to β 600		i	i	s alk; v sl s h a
g43	oxide, sesqui-(β)	Ga_2O_3	187.44	monocl, rhomb	5.88	1795 ± 15		i	i	s alk; v sl s ha a
g44	oxide, sesqui-, monohydrate	$Ga_2O_3.H_2O$	205.45	wh micr cr, orthorhom, 1.84	5.2	$-H_2O$, 400 tr to Ga_2O_3		i	i	sl s a; s alk
g45	oxide, sub-	Ga_2O	155.44	blk brn powd	4.77^{25}_{4}	>660	subl >500	i	i	s a, alk
g46	oxalate	$Ga_2(C_2O_4)_3.4H_2O$	475.56	wh micro cr, hygr		$-4H_2O$, 180	d 200	0.4		
g47	oxychloride	$6GaOCl.14H_2O$	979.25	oct				i		v s KOH; i dil HNO_3; s acet
g48	selenate	$Ga_2(SeO_4)_3.16H_2O$	856.56	col, monocl or tricl cr				57.5[25]	v s	
g49	selenide, mono-	GaSe	148.68	dk red-br greasy leaf	5.03^{25}_{4}	960 ± 10				
g50	selenide, sesqui-	Ga_2Se_3	376.32	rdsh-bl brittle, hard	4.92^{25}_{4}	>1020 ± 10				
g51	selenide, sub-	Ga_2Se	218.40	bl	5.02^{25}_{4}					
g52	sulfate	$Ga_2(SO_4)_3$	427.62	wh powd		diss 690[750]		v s	v s	s al; i eth
g53	sulfate, hydrate	$Ga_2(SO_4)_3.18H_2O$	751.90	oct cr				v s	v s	s 60 % al; i eth
g54	sulfide, mono-	GaS	101.78	yel cr	3.86^{25}_{4}	965 ± 10		i	d	s a, alk
g55	sulfide, sesqui-	Ga_2S_3	235.63	yel cr, or wh amorph	3.65[25]	1255 ± 10		d	d	s a, alk
g56	sulfide, sub-	Ga_2S	171.50	dk gray	4.18^{25}_{4}	d vac 800		d	d	s a, alk
g57	telluride, mono-	GaTe	197.32	blk soft cr	5.44^{25}_{4}	824 ± 2				
g58	telluride, sesqui-	Ga_2Te_3	522.24	blk brittle cr	5.57^{25}_{4}	790 ± 2				
	Germane									
g59	bromo-	GeH_3Br	155.52	col liq	2.34[29.5]	−32	52	d	d	i al; d alk
g60	chloro-	GeH_3Cl	111.07	col liq	1.75[−26]	−52	28.0	d	d	i al; d alk
g61	chloro trifluoro-	GeF_3Cl	165.04	col gas		−66.2	−20.63	d	d	s abs al
g62	dibromo-	GeH_2Br_2	234.42		2.80[0]	−15.0	89.0	d	d	i al; d alk
g63	dichloro-	GeH_2Cl_2	145.51	col liq	1.90[−68]	−68.0	69.5	d	d	i al; d alk
g64	dichlorodifluoro-	$GeCl_2F_2$	181.49	col gas		−51.8	−2.8	d		s abs al
g65	tribromo-	Germanium bromoform $GeHBr_3$	313.33	col liq		−24	d	d	d	d alk
g66	trichloro-	Germanium chloroform. $GeHCl_3$	179.96	col liq	1.93[0]	−71	75.2 d	d	d	d alk
g67	trichlorofluoro-	$GeCl_3F$	197.95	col liq		−49.8	37.5	d		s abs al
g68	**Germanium**	Ge	72.59	gray-wh met-cub	5.35^{20}_{20}	937.4	2830	i	i	s h H_2SO_4, aq reg; i alk

No.	Name	Synonyms and Formulae	Mol. wt.	Crystalline form, properties and index of refraction	Density or spec. gravtiy	Melting point, °C	Boiling point, °C	Solubility, in grams per 100 cc: Cold water	Hot water	Other solvents
	Germanium									
g69	bromide, di-	$GeBr_2$	232.41	col need or pl	...	122	d	d	d	s a, $GeBr_4$, al; i bz
g70	bromide, tetra-	$GeBr_4$	392.23	gray-wh oct, 1.6269	3.132^{29}_{29}	26.1	186.5	d	d	s abs al, eth, bz; i conc H_2SO_4
g71	chloride, di-	$GeCl_2$	143.50	wh powd	...	d to Ge+ $GeCl_4$	...	d	d	s $GeCl_4$; i al, chl
g72	chloride, tetra-	$GeCl_4$	214.41	col liq, 1.464	1.8443^{30}	−49.5	84	d	d	s al, eth; v s dil HCl; i conc HCl conc H_2SO_4
g73	fluoride, di-	GeF_2	110.59	wh cr, hygr	...	d>350	subl	s	v s	...
g74	fluoride, tetra-	GeF_4	148.58	col gas or liq, not liq at atm press	$2.46^{-36.5}$	subl −37		d to GeO+ H_2GeF_6	...	...
g75	fluoride, tetra-	$GeF_4.3H_2O$	202.63	wh cr, deliq	...	d	...	s	s	...
g76	hydride	Digermane. Ge_2H_6	151.23	liq	1.98^{-100}	−109	29; d 215	d	...	s liq NH_3
g77	hydride	Trigermane. Ge_3H_8	225.83	col liq	2.2^{20}	−105.6	110.5; d 195	i	i	s CCl_4
g78	hydride, tetra-	Germane. GeH_4	76.62	col gas	1.523^{-142}	−165	−88.5; d 350	i	i	s liq NH_3, NaOCl; sl s h HCl
g79	imide	$Ge(NH)_2$	102.62	wh amorph powd	...	d 150	...	...	d to NH_3 +GeO_2	...
g80	iodide, di-	GeI_2	326.40	or hex pl	5.37	d	subl vac 240	s	s d	s conc HI, dil a; sl s CCl_4, chl; i CS_2
g81	iodide, tetra-	GeI_4	580.21	red-or, cub	4.322^{26}_{26}	144	d 440	s d	...	d al, acet; s CS_2, CCl_4, bz, MeOH
g82	(tri-) nitride, di-	Ge_3N_2	245.78	blk cr	...	...	subl 650	...	...	...
g83	(tri-) nitride, tetra-	Ge_3N_4	273.80	wh-lt brn powd	5.25^{25}_{4}	d 450	...	i	i	i a, alk
g84	oxide, d-(insoluble)	GeO_2	104.59	tetr	6.239	1086±5	...	i	...	sl s NaOH; i HCl
g85	oxide, di-(soluble)	GeO_2	104.59	col, hex, 1.650	4.228^{25}	1115.0±4	...	0.447^{25}	1.07^{100}	s a, alk; i HCl, HF; one form NaOH, NH_4OH; one form
g86	oxide, mono-	GeO	88.59	blk cr powd, 1.607	...	subl 710	...	i	i	s Cl_2 water, H_2O_2+NH_4OH; i a, alk
g87	oxychloride	$GeOCl_2$	159.50	col liq	...	−56	d>20	d	d	i all solv
g88	selenide	$GeSe_2$	230.51	orange, rhomb(?)	4.56^{25}	707±3	d	i	i	v sl s a; sl s alk
g89	sulfide, di-	GeS_2	136.72	wh powd, or wh, orthorhomb	2.94^{14}	*ca* 800	subl >600	0.45 d	d to GeO_2 +H_2S	s alk, alk sulf; i al, eth, a; 3.112 liq NH_3
g90	sulfide, mono-	GeS	104.65	yel-red amorph, or rhomb bipyram, blk	amorph: 3.31 rhomb:4.01	530	subl 430	0.24	i	s HCl, alk or alk sulf; sl s NH_4OH; 0.0473 liq NH_3
	Glucinum	**See Beryllium**								
g91	**Gold**	Au	196.967	yel duct met, cub, coll blue-viol	19.3 liq 17.0^{1063}	1064.43	2807	i	i	s aq reg, KCN, h H_2SeO_4; ia
g92	(I) bromide	AuBr	276.88	yel-gray mass, or cr powd	7.9	d 115	...	i	i	d a; s NaCN
g93	(III) bromide	$AuBr_3$	436.69	gray powd, or br cr	...	97.5–Br, 160	...	sl s	...	s eth, al
g94	(I) chloride	AuCl	232.42	yel cr	7.4	170 d to $AuCl_3$	d 289.5	v sl s d	d	s HCl, HBr
g95	(III) chloride	$AuCl_3$ or Au_2Cl_6	303.33	claret-red cr pr	3.9	d 254	subl 265	68	v s	s al, eth; sl s NH_3; i CS_2
g96	(I) cyanide	AuCN	222.98	lt yel cr powd	7.12^{25}	d	...	v sl s	v sl s	s KCN, NH_4OH; i eth, alk
g97	(III) cyanide	Cyanoauric acid. $Au(CN)_3.3H_2O$ or $HAu(CN_4).3H_2O$	329.07	col pl, hygr	...	d 50	...	v s	d, v s	s al, eth
g98	(I) iodide	AuI	323.87	grnsh-yel powd	8.25	d 120	...	v sl s	sl s d	s KI
g99	(III) iodide	AuI_3	577.68	dk grn	...	...	...	i	d	s iodides
g100	(III) nitrate, hydrogen	Nitratoauric acid. $AuH(NO_3)_4.3H_2O$ or $H[Au(NO_3)_4].3H_2O$	500.04	yel, tricl, oct	2.84	d 72	...	s, d	...	s HNO_3
g101	(III) oxide	Au_2O_3	441.93	...	...	−O, 160	−3O, 250	i	i	s HCl, conc HNO_3, NaCN
g102	(III) oxide	$Au_2O_3.xH_2O$	...	...	...	−1½H_2O, 250	...	5.7×10^{-11} 25	...	s HCl, NaCN, conc HNO_3
g103	phosphide	Au_2P_3	486.86	gray	6.67	...	d	...	...	i HCl, dil HNO_3
g104	selenide	Au_2Se_3	630.81	...	4.65^{22}	...	...	...	...	...
g105	(I) sulfide	Au_2S	426.00	br blk powd	...	d 240	...	i fresh sol	ppt coll	s aq reg, KCN; i a
g106	(III) sulfide	Au_2S_3	490.13	br blk powd	8.754	d 197	...	i	...	i al, eth, s Na_2S

No.	Name	Synonyms and Formulae	Mol. wt.	Crystalline form, properties and index of refraction	Density or spec. gravity	Melting point, °C	Boiling point, °C	Solubility, in grams per 100 cc		
								Cold water	Hot water	Other solvents
	Gold									
g107	telluride, di-	Nat. krennerite. $AuTe_2$	452.16	1) rhomb, 2) monocl, 3) tricl yel earthy to massive	8.2–9.3	d 472	......	i	i	......
h1	**Hafnium**	Hf	178.49	hex	13.31^{20}	2227±20	4602	i	i	s HF
h2	bromide	$HfBr_4$	498.13	wh	......	subl 420	......	......	......	......
h3	carbide	HfC	190.54	......	12.20	ca 3890	......	i	......	......
h4	chloride	$HfCl_4$	320.30	wh	......	subl 319	......	d	......	s MeOH, acet
h5	fluoride	HfF_4	254.48	monocl, 1.56	......	......	......	......	......	......
h6	iodide	HfI_4	686.11	......	......	......	400 subl vac	......	......	......
h7	nitride	HfN	192.50	yel-brn, cub	......	3305	......	......	......	......
h8	oxide	Hafnia. HfO_2	210.49	wh, cub	9.68^{20}	2758 ± 25	~5400(?)	i	i	......
h9	oxychloride	$HfOCl_2.8H_2O$	409.52	col	......	......	......	s	......	......
h10	**Helium**	He	4.0026	col gas, inert odorless	0.1785^{0} g/l liq $0.147^{-270.8}$	$-272.2^{26 atm}$	-268.934	0.94^{0} cm³ 0.94^{25} cm³	1.05^{50} cm³ 1.21^{75} cm³	i al; absorbed by Pt
h11	**Holmium**	Ho	164.9304	met, hex	8.7947	1474	2695	i	i	......
h12	bromide	$HoBr_3$	404.66	lt yel	......	914	1470	s	s	......
h13	chloride	$HoCl_3$	271.29	lt yel	......	718	1500	s	s	......
h14	iodide	HoI_3	545.64	lt yel	......	989	1300	s	s	......
h15	fluoride	HoF_3	221.93	lt yel	......	1143	>2200	i	i	i dil a
h16	oxalate	$Ho_2(C_2O_4)_3.10H_2O$	774.10	pa tan	......	$-H_2O$, 40	d	i	i	i dil a
h17	oxide	Holmia. Ho_2O_3	377.86	tan	......	......	......	i	i	s a
h18	**Hydrazine**	NH_2NH_2	32.05	col liq or wh cr, 1.470^{22}	liq 1.011^{15}	1.4	113.5	v s	......	s al
h19	azide	$N_2H_4.HN_3$	75.07	wh pr, deliq, 1.53, 1.76	......	75.4	......	v s	v s	1.2^{20} al; 6.1^{20} MeOH; i CS_2 bz
h20	fluogermanate	$2N_2H_4.H_2GeF_6$	252.80	monocl pr, 1.452, 1.460, 1.464	2.406^{25}_{25}	......	......	s	......	......
h21	fluosilicate	$N_2H_4.H_2SiF_6$	176.14	cr	......	d 186	......	v s	......	sl s al
h22	formate	$N_2H_4.2CH_2O_2$	124.10	......	......	128	......	s	......	......
h23	hydrate	$N_2H_4.H_2O$	50.07	col fum liq or cub cr, 1.42842	1.03^{21}	−40	118.5^{740}	∞	∞	s al; i eth, chl
h24	hydrochloride, di-	$N_2H_4.2HCl$	104.97	col vitr, oct	1.42	198, −HCl	d 200	27.2^{20}	v s	sl s al
h25	hydrochloride, mono-	$N_2H_4.HCl$	68.51	wh need	......	89	d 240	v s	......	v sl s al; v s liq NH_3
h26	hydroiodide	Hydrazine monoiodide. $N_2H_4.HI$	159.96	col pr	......	124–126	......	s	......	......
h27	nitrate, di-	$N_2H_4.2HNO_3$	158.07	col cr	......	104 (rapid heat); d 80 (slow heat)	......	v s	d	......
h28	nitrate, mono-	$N_2H_4.HNO_3$	95.06	col dimorph need (α, β)	......	α70.71; β62.09	subl 140	174.9^{10}	2127^{60}	sl s al
h29	oxalate	$2N_2H_4.H_2C_2O_4$	154.14	wh need	......	148	......	200^{25}	......	0.0003^{20} al; i eth
h30	perchlorate	$N_2H_4.HClO_4.\frac{1}{2}H_2O$	141.51	exp	1.939	137	d 145	d	......	s al; i eth, bz, chl CS_2
h31	*hypo*phosphate	$N_2H_4.2H_2PO_3$	194.03	......	......	152	......	......	......	......
h32	*ortho*phosphate	$N_2H_4.H_3PO_4$	130.05	cr, hygr	......	82	......	v s	......	......
h33	*ortho*phosphite	$N_2H_4.2H_3PO_3$	196.06	......	......	82	......	......	......	......
h34	*ortho*phosphite	$N_2H_4.H_3PO_3$	114.05	......	......	36	......	v s	......	......
h35	picrate	$N_2H_4.HC_6H_2N_3O_7.\frac{1}{2}H_2O$	270.16	......	......	201.3	......	s	s	......
h36	selenate	$N_2H_4.H_2SeO_4$	177.02	col cr powd, unstab	......	exp	......	v sl s	v s	......
h37	sulfate	$N_2H_4H_2SO_4$	130.13	col, rhomb	1.37	254	d	3.415^{25}	14.39^{80}	i al
h38	sulfate	$(N_2H_4)_2.H_2SO_4$	162.18	col cr, hygr	......	85	......	202.2^{25}	554.4^{60}	i al
h39	tartrate	$(N_2H_4)C_4H_6O_6$	182.13	col cr; $[\alpha]^{20}_{4}$ +22.5	......	182–183	......	6.0^{0}	......	......
h40	**Hydrazoic acid**	Azoimide. HN_3	43.03	col liq	1.09^{25}_{4}	−80	37	∞	∞	s al, alk, eth
h41	**Hydrogen**	H_2	2.0158	col gas, cub sol	gas 0.0899 g/l liq 0.070	−259.14	−252.87	2.14^{0} cm³ 1.91^{25} cm³	0.85^{80} cm³ 1.80^{50} cm³	6.925^{0} cm³ al
h42	antimonide	Stibine H_3Sb	124.77	col gas, pois	gas 5.30^{0} g/l liq 2.26^{-25}	−88.5	−17	20 cm³	4 cm³	1500 cm³ al; 2500 cm³ CS_2
h43	arsenide	Arsine H_3As	77.95	col gas, pois	3.484 g/l	−113.5	−55, d 230	20 cm³	sl s	sl s al, alk
h44	arsenide (solid)	H_2As_2	151.86	br powd	......	d 200	......	i	d	i al, eth, CS_2, alk; s HNO_3
h45	bismuthide	Bismuthine. H_3Bi	212.00	liq, v unstab	......	......	22	......	......	......
h46	bromide	Hydrobromic acid. HBr	80.92	col gas or pa yel liq, 1.325	gas 3.5^{0} g/l liq 2.77^{-67}	−88.5	−67.0	221^{0}	130^{100}	s al
h47	bromide (const. boiling)	HBr(47%) + H_2O	......	col liq	1.49	−11	126	......	......	......
h48	bromide, dihydrate	$HBr.2H_2O$	116.95	wh cr, col liq	2.11^{-15}	−11	......	s	s	......

No.	Name	Synonyms and Formulae	Mol. wt.	Crystalline form, properties and index of refraction	Density or spec. gravity	Melting point, °C	Boiling point, °C	Solubility, in grams per 100 cc		
								Cold water	Hot water	Other solvents
	Hydrogen									
h49	bromide, monohydrate	$HBr.H_2O$	98.93	col liq	1.78	stab	−3.6 to −15.5 between 1–2.5 atm			
h50	chloride	Hydrochloric acid. HCl	36.46	col gas or col liq, pois	$1.187^{-84.9}$ gas 1.00045 g/l	−114.8	−84.9	82.3^{0}	56.1^{60}	327 cm^3 al; s eth, bz
h51	chloride (const. boiling)	$HCl(20.24\,\%)+H_2O$		col liq	1.097		110			
h52	chloride, dihydrate	$HCl.2H_2O$	72.49	col liq	$1.46^{18.3}$	−17.7	d	d	∞	s al
h53	chloride, monohydrate	$HCl.H_2O$	54.48	col liq	1.48	−15.35		∞	∞	s al
h54	chloride, trihydrate	$HCl.3H_2O$	90.51	col liq		−24.4	d	∞		s al
h55	cyanide	Hydrocyanic acid. HCN	27.03	col liq or gas, pois, liq 1.2675^{10}	gas 0.901 g/l liq 0.699^{22}	−14	26	∞	∞	∞ al; s eth
h56	fluoride	Hydrofluoric acid. HF	20.01	col fum cor liq, or gas; gas 1.90	$0.991^{19.54}$ liq 0.987	−83.1	19.54	∞	v s	
h57	fluoride (const. boiling)	$HF(35.35\,\%)+H_2O$		col liq			120			
h58	iodide	Hydroiodic acid. HI	127.91	col gas, or pa yel liq, $n_D^{16.5}$ 1.466	gas 5.66^{0} g/l liq $2.85^{-4.7}$	−50.8	-35.38^{4} atm	42.5^{0} cm^3	v s	s al
h59	iodide (const. boiling)	$HI(57\,\%)+H_2O$		col or pa yel fum liq	1.70^{15}		127^{774}			
h60	iodide, dihydrate	$HI.2H_2O$	163.94	col liq		−43		∞		
h61	iodide, tetrahydrate	$HI.4H_2O$	199.97	col liq		−36.5		∞		
h62	iodide, trihydrate	$HI.3H_2O$	181.96	col liq		−48		∞		
h63	oxide	Water. H_2O	18.0153	col liq or hex cr, liq 1.333, sol 1.309, 1.313	1.000_4^4	0.000	100.000			∞ al
h64	oxide, per-	H_2O_2	34.01	col liq; 1.414^{22}	1.4422^{25}	−0.41	150.2^{760}	∞		s al, eth; i pet eth
h65	phosphide	Phosphine. H_3P	34.00	col pois inflam gas or col liq, 1.317 liq	gas 1.529 g/l liq 0.746^{-90}	−133.5	−87.4	26^{17} cm^3	i	s al, eth, Cu_2Cl_2
h66	phosphide	H_4P_2	65.98	col liq	1.012	−90	57.5^{735}	i	i	s al, turp
h67	phosphide	$(H_2P_4)_3$	377.73	yel solid	1.83^{19}	ign 160	d	i	i	i al; s P, P_2H_4
h68	sulfide	H_2S	34.08	col gas, infl, liq 1.374	1.539^{0} g/l	−85.5	−60.7	437^{0} cm^3	186^{40} cm^3	9.54^{20} cm^3 al; s CS_2
h69	sulfide, di-	H_2S_2	66.14	yel oil, 1.885	1.334^{20}	−89.6	70.7^{760}	d		s bz, eth, CS_2; i al
h70	sulfide, penta-	H_2S_5	162.34	clear yel oil	1.67^{16}	−50	50^{4}			s bz, eth, CS_2; i al
h71	sulfide, tetra-	H_2S_4	130.27	lt yel liq	1.588^{15}	−85				
h72	sulfide, tri-	H_2S_3	98.21	brt yel liq, 1.705^{15}	1.496^{15}	−52	d 90			s bz, eth, CS_2; i al
h73	telluride	H_2Te	129.62	col gas or yel need	gas 5.81 g/l liq 2.57^{-20}	−49	−2	s unstab		s al, alk
h74	**Hydroxylamine**	NH_2OH	33.03	wh need or col liq, deliq	1.204	33.05	56.5	s	d	s a, al, MeOH; v sl s eth, chl, bz, CS_2
h75	acetate	$NH_2OH.CH_3CO_2$	92.08	col cr		87	subl 90	v s		
h76	bromide	$NH_2OH.HBr$	113.95	wh, monocl	2.35_4^{22}			v s	v s	i eth
h77	fluogermanate	$(NH_2OH)_2.H_2GeF_6.2H_2O$	290.69	monocl pr, 1.418, 1.438, 1.443	2.229^{25}			s		s abs al
h78	fluosilicate	$(NH_2OH)_2H_2SiF_6.2H_2O$	246.18	scales				v s		i al
h79	formate	$NH_2OH.HCO_2$	78.05	col need		76	d 80	v s	d	s h al; i eth
h80	hydrochloride	$NH_2OH.HCl$	69.49	col, monocl	1.67^{17}	151	d	83^{17}	v s	4.43^{20} al; 16.4^{20} MeOH; s gluc; i eth
h81	iodide	$NH_2OH.HI$	160.94	col need, hygr		83–84 exp		v s	d	v s MeOH; sl s eth
h82	nitrate	$NH_2OH.HNO_3$	96.04	wh		48	d <100	v s	d	v s al
h83	*ortho*phosphate	$(NH_2OH)_3.H_3PO_4$	197.09			148 exp		1.9^{20}	16.8^{90}	
h84	sulfate	$(NH_2OH)_2.H_2SO_4$	164.14	col, monocl		d 170	d	32.9^{0}	68.5^{20}	sl s al; s eth
i1	**Indium**	In	114.82	soft silv wh met, tetr	7.30^{20}	156.61	2080	i	i	s a; v sl s NaOH
i2	antimonide	InSb	236.57	cr		535				
i3	arsenide	InAs	189.74	met cr		943				i a
i4	bromide, di-	$InBr_2$	274.64	pa yel solid	4.22^{25}	235	632 subl	d		s a
i5	bromide, mono-	InBr	194.73	red br solid	4.96^{25}	220	662 subl	d		s a
i6	bromide, tri-	$InBr_3$	354.55	wh to yel need, deliq	4.74^{25}	436 ± 2	subl	v s		
i7	*per*chlorate	$In(ClO_4)_3.8H_2O$	557.29	col cr, deliq		*ca* 800	d 200	v s	d	s abs al; sl s eth
i8	chloride, di-	$InCl_2$	185.73	wh rhomb, deliq	3.655^{25}	235	550–570	d	d	
i9	chloride, mono-	InCl	150.27	1) yel or 2) dk red, deliq	4.19^{25} yel 4.18^{25} red	225 ± 1	608	d	d	s a

No.	Name	Synonyms and Formulae	Mol. wt.	Crystalline form, properties and index of refraction	Density or spec. gravity	Melting point, °C	Boiling point, °C	Solubility, in grams per 100 cc		
								Cold water	Hot water	Other solvents
	Indium									
i10	chloride, tri-	$InCl_3$	221.18	wh pl, deliq	3.46_4^{25}	586 subl 300	volat 600	v s	v s	sl s al, eth
i11	cyanide	$In(CN)_3$	192.87	wh ppt				unstab		i dil a; s HCN; v sl s NaOH
i12	fluoride	InF_3	171.82	col	$4.39^{25} \pm 1$	1170 ± 10	>1200	0.040^{25}		
i13	fluoride, trihydrate	$InF_3.3H_2O$	225.86	cr		$-3H_2O$, 100		8.49^{22}		s a; i al, eth
i14	fluoride, nonahydrate	$InF_3.9H_2O$	333.95	wh need		d		sl s	d	s HCl, HNO_3; i al, eth
i15	hydroxide	$In(OH)_3$	165.84	wh ppt		$-H_2O<150$		i		s a; v sl s NaOH; i NH_4OH
i16	iodate	$In(IO_3)_3$	639.53	wh cr			d	0.067^{20}		s dil HNO_3, dil H_2SO_4; s d HCl
i17	iodide, di-	InI_2	368.63		4.71^{25}	212				s a
i18	iodide, mono-	InI	241.72	br red solid	5.31	351	711–715		sl d	i al, eth, chl; s dil a
i19	iodide, tri-	InI_3	495.53	carmine red, or yel cr	4.69	210		v unstable	s	s a, chl, bz yxl
i20	methylate	$In(CH_3)_3$	159.93	col cr	1.568_{19}^{19}			d		d al, MeOH; s liq NH_3, eth; v s acet, bz
i21	nitrate	$In(NO_3)_3.3H_2O$	354.88	pl, deliq		$-2H_2O$, 100	d	v s		s al
i22	nitrate	$In(NO_3)_3.4\frac{1}{2}H_2O$	381.90	need, deliq		$-4\frac{1}{2}H_2O$	d	v s		s al
i23	oxide, mon-	InO	130.81	wh gray				i		s a
i24	oxide, sesqui-	In_2O_3	277.64	red brn, (h) pa yel (c) amorph and trig	7.179	1910 ± 10	volat 850	i		amorph s a; cr i a
i25	oxide, sub-	In_2O	245.64	blk cr	6.99_4^{25}	subl vac 565–700				s HCl
i26	phosphide	InP	145.79	brittle mass met.		1070				v sl s min a
i27	selenate	$In_2(SeO_4)_3.10H_2O$	838.67	cr, deliq				v s		
i28	selenide, sesqui-	In_2Se_3	466.52	blk cr, or soft dk scales	5.67_4^{25}	890 ± 10				s, d conc a
i29	sulfate	$In_2(SO_4)_3$	517.83	wh gray powd, monocl pr, hygr	3.438			s	v s	
i30	sulfate	$In_2(SO_4)_3.9H_2O$	679.96	wh powd, hygr	3.44	d 250		v s		
i31	sulfate, dihydrate	$In_2(SO_4)_3.H_2SO_4.7H_2O$	742.01	rhomb cr		$-7H_2O$, $-H_2SO_4$, *ca* 250				
i32	sulfide, mono-	InS	146.88	dk	5.18^{25}	692 ± 5	subl vac 850			s HCl, HNO_3
i33	sulfide, sesqui-	In_2S_3	325.83	red cr or yel ppt	4.90	1050	subl *ca* 850 in high vac	i		s a; sl s Na_2S
i34	sulfide, sub-	In_2S	261.70	yel or blk need	5.87^{25}	653 ± 5				
i35	sulfite, basic	$2In_2O_3.3SO_2.8H_2O$	891.59	wh cr		$-3H_2O$, 100	$-8H_2O$, 260	i		
i36	telluride, sesqui-	InTe	242.42	dk met, shiny	6.29_4^{25}	696 ± 2				i HCl, s HNO_3
i37	telluride	In_2Te_3	612.44	bl brittle cr	5.78	667				
i38	**Iodic acid**	HIO_3	175.91	col or pa yel cr powd, rhomb	4.629^0	d 110		286^0 310^{16}	473^{80} 576^{101}	v s 87 % al; sl s HNO_3; i abs al, eth, chl
i39	metaper-	HIO_4	191.91	col		subl 110	d 138	v s, d		
i40	*ortho*paraper-	H_5IO_6 or $HIO_4.2H_2O$	227.94	wh monocl, deliq			d 140	113	v s	s al, eth
i41	**Iodine**	I_2	253.809	vlt blk met lust, rhomb, 3.34	4.93	113.5	184.35^{atm}	0.029^{20} 0.030^{25}	0.078^{50}	20.5^{15} al; 16.46^{25} bz 20.6^{17} eth; s chl glyc, KI; 24^{25} eth 23^{25} MeOH 20.15^{25} CS_2; 2.91^{25} CCl_4
i42	azide	Iod(o)azide. IN_3	168.92	yel, exp				s d		s $Na_2S_2O_3$
i43	bromide, mono-	IBr	206.81	dk gray cr	4.4157^0	(42) subl 50	d 116	s d		s al, eth, chl, CS_2
i44	bromide, tri-	IBr_3	366.63	br liq				s		s al
i45	chloride, mono-(α)	ICl	162.36	dk-red need, cub, red br oily liq	3.1822^0	27.2	97.4	d to HIO_3 +Cl		s al, eth, CS_2, HCl
i46	chloride, mono-(β)	ICl	162.36	brn red, rhombic 6 sided pl	liq 3.24^{34}	13.92	97.4; d 100	d		s al, eth, HCl
i47	chloride, tri-	ICl_3	233.26	yel brn, rhomb, red liq	3.117^{15}	101^{16atm}	d 77	s d		s al, eth, CCl_4, ac a, bz
i48	cyanide	ICN	152.92	wh cr				sl s	sl s	s al, eth
i49	fluoride, hepta-	IF_7	259.89	col cr or liq	liq 2.8^6	5.5	4.5 subl	v s, d	d	d a, alk
i50	fluoride, penta-	IF_5	221.90	col liq	3.75	9.6	98	d	d	d a, ak
i51	oxide, di- (or tetra-)	IO_2 or I_2O_4	158.90	lem-yel cr	4.2_{10}^{10}	d slow 75 rap 130		d to HIO_3+I_2		s H_2SO_4; sl s acet; i al, eth

No.	Name	Synonyms and Formulae	Mol. wt.	Crystalline form, properties and index of refraction	Density or spec. gravity	Melting point, °C	Boiling point, °C	Solubility, in grams per 100 cc Cold water	Hot water	Other solvents
	Iodine									
i52	oxide, pent-	I_2O_5	333.81	wh trim	4.799^{25}_{4}	d 300–350		187.4^{13}	v s	i abs al, eth, chl, CS_2; sl s dil a
i53	(tetra-) oxide non-	I_4O_9	651.61	yel powd, hygr			d 75			
i54	**Iodo platinic acid**	$H_2PtI_6.9H_2O$	1120.67	blk, monocl, deliq		>100		v s, d	d	
i55	**Iodous acid, hypo-**	Iodine hydroxide. HOI	143.91	only in sol, yel to grayish				d	d	
i56	**Iridium**	Ir	192.20	silv wh met, cub	22.421	2410	4130	i	i	sl s aq reg; i a, alk
i57	bromide, tetra-	$IrBr_4$	511.84	blk, deliq		d		s d		s al
i58	bromide, tri-	$IrBr_3.4H_2O$	503.99	olv-grn cr		$-3H_2O$, 100		v s		i al
i59	carbonyl	$Ir_2(CO)_8$	608.48	yel cr		subl 160 (in CO_2)				s CCl_4
i60	carbonyl	$Ir_4(CO)_{12}$	1104.93	yel cr		subl 250 (in CO_2)				i CCl_4
i61	carbonyl chloride	$Ir(CO)_2Cl_2$	319.13	col need		d 140		d		d HCl, KOH
i62	chloride, di-	$IrCl_2$	263.11	blk gray cr (exist ?)		d 773		s		i a, alk
i63	chloride, tetra-	$IrCl_4$	334.01	dk-brn, amorph, hygr		d		s	d	s al, dil HCl
i64	chloride, tri-	$IrCl_3$	298.56	olv-grn hex, or trig	5.30	d 763		i	i	i a, alk
i65	fluoride, hexa-	IrF_6	306.19	yel glass or tetr	6.0	44.4^{20}	53	d	d	
i66	iodide, tetra-	IrI_4	699.82	blk		d 100		i	i	i al, s KI
i67	iodide, tri-	IrI_3	572.91	grn		d		sl s	s	sl s al
i68	oxalic acid	$H_3[Ir(C_2O_4)_3.xH_2O$		pa yel cr				v s	v s	sl s al; i eth
i69	oxide, di-	IrO_2	224.20	blk tetr or bl cr	11.665	d 1100		0.0002^{20}		i a, alk
i70	oxide, di- hydrate	$IrO_2.2H_2O$ or $Ir(OH)_4$	260.23	indigo bl cr		$-2H_2O$, 350		i	i	s HCl
i71	oxide, sesqui-	Ir_2O_3	432.40	bl-blk (exist ?)		−O, 400		i		s H_2SO_4, h HCl; i alk
i72	oxide, sesqui-	$Ir_2O_3.xH_2O$		olive green		d		i		s a, alk
i73	phosphor chloride	IrP_3Cl_{12} or $IrCl_3.3PCl_3$	710.56	yel pr		d 250		sl s	d 100	sl s, d al
i74	selenide	$IrSe_2$	350.12	dk gray cr powd		d 600–700 (in CO_2)				sl s aq reg; i a
i75	sulfate	$Ir_2(SO_4)_3.xH_2O$		yel pr		d		s		
i76	sulfide, di	IrS_2	256.33	br-blk	8.43^{25}_{4}	d 300		i		s aq reg; i a
i77	sulfide, hydro-	$Ir(HS)_3.2H_2O$	327.45	choc br		d		i		s HNO_3
i78	sulfide, mono-	IrS	224.26	bl-blk			d	i		s K_2S, i a
i79	sulfide, sesqui-	Ir_2S_3	480.59	br-blk	9.64^{25}_{4}	d		sl s		s K_2S, HNO_3
i80	telluride	$IrTe_3$	575.00	dk gray cr	9.5^{25}_{4}			i	i	i a; s h aq reg
	Iridium complexes									
i81	aquopentammine iridium-chloride	$[Ir(NH_3)_5H_5(H_2O)]Cl_3$	402.54	wh micro cr	2.474^{15}_{4}	$-H_2O$, 100		ca 75^{26}	d	i al, eth
i82	chloropentammine iridiumchloride	$[Ir(NH_3)_5Cl]Cl_2$	383.71	pa yel, rhomb	$2.681^{15.5}_{4}$	d		sl s	s	i HCl
i83	hexammine iridium chloride	$[Ir(NH_3)_6]Cl_3$	400.74	col, rhomb,	$2.434^{15.5}_{4}$			20		
i84	hexammine iridium nitrate	$[Ir(NH_3)_6](NO_3)_3$	408.40	col, tetr micro cr	2.395^{15}_{4}			1.7^{14}		
i85	nitratopentammine iridium nitrate	$[Ir(NH_3)_5(NO_3)](NO_3)_2$	463.37	wh micro cr	2.515^{18}_{4}	heat exp		0.28	2.5^{100}	
i86	**Iron**	Fe	55.847	silv met, cub	7.86	1535	2750	i	i	s a; i alk, al, eth
i87	(II) acetate	$Fe(C_2H_3O_2)_2.4H_2O$	246.00	lt grn need, monocl		d		v s		
i88	(III) acetate, basic	$FeOH(C_2H_3O_2)_2$	190.94	br-red powd				i		s a, al
i89	(III) acetylacetonate	$Fe(C_5O_2H_7)_3$	353.18	rubyred, rhomb	1.33	184		sl s	sl s	s al, acet, bz, chl
i90	(II) *ortho*arsenate	$Fe_3(AsO_4)_2.6H_2O$	553.47	grn amorph powd		d		i	i	s dil HCl; sl s NH_4OH
i91	(III) *ortho*arsenate	Nat. scorodite. $FeAsO_4.2H_2O$	230.80	grn, rhomb, 1.765, 1.774, 1.797	3.18	d		i	i	s HCl; i HNO_3
i92	(III) *ortho*arsenite, basic	$2FeAsO_3.Fe_2O_3.5H_2O$	607.30	br-yel powd		d		sl s		s a, alk
i93	(II) *pyro*arsenite	$Fe_2As_2O_5$	341.53	grn-wh				i		s NH_4OH
i94	arsenide	FeAs	130.77	wh	7.83	1020		v sl s		
i95	arsenide, di-	Arsenoferrite. $FeAs_2$	205.69	silv gray, cub	7.4	990		i		sl s HNO_3; i HCl
i96	(III) benzoate	$Fe(C_7H_5O_2)_3$	419.20	br powd				i		s h al, h eth
i97	boride	FeB	66.66	gray cr	7.15^{18}			i		s HNO_3, h conc H_2SO_4

No.	Name	Synonyms and Formulae	Mol. wt.	Crystalline form, properties and index of refraction	Density or spec. gravity	Melting point, °C	Boiling point, °C	Solubility, in grams per 100 cc		
								Cold water	Hot water	Other solvents
	Iron									
i98	(II) bromide	$FeBr_2$	215.67	grn-yel, hex	4.636^{25}	d 684 (?)		109^{10}	170^{95}	s al; sl s bz
i99	(III) bromide	$FeBr_3$ or Fe_2Br_6	295.57	dk red-brn, rhomb (?) deliq		subl d		s	s	s al, eth, sl s NH_3
i100	(III) bromide-, hexahydrate	$FeBr_3.6H_2O$	403.68	dk grn		27		v s	v s	s al, eth
i101	(III) cacodylate	$Fe[(CH_3)_2AsO_2]_3$	466.82	yelsh amorph powd				6.67		sl s al
i102	carbide	Fe_3C	179.55	gray, cub	7.694	1837		i	i	s a
i103	(II) carbonate	Nat. siderite. $FeCO_3$	115.85	gray, trig, 1.875, 1.633	3.8	d		0.0067^{25}		s CO_2 sol
i104	(II) carbonate, hydrate	$FeCO_3.H_2O$	133.86	amorph		d		sl s		s a, CO_2 sol
i105	carbonyl, ennea-	$Fe_2(CO)_9$	363.79	yel met cr, hex	2.085^{18}	d 80		i	i	v sl s al, MeOH; d HNO_3; i a
i106	carbonyl, penta-	$Fe(CO)_5$	195.90	visc yel liq	liq 1.457^{21}	−21	102.8^{749}	i		s al, eth, bz, alk, conc H_2SO_4
i107	carbonyl, tetra-	$Fe(CO)_4$	167.89	dk grn lust cr, tetr	1.996^{18}	d 140–150		i		s org solv, conc HNO_3 h H_2SO_4
i108	(II) *per*chlorate	$Fe(ClO_4)_2$	254.75	wh or grnsh-wh need, hygr		d>100		v s		
i109	(II) *per*chlorate hexahydrate	$Fe(ClO_4)_2.6H_2O$	362.84	grn, 1.493, 1.478		d>100		97.8^{0}	116.1^{60}	86.5^{20} al; s $HClO_4$
i110	*oxy*chloride	FeOCl	107.30	brn, rhomb	3.1	d 200				
i111	(II) chloride	Nat. lawrencite. $FeCl_2$	126.75	grn to yel, hex deliq, 1.567	3.16^{25}_{4}	670–674	subl	64.4^{10}	105.7^{100}	100 al; s acet; i eth
i112	(II) chloride, dihydrate	$FeCl_2.2H_2O$	162.78	grn, monocl	2.358					
i113	(II) chloride, tetrahydrate	$FeCl_2.4H_2O$	198.81	bl-grn, monocl, deliq	1.93			160.1^{10}	415.5^{100}	s al; sl s acet
i114	(III) chloride	Nat. molysite. $FeCl_3$ or Fe_2Cl_6	162.21	blk-brn hex	2.898^{25}_{4}	306	d 315	74.4^{0}	535.7^{100}	v s al, MeOH, eth; 63^{18} acet
i115	(III) chloride, hydrate	$FeCl_3.2\frac{1}{2}H_2O$	207.24	dk yel-red, rhomb, deliq		56		v s	v s	v s al, eth
i116	(III) chloride, hexahydrate	$FeCl_3.6H_2O$	270.30	br yel cr mass, v deliq		37	280–285	91.9^{20}	∞	s al, eth
i117	(II) chloride, hexammine	$FeCl_2.6NH_3$	228.94	wh powd	1.928^{25}_{4}					
i118	(II) chloroplatinate	$FePtCl_6.6H_2O$	571.75	yel, hex	2.714	d		v s	v s	
i119	(III) *di*chromate	$Fe_2(Cr_2O_7)_3$	759.66	red-brn gran				s		s a
i120	(II) chromite	$FeCr_2O_4$	223.84	brn-blk, cub	4.97^{20}			i	i	sl s a
i121	(II) citrate	$FeC_6H_6O_7.H_2O$	263.97	wh micr, rhomb		d 350 (in H_2)		sl s		s NH_4OH
i122	(III) citrate	$FeC_6H_5O_7.5H_2O$	335.03	red-brn scales				sl s	s	i al
i123	hydroxide	Nat. goethite. FeO(OH)	88.85	brn, blksh, rhomb, 2.260, 2.394, 2.400	4.28	$-\frac{1}{2}H_2O$, 136				s HCl
i124	(II) ferricyanide	$Fe_3[Fe(CN)_6]_2$(?)	591.45	deep bl		d		i		i al, dil a
i125	(III) ferricyanide	Berlin green. $Fe[Fe(CN)_6]$	267.80	cub						
i126	(II) ferrocyanide	$Fe_2[Fe(CN)_6]$	323.65	bl-wh, amorph	1.601^{25}_{4}	d 100	d 430 (vac)	i		
i127	(III) ferrocyanide	$Fe_4[Fe(CN)_6]_3$	859.25	dk bl cr		d		i	i	s HCl, H_2SO_4; i al, eth
i128	(II) fluoride	FeF_2	93.84	wh, rhomb	4.09^{25}_{4}	>1000 (?)		sl s		s a; i al, eth
i129	(II) fluoride, octahydrate	$FeF_2.8H_2O$	237.97	grn-bl	4.20 (anh)	$-8H_2O$, 100		sl s	s	s HF, a; i al, eth
i130	(II) fluoride, tetrahydrate	$FeF_2.4H_2O$	165.91	wh, rhomb	2.095	d		v sl s		s a; sl s al, eth
i131	(III) fluoride	FeF_3 or Fe_2F_6	112.84	grn, rhomb	3.52	>1000		sl s	s	s a; i al, eth
i132	(III) fluoride, tetrahydrate	$FeF_3.4\frac{1}{2}H_2O$	193.91	yel cr		$-3H_2O$, 100	d	sl s	s	i al
i133	(II) fluosilicate	$FeSiF_6.6H_2O$	306.01	col, trig, 1.361, 1.385	1.961			128.2		
i134	(III) fluosilicate	$Fe_2(SiF_6)_3$ (exist ?)	537.92	flesh col, gel				s	s, d	
i135	(III) formate	$Fe(CHO_2)_3.H_2O$	208.92	red cr or powd				s		v sl s al
i136	(III) glycerophosphate	$Fe_2[C_3H_5(OH)_2.PO_4]_3$	621.39	yelsh-grn sc or powd				50^{25}		i al
i137	(II) hydrogen cyanide	$H_4[Fe(CN)_6]$	215.99	wh, rhomb	1.536^{25}_{4}	d 190		s	s	v s al; s a; i acet
i138	(III) hydrogen cyanide	$H_3[Fe(CN)_6]$	214.98	brn-yel need		d 50–60		s	s	v s al; i eth

No.	Name	Synonyms and Formulae	Mol. wt.	Crystalline form, properties and index of refraction	Density or spec. gravity	Melting point, °C	Boiling point, °C	Solubility, in grams per 100 cc		
								Cold water	Hot water	Other solvents
	Iron									
i139	(II) hydroxide	$Fe(OH)_2$	89.86	pa grn hex, or wh amorph	3.4	d		0.00015^{18}		s a, NH_4Cl; i alk
i140	(III) hydrosulfate	Iron(ic) tetrasulfate. Nat. rhomboklas. $Fe_2O_3.4SO_3.9H_2O$	642.08	wh to pink powd, 1.533, 1.550, 1.635	2.172	$-6H_2O$, 80		s		sl s abs al
i141	(III) iodate	$Fe(IO_3)_3$	580.55	grn yel powd	4.80_4^{20}	d 130		sl s		i dil HNO_3
i142	(II) iodide	FeI_2	309.66	gray hex, hygr	5.315	red heat		s	s	s al, acet
i143	(II) iodide, tetrahydrate	$FeI_2.4H_2O$	381.72	gray blk cr, deliq	2.873	d 90–98		v s	d	s al, eth
i144	(II) lactate	$Fe(C_3H_5O_3)_2.3H_2O$	287.96	grn-wh cr or powd		d		2.1^{10}	8.5^{100}	s alk citrate; v sl s al; i eth
i145	(III) lactate	$Fe(C_3H_5O_3)_3$	323.06	br, amorph, deliq				s	v s	i eth
i146	(III) malate	$Fe_2(C_4H_4O_5)_3$	507.91	br scales, hygr				s		s al
i147	methanoarsenate	$Fe(CH_3AsO_3)_3$	525.60	redsh br lustr scales				50		i al, eth
i148	(II) nitrate	$Fe(NO_3)_2.6H_2O$	287.95	grn, rhomb		60.5		83.5^{20}	166.7^{61}	
i149	(III) nitrate	$Fe(NO_3)_3.6H_2O$	349.95	cub		35	d	150^{0}	∞	
i150	(III) nitrate	$Fe(NO_3)_3.9H_2O$	404.02	col-pa vlt, monocl, deliq	1.684	47.2	d 125	s	s	s al, acet; sl s HNO_3
i151	nitride	Fe_4N	237.39		6.57(?)					
i152	nitride	Fe_2N or Fe_4N_2	125.70	gray	6.35	d 200		i		s HCl, H_2SO_4
i153	nitrosyl carbonyl	$Fe(NO)_2(CO)_2$	171.88	dk red cr	1.56	18.5	d 50, 110	i		s org solv
i154	(III) oleate	$Fe(C_{18}H_{33}O_2)_3$	900.23	br-red fatty lumps				i		s a, al, eth
i155	(II) oxalate	$FeC_2O_4.2H_2O$	179.90	pa yel, rhomb	2.28	d 190		0.022	0.026	s a
i156	(III) oxalate	$Fe_2(C_2O_4)_3.5H_2O$	465.83	yel micro cr powd		d 100		v s	v s	s a; i al
i157	(II) oxide	Nat. wuestite. FeO	71.85	blk, cub, 2.32	5.7	1369 ± 1		i	i	s a; i al, alk
i158	(III) oxide	Nat. hematite. Fe_2O_3	159.69	red-brn to blk, trig, 3.01, 2.94(Li)	5.24	1565		i	i	s HCl, H_2SO_4; sl s HNO_3
i159	oxide	Iron ferrosoferric, nat. magnetite. Fe_3O_4	231.54	blk, cub or red-blk powd, 2.42	5.18	1594 ± 5		i	i	s conc a; i al, eth
i160	(III) oxide, hydrate	$Fe_2O_3.xH_2O$		red-brn amorph powd or gelat	2.44–3.60	all H_2O, 350–400		i	i	s a; i al, eth
i161	(II) *ortho*phosphate	Nat. vivianite. $Fe_3(PO_4)_2.8H_2O$	501.61	wh-bl, monocl, 1.579, 1.603, 1.633	2.58			i	i	s a; i ac a
i162	(III) *ortho*phosphate	$FePO_4.2H_2O$	186.85	pink, monocl	2.74	d		v sl s	0.67^{100}	s HCl, H_2SO_4; i HNO_3
i163	(III) *pyro*phosphate	$Fe_4(P_2O_7)_3.9H_2O$	907.36	yel-wh powd				i		s a, alk citr
i164	phosphide, mono-	FeP	86.82	rhomb	6.07 (5.2^{20})					
i165	(di-)phosphide	Fe_2P	142.67	bl-gray cr or powd	6.56	1290		i	i	s aq reg, HNO_3+HF; i dil a
i166	(tri-)phosphide	Fe_3P	198.51	gray	6.74	1100		i		
i167	(III) *hypo*phosphite	$Fe(H_2PO_2)_3$	250.81	wh or gray-wh powd		d		0.043^{25}	0.083^{100}	s alk citr
i168	(II) *meta*silicate	Nat. gruenerite. $FeSiO_3$	131.93	gray-grn, rhomb, 1.672, 1.697, 1.717	3.5	1146				
i169	*ortho*silicate	Nat. fayalite. Fe_2SiO_4	203.78	col, rhomb	4.34	1503 (?)		i	i	d HCl
i170	silicide	FeSi	83.93	yel-gray, oct	6.1			i	i	i aq reg
i171	(II) sulfate	Nat. szomolnikite. $FeSO_4.H_2O$	169.96	off-wh, monocl	2.970^{25}			sl s	s	
i172	(III) sulfate	$Fe_2(SO_4)_3$	399.87	yel rhomb, hygr, 1.814	3.097^{18}	d 480		sl s	d	i H_2SO_4, NH_3
i173	(II) sulfate, heptahydrate	Nat. melanterite. $FeSO_4.7H_2O$	278.05	bl-grn, monocl, 1.471, 1.478, 1.486	1.898	64 $-6H_2O$, 90	$-7H_2O$, 300	15.65	48.6^{50}	sl s al; s abs MeOH
i174	(II) sulfate, pentahydrate	Nat siderotil. $FeSO_4.5H_2O$	242.02	wh, tricl, 1.526, 1.536, 1.542	2.2	$-5H_2O$, 300		s	s	i al
i175	(II) sulfate, tetrahydrate	$FeSO_4.4H_2O$	224.01	grn monocl pr, 1.533, 1.535	2.23–2.29					
i176	(III) sulfate, enneahydrate	Nat. coquimbite. $Fe_2(SO_4)_3.9H_2O$	562.01	rhomb, deliq, 1.552, 1.558	2.1	$-7H_2O$, 175		440	d	s abs al
i177	sulfide, di-	Nat. pyrite. FeS_2	119.98	yel, cub	5.0	1171		0.00049		d HNO_3. dil a
i178	sulfide, di-	Nat. marcasite. FeS_2	119.98	yel, rhomb	4.87	tr 450	d	0.00049		d HNO_3, i dil a
i179	(II) sulfide	Nat. troilite. FeS	87.91	blk-brn, hex	4.74	1193–1199	d	0.00062^{18}	d	s d a; i NH_3
i180	(III) sulfide	Fe_2S_3	207.87	yel-grn	4.3	d		sl d	d FeS+S	d a
i181	(II) sulfite	$FeSO_3.3H_2O$	189.96	grnsh or wh cr		d 250		v sl s		s SO_2 sol; i al
i182	tantalate	Nat. tapiolite. $Fe(TaO_3)_2$	513.73	lt brn, tetr, 2.27, 2.42 (Li)	7.33					
i183	*d*-tartrate	$FeC_4H_4O_6$	203.92	wh cr				0.877^{16}	v sl s	v s min a; s NH_4OH
i184	(II) thiocyanate	$Fe(SCN)_2.3H_2O$	226.06	grn, rhomb		d		v s		s al, eth, acet
i185	(III) thiocyanate	$Fe(SCN)_3$ or $Fe_2(SCN)_6$	230.09	blk-red, cub, deliq				v s	d	s al, eth, acet

No.	Name	Synonyms and Formulae	Mol. wt.	Crystalline form, properties and index of refraction	Density or spec. gravity	Melting point, °C	Boiling point, °C	Solubility, in grams per 100 cc: Cold water	Hot water	Other solvents
	Iron									
i186	(II) thiosulfate	$FeS_2O_3.5H_2O$	258.05	grn cr, deliq				v s	d	v s al
i187	tungstate	Nat. ferberite. $FeWO_4$	303.69	tetr, 2.40 (Li)	6.64					
i188	*meta*vanadate	$Fe(VO_3)_3$	352.67	grayish-brn powd.				i		s a; i al
k1	**Krypton**	Kr	83.80	inert gas	gas 3.736g/l liq 2.155 *at* −152.9	−156.6	−152.30±0.10	11.0^{0} cm³ 6.0^{25} cm³	4.67^{50} cm³	
l1	**Lanthanum**	La	138.91	wh met, tarnish in air, α: hex, β: cub above 350	α 6.1453, β 6.17	921	3457	d	d	s min a; i conc H_2SO_4
l2	acetate	$La(C_2H_3O_2)_3.1\frac{1}{2}H_2O$	343.07					16.88^{25}		
l3	boride, hexa-	LaB_6	203.78	purp met cub	2.61	2210	d	i	i	i HCl
l4	bromate	$La(BrO_3)_3.9H_2O$	684.77	hex pr		37.5	$-7H_2O$, 100	28.5^{15}		i al
l5	bromide	$LaBr_3.7H_2O$	504.74	col cr	5.057^{25} anh	783±3 anh	1577	v s		v s al; i eth
l6	carbide	LaC_2	162.93	yel cr	5.02			d	d	s H_2SO_4; i conc HNO_3
l7	carbonate	$La_2(CO_3)_3.8H_2O$	601.97	wh	2.6–2.7			i	i	s dil a; sl s aq CO_2; i acet
l8	chloride	$LaCl_3$	245.27	wh cr, deliq	3.842^{25}	860	>1000	v s	d	v s al, pyr; i eth, bz
l9	chloride, hepta-hydrate	$LaCl_3.7H_2O$	371.38	wh, tricl, hygr.		d 91		v s	v s	v s al
l10	hexaantipyrin *per*-chlorate	$[La(C_{11}H_{12}N_2O)].(ClO_4)_3$	1566.65	col, hex cr		d 290–295		1.50^{20}		
l11	hydroxide	$La(OH)_3$	189.93	wh powd		d		i		s a
l12	iodate	$La(IO_3)_3$	663.62	col, cr.				1.7^{25}		
l13	iodide	LaI_3	519.62	gray-wh, rhomb, hygr	5.63	772		v s		s acet
l14	molybdate	$La_2(MoO_4)_3$	757.63	tetr	4.77^{18}	1181		0.00179^{25}	0.0033^{85}	
l15	nitrate	$La(NO_3)_3.6H_2O$	433.02	col cr, deliq, tricl		40	d 126	151.1^{25}	v s	v s al; s acet
l16	oxalate	$La_2(C_2O_4)_3.9H_2O$	704.02	wh cr		d		0.00008^{25}		s min a
l17	oxide	Lanthana. La_2O_3	325.82	wh rhomb, or amorph	6.51^{15}	2307	4200	0.0004^{29}	d	s a, NH_4Cl; i acet
l18	sulfate	$La_2(SO_4)_3$	566.00	wh powd, hygr	3.60^{15}	d 1150		3.0	0.69^{100}	sl s al; i acet
l19	sulfate, hydrate	$La_2(SO_4)_3.9H_2O$	728.14	col hex, 1.564	2.821	d white heat		3.8^{0}	0.87^{100}	sl s HCl; i al
l20	sulfide	La_2S_3	374.01	red-yel cr, hex	4.911^{11}	2100–2150 vac		d	d	s a
l21	**Lead**	Pb	207.19	silv-blsh wh soft met, cub, 2.01	11.3437^{16} Ra-Pb 11.288^{20}_{20} UPb 11.2960^{16}	327.502	1740	i	i	s HNO_3, h conc H_2SO_4
l22	abietate	$Pb(C_{20}H_{29}O_2)_2$	810.10	brn lumps or yelsh-wh powd				i		
l23	acetate	$Pb(C_2H_3O_2)_2$	325.28	wh cr	3.25^{20}_{4}	280		44.3^{20}	221^{50}	s glyc; v sl s al
l24	acetate, basic	$Pb_2(OH)(C_2H_3O_2)_3$	608.52	wh				v s		sl s al
l25	acetate, basic	$Pb(C_2H_3O_2)_2.3PbO.H_2O$	1012.86	wh powd						
l26	acetate, basic	$Pb(C_2H_3O_2)_2.Pb(OH)_2.H_2O$	584.50	wh, monocl				v s		v s al
l27	acetate, decahydrate	$Pb(C_2H_3O_2)_2.10H_2O$	505.44	wh, rhomb cr	1.69	22		s	s	i al
l28	acetate, trihydrate	Sugar of lead. $Pb(C_2H_3O_2)_2.3H_2O$	379.33	wh, monocl, β 1.567	2.55	$-H_2O$, 75	d 200	45.61^{15}	200^{100}	i al
l29	acetate, tetra-	$Pb(C_2H_3O_2)_4$	443.37	col, monocl	2.228^{17}	175		d		d al; s chl, h ac a
l30	*di*antimonate	$Pb_2Sb_2O_7$	769.88	dk yel powd	6.72			i	i	v sl s HCl
l31	*ortho*antimonate	$Pb_3(SbO_4)_2$	993.07	or-yel powd	6.58^{20}_{4}			i	i	v sl s HCl
l32	*ortho*antimonate	Nat. monimolite. $Pb_3(SbO_4)_2$	993.07	orange powd	6.58^{20}_{4}			i		i dil a
l33	*meta*arsenate	$Pb(AsO_3)_2$	453.03	hex tabl	6.42^{15}			d	d	s HNO_3
l34	*ortho*arsenate	$Pb_3(AsO_4)_2$	899.41	wh cr, v pois	7.80	1042, sl d 1000		v sl s		s HNO_3
l35	*ortho*arsenate, di-	Nat. schultenite. $PbHAsO_4$	347.12	monocl leaf, α 1.90, γ 1.97	5.79	d 720	$-H_2O$, 220	i	sl s	s HNO_3, caust alk
l36	*ortho*arsenate, mono-	$Pb(H_2AsO_4)_2$	489.06	tricl, 1.74, 1.82	4.46^{15}	d 140		d		s HNO_3
l37	*pyro*arsenate	$Pb_2As_2O_7$	676.22	rhomb, β 2.03	6.85^{15}_{15}	802		i	d	s HCl, HNO_3; i ac a
l38	*meta*arsenite	$Pb(AsO_2)_2$	421.03	wh powd	5.85			i		s HNO_3
l39	*ortho*arsenite	$Pb_3(AsO_3)_2.xH_2O$		wh powd	5.85			i		s alk, HNO_3
l40	azide	$Pb(N_3)_2$	291.23	col need, or powd.			expl 350	0.023^{18}	0.09^{70}	v s ac a; i NH_4OH
l41	*meta*borate	$Pb(BO_2)_2.H_2O$	310.82	wh cr powd	5.598, anhydr	$-H_2O$, 160		i	i	s a; i alk

No.	Name	Synonyms and Formulae	Mol. wt.	Crystalline form, properties and index of refraction	Density or spec. gravity	Melting point, °C	Boiling point, °C	Solubility, in grams per 100 cc		
								Cold water	Hot water	Other solvents
	Lead									
142	borofluoride	$Pb(BF_4)_2$	380.80	cr pr				d		d al
143	bromate	$Pb(BrO_3)_2.H_2O$	481.02	col, monocl	5.53	d 180		1.38^{20}	sl s	
144	bromide	$PbBr_2$	367.01	wh, rhomb	6.66	373	916	0.4554^{0} 0.8441^{20}	4.71^{100}	s a, KBr; sl s NH_3; i al
145	butyrate	$Pb(C_4H_7O_2)_2$	381.39	col scales, pois		90		i	i	s dil HNO_3
146	caprate	$Pb(C_{10}H_{19}O_2)_2$	549.71			103–104		i	i	0.0029^{20} eth
147	caproate	$Pb(C_6H_{11}O_2)_2$	437.50			73–74				1.09^{25} eth
148	caprylate	Lead octoate. $Pb(C_8H_{15}O_2)_2$	493.60	wh leaf		83.5–84.5		i	i	s al; 0.0938 eth
149	carbonate	Nat. cerussite. $PbCO_3$	267.20	col, rhomb, 1.804, 2.076, 2.078	6.6	d 315		0.00011^{20}	d	s a, alk; i NH_3, al
150	carbonate, basic	White lead, hydro-cerussite. $2PbCO_3.Pb(OH)_2$	775.60	wh powd, or hex.	6.14	d 400		i	i	sl s aq CO_2; s HNO_3; i al
151	cerotate	$Pb(C_{26}H_{51}O_2)_2$	998.57	wh need		113		i		i al, eth; s bz
152	chlorate	$Pb(ClO_3)_2$	374.09	wh monocl, deliq	3.89	d 230		v s		s al
153	chlorate, hydrate	$Pb(ClO_3)_2.H_2O$	392.11	wh, monocl, deliq.	4.037	d 110		151.3^{18}	171^{80}	s al
154	*per*chlorate	$Pb(ClO_4)_2.3H_2O$	460.14	wh, rhomb	2.6	d 100		499.7^{25}		s al
155	chloride	Nat. cotunite. $PbCl_2$	278.10	wh, rhomb, 2.199, 2.217, 2.260	5.85	501	950^{760}	0.99^{20}	3.34^{100}	sl s dil HCl,NH_3; i al; s NH_4 salts
156	chloride, tetra-	$PbCl_4$	349.00	yel oily liq	3.18^{0}	−15	expl 105	d (Cl_2)	d	s conc HCl
157	chloride, sulfide	$PbCl_2.3PbS$	995.86	red				i	d	d a, alk; i dil a
158	chlorite	$Pb(ClO_2)_2$	342.09	yel, monocl		expl 126		0.095^{20}	0.42^{100}	s KOH
159	chromate	Nat. crocoite, chrome yellow. $PbCrO_4$	323.18	yel, monocl, 2.31, 2.37(Li), 2.66	6.12^{15}	844	d	0.0000058^{25}	i	s a, alk; i ac a, NH_3
160	chromate, basic	Chrome red. $PbCrO_4.PbO$	546.37	red cr powd	6.63			i	i	s a, alk
161	chromate, basic	$Pb_2(OH)_2CrO_4$	564.39	red amorph or cr	6.63	920		i	i	s KOH
162	*di*chromate	$PbCr_2O_7$	423.18	red cr				d		s a, alk
163	citrate	$Pb_3(C_6H_5O_7)_2.3H_2O$	1053.82	wh cr powd				s		v sl s al
164	cyanate	$Pb(OCN)_2$	291.22	wh need		d		i	sl s	
165	cyanide	$Pb(CN)_2$	259.23	yelsh-wh powd, pois				sl s	s	s KCN
166	enanthate	$Pb(C_7H_{13}O_2)_2$	465.55	wh leaf		91.5		sl s		i al
167	ethylsulfate	$Pb(C_2H_5SO_4)_2.2H_2O$	493.57	col liq, pois				s		
168	ferricyanide	$Pb_3[Fe(CN)_6]_2$.5 (or 6) H_2O	1135.55	blk-brn to red, monocl pr		$-H_2O$, 110–120 d		sl s	s, d 100	s alk, HNO_3
169	ferrite	$PbFe_2O_4$	382.88	hex		1530 d, 725				
170	ferrocyanide	$Pb_2Fe(CN)_6$ $3H_2O$	680.38	yelsh-wh powd		$-H_2O$, 100		i		sl s H_2SO_4
171	fluoride	PbF_2	245.19	col, rhomb, pois	8.24	855	1290	0.064^{20}		s HNO_3; i acet, NH_3
172	fluorochloride	Nat. matlockite. PbFCl	261.64	wh, tetr, 2.145, 2.006	7.05	601		0.037^{25}	0.1081^{100}	
173	fluosilicate	$PbSiF_6.2H_2O$	385.30	col, monocl		d		s	v s	
174	fluosilicate, tetra-hydrate	$PbSiF_6.4H_2O$	421.33	col, monocl		d<100				
175	formate	$Pb(CHO_2)_2$	297.23	wh, rhomb, lust, 1.789, 1.852, 1.877	4.63	d 190		1.6^{15}	20^{100}	i al
176	hydride, di-	PbH_2	209.21	gray powd		d				
177	hydroxide	$Pb(OH)_2$	241.20	wh, amorph		d 145		0.0155^{20}	sl s	s a, alk; i ac a
178	hydroxide	$Pb_2O(OH)_2$ or $2PbO.H_2O$	464.39	wh cub, or amorph powd, pois	7.592	d 145		0.014	sl s	s alk, ac a, HNO_3
179	iodate	$Pb(IO_3)_2$	557.00	wh	6.155^{20}	d 300		0.0012^{2}	0.003^{25}	sl s HNO_3; i NH_3
180	*paraper*iodate	$PbHIO_5$	415.10	wh cr		d 130		i	i	s dil HNO_3
181	*paraper*iodate, hydrate	$PbHIO_5.H_2O$	433.11	amorph		$-H_2O$, 110		i	i	sl s dil HNO_3
182	iodide, basic	$PbI_2.PbO.H_2O$	702.20	rhomb cr	6.83^{20}	d 100				
183	iodide,di-	PbI_2	461.00	yel hex powd, pois	6.16	402	954	0.044^{0} 0.063^{20}	0.41^{100}	s alk, KI; i al
184	iodide, mono-	PbI	334.09	pa yel		d 300		0.1		
185	isobutyrate	$Pb(C_4H_7O_2)_2$	381.39	wh pr		<100		9.1^{15}		
186	lactate	$Pb(C_3H_5O_3)_2$	385.33	wh cr powd				s		s h al
187	laurate	$Pb(C_{12}H_{23}O_2)_2$	605.82	chalky wh powd		104.7		0.009^{35}		0.008^{25} al; $0.007^{14.5}$ eth
188	lignocenate	$Pb(C_{24}H_{47}O_2)_2$	942.47	wh powd		117		i		v s h bz; sl s al; i eth
189	malate	$Pb(C_4H_4O_5).3H_2O$	393.31	wh powd				sl s		v sl s al
190	melissate	$Pb(C_{31}H_{61}O_2)_2$	1138.85	wh powd		115–116		i	i	s boil tol, ac a; sl s h bz, chl; i al, eth
191	molybdate	Nat. wulfenite. $PbMoO_4$	367.13	col-lt yel, tetr pl	6.92^{25}_{4}	1060–1070			i	d conc H_2SO_4; s a, KOH; i al
192	myristate	$Pb(C_{14}H_{27}O_2)_2$	661.93	wh powd		107		0.005^{35}	0.006^{50}	0.004^{25} al; $0.010^{14.5}$ eth

No.	Name	Synonyms and Formulae	Mol. wt.	Crystalline form, properties and index of refraction	Density or spec. gravity	Melting point, °C	Boiling point, °C	Solubility, in grams per 100 cc: Cold water	Hot water	Other solvents
	Lead									
l93	2-naphthalene-sulfonate	$Pb(C_{10}H_7SO_3)_2$	621.65	wh cr powd, pois				i		s al
l94	nitrate	$Pb(NO_3)_2$	331.20	col, cub or monocl, pois, 1.782	4.53^{20}	d 470		37.65^{0} 56.5°	127^{100}	8.77^{22} 43 % al; s alk, NH_3
l95	nitrate, basic	$Pb(OH)NO_3$	286.20	wh rhomb cr	5.93	d 180		19.4^{19}	s	s a
l96	nitrite	$3PbO.N_2O_3.H_2O$	763.60	lt yel powd				v s		s dil HNO_3
l97	oleate	$Pb(C_{18}H_{33}O_2)_2$	770.12					i		6.46^{20} eth; s pet eth; sl s al
l98	oxalate	PbC_2O_4	295.21	wh powd	5.28	d 300		0.00016^{18}		s HNO_3; i al
l99	oxide-, di	Plattnerite. PbO_2	239.19	bz, tetr, ω 2.3(Li)	9.375	d 290		i	i	s dil HCl; sl s ac a; i al
l100	oxide, mono-	Litharge. PbO	223.19	yel, tetr	9.53	886		0.0017^{20}		s HNO_3. alk, Pb acet, NH_4Cl, $CaCl_2$, $SrCl_2$
l101	oxide, mono-	Massicot. PbO	223.19	yel, rhomb, 2.51, 2.61(Li), 2.71	8.0	886		0.0023^{22}	i	s alk
l102	oxide, red	Minium. Pb_3O_4	685.57	red cr sc, or amorph powd	9.1	d 500		i	i	s HCl, acet a; i al
l103	oxide, sesqui-	Pb_2O_3	462.38	or-yel powd, amorph		d 370		i	d	d a
l104	oxide, sub	Pb_2O	430.38	blk, amorph	8.342	d		i	i	s a, alk
l105	oxychloride	$PbCl_2.3PbO$	947.66	yel,				0.0056^{18}	0.07^{74}	
l106	oxychloride	Cassel yellow. $PbCl_2.7PbO$	1840.42	yel cr, or powd				i		
l107	oxychloride	Fiedlerite. $2PbCl_2.PbO.H_2O$	797.40	monocl, 1.816, 2.1023, 2.026	5.88^{20}	d 150				s HNO_3
l108	oxychloride	Nat. laurionite. $PbCl_2.Pb(OH)_2$	519.29	rhomb	6.24	d 142				
l109	oxychloride	Matlockite. $PbCl_2.PbO$	501.29						i	s alkalies, hot conc HCl
l110	oxychloride	Nat. matlockite. $PbCl_2.Pb(OH)_2$	519.29	wh, tetrag, 2.04, 2.15, 2.15	7.21	d 524		0.0095^{18}		s alk
l111	oxychloride	Nat. mendipite. $PbCl_2.2PbO$	724.47	yel, rhomb, 2.24, 2.27, 2.31	7.08	693		i	i	s alk
l112	oxychloride	Nat. paralaurionite. $PbCl_2.PbO.H_2O$	519.29	col to wh, monocl pr, 2.146	6.05^{15}	d 150				
l113	palmitate	$Pb(C_{16}H_{31}O_2)_2$	718.04	wh powd		112.3		0.005^{35}	0.007^{50}	s al; 0.148^{20} eth
l114	phenolate	Lead phenate, lead carbolate. $Pb(OH)OC_6H_5$	317.30	yelsh-wh powd				i		
l115	phenolsulfonate	Lead sulfocarbolate. $Pb[C_6H_4(OH)SO_3]_2.5H_2O$	643.60	wh lustr need				s		s al
l116	*meta*phosphate	$Pb(PO_3)_2$	365.13	col cr		800(?)			v sl s	
l117	*ortho*phosphate	$Pb_3(PO_4)_2$	811.51	col or wh powd, hex, 1.970, 1.936	6.9–7.3	1014		0.000014^{20}	i	s HNO_3, alk; i ac a, al
l118	*ortho*phosphate, di-	$PbHPO_4$	303.17	rhomb	5.661^{15}	d				s HNO_3, alk, NH_4Cl
l119	*ortho*phosphate, mono-	$Pb(H_2PO_4)_2$	401.16	need						s alk, dil HNO_3, h conc HCl; i acet a
l120	phosphide	PbP_5	362.06	blk unstable, inflam		d 400 (vac)		d	d	d dil a
l122	*ortho*phosphite	$PbHPO_3$	287.17	wh, powd		d		i		s HNO_3
l123	picrate	$Pb(C_6H_2N_3O_7)_2.H_2O$	681.45	yel, need	2.831^{20}	$-H_2O$, 130	expl	0.88^{15}		
l124	proprionate, tetra-	$Pb(C_3H_5O_2)_4$	499.48	solid		132				
l125	*pyro*phosphate	$Pb_2P_2O_7$	588.32	wh, rhomb	5.8^{20}	824		i	i	s HNO_3, KOH
l126	*pyro*phosphate, hydrate	$Pb_2P_2O_7.H_2O$	606.34	wh, rhomb		806 anhydr		i	d	s HNO_3, KOH, $Na_4P_2O_7$
l127	selenate	$PbSeO_4$	350.17	wh, rhomb	6.37^{20}_{4}	d		i	i	s conc a
l128	selenide	Nat. clausthalite. PbSe	286.15	gray, cub	8.10^{15}	1065		i		s HNO_3
l129	metasilicate	Nat. alamosite. $PbSiO_3$	283.27	col or wh, monocl	6.49	766		i		d a
l130	*ortho*silicate, di-	Nat. barysilite. $Pb_3Si_2O_7$	582.55	wh, trig, 2.070, 2.050	6.707			i	i	
l131	stearate	$Pb(C_{18}H_{35}O_2)_2$	774.15	wh powd		115.7		0.05^{35}	0.06^{50}	$0.005^{14.5}$ eth; i al
l132	sulfate	Nat. anglesite. $PbSO_4$	303.25	wh, monocl, or rhomb, 1.877, 1.822, 1.894	6.2	1170		0.00425^{25}	0.0056^{40}	s NH_4 salts; sl s conc H_2SO_4; i a
l133	sulfate, basic	Nat. lanarkite. $PbSO_4.PbO$	526.44	wh, monocl, 1.93, 1.99, 2.02	6.92	977		0.0044^{0}	v sl s	sl s H_2SO_4
l134	sulfate, hydrogen	$Pb(HSO_4)_2.H_2O$	419.34	wh cr		d		0.0001^{18} d		sl s H_2SO_4
l135	*peroxydi*sulfate	$PbS_2O_8.3H_2O$	453.36	deliq				v s		
l136	sulfide	Nat. galena. PbS	239.25	bl met cub, 3.921	7.5	1114		0.000086^{13}		s a; i al, KOH
l137	sulfite	$PbSO_3$	287.25	wh powd		d		i	i	s HNO_3
l138	tartrate, *dl*-	$PbC_4H_4O_6$	355.26	wh cr powd	2.53^{19}			0.0025^{20}	0.0074^{100}	s HNO_3, KOH; i al, ac a, NH_4 ac

No.	Name	Synonyms and Formulae	Mol. wt.	Crystalline form, properties and index of refraction	Density or spec. gravity	Melting point, °C	Boiling point, °C	Solubility, in grams per 100 cc		
								Cold water	Hot water	Other solvents
	Lead									
1139	*di*thionate	$PbS_2O_6.4H_2O$	439.38	trig, 1.635, 1.653	3.22	d		115.0^{20}		
1140	thiosulfate	PbS_2O_3	319.32	wh cr	5.18	d		0.03		s a, $Na_2S_2O_3$
1141	*meta*titanate	$PbTiO_3$	303.09	yel, rhomb-pyr	7.52			i	i	
1142	telluride	Nat. altaite. PbTe	334.79	wh, cub	8.164$^{20}_{4}$	917				i a
1143	thiocyanate	$Pb(SCN)_2$	323.35	wh, monocl	3.82	d 190		0.05^{20}	0.2^{100}	s KCNS, HNO_3
1144	tungstate	Nat. stolzite. $PbWO_4$	455.04	tetr, 2.269, 2.182	8.23			i		i HNO_3, s KOH
1145	tungstate	Nat. raspite. $PbWO_4$	455.04	col, monocl, 2.27, 2.27, 2.30		1123		0.03		d a; i al
1146	*meta*vanadate	$Pb(VO_3)_2$	405.07	yel powd				sl s		d HCl; s dil HNO_3
1147	**Lithium**	Li	6.939	silver white, soft	0.534^{20}	180.54	1347	d		
1148	acetate	$LiC_2H_3O_2.2H_2O$	102.01	wh, rhomb, α 1.40, β 1.50		70	d	300^{15}	v s	21.5 al
1149	acetylsalicylate	$LiC_9H_7O_4$	186.09	wh powd hygr, d in moist air				100		25 al
1150	*meta*aluminate	$LiAlO_2$ (or $Li_2Al_2O_4$)	65.92	wh, rhomb, 1.604, 1.614	2.55$^{25}_{4}$	1900–2000		i		
1151	aluminum hydride	$LiAlH_4$	37.95	wh cr powd	0.917	d 125		d		*ca* 30 eth
1152	amide	$LiNH_2$	22.96	col need, cub	1.178$^{17.5}$	380–400	d 750–200 subl	s	d	sl s liq NH_3, al; i eth, bz
1153	antimonide	Li_3Sb	142.57		3.2^{17}	>950		d	d	d a
1154	*ortho*arsenate	Li_3AsO_4	159.74	wh powd, rhomb	3.07^{15}			v sl s		s dil ac a; i pyr
1155	azide	LiN_3	48.96	col cr, hygr		d 115–298		66.41^{16}		20.26^{16} abs al; i eth
1156	benzoate	$LiC_7H_5O_2$	128.06	wh cr or powd				33^{25}	40^{100}	7.7^{25} al, 10^{78} al
1157	*meta*borate	$LiBO_2$	49.75	wh, tricl	1.397$^{41.7}$	845		2.57^{20}	11.83^{80}	
1158	*meta*borate	$LiBO_2.8H_2O$	193.87	col, trig	1.38$^{14.9}$	47				
1159	*penta*borate	$Li_2B_{10}O_{16}.8H_2O$	522.10	wh	1.72	300–350 −8H_2O		36.3^{45}	194^{100}	3.9^{20} al; 22^{23} glycerine; i bz
1160	*tetra*borate	$Li_2B_4O_7$	169.11	wh cr		930		2.89^{20}	5.45^{100}	i org solv
1161	borohydrate	$LiBH_4$	21.78	rhomb cr	0.66	d 279		s d		s eth
1162	borohydrate	$LiBH_4$	21.78	wh, orthorhomb	0.666	275 d		v sl s		d al; 2.5 eth
1163	bromide	LiBr	86.85	wh, cub, deliq, 1.784	3.464^{25}	550	1265	145^{4}	254^{90}	73^{40} al; 8 MeOH; s al, eth; sl s pyrid
1164	bromide, dihydrate	$LiBr.2H_2O$	122.28	wh cr		−1H_2O; 44		246.0^{20}	v s	s al
1165	carbide	Li_2C_2	37.90	wh cr or powd	1.65^{18}			d	d	s a
1166	carbonate	Li_2CO_3	73.89	wh, monocl, 1.428, 1.567, 1.572	2.11	723	d 1310^{760}	1.54^{0}	0.72^{100}	i al; acet
1167	carbonate, acid	Lithium bicarbonate. $LiHCO_3$	67.96	wh				5.5^{13}		
1168	chlorate	$LiClO_3$	90.39	col, rhomb need, deliq, α 1.63, γ 1.64	1.1190$^{18}_{4}$ (18 %Soln)	127.6	300 d	500^{27}		v s al; 0.142^{25} acetone
1169	chlorate	$LiClO_3.\frac{1}{2}H_2O$ (or $\frac{1}{3}H_2O$)	99.39	wh, tetr, deliq		65(?)	−½H_2O, 90 d 290	v s	v s	v s al
1170	perchlorate	$LiClO_4$	106.39	wh	2.428	236	430 d	60.0^{25}	150^{89}	152^{25} al; 182^{25} MeOH; 114^{25} eth; 137^{25} acetone
1171	perchlorate, trihydrate	$LiClO_4.3H_2O$	160.44	wh, hex	1.841	95 deliq 236 (anhydr)	d 100 −2H_2O	130^{25}		72.9^{25} al; 156^{25} MeOH; 96.2^{25} acetone; 0.096^{25} eth
1172	chloride	LiCl	42.39	wh, cub, 1.662	2.068^{25}	605	1325–1360	63.7^{0}	130^{95}	25.10^{30} al; 42.36^{25} MeOH; 4.11^{25} acetone; 0.538$^{33.9}$ NH_4OH
1173	chloride, monohydrate	$LiCl.H_2O$	60.41	wh cr, hygr	1.78	−H_2O>98		86.2^{20}	s	s HCl
1174	chloroplatinate	$Li_2PtCl_6.6H_2O$	529.78	or prism		−6H_2O, 180		v s	v s	v s al; i eth
1175	dichromate, dihydrate	$Li_2Cr_2O_7.2H_2O$	265.90	orange-red cr, deliq	2.34^{30}	187 d	110 −2H_2O	187^{30}	278^{100}	s reacts al
1177	citrate	$Li_3C_6H_5O_7.4H_2O$	281.98	col cr or powd, deliq		−4H_2O, 105		74.5^{25}	66.7^{100}	sl s al, eth
1178	fluoride	LiF	25.94	wh, cub, 1.3915	2.635^{20}	845	1676	0.27^{18}		i al; s HF
1179	fluosilicate	$Li_2SiF_6.2H_2O$	191.99	wh, monocl, 1.300, 1.296	2.33^{12}	−2H_2O, 100	d	73^{17}		s al; i eth, acet
1180	fluosulfonate	$LiSO_3F$	106.00	wh powd		360		v s	s	v s al, eth, acet; i ligorin
1181	formate, monohydrate	$H.COOLi.H_2O$	69.97	wh, rhomb	1.46	−H_2O, 94	d 230	27.85^{18}	57.05^{98}	sl s al, acet; i bz
1182	gallium hydride	$LiGaH_4$	80.69	wh cr				d	d	s eth
1183	gallium nitride	$LiGaN_2$	118.55	lt gr powd	3.35	d 800		d	d	s a, alk
1184	*meta*germanate	Li_2GeO_3	134.47	monocl, 1.7	3.53^{21}	1239		0.85^{25}		s a

No.	Name	Synonyms and Formulae	Mol. wt.	Crystalline form, properties and index of refraction	Density or spec. gravity	Melting point, °C	Boiling point, °C	Solubility, in grams per 100 cc: Cold water	Hot water	Other solvents
	Lithium									
1185	hydride	LiH	7.95	wh cr	0.82	680		d		v al s a
1186	hydroxide	LiOH	23.95	wh tetr, 1.464, 1.452	1.46	450	d 924	12.8^{20}	17.5^{100}	al s al
1187	hydroxide, monohydrate	$LiOH.H_2O$	41.96	wh monocl, 1.460, 1.524	1.51			22.3^{10}	26.8^{80}	al s al; i eth
1188	iodate	$LiIO_3$	181.84	wh, hex, hygr	4.502^{22}_4			80.3^{18}		i al
1189	iodide	LiI	133.84	wh, cub, 1.955 ± 0.003	3.494 ± 0.015	449	1180 ± 10	165^{20}	433^{80}	250.8^{20} al; 42.6^{18} acet 343.4^{20} MeOH; v s NH_4OH
1190	iodide, trihydrate	$LiI.3H_2O$	187.89	col-yelsh, hex, hygr	3.48	73 − H_2O	−$2H_2O$, 80 −H_2O, 300	151^0	201.2^{80}	s abs al, acet
1191	laurate	$LiC_{12}H_{23}O_2$	206.25	wh powd		229.2–229.8		$0.154^{18.3}$	0.178^{25}	0.322^{25} al; $0.008^{15.3}$ eth; 0.240^{25} acet
1192	*per*manganate	$LiMnO_4.3H_2O$	179.92	cub	2.06	d 190		71.43^{16}		d alk
1193	molybdate	Li_2MoO_4	173.82	wh trig, hygr	2.66	705		v s		
1194	myristate	$LiC_{14}H_{27}O_2$	234.31			223.6–224.2		$0.027^{18.3}$ 0.036^{25}	0.062^{50}	$0.010^{15.3}$ eth; 0.331^{15} acet; 0.155^{20} al
1195	nitrate	$LiNO_3$	68.94	wh, trig, 1.735, 1.735	2.38	264	d 600	$89.8^{27.86}$	234^{100}	s NH_4OH, al; 37.15 pyridine
1196	nitrate, trihydrate	$LiNO_3.3H_2O$	122.99	col need		−2½H_2O, 29.9	−$3H_2O$, 61.1	34.8^0	$57.48^{29.6}$	s al, MeOH, acet
1197	nitride	Li_3N	34.82	red-brn amorph, or blk-gray cr, cub		tr 840–850 (in N_2)				
1198	nitrite	$LiNO_2.H_2O$	70.96	col flat need	1.615^0	>100	d	125^0	459^{80}	v s abs al
1199	oxalate	$Li_2C_2O_4$	101.90	col, rhomb, 1.465, 1.53, 1.696	$2.121^{17.5}$	d		$8^{19.5}$		i al, eth
1200	oxalate, acid	$LiHC_2O_4.H_2O$	113.99			d		8^{17}		
1201	oxide	Li_2O	29.88	wh cr, cub, n_D 1.644	$2.013^{25.2}$	>1700	1200^{800}	6.67^0 d	10.02^{100}	
1202	palmitate	$LiC_{16}H_{31}O_2$	262.36	wh powd		224.5		0.01^{18}	0.015^{25}	0.347^{15} acet; 0.077^{20} al; $0.005^{18.8}$ eth
1203	*meta*phosphate	$LiPO_3$	85.91	col pl	2.461	red heat		i	i	s a
1204	*ortho*phosphate	Li_3PO_4	115.79	col, rhomb	$2.537^{17.5}$	837		0.039^{18}		s a, NH_4OH; i acet
1205	*ortho*phosphate	Li_3PO_4.½H_2O	124.80	wh cr powd	2.41	−½H_2O, 100		0.04^{25}		s a
1206	phosphate, di- H	LiH_2PO_4	103.93	col cr, hygr	2.461	>100				
1207	potassium sulfate	$LiKSO_4$	142.10	col, hex; n_D 1.4723, 1.4717	2.393^{20}			s	s	
1208	potassium *dl*-tartrate	$LiKC_4H_4O_6.H_2O$	212.13	col, monocl, β 1.523 (red)	1.610			s		
1209	salicylate	$LiC_7H_5O_3$	144.06	wh, powd, deliq		d		133.3		50 al
1210	selenide	$Li_2Se.9H_2O$	254.98	col, rhomb, deliq				d		
1211	*meta*silicate	Li_2SiO_3	89.96	col, rhomb; α 1.584, γ 1.604	2.52^{25}_4	1204		i	s d	s dil HCl
1212	*ortho*silicate	Li_4SiO_4	119.84	col, rhomb; α 1.594, γ 1.614	2.392^{25}_4	1256		i	d	d a
1213	silicide	Li_6Si_2	97.81	bl cr, hygr	*ca.* 1.12	d 600 (vac)		d	d	d a; i NH_3 turp
1214	sodium fluoaluminate	$Li_3Na_3(AlF_6)_2$	371.73	cub cr, 1.3395	2.774	710		0.074^{18}		
1215	stearate	$LiC_{18}H_{35}O_2$	290.41	wh cr		220.5–221.5		0.010^{18}		0.010^{25} al; 0.040^{18} eth; 0.457^{15} acet
1216	sulfate	Li_2SO_4	109.94	α monocl; β hex or rhomb, γ cub 500°C; β 1.465	2.221	845		26.1^0	23^{100}	i abs al, acet
1217	sulfate, hydrogen	$LiHSO_4$	104.01	col pr	2.123^{13}	120		d		
1218	sulfate, monohydrate	$Li_2SO_4.H_2O$	127.95	col cr, monocl, 1.465, 1.477, 1.488		880		34.9^{25}	29.2^{100}	11.5^{20} al + H_2O (23.9 % alco); i acet, pryidine
1219	sulfide	Li_2S	45.94	wh-yel, cub, deliq	1.66	900–975		v s	v s	v s al
1220	sulfide, hydro-	LiHS	40.01	wh powd, hygr				s	s	s al
1221	sulfite, monohydrate	$Li_2SO_3.H_2O$	111.96	wh need, α 1.53, γ 1.59		455 d	140 − H_2O	24.9^{20}	22^{80}	i org solv
1222	tartrate	$Li_2C_4H_4O_6.H_2O$	179.97	wh cr powd				s		
1223	thallium *dl*-tartrate	$LiTlC_4H_4O_6.2H_2O$	395.41	tricl	3.144					
1224	thiocyanate	LiSCN	65.02	wh cr, deliq, n_D 1.333				v s		s methylacet
1225	dithionate	$Li_2S_2O_6.2H_2O$	210.03	col, rhomb, 1.5602	2.158	d		v s		
1226	tungstate	Li_2WO_4	261.73	col, trig	3.71	742		v s	v s	d a; i al
1227	**Lutetium**	Cassiopeium. Lu	174.967	met, hex	9.8404	1663	3395			

No.	Name	Synonyms and Formulae	Mol. wt.	Crystalline form, properties and index of refraction	Density or spec. gravity	Melting point, °C	Boiling point, °C	Solubility, in grams per 100 cc: Cold water	Hot water	Other solvents
	Lutetium									
l228	bromide	$LuBr_3$	414.70	...	...	1025	1400	s	s	...
l229	chloride	$LuCl_3$	281.33	col cr	3.98	905	subl 750	s	...	...
l230	fluoride	LuF_3	231.97	...	...	1182	2200	i	i	...
l231	iodide	LuI_3	555.68	...	...	1050	1200	s	s	...
l232	oxalate	$Lu_2(C_2O_4)_3.6H_2O$	722.09	wh cr	...	50 ($-H_2O$)	...	i	i	i dil a
l233	oxide	Lu_2O_3	397.94	cub cr	9.42	...	...	...	...	...
l234	sulfate	$Lu_2(SO_4)_3.8H_2O$	782.25	col cr	...	...	...	42.27^{20}	16.93^{40}	...
m1	**Magnesium**	Mg	24.312	silv wh met, hex	1.74^{5}	648.8	1090	i	d to $Mg(OH)_2$	s min a, conc. HF, NH_4 salts; i CrO_3, alk.
m2	acetate	$Mg(C_2H_3O_2)_2$	142.40	wh cr	1.42	323 d	...	v s	v s	5.25^{15} MeOH
m3	acetate, tetrahydrate	$Mg(C_2H_3O_2)_2.4H_2O$	214.46	col, monocl deliq, β 1.491	1.454	80	...	120^{15}	∞	v s al
m4	aluminate	Nat. spinel. $MgAl_2O_4$	142.27	col, cub, 1.723	3.6	2135	...	...	...	sl s H_2SO_4; v sl s dil HCl; i HNO_3
m5	amide	$Mg(NH_2)_2$	56.36	gray powd	...	d 350–400	d	d	...	v sl s liq NH_3; d al
m6	antimonate	$MgO.Sb_2O_5.12H_2O$	579.99	hex or tricl cr	2.60 (hex)	$-12H_2O$, 200	...	v sl s	...	d HCl
m7	antimonide	Mg_3Sb_2	316.44	met hex pl	4.088^{25}_{4}	961	...	i	...	...
m8	*ortho*arsenate	Nat. hoernesite. $Mg_3(AsO_4)_2.8H_2O$	494.90	wh monocl	2.60–2.61	...	...	...	...	...
m9	*ortho*arsenate	$Mg_3(AsO_4)_2.22H_2O$	747.11	wh cr	1.788	$-17H_2O$, 100	$-21H_2O$, 220	i	i	s a, NH_4Cl
m10	*ortho*arsenate, mono-H	Nat. roesslerite. $MgHAsO_4.7H_2O$	290.35	monocl	1.943^{15}	$-5H_2O$, 100	...	d	...	...
m11	arsenide	Mg_3As_2	222.78	brn-red, cub	3.148^{25}_{4}	800	...	d	d	d dil a, al
m12	*ortho*arsenite	$Mg_3(AsO_3)_2$	318.78	wh	...	...	...	s	v s	s a, NH_4Cl; i NH_4OH
m13	benzoate	$Mg(C_7H_5O_2)_2.3H_2O$	320.59	wh powd	...	$-3H_2O$, 110	d 200	6.16^{15}	19.6^{100}	s al
m14	bismuthide	Mg_3Bi_2	490.90	met, hex	5.945^{25}_{4}	823	...	...	...	...
m15	bismuth nitrate	$3Mg(NO_3)_2.2Bi(NO_3)_3.24H_2O$	1543.31	col cr, deliq	2.32^{16}_{16}	71	...	d	d	s HNO_3
m16	*di*borate	Nat. ascharite. $Mg_2B_2O_5.H_2O$	168.26	orthorhmb, 1.54	2.60–2.70	...	...	...	...	...
m17	*meta*borate	Nat. pinnoite. $Mg(BO_2)_2.3H_2O$	163.98	yel, tetr, pyram, 1.565, 1.575	2.27–2.30	...	...	...	...	...
m18	metaborate, octahydrate	$Mg(BO_2)_2.8H_2O$	254.05	col, tetr, 1.565, 1.575	2.30	...	...	i	v sl s	s a
m19	*ortho*borate	$Mg_3(BO_3)_2$	190.55	col, rhomb, 1.6527, 1.6537, 1.6548	2.99^{21}	...	...	i	i	s min a; i ac a
m20	boride	MgB_2	89.18	bl	...	d 1200 (vac)	...	d	...	sl s a
m21	bromate	$Mg(BrO_3)_2.6H_2O$	388.22	col, cub, 1.514	2.29	$-6H_2O$, 200	d	42^{18}	v s	i al
m22	bromide	$MgBr_2$	184.13	wh hex cr, deliq	3.72^{25}_{4}	711	...	101.50^{20}	125.6^{100}	6.9 al; 21.8^{20} MeOH
m23	bromide, hexahydrate	$MgBr_2.6H_2O$	292.22	col, hex pr or need, hygr, fluo in x-rays	2.00	172.4	...	316^{0}	v s	s al, acet; sl s NH_3
m24	bromoplatinate	$MgPtBr_6.12H_2O$	915.04	trig	2.802	...	...	...	...	...
m25	carbonate	Nat. magnesite. $MgCO_3$	84.32	wh, trig, 1.717, 1.515	2.958	d 350	$-CO_2$, 900	0.0106	...	s a, aq + CO_2; i acet, NH_3
m26	carbonate, basic artinite	Nat. artinite. $MgCO_3.Mg(OH_2).3H_2O$	196.69	wh, rhomb, 1.489, 1.534, 1.557	2.02^{20}	...	...	...	...	...
m27	carbonate, basic	Nat. hydromagnesite. $3MgCO_3.Mg(OH)_2.3H_2O$	365.34	wh, rhomb, 1.527, 1.530, 1.540	2.16	d	...	0.04	0.011	s a, NH_4 salts
m28	carbonate, pentahydrate	Nat. lansfordite. $MgCO_3.5H_2O$	174.40	wh, monocl, 1.456 1.476, 1.502	1.73	d in air	...	0.176^{7}	0.375^{20}	s HCl, $MgSO_4$ soln
m29	carbonate, trihydrate	Nat. nesquehonite. $MgCO_3.3H_2O$	138.37	col, rhomb need, 1.495, 1.501, 1.526	1.850	165	...	0.179^{16}	d	s a; 1.4 aq + CO_2
m30	chlorate	$Mg(ClO_3)_2.6H_2O$	299.31	wh, rhomb need, deliq	1.80^{25}	35	d 120	128.6^{18}	v s	s al
m31	*per*chlorate	$Mg(ClO_4)_2$	223.21	deliq	2.21^{18}	d 251	...	99.3^{25}	v s	23.96^{25} al
m32	*per*chlorate, hexahydrate	$Mg(ClO_4)_2.6H_2O$	331.31	wh, rhomb cr, 1.482, 1.458	1.98	185–190	...	v s	v s	...
m33	*per*chlorate, hexammine	$Mg(ClO_4)_2.6NH_3$	325.40	wh, cub	1.41^{20}_{4}	...	...	...	...	s liq NH_3; s d al
m34	chloride	$MgCl_2$	95.22	wh, lustr hex cr, 1.675, 1.59	2.316–2.33	714	1412	54.25^{20}	72.7^{100}	7.40^{20} al

No.	Name	Synonyms and Formulae	Mol. wt.	Crystalline form, properties and index of refraction	Density or spec. gravity	Melting point, °C	Boiling point, °C	Solubility, in grams per 100 cc		
								Cold water	Hot water	Other solvents
	Magnesium									
m35	chloride, hexahydrate	Nat. bischofite. $MgCl_2.6H_2O$	203.31	col, monocl, deliq, 1.495, 1.507 1.528	1.569	d 116–118	d	167	367	s al
m36	chloropalladate	$MgPdCl_6.6H_2O$	451.52	hex	2.12	d				
m37	chloroplatinate	$MgPtCl_6.6H_2O$	540.21	yel, trig	2.692	$-H_2O$, 180		s	s	
m38	chlorostannate	$MgSnCl_6.6H_2O$	463.81	tricl	2.08	d 100				
m39	chromate	$MgCrO_4.7H_2O$	266.41	yel, rhomb, 1.521, 1.550, 1.568	1.695		211.5^{18}	v s		
m40	chromite	$MgCr_2O_4$	192.30	dk-grn or red, cub	4.6^{20}			i	i	s conc H_2SO_4; i dl a, dil alk
m41	citrate, nono-H	$MgHC_6H_5O_7.5H_2O$	304.50	wh gran powd				20^{25}	s	s a; i al
m42	cyanide	$Mg(CN)_2$	76.35			d 300 to $MgCN_2$	d 600	s	d	
m43	cyanoplatinite	$MgPt(CN)_4.7H_2O$	449.58	red, pr	2.185^{16}	$-H_2O$, 45		v s	v s	i al, eth
m44	ferrite	$MgFeO_4$	200.00	blk, oct, 2.35	4.44–4.60	1750 ± 25				s conc HCl; i dil a, h HNO_3, al
m45	ferrocyanide	$Mg_2Fe(CN)_6.12H_2O$	476.76	pa yel cr		d *ca* 200		33		i al
m46	fluoride	Nat. sellaite. MgF_2	62.31	col, tetr, faint vlt, lumin, 1.378, 1.390; 3.14		1261	2239	0.0076^{18}	i	s HNO_3; sl s a; i al
m47	fluosilicate	$MgSiF_6$	166.39	wh, cr or powd				65		
m48	fluosilicate, hexahydrate	$MgSiF_6.6H_2O$	274.48	wh, trig	1.788	d 120		$64.8^{17.5}$		i al
m49	formate	$Mg(CHO_2)_2.2H_2O$	150.38	col, rhomb		$-2H_2O$, 100		14^{0} (anh)	24^{100} (anh)	i al, eth
m50	*ortho*germanate	Mg_2GeO_4	185.21	wh ppt				0.0016^{25}		s a; i alk
m51	germanide	Mg_2Ge	121.21			1115				
m52	hydride	MgH_2	26.33	wh tetr cr or mass		d 280 (vac)		d viol		i eth
m53	hydroxide	Nat. brucite. $Mg(OH)_2$	58.33	col, hex pl, 1.559, 1.580	2.36	$-H_2O$, 350		0.0009^{18}	0.004^{100}	s a, NH_4 salts
m54	iodate	$Mg(IO_3)_2.4H_2O$	446.18	wh, monocl	$3.3^{13.5}$	$-4H_2O$, 210	d	10.2^{20}	19.3^{100}	
m55	iodide	MgI_2	278.12	wh, hex, deliq	4.43^{25}_{4}	d < 637		148^{18}	164.9^{110}	s al, eth, NH_3
m56	iodide, octahydrate	$MgI_2.8H_2O$	422.24	wh powd, deliq		d 41		81^{20}	90.3^{20}	s al, eth
m57	lactate	$Mg(C_3H_5O_3).3H_2O$	256.50	wh cr powd, v bitter taste				3.3	16.7^{100}	i al, eth
m58	laurate	$Mg(C_{12}H_{23}O_2)_2.2H_2O$	458.97	wh lumps		150.4		0.007^{25}	0.041^{100}	0.415^{15} al: 0.012^{25} eth
m59	*per*manganate	$Mg(MnO_4)_2.6H_2O$	370.27	dk purp need, deliq	2.18(?)	d		v s	d	s MeOH, ac a
m60	molybdate	$MgMoO_4$	184.25	rhomb, tricl	2.208			13.7^{25}		
m61	myristate	$Mg(C_{14}H_{27}O_2)_2$	479.05	wh powd		131.6		$0,006^{15}$	0.014^{50}	0.189^{25} al; 0.007^{25} eth
m62	nitrate, dihydrate	$Mg(NO_3)_2.2H_2O$	184.35	col pr	2.0256^{25}	129		s	s	s al, liq NH_3; al s conc HNO_3
m63	nitrate, hexahydrate	$Mg(NO_3)_2.6H_2O$	256.41	col, monocl, deliq	1.6363^{25}_{4}	89	d 330	125	v s	s al, liq NH_3
m64	nitride	Mg_3N_2	100.95	grn-yel, powd or mass	2.712^{25}_{4}	d 800	subl 700 (vac)	d	d	s a; i al
m65	nitrite, trihydrate	$Mg(NO_2)_2.3H_2O$	170.37	wh pr, hygr		d 100		s		s al
m66	oleate	$Mg(C_{18}H_{33}O_2)_2$	587.24	yel powd or mass				0.024^{25}		6.64^{20} al; s linseed oil; al s eth
m67	oxalate	$MgC_2O_4.2H_2O$	148.36	wh powd	2.45	d 150		0.07^{16}	0.08^{100}	s alk, a, oxalate
m68	oxide	Nat. periclase. MgO	40.31	col, cub, 1.736	$3,58^{25}$	2852	3600	0.00062	0.0086^{30}	s a, NH_4 salts; i al
m69	oxide, per-	MgO_2	56.31	wh powd				i	i	s a
m70	palmitate	$Mg(C_{16}H_{31}O_2)_2$	535.16	wh cr need or lumps		121.5		0.008^{25}	0.009^{50}	0.047^{25} al; 0.003^{25} eth
m71	*ortho*phosphate	$Mg_3(PO_4)_2$	262.88	rhomb pl, iridisc		1184		i	i	s NH_4 salts, i liq NH_3
m72	*ortho*phosphate	$Mg_3(PO_4)_2.22H_2O$	659.22	col, monocl pr	1.640^{15}	$-18H_2O$, 100	d 200	v sl s		d a
m73	*ortho*phosphate, mono-H	Nat. newberyite. $MgHPO_4.3H_2O$	174.34	wh, rhomb, 1.514, 1.518, 1.533	2.123^{15}	$-H_2O$, 205	d 550–650	sl s		s a
m74	*ortho*phosphate, mono-H, heptahydrate	$MgHPO_4.7H_2O$	246.40	wh, monocl need	1.728^{15}	$-4H_2O$, 100	d 550–650	0.3	0.2	s a; i al
m75	*ortho*phosphate, octahydrate.	Nat. bobierite. $Mg_3(PO_4)_2.8H_2O$	407.00	wh, monocl pl, 1.510, 1.520, 1.543	2.195^{15}	$-5H_2O$, 150	$-8H_2O$, 400	v sl s		s NH_4 citrate
m76	*ortho*phosphate, tetrahydrate	$Mg_3(PO_4)_2.4H_2O$	334.97	monocl	1.64^{15}			0.0205		s a; i NH_4 salts
m77	phosphide	Mg_3P_2	134.88	yel-grn cub cr	2.055			d	d	d dil min a; sl d conc H_2SO_4

No.	Name	Synonyms and Formulae	Mol. wt.	Crystalline form, properties and index of refraction	Density or spec. gravity	Melting point, °C	Boiling point, °C	Solubility, in grams per 100 cc: Cold water	Hot water	Other solvents
	Magnesium									
m78	*hypo*phosphite	$Mg(H_2PO_2)_2.6H_2O$	262.38	wh, ditetrag	$1.59_4^{12.5}$	$-5H_2O$, 100	$-6H_2O$, 180	20^{25}		i al, eth
m79	*ortho*phosphite	$MgHPO_3.3H_2O$	158.34						0.25	s a
m80	*pyro*phosphate	$Mg_2P_2O_7$	222.57	col, tab monocl, 1.602, 1.604, 1.615	2.559, (3.06)	1383		i	i	s a; i al
m81	platinocyanide	$MgPt(CN)_4.7H_2O$	449.58	red cr, 1.561	2.185^{16}	$-2H_2O$, 45		s	s	s al; i eth
m82	salicylate	$Mg(C_7H_5O_3)_2.4H_2O$	370.61	col or sl redsh cr powd, effl				s		s al
m83	selenate	$MgSeO_4.6H_2O$	275.36	col, monocl, 1.468, 1.489, 1.491	1.928			v s	v s	
m84	selenide	MgSe	103.27	lght gray powd or cr, 2.44	4.21			d	d	d a
m85	*meta*silicate	Nat. clinoenstatite. $MgSiO_3$	100.40	wh, monocl, α 1.651, γ 1.660	3.192_4^{25}	d 1557		i	i	v sl s HF
m86	*ortho*silicate	Nat. forstérite. Mg_2SiO_4	140.71	wh, orthorhmb, 1.65, 1.66, 1.67	3.21	1910		i		d h HCl
m87	(di-) silicide	Mg_2Si	76.71	blue cub	1.94	1102		i	d	s a, NH_4Cl, HCl
m88	silicofluoride	$MgSiF_6.6H_2O$	274.48	wh hex-rhomb, 1.3439, 1.3602	1.788	d 100		60^{25}	5	s dil a, v sl s HF; i al
m89	stannide	Mg_2Sn	167.31	blsh-wh met		778		s		s dil HCl
m90	stearate	$Mg(C_{18}H_{35}O_2)_2$	591.27	wh powd or lumps		86-88		0.003^{15} 0.004^{25}	0.008^{50}	0.020^{25} al; 0.003^{25} eth
m91	sulfate	$MgSO_4$	120.37	col, rhomb cr, 1.56	2.66	d 1124		26^{0}	73.8^{100}	s al, glyc; 1.16^{18} eth; i acet
m92	sulfate, heptahydrate	Epsom salt, nat. epsomite. $MgSO_4.7H_2O$	246.48	col, rhomb or monocl, 1.433, 1.455, 1.461	1.68	$-6H_2O$, 150	$-7H_2O$, 200	71^{20}	91^{40}	sl s al, glyc
m93	sulfate, monohydrate	Nat. kieserite, $MgSO_4.H_2O$	138.39	col, monocl pr, 1.523, 1.535, 1.586	2.445				68.4^{100}	
m94	sulfide	MgS	56.38	pa red-brn, cub, phosph, 2.271	2.84	d >2000		d	d	s a, PCl_3
m95	sulfite	$MgSO_3.6H_2O$	212.47	wh, rhomb or hex, 1.511, 1.464 (hex)	1.725	$-6H_2O$, 200	d	1.25	s	i al, NH_3
m96	*d*-tartrate	$MgC_4H_4O_6.5H_2O$	262.46	wh, rhomb	1.67	$-4H_2O$, 100	$-5H_2O$, 200	0.8^{15}	1.44^{90}	s min a; i al, NH_3
m97	*d*-tartrate, hydrogen	$Mg(HC_4H_4O_6)_2.4H_2O$	394.54	wh, rhomb	1.72				1.893^{100}	
m98	telluride	MgTe	151.91	wh, hex cr	3.86			d		d a
m99	thiosulfate	$MgS_2O_3.6H_2O$	244.53	col, rhomb pr	1.818^{24}	$-3H_2O$, 170	d	v s	v s	i al
m100	thiotellurite	Mg_3TeS_6	360.86	pa yel cr mass				s	s	s al
m101	tungstate	$MgWO_4$	272.16	col, monocl	5.66			i		d a; i al
m102	**Manganese**	Mn	54.938	gray-pink met, cub or tetr	7.20	1244 ± 3	1962	d	d	s dil a
m103	(II) acetate	$Mn(C_2H_3O_2)_2$	173.02	brn cr	1.74			s, d		s al
m104	acetate, tetrahydrate	$Mn(C_2H_3O_2)_2.4H_2O$	245.08	pa red, monocl	1.589			s		s al
m105	arsenide, mono-	MnAs	129.86	blk, hex	6.17–6.20 (5.55)	d 400		i	i	s HCl, aq reg
m106	arsenide, di-	Mn_2As	184.80			1400		i	i	s aq reg
m107	arsenide, tri-	Mn_3As_2	314.66	magnetic, (exist ?)				i	i	s aq reg
m108	(II) benzoate	$Mn(C_7H_5O_2)_2.4H_2O$	369.23	pa red pr				6.55^{15}		
m109	boride, di-	MnB_2	76.56	gray-vlt cr	6.9			d	d	s a
m110	boride, mono-	MnB	65.75	cr powd	6.2^{15}					
m111	bromide, di-	$MnBr_2$	214.76	rose cr	4.385_4^{25}	d		127.3^{0}	228^{100}	i NH_3
m112	bromide, di-, tetrahydrate	$MnBr_2.4H_2O$	286.82	α stable, rose monocl, deliq β labile, col, rhomb		d 64.3		296.7^{0}		
m113	carbide	Mn_3C	176.83	tetr	6.89^{17}			d	d	s a
m114	(II) carbonate	Nat rodochrosite. $MnCO_3$	114.95	rose, rhomb, lt brn in air	3.125	d		0.0065^{25}		s dil a, aq CO_2; i al, NH_3
m115	chloride, di-	Scacchite. $MnCl_2$	125.84	pink, cub cr, deliq	2.977_4^{25}	650	1190	72.3^{25}	123.8^{100}	s al; i eth, NH_3
m116	chloride, di-, tetrahydrate	$MnCl_2.4H_2O$	197.91	rose, monocl, deliq	2.01	58	$-H_2O$, 106; $-4H_2O$, 198	151^{8}	656^{100}	s al; i eth
m117	chloride, tri-	$MnCl_3$	161.30	brn cr or grnsh-blk		d sl				s abs al
m118	chloroplatinate	$MnPtCl_6.6H_2O$	570.84	trig	2.692					
m119	chromite	$MnCr_2O_4$	222.93	gray-blk, cub	4.97^{20}			i		i a
m120	(II) citrate	$Mn_3(C_6H_5O_7)_2$	543.02	wh-redsh powd				v sl s		s Na-citr sol
m121	(II) ferrocyanide	$Mn_2Fe(CN)_6.7H_2O$	447.94	grnsh-wh powd				i		s HCl; i NH_4 salts
m122	fluogallate	$[Mn(H_2O)_6][GaF_5.H_2O]$	345.76	pink, orthorhomb, 1.45	2.22	d 230		v s		s HF
m123	fluosilicate	$MnSiF_6.6H_2O$	305.11	rose, hex pr, 1.357, 1.374	1.903	d		140^{25}	v s	s al

No.	Name	Synonyms and Formulae	Mol. wt.	Crystalline form, properties and index of refraction	Density or spec. gravity	Melting point, °C	Boiling point, °C	Solubility, in grams per 100 cc: Cold water	Hot water	Other solvents
	Manganese									
m124	fluoride, di-	MnF_2	92.93	red, tetr, or redsh powd	3.98	856		0.66^{40}	0.48^{100}	s a; i al, eth
m125	fluoride, tri-	MnF_3	111.93	red cr	3.54	d		d	d	s a
m126	formate	$Mn(CHO_2)_2.2H_2O$	181.00	rhomb	1.953	d		s	s	
m127	(II) glycerophosphate	$MnC_3H_7O_6P$	225.00	wh or sl red powd				sl s		s a, citr a; i al
m128	hydroxide	$MnO(OH)_2$	104.95	blk-brn, amorph (exist ?)	2.58			v sl s		
m129	(II) hydroxide	Nat. pyrochroite. $Mn(OH)_2$	88.95	wh-pink, trig 1.723, 1.681	3.258^{13}	d		0.0002^{18}		s a, NH_4 salts; i alk
m130	(III) hydroxide	Magnanite. $MnO(OH)$	87.94	br-blk, rhomb. 2.24, 2.24, 2.53 (Li)	4.2–4.4	d		i	i	s HCl, h H_2SO_4
m131	iodide, di-	MnI_2	308.75	pink, hex cr, deliq, br. in air	5.0^{4}	638 (vac) d ca 80	subl vac 500	s	s	0.02^{25} NH_3
m132	iodide, di-, tetrahydrate	$MnI_2.4H_2O$	380.81	rose, monocl, deliq		d		s	v s	
m133	*hexa*iodoplatinate	$MnPtI_6.9H_2O$	1173.59	trig	3.604^{20}_{4}	d				
m134	(II) nitrate	$Mn(NO_3)_2.4H_2O$	251.01	col, or rose, monocl	1.82	25.8	129.4	426.4^{0}	∞	v s al
m135	(II) lactate	$Mn(C_3H_5O_3)_2.3H_2O$	287.04	pa red, monocl		d		10	v s	s al
m136	(II) oxalate	MnC_2O_4	142.96	wh cr powd	$2.43^{21.7}$	d 150		i	i	s a, NH_4Cl
m137	(II) oxalate, dihydrate	$MnC_2O_4.2H_2O$	178.98	redsh-wh oct cr powd		$-2H_2O$, 100	d	0.0312^{25}	0.037^{25}	
m138	(II) oxalate, trihydrate	$MnC_2O_4.3H_2O$	197.00	pink, tricl		$-H_2O$, 25				
m139	(II, III) oxide	Nat. hausmannite. Mn_3O_4	228.81	blk, tetr (rhomb), 2.46, (Li) 2.15 (Li)	4.856	1564		i	i	s HCl
m140	oxide, di-	Nat. pyrolusite. MnO_2	86.94	blk, rhomb, or brn-blk powd	5.026	−O, 535		i	i	s HCl; i HNO_3, acet
m141	oxide, hept-	Mn_2O_7	221.87	dk red oil, hyg, exp	2.396^{20}_{4}	5.9	d 55, exp 95	v s	d	s H_2SO_4
m142	(II) oxide, mon-	Nat. manganosite, MnO	70.94	grn, cub, 2.16	5.43–5.46 (3.7–3.9)			i	i	s a, NH_4Cl
m143	(III) oxide, sesqui-	Nat. braunite. Mn_2O_3	157.87	blk, cub (tetr)	4.50	−O, 1080		i	i	s a; i ac a
m144	oxide, tri-	MnO_3	102.94	redsh, deliq (exist ?)		d		s	d	s alk, H_2SO_4
m145	(III) *meta*phosphate	$Mn_2(PO_3)_6.2H_2O$	619.74					sl s	s	
m146	(II) *ortho*phosphate	Nat. reddingite. $Mn_3(PO_4)_2.3H_2O$	408.80	rose or yelsh-wh rhomb, 1.651, 1.656, 1.683	3.102					
m147	(III) *ortho*phosphate	$MnPO_4.H_2O$	167.92	gray cr powd		$-H_2O$, 300	d	i		s h conc H_2SO_4, conc HCl, molten H_3PO_4
m148	(II) *ortho*phosphate, di-H	$Mn(H_2PO_4)_2.2H_2O$	284.94			$-H_2O$, >100		s		i al
m149	(II) *ortho*phosphate mono-H	$MnHPO_4.3H_2O$	204.97	red, rhomb or pink powd, 1.656				sl s	d	s a; i al
m150	(II) *pyro*phosphate	$Mn_2P_2O_7$	283.82	br-pink, monocl, 1.695, 1.704, 1.710	3.707^{25}	1196		i		s a
m151	(II) *pyro*phosphate, trihydrate	$Mn_2P_2O_7.3H_2O$	337.87	wh, amorph powd				i		s $K_4P_2O_7$ sol, H_2SO_3 i acet
m152	phosphide, mono-	MnP	85.91	dk gray	5.39^{21}	1190		i	i	sl s HNO_3
m153	(tri-)phosphide, di-	Mn_3P_2	226.76	dk gray	5.12^{18}	1095		i	i	sl s dil HNO_3
m154	(II) *hypo*phosphite	$Mn(H_2PO_2)_2.H_2O$	202.93	rose cr or powd		$-H_2O$, >150		12.5	16.7	i al
m155	(II) *ortho*phosphite	$MnHPO_3.H_2O$	152.93	redsh		$-H_2O$, 200		sl s		s $MnSO_4$, $MnCl_2$
m156	selenate, dihydrate	$MnSeO_4.2H_2O$	233.93	rhomb	2.95–3.01			s	s	
m157	selenate, pentahydrate	$MnSeO_4.5H_2O$	287.97	pa red, trig	2.33–2.39					
m158	selenide	MnSe	133.90	gray, cub	5.55^{15}			i		d dil a
m159	selenite	$MnSeO_3.2H_2O$	217.93	monocl cr				v sl s	v sl s	
m160	(II) *meta*silicate	Nat. rhodonite. $MnSiO_3$	131.02	red, tricl, 1.733, 1.740, 1.744	3.72^{25}	1323		i		i HCl
m161	silicide, di-	$MnSi_2$	111.11	gray, oct	5.24^{13}			i	i	s HF, alk; i HNO_3, H_2SO_4
m162	silicide, mono-	MnSi	83.02	tetrah	5.90^{15}	1280		i	i	s HF; v sl s a
m163	(di-)silicide,	Mn_2Si	137.96	quadr pr	6.20^{15}	1316		i	i	s HCl, NaOH; i HNO_3
m164	(II) sulfate	$MnSO_4$	151.00	redsh	3.25	700	d 850	52^{5}	70^{70}	s al; i eth
m165	(III) sulfate	$Mn_2(SO_4)_3$	398.06	grn cr, deliq, hex	3.24	d 160		d	d	s HCl, dil H_2SO_4; i conc. H_2SO_4, HNO_3

No.	Name	Synonyms and Formulae	Mol. wt.	Crystalline form, properties and index of refraction	Density or spec. gravity	Melting point, °C	Boiling point, °C	Solubility, in grams per 100 cc: Cold water	Hot water	Other solvents
	Manganese									
m166	(II) sulfate, dihydrate	$MnSO_4.2H_2O$	187.03	(exist ?)	2.526^{15}	stab 57–117	...	85.27^{35}	106.8^{55}	...
m167	(II) sulfate, heptahydrate	$MnSO_4.7H_2O$	277.11	red monocl or rhomb	2.09	$-7H_2O$, 280; stab +9	...	172	118^{13}	i al
m168	(II) sulfate, hexahydrate	$MnSO_4.6H_2O$	259.09	(exist?)	...	stab +5 to +8	...	147.4	1.345^{38}	...
m169	(II) sulfate, monohydrate	Nat. szmikite. $MnSO_4.H_2O$	169.01	pa pink monocl, 1.562, 1.595, 1.632	2.95	stab 57–117	...	98.47^{48}	79.8^{100}	
m170	(II) sulfate, pentahydrate	$MnSO_4.5H_2O$	241.08	rose, tricl, 1.495, 1.508, 1.514	2.103^{15}	stab 9–26	...	124^{0}	142^{54}	...
m171	(II) sulfate, tetrahydrate	Common form. $MnSO_4.4H_2O$	223.06	pink, monocl or rhomb effl, 1.508, 1.522	2.107	stab 26–27	...	105.3^{0}	111.2^{54}	i al
m172	(II) sulfate, trihydrate	$MnSO_4.3H_2O$	205.05	(exist?)	2.356^{15}	stab 30–40	...	74.22^{5}	99.31^{57}	...
m173	(II) sulfide	Nat. alabandite. MnS	87.00	grn cub or pink amorph, 2.70 (Li)	3.99	d	...	0.00047^{18}	...	s dil a, al; i $(NH_4)_2S$
m174	(II) sulfide	$3MnS.H_2O$	279.05	gray-pink	...	d	...	0.0006	i	s dil a; i $(NH_4)_2S$
m175	(IV) sulfide	Nat. hauerite. MnS_2	119.07	blk cub, 2.69 (Li)	3.463	d	...	i	i	d HCl
m176	(II) tantalate	$Mn(TaO_3)_2$	512.83	blk, rhomb, 2.22, 2.25, 2.29	7.03	...	...	...	...	...
m177	(II) tartrate	$MnC_4H_4O_6$	203.01	wh powd	...	...	...	v sl s	...	s dil a
m178	(II) thiocyanate	$Mn(SCN)_2.3H_2O$	225.16	deliq	...	$-3H_2O$, 160–170	...	s	v s	v s al
m179	(II) *di*thionate	$Mn(SO_3)_2$	215.06	tricl cr	1.757	...	...	s	v s	...
m180	(II) titanate	Nat. pyrophanite. $MnTiO_3$	150.84	yel, trig, 2.481, 2.210	4.54	1360	...	...	...	...
m181	valerate	$Mn(C_5H_9O_2)_2.2H_2O$	293.22	br powd	...	...	...	s	...	...
m182	**Manganic acid, per-**	$HMnO_4$	119.94	...	...	...	...	v s	d	...
m183	**Manganocyanic acid**	$H_4Mn(CN)_6$	215.08	...	...	d	...	i	...	v s al; i eth
m184	**Mercury**	Quicksilver. Hg	200.59	silv wh met, liq.	13.5939^{20}_{4}	−38.87	356.58	i	i	s HNO_3; i dil HCl, HBr, HI cold H_2SO_4
m185	(I) acetate	$Hg_2(C_2H_3O_2)_2$	519.27	micaceous plates	...	d	...	0.75^{12}	...	s HNO_3, H_2SO_4; i al, eth
m186	(II) acetate	$Hg(C_2H_3O_2)_2$	318.76	wh, sc or powd	3.270	d	...	25^{10}	100^{100}	s al, ac a
m187	(II) acetylide	$3HgC_2.H_2O$	691.85	wh powd	5.3	expl	...	i	i	i al
m188	(II) *ortho*arsenate	$Hg_3(AsO_4)_2$	879.61	yel	...	...	...	v sl s	...	s HCl, HNO_3
m189	(I) *ortho*arsenate mono-H	Hg_2HAsO_4	541.11	...	...	...	...	i	...	s HNO_3; i ac a, NH_4OH
m190	(I) azide	$Hg_2(N_3)_2$	485.22	wh cr	...	expl, d by light	...	0.025	...	...
m191	(II) benzoate	$Hg(C_7H_5O_2)_2.H_2O$	460.84	wh cr powd	...	165	...	1.2^{15}	2.5^{100}	s al, NaCl, NH_4Cl, bz
m192	(I) bromate	$Hg_2(BrO_3)_2$	656.99	cr	...	d	...	d	...	sl s HNO_3
m193	(II) bromate	$Hg(BrO_3)_2.2H_2O$	492.44	cr	...	d 130–140	...	0.15	1.6	s HCl, HNO_3, $Hg(NO_3)_2$
m194	(I) bromide	Hg_2Br_2	561.00	wh, yel, tetr	7.307	subl 345	...	0.000004^{25}	...	s a; i al, acet
m195	(II) bromide	$HgBr_2$	360.41	col rhomb	6.109^{25} 5.12^{240}	236	322	0.61^{25}	4.0^{100}	15^{0} al; s MeOH; v sl s eth
m196	bromide iodide	HgBrI	407.40	yel, rhomb	...	229	360	...	...	s al, eth
m197	(I) carbonate	Hg_2CO_3	461.19	yel br cr	...	d 130	...	0.0000045	d	s NH_4Cl; i al
m198	(II) carbonate, basic	$HgCO_3.2HgO$	693.78	br red	...	...	...	i	...	s NH_4Cl, aq CO_2
m199	(I) chlorate	$Hg_2(ClO_3)_2$	568.08	wh, rhomb	6.409	d 250	...	s	d	s al, ac a
m200	(II) chlorate	$Hg(ClO_3)_2$	367.49	wh need	4.998	d	...	25	...	...
m201	(I) chloride	Calomel. Hg_2Cl_2	472.09	wh, tetr, 1.973, 2.656	7.150	subl 400	...	0.00020^{25}	0.001^{43}	s aq reg, $Hg(NO_3)_2$; sl s HCl, h HNO_3; i al, eth
m202	(II) chloride	Corrosive sublimate. $HgCl_2$	271.50	col, rhomb or wh powd pois, 1.859	5.44^{25}, liq 4.44^{280}	276	302	6.9^{20}	48^{100}	33^{25} al; 4 eth; s ac a, pyr
m203	(I) chromate	Hg_2CrO_4	517.17	red need, or powd	...	d	...	v sl s	sl s	s HCN, HNO_3; i al, ac
m204	(II) chromate	$HgCrO_4$	316.58	red, rhomb	...	d	...	sl s, d	d	s HN_4Cl; d a; i acet
m205	(II) cyanide	$Hg(CN)_2$	252.63	col, tetr, or wh powd pois, 1.645	3.996	d	...	9.3^{14}	33^{100}	$25^{19.5}$ MeOH; 8 al; s NH_3, glyc; i bz
m206	(I) fluoride	Hg_2F_2	439.18	yel, cub	8.73^{15}_{4}	570	d	d to Hg_2O	...	...
m207	(II) fluoride	HgF_2	238.59	col, cub	8.95^{15}	d 645	650	d	...	s HF, dil HNO_3
m208	(I) fluosilicate	$Hg_2SiF_6.2H_2O$	579.29	col pr	2.134	...	...	sl s		i HCl

No.	Name	Synonyms and Formulae	Mol. wt.	Crystalline form, properties and index of refraction	Density or spec. gravity	Melting point, °C	Boiling point, °C	Solubility, in grams per 100 cc: Cold water	Hot water	Other solvents
	Mercury									
m209	(II) fluosilicate	$HgSiF_6.6H_2O$	450.76	col, rhomb, deliq		d easily				
m210	(I) formate	$Hg_2(CHO_2)_2$	491.22	glist scales		d		0.4^{17}	d	i al
m211	(II) fulminate	$Hg(CNO)_2$	284.62	wh, cub	4.42	expl		sl s	s	s al, NH_4OH
m212	(I) iodate	$Hg_2(IO_3)_2$	750.99	yelsh		d 250		i	i	s dil HCl, conc HNO_3
m213	(II) iodate	$Hg(IO_3)_2$	550.48	wh, amorph powd				i		s HCl, NH_4Cl, NaCl, KI; i HNO_3
m214	(I) iodide	Hg_2I_2	654.99	yel, tetr or amorph powd	7.70	subl 140	d 290	v sl s		s KI, NH_4OH; i al, eth
m215	(II) iodide (α)	HgI_2	454.90	red, tetr	6.36^{25}_{4}	tr 127		0.01^{25}		3.18^{25} acet; 2.23^{25} al; s chl; d NH_4OH
m216	(II) iodide (β)	HgI_2	454.90	yel, rhomb cr or powd	6.094^{127}_{4}	259	354	v sl s	sl s	s eth, KI, $Na_2S_2O_3$; v sl s al
m217	(I) nitrate	$Hg_2(NO_3)_2.2H_2O$	561.22	col, monocl, effl.	4.79^{4}	70		d	s, d	s dil HNO_3; i NH_4OH
m218	(II) nitrate	$Hg(NO_3)_2.\frac{1}{2}H_2O$	333.61	wh-yelsh cr or powd, deliq	4.39	79	d	v s	d	s acet, HNO_3, NH_3; i al
m219	(II) nitrate	$Hg(NO_3)_2.H_2O$	342.61	col cr or wh powd, deliq	4.3			s		s HNO_3; i al
m220	(I) nitrite	$Hg_2(NO_2)_2$	493.19	yel	7.33	d 100		d		
m221	nitride	Hg_3N_2	629.78	br powd		expl		d		s NH_4OH, NH_4 salts; d a
m222	(I) oxalate	$Hg_2C_2O_4$	489.20					i	i	sl s HNO_3
m223	(II) oxalate	HgC_2O_4	288.61			d		0.0107^{20}		s HCl; sl s HNO_3
m224	(I) oxide	Hg_2O	417.18	blk or brnish-blk powd	9.8	d 100		i	i	s HNO_3
m225	(II) oxide	Nat. montroydite. HgO	216.59	yel or red, rhomb, 2.37, 2.5, 2.65	11.1^{4}	d 500		0.0053^{25}	0.0395^{100}	s a; i al, eth, acet, alk, NH_3
m226	(II) oxybromide	$HgBr_2.3HgO$	1010.17	yel cr				i	sl s	v s al
m227	(II) oxychloride	$HgCl_2.2HgO$	704.67	red hex, or blk monocl	red 8.16–8.43 blk 8.53					
m228	(II) oxychloride	$HgCl_2.3HgO$	921.26	yel, hex	7.93	d 260		i	d	
m229	(II) oxycyanide	$Hg(CN)_2.HgO$	469.22	wh need or cr powd	4.437^{19}	expl		1.25	s	
m230	(II) oxyfluoride	$HgF_2.HgO.H_2O$	473.19	yel cr		d 100		d		s dil HNO_3
m231	(II) oxyiodide	$HgI_2.3HgO$	1104.17	yel br					d	s HI
m232	(II) selenide	Nat. tiemannite. HgSe	279.55	gray plates	8.266	vac subl		i		s aq reg
m233	(I) sulfate	Hg_2SO_4	497.24	col monocl, wh-yelsh powd	7.56	d	d	0.06^{25}	0.09^{100}	s HNO_3, H_2SO_4
m234	(II) sulfate	$HgSO_4$	296.65	col rhomb or wh powd	6.47	d		d		s a, NaCl; i al, acet, NH_3
m235	(II) sulfate, basic	$HgSO_4.2HgO$	729.83	lem yel powd	6.44		volat	0.003^{16}	sl s	s a; i al
m236	(I) sulfide	Hg_2S	433.24	blk		d		i		i al, $(NH_4)_2S$
m237	(II) sulfide (α)	Cinnabar, vermillion. HgS	232.65	red cr hex, or powd, 2.854, 3.201	8.10	subl 583.5		0.000001^{18}		s aq reg, Na_2S; i al, HNO_3
m238	(II) sulfide (β)	Metacinnabar. HgS	232.65	blk, cub or amorph powd	7.73	583.5		i		s aq reg, Na_2S, alk; i al, HNO_3
m239	(I) tartrate	$Hg_2C_4H_4O_6$	549.25	yelsh-wh cr powd.				i		i a
m240	(I) *ortho*tellurate	HgH_4TeO_6	428.22	trans, orthorhomb		d 20		slow d	rapid d	
m241	(II) *ortho*tellurate	Hg_3TeO_6	825.37	amber, cub		unalt at 140		i	i	s HCl, HNO_3
m242	(I) thiocyanate	$Hg_2(SCN)_2$	517.34			d		i		s HCl, KCNS
m243	(II) thiocyanate	$Hg(SCN)_2$	316.75	wh powd, pois.		d		0.07^{25}	s	s NH_4 salts, HCl, NH_3, KCN; sl s al, eth
m244	(I) tungstate	Hg_2WO_4	649.03	yel, amorph		d		i	i	d a; i al
m245	(II) tungstate	$HgWO_4$	448.44	yel		d		i	d	d a; i al
	Mercury nitrogen compounds									
m246	mercury (II) bromide, ammonobasic	$Hg(NH_2)Br$	296.52	wh powd		d		d		s NH_4OH; i al
m247	mercury (II) bromide, diammine	$Hg(NH_3)_2Br_2$	394.47	wh powd		180		d		s NH_4Cl, NH_4Br, NH_4I
m248	mercury (II) chloride ammonobasic	Infusible ppt. $Hg(NH_2)Cl$	252.07	wh powd or sm pr	5.70	infus		0.14	d 100	d a; i al
m249	mercury (II) chloride, aquobasic ammonobasic	Chloride of Millon's base. OHg_2NH_2Cl	468.66	pa yel or wh powd		d >120		sl s		s HCl, HNO_3
m250	mercury (II) chloride, diammine	Fusible white ppt. $Hg(NH_3)_2Cl_2$	305.56	rhombd		300		i	d	s a, KI

No.	Name	Synonyms and Formulae	Mol. wt.	Crystalline form, properties and index of refraction	Density or spec. gravity	Melting point, °C	Boiling point, °C	Solubility, in grams per 100 cc: Cold water	Hot water	Other solvents
	Mercury nitrogen compounds									
m251	mercury (II) iodide, ammonobasic	$Hg(NH_2)I$	353.52							i eth
m252	mercury (II) iodide aquobasic ammonobasic	Iodide of Millon's base. OHg_2NH_2I	560.11	yel to brn		>128	expl	i		s d, HCl, KI soln
m253	mercury (II) iodide, diammine	$Hg(NH_3)_2I_2$	488.48	col or pa yel powd or need				d		s NH_4OH
m254	**Millons's base**	$(HO)_2Hg_2NH_2OH$	468.22		4.083^{18}					
m255	**Molybdenum**	Mo	95.94	silv-wh met, or gray-blk powd, cub	10.2	2617	4612	i	i	s h conc HNO_3, h conc H_2SO_4, aq reg; sl s HCl; i HF, NH_3
m256	boride, (di-)	MoB_2	117.56	rhomb	7.12					
m257	boride, (mono-)	MoB	106.75	tetr	8.65					
m258	(di-) boride	Mo_2B	202.69	tetr	9.26					
m259	bromide, di-	$MoBr_2$ (or Mo_2Br_4)	255.76	yel red, amorph	$4.88^{17.5}$			i	i	s alk; i a, aq reg
m260	bromide, tetra-	$MoBr_4$	415.58	blk need, deliq		d	volat	v s		d alk
m261	bromide, tri-	$MoBr_3$	335.67	dk-grn need		d		i	i	d alk, NH_3; i a
m262	carbide, mono-	MoC	107.95	gray, hex	8.20^{20}	2692		i	i	sl s HNO_3, HF, h H_2SO_4, HCl; i alk hydr
m263	carbide(di-)	Mo_2C	203.89	wh, hex pr	8.9	2687		i		sl s HNO_3, HF h H_2SO_4, aq reg, HCl; i alk
m264	carbonyl	$Mo(CO)_6$	264.00	wh cr, rhomb diamagnet	1.96	d 150, without meltg	156.4^{766}	i	i	s bz; sl s eth
m265	carbonyl, tri-pyridine, tri-	$Mo(CO)_3(C_5H_5N)_3$	287.08	yel-brn cr		d				
m266	chloride, di-	$MoCl_2$ (or Mo_3Cl_6)	166.85	yel, amorph	3.714^{25}	d		i	i	s HCl, H_2SO_4, alk, NH_4OH, al, acet
m267	chloride, penta-	$MoCl_5$	273.21	grn-blk cr, trig, deliq	2.928	194	268	d	d	s conc min a, liq NH_3, CCl_4, chl; s d al, eth
m268	chloride, tetra-	$MoCl_4$	237.75	brn powd or cr, deliq		d	vol	d	d	s conc min a; d al, eth
m269	chloride, tri-	$MoCl_3$	202.30	dk red need or powd	3.578^{25}_{4}	d		i	sl d	s conc H_2SO_4, conc HNO_3; v sl s al, eth; i HCl; d alk
m270	fluoride, hexa-	MoF_6	209.93	col cr	liq $2.55^{17.5}$	17.5^{406}	35^{760}	s d	d	s NH_4OH, alk
m271	hydro*tetra*chloro-hydroxide, di-	$[Mo_3Cl_4(H_2O)_2](OH)_2.6H_2O$	607.77	lt yel cr		$-H_2O$, 35–300		i	i	i a, al
m272	hydroxide	$Mo(OH)_3$ (or $Mo_2O_3.3H_2O$)	146.96	blk powd		d		0.2		sl s H_2SO_4, HCl; s 30% H_2O_2
m273	hydroxide	$MoO(OH)_3$ (or $Mo_2O_5.3H_2O$)	162.96	br to blk powd				0.2 (coll)		s a, alk carb; i alk hydr
m274	(VI) hydroxide	$MoO_3.2H_2O$	179.97	lt yel, monocl pr.	3.124^{15}			0.05^{15}		s dil H_2O_2, alk hydr; sl s a
m275	hydroxy*tetra*-bromide, di-	$Mo_3Br_4(OH)_2$	641.47	red powd						s alk
m276	hydroxy*tetra*-bromide, diocta-hydrate	$Mo_3Br_4(OH)_2.8H_2O$	785.59	golden yel cr		d	d			s HCl; d alk, HNO_3
m277	hydroxy*tetra*-chloride, di-	$Mo_3Cl_4(OH)_2.2H_2O$	499.68	pa yel, amorph					i	s conc a; i al
m278	iodide, di-	MoI_2	349.75	brn powd	$5.278^{25.4}$			i	d	v sl s a
m279	iodide, tetra-	MoI_4	603.56	blk cr		d 100				
m280	oxide, di-	MoO_2	127.94	lead gray, tetr or monocl	6.47			i	i	sl s h conc H_2SO_4; i alk, HCl, HF
m281	oxide, pent-	Mo_2O_5	271.88	vlt-bl powd (exist?)						s h H_2SO_4, h HCl
m282	oxide, pent-	"Molybd. blue" $Mo_2O_5.xH_2O$ (variations in Mo and O)		dk blue coll or powd	3.6^{18}	d		s		s a, MeOH; i acet, bz, chl
m283	oxide, sesqui-	Mo_2O_3	239.88	blk, opaque (exist?)				i	i	i a, alk, NH_4OH
m284	oxide, tri-	Molybdic anhydride. Nat. Molybdite. MoO_3	143.94	col or wh-yel, rhomb	4.692^{21}	795	subl 1155^{760}	0.1066^{18}	2.055^{70}	s a, alk sulf, NH_4OH
m285	oxy*di*bromide, di-	MoO_2Br_2	287.76	yel-red tabl, deliq				s		

No.	Name	Synonyms and Formulae	Mol. wt.	Crystalline form, properties and index of refraction	Density or spec. gravity	Melting point, °C	Boiling point, °C	Solubility, in grams per 100 cc		
								Cold water	Hot water	Other solvents
	Molybdenum									
m286	oxy*tetra*chloride	$MoOCl_4$	253.75	grn cr, deliq		subl		s		
m287	oxy*tri*chloride	$MoOCl_3$	218.30	grn cr		subl 100		d		
m288	oxy*di*chloride, di-	MoO_2Cl_2	198.84	yelsh wh scaly cr	3.31^{17}	subl		s	s	s al, eth
m289	oxy*di*chloride, dihydrate	$MoO_2Cl_2.H_2O$	216.86	pa yel cr		subl		s	s	sl s al, acet, eth
m290	oxy*hexa*chloride, tri-	$Mo_2O_3Cl_6$	452.60	rubyred or dk vlt cr		d		d		s eth
m291	oxy*penta*chloride, tri-	$Mo_2O_3Cl_5$	417.14	dk brn-blk cr, deliq		melts easily	subl	s	s	
m292	oxychloride acid	$MoO(OH)_2Cl_2$	216.63	wh need, deliq		d 160		v s		s al, eth, acet
m293	oxy*di*fluoride, di-	MoO_2F_2	165.94	wh cr, hygr	3.494^{25}	subl 270		v s	v s	s al, MeOH; i eth, chl, tol
m294	oxy*tetra*fluoride	$MoOF_4$	187.93	col-wh, deliq	3.001^{25}	98	180	s		s al, eth, CCl_4, s d H_2SO_4; v sl s bz
m295	*meta*phosphate	$Mo(PO_3)_6$	569.77	yel powd	3.28^{0}			i	i	sl s h aq reg; i HCL, HNO_3, H_2SO_4
m296	phosphide	MoP (or Mo_2P_2)	126.93	gray-grn cr powd	6.167					s h HNO_3
m297	phosphide	MoP_2	157.89	blk powd	5.35^{25}					s HNO_3, h conc H_2SO_4, aq reg; i conc HCl
m298	silicide	$MoSi_2$	152.11	gray, met, tetr	$6.31^{20.5}$					i a, aq reg; v s HF+HNO_3
m299	sulfide, di-	Nat. molybdenite. MoS_2	160.07	blk luster, hex.	4.80^{14}	1185	subl 450, d in air	i	i	s h H_2SO_4, aq reg, HNO_3; i dil a, conc H_2SO_4
m300	sulfide, penta	$Mo_2S_5.3H_2O$	406.25	dk br powd		$-H_2O$, 135	d	i	i	s NH_4OH, alk, sulfides
m301	sulfide, sesqui-	Mo_2S_3	288.07	steel gray need	5.91^{15}	d 1100	vol 1200			i conc HCl; d h HNO_3
m302	sulfide, tetra-	MoS_4	224.20	brn powd		d		i	i	i a; s h H_2SO_4, alk sulfide
m303	sulfide, tri-	MoS_3	192.13	blk pl		d	d	sl s	s	s alk sulf, conc KOH
m304	**Molybdic acid**	$H_2MoO_4.H_2O$ (or $MoO_3.2H_2O$)	179.97	yel, monocl	3.124^{15}	$-H_2O$, 70	d	0.133^{18}	2.568^{70}	s alk hydr, alk carb; sl s a
m305	anhydrous	H_2MoO_4 (or $MoO_3.H_2O$)	161.95	wh or sl yelsh, hex	3.112	$-H_2O$, 70		sl s	sl s	s alk, NH_4OH, H_2SO_4; i NH_3
m306	**Molybdic arsenic acid**	$As_2O_5.6MoO_3.18H_2O$	1475.75	col, trig	$2.493^{19.8}$	$-15H_2O$, 150		v s		s abs al; i chl, CS_2
m307	**Molybdic phosphoric acid**	$H_7[P(Mo_2O_7)_6].28H_2O$	2365.71	yel, oct	2.53	78		d		
m308	**Molybdic silicic acid**	$H_8[Si(Mo_2O_7)_6].28H_2O$	2363.83	yel, tetr		45	d 100	600^{14}		s dil a; i bz, chl, CS_2
n1	**Neodymium**	Nd	144.24	silv-wh to yelsh met, hex to m.p. 868, cub from 868	hex 7.007 cub 6.80	1021	3068	d		
n2	acetate	$Nd(C_2H_3O_2)_3.H_2O$	339.39					26.2		
n3	acetylacetonate	$Nd[CH(COCH_3)_2]_3$	441.57	pink cr	1.618	150–152				
n4	*hexa*antipyrin *per*chlorate	$[Nd(C_{11}H_{12}N_2O)_6].(ClO_4)_3$	1571.99	rose, hex cr		d 285–289		0.99^{20}		
n5	bromate	$Nd(BrO_3)_3.9H_2O$	690.10	red, hex		66.7	$-9H_2O$, 150	151^{25}		
n6	bromide	$NdBr_3$	383.97	grn cr		684	1540	sl s		
n7	carbide	NdC_2	168.26	yel, hex leaf	5.15	d		d	d	s dil a; i conc HNO_3
n8	chloride	$NdCl_3$	250.60	rose-vlt pr	4.134^{25}	784	1600	96.7^{13}	140^{100}	44.5 al; i eth, chl
n9	chloride, hexahydrate	$NdCl_3.6H_2O$	358.69	red, rhomb		124	$-6H_2O$, 160	246^{13}	511^{100}	v s al
n10	chromate	$Nd_2(CrO_4)_3.8H_2O$	780.58	yel cr				0.027		
n11	fluoride	NdF_3	201.24	pa lilac		1410	2300	i	i	
n12	iodide	NdI_3	524.95	blk cr powd		775 ± 3	1370	s	s	
n13	kojate	$Nd(C_6H_5O_4)_3$	567.60	lt choc		d 275		i		
n14	manganous nitrate	$2Nd(NO_3)_3.3Mn(NO_3)_2.24H_2O$	1629.72	vlt-red	2.114	77		77.4^{30}		
n15	magnesium nitrate	$2Nd(NO_3)_3.3Mg(NO_3)_2.24H_2O$	1537.84	vlt-red	2.020	109		69.5^{30}		
n16	dimethylphosphate	$Nd[(CH_3)_2PO_4]_3$	519.35	pa lilac, hex pl				56.1^{25}	22.3^{100}	
n17	molybdate	$Nd_2(MoO_4)_3$	768.29	tetr, 2.005	5.14^{18}	1176				
n18	nickel nitrate	$2Nd(NO_3)_3.3Ni(NO_3)_2.24H_2O$	1641.04	blsh-grn	2.202	105.6		71.5^{30}		
n19	nitrate	$Nd(NO_3)_3.6H_2O$	438.35	tricl				152.9^{25}		s al, acet
n20	nitride	NdN	158.25	blk powd				d		
n21	oxalate	$Nd_2(C_2O_4)_3.10H_2O$	732.69	rose cr				0.000074^{25}		
n22	oxide	Neodymia. Nd_2O_3	336.48	lt bl powd, red fluores	7.24	2272 ± 20		0.00019^{20}	0.003^{75}	s a

No.	Name	Synonyms and Formulae	Mol. wt.	Crystalline form, properties and index of refraction	Density or spec. gravity	Melting point, °C	Boiling point, °C	Solubility, in grams per 100 cc: Cold water	Hot water	Other solvents
	Neodymium									
n23	sulfate	$Nd_2(SO_4)_3.8H_2O$	720.79	red, monocl, 1.41, 1.551, 1.562	2.85	1176		8^{20}	5.4^{40}	
n24	sulfide	Nd_2S_3	384.67	oliv grn powd	5.179^{11}	d		i	d	s dil a
n25	Neon	Ne	20.183	inert gas col sol, cub	gas: 0.9002^{0} g/1; liq: $1.204^{-245.9}$	−248.67	−246.048	1.47^{20} cm^3		s liq O_2
n26	Neptunium	Np	237.00	α: orthorhomb silvery β: tetr (above 278) γ: cub (above 500)	α; 20.45 β: 19.36^{313} γ: 18.0^{600}	640 ± 1 278 ± 5 stab 278–570	3902			s HCl
n27	bromide, tri-	$NpBr_3$	476.73	α: hex β: grn orthorhomb	α 6.92	subl *ca* 800		s		
n28	chloride, tetra-	$NpCl_4$	378.81	red-brn tetr	4.92	538		s		
n29	chloride, tri-	$NpCl_3$	343.36	wh, hex	5.38	*ca* 800		s		
n30	fluoride, hexa-	NpF_6	350.99	brn, orthorhomb.		53	d	d		
n31	fluoride, tetra-	NpF_4	312.99	grn, monocl	6.8			i		i conc HNO_3
n32	fluoride, tri-	NpF_3	294.00	purple, hex	9.12			i		
n33	iodide, tri-	NpI_3	617.71	brn, orthorhomb.	6.82			s		
n34	oxide, di-	NpO_2	269.00	apple grn, cub.	11.11			i		s conc a
n35	(tri-) oxide, octa-	Np_3O_8	839.00	brn, cub		d 500				s HNO_3
n36	Nickel	Ni	58.71	silv met, cub	8.90	1453	2732	i	i	s dil HNO_3; sl s HCl, H_2SO_4; i NH_3
n37	acetate	$Ni(C_2H_3O_2)_2$	176.80	grn pr	1.798	d				i al
n38	acetate, tetrahydrate	$Ni(C_2H_3O_2)_2.4H_2O$	248.86	grn pr	1.744	d	16			s dil al
n39	antimonide	Nat. breithauptite. NiSb	180.46	lt copper red, hex	7.54	1158	d 1400			
n40	*ortho*arsenate, octahydrate	$Ni_3(AsO_4)_2.8H_2O$	598.09	yelsh-grn powd	4.98			i		s a
n41	arsenide	Nat. niccolite. NiAs,	133.63	hex	7.57^{0}	968		i	i	s aq reg
n42	*ortho*arsenite, acid	$Ni_3H_6(AsO_3)_4.H_2O$	691.87	grn-wh		d		i		s a, alk
n43	benzenesulfonate	$Ni(C_6H_5SO_3)_2.6H_2O$	481.10	grn, monocl	1.628^{25}	$-H_2O$	d	14.3^{18}	51.5^{82}	5.9 al; 4.5 eth
n44	boride	NiB	69.52	pr	7.39^{18}			d	d	s aq reg, HNO_3
n45	bromate	$Ni(BrO_3)_2.6H_2O$	422.62	monocl	2.575	d		28		
n46	bromide	$NiBr_2$	218.53	yel brn, deliq	5.098^{27}	963		112.8^{0}	155.1^{100}	s al, eth, NH_4OH
n47	bromide, trihydrate	$NiBr_2.3H_2O$	272.57	yelsh-grn need, deliq		$-3H_2O$, 300		199^{0}	315.7^{100}	s al, eth, NH_4OH
n48	bromoplatinate	$NiPtBr_6.6H_2O$	841.35	trig	3.715					
n49	di-*N*-butyldithiocarbamate	(NBC).$Ni[(C_4H_9)_2NCSS]_2$	467.47	dk olive grn powd	1.29	89–90		i	i	sl s bz, pet comp; i al
n50	carbide	Ni_3C	188.14	dk gray powd	7.957^{25}					
n51	carbonate	$NiCO_3$	118.72	lt grn, rhomb		d		0.0093^{25}	i	s a
n52	carbonate, basic	$2NiCO_3.3Ni(OH)_2.4H_2O$	587.67	lt grn cr or brn powd		d		i	d	s a, NH_4 salts
n53	carbonate, basic	Zaratite. $NiCO_3$. $2Ni(OH)_2.4H_2O$(?)	376.23	emerald grn, cub, 1.56–1.61	2.6			i	i	s h dil HCl, NH_4OH
n54	carbonyl	$Ni(CO)_4$	170.75	col, volat, inflamm, liq, or need	1.32^{17}	−25	43	$0.018^{9.8}$		s aq reg, al, eth, bz, HNO_3; i dil a, dil alk
n55	chlorate	$Ni(ClO_3)_2.6H_2O$	333.70	dk red	2.07	d 80		0.9^{27}		
n56	*per*chlorate	$Ni(ClO_4)_2.6H_2O$	365.70	grn, hex need, 1.518, 1.498		140		222.5^{0}	273.7^{45}	s al, acet, chl
n57	chloride	$NiCl_2$	129.62	yel sc, deliq	3.55	1001	subl 973	64.2^{20}	87.6^{100}	s al, NH_4OH; i NH_3
n58	chloride, hexahydrate	$NiCl_2.6H_2O$	237.70	grn, monocl, deliq ~1.57				254^{20}	599^{100}	v s al
n59	chloropalladate	$NiPdCl_6.6H_2O$	485.92	hex	2.353					
n60	chloroplatinate	$NiPtCl_6.6H_2O$	574.61	trig	2.798					
n61	cyanide	$Ni(CN)_2$	110.75	yel-brn				i	i	s KCN
n62	cyanide, tetrahydrate	$Ni(CN)_2.4H_2O$	182.81	lt grn pl or powd pois		$-4H_2O$, 200	d	i	i	s KCN, NH_4OH, alk; sl s dil a
n63	ferrocyanide	$Ni_2Fe(CN)_6.xH_2O$		grn-wh	1.892(?)			i		s KCN, NH_4OH; i HCl
n64	fluogallate	$[Ni(H_2O)_6][GaF_5.H_2O]$	349.53	pa grn, monocl(?), 1.45	2.45	$-5H_2O$, 110		sl s		s HF
n65	fluoride	NiF_2	96.71	grn, tetr	4.63	subl 1000 (in HF)		4^{25}		s a, alk, eth, NH_3
n66	fluosilicate	$NiSiF_6.6H_2O$	308.88	grn, trig, 1.391, 1.407	2.134	d				
n67	formate, dihydrate	$Ni(CHO_2)_2.2H_2O$	184.78	grn cr	2.154	d		s		
n68	(II) hydroxide	$Ni(OH)_2$ (or $NiO.xH_2O$)	92.72	grn cr, or amorph	4.15(3.65)	d 230		0.013		s a, NH_4OH
n69	iodate	$Ni(IO_3)_2$	408.52	yel need	5.07			1.1^{30}	1.0^{90}	
n70	iodate, tetrahydrate	$Ni(IO_3)_2.4H_2O$	480.59	hex		d 100		1.4^{30}	1.1^{90}	
n71	iodide	NiI_2	312.52	blk cr, deliq	5.834	797		124.2^{0}	188.2^{100}	s al

No.	Name	Synonyms and Formulae	Mol. wt.	Crystalline form, properties and index of refraction	Density or spec. gravity	Melting point, °C	Boiling point, °C	Solubility, in grams per 100 cc: Cold water	Hot water	Other solvents
	Nickel									
n72	dimethylglyoxime	$Ni(HC_4H_6N_2O_2)_2$	288.94	scarlet red cr		subl 250		i	i	s a, abs al; i a ac, NH_4OH
n73	nitrate, hexahydrate	$Ni(NO_3)_2.6H_2O$	290.81,	grn, monocl, deliq	2.05	56.7	136.7	238.5^{0}	v s	s al, NH_4OH
n74	oleate	$Ni(C_{18}H_{33}O_2)_2$	621.64	grn oil		18–20				
n75	oxalate	$NiC_2O_4.2H_2O$	182.76	lt grn powd				i		s a, NH_4 salts; v sl s oxal a
n76	oxide, mono-	Nat. bunsenite. NiO	74.71	grn-blk, cub, 2.1818(red)	6.67	1984		i	i	s a, NH_4OH
n77	*ortho*phosphate	$Ni_3(PO_4)_2.8H_2O$	510.20	apple grn pl or emerald cr gran		d		i	i	s a, NH_4 salts; i me acet. et acet
n78	*pyro*phosphate	$Ni_2P_2O_7.xH_2O$		grn	3.93 (anhydr)					s a, NH_4OH
n79	(di-)phosphide	Ni_2P	148.39	gray cr	6.31^{15}	1112		i		s HNO_3+HF; i a
n80	(penta)phosphide, (di)-	Ni_5P_2	355.50	need or tabl cr		1185				
n81	(tri-)phosphide, (di-)	Ni_3P_2	238.08	dk grn-blk	5.99			i	i	s HNO_3; i HCl
n82	*hypo*phosphite	$Ni(H_2PO_2)_2.6H_2O$	296.78	grn	$1.82^{19.5}$	d 100		s		
n83	selenate	$NiSeO_4.6H_2O$	309.76	grn, tetr, 1.5393	2.314			s		
n84	selenide	NiSe	137.67	wh or gray, cub	8.46	red heat		i		s aq reg, HNO_3; i a, HCl
n85	silicide	Ni_2Si	145.51		7.2^{17}	1309		i		i a
n86	stearate	$Ni(C_{18}H_{35}O_2)_2$	625.67	grn powd		100		i		s CCl_4, pyr; sl s acet; i MeOH, eth
n87	sulfate	$NiSO_4$	154.78	yel, cub	3.68	d 848^{700}		29.3^{0}	83.7^{100}	i al, eth, acet
n88	sulfate, heptahydrate	Morenosite. $NiSO_4.7H_2O$	280.88	grn, rhomb, 1.467, 1.489, 1.492	1.948	99; $-H_2O$ 31.5	$-6H_2O$, 103	$75.6^{15.5}$	475.8^{100}	s al
n89	sulfate, hexahydrate	Single nickel salt. $NiSO_4.6H_2O$	262.86	α: bl, tetr β: grn, monocl, 1.511, 1.487	2.07	tr: 53.3	$-6H_2O$, 103	62.52^{0}	340.7^{100}	12.5 MeOH; v s al, NH_4OH
n90	sulfide, mono-	Nat. millerite. NiS	90.77	blk; trig or amorph	5.3–5.65	797		0.00036^{18}		s aq reg, HNO_3, KHS; sl s a
n91	sulfide, sub-	Heazlewoodite. Ni_3S_2	240.26	pa yelsh bronze met, lust	5.82	790		i		s HNO_3
n92	(II, III) sulfide	Poydomite. Ni_3S_4	304.39	gray-blk, cub	4.7			i		s HNO_3
n93	sulfite	$NiSO_3.6H_2O$	246.86	grn, tetrah				i		s HCl, H_2SO_4
n94	dithionate	$NiS_2O_6.6H_2O$	326.93	grn, tricl	1.908	d				
	Nickel Complexes									
n95	diaquotetrammine nickel (II) nitrate	$[Ni(NH_3)_4(H_2O)_2].(NO_3)_2$	286.87	grn cr				s		i al
n96	hexamminenickel (II) bromide	$[Ni(NH_3)_6]Br_2$	320.71	vlt powd	1.837			v s	d	
n97	hexamminenickel (II) chlorate	$[Ni(NH_3)_6](ClO_3)_2$	327.80		1.52	180		d to $Ni(NH_3)_4$		
n98	hexamminenickel (II) chloride	$[Ni(NH_3)_6]Cl_2$	231.80	blsh, cub	1.468^{25}			s	d	s NH_4OH; i al
n99	hexamminenickel (II) iodide	$[Ni(NH_3)_6]I_2$	414.70	pa bl, cub	2.101	d		d		s NH_4OH
n100	hexamminenickel (II) nitrate	$[Ni(NH_3)_6](NO_3)_2$	284.90	bl, oct or cub				4.46		
n101	tetrapyridinnickel (II) fluosilic	$[Ni(C_5H_5N)_4]SiF_6$	517.20	bl grn, rhomb	2.307					
n102	**Niobium**	Columbium. Nb	92.906	steel gray, lustr met cub, 1.80	8.57	2468 ± 10	4742	i	i	s fus alk; i HCl, HNO_3, aq reg
n103	boride	NbB_2	114.53	hex	6.97	2900(?)				
n104	bromide, penta-	$NbBr_5$	492.46	purp red		265.2	361.6	d		s al, ethyl bromide
n105	carbide	NbC	104.92	blk, cub or lav-ender-gray powd	7.6	3500		i		s HNO_3, HF
n106	chloride, penta-	$NbCl_5$	270.17	yel-wh, deliq	2.75	204.7	254	d		s al, HCl, CCl_4
n107	fluoride, penta-	NbF_5	187.90	col, monocl pr, hygr	3.293	72–73	236	d		s al; sl s chl, CS_2, H_2SO_4
n108	hydride	NbH	93.91	gray powd	6.6	infus				s HF, conc H_2SO_4; i HCl, alk, HNO_3
n109	nitride	NbN	106.91	blk, cub	8.4	2573		i		s $HF+HNO_3$; i HNO_3
n110	oxalate, hydrogen	$Nb(HC_2O_4)_5$	538.05	col, monocl				d	d	s $H_2C_2O_4$; d al
n111	oxide, di-	NbO_2	124.90	blk	5.9			i	i	sl s alk; i a
n112	oxide, mon- (or di-)	NbO (or Nb_2O_2)	108.91	blk, cub	7.30			i	i	s a, alk; i al, NHO_3
n113	oxide, pent-	Nb_2O_5	265.81	wh, rhomb	4.47	1485 ± 5		i	i	s HF, alk; i a
n114	oxide, pent-, hydrate	$Nb_2O_5.xH_2O$				d		i		s conc H_2SO_4, conc HCl, HF, alk; i NH_3
n115	oxide, tri-(sesqui)	Nb_2O_3	233.81	bl-blk		1780				

No.	Name	Synonyms and Formulae	Mol. wt.	Crystalline form, properties and index of refraction	Density or spec. gravity	Melting point, °C	Boiling point, °C	Solubility, in grams per 100 cc: Cold water	Hot water	Other solvents
	Niobium									
n116	oxybromide	$NbOBr_3$	348.63	yel cr		subl		d		s a
n117	oxychloride	$NbOCl_3$	215.26	col need		subl 400		s, d	d	s al, H_2SO_4; i HCl
n118	potassium fluoride	$NbOF_3.2KF.H_2O$	300.12	monocl leaf, lustr fatty				7.8	v s	
n119	**Nitric acid**	HNO_3	63.01	col liq, corr, pois, $1.397^{16.4}$	1.5027^{25}_{4}	−42	83	∞	∞	d al, viol; s eth
n120	const boil	68 % HNO_3+32 % H_2O		col liq	1.41^{20}_{4}		120.5	∞	∞	
n121	**Nitrogen**	N_2	28.0134	col gas, col liq, sol cub cr	gas 1.2506 g/l liq $0.8081^{-195.8}$ sol $1.026^{-252.5}$	−209.86	−195.8	2.33^{0} cm^3	1.42^{40} cm^3	sl s al
n122	chloride, tri-	NCl_3	120.37	yel oil or rhomb cr	1.653	<−40	<71, expl 95	i	d	s chl, bz, CCl_4, CS_2, PCl_3
n123	fluoride, tri-	NF_3	71.00	col gas	liq: 1.537^{-129}	−206.60	−128.8	v sl s		
n124	iodide, tri-	NI_3	394.72	blk, expl		expl	subl vac	i	d	s KCNS, $Na_2S_2O_3$
n125	iodide, tri-, monoammine	$NI_3.NH_3$	411.75	dk red, rhomb	3.5	d>20	expl	i	d	s HCl, KCNS, $Na_2S_2O_3$; i abs al
n126	oxide(ic)	NO	30.01	col gas, bl liq, sol liq 1.330^{-90}	gas 1.3402 g/l liq; $1.269^{-150.2}$	−163.6	−151.8	7.34^{0} cm^3	2.37^{60} cm^3	3.4 cm^3 H_2SO_4; 26.6 cm^3 al; s $FeSO_4$, CS_2
n127	oxide(ous)	N_2O	44.01	col gas or liq or cub cr, 1.0005^{20}_{760}	1.977^{0}_{760} g/l	−90.8	−88.5	130^{0} cm^3	56.7^{25} cm^3	s al, eth, H_2SO_4
n128	oxide, pent-	Nitric anhydride. N_2O_5	108.01	wh, rhomb or hex	1.642^{18}	30	d 47	s	d to HNO_3	s chl
n129	oxide, tri-	NO_3	62.00	blsh gas		d at ord temp				s eth
n130	peroxide	Nitrogen oxide, di-. NO_2	46.01	col sol, yel liq or brn gas, 1.40^{20}	1.4494^{20}_{20}	−11.20	21.2	s, d		s alk, CS_2, chl
n131	(di-) oxide(tri-)	Nitrous anhydride. N_2O_3	76.01	red-brn gas, bl sol or liq	1.447^{2}	−102	d 3.5	s	d	s eth, a, alk
n132	oxi (tri-) fluoride	NO_3F	81.06	col gas expl	liq; $1.507^{-45.9}$ sol: $1.951^{-193.2}$	−175	−45.9	d		s acet; expl al, eth
n133	sulfide, penta-	N_2S_5	188.33	gray cr		10–11	d	d	d	s CS_2, eth; i bz, al
n134	sulfide, tetra-	Tetranitrogen tetrasulfide sulfurnitride. N_4S_4	184.28	yel cr, 2.046	2.24^{18}	d 178				s al, bz, CS_2
n135	**Nitrosyl bromide**	NOBr	109.92	br gas or dk br liq	>1.0	−55.5	−2	d	d	s alk
n136	perchlorate	$NOClO_4.H_2O$	147.47	rhomb, deliq	2.169	d 100		d		expl al, eth
n137	chloride	NOCl	65.46	yel gas or yel-red liq or cr	gas: 2.99 g/l liq: 1.417^{-12}	−64.5	−5.5	d	d	s fum H_2SO_4
n138	fluoborate	$NOBF_4$	116.81	col, rhomb cr, hygr	2.185^{25}_{4}	subl $250^{0.01}$		d		
n139	fluoride	NOF	49.00	col gas	2.176 g/l	−134	−56			d to HNO_2+HF
n140	**Nitrosylsulfuric acid**	Chamber crystals. $NOHSO_4$	127.08	col, rhomb		d 73.5		d		s H_2SO_4
n141	**Nitrosylsulfuric anhydride**	$(NOSO_3)_2O$	236.14	tetr		217	360	d		s H_2SO_4
n142	**Nitrous acid**	HNO_2	47.01	only in sol (pa bl)				d		
n143	hypo-	$H_2N_2O_2$	62.03	wh, sol		expl		s		
n144	**Nitryl chloride**	NO_2Cl	81.46	pa yel-br gas	gas: 2.57 g/l liq: 1.32^{14}	<−31	5	d		
n145	fluoride	NO_2F	65.00	col gas, col sol	2.90 g/l	−139	−63.5	d		d al, eth, chl
o1	**Osmium**	Os	190.20	gray-blsh met, hex	22.48^{20}	3045	5027±100	i	i	sl s aq reg, HNO_3; i NH_3
o2	carbonyl chloride	$Os(CO)_3Cl_2$	345.14	col pr		269–273	d 280	i	i	s NaOH; i a
o3	chloride, di-	$OsCl_2$	261.11	dk brn, deliq	d			i	sl d	s al, eth, HNO_3; sl s alk
o4	chloride, tetra-	$OsCl_4$	332.01	red br need		subl		sl s, d		i al
o5	chloride, tri-	$OsCl_3$	296.56	br, cub		d 500–600		v s		s alk, al, a; sl s eth
o6	chloride, tri-, trihydrate	$OsCl_3.3H_2O$	350.61	dk gr cr		d		v s		s al
o7	fluoride, hexa-	OsF_6	304.19	grn cr		32.1	45.9	d	d	
o8	fluoride, tetra-	OsF_4	266.19	br powd				d	d	
o9	iodide	OsI_4	697.82	vlt-blk, hygr met, lust				v s	d	s al

No.	Name	Synonyms and Formulae	Mol. wt.	Crystalline form, properties and index of refraction	Density or spec. gravity	Melting point, °C	Boiling point, °C	Solubility, in grams per 100 cc		
								Cold water	Hot water	Other solvents
	Osmium									
o10	oxide, di-, brown	OsO_2	222.20	brn cr	$11.37^{21.4}$	30 % tr to OsO_4, 500		i	i	i a
o11	oxide, di-, black	OsO_2	222.20	blk powd	7.71^{21}	tr to br 350–400		i	i	s dil HCl
o12	oxide, mon-	OsO	206.20	blk				i	i	
o13	oxide, sesqui-	Os_2O_3	428.40	dk brn		d		i	i	i a
o14	oxide, tetra-	OsO_4	254.10	a) col, monocl b) yel mass	4.906^{22}	a) 39.5 b) 41.0	130	5.70^{10}	6.23^{25}	250 ± 10^{20} CCl_4; s al, eth, NH_4OH, $POCl_3$
o15	sulfide, di-	OsS_2	254.33	blk, cub	9.47	d		i	i	s HNO_3; i alk
o16	sulfide, tetra-	OsS_4	318.46	br blk (exist?)		d		i		s dil HNO_3; i $(NH_4)_2S$
o17	sulfite	$OsSO_3$	270.26	bl blk		d		i		s dil HCl, alk
o18	telluride	$OsTe_2$	445.40	gray-blk cr		*ca* 600				i a; d dil HNO_3
o19	**Oxygen**	O_2	31.9988	col gas, sol hex cr	gas: 1.429^0 g/l, liq: 1.149^{-183} sol: $1.426^{-252.5}$	−218.4	−182.962	4.89^0 cm^3 3.16^{25} cm^3	2.46^{50} cm^3 2.30^{100} cm^3 3	2.78^{25} al
o20	fluoride	OF_2	54.00	col gas, unst	liq: $1.90^{-223.8}$	−223.8	−144.8	sl s, d	i	sl s a, alk
o21	**Ozone**	O_3	47.9982	col gas, or dk bl liq, or bl-blk cr, liq: 1.2226	gas: 2.144^0 g/l liq: $1.614^{-195.4}$ g/l	−192.7 ± 2 1.0	−111.9	49^0 cm^3		s alk sol, oils
p1	**Palladium**	Pd	106.40	silv wh, met, cub	12.02^{20} $11.40^{22.5}$	1552	3140	i	i	s aq reg, h HNO_3, H_2SO_4; sl s HCl
p2	bromide	$PbBr_2$	266.22	red br	5.173^{16}	d		i	i	s HBr
p3[1]	chloride	$PdCl_2$	177.31	dk red, cub need, deliq	4.0^{18}	d 500		s	s	s HBr; acet
p3[2]	chloride, dihydrate	$PdCl_2.2H_2O$	213.34	br pr, deliq		d		v s	v s	s HCl, acet
p4	cyanide	$Pd(CN)_2$	158.44	yelsh-wh		d		i	i	s KCN, NH_4OH; i dil a
p5	fluoride, di-	PdF_2	144.40	br, tetr	5.80	volat	d red heat	sl s, d		s HF
p6	fluoride, tri-	PdF_3	163.40	blk, rhomb	5.06	d	d	d	d	s HF
p7	hydride	Pd_2H (or Pd_4H_2)	213.81	silv met (exist?)	10.76	d				
p8	iodide	PdI_2	360.21	blk powd	6.003^{18}	d 350		i	i	s KI; i al, eth, dil HCl
p9	nitrate	$Pd(NO_3)_2$	230.41	br yel, rhomb, deliq		d		s, d		s HNO_3
p10	oxide, di-	$PdO_2.xH_2O$		dull red		d − H_2O, −O		i	i,d	s a, alk
p11	oxide, mon-	PdO	122.40	grnsh-bl or amber mass, or blk powd	8.70_4^{20}	870		i	i	i aq reg
p12	oxide, mon-, hydrate	$PdO.xH_2O$		yel to brn		d		i	i	s a, NH_3, NH_4Cl
p13	selenate	$PdSeO_4$	249.36	dk brn-red, rhomb, deliq	6.5	d red heat		v s	v s	i al, eth, alk; s NH_3
p14	selenide	PdSe	185.36	dk gray		<960		i		s aq reg
p15	selenide, di-	$PdSe_2$	264.32	olive gray, hex		<1000		i	i	v s aq reg; v sl s HNO_3; i alk
p16	silicide	PdSi	134.49	cr	7.31^{15}			i		
p17	sulfate	$PdSO_4.2H_2O$	238.50	red-br cr, deliq		d		v s	d	
p18	sulfide, di-	PdS_2	170.53	dk br cr	$4.7–4.8_4^{25}$	d		i	i	s aq reg, $(NH_4)_2S$
p19	sulfide, mono-	PdS	138.46	brn-blk tetr	6.6_4^{25}	d 950		i	i	sl s HNO_3, aq reg; i HCl, $(NH_4)_2S$
p20	sulfide, sub-	Pd_2S	244.86	grn gray (exist?)	7.303^{15}	d 800		i	i	sl s a, aq reg
p21	telluride, di-	$PdTe_2$	361.60	silvery cr, hex				i	i	v s aq reg; s HNO_3; i alk
	Palladium Complexes									
p22	diamminepalladium (II) hydroxide	$Pd(NH_3)_2(OH)_2$	174.48	yel micr cr		>105		v s	d	
p23	dichlorodiammine-palladium (II) *trans* (or *α*)	$Pd(NH_3)_2Cl_2$	211.37	yel, tetr	2.5	d		0.304^{10}	s, d	s, d a; s NH_4OH; i chl, acet
p24	tetramminepalladium (II) chloride	$Pd(NH_3)_4Cl_2.H_2O$	263.44	col, tetr	1.91^{18}	d 120		v s		
p25	tetramminepalladium tetrachloropalladate	Vauquelin's salt. $Pd(NH_3)_4.PdCl_4$	422.73	pink powd or need	2.489^{21}	tr yel 184 d above 192		i	sl s	sl s dil HCl; s KOH
p26	**Phospham**	PN_2H	60.00	wh, amorph		infus		i	d	s conc H_2SO_4, alk; i a
p27	**Phosphomolybdic acid**	Molybdophosphoric acid $H_3[P(Mo_3D_{10})_4]$		yel, tetr		78–90		s	s	

PHYSICAL CONSTANTS OF INORGANIC COMPOUNDS (Continued)

No.	Name	Synonyms and Formulae	Mol. wt.	Crystalline form, properties and index of refraction	Density or spec. gravity	Melting point, °C	Boiling point, °C	Solubility, in grams per 100 cc		
								Cold water	Hot water	Other solvents
	Phosphonium									
	Phosphonium									
p28	bromide	PH_4Br	114.91	col, cub	gas: 2.464 g/l	subl ca 30	38.8^{764}	d	d	
p29	chloride	PH_4Cl	70.46	col, cub		$28^{46\ atm}$	subl	d		
p30	iodide	PH_4I	161.91	col, tetr, deliq	2.86	18.5, subl 61.8	80	d		d, s a, alk
p31	sulfate	$(PH_4)_2SO_4$	166.07					d		
p32	**Phosphoramide**	Phosphorylamide. $PO(NH_2)_3$	95.04	wh, cr		d		40.66^{15}	i	sl, me
	Phosphoric acid									
p33	difluoro-	$H_2PO_2F_2$	102.99	col, fum liq	1.583_4^{25}	-96.5 ± 0.1	115.9 sl d			
p34	hypo-	$H_4P_2O_6.2H_2O$	198.01	col, rhomb deliq		70	d 100	d to $H_3PO_3+HPO_3$		
p35	meta-	HPO_3	79.98	col, vitrous, deliq	2.2–2.5	subl		d to H_3PO_4	d	s al; i liq CO_2
p36	monofluo-	H_2PO_3F	99.99	oily, col liq	1.818	−80				
p37	ortho-	H_3PO_4	98.00	col, liq, or rhomb cr, deliq	1.834^{18}	42.35	$-\frac{1}{2}H_2O$, 213	548	v s	s al
p38	ortho-	$2H_3PO_4.H_2O$	214.01	col, hex pr deliq		29.32	d	v s		
p39	pyro-	$H_4P_2O_7$	177.98	col, need or liq, hygr		61		709^{23}	d to H_3PO_4	v s al, eth
p40	**Phosphorus**, black	P_4	123.89504	blk, incombust	2.70					i CS_2, conc H_2SO_4
p41	red	P_4	123.89504	redsh-brn, cub, or amorph powd, (mix of col and vlt?)	2.34	$590^{43\ atm}$	ign 200 280	v sl s	i	s abs al; i CS_2, eth, NH_3
p42	violet	P_4	123.89504	vlt, monocl	2.36	590				i org solv
p43	yellow	Phosphorus, white. P_4	123.89504	yel (or wh) cub or wax like solid, 2.144	1.82^{20}	44.1	280	0.0003^{15}	sl s	0.3 al; $880^{10}CS_2$; s bs, NH_3, alk, eth, chl, tol
p44	bromide, penta-	PBr_5	430.52	yel, rhomb		d < 100	d 106	d		s CS_2, CCl_4, bs
p45	bromide, tri-	PBr_3	270.70	col, fum liq, $1.697^{19.5}$	2.852^{15}	−40	172.9	d		d al; s eth, chl, CS_2, CCl_4
p46	bromide(di-) chloride, tri-	PBr_2Cl_3	297.15	or cr		d 35		d		
p47	bromide(hepta-) chloride, di-	PBr_7Cl_2	661.24	pr				d		s PCl_3, PCl_5
p48	bromide(mono-) chloride, tetra-	$PBrCl_4$	252.69	yel cr				d		
p49	bromide(octa-) chloride, tri-	PBr_8Cl_3	776.60	brn need		25		d		
p50	bromide(di-) fluoride, tri-	PBr_2F_3	247.79	pa yel liq		−20	d 15	d		d glass
p51	bromide nitride	$(PNBr_2)_3$	614.40	col, rhomb		190	subl v 150	i		s eth; sl s chl, CS_2
p52	chloride, di-	PCl_2 (or P_2Cl_4 ?)	101.88	col liq		−28	180	hydr		
p53	chloride, penta-	PCl_5	208.24	yelsh-wh, tetr, fum	gas: 4.65^{296} g/l	d 166.8 (press)	subl 162	d		d a; s CS_2, CCl_4
p54	chloride, tri-	PCl_3	137.33	col, fum liq, 1.516^{14}	1.574^{21}	−112	75.5^{760}	d	d	s eth, bs, chl, CS_2, CCl_4
p55	chloride(di-) fluoride, tri-	PCl_2F_3	158.88	col liq	5.4 g/l	−8	10			
p56	chloride(tri-)iodide, di-	PCl_3I_2	391.14	red, hex		d 259		d		s CS_2
p57	chloride(di-)nitride	$(PNCl_2)_3$	347.66	rhomb	1.98	114	256.5	i	d	s al, eth, bs, chl, a ac, CS_2
p58	chloride(di-)nitride	$(PNCl_2)_4$	463.55		2.18_{24}^{24}	123.5	328.5			
p59	chloride(di-)nitride	$(PNCl_2)_5$	579.43			41	224^{13}, polym >250			
p60	chloride(di-)nitride	$(PNCl_2)_6$	695.32			90	262^{13}, polym >250			
p61	cyanide	$P(CN)_3$	109.03	wh need		subl 130		d		v s eth; sl s h bs
p62	fluoride, penta-	PF_5	125.97	col gas	5.805 g/l	−83	−75	d		
p63	fluoride, tri-	PF_3	87.97	col gas	3.907 g/l	−151.5	−101.5	d		s al; d alk
p64	hydride, tri-	Phosphine. PH_3	34.00	col gas, pois		−133	−87.7	0.26 vol 20		
p65	iodide, di	P_2I_4	569.57	or, tricl		110		d	d	s CS_2
p66	iodide, tri-	PI_3	411.68	red, hex, deliq	4.18	61	d	d	d	v s CS_2
p67	oxide, pent-	Phosphoric anhydride. P_2O_5 (or P_4O_{10})	141.94	wh, monocl or powd, v deliq	2.39	569	subl 300	d to H_3PO_4	d	s H_2SO_4; i acet, NH_3
p68	oxide, sesqui-	Phosphorus trioxide. P_4O_6 (or P_2O_3)	219.89	col, or wh powd, monocl cr, deliq	2.135^{21}	23.8	175.4	d to H_3PO_3	d	s chl, bs, eth, CS_2
p69	oxide, tetra-	P_2O_4	125.95	col, rhomb, deliq	2.54^{20}	>100	180 vac	v s to H_3PO_3	d	
p70	oxide, tri-	P_2O_3 (or P_4O_6)	109.95	col, or wh powd, or monocl, deliq	2.135^{21}	23.8	173.8 (in N_2)	d to H_3PO_3	d	s CS_2, eth, chl, bs,

No.	Name	Synonyms and Formulae	Mol. wt.	Crystalline form, properties and index of refraction	Density or spec. gravity	Melting point, °C	Boiling point, °C	Solubility, in grams per 100 cc		
								Cold water	Hot water	Other solvents
	Phosphorus									
p71	oxybromide	$POBr_3$	286.70	col pl	2.822	56	189.5	d		s H_2SO_4, CS_2, eth, bz, chl
p72	oxy*di*bromide chloride	$POBr_2Cl$	242.27	sol or liq	liq: 2.45^{50}	30	165	d		
p73	oxybromide chloride, di	$POBrCl_2$	197.79	tabl or liq	liq: 2.104^{14}	13	137.6	d		
p74	oxychloride	$POCl_3$	153.33	col, fum liq, $1.460^{25.1}$	1.675	2	105.3	d	d	d al, a
p75	oxychloride	$P_2O_3Cl_4$	251.76	col fum liq	liq: 1.58^{7}	<−50	212	d		
p76	oxyfluoride	POF_3	103.97	col gas	4.69 g/l	−68	−39.8	d		d al
p77	oxynitride	PON	60.98	wh, amorph		red heat		i	i	i a, alk
p78	oxysulfide	$P_4O_6S_4$	348.15	wh, tetr, deliq		102	295	d		50 CS_2
p79	selenide, penta-	P_2Se_5	456.75	dk red-blk need		d		d		s CCl_4; i CS_2
p80	(tetra-)selenide, tri-	P_4Se_3	360.78	or red cr	1.31	242	360–400			
p81	(tetra-)sulfide, hepta-	P_4S_7	348.34	lt yel cr	2.19^{17}	310	523			sl s CS_2
p82	sulfide, penta-	P_2S_5 (or P_4S_{10})	222.27	gray-yel cr, deliq	2.03	286–90	514	i	d	0.22 CS_2; s alk
p83	sesquisulfide	Tetraphosphorus trisulfide. P_4S_3	220.09	yel, rhomb	2.03^{17}	174	408	i	d	100^{17} CS_2; 11.1^{80} bz
p84	thiobromide	$PSBr_3$	302.76	yel, oct	2.85^{17}	38	d 212	s		s eth, CS_2, PCl_3
p85	thiochloride	$PSCl_3$	169.40	col, fum liq	1.668	−35	125	sl d	d	s CS_2
p86	thiocyanate	$P(SCN)_3$	205.22	liq	1.625^{18}	*ca* −4	265	d		s al, eth, bz, CS_2
	Phosphorous acid									
p87	hypo-	$H(H_2PO_2)$	66.00	col oily liq or deliq cr	1.493^{19}	26.5	d 130	s	v s	v s al, eth
p88	meta-	HPO_2	63.98	feather like cr				d		
p89	ortho-	$H_2(HPO_3)$	82.00	col-yel cr, deliq	$1.651^{21.2}$	73.6	d 200	309^{0}	694^{40}	s al
p90	pyro-	$H_4P_2O_5$	145.98	need		38	d 120	d		
p91	**Phosphotungstic acid**	Tungstophosphoric acid $H_3[P(W_3O_{10})_4].14H_2O$	3132.39	yel-grn cr, tricl		d		s		s al, eth
p92	phosphotungstic acid.	Dodecatungtophosphoric acid. $H_3[P(W_3O_{10})_4].24H_2O$	3312.54	trig		89		s		
p93	**Platinic acid,** *hexa*chloro	$H_2PtCl_6.6H_2O$	517.92	red, brn, deliq	2.431	60		v s	v s	s al, eth
p94	*tetra*cyano	$H_2Pt(CN)_4$	301.18			d 100		v s	v s	v s al, eth, chl
p95	*hexa*hydroxy-	$H_2Pt(OH)_6$	299.15	yel need		$-2H_2O$, 100	$-3H_2O$, 120	i	v sl s	s H_2SiF_6, dil a, alk
p96	*hexa*iodo-	$H_2PtI_6.9H_2O$	1120.67	blk-red, deliq				d		
p97	**Platinum**	Pt	195.09	silv met, cub	21.45^{20}	1772	3827 ± 100			s aq reg, fus alk
p98	arsenide	Nat. sperrylith. $PtAs_2$	344.93	gray, cub	11.8	d >800		sl d	sl d	v sl s a
p99	bromic acid	$H_2PtBr_6.9H_2O$	838.70	br-red, monocl, hygr				v s	v s	v s al, eth
p100	(II) bromide, di-	$PtBr_2$	354.91	br	6.65_4^{25}	d 250		i	i	i al; s HBr, KBr, aq Br
p101	(IV) bromide, tetra-	$PtBr_4$	514.73	dk cr	5.69_4^{25}	d 180		0.41^{20}	sl s	v s al, eth, HBr
p102	carbonyl bromide	$[Pt_2(CO)_2]Br_4$	765.84	lt red, need, hygr.	5.115_4^{25}	d 180		s d		s abs al, CCl_4, bz
p103	carbonyl chloride, di-	$Pt(CO)Cl_2$	294.00	yel need	4.2346_4^{25}	195, subl 240 in CO_2	d 300	d	d	s conc HCl, H_2SO_4, al
p104	*di*carbonyl chloride, di-	$Pt(CO)_2Cl_2$	322.02	lt yel need	3.4882_4^{25}	142	−CO, 210	d	d	d HCl; s CCl_4
p105	*di*platinum dicarbonyl tetrachloride	$Pt_2(CO)_2Cl_4$	588.01	or-yel need	4.235_4^{25}	195	subl 240 (in CO_2)	d		d HCl
p106	*di*platinum tricarbonyl tetrachloride	$Pt_2(CO)_3Cl_4$	616.02	or-yel need		130	d 250	d		s h CCl_4, d al
p107	carbonyl diiodide	PtI_2CO	476.91	red cr	5.257	d 140–150		d		s d, al; s bz
p108	carbonyl sulfide	$Pt(CO)S$	255.16	br-blk		d 300–400		d		d alk, al
p109	chloric acid	$H_2PtCl_6.6H_2O$	517.92	brn-red cr, hygr	2.431	60	d >115	s	s	s abs al, acet; v s eth
p110	(II) chloride, di-	$PtCl_2$	266.00	olive grn, hex	6.05	d 581 (in Cl_2)		v sl s		i al, eth; s HCl, NH_4OH
p111	(IV) chloride, tetra-	$PtCl_4$	336.90	br-red cr	4.303_4^{25}	d 370 (in Cl_2)		58.7^{25}	v s	sl s al, NH_3; s acet; i eth
p112	(IV) chloride, tetra-, hydrate	$PtCl_4.5H_2O$	426.98	red, monocl	2.43	$-H_2O$, 100		v s	v s	s al, eth.
p113	chloride, tri-	$PtCl_3$	301.45	grnsh-blk	5.256^{25}	435		sl s	s	s h HCl; v sl s conc HCl

No.	Name	Synonyms and Formulae	Mol. wt.	Crystalline form, properties and index of refraction	Density or spec. gravity	Melting point, °C	Boiling point, °C	Solubility, in grams per 100 cc		
								Cold water	Hot water	Other solvents
	Platinum									
p114	*di*chlorocarbonyl, dichloride	$Pt(COCl_2)Cl_2$	364.91	yel cr		d		v s		sl s al; v sl s CCl_4
p115	(II) cyanide, di-	$Pt(CN)_2$	247.13	yel-br cr				i	i	i al, a, alk; s KCN
p116	(II) fluoride, di-	PtF_2	233.09	yelsh-grn				i	i	
p117	fluoride, hexa-	PtF_6	309.08	dk red solid, very unstable		57.6				
p118	(IV) fluoride, tetra-	PtF_4	271.08	deep red, fused, or yel-lt brn cr, deliq		d red heat		s d	v s	s a, alk
p119	(II) hydroxide	$Pt(OH)_2$	229.10	blk		d		i	i	s HCl, HBr, alk; i H_2SO_4, dil HNO_3
p120	(II) hydroxide, hydrate	$Pt(OH)_2.2H_2O$	265.14			$-2H_2O$, 100		i	i	s conc a
p121	*mono*hydroxy chloric acid	$H_2[PtCl_5(OH)].H_2O$	409.39	red-brn cr, hygr						
p122	(II) iodide, di-	PtI_2	448.90	blk powd	6.403^{25}_{4}	d 360		i	i	i eth, a; s HI; sl s Na_2SO_3
p123	(IV) iodide, tetra-	PtI_4	702.71	brn, amorph, or blk cr	6.064^{25}_{4}	d 130		s d		s al, acet, alk, HI, KI, liq NH_3
p124	iodide, tri-	PtI_3	575.80	blk, like graphite	7.414^{25}_{4}	d 270		i	i	i al, eth; s KI
p125	(II) oxide, mon-	PtO	211.09	vlt-blk	14.9^{15}	d 550		i	i	s HCl; i a, aq reg
p126	(II) oxide, mon-, dihydrate	$PtO.2H_2O$	247.12			$-2H_2O$, 140–150				s conc HCl, conc H_2SO_4, conc HNO_3
p127	(IV) oxide, di-	PtO_2	227.03	blk	10.2	450		i	i	i a, aq reg
p128	(IV) oxide, di-dihydrate	Platinic hydroxide. $PtO_2.2H_2O$ or $Pt(OH)_4$	263.12			$-2H_2O$, 100		i	i	s HCl, aq reg, KOH
p129	(IV) oxide, di-, hydrate	$PtO_2.H_2O$	245.10					i	i	i aq reg, ac a, HCl; sl s NaOH
p130	(IV) oxide, di-trihydrate	$PtO_2.3H_2O$	281.13	ochre		d 300		i	i	i aq reg, HCl
p131	(IV) oxide, di-, tetrahydrate	Hydroxoplatinic acid. $PtO_2.4H_2O$ or $H_2Pt(OH)_6$	299.15	yel need		$-2H_2O$, 100	$-3H_2O$, 120	i		s a, dil alk
p132	(II, IV) oxide,	Pt_3O_4	649.27			d		i		i a, aq reg
p133	oxide, sesqui,	$Pt_2O_3.3H_2O$	492.22			$-H_2O$, 100		i	i	s conc H_2SO_4, caust alk
p134	oxide, tri-	PtO_3	243.09	redsh-brn powd						s HCl, H_2SO_3; sl s HNO_3, H_2SO_4
p135	*pyro*phosphate	PtP_2O_7	369.03	grn-yel	4.85	d 600		v sl s		
p136	phosphide	PtP_2	257.04	met shine	9.01^{25}_{4}	*ca* 1500		i	i	i a; v sl s aq reg
p137	selenide, di-	$PtSe_2$	353.01	blk or gray cr or amorph	7.65	d when dry				s aq reg; sl s HNO_3, H_2SO_4
p138	selenide, tri-	$PtSe_3$	431.97	bl flakes	7.15	d 140		i	i	i conc a, CS_2; s aq reg
p139	sulfate	$Pt(SO_4)_2.4H_2O$	459.27	yel pl				s	d	s al, eth, a
p140	(IV) sulfide, di-	PtS_2	259.22	blk-brn powd	7.66^{25}_{4}	d 225–250		i		s HCl, HNO_3; i $(NH_4)_2S$
p141	(II) sulfide, mono-	PtS	227.15	blk, tetr	10.04^{25}_{4}	d		i	i	s $(NH_4)_2$ S; i a, alk
p142	sulfide, sesqui-	Pt_2S_3	486.37	gray (exist ?)	5.52	d		i	i	sl s aq reg; i a
p143	(III) sulfuric acid	$H_2[Pt_2(SO_4)_4(H_2O)_2].9\frac{1}{2}H_2O$	983.62	or, red, tricl		d 150		s	s	d alk
p144	telluride	$PtTe_2$	450.29	gray, hex		1200–1300				sl s Na_2S, $(NH_4)_2S$
	Platinum complexes									
p145	tetramine platinum (II) chloride	$[Pt(NH_3)_4]Cl_2.H_2O$	352.13	col, tetr, 1.672, 1.667	2.737	250, $-H_2O$, 100				
p146	tetrammineplatinum (II) chloroplatinite	Magnus, salt. $[Pt(NH_3)_4]PtCl_4$	600.11	grn or red, tetr	<4.1	d		sl s	sl s	
p147	tetrachlorodiammine platinum (IV), *trans-*	$[Pt(NH_3)_2]Cl_4$	370.96		3.3	200–216				
p148	tetrachlorodiammine platinum (IV), *cis-*	$[Pt(NH_3)_2]Cl_4$	370.96	or-yel, rhomb or hexag pl or need		240				
p149	**Plumbous, plumbic**	see Lead								
p150	**Plutonium α**	Pu	242.00	sil wh, monocl	19.84	641	3232			s HCl; i HNO_3, conc H_2SO_4

No.	Name	Synonyms and Formulae	Mol. wt.	Crystalline form, properties and index of refraction	Density or spec. gravity	Melting point, °C	Boiling point, °C	Solubility, in grams per 100 cc: Cold water	Hot water	Other solvents
	Plutonium									
p151	β	Pu	239.05	monocl	17.70	stab 122 ± 2 to 206 ± 3				
p152	γ	Pu	239.05	orthorhomb	17.14	stab 206 ± 3 to 319 ± 5				
p153	δ	Pu	239.05	cub	15.92	stab 319 ± 5 to 451 ± 4				
p154	δ'	Pu	239.05	tetrag	16.00	stab 451 ± 4 to 476 ± 5				
p155	ε	Pu	239.05	cub	16.51	stab 476 ± 5 to 639.5 ± 2				
p156	bromide, tri-	$PuBr_3$	481.73	grn, orthorhomb	6.69	681		s		
p157	chloride, tri-	$PuCl_3$	348.36	emerald grn, hex	5.70	760		s		s dil a
p158	fluoride, hexa-	PuF_6	355.99	redsh-brn, orthorhomb		50.75	62.3	d		
p159	fluoride, tetra-	PuF_4	317.99	pa brn, monocl	7.0 ± 0.2	1037				
p160	fluoride, tri-	PuF_3	299.00	purple, hex	9.32	1425(± 3)		i		
p161	iodide, tri-	PuI_3	622.71	bright grn, orthorhomb	6.92	777		s		
p162	nitride	PuN	256.01	blk, cub	14.25			hydrol		s HCl, H_2SO_2
p163	oxalate	$Pu(C_2O_4)_2.6H_2O$	526.13	yel-grn				i		
p164	oxide, di-	PuO_2	274.00	yelsh-grn, cub	11.46					sl s h conc H_2SO_4, HNO_3, HF
p165	sulfate	$Pu(SO_4)_2$	434.12	light pink						s dil min a
p166	sulfate, tetrahydrate	$Pu(SO_4)_2.4H_2O$	506.18	coral pink		d 280				s dil min a
p167	**Polonium**	Po	210.05	α-Po: simple cub; β-Po: rhbr	9.4 (for β-Po)	254	962	sl s		s dil min a; v sl s dil KOH
p168	ammonium chloride	$(NH_4)_2PoCl_6$	458.85		2.76					
p169	*tetra*bromide	$PoBr_4$	529.67	bright red, cub		330 (in Br atm)	360^{200}			s al, acet; i bz, CCl_4
p170	*di*chloride	$PoCl_2$	280.96	ruby red, orthorhomb	6.50	subl 190				s dil HNO_3
p171	*tetra*chloride	$PoCl_4$	351.86	yel, monocl or tric		300 (in Cl atm)	390	s, d		s HCl; sl s al, acet
p172	*tetra*iodide	PoI_4	717.67	blk cr		200 (in N atm subl)				sl s al, acet; i bz, CCl_4
p173	*di*oxide	PoO_2	242.05	red, tetr		d 500				
p174	selenate	$2PoO_2.SeO_3$	611.06	wh powd		>400				s dil HCl
p175	sulfate, basic	$2PoO_2.SO_3$	564.16	wh powd		>400, d 550				s dil HCl
p176	*di*sulfate	$Po(SO_4)_2$	402.17	purp		d 550				i al; v s dil HCl
p177	monosulfide	PoS	242.11	blk		d 275				i al; sl s dil HCl
p178	**Potassium**	Kalium. K	39.0983	cub silv met	0.86^{20}	63.65	774	d to KOH	d	d al; s a, Hg, NH_3
p179	acetate	$KC_2H_3O_2$	98.15	wh, lust powd, deliq	1.57^{25}	292		253^{20}	492^{62}	33 al; 24.24^{15} MeOH; s liq NH_3; i eth, acet
p180	acetate, acid	$K_2C_2H_3O_2.HC_2H_3O_2$	158.20	col, need or pl, hygr		148	d 200	d	d	s al, acet
p181	acetyl salicylate	$KC_9H_7O_4.2H_2O$	254.29			65				
p182	*meta*aluminate	$K_2Al_2O_4.3H_2O$	250.22	col cr				v s, d	v s, d	s alk; i al
p183	aluminosilicate	Nat. orthoclase. $KAlSi_3O_8$ (or $K_2O.Al_2O_3.6SiO_2$)	278.34	wh, monocl, 1.518, 1.524, 1.526	2.56	*ca* 1200				
p184	aluminosilicate	Nat. microcline. $KAlSi_3O_8$ (or $K_2O.Al_2O_3.6SiO_2$)	278.34	wh, tricl, 1.522, 1.526, 1.530	2.54–2.57	1140–1300				
p185	aluminosilicate	Nat. muscovite, white mica. $KAl_3Si_3O_{10}.(OH)_2$ (or $K_2O.3Al_2O_3.6SiO_2.2H_2O$)	398.31	col, monocl, 1.551, 1.587, 1.581	2.76–2.80	d		i		
p186	aluminum *meta*silicate	Nat. leucite. $KAlSi_2O_6$	218.25	col cr, 1.508	2.47	1686 ± 5		i	i	d a
p187	aluminum *ortho*silicate	Nat. kaliophilite. $KAlSiO_4$	158.17	col, hex or rhomb (hex→rhomb 1540°) hex: 1.532, 1.572; rhomb: 1.528, 1.536	2.5	*ca* 1800 (rhomb)				
p188	aluminum sulfate	Nat. kalinite. $KAl(SO_4)_2.12H_2O$	474.39	col, cub, oct or monocl, cub: 1.454, 1.4564; hex: 1.456, 1.429	1.757^{20}_{4}	92.5 $-9H_2O$, 64.5	$-12H_2O$, 200	11.4^{20}	v s	i al, acet; s dil a
p189	amide	Potassamide. KNH_2	55.12	col-wh, or yel-grn, hygr		335	subl 400	d	d	d al; s liq NH_3
p190	*peroxyl*amine sulfonate	$(KSO_3)_2NO$	268.39	yel cr, expl				0.62^{3}, d	6.6^{29}, d	i al

No.	Name	Synonyms and Formulae	Mol. wt.	Crystalline form, properties and index of refraction	Density or spec. gravity	Melting point, °C	Boiling point, °C	Solubility, in grams per 100 cc: Cold water	Hot water	Other solvents
	Potassium									
p191	ammonium tartrate	$KNH_4C_4H_4O_6$	205.21	wh, cr powd	...	...	...	v s	...	...
p192	antimonate, hydroxo-	"*Pyro*"-antimonate. $KSb(OH)_6.\frac{1}{2}H_2O$	271.90	wh gran or cr powd	...	...	...	2.82^{20}	s	...
p193	antimonide	K_3Sb	239.06	yel-grn	...	812	...	d	...	d air
p194	antimony tartrate	$KSbC_4H_4O_7.\frac{1}{2}H_2O$	333.93	col, rhomb, 1.620, 1.636, 1.638	2.607	$-H_2O$, 100	...	$5.26^{8.7}$	35.7^{100}	i al; 6.67^{25} glyc
p195	*ortho*arsenate	K_3AsO_4	256.23	col need, deliq	...	1310	...	18.87	v s	4 al
p196	*ortho*arsenate, di-H	KH_2AsO_4	180.04	col, tetr, 1.567, 1.518	2.867	288	...	19^{6}	v s	i al; s NH_3, a; 52.5 glyc
p197	*ortho*arsenate, mono H	K_2HAsO_4	218.13	col monocl pr	...	d 300	...	18.86^{6}	s	i al
p198	*meta*arsenite	$KAsO_2$	146.02	wh powd, hygr	...	...	...	s	s	sl s al
p199	*ortho*arsenite	K_3AsO_3	240.23	col need	...	...	...	v s	...	s al
p200	*meta*arsenite, acid	$KH(AsO_2)_2.H_2O$	271.97	...	...	...	...	s	...	sl s al
p201	aurate	$KAuO_2.3H_2O$ (or $2H_2O$)	322.11	lt yel need	...	d	...	s	d	s al
p202	azide	KN_3	81.12	col, tetr	2.04	350 (vac)	...	49.6^{17}	105.7^{100}	s al; i eth
p203	benzoate	$KC_7H_5O_2.3H_2O$	214.26	wh cr powd	...	$-3H_2O$, 110	d	52^{25}	112^{100}	s al
p204	*di*borane	Diboranidex. $K_2B_2H_6$	105.87	wh, cub cr, 1.493	1.18	subl 400, vac	...	d	...	...
p205	*penta*borate	Pentaboranidex. $K_2B_5H_9$	141.32	wh powd	...	d <180	...	s, d	d	...
p206	*di*borane, dihydroxy	$K_2B_2H_6O_2$	137.87	col, cub cr	1.39	d→K, 400–500	...	s, d	...	s al; d a
p207	*meta*borate	KBO_2 (or $K_2B_2O_4$)	81.91	col, hex 1.526, 1.450	...	950	...	71^{30}	v s	i al, eth
p208	*penta*borate	$KB_5O_8.4H_2O$	293.21	col, rhomb	...	780	...	...	0.007°	...
p209	*peroxy*borate	$KBO_3.\frac{1}{2}H_2O$	106.92	wh cr	...	$-O_2$, 100	d 150	1.22^{0}	...	i al, eth
p210	*tetra*borate	$K_2B_4O_7.8H_2O$	377.57	col, monocl	1.74 (anhydr)	d	...	26.7^{30}	v s	...
p211	borohydride	KBH_4	53.94	wh, cub, 1.494	1.178	d 500	...	19.3^{20}	v s	0.25 al; 0.56 MeOH; i eth
p213	boroxalate	$KHC_2O_4.HBO_2.2H_2O$	207.98	...	...	$-H_2O$, 110	...	v s	v s	i, d al
p214	borotartrate	Solution: cream of tartar. $KC_4H_4BO_7$	213.99	wh cr powd	1.832	...	...	v s	...	i al, eth, chl
p215	bromate	$KBrO_3$	167.01	col, trig	$3.27^{17.5}$	434 d 370	...	13.3^{40}	49.75^{100}	sl s al; i acet
p216	bromide	KBr	119.01	col, cub, sl hygr, 1.559	2.75^{25}	734	1435	53.48^{0}	102^{100}	0.142^{25} al; sl s eth; s glyc
p217	bromoaurate	$K[AuBr_4]$	555.71	red-brn, rhomb	...	d 120	...	sl s	...	s al
p218	bromoaurate, dihydrate	$K[AuBr_4].2H_2O$	591.74	vlt, monocl cr	4.08	...	...	19.5^{15}	204^{67}	s al, KBr; d eth
p219	bromoiodide, di-	$KIBr_2$	325.82	red, rhomb	...	60	d 180	v s	...	...
p220	*hexa*bromoplatinate	$K_2[PtBr_6]$	752.75	dk red-brn, cub	4.66^{24}	d>400	...	2.02^{20}	10^{100}	i al
p221	*tetra*bromoplatinite	$K_2[PtBr_4]$	592.93	br, rhomb	...	...	...	v s	v s	...
p222	bromoplatinite, dihydrate	$K_2[PtBr_4].2H_2O$	628.96	blk, rhomb	3.747^{25}_{4}	$-H_2O$, vac	...	v s	v s	...
p223	bromostannate	$K_2[SnBr_6]$	676.35	wh cr	3.783	...	...	...	...	...
p224	cacodylate	$K[(CN_3)_2AsO_2].H_2O$	194.10	wh cr	...	...	...	s	...	sl s al; i eth
p225	cadmium cyanide	$K_2[Cd(CN)_4]$	294.68	col, oct	1.846	450	...	33.3	100^{100}	s al
p226	cadmium iodide	$2KI.CdI_2.2H_2O$	734.25	wh-yelsh cr powd, deliq	3.359	...	...	137^{15}	...	s a, al, eth
p227	calcium chloride	Chlorocalcite. $KCl.CaCl_2$	185.54	col cub, β 1.52	754	...	...	s	...	...
p228	calcium magnesium sulfate	Polyhalite. $K_2Ca_2Mg(SO_4)_4.2H_2O$	602.95	wh, trig, 1.548, 1.562, 1.567	2.775	...	...	...	...	...
p228	calcium magnesium sulfate	Krugite $K_2Ca_4Mg(SO_4)_6.2H_2O$	875.24	gray cr	2.801	...	...	...	...	...
p229	calcium sulfate	Kalussite, syngenite. $K_2Ca(SO_4)_2.H_2O$	328.42	col, monocl, 1.500, 1.517, 1.518	2.60	1004	...	0.25	d	i al; s a
p230	*d*-camphorate	$K_2C_{10}H_{14}O_4.5H_2O$	366.50	col, need cluster, hygr	...	$-5H_2O$, 110	...	260^{14}	...	s al
p231	carbide	KHC_2	64.13	col, rhomb cr	1.37	...	...	...	...	...
p232	carbonate	K_2CO_3	138.21	col, monocl, hygr, 1.531	2.428^{19}	891	d	112^{20}	156^{100}	i al, acet
p233	*peroxy*carbonate	$K_2C_2O_6.H_2O$	216.22	...	...	200–300	...	s	...	...
p234	carbonate, dihydrate	$K_2CO_3.2H_2O$	174.24	col, monocl, hygr, 1.380, 1.432, 1.573	2.043	$-H_2O$, 130	...	146.9	331^{100}	...
p235	carbonate, hydrogen	$KHCO_3$	100.12	col, monocl, 1.482	2.17	d 100–200	...	22.4	60^{60}	i al
p236	carbonate, trihydrate	$2K_2CO_3.3H_2O$	330.47	col, monocl, 1.380, 1.482, 1.573	2.043	...	...	129.4	268.3^{100}	i al, conc HN_4OH
p237	carbonyl	$(KCO)_6$	402.68	gray-red	...	expl	...	expl	...	d al
p238	chlorate	$KClO_3$	122.55	col, monocl, 1.409, 1.517, 1.524	2.32	356	d 400	7.1^{20}	57^{100}	14.1^{100} 50 % al; sl s glyc, liq NH_3; i acet; s alk
p239	*per*chlorate	$KClO_4$	138.55	col, rhomb, 1.4717, 1.4724, 1.476	2.52^{10}	610 ± 10	d 400	0.75^{0}	21.8^{100}	v sl s al; i eth

No.	Name	Synonyms and Formulae	Mol. wt.	Crystalline form, properties and index of refraction	Density or spec. gravity	Melting point, °C	Boiling point, °C	Solubility, in grams per 100 cc		
								Cold water	Hot water	Other solvents
	Potassium									
p240	chloride	Nat. sylvite. KCl	74.56	cub, col 1.490	1.984	770	subl 1500	34.7^{20}	56.7^{100}	sl s al; s eth, glyc, alk
p241	*hypo*chlorite	KClO	90.55	in sol only	...	d	...	v s	v s	...
p242	chloroaquoruthenate (III) penta-	$K_2[Ru(H_2O)Cl_5]$	374.55	rose pr	...	$-H_2O$, 200	...	s	s	sl s al
p243[1]	chloroaurate	K[AuCl]	377.88	yel, monocl	3.75	d 357	...	61.8^{20}	80.2^{60}	25 al; s a
p243[2]	chloroaurate, dihydrate	$K[AuCl_4].2H_2O$	413.91	yel , rhomb pl	...	...	...	s	s	s al, eth
p244	chlorochromate	Peligot's salt. $KCrO_3Cl$.	174.55	red, monocl	2.497	d	...	s, d	...	s a
p245	chlorohydroxoruthenate	$K_2[Ru(OH)Cl_5]$	373.55	brn-red cr	...	d	...	s, d	d	i al
p246	chloroiodate (III)	$KICl_4$	307.82	yel, rhomb	1.76^{45}	d	...	d	...	d eth
p247	chloroiodide, di-	$KICl_2$	236.91	col, monocl	...	60	d 215	d	...	...
p248	chloroiridate	$K_4[IrCl_6]$	483.22	blk, cub	3.546	d	...	125^{19}	6.67	i al, KCl, HN_4OH
p249	chloronitrosylruthenate (III) penta-	$K_2Ru(NO)Cl_5$	386.55	dk red, rhomb	...	d	...	12^{25}	80^{80}	i al
p250	chloroösmate (III)	$K_3[OsCl_6].3H_2O$	574.27	dk red cr	...	$-3H_2O$, 150	...	v s	...	s a; i eth
p251	chloroösmate (IV)	K_2OsCl_6	481.12	red, cub	...	d	...	sl s	s	i al; s dil HCl
p252	chloropalladate	$K_2[PdCl_6]$	397.32	red, cub	2.738	d	...	sl s, d	d	i al; sl s HCl
p253	chloropalladite	K_2PdCl_4	326.42	red-brn, tetr (yel cub)	2.67	d 105	...	s	v s	s KCl, HN_4OH; i al
p254	*hexa*chloroplatinate	$K_2[PtCl_6]$	486.01	yel, cub, ~1.825	3.499^{24}	d 250	...	0.481^{2}	5.22^{100}	i al, eth
p255	*tetra*chloroplatinite	$K_2[PtCl_4]$	415.11	red-brn tetr, 1.64, 1.67	3.38	d	...	0.93^{16}	5.3^{100}	i al
p256	chloroplumbate	K_2PbCl_6	498.11	lt yel, cub	...	d 190	...	d	...	s h HCl
p257	chlororhenate (IV)	K_2ReCl_6	477.12	yel-grn, oct	3.34	...	...	0.8	d	d alk; sl s HCl
p258	chlororhenate (V)	K_2ReOCl_5	457.67	grn hex pl	...	...	...	d	d	s a; i al, eth
p259	chlororhodate, hexa-	$K_3RhCl_6.3H_2O$	486.98	red, tricl	3.291	d	...	d	...	sl s al, KCl
p260	chlororhodite, penta-	K_2RhCl_5	358.37	red, rhomb	...	d	...	sl s	d	i al
p261	chlororuthenate (IV)	K_2RuCl_6	391.99	blk, cub	...	d	...	s, d	...	i al
p262	chlorostannate	K_2SnCl_6	409.63	col, cub, 1.657	2.71	...	...	s	s	...
p263	chlorotellurate	K_2TeCl_6	418.52	pale yel, octahedral	...	...	...	d	d	s HCl
p264	chromate	Nat. tarapacaite. K_2CrO_4	194.20	yel, rhomb, β 1.74	2.732^{18}	968.3	...	62.9^{20}	79.2^{100}	i al
p265	*di*chromate	$K_2Cr_2O_7$	294.19	red, monocl or tricl, 1.738	2.676^{25}_{4}	tricl→ monocl 241.6 m.p. 398	d 500	4.9^{0}	102^{100}	i al
p266	*peroxy*chromate	K_3CrO_8	297.30	brn-red, cub	...	d 170	...	sl s	...	i a, al, eth
p267	chromium sulfate	Potassium chromium alum. $K[Cr(SO_4)_2].12H_2O$	499.41	vlt-ruby red, cub, oct, 1.4814	1.826^{25}	89	$-10H_2O$, 100 $-12H_2O$, 400	24.39^{25}	50	i al; s dil a
p268	chromium chromate, basic	$K_2CrO_4.2[Cr(OH).CrO_4]$	564.19	vlt brn amorph powd	2.28^{14}	300		i	...	i al, acet a
p269	citrate	$K_3C_6H_5O_7.H_2O$	324.42	wh cr	1.98	d 230	...	167^{15}	199.7^{31}	sl s al; s glyc
p270	citrate, monobasic	$KH_2C_6H_5O_7$	230.22	wh cr powd	...	...	...	s	...	...
p271	cobalt carbonate, hydrogen(ous)	$KHCO_3.CoCO_3.4H_2O$	291.12	rose need	...	...	...	d	...	...
p272	cobalt (II) cyanide	$K_4[Co(CN)_6]$	371.45	redsh-brn cr, deliq	2.039^{25}_{4}	...	...	v s	v s	d a; i al, eth, $CHCl_3$
p273	cobalt (III) cyanide	$K_3[CO(CN)_6]$	332.35	wh-yel, monocl pr	1.878^{25}_{4}	...	...	sl s	sl s	s dil HCl, dil HNO_3; sl s al
p274	cobaltinitrite	Fischer's salt. $K_3[Co(NO_2)_6]$	452.27	yel pr, cub	...	...	...	0.9^{17}	d	i al
p275	cobaltinitrite, hydrate	$K_3[Co(NO_2)_6].H_2O$	470.29	yel cr powd	...	...	...	i	s, d	s min a; sl s ac a; i al, eth
p276	cobaltinitrite, hydrate	$K_3[Co(NO_2)_6].1\frac{1}{2}H_2O$	479.30	yel, tetr	...	d 200	...	0.089^{17}	sl s	i al, meth
p277	cobaltmalonate (II)	$K_2[Co(C_3H_2O_4)_2]$	341.23	...	2.234	...	...	...	...	...
p278	cobalt sulfate (II)	$K_2SO_4.CoSO_4.6H_2O$	437.36	red pr, monocl, 1.481, 1.487, 1.500	2.218	...	...	25.5^{0}	108.4^{40}	...
p279	copperchloride	$KCl.CuCl_2$	209.00	red need	2.86	...	...	...	...	...
p280	cyanate	KOCN	81.12	col, tetrag	2.056^{20}	d 700–900	...	75^{25}	s	i al
p281	cyanide	KCN	65.12	col, cub, wh gran, deliq, very pois, 1.410	1.52^{16}	634.5	...	50	100	$0.88^{19.5}$ al; $4.91^{19.5}$ MeOH; s glyc
p282	cyanoargentate (I)	Potassium argentocyanide. $K[Ag(CN)_2]$	199.01	cub, 1.625	2.36		...	25^{20}	100	4.85 % al; i a
p283	cyanoaurate	$K[Au(CN)_2]$	288.10	col, rhomb	3.45	...	...	14.3	200	sl s al; i eth
p284	cyanoaurate (III)	$K[Au(CN)_4].1\frac{1}{2}H_2O$	367.16	col tabl	...	d 200	...	s	v s	s al

No.	Name	Synonyms and Formulae	Mol. wt.	Crystalline form, properties and index of refraction	Density or spec. gravity	Melting point, °C	Boiling point, °C	Solubility, in grams per 100 cc		
								Cold water	Hot water	Other solvents
	Potassium									
p285	cyanocadmate	$K_2[Cd(CN)_4]$	294.68	col, cub	1.85			33	100^{100}	sl s al
p286	cyanochromate (III)	$K_3[Cr(CN)_6]$	325.41	yel, rhomb	1.71			30.9^{20}		i al
p287	cyanocobaltate (II)	$K_4[Co(CN)_6]$	371.42					s	s	i al, eth
p288	cyanocobaltate (III)	$K_3[Co(CN)_6]$	332.32	yel, monocl	1.906	d		s	s	i al; sl s NH_3
p289	cyanocuprate (I)	$K_3[Cu(CN)_4]$	284.92	col, rhbdr		d		v s		
p290	cyanomanganate (II)	$K_4[Mn(CN)_6].3H_2O$	421.50	deep bl, tetr				s	d	
p291	cyanomanganate (III)	$K_3[Mn(CN)_6]$	328.35	red, rhomb, 1.553, 1.555, 1.571 (Li)				s		
p292	cyanomercurate	$K_2[Hg(CN)_4]$	382.87	col, cr pois				s		s al
p293	cyanomolybdate	$K_4[Mo(CN)_8].2H_2O$	496.52	yel, rhomb	2.337^{25}_{4} (anhyd)	$-H_2O$, 105–110		v s	v s	i eth; 0.0017^{20} al
p294	cyanonickelate (II)	$K_2[Ni(CN)_4].H_2O$	259.00	red-yel, monocl cr or powd	1.875^{11}	$-H_2O$, 100		s		d a
p295	cyanoosmite	$K_4[Os(CN)_6].6H_2O$	556.76	col, yel, monocl, β 1.607		d		sl s	s	i al, eth
p296	cyanoplatinite	$K_2[Pt(CN)_4].3H_2O$	431.41	col, yel, rhomb, blue fluor, deliq	2.455^{16}	$-3H_2O$, 100	d 400–600	sl s	s	sl s al, eth, H_2SO_4
p297	cyanotungstate (IV)	$K_4[W(CN)_8].2H_2O$	584.43	lt yel- grn cr powd	1.989^{25}_{4} (anhydr)	$-2H_2O$, 115		130^{18}	s	i al, eth
p298	ethylsulfate	$KC_2H_5SO_4$	164.23	wh, monocl	1.843			s		s al
p299	ferricyanide	$K_3Fe(CN)_6$	329.26	red, monocl, 1.566, 1.569, 1.583	1.85^{25}	d		33^{4}	77.5^{100}	i al; s acet
p300	ferrocyanide	Yellow prussiate of **potash.** $K_4Fe(CN)_6.3H_2O$	422.41	lem yel, monocl, β 1.577	1.85^{17}	$-3H_2O$, 70	d	27.8^{12} anhydr 14.5^{0}	$90.6^{96.3}$ anhydr 74^{96}	s acet; i al, eth, NH_3
p301	fluoberyllate	K_2BeF_4	163.21	col, rhomb		red heat		2^{20}	5.26^{100}	
p302	fluoborate	Nat. avogadrite. KBF_4	125.91	col, rhomb or cub, 1.324, 1.325, 1.325	2.498^{20}	d 350	d	0.44^{20}	6.27^{100}	sl s al, eth; i alk
p303	fluogermanate	K_2GeF_6	264.78	wh, hex		730	*ca* 835	0.542^{18}	2.58^{100}	
p304	fluomanganate (IV)	K_2MnF_6	247.13	yel, hex, tab		d		d	d	s conc HCl
p305	fluoniobate, penta-	Potassium oxyniobate. $K_2NbOF_5.H_2O$	300.12	col, monocl pl or leaf				7.69	s	s d conc HF
p306	fluorescein deriv	$K_2C_{20}H_{10}O_5$	408.50	yelsh-red powd				s		
p307	fluoride	KF	58.10	col, cub deliq, 1.363	2.48	858	1505	92.3^{18}	v s	s HF, NH_3; i al
p308	fluoride, acid	KHF_2	78.11	col, cub, deliq	2.37	d *ca* 225	d	41^{21}	v s	s $KC_2H_3O_2$; i al
p309	fluoride, dihydrate	$KF.2H_2O$	94.13	col, monocl pr, deliq, 1.352	2.454	41	156	349.3^{18}	v s	s HF; i al
p310	*hexa*fluorophosphate	KPF_6	184.07			*ca* 575	d	9.3^{25}	20.6^{80}	
p311	fluorotungstate	$2KF.WO_2F_2.H_2O$	388.06	monocl		$-H_2O$, red heat		6^{17}	s	
p312	fluosilicate	Nat, hieratite. K_2SiF_6	220.25	col, cub or hex 1.3991	hex 3.08 cub 2.665^{17}	d		$0.12^{17.5}$ 6.9^{19}	0.954^{100}	s HCl; i NH_3; v sl s al
p313	fluostannate	$K_2SnF_6.H_2O$	328.90	monocl pr	3.053			3.7^{18}	33.3^{100}	i al, NH_3
p314	fluosulfonate	$KFSO_3$	138.16	short, thick pr		311		6.9^{19}		
p315	fluotantalate	K_2TaF_7	392.14	col, rhomb	4.56; 5.24			sl s, d		sl s HF
p316	fluotellurate, di-	$K_2TeO_2F_2.3H_2O$	345.84	micros oct, monocl		d		sl s	sl s	s HF
p317	fluothorate	$K_2ThF_6.4H_2O$	496.29	col		d		2.15	6.6	d a; i al
p318	fluotitanate	$K_2TiF_6.H_2O$	258.11	col, monocl lust leaf		$-H_2O$, 32 m.p. 780	d	0.556^{0}	1.27^{21}	sl s min a; i NH_3
p319	fluozirconate	K_2ZrF_6	283.41	col, monocl, 1.466, 1.455	3.48			0.781^{2}	25^{100}	i NH_3
p320	formate	$KCHO_2$	84.12	col, rhomb deliq	1.91	167.5	d	331^{18}	657^{80}	s al; i eth
p321	gadolinium sulfate	$K_2SO_4.Gd_2(SO_4)_3.2H_2O$	812.98	cr	3.503^{16}			s	s	s K_2SO_4
p322	gallium sulfate	$KGa(SO_4)_2.12H_2O$	517.13	col cr	1.895			s		
p323	*di*germanate	$K_2Ge_2O_5$	303.38	wh cr	$4.31^{21.5}$	>83		s		s a
p324	*meta*germanate	K_2GeO_3	198.79	wh cr	$3.40^{21.5}$	823		s		s a
p325	*tetra*germanate	$K_2Ge_4O_9$	512.55	wh cr	$4.12^{21.5}$	1083		s		s a
p326	glycerophosphate	$K_2C_3H_7PO_6$	248.26	col-sl yelsh mass, hygr				v s	v s	s al
p327	hydride	KH	40.11	wh need 1.453	1.47	d		d	d	i CS_2, eth, bz
p328	hydroxide	KOH	56.11	wh, rhomb deliq	2.044	360.4 ± 0.7	1320–1324	107^{15}	178^{100}	v s al; i eth, NH_3
p329	(tri-)hydroxyl-ammine trisulfonate	$(KSO_3)_3.NO.1\frac{1}{2}H_2O$	414.52	col, monocl pr		$-H_2O$, 100–200		4^{18}	sl d	
p330	*hexa*hydroxyplatinate	$K_2[Pt(OH)_6]$	375.34	yel, rhomb	5.18	d 160		s	s	i al
p331	imidolsulfonate	$(KSO_3)_2NH$	253.34	col, monocl	2.515	d 170–180	d 360–440, vac	1.3^{23}	d	i HNO_3
p332	iodate	KIO_3	214.00	col, monocl	3.93^{32}_{4}	560	d>100	4.74^{0}	32.3^{100}	s KI; i al, NH_3
p333	iodate, acid	$KIO_3.HIO_3$	389.92	col, monocl				1.33^{15}		i al

No.	Name	Synonyms and Formulae	Mol. wt.	Crystalline form, properties and index of refraction	Density or spec. gravity	Melting point, °C	Boiling point, °C	Solubility, in grams per 100 cc: Cold water	Hot water	Other solvents
	Potassium									
p334	iodate, acid	$KIO_3.2HIO_3$	565.83	col, tricl				4.15		
p335	*meta*periodate	KIO_4	230.00	col, tetr, 1.6205	3.618^{15}_{4}	582	−O, 300	0.66^{13}	s	v sl s KOH
p336	iodide	KI	166.01	col or wh, cub or gran, 1.677	3.13	681	1330	127.5^{0}	208^{100}	1.88^{25} al; 1.31^{25} acet; sl s eth; s NH_3
p337	iodide, tri-	$KI_3.\frac{1}{2}H_2O$	428.82	dk bl, monocl, deliq	3.498	31	d 225	v s		s al, KI
p338	iodoaurate	$KAuI_4$	743.69	blk lust cr		d 150		s, d		s dil KI sol
p339	iodoiridite	K_3IrI_6	1070.93	gr cr		d		i	i	i al
p340	iodomercurate (II) tetra-	K_2HgI_4 (or $KI.HgI_2$)	786.41	yel cr, deliq				v s		i al
p341	iodomercurate (II) tri-	Potassium mercury iodide. $KHgI_3$ (or $KI.HgI_2$)	620.47	yel pr, deliq		105		v s		341^{24} al; s KI sol, ac a, eth
p342	iodoplatinate	K_2PtI_6	1034.72	blk, rect	4.96^{25}_{4}			s	s, d	i al
p343	iridium chloride	Potassium hexachloro iridate. K_2IrCl_6	483.12	redsh-blk, cub	3.549	d		1.12^{20}	s	i al
p344	iridium oxalate	$K_3[Ir(C_2O_4)_3].4H_2O$	645.63	orange, tricl cr	2.510^{19}	$-H_2O$, 120	d 160	s	v s	i al, eth
p345	iron chloride, (III)	Nat. erythrosiderite. $2KCl.FeCl_3.H_2O$	329.33	red, orthorhomb.	2.372					
p346	iron chloride (II)	Rinneite. $3KCl.NaCl.FeCl_2$	408.86	rhbdr, 1.589, 1.590	2.3					
p347	iron (III) oxalate	$K_3Fe(C_2O_4)_3.3H_2O$	491.26	emerald grn, monocl, 1.5019, 1.5558, 1.5960	2.133^{20}_{4}	$-3H_2O$, 100	d 230	4.7^{0}	118^{100}	i al
p348	iron sulfate (III)	$KFe(SO_4)_2.12H_2O$	599.32	vlt, cub oct, 1.452	1.83	33		$20^{12.5}$	v s	i al
p349	iron sulfate (III)	Nat. krausite. $K_2SO_4.Fe_2(SO_4)_3.24H_2O$	1006.51	pa yel-grn, monocl, 1.482	1.806	28	d 33			
p350	iron sulfate (II)	$K_2SO_4.FeSO_4.6H_2O$	434.27	grn pr, monocl, 1.476, 1.482, 1.497	2.169	d		s	s	
p351	iron sulfide	$KFeS_2$	159.08	purp, hex	2.563			d		
p352	lactate	$KC_3H_5O_3.xH_2O$						s		s al; i eth
p353	laurate	$KC_{12}H_{23}O_2$	238.42	amorph						4.5^{15} al
p354	laurate, acid (mixt)	$KC_{12}H_{23}O_2.C_{12}H_{24}O_2$	438.74	wh, wax like sol		160				$0.904^{13.5}$ al
p355	lead chloride	Nat. pseudocotunnite. $2KCl.PbCl_2$	427.21	yel		490		s		
p356	magnesium carbonate, hydrogen	$KHCO_3.MgCO_3.4H_2O$	256.50	col, tricl or rhomb	2.98			d		
p357	magnesium chloride	Nat. carnalite. $KCl.MgCl_2.6H_2O$	277.86	col, rhomb, deliq, 1.466, 1.475, 1.494	1.61	265		64.5^{19}	d	d al
p358	magnesium chloride sulfate	Nat. kainite. $KCl.MgSO_4.3H_2O$	248.98	col, monocl	2.131			79.56^{18}		i al, eth
p359	magnesium chromate	$K_2CrO_4.MgCrO_4.2H_2O$	370.53	tricl	2.59					
p360	magnesium phosphate, hexahydrate	$KMgPO_4.6H_2O$	266.48	wh, rhomb cr		$-5H_2O$, 110		d		
p361	magnesium selenate	$K_2SeO_4.MgSeO_4.6H_2O$	496.52	col, monocl, 1.497, 1.499, 1.514	2.3645^{20}_{4}	$-2H_2O$, 33		s	s	
p362	magnesium sulfate	Nat. langbelinite. $K_2SO_4.2MgSO_4$	415.01	tetrah	2.829	927				
p363	magnesium sulfate	Nat. leonite. $K_2SO_4.MgSO_4.4H_2O$	366.70	col, monocl, 1.483, 1.487, 1.490	2.201^{20}			v s		
p364	magnesium sulfate	$K_2SO_4.MgSO_4.6H_2O$	402.73	col, monocl, 1.461, 1.463, 1.476	2.15	d 72		19.26^{0} 25^{20}	59.8^{75}	
p365	malate	$K_2C_4H_4O_5$	210.28	col, viscid mass				s		
p366	manganate	K_2MnO_4	197.14	grn, rhomb		d 190		d	d	s KOH
p367	*per*manganate	$KMnO_4$	158.04	purple rhomb, 1.59	2.703	d < 240		6.38^{20}	25^{65}	d al; v s MeOH, acet; s H_2SO_4
p368	manganese chloride(ous)	Chloromanganokalite. $4KCl.MnCl_2$	424.06	trig. 1.50	2.31			s	s	
p369	manganese sulfate(ic)	$KMn(SO_4)_2.12H_2O$	502.35	vlt, cub, (oct)				d		
p370	manganese sulfate(ous)	Manganolongbeinite. $K_2SO_4.2MnSO_4$	476.20	rose, tetrah, 1.572	3.02	850				
p371	mercury tartrate(ous)	$KHgC_4O_6$	387.76	wh cr powd				i		i al
p372	methionate	Potassium methane disulfonate. $K_2CH_2(SO_3)_2$	252.36	monocl, β 1.539	2.376			s		
p373	methylsulfate	$2KCH_3SO_4.H_2O$	318.41	wh cr				s		s al

No.	Name	Synonyms and Formulae	Mol. wt.	Crystalline form, properties and index of refraction	Density or spec. gravity	Melting point, °C	Boiling point, °C	Solubility, in grams per 100 cc		
								Cold water	Hot water	Other solvents
	Potassium									
p374	molybdate	K_2MoO_4	238.14	wh powd or 4-sid pr, deliq	2.91^{18}	919	d 1400	184.6^{25}	v s	i al
p375	*per*molybdate	$K_2O.3MoO_4.3H_2O$	596.06	lt yel cr, monocl.		d 180		sl s	s	v sl s al
p376	*tri*molybdate	$K_2O.3MoO_3.3H_2O$	580.06	wh need		571 (anhydr)		0.22^{15}	s	
p377	molybdenum cyanate	$K_4[Mo(CN)_8].2H_2O$	496.52	yel, rhomb	2.337^{25}_{4} (anhydr)	$-2H_2O$, 105–110		v s	v s	0.0017^{20} abs al; i eth; d HCl, H_2SO_4
p378	myristate, acid (mixt)	$KC_{14}H_{27}O_2.C_{14}H_{28}O_2$	494.85	wh, waxlike sol		153				$0.453^{13.5}$ al
p379	naphthalene − 1,5-disulfonate	$K_2C_{10}H_6(SO_3)_2.2H_2O$	400.53	monocl, 1.485, 1.669, 1.697	1.797			s		
p380	nickelsulfate	$K_2SO_4.NiSO_4.6H_2O$	437.13	bl, monocl, 1.484, 1.492, 1.505	2.124	d < 100		7^{0}	60.8^{75}	
p381	nitrate	Saltpeter. KNO_3	101.11	col, rhomb or trig, 1.335, 1.5056, 1.5064	2.109^{16}	tr-trig 129 m.p. 334	d 400	13.3^{0}	247^{100}	i dil al, eth; s liq NH_3, glyc
p382	nitride	K_3N	131.31	grnsh-blk		d		d		
p383	nitrite	KNO_2	85.11	wh-yelsh pr, deliq	1.915	440	d	281^{0}	413^{100}	s hot al; v s liq NH_3
p384	*m*-nitrophenoxide	$KOC_6H_4NO_2.2H_2O$	213.24	flat or need	1.691^{20}	$-H_2O$, 130	d	16.3^{15}		s al
p385	*p*-nitrophenoxide	$KOC_6H_4NO_2.2H_2O$	213.24	yel leaf	1.652^{20}	$-2H_2O$, 130	d	7.5^{15}		sl s al
p386	nitroplatinite	$K_2Pt(NO_2)_4$	457.32	col, monocl			d	3.8^{15}	s	
p387	nitroprusside	$K_2[Fe(NO)(CN)_5].2H_2O$	330.18	red, monocl, hygr.				100^{16}		s al
p388	nitrososulfate	$K_2SO_3(NO)_2$	218.28	col need		d 127, expl		$12^{14.5}$, d		i al
p389	oleate	$KC_{18}H_{33}O_2$	320.56	yelsh or brnsh soft mass or cr, 1.452				25	s	$4.315^{13.5}$ al; 100^{50} al; 3.5^{25} eth
p390	oleate, acid (mixt)	$KC_{18}H_{33}O_2.C_{18}H_{34}O_2$	603.03	wh, wax-like solid		95		s	s	$5.2^{13.5}$ al
p391	osmate	$K_2OsO_4.2H_2O$	368.43	vlt, cub, hygr		$-H_2O$, >100		sl s	s, d	i al, eth
p392	osmiumchloride	K_2OsCl_6	481.12	blk, oct	3.42^{16}	d 600		s	s	i al; s HCl
p393	osmylchloride	$K_2OsO_2Cl_4$	442.21	red, tetr	3.42	d 200 (in H atm)		s	s	
p394	osmyloxalate	$K_2[OsO_2(C_2O_4)_2].2H_2O$	512.47	brn need, tricl		$-H_2O$, 80	d 180	0.75^{15}	3.0^{15}	
p395	oxalate	$K_2C_2O_4.H_2O$	184.24	wh, monocl, 1.440, 1.485 1.550	$2.127^{3.9}$	$-H_2O$, 100			33^{16}	
p396	oxalate, hydrogen	KHC_2O_4	128.11	col, monocl, 1.382, 1.553, 1.573	2.044	d		2.5	16.7^{100}	i al, eth
p396	oxalate, hydrogen. monohydrate	$KHC_2O_4.H_2O$	146.13	rhomb	$2.044^{3.9}$					
p398	oxalate, tetra-	$KHC_2O_4.H_2C_2O_4.2H_2O$	254.20	col, tricl, 1.415, 1.536, 1.560	1.836	d		1.8^{13}		d al
p399	oxaloferrate (II)	$K_2[Fe(C_2O_4)_2].2H_2O$	346.12	gold need		d		s	s	
p400	oxaloferrate (II)	$K[Fe(C_2O_4)_2].2\frac{1}{2}H_2O$	316.03	br cr		d		92^{21}	d	i al
p401	oxaloferrate (III)	$K_3[Fe(C_2O_4)_3].3H_2O$	491.25	grn, monocl		$-3H_2O$, 100	d 230	4.7^{0}	117.7^{100}	i al, NH_3; s acet
p402	oxalatoplatinate	$K_2[Pt(C_2O_4)_2].2H_2O$	485.36	col, monocl pr	3.04^{12}	$-H_2O$, 100		sl s	s	
p403	oxalatouranate (IV)	$K_4[U(C_2O_4)_4].5H_2O$	836.59	yel, monocl.	2.563					
p404	oxide, mon-	K_2O	94.20	col, cub, hygr	2.32^{0}	d 350		v s	v s	s al, eth
p405	*per*oxide	K_2O_2	110.20	wh, amorph, deliq		490	d			
p406	oxide, super-	KO_2	71.10	yel, cub leaf	2.14	380	d	v s, d		d al
p407	oxide, tri- (sesqui-)	K_2O_3	126.20	red		430		ev O_2		d dil H_2SO_4
p408	palladium chloride	$K_2(PdCl_4)$	326.42	yel-grnsh-br cr, tetr, 1.710, 1.523	2.67	524		s	s	sl s^{80} al
p409	palladium chloride	$K_2(PdCl_6)$	397.32	lt red, oct	2.738	d 170		v sl s	sl s	s HCl; i al
p410	palladium oxalate	$K_2[Pd(C_2O_4)_2].4H_2O$	432.71	yel need		dec in air, $-4H_2O$, 80		0.833^{27}	$9.98^{82.9}$	
p411	palmitate, acid (mixt)	$KC_{16}H_{31}O_2.C_{16}H_{32}O_2$	550.96	wh, fatty sol		138				0.198^{13} al
p412	1-phenol-2-sulfonate(*o*-)	$KC_6H_4(OH)SO_3.H_2O$	230.29	rhomb, 1.527, 1.568, 1.647	1.87	400		s		s al
p413	1-phenol-4-sulfonate(*p*-)	$KC_6H_4(OH)SO_3$	212.27	rhomb, 1.571, 1.608, 1.694	1.87	>260				
p414	phenylsulfate	$KC_6H_5SO_4$	212.27	rhomb leaf		d 150–160	d	14^{15}		v sl s al
p415	*meta*phosphate, hexa-	$(KPO_3)_6$	708.44	col mass, hygr	2.107	810	1320	s	s	i al
p416	*meta*phosphate, tetra-	$(KPO_3)_4.2H_2O$	508.33	col cr		$-2H_2O$, 100		100^{15}	s	
p417	*ortho*phosphate	K_3PO_4	212.28	col, rhomb, deliq	2.564^{17}	1340		90^{20}	s	i al
p418	*ortho*phosphate, di-H	KH_2PO_4	136.09	col, tetr, deliq, 1.510, 1.4864	2.338	252.6		33^{25}	83.5^{90}	i al
p419	*ortho*phosphate, mono-H	K_2HPO_4	174.18	wh, amorph, deliq		d		167^{20}	v s	v s al
p420	*pyro*phosphate	$K_4P_2O_7.3H_2O$	384.40	col, deliq	2.33	$-2H_2O$, 180	$-3H_2O$, 300	s	v s	i al
p421	*sub*phosphate	$K_2PO_3.4H_2O$	229.24	col, rhomb		40	$-4H_2O$, 150	v s	v s	

No.	Name	Synonyms and Formulae	Mol. wt.	Crystalline form, properties and index of refraction	Density or spec. gravity	Melting point, °C	Boiling point, °C	Solubility, in grams per 100 cc		
								Cold water	Hot water	Other solvents
	Potassium									
p422	*hypo*phosphite	KH_2PO_2	104.09	wh, hex, deliq		d		200^{25}	330	v sl s abs al, NH_3; i eth; 11.1^{25} chl
p423	*ortho*phosphite, di-H	KH_2PO_3	120.09	wh cr, deliq		d		220^{20}	v s	i al
p424	phthalate, hydrogen	$KHC_8H_4O_4$	204.23	col, rhomb	1.636			10^{25}	33^{100}	
p425	picrate	$KC_6H_2N_3O_7$	267.20	yel-redsh or grnsh, rhomb, 1.527, 1.903, 1.952	1.852	expl 310		0.5^{15}	25^{100}	0.184^{25} al
p426	piperate	$KC_{12}H_9O_4$	256.31	lt yel cr powd				sl s	v s	
p427	platinate, hydroxo-	$K_2Pt(OH)_6$	375.34	yel, rhomb		d 160		s		i al
p428	platinorhodanide	$K_2[Pt(CNS)_6].2H_2O$	657.82	red, rhomb				s	8^{60}	s h al
p429	platinum iodide, hexa-	K_2PtI_6	1034.72	blk, cub	4.963^{25}_{4}			s		sl s al
p430	plumbate, hydroxo-	$K_2Pb(OH)_6$	387.44	col, rhomb				d	d	s KOH
p431	praseodymium sulfate	$3K_2SO_4.Pr_2(SO_4)_3.H_2O$	1110.81	cr	3.275^{16}			sl s		s HCl, HNO_3
p432	propionate	$KC_3H_5O_2.H_2O$	130.19	wh cr, hygr, or wh leaf, deliq		$-H_2O$, 120		207^{16}	359	22.2^{13} 95 % al
p433	propyl sulfate	$KC_3H_7SO_4$	178.25	wh cr powd				v s		
p434	*per*rhenate	$KReO_4$	289.30	wh, tetr, 1.643	4.887	550	1360–1370	1.21^{20}	14.0^{100}	v sl s al
p435	rhenium (IV) chloride	$K_2[ReCl_6]$	477.12	yel-grn cr, oct	3.34	d		s	s	s a; i conc H_2SO_4; d h H_2SO_4
p437	rhenium (V) oxychloride	$K_2[ReOCl_5]$	457.67	yel-grn cr, rhomb or monocl, 1.52				s		sl s HCl; s H_2SO_4; i al, eth
p438	rhenium oxycyanide	$K_3[ReO_2(CN)_4]$	439.58	red cr, monocl	2.70^{25}_{4}	d 300–400, vac		s	s	v sl s al; i alk
p439	rhodium cyanide	$K_3[Rh(CN)_6]$	376.32	pa yel, monocl, 1.5498, 1.5513, 1.5634				v s	v s	
p440	rhodium oxalate	$K_3[Rh(C_2O_4)_3].4\frac{1}{2}H_2O$	565.34	col, tricl	2.171^{20}_{4}	$-4\frac{1}{2}H_2O$, 190		v s	v s	
p441	rhodium sulfate	$KRh(SO_4)_2.12H_2O$	646.38	yel, cub	2.23			s		
p442	ruthenate	$K_2RuO_4.H_2O$	261.29	blk, tetr		$-H_2O$, 200	d 400 vac	v s	d	d a, al
p443	*per*ruthenate	$KRuO_4$	204.17	blk, tetr		d 44		sl s	s, d	
p444	*d*-saccharate, acid	$KHC_6H_8O_8$	248.24	rhomb need				1.1^{6}	s	
p445	salicylate	$KC_7H_5O_3$	176.22	wh powd				s		s al
p446	santoninate	$KC_{15}H_{19}O_4$	302.42	wh cr powd, deliq				s		s al
p447	selenate	K_2SeO_4	221.16	col, rhomb, hygr, 1.535, 1.539, 1.545	3.066			110.5^{0}	122.2^{100}	
p448	selenide	K_2Se	157.16	wh cub, reddens in air, hygr	2.851^{15}			s d	s	
p449	selenite	K_2SeO_3	205.16	wh, deliq		d 875		s		sl s al
p450	selenocyanate	KSeCN	144.08	need, deliq	2.347	d 100		s	s	d a; s al
p451	selenocyanoplatinate	$K_2Pt(SeCN)_6$	903.16	rhomb	$3.378^{12.5}$	d 80				
p452	selenothionate	$K_2SeS_2O_6$	317.29	col, monocl pr		d 250		s, d		
p453	*di*silicate	$K_2Si_2O_5$	214.37	col, rhomb, 1.502, 1.513	2.456^{25}_{4}	1015 ± 10				
p454	*meta*silicate	K_2SiO_3	154.29	col, rhomb (?), 1.520, 1.528		976		s	s	i al
p455	*tetra*silicate	$K_2Si_4O_9.H_2O$	352.56	wh, rhomb, α 1.495, β 1.535	2.417	d 400		s	s	i al
p456	*di*silicate, hydrogen	$KHSi_2O_5$	176.28	wh, rhomb, 1.480, 1.530	2.417^{15}_{4}	515				d HCl
p457	silicotungstate	$K_4SiW_{12}O_{40}.18H_2O$	3354.95	col, hex		$-17H_2O$, 100		33.3^{20}	v s	v s acet; s MeOH; sl s al; eth, bz
p458	silver carbonate	$KAgCO_3$	206.98	rect pl	3.769	d		d	d	
p459	silver nitrate	$KNO_3.AgNO_3$	270.98	monocl	3.219	125		v s	v s	
p460	sodium antimony tartrate	$KNaSbC_4H_2O_7$	346.97	wh, scales or powd				s		
p461	sodium carbonate	$KNaCO_3.6H_2O$	230.19	monocl, hygr, effl	$1.61–1.63^{14}$	$-6H_2O$, 100		185.2^{15}		
p462	sodium ferricyanide	$K_2Na[Fe(CN)_6]$	313.15	or-red, monocl		d		50^{25}	80^{80}	
p463	sodium nitrocobaltate (III)	$K_2Na[Co(NO_2)_6].H_2O$	454.18	yel cr	1.633	135		0.07^{25}		i al
p464	sodium sulfate	$3K_2SO_4.Na_2SO_4$	664.84	col, rhbdr	2.7			s	s	
p465	sodium tartrate	Rochelle salt, seignette salt. $KNaC_4H_4O_6.4H_2O$	282.23	col, rhomb, 1.492, 1.493, 1.496	1.790	70–80	$-4H_2O$, 215	26^{0}	66^{26}	v sl s al
p466	sorbate	$KC_6H_7O_2$	150.22	col cr	1.363^{25}_{20}	d 270		58.5^{25}		sl s MeOH

No.	Name	Synonyms and Formulae	Mol. wt.	Crystalline form, properties and index of refraction	Density or spec. gravity	Melting point, °C	Boiling point, °C	Solubility, in grams per 100 cc: Cold water	Hot water	Other solvents
	Potassium									
p467	stannate, hydroxo-	$K_2Sn(OH)_6$	298.94	col, trig	3.197			85^{10}	110.5^{20}	sl s KOH; i al, acet
p468	stearate	$KC_{18}H_{35}O_2$	322.58	wh cr powd					s	$0.145^{13.5}$ al; i eth, chl, CS_2
p469	stearate, acid (mixt)	$KC_{18}H_{35}O_2.C_{18}H_{36}O_2$	607.07	wh powd		153		s	s	$0.091^{13.5}$ al
p470	strontium chromium oxalate(ic)	$KSrCr(C_2O_4).6H_2O$	550.87	grnsh blk cr	2.155^{13}					
p471	styphnate	$KC_6H_2N_3O_8.H_2O$	301.22	yel, monocl pr		$-H_2O$, 120	expl	1.54^{20}		v sl s al
p472	succinate	$K_2C_4H_4O_4.3H_2O$	248.32	rhomb, hygr	1.564			s		
p473	succinate, hydrogen	$KHC_4H_4O_4$	156.18	monocl	1.767	d 242				
p474	succinate, hydrogen, dihydrate	$KHC_4H_4O_4.2H_2O$	192.21	rhomb, 1.417, 1.530, 1.533	1.616			s		s al
p475	succinate, hydrogen	$KHC_4H_4O_4.C_4H_6O_4$	274.27	monocl	1.56	162				
p476	sulfate	Nat. arcanite. K_2SO_4	174.27	col, rhomb or hex, 1.494, 1.495, 1.497	2.662	tr 588 m. p. 1069	1689	12^{25}	24.1^{100}	i al, acet, CS_2
p477	*peroxydi*sulfate	$K_2S_2O_8$	270.33	col, tricl, 1.461, 1.467, 1.566	2.477	d<100		1.75^{0}	5.3^{20}	i al
p478	*pyro*sulfate	$K_2S_2O_7$	254.33	col, need	2.512^{25}_{4}	>300	d	s	d	
p479	sulfate, hydrogen	Nat. mercallite, misenite. $KHSO_4$	136.17	col, rhomb, deliq 1.480	2.322	214	d	36.3^{0}	121.6^{100}	i al, acet
p480	sulfide, di-	K_2S_2	142.33	red-yel cr		470		s	d	s al
p481	sulfide, hydro-	KHS	72.17	yel, rhomb deliq	1.68–1.70	455		d	d	s al
p482	sulfide, mono-	K_2S	110.27	yel-br, cub, deliq	1.805^{14}	840		s	v s	s al, glyc; i eth
p483	sulfide, mono-, pentahydrate	$K_2S.5H_2O$	200.34	col, rhomb		60	$-3H_2O$, 150	s		s al, glyc; i eth
p484	sulfide, penta-	K_2S_5	238.52	or cr, hygr		206	d 300	v s	v s	sl s al
p485	sulfide, tetra-	K_2S_4	206.46	red-brn cr		145	d 850	s		s al
p486	sulfide, tetra- dihydrate	$K_2S_4.2H_2O$	242.49	yel				s	s	sl s al
p487	sulfide, tri-	K_2S_3	174.40	br-yel cr		252		s	d	s al
p488	sulfide, di-, trihydrate	$K_2S_2.3H_2O$	196.38	yel				v s	v s	s al
p489	sulfite	$K_2SO_3.2H_2O$	194.30	wh-yelsh, hex		d		100	<100	sl s al; i NH_3; d dil a
p490	*pyro*sulfite	Potassium *meta*bisulphite. $K_2S_2O_5$	222.33	col, monocl pl	2.34	d 190		sl s		sl s al; i eth
p491	sulfite, hydrogen	$KHSO_3$	120.17	col cr		d 190		s	s	i al
p492	*d*-tartrate	$K_2C_4H_4O_6.\frac{1}{2}H_2O$	235.28	col, monocl, β 1.526	1.98^{20}_{4}	$-H_2O$, 155	d 200–220	150^{14}	278^{100}	sl s al
p493	*dl*-tartrate	$K_2C_4H_4O_6.2H_2O$	262.31	col, monocl	1.984	$-2H_2O$, 100		100^{25}		
p494	*d*-tartrate, hydrogen	$KHC_4H_4O_6$	188.18	col, rhomb, 1.511, 1.550, 1.590	1.984^{18}			0.37	6.1^{100}	s a, alk; i al, ac a
p495	*dl*-tartrate, hydrogen	$KHC_4H_4O_6$	188.18	col, monocl	1.954			0.42^{15}	7.0^{100}	i al; s min a
p496	*meta*tellurate	K_2TeO_4	269.80	soft glutinous mass		d 200		d		
p497	*ortho*tellurate	$K_2H_4TeO_6.3H_2O$	359.88	col, rhomb, deliq		$-H_2O$	−O, 300	sl s	s	i al; sl s KOH
p498	telluride	K_2Te	205.80	col, cub, hygr	2.51			s, d	s	
p499	tellurite	K_2TeO_3	253.80	wh cr, deliq		d 460–470		v s	v s	s h K_2CO_3, KOH
p500	tellurium chloride	K_2TeCl_6	418.52	yel, oct, hygr	2.645			s d,		s, d al; s dil HCl
p501	thioantimonate	$2K_3SbS_4.4\frac{1}{2}H_2O$	815.68	yel cr, deliq				300^{0}	400^{80}	i al
p502	thioarsenate	K_3AsS_4	320.48	wh cr, deliq			d	v s		i al
p503	thioarsenite	K_3AsS_3	288.42			d		s		i al
p504	thiocarbonate, tri-	K_2CS_3	186.41	yel-red-brn cr, deliq		d		v s	s	s NH_3; sl s al; i eth
p505	thiocyanate	KNCS	97.18	col, rhomb pr, deliq	1.886^{14}	173.2	d 500	177.2^{0}	217^{20}	s al; 20.75^{22} acet; 0.18^{13} amyl al; v s liq NH_3
p506	*di*thionate	$K_2S_2O_6$	238.32	col, trig, 1.455, 1.515	2.278	d		6	66^{100}	i al
p507	*penta*thionate	$K_2S_5O_6.1\frac{1}{2}H_2O$	361.55	col, rhomb, —1.63	2.112	d		s	d	i al
p508	*tetra*thionate	$K_2S_4O_6$	302.46	col, monocl, 1.6057	2.296			v s		i al
p509	*tri*thionate	$K_2S_3O_6$	270.39	col, rhomb, 1.475, 1.480, 1.487	2.304	d 30–40		s	d	i al
p510	thioplatinate	$K_2Pt_4S_6$	1050.95	bl-gray cr	6.44^{15}	d, ign		i		d HCl
p511	thiosulfate, pentahydrate	$3K_2S_2O_3.5H_2O$	661.07	col, rhomb		d		$150.2^{17.2}$		
p512	*meta*thiostannate	$K_2SnS_3.3H_2O$	347.13	dk brn oil	1.847^{18}	$-3H_2O$, 100		s		i al
p513	thiosulfate	$K_2S_2O_3\frac{1}{3}.H_2O$	196.34	col, monocl, deliq	2.590 anhydr 2.23	$-H_2O$, 200	d	96.1^{0}	312^{90}	i al
p514	tungstate	$K_2WO_4.2H_2O$	362.08	col, monocl, deliq	3.113	tr 388, 921		51.5	151.5	d a; i al
p515	*meta*tungstate	$K_6[H_2W_{12}O_{40}].18H_2O$	3407.08	hex		*ca* 930		s	v s	d a
p516	*meta*uranate	K_2UO_4	380.23	or-yel, rhomb				i		v s a
p517	uranyl acetate	$KUO_2(C_2H_3O_2)_3.H_2O$	504.28	tetr	3.296^{15} ($\frac{1}{2}H_2O$)	$-H_2O$, 275		s		
p518	uranyl carbonate	$2K_2CO_3.UO_2CO_3$	606.46	yel, hex		$-CO_2$		7.4^{15}	d	s K_2CO_3 sol; i a

No.	Name	Synonyms and Formulae	Mol. wt.	Crystalline form, properties and index of refraction	Density or spec. gravity	Melting point, °C	Boiling point, °C	Solubility, in grams per 100 cc: Cold water	Hot water	Other solvents
	Potassium									
p519	uranyl sulfate	$K_2SO_4.UO_2SO_4.2H_2O$	576.39	yel, monocl	$3.363^{19.1}$	$-2H_2O$, 120		s		
p520	urate, acid	$KHC_5H_2N_4O_3$	206.21	wh powd				sl s		
p521	*meta*vanadate	KVO_3	138.04	col cr				sl s	s	sl s KOH; i al
p522	vanadium sulfate	Potassium vanadium alum. $KV(SO_4)_2.12H_2O$	498.35	violet cubic	1.783^{20}_{4}	20	$-12H_2O$, 230	1984^{10}		
p523	*ethyl*xanthate	KC_2H_5OCSS	160.30	wh to pa yel cr or cr powd	$1.558^{21.5}$	d>200		v s	d	s al; i eth
p524	**Praseodymium** (α form)	Pr	140.907	pa yel, met, hex up to 798	6.773	931	3512	d		s a
p525	(β form)	Pr	140.907	cub	6.64	935	3127	d		s a
p526	acetate	$Pr(C_2H_3O_2)_3.3H_2O$	372.09	grn need				v s		
p527	acetylacetonate	$Pr(C_5H_7O_2)_3$	438.24	cr ppt		146				s CS_2
p528	bromate	$Pr(BrO_3)_3.9H_2O$	686.77	grn, hex		56.5	$-7H_2O$, 170	196^{25}		
p529	bromide	$PrBr_3$	380.63	grn cr powd		691	1547	d, sl s		
p530	carbide	PrC_2	164.93	yel cr	5.10	d		d	d	s dil a
p531	carbonate	$Pr_2(CO_3)_3.8H_2O$	605.96	grn silky pl		$-6H_2O$, 100		i		s a
p532	chloride	$PrCl_3$	247.27	bl grn need	4.02^{25}	786	1700	103.9^{13}	∞^{100}	v s al; 2.4 pyr; i chl, eth
p533	chloride, heptahydrate	$PrCl_3.7H_2O$	373.37	grn, tricl	2.25^{17}	115		334^{13}	∞^{100}	s al, HCl
p534	hexantipyrine *per*chlorate	$[Pr(C_{11}H_{12}N_2O)_6].(ClO_4)_3$	1568.65	grn, hex leaf		d 286–291				
p535	iodide	PrI_3	521.62	gr cr, hygr		737		v s		
p536	molybdate	$Pr_2(MoO_4)_3$	761.63	grass-green tetr	4.84	1030				
p537	oxalate	$Pr_2(C_2O_4)_3.10H_2O$	726.03	lt grn cr				i		s a
p538	oxide, di-	PrO_2	172.91	br-bl powd	6.82	>350 tr to Pr_6O_{11}				
p539	oxide, sesqui-	Praseodymia. Pr_2O_3	329.81	yel-grn, amorph	7.07	d		0.000020^{20}		s a
p540	selenate	$Pr_2(SeO_4)_3$	710.69		4.30^{15}			36^{0}	3^{90}	
p541	sulfate	$Pr_2(SO_4)_3$	570.00	lt grn powd	3.72^{16}			23.7^{0} 17.7^{20}	1.02^{95}	
p542	sulfate, octahydrate	$Pr_2(SO_4)_3.8H_2O$	714.12	grn, monocl, 1.540, 1.549, 1.561	$2.827^{15.2}$			17.4^{20}	sl s	
p543	sulfate, pentahydrate	$Pr_2(SO_4)_3.5H_2O$	660.08	monocl pr	3.176^{16}				1.85^{25}	
p544	sulfide	Pr_2S_3	378.01	br powd	5.042^{11}	d		i	d	s dil a
p545	**Protactinium**	Pa	231.10	gray met, tetrag	15.37	<1600				
p546	chloride	$PaCl_4$	372.91	yel-grn, tetrag		subl 400 in vac		s		s dil HCl
p547	fluoride	PaF_4	307.09	monocl				i		
p548	oxide, di-	PaO_2	263.10	blk, cub						
p549	oxide, pent-	Pa_2O_5	542.20	wh, cub						s dil HF
r1	**Radium**	Ra	226.0254	silver wh met	5(?)	700	<1140	d, ev H_2		d a
r2	bromide	$RaBr_2$	385.82	col-yelsh, monocl	5.79	728	subl 900	s	s	s al
r3	bromide dihydrate	$RaBr_2.2H_2O$	421.85	wh, monocl		$-2H_2O$, 100		s	s	
r4	carbonate	$RaCO_3$	286.01	wh, or sl brnsh, monocl				i		d a
r5	chloride	$RaCl_2$	296.91	col-yelsh, monocl	4.91	1000		s	s	s al
r6	chloride, dihydrate	$RaCl_2.2H_2O$	332.94	wh, monocl, discol		$-2H_2O$, 100		s	s	s HCl
r7	iodate	$Ra(IO_3)_2$	575.81					0.0175^{0}	0.170^{100}	
r8	nitrate	$Ra(NO_3)_2$	350.01	cr				13.9^{20}		
r9	sulfate	$RaSO_4$	322.06	col, rhomb				0.000002^{20}	0.000005^{40}	i a
r10	**Radon**	Niton, Radium emanation. Rn	222.00	col gas, opaque cr	gas 9.73 g/l liq 4.4^{-62} sol 4.0	−71	−61.8	51.0^{0} cm^3 22.4^{25} cm^3	13.0^{50} cm^3	sl s al, org liqu
r11	**Rhenium**	Re	186.2	met lust, hex	20.53	3180	5627 (est)	i	i	s dil HNO_3, H_2O_2; sl s H_2SO_4; i HCl
r12	bromide	$ReBr_3$	425.93	grn-blk cr		subl 500, vac				s dil H_2SO_4, HBr, liq NH_3
r13	carbonyl, penta-	$[Re(CO)_5]_2$	652.51	col, cub cr		d 250				v sl s org solv
r14	chloride, penta-	$ReCl_5$	363.47	dk grn to blk	4.9	d	d	d	d	s HCl, alk
r15	chloride, tetra-	$ReCl_4$	328.01	blk (exist. ?)			500	s, d	s, d	s HCl
r16	chloride, tri-	$ReCl_3$	292.56	dk red, hex			>550	s	s	s a, alk, liq NH_3, al; sl s eth
r17	fluoride	ReF_4	262.19	dk grn	5.383^{25}	124.5	d 500	d		s a
r18	fluoride, hexa-	ReF_6	300.19	pa yel, v hygr	liq 6.1573, sld $3.616^{18.8}$	18.8	47.6	s, d	s, d	d HNO_3, H_2SO_4
r19	iodide pentacarbonyl	$ReI.5CO$	453.16	yel, rhombd		200	d 400, subl vac 90	i		s bz
r20	oxide, di-	ReO_2	218.20	blk	11.4^{25}_{4}	d 1000		i	i	s conc HCl, H_2O_2
r21	oxide, hept-	Re_2O_7	484.40	yel pl or hex or powd, hygr	6.103	*ca* 297	subl 250	v s	v s	v s al; s alk, a
r22	oxide, per-	Re_2O_8(?)	500.40	wh	8.4	145		v s	v s	s alk; sl s eth

No.	Name	Synonyms and Formulae	Mol. wt.	Crystalline form, properties and index of refraction	Density or spec. gravity	Melting point, °C	Boiling point, °C	Solubility, in grams per 100 cc		
								Cold water	Hot water	Other solvents
	Rhenium									
r23	oxide, sesqui-	Dirhenium trioxide $Re_2O_3.xH_2O$		unstable				d, ev H_2		
r24	oxide, tri-	ReO_3	234.20	red, blue, cub	6.9–7.4		d 400	i	i	s H_2O_2, HNO_3
r25	oxybromide, tri-	ReO_3Br	314.11	wh		39.5	163			
r26	oxychloride, tri-	ReO_3Cl	269.65	col liq	3.867^{20}_{4}	4.5	131^{760}	d	d	s CCl_4
r27	oxy*tetra*chloride	$ReOCl_4$	344.01	or need		29.3	223.00	d	d	
r28	oxy*tetra*fluoride	$ReOF_4$	278.19	wh	lq 3.717 sol 4.032	39.7	62.7			
r29	oxy*di*fluoride, di-	ReO_2F_2	256.20	col		156				
r30	sulfide, di-	ReS_2	250.33	blk tr leaf	7.506^{20}_{4}		d	i	i	s HNO_3; i al, alk, HCl
r31	sulfide, hepta-	Re_2S_7	596.85	blk powd	4.866		d	i	i	s HNO_3, H_2O_2, alk; i HCl
r32	**Rhodium**	Rh	102.905	gray-wh, cub	12.4	1966 ± 3	3727 ± 100	i	i	s H_2SO_4+HCl, conc H_2SO_4; sl s a, aq reg
r33	amminechloride, hexa-	$[Rh(NH_3)_6]Cl_3$	311.45	rhomb pl	2.008^{25}_{4}	$-NH_3$ 210, d		12.5^{8}	s	
r34	carbonylchloride, basic	$RhCl_2.RhO.3CO$	376.75	ruby red need		subl 125.5		sl s	d	s CCl_4, ac a, bz
r35	chloride, tri-,	$RhCl_3$	209.26	br-red powd, deliq		d 450–500	subl 800	i	i	i a, aq reg
r36	chloride, tri-	$RhCl_3.xH_2O$		dk red, deliq		d 100		v s		s al, HCl; i eth
r37	fluoride, tri-	RhF_3	159.90	red, rhomb	5.38	>600 subl		i	i	i a, alk
r38	iodide, tri-	RhI_3	483.62	blk				i		
r39	nitrate	$Rh(NO_3)_3.2H_2O$	324.93	red, deliq				s	s	i al
r40	oxide, di-	RhO_2	134.90	br				i	i	i a, alk
r41	oxide, di-, dihydrate	$RhO_2.2H_2O$	170.93	olive grn		d		i		s HCl, acet a, alk
r42	oxide, sesqui-	Rh_2O_3	253.81	gray cr or amorph	8.20	d 1100–1150		i	i	i a, aq reg, KOH
r43	oxide, sesqui-, pentahydrate	$Rh_2O_3.5H_2O$	343.88	yel ppt		d		i	s	s a
r44	sulfate	$Rh_2(SO_4)_3.4H_2O$	566.05	red		d		s	s	
r45	sulfate	$Rh_2(SO_4)_3.12H_2O$	710.18	lt yel cr				v s	d	i al
r46	sulfate	$Rh_2(SO_4)_3.15H_2O$	764.22	pa yel cr		d		v s	d	i al, eth
r47	sulfide, hydro-	$Rh(HS)_3$	202.12	blk		d		i	d	s aq reg, aq Br; i Na_2S
r48	sulfide, mono-	RhS	134.97	gray-blk cr		d		i	i	i a, aq reg
r49	sulfide, sesqui-	Rh_2S_3	302.00	blk	6.40^{25}_{4}	d		i	i	i a, aq reg, aq Br
r50	sulfite	$Rh_2(SO_3)_3.6H_2O$	554.09	yel cr		d		s		i al
r51	**Rubidium**	Rb	85.47	soft, silver wh met	1.532 liq: $1.475^{38.5}$	38.89	688	d	d	d al; s a
r52	acetate	$RbC_2H_3O_2$	144.52	col, nacreous leaf, hygr		246		$86^{44.7}$	$89.3^{99.4}$	
r53	aluminium sulfate	$RbAl(SO_4)_2.12H_2O$	520.76	col, cub, oct, 1.457, 1.45232, 1.46618	1.867^{0}	99		2.59^{20}	43.25^{80}	
r54	azide	RbN_3	127.49	col need or plates	2.7876	d *ca* 310		107.1^{16}		0.182^{16} al; i eth
r55	borofluoride	$RbBF_4$	172.27	very sm rhomb cr, 1.333	2.820^{20}	590	d 500	0.6^{17}	10^{100}	
r56	borohydride	$RbBH_4$	100.31	white, cubic	1.920	1.487		v s		sl s al, i ether, bz
r57	bromate	$RbBrO_3$	213.38	cub	3.68	430		2.93^{25}	5.08^{40}	
r58	bromide	RbBr	165.38	col, cub, 1.5530	3.35 liq: 2.79^{730}	693	1340	98^{5}	$205.2^{113.5}$	i al, sl s acet
r59	bromide, tri-	$RbBr_3$	325.20	red, rhomb		d 140				
r60	bromochloroiodide	RbIBrCl	327.74	rhomb		d 200				s al, d eth
r61	bromoiodide, di-	$RbIBr_2$	372.13	red, rhomb	3.84	225	d 265	s		s al, d eth
r62	carbonate	Rb_2CO_3	230.95	col cr, deliq		837	d 740	450^{20}	s	0.7 abs al
r63	carbonate, acid	$RbHCO_3$	146.49	wh, rhomb		d 175		53.73^{20}		2.0 al
r64	chlorate	$RbClO_3$	168.92	trim	3.19			5.0^{19}	62.8^{100}	
r65	perchlorate	$RbClO_4$	184.92	rhomb, 1.4701	2.80	fus	d	0.5^{0}	18^{100}	0.009^{25} al, 0.06^{25} MeOH
r66	chloride	RbCl	120.92	col, cub, 1.493^{25}	2.80: liq: 2.088^{750}	718	1390	77^{0}	138.9^{100}	0.08^{25} al, 1.41^{25} MeOH; v sl s NH_3
r67	chlorobromide, di-	$RbBrCl_2$	236.29	rhomb		d 110				
r68	chloro*di*bromide	$RbBr_2Cl$	280.74	rhomb		76				
r69	chloroiodide, di-	$RbICl_2$	283.28	dk orange, rhomb		180–200	d 265	v s	v s	
r70	chloroplatinate	Rb_2PtCl_6	578.75	yel, cub	$3.94^{17.5}$	d		0.184^{0}	0.634^{100}	i al
r71	chloroplatinate, hexa-	$Rb_2[PtCl_6]$	578.75	yel, cub	$3.94^{17.5}$	d		0.014^{0}	0.33^{100}	i al
r72	chromate	Rb_2CrO_4	286.93	yel, rhomb, ~1.71	3.518			62^{0}	95.7^{60}	
r73	*di*chromate	$Rb_2Cr_2O_7$	386.93	tricl or monocl, >1.95, 1.70	3.02 monocl 3.125 tricl			tricl: 4.96^{18} monocl: 5.42^{18}	27.3^{60} 28.1^{60}	
r74	chromiumsulfate	$RbCr(SO_4)_2.12H_2O$	545.77	vlt cub, 1.482	1.946	107		43.4^{25}		

No.	Name	Synonyms and Formulae	Mol. wt.	Crystalline form, properties and index of refraction	Density or spec. gravity	Melting point, °C	Boiling point, °C	Solubility, in grams per 100 cc: Cold water	Hot water	Other solvents
	Rubidium									
r75	cobalt (II) sulfate	$RbSO_4.CoSO_4.6H_2O$	449.22	rubyred, monocl, 1.486, 1.491, 1.501	2.56^{15}			9.3^{25}	s	
r76	coppersulfate	$Rb_2SO_4.CuSO_4.6H_2O$	534.70	monocl, 1.489, 1.491, 1.504	2.57			10.28^{25}		
r77	cyanide	RbCN	111.49	col cr powder	2.32			s	s	i al, eth
r78	galliumsulfate	$RbGa(SO_4)_2.12H_2O$	563.50	col cr, 1.46579	1.962			s		
r79	fluoride	RbF	104.47	col, cubic, 1.398	3.557	795	1410	130.6^{18}		s dil HF; i al, eth, NH_3
r80	fluorogermanate	Rb_2GeF_6	357.52	wh cr				sl s	v s	
r81	fluosilicate	Rb_2SiF_6	313.02	cub, oct	3.332^{20}			0.16^{20}	1.35^{100}	s a, i al
r82	fluosulfonate	$RbFSO_3$	184.53	need		304				
r83	iodate	$RbIO_3$	260.37	monocl or cub	$4.33^{19.5}$	d		2.1^{23}		v s HCl
r84	*meta*periodate	$RbIO_4$	276.37	tetr	3.918^{16}			0.65^{13}		
r85	iodide	RbI	212.37	col, cub, 1.6474	3.55; liq: 2.87^{825}	647	1300	152^{17}	163^{25}	s liq NH_3; 0.674^{25} acet
r86	iodide, tri-	RbI_3	466.18	blk, rhomb	4.03^{22}	190		s		
r87	iodide, cmpd. with SO_2	$RbI.4SO_2$	468.63	lemon yel		13.5				
r88	iron (II) selenate	$Rb_2SeO_4.FeSeO_4.6H_2O$	620.79	blue-grn monocl, prism, 1.513, 1.520, 1.532	2.819					
r89	iron (III) selenate	$RbFe(SeO_4)_2.12H_2O$	643.42	cub, 1.507^{18}	2.31^{11}	45	$-12H_2O$, 100			
r90	iron (II) sulfate	$Rb_2SO_4.FeSO_4.6H_2O$	527.00	gr monocl, prism, 1.4815, 1.4874, 1.4977	2.516	d 60		24.2^{25} (anhydr)		
r91	iron (III) sulfate	$RbFe(SO_4)_2.12H_2O$	549.62	cub, 1.4823	1.91–1.95	48–53	$4.55^{6.6}$	52.6^{90}		
r92	hydride	RbH	86.48	col need	2.60	d 300		d	d	d a
r93	hydroxide	RbOH	102.48	gray-wh, deliq	3.203^{11}	301 ± 0.9		180^{15}	v s	s al
r94	neodymium nitrate	$2(?)RbNO_3.Nd(NO_3)_3.4H_2O$	697.26	redsh-vlt pl	2.56	47	$-4H_2O$, 60			
r95	nitrate	$RbNO_3$	147.47	col, hex cub, rhomb or tricl, hygr, 1.51, 1.52, 1.524	3.11; liq: 2.395^{100}	tr cub 161.4 m.p. 310	tricl rhom 219	44.28^{16}	452^{100}	sl s acet; v s HNO_3
r96	nitrate, hydrogen	$RbNO_3.HNO_3$	210.49	tetr		62				
r97	nitrate, hydrogen	$RbNO_3.2HNO_3$	273.50	col need		45				
r98	magnesium sulfate	$Rb_2SO_4.MgSO_4.6H_2O$	495.47	col, monocl, 1.467, 1.469, 1.478	2.386^{20}_{4}			20.2^{25} (anhydr)		
r99	permanganate	$RbMnO_4$	204.41	cr	$3.235^{10.4}$	d 295		0.5^{0}	4.7^{60}	
r100	oxide, mon-	Rb_2O	186.94	col-yel, cub	3.72	d 400		s d	s d	s liq NH_3
r101	oxide, per-	Rb_2O_2	202.94	yel, cub	3.65^{0}	570	d 1011^{760} mm	dec to RbOH $+H_2$		
r102	oxide, super	RbO_2, unstable	117.47	yel plates	3.80	432	d 1157^{1} atm			
r103	oxide (tetr-)	Rb_2O_4	234.94	dk orange cr, deliq		dec 500 vac				
r104	oxide, tri- (sesqui)	Rb_2O_3 (or Rb_4O_6)	218.94	blk cub	3.53^{0}	489		s d		
r105	praseodymium nitrate	$2RbNO_3.Pr(NO_3)_3.4H_2O$	693.93	grnsh, monocl, need hygr	2.50	63.5	$-4H_2O$, 60			
r106	selenate	Rb_2SeO_4	313.90	col, rhomb, 1.5515, 1.5537, 1.5582	3.90			159^{12}		
r107	silicofluoride	Rb_2SiF_6	313.02	col, oct or hex	3.3383^{20}_{4}			sl s	s	s a, i al
r108	sulfate	Rb_2SO_4	267.00	col, rhomb hex, 1.513, 1.513, 1.514	3.613^{20}_{4} liq 2.53^{1100}	1060 trig 653	*ca* 1700	42.4^{10}	81.8^{100}	i acet; sl s NH_3
r109	sulfate, hydrogen	$RbHSO_4$	182.54	rhomb, 1.473	2.892^{16}	< red heat				
r110	sulfide, di-	Rb_2S_2	235.07	dk red		420	volat >850			
r111	sulfide, hexa-	Rb_2S_6	363.32	brn-red		201				
r112	sulfide, mono-	Rb_2S	203.00	wh-pale yel	2.912	530 d vac		v s	v s	
r113	sulfide, mono-tetrahydrate	$Rb_2S.4H_2O$	275.07	cr, deliq				v s	v s	
r114	sulfide, penta-	Rb_2S_5	331.26	red, rhomb, deliq	2.618^{15}	225		d		s 70 % al; i eth, chl
r115	sulfide, tri-	Rb_2S_3	267.13	redsh yel		213				
r116	tartrate, d & l	$Rb_2C_4H_4O_6$	319.01	trig	2.658^{20}_{4}			200^{25}		i toluol
r117	*dl*-tartrate, hydrogen	$RbHC_4H_4O_6$	234.55	trim pr	2.282	201 d		1.18^{25}	11.7^{100}	
r118	vanadium sulfate	Rubidium vanadium alum. $RbV(SO_4)_2.12H_2O$	544.72	yellow, cubic, 1.4689	1.915^{20}_{4}	64	230 $-12H_2O$, 300 dec	2.56^{10}		

No.	Name	Synonyms and Formulae	Mol. wt.	Crystalline form, properties and index of refraction	Density or spec. gravity	Melting point, °C	Boiling point, °C	Solubility, in grams per 100 cc: Cold water	Hot water	Other solvents
	Ruthenium									
r119	**Ruthenium**	Ru	101.07	gray-wh or silv brittle met, hex	12.30	2310	3900	i	i	i aq reg, a, al; s fus alk
r120	carbonyl, penta-	$Ru(CO)_5$	241.12	col liq		−22				s al, bz
r121	chloride, tetra-	$RuCl_4.5H_2O$	332.96	rdsh-br cr, hygr		d		s		s al
r122	chloride, tri-	$RuCl_3$	207.43	br cr, deliq	3.11	d >500		i	d	sl s al; s HCl; i CS_2
r123	fluoride, penta-	RuF_5	196.06	dk grn cr	$2.963^{18.5}$	101	250	d	d	
r124	hydroxide	$Ru(OH)_3$	152.09	blk powd				v sl s		s a; i alk
r125	oxide, di-	RuO_2	133.07	dk bl, tetr	6.97	d		i	i	i a; s fus alk
r126	oxide, tetr-	RuO_4	165.07	yel, rhomb need	3.29^{21}	25.5	d 108	2.033^{20}	2.249^{74}	s al, a, alk, CCl_4
r127	oxychloride ammoniated	Ruthenium red. $Ru_2(OH)_2Cl_4.7NH_3.3H_2O$	551.19	brn-red powd				s		
r128	silicide	RuSi	129.16	met pr	5.40^4			i	i	s HNO_3+HF
r129	sulfide	Nat. laurite. RuS_2	165.20	gray-blk, cub	6.99	d 1000		i	i	i a; s fus alk
s1	**Samarium**	Sm	150.35	wh-gray met, hex	7.520	1077	1791	i	i	s a
s2	acetate	$Sm(C_2H_3O_2)_3.3H_2O$	381.53		1.94			15^{25}		
s3	acetylacetonate	$Sm(C_5H_7O_2)_3$	447.68	cr mass		146–147		i		
s4	benzylacetonate	$Sm(C_{10}H_9O_2)_3.2H_2O$	669.92	straw color		103–105		i		s org solv
s5	bromate	$Sm(BrO_3)_3.9H_2O$	696.21	yel, hex		75	$-9H_2O$, 150	114^{25}		v sl s al
s6	(II) bromide	$SmBr_2$	310.17	dk brn	5.1	508	1880	d		
s7	(III) bromide	$SmBr_3.6H_2O$	498.17	yel cr, deliq	2.971^{22}	640				
s8	carbide	SmC_2	174.37	yel, hex	5.86			d	d	s, d a
s9	(II) chloride	$SmCl_2$	221.26	red-brn cr	4.56^{25}	740		s, d		i al, CS_2
s10	(III) chloride	$SmCl_3$	256.71	yelsh-wh cr, hygr	4.46^{18}	678±2	d	92.4^{10}	99.9^{60}	v s al; 6.4^{25} pyr
s11	(III) chloride, hexahydrate	$SmCl_3.6H_2O$	364.80	grn-yel, tricl, hygr	2.383	$-5H_2O$, 110				
s12	chromate	$Sm_2(CrO_4)_3.8H_2O$	792.80	yel				0.043^{25}		
s13	(II) fluoride	SmF_2	188.35			1306	>2400			
s14	(III) fluoride	SmF_3	207.35			1306	2323	i	i	
s15	hydroxide	$Sm(OH)_3$	201.37	pa yel powd				i		s a; i alk
s16	(II) iodide	SmI_2	404.16	dk brn		527	1580	d		
s17	(III) iodide	SmI_3	531.06	or-yel cr		850				
s18	kojate	$Sm(C_6H_5O_4)_3$	573.66			d 275		i		
s19	(III) methyl-phosphate, di	$Sm[(CH_3)_2PO_4]_3$	525.47	cream col, hex pr				35.2^{25}	10.8^{95}	
s20	(III) molybdate	$Sm_2(MoO_4)_3$	780.51	vlt, rhomb oct	5.36					
s21	(III) nitrate	$Sm(NO_3)_3.6H_2O$	444.46	pa yel, tricl	2.375	78–79		v s		
s22	(III) oxalate	$Sm_2(C_2O_4)_3.10H_2O$	744.91	wh cr				0.000054		s H_2SO_4
s23	oxide, sesqui-	Samaria. Sm_2O_3	348.70	wh-yelsh powd	8.347			i		v s a
s24	(III) sulfate	$Sm_2(SO_4)_3.8H_2O$	733.01	lt yel, monocl, 1.543, 1.552, 1.563	2.930	$-8H_2O$, 450		2.67^{20} 4.4^{25}	1.99^{40}	
s25	(III) sulfide	Sm_2S_3	396.89	yelsh-pink	5.729	1900			d	d dil a
s26	sulfate, basic	$Sm_2O_2SO_4$	428.76	yel powd		d 1100		i		i dil H_2SO_4
s27	**Scandium**	Sc	44.956	silv met, cubic or hex	2.9890	1541	2831	d, ev H_2		
s28	acetylacetonate	$Sc(C_5H_7O_2)_3$	342.29	col pl		187.5	subl 210–215			s al, bz, chl
s29	bromide	$ScBr_3$	284.68		3.914	subl >1000				
s30	chloride	$ScCl_3$	151.32	col cr	2.39^{25}_{4}	939	subl 800–850	v s	v s	i abs al
s31	hydroxide	$Sc(OH)_3$	95.98	col amorph				i		s dil a
s32	nitrate	$Sc(NO_3)_3$	230.97	col, deliq		150		s		s al
s33	nitrate, tetrahydrate	$Sc(NO_3)_3.4H_2O$	303.03	col, pr, deliq		$-4H_2O$, 100		v s		
s34	oxalate	$Sc_2(C_2O_4)_3.5H_2O$	444.05	cr		$-4H_2O$, 140				
s35	oxide	Scandia. Sc_2O_3	137.91	wh powd	3.864			i	i	s h a
s36	sulfate	$Sc_2(SO_4)_3$	378.10	col cr	2.579	d		10.3^{25}	v s	
s37	sulfate, hexahydrate	$Sc_2(SO_4)_3.6H_2O$	486.19			$-4H_2O$, 100 $-6H_2O$, 250		v s		
s38	sulfate, pentahydrate	$Sc_2(SO_4)_3.5H_2O$	468.17		2.519	$-3H_2O$, 100		54.6^{25}		
s39	**Selenic acid**	H_2SeO_4	144.97	wh, hex prism, hygr	3.004^{15}_{4}	58 eas undercools	d 260	1300^{30}	∞^{60}	s H_2SO_4; d al; i NH_3
s40	monohydrate	$H_2SeO_4.H_2O$	162.99	wh, need	2.627^{15}_{4} liq 2.3564^{15}_{4}	26 eas undercools	205	v s	v s	
s41	tetrahydrate	$H_2SeO_4.4H_2O$	217.03	col liq		51.7 eas undercools	$-H_2O$, 172^{20}	∞		s H_2SO_4; d org solv
s42	**Selenium**	Se	78.96	blsh-gray, met hex	4.81^{20}_{4}	217	684.9±1.0	i	i	s H_2SO_4, $CHCl_3$; i al; v sl s CS_2
s43	"	Se	78.96	red, monocl prism	4.50	170–180 trsf to hex	684.8	i	i	$0.1^{46.6}$ CS_2; s H_2SO_4, HNO_3
s44	"	Se	78.96	red amorph, blk vitr	red 4.26 blk 4.28	tr to hex, 60–80	684.8	i	i	s H_2SO_4, CS_2, bz

No.	Name	Synonyms and Formulae	Mol. wt.	Crystalline form, properties and index of refraction	Density or spec. gravity	Melting point, °C	Boiling point, °C	Solubility, in grams per 100 cc		
								Cold water	Hot water	Other solvents
	Selenium									
s45	bromide, "mono-"	Diselenium dibromide. Se_2Br_2	317.74	dk red liq	3.604^{15}		227 d	d	d	d al; s CS_2, $CHCl_3$, C_2H_5Br
s46	bromide, tetra-	$SeBr_4$	398.60	or-red-brn cr		d 75		d	d	s CS_2, chl, C_2H_5Br, HCl
s47	bromide (mono-) chloride, tri-,	$SeBrCl_3$	265.23	yel br cr		190				i CS_2
s48	bromide (tri-) chloride	$SeBr_3Cl$	354.14	or cr, hygr		d				v sl s CS_2
s49	carbide	SeC_2	102.98	yel liq, 1.845	2.682^{20}_{4}	45.5	$125–126^{760}$	i		s CS_2, eth, CCl_4, bz, al
s50	chloride "mono"-	Diselenium dichloride. Se_2Cl_2	228.83	br red liq, 1.596	2.77^{25}_{4}	−85	d 130	d	d	d al, eth; s CS_2, chl, CCl_4, bz
s51	chloride, tetra-	$SeCl_4$	220.77	wh-yel, cub, deliq, 1.807	$3.78–3.85^{360}$	305, subl 170–196	d 288	d	d	i al, eth, CS_2; s $POCl_3$; d a, alk
s52	fluoride, hexa-	SeF_6	192.95	col gas, 1.895	3.25^{-28} g/l	−39, subl −46.6	−34.5	s d		
s53	fluoride, tetra-	SeF_4	154.95	col liq or wh cr		m.p. −13.8 frz −90	>100	d	d	
s54	hydride	H_2Se	80.98	col gas, pois	gas 3.664^{760} air; liq: $2.004^{-41.5}$	-60.4	−41.5	3.77^{4}	$270^{22.5}$	s CS_2, $COCl_2$
s55	iodide, "mono"-	Diselenium diiodide. Se_2I_2	411.73	steelgray cr (exist ?)		68–70	d 100	d	d	
s56	nitride	Se_4N_4	371.87	amorph, or yel-brickred, hygr		expl 160–200	d	i	sl d	i al, eth; v sl s acet ac, bz
s57	oxide, di-	SeO_2	110.96	wh, monocl, col, tetr, pois, >1.76	3.95^{15}_{15}	340–350 subl 315–317		38.4^{14}	82.5^{65}	6.67^{14} al; $4.35^{15.3}$ acet; $1.11^{13.9}$ ac a; s bz
s58	oxide, tri-	SeO_3	126.96	pa yel cub or fibre, deliq	3.6	118	d 180	d, v s	d, v s	s al, conc H_2SO_4; i eth, bz, chl, CCl_4
s59	oxybromide	Selenyl bromide. $SeOBr_2$	254.78	red yel cr	liq 3.38^{50}	41.6	217^{710} d	d		s CS_2, CCl_4, chl, H_2SO_4, bz
s60	oxychloride	$SeOCl_2$	165.87	col-yel, liq 1.651^{20}	2.42^{22}	8.5	176.4	d		s CS_2, CCl_4, chl, bz
s61	sulfur oxy*tetra*-chloride	$SeSO_3Cl_4$	300.83	hex pr		165	183	d		
s62	oxyfluoride	Selenyl fluoride. $SeOF_2$	132.96	col liq	2.67	4.6	124	d		s al, CCl_4
s63	sulfide, di-	SeS_2	143.09	br red-yel		<100	d	i		d aq reg, HNO_3; s $(NH_4)_2S$
s64	sulfide, "mono"-	SeS	111.02	or-yel tabl or powd	3.056^{0}	d 118–119		i	i	s CS_2; i eth; d al
s65	sulfur oxide	$SeSO_3$	159.02	grn pr or yel powd		−SO_2, 40		d, 118		s H_2SO_4; i SO_3
s67	**Selenious acid**	H_2SeO_3	128.97	col, hex, deliq	3.004^{15}_{4}	d 70	−H_2O	167^{20}	v s	v s al; i NH_3
s68	**Silane,** bromo-	SiH_3Br	111.02	col gas, expl in air	1.72^{-80} 1.533^{0}	−94	1.9			
s69	bromotrichloro-	$SiBrCl_3$	214.35	col liq	1.826	−62	80.3	d	d	
s70	chloro-	SiH_3Cl	66.56	col gas	gas: 3.033 g/l liq: 1.145^{-113}	−118.1	−30.4			
s71	dibromo-	SiH_2Br_2	189.92	col liq, inflam	2.17^{0}	−70.1	66	d		d alk
s72	dibromo-	Silicobromoform. $SiHBr_3$	268.82	col liq, inflam	2.7^{17}	−73	109	d	d	d NH_3
s73	dibromodichloro-	$SiBr_2Cl_2$	258.81	col liq	2.172^{25}_{4}	−45.5	104	d	d	
s74	dichloro-	SiH_2Cl_2	101.01	gas	gas 4.599 g/l liq 1.42^{-122}	−122	8.3	d	d	
s75	dichloro difluoro-	$SiCl_2F_2$	136.99	gas	6.2784 g/l	−144 ± 2	−31.7 ± 0.2	d	d	
s76	(hexa-)hexaoxocyclo-	Siloxane. $Si_6O_3H_6$	222.56	wh, pl	1.32^{20}	d 140, inflam		sl d	d	
s77	monochloro trifluoro-	$SiClF_3$	120.53	gas	5.455 g/l	−138.0 ± 2	−70.0 ± 0.2	d	d	
s78	monoiodo-	SiH_3I	158.01	col liq	$2.035^{14.8}$	−57.0	45.5	d		
s79	(tri-)nitrilo-	Silicylamine, tri-$(SiH_3)_3N$	107.34	col inflam liq	0.895^{-106}	−105.6				
s81	tribromochloro-	$SiBr_3Cl$	303.27	col liq	2.497^{25}_{4}	−20.8 ± 1	126–128	d	d	
s82	trichloro-	Silicochloroform. $SiHCl_3$	135.45	col liq	1.34	−126.5	33^{758} mm	d	d	s CS_2, CCl_4, chl, bz
s83	trichloroiodo-	$SiCl_3I$	261.35	col liq		>−60	113.5	d		
s84	trifluoro-	Silicofluoroform. $SiHF_3$	86.09	col gas	3.86^{0} g/l	−131.4	*ca* −95	d	d	d al, eth, alk; s tol
s85	triiodo-	Silicoiodoform. $SiHI_3$	409.81	red liq	3.314^{20}	8	220	d	d	s CS_2, bz

No.	Name	Synonyms and Formulae	Mol. wt.	Crystalline form, properties and index of refraction	Density or spec. gravity	Melting point, °C	Boiling point, °C	Solubility, in grams per 100 cc: Cold water	Hot water	Other solvents
	Silicane cyanate									
s86	**Silicane cyanate**	$Si(OCN)_4$	196.16	sol or liq	1.414^{20}_{4}	34.5 ± 0.5	247.2 ± 0.5 0.5^{760}	d	...	...
s87	diimide	$Si(NH)_2$	58.12	wh powd		d 900	...	...	...	...
s88	isocyanate	$Si(NCO)_4$	196.16	sol or liq	1.434^{25}_{4}	26.0 ± 0.5	185.6 ± 0.3^{760}	d	...	i acet; s bz, CCl_4, CS_2
s89	**Silicic acid, di-**	$H_2Si_2O_5$	138.18	col cr		d 150	...	i	i	s NH_3,HF
s90	meta-	H_2SiO_3	78.10	col, amorph		d room temp	...	i	i	s NH_3, HF, h alk
s91	**Silicon**	Si	28.0855	steel gray, large to micr cr, cub	2.32–2.34	1410	2355	i	i	s HF+HNO_3; i HF
s92	acetate, tetra-	$Si(C_2H_3O_2)_4$	264.27	col cr, hygr		subl 110 d 160–170	148^{6} mm	d	...	d al; sl s acet, bz
s93	bromide, tetra-	Tetrabromosilane. $SiBr_4$	347.72	col fum liq, sol cub	liq: 2.7715^{25}_{4} sol: 3.292^{-79}	5.4	154	d	d	d H_2SO_4
s94	(di-)bromide, hexa-	Si_2Br_6	535.62	wh, rhomb	...	95	240	d	d	s CS_2; d KOH
s95	bromide(di-) sulfide	$SiSBr_2$	219.97	col pl	...	93	150$^{18.3}$ mm	d	d	s bz, CS_2
s96	carbide	SiC	40.10	col-blk, hex or cub, 2.654, 2.697	3.217	~2700, subl, d	...	i	i	i a; s fus KOH
s97	chloride, tetra-	Tetrachlorosilane. $SiCl_4$	169.90	col fum liq	liq: 1.483^{20}, sol: 1.90^{-97} gas: 7.59 g/l	−70	57.57	d	d	d al
s98	(di-)chloride, hexa-	Hexachlorodisilane. Si_2Cl_6	268.89	col liq, 1.4748^{18}	1.58^{0}	−1	145^{760}	d	d	d al
s99	chloride(di-) sulfide	$SiSCl_2$	131.06	col pr	...	75	92$^{22.5}$	d	d	s CCl_4, CS_2 bz
s100	chloride(tri-) sulfide, hydro-	$SiCl_3HS$	167.52	col liq	1.45	...	96–100	d	d	d al
s101	fluoride, tetra-	Tetrafluorosilane. SiF_4	104.08	col gas	gas: 4.69 g/l^{760} liq: 1.66^{-95}	−90.2	−86	d	d	s abs al, HF; i eth
s102	(di-)fluoride, hexa	Hexa-fluorodisilane. Si_2F_6	170.16	gas	7.759 g/l	−18.7	−18.5	d	d	...
s103	hydride	Silane, silicane. SiH_4	32.12	col gas	liq: 0.68^{-185} gas: 1.44 g/l	−185	−111.8^{760} mm	i	...	d KOH
s104	hydride	Disilane, disilicane. Si_2H_6	62.22	col gas	gas: 2.865 g/l liq: 0.686^{-20}	−132.5	−14.5	sl d	...	s al, bz, CS_2
s105	hydride	Trisilane, trisilanepropane. Si_3H_8	92.32	col liq	liq: 0.743^{0}; gas: 4.15 g/l^{0}_{760}	−117.4	52.9	d	d	d CCl_4
s106	hydride	Tetrasilane, tetrasilane butane. Si_4H_{10}	122.42	col liq	liq: 0.79^{0} gas 5.48 g/l^{0}_{760}	−108	84.3	d	...	...
s107	iodide, tetra-	Tetraiodosilane. SiI_4	535.70	col, cub	4.198	120.5	287.5	d	...	2.2^{27} CS_2
s108	(di-)iodide, hexa-	Hexaiodosilane. Si_2I_6	817.60	col, hex	...	d 250	d	d	d	19^{19} CS_2
s109	nitride	Si_3N_4	140.28	gray-wh amorph powd	3.44	1900 press	...	i	i	s HF
s110	oxide, di-	Nat. cristobalite. SiO_2	60.08	col, cub or tetr, 1.487, 1.484	2.32	1723 ± 5	2230 (2590)	i	i	s HF; v sl s alk
s111	oxide, di-	Nat. lechatelierite. SiO_2	60.08	col, amorph, vitr, 1.4588	2.19	...	2230 (2590)	i	i	s HF; v sl s alk
s112	oxide, di-	Nat. opal. $SiO_2.xH_2O$	...	col, amorph 1.41–1.46	2.17–2.20	>1600	...	i	i	s HF; v sl s alk
s113	oxide, di-	Nat. tridymite. SiO_2	60.08	col, rhomb, 1.469, 1.470, 1.471	2.26^{25}_{4}	1703	2230 (2590)	i	i	s HF; v sl s alk
s114	oxide, di-	Nat. quartz. SiO_2	60.08	col, hex, 1.544, 1.553	2.635–2.660	1610	2230 (2590)	i	i	s HF; v sl s alk
s115	oxide, mon-	SiO	44.09	wh, cub	2.13	>1702	1880	i	i	s dil HF+HNO_3
s116	oxychloride	Chlorosiloxane. Si_2OCl_6	284.89	col liq	...	28.1 ± 0.2	137	d	...	∞ CS_2, CCl_4
s117	oxyfluoride	Si_2OF_6	186.16	col gas	1.358 liq	−47.8 ± 0.5	−23.3	d	d	d alk
s118	sulfide, di-	SiS_2	92.21	wh need, rhomb	2.02	subl 1090	white heat	d	...	d al, liq NH_3; s dil alk; i bz
s119	sulfide, mono-	SiS	60.15	yel need	1.853^{16}	subl 940^{20}	...	d	d	d al, alk
s120	thiocyanate	$Si(CNS)_4$	260.41	wh, rhomb need	1.409^{20}_{4}	143.8	314.2	d	...	d al, a, alk; i eth, CS_2, $CHCl_3$
s121	**Silicotungstic acid**	$H_4[Si(W_3O_{10})_4]\cdot 26H_2O$	3346.47	wh-sl yel cr, deliq	...			v s	v s	v s al
s122	**Silicyl oxide**	Disiloxane. $(SiH_3)_2O$	78.22	col gas	gas: 3.491 g/l liq: 0.881^{-80}	−144	−15.2	v sl s	sl d	...

No.	Name	Synonyms and Formulae	Mol. wt.	Crystalline form, properties and index of refraction	Density or spec. gravity	Melting point, °C	Boiling point, °C	Solubility, in grams per 100 cc: Cold water	Hot water	Other solvents
	Siloxane									
s123	**Siloxane, (di-), oxide.**	$[H(O)Si]_2.O$	106.19	wh volum subst	...	expl *ca* 300	...	sl s	...	s, d HF; d al
s124	**Silver**	Ag	107.868	wh met, cub 0.54.	10.5^{20}	961.93	2212	i	i	s HNO_3, h H_2SO_4, KCN; i alk
s125	acetate	$AgC_2H_3O_2$	166.92	wh pl	3.259^{15}	d	...	1.02^{20}	2.52^{80}	s dil HNO_3
s126	acetylide	Ag_2C_2	239.76	wh ppt	...	expl	...	i	...	s a; sl s al
s127	*ortho*arsenate	Ag_3AsO_4	462.53	dk red, cub	6.657^{25}	d	...	0.00085^{20}	...	s NH_4OH, ac a
s128	*ortho*arsenite	Ag_3AsO_3	446.53	yel, powd	...	d 150	...	0.00115^{20}	i	s ac a, NH_4OH, HNO_3; i al
s129	azide	AgN_3	149.89	wh rhomb pr, expl	...	252	297	i	0.01^{100}	s KCN, dil HNO_3; sl s NH_4OH
s130	benzoate	$AgC_7H_5O_2$	228.99	wh powd	...	...	...	0.262^{25}	s	0.017 al
s131	*tetra*borate	$Ag_2B_4O_7.2H_2O$	407.01	wh cr	...	...	...	sl s	...	s a
s132	bromate	$AgBrO_3$	235.78	col, tetr, 1.874, 1.920	5.206	d	...	0.196^{25}	1.33^{90}	s NH_4OH; sl s HNO_3
s133	bromide	Bromyrite: AgBr	187.78	pa yel, 2.253	6.473^{25}	432	d>1300	8.4×10^{-6}	0.00037^{100}	s KCN, $Na_2S_2O_3$, NaCl sol; sl s NH_4OH; i al
s134	carbonate	Ag_2CO_3	275.75	yel powd	6.077	d 218	...	0.0032^{20}	0.05^{100}	s NH_4OH, $Na_2S_2O_3$; i al
s135	chlorate	$AgClO_3$	191.32	wh, tetr	4.430^{20}_{4}	230	d 270	10^{15}	50^{80}	sl s al
s136	*per*chlorate	$AgClO_4$	207.32	wh, cr, deliq	2.806^{25}	d 486	...	557^{25}	s	s al; 101 tol; 5.28 bz
s137	chloride	Nat. cerargyrite. AgCl	143.32	wh, cub, 2.071	5.56	455	1550	0.000089^{10}	0.0021^{100}	s NH_4OH, $Na_2S_2O_3$, KCN
s138	chlorite	$AgClO_2$	175.32	yel cr	...	105 expl	...	0.45^{25}	2.13^{100}	...
s139	chromate	Ag_2CrO_4	331.73	red, monocl	5.625	...	...	0.0014^{0}	0.008^{70}	s NH_4OH, KCN
s140	*di*chromate	$Ag_2Cr_2O_7$	431.72	red, tricl	4.770	d	...	0.0083^{15}	d	s a, NH_4OH, KCN
s141	citrate	$Ag_3C_6H_5O_7$	512.71	wh need	...	d	...	0.028^{18}	sl s	s a, NH_4OH, KCN, $Na_2S_2O_3$
s142	cyanate	AgOCN	149.89	col	4.00	d	...	sl s	s	s KCN, HNO_3, NH_4OH
s143	cyanide	AgCN	133.84	wh, hex	3.95	d 320	...	0.000023^{20}	...	s HNO_3, NH_4OH, KCN, $Na_2S_2O_3$
s144	ferricyanide	$Ag_3Fe(CN)_6$	535.56	...	...	...	...	0.000066^{20}	...	i a; s NH_4OH, h $(NH_4)_2CO_3$
s145	ferrocyanide	$Ag_4Fe(CN)_6.H_2O$	661.45	wh	...	...	...	i	i	s KCN; i a, NH_4 salts, NH_4OH
s146	fluogallate	$Ag_3[GaF_6].10H_2O$	687.47	col, orthorhomb cr, 1.493	2.90	...	...	v s	...	i al
s147	fluoride	AgF	126.87	yel, cub, deliq	$5.852^{15.5}$	435	*ca* 1159	$182^{15.5}$	205^{108}	sl s NH_4OH
s148	fluoride, di-	AgF_2	145.87	brn, rhomb	4.57–4.58	690	d 700	d	d	...
s149	(di-)fluoride	Ag_2F	234.74	yel, hex	8.57	d 90	...	d	...	...
s150	fluosilicate	$Ag_2SiF_6.4H_2O$	429.88	wh powd or col cr, deliq	...	>100	d	v s	...	...
s151	fulminate	$Ag_2C_2N_2O_2$	299.77	need	...	expl	...	0.075^{13}	s	i NHO_3; s NH_4OH
s152	iodate	$AgIO_3$	282.77	col, rhomb	$5.525^{16.5}$	>200	d	0.003^{10}	0.019^{60}	s HNO_3, NH_4OH, KI
s153	*per*iodate	$AgIO_4$	298.77	or yel, tetrag	5.57	d 180	...	d	...	s HNO_3
s154	iodide(α)	Nat. iodyrite. AgI	234.77	yel, hex 2.21, 2.22	5.683^{30}_{4}	tr 146 to β	...	$2.8 \times 10^{-7.25}$	$2.5 \times 10^{-6.60}$	s KCN, $Na_2S_2O_3$, KI; sl s NH_4OH
s155	iodide (β)	AgI	234.77	or, cub	$6.010^{14.6}_{4}$	558	1506	...	...	...
s156	iodomercurate (α)	Ag_2HgI_4	923.95	yel, tetrag	6.02	tr to β 50.7	...	i	...	s KI, KCN; i dil a
s157	iodomercurate (β)	Ag_2HgI_4	923.95	red, cub	5.90	d 158	...	i	...	s KI, KCN; i dil a
s158	hydrogen(tri-) *paraperiodate*	$Ag_2H_3IO_6$	441.69	yel, rhomb	5.68^{25}	60 d	...	1.68^{25}	...	s HNO_3
s159	hyponitrite	$Ag_2N_2O_2$	275.75	yel	5.75^{30}	d 110	...	v sl s	...	d HNO_3, H_2SO_4
s160	lactate	$AgC_3H_5O_3.H_2O$	214.96	wh or sl gray cr, powd	...	...	...	*ca* 7.7	...	...
s161	laurate	$AgC_{12}H_{23}O_2$	307.19	wh, greasy powd	...	212.5	...	...	...	0.007^{25} al; 0.008^{15} eth
s162	levunilate	$AgC_5H_7O_3$	222.98	leaf	...	...	...	0.67^{17}	d	...
s163	*per*manganate	$AgMnO_4$	226.81	dk vlt, monocl	4.27^{25}	d	...	0.55^{0}	$1.69^{28.5}$	d al
s164	mercury iodide (α)	Ag_2HgI_4	923.98	yel, tetrag	6.02	trst 50.7	...	i	...	...
s165	mercury iodide (β)	Ag_2HgI_4	923.98	red, cub	5.90	158 d	...	i	...	...
s166	myristate	$AgC_{14}H_{27}O_2$	335.24	...	...	211	...	0.007^{25}	...	0.006^{25} al; 0.007^{15} eth
s167	nitrate	$AgNO_3$	169.87	col, rhomb, 1.729, 1.744, 1.788	4.352^{19}	212	d 444	122^{0}	952^{190}	s eth, glyc; v sl s abs al

No.	Name	Synonyms and Formulae	Mol. wt.	Crystalline form, properties and index of refraction	Density or spec. gravity	Melting point, °C	Boiling point, °C	Solubility, in grams per 100 cc: Cold water	Hot water	Other solvents
	Silver									
a168	nitrite	$AgNO_2$	153.88	wh, rhomb	4.453^{26}	d 140		0.155^{0}	1.363^{60}	s ac a, NH_4OH; i al
a169	nitroplatinite	$Ag_2[Pt(NO_2)_4]$	594.85	yel-brn monocl pr		d 100		sl s	s	
a170	nitroprusside	$Ag_2[FeNO(CN)_5]$	431.68	lt pink				i		s NH_4OH; i al, HNO_3
a171	oxalate	$Ag_2C_2O_4$	303.76	col cr	5.029^{4}	expl 140		0.00339^{18}		s KCN, NH_4OH, a
a172	oxide	Ag_2O	231.74	br-blk, cub	$7.143^{16.6}$	d 230		0.0013^{20}	0.0053^{80}	s a, KCN, NH_4OH, al
a173	oxide, per	Ag_2O_2 (or AgO)	247.74	gray-blk, cub	7.44	d>100		i		s H_2SO_4, HNO_3, NH_4OH
a174	palmitate	$AgC_{16}H_{31}O_2$	363.29	wh, greasy powd		209		0.0012^{20}	0.006^{25}	0.007^{15} eth; 0.006^{25} al
a175	*meta*phosphate	$AgPO_3$	186.84	wh, amorph	6.37	*ca* 482		i		s HNO_3, NH_4OH
a176	*ortho*phosphate	Ag_3PO_4	418.58	yel, cub	6.370^{25}	849		$0.00065^{19.5}$		s a, KCN, NH_4OH, NH_3
a177	*ortho*phosphate, mono-H	Ag_2HPO_4	311.75	wh, trig	1.8036	d 110				
a178	*pyro*phosphate	$Ag_4P_2O_7$	605.42	wh	$5.306^{7.5}$	585		i	i	s a, NH_4OH, KCN, ac a
a179	propionate	$AgC_3H_5O_2$	180.94	wh leaf or need.	2.687^{25}_{4}			0.842^{20}	2.04^{80}	
a180	*per*rhenate	$AgReO_4$	358.07	wh cr, tetrag or rhomb	7.05	430		0.32^{20}		
a181	salicylate	$AgC_7H_5O_3$	244.99	wh to redsh-wh cr				sl s		s al
a182	selenate	Ag_2SeO_4	358.73	wh, orthorhomb cr	5.72			0.118^{20}		
a183	selenide	Ag_2Se	294.70	thin gray pl, cub	8.0	880	d	i		s NH_4OH, h HNO_3
a184	stearate	$AgC_{18}H_{35}O_2$	391.35	wh powd amorph		205		0.006^{20}		0.006^{25} al; 0.006^{25} eth
a185	sulfate	Ag_2SO_4	311.80	wh, rhomb, 1.7583, 1.7748, 1.7852	$5.45^{29.2}$	652	d 1085	0.57^{0}	1.41^{100}	s a, NH_4OH; i al
a186	sulfide	Nat. acanthite. Ag_2S	247.80	gray-blk, rhomb	7.326	tr 175	d		v sl s	s KCN, conc H_2SO_4, HNO_3
a187	sulfide	Nat. argentite. Ag_2S	247.80	blk, cub	7.317	825	d	8.4×10^{-15}		s KCN, a
a188	sulfite	Ag_2SO_3	295.80	wh cr		d 100		v sl s		s a, NH_4OH, KCN; i HNO_3
a189	*d*-tartrate	$Ag_2C_4H_4O_6$	363.81	wh, scales	3.423^{15}	d		0.2^{18}	0.203^{25}	s a, KCN, NH_4OH
a190	*ortho*tellurate, tetra-H	$Ag_2H_4TeO_6$	443.40	straw yel, rhomb bipyr		d>200		i	i	s KCN, NH_4OH
a191	telluride	Nat. hessite, Ag_2Te	343.34	gray, cub	8.5	955		i	i	s KCN, NH_4OH
a192	tellurite	Ag_2TeO_3	391.36	yel-wh ppt		250-bl 450-pa yel		i	i	s KCN, NH_3
a193[1]	thioantimonite	Nat. pyrargyrite. Ag_3SbS_3	541.55	red, trig, 3.084 2.881 (Li)	5.76	486		i	i	s HNO_3
a193[2]	thioarsenite	Nat. proustite. Ag_3AsS_3	494.72	scarlet red, trig, 3.088, 2.792	5.49	490		i	i	s HNO_3
a194	thiocyanate	AgSCN	165.95	col cr		d		0.000021^{25}	0.00064^{100}	s NH_4OH; i a
a195	*di*-thionate	$Ag_2S_2O_6.2H_2O$	411.90	rhomb cr, ~1.662	3.61					
a196	thiosulfate	$Ag_2S_2O_3$	327.87	wh cr		d		sl s		s $Na_2S_2O_3$, NH_4OH
a197	tungstate	Ag_2WO_4	463.59	pa yel cr				0.05^{15}		s KCN, NH_4OH, HNO_3
	Silver complex									
a198	diamminesilver *per*rhenate	$[Ag(NH_3)_2]ReO_4$	392.13	col monocl cr	3.901					1.618 conc NH_4OH
a199	**Sodium**	Na	22.9898	silv, met cub, 4.22	0.97	97.81±0.03	882.9	d to $NaOH+H_2$		d al; i eth, bz
a200	acetate	$NaC_2H_3O_2$	82.03	wh gr powd, monocl, 1.464	1.528	324		119^{0}	170.15^{100}	sl s al
a201	acetate trihydrate	$NaC_2H_3O_2.3H_2O$	136.08	col, monocl pr, effl, β 1.464	1.45	58	123, $-3H_2O$, 120	76.2^{0}	138.8^{50}	2.1^{18} al; s
a202	alumina trisilicate	Nat. albite. $NaAlSi_3O_8$. (or $Na_2O.Al_2O_3.6SiO_2$)	262.22	col, tricl, 1.525, 1.529, 1.536	2.61	1100			sl d	s HCl; d dil al
a203	*meta*aluminate	$NaAlO_2$	81.97	wh amorph powd, hygr, 1.566, 1.595, 1.580		1800		s	v s	i al
a204	aluminum chloride	$NaCl.AlCl_3$	191.78	wh-yelsh cr powd, hygr		185		s	s	
a205	aluminum *meta*-silicate	Nat. jadeite. $Na_2O.Al_2O_3.4SiO_2$	404.28	col, monocl	3.3	1000–1060		i	i	d HCl

No.	Name	Synonyms and Formulae	Mol. wt.	Crystalline form, properties and index of refraction	Density or spec. gravity	Melting point, °C	Boiling point, °C	Solubility, in grams per 100 cc: Cold water	Hot water	Other solvents
	Sodium									
s206	aluminum *ortho*silicate	Nat. nephelite. $Na_2O.Al_2O_3.2SiO_2$	284.11	col, hex, 1.537 ± 0.002	2.619^{21}	1526		i	d	d a
s207	aluminum sulfate	$NaAl(SO_4)_2.12H_2O$	458.28	col, cub oct, 1.4388	1.6754^{20}	61		110^{15} (anhydr)	146^{80} (anhydr)	
s208	amide	Sodamide. $NaNH_2$	39.01	wh, conchoid fract		210	400	d	d	d hot al; 0.1 liq NH_3
s209	ammonium phosphate	Microcosmic salt, stercorite. $NaNH_4HPO_4.4H_2O$	209.07	col, monoc, 1.439, 1.441, 1.469	1.554	d 79		16.7	100	i al, acet
s210	ammonmium sulfate	$NaNH_4SO_4.2H_2O$	173.12	wh, rhomb	1.63^{15}	d 80		s	s	
s211	ammonium tartrate	$NaNH_4C_4H_4O_6.4H_2O$	261.16	wh, rhomb	1.590			21.09^{0}		
s212	*meta*antimonate	Leuconine. $NaSbO_3$	192.74	wh powd				i	s	s Na_2S sol
s213	antimonate, hydroxy	"Pyroantimonate". $NaSb(OH)_6$	246.78	pseudo cub				$0.03^{12.3}$	0.3^{100}	sl s al
s214	*pyro*antimonate, dihydro-	$Na_2H_2Sb_2O_7.6H_2O(?)$	511.58	wh, tetrag		d 280		i	0.28^{100}, d	
s215	antimonide	Na_3Sb	190.72	blk powd or bl cr, inflamm		856		d		sl s NH_3
s216	*meta*antimonite	$NaSbO_2.3H_2O$	230.78	col, rhomb	2.864	d		d		
s217	*meta*arsenate	$NaAsO_3$	145.91	rhomb, effl, 1.479, 1.502, 1.527	2.301	615		v s		
s218	*ortho*arsenate	$Na_3AsO_4.12H_2O$	423.93	col, trig or hex prism, 1.457, 1.466	1.752–1.804	86.3		$38.9^{15.5}$		1.67 al; 50^{15} glyc
s219	*ortho*arsenate, di-H	$NaH_2AsO_4.H_2O$	181.94	col, rhomb or monocl, 1.583, 1.553, 1.507	2.53	130, $-H_2O$, 100	d 200–280	s		
s220	*ortho*arsenate, mono-H	$Na_2HAsO_4.7H_2O$	312.01	col, monocl, pois, 1.462, 1.466, 1.478	1.88	130, $-5H_2O$, 50	d 180	5.46^{0}	100^{100}	s glyc; sl s al
s221	*ortho*arsenate, mono-H	$Na_2HAsO_4.12H_2O$	402.09	col, monocl, effl, 1.445, 1.466, 1.451	1.736	28	$-12H_2O$, 100	56^{14}	140.7^{30}	sl s al; i liq Cl
s222	*pyro*arsenate	$Na_4As_2O_7$	353.79	wh cr	2.205	850	d 1000	v s		
s223	arsenate fluoride	$2Na_3AsO_4.NaF.19H_2O$	800.06	wh, cub, 1.4657, 1.4693, 1.4726	2.849^{25}			10^{75}		
s224	arsenite	Sodium metaarsenite (?) (com'l) $NaAsO_2$ (or mixt with Na_3AsO_3)	129.91	gray-wh powd, pois	1.87			v s	v s	sl s al
s225	arsenotartrate	$Na(AsO)C_4H_4O_6.2\frac{1}{2}H_2O$	307.02	shiny cr, pois		$-2\frac{1}{2}H_2O$, 275	d 275	6.5^{19}		i al
s226	azide	NaN_3	65.01	col, hex	1.846^{20}	d Na+N	d in vac	41.7^{17}		0.314^{16} al; s liq NH_3, i eth
s227	barbital	$NaC_8H_{11}N_2O_3$	206.18	wh powd				20^{25}	40^{100}	sl s al; i eth
s228	benzenesulfonate	$NaC_6H_5SO_3$	180.16	wh cr				35.8^{30}	v s	
s229	benzoate	$NaC_7H_5O_2$	144.11	col cr, or wh amorph, or gran powd				66^{20}	74.2^{100}	1.64^{25} al
s230	*meta*bismuthate	$NaBiO_3$	279.97	yel-brn powd (com'l), yel (pure)				i	d	d a
s231	*meta*borate	$NaBO_2$	65.80	col, hex pr	2.464	966	1434	26^{20}	36^{35}	
s232	*meta*borate, tetrahydrate	$NaBO_2.4H_2O$	137.86	tricl, coll		57	$-H_2O$, 120	v s	v s	
s233	*meta*borate, peroxyhydrate	Sodium perborate (com'l). $NaBO_2.H_2O_2.3H_2O$	153.86	col, monocl		63.0	$-H_2O$, 130–150	2.55^{15}	3.75^{32}	s a, al, glyc
s234	*tetra*borate	$Na_2B_4O_7$	201.22	cr, 1.5010	2.367	741	d 1575	1.06^{0}	8.79^{40}	i al
s235	*tetra*borate, decahydrate	Borax. $Na_2B_4O_7.10H_2O$	381.37	col, monocl, effl, 1.447, 1.469, 1.472	1.73	75, $-8H_2O$, 60	$-10H_2O$, 320	2.01^{0}	170^{100}	v sl s al; s glyc; i a
s236	*tetra*borate, pentahydrate	$Na_2B_4O_7.5H_2O$	291.30	col, cub or hex, deliq, 1.461	1.815	$-H_2O$, 120		22.65^{65} (anhydr)	52.3^{100}	
s237	borohydride	$NaBH_4$	37.83	white cub, 1.542	1.074	400 dec		55^{25}	v s	4 al; 16.4 MeOH; s pyr; i eth
s238	bromate	$NaBrO_3$	150.90	col, cub, 1.594	$3.339^{17.5}$	381		27.5^{0}	90.9^{100}	i al
s239	bromide	NaBr	102.90	col, cub, hygr, 1.6412	3.203^{25}_{4}	747	1390	116.0^{50}	121^{100}	sl s al
s240	bromide, dihydrate	$NaBr.2H_2O$	138.93	col, monocl pr	2.176	$-2H_2O$, 51		79.5^{0}	$118.6^{80.5}$	2.31^{25} al; s liq NH_3; 17.42^{15} MeOH
s241	bromoaurate	$NaAuBr_4.2H_2O$	575.62	br-blk cr				s		
s242	bromoiridite	$Na_3IrBr_6.12H_2O$	956.81	dk grn, rhomb, effl		100	$-H_2O$, 150			s NH_4OH

No.	Name	Synonyms and Formulae	Mol. wt.	Crystalline form, properties and index of refraction	Density or spec. gravity	Melting point, °C	Boiling point, °C	Solubility, in grams per 100 cc: Cold water	Hot water	Other solvents
	Sodium									
s243	bromoplatinate	$Na_2PtBr_6.6H_2O$	828.62	dk red, tricl	3.323	d 150		v s	v s	v s al
s244	cacodylate	$Na[(CH_3)_2AsO_2].3H_2O$	214.03	wh		*ca* 60	$-H_2O$, 120	200^{15-20}		40^{25} al; 100^{15-20} 90 % al
s245	calcium sulfate	$Na_2Ca(SO_4)_2.2H_2O$	314.21	col, monocl need.	2.64	$-2H_2O$, 80		d	d	
s246	*d*-camphorate	$Na_2C_{10}H_{14}O_4.3H_2O$	298.25	wh need, hygr		$-3H_2O$, 100		122^{14}		s al
s247	carbide	Na_2C_2	70.00	wh powd	1.575^{15}	*ca* 700		d	d	s a; d al
s248	carbonate	Na_2CO_3	105.99	wh powd, hygr, 1.535	2.532	851	d	7.1^{0}	45.5^{100}	sl s abs al; i acet
s249	carbonate, decahydrate	Washing soda. $Na_2CO_3.10H_2O$	286.14	wh, monocl, 1.405, 1.425, 1.440	1.44^{15}	32.5–34.5	$-H_2O$, 33.5	21.52^{0}	421^{104}	i al
s250	carbonate, heptahydrate	$Na_2CO_3.7H_2O$	232.10	rhomb bipyr, effl	1.51	$-H_2O$, 32		16.90	33.9^{35}	
s251	carbonate, monohydrate	Crystal carbonate, thermonatrite. $Na_2CO_3.H_2O$	124.00	col, rhomb, deliq, 1.506, 1.509	2.25	$-H_2O$, 100		33	52.08	14^{25} glyc; i al, eth
s252	carbonate, sesqui-	$Na_2CO_3.NaHCO_3.2H_2O$	226.03	col, monocl, 1.5073	2.112	d		13^{0}	42^{100}	
s253	carbonate hydrogen	$NaHCO_3$	84.00	wh, monocl pr, 1.500	2.159	$-CO_2$,270		6.9^{0}	16.4^{60}	sl s al
s254	chlorate	$NaClO_3$	106.44	col, cub or trig, 1.513	2.490^{15}	248–261	d	79^{0}	230^{100}	s al, liq NH_3, glyc
s255	*per*chlorate	$NaClO_4$	122.44	wh, rhomb, deliq, 1.4606, 1.4617, 1.4731		d 482	d	s	v s	s al
s256	*per*chlorate, hydrate	$NaClO_4.H_2O$	140.46	col rhbdr, deliq	2.02	130	d 482	209^{15}	284^{50}	s al
s257	chloride	Common salt, nat. halite. NaCl	58.44	col, cub, 1.5442	2.165_4^{25}	801	1413	35.7^{0}	39.12^{100}	sl s al, liq, NH_3; s glyc; i HCl
s258	chlorite	$NaClO_2$	90.44	wh, cr, hygr		d 180–200		39^{17}	55^{60}	
s259	chlorite, pentahydrate	$NaOCl.5H_2O$	164.52	col		24.5		29.3^{0}	94.2^{23}	
s260	*hypo*chlorite	NaOCl	74.44	in solution only						
s261	*hypo*chlorite, dihydrate	$NaOCl.2\frac{1}{2}H_2O$	119.48	col, hygr		57.5		v s		
s262	chloroaurate	$NaAuCl_4.2H_2O$	397.80	yel, rhomb, ω 1.545 ϵ>1.75		d 100		150^{10}	990^{60}	v s al, eth
s263	chloroiridate	$Na_2IrCl_6.6H_2O$	558.99	dull red-blk, tricl		d 600		v s	v s	sl s al
s264	chloroiridite	$Na_3IrCl_6.12H_2O$	690.07	dk grn cr		$-H_2O$, 50		31.46^{15}	307.26^{85}	
s265	chloroosmate	$Na_2OsCl_6.2H_2O$	484.93	or-red, rhomb pr				v s		s al
s266	chloropalladite	$Na_2PdCl_4.3H_2O$	348.24	br-red cr, deliq				v s		s al
s267	chloroplatinate	Na_2PtCl_6	453.79	or-yel powd, hygr		tr 150–160		s	v s	s al
s268	*hexa*chloroplatinate	$Na_2PtCl_6.6H_2O$	561.88	or-red, tricl	2.500	$-6H_2O$, 100		66^{15}	v s	11.9 al, MeOH; i eth
s269	chloroplatinite	$Na_2PtCl_4.4H_2O$	454.98	red pr		100	$-H_2O$, 150	s		s al
s270	chlororhodite, hexa-	Na_3RhCl_6	384.59	red, tricl		d>550		v s		
s271	chlororhodite, hexa-, hydrate	$Na_3RhCl_6.18H_2O$	708.87	garnet red, oct, effl		d 904, effl		v s		i al
s272	chromate	Na_2CrO_4	161.97	yel, rhomb bipyram	2.710–2.736			87.3^{30}		sl s al; s MeOH
s273	chromate decahydrate	$Na_2CrO_4.10H_2O$	342.13	yel, monocl, deliq	1.483	19.92		50^{10}	126^{100}	sl s al; i ac a
s274	*di*chromate	$Na_2Cr_2O_7.2H_2O$	298.00	red, monocl pr, deliq, 1.661, 1.699, 1.751	2.52^{13}	$-2H_2O$, 100 356.7 (anhydr)	d 400 (anhydr)	238^{0} (anhydr) 180^{20}	508^{80} (anhydr) 433^{98}	i al
s275	*peroxy*chromate	Na_3CrO_8	248.96	or pl		d 115		sl s		i al, eth
s276	cinnamate	$NaC_9H_7O_2$	170.14	wh cr powd				9.1	5^{100}	0.625 90 % al; s glyc
s277	citrate, dihydrate	$Na_3C_6H_5O_7.2H_2O$	294.10	wh cr, gran or powd		$-2H_2O$, 150		72^{25}	167^{100}	0.625 90 % al; s glyc
s278	citrate, pentahydrate	$Na_3C_6H_5O_7.5(\text{or } 5\frac{1}{2})H_2O$	348.15	wh, rhomb	$1.857^{23.5}$	$-5H_2O$, 150	d	92.6^{25}	250^{100}	sl s al
s279	cobaltinitrite	$Na_3Co(NO_2)_6$	403.94	yelsh-brnsh cr powd				v s, d		sl s al; d min a; i dil ac a
s280	cyanamide, mono-	$NaHCN_2$	64.02	wh cr powd, hygr				v s		
s281	cyanate	NaOCN	65.01	col need	1.937^{20}	d 700 vac		s	s	v sl s eth, bz
s282	cyanide	NaCN	49.01	col, cub, deliq, pois, 1.452		563.7	1496	48^{10}	82^{35}	sl s al; s NH_3
s283	cyanoaurite	Sodium aurocyanide. $NaAu(CN)_2$	271.99					s		
s284	cyanocuprate (I)	$NaCu(CN)_2$	138.57		1.013^{20}	d 100		s		
s285	cyanoplatinite	$Na_2[Pt(CN)_4].3H_2O$	399.19	col, tricl	2.646	$-3H_2O$, 120–125		s	s	s al
s286	enanthate	Sodium heptanoate. $NaC_7H_{13}O_2$	152.17	wh cr powd or leaf		240–350		s		s al
s287	ethyl acetoacetate	$NaC_6H_9O_3$	152.13	need		d		d		s eth
s288	ethyl sulfate	$NaC_2H_5SO_4.H_2O$	166.13	wh, hex pl, deliq		d		164^{17}		d alk, H_2SO_4; 142 al

No.	Name	Synonyms and Formulae	Mol. wt.	Crystalline form, properties and index of refraction	Density or spec. gravity	Melting point, °C	Boiling point, °C	Solubility, in grams per 100 cc		
								Cold water	Hot water	Other solvents
	Sodium									
s289	ferrate (III)	Ferrite. $Na_2Fe_2O_4$	221.67	br, hex pl or need	4.05			d		v s dil HCl
s290	ferricyanide	$Na_3Fe(CN)_6.H_2O$	298.92	red cr, deliq				18.9^{0}	67^{100}	i al
s291	ferrocyanide	Yellow prussiate of soda. $Na_4Fe(CN)_6.10H_2O$ Sodium hexacyanoferrate (II).	484.04	pa yel, monocl, 1.519, 1.530 1.544	1.458			31.85^{20}	156.5^{98}	i al
s292	fluoaluminate	Na_3AlF_6	209.94	col, monocl, β 1.364	2.90	1000		sl s		i HCl,; d alk
s293	fluoantimonate	$NaSbF_6$	258.73	rhomb	3.375^{18}	<1360		128.6^{20}		s al, acet
s294	fluoberyllate	Na_2BeF_4	130.99	wh, rhomb or monocl		d		1.47^{18}	2.94^{100}	
s295	fluoborate	$NaBF_4$	109.79	wh, rhomb	2.47^{20}	sl d 384	d	108^{26}	210^{100}	sl s al; d H_2SO_4
s296	fluoride	Nat. villiaumite. NaF	41.99	col, cub or tetr, 1.336	2.558^{41}	993	1695	4.22^{18}		s HF; v sl s al
s297	fluoride, hydrogen	NaF.HF	61.99	col, or wh cr powd, rhdr	2.08			s	s	
s298	fluoride *ortho*phosphate	$NaF.Na_3PO_4.12H_2O$	422.11		2.2165			12^{25}	57.5^{50}	
s299	fluoroacetate, mono-	$NaC_2H_2FO_2$	100.02	wh powd		200		111^{25}		1.4^{25} al; 5^{25} MeOH; 0.04^{25} acet; 0.0049^{25} CCl_4
s300	fluorophosphate, hexa-	$NaPF_6.H_2O$	185.97		2.369^{19}			103.2^{0}		
s301	fluorophosphate, mono-	Na_2PO_3F	143.95	col		*ca* 625		25		
s302	fluosilicate	Na_2SiF_6	188.06	col, hex, 1.312, 1.309	2.679	d		0.652^{17}	2.46^{100}	i al
s303	fluosulfonate	$NaSO_3F$	122.05	shiny leaf, hygr		d red heat		s		s al, acet; i eth
s304	formaldehyde-sulfoxylate	$NaHSO_2.CH_2O.2H_2O$	154.12	rhomb pr, hygr		64	d	v s		d a; s al, alk
s305	formate	$NaCHO_2$	68.01	col, monocl, deliq	1.92^{20}	253	d	97.2^{20}	160^{100}	sl s al; i eth
s306	2-furanacrylate	$NaC_7H_5O_3$	160.10	lt brn powd	1.919	d		s	s	sl s al; i eth
s307	*meta*germanate	Na_2GeO_3	166.57	wh, monocl, deliq, 1.59	3.31^{22}	1083			d	s a
s308	*meta*germanate, heptahydrate	$Na_2GeO_3.7H_2O$	292.68	col, rhomb		83		24.6^{0} 45.5^{25}		s a
s309	(mono-) *d*-glutamate	$NaC_5H_8NO_4$	169.11	wh cr		d		v s		sl s al
s310	glycerophosphate, monohydrate	$Na_2C_3H_7O_6P.H_2O$	234.05	yelsh visc liq; wh cr or powd				s		s al
s311	glycerophosphate, pentahydrate	$Na_2C_3H_7O_6P.5\frac{1}{2}H_2O$	315.12	wh pl, sc or powd		>130		v s		i al
s312	gold sulfide	$NaAuS.4H_2O$	324.08	col, monocl		d		s		s al
s313	hydride	NaH	24.00	silver need, 1.470	0.92	d 800		d	d	s molten Na; i CS_2, CCl_4, NH_3, bz
s314	hydroxide	NaOH	40.00	wh, deliq, 1.3576	2.130	318.4	1390	42^{0}	347^{100}	v s al, glyc; i acet, eth
s315	iodate	$NaIO_3$	197.89	wh, rhomb	$4.277^{17.5}$	d		9^{20}	34^{100}	i al; s ac a
s316	*meta*periodate	$NaIO_4$	213.89	col, tetr	4.174	d 300		14.44^{25}	$38.9^{51.5}$	s H_2SO_4, HNO_3, ac a
s317	*meta*periodate, trihydrate	$NaIO_4.3H_2O$	267.94	col, rhombdsh, effl	3.219^{18}_{4}	d 175		18.78^{25}	$36.4^{24.5}$	
s318	*para*periodate	Na_3IO_5	337.85	wh		800 d		d		
s319	(tri-)*para*periodate	$Na_3H_2IO_6$	293.88	col, hexag				sl s		s con NaOH sol
s320	iodide	NaI	149.89	col, cub, 1.7745	3.667^{25}_{4}	661	1304	184^{25}	302^{100}	42.57^{25} al; 39.9^{25} acet; s glyc
s321	iodide, dihydrate	$NaI.2H_2O$	185.92	col, monocl	$2.448^{20.8}$	752		317.9^{0}	1550^{100}	v s NH_3
s322	iodoplatinate	$Na_2PtI_6.6H_2O$	1110.59	brn, monocl	3.707			v s		s al
s323	iridium chloride	Sodium hexachloroiridate. $Na_3IrCl_6.12H_2O$	690.07	olive cr, rhomb or trig-rhomb		50		s	s	i al
s324	iron (III) nitrosopentacyanide	$Na_2[Fe(CN)_5NO].2H_2O$	297.95	ruby red, rhomb, 1.605, 1.575, 1.56	1.687^{25}	$-H_2O$, 100	d 160	40^{16}	s	
s325	iron (III) oxalate	$Na_3[Fe(C_2O_4)_3].5\frac{1}{2}H_2O$	487.96	grn, monocl	$1.973^{17.5}$	$-4H_2O$	d 300	32^{0}	182^{100}	
s326	iron (III) sulfate	$3Na_2SO_4.Fe_2(SO_4)_3.6H_2O$	934.09	wh, trig, 1.558, 1.613	2.5	$-6H_2O$, 100		d v sl		i al
s327	lactate	$NaC_3H_5O_3$	112.06	col or yelsh liq, very hygr		17	d 140	v s		s al; i eth
s328	lithium sulfate	$Na_3Li(SO_4)_2.6H_2O$	376.12	col, ditrig	2.009	$-6H_2O$, 50		s	s	
s329	magnesium carbonate	$Na_2CO_3.MgCO_3$	190.31	wh, rhomb	2.729^{15}	677 CO_2 1240/atm		d	d	
s330	magnesium sulfate	Nat. bloedite. $Na_2SO_4.MgSO_4.4H_2O$	334.48	col, monocl, 1.486, 1.488, 1.489	2.23			s		
s331	magnesium tartrate	$Na_2Mg(C_4H_4O_6)_2.10H_2O$	546.59	wh, monocl pr or powd				s	s	

No.	Name	Synonyms and Formulae	Mol. wt.	Crystalline form, properties and index of refraction	Density or spec. gravity	Melting point, °C	Boiling point, °C	Solubility, in grams per 100 cc Cold water	Hot water	Other solvents
	Sodium									
s332	manganate	$Na_2MnO_4.10H_2O$	345.07	grn, monocl		17		s	d	
s333	*per*manganate	$NaMnO_4$	141.93	red cr, deliq		d		v s	v s	
s334	*per*manganate, trihydrate	$NaMnO_4.3H_2O$	195.97	purp cr, deliq	2.47	d 170		v s	v s	s NH_3; d alk
s335	methanearsenate	$Na_2CH_3AsO_3.6H_2O$	292.03	wh cr powd		130–140		*ca* 100		sl s al; i bz, eth, oils
s336	methoxide	$CH_3ONa.2CH_3OH$	118.11	wh powd		d, $-CH_3OH$		s, d		s CH_3OH
s337	methylsulfate	$NaCH_3SO_4.H_2O$	152.10	col cr, hygr				s		s al
s338	molybdate	Na_2MoO_4	205.92	opaque wh	3.28^{18}	687		s 44.3'	84^{100}	
s339	molybdate, dihydrate	$Na_2MoO_4.2H_2O$	241.95	wh, rhbdr	3.28(?)	$-2H_2O$, 100		56.2^{0}	115.5^{100}	i meth acet
s340	*deca*molybdate	$Na_2Mo_{10}O_{31}.21H_2O$	1879.68	wh, monocl pr				sl s	0.842^{100}	
s341	*di*molybdate	$Na_2Mo_2O_7$	349.86	wh need		612		sl s	sl s	
s342	*octa*molybdate	$Na_2MoO_{25}.17H_2O$	1519.75	monocl cr		$-H_2O$, 20		v s	v s	
s343	*para*molybdate	$Na_6Mo_7O_{24}.22H_2O$	1589.84	col, monocl, effl		700 $-H_2O$, 100–120		117.9^{30} (anhydr)	s	
s344	*tetra*molybdate	$Na_2Mo_4O_{13}.6H_2O$	745.82	yel need				39.8^{21}	v s	
s345	*tri*molybdate	$Na_2Mo_3O_{10}.7H_2O$	619.96	need acicular		528 $-6H_2O$, 100–120		3.878^{20}	13.7^{100}	
s346	*meta*niobate	$Na_2Nb_2O_6.7H_2O$	453.90	pseudo-cub	4.512–4.559	$-H_2O$, 100		s		
s347	nitrate	Soda niter. $NaNO_3$	84.99	col, trig or rhbdr, 1.587, 1.336	2.261	306.8	d 380	92.1^{25}	180^{100}	s al, MeOH; v s NH_3; v sl s acet; sl s glyc
s348	nitride	Na_3N	82.98	dk gray		d 300		d		
s349	nitrite	$NaNO_2$	69.00	col-yel, rhomb pr, hygr	2.168^{0}	271	d 320	81.5^{15}	163^{100}	0.3^{20} eth; 4.4^{20} MeOH; 3 abs al; v s NH_3
s350	*hypo*nitrite	$Na_2N_2O_2$	105.99		1.728^{25}	d 300		d		i al
s351	*p*-nitrophenoxide	$NaOC_6H_4NO_2.4H_2O$	233.15	yel, monocl pr		$-2H_2O$, 36 $-4H_2O$, 120	d	5.97^{25}		sl s al
s352	nitroplatinite	$Na_2Pt(NO_2)_4$	425.09	pa yel rhomb or monocl pr, effl				s	s	
s353	nitroprusside	$Na_2[Fe(NO)(CN)_5].2H_2O$	297.95	red, rhomb	1.72			40^{16}		s al
s354	oleate	$NaC_{18}H_{33}O_2$	304.45	wh cr, or yel amorph gran		232–235		10^{12}		s al; sl s eth
s355	oxalate	$Na_2C_2O_4$	134.00	col cr, or wh powd	2.34	d 250–270		3.7^{20}	6.33^{100}	i al, eth
s356	oxalate, hydrogen	$NaHC_2O_4.H_2O$	130.03	wh, monocl		$-H_2O$, 100	d 200	1.7^{15}	21^{100}	
s357	oxalatoferrate (III)	$Na_3[Fe(C_2O_4)_3].xH_2O$	365.89 $+xH_2O$	grn, monocl cr	$1.973^{17.5}$	$-H_2O$, 100–120		32.5	182^{100}	
s358	oxide, mon-	Na_2O	61.98	wh-gray, deliq	2.27	subl 1275		d	d	d al
s359	oxide, per-	Na_2O_2	77.98	yel-wh powd	2.805	d 460	d 657	s	d	d al, NH_3; s dil a; i alk
s360	oxide, per-, octahydrate	$Na_2O_2.8H_2O$	222.10	wh, hex		d 30	d	s	d	i al
s361	palmitate	$NaC_{16}H_{31}O_2$	278.41	wh cr		270				
s362	pentobarbital	$NaC_{11}H_{17}N_2O_3$	248.26					s	s, d	s al
s363	phenobarbital	$NaC_{12}H_{11}N_2O_3$	254.22	wh				v s		s al; i eth, chl
s364	1-phenol-4-sulfonate(*p*-)	$NaC_6H_4(OH)SO_3.2H_2O$	232.19	col, monocl or gran, sl effl		d		23.8^{25}	125^{100}	0.75^{25} al; 20^{25} glyc
s365	phenoxide	$NaOC_6H_5$	116.10	wh cr need, deliq				v s		s al, acet; d a
s366	phenylcarbonate	$NaC_7H_5O_3$	160.11	col powd		d 120		d		d acet
s367	*hypo*phosphate	$Na_4P_2O_6.10H_2O$	430.06	col, monocl, 1.477, 1.482, 1.504	1.823	d		1.49^{25}	5.46^{80}	
s368	*hypo*phosphate, di-H	$Na_2H_2P_2O_6.6H_2O$	314.03	col, monocl, 1.468, 1.490, 1.504	1.849	250 (anhydr)	$-6H_2O$, 100	2.35	25	s dil H_2SO_4, NH_4OH; i al
s369	*meta*phosphate, hexa-	Graham's salt. $(NaPO_3)_6$	611.17	col glass, 1.482 ± 0.002				v s		
s370	*meta*phosphate, tri-, hexahydrate	Knorre's salt. $(NaPO_3)_3.6H_2O$	413.98	col, tricl, effl, 1.433, 1.442, 1.446		53; $-6H_2O$, 50		s		
s371	*ortho*phosphate	$Na_3PO_4.10H_2O$	344.09	col, oct	$2.536^{17.5}$ (anhydr)	100		8.8 (anhydr)		
s372	*ortho*phosphate	$Na_3PO_4.12H_2O$	380.12	col, trig, 1.446, 1.452	1.62^{20}	d 73.3–76.7	$-12H_2O$, 100	1.5^{0}	157^{70}	i CS_2, al
s373	*ortho*phosphate, di-H	$NaH_2PO_4.H_2O$	137.99	col, rhomb, 1.456, 1.458, 1.487	2.040	$-H_2O$, 100	d 204	59.9^{0}	427^{100}	v sl s eth, chl, tol; i al
s374	*ortho*phosphate, di-H	$NaH_2PO_4.2H_2O$	156.01	col, rhomb, 1.4629	1.91	60		v s	v s	—
s375	*ortho*phosphate, mono-H	Sŏrensen's sodium phosphate. $Na_2HPO_4.2H_2O$	177.99	rhomb bispheroidal, 1.463	2.066^{16}	$-2H_2O$, 95		100^{50}	117^{80}	
s376	*ortho*phosphate, mono-H	$Na_2HPO_4.7H_2O$	268.07	col, monocl pr, 1.442	1.679	$-5H_2O$, 48.1		104^{40}		i al
s377	*ortho*phosphate, mono-H	$Na_2HPO_4.12H_2O$	358.14	col, rhomb or monocl, eff, wh powd, 1.432, 1.436, 1.437	1.52	$-5H_2O$, 35.1	$-12H_2O$, 100	4.15	87.4^{34}	i al

No.	Name	Synonyms and Formulae	Mol. wt.	Crystalline form, properties and index of refraction	Density or spec. gravity	Melting point, °C	Boiling point, °C	Solubility, in grams per 100 cc Cold water	Hot water	Other solvents
	Sodium									
s378	*pyro*phosphate	$Na_4P_2O_7$	265.90	wh cr, 1.425	2.534	880		3.16^{0}	40.26^{100}	
s379	*pyro*phosphate	$Na_4P_2O_7.10H_2O$	446.06	col, monocl, 1.450, 1.453, 1.460	1.815–1.836	$-H_2O$, 93.8 m.p. 880		5.41^{0}	93.11^{100}	i al, NH_3
s380	*pyro*phosphate, di-H	$Na_2H_2P_2O_7.6H_2O$	330.03	monocl, 1.4599, 1.4645, 1.4649	1.85	$-H_2O$, 220		6.9^{0}	35^{40}	
s381	phosphide	Na_3P	99.94	red		d		d, PH_3		
s382	*hypo*phosphite	$NaH_2PO_2.H_2O$	105.99	col, monocl, deliq		d viol		100^{25}	667^{100}	v s al; s glyc; sl s NH_3, NH_4OH
s383	*ortho*phosphite, di-H	$NaH_2PO_3.2\frac{1}{2}H_2O$	149.01	col, monocl, 1.419, 1.431, 1.449		42	$-2\frac{1}{2}H_2O$, 100	56^{0}	193^{42}	
s384	*ortho*phosphite, mono-H	$Na_2HPO_3.5H_2O$	216.04	wh, rhomb deliq, β, 1.443		53	d 200−250	s	v s	i al, NH_4OH
s385	*tri*phosphate	Sodium tripolyphosphate. $Na_5P_3O_{10}$	367.86	powd and gran				14.5^{25}	32.5^{100}	
s386	phthalate	$Na_2C_8H_4O_4$	210.10	wh powd or pearly pl		$-H_2O$, 150				
s387	platinate, hydroxo-	$Na_2Pt(OH)_6$	343.11	yel or red-brn, hex		$-3H_2O$, 150–170	d	s		i al; sl s HCl
s388	platinum cyanide	$Na_2[Pt(CN)_4].3H_2O$	399.19	col, tricl	2.646	$-H_2O$, 120–125		s		s al
s389	plumbate, hydroxo-	$Na_2Pb(OH)_6$	355.21	yel-wh lumps, hygr				d to PbO_2		d a; s alk
s390	potassium(*dl*)-tartrate	$NaKC_4H_4O_6$	210.14	col, tricl		90–100	d 200	47.4^{6} (anhydr)	v s	
s391	propionate	$NaC_3H_5O_2$	96.06	wh, gran powd				s		s al
s392	*per*rhenate	$NaReO_4$	273.19	col, hex pl, hygr	5.39	300 (in O_2) d 440 (vac)		100^{20}		s al
s393	*pyrohypo*rhenate	$Na_4Re_2O_7.H_2O$	594.37	sandy yel cr				0.004		
s394	rhodiumchloride	$Na_3RhCl_6.12H_2O$	600.78	dk red cr, monocl pr		$-12H_2O$, 120		v s	v s	i al
s395	rhodiumnitrite	$Na_3[Rh(NO_2)_6]$	447.91	wh cr		d 360		40^{17}	s	i al; d a
s396	*per*ruthenate	$NaRuO_4.H_2O$	206.07	blk cr, lamellar		d 440 vac		v s	d	
s397	salicylate	$NaC_7H_5O_3$	160.11	wh cr powd				111^{15}	125^{25}	17^{15} al; 25 glyc
s398	selenate	Na_2SeO_4	188.94	col, rhomb	$3.213^{17.4}$			84^{35}	72.8^{100}	
s399	selenate, decahydrate	$Na_2SeO_4.10H_2O$	369.09	col, monocl	1.603–1.620	*ca* 32 trans		43.5^{20}	340^{100}	
s400	selenide	Na_2Se	124.94	wh to red, cr, deliq	2.625^{10}	>875		d		i NH_3
s401	selenite	$Na_2SeO_3.5H_2O$	263.01	wh cr, tetrag				s	s	i al
s402	silicate	Waterglass. $Na_2O.xSiO_2(x=3-5)$		col, amorph, deliq				s	s	i al, K and Na salts
s403	*di*silicate	$Na_2Si_2O_5$	182.15	rhomb pearly luster 1.500, 1.510		874		s	s	
s404	*meta*silicate	Na_2SiO_3	122.06	col, monocl, α 1.518, γ 1.527	2.4	1088		s	s, d	i al, K and Na salts
s405	*meta*silicate	$Na_2SiO_3·9H_2O$	284.20	col, rhomb bipyramid, effl		40–48	$-6H_2O$, 100	v s	v s	s dil NaOH; i al, a
s406	*ortho*silicate	Na_4SiO_4	184.04	col, hex, 1.530		1018		s	s	
s407	silicotungstate, dodeca-	$Na_4[Si(W_3O_{10})_4].20H_2O$ **Sodium tungstosilicate**	3326.53	col, tricl		$-7H_2O$, 100	d	v s	v s	s a, sl s al
s408	stannate, hydroxo-	$Na_2Sn(OH)_6$	266.71	col, hex or wh powd, or lumps		$-3H_2O$, 140		$61.3^{15.5}$	50^{100}	i al, acet
s409	stearate	$NaC_{18}H_{35}O_2$	306.47	wh fatty powd				s	s	s h al
s410	succinate	$Na_2C_4H_4O_4.6H_2O$	270.15	wh, gran or powd		$-6H_2O$, 120		21.45^{0}	86.63^{75}	v sl s al
s411	succinate, tetrahydroxy-	Sodium dihydroxy tartrate. $Na_2C_4H_4O_8.3H_2O$	280.10			d		0.032^{0}	d	d min a; i al, eth
s412	sulfanilate	$NaC_6H_4(NH_2)SO_3$	195.17	wh lust cr leaf				s		
s413	sulfate, anhydr	Na_2SO_4	142.04	monocl (between *ca* 160–185), 1.480		884; tr to hex *ca* 241		s	$42-5^{100}$	s HI
s414	sulfate, anhydr	Nat. thenardite. Na_2SO_4	142.04	orthorhomb, 1.484, 1.477, 1.471	2.68			4.76^{0}	42.7^{100}	s glyc; i al
s415	sulfate, decahydrate	Glauber's salt, mirabilite. $Na_2SO_4.10H_2O$	322.19	col, monocl, effl, 1.394, 1.396, 1.398	1.464	32.38	$-10H_2O$, 100	11^{0}	92.7^{30}	i al
s416	sulfate, heptahydrate	$Na_2SO_4.7H_2O$	268.15	wh, rhomb or tetrag		tr to anhydr 24.4		19.5^{0}	44^{20}	i al
s417	*pyro*sulfite	Sodium metabisulfite. $Na_2S_2O_5$	190.10	wh powd or cr $(+7H_2O)$	1.4	>d 150		54^{20}	81.7^{100}	sl s al; s glyc
s418	*pyro*sulfate	$Na_2S_2O_7$	222.16	wh, transluc cr, deliq	2.658^{25}	400.9	d 460	s		s fum H_2SO_4
s419	sulfate hydrogen	$NaHSO_4$	120.06	col, tricl	2.435^{13}	>315	d	28.6^{25}	100^{100}	sl s al; i NH_3
s420	sulfate hydrogen, monohydrate	$NaHSO_4.H_2O$	138.07	col, monocl, deliq ~1.46	$2.103^{13.5}_{4}$	58.54 ± 0.5		*ca* 67, d	d	d al
s421	sulfide, hydro-	NaHS	56.06	col, rhomb or wh gran cr, deliq		350		vs		s al

No.	Name	Synonyms and Formulae	Mol. wt.	Crystalline form, properties and index of refraction	Density or spec. gravity	Melting point, °C	Boiling point, °C	Solubility, in grams per 100 cc: Cold water	Hot water	Other solvents
	Sodium									
s422	sulfide, hydrodihydrate	$NaHS.2H_2O$	92.09	col need, deliq		d		s	s	d a; s al
s423	sulfide, mono-	Na_2S	78.04	wh cr, deliq	1.856^{14}	1180		15.4^{10}	57.2^{90}	d a; sl s al; i eth
s424	sulfide, mono-hydrate	$Na_2S.9H_2O$	240.18	col, tetr, deliq	1.427^{16}_{4}	d 920		47.5^{10}	96.7^{10}	d, sl s al
s425	sulfide, penta-	Na_2S_5	206.30	yel (exist ?)		251.8		s	s	s al
s426	sulfide, tetra-	Na_2S_4	174.24	yel, cub, hygr		275	d	s		s al
s427	sulfide, hydrotrihydrate	$NaHS.3H_2O$	110.11	col, lust rhomb cr		22	d	s	s	s al
s428	sulfite	Na_2SO_3	126.04	wh powd or hex, prism 1.565, 1.515	$2.633^{15.4}$	d red heat	d	12.54^{0}	28.3^{80}	sl s al; i liq Cl_2, NH_3
s429	sulfite hydrate	$Na_2SO_3.7H_2O$	252.15	col, monocl, effl	1.539^{15}	$-7H_2O$, 150	d	32.8^{0}	196^{40}	sl s al
s430	*hydro*sulfite	Dithionite, hyposulfate $Na_2S_2O_4.2H_2O$	210.14	col, monocl(?) cr, or yel-wh powd		d 52		25.4^{20}	d	d a; s alk; i al
s431	sulfite, hydrogen	$NaHSO_3$	104.06	wh, monocl, yel in sol, 1.526	1.48	d		v s	v s	sl s al
s432	*d*(& *l*)-tartrate	$Na_2C_4H_4O_6.2H_2O$	230.08	col, rhomb, 1.545, 1.49	1.818	$-2H_2O$, 150		29^{6}	66^{43}	i al
s433	*d*-tartrate, hydrogen	$NaHC_4H_4O_6.H_2O$	190.09	wh cr powd, rhomb, 1.53, 1.54, 1.60		$-H_2O$, 100	d 234	6.7^{18}	9.2^{80}	
s434	*dl*-tartrate hydrogen	$NaHC_4H_4O_6.H_2O$	190.09	col, monocl or tricl, 1.53, 1.54, 1.60		$-H_2O$, 100	d 219	8.9^{19}		
s435	*ortho*tellurate, tetra-H	$Na_2H_4TeO_6$	273.61	col, hex pl		d	d	0.77^{18}	2^{100}	s h dil HNO_3: i NaOH
s436	telluride	Na_2Te	173.58	wh cr powd very hygr, d in air	2.90	953		v s, d	v s, d	
s437	tellurite	Na_2TeO_3	221.58	wh, rhomb pr				sl s	s	
s438	thioantimonate	Schlippe's salt. $Na_3SbS_4.9H_2O$	481.11	pa yel, cub	1.806	87	d 234	20.15^{0}	100^{100}	i al, eth
s439	thioarsenate	$Na_3AsS_4.8H_2O$	416.27	yel, monocl, β 1.6802		d		v s	d	i al
s440	thiocarbonate, tri	$Na_2CS_3.H_2O$	172.20	yel need, deliq		d 75		s	d	s al; i eth, bz
s441	thiocyanate	$NaSCN$	81.07	col, rhomb deliq pois, ~1.625		287		$139.31^{21.3}$	225^{100}	v s al, acet
s442	*di*thionate	$Na_2S_2O_6.2H_2O$	242.13	col, rhomb, 1.482, 1.495, 1.519	2.189	$-2H_2O$, 110	$-SO_2$, 267	47.6^{16}	90.9^{100}	s HCl; i al
s443	thiosulfate	$Na_2S_2O_3$	158.11	col, monocl	1.667			50	231^{100}	i al
s444	thiosulfate, pentahydrate	"Hypo", sodium hyposulfite. $Na_2S_2O_3.5H_2O$	248.18	col, monocl, effle, 1.489, 1.508, 1.536	1.729^{17}	40–45 d 48	$-5H_2O$, 100	79.4^{0}	291.1^{45}	s NH_3; i al
s445	thiosulfoaurate (I)	$Na_3[Au(S_2O_3)_2].2H_2O$	526.22	wh cr, monocl	3.09	$-H_2O$, 150	d	50		i al
s446	*tri*titanate	$Na_2Ti_3O_7$	301.68	wh need, monocl	3.35–3.50	1128		i		s h HCl
s447	tungstate	Na_2WO_4	293.83	wh, rhomb	4.179	698		57.5^{0} 73.2^{21}	96.9^{100}	
s448	tungstate, dihydrate	$Na_2WO_4.2H_2O$	329.86	col pl, rhomb, 1.5533	3.23–3.25	$-2H_2O$, 100 anhydr 698		41^{0}	123.5^{100}	sl s NH_3; i al, a
s449	*meta*tungstate	$Na_2O.4WO_3.10H_2O$	1169.53	col, oct.		706.6		s	v s	i a
s450	*para*tungstate	$Na_6W_7O_{24}.16H_2O$	2097.12	col, tricl.	3.987	$-12H_2O$, 100; $-16H_2O$, 300		8	d	
s451	*meta*uranate	Na_2UO_4	348.01	gr yel or red pl, rhomb pr				i	i	s dil a, alk carb
s452	uranyl acetate	$NaUO_2(C_2H_3O_2)_3$	470.15	yel, tetr pr, 1.501	2.56					
s453	uranyl carbonate	$2Na_2CO_3.UO_2CO_3$	542.02	yel cr.		d 400		sl s		i al
s454	urate	$Na_2C_5H_2N_4O_3.H_2O$	230.09	wh gran powd or hard cr nodules					1.3^{100}	v sl s 90 % al
s455	urate, acid	$NaHC_5H_2N_4O_3$	190.09	wh gran powd				0.083	0.8^{100}	
s456	valerate	$NaC_5H_9O_2$	124.12	wh cr or mass, hygr		140		s		s al
s457	*meta*vanadate	$NaVO_3$	121.93	col, monocl pr		630		21.1^{25}	38.8^{75}	
s458	*ortho*vanadate	Na_3VO_4	183.94	col, hex pr		850–866		s		i al
s459	*ortho*vanadate, decahydrate	$Na_3VO_4.10H_2O$	364.06	wh, cub or hex cr, 1.5305, 1.5398, 1.5475				s		
s460	*ortho*vanadate, hexadecylhydrate	$Na_3VO_4.16H_2O$	472.15	col need		866 (anhydr)		v s	d	i al
s461	*pyro*vanadate	$Na_4V_2O_7$	305.84	col, hex		632–654		s	s	i al
s462	ethylxanthate	NaC_2H_5OCSS	144.19	yelsh powd				s		s al
s463	zinc uranyl acetate	$NaZn(UO_2)_3(C_2H_3O_2)_9.9H_2O$	1591.98	monocl cr, α 1.475, γ 1.480				i		s al
	Stannous	See under tin								
	Stannic	See under tin								

No.	Name	Synonyms and Formulae	Mol. wt.	Crystalline form, properties and index of refraction	Density or spec. gravity	Melting point, °C	Boiling point, °C	Solubility, in grams per 100 cc: Cold water	Hot water	Other solvents
	Strontium									
s464	**Strontium**	Sr	87.62	silv wh to pa yel met	2.6^{20}	769	1384	d	d	s a, al, liq NH_3
s465	acetate	$Sr(C_2H_3O_2)_2$	205.71	wh cr	2.099	d		36.9	36.4^{97}	0.26^{15} MeOH
s466	acetate	$Sr(C_2H_3O_2)_2.\frac{1}{2}H_2O$	214.72	wh cr powd		$-\frac{1}{2}H_2O$, 150		s		sl s al
s467	*ortho*arsenate, acid	$SrHAsO_4.H_2O$	245.56	rhomb, need	3.606^{15}; 4.035 (anhydr)	$-H_2O$, 125		$0.284^{15.5}$	d	s a
s468	*ortho*arsenite	$Sr_3(AsO_3)_2.4H_2O$	580.76	cr, or wh powd				sl s		s a; sl s al
s469	borate, *tetra-*	$SrB_4O_7.4H_2O$	314.92						77^{100}	s HNO_3; NH_4 salts
s470	boride, hexa-	SrB_6	152.49	blk, cub	3.39^{15}	2235		i	i	s HNO_3; i HCl
s471	bromate	$Sr(BrO_3)_2.H_2O$	361.45	col yelsh, monocl, hygr	3.773	$-H_2O$, 120	d 240	33^{16}		
s472	bromide	$SrBr_2$	247.44	wh, hex need, hygr, 1.575	4.216^{24}	643	d	100^{20}	222.5^{100}	s al, amyl al
s473	bromide, hexahydrate	$SrBr_2.6H_2O$	355.53	col, hex, hygr	2.386^{25}_{4}	tr to $2H_2O$, 88.6	$-6H_2O$, >180	204.2^{0}	∞	63.9^{20} al; 113.4^{20} MeOH; 0.6^{20} acet; i eth
s474	carbide	SrC_2	111.64	blk, tetr	3.2	>1700		d	d	d a
s475	carbonate	Nat. strontianite. $SrCO_3$	147.63	col, rhomb, or wh powd trfrs to −hex at 926, 1.516, 1.664, 1.666	3.70	$1497^{69 atm}$	$-CO_2$, 1340	0.0011^{18}	0.065^{100}	0.12 aq CO_2; s a, NH_4 salts
s476	chlorate	$Sr(ClO_3)_2$	254.52	col, rhomb, or wh powd, 1.516, 1.605, 1.626	3.152	d 120		174.9^{18}	v s	s dil al; i abs al
s477	*per*chlorate	$Sr(ClO_4)_2$	286.52	col cr, hygr				310^{25}	v s	212 MeOH; 181 al; i eth
s478	chloride	$SrCl_2$	158.53	col, cub 1.650^{25}	3.052	875	1250	53.8^{20}	100.8^{100}	v sl s abs al, acet; i NH_3
s479	chloride, dihydrate	$SrCl_2.2H_2O$	194.56	transp leaf, 1.594, 1.595, 1.617	2.6715^{25}					
s480	chloride, fluoride	$SrCl_2.SrF_2$	284.14	col, tetr, 1.651, 1.627	4.18	962		d	d	s conc HNO_3, conc HCl, i al
s481	chloride, hexahydrate	$SrCl_2.6H_2O$	266.62	col, trig, 1.536, 1.487	1.93	115, $-4H_2O$, 60	$-6H_2O$, 100	106.2^{0}	205.8^{40}	3.8^{6} al
s482	chromate	$SrCrO_4$	203.61	yel, monocl	3.895^{15}			0.12^{15}	3^{100}	s HCl, HNO_3, ac a, NH_4 salts
s483	cyanide	$Sr(CN)_2.4H_2O$	211.72	wh, rhomb, deliq		d		v s		
s484	cyanoplatinite	$Sr[Pt(CN)_4].5H_2O$	476.86	col, monocl pr, 1.696		$-5H_2O$, 150				s abs al
s485	glycerophosphate	$SrC_3H_7O_6P$	257.68	wh powd				sl s		i al
s486	ferrocyanide	$Sr_2Fe(CN)_6.15H_2O$	657.42	yel, monocl				50	100	
s487	fluoride	SrF_2	125.62	col, cub or wh powd, 1.442	4.24	1473	2489^{760}	0.011^{0}	0.012^{27}	s hot HCl; i HF, al, acet
s488	fluosilicate	$SrSiF_6.2H_2O$	265.73	monocl	$2.99^{17.5}$	d		3.2^{15}	v s	s HCl; 0.065^{15} 50 % al
s489	formate	$Sr(CHO_2)_2$	177.66	col, rhomb, 1.559, 1.547, 1.598	2.693	71.9		9.1^{0}	34.4^{100}	
s490	formate, dihydrate	$Sr(CHO_2)_2.2H_2O$	213.69	col, rhomb, 1.484, 1.521, 1.538	2.25	d, $-2H_2O$, 100		$11.62^{28.6}$	26.57^{100}	i al, eth
s491	hydride	SrH_2(?)	89.64	wh, rhomb, hygr	3.72	d 675	subl 1000 (in H_2)	d	d	d al
s492	hydroxide	$Sr(OH)_2$	121.63	wh, deliq	3.625	375 (in H_2)	$-H_2O$, 710	0.41^{0}	21.83^{100}	s a, NH_4Cl
s493	hydroxide, octahydrate	$Sr(OH)_2.8H_2O$	265.76	col, tetr, deliq, 1.499, 1.476	1.90	$-8H_2O$, 100		0.90^{0}	47.71^{100}	s a, NH_4Cl; i acet
s494	iodate	$Sr(IO_3)_2$	437.43	tricl	5.045^{15}			0.03^{15}	0.8^{100}	
s495	iodide	SrI_2	341.43	col pl	4.549^{25}_{4}	515	d	165.3^{0}	383^{100}	4.5^{20} al; 0.31^{0} NH_4OH; s MeOH
s496	iodide, hexahydrate	$SrI_2.6H_2O$	449.52	col-yelsh, hex, deliq	2.672^{25}	d 90		448.9^{0}	∞	s al; i eth
s497	lactate	$Sr(C_3H_5O_3)_2.3H_2O$	319.81	wh cr or gran powd		$-3H_2O$, 120		25	200^{100}	sl s al
s498	*per*manganate	$Sr(MnO_4)_2.3H_2O$	379.54	purpl, cub	2.75	d 175		270^{0}	291^{18}	
s499	molybdate	$SrMoO_4$	247.56	col, tetr ~1.91	4.54^{25}_{25}	d		0.0104^{17}		s a
s500	nitrate	$Sr(NO_3)_2$	211.63	col, cub	2.986	570		70.9^{18}	100^{90}	0.012 abs al; v s NH_3; sl s acet
s501	nitrate, tetrahydrate	$Sr(NO_3)_2.4H_2O$	283.69	col, monocl	2.2	$-4H_2O$, 100	1100 tr SrO	60.43^{0}	206.5^{100}	s liq NH_3; v sl s abs al, acet; i HNO_3

No.	Name	Synonyms and Formulae	Mol. wt.	Crystalline form, properties and index of refraction	Density or spec. gravity	Melting point, °C	Boiling point, °C	Solubility, in grams per 100 cc: Cold water	Hot water	Other solvents
	Strontium									
s502	nitride	Sr_3N_2	290.87					d	d	s HCl
s503	nitrite	$Sr(NO_2)_2.H_2O$	197.65	col, hex, 1.588	2.408^{0}_{0}	$-H_2O>100$	d 240	58.9^{0}	182^{100}	0.42^{20} 90 % al
s504	*hypo*nitrite	$SrN_2O_2.5H_2O$	237.71	wh need	2.173^{25}_{4}			v al s	al s	v al s NH_3
s505	oxalate	$SrC_2O_4.H_2O$	193.64	col cr		$-H_2O$, 150		0.0051^{18}	0.15^{100}	s HCl, HNO_3
s506	oxide	SrO	103.62	gray-wh, cub, 1.810	4.7	2420	~3000	0.69^{20}	22.85^{100}	30 fus KOH; sl s al; i eth, acet
s507	oxide, per-	SrO_2	119.62	wh powd	4.56	d 215^{760}		0.018^{20}	d	v s al, NH_4Cl; i acet
s508	oxide, per-, octahydrate	$SrO_2.8H_2O$	263.74	col cr	1.951	$-8H_2O$, 100	d	0.018^{20}	d	s NH_4Cl; i al, acet, NH_4OH
s509	*ortho*phosphate, di-	$SrHPO_4$	183.60	col, rhomb	3.544^{15}	1.62		i	i	s a, NH_4 salts
s510	salicylate	$Sr(C_7H_5O_3)_2.2H_2O$	397.88	col cr		d		5.6^{25}	28.6^{100}	1.5^{25}, 9.5^{78} al
s511	selenate	$SrSeO_4$	230.58	col, rhomb	4.23			i	i	s hot HCl; i HNO_3
s512	selenide	SrSe	166.58	wh, cub, 2.220	4.38			d	d	s HCl
s513	*meta*silicate	$SrSiO_3$	163.70	col, pr monocl, 1.599, 1.637	3.65	1580		i	i	
s514	*ortho*silicate	$SrSiO_4$	179.70	monocl, 1.728, 1.732, 1.758	3.84	>1750				
s515	sulfate	Nat. celestite. $SrSO_4$	183.68	col, rhomb, 1.622, 1.624, 1.631	3.96	1605		0.0113^{0}	0.014^{30}	sl s a; i al, dil H_2SO_4
s516	sulfate, hydrogen	$Sr(HSO_4)_2$	281.76	col		d		d		14^{70} H_2SO_4
s517	sulfide, hydro	$Sr(HS)_2$	153.76	col, cub need, 2.107		d		s	d	
s518	sulfide, mono	SrS	119.68	col, lt gray, cub, 2.107	3.70^{15}	>2000		i	d	d a
s519	sulfide, tetra-	$SrS_4.6H_2O$	323.97	redsh cr, hygr		25	$-4H_2O$, 100	s	s	s al
s520	sulfite	$SrSO_3$	167.68	col cr		d		0.0033^{17}		v s H_2SO_4; s a, HCl
s521	tartrate	$SrC_4H_4O_6.4H_2O$	307.75	wh, monocl	1.966			0.112^{0}	0.755^{85}	s dil HCl, dil HNO_3
s522	telluride	SrTe	215.22	wh, cub, 2.408	4.83					
s523	thiocyanate	$Sr(SCN)_2.3H_2O$	257.83	deliq		$-3H_2O$, 100	d 160–170	v s		v s al
s524	thiosulfate	$SrS_2O_3.5H_2O$	289.82	monocl need	2.17^{17}	$-4H_2O$, 100		2.5^{13}	57^{100}	i al
s525	*di*thionate	$SrS_2O_6.4H_2O$	319.81	trig, 1.530, 1.525	2.373	$-4H_2O$, 78		22^{16}	67^{100}	i al
s526	tungstate	$SrWO_4$	335.47	col, tetr	6.187	d		0.14^{15}		id a; i al
s527	**Sulfamic acid**	Amidosulfuric, aminosulfonic acid. NH_2SO_3H	97.09	col, rhomb	2.126^{25}	200 d	d	14.68	47.08^{80}	v sl s al, eth, acet; i CS_2, CCl_4
s528	**Sulfamide**	Sulfuryl amide. $SO_2(NH_2)_2$	96.11	rhomb pl	1.611	91.5	d 250	s		s al
s529	**Sulfur(α)**	S_8	256.512	yel, rhomb, 1.957	2.07^{20}	112.8; 95.5 (revers.); 444.6	444.674	i	i	23^{0} CS_2; sl s tol, al, bz, eth, liq NH_3; s CCl_4
s530	(β)	S_8	256.512	pa yel, monocl	1.96	119.0	444.674	i	i	70 CS_2; s al, bz
s531	(γ)	S_8	256.512	pa yel, amorph	1.92	*ca* 120	444.6	i	i	i CS_2
s532	bromide, mono-	S_2Br_2	223.95	red liq, 1.730	2.63	−40	$54^{0.2}$	d	d	s CS_2
s533	chloride, di-	SCl_2	102.97	dk red liq, 1.557^{11}	1.621^{15}_{15}	−78	d 59			s CCl_4, bz; d al, eth
s534	chloride, mono-	S_2Cl_2	135.03	yel-red liq, 1.666^{14}	1.678	−80	135.6	d	d	s bz, eth, CS_2
s535	chloride, tetra-	SCl_4	173.88	yel-br liq		−30	d −15	d	d	
s536	fluoride, hexa-	SF_6	146.05	col gas	gas 6.602 g/l; liq $1.88^{-50.5}$	−50.5	−63.8 (subl)	sl s	sl s	s al, KOH
s537	fluoride, mono-	S_2F_2	102.12	col gas	liq 1.5^{-100}	−120.5	−38.4	d	d	d KOH
s538	fluoride, tetra-	SF_4	108.06	gas (exist ?)		−124	−40	d	d	
s539	(di) fluoride, deca-	S_2F_{10}	254.11	col liq	$2.08^{0}_{}$	−92	29			d fus caust
s540	(tetra-) nitride, di-	S_4N_2	156.27	red liq or gray solid	1.901^{18}	23	d 100 expl	i		s eth; sl s al, CS_2
s541	(tetra-) nitride, tetra-	S_4N_4	184.28	or red, monocl	2.22^{15}	subl 179	expl 160	d		s CS_2, chl, bz, NH_3; sl s al, eth
s542	(tri-)*di*nitrogen dioxide	$S_3N_2O_2$	156.20	pa yel cr		100.7	d	i		s al, bz
s543	oxide, di-	SO_2	64.06	col gas or liq suffoc odor	gas 2.927 g/l; liq 1.434	−72.7	−10	22.8^{0}	0.58^{90}	s al, ac a, H_2SO_4
s544	oxide, hept-	Sulfur oxide, per- S_2O_7	176.12	visc liq, or need		0	subl 10	d	d	s H_2SO_4
s545	oxide, mono-	SO (or S_2O_2)	48.06	col gas		d	d	d		
s546	oxide, sesqui-	S_2O_3	112.13	bl-grn cr		d 70–95		d	d	s al, eth, fum H_2SO_4
s547	oxide, tetra-	Sulfurperoxide. SO_4	96.06	wh		d 0–3		s, d		d dil H_2SO_4
s548	oxide, tri-(α)	SO_3	80.06	silky fibr need, stable modific	1.97^{20}	16.83	44.8	d	d	forms fum H_2SO_4
s549	oxide, tri-(β)	$(SO_3)_2$	160.12	asbestos like fibre, metastable		62.4	50 (subl)	d	d	forms fum H_2SO_4

No.	Name	Synonyms and Formulae	Mol. wt.	Crystalline form, properties and index of refraction	Density or spec. gravity	Melting point, °C	Boiling point, °C	Solubility, in grams per 100 cc: Cold water	Hot water	Other solvents
	Sulfur(α)									
s550	oxide, tri-(γ)	SO_3	80.06	vitreous, orthorhomb, metastable	liq 1.920^{20}_{4} sld 2.29^{-10}	16.8	44.8	d	d	forms fum H_2SO_4
s551	oxy*tetra*chloride, mono-	S_2OCl_4	221.94	dk red liq	1.656^{0}	...	60	d	d	d al
s552	oxy*tetra*chloride, tri-	$S_2O_3Cl_4$	253.94	wh, rhomb need or pl	...	d 57	...	d	d	d al
s553	*tri*thiazyl chloride	S_4N_3Cl	205.73	pa yel cr	...	d 170 (vac)	...	d	d	...
s554	**Sulfuric acid**	H_2SO_4	98.08	col liq	1.841 (96–98 %)	10.36 (100 %) 3.0 (98 %)	338 (98.3 %)	∞ ev heat	∞	d al
s555	dihydrate	$H_2SO_4.2H_2O$	134.11	col liq, 1.405	1.650^{0}	−38.9	167	∞	∞	d al, eth
s556	hexahydrate	$H_2SO_4.6H_2O$	206.17	liq	...	−54	...	v s	v s	...
s557	monohydrate	$H_2SO_4.H_2O$	116.09	col liq or monocl cr, 1.438	1.788	8.62	290	∞	∞	d al
s558	octahydrate	$H_2SO_4.8H_2O$	242.20	liq	...	−62	...	v s	v s	...
s559	peroxidi-	Per(di-)sulfuric acid. $H_2S_2O_8$	194.14	hygr cr	...	d 65	d	d	d	s al, eth, H_2SO_4
s560	peroximono-	Permonosulfuric acid. Caro's acid H_2SO_5	114.08	...	...	d 45	...	d	d	s H_3PO_4
s561	pyro-	$H_2S_2O_7$	178.14	col cr, hygr	1.9^{20}	35	d	d	d	d al
s562	tetrahydrate	$H_2SO_4.4H_2O$	170.14	...	...	−27	...	∞	∞	d al, eth
s563	**Sulfurous acid**	H_2SO_3	82.08	in sol only	*ca* 1.03	...	...	s	...	s al, eth, ac a
s564	**Sulfuryl** chloride	SO_2Cl_2	134.97	col liq, 1.444	1.6674^{20}_{4}	−54.1	69.1	d	d	s bz, ac a
s565	chloride fluoride	SO_2ClF	118.52	col gas	1.623^{0} g/l	−124.7	7.1	d	...	...
s566	fluoride	SO_2F_2	102.06	col gas	gas 3.72 g/l liq 1.7	−136.7	−55.4	10^{0}	...	s al, CCl_4; al s alk
s567	pyro-, chloride	$S_2O_5Cl_2$	215.03	col liq, 1.937^{20}	gas 9.6 g/l liq 1.818^{11}_{4}	−39 to −37	152.5	d	d	d a
t1	**Tantalum**	Ta	180.948	gray black hard metal, cub or powd	met 16.6 powd 14.401	2996	5425 ± 100	i	i	s HF, fus alk; i a
t2	boride, di-	TaB_2	202.57	...	11.15	3000(?)	...	...	...	...
t3	bromide	$TaBr_5$	580.49	yel cr	4.67	265	348.8	d	d	s abs al, eth
t4	carbide	TaC	192.96	blk, cub	13.9	3880	5500	i	i	sl s H_2SO_4. HF
t5	chloride, penta-	$TaCl_5$	358.21	lt yel, vitr cr powd	3.68^{27}	216	242	d	...	s abs al, H_2SO_4
t6	fluoride	TaF_5	275.94	col, tetrag, deliq	4.74	96.8	229.5	s	...	s HF, eth
t7	nitride	TaN	194.95	br bronze or blk, hex	16.30	3360 ± 50	...	i	i	sl s aq reg, HF, HNO_3
t8	oxide, pent-	Ta_2O_5	441.89	col, rhomb	8.2	1872 ± 10	...	i	i	s fus $KHSO_4$, HF; i a
t9	oxide, pent- hydrate	Tantalic acid $Ta_2O_5.xH_2O$	...	col gel	...	...	...	s	...	s alk, exc conc HNO_3; i a
t10	oxide, tetr-	Ta_2O_4 (or TaO_2)	425.89	dk gray powd	...	oxidizes	...	i	i	i a
t11	sulfide	Ta_2S_4 (or TaS_2)	490.15	blk powd or cr	...	>1300	...	i	i	sl s HF+HNO_3; i HCl
t12	**Telluric acid, ortho-**	$Te(OH)_6$ or $H_2TeO_4.2H_2O$	229.64	wh, monocl pr	3.071	136	...	s	s	sl s dil a, HNO_3; i abs al, acet, eth
t13	**Telluric acid,**	$Te(OH)_6$ or H_6TeO_6	229.64	wh, cub	3.158	136	...	s	s	sl s dil a, HNO_3; i abs al, acet, eth
t14	**Tellurium**	Te	127.60	br blk, amorph, 1.0025	6.00	449.5 ± 0.3	989.8 ± 3.8	i	i	s H_2SO_4, HNO_3, aq reg, KCN, KOH; i HCl, CS_2
t15	**Tellurium**	Te	127.60	rhomb silv wh met, 1.0025	6.25	452	1390	i	i	s H_2SO_4, HNO_3, aq reg, KCN KOH; i HCl, CS_2
t16	bromide, di-	$TeBr_2$	287.42	brn to gray grn, need, unstable	...	210	339	d	...	s eth; sl s a; d NaOH
t17	bromide, tetra-	$TeBr_4$	447.27	or cr	4.31^{15}_{4}	380 ± 6	d 421	sl s	d	s eth, a, tart a, NaOH
t18	chloride, di-	$TeCl_2$	198.50	blk cr or amorph, unstable	7.05	209 ± 5	327	d	d	s min a, tart a; d NaOH
t19	chloride, tetra	$TeCl_4$	269.41	wh to yel cr, deliq	3.26^{18} 2.559^{232}	224	380^{760}	s d	s d	s HCl, bz, al, chl, CCl_4; i CS_2
t20	ethoxide	$Te(OC_2H_5)_4$	307.85	...	...	20	$107–107.5^{5.5}$	...	...	...
t21	fluoride, hexa-	TeF_6	241.59	col gas unpleas odor	sol 4.006^{-191} liq 2.56^{-25}	−36	+35.5	d	d	d a, alk
t22	fluoride, tetra-	TeF_4	203.59	wh cr hygr	...	subl	>97	d	d	...
t23	hydride	H_2Te	129.62	col gas pois	4.49	−48.9	-2.2^{760}	v s	s	d al

No.	Name	Synonyms and Formulae	Mol. wt.	Crystalline form, properties and index of refraction	Density or spec. gravity	Melting point, °C	Boiling point, °C	Solubility, in grams per 100 cc: Cold water	Hot water	Other solvents
	Tellurium									
t24	iodide, di-	TeI_2	381.41	blk cr (exist ?)		subl		i	i	
t25	iodide, tetra,-	TeI_4	635.22	blk cr	5.403^{15}_{4}	280	d	sl s	d	s alk, aq NH_3, HI
t26	methoxide	$Te(OCH_3)_4$	251.74	solid			123–124			
t27	oxide, di-	Tellurite. TeO_2	159.60	wh, tetr or rhomb, 2.00, 2.18(Li), 2.35	tetr 5.67^{15} rhomb 5.91^{0}	733	1245	i	i	s HCl, hot HNO_3, alk; i NH_4OH
t28	oxide, mon-	TeO	143.60	blk, amorph (exist ?)	5.682	d 370 (in CO_2)	d	i	i	s dil a, H_2SO_4, KOH
t29	oxide, tri-	TeO_3	175.60	α yel amorph β gray cr	α 5.075^{16}_{4} β 6.21	d 395		i	i	d conc HCl; s hot KOH; i a, al
t30	sulfide	TeS_2	191.72	red-blk powd amorph (exist ?)				i		i a; s alk sulf
t31	sulfoxide	$TeSO_3$	207.66	deep red amorph		d 30	d	d	d	s H_2SO_4
t32	**Tellurous acid**	H_2TeO_3(?)	177.61	wh flocks, indef not isolated	3.05	d 40		0.00067	d	s a, NaOH; sl s NH_4OH; i al
t33	**Terbium**	Tb	158.924	silv-gray met, hex	8.2294	1360±4	3123	i	i	s a
t34	bromide	$TbBr_3$	398.65			827	1490	s	s	
t35	chloride hexa-hydrate	$TbCl_3.6H_2O$	373.38	col pr cr, deliq	4.35 (anhydr)	588 (anhydr)	$-H_2O$, 180–200 (in HCl gas)	v s		
t36	fluoride	TbF_3	215.92			1172	2280(?)	i		i dil a
t37	iodide	TbI_3	539.64			946	>1300	s	s	
t38	*di*methylphosphate	$Tb[(CH_3)_2PO_4]_3$	534.05					12.6^{25}	8.07^{40}	
t39	nitrate	$Tb(NO_3)_3.6H_2O$	453.03	col, monocl cr		893		s		
t40	oxalate	$Tb_2(C_2O_4)_3.10H_2O$	762.06	wh cr	2.60	$-H_2O$, 40		i	i	i dil a
t41	oxide	Terbia. Tb_2O_3	365.85	wh solid						s dil a
t42	oxide, per-	Tb_4O_7	747.69	dk-brn or blk solid		$-O_2$		i	i	s hot conc a
t43	sulfate	$Tb_2(SO_4)_3.8H_2O$	750.16	wh cr		$-8H_2O$, 360		3.561^{20}	2.51^{40}	
t44	**Thallium**	Tl	204.37	bl-wh met, tetr	11.85	303.5	1457 ± 10	i	i	s HNO_3. H_2SO_4; sl s HCl
t45	acetate	$TlC_2H_3O_2$	263.42	silk wh cr, deliq	3.765^{127}	131		v s		v s al, $CHCl_3$; i acet
t46	aluminium sulfate	$TlAl(SO_4)_2.12H_2O$	639.66	oct, 1.488	2.306^{20}	91		11.78^{25}		
t47	azide	TlN_3	246.39	yel, tetr		330 (vac)		0.1712^{0}	0.3^{16}	i al, eth
t48	bromate	$TlBrO_3$	332.28	col, need				0.35^{20}	s	s dil al
t49	bromide, di-	Bromothallate(ous). Tl_2Br_4 or $Tl^{I}_{3}[Tl^{III}Br_6]$	728.38	yel need				d	d	
t50	bromide, mono-	TlBr	284.28	yel-wh, cub, 2.4–2.8	$7.557^{17.3}$	480	815	0.05^{25}	0.25^{68}	s al; i HB_2, acet
t51	bromide, tri-	$TlBr_3$	444.10	yel, deliq, unstable		d		s	v s	v s al
t52	carbonate	Tl_2CO_3	468.75	col, monocl	7.11	273		$4.03^{15.5}$	27.2^{100}	i abs al, eth, acet
t53	chlorate	$TlClO_3$	287.82	need (rhomb ?)	5.047^{9}			2^{0}	57.31^{100}	
t54	*per*chlorate	$TlClO_4$	303.82	col, rhomb	4.89	501	d	20.5^{30}	167^{100}	sl s al
t55	chloride	TlCl	239.82	wh reg discol in air, 2.247	7.004^{30}_{4}	430	720	$0.29^{15.6}$	$2.41^{99.35}$	i al, acet; d a
t56	chloride, tri-	$TlCl_3$	310.73	hex pl, hygr		25	d	v s		s al, eth
t57	chloride, tri-	$TlCl_3.H_2O$	328.74	col, need		$-H_2O$, 60	d 100	v s	d	v s al, eth
t58	chloride, tri-	$TlCl_3.4H_2O$	382.79	col, need		37	$-4H_2O$, 100	86.2^{17}	d	s al, eth
t59	chloroplatinate	Tl_2PtCl_6	816.55	pale or cr	5.76^{17}			0.0064^{15}	0.05^{100}	
t60	chromate	Tl_2CrO_4	524.73	yel				0.03^{60}	0.2^{100}	sl s a, alk; i ac a
t61	*di*chromate	$Tl_2Cr_2O_7$	624.73	red				i		d a
t62	chromium sulfate	Thallium chromium alum. $Tl[Cr(H_2O)_6](SO_4)_2.6H_2O$	664.67	vlt cr	2.394	92		163.8^{25}		
t63	cyanate	TlCNO	246.39	col, need	5.487^{20}_{4}			s	v s	sl s al
t64	cyanide	TlCN	230.39	tabl	6.523	d		$16.8^{28.5}$		s a
t65	ethoxide	$(TlOC_2H_5)_4$	997.73	col liq	3.522	−3	d 80	s d		9.11^{25} al; s bz; i liq NH_3
t66	ethylate	$TlOC_2H_5$	249.43	liq, 1.6714^{20}	3.493^{20}_{4}	−3	d 130			sl s al; s eth
t67	ferrocyanide	$Tl_4Fe(CN)_6.2H_2O$	1065.46	yel, tricl	4.641			0.37^{18}	3.93^{101}	
t68	fluogallate	$Tl_2(GaF_5H_2O)$	591.47	col, orthorhomb	6.44			s		
t69	fluoride, mono-	TlF	223.37	col, cub, oct	8.23^{4}	327	655	78.6^{15}	d	sl s al
t70	fluoride, tri-	TlF_3	261.37	olive grn	8.36^{25}	d 550		d		i conc HCl
t71	fluosilicate	$Tl_2SiF_6.2H_2O$	586.85	hex pl	5.72			v s		
t72	formate	$TlHCO_2$	249.39	col, need, hygr	4.967^{104}	101		500^{10}		v s MeOH, sl s al; i $CHCl_3$

No.	Name	Synonyms and Formulae	Mol. wt.	Crystalline form, properties and index of refraction	Density or spec. gravity	Melting point, °C	Boiling point, °C	Solubility, in grams per 100 cc: Cold water	Hot water	Other solvents
	Thallium									
t73	(I) hydroxide	TlOH	221.38	pa yel, need		d 139		25.9^{0}	52^{40}	s al
t74	iodate	$TlIO_3$	379.27	wh need				0.058^{20}	sl s	sl s HNO_3
t75	iodide (α)	TlI	331.27	yel, rhomb	7.29	tr to (β) 170		0.0006^{20}	0.12^{100}	s liq NH_3
t76	iodide (β)	TlI	331.27	red, cub	$7.098^{14.7}$	440	823	i	i	i al
t77	iodide, tri-	TlI_3	585.08	blk, lust rhomb		d		s		
t78	iron (III) sulfate	$TlFe(SO_4)_2.12H_2O$	668.52	pink, oct, n_D^{17} 1.524	2.351^{15}	$-H_2O$, *ca* 100		36.15^{25} (anhydr)		
t79	magnesium sulfate	$Tl_2SO_4.MgSO_4.6H_2O$	733.26	wh dull cr, 1.5660, 1.5836, 1.5900	3.573^{20}_{4}	$-6H_2O$, 40		d 0		
t80	methoxide	$TlOCH_3$	235.40	wh cr powd		d>120		s d		1.70^{25} CH_3OH; 3.16^{25} bz
t81	molybdate	Tl_2MoO_4	568.68	wh powd or cr		vol red heat		i	v sl s	i al; s alk carb, conc NH_4OH, HF
t82	myristate	$TlC_{14}H_{27}O_2$	431.74	wh powd		120–3				0.52^{25} 50 % al
t83	(I) nitrate (α)	$TlNO_3$	266.37	cubic		206	430	9.55^{20}	4.13^{100}	i al, s acet
t84	(I) nitrate (β)	$TlNO_3$	266.37	trig		tr 145 to (α)				
t85	(I) nitrate (γ)	$TlNO_3$	266.37	rhomb, α 1.817	$5.556^{21.4}$	tr 75 to (β)		3.91^{0}	414^{100}	i al; s acet
t86	(III) nitrate	$Tl(NO_3)_3$	390.38	cr				s		
t87	(III) nitrate	$Tl(NO_3)_3.3H_2O$	444.43	col, rhomb, deliq			s 100	d	d	
t88	nitrite	$TlNO_2$	250.38	yel micro cr		182		32.10^{25}	95.78^{98}	i a
t89	oleate	$TlC_{18}H_{33}O_2$	485.83	wh cr clusters		131–2		0.05^{15}	0.3^{80}	3.0^{25} al
t90	oxalate	$Tl_2C_2O_4$	496.76	monocl pr	6.31			1.48^{15}	9.02^{100}	
t91	oxalate, tetra-	$TlH_3(C_2O_4)_2.2H_2O$	419.46	tricl, leaf, 1.5097, 1.6319, 1.6538	2.992^{17}	d 100		76.9^{23}	v s	i cold al; s hot al
t92	(I) oxide	Tl_2O	424.74	blk, deliq	9.52^{16}	300	1080^{760} −O, 1865	v s d to TlOH		s a, al
t93	(III) oxide	Tl_2O_3	456.74	col, amorph pr, hex	hex 10.19^{22} am 9.65^{21}	717 ± 5	−2O, 875	i	i	s a; i alk
t94	palmitate	$TlC_{16}H_{31}O_2$	459.79	cr need		115–117		0.01^{15}	0.07^{60}	1.04^{45} al
t95	phenoxide	$TlOC_6H_5$	297.48	wh cr		233–5		d		s hot bz; sl s lgr
t96	*ortho*phosphate	Tl_3PO_4	708.08	col, need	6.89^{10}			0.5^{15}	0.67^{100}	i al, s NH_4 salts
t97	*ortho*phosphate, (di)-β	TlH_2PO_4	301.36	monocl	4.726	*ca* 190		sl s	sl s	i al
t98	*pyro*phosphate	$Tl_4P_2O_7$	991.42	monocl pr	6.786^{20}	>120		40		
t99	picrate	$TlC_6H_2N_3O_7$	432.47	red, monocl or yel tricl	red 3.164^{17} yel 2.993^{17}	expl 723–725		0.135^{0}	2.43^{70}	0.40 CH_3OH
t100	rhodanide	TlCNS	262.45	glossy leaflets, rhomb, tetr cr	4.954^{20}_{4}	d low temp		0.393^{25}		0.024^{0} liq SO_2; i acet; s MeOH
t101	selenate	Tl_2SeO_4	551.70	rhomb need, 1.949, 1.959, 1.964		>400		2.13^{10}	8.5^{50}	i al, eth
t102	selenide	Tl_2Se	487.70	gray leaf	9.05^{25}_{4}	340		i		s a; i acet a
t103	silver nitrate	$TlNO_3.AgNO_3$	436.25	wh cr powd		75		s		
t104	stearate	$TlC_{18}H_{35}O_2$	487.85	need		119		0.005^{15}	0.095^{75}	0.18^{15} al, 0.60^{60} al
t105	(I) sulfate	Tl_2SO_4	504.80	col, rhomb, 1.860, 1.867, 1.885	6.77	632	d	4.87^{20}	19.14^{100}	
t106	(I) sulfate hydrogen	$TlHSO_4$	301.44	pr need		120 d				v sl s dil H_2SO_4
t107	(III) sulfate	$Tl_2(SO_4)_3.7H_2O$	823.03	col leaf		$-6H_2O$, 220		d	d	s dil H_2SO_4
t108	(I) sulfide	Tl_2S	440.80	bl-blk tetr	8.46	448.5	d	0.02^{20}	sl s	s a; i alk, acet
t109	(III) sulfide	Tl_2S_3	504.93	blk, amorph		260 (in N_2)	d	i	i	s hot H_2SO_4
t110	sulfite	Tl_2SO_3	488.80	wh cr	6.427			3.34^{15}	v s	i al
t111	tartrate(dl)	$Tl_2C_4H_4O_6$	556.81	monocl	4.659	d 165		13.3^{15}		
t112	*meta*tellurate	Tl_2TeO_4	600.34	heavy wh ppt	$6.760^{17.6}$	red heat		sl s	sl s	
t113	thiocyanate	TlSCN	262.45	col, tetr	4.956^{20}			0.315^{20}	0.727^{40}	i al
t114	*di*thionate	$Tl_2S_2O_6$	568.86	monocl	5.57^{20}_{4}	d		41.8^{19}		
t115	thiosulfate	$Tl_2S_2O_3$	520.87	wh rhomb cr		d 130		sl s	v s	
t116	*meta*vanadate	$TlVO_3$	303.41	gray cr	6.09^{17}	424		0.87^{11}	0.21^{100}	
t117	*pyro*vanadate	$Tl_4V_2O_7$	1031.36	light yel	8.21^{19}	454		0.2^{14}	0.26^{100}	
t118	**Thiocarbonyl chloride**	Thiophosgene. $CSCl_2$	114.98	red yel liq, 1.5442	1.509^{15}		73.5	d		d al; s eth
t119	**Thiocarbonyl chloride, tetra-**	$CSCl_4$	185.89	yel	1.712^{13}		146–147	d		
t120	**Thiocyanic acid(iso)**	HSCN(HNCS)	59.09	col mass or gas		>−110	polym to solid −90	v s		v s al, eth, bz
t121	**Thiocyanogen**	$(SCN)_2$	116.16	liq, or yel solid		−2 to −3		d		s al, eth, CS_2, CCl_4

No.	Name	Synonyms and Formulae	Mol. wt.	Crystalline form, properties and index of refraction	Density or spec. gravity	Melting point, °C	Boiling point, °C	Solubility, in grams per 100 cc: Cold water	Hot water	Other solvents
	Thionyl bromide									
t122	**Thionyl bromide**	$SOBr_2$	207.88	or, yel liq	2.68^{18}	−52	138^{773}, 68^{40}	d	d	s bz, chl, CS_2, CCl_4
t123	chloride	$SOCl_2$	118.97	col, or yel liq; 1.527^{10}	1.655^{10}_{4}	−105	78.8^{746} d 140	d	d	d a, al, alk; s bz, chl
s124	chloride fluoride	SOClF	102.51	gas	...	−139.5	12.2	...	...	...
t125	fluoride	SOF_2	86.06	col, gas	gas 2.93 g/l liq 1.780^{-100}	−110.5	−43.8	d	d	s eth, bz, chl, acet, $AsCl_3$
t126	**Thiophosphoramide**	Thiophosphorylamide. $PS(NH_2)_3$	111.11	yel wh cr	1.7^{13}	d 200	...	14.3^{25}	d	sl, me
t127	**Thiophosphoryl** bromide	$PSBr_3$	302.76	yel, cub	2.85^{17}	37.8	125^{25}	d	...	s eth, CS_2, PCl_3
t128	thiophosphoryl bromide, hydrate	$PSBr_3.H_2O$	320.78	yel cr	2.794^{18}	35	...	...	...	...
t129	thiophosphoryl bromide (mono-) chloride, di-	$PSBrCl_2$	213.85	yel liq	2.12^{6}	−30	d 150	d	...	...
t130	thiophosphoryl bromide (di-) chloride	$PSBr_2Cl$	258.31	pa grn fum liq	2.48^{0}	−60	95^{60}	d	...	...
t131	thiophosphoryl chloride	$PSCl_3$	169.40	col liq, 1.563 (c)	1.635	−35	125	d	...	s bz, CS_2, CCl_4
t132	thiophosphoryl fluoride	PSF_3	120.03	gas	...	$3.8^{7.64 atm}$	d	sl s d	...	s eth; i bz, CS_2
t133	**Thiosulfuric acid**	$H_2S_2O_3$	114.14	in sol only	...	...	...	s	...	...
t134	**Thorium**	Th	232.038	gray, cub radioactive	11.7			i	i	s HCl, H_2SO_4, aq reg; sl s HNO_3
t135	boride, hexa-	ThB_6	296.92	dk viol-blk met, cub	6.4^{15}	2195	...	i	i	s HNO_3; i H_2SO_4, HCl, HF, aq alk
t136	boride, tetra	ThB_4	275.28	tetr pr	7.5^{15}	...	...	i	i	s HNO_3 HCl, hot H_2SO_4
t137	bromide	$ThBr_4$	551.67	col cr, hygr	5.67	subl 610	725	s	s	...
t138	carbide	ThC_2	256.06	yel, tetr	8.96^{18}	2655 ± 25	*ca* 5000 (?)	d	...	v sl s conc a
t139	carbonate	$Th(CO_3)_2$	352.06	exist?	...	...	...	i	d	s conc Na_2CO_3
t140	chloride	$ThCl_4$	373.85	wh, rhomb, deliq	4.59	770 ± 2 subl 820	d 928	v s	v s	s al, a, KCl; al s eth
t142	*tetra*cyanoplatinate	$Th[Pt(CN)_4]_2.16H_2O$	1118.61	yel-grn, rhomb	2.460	...	...	sl s	s	...
t143	fluoride	ThF_4	308.03	wh cub powd	6.32^{24}	>900	...	...	...	sl d dil H_2SO_4, HCl; i conc H_2SO_4
t144	fluoride	$ThF_4.4H_2O$	380.09	cr	...	$-H_2O$, 100	$-2H_2O$, 140–200	0.017^{25}	...	i HF
t145	hydroxide	$Th(OH)_4$	300.02	wh gelat	...	d	...	i	i	s a; i alk, HF
t146	iodate	$Th(IO_3)_4$	931.65	...	...	...	...	i	i d	s dil H_2SO_4; i dil HNO_3
t147	iodide, tetra-	ThI_4	739.66	yel	...	566	839	s	...	...
t148	nitrate	$Th(NO_3)_4$	480.06	plates, deliq	...	d 500	...	v s	...	s al
t149	nitrate	$Th(NO_3)_4.4H_2O$	552.12	col cr	...	swells	...	v s	...	v s al; sl s acet; 36.9 eth
t150	nitrate	$Th(NO_3)_4.12H_2O$	696.24	col leaf, deliq	...	d	...	v s	...	v s al, a
t151	nitride	Th_3N_4	752.14	dk brn powd, or blk cr	...	...	...	sl d	d	s HCl
t152	oxalate	$Th(C_2O_4)_2$	408.08	wh cr	4.637^{16}	d	...	0.0017^{17}	0.0017^{60}	s h aq $(NH_4)_2C_2O_4$; al s a
t153	oxalate	$Th(C_2O_4)_2.6H_2O$	516.17	wh amorph powd	...	...	...	i	...	s Na_2CO_3, $(NH_4)_2C_2O_4$ sol; i HNO_3
t154	oxide, di-	Thorianite. ThO_2	264.04	wh cub, 2.20 (liq)	9.86	3220 ± 50	4400	i	i	s hot H_2SO_4; i dil a, alk
t155	oxysulfide	ThOS	280.10	yel cr	6.44^{0}	d	...	i	...	s aq reg; sl s HNO_3
t156	2,4-pentanedione	Thorium acetylacetonate. $Th(C_5H_7O_2)_4$	628.48	col cr	...	171 subl 160^{10}	$260–270^{10}$	sl s	...	v s al, chl; s eth
t157	*hypo*phosphate	$ThP_2O_6.11H_2O$	588.15	wh amorph ppt.	...	$-11H_2O$, 160	...	i	i	i a, alk
t158	*meta*phosphate	$Th(PO_3)_4$	547.93	col rhomb pr	$4.08^{16.4}$	...	...	...	...	...
t159	*ortho*phosphate	$Th_3(PO_4)_4.4H_2O$	1148.06	wh gelat	...	...	...	i	i	s 30° HCl; i a
t160	picrate	$Th(C_6H_2N_3O_7)_4.10H_2O$	1324.59	...	...	...	...	0.305^{25}	...	...
t161	selenate	$Th(SeO_4)_2.9H_2O$	680.09	col, monocl	3.026	$-8H_2O$, 200	d 1500	0.5^{0}	2.0^{100}	...
t162	*ortho*silicate	Thorite $ThSiO_4$	324.12	col, tetr, 1.80, 1.81	6.82^{16}	...	...	v sl s	...	i a
t163	silicide	$ThSi_2$	288.21	blk, tetr	7.96^{16}	...	...	...	...	s hot HCl; sl s H_2SO_4
t164	sulfate	$Th(SO_4)_2$	424.16	wh cr, hygr	4.225^{17}	...	...	s	s	i a; v s $NH_4C_2H_3O_2$
t165	sulfate	$Th(SO_4)_2.4H_2O$	496.22	wh need, or cr powd	...	$-4H_2O$, 400	...	9.41^{17} (anhydr)	2.54^{50} (anhydr)	i a

No.	Name	Synonyms and Formulae	Mol. wt.	Crystalline form, properties and index of refraction	Density or spec. gravity	Melting point, °C	Boiling point, °C	Solubility, in grams per 100 cc		
								Cold water	Hot water	Other solvents
	Thorium									
t166	sulfate	$Th(SO_4)_2.6H_2O$	532.26					1.63^{15}	6.64^{60}	i a
t167	sulfate	$Th(SO_4)_2.8H_2O$	568.29	monocl, prism, 1.5168		$-4H_2O$, 42		1.88^{25}	3.71^{44}	i a
t168	sulfate	$Th(SO_4)_2.9H_2O$	586.31	wh monocl	2.77	$-9H_2O$, 400		1.57^{20}	6.67^{55}	i a
t169	sulfide	ThS_2	296.17	dk brn-blk cr	7.30^{25}_{4}	1925 ± 50 (vac)		i	d 200	s hot aq reg; sl s a
t170	*pyro*vanadate	$ThV_2O_7.6H_2O$	554.01	yel				i	i	s conc a
t171	**Thulium**	Tm	168.934	silv wh met, hex.	9.3208	1545	1947	i	i	
t172	bromide	$TmBr_3$	408.66			952	1440	s	s	
t173	chloride	$TmCl_3.7H_2O$	401.40	grn cr, deliq		824	1440	v s		v s al
t174	fluoride	TmF_3	225.93			1158	>2200	i	i	s dil a
t175	iodide	TmI_3	549.65	brt yel cr		1015	1260	s	s	
t176	oxalate	$Tm_2(C_2O_4)_3.6H_2O$	710.02	grn-wh ppt		$-H_2O$, 50		i		s alk oxal sol; i dil a
t177	oxide	Thulia. Tm_2O_3	385.87	grn-wh powd						sl s min a
t178	**Tin gray**	Sn	118.69	gray, cub	5.75	231.9681	2270	i	i	s HCl, H_2SO_4, aq reg, alk; sl s dil HNO_3
t179	**Tin white**	Sn	118.69	wh met, tetr	7.28	231.88 stable 13.2 − 161	2260	i	i	s HCl, H_2SO_4, aq reg, alk; sl s dil HNO_3
t180	**Tin brittle**	Sn	118.69	wh, rhomb	6.52–56	231.89 stable >161	2260	i	i	s HCl, H_2SO_4, aq reg, alk; sl s dil HNO_3
t181	(II) acetate	$Sn(C_2H_3O_2)_2$	236.78	yelsh powd		182	240	d		s dil HCl
t182	*pyro*arsenate	$Sn_2As_2O_7$	499.22	flocculent ppt		d AS_2O_3 + SnO_2		i	i	i conc ac a
t183	(II) bromide	$SnBr_2$	278.51	pa yel, rhomb	5.117^{17}	215.5	620	85.2^{0}	222.5^{100}	s al, eth, acet
t184	(IV) bromide	$SnBr_4$	438.33	col, rhomb pyr, deliq	liq 3.34^{35}	31	202^{734}	s d	d	s acet, PCl_3, $AsBr_3$
t185	bromide chloride(tri-)	$SnBrCl_3$	304.96	col liq	2.51^{13}	−31	50^{20}			
t186	bromide(di-) chloride(di-)	$SnBr_2Cl_2$	349.41		2.82^{13}	−20	65^{20} d 191	d	d	
t187	bromide(tri-) chloride	$SnBr_3Cl$	393.87	liq	3.12^{13}	1	73^{20}			
t188	bromide(di-) iodode(di-)	$SnBr_2I_2$	532.32	or-red, hex pl	3.631^{15}	50	225	s	d<80	
t189	(II) chloride	$SnCl_2$	189.60	wh, rhomb	3.95^{25}_{4}	246	652	83.9^{0}	269.8^{15} d	s al, eth, acet, et acet, me acet, pyr
t190	(II) chloride dihydrate	$SnCl_2.2H_2O$	225.63	wh, monocl	$2.710^{15.5}$	37.7	d	d	d	s al, eth, acet, glac ac a
t191	(IV) chloride	$SnCl_4$	260.50	col liq, solid cub, 1.512	liq 2.226	−33	114.1	s	d	s eth
t192	(IV) chloride pentahydrate	$SnCl_4.5H_2O$	350.58	monocl cr		stable 19–56		s		
t193	(IV) chloride tetrahydrate	$SnCl_4.4H_2O$	332.56	opaque			stable 56–83	s		
t194	(IV) chloride trihydrate	$SnCl_4.3H_2O$	314.55	col, monocl cr		80	stable 64–83	s		
t195	(IV) chloride *di*ammine	$SnCl_4.2NH_3$	294.56	cr				s		d HCl
t196	chloride(tri-) bromide	$SnCl_3Br$	304.96	col liq	2.51^{13}	−31	50^{20}			
t197	chloride (di-) iodide(di-)	$SnCl_2I_2$	443.40	red mobile liq	3.287^{15}		297	s	d	s chl, bz CS_2
t198	(IV) chloride nitrosyl-chloride	$SnCl_4.2NOCl$	391.42	pa yel, oct cr	2.60	180		d	d	
t199	(IV) chromate	$Sn(CrO_4)_2$	350.68	br yel cr powd		d		s		
t200	(II) ferricyanide	$Sn_3[Fe(CN)_6]_2$	779.98	wh		d		i	i	s HCl
t201	(II) ferrocyanide	$Sn_2Fe(CN)_6$	449.33	wh gel				i	i	d HCl
t202	(IV) ferrocyanide	$SnFe(CN)_6$	330.64					i	i	d h HCl
t203	(II) fluoride	Fluoristan. SnF_2	156.69	wh, monocl cr				s		
t204	(IV) fluoride	SnF_4	194.68	wh, monocl cr, hygr	4.780^{19}	705 subl		v s	d	
t205	hydride	Stannane. SnH_4	122.72	gas		d − 150	−52			s $AgNO_3$, $HgCl_2$, conc alk, conc H_2SO_4
t206	iodide	SnI_2	372.50	yelsh-red to red monocl need	5.285	320	717	0.98^{20}	4.03^{100}	v s NH_4OH, HI soln
t207	(IV) iodide	SnI_4	626.31	or red cub, 2.106	4.473^{0}	144.5	364.5	s	d	141.1^{25} CS_2; 6.03^{15} CCl_4; 17.88^{25} bz

No.	Name	Synonyms and Formulae	Mol. wt.	Crystalline form, properties and index of refraction	Density or spec. gravity	Melting point, °C	Boiling point, °C	Solubility, in grams per 100 cc		
								Cold water	Hot water	Other solvents
	Tin									
t208	(II) nitrate	$Sn(NO_3)_2.20H_2O$	603.07	col leaf		−20		d	d	d HNO_3
t209	(II) nitrate, basic	$SnO.Sn(NO_3)_2$	377.39	wh cr mass		d >100 expl		d	d	
t210	(IV) nitrate	$Sn(NO_3)_4$	366.71	silky need		d 50		d		
t211	(II) oxide, mon-	SnO	134.69	blk, cub (tetr)	6.446^{0}	d 1080^{600}		i	i	s a, alk; sl s NH_4Cl
t212	oxide, mon-hydrate	$SnO.xH_2O$		wh powd or yellow-brn cr					d to SnO	d a; alk; s alk carb; i NH_4OH
t213	(IV) oxide, di-	Nat. cassiterite. SnO_2	150.69	wh, tetr, (also hex or rhomb), 1.997, 2.093	6.95	1630	subl 1800–1900	i	i	d KOH, NaOH; i aq reg
t214	oxide, di-hydrate	α-Stannic acid or "ordinary" stannic acid. $SnO_2.xH_2O$		amorph or gel				i	i	s a, alk, K_2CO_3
t215	oxide, di-hydrate	β-Stannic acid or "meta" stannic acid. $SnO_2.xH_2O$		wh, amorph or gel				i	i	i a, K_2CO_3; sol alk
t216	(II) *meta*phosphate	$Sn(PO_3)_2$	276.63	amorph mass	$3.380^{22.8}$					
t217	(II) *ortho*phosphate	$Sn_3(PO_4)_2$	546.01	wh, amorph	3.823^{17}			i	i	d a, alk
t218	(II) *ortho*phosphate, di-H	$Sn(H_2PO_4)_2$	312.66	wh, rhomb cr	$3.167^{22.8}$	d	d		d	
t219	(II) *ortho*phosphate, mono-H	$SnHPO_4$	214.67	cr	$3.476^{15.5}$	stabl >100	d	i	i	s dil min a
t220	(II) *pyro*phosphate	$Sn_2P_2O_7$	411.32	amorph powd	$4.009^{16.4}$					s conc a
t221	phosphide, mono-	SnP	149.66	silv wh	6.56	d	d	i	i	s HCl; i HNO_3
t222	phosphide, tri-	SnP_3	211.61	cr	4.10^{0}	<415 d to Sn_4P_3		i	i	d HNO_3; i HCl
t223	*tetra*phosphide, tri-	Sn_4P_3	567.68	wh cr	5.181	d <480		i	i	d fixed alk hydr, HCl
t224	phosphorus chloride	$SnCl_4.PCl_5$	468.74	col cr		subl 200		d	d	
t225	(II) selenide	SnSe	197.65	steelgray cr	6.179^{0}	861		i	i	d HCl, HNO_3, aq reg, alk sulf
t226	(II) sulfate	$SnSO_4$	214.75	wh-yelsh cr powd		>360 (SO_2)		33^{25}		s H_2SO_4
t227	(IV) sulfate	$Sn(SO_4)_2.2H_2O$	346.84	wh, hex pr, deliq				v s	d	s eth, dil H_2SO_4, HCl
t228	(II) sulfide	SnS	150.75	gray-blk cub, monocl	5.22^{25}	882	1230	0.000002^{18}		d HCl, alk, $(NH_4)_2S$
t229	(IV) sulfide	Mosaic gold. SnS_2	182.82	gold yel, hex	4.5	d 600		0.0002^{18}		d alk sulf, aq reg, alk hydr, PCl_5, $SnCl_2$; i a
t230	(IV) sulfur chloride	$SnCl_4.2SCl_4$	608.25	yel cr		37	d <40	d	d	s eth, bz, CS_2, ethyl acet; d HNO_3
t231	tartrate	$SnC_4H_4O_6$	266.76	heavy wh powd				s		v s dil HCl
t232	(II) telluride	SnTe	246.29	gray cr	6.48	780	d	i	i	d alk sulf
t233	(IV) telluride	$SnTe_2$	373.89	blk, flocc ppt				i	i	d dil a, alk
t234	**Titanic acid, ortho-**	α Titanic acid. H_2TiO_4	113.91	wh		d		v sl s d		s dil HCl, dil H_2SO_4, conc alk
t235	**Titanium**	Ti	47.90	α hex, tr β cub 838, silv gray	4.5^{20}	1660±10	3287	i	i	s dil a
t236	boride, di-	TiB_2	69.52	hex	4.50	2900				
t237	bromide, di-	$TiBr_2$	207.72	blk powd	4.31	d >500		s ev H_2		
t238	bromide, tetra-	$TiBr_4$	367.54	or yel, deliq	2.6	39	230	d		s abs al, abs eth
t239	bromide, tri-	$TiBr_3.6H_2O$	395.72	redsh-viol or dk blue cr, deliq		115	d 400	v s		v s al, acet
t240	carbide	TiC	59.91	gr met, cub	4.93	3140±90	4820	i	i	s aq reg, HNO_3
t241	chloride, di-	$TiCl_2$	118.81	lt br-blk, hex, deliq	3.13	subl H_2	d 475 vac	d		s al, i eth, chl, CS_2
t242	chloride, tetra-	$TiCl_4$	189.71	lt yel liq, $1.61^{10.5}$	liq 1.726 sol 2.06^{-79}	−25	136.4	s	d	s dil HCl, al
t243	chloride, tri-	$TiCl_3$	154.26	dk viol, deliq	2.64	d 440	660^{108}	s	s	v s al; s HCl; i eth
t244	fluoride, tetra	TiF_4	123.89	wh powd, hygr	$2.798^{20.5}$	>400 (pressure)	284 (subl.)	s d		s H_2SO_4, al, C_5H_5N; i eth
t245	fluoride, tri-	TiF_3	104.90	purp-red or vlt	3.40	1200	1400	red s vlt i		
t246	hydride	TiH_2	49.92	gray powd	3.9^{12}	d 400				
t247	iodide, di-	TiI_2	301.71	blk, hygr	4.99	600	1000	d		d alk; s conc HF, conc HCl
t248	iodide, tetra-	TiI_4	555.52	red, cub	4.3	150	377.1	v s	d	
t249	nitride	TiN	61.91	yel-bronze, cub	5.22	2930		i	i	sl s hot aq reg +HF
t250	oxalate	$Ti_2(C_2O_4)_3.10H_2O$	540.01	yel pr				s	s	i al, eth
t251	oxide, di-	Nat. brookite. TiO_2	79.90	wh, rhomb, 2.583, 2.586, 2.741	4.17	1825		i	i	s H_2SO_4, alk; i a
t252	oxide, di-	Nat. octahedrite, anatase. TiO_2	79.90	br-blk, tetr, 2.554, 2.493	3.84			i	i	s H_2SO_4, alk; i a

No.	Name	Synonyms and Formulae	Mol. wt.	Crystalline form, properties and index of refraction	Density or spec. gravity	Melting point, °C	Boiling point, °C	Solubility, in grams per 100 cc: Cold water	Hot water	Other solvents
	Titanium									
t253	**oxide, di-**	Nat. rutile. TiO_2	79.90	col, tetr, 2.616, 2.903	4.26	1830–1850	2500–3000	i	i	s H_2SO_4, alk; i a
t254	**oxide, mon-**	TiO	63.90	yel blk, pr	4.93	1750	>3000			s dil H_2SO_4; i HNO_3
t255	**oxide, sesqui-**	Ti_2O_3	143.80	vlt blk, trig	4.6	2130 d		i	i	s H_2SO_4; i HCl, HNO_3
t256	**phosphide**	TiP	78.87	gray, met	3.95^{25}			i	i	i a
t257	**sulfate**	$Ti_2(SO_4)_3$	383.98	green powd				i	i	s dil a; i al, eth, conc H_2SO_4
t258	**sulfate, basic**	$TiOSO_4$	159.96	wh or sl yelsh powd, 1.80–1.89				d		
t259	**sulfide, di-**	TiS_2	112.03	yel sc	3.22^{20}			hyd sl	d in steam	d HCl; s dil HNO_3, H_2SO_4
t260	**sulfide, mono-**	TiS	79.96	redsh solid	4.12			i		s conc H_2SO_4; i HCl, HF, dil H_2SO_4
t261	**sulfide, sesqui**	Ti_2S_3	191.99	grayish-blk cr	3.584			i	i	s conc H_2SO_4, conc HNO_3; i dil H_2SO_4, dil HCl
t262	**Tungsten**	Wolfram. W	183.85	gray-blk, cub	19.35^{20}_{4}	3410±20	5660	i	i	v sl s HNO_3, H_2SO_4, aq reg; s HNO_3+HF, fus NaOH+$NaNO_3$; i HF, KOH
t263	**arsenide**	WAs_2	333.69	blk cr	6.9^{18}	d red heat		i		d hot HNO_3, hot H_2SO_4
t264	**boride, di-**	WB_2	205.47	silvery, oct	10.77	*ca* 2900		i	i	s aq reg
t265	**bromide, di-**	WBr_2	343.67	bl-blk need		d 400		d		
t266	**bromide, penta-**	WBr_5	583.40	vlt-brn need, hygr		276	333	d		s abs al, chl, eth, alk
t267	**bromide, hexa-**	WBr_6	663.30	bl-blk, need	6.9	232		i	d	s abs a, eth, CS_2, NH_4OH
t268	**carbide**	WC	195.86	blk, hex	15.63^{18}	2870±50	6000	i		s HNO_3+HF, aq reg
t269	**(di-)carbide**	W_2C	379.71	blk, hex	17.15	2860	6000	i		s HNO_3+HCl
t270	**carbonyl**	$W(CO)_6$	351.91	col, rhomb cr	2.65	d~150	175^{766}	i	i	s fum HNO_3; v sl s al, eth, bz
t271	**chloride, di-**	WCl_2	254.76	gray, amorph	5.436			d		
t272	**chloride, hexa-**	WCl_6	396.57	dk bl, cub	3.52^{25}_{4}	275	346.7		d^{60}	s al, eth, bz, CCl_4; v s CS_2, POCl
t273	**chloride, penta-**	WCl_5	361.12	blk, deliq	3.875^{25}_{4}	248	275.6		d to W_2O_5	v sl s CS_2
t274	**chloride, tetra-**	WCl_4	325.66	gray, deliq	4.624^{25}_{4}	d		d		
t275	**fluoride, hexa-**	WF_6	297.84	col gas, or lt yel liq	liq 3.44 gas 12.9 g/l	2.5^{420}	17.5	d	d	s alk
t276	**iodide, di-**	WI_2	437.66	br-gr, amorph	6.799^{25}_{4}	d		i	d	s alk; i al CS_2
t277	**iodide, tetra-**	WI_4	691.47	blk, cr	5.2^{18}	d		i	d	s abs al; i eth, chl, turp
t278	**nitride, di-**	WN_2	211.86	brn, cub		above 400 (vac)		d	d	
t279	**oxide, di-**	WO_2	215.85	br, cub	12.11	1500–1600 (in N_2)	*ca* 1430 subl 800	i	i	s a, KOH
t280	**oxide, pent-**	Mineral blue. W_2O_5 or W_4O_{11}	447.70 or 911.39	blue-vlt, tricl		subl 800–900	*ca* 1530 d 2000	i		i a
t281	**oxide, tri-**	WO_2	231.85	yel, rhomb, or yel-or powd	7.16	1473		i		s hot alk; sl s HF; i a
t282	**oxy*di*bromide, di-**	WO_2Br_2	375.67	red, prism		d				
t283	**oxy*tetra*bromide**	$WOBr_4$	519.49	blk, deliq		277	327	d	d	
t284	**oxy*tetra*chloride**	$WOCl_4$	341.66	red, need		211	227.5	d	d	s CS_2, S_2Cl_2, bz
t285	**oxy*di*chloride, di-**	WO_2Cl_2	286.75	lt yel tabl		266		s	d	i al; s NH_4OH, alk
t286	**oxy*tetra*fluoride-**	WOF_4	75.84	col pl, hygr		110	187.5	d		sl s CS_2; i CCl_4
t287	**phosphide**	WP	214.82	gray, prism	8.5			i		s HNO_3+HF; i alk, HCl
t288	**phosphide**	WP_2	245.80	blk cr	5.8	d		i	i	s HNO_3+HF, aq reg; i al, eth
t289	**phosphide**	W_2P	398.67	dk gray prism	5.21	d				s fus Na_2CO_3+$NaNO_3$; i a, aq reg

No.	Name	Synonyms and Formulae	Mol. wt.	Crystalline form, properties and index of refraction	Density or spec. gravity	Melting point, °C	Boiling point, °C	Solubility, in grams per 100 cc		
								Cold water	Hot water	Other solvents
	Tungsten									
t290	silicide	WSi_2	240.02	blue, gray, tetrag	9.4	above 900		i	i	s HNO_3+HF; i aq reg
t291	sulfide, di-	Nat. tungstenite. WS_2	247.98	dk gray, hexag	7.5^{10}	d 1250		i		s HNO_3+HF, fus alk; i al
t292	sulfide, tri-	WS_3	280.04	choc brn powd				sl s	s	s alk
t293	**Tungstic acid, meta-**	$H_2W_4O_{13}.9H_2O$	1107.55	col, tetrag	3.93	d 50		88.57^{22}	$111.87^{43.5}$	$110.76^{24.3}$ eth; s al
t294	**Tungstic acid, ortho-**	H_2WO_4	249.86	yel powd, 2.24	5.5	$-H_2O$, 100	1473	i	sl s	s alk, HF, NH_3; i most a
t295	**Tungstic acid, ortho-**	$H_2WO_4.H_2O$	267.88	wh		$H_2W_2O_7$ *at* 100		sl s		s alk
u1	**Uranic acid meta-**	Uranyl hydroxide. H_2UO_4(or $UO_2(OH)_2$)	304.04	yel, rhomb, or powd	5.926	$-H_2O$ 250–300		i	i	s a, alk carb
u2	**Uranium**	U	238.03	silvery, cubic, radioactive	19.05 ± 0.02^{25}	1132.3 ± 0.8	3818	i	i	s a; i alk, al
u3	boride, di-	UB_2	259.65	hex	12.70	2365				
u4	bromide tetra-	UBr_4	557.67	br leaf, deliq	5.35	516	792^{760}	v s	v s	s d al, MeOH; i bz; s liq NH_3
u5	bromide tri-	UBr_3	477.76	dk brn need, hygr	6.53	730	volat	s		d al
u6	*di*carbide	UC_2	262.05	met cr	11.28^{16}	2350–2400	4370^{760}	d	d	i al; d dil inorg a
u7	chloride, penta-	UCl_5	415.30	dk green, gray need, red by trans light, hydr	3.81(?)	d 300		d		s abs als, a acet, NH_4Cl; d ac a; i bz, eth
u8	chloride, tetra-	UCl_4	379.84	dk grn met, cub oct, hygr	4.87	590 ± 1	792^{760}	v s	s	s al, acet, ac a; i eth, $CHCl_3$
u9	chloride, tri-	UCl_3	344.39	dk red need, hygr	5.44^{25}_{4}	842 ± 5		s	s	s MeOH, acet, glac acet a; i eth
u10	fluoride, hexa-	UF_6	352.02	col cr, deliq, monocl	4.68^{21}	64.5–64.8	56.2^{765}	d		d al, eth; s CCl_4, chl; i CS_2
u11	fluoride, tetra-	UF_4	314.02	green, tricl need	6.70 ± 0.10	960 ± 5		v sl s		i dil a, alk; s conc a, conc alk
u12	fluoride, tri-	UF_3	295.03	blk cr or fused		d above 1000		sl d		v sl s dil inorg a
u13	hydride	UH_3	241.05	blk-brn powd	10.95			i	i	i al, acet, liq NH_3; sl s dil HCl; d HNO_3
u14	hydride	UH_3	241.05	blk powd, cub	11.4					
u15	iodide, tetra-	UI_4	745.65	blk, need	5.6^{15}	506	759	s	s d	
u16	nitride, mono-	UN	252.04	br powd	14.31	*ca* 2630 ± 50				i HCl, H_2SO_4
u17	oxide, di-	UO_2	270.03	br-blk rhomb, or cub	10.96	2878 ± 20		i	i	s HNO_3, conc H_2SO_4
u18	oxide, per-	$UO_4.2H_2O$	338.06	pa yel cr, hygr		d 115		0.0006^{20}	0.008^{20}	d HCl
u19	oxide, tri-	Uranyl oxide. UO_3	286.03	yel-red powd	7.29	d		i		s HNO_3, HCl
u20	*tri*-oxide, oct-	U_3O_8	842.09	olive green-blk	8.30	d 1300 to UO_2		i	i	s HNO_3, H_2SO_4
u21	(IV) sulfate	$U(SO_4)_2.4H_2O$	502.21	grn, rhomb		$-4H_2O$, 300		23^{11}	9^{63} (anhydr)	s dil a
u22	(IV) sulfate	$U(SO_4)_2.8H_2O$	574.28			d 90		11.3^{18}	58.2^{62}	i al; s dil a
u23	(IV) sulfate	$U(SO_4)_2.9H_2O$	592.29	grnsh, monocl		$-7H_2O$, 230	$-9H_2O$ red heat			s dil H_2SO_4
u24	sulfide, di-	US_2	302.16	gray-blk, tetr	7.96^{26}	>1100	oxidizes	sl d		s conc HCl; d HNO_3 v s al
u25	sulfide, mono-	US	270.09	blk amorph powd	10.87	above 2000				i HCl, HNO_3
u26	sulfide, sesqui-	U_2S_3	572.25	gray blk, rhomb need		ign				s+O aq reg conc HNO_3; i dil a
u27	**Uranyl acetate**	$UO_2(C_2H_3O_2)_2.2H_2O$	422.19	yel, rhomb	2.893^{15}	$-2H_2O$, 110	d 275	7.694^{15}	d	v s al
u28	benzoate	$UO_2(C_7H_5O_2)_2$	512.26	yel powd				sl s		sl s al
u29	bromide	UO_2Br_2	429.85	grn-yel need, hygr				s d		s al, eth
u30	*per*chlorate	$UO_2(ClO_4)_2.6H_2O$	577.02	yel cr, deliq, rhomb		90 d 110				
u31	chloride	UO_2Cl_2	340.93	yel, deliq		578	d	320^{18}	v s	s al, amyl al, eth
u32	formate	$UO_2(CHO_2)_2.H_2O$	378.08	yel, oct	3.695^{19}	$-H_2O$, 110		7.2^{15}		sl s form a; 0.74^{15} MeOH, 2.37^{15} acet
u33	iodate	$UO_2(IO_3)_2$	619.83	yel, rhomb	5.2	d 250		s	s	i HNO_3
u34	iodate	$UO_2(IO_3)_2.H_2O$	637.85	α prismatic, stable, β pyramidal	α 5.220^{18} β 5.052^{18}			α 0.1049^{18} β 0.1214^{18}		
u35	iodide	UO_2I_2	523.84	red, deliq		d in air				s al, eth, bz
u36	nitrate	$UO_2(NO_3)_2.6H_2O$	502.13	yel, rhomb, deliq, 1.4967	2.807^{13}	60.2 d 100	118		∞ 60	v s al, eth, ac a, acet, MeOH
u37	oxalate	$UO_2C_2O_4.3H_2O$	412.09	yel cr		$-H_2O$, 110		0.8^{14}	3.3^{100}	s inorg a, alk, oxal a
u38	phosphate, mono-H	$UO_2HPO_4.4H_2O$	438.07	yel pl, tetr				i	i	s HNO_3, aq Na_2CO_3; i ac a
u39	potassium carbonate	$UO_2CO_3.2K_2CO_3$	606.46	yel cr		$-CO_2$, 300		7.4^{15}	d	i al

PHYSICAL CONSTANTS OF INORGANIC COMPOUNDS (Continued)

No.	Name	Synonyms and Formulae	Mol. wt.	Crystalline form, properties and index of refraction	Density or spec. gravity	Melting point, °C	Boiling point, °C	Solubility, in grams per 100 cc: Cold water	Hot water	Other solvents
	Uranyl									
u40	sodium carbonate	$UO_2CO_3.2Na_2CO_3$	542.02	yel cr				al s		i al
u41	sulfate	$UO_2SO_4.3H_2O$	420.14	yel-grn cr	$3.28^{16.5}$	d 100		$20.5^{15.5}$	22.2^{100}	24.3^{15} conc H_2SO_4; 30^{15} conc HCl
u42	sulfate	$2(UO_2SO_4).7H_2O$	858.29	yel		anh 300		al s		s H_2SO_3
u43	sulfide	UO_2S	302.09	brn-blk, tetr		d 40–50		al s	s d	s dil a, dil al, $(NH_4)_2CO_3$; i abs al
u44	sulfite	$UO_2SO_3.4H_2O$	422.15	pa-gr cr						s H_2SO_3
v1	**Vanadic acid, meta**	HVO_3	99.95	yel sc				i		s a, alk; i NH_4OH
v2	tetra-	$H_2V_4O_{11}$	381.78	br amorph				i		s a, alk, NH_4OH
v3	**Vanadium**	V	50.9415	lt gray met, cub, 3.03	5.96	1890 ± 10	3380	i	i	s aq reg, HNO_3, H_2SO_4, HF; i HCl, alk
v4	boride, di-	VB_2	72.56	hex	5.10					
v5[1]	bromide, tri-	VBr_3	290.67	grn-blk, deliq	4.00^{18}	d		s		s al, eth; i HBr
v5[2]	carbide	VC	62.95	blk. cub	5.77	2810	3900	i		s HNO_3, fus KNO_3; i HCl, H_2SO_4
v6	chloride, di-	VCl_2	121.85	grn, hex, deliq	3.23^{18}			s d	s d	s al, eth
v7	chloride, tetra-	VCl_4	192.75	red-br liq	1.816^{30}	−28 ± 2	148.5^{755}	s d		s abs al, eth, chl, acet a
v8	chloride, tri-	VCl_3	157.30	pink cr, deliq	3.00^{18}	d		s d	s d	s abs al, eth
v9	fluoride, penta	VF_5	145.93		2.177^{19}		111.2^{758}			s al
v10	fluoride, tetra-	VF_4	126.94	br yel	2.975^{23}	d 325		s		s acet; sl s al, chl
v11	fluoride, tri-	VF_3	107.94	grn, rhomb	3.363^{19}	>800	subl	i		s al, chl, CS_2.
v12	fluoride, tri-	$VF_3.3H_2O$	161.98	dk gr, rhomb		$-3H_2O$, 100		s	v s d	i abs al
v13	iodide, di-	VI_2	304.71	vlt-rose, hex	5.44	750–800 subl vac		s		i al, CCl_4, CS_2, bz
v14	iodide, tri-	$VI_3.6H_2O$	539.75	gr cr, deliq		d		v s		s al
v15	nitride	VN	64.95	blk, cub	6.13	2320		i		al s aq reg
v16	oxide	VO (or V_2O_2)	66.94	lt gray cr	5.758^{14}	ign		i	i	s a
v17	oxide, di- (or tetr)-	VO_2 (or V_2O_4)	82.94	bl cr	4.339	1967		i	i	s a, alk
v18	oxide, pent-	V_2O_5	181.88	yel-red, rhomb, 1.46, 1.52, 1.76	3.357^{18}	690	d 1750	0.8^{20}		s a, alk; i abs al
v19	oxide, sesqui	Vanadium trioxide. V_2O_3	149.88	blk cr	4.87^{18}_{4}	1970		al s	s	s HNO_3, HF, alk
v20	oxybromide	VOBr	146.85	vlt, oct	4.00^{18}	d 480		v sl s		s acet, anhydr eth, acet
v21	oxy di-bromide	$VOBr_2$	226.76	br powd, deliq		d 180		s		
v22	oxy*tri*bromide	$VOBr_3$	306.67	red liq	$2.933^{14.5}$	d 180	130^{100}	s		
v23	oxychloride	VOCl	102.39	yel brn powd	2.824, 3.64^{20}		127	i		v s HNO_3
v24	oxy*di*chloride	$VOCl_2$	137.85	grn, deliq	2.88^{13}			d		s dil HNO_3
v25	oxy*tri*chloride	$VOCl_3$	173.30	yel liq	1.829	−77 ± 2	126.7	s d		s al, eth, ac a
v26	oxy*di*fluoride	VOF_2	104.94	yel	3.396^{19}	d				al s acet
v27	oxy*tri*fluoride	VOF_3	123.94	yel-wh, hygr	2.459^{19}	300	480			
v28	silicide, di-	VSi_2	107.11	met pr	4.42			i	i	s HF; i al, eth, a
v29	(*di*-)silicide	V_2Si	129.97	silv wh pr	5.48^{17}			i	i	s HF; i al, eth, a
v30	sulfate (hypovanadous)	$VSO_4.7H_2O$	273.11	vlt, monocl		d in air				
v31	sulfide, mono- or (di-)	VS (or V_2S_2)	83.01	blk pl (exist ?)	4.20	d				s hot H_2SO_4, HNO_3; sl s KSH; i HCl, alk
v32	sulfide, penta-	V_2S_5	262.20	blk-grn powd	3.0	d		i		s HNO_3, alk sulf, alk
v33	sulfide, sesqui- or (tri-)	V_2S_3	198.08	grn-blk pl, or powd	4.72^{21}	d>600		i		s alk sulf; sl s, alk, HCl, HNO_3, H_2SO_4
v34	**Vanadyl sulfate**	$VOSO_4$	163.00	bl				v s		
w1	**Water**	H_2O	18.01534	col liq, or col hex cr	liq 1.000^{4} sld 0.9168^{0}	0.00	100.00			s al
w2	**Water heavy**	Deuterium oxide. D_2O	20.03	col liq or hex cr, 1.33844^{20}	1.105^{20}_{4}	3.82	101.42	∞	∞	∞ al; sl s eth
w3	**Wolfram**	See tungsten								
x1	**Xenon**	Xe	131.30	col inert gas	gas 5.887 g/l ± 0.009 liq 3.52^{-109} solid 2.7^{-140}	−111.9	−107.1 ± 3	24.1^{0} cm^3, 11.9^{25} cm^3	8.4^{50}, 7.12^{60}	
y1	**Ytterbium**	Yb	173.04	cub	6.9654 up to 789 6.54 above 789	819 ± 5	1194	i		s a
y2	(III) acetate	$Yb(C_2H_3O_2)_3.4H_2O$	422.24	hex pl	2.09	$-4H_2O$, 100		v s	v s	
y3	(II) bromide	$YbBr_2$	332.86		5.91^{25}_{4}	677	1800	s	s	s dil a

No.	Name	Synonyms and Formulae	Mol. wt.	Crystalline form, properties and index of refraction	Density or spec. gravity	Melting point, °C	Boiling point, °C	Solubility, in grams per 100 cc: Cold water	Hot water	Other solvents
	Ytterbium									
y4	(III) bromide	$YbBr_3$	412.77	col cr		956	d	s	s	
y5	(II) chloride	$YbCl_2$	243.95	grn-yel cr	5.08	702	1900	s	s	s dil a
y6	(III) chloride	$YbCl_3.6H_2O$	387.49	grn, rhomb cr, deliq	2.575	865 $-6H_2O$, 180		v s	v s	s abs al
y7	(II) fluoride	YbF_2	211.04			1052	2380	i	i	
y8	fluoride	YbF_3	230.04			1157	2200	i	i	i dil a
y9	(II) iodide	YbI_2	426.85	lt yel, hex cr	5.40^{25}_{4}	780 ± 4	1300 d(700) vac	s	s	s dil a
y10	(III) iodide	YbI_3	553.75	gold yel cr		d 700	d	s	s	s dil a
y11	(III) oxalate	$Yb_2(C_2O_4)_3.10H_2O$	790.29	col cr	2.644			0.00033^{25}		sl s dil a
y12	(III) oxide	Ytterbia. Yb_2O_3	394.08	col	9.17			i	i	s h dil a
y13	(III) selenate	$Yb_2(SeO_4)_3.8H_2O$	919.08	hex pl	3.30			s d	s	
y14	(III) selenite	$Yb_2(SeO_3)_3$	726.95					i		
y15	(III) sulfate	$Yb_2(SO_4)_3$	634.26	col cr	3.793	d 900		44.2^{0}	4.7^{100}	
y16	(III) sulfate, octohydrate	$Yb_2(SO_4)_3.8H_2O$	778.39	prism	3.286			35.9^{25}	21.1^{40}	
y17	**Yttrium**	Y	88.905	gray-blk met, hex	4.4689	1522	3338	sl d	d	v s dil a; s h KOH
y18	acetate	$Y(C_2H_3O_2)_3.4H_2O$	338.10	col, tricl					9.03^{25}	
y19	bromate	$Y(BrO_3)_3.9H_2O$	634.76	hex pr		74	$-6H_2O$, 100	168^{25}		sl s al; i eth
y20	bromide	YBr_3	328.63	deliq		904		v s		s al; i eth
y21	bromide hydrate	$YBr_3.9H_2O$	490.77	col tabl, deliq				v s		sl s al; i eth
y22	carbide	YC_2	112.93	yel., microcr	4.13^{18}			d		
y23	carbonate	$Y_2(CO_3)_3.3H_2O$	411.88	wh-redsh powd						s dil min a, $(NH_4)_2CO_3$; sl s aq CO_2; i al, eth
y24	chloride	YCl_3	195.26	shiny wh leaf	2.67	721	1507	78^{10}	82^{50}	60.1^{15} al; 60.6^{15} pyr
y25	chloride, hexahydrate	$YCl_3.6H_2O$	303.36	redsh-wh, rhomb, deliq	2.18^{18}	$-5H_2O$, 100		217^{20}	235^{80}	s al; i eth
y26	chloride, monohydrate	$YCl_3.H_2O$	213.28	col cr		$-H_2O$, 160		v s		
y27	fluoride	YF_3	145.90	gelat	4.01	1387		i		v sl s dil a
y28	hydroxide	$Y(OH)_3$	139.93	wh-yel gelat or powd		d		i	i	s a, NH_4Cl; i alk
y29	iodide	YI_3	469.62	wh, cr, deliq		1004	$650–700^{0.02}$	v s		s al, acet; sl s eth
y30	molybdate	$Y_2(MoO_4)_3.4H_2O$	729.68	grayish or yelsh, tetr pl, 2.03	4.79^{16}_{16}	1347				
y31	nitrate, hexahydrate	$Y(NO_3)_3.6H_2O$	383.01	col, redsh cr, deliq	2.68	$-3H_2O$, 100		$134.7^{22.5}$		v s al, eth, HNO_3
y32	nitrate, tetrahydrate	$Y(NO_3)_3.4H_2O$	346.98	redsh-wh pr	2.682			s		s al, HNO_3
y33	oxalate	$Y_2(C_2O_4)_3.9H_2O$	604.01	wh cr powd		d		0.0001		sl s HCl
y34	oxide	Yttria. Y_2O_3	225.81	col-yelsh, cub or powd	5.01	2410		0.00018^{29}		s a; i alk
y35	sulfate	$Y_2(SO_4)_3$	465.99	wh powd	2.52	d 1000		5.38^{25}	s	s sat K_2SO_4 sol
y36	sulfate, octahydrate	$Y_2(SO_4)_3.8H_2O$	610.12	col-redsh, monocl, 1.543, 1.549, 1.576	2.558	$-8H_2O$, 120	d 700	7.47^{16} (anhydr)	1.99^{95} (anhydr)	i al, alk; s conc H_2SO_4
y37	sulfide	Y_2S_3	273.99	yel-gr powd						d a
y38	**Yttrium hexaantipyrine** *per***chlorate**	$[Y(C_{11}H_{12}N_2O)_6](ClO_4)_3$	1516.60	col, hex cr		d 293–296		0.55^{20}		
y39	hexaantipyrine iodide	$[Y(C_{11}H_{12}N_2O)_6]I_3$	1598.96	col cr		280–282		4.65^{20}		
z1	**Zinc**	Zn	65.38	bluish-wh met, hex	7.14	419.58	907	i	i	s a, alk, ac a
z2	acetate	$Zn(C_2H_3O_2)_2$	183.46	col, monocl	1.84	d 200	subl vac	30^{20}	44.6^{100}	2.8^{25} al; 166.7^{79} al
z3	acetate, dihydrate	$Zn(C_2H_3O_2)_2.2H_2O$	219.49	col, monocl, β 1.494	1.735	237	$-2H_2O$, 100	31.1^{20}	66.6^{100}	2 al
z4	acetylacetonate	$Zn(C_5H_7O_2)_2$	263.59	need		138	subl	v s d		v s bz, acet; s al
z5	aluminate	Nat. gahnite. $ZnAl_2O_4$	183.33	cub, grn 1.78	4.58			i	i	i a; sl s alk
z6	amide	$Zn(NH_2)_2$	97.42	wh powd, amorph	2.13^{25}	d 200 vac		d	d	i al, eth
z7	antimonide	Zn_3Sb_2	439.61	silv wh, rhomb pr	6.33	570		d		
z8	*ortho*arsenate	Nat. koettigite. $Zn_3(AsO_4)_2.8H_2O$	618.08	monocl, 1.662, 1.683, 1.717	3.309^{15}	$-1H_2O$, 100		i	i	s HNO_3, H_3PO_4, alk
z9	*ortho*arsenate, basic	Nat. adamite. $Zn_3(AsO_4)_2.Zn(OH)_2$	573.34	col, rhomb	4.475^{15}	d 250				
z10	*ortho*arsenate, hydrogen	$ZnHAsO_4.4H_2O$	277.36	wh, rhomb		$-H_2O$, 327		d	d	
z11	arsenide	Zn_3As_2	345.95	met-gray, tetr	5.528	1015		i		d a
z12	benzoate	$Zn(C_7H_5O_2)_2$	307.60	wh powd				2.46^{20}	1.44^{20}	
z13	borate	$3ZnO.2B_2O_3$	383.35	wh tricl cr, or amorph powd	cr 4.22 powd 3.64	980		s		cr i HCl; amorph; s HCl
z14	bromate	$Zn(BrO_3)_2.6H_2O$	429.28	wh, cub, 1.5452	2.566	100	$-6H_2O$, 200	v s	∞	
z15	bromide	$ZnBr_2$	225.19	col, rhomb, hygr n_D^{18} 1.5452	4.201^{25}_{4}	394	650	447^{20}	675^{100}	v s al, eth, acet; s NH_4OH

No.	Name	Synonyms and Formulae	Mol. wt.	Crystalline form, properties and index of refraction	Density or spec. gravity	Melting point, °C	Boiling point, °C	Solubility, in grams per 100 cc: Cold water*	Hot water	Other solvents
	Zinc									
z16	butyrate	$Zn(C_4H_7O_2)_2.2H_2O$	275.60	wh pr				10.7^{16}	d	
z17	caproate	$Zn(C_6H_{11}O_2)_2$	295.68					$1.03^{24.5}$		
z18	carbonate	Nat. smithsonite. $ZnCO_3$	125.39	col, trig, 1.818, 1.618	4.398	$-CO_2$, 300		0.001^{15}		s a, alk, NH_4 salts; i NH_3, acet, pyr
z19	chlorate	$Zn(ClO_3)_2.4H_2O$	304.33	col yelsh, cub, deliq	2.15	d 60	d	262^{20}	v s	167 al; s acet, eth, glyc
z20	chlorate, per-	$Zn(ClO_4)_2.6H_2O$	372.36	wh, rhomb, deliq, 1.508, 1.480	2.252 ± 0.01	105–107	d 200	s		s al
z21	chloride	$ZnCl_2$	136.28	wh, hex, deliq, 1.681, 1.713	2.91^{25}	283	732	432^{25}	615^{100}	$100^{12.5}$ al; v s eth; i NH_3
z22	chloroplatinate	$ZnPtCl_6.6H_2O$	581.27	yel, trig, hygr	2.717^{12}	d 160		v s	v s	v s al; d H_2SO_4
z23	chromate	$ZnCrO_4$	181.36	lem-yel pr	3.40			i	d	s a, liq NH_3; i acet
z24	chromate	$ZnCr_2O_4$	233.36	dk grn to black, cub	5.30^{15}					
z25	*di*chromate	$ZnCr_2O_7.3H_2O$	335.40	redsh-brn cr, or or-yel powd, hygr				v s	d	i al, eth; s a
z26	citrate	$Zn_3(C_6H_5O_7)_2.2H_2O$	610.35					sl s		
z27	cyanide	$Zn(CN)_2$	117.41	col, rhomb	1.852	d 800		0.0005^{20}		s alk, KCN, NH_3; i al
z28	ferrate (III)	Ferrite. $ZnFe_2O_4$	241.06	blk, oct	5.33^{20}	1590				s conc HCl; i dil a, alk
z29	ferrocyanide	$Zn_2Fe(CN)_6$	342.69	wh powd	1.85^{25}_{4}			i		s excess alk; i dil a
z30	ferrocyanide, trihydrate	$Zn_2Fe(CN)_6.3H_2O$	396.74	wh powd		d		i	i	i al, HCl; d NaOH; s NH_4OH; v sl s NH_3
z31	fluoride	ZnF_2	103.37	col, monocl or tricl	4.95^{25}_{4}	872	*ca* 1500	1.62^{20}	s	s hot a, NH_4OH; i al, NH_3
z32	fluoride, tetrahydrate	$ZnF_2.4H_2O$	175.43	col, rhomb	2.255	$-4H_2O$, 100	tr to ZNO, 3000	1.6^{18}	s	s a, alk, NH_4OH
z33	fluosilicate	$ZnSiF_6.6H_2O$	315.54	col, hex pr, 1.3824, 1.3956	2.104	d 100		v s		
z34	formaldehyde-sulfoxylate	$Zn(HSO_2.CH_2O)_2$	255.56	rhomb pr		d		v s	v s	d a; i al
z35	formaldehyde-sulfoxylate, basic	$Zn(OH)HSO_2.CH_2O$	177.47	rhomb pr		d		i	i	d a; i al
z36	formate	$Zn(CHO_2)_2$	155.41	col, cr	2.368	d		3.80	62^{100}	
z37	formate	$Zn(CHO)_2.2H_2O$	191.44	wh, monocl, 1.513, 1.526, 1.566	2.207^{20}	$-2H_2O$, 140	d	5.2^{20}	38^{100}	i al
z38	gallate	$ZnGa_2O_4$	268.81	wh fine cr, 1.74	6.15 calc	<800		i	i	i org solv; s dil a, NH_4OH
z39	glycerophosphate	$ZnC_3H_7O_6P$	235.43	wh amorph powd				s		i al, eth
z40	hydroxide(ε)	$Zn(OH)_2$	99.38	col, rhomb	3.053	d 125		v sl s		s a, alk
z41	iodate	$Zn(IO_3)_2$	415.18	wh, need	5.0632^{5}	d		0.87	1.31	s alk, HNO_3
z42	iodate, dihydrate	$Zn(IO_3)_2.2H_2O$	451.21	wh, cr powd	4.223^{25}_{4}	$-H_2O$, 200		0.877	1.32	s HNO_3, NH_4OH
z43	iodide	ZnI_2	319.18	col, hexag	4.7364^{25}_{4}	446	d 624	432^{18}	511^{100}	s a, al, eth, NH_3, $(NH_4)_2CO_3$
z44	*d*-lactate	$Zn(C_3H_5O_3)_2.2H_2O$	279.45					5.7^{15}	9^{33}	0.104 h 98 % al
z45	*dl*-lactate	$Zn(C_3H_5O_3)_2.3H_2O$	297.47	wh, rhomb cr				1.67^{105}	16.7^{100}	v sl s al
z46	laurate	$Zn(C_{12}H_{23}O_2)_2$	464.00	wh powd		128		0.01^{15}	0.019^{100}	0.010^{15} al
z47	*per*manganate	$Zn(MnO_4)_2.6H_2O$	411.33	vlt-br or bl, deliq.	2.47	$-5H_2O$, 100		33.3	v s	d al, a
z48	nitrate, trihydrate	$Zn(NO_3)_2.3H_2O$	243.43	col, need		45.5		327.3^{40}		
z49	nitrate, hexahydrate	$Zn(NO_3)_2.6H_2O$	297.47	col, tetrag	2.065^{14}	36.4	$-6H_2O$, 105–131	184.3^{20}	∞	v s al
z50	nitride	Zn_3N_2	224.12	gray	6.22^{25}_{4}			d		s HCl
z51	oleate	$Zn(C_{18}H_{33}O_2)_2$	628.30	wax-like solid		70		i		s al, eth, bz, CS_2; sl s acet
z52	oxalate	$ZnC_2O_4.2H_2O$	189.42	wh powd	3.28^{25}_{4}	d 100		0.00079^{18}		s a, alk
z53	oxide	Nat. zincite. ZnO	81.37	wh, hex, 2.008, 2.029	5.606	1975		0.00016^{29}		s a, alk, NH_4Cl; i al, NH_3
z54	oxide, per-	$ZnO_2.\frac{1}{2}H_2O$	106.38	yelsh, powd	3.00 ± 0.08	$-O_2$, vac		sl d	d	d al, eth, acet
z55	1-phenol-4-sulfonate(p)	$Zn(C_6H_5SO_4)_2.8H_2O$	555.83	col cr or fine wh powd, effl		$-8H_2O$, 125		62.5	250^{100}	55.6^{25} al
z56	*ortho*phosphate	$Zn_3(PO_4)_2$	386.05	col, rhomb	3.998^{15}	900		i	i	s a, NH_4OH; i al
z57	*ortho*phosphate, dihydrogen	$Zn(H_2PO_4)_2.2H_2O$	295.38	tricl		d 100		d		
z58	*ortho*phosphate, octahydrate	$Zn_3(PO_4)_2.8H_2O$	530.18	rhomb pl	3.109^{15}			i		s alk

PHYSICAL CONSTANTS OF INORGANIC COMPOUNDS (Continued)

No.	Name	Synonyms and Formulae	Mol. wt.	Crystalline form, properties and index of refraction	Density or spec. gravity	Melting point, °C	Boiling point, °C	Solubility, in grams per 100 cc		
								Cold water	Hot water	Other solvents
	Zinc									
z59	*ortho*phosphate, tetrahydrate	α—Hopeite. $Zn_3(PO_4)_2.4H_2O$	458.11	col, rhomb, 1.572, 1.591, 1.59	3.04	tr >105		i	i	v s a, NH_4OH, NH_4 salts
z60	*ortho*phosphate tetrahydrate	β-Hopeite. $Zn_3(PO_4)_2.4H_2O$	458.11	col, rhomb, 1.574, 1.582, 1.582	3.03	tr >140		i	i	v s a, NH_4OH, NH_4 salts
z61	*ortho*phosphate tetrahydrate	Parahopeite. $Zn_3(PO_4)_2.4H_2O$	458.11	col, tricl, 1.614, 1.625, 1.665	3.75	tr >163		i	i	v s a, NH_4OH, NH_4 salts
z62	*pyro*phosphate	$Zn_2P_2O_7$	304.68	wh powd	3.75^{23}			i	i	s a, alk, NH_4OH
z63	phosphide	Zn_3P_2	258.06	dk gray, tetrag, pois	4.55^{13}	>420	1100; subl in H_2	d		d H_2SO_4 ev H_3P s HNO_3; s (viol) dil a; i al
z64	hypophosphite	$Zn(H_2PO_2)_2.H_2O$	213.36	col, cr powd, hygr				s		s alk
z65	picrate	$Zn(C_6H_2N_3O_7)_2.8H_2O$	665.69	yel cr powd, expl		expl		s		
z66	salicylate	$Zn(C_7H_5O_3)_2.3H_2O$	393.65	need				5^{20}		s al
z67	selenate	$ZnSeO_4.5H_2O$	298.40	wh, tricl	2.591_4^{20}	d >50		s		
z68	selenide	ZnSe	144.33	yelsh to redsh, cub, 2.89	5.42_4^{15}	>1100		i		s a; d HNO_3
z69	silicate	Nat. hemimorphite. $2ZnO.SiO_2.H_2O$	240.84	rhomb, or trigon 1.614, 1.617, 1.636	3.45			i	i	
z70	*meta*silicate	$ZnSiO_3$	141.45	col, rhomb	3.42	1437		i		i a
z71	*ortho*silicate	Nat. willemite. Zn_2SiO_4	222.82	trig, 1.694, 1.723	4.103	1509		i	i	s acet a
z72	stearate	$Zn(C_{18}H_{35}O_2)_2$	632.33	light powd		130		i		i al, eth
z73	sulfate	Nat. zinkosite. $ZnSO_4$	161.43	col, rhomb, 1.658, 1.669, 1.670	3.54_4^{25}	d 600		s	s	sl s al; s MeOH, glyc
z74	sulfate, heptahydrate	Nat. goslarite. $ZnSO_4.7H_2O$	287.54	col, rhomb, effl, 1.457, 1.480, 1.484	1.957_4^{25}	100	$-7H_2O$, 280	96.5^{20}	663.6^{100}	sl s al, glyc
z75	sulfate, hexahydrate	$ZnSO_4.6H_2O$	269.52	col, monocl or tetrag	2.072^{15}	$-5H_2O$, 70		s	117.5^{40}	
z76	sulfide,(α)	Nat. wurtzite. ZnS	97.43	col, hex, 2.356, 2.378	3.98	1700 ± 20 $^{50\ atm}$	1185	0.00069^{18}		v s a; i ac a
z77	sulfide,(β)	Nat. sphalerite. ZnS	97.43	col, cub, 2.368	4.102^{25}	tr 1020		0.000065^{18}		v s a
z78	sulfide, monohydrate	$ZnS.H_2O$	115.45	yelsh-wh powd	3.98	1049		i		s a
z79	sulfite	$ZnSO_3.2H_2O$	181.46	wh, cr powd		$-2H_2O$, 100	d 200	0.16	d	i al; s H_2SO_3
z80	tartrate	$ZnC_4H_4O_6.H_2O$ (or $2H_2O$)	231.46	wh powd				0.055^{20}		s KOH, NaOH
z81	tellurate	Zn_3TeO_6	419.71	wh, gran ppt				i	i	s a
z82	telluride	ZnTe	192.97	red, cub, 3.56	6.34^{15}	1238.5		d		s d a
z83	thiocyanate	$Zn(SCN)_2$	181.53	wh powd, deliq				s		s al, NH_4OH
z84	valerate	$Zn(C_5H_9O_2)_2.2H_2O$	303.65	wh glist sc or powd				2.6^{24-25}	s	*ca* 2.5 al; v sl s eth
	Zinc complexes									
z85	diamminezinc chloride	$[Zn(NH_3)_2]Cl_2$	170.34	col, rhomb, 1.625, 1.590	2.10	210.8	d 271	d		
z86	tetrammine perrhenate	$[Zn(NH_3)_4](ReO_4)_2$	633.89	wh, cub cr	3.608_4^{25}					0.1852 conc NH_4OH
z87	tetrapyridine fluosilicate	$[Zn(C_5H_5N)_4]SiF_6$	523.86	wh, rhomb	2.197					
z88	**Zirconium**	Zr	91.22	silver gray, met	6.49	1852 ± 2	4377	i	i	s HF, aq reg; sl s a
z89	bromide, di-	$ZrBr_2$	112.84	blk powd, ign in air		d >350		d ev H_2		
z90	boride, di-	ZrB_2	251.04	hex	6.085	*ca* 3000				
z93	bromide, tetra-	$ZrBr_4$	410.86	wh cr powd, deliq		450 ± $1^{15 atm}$	357 subl	i d		s liq NH_3, acetone; i bz, CCl_4
z94	bromide, tri-	$ZrBr_3$	330.95	bl-blk powd		d 350		d ev H_2		
z95	carbide	ZrC	103.23	gray met, cub	6.73	3540	5100	i		sl s conc H_2SO_4
z96	carbonate, basic	$3ZrO_2.CO_2.H_2O$	431.68	wh, amorph powd				i		s a
z97	chloride, di-	$ZrCl_2$	162.13	blk	3.6^{16}	d 350		d ev H_2		
z98	chloride, tetra-	$ZrCl_4$	233.03	wh cr	2.803^{15}	$437^{25 atm}$	subl 331	s	d	s al, eth, conc HCl
z99	chloride, tri-	$ZrCl_3$	197.58	br cr	3.00^{16}	d 350		d ev H_2		s $-H_2$ conc al; i org cpd
z100	fluoride	ZrF_4	167.21	wh hex, 1.59	4.43	subl ~ 600		1.388^{25}	d	sl s HF
z101	hydride	ZrH_2	93.24	gray-blk powd						s dif HF, conc a
z102	hydroxide	$Zr(OH)_4$	159.25	wh amorp powd	3.25	$-2H_2O$, 500		0.02	i	s min a
z103	iodide	ZrI_4	598.84	wh need, hygr		499 ± 2 6.3^{atm}	d ~600	s d	s	d al; s eth; v sl s CS_2, bz; i liq NH_3
z104	nitrate	$Zr(NO_3)_4.5H_2O$	429.32	col cr, deliq, 1.60, 1.61				v s		s al

No.	Name	Synonyms and Formulae	Mol. wt.	Crystalline form, properties and index of refraction	Density or spec. gravity	Melting point, °C	Boiling point, °C	Solubility, in grams per 100 cc: Cold water	Hot water	Other solvents
	Zirconium									
z105	nitride	ZrN	105.23	yel-brn cr	7.09	2980 ± 50		i	i	sl s inorg ac; s conc H_2SO_4, HF, aq reg
z106	oxide	Nat. baddeleyite. ZrO_2	123.22	col-yel-brn, monocl, 2.13, 2.19, 2.20	5.89	*ca* 2700	*ca* 5000	i	i	s H_2SO_4, HF
z107	oxide	Zirconia. ZrO_2 $HfO_2 < 2\%$	123.22	wh, monocl below 1000°, cub, above	5.6	2715		i	i	s H_2SO_4, HF
z108	oxide	Zirconium hydroxide, zirconic acid. $ZrO_2.xH_2O$		gel or wh amorph powd	3.25	$-2H_2O$, 550		0.02		s acids; i al, alk
z109	phosphide	ZrP_2	153.17	gray, brittle	4.77^{25}_{4}			i		v s conc hot H_2SO_4
z110	selenate	$Zr(SeO_4)_2.4H_2O$	449.20	hex trsp cr				s		sl s al, conc a
z111	selenite	$Zr(SeO_3)_2$	345.14	wh sm cr	4.3	d ~ 400		i	i	sl s H_2SO_4
z112	*ortho*silicate	Zircon, hyacinth. $ZrSiO_4$	183.30	tetr, var colors, 1.92–96; 1.97–2.02	4.56	2550		i	i	i a, aq reg, alk
z113	silicide	$ZrSi_2$	147.39	steel gray rhomb, lust met	4.88^{22}			i	i	s HF; i inorg a, aq reg
z114	sulfate	$Zr(SO_4)_2$	283.35	microcr powd, hygr	3.22^{16}	410 d		s	v s	
z115	sulfate	$Zr(SO_4)_2.4H_2O$	355.41	wh cr powd, rhomb	3.22^{16}	$-3H_2O$, 135–150		v s		i al
z116	sulfide	ZrS_2	155.35	steelgray cr, hexag	3.87	~ 1550		i	i	i a
z117	**Zirconyl bromide**	$ZrOBr_2.xH_2O$		brill need, deliq		$-H_2O$, 120		s		s hot conc HBr
z118	chloride	$ZrOCl_2.8H_2O$	322.25	wh, need, tetr, effl, 1.552, 1.563		$-6H_2O$, 150	$-8H_2O$, 210	s	d	s al, eth; sl s HCl
z119	iodide	$ZrOI_2.8H_2O$	505.15	col, need, hygr		d		v s	v s	s al; v s eth
z120	sulfide	Zirconium sulfoxide. ZrOS	139.28	yel powd	4.87	ign in air		i	i	

GRAVIMETRIC FACTORS AND THEIR LOGARITHMS

Rudolf Loebel

Compiled from International atomic weights of 1964. To facilitate use of this table the group of substances weighed under each element as well as the substance sought under each substance weighed are arranged in alphabetical order of their formulas.

Weighed	Sought	Factor	Log of Factor +10	Reciprocal of Factor	Log of Reciprocal of Factor +10
Aluminum					
Al	Al_2O_3	1.88946	10.27634	0.52925	9.72366
	$AlPO_4$	4.51987	10.65513	0.22125	9.34488
Al_4C_3	Al_2O_3	1.41653	10.15122	0.70595	9.84877
$Al(C_9H_6ON)_3$	Al	0.05873	8.76886	17.02811	11.23114
	Al_2O_3	0.11096	9.04516	9.01226	10.95483
$AlCl_3$	Al_2O_3	0.38233	9.58244	2.61554	10.41756
AlF_3	CaF_2	1.39464	10.14446	0.71703	9.85554
Al_2O_3	Al	0.52925	9.72366	1.88946	10.27634
	Al_4C_3	0.70595	9.84877	1.41653	10.15122
	$AlCl_3$	2.61552	10.41756	0.38233	9.58244
	$AlPO_4$	2.39214	10.37879	0.41804	9.62122
	$Al_2(SO_4)_3$	3.35567	10.52578	0.29800	9.47422
	$Al_2(SO_4)_3.18H_2O$	6.53605	10.81531	0.15300	9.18469
	$K_2SO_4.Al_2(SO_4)_3.24H_2O$	9.30532	10.96873	0.10747	9.03129
	$(NH_4)_2SO_4.Al_2(SO_4)_3.24H_2O$	8.89216	10.94901	0.11246	9.05100
$AlPO_4$	Al	0.22125	9.34488	4.51977	10.65512
	Al_2O_3	0.41804	9.62122	2.39211	10.37878
	P_2O_5	0.58196	9.76489	1.71833	10.23511
$Al_2(SO_4)_3$	Al_2O_3	0.29800	9.47422	3.35570	10.52578
$Al_2(SO_4)_3.18H_2O$	Al_2O_3	0.15300	9.18469	6.53595	10.81531
CaF_2	AlF_3	0.71703	9.85554	1.39464	10.14446
$K_2SO_4.Al_2(SO_4)_3.24H_2O$	Al_2O_3	0.10747	9.03129	9.30492	10.96872
$(NH_4)_2SO_4.Al_2(SO_4)_3.24H_2O$	Al_2O_3	0.11246	9.05100	8.89205	10.94900
P_2O_5	$AlPO_4$	1.71831	10.23510	0.58196	9.76489
Ammonium					
Ag	NH_4Br	0.90802	9.95810	1.10130	10.04191
	NH_4Cl	0.49589	9.69538	2.01657	10.30462
	NH_4I	1.34366	10.12829	0.74424	9.87171
AgBr	NH_4Br	0.52161	9.71735	1.91714	10.28265
	NH_4Cl	0.37323	9.57198	2.67931	10.42802
AgI	NH_4I	0.61737	9.79055	1.61977	10.20945
$BaSO_4$	$(NH_4)_2SO_4$	0.56615	9.75293	1.76632	10.24707
Br	NH_4Br	1.22574	10.08840	0.81583	9.91160
Cl	NH_4	0.50880	9.70655	1.96541	10.29345
	NH_4Cl	1.50881	10.17864	0.66277	9.82136
HCl	NH_4Cl	1.46710	10.16646	0.68162	9.83354
I	NH_4I	1.14214	10.05772	0.87555	9.94229
$MgNH_4PO_4.6H_2O$	NH_3	0.06941	8.84142	14.40714	11.15857
	NH_4	0.07352	8.86641	13.60174	11.13360
	$(NH_4)_2O$	0.10613	9.02584	9.42241	10.97416
N	NH_3	1.21589	10.08489	0.82244	9.91510
	NH_4	1.28785	10.10987	0.77649	9.89014
	NH_4Cl	3.81903	10.58195	0.26184	9.41804
	NH_4NO_3	5.71466	10.75699	0.17499	9.24302
	$(NH_4)_2O$	1.85899	10.26928	0.53793	9.73072
	$(NH_4)_2SO_4$	4.71699	10.67367	0.21200	9.32634
NH_3	$MgNH_4PO_4.6H_2O$	14.40648	11.15855	0.06941	8.84142
	N	0.82244	9.91510	1.21589	10.08489
	NH_4	1.05919	10.02498	0.94412	9.97503
	NH_4Cl	3.14093	10.49706	0.31837	9.50293
	$(NH_4)_2CO_3$	2.82099	10.45040	0.35448	9.54959
	NH_4HCO_3	4.64199	10.66671	0.21542	9.33329
	NH_4NO_3	4.69998	10.67210	0.21277	9.32791
	$(NH_4)_2O$	1.52891	10.18438	0.65406	9.81562
	NH_4OH	2.05783	10.31341	0.48595	9.68659
	$(NH_4)_2PtCl_6$	13.03213	11.11501	0.07673	8.88497
	$(NH_4)_2SO_4$	3.87945	10.58877	0.25777	9.41123
	N_2O_5	3.17106	10.50121	0.31535	9.49879
	Pt	5.72763	10.75797	0.17459	9.24202
	SO_3	2.35053	10.37117	0.42544	9.62884
NH_4	Cl	1.96540	10.29345	0.50880	9.70655
	$MgNH_4PO_4.6H_2O$	13.60144	11.13359	0.07352	8.86641
	N	0.77648	9.89013	1.28786	10.10986
	NH_3	0.94412	9.97503	1.05919	10.02498

Weighed	Sought	Factor	Log of Factor +10	Reciprocal of Factor	Log of Reciprocal of Factor +10
Ammonium (contd.)					
NH_3	NH_4Cl	2.96542	10.47208	0.33722	9.52792
	$(NH_4)_2PtCl_6$	12.30389	11.09005	0.08128	8.90998
	Pt	5.40757	10.73301	0.18493	9.26701
NH_4Br	Ag	1.10130	10.04191	0.90802	9.95810
	AgBr	1.91713	10.28265	0.52161	9.71735
	Br	0.81583	9.91160	1.22575	10.08840
NH_4Cl	Ag	2.01656	10.30461	0.49589	9.69538
	AgCl	2.67934	10.42802	0.37323	9.57198
	Cl	0.66277	9.82136	1.50882	10.17864
	HCl	0.68162	9.83354	1.46709	10.16646
	N	0.26185	9.41805	3.81898	10.58194
	NH_3	0.31838	9.50294	3.14090	10.49706
	NH_4	0.33722	9.52791	2.96542	10.47208
	$(NH_4)_2O$	0.48677	9.68732	2.05436	10.31268
	NH_4OH	0.65516	9.81635	1.52634	10.18365
	$(NH_4)_2PtCl_6$	4.14913	10.61795	0.24101	9.38204
	Pt	1.82354	10.26091	0.54838	9.73908
$(NH_4)_2CO_3$	NH_3	0.35448	9.54959	2.82103	10.45040
NH_4HCO_3	NH_3	0.21542	9.33329	4.64209	10.66672
NH_4I	Ag	0.74422	9.87170	1.34369	10.12830
	AgI	1.61977	10.20946	0.61737	9.79055
	I	0.87555	9.94228	1.14214	10.05772
NH_4NO_3	NH_3	0.21277	9.32791	4.69991	10.67209
	$(NH_4)_2PtCl_6$	2.77280	10.44292	0.36064	9.55709
	N_2O_5	0.67470	9.82911	1.48214	10.17089
	Pt	1.21865	10.08588	0.82058	9.91412
$(NH_4)_2O$	$MgNH_4PO_4.6H_2O$	9.42281	10.97418	0.10613	9.02584
	N	0.53733	9.73069	1.86105	10.26931
	NH_3	0.65407	9.81562	1.52889	10.18438
	NH_4Cl	2.05437	10.31268	0.48677	9.68732
	$(NH_4)_2PtCl_6$	8.52378	10.93063	0.11732	9.06937
	N_2O_5	2.07406	10.31682	0.48215	9.68319
	Pt	3.74621	10.57359	0.26694	9.42641
NH_4OH	N	0.39967	9.60170	2.50206	10.39830
	NH_3	0.48595	9.68660	2.05782	10.31341
	NH_4	0.51471	9.71156	1.94284	10.28843
	NH_4Cl	1.52633	10.18365	0.65517	9.81635
	$(NH_4)_2PtCl_6$	6.33297	10.80158	0.15790	9.19841
	Pt	2.78334	10.44456	0.35928	9.55544
$(NH_4)_2PtCl_6$	NH_3	0.07674	8.88502	13.03101	11.11497
	NH_4	0.08128	8.90998	12.30315	11.09002
	NH_4Cl	0.24102	9.38206	4.14903	10.61794
	NH_4NO_3	0.36065	9.55709	2.77277	10.44291
	$(NH_4)_2O$	0.11732	9.06937	8.52370	10.93063
	NH_4OH	0.15791	9.19841	6.33272	10.80159
	$(NH_4)_2SO_4$	0.29768	9.47375	3.35931	10.52625
$(NH_4)_2SO_4$	$BaSO_4$	1.76632	10.24709	0.56615	9.75291
	H_2SO_4	0.74223	9.87054	1.34729	10.12947
	N	0.21200	9.32634	4.71698	10.67367
	NH_3	0.25777	9.41123	3.87943	10.58877
	$(NH_4)_2PtCl_6$	3.35927	10.52625	0.29768	9.47375
	$(NH_4)_2SO_4$	1.47640	10.16921	0.67732	9.83079
	SO_3	0.60589	9.78239	1.65046	10.21761
N_2O_5	NH_3	0.31535	9.49879	3.17108	10.50121
	NH_4NO_3	1.48214	10.17089	0.67470	9.82911
	$(NH_4)_2O$	0.48214	9.68318	2.07409	10.31682
Pt	NH_3	0.17459	9.24202	5.72770	10.75798
	NH_4	0.18493	9.26701	5.40745	10.73300
	NH_4Cl	0.54838	9.73908	1.82355	10.26092
	NH_4NO_3	0.82058	9.91412	1.21865	10.08588
	$(NH_4)_2O$	0.26694	9.42641	3.74616	10.57364
	NH_4OH	0.35928	9.55544	2.78334	10.44456
	$(NH_4)_2SO_4$	0.67732	9.83079	1.47641	10.16921
SO_3	NH_3	0.42543	9.62883	2.35056	10.37118
	$(NH_4)_2SO_4$	1.65046	10.21760	0.60589	9.78239
Antimony					
$K(SbO)C_4H_4O_6.\frac{1}{2}H_2O$	Sb	0.36460	9.56182	2.74273	10.43819
	Sb_2O_3	0.43647	9.63995	2.29111	10.36005

Weighed	Sought	Factor	Log of Factor +10	Reciprocal of Factor	Log of Reciprocal of Factor +10
Antimony (contd.)					
$K(SbO)C_4H_4O_6\cdot\frac{1}{2}H_2O$	Sb_2O_4	0.46043	9.66317	2.17188	10.33684
	Sb_2S_3	0.50862	9.70640	1.96610	10.29360
Sb	$K(SbO)C_4H_4O_6\cdot\frac{1}{2}H_2O$	2.74275	10.43819	0.36460	9.56182
	Sb_2O_3	1.19713	10.07814	0.83533	9.92186
	Sb_2O_4	1.26283	10.10137	0.79187	9.89866
	Sb_2O_5	1.32854	10.12340	0.75271	9.87659
	Sb_2S_3	1.39503	10.14458	0.71683	9.85540
	Sb_2S_5	1.65840	10.21969	0.60299	9.78031
Sb_2O_3	$K(SbO)C_4H_4O_6\cdot\frac{1}{2}H_2O$	2.29111	10.36005	0.43647	9.63995
	Sb	0.83533	9.92186	1.19713	10.07814
	Sb_2O_4	1.05489	10.02320	0.94797	9.97680
	Sb_2O_5	1.10978	10.04523	0.90108	9.95476
	Sb_2S_3	1.16532	10.06645	0.85813	9.93356
	Sb_2S_5	1.38532	10.14155	0.72185	9.85845
Sb_2O_4	$K(SbO)C_4H_4O_6\cdot\frac{1}{2}H_2O$	2.17190	10.33684	0.46043	9.66317
	Sb	0.79187	9.89866	1.26283	10.10137
	Sb_2O_3	0.94797	9.97680	1.05489	10.02300
	Sb_2O_5	1.05203	10.02203	0.95054	9.97797
	Sb_2S_3	1.10468	10.04324	0.90523	9.95671
	Sb_2S_5	1.31324	10.11834	0.76148	9.88166
Sb_2O_5	Sb	0.75270	9.87663	1.32853	10.12337
	Sb_2O_3	0.90108	9.95477	1.10977	10.04523
	Sb_2O_4	0.95054	9.97798	1.05202	10.02202
	Sb_2S_5	1.24828	10.09632	0.80109	9.90368
Sb_2S_3	$K(SbO)C_4H_4O_6\cdot\frac{1}{2}H_2O)$	1.96603	10.29360	0.50863	9.70640
	Sb	0.71681	9.85542	1.39503	10.14458
	Sb_2O_3	0.85814	9.93356	1.16532	10.06645
	Sb_2O_4	0.90524	9.95676	1.10469	10.04332
	Sb_2O_5	0.95234	9.97879	1.05006	10.02121
Sb_2S_5	Sb	0.60299	9.78031	1.65840	10.21969
	Sb_2O_3	0.72186	9.85834	1.38570	10.14167
	Sb_2O_4	0.76148	9.88166	1.31323	10.11834
	Sb_2O_5	0.80110	9.90369	1.24828	10.09631
Arsenic					
As	As_2O_3	1.32032	10.12068	0.75739	9.87932
	As_2O_5	1.53387	10.18579	0.65195	9.81422
	As_2S_3	1.64194	10.21535	0.60904	9.78465
	As_2S_5	2.06991	10.31597	0.48311	9.68405
	$BaSO_4$	4.67291	10.66959	0.21400	9.33041
	$Mg_2As_2O_7$	2.07191	10.31637	0.48265	9.68364
	$MgNH_4AsO_4\cdot\frac{1}{2}H_2O$	2.53968	10.40478	0.39375	9.59523
AsO_3	$BaSO_4$	2.84821	10.45457	0.35110	9.54543
	$Mg_2As_2O_7$	1.26286	10.10135	0.79185	9.89865
	$MgNH_4AsO_4\cdot\frac{1}{2}H_2O$	1.54797	10.18976	0.64601	9.81024
AsO_4	$BaSO_4$	2.52018	10.40143	0.39680	9.59857
	$Mg_2As_2O_7$	1.11742	10.04821	0.89492	9.95178
	$MgNH_4AsO_4\cdot\frac{1}{2}H_2O$	1.36969	10.13662	0.73009	9.86337
As_2O_3	As	0.75739	9.87932	1.32032	10.12068
	As_2O_5	1.16173	10.06511	0.86079	9.93490
	As_2S_3	1.24359	10.09468	0.80412	9.90532
	As_2S_5	1.56773	10.19527	0.63781	9.80472
	$BaSO_4$	3.53922	10.54891	0.28255	9.45110
	$Mg_2As_2O_7$	1.56925	10.19569	0.63725	9.80431
	$MgNH_4AsO_4\cdot\frac{1}{2}H_2O$	1.92353	10.28410	0.51988	9.71590
As_2O_5	As	0.65195	9.81422	1.53386	10.18579
	As_2O_3	0.86077	9.93489	1.16175	10.06511
	As_2S_3	1.07046	10.02957	0.93418	9.97043
	As_2S_5	1.34947	10.13016	0.74103	9.86984
	$BaSO_4$	3.04648	10.48380	0.32825	9.51621
	$Mg_2As_2O_7$	1.35077	10.13058	0.74032	9.86942
	$MgNH_4AsO_4\cdot\frac{1}{2}H_2O$	1.65573	10.21899	0.60396	9.78101
As_2S_3	As	0.60903	9.78464	1.64196	10.21536
	As_2O_3	0.80412	9.90532	1.24360	10.09468
	As_2O_5	0.93418	9.97043	1.07046	10.02957
	As_2S_5	1.26065	10.10060	0.79324	9.89939
	$Mg_2As_2O_7$	1.26187	10.10102	0.79247	9.89898
As_2S_5	As	0.48311	9.68405	2.06992	10.31595
	As_2O_3	0.63786	9.80472	1.56774	10.19527

Weighed	Sought	Factor	Log of Factor +10	Reciprocal of Factor	Log of Reciprocal of Factor +10
Arsenic (contd.)					
As_2S_5	As_2O_5	0.74103	9.86984	1.34947	10.13016
	As_2S_3	0.79324	9.89940	1.26065	10.10060
$BaSO_4$	As	0.21399	9.33039	4.67312	10.66961
	AsO_3	0.35110	9.54543	2.84819	10.45457
	AsO_4	0.39679	9.59856	2.52022	10.40143
	As_2O_3	0.28255	9.45110	3.53920	10.54890
	As_2O_5	0.32825	9.51621	3.04646	10.48379
$Ca_2As_2O_7$	As	0.43814	9.64161	2.28238	10.35839
	As_2O_3	0.57849	9.76230	1.72864	10.23770
$Mg_2As_2O_7$	As	0.48264	9.68363	2.07194	10.31638
	AsO_3	0.79184	9.89864	1.26288	10.10136
	AsO_4	0.89491	9.95178	1.11743	10.04822
	As_2O_3	0.63725	9.80431	1.56924	10.19569
	As_2O_5	0.74031	9.86942	1.35079	10.13059
	As_2S_3	0.79248	9.89899	1.26186	10.10102
$MgNH_4AsO_4.\frac{1}{2}H_2O$	As	0.39375	9.59523	2.53968	10.40478
	AsO_3	0.64600	9.81023	1.54799	10.18977
	AsO_4	0.73009	9.86337	1.36969	10.13662
	As_2O_3	0.51988	9.71590	1.92352	10.28410
	As_2O_5	0.60396	9.78101	1.65574	10.21899
Barium					
Ba	$BaCO_3$	1.43694	10.15744	0.69592	9.84256
	$BaCrO_4$	1.84455	10.26589	0.54214	9.73411
	$BaSiF_6$	2.03444	10.30844	0.49154	9.69156
	$BaSO_4$	1.69943	10.23030	0.58843	9.76969
$BaCl_2$	$BaCO_3$	0.94766	9.97665	1.05523	10.02334
	$BaCrO_4$	1.21647	10.08510	0.82205	9.91490
	$BaSO_4$	1.12077	10.04952	0.89224	9.95048
$BaCl_2.2H_2O$	$BaSO_4$	0.95546	9.98021	1.04662	10.01979
$BaCO_3$	Ba	0.69592	9.84256	1.43695	10.15747
	$BaCl_2$	1.05523	10.02334	0.94766	9.97665
	$BaCrO_4$	1.28366	10.10845	0.77902	9.89155
	$Ba(HCO_3)_2$	1.31426	10.11869	0.76088	9.88132
	BaO	0.77700	9.89042	1.28700	10.10958
	$BaSO_4$	1.18267	10.07286	0.84554	9.92713
	CO_2	0.22300	9.34830	4.48430	10.65170
$BaCrO_4$	Ba	0.54214	9.73411	1.84455	10.26589
	$BaCl_2$	0.82205	9.91490	1.21647	10.08510
	$BaCO_3$	0.77902	9.89155	1.28366	10.10846
	BaO	0.60530	9.78197	1.65207	10.21803
BaF_2	$BaSiF_6$	1.59359	10.20238	0.62751	9.79762
$Ba(HCO_3)_2$	$BaCO_3$	0.76088	9.88132	1.31427	10.11869
$Ba(NO_3)_2$	$BaSO_4$	0.89306	9.95088	1.11975	10.04912
BaO	$BaCO_3$	1.28701	10.10958	0.77699	9.89042
	$BaCrO_4$	1.65208	10.21803	0.60530	9.78197
	$BaSiF_6$	1.82233	10.26061	0.54878	9.73934
	$BaSO_4$	1.52211	10.18244	0.65698	9.81756
	CO_2	0.28701	9.45790	3.48420	10.54210
BaO_2	$BaSO_4$	1.37829	10.13934	0.72554	9.86066
BaS	$BaSO_4$	1.37780	10.13919	0.72579	9.86081
$BaSiF_6$	Ba	0.49152	9.69154	2.03451	10.30846
	BaF_2	0.62751	9.79762	1.59360	10.20238
	BaO	0.54878	9.73939	1.82222	10.26060
$BaSO_4$	Ba	0.58843	9.76969	1.69944	10.23030
	$BaCl_2$	0.89225	9.95049	1.12076	10.04952
	$BaCl_2.2H_2O$	1.04662	10.01979	0.95546	9.98021
	$BaCO_3$	0.84554	9.92713	1.18268	10.07286
	$Ba(NO_3)_2$	1.11975	10.04912	0.89306	9.95088
	BaO	0.65698	9.81756	1.52212	10.18244
	BaO_2	0.72554	9.86066	1.37828	10.13934
	BaS	0.72579	9.86081	1.37781	10.13911
CO_2	$BaCO_3$	4.48421	10.65169	0.22300	9.34830
	BaO	3.48421	10.54211	0.28701	9.45790
Beryllium					
Be	BeO	2.77530	10.44331	0.36032	9.55668
$BeCl_2$	BeO	0.31297	9.49551	3.19519	10.50450
BeO	Be	0.36032	9.55668	2.77531	10.44331
	$BeCl_2$	3.19525	10.50451	0.31296	9.49549

Weighed	Sought	Factor	Log of Factor +10	Reciprocal of Factor	Log of Reciprocal of Factor +10
Beryllium (contd.)					
BeO	$BeSO_4.4H_2O$	7.08211	10.85017	0.14120	9.14983
$Be_2P_2O_7$	Be	0.09389	8.97262	10.65076	11.02738
$BeSO_4.4H_2O$	BeO	0.14120	9.14983	7.08215	10.85017
Bismuth					
Bi	$BiAsO_4$	1.66475	10.22135	0.60069	9.77865
	Bi_2O_3	1.11484	10.04721	0.89699	9.95279
	BiOCl	1.24620	10.09559	0.80244	9.90441
	Bi_2S_3	1.23014	10.08996	0.81292	9.91005
$BiAsO_4$	Bi	0.60069	9.77865	1.66475	10.22135
	Bi_2O_3	0.66968	9.82587	1.49325	10.17413
$Bi(NO_3)_3.5H_2O$	Bi_2O_3	0.48030	9.68151	2.08203	10.31849
	BiOCl	0.53689	9.72988	1.86258	10.27011
Bi_2O_3	Bi	0.89699	9.95279	1.11484	10.04722
	$BiAsO_4$	1.49326	10.17414	0.66968	9.82587
	$Bi(NO_3)_3.5H_2O$	2.08204	10.31849	0.48030	9.68151
	BiOCl	1.11783	10.04837	0.89459	9.95163
	$BiONO_3$	1.23180	10.09054	0.81182	9.90946
	Bi_2S_3	1.10343	10.04275	0.90627	9.95726
BiOCl	Bi	0.80244	9.90441	1.24619	10.09559
	$Bi(NO_3)_3.5H_2O$	1.86256	10.27011	0.53690	9.72989
	Bi_2O_3	0.89458	9.95162	1.11784	10.04838
	$BiONO_3$	1.10195	10.04215	0.90748	9.95784
$BiONO_3$	Bi_2O_3	0.81182	9.90946	1.23180	10.09054
	BiOCl	0.90748	9.95784	1.10195	10.04216
$BiPO_4$	Bi	0.68754	9.83730	1.45446	10.16270
	Bi_2O_3	0.76651	9.88452	1.30461	10.11548
Bi_2S_3	Bi	0.81291	9.91005	1.23015	10.08996
	Bi_2O_3	0.90627	9.95726	1.10342	10.04274
Boron					
B	B_2O_3	3.21987	10.50784	0.31057	9.49216
	KBF_4	11.64619	11.06618	0.08586	8.93379
BO_2	B_2O_3	0.81314	9.91016	1.22981	10.08984
BO_3	B_2O_3	0.59192	9.77226	1.68942	10.22774
B_2O_3	B	0.31057	9.49216	3.21990	10.50785
	BO_2	1.22982	10.08985	0.81313	9.91016
	BO_3	1.68943	10.22774	0.59192	9.77226
	B_4O_7	1.11491	10.04724	0.89693	9.95276
	H_3BO_3	1.77630	10.24952	0.56297	9.75049
	KBF_4	3.61678	10.55832	0.27647	9.44165
	$Na_2B_4O_7.10H_2O$	2.73896	10.43758	0.36510	9.56241
B_4O_7	B_2O_3	0.89693	9.95276	1.11491	10.04724
$C_{20}H_{16}N_4.HBF_4$	B	0.02702	8.43169	37.00962	11.56832
	B_2O_3	0.08698	8.93942	11.49690	11.06058
H_3BO_3	B_2O_3	0.56297	9.75049	1.77629	10.24951
	KBF_4	2.03624	10.30883	0.49110	9.69117
KBF_4	B	0.08587	8.93384	11.64551	11.06616
	B_2O_3	0.27648	9.44167	3.61690	10.55834
	H_3BO_3	0.49110	9.69117	2.03625	10.30885
	$Na_2B_4O_7.10H_2O$	0.75725	9.87924	1.32057	10.12076
$Na_2B_4O_7.10H_2O$	B_2O_3	0.36510	9.56241	2.73898	10.43759
	KBF_4	1.32056	10.12075	0.75725	9.87924
Bromine					
Ag	Br	0.74079	9.86969	1.34991	10.13030
	BrO_3	1.18575	10.07399	0.84335	9.92601
	HBr	0.75013	9.87514	1.33310	10.12487
AgBr	Br	0.42555	9.62895	2.34990	10.37105
	BrO_3	0.68116	9.83325	1.46808	10.16675
	HBr	0.43092	9.63440	2.32062	10.36561
AgCl	Br	0.55754	9.74627	1.79359	10.25372
Br	Ag	1.34991	10.13030	0.74079	9.86969
	AgBr	2.34991	10.37105	0.42555	9.62895
	AgCl	1.79358	10.25372	0.55754	9.74627
	O	0.10011	9.00047	9.98901	10.99952
BrO_3	Ag	0.84335	9.92601	1.18575	10.07399
	AgBr	1.46809	10.16676	0.68116	9.83325
O	Br	9.98899	10.99952	0.10011	9.00047
HBr	Ag	1.33309	10.12486	0.75014	9.87514
	AgBr	2.32064	10.36561	0.43092	9.63440

Weighed	Sought	Factor	Log of Factor +10	Reciprocal of Factor	Log of Reciprocal of Factor +10
Cadmium					
Cd	$CdCl_2$	1.63087	10.21242	0.61317	9.78758
	$Cd(NO_3)_2$	2.10329	10.32290	0.47545	9.67711
	CdO	1.14235	10.05780	0.87539	9.94221
	CdS	1.28523	10.10898	0.77807	9.89102
	$CdSO_4$	1.85463	10.26825	0.53919	9.73174
$Cd(C_{13}H_8O_2N)_2$	Cd	0.21095	9.32418	4.74046	10.67582
	CdO	0.24098	9.38198	4.14972	10.61802
$CdCl_2$	Cd	0.61317	9.78758	1.63087	10.21242
	CdO	0.70045	9.84538	1.42765	10.15462
	CdS	0.78807	9.89657	1.26892	10.10343
	$CdSO_4$	1.13720	10.05584	0.87935	9.94417
$Cd(C_9H_6NO)_2$	Cd	0.28050	9.44793	3.56506	10.55207
	CdO	0.32043	9.50573	3.12081	10.49426
CdMoO4	Cd	0.41272	9.61566	2.42295	10.38434
	CdO	0.47147	9.67345	2.12103	10.32655
$Cd(NO_3)_2$	Cd	0.47545	9.67711	2.10327	10.32290
	CdO	0.54312	9.73490	1.84121	10.26510
	CdS	0.61106	9.78608	1.63650	10.21392
	$CdSO_4$	0.88177	9.94536	1.13408	10.05464
CdO	Cd	0.87539	9.94221	1.14235	10.05781
	$CdCl_2$	1.42765	10.15462	0.70045	9.85438
	$Cd(NO_3)_2$	1.84120	10.26510	0.54312	9.73490
	CdS	1.12508	10.05118	0.88883	9.94882
	$CdSO_4$	1.62352	10.21046	0.61595	9.78955
CdS	Cd	0.77807	9.89102	1.28523	10.10898
	$CdCl_2$	1.26893	10.10343	0.78807	9.89657
	$Cd(NO_3)_2$	1.63651	10.21392	0.61106	9.78608
	CdO	0.88883	9.94882	1.12507	10.05118
	$CdSO_4$	1.44303	10.15928	0.69299	9.84072
$CdSO_4$	Cd	0.53919	9.73174	1.85463	10.26825
	$CdCl_2$	0.87935	9.94417	1.13720	10.05584
	$Cd(NO_3)_2$	1.13408	10.05464	0.88177	9.94536
	CdO	0.61595	9.78955	1.62351	10.21046
	CdS	0.69299	9.84072	1.44302	10.15928
Calcium					
$BaSO_4$	CaS	0.30908	9.49007	3.23541	10.50993
	$CaSO_4$	0.58329	9.76588	1.71441	10.23411
	$CaSO_4.2H_2O$	0.73766	9.86786	1.35564	10.13214
Ca	$CaCl_2$	2.76921	10.44235	0.36111	9.55764
	$CaCO_3$	2.49726	10.39746	0.40044	9.60253
	CaF_2	1.94810	10.28961	0.51332	9.71039
	CaO	1.39920	10.14588	0.71469	9.85411
	$CaSO_4$	3.39671	10.53106	0.29440	9.46894
	Cl	1.76911	10.24776	0.56526	9.75225
$Ca_3(AsO_4)_2$	$Mg_2As_2O_7$	0.77990	9.89204	1.28222	10.10798
$Ca_2As_2O_7$	Ca	0.23439	9.36994	4.26639	10.63006
	CaO	0.32795	9.51581	3.04925	10.48420
$CaC_2O_4.H_2O$	Ca	0.27430	9.43823	3.64564	10.56178
	CaO	0.38379	9.58410	2.60550	10.41589
$CaCl_2$	Ca	0.36111	9.55764	2.76924	10.44236
	$CaCO_3$	0.90179	9.95511	1.10891	10.04489
	CaO	0.50527	9.70352	1.97914	10.29648
	$CaSO_4$	1.22660	10.08870	0.81526	9.91130
	Cl	0.63885	9.80540	1.56531	10.19460
$CaCO_3$	Ca	0.40044	9.60253	2.49725	10.39751
	$CaCl_2$	1.10890	10.04489	0.90179	9.95511
	$Ca(HCO_3)_2$	1.61964	10.20942	0.61742	9.79058
	CaO	0.56030	9.74842	1.78476	10.25158
	$CaSO_4$	1.36018	10.13360	0.73520	9.86641
	$CaSO_4.2H_2O$	1.72015	10.23557	0.58134	9.76442
	CO_2	0.43970	9.64316	2.27428	10.35684
	HCl	0.72854	9.86245	1.37261	10.13755
CaF_2	Ca	0.51332	9.71039	1.94810	10.28961
	$CaSO_4$	1.74360	10.24145	0.57353	9.75855
$Ca(HCO_3)$	$CaCO_3$	0.61742	9.79058	1.61964	10.20942
	CaO	0.34594	9.53900	2.89067	10.46100
$Ca(IO_3)_2$	Ca	0.10280	9.01199	9.72763	10.98801
	CaO	0.14384	9.15788	6.95217	10.84212

Weighed	Sought	Factor	Log of Factor +10	Reciprocal of Factor	Log of Reciprocal of Factor +10
Calcium (contd.)					
$Ca(NO_3)_2$	N_2O_5	0.65824	9.81838	1.51920	10.18162
CaO	Ca	0.71470	9.85412	1.39919	10.14588
	$CaCl_2$	1.97914	10.29648	0.50527	9.70352
	$CaCO_3$	1.78477	10.25158	0.56030	9.74842
	CaF_2	1.39230	10.14373	0.71824	9.85627
	$Ca(HCO_3)_2$	2.89069	10.46100	0.34594	9.53900
	$Ca_3(PO_4)_2$	1.84368	10.26569	0.54239	9.73431
	$CaSO_4$	2.42760	10.38518	0.41193	9.61482
	$CaSO_4.2H_2O$	3.07008	10.48715	0.32572	9.51285
	Cl	1.26437	10.10187	0.79091	9.89813
	CO_2	0.78477	9.89474	1.27426	10.10527
	MgO	0.71879	9.85660	1.39123	10.14340
	SO_3	1.42760	10.15461	0.70048	9.84540
$Ca_3(PO_4)_2$	CaO	0.54239	9.73431	1.84369	10.26569
	$CaSO_4$	1.31672	10.11950	0.75946	9.88051
	$Mg_2P_2O_7$	0.71755	9.85585	1.39363	10.14415
	$(NH_4)_3PO_4.12MoO_3$	12.09843	11.08273	0.08266	8.91730
	P_2O_5	0.45761	9.66050	2.18527	10.33950
CaS	$BaSO_4$	3.23538	10.50992	0.30908	9.49007
$CaSO_4$	$BaSO_4$	1.71441	10.23411	0.58329	9.76581
	Ca	0.29440	9.46894	3.39674	10.53107
	$CaCl_2$	0.81526	9.91130	1.22660	10.08870
	$CaCO_3$	0.73520	9.86641	1.36017	10.13360
	CaF_2	0.57353	9.75855	1.74359	10.24145
	CaO	0.41193	9.61482	2.42760	10.38518
	$Ca_3(PO_4)_2$	0.75946	9.88051	1.31673	10.11950
	SO_3	0.58809	9.76944	1.70042	10.23056
$CaSO_4.2H_2O$	$BaSO_4$	1.35564	10.13213	0.73766	9.86786
	$CaCO_3$	0.58134	9.76443	1.72016	10.23557
	CaO	0.32572	9.51285	3.07012	10.48716
	SO_3	0.46502	9.66747	2.15045	10.33252
$CaWO_4$	WO_3	0.80523	9.90592	1.24188	10.09408
Cl	Ca	0.56526	9.75225	1.76910	10.24775
	$CaCl_2$	1.56531	10.19460	0.63885	9.80540
	CaO	0.79091	9.89813	1.26437	10.10188
CO_2	$CaCO_3$	2.27428	10.35684	0.43970	9.64316
	CaO	1.27427	10.10526	0.78476	9.89474
HCl	$CaCO_3$	1.37256	10.13753	0.72857	9.86247
$Mg_2As_2O_7$	$Ca_3(AsO_4)_2$	1.28221	10.10796	0.77990	9.89204
MgO	CaO	1.39118	10.14339	0.71881	9.85661
$Mg_2P_2O_7$	$Ca_3(PO_4)_2$	1.39365	10.14415	0.71754	9.85585
$(NH_4)_3PO_4.12MoO_3$	$Ca_3(PO_4)_2$	0.08265	8.91725	12.09892	11.08275
N_2O_5	$Ca(NO_3)_2$	1.51920	10.18162	0.65824	9.81838
P_2O_5	$Ca_3(PO_4)_2$	2.18521	10.33949	0.45762	9.66051
SO_3	CaO	0.70046	9.84539	1.42763	10.15462
	$CaSO_4$	1.70043	10.23056	0.58809	9.76944
	$CaSO_4.2H_2O$	2.15046	10.33253	0.46502	9.66747
WO_3	$CaWO_4$	1.24188	10.09408	0.80523	9.90592
Carbon					
Ag	CN	0.24120	9.38238	4.14594	10.61762
	HCN	0.25054	9.39888	3.99138	10.60112
	KCN	0.60369	9.78081	1.65648	10.21918
$AgCN$	CN	0.19433	9.28854	5.14589	10.71146
	HCN	0.20185	9.30503	4.95417	10.69497
	KCN	0.48638	9.68698	2.05601	10.31302
$AgCNS$	CNS	0.34999	9.54406	2.85722	10.45594
$Ag_4Fe(CN)_6$	C	0.11200	9.04922	8.92857	10.95078
$AgOCN$	OCN	0.28033	9.44767	3.56722	10.55233
$BaCO_3$	C	0.06086	8.78433	16.43115	11.21567
	CO_2	0.22300	9.34830	4.48431	10.65170
	CO_3	0.30408	9.48299	3.28861	10.51701
BaO	CO_2	0.28701	9.45790	3.48420	10.54210
	CO_2 (bicarbonate)	0.57402	9.75893	1.74210	10.24107
$BaSO_4$	CNS	0.24885	9.39594	4.01849	10.60406
C	$BaCO_3$	16.4305	11.21565	0.06086	8.78433
	CO_2	3.66409	10.56397	0.27292	9.43604
$CaCO_3$	CO_2	0.43970	9.64316	2.27428	10.35684
$Ca(HCO_3)_2$	CO_2	0.54295	9.73476	1.84179	10.26524

Weighed	Sought	Factor	Log of Factor +10	Reciprocal of Factor	Log of Reciprocal of Factor +10
Carbon (contd.)					
CaO	CO_2	0.78477	9.89474	1.27426	10.10526
	CO_2 (bicarbonate)	1.56954	10.19577	0.63713	9.80423
CN	Ag	4.14599	10.61762	0.24120	9.38238
	AgCN	5.14599	10.71147	0.19433	9.28854
CNS	AgCNS	2.85720	10.45594	0.34999	9.54406
	$BaSO_4$	4.01846	10.60406	0.24885	9.39594
	CuCNS	2.09394	10.32097	0.47757	9.67904
CO_2	$BaCO_3$	4.48420	10.65169	0.22301	9.34832
	$Ba(HCO_3)_2$	2.94672	10.46934	0.33963	9.53066
	BaO	3.48421	10.54211	0.28701	9.45790
	C	0.27292	9.43603	3.66408	10.56397
	$CaCO_3$	2.27426	10.35684	0.43970	9.64316
	$Ca(HCO_3)_2$	1.84174	10.26523	0.54296	9.73477
	CaO	1.27426	10.10526	0.78477	9.89474
	CO_3	1.36354	10.13467	0.73339	9.86534
	Cs_2CO_3	7.40329	10.86943	0.13508	9.13059
	$CsHCO_3$	4.40632	10.64407	0.22695	9.35593
	$FeCO_3$	2.63249	10.42037	0.37987	9.57964
	$Fe(HCO_3)_2$	2.02092	10.30555	0.49482	9.69445
	K_2CO_3	3.14049	10.49700	0.31842	9.50300
	$KHCO_3$	2.27491	10.35696	0.43958	9.64304
	K_2O	2.14049	10.33050	0.46718	9.66948
	Li_2CO_3	1.67887	10.22502	0.59564	9.77498
	$LiHCO_3$	1.54410	10.18868	0.64763	9.81133
	Li_2O	0.67887	9.83179	1.47304	10.16821
	$MgCO_3$	1.91595	10.28239	0.52193	9.71761
	$Mg(HCO_3)_2$	1.66265	10.22080	0.60145	9.77920
	MgO	0.91595	9.96188	1.09176	10.03812
	$MnCO_3$	2.61185	10.41695	0.38287	9.58305
	$Mn(HCO_3)_2$	2.01059	10.30332	0.49737	9.69668
	MnO	1.61185	10.20733	0.62041	9.79268
	Na_2CO_3	2.40829	10.38171	0.41523	9.61829
	$NaHCO_3$	1.90882	10.28077	0.52388	9.71923
	Na_2O	1.40829	10.14869	0.71008	9.85131
	$(NH_4)_2CO_3$	2.18329	10.33911	0.45802	9.66088
	NH_4HCO_3	1.79632	10.25439	0.55669	9.74561
	$PbCO_3$	6.07135	10.78328	0.16471	9.21672
	Rb_2CO_3	5.24767	10.71996	0.19056	9.28003
	$RbHCO_3$	3.32856	10.52225	0.30043	9.47774
	Rb_2O	4.24767	10.62815	0.23542	9.37184
	$SrCO_3$	3.35446	10.52562	0.29811	9.47438
	$Sr(HCO_3)_2$	2.3818	10.37690	0.41985	9.62309
	SrO	2.35446	10.37189	0.42473	9.62811
CO_3	$BaCO_3$	3.2886	10.51701	0.30408	9.48299
	CO_2	0.73339	9.86533	1.36353	10.13466
Cs_2CO_3	CO_2	0.13507	9.13056	7.40357	10.86944
$CsHCO_3$	CO_2	0.22695	9.35593	4.40626	10.64407
CuCNS	CNS	0.47757	9.67904	2.09393	10.32097
$FeCO_3$	CO_2	0.37987	9.57964	2.63248	10.42037
$Fe(HCO_3)_2$	CO_2	0.49482	9.69445	2.02094	10.30556
HCN	Ag	3.99137	10.60112	0.25054	9.39888
	AgCN	4.95408	10.69497	0.20185	9.30503
KCN	Ag	1.65648	10.21918	0.60369	9.78081
	AgCN	2.05602	10.31302	0.48638	9.68698
K_2CO_3	CO_2	0.31842	9.50300	3.14051	10.49700
$KHCO_3$	CO_2	0.43958	9.64304	2.27490	10.35696
K_2O	CO_2	0.46718	9.66948	2.14050	10.33052
Li_2CO_3	CO_2	0.59564	9.77498	1.67887	10.22502
Li_2O	CO_2	1.47304	10.16821	0.67887	9.83179
$MgCO_3$	CO_2	0.52193	9.71761	1.91597	10.28239
$Mg(HCO_3)_2$	CO_2	0.60145	9.77920	1.66265	10.22080
MgO	CO_2	1.09176	10.03812	0.91595	9.96188
$MnCO_3$	CO_2	0.38287	9.58305	2.61185	10.41695
MnO	CO_2	0.62041	9.79268	1.61184	10.20733
Na_2CO_3	CO_2	0.41523	9.61829	2.40830	10.38171
$NaHCO_3$	CO_2	0.52388	9.71923	1.90883	10.28077
Na_2O	CO_2	0.71008	9.85131	1.40829	10.14869
$(NH_4)_2CO_3$	CO_2	0.45802	9.66088	2.18331	10.33911

Weighed	Sought	Factor	Log of Factor +10	Reciprocal of Factor	Log of Reciprocal of Factor +10
Carbon (contd.)					
NH_4HCO_3	CO_2	0.55669	9.74561	1.79633	10.25439
$PbCO_3$	CO_2	0.16471	9.21672	6.07128	10.78328
Rb_2CO_3	CO_2	0.19056	9.28003	5.24769	10.71996
$RbHCO_3$	CO_2	0.30043	9.47774	3.32856	10.52225
Rb_2O	CO_2	0.23542	9.37184	4.24773	10.62815
$SrCO_3$	CO_2	0.29811	9.47438	3.35447	10.52562
$Sr(HCO_3)_2$	CO_2	0.41984	9.62308	2.38186	10.37691
SrO	CO_2	0.42472	9.62810	2.35449	10.37190
Cerium					
Ce	$Ce_2(C_2O_4)_3.3H_2O$	2.13513	10.32943	0.46836	9.67058
	$Ce(NO_3)_4$	2.77005	10.44249	0.36100	9.55751
	$Ce(NO_3)_4(NH_4NO_3)_2.H_2O$	4.04111	10.60650	0.24746	9.39351
	CeO_2	1.22838	10.08933	0.81408	9.91067
	Ce_2O_3	1.17128	10.06866	0.85377	9.93134
	$Ce_2(SO_4)_3$	2.02833	10.30714	0.49302	9.69286
$Ce_2(C_2O_4)_3.3H_2O$	Ce	0.46835	9.67057	2.13516	10.32943
	$Ce_2(SO_4)_3$	0.95000	9.97772	1.05263	10.02228
$Ce(NO_3)_4$	Ce	0.36100	9.55751	2.77008	10.44249
	CeO_2	0.44345	9.64684	2.25505	10.35316
	Ce_2O_3	0.42284	9.62618	2.36496	10.37382
$Ce(NO_3)_4(NH_4NO_3)_2.$ H_2O	Ce	0.24745	9.39349	4.04122	10.60651
	CeO_2	0.30396	9.48282	3.28991	10.51719
	Ce_2O_3	0.28984	9.46216	3.45018	10.53784
CeO_2	Ce	0.81408	9.91067	1.22838	10.08933
	$Ce(NO_3)_4$	2.25505	10.35316	0.44345	9.64685
	$Ce(NO_3)_4(NH_4NO_3)_2.H_2O$	3.28986	10.51718	0.30396	9.48281
	Ce_2O_3	0.95352	9.97933	1.04875	10.02068
Ce_2O_3	Ce	0.85377	9.93134	1.17128	10.06866
	$Ce(NO_3)_4$	2.36498	10.37383	0.42284	9.62618
	$Ce(NO_3)_4(NH_4NO_3)_2.H_2O$	3.45016	10.53784	0.28984	9.46216
	CeO_2	1.04874	10.02067	0.95353	9.97933
	$Ce_2(SO_4)_3$	1.73172	10.23848	0.57746	9.76152
$Ce_2(SO_4)_3$	Ce	0.49302	9.69286	2.02832	10.30714
	$Ce_2(C_2O_4)_3.3H_2O$	1.05265	10.02220	0.94998	9.97771
	Ce_2O_3	0.57746	9.76152	1.73172	10.23848
Cesium					
AgCl	CsCl	1.17468	10.06992	0.85130	9.93008
Cl	Cs	3.74877	10.57389	0.26675	9.42610
	CsCl	4.74877	10.67658	0.21058	9.32342
Cs	Cl	0.26675	9.42610	3.74883	10.57390
	CsCl	1.26675	10.10269	0.78942	9.89731
	Cs_2O_3	1.22576	10.08840	0.81582	9.91159
	Cs_2O	1.06019	10.02539	0.94323	9.97462
	Cs_2PtCl_6	2.53422	10.40385	0.39460	9.59616
	Cs_2SO_4	1.36139	10.13398	0.73454	9.86602
$CsB(C_6H_5)_4$	Cs	0.29394	9.46826	3.40205	10.53174
	Cs_2O	0.31164	9.49365	3.20883	10.50635
CsCl	AgCl	0.85130	9.93008	1.17467	10.06993
	Cl	0.21058	9.32342	4.74879	10.67658
	Cs	0.78942	9.89731	1.26675	10.10269
	Cs_2O	0.83693	9.92269	1.19484	10.07731
	Cs_2PtCl_6	2.00055	10.30114	0.49986	9.69885
	Cs_2SO_4	1.07470	10.03129	0.93049	9.96871
Cs_2CO_3	Cs	0.81582	9.91159	1.22576	10.08841
	Cs_2PtCl_6	2.06746	10.31544	0.48369	9.68457
	Cs_2SO_4	1.11065	10.04557	0.90037	9.95442
$CsClO_4$	Cs	0.57199	9.75739	1.74828	10.24261
	CsCl	0.72457	9.86008	1.38013	10.13992
	Cs_2O_3	0.70110	9.84578	1.42631	10.15421
	Cs_2O	0.60642	9.78277	1.64902	10.21723
	Cs_2SO_4	0.77870	9.89137	1.28419	10.10863
Cs_2O	Cs	0.94323	9.97462	1.06019	10.02539
	CsCl	1.19483	10.07731	0.83694	9.92269
	Cs_2PtCl_6	2.39034	10.37846	0.41835	9.62154
	Cs_2SO_4	1.28410	10.10860	0.77876	9.89140
	SO_3	0.28410	9.45347	3.51989	10.54653
Cs_2PtCl_6	Cs	0.39460	9.59616	2.53421	10.40385
	CsCl	0.49986	9.69885	2.00056	10.30115

Weighed	Sought	Factor	Log of Factor +10	Reciprocal of Factor	Log of Reciprocal of Factor +10
Cesium (contd.)					
Cs_2PtCl_6	Cs_2CO_3	0.48368	9.68456	2.06748	10.31544
	Cs_2O	0.41835	9.62154	2.39034	10.37846
Cs_2SO_4	Cs	0.73454	9.86602	1.36140	10.13399
	CsCl	0.93048	9.96871	1.07471	10.03129
	Cs_2CO_3	0.90037	9.95442	1.11065	10.04557
	Cs_2O	0.77876	9.89140	1.28409	10.10860
SO_3	Cs_2O	3.51987	10.54652	0.28410	9.45347
	Cs_2SO_4	4.51988	10.65513	0.22124	9.34486
Chlorine					
Ag	Cl	0.32866	9.51675	3.04266	10.48325
	HCl	0.33801	9.52893	2.95850	10.47107
AgCl	Cl	0.24736	9.39333	4.04269	10.60667
	ClO_3	0.58226	9.76512	1.71745	10.23488
	ClO_4	0.69389	9.84129	1.44115	10.15871
	HCl	0.25440	9.40552	3.93082	10.59448
$BaCrO_4$	Cl	0.27990	9.44700	3.57270	10.55299
Ca	Cl	1.76911	10.24776	0.56526	9.75225
Cl	Ag	3.04262	10.48325	0.32866	9.51675
	AgCl	4.04262	10.60666	0.24736	9.39333
	$BaCrO_4$	3.57276	10.55300	0.27990	9.44700
	Ca	0.56526	9.75225	1.76910	10.24775
	HCl	1.02843	10.01217	0.97236	9.98783
	K	1.10292	10.04255	0.90668	9.95745
	KCl	2.10292	10.32282	0.47553	9.67718
	Li	0.19572	9.29164	5.10934	10.70837
	Mg	0.34288	9.53514	2.91647	10.46486
	$MgCl_2$	1.34288	10.12804	0.74467	9.87196
	MnO_2	1.22609	10.08852	0.81560	9.91148
	Na	0.64846	9.81188	1.54212	10.18811
	NaCl	1.64846	10.21708	0.60663	9.21708
Cl	NH_4	0.50880	9.70655	1.96541	10.29346
	$PbCrO_4$	4.55787	10.65876	0.21940	9.34124
ClO_3	AgCl	1.71745	10.23488	0.58226	9.76512
	KCl	0.89340	9.95105	1.11932	10.04895
	NaCl	0.70032	9.84530	1.42792	10.15471
ClO_4	AgCl	1.44114	10.15870	0.69390	9.84130
	KCl	0.74967	9.87487	1.33392	10.12513
	NaCl	0.58765	9.76912	1.70169	10.23088
HCl	Ag	2.95850	10.47107	0.33801	9.52893
	AgCl	3.93086	10.59448	0.25440	9.40552
	NH_4Cl	1.46710	10.16646	0.68162	9.83354
	$(NH_4)_2SO_4$	1.81206	10.25817	0.55186	9.74183
K	Cl	0.90668	9.95745	1.10292	10.04255
KCl	Cl	0.47553	9.67718	2.10292	10.32282
	ClO_3	1.11932	10.04895	0.89340	9.95105
	ClO_4	1.33393	10.12513	0.74966	9.87486
Li	Cl	5.10924	10.70836	0.19572	9.29164
Mg	Cl	2.91650	10.46487	0.34288	9.53514
$MgCl_2$	Cl	0.74467	9.87196	1.34288	10.12804
MnO_2	Cl	0.81560	9.91148	1.22610	10.08853
Na	Cl	1.54212	10.18811	0.64896	9.81188
NaCl	Cl	0.60663	9.78293	1.64845	10.21708
	ClO_3	1.42791	10.15470	0.70032	9.84530
	ClO_4	1.70168	10.23088	0.58765	9.76912
NH_4	Cl	1.96539	10.29345	0.50880	9.70655
NH_4Cl	HCl	0.68162	9.83354	1.46709	10.16646
$(NH_4)_2SO_4$	HCl	0.55186	9.74184	1.81205	10.25817
$PbCrO_4$	Cl	0.21940	9.34124	4.55789	10.65876
Chromium					
Ag_2CrO_4	Cr	0.15674	9.19518	6.37999	10.80482
	Cr_2O_3	0.22895	9.35974	4.36777	10.64026
	CrO_3	0.30143	9.47919	3.31752	10.52082
	CrO_4	0.34966	9.54365	2.85992	10.45635
$BaCrO_4$	Cr	0.20525	9.31228	4.87211	10.68772
	CrO_3	0.39472	9.59629	2.53344	10.40371
	CrO_4	0.45788	9.66075	2.18398	10.33924
	Cr_2O_3	0.29998	9.47709	3.33356	10.52291
	$Cr_2(SO_4)_3.18H_2O$	1.41404	10.15046	0.70719	9.84954

Weighed	Sought	Factor	Log of Factor +10	Reciprocal of Factor	Log of Reciprocal of Factor +10
Chromium (contd.)					
$BaCrO_4$	$PbCrO_4$	1.27573	10.10576	0.78386	9.89424
Cr	$BaCrO_4$	4.87210	10.68772	0.20525	9.31228
	Cr_2O_3	1.46155	10.16482	0.68421	9.83519
	$PbCrO_4$	6.21547	10.79347	0.16089	9.20653
$Cr(C_9H_6NO)_3$	Cr	0.10733	9.03072	9.31706	10.96928
	Cr_2O_3	0.15677	9.19526	6.37877	10.80473
CrO_3	$BaCrO_4$	2.53345	10.40372	0.39472	9.59629
	Cr	0.51999	9.71600	1.92311	10.28401
	Cr_2O_3	0.75999	9.88081	1.31581	10.11920
	K_2CrO_4	1.94210	10.28827	0.51491	9.71173
	$K_2Cr_2O_7$	1.47105	10.16762	0.67979	9.83237
	$PbCrO_4$	3.23199	10.50947	0.30941	9.49053
CrO_4	$BaCrO_4$	2.18399	10.33925	0.45788	9.66075
	$PbCrO_4$	2.78618	10.44501	0.35891	9.55499
Cr_2O_3	$BaCrO_4$	3.33351	10.52291	0.29998	9.47709
	Cr	0.68420	9.83518	1.46156	10.16482
	CrO_3	1.31580	10.11919	0.75999	9.88081
	CrO_4	1.52634	10.18364	0.65516	9.81635
	$PbCrO_4$	4.25265	10.62866	0.23515	9.37135
$Cr_2(SO_4)_3.18H_2O$	$BaCrO_4$	0.70718	9.84953	1.41407	10.15047
	$PbCrO_4$	0.90217	9.95529	1.10844	10.04471
K_2CrO_4	CrO_3	0.51491	9.71173	1.94209	10.28827
	$PbCrO_4$	1.66418	10.22120	0.60090	9.77880
$K_2Cr_2O_7$	CrO_3	0.67979	9.83237	1.47104	10.16762
	$PbCrO_4$	2.19707	10.34184	0.45515	9.65815
$PbCrO_4$	Cr	0.16089	9.20652	6.21543	10.79347
	CrO_3	0.30941	9.49053	3.23196	10.50946
	CrO_4	0.35891	9.55499	2.78621	10.44501
	Cr_2O_3	0.23514	9.37133	4.25279	10.62868
	$Cr_2(SO_4)_3.18H_2O$	1.10843	10.04471	0.90218	9.95529
	K_2CrO_4	0.60090	9.77880	1.66417	10.22119
	$K_2Cr_2O_7$	0.45514	9.65815	2.19713	10.34186
Cobalt					
Co	$Co(NO_3)_2.6H_2O$	4.93731	10.69349	0.20254	9.30651
	$Co(NO_2)_3.(KNO_2)_3$	7.67433	10.88504	0.13030	9.11494
	CoO	1.27148	10.10431	0.78649	9.89569
	Co_3O_4	1.36197	10.13417	0.73423	9.86583
	$CoSO_4$	2.63001	10.41996	0.38023	9.58005
	$CoSO_4.7H_2O$	4.76984	10.67851	0.20965	9.32149
$Co(C_{10}H_6O_2N).2H_2O$	$(CoSO_4)_2.(K_2SO_4)_3$	7.06550	10.84914	0.14153	9.15085
	Co	0.09638	8.98399	10.37560	11.01602
	CoO	0.12255	9.08831	8.15993	10.91169
$(Co[C_5H_5N]_4).(SCN)_2$	Co	0.11990	9.07882	8.34028	10.92118
	CoO	0.15246	9.18316	6.55910	10.81684
$Co(C_9H_6ON)_2$	Co	0.16972	9.22973	5.89206	10.77027
	Co_2O_3	0.23883	9.37809	4.18708	10.62191
$Co(NO_3)_2.6H_2O$	Co	0.20254	9.30651	4.93730	10.69349
$Co(NO_2)_3.(KNO_2)_3$	Co	0.13030	9.11494	7.67460	10.88506
	CoO	0.16568	9.21927	6.03573	10.78073
CoO	Co	0.78648	9.89569	1.27149	10.10432
	$Co(NO_2)_3.(KNO_2)_3$	6.03571	10.78073	0.16568	9.21927
	Co_3O_4	1.07117	10.02986	0.93356	9.97014
	$CoSO_4$	2.06845	10.31564	0.48345	9.68435
	$(CoSO_4)_2.(K_2SO_4)_3$	5.55690	10.74483	0.17996	9.25518
Co_3O_4	Co	0.73423	9.86583	1.36197	10.13417
	CoO	0.93356	9.97014	1.07117	10.02986
$Co_2P_2O_7$	Co	0.40392	9.60630	2.47574	10.39371
	CoO	0.51357	9.71060	1.94715	10.28940
	Co_3O_4	0.55012	9.74046	1.81779	10.25953
	$CoSO_4$	1.06230	10.02624	0.94135	9.97375
	$CoSO_4.7H_2O$	1.92661	10.28479	0.51905	9.71521
$CoSO_4$	Co	0.38023	9.58005	2.62999	10.41996
	CoO	0.48345	9.68435	2.06847	10.31565
$CoSO_4.7H_2O$	Co	0.20965	9.32150	4.76985	10.67851
	CoO	0.26657	9.42581	3.75136	10.57419
$(CoSO_4)_2.(K_2SO_4)_3$	Co	0.14153	9.15085	7.06553	10.84914
	CoO	0.17996	9.25518	5.55679	10.74482
Columbium	See Niobium				

Weighed	Sought	Factor	Log of Factor +10	Reciprocal of Factor	Log of Reciprocal of Factor +10
Copper					
Cu	$Cu_2C_2H_3O_2.(AsO_2)_3$	3.98875	10.60084	0.25071	9.39917
	CuCNS	1.91407	10.28196	0.52245	9.71804
	CuO	1.25181	10.09754	0.79884	9.90246
	Cu_2O	1.12590	10.05150	0.88818	9.94850
	Cu_2S	1.25228	10.09770	0.79854	9.90230
	$CuSO_4.5H_2O$	3.92949	10.59433	0.25449	9.40567
$Cu_2C_2H_3O_2.(AsO_2)_3$	Cu	0.25071	9.39917	3.98867	10.60083
CuCNS	Cu	0.52245	9.71804	1.91406	10.28195
	CuO	0.65400	9.81558	1.52905	10.18443
$(Cu[C_5H_5N]_2.(SCN)_2$	Cu	0.18804	9.27425	5.31802	10.72575
	CuO	0.23539	9.37179	4.24827	10.62821
$Cu(C_7H_6NO_2)_2$	Cu	0.18922	9.27697	5.28485	10.72303
	CuO	0.23687	9.37451	4.22172	10.62549
$Cu(C_9H_6NO)_2$	Cu	0.18059	9.25669	5.53741	10.74330
	CuO	0.22606	9.35422	4.42360	10.64578
$Cu(C_{12}H_{10}ONS)_2$	Cu	0.12795	9.10704	7.81555	10.89296
	CuO	0.16017	9.20458	6.24337	10.79542
$Cu(NH_2C_6H_4CO_2)_2$	Cu	0.18922	9.27697	5.28485	10.72303
	CuO	0.23687	9.37451	4.22172	10.62549
CuO	Cu	0.79884	9.90246	1.25182	10.09754
	CuCNS	1.52904	10.18442	0.65401	9.81558
	Cu_2S	1.00038	10.00016	0.99962	9.99983
	$CuSO_4.5H_2O$	3.13905	10.49680	0.31857	9.50320
Cu_2O	Cu	0.88817	9.94850	1.12591	10.05150
	CuO	1.11183	10.04603	0.89942	9.95396
	Cu_2S	1.11224	10.04620	0.89909	9.95380
Cu_2S	Cu	0.79854	9.90230	1.25229	10.09770
	CuO	0.99962	9.99983	1.00004	10.00002
	Cu_2O	0.89908	9.95380	1.11225	10.04620
	$CuSO_4.5H_2O$	3.13787	10.49663	0.31869	9.50337
$CuSO_4.5H_2O$	Cu	0.25449	9.40567	3.92943	10.59433
	CuO	0.31857	9.50320	3.13903	10.49679
	Cu_2S	0.31869	9.50337	3.13785	10.49663
$Mg_2As_2O_7$	$Cu_2C_2H_3O_2.(AsO_2)_3$	1.08844	10.03681	0.91875	9.96320
Dysprosium					
Dy_2O_3	Dy	0.87131	9.94017	1.14770	10.05983
Erbium					
Er	Er_2O_3	1.14349	10.05824	0.87452	9.94177
Er_2O_3	Er	0.87452	9.94177	1.14349	10.05824
Europium					
Eu_2O_3	Eu	0.86361	9.93632	1.15793	10.06368
Fluorine					
BaF_2	$BaSiF_6$	1.59359	10.20238	0.62751	9.79762
$BaSiF_6$	BaF_2	0.62751	9.79762	1.59360	10.20238
	F	0.40795	9.61061	2.45128	10.38939
	HF	0.42960	9.63306	2.32775	10.36694
	H_2SiF_6	0.51568	9.71238	1.93919	10.28762
	SiF_4	0.37248	9.57110	2.68471	10.42889
	SiF_6	0.50847	9.70627	1.96668	10.29373
CaF_2	F	0.48664	9.68721	2.05491	10.31279
	6HF	0.51248	9.70968	1.95130	10.29033
	H_2SiF_6	0.61517	9.78900	1.62557	10.21100
	SiF_6	0.60657	9.78288	1.64861	10.21712
$CaSO_4$	F	0.27910	9.44576	3.58295	10.55424
	HF	0.29391	9.46821	3.40240	10.53178
F	$BaSiF_6$	2.45125	10.38939	0.40796	9.61062
	CaF_2	2.05491	10.31279	0.48664	9.68721
	$CaSO_4$	3.58293	10.55433	0.27910	9.44576
	H_2SiF_6	1.26407	10.10177	0.79110	9.89823
	K_2SiF_6	1.93244	10.28611	0.51748	9.71389
2HF	H_2SiF_6	3.60115	10.55644	0.27769	9.44356
6HF	H_2SiF_6	1.20038	10.07932	0.83307	9.92068
H_2SiF_6	$BaSiF_6$	1.93917	10.28762	0.51568	9.71237
	CaF_2	1.62555	10.21100	0.61518	9.78900
	F	0.79109	9.89823	1.26408	10.10178
	2HF	0.27769	9.44356	3.60114	10.55644
	6HF	0.83307	9.92068	1.20038	10.07932
	K_2SiF_6	1.52875	10.18434	0.65413	9.81566

Weighed	Sought	Factor	Log of Factor +10	Reciprocal of Factor	Log of Reciprocal of Factor +10
Fluorine (contd.)					
H_2SiF_6	SiF_4	0.72232	9.85873	1.38443	10.14129
	SiF_6	0.98601	9.99388	1.01419	10.00612
2KF	K_2SiF_6	1.89568	10.27777	0.52752	9.72224
K_2SiF_6	F	0.51748	9.71389	1.93244	10.28611
	6HF	0.54493	9.73634	1.83510	10.26366
	H_2SiF_6	0.65413	9.81566	1.52875	10.18434
	KF	0.52751	9.72223	1.89570	10.27777
	SiF_6	0.64498	9.80955	1.55044	10.19045
Na_2SiF_6	F	0.60615	9.78258	1.64976	10.21742
	6HF	0.63831	9.80503	1.56664	10.19498
	H_2SiF_6	0.76622	9.88435	1.30511	10.11565
	2NaF	0.44655	9.64987	2.23939	10.35013
	SiF_4	0.55344	9.74307	1.80688	10.25693
	SiF_6	0.75550	9.87823	1.32363	10.12177
PbClF	F	0.07261	8.86100	13.77221	11.13900
SiF_4	$BaSiF_6$	2.68467	10.42889	0.37249	9.57111
	H_2SiF_6	1.38444	10.14128	0.72231	9.85872
SiF_6	$BaSiF_6$	1.96666	10.29373	0.50848	9.70627
	CaF_2	1.64862	10.21712	0.60657	9.78288
	H_2SiF_6	1.01419	10.00612	0.98601	9.99388
	K_2SiF_6	1.55043	10.19045	0.64498	9.80955
Gallium					
Ga	Ga_2O_3	1.34423	10.12847	0.74392	9.87153
$Ga(C_9H_4NOBr_2)_3$	Ga	0.07146	8.85406	13.99384	11.14594
	Ga_2O_3	0.09606	8.98254	10.41016	11.01746
Ga_2O_3	Ga	0.74392	9.87153	1.34423	10.12847
Ga_2S_3	Ga	0.59178	9.77216	1.68982	10.22785
Germanium					
Ge	GeO_2	1.44083	10.15861	0.69404	9.84138
$(C_9H_6NOH)_4.(GeO_2.12MoO_3)$	Ge	0.03009	8.47842	33.23363	11.52158
	GeO_2	0.04335	8.63699	23.06805	11.36301
GeO_2	Ge	0.69404	9.84139	1.44084	10.15861
K_2GeF_6	Ge	0.27415	9.43799	3.64764	10.56201
Mg_2GeO_4	Ge	0.39193	9.59321	2.55148	10.40679
	GeO_2	0.56471	9.75183	1.77082	10.24817
Gold					
Au	$AuCl_3$	1.53998	10.18749	0.64936	9.81249
	$HAuCl_4.4H_2O$	2.09095	10.32034	0.47825	9.67966
	$KAu(CN)_4.H_2O$	1.81836	10.25968	0.54995	9.74032
$AuCl_3$	Au	0.64938	9.81250	1.53993	10.18750
C_6H_5SAu	Au	0.64339	9.80847	1.55427	10.19153
$HAuCl_4.4H_2O$	Au	0.47825	9.67966	2.09096	10.32034
$KAu(CN)_4.H_2O$	Au	0.54995	9.74032	1.81835	10.25967
Hafnium					
HfO_2	Hf	0.84797	9.92838	1.17929	10.07162
$Hf(C_6H_5CHOHCO_2)_4$	Hf	0.22794	9.35782	4.38712	10.64218
	HfO_2	0.26880	9.42943	3.72024	10.57057
Holmium					
Ho_2O_3	Ho	0.87297	9.94100	1.14551	10.05900
Hydrogen					
AgCNS	HCNS	0.35606	9.55152	2.80852	10.44848
$BaSO_4$	HCNS	0.25317	9.40341	3.94992	10.59659
CuCNS	HCNS	0.48586	9.68651	2.05821	10.31349
H	H_2O	8.93644	10.95116	0.11190	9.04883
	O	7.93644	10.89963	0.12600	9.10037
HCNS	AgCNS	2.80846	10.44847	0.35607	9.55154
	$BaSO_4$	3.94991	10.59659	0.25317	9.40341
	CuCNS	2.05822	10.31350	0.48586	9.68651
H_2O	H	0.11190	9.04883	8.93655	10.95117
O	H	0.12600	9.10037	7.93651	10.89963
Indium					
In	In_2O_3	1.20902	10.08244	0.92712	9.91757
	In_2S_3	1.41888	10.15194	0.70478	9.84805
$In(C_9H_6ON)_3$	In	0.20980	9.32181	4.76644	10.67819
	In_2O_3	0.25365	9.40423	3.94244	10.59577
In_2O_3	In	0.82711	9.91756	1.20903	10.08244
In_2S_3	In	0.70478	9.84805	1.41888	10.15194

Weighed	Sought	Factor	Log of Factor +10	Reciprocal of Factor	Log of Reciprocal of Factor +10
Iodine					
Ag	HI	1.18580	10.07401	0.84331	9.92599
	I	1.17646	10.07058	0.85001	9.92942
AgCl	I	0.88544	9.94716	1.12938	10.05284
AgI	HI	0.54483	9.73626	1.83543	10.26374
	I	0.54054	9.73283	1.85000	10.26717
	IO_3	0.74498	9.87214	1.34232	10.12786
	IO_4	0.81313	9.91016	1.22982	10.08985
	I_2O_5	0.71091	9.85181	1.40665	10.14819
	I_2O_7	0.77906	9.89157	1.28360	10.10843
HI	Ag	0.84331	9.92599	1.18580	10.07401
	AgI	1.83542	10.26374	0.54483	9.73626
	Pd	0.41590	9.61899	2.40442	10.38101
	PdI_2	1.40799	10.14860	0.71023	9.85140
	TlI	2.58981	10.41327	0.38613	9.58673
I	Ag	0.85000	9.92942	1.17647	10.07058
	AgCl	1.12937	10.05283	0.88545	9.94716
	AgI	1.85000	10.26717	0.54054	9.73283
	Pd	0.41921	9.62243	2.38544	10.37757
	PdI_2	1.41917	10.15203	0.70464	9.84797
	TlI	2.61039	10.41671	0.38308	9.58329
IO_3	AgI	1.34231	10.12785	0.74498	9.87214
	PdI_2	1.02971	10.01272	0.97115	9.98729
	TlI	1.89402	10.27738	0.52798	9.72262
IO_4	AgI	1.22981	10.08984	0.81313	9.91016
	PdI_2	0.94341	9.97470	1.05998	10.02530
	TlI	1.73528	10.23937	0.57628	9.76063
I_2O_5	AgI	1.40665	10.14819	0.71091	9.95181
	PdI_2	1.07907	10.03305	0.92672	9.96695
	TlI	1.98481	10.29772	0.50383	9.70228
I_2O_7	AgI	1.28360	10.10843	0.77906	9.89157
	PdI_2	0.98467	9.99329	1.01557	10.00671
	TlI	1.81120	10.25797	0.55212	9.74203
Pd	HI	2.40437	10.38100	0.41591	9.61900
	I	2.38542	10.37757	0.41921	9.62243
PdI_2	HI	0.71023	9.85140	1.40799	10.14860
	I	0.70463	9.70463	1.41918	10.15204
	IO_3	0.97114	9.98728	1.02972	10.01272
	IO_4	1.05998	10.02530	0.94341	9.97470
	I_2O_5	0.92672	9.96695	1.07907	10.03305
	I_2O_7	1.01556	10.00671	0.98468	9.99330
TlI	HI	0.38613	9.58673	2.58980	10.41327
	I	0.38308	9.58329	2.61042	10.41671
	IO_3	0.52798	9.72262	1.89401	10.27738
	IO_4	0.57627	9.76063	1.73530	10.23937
	I_2O_5	0.50382	9.70228	1.98484	10.29772
	I_2O_7	0.55212	9.74203	1.81120	10.25797
Iron					
Ag	$Fe_7(CN)_{18}$ (Prussian blue)	0.44253	9.64594	2.25973	10.35406
CN	$Fe_7(CN)_{18}$	1.83474	10.26358	0.54504	9.73643
CO_2	$FeCO_3$	2.63249	10.42037	0.37987	9.57964
	$Fe(HCO_3)_2$	2.02092	10.30555	0.49482	9.69445
	FeO	1.63249	10.21287	0.61256	9.78715
Fe	$Fe(HCO_3)_2$	3.18517	10.50313	0.31395	9.49686
	FeO	1.28648	10.10940	0.77731	9.89059
	Fe_2O_3	1.42973	10.15526	0.69943	9.84474
	$FePO_4$	2.70056	10.43145	0.37029	9.56854
	FeS	1.57414	10.19704	0.63527	9.80296
	$FeSO_4$	2.72009	10.43458	0.36763	9.56541
	$FeSO_4.7H_2O$	4.97817	10.69707	0.20088	9.30294
	$FeSO_4.(NH_4)_2SO_4.6H_2O$	7.02054	10.84637	0.14244	9.15363
$FeAsO_4$	$Mg_2As_2O_7$	0.79702	9.90147	1.25467	10.09853
$Fe(C_6H_5N[NO]O)_3$	Fe	0.11953	9.07748	8.36610	10.92252
	Fe_2O_3	0.17090	9.23274	5.85138	10.76726
$Fe(C_9H_6NO)_3$	Fe	0.11437	9.05821	8.74355	10.94169
	Fe_2O_3	0.16351	9.21354	6.11583	10.78646
$FeCl_3$	Fe_2O_3	0.49225	9.69219	2.03149	10.30781
$Fe_7(CN)_{18}$	Ag	2.25971	10.35405	0.44253	9.64594
	CN	0.54503	9.73642	1.83476	10.26358

GRAVIMETRIC FACTORS AND THEIR LOGARITHMS (Continued)

Weighed	Sought	Factor	Log of Factor +10	Reciprocal of Factor	Log of Reciprocal of Factor +10
Iron (contd.)					
$FeCO_3$	CO_2	0.37986	9.57962	2.63255	10.42038
	FeO	0.62013	9.79248	1.61257	10.20752
	Fe_2O_3	0.68918	9.83833	1.45100	10.16167
$Fe(HCO_3)_2$	CO_2	0.49482	9.69445	2.02094	10.30555
	Fe	0.31396	9.49687	3.18512	10.50313
	FeO	0.40390	9.60627	2.47586	10.39373
	Fe_2O_3	0.44887	9.65212	2.22782	10.34788
FeO	CO_2	0.61256	9.78715	1.63249	10.21285
	Fe	0.77732	9.89060	1.28647	10.10940
	$FeCO_3$	1.61256	10.20751	0.62013	9.79248
	$Fe(HCO_3)_2$	2.47588	10.39373	0.40390	9.60627
	Fe_2O_3	1.11134	10.04584	0.89981	9.95415
	$FePO_4$	2.09918	10.32205	0.47638	9.67795
	FeS	1.22360	10.08764	0.81726	9.91236
	SO_3	1.11436	10.04703	0.89738	9.95298
Fe_2O_3	Fe	0.69943	9.84474	1.42974	10.15524
	$FeCl_3$	2.03149	10.30781	0.49225	9.69219
	$FeCO_3$	1.45099	10.16167	0.68918	9.83833
	$Fe(HCO_3)_2$	2.22781	10.34788	0.44887	9.65212
	$Fe(HCO_3)_3$	2.99200	10.47596	0.33422	9.52403
	FeO	0.89981	9.95415	1.11135	10.04584
	Fe_3O_4	0.96660	9.98525	1.03455	10.01475
	$FePO_4$	1.88886	10.27620	0.52942	9.72380
	FeS	1.10101	10.04113	0.90826	9.95821
	$FeSO_4$	1.90252	10.27933	0.52562	9.72067
	$FeSO_4.7H_2O$	3.48190	10.54182	0.28720	9.45818
	$Fe_2(SO_4)_3$	2.50406	10.39864	0.39935	9.60135
	$FeSO_4.(NH_4)_2SO_4.6H_2O$	4.91040	10.69112	0.20365	9.30888
Fe_3O_4	Fe_2O_3	1.03455	10.01475	0.96660	9.98525
$FePO_4$	Fe	0.37029	9.56854	2.70059	10.43145
	FeO	0.47638	9.67795	2.09916	10.32204
	Fe_2O_3	0.52942	9.72380	1.88886	10.27620
FeS	Fe	0.63527	9.80296	1.57413	10.19704
	FeO	0.81726	9.91236	1.22360	10.08764
	Fe_2O_3	0.90826	9.95821	1.10101	10.04179
$FeSO_4$	Fe	0.36763	9.56541	2.72013	10.43459
	Fe_2O_3	0.52562	9.72067	1.90252	10.27933
	SO_3	0.52704	9.72184	1.89739	10.27817
$FeSO_4.7H_2O$	Fe	0.20088	9.30294	4.97810	10.69706
	Fe_2O_3	0.28720	9.45818	3.48189	10.54182
$FeSO_4.(NH_4)_2SO_4.6H_2O$	Fe	0.14244	9.15363	7.02050	10.84637
	Fe_2O_3	0.20365	9.30888	4.91039	10.69112
$Fe_2(SO_4)_3$	Fe_2O_3	0.39935	9.60135	2.50407	10.39864
$Mg_2As_2O_7$	$FeAsO_4$	1.25468	10.09853	0.79702	9.90147
SO_3	FeO	0.89738	9.95298	1.11436	10.04703
	$FeSO_4$	1.89739	10.27816	0.52704	9.72184
Lanthanum					
La	La_2O_3	1.17277	10.06921	0.85268	9.93079
La_2O_3	La	0.85268	9.93079	1.17277	10.06921
Lead					
$BaSO_4$	$PbSO_4$	1.29927	10.11370	0.76966	9.88630
Pb	$PbCl_2$	1.34225	10.12783	0.74502	9.87217
	$PbCO_3$	1.28958	10.11044	0.77545	9.88955
	$(PbCO_3)_2.Pb(OH)_2$	1.24781	10.09615	0.80140	9.90385
	$PbCrO_4$	1.55982	10.19307	0.64110	9.80693
	PbO	1.07722	10.03231	0.92832	9.96770
	PbO_2	1.15445	10.06238	0.86621	9.93762
	$Pb(OH)_2$	1.16415	10.06601	0.85900	9.93399
	PbS	1.15474	10.06248	0.86600	9.93752
	$PbSO_4$	1.46363	10.16543	0.68323	9.83457
$Pb(C_2H_3O_2).3H_2O$	$PbCrO_4$	0.85198	9.93043	1.17374	10.06957
	$PbSO_4$	0.79944	9.90279	1.25088	10.09722
$Pb(C_7H_5NO_2)_2$	Pb	0.43396	9.63746	2.30436	10.36255
	PbO	0.46747	9.66975	2.13971	10.33024
$PbCl_2$	Pb	0.74502	9.87218	1.34225	10.12783
	PbO	0.80255	9.90448	1.24603	10.09553
$PbCO_3$	Pb	0.77541	9.88954	1.28964	10.11047
	PbO	0.83529	9.92184	1.19719	10.07816

Weighed	Sought	Factor	Log of Factor +10	Reciprocal of Factor	Log of Reciprocal of Factor +10
Lead (contd.)					
$PbCO_3$	$PbSO_4$	1.13492	10.05497	0.88112	9.94504
$(PbCO_3)_2.Pb(OH)_2$	Pb	0.80141	9.90386	1.24780	10.09614
	$PbCrO_4$	1.25007	10.09693	0.79996	9.90307
	$PbSO_4$	1.17298	10.06929	0.85253	9.93017
$PbCrO_4$	Pb	0.64110	9.80692	1.55982	10.19307
	$Pb(C_2H_3O_2)_2.3H_2O$	1.17371	10.06956	0.85200	9.93044
	$(PbCO_3)_2.Pb(OH)_2$	0.79996	9.90307	1.25006	10.09693
	PbO	0.69061	9.83922	1.44800	10.16077
	Pb_3O_4	0.70710	9.84947	1.41423	10.15052
	$PbSO_4$	0.93833	9.97235	1.06572	10.02765
$Pb(NO_3)_2$	PbO	0.67388	9.82859	1.48394	10.17141
	PbO_2	0.72219	9.85865	1.38468	10.14135
	$PbSO_4$	0.91561	9.96171	1.09217	10.03829
PbO	Pb	0.92831	9.96770	1.07723	10.03231
	$PbCl_2$	1.24602	10.09553	0.80256	9.90448
	$PbCO_3$	1.19718	10.07816	0.83530	9.92184
	$PbCrO_4$	1.44800	10.16077	0.69061	9.83923
	$Pb(NO_3)_2$	1.48393	10.17141	0.67389	9.82859
	PbO_2	1.07169	10.03007	0.93311	9.96993
	PbS	1.07196	10.03018	0.93287	9.96982
	$PbSO_4$	1.35871	10.13313	0.73599	9.86687
PbO_2	Pb	0.86622	9.93763	1.15444	10.06237
	$Pb(NO_3)_2$	1.38467	10.14135	0.72219	9.85865
	PbO	0.93311	9.96993	1.07169	10.03007
	$PbSO_4$	1.26782	10.10306	0.78876	9.89694
Pb_3O_4	$PbCrO_4$	1.41423	10.15052	0.70710	9.84948
	$PbSO_4$	1.32702	10.12288	0.75357	9.87712
$Pb(OH)_2$	Pb	0.85900	9.93399	1.16414	10.06601
$Pb_2P_2O_7$	Pb	0.70434	9.84778	1.41977	10.15222
	PbO	0.75874	9.88009	1.31797	10.11991
PbS	Pb	0.86600	9.93752	1.15473	10.06249
	PbO	0.93287	9.96982	1.07196	10.03017
	$PbSO_4$	1.26750	10.10296	0.78895	9.89705
$PbSO_4$	$BaSO_4$	0.76966	9.88630	1.29928	10.11370
	Pb	0.68323	9.83457	1.46364	10.16543
	$Pb(C_2H_3O_2)_2.3H_2O$	1.25088	10.09722	0.79944	9.90279
	$PbCO_3$	0.88112	9.94504	1.13492	10.05497
	$(PbCO_3)_2.Pb(OH)_2$	0.85253	9.93071	1.17298	10.06929
	$PbCrO_4$	1.06572	10.02765	0.93833	9.97236
	$Pb(NO_3)_2$	1.09216	10.03828	0.91562	9.96172
	PbO	0.73599	9.86687	1.35871	10.13313
	PbO_2	0.78876	9.89694	1.26781	10.10305
	Pb_3O_4	0.75357	9.87712	1.32702	10.12288
	PbS	0.78895	9.89705	1.26751	10.10295
Lithium					
CO_2	Li_2CO_3	1.67887	10.22504	0.59564	9.77498
	$LiHCO_3$	1.54410	10.18868	0.64763	9.81133
	Li_2O	0.67887	9.83179	1.47304	10.16821
Li	LiCl	6.10924	10.78599	0.16369	9.21402
	Li_2CO_3	5.32404	10.72624	0.18783	9.27377
	Li_2O	2.15283	10.33201	0.46450	9.66699
	Li_3PO_4	5.56218	10.74524	0.17979	9.25477
	Li_2SO_4	7.92189	10.89883	0.12623	9.10115
LiCl	Li	0.16369	9.21402	6.10911	10.78598
	Li_2CO_3	0.87147	9.94025	1.14749	10.05975
	Li_2O	0.35239	9.54702	2.83776	10.45297
	Li_3PO_4	0.91045	9.95926	1.09836	10.04074
	Li_2SO_4	1.29670	10.11284	0.77119	9.88716
Li_2CO_3	CO_2	0.59564	9.77498	1.67887	10.22502
	Li	0.18782	9.27374	5.32425	10.72626
	LiCl	1.14747	10.05974	0.87148	9.94026
	$LiHCO_3$	1.83963	10.26473	0.54359	9.73527
	Li_2O	0.40436	9.60677	2.47304	10.39323
	Li_3PO_4	1.04473	10.01901	0.95719	9.98100
$LiHCO_3$	CO_2	0.64762	9.81132	1.54412	10.18868
	Li_2CO_3	0.54363	9.73530	1.83949	10.26470
	Li_2O	0.21982	9.34207	4.54918	10.65794
	Li_3PO_4	0.56795	9.75431	1.76072	10.24569

Weighed	Sought	Factor	Log of Factor +10	Reciprocal of Factor	Log of Reciprocal of Factor +10
Lithium (contd.)					
Li_2O	CO_2	1.47304	10.16821	0.67887	9.83179
	Li	0.46449	9.66698	2.15290	10.33303
	LiCl	2.83767	10.45296	0.35240	9.54704
	Li_2CO_3	2.47304	10.39323	0.40436	9.60677
	$LiHCO_3$	4.54890	10.65791	0.21983	9.34209
	Li_3PO_4	2.58364	10.41223	0.38705	9.58777
	Li_2SO_4	3.67975	10.56582	0.27175	9.43417
	SO_3	2.67872	10.42809	0.37317	9.57191
Li_3PO_4	Li	0.17979	9.25477	5.56204	10.74523
	LiCl	1.09835	10.04074	0.91046	9.95926
	Li_2CO_3	0.95718	9.98099	1.04474	10.01901
	$LiHCO_3$	1.76070	10.24569	0.56796	9.75432
	Li_2O	0.38705	9.58777	2.58365	10.41223
	$Li_2.SO_4.H_2O$	1.65761	10.21949	0.60328	9.78052
Li_2SO_4	Li	0.12623	9.10116	7.92205	10.89883
	LiCl	0.77118	9.88716	1.29671	10.11284
	Li_2O	0.27175	9.43417	3.67975	10.56582
	Li_3PO_4	0.70213	9.84642	1.42424	10.15358
	SO_3	0.72823	9.86227	1.37319	10.13773
$Li_2SO_4.H_2O$	Li_3PO_4	0.60328	9.78052	1.65761	10.22948
SO_3	Li_2O	0.37317	9.57191	2.67974	10.42809
	Li_2SO_4	1.37318	10.13773	0.72824	9.86227
Lutetium					
Lu_2O_3	Lu	0.87938	9.94418	1.13716	10.05582
Magnesium					
$BaSO_4$	$MgSO_4$	0.51574	9.71243	1.93896	10.28757
	$MgSO_4.7H_2O$	1.05605	10.02368	0.94692	9.97631
Br	Mg	0.15212	9.18219	6.57376	10.81781
	$MgBr_2$	1.15212	10.06150	0.86797	9.93850
	$MgBr_2.6H_2O$	1.82807	10.26200	0.54703	9.73801
Cl	Mg	0.34288	9.53514	2.91647	10.46486
	$MgCl_2$	1.34288	10.12804	0.74467	9.87196
	$MgCl_2.6H_2O$	2.86643	10.45734	0.34887	9.54266
CO_2	$MgCO_3$	1.91595	10.28239	0.52193	9.71761
	MgO	0.91595	9.96187	1.09176	10.03812
I	Mg	0.09579	8.98132	10.43950	11.01868
	MgI_2	1.09578	10.03972	0.91259	9.96028
Mg	Br	6.57262	10.81780	0.15212	9.18219
	Cl	2.91650	10.46487	0.34288	9.53514
	I	10.43965	11.01869	0.09579	8.98132
	$MgCO_3$	3.46829	10.54011	0.28833	9.45989
	MgO	1.65807	10.21960	0.60311	9.78040
	$Mg_2P_2O_7$	4.57731	10.66061	0.21847	9.33939
	$MgSO_4$	4.95122	10.69471	0.20197	9.30529
$MgBr_2$	Br	0.86796	9.93850	1.15213	10.06150
$MgBr.6H_2O$	Br	0.54702	9.73800	1.82809	10.26200
MgC_2O_4	Mg	0.21643	9.33532	4.62043	10.66468
	MgO	0.35886	9.55493	2.78660	10.44508
$Mg(C_9H_6NO)_2$	Mg	0.07777	8.89081	12.85843	11.10919
	$MgBr_2$	0.58898	9.77010	1.69785	10.22990
	$MgBr_2.6H_2O$	0.93455	9.97060	1.07003	10.02939
	$MgCO_3$	0.26972	9.43091	3.70755	10.56909
	$MgCl_2$	0.30458	9.48370	3.28321	10.51630
	$MgCl_2.6H_2O$	0.65014	9.81301	1.53813	10.18699
	$Mg(HCO_3)_2$	0.46813	9.67037	2.13616	10.32963
	MgO	0.12895	9.11042	7.75494	10.88958
	$MgSO_4$	0.38505	9.58552	2.59707	10.41448
	$MgSO_4.7H_2O$	0.78842	9.89676	1.26834	10.10324
$MgCl_2$	Cl	0.74467	9.87196	1.34288	10.12804
	$Mg_2P_2O_7$	1.16872	10.06771	0.85564	9.93229
$MgCl_2.6H_2O$	Cl	0.34887	9.54266	2.86640	10.45723
	$Mg_2P_2O_7$	0.54753	9.73841	1.82638	10.26159
$MgCl_2.KCl.6H_2O$	$Mg_2P_2O_7$	0.40059	9.60270	2.49632	10.39730
$MgCO_3$	CO_2	0.52193	9.71761	1.91597	10.28239
	Mg	0.28833	9.45989	3.46825	10.54011
	$Mg(HCO_3)_2$	1.73559	10.23945	0.57617	9.76055
	MgO	0.47807	9.67949	2.09174	10.32051
	$Mg_2P_2O_7$	1.31977	10.12049	0.75771	9.87950

GRAVIMETRIC FACTORS AND THEIR LOGARITHMS (Continued)

Weighed	Sought	Factor	Log of Factor +10	Reciprocal of Factor	Log of Reciprocal of Factor +10
Magnesium (contd.)					
$Mg(HCO_3)_2$	$MgCO_3$	0.57617	9.76055	1.73560	10.23945
	MgO	0.27545	9.44004	3.63042	10.55996
	$Mg_2P_2O_7$	0.76041	9.88105	1.31508	10.11895
MgI_2	I	0.91258	9.96027	1.09579	10.03973
MgO	CO_2	1.09176	10.03812	0.91595	9.96187
	Mg	0.60311	9.78040	1.65807	10.21960
	$MgCO_3$	2.09176	10.32051	0.47807	9.67949
	$Mg(HCO_3)_2$	3.63045	10.55996	0.27545	9.44004
	$Mg_2P_2O_7$	2.76064	10.44101	0.36223	9.55898
	$MgSO_4$	2.98613	10.47511	0.33488	9.52489
	SO_3	1.98611	10.29800	0.50350	9.70200
$Mg_2P_2O_7$	Mg	0.21847	9.33939	4.57729	10.66061
	$MgCl_2$	0.85563	9.93229	1.16873	10.06771
	$MgCl_2.6H_2O$	1.82638	10.26159	0.54753	9.73841
	$MgCl_2.KCl.6H_2O$	2.49633	10.39730	0.40059	9.60270
	$MgCO_3$	0.75771	9.87950	1.31977	10.12049
	$Mg(HCO_3)_2$	1.31508	10.11896	0.76041	9.88105
	MgO	0.36224	9.55900	2.76060	10.44101
	$MgSO_4$	1.08168	10.03410	0.92449	9.96590
	$MgSO_4.7H_2O$	2.21488	10.34535	0.45149	9.65465
$MgSO_4$	$BaSO_4$	1.93896	10.28757	0.51574	9.71243
	Mg	0.20197	9.30529	4.85123	10.69471
	MgO	0.33488	9.52489	2.98614	10.47511
	$Mg_2P_2O_7$	0.92449	9.96590	1.08168	10.03410
	SO_3	0.66511	9.82289	1.50351	10.17711
$MgSO_4.7H_2O$	$BaSO_4$	0.94693	9.97632	1.05604	10.02368
	$Mg_2P_2O_7$	0.45149	9.65465	2.21489	10.34535
	SO_3	0.32482	9.51164	3.07863	10.48836
SO_3	MgO	0.50350	9.70200	1.98610	10.29800
	$MgSO_4$	1.50351	10.17711	0.66511	9.82289
	$MgSO_4.7H_2O$	3.07863	10.48836	0.32482	9.51164
Manganese					
$BaSO_4$	$MnSO_4$	0.64696	9.81088	1.54569	10.18912
CO_2	$MnCO_3$	2.61184	10.41694	0.38287	9.58305
	MnO	1.61184	10.20733	0.62041	9.79268
Mn	$MnCO_3$	2.09230	10.32062	0.47794	9.67937
	MnO	1.29122	10.11100	0.77446	9.88900
	MnO_2	1.58246	10.19933	0.63193	9.80067
	Mn_2O_3	1.43684	10.15741	0.69597	9.84259
	Mn_3O_4	1.38830	10.14248	0.72031	9.85752
	$Mn_2P_2O_7$	2.58308	10.41213	0.38713	9.58786
MnC_2O_4	Mn	0.38429	9.58466	2.60220	10.41534
	Mn_3O_4	0.53352	9.72715	1.87434	10.27285
$Mn(C_9H_6NO)_2$	Mn	0.16005	9.20426	6.24805	10.79574
	Mn_3O_4	0.22220	9.34674	4.50045	10.65326
$MnCO_3$	CO_2	0.38287	9.58305	2.61185	10.41695
	Mn	0.47794	9.67937	2.09231	10.32063
	$Mn(HCO_3)_2$	1.53961	10.18741	0.64952	9.81259
	MnO	0.61713	9.79038	1.62040	10.20962
	Mn_3O_4	0.66353	9.82186	1.50709	10.17814
	$Mn_2P_2O_7$	1.23456	10.09152	0.81001	9.90849
	MnS	0.75689	9.87903	1.32120	10.12097
	$MnSO_4$	1.31365	10.11848	0.76124	9.88152
$Mn(HCO_3)_2$	$MnCO_3$	0.64952	9.81259	1.53960	10.18741
	MnO	0.40084	9.60297	2.49476	10.39703
	Mn_3O_4	0.43098	9.63446	2.32029	10.36555
MnO	CO_2	0.62041	9.79268	1.61184	10.20733
	Mn	0.77446	9.88900	1.29122	10.11099
	$MnCO_3$	1.62041	10.20963	0.61713	9.79038
	$Mn(HCO_3)_2$	2.49479	10.39703	0.40084	9.60297
	Mn_2O_3	1.11277	10.04641	0.89866	9.95360
	Mn_3O_4	1.07518	10.03148	0.93008	9.96852
	$Mn_2P_2O_7$	2.00049	10.30114	0.49988	9.69887
	MnS	1.22647	10.08865	0.81535	9.91134
	$MnSO_4$	2.12865	10.32811	0.46978	9.67189
	SO_3	1.12864	10.05255	0.88602	9.94744
MnO_2	Mn	0.63193	9.80067	1.58245	10.19933
	Mn_3O_4	0.87731	9.94315	1.13985	10.05684

Weighed	Sought	Factor	Log of Factor +10	Reciprocal of Factor	Log of Reciprocal of Factor +10
Manganese (contd.)					
MnO_2	$Mn_2P_2O_7$	1.63233	10.21281	0.61262	9.78719
Mn_2O_3	Mn	0.69597	9.84259	1.43684	10.15741
	MnO	0.89865	9.95359	1.11278	10.04641
	Mn_3O_4	0.96622	9.98508	1.03496	10.01492
Mn_3O_4	Mn	0.72030	9.85751	1.38831	10.14249
	$MnCO_3$	1.50709	10.17814	0.66353	9.82186
	$Mn(HCO_3)_2$	2.32032	10.36555	0.43098	9.63446
	MnO	0.93007	9.96852	1.07519	10.03149
	MnO_2	1.13984	10.05684	0.87732	9.94316
	Mn_2O_3	1.03496	10.01492	0.96622	9.98508
	$MnSO_4$	1.97978	10.29662	0.50511	9.70339
$Mn_2P_2O_7$	Mn	0.38713	9.58786	2.58311	10.41214
	$MnCO_3$	0.81000	9.90849	1.23457	10.09152
	MnO	0.49987	9.69886	2.00052	10.30114
	MnO_2	0.61262	9.78719	1.63233	10.21281
	$MnSO_4$	1.06405	10.02696	0.93981	9.97304
MnS	Mn	0.63146	9.80035	1.58363	10.19966
	$MnCO_3$	1.32120	10.12097	0.75689	9.87903
	MnO	0.81535	9.91134	1.22647	10.08865
	$MnSO_4$	1.73559	10.23945	0.57617	9.76055
$MnSO_4$	$BaSO_4$	1.54570	10.18913	0.64696	9.81088
	Mn	0.36383	9.56090	2.74854	10.43910
	MnO	0.46978	9.67189	2.12866	10.32811
	Mn_3O_4	0.50510	9.70338	1.97981	10.29663
	$Mn_2P_2O_7$	0.93980	9.97304	1.06406	10.02696
	MnS	0.57617	9.76055	1.73560	10.23945
	SO_3	0.53021	9.72445	1.88605	10.27555
SO_3	MnO	0.88603	9.94745	1.12863	10.05255
	$MnSO_4$	1.88604	10.27555	0.53021	9.72445
Mercury					
Hg	HgCl	1.17673	10.07068	0.84981	9.92932
	$HgCl_2$	1.35351	10.13146	0.73882	9.86854
	HgO	1.07976	10.03332	0.92613	9.96667
	HgS	1.15983	10.06440	0.86220	9.93561
$Hg_3(AsO_4)_2$	Hg	0.68413	9.83514	1.46171	10.16486
$Hg(C_{12}H_{10}ONS)_2$	Hg	0.31681	9.50080	3.15647	10.49920
HgCl	Hg	0.84981	9.92932	1.17673	10.07068
	$HgCl_2$	1.15023	10.06079	0.86939	9.93921
	$HgNO_3$	1.11248	10.04629	0.89889	9.95371
	HgO	0.91760	9.96265	1.08980	10.03735
	Hg_2O	0.88369	9.94630	1.13162	10.05370
	HgS	0.98564	9.99372	1.01457	10.00629
$HgCl_2$	Hg	0.73882	9.86854	1.35351	10.13146
	HgCl	0.86939	9.93921	1.15023	10.06079
	HgS	0.85691	9.93294	1.16698	10.06706
$Hg(CN)_2$	HgS	0.92091	9.96422	1.08588	10.03578
Hg_2CrO_4	Hg	0.77572	9.88971	1.28912	10.11029
$HgNO_3$	HgCl	0.89889	9.95371	1.11248	10.04629
	HgS	0.88598	9.94742	1.12869	10.05257
$Hg(NO_3)_2$	HgS	0.71673	9.85536	1.39523	10.14464
$Hg(NO_3)_2.H_2O$	HgS	0.67903	9.83189	1.47269	10.16812
HgO	Hg	0.92613	9.96667	1.07977	10.03333
	HgCl	1.08980	10.03735	0.91760	9.96265
	HgS	1.07415	10.03106	0.93097	9.96894
Hg_2O	HgCl	1.13160	10.05369	0.88370	9.94630
	HgS	1.11535	10.04741	0.89658	9.95259
HgS	Hg	0.86220	9.93561	1.15982	10.06440
	HgCl	1.01457	10.00629	0.98564	9.99372
	$HgCl_2$	1.16698	10.06706	0.85691	9.93294
	$Hg(CN)_2$	1.08588	10.03578	0.92091	9.96422
	$HgNO_3$	1.12869	10.05257	0.88598	9.94743
	$Hg(NO_3)_2$	1.39522	10.14464	0.71673	9.85536
	$Hg(NO_3)_2.H_2O$	1.47268	10.16811	0.67903	9.83189
	HgO	0.93097	9.96894	1.07415	10.03106
	Hg_2O	0.89656	9.95258	1.11537	10.04741
	$HgSO_4$	1.27509	10.10554	0.78426	9.89446
$HgSO_4$	HgS	0.78426	9.89446	1.27509	10.10554
Molybdenum					
Mo	MoO_3	1.50031	10.17618	0.66653	9.82382

Weighed	Sought	Factor	Log of Factor +10	Reciprocal of Factor	Log of Reciprocal of Factor +10
Molybdenum					
Mo	MoC	1.12518	10.05122	0.88875	9.94878
	MoS_3	2.00261	10.30159	0.49935	9.69841
	$PbMoO_4$	3.82666	10.58282	0.26132	9.41717
MoC	C	0.11127	9.04638	8.89715	10.95362
	Mo	0.88874	9.94877	1.12519	10.05122
$MoO_2(C_9H_6NO)_2$	Mo	0.23049	9.36265	4.33858	10.63735
	MoO_3	0.34580	9.53883	2.89184	10.46118
MoO_3	Mo	0.66653	9.82382	1.50031	10.17617
	MoS_3	1.33479	10.12541	0.74918	9.87459
	$(NH_4)_2MoO_4$	1.36175	10.13410	0.73435	9.86590
	$(NH_4)_3PO_4.12MoO_3$	1.08630	10.03595	0.92056	9.96405
	$PbMoO_4$	2.55058	10.40664	0.39207	9.59336
MoS_3	Mo	0.49935	9.69841	2.00260	10.30159
	MoO_3	0.74918	9.87459	1.33479	10.12541
	$(NH_4)_2MoO_4$	1.02019	10.00868	0.98021	9.99132
MoSi	Mo	0.77352	9.88847	1.29279	10.11152
	Si	0.22645	9.35497	4.41599	10.64503
$(NH_4)_2MoO_4$	MoO_3	0.73435	9.86590	1.36175	10.13410
	MoS_3	0.98021	9.99132	1.02019	10.00868
	$(NH_4)_3PO_4.12MoO_3$	0.79771	9.90185	1.25359	10.09816
	$PbMoO_4$	1.87302	10.27254	0.53390	9.72746
$(NH_4)_3PO_4.12MoO_3$	MoO_3	0.92054	9.96404	1.08632	10.03596
	$(NH_4)_2MoO_4$	1.25359	10.09816	0.79771	9.90185
$PbMoO_4$	Mo	0.26132	9.41717	3.82673	10.58283
	MoO_3	0.39207	9.59336	2.55056	10.40664
	$(NH_4)_2MoO_4$	0.53390	9.72746	1.87301	10.27254
Neodymium					
Nd	Nd_2O_3	1.16639	10.06684	0.85735	9.93315
Nd_2O_3	Nd	0.85735	9.93316	1.16638	10.06684
Nickel					
Ni	$Ni(C_4H_7N_2O_2)_2$	4.92148	10.69209	0.20319	9.30790
	$Ni(NO_3)_2.6H_2O$	4.95231	10.69481	0.20193	9.30520
	NiO	1.27254	10.10467	0.78584	9.89533
	$NiSO_4$	2.63618	10.42099	0.37934	9.57903
Ni	$NiSO_4.7H_2O$	4.78419	10.67981	0.20902	9.32019
$Ni(C_4H_7N_2O_2)_2$	Ni	0.20319	9.30790	4.92150	10.69210
	NiO	0.25857	9.41258	3.86742	10.58742
$(Ni[C_5H_5N]_4)(SCN)_2$	Ni	0.11950	9.07737	8.36820	10.92263
	NiO	0.15207	9.18204	6.57592	10.81776
$Ni(C_7H_6O_2N)_2$	Ni	0.17736	9.24886	5.63825	10.75114
	NiO	0.22573	9.35359	4.43007	10.64641
$Ni(C_9H_6NO)_2$	Ni	0.16918	9.22835	5.91086	10.77165
	NiO	0.21529	9.33303	4.64490	10.66698
$Ni(NO_3)_2.6H_2O$	Ni	0.20193	9.30520	4.95221	10.69480
	NiO	0.25696	9.40987	3.89166	10.59013
	$NiSO_4$	0.53231	9.72616	1.87860	10.27383
NiO	Ni	0.78584	9.89533	1.27252	10.10467
	$Ni(C_4H_7N_2O_2)_2$	3.86749	10.58743	0.25857	9.41258
	$Ni(NO_3)_2.6H_2O$	3.89171	10.59014	0.25696	9.40987
	$NiSO_4$	2.07161	10.31631	0.48272	9.68370
	$NiSO_4.7H_2O$	3.75960	10.57514	0.26599	9.42487
$NiSO_4$	Ni	0.37934	9.57903	2.63616	10.42098
	$Ni(NO_3)_2.6H_2O$	1.87859	10.27384	0.53231	9.72616
	NiO	0.48272	9.68370	2.07159	10.31630
	$NiSO_4.7H_2O$	1.81482	10.25884	0.55102	9.74117
$NiSO_4.7H_2O$	Ni	0.20902	9.32019	4.78423	10.67981
	NiO	0.26599	9.42487	3.75954	10.57514
	$NiSO_4$	0.55102	9.74117	1.81482	10.25884
Niobium					
C	Nb	7.73497	10.88846	0.12928	9.11153
	NbC	8.73493	10.94126	0.11448	9.05873
	$\frac{1}{2}Nb_2O_5$	11.06508	11.04396	0.09038	8.95607
Nb	C	0.12928	9.11153	7.73515	10.88848
	Nb_2O_5	1.43053	10.15550	0.69904	9.84450
NbC	C	0.11448	9.05873	8.73515	10.94127
	Nb	0.88552	9.94720	1.12928	10.05280
Nb_2O_5	Nb	0.69904	9.84450	1.43053	10.15550
Nitrogen					
$AgNO_2$	HNO_2	0.30552	9.48504	3.27311	10.51496

Weighed	Sought	Factor	Log of Factor +10	Reciprocal of Factor	Log of Reciprocal of Factor +10
Nitrogen (contd.)					
$AgNO_2$	N_2O_3	0.24698	9.39266	4.04891	10.60733
$C_{20}H_{16}N_4.HNO_3$	N	0.37312	9.57185	2.68010	10.42815
	NO_3	0.16517	9.21793	6.05437	10.78207
HNO_2	$AgNO_2$	3.27310	10.51496	0.30552	9.48504
HNO_3	N	0.22228	9.34690	4.49883	10.65310
	NH_3	0.27027	9.43180	3.70000	10.56820
	NH_4Cl	0.84889	9.92885	1.17801	10.07115
	$(NH_4)_2PtCl_6$	3.52221	10.54682	0.28391	9.45318
	NO	0.47619	9.67778	2.10000	10.32222
	Pt	1.54801	10.18977	0.64599	9.81023
	SO_3	0.63528	9.80297	1.57411	10.19703
	N_2O_5	0.53413	9.72765	1.87220	10.27235
N	HNO_3	4.49877	10.65310	0.22229	9.34692
	$NaNO_3$	6.06816	10.78306	0.16479	9.21693
	NH_3	1.21589	10.08489	0.82244	9.91510
	NH_4Cl	3.81902	10.58195	0.26185	9.41805
	$(NH_4)_2PtCl_6$	15.84563	11.19991	0.06311	8.80010
	$(NH_4)_2SO_4$	4.71699	10.67367	0.21200	9.32634
	NO_2	3.28454	10.51648	0.30446	9.48353
	NO_3	4.42680	10.64609	0.22590	9.35392
	N_2O_3	2.71341	10.43352	0.36854	9.56648
	N_2O_5	3.85566	10.58610	0.25936	9.41390
	Pt	6.96416	10.84287	0.14359	9.15712
	SO_3	2.85798	10.45606	0.34990	9.54394
$NaNO_3$	N	0.16479	9.21693	6.06833	10.78307
	N_2O_5	0.63539	9.80304	1.57384	10.19696
NH_3	HNO_3	3.69998	10.56820	0.27027	9.43180
	N	0.82244	9.91510	1.21589	10.08489
	NO_3	3.64079	10.56119	0.27467	9.43881
	N_2O_5	3.17107	10.50121	0.31535	9.49879
$NH_4B(C_6H_5)_4$	N	0.04153	8.61836	24.07898	11.38164
	NH_3	0.05049	8.70320	19.80590	11.29679
	NH_4	0.05348	8.72819	18.69858	11.27181
NH_4Cl	HNO_3	1.17799	10.07115	0.84890	9.92886
	N	0.26185	9.41805	3.81898	10.58195
	NO_3	1.15914	10.06413	0.86271	9.93586
	N_2O_5	1.00959	10.00415	0.99050	9.99585
$(NH_4)_2PtCl_6$	HNO_3	0.28392	9.45320	3.52212	10.54680
	N	0.06311	8.80010	15.84535	11.19990
	NO_3	0.27938	9.44620	3.57935	10.55380
	N_2O_5	0.24333	9.38620	4.10965	10.61380
$(NH_4)_2SO_4$	N	0.21200	9.32634	4.71698	10.67367
	N_2O_5	0.81740	9.91243	1.22339	10.08757
NO	HNO_3	2.10000	10.32222	0.47619	9.67778
	NO_2	1.53320	10.18560	0.65223	9.81440
	NO_3	2.06641	10.31522	0.48393	9.68478
	N_2O_3	1.26660	10.10264	0.78952	9.89736
	N_2O_5	1.79980	10.25522	0.55562	9.74478
NO_2	N	0.30446	9.48353	3.28450	10.51647
	NO	0.65223	9.81440	1.53320	10.18560
NO_3	N	0.22590	9.35392	4.42674	10.64608
	NH_3	0.27467	9.43881	3.64073	10.56119
	NH_4Cl	0.86270	9.93586	1.15915	10.06413
	NO	0.48393	9.68478	2.06641	10.31522
	Pt	1.57318	10.19678	0.63566	9.80322
N_2O_3	$AgNO_2$	4.04875	10.60732	0.24699	9.39268
	N	0.36854	9.56648	2.71341	10.43352
	NO	0.78951	9.89736	1.26661	10.10264
N_2O_5	KNO_3	1.87217	10.27235	0.53414	9.72766
	N	0.25936	9.41390	3.85564	10.58610
	$NaNO_3$	1.57382	10.19695	0.63540	9.80305
	NH_3	0.31535	9.49879	3.17108	10.50121
	NH_4Cl	0.99049	9.99585	1.00960	10.00415
	$(NH_4)_2PtCl_6$	4.10965	10.61380	0.24333	9.38620
	$(NH_4)_2SO_4$	1.22339	10.08757	0.81740	9.91243
	NO	0.55562	9.74478	1.79980	10.25522
	Pt	1.80621	10.25677	0.55365	9.74324
	SO_3	0.74124	9.86996	1.34909	10.13004

Weighed	Sought	Factor	Log of Factor +10	Reciprocal of Factor	Log of Reciprocal of Factor +10
Nitrogen (contd.)					
Pt	HNO_3	0.64595	9.81020	1.54811	10.18980
	N	0.14358	9.15709	6.96476	10.84291
	NO_3	0.63562	9.80320	1.57327	10.19680
	N_2O_5	0.55364	9.74323	1.80623	10.25678
SO_3	HNO_3	1.57410	10.19703	0.63528	9.80297
	N	0.34990	9.54394	2.85796	10.45605
	N_2O_5	1.34908	10.13004	0.74125	9.86996
Osmium					
Os	OsO_4	1.33649	10.12597	0.74823	9.87404
OsO_4	Os	0.74823	9.87404	1.33649	10.12597
Palladium					
K_2PdCl_6	Pd	0.26781	9.42783	3.73399	10.57217
	$PdCl_2.2H_2O$	0.53687	9.72987	1.86265	10.27013
Pd	K_2PdCl_6	3.73402	10.57217	0.26781	9.42783
	$PdCl_2.2H_2O$	2.00460	10.30205	0.49883	9.69795
	PdI_2	3.38534	10.52960	0.29539	9.47040
	$Pd(NO_3)_2$	2.16541	10.33554	0.46181	9.66446
$Pd(C_4H_7N_2O_2)_2$	Pd	0.31610	9.49982	3.16356	10.50018
$PdCl_2.2H_2O$	K_2PdCl_6	1.86263	10.27012	0.53688	9.72988
	Pd	0.49883	9.69795	2.00469	10.30204
$PdCl_2.C_{12}H_8N_2$	Pd	0.29762	9.47366	3.35999	10.52634
PdI_2	Pd	0.29539	9.47040	3.38536	10.52960
$Pd(NO_3)_2$	Pd	0.46181	9.66446	2.16539	10.33554
Phosphorus					
Ag_3PO_4	P	0.07400	8.86923	13.51351	11.13077
	PO_4	0.22689	9.35582	4.40742	10.64418
	P_2O_5	0.16954	9.22927	5.89831	10.77072
$Ag_4P_2O_7$	P	0.10232	9.00996	9.77326	10.99004
	PO_4	0.31374	9.49657	3.18735	10.50343
	P_2O_5	0.23446	9.37007	4.26512	10.62993
Al_2O_3	P_2O_5	1.39214	10.14368	0.71832	9.85632
$AlPO_4$	PO_4	0.77875	9.89140	1.28411	10.10860
	P_2O_5	0.58196	9.76489	1.71833	10.23511
$Ca_3(PO_4)_2$	P_2O_5	0.45762	9.66051	2.18522	10.33949
$FePO_4$	PO_4	0.62971	9.79914	1.58803	10.20086
	P_2O_5	0.47058	9.67263	2.12504	10.32737
$Mg_2P_2O_7$	H_3PO_4	0.88059	9.94471	1.13560	10.05523
	Na_2HPO_4	1.27565	10.10573	0.78391	9.89427
	$Na_2HPO_4.12H_2O$	3.21828	10.50763	0.31073	9.49238
	$NaNH_4.HPO_4.4H_2O$	1.87870	10.27386	0.53228	9.72614
	P	0.27833	9.44456	3.59286	10.55544
	PO_4	0.85341	9.93116	1.17177	10.06884
	P_2O_5	0.63776	9.80466	1.56799	10.19535
Na_2HPO_4	$Mg_2P_2O_7$	0.78392	9.89427	1.27564	10.10573
	P_2O_5	0.49995	9.69893	2.00020	10.30105
$Na_2HPO_4.12H_2O$	$Mg_2P_2O_7$	0.31072	9.49237	3.21833	10.50763
	P_2O_5	0.19816	9.29702	5.04643	10.70299
$NaNH_4HPO_4.4H_2O$	$Mg_2P_2O_7$	0.53228	9.72614	1.87871	10.27386
	P_2O_5	0.33946	9.53079	2.94586	10.46921
$(NH_4)_3PO_4.12MoO_3$	P	0.01651	8.21775	60.56935	11.78223
	PO_4	0.05061	8.70424	19.75894	11.29577
	P_2O_5	0.03782	8.57772	26.44104	11.42227
P	Ag_3PO_4	13.51403	11.13078	0.07400	8.86923
	$Ag_4P_2O_7$	9.77314	10.99004	0.10232	9.00996
	$Mg_2P_2O_7$	3.59282	10.55544	0.27833	9.44456
	$(NH_4)_3PO_4.12MoO_3$	60.5786	11.78234	0.01651	8.21775
	P_2O_5	2.29136	10.36009	0.43642	9.63990
	$P_2O_5.24MoO_3$	58.0564	11.76385	0.01723	8.23616
	$U_2P_2O_{11}$	11.52587	11.06167	0.08676	8.93832
PO_4	Ag_3PO_4	4.40744	10.64418	0.22689	9.35582
	$Ag_4P_2O_7$	3.18739	10.50343	0.31374	9.49657
	$AlPO_4$	1.28410	10.10860	0.77876	9.89140
	$FePO_4$	1.58803	10.20086	0.62971	9.79914
	$Mg_2P_2O_7$	1.17175	10.06884	0.85342	9.93116
	$(NH_4)_3PO_4.12MoO_3$	19.7570	11.29572	0.05062	8.70432
	$P_2O_5.24MoO_3$	18.9344	11.27725	0.05281	8.72272
	$U_2P_2O_{11}$	3.75902	10.57507	0.26603	9.42493
P_2O_5	Ag_3PO_4	5.89780	10.77069	0.16955	9.22930

Weighed	Sought	Factor	Log of Factor +10	Reciprocal of Factor	Log of Reciprocal of Factor +10
Phosphorus (contd.)					
P_2O_5	$Ag_4P_2O_7$	4.26520	10.62994	0.23446	9.37007
	Al_2O_3	0.71831	9.85631	1.39216	10.14369
	$AlPO_4$	1.71831	10.23510	0.58197	9.76440
	$Ca_3(PO_4)_2$	2.18521	10.33949	0.45762	9.66051
	$FePO_4$	2.12502	10.32736	0.47058	9.67263
	$Mg_2P_2O_7$	1.56798	10.19534	0.63776	9.80466
	Na_2HPO_4	2.00020	10.30107	0.49995	9.69893
	$Na_2HPO_4.12H_2O$	5.04623	10.70297	0.19817	9.29704
	$NaNH_4HPO_4.4H_2O$	2.94578	10.46920	0.33947	9.53080
	$(NH_4)_3PO_4.12MoO_3$	26.4377	11.42222	0.03783	8.57784
	P	0.43642	9.63990	2.29137	10.36010
	$P_2O_5.24MoO_3$	25.3370	11.40376	0.03947	8.59622
	$U_2P_2O_{11}$	5.03013	10.70158	0.19880	9.29842
$P_2O_5.24MoO_3$	P	0.01722	8.23603	58.07201	11.76397
	PO_4	0.05281	8.72272	18.93581	11.27728
	P_2O_5	0.03947	8.59627	25.33570	11.40373
$U_2P_2O_{11}$	P	0.08676	8.93832	11.52605	11.06176
	PO_4	0.26603	9.42493	3.75897	10.57507
	P_2O_5	0.19880	9.29842	5.03018	10.70158
Platinum					
$H_2PtCl_6.6H_2O$	K_2PtCl_6	0.93852	9.97244	1.06551	10.02756
	Pt	0.37673	9.57603	2.65442	10.42397
K_2PtCl_6	$H_2PtCl_6.6H_2O$	1.06551	10.02756	0.93852	9.97244
	Pt	0.40141	9.60359	2.49122	10.39541
	$PtCl_4$	0.69320	9.84086	1.44259	10.15915
	$PtCl_4.5H_2O$	0.87854	9.94376	1.13825	10.05624
$(NH_4)_2PtCl_6$	Pt	0.43950	9.64296	2.27531	10.35704
	$PtCl_4$	0.75897	9.88022	1.31758	10.11978
	$PtCl_6$	0.91854	9.96310	1.08868	10.03690
Pt	$H_2PtCl_6.6H_2O$	2.65442	10.42397	0.37673	9.57603
	K_2O	0.48287	9.68383	2.07095	10.31617
	K_2PtCl_6	2.49121	10.39641	0.40141	9.60359
	$(NH_4)_2PtCl_6$	2.27531	10.35704	0.43950	9.64296
	$PtCl_4$	1.72690	10.23727	0.57907	9.76273
	$PtCl_4.5H_2O$	2.18863	10.34017	0.45691	9.65983
$PtCl_4$	K_2PtCl_6	1.44259	10.15915	0.69320	9.84086
	$(NH_4)_2PtCl_6$	1.31757	10.11978	0.75897	9.88022
	Pt	0.57907	9.76273	1.72691	10.23727
$PtCl_4.5H_2O$	K_2PtCl_6	1.13825	10.05624	0.87854	9.94376
	Pt	0.45691	9.65983	2.18861	10.34017
$PtCl_6$	K_2PtCl_6	1.19199	10.07628	0.83893	9.92373
	$(NH_4)_2PtCl_6$	1.08869	10.03691	0.91854	9.96310
Potassium					
Ag	KBr	1.10328	10.04269	0.90639	9.95732
	KCl	0.69116	9.83958	1.44684	10.16042
	$KClO_3$	1.13612	10.05543	0.88019	9.94458
	$KClO_4$	1.28444	10.10872	0.77855	9.89129
	KCN	0.60369	9.78081	1.65648	10.21919
	KI	1.53895	10.18723	0.64979	9.81277
AgBr	KBr	0.63378	9.80194	1.57783	10.19806
	$KBrO_3$	0.88939	9.94909	1.12437	10.05091
AgCl	KCl	0.52019	9.71616	1.92237	10.28384
	$KClO_3$	0.85508	9.93201	1.16948	10.06799
	$KClO_4$	0.96672	9.98530	1.03443	10.01470
AgCN	KCN	0.48638	9.68698	2.05601	10.31302
AgI	KI	0.70709	9.84947	1.41425	10.15053
	KIO_3	0.91154	9.95978	1.09704	10.04022
$BaCrO_4$	K_2CrO_4	0.76658	9.88456	1.30450	10.11544
	$K_2Cr_2O_7$	0.58064	9.76391	1.72224	10.23609
$BaSO_4$	$KHSO_4$	0.58343	9.76599	1.71400	10.23401
	K_2S	0.47244	9.67435	2.11667	10.32565
	K_2SO_4	0.74664	9.87311	1.33933	10.12689
Br	K	0.48933	9.68960	2.04361	10.31040
	KBr	1.48933	10.17299	0.67144	9.82701
CaF_2	$KF.2H_2O$	2.41114	10.38223	0.41474	9.61778
$CaSO_4$	$KF.2H_2O$	1.38286	10.14078	0.72314	9.85922
Cl	K	1.10292	10.04255	0.90668	9.95745
	KCl	2.10292	10.32282	0.47553	9.67718

Weighed	Sought	Factor	Log of Factor +10	Reciprocal of Factor	Log of Reciprocal of Factor +10
Potassium (contd.)					
Cl	$KClO_3$	3.45677	10.53867	0.28929	9.46133
	$KClO_4$	3.90808	10.59196	0.25588	9.40804
	K_2O	1.32856	10.12338	0.75269	9.87662
CO_2	$KHCO_3$	2.27491	10.35696	0.43958	9.64304
	K_2CO_3	3.14049	10.49700	0.31842	9.50300
	K_2O	2.14049	10.33051	0.46718	9.66948
I	KI	1.30811	10.11665	0.76446	9.88335
	KIO_3	1.68634	10.22695	0.59300	9.77306
K	Br	2.04360	10.31040	0.48933	9.68960
	Cl	0.90668	9.95745	1.10292	10.04256
	KBr	3.04360	10.48339	0.32856	9.51661
	KCl	1.90668	10.28028	0.52447	9.71972
	$KClO_3$	3.13419	10.49614	0.31906	9.50387
	$KClO_4$	3.54337	10.54941	0.28222	9.45059
	KI	4.24546	10.62793	0.23555	9.37208
	KNO_3	2.58572	10.41259	0.38674	9.58742
	K_2O	1.20458	10.08084	0.83016	9.91916
	K_2PtCl_6	6.21467	10.79342	0.16091	9.20658
	K_2SO_4	2.22835	10.34799	0.44876	9.65201
	Pt	2.49463	10.39701	0.40086	9.60299
K_3AsO_4	$Mg_2As_2O_7$	0.60584	9.78236	1.65060	10.21764
$K(B[C_6H_5]_4)$	K	0.10912	9.03790	9.16422	10.96210
	K_2O	0.13144	9.11873	7.60803	10.88127
KBr	Ag	0.90639	9.95732	1.10328	10.04269
	AgBr	1.57783	10.19806	0.63378	9.80194
	Br	0.67144	9.82701	1.48934	10.17299
	K	0.32856	9.51662	3.04358	10.48338
	K_2O	0.39580	9.59748	2.52653	10.40253
$KBrO_3$	AgBr	1.12436	10.05091	0.88939	9.94909
$KC_{12}H_4O_{12}N_7$	K	0.08192	8.91339	12.20703	11.08661
	K_2O	0.09868	8.99423	10.13377	11.00577
KCl	Ag	1.44685	10.16043	0.69116	9.83958
	AgCl	1.92238	10.28384	0.52019	9.71616
	Cl	0.47553	9.67718	2.10292	10.32283
	K	0.52447	9.71972	1.90669	10.28028
	$KClO_3$	1.64379	10.21585	0.60835	9.78415
	$KClO_4$	1.85840	10.26914	0.53810	9.73086
	K_2CO_3	0.92692	9.96704	1.07884	10.03296
	$K_2Cr_2O_7$	1.97299	10.29512	0.50684	9.70487
	$KHCO_3$	1.34289	10.12804	0.74466	9.87196
	KNO_3	1.35614	10.13230	0.73739	9.86770
	K_2O	0.63180	9.80058	1.58278	10.19942
	K_2PtCl_6	3.25941	10.51314	0.30680	9.48686
	K_2SO_4	1.16871	10.06770	0.85564	9.93229
	Pt	1.30836	10.11673	0.76432	9.88328
$KClO_3$	Ag	0.88019	9.94458	1.13612	10.05543
	AgCl	1.16948	10.06799	0.85508	9.93201
	Cl	0.28929	9.46133	3.45674	10.53867
	KCl	0.60835	9.78415	1.64379	10.21584
$KClO_4$	Ag	0.77855	9.89129	1.28444	10.10872
	AgCl	1.03443	10.01470	0.96672	9.98530
	Cl	0.25588	9.40804	3.90808	10.59196
	K	0.28222	9.45059	3.54333	10.54941
	KCl	0.53810	9.73086	1.85839	10.26914
	K_2O	0.33995	9.53142	2.94161	10.46858
KCN	Ag	1.65648	10.21918	0.60369	9.78081
	AgCN	2.05602	10.31302	0.48638	9.68698
K_2CO_3	CO_2	0.31842	9.50300	3.14051	10.49700
	KCl	1.07883	10.03295	0.92693	9.96705
	K_2O	0.68158	9.83352	1.46718	10.16648
	KOH	0.81191	9.90951	1.23166	10.09049
	K_2PtCl_6	3.51638	10.54610	0.28438	9.45390
	K_2SO_4	1.26085	10.10067	0.79312	9.89934
K_2CrO_4	$BaCrO_4$	1.30449	10.11544	0.76658	9.88456
$K_2Cr_2O_7$	$BaCrO_4$	1.72220	10.23608	0.58065	9.76391
	KCl	0.50685	9.70488	1.97297	10.29512
	K_2O	0.32021	9.50543	3.12295	10.49456
$KF.2H_2O$	CaF_2	0.41474	9.61778	2.41115	10.38222

Weighed	Sought	Factor	Log of Factor +10	Reciprocal of Factor	Log of Reciprocal of Factor +10
Potassium (contd.)					
$KF.2H_2O$	$CaSO_4$	0.72314	9.85922	1.38286	10.14078
K_2HAsO_4	$Mg_2As_2O_7$	0.71165	9.85227	1.40519	10.14774
$KHCO_3$	KCl	0.74466	9.87196	1.34289	10.12804
	K_2O	0.47046	9.67252	2.12558	10.32748
	K_2PtCl_6	2.42716	10.38510	0.41200	9.61490
	K_2SO_4	0.87029	9.93966	1.14904	10.06034
$KHSO_4$	$BaSO_4$	1.71401	10.23401	0.58343	9.76599
	K_2SO_4	0.63988	9.80610	1.56279	10.19389
KI	Ag	0.64980	9.81278	1.53894	10.18722
	AgI	1.41425	10.15053	0.70709	9.84947
	I	0.76446	9.88335	1.30811	10.11665
	K	0.23555	9.37208	4.24538	10.62792
	K_2O	0.28374	9.45292	3.52435	10.54708
KIO_3	AgI	1.09705	10.04023	0.91154	9.95978
	I	0.59300	9.77305	1.68634	10.22695
$KMnO_4$	Mn_2O_3	0.49948	9.69852	2.00208	10.30148
	MnS	0.55051	9.74007	1.81650	10.25924
K_2MnO_4	Mn_2O_3	0.40041	9.60250	2.49744	10.39749
	MnS	0.44132	9.64475	2.26593	10.35525
KNO_2	K_2SO_4	1.02380	10.01022	0.97675	9.98978
	N_2O_3	0.44656	9.64988	2.23934	10.35012
KNO_3	K	0.38674	9.58742	2.58572	10.41260
	KCl	0.73739	9.86770	1.35613	10.13230
	K_2O	0.46586	9.66826	2.14657	10.33174
	K_2PtCl_6	2.40344	10.38083	0.41607	9.61917
	K_2SO_4	0.86179	9.93540	1.16038	10.06460
	N	0.13853	9.14154	7.21865	10.85846
	NH_3	0.16844	9.22645	5.93683	10.77356
	NO	0.29678	9.47243	3.36950	10.52757
	N_2O_5	0.53414	9.72766	1.87217	10.27235
K_2O	Cl	0.75269	9.87662	1.32857	10.12338
	CO_2	0.46718	9.66948	2.14050	10.33052
	K	0.83016	9.91916	1.20459	10.08085
	KBr	2.52667	10.40255	0.39578	9.59745
	KCl	1.58284	10.19944	0.63178	9.80057
	$KClO_3$	2.60186	10.41529	0.38434	9.58472
	$KClO_4$	2.94155	10.46858	0.33996	9.53143
	K_2CO_3	1.46718	10.16648	0.68158	9.83352
	$K_2Cr_2O_7$	3.12294	10.49456	0.32021	9.50543
	$KHCO_3$	2.12558	10.32748	0.47046	9.67252
	KI	3.52439	10.54709	0.28374	9.45292
	KNO_3	2.14655	10.33174	0.46586	9.66826
	KOH	1.19122	10.07599	0.83948	9.92401
	K_2PtCl_6	5.15918	10.71258	0.19383	9.28742
	K_2SO_4	1.84990	10.26715	0.54057	9.73285
	N_2O_5	1.14657	10.05940	0.87217	9.94060
KOH	K_2CO_3	1.23164	10.09048	0.81193	9.90952
	K_2O	0.83946	9.92400	1.19124	10.07600
K_2PtCl_6	K	0.16091	9.20658	6.21465	10.79342
	KCl	0.30680	9.48686	3.25948	10.51315
	K_2CO_3	0.28438	9.45390	3.51642	10.54610
	$KHCO_3$	0.41200	9.61490	2.42718	10.38510
	KNO_3	0.41606	9.61916	2.40350	10.38084
	K_2O	0.19383	9.28742	5.15916	10.71258
	K_2SO_4	0.35856	9.55456	2.78893	10.44544
	$K_2SO_4.Al_2(SO_4)_3.24H_2O$	1.95218	10.29052	0.51225	9.70948
	$K_2SO_4.Cr_2(SO_4)_3.24H_2O$	2.05512	10.31283	0.48659	9.68716
K_2S	$BaSO_4$	2.11666	10.32565	0.47244	9.67435
	K_2SO_4	1.58039	10.19877	0.63276	9.80124
K_2SiO_3	SiO_2	0.38943	9.59043	2.56786	10.40958
K_2SO_4	$BaSO_4$	1.33933	10.12689	0.74664	9.87311
	K	0.44876	9.65201	2.22836	10.34799
	KCl	0.85565	9.93230	1.16870	10.06770
	K_2CO_3	0.79312	9.89934	1.26084	10.10066
	$KHCO_3$	1.14904	10.06034	0.87029	9.93966
	$KHSO_4$	1.56281	10.19390	0.63987	9.80609
	KNO_2	0.97676	9.98979	1.02379	10.01021
	KNO_3	1.16038	10.06460	0.86179	9.93540

GRAVIMETRIC FACTORS AND THEIR LOGARITHMS (Continued)

Weighed	Sought	Factor	Log of Factor +10	Reciprocal of Factor	Log of Reciprocal of Factor +10
Potassium (contd.)					
K_2SO_4	K_2O	0.54057	9.73285	1.84990	10.26715
	K_2PtCl_6	2.78889	10.44543	0.35857	9.55457
	K_2S	0.63276	9.80124	1.58038	10.19876
	SO_3	0.45942	9.66221	2.17666	10.33779
$K_2SO_4.Al_2(SO_4)_3.24H_2O$	K_2PtCl_6	0.51224	9.70947	1.95221	10.29053
$K_2SO_4.Cr_2(SO_4)_3.24H_2O$	K_2PtCl_6	0.48658	9.68715	2.05516	10.31284
$Mg_2As_2O_7$	K_3AsO_4	1.65059	10.21764	0.60584	9.78236
	K_2HAsO_4	1.40519	10.14774	0.71165	9.85227
Mn_2O_3	$KMnO_4$	2.00207	10.30148	0.49948	9.69852
	K_2MnO_4	2.49731	10.39747	0.40043	9.60253
MnS	$KMnO_4$	1.81649	10.25923	0.55051	9.74077
	K_2MnO_4	2.26592	10.35524	0.44132	9.64475
N	KNO_3	7.21847	10.85845	0.13853	9.14154
NH_3	KNO_3	5.93678	10.77355	0.16844	9.22645
NO	KNO_3	3.36954	10.52757	0.29678	9.47243
N_2O_3	KNO_2	2.23934	10.35012	0.44656	9.64988
N_2O_5	KNO_3	1.87217	10.27235	0.53414	9.72765
	K_2O	0.87217	9.94060	1.14657	10.05940
Pt	K	0.40086	9.60299	2.49464	10.39701
	KCl	0.76431	9.88327	1.30837	10.11673
SiO_2	K_2SiO_3	2.56783	10.40957	0.38943	9.59043
SO_3	K_2SO_4	2.17664	10.33779	0.45942	9.66221
Praseodymium					
Pr	Pr_2O_3	1.17032	10.06832	0.85447	9.93169
Pr_2O_3	Pr	0.85447	9.93169	1.17032	10.06831
Pr_6O_{11}	Pr	0.82770	9.91787	1.20817	10.08213
Rhenium					
$C_{20}H_{16}N_4.HReO_4$	Re	0.33038	9.51901	3.02682	10.48098
$(C_6H_5)_4AsReO_4$	Re	0.29392	9.46823	3.40229	10.53177
Rhodium					
Na_3RhCl_6	Rh	0.26757	9.42744	3.73734	10.57256
Rh	Na_3RhCl_6	3.73735	10.57256	0.26757	9.42744
$RhCl_3$	Rh	0.49175	9.69174	2.03355	10.30826
Rubidium					
AgCl	Rb	0.59635	9.77550	1.67687	10.22450
	RbCl	0.84368	9.92618	1.18528	10.07382
Cl	Rb	2.41080	10.38216	0.41480	9.61784
	RbCl	3.41071	10.53284	0.29319	9.46715
Rb	AgCl	1.67688	10.22450	0.59635	9.77550
	Cl	0.41480	9.61784	2.41080	10.38216
	RbCl	1.41477	10.15069	0.70683	9.84932
	Rb_2CO_3	1.35106	10.13068	0.74016	9.86933
	Rb_2O	1.09360	10.03886	0.91441	9.96114
	Rb_2PtCl_6	3.38569	10.52965	0.29536	9.47035
	Rb_2SO_4	1.56195	10.19367	0.64023	9.80634
$RbB(C_6H_5)_4$	Rb	0.21119	9.32467	4.73507	10.67533
	Rb_2O	0.23096	9.36354	4.32975	10.63647
RbCl	AgCl	1.18527	10.07382	0.84369	9.92618
	Cl	0.29319.	9.46715	3.41076	10.53285
	Rb	0.70683	9.84932	1.41477	10.15069
	Rb_2CO_3	0.95493	9.97997	1.04720	10.02003
	Rb_2O	0.77296	9.08816	1.29373	10.11184
	Rb_2PtCl_6	2.39301	10.37896	0.41788	9.62105
	Rb_2SO_4	1.10399	10.04297	0.90581	9.95704
Rb_2CO_3	Rb	0.74019	9.86934	1.35100	10.12066
	RbCl	1.04720	10.02003	0.45493	9.97997
	$RbHCO_3$	1.26864	10.10335	0.78825	9.89666
	Rb_2PtCl_6	2.50595	10.39897	0.39905	9.60103
	Rb_2SO_4	1.15609	10.06299	0.86498	9.93701
$RbClO_4$	Rb	0.46220	9.66483	2.16357	10.33517
	Rb_2CO_3	0.62446	9.79550	1.60138	10.20449
	RbCl	0.65390	9.81551	1.52929	10.18448
	$RbHCO_3$	0.79218	9.89883	1.26234	10.10118
	Rb_2O	0.50546	9.70369	1.97840	10.29632
	Rb_2SO_4	0.72193	9.85850	1.38518	10.14151
$RbHCO_3$	Rb_2CO_3	0.78831	9.89670	1.26854	10.10330
	Rb_2PtCl_6	1.97546	10.29567	0.50621	9.70433
	Rb_2SO_4	0.91136	9.95969	1.09726	10.04031

Weighed	Sought	Factor	Log of Factor +10	Reciprocal of Factor	Log of Reciprocal of Factor +10
Rubidium (contd.)					
Rb_2O	Rb	0.91441	9.96114	1.09360	10.03887
	RbCl	1.29368	10.11182	0.77299	9.88817
	Rb_2PtCl_6	3.09591	10.49079	0.32301	9.50922
	Rb_2SO_4	1.42827	10.15481	0.70015	9.84519
Rb_2PtCl_6	Rb	0.29537	9.47037	3.38558	10.52963
	RbCl	0.41787	9.62103	2.39309	10.37896
	Rb_2CO_3	0.39905	9.60103	2.50595	10.39898
	$RbHCO_3$	0.50624	9.70436	1.97535	10.29565
	Rb_2O	0.32301	9.50922	3.09588	10.49079
Rb_2SO_4	Rb	0.64022	9.80633	1.56196	10.19367
	RbCl	0.90577	9.95701	1.10403	10.04298
	Rb_2CO_3	0.86498	9.93701	1.15610	10.06300
	$RbHCO_3$	1.09730	10.04033	0.91133	9.95968
	Rb_2O	0.70015	9.84520	. 1.42827	10.15481
Samarium					
Sm_2O_3	Sm	0.86235	9.93568	1.15962	10.06431
Scandium					
Sc_2O_3	Sc	0.65196	9.81422	1.53384	10.18578
$Sc(C_9H_6NO)_3.(C_9H_6NOH)$	Sc	0.07221	8.85860	13.84850	11.14140
	Sc_2O_3	0.11076	9.04438	9.02853	10.95562
Selenium					
H_2SeO_3	Se	0.61224	9.78692	1.63335	10.21308
H_2SeO_4	Se	0.54466	9.73613	1.83601	10.26389
$PbSeO_4$	Se	0.22550	9.35315	4.43459	10.64685
	SeO_2	0.31689	9.50091	3.15567	10.49909
Se	H_2SeO_3	1.63336	10.21308	0.61223	9.78691
	H_2SeO_4	1.83599	10.26387	0.54467	9.73613
	SeO_2	1.40527	10.14776	0.71161	9.85224
	SeO_3	1.60790	10.20626	0.62193	9.79374
SeO_2	Se	0.71161	9.85224	1.40526	10.14776
SeO_3	Se	0.62193	9.79374	1.60790	10.20626
Silicon					
$BaSiF_6$	SiF_4	0.37249	9.57111	2.68464	10.42888
	SiO_2	0.21503	9.33250	4.65051	10.66750
H_2SiO_3	SiO_2	0.76933	9.88611	1.29983	10.11388
K_2SiF_6	SiF_4	0.47249	9.67439	2.11645	10.32561
	SiO_2	0.27277	9.43580	3.66609	10.56420
Si	SiO_2	2.13932	10.33027	0.46744	9.66973
SiC	C	0.29955	9.47647	3.33834	10.52353
	Si	0.70045	9.84538	1.42765	10.15462
SiF_4	$BaSiF_6$	0.26847	9.42890	3.72481	10.57110
	K_2SiF_6	2.11645	10.32561	0.47249	9.67439
SiF_4	SiO_2	0.57730	9.76140	1.73220	10.23860
SiO_2	$BaSiF_6$	4.65041	10.66749	0.21503	9.33250
	H_2SiO_3	1.29983	10.11388	0.76933	9.88611
	K_2SiF_6	3.66614	10.56421	0.27277	9.43580
	Si	0.46744	9.66973	2.13931	10.33027
	SiF_4	1.73221	10.23861	0.57730	9.76140
	SiO_3	1.26627	10.10252	0.78972	9.89747
	SiO_4	1.53256	10.18542	0.65250	9.81458
	Si_2O	0.60058	9.77857	1.66506	10.22182
	$Si(OH)_4$	1.59965	10.20403	0.62514	9.79597
$SiO_2.12MoO_3$	Si	0.01571	8.19618	63.65372	11.80382
	SiO_2	0.03362	8.52660	29.74420	11.47340
SiO_3	SiO_2	0.78972	9.89747	1.26627	10.10252
SiO_4	SiO_2	0.65250	9.81458	1.53257	10.18542
Si_2O	SiO_2	1.66505	10.22142	0.60058	9.77857
$Si(OH)_4$	SiO_2	0.62514	9.79598	1.59964	10.20402
Silver					
Ag	AgBr	1.74079	10.24075	0.57445	9.75925
	AgCl	1.32866	10.12341	0.75264	9.87659
	AgCN	1.24120	10.09384	0.80567	9.90616
	AgI	2.17645	10.33775	0.45946	9.66225
	$AgNO_3$	1.57481	10.19723	0.63500	9.80277
	Ag_2O	1.07416	10.03107	0.93096	9.96893
	Ag_3PO_4	1.29347	10.11176	0.77311	9.88824

Weighed	Sought	Factor	Log of Factor +10	Reciprocal of Factor	Log of Reciprocal of Factor +10
Silver (contd.)					
Ag	$Ag_4P_2O_7$	1.40313	10.14710	0.71269	9.85290
	Br	0.74079	9.86970	1.34991	10.13030
	Cl	0.32866	9.51675	3.04266	10.48325
	I	1.17646	10.07058	0.85001	9.92942
AgBr	Ag	0.57445	9.75925	1.74080	10.24075
	Br	0.42555	9.62895	2.34990	10.37105
AgCl	Ag	0.75264	9.87659	1.32866	10.12342
	$AgNO_3$	1.18526	10.07382	0.84370	9.92619
	Ag_2O	0.80845	9.90765	1.23693	10.09235
	Br	0.55754	9.74628	1.79359	10.25372
	Cl	0.24736	9.39333	4.04269	10.60667
AgCN	Ag	0.80567	9.90616	1.24120	10.09384
Ag_2CrO_4	Ag	0.65034	9.81314	1.53766	10.18686
AgI	Ag	0.45946	9.66225	2.17647	10.33775
	I	0.54054	9.73283	1.85000	10.26717
$AgNO_3$	Ag	0.63499	9.80277	1.57483	10.19723
	AgCl	0.84370	9.92619	1.18526	10.07381
Ag_2O	Ag	0.93096	9.96893	1.07416	10.03107
	AgCl	1.23693	10.09235	0.80845	9.90765
Ag_3PO_4	Ag	0.77311	9.88824	1.29348	10.11176
$Ag_4P_2O_7$	Ag	0.71269	9.85290	1.40313	10.14710
Br	Ag	1.34991	10.13030	0.74079	9.86970
	AgBr	2.34991	10.37105	0.42555	9.62895
	AgCl	1.79358	10.25372	0.55754	9.74628
Cl	Ag	3.04261	10.48325	0.32867	9.51676
	AgCl	4.04262	10.60666	0.24736	9.39333
I	Ag	0.85000	9.92942	1.17647	10.07058
	AgI	1.85000	10.26717	0.54054	9.73283
Sodium					
Ag	NaBr	0.95392	9.97951	1.04831	10.02049
	NaCl	0.54179	9.73383	1.84573	10.26617
	NaI	1.38958	10.14288	0.71964	9.85712
AgBr	NaBr	0.54798	9.73876	1.82488	10.26123
AgCl	NaCl	0.40777	9.61042	2.45236	10.38959
	$NaClO_3$	0.74267	9.87080	1.34649	10.21921
	$NaClO_4$	0.85429	9.93161	1.17056	10.06839
AgI	NaI	0.63846	9.80513	1.56627	10.19487
$BaSO_4$	$NaHSO_4$	0.51439	9.71129	1.94405	10.28871
	$NaHSO_4.H_2O$	0.59157	9.77201	1.69042	10.22799
	Na_2S	0.33438	9.52424	2.99061	10.47577
	Na_2SO_3	0.54002	9.73241	1.85178	10.26759
	$Na_2SO_3.7H_2O$	1.08032	10.03355	0.92565	9.96645
	Na_2SO_4	0.60857	9.78431	1.64320	10.21569
	$Na_2SO_4.10H_2O$	1.38043	10.14001	0.72441	9.85998
B_2O_3	$Na_2B_4O_7$	1.44511	10.15990	0.69199	9.84010
	$Na_2B_4O_7.10H_2O$	2.73895	10.43758	0.36510	9.56241
Br	Na	0.28770	9.45894	3.47584	10.54106
	NaBr	1.28770	10.10981	0.77658	9.89019
	Na_2O	0.38781	9.58862	2.57858	10.41138
$CaCl_2$	NaCl	1.05321	10.02252	0.94948	9.97749
$CaCO_3$	Na_2CO_3	1.05894	10.02488	0.94434	9.97513
CaF_2	NaF	1.07551	10.03161	0.92979	9.96838
CaO	Na_2CO_3	1.88996	10.27645	0.52911	9.72355
$CaSO_4$	Na_2CO_3	0.77853	9.89128	1.28447	10.10873
Cl	Na	0.64846	9.81188	1.54212	10.18812
	NaCl	1.64846	10.21708	0.60663	9.78292
	Na_2O	0.87410	9.94156	1.14403	10.05844
CO_2	Na_2CO_3	2.40829	10.38171	0.41523	9.61829
	Na_2O	1.40829	10.14869	0.71008	9.85131
H_3BO_3	$Na_2B_4O_7$	0.81356	9.91039	1.22917	10.08961
	$Na_2B_4O_7.10H_2O$	1.54195	10.18807	0.64853	9.81193
I	Na	0.18116	9.25806	5.51998	10.74194
	NaI	1.18115	10.07231	0.84663	9.92769
	Na_2O	0.24419	9.38773	4.09517	10.61227
KBF_4	$Na_2B_4O_7$	0.39954	9.60156	2.50288	10.39844
	$Na_2B_4O_7.10H_2O$	0.75725	9.87924	1.32057	10.12076
$Mg_2As_2O_7$	Na_2HAsO_3	1.09454	10.03923	0.91363	9.96077
	Na_2HAsO_4	1.19761	10.07832	0.83500	9.92169

Weighed	Sought	Factor	Log of Factor +10	Reciprocal of Factor	Log of Reciprocal of Factor +10
Sodium (contd.)					
$MgCl_2$	$NaCl$	1.22756	10.08904	0.81462	9.91096
$Mg_2P_2O_7$	Na_2HPO_4	1.27565	10.10573	0.78391	9.89427
	$Na_2HPO_4.12H_2O$	3.21828	10.50763	0.31072	9.49237
	$NaNH_4HPO_4.4H_2O$	1.87870	10.27386	0.53228	9.72614
	Na_3PO_4	1.47318	10.16825	0.67880	9.83174
	$Na_4P_2O_7.10H_2O$	2.00414	10.30193	0.49897	9.69807
N	$NaNO_3$	6.06815	10.78306	0.16479	9.21693
Na	Br	3.47585	10.54106	0.28770	9.45894
	Cl	1.54212	10.18811	0.64846	9.81188
	I	5.52003	10.74194	0.18116	9.25806
	NaBr	4.47585	10.65088	0.22342	9.34912
	NaCl	2.54213	10.40520	0.39337	9.59480
	Na_2CO_3	2.30513	10.36269	0.43382	9.63731
	$NaHCO_3$	3.65410	10.56278	0.27367	9.43723
	NaI	6.52003	10.81425	0.15337	9.18574
	Na_2O	1.34797	10.12968	0.74186	9.87032
	Na_2SO_4	3.08921	10.48985	0.32371	9.51016
$Na_2B_4O_7$	B_2O_3	0.69198	9.84009	1.44513	10.15991
	H_3BO_3	1.22916	10.08961	0.81356	9.91039
	KBF_4	2.50287	10.39844	0.39954	9.60156
$Na_2B_4O_7.10H_2O$	B_2O_3	0.36510	9.56241	2.73898	10.43759
	H_3BO_3	0.64853	9.81193	1.54195	10.18807
	KBF_4	1.32057	10.12076	0.75725	9.87924
NaBr	Ag	1.04831	10.02048	0.95392	9.97951
	AgBr	1.82489	10.26123	0.54798	9.73876
	Br	0.77658	9.89019	1.28770	10.10981
	Na	0.22342	9.34912	4.47588	10.65088
	Na_2O	0.30116	9.47880	3.32049	10.52120
$NaC_2H_3O_2.Mg(C_2H_3O_2)_2.3UO_2(C_2H_3O_2)_2.6\frac{1}{2}H_2O$	Na	0.01527	8.18375	65.50075	11.81625
	NaBr	0.06833	8.83464	14.63400	11.16536
	Na_2CO_3	0.03519	8.54646	28.41474	11.45354
	NaCl	0.03881	8.58895	25.76589	11.41104
	NaF	0.02788	8.44536	35.86286	11.55465
	$NaHCO_3$	0.05579	8.74654	17.92500	11.25346
	NaI	0.09954	8.99801	10.04590	11.00199
	$NaNO_3$	0.05644	8.75162	17.71667	11.24837
	Na_2O	0.02058	8.31345	48.59086	11.68655
	NaOH	0.02656	8.42426	37.64746	11.57574
	Na_2SO_4	0.04716	8.67361	21.20261	11.32638
$NaC_2H_3O_2.Zn(C_2H_3O_2)_2.3UO_2(C_2H_3O_2)_2.6\frac{1}{2}H_2O$	Na	0.01495	8.17461	66.89410	11.85239
	NaBr	0.06691	8.82549	14.94545	11.17450
	Na_2CO_3	0.03446	8.53730	29.01999	11.46270
	NaCl	0.03800	8.57981	26.31440	11.42019
	NaF	0.02733	8.43621	36.62601	11.56379
	$NaHCO_3$	0.05466	8.73739	18.30663	11.26261
	NaI	0.09747	8.98886	10.25977	11.01114
	$NaNO_3$	0.05527	8.74247	18.09398	11.25753
	NaOH	0.02601	8.41511	38.44970	11.58489
	Na_2O	0.02015	8.30430	49.62532	11.69570
	Na_2SO_4	0.04618	8.66446	21.65392	11,33554
Na_2CO_3	$CaCO_3$	0.94434	9.97513	1.05894	10.02488
	CaO	0.52911	9.72355	1.88997	10.27645
	$CaSO_4$	1.28447	10.10873	0.77853	9.89128
	CO_2	0.41523	9.61829	2.40830	10.38171
	Na	0.43381	9.63730	2.30516	10.36270
	NaCl	1.10281	10.04250	0.90677	9.95750
	$NaHCO_3$	1.58520	10.20008	0.63084	9.79992
	NaOH	0.75474	9.87780	1.32496	10.12220
	Na_2O	0.58477	9.76699	1.71007	10.23301
	Na_2SO_4	1.34015	10.12715	0.74619	9.87285
$Na_2CO_3.10H_2O$	Na_2SO_4	0.49639	9.69582	2.01455	10.30417
NaCl	Ag	1.84573	10.26617	0.54179	9.73383
	AgCl	2.45236	10.38958	0.40777	9.61042
	Cl	0.60663	9.78292	1.64845	10.21707
	Na	0.39337	9.59480	2.54214	10.40520
	$NaClO_3$	1.82129	10.26038	0.54906	9.73962

Weighed	Sought	Factor	Log of Factor +10	Reciprocal of Factor	Log of Reciprocal of Factor +10
Sodium (contd.)					
	$NaClO_4$	2.09503	10.32118	0.47732	9.67881
	Na_2CO_3	0.90677	9.95750	1.10282	10.04251
	$NaHCO_3$	1.43742	10.15759	0.69569	9.84242
	Na_2HPO_4	1.21451	10.08440	0.82338	9.91560
	Na_2O	0.53025	9.72448	1.88590	10.27552
NaCl	Na_2SO_4	1.21521	10.08465	0.82290	9.91535
$NaClO_3$	AgCl	1.34650	10.12921	0.74267	9.87080
	NaCl	0.54906	9.73961	1.82129	10.26038
$NaClO_4$	AgCl	1.17055	10.06839	0.85430	9.93161
	NaCl	0.47732	9.67881	2.09503	10.32119
NaF	CaF_2	0.92979	9.96838	1.07551	10.03161
Na_2HAsO_3	$Mg_2As_2O_7$	0.91362	9.96077	1.09455	10.03923
Na_2HAsO_4	$Mg_2As_2O_7$	0.83499	9.92168	1.19762	10.07832
$NaHCO_3$	Na	0.27366	9.43721	3.65417	10.56279
	NaCl	0.69569	9.84242	1.43742	10.15759
	Na_2CO_3	0.63083	9.79991	1.58521	10.20009
	Na_2O	0.36889	9.56690	2.71084	10.43310
Na_2HPO_4	$Mg_2P_2O_7$	0.78392	9.89427	1.27564	10.10573
	NaCl	0.82338	9.91560	1.21451	10.08440
	Na_2O	0.43660	9.64008	2.29043	10.35992
	$Na_4P_2O_7$	0.93654	9.97153	1.06776	10.02847
	P_2O_5	0.49994	9.69892	2.00024	10.30108
$Na_2HPO_4.12H_2O$	$Mg_2P_2O_7$	0.31072	9.49237	3.21833	10.50763
	$Na_4P_2O_7$	0.37122	9.56963	2.69382	10.43037
	P_2O_5	0.19816	9.29702	5.04643	10.70298
$NaHSO_3$	SO_2	0.61564	9.78933	1.62433	10.21076
$NaHSO_4$	$BaSO_4$	1.94404	10.28871	0.51439	9.71129
$NaHSO_4.H_2O$	$BaSO_4$	1.69039	10.22799	0.59158	9.77201
NaI	Ag	0.71964	9.85712	1.38958	10.14288
	AgI	1.56626	10.19486	0.63846	9.80513
	I	0.84662	9.92769	1.18117	10.07231
	Na	0.15337	9.18574	6.52018	10.81426
	Na_2O	0.20674	9.31542	4.83699	10.68458
$NaNH_4HPO_4.4H_2O$	$Mg_2P_2O_7$	0.53228	9.72614	1.87871	10.27386
	NH_3	0.08146	8.91094	12.27596	11.08905
	P_2O_5	0.33946	9.53079	2.94586	10.46921
$NaNO_3$	N	0.16479	9.21693	6.06833	10.78307
	Na_2O	0.36461	9.56183	2.74266	10.43818
	NH_3	0.20037	9.30183	4.99077	10.69817
	NO	0.35303	9.54781	2.83262	10.45219
	N_2O_5	0.63539	9.80304	1.57384	10.19696
Na_2O	Br	2.57854	10.41137	0.38782	9.58863
	Cl	1.14401	10.05843	0.87412	9.94157
	CO_2	0.71008	9.85131	1.40829	10.14869
	I	4.09501	10.61225	0.24420	9.38775
	Na	0.74185	9.87032	1.34798	10.12968
	NaBr	3.32039	10.52119	0.30117	9.47881
	NaCl	1.88587	10.27551	0.53026	9.72449
	Na_2CO_3	1.71008	10.23302	0.58477	9.76699
	$NaHCO_3$	2.71078	10.43310	0.36890	9.56691
	Na_2HPO_4	2.29044	10.35992	0.43660	9.64008
	NaI	4.83686	10.68457	0.20675	9.31545
	$NaNO_3$	2.74265	10.43817	0.36461	9.56183
	NaOH	1.29065	10.11081	0.77480	9.88919
	Na_2SO_4	2.29176	10.36017	0.43635	9.63984
	N_2O_5	1.74269	10.24122	0.57383	9.75878
	SO_3	1.29176	10.11118	0.77414	9.88882
NaOH	Na_2CO_3	1.32495	10.12220	0.75475	9.87780
	Na_2O	0.77479	9.88918	1.29067	10.11082
$Na_4P_2O_7$	Na_2HPO_4	1.06775	10.02847	0.93655	9.97153
	$Na_2HPO_4.12H_2O$	0.26938	9.43037	3.71223	10.56963
$Na_4P_2O_7.10H_2O$	$Mg_2P_2O_7$	0.49897	9.69807	2.00413	10.30193
Na_2S	$BaSO_4$	2.99062	10.47576	0.33438	9.52424
Na_2SO_3	$BaSO_4$	1.85176	10.26758	0.54003	9.73242
	SO_2	0.50827	9.70609	1.96746	10.29391
$Na_2SO_3.7H_2O$	$BaSO_4$	0.92564	9.96644	1.08033	10.03356
	SO_2	0.25407	9.40495	3.93592	10.59505
Na_2SO_4	$BaSO_4$	1.64319	10.21569	0.60857	9.78431
	Na	0.32370	9.51014	3.08928	10.48986

Weighed	Sought	Factor	Log of Factor +10	Reciprocal of Factor	Log of Reciprocal of Factor +10
Sodium (contd.)					
	$NaCl$	0.82289	9.91534	1.21523	10.08466
	Na_2CO_3	0.74619	9.87285	1.34014	10.21715
	$Na_2CO_3.10H_2O$	2.01451	10.30417	0.49640	9.69583
	Na_2O	0.43635	9.63984	2.29174	10.36016
	SO_3	0.56365	9.75101	1.77415	10.24899
$Na_2SO_4.10H_2O$	$BaSO_4$	0.72441	9.85998	1.38043	10.14001
$Na_2U_2O_7.2ZnU_2O_7$	Na	0.02369	8.37457	42.21190	11.62544
	Na_2O	0.03193	8.50420	31.31851	11.49579
NH_3	$NaNH_4HPO_4.4H_2O$	12.27607	11.08906	0.08146	8.91094
	$NaNO_3$	4.99070	10.69816	0.20037	9.30183
NO	$NaNO_3$	2.83258	10.45218	0.35304	9.54782
N_2O_5	$NaNO_3$	1.57382	10.19695	0.63540	9.80305
	Na_2O	0.57382	9.75878	1.74271	10.24123
P_2O_5	Na_2HPO_4	2.00020	10.30107	0.49995	9.69893
	$Na_2HPO_4.12H_2O$	5.04623	10.70297	0.19817	9.29704
	$NaNH_4HPO_4.4H_2O$	2.94578	10.46920	0.33947	9.53080
SO_2	$NaHSO_3$	1.62442	10.21070	0.61560	9.78930
	Na_2SO_3	1.96756	10.29393	0.50824	9.70607
SO_3	Na_2O	0.77414	9.88882	1.29176	10.11118
	Na_2SO_4	1.77414	10.24899	0.56365	9.75101
Strontium					
CO_2	$SrCO_3$	3.35446	10.52562	0.29811	9.47438
SO_3	SrO	1.29424	10.11202	0.77265	9.88798
	$SrSO_4$	2.29422	10.36063	0.43588	9.63937
Sr	$SrCO_3$	1.68489	10.22657	0.59351	9.77343
	$Sr(NO_3)_2$	2.41532	10.38298	0.41402	9.61702
	SrO	1.18261	10.07284	0.84559	9.92716
	$SrSO_4$	2.09633	10.32146	0.47702	9.67854
$SrCl_2$	$SrCO_3$	0.93124	9.96906	1.07384	10.03094
	SrO	0.65363	9.81533	1.52992	10.18467
	$SrSO_4$	1.15865	10.06395	0.86307	9.93605
$SrCO_3$	CO_2	0.29811	9.47438	3.35447	10.52562
	Sr	0.59351	9.77343	1.68489	10.22657
	$SrCl_2$	1.07383	10.03095	0.93125	9.96907
	$Sr(HCO_3)_2$	1.42010	10.15232	0.70418	9.84768
	$Sr(NO_3)_2$	1.43352	10.15640	0.69758	9.84359
	SrO	0.70189	9.84627	1.42472	10.15373
	$SrSO_4$	1.24419	10.09489	0.80374	9.90511
SrC_2O_4	Sr	0.49886	9.69798	2.00457	10.30202
	SrO	0.58996	9.77082	1.69503	10.22918
$Sr(HCO_3)_2$	$SrCO_3$	0.70417	9.84768	1.42011	10.15232
	SrO	0.49425	9.69395	2.02327	10.30602
$Sr(IO_3)_2$	Sr	0.20031	9.30170	4.99226	10.69829
	SrO	0.23688	9.37453	4.22155	10.69829
$Sr(NO_3)_2$	Sr	0.41402	9.61702	2.41434	10.38298
	$SrCO_3$	0.69759	9.84360	1.43351	10.15640
	SrO	0.48963	9.68987	2.04236	10.31014
	$SrSO_4$	0.86793	9.93849	1.15217	10.06151
SrO	SO_3	0.77265	9.88798	1.29425	10.11202
	Sr	0.84559	9.92716	1.18261	10.07284
	$SrCl_2$	1.52992	10.18467	0.65363	9.81533
	$SrCO_3$	1.42472	10.15373	0.70189	9.84627
	$Sr(HCO_3)_2$	2.02326	10.30605	0.49425	9.69395
	$Sr(NO_3)_2$	2.04237	10.31014	0.48963	9.68987
	$SrSO_4$	1.77263	10.24862	0.56413	9.75138
$SrSO_4$	SO_3	0.43588	9.63937	2.29421	10.36063
	Sr	0.47703	9.67855	2.09630	10.32145
	$SrCl_2$	0.86308	9.93605	1.15864	10.06395
	$SrCO_3$	0.80373	9.90511	1.24420	10.09489
	$Sr(NO_3)_2$	1.15217	10.06151	0.86793	9.93848
	SrO	0.56413	9.75138	1.77264	10.24862
Sulfur					
As_2S_3	H_2S	0.41555	9.61862	2.40645	10.38138
	S	0.39097	9.59214	2.55774	10.40786
$BaSO_4$	FeS_2	0.25702	9.40997	3.89077	10.59003
	H_2S	0.14602	9.16441	6.84838	10.83559
	H_2SO_3	0.35166	9.54612	2.84366	10.45388
	H_2SO_4	0.42021	9.62347	2.37976	10.37653

Weighed	Sought	Factor	Log of Factor +10	Reciprocal of Factor	Log of Reciprocal of Factor +10
Sulfur (contd.)					
	S	0.13738	9.13792	7.27908	10.86208
	SO_2	0.27448	9.43851	3.64325	10.56149
	SO_3	0.34302	9.53532	2.91528	10.46468
	SO_4	0.41158	9.61445	2.42966	10.38555
$C_{12}H_{12}N_2.H_2SO_4$	S	0.11357	9.05526	8.80514	10.94474
	SO_4	0.34026	9.53181	2.93893	10.46819
CdS	H_2S	0.23591	9.37275	4.23890	10.62725
	S	0.22196	9.34627	4.50532	10.65372
FeS_2	$BaSO_4$	3.89081	10.59004	0.25702	9.40997
H_2S	As_2S_3	2.40646	10.38138	0.41555	9.61862
	$BaSO_4$	6.84859	10.83560	0.14602	9.16441
	CdS	4.23885	10.62725	0.23591	9.37275
	SO_3	2.34924	10.37093	0.42567	9.62907
H_2SO_3	$BaSO_4$	2.84363	10.45287	0.35166	9.54612
H_2SO_4	$BaSO_4$	2.37974	10.37653	0.42021	9.62347
	$(NH_4)_2SO_4$	1.34728	10.12946	0.74224	9.87054
	SO_3	0.81631	9.91186	1.22502	10.08814
$(NH_4)_2SO_4$	H_2SO_4	0.74223	9.87054	1.34729	10.12946
	SO_3	0.60589	9.78239	1.65046	10.21760
S	As_2S_3	2.55775	10.40786	0.39097	9.59214
	$BaSO_4$	7.27919	10.86208	0.13738	9.13792
	CdS	4.50536	10.65373	0.22196	9.34627
SO_2	$BaSO_4$	3.64329	10.56149	0.27448	9.43851
SO_3	$BaSO_4$	2.91524	10.46468	0.34302	9.53532
	H_2S	0.42567	9.62907	2.34924	10.37093
	$(NH_4)_2SO_4$	1.65047	10.21761	0.60589	9.78239
SO_4	$BaSO_4$	2.42968	10.38555	0.41158	9.61445
Tantalum					
Ta	$TaCl_5$	1.97965	10.29659	0.50514	9.70341
	Ta_2O_5	1.22105	10.08674	0.81897	9.91327
TaC	C	0.06225	8.79413	16.06451	11.20587
	Ta	0.93775	9.97209	1.06638	10.02791
$TaCl_5$	Ta	0.50514	9.70341	1.97965	10.29659
	Ta_2O_5	0.61680	9.79041	1.62127	10.20985
Ta_2O_4	Ta_2O_5	1.03757	10.01602	0.96379	9.98398
Ta_2O_5	Ta	0.81897	9.91327	1.22105	10.08673
	$TaCl_5$	1.62126	10.20985	0.61680	9.79014
	Ta_2O_4	0.96379	9.98398	1.03757	10.01602
Tellurium					
H_2TeO_4	Te	0.65906	9.81893	1.51731	10.18108
$H_2TeO_4.2H_2O$	Te	0.55565	9.74480	1.79969	10.25520
Te	H_2TeO_4	1.51732	10.18108	0.65906	9.81893
	$H_2TeO_4.2H_2O$	1.79969	10.25520	0.55565	9.74480
	TeO_2	1.25078	10.09718	0.79950	9.90282
	TeO_3	1.37618	10.13868	0.72665	9.86133
	$(TeO_2)_2SO_3$	1.56450	10.19438	0.63918	9.80562
TeO_2	Te	0.79950	9.90282	1.25078	10.09718
TeO_3	Te	0.72665	9.86133	1.37618	10.13868
$(TeO_2)_2SO_3$	Te	0.63918	9.80562	1.56450	10.19438
Terbium					
Tb_4O_7	Tb	0.85021	9.92953	1.17618	10.07048
Thallium					
$(C_6H_5)_4AsTlCl_4$	Tl	0.28014	9.44738	3.56964	10.55263
Tl	TlCl	1.17346	10.06947	0.85218	9.93053
	Tl_2CO_3	1.14682	10.05949	0.87198	9.94051
	Tl_2CrO_4	1.28377	10.10849	0.77896	9.89151
	$TlHSO_4$	1.47497	10.16878	0.67798	9.83122
	TlI	1.62093	10.20976	0.61693	9.79024
	$TlNO_3$	1.30337	10.11507	0.76724	9.88493
	Tl_2O	1.03914	10.01668	0.96233	9.98332
	Tl_2PtCl_6	1.99772	10.30055	0.50057	9.69946
	Tl_2SO_4	1.23501	10.09167	0.80971	9.90833
TlCl	Tl	0.85218	9.93053	1.17346	10.06947
	Tl_2PtCl_6	1.70240	10.23106	0.58741	9.76894
Tl_2CO_3	Tl	0.87198	9.94051	1.14682	10.05951
	Tl_2PtCl_6	1.74197	10.24104	0.57406	9.75896
Tl_2CrO_4	Tl	0.77895	9.89151	1.28378	10.10850
$TlHSO_4$	Tl	0.67798	9.83122	1.47497	10.16878

Weighed	Sought	Factor	Log of Factor +10	Reciprocal of Factor	Log of Reciprocal of Factor +10
Thallium (contd.)					
TlI	Tl	0.61693	9.79024	1.62093	10.20976
	Tl_2PtCl_6	1.23244	10.09076	0.81140	9.90924
$TlNO_3$	Tl	0.76724	9.88493	1.30338	10.11507
	Tl_2PtCl_6	1.53272	10.18546	0.65243	9.81453
Tl_2O	Tl	0.96223	9.98332	1.03914	10.01668
	Tl_2PtCl_6	1.92247	10.28386	0.52016	9.71614
Tl_2PtCl_6	Tl	0.50057	9.69946	1.99772	10.30054
	TlCl	0.58741	9.76894	1.70239	10.23106
	Tl_2CO_3	0.57406	9.75896	1.74198	10.24105
	TlI	0.81140	9.90924	1.23244	10.09076
	$TlNO_3$	0.65244	9.81454	1.53271	10.18546
	Tl_2O	0.52016	9.71614	1.92249	10.28386
	Tl_2SO_4	0.61821	9.79114	1.61757	10.20886
Tl_2SO_4	Tl	0.80971	9.90833	1.23501	10.09167
	Tl_2PtCl_6	1.61757	10.20886	0.61821	9.79114
Thorium					
Th	ThO_2	1.13790	10.05610	0.87881	9.94390
$Th(C_9H_6NO)_4.(C_9H_6NOH)$	Th	0.24327	9.38609	4.11066	10.61391
	ThO_2	0.27682	9.44220	3.61246	10.55781
$ThCl_4$	ThO_2	0.70626	9.84896	1.41590	10.15103
$Th(NO_3)_4.6H_2O$	ThO_2	0.44898	9.65223	2.22727	10.34777
ThO_2	Th	0.87881	9.94390	1.13790	10.05610
	$ThCl_4$	1.41590	10.15103	0.70626	9.84896
	$Th(NO_3)_4.6H_2O$	2.22729	10.34778	0.44898	9.65223
Thulium					
Tm_2O_3	Tm	0.87561	9.94231	1.14206	10.05769
Tin					
Sn	$SnCl_2$	1.59744	10.20342	0.62600	9.79657
	$SnCl_2.2H_2O$	1.90100	10.27898	0.52604	9.72102
	$SnCl_4$	2.19479	10.34140	0.45562	9.65860
	$SnCl_4.(NH_4Cl)_2$	3.09622	10.49083	0.32297	9.50916
	SnO	1.13480	10.05492	0.88121	9.94508
	SnO_2	1.26961	10.10367	0.78764	9.89633
$SnCl_2$	Sn	0.62600	9.79657	1.59744	10.20342
	SnO_2	0.79478	9.90025	1.25821	10.09975
$SnCl_2.2H_2O$	Sn	0.52604	9.72102	1.90100	10.27898
	SnO_2	0.66786	9.82469	1.49732	10.17531
$SnCl_4$	Sn	0.45562	9.65860	2.19481	10.34139
	SnO_2	0.57846	9.76227	1.72873	10.23772
$SnCl_4.(NH_4Cl)_2$	Sn	0.32297	9.50916	3.09626	10.49084
	SnO_2	0.41005	9.61284	2.43873	10.38716
SnO	Sn	0.88121	9.94508	1.13480	10.05491
	SnO_2	1.11879	10.04875	0.89382	9.95125
SnO_2	Sn	0.78764	9.89633	1.26962	10.10367
	$SnCl_2$	1.25821	10.09976	0.79474	9.90023
	$SnCl_2.2H_2O$	1.49731	10.17531	0.66786	9.82469
	$SnCl_4$	1.72871	10.23772	0.57847	9.76228
	$SnCl_4.(NH_4Cl)_2$	2.43872	10.38716	0.41005	9.61284
	SnO	0.89382	9.95125	1.11879	10.04875
Titanium					
K_2TiF_6	F	0.47472	9.67644	2.10650	10.32356
	K	0.32573	9.51286	3.07003	10.48714
	Ti	0.19951	9.29996	5.01228	10.70004
	TiO_2	0.33279	9.52217	3.00490	10.47783
Ti	K_2TiF_6	5.01232	10.70004	0.19951	9.29996
	TiC	1.25073	10.09717	0.79953	9.90283
	TiO_2	1.66806	10.22222	0.59950	9.77779
TiC	C	0.20049	9.30210	4.98778	10.69791
	Ti	0.79953	9.90283	1.25073	10.09717
$TiO(C_9H_4NOCl_2)_2$	Ti	0.09776	8.99016	10.22913	11.00984
	TiO_2	0.16306	9.21235	6.13271	10.78765
$TiO(C_9H_6ON)_2$	Ti	0.13600	9.13354	7.35294	10.86646
	TiO_2	0.22685	9.35574	4.40820	10.64426
TiO_2	K_2TiF_6	3.00488	10.47782	0.33279	9.52217
	Ti	0.59950	9.77779	1.66806	10.22221
	TiC	0.74969	9.87488	1.33388	10.12512

Weighed	Sought	Factor	Log of Factor +10	Reciprocal of Factor	Log of Reciprocal of Factor +10
Tungsten					
$FeWO_4$	W	0.60539	9.78203	1.65183	10.21797
	WO_3	0.76344	9.88277	1.30986	10.11722
$MgWO_4$	W	0.67552	9.82964	1.48034	10.17036
	WO_3	0.85189	9.93038	1.17386	10.06962
$MnWO_4$	W	0.60719	0.78332	1.64693	10.21668
	WO_3	0.76571	9.88406	1.30598	10.11593
$PbWO_4$	W	0.40403	9.60641	2.47506	10.39359
	WO_3	0.50952	9.70716	1.96263	10.29284
W	W_2C	1.03266	10.01396	0.96837	9.98604
	WC	1.06532	10.02748	0.93869	9.97252
	WO_2	1.17405	10.06969	0.85175	9.93031
	WO_3	1.26108	10.10075	0.79297	9.89925
W_2C	C	0.03163	8.50010	31.61555	11.49990
	W	0.96837	9.98604	1.03266	10.01396
WC	C	0.06133	8.78767	16.30523	11.21233
	W	0.93868	9.97252	1.06533	10.02749
WO_2	W	0.85175	9.93031	1.17405	10.06969
$WO_2(C_9H_6ON)_2$	W	0.36467	9.56190	2.74221	10.43810
	WO_3	0.45987	9.66264	2.17453	10.33736
WO_3	W	0.79297	9.89926	1.26108	10.10074
Uranium					
U	UO_2	1.13444	10.05478	0.88149	9.94522
	U_3O_8	1.17925	10.07160	0.84800	9.92840
	$U_2P_2O_{11}$	1.49981	10.17604	0.66675	9.82396
UO_2	U	0.88149	9.94522	1.13444	10.05478
	U_3O_8	1.03950	10.01682	0.96200	9.98318
	$U_2P_2O_{11}$	1.32268	10.12145	0.75604	9.87854
$UO_2(C_9H_6ON)_2.(C_9H_7ON)$	U	0.33835	9.52937	2.95552	10.47063
	UO_2	0.38384	9.58415	2.60525	10.41585
U_3O_8	U	0.84799	9.92839	1.17926	10.07161
	UO_2	0.96199	9.98317	1.03951	10.01683
	$UO_2(NO_3)_2.6H_2O$	1.78864	10.25252	0.55908	9.74747
$UO_2(NO_3)_2.6H_2O$	U_3O_8	0.55909	9.74748	1.78862	10.25252
$U_2P_2O_{11}$	U	0.66675	9.82396	1.49981	10.17603
	UO_2	0.75639	9.87875	1.32207	10.12125
Vanadium					
V	VC	1.23578	10.09194	0.80921	9.90806
	V_2O_5	1.78518	10.25168	0.56017	9.74832
VC	C	0.19080	9.28058	5.24109	10.71942
	V	0.80921	9.90806	1.23577	10.09194
VO_4	V_2O_5	0.79120	9.89829	1.26390	10.10171
V_2O_5	V	0.56017	9.74832	1.78517	10.25168
	VC	0.69225	9.84026	1.44456	10.15974
	VO_4	1.26390	10.10171	0.79120	9.89829
Ytterbium					
Yb	YbO_3	1.13870	10.05641	0.87819	9.94359
Yb_2O_3	Yb	0.87820	9.94359	1.13869	10.05640
Yttrium					
Y	Y_2O_3	1.26994	10.10378	0.78744	9.89622
Y_2O_3	Y	0.78744	9.89622	1.26994	10.10378
Zinc					
$BaSO_4$	ZnS	0.41744	9.62059	2.39555	10.37941
	$ZnSO_4.7H_2O$	1.23196	10.09059	0.81171	9.90940
Zn	$ZnNH_4PO_4$	2.72877	10.43596	0.36647	9.56404
	ZnO	1.24476	10.09508	0.80337	9.90492
	$Zn_2P_2O_7$	2.33043	10.36744	0.42911	9.63257
	ZnS	1.49044	10.17332	0.67094	9.82668
$Zn(C_9H_6NO)_2$	Zn	0.18483	9.26677	5.41038	10.73323
	ZnO	0.23007	9.36186	4.34650	10.63814
$ZnCl_2$	ZnO	0.59708	9.77603	1.67482	10.22397
$ZnCO_3$	ZnO	0.64899	9.81224	1.54086	10.18777
$(Zn[C_5H_5N]_2).(SCN)_2$	Zn	0.19241	9.28423	5.19724	10.71577
	ZnO	0.23951	9.37932	4.17519	10.62068
$ZnNH_4PO_4$	Zn	0.36646	9.56403	2.72881	10.43597
	ZnO	0.45616	9.65912	2.19221	10.34089
ZnO	Zn	0.80337	9.90492	1.24476	10.09509
	$ZnCl_2$	1.67482	10.22397	0.59708	9.77603

Weighed	Sought	Factor	Log of Factor +10	Reciprocal of Factor	Log of Reciprocal of Factor +10
Zinc (contd.)					
	$ZnCO_3$	1.54086	10.18776	0.64899	9.81224
	$ZnNH_4PO_4$	2.19221	10.34088	0.45616	9.65912
	$Zn_2P_2O_7$	1.87219	10.27235	0.53413	9.72765
	ZnS	1.19737	10.07823	0.83516	9.92177
	$ZnSO_4.7H_2O$	3.53373	10.54823	0.28299	9.45177
$Zn_2P_2O_7$	Zn	0.42911	9.63257	2.33040	10.36743
	ZnO	0.53413	9.72765	1.87220	10.27235
ZnS	$BaSO_4$	2.39556	10.37941	0.41744	9.62059
	Zn	0.67094	9.82668	1.49045	10.17332
	ZnO	0.83516	9.92177	1.19738	10.07822
	$ZnSO_4.7H_2O$	2.95125	10.47001	0.33884	9.52999
$ZnSO_4.7H_2O$	$BaSO_4$	0.81171	9.90940	1.23197	10.09060
	ZnO	0.28299	9.45177	3.53369	10.54823
	ZnS	0.33884	9.52999	2.95125	10.47001
Zirconium					
K_2ZrF_6	Zr	0.32187	9.50768	3.10684	10.49232
	ZrF_4	0.58999	9.77084	1.69494	10.22915
	ZrF_6	0.72407	9.85978	1.38108	10.14022
	ZrO_2	0.43478	9.63827	2.30001	10.36173
Zr	ZrO_2	1.35080	10.13059	0.74030	9.86941
	ZrC	1.13166	10.05372	0.88366	9.94629
$Zr(BrC_8H_6O_3)_4$	Zr	0.09020	8.95521	11.08647	11.04479
	ZrO_2	0.12184	9.08579	8.20749	10.91421
$Zr(C_8H_7O_3)_4$	Zr	0.13110	9.11760	7.62777	10.88240
	ZrO_2	0.17709	9.24819	5.64685	10.75181
ZrC	C	0.11635	9.06577	8.59476	10.93424
	Zr	0.88366	9.94629	1.13166	10.05372
ZrO_2	Zr	0.74030	9.86941	1.35080	10.13059
	ZrC	0.83777	9.92312	1.19365	10.07687
$Zr_2P_2O_7$	Zr	0.34402	9.53658	2.90681	10.46342
	ZrO_2	0.46470	9.66717	2.15193	10.33283

PHYSICAL CONSTANTS OF MINERALS

Compiled by Ralph Kretz

The following table presents data for many of the more common minerals.

In order to avoid duplication and save space, very few cross references are given in the body of the table. If the name sought is not found in the table, consult the **synonym index** given below.

Specific gravities are given at normal atmospheric temperatures, a more precise statement being valueless considering the large variations in natural minerals.

Hardness is given in terms of Mohs' scale. (See under Hardness.)

Indices of refraction for the sodium line, λ = 5893 Å, unless otherwise indicated. Li, λ = 6708 Å. Indices will invariably be given in the order ω, ϵ or α, β, γ. Uniaxial crystals are considered positive if $\epsilon > \omega$, negative if $\omega > \epsilon$. Biaxial crystals are considered positive if β is nearer α in value than it is γ and negative if β is nearer γ than α.

ABBREVIATIONS

Abbreviation	Meaning of abbreviation	Abbreviation	Meaning of abbreviation	Abbreviation	Meaning of abbreviation
bl	blue	grn	green	rhbdr	rhombohedral
blk	black	grnsh	greenish	rhomb	rhombic
blksh	blackish	hex	hexagonal	somet	sometimes
blsh	bluish	iridesc	iridescent	tarn	tarnishes
br	brown	monocl	monoclinic	tetr	tetragonal
brnsh	brownish	oft	often	tricl	triclinic
col	colorless	pa	pale	vlt	violet
cub	cubic	purp	purple	wh	white
dk	dark	(R)	radioactive	yel	yellow
Fe	Fe, ferrous iron	redsh	redish	yelsh	yellowish
Fe^{+3}	Fe, ferric iron				

SYNONYM INDEX

Compound sought	Listed
Acmite	Aegirine
Agate	Quartz (impure)
Aluminum hydroxide	Boehmite, Diaspore, Gibbsite
Amphibole	Actinolite, Anthophyllite, Cummingtonite, Glaucophane, Hornblende, Riebeckite, Tremolite
Antimony oxide	Senarmontite, Valentinite
Antimony sulfide	Stibnite
Arsenic oxide	Arsenolite, Claudetite
Arsenic sulfide	Orpiment, Realgar
Barium carbonate	Witherite
Barium sulfate	Barite
Barytes	Barite
Bauxite	Gibbsite, Boehmite, Diaspore
Brimstone	Sulfur
Bronzite	Orthopyroxene
Cadmium sulfide	Greenockite
Calamine	Hemimorphite
Calcium carbonate	Aragonite, Calcite, Vaterite
Calcium sulfate	Anhydrite, Gypsum
Calcium sulfide	Oldhamite
Carborundum	Moissanite
Chalcedony	Quartz (impure, fibrous)
Chinaclay	Kaolinite
Chloanthite	Skutterodite
Chromespinel	Chromite
Chrysolite	Serpentine
Clinoptolite	Heulandite
Clayminerals	Illite, Kaolinite, Montmorillonite
Clinochlore	Chlorite
Cobaltbloom	Erythrite
Copper chloride	Nantokite
Copper oxide	Cuprite
Copper sulfide	Chalcocite, Covellite, Digenite
Emerald	Beryl
Emery	Mixture of Corundum, Magnetite and other minerals
Epsom salt	Epsomite
Feldspar	Orthoclase, Microcline, Anorthoclase, Albite, Oligoclase, Andesine, Anorthite
Fibrolite	Sillimanite
Flint	Quartz (impure)
Fluorapatite	Apatite
Fluorspar	Fluorite
Garnet	Almandine, Pyrope, Spessartite, Andradite, Grossularite, Uvarovite, Hydrogrossularite
Garnierite	Serpentine (Ni-bearing)
Glauber salt	Mirabilite
Hyacinth	Zircon
Iceland spar	Calcite
Idocrase	Vesuvianite
Iron carbonate	Siderite
Iron hydroxide	Goethite, Lepidocrocite
Iron oxide	Hematite, Magnetite
Iron spinel	Hercynite
Iron sulfide	Marcasite, Pyrite, Pyrrhotite
Lapis lazuli	Lazurite
Lead carbonate	Cerussite
Lead chloride	Cotunnite
Lead chromate	Crocoite
Lead oxide	Litharge, Minium
Lead sulfate	Anglesite
Lead sulfide	Galena
Limonite	Goethite (impure)
Lithiophyllite	Triphylite
Lithium mica	Lepidolite
Lodestone	Magnetite
Magnesium carbonate	Magnesite
Magnesium hydroxide	Brucite
Magnesium oxide	Periclase
Magnesium sulfate	Kieserite
Manganese carbonate	Rhodochrosite
Manganese hydroxide	Pyrochroite
Manganese oxide	Hausmannite, Manganosite, Pyrolusite
Manganese sulfide	Alabandite
Meerschaum	Serpentine
Mica	Muscovite, Paragonite, Phlogopite, Biotite, Lepidolite
Native copper	Copper
Native gold	Gold
Nickel oxide	Bunsenite
Nickel sulfide	Millerite
Orthite	Allanite
Penninite	Chlorite
Peridote	Olivine
Pistacite	Epidote
Pitchblende	Uraninite
Plagioclase	Albite, Oligoclase, Andesine, Anorthite
Potassium chloride	Sylvite
Potassium sulfate	Arcanite
Pyroxene	Diopside, Angite, Aegirine, Jadeite, Pigeonite, Eustatite, Orthopyroxene
Rocksalt	Halite
Ruby	Corundum
Sapphire	Corundum
Silica	Christobalite, Quartz, Tridymite
Silver chloride	Cerargyrite
Silver iodide	Jodyrite, Miersite
Silver sulfide	Acanthite, Argentite
Smalltite	Skutterotite
Soapstone	Mixture of Talc and other minerals
Sodium chloride	Halite
Sodium sulfate	Thenardite
Strontium carbonate	Strontianite
Strontium sulfate	Celestite
Thorium oxide	Thorianite
Tin oxide	Cassiterite
Titanite	Sphene
Titanium oxide	Anatase, Brookite, Rutile
Uranium oxide	Uraninite
Zeolite	Natrolite, Mesolite, Scolecite, Thomasonite, Harmatome, Eddingtonite, Heulandite, Stilbite, Phillipsite, Chabazite, Gmelinite, Levyn, Laumontite, Mordenite
Zincblende	Sphalerite
Zinc carbonate	Smithsonite
Zinc oxide	Zincite
Zinc spinel	Gahnite
Zinc sulfide	Sphalerite, Wurtzite
Zirconium oxide	Baddeleyite

PHYSICAL CONSTANTS OF MINERALS

Name	Formula	Sp. gr.	Hardness	Crystalline form and color	Index of refraction (Na) η; ω ϵ α β γ
Acanthite	AgS	7.2–7.3	2–2.5	rhomb.(?), iron-blk.	....
Actinolite	$Ca_2((Mg,Fe)_5Si_8O_{22}(OH,F)_2$	3.02–3.44	5–6	monocl., pa. to dk. grn.	1.599–1.688, 1.612–1.697, 1.622–1.705
Aegirine	$NaFe^{+3}Si_2O_6$	3.55–3.60	6	monocl., dk. grn. to grnsh. blk.	1.750–1.776, 1.780–1.820, 1.800–1.836
Åkermanite	$Ca_2MgSi_2O_7$	2.944	5–6	tetr., col., gray-grn., br.	1.632, 1.640
Alabandite	MnS	4.050	3.5–4	cub., iron-blk., tarn., br.	....
Albite	$NaAlSi_3O_8$	2.63	6–6.5	tricl., col., wh., somet. yel., pink, grn.	1.527, 1.531, 1.538
Allanite	$(Ca,Mn,Ce,La,Y,Th)_2(Fe,Fe^{+3},Ti)(Al,Fe^{+3})_2Si_3O_{12}(OH)$	3.4–4.2	5–6.5	monocl., pa. br. to blk.	1.690–1.791, 1.700–1.815, 1.706–1.828
Allemontite	$AsSb$	5.8–6.2	3–4	hex., tin-wh. to redsh., gray, tarn. gray–brnsh. blk.	....
Almandine	$Fe_3Al_2Si_3O_{12}$	4.318	6–7.5	cub., red, dk. red, blk.	1.830
Altaite	$PbTe$	8.15	3	cub., tin-wh., yelsh., tarn. bronze-yel.	....
Aluminite	$Al_2(SO_4)(OH)_4.7H_2O$	1.66–1.82	1–2	monocl.(?), wh.	1.459, 1.464, 1.470
Alunite	$(K,Na)Al_3(SO_4)_2(OH)_6$	2.6–2.9	3.5–4	rhbdr., wh., gray, yel., redsh., br.	1.572, 1.592
Alunogen	$Al_2(SO_4)_3.18H_2O$	1.77	1.5–2	tricl., col., wh., yelsh. wh., redsh. wh.	1.459–1.475, 1.461–1.478, 1.470–1.485
Amblygonite	$(Li,Na)Al(PO_4)(F,OH)$	3.0–3.1	5.5–6	tricl., wh., yelsh. wh., grnsh. wh., blsh. wh., gray	1.591, 1.604, 1.613
Analcite	$NaAlSi_2O_6.H_2O$	2.24–2.29	5.5	cub., wh., pink, gray	1.479–1.493
Anatase	TiO_2	3.90	5.5–6	tetr., br., yelsh. br., redsh. br., bl., blk., grn., gray	2.5612, 2.4880
Andalusite	Al_2OSiO_4	3.13–3.16	6.5–7.5	rhomb., pink, wh., red	1.629–1.640, 1.633–1.644, 1.638–1.650
Andesine	$([NaSi]_{0.7-0.5}[CaAl]_{0.3-0.5})AlSi_2O_8$	2.65–2.68	6–6.5	tricl., wh., gray, grn.	1.544–1.555, 1.548–1.558, 1.551–1.563
Andorite	$PbAgSb_3S_6$	5.33–5.37	3–3.5	rhomb., dk. steel gray, somet. tarn. yel. or iridesc.	....
Andradite	$Ca_3Fe_2{}^{+3}Si_3O_{12}$	3.859	6–7.5	cub., brnsh. red, blk., somet. yel., grn.	1.887
Anglesite	$PbSO_4$	6.37–6.39	2.5–3	rhomb., col., wh., somet. gray, yelsh., grn. tinge	1.8771, 1.8826, 1.8937
Anhydrite	$CaSO_4$	2.98	3.5	rhomb., col., blsh. wh., vlt.	1.5698, 1.5754, 1.6136
Ankerite	$Ca(Fe,Mg,Mn)(CO_3)_2$	2.8–3.1	3.5–4	rhbdr., br., yelsh. br., grnsh. br., pink	1.690–1.750, 1.510–1.548
Anorthite	$CaAl_2Si_2O_8$	2.76	6–6.5	tricl., wh., yel., grn., blk.	1.577, 1.585, 1.590
Anorthoclase	$(Na,K)AlSi_3O_8$	2.56–2.60	6	tricl., col., wh.	1.523, 1.528, 1.529
Anthophyllite	$(Mg,Fe)_7Si_8O_{22}(OH,F)_2$	2.85–3.57	5.5–6	rhomb., wh., gray, grn., br., yelsh. br., dk. br.	1.596–1.694, 1.605–1.710, 1.615–1.722
Antimony	Sb	6.61–6.72	3–3.5	hex., tin-wh.	....
Apatite	$Ca_5(PO_4)_3(OH,F,Cl)$	3.1–3.35	5	hex., grn., wh., yel., br. red, bl.	1.629–1.667, 1.624–1.666
Apophyllite	$KFCa_4Si_8O_{20}.8H_2O$	2.33–2.37	4.5–5	tetr., col., wh., pink, pa. yel., pa. grn.	1.534–1.535, 1.535–1.537
Aragonite	$CaCO_3$	2.94–2.95	3.5–4	rhomb., col., wh.	1.530–1.531, 1.680–1.681, 1.685–1.686
Arcanite	K_2SO_4	2.663	....	rhom., col., wh.	1.4935, 1.4947, 1.4973
Argentite	Ag_2S	7.2–7.4	2–2.5	cub., blksh. lead gray	....
Arsenic	As	5.63–5.78	3.5	hex., tin-wh., tarn. dk. gray	....
Arsenolite	As_2O_3	3.86–3.88	1.5	cub., wh., somet. blsh., yelsh., redsh. tinge	1.755
Arsenopyrite	$FeAsS$	5.9–6.2	5.5–6	monocl., silver-wh., to steel gray	....
Atacamite	$Cu_2(OH)_3Cl$	3.74–3.78	3–3.5	rhomb., grn., dk. grn., blksh. grn.	1.831, 1.861, 1.880
Augelite	$Al_2(PO_4)(OH)_3$	2.696	4.5–5	monocl., col., wh., yelsh. wh., rose	1.5736, 1.5759, 1.5877
Augite	$(Ca,Mg,Fe,Fe^{+3},Ti,Al)_2(Si,Al)_2O_6$	3.23–3.52	5.5–6	monocl., pa. br., br., purp. br., grn., blk.	1.671–1.735, 1.672–1.741, 1.703–1.761
Autunite	$Ca(UO_2)_2(PO_4)_2.10\text{–}12H_2O$	3.1–3.2	2–2.5	tetr., yel., somet. grnsh. yel. to pa. grn.	1.577, 1.553
Axinite	$(Ca,Mn,Fe)_3Al_2BO_3Si_4O_{12}(OH)$	3.26–3.36	6.5–7	tricl., br., yelsh.	1.674–1.693, 1.681–1.701, 1.684–1.704
Azurite	$Cu_3(OH)_2(CO_3)_2$	3.77	3.5–4	monocl., azure bl., dk. bl., pa. bl.	1.730, 1.758, 1.838
Baddeleyite	ZrO_2	5.4–6.02	6.5	monocl., col., yel., gr., redsh. br., br., blk.	2.13, 2.19, 2.20
Barite	$BaSO_4$	4.50	3–3.5	rhomb., col., wh., somet. br., dk. br., gray	1.6362, 1.6373, 1.6482
Benitoite	$BaTi(SiO_3)_3$	3.65	6–6.5	rhbdr., bl., purp., col.	1.757, 1.804
Bertrandite	$Be_4Si_2O_7(OH)_2$	2.6	6	rhomb., col.	1.589, 1.602, 1.613
Beryl	$Be_3Al_2Si_6O_{18}$	2.66–2.83	7.5–8	hex., col., wh., blsh. grn., grnsh. yel., yel., bl.	1.565–1.590, 1.567–1.598
Beryllonite	$NaBe(PO)_4$	2.81	5.5–6	monocl., col., wh., pa. yel.	1.5520, 1.5579. 1.561
Biotite	$K(Mg,Fe)_3AlSi_3O_{10}(OH,F)_2$	2.7–3.3	2.5–3	monocl., blk., dk. br., redsh. br.	1.565–1.625–1.605–1.696, 1.605–1.696
Bismuth	Bi	9.70–9.83	2–2.5	rhbdr., silver-wh. to redsh wh.	....
Bismuthinite	Bi_2S_3	6.75–6.81	2	rhomb., lead gray to tin-wh., tarn. yel. or iridesc.	....
Bixbyite	$(Mn,Fe)_2O_3$	4.945	6–6.5	cub., blk.	....
Bloedite	$Na_2Mg(SO_4)_2.4N_2O$	2.22–2.28	2.5–3	monocl., col., somet. blsh.-grn. or redsh.	1.483, 1.486, 1.487
Boehmite	$AlO(OH)$	3.01–3.06	3.5–4	rhomb., wh.	1.64–1.65, 1.65–1.66, 1.65–1.67
Boracite	$Mg_3B_7O_{13}Cl$	2.91–2.97	7–7.5	rhomb., col., wh., gray, yel., blsh.-grn., grn.	1.66, 1.66, 1.67
Borax	$Na_2B_4O_7.10H_2O$	1.715	2–2.5	monocl., col., wh., gray, blsh. or grnsh-wh.	1.4466, 1.4687, 1.4717
Bornite	Cu_5FeS_4	5.06–5.08	3	cub., copper red to pinchbeck br., tarn. purp., iridesc.	....
Boulangerite	$Pb_5Sb_4S_{11}$	6.0–6.2	2.5–3	monocl., blsh. lead gray, oft, with yel. spots	....
Bournonite	$PbCuSbS_3$	5.80–5.86	2.5–3	rhomb., steel gray to blk.	....
Braggite	PtS	10.0	....	tetr., steel gray	....
Braunite	$(Mn,Si)_2O_3$	4.72–4.83	6–6.5	tetr., brns. blk. to steel gray	....
Bravoite	$(Ni,Fe)S_2$	4.62	5.5–6	cub., steel gray	....
Breithauptite	$NiSb$	8.23	5.5	hex., pa. copper red to vlt., tarn.	....
Brochantite	$Cu_4(SO_4)(OH)_6$	3.79	3.5–4	monocl., emerald-grn. to blksh. grn., pa. grn.	1.728, 1.771, 1.800
Bromyrite	$AgBr$	6.47	2.5	cub., col., gray, yelsh., grnsh.-br.	2.253
Brookite	TiO_2	4.08–4.20	5.5–6	rhomb., br., yelsh. br., redsh. br., blk.	2.5831, 2.5843, 2.7004
Brucite	$Mg(OH)_2$	2.38–3.40	2.5	hex., wh., pa. grn., gray, bl., yel., br.	1.560–1.590, 1.580–1.600
Bunsenite	NiO	6.898	5.5	cub., dk. pistachio-grn.	(Li) 2.37
Cacoxenite	$Fe_4(PO_4)_3(OH)_3.12H_2O$	2.2–2.4	3–4	hex., yel. to brnsh.-yel., redsh. yel., somet. grnsh.	1.575–1.585, 1.635–1.656
Calcite	$CaCO_3$	2.715–2.94	3	rhbdr., col., wh., somet. gray, yel., pink, bl.	1.658–1.740, 1.486–1.550
Caledonite	$Cu_2Pb_5(SO_4)_3(CO_3)(OH)_6$	5.75–5.77	2.5–3	rhomb., dk. grn., blsh. grn.	1.815–1.821, 1.863–1.869, 1.906–1.912
Calomel	$HgCl$	7.15	1.5	tetr., col., wh., gray, yelsh. wh., br.	1.973, 2.656
Cancrinite	$(Na,Ca)_{7-8}Al_6Si_6O_{24}(CO_3SO_4Cl)_{1.5-2}.1\text{–}5H_2O$	2.51–2.42	5–6	hex., col., wh., pa. bl., pa. grn., yel., redsh.	1.528–1.507, 1.503–1.495
Carnalite	$KMgCl_3.6H_2O$	1.602	2.5	rhomb., col., wh., oft. redsh., somet. yel., bl.	1.466, 1.475, 1.494
Carnotite	$K_2(UO_2)_2(VO_4)_2.3H_2O$	....	1–2	rhomb. or monocl., bright yel., yel., grnsh. yel.	1.75, 1.92, 1.95
Cassiterite	SnO_2	6.99	6–7	tetr., yelsh. or redsh. br., brnsh.-blk.	2.006, 2.0972
Celestite	$SrSO_4$	3.96	3–3.5	rhomb., col., wh.. pa. bl., redsh., grnsh., brnsh.	1.621–1.622, 1.623–1.624, 1.630–1.631
Celsian	$BaAl_2Si_2O_8$	3.10–3.39	6–6.5	monocl., col., wh., yel.	1.579–1.587, 1.583–1.593, 1.588–1.600
Cervantite	$Sb_2O_4(?)$	6.64	4–5	rhomb.(?)., yel., wh., somet. redsh.-wh.	....
Cerargyrite	$AgCl$	5.55	2.5	cub., col., gray, grnsh.-br., tarn. purp., yelsh.	2.071
Cerussite	$PbCO_3$	6.53–6.57	3–3.5	rhomb., col., wh., gray, somet. bl., blk., grn.	1.8036, 2.0765, 2.0786
Chabazite	$(Ca,Na_2)Al_2Si_4O_{12}.6H_2O$	2.05–2.10	4.5	rhbdr., redsh.-wh., wh., yelsh., grnsh.	1.470–1.494
Chalcocite	Cu_2S	5.5–5.8	2.5–3	rhomb., blksh., lead gray	....
Chalcanthite	$CuSO_4.5H_2O$	2.28	2.5	tricl., dk. bl. to sky bl., somet. grnsh.	1.514, 1.537, 1.543
Chalcopyrite	$CuFeS_2$	4.1–4.3	3.5–4	tetr., brass-yel., tarn., iridisc.	....
Chiolite	$Na_5Al_3F_{14}$	3.00	3.5–4	tetr., wh. to col.	1.349, 1.342
Chlorite	$(Mg,Al,Fe)_{12}(Si,Al)_8O_{20}(OH)_{16}$	2.6–3.3	2–3	monocl., grn., wh., yel., pink, br., red	1.57–1.66, 1.57–1.67, 1.57–1.67
Chloritoid	$(Fe,Mg,Mn)_2(AlFe^{+3})Al_3O_2SiO_4(OH)_4$	3.51–3.80	6.5	monocl., tricl., dk. grn.	1.713–1.730, 1.719–1.734, 1.723–1.740

PHYSICAL CONSTANTS OF MINERALS (Continued)

Name	Formula	Sp. gr.	Hardness	Crystalline form and color	Index of refraction (Na) η; ω ϵ / α β γ
Chondrodite	$Mg(OH,F)_2.2Mg_2SiO_4$	3.16–3.26	6.5	monocl., yel., br., red	1.592–1.615, 1.602–1.627, 1.621–1.646
Chromite	$FeCr_2O_4$	4.5–5.1	5.5	cub., blk.	2.16
Chrysoberyl	$BeAl_2O_4$	3.65–3.85	8.5	rhomb., grn., yel.	1.746, 1.748, 1.756
Chrysocolla	$CuSiO_3.2H_2O$	~2.4	2	rhomb., (?)., grn., bl., br., blk.	1.575, 1.597, 1.598
Cinnabar	HgS	8.090	2–2.5	hex., red, brnsh. red., gray	(Li) 2.814, 3.143
Claudetite	As_2O_3	4.15	2.5	monocl., col. to wh.	1.87, 1.92, 2.01
Clinozoisite	$Ca_2Al_3Si_3O_{12}(OH)$	3.21–3.38	6.5	monocl., col., pa. yel., gray, grn.	1.670–1.715, 1.674–1.725, 1.690–1.734
Cobaltite	$CoAsS$	6.33	5.5	cub., silver wh., redsh., steel gray, blk.	....
Colemanite	$Ca_2B_6O_{11}.5H_2O$	2.42–2.43	4.5	monocl., col., wh., yelsh. wh., gray	1.586, 1.592, 1.614
Columbite	$(Fe,Mn)(Cb,Ta)_2O_6$	5.15–5.25	6	rhomb., iron blk. to br. blk.	....
Connellite	$Cu_{19}(SO_4)Cl_4(OH)_{32}.3H_2O(?)$	3.36	3	hex., azure bl.	....
Copiapite	$(Fe,Mg)Fe_4^{+3}(SO_4)_6(OH)_2.20H_2O$	2.08–2.17	2.5–3	tricl., yel., grnsh. yel.	1.724–1.738, 1.746–1.758
Copper	Cu	8.95	2.5–3	cub., red	1.51–1.53, 1.53–1.55, 1.58–1.60
Coquimbite	$Fe_2(SO_4)_3.9H_2O$	2.10–2.12	2.5	hex., pa. vlt. to dk. amethystine, yelsh., grnsh.	....
Cordierite	$Al_3(Mg,Fe)_2Si_5AlO_{18}$	2.53–2.78	7	rhomb., gray-bl., bl., dk. bl.	1.53–1.55, 1.55–1.57
Corundum	Al_2O_3	4.022	9	hex., col., bl., yel., purp., grn., pink, red	1.522–1.558, 1.524–1.574, 1.527–1.578
Cotunnite	$PbCl_2$	5.80	2.5	rhomb., col. to wh., somet. yelsh., grnsh.	1.767–1.772, 1.759–1.763
Covellite	CuS	4.6–4.76	1.5–2	hex., indigo bl., dk. bl., iridesc. brass yel. to red	2.199, 2.217, 2.260
Cristobalite	SiO_2	2.33	6–7	tetr.(?)., col., wh., yel.	....
Crocoite	$PbCrO_4$	5.96–6.02	2.5–3	monocl., red, orange red, orange yel.	1.487, 1.484
Cryolite	Na_3AlF_6	2.96–2.98	2.5	monocl., col. to wh., brnsh., redsh., blk.	2.29, 2.36, 2.66
Cryolithionite	$Na_3Li_3Al_2F_{12}$	2.77	2.5–3	cub., col. to wh.	1.338, 1.338, 1.339
Cubanite	$CuFe_2S_3$	4.03–4.18	3.5	rhomb., brass to bronze yel.	1.3395
Cummingtonite	$(Mg,Fe)_7Si_8O_{22}(OH)_2$	3.2–3.5	5–6	monocl., dk. grn., br.	....
Cuprite	Cu_2O	6.14	3.5–4	cub., red, somet. blk.	1.635–1.665, 1.644–1.675, 1.655–1.698
Danburite	$CaSi_2B_2O_8$	3.0	7	rhomb., pa. yel., col., dk. yel., yelsh. br.	....
Datolite	$CaBSiO_4(OH)$	2.96–3.00	5–5.5	monocl. col., wh., yelsh., grnsh., pinksh.	1.63, 1.63–1.64, 1.63–1.64
Daubreelite	Cr_2FeS_4	3.80–3.82	?	cub., blk.	1.622–1.626, 1.649–1.654, 1.666–1.670
Derbylite	$Fe_6Ti_6Sb_2O_{23}(?)$	4.53	5	rhomb., pitch blk.	....
Diamond	C	3.50–3.53	10	cub., col., pa. yel. to dk. yel., pa. br. to dk. br., wh., blsh. wh.	2.45, 2.45, 2.51
Diaspore	$AlO(OH)$	3.3–3.5	6.5–7	rhomb., wh., graysh. wh., col.	2.4175
Digenite	$Cu_{2x}S$	5.546	2.5–3	cub., bl. to blk.	1.682–1.706, 1.705–1.725, 1.730–1.752
Diopside	$CaMgSi_2O_6$	3.22–3.38	5.5–6.5	monocl., wh., pa. grn., dk. grn.	....
Dioptase	$Cu_6Si_6O_{18}.6H_2O$	3.5	5	rhbdr., emerald grn.	1.664–1.695, 1.672–1.701, 1.695–1.721
Dolomite	$CaMg(CO_3)_2$	2.86	3.5–4	rhbdr., wh., oft. yel. or br. tinge, col.	1.64–1.66, 1.70–1.71
Douglasite	$K_2FeCl_4.2H_2O(?)$	2.16	......	pa. grn., tarn. brnsh. red	1.679, 1.500
Dyscrasite	Ag_3Sb	9.67–9.81	3.5–4	rhomb., silver wh., tarn. gray, yelsh. or blksh.	1.485–1.491, 1.497–1.503
Eddingtonite	$BaAl_2Si_3O_{10}.4H_2O$	2.7–2.8	......	rhomb. or monocl., col., pink, br. wh.	....
Eglestonite	Hg_4OCl_2	8.4	2.5	cub., yel., orange-yel. to dk. brnsh., tarn. bl.	1.541, 1.553, 1.557
Emplectite	$CuBiS_2$	6.38	2	rhomb., gray to tin wh.	2.47–2.51
Empressite	$AgTe$	7.510	3–3.5	pa. bronze	....
Enargite	Cu_3AsS_4	4.4–4.5	3	rhomb., gray-blk. to iron-blk.	....
Enstatite	$MgSiO_3$	3.209	5–6	rhomb., col., gray, grn., yel., brn.	....
Epidote	$Ca_2Fe^{+3}Al_2Si_3O_{12}(OH)$	3.38–3.49	6	monocl., grn., yel., gray	1.650–1.662, 1.653–1.671, 1.658–1.680
Epsomite	$MgSO_4.7H_2O$	1.675–1.679	2–2.5	rhomb., col., wh. pink, grn.	1.715–1.751, 1.725–1.784, 1.734–1.797
Erythrite	$(Co,Ni)_3(AsO_4)_2.8H_2O$	3.06	1.5–2.5	monocl., crimson-red, red, pa. pink	1.4325, 1.4554, 1.4609
Eucairite	$CuAgSe$	7.6–7.8	2.5	silver wh. to lead gray	1.626, 1.661, 1.699
Euclasite	$BeAlSiO_4(OH)$	3.0–3.1	7.5	monocl., col., pa. grn., bl.	....
Eudialite	$(Na,Ca,Fe)_6ZrSi_6O_{18}(OH,Cl)(?)$	2.8–3.1	5–6	hex., pa. pink, red, br.	1.651, 1.655, 1.671
Eulytite	$Bi_4Si_3O_{12}$	6.6	4.5	cub., br., yel., gray	1.59–1.61, 1.59–1.61
Euxenite	$(Y,Ca,Ce,U,Th)(Cb,Ta,Ti)_2O_6$	5.0–5.9	5.5–6.5	rhomb., blk., grnsh. or brnsh. tint.	2.05
Fayalite	Fe_2SiO_4	4.392	6.5	rhomb., grnsh., yelsh.	~2.2
Ferberite	$FeWO_4$	7.51	4–4.5	monocl., br. to blk.	1.827, 1.869, 1.879
Fergussonite	$(Y,Er,Ce,Fe)(Cb,Ta,Ti)O_4$	5.6–5.8	5.5–6.5	tetr., gray, yel., br., dk. br.	(Li)2.37–2.43
Fluorite	CaF_2	3.18	4	cub., bl., purp., wh., col., yel., grn.	2.1
Forsterite	Mg_2SiO_4	3.222	7	rhomb., wh., grnsh., yelsh.	1.433–1.435
Franklinite	$ZnFe_2^{+3}O_4$	5.07–5.34	5.5–6.5	Cub., blk. to br.-blk.	1.635, 1.651, 1.670
Gahnite	$ZnAl_2O_4$	4.62	7.5–8	cub., dk. bl.-grn., somet. yelsh. or brnsh.	(Li) ~2.36
Galena	PbS	7.57–7.59	2.5–2.75	cub. lead gray	1.79–1.81
Galenabismuthite	$PbBi_2S_4$	7.04	2.5–3.5	rhomb., pa. gray to tin-wh., lead gray, somet. tarn., yel. or irid.	....
Ganomalite	$(Ca,Pb)_{10}(OH,Cl)_2(Si_2O_7)_3$	5.4–5.7	3–4	hex., col., gray	....
Gaylussite	$Na_2Ca(CO_3)_2.5H_2O$	1.991	2.5–3	monocl., col. to yelsh. wh., graysh. wh., wh.	1.910, 1.945
Gehlenite	$Ca_2Al_2SiO_7$	3.038	5–6	tetr., col., gray-grn., br.	1.4435, 1.5156, 1.5233
Geikielite	$MgTiO_3$	4.05	5–6	rhbdr., brnsh blk., blsh.	1.669, 1.658
Gibbsite	$Al(OH)_3$	2.38–2.42	2.5–3.5	monocl., wh., graysh., grnsh. or redsh.-wh.	2.31, 1.95
Glauberite	$Na_2Ca(SO_4)_2$	2.75–2.85	2.5–3	monocl., gray, yelsh., somet. col., redsh.	1.56–1.58, 1.56–1.58, 1.58–1.60
Glauconite	$(K,Na,Ca)_{1.2-2}(Fe^{+3},Al,Fe,Mg)_4Si_{7-7.6}Al_{1-0.4}O_{20}(OH)_4.nH_2O$	2.4–2.95	2	monocl., col., ye.lsh. grn., grn., blsh. gray	1.515, 1.535, 1.536
Glaucophane	$Na_2Mg_3Al_2Si_8O_{22}(OH)_2$	3.08–3.30	6	monocl., gray, lavender bl.	1.592–1.610, 1.614–1.641, 1.614–1.641
Gmelinite	$(Ca,Na_2)Al_2Si_4O_{12}.6H_2O$	~2.1	4.5	rhbdr., wh., redsh.-wh., yelsh., grnsh.	1.606–1.661, 1.622–1.667, 1.627–1.670
Goethite	$FeO(OH)$	3.3–4.3	5–5.5	rhomb., blksh.-br., yelsh. or redsh.-br., yel.	1.476–1.494, 1.474–1.480
Gold	Au	19.3	2.5–3	cub., yel.	2.260–2.275, 2.393–2.409, 2.398–2.515
Goslarite	$ZnSO_4.7H_2O$	1.978	2–2.5	rhomb., col., wh., somet. br., grn., bl.	....
Graphite	C	2.09–2.23	1–2	hex., iron-blk. to steel gray	1.4568, 1.4801, 1.4844
Greenockite	CdS	4.9	3–3.5	hex., yel. to orange	....
Grossularite	$Ca_3Al_2Si_3O_{12}$	3.594	6–7.5	cub., wh., yel., grn., br., red	2.506, 2.529
Gummite (R)	$UO_3.H_2O$	3.9–6.4	2.5–5	yel., orange, redsh.-yel., red, br. blk.	1.734
Gypsum	$CaSO_4.2H_2O$	2.30–2.37	2	monocl., wh., col., somet. gray, red, yel., br.	....
Halite	$NaCl$	2.16–2.17	2.5	cub., col., wh., orange, red	1.519–1.521, 1.523–1.526, 1.529–1.531
Hambergite	$Be_2(OH)(BO_3)$	2.36	7.5	rhomb., col. to gray, wh., yel.	1.544
Hanksite	$Na_{22}K(SO_4)_9(CO_3)_2Cl$	2.562	3–3.5	hex., col., somet. pa.-yelsh. or gray	1.56, 1.59, 1.63
Harmotome	$BaAl_2Si_6O_{16}.6H_2O$	2.41–2.47	4.5	monocl., or rhomb., col., wh., pink, gray, yel.	1.481, 1.461
Hausmannite	Mn_3O_4	4.83–4.85	5.5	tetr., brnsh.-blk.	1.503–1.508, 1.505–1.509, 1.508–1.514
Haüyne	$(Na,Ca)_{4-8}Al_6Si_6O_{24}(SO_4,S)_{1-2}$	2.44–2.50	5.5–6	cub., wh., gray, grn., bl.	(Li) 2.46, 2.15
Hedenbergite	$CaFeSi_2O_6$	3.50–3.56	6	monocl., brnsh.-grn., dk. grn., blk.	1.496–1.505
Helvite	$Mn_4Be_3Si_3O_{12}S$	3.20–3.44	6	cub., yel., br., redsh.-brn.	1.716–1.726, 1.723–1.730, 1.741–1.751
Hematite	Fe_2O_3	5.26	5–6	rhbdr., steel gray, dull red to bright red	1.728–1.749
Hemimorphite	$Zn_4Si_2O_7(OH)_2.H_2O$	3.45	5	rhomb., col., wh., pa. bl., pa. grn., br.	3.22, 2.94
Hercynite	$FeAl_2O_4$	4.40	7.5–8	cub., blk.	1.614, 1.617, 1.636
Herderite	$CaBe(PO_4)(Fe,OH)$	2.95–3.01	5–5.5	monocl., col. to pa. yel. or grnsh.-wh.	1.835
Hessite	Ag_2Te	8.24–8.45	2–3	monocl., (<149.5°), cub. (>149.5°)., gray	1.592, 1.612, 1.621
Heulandite	$(Ca,Na_2)Al_2Si_7O_{18}.6H_2O$	2.1–2.2	3.5–4	pseudo-monocl., col., wh., yel., pink, red, gray, br.	1.491–1.505, 1.493–1.503, 1.500–1.512

PHYSICAL CONSTANTS OF MINERALS (Continued)

Name	Formula	Sp. gr.	Hardness	Crystalline form and color	Index of refraction (Na) η; ω ϵ ω β γ
Hopeite	$Zn_3(PO_4)_2.4H_2O$	3.0–3.1	3.25	rhomb., col. to grayish-wh., pa. yel.	1.57–1.59, 1.58–1.60, 1.58–1.60
Hornblende	$(Ca,Na,K)_{2-3}(Mg,Fe,Fe^{+3}Al)_5Si_6(Si,Al)_2O_{22}(OH,F)_2$	3.02–3.45	5–6	monocl., grn., dk. grn., blk.	1.615–1.705, 1.618–1.714, 1.632–1.730
Huebnerite	$MnWO_4$	7.12	4–4.5	monocl., yel.-br. to red br., somet. br., blk.	2.17, 2.22, 2.32
Humite	$Mg(OH,F)_2.3Mg_2SiO_4$	3.2–3.32	6	rhomb., yel., orange	1.607–1.643, 1.619–1.653, 1.639–1.675
Huntite	$Mg_3Ca(CO_3)_4$	2.696	...	rhomb.(?)., wh.	...
Hydrogrossularite	$Ca_3Al_2Si_2O_8(SiO_4)_{1-m}(OH)_{4m}$	3.594–3.13	6–7.5	cub., wh., buff, pa. grn., gray, pink	1.734–1.675
Hydromagnesite	$Mg_4(OH)_2(CO_3.3H_2O$	2.236	3.5	monocl., col. to wh.	1.520–1.526, 1.524–1.530, 1.544–1.546
Illite	$K_{1-1.5}Al_4Si_{7-6.5}Al_{1-1.5}O_{20}(OH)_4$	2.6–2.9	1–2	monocl., wh.	1.54–1.57, 1.57–1.61, 1.57–1.61
Ilmenite	$FeTiO_3$	4.68–4.76	5–6	rhbdr., iron-blk.	...
Iodyrite	AgI	5.69	1.5	hex., col. on exposure to light, yel., br.	2.21, 2.22
Jadeite	$NaAlSi_2O_6$	3.24–3.43	6	monocl., col., wh., grn., grnsh. bl.	1.640–1.658, 1.645–1.663, 1.652–1.673
Jamesonite	$Pb_4FeSb_6S_{14}$	5.63	2.5	monocl., gray-blk., somet. tarn. iridesc.	...
Jarosite	$KFe_3(SO_4)_2(OH)_6$	2.91–3.26	2.5–3.5	rhbdr., ocherous, amber yel. to dk. br.	1.820, 1.715
Kainite	$KMg(SO_4)Cl{\cdot}3H_2O$	2.15	2.5–3	monocl., col., gray bl., vlt., yelsh., redsh.	1.494, 1.505, 1.516
Kaliophyllite	$KAlSiO_4$	2.61	6	hex., col.	1.532, 1.537
Kaolinite	$Al_4Si_4O_{10}(OH)_8$	2.61–2.68	2–2.5	tricl. or monocl., wh., redsh.-wh., grnsh.-wh.	1.533–1.565, 1.559–1.569, 1.560–1.570
Kernite	$Na_2B_4O_7.4H_2O$	1.908	2.5	monocl., col., wh.	1.454, 1.472, 1.488
Kieserite	$MgSO_4.H_2O$	2.571	3.5	monocl., col., gray, wh., yelsh.	1.520, 1.533, 1.584
Kyanite	Al_2OSiO_4	3.53–3.65	5.5–7	tricl., bl., wh., gray, grn., yel., pink	1.712–1.718, 1.721–1.723, 1.727–1.734
Lanarkite	$Pb_2(SO_4)O$	6.92	2–2.5	monocl., gray to grnsh. wh., pa. yel.	1.925–1.931, 2.004–2.010, 2.033–2.039
Lanthanite	$(La,Ce)_2(CO_3)_38H_2O$	2.69–2.74	2.5–3	rhomb., col. to wh., pink, yelsh.	1.51–1.53, 1.584–1.590, 1.610–1.616
Laumontite	$CaAl_2Si_4O_{12}.4–3.5H_2O$	2.2–2.3	3–3.5	monocl., col., wh., red, yel., brn.	1.502–1.514, 1.512–1.522, 1.514–1.525
Laurionite	$Pb(OH)Cl$	6.24	3–3.5	rhomb., col. to wh.	2.08, 2.12, 2.16
Lawsonite	$CaAl_2(OH)_2Si_2O_7.H_2O$	3.05–3.10	6	rhomb., col., wh.	1.655, 1.674–1.675, 1.684–1.686
Lazulite	$(Mg,Fe)Al_2(PO_4)_2(OH)_2$	3.08–3.38	5.5–6	monocl., bl., blsh. wh., dk. bl., blsh. grn.	1.604–1.626, 1.626–1.654, 1.637–1.663
Lazurite	$Na_4SSi_3Al_3O_{12}$	2.38–2.45	5–5.5	cub., berlin bl., azure bl., grnsh. bl., vlt.	1.500
Leadhillite	$Pb_4(SO_4)(CO_3)_2(OH)_2$	6.55	2.5–3	monocl., col. to wh., gray, pa. grn., pa. bl., yelsh.	1.87, 2.00, 2.01
Lepidocrocite	$FeO(OH)$	4.05–4.31	5	rhomb., ruby-red to red-br.	1.94, 2.20, 2.51
Lepidolite	$K_2(Li,Al)_{5-6}Si_{6-7}Al_{2-1}O_{20}(OH,F)_4$	2.80–2.90	2.5–4	monocl., col., pa. pink, pa. purp.	1.525–1.548, 1.551–1.585, 1.554–1.587
Leucite	$KAlSi_2O_6$	2.47–2.50	5.5–6	tetr., (pseudo-cub.) wh., gray	1.508–1.511
Levyne	$(Ca,Na_2)Al_2Si_4O_{12}.6H_2O$	~2.1	4.5	rhbdr., wh., redsh. wh., yelsh., grnsh.	1–496–1.505, 1.491–1.500
Litharge	PbO	9.14	2	tetr., red	(Li) 2.665, 2.535
Loellingite	$FeAs_2$	7.39–7.41	5–5.5	rhomb., silver wh. to steel-gray	...
Magnesite	$MgCO_3$	2.98–3.44	3.5–4.5	rhbdr., wh., col., somet. yel., br.	1.700–1.782, 1.509–1.563
Magnetite	Fe_3O_4	5.175	5.5–6.5	cub., blk. to br.-blk.	2.42
Malachite	$Cu_2(OH)_2(CO_3)$	4.03–4.07	3.5–4	monocl., bright grn. to dk. grn., blksh. grn.	1.652–1.658, 1.872–1.878, 1.906–1.912
Manganite	$MnO(OH)$	4.32–4.43	4	monocl., dk. steel-gray to iron-blk.	(Li) 2.25, 2.25, 2.53
Manganosite	MnO	5.364	5.5	cub., emerald grn., tarn. bl.	...
Marcasite	FeS_2	4.887	6–6.5	rhomb., pa. bronze-yel., tin-wh.	...
Marialite	$Na_4Al_3Si_9O_{24}Cl$	2.50–2.62	5–6	tetr., col., wh., pa. grnsh. yel., gray, br.	1.546–1.550, 1.540–1.541
Marshite	CuI	5.68	2.5	cub., col. to pa. yel., on exposure to light, red	2.346
Mascagnite	$(NH_4)_2SO_4$	1.768	2–2.5	rhomb., col., gray, yelsh.	1.5202, 1.5230, 1.5330
Matlockite	$PbFCl$	7.12	2.5–3	tetr., col. or yel. to pa. amber, grnsh.	2.145, 2.006
Meionite	$Ca_4Al_6Si_6O_{24}CO_3$	2.78	5–6	tetr., col., wh., pa. grnsh. yel., gray, br.	1.590–1.600, 1.556–1.562
Melanterite	$FeSO_4.7H_2O$	1.898	2	monocl., grn., grnsh. bl., grnsh. wh.	1.47, 1.48, 1.49
Melilite	$(Ca,Na,K)_2(Mg,Fe,Fe^{+3},Al,Si)_3O_7$	2.95–3.05	5–6	tetr., yelsh., br., grn.-br.	1.624–1.666, 1.616–1.661
Mellite	$Al_2C_{12}O_{12}.18H_2O$	1.64	2–2.5	tetr., yel., redsh., brnsh., somet. wh.	1.5393, 1.5110
Mendipite	$Pb_3O_2Cl_2$	7.24	2.5	rhomb., col. to wh., gray, oft. yel., red, bl. tinge	2.22–2.26, 2.25–2.29, 2.29–2.33
Mesolite	$Na_2Ca_2(Al_2Si_3O_{10}).8H_2O$	~2.26	5	monocl., col., wh., gray, yel., pink, red	β = 1.504–1.508
Metacinnabar	HgS	7.65	3	cub., graysh.-blk.	...
Microcline	$KAlSi_3O_8$	2.56–2.63	6–6.5	tricl., col., wh., pink, red, yel., grn.	1.514–1.529, 1.518–1.533, 1.521–1.539
Microlite	$(Na,Ca)_2Ta_2O_6(O,OH,F)$	4.2–6.4	5–5.5	cub., pa. yel. to br., somet. red, grn.	~2.0
Miersite	AgI	5.64–5.68	2.5	cub., canary-yel.	2.18–2.22
Millerite	NiS	5.3–5.7	3–3.5	hex., pa. brass-yel. to bronze-yel., gray, tarn. iridesc.	...
Mimetite	$Pb_5(AsO_4,PO_4)_3Cl$	7.24	3.5–4	hex. pa. yel. to yelsh. br., orange-yel., wh.	2.147, 2.128
Minium	Pb_3O_4	8.9–9.2	2.5	scarlet red, bl. red, somet. yel. tint.	(Li) 2.40–2.44
Mirabilite	$Na_2SO_4.10H_2O$	1.490	1.5–2	monocl., col. to wh.	1.391–1.397, 1.393–1.399, 1.395–1.401
Moissanite	SiC	3.218	9.5	hex., grn. to blk., somet. blsh., red	2.647–2.649, 2.689–2.693
Molybdenite	MoS_2	4.62–4.73	1–1.5	hex., lead-gray	...
Monazite	$(Ce,La,Th)PO_4$	5.0–5.3	5	monocl., yel., br., redsh. br.	1.774–1.800, 1.777–1–801, 1.828–1.851
Monetite	$CaH(PO_4)$	2.929	3.5	tricl., wh., pa. yelsh.-wh.	1.587, ~1.615, 1.640
Monticellite	$CaMgSiO_4$	3.08–3.27	5.5	rhomb., col.	1.639–1.654, 1.646–1.664, 1.653–1.674
Montmorillonite	$(0.5Ca,Na)_{0.7}(Al,Mh,Fe)_4(Si,Al)_8O_{20}(OH)_4.nH_2O$	2–3	1–2	monocl., wh., yel., grn.	1.48–1.61, 1.50–1.64, 1.50–1.64
Montroydite	HgO	11.23	2.5	rhomb., dk. red to brnsh. red, br.	(Li) 2.37, 2.5, 2.65
Mordenite	$(Na_2,K_2Ca)Al_2Si_{10}O_{24}.7H_2O$	2.12–2.15	3–4	rhomb., col., wh., red, yel., br.	1.472–1.483, 1.475–1.485, 1.477–1.487
Muscovite	$KAl_2Si_3AlO_{10}(OH,F)_2$	2.77–2.88	2.5–3	monocl., col., pa. grn., pa. red, pa. br.	1.552–1.574, 1.582–1.610, 1.587–1.616
Nantokite	$CuCl$	4.136	2.5	cub., col. to wh., grayish, grn.	1.925–1.935
Natrolite	$Na_2Al_2Si_3O_{10}.2H_2O$	2.20–2.26	5	rhomb., col., wh., gray, yel., pink, red	1.473–1.483, 1.476–1.486, 1.485–1.496
Nepheline	$Na_3KAl_4Si_4O_{16}$	2.56–2.665	5.5–6	hex., col., wh., gray	1.529–1.546, 1.526–1.542
Newberyite	$MgH(PO_4).3H_2O$	2.10	3.0–3.5	rhomb., col.	1.511–1.517, 1.514–1.520. 1.530–1.536
Niccolite	$NiAs$	7.784	5–5.5	hex., pa. copper-red, tarn. gray to blk.	...
Nosean	$Na_8Al_6Si_6O_{24}SO_4$	2.30–2.40	5.5	cub., gray, bl., br.	1.495
Oldhamite	CaS	2.58	4	cub., pa. chestnut-br.	2.137
Oligoclase	$([NaSi]_{0.9-0.7}[CaAl]_{0.1-0.3})AlSi_2O_8$	2.63–2.65	6–6.5	tricl., col., wh., gray, grnsh., pink	1.533–1.544, 1.537–1.548, 1.543–1.552
Olivenite	$Cu_2(AsO_4)(OH)$	3.9–4.5	3	rhomb., olive grn., grnsh.-br., br., gray	1.75–1.78, 1.79–1.82, 1.83–1.87
Olivine	$(Mg,Fe)SiO_4$	3.22–4.39	6.5–7	rhomb., olive grn., grayish grn. to yelsh. br.	1.63–1.83, 1.65–1.87, 1.67–1.88
Opal	$SiO_2.nH_2O$	1.73–2.16	~6	col., wh., yel., br., red, grn., bl., blk., amorp.	1.41–1.46
Orpiment	As_2S_3	3.49	1.5–2	monocl., yel., brnsh. yel.	(Li) 2.4, 2.81, 3.02
Orthoclase	$KAlSi_3O_8$	2.55–2.63	6–6.5	monocl., col., wh., pink, red, yel., grn.	1.518–1.529, 1.522–1.533, 1.522–1.539
Orthopyroxene	$(Mg,Fe)SiO_3$	3.209–3.96	5–6	rhomb., col., gray, grn., yel., dk. brn.	1.650–1.768, 1.653–1.770, 1.658–1.788
Paragonite	$NaAl_2Si_3AlO_{10}(OH)_2$	2.85	2.5	monocl., col., pa. yel.	1.564–1.580, 1.594–1.609, 1.600–1.609
Parisite	$(Ce,La,Na)FCO_3.CaCO_3$	4.42	4.5	hex., brnsh., yel.	1.672, 1.771
Pectolite	$Ca_2NaH(SiO_3)_3$	2.86–2.90	4.5–5	tricl., col., wh.	1.595–1.610, 1.605–1.615, 1.632–1.645
Penfieldite	$Pb_4Cl_6(OH)_2$	6.6	...	hex., wh.	2.13, 2.21
Pentlandite	$(Fe,Ni)_9S_8$	4.6–5.0	3.5–4	cub., pa. bronze-yel.	...
Percylite	$PbCuCl_2(OH)_2(?)$	?	2.5	cub(?), sky bl.	2.04–2.06
Periclase	MgO	3.55–3.68	5.5	cub., col. to gray-wh., yel., brnsh. yel., grn., bl.	1.7350
Pekovskite	$CaTiO_3$	3.97–4.26	5.5	pseudo cub., blk., gray-blk., brnsh. bl., redsh. br., br., yel.	2.30–2.38
Petalite	$LiAlSi_4O_{10}$	2.412–2.422	6.5	monocl., wh., gray, somet. pink, grn.	1.504–1.507, 1.510–1.513, 1.516–1.523

Name	Formula	Sp. gr.	Hardness	Crystalline form and color	Index of refraction (Na) η; ω ϵ / α β γ
Pharmacosiderite	$Fe_3(AsO_4)_2(OH)_3.5H_2O$	2.797	2.5	cub., olive-grn. to yel., br., redsh.	1.676–1.704
Phenakite	Be_2SiO_4	2.98	7.5	rhbder., col., rose, yel., br.	1.654, 1.670
Phillipsite	$(0.5Ca,Na,K)_3Al_3Si_5O_{16}.6H_2O$	2.2	4–4.5	monocl. or rhomb., col., wh., pink, gray, yel.	1.483–1.504, 1.484–1.509, 1.496–1.514
Phlogopite	$KMg_3AlSi_3O_{10}(OH,F)_2$	2.76–2.90	2–2.5	monocl., col., yelsh.-br., grn., redsh.-br., br.	1.530–1.590, 1.557–1.637, 1.558–1.637
Phosgenite	$Pb_2(CO_3)Cl_2$	6.133	2–3	tetr., yelsh. wh. to yelsh. br., br., somet. wh., rose, gray	2.1181, 2.1446
Piemontite	$Ca_2(Mn,Fe^{+3},A^{1})_2AlSi_3O_{12}(OH)$	3.45–3.52	6	monocl., redsh. brn., blk.	1.732–1.794, 1.750–1.807, 1.762–1.829
Pigeonite	$(Mg,Fe,Ca)(Mg,Fe)Si_2O_6$	3.30–3.46	6	monocl., br., grnsh. br., blk.	1.682–1.722, 1.684–1.722, 1.705–1.751
Platinum	Pt	14–19	4–4.5	cub., whitish, steel gray to dk. gray	......
Pollucite	$CsAlSi_2O_6$	2.9	6.5	tetr., (pseudo-cub.) col.	1.507–1.527
Polybasite	$(Ag,Cu)_{16}Sb_2S_{11}$	6.0–6.2	2–3	monocl., iron-blk.	......
Powellite	$Ca(Mo,W)O_4$	4.21–4.25	3.5–4	tetr., straw-yel., br., grnsh., somet. gray, bl., blk.	1.959–1.982, 1.967–1.993
Prehnite	$Ca_2Al_2Si_3O_{10}(OH)_2$	2.90–2.95	6–6.5	rhomb., pa. grn., yel., gray, wh.	1.611–1.632, 1.615–1.642, 1.632–1.665
Proustite	Ag_3AsS_3	5.57	2–2.5	rhbdr., scarlet-vermillion	3.0877, 2.7924
Pseudobrookite	Fe_2TiO_5	4.33–4.39	6	rhomb., dk. red-br. to brnsh. blk. and blk.	(Li) 2.38, 2.39, 2.42
Psilomelane	$BaMn^{+2}Mn_8^{+4}O_{16}(OH)_4$	4.71	5–6	rhomb., iron-blk. to steel-gray	
Pumpellyite	$Ca_4(Mg,Fe,Mn)(Al,Fe^{+3},Ti)_5(OH)_3Si_6O_{23}.2H_2O$	3.18–3.23	6	monocl., grn., blsh. grn., br.	1.674–1.702, 1.675–1.715, 1.688–1.722
Pyrargyrite	Ag_3SbS_3	5.85	2.5	rhbdr., deep red	(Li) 3.084, 2.881
Pyrite	FeS_2	5.018	6–6.5	cub., pa. brass-yel., tarn. iridesc.	......
Pyrochlore	$NaCaCb_2O_6F$	4.2–6.4	5–5.5	cub., br. to blk., yelsh., redsh. or blksh. br.	......
Pyrochroite	$Mn(OH)_2$	3.23–3.27	2.5	hex., col. to pa. grn. or bl., tarn. br. to blk.	1.72, 1.68
Pyrolusite	MnO_2	5.04–5.08	6–6.5	tetr., pa. steel-gray, iron-gray, blk., blsh.	......
Pyromorphite	$Pb_5(PO_4,AsO_4)_3Cl$	7.00–7.08	3.5–4	hex., grn., yel., br., orange, brnsh. red., gray	2.058, 2.048
Pyrope	$Mg_3Al_2Si_3O_{12}$	3.582	6–7.5	cub., red, pink	1.714
Pyrophyllite	$Al_2Si_4O_{10}(OH)_2$	2.65–2.90	1–2	monocl., wh., yel., pa. bl., gray-grn., brnsh.-grn.	1.534–1.556, 1.568–1.589, 1.596–1.601
Pyrrhotite	$Fe_{1-0.8}S$	4.58–4.65	3.5–4.5	hex., bronze-yel. to br., tarn., somet. iridesc.	......
Quartz	SiO_2	2.65	7	rhbdr., col., wh., blk., purp., grn., bl., rose	1.544, 1.553
Rammelsbergite	$NiAs_2$	7.0–7.2	5.5–6	tin. wh., redsh. tinge	
Raspite	$PbWO_4$	8.46	2.5–3	monocl., yelsh. br., pa. yel., gray	1.25–1.29, 1.25–1.29, 1.28–1.32
Realgar	AsS	3.56	1.5–2	monocl., aurora-red to orange-yel.	2.538, 2.684, 2.704
Riebeckite	$Na_2Fe_3Fe_2^{+3}Si_8O_{22}(OH,F)_2$	3.02–3.42	5	monocl., dk. bl., bl.	1.654–1.701, 1.662–1.711, 1.668–1.717
Rhodochrosite	$MnCO_3$	3.70	3.5–4	rhbdr., pink, red, br., brnsh.-yel.	1.816, 1.597
Rhodonite	$(Mn,Fe,Ca)SiO_3$	3.57–3.76	5.5–6.5	tricl., pink to brnsh. red	1.711–1.738, 1.716–1.741, 1.724–1.751
Rutile	TiO_2	4.23–5.5	6–6.5	tetr., redsh. brn. to red, somet. yelsh., blsh.	2.605–2.613, 2.899–2.901
Safflorite	$(Co,Fe)As_2$	7.0–7.5	4.5–5	rhomb., tin-wh., tarn. dk. gray	......
Samarskite	$(Y,Er,Ce,U,Ca,Fe,Pb,Th)(Cb,Ta,Ti,Sn)_2O_6$	5.69	5–6	rhomb., velvet blk., somet. brnsh. tint	~2.20
Sapphirine	$(Mg,Fe)_2Al_4O_6SiO_4$	3.40–3.58	7.5	monocl., pa. bl., pa. grn.	1.701–1.717, 1.703–1.720, 1.705–1.724
Scapolite	$(Na,Ca)_4Al_3(Al,Si)_3Si_6O_{24}(Cl,F,OH,CO_3,SO_4)$	2.50–2.78	5–6	tetr., col., wh., pa. grnsh. yel., gray, bl.	1.546–1.600, 1.540–1.562
Scheelite	$CaWO_4$	6.08–6.12	4.5–5	tetr., yelsh. wh., pa. yel., brnsh., col., wh., gray	1.920, 1.936
Scolecite	$CaAl_2Si_3O_{10}.3H_2O$	2.25–2.29	5	monocl., col., wh., gray, yel., pink, red	1.507–1.513, 1.516–1.520, 1.517–1.521
Scorodite	$Fe^{+3}(AsO_4).2H_2O$	3.28	3.5–4	rhomb., pa. grn., gray grn., br. somet. col., blsh., yel.	1.784, 1.795, 1.814
Sellaite	MgF_2	3.15	5	tetr., col. to wh.	1.378, 1.390
Senarmontite	Sb_2O_3	5.50	2–2.5	pseudo-cub., col., gray-wh.	2.087
Serpentine	$Mg_3Si_2O_5(OH)_4$	~2.55	2.5–3.5	monocl., wh., yel., gray, grn., blsh. grn.	1.53–1.57, 1.56, 1.54–1.57
Siderite	$FeCO_3$	3.96	4–4.5	rhbdr., yelsh. br., br., dk. br.	1.875. 1.635
Sillimanite	Al_2OSiO_4	3.23–3.27	6.5–7.5	rhomb., col., wh., yelsh., br., grnsh.	1.654–1.661, 1.658–1.662, 1.637–1.683
Silver	Ag	10.1–11.1	2.5–3	cub., wh., tarn. gray or blk.	......
Skutterudite	$(Co,Ni)As_3$	6.1–6.9	5.5–6	cub., between tin-wh. and silver-gray, tarn. gray or iridesc.	......
Smithsonite	$ZnCO_3$	4.42–4.44	4–4.5	rhbdr., grayish wh. to dk. gray, grnsh., brnsh. wh.	1.848, 1.621
Sodalite	$Na_8Al_6Si_6O_{24}Cl_2$	2.27–2.33	5.5–6	cub., bl., grn., yel., gray, pink	1.483–1.487
Sperrylite	$PtAs_2$	10.58	6–7	cub , tin-wh.	......
Spessartite	$Mn_3Al_2Si_3O_{12}$	4.190	6–7.5	cub., blk., dk. red, brnsh. red., bl., yelsh. orange	1.800
Sphalerite	ZnS	3.9–4.1	3.5–4	cub., br., blk., yel., red, wh.	2.369
Sphene	$CaTiSiO_4(O,OH,F)$	3.45–3.55	5	monocl., col., yel., grn., br., blk.	1.843–1.950, 1.870–2.034, 1.943–2.110
Spinel	$MgAl_2O_4$	3.55	7.5–8	cub., grn., red, bl., br. to col.	1.719
Spodumene	$LiAlSi_2O_6$	3.03–3.22	6.5–7	monocl., col., gray-wh., pa. bl., pa. grn., yelsh.	1.648–1.663, 1.655–1.669, 1.662–1.679
Stannite	Cu_2FeSn_4	4.3–4.5	4	tetr., steel gray to iron blk.	......
Staurolite	$(Fe,Mg)_2(AlFe^{+3})_9O_6SiO_4(O,OH)_2$	3.74–3.83	7.5	monocl., brn., redsh., yelsh.	1.739–1.747, 1.745–1.753, 1.752–1.761
Stercorite	$Na(NH_4)H(PO_4).4H_2O$	1.615	2	tricl., wh., yelsh., brnsh.	1.439, 1.442, 1.469
Stibiotantalite	$Sb(Ta,Cb)O_4$	5.7–7.5	5.5	rhomb., dk. br. to pa. yel.-br., red-br., grnsh.-yel.	2.38, 2.41, 2.46
Stibnite	Sb_2S_3	4.61–4.65	2	rhomb., lead-gray to steel-gray	......
Stilbite	$(Ca,Na_2K_2)Al_2Si_7O_{18}.7H_2O$	2.1–2.2	3.5–4	monocl., col., wh., yel., pink, red, gray, br.	1.484–1.500, 1.492–1.507, 1.494–1.513
Stilpnomelane	$(K,Na,Ca)_{0-1.4}(Fe^{+3}Fe,Mg,Al)_{6-8}Si_8O_{20}(OH)_4(O,OH,H_2O)_{4-8}$	2.59–2.96	3–4	monocl., br., dk. br., redsh. br., blk., dk. grn.	1.543–1.634, 1.576–1.745, 1.576–1.745
Stolzite	$PbWO_4$	7.9–8.4	2.5–3	tetr., redsh. br., yelsh. gray, straw-yel., grnsh.	2.26–2.28, 2.18–2.20
Strengite	$Fe^{+3}(PO_4).2H_2O$	2.90	3.5–4.5	rhomb., red, carmine, vlt., near col.	1.707, 1.719, 1.741
Strontianite	$SrCO_3$	3.72	3.5	rhomb., col., wh., yel., grnsh., brnsh.	1.516–1.520, 1.664–1.667, 1.666–1.669
Struvite	$Mg(NH_4)(PO_4).6H_2O$	1.71	2	rhomb., col., somet. yelsh., brnsh.	1.495, 1.496, 1.504
Sulfur	S	2.07	1.5–2.5	rhomb., yel., brnsh., grnsh., redsh., gray	1.9579, 2.0377, 2.2452
Sylvanite	$(Ag,Au)Te_2$	8.161	1.5–2	monocl., steel-gray to silver-wh.	......
Sylvite	KCl	1.99	2	cub., col., wh., somet. grayish, blsh., yelsh., red	1.49031
Talc	$Mg_3Si_4O_{10}(OH)_2$	2.58–2.83	1	monocl., col., wh., pa. grn., dk. grn., br.	1.539–1.550, 1.589–1.594, 1.589–1.600
Tantalite	$(Fe,Mn)(Ta,Cb)_2O_6$	7.90–8.00	6.5	rhomb., iron-bl. to br.-blk.	2.26, 2.32, 2.43
Tapiolite	$FeTa_2O_6$	7.9	6–6.5	tetr., blk.	(Li) 2.27, 2.42
Tellurobismuthite	Bi_2Te_3	7.800–7.830	1.5–2	rhbdr., pa. lead-gray	......
Terlinguaite	Hg_2OCl	8.725	2.5	monocl., yel. to grnsh.-yel., somet. br.	(Li) 2.33–2.37, 2.62–2.66, 2.64–2.68
Tetrahedrite	$(Cu,Fe)_{12}Sb_4S_{13}$	4.6–5.1	3–4.5	cub., flint-gray to iron-blk. to dull-blk.	......
Thenardite	Na_2SO_4	2.664	2.5–3	rhomb., col., grayish-wh., yelsh., yelsh. br., redsh.	1.464–1.471, 1.473–1.477, 1.481–1.485
Thermonatrite	$Na_2CO_3.H_2O$	2.255	1–1.5	rhomb., col. to wh., grayish, yelsh.	1.420, 1.506, 1.524
Thomsenolite	$NaCaAlF_6.H_2O$	2.981	2	monocl., col. to wh., somet. brnsh., redsh.	1.4072, 1.4136, 1.4150
Thomsonite	$NaCa_2([Al,Si]_5O_{10})_2.6H_2O$	2.10–2.39	5–5.5	rhomb., col., wh., pink, br.	1.497–1.530, 1.513–1.533, 1.518–1.544
Thorianite (R)	ThO_2	9.7	6.5	cub., dk. gray to brnsh.-blk., blk.	~2.20
Thorite (R)	$ThSiO_4$	5.2–5.4	4.5–5	tetr., orange-yel., brnsh. to blk.	~1.8
Topaz	$Al_2SiO_3(OH,F)_2$	3.49–3.57	8	rhomb., col., wh., yel., gray, grn., red, bl.	1.606–1.629, 1.609–1.631, 1.616–1.638

PHYSICAL CONSTANTS OF MINERALS (Continued)

Name	Formula	Sp. gr.	Hardness	Crystalline form and color	Index of refraction (Na) η; ω ϵ / ω β γ
Torbernite (R)	$Cu(UO_2)_2(PO_4)_2.8\text{–}12H_2O$	3.22	2–2.5	tetr., various shades of grn.	1.592, 1.582
Tourmaline	$Na(Mg,Fe,Mn,Li,Al)_3Al_6Si_6O_{18}(BO_3)_3(OH,F)_4$	3.03–3.25	7	rhbdr., blk., bl., grn., yel., red, col., br.	1.635–1.675, 1.610–1.650
Tremolite	$Ca_2Mg_5Si_8O_{22}(OH,F)_2$	3.0	5–6	monocl., col., gray, wh.	1.599, 1.612, 1.622
Tridymite	SiO_2	2.27	7	rhomb., col., wh.	1.471–1.479, 1.472–1.480, 1.474–1.483
Triphyllite-Lithiophyllite	$Li(Fe,Mn)PO_4$	3.34–3.58	4–5	rhomb., blsh. or grnsh. gray to yelsh. br., br.	1.66–1.70, 1.67–1.70, 1.68–1.71
Troegerite (R)	$(UO_2)_3(AsO_4)_2.12H_2O$	....	2–3	tetr., lemon-yel.	1.58–1.59, 1.625–1.635
Trona	$Na_3H(CO_3)_2.2H_2O$	2.14	2.5–3	monocl.. gray or yelsh. wh., col.	1.412, 1.492, 1.540
Turquois	$Cu(Al,Fe^{+3})_6(PO_4)_4(OH)_8.4H_2O$	2.6–3.2	4.5–6	tricl., bl., grn., grnsh.-gray	1.61–1.78, 1.62–1.84, 1.65–1.84
Ullmannite	$NiSbS$	6.61–6.69	5–5.5	cub., steel-gray to silver-wh.	....
Uraninite (R)	UO_2	8.0–11	5–6	cub., steel-blk., brnsh.-blk., grayish, grn.	....
Uvarovite	$Ca_3Cr_2Si_3O_{12}$	3.90	6–7.5	cub., emerald-grn.	1.86
Valentinite	Sb_2O_3	5.76	2.5–3	rhomb., col. to wh., somet. yelsh., redsh., gray, br.	2.18, 2.35, 2.35
Vanadinite	$Pb_5(VO_4)_3Cl$	6.5–7.1	2.75–3	hex., orange-red, red, brnsh.-red, br., brnsh.-yel., yel.	2.416, 2.350
Variscite-Strengite	$(AlFe^{+3})(PO_4).2H_2O$	2.57–2.87	3.5–4.5	rhomb., pa. grn., grn., blsh.-grn., red, vlt., col.	1.563–1.707, 1.588–1.719, 1.594–1.741
Vaterite	$CaCO_3$	2.645	....	hex., col.	1.550, 1.640–1.650
Vermiculite	$(Mg,Ca)_{0.7}(Mg,Fe^{+3}Al)_6(Al,Si)_8O_{20}(OH)_4.8H_2O$	~2.3	~1.5	monocl., col., yel., grn., br.	1.525–1.564, 1.545–1.583, 1.545–1.583
Vesuvianite	$Ca_{10}(Mg,Fe)_2Al_4(Si_2O_7)_2(SiO_4)_5(OH,F)_4$	3.33–3.43	6–7	tetr., yel., grn., br.	1.700–1.746, 1.703–1.752
Villiaumite	NaF	2.79	2–2.5	cub., carmine, (nat.), col. (artif.)	1.327
Vivianite	$Fe_3(PO_4)_2.8H_2O$	2.67–2.69	1.5–2	monocl., col., tarn. pa. bl., grnsh. bl., dk. bl., blsh. blk.	1.579–1.616, 1.602–1.656, 1.629–1.675
Wagnerite	$Mg_2(PO_4)F$	3.15	5–5.5	monocl., yel., gray, somet. red, grn.	1.568, 1.572, 1.582
Wavellite	$Al_3(OH)_3(PO_4)_2.5H_2O$	2 36	3.25–4	rhomb., grnsh. wh., grn. to yel., somet. br., bl., wh.	1.520–1.535, 1.526–1.543, 1.545–1.561
Whewellite	$Ca(C_2O_4).H_2O$	2.23	2.5–3	monocl., col., somet. yelsh., brnsh.	1.491, 1.554, 1.650
Willemite	Zn_2SiO_4	3.9–4.1	5.5	rhbdr., wh., yel., grn., red, gray, br.	1.691, 1.719
Witherite	$BaCO_3$	4.29–4.30	3.5	rhomb., col., wh., gray, yelsh. br.	1.529, 1.676, 1.677
Wolframite	$(Fe,Mn)WO_4$	7.12–7.51	4–4.5	monocl., dk. gray, brnsh. blk. to iron blk.	(Li) ~2.26, 2.32, 2.42
Wollastonite	$CaSiO_3$	2.87–3.09	4.5–5	tricl., wh., col., gray, pa. grn.	1.616–1.640, 1.628–1.650, 1.631–1.653
Wulfenite	$PbMoO_4$	6.5–7.0	2.75–3	tetr., orange-yel. to yel., gray, grn., br., red	2.403, 2.283
Wurtzite	ZnS	3.98	3.5–4	hex., brnsh. blk.	2.356, 2.378
Xenotime	$Y(PO_4)$	4.4–5.1	4–5	tetr., yelsh. br. to redsh. br., somet. gray, wh., pa. yel., grnsh.	1.721, 1.816
Zeunerite (R)	$Cu(UO_2)_2(AsO_4)_2.10\text{–}16H_2O$	....	....	tetr.	1.602–1.610
Zincite	ZnO	5.64–5.68	4	hex., orange-yel. to dk. red, somet. yel.	2.013, 2.029
Zircon	$ZrSiO_4$	4.6–4.7	7.5	tetr., redsh. br., yel., gray, grn., col.	1.923–1.960, 1.968–2.015
Zoisite	$Ca_2Al_3Si_3O_{12}(OH)$	3.15–3.365	6	rhomb., gray, grnsh., brnsh.	1.685–1.705, 1.688–1.710, 1.697–1.725

X-Ray Crystallographic Data, Molar Volumes, and Densities of Minerals and Related Substances

From U.S. Geological Survey Bulletin 1248 by
Richard A. Robie, Philip M. Bethke and Keith M. Beardsley

An extensive list of references and the bases for the calculations and the selection of data are given in the above referenced Bulletin. Bulletin 1248 may be obtained from the Superintendent of Documents, U.S. Government Printing Office, Washington, D.C., 20402.

X-Ray Crystallographic Data of Minerals

	Name and formula	Crystal system	Space group	Structure type	Z	a_o	b_o
	Elements						
1	Silver Ag	cubic	Fm3m(225)	face-centered cubic	4	4.0862 ±.0002	
2	Arsenic As	hex-R	R$\bar{3}$m(166)	arsenic	6	3.760 ±.001	
3	Gold Au	cubic	Fm3m(225)	face-centered cubic	4	4.0786 ±.0002	
4	Bismuth Bi	hex-R	R$\bar{3}$m(166)	arsenic	6	4.5459 ±.0010	
5	Diamond C*	cubic	Fd3m(227)	diamond	8	3.5670 ±.0001	
6	Graphite C*	hex.	C6/mmc(194)	graphite	4	2.4612 ±.0001	
7	Copper Cu	cubic	Fm3m(225)	face-centered cubic	4	3.6150 ±.0005	
8	α-Iron Fe	cubic	Im3m(229)	body-centered cubic	2	2.8664 .0005	
9	Nickel Ni	cubic	Fm3m(225)	face-centered cubic	4	3.5238 ±.0005	
10	Lead Pb	cubic	Fm3m(225)	face-centered cubic	4	4.9505 ±.0005	
11	Platinum Pt	cubic	Fm3m(225)	face-centered cubic	4	3.9231 ±.0005	
12	orthorhombic Sulfur S	orth.	Fddd(70)	S_8 ring molecules	128	10.4646 ±.0020	12.8660 ±.0020
13	monoclinic Sulfur S	mon.	$P2_1/c$(14)	S_8 ring molecules	48	11.04 ±.03	10.98 ±.03
14	rhombohedral Sulfur S	hex-R	R$\bar{3}$(148)	S_6 ring molecules	18	10.818 ±.002	
15	Antimony Sb	hex-R	R$\bar{3}$m(166)	arsenic	6	4.310 ±.001	
16	Selenium Se	hex.	$P3_121$(152) $P3_221$(154)		3	4.3642 ±.0008	
17	Silicon Si	cubic	Fd3m(227)	diamond	8	5.4305 ±.0003	
18	β-Tin (white) Sn	tet.	14_1/amd(141)		4	5.8315 ±.0008	
19	Tellurium Te	hex.	$P3_121$(152) $P3_221$(154)		3	4.4570 ±.0008	
20	Zinc Zn	hex.	$P6_3$/mmc(194)	hexagonal close packed	2	2.665 ±.001	
	Sulfides, arsenides, tellurides, selenides, and sulfosalts						
21	Shandite β-$Ni_3Pb_2S_2$*	hex-R	R$\bar{3}$m(166)		3	5.576 ±.010	
22	High-Argentite Ag_2S I	cubic			4	6.269 ±.020	

X-Ray Crystallographic Data of Minerals

Z; The number of gram formula weights per unit cell.

r; Indicates the data were obtained at an unspecified room temperature and may be taken as 25°+5°C.

*; Indicates the measurements were made on a natural specimen which may have deviated slightly from the listed formula. Densities for these minerals were calculated using the formula weight for the stoichiometric phase.

hex-R; Rhombohedral symmetry. To distinguish from true hexagonal symmetry.

X-Ray Crystallographic Data of Minerals

c_o	α_o	β_o	γ_o	Cell volume 10^{-24} cm^3	Molar volume cm^3	Molar volume cal bar^{-1}	X-Ray density grams cm^{-3}	Temp. °C	
				Elements					
				68.227 ± .010	10.272 ± .002	.24556 ±.00008	10.501 ±.002	25	1
10.555 ±.003				129.23 ± .08	12.972 ± .002	.31007 ±.00023	5.776 ±.004	26	2
				67.847 ± .010	10.215 ± .002	.24420 ±.00008	19.282 ±.003	25	3
11.8622 ±.0030				212.29 ± .11	21.309 ± .011	.50934 ±.00030	9.8071 ±.0050	26	4
				45.385 ± .004	3.4166 ± .0003	.08170 ±.00005	3.5155 ±.0003	25	5
6.7079 ±.0010				35.189 ± .006	5.2982 ± .0009	.12668 ±.00007	2.2670 ±.0004	15	6
				47.242 ± .020	7.1128 ± .0030	.17005 ±.00012	8.9331 ±.0037	25	7
				23.551 ± .012	7.0918 ± .0037	.16954 ±.00013	7.8748 ±.0041	25	8
				43.756 ± .019	6.5880 ± .0028	.15750 ±.00011	8.9117 ±.0038	25	9
				121.32 ± .04	18.267 ± .006	.43663 ±.00018	11.342 ±.003	25	10
				60.379 ± .023	9.0909 ± .0035	.21732 ±.00013	21.460 ±.008	25	11
24.4860 ±.0040				3296.73 ± .97	15.511 ± .005	.37078 ±.00015	2.0671 ±.0006	25	12
10.92 ±.03		96.73 ±.50		1314.6 ± 6.4	16.49 ± .08	.3943 ±.0020	1.944 ±.009	103	13
4.280 ±.001				433.78 ± .19	14.514 ± .006	.34693 ±.00020	2.2092 ±.0010	r	14
11.279 ±.003				181.45 ± .09	18.213 ± .010	.43535 ±.00028	6.685 ±.004	26	15
4.9588 ±.0008				81.793 ± .033	16.420 ± .007	.39249 ±.00020	4.8088 ±.0019	26	16
				160.15 ± .03	12.056 ± .002	.28819 ±.00009	2.3296 ±.0004	25	17
3.1813 ±.0006				108.18 ± .04	16.289 ± .005	.38935 ±.00017	7.2867 ±.0024	26	18
5.9290 ±.0010				102.00 ± .04	20.476 ± .008	.48944 ±.00024	6.2316 ±.0025	25	19
4.947 ±.001				30.428 ± .024	9.162 ± .007	.2190 ±.0002	7.134 ±.006	25	20
				Sulfides, arsenides, tellurides, selenides, and sulfosalts					
13.658 ±.010				367.76 ± 1.35	73.83 ± .27	1.765 ±.007	8.867 ±.033	r	21
				246.4 ± 2.4	37.09 ± .36	.8866 ±.0085	6.680 ±.064	600	22

	Name and formula	Crystal system	Space group	Structure type	Z	a_o	b_o
	Sulfides, arsenides, tellurides, selenides, and sulfosalts—Continued						
23	Argentite Ag_2S II	cubic			2	4.870 ±.008	
24	Acanthite Ag_2S III	mon.	$P2_1/c$(14)		4	4.228 ±.002	6.928 ±.005
25	High-Naumanite Ag_2Se	cubic			2	4.993 ±.016	
26	Ag_2Te I	cubic			2	5.29 ±.01	
27	Ag_2Te II	cubic			4	6.585 ±.010	
28	Hessite Ag_2 Te III	mon.	$P2_1/c$(14)		4	8.09 ±.02	4.48 ±.01
29	$Ag_{1.55}Cu_{.45}S$ I	cubic			4	6.110 ±.010	
30	$Ag_{1.55}Cu_{.45}S$ II	cubic			2	4.825 ±.005	
31	Jalpaite $Ag_{1.55}Cu_{.45}S$ III	tet.			16	8.673 ±.004	
32	$Ag_{.93}Cu_{1.07}S$ I	cubic			4	5.961 ±.009	
33	$Ag_{.93}Cu_{1.07}S$ II	hex.			2	4.138 ±.004	
34	Stromeyerite $Ag_{.93}Cu_{1.07}S$ III	orth.	Cmcm(63)		4	4.066 ±.002	6.628 ±.003
35	Eucairite AgCuSe	orth.	pseudo P4/nmm(129)		10	4.105 ±.010	20.35 ±.02
36	Petzite Ag_3AuTe_2*	cubic	$I4_132$(214)		8	10.38 ±.02	
37	Maldonite Au_2Bi	cubic	Fd3m(227)	Cu_2Mg	8	7.958 ±.002	
38	High-Digenite Cu_2S I	cubic			4	5.725 ±.010	
39	High-Chalcocite Cu_2S II	hex.			2	3.961 ±.004	
40	Chalcocite Cu_2S III	orth.	Ab2m(39)		96	11.881 ±.004	27.323 ±.010
41	Digenite $Cu_{1.79}S$ (Cu rich side)	cubic		deformed fluorite	4	5.5695 ±.0010	
42	Digenite $Cu_{1.77}S$ (S rich side)	cubic		deformed fluorite	4	5.5542 ±.0010	
43	Berzelianite Cu_2Se	cubic			4	5.85 ±.01	
44	High-Bornite Cu_5FeS_4*	cubic			1	5.50 ±.01	
45	Metastable Bornite Cu_5FeS_4	cubic			8	10.94 ±.02	
46	Low-Bornite Cu_5FeS_4*	tet.	$P\bar{4}2_1c$(144)		16	10.94 ±.02	
47	Umangite Cu_3Se_2	tet.	P4/mmm(123)		2	6.402 ±.010	
48	Heazelwoodite Ni_3S_2	hex-R	R32(155)		3	5.746 ±.001	
49	Maucherite $Ni_{11}As_8$	tet.	$P4_12_12$(92)		4	6.870 ±.001	

c_o	α_o	β_o	γ_o	Cell volume 10^{-24} cm^3	Molar volume cm^3	Molar volume cal bar^{-1}	X-Ray density grams cm^{-3}	Temp. °C	
				Sulfides, arsenides, tellurides, selenides, and sulfosalts—Continued					
				115.5 ± .6	34.78 ± .17	.8313 ±.0041	7.125 ±.035	189	23
7.862 ±.003		99.58 ±.30		227.08 ± .29	34.19 ± .04	.8172 ±.0011	7.248 ±.009	25	24
				124.48 ± 1.20	37.48 ± .36	.8959 ±.0087	7.862 ±.076	170	25
				148.0 ± .8	44.58 ± .26	1.065 ±.006	7.702 ±.044	825	26
				285.54 ± 1.30	42.99 ± .20	1.028 ±.005	7.986 ±.036	250	27
8.96 ±.02		123.33 ±.30		271.33 ± 1.43	40.85 ± .22	.9764 ±.0052	8.405 ±.044	r	28
				228.10 ± 1.12	34.34 ± .17	.8209 ±.0041	6.635 ±.033	300	29
				112.33 ± .35	33.83 ± .11	.8085 ±.0026	6.736 ±.021	116	30
11.756 ±.006				884.30 ± .93	33.286 ± .035	.79559 ±.00088	6.8455 ±.0072	r	31
				211.82 ± .96	31.89 ± .14	.7623 ±.0035	6.283 ±.029	196	32
7.105 ±.007				105.36 ± .23	31.73 ± .07	.7583 ±.0017	6.316 ±.014	100	33
7.972 ±.004				214.84 ± .18	32.35 ± .03	.7732 ±.0007	6.194 ±.005	r	34
6.31 ±.01				527.12 ± 1.62	31.75 ± .10	.7588 ±.0024	7.887 ±.024	r	35
				1118.4 ± 6.5	84.19 ± .49	2.012 ±.012	9.214 ±.053	r	36
				503.98 ± .38	37.94 ± .03	.9068 ±.0007	15.891 ±.012	r	37
				187.64 ± .98	28.25 ± .15	.6753 ±.0036	5.633 ±.030	465	38
6.722 ±.007				91.34 ± .21	27.50 ± .06	.6574 ±.0015	5.786 ±.013	152	39
13.491 ±.004				4379.5 ± 2.5	27.475 ± .016	.65671 ±.00043	5.7924 ±.0034	r	40
				172.76 ± .09	26.012 ± .014	.6217 ±.0004	5.605 ±.003	25	41
				171.34 ± .09	25.798 ± .014	.6166 ±.0004	5.602 ±.003	25	42
				200.2 ± 1.0	30.14 ± .15	.7205 ±.0037	6.835 ±.035	170	43
				166.4 ± .9	100.2 ± .5	2.395 ±.013	5.008 ±.027	240	44
				1309.34 ± 7.18	98.57 ± .54	2.356 ±.013	5.091 ±.028	r	45
21.88 ±.04				2618.7 ±10.7	98.57 ± .40	2.356 ±.010	5.091 ±.021	r	46
4.276 ±.010				175.25 ± .68	52.77 ± .21	1.261 ±.005	6.604 ±.026	r	47
7.134 ±.002				203.98 ± .09	40.95 ± .02	.9788 ±.0005	5.867 ±.003	r	48
21.81 ±.01				1029.36 ± .56	154.98 ± .08	3.7043 ±.0021	8.0343 ±.0044	r	49

	Name and formula	Crystal system	Space group	Structure type	Z	a_o	b_o
	Sulfides, arsenides, tellurides, selenides, and sulfosalts—Continued						
50	Pentlandite	cubic	Fm3m(225)		4	10.196	
	$Fe_{5.25}Ni_{3.75}S_8$					±.010	
51	Pentlandite	cubic	Fm3m(225)		4	10.095	
	$Fe_{4.75}Ni_{5.25}S_8$					±.010	
52	Sternbergite	orth.	Ccmm(63)		8	11.60	12.675
	$AgFe_2S_3$*					±.02	±.020
53	Argentopyrite	orth.	Pmmm(47)		4	6.64	11.47
	$AgFe_2S_3$*					±.01	±.02
54	Realgar	mon.	$P2_1/m(11)$		16	9.29	13.53
	AsS*					±.05	±.05
55	Oldhamite	cubic	Fm3m(225)	rock salt	4	5.689	
	CaS					±.006	
56	Greenockite	hex.	$P6_3mc(186)$	zincite	2	4.1354	
	CdS					±.0010	
57	Hawleyite	cubic	$F\bar{4}3m(216)$	sphalerite	4	5.833	
	CdS					±.002	
58	(hypothetical)	cubic	Fm3m(225)	rock salt	4	5.516	
	CdS					±.002	
59	Cadmoselite	hex.	$P6_3mc(186)$	zincite	2	4.2977	
	CdSe					±.0010	
60		cubic	$F\bar{4}3m(216)$	sphalerite	4	6.4805	
	CdTe					±.0006	
61	(hypothetical)	cubic	$F\bar{4}3m(216)$	sphalerite	4	5.339	
	CoS					±.001	
62	Chalcopyrite	tet.	I42d(122)		4	5.2988	
	$(CuFeS_2)$ $CuFeS_{1.90}$					±.0010	
63	Cubanite	orth.	Pcmn(62)		4	6.46	11.12
	$CuFe_2S_3$*					±.01	±.01
64	Covellite	hex.	$P6_3/mmc(194)$		6	3.792	
	CuS					±.001	
65	Klockmannite	hex.		deformed	78	14.206	
	CuSe			covellite		±.010	
66	Troilite	hex.	$P6_3/mmc(194)$	niccolite	2	3.446	
	FeS					±.003	
67	Pyrrhotite	hex.	$P6_3/mmc(194)$	defect	2	3.446	
	$Fe_{.980}S$			niccolite		±.001	
68	Pyrrhotite	hex.	$P6_3/mmc(194)$	defect	2	3.440	
	$Fe_{.885}S$			niccolite		±.001	
69	(hypothetical)	cubic	$F\bar{4}3m(216)$	sphalerite	4	5.455	
	FeS					±.001	
70	(hypothetical)	hex.	$P6_3mc(186)$	zincite	2	3.872	
	FeS					±.001	
71	Cinnabar	hex.	$P3_121(152)$	cinnabar	3	4.149	
	HgS		$P3_21(154)$			±.001	
72	Metacinnabar	cubic	$F\bar{4}3m(216)$	sphalerite	4	5.8517	
	HgS					±.0010	
73	Tiemannite	cubic	$F\bar{4}3m(216)$	sphalerite	4	6.0853	
	HgSe					±.0050	
74	Coloradoite	cubic	$F\bar{4}3m(216)$	sphalerite	4	6.4600	
	HgTe					±.0006	
75	Alabandite	cubic	Fm3m(225)	rock salt	4	5.2234	
	MnS					±.0005	
76	(hypothetical)	cubic	$F\bar{4}3m(216)$	sphalerite	4	5.611	
	MnS					±.002	

c_o	α_o	β_o	γ_o	Cell volume 10^{-24} cm^3	Molar volume cm^3	Molar volume cal bar^{-1}	X-Ray density grams cm^{-3}	Temp. °C	
Sulfides, arsenides, tellurides, selenides, and sulfosalts—Continued									
				1059.96 ± 3.12	159.59 ± .47	3.8144 ±.0113	4.823 ±.014	r	50
				1028.77 ± 3.06	154.89 ± .46	3.702 ±.011	4.998 ±.015	r	51
6.63 ±.01				974.81 ± 2.71	73.39 ± .20	1.754 ±.005	4.303 ±.012	r	52
6.45 ±.02				491.2 ± 1.9	73.96 ± .29	1.768 ±.007	4.269 ±.017	r	53
6.57 ±.03		106.55 ±.30		791.6 ± 6.4	29.80 ± .24	.7122 ±.0058	3.591 ±.029	r	54
				184.12 ± .58	27.722 ± .088	.6626 ±.0021	2.602 ±.008	r	55
6.7120 ±.0010				99.407 ± .050	29.934 ± .015	.71549 ±.00041	4.8261 ±.0024	r	56
				198.46 ± .20	29.88 ± .03	.7142 ±.0008	4.835 ±.005	r	57
				167.83 ± .18	25.27 ± .03	.6040 ±.0007	5.717 ±.006	r	58
7.0021 ±.0010				112.00 ± .05	33.727 ± .016	.80614 ±.00044	5.6738 ±.0028	r	59
				272.16 ± .08	40.977 ± .012	.97943 ±.00032	5.8569 ±.0016	25	60
				152.19 ± .09	22.91 ± .02	.5477 ±.0004	3.971 ±.002	r	61
10.434 ±.005				292.96 ± .18	44.109 ± .027	1.0543 ±.0007	4.0878 ±.0025	r	62
6.23 ±.01				447.53 ± 1.08	67.38 ± .16	1.611 ±.004	4.026 ±.010	r	63
16.34 ±.01				203.48 ± .16	20.42 ± .02	.4882 ±.0005	4.682 ±.001	r	64
17.25 ±.05				3014.8 ± 9.7	23.28 ± .08	.5564 ±.0018	6.122 ±.020	r	65
5.877 ±.001				60.439 ± .106	18.20 ± .03	.4350 ±.0008	4.830 ±.009	28	66
5.848 ±.002				60.14 ± .04	18.11 ± .01	.4329 ±.0003	4.793 ±.003	28	67
5.709 ±.003				58.507 ± .046	17.62 ± .02	.4211 ±.0004	4.625 ±.004	28	68
				162.32 ± .09	24.44 ± .01	.5842 ±.0004	3.597 ±.002	r	69
6.345 ±.002				82.38 ± .05	24.81 ± .02	.5930 ±.0004	3.544 ±.002	r	70
9.495 ±.002				141.55 ± .07	28.416 ± .015	.6792 ±.0004	8.187 ±.004	r	71
				200.38 ± .10	30.169 ± .016	.7211 ±.0004	7.712 ±.004	r	72
				225.34 ± .56	33.928 ± .084	.8110 ±.0020	8.239 ±.020	r	73
				269.59 ± .08	40.590 ± .011	.97016 ±.00032	8.0855 ±.0023	r	74
				142.51 ± .04	21.457 ± .006	.51289 ±.00019	4.0546 ±.0012	r	75
				176.65 ± .19	26.60 ± .03	.6357 ±.0007	3.271 ±.004	r	76

	Name and formula	Crystal system	Space group	Structure type	Z	a_o	b_o
	Sulfides, arsenides, tellurides, selenides, and sulfosalts—Continued						
77	(hypothetical) MnS	hex.	$P6_3mc(186)$	zincite	2	3.986 ±.001	
78	Niccolite $NiAs$	hex.	$P6_3/mmc(194)$	niccolite	2	3.618 ±.001	
79	Millerite NiS	hex-R	$R3m(160)$		9	9.616 ±.001	
80	Breithauptite $NiSb$	hex.	$P6_3/mmc(194)$	niccolite	2	3.942 ±.001	
81	Galena PbS	cubic	$Fm3m(225)$	rock salt	4	5.9360 ±.0005	
82	Clausthalite $PbSe$	cubic	$Fm3m(225)$	rock salt	4	6.1255 ±.0005	
83	Teallite $PbSnS_2$	orth.	$Pbnm(62)$	GeS	2	4.266 ±.003	11.419 ±.007
84	Altaite $PbTe$	cubic	$Fm3m(225)$	rock salt	4	6.4606 ±.0005	
85	Cooperite PtS	tet.	$P4_2/mmc(131)$		2	3 4699 ±.0006	
86	Herzenbergite SnS	orth.	$Pbnm(62)$	GeS	4	4.328 ±.002	11.190 ±.004
87	Sphalerite ZnS	cubic	$F\bar{4}3m(216)$	sphalerite	4	5.4093 ±.0005	
88	Wurtzite ZnS	hex.	$P6_3mc(186)$	zincite	2	3.8230 ±.0010	
89	Stilleite $ZnSe$	cubic	$F\bar{4}3m(216)$	sphalerite	4	5.6685 ±.0005	
90	$ZnTe$	cubic	$F\bar{4}3m(216)$	sphalerite	4	6.1020 ±.0006	
91	Orpiment As_2S_3*	mon.	$P2_1/n(14)$		4	11.49 ±.02	9.59 ±.02
92	Bismuthinite Bi_2S_3	orth.	$Pbnm(62)$	stibnite	4	11.150 ±.004	11.300 ±.004
93	Tellurobismuthite Bi_2Te_3	hex-R	$R\bar{3}m(166)$	Bi_2Te_2S	3	4.3835 ±.0020	
94	Stibnite Sb_2S_3	orth.	$Pbnm(62)$	stibnite	4	11.229 ±.004	11.310 ±.004
95	Linnaeite Co_3S_4	cubic	$Fd3m(227)$	spinel	8	9.401 ±.001	
96	Greigite Fe_3S_4	cubic	$Fd3m(227)$	spinel	8	9.876 ±.002	
97	Daubreeite $FeCr_2S_4$	cubic	$Fd3m(227)$	spinel	8	9.966 ±.005	
98	Violarite $FeNi_2S_4$	cubic	$Fd3m(227)$	spinel	8	9.464 ±.005	
99	Polymidite Ni_3S_4	cubic	$Fd3m(227)$	spinel	8	9.480 ±.001	
100	Co-Safflorite $CoAs_2$	mon.		deformed marcasite	2	5.049 ±.002	5.872 ±.002
101	Safflorite $(CO_{.5}Fe_{.5})As_2$	orth.	$Pnnm(58)$	marcasite	2	5.231 ±.002	5.953 ±.002
102	Cobaltite $CoAsS$*	cubic	$P2_13(198)$	NiSbS	4	5.60 ±.05	
103	Glaucodot $(Co,Fe)AsS$*	orth.	$Cmmm(65)$		24	6.64 ±.05	28.39 ±.10

c_o	α_o	β_o	γ_o	Cell volume 10^{-24} cm^3	Molar volume cm^3	Molar volume cal bar^{-1}	X-Ray density grams cm^{-3}	Temp. °C	
Sulfides, arsenides, tellurides, selenides, and sulfosalts—Continued									
6.465 ±.002				88.96 ± .05	26.79 ± .02	.6403 ±.0004	3.248 ±.002	r	77
5.034 ±.001				57.07 ± .03	17.18 ± .01	.4108 ±.0003	7.776 ±.005	r	78
3.152 ±.001				252.41 ± .10	16.891 ± .006	.40374 ±.00020	5.3743 ±.0020	r	79
5.155 ±.001				69.37 ± .04	20.89 ± .01	.4994 ±.0004	8.639 ±.005	r	80
				209.16 ± .05	31.492 ± .008	.75272 ±.00024	7.5973 ±.0019	26	81
				229.84 ± .06	34.605 ± .009	.82713 ±.00025	8.2690 ±.0020	r	82
4.090 ±.002				199.24 ± .21	59.996 ± .063	1.4340 ±.0016	6.501 ±.007	r	83
				269.66 ± .06	40.601 ± .009	.97043 ±.00027	8.2459 ±.0019	r	84
6.1098 ±.0010				73.563 ± .028	22.152 ± .008	.5295 ±.0003	10.254 ±.004	r	85
3.978 ±.001				192.66 ± .12	29.01 ± .02	.6933 ±.0005	5.197 ±.003	r	86
				158.28 ± .04	23.831 ± .007	.56962 ±.00020	4.0885 ±.0011	r	87
6.2565 ±.0010				79.190 ± .043	23.846 ± .013	.56998 ±.00036	4.0859 ±.0022	r	88
				182.14 ± .05	27.424 ± .007	.65548 ±.00022	5.2630 ±.0014	r	89
				227.20 ± .07	34.209 ± .010	.81765 ±.00029	5.6410 ±.0017	r	90
4.25 ±.01		90.45 ±.30		468.3 ± 1.7	70.51 ± .25	1.685 ±.006	3.490 ±.013	r	91
3.981 ±.001				501.59 ± .28	75.520 ± .043	1.8050 ±.0011	6.8081 ±.0038	26	92
30.487 ±.003				507.33 ± .47	101.85 ± .09	2.4342 ±.0023	7.862 ±.007	25	93
3.8389 ±.0010				487.54 ± .28	73.406 ± .042	1.7545 ±.0010	4.6276 ±.0026	25	94
				830.85 ± .27	62.548 ± .020	1.4950 ±.0005	4.8772 ±.0016	r	95
				963.26 ± .59	72.52 ± .04	1.733 ±.001	4.079 ±.003	r	96
				989.83 ± 1.49	74.52 ± .11	1.781 ±.003	3.866 ±.006	r	97
				847.66 ± 1.34	63.81 ± .10	1.525 ±.002	4.725 ±.008	r	98
				851.97 ± .27	64.138 ± .020	1.5330 ±.0005	4.7458 ±.0015	r	99
3.127 ±.001		90.45 ±.20		92.706 ± .057	27.92 ± .02	.6672 ±.0005	7.479 ±.005	26	100
2.962 ±.002				92.237 ± .078	27.775 ± .024	.6639 ±.0006	7.461 ±.006	26	101
				175.62 ± 4.70	26.44 ± .71	.6320 ±.0170	6.275 ±.168	r	102
5.64 ±.05				1063.2 ±12.9	26.68 ± .32	.6377 ±.0078	6.161 ±.075	r	103

	Name and formula	Crystal system	Space group	Structure type	Z	a_o	b_o
	Sulfides, arsenides, tellurides, selenides, and sulfosalts—Continued						
104	Cattierite CoS_2	cubic	Pa3(205)	pyrite	4	5.5345 ±.0005	
105	Trogtalite $CoSe_2$	cubic	Pa3(205)	pyrite	4	5.8588 ±.0010	
106	Loellingite $FeAs_2$	orth.	Pnnm(58)	marcasite	2	5.300 ±.002	5.981 ±.002
107	Arsenopyrite FeAsS*	tri.	P$\bar{1}$(2)		4	5.760 ±.010	5.690 ±.005
108	Gudmundite FeSbS*	mon.	$B2_1$/d(14)		8	10.00 ±.05	5.93 ±.03
109	Pyrite FeS_2	cubic	Pa3(205)	pyrite	4	5.4175 ±.0005	
110	Marcasite FeS_2*	orth.	Pnnm(58)	marcasite	2	4.443 ±.002	5.423 ±.002
111	Ferroselite $FeSe_2$	orth.	Pnnm(58)	marcasite	2	4.801 ±.005	5.778 ±.005
112	Frohbergite $FeTe_2$	orth.	Pnnm(58)	marcasite	2	5.265 ±.005	6.265 ±.005
113	Hauerite MnS_2	cubic	Pa3(205)	pyrite	4	6.1014 ±.0006	
114	Molybdenite MoS_2	hex.	$P6_3$/mmc(194)	molybdenite	2	3.1604 ±.0010	
115	Rammelsbergite $NiAs_2$	orth.	Pnnm(58)	marcasite	2	4.757 ±.002	5.797 ±.004
116	Pararammelsbergite $NiAs_2$	orth.	Pbca(61)		8	5.75 ±.01	5.82 ±.01
117	Gersdorfite NiAsS	cubic	$P2_13$(198)		4	5.693 ±.001	
118	Vaesite NiS_2	cubic	Pa3(205)	pyrite	4	5.6873 ±.0005	
119	$NiSe_2$	cubic	Pa3(205)	pyrite	4	5.9604 ±.0010	
120	Melonite $NiTe_2$	hex.	P$\bar{3}$ml(164)	cadmium iodide	1	3.869 ±.010	
121	Sperrylite $PtAs_2$	cubic	Pa3(205)	pyrite	4	5.968 ±.005	
122	Laurite RuS_2	cubic	Pa3(205)	pyrite	4	5.60 ±.02	
123	Tungstenite WS_2	hex.	$P6_3$/mmc(194)	molybdenite	2	3.154 ±.001	
124	Co-Skutterudite $CoAs_{3-x}$ $CoAs_{2.95}$	cubic	Im3(204)		8	8.2060 ±.0010	
125	Fe-Skutterudite $FeAs_{3-x}$ $FeAs_{2.95}$	cubic	Im3(204)		8	8.1814 ±.0010	
126	Ni-Skutterudite $NiAs_{3-x}$ $NiAs_{2.95}$	cubic	Im3(204)		8	8.3300 ±.0010	
127	Tennantite $Cu_{12}As_4S_{13}$	cubic	I$\bar{4}$3m(217)	tetrahedrite	2 2	10.190 ±.004	
128	Tetrahedrite $Cu_{12}Sb_4S_{13}$	cubic	I$\bar{4}$3m(217)	tetrahedrite	2 2	10.327 ±.004	
129	Enargite Cu_3AsS_4	orth.	$Pnn2$(34)		2	6.426 ±.005	7.422 ±.005
130	Luzonite Cu_3AsS_4*	tet.	I$\bar{4}$2m(121)		2	5.289 ±.005	

X-Ray Crystallographic Data of Minerals

c_o	α_o	β_o	γ_o	Cell volume 10^{-24} cm³	Molar volume cm³	Molar volume cal bar⁻¹	X-Ray density grams cm⁻³	Temp. °C	
Sulfides, arsenides, tellurides, selenides, and sulfosalts—Continued									
				169.53 ± .05	25.524 ± .007	.61009 ±.00021	4.8213 ±.0013	r	104
				201.11 ± .10	30.279 ± .016	.72374 ±.00042	7.1618 ±.0037	r	105
2.882 ±.001				91.357 ± .056	27.51 ± .02	.6576 ±.0005	7.477 ±.005	26	106
5.785 ±.005	90.00 ±.20	112.23 ±.20	90.00 ±.20	175.51 ± .44	26.42 ± .07	.6316 ±.0016	6.162 ±.015	r	107
6.73 ±.03		90.00 ±.50		399.09 ± 3.35	30.04 ± .25	.7181 ±.0061	6.978 ±.059	r	108
				159.00 ± .04	23.940 ± .007	.57221 ±.00020	5.0116 ±.0014	r	109
3.3876 ±.0015				81.622 ± .060	24.579 ± .018	.58749 ±.00047	4.8813 ±.0036	25	110
3.587 ±.004				99.50 ± .17	29.96 ± .05	.7162 ±.0013	7.134 ±.013	r	111
3.869 ±.002				127.62 ± .17	38.43 ± .05	.9185 ±.0013	8.094 ±.011	r	112
				227.14 ± .07	34.198 ± .010	.81741 ±.00029	3.4816 ±.0010	28	113
12.295 ±.002				106.35 ± .07	32.025 ± .021	.76547 ±.00055	4.9982 ±.0033	26	114
3.542 ±.002				97.645 ± .096	29.41 ± .03	.7030 ±.0007	7.091 ±.007	26	115
11.428 ±.02				382.42 ± 1.15	28.79 ± .09	.6882 ±.0021	7.244 ±.022	r	116
				184.51 ± .10	27.78 ± .01	.6640 ±.0004	5.964 ±.003	26	117
				183.96 ± .05	27.697 ± .007	.66203 ±.00022	4.4350 ±.0012	r	118
				211.75 ± .11	31.882 ± .016	.76204 ±.00043	6.7948 ±.0034	20	119
5.308 ±.010				68.81 ± .38	41.44 ± .23	.9905 ±.0055	7.575 ±.042	84	120
				212.56 ± .53	32.00 ± .08	.7650 ±.0020	10.778 ±.027	r	121
				175.6 ± 1.9	26.44 ± .28	.6320 ±.0068	6.248 ±.067	r	122
12.362 ±.004				106.50 ± .08	32.069 ± .023	.76652 ±.00059	7.7325 ±.0055	26	123
				552.58 ± .20	41.599 ± .015	.99428 ±.00041	6.7298 ±.0025	r	124
				547.62 ± .20	41.226 ± .015	.98537 ±.00041	6.7158 ±.0025	r	125
				578.01 ± .21	43.513 ± .016	1.0400 ±.0004	6.4286 ±.0023	r	126
				1058.09 ± 1.25	318.62 ± .38	7.1652 ±.0090	4.642 ±.006	r	127
				1101.3 ± 1.3	331.64 ± .39	7.9266 ±.0094	5.024 ±.006	r	128
6.144 ±.005				293.03 ± .38	88.24 ± .12	2.109 ±.003	4.463 ±.006	26	129
10.440 ±.008				292.04 ± .60	87.94 ± .18	2.1019 ±.0043	4.478 ±.009	26	130

	Name and formula	Crystal system	Space group	Structure type	Z	a_o	b_o
	Sulfides, arsenides, tellurides, selenides, and sulfosalts—Continued						
131	Famatimite Cu_3SbS_4*	tet.	$I\bar{4}m(121)$		2	5.384 ±.005	
132	Proustite Ag_3AsS_3	hex-R	R3c(161)		6	10.816 ±.001	
133	Pyrargyrite Ag_3SbS_3	hex-R	R3c(161)		6	11.052 ±.002	
134	Miargyrite $AgSbS_2$*	mon.	Cc(9)		8	12.862 ±.013	4.111 ±.004
	Oxides and hydroxides						
135	Corundum Al_2O_3	hex-R	$R\bar{3}c(167)$	corundum	6	4.7591 ±.0004	
136	Boehmite $AlO(OH)$*	orth.	Cmcm(63)	lepidocrocite	4	2.868 ±.003	12.227 ±.003
137	Diaspore $AlO(OH)$*	orth.	Pbnm(62)		4	4.401 ±.005	9.421 ±.005
138	Gibbsite $Al(OH)_3$	mon.	$P2_1/n(14)$		8	9.719 ±.002	5.0705 ±.0010
139	Arsenolite As_2O_3	cubic	Fd3m(227)	diamond	16	11.074 ±.005	
140	Claudetite As_2O_3	mon.	$P2_1/n(14)$		4	5.339 ±.002	12.984 ±.005
141	Bromellite BeO	hex.	$P6_3mc(186)$	zincite	2	2.6979 ±.0005	
142	Bismite α-Bi_2O_3	mon.	$P2_1/c(14)$	pseudo orthorhombic	8	8.166 ±.005	13.827 ±.010
143	Lime CaO	cubic	Fm3m(225)	rock salt	4	4.8108 ±.0005	
144	Portlandite $Ca(OH)_2$	hex.	$P\bar{3}ml(164)$	CdI_2	1	3.5933 ±.0005	
145	Monteponite CdO	cubic	Fm3m(225)	rock salt	4	4.6953 ±.0010	
146	Cerianite CeO_2	cubic	Fm3m(225)	fluorite	4	5.4110 ±.0020	
147	CoO	cubic	Fm3m(225)	rock salt	4	4.260 ±.002	
148	Eskolaite Cr_2O_3	hex-R	$R\bar{3}c(167)$	corundum	6	4.9607 ±.0020	
149	Tenorite CuO	mon.	C2/c(15)		4	4.684 ±.005	3.425 ±.005
150	Cuprite Cu_2O	cubic	Pn3m(224)		2	4.2696 ±.0010	
151	Wustite $Fe_{.953}O$	cubic	Fm3m(225)	defect rock salt	4	4.3088 ±.0003	
152	Hematite Fe_2O_3	hex-R	$R\bar{3}c(167)$	corundum	6	5.0329 ±.0010	
153	Magnetite Fe_3O_4	cubic	Fd3m(227)	spinel	8	8.3940 ±.0005	
154	Goethite α-$FeO(OH)$*	orth.	Pbnm(62)		4	4.596 ±.005	9.957 ±.010
155	Lepidocrocite γ-$FeO(OH)$*	orth.	Amam(63)		4	3.868 ±.010	12.525 ±.010
156	α-Ga_2O_3	hex-R	$R\bar{3}c(167)$	corundum	6	4.9793 ±.0010	

c_o	α_o	β_o	γ_o	Cell volume 10^{-24} cm³	Molar volume cm³	Molar volume cal bar⁻¹	X-Ray density grams cm⁻³	Temp. °C	
				Sulfides, arsenides, tellurides, selenides, and sulfosalts—Continued					
10.770				312.19	94.01	2.2469	4.687	26	131
±.008				± .62	± .19	±.0045	±.009		
8.6948				880.89	88.420	2.1133	5.595	26	132
±.0013				± .21	± .021	±.0006	±.001		
8.7177				922.18	92.564	2.2124	5.8506	26	133
±.0020				± .40	± .040	±.0010	±.0025		
13.220		98.63		691.10	52.027	1.244	5.646	r	134
±.010		±.15		± 1.14	± .086	±.002	±.009		
				Oxides and hydroxides					
12.9894				254.78	25.575	.61128	3.9869	25	135
±.0030				± .07	± .007	±.00022	±.0011		
3.700				129.75	19.535	.46695	3.071	26	136
±.003				± .17	± .026	±.00067	±.004		
2.845				117.96	17.760	.4245	3.378	r	137
±.002				± .17	± .026	±.0007	±.005		
8.6412		94.57		424.49	31.956	.7638	2.441	r	138
±.0010		±.25		± .20	± .015	±.0004	±.001		
				1358.0	51.118	1.2218	3.870	25	139
				± 1.8	± .069	±.0017	±.005		
4.5405		94.27		313.88	47.259	1.1296	4.1863	25	140
±.0010		±.10		± .19	± .028	±.0007	±.0025		
4.3772				27.592	8.3086	.19862	3.0104	26	141
±.0005				± .011	± .0032	±.00012	±.0012		
5.850		90.00		660.53	49.73	1.1885	9.371	25	142
±.004		±.20		± .77	± .06	±.0014	±.011		
				111.34	16.764	.40071	3.3453	26	143
				± .03	± .005	±.00017	±.0010		
4.9086				54.888	33.056	.79011	2.2415	26	144
±.0020				± .027	± .016	±.00043	±.0011		
				103.51	15.585	.37254	8.2386	27	145
				± .07	± .010	±.00028	±.0053		
				158.43	23.853	.57016	7.216	26	146
				± .18	± .026	±.00068	±.008		
				77.31	11.64	.2782	6.438	26	147
				± .11	± .02	±.0004	±.009		
13.599				289.82	29.090	.6953	5.225	r	148
±.010				± .32	± .032	±.0008	±.006		
5.129		99.47		81.16	12.22	.2921	6.509	26	149
±.005		±.17		± .17	± .03	±.0007	±.014		
				77.833	23.437	.56021	6.1047	26	150
				± .055	± .016	±.00044	±.0043		
				79.996	12.044	.28791	5.7471	17	151
				± .017	± .003	±.00011	±.0012		
		13.7492		301.61	30.274	.72361	5.2749	25	152
		±.0010		± .12	± .012	±.00034	±.0021		
				591.43	44.524	1.0642	5.2003	22	153
				± .11	± .008	±.0002	±.0009		
3.021				138.2	20.82	.4975	4.269	r	154
±.003				± .2	± .04	±.0009	±.008		
3.066				148.54	22.364	.5346	3.973	r	155
±.003				± .43	± .064	±.0016	±.011		
13.429				288.34	28.943	.69179	6.4762	24	156
±.003				± .13	± .013	±.00036	±.0030		

	Name and formula	Crystal system	Space group	Structure type	Z	a_o	b_o
	Oxides and hydroxides—Continued						
157	Low-germania GeO_2	tet.	P4/mnm(136)	rutile	2	4.3963 ±.0010	
158	High-germania GeO_2	hex.	$P3_121(152)$ $P3_221(154)$	α-quartz	3	4.987 ±.002	
159	Ice H_2O	hex.	$P6_3/mmc(194)$		4	4.5212 ±.0010	
160	Hafnia HfO_2	mon.	$P2_1/c(14)$	baddeleyite	4	5.1156 ±.0010	5.1722 ±.0010
161	Montroydite HgO	orth.	Pnma(62)		4	6.608 ±.003	5.518 ±.003
162	Periclase MgO	cubic	Fm3m(225)	rock salt	4	4.2117 ±.0005	
163	Brucite $Mg(OH)_2$	hex.	$P\bar{3}ml(164)$	CdI_2	1	3.147 ±.004	
164	Manganosite MnO	cubic	Fm3m(225)	rock salt	4	4.4448 ±.0005	
165	Pyrolusite MnO_2	tet.	P4/mnm(136)	rutile	2	4.388 ±.003	
166	Bixbyite Mn_2O_3	cubic	Ia3(206)	Tl_2O_3	16	9.411 ±.005	
167	Hausmanite Mn_3O_4	tet.	$I4_1/amd(141)$		8	8.136 ±.005	
168	Molybdite MoO_3	orth.	Pbnm(62)		4	3.962 ±.002	13.858 ±.005
169	Bunsenite NiO	cubic	Fm3m(225)	rock salt	4	4.177 ±.002	
170	Litharge PbO red	tet.	P4/nmm(129)		2	3.9759 ±.0040	
171	Massicot PbO yellow	orth.	Pb2a(32)		4	5.489 ±.003	4.755 ±.004
172	Minium Pb_3O_4	tet.	$P4_2/mbc(135)$		4	8.815 ±.005	
173	Senarmontite Sb_2O_3	cubic	Fm3m(225)	arsenic trioxide	16	11.152 ±.003	
174	Valentinite Sb_2O_3	orth.	Pccn(56)	antimony trioxide	4	4.914 ±.002	12.468 ±.005
175	Cervantite Sb_2O_4	cubic	Fd3m(227)		8	10.305 ±.005	
176	Selenolite SeO_2	tet.	$P4_2/mbc(135)$ $P4_2bc(106)$		8	8.35 ±.01	
177	α-Quartz SiO_2*	hex.	$P3_121(152)$ $P3_221(154)$		3	4.9136 ±.0001	
178	β-Quartz SiO_2*	hex.	$P6_422(181)$ $P6_222(180)$		3	4.999 ±.001	
179	α-Cristobalite SiO_2	tet.	$P4_12_12(92)$ $P4_32_12(96)$		4	4.971 ±.003	
180	β-Cristobalite SiO_2	cubic	Fd3m(227)		8	7.1382 ±.0010	
181	Keatite SiO_2	tet.	$P4_12_12(92)$ $P4_32_12(96)$		12	7.456 ±.003	
182	β-Tridymite SiO_2	hex.	$P\bar{6}2c(172)$ $P6_3/mmc(194)$		4	5.0463 ±.0020	
183	Coesite SiO_2*	mon.	B2/b(15)		16	7.152 ±.001	12.379 ±.002

c_o	a_o	β_o	γ_o	Cell volume 10^{-24} cm³	Molar volume cm³	Molar volume cal bar⁻¹	X-Ray density grams cm⁻³	Temp. °C	
				Oxides and hydroxides—Continued					
2.8626 ±.0010				55.327 ± .032	16.660 ± .010	.39824 ±.00027	6.2777 ±.0036	25	157
5.652 ±.002				121.73 ± .11	24.438 ± .021	.58413 ±.00056	4.2797 ±.0038	26	158
7.3666 ±.0010				130.41 ± .06	19.635 ± .009	.46932 ±.00026	.9175 ±.0004	0	159
5.2948 ±.0010		99.18 ±.08		138.30 ± .06	20.823 ± .008	.49772 ±.00025	10.108 ±.004	r	160
3.519 ±.003				128.3 ± .1	19.32 ± .02	.4618 ±.0006	11.21 ±.01	25	161
				74.709 ± .027	11.248 ± .004	.26889 ±.00014	3.5837 ±.0013	25	162
4.769 ±.004				40.90 ± .11	24.63 ± .07	.5888 ±.0016	2.368 ±.006	26	163
				87.813 ± .030	13.221 ± .004	.31604 ±.00015	5.3653 ±.0018	26	164
2.865 ±.002				55.16 ± .08	16.61 ± .02	.3971 ±.0007	5.234 ±.008	r	165
				833.5 ± 1.3	31.37 ± .05	.7499 ±.0012	5.032 ±.008	25	166
9.422 ±.005				623.68 ± .84	46.95 ± .06	1.1222 ±.0016	4.873 ±.007	20	167
3.697 ±.004				202.98 ± .25	30.56 ± .04	.7305 ±.0010	4.710 ±.006	26	168
				72.88 ± .10	10.97 ± .02	.2623 ±.0004	6.809 ±.010	26	169
5.023 ±.004				79.40 ± .17	23.91 ± .05	.5715 ±.0013	9.334 ±.020	27	170
5.891 ±.004				153.8 ± .2	23.15 ± .03	.5533 ±.0007	9.641 ±.012	27	171
6.565 ±.003				510.13 ± .62	76.81 ± .09	1.836 ±.002	8.926 ±.009	25	172
				1386.9 ± 1.1	52.206 ± .042	1.2478 ±.0011	5.5837 ±.0045	26	173
5.421 ±.004				332.13 ± .31	50.007 ± .047	1.1952 ±.0012	5.8292 ±.0054	25	174
				1094.3 ± 1.6	82.38 ± .12	1.9690 ±.0029	3.733 ±.005	26	175
5.08 ±.01				354.2 ± 1.1	26.66 ± .08	.6373 ±.0020	4.161 ±.013	26	176
5.4051 ±.0001				113.01 ± .01	22.688 ± .001	.54229 ±.00007	2.6483 ±.0001	25	177
5.4592 ±.0020				118.15 ± .06	23.718 ± .013	.5669 ±.0004	2.533 ±.002	575	178
6.918 ±.003				170.95 ± .22	25.739 ± .033	.61521 ±.00083	2.3344 ±.0030	25	179
				363.72 ± .15	27.381 ± .012	.65447 ±.00032	2.1944 ±.0009	405	180
8.604 ±.005				478.3 ± .5	24.01 ± .02	.5738 ±.0006	2.503 ±.003	r	181
8.2563 ±.0030				182.08 ± .16	27.414 ± .024	.65527 ±.00062	2.1917 ±.0019	405	182
7.152 ±.001		120.00 ±.17		548.37 ± .95	20.641 ± .036	.49338 ±.00090	2.9110 ±.0050	25	183

	Name and formula	Crystal system	Space group	Structure type	Z	a_o	b_o
			Oxides and hydroxides—Continued				
184	Stishovite SiO_2*	tet.	P4/mnm(136)	rutile	2	4.1790 ±.0010	
185	Melanophlogite SiO_2*	cubic	Pm3n(223)	clathrate type	46	13.402 ±.004	
186	Cassiterite SnO_2	tet.	P4/mnm(136)	rutile	2	4.738 ±.003	
187	Tellurite TeO_2*	orth.	Pbca(61)	tellurite	8	5.607 ±.003	12.034 ±.005
188	Paratellurite TeO_2	tet.	$P4_12_12(92)$ $P4_32_12(96)$		4	4.810 ±.002	
189	Thorianite ThO_2	cubic	Fm3m(225)	fluorite	4	5.5952 ±.0005	
190	Rutile TiO_2	tet.	P4/mnm(136)		2	4.5937 ±.0005	
191	Anatase TiO_2	tet.	$I4_1/amd(141)$		4	3.785 ±.002	
192	Brookite TiO_2*	orth.	Pcab(61)		8	5.456 ±.002	9.182 ±.005
193	Titanium sesquioxide Ti_2O_3	hex-R	$R\bar{3}c(167)$	corundum	6	5.149 ±.002	
194	Uraninite UO_2	cubic	Fm3m(225)	fluorite	4	5.4682 ±.0010	
195	Karelianite V_2O_3	hex-R	$R\bar{3}c(167)$	corundum	6	4.952 ±.002	
196	Zincite ZnO	hex.	$P6_3mc(186)$	zincite	2	3.2495 ±.0005	
197	Baddeleyite ZrO_2	mon.	$P2_1/c(14)$	baddeleyite	4	5.1454 ±.0010	5.2075 ±.0010
			Multiple oxides				
198	Spinel $MgAl_2O_4$	cubic	Fd3m(227)	spinel	8	8.080 ±.002	
199	Hercynite $FeAl_2O_4$	cubic	Fd3m(227)	spinel	8	8.150 ±.004	
200	Galaxite $MnAl_2O_4$	cubic	Fd3m(227)	spinel	8	8.258 ±.002	
201	Gahnite $ZnAl_2O_4$	cubic	Fd3m(227)	spinel	8	8.0848 ±.0020	
202	Magnetite $FeFe_2O_4$	cubic	Fd3m(227)	spinel	8	8.3940 ±.0005	
203	Jacobsite $MnFe_2O_4$	cubic	Fd3m(227)	spinel	8	8.499 ±.002	
204	Trevorite $NiFe_2O_4$	cubic	Fd3m(227)	spinel	8	8.339 ±.003	
205	Picrochromite $MgCr_2O_4$	cubic	Fd3m(227)	spinel	8	8.333 ±.003	
206	Ilmenite $FeTiO_3$	hex-R	$R\bar{3}(148)$	ilmenite	6	5.093 ±.005	
207	Geikielite $MgTiO_3$	hex-R	$R\bar{3}(148)$	ilmenite	6	5.054 ±.005	
208	Pyrophanite $MnTiO_3$	hex-R	$R\bar{3}(148)$	ilmenite	6	5.155 ±.005	
209	Cobalt Titanate $CoTiO_3$	hex-R	$R\bar{3}(148)$	ilmenite	6	5.066 ±.001	

c_o	α_o	β_o	γ_o	Cell volume 10^{-24} cm^3	Molar volume cm^3	Molar volume cal bar^{-1}	X-Ray density grams cm^{-3}	Temp. °C	
Oxides and hydroxides—Continued									
2.6649 ±.0010				46.540 ± .028	14.014 ± .009	.33500 ±.00025	4.2874 ±.0026	r	184
				2407.2 ± 2.2	31.516 ± .028	.75325 ±.00072	1.9065 ±.0017	r	185
3.188 ±.003				71.57 ± .11	21.55 ± .03	.5151 ±.0009	6.992 ±.011	26	186
5.463 ±.003				368.61 ± .32	27.750 ± .024	.66328 ±.00062	5.7514 ±.0050	25	187
7.613 ±.002				176.14 ± .15	26.52 ± .02	.6339 ±.0006	6.018 ±.005	25	188
				175.16 ± .05	26.373 ± .007	.63038 ±.00021	10.012 ±.003	25	189
2.9618 ±.0010				62.500 ± .025	18.820 ± .008	.44986 ±.00023	4.2453 ±.0017	25	190
9.514 ±.006				136.30 ± .17	20.522 ± .025	.4905 ±.0007	3.893 ±.005	r	191
5.143 ±.003				257.6 ± .2	19.40 ± .02	.4636 ±.0005	4.119 ±.004	r	192
13.642 ±.010				313.2 ± .3	31.44 ± .03	.7515 ±.0009	4.574 ±.005	r	193
				163.51 ± .09	24.618 ± .014	.58843 ±.00037	10.969 ±.006	26	194
14.002 ±.010				297.36 ± .32	29.848 ± .032	.71342 ±.00081	5.0216 ±.0054	r	195
5.2069 ±.0005				47.615 ± .015	14.338 ± .005	.34273 ±.00016	5.6750 ±.0018	25	196
5.3107 ±.0010		99.23 ±.08		140.46 ± .06	21.148 ± .009	.50548 ±.00025	5.8267 ±.0023	r	197
Multiple oxides									
				527.5 ± .4	39.71 ± .03	.9492 ±.0008	3.583 ±.003	26	198
				541.3 ± .6	40.75 ± .05	.9740 ±.0011	4.265 ±.005	25	199
				563.2 ± .4	42.39 ± .03	1.013 ±.001	4.078 ±.003	25	200
				528.45 ± .39	39.783 ± .030	.95088 ±.00075	4.6083 ±.0034	26	201
				591.43 ± .11	44.524 ± .008	1.0642 ±.0002	5.2003 ±.0009	22	202
				613.9 ± .4	46.22 ± .03	1.105 ±.001	4.990 ±.004	25	203
				579.9 ± .6	43.65 ± .05	1.043 ±.001	5.370 ±.006	25	204
				578.6 ± .6	43.56 ± .05	1.041 ±.001	4.415 ±.005	26	205
14.055 ±.020				315.73 ± .75	31.69 ± .08	.7574 ±.0019	4.788 ±.012	r	206
13.898 ±.010				307.44 ± .65	30.86 ± .07	.7376 ±.0016	3.896 ±.008	26	207
14.18 ±.01				326.3 ± .7	32.76 ± .07	.7829 ±.0017	4.605 ±.010	r	208
13.918 ±.005				309.34 ± .17	31.05 ± .02	.7422 ±.0004	4.986 ±.003	r	209

	Name and formula	Crystal system	Space group	Structure type	Z	a_o	b_o
	Multiple oxides—Continued						
210	Perovskite $CaTiO_3$	orth.	Pcmn(62)	perovskite	4	5.3670 ±.0010	7.6438 ±.0010
211	Chrysoberyl $BeAl_2O_4$	orth.	Pmnb(62)	olivine	4	5.4756 ±.0020	9.4041 ±.0030
	Halides						
212	Halite NaCl	cubic	Fm3m(225)	rock salt	4	5.6402 ±.0002	
213	Sylvite KCl	cubic	Fm3m(225)	rock salt	4	6.2931 ±.0002	
214	Villiaumite NaF	cubic	Fm3m(225)	rock salt	4	4.6342 ±.0005	
215	Chlorargyrite AgCl	cubic	Fm3m(225)	rock salt	4	5.5491 ±.0005	
216	Bromargyrite AgBr	cubic	Fm3m(225)	rock salt	4	5.7745 ±.0005	
217	Nantockite CuCl	cubic	$F\bar{4}3m(216)$	sphalerite	4	5.416 ±.003	
218	Marshite CuI	cubic	$F\bar{4}3m(216)$	sphalerite	4	6.0507 ±.0010	
219	Miersite AgI	cubic	$F\bar{4}3m(216)$	sphalerite	4	6.4963 ±.0010	
220	Iodargyrite AgI	hex.	$P6_3mc(186)$	zincite	2	4.5955 ±.0010	
221	Calomel HgCl	tet.	14/mm(139)		4	4.478 ±.005	
222	Fluorite CaF_2	cubic	Fm3m(225)	fluorite	4	5.4638 ±.0004	
223	Sellaite MgF_2	tet.	$P4_2/mnm(136)$	rutile	2	4.621 ±.001	
224	Chloromagnesite $MgCl_2$	hex-R	$R\bar{3}m(166)$		3	3.632 ±.004	
225	Lawrencite $FeCl_2$	hex-R	$R\bar{3}m(166)$		3	3.593 ±.003	
226	Scacchite $MnCl_2$	hex-R	$R\bar{3}m(166)$		3	3.711 ±.002	
227	Cotunnite $PbCl_2$	orth.	Pnmb(62)		4	4.535 ±.005	7.62 ±.01
228	Matlockite PbFCl	tet.	P4/nmm(129)		2	4.106 ±.005	
229	Cryolite Na_3AlF_6*	mon.	$P2_1/n(14)$		2	5.40 ±.01	5.60 ±.01
230	Neighborite $NaMgF_3$	orth.	Pcmn(62)	perovskite	4	5.363 ±.001	7.676 ±.001
	Carbonates and nitrates						
231	Calcite $CaCO_3$	hex-R	$R\bar{3}c(167)$	calcite	6	4.9899 ±.0010	
232	Otavite $CdCO_3$	hex-R	$R\bar{3}c(167)$	calcite	6	4.9204 ±.0010	
233	Cobalticalcite $CoCO_3$	hex-R	$R\bar{3}c(167)$	calcite	6	4.6581 ±.0010	
234	Siderite $FeCO_3$	hex-R	$R\bar{3}c(167)$	calcite	6	4.6887 ±.0010	

c_o	α_o	β_o	γ_o	Cell volume 10^{-24} cm^3	Molar volume cm^3	Molar volume cal bar^{-1}	X-Ray density grams cm^{-3}	Temp. °C	
Multiple oxides—Continued									
5.4439 ±.0010				223.33 ± .07	33.626 ± .010	.80371 ±.00028	4.0439 ±.0012	r	210
4.4267 ±.0020				227.94 ± .15	34.320 ± .023	.82031 ±.00059	3.6997 ±.0025	25	211
Halides									
				179.43 ± .02	27.015 ± .003	.64571 ±.00011	2.1634 ±.0002	26	212
				249.23 ± .02	37.524 ± .004	.89690 ±.00013	1.9868 ±.0002	25	213
				99.523 ± .032	14.984 ± .005	.35818 ±.00016	2.8021 ±.0009	25	214
				170.87 ± .05	25.727 ± .007	.61493 ±.00021	5.5710 ±.0015	26	215
				192.55 ± .05	28.991 ± .008	.69294 ±.00022	6.4772 ±.0017	26	216
				158.87 ± .26	23.92 ± .04	.5717 ±.0010	4.139 ±.007	25	217
				221.52 ± .11	33.353 ± .017	.7972 ±.0004	5.710 ±.003	26	218
				274.16 ± .10	41.278 ± .020	.9866 ±.0004	5.688 ±.003	r	219
7.5005 ±.0033				137.18 ± .10	41.308 ± .030	.9873 ±.0009	5.683 ±.004	25	220
10.910 ±.005				218.77 ± .50	32.939 ± .075	.7873 ±.0018	7.166 ±.016	26	221
				163.11 ± .04	24.558 ± .005	.58701 ±.00017	3.1792 ±.0007	25	222
3.050 ±.001				65.13 ± .04	19.61 ± .01	.4688 ±.0003	3.177 ±.002	18	223
17.795 ±.016				203.29 ± .48	40.81 ± .10	.9754 ±.0024	2.333 ±.006	r	224
17.58 ±.09				196.55 ± 1.06	39.46 ± .21	.9431 ±.0051	3.212 ±.017	r	225
17.59 ±.07				209.79 ± .86	42.11 ± .17	1.007 ±.004	2.988 ±.012	r	226
9.05 ±.01				312.74 ± .64	47.09 ± .10	1.1254 ±.0023	5.906 ±.012	26	227
7.23 ±.01				121.89 ± .34	36.70 ± .10	.8773 ±.0025	9.853 ±.028	26	228
7.776 ±.010		90.18 ±.25		235.1 ± .7	70.81 ± .20	1.692 ±.005	2.965 ±.009	r	229
5.503 ±.001				226.54 ± .07	34.11 ± .01	.8152 ±.0003	3.058 ±.001	18	230
Carbonates and nitrates									
17.064 ±.002				367.96 ± .15	36.934 ± .015	.88278 ±.00041	2.7100 ±.0011	26	231
16.298 ±.003				341.72 ± .15	34.300 ± .015	.81983 ±.00041	5.0265 ±.0022	26	232
14.958 ±.003				281.07 ± .13	28.213 ± .013	.67435 ±.00036	4.2159 ±.0020	26	233
15.373 ±.003				292.68 ± .14	29.378 ± .014	.70219 ±.00037	3.9436 ±.0018	26	234

	Name and formula	Crystal system	Space group	Structure type	Z	a_o	b_o
	Carbonates and nitrates—Continued						
235	Magnesite $MgCO_3$	hex-R	$R\bar{3}c(167)$	calcite	6	4.6330 ±.0010	
236	Rhodochrosite $MnCO_3$	hex-R	$R\bar{3}c(167)$	calcite	6	4.7771 ±.0010	
237	Nickelous Carbonate $NiCO_3$	hex-R	$R\bar{3}c(167)$	calcite	6	4.5975 ±.0010	
238	Smithsonite $ZnCO_3$	hex-R	$R\bar{3}(167)$	calcite	6	4.6528 ±.0010	
239	Dolomite $CaMg(CO_3)_2$*	hex-R	$R\bar{3}(148)$	calcite	3	4.8079 ±.0010	
240	Huntite $Mg_3Ca(CO_3)_4$*	hex-R	R32(155)	calcite	3	9.498 ±.003	
241	Norsethite $BaMg(CO_3)_2$*	hex-R	R32(155)	calcite	3	5.020 ±.005	
242	Vaterite $CaCO_3$	hex.			6	7.135 ±.005	
243	Witherite $BaCO_3$	orth.	Pnam(62)	aragonite	4	6.430 ±.005	8.904 ±.005
244	Aragonite $CaCO_3$	orth.	Pnam(62)	aragonite	4	5.741 ±.005	7.968 ±.005
245	Cerussite $PbCO_3$	orth.	Pnam(62)	aragonite	4	6.152 ±.005	8.436 ±.005
246	Strontianite $SrCO_3$	orth.	Pnam(62)	aragonite	4	6.029 ±.005	8.414 ±.005
247	Shortite $Na_2Ca_2(CO_3)_3$	orth.	Amm2(38)		2	4.961 ±.005	11.03 ±.02
248	Malachite $Cu_2(OH)_2CO_3$	mon.	$P2_1/a(14)$		4	9.502 ±.007	11.974 ±.007
249	Azurite $Cu_3(OH)_2(CO_3)_2$	mon.	$P2_1/a(14)$		2	5.008 ±.005	5.844 ±.005
250	Niter KNO_3	orth.	Pnam(62)	aragonite	4	6.431 ±.005	9.164 ±.005
251	Soda Niter $NaNO_3$	hex-R	$R\bar{3}c(167)$	calcite	6	5.0696 ±.0010	
252	Gerhardite $Cu_2(NO_3)(OH)_3$	orth.	$P2_12_12_1(19)$		4	6.075 ±.004	13.812 ±.008
	Sulfates and borates						
253	Barite $BaSO_4$	orth.	Pnma(62)	barite	4	8.878 ±.005	5.450 ±.005
254	Anhydrite $CaSO_4$	orth.	Amma(63) Ccmm(63)	anhydrite	4	6.991 ±.005	6.996 ±.005
255	Anglesite $PbSO_4$	orth.	Pnma(62)	barite	4	8.480 ±.005	5.398 ±.005
256	Celestite $SrSO_4$	orth.	Pnma(62)	barite	4	8.359 ±.005	5.352 ±.005
257	Zinkosite $ZnSO_4$	orth.	Pnma(62)	barite	4	8.588 ±.008	6.740 ±.006
258	Arcanite KS_2SO_4	orth.	Pnma(62)	arcanite	4	5.772 ±.005	10.072 ±.005
259	Mascagnite $(NH_4)_2SO_4$	orth.	Pnma(62)	arcanite	4	7.782 ±.005	5.993 ±.005
260	Thenardite Na_2SO_4	orth.	Fddd(70)	thenardite	8	5.863 ±.005	12.304 ±.005

X-Ray Crystallographic Data of Minerals

c_o	a_o	β_o	γ_o	Cell volume 10^{-24} cm³	Molar volume cm³	Molar volume cal bar⁻¹	X-Ray density grams cm⁻³	Temp. °C	
Carbonates and nitrates—Continued									
15.016 ±.003				279.13 ± .13	28.018 ± .013	.66969 ±.00036	3.0095 ±.0014	26	235
15.664 ±.003				309.57 ± .14	31.073 ± .014	.74272 ±.00039	3.6992 ±.0017	26	236
14.723 ±.002				269.51 ± .12	27.052 ± .012	.64660 ±.00034	4.3886 ±.0020	26	237
15.025 ±.003				281.69 ± .13	28.275 ± .013	.67583 ±.00037	4.4343 ±.0021	26	238
16.010 ±.003				320.50 ± .15	64.341 ± .029	1.5378 ±.0008	2.8661 ±.0013	26	239
7.816 ±.004				610.63 ± .50	122.58 ± .10	2.9299 ±.0024	2.880 ±.002	26	240
16.75 ±.02				365.6 ± .8	73.39 ± .17	1.754 ±.004	3.838 ±.009	r	241
8.524 ±.007				375.80 ± .61	37.72 ± .06	.9016 ±.0015	2.653 ±.004	r	242
5.314 ±.005				304.24 ± .41	45.81 ± .06	1.095 ±.002	4.308 ±.006	26	243
4.959 ±.005				226.85 ± .33	34.15 ± .05	.8164 ±.0012	2.930 ±.004	26	244
5.195 ±.005				269.61 ± .38	40.59 ± .06	.9702 ±.0014	6.582 ±.009	26	245
5.107 ±.005				259.07 ± .37	39.01 ± .06	.9323 ±.0014	3.785 ±.005	26	246
7.12 ±.01				389.6 ± 1.0	117.3 ± .3	2.804 ±.007	2.610 ±.007	r	247
3.240 ±.003		98.75 ±.25		364.35 ± .54	54.86 ± .08	1.311 ±.002	4.030 ±.006	25	248
10.336 ±.005		92.45 ±.25		302.22 ± .43	91.01 ± .13	2.1752 ±.0031	3.787 ±.005	25	249
5.414 ±.005				319.07 ± .42	48.04 ± .06	1.148 ±.002	2.105 ±.003	26	250
16.829 ±.005				374.57 ± .19	37.598 ± .019	.89866 ±.00049	2.2606 ±.0011	25	251
5.592 ±.004				469.21 ± .53	70.65 ± .08	1.689 ±.002	3.399 ±.004	r	252
Sulfates and borates									
7.152 ±.003				346.05 ± .40	52.10 ± .06	1.245 ±.002	4.480 ±.005	26	253
6.238 ±.005				305.09 ± .39	45.94 ± .06	1.098 ±.002	2.964 ±.004	26	254
6.958 ±.003				318.50 ± .38	47.95 ± .06	1.146 ±.002	6.324 ±.008	25	255
6.866 ±.005				307.17 ± .41	46.25 ± .06	1.105 ±.002	3.972 ±.005	26	256
4.770 ±.005				276.10 ± .46	41.57 ± .07	.9936 ±.0017	3.883 ±.006	25	257
7.483 ±.004				435.03 ± .49	65.50 ± .07	1.566 ±.002	2.661 ±.003	25	258
10.636 ±.005				496.04 ± .57	74.68 ± .09	1.7851 ±.0021	1.7693 ±.0020	25	259
9.821 ±.005				708.47 ± .76	53.33 ± .06	1.275 ±.002	2.663 ±.003	25	260

	Name and formula	Crystal system	Space group	Structure type	Z	a_o	b_o
	Sulfates and borates—Continued						
261	Gypsum $CaSO_4.2H_2O$*	mon.	$C2/c(15)$		4	5.68 ±.01	15.18 ±.01
262	Epsomite $MgSO_4.7H_2O$	orth.	$P2_12_12_1(19)$		4	11.86 ±.01	11.99 ±.01
263	Goslarite $ZnSO_4.7H_2O$	orth.	$P2_12_12_1(19)$	epsomite	4	11.779 ±.005	12.050 ±.005
264	Mirabilite $Na_2SO_4.10H_2O$	mon.	$P2_1/c(14)$		4	11.51 ±.01	10.38 ±.01
265	Chalcanthite $CuSO_4.5H_2O$	tri.	$P\bar{1}(2)$		2	6.1045 ±.0050	10.72 ±.01
266	Brochantite $Cu_4SO_4(OH)_6$*	mon.	$P2_1/c(14)$		4	13.066 ±.010	9.85 ±.01
267	Syngenite $K_2Ca(SO_4)_2.H_2O$	mon.	$P2_1/m(11)$		2	9.775 ±.005	7.156 ±.005
268	Alunite $KAl_3(SO_4)_2(OH)_6$	hex-R	$R3m(160)$		3	6.982 ±.005	
269	Natroalunite $NaAl_3(SO_4)_2(OH)_6$	hex-R	$R3m(160)$		3	6.974 ±.005	
270	Hexahydrite $MgSO_4.6H_2O$	mon.	$C2/c(15)$		8	10.110 ±.005	7.212 ±.004
271	Leonhardtite $MgSO_4.4H_2O$	mon.	$P2_1/n(14)$		4	5.922 ±.006	13.604 ±.004
272	Melanterite $FeSO_4.7H_2O$	mon.	$P2_1/c(14)$		4	14.072 ±.010	6.503 ±.007
273	Vanthoffite $MgSO_4.3Na_2SO_4$	mon.	$P2_1/c(14)$		2	9.797 ±.003	9.217 ±.003
274	Dolerophanite $Cu_2O(SO_4)$	mon.	$C2/m(15)$		4	9.355 ±.010	6.312 ±.005
275	Retgersite $NiSO_4.4H_2O$	tet.	$P4_12_12(92)$ $P4_32_1(96)$		4	6.782 ±.004	
276	Colemanite $CaB_3O_4(OH)_3.H_2O$*	mon.	$P2_1/a(14)$		4	8.743 ±.004	11.264 ±.002
277	Borax $Na_2B_4O_7.10H_2O$	mon.	$C2/c(15)$		4	11.858 ±.005	10.674 ±.005
278	Kernite $Na_2B_4O_7.4H_2O$	mon.	$P2_1/c(14)$		4	7.022 ±.003	9.151 ±.004
279	Hambergite $Be_2BO_3.(OH,F)$*	orth.	$Pbca(61)$		8	9.755 ±.001	12.201 ±.001
	Phosphates, molybdates, and tungstates						
280	Berlinite $AlPO_4$	hex.	$P3_121(152)$ $P3_221(154)$	*a*-quartz	3	4.942 ±.005	
281	Xenotime YPO_4	tet.	$I4_1/amd(141)$	zircon	4	6.885 ±.005	
282	Hydroxylapatite $Ca_5(PO_4)_3OH$	hex.	$P6_3/m(176)$	apatite	2	9.418 ±.003	
283	Fluorapatite $Ca_5(PO_4)_3F$	hex.	$P6_3/m(176)$	apatite	2	9.3684 ±.0030	
284	Chlorapatite $Ca_5(PO_4)_3Cl$	hex.	$P6/_3m(176)$	apatite	2	9.629 ±.005	
285	Carbonate-apatite $Ca_{10}(PO_4)_6CO_3H_2O$	hex.	$P6_3/m(176)$	apatite	1	9.436 ±.010	
286	Turquois $CuAl_6(PO_4)_4(OH)_8.4H_2O$*	tri.	$P\bar{1}(2)$		1	7.424 ±.008	7.629 ±.008

c_o	α_o	β_o	γ_o	Cell volume 10^{-24} cm³	Molar volume cm³	Molar volume cal bar⁻¹	X-Ray density grams cm⁻³	Temp. °C	
Sulfates and borates—Continued									
6.29		113.83		496.1	74.69	1.785	2.305	r	261
±.01		±.22		± 1.5	± .22	±.005	±.007		
6.858				975.22	146.83	3.5094	1.679	25	262
±.007				± 1.53	± .23	±.0055	±.003		
6.822				968.29	145.79	3.4845	1.9723	25	263
±.003				± .72	± .11	±.0026	±.0015		
12.83		107.75		1459.9	219.8	5.253	1.466	24	264
±.01		±.17		± 2.6	± .4	±.009	±.003		
5.949	97.57	107.28	77.43	361.88	108.97	2.6045	2.2912	r	265
±.007	±.17	±.17	±.17	± .72	± .22	±.0052	±.0046		
6.022		103.27		754.3	113.6	2.715	3.982	r	266
±.010		±.25		± 1.8	± .2	±.006	±.009		
6.251		104.00		424.27	127.76	3.0535	2.5707	r	267
±.005		±.25		± .68	± .20	±.0049	±.0041		
17.32				731.2	146.8	3.508	2.822	r	268
±.01				± 1.1	± .2	±.005	±.004		
16.69				702.99	141.1	3.373	2.821	r	269
±.01				± 1.09	± .2	±.005	±.004		
24.41		98.30		1761.2	132.58	3.1689	1.7232	r	270
±.01		±.10		± 1.6	± .12	±.0029	±.0015		
7.905		90.85		636.78	95.88	2.2915	2.0071	r	271
±.005		±.20		± .78	± .12	±.0029	±.0025		
11.041		105.57		973.29	146.54	3.5025	1.8972	r	272
±.010		±.15		± 1.69	± .25	±.0061	±.0033		
8.199		113.50		678.96	204.45	4.8866	2.6730	r	273
±.003		±.10		± .65	± .20	±.0047	±.0025		
7.628		122.29		380.77	57.33	1.3703	4.171	r	274
±.005		±.10		± .70	± .11	±.0026	±.008		
18.28				840.80	126.59	3.0257	2.076	25	275
±.01				± 1.09	± .16	±.0040	±.003		
6.102		110.12		564.26	84.957	2.0306	2.4194	r	276
±.003		±.08		± .49	± .073	±.0018	±.0021		
12.197		106.68		1478.8	222.66	5.3217	1.7128	r	277
±.005		±.03		± 1.1	± .17	±.0041	±.0013		
15.676		108.83		953.40	143.55	3.4309	1.9038	r	278
±.008		±.25		± 1.61	± .24	±.0058	±.0032		
4.426				526.79	39.658	.9479	2.3663	r	279
±.001				± .14	± .011	±.0003	±.0006		
Phosphates, molybdates, and tungstates									
10.97				232.03	46.58	1.113	2.618	25	280
±.007				± .50	± .10	±.002	±.006		
5.982				283.57	42.69	1.020	4.307	26	281
±.005				± .48	± .07	±.002	±.008		
6.883				528.7	159.2	3.805	3.155	r	282
±.003				± .5	± .2	±.004	±.004		
6.8841				523.25	157.56	3.7659	3.2007	25	283
±.0030				± .41	± .12	±.0030	±.0025		
6.777				544.16	163.86	3.916	3.178	r	284
±.003				± .61	± .19	±.004	±.004		
6.883				530.74	319.6	7.640	3.281	r	285
±.010				± 1.36	± .8	±.020	±.008		
9.910	68.61	69.71	65.08	461.40	277.9	6.6416	2.927	r	286
±.010	±.20	±.20	±.20	± 1.12	± .7	±.0162	±.007		

X-Ray Crystallographic Data of Minerals

	Name and formula	Crystal system	Space group	Structure type	Z	a_o	b_o
	Phosphates, molybdates, and tungstates—Continued						
287	Powellite $CaMoO_4$	tet.	$I4_1/a$(100)	scheelite	4	5.226 ±.005	
288	Wulfenite $PbMoO_4$	tet.	$I4_1/a$(100)	scheelite	4	5.435 ±.005	
289	Scheelite $CaWO_4$	tet.	$I4_1/a$(100)	scheelite	4	5.242 ±.005	
290	Stolzite $PbWO_4$	tet.	$I4_1/a$(100)	scheelite	4	5.4616 ±.0030	
291	Ferberite $FeWO_4$	mon.	P2/c(13)	wolframite	2	4.732 ±.004	5.708 ±.003
292	Huebnerite $MnWO_4$	mon.	P2/c(13)	wolframite	2	4.834 ±.004	5.758 ±.005
293	Wolframite $Fe_{.5}Mn_{.5}WO_4$	mon.	P2/c(13)	wolframite	2	4.782 ±.004	5.731 ±.004
294	Sanmartinite $ZnWO_4$	mon.	P2/c(13)	wolframite	2	4.691 ±.003	5.720 ±.003
	Ortho and ring structure silicates						
295	Forsterite Mg_2SiO_4	orth.	Pbnm(62)	olivine	4	4.758 ±.002	10.214 ±.003
296	Fayalite Fe_2SiO_4	orth.	Pbnm(62)	olivine	4	4.817 ±.005	10.477 ±.005
297	Tephroite Mn_2SiO_4*	orth.	Pbnm(62)	olivine	4	4.871 ±.005	10.636 ±.005
298	Lime Olivine γCa_2SiO_4	orth.	Pbnm(62)	olivine	4	5.091 ±.010	11.371 ±.020
299	Nickel Olivine Ni_2SiO_4	orth.	Pbnm(62)	olivine	4	4.727 ±.002	10.121 ±.005
300	Cobalt Olivine Co_2SiO_4	orth.	Pbnm(62)	olivine	4	4.782 ±.002	10.301 ±.005
301	Monticellite $CaMgSiO_4$	orth.	Pbnm(62)	olivine	4	4.827 ±.005	11.084 ±.005
302	Kerschsteinite $CaFeSiO_4$	orth.	Pbnm(62)	olivine	4	4.886 ±.005	11.146 ±.005
303	Knebelite $MnFeSiO_4$*	orth.	Pbnm(62)	olivine	4	4.854 ±.010	10.602 ±.010
304	Glauchroite $CaMnSiO_4$	orth.	Pbnm(62)	olivine	4	4.944 ±.004	11.19 ±.01
305	Fluor-Norbergite $Mg_2SiO_4.MgF_2$	orth.	Pnmb(62)		4	8.727 ±.005	10.271 ±.010
306	Chondrodite $2Mg_2SiO_4.MgF_2$*	mon.	$P2_1/c$(14)		2	7.89 ±.03	4.743 ±.020
307	Fluor-Humite $3Mg_2SiO_4.MgF_2$	orth.	Pnma(62)		4	10.243 ±.005	20.72 ±.02
308	Clinohumite $4Mg_2SiO_4.MgF_2$*	mon.	$P2_1/c$(14)		2	13.68 ±.04	4.75 ±.02
309	Grossularite $Ca_3Al_2Si_3O_{12}$	cubic	Ia3d(230)	garnet	8	11.851 ±.001	
310	Uvarovite $Ca_3Cr_2Si_3O_{12}$	cubic	Ia3d(230)	garnet	8	11.999 ±.002	
311	Andradite $Ca_3Fe_2Si_3O_{12}$	cubic	Ia3d(230)	garnet	8	12.048 ±.001	
312	Goldmanite $Ca_3V_2Si_3O_{12}$	cubic	Ia3d(230)	garnet	8	12.070 ±.005	

X-Ray Crystallographic Data of Minerals

c_o	a_o	β_o	γ_o	Cell volume 10^{-24} cm³	Molar volume cm³	Molar volume cal bar⁻¹	X-Ray density grams cm⁻³	Temp. °C	
Phosphates, molybdates, and tungstates—Continued									
11.43				312.17	47.00	1.1234	4.256	25	287
±.007				± .63	± .09	±.0023	±.009		
12.110				357.72	53.859	1.2873	6.816	25	288
±.007				± .69	± .104	±.0025	±.013		
11.372				312.49	47.049	1.1245	6.120	25	289
±.005				± .61	± .092	±.0023	±.012		
12.046				359.32	54.100	1.2931	8.4110	25	290
±.005				± .42	± .064	±.0016	±.0099		
4.965		90.00		134.11	40.38	.9652	7.520	r	291
±.004		±.05		± .17	± .05	±.0013	±.010		
4.999		91.18		139.11	41.89	1.001	7.228	r	292
±.004		±.10		± .20	± .06	±.002	±.010		
4.982		90.57		136.53	41.11	.9826	7.376	r	293
±.004		±.10		± .18	± .06	±.0014	±.010		
4 925		89.36		132.14	39.79	.9511	7.872	25	294
±.003		±.20		± .14	± .04	±.0010	±.008		
Ortho and ring structure silicates									
5.984				290.81	43.786	1.0465	3.2136	25	295
±.002				± .18	± .027	±.0007	±.0020		
6.105				308.11	46.389	1.1088	4.3928	r	296
±.010				± .62	± .093	±.0023	±.0088		
6.232				322.87	48.612	1.1619	4.1545	r	297
±.005				± .45	± .067	±.0017	±.0058		
6.782				392.61	59.11	1.4129	2.914	r	298
±.010				± 1.19	± .18	±.0043	±.009		
5.915				282.98	42.61	1.0184	4.917	r	299
±.002				± .21	± .03	±.0008	±.004		
6.003				295.70	44.52	1.0642	4.716	r	300
±.002				± .21	± .03	±.0008	±.003		
6.376				341.13	51.362	1.2276	3.046	r	301
±.005				± .47	± .071	±.0017	±.004		
6.434				350.39	52.756	1.2609	3.564	r	302
±.010				± .67	± .101	±.0025	±.007		
6.162				317.11	47.74	1.1412	4.249	r	303
±.010				± .88	± .13	±.0032	±.012		
6.529				361.2	54.38	1.2997	3.441	r	304
±.005				± .9	± .14	±.0032	±.009		
4.709				422.09	63.551	1.5190	3.194	25	305
±.002				± .51	± .077	±.0019	±.004		
10.29		109.03		364.0	109.6	2.620	3.136	r	306
±.03		±.30		± 2.4	± .7	±.017	±.021		
4.735				1004.9	151.31	3.6163	3.2017	25	307
±.002				± 1.2	± .18	±.0042	±.0037		
10.27		100.83		655.5	197.4	4.717	3.167	r	308
±.02		±.50		± 3.8	± 1.1	±.027	±.018		
				1664.43	125.30	2.9948	3.595	25	309
				± .42	± .03	±.0008	±.001		
				1727.57	130.05	3.1084	3.848	26	310
				± .86	± .07	±.0016	±.002		
				1748.82	131.65	3.1466	3.860	25	311
				± .44	± .03	±.0008	±.001		
				1758.42	132.38	3.1639	3.765	r	312
				2.19	± .16	±.0040	±.005		

	Name and formula	Crystal system	Space group	Structure type	Z	a_o	b_o
	Ortho and ring structure silicates—Continued						
313	Almandite $Fe_3Al_2Si_3O_{12}$	cubic	Ia3d(230)	garnet	8	11.526 ±.001	
314	Pyrope $Mg_3Al_2Si_3O_{12}$	cubic	Ia3d(230)	garnet	8	11.459 ±.001	
315	Spessartite $Mn_3Al_2Si_3O_{12}$	cubic	Ia3d(230)	garnet	8	11.621 ±.001	
316	Zircon $ZrSiO_4$*	tet.	I4/amd(141)	zircon	4	6.604 ±.005	
317	Thorite $ThSiO_4$	tet.	I4/amd(141)	zircon	4	7.143 ±.004	
318	Coffinite $USiO_4$	tet.	I4/amd(141)	zircon	4	6.995 ±.004	
319	Kyanite Al_2SiO_5*	tri.	P$\bar{1}$(2)		4	7.123 ±.001	7.848 ±.002
320	Andalusite Al_2SiO_5*	orth.	Pnnm(58)		4	7.7959 ±.0050	7.8983 ±.0020
321	Sillimanite Al_2SiO_5*	orth.	Pbnm(62) Pnma(62)		4	7.4843 ±.0030	7.6730 ±.0030
322	3.2 Mullite $3AL_2O_3.2SiO_2$	orth.			3/4	7.557 ±.002	7.6876 ±.0020
323	2.1 Mullite $2Al_2O_3.SiO_2$	orth.	Pbam(55)		6/5	7.5788 ±.0020	7.6909 ±.0020
324	Staurolite $Fe_2Al_9Si_4O_{22}(OH)_2$*	mon.	C2/m(15)		2	7.90 ±.10	16.65 ±.15
325	Topaz $Al_2(SiO_4)(OH)$*	orth.	Pmnb(62)		4	8.394 ±.005	8.792 ±.007
326	Phenacite Be_2SiO_4*	hex-R	R$\bar{3}$(148)	phenacite	18	12.472 ±.005	
327	Willemite Zn_2SiO_4	hex-R	R$\bar{3}$(148)	phenacite	18	13.94 ±.01	
328	Dioptase CuH_2SiO_4*	hex-R	R$\bar{3}$(148)	phenacite	18	14.61 ±.02	
329	Larnite β-Ca_2SiO_4*	mon.	P2$_1$/n(14)		4	5.48 ±.02	6.76 ±.02
330	Akermanite $Ca_2MgSi_2O_7$	tet.	P$\bar{4}$2$_1$m(113)	melilite	2	7.8435 ±.0030	
331	Gehlenite $Ca_2Al_2SiO_7$	tet.	P$\bar{4}$2$_1$m(113)	melilite	2	7.690 ±.003	
332	Fe-Gehlenite $Ca_2Fe_2SiO_7$	tet.	P$\bar{4}$2$_1$m(113)	melilite	2	7.54 ±.01	
333	Hardystonite $Ca_2ZnSi_2O_7$*	tet.	P$\bar{4}$2$_1$m(113)	melilite	2	7.87 ±.03	
334	Sodium Melilite $NaCaAlSi_2O_7$	tet.	P$\bar{4}$2$_1$m(113)	melilite	2	8.511 ±.005	
335	Beryl $Be_3Al_2(Si_6O_{18})$*	hex.	P6/mmc(192)	beryl	2	9.215 ±.005	
336	Indialite high Cordierite $Mg_2Al_3(AlSi_5O_{18})$	hex.	P6/mmc(192)	beryl	2	9.7698 ±.0030	
337	Low Cordierite $Mg_2Al_3(AlSi_5O_{18})$	orth.	Cccm(66)		4	9.721 ±.003	17.062 ±.006
338	Fe-Indialite $Fe_2Al_3(AlSi_5O_{18})$	hex.	P6/mmc(192)	beryl	2	9.860 ±.010	
339	Fe-Cordierite $Fe_2Al_3(AlSi_5O_{18})$	orth.	Cccm(66)	cordierite	4	9.726 ±.010	17.065 ±.010

X-Ray Crystallographic Data of Minerals

c_o	α_o	β_o	γ_o	Cell volume 10^{-24} cm³	Molar volume cm³	Molar volume cal bar⁻¹	X-Ray density grams cm⁻³	Temp. °C	
Ortho and ring structure silicates—Continued									
				1531.21 ± .40	115.27 ± .04	2.7551 ±.0008	4.318 ±.001	25	313
				1504.67 ± .39	113.27 ± .03	2.7074 ±.0008	3.559 ±.001	25	314
				1569.39 ± .41	118.15 ± .03	2.8238 ±.0008	4.190 ±.001	25	315
5.979 ±.005				260.76 ± .45	39.261 ± .068	.9384 ±.0017	4.669 ±.008	25	316
6.327 ±.003				322.82 ± .39	48.60 ± .06	1.1617 ±.0015	6.668 ±.008	r	317
6.263 ±.005				306.45 ± .43	46.140 ± .064	1.103 ±.002	7.155 ±.010	r	318
5.564 ±.008	89.92 ±.15	101.25 ±.08	105.97 ±.08	292.83 ± .45	44.09 ± .07	1.054 ±.002	3.675 ±.006	25	319
5.5583 ±.0020				342.25 ± .27	51.530 ± .040	1.2316 ±.0010	3.145 ±.002	25	320
5.7711 ±.0040				331.42 ± .30	49.899 ± .044	1.1927 ±.0011	3.248 ±.003	25	321
2.8842 ±.0010				167.56 ± .09	134.55 ± .07	3.2159 ±.0016	3.166 ±.002	r	322
2.8883 ±.0010				168.35 ± .09	84.492 ± .043	2.0195 ±.0011	3.125 ±.002	r	323
5.63 ±.10		90.00 ±.25		740.5 ±17.5	223.0 ± 5.3	5.330 ±.126	3.825 ±.090	r	324
4.649 ±.003				343.10 ± .41	51.66 ± .06	1.2347 ±.0015	3.563 ±.005	26	325
8.252 ±.005				1111.6 ± 1.1	37.194 ± .037	.8890 ±.0009	2.960 ±.003	25	326
9.309 ±.003				1566.6 ± 2.3	52.42 ± .08	1.253 ±.002	4.251 ±.006	25	327
7.80 ±.01				1441.9 ± 4.4	48.24 ± .15	1.153 ±.004	3.247 ±.010	r	328
9.28 ±.02		94.55 ±.33		342.7 ± 1.8	51.60 ± .27	1.233 ±.006	3.338 ±.017	r	329
5.010 ±.003				308.22 ± .30	92.812 ± .090	2.2183 ±.0022	2.9375 ±.0029	r	330
5.0675 ±.0030				299.67 ± .29	90.239 ± .088	2.1568 ±.0022	3.0387 ±.0030	r	331
4.855 ±.005				276.01 ± .79	83.12 ± .24	1.9865 ±.0057	3.994 ±.011	r	332
5.01 ±.02				310.3 ± 2.7	93.44 ± .80	2.233 ±.019	3.357 ±.029	r	333
4.809 ±.003				348.35 ± .46	104.90 ± .14	2.507 ±.003	2.462 ±.003	r	334
9.192 ±.005				675.98 ± .82	203.55 ± .25	4.8651 ±.0060	2.641 ±.003	25	335
9.3517 ±.0030				773.02 ± .54	232.78 ± .16	5.5636 ±.0039	2.513 ±.002	25	336
9.339 ±.003				1548.96 ± .88	233.22 ± .13	5.5741 ±.0032	2.508 ±.001	25	337
9.285 ±.010				781.75 ± 1.80	235.40 ± .54	5.6264 ±.0130	2.753 ±.006	r	338
9.287 ±.010				1541.40 ± 2.47	232.08 ± .37	5.5468 ±.0089	2.792 ±.005	r	339

	Name and formula	Crystal system	Space group	Structure type	Z	a_o	b_o
	Ortho and ring structure silicates—Continued						
340	Mn-Indialite $Mn_2Al_3(AlSi_5O_{18})$	hex.	P6/mmc(192)	beryl	2	9.925 ±.010	
341	Sapphirine $Mg_2Al_4O_6SiO_4$*	mon.	$P2_1/c(14)$		8	11.26 ±.03	14.46 ±.03
342	Elbaite $NaLiAl_{7.67}B_3Si_6O_{27}(OH)_4$*	hex-R	R3m(160)	tourmaline	3	15.842 ±.010	
343	Schorl $NaFe_3Al_6B_3Si_6O_{27}(OH)_4$*	hex-R	R3m(160)	tourmaline	3	16.032 ±.010	
344	Dravite $NaMg_3Al_6B_3Si_6O_{27}(OH)_4$	hex-R	R3m(160)	tourmaline	3	15.942 ±.010	
345	Uvite $CaMg_4Al_5B_3Si_6O_{27}(OH)_4$	hex-R	R3m(160)	tourmaline	3	15.86 ±.01	
346	Sphene $CaTiSiO_5$*	mon.	A2/a(15)		4	7.07 ±.01	8.72 ±.01
347	Datolite $CaBSiO_4(OH)$*	mon.	$P2_1/c(14)$		4	9.62 ±.03	7.60 ±.03
348	Euclase $AlBeSiO_4(OH)$*	mon.	$P2_1/a(14)$		4	4.763 ±.005	14.29 ±.02
349	Chloritoid $H_2FeAl_2SiO_7$*	mon.	C2/c(15)		8	9.48 ±.01	5.48 ±.01
350	Hemimorphite $Zn_4(OH)_2Si_2O_7.H_2O$*	orth.	Imm2(35)		2	8.370 ±.005	10.719 ±.005
351	Zoisite $Ca_2Al_3(SiO_4)_3OH$	orth.	Pnma(62)		4	16.15 ±.01	5.581 ±.005
352	Clinozoisite $Ca_2Al_3(SiO_4)_3OH$	mon.	$P2_1/m(11)$		2	8.887 ±.007	5.581 ±.005
353	Epidote $Ca_2Al_{1.5}Fe_{1.5}(SiO_4)_3OH$*	mon.	$P2_1/m(11)$		2	8.89 ±.02	5.63 ±.01
354	Piemontite $Ca_2Al_{1.5}Mn_{1.5}(SiO_4)_3OH$*	mon.	$P2_1/m(11)$		2	8.95 ±.02	5.70 ±.01
355	Lawsonite $CaAl_2Si_2O_7(OH)_2.H_2O$	orth.	Cccm(63)		4	8.787 ±.005	5.836 ±.005
	Chain and band structure silicates						
356	Enstatite $MgSiO_3$*	orth.	Pcab(61)		16	8.829 ±.010	18.22 ±.01
357	Clinoenstatite $MgSiO_3$	mon.	$P2_1/c(15)$		8	9.620 ±.005	8.825 ±.005
358	Protoenstatite $MgSiO_3$	orth.	Pbcn(60)		8	9.25 ±.01	8.74 ±.01
359	High Clinoenstatite $MgSiO_3$	tri.			8	10.000 ±.005	8.934 ±.004
360	Clinoferrosilite $FeSiO_3$	mon.	$P2_1/c(14)$		8	9.7085 ±.0010	9.0872 ±.0011
361	Orthoferrosilite $FeSiO_3$	orth.	Pcab(61)	enstatite	16	9.080 ±.002	18.431 ±.004
362	Diopside $CaMg(SiO_3)_2$	mon.	C2/c(15)	diopside	4	9.743 ±.005	8.923 ±.005
363	Hedenbergite $CaFe(SiO_3)_2$*	mon.	C2/c(15)	diopside	4	9.854 ±.010	9.024 ±.010
364	Johannsenite $CaMn(SiO_3)_2$*	mon.	C2/c(15)	diopside	4	9.83 ±.03	9.04 ±.03
365	Ureyite $NaCr(SiO_3)_2$	mon.	C2/c(15)	diopside	4	9.550 ±.016	8.712 ±.007

X-Ray Crystallographic Data of Minerals

c_o	α_o	β_o	γ_o	Cell volume 10^{-24} cm^3	Molar volume cm^3	Molar volume cal bar^{-1}	X-Ray density grams cm^{-3}	Temp. °C	
Ortho and ring structure silicates—Continued									
9.297 ±.010				793.11 ± .81	238.8 ± .5	5.708 ±.013	2.706 ±.006	r	340
9.95 ±.02		125.33 ±.50		1321.7 ± 9.7	99.50 ± .73	2.378 ±.017	3.464 ±.025	r	341
7.009 ±.010				1523.4 ± 2.9	305.82 ± .58	7.3093 ±.0140	3.271 ±.006	r	342
7.149 ±.010				1591.3 ± 3.0	319.45 ± .60	7.635 ±.014	3.297 ±.006	r	343
7.224 ±.010				1589.99 ± 2.97	319.19 ± .60	7.629 ±.014	3.004 ±.006	r	344
7.19 ±.01				1566.3 ± 2.9	314.4 ± .6	7.515 ±.014	3.095 ±.006	r	345
6.56 ±.01		113.95 ±.25		369.61 ± 1.13	55.65 ± .17	1.330 ±.004	3.523 ±.011	r	346
4.84 ±.02		90.15 ±.25		353.9 ± 2.3	53.28 ± .35	1.273 ±.008	3.003 ±.020	r	347
4.618 ±.005		100.25 ±.10		309.30 ± .64	46.57 ± .10	1.113 ±.002	3.116 ±.007	r	348
18.18 ±.01		101.77 ±.25		924.6 ± 2.2	69.61 ± .16	1.664 ±.004	3.619 ±.008	r	349
5.120 ±.005				459.36 ± .57	138.32 ± .17	3.306 ±.004	3.482 ±.004	25	350
10.06 ±.01				906.74 ± 1.34	136.52 ± .20	3.263 ±.005	3.328 ±.005	r	351
10.14 ±.01		115.93 ±.33		452.30 ± 1.45	136.20 ± .44	3.255 ±.010	3.336 ±.011	r	352
10.19 ±.02		115.40 ±.30		460.72 ± 1.97	138.7 ± .6	3.316 ±.014	3.587 ±.015	r	353
9.41 ±.02		115.70 ±.50		432.56 ± 2.38	130.3 ± .7	3.113 ±.017	3.810 ±.021	r	354
13.123 ±.008				672.96 ± .80	101.32 ± .12	2.4217 ±.0029	3.101 ±.004	r	355
Chain and band structure silicates									
5.192 ±.005				835.21 ± 1.32	31.44 ± .05	.7514 ±.0012	3.194 ±.005	r	356
5.188 ±.005		108.33 ±.17		418.10 ± .66	31.47 ± .05	.7523 ±.0012	3.190 ±.005	r	357
5.32 ±.01				430.10 ± 1.05	32.38 ± .08	.7739 ±.0019	3.101 ±.008	r	358
5.170 ±.003	88.27 ±.05	70.03 ±.04	91.01 ±.04	433.72 ± .40	32.65 ± .03	.7804 ±.0008	3.075 ±.003		359
5.2284 ±.004		108.43 ±.05		437.60 ± .15	32.943 ± .011	.7874 ±.0003	4.005 ±.002		360
5.238 ±.001				876.6 ± .54	33.00 ± .02	.7887 ±.0008	3.998 ±.004	r	361
5.251 ±.003		105.93 ±.25		438.97 ± .69	66.09 ± .10	1.580 ±.003	3.277 ±.005	r	362
5.263 ±.010		104.23 ±.33		453.64 ± 1.28	68.30 ± .19	1.632 ±.005	3.632 ±.010	r	363
5.27 ±.02		105.00 ±.50		452.35 ± 2.87	68.11 ± .43	1.628 ±.010	3.629 ±.023	r	364
5.273 ±.008		107.44 ±.16		418.6 ± 1.1	63.02 ± .16	1.506 ±.004	3.605 ±.009	r	365

	Name and formula	Crystal system	Space group	Structure type	Z	a_o	b_o
	Chain and band structure silicates—Continued						
366	Jadeite $NaAl(SiO_3)_2$*	mon.	C2/c(15)	diopside	4	9.409 ±.005	8.564 ±.005
367	Acmite (Aegirine) $NaFe(SiO_3)_2$	mon.	C2/c(15)	diopside	4	9.658 ±.005	8.795 ±.005
368	Ca Tschermak Molecule $CaAl_2SiO_6$	mon.	C2/c(15)	diopside	4	9.615 ±.005	8.661 ±.005
369	Spodumene $LiAl(SiO_3)_2$	mon.	C2/c(15)	diopside	4	9.451 ±.002	8.387 ±.002
370	β-Spodumene $LiAl(SiO_3)_2$	tet.	$P4_32_12$(96) $P4_12_12$(92)		4	7.5332 ±.0008	
371	Pectolite $Ca_2NaH(SiO_3)_3$*	tri.	$P\bar{1}$(2)		2	7.99 ±.01	7.04 ±.01
372	Wollastonite $CaSiO_3$*	tri.	$P\bar{1}$(2)		6	7.94 ±.01	7.32 ±.01
373	Parawollastonite $CaSiO_3$*	mon.	$P2_1$(4)		12	15.417 ±.004	7.321 ±.002
374	Pseudowollastonite $CaSiO_3$*	tri.			24	6.90 ±.02	11.78 ±.02
375	Rhodonite $MnSiO_3$*	tri.	$P\bar{1}$(2)		10	7.682 ±.002	11.818 ±.003
376	Bustamite $CaMn(SiO_3)_2$*	tri.	$A\bar{1}$(2)		6	7.736 ±.003	7.157 ±.003
377	Pyroxmangite $MnFe(SiO_3)_2$*	tri.	$P\bar{1}$(2)		7	7.56 ±.02	17.45 ±.05
378	Tremolite $Ca_2Mg_5[Si_8O_{22}](OH)_2$*	mon.	C2/m(12)	tremolite	2	9.840 ±.010	18.052 ±.020
379	Fluor-tremolite $Ca_2Mg_5[Si_8O_{22}]F_2$	mon.	C2/m(12)	tremolite	2	9.781 ±.007	18.01 ±.01
380	Ferrotremolite $Ca_2Fe_5[Si_8O_{22}](OH)_2$	mon.	C2/m(12)	tremolite	2	9.97 ±.01	18.34 ±.02
381	Grunerite $Fe_7[Si_8O_{22}](OH)_2$	mon.	C2/m(12)	tremolite	2	9.572 ±.005	18.44 ±.01
382	Cummingtonite (hypo.) $Mg_7[Si_8O_{22}](OH)_2$	mon.	C2/m(12)	tremolite	2	9.476 ±.010	17.935 ±.010
383	Riebeckite $Na_2Fe_3Fe_2[Si_8O_{22}](OH)_2$	mon.	C2/m(12)	tremolite	2	9.729 ±.020	18.065 ±.020
384	Magnesioriebeckite $Na_2Mg_3Fe_2[Si_8O_{22}](OH)_2$	mon.	C2/m(12)	tremolite	2	9.733 ±.010	17.946 ±.020
385	Gaucophane I $Na_2Mg_3Al_2[Si_8O_{22}](OH)_2$	mon.	C2/m(12)	tremolite	2	9.748 ±.010	17.915 ±.020
386	Glaucophane II $Na_2Mg_3Al_2[Si_8O_{22}](OH)_2$	mon.	C2/m(12)	tremolite	2	9.663 ±.010	17.696 ±.020
387	Fluor-edenite $NaCa_2Mg_5[AlSi_7O_{22}]F_2$	mon.	C2/m(12)	tremolite	2	9.847 ±.005	18.00 ±.01
388	Fluor-richterite $Na_2CaMg_5[Si_8O_{22}]F_2$	mon.	C2/m(12)	tremolite	2	9.823 ±.005	17.96 ±.01
389	Anthophyllite $Mg_7[Si_8O_{22}](OH)_2$	orth.	Pnma(62)		4	18.61 ±.02	18.01 ±.06
	Framework structure silicates						
390	Microcline $KAlSi_3O_8$	tri.	$C\bar{1}$(2)		4	8.582 ±.002	12.964 ±.005
391	High Sanidine $KAlSi_3O_8$	mon.	C2/m(12)		4	8.615 ±.002	13.031 ±.003

X-Ray Crystallographic Data of Minerals

c_o	α_o	β_o	γ_o	Cell volume 10^{-24} cm^3	Molar volume cm^3	Molar volume cal bar^{-1}	X-Ray density grams cm^{-3}	Temp. °C	
				Chain and band structure silicates—Continued					
5.220 ±.005		107.50 ±.20		401.15 ± .67	60.40 ± .10	1.444 ±.002	3.347 ±.006	r	366
5.294 ±.005		107.42 ±.20		429.06 ± .70	64.60 ± .11	1.544 ±.003	4.411 ±.007	r	367
5.272 ±.003		106.12 ±.20		421.77 ± .59	63.50 ± .09	1.518 ±.002	3.435 ±.005	r	368
5.208 ±.001		110.07 ±.03		387.7 ± .1	58.37 ± .02	1.395 ±.001	3.188 ±.001	r	369
9.1540 ±.0008				519.48 ± .12	78.215 ±.018	1.8694 ±.0005	2.379 ±.001	r	370
7.02 ±.01	90.05 ±.25	95.27 ±.25	102.47 ±.25	383.84 ± .99	115.58 ± .30	2.763 ±.007	2.876 ±.007	r	371
7.07 ±.01	90.03 ±.25	95.37 ±.25	103.43 ±.25	397.82 ± 1.03	39.93 ± .10	.9544 ±.0025	2.909 ±.008	r	372
7.066 ±.003		95.40 ±.10		793.98 ± .47	39.85 ± .02	.9524 ±.0006	2.915 ±.002	r	373
19.65 ±.02	90.00 ±.30	90.80 ±.30	90.00 ±.30	1597.0 ± 5.6	40.08 ± .14	.9579 ±.0034	2.899 ±.010	r	374
6.707 ±.002	92.36 ±.05	93.95 ±.05	105.66 ±.05	583.77 ± .31	35.158 ± .019	.8403 ±.0005	3.727 ±.002	r	375
13.824 ±.010	90.52 ±.25	94.58 ±.25	103.87 ±.25	740.38 ± 1.08	74.32 ± .11	1.776 ±.003	3.326 ±.005	r	376
6.67 ±.02	84.00 ±.30	94.30 ±.30	113.70 ±.30	800.77 ± 4.29	68.90 ± .36	1.647 ±.009	3.817 ±.020	r	377
5.275 ±.010		104.70 ±.25		906.34 ± 2.43	272.92 ± .73	6.523 ±.018	2.977 ±.008	r	378
5.267 ±.005		104.52 ±.25		898.18 ± 1.56	270.46 ± .47	6.464 ±.011	3.018 ±.005	20	379
5.30 ±.01		104.50 ±.10		938.24 ± 2.92	282.53 ± .69	6.753 ±.017	3.434 ±.008	r	380
5.342 ±.007		101.77 ±.25		923.08 ± 1.63	277.96 ± .49	6.644 ±.012	3.603 ±.006	r	381
5.292 ±.005		102.23 ±.25		878.97 ± 1.58	264.68 ± .47	6.326 ±.011	2.950 ±.005	r	382
5.334 ±.010		103.31 ±.25		912.29 ± 2.89	274.71 ± .87	6.566 ±.021	3.407 ±.011	r	383
5.299 ±.010		103.30 ±.25		900.74 ± 2.37	271.24 ± .71	6.483 ±.017	3.102 ±.008	r	384
5.273 ±.010		102.78 ±.25		898.04 ± 2.35	270.42 ± .71	6.463 ±.017	2.898 ±.008	r	385
5.277 ±.010		103.67 ±.10		876.79 ± 2.17	264.02 ± .65	6.310 ±.016	2.968 ±.007	r	386
5.282 ±.005		104.83 ±.25		905.03 ± 1.51	272.53 ± .46	6.514 ±.011	3.076 ±.005	r	387
5.268 ±.005		104.33 ±.25		900.47 ± 1.48	271.15 ± .45	6.481 ±.011	3.033 ±.005		388
5.24 ±.01				1756.3 ± 7.0	264.4 ± 1.1	6.320 ±.025	2.953 ±.012	r	389
				Framework structure silicates					
7.222 ±.002	90.62 ±.10	115.92 ±.10	87.68 ±.10	722.06 ± .67	108.72 ± .10	2.5984 ±.0025	2.560 ±.002	r	390
7.177 ±.002		115.98 ±.10		724.28 ± .69	109.05 ± .10	2.6064 ±.0025	2.552 ±.002	r	391

X-Ray Crystallographic Data of Minerals

	Name and formula	Crystal system	Space group	Structure type	Z	a_o	b_o
	Framework structure silicates—Continued						
392	Orthoclase $KAlSi_3O_8$*	mon.	C2/m(12)		4	8.562 ±.003	12.996 ±.004
393	Fe-Sanidine $KFeSi_3O_8$	mon.	C2/m(12)		4	8.689 ±.008	13.12 ±.01
394	Fe-Microcline $KFeSi_3O_8$	tri.	$C\bar{1}(2)$		4	8.68 ±.01	13.10 ±.01
395	Low Albite $NaAlSi_3O_8$	tri.	$C\bar{1}(2)$		4	8.139 ±.002	12.788 ±.003
396	High Albite (Analbite) $NaAlSi_3O_8$	tri.	$C\bar{1}(2)$		4	8.160 ±.002	12.870 ±.003
397	Anorthite $CaAl_2Si_2O_8$	tri.	$P\bar{1}(2)$	primitive cell	8	8.177 ±.002	12.877 ±.003
398	Synthetic $CaAl_2Si_2O_8$	hex.	$P6_3/mcm(193)$		2	5.10 ±.02	
399	Synthetic $CaAl_2Si_2O_8$	orth.	$P2_12_12(18)$		2	8.22 ±.02	8.60 ±.02
400	Celsian $BaAl_2Si_2O_8$*	mon.	$I2_1/c(15)$		8	8.627 ±.010	13.045 ±.010
401	Paracelsian $BaAl_2Si_2O_8$*	mon.	$P2_1/a(14)$		4	8.58 ±.02	9.583 ±.020
402	Banalsite $BaNa_2Al_4Si_4O_{16}$*	orth.			4	8.50 ±.02	9.97 ±.02
403	Danburite $CaB_2Si_2O_8$*	orth.	Pnam(62)		4	8.04 ±.02	8.77 ±.02
404	Low Nepheline $NaAlSiO_4$	hex.	$C6_3(178)$		8	9.986 ±.005	
405	High Carnegeite $NaAlSiO_4$	cubic			4	7.325 ±.004	
406	Kaliophilite natural $KAlSiO_4$*	hex.	$P6_322(182)$		54	26.930 ±.010	
407	Kaliophilite synthetic $KAlSiO_4$	hex.	$P6_3(173)$ $P6_322(182)$		2	5.180 ±.002	
408	Kalsilite $KAlSiO_4$	hex.	$P6_3(173)$		2	5.1597 ±.0020	
409	Leucite $KAlSi_2O_6$	tet.	$I4_1/a(100)$		16	13.074 ±.003	
410	High Leucite $KAlSi_2O_6$*	cubic	Ia3d(230)		16	13.43 ±.05	
411	Fe-Leucite $KFeSi_2O_6$	tet.	$I4_1/a(100)$		16	13.205 ±.002	
412	Petalite $LiAlSi_4O_{10}$*	mon.	$P2_1/n(14)$		2	11.32 ±.03	5.14 ±.01
413	Marialite $Na_4Al_3Si_9O_{24}Cl$	tet.	I4/m(87) P4/m(83)		2	12.064 ±.008	
414	Meionite $Ca_4Al_6Si_6O_{24}CO_3$	tet.	I4/m(87) P4/m(83)		2	12.174 ±.008	
	Sheet structure silicates						
415	Muscovite $KAl_2[AlSi_3O_{10}](OH)_2$*	mon.	C2/c(15)	$2M_2$ mica	4	5.203 ±.005	8.995 ±.005
416	Paragonite $NaAl_2[AlSi_3O_{10}](OH)_2$*	mon.	C2/c(15)	$2M_1$ mica	4	5.13 ±.03	8.89 ±.05
417	Lepidolite $K_2Al_3Li_2[AlSi_7O_{20}](OH)_4$*	mon.	C2/c(15)	$2M_2$ mica	2	9.2 ±.1	5.3 ±.1

X-Ray Crystallographic Data of Minerals

c_o	α_o	β_o	γ_o	Cell volume 10^{-24} cm³	Molar volume cm³	Molar volume cal bar⁻¹	X-Ray density grams cm⁻³	Temp. °C	
				Framework structure silicates—Continued					
7.193 ±.003		116.02 ±.15		719.25 ± 1.02	108.29 ± .15	2.5883 ±.0037	2.570 ±.004	r	392
7.319 ±.007		116.10 ±.30		749.28 ± 2.24	112.81 ± .34	2.6964 ±.0081	2.723 ±.008	r	393
7.340 ±.007	90.75 ±.25	116.05 ±.25	86.23 ±.25	748.09 ± 1.92	112.63 ± .29	2.692 ±.007	2.727 ±.007	r	394
7.160 ±.002	94.27 ±.10	116.57 ±.10	87.68 ±.10	664.65 ± .60	100.07 ± .09	2.3918 ±.0022	2.620 ±.002	26	395
7.106 ±.002	93.54 ±.10	116.36 ±.10	90.19 ±.10	667.00 ± .60	100.43 ± .09	2.4003 ±.0022	2.611 ±.002	r	396
14.169 ±.003	93.17 ±.02	115.85 ±.02	91.22 ±.02	1338.9 ± .6	100.79 ± .04	2.4090 ±.0011	2.760 ±.001	r	397
14.72 ±.02				331.57 ± 2.64	99.85 ± .79	2.386 ±.019	2.786 ±.022	r	398
4.83 ±.01				341.44 ± 1.35	102.82 ± .41	2.457 ±.010	2.706 ±.011	r	399
14.408 ±.020		115.20 ±.25		1467.1 ± 4.2	110.45 ± .31	2.640 ±.008	3.400 ±.010	r	400
9.08 ±.02		90.00 ±.50		746.6 ± 2.9	112.4 ± .4	2.687 ±.010	3.340 ±.013	r	401
16.72 ±.03				1416.9 ± 5.1	213.3 ± .8	5.099 ±.018	3.092 ±.011	r	402
7.74 ±.02				545.8 ± 2.3	82.17 ± .35	1.964 ±.008	2.992 ±.013	r	403
8.330 ±.004				719.38 ± .80	54.16 ± .06	1.294 ±.002	2.623 ±.003	r	404
				393.03 ± .64	59.18 ± .10	1.414 ±.002	2.401 ±.004	750	405
8.522 ±.004				5352.4 ± 4.7	59.69 ± .05	1.427 ±.001	2.650 ±.002	r	406
8.559 ±.004				198.89 ± .18	59.89 ± .05	1.431 ±.001	2.641 ±.002	r	407
8.7032 ±.0030				200.66 ± .17	60.424 ± .051	1.4442 ±.0031	2.618 ±.002	r	408
13.738 ±.003				2348.23 ± 1.19	88.389 ± .045	2.1126 ±.0011	2.469 ±.001	25	409
				2422.3 ±27.1	91.18 ± 1.02	2.179 ±.024	2.394 ±.027	625	410
13.970 ±.003				2435.98 ± .91	91.692 ± .034	2.1915 ±.0009	2.695 ±.001	25	411
7.62 ±.01		105.90 ±.20		426.41 ± 1.57	128.4 ± .5	3.069 ±.011	2.385 ±.009	r	412
7.514 ±.004				1093.6 ± 1.6	329.3 ± .5	7.871 ±.011	2.566 ±.004	r	413
7.652 ±.015				1134.07 ± 2.68	341.5 ± .8	8.162 ±.019	2.737 ±.007	r	414
				Sheet structure silicates					
20.030 ±.010		94.47 ±.33		934.57 ± 1.21	140.71 ± .18	3.363 ±.004	2.831 ±.004	r	415
19.32 ±.10		95.17 ±.50		877.52 ± 8.47	132.1 ± 1.3	3.158 ±.031	2.893 ±.028	r	416
20.0 ±.2		98.00 ±.50		965.7 ±23.2	290.8 ± 7.0	6.950 ±.167	2.698 ±.065	r	417

	Name and formula	Crystal system	Space group	Structure type	Z	a_o	b_o
	Sheet structure silicates—Continued						
418	Phlogopite $KMg_3[AlSi_3O_{10}](OH)_2$	mon.	Cm(8)	1M mica	2	5.326 ±.010	9.210 ±.010
419	Fluor-phlogopite $KMg_3[AlSi_3O_{10}]F_2$	mon.	Cm(8)	1M mica	2	5.299 ±.005	9.188 ±.005
420	Annite $KFe_3[AlSi_3O_{10}](OH)_2$	mon.	Cm(8)	1M mica	2	5.391 ±.010	9.350 ±.005
421	Ferriannite $KFe_3[FeSi_3O_{10}](OH)_2$	mon.	C2/m(12)		2	5.430 ±.002	9.404 ±.003
422	Margarite $CaAl_2[Al_2Si_2O_{10}](OH)_2$*	mon.	C2/c(15)	2M mica	4	5.13 ±.02	8.92 ±.03
423	Talc $Mg_3Si_4O_{10}(OH)_2$*	mon.	C2/c(15)	$2M_1$	4	5.287 ±.007	9.158 ±.010
424	Pyrophyllite $Al_2Si_4O_{10}(OH)_2$*	mon.	C2/c(15)	$2M_1$	4	5.14 ±.02	8.90 ±.02
425	Minnesotaite $Fe_3Si_4O_{10}(OH)_2$*	mon.	C2/c(15)		4	5.4 ±.1	9.42 ±.04
426	Dickite $Al_2Si_2O_5(OH)_4$*	mon.	Cc(9)		4	5.150 ±.002	8.940 ±.003
427	Kaolinite $Al_2Si_2O_5(OH)_4$*	tri.	P1(1)		2	5.155 ±.007	8.959 ±.010
428	Nacrite $Al_2Si_2O_5(OH)_4$*	mon.	Cc(9)		4	8.909 ±.010	5.146 ±.010
	Zeolites						
429	Analcite $NaAlSi_2O_6.H_2O$	cubic	Ia3d(230)		16	13.733 ±.005	
430	Natrolite $Na_2Al_2Si_3O_{10}.2H_2O$*	orth.	Fdd2(43)		8	18.30 ±.02	18.63 ±.02

HEAT CAPACITY OF ROCK FORMING MINERALS

The units of heat capacity at constant pressure, C_p, in this table are cal $g^{-1} deg^{-1}$. Values of these units are given for several temperatures with the temperatures being in degrees C.

Heat capacity at other temperatures may be calculated by use of the equation $C_p = a + bT - cT^{-5}$. The units of these constants are cal $g^{-1} deg^{-1}$.

	Heat capacity at various temperatures						Constants for heat capacity equation		
Mineral	**-200**	**0**	**200**	**400**	**800**	**1200**	**a**	**10^3b**	**10^5c**
Albite	–	0.1695	0.236	0.26	0.286	–	1.018	0.187	0.268
Amphibole	–	0.177	0.246	0.27	0.296	–	1.067	1.183	0.281
Apatite	–	0.24	–	–	–	–	–	–	–
Arsenopyrite	–	0.103 at 55°C	–	–	–	–	–	–	–
Asbestos	–	0.195	–	–	–	–	–	–	–
Barite	0.047	0.1076	0.12	0.1315	0.1555	–	0.383	0.253	0
Cassiterite	–	0.0814	0.103	0.115	0.132	–	0.387	0.157	0.007
Chalcopyrite	–	0.129 at 50°C	–	–	–	–	–	–	–
Diamond	–	0.104	0.253	0.328	0.445	–	0.754	1.067	0.454
Dolomite	–	0.222 at 60°C	–	–	–	–	–	–	–
Fluorite	0.0525	0.203	0.213	0.222	0.24	0.263	0.798	0.204	0
Galena	0.034	0.0496	0.0528	0.0562	–	–	0.188	0.007	0
Garnet	–	0.177 at 58°C	–	–	–	–	–	–	–
β-Graphite	–	0.152	0.282	0.348	0.45	–	0.932	0.913	0.4077
Hematite	–	0.146	0.189	0.215	0.258	–	0.640	0.420	0.111
Ice	0.156	0.492	–	–	–	–	–	–	–
Kaolinite	–	0.222	0.244	–	–	–	0.806	0.463	0.0

c_o	a_o	β_o	γ_o	Cell volume 10^{-24} cm³	Molar volume cm³	Molar volume cal bar⁻¹	X-Ray density grams cm⁻³	Temp. °C	
Sheet structure silicates—Continued									
10.311 ±.010		100.17 ±.10		497.83 ± 1.19	149.91 ± .36	3.5830 ±.0086	2.784 ±.007	r	418
10.135 ±.005		99.92 ±.10		486.07 ± .60	146.37 ± .18	3.498 ±.004	2.878 ±.004	r	419
10.313 ±.020		99.70 ±.25		512.40 ± 1.45	154.30 ± .44	3.688 ±.010	3.318 ±.009	r	420
10.341 ±.006		100.07 ±.20		519.92 ± .51	156.56 ± .15	3.7419 ±.0037	3.454 ±.003	r	421
19.50 ±.05		95.00 ±.50		888.9 ± 5.2	133.8 ± .8	3.199 ±.019	2.975 ±.017	r	422
18.95 ±.01		99.50 ±.20		904.94 ± 1.71	136.25 ± .26	3.2565 ±.0062	2.784 ±.005	r	423
18.55 ±.03		99.92 ±.20		835.9 ± 4.0	125.9 ± .6	3.008 ±.015	2.863 ±.014	r	424
19.4 ±.1		100.00 ±.50		971.8 ±19.2	146.3 ± 2.9	3.497 ±.069	3.239 ±.064	r	425
14.736 ±.005		103.58 ±.10		659.49 ± .49	99.30 ± .07	2.3733 ±.0018	2.600 ±.002	r	426
7.407 ±.008	91.68 ±.35	104.87 ±.35	89.93 ±.35	330.48 ± .86	99.52 ± .26	2.3785 ±.0062	2.594 ±.007	r	427
15.697 ±.020		113.70 ±.25		658.9 ± 2.1	99.21 ± .32	2.3713 ±.0076	2.602 ±.008	r	428
Zeolites									
				2589.98 ± 2.83	97.49 ± .11	2.3301 ±.0026	2.258 ±.003	r	429
6.60 ±.01				2250.1 ± 4.8	169.39 ± .37	4.049 ±.009	2.245 ±.005	r	430

HEAT CAPACITY OF ROCK FORMING MINERALS (continued)

Mineral	Heat capacity at various temperatures: −200	0	200	400	800	1200	Constants for heat capacity equation: a	10^3 b	10^5 c
Labradorite	–	0.196 at 60°C	–	–	–	–	–	–	–
Magnesite	0.0385	0.207							
Magnetite	–	0.1435	0.1985	0.222		–	0.744	0.340	0.177
Mica (mono-crystal)	–	0.208							
Microcline	–	0.163	0.227	0.248	0.342	–	0.988	0.166	0.263
Oligoclase	–	0.2035 at 60°C	–	–	–	–	–	–	–
Olivine	–	0.189 at 36°C	–	–	–	–	–	–	–
Orthoclase	–	0.146	0.226	0.251	0.347	–	0.043	0.124	0.351
Pyrite	0.0179	0.1195	0.142	0.165		–	0.373	0.466	0
Pyroxene	–	0.18	0.246	0.275	–	–	0.973	0.336	0.233
α-Quartz	0.0414	0.167	0.232	0.2695		–	0.7574	0.607	0.168
β-Quartz	–	–	–	–	0.28	0.3165	0.763	0.383	0
Serpentine		0.227	–	–	–	–	–	–	–
Siderite	0.056	0.163	–	–	–	–	–	–	–
Talc	–	0.208 at 59°C	–	–	–	–	–	–	–
Zircon	–	0.146 at 60°C	–	–	–	–	–	–	–

SOLUBILITY PRODUCT

The solubility product (or ion product constant) is the product of the concentrations of the ions in the saturated solution of a difficultly soluble salt. The concentrations are expressed as moles per liter of solution. The number of cations (or anions) resulting from the dissociation of one molecule of the salt, appears in the formula for calculations of the solubility product as the exponent of the concentration of the cation (or anion).

If two solutions, each containing one of the ions of a difficultly soluble salt, are mixed, no precipitation takes place unless the product of the ion concentrations in the mixture is greater than the solubility product.

In a solution containing two salts which yield a common ion the ratio of solubilities of the two salts is the ratio of the solubility products.

Substance	Solubility product at temperature noted
Aluminum hydroxide	4×10^{-13} (15°)
Aluminum hydroxide	1.1×10^{-15} (18°)
Aluminum hydroxide	3.7×10^{-15} (25°)
Barium carbonate	7×10^{-9} (16°)
Barium carbonate	8.1×10^{-9} (25°)
Barium chromate	1.6×10^{-10} (18°)
Barium chromate	2.4×10^{-10} (28°)
Barium fluoride	1.6×10^{-6} (9.5°)
Barium fluoride	1.7×10^{-6} (18°)
Barium fluoride	1.73×10^{-6} (25.8°)
Barium iodate, $Ba(IO_3)..2H_2O$	8.4×10^{-11} (10°)
Barium iodate, $Ba(IO_3)..2H_2O$	6.5×10^{-10} (25°)
Barium oxalate, $BaC_2O_4.3\frac{1}{2}H_2O$	1.62×10^{-7} (18°)
Barium oxalate, $BaC_2O_4.2H_2O$	1.2×10^{-7} (18°)
Barium oxalate, $BaC_2O_4.\frac{1}{2}H_2O$	2.18×10^{-7} (18°)
Barium sulfate	0.87×10^{-10} (18°)
Barium sulfate	1.08×10^{-10} (25°)
Barium sulfate	1.98×10^{-10} (50°)
Cadmium oxalate $CdC_2O_4.3H_2O$	1.53×10^{-8} (18°)
Cadmium sulfide	3.6×10^{-29} (18°)
Calcium carbonate (calcite)	0.99×10^{-8} (15°)
Calcium carbonate (calcite)	0.87×10^{-8} (25°)
Calcium fluoride	3.4×10^{-11} (18°)
Calcium fluoride	3.95×10^{-11} (26°)
Calcium iodate, $Ca(IO_3)_2. 6H_2O$	22.2×10^{-8} (10°)
Calcium iodate, $Ca(IO_3)_2.6H_2O$	64.4×10^{-8} (18°)
Calcium oxalate, $CaC_2O_4.H_2O$	1.78×10^{-9} (18°)
Calcium oxalate, $CaC_2O_4.H_2O$	2.57×10^{-9} (25°)
Calcium sulfate	2.45×10^{-5} (25°C)
Calcium tartrate, $CaC_4H_4O_6.2H_2O$	0.77×10^{-6} (18°)
Cobalt sulfide	3×10^{-26} (18°)
Cupric iodate	1.4×10^{-7} (25°)
Cupric oxalate	2.87×10^{-8} (25°)
Cupric sulfide	8.5×10^{-45} (18°)
Cuprous bromide	4.15×10^{-8} (18–20°)
Cuprous chloride	1.02×10^{-6} (18–20°)
Cuprous iodide	5.06×10^{-12} (18–20°)
Cuprous sulfide	2×10^{-47} (16–18°)
Cuprous thiocyanate	1.6×10^{-11} (18°)
Ferric hydroxide	1.1×10^{-36} (18°)
Ferrous hydroxide	1.64×10^{-14} (18°)
Ferrous oxalate	2.1×10^{-7} (25°)
Ferrous sulfide	3.7×10^{-19} (18°)
Lead carbonate	3.3×10^{-14} (18°)
Lead chromate	1.77×10^{-14} (18°)
Lead fluoride	2.7×10^{-8} (9°)
Lead fluoride	3.2×10^{-8} (18°)
Lead fluoride	3.7×10^{-8} (26.6°)
Lead iodate	5.3×10^{-14} (9.2°)
Lead iodate	1.2×10^{-13} (18°)
Lead iodate	2.6×10^{-13} (25.8°)
Lead iodide	7.47×10^{-9} (15°)
Lead iodide	1.39×10^{-8} (25°)
Lead oxalate	2.74×10^{-11} (18°)
Lead sulfate	1.06×10^{-8} (18°)
Lead sulfide	3.4×10^{-28} (18°)
Lithium carbonate	1.7×10^{-3} (25°)
Magnesium ammonium phosphate	2.5×10^{-13} (25°)
Magnesium carbonate	2.6×10^{-5} (12°)
Magnesium fluoride	7.1×10^{-9} (18°)
Magnesium fluoride	6.4×10^{-9} (27°)
Magnesium hydroxide	1.2×10^{-11} (18°)
Magnesium oxalate	8.57×10^{-5} (18°)
Manganese hydroxide	4×10^{-14} (18°)
Manganese sulfide	1.4×10^{-15} (18°)
Mercuric sulfide	4×10^{-53} to 2×10^{-49} (18°)
Mercurous bromide	1.3×10^{-21} (25°)
Mercurous chloride	2×10^{-18} (25°)
Mercurous iodide	1.2×10^{-28} (25°)
Nickel sulfide	1.4×10^{-24} (18°)
Potassium acid tartrate $[K^+][HC_4H_4O_6^-]$	3.8×10^{-4} (18°)
Silver bromate	3.97×10^{-5} (20°)
Silver bromate	5.77×10^{-5} (25°)
Silver bromide	4.1×10^{-13} (18°)
Silver bromide	7.7×10^{-13} (25°)
Silver carbonate	6.15×10^{-12} (25°)
Silver chloride	0.21×10^{-10} (4.7°)
Silver chloride	0.37×10^{-10} (9.7°)
Silver chloride	1.56×10^{-10} (25°)
Silver chloride	13.2×10^{-10} (50°)
Silver chloride	215×10^{-10} (100°)
Silver chromate	1.2×10^{-12} (14.8°)
Silver chromate	9×10^{-12} (25°)
Silver cyanide $[Ag^+][Ag(CN)^-_2]$	2.2×10^{-12} (20°)
Silver dichromate	2×10^{-7} (25°)
Silver hydroxide	1.52×10^{-8} (20°)
Silver iodate	0.92×10^{-8} (9.4°)
Silver iodide	0.32×10^{-16} (13°)
Silver iodide	1.5×10^{-16} (25°)
Silver sulfide	1.6×10^{-49} (18°)
Silver thiocyanate	0.49×10^{-12} (18°)
Silver thiocyanate	1.16×10^{-12} (25°)
Strontium carbonate	1.6×10^{-9} (25°)
Strontium fluoride	2.8×10^{-9} (18°)
Strontium oxalate	5.61×10^{-8} (18°)
Strontium sulfate	2.77×10^{-7} (2.9°)
Strontium sulfate	3.81×10^{-7} (17.4°)
Zinc hydroxide	1.8×10^{-14} (18–20°)
Zinc oxalate, $ZnC_2O_4.2H_2O$	1.35×10^{-9} (18°)
Zinc sulfide	1.2×10^{-23} (18°)

PROPERTIES OF RARE EARTH METALS

F. H. Spedding

Symbol	M. P. °C[1]	B. P. °C[1]	Heat of Vaporization $\Delta H_{v,o}$ Kcal/g atm[2]	Density	Atomic Vol. (cm³/mole)	Metallic Radius Å	Electrical Resistivity Polycrystalline Wire 298°K (ohm-cm x 10^{-6})[3]	Residual Resistivity Wire 4.2°K (ohm-cm x 10^{-6})[3]	Compressibility cm²/kg x 10^{-6} **
Sc	1541	2831	89.9	2.989	15.041	1.640	52	3	2.26
Y	1522	3338	101.3	4.469	19.894	1.801	59	2	2.68
La	921	3457	103.1	6.145	22.603	1.879	61-80*	SC	4.04
Ce	799	3426	101.1	γ=6.767	20.400	1.820	70-80	10	4.10
				β=6.657	21.049				
Pr	931	3512	85.3	6.773	20.805	1.828	68	1	3.21
Nd	1021	3068	78.5	7.007	20.585	1.821	65	7	3.0
Pm	1168	2700 est.							
Sm	1077	1791	49.2	7.520	20.001	1.804	91	7	3.34
Eu	822	1597	41.9	5.243	28.981	1.984	91	1	8.29
			$\Delta H°298$						
Gd	1313	3266	95.3	7.900	19.904	1.801	127	1	2.56
Tb	1356	3123	93.4	8.229	19.312	1.783	114	4	2.45
Dy	1412	2562	70.0	8.550	19.006	1.774	100	5	2.55
Ho	1474	2695	72.3	8.795	18.753	1.766	88	3	2.47
Er	1529	2863	76.1	9.066	18.450	1.757	71	3	2.39
Tm	1545	1947	55.8	9.321	18.124	1.746	74	3	2.47
Yb	819	1194	36.5	6.965	24.843	1.939	28	2	7.39
Lu	1663	3395	102.2	9.840	17.781	1.735	60	2	2.38

* Crystal usually mixture of α hcp and fcc lattice. SC - Superconductor

** Best values in author's opinion. Many numbers are taken from P. W. Bridgman's publication.

[1] Corrected for new temperature scale. Best values Ames Laboratory 2/1/72.

[2] Values from Thermodynamic Properties of Metals and Alloys (Review) R. Hultgren, R. Orr and K. Kelley. Supplements 1967.

[3] Weighted average of original papers 1/1/72.

PROPERTIES OF RARE EARTH METALS

F. H. Spedding

Symbol	Crystal Structure at Room Temperature 25°C		Allotropic Forms Transition Temperatures Expressed in °C	
Sc	Hex. (to 1335°)	a=3.3088Å, c=5.2680Å	bcc(above 1335°)	
Y	Hex. (to 1478°)	a=3.6482Å, c=5.7318Å	bcc(above 1478°)	a=4.08Å
La	Hex. (to 310°) (usually contains fcc also)	a=3.7740Å, c=12.171Å	fcc(310°-865°) bcc(above 865°)	 a=4.26Å
Ce	fcc(~ 0° to 726°)	a=5.160Å	fcc(below-157° on cooling) (up to -94° on heating) Hex. (below-23° on cooling) (up to 168° on heating)	a=4.85Å a=3.68Å c=11.92Å
Pr	Hex. (to 795°)	a=3.6721Å, c=11.832Å	bcc(above 795°)	a=4.12Å
Nd	Hex. (to 863°)	a=3.6583Å, c=11.7966Å	bcc(above 863°)	a=4.13Å
Sm	*Rhom (to 926°)	a=8.9834Å, α=23° 49.5′	bcc(above 926°)	a=4.07Å
Eu	bcc	a=4.5827Å		
Gd	Hex. (to 1235°)	a=3.6336Å, c=5.7810Å	bcc(above 1235°)	a=4.05Å†
Tb	Hex. (to 1289°)	a=3.6055Å, c=5.6966Å	bcc(above 1289°)	a=4.02Å †
Dy	Hex. (to 1381°)	a=3.5915Å, c=5.6501Å	bcc(above 1381°)	a=3.98Å †
Ho	Hex.	a=3.5778Å, c=5.6178Å	Not present pure metals	
Er	Hex.	a=3.5592Å, c=5.5850Å	Not present pure metals	
Tm	Hex.	a=3.5375Å, c=5.5540Å	Not present pure metals	
Yb	Hex. (to 795°)	a=5.4848Å,	bcc(above 795°)	a=4.44Å
Lu	Hex.	a=3.5052Å, c=5.5494Å	Not present pure metals	

† Extrapolated from magnesium alloy studies. Hex. refers to close packed hexagonal. La, Ce, Pr, and Nd have a stacking order ABAC, the other rare earths ABAB. fcc refers to close packed face centered cubic with a stacking order ABC, ABC, bcc refers to body centered cubic. These high temperature forms are very soft and deform very easily.

* The rhombic Sm cell can be expressed as hexagonal with a=3.6290Å, c=26.207 and possess a stacking order ABABCBCAC

DENSITY OF LIQUID ELEMENTS

Gernot Lang

Data in this table consist of those published in various places in the literature. Generally, there has been excellent agreement among values obtained by different methods of measurement. Temperatures are in degrees C and densities in grams per cubic centimeter. Grams per cc × 62.427961 = pounds per cu ft.

Element:	Purity:	ρ_{mp}	ρ_{t_1}		ρ_{t_2}		ρ_{t_3}		Atm.:	Method:	Ref.
			t	ρ	t	ρ	t	ρ			
Ag	—	9.32	1100	9.17	1200	9.07	—	—	—	IBF	74
Ag	—	9.33	1100	9.20	1200	9.10	1300	9.00	—	DBF	33
Ag	99.9	—	1100	9.170	1200	9.070	1300	8.980	—	IBF	Q
Ag	—	9.285	1100	9.15	1200	9.07	1300	8.97	—	IBF	24
Ag	—	9.345	1100	9.20	—	—	—	—	H_2	Bubble pressure	46
Ag	—	9.33 ±0.01	1100	9.18	1200	9.08	1300	8.97	Ar	Bubble pressure	47
Ag	—	9.346	1100	9.22	1200	9.13	1300	9.04	—	DBF	39
Al	99.4	2.41	—	—	—	—	—	—	—	BF	62
Al	99.4	2.38_4	—	—	—	—	—	—	—	Pycnometer	15
Al	99.996	2.368	700	2.357	800	2.332	900	2.304	—	BF	25
Al	99.99	2.39	—	—	—	—	—	—	—	Pycnometer	60
Al	99.997	2.39	677	2.380	807	2.334	912	2.292	Ar	Bubble pressure	11
Au	99.96	17.28	1100	17.24	1200	17.12	1300	16.99	H_2	IBF	24
Au	—	17.361	1100	17.221	1200	17.099	1300	16.950	—	IBF	P
B	99.8	2.08 ±0.03	—	—	—	—	—	—	vac.	Weighing and vol. determin.	80
Ba	—	3.325	$\rho_t = 3.476 - (2.14 \cdot 10^{-4})t$ (t°C)						Ar	DBF	1
Ba	—	3.320	$\rho_t = 3.847 - 5.26 \cdot 10^{-4}T$ (T°K)						—	calculated	K
Be	—	—	1500	1.42 ±0.04	—	—	—	—	—	calculated	22
Bi	—	10.07	300	10.03	400	9.91	500	9.78	—	Manometer pressure	31
Bi	—	10.02	400	9.87	600	9.62	800	9.40	—	Dilatometer	7
Bi	—	10.07	400	9.90	500	9.78	600	9.66	—	Dilatometer	53
Bi	—	10.04	400	9.91	600	9.67	800	9.44	—	DBF	34
Bi	—	—	400	9.86	600	9.62	700	9.49	N_2	Bubble pressure	78
Bi	99.94	10.03	400	9.88	600	9.62	700	9.51	—	DBF	66
Bi	99.90	—	800	9.43	900	9.28	1000	9.15	H_2	Bubble pressure	57
Bi	—	10.02	300	9.98	400	9.85	500	9.73	—	Dilatometer	30
Bi	99.98	10.07	500	9.78	700	9.53	900	9.30	Ar	Bubble pressure	49
Bi	—	10.057	700	9.51 ±0.03	—	—	—	—	—	—	L
Ca	—	1.365	$\rho_t = 1.613 - 2.21 \cdot 10^{-4}T$ (T°K)						—	calculated	K
Cd	—	8.02	$\rho_t = 8.02 - 0.00110\,(t - 320)$ (t°C)						—	Manometer pressure	31
Cd	99.97	—	340	8.009	400	7.942	500	7.821	N_2	Bubble pressure	28
Cd	—	—	385	7.94	479	7.83	560	7.74	Ar	Bubble pressure	76
Co	—	—	1600	8.08 – 8.13	—	—	—	—	—	Drop volume	59
Co	—	—	1500	ca. 8.05	—	—	—	—	—	BF	26
Co	99.99	7.67	1500	7.66	1600	7.54	1700	7.42	Ar	Bubble pressure	50
Co	—	—	1500	7.70	—	—	—	—	—	—	81
Co	99.53	7.76	$\rho_t = 9.51 - 9.88 \cdot 10^{-4}T$ (T°K)						vac.	Sessile drop (mp – 2200)	U
Cr	—	—	1950	6.00 ±0.13	—	—	—	—	vac.	Sessile drop	18
Cr	—	6.46	—	—	—	—	—	—	—	calculated	3
Cs	—	1.843	110	1.800	310	1.691	510	1.575	—	Pycnometer	H
Cs	—	1.851	$\rho_t = 1.853 - 5.71 \cdot 10^{-4}t$ (t°C)						—	Pycnometer (100–750°C)	F
Cu	—	7.99	1100	7.97	1200	7.81	1300	7.66	—	IBF	6
Cu	—	7.96_2	1100	7.94	1200	7.77	1300	7.62	—	Dilatometer	85
Cu	—	7.92_4	1100	7.91	1300	7.76	1500	7.61	—	DBF	84
Cu	99.99	7.940	1100	7.924	1200	7.846	1300	7.764	—	IBF	Q
Cu	—	—	1100	8.10	—	—	—	—	—	Drop volume	43
Cu	—	7.87_5	1100	7.86	1300	7.70	1500	7.55	—	DBF	45
Cu	—	—	1100	7.90	—	—	—	—	—	Pycnometer	52
Cu	—	7.99_2	1100	7.98	1300	7.81	1500	7.66	Ar	DBF	8
Cu	—	—	1100	8.07	—	—	—	—	—	Bubble pressure	Beer in 49
Cu	—	8.090	$\rho_t = 9.370 - 9.442 \cdot 10^{-4}T$ (T°K)						—	Drop volume	Mehairy in 49
Cu	—	8.03	1100	8.02	1300	7.86	1500	7.70	Ar	Bubble pressure	49
Fe	—	7.13	—	—	—	—	—	—	—	DBF	75
Fe	—	7.24	—	—	—	—	—	—	vac.	Drop volume	35
Fe	—	—	1550	7.01	—	—	—	—	air	Casting	79
Fe	—	7.15	—	—	—	—	—	—	—	DBF	42
Fe	—	6.99	$\rho_t = 8.523 - 8.358 \cdot 10^{-4}T$ (T°K)						Ar	DBF	38
Fe	Carbonyl	—	1550	7.189	—	—	—	—	—	—	27
Fe	—	7.015	$\rho_t = 8.618 - 8.83 \cdot 10^{-4}T$ (T°K)						Ar	DBF	C
Fe	—	—	1555	6.98	—	—	—	—	—	Bubble pressure	Beer in 49
Fe	99.96	7.03_5	1550	7.01	1600	6.94	1700	6.80	Ar	Bubble pressure	49
Fe	—	—	1550	7.13	—	—	—	—	—	—	70
Fe	99.9	7.02	$\rho_t = 8.50 - 8.17 \cdot 10^{-4}T$ (T°K)						vac.	Sessile drop	U

DENSITY OF LIQUID ELEMENTS (Continued)

Element:	Purity:	ρ_{mp}	ρ_{t_1}		ρ_{t_2}		ρ_{t_3}		Atm.:	Method:	Ref.
			t	ρ	t	ρ	t	ρ			
Ga	—	6.20 ±0.01	—	—	—	—	—	—	—	—	E
Ge	—	5.52	1000	5.50	1100	5.44	1200	5.39	—	Volumetric measurement	58
Ge	—	5.57_5	1000	5.56	1100	5.51	1200	5.45	N_2	Pycnometer	40
Ge	99.990	5.49	1000	5.46	1200	5.36	1500	5.21	Ar	Bubble pressure	49
Hf	—	12.0	—	—	—	—	—	—	—	calculated	3
Hg	—	—	0	13.5951	20	13.5457	100	13.3514	—	—	—
In	—	—	231	6.99	302	6.93	421	6.84	Ar	Bubble pressure	76
Ir	—	20.0	—	—	—	—	—	—	—	Calculated	3
K	—	0.826	$\rho_t = 0.826 - 0.000222\,(t - 62.4)$ (t°C)						—	—	72
K	—	0.819	$\rho_t = 0.819 - 0.238 \cdot 10^{-3}\,(t - 64)$ (t°C)						—	Pycnometer (64–1400°C)	D
K	—	0.828	110	0.817	310	0.772	510	0.724	—	Pycnometer	H
Li	—	0.515	$\rho_t = 0.515 - 0.101 \cdot 10^{-3}\,(t - 200)$ (t°C)						—	Pycnometer (200–1600°C)	D
Li	—	0.518	$\rho_t = 0.5368 - 1.021 \cdot 10^{-4}\,t$ (t°C)						—	Pycnometer (400–1125°C)	F
Mg	—	—	700	1.575	800	1.555	—	—	—	DBF	67
Mg	99.5	1.585	700	1.570	800	1.550	900	1.525	—	IBF	N
Mg	—	1.57	—	—	—	—	—	—	—	Estimated	17
Mg	—	1.590	$\rho_t = 1.834 - 2.67 \cdot 10^{-4}\,T$ (T°K)						Ar	DBF	B
Mn	—	—	1440	5.84 ±2 %	—	—	—	—	He	Volumetric method	M
Mn	—	6.43	—	—	—	—	—	—	—	—	71
Mo	99.7	9.33	—	—	—	—	—	—	vac.	Drop volume	64
Mo	—	9.35	—	—	—	—	—	—	—	calculated	3
Na	—	0.938_5	$\rho_t = 0.938_5 - 0.000260\,(t - 96.5)$ (t°C)						—	—	72
Na	—	—	$\rho_t = 0.938 - 1.9 \cdot 10^{-4}\,T$ (T°K)						—	Bubble pressure	82
Na	—	0.927	$\rho_t = 0.927 - 0.238 \cdot 10^{-3}\,(t - 100)$ (t°C)						—	Pycnometer	D
Na	—	0.928	410	0.854	610	0.806	810	0.754	—	Pycnometer	H
Nb, Cb	—	7.83	—	—	—	—	—	—	—	calculated	3
Ni	—	—	1500	ca. 8.04	—	—	—	—	—	BF	26
Ni	—	7.905	$\rho_t = 9.908 - 11.598 \cdot 10^{-4}\,T$ (T°K)						Ar	DBF	C
Ni	99.85	7.77	1500	7.71_5	1600	7.59_5	1700	7.48	Ar	Bubble pressure	49
Ni	99.99	7.78	$\rho_t = 7.78 - 0.006\,(t - 1453)$ (t°C)						vac.	Sessile drop	21
Ni	—	—	1500	7.78	—	—	—	—	—	—	81
Ni	99.95	7.91	$\rho_t = 9.51 - 10.00 \cdot 10^{-4}\,T$ (T°K)						vac.	Sessile drop	U
Os	—	20.1	—	—	—	—	—	—	—	calculated	3
Pb	—	10.71	$\rho_t = 10.71 - 0.00139\,(t - 327)$ (t°C)						—	Manometer pressure	31
Pb	99.98	—	340	10.57	400	10.49	440	10.43	N_2	Bubble pressure	28
Pb	—	—	568	10.380	640	10.294	720	10.201	—	DBF	O
Pb	—	—	400	10.56	500	10.43	700	10.17	N_2	Bubble pressure	78
Pb	—	—	437	10.52	623	10.28	705	10.16	Ar	Bubble pressure	76
Pb	99.9923	10.678	$\rho_t = 10.678 - 13.174 \cdot 10^{-4}\,T$ (T°K)						Ar	DBF	A
Pd	—	10.7	—	—	—	—	—	—	—	calculated	20
Pd	—	10.7	—	—	—	—	—	—	—	calculated	3
Pd	99.95	—	1600	10.43	1700	10.31	1800	10.18_5	Ar	Bubble pressure	49
Pt	99.84	19.7 ±0.25	—	—	—	—	—	—	vac.	Drop volume	19
Pt	99.999	—	1800	18.82 ±0.02	—	—	—	—	Ar	Sessile drop	44
Pt	—	18.91	—	—	—	—	—	—	Ar	Bubble pressure	48
Pt	—	—	1800	18.82	1850	18.67_5	1875	18.60_5	Ar	Bubble pressure	49
Pu	—	—	$\rho_t = 17.57 - 1.45 \cdot 10^{-3}\,t$ (t°C) ±0.21						—	(655–P60°C)	T
Pu	99.95	16.623	699	16.548	824	16.370	950	16.185	—	Pycnometer	32
Rb	—	1.48	—	—	—	—	—	—	—	estimated	56
Rb	—	1.463	$\rho_t = 1.481 - 4.51 \cdot 10^{-4}\,t$ (t°C)						—	Pycnometer	F
Rb	—	—	45.5	1.4505	47.4	1.4498	49.3	1.4469	vac.	Pycnometer	G
Re	99.4	18.9	—	—	—	—	—	—	vac.	Drop volume	64
Re	—	18.7	—	—	—	—	—	—	—	calculated	3
Rh	—	10.65	—	—	—	—	—	—	—	calculated	20
Rh	—	11.1	—	—	—	—	—	—	—	calculated	3
Ru	—	10.9	—	—	—	—	—	—	—	calculated	3
S	—	1.819	120–160°C: $\rho_t = 1.901 - 8.00 \cdot 10^{-4}\,t$ (t°C)						N_2	Bubble pressure	61
Sb	—	6.49	700	6.45	800	6.39	1000	6.27	—	IBF	7
Sb	99.52	—	650	6.530	700	6.509	800	6.424	N_2	Bubble pressure	28
Sb	—	—	700	6.41	800	6.38	900	6.35	H_2	Bubble pressure	78

Element:	Purity:	ρ_{mp}	ρ_{t_1} t	ρ_{t_1} ρ	ρ_{t_2} t	ρ_{t_2} ρ	ρ_{t_3} t	ρ_{t_3} ρ	Atm.:	Method:	Ref.
Sb	—	—	650	6.52	700	6.50	—	—	H_2	Bubble pressure	23
Sb	—	—	728	6.44	827	6.38	917	6.32	Ar	Bubble pressure	76
Sb	99.9	6.50	650	6.49	700	6.45	—	—	—	DBF	66
Sb	99.992	6.483	$\rho_t = 6.596 + 2.022 \cdot 10^{-4}\,T - 3.629 \cdot 10^{-7}\,T^2$ (T°K)							DBF, Bubble pressure	37
Sb	99.6	6.46_5	700	6.42	900	6.30	1200	6.13	Ar	Bubble pressure	49
Se	—	3.987	$\rho_t = 3.987 - 16 \cdot 10^{-4}\,(T - 493)$ (T°K)						—	—	13
Se	—	3.985	$\rho_t = 3.985 - 15.5 \cdot 10^{-4}\,(T - 490)$ (T°K)						—	Pycnometer	10
Se	—	4.06	$\rho_t = 4.06 - 5 \cdot 10^{-4}\,(T - 491)$ (T°K)						—	—	4
Se	—	3.984	—	—	—	—	—	—	—	—	1
Se	—	4.01	250	3.97	300	3.91	400	3.79	—	BF	51
Si	—	—	1550	2.54	—	—	—	—	—	—	14
Si	99.9	2.52_5	1450	2.51	1500	2.49	1600	2.46	Ar	Bubble pressure	49
Sn	—	6.988	300	6.94	500	6.81	800	6.64	—	DBF	12
Sn	—	6.98	—						—	DBF	63
Sn	—	7.01	$\rho_t = 7.01 - 0.00074\,(t - 232)$ (t°C)						—	Manometer pressure	31
Sn	—	6.97	300	6.92	500	6.78	800	6.57	—	IBF	74
Sn	—	6.896	300	6.93	500	6.78	600	6.71	—	Dilatometer	53
Sn	—	6.966	300	6.92	600	6.72	1200	6.34	—	DBF	84
Sn	—	6.983	300	6.93	400	6.86	500	6.79	—	Pycnometer	73
Sn	—	7.00	300	6.95	500	6.80	600	6.74	N_2	Bubble pressure	65
Sn	—		625	6.67	—	—	—	—	N_2	Bubble pressure	78
Sn	—	6.99	300	6.94	400	6.87	—	—	H_2	Bubble pressure	23
Sn	—	6.993	$\rho_t = 6.98 - 0.00074\,(t - 250)$ (t°C)						—	Manometer pressure	69
Sn	—	6.968 ±0.005	300	6.93	400	6.87	500	6.79	—	Dilatometer	30
Sn	—	6.968	300	6.96	400	6.84	500	6.77	Ar	Bubble pressure	83
Sn	—	6.978 ±0.022	300	6.93	600	6.70	1200	6.29	Ar	Bubble pressure	49
Ta	—	15.0	—	—	—	—	—	—	—	calculated	3
Te	—	5.75	460	5.75	500	5.74	600	5.70	—	Pycnometer	40
Te	—	—	460	5.58	600	5.53	—	—	N_2	Bubble pressure	77
Te	99.9999	5.797	460	5.79	600	5.72	700	5.66	Ar	Bubble pressure	49
Ti	98.7	4.11 ±0.08	—	—	—	—	—	—	—	Capillary method	16
Ti	—	4.15	—	—	—	—	—	—	—	calculated	3
Tl	—	11.29	412	11.13	504	11.01	651	10.77	Ar	Bubble pressure	76
U	—	17.907	$\rho_t = 19.356 - 10.328 \cdot 10^{-4}\,T$ (T°K)						Ar	DBF	29
V	—	5.55	—	—	—	—	—	—	—	calculated	3
W	—	17.6	—	—	—	—	—	—	—	estimated	9
W	99.8	17.7	—	—	—	—	—	—	vac.	Drop volume	64
W	—	17.5	—	—	—	—	—	—	—	calculated	3
Zn	—	6.59	$\rho_t = 6.59 - 0.00097\,(t - 419)$ (t°C)						—	Manometer pressure	31
Zn	—		600	6.35	—	—	—	—	—	IBF	7
Zn	—	6.56_2	500	6.47	600	6.37	700	6.29	—	Dilatometer	53
Zn	—	6.55_1	500	6.47	600	6.38	700	6.29	—	Pycnometer	73
Zn	—	6.64_5	500	6.56	600	6.45	800	6.25	—	DBF	67
Zn	—	6.64_4	500	6.58	600	6.49	800	6.22	—	DBF	68
Zn	—	6.64	500	6.57	600	6.50	700	6.43	N_2	Bubble pressure	65
Zn	—		500	6.55	—	—	—	—	H_2	Bubble pressure	78
Zn	—	6.64_5	—	—	—	—	—	—	—	Pycnometer	41
Zn	99.995	6.562	500	6.417	600	6.370	700	6.259	—	IBF	25
Zn	—	6.57_5	500	6.49	600	6.37	700	6.26	Ar	Bubble pressure	83
Zn	99.999	6.57_7 ±0.012	500	6.48	600	6.37	700	6.27	Ar	Bubble pressure	49
Zr	—	5.80	—	—	—	—	—	—	—	calculated	3

References

1. Addison and Pulham, J. Chem. Soc., 3873 (1962).
2. Allen, Trans, AIME, **227,** 1175 (1963).
3. Allen, Trans. AIME, **230,** 1537 (1964).
4. Astachov, Penin and Dobkina, Zh. Fiz. Chim., **20,** 403 (1946).
5. Beer (in Lucas), Mém. Sci. Rev. Mét., **61,** 1, 97 (1964).
6. Bornemann and Sauerwald, Z. Metallkunde, **14,** 145 (1922).
7. Bornemann and Sauerwald, Z. Metallkunde, **14,** 254 (1922).
8. Cahill and Kirshenbaum, J. Phys. Chem., **66,** 1080 (1962).
9. Calverley, Proc. Phys. Soc., **70,** 1040 (1957).
10. Campbell and Epstein, J. Amer. Chem. Soc., **64,** 2679 (1942).
11. Coy and Mateer, Trans. ASM, **58,** 99 (1965).
12. Day, Sosman and Hostetter, Am. J. Sci., **187,** 1 (1914).
13. Dobinsky and Weselowsky, Bull. Int. Acad. Polon., **A, 446** (1936).

References (Continued)

14. Dshemilev, Popel and Zarevski, Fiz. Met. i Met., **18,** 83 (1964).
15. Edwards and Moorman, Chem & Met. Eng., **24,** 61 (1921).
16. Eljutin and Maurach, Izv. A. N., OTN, **4,** 129 (1956).
17. Eremenko, Ukr. Chim. Zh., **28,** 427 (1962).
18. Eremenko and Naidich, Izv. A. N., OTN, **2,** 111 (1959).
19. Eremenko and Naidich, Izv. A. N., OTN, **6,** 129 (1959).
20. Eremenko and Naidich, Izv. A. N., OTN, **6,** 100 (1961).
21. Eremenko and Nishenko, Ukr. Chim. Zh., **30,** 125 (1964).
22. Eremenko, Nishenko and Taj-Shou-Wej, Izv. A. N., OTN, **3,** 116 (1960).
23. Fisher and Phillips, J. Metals, **6,** 1060 (1954).
24. Gebhardt and Becker, Z. Metallkunde, **42,** 111 (1951).
25. Gebhardt, Becker and Dorner, Aluminium, **31,** 315 (1955).
26. Geld and Vertman, Fiz. Met i Met., **10,** 793 (1960).
27. Gogiberidse and Kekelidse, Izv. A. N., **3,** 125 (1963).
28. Greenaway, J. Inst. Met., **74,** 133 (1947).
29. Grosse, Cahill and Kirshenbaum, J. Amer. Chem. Soc., **83,** 4665 (1961).
30. Herczynska, Naturwiss., **47,** 200 (1960).
31. Hogness, J. Amer. Chem. Soc., **43,** 1621 (1921).
32. Jones, Ofte, Rohr and Wittenberg, Trans. ASM, **55,** 819 (1962).
33. Jouniaux, Bull. Soc. Chim. France, **47,** 524 (1930).
34. Jouniaux, Bull. Soc. Chim. France, **51,** 677 (1932).
35. Kingery and Humenik, J. Phys. Chem., **57,** 359 (1953).
36. Kirshenbaum and Cahill, Trans. ASM, **55,** 844 (1962).
37. Kirshenbaum and Cahill, Trans. ASM, **55,** 849 (1962).
38. Kirshenbaum and Cahill, Trans. AME, **224,** 816 (1962).
39. Kirshenbaum, Cahill and Grosse, J. Inorg. Nucl. Chem., **24,** 333 (1962).
40. Klemm et al., Monatsh. Chem., **83,** 629 (1952).
41. Knappwost and Restle, Z. Elektrochemie, **58,** 112 (1954).
42. Königer and Nagel, GieBerei TWB, **13,** 57 (1960).
43. Kozakevitch, Châtel, Urbain and Sage, Rev. de Mét., **52,** 139 (1955).
44. Kozakevitch and Urbain, C. R., Paris, **253,** 2229 (1961).
45. Lang, WADC-Techn. Rep., 57–488 (1957).
46. Lauermann and Metzger, Z. Phys. Chem., **216,** 37 (1961).
47. Lucas, C. R., Paris, **250,** 1850 (1960).
48. Lucas, C. R., Paris, **253,** 2526 (1961).
49. Lucas, Mém. Sci. Rev. Mét., **61,** 1 (1964).
50. Lucas, Mém. Sci. Rev. Mét., **61,** 97 (1964).
51. Lucas and Urbain, C. R., Paris, **258,** 6403 (1964).
52. Malmberg, J. Inst. Met., **89,** 137 (1960).
53. Matuyama, Sci. Rep. RITU, **18,** 19 (1929).
54. Mehairy and Wood (in Lucas), Mém. Sci. Rev. Mét., **61,** 1 (1964).
55. Metals Handbook: Properties and Selection of Metals, 8th edit., **1,** (1961).
56. Metals Reference Book, C. J. Smithells, 4th edit., **1,** 688 (1967).
57. Metzger, Z. Phys. Chem., **211,** 1 (1959).
58. Mokrovski and Regel, J. Phys. Techn., **22,** 1281 (1952).
59. Monma and Suto, J. Jap. Inst. Met., **1,** 69 (1960).
60. Naidich and Eremenko, Fiz. Met. i Met., **6,** 62 (1961).
61. Ono and Matsushima, Sci. Rep. RITU, **9,** 309 (1957).
62. Pascal and Jouniaux, C. R., Paris, **158,** 414 (1914).
63. Pascal and Jouniaux, Z. Elektrochemie, **22,** 72 (1916).
64. Pekarev, Izv. Vyss. Utch. Saved., Tsvetn. Met., **6,** 111 (1963).
65. Pelzel, Berg-u. Hütt. Mon. Hefte, Leoben, **93,** 248 (1948).
66. Pelzel, Z. Metallkunde, **50,** 392 (1959).
67. Pelzel and Sauerwald, Z. Metallkunde, **33,** 229 (1941).
68. Pelzel and Schneider, Z. Metallkunde, **35,** 121 (1943).
69. Pokrovski and Saidov, Zh. Fiz. Chim., **29,** 1601 (1955).
70. Popel, Smirnov, Zarevski, Dshemilev and Pastuchov, Izv. A. N., **1,** 62 (1965).
71. Popel, Zarevski and Dshemilev, Fiz. Met. i Met., **18,** 468 (1964).
72. Rink, C. R., Paris, **189,** 39 (1929).
73. Saeger and Ash, J. Res. Nat. Bur. Stand., **8,** 37 (1932).
74. Sauerwald, Z. Metallkunde, **14,** 457 (1922).
75. Sauerwald and Widawski, Z. anorg. allg. Chem., **155.** 1 (1926).
76. Schneider and Heymer, Z. anorg. allg. Chem., **286,** 118 (1956).
77. Smith and Spitzer, J. Phys. Chem., **66,** 946 (1962).
78. Stauffer, Thesis Göttingen (1953).
79. Stott and Rendall, J. Iron & Steel Inst., **175,** 374 (1953).
80. Tavadse, Bairamashvili, Chantadse and Zagareishvili, Doklady A. N., **150,** 544 (1963).

DENSITY OF LIQUID ELEMENTS (Continued)

References (Continued)

81. Tavadse, Bairamashvili and Chantadse, Doklady A. N., **162,** 67 (1965).
82. Taylor, J. Inst. Met., **83,** 143 (1954).
83. Übelacker and Lucas, C. R., Paris, **254,** 1622 (1962).
84. Widawski and Sauerwald, Z. anorg. allg. Chem., **192,** 145 (1930).
85. Zimmermann and Esser, Arch. f. Eisenh., **2,** 867 (1929).

A) Kirshenbaum, Cahill and Grosse, J. Inorg. Nucl. Chem., **22,** 33 (1961).
B) McGonigal, Kirshenbaum and Grosse, J. Phys. Chem., **66,** 737 (1962).
C) Grosse and Kirshenbaum, J. Inorg. Nucl. Chem., **25,** 331 (1963).
D) Golchova, Teplofiz. Vysok. Temp., **4,** 360 (1966).
E) Bosio, C. R., Paris, **259,** 4545 (1964).
F) Spilrajn and Jakimovich, Teplofiz. Vysok. Temp., **5,** 239 (1967).
G) Jakimovich and Saars, Teplofiz. Vysok. Temp., **5,** 532 (1967).
H) Stone, Ewing, Spann, Steinkuller, Williams and Miller, J. Chem. Engng. Data, **11,** 320 (1966).
I) Shirai, Hamada and Kobayashi, J. Chem. Soc. Japan, **84,** 968 (1963).
K) Grosse and McGonigal, J. Phys. Chem., **68,** 414 (1964).
L) Cubicciotti, J. Phys. Chem., **68,** 537 (1964).
M) Watolin and Esin, Fiz. Met. i Met., **16,** 936 (1963).
N) Gebhardt, Becker and Trägner, Z. Metallkunde, **46,** 90 (1955).
O) Kubaschewski and Hörnle, Z. Metallkunde, **42,** 129 (1951).
P) Gebhardt and Dorner, Z. Metallkunde, **42,** 353 (1951).
Q) Gebhardt and Wörwag, Z. Metallkunde, **42,** 358 (1951).
R) Been, Edwards, Teeter and Chalkins, NEPA-1585, US-AEC (1950).
S) da C. Andrade and Dobbs, Proc. Roy. Soc., **211,** 12 (1952).
T) Wilkinson: Extractive and Physical Metallurgy of Plutonium and its Alloys, New York-London, 1960.
U) Saito and Sakuma, J. Jap. Inst. Met., **31,** 1140 (1967).

DENSITY OF LIQUID ELEMENTS

SUPPLEMENTARY TABLE

Gernot Lang

Element	Purity	ρ_{mp}	ρt_1 t°C	ρt_1 ρ	ρt_2 t°C	ρt_2 ρ	ρt_3 t°C	ρt_3 ρ	Atm.	Method	Ref.
Ag		9.36		$\rho_t = 9.36-0.00114\ (t-t_{mp})$				(t°C)		Bubble pressure	110
Ag	99.95			$\rho_t = 9.318-1.08 \cdot 10^{-3}\ (t-1000) \pm 0.006$ (1000–1200°C)					H_2	Bubble pressure	81
Ag	99.999	9.31		$\rho_t = 10.31-1.05 \cdot 10^{-3} \cdot t$(t°C) (960–1150)					Ar	Sessile drop	88
Al	99.99	2.39		$\rho_t = 2.60-3.2 \cdot 10^{-4}\,t$ (t°C) (900–1750°C)					He	Drop volume	87
Al	99.999	2.365		$\rho_t = 2.576-3.2 \cdot 10^{-4}\,t$ (t°C)					He	Sessile drop	108
Al	99.998	2.375		$\rho_t = 2.375-5.4 \cdot 10^{-4}\ (t-660)$				(t°C)	Ar	Sessile drop	97
Ba	99.5	3.32		$\rho_t = 3.59-2.74 \cdot 10^{-4}\,T$				(T°K)		Bubble pressure	90
Bi	99.9999	10.05		$\rho_t = 10.05-1.41 \cdot 10^{-3}\ (t-t_{mp})$				(t°C)	N_2	Volume determin.	96
Bi	99.98	10.04		$\rho_t = 10.04-1.44 \cdot 10^{-3}\ (t-t_{mp})$				(t°C)	N_2	Volume determin.	96
Bi	99.999	10.114		$\rho_t = 10.406-1.078 \cdot 10^{-3}\,t$ (271–414)				(t°C)	N_2	DBF	89
Bi	–	10.031		$\rho_t = 10.336-12.367 \cdot 10^{-4}\,t$				(t°C)	vac.	Pycnometer	92
Ca	–	1.365		$\rho_t = 1.613-2.21 \cdot 10^{-4}\,T$				(T°K)		Bubble pressure	90
Cd	99.999	7.996		$\rho_t = 8.388-12.205 \cdot 10^{-4}\,t$				(t°C)	Ar	Pycnometer	91
Co	99.53	7.78		$\rho_t = 9.5_7-10.1_7 \cdot 10^{-4}\,T$ (1780–2200)				(T°K)	Ar	Drop volume	103
Co	99.67	7.75		$\rho_t = 9.7_1-1.1_1 \cdot 10^{-3}\,T$				(T°K)	Ar	Bubble pressure	109
Cs	–	1.86		$\rho_t = 1.84/1+1.1755 \cdot 10^{-4}\ (t-82.4)+7.656 \cdot 10^{-8}\ (t-82.4)^2$ (t°F, 29–704°C)					Ar	Pycnometer	86
Cs	–	–	40	1.840	271	1.711	376	1.655	Ar	Dilatometer	106
Fe	–	7.06		$\rho_t = 7.05-7.3 \cdot 10^{-4}\ (t-1550)$				(t°C)	H_2, He	Sessile drop	101
Fe	99.9	7.03		$\rho_t = 8.5_7-8.5_3 \cdot 10^{-4}\,T$ (1800–2150)				(T°K)	Ar	Drop volume	103
Fe	99.98	7.05		$\rho_t = 8.7_8-0.95_8 \cdot 10^{-3}\,T$				(T°K)	Ar	Bubble pressure	109
Fr	–		100	2.29						calculated	100
Ga	–	6.1136		$\rho_t = 6.11564-7.37437 \cdot 10^{-4}\,t + 1.37767 \cdot 10^{-7}\,t^2$ (t°C) (50–600°C)						Pycnometer	98

DENSITY OF LIQUID ELEMENTS (*Continued*)

Element	Purity	ρ_{mp}	ρt_1 t°C	ρt_1 ρ	ρt_2 t°C	ρt_2 ρ	ρt_3 t°C	ρt_3 ρ	Atm.	Method	Ref.
Hg	–	–	0	13.5951	20	13.5457	100	13.3514		–	105
In	99.999	7.032		$\rho_t = 7.153–0759 \cdot 10^{-3} t$				(t°C)	N_2	DBF (160–532)	89
In	99.999	7.016		$\rho_t = 7.146–8.362 \cdot 10^{-4} t$				(t°C)	Ar	Pycnometer	91
Mg	99.5		959	1.52	1053	1.47	–	–		Bubble pressure	90
Mo	–	9.1							vac.	Pendant drop and drop weight	95
Nb, Cb	–	7.6							vac.	Pendant drop and drop weight	95
Ni	–	7.82		$\rho_t = 7.81–8.7 \cdot 10^{-4}$ (t–1460)				(t°C)	H_2, He	Sessile drop	U
Ni	–	7.65							vac.	Pendant drop and drop weight	95
Ni	99.95	7.95		$\rho_t = 9.8_1–10.8_0 \cdot 10^{-4} T$ (1890–2150)				(T°K)	Ar	Drop volume	103
Pb	99.999	10.330		$\rho_t = 10.650–9.8 \cdot 10^{-4} t$ (1000–1600)				(t°C)	He	Sessile drop	108
Pb	99.997	10.660		$\rho_t = 11.060–12.220 \cdot 10^{-4} t$				(t°C)		Pycnometer	107
Pd	99.998	10.379		$\rho_t = 12.193–11.69 \cdot 10^{-4} t$ (MP–1700)				(t°C)	He	Sessile drop	108
Rb	99.4	1.385		$\rho_t = 1.55643–2.6511 \cdot 10^{-4} t–6.26779/t$ (t°F, 66–1076°C)					Ar	Dilatometer	106
Rb	–	1.484		$\rho_t = 1.472/1+1.3309 \cdot 10^{-4} (t–102) + 5.2106 \cdot 10^{-8} (t–102)^2$ (t°F, 39–730°C)					Ar	Pycnometer	86
Sb	99.999	6.452		$\rho_t = 6.818–5.8 \cdot 10^{-4} t$ (1000–1600°C)				(t°C)	He	Sessile drop	108
Sb	99.999	6.535		$\rho_t = 6.962–0.673 \cdot 10^{-3} t$ (635–745°C)				(t°C)	N_2	D B F	89
Sb	–	6.493		$\rho_t = 6.902–6.486 \cdot 10^{-4} t$ (MP–746°C)				(t°C)	Ar	Pycnometer	93
Si	99.9999		1500	2.46					Ar		94
Sn	99.999		246	6.964	336	6.900	400	6.854	vac.	Pycnometer	111
Sn	99.999	7.01		$\rho_t = 7.16–6.3 \cdot 10^{-4} t$				(t°C)		Bubble pressure	104
Sn	99.999	6.981		$\rho_t = 7.135–0.663 \cdot 10^{-3} t$ (MP–438°C)				(t°C)	N_2	D B F	89
Sn	99.999	6.973		$\rho_t = 7.139–7.125 \cdot 10^{-4} t$				(t°C)		Pycnometer	107
Sr	99.5	2.375		$\rho_t = 2.648–2.62 \cdot 10^{-4} T$				(T°K)		Bubble pressure	90

DENSITY OF LIQUID ELEMENTS (*Continued*)

Element	Purity	ρ_{mp}	ρt_1 t°C	ρt_1 ρ	ρt_2 t°C	ρt_2 ρ	ρt_3 t°C	ρt_3 ρ	Atm.	Method	Ref.
Te	–	5.71	$\rho_t = 5.71 - 3.6 \cdot 10^{-4} (t - t_{mp})$					(t°C)		Bubble pressure	110
Te	99.7	5.86	$\rho_t = 5.86 - 0.73 \cdot 10^{-3} (t - t_{mp})$					(t°C)	N_2	Volume determin.	96
Ti	–	4.10							vac.	Pendant drop and drop weight	95
U		17.27	$\rho_t = 19.520 - 16.01 \cdot 10^{-4} T$					(T°K)	vac.	Pycnometer	102
V	–	5.3							vac.	Pendant drop and drop weight	95
Zr	–	5.60							vac.	Pendant drop and drop weight	95

REFERENCES

86. Achener, HTLMHTTM., Vol. I, Oak Ridge, p. 5, November 1964.
87. Ayushina, Lewin and Geld, Zh. Fiz. Chim., **42**, 2799 (1968).
88. Bernard and Lupis, Met. Trans., **2**, 555 (1971).
89. Berthou and Tougas, Met. Trans., **1**, 2978 (1970).
90. Bohdansky and Schins, J. Inorg. Nucl. Chem., **30**, 2331 (1968).
91. Crawley, Trans. AIME, **242**, 2237 (1968).
92. Crawley and Kiff, Met. Trans., **2**, 609 (1971).
93. Crawley and Kiff, Met. Trans., **3**, 158 (1972).
94. Eljutin, Kostikow and Lewin, Izv. Vys. Uch. Sav., Tsvetn. Met., **2**, 131 (1970).
95. Eljutin, Kostikow and Penkow, Poroshk. Met., **9**, 46 (1970).
96. Keskar and Hruska, Met. Trans., **1**, 2357 (1970).
97. Körber and Löhberg, Gießereiforschung, **23**, 173 (1971).
98. Koster, Hensel and Franck, Ber. Bunsenges., **74**, 43 (1970).
99. Nagamori, Trans. AIME, **245**, 1897 (1969).
100. Osminin, Zh. Fiz. Chim., **43**, 2610 (1969).
101. Popel, Shergin and Zarewski, Zh. Fiz. Chim., **43**, 2365 (1969).
102. Rohr and Wittenberg, J. Phys. Chem., **74**, 1151 (1970).
103. Saito and Sakuma, Sci. Rep. RITU, **22**, 57 (1970).
104. Schwaneke and Folke, Us-Bur. Min., Invest. Rep. No. **7372** (1970).
105. Stoffhütte. Taschenbuch der Werkstoffkunde, 4th ed. p. 1059, 1967.
106. Tepper, Murchison, Zelenak and Roehlich, HTLMHTTM, Vol. I, Oak Ridge, p. 26, November 1964.
107. Thresh, Crawley and White, Trans. AIME, **242**, 819 (1968).
108. Watolin, Esin, Uchow and Dubinin, Trudy Inst. Met. Sverdlovsk, **18**, 73 (1969).
109. Watanabe, Trans. Jap. Inst. Met., **12**, 17 (1971).
110. Wobst and Rentzsch, Z. Phys. Chem., **240**, 36 (1969).
111. Crawley, Trans. AIME, **245**, 1655 (1969).

HEAT OF FUSION OF SOME INORGANIC COMPOUNDS

RUDOLF LOEBEL

Values in parentheses are of uncertain reliability

Compound	Formula	M.P., °C	H_f cal/g mole	H_f cal/g
Actinium^{227}	Ac	1050 ± 50	(3400)	(11.0)
Aluminum	Al	658.5	2550	94.5
Aluminum bromide	Al_2Br_6	87.4	5420	10.1
Aluminum chloride	Al_2Cl_6	192.4	19600	63.6
Aluminum iodide	Al_2I_6	190.9	7960	9.8
Aluminum oxide	Al_2O_3	2045.0	(26000)	(256.0)
Antimony	Sb	630	4770	39.1
Antimony tribromide	$SbBr_3$	96.8	3510	9.7
Antimony trichloride	$SbCl_3$	73.3	3030	13.3
Antimony pentachloride	$SbCl_5$	4.0	2400	8.0
Antimony trioxide	Sb_4O_6	655.0	(26990)	(46.3)
Antimony trisulfide	Sb_4S_6	546.0	11200	33.0
Argon	Ar	−190.2	290	7.25
Arsenic	As	816.8	(6620)	(22.0)
Arsenic pentafluoride	AsF_4	−80.8	2800	16.5
Arsenic tribromide	$AsBr_3$	30.0	2810	8.9
Arsenic trichloride	$AsCl_3$	−16.0	2420	13.3
Arsenic trifluoride	AsF_3	−6.0	2486	18.9
Arsenic trioxide	As_4O_6	312.8	8000	22.2
Barium	Ba	725	1830	13.3
Barium bromide	$BaBr_2$	846.8	6000	21.9
Barium chloride	$BaCl_2$	959.8	5370	25.9
Barium fluoride	BaF_2	1286.8	3000	17.1
Barium iodide	BaI_2	710.8	(6800)	(17.3)
Barium nitrate	$Ba(NO_3)_2$	594.8	(5900)	(22.6)
Barium oxide	BaO	1922.8	13800	93.2
Barium phosphate	$Ba_3(PO_4)_2$	1727	18600	30.9
Barium sulfate	$BaSO_4$	1350	9700	41.6
Beryllium	Be	1278	—	260.0
Beryllium bromide	$BeBr_2$	487.8	(4500)	(26.6)
Beryllium chloride	$BeCl_2$	404.8	(3000)	(30)
Beryllium fluoride	BeF_2	796.8	(6000)	(127.6)
Beryllium iodide	BeI_2	479.8	(4500)	(17.1)
Beryllium oxide	BeO	2550.0	17000	679.7
Bismuth	Bi	271	2505	12.0
Bismuth tribromide	$BiBr_3$	216.8	(4000)	(8.9)
Bismuth trichloride	$BiCl_3$	223.8	2600	8.2
Bismuth trifluoride	BiF_3	726.0	(6200)	(23.3)
Bismuth trioxide	Bi_2O_3	815.8	6800	14.6
Boron	B	2300	(5300)	(490)
Boron tribromide	BBr_3	−48.8	(700)	(2.9)
Boron trichloride	BCl_3	−107.8	(500)	(4.3)
Boron trifluoride	BF_3	−128.0	480	7.0
Boron triiodide	BI_3	31.8	(1000)	(2.5)
Boron trioxide	B_2O_3	448.8	5500	78.9
Bromine	Br_2	−7.2	2580	16.1
Bromine pentafluoride	BrF_5	−61.4	1355	7.07
Cadmium	Cd	320.8	1460	12.9
Cadmium bromide	$CdBr_2$	567.8	(5000)	(18.4)
Cadmium chloride	$CdCl_2$	567.8	5300	28.8
Cadmium fluoride	CdF_2	1110	(5400)	(35.9)
Cadmium iodide	CdI_2	386.8	3660	10.0
Cadmium sulfate	$CdSO_4$	1000	4790	22.9
Calcium	Ca	851	2230	55.7
Calcium bromide	$CaBr_2$	729.8	4180	20.9
Calcium carbonate	$CaCO_3$	1282	(12700)	(126)
Calcium chloride	$CaCl_2$	782	6100	55
Calcium fluoride	CaF_2	1382	4100	52.5
Calcium nitrate	$Ca(NO_3)_2$	560.8	5120	31.2
Calcium oxide	CaO	2707	(12240)	(218.1)
Calcium metasilicate	$CaSiO_3$	1512	13400	115.4
Calcium sulfate	$CaSO_4$	1297	6700	49.2
Carbon dioxide	CO_2	−57.6	1900	43.2
Carbon monoxide	CO	−205	199.7	7.13
Cyanogen	C_2N_2	−27.2	2060	39.6
Cyanogen chloride	CNCl	−5.2	2240	36.4
Cerium	Ce	775	2120	15.1
Cerium (III) chloride	$CeCl_3$	820.8	(8000)	(32.4)
Cerium (IV) fluoride	CeF_4	976	(10000)	(46.3)

HEAT OF FUSION OF SOME INORGANIC COMPOUNDS (Continued)

Compound	Formula	M.P.,°C	H_f cal/g mole	H_f cal/g
Cesium	Cs	28.3	500	3.7
Cesium chloride	CsCl	641.8	3600	21.4
Cesium fluoride	CsF	704.8	(2450)	(16.1)
Cesium nitrate	$CsNO_3$	406.8	3250	16.6
Chlorine	Cl_2	−103±5	1531	22.8
Chromium	Cr	1890	3660	62.1
Chromium (II) chloride	$CrCl_2$	814	7700	65.9
Chromium (III) sequioxide	Cr_2O_3	2279	4200	27.6
Chromium trioxide	CrO_3	197	3770	37.7
Cobalt	Co	1490	3640	62.1
Cobalt (II) chloride	$CoCl_2$	727	7390	56.9
Cobalt (II) fluoride	CoF_2	1201	(9000)	(92.9)
Copper	Cu	1083	3110	49.0
Copper (I) bromide	CuBr	487	(2300)	(16.03)
Copper (II) chloride	$CuCl_2$	430	4890	24.7
Copper (I) chloride	CuCl	429	2620	26.4
Copper (I) cyanide	$Cu_2(CN)_2$	473	(5400)	(30.1)
Copper (I) iodide	CuI	587	(2600)	(13.6)
Copper (II) oxide	CuO	1446	2820	35.4
Copper (I) oxide	Cu_2O	1230	(13400)	(93.6)
Copper (I) sulfide	Cu_2S	1129	5500	34.6
Dysprosium	Dy	1407	4100	25.2
Dysprosium chloride	$DyCl_3$	646	(7000)	(26.0)
Erbium	Er	1496	4100	24.5
Erbium trichloride	$ErCl_3$	775	(7000)	(26.0)
Europium	Eu	826	2500	16.4
Europium trichloride	$EuCl_3$	622	(8000)	(20.9)
Fluorine	F_2	−219.6	244.0	6.4
Gadolinium	Gd	1312	3700	23.8
Gadolinium trichloride	$GdCl_3$	608	(7000)	(26.6)
Gallium	Ga	29	1336	19.1
Germanium	Ge	959	(8300)	(114.3)
Gold	Au	1063	3030	15.3
Hafnium	Hf	2214	(6000)	(34.1)
Holmium	Ho	1461	4100	24.8
Holmium trichloride	$HoCl_3$	717	(8000)	(29.5)
Hydrogen	H_2	−259.25	28	13.8
Hydrogen bromide	HBr	−86.96	575.1	7.1
Hydrogen chloride	HCl	−114.3	476.0	13.0
Hydrogen fluoride	HF	−83.11	1094	54.7
Hydrogen iodide	HI	−50.91	686.3	5.4
Hydrogen nitrate	HNO_3	−47.2	601	9.5
Hydrogen oxide (water)	H_2O	0	1436	79.72
Deuterium oxide	D_2O	3.78	1516	75.8
Hydrogen peroxide	H_2O_2	−0.7	2920	8.58
Hydrogen selenate	H_2SeO_4	57.8	3450	23.8
Hydrogen sulfate	H_2SO_4	10.4	2360	24.0
Hydrogen sulfide	H_2S	−85.6	5683	16.8
Hydrogen sulfide (di-)	H_2S_2	−89.7	1805	27.3
Hydrogen telluride	H_2Te	−49.0	1670	12.9
Indium	In	156.3	781	6.8
Iodine	I_2	112.9	3650	14.3
Iodine chloride (α)	ICl	17.1	2660	16.4
Iodine chloride (β)	ICl	13.8	2270	13.3
Iron	Fe	1530.0	3560	63.7
Iron carbide	Fe_3C	1226.8	12330	68.6
Iron (III) chloride	Fe_2Cl_6	303.8	20500	63.2
Iron (II) chloride	$FeCl_2$	677	7800	61.5
Iron (II) oxide	FeO	1380	(7700)	(107.2)
Iron oxide	Fe_3O_4	1596	33000	142.5
Iron pentacarbonyl	$Fe(CO)_5$	−21.2	3250	16.5
Iron (II) sulfide	FeS	1195	5000	56.9
Lanthanum	La	920	2400	17.4
Lanthanum chloride	$LaCl_3$	861	(9000)	(36.6)
Lead	Pb	327.3	1224	5.9
Lead bromide	$PbBr_2$	487.8	4290	11.7
Lead chloride	$PbCl_2$	497.8	5650	20.3
Lead fluoride	PbF_2	823	1860	7.6
Lead iodide	PbI_2	412	5970	17.9
Lead molybdate	$PbMoO_4$	1065	(25800)	70.8
Lead oxide	PbO	890	2820	12.6
Lead sulfide	PbS	1114	4150	17.3
Lead sulfate	$PbSO_4$	1087	9600	31.6
Lead tungstate	$PbWO_4$	1123	(15200)	(33.4)
Lithium	Li	178.8	1100	158.5
Lithium borate, meta-	$LiBO_2$	845	(5570)	(111.9)
Lithium bromide	LiBr	552	2900	33.4
Lithium chloride	LiCl	614	3200	75.5

HEAT OF FUSION OF SOME INORGANIC COMPOUNDS (Continued)

Compound	Formula	M.P.,°C	H_f cal/g mole	H_f cal/g
Lithium fluoride	LiF	896	(2360)	(91.1)
Lithium hydroxide	LiOH	462	2480	103.3
Lithium iodide	LiI	440	(1420)	(10.6)
Lithium metasilicate	Li_2SiO_3	1177	7210	80.2
Lithium molybdate	Li_2MoO_4	705	4200	24.1
Lithium nitrate	$LiNO_3$	250	6060	87.8
Lithium orthosilicate	Li_4SiO_4	1249	7430	60.5
Lithium sulfate	Li_2SO_4	857	3040	27.6
Lithium tungstate	Li_2WO_4	742	(6700)	(25.6)
Lutetium	Lu	1651	4600	26.3
Lutetium chloride	$LuCl_3$	904	(9000)	(32)
Magnesium	Mg	650	2160	88.9
Magnesium bromide	$MgBr_2$	711	8300	45.0
Magnesium chloride	$MgCl_2$	712	8100	82.9
Magnesium fluoride	MgF_2	1221	5900	94.7
Magnesium oxide	MgO	2642	18500	459.0
Magnesium phosphate	$Mg_3(PO_4)_2$	1184	(11300)	(42.9)
Magnesium silicate	$MgSiO_3$	1524	14700	146.4
Magnesium sulfate	$MgSO_4$	1327	3500	28.9
Manganese	Mn	1220	3450	62.7
Manganese dichloride	$MnCl_2$	650	7340	58.4
Manganese metasilicate	$MnSiO_3$	1274	(8200)	(62.6)
Manganese (II) oxide	MnO	1784	13000	183.3
Manganese oxide	Mn_3O_4	1590	(39000)	(170.4)
Mercury	Hg	−39	557.2	2.7
Mercury bromide	$HgBr_2$	241	3960	10.9
Mercury chloride	$HgCl_2$	276.8	4150	15.3
Mercury iodide	HgI_2	250	4500	9.9
Mercury sulfate	$HgSO_4$	850	(1440)	(4.8)
Molybdenum	Mo	2622	(6600)	(68.4)
Molybdenum dichloride	$MoCl_2$	726.8	6000	3.58
Molybdenum hexafluoride	MoF_6	17	2500	11.9
Molybdenum trioxide	MoO_3	795	(2500)	(17.3)
Neodymium	Nd	1020	1700	11.8
Neodymium trichloride	$NdCl_3$	758	(8000)	(31.9)
Neon	Ne	−248.6	77.4	3.83
Nickel	Ni	1452	4200	71.5
Nickel chloride	$NiCl_2$	1030	18470	142.5
Nickel subsulfide	Ni_3S_2	790	5800	25.8
Niobium	Nb	2496	(6500)	(68.9)
Niobium pentachloride	$NbCl_5$	211	8400	30.8
Niobium pentoxide	Nb_2O_5	1511	24200	91.0
Nitrogen	N_2	−210	172.3	6.15
Nitrogen oxide (ic)	NO	−163.7	549.5	18.3
Nitrogen oxide (ous)	N_2O	−90.9	1563	35.5
Nitrogen tetroxide	N_2O_4	−13.2	5540	60.2
Osmium	Os	2700	(7000)	(36.7)
Osmium tetroxide (white)	OsO_4	41.8	2340	9.2
Osmium tetroxide (yellow)	OsO_4	55.8	4060	15.5
Oxygen	O_2	−218.8	106.3	3.3
Palladium	Pd	1555	4120	38.6
Phosphoric acid	H_3PO_4	42.3	2520	25.8
Phosphoric acid, hypo-	$H_4P_2O_6$	54.8	8300	51.2
Phosphorous acid, hypo-	H_3PO_2	17.3	2310	35.0
Phosphorous acid, ortho-	H_3PO_3	73.8	3070	37.4
Phosphorus, yellow	P_4	44.1	600	4.8
Phosphorus oxychloride	$POCl_3$	1.0	3110	20.3
Phosphorus pentoxide	P_4O_{10}	569.0	17080	60.1
Phosphorus trioxide	P_4O_6	23.7	3360	15.3
Platinum	Pt	1770	4700	24.1
Potassium	K	63.4	574	14.6
Potassium borate, meta-	KBO_2	947	(5660)	(69.1)
Potassium bromide	KBr	742	5000	42.0
Potassium carbonate	K_2CO_3	897	7800	56.4
Potassium chloride	KCl	770	6410	85.9
Potassium chromate	K_2CrO_4	984	6920	35.6
Potassium cyanide	KCN	623	(3500)	(53.7)
Potassium dichromate	$K_2Cr_2O_7$	398	8770	29.8
Potassium fluoride	KF	875	6500	111.9

HEAT OF FUSION OF SOME INORGANIC COMPOUNDS (Continued)

Compound	Formula	M.P., °C	H_f cal/g mole	H_f cal/g
Potassium hydroxide	KOH	360	(1980)	(35.3)
Potassium iodide	KI	682	4100	24.7
Potassium nitrate	KNO_3	338	2840	28.1
Potassium peroxide	K_2O_2	490	6100	55.3
Potassium phosphate	K_3PO_4	1340	8900	41.9
Potassium pyro-phosphate	$K_4P_2O_7$	1092	14000	42.4
Potassium sulfate	K_2SO_4	1074	8100	46.4
Potassium sulfocyanide	KCNS	179	2250	23.1
Potassium tungstate	K_2WO_4	929	(4400)	(13.6)
Praseodymium	Pr	931	2700	19.0
Praseodymium chloride	$PrCl_3$	786	(8000)	(32.3)
Promethium	Pm	(1027)	(3000)	(21)
Rhenium	Re	3167 ± 60	(7900)	(42.4)
Rhenium heptoxide	Re_2O_7	296	15340	30.1
Rhenium hexafluoride	ReF_6	19.0	5000	16.6
Rubidium	Rb	38.9	525	6.1
Rubidium bromide	RbBr	677	3700	22.4
Rubidium chloride	RbCl	717	4400	36.4
Rubidium fluoride	RbF	833	4130	39.5
Rubidium iodide	RbI	638	2990	14.0
Rubidium nitrate	$RbNO_3$	305	1340	9.1
Samarium	Sm	1072	2600	17.3
Samarium chloride	$SmCl_2$	562	(6000)	(27.1)
Scandium	Sc	1538	3800	84.4
Scandium trichloride	$ScCl_3$	940	(19000)	(125.4)
Selenium	Se	217	1220	15.4
Selenium oxychloride	$SeOCL_3$	9.8	1010	6.1
Silicon	Si	1427	9470	337.0
Silicondioxide (Cristobalite)	SiO_2	2100	2100	35.0
Silicondioxide (Quartz)	SiO_2	1470	3400	56.7
Disilane, hexafluoro-	Si_2F_6	−28.6	3900	22.9
Silicon tetrachloride	$SiCl_4$	−67.7	1845	10.8
Silver	Ag	961	2700	25.0
Silver bromide	AgBr	430	2180	11.6
Silver chloride	AgCl	455	3155	22.0

Compound	Formula	M.P., °C	H_f cal/g mole	H_f cal/g
Silver cyanide	AgCN	350	2750	20.5
Silver iodide	AgI	557	2250	9.5
Silver nitrate	$AgNO_3$	209	2755	16.2
Silver sulfide	Ag_2S	841	3360	13.5
Silver sulfate	Ag_2SO_4	657	(4280)	(13.7)
Sodium	Na	97.8	630	27.4
Sodium borate, meta-	$NaBO_2$	966	8660	134.6
Sodium bromide	NaBr	747	6140	59.7
Sodium carbonate	Na_2CO_3	854	7000	66.0
Sodium chloride	NaCl	800	7220	123.5
Sodium chlorate	$NaClO_3$	255	5290	49.7
Sodium cyanide	NaCN	562	(4360)	(88.9)
Sodium fluoride	NaF	992	7000	166.7
Sodium hydroxide	NaOH	322	2000	50.0
Sodium iodide	NaI	662	5340	35.1
Sodium molybdate	Na_2MoO_4	687	3600	17.5
Sodium nitrate	$NaNO_3$	310	3760	44.2
Sodium phosphate, meta-	$NaPO_3$	988	(4960)	(48.6)
Sodium peroxide	Na_2O_2	460	5860	75.1
Sodium pyro-phosphate	$Na_4P_2O_7$	970	(13700)	(51.5)
Sodium silicate, meta-	Na_2SiO_3	1087	10300	84.4
Sodium silicate, di-	$Na_2Si_2O_5$	884	8460	46.4
Sodium silicate, aluminum-	$NaAlSi_3O_8$	1107	13150	50.1
Sodium sulfide	Na_2S	920	(1200)	15.4
Sodium sulfate	Na_2SO_4	884	5830	41.0
Sodium thiocyanate	NaCNS	323	4450	54.8
Sodium tungstate	Na_2WO_4	702	5800	19.6
Strontium	Sr	757	2190	25.0
Strontium bromide	$SrBr_2$	643	4780	19.3
Strontium chloride	$SrCl_2$	872	4100	26.5
Strontium fluoride	SrF_2	1400	4260	34.0

HEAT OF FUSION OF SOME INORGANIC COMPOUNDS (Continued)

Compound	Formula	M.P.,°C	H_f cal/g mole	H_f cal/g
Strontium oxide	SrO	2430	16700	161.2
Sulfur (monatomic)	S	119	295	9.2
Sulfur dioxide	SO_2	−73.2	2060	32.2
Sulfur trioxide (α)	SO_3	16.8	2060	25.8
Sulfur trioxide (β)	SO_3	32.3	2890	36.1
Sulfur trioxide (γ)	SO_3	62.1	6310	79.0
Tantalum	Ta	2996 ± 50	(7500)	34.6 to 41.5
Tantalum pentachloride	$TaCl_5$	206.8	9000	25.1
Tantalum pentoxide	Ta_2O_5	1877	48000	108.6
Tellurium	Te	453	3230	25.3
Terbium	Tb	1356	3900	24.6
Thallium	Tl	302.4	1030	5.0
Thallium bromide, mone-	TlBr	460	5990	21.0
Thallium carbonate	Tl_2CO_3	273	4400	9.5
Thallium chloride, mono-	TlCl	427	4260	17.7
Thallium iodide, mono-	TlI	440	3125	9.4
Thallium nitrate	$TlNO_3$	207	2290	8.6
Thallium sulfide	Tl_2S	449	3000	6.8
Thallium sulfate	Tl_2SO_4	632	5500	10.9
Thorium	Th	1845	(<4600)	(<19.8)
Thorium dioxide	ThO_2	2952	291100	1102.0
Thorium chloride	$ThCl_4$	765	22500	61.6
Thulium	Tm	1545	4400	26.0
Thulium chloride	$TmCl_3$	821	(9000)	(32.6)
Tin	Sn	231.7	1720	14.4
Tin bromide, ic-	$SnBr_4$	29.8	3000	6.8
Tin bromide, ous-	$SnBr_2$	231.8	(1720)	(6.1)
Tin chloride, ic-	$SnCl_4$	−33.3	2190	8.4
Tin chloride, ous-	$SnCl_2$	247	3050	16.0
Tin iodide, ic-	SnI_4	143.4	(4330)	(6.9)
Tin oxide	SnO	1042	(6400)	(46.8)
Titanium	Ti	1800	(5000)	(104.4)
Titanium bromide, tetra-	$TiBr_4$	38	(2060)	(5.6)
Titanium chloride, tetra-	$TiCl_4$	−23.2	2240	11.9
Titanium dioxide	TiO_2	1825	(11400)	(142.7)
Titanium oxide	TiO	991	14000	219
Tungsten	W	3387	(8420)	(45.8)
Tungsten dioxide	WO_2	1270	13940	60.1
Tungsten hexafluoride	WF_6	−0.5	1800	6.0
Tungsten tetrachloride	WCl_4	327	6000	18.4
Tungsten trioxide	WO_3	1470	13940	60.1
Uranium^{235}	U	~1133	3700	20
Uranium tetrachloride	UCl_4	590	10300	27.1
Vanadium	V	1917	(4200)	(70)
Vanadium dichloride	VCl_2	1027	8000	65.6
Vanadium oxide	VO	2077	15000	224.0
Vanadium pentoxide	V_2O_5	670	15560	85.5
Xenon	Xe	−111.6	740	5.6
Ytterbium	Yb	823	2200	12.7
Yttrium	Y	1504	4100	46.1
Yttrium oxide	Y_2O_3	2227	25000	110.7
Yttrium trichloride	YCl_3	709	(9000)	(46)
Zinc	Zn	419.4	1595	24.4
Zinc chloride	$ZnCl_2$	283	(5540)	(40.6)
Zinc oxide	ZnO	1975	4470	54.9
Zinc sulfide	ZnS	1700 ± 20	(9100)	(93.3)
Zirconium	Zr	1857	(5500)	(60)
Zirconium dichloride	$ZrCl_2$	727	7300	45.0
Zirconium oxide	ZrO_2	2715	20800	168.8

Table of the Isotopes

Compiled by
Russell L. Heath
National Reactor Testing Station
Idaho Nuclear Corporation
Idaho Falls, Idaho

The following describes the information contained in each column of this table:

Column 1. This column lists the isotopes with the atomic number as a subscript and the atomic weight as a superscript. Isomers are indicated by the addition of the letter m to the superscript. Although they are not isotopes, the elements and their corresponding atomic numbers are also included in this column.

Column 2. % Natural Abundance. This column lists isotopic abundance in percent.

Column 3. Atomic Mass. This column lists atomic mass of isotopes in the physical scale. The atomic masses are all relative to the mass of the Carbon 12 isotope which has arbitrarily been assigned 12.00000 atomic mass units as recommended by the International Union of Pure and Applied Chemistry and the International Commission of Atomic Weights.

Column 4. Lifetime. This column lists the half-life of the radioactive isotopes. The notation used is: μs = microseconds, ms = milliseconds, s = seconds, m = minutes, d = days, and y = years.

Column 5. This column lists the observed modes of decay for all radioactive isotopes. Symbols used are: β^- = negative beta emission, β^+ = positron emission, α = alpha particle decay, E.C. = orbital electron capture, I.T. = isomeric transition from upper to lower isomeric state, n = neutron emission, and S.F. = spontaneous fission.

Column 6. Decay Energy. This lists the currently adopted values for the total disintegration energies (Q) in MeV. Where more than one mode of decay exists separate values are given.

Column 7. Particle Energies. This column lists the end-point energies of beta particle transitions or discrete energies of alpha particles in MeV.

Column 8. Particle Intensities. This column lists the intensities of beta groups or alpha particle transitions in percent.

Column 9. Thermal Neutron Capture Cross Section. This column lists cross sections for thermal-neutron capture in barns (10^{-24} cm^2) or in millibarns (mb). Where neutron capture for a given nucleus may produce two isomers, the separate cross sections are listed if known. Fission cross sections are listed for the heavy elements and are designated by the symbol σ_f.

Column 10. I. This table lists the Nuclear Spin or Angular Momentum in units of $h/2\pi$.

Column 11. μ. This column lists values of the Nuclear Magnetic Moments of the isotope in nuclear magnetons, with diamagnetic correction.

Isotope	% nat. abundance	Atomic mass	Lifetime	Modes of decay	Decay energy (MeV)	Particle energies (MeV)	Particle intensities	Thermal neutron capture cross-section	I	μ
n^1		1.008665	12m	β^-	0.7825	0.7825	100%		1/2	−1.9131
H		1.00797						0.332 ± 2mb		
$_1H$	99.985	1.007825							1/2	+2.79278
$_1H^2$	0.015	2.0140						0.51 ± 0.01mb	1	+0.85742
$_1H^3$		3.01605	12.26y	β^-	0.01861	0.01861	100%	<6μb	1/2	
He		4.0026						≥7mb		
$_2He^3$	0.00013	3.01603						<0.1mb	1/2	−2.1275
$_2He^4$	100	4.00260						≈0	0	
$_2He^5$		5.0123								
$_2He^6$		6.01888	0 81s	β^-	3.5098	3.5098	100%		0	<0.16
$_2He^8$		8.0375	0.122s	β^- η	13	13	88% 12%			
Li		6.941								
$_3Li^5$		5.0125								
$_3Li^6$	7.42	6.01512						45 ± 10mb	1	+0.82202
$_3Li^7$	92.58	7.01600						37 ± 4mb	3/2	+3.2564
$_3Li^8$			0.855s	β^-, α	13	13			(2)	+1.6532
$_3Li^9$			0.175s	β^-, η	13.5	13.5 11	75% 25%			
Be		9.0122						9.2 ± 0.5mb		
$_4Be^6$		6.0197								
$_4Be^7$		7.0169	53.37d	EC	0.861					
$_4Be^8$		8.0053	2×10^{-16}s	α						
$_4Be^9$	100%	9.01218						9.2 ± 0.5mb	3/2	−1.1776
$_4Be^{10}$		10.0135	2.5×10^6y	β^-	0.55	0.55	100%	≤1mb		
$_4Be^{11}$		11.0216	13.6s	β^-	11.5	11.5 9.3 4.7 3.6	61% 29% 6% 4%			
B		10.811								
$_5B^8$		8.0246	0.77s	β^+	18	13.7				
$_5B^9$		9.0133	8×10^{-19}s	P						
$_5B^{10}$	19.78	10.0129						($\sigma[\eta, \alpha]$ 3836b) 0.5 ± 0.2b	3	+1.8007
$_5B^{11}$	80.22	11.00931						5 ± 3mb	3/2	+2.6885
$_5B^{12}$		12.0143	0.02s	β^-, α	13.4	13.4 9.0	~100% 1.3%		(1)	+1.003
$_5B^{13}$		13.0178	0.019s	β^-	13.4	13.4	93%			

Isotope	% nat. abundance	Atomic mass	Lifetime	Modes of decay	Decay energy (MeV)	Particle energies (MeV)	Particle intensities	Thermal neutron capture cross-section	I	μ
C		12.01115						3.4 ± 0.2mb		
$_6C^{10}$			19.45s	β^+	3.61	1.87				
$_6C^{11}$			20.3min	β^+, EC	1.98	0.98			3/2	±1.03
$_6C^{12}$	98.89	12.0000						3.4 ± 0.3mb	0	
$_6C^{13}$	1.11	13.00335						0.9 ± 0.2mb	1/2	+0.7024
$_6C^{14}$			5730y	β^-	0.156	0.156		$<10^{-6}$b	0	
$_6C^{15}$			2.4s	β^-	9.8	9.82 4.51	32% 68%			
$_6C^{16}$			0.74s	β^-, η						
N		14.0067								
$_7N^{12}$			0.011s	β^+	16.4	16.38				
$_7N^{13}$										
$_7N^{13}$			9.97m	β^+	1.19	1.19			1/2	±0.3221
$_7N^{14}$	99.63	14.00307						75 ± 7.5mb	1	+0.4036
$_7N^{15}$	0.37	15.00011						24 ± 8μb	1/2	−0.2831
$_7N^{16}$			7.2s	β^-	10.44	10.44 4.27	26% 68%			
$_7N^{17}$			4.16s	$\beta^-, (\eta)$	8.68	8.7 7.8 4.12 3.3 2.7	1.6% 2.6% 45% 45% 5%			
$_7N^{18}$			0.63s	β^-	13.9	9.4	100%			
O		15.9994								
$_8O^{13}$			0.0087s	β^+, (P)	17.8	6.40 (proton) 6.79 (proton)	100† 24			
$_8O^{14}$			71.0s	β^+	5.14	1.813 4.09	99.35% 0.65%			
$_8O^{15}$			124s	β^+	2.76	1.70	100%		1/2	±0.7189
$_8O^{16}$	99.759	15.99491						0.178 ± 0.025mb	0	
$_8O^{17}$	0.037							235 ± 10mb	5/2	−1.8937
$_8O^{18}$	0.204							0.21 ± 0.04mb	0	
$_8O^{19}$			29s	β^-	4.819	4.60 3.25	42% 58%		5/2	
$_8O^{20}$			14s	β^-	3.81	2.7				
F		18.9984						9.8 ± 0.7mb		
$_9F^{17}$			66s	β^+	2.76	1.73	100%		(5/2)	±4.722
$_9F^{18}$			109.7m	β^+, EC	1.65	0.635 EC	97% 3%			
$_9F^{19}$	100%	18.99840						9.8 ± 0.7mb	1/2	+2.6288

Isotope	% nat. abundance	Atomic mass	Lifetime	Modes of decay	Decay energy (MeV)	Particle energies (MeV)	Particle intensities	Thermal neutron capture cross-section	I	μ
$_9F^{20}$			11.4s	β^-					(2)	+2.094
$_9F^{21}$			4.4s	β^-	5.68	5.4 4.0				
$_9F^{22}$			4.0s	β^-	12	11				
Ne		20.183						38 ± 10mb		
$_{10}Ne^{17}$			0.10s	β^+, (P)						
$_{10}Ne^{18}$			1.5s	β^+	4.45	3.42 2.38	93% 7%			
$_{10}Ne^{19}$			17.5s	β^+	3.24	2.24			1/2	−1.887
$_{10}Ne^{20}$	90.92%	19.99244						38 ± 10mb	0	
$_{10}Ne^{21}$	0.257%	20.99395							3/2	−0.6618
$_{10}Ne^{22}$	8.82%	21.99138						36 ± 10mb	0	
$_{10}Ne^{23}$			37.6s	β^-	4.38	4.38 3.95 2.4	67% 32% 1.0%			
$_{10}Ne^{24}$			3.38m	β^-	2.47	1.98 1.10	92% 8%			
Na		22.9898						534 ± 5mb		
$_{11}Na^{20}$			0.39s	$\beta^+(\alpha)$	14					
$_{11}Na^{21}$			23s	β^+	3.54	2.51	98%		3/2	+2.386
$_{11}Na^{22}$			2.602y	β^+, EC				40,000 ± 5000b	3	+1.746
$_{11}Na^{23}$	100%	22.9898						400 ± 30mb	3/2	+2.2175
$_{11}Na^{24}$			15.0h	β^-	5.51	4.17 1.389	0.003% 99+%		4	+1.690
$_{11}Na^{25}$			60s	β^-	3.83	4.0 3.1 2.6	65% 25% 7%			
$_{11}Na^{26}$			1.0s	β^-	8.5	6.7	>80%			
Mg		24.312						64 ± 2mb		
$_{12}Mg^{20}$			0.6s	β^+	9					
$_{12}Mg^{21}$			0.12s	β^+(P)						
$_{12}Mg^{23}$			12.1s	β^+	4.06	3.10				
$_{12}Mg^{24}$	78.70%	23.98504						52 ± 15mb	0	
$_{12}Mg^{25}$	10.13%	24.98584						0.18 ± 0.05b	5/2	−0.8553
$_{12}Mg^{26}$	11.17%	25.98259						0.03 ± 0.005b	0	
$_{12}Mg^{27}$			9.5min	β^-	2.61	1.75 1.59	58% 42%			

TABLE OF THE ISOTOPES (*Continued*)

Isotope	% nat. abundance	Atomic mass	Lifetime	Modes of decay	Decay energy (MeV)	Particle energies (MeV)	Particle intensities	Thermal neutron capture cross-section	I	μ
$_{12}Mg^{28}$			21h	β^-	1.84	0.46				
Al		26.98153						232 ± 3mb		
$_{13}Al^{24}$			2.1s	$\beta^+(\alpha)$	14	8.5				
$_{13}Al^{25}$			7.2s	β^+	4.26	3.24	100%			
$_{13}Al^{26m}$			6.4s	β^+		3.21				
$_{13}Al^{26}$			7.4×10^5y	β^+, EC	4.003	1.16				
$_{13}Al^{27}$	100%	26.98153						232 ± 3mb	5/2	+3.6414
$_{13}Al^{28}$			2.31m	β^-	4.635	2.82	100%			
$_{13}Al^{29}$			6.6min	β^-	3.68	2.5 1.4	70% 30%			
$_{13}Al^{30}$			3.3s	β^-	7.3	5.0				
Si		28.086						160 ± 20mb		
$_{14}Si^{25}$			0.23s	β^+(P)						
$_{14}Si^{26}$			2.1s	β^+	5.08	3.83 3.00	66† 34			
$_{14}Si^{27}$			4.2s	β^+	4.810	3.85				
$_{14}Si^{28}$	92.21%	27.97693						160 ± 40mb	0	
$_{14}Si^{29}$	4.70%	28.97649						280 ± 90mb	1/2	−0.5553
$_{14}Si^{30}$	3.09%	29.97376						100 ± 10mb	0	
$_{14}Si^{31}$			2.62h	β^-	1.48	1.49	99+%			
$_{14}Si^{32}$			~650y	β^-	0.21	0.21				
P		30.9738						190 ± 10mb		
$_{15}P^{28}$			0.28s	β^+	13.8	10.6	50%			
$_{15}P^{29}$			4.4s	β^+	4.95	3.96				
$_{15}P^{30}$			2.50m	β^+	4.24	3.27				
$_{15}P^{31}$	100%	30.97376						190 ± 10mb	1/2	+1.1317
$_{15}P^{32}$			14.3d	β^-	1.710	1.710	100%		1	−0.2523
$_{15}P^{33}$			25d	β^-	0.248	0.249	100%			
$_{15}P^{34}$			12.4s	β^-	5.1	5.1 3.2	75% 25%			
S		32.064								
$_{16}S^{29}$			0.19s	β^+(P)						
$_{16}S^{30}$			1.4s	β^+	6.11	5.09 4.42	20† 80			
$_{16}S^{31}$			2.7s	β^+	5.44	4.39				

TABLE OF THE ISOTOPES (*Continued*)

Isotope	% nat. abundance	Atomic mass	Lifetime	Modes of decay	Decay energy (MeV)	Particle energies (MeV)	Particle intensities	Thermal neutron capture cross-section	I	μ
$_{16}S^{32}$	95.0%	31.97207							0	
$_{16}S^{33}$	0.76%	32.97146						15 ± 10mb	3/2	+0.6433
$_{16}S^{34}$	4.22%	33.96786						200 ± 100mb	0	
$_{16}S^{35}$			88d	β^-	0.1674	0.1674	100%		3/2	±1.0
$_{16}S^{36}$	0.014%	35.96709						140 ± 40mb	0	
$_{16}S^{37}$			5.06m	β^-	4.8	4.3 1.6	10% 90%			
$_{16}S^{38}$			2.87h	β^-	3.0	3.0 1.1	5% 95%			
Cl		35.453								
$_{17}Cl^{32}$			0.31s	$\beta^+(\alpha)$	13.2	9.5	~50%			
$_{17}Cl^{33}$			2.5s	β^+	5.57	4.51				
$_{17}Cl^{34m}$			32.0m	β^+, IT		2.5 1.3 4.5	26† 26 47			
$_{17}Cl^{34}$			1.56s	β^+	5.482	4.50				
$_{17}Cl^{35}$	75.53%	34.96885						44 ± 2b	3/2	+0.82183
$_{17}Cl^{36}$			3.1×10^5y	β^- β^+, EC	0.712 1.14	0.714 (β^-)	98.1% 1.9%	100 ± 30b	2	+1.285
$_{17}Cl^{37}$	24.47%							430 ± 100mb	3/2	+0.68411
$_{17}Cl^{38m}$			0.74s	IT						
$_{17}Cl^{38}$			37.3m	β^-	4.91	4.81 2.77 1.11	53% 16% 31%			
$_{17}Cl^{39}$			55.5m	β^-	3.44	3.45 2.18 1.91	7% 8% 85%			
$_{17}Cl^{40}$			1.4m	β^-	7.5	~7.5 ~3.2				
Ar		39.948								
$_{18}Ar^{35}$			1.83s	β^+	5.96	4.93	93%		(3/2)	+0.632
$_{18}Ar^{36}$	0.337%	35.96755						6 ± 2b	0	
$_{18}Ar^{37}$			35d	EC	0.814	EC	100%		3/2	+0.95
$_{18}Ar^{38}$	0.063%	37.96272						0.8 ± 0.2b	0	
$_{18}Ar^{39}$			265y	β^-	3.44	0.565	100%	500 ± 200b	7/2	−1.3
$_{18}Ar^{40}$	99.60%							650 ± 30mb	0	
$_{18}Ar^{41}$			1.83h	β^-	2.491	1.198 2.49	99% 0.78%	0.5 ± 0.1b		
$_{18}Ar^{42}$			33y	β^-	0.60					
K		39.0983								
$_{19}K^{37}$			1.23s	β^+	6.16	~5.1	98%			

TABLE OF THE ISOTOPES (*Continued*)

Isotope	% nat. abundance	Atomic mass	Lifetime	Modes of decay	Decay energy (MeV)	Particle energies (MeV)	Particle intensities	Thermal neutron capture cross-section	I	μ
$_{19}K^{38m}$			0.95s	β^+		5.0	100%			
$_{19}K^{38}$			7.71m	β^+, EC	5.93	2.68			3	+1.374
$_{19}K^{39}$	93.1%	38.96371						2.2 ± 0.2b	3/2	0.3914
$_{19}K^{40}$	0.0118%	39.974	1.28 × 10^9y	β^-, β^+, EC	1.35 1.505	β^- β^+	1.35% 0.49%	70 ± 20b	4	−1.298
$_{19}K^{41}$	6.88%							1.3 ± 0.2b	3/2	+0.2149
$_{19}K^{42}$			12.4h	β^-	0.60	3.52 1.97	82% 18%		2	−1.140
$_{19}K^{43}$			22.4h	β^-	1.82	1.81 1.24 0.84 0.46	1.3% 3.5% 87% 8%		3/2	±0.163
$_{19}K^{44}$			22.0m	β^-	5.2	5.3 4.0 2.6	~35% 9% 38%			
$_{19}K^{45}$			16m	β^-	4.2	4.0 2.1 1.1	15† 60 25		3/2	±0.173
Ca		40.08						0.44 ± 0.02b		
$_{20}Ca^{37}$			0.173s	β^+(P)	11.5					
$_{20}Ca^{38}$			0.66s	β^+						
$_{20}Ca^{39}$			0.87s	β^+	6.50	5.49				
$_{20}Ca^{40}$	96.947%	39.96259						430 ± 40mb	0	
$_{20}Ca^{41}$			8 × 10^4y	EC	0.427				7/2	−1.595
$_{20}Ca^{42}$	0.646%	41.95863						700 ± 160mb		
$_{20}Ca^{43}$	0.135%	42.95878						6.2 ± 1.1b	7/2	−1.317
$_{20}Ca^{44}$	2.083%	43.95549						1.1 ± 0.3b		
$_{20}Ca^{45}$			165d	β^-	0.252	0.258	100%			
$_{20}Ca^{46}$	0.186%	45.9537						250 ± 100mb		
$_{20}Ca^{47}$			4.53d	β^-	1.979	1.979 1.48 0.67	16% 2% 82%			
$_{20}Ca^{48}$	0.18%	47.9524						1.1 ± 0.1b		
$_{20}Ca^{49}$			8.8m	β^-	5.26	1.95 0.9	88† 12			
$_{20}Ca^{50}$			9s	β^-	3.9					
Sc		44.95592						25 ± 2b		
$_{21}Sc^{40}$			0.179s	β^+	14.5	9.1				
$_{21}Sc^{41}$			0.60s	β^+	6.50	5.6	100%			
$_{21}Sc^{42m}$			61s	β^+		2.82				

TABLE OF THE ISOTOPES (*Continued*)

Isotope	% nat. abundance	Atomic mass	Lifetime	Modes of decay	Decay energy (MeV)	Particle energies (MeV)	Particle intensities	Thermal neutron capture cross-section	I	μ
$_{21}Sc^{42}$			0.68s	β^+	6.431	5.3	100%			
$_{21}Sc^{43}$			3.92h	β^+, EC	2.2	1.20 0.82 0.50	79† 17 4		7/2	+4.62
$_{21}Sc^{44m}$			2.44d	IT EC		98.6% 1.4%			6	+3.88
$_{21}Sc^{44}$			3.92h	β^+	3.647	1.471	99%		2	+2.56
$_{21}Sc^{45}$	100%	44.95592						25 ± 2b 11 ± 4b (Sc^{46m})	7/2	+4.7564
$_{21}Sc^{46m}$			20s	IT	0.142					
$_{21}Sc^{46}$			83.80d	β^-	2.367	1.48 0.357	0.004% 100%	8.0 ± 1b	4	+3.03
$_{21}Sc^{47}$			3.43d	β^-	0.60	0.600 0.439	40% 60%		7/2	+5.34
$_{21}Sc^{48}$			1.83d	β^-	3.98	0.65 0.47	94% 6%		6	
$_{21}Sc^{49}$			57.5m	β^-	2.008	2.01	100%			
$_{21}Sc^{50m}$			0.35s	IT	0.258					
$_{21}Sc^{50}$			1.72m	β^-	6.5	3.6				
Ti		47.90						6.1 ± 0.2b		
$_{22}Ti^{43}$			0.56s	β^+	6.8	5.5	100%			
$_{22}Ti^{44}$			48y	EC	0.16					
$_{22}Ti^{45}$			3.09h	β^+, EC	2.059	1.02	99+%		7/2	±0.095
$_{22}Ti^{46}$	7.93%	45.95263						0.6 ± 0.2b		
$_{22}Ti^{47}$	7.28%	46.9518						1.7 ± 0.3b	5/2	−0.7883
$_{22}Ti^{48}$	73.94%							8.3 ± 0.6b		
$_{22}Ti^{49}$	5.51%	48.94787						1.9 ± 0.5b	7/2	−1.1039
$_{22}Ti^{50}$	5.34%	49.9448						140 ± 30mb		
$_{22}Ti^{51}$			5.8m	β^-	2.46	2.13 1.50	94% 6%			
V		50.942						5.06 ± 0.06b		
$_{23}V^{46}$			0.426s	β^+	7.054	6.0	100%			
$_{23}V^{47}$			33m	β^+, EC	2.92	1.89				
$_{23}V^{48}$			16.0d	β^+, EC	4.013	0.698				
$_{23}V^{49}$			330d	EC	0.601					
$_{23}V^{50}$	0.24%	49.9472	6×10^{15}y	EC				100 ± 60b	6	+3.3470
$_{23}V^{51}$	99.76%	50.9440						4.8 ± 0.2b	7/2	+5.149

TABLE OF THE ISOTOPES (*Continued*)

Isotope	% nat. abundance	Atomic mass	Lifetime	Modes of decay	Decay energy (MeV)	Particle energies (MeV)	Particle intensities	Thermal neutron capture cross-section	I	μ
$_{23}V^{52}$			3.76m	β^-	3.97	2.47				
$_{23}V^{53}$			2.0m	β^-	3.5	2.50	100%			
$_{23}V^{54}$			55s	β^-	7.3	3.3	100%			
Cr		51.996						3.1 ± 0.2b		
$_{24}Cr^{48}$			23h	EC	1.4					
$_{24}Cr^{49}$			41.9m	β^-	5.26	1.54 1.39 0.73	50† 35 15		5/2	±0.48
$_{24}Cr^{50}$	4.31%	49.9461						16.0 ± 0.5b		
$_{24}Cr^{51}$			27.8d	EC	0.752				7/2	±0.94
$_{24}Cr^{52}$	83.76%	51.9405						0.76 ± 0.06b		
$_{24}Cr^{53}$	9.55%	52.9407						18.2 ± 1.5b	3/2	−0.4744
$_{24}Cr^{54}$	2.38%	53.9389						380 ± 40mb		
$_{24}Cr^{55}$			3.5m	β^-	2.59	2.50	100%			
$_{24}Cr^{56}$			5.9m	β^-	1.6	1.5				
Mn								13.3 ± 0.1b		
$_{25}Mn^{50m}$			2m	β^+, EC						
$_{25}Mn^{50}$			0.286s	β^+	7.631	6.6	100%			
$_{25}Mn^{51}$			45m	β^+, EC	3.19	2.17	97%		5/2	±3.60
$_{25}Mn^{52m}$			21m	IT, β^+, EC	0.383	2% 98%			2	±0.0077
$_{25}Mn^{52}$			5.7d	β^+, EC	4.708	0.575			6	±3.075
$_{25}Mn^{53}$			2 × 10^6y	EC	0.598		100%	~170b	7/2	±5.05
$_{25}Mn^{54}$			303d	EC	1.379			<10b	3	±3.30
$_{25}Mn^{55}$	100%	54.9381						13.3 ± 0.1b	5/2	±3.444
$_{25}Mn^{56}$			2.576h	β^-	3.702	2.84 1.03 0.72 0.30	47% 34% 18% 1%		3	+3.218
$_{25}Mn^{57}$			1.7m	β^-	2.7	2.55 1.10	82% 18%			
$_{25}Mn^{58}$			1.1m	β^-	6.5					
Fe		55.847						2.56 ± 0.05b		
$_{26}Fe^{52}$			8.2h	β^+, EC	2.37	0.80				
$_{26}Fe^{53}$			8.5m	β^+, EC	3.98	2.8 2.4 1.6	50† 38 12			
$_{26}Fe^{54}$	5.82%	53.9396						2.8 ± 0.4b		

TABLE OF THE ISOTOPES (*Continued*)

Isotope	% nat. abundance	Atomic mass	Lifetime	Modes of decay	Decay energy (MeV)	Particle energies (MeV)	Particle intensities	Thermal neutron capture cross-section	I	μ
$_{26}Fe^{55}$			2.6y	EC	0.232					
$_{26}Fe^{56}$	91.66%	55.9349						2.5 ± 0.3b		
$_{26}Fe^{57}$	2.19%	56.9354						2.5 ± 0.2b	1/2	+0.0902
$_{26}Fe^{58}$	0.33%	57.9333						1.23 ± 0.05b		
$_{26}Fe^{59}$			45.1 ± 0.5d	β^-		1.573 0.475 0.273	0.3% 51% 48%		3/2	
$_{26}Fe^{60}$			3×10^5y	β^-	0.14					
$_{26}Fe^{61}$			6.0m	β^-	3.8	2.8				
Co		58.9332						37.5 ± 0.2b		
$_{27}Co^{54m}$			1.5m	β^+		4.3	100%			
$_{27}Co^{54}$			0.1945 s	β^+	8.252	>7.4				
$_{27}Co^{55}$			18.2h	β^+, EC	3.46	1.51 1.04 0.79	45† 33 2		(7/2)	±4.3
$_{27}Co^{56}$			77d	β^+, EC	4.57	1.50 0.44	96† 4		4	±3.85
$_{27}Co^{57}$			270d	EC	0.837				7/2	±4.65
$_{27}Co^{58m}$			9.0h	IT	0.025			1.4×10^5 $\pm 1 \times 10^4$b		
$_{27}Co^{58}$			71.3d	β^+, EC	2.309	1.3 0.474	0.0006% 83% EC 15% β^+	1700 ± 200b	2	±4.06
$_{27}Co^{59}$	100%	58.9332						19.9 ± 0.91b (Co^{60m}) 17 ± 2b (Co^{60})	7/2	+4.649
$_{27}Co^{60m}$			10.5m	IT β^-	0.05860	 1.56	99% 1%	58 ± 8b		
$_{27}Co^{60}$			5.26 ± 0.01y		2.819	0.315 0.663 1.488	99.87% 0.008% 0.12%	2.0 ± 0.2b	5	+3.81
$_{27}Co^{61}$			1.650h ± 0.05	β^-	1.29	1.22 ± 0.04	100%			
$_{27}Co^{62m}$			1.6m	β^-						
$_{27}Co^{62}$			13.9m	β^-	5.22	2.88 0.88	75% 25%			
$_{27}Co^{63}$			52s	β^-	3.6	3.6	100%			
Ni		58.71						4.54 ± 0.1b		
$_{28}Ni^{56}$			6.10d	EC	2.11					
$_{28}Ni^{57}$			36.0h	β^+, EC	3.24	0.85 0.72 0.35	87† 11 2			
$_{28}Ni^{58}$	68.274%	57.9353						4.4 ± 0.4b		

TABLE OF THE ISOTOPES (*Continued*)

Isotope	% nat. abundance	Atomic mass	Lifetime	Modes of decay	Decay energy (MeV)	Particle energies (MeV)	Particle intensities	Thermal neutron capture cross-section	I	μ
$_{28}Ni^{59}$			8×10^4y	EC	1.072		100%			
$_{28}Ni^{60}$	26.095%	58.9332						2.6 ± 0.2b		
$_{28}Ni^{61}$	1.134%	60.9310						2.5 ± 0.9b	3/2	−0.7487
$_{28}Ni^{62}$	3.593%	61.9283						15 ± 2b		
$_{28}Ni^{63}$			92y	β^-	0.067	0.067	100%			
$_{28}Ni^{64}$	0.904%	63.9280						1.52 ± 0.24b		
$_{28}Ni^{65}$			2.521h ± 0.005	β^-	2.139	0.412 0.513 0.655 1.022 2.139	0.7% 1.2% 30.2% 9.9% 58%	20 ± 2b		
$_{28}Ni^{66}$			54.6h ± 0.9	β^-	0.20	0.20	100%			
$_{28}Ni^{67}$			50s	β	4.1	~2 2.3 3.2 4.1	22% 24% 4% 50%			
Cu		63.54						3.8 ± 0.1b		
$_{29}Cu^{58}$			3.20s	β^+	8.569	8.2				
$_{29}Cu^{59}$			82.0s ± 0.4	β^+, EC	4.8	2.10 2.48 2.90 3.32 3.44 3.75	1.4% 10.9% 7.8% 4.1% 3.6% 70.8%			
$_{29}Cu^{60}$			23.0m ± 0.3	β^+, EC	6.12	0.52 0.55 0.61 0.77 1.02 1.08 1.71 1.83 1.91 2.00 2.48 3.00 3.92	0.1% 0.2% 0.2% 0.3% 1.0% 1.0% 0.8% 2.4% 11.8% 47.0% 4.2% 14.4% 8.4%		2	
$_{29}Cu^{61}$			3.41h	β^+, EC	2.242	0.307 0.560 0.933 1.148 1.216	0.03% 2.5% 6.3% 2.4% 52%		3/2	+2.13
$_{29}Cu^{62}$			9.8m	β^+, EC	3.939	0.87 1.75 2.923	1.5% 1.8% 93.9%		1	
$_{29}Cu^{63}$	69.09%	62.9298						4.5 ± 0.2b	3/2	+2.226
$_{29}Cu^{64}$			12.9h	β^- β^+, EC	0.573 1.677	0.573 0.654	39.6% 19.3%	<6000b	1	−0.216

TABLE OF THE ISOTOPES (*Continued*)

Isotope	% nat. abundance	Atomic mass	Lifetime	Modes of decay	Decay energy (MeV)	Particle energies (MeV)	Particle intensities	Thermal neutron capture cross-section	I	μ
${}_{29}Cu^{65}$	30.91%	64.9278						2.3 ± 0.3b	3/2	+2.385
${}_{29}Cu^{66}$			5.10m ±0.02	β^-	2.633	0.760 1.64 2.63	0.25% 9% 91%	130 ± 30b	1	±0.283
${}_{29}Cu^{67}$			61.88h ±0.14	β^-	0.576	0.183 0.395 0.484 0.577	0.9% 51% 28% 20%			
${}_{29}Cu^{68}$			30s	β^-	4.6	3.5 2.7 2.3	75% 22% 3%			
Zn		65.37						1.10 ± 0.04b		
${}_{30}Zn^{60}$			2.1m	β^+, EC	4.16					
${}_{30}Zn^{61}$			88 ± 1s	β^+, EC	5.91	4.27 4.94 5.44 5.91	7.8% 1.1% 10.8% 79.6%			
${}_{30}Zn^{62}$			9.15 ± 0.03h	β^+, EC	1.679	0.66	15.2%			
${}_{30}Zn^{63}$			38.40 ± 0.04m	β^+, EC	3.365	0.26 0.79 0.932 1.40 1.69 2.34	0.002% 0.08% 0.3% 6.9% 9.7% 76%		3/2	−0.282
${}_{30}Zn^{64}$	48.89%	63.9291						0.82 ± 0.01b	0	
${}_{30}Zn^{65}$			243.6d ±0.1	β^+, EC	1.353	0.325	1.54%		5/2	+0.769
${}_{30}Zn^{66}$	27.81%	65.9260							0	
${}_{30}Zn^{67}$	4.11%	65.9271							5/2	+0.8755
${}_{30}Zn^{68}$	18.57%	67.9249						0.09 ± 0.01b (Zn^{69m}) 1.0 ± 0.1b (Zn^{69})	0	
${}_{30}Zn^{69m}$			13.9h	IT	0.4387					
${}_{30}Zn^{69}$			58 ± 2m	β^-	0.925	0.925	100%			
${}_{30}Zn^{70}$	0.62%	69.9253						9 ± 1mb (Zn^{71m}) 90 ± 10mb (Zn^{71})		
${}_{30}Zn^{71m}$			4.0h	β^-		1.45				
${}_{30}Zn^{71}$			2.4m	β^-	2.61	2.61 2.1	82% 14%			
${}_{30}Zn^{72}$			46.5h	β^-	0.45	0.30 0.25	90% 10%			

TABLE OF THE ISOTOPES (*Continued*)

Isotope	% nat. abundance	Atomic mass	Lifetime	Modes of decay	Decay energy (MeV)	Particle energies (MeV)	Particle intensities	Thermal neutron capture cross-section	I	μ
Ga	69.72							3.1 ± 0.3b		
$_{31}Ga^{63}$			31s	β^+, EC	5.5	5.5	100%			
$_{31}Ga^{64}$			2.6m	β^+, EC	7.08	0.85 1.31 1.60 2.27 2.64 2.69 2.80 2.87	0.05% 0.04% 1.6% 5.3% 6.9% 31% 2.5% 31%			
$_{31}Ga^{65}$			15.2m	β^+, EC	3.26	0.266 0.769 0.895 1.190 1.327 1.370 1.468 2.030 2.122 2.183 2.237	0.003% 0.64% 0.8% 2.1% 0.4% 6.3% 1.5% 10.3% 53.1% 6.9% 7.2%			
$_{31}Ga^{66}$			9.4h	β^+, EC	5.175	0.367 0.720 0.770 0.935 1.373 1.781 4.153	1.1% 0.16% 0.8% 3.8% 0.1% 0.56% 48.6%		0	
$_{31}Ga^{67}$			78.1h	EC	1.0003				3/2	+1.850
$_{31}Ga^{68}$			68.3m	β^+, EC	2.919	0.82 1.895	1.2% 88%		1	±0.0117
$_{31}Ga^{69}$	60.4%	68.9257						1.8 ± 0.4b	3/2	+2.016
$_{31}Ga^{70}$			21.1m	β^-	1.66	1.65 0.61 0.44	99+% 0.3% 0.5%		1	
$_{31}Ga^{71}$	39.6%	70.9249						0.15 ± 0.05b	3/2	+2.562
$_{31}Ga^{72}$			14.10h	β^-	4.00	1.51 1.97 2.52 3.15	6% 4% 8% 8%		3	−0.1322
$_{31}Ga^{73}$			4.9h	β^-	1.55	0.4 1.19	5% 95%			
$_{31}Ga^{74}$			7.9m	β^-	5.60	2.6 4.0?				
$_{31}Ga^{75}$			2m	β^-	3.3	5.4	96%			
$_{31}Ga^{76}$			32s	β^-		~6				
Ge		72.59						2.3 ± 0.26		

TABLE OF THE ISOTOPES (*Continued*)

Isotope	% nat. abundance	Atomic mass	Lifetime	Modes of decay	Decay energy (MeV)	Particle energies (MeV)	Particle intensities	Thermal neutron capture cross-section	I	μ
$_{32}Ge^{65}$			1.5m	β^+, EC	~6.5	3.8 4.8 5.5	2% 3% 94%			
$_{32}Ge^{66}$			9.4h	β^+, EC	3.0	1.2 1.44 1.60 1.69 1.74 1.87 1.93	2.7% 12.6% 43.1% 1.1% 4.0% 3.8% 16.5%			
$_{32}Ge^{67}$			19m	β^+, EC	4.43	0.68 1.24 1.60 1.96 2.33 2.58 3.08 3.24	0.08% 4.6% 11.4% 2.6% 7.7% 2.7% 13% 50.8%			
$_{32}Ge^{68}$			287d	EC	~0.7					
$_{32}Ge^{69}$			39h	β^+, EC	2.227	0.333 0.631 1.213	0.2% 2.3% 32.5%			
$_{32}Ge^{70}$	20.52%	69.9243						0.28 ± 0.07b	0	
$_{32}Ge^{71m}$			20ms	IT	0.198					
$_{32}Ge^{71}$			11.4d	EC	0.235		100%		1/2	+0.546
$_{32}Ge^{72}$	27.43%	71.9217						0.98 ± 0.09b	0	
$_{32}Ge^{73m}$			0.53s	IT	0.067		100%			
$_{32}Ge^{73}$	7.76%	72.9234						14 ± 1b	9/2	−0.8792
$_{32}Ge^{74}$	36.54%	73.9219						0.14 ± 0.3b	0	
$_{32}Ge^{75m}$			? 48.9s	IT	0.139		100%			
$_{32}Ge^{75}$			82.78m ±0.04	β^-	1.20	0.58 0.72 0.92 0.98 1.19	0.4% 0.3% 11% 1.4% 87%			
$_{32}Ge^{76}$	7.76%	75.9214						0.09 ± 0.02b	0	
$_{32}Ge^{77m}$			54s	IT β^-	0.159 2.75	2.7 2.9	1† 10			
$_{32}Ge^{77}$			11.3 ± 0.01h	β^-	2.75	0.71 1.38 2.20	23% 35% 42%			
$_{32}Ge^{78}$			1.47h	β^-	0.99	0.69 0.71	6% 94%			
As		74.9216						4.30 ± 0.10b		
$_{33}As^{69}$			15m	β^+, EC	3.9	2.9				

TABLE OF THE ISOTOPES (*Continued*)

Isotope	% nat. abundance	Atomic mass	Lifetime	Modes of decay	Decay energy (MeV)	Particle energies (MeV)	Particle intensities	Thermal neutron capture cross-section	I	μ
$_{33}As^{70}$			53.0m $\pm$0.6	β^+, EC	6.222	1.44 2.14 2.89	10† 75 6			
$_{33}As^{71}$			62h	β^+, EC	2.01	0.25 0.81	 30%			
$_{33}As^{72}$			26h	β^+, EC	4.36	0.62 1.027 1.33 1.45 1.84 2.50 3.34	4% 3% 3% 5% 2% 56% 17%			
$_{33}As^{73}$			80.3d	EC	0.37					
$_{33}As^{74m}$			8.0s	IT	0.283					
$_{33}As^{74}$			17.9d	β^- β^+, EC	1.36	0.72 1.35 0.91 1.51	14% 18% 26% 3.5%			
$_{33}As^{75}$	100%	74.9216						4.30 $\pm$ 0.10b	3/2	+1.439
$_{33}As^{76}$			26.5h	β^-	2.97	0.35 1.20 1.75 2.40 2.96	3% 6% 6% 32% 53%		2	−0.905
$_{33}As^{77}$			38.83h $\pm$0.05	β^-	0.686	0.165 0.437 0.477 0.684	0.68% 0.6% 1.6% 97%			
$_{33}As^{78}$			91m	β^-	4.3	1.4 4.1	30% 70%			
$_{33}As^{79}$			9.0m	β^-	2.2	1.25 1.43 1.70 1.80 2.15	1.5% 0.5% 2% 1.5% 95%			
$_{33}As^{80}$			15.3s	β^-	6.0	3.0 3.5 3.7 4.1 4.2 4.5 5.3 6.0	0.4% 0.5% 4.3% 3.6% 1.7% 1.4% 32% 56%			
$_{33}As^{81}$			33s	β^-	3.8	3.8				
Se		78.96						12.2 $\pm$ 0.6b		
$_{34}Se^{70}$			44m	β^+, EC						
$_{34}Se^{71}$			5m	β^+, EC	4.4	3.4				
$_{34}Se^{72}$			8.4d	EC	0.6		100%			

Isotope	% nat. abundance	Atomic mass	Lifetime	Modes of decay	Decay energy (MeV)	Particle energies (MeV)	Particle intensities	Thermal neutron capture cross-section	I	μ
$_{34}Se^{73}$			7.1h	β^+, EC	2.75	0.25 0.75 1.32 1.68	1† 10 88 1			
$_{34}Se^{73}$			42m	β^+, EC		1.8				
$_{34}Se^{74}$	0.87%	73.9225						55 ± 5b	0	
$_{34}Se^{75}$			120.4d	β^-	0.865	0.293 0.465 0.625 0.601	0.1% 94% <2% 3%		5/2	
$_{34}Se^{76}$	9.02%	75.9192						21 ± 2b	0	
$_{34}Se^{77m}$			17.5s	IT	0.161		100%			
$_{34}Se^{77}$	7.58%	76.9199						42 ± 4b	1/2	+0.534
$_{34}Se^{78}$	23.52%	77.9173						0.33 ± 0.04b (Se^{79m}) 0.20 ± 0.106 (Se^{79})	0	
$_{34}Se^{79}$			6.5×10^4y	β^-	0.154	0.154	100%		7/2	−1.02
$_{34}Se^{80}$	49.82%	79.9165						0.08 ± 0.01b	0	
$_{34}Se^{81m}$			57m	IT	0.103		100%			
$_{34}Se^{81}$			18.6m	β^-	1.576	0.74 1.01 1.51	2% 2% ~93%			
$_{34}Se^{82}$	9.19%	81.9167						40 ± 10mb (Se^{83m}) 6 ± 1mb (Se^{83})	0	
$_{34}Se^{83m}$			70s	β^-		1.5 2.4 3.5	25% 25% 50%			
$_{34}Se^{83}$			23m	β^-	3.6	0.83 0.87 0.91 0.96 1.55 2.52 3.25	1% 13% 29% 37% 5% 10% 5%			
$_{34}Se^{84}$			3.3m	β^-						
$_{34}Se^{85}$			39s	β^-						
Br		79.909						6.8 ± 0.1b		
$_{35}Br^{74}$			36m	β^+, EC	~6.8	4.7				
$_{35}Br^{75}$			1.7h	β^+, EC	2.72	0.3 0.6 0.8 1.70	19† 15 20 46			
$_{35}Br^{76}$			16.1h	4.6	β^+, EC	1.2 1.7 2.4 3.1 3.6	~3% 9% 12% 33% ~8%		1	±0.548

TABLE OF THE ISOTOPES (*Continued*)

Isotope	% nat. abundance	Atomic mass	Lifetime	Modes of decay	Decay energy (MeV)	Particle energies (MeV)	Particle intensities	Thermal neutron capture cross-section	I	μ
$_{35}Br^{77}$			57h	β^+, EC	1.365	0.361	0.9%		3/2	
$_{35}Br^{78}$			6.4m	β^+, EC	3.573	1.937 2.52	11% 82%			
$_{35}Br^{79}$	50.54%	78.9183						2.6 ± 0.2b	3/2	+2.106
$_{35}Br^{80m}$			4.4h	IT	0.085				5	+1.317
$_{35}Br^{80}$			17.6m	β^-	β^+, EC β^-	1.871 2.01	β^+ 0.866 β^- 0.70 0.76 1.38 2.05	7.2% 0.23% 1.1% 0.2% 0.004% 0.08%	1	±0.514
$_{35}Br^{81}$	49.46%	80.9163						3.0 ± 0.2b (Br^{82m}) [c] 0.26b (Br^{82})	3/2	+2.270
$_{35}Br^{82m}$			? 6.1m	IT β^-	0.046	1.659 2.357	0.04% 0.2%			
$_{35}Br^{82}$			35.5h	β^-	3.092	0.257 0.440	~2% ~98%		5	±1.626
$_{35}Br^{83}$			2.41h	β^-	0.97	0.395 0.925	1.4% 98.6%			
$_{35}Br^{84}$			31.8m	β^-	4.8	0.89 1.45 2.9 3.9 4.8	19% 14% 15% 14% 32%			
$_{35}Br^{85}$			3.0m	β^-	2.8	2.8	100%			
$_{35}Br^{86}$			54s	β^-	7.1					
$_{35}Br^{87}$			55s	β^-, η	6.1	8.0 2.6	30† 70			
$_{35}Br^{88}$			16s	β^-, η						
$_{35}Br^{89}$			45s	β^-, η						
$_{35}Br^{90}$			1.6s	β^-, η						
Kr		83.80						24.5 ± 1.0b		
$_{36}Kr^{74}$			20m	β^+, EC	4.1	3.1				
$_{36}Kr^{75}$			5m	β^+, EC	~5.1					
$_{36}Kr^{76}$			14.8h	EC	~1.2					
$_{36}Kr^{77}$			1.19h	β^+, EC	2.99	1.64 1.68 1.85 1.86	2.3% 34.4% 40.7% 4.2%			
$_{36}Kr^{78}$	0.35%	77.9204						2.0 ± 0.5b		
$_{36}Kr^{79}$			34.9h	β^+, EC	1.628	0.209 0.299 0.325 0.604	0.03% 0.01% 0.2% 7.3%			

TABLE OF THE ISOTOPES (*Continued*)

Isotope	% nat. abundance	Atomic mass	Lifetime	Modes of decay	Decay energy (MeV)	Particle energies (MeV)	Particle intensities	Thermal neutron capture cross-section	I	μ
$_{36}Kr^{80}$	2.27%	79.9164						14 ± 2b		
$_{36}Kr^{81}$			2.1×10^5y	EC	0.29		100%			
$_{36}Kr^{82}$	11.56%	81.9135						3 ± 1b (Kr^{83m}) 32 ± 15b (Kr^{83})	0	
$_{36}Kr^{83m}$			1.86h	IT						
$_{36}Kr^{83}$	11.55%							180 ± 40b	9/2	−0.970
$_{36}Kr^{84}$	56.90%							0.10 ± 0.03b	0	
$_{36}Kr^{85m}$			4.39hr	IT β^-	0.305	0.82				
$_{36}Kr^{85}$			10.76y	β^-	0.67				9/2	±1.005
$_{36}Kr^{86}$	17.37%							0.06 ± 0.02b	0	
$_{36}Kr^{87}$			76m	β^-	3.89	3.8 3.3 1.3	~70% ~5% 25%			
$_{36}Kr^{88}$			2.80h	β^-	2.8	0.52 0.90 2.8	68% 12% 20%			
$_{36}Kr^{89}$			3.2m	β^-, η	4.6	4.0				
$_{36}Kr^{90}$			33s	β^-, η		~1 2.0 2.80	105 120 236			
$_{36}Kr^{91}$			10s	β^-		~3.6				
$_{36}Kr^{92}$			3.0s	β^-						
$_{36}Kr^{93}$			2.0s	β^-						
$_{36}Kr^{94}$			1.4s	β^-						
Rb		85.47						0.5 ± 0.1b		
$_{37}Rb^{79}$			21m	β^+, EC	~3.6					
$_{37}Rb^{80}$			34s	β^+, EC	5.1	3.5 4.1	40% 60%			
$_{37}Rb^{81m}$			32m	IT	0.085				9/2	
$_{37}Rb^{81}$			4.7h	β^+, EC	2.26	0.575 1.05 1.4	2.3% 30% ~50%		3/2	+2.05
$_{37}Rb^{82m}$			6.4h	β^+		0.80	100%		5	+1.5
$_{37}Rb^{83}$			83d	EC	0.83				5/2	+1.4
$_{37}Rb^{84m}$			20m	EC, IT						
$_{37}Rb^{84}$			33.0d	β^+, EC	2.68	0.78 1.657	53† 47		2	−1.32

Isotope	% nat. abundance	Atomic mass	Lifetime	Modes of decay	Decay energy (MeV)	Particle energies (MeV)	Particle intensities	Thermal neutron capture cross-section	I	μ
$_{37}Rb^{85}$	72.15%	84.9117						55 ± 5mb (Rb^{86m}) 0.45 ± 0.04b (Rb^{86})	5/2	+1.352
$_{37}Rb^{86m}$			1.04m	IT	0.56		100%			
$_{37}Rb^{86}$			18.66d	β^-	1.78	0.71 1.78	8.8% 91.2%		2	−1.691
$_{37}Rb^{87}$	27.85%		5 × 10^{11}y	β^-	0.274	0.274	100%	0.12 ± 0.03b	3/2	+2.750
$_{37}Rb^{88}$			17.8m	β^-	5.2	2.5 3.6 5.3	9% 13% 78%	1.0 ± 0.2b	2	±0.51
$_{37}Rb^{89}$			15.4m	β^-	3.92	0.67 1.61 2.8 3.92	28% 53% 5% 7%			
$_{37}Rb^{90}$			2.9m	β^-	6.6	1.3 2.0 2.2 2.5 4.4 5.8 6.6	10% 5% 17% 23% 15% 10%			
$_{37}Rb^{91}$			1.2m	β^-	~5.5	4.6				
$_{37}Rb^{92}$			5.3s	β^-	~7.6					
$_{37}Rb^{93}$			5.6s	β^-						
$_{37}Rb^{94}$			2.9s	β^-						
$_{37}Rb^{95}$			<2.5s	β^-						
Sr		87.62						1.21 ± 0.06b		
$_{38}Sr^{80}$			1.7h	EC	2.0					
$_{38}Sr^{81}$			29m	β^+, EC	~3.8					
$_{38}Sr^{82}$			25d	EC	~0.4					
$_{38}Sr^{83}$			33h	β^+	2.21	0.44 0.81 1.15	1.3% 7.4% 11%			
$_{38}Sr^{84}$	0.56%	83.9134						0.57 ± 0.5 (Sr^{85m}) 0.31 ± 0.03 (Sr^{85})		
$_{38}Sr^{85m}$			70m	IT, EC	0.238					
$_{38}Sr^{85}$			64d	EC	1.11					
$_{38}Sr^{86}$	9.86%	85.9094						0.8 ± 0.1b (Sr^{87m})	0	
$_{38}Sr^{87m}$			2.83h	IT, EC			~100%			
$_{38}Sr^{87}$	7.02%	86.9089							9/2	−1.093
$_{38}Sr^{88}$	82.56%	87.9056						5 ± 1mb	0	
$_{38}Sr^{89}$			52d	β^-	1.463	1.463 0.55	>99% <0.01%	0.5 ± 0.1b		

TABLE OF THE ISOTOPES (*Continued*)

Isotope	% nat. abundance	Atomic mass	Lifetime	Modes of decay	Decay energy (MeV)	Particle energies (MeV)	Particle intensities	Thermal neutron capture cross-section	I	μ
$_{38}Sr^{90}$			28.1y	β^-	0.546	0.546	100%	0.9 ± 0.5b		
$_{38}Sr^{91}$			9.67h	β^-	2.67	0.62 1.09 1.36 2.03 2.67	7% 33% 29% 4% 26%			
$_{38}Sr^{92}$			2.71h	β^-	1.9	0.55 1.5	90% 10%			
$_{38}Sr^{93}$			8m	β^-	4.8	2.2 2.6 2.8	10% 25% 65%			
$_{38}Sr^{94}$						2.1				
$_{38}Sr^{95}$			0.8m	β^-						
Y		88.905						1.3 ± 0.1b		
$_{39}Y^{82}$			12m	β^+, EC	~7.5					
$_{39}Y^{83}$			7.5m	β^+, EC						
$_{39}Y^{84}$			41m	β^+, EC	6.3	2.9 3.5				
$_{39}Y^{85m}$			2.7h	β^+, EC		1.54	50%			
$_{39}Y^{85}$			5.0h	β^+, EC	3.26	1.1 2.1 2.24	4% 10% 55%			
$_{39}Y^{86m}$			48m	IT						
$_{39}Y^{86}$			14.6h	β^+, EC	5.27	1.04 1.25 1.60 2.02 2.34 3.15	4% 11% 5% 4% 1.1% 0.5%			
$_{39}Y^{87m}$			2.83h	IT	0.388					
$_{39}Y^{87}$			80h	β^+, EC	1.7	0.7	~0.3%			
$_{39}Y^{88}$			106.6d ±0.1	β^+, EC	3.621	0.76				
$_{39}Y^{89m}$			16s	IT	0.909					
$_{39}Y^{89}$	100%	88.9054						1.3 ± 0.1b (Y^{90}) 1.0 ± 0.2mb (Y^{90m})	1/2	−0.1373
$_{39}Y^{90m}$			3.1h	IT β^-	0.685 2.95					
$_{39}Y^{90}$			64h	β^-	2.27	2.273	~100%		2	−1.63
$_{39}Y^{91m}$			50m	IT	0.555					
$_{39}Y^{91}$			58.8d	β^-	1.545	1.545 0.33	99.7% 0.3%	1.4 ± 0.3b	1/2	±0.164

Isotope	% nat. abundance	Atomic mass	Lifetime	Modes of decay	Decay energy (MeV)	Particle energies (MeV)	Particle intensities	Thermal neutron capture cross-section	I	μ
$_{39}Y^{92}$			3.53h	β^-	3.63	1.80 2.14 2.25 2.70 3.63	0.8% 1.5% 2.3% 3.1% 86%			
$_{39}Y^{93}$			10.2h	β^-	2.89	0.71 1.47 1.95 2.62 2.89	1.8% 0.9% 3.0% 4% 90%			
$_{39}Y^{94}$			20.3m	β^-	5.0	5.0				
$_{39}Y^{95}$			10.9m	β^-	~4.2					
$_{39}Y^{96}$			2.3m	β^-	~6.9	3.5				
Zr†		91.22						0.182 ± 0.005b		
$_{40}Zr^{81}$			~10m	β^+, EC						
$_{40}Zr^{82}$			10m	β^+, EC						
$_{40}Zr^{83}$			~7m	β^+, EC						
$_{40}Zr^{84}$			16m	β^+, EC						
$_{40}Zr^{85}$			1.4h	β^+, EC						
$_{40}Zr^{86}$			16.5h	β^+, EC	1.2					
$_{40}Zr^{87}$			1.6h	β^+, EC	3.50	2.10				
$_{40}Zr^{88}$			85d	EC	~0.5					
$_{40}Zr^{89m}$			4.18m	IT, β^+		2.4				
$_{40}Zr^{89}$			78.4h	β^+, EC	2.834					
$_{40}Zr^{90m}$			0.81s	IT	2318.7					
$_{40}Zr^{90}$	51.46%	89.9043						0.10 ± 0.07b		
$_{40}Zr^{91}$	11.23%	90.9053						1.58 ± 0.12b	5/2	−1.303
$_{40}Zr^{92}$	17.11%	91.9046						0.25 ± 0.12b		
$_{40}Zr^{93}$			1.5×10^6y	β^-	0.090	0.060	≥95%	<4b		
$_{40}Zr^{94}$	17.40%	93.9061						75 ± 8mb		
$_{40}Zr^{95}$			65d	β^-	1.121	0.360 0.396 0.890 1.130	43% 55% 2% 0.4%			
$_{40}Zr^{96}$	2.80%	95.9082	$>3.6 \times 10^{17}$y					0.05 ± 0.01b		
$_{40}Zr^{97}$			17h	β^-	2.67	0.46 1.91				
$_{40}Zr^{98}$			1m	β^-	~1.5					
Nb		92.906						1.15 ± 0.05b		

Isotope	% nat. abundance	Atomic mass	Lifetime	Modes of decay	Decay energy (MeV)	Particle energies (MeV)	Particle intensities	Thermal neutron capture cross-section	I	μ
$_{41}Nb^{88}$			14m	β^+, EC	~7.2	3.2				
$_{41}Nb^{89m}$			42m	β^+, EC		3.1				
$_{41}Nb^{89}$			1.9h	β^+, EC	3.9	2.9				
$_{41}Nb^{90m}$			24s	IT						
$_{41}Nb^{90}$			14.6h	β^+, EC	6.11	0.55 0.86 1.50	4† 9 87			
$_{41}Nb^{91m}$			62d	EC IT						
$_{41}Nb^{91}$				EC	1.1					
$_{41}Nb^{92m}$			10.14 ± 0.03d	β^+, EC						
$_{41}Nb^{92}$			>350y	γ						
$_{41}Nb^{93}$	100%	92.9060						0.15 ± 0.1b (Nb^{94m}) 1.15 ± 0.05b (Nb^{94})	9/2	+6.167
$_{41}Nb^{94m}$			6.26 ± 0.01m	IT β^-	0.0415 2.1	1.15				
$_{41}Nb^{94}$			2.0 × 10^4y	β^-	2.06	0.60	100%	11 ± 2b		
$_{41}Nb^{95m}$				IT	0.2357		100%			
$_{41}Nb^{95}$			35.15 ± 0.03d	β^-	0.925	0.1597 0.924	~100% ≤0.07%	<7b		
$_{41}Nb^{96}$			23.4h	β^-	3.15	0.71 0.70 0.80	27% 68% 5%			
$_{41}Nb^{97m}$			1.0m	IT	0.743					
$_{41}Nb^{97}$			72m	β^-	1.93	0.97 1.27	2% 98%			
$_{41}Nb^{98m}$			<2m	β^-						
$_{41}Nb^{98}$			51m	β^-	4.6	3.1				
$_{41}Nb^{99}$			10s	β^-						
$_{41}Nb^{99}$			2.4 m	β^-	3.2					
$_{41}Nb^{100}$			3.0m	β^-	~6.1					
$_{41}Nb^{100}$			11m	β^-	~6.1	3.1 3.5 42	~45% ~45% ~10%			
$_{41}Nb^{101}$			1.0m	β^-						
Mo		95.94						2.65 ± 0.05b		

Isotope	% nat. abundance	Atomic mass	Lifetime	Modes of decay	Decay energy (MeV)	Particle energies (MeV)	Particle intensities	Thermal neutron capture cross-section	I	μ
$_{42}Mo^{88}$			27m	β^+, EC						
$_{42}Mo^{89}$			7m	β^+, EC		4.0 4.9				
$_{42}Mo^{90}$			5.7h	β^+, EC	1.465	1.085	25%			
$_{42}Mo^{91m}$			66s	IT β^+, EC	0.658 6.1					
$_{42}Mo^{91}$			15.49m	IT, β^+	4.46	3.44				
$_{42}Mo^{92}$	15.84%	91.9063						<0.006b (Mo^{93m}) <0.3b (Mo^{93})	0	
$_{42}Mo^{93m}$			6.9h	IT	2.428					
$_{42}Mo^{93}$			>100y	EC						
$_{42}Mo^{94}$	9.04%	93.9047							0	
$_{42}Mo^{95}$	15.72%	94.90584						14.4 ± 0.5b	5/2	−0.9133
$_{42}Mo^{96}$	16.53%	95.9046						1.2 ± 0.6b	0	
$_{42}Mo^{97}$	9.46%	96.9058						2.2 ± 0.7b	5/2	−0.9325
$_{42}Mo^{98}$	23.78%	97.9055						0.14 ± 0.02b	0	
$_{42}Mo^{99}$			66.69 ± 0.06h	β^-	1.37	0.26 0.45 0.86 1.19	0.3% 17% 1% 82%			
$_{42}Mo^{100}$	9.63%	99.9076						0.20 ± 0.05b	0	
$_{42}Mo^{101}$			14.6m	β^-	2.82	0.84 1.23 1.61 2.23	38% 13% 25% 10%			
$_{42}Mo^{102}$			11m	β^-	1.2	1.2				
$_{42}Mo^{103}$			62s	β^-						
$_{42}Mo^{104}$			1.3m	β^-		2.2 4.8				
$_{42}Mo^{105}$			40s	β^-						
Tc										
$_{43}Tc^{92}$			4.4m	β^+, EC	7.9	4.1	92%			
$_{43}Tc^{93m}$			43m	IT EC	0.39 ~3.6					
$_{43}Tc^{93}$			2.7h	β^+, EC	3.19	0.64 0.80	32% 11%			
$_{43}Tc^{94m}$			53m	β^+, EC	~4.3	2.47				
$_{43}Tc^{94}$			293m	β^+, EC	4.26	0.816	11%			
$_{43}Tc^{95m}$			61d	IT β^+, EC	0.0389 ~1.7	0.49 0.71	10† 25			

TABLE OF THE ISOTOPES (*Continued*)

Isotope	% nat. abundance	Atomic mass	Lifetime	Modes of decay	Decay energy (MeV)	Particle energies (MeV)	Particle intensities	Thermal neutron capture cross-section	I	μ
$_{43}Tc^{95}$			20.0h	EC	1.66					
$_{43}Tc^{96m}$			52m	IT	0.0344					
$_{43}Tc^{96}$					2.9					
$_{43}Tc^{97m}$			90d	IT	0.0965					
$_{43}Tc^{97}$			2.6×10^6y	EC	~0.3					
$_{43}Tc^{98}$			1.5×10^6y	β^-	1.7	0.3				
$_{43}Tc^{99m}$			6.0h	IT	0.1427					
$_{43}Tc^{99}$			2.12×10^5y	β^-	0.292	0.292		22 ± 3b	9/2	+5.68
$_{43}Tc^{100}$			17s	β^-		2.2 2.88 3.38				
$_{43}Tc^{101}$			14m	β^-	1.63	1.07 1.32				
$_{43}Tc^{102}$			4.5m	β^-	4.4	~2				
$_{43}Tc^{102}$			5s	β^-	4.4	4.4				
$_{43}Tc^{103}$			50s	β^-	2.4	2.0 2.2				
$_{43}Tc^{104}$			18m	β^-	5.8	2.4 3.3 4.6 >5.3				
$_{43}Tc^{105}$			8m	β^-	3.4					
$_{43}Tc^{106}$			37s	β^-						
$_{43}Tc^{107}$			29 s	β^-						
Ru		101.07						3.0 ± 0.8b		
$_{44}Ru^{93}$			50s	β^+						
$_{44}Ru^{94}$			57m	EC						
$_{44}Ru^{95}$			1.7h	β^+, EC	2.03	0.7 1.01 1.33	1.3† 12 4			
$_{44}Ru^{96}$	5.51%	95.9076						0.21 ± 0.02b		
$_{44}Ru^{97}$			2.9d	EC	~1.2					
$_{44}Ru^{98}$	1.87%	97.9055						<8b		
$_{44}Ru^{99}$	12.72%	98.9061						10.6 ± 0.6b	5/2	−0.63
$_{44}Ru^{100}$	12.62%	99.9030						10.4 ± 0.7b		
$_{44}Ru^{101}$	17.07%							3.1 ± 0.9b	5/2	−0.69
$_{44}Ru^{102}$	31.61%	101.9037								
$_{44}Ru^{103}$			39.6d	β^-	0.74	0.90 0.203 0.382 0.445	7% 89% 1% 1%			

Isotope	% nat. abundance	Atomic mass	Lifetime	Modes of decay	Decay energy (MeV)	Particle energies (MeV)	Particle intensities	Thermal neutron capture cross-section	I	μ
						0.700	3%			
$_{44}Ru^{104}$	18.58%	103.9055						0.47 ± 0.2b		
$_{44}Ru^{105}$			4.44h	β^-	1.871	0.49 1.06 1.08 1.15 1.52 1.75 1.87	5% 14% 13% 44% ~2% ≤10% 11%	0.2 ± 0.02b		
$_{44}Ru^{106}$			367d	β^-	0.0394	0.0392	100%	146 ± 45mb		
$_{44}Ru^{107}$			4.2m	β^-	3.2	2.1 3.2				
$_{44}Ru^{108}$			4.5m	β^-	1.3	1.1 1.3	28% 72%			
Rh		102.905						150 ± 5b		
$_{45}Rh^{97}$			32m	β^+, EC	3.49	1.8 2.06 2.47				
$_{45}Rh^{98}$			8.7m	β^+, EC	4.2	2.5				
$_{45}Rh^{99m}$			4.7h	β^+, EC	~2.05	0.74				
$_{45}Rh^{99}$			16d	β^+, EC	~2.05	0.44 0.59 0.71 1.03				
$_{45}Rh^{100}$			20h	β^+, EC	3.64	0.20 0.58 0.8 1.30 1.6 2.12 2.68	~2% 3%			
$_{45}Rh^{101m}$			4.5d	IT EC	0.157 ~0.7		10%			
$_{45}Rh^{101}$			3.1y	EC	0.56					
$_{45}Rh^{102}$			2.9y	EC	2.3					
$_{45}Rh^{102}$			206d	β^+, EC β^-	2.32 1.15	β^- 0.59 1.15 β^+0.20 0.36 0.83 1.30	2% 20% 7% 3% 37% 29%			
$_{45}Rh^{103}$	100%	102.9048						11 ± 1b (Rh^{104m}) 139 ± 5b (Rh^{104})	1/2	−0.0883
$_{45}Rh^{104m}$			4.4m	IT β^-	0.129 ~2.5			800 ± 100b	5	

TABLE OF THE ISOTOPES (*Continued*)

Isotope	% nat. abundance	Atomic mass	Lifetime	Modes of decay	Decay energy (MeV)	Particle energies (MeV)	Particle intensities	Thermal neutron capture cross-section	I	μ
$_{45}Rh^{104}$			43s	EC β^-	1.15 2.47	0.7 1.9 2.5	0.08% 1.4% 98.5%	40 ± 30b		
$_{45}Rh^{105m}$			45s	IT	0.129					
$_{45}Rh^{105}$			35.9h	β^-	0.565	0.249 0.262 0.568	20% 5% 75%	5700 ± 1200b (Rh^{106m}) 13,000 ± 2000b (Rh^{106})		
$_{45}Rh^{106}$			30s	β^-	3.54	2.0 2.4 3.1 3.53	3% 12% 11% 68%			
$_{45}Rh^{106}$			130m	β^-	~3.63	0.79 0.95 1.18 1.62	40% 38% 11% 10%			
$_{45}Rh^{107}$			22m	β^-	1.51	0.84 0.94 1.14 1.20				
$_{45}Rh^{108}$			17s	β^-	~4.5	3.5 4.0 4.5	22% 17% 51%			
$_{45}Rh^{109}$			<1h	β^-	~2.5					
$_{45}Rh^{110}$			5s	β^-	5.5	5.5				
Pd		106.4						6.0 ± 1.0b		
$_{46}Pd^{98}$			17m	EC β^+						
$_{46}Pd^{99}$			22m	β^+, EC	3.8	2.0				
$_{46}Pd^{100}$			4.0d	EC	~0.6					
$_{46}Pd^{101}$			8.4h	β^+, EC	1.99	0.49 0.78	17† 83			
$_{46}Pd^{102}$	0.96%	101.9049						4.8 ± 1.5b		
$_{46}Pd^{103}$			17d	EC	0.56					
$_{46}Pd^{104}$	10.97%	103.9036								
$_{46}Pd^{105}$	22.23%	104.9046							5/2	−0.642
$_{46}Pd^{106}$	27.33%	105.9032						13 ± 2mb (Pd^{107m}) 279 ± 29mb (Pd^{107})		
$_{46}Pd^{107m}$			22s	IT	0.21					
$_{46}Pd^{107}$			~7 × 10^6y	β^-	0.035	0.035				
$_{46}Pd^{108}$	26.71%	107.90389						0.20 ± 0.04b (Pd^{109m}) 12 ± 2b (Pd^{109})	0	
$_{46}Pd^{109m}$			4.7m	IT	0.188					
$_{46}Pd^{109}$			13.47h	β^-	1.115	0.248 0.255 0.278	0.023% 0.002% 0.002%			

TABLE OF THE ISOTOPES (*Continued*)

Isotope	% nat. abundance	Atomic mass	Lifetime	Modes of decay	Decay energy (MeV)	Particle energies (MeV)	Particle intensities	Thermal neutron capture cross-section	I	μ
						0.383 0.394 0.416 0.703 0.806 1.030	0.037% 0.022% 0.005% 0.010% 0.027% 99.87%			
$_{46}Pd^{110}$	11.81%							20 ± 15mb (Pd^{111m}) 0.4 ± 0.1b (Pd^{111})		
$_{46}Pd^{111m}$			5.5h	IT β^-	0.17		75% 25%			
$_{46}Pd^{111}$			22m	β^-	2.2	0.6 0.7 1.1 1.2 2.2				
$_{46}Pd^{112}$			21h	β^-	0.30	0.30	100%			
$_{46}Pd^{113}$			1.4m	β^-						
$_{46}Pd^{114}$			2.4m	β^-						
$_{46}Pd^{115}$			45s	β^-						
Ag		107.870						63.8 ± 0.6b		
$_{47}Ag^{102}$			15m	β^+, EC	~5.3					
$_{47}Ag^{103m}$			5.7s	IT	0.138					
$_{47}Ag^{103}$			66m	β^+,EC	2.6	1.3 1.6			7/2	
$_{47}Ag^{104m}$			30m	IT β^+, EC	0.02	2.7			2	+3.7
$_{47}Ag^{104}$			67m	β^+, EC	4.27				5	+4.0
$_{47}Ag^{105}$			40d	EC					1/2	±1.01
$_{47}Ag^{106m}$			8.4d	EC					6	
$_{47}Ag^{106}$			24m	β^+, EC	2.96	1.45 1.96	11† 89		1	
$_{47}Ag^{107m}$			44.3s	IT	0.093					
$_{47}Ag^{107}$	51.82%	106.90509						35 ± 5b	1/2	−0.1135
$_{47}Ag^{108m}$			> 5y	IT EC	0.110		10% 90%			
$_{47}Ag^{108}$			2.42m	β^+, EC β^-	1.92 1.64	β^- 1.8 1.2 0.47 β^+ 0.88	96% 1.8% ~0.02%		1	+2.8
$_{47}Ag^{109m}$			40s	IT	0.0877				7/2	±4.3
$_{47}Ag^{109}$	48.18%	108.9047						4.2 ± 0.7b (Ag^{110m}) 89 ± 4b (Ag^{110})	1/2	−0.1305
$_{47}Ag^{110m}$			253d	β^- IT	~2.9 0.12	0.087 0.529 1.5	61% 36% 0.6%	82 ± 11b	6	+3.604

Isotope	% nat. abundance	Atomic mass	Lifetime	Modes of decay	Decay energy (MeV)	Particle energies (MeV)	Particle intensities	Thermal neutron capture cross-section	I	μ
$_{47}Ag^{110}$			24.4s	β^-	~2.9	1.40 2.14 2.87	0.05% 12% 88%			
$_{47}Ag^{111m}$			74s	IT	0.07					
$_{47}Ag^{111}$			7.5d	β^-	1.05	0.69 0.79 1.04	6% 1% 93%		1/2	−0.146
$_{47}Ag^{112}$			3.2h	β^-	4.01	1.96 3.35 3.94	10% 22% 54%		2	±0.054
$_{47}Ag^{113m}$			1.2m	β^-						
$_{47}Ag^{113}$			5.3h	β^-	2.00	2.0			1/2	±0.159
$_{47}Ag^{114}$			4.5s	β^-	4.6	4.6				
$_{47}Ag^{114}$			2m	β^-	4.6					
$_{47}Ag^{115}$			20s	β^-						
$_{47}Ag^{115}$			20m	β^-	3.3	1.0 2.1 2.7 3.0 3.2				
$_{47}Ag^{116}$			2.5m	β^-	~6.1					
$_{47}Ag^{117}$			1.1m	β^-						
Cd		112.40						2450 ± 20b		
$_{48}Cd^{103}$			10m	β^+, EC						
$_{48}Cd^{104}$			57m	EC	~1.2					
$_{48}Cd^{105}$			55m	β^+, EC	~2.8	0.8 1.69				
$_{48}Cd^{106}$	1.22%	105.907						1.0 ± 0.5b		
$_{48}Cd^{107}$			6.5h	β^+, EC	1.417	0.302			5/2	−0.6144
$_{48}Cd^{108}$	0.88%	107.9040						1.5 ± 0.5b	0	
$_{48}Cd^{109}$			450 ± 5d	EC	0.16			700 ± 100b	5/2	−0.8270
$_{48}Cd^{110}$	12.39%	109.9030						0.10 ± 0.03b (Cd^{111m})	0	
$_{48}Cd^{111m}$			48.6m	IT						
$_{48}Cd^{111}$	12.75%	110.9042							1/2	−0.5943
$_{48}Cd^{112}$	24.07%	111.9028						0.06 ± 0.02b (Cd^{113m})	0	
$_{48}Cd^{113m}$			14y	β^-	0.58	0.58	99+%		11/2	−1.087
$_{48}Cd^{113}$	12.26%	112.9046						20,000 ± 300b	1/2	−0.6217
$_{48}Cd^{114}$	28.86%	113.9036						0.36 ± 0.007b (Cd^{115m}) 0.30 ± 0.015b (Cd^{115})	0	

Isotope	% nat. abundance	Atomic mass	Lifetime	Modes of decay	Decay energy (MeV)	Particle energies (MeV)	Particle intensities	Thermal neutron capture cross-section	I	μ
$_{48}Cd^{115m}$			43d	IT β^-	0.18 1.62	 0.34 0.68 1.62	 1.6% 97%		11/2	−1.042
$_{48}Cd^{115}$			53.5h	β^-	1.45	0.58} 0.86} 1.11	 42% 58%		1/2	−0.648
$_{48}Cd^{116}$	7.58%	115.9050						27 ± 5mb (Cd^{117m}) 50 ± 8mb (Cd^{117})	0	
$_{48}Cd^{117m}$			3.4h	β^-	2.65	0.41 0.70				
$_{48}Cd^{117}$			2.4h	β^-	2.52	0.65 1.79 2.23				
$_{48}Cd^{118}$			49m	β^-	~0.8					
$_{48}Cd^{119}$			2.7m	β^-	3.5	~3.5				
$_{48}Cd^{119}$			10m	β^-	3.5	~3.5				
In		114.82						194 ± 2b		
$_{49}In^{106}$			5.3m	β^+, EC	6.5	3.1 4.85				
$_{49}In^{107}$			32m	β^+, EC	3.5	2.2				
$_{49}In^{108}$			39m	β^+, EC		2.28 2.66 3.50				
$_{49}In^{108}$			56m	β^+, EC		1.29				
$_{49}In^{109}$			4.3h	β^+, EC	2.02				9/2	+5.53
$_{49}In^{110m}$			4.9h	EC β^+					7	±10.5
$_{49}In^{110}$			67m	β^+, EC	3.93	2.25			2	
$_{49}In^{111}$			2.81d						9/2	+5.53
$_{49}In^{112m}$			20.7m	IT	0.156					
$_{49}In^{112}$			14m	β^- β^+, EC	0.66	β^- 0.66 β^+ 1.0 1.52	44% 6% 28%			
$_{49}In^{113m}$			100m	IT	0.3916				1/2	−0.210
$_{49}In^{113}$	4.28%	112.9043						3.2 ± 0.7b (In^{114m_2}) 4.5 ± 0.7b (In^{114m_1}) 3 ± 1b (In^{114})	9/2	+5.523
$_{49}In^{114m_2}$			0.0425	IT	0.502					

TABLE OF THE ISOTOPES (*Continued*)

Isotope	% nat. abundance	Atomic mass	Lifetime	Modes of decay	Decay energy (MeV)	Particle energies (MeV)	Particle intensities	Thermal neutron capture cross-section	I	μ
$_{49}In^{114m_1}$			50.0d	IT EC	0.191		96.5% 3.5%		5	+4.7
$_{49}In^{114}$			72s	β^+, EC β^-	1.44 1.986	β^- 1.989 β^+ 0.40	98+% 0.039%			+1.7
$_{49}In^{115m}$			4.50h	IT β^-	0.335 0.82		95% 5%		1/2	−0.244
$_{49}In^{115}$	95.72%	114.9041						85 ± 10b (In^{116m_2}) 72 ± 4b (In^{116m_1}) 42 ± 4b (In^{116})	9/2	+5.534
$_{49}In^{116m_2}$			2.16s	IT	0.22					
$_{49}In^{116m_1}$			54m	β^-	3.39	0.60 0.87 1.00	21% 28% 51%		5	+4.3
$_{49}In^{116}$			14s	β^-	3.33	3.3	99%			
$_{49}In^{117m}$			1.93h	IT β^-	0.314 ~1.77	 1.62 1.77	47% 16% 37%		1/2	±0.25
$_{49}In^{117}$			44m	β^-	1.47	0.74			9/2	
$_{49}In^{118}$			5s	β^-	4.2	4.2				
$_{49}In^{118}$			4.4m	β^-	4.3	1.3 2.0				
$_{49}In^{119m}$			18m	IT β^-	0.30 2.8	2.7	5%			
$_{49}In^{119}$			2.1m	β^-	2.5	1.6				
$_{49}In^{120}$			3.2s	β^-	~5.6	4.4 5.6	15% 85%			
$_{49}In^{120}$			46s	β^-	~5.3	2.2 3.1	~40% ~27%			
$_{49}In^{121}$			3.1m	β^-	~3.6	3.7				
$_{49}In^{121}$			30s	β^-						
$_{49}In^{122}$			8s	β^-	~6.7					
$_{49}In^{123}$			36s	β^-	~4.5					
$_{49}In^{123}$			10s	β^-						
$_{49}In^{124}$			4s	β^-	7.4	5				
Sn		118.69						0.63 ± 0.01b		
$_{50}Sn^{108}$			9m	EC						
$_{50}Sn^{109}$			18.1m	β^+, EC		~1.6				
$_{50}Sn^{110}$			4.0h	EC						
$_{50}Sn^{111}$			35m	β^+, EC	2.52	1.51	35.3%			
$_{50}Sn^{112}$	0.96%	111.9040						0.35 ± 0.08b (Sn^{113m}) 0.8 ± 0.2 (Sn^{113})		

TABLE OF THE ISOTOPES (*Continued*)

Isotope	% nat. abundance	Atomic mass	Lifetime	Modes of decay	Decay energy (MeV)	Particle energies (MeV)	Particle intensities	Thermal neutron capture cross-section	I	μ
$_{50}Sn^{113m}$			20m	IT EC	0.079 1.1		91% 9%			
$_{50}Sn^{113}$			115d	EC	1.02					
$_{50}Sn^{114}$	0.66%	113.9030								
$_{50}Sn^{115}$	0.35%	114.9035							1/2	−0.918
$_{50}Sn^{116}$	14.30%	115.9021						6 ± 2mb (Sn^{117m})	0	
$_{50}Sn^{117m}$			14d	IT	0.317					
$_{50}Sn^{117}$	7.61%	116.9031							1/2	−1.000
$_{50}Sn^{118}$	24.03%	117.9018						10 ± 6mb (Sn^{119m})	0	
$_{50}Sn^{119m}$			250d	IT	0.089				(3/2)	+0.67
$_{50}Sn^{119}$	8.58%	118.9034							1/2	−1.046
$_{50}Sn^{120}$	32.85%							1 ± 1mb (Sn^{121m}) 140 ± 30mb (Sn^{121})	0	
$_{50}Sn^{121m}$			76y	β^-	0.45	0.35	100%			
$_{50}Sn^{121}$			27h	β^-	0.383	0.42	100%		3/2	±0.70
$_{50}Sn^{122}$	4.72%	121.9034						1.0 ± 0.5 mb (Sn^{123m}) 0.15 ± 0.02b (Sn^{123})		
$_{50}Sn^{123m}$			125d	β^-	1.42	1.42	100%			
$_{50}Sn^{123}$			42m	β^-	1.46	1.46				
$_{50}Sn^{124}$	5.94%	123.9052						0.14 ± 0.02b (Sn^{125m}) 4 ± 2mb (Sn^{125})		
$_{50}Sn^{125m}$			9.7m	β^-	2.39	0.281 0.447 0.480 0.658 0.694 0.910 1.045 1.473 1.751 2.062	0.002% 0.2% 0.03% 0.9% 0.1% 0.3% <0.02% 0.2% <0.1% 98.3%			
$_{50}Sn^{125}$			9.4d	β^-	2.34	0.34 0.46 0.95 1.3 2.37	1.9% 1.9% <0.25% ~0.8% 95%			
$_{50}Sn^{126}$			~10^5y	β^-	~0.3					
$_{50}Sn^{127}$			2.1h	β^-		1.5				
$_{50}Sn^{127}$			4m	β^-	~3.1	2.7				

Isotope	% nat. abundance	Atomic mass	Lifetime	Modes of decay	Decay energy (MeV)	Particle energies (MeV)	Particle intensities	Thermal neutron capture cross-section	I	μ
${}_{50}Sn^{128}$			59m	β^-	1.3	0.8				
Sb		121.75						5 ± 1b		
${}_{51}Sb^{112}$			0.9m	β^+, EC						
${}_{51}Sb^{113}$			6.7m	β^+, EC	4.47	1.85 2.42				
${}_{51}Sb^{114}$			3.3m	β^+, EC	6.3	2.7				
${}_{51}Sb^{115}$			31m	β^+, EC	3.03	1.51				
${}_{51}Sb^{116m}$			60m	β^+, EC		1.16				
${}_{51}Sb^{116}$			15m	β^+, EC	4.6	1.5 2.4				
${}_{51}Sb^{117}$			2.8h	β^+, EC	1.82	0.57			5/2	+2.67
${}_{51}Sb^{118m_2}$			0.87s	IT						
${}_{51}Sb^{118m_1}$			5.1h	β^+, EC	3.9					
${}_{51}Sb^{118}$			3.5m	β^+, EC	3.7	1.5 2.7			1	±2.4
${}_{51}Sb^{119}$			38hr	EC			100%		5/2	+3.45
${}_{51}Sb^{120}$			15.9m	β^+, EC	2.69	1.70			1	+2.3
${}_{51}Sb^{121}$	57.25%	120.9038						55 ± 10mb (Sb^{122m}) 6.2 ± 0.3b (Sb^{122})	5/2	+3.359
${}_{51}Sb^{122m}$			4.2m	IT	0.162					
${}_{51}Sb^{122}$			2.8d	β^- β^+, EC	1.972 1.62	β^+ 0.56 β^- 0.74 1.40 1.97	 4% 63% 30%		2	−1.90
${}_{51}Sb^{123}$	42.75%	122.9041						15 ± 4mb (Sb^{124m_2}) 30 ± 8mb (Sb^{124m_1}) 3.4 ± 0.8b (Sb^{124})	7/2	+2.547
${}_{51}Sb^{124m_2}$			21m	IT	0.035					
${}_{51}Sb^{124m_1}$			93s	IT β^-	0.010					
${}_{51}Sb^{124}$			60.3 ±0.2d	β^-	2.916	0.06 0.23 0.621 0.950 1.01 1.59 1.67 2.317	2% 11% 50% 5% 1.5% 5% 3% 22%	6.5 ± 1.5b	3	

TABLE OF THE ISOTOPES (*Continued*)

Isotope	% nat. abundance	Atomic mass	Lifetime	Modes of decay	Decay energy (MeV)	Particle energies (MeV)	Particle intensities	Thermal neutron capture cross-section	I	μ
$_{51}Sb^{125}$			2.7y	β^-	0.764	0.1 0.619 0.45 0.299 0.240 0.125	2% 13% 6% 43% 1% 28%			
$_{51}Sb^{126m}$			19m	IT β^-		1.9				
$_{51}Sb^{126}$			12.5d	β^-	3.7	1.9				
$_{51}Sb^{127}$			93h	β^-	1.60	0.86 1.11 1.57	50% 20% 30%			
$_{51}Sb^{128}$			11m	β^-	4.3	2.5				
$_{51}Sb^{129}$			4.3h	β^-	2.5	1.87				
$_{51}Sb^{130}$			2.6m	β^-						
$_{51}Sb^{131}$			25m	β^-						
$_{51}Sb^{132}$			2.1m	β^-						
$_{51}Sb^{133}$			4.2m	β^-						
Te		127.60						4.7 ± 0.1b		
$_{52}Te^{115}$			6.0m	β^+, EC	4.5	2.8				
$_{52}Te^{116}$			2.50h	β^+, EC	1.6	0.44			0	
$_{52}Te^{117}$			61m	β^+, EC	3.50	1.7			1/2	
$_{52}Te^{118}$			6.00d	EC	~0.3					
$_{52}Te^{119m}$			4.7d	EC	~2.6				11/2	
$_{52}Te^{119}$			15.9h	β^+, EC	2.294	0.627	5%		1/2	±0.25
$_{52}Te^{120}$	0.089%	119.9045						0.34 ± 0.06b (Te^{121m}) 2.0 ± 0.3b (Te^{121})		
$_{52}Te^{121m}$			154d	IT EC, β^+	0.293 ~1.6		90%			
$_{52}Te^{121}$			17d	EC	1.29					
$_{52}Te^{122}$	2.46%	121.9030						1.1 ± 0.5b (Te^{123m}) 1.7 ± 0.8b (Te^{123})		
$_{52}Te^{123m}$			117d	IT	0.247					
$_{52}Te^{123}$	0.87%	122.9042	1.2 × 10^{13}y	EC	~0.06			410 ± 30b	1/2	−0.7359
$_{52}Te^{124}$	4.61%	123.9028						40 ± 25mb (Te^{125m}) 6.8 ± 1.3b (Te^{125})		
$_{52}Te^{125m}$			58d	IT					3/2	~+0.7
$_{52}Te^{125}$	6.99%	124.9044						1.56 ± 0.16b	1/2	−0.8871

Isotope	% nat. abundance	Atomic mass	Lifetime	Modes of decay	Decay energy (MeV)	Particle energies (MeV)	Particle intensities	Thermal neutron capture cross-section	I	μ
${}_{52}Te^{126}$	18.71%	125.9032						125 ± 23mb (Te^{127m}) 0.90 ± 0.15b (Te^{127})	0	
${}_{52}Te^{127m}$			109d	IT β^-	0.0887 0.77		99.2%			
${}_{52}Te^{127}$			9.4h	β^-	0.69	0.28 0.50 0.70	0.32% 0.005% 99.7%			
${}_{52}Te^{128}$	31.79%	127.9047						14 ± 4mb (Te^{129m}) 155 ± 40mb (Te^{129})	0	
${}_{52}Te^{129m}$			34d	β^- IT	1.48 0.105	0.91 1.60				
${}_{52}Te^{129}$			69m	β^-	1.48	0.39 0.67 1.00 1.45				
${}_{52}Te^{130}$	34.48%	129.9067						0.02 ± 0.01b (Te^{131m}) 0.2 ± 0.1b (Te^{131})	0	
${}_{52}Te^{131m}$			30h	β^- IT	2.46 0.1817 (18%)	0.45 0.48 0.50 0.53 0.56 0.63 0.84 2.46	10% 11% 12% 3% 13% 7% 16% 5%			
${}_{52}Te^{131}$			25m	β^-		0.80 0.86 0.88 1.32 1.80 2.14	1.4% 1.6% 1.5% 1.0% 20% 62%			
${}_{52}Te^{132}$			78h	β^-	0.50	0.22	100%			
${}_{52}Te^{133m}$			50m	IT β^-	0.334		13%			
${}_{52}Te^{133}$			12.5m	β^-	2.7	2.4 1.3	~30% ~70%			
${}_{52}Te^{134}$			42m	β^-						
${}_{52}Te^{135}$			<2m	β^-						
I		126.9044						6.2 ± 0.2b		
${}_{53}I^{117}$			7m	β^+, EC						
${}_{53}I^{118}$			14m	β^+, EC	7	5.5				
${}_{53}I^{119}$			19m	β^+, EC						
${}_{53}I^{120}$			1.3h	β^+, EC	5.6	2.1 4.0				

Isotope	% nat. abundance	Atomic mass	Lifetime	Modes of decay	Decay energy (MeV)	Particle energies (MeV)	Particle intensities	Thermal neutron capture cross-section	I	μ
$_{53}I^{121}$			2.1h	β^+, EC	2.36	1.2				
$_{53}I^{122}$			3.5m	β^+, EC	4.14	1.8 2.6 3.1				
$_{53}I^{123}$			13.3h	EC	~1.4				5/2	
$_{53}I^{124}$			4.2d	β^+, EC	3.17	0.79 1.53 2.13	8† 46 46		2	
$_{53}I^{125}$			60d	EC	0.149			900 ± 90b	5/2	+3.0
$_{53}I^{126}$			13d	β^+, EC β^-	2.150 1.251	β^+ 1.129 β^- 0.385 0.865 1.25	 6† 29 12		2	
$_{53}I^{127}$	100%	126.9004						6.2 ± 0.2b	5/2	+2.808
$_{53}I^{128}$			25.08 ±0.05m	β^+, EC β^-	1.27 2.14	β^- 1.13 1.67 2.12	1.6% 13% 79%		1	
$_{53}I^{129}$			1.7 × 10^7y	β^-	0.189	0.189	100%	19 ± 2b (I^{130m}) 9 ± 1b (I^{130})	7/2	+2.617
$_{53}I^{130m}$			8.82 ± 0.04m	IT						
$_{53}I^{130}$			12.3 ± 0.1h	β^-	2.99	0.62 1.04 1.7	52% 48% 0.4%	18 ± 3b	5	
$_{53}I^{131}$			8.070 ± 0.009d	β^-	0.970	0.257 0.333 0.487 0.606 0.806	1.6% 6.9% 0.5% 90.4% 0.6%	~0.7b	7/2	+2.74
$_{53}I^{132}$			2.3h	β^-	3.56	0.72 0.80 0.91 0.98 1.16 1.60 1.75 2.12	16% 8% 5% 18% 20% 9% 6% 18%		4	±3.08
$_{53}I^{133}$			20.9 ± 0.1h	β^-	1.80	0.7 0.94 1.27			7/2	+2.84
$_{53}I^{134}$			52m	β^-	4.2	1.10 1.32 1.56 1.72 1.86 2.28 2.46	1.0% 23% 15% <4% 9% 12% 25%			
$_{53}I^{135}$			6.7h	β^-	~2.8	0.5 1.0 1.4	35% 40% 25%		7/2	
$_{53}I^{136}$			83s	β^-	7.0	2.7 4.3 5.6	9% 23% 16%			

TABLE OF THE ISOTOPES (*Continued*)

Isotope	% nat. abundance	Atomic mass	Lifetime	Modes of decay	Decay energy (MeV)	Particle energies (MeV)	Particle intensities	Thermal neutron capture cross-section	I	μ
						7.0	~6%			
$_{53}I^{137}$			23s	β^-, η						
$_{53}I^{138}$			5.9s	β^-, η						
$_{53}I^{139}$			2s	β^-, η						
Xe		131.30						24.5 ± 1b		
$_{54}Xe^{118}$			6m	β^+, EC						
$_{54}Xe^{119}$			6m	β^+, EC						
$_{54}Xe^{120}$			40m	β^+, EC						
$_{54}Xe^{121}$			39m	β^+, EC	3.0	2.8				
$_{54}Xe^{122}$			20h	EC	~0.5					
$_{54}Xe^{123}$			2.1h	β^+, EC	2.8	1.6				
$_{54}Xe^{124}$	0.096%	123.9061						100 ± 20b		
$_{54}Xe^{125m}$			55s	IT						
$_{54}Xe^{125}$			17h	EC						
$_{54}Xe^{126}$	0.090%	125.9042						1.5 ± 1.0b		
$_{54}Xe^{127m}$			75s	IT						
$_{54}Xe^{127}$			36.4d	EC	0.44					
$_{54}Xe^{128}$	1.92%	127.9035						0.43 ± 0.1b (Xe^{129m}) <8b (Xe^{129})		
$_{54}Xe^{129m}$			8.0d	IT	0.236					
$_{54}Xe^{129}$	26.44%	128.9048						21 ± 7b	1/2	−0.7768
$_{54}Xe^{130}$	4.08%	129.9035						0.34 ± 0.08b (Xe^{131m}) <26b (Xe^{131})		
$_{54}Xe^{131m}$			11.8d	IT	0.163				3/2	+0.6908
$_{54}Xe^{131}$	21.18%	130.9051						110 ± 20b		
$_{54}Xe^{132}$	26.89%	131.9042						0.53 ± 0.10b (Xe^{133m}) 50 ± 20mb (Xe^{133})	0	
$_{54}Xe^{133m}$			2.26d	IT	0.233					
$_{54}Xe^{133}$			5.27d	β^-	0.427	0.267 0.347	0.7% 99.3%			
$_{54}Xe^{134}$	10.44%	133.9054						0.23 ± 0.02b	0	
$_{54}Xe^{135m}$			15.6m	0.526						

Isotope	% nat. abundance	Atomic mass	Lifetime	Modes of decay	Decay energy (MeV)	Particle energies (MeV)	Particle intensities	Thermal neutron capture cross-section	I	μ
$_{54}Xe^{135}$			9.2h	β^-	1.16	0.55 0.92	3% 97%	2.64×10^6b $\pm 0.1 \times 10^6$		
$_{54}Xe^{136}$	8.87%	135.9072						280 ± 28mb	0	
$_{54}Xe^{137}$			3.9m	β^-	4.0	3.6 4.1	33% 67%			
$_{54}Xe^{138}$			17m	β^-	2.8					
$_{54}Xe^{139}$			42s	β^-						
$_{54}Xe^{140}$			16s	β^-						
$_{54}Xe^{141}$			2s	β^-						
$_{54}Xe^{142}$			~1.5s	β^-						
Cs		132.905						30.0 ± 1.5b		
$_{55}Cs^{123}$			8m	β^+, EC						
$_{55}Cs^{125}$			45m	β^+, EC		2.05			1/2	+1.41
$_{55}Cs^{126}$			1.6m	β^+, EC		3.8				
$_{55}Cs^{127}$			6.2h	β^+, EC	2.1	1.02 0.677	2.5† 3		1/2	+1.46
$_{55}Cs^{128}$			3.8m	β^+, EC	3.93	1.30 1.90 2.44 2.88	8† 16 39 100			
$_{55}Cs^{129}$			32h	EC	~1.1				1/2	+1.479
$_{55}Cs^{130}$			30m	β^+, EC β^-	2.99 0.442	1.97 0.442	 2%		1	±1.4
$_{55}Cs^{131}$			9.70d	EC	0.35				5/2	+3.54
$_{55}Cs^{132}$			6.5d	β^+, EC	2.09				2	+2.22
$_{55}Cs^{133}$	100%	133.9051						2.6 ± 0.2b (Cs^{134m}) 27.4 ± 1.5b (Cs^{134})	7/2	+2.579
$_{55}Cs^{134m}$			2.90 ±0.01h	IT	0.138				8	+1.096
$_{55}Cs^{134}$			2.05y	β^-	2.062	0.089 0.410 0.662	28% 1% 71%	134 ± 12b	4	+2.990
$_{55}Cs^{135}$			3×10^6y	β^-	0.210	0.210	100%	8.7 ± 0.5b	7/2	+2.729
$_{55}Cs^{136}$			13d	β^-	2.54	0.341 0.56 0.657	93% 2% 7%		5	±3.70
$_{55}Cs^{137}$			30.23 ±0.16y	β^-	1.176	0.511 1.176	94% 6%	110 ± 33 mb	7/2	+2.838

Isotope	% nat. abundance	Atomic mass	Lifetime	Modes of decay	Decay energy (MeV)	Particle energies (MeV)	Particle intensities	Thermal neutron capture cross-section	I	μ
$_{55}Cs^{138}$			32.2m	β^-	4.83	1.49 2.20 2.39 2.53 2.62 2.94 3.4	0.5% 10% 36% 5% 16% 12% 21%		3	±0.5
$_{55}Cs^{139}$			9.5m	β^-	4.0					
$_{55}Cs^{140}$			66s	β^-	6.1					
$_{55}Cs^{141}$			24s	β^-						
$_{55}Cs^{142}$			2.3s	β^-						
$_{55}Cs^{143}$			2.0s	β^-						
$_{55}Cs^{144}$				β^-						
Ba		137.34						1.2 ± 0.1b		
$_{56}Ba^{123}$			2m	β^+, EC						
$_{56}Ba^{125}$			6m	β^+, EC						
$_{56}Ba^{126}$			97m	EC	~1.3					
$_{56}Ba^{127}$			10m	β^+, EC	~3.6					
$_{56}Ba^{128}$			2.4d	EC	~0.7					
$_{56}Ba^{129+129m}$			~2.5h	β^+, EC	2.45	1.0 1.24 1.43	6† 24 78			
$_{56}Ba^{130}$	0.101%	129.9062						2.5 ± 0.3b (Ba^{131m}) 11 ± 3b (Ba^{131})		
$_{56}Ba^{131m}$			15m	IT	0.18					
$_{56}Ba^{131}$			12d	EC	1.16					
$_{56}Ba^{132}$	0.097%	131.9057						8.5 ± 1.0b		
$_{56}Ba^{133m}$			38.9h	IT	0.288					
$_{56}Ba^{133}$				EC	0.488					
$_{56}Ba^{134}$	2.42%	133.9043						0.16 ± 0.02b (Ba^{135m}) 2 ± 2b (Ba^{135})	0	
$_{56}Ba^{135m}$			28.7h	IT	0.268					
$_{56}Ba^{135}$	6.59%	134.9056						5.8 ± 0.9b	3/2	+0.8365
$_{56}Ba^{136m}$			0.32s	IT	2.04					
$_{56}Ba^{136}$	7.81%	135.9044						10 ± 1mb (Ba^{137m}) 0.4 ± 0.4b (Ba^{137})	0*	
$_{56}Ba^{137m}$			2.55m	IT	0.6616					
$_{56}Ba^{137}$	11.32%							5.1 ± 0.4b	3/2	+0.9357

TABLE OF THE ISOTOPES (*Continued*)

Isotope	% nat. abundance	Atomic mass	Lifetime	Modes of decay	Decay energy (MeV)	Particle energies (MeV)	Particle intensities	Thermal neutron capture cross-section	I	μ
$_{56}Ba^{138}$	71.66%	137.9050						0.35 ± 0.15b	0	
$_{56}Ba^{139}$			82.9m	β^-	2.34	0.95 2.23 2.38	<1% 27% 72%	4 ± 1b		
$_{56}Ba^{140}$			12.8d	β^-	1.05	0.47 0.58 0.89 1.01 1.02	34% 11% ~3% 33% 19%	1.6 ± 0.3b		
$_{56}Ba^{141}$			18m	β^-	3.0	2.0 2.4 2.6 2.8 3.0				
$_{56}Ba^{142}$			11m	β^-	2.2	1.0 ~1.7				
$_{56}Ba^{143}$			12s	β^-						
La		138.91						8.9 ± 0.2b		
$_{57}La^{126}$			1.0m	β^+, EC						
$_{57}La^{127}$			3.5m	β^+, EC						
$_{57}La^{128}$			4.4m	β^+, EC						
$_{57}La^{129}$			10m	β^+, EC	~4.0					
$_{57}La^{130}$			8.7m	β^+, EC	~5					
$_{57}La^{131}$			59m	β^+, EC	2.96	0.70 1.42 1.94	17† 56 27			
$_{57}La^{132}$			4.5h	β^+, EC	5.3	3.8				
$_{57}La^{133}$			4.0h	β^+, EC	2.2	~1.2				
$_{57}La^{134}$			6.7m	β^+, EC	3.77	2.7				
$_{57}La^{135}$			19.5h	EC	~1.2	0.81				
$_{57}La^{136}$			9.5m	β^+, EC	2.9	1.8				
$_{57}La^{137}$			6×10^4y	EC	~0.5					
$_{57}La^{138}$	0.089%	137.9068							5	+3.707
$_{57}La^{139}$	99.911%	138.9061						8.8 ± 0.7b	7/2	
$_{57}La^{140}$			40.22 ± 0.1h	β^-	3.769	1.25 1.36 1.42 1.69 2.17	15% 35% 4% 18% 27%	2.8 ± 0.3b	3	
$_{57}La^{141}$			3.9h	β^-	2.43	0.9 2.43	~5% ~95%			
$_{57}La^{142}$			92m	β^-	4.51	0.86 1.06	12% 16%			

TABLE OF THE ISOTOPES (*Continued*)

Isotope	% nat. abundance	Atomic mass	Lifetime	Modes of decay	Decay energy (MeV)	Particle energies (MeV)	Particle intensities	Thermal neutron capture cross-section	I	μ
						1.20 1.79 1.98 2.11 2.31 2.33 2.98 3.85 4.52	5% 11% 19% 24% 6% 1% 1.7% 2.4% 12%			
$_{57}La^{143}$			14m	β^-	3.3	3.3				
$_{57}La^{144}$			40s	β^-	5.6					
Ce		140.12						0.73 ± 0.08b		
$_{58}Ce^{132}$			4.2h	EC	~1.2					
$_{58}Ce^{133}$			6.3h	β^+, EC	~2.8	1.3				
$_{58}Ce^{134}$			72h	EC	~0.16					
$_{58}Ce^{135}$			17.2h	β^+, EC	~2.1	0.81				
$_{58}Ce^{136}$	0.193%							0.95 ± 0.25b (Ce^{137m}) 6.3 ± 1.5b (Ce^{137})		
$_{58}Ce^{137m}$			34.4h	IT EC	0.255				(11/2)	±0.89
$_{58}Ce^{137}$			9.0h	β^+, EC						
$_{58}Ce^{138}$	0.250%	137.9057						15 ± 5mb (Ce^{139m}) 1.1 ± 0.3b (Ce^{139})		
$_{58}Ce^{139m}$			55s	IT	0.746					
$_{58}Ce^{139}$			140d	EC	0.27				(3/2)	±0.9
$_{58}Ce^{140}$	88.48%	139.9053						0.58 ± 0.06b		
$_{58}Ce^{141}$			33d	β^-	0.581	0.444 0.582	60% 40%	29 ± 3b	7/2	±0.97
$_{58}Ce^{142}$	11.07%	141.9090						0.95 ± 0.05b		
$_{58}Ce^{143}$			33h	β^-	1.44	0.28 0.50 0.72 1.09 1.38	1% 3% 16% 42% 38%	6.0 ± 0.7b	3/2	
$_{58}Ce^{144}$			284.9 ± 0.8d	β^-		0.175 0.24 0.309	24% 4.5% 76%	1.0 ± 0.1b		
$_{58}Ce^{145}$			3.0m	β^-	~2.6	~2.0				
$_{58}Ce^{146}$			14m	β^-	~1.0	0.7				
$_{58}Ce^{147}$			65s	β^-						
$_{58}Ce^{148}$			43s	β^-						
Pr		140.907						11.5 ± 1.0b		
$_{59}Pr^{134}$			17m	β^+, EC						

Isotope	% nat. abundance	Atomic mass	Lifetime	Modes of decay	Decay energy (MeV)	Particle energies (MeV)	Particle intensities	Thermal neutron capture cross-section	I	μ
$_{59}Pr^{135}$			22m	β^+, EC		2.5				
$_{59}Pr^{136}$			1.1h	β^+, EC		~2.0				
$_{59}Pr^{137}$			1.5h	β^+, EC	2.7	1.7				
$_{59}Pr^{138}$			2.1h	β^+, EC	4.79	1.65				
$_{59}Pr^{139}$			4.5h	β^+, EC	2.11	1.09				
$_{59}Pr^{140}$			3.39m	β^+, EC	3.34	2.32				
$_{59}Pr^{141}$	100%	140.9074						3.9 ± 0.5b	5/2	+4.3
$_{59}Pr^{142}$			19.2h	β^-	2.16	0.64 2.15	4% 96%	18 ± 3b	2	±0.26
$_{59}Pr^{143}$			13.7d	β^-	0.933	0.933	100%	89 ± 10b	7/2	
$_{59}Pr^{144}$			17.3m	β^-	2.989	2.99 0.80 2.29	98% 1% 1.2%			
$_{59}Pr^{145}$			5.98h	β^-	1.80	1.80	95%			
$_{59}Pr^{146}$			24m	β^-	4.2	3.7 2.3	56% 44%			
$_{59}Pr^{147}$			12.0m	β^-	2.7	1.0 1.4 2.1 2.7	≤15% 30% 50% ≤ 5%			
$_{59}Pr^{148}$			2.0m	β^-	4.5					
Nd		144.24						49 ± 2b		
$_{60}Nd^{138}$			22m	β^+	3.4	~2.4				
$_{60}Nd^{139m}$			5.5h	IT β^+	0.232					
$_{60}Nd^{139}$			5.2h	EC β^+	4.1	1.0 3.3	8† 100			
$_{60}Nd^{140}$			3.3d	EC						
$_{60}Nd^{141}$			2.5h	EC β^+	 1.80	EC β^+ 0.78	95% 5%			
$_{60}Nd^{141m}$			64s	IT	0.755				3/2	
$_{60}Nd^{142}$	27.11%	141.9075						18.8 ± 0.7b		
$_{60}Nd^{143}$	12.17%	142.9096						330 ± 10b 20 ± 2mb (η, α)	7/2	−1.08
$_{60}Nd^{144}$	23.85%	143.9099	≈5 × 10^{15}y	α		1.8		4.0 ± 0.5b		
$_{60}Nd^{145}$	8.30%	144.9122						50 ± 4b	7/2	−0.66
$_{69}Nd^{146}$	17.62%	145.9127						1.4 ± 0.2b		
$_{60}Nd^{147}$			11.1d	β^-	0.91	0.82 0.38	77% 17%		5/2	±0.59

TABLE OF THE ISOTOPES (*Continued*)

Isotope	% nat. abundance	Atomic mass	Lifetime	Modes of decay	Decay energy (MeV)	Particle energies (MeV)	Particle intensities	Thermal neutron capture cross-section	I	μ
$_{60}Nd^{148}$	5.73%	147.9165						2.5 ± 0.2b		
$_{60}Nd^{149}$			1.73h	β^-	1.66	1.55 1.01 1.45 1.12 1.40 1.40 1.12 1.45 1.01 1.55	17% 18% 18% 25% 11%		5/2	
$_{60}Nd^{150}$	5.62%	149.9207						1.3 ± 0.3b		
$_{60}Nd^{151}$			10m	β^-	2.31	2.14 2.06 1.99 1.78 1.47	17% 40% 9% 11% 16%			
Pm										
$_{61}Pm^{141}$			22m	β^+	3.6	2.6				
$_{61}Pm^{142}$			34s	β^+	4.82	β^+, 3.8 EC	69% 31%			
$_{61}Pm^{143}$			265d	EC	1.1	EC				
$_{61}Pm^{144}$			≈400d	EC		EC				
$_{61}Pm^{145}$			17.7y	EC	0.14	EC				
$_{61}Pm^{146}$			≈710d	EC β^-	1.72 1.53	EC β^-, 0.78	 35%	8400 ± 1680b		
$_{61}Pm^{147}$			2.5y	β^-	0.225	0.225	100%	90 ± 30b (Pm^{148m}) 100 ± 30b (Pm^{148})	7/2	±2.7
$_{61}Pm^{148m}$			42d	IT β^-	2.59	β^- 0.5 0.6 0.68 0.8 IT	37% 19% 8% 24% 6%	25,000 ± 4000b		
$_{61}Pm^{148}$			5.39d	β^-	2.46	1.1 2.0 2.6	41% 14% 45%		1	+2.0
$_{61}Pm^{149}$			53.1h	β^-	1.06	0.19 0.47 0.78 1.06	0.6% 0.3% 10% 89%	1350 ± 400b	7/2	
$_{61}Pm^{150}$			2.7h	β^-	3.43	1.5 1.8 2.3 2.8 3.2				
$_{61}Pm^{151}$			28h	β^-	1.19	0.35 0.50 0.73 0.84 0.98 1.05 1.13 1.19	11% 6% 10% 43% 3% 11% 6% 10%	<700b	5/2	±1.8
$_{61}Pm^{152}$			6m	β^-		2.2				
$_{61}Pm^{154}$			2.5m	β^-		2.5				
Sm		150.35						5820 ± 100b		

Isotope	% nat. abundance	Atomic mass	Lifetime	Modes of decay	Decay energy (MeV)	Particle energies (MeV)	Particle intensities	Thermal neutron capture cross-section	I	μ
$_{62}Sm^{142}$			72m	EC β^+		EC β^+ 1.03	90% 10%			
$_{62}Sm^{143m}$			64s	IT	0.748					
$_{62}Sm^{143}$			8.9m	EC β^+	3.3	EC 2.5	50% 50%			
$_{62}Sm^{144}$	3.09%	143.9117						≈0.7b		
$_{62}Sm^{145}$			340d	EC	0.65			≈110b		
$_{62}Sm^{146}$		146.9129	7×10^7y	α		2.55				
$_{62}Sm^{147}$	14.97%	146.9146	1.06×10^{11}y	α	2.314	2.314		75 ± 11b	7/2	−0.813
$_{62}Sm^{148}$	11.24%	147.9146	1.2×10^{13}y	α	2.001	2.04		2.4 ± 0.3b		
$_{62}Sm^{149}$	13.83%	148.9169	$\sim 4 \times 10^{14}$y	α	1.90	1.84		41,000 ± 2000b 50 ± 10mb (η, α)	7/2	−0.66
$_{62}Sm^{150}$	7.44%	149.9170						102 ± 5b		
$_{62}Sm^{151}$			93 ± 8y	β^-	0.076	0.055 0.076	1.7% 98.3%	15,000 ± 1800b		
$_{62}Sm^{152}$	26.72%	151.9195						210 ± 10b		
$_{62}Sm^{153}$			46.8 ± 0.1h	β^-	0.801	0.679 0.698 0.801	32% 48% 20%		3/2	−0.03
$_{62}Sm^{154}$	22.71%	153.9220						5.5 ± 1.1b		
$_{62}Sm^{155}$			22m	β^-	1.65	1.65 1.50	93% 7%		3/2	
$_{62}Sm^{156}$			9.4h	β^-	0.72	0.43 0.70	44% 51%			
$_{62}Sm^{157}$			0.5m	β^-						
Eu		151.96						4100 ± 100b		
$_{63}Eu^{144}$			10s	β^+, EC	6.32					
$_{63}Eu^{145}$			5.9d	β^+, EC						
$_{63}Eu^{146}$			4.6d	β^+, EC	3.86	0.80 1.47 2.1	0.4% 3.3% 0.14%			
$_{63}Eu^{147}$			22d	β^+, EC	1.8					
$_{63}Eu^{148}$			54d	β^+, EC	3.11	0.92	0.13%			
$_{63}Eu^{149}$			106d	EC	0.80	0.25 0.52 0.78 0.80	5% 4% 5% 85%			
$_{63}Eu^{150m}$			12.8h	β^+, EC β^-	2.25 1.01	β^- 1.01 β^+ 1.24				
$_{63}Eu^{150}$			5y	EC	2.25					

TABLE OF THE ISOTOPES (*Continued*)

Isotope	% nat. abundance	Atomic mass	Lifetime	Modes of decay	Decay energy (MeV)	Particle energies (MeV)	Particle intensities	Thermal neutron capture cross-section	I	μ
${}_{63}Eu^{151m}$			58μs	IT	0.196				7/2	+2.57
${}_{63}Eu^{151}$	47.82%	150.9196						3.8 ± 1.9b (Eu^{152m_2}) 3000 ± 200b (Eu^{152m_1}) 5000 ± 300b (Eu^{152})	5/2	+3.464
${}_{63}Eu^{152m_2}$			96m	IT	0.147					
${}_{63}Eu^{152m_1}$			9.2h	β^- β^+, EC	1.82 1.857	β^- 0.55 1.88 β^+ 0.77 0.895	10† 0.004% 0.007%			
${}_{63}Eu^{152}$			13y	β^+, EC β^-	1.86 1.82	β^- 0.22 0.36 0.68 1.05 1.46 β^+ 0.47 0.713	9† 13 51 6 21 0.005% 0.014%		3	±1.924
${}_{63}Eu^{153}$	52.18%	152.9209						450 ± 100b	5/2	+1.530
${}_{63}Eu^{154}$			16y	β^-	1.97	β^- 0.27 0.59 0.89 1.86	20% 45% 23% 12%	1500 ± 400b	3	±2.000
${}_{63}Eu^{155}$			1.81y	β^-	0.248	0.10 0.15 0.19 0.25	34% 40% 10% 16%	14,000 ± 4000b		
${}_{63}Eu^{156}$			15d	β^-	2.45	0.25 0.26 0.27 0.43 0.49 1.08 1.21 1.28 2.45	8% 9% 2.2% 8% 33% 2.3% 1.3% 6% 31%			
${}_{63}Eu^{157}$			15.2h	β^-	1.34	0.55 0.65 0.86 0.90 1.28 1.35	2% 10% 35% 15% 30% 3%			
${}_{63}Eu^{158}$			46m	β^-	3.5	1.11 1.9 2.5				
${}_{63}Eu^{159}$			18m	β^-	2.57	1.0 1.5 1.75 1.90 2.35 2.57	10% 11% 11% 21% 21% 25%			
${}_{63}Eu^{160}$			≈2.5m	β^-		3.6				
Gd		157.25						49,000 ± 2000b		

TABLE OF THE ISOTOPES (*Continued*)

Isotope	% nat. abundance	Atomic mass	Lifetime	Modes of decay	Decay energy (MeV)	Particle energies (MeV)	Particle intensities	Thermal neutron capture cross-section	I	μ
$_{64}Gd^{145}$			25m	EC β^+	5.2	EC β^+ 2.4	55% 45%			
$_{64}Gd^{146}$			4.6d	EC	~1.1					
$_{64}Gd^{147}$			35h	β^+, EC	~2.2					
$_{64}Gd^{148}$		147.9177	≈130y	α	3.27	3.18				
$_{64}Gd^{149}$		148.9189	9d	EC, α	3.17	α 3.01	0.0005%			
$_{64}Gd^{150}$		149.9185	2.1×10^6y	α	2.80	2.73	100%			
$_{64}Gd^{151}$			120d	EC, α	~0.4 2.61	2.60	~8 × 10^{-7}%			
$_{64}Gd^{152}$	0.200%	151.9195	1.1×10^{14}y	α	2.24	2.15	100%	<125b		
$_{64}Gd^{153}$			242d	EC	0.243					
$_{64}Gd^{154}$	2.15%	153.9207						102 ± 7b		
$_{64}Gd^{155}$	14.73%	154.9226						61,000 ± 1000b	3/2	−0.27
$_{64}Gd^{156}$	20.47%	155.9221								
$_{64}Gd^{157}$	15.68%	156.9339						254,000 ± 2000b	3/2	−0.36
$_{64}Gd^{158}$	24.87%	157.9241						3.5 ± 1.0b		
$_{64}Gd^{159}$			18h	β^-	0.95	0.59 0.89 0.95	13% 24% 63%		3/2	
$_{64}Gd^{160}$	21.90%	159.9271						0.77 ± 0.01b		
$_{64}Gd^{161}$			3.7m	β^-	2.02	1.5 1.6		(9.6 ± 5) × 10^4b		
Tb		158.924						30 ± 10b		
$_{65}Tb^{147}$			24m	β^+, EC						
$_{65}Tb^{148}$			70m	β^+, EC	5.6	2.6 4.6				
$_{65}Tb^{149m}$			4.3m	EC α		3.99	0.025%			
$_{65}Tb^{149}$			4.1h	EC α	4.06	EC α 3.95	84% 16%			
$_{65}Tb^{150}$			3.1h	β^+, EC	4.79	1.5 2.8 3.6	5† 5 10			
$_{63}Tb^{151}$		150.9230	18h	EC α	~2.7 3.49	EC α 3.44	99+% 0.0005%			
$_{65}Tb^{152m}$			4m	β^+, EC α		α	0.002%			
$_{65}Tb^{152}$			18h	β^+, EC	4.2	0.42 0.85 1.30 1.93 2.82	1† 4 14 45 100			
$_{65}Tb^{153}$			55h	EC	~1.6					
$_{65}Tb^{154}$			8.5h	EC						

TABLE OF THE ISOTOPES (*Continued*)

Isotope	% nat. abundance	Atomic mass	Lifetime	Modes of decay	Decay energy (MeV)	Particle energies (MeV)	Particle intensities	Thermal neutron capture cross-section	I	μ
$_{65}Tb^{154}$			21h	β^+, EC		EC	99+ %			
$_{65}Tb^{155}$			5.6d	EC	0.9					
$_{65}Tb^{156m}$			5.5h	IT EC						
$_{65}Tb^{156}$			5.4d	EC					(3)	±1.4
$_{65}Tb^{157}$			150y	EC	0.06		100%			
$_{65}Tb^{158m}$			11s	IT	0.111					
$_{65}Tb^{158}$			1.2 × 10^3y	EC β^-	1.20 0.95	0.64 0.86	1% 13%		3	±1.74
$_{65}Tb^{159}$	100%	159.9250						30 ± 10b	3/2	+1.95
$_{65}Tb^{160}$			73d	β^-	1.72	0.32 0.36 0.43 0.46 0.68 0.95 1.55 1.74	5% 10% 4% 44% 7% 29% 0.12% 0.34%	525 ± 100b	3	
$_{65}Tb^{161}$			6.9d	β^-	0.58	0.04 0.17 0.21 0.45 0.57 0.58	0.07% 0.02% 0.11% 26% 64% 10%		3/2	
$_{65}Tb^{162}$			7.5m	β^-	~2.8					
$_{65}Tb^{163m}$			7m	β^-						
$_{65}Tb^{163}$			6.5h	β^-	1.68	1.1 1.4 1.5 1.65	15% 40% 15% 30%			
$_{65}Tb^{164}$			23h	β^-	~3.8					
Dy		162.50						930 ± 20b		
$_{66}Dy^{149}$			≈15m	EC						
$_{66}Dy^{150}$			7.2m	α β^+, EC		α 4.2	18% 82%			
$_{66}Dy^{151}$			18m	β^+, EC α	~2.9 4.1	4.06	6%			
$_{66}Dy^{152}$		151.9244	2.5h	β^+, EC α	0.4 3.75	α 3.7	0.05%			
$_{66}Dy^{153}$		152.9254	6.4h	β^+, EC α	~2.1 3.57	α 3.48				
$_{66}Dy^{154m}$			13h	α		3.37				
$_{66}Dy^{154}$		153.9248	≈10^6y	α	2.93	2.85				
$_{66}Dy^{155}$			10.2h	β^+, EC β^+, EC		0.85 1.08	2% 0.14%			
$_{66}Dy^{156}$	0.052%	155.9238								
$_{66}Dy^{157}$			8.1h	EC						

TABLE OF THE ISOTOPES (*Continued*)

Isotope	% nat. abundance	Atomic mass	Lifetime	Modes of decay	Decay energy (MeV)	Particle energies (MeV)	Particle intensities	Thermal neutron capture cross-section	I	μ
$_{66}Dy^{158}$	0.090%	157.9240						96 ± 20b		
$_{66}Dy^{159}$			144d	EC	0.38				3/2	
$_{66}Dy^{160}$	2.29%	159.9248						55 ± 9b		
$_{66}Dy^{161}$	18.88%	160.9266						585 ± 50b	5/2	−0.46
$_{66}Dy^{162}$	25.53%	161.9265						200 ± 50b		
$_{66}Dy^{163}$	24.97%	162.9284						140 ± 30b	5/2	+0.64
$_{66}Dy^{164}$	28.18%	163.9288						2100 ± 400b (Dy^{165m},4) 2600 ± 200b ($Dy^{165m}+Dy^{165}$,4)		
$_{66}Dy^{165m}$			75s	IT β^-	0.108 1.30	β^- 0.89 1.0	87† 13			
$_{66}Dy^{165}$			2.3h	β^-	1.30	0.22 0.24 0.3 1.25 1.30	0.1% 0.03% 1.3% 15% 83%		7/2	
$_{66}Dy^{166}$			81.5h	β^-	0.482	0.05 0.40 0.482	1% 94% 5%			
$_{66}Dy^{167}$			4.4m	β^-						
Ho		164.930						65 ± 2b		
$_{67}Ho^{151}$			42s	α β^+, EC		4.60	30%			
$_{67}Ho^{151}$			36s	α β^+, EC		4.51	20%			
$_{67}Ho^{152}$			2.4m	α β^+, EC	4.50 6.4	4.38	30%			
$_{67}Ho^{152}$			52s	α β^+, EC	4.57	4.45	19%			
$_{67}Ho^{153}$			9m	α	4.02	3.92	0.3%			
$_{67}Ho^{154}$			7m	β^+, EC	~5.5					
$_{67}Ho^{155}$			50m	β^+, EC		2.1				
$_{67}Ho^{156}$			55m	β^+, EC	≈5	1.3 1.80 2.9	4† 14 1			
$_{67}Ho^{157}$			14m	β^+, EC						
$_{67}Ho^{158m}$			29m	IT β^+, EC	0.0673 4.1	1.32				
$_{67}Ho^{158}$			11m	β^+, EC	4.04					
$_{67}Ho^{159m}$			6.9s	IT	~2.0					
$_{67}Ho^{159}$			33m	EC	~1.8					
$_{67}Ho^{160m}$			5.0h	IT	3.36		66%			

TABLE OF THE ISOTOPES (*Continued*)

Isotope	% nat. abundance	Atomic mass	Lifetime	Modes of decay	Decay energy (MeV)	Particle energies (MeV)	Particle intensities	Thermal neutron capture cross-section	I	μ
				β^+, EC		0.31 0.57 1.01	0.01% 0.2% 0.08%			
$_{67}Ho^{160}$			23m	β^+, EC	3.30	1.9	0.02%			
$_{67}Ho^{161m}$			6s	IT	0.211					
$_{67}Ho^{161}$			2.5h	EC	~0.8				7/2	
$_{67}Ho^{162m}$			68m	IT EC	~2.26		63%			
$_{67}Ho^{162}$			15m	β^+, EC	2.16	~1.06 ~1.14	~56% ~44%			
$_{67}Ho^{163m}$			1.1s	IT	0.305					
$_{67}Ho^{163}$			$>10^3$y	EC	~0.01					
$_{67}Ho^{164}$			37m	β^- EC	1.03 1.11	0.99	53% 47%			
$_{67}Ho^{165}$	100%	164.9303						3.45 ± 0.35b (Ho^{166m}) 61.2 ± 2.0b (Ho^{166})	7/2	+4.08
$_{67}Ho^{166}$			26.9h	β^-	1.847	0.18 0.38 1.76 1.84	0.3% 1% 47% 52%		0*	
$_{67}Ho^{166m}$			1.2×10^3y	β^-	1.847	≤0.067				
$_{67}Ho^{167}$			3.1h	β^-	1.0	0.28 0.96	50% 50%			
$_{67}Ho^{168}$			3.3m	β^-	3.3	≈2.2				
$_{67}Ho^{169}$			4.8m	β^-	2.1	1.20 1.95	75% 25%			
$_{67}Ho^{170}$			45s	β^-	4.2	≈3.1				
Er		167.26						160 ± 30b		
$_{68}Er^{158}$			2.5h	β^+, EC		0.8				
$_{68}Er^{159}$			36m	β^+, EC						
$_{68}Er^{160}$			29h	EC	~0.8					
$_{68}Er^{161}$			3.1h	β^+, EC	2.4	1.2				
$_{68}Er^{162}$	0.136%	161.9288						160 ± 30b		
$_{68}Er^{163}$			75m	β^+, EC	1.21	0.19	0.004%			
$_{68}Er^{164}$	1.56%	163.9293						13 ± 5b		
$_{68}Er^{165}$			10.3h	EC	0.37				5/2	±0.65
$_{68}Er^{166}$	33.41%	165.9304						10 ± 5b (Er^{167m}) 30 ± 5b (Er^{167m} + Er^{167})		
$_{68}Er^{167m}$			2.3s	IT	0.208					

Isotope	% nat. abundance	Atomic mass	Lifetime	Modes of decay	Decay energy (MeV)	Particle energies (MeV)	Particle intensities	Thermal neutron capture cross-section	I	μ
$_{68}Er^{167}$	22.94%	166.9320						700 ± 50b	7/2	−0.564
$_{68}Er^{168}$	27.07%	167.9324						1.9 ± 0.2b		
$_{68}Er^{169}$			9.4d	β^-	0.34	0.332 0.340	42% 58%		1/2	+0.513
$_{68}Er^{170}$	14.88%	169.9355						6 ± 1b		
$_{68}Er^{171}$			7.52h	β^-	1.490	0.50 0.54 0.59 0.77 0.83 1.065 1.49	0.4% 0.5% 3.6% 0.4% 0.2% 91% 2.3%		5/2	±0.70
$_{68}Er^{172}$			49h	β^-	0.91	0.30 0.38 0.43 0.91	42% 44% 12% <10%			
Tm		168.934						115 ± 15b		
$_{69}Tm^{161}$			30m	EC	3.5					
$_{69}Tm^{162}$			77m	β^+, EC	4.89					
$_{69}Tm^{162}$			22m	β^+, EC	4.89	0.9 2.3 3.82	3† 12 10			
$_{69}Tm^{163}$			1.8h	β^+, EC	2.27	0.40 1.1	7† 100		1/2	
$_{69}Tm^{164}$			1.9m	β^+, EC	~3.8	1.3 2.94	4† 100			
$_{69}Tm^{165}$			30.1h	β^+, EC	1.57	0.1	~0.004%			
$_{69}Tm^{166}$			7.7h	β^+, EC	3.04	1.10 1.94	0.3% 1.5%		2	±0.05
$_{69}Tm^{167}$			9.6d	EC	1.16				1/2	
$_{69}Tm^{168}$			87d	EC	1.72					
$_{69}Tm^{169}$	100%	168.9344						115 ± 15b	1/2	−0.230
$_{69}Tm^{170}$			128.6 ± 0.3d	β^-	0.967	0.886 0.97	23% 77%	92 ± 4b	1	±0.25
$_{69}Tm^{171}$			1.92y	β^-	0.098	0.097 0.030	98% 2%	4.5 ± 0.2b	1/2	±0.23
$_{69}Tm^{172}$			63.6h	β^-	1.88	0.13 0.22 0.28 0.33 0.41 0.71 1.62 1.80 1.88	~0.05% 1% 10% 4% 14% 8% 1% 39% 23%			
$_{69}Tm^{173}$			8.2h	β^-	1.32	0.80 0.84 1.26	23% 75% 2%			

Isotope	% nat. abundance	Atomic mass	Lifetime	Modes of decay	Decay energy (MeV)	Particle energies (MeV)	Particle intensities	Thermal neutron capture cross-section	I	μ
$_{69}Tm^{174m}$			5.2m	β^-	3.0	0.7 1.2	~20% ~80%			
$_{69}Tm^{174}$			5.5m	β^-	2.5	2.5	100%			
$_{69}Tm^{175}$			20m	β^-	2.5	2.0				
$_{69}Tm^{176}$			1.5m	β^-	4.2	4.2				
Yb		173.04						37 ± 3b		
$_{70}Yb^{164}$			76m	EC						
$_{70}Yb^{165}$			10m	β^+, EC	~2.7					
$_{70}Yb^{166}$			57.5h	EC	0.3					
$_{70}Yb^{167}$			18m	β^+, EC	1.96	0.65	0.4%			
$_{70}Yb^{168}$	0.135%	167.9339						3200 ± 400b		
$_{70}Yb^{169}$			32d	EC	~1.2					
$_{70}Yb^{170}$	3.03%	169.9349						9.4 ± 0.9b		
$_{70}Yb^{171}$	14.31%	170.9365						50 ± 5b	1/2	+0.4919
$_{70}Yb^{172}$	21.82%	171.9366						0.4 ± 0.1b		
$_{70}Yb^{173}$	16.13%	172.9383						19 ± 2b	5/2	−0.6776
$_{70}Yb^{174}$	31.84%	173.9390						46 ± 4b (Yb^{175m}) 65 ± 5b (Yb^{175m} + Yb^{175})		
$_{70}Yb^{175m}$			0.067s	IT	0.513					
$_{70}Yb^{175}$			101h	β^-	0.467	0.073 0.22 0.35 0.466	11% <0.1% 2.1% 87%		(7/2)	±0.13
$_{70}Yb^{176}$	12.73%	175.9427						5.5 ± 1.0b		
$_{70}Yb^{177m}$			6.5s	IT	0.33					
$_{70}Yb^{177}$			1.9h	β^-	1.40	0.17 0.18 0.25 1.1 1.25 1.3 1.40	3% 6% 1% 2% 17% 9% 60%			
Lu		174.97						75 ± 2b		
$_{71}Lu^{167}$			54m	β^+, EC	3.0	1.1 1.5	1† 1			
$_{71}Lu^{168}$			7.1m	β^+, EC	~4.6	1.20				
$_{71}Lu^{169m}$			2.7m	IT	0.02%					
$_{71}Lu^{169}$			34h	β^+, EC	2.26	0.4 1.2	17† 100			
$_{71}Lu^{170m}$			0.7s	IT	0.093					

TABLE OF THE ISOTOPES (*Continued*)

Isotope	% nat. abundance	Atomic mass	Lifetime	Modes of decay	Decay energy (MeV)	Particle energies (MeV)	Particle intensities	Thermal neutron capture cross-section	I	μ
$_{71}Lu^{170}$			2.0d	β^+, EC	3.41	2.39	0.19%			
$_{71}Lu^{171m}$			76s	IT	0.071					
$_{71}Lu^{171}$			8.3d	β^+, EC	~1.6					
$_{71}Lu^{172m}$			3.7m	IT	0.041	γ Energies 41.86				
$_{71}Lu^{172}$			6.70d	EC	~2.7					
$_{71}Lu^{173}$			≈1.37y	EC	0.69					
$_{71}Lu^{174m}$			140d	IT EC	0.171 ~1.7					
$_{71}Lu^{174}$			3.6y	EC	1.5					
$_{71}Lu^{175}$	97.41%	174.9409						21 ± 3b	7/2	+2.23
$_{71}Lu^{176m}$			3.7h	β^-	1.3	1.22 1.31	~65% ~35%		1	+0.318
$_{71}Lu^{176}$	2.59%		3×10^{10}y	β^-	1.02	0.42	100%	7 ± 2b (Lu^{177m}) 2050 ± 50b (Lu^{177})	7	+3.18
$_{71}Lu^{177m}$			155d	IT β^-	0.97		22% 78%	gamma rays in equilibrium with Lu^{177} (* with IT)		
$_{71}Lu^{177}$			6.7d	β^-	0.497	0.17 0.38 0.497	6.7% 7% 86%		7/2	+2.24
$_{71}Lu^{178}$			20m	β^-		1.5				
$_{71}Lu^{178}$			30m	β^-	2.25	2.25				
$_{71}Lu^{179}$			4.6h	β^-	1.34	1.1 1.35	13% 87%			
$_{71}Lu^{180}$			2.5m	β^-	3.3	3.3				
Hf		178.49						103 ± 3b		
$_{72}Hf^{168}$			22m	β^+, EC		EC β^+ 1.7				
$_{72}Hf^{169}$			1.5h	β^+, EC		~1.3				
$_{72}Hf^{170}$			12.2h	β^+, EC						
$_{72}Hf^{171}$			11h	EC						
$_{72}Hf^{172}$			≈5y	EC	~1					
$_{72}Hf^{173}$			23.6h	EC	>2					
$_{72}Hf^{174}$	0.18%	173.9403	2.0×10^{15}y	α	2.55	≈2.55		390 ± 55b		
$_{72}Hf^{175}$			70d	EC	0.59					
$_{72}Hf^{176}$	5.20%	175.94165						15 ± 15b		
$_{72}Hf^{177}$	18.50%	176.9435						1.1 ± 0.1b (Hf^{178m})	7/2	+0.61

Isotope	% nat. abundance	Atomic mass	Lifetime	Modes of decay	Decay energy (MeV)	Particle energies (MeV)	Particle intensities	Thermal neutron capture cross-section	I	μ
								380 ± 30b (Hf^{178})		
$_{72}Hf^{178m_2}$			>10y	IT	2.5					
$_{72}Hf^{178m_1}$			5s	IT	1.148					
$_{72}Hf^{178}$	27.14%	177.9439						52 ± 6b (Hf^{179m}) 75 ± 10b (Hf^{179m} + Hf^{179})	0	
$_{72}Hf^{179m}$			18.6s	IT	0.375					
$_{72}Hf^{179}$	13.75%	178.9460						0.34 ± 0.03b (Hf^{180m}) 65 ± 15b (Hf^{180})	9/2	−0.47
$_{72}Hf^{180m}$			5.5h	IT	1.143					
$_{72}Hf^{180}$	35.24%	179.9468						12.6 ± 0.7b	0	
$_{72}Hf^{181}$			42.4 ± 0.2d	β^-	1.023	0.4 0.408	7% 93%	40^{+40}_{-20}b		
$_{72}Hf^{182}$			9 × 10^6y	β^-	~0.5	~0.5				
$_{72}Hf^{183}$			1.1h	β^-	2.2	0.7 1.2 1.6	25% 50% 25%			
Ta		180.948						22 ± 1b		
$_{73}Ta^{172}$			44m	β^+, EC	~5					
$_{73}Ta^{173}$			3.7h	β^+, EC						
$_{73}Ta^{174}$			1.3h	β^+, EC						
$_{73}Ta^{175}$			10.5h	EC	~2.3					
$_{73}Ta^{176}$			8.0h	EC	~3					
$_{73}Ta^{177}$			56.6h	EC	1.17					
$_{73}Ta^{178}$			9.4m	β^+, EC	1.9	0.80 0.89	7† 10			
$_{73}Ta^{178}$			2.1h	β^+, EC	1.9	1				
$_{73}Ta^{179}$			~600y	EC	0.115					
$_{73}Ta^{180m}$			8.1h	EC β^-	0.7 0.50	0.61 0.71	3% 10%			
$_{73}Ta^{180}$	0.0123%	179.9415	>10^{13}y	EC, β^-						
$_{73}Ta^{181m}$			6.8μs	IT	0.0063					
$_{73}Ta^{181}$	99.988%	180.9480						10 ± 2mb (Ta^{182m}) 22 ± 1b (Ta^{182})	7/2	+2.36

Isotope	% nat. abundance	Atomic mass	Lifetime	Modes of decay	Decay energy (MeV)	Particle energies (MeV)	Particle intensities	Thermal neutron capture cross-section	I	μ
$_{73}Ta^{182m}$			16.5m	IT	0.503					
$_{73}Ta^{182}$			115d	β^-	1.811	0.17 0.23 0.35 0.37 0.43 0.50 1.48 1.71	29% 3% 2.2% ~2% 42% 0.9% 0.24% 0.05%	8200 ± 600b		
$_{73}Ta^{183}$			5.1d	β^-	1.068	0.48 0.62 0.66 0.76 0.78	0.7% 84% 6% 7% ~3%		7/2	
$_{73}Ta^{184}$			8.7h	β^-	2.75	1.19 1.45 1.76 2.22 2.64	93% 6% 0.9% 0.1% 0.15%			
$_{73}Ta^{185}$			50m	β^-	2.0	0.15 1.7	~30% ~70%			
$_{73}Ta^{186}$			10.5m	β^-	3.7	2.2				
W		183.85						18.5 ± 0.5b		
$_{74}W^{173}$			16m	EC						
$_{74}W^{174}$			31m	EC						
$_{74}W^{175}$			34m	EC						
$_{74}W^{176}$			2.5h	β^+, EC	≈1.5	~2				
$_{74}W^{177}$			135m	EC	≈2					
$_{74}W^{178}$			21.5d	EC	≈0.3					
$_{74}W^{179m}$			5.2m	IT	0.222					
$_{74}W^{179}$			38m	EC	≈1.2					
$_{74}W^{180m}$			0.005s	IT						
$_{74}W^{180}$	0.14%	179.9470						10 ± 10b		
$_{74}W^{181m}$			14μs	IT						
$_{74}W^{181}$			140d	EC	0.19					
$_{74}W^{182}$	26.41	181.9483						20.7 ± 0.5b	0	
$_{74}W^{183m}$			5.3s	IT	0.309					
$_{74}W^{183}$	14.40%	182.9503						10.2 ± 0.3b	1/2	+0.1172
$_{74}W^{184}$	30.64%	183.9510						2.4 ± 0.4mb (W^{185m}) 1.8 ± 0.2b (W^{185})	0	
$_{74}W^{185m}$			1.62m	IT	0.368					
$_{74}W^{185}$			75.8d	β^-	0.432	0.432	100%		3/2	
$_{74}W^{186}$	28.41%	185.9543						37 ± 2b	0	

Isotope	% nat. abundance	Atomic mass	Lifetime	Modes of decay	Decay energy (MeV)	Particle energies (MeV)	Particle intensities	Thermal neutron capture cross-section	I	μ
$_{74}W^{187}$			24h	β^-	1.315	0.43 0.47 0.55 0.63 ~0.71 1.316	0.5% ~1% 4.3% 59% 8% 19%	90 ± 40b	3/2	
$_{74}W^{188}$			69d	β^-	0.349	0.059 0.285 0.349	0.9% 0.6% 98.6%			
$_{74}W^{189}$			11.5m	β^-	2.5	2.0 2.5				
Re		186.2						85 ± 5b		
$_{75}Re^{177}$			17m	β^+, EC	≈3	>0.4				
$_{75}Re^{178}$			15m	β^+, EC	≥4.1	3.1				
$_{75}Re^{179}$			20m	EC	≈2.5					
$_{75}Re^{180}$			2.4m	β^+, EC		1.1				
$_{75}Re^{180}$			20h	β^+, EC	2.9	≈1.9				
$_{75}Re^{181}$			18h	EC	≈1.7					
$_{75}Re^{182}$			12.7h	β^+, EC	2.86	0.55 1.74	0.06% 0.2%			
$_{75}Re^{182}$			64h	EC						
$_{75}Re^{183}$			71d	EC	~0.9					
$_{75}Re^{184m}$			169d	IT	0.188					
$_{75}Re^{184}$			3.8d	EC	~1.6					
$_{75}Re^{185}$	37.500%	184.9530						110 ± 5b	5/2	+3.172
$_{75}Re^{186}$			90h	β^-	1.071	0.30 0.933 1.071	0.1% 21% 74%		1	+1.73
$_{75}Re^{187}$	62.500%	186.9560	7×10^{10}y	β^-	<0.01	≤0.008		2.0 ± 0.5b (Re^{184m}) 75 ± 4b (Re^{184})	5/2	+3.204
$_{75}Re^{188m}$			18.7m	IT	0.172					
$_{75}Re^{188}$			16.7h	β^-	2.116	0.21 0.23 0.30 0.37 0.41 0.56 0.71 0.715 0.77 1.09 1.54 1.96 2.116	0.05% 0.06% 0.02% 0.026% 0.14% 0.012% 0.35% 0.026% 0.025% 0.5% 1.2% 18% 80%	<2b	1	+1.78
$_{75}Re^{189}$			24h	β^-	1.00	0.72 0.80 1.02				

TABLE OF THE ISOTOPES (*Continued*)

Isotope	% nat. abundance	Atomic mass	Lifetime	Modes of decay	Decay energy (MeV)	Particle energies (MeV)	Particle intensities	Thermal neutron capture cross-section	I	μ
$_{75}Re^{190}$			2.8m	β^-	3.1	1.7				
$_{75}Re^{190m}$			2.8h	IT						
$_{75}Re^{191}$			9.8m	β^-	1.8	1.8				
$_{75}Re^{192}$			6s	β^-		2.5				
Os		190.2						15.3 ± 0.7b		
$_{76}Os^{181}$			23m	EC	≈2.8					
$_{76}Os^{182}$			22h	EC	≈1.1					
$_{76}Os^{183m}$			9.9h	IT EC	0.171 ≈2.2	IT EC	46% 54%			
$_{76}Os^{183}$			12h	EC	≈2.0					
$_{76}Os^{184}$	0.018%	183.9526						<200b		
$_{76}Os^{185}$			94d	EC	0.982					
$_{76}Os^{186}$	1.59%	185.9539								
$_{76}Os^{187}$	1.64%	186.9560							1/2	+0.0643
$_{76}Os^{188}$	13.3%	187.9560								
$_{76}Os^{189m}$			5.7h	IT	0.03					
$_{76}Os^{189}$	16.1%	188.9586						0.26 ± 0.03mb (Os^{190m})	3/2	+0.6566
$_{76}Os^{190m}$			9.9m	IT	1.706					
$_{76}Os^{190}$	26.4%	189.9586						12 ± 6b (Os^{191m}) 4 ± 2b (Os^{191})		
$_{76}Os^{191m}$			13h	IT	0.074					
$_{76}Os^{191}$			15d	β^-	0.310	0.13				
$_{76}Os^{192}$	39.952%									
$_{76}Os^{193}$			31h	β^-	1.132	0.57 0.67 0.85 0.99 1.059	4% 7% 1% 9% 6%			
$_{76}Os^{194}$			6.0y	β^-	0.097	0.009 0.054 0.097	2% 98% <2%			
$_{76}Os^{195}$			6.5m	β^-	2.0	~2				
Ir		192.2						425 ± 15b		
$_{77}Ir^{182}$			15m	β^+, EC	≈5.5					

Isotope	% nat. abundance	Atomic mass	Lifetime	Modes of decay	Decay energy (MeV)	Particle energies (MeV)	Particle intensities	Thermal neutron capture cross-section	I	μ
$_{77}Ir^{183}$			0.9h	EC						
$_{77}Ir^{184}$			3.2h	β^+, EC	>4.3					
$_{77}Ir^{185}$			14h	EC	≈2.5					
$_{77}Ir^{186}$			16h	β^+, EC	3.83	1.06 1.37 1.94	 0.5% 1.9%			
$_{77}Ir^{186}$			1.7h	β^+, EC		0.8 1.30 1.93 2.6 3.4	22† 12 44 20 1			
$_{77}Ir^{187}$			10.5h	EC	≈1.6					
$_{77}Ir^{188}$			41h	β^+, EC		1.21 1.66				
$_{77}Ir^{189}$			13.3d	EC	≈0.6					
$_{77}Ir^{190m_2}$			3.2h	IT β^+, EC	0.175 ~2.3		6% 94%			
$_{77}Ir^{190m_1}$			1.2h	IT	0.026					
$_{77}Ir^{190}$			11d	EC	2.1					
$_{77}Ir^{191m}$			4.9s	IT	0.042				11/2	±6
$_{77}Ir^{191}$	37.4%	190.9609						0.4 ± 0.2b (Ir^{192m_2}) 610 ± 60b (Ir^{192m_1}) 300 ± 30b (Ir^{192})	3/2	+0.18
$_{77}Ir^{192m_2}$			>5y	IT	0.161					
$_{77}Ir^{192m_1}$			1.4m	IT β^-	0.058 1.50	0.9 1.2 1.5	0.0025% 0.008% 0.007%			
$_{77}Ir^{192}$			74d	β^- β^+, EC	1.453 1.2	β^-, 0.24 0.536 0.672 β^+, 0.24	8% 41% 46% 1.5×10^{-5}%	1100 ± 400b (Ir^{193m})	4	
$_{77}Ir^{193m}$			12d	IT	0.08			0.05 ± 0.02b (Ir^{194m_2})	[1/2?]	<2
$_{77}Ir^{193}$	62.6%	192.9633						110 ± 15b	3/2	+0.18
$_{77}Ir^{194}$			17.4h	β^-	2.24	0.98 1.62	2.0% 1.2%		1	
$_{77}Ir^{195}$			4.2h	β^-	1.0	0.6 1.0	≈25% ≈75%			
$_{77}Ir^{196m}$			84m	β^-	3.40	0.24 0.39 0.66 1.08	10% 4.7% 5.4% 80%			
$_{77}Ir^{196}$			52s	β^-	3.22	0.3 0.4	~1% ~1%			

Isotope	% nat. abundance	Atomic mass	Lifetime	Modes of decay	Decay energy (MeV)	Particle energies (MeV)	Particle intensities	Thermal neutron capture cross-section	I	μ
						0.82 1.09 3.22	~1% 16% 80%			
$_{77}Ir^{197}$			7m	β^-	2.0	1.5 2.0	~50% ~50%			
$_{77}Ir^{198}$			50s	β^-	4.4	~3.6				
Pt		195.09						9 ± 1b		
$_{78}Pt^{173}$			short	α	6.34	α 6.19				
$_{78}Pt^{174}$			0.7s	β^+, EC α	6.17	 6.03	20% 80%			
$_{78}Pt^{175}$			2.1s	α	6.09	5.95				
$_{78}Pt^{176}$			6s	α β^+, EC	5.87	5.74	1% 99%			
$_{78}Pt^{177}$			6.6s	α β^+, EC	5.64	5.51	0.3% 99+%			
$_{78}Pt^{178}$			21s	α β^+, EC	5.56	5.28 5.44	0.07% 1.3%			
$_{78}Pt^{179}$			33s	α β^+, EC	5.27	5.15	0.1%			
$_{78}Pt^{180}$			50s	α β^+, EC	5.26	5.14	0.3%			
$_{78}Pt^{181}$			51s	α β^+, EC	5.13	5.02	0.0006%			
$_{78}Pt^{182}$			3.0m	α β^+, EC	4.95 ≈3		≈0.02%			
$_{78}Pt^{183}$			7m	β^+, EC α	4.84	4.74				
$_{78}Pt^{184}$			20m	α β^+, EC	4.60	4.50				
$_{78}Pt^{185}$			1.2h	EC						
$_{78}Pt^{186}$			3h	α EC	4.32	4.23				
$_{78}Pt^{187}$			2h	EC						
$_{78}Pt^{188}$			10.2d	EC	0.51					
$_{78}Pt^{189}$			10.9h	EC						
$_{78}Pt^{190}$	0.0127%	189.960	6×10^{11}y	α		3.18				
$_{78}Pt^{191}$			3.0d	EC	~2.0					
$_{78}Pt^{192}$	0.78%	191.9614	~10^{15}y	α		2.6		2 ± 1b (Pt^{193m})		
$_{78}Pt^{193m}$			4.3d	IT	0.148					
$_{78}Pt^{193}$			<500y	EC	0.045					
$_{78}Pt^{194}$	32.9%	193.9628						87 ± 13mb		

Isotope	% nat. abundance	Atomic mass	Lifetime	Modes of decay	Decay energy (MeV)	Particle energies (MeV)	Particle intensities	Thermal neutron capture cross-section	I	μ
								(Pt^{195m}) 1.2 ± 0.9b (Pt^{195})	0	
$_{78}Pt^{195m}$			4.1d	IT	0.2593					
$_{78}Pt^{195}$	33.8%	194.9648						27 ± 2b	1/2	+0.6060
$_{78}Pt^{196}$	25.3%	195.9650						0.06 ± 0.02b (Pt^{197m}) 0.9 ± 0.1b (Pt^{197})	0	
$_{78}Pt^{197m}$			80m	IT β^-	0.399 1.15		97%			
$_{78}Pt^{197}$			18h	β^-	0.75	0.48 0.67	10% 90%			
$_{78}Pt^{198}$	7.21%	197.9675						4.0 ± 0.5b		
$_{78}Pt^{199m}$			14.1s	IT	0.425					
$_{78}Pt^{199}$			30m	β^-	1.68	0.72 0.89 0.94 1.14 1.38 1.69	1% 10% 8% 14% 4% 63%	15 ± 10b		
$_{78}Pt^{200}$			11.5h	β^-	≈0.7					
Au		196.967						98.8 ± 0.3b		
$_{79}Au^{185}$			7m	α	5.18	5.07				
$_{79}Au^{186}$			12m	EC						
$_{79}Au^{187}$			8m	EC						
$_{79}Au^{188}$			8m	β^+, EC	5.2					
$_{79}Au^{189}$			30m	EC						
$_{79}Au^{190}$			39m	β^+, EC	≈4.4				1	±0.066
$_{79}Au^{191}$			3.2h	EC	≈2.0				3/2	±0.137
$_{79}Au^{192}$			4.1h	β^+, EC	3.24	2.22	1%		1	±0.00785
$_{79}Au^{193m}$			3.9s	IT EC	0.290	I.T	99%			
$_{79}Au^{193}$			16h	EC	≈1.1				3/2	±0.139
$_{79}Au^{194}$			39.5h	β^+, EC	2.51	1.21 1.55	1.3† 1.7		1	±0.074
$_{79}Au^{195m}$			31s	IT	0.318					
$_{79}Au^{195}$			183d	EC	0.227				3/2	±0.147
$_{79}Au^{196}$			6.18d	β^+, EC β^-	1.48 0.684	β^-, 0.259			2	±0.6
$_{79}Au^{197m}$			7.2s	IT	0.409				[1/2]†	+0.37†

TABLE OF THE ISOTOPES (*Continued*)

Isotope	% nat. abundance	Atomic mass	Lifetime	Modes of decay	Decay energy (MeV)	Particle energies (MeV)	Particle intensities	Thermal neutron capture cross-section	I	μ
$_{79}Au^{197}$	100%	196.9666						98.8 ± 0.3b	3/2	+0.14486
$_{79}Au^{198}$			2.693 ±0.005d	β^-	1.374	0.28 0.961 1.374	1.1% 98.9% 0.025%	25,800 ± 1200b	2	+0.590
$_{79}Au^{199}$			3.15d	β^-	0.46	0.250 0.296 0.462	22.4% 71.6% 6.0%	30 ± 15b	3/2	+0.270
$_{79}Au^{200}$			48.4m	β^-	2.21	0.61 1.84 2.21	24% 6% 70%			
$_{79}Au^{201}$			26m	β^-	1.5	1.0 1.5	5% 95%			
$_{79}Au^{203}$			5.5s	β^-	≈2.1	1.9				
Hg		200.59						375 ± 5b		
$_{80}Hg^{185}$			50s	EC						
$_{80}Hg^{186}$			1.5m	EC						
$_{80}Hg^{187}$			3m	EC α		 5.14?				
$_{80}Hg^{188}$			3.7m	EC						
$_{80}Hg^{189}$			10m	β^+, EC						
$_{80}Hg^{190}$			20m	EC						
$_{80}Hg^{191}$			55m	β^+, EC						
$_{80}Hg^{192}$			4.8h	EC	≈1.4					
$_{80}Hg^{193m}$			10h	IT β^+, EC	0.141	1.17	16%		13/2	−1.063
$_{80}Hg^{193}$			~6h	EC	2.34				3/2	−0.62
$_{80}Hg^{194}$			1.9y?	EC	≈0.4					
$_{80}Hg^{195m}$			40h	IT EC	0.176 ~1.7				13/2	−1.049
$_{80}Hg^{195}$			9.5h	EC	~1.5				1/2	+0.538
$_{80}Hg^{196}$	0.146%	195.9658						120 ± 15b (Hg^{197m}) 3000 ± 100b (Hg^{197})		
$_{80}Hg^{197m}$			24h	IT EC	0.72 0.42	IT EC	94% 6%		13/2	−1.032
$_{80}Hg^{197}$			65h	EC	0.42				1/2	+0.5241
$_{80}Hg^{198}$	10.02%	197.9668						0.02 ± 0.01b (Hg^{199m})	0	
$_{80}Hg^{199m}$			43m	IT	0.533					
$_{80}Hg^{199}$	16.84%	198.9683						2000 ± 1000b	1/2	+0.5027
$_{80}Hg^{200}$	23.13%	199.9683						<60b	0	
$_{80}Hg^{201}$	13.22%	200.9703						<60b	3/2	−0.5567

TABLE OF THE ISOTOPES (*Continued*)

Isotope	% nat. abundance	Atomic mass	Lifetime	Modes of decay	Decay energy (MeV)	Particle energies (MeV)	Particle intensities	Thermal neutron capture cross-section	I	μ
$_{80}Hg^{202}$	29.80%	201.9706						4.9 ± 0.2b	0	
$_{80}Hg^{203}$			46.57 ±0.03d	β^-	0.492	0.210 0.49	100% <0.004%		5/2	+0.84
$_{80}Hg^{204}$	6.85%	203.9735						0.43 ± 0.10b	0	
$_{80}Hg^{205}$			5.5m	β^-	1.6	~1.4 ~1.6				
$_{80}Hg^{206}$			7.5m	β^-	1.31					
Tl		204.37						3.4 ± 0.5b		
$_{81}Tl^{191}$			10m	β^+, EC						
$_{81}Tl^{192}$			11m	β^+, EC						
$_{81}Tl^{193m}$			2.1m	IT						
$_{81}Tl^{193}$			23m	EC						
$_{81}Tl^{194m}$			32.8m	EC						
$_{81}Tl^{194}$			33m	EC	≈5.4					
$_{81}Tl^{195m}$			3.5s	IT	0.482					
$_{81}Tl^{195}$			1.2h	β^+, EC		1.8			1/2	
$_{81}Tl^{196m}$			1.41h	IT EC	0.395	IT EC	4% 96%			
$_{81}Tl^{196}$			1.84h	EC	4.6					
$_{81}Tl^{197m}$			0.54s	IT	0.609					
$_{81}Tl^{197}$			2.8h	EC	2.2				1/2	
$_{81}Tl^{198m}$			1.87h	IT EC	0.543 ~4.0		55% 45%		7	
$_{81}Tl^{198}$			5.3h	β^+, EC	3.5	0.7 1.4 2.1 2.4	4† 10 12 14		2	± <0.002
$_{81}Tl^{199}$			7.4h	EC	1.1				1/2	+1.59
$_{81}Tl^{200}$			26.1h	β^+, EC	2.454	1.07 1.44	5.6† 1.0		2	± ≤0.15
$_{81}Tl^{201m}$			2.1ms	IT	0.93					
$_{81}Tl^{201}$			73h	EC	≈0.41				1/2	+1.60
$_{81}Tl^{202}$			12.0d	EC	1.1				2	±≤0.15
$_{81}Tl^{203}$	29.50%	202.9723						10 ± 1b	1/2	+1.6115
$_{81}Tl^{204}$			3.8y	β^- EC	0.763 0.34	0.7635 ± 0.0002 EC	97.9% 2%		2	±0.089
$_{81}Tl^{205}$	70.50%	204.9745						0.5 ± 0.2b	1/2	+1.6274
$_{81}Tl^{206}$			4.19m	β^-	1.524	1.524	100%		0*	
$_{81}Tl^{207m}$			1.3s	IT	1.34					

Isotope	% nat. abundance	Atomic mass	Lifetime	Modes of decay	Decay energy (MeV)	Particle energies (MeV)	Particle intensities	Thermal neutron capture cross-section	I	μ
$_{81}Tl^{207}$			4.78m	β^-	1.44	0.53 1.44	0.16% 99+%			
$_{81}Tl^{208}$			3.10m	β^-	4.994	1.04 1.29 1.52 1.792 2.38	4% 24% 23% 49% 0.03%			
$_{81}Tl^{209}$			2.2m	β^-	3.98	1.99	100%			
$_{81}Tl^{210}$			1.3m	β^-	5.50	1.3 1.9 2.3	25% 56% 19%			
Pb		207.2						180 ± 10mb		
$_{82}Pb^{194}$			11m	EC						
$_{82}Pb^{195}$			17m	EC						
$_{82}Pb^{196}$			37m	EC	≈2.8					
$_{82}Pb^{197m}$			42m	IT EC	0.319 ≈4.5	IT EC	20% 80%			
$_{82}Pb^{197}$				EC	≈4.1					
$_{82}Pb^{198}$			2.4h	EC	≈1.8					
$_{82}Pb^{199m}$			12.2m	IT	0.424					
$_{82}Pb^{199}$			1.5h	β^+, EC	≈3.2	2.8	17%			
$_{82}Pb^{200}$			21.5h	EC	≈0.9					
$_{82}Pb^{201m}$			61s	IT	0.629					
$_{82}Pb^{201}$			9.4h	β^+, EC	≈2.0	0.55				
$_{82}Pb^{202m}$			3.62h	IT EC	2.17 2.17	IT EC	90% 10%			
$_{82}Pb^{202}$			~3 × 10^5y	EC	0.05					
$_{82}Pb^{203m}$			6.1s	IT	0.825					
$_{82}Pb^{203}$			52.1h	EC	0.96					
$_{82}Pb^{204m}$			66.9m	IT	2.186					
$_{82}Pb^{204}$	1.48%	203.973						0.66 ± 0.07b		
$_{82}Pb^{205m}$			4ms	IT	1.01					
$_{82}Pb^{205}$			3 × 10^7y	EC	≈0.035					
$_{82}Pb^{206}$	23.6%	205.9745						30 ± 1mb	0	
$_{82}Pb^{207m}$			0.8s	IT	1.633					
$_{82}Pb^{207}$	22.6%	206.9759						0.71 ± 0.01b	1/2	+0.5895
$_{82}Pb^{208}$	52.3%	207.9766						0.02 ± 0.01b	0	
$_{82}Pb^{209}$			3.30h	β^-	0.635	0.635	100%			

Isotope	% nat. abundance	Atomic mass	Lifetime	Modes of decay	Decay energy (MeV)	Particle energies (MeV)	Particle intensities	Thermal neutron capture cross-section	I	μ
$_{82}Pb^{210}$			21y	β^- α	0.061 3.72	β^-0.015 0.061 α 3.72	81% 19% 1.7×10^{-6}%			
$_{82}Pb^{211}$			36.1m	β^-	1.37	0.10 0.26 0.34 0.53 0.59 0.96 1.36	0.01% 0.7% 0.1% 6% 0.4% 1% 92%			
$_{82}Pb^{212}$			10.64h	β^-	0.58	0.17 0.35 0.589	5% 81% 14%			
$_{82}Pb^{213}$			10m	β^-	≈1.8					
$_{82}Pb^{214}$			26.8m	β^-	1.04	0.59 0.65 1.03	6%			
Bi		208.980						34 ± 2mb		
$_{83}Bi^{199}$			24m	EC α		 5.47	99+% 0.01%		9/2	
$_{83}Bi^{200}$			35m	EC	≈6.5				7	
$_{83}Bi^{201m}$			52m	EC α	≈4.2 ≈4.7	5.28				
$_{83}Bi^{201}$			1.8h	EC	≈4.2				9/2	
$_{83}Bi^{202}$			95m	EC	≈5.5				5	
$_{83}Bi^{203}$			11.8h	β^+, EC α	≈4.4 4.3	β^+ 0.74 1.35 α 4.85	33† 67		9/2	+4.59
$_{83}Bi^{204}$			11.2h	EC	≈4.4				6	+4.25
$_{83}Bi^{205}$			15.3d	β^+, EC	2.70	0.98	0.06%		9/2	+5.5
$_{83}Bi^{206}$			6.24d	β^+, EC	3.65	0.98	8×10^{-4}%		6	+4.56
$_{83}Bi^{207}$			30y	EC	2.40					
$_{83}Bi^{208}$			3.7×10^5y	EC	2.87					
$_{83}Bi^{209}$	100%	208.9804						19 ± 2mb	9/2	+4.080
$_{83}Bi^{210m}$			3×10^6y	α		4.43 4.57 4.92 4.96	0.4% 6% 36% 58%			
$_{83}Bi^{210}$			5.01d	β^- α	1.16 5.044	β^- 1.160 α 4.654 4.691	99+% 1.3×10^{-4}% (4.654, 4.691)	54 ± 5mb	1	±0.0442
$_{83}Bi^{211}$		210.9873	2.15m	α β^-	6.750 0.60	α 5.946 6.278 6.622 β^- 0.060	0.0037% 15.9% 84.1% 0.28%			
$_{83}Bi^{212}$			60.6m	β^- α	2.25 6.206	β^- 0.08 0.45 0.67	0.7% 0.1% 2.2%			

Isotope	% nat. abundance	Atomic mass	Lifetime	Modes of decay	Decay energy (MeV)	Particle energies (MeV)	Particle intensities	Thermal neutron capture cross-section	I	μ
						0.93	1.4%			
						1.55	5%			
						2.27	54%			
						α 5.178	5×10^{-5} †			
							$\alpha_T = 36\%$			
						5.291	1.3×10^{-4}			
						5.334	1×10^{-3}			
						5.473	0.014			
						5.547	1.1			
						5.617	1.7			
						6.051	69.9			
						6.090	27.2			
$_{83}Bi^{213}$			47m	β^-	1.42	β^- 0.96 1.39	97.8%			
				α	5.98	α 5.55 5.87	2.2%			
$_{83}Bi^{214}$			19.7m	β^-	3.28	β^- 0.82	2%			
				α	5.616	1.06	8%			
						1.14	5%			
						1.41	8%			
						1.50	19%			
						1.51	19%			
						1.88	10%			
						1.98	2%			
						2.65	5%			
						3.26	19%			
						α 4.941	0.25†			
						5.023	0.21			
						5.184	0.6			
						5.268	5.8			
						5.448	53.9			
						5.512	39.2			
$_{83}Bi^{215}$			7m	β^-	2.2					
Po										
$_{84}Po^{192}$										
$_{84}Po^{193}$			Short	α	7.1	7.0				
$_{84}Po^{194}$			0.5s	α	6.97	6.85				
$_{84}Po^{195m}$			1.4s	α		6.72				
$_{84}Po^{195}$			3s	α	6.77	6.63				
$_{84}Po^{196}$			6s	α	6.67	6.53				
$_{84}Po^{197m}$			25s	α		6.39				
$_{84}Po^{197}$			54s	α	6.43	6.30				
$_{84}Po^{198}$			1.7m	α	6.29	6.16				
$_{84}Po^{199m}$			4.2m	EC α		EC α 6.047	74% 26%			
$_{84}Po^{199}$			5.0m	EC α	6.06	EC α 5.942	97% 3%			
$_{84}Po^{200}$			11m	EC α	≈3.6 est. 5.97	EC α 5.850	88% 12%			
$_{84}Po^{201m}$			9m	EC α		EC α 5.776	97% 3%			

Isotope	% nat. abundance	Atomic mass	Lifetime	Modes of decay	Decay energy (MeV)	Particle energies (MeV)	Particle intensities	Thermal neutron capture cross-section	I	μ
$_{84}Po^{201}$			15.1m	EC α	≈5.3 est. 5.79	EC α 5.677	99% 1.1%		3/2	
$_{84}Po^{202}$			45m	EC α	≈3.0 est. 5.69	EC α 5.581	98% 2%		0	
$_{84}Po^{203}$			42m	EC α	≈4.4 est. 5.59	EC α 5.49	99% 0.02%		5/2	
$_{84}Po^{204}$			3.6h	EC α	≈2.5 est. 5.48	EC α 5.376	99% 0.6%		0	
$_{84}Po^{205}$			1.8h	EC α	≈3.6 est. 5.34	EC α 5.25	99+% 0.07%		5/2	+0.26
$_{84}Po^{206}$		205.9805	8.8d	EC α	1.80 calc. 5.33	EC α 5.224	95% 5%		0	
$_{84}Po^{207m}$			2.8s	IT						
$_{84}Po^{207}$		206.9816	5.7h	EC β^+ α	2.91 5.33	EC β^+, 1.14, 0.89 α 5.11	99+% 0.5% ≈0.01%		5/2	≈0.27
$_{84}Po^{208}$		207.9813	2.93y	α EC	5.21 1.41 calc.	5.114 EC	99+% 0.006%			
$_{84}Po^{209}$		208.9825	103y	α EC	4.98 1.89 calc.	4.882 EC	99+% 0.5%		1/2	
$_{84}Po^{210}$		209.9829	138.4d	α	5.408	5.305		<0.5mb (25s Po^{211}) <30 mb (0.5s Po^{211})	0	
$_{84}Po^{211m}$			25s	α		8.88 8.31 8.00 7.28	7% 0.25 1.66 91%			
$_{84}Po^{211}$		210.9866	0.52s	α	7.592	7.448 6.891 6.551	99% 0.50 0.53			
$_{84}Po^{212m}$			45s	α		11.65 14.27 14.85	97% 1.00% 2.0%			
$_{84}Po^{212}$		211.9889	3.04×10^{-7}s	α		8.785 + long range α's				
$_{84}Po^{213}$		212.9928	4.2×10^{-6}s	α	8.54	8.377 7.61	99+% 0.003			
$_{84}Po^{214}$		213.9952	1.64×10^{-4}s	α	7.835	7.687 6.905 + long range α's	99+% 0.01			
$_{84}Po^{215}$		214.9995	1.78×10^{-3}s	α	7.524	7.384 6.954 6.948	100% 0.034 0.022			

Isotope	% nat. abundance	Atomic mass	Lifetime	Modes of decay	Decay energy (MeV)	Particle energies (MeV)	Particle intensities	Thermal neutron capture cross-section	I	μ
$_{84}Po^{216}$		216.0019	0.15s	α	6.906	6.777 5.984	100% 2.1×10^{-3}			
$_{84}Po^{217}$			<10s	α β^-	6.67 ≈1.3 est.	6.55 β^-	80% <20%			
$_{84}Po^{218}$		218.0089	3.05m	α β^-	6.111 0.278	α 6.002 5.181 β^-	99.98% 99.9+ 0.0011 0.019%			
At										
$_{85}At^{200}$			0.9m	EC α	6.55	6.47 6.42				
$_{85}At^{201}$			1.5m	EC α	6.48	6.35				
$_{85}At^{202}$			3.0m	EC α	≈8.1 est. 6.359	EC α 6.234 6.120	88% 12% 36† 64			
$_{85}At^{203}$			7.4m	EC α	≈5.8 est. 6.211	EC α 6.092	86% 14%			
$_{85}At^{204}$			9.3m	EC α	7.1 est. 6.071	EC α 5.955	95.5% 4.52%			
$_{85}At^{205}$			26m	EC α	≈4.9 est. 6.018	EC α 5.902	82% 18%			
$_{85}At^{206}$			32m	EC α	≈6.0 est. 5.88	EC α 5.70	≈12% ≈88%			
$_{85}At^{207}$			1.79h	EC α	3.73 calc. 5.87	EC α 5.76	≈90% ≈10%			
$_{85}At^{208}$			1.6h	EC α	≈5.0 est. 5.77	EC α 5.65	99+% 0.5%			
$_{85}At^{209}$			5.5h	EC α	3.49 calc. 5.758	EC α 5.648	95% ≈5%			
$_{85}At^{210}$			8.3h	EC α	3.87 calc. 5.631	EC α 5.525 5.443 5.361	99+% 0.17% 32† 31 37			
$_{85}At^{211}$		210.9875	7.21h	EC α	0.79 calc. 5.981	EC α 5.868	59.1% 40.9%		9/2	
$_{85}At^{212m}$			0.12s	α		7.899 7.837	34% 66			
$_{85}At^{212}$			0.30s	α	7.81	7.678 7.616	80% 20			
$_{85}At^{213}$		212.993090	Short	α	9.4	9.06				

TABLE OF THE ISOTOPES (*Continued*)

Isotope	% nat. abundance	Atomic mass	Lifetime	Modes of decay	Decay energy (MeV)	Particle energies (MeV)	Particle intensities	Thermal neutron capture cross-section	I	μ
$_{85}At^{214}$		213.9963	Short	α	8.94	8.78 8.44 8.23	100% <0.2 <0.1			
$_{85}At^{215}$		214.9987	≈10^{-4}s	α	8.16	8.01 7.60	100% 0.05			
$_{85}At^{216}$		216.0024	≈3 × 10^{-4}s	α	7.95	7.81 7.71 7.60 } 7.58 } 7.49 7.40 7.33 7.25	97% 2.1 0.2 0.5 0.23 0.09 0.06			
$_{85}At^{217}$		217.0046	0.032s	α	7.199	7.066 6.820 6.623 6.486	100% 0.025 0.003 0.040			
$_{85}At^{218}$		218.0086	≈2s	α β^-	6.816 2.83 calc.	6.697 6.653	94% ≈6 0.1%			
$_{85}At^{219}$		219.0114	0.9m	α β^-	6.39 1.7 calc.	α 6.28	97% ≈3%			
Rn										
$_{86}Rn^{204}$			75s	α	6.545	6.416				
$_{86}Rn^{205}$			1.8m	α	6.387	6.262				
$_{86}Rn^{206}$			6.5 m	α EC	6.382 ≈3.5 est.	α 6.258 EC	65% 35%			
$_{86}Rn^{207}$			11m	EC α	≈4.8 est. 6.23	EC α 6.15	96% 4%			
$_{86}Rn^{208}$			23m	EC α	≈3.0 est. 6.268	EC α 6.148	≈80% ≈20%			
$_{86}Rn^{209}$			30m	EC α	≈4.0 est. 6.161	EC α 6.044	83% 17%			
$_{86}Rn^{210}$		209.9897	2.42h	α EC	6.160 2.33 calc.	α EC 6.044	≈96% ≈4%			
$_{86}Rn^{211}$		210.9906	15h	EC α	2.89 calc. 5.966	EC α 5.853 5.785 5.619	74% 26% 33.5† 64.5 2.0			
$_{86}Rn^{212}$		211.9907	25m	α	6.39	6.271				
$_{86}Rn^{213}$			19ms	α	8.296	8.14				
$_{86}Rn^{215}$		214.9987	Short	α	8.78	8.6				
$_{86}Rn^{216}$		216.0002	45μs	α	8.201	8.05				
$_{86}Rn^{217}$		217.0039	540μs	α	7.888	7.742				
$_{86}Rn^{218}$		218.0056	0.035s	α	7.26	7.14 6.54	99.8% 0.16			

TABLE OF THE ISOTOPES (*Continued*)

Isotope	% nat. abundance	Atomic mass	Lifetime	Modes of decay	Decay energy (MeV)	Particle energies (MeV)	Particle intensities	Thermal neutron capture cross-section	I	μ
$_{86}Rn^{219}$		219.0095	4.0s	α	6.944	6.817 6.551 6.527 6.423 6.310 6.222 6.157 6.146 6.101 5.999 5.946 5.785	81% 11.5 0.12 7.5 0.054 0.0026 0.017 0.0026 0.003 0.0044 0.0037 ≈0.001			
$_{86}Rn^{220}$		220.0114	55s	α	6.405	6.287 5.747	100% 0.07	<0.2b		
$_{86}Rn^{221}$			25m	β^- α	≈1.0 est. 6.1	β^- 6.0 est.	≈80% ≈20%			
$_{86}Rn^{222}$		222.0175	3.823d	α	5.587	5.486 4.984 4.824	100% 8×10^{-2} 5×10^{-4}	0.72 ± 0.07b		
$_{86}Rn^{223}$			43m	β^-						
$_{86}Rn^{224}$			1.9h	β^-						
Fr										
$_{87}Fr^{204}$			2.0s	α	7.17	7.03				
$_{87}Fr^{205}$			3.7s	α	7.05	6.92				
$_{87}Fr^{206}$			15.8s	α	6.93	6.80				
$_{87}Fr^{207}$			19s	α	6.91	6.78				
$_{87}Fr^{208}$			37s	α	6.79	6.66				
$_{87}Fr^{209}$			55s	α	6.79	6.66				
$_{87}Fr^{210}$			2.6m	α	6.68	6.56				
$_{87}Fr^{211}$			3.1m	α	6.68	6.56				
$_{87}Fr^{212}$		211.996	19m	EC α	≈5.2 est. 6.530	EC α 6.407 6.383 6.338 6.261 6.179 6.127 6.077 5.983 5.83	56% 44% 21† 23 12 40 2.1 1.0 0.4 0.1 0.05			
$_{87}Fr^{213}$			34s	α EC	6.91 2.10 calc.	6.78 EC	99.48% 0.52%			
$_{87}Fr^{214m}$			3.35ms	α	8.59	8.549 8.477	51% 49			
$_{87}Fr^{214}$			5.0ms	α	8.70	8.426 8.353	94.5% 5.5			
$_{87}Fr^{215}$			≪1ms	α	9.59	9.4				
$_{87}Fr^{217}$		217.0048	Short	α	8.5	8.31				
$_{87}Fr^{218}$		218.0095	Short	α	8.00	7.85 7.70	93% <0.5			

TABLE OF THE ISOTOPES (*Continued*)

Isotope	% nat. abundance	Atomic mass	Lifetime	Modes of decay	Decay energy (MeV)	Particle energies (MeV)	Particle intensities	Thermal neutron capture cross-section	I	μ
						7.55 7.52 7.36	5 1 <0.5			
${}_{87}Fr^{219}$		219.0092	20ms	α	7.44	7.30 7.14 6.95 6.72 6.68	98.4% 0.3 0.8 0.2 0.3			
${}_{87}Fr^{220}$		220.0123	27.5s	α	6.80	6.68 6.64 6.58 6.53 6.48 6.40	85% 13 1 0.5 <0.1 <0.1			
${}_{87}Fr^{221}$		221.0142	4.8m	α	6.457	6.340 6.242 6.126 6.075 6.036 5.979 5.965 5.938 5.925 5.813 5.782 5.775 5.696? 5.688	83.4% 1.34 15.1 0.15 0.003 0.49 0.08 0.17 0.03 0.004 0.005 0.06 ~0.001 0.002			
${}_{87}Fr^{222}$			14.8m	β^- α	2.03 calc. 5.87 calc.	β^- α	99+% 0.01–0.1%			
${}_{87}Fr^{223}$		223.0198	22m	β^- α	1.15 ≈5.44 calc.	1.15 5.34	99.994% 0.006%			
${}_{87}Fr^{224}$			≈2m	β^-	≈2.9 est.					
Ra										
${}_{88}Ra^{213}$			2.7m	α	6.89	6.74 6.61	50% 50			
${}_{88}Ra^{214}$			2.6s	α	7.31	7.17				
${}_{88}Ra^{215}$			1.6ms	α	8.90	8.73				
${}_{88}Ra^{216}$			<1ms	α	9.48	9.3				
${}_{88}Ra^{217}$			Short	α	9.18	9.0				
${}_{88}Ra^{219}$		219.0100	Short	α	8.16	8.0				
${}_{88}Ra^{220}$		220.0110	23ms	α	7.59	7.45 6.90	≈99% ≈1			
${}_{88}Ra^{221}$		221.0139	30s	α	6.883	6.758 6.665 6.610 6.588 6.578 6.46? 6.40? 6.25 6.16?	31% 21 35 8 3 0.4 0.3 0.7 0.3			
${}_{88}Ra^{222}$		222.0154	38s	α	6.68	6.56 (6.23) (5.36) (4.31)	96% 4 7×10^{-3} 6×10^{-3}			

Isotope	% nat. abundance	Atomic mass	Lifetime	Modes of decay	Decay energy (MeV)	Particle energies (MeV)	Particle intensities	Thermal neutron capture cross-section	I	μ
$_{88}Ra^{223}$		223.0186	11.43d	α	5.977	5.870	0.87%	σ_f		
						5.865?	<0.02	1.0 ± 0.5b		
						5.856	0.32	$\sigma(\eta, \gamma)$		
						5.833?		130 ± 20b		
						5.745	9.1			
						5.714	53.7			
						5.605	26			
						5.538	9.1			
						5.500	0.8			
						5.479	≈0.008			
						5.432	2.3			
						5.364	0.11			
						5.337	0.10			
						5.285	0.13			
						5.281	0.10			
						5.257	0.043			
						5.234	0.042			
						5.210	0.0054			
						5.171	0.026			
						5.150	0.021			
						5.133	≈0.0017			
						5.110	$\approx 6 \times 10^{-4}$			
						5.084	$\approx 6 \times 10^{-4}$			
						5.054	$\approx 2 \times 10^{-4}$			
						5.034	$\approx 4 \times 10^{-4}$			
						5.023	$\approx 6 \times 10^{-4}$			
						5.012	$\approx 4 \times 10^{-4}$			
$_{88}Ra^{224}$		224.0202	3.64d	α	5.787	5.684	94.5%	12.0 ± 0.5b		
						5.447	5.5			
						5.159	0.007			
						5.049	0.007			
						5.032	0.003			
$_{88}Ra^{225}$			14.8d	β^-	0.39 calc.	0.32				
$_{88}Ra^{226}$		226.0254	1600y	α		4.781	94.5%	20 ± 3b		
						4.598	5.5			
						4.340	6×10^{-3}			
						4.194	9×10^{-4}			
						4.160	2.7×10^{-4}			
$_{88}Ra^{227}$			41.2m	β^-	1.31	1.31				
$_{88}Ra^{228}$			5.77 ± 0.02y	β^-	0.055	0.024	30%	36 ± 5b		
						0.048	70			
$_{88}Ra^{230}$			1h	β^-	0.9 est.	1.2				
Ac										
$_{89}Ac^{221}$		221.0157	Short	α	7.75	7.63				
$_{89}Ac^{222}$		222.0178	5s	α	7.13	7.00	93†			
						6.96	6			
$_{89}Ac^{223}$		223.0191	2.2m	α	6.781	6.659	37.6%			
						6.648	42.1			
						6.56	13.3			
						6.52	3.8			
						6.47	3.2			
				EC	0.56	EC	1%			
$_{89}Ac^{224}$		224.0217	2.9h	EC	1.38	EC	90%			
				α	6.32	α	≈10%			
						6.210	20.4†			
						6.203	11.9			
						6.155	1.03			
						6.138	25.6			
						6.056	21.9			
						6.014	1.4			
						6.000	6.7			

Isotope	% nat. abundance	Atomic mass	Lifetime	Modes of decay	Decay energy (MeV)	Particle energies (MeV)	Particle intensities	Thermal neutron capture cross-section	I	μ
						5.968	0.2			
						5.958	~0.04			
						5.942	4.4			
						5.916	0.94			
						5.909	0.15			
						5.902	0.15			
						5.877	1.7			
						5.867	0.1			
						5.860	0.75			
						5.854	0.25			
						5.851	~0.07			
						5.841	0.55			
						5.837	0.26			
						5.803	~0.02			
						5.776	0.06			
						5.768	0.24			
						5.739	~0.05			
						5.718	0.12			
						5.709	0.12			
						5.641	0.06			
$_{89}Ac^{225}$		225.0231	10.0d	α	5.931	5.829	52%			
						5.793	29			
						5.732	9.5			
						5.724	2.6			
						5.683	1.0			
						5.638	4.0			
						5.610	0.8			
						5.58	0.8			
						5.55	0.07			
						5.49	≈0.02			
						5.45	0.1			
						5.40	≈0.01			
						5.37	≈0.01			
						5.33	0.07			
						5.30	0.1			
						5.23	≈0.02			
$_{89}Ac^{226}$			29h	β^-	1.12	β^-	77%			
						1.1	41†			
						0.89	59			
				EC	0.77	EC	23%			
				α	5.54	5.44	Weak			
$_{89}Ac^{227}$		227.0278	21.6y	β^-	0.043	β^- 0.044	98.6%	810 ± 20b	3/2	+1.1
				α	5.040	α	1.4%			
						4.950	47†			
						4.938	38			
						4.897	0.11			
						4.870	6.5			
						4.853	4.0			
						4.819	0.07			
						4.793	0.9			
						4.782	0.08			
						4.766	1.6			
						4.735	0.09			
						4.712	0.31			
						4.591	0.02			
						4.586	0.01			
						4.578	≈0.003			
						4.509	≈0.003			
						4.456	≈0.005			
						4.442	0.05			
						4.420	0.006			
						4.360	≈0.003			
$_{89}Ac^{228}$			6.13h	β^-	2.14	0.45	13%			
						0.62	6			
						1.18	53			
						1.70	6.7			

TABLE OF THE ISOTOPES (*Continued*)

Isotope	% nat. abundance	Atomic mass	Lifetime	Modes of decay	Decay energy (MeV)	Particle energies (MeV)	Particle intensities	Thermal neutron capture cross-section	I	μ
						1.76	20			
						2.10	13			
$_{89}Ac^{229}$			66m	β^-	1.1	≈ 1.0				
$_{89}Ac^{230}$			<1m	β^-	≈ 2.9	≈ 2.2				
$_{89}Ac^{231}$			15m	β^-	2.1	2.1				
Th								7.4 ± 0.1b		
$_{90}Th^{223}$		223.0209	0.9s	α	7.7	7.55				
$_{90}Th^{224}$		224.0214	1.05s	α	7.30	7.17	79%			
						6.99	19			
						6.77	1.2			
						6.70	0.4			
$_{90}Th^{225}$		225.0237	8.0m	α	6.92	α	$\geq 95\%$			
						6.797	9†			
						6.743	7			
						6.699	2			
						6.649	3			
						6.626	3			
						6.500	14			
						6.477	43			
						6.440	15			
						6.344	2			
						6.311	2			
				EC	0.68	EC	$\leq 5\%$			
$_{90}Th^{226}$		226.0249	30.9m	α	6.45	6.33	79%			
						6.22	19			
						6.10	1.7			
						6.03	0.6			
$_{90}Th^{227}$		227.0278	18.5d	α	6.145	6.037	24.5%	$\sigma_f = 1500 \pm 1000$b		
						6.007	2.9			
						5.988	0.002			
						5.976	23.4			
						5.958	3.0			
						5.914	0.8			
						5.908	0.17			
						5.865	2.4			
						5.806	1.3			
						5.794	0.31			
						5.761	0.23			
						5.755	20.3			
						5.726	0.034			
						5.712	4.89			
						5.707	8.20			
						5.669	3.63			
						5.691	1.50			
						5.673	0.057			
						5.666	2.06			
						5.639	0.018			
						5.620	0.0070			
						5.612	0.22			
						5.599	0.17			
						5.584	0.18			
						5.531	0.021			
						5.508	0.0166			
						5.479	0.0012			
						5.457	0.0027			
						5.407	0.0004			
						5.363	0.00066			
						5.334	0.0002			
						5.320	0.0002			
						5.263	0.0026			
						5.247	0.0032			
						5.227	0.0098			
						5.208	0.0070			

Isotope	% nat. abundance	Atomic mass	Lifetime	Modes of decay	Decay energy (MeV)	Particle energies (MeV)	Particle intensities	Thermal neutron capture cross-section	I	μ
						5.192 5.179 5.169 5.145 5.127 5.109 5.082 5.054 5.032	0.0038 0.0012 0.0017 0.004 0.00062 0.00028 0.00015 0.00023 0.00031			
$_{90}Th^{228}$		228.0287	1.913y	α	5.521	5.424 5.342 5.211 5.176 5.140	71% 28 0.4 0.2 0.03	120 ± 15b σ_f = <0.3b		
$_{90}Th^{229}$		229.0316	7340y	α	5.167	5.054 5.034 5.009 4.977 4.967 4.930 4.899 4.842 4.811 4.793 4.756 4.683	6.7 ≈0.2 ≈0.1 3.4 6.0 0.25 10.7 58.2 11.4 1.0 1.5 0.4	σ_f = 32 ± 3b	5/2	+0.38
$_{90}Th^{230}$		230.0331	8.0 × 10^4y	α	4.767	4.684 4.617 4.474 4.431	76% 24 0.12 0.03	33 ± 10b		
$_{90}Th^{231}$			25.5h	β^-	0.381	0.299 0.218 0.134 0.090	39% 33 20 8			
$_{90}Th^{232}$		232.0382	1.41 × 10^{10}y	α	4.08	3.994 3.935 3.809	77% 23 0.2	7.4 ± 0.1b		
$_{90}Th^{233}$			22.2m	β^-	1.246	1.245 1.158 1.073 0.88 0.79 0.58	<87%	1500 ± 100b σ_f = 15 ± 2b		
$_{90}Th^{234}$		234.0436	24.1d	β^-	0.263	0.191 0.100	65% 35	1.8 ± 0.5b		
Pa										
$_{91}Pa^{225}$			2.0s	α	≈7.5est.	7.25				
$_{91}Pa^{226}$		226.0278	1.8m	α EC	6.99 2.78	α 6.86 6.82 6.73 EC	74% 52† 46 1 26%			
$_{91}Pa^{227}$		227.0289	38.3m	α	6.581	α 6.465 6.423 6.415 6.401 6.376 6.356 6.336	≈85% 50.7 11.8 15.2 9.6 2.6 8.0 0.7			

Isotope	% nat. abundance	Atomic mass	Lifetime	Modes of decay	Decay energy (MeV)	Particle energies (MeV)	Particle intensities	Thermal neutron capture cross-section	I	μ
						6.326	0.4			
						6.299	0.8			
				EC	1.00	EC	15%			
$_{91}Pa^{228}$		228.0310	26h	EC	2.10	EC	≈ 98%			
				α	6.227	α	≈2			
						6.118	10.5†			
						6.105	12.0			
						6.091	2.3			
						6.078	20.7			
						6.066	1.0			
						6.041	2.3			
						6.028	9.0			
						6.011	0.8			
						5.998	0.3			
						5.989	1.1			
						5.982	2.8			
						5.975	2.7			
						5.947	0.6			
						5.941	0.5			
						5.922	0.8			
						5.907	1.1			
						5.874	1.4			
						5.858	0.3			
						5.843	0.4			
						5.805	7.3			
						5.799	11.3			
						5.779	1.4			
						5.765	2.0			
						5.760	1.4			
						5.756	2.5			
						5.711	1.0			
$_{91}Pa^{229}$			1.5d	EC	0.22	EC	99 +%			
				α	5.837	α	0.25			
						5.74	0.5†			
						5.70	1.5			
						5.671	18.5			
						5.631	9.7			
						5.616	13.3			
						5.592	4.6			
						5.581	36.5			
						5.566	3.9			
						5.537	8.8			
						5.518	0.6			
						5.502	0.7			
						5.480	1.7			
						5.423	0.07			
						5.414	0.15			
						5.321	0.05			
$_{91}Pa^{230}$								$\sigma_f = 1500 \pm 250b$		
			17.7d	EC	1.30	EC	89.6%			
				β^-	0.56	β^- 0.41	10.4%			
				α	5.438	α	0.0032%			
						5.343	23†			
						5.338	15			
						5.325	18			
						5.310	13			
						5.299	17			
						5.286	3			
						5.274	3			
						5.267	3.5			
						5.215	0.5			
						5.181	0.5			
						5.151	0.4			
						5.117	0.6			
						5.082	0.7			
						5.058	0.4			
						4.971	0.7			
						4.932	0.4			
						4.796	≈0.03			
						4.764	0.2			

Isotope	% nat. abundance	Atomic mass	Lifetime	Modes of decay	Decay energy (MeV)	Particle energies (MeV)	Particle intensities	Thermal neutron capture cross-section	I	μ
$_{91}Pa^{231}$		231.0359	3.25×10^4y	α	5.148	5.051	11.0%	200 ± 10b	3/2	±1.98
						5.024	? ≈ 2.5	σ_f = 10 ± 5mb		
						5.022	22			
						5.017	?			
						5.005	25.4			
						4.977	1.4			
						4.967	0.4			
						4.943	22.8			
						4.926	3.0			
						4.892	0.002			
						4.845	1.4			
						4.787	0.04			
						4.729	8.4			
						4.705	≈ 1.0			
						4.673	1.5			
						4.635	0.1			
						4.624	0.1			
						4.591	0.015			
						4.558	0.008			
						4.500	0.003			
						4.14				
$_{91}Pa^{232}$			1.31d	β^-	1.34	0.32	98%	760 ± 100b		
						1.19	0.8	σ_f = 700 ± 100b		
						1.30	0.7			
$_{91}Pa^{233}$			27.0d	β^-	0.571	0.58	12%	22 ± 4b	3/2	+3.4
						0.254	56	(Pa^{234m})		
						0.154	32	21 ± 3b		
								(Pa^{234})		
$_{91}Pa^{234m}$			1.17m	β^-		2.29	98+%			
						1.53				
						1.25				
				IT			0.13%			
$_{91}Pa^{234}$		234.043	6.75h	β^-	2.23	0.23	14%			
						0.51	66			
						0.73	11			
						1.02	7			
						1.35	≤2			
$_{91}Pa^{235}$			23.7m	β^-	1.4	1.4				
$_{91}Pa^{236}$			12m	β^-	≈ 2.9	3.3				
$_{91}Pa^{237}$			39m	β^-	2.30	β^- 2.30	60%			
						1.3	30			
						0.8	10			
U								σ(abs) = 7.595		
								± 0.07b		
								σ_f = 4.172		
								± 0.015b		
$_{92}U^{227}$		227.0309	1.3m	α	7.1	6.8				
$_{92}U^{228}$		228.0313	9.3m	α	6.80	α	≥95%			
						6.68	70†			
						6.59	29			
						6.44	0.7			
						6.40	0.6			
				EC	0.36	EC	≤5%			
$_{92}U^{229}$		229.0332	58m	EC	1.32	EC	≈ 80%			
				α	6.473	α	≈ 20%			
						6.362	64†			
						6.334	20			
						6.299	11			
						6.262	1			
						6.225	3			
						6.187	1			

TABLE OF THE ISOTOPES (*Continued*)

Isotope	% nat. abundance	Atomic mass	Lifetime	Modes of decay	Decay energy (MeV)	Particle energies (MeV)	Particle intensities	Thermal neutron capture cross-section	I	μ
$_{92}U^{230}$		230.0339	20.8d	α	5.991	5.887	67.5%	$\sigma_f = 25 \pm 10b$		
						5.816	31.9			
						5.666	0.038			
						5.661	0.26			
						5.585	0.012			
						5.542	0.0005			
						5.532	0.0001			
$_{92}U^{231}$			4.3d	EC	0.36	EC	99+%	$\sigma_f = 400 \pm 300b$		
				α	5.55	α	0.0055%			
						5.46				
$_{92}U^{232}$		232.0372	73.6y	α	5.414	5.320	68.6%	$75 \pm 10b$		
				SF		5.263	31.2	$\sigma_f = 75 \pm 10b$		
						5.137	0.28	σ(abs.) = 150		
						4.997	2.9×10^{-3}	$\pm 10b$		
						4.946	1.7×10^{-4}			
						4.929	2.1×10^{-4}			
						4.51	2.4×10^{-5}			
$_{92}U^{233}$		233.0395	1.62×10^5y	α	4.909	4.824	84.4%		5/2	+0.54
						4.804	0.051			
						4.796	0.28			
						4.783	13.23			
						4.758	0.016			
						4.754	0.163			
						4.751	0.01			
						4.729	1.61			
						4.701	0.06			
						4.687	0.0028			
						4.681	0.01			
						4.664	0.042			
						4.656	0.005			
						4.641	0.003			
						4.634	0.01			
						4.626	≤0.004			
						4.615	0.004			
						4.611	0.006			
						4.590	0.007			
						4.572	0.0023			
						4.565	0.0028			
						4.538	0.004			
						4.513	0.018			
						4.507	0.012			
						4.503	0.001			
						4.483	0.0014			
						4.465	0.003			
						4.457	0.0028			
						4.411	0.0004			
						4.404	0.0003			
						4.309	0.0009			
$_{92}U^{234}$		234.0409	2.47×10^5y	α	4.856	4.773	72%	$\sigma_{(abs)} = 95 \pm 7b$		
						4.722	28	$\sigma_f \leq 0.65b$		
						4.599	0.35			
						4.265	4.5×10^{-5}			
						4.139	3×10^{-5}			
$_{92}U^{235}$		235.0439	7.1×10^8y	α	4.681	4.597	4.6%	$100.5 \pm 1.4b$	7/2	−0.35
				SF		4.556	3.7	$\sigma_f = 579.5 \pm 2.0b$		
						4.502	1.2	$\sigma_{(abs)} = 679.9 \pm 2.3b$		
						4.445?	0.6			
						4.415	4			
						4.396	57			
						4.366	18			
						4.344?	1.5			
						4.323	3			
						4.266	0.6			
						4.216	5.7			
						4.157	≈0.5			
$_{92}U^{235m}$			26.1m	IT	<0.001					

TABLE OF THE ISOTOPES (*Continued*)

Isotope	% nat. abundance	Atomic mass	Lifetime	Modes of decay	Decay energy (MeV)	Particle energies (MeV)	Particle intensities	Thermal neutron capture cross-section	I	μ
$_{92}U^{236}$		236.0457	2.39×10^7y	α SF	4.573	4.493 4.443 4.330	74% 26 0.3	6.0 ± 0.5b $\sigma_f = 0$		
$_{92}U^{237}$			6.75d	β^-	0.517	0.248 ≈0.09	>80% ≈12	480 ± 160b σ_f = 2b		
$_{92}U^{238}$		238.0508	4.51×10^9y	α SF	4.268	4.195 4.147 4.135	77% 23 0.23	2.720 ± 0.025b		
$_{92}U^{239}$			23.5m	β^-	1.28	1.29 1.21	20% 80	22 ± 5b σ_f = 14 ± 3b		
$_{92}U^{240}$			14.1h	β^-	0.5	0.36 0.32	75% 25			
Np										
$_{93}Np^{229}$			4.0m	α		6.89				
$_{93}Np^{230}$			4.6m	α		6.66				
$_{93}Np^{231}$		231.0383	~50m	α	6.40	6.29				
$_{93}Np^{232}$			≈13m	EC	≈2.6					
$_{93}Np^{233}$		233.0406	35m	EC α	≈1.0 5.7	EC α 5.54	99+% 0.001%			
$_{93}Np^{234}$			4.40d	EC β^+	1.80	EC 0.8	99+% ≈0.05%	σ_f = 900 ± 300b		
$_{93}Np^{235}$			410d	EC α	1.23 5.187	EC α 5.10 5.02 4.93 4.87	99+% 0.0016 3.8† 84 12 1			
$_{93}Np^{236m}$			>5000y	EC, β^-				σ_f = 2800 ± 800b		
$_{93}Np^{236}$		236.0466	22h	β^- EC	0.52 0.91	0.51 0.36 EC	≈40% ≈10 ≈50%			
$_{93}Np^{237}$		237.0480	2.14×10^6y	α	4.956	4.872 4.870 4.862 4.816 4.802 4.787 4.770 4.765 4.740 4.711 4.707 4.698? 4.693 4.663 4.658 4.638 4.598 4.594 4.580 4.573 4.514 4.385	0.4% 0.9 0.2 1.5 1.6 51 19 17 0.02 0.13 0.29 0.07 0.18 1.6 0.57 4.6 0.06 0.08 0.02 0.05 0.01 0.02	170 ± 5b σ_f = 19 ± 3mb	5/2	+5
$_{93}Np^{238}$			2.1d	β^-	1.29	1.24 1.13?	38% 3	$\sigma(\eta, \gamma)$ 0 σ_f = 2000 ± 300b	2	

TABLE OF THE ISOTOPES (*Continued*)

Isotope	% nat. abundance	Atomic mass	Lifetime	Modes of decay	Decay energy (MeV)	Particle energies (MeV)	Particle intensities	Thermal neutron capture cross-section	I	μ
						0.28 0.25 0.20?	20 31 8			
$_{93}Np^{239}$			2.35d	β^-	0.723	0.713 0.654 0.437 0.393 0.332	7% 4 48 13 28	35 ± 10b (Np^{240m}) 25 ± 15b (Np^{240}) σ_f = <1b	5/2	
$_{93}Np^{240m}$			7.3m	β^-	2.18	0.7 1.30 1.60 2.18	7% 10 31 52			
$_{93}Np^{240}$			63m	β^-	2.0	0.89				
$_{93}Np^{241}$			16m	β^-	1.4	1.4				
Pu										
$_{94}Pu^{232}$		232.0411	36m	EC α	1.0 6.70	EC α 6.70	≤98% ≥2%			
$_{94}Pu^{233}$			20m	EC α	≈2.0 6.42	EC 6.31	99+% 0.1%			
$_{94}Pu^{234}$		234.0433	9.0h	EC α	0.4 6.30	EC α 6.203 6.152 6.032	94% 6% 68† 32 0.4			
$_{94}Pu^{235}$			26m	EC α	1.1 5.96	EC α 5.86	99+% 0.003%			
$_{94}Pu^{236}$		236.0461	2.85y	α SF	5.868	5.769 5.722 5.616 5.21 5.08	69% 31 0.18 2.7 × 10^{-4} 6 × 10^{-4}	σ_f = 170 ± 35b		
$_{94}Pu^{237m}$			0.18s	IT SF	0.145					
$_{94}Pu^{237}$		237.0483	45.6d	EC α SF	0.22 5.74	EC α 5.66 5.37	99+% 3.3 × 10^{-3}% 21† 79	σ_f = 2500 ± 500b		
$_{94}Pu^{238}$		238.0495	86y	α SF	5.592	5.499 5.456 5.358 5.214 4.70	72% 28 0.09 0.005 1.1 × 10^{-4}	560 ± 25b σ_f = 16.5 ± 1.0b		
$_{94}Pu^{239}$		239.0522	24,400y	α	5.243	5.157 5.145 5.112? 5.107 5.078 5.066 5.056 5.031 5.010 5.001 4.988 4.963 4.957 4.937 4.914	73.3% 15.1 ≤0.03 11.5 0.032 0.0009 0.021 0.005 0.008 0.0006 0.005 0.003 0.0005 0.003 0.0008	265.7 ± 3.7b σ_f = 742.4 ± 3.5b $\sigma_{(abs)}$ = 1008.1 ± 4.9b	1/2	+0.200

TABLE OF THE ISOTOPES (*Continued*)

Isotope	% nat. abundance	Atomic mass	Lifetime	Modes of decay	Decay energy (MeV)	Particle energies (MeV)	Particle intensities	Thermal neutron capture cross-section	I	μ
						4.873	0.0007			
						4.863	0.0008			
						4.830	0.0015			
						4.801	0.0006			
						4.775	0.0006			
						4.754?	0.0004			
						4.743 }	0.0026			
						4.739 }				
						4.695	0.0004			
						4.636	0.0002			
$_{94}Pu^{240}$		240.0540	6580y	α	5.255	5.168	76%	290 ± 15b		
				SF		5.123	24			
						5.020	0.09			
						4.856	3×10^{-3}			
						4.49	2.1×10^{-5}			
$_{94}Pu^{241}$			13.2y	β^-	0.0208	0.0208	99 + %	360 ± 15b	5/2	−0.73
				α	5.144	α	0.0023%	σ_f = 1011 ± 9b		
						5.051	0.35†	$\sigma_{(abs)}$ = 1371 ± 18b		
						5.041	1.0			
						4.998?				
						4.995	0.36			
						4.971	1.1			
						4.968?				
						4.896	83			
						4.853	12			
						4.798	1.2			
						4.73	0.03			
$_{94}Pu^{242}$		242.0587	3.79×10^5y	α	4.98	4.903	76%	20 ± 3b		
				SF		4.863	24			
$_{94}Pu^{243}$			4.98h	β^-	0.59	0.579	62%	75 ± 50b		
						0.490	38	σ_f = 196 ± 16b.		
$_{94}Pu^{244}$			8×10^7y	α	4.66	α		1.8 ± 0.3b		
				SF						
$_{94}Pu^{245}$			10h	β^-	≈1.2			260 ± 150b		
$_{94}Pu^{246}$			10.9d	β^-	0.37	0.15	≈73%			
						0.33	≈27			
Am										
$_{95}Am^{237}$			≈1.3h	EC	≈1.5	EC	99 + %			
				α	≈6.2	6.02	0.005%			
$_{95}Am^{238}$			1.9h	EC	≈2.3					
$_{95}Am^{239}$			12.1h	EC	0.81	EC	99 + %			
				α	5.93	5.78	0.005%			
$_{95}Am^{240}$			51h	EC	≈1.3					
$_{95}Am^{241}$		241.0567	458y	α	5.640	5.545	0.25%	70 ± 5b	5/2	+1.59
						5.313	0.12	(Am^{242m})		
						5.486	86	670 ± 60b		
						5.469?	<0.04	(Am^{242})		
						5.443	12.7	σ_f = 3.15 ± 0.1b		
						5.417	≈0.01			
						5.389	1.3			
						5.322	0.015			
						5.279	0.0005			
						5.244	0.0024			
						5.223	0.0013			
						5.194	0.0006			
						5.182	0.0009			
						5.178	0.0003			
						5.156	0.0007			
						5.114	0.0004			

Isotope	% nat. abundance	Atomic mass	Lifetime	Modes of decay	Decay energy (MeV)	Particle energies (MeV)	Particle intensities	Thermal neutron capture cross-section	I	μ
						5.096	0.0004			
						5.089	0.0004			
						5.068	0.00014			
						5.004	0.001			
						4.834	0.0007			
						4.800	0.00009			
$_{95}Am^{242m}$			152y	IT	0.048	IT	99.52%	2400 ± 500b		
				α	5.588	α	0.48	σ_f = 7200 ± 300b		
						5.411	1.2†			
						5.367	1.5			
						5.315	0.8			
						5.207	89			
						5.142	6.1			
						5.088	0.3			
						5.067	0.2			
$_{95}Am^{242}$			16.0h	β^-	0.66	0.667	≈33%	<200b		
						0.625	≈49	σ_f = 2100 ± 200b	1	±0.382
				EC	0.73	EC	≈18			
$_{95}Am^{243}$		243.0614						180 ± 20b	5/2	+1.4
			7.37×10^3y	α	5.439	5.350	0.16%			
						5.321	0.12			
						5.276	87.9			
						5.234	10.6			
						5.181	1.1			
						5.113	5.4×10^{-3}			
						5.088	4×10^{-3}			
						5.035 } 5.029 }	2.2×10^{-3}			
						5.008	1.6×10^{-3}			
						4.946	3.4×10^{-4}			
						4.930	1.8×10^{-4}			
						4.919	8×10^{-5}			
						4.695	6×10^{-4}			
$_{95}Am^{244m}$			26m	β^-		1.50	99+%			
				EC		EC	0.039%			
$_{95}Am^{244}$			10.1h	β^-	1.429	0.387	100%	2300 ± 300b		
$_{95}Am^{245}$			2.1h	β^-	0.905	0.905	78%			
						(0.709)	≈5			
						(0.653)	≈17			
$_{95}Am^{246}$			25.0m	β^-	2.29	1.31	79%			
						1.60	14			
						2.10	7			
$_{95}Am^{247}$			20m	β^-		227				
						285				
Cm										
$_{96}Cm^{238}$		238.0530	2.5h	EC	≈1.0	EC	<90%			
				α	6.63	6.51	>10			
$_{96}Cm^{239}$			2.9h	EC	≈1.7	EC	100%			
$_{96}Cm^{240}$		240.0555	26.8d	α	6.441	6.292	71.1%			
						6.249	28.9			
						6.150	0.052			
				SF		5.992	0.014			
$_{96}Cm^{241}$			35d	EC	0.77	EC	≈99%			
				α	6.184	α	1%			
						6.083	0.64%			
						6.039	0.10			
						5.941	71.5			
						5.929	16.3			
						5.887	11.5			
						5.831	2			

TABLE OF THE ISOTOPES (*Continued*)

Isotope	% nat. abundance	Atomic mass	Lifetime	Modes of decay	Decay energy (MeV)	Particle energies (MeV)	Particle intensities	Thermal neutron capture cross-section	I	μ
						5.824	1.4			
						5.810	1.2			
						5.789	1.8			
						5.784	0.6			
						5.781	3.0			
						5.732	0.9			
$_{96}Cm^{242}$		242.0588	163d	α		6.115	74.2%	20 ± 10b		
						6.071	25.8	σ_f = <5b	0	
						5.974	0.036			
						5.819	4.6×10^{-3}			
						5.611	2×10^{-5}			
						5.516	2.5×10^{-4}			
						5.191	2.5×10^{-5}			
				SF		5.148	$\leq 5 \times 10^{-6}$			
$_{96}Cm^{243}$		243.0614	32y	α	6.168	α	99.7%	250 ± 50b		
						6.067	1.5	σ_f = 700 ± 50b		
						6.057	4.7			
						6.010	1.1			
						5.993	5.6			
						5.907	0.1			
						5.876	0.7			
						5.786	73.5			
						5.742	10.6			
						5.686	1.6			
						5.682	0.2			
						5.646	0.03			
						5.639	0.14			
						5.622	0.06			
						5.612	≈0.04			
						5.609 } 5.607?}	0.01			
						5.593	0.01			
						5.587	≈0.02			
						5.582	≈0.009			
						5.575	0.007			
						5.568	0.007			
						5.537	0.002			
						5.532	0.006			
						5.523	0.002			
						5.332	0.003			
						5.323	0.003			
						5.316	0.001			
						5.267	0.0015			
				EC	0.006	EC	0.3%			
$_{96}Cm^{244}$		244.0629	17.6y	α	5.902	5.806	76.4%	13 ± 5b		
						5.765	23.6	σ_f = 1.0 ± 0.5b		
						6.664	0.02			
						5.513	3.4×10^{-3}			
						5.313	$\sim 4 \times 10^{-5}$			
						5.215	$\sim 1 \times 10^{-4}$			
						4.960	3×10^{-4}			
				SF		4.920	1.3×10^{-4}			
$_{96}Cm^{245}$		245.0653	9.3×10^3y	α	5.624	5.533	0.5%	250 ± 50b		
						5.492		σ_f = 2000 ± 80b		
						5.468	2.7			
						5.464	2.0			
						5.359	87.6			
						5.305	4.5			
						5.255	0.3			
						5.245	0.7			
						5.240	0.5			
						5.231	0.6			
						5.199	0.35			
						5.175	0.25			
						5.158	0.3			
$_{96}Cm^{246}$		246.0674	5.5×10^3y	α	5.476	5.385	79%	8.4 ± 2.0b		
				SF		5.342	21			

TABLE OF THE ISOTOPES (*Continued*)

Isotope	% nat. abundance	Atomic mass	Lifetime	Modes of decay	Decay energy (MeV)	Particle energies (MeV)	Particle intensities	Thermal neutron capture cross-section	I	μ
$_{96}Cm^{247}$			1.6×10^7y	α	≈5.3			180b $\sigma_f = 108 \pm 5$b		
$_{96}Cm^{248}$			4.7×10^5y	α SF	5.161	α 5.080 5.036 SF	89% 82† 18 11%	6 ± 2b		
$_{96}Cm^{249}$		249.0758	64m	β^-	0.9	0.9		1.6 ± 1.0b		
$_{96}Cm^{250}$			1.7×10^4y	SF		SF	100%			
Bk										
$_{97}Bk^{243}$			4.6h	EC α	1.51 6.871	EC α 6.758 6.718 6.666 6.605 6.574 6.542 6.502 6.446 6.394 6.210 6.182	99+% 0.15% 15† 12 1.2 0.7 26 19 7 0.7 0.3 14 3.9			
$_{97}Bk^{244}$			4.4h	EC α	≈2.2 6.79	EC α 6.666 6.624	99+% 0.006% ≈50† ≈50			
$_{97}Bk^{245}$			4.98d	EC α	0.82 6.464	EC α 6.358 6.317 6.265 6.200 6.153 6.124 6.087 6.038 5.985 5.889 5.858	99+% 0.11% 16† 15 1.4 1.4 19 15 5.6 0.6 0.2 22 4.0			
$_{97}Bk^{246}$			1.8d	EC	≈1.5	EC	100%			
$_{97}Bk^{247}$		247.0702	1.4×10^3y	α	5.86	5.68 5.52 5.31	37% 58 5			
$_{97}Bk^{248}$			>9y	β^- EC	0.65 0.5	0.65 EC	70% 30%			
$_{97}Bk^{249}$			314d	β^- α SF	0.125 5.520	β^- 0.125 α 5.431 5.412 5.384 5.345 5.109 5.046	99+% 0.0022% 6.7† 69 18 2.6 2.7 0.12	1000 ± 500b		
$_{97}Bk^{250}$			3.22h	β^-	1.76	0.73 1.72 1.76	89% 6 5	$\sigma_f = 960 \pm 150$b		
Cf										
$_{98}Cf^{242}$			3.7m	α	7.49	7.37				

TABLE OF THE ISOTOPES (*Continued*)

Isotope	% nat. abundance	Atomic mass	Lifetime	Modes of decay	Decay energy (MeV)	Particle energies (MeV)	Particle intensities	Thermal neutron capture cross-section	I	μ
$_{98}Cf^{244}$		244.0659	25m	α	7.59	7.18				
$_{98}Cf^{245}$			44m	EC	1.52	EC	70%			
				α	7.23	7.12	30%			
$_{98}Cf^{246}$		246.0688	36h	α	6.87	6.760	78%			
						6.718	22			
						6.63	0.18			
				SF		6.47				
$_{98}Cf^{247}$			2.5h	EC	≈0.8					
$_{98}Cf^{248}$		248.0724	360d	α	6.37	6.27	82%			
				SF		5.84	18			
$_{98}Cf^{249}$		249.0748	360y	α	6.295	6.201	1.9%	300 ± 200b		
						6.146	1.1	σ_f = 1735 ± 70b		
						6.079	0.4			
						5.997?	0.08			
						5.948	3.3			
						5.905	3.0			
						5.848	1.2			
						5.812	84			
						5.784?	0.5			
						5.755	4.4			
				SF		5.693	0.4			
$_{98}Cf^{250}$		250.0766	13y	α	6.128	6.031	83%	1500 ± 500b		
						5.987	17	σ_f = <350b		
				SF		5.889	0.32			
$_{98}Cf^{251}$			≈800y	α	5.94	6.072	2.7%	2100 ± 1000b		
						6.012	12.3	σ_f = 4000		
						5.937	0.5	± 1000b		
						5.849	27.8			
						5.813	4.1			
						5.797	2.8			
						5.762	3.9			
						5.726	1.1			
						5.672	34.7			
						5.645	2.8			
						5.630	4.8			
						5.600	0.5			
						5.564	1.8			
						5.501	0.3			
$_{98}Cf^{252}$			2.65y	α	6.217	α	97%	10 ± 5b		
						6.119	84.3†			
						6.076	15.5			
						5.975	0.28			
				SF		SF	3%			
$_{98}Cf^{253}$		253.0850	17.6d	β^-	0.27	0.27		≈165b		
						0.17? }	99+ %			
				α	6.2	α	0.31%			
						5.979	94.7†			
						5.921	5.3			
$_{98}Cf^{254}$			60.5d	α	5.929	α	≈0.2%	<2b		
						5.834	83†			
						5.792	17			
				SF		SF	99+ %			
Es										
$_{99}Es^{246}$			7.3m	EC	≈3.6	EC	90%			
				α	7.48	α	10%			
						7.33				
$_{99}Es^{247}$			5.0m	EC	≈2.2	EC	≈93%			
				α	≈7.3	7.33	≈7%			

Isotope	% nat. abundance	Atomic mass	Lifetime	Modes of decay	Decay energy (MeV)	Particle energies (MeV)	Particle intensities	Thermal neutron capture cross-section	I	μ
$_{99}Es^{248}$			25m	EC α	$\approx$2.8 6.99	EC 6.88	99+ % $\approx$0.3%			
$_{99}Es^{249}$			2h	EC α	1.40 6.88	EC 6.77	99+ % 0.13%			
$_{99}Es^{250}$			8h	EC	2.1	EC	100%			
$_{99}Es^{251}$			1.5d	EC α	$\approx$0.4 6.59	EC α 6.48	99+ % 0.53%			
$_{99}Es^{252}$		252.0829	$\approx$140d	α	6.75	6.639 6.58 6.493 6.26 6.23 6.09 6.07? 6.02 5.99 5.94?	82† 13 2.3 0.8 0.33 0.6 0.11 1.1 0.07 0.09			
$_{99}Es^{253}$		253.0847	20.47d	α SF	6.747	6.640 6.631 6.601 6.597 6.559 6.547 6.504 6.486 6.436 6.256 6.216 6.165 6.03 5.93–6.04 5.91 5.73	90% 0.8 0.7 6.6 0.75 0.85 0.26 0.08 0.1 0.04 0.04 0.015 2×10^{-4} 1×10^{-4} 3×10^{-5} 8×10^{-5}	150 ± 5db		
$_{99}Es^{254m}$			39.3h	β^- EC SF		β^- 1.127 0.48 EC	99+ % 25† 75 0.08%	$\sigma_f = 1840 \pm 80$b		
$_{99}Es^{254}$		254.0881	276d	α	6.623	6.486 6.437 6.424 6.392 6.367 6.355 6.331 6.284 6.276 6.20 6.19 6.113 6.056	0.27% 93 1.7 0.13 2.9 0.74 0.05 0.16 0.22 0.05 0.05 0.33 0.16	<40b $\sigma_f = 3060 \pm 180$b		
$_{99}Es^{255}$			38.3d	β^- α	$\approx$0.5 $\approx$6.8	 6.307	91.5% 8.5%	$\approx$40b		
$_{99}Es^{256}$			Short	β^-						
Fm										
$_{100}Fm^{248}$		248.0772	0.6m	α						
$_{100}Fm^{249}$			$\approx$2.5m	α	8.0	7.9				

Isotope	% nat. abundance	Atomic mass	Lifetime	Modes of decay	Decay energy (MeV)	Particle energies (MeV)	Particle intensities	Thermal neutron capture cross-section	I	μ
$_{100}Fm^{250}$		250.0795	30m	α	7.56	7.44	100%			
$_{100}Fm^{251}$			7h	EC α	≈1.6 ≈7.4	EC 6.90	≈99% ≈1%			
$_{100}Fm^{252}$		252.0827	23h	α	7.17	7.05	100%			
$_{100}Fm^{253}$			3d	EC α	0.4 7.3	EC α 6.96 6.91	89% 11% 4† 1			
$_{100}Fm^{254}$		254.0870	3.24h	α SF	7.310	7.200 7.158 7.061 SF	85% 14 0.9 0.055%			
$_{100}Fm^{255}$			20.1h	α	7.244	7.130 7.106 7.084 7.027 6.985 6.968 6.956 6.923 6.896 6.895 6.880 6.842 6.811 6.710 6.59 6.54 6.49 6.41	0.09% 0.10 0.43 93.4 0.11 5.3 0.024 0.017 0.01 0.60 0.013 0.002 0.12 0.03 0.020 0.015 0.0035 0.0003	26 ± 3b		
$_{100}Fm^{256}$			2.7h	SF α	 6.96	SF 6.86	97% 3%			
$_{100}Fm^{257}$			80d	α	6.871	6.76? 6.703 6.526 6.450 6.36?	0.4% 3.2 94 2.2 0.5			
Md										
$_{101}Md^{255}$		255.0906	0.6h	EC α	≈0.6 7.46	EC 7.35	90% 10%			
$_{101}Md^{256}$			1.5h	EC α	1.4 7.28	EC 7.18	97% 3%			
$_{101}Md^{257}$			3h	EC α	 7.18?	EC α 7.25? 7.08	≈92% ≈8%			
No										
$_{102}No^{251}$			1s	α	8.8	8.68? 8.58	≈20% ≈80			
$_{102}No^{252}$			2.1s	α SF	8.55	8.41 SF	≈70% 30%			
$_{102}No^{253}$			95s	α	8.2	8.01				
$_{102}No^{254}$			55s	α	8.23	8.10				
$_{102}No^{255}$			3m	α	8.24	8.11				

Isotope	% nat. abundance	Atomic mass	Lifetime	Modes of decay	Decay energy (MeV)	Particle energies (MeV)	Particle intensities	Thermal neutron capture cross-section	I	μ
$_{102}No^{256}$			2.7s	α SF	8.56	8.43 SF	99 + % 0.5%			
$_{102}No^{257}$			20s	α	8.4	8.27 8.23	≈50% ≈50			
Lr										
$_{103}Lr^{256}$			8s	α		8.6				
$_{103}Lr^{257}$										
Ku										
$_{104}Ku^{260}$			0.3s	SF						
$_{105}?^{260}$				α		9.4				
$_{105}?^{261}$			1.8s	α		8.93				
$_{105}?^{262}$			40s	α		8.45				

Gamma Energies and Intensities of Radionuclides

Compiled by
Russell L. Heath

National Reactor Testing Station
Idaho Nuclear Corporation
Idaho Falls, Idaho

The following table lists the Gamma Energies and Intensities of Radionuclides for a large number of the radioactive isotopes. The values of the gamma-ray energies are given in keV. The abbreviation Ann. Rad. refers to 511.006 keV photon associated with the annihilation of positrons in matter. The values for the gamma-ray intensities are given in percent of decays for a given gamma-ray transition and not the percent of transitions between two given energy levels in the daughter nucleus. Relative intensities for gamma rays are indicated by the symbol †.

GAMMA ENERGIES AND INTENSITIES OF RADIONUCLIDES (*Continued*)

Gamma energies (KeV)	Gamma intensities
$_{2}He^{8}$	
99	88%
$_{4}Be^{7}$	
477.575 ± 0.02	
$_{4}Be^{11}$	
2120	32%
4640	2.1%
5860	2.4%
6790	4.4%
7970	1.7%
$_{5}B^{12}$	
3230	0.0005%
4430	1·3%
$_{5}B^{13}$	
3670	7%
$_{6}C^{10}$	
717	100%
1030	1.7%
$_{6}C^{15}$	
5299	68%
$_{7}N^{12}$	
4430	2.4%
$_{7}N^{13}$	
Ann. Rad.	
$_{7}N^{16}$	
6128.9 ± 0.4	100†
7117.0 ± 0.5	7
$_{7}N^{17}$	
870	7†
2190	1
$_{7}N^{18}$	
820	60†
1650	63
1980	100
2470	43
$_{8}O^{14}$	
2311	99%
Ann. Rad.	
$_{8}O^{15}$	
Ann. Rad.	
$_{8}O^{19}$	
112	2.7†
200	97

Gamma energies (KeV)	Gamma intensities
1370	59
1440	2.7
$_{8}O^{20}$	
1070	
$_{9}F^{17}$	
Ann. Rad.	
$_{9}F^{18}$	
Ann. Rad.	
$_{9}F^{21}$	
345	100†
1380	13
$_{9}F^{22}$	
1274.52 ± 0.07	15†
2060	10
$_{10}Ne^{17}$	
Ann. Rad.	
$_{10}Ne^{18}$	
Ann. Rad.	
1041	7%
$_{10}Ne^{19}$	
Ann. Rad.	
$_{10}Ne^{23}$	
440	1000†
1630	28
1960	0.4
2070	3.0
2270	0.2
2550	0.8
2870	0.1
2990	0.9
$_{10}Ne^{24}$	
472	100†
880	8
$_{11}Na^{20}$	
Ann. Rad.	
$_{11}Na^{21}$	
Ann. Rad.	
347	2.3%
$_{11}Na^{22}$	
Ann. Rad.	
1274.52 ± 0.07	
$_{11}Na^{24}$	
1368.650 ± 0.050	100%

Gamma energies (KeV)	Gamma intensities
2754.10 ± 0.07	100%
3850	0.09%
4230	0.0015%
$_{11}Na^{25}$	
400	10†
580	9
980	10
1610	3
$_{11}Na^{26}$	
1820	
$_{12}Mg^{20}$	
Ann. Rad.	
$_{12}Mg^{21}$	
Ann. Rad.	
$_{12}Mg^{23}$	
Ann. Rad.	
440	9.1%
$_{12}Mg^{27}$	
80	0.7%
840	70%
1013	30%
$_{12}Mg^{28}$	
32	96†
391	31
950	29
350	70
$_{13}Al^{24}$	
Ann. Rad.	
1368	40†
2754	32
4200	15
5400	6
7100	7
$_{13}Al^{25}$	
Ann. Rad.	
$_{13}Al^{26m}$	
Ann. Rad	
$_{13}Al^{26}$	
Ann. Rad.	
1830	99.7%
1120	3.7%
2960	0.3%
$_{13}Al^{27}$	
1780	100%
$_{13}Al^{28}$	
1280	85%
2430	15%

Gamma energies (KeV)	Gamma intensities
$_{13}Al^{30}$	
2240	10†
3520	6
$_{14}Si^{26}$	
Ann. Rad.	
820	
$_{14}Si^{27}$	
Ann. Rad.	
$_{14}Si^{31}$	
1260	<1%
$_{15}P^{28}$	
1790	75%
2600	
4440	10%
4900	
6100	
6700	
7000	
7600	5%
$_{15}P^{29}$	
1280	0.8%
2430	0.2%
$_{15}P^{30}$	
2160	0.5%
$_{15}P^{34}$	
2100	
4000	0.2%
$_{16}S^{29}$	
Ann. Rad.	
$_{16}S^{30}$	
Ann. Rad.	
687	80%
$_{16}S^{31}$	
Ann. Rad.	
1270	1.1%
$_{16}S^{37}$	
3090	90%
$_{16}S^{38}$	
1880	95%
$_{17}Cl^{32}$	
Ann. Rad.	

Gamma energies (KeV)	Gamma intensities
2210	70%
2770	
4270	7%
4770	14%
$_{17}Cl^{33}$	
2900	0.3%
$_{17}Cl^{34m}$	
Ann. Rad.	
640	
770	
1176.9 ± 1.0	32%†
2128.2 ± 1.0	100%
3303.9 ± 1.0	32%
4100	1%
146.5 ± 0.3	IT (50%)
$_{17}Cl^{34}$	
Ann. Rad.	
$_{17}Cl^{38m}$	
660	
$_{17}Cl^{38}$	
1600	85†
2167.6 ± 0.7	100
3760	0.05
$_{17}Cl^{39}$	
246	90†
1270	100
1520	85
$_{17}Cl^{40}$	
1460	
2830	
3100	
5800	
$_{18}Ar^{35}$	
Ann. Rad.	
1190	5%
1730	2%
$_{18}Ar^{37}$	
Cl X-ray	
$_{18}Ar^{41}$	
1293	99.2%
1660	0.05%
$_{19}K^{37}$	
Ann. Rad.	
2790	2%
$_{19}K^{38m}$	
Ann. Rad.	
$_{19}K^{38}$	
Ann. Rad.	
1650	weak
2167.6 ± 0.7	99^+%

Gamma energies (KeV)	Gamma intensities
$_{19}K^{40}$	
Ann. Rad.	
1460	11%
$_{19}K^{42}$	
310	1.1†
600	0.1
900	0.1
1020	0.1
1524	100
1924	0.3
2440	0.2
$_{19}K^{43}$	
220.4 ± 0.8	3%
372.6 ± 1.0	85%
396.9 ± 0.8	11%
593.6 ± 1.0	13%
617.5 ± 1.0	81%
990.2 ± 2.0 } 1021.9 ± 2.0 }	~2%
$_{19}K^{44}$	
480	
630	
740	
900	1†
1060	1
1157.02 ± 0.1	100
1499.63 ± 0.25	
1740	13†
2080	
2170	
2550	12
3400	
3660	
4400	6
4600	
5000	0.6
$_{19}K^{45}$	
175	
500	
900	
1230	
1710	
1900	
2100	
2350	
2600	
3100	
$_{20}Ca^{37}$	
Ann. Rad.	
$_{20}Ca^{38}$	
Ann. Rad.	
3.5	
$_{20}Ca^{39}$	
Ann. Rad.	
$_{20}Ca^{45}$	
12.5	

Gamma energies (KeV)	Gamma intensities
$_{20}Ca^{49}$	
490	5%
520	0.05%
760	0.05%
810	5%
1290	71%
$_{20}Ca^{49}$	
3100	88%
4050	10%
4680	0.8%
$_{20}Ca^{50}$	
72	
258	
$_{21}Sc^{40}$	
Ann. Rad.	
3750	
$_{21}Sc^{41}$	
Ann. Rad.	
$_{21}Sc^{42m}$	
Ann. Rad.	
438	100†
1230	
1520	100
$_{21}Sc^{42}$	
Ann. Rad.	
$_{21}Sc^{43}$	
220	
370	
620	
960	
$_{21}Sc^{44m}$	
271.19 ± 0.8 (IT)	98.6%
1001.99 ± 0.1	
1126.16 + 0.15	
1157.02 ± 0.1	
$_{21}Sc^{44}$	
Ann. Rad.	
1126.16 ± 0.15	
1157.02 ± 0.1	
1499.63 ± 0.25	
$_{21}Sc^{46m}$	
142	
$_{21}Sc^{46}$	
889.25 ± 0.07	100%
1120.50 ± 0.07	100%

Gamma energies (KeV)	Gamma intensities
$_{21}Sc^{47}$	
159.38 ± 0.08	40%
$_{21}Sc^{48}$	
175.4 ± 0.2	6%
983.46 ± 0.15	100%
1037.5 ± 0.2	100%
1311.9 ± 0.2	100%
$_{21}Sc^{49}$	
1780	0.03%
$_{21}Sc^{50m}$	
258	
$_{21}Sc^{50}$	
520	85†
1120	100
1570	100
$_{22}Ti^{44}$	
67.85 ± 0.07	10†
78.40 ± 0.10	14
$_{22}Ti^{45}$	
718	
1238	
1408	
1665	
$_{22}Ti^{51}$	
320.07 ± 0.05	100†
605	1.5
928	4.4
$_{23}V^{46}$	
Ann. Rad.	
$_{23}V^{47}$	
1500	
1800	0.8%
2160	0.2%
$_{23}V^{48}$	
Ann. Rad.	
290.1 ± 0.4	
928.3 ± 0.3	
944.1 ± 0.2	10†
983.46 ± 0.15	100
1311.9 ± 0.2	98
1437.6 ± 0.4	
2240.1 ± 0.2	2.7
$_{23}V^{52}$	
935	0.14%
1331	0.9%
1433	100%
1531	0.16%

GAMMA ENERGIES AND INTENSITIES OF RADIONUCLIDES (*Continued*)

Gamma energies (KeV)	Gamma intensities
$_{23}V^{53}$	
1000	100%
$_{23}V^{54}$	
834.795 ± 0.04	100%
990	100%
2210	100%
$_{24}Cr^{48}$	
116	95
310	100%
$_{24}Cr^{49}$	
62.31 ± 0.1	15†
90.7 ± 0.1	30
152.85 ± 0.1	14
610	
$_{24}Cr^{51}$	
320.080 ± 0.013	9%
$_{24}Cr^{56}$	
26	
83	
$_{25}Mn^{50m}$	
Ann. Rad.	
660	<1†
790	3
1110	3
1280	<1
1450	1
$_{25}Mn^{50}$	
Ann. Rad.	
$_{25}Mn^{51}$	
1560	
2030	
$_{25}Mn^{52m}$	
3.83	2%
Ann. Rad.	
$_{25}Mn^{52}$	
Ann. Rad.	
345.7	1%
630	
744	100%
846.8	100%
938.1	84%
1070	
1214	
1246	
1332	6%
1434	100%
1463	ω
2620	0.08%
$_{25}Mn^{54}$	
834.795 ± 0.04	100%
$_{25}Mn^{56}$	
846.78 ± 0.06	100†
1810.96 ± 0.2	30
2113.2 ± 0.2	15.3
2522.6 ± 0.3	1.2
2657.24 ± 0.2	0.7
2959.8 ± 0.2	0.4
3367 ± 1.0	0.21
$_{25}Mn^{57}$	
117	
134	
220	
350	
690	
$_{25}Mn^{58}$	
360	
410	
520	
570	
820	
1000	
1250	
1400	
1600	
2200	
2800	
$_{26}Fe^{52}$	
165	
Ann. Rad.	
383	
$_{26}Fe^{53}$	
Ann. Rad.	38%
377.5 ± 0.5	
$_{26}Fe^{55}$	
Mn X-rays	
$_{26}Fe^{59}$	
142.45 ± 0.08	0.8%
192.23 ± 0.08	2.5%
334.81 ± 0.10	0.3%
1099.27 ± 0.08	56%
1291.58 ± 0.10	44%
$_{26}Fe^{61}$	
130	11†
180	3
230	0.3
300	48
1030	98
1200	100
1650	1.4
1980	0.7
$_{27}Co^{54m}$	
Ann. Rad.	
410	1†
1140	1
1410	1
$_{27}Co^{54}$	
Ann. Rad.	
$_{27}Co^{55}$	
Ann. Rad.	
95	0.5%
245	<0.06%
390	
415	
480	12%
650	0.5%
750	0.2%
805	2.5%
892	2.4%
930	80%
985	0.9%
1060	0.4%
1210	1.0%
1320	6%
1370	4.2%
1410	13%
1580	0.09%
1800	0.06%
2190	0.11%
2310	
$_{27}Co^{56}$	
Ann. Rad.	
263	
788.0 ± 0.2	0.36%
846.78 ± 0.06	100%
977.4 ± 0.1	1.5%
1037.85 ± 0.06	14%
1175.08 ± 0.07	1.6%
1140.18 ± 0.1	
1175.1 ± 0.1	
1238.29 ± 0.04	64%
1360.219 ± 0.04	4%
1771.33 ± 0.06	14%
1810.46 ± 0.07	
1963.64 ± 0.1	0.68%
2015.33 ± 0.07	2.6%
2034.90 ± 0.07	6.6%
2112.95 ± 0.1	0.56%
2212.8 ± 0.1	0.60%
2598.52 ± 0.1	14%
3009.56 ± 0.1	0.6%
3202.18 ± 0.1	2.9%
3253.61 ± 0.1	7.2%
3273.16 ± 0.1	1.6%
3451.68 ± 0.1	0.72%
3547.5 ± 0.2	0.2%
$_{27}Co^{57}$	
Fe X-rays	
14.41 ± 0.020	8.4%
122.060 ± 0.010	85%
136.471 ± 0.010	11%
231	0.0005%
339.7	0.0048%
352.4	0.0037%
366.7	0.0007%
570.3	0.014%
692.1	0.16%
706.8	0.0067%
$_{27}Co^{58}$	
810.81 ± 0.12	100†
864.02 ± 0.17	1.38
1674.94 ± 0.2	0.61
$_{27}Co^{60m}$	
58.60 ± 0.010	
830	0.008%
1332.483 ± 0.040	0.25%
2160	0.0008%
$_{27}Co^{60}$	
1173.226 ± 0.040	99.88%
1332.483 ± 0.046	100%
2158	0.001%
$_{27}Co^{61}$	
67.4 ± 0.4	100%
$_{27}Co^{62}$	
1172	100†
1470	11
1740	11
2030	4
$_{28}Ni^{56}$	
163	99%
276	35%
472	35%
748	48%
812	87%
1560	14%
$_{28}Ni^{57}$	
Ann. Rad.	
129.3 ± 1.0	14%
1377.6 ± 1.0	86%
1757.8 ± 1.0	6%
1783.0 ± 1.5	1%
1882.4 ± 1.0	8%
$_{28}Ni^{59}$	
Co X-ray	
$_{28}Ni^{65}$	
344	
366	4.8%
507	0.35%
610	0.15%
771	0.07%
855	0.07%
1115.51 ± 0.07	15.2%
1480	
1624	0.76%
1725	0.53%

Gamma energies (KeV)	Gamma intensities
$_{28}Ni^{67}$	
890	24%
900	50%
1260	22%
$_{29}Cu^{58}$	
Ann. Rad.	
$_{29}Cu^{59}$	
Ann. Rad.	
340	3.6%
425	1.4%
465	4.1%
878	9.4%
1301	9.9%
1680	1.5%
$_{29}Cu^{60}$	
Ann. Rad.	
467	2.8%
826	19.2%
910	3.4%
1037	3.3%
1292	1.5%
1332.483 ± 0.04	87.3%
1792	44.9%
1863	4.7%
1920	0.6%
1938	1.7%
2061 ± 3	0.9%
2158 ± 2	3.5%
2693	0.5%
2747	0.9%
2268	0.09%
2399	0.7%
2428	0.6%
3124	5.1%
3164 ± 2	0.7%
3195 ± 2	1.9%
3222	0.2%
3252	0.3%
3271 ± 3	0.9%
4022 ± 3	1.0%
4080	0.1%
4332	0.09%
4503	0.09%
4550	0.06%
4584	0.02%
$_{29}Cu^{61}$	
Ann. Rad.	
23.2 ± 0.3	
44.8 ± 0.5	
55.4 ± 0.5	
67.6 ± 0.2	4.9%
283.0 ± 0.2	13%
373.6 ± 0.3	1.9%
405.2 ± 0.3	
529.3 ± 0.3	0.3%
588.6 ± 0.3	1.2%
656.01 ± 0.2	9.6%
1074.5	0.03%
1185.0 ± 0.3	3.6%
1446.7	0.04%
1543.6	0.03%

Gamma energies (KeV)	Gamma intensities
1611.1	0.025%
1662.9	0.04%
1730.4	0.04%
1997	0.004%
2122.4	0.04
$_{29}Cu^{62}$	
Ann. Rad.	
875.4	1%
1170	3%
$_{29}Cu^{64}$	
Ann. Rad.	
1348	0.6%
$_{29}Cu^{66}$	
833.6	0.25%
1039.0 ± 1	9%
$_{29}Cu^{67}$	
91.26 ± 0.02	6.2%
93.31 ± 0.02	35%
184.6	45%
209.0	0.09%
300.2	0.6%
393.6	0.2%
$_{29}Cu^{68}$	
810	18†
1080	100
1240	3
188	5
$_{30}Zn^{60}$	
Ann. Rad.	
$_{30}Zn^{61}$	
Ann. Rad.	
470	10.9%
690	1.8%
966	2.9%
1640	6%
$_{30}Zn^{62}$	
Ann. Rad.	
40.88	
243.47 ± 0.2	1.6%
246.99 ± 0.2	1.0%
305.2 ± 0.3	0.2%
349.7 ± 0.3	0.3%
393.94 ± 0.25	1.4%
507.35 ± 0.18	12.8%
548.38 ± 0.2	12.6%
596.76 ± 0.2	20.3%
$_{30}Zn^{63}$	
Ann. Rad.	
669.8 ± 0.4	11.2%
961.2 ± 0.4	8.4%
1411.3 ± 0.4	0.9%
1555	0.2%

Gamma energies (KeV)	Gamma intensities
$_{30}Zn^{65}$	
1115.51 ± 0.07	50.6%
344	0.003%
771	0.003%
$_{30}Zn^{69m}$	
438.7 ± 0.2	100%
$_{30}Zn^{71m}$	
130	10†
210	
385	100
495	73
609	68
760	6
880	
920	3
990	8
1120	4
1310	1
1370	0.7
1490	1.0
1640	0.11
1740	0.27
1850	0.08
1960	0.06
2400	0.08
$_{30}Zn^{71}$	
120	7†
385	10
510	100
680	2
920	24
1120	10
1630	1.1
$_{30}Zn^{72}$	
18	
46	
145	90%
190	10%
$_{31}Ga^{63}$	
Ann. Rad.	
$_{31}Ga^{64}$	
Ann. Rad.	
427.1	0.9%
756.7	1.6%
808.8	14%
991.6	46%
918.9	8.3%
1276.4	6.9%
1386.9	14%
1455.2	1.9%
1566.6	4.3%
1617.2	1.7%
1626.2	1.4%
1799.3	4.6%
1995.2	2.9%
2195.2	11%
2270.5	2.3%
2374.8	7.8%

Gamma energies (KeV)	Gamma intensities
2433.5	0.6%
2550	0.9%
2804.3	0.7%
3262.9	0.3%
3366.1	1.7%
3425.1	5.1%
4215	0.1%
4454.7	0.9%
4749.0	0.05%
$_{31}Ga^{65}$	
Ann. Rad.	
53.9 ± 0.4	31%
61.2 ± 0.4	15%
90.1 ± 1	1%
115.2 ± 0.4	55%
153.2 ± 0.4	8.5%
207.1 ± 0.4	1.8%
751.3 ± 0.4	7.3%
769	1.2%
932.4 ± 0.4	1.7%
1047	1.1%
1227	0.8%
1342	0.3%
1353	0.8%
1414	0.27%
1480	0.7%
1764	0.1%
1855	0.15%
1970	0.04%
$_{31}Ga^{66}$	
Ann. Rad.	
833.6	6%
1039.0	39%
1232.6	0.63%
1332	1.3%
1356.9	0.67%
1419.0	0.63%
1508.5	0.59%
1918.7	2.4%
2190.3	5.9%
2215.4	0.16%
2394	0.24%
2423.0	2.0%
2752.1	25%
2780.5	0.13%
2932	0.34%
3229.5	1.5%
3257	0.13%
3381.2	1.6%
3423.0	0.94%
3433.0	0.29%
3766	0.12%
3791.3	1.0%
4087.5	1.1%
4295.5	3.6%
4462.0	0.71%
4806.0	1.6%
$_{31}Ga^{67}$	
91.26	2.9%
93.31	70%
184.46 ± 0.2	20.7%
200.96 ± 0.2	2.3%
300.17 ± 0.2	15%

Gamma energies (KeV)	Gamma intensities
353.0 ± 0.4	
393.43 ± 0.2	4.1%
494.4 ± 0.3	0.08%
595.9 ± 0.4	
703.6	0.012%
794.7	0.05%
887.9 ± 0.4	0.12%
$_{31}Ga^{68}$	
Ann. Rad.	
578.3	0.04%
805.8	0.09%
1077.1 ± 1	3.2%
1261.3	0.1%
1764.5	0.01%
1884.5 ± 1	0.13%
2338	0.001%
$_{31}Ga^{70}$	
173	0.15%
1042	0.5%
1215	
$_{31}Ga^{72}$	
601	8†
630	24
690	
786	4
812	4
834	100
894	11
1050	7
1231	1.4
1263	1.3
1278	1.8
1377	0.4
1400	0.3
1465	5
1602	6
1685	1.0
1718	0.5
1866	7
2114	1.1
2201	35
2460	0.7
2491	11
2508	19
2620	0.2
2849	0.4
3050	
3050	0.01
3100	0.01
3340	0.007
3680	0.0005
$_{31}Ga^{73}$	
13.5	
53.39 ± 0.2	9†
68.66 ± 0.3	1
285.0 ± 0.3	3
297.37 ± 0.2	97
325.74 ± 0.2	10
379.28 ± 0.4	
739.37 ± 0.25	
767.87 ± 0.3	
833.3 ± 0.3	

Gamma energies (KeV)	Gamma intensities
$_{31}Ga^{74}$	
53.92 ± 0.4	
380	2†
500	11
595.6 ± 0.2	87
600	13
720	2
870	5
980	4
1110	5
1200	6
1330	5
1460	6
1560	2
1700	3
1760	3
1930	6
2350	45
2550	3
2730	3
2970	2
3230	3
3410	2
$_{31}Ga^{75}$	
360?	1%
580	3%
$_{31}Ga^{76}$	
560	
960	
1120	
$_{32}Ge^{65}$	
655	
1670	
$_{32}Ge^{66}$	
Ann. Rad.	
43.83	43.8%
65.11	7.8%
90.9	0.5%
108.93	16.5%
120.05	0.4%
147.87	1.0%
154.85	0.6%
181.96	7.5%
190.17	7.5%
245.75	6.1%
272.90	11.9%
302.51	2.4%
338.01	9.9%
381.85	29%
427.51	0.3%
492.76	0.4%
536.58	6.1%
706.04	3.8%
$_{32}Ge^{67}$	
Ann. Rad.	
166	89.7%
166.5	24.2%
360	5.3%
558	0.09%
661	0.26%

Gamma energies (KeV)	Gamma intensities
720	0.14%
728	3.9%
828	5.0%
915	9.8%
981	2.4%
1082	2.3%
1162	1.7%
1283	0.35%
1450	0.9%
1477	8.5%
1644	2.1%
1810	1.7%
1837	0.4%
2004	0.8%
2170	0.09%
2559	0.13%
2726	0.1%
2991	0.1%
3065	0.1%
3157	0.1%
3398	
$_{32}Ge^{69}$	
Ann. Rad.	
234.8 ± 0.3	0.29%
318.7 ± 0.3	1.33%
532.8 ± 0.3	0.2%
553.3 ± 0.4	0.61%
574.0 ± 0.3	11.8%
587.8 ± 0.4	0.27%
762.0 ± 0.4	0.18%
788.0 ± 0.4	0.32%
872.0 ± 0.3	9.6%
1001.7 ± 0.4	
1051.8 ± 0.4	0.32%
1106.5 ± 0.3	26%
1206.3 ± 0.4	0.26%
1336.2 ± 0.3	2.94%
1349.5 ± 0.4	0.27%
1487.8 ± 1	0.06%
1525.6 ± 1	0.15%
1572.3 ± 1	0.13%
1891.8 ± 0.4	0.27%
1924.0 ± 0.5	0.08%
2023.9 ± 0.5	0.34%
$_{32}Ge^{71m}$	
23.0	
175.0	
$_{32}Ge^{73m}$	
13.5	
54	
$_{32}Ge^{75m}$	
139	~100%
$_{32}Ge^{75}$	
66	2.5†
200	13
264	100
427	2.6

Gamma energies (KeV)	Gamma intensities
477	2.6
628	1.3
$_{32}Ge^{77m}$	
159 (IT)	55†
215	100
$_{32}Ge^{77}$	
156.3 ± 0.3	16.3 ± 2.2†
177.4 ± 0.4	1.0 ± 0.4
194.9 ± 0.2	48.5 ± 4.3
211.03 ± 0.04	998.0 ± 90.0
215.51 ± 0.04	871.0 ± 78.0
255.0 ± 0.25	2.6 ± 2.1
264.45 ± 0.025	1700.0 ± 116.0
338.5 ± 0.15	30.8 ± 2.8
367.49 ± 0.04	485.0 ± 32.0
416.35 ± 0.04	815.0 ± 51.0
439.5 ± 0.23	10.3 ± 5.1
461.4 ± 0.22	41.9 ± 3.7
475.5 ± 0.15	35.9 ± 3.3
557.7 ± 0.08	569.0 ± 28.0
582.5 ± 0.18	25.5 ± 3.9
613.6 ± 0.4	23.9 ± 5.8
624.6 ± 0.4	2.0 ± 0.6
632.3 ± 0.23	314.0 ± 16.0
673.1 ± 0.25	21.9 ± 1.9
698.8 ± 0.35	8.35 ± 0.69
714.1 ± 0.09	269.0 ± 14.0
743.2 ± 0.40	4.8 ± 1.4
745.6 ± 0.28	35.2 ± 4.9
749.9 ± 0.28	32.6 ± 4.6
766.8 ± 0.25	26.0 ± 2.6
781.3 ± 0.25	33.5 ± 4.0
784.8 ± 0.25	40.9 ± 4.9
794.7 ± 0.35	4.9 ± 1.4
810.6 ± 0.20	78.1 ± 5.8
823.6 ± 0.26	20.9 ± 3.6
843.7 ± 0.30	7.9 ± 2.1
875.3 ± 0.30	26.0 ± 10.0
907.2 ± 0.35	22.3 ± 4.1
914.2 ± 0.5	4.2 ± 1.0
922.9 ± 0.35	16.8 ± 2.5
925.5 ± 0.30	26.2 ± 3.9
929.1 ± 0.30	29.2 ± 4.4
939.6 ± 0.35	6.5 ± 1.6
946.9 ± 1.0	0.4 ± 0.25
959.1 ± 0.8	2.6 ± 0.7
968.1 ± 0.8	2.4 ± 0.6
986.3 ± 0.8	2.3 ± 0.6
997.1 ± 0.5	2.3 ± 0.5
1062.1 ± 1.5	1.4 ± 0.5
1085.0 ± 0.15	216.0 ± 14.0
1115.4 ± 0.4	3.1 ± 0.6
1125.7 ± 0.4	3.0 ± 0.6
1151.8 ± 0.4	6.2 ± 1.4
1193.3 ± 0.2	86.2 ± 4.3
1202.7 ± 0.6	2.0 ± 1.0
1216.0 ± 0.4	3.5 ± 1.3
1242.4 ± 0.3	10.4 ± 2.8
1264.2 ± 0.3	26.7 ± 3.2
1280.8 ± 0.4	3.4 ± 0.9
1296.3 ± 0.4	2.9 ± 0.7
1309.5 ± 0.5	16.1 ± 2.9
1313.1 ± 0.5	9.2 ± 1.7
1319.8 ± 0.5	13.5 ± 2.2
1368.5 ± 0.5	90.5 ± 9.1
1452.6 ± 0.7	5.9 ± 1.4
1464.8 ± 0.8	5.5 ± 1.3

Gamma energies (KeV)	Gamma intensities
1477.1 ± 0.6	11.1 ± 1.9
1479.9 ± 0.6	8.3 ± 1.4
1495.1 ± 0.5	18.6 ± 3.0
1529.0 ± 0.5	2.1 ± 0.6
1539.4 ± 0.5	4.9 ± 1.4
1573.3 ± 0.25	22.7 ± 1.4
1709.6 ± 0.6	10.4 ± 2.5
1719.7 ± 0.6	14.7 ± 2.9
1726.9 ± 0.9	4.2 ± 1.3
1830.9 ± 0.8	1.5 ± 0.5
1846.5 ± 0.5	5.4 ± 1.6
1879.9 ± 0.8	1.0 ± 0.4
1928.8 ± 0.8	0.7 ± 0.3
2000.1 ± 0.3	19.2 ± 3.7
2038.5 ± 0.5	1.2 ± 0.3
2077.4 ± 0.5	8.0 ± 2.3
2089.7 ± 0.5	7.3 ± 2.2
2126.5 ± 0.4	4.8 ± 1.5
2248.0 ± 1.0	0.7 ± 0.3
2329.4 ± 1.0	0.8 ± 0.2
2341.5 ± 0.4	16.4 ± 3.2
$_{32}Ge^{78}$	
277	94%
294	6%
$_{33}As^{69}$	
Ann. Rad.	
86.8 ± 1.0	
146.0 ± 0.5	
232.8 ± 0.4	
$_{33}As^{70}$	
Ann. Rad.	1.98†
0.176 ± 0.2	3.2
0.2405	0.26
0.2523	3.6
0.2942	0.23
0.2988	0.47
0.4480 ± 0.0010	0.20
0.4509 ± 0.0010	0.13
0.4922	1.2
0.4970	3.1
0.5952	20.0
0.6076	4.8
0.6530	0.73
0.6684	25.9
0.7448	25.5
0.7602	0.30 ± 0.15
0.8281	0.43
0.8893	3.8
0.8931	2.4
0.9019	1.7
0.9057	14.9
0.9421	1.7
0.9538	0.58
1.0400	100
1.0993	5.4
1.1143	25.9
1.1181	3.9
1.2183	0.22
1.2961	0.21 ± 0.10
1.3322	0.75
1.3360	0.75
1.3394	10.9
1.3518	0.72

Gamma energies (KeV)	Gamma intensities
1.4125	10.5
1.4183	0.61 ± 0.20
1.4961	1.9
1.5071	0.48 ± 0.30
1.5121	0.34 ± 0.20
1.5233	6.2
1.5666	0.34 ± 0.20
1.5879	0.54
1.7079	21.9
1.7813	4.8
1.8831	0.63
1.9450	0.10
1.9490	0.20
2.0077	3.6
2.0200	20.4
2.0647 ± 0.0030	≦0.15
2.0955 ± 0.0030	0.22 ± 0.18
2.1576	0.45
2.2193	0.13 ± 0.07
2.2561	0.16 ± 0.08
2.3266	0.14 ± 0.07
2.3334	0.10 ± 0.07
2.4212	0.10
2.4250	0.10
2.4493	0.43
2.5195 ± 0.0030	0.09
2.6372	0.37
2.7804 ± 0.0030	0.09
2.8523 ± 0.0030	0.05
2.9649 ± 0.0030	0.09
3.1256	0.08
4.0906 ± 0.0030	0.02
4.3279 ± 0.0030	0.02
4.4345 ± 0.0030	0.01
$_{33}As^{71}$	
Ann. Rad.	
23.0	
174.70 ± 0.2	
247.14 ± 0.5	
279.22 ± 0.5	
326.53 ± 0.3	
349.87 ± 0.4	
373.57 ± 0.5	
391.22 ± 0.3	
465.00 ± 0.6	
499.85 ± 0.17	
526.57 ± 0.2	
572.15 ± 0.3	
615.26 ± 0.3	
622.72 ± 0.5	
679.71 ± 0.5	
685.41 ± 0.8	
707.0 ± 0.3	
712.5 ± 0.3	
747.0 ± 0.3	
851.2 ± 0.3	
920.6 ± 0.2	
1026.7 ± 0.2	
1033.54 ± 0.2	
1037.87 ± 1.0	
1095.48 ± 0.15	
1139.3 ± 0.2	
1231.28 ± 0.4	
1298.43 ± 0.2	
1331.5 ± 1.5	
$_{33}As^{72}$	

Gamma energies (KeV)	Gamma intensities
Ann. Rad.	
595.79 ± 0.2	
600.84 ± 0.17	
629.88 ± 0.15	10†
634.6 ± 0.4	
756.7 ± 0.3	
765.7 ± 0.3	
786.42 ± 0.17	
834.03 ± 0.13	100
894.17 ± 0.15	2
1050.65 ± 0.16	3
1212.6 ± 0.5	
1215.3 ± 0.5	
1379.2 ± 0.4	
1390.3 ± 0.3	
1464.08 ± 0.18	
1475.8 ± 0.2	
1710.7 ± 0.3	
1917.8 ± 0.3	1
1991.1 ± 0.4	
2071.7 ± 0.4	
2090.6 ± 0.5	
2105.2 ± 0.5	
2109.6 ± 0.5	
2201.57 ± 0.1	4
2248.3 ± 0.3	0.9
2507.7 ± 0.16	
2523.4 ± 0.4	
2528.8 ± 0.4	
2620.9 ± 0.4	
2780.0 ± 0.5	0.6
3111.2 ± 1.0	
3481.3 ± 1.0	
3994.3 ± 1.0	
$_{33}As^{73}$	
13.3 ± 0.2	0.44†
53.415 ± 0.035	100
$_{33}As^{74m}$	
283	100%
$_{33}As^{74}$	
Ann. Rad.	
595.86 ± 0.14	100†
608.4 ± 0.2	0.2
634.73 ± 0.15	25.6
887.2 ± 0.7	0.048
993.6 ± 0.8	0.021
1204.0 ± 0.3	0.47
1604.0 ± 1.0	0.013
2198.8 ± 1.0	0.019
$_{33}As^{76}$	
510	0.5†
559.1	100
562.8	1.0
657.04	15
665.4	
708	0.3
740	0.5
775	0.4
858	0.3
869	0.24
1213.3	

Gamma energies (KeV)	Gamma intensities
1216.25	10
1220	2
1228.63	2.3
1439.4	1.2
1453	0.54
1550	0.2
1789.8	0.77
1880	0.2
2096.6	1.3
2111.1	0.73
2434	0.05
2656	0.08
$_{33}As^{77}$	
87.86 ± 0.12	0.26%
161.86 ± 0.12	0.34%
238.96 ± 0.12	1.65%
270.74 ± 0.15	0.008%
281.63 ± 0.12	0.06%
520.63 ± 0.15	0.61%
$_{33}As^{78}$	
620	42†
700	15
830	8
990	2.5
1110	2.9
1210	4.2
1310	11
1490	2.1
1700	1.7
1820	1.4
1940	1.5
2050	1.0
2240	1.0
2650	1.4
$_{33}As^{79}$	
96	100%
360	2%
430	2%
540	0.5%
730	0.5%
890	1%
$_{33}As^{80}$	
666	42%
785 } 812 }	1.4%
1220	3.6%
1640	4%
1770	1.7%
1840	0.5%
2300	0.3%
2350	0.4%
$_{34}Se^{70}$	
Ann. Rad.	
$_{34}Se^{71}$	
Ann. Rad.	
160	

GAMMA ENERGIES AND INTENSITIES OF RADIONUCLIDES (*Continued*)

Gamma energies (KeV)	Gamma intensities
$_{34}Se^{72}$	
46.0	
$_{34}Se^{73}$	
Ann. Rad.	
67.5 ± 0.5	
360.7 ± 0.5	
427.9 ± 0.5	
$_{34}Se^{73}$	
Ann. Rad.	
88	
251	
580	
1080	
$_{34}Se^{75}$	
24.3	
66.05	1.6†
96.731 ± 0.007	5.6
121.113 ± 0.010	28
135.998 ± 0.010	96
198.600 ± 0.020	2.4
264.648 ± 0.015	100
279.522 ± 0.012	42
303.892 ± 0.020	2.3
400.641 ± 0.015	20
$_{34}Se^{77m}$	
161	
$_{34}Se^{81m}$	
103	
$_{34}Se^{81}$	
270	0.3%
276	2.0%
290	1.3%
561	0.3%
565	1%
836	1%
$_{34}Se^{83m}$	
356	15%
676	13%
989	14%
1031	13%
1063	4%
1664	2%
2054	10%
2147	0.2%
$_{34}Se^{83}$	
226	31%
356	73%
357	3.5%
512	45%
554	3%
676	13%
720	22%
801	15%
837	14%
866	9%
1065	6%
1082	2%
1192	4%
1299	9%
1319	4%
1344	6%
1355	3%
1421	0.7%
1558	3%
1784	4%
1830	1.3%
1855	3%
1897	9%
2291	12%
2338	4%
2421	1%
$_{35}Br^{74}$	
Ann. Rad.	
640	
$_{35}Br^{75}$	
Ann. Rad.	
112.51 ± 0.5	
141.50 ± 0.25	
236.0 ± 0.2	
286.70 ± 0.18	
293.1 ± 0.2	
316.0 ± 0.4	
377.6 ± 0.3	
428.2 ± 0.3	
431.8 ± 0.3	
566.1 ± 0.3	
598.5 ± 0.4	
608.8 ± 0.3	
657.0 ± 0.4	
771.0 ± 0.6	
796.4 ± 0.8	
860.0 ± 1.0	
897.6 ± 1.0	
912.1 ± 0.6	
949.9 ± 1.5	
952.8 ± 1.0	
962.1 ± 1.0	
$_{35}Br^{76}$	
Ann. Rad.	
358.4 ± 0.5	
399.31 ± 0.5	
472.7 ± 0.3	
490.1 ± 0.5	
558.99 ± 0.25 }	
562.4 ± 0.5 }	95†
598.8 ± 0.4	
604.2 ± 0.5	
657.2 ± 0.25	28
665.3 ± 0.5	
681.6 ± 0.5	
696.0 ± 0.5	
727.1 ± 0.5	
730.7 ± 0.5	
789.2 ± 0.5	
803.5 ± 0.5	
836.6 ± 0.5	
867.7 ± 0.5	
882.2 ± 0.5	
886.6 ± 0.5	
901.0 ± 0.5	
942.7 ± 0.5	
981.2 ± 0.5	
1029.7 ± 0.5	
1032.2 ± 0.5	
1129.9 ± 0.4	
1212.9 ± 0.5	
1216.2 ± 0.3	
1228.7 ± 0.5	
1369.0 ± 0.4	
1380.4 ± 0.4	
1428.8 ± 0.6	
1439.2 ± 0.6	
1453.7 ± 0.5	
1470.9 ± 0.5	
1488.5 ± 0.4	
1560.1 ± 0.6	
1568.2 ± 0.5	
1578.7 ± 0.6	
1787.6 ± 0.5	
1853.1 ± 0.25	
2096.2 ± 0.4	
2110.8 ± 0.4	
2135.0 ± 0.4	
2281.2 ± 0.4	
2329.2 ± 0.5	
2334.0 ± 0.7	
2348.7 ± 0.5	
2390.50 ± 0.3	
2439.0 ± 0.6	
2510.2 ± 0.5	
2600.7 ± 0.6	
2617.6 ± 0.7	
2792.15 ± 0.25	8
2900.3 ± 0.5	
2950.0 ± 0.25	12
2996.8 ± 0.3	
3023.4 ± 0.7	
3092.3 ± 0.7	
3603.5 ± 0.5	3
3929.5 ± 0.7	0.1
4436.7 ± 1.0	0.07
$_{35}Br^{77}$	
87.86 ± 0.12	1.6%
161.86 ± 0.12	1.9%
180.74 ± 0.15	
187.34 ± 0.2	
200.4 ± 0.12	0.7%
231.3 ± 0.2	
238.96 ± 0.12	27%
243.3 ± 0.2	
249.77 ± 0.12	2.5%
270.74 ± 0.15	1%
281.63 ± 0.12	3%
297.15 ± 0.13	8%
303.76 ± 0.15	
384.98 ± 0.18	0.6%
439.37 ± 0.15	1.6%
484.52 ± 0.18	3%
510.8 ± 0.2	
Ann. Rad.	
520.63 ± 0.15	23%
565.82 ± 0.25	
567.94 ± 0.2	
574.64 ± 0.2	
578.85 ± 0.18	7%
585.39 ± 0.18	
662.5 ± 0.5	
757.3 ± 0.25	2%
817.8 ± 0.25	3%
1005.2 ± 0.3	2%
$_{35}Br^{78}$	
Ann. Rad.	13%
614.1	
$_{35}Br^{80m}$	
85	
37	
$_{35}Br^{80}$	
	616†
2.8%	639
	666
0.2%	704
0.3%	812
6.8%	1257
83.6%	
$_{35}Br^{82m}$	
46(IT)	
698.4	
776.8	
1474.8	
$_{35}Br^{82}$	
92.3 ± 0.1	0.4%
100.9 ± 0.1	
137.1 ± 0.1	
221.28 ± 0.05	2.3%
273.22 ± 0.05	1.2%
295.5 ± 0.1	
452.9 ± 0.1	
554.24 ± 0.05	73%
606.23 ± 0.1	
619.02 ± 0.05	43%
698.30 ± 0.05	27%
776.45 ± 0.05	83%
827.80 ± 0.05	24%
1007.55 ± 0.1	
1044.02 ± 0.05	29%
1084	0.4%
1317.52 ± 0.07	28%
1474.93 ± 0.1	17%
1650.5 ± 0.2	0.8%
1777.6 ± 0.5	0.12%
1874	0.05%
1959	0.05%
2056	0.02%
$_{35}Br^{83}$	
32	100%
93	100%
521	1.4%

Gamma energies (KeV)	Gamma intensities
$_{35}Br^{84}$	
270	1†
350	3
430	5
470	2
520	6
610	5
740	7
810	18
880	100
1010	20
1210	8
1470	4
1570	2
1740	4
1900	36
2050	4
2170	4
2470	16
2820	4
3030	8
3280	6
3930	25
$_{35}Br^{85}$	
305.0 (Kr^{85m})	23%
$_{35}Br^{87}$	
1440	100†
1850	18
2480	18
2640	16
2980	25
3180	16
3800	11
4000	7
4190	21
4800	17
5000	17
5200	12
$_{35}Br^{88}$	
760	
$_{36}Kr^{74}$	
Ann. Rad.	
$_{36}Kr^{76}$	
39	
73	
104	
135	
197	
267	
316	
360	
407	
452	
$_{36}Kr^{77}$	

Gamma energies (KeV)	Gamma intensities
24	4%
107.6	9%
131	84%
148	40%
279	3%
313	3%
607	
734	
$_{36}Kr^{79}$	
44	
136.0	0.7%
181	0.08%
208.5	0.7%
217.3	2%
261.3	11.5%
299.7	1.3%
306.7	2.5%
308	
345	0.1%
389.1	1.6%
397.4	10.1%
523	0.3%
526	0.2%
616	0.1%
726	0.05%
810	0.14%
833	1.7%
860	0.08%
935	0.1%
1026	0.13%
1072	0.09%
1115	0.4%
1165	0.09%
1332	0.5%
$_{36}Kr^{81}$	
Br X-rays	
$_{36}Kr^{83m}$	
32	
93	
$_{36}Kr^{85m}$	
151.2	
305.0	
$_{36}Kr^{85}$	
514	
$_{36}Kr^{87}$	
402.4 ± 0.2	100†
850	19
2050	~6
2570	42
$_{36}Kr^{88}$	
28	
166	20†
196.1 ± 0.2	100

Gamma energies (KeV)	Gamma intensities
360	14
850	65
1369.6 ± 0.2	
1550	40
2391.6 ± 0.3	100
$_{36}Kr^{89}$	
230	85†
360	28
430	29
510	42
600	100
740	32
880	65
1120	45
1290	31
1510	88
1710	34
1930	10
2040	16
2230	10
2420	22
2570	10
2840	25
$_{36}Kr^{90}$	
105	15†
120	65
236	16
495	12
536	48
640	7
670	4
720	4
770	3
890	2
970	4
1110	48
1230	3
1340	4
1400	5
1540	17
1630	4
1700	5
1790	11
1940	2
2030	1
2120	2
2480	4
2580	2
2700	2
2940	3
3080	2
3170	1
3600	1
$_{37}Rb^{79}$	
Ann. Rad.	
$_{37}Rb^{80}$	
Ann. Rad.	
616	
704?	

Gamma energies (KeV)	Gamma intensities
$_{37}Rb^{81m}$	
85	
$_{37}Rb^{81}$	
190	100%
370	1.1%
450	29%
560	2.9%
820	1.1%
1040	0.9%
$_{37}Rb^{82m}$	
554.3	70%
619.1	41%
698.4	27%
776.8	83%
827.6	26%
1044.1	31%
1317.2	30%
1474.8	17%
$_{37}Rb^{83}$	
93	
520.43 ± 0.2	46%
529.65 ± 0.2	16%
552.63 ± 0.2	16%
651	0.09%
682	0.03%
790.1 ± 0.2	0.7%
801	0.3%
$_{37}Rb^{84m}$	
215.6 ± 0.3	
248.3 ± 0.4	
445.5	
463.6 ± 0.4	
$_{37}Rb^{84}$	
Ann. Rad.	
882.9 ± 0.2	100†
1016	0.53
1892	1.1
$_{37}Rb^{86m}$	
560	100%
$_{37}Rb^{86}$	
1078	8.8%
$_{37}Rb^{88}$	
898.0 ± 0.2	63†
1390	6
1835.9 ± 0.2	100
2110	4.5
2680	11
3010	1.4

GAMMA ENERGIES AND INTENSITIES OF RADIONUCLIDES (*Continued*)

Gamma energies (KeV)	Gamma intensities
3240	1.4
3520	1.1
3680	0.4
4780	1.4
$_{37}Rb^{89}$	
272.69 ± 0.1	2.3†
287.97 ± 0.1	1.0
657.76 ± 0.05	16.9
947.72 ± 0.05	16.4
1031.89 ± 0.05	100
1248.07 ± 0.05	75.4
1538.41 ± 0.1	3.2
2008.21 ± 0.1	3.4
2196.13 ± 0.1	25.4
2569.96 ± 0.1	16.6
2706.76 ± 0.2	2.1
3508.43 ± 0.5	2.3
$_{37}Rb^{90}$	
530	4%
720	4%
830	56%
860	6%
1030	5%
1110	7%
1240	2%
1400	5%
1700	3%
1820	3%
2200	2%
2510	2%
2710	1%
2840	2%
3070	5%
3340	15%
3540	5%
4130	11%
4340	13%
4370	4%
4600	5%
5100	2%
5200	4%
$_{38}Sr^{80}$	
580	
$_{38}Sr^{81}$	
Ann. Rad.	
$_{38}Sr^{82}$	
Rb X-rays	
$_{38}Sr^{83}$	
41.9 ± 0.5	
94.1 ± 0.2	
159.7 ± 0.25	
289.9 ± 0.25	
381.56 ± 0.15	
389.4 ± 0.2	
418.4 ± 0.2	
423.5 ± 0.2	
438.2 ± 0.2	
Ann. Rad.	
658.8 ± .0.5	
736.9 ± 1.0	
762.51 ± 0.15	
778.35 ± 0.2	
804.9 ± 1.0	
818.92 ± 0.3	
831.3 ± 2.0	
818.3 ± 2.0	
853.9 ± 1.5	
890.1 ± 2.0	
908.0 ± 1.0	
944.6 ± 1.5	
994.1 ± 0.5	
1043.6 ± 0.4	
1054.8 ± 0.3	
1098.1 ± 2.0	
1147.09 ± 0.2	
1159.81 ± 0.2	
1214.8 ± 1.5	
1237.6 ± 2.0	
1324.7 ± 3.0	
1562.63 ± 0.2	
1951.81 ± 0.3	
2089.84 ± 1.0	
2146.9 ± 1.0	
$_{38}Sr^{85m}$	
7 ± 0.1	
151.28 ± 0.1	14% (EC)
231.69 ± 0.1	85% (IT)
238.65 ± 0.15	~4% (IT)
$_{38}Sr^{85}$	
360	0.002%
513.998 ± 0.02	~100%
880	0.017%
$_{38}Sr^{87m}$	
388.40 ± 0.08	~100%
$_{38}Sr^{89}$	
910	<0.01%
$_{38}Sr^{91}$	
118.31 ± 0.1	0.21†
261.00 ± 0.1	1.25
272.30 ± 0.1	0.29
274.29 ± 0.05	3.0
555.57 ± 0.05	94.7
620.13 ± 0.1	5.1
631.29 ± 0.1	1.75
652.91 ± 0.05	36.9
749.84 ± 0.05	69.7
761.29 ± 0.1	2.08
925.83 ± 0.05	12.2
1024.29 ± 0.05	100
1054.70 ± 0.2	0.5
1281.09 ± 0.1	2.75
1413.58 ± 0.1	3.39
1473.83 ± 0.2	0.47
1546.53 ± 0.3	0.13
1723.63 ± 0.3	0.40
$_{38}Sr^{92}$	
241.53 ± 0.1	3.36†
430.45 ± 0.1	4.64
953.32 ± 0.1	4.30
1384.00 ± 0.05	100
$_{38}Sr^{93}$	
178	17†
210	6
255	13
360	7
400	5
450	9
530	7
600	100
710	35
880	73
1080	9
1180	19
1260	19
1660	16
1860	4
2190	4
2450	6
2580	5
3000	
$_{38}Sr^{94}$	
1420	
$_{39}Y^{82}$	
Ann. Rad.	
$_{39}Y^{83}$	
Ann. Rad.	
$_{39}Y^{84}$	
200	
453	
590	15†
795	100
982	90
1041	50
1270	9
1470	6
$_{39}Y^{85m}$	
503	
700	
770	
920	
$_{39}Y^{85}$	
Ann. Rad.	
230.86 ± 0.2	
627.4 ± 0.2	
699.06 ± 0.2	
768.0 ± 0.2	
1030.1 ± 0.2	
1101.9 ± 0.2	
1122.8 ± 0.2	
1151.6 ± 0.2	
1220.8 ± 0.2	
1403.8 ± 0.7	
1584.1 ± 0.7	
1889.7 ± 0.7	
2123.5 ± 1.0	
2172.4 ± 1.0	
2351.2 ± 1.0	
2750	
$_{39}Y^{86m}$	
10.2	
208.0	
152.2 ± 0.5	
188.8 ± 0.5	
191.8 ± 0.5	
252.0 ± 0.3	
265.2 ± 0.3	
307.3 ± 0.3	
331.4 ± 0.3	
370.7 ± 0.3	
443.2 ± 0.25	
514 ± 2	
580.6 ± 0.3	
608.1 ± 0.6	
627.7 ± 0.15	
645.6 ± 0.18	
703.3 ± 0.2	
709.7 ± 0.3	
767.7 ± 0.4	
777.3 ± 0.2	
826.2 ± 0.5	
833.5 ± 1.0	
896.6 ± 2	
1024.0 ± 0.25	
1076.8 ± 0.2	
1153.1 ± 0.3	
1253.6 ± 0.8	
1349.5 ± 0.5	
1407	
1854.1 ± 0.3	
1920.4 ± 0.3	
2567.2 ± 0.6	
2610.7 ± 1.0	
2830	
3060	
3340	
3660	
3860	
$_{39}Y^{87m}$	
381.1 ± 0.5	99%
$_{39}Y^{87}$	
388.4 ± 0.08	~100%
484.8 ± 0.5	~100%
Ann. Rad.	
$_{39}Y^{88}$	

Gamma energies (KeV)	Gamma intensities
898.00 ± 0.03	92%
1836.075 ± 0.05	~100%
2734.07 ± 0.08	<1%
$_{39}Y^{89m}$	
909.07 ± 0.1	100%
$_{39}Y^{90m}$	
202.4	100†
482	95.7
2320	0.4
$_{39}Y^{90}$	
1.734	0.4%
$_{39}Y^{91m}$	
555.59 ± 0.07	~100%
$_{39}Y^{91}$	
1.21	0.3%
$_{39}Y^{92}$	
447.99 ± 0.05	18.2†
492.17 ± 0.1	3.6
560.81 ± 0.05	18.5
844.12 ± 0.1	9.1
912.66 ± 0.1	4.8
934.44 ± 0.05	100
972.35 ± 0.2	0.6
1132.29 ± 0.1	1.5
1405.44 ± 0.05	32.3
1848.07 ± 0.2	2.1
1885.9 ± 0.5	0.2
2066	
$_{39}Y^{93}$	
266.75 ± 0.05	100†
428.31 ± 0.1	0.1
478.15 ± 0.1	0.2
658.21 ± 0.1	1.8
680.27 ± 0.05	9.7
714.42 ± 0.1	0.3
743.51 ± 0.1	1.9
947.12 ± 0.05	29.6
1158.73 ± 0.1	0.4
1182.5 ± 0.1	0.5
1203.28 ± 0.1	1.8
1425.39 ± 0.1	3.6
1470.04 ± 0.2	1.0
1827.15 ± 0.1	1.4
1917.71 ± 0.07	22.3
2184.2 ± 0.2	1.4
2190.8 ± 0.2	2.6
2472.5 ± 0.3	0.2
$_{39}Y^{94}$	

Gamma energies (KeV)	Gamma intensities
560	
920	
1130	6%
1420	43%
1650	5%
1900	
2130	2.4%
2570	1.6%
2840	2.4%
3060	1.5%
3530	0.7%
	1.3%
$_{39}Y^{96}$	1.1%
700	
1000	
1500	
$_{40}Zr^{81}$	
Ann. Rad.	
$_{40}Zr^{82}$	
Ann. Rad.	
$_{40}Zr^{83}$	
Ann. Rad.	
$_{40}Zr^{84}$	
Ann. Rad.	
$_{40}Zr^{85}$	
Ann. Rad.	
$_{40}Zr^{86}$	
Ann. Rad.	
28	
243	
612	
	21†
	100
$_{40}Zr^{87}$	5.4
1200	
2200	
Ann. Rad.	
$_{40}Zr^{88}$	
392.6 ± 0.3	
$_{40}Zr^{89m}$	100%
588	
1506	
	94%
$_{40}Zr^{89}$	6%
Ann. Rad.	
909.07 ± 0.1	
1621.4 ± 0.6	
1713.3 ± 0.4	

Gamma energies (KeV)	Gamma intensities
$_{40}Zr^{90m}$	
132.5 ± 0.4	
2186.1 ± 2.0	
2318.7 ± 2.0	
$_{40}Zr^{95}$	
235.7 ± 0.2	
724.24 ± 0.06	
756.87 ± 0.07	
$_{40}Zr^{97}$	
254.1 ± 0.2	
355.6 ± 0.1	
602.5 ± 0.1	
703.8 ± 0.1	
743.2 ± 0.2	
971.5 ± 0.4	
1147.9 ± 0.2	
1276.1 ± 0.2	
1362.7 ± 0.2	
1712.6 ± 0.4	
1750.6 ± 0.3	
1852.0 ± 0.3	
$_{41}Nb^{88}$	
76	
141	
272	
399	
671	
1058	
1083	
$_{41}Nb^{89m}$	
588	
(Zr^{89m})	
1506	
$_{41}Nb^{89}$	
1510	
1626	
1713	
1832	
1911	
2220	
2572	
2574	
2961	
3018	
3093	
3282	
3333	
3513	
3577	
3838	
3908	
$_{41}Nb^{90m}$	
3	
122.4	

Gamma energies (KeV)	Gamma intensities
$_{41}Nb^{90}$	
132.5 ± 0.3	5†
141.5 ± 0.3	75
371.5 ± 1.0	3
Ann. Rad.	
890.0 ± 1.0	
1128.7 ± 1.5	97
1269.5 ± 2.0	
2186.27 ± 1.0	14
2318.7 ± 1.0	82
$_{41}Nb^{91m}$	
104.5	
1210	
$_{41}Nb^{92m}$	
Ann. Rad.	
912.66 ± 0.1	1.8%
934.44 ± 0.05	99%
1848.07 ± 0.2	0.8%
$_{41}Nb^{94m}$	
41.5	
703	0.002%
872	0.2%
$_{41}Nb^{94}$	
702.59 ± 0.15	100%
871.16 ± 0.15	100%
$_{41}Nb^{95m}$	
235.7 ± 0.2	100%
$_{41}Nb^{95}$	
765.87 ±	~100%
$_{41}Nb^{96}$	
219.13 ± 0.3	
241.46 ± 0.1	
350.13 ± 0.2	
372.20 ± 0.1	
460.16 ± 0.15	
480.43 ± 0.15	
568.69 ± 0.08	
719.38 ± 0.08	
778.18 ± 0.15	
810.53 ± 0.15	
849.96 ± 0.15	
1091.45 ± 0.15	
1199.97 ± 0.15	
1497.6 ± 0.2	
$_{41}Nb^{97m}$	
743.20 ± 0.15	100%

GAMMA ENERGIES AND INTENSITIES OF RADIONUCLIDES (*Continued*)

Gamma energies (KeV)	Gamma intensities
$_{41}Nb^{97}$	
658.0 ± 0.2	98%
743.2 ± 0.2	2%
1021.66 ± 0.2	
1148.0 ± 0.2	
$_{41}Nb^{98}$	
330	9†
720	75
780	100
1160	30
1440	10
1520	4
1680	10
1880	4
1930	8
2240	
2440	
2700	
$_{41}Nb^{99}$	
100	1†
260	1
$_{41}Nb^{100}$	
140	10†
360	55
450	40
530	100
650	
2200	
2300	
2650	
2850	
$_{41}Nb^{100}$	
530	100†
620	60
1040	10
1150	10
1470	5
$_{42}Mo^{88}$	
Ann. Rad.	
$_{42}Mo^{89}$	
Ann. Rad.	
$_{42}Mo^{90}$	
42.70	6†
122.4	36
162.9	0.43
203.1	0.26
257.34	14.2
323.2	0.08
421.0	0.001
445.4	0.036
472.2	0.005
489.8	0.003
941.5	0.004
990	0.001
1271	0.002
1387	0.001
1455	0.0008
1463	0.0004
1482	0.0004
$_{42}Mo^{91m}$	
658	57%
Ann. Rad.	
1210	16%
1530	22%
$_{42}Mo^{91}$	
Ann. Rad.	
$_{42}Mo^{93m}$	
264	100%
685	100%
1479	100%
$_{42}Mo^{93}$	
30.4	85%
$_{42}Mo^{99}$	
40.6	1†
140.3 ± 0.2	
180.9 ± 0.2	7
366.3 ± 0.2	1.8
410	0.15
739.3 ± 0.2	15
777.6 ± 0.2	4
940	0.14
$_{42}Mo^{101}$	
80	3%
191	25%
193	2%
300	7%
400	2%
510	15%
590	21%
700	1%
704	11%
840	1%
890	15%
950	2%
1020	25%
1140	1%
1180	11%
1280	3%
1380	9%
1460	1%
1560	11%
1660	3%
2080	16%
$_{42}Mo^{104}$	
70	
$_{43}Tc^{92}$	
90	~20†
140	67
240	30
330	90
790	95
1540	100
$_{43}Tc^{93m}$	
390	63%
2660	18%
$_{43}Tc^{93}$	
860	3%
1350	65%
1490	33%
2030	0.4%
2440	0.4%
$_{43}Tc^{94m}$	
871.01 ± 0.08	100 ± 5
875.45 ± 0.12	1.6 ± 0.1
993.01 ± 0.10	2.3 ± 0.8
1195.8 ± 0.4	0.9 ± 0.2
1521.52 ± 0.17	5.6 ± 1.1
1868.50 ± 0.25	6.2 ± 0.4
2393 ± 1	0.5 ± 0.2
2739.7 ± 0.3	4.5 ± 0.5
3128.2 ± 0.4	2.7 ± 0.7
$_{43}Tc^{94}$	
449.07 ± 0.06	2.6 ± 1.0†
532.13 ± 0.10	2.6 ± 1.0
702.60 ± 0.08	99.8 ± 6
742.2 ± 0.3	1.2 ± 0.6
849.65 ± 0.08	97.7 ± 6
871.01 ± 0.08	100 ± 5
916.12 ± 0.10	7.4 ± 1.5
1362.7 ± 0.6	0.9 ± 0.4
1508.6 ± 0.7	0.66 ± 0.3
1591.7 ± 0.3	2.4 ± 1.0
$_{43}Tc^{95m}$	
Ann. Rad.	
38.9	
204.2	100†
584	55
763	1
784	2
788	15
823	12
837	39
1040	6
1630	0.1
$_{43}Tc^{95}$	
204	1%
680	2%
768	82%
840	11%
930	1.7%
1060	4%
$_{43}Tc^{96m}$	
34.4	
$_{43}Tc^{96}$	
312	0.5†
778.18 ± 0.15	100
810.53 ± 0.15	82
849.96 ± 0.15	100
1119	17
$_{43}Tc^{97m}$	
96.5	100%
$_{43}Tc^{97}$	
Mo X-rays	
$_{43}Tc^{98}$	
669	100†
770	100
$_{43}Tc^{99m}$	
142.7	100%
$_{43}Tc^{100}$	
542	
600	
710	
810	
890	
1010	
1310	
1490	
1800	
$_{43}Tc^{101}$	
127	0.8†
130	2.6
183	0.4
186	2.3
235	0.8
307	100
380	1.8
410	0.4
545	8
640	1.0
720	1.2
850	0.36
940	0.24
$_{43}Tc^{102}$	
470	
$_{43}Tc^{103}$	
135	17†
210	10
350	
$_{43}Tc^{104}$	
360	
530	

Gamma energies (KeV)	Gamma intensities
630	
890	
1150	
1250	
1370	
1580	
1630	
1900	
2200	
2500	
2700	
3200	
3400	
3700	
4000	
4400	
4700	
$_{43}Tc^{105}$	
110	
$_{44}Ru^{95}$	
Ann. Rad.	
254.6 ± 0.4	
290.3 ± 0.3	
301.0 ± 0.3	
336.5 ± 0.2	
367.2 ± 0.3	
511.006	
551.6 ± 0.3	
591.1 ± 0.3	
627.0 ± 0.3	
653.0 ± 0.5	
749.4 ± 0.5	
807.6 ± 0.3	
843.2 ± 0.5	
1051.7 ± 0.3	
1097.5 ± 0.2	
1159.5 ± 0.3	
1179.6 ± 0.3	
1218.8 ± 0.5	
1230.2 ± 0.4	
1267.5 ± 0.5	
1303.1 ± 0.3	
1353.9 ± 0.5	
1410.8 ± 0.4	
1434.0 ± 0.5	
1459.5 ± 0.4	
2323.8 ± 0.5	
$_{44}Ru^{97}$	
96.5	0.04†
215.4	100
324.4	9
565	<1
$_{44}Ru^{103}$	
39.8 ± 0.1	0.072 ± 0.008
53.3 ± 0.1	0.30 ± 0.03
65	<0.01
295.2 ± 0.1	0.21 ± 0.02
358	<0.03
443.0 ± 0.1	0.31 ± 0.03
497.1 ± 0.1	100 ± 10
557.1 ± 0.1	0.76 ± 0.07
610.3 ± 0.1	4.8 ± 0.4

Gamma energies (KeV)	Gamma intensities
$_{44}Ru^{105}$	
129.53	27%
149.04	2%
150	2%
188	2%
210	2%
262.84	9%
315.50	10%
326.1	3%
330.9	
393.36	3%
413.51	2%
469.38	19%
499.28	3%
575.19	1%
656.15	2%
676.32	6%
724.2	44%
875.8	2%
969.4	1.0%
1350	0.1%
1380	<0.1%
1580	<0.1%
1730	<0.1%
$_{44}Ru^{107}$	
195	14%
370	
480	
860	7%
930	4%
1030	4%
1290	4%
$_{44}Ru^{108}$	
165	28%
$_{45}Rh^{97}$	
Ann. Rad.	
190	
260	
430	
860	
1180	
1570	
1760	
1960	
2160	
2540	
$_{45}Rh^{98}$	
Ann. Rad.	
650	
$_{45}Rh^{99m}$	
Ann. Rad.	
89.3 ± 0.6	
250.6 ± 0.5	
276.6 ± 0.5	
321.7 ± 0.4	
340.3 ± 0.25	
446.0 ± 0.3	
558.2 ± 0.4	
575.7 ± 0.4	

Gamma energies (KeV)	Gamma intensities
921.0 ± 0.7	
938.0 ± 0.6	
1261.27 ± 0.2	
$_{45}Rh^{99}$	
Ann. Rad.	
89.7 ± 0.3†	100 ± 4†
120.3 ± 0.2	0.22 ± 0.05
175.45 ± 0.05	6.58 ± 0.3
232.75 ± 0.05	1.64 ± 0.25
279.00 ± 0.15	0.78 ± 0.08
295.76 ± 0.08	4.04 ± 0.25
322.32 ± 0.05	20.4 ± 0.9
353.06 ± 0.05	115 ± 9
442.59 ± 0.05	6.42 ± 0.4
486.28 ± 0.08	1.44 ± 0.08
528.53 ± 0.05	132 ± 10
575.6 ± 0.3	0.65 ± 0.1
618.08 ± 0.05	14.4 ± 0.8
733.93 ± 0.06	0.93 ± 0.15
764.90 ± 0.07	1.25 ± 0.09
807.07 ± 0.05	4.1 ± 0.3
896.65 ± 0.15	2.7 ± 0.3
940.38 ± 0.12	4.9 ± 0.8
998.51 ± 0.10	2.8 ± 0.2
1009.6 ± 0.2	0.19 ± 0.05
1060.6 ± 0.2	0.86 ± 0.08
1088.88 ± 0.15	1.0 ± 0.2
1209.1 ± 0.4	0.80 ± 0.08
1293.34 ± 0.15	0.95 ± 0.05
1325.2 ± 0.4	0.60 ± 0.06
1482.8 ± 0.3	0.42 ± 0.05
1531.5 ± 0.4	1.62 ± 0.12
1571.9 ± 0.6	0.74 ± 0.07
1616.5 ± 0.8	0.77 ± 0.07
1660.9 ± 0.4	0.20 ± 0.05
1970.2 ± 0.7	0.34 ± 0.03
$_{45}Rh^{100}$	
Ann. Rad.	†
302.2 ± 0.3	
340.9 ± 0.4	
370.7 ± 1.5	
446.1 ± 0.4	
539.5 ± 0.3	
588.3 ± 0.4	
591.0 ± 1.0	
686.6 ± 0.6	
822.5 ± 0.4	
1033.7 ± 0.6	
1107.6 ± 0.4	
1341.4 ± 0.4	
1362.2 ± 0.3	
1553.7 ± 0.3	
1865.5 ± 0.5	
1930.2 ± 0.4	
2376.2 ± 0.3	
2530.3 ± 0.5	
$_{45}Rh^{101m}$	
127.23 ± 0.2	
157.35 ± 0.2	IT
179.40 ± 0.2	
183.91 ± 0.2	
197.80 ± 0.2	

Gamma energies (KeV)	Gamma intensities
233.22 ± 0.2	
237.60 ± 0.2	
306.52 ± 0.2	
545.15 ± 0.2	
595.73 ± 0.3	
$_{45}Rh^{101}$	
127.23 ± 0.04	117†
157.44 ± 0.08 IT	
179.64 ± 0.06	
184.09 ± 0.08	
198.02 ± 0.04	100
233.72 ± 0.08	
238.20 ± 0.15	
306.87 ± 0.03	1
325.24 ± 0.06	15
545.00 ± 0.10	
$_{45}Rh^{102}$	
345.89 ± 0.12	0.87 ± 0.10†
415.25 ± 0.15	2.1 ± 0.3
418.52 ± 0.18	9.4 ± 0.1
420.4 ± 0.2	3.2 ± 0.3
475.06 ± 0.04	95 ± 4
628.05 ± 0.05	8.3 ± 0.4
631.29 ± 0.05	56 ± 2
692.4 ± 0.2	1.6 ± 0.2
695.6 ± 0.3	2.9 ± 0.4
697.49 ± 0.08	44 ± 2
766.84 ± 0.06	34 ± 2
1046.59 ± 0.07	34 ± 2
1103.16 ± 0.06	4.6 ± 0.3
1112.84 ± 0.07	19 ± 1
1323.6 ± 0.5	0.46 ± 0.08
$_{45}Rh^{102}$	
Ann. Rad.	
415.25 ± 0.15	0.03 ± 0.02†
418.52 ± 0.18	0.12 ± 0.02
456.42 ± 0.15	0.08 ± 0.02
468.58 ± 0.04	2.9 ± 0.2
475.06 ± 0.04	46 ± 3
556.60 ± 0.04	2.0 ± 0.2
628.05 ± 0.05	4.5 ± 0.4
631.29 ± 0.05	0.10 ± 0.03
636.81 ± 0.10	0.23 ± 0.03
680.66 ± 0.05	0.58 ± 0.04
733.93 ± 0.08	0.10 ± 0.02
739.58 ± 0.07	0.53 ± 0.08
930.5 ± 0.3	0.03 ± 0.02
1046.59 ± 0.07	0.43 ± 0.03
1103.16 ± 0.06	2.9 ± 0.1
1105.7 ± 0.3	0.39 ± 0.03
1158.10 ± 0.06	0.58 ± 0.04
1362.08 ± 0.20	0.39 ± 0.05
1562.2 ± 0.4	0.11 ± 0.03
1568.7 ± 0.6	0.01 ± 0.01
1580.5 ± 0.3	0.05 ± 0.01
1786.4 ± 0.4	0.01 ± 0.01
2037.0 ± 0.3	0.03 ± 0.02
2261.3 ± 0.4	0.02 ± 0.02
$_{45}Rh^{104m}$	

Gamma energies (KeV)	Gamma intensities
31.86 51.43 77.55 97.11	} IT
560	
760	0.1%
770	0.1%
930	0.03%
1340	0.02%
1530	0.03%
1700	<0.03%
$_{45}Rh^{104}$	
555.5	2%
1240	0.12%
1800	<0.002%
$_{45}Rh^{105m}$	
129.4	100%
$_{45}Rh^{105}$	
38.8	
215.8	
280.54	0.17%
306.31	5%
319.24	19%
442.3	
497.0	
$_{45}Rh^{106}$	
511.9 ± 0.2	100†
616.30 ± 0.2	45
622.1 ± 0.2	6
710	0.2
873.8	1.8
1050.4 ± 0.2	6.8
1130	2.3
1490	0.16
1550	0.64
1770	0.19
1950	0.08
2090	0.13
2360	0.17
2640	0.03
$_{45}Rh^{106}$	
220	20†
407	20
450	35
511.9 ± 0.2	100
620	33
735	47
820	40
1050.4 ± 0.2	29
1140	14
1220	19
1560	2.1
1740	2.0
1860	2.0
2120	1.0
2280	0.5

Gamma energies (KeV)	Gamma intensities
$_{45}Rh^{107}$	
115	0.5%
285	3%
307	73%
570	1.8%
680	2.7%
$_{45}Rh^{108}$	
433.99 ± 0.2	43%
614.29 ± 0.2	22%
510	40%
1520	5%
$_{45}Rh^{110}$	
375	
$_{46}Pd^{98}$	
132	
Ann. Rad.	
$_{46}Pd^{99}$	
Ann. Rad.	
140	
275	
420	
670	
$_{46}Pd^{100}$	
32.7	
42.1	51†
55.8	3
74.8	70
84.0	100
126.1	33
139.7	1.2
151.7	2.5
158.8	8
$_{46}Pd^{101}$	
24	†
269.6 ± 0.2	27
296.1 ± 0.2	100
320.6 ± 0.2	3.6
355.2 ± 0.3	
428.2 ± 0.4	
453.5 ± 0.25	4.0
556.0 ± 0.2	2.2
590.54 ± 0.15	80
611.7 ± 0.4	
723.8 ± 0.4	16
748.6 ± 0.4	3.9
854.0 ± 1.0	
881.0 ± 1.0	
992.8 ± 0.4	5.5
1177.7 ± 0.4	1.6
1201.8 ± 0.4	7
1218.3 ± 0.4	2.5
1288.7 ± 0.4	10
1311.4 ± 0.6	1.5
1638.0 ± 2.5	

Gamma energies (KeV)	Gamma intensities
$_{46}Pd^{103}$	
53	†
65	0.001
298	0.009
324	0.003
362	0.06
498	0.013
$_{46}Pd^{107m}$	
210	
$_{46}Pd^{109m}$	
188	
$_{46}Pd^{109}$	
44.8 ± 0.6†	3.6
88.036 ± 0.008	8900 ± 800
103.6 ± 0.7	2.2
134.7 ± 0.7	3.2
145.9 ± 0.7	2.7
311.5 ± 0.5	(100)
390.9 ± 0.8	2.5 ± 0.5
413.5 ± 1	49 { 26[b]
415.2 ± 0.6	23
424.7 ± 0.8	1.8 ± 0.4
448.2 ± 0.8	2.6 ± 0.6
551.3 ± 0.8	1.5 ± 0.5
557.8 ± 0.5	6.2 ± 0.8
602.4 ± 0.5	21.5 ± 2
636.1 ± 0.5	27 ± 3
647.3 ± 0.6	65 ± 5
701.8 ± 0.8	9.2 ± 1
707.3 ± 0.8	4.5 ± 0.5
736.7 ± 0.7	5.0 ± 0.6
781.8 ± 0.7	33 ± 3
(863)	<0.5
$_{46}Pd^{111m}$	
170	
others (see below)*	
$_{46}Pd^{111}$	
50	
70	
160	620†
280 290	} 71
380 400	} 177
450	36
500	32
560	68
630	100
750	39
830	10
900	9
960	11
*1080	18
1130	17
*1250	16
1380	9
1440	5
*1640	23
*1690	11
*1900	9

Gamma energies (KeV)	Gamma intensities
$_{46}Pd^{112}$	
18.5	20%
$_{47}Ag^{102}$	
Ann. Rad.	
$_{47}Ag^{103m}$	
138	
$_{47}Ag^{103}$	
Ann. Rad	
120	26†
150	23
240	10
270	34
380	
760	
920	
1010	12
1160	9
1280	13
$_{47}Ag^{104m}$	
Ann. Rad.	
555	
$_{47}Ag^{104}$	
Ann. Rad.	
167	
262	
355	
443	
478	
556	100†
750	20
767	57
854	15
938	35
1260	4
1340	10
1530	8
1620	9
1810	8
$_{47}Ag^{105}$	
64.0 ± 0.1	18†
74.1	
89.9 ± 0.2	
112.7 ± 0.2	0.013
155.7 ± 0.3	0.9
183.0 ± 0.3	0.7
280.4 ± 0.3	62
284.8 ± 0.4	0.003
289.4 ± 0.4	
306.3 ± 0.3	1.6
311.8 ± 0.4	
319.4 ± 0.3	0.11
325.8 ± 0.4	
329.0 ± 0.4	
331.8 ± 0.3	
344.7 ± 0.3	69
360.8 ± 0.3	
370.3 ± 0.3	
392.8 ± 0.3	

Gamma energies (KeV)	Gamma intensities
401.7 ± 0.4	
415.0 ± 0.4	
421.1 ± 0.4	
437.4 ± 0.4	
443.5 ± 0.2	18
527.4	
560.9 ± 0.4	
618.1 ± 0.4	
644.8 ± 0.3	16
650.9 ± 0.4	
673.3 ± 0.4	
681.4 ± 0.5	
727.6 ± 0.4	
743.4 ± 0.4	
807.6 ± 0.4	
962.8 ± 0.5	
1088.2 ± 0.3	4.2
$_{47}Ag^{106m}$	
194.9	
221.5	
228.5	
328.3	
309.9	
406.0	
429.5	
450.8	
474.2	
511.8	
600.9	
616.0	
680	
703.3	
716.2	
717.1	
748.2	
792.8	
803.9	
807.5	
824.5	
847.5	
1045.7	
1127.8	
1199	
1223	
$_{47}Ag^{106}$	
Ann. Rad.	
295	
450	
616.2 ± 0.6	
621.4 ± 0.6	
870.3 ± 0.6	
1045.9 ± 0.6	
1124.6 ± 0.6	
1520	
1730	
1880	
2170	
$_{47}Ag^{107m}$	
93.1	100%
$_{47}Ag^{108m}$	
30.2	
79.49 ± 0.18	

Gamma energies (KeV)	Gamma intensities
433.99 ± 0.13	10†
614.29 ± 0.15	10
722.81 ± 0.18	10
$_{47}Ag^{108}$	
433.8	0.45%
510	
614.29 ± 0.15	0.26%
633.2	1.7%
841	0.02%
1010	
$_{47}Ag^{109m}$	
88.036 ± 0.008	
$_{47}Ag^{110m}$	
116.41 (IT)	
446.2	
620.10	
657.6	100†
677.5	10
686.8	7
706.6	10
744.2	
763.8	24
817.9	8
884.5	74
937.3	33
1384.3	22
1475.9	4
1505.2	11
1562.5	1.2
$_{47}Ag^{110}$	
657.6	12%
817.9	
1475.9	0.02%
$_{47}Ag^{111m}$	
70	100%
$_{47}Ag^{111}$	
95	
247	1%
342	6%
610	
$_{47}Ag^{112}$	
606.2 ± 0.5	41%
617.0 ± 0.4	
694.1 ± 0.5	
716.8 ± 0.8	
775.8 ± 1.0	1%
850.8 ± 0.6	2%
1103.4 ± 1.0	1%
1252.2 ± 1.0	0.5%
1311.5 ± 1.0	1%
1386.7 ± 0.5	4%
1539.0 ± 1.0	1%
1612.9 ± 0.7	2%
2055.8 ± 0.8	

Gamma energies (KeV)	Gamma intensities
2105.7 ± 0.6	2%
2211.0 ± 1.5	
2359.8 ± 2.5	0.5%
2506.4 ± 1.0	
2685.0 ± 1.0	0.6%
2827.2 ± 1.0	0.7%
2749.5 ± 2.0	0.6%
$_{47}Ag^{113}$	
97.2 ± 0.3	
258.8 ± 0.3	20
298.6 ± 0.2	100†
316.2 ± 0.3	10
333.1 ± 0.3	5
339.3 ± 0.3	5
364.6 ± 0.3	1
382.0 ± 0.3	1
583.7 ± 0.3	5
672.4 ± 0.3	9
680.9 ± 0.3	9
884.0 ± 0.3	4
988.7 ± 0.3	5
1194.8 ± 0.3	4
$_{47}Ag^{114}$	
570	
$_{47}Ag^{115}$	
110	1%
140	9%
170	2%
220	44%
240	7%
280	13%
360	10%
420	7%
450	1%
470	10%
620	1%
640	3%
1040	2%
1080	2%
1480	11%
1660	8%
1760	2%
1860	3%
1900	7%
2030	2%
2120	13%
2500	1%
$_{48}Cd^{103}$	
Ann. Rad.	
220	
630	
850	
$_{48}Cd^{104}$	
66.7	

Gamma energies (KeV)	Gamma intensities
83.6	
123.6	
134.2	
$_{48}Cd^{105}$	
Ann. Rad.	
25.5	
27.7	
263	
292.5	
308	
312	
317	
321	
325	
336	
341	
347	
433	
607	
1901	
1960	
2000	
2050	
2280	
2320	
$_{48}Cd^{107}$	
Ann. Rad.	
32.6	
93.1(IT)	
300.7	
324.6	
422.6	
526.4	
596.8	
624.5	
719.8	
786.7	
796.0	
799.0	
818	
828.9	
897	
949.0	
1049	
1130	
1222.4	
$_{48}Cd^{109}$	
88.036 ±	
$_{48}Cd^{111m}$	
149.6	30†
246	94
$_{48}Cd^{113m}$	
265	~0.1%

GAMMA ENERGIES AND INTENSITIES OF RADIONUCLIDES (*Continued*)

Gamma energies (KeV)	Gamma intensities
$_{48}Cd^{115m}$	
130	0.07†
162	1.3
210	0.15
292	0.2
485	16
935	100
1130	5
1240	45
1420	2
$_{48}Cd^{115}$	
35	0.5†
230	0.6
262	2.0
485	10
530	26
$_{48}Cd^{117m}$	
89	
161	
222	
273	
293	
314	
345	
366	6†
434	
462	
565	
631	2
702	3
715	5
748	3
762	
832	
862	3
880	
931	2
1029	4
1052	
1065	11
1117	5
1165	
1233	8
1248	3
1260	2
1338	9
1371	
1408	9
1433	12
1562	7
1652	
1659	
1682	3
1723	9
1939	
1998	17
2095	2
2311	
2319	4
2394	
2407	
$_{48}Cd^{117}$	
89	
161	
222	
273	
293	
314	
345	
434	
462	
565	
832	11†
880	
947	8
964	3
1009	0.7
1052	
1142	
1274	1
1303	54
1381	2
1397	6†
1577	46
1708	6
1835	
$_{49}In^{106}$	
Ann. Rad.	
530	17†
630	10
1650	
1850	
$_{49}In^{107}$	
Ann. Rad.	
220	46%
320	
730	
840	
940	
1050	
1250	
1520	
1720	
1940	
2260	
$_{49}In^{108}$	
Ann. Rad.	
383	
633	
842	
$_{49}In^{108}$	
75	
120	
150	
175	
243	
264	
325	
330	
383	
391	
411	
570	
633	
731	
753	
781	
842	
872	
924	
1013	
1033	
1054	
$_{49}In^{109}$	
Ann. Rad.	
60	
89	
94	
205	
228	
262	
289	
301	
326	
329	
340	
349	
410	
422	
428	
446	
456	
519	
533	
546	
578	
615	
623	
627	
651	
678	
725	
821	
825	
862	
920	
947	
1050	
1190	
1550	
1610	
1680	
$_{49}In^{110m}$	
121	
399	
411	
463	
562	
584	
642	
658	
707	
745	
761	
816	
844	
884	
937	
995	
1118	
1500	
$_{49}In^{110}$	
658	100†
817	
1180	3
1520	2
1780	1.3
2200	2
2800	0.6
3000	
3460	
3650	
$_{49}In^{111}$	
172.5	93†
247	100
$_{49}In^{112m}$	
156	
$_{49}In^{112}$	
Ann. Rad.	
621	6%
860	0.4%
1240	0.5%
1480	0.3%
$_{49}In^{113m}$	
391.689 ± 0.010	100%
$_{49}In^{114m_2}$	
311	
191.6	
$_{49}In^{114m_1}$	
191.6	
358	
724	
$_{49}In^{114}$	
Ann. Rad.	
1299	

Gamma energies (KeV)	Gamma intensities
$_{49}In^{115m}$	
335	
$_{49}In^{116m_2}$	
164	
$_{49}In^{116m_1}$	
138.3	3%
385	1%
417.0	36%
818.7	17%
1097.2	53%
1293.3	80%
1508	11%
1753	<0.4%
1770	1.5%
2111	20%
$_{49}In^{116}$	
434	0.12%
950	0.1%
1293	1.2%
2200	0.01%
$_{49}In^{117m}$	
158.63 ± 0.15	16%
315.33 ± 0.18	
$_{49}In^{117}$	
158.63 ± 0.15	
552.49 ± 0.15	
$_{49}In^{118}$	
1220	
$_{49}In^{118}$	
210	3†
440	7
640	10
690	42
810	7
1050	83
1230	100
1250	6
1490	1.4
1540	1.0
2040	3
$_{49}In^{119m}$	
24	
311 ± 0.2	
762.91 ± 0.4	
$_{49}In^{119}$	

Gamma energies (KeV)	Gamma intensities
730	5†
820	100
$_{49}In^{120}$	
1171	
$_{49}In^{120}$	
90	12†
198	9
260	1
710	12
860	34
940	12
1020	61
1170	100
1190	7
1280	14
1470	6
1870	7
2010	6
2220	3
2630	2
$_{49}In^{121}$	
940	
$_{49}In^{124}$	
990	3†
1130	10
3210	3
$_{50}Sn^{109}$	
Ann. Rad.	
335	
521	
890	
1120	
$_{50}Sn^{110}$	
283	
$_{50}Sn^{111}$	
Ann. Rad.	
372.9 ± 0.8	16.5 ± 2†
458.0 ± 0.6	11.8 ± 1
536.2 ± 0.6	6.5 ± 1
563.6 ± 0.8	10.5 ± 1.5
616 ± 1	≦ 4
635 ± 1	≦ 4
762.0 ± 0.6	46 ± 3
953.6 ± 0.8	18 ± 1.5
1026.1 ± 0.8	10 ± 1.5
1101.2 ± 0.8	28 ± 3
1153.0 ± 0.6	(100)
1542.5 ± 0.8	27 ± 3
1610.3 ± 0.8	40 ± 3
1915.1 ± 0.8	56 ± 4
2106.6 ± 1	12 ± 1
2178.2 ± 1	9 ± 1
2212.9 ± 1	5.8 ± 0.7
2324.8 ± 1	9 ± 1

Gamma energies (KeV)	Gamma intensities
$_{50}Sn^{113}$	
391.689 ± 0.010	98.2%
255	1.8%
$_{50}Sn^{117m}$	
158	
159	
$_{50}Sn^{119m}$	
23.83	~20%
65	100%
$_{50}Sn^{121m}$	
37.15	
$_{50}Sn^{123}$	
159.7 ± 0.	100†
381 ± 1	0.04 ± 0.004
542 ± 1	0.02 ± 0.004
552 ± 1	0.007 ± 0.002
$_{50}Sn$	
331.7	100†
386.1	0.082
589.6	0.20
643.0	0.15
662 ± 1	0.046
776 ± 1	0.018
840.8	0.075
1017.5	0.093
1093 ± 1	0.050
1152 ± 1	0.015
1294 ± 1	0.013
1305 ± 1	0.012
1349 ± 1	0.016
1368.5	0.10
1404.1	0.71
1484.2	0.22
1582 ± 1	0.0058
1615.3	0.13
1634 ± 1	0.024
1735.5	0.033
1913.7	0.020
1947 ± 1	0.013
2113 ± 1	0.0016
$_{50}Sn^{12}$	
342	10†
470	14
810	30
910	30
1070	100
1160	3.3
1410	3.3
1990	14
2230	1.1
$_{50}Sn^{126}$	

Gamma energies (KeV)	Gamma intensities
60	
67	
92	
$_{50}Sn^{127}$	
270	
440	
490	
580	
820	
990	
1100	
1480	
1600	
2000	
2140	
2320	
2580	
2680	
2820	
$_{50}Sn^{127}$	
495	
$_{50}Sn^{128}$	
44	7†
72	19
500	61
570	22
$_{51}Sb^{112}$	
Ann. Rad.	
1257	
$_{51}Sb^{113}$	
79 (Sn^{113m})	
320	
640	
700	
760	
810	
950	
1030	
1200	
1520	
$_{51}Sb^{114}$	
Ann. Rad.	
900	
1299	
$_{51}Sb^{115}$	
Ann. Rad.	
499	100†
980	5
1240	5
2220	1

GAMMA ENERGIES AND INTENSITIES OF RADIONUCLIDES (*Continued*)

Gamma energies (KeV)	Gamma intensities
$_{51}Sb^{116m}$	
Ann. Rad.	
99	30%
140	30%
406	36%
545	68%
960	75%
1060	27%
1290	100%
$_{51}Sb^{116}$	
Ann. Rad.	
99	
930	26†
1290	85
2230	14
$_{51}Sb^{117}$	
Ann. Rad.	
158	100%
$_{51}Sb^{118m_2}$	
140	4†
300	10
380	10
$_{51}Sb^{118m_1}$	
Ann. Rad.	
40.7	29%
253.5	93%
1049	100%
1090	
1229	100%
$_{51}Sb^{118}$	
Ann. Rad.	
830	
1229	
1250	
$_{51}Sb^{119}$	
23.83	
$_{51}Sb^{120}$	
Ann. Rad.	
1171	
$_{51}Sb^{122m}$	
26	
61	
75	
$_{51}Sb^{122}$	
Ann. Rad.	
563.9	70%
686	3.5%
1137	0.7%
1258	0.7%

Gamma energies (KeV)	Gamma intensities
$_{51}Sb^{124m_2}$	
25	
$_{51}Sb^{124m_1}$	
505	100†
602.70 ± 0.18	100
645.76 ± 0.18	100
$_{51}Sb^{124}$	
443.86 ± 0.3	
602.70 ± 0.18	100†
645.76 ± 0.18	7.5
709.3 ± 0.2	
713.8 ± 0.2	4
722.72 ± 0.18	
735.2 ± 0.5	
790.7 ± 0.25	
968.2 ± 0.2	2.5
1045.2 ± 0.2	2.4
1325.6 ± 0.2	2
1355.4 ± 0.5	
1368.5 ± 0.2	
1489.08 ± 0.3	2.1
1526.4 ± 0.3	1.1
1580.1 ± 0.2	0.6
1691.0 ± 0.07	51
2091.1 ± 0.2	7
2183.3 ± 0.5	
2293.6 ± 0.4	0.6
$_{51}Sb^{125}$	
35.6 } (Te^{125m})	
109.5 } (Te^{125m})	1†
117.1	
172.8	~1.5
176.43	20
204.2	<1
208.2	<1
227.9	2.6
321.1	5
380.4	4
407.8	100
427.9	
443.6	
463.4	33
600.6	62
606.7	17
635.9	36
671.4	6
$_{51}Sb^{126m}$	
410	
670	
$_{51}Sb^{126}$	
290	
410	
580	
690	
850	
990	

Gamma energies (KeV)	Gamma intensities
$_{51}Sb^{127}$	
62	
252.7	
290.6	
392.1	
412.0	
445.2	
473.2	
543.2	
603.8	
684.9	
697.9	
721.4	
782.6	
$_{51}Sb^{128}$	
320	
743	
750	
1070	
$_{51}Sb^{129}$	
73	
340	
460	
540	
660	
810	
910	
1024	
1240	
1740	
$_{51}Sb^{130}$	
190	
330	
820	
940	
$_{51}Sb^{131}$	
640	37%
950	~48%
$_{52}Te^{115}$	
Ann. Rad.	
720	34†
960	6
1080	24
1280	32
1380	32
1580	6
2200	2
$_{52}Te^{116}$	
	93.9
$_{52}Te^{117}$	
	720

Gamma energies (KeV)	Gamma intensities
$_{52}Te^{118}$	
Sb X-rays	
$_{52}Te^{119m}$	
153.0	94†
164	6
207.6	38
400	
760	
917	6
946	10
984	12
1008	
1050	8
1099	7
1140	10
1221	100
1370	2.2
2090	7
$_{52}Te^{119}$	
645	100†
700	13
1760	4.2
$_{52}Te^{121m}$	
81.78 } IT	
212.21 } IT	
37.15	
103	0.02†
910	0.2
1000	0.2
1100	3.4
$_{52}Te^{121}$	
37.15	
65.58	0.4
470.4	1.4
507.5	18
573.1	80†
$_{52}Te^{123m}$	
88.46	100%
159.0	100%
$_{52}Te^{125m}$	
35.3	
109.6	
$_{52}Te^{127m}$	
88.7 (IT)	
57.6	0.19†
591	0.0006
657	0.0043
$_{52}Te^{127}$	
57.6	
145	0.0016†

GAMMA ENERGIES AND INTENSITIES OF RADIONUCLIDES (*Continued*)

Gamma energies (KeV)	Gamma intensities
203	0.017
214	0.012
360	0.046
417	0.31
$_{52}Te^{129m}$	
105.7 (IT)	2.3% (equilib.)
158.91	0.15%
208.9	0.3%
250.57 ± 0.2	0.46%
278.4	0.7%
281.15	0.22%
342.90	0.05%
459.4 ± 0.3	0.90%
487.4	1.7%
556.7	0.22%
624.4	0.1%
671.9	0.05%
696.0	5%
729.7	1.36%
741.1	0.09%
802.2	0.24%
817.2	0.17%
833.5	0.05%
844.9	0.07%
982.2	0.03%
1022.6	0.04%
1084.0	0.71%
1111.8	0.27%
$_{52}Te^{129}$	
27	
208.95	2.9†
250.6	5
278.4	9
281.1	1
342.9	0.3
345	0.1
459.6	100
487.4	19
532	2.3
556.7	2.1
573	0.6
624.4	1.0
671.9	0.5
696.0	73
729.7	16
741.1	0.8
802.2	2.8
817.2	1.8
833.5	0.7
844.9	0.8
982.2	0.3
1022.6	0.5
1050	0.5
1084.0	10
1112	3.5
1233	0.3
1262	0.7
1400	0.3
$_{52}Te^{131m}$	
181.7 (IT)	68% (equilibrium)
102.2	7.5%
149.8	24%
182.4	2.1%
200.7	7.6%
240.9	8.4%
334.3	11%
452.4	6.5%
665.1	5%
773.7	46%
782.5	7.6%
793.8	16%
822.8	6.7%
852.3	26%
1125.5	15%
1206.6	12%
1645.8	1.5%
1887.7	1.7%
2001.0	2.7%
2240	0.7%
2330	0.3%
$_{52}Te^{131}$	
149.8	100†
279	1.0
343	1.1
384	1.1
452.4	24.0
493	7.0
544	0.4
603	5.9
654	1.8
695	0.5
727	0.4
842	0.2
898	0.3
933	1.2
948	3.0
952	0.6
997	5.1
$_{52}Te^{132}$	
49.3 ± 0.3	
111.81 ± 0.2	
116.48 ± 0.2	
228.3 ± 0.2	
$_{52}Te^{133m}$	
74.1 ± 0.2	0.8†
81.5 ± 0.2	0.8
88.0 ± 0.2	3.0
94.9 ± 0.2	6.0
164.34 ± 0.09	1.6
168.87 ± 0.09	12
177.1 ± 0.2	1.5
178.2 ± 0.2	1.0
184.45 ± 0.10	0.4
193.22 ± 0.10	1.2
198.2 ± 0.2	0.6
213.36 ± 0.08	4.2
220.94 ± 0.13	0,5
224.03 ± 0.13	0.4
244.28 ± 0.10	0.7
251.49 ± 0.10	0.6
257.64 ± 0.09	1.0
261.55 ± 0.07	14
(IT) 334.14 ± 0.07	15
344.5 ± 0.2	1.5
347.22 ± 0.09	1.3
355.57 ± 0.14	0.6
362.81 ± 0.15	1.1
376.83 ± 0.14	0.6
396.96 ± 0.09	1.7
429.02 ± 0.11	3.6
444.90 ± 0.09	4.4
462.11 ± 0.16	3.4
471.85 ± 0.09	2.3
478.59 ± 0.10	1.8
519.6 ± 0.2	0.5
534.85 ± 0.11	2.0
574.04 ± 0.10	3.6
622.03 ± 0.16	1.6
647.40 ± 0.08	34
702.75 ± 0.12	4.3
731.69 ± 0.15	1.7
733.89 ± 0.10	3.3
779.75 ± 0.10	3.9
795.7 ± 0.4	1.5
800.51 ± 0.12	2.2
863.91 ± 0.13	29
882.83 ± 0.12	4.8
897.7 ± 0.4	0.5
912.58 ± 0.10	100
914.72 ± 0.13	19
934.4 ± 0.3	1.5
978.19 ± 0.09	9.3
980.4 ± 0.2	2.7
982.9 ± 0.2	1.3
1029.8 ± 0.2	1.8
1061.83 ± 0.11	3.1
1348.9 ± 0.2	2.9
1459.1 ± 0.2	2.5
1516.1 ± 0.3	1.1
1531.6 ± 0.4	1.0
1587.4 ± 0.2	2.2
1683.3 ± 0.2	6.7
1704.4 ± 0.3	1.1
1855.7 ± 0.3	1.3
2004.9 ± 0.3	5.4
2027.7 ± 0.4	1.4
2049.2 ± 0.4	1.7
$_{52}Te^{133}$	
311.99 ± 0.08	70%
407.63 ± 0.07	31%
474.72 ± 0.13	1.1%
719.65 ± 0.10	8.3%
786.77 ± 0.10	7.2%
844.39 ± 0.07	4.6%
930.67 ± 0.10	5.4%
1000.77 ± 0.11	4.5%
1021.07 ± 0.15	3.4%
1252.2 ± 0.2	1.4%
1333.23 ± 0.12	11%
1405.7 ± 0.2	0.8%
1717.65 ± 0.15	3.2%
1881.5 ± 0.4	1.3%
$_{52}Te^{134}$	
80	13%
170	16%
204	21%
262	19%
$_{53}I^{117}$	
Ann. Rad.	
160	
340	
$_{53}I^{118}$	
Ann. Rad.	
550	1†
600	4
1150	
$_{53}I^{119}$	
Ann. Rad.	
260	
780	
$_{53}I^{120}$	
Ann. Rad.	
560	5†
620	1
1520	
$_{53}I^{121}$	
Ann. Rad.	
213	100†
270	3
320	7
600	
740	
$_{53}I^{122}$	
Ann. Rad.	
560	100†
690	7
780	5
1250	
1420	1
1750	2
2180	1
2540	0.2
2740	
2920	
3170	
3450	
$_{53}I^{123}$	
159.1 ± 0.1	100
183.7 ± 1.0	0.03 ± 0.02
192.7 ± 1.0	0.03 ± 0.02
248.3 ± 0.5	0.08 ± 0.01
281.0 ± 0.5	0.08 ± 0.01
346.6 ± 0.5	0.12 ± 0.02
440.4 ± 0.5	0.42 ± 0.02
505.6 ± 0.6	0.31 ± 0.05
529.0 ± 0.4	1.27 ± 0.11
538.5 ± 0.5	0.32 ± 0.02
624.9 ± 0.5	0.08 ± 0.01
687.7 ± 0.6	0.03 ± 0.01
736.1 ± 0.06	0.04 ± 0.01
784.4 ± 0.6	0.05 ± 0.01

GAMMA ENERGIES AND INTENSITIES OF RADIONUCLIDES (*Continued*)

Gamma energies (KeV)	Gamma intensities
$_{53}I^{124}$	
Ann. Rad.	
602.7 ± 0.18	100†
645.76 ± 0.18	18
722.72 ± 0.18	21
890	0.4
968.2 ± 0.2	0.8
1045.2 ± 0.2	0.8
1325.6 ± 0.2	1.5
1368.5 ± 0.2	5
1450	1.1
1500	6
1691.0 ± 0.07	21
1900	0.8
2091.1 ± 0.2	3
2293.6 ± 0.4	2.3
2740	0.9
$_{53}I^{125}$	
35.48	
$_{53}I^{126}$	
Ann. Rad.	
388.7 ± 0.2	
491.3 ± 0.2	
666.2 ± 0.2	
753.8 ± 0.2	
879.9 ± 0.2	
1420.0 ± 0.2	
$_{53}I^{128}$	
442.9 ± 0.2	17%
526.6 ± 0.2	1.8%
743.3 ± 0.15	0.3%
969.5 ± 0.2	0.3%
$_{53}I^{129}$	
39.58	
$_{53}I^{130}$	
419	36†
538	100
669	100
743	87
1150	12
$_{53}I^{131}$	
80.164 ± 0.009	2.6%
177.19 ± 0.08	0.3%
272.45 ± 0.2	0.07%
284.31 ± 0.08	5.9%
318.14 ± 0.15	0.09%
325.80 ± 0.15	0.24%
364.49 ± 0.08	79%
503.13 ± 0.15	0.35%
637.01 ± 0.15	6.7%
642.80 ± 0.17	0.2%
722.92 ± 0.15	1.8%
$_{53}I^{132}$	
147.10 ± 0.3	5.0†
254.9	2.0
262.75 ± 0.3	8.42
505.80 ± 0.3	7.87
522.48 ± 0.3	24.55
546.95 ± 0.4	1.70
620.95 ± 0.3	2.49
630.16 ± 0.3	15.44
650.51 ± 0.3	2.80
727.04 ± 0.3	4.96
729.08 ± 0.25	1.44
772.49 ± 0.25	59.77
780.24 ± 0.25	1.0
784.92 ± 0.4	0.35
809.77 ± 0.2	2.04
812.30 ± 0.3	4.19
863.26 ± 0.4	0.40
876.96 ± 0.4	0.63
910.44 ± 0.3	0.57
927.82 ± 0.4	0.37
954.66 ± 0.25	9.72
984.38 ± 0.4	0.40
1035.29 ± 0.4	0.29
1136.11 ± 0.3	1.29
1143.64 ± 0.4	0.62
1173.22 ± 0.4	0.48
1290.74 ± 0.4	0.39
1295.34 ± 0.5	0.54
1298.0 ± 0.4	0.25
1372.0 ± 0.3	0.70
1398.71 ± 0.4	2.21
1442.46 ± 0.3	0.43
1920.84 ± 0.4	0.24
2002.08 ± 0.4	0.21
2088.5 ± 0.5	0.07
2175.0 ± 1	0.02
2226.0 ± 1	0.03
2395.0 ± 1	0.01
$_{53}I^{133}$	
151.1 ± 0.3	0.029
234.3 ± 0.4	†
245.9 ± 0.3	0.081
262.7 ± 0.3	0.38
267.3 ± 0.4	0.10
386.4 ± 0.3	0.088
417.1 ± 0.3	0.18
422.9 ± 0.3	0.43
427.7 ± 0.4	0.099
438.9 ± 0.4	0.083
510.5 ± 0.3	2.08
529.90 ± 0.15	89% 100
600.4 ± 0.3	0.046
608.1 ± 0.3	0.086
618.0 ± 0.25	0.64
680.3 ± 0.2	0.79
706.65 ± 0.2	1.77
768.46 ± 0.25	0.51
789.5 ± 0.3	0.058
820.63 ± 0.3	0.187
835.3 ± 0.35	
856.37 ± 0.25	1.39
875.40 ± 0.25	5.26
910.1 ± 0.3	0.26
1052.3 ± 0.3	0.64
1060.3 ± 0.4	0.20
1236.56 ± 0.2	1.78
1293.9 ± 0.3	0.30
1298.32 ± 0.2	2.72
1350.6 ± 0.3	0.18
1460.9 ± 0.3	0.18
1592.5 ± 0.3	0.08
$_{53}I^{134}$	
135.4 ± 0.3	3.51†
152.2 ± 0.4	0.12
162.6 ± 0.3	0.26
188.5 ± 0.3	
216.9 ± 0.4	0.29
235.43 ± 0.3	1.85
319.68 ± 0.3	0.44
356.88 ± 0.3	
405.35 ± 0.15	6.84
433.24 ± 0.17	4.13
458.88 ± 0.25	1.04
488.90 ± 0.25	1.63
514.48 ± 0.2	2.30
540.81 ± 0.15	7.66
565.0 ± 0.4	
595.29 ± 0.15	11.24
621.77 ± 0.17	9.93
628.12 ± 0.25	2.60
667.58 ± 0.25	0.74
677.26 ± 0.17	7.70
730.77 ± 0.2	1.79
766.67 ± 0.18	3.98
772.2 ± 0.4	0.60
816.7 ± 0.3	
847.04 ± 0.08	96% 100
857.29 ± 0.15	7.51
875.3 ± 0.3	0.35
884.18 ± 0.1	67.81
947.69 ± 0.15	4.19
967.4 ± 0.25	0.6
974.72 ± 0.17	4.90
1040.1 ± 0.2	1.52
	0.5
1072.52 ± 0.17	15.91
1100.3 ± 0.3	0.75
1103.5 ± 0.3	0.66
1136.15 ± 0.18	10.68
1353.5 ± 0.3	0.33
1455.2 ± 0.3	2.48
1469.5 ± 0.5	0.81
1613.54 ± 0.3	4.61
1741.4 ± 0.35	2.41
1806.7 ± 0.35	5.62
2311.9 ± 1.0	0.4
2467.1 ± 1.0	0.21
$_{53}I^{135}$	
135.4 ± 0.5	
197.0 ± 0.4	
220.45 ± 0.08	5.37†
229.64 ± 0.18	0.64
249.56 ± 0.08	6.24
264.16 ± 0.08	
288.26 ± 0.08	9.03
289.6 ± 0.4	1.24
305.3 ± 0.2	0.43
361.6 ± 0.2	
403.0 ± 0.3	
415.0 ± 0.2	1.44
417.54 ± 0.1	10.79
429.9 ± 0.2	0.97
433.69 ± 0.15	1.93
451.7 ± 0.2	0.83
510.60 ± 0.17	0.94
526.43 ± 0.10 (Xe^{135m})	21.16
546.54 ± 0.10	22.75
559.0 ± 0.5	
648.8 ± 0.3	
650.2 ± 0.3	2.56
667.8 ± 0.3	1.25
707.9 ± 0.6	2.92
836.84 ± 0.12	20.10
884.1 ± 0.3	4.07
972.42 ± 0.18	7.41
1038.74 ± 0.15	29.86
1101.5 ± 0.4	5.79
1123.8 ± 0.2	13.18
1131.56 ± 0.18	79.73
1168.9 ± 0.4	3.02
1260.45 ± 0.15	100 34.9%
1281.8 ± 0.4	
1298.3 ± 0.4	
1367.5 ± 0.3	
1398.0 ± 0.6	
1457.65 ± 0.15	31.36
1470.0 ± 0.6	
1503.0 ± 0.5	3.92
1566.6 ± 0.4	3.70
1678.20 ± 0.17	30.10
1706.6 ± 0.2	13.03
1791.36 ± 0.2	25.16
1830.7 ± 0.4	1.89
1927.1 ± 0.5	
2046.1 ± 0.5	2.76
2256.6 ± 0.8	
2268.6 ± 1	
2408.6 ± 1	2.78
2465.9 ± 1.5	
$_{53}I^{136}$	
200	12%
270	18%
390	19%
460	~2%
530	~2%
710	3%
1000	6%
1320	70%
1325	25%
1550	4%
1720	2%
1910	5%
2200	7%
2400	12%
2630	10%
2800	8%
3200	5%
$_{54}Xe^{118}$	
Ann. Rad.	
$_{54}Xe^{119}$	
Ann. Rad.	
100	

GAMMA ENERGIES AND INTENSITIES OF RADIONUCLIDES (*Continued*)

Gamma energies (KeV)	Gamma intensities
$_{54}Xe^{120}$	
Ann. Rad	
55	
73	
176	
760	
$_{54}Xe^{121}$	
Ann. Rad.	
80	
95.5	
132.5	
437	
$_{54}Xe^{122}$	
90.3	
148.2	
239	
$_{54}Xe^{123}$	
Ann. Rad.	
148.6	
177.6	
328.6	
680	
900	
1100	
$_{54}Xe^{125m}$	
75	
111	
$_{54}Xe^{125}$	
54.7	
74.6	
113.1	
187.6	
242.2	
460	
$_{54}Xe^{127m}$	
125	
175	
$_{54}Xe^{127}$	
57.60	
145.22	3.9%
172.10	24%
202.84	67%
374.96	20%
$_{54}Xe^{129m}$	
39.58	
196.56	
$_{54}Xe^{131m}$	
163.98 ± 0.10	
$_{54}Xe^{133m}$	
232.8	
$_{54}Xe^{133}$	
79.55	0.25%
80.97	35%
161	0.03%
220	
302	0.003%
382	0.001%
$_{54}Xe^{135m}$	
526.43 ± 0.10	
$_{54}Xe^{135}$	
249.56 ± 0.1	9.2%
360	0.009%
610	0.3%
$_{54}Xe^{137}$	
455	33%
$_{54}Xe^{138}$	
160	33†
260	100
420	63
1730	44
1990	30
$_{54}Xe^{139}$	
180	4†
220	100
300	57
1150	23
$_{54}Xe^{140}$	
130	
$_{55}Cs^{123}$	
Ann. Rad.	
$_{55}Cs^{125}$	
Ann. Ra	
112	
$_{55}Cs^{126}$	
Ann. Rad.	
385	38%
$_{55}Cs^{127}$	
Ann. Rad.	
125	13†
286	
406	100
$_{55}Cs^{128}$	
Ann. Rad.	
441	50†
528	
576	
940	1.5
1200	2.1
1660	0.4
2180	0.4
2420	0.2
$_{55}Cs^{129}$	
94	1.3†
174	1.0
270	0.8
278	6
321	8
375	100
416	52
550	9
591	1.7
900	0.5
950	0.18
$_{55}Cs^{130}$	
Ann. Rad.	
$_{55}Cs^{132}$	
Ann. Rad.	
464.58 ± 0.4	
505.55 ± 0.4	
510.58 ± 0.4	
566.91 ± 0.3	
630.12 ± 0.3	
667.60 ± 0.2	
772.53 ± 0.3	
796.09 ± 0.3	
1031.87 ± 0.3	
1136.00 ± 0.3	
1298.11 ± 0.4	
1317.90 ± 0.4	
1461.09 ± 0.5	
1985.68 ± 0.4	
$_{55}Cs^{134m}$	
~10	
127.6	
137.4	
$_{55}Cs^{134}$	
475.34 ± 0.1	1.8%
563.20	8%
569.33	14%
604.7	98%
795.8	88%
801.9	9%
1038.6	1.1%
1167.9	1.9%
1365.1	3.4%
1400.5	0.08%
$_{55}Cs^{136}$	
66.9 ± 0.2	13.7%
86.4	5.6%
153.38 ± 0.2	8.2%
164.0 ± 0.2	4.5%
176.7 ± 0.2	13%
273.8 ± 0.2	12%
340.6 ± 0.15	44%
818.5 ± 0.2	100%
1048.1 ± 0.2	80%
1235.4 ± 0.2	20%
$_{55}Cs^{137}$	
661.630 ± 0.03	84.8 ± 0.5%
$_{55}Cs^{138}$	
66.7 ± 0.3	
75.41 ± 0.4	0.08†
107.6 ± 0.3	0.06
112.27 ± 0.25	0.16
137.89 ± 0.15	1.58
165.75 ± 0.2	3.17
191.83 ± 0.3	0.5
212.04 ± 0.3	
227.60 ± 0.18	1.56
324.53 ± 0.2	0.43
364.1 ± 0.3	
408.67 ± 0.2	5.3
421.39 ± 0.3	0.43
462.51 ± 0.18	35.49
516.53 ± 0.2	0.65
546.71 ± 0.15	12.69
871.61 ± 0.18	6.07
1009.65 ± 0.18	37.86
1146.88 ± 0.2	1.49
1283.4 ± 0.8	
1343.33 ± 0.3	1.64
1435.74 ± 0.18	73% 100
1444.59 ± 0.25	1.32
2217.75 ± 0.2	18.52
2316.5 ± 1.5	
2345.0 ± 0.5	0.81
2582.6 ± 0.4	
2639.4 ± 0.3	9.97
3340	0.5
$_{55}Cs^{139}$	
500	
630	
800	
1280	
1900	
2080	
$_{55}Cs^{140}$	
590	
880	
1140	
1620	
1850	
2060	
2320	
2720	
3150	
$_{56}Ba^{123}$	

Gamma energies (KeV)	Gamma intensities
Ann. Rad.	
$_{56}Ba^{125}$	
Ann. Rad.	
$_{56}Ba^{126}$	
230	100†
700	33
$_{56}Ba^{128}$	
134	
278	
$_{56}Ba^{+129m}$	
Ann. Rad.	
53.2	
73.2	
129	
135.7	
176.9	
182.3	
202.3	
214.2	
220.6	
237.4	
263.6	
286.0	
293.5	
328.2	
475	
481	
501	
534	
543	
547	
554	
557	
561	
597	
679	
690	
700	
713	
738	
748	
759	
768	
779	
802	
820	
829	
833	
872	
892	
933	
957	
963	
1000	
1035	
1045	
1122	
1163	
1179	
1208	
1221	
1252	
1458	
1624	
$_{56}Ba^{131m}$	
78	2†
110	100
$_{56}Ba^{131}$	
54.84	
78.69	
82.24	
92.25	
123.73	
133.54	
137.3	
157.0	
216.0	
239.6	
246.8	
249.4	
294.5	
351.2	
373.2	
404.3	
427.7	
451.7	
461.3	
462.9	
480.6	
486.6	
496.3	
573	
585	
620.2	
674	
696	
832	
914	
924.4	
969.1	
1048.2	
$_{56}Ba^{133m}$	
275.7	
12.29	
$_{56}Ba^{133}$	†
53.17 ± 0.04	3.68 ± 0.09
79.60 ± 0.05	4.9 ± 0.59
80.997 ± 0.006	55.3 ± 4.2
160.66 ± 0.06	1.01 ± 0.05
223.43 ± 0.26	0.75 ± 0.04
276.43 ± 0.26	11.6 ± 0.17
303.09 ± 0.21	29.9 ± 0.29
356.26 ± 0.15	≡100
384.09 ± 0.20	14.9 ± 0.26
$_{56}Ba^{135m}$	
268	100%
$_{56}Ba^{136m}$	
163.7	
818	
1050	
$_{56}Ba^{137m}$	
661.630 ± 0.030	
$_{56}Ba^{139}$	
165.85 ± 0.010	27%
1270	
1430	
$_{56}Ba^{140}$	
13.85	
29.97	30%
132.7	1.7%
162.7	8%
304.9	7%
423.8	3%
437.6	5%
537.3	29%
$_{56}Ba^{141}$	
113.08 ± 0.05	2†
180.70 ± 0.09	0.7
190.33 ± 0.05	100
276.94 ± 0.04	52
304.18 ± 0.04	57
343.66 ± 0.04	31
389.66 ± 0.07	3.1
457.52 ± 0.05	12
462.13 ± 0.06	10
467.26 ± 0.06	13
625.21 ± 0.04	7
647.78 ± 0.04	13
739.02 ± 0.1	10
831.70 ± 0.04	4.1
876.01 ± 0.14	8
929.28 ± 0.08	1.5
981.40 ± 0.15	1.7
1160.55 ± 0.1	3
1197.3 ± 0.2	13
1273.4 ± 0.2	1.3
1310.7 ± 0.2	2.1
1436.64 ± 0.1	2.0
1501.28 ± 0.18	0.7
1551.0 ± 0.3	0.9
1740.8 ± 0.2	7
$_{56}Ba^{142}$	
77 ± 1	41†
231.57 ± 0.09	59
255.33 ± 0.07	100
286.13 ± 0.16	7
309.5 ± 0.3	11
364.03 ± 0.07	28
379.42 ± 0.16	3
425.05 ± 0.07	30
432.42 ± 0.16	5.6
589.95 ± 0.2	3.5
599.61 ± 0.1	13
769.71 ± 0.17	4.2
840.07 ± 0.12	19
894.86 ± 0.1	82
944.02 ± 0.07	47
1000.90 ± 0.1	53
1078.52 ± 0.07	56
1093.95 ± 0.09	17
1126.1 ± 0.1	10
1203.81 ± 0.1	97
1380.3 ± 0.2	14
$_{57}La^{126}$	
Ann. Rad.	
$_{57}La^{127}$	
Ann. Rad.	
$_{57}La^{128}$	
Ann. Rad.	
$_{57}La^{129}$	
Ann. Rad.	
$_{57}La^{130}$	
Ann. Rad.	
356	
450	
550	
720	
810	
910	
1010	
1190	
1450	
1550	
$_{57}La^{131}$	
Ann. Rad.	
115	23†
169	5
214	8
254	8
285	17
364	20
417	20
450	8
600	7
878	1
$_{57}La^{132}$	
Ann. Rad.	
470	
560	
660	
900	
1030	
1220	
1450	
1580	
1920	
$_{57}La^{133}$	
Ann. Rad.	
260	
$_{57}La^{134}$	
Ann. Rad.	
560	

GAMMA ENERGIES AND INTENSITIES OF RADIONUCLIDES (*Continued*)

Gamma energies (KeV)	Gamma intensities
604.6	
760	
820	
900	
980	
1170	
1300	
1470	
1550	
1760	
1980	
2150	
2240	
2590	
$_{57}La^{135}$	
52.4	
87.4	
117.0	
118	
132	
205	
265	
300	
379	
515	
544	
549	
569	
571	
602	
661	
679	
777	
821	
865	
901	
$_{57}La^{136}$	
Ann. Rad.	
830	
$_{57}La^{140}$	
109.6 ± 0.2	0.22%
131.15 ± 0.2	0.45%
241.9	0.55%
266.5	0.5%
328.7	21%
432.5	3%
487.0	46%
510.9	0.35%
751.7	0.45%
815.8	24%
867.9	5.7%
919.6	2.5%
925.2	6.7%
951.0	0.6%
1085.3	1%
1596.2	97%
2010.4	0.4%
2348.2	0.8%
2521.8	3.2%
2547.7	0.09%
$_{57}La^{141}$	
1370	2%
$_{57}La^{142}$	
106.47 ± 0.1	0.25
178.32 ± 0.1	0.23
418.2 ± 0.3	0.1
430.3 ± 0.3	0.1
510.11 ± 0.1	0.4
578.25 ± 0.08	2.8
641.27 ± 0.05	100 (50%)
861.68 ± 0.1	3.5
894.94 ± 0.05	18.4
1011.42 ± 0.07	7.8
1043.89 ± 0.07	4.6
1160.40 ± 0.1	2.4
1233.07 ± 0.1	3.6
1363.55 ± 0.2	3.4
1545.93 ± 0.2	6.5
1723.23 ± 0.3	2.4
1756.60 ± 0.1	5.4
1901.56 ± 0.1	11.3
2004.87 ± 0.3	2.0
2055.6 ± 0.2	3.5
2187.49 ± 0.1	8.8
2397.95 ± 0.1	24.4
2543.1 ± 0.3	15.7
2666.5 ± 0.3	2.0
2971.8 ± 0.3	6.5
3314.1 ± 0.3	0.9
3460.1 ± 0.3	0.6
3611 ± 1	1.2
3632.6 ± 0.4	1.8
3719.7 ± 0.5	0.3
3850.1 ± 0.5	0.3
$_{57}La^{143}$	
200	8†
440	13
620	100
800	44
920	8
1070	26
1170	57
1580	28
1700	19
1980	35
2220	6
2460	13
2560	27
2850	15
$_{58}Ce^{132}$	
180	
$_{58}Ce^{133}$	
1800	
$_{58}Ce^{135}$	
52.4	
87.4	
117.0	
118.0	
132	
205	
265	
300	
379	
515	
544	
549	
569	
571	
602	
661	
679	
777	
821	
865	
901	
$_{58}Ce^{137m}$	
254.5 (IT)	99.4%
168	45†
762	16
825	38
836	8.5
994	0.15
1004	2.1
1160	0.2
$_{58}Ce^{137}$	
Ann. Rad.	
10	†
433	5
436	31
446	204
479	1.7
481	~4.0
492	1.0
698	3.5
771	0.9
781	0.4
916	6.7
926	3.4
$_{58}Ce^{139m}$	
746	
$_{58}Ce^{139}$	
165.854 ± 0.010	100%
$_{58}Ce^{141}$	
145.443 ± 0.006	49%
$_{58}Ce^{143}$	
57.37	11.5%
142	0.4%
231.60	2%
293.22	43%
350.6 ± 0.1	3%
374	0.4%
490.42	2%
587.38	0.5%
664.59 ± 0.1	5.5%
721.96 ± 0.1	6%
880.4 ± 0.2	1%
940	0.2%
1040	0.2%
1110	0.4%
$_{58}Ce^{144}$	
33.57	
40.93	
53.41	
59.03	
80.12	
99.95	
133.53	11%
$_{58}Ce^{146}$	
50	
110	20†
142	42
220	50
270	12
320	100
$_{59}Pr^{134}$	
Ann. Rad.	
220	
300	
410	
640	
960	
$_{59}Pr^{135}$	
Ann. Rad.	
80	
220	
300	
$_{59}Pr^{136}$	
Ann. Rad.	
170	
~800	
~1100	
$_{59}Pr^{137}$	
Ann. Rad.	
$_{59}Pr^{138}$	
Ann. Rad.	
298	77†
400	9
550	~5
790	100
1040	98
1200	1
1830	0.4
$_{59}Pr^{139}$	
Ann. Rad.	
270	12†
1350	27
1610	16
$_{59}Pr^{140}$	
Ann. Rad.	
306	
1596	
1902	

Gamma energies (KeV)	Gamma intensities
$_{59}Pr^{142}$	
1575.43 ± 0.3	4%
$_{59}Pr^{143}$	
no γ	
$_{59}Pr^{144}$	
696.48 ± 0.09	1.5%
1489.14 ± 0.07	0.35%
2185.32 ± 0.05	0.9%
$_{59}Pr^{145}$	
72	
680	
750	
920	
980	
1050	
1160	
$_{59}Pr^{146}$	
455	77%
597	10%
750	16%
780	15%
920	6%
1020	5%
1370	6%
1510	27%
1720	4%
1970	2%
2230	4%
2390	3%
2730	1.7%
$_{59}Pr^{147}$	
78	17%
105	
127	
310	
340	25%
560	39%
610	10%
650	24%
1260	11%
$_{59}Pr^{148}$	
300	
$_{60}Nd^{138}$	
Ann. Rad.	
$_{60}Nd^{139m}$	
232	
$_{60}Nd^{139}$	
Ann. Rad.	
36.5	
114	
132.7	
200	
210	
217	
708	100
737	42
773	68†
821	
822	
828	
900	25
983	70
1030	30
1100	30
1240	20
1340	20
1480	10
1580	8
1720	4
1850	3
2050	10
$_{60}Nd^{141}$	
Ann. Rad.	
145.443 ± 0.0006	
1126.8 ± 0.3	
1147.5 ± 0.3	
1292.6 ± 0.3	
1298.6 ± 0.3	
$_{60}Nd^{141m}$	
0.755	100%
$_{60}Nd^{147}$	
91.0 ± 0.2	27%
120.5	0.5%
196.6	0.4%
275.4	0.8%
319.4	2.0%
398.1	1.6%
410.4	0.7%
439.8	1.2%
489.3	0.2%
531.0	13%
594.7	0.25%
685.8	0.8%
$_{60}Nd^{149}$	
30.00 ± 0.10	
58.90 ± 0.03	1.3%
65.37 ± 0.05	0.041%
74.3 ± 0.2	
74.7 ± 0.2	
75.55 ± 0.2	
91.12 ± 0.05	0.063%
97.00 ± 0.03	1.3%
107.76 ± 0.10	
114.31 ± 0.03	16.1%
117.0 ± 0.2	
122.41 ± 0.05	0.20%
126.62 ± 0.05	0.10%
137.02 ± 0.08	0.040%
139.21 ± 0.05	0.41%
155.87 ± 0.03	5.2%
177.82 ± 0.05	0.14%
185.50 ± 0.08	0.079%
188.64 ± 0.03	1.7%
192.03 ± 0.05	0.51%
198.93 ± 0.03	1.25%
208.15 ± 0.05	2.5%
211.32 ± 0.03	23.4%
214.02 ± 0.07	
226.85 ± 0.05	0.14%
229.57 ± 0.05	0.43%
240.23 ± 0.03	3.4%
245.71 ± 0.05	0.89%
250.84 ± 0.10	0.032%
254.20 ± 0.07	0.075%
258.07 ± 0.05	0.33%
267.70 ± 0.03	5.2%
270.18 ± 0.03	9.2%
273.28 ± 0.08	0.20%
275.46 ± 0.05	0.51%
276.96 ± 0.10	0.28%
282.47 ± 0.05	0.52%
288.21 ± 0.05	0.59%
294.82 ± 0.05	0.50%
301.14 ± 0.05	0.33%
310.99 ± 0.05	0.44%
326.57 ± 0.03	4.0%
342.85 ± 0.10	0.029%
347.8 ± 0.2	
349.23 ± 0.03	1.29%
352.75 ± 0.10	0.056%
357.05 ± 0.10	0.047%
360.05 ± 0.05	0.14%
366.64 ± 0.05	0.57%
380.94 ± 0.10	0.058%
384.69 ± 0.05	0.29%
396.77 ± 0.10	0.085%
423.54 ± 0.03	8.2%
425.7 ± 0.3	
443.53 ± 0.03	1.29%
480.21 ± 0.15	0.051%
483.53 ± 0.15	0.073%
493.86 ± 0.15	0.044%
497.81 ± 0.15	0.047%
510.29 ± 0.15	0.051%
515.45 ± 0.20	
533.17 ± 0.15	0.074%
538.00 ± 0.20	
540.49 ± 0.03	6.6%
556.40 ± 0.10	1.00%
579.25 ± 0.07	0.077%
583.07 ± 0.07	0.072%
594.48 ± 0.10	0.028%
598.06 ± 0.15	0.033%
630.24 ± 0.07	0.19%
635.47 ± 0.07	0.096%
654.82 ± 0.03	8.4%
661.8 ± 0.3	0.016%
671.6 ± 1.0	0.012%
673.6 ± 0.5	0.014%
675.8 ± 0.2	0.028%
681.29 ± 0.2	0.011%
686.96 ± 0.07	0.092%
696.29 ± 0.05	0.17%
712.62 ± 0.07	0.072%
718.40 ± 0.10	0.026%
727.90 ± 0.15	0.019%
740.61 ± 0.10	0.017%
749.68 ± 0.15	0.012%
754.33 ± 0.10	0.030%
758.6 ± 0.3	0.014%
761.45 ± 0.15	0.020%
768.17 ± 0.10	0.059%
781.42 ± 0.20	0.006%
786.74 ± 0.10	0.011%
793.43 ± 0.10	0.022%
795.84 ± 0.20	0.006%
808.80 ± 0.10	0.146%
809.8 ± 0.5	0.025%
813.2 ± 0.5	0.008%
829.4 ± 0.4	0.006%
832.20 ± 0.10	0.019%
837.41 ± 0.10	0.030%
842.86 ± 0.10	0.048%
849.93 ± 0.10	0.022%
859.43 ± 0.20	0.013%
861.54 ± 0.10	0.019%
865.03 ± 0.15	0.018%
871.40 ± 0.10	0.032%
873.8 ± 0.2	0.0052%
886.59 ± 0.3	0.0073%
907.70 ± 0.2	0.0063%
915.3 ± 0.3	0.0045%
923.88 ± 0.08	0.099%
929.44 ± 0.15	0.012%
935.94 ± 0.2	0.0056%
938.77 ± 0.2	0.0060%
945.82 ± 0.10	0.021%
951.96 ± 0.15	0.0086%
964.00 ± 0.15	0.018%
967.44 ± 0.15	0.0094%
971.7 ± 0.3	0.0033%
979.03 ± 0.08	0.096%
986.7 ± 0.3	0.0033%
992.9 ± 0.2	0.013%
1022.79 ± 0.08	0.105%
1027.19 ± 0.10	0.0098%
1031.8 ± 0.2	0.0041%
1041.97 ± 0.10	0.023%
1051.9 ± 0.3	0.0037%
1076.05 ± 0.15	0.021%
1078.78 ± 0.10	0.058%
1100.78 ± 0.10	0.052%
1123.6 ± 0.4	0.014%
1125.41 ± 0.15	0.025%
1128.5 ± 0.3	0.0034%
1135.9 ± 0.2	0.0021%
1141.8 ± 0.2	0.0034%
1150.1 ± 0.2	0.0030%
1172.1 ± 0.3	0.0044%
1175.7 ± 0.2	0.0044%
1190.4 ± 0.2	0.0029%
1197.8 ± 0.2	0.0066%
1202.3 ± 0.2	0.0019%
1225.7 ± 0.3	0.0014%
1234.11 ± 0.10	0.0250%
1239.0 ± 0.6	0.0014%
1259.6 ± 0.2	0.0033%
1264.0 ± 0.2	0.0060%
1280.3 ± 0.3	0.0010%
1284.3 ± 0.3	0.0007%
1290.10 ± 0.15	0.0039%
1298.4 ± 0.3	0.0015%
1312.13 ± 0.15	0.0068%
1357.2 ± 0.40	0.0021%
1381.5 ± 0.3	0.0021%
1448.1 ± 0.5	0.0004%
1454.3 ± 0.4	0.0011%
1473.9 ± 0.5	0.0003%
1495.8 ± 0.3	0.0008%
1568.4 ± 0.4	0.0005%
$_{60}Nd^{151}$	

Gamma energies (KeV)	Gamma intensities
58.62 ± 0.10	0.35%
69.19 ± 0.05	1.27%
81.85 ± 0.25	0.49%
85.18 ± 0.06	1.98%
90.04 ± 0.05	1.60%
102.89 ± 0.13	0.48%
109.07 ± 0.55	<0.001%
116.79 ± 0.04	32.56%
138.88 ± 0.04	6.31%
149.49 ± 0.07	0.33%
158.74 ± 0.20	0.50%
170.75 ± 0.04	3.39%
175.07 ± 0.05	6.12%
183.14 ± 0.08	0.36%
197.23 ± 0.13	0.27%
199.73 ± 0.12	0.30%
208.10 ± 0.14	0.31%
239.29 ± 0.30	0.83%
249.27 ± 0.30	0.16%
255.69 ± 0.04	13.3%
263.56 ± 0.07	0.67%
300.58 ± 0.08	1.58%
312.49 ± 0.35	0.25%
320.13 ± 0.11	0.63%
324.56 ± 0.35	0.48%
332.76 ± 0.11	0.63%
347.11 ± 0.11	0.39%
356.91 ± 0.11	0.38%
402.33 ± 0.06	1.41%
407.53 ± 0.10	0.45%
414.49 ± 0.35	0.24%
423.54 ± 0.03	5.31%
426.60 ± 0.30	0.50%
439.37 ± 0.20	0.35%
460.58 ± 0.08	0.74%
524.35 ± 0.11	0.42%
531.99 ± 0.15	0.16%
542.19 ± 0.35	0.45%
549.99 ± 0.11	0.70%
562.72 ± 0.17	0.21%
577.41 ± 0.11	0.26%
585.20 ± 0.16	1.09%
589.61 ± 0.10	0.30%
596.93 ± 0.09	0.60%
619.02 ± 0.12	0.35%
629.78 ± 0.31	0.14%
658.58 ± 0.20	0.66%
670.50 ± 0.18	0.22%
677.95 ± 0.06	2.22%
724.17 ± 0.18	0.28%
736.21 ± 0.06	5.92%
739.17 ± 0.12	1.71%
755.56 ± 0.11	1.08%
765.46 ± 0.70	0.16%
768.09 ± 0.70	0.22%
773.60 ± 0.18	0.28%
797.48 ± 0.06	4.65%
829.05 ± 0.18	0.22%
841.05 ± 0.11	0.96%
852.98 ± 0.14	0.42%
866.44 ± 0.26	0.32%
870.65 ± 0.24	0.32%
876.33 ± 0.14	0.48%
897.72 ± 0.24	0.41%
904.69 ± 0.16	0.35%
914.15 ± 0.11	1.21%
925.53 ± 0.22	0.16%
943.11 ± 0.22	0.37%
958.09 ± 0.11	0.46%
1016.30 ± 0.10	2.80%
1020.21 ± 0.50	0.19%

Gamma energies (KeV)	Gamma intensities
1032.47 ± 0.18	0.11%
1035.98 ± 0.14	0.19%
1041.76 ± 0.11	0.35%
1048.05 ± 0.09	0.67%
1066.60 ± 0.14	0.17%
1073.00 ± 0.22	0.085%
1077.49 ± 0.37	0.15%
1080.09 ± 0.21	0.30%
1100.11 ± 0.20	0.093%
1107.20 ± 0.11	0.44%
1122.59 ± 0.06	4.46%
1126.56 ± 0.25	0.28%
1132.53 ± 0.18	0.18%
1136.54 ± 0.15	0.23%
1145.84 ± 0.17	0.25%
1157.02 ± 0.13	0.21%
1180.90 ± 0.06	14.65%
1189.33 ± 0.10	0.31%
1201.15 ± 0.13	0.20%
1213.34 ± 0.14	0.093%
1217.44 ± 0.25	0.046%
1224.39 ± 0.27	0.051%
1232.79 ± 0.14	0.11%
1238.30 ± 0.18	0.065%
1250.84 ± 1.12	0.051%
1254.53 ± 1.12	0.051%
1270.64 ± 0.21	0.19%
1286.17 ± 0.12	0.26%
1293.94 ± 0.33	0.11%
1297.75 ± 0.21	0.19%
1314.61 ± 0.12	0.31%
1328.36 ± 0.12	0.31%
1333.33 ± 0.14	0.12%
1341.61 ± 0.23	0.11%
1346.45 ± 0.2	0.033%
1360.20 ± ?	0.15%
1362.87 ± 0.20	0.31%
1379.25 ± 0.17	0.16%
1363.48 ± 0.21	0.078%
1395.41 ± 1.09	0.13%
1425.30 ± 0.27	0.068%
1444.41 ± 0.60	0.055%
1446.97 ± 0.90	0.035%
1465.62 ± 0.24	0.077%
1476.06 ± 0.19	0.11%
1485.66 ± 0.14	0.28%
1498.74 ± 1.14	0.029%
1501.58 ± 1.14	0.029%
1507.58 ± 0.41	0.032%
1533.61 ± 0.28	0.049%
1550.03 ± 0.14	0.36%
1554.05 ± 0.25	0.072%
1566.73 ± 0.24	0.085%
1572.02 ± 0.25	0.10%
1578.63 ± 0.18	0.21%
1592.75 ± 0.25	0.044%
1598.20 ± 0.26	0.12%
1618.17 ± 0.16	0.41%
1636.75 ± 0.24	0.12%
1640.44 ± 0.37	0.032%
1647.45 ± 0.30	0.051%
1686.86 ± 0.26	0.015%
1717.27 ± 0.20	0.091%
1732.31 ± 0.27	0.048%
1753.48 ± 0.25	0.045%
1757.10 ± 0.31	0.053%
1762.06 ± 0.29	0.025%
1775.61 ± 0.19	0.25%
1782.81 ± 0.31	0.051%
1786.86 ± 0.26	0.078%
1794.20 ± 0.26	0.056%

Gamma energies (KeV)	Gamma intensities
1807.24 ± 0.35	0.068%
1811.34 ± 0.37	0.074%
1819.23 ± 0.24	0.062%
1830.21 ± 0.50	0.012%
1836.12 ± 0.43	0.014%
1849.02 ± 0.31	0.023%
1855.24 ± 0.63	0.011%
1863.59 ± 0.23	0.043%
1873.68 ± 1.19	0.0074%
1877.58 ± 0.36	0.011%
1892.45 ± 0.19	0.018%
1903.89 ± 0.44	0.021%
1909.40 ± 0.35	0.037%
1912.38 ± 1.21	0.0040%
1926.15 ± 0.28	0.042%
1932.83 ± 0.30	0.024%
2019.32 ± 0.29	0.037%
2105.83 ± 0.57	0.0058%
2254.82 ± 0.42	0.0065%
$_{61}Pm^{141}$	
Ann. Rad.	
$_{61}Pm^{142}$	
1570	0.2%
$_{61}Pm^{143}$	
742	47%
$_{61}Pm^{144}$	
476	45%
616	100%
696.48 ± 0.09	100%
$_{61}Pm^{145}$	
67	12%
725	8%
$_{61}Pm^{146}$	
453	65%
742	30%
749	35%
$_{61}Pm^{148m}$	
61.5 ± 0.2 (IT)	
75.6 ± 0.2 (IT)	
98.5 ± 0.2	3.3%
189.5 ± 0.3	
288.0 ± 0.2	13%
311.7 ± 0.3	2.8%
414.1 ± 0.3	17%
432.7 ± 0.3	6%
443.1 ± 0.4	
501.1 ± 0.4	7%
550.1 ± 0.2	95%
592.4 ± 0.4	
599.5 ± 0.3	8%
611.1 ± 0.3	6%
629.9 ± 0.2	87%
725.6 ± 0.2	36%
896.2 ± 0.3	

GAMMA ENERGIES AND INTENSITIES OF RADIONUCLIDES (*Continued*)

Gamma energies (KeV)	Gamma intensities
914.9 ± 0.2	21%
1013.7 ± 0.3	20%
1465.1 ± 0.2	27%
$_{61}Pm^{148}$	
550.1 ± 0.2	50%
914.9 ± 0.2	16%
1465.1 ± 0.2	24%
$_{61}Pm^{149}$	
K X-ray 22.5	2 ± 0.7
208.1 ± 1	0.008 ± 0.006
242.4 ± 1	0.009 ± 0.006
254.4 ± 0.5	0.045 ± 0.02
263.3 ± 0.4	0.091 ± 0.04
276.9 ± 0.4	0.27 ± 0.10
285.90 ± 0.05	31.0 ± 2
304.9 ± 0.4	0.022 ± 0.01
327.4 ± 0.4	0.034 ± 0.01
350.6 ± 0.4(doublet)	0.023 ± 0.01
359.4 ± 0.4	0.022 ± 0.01
531 ± 1	0.012 ± 0.01
535.8 ± 0.4	0.13 ± 0.03
545 ± 1	0.025 ± 0.01
548 ± 1	0.019 ± 0.01
558.4 ± 0.4	0.14 ± 0.03
568.5 ± 0.4	0.18 ± 0.03
590.9 ± 0.3	0.68 ± 0.08
613.9 ± 0.4	0.13 ± 0.02
636.6 ± 0.4	0.062 ± 0.02
807.8 ± 0.5	0.15 ± 0.02
830.5 ± 0.5	0.30 ± 0.04
833.2 ± 0.5	0.30 ± 0.04
859.4 ± 0.5	1.02 ± 0.15
881.9 ± 0.5	0.23 ± 0.03
$_{61}Pm^{150}$	
334	100†
410	10
590	5
720	11
840	25
880	17
1170	32
1230	6
1330	31
1750	14
1960	3.5
2060	1.7
2530	1.3
2750	0.4
2910	0.5
3080	0.16
$_{61}Pm^{151}$	
100.00	2.7%
101.80	1.6%
104.80	3.8%
139.30	0.5%
156.20	0.2%
163.37	2.5%
167.84	9.9%
177.05	4.8%

Gamma energies (KeV)	Gamma intensities
186.60	0.2%
201.94	0.1%
208.97	2.0%
227.21	0.3%
232.36	1.2%
236.73	1.0%
240.00	3.9%
254.07	0.2%
270.41	0.1%
275.10	7.4%
280.11	0.2%
290.63	0.8%
306.60	0.2%
323.79	1.5%
329.60	0.25%
339.93	24%
344.80	2.2%
349.60	0.1%
353.10	0.1%
379.67	1.0%
406.80	0.2%
440.66	1.6%
445.51	4.3%
451.24	0.3%
490.10	0.1%
516.00	0.2%
564.90	0.4%
636.20	1.5%
654.30	0.3%
669.10	0.8%
671.40	1.0%
704.40	0.3%
717.85	4.4%
736.30	4.8%
753.00	1.4%
773.00	0.8%
808.10	0.6%
817.90	0.3%
848.90	0.3%
948.90	0.4%
953.90	0.1%
$_{61}Pm^{152}$	
120	
240	
1000	
$_{62}Sm^{142}$	
Ann. Rad.	
$_{62}Sm^{143m}$	
748	
$_{62}Sm^{143}$	
Ann. Rad.	
1050	
$_{62}Sm^{145}$	
61.3	92%(78%)
485	0.003%
$_{62}Sm^{151}$	
21.6	1.7%

Gamma energies (KeV)	Gamma intensities
$_{62}Sm^{153}$	
54.19	0.006†
68.23	0.004
69.68	17
75.43	0.61
83.4	0.75
89.5	0.58
103.23	100
151.6	0.03
172.9	0.31
411	0.01
424	0.013
438	0.01
450	0.01
463	0.05
521	0.025
531	0.23
533	0.12
539	0.06
578	0.01
603	0.014
609	0.04
636	0.018
$_{62}Sm^{155}$	
75	0.3†
104.35	73
140	1.7
246	4
450	0.18
510	0.3
650	0.2
920	0.03
990	0.04
1180	0.06
1270	0.13
$_{62}Sm^{156}$	
22.6	
25.3	
38.1	
65.0	
87.6	40%
103	
166	
204	
219	
244	
269	
291	
$_{62}Sm^{157}$	
570	
$_{63}Eu^{144}$	
Ann. Rad.	
$_{63}Eu^{145}$	
Ann. Rad.	
110.94 ± 0.15	2.4†
191.25 ± 0.15	0.9
542.51 ± 0.15	7.5
653.46 ± 0.15	24

Gamma energies (KeV)	Gamma intensities
713.7 ± 0.3	0.5
764.65 ± 0.15	3.1
893.57 ± 0.20	75
974.90 ± 0.15	3.8
1658.38 ± 0.15	1.9
1803.90 ± 0.15	1.9
1876.84 ± 0.15	2.0
1996.61 ± 0.15	12.3
$_{63}Eu^{146}$	
Ann. Rad.	†
27.18	
146.3 ± 0.2	
267.6 ± 0.2	
271.78 ± 0.20	
397.32 ± 0.20	
410.80 ± 0.20	
430.29 ± 0.20	
621.68 ± 0.20	
628.7 ± 0.20	
633.6 ± 0.20	
665.31 ± 0.20	
667.84 ± 0.20	12†
743.1 ± 0.2	
747.1 ± 0.2	100
791.14 ± 0.20	
865.0 ± 0.20	
888.60 ± 0.20	
893.61 ± 0.20	
900.23 ± 0.20	
909.33 ± 0.20	
913.93 ± 0.20	
702.50 ± 0.20	
704.77 ± 0.20	
1058.28 ± 0.20	7
1133.02 ± 0.20	
1150.53 ± 0.20	
1176.26 ± 0.20	
1296.88 ± 0.20	6
1378.26 ± 0.20	
1407.55 ± 0.20	
1414.89 ± 0.20	5
1516.96 ± 0.20	
1522.67 ± 0.20	
1533.55 ± 0.20	8
1685.95 ± 0.20	2.4
1690.94 ± 0.20	
1756.02 ± 0.20	3
1880	1.1
1930	2
2080	3
2160	1.8
2300	1.3
2410	1.4
2510	0.7
$_{63}Eu^{147}$	
Ann. Rad.	
76.4	1.1†
121.23 ± 0.20	10.9
160.87 ± 0.20	0.04
166.3 ± 0.20	0.04
197.23 ± 0.20	21.4
327.92 ± 0.20	0.1
601.49 ± 0.20	5.6
677.54 ± 0.20	9.3
798.78 ± 0.20	4.8

Gamma energies (KeV)	Gamma intensities
857.02 ± 0.20	2.8
879.5 ± 0.20	0.14
933.2 ± 0.20	3.6
943.2 ± 0.20	0.4
955.97 ± 0.20	3.6
1064.0 ± 0.20	0.7
1077.11 ± 0.20	6.0
1119.87 ± 0.20	0.46
1179.95 ± 0.20	0.1
1197.0 ± 0.20	0.1
1255.90 ± 0.20	0.82
1318.0 ± 0.20	0.18
1331.50 ± 0.20	0.34
1349.93 ± 0.20	0.16
1448.74 ± 0.20	0.18
1459.92 ± 0.20	0.22
$_{63}Eu^{148}$	
Ann. Rad.	% β decays
22.41 ± 0.10	
67.63 ± 0.10	
98.40 ± 0.10	0.09 ± 0.03
116.3 ± 0.2	0.10 ± 0.03
121.18 ± 0.12	0.006 ± 0.003
166.2 ± 0.3	0.05 ± 0.02
182.8 ± 0.2	0.11 ± 0.3
189.6 ± 0.2	0.10 ± 0.3
215.9 ± 0.2	
241.5 ± 0.1	1.04 ± 0.10
243.6 ± 0.2	0.23 ± 0.05
252.3 ± 0.3	0.05 ± 0.02
279.1 ± 0.5	0.06 ± 0.02
287.9 ± 0.2	0.22 ± 0.05
296.0 ± 0.3	
310.0 ± 0.3	0.22 ± 0.05
311.4 ± 0.1	1.14 ± 0.10
319.1 ± 0.3	0.05 ± 0.02
377.5 ± 0.5	
413.9, 413.9 } ± 0.1	{ 18.6 ± 0.5
432.6 ± 0.1	2.79 ± 0.20
437.2 ± 0.3	
468.4 ± 0.3	0.49 ± 0.08
481.2 ± 0.5	0.16 ± 0.05
495.2 ± 0.3	0.25 ± 0.06
501.3 ± 0.1	0.75 ± 0.13
505.2 ± 0.5	0.28 ± 0.07
516.7 ± 0.3	0.40 ± 0.10
528.5 ± 0.3	0.19 ± 0.05
550.20 ± 0.5	99.00 ± 0
553.2 ± 0.1	17.1 ± 0.5
571.90 ± 0.05	9.1 ± 0.3
587.6 ± 0.5	0.08 ± 0.03
590.1 ± 0.2	0.32 ± 0.08
595.2 ± 0.3	0.20 ± 0.06
599.5 ± 0.2	0.51 ± 0.10
602.6 ± 0.3	0.23 ± 0.16
611.26 ± 0.05	19.3 ± 0.5
620.0 ± 0.1	1.0 ± 0.1
629.90 ± 0.05	70.9 ± 3.0
654.3 ± 0.1	1.31 ± 0.15
657.0 ± 0.5	0.17 ± 0.06
667.7 ± 0.5	0.13 ± 0.05
669.9 ± 0.2	0.34 ± 0.08
675.4 ± 0.5	0.08 ± 0.03
683.2 ± 0.1	1.06 ± 0.10
690.7 ± 0.5	0.08 ± 0.03
714.8 ± 0.1	1.54 ± 0.12

Gamma energies (KeV)	Gamma intensities
719.6 ± 0.3	0.11 ± 0.03
725.7 ± 0.1	12.2 ± 1.0
734.8 ± 1.0	0.06 ± 0.03
756.6 ± 0.5	0.19 ± 0.05
770.4 ± 0.2	0.38 ± 0.07
780.2 ± 0.5	0.08 ± 0.03
799.2 ± 0.2	0.34 ± 0.07
870.0 ± 0.1	4.9 ± 0.5
895.8 ± 0.1	0.55 ± 0.10
903.9 ± 0.2	0.34 ± 0.07
906.9 ± 0.3	0.15 ± 0.04
915.3 ± 0.1	2.4 ± 0.2
924.9 ± 0.2	0.27 ± 0.06
930.5 ± 0.1	2.4 ± 0.2
938.2 ± 1.0	0.11 ± 0.04
949.7 ± 0.2	0.21 ± 0.06
964.2 ± 0.4	0.20 ± 0.07
967.3 ± 0.1	2.9 ± 0.3
980.0 ± 0.5	0.13 ± 0.04
989.7 ± 0.2	0.39 ± 0.08
1013.9 ± 0.2	0.41 ± 0.08
1034.1 ± 0.1	7.9 ± 0.7
1043.9 ± 1.0	0.04 ± 0.02
1047.5 ± 0.3	0.20 ± 0.07
1066.8 ± 0.2	0.29 ± 0.08
1069.2 ± 0.4	0.22 ± 0.08
1082.2 ± 0.5	0.17 ± 0.07
1089.3 ± 0.5	0.20 ± 0.07
1097.5 ± 0.7	0.10 ± 0.04
1104.3 ± 0.2	0.45 ± 0.09
1107.9 ± 0.3	0.10 ± 0.03
1113.8 ± 0.3	0.13 ± 0.04
1126.9 ± 0.5	0.10 ± 0.03
1146.9 ± 0.1	1.9 ± 0.2
1151.7 ± 0.5	0.08 ± 0.03
1156.4 ± 0.5	0.04 ± 0.02
1165.3 ± 0.5	0.08 ± 0.03
1180.5 ± 0.5	0.30 ± 0.08
1183.3 ± 0.1	1.7 ± 0.2
1194.1 ± 0.5	0.13 ± 0.04
1207.4 ± 0.2	0.59 ± 0.12
1219.7 ± 0.5	0.08 ± 0.03
1222.0 ± 0.5	0.09 ± 0.03
1230.0 ± 0.5	0.03 ± 0.02
1236.6 ± 0.2	0.39 ± 0.09
1267.1 ± 0.5	0.09 ± 0.03
1276.5 ± 0.5	0.07 ± 0.03
1309.8 ± 0.2	0.46 ± 0.09
1328.5 ± 0.2	1.22 ± 0.15
1344.6 ± 0.1	3.3 ± 0.3
1353.6 ± 0.2	0.47 ± 0.09
1362.6 ± 0.2	0.55 ± 0.10
1409.0 ± 0.6	0.10 ± 0.03
1454.3 ± 0.3	0.25 ± 0.07
1460.5 ± 0.2	1.12 ± 0.12
1493.2 ± 0.6	0.07 ± 0.03
1503.0 ± 0.3	0.13 ± 0.04
1512.0 ± 0.5	0.04 ± 0.02
1521.9 ± 0.3	0.11 ± 0.03
1536.1 ± 0.3	0.08 ± 0.03
1543.1 ± 0.2	0.71 ± 0.13
1547.3 ± 0.4	0.08 ± 0.03
1551.7 ± 0.5	0.02 ± 0.01
1560.7 ± 0.2	0.87 ± 0.10
1565.0 ± 0.6	0.03 ± 0.02
1621.5 ± 0.1	4.6 ± 0.5
1635.3 ± 0.3	0.18 ± 0.4
1650.4 ± 0.1	3.7 ± 0.5
1664.5 ± 0.5	0.08 ± 0.03
1677.8 ± 0.1	0.42 ± 0.08
1748.1 ± 0.5	0.02 ± 0.01

Gamma energies (KeV)	Gamma intensities
1776.9 ± 0.3	0.04 ± 0.02
1940.0 ± 0.5	0.05 ± 0.03
1974.2 ± 0.5	0.03 ± 0.02
2173.2 ± 0.3	0.22 ± 0.06
$_{63}Eu^{149}$	
22.5	760 ± 40†
72.9	0.08 ± 0.04
178.4 ± 0.4	0.17 ± 0.08
207.9 ± 0.4	0.06 ± 0.03
251.0 ± 1	0.14 ± 0.07
254.5 ± 0.3	5.9 ± 0.6
277.0 ± 0.3	33 ± 2
281.1 ± 0.5	0.19 ± 0.07
327.5 ± 0.3	39 ± 2
350.0 ± 0.3	3.4 ± 0.3
505.9 ± 0.3	5.5 ± 0.5
528.5 ± 0.3	5.3 ± 0.5
535.9 ± 0.3	0.46 ± 0.1
558.4 ± 0.3	0.62 ± 0.1
$_{63}Eu^{150m}$	
209.27 ± 0.35	0.55 ± 0.08†
285.23 ± 0.35	
334.04 ± 0.20	100 ± 8.0
372.96 ± 0.35	
406.47 ± 0.20	70.9 ± 5.7
424.96 ± 0.35	0.2 ± 0.04
438.67 ± 0.30	
446.22 ± 0.35	
620.48 ± 0.35	0.80 ± 0.12
632.57 ± 0.25	
712.13 ± 0.25	3.34 ± 0.30
747.12 ± 0.30	
831.77 ± 0.30	5.0 ± 0.4
860.2 ± 0.6	0.23 ± 0.06
893.6 ± 0.4	
921.22 ± 0.30	5.6 ± 0.4
1046.2 ± 0.8	0.21 ± 0.05
1057.3 ± 0.8	
1165.63 ± 0.30	6.45 ± 0.58
1193.89 ± 0.7	0.46 ± 0.1
1223.0 ± 0.3	5.0 ± 0.4
1268.9 ± 0.5	
1452.0 ± 0.5	0.37 ± 0.13
1462.9 ± 0.4	
1533.3 ± 1.0	
1592.5 ± 1.0	
1629.8 ± 0.7	1.45 ± 0.19
1963.90 ± 0.30	2.91 ± 0.30
1998.3 ± 0.7	
$_{63}Eu^{150}$	
61.278 ± 0.02	
78.830 ± 0.02	
205.101 ± 0.09	
284.760 ± 0.13	
298.031 ± 0.03	
305.438 ± 0.15	
310.826 ± 0.23	
315.132 ± 0.11	
333.965 ± 0.02	
340.369 ± 0.04	
345.968 ± 0.13	
382.063 ± 0.05	

GAMMA ENERGIES AND INTENSITIES OF RADIONUCLIDES (*Continued*)

Gamma energies (KeV)	Gamma intensities
402.252 ± 0.07	
439.402 ± 0.01	
448.744 ± 0.03	
461.788 ± 0.02	
464.497 ± 0.08	
474.544 ± 0.14	
496.298 ± 0.23	
505.480 ± 0.02	
510.078 ± 0.10	
515.776 ± 0.03	
520.042 ± 0.02	
542.989 ± 0.10	
571.244 ± 0.02	
584.253 ± 0.07	
591.026 ± 0.19	
596.334 ± 0.09	
607.237 ± 0.12	
612.446 ± 0.10	
614.908 ± 0.09	
667.031 ± 0.06	
675.804 ± 0.03	
712.139 ± 0.02	
731.212 ± 0.03	
737.405 ± 0.02	
741.475 ± 0.02	
748.062 ± 0.02	
750.890 ± 0.03	
756.473 ± 0.11	
759.336 ± 0.13	
828.486 ± 0.04	
830.860 ± 0.05	
836.574 ± 0.09	
859.885 ± 0.04	
869.250 ± 0.03	
899.079 ± 0.02	
910.989 ± 0.16	
923.371 ± 0.06	
1045.975 ± 0.06	
1049.065 ± 0.02	
1071.251 ± 0.10	
1083.081 ± 0.10	
1122.762 ± 0.05	
1165.546 ± 0.17	
1170.605 ± 0.02	
1193.801 ± 0.02	
1197.125 ± 0.02	
1212.956 ± 0.12	
1246.984 ± 0.01	
1251.077 ± 0.11	
1261.831 ± 0.04	
1274.1 ± 0.4	
1308.755 ± 0.03	
1322.037 ± 0.07	
1334.064 ± 0.04	
1343.877 ± 0.02	
1350.386 ± 0.08	
1379.064 ± 0.20	
1383.431 ± 0.30	
1390.0 ± 0.4	
1401.509 ± 0.31	
1413.507 ± 0.39	
1460.913 ± 0.15	
1465.365 ± 0.81	
1476.686 ± 0.59	
1485.557 ± 0.03	
1592.893 ± 0.20	
1636.623 ± 0.07	
1677.832 ± 0.35	
1690.696 ± 0.10	
1783.269 ± 0.10	
$_{63}Eu^{151m}$	
196	
$_{63}Eu^{152m_2}$	
18.3	
39.7	
89.8	
$_{63}Eu^{152m_1}$	
Ann. Rad.	
73.2 ± 0.3	†
76.8 ± 0.3	
85.15 ± 0.10	0.13
90.14 ± 0.10	0.23
121.92 ± 0.10	5.0
244.3 ± 0.3	
270.97 ± 0.15	0.07
344.08 ± 0.15	1.7
562.66 ± 0.15	0.3
688.9 ± 0.15	
841.39 ± 0.15	11.1
869.72 ± 0.15	0.09
963.10 ± 0.15	9.3
970.03 ± 0.15	0.5
995.03 ± 0.15	0.07
1314.33 ± 0.15	0.8
1388.79 ± 0.15	0.6
1411.26 ± 0.15	0.04
$_{63}Eu^{152}$	
Ann. Rad.	
121.83 ± 0.10	0.3†
147.79 ± 0.10	0.2
161.41 ± 0.10	
166.89 ± 0.15	0.1
167.9 ± 0.3	0.2
244.54 ± 0.15	7.6
251.40 ± 0.15	0.14
271.01 ± 0.15	0.09
275.07 ± 0.15	0.04
284.42 ± 0.15	0.08
295.84 ± 0.15	0.44
315.37 ± 0.15	0.05
324.81 ± 0.15	0.07
329.23 ± 0.15	0.13
344.19 ± 0.15	28.5
366.25 ± 0.15	0.22
367.71 ± 0.15	1.0
385.91 ± 0.15	0.31
411.06 ± 0.15	2.8
415.94 ± 0.15	0.13
443.96 ± 0.15	3.6
488.71 ± 0.15	0.44
493.51 ± 0.15	0.05
503.41 ± 0.15	0.13
563.96 ± 0.15	0.47
566.25 ± 0.15	0.15
586.17 ± 0.15	0.41
674.78 ± 0.15	0.14
678.62 ± 0.15	0.61
688.88 ± 0.15	0.90
719.11 ± 0.15	0.40
779.08 ± 0.15	14.5
810.54 ± 0.15	0.32
841.44 ± 0.15	0.20
867.53 ± 0.15	4.4
919.40 ± 0.15	0.57
926.98 ± 0.15	0.49
964.21 ± 0.15	16.2
1005.15 ± 0.15	1.0
1086.00 ± 0.15	11.8
1089.91 ± 0.15	1.9
1112.20 ± 0.15	15.4
1213.02 ± 0.15	1.7
1234.0 ± 0.3	0.25
1249.95 ± 0.15	0.3
1292.62 ± 0.15	0.2
1299.18 ± 0.15	1.9
1408.11 ± 0.15	22.9
1447.93 ± 0.15	0.14
1457.76 ± 0.15	0.6
1528.94 ± 0.15	0.7
$_{63}Eu^{154}$	
59.75 ± 0.10	4.0%
122.91 ± 0.10	40.0%
145.60 ± 0.10	0.2%
188.04 ± 0.15	0.3%
238.34 ± 0.15	0.1%
247.73 ± 0.15	6.7%
444.22 ± 0.15	0.7%
478.23 ± 0.15	0.3%
581.91 ± 0.15	1.0%
591.70 ± 0.15	5.4%
692.38 ± 0.15	1.9%
723.29 ± 0.15	21%
756.78 ± 0.15	4.6%
765.74 ± 0.15	1.0%
815.71 ± 0.15	0.6%
845.57 ± 0.15	0.5%
873.24 ± 0.15	12.3%
904.21 ± 0.15	1.0%
996.36 ± 0.15	11.3%
1004.84 ± 0.15	20%
1274.53 ± 0.15	37%
1396.00 ± 0.15	1.9%
1460.97 ± 0.15	1.2%
1494.14 ± 0.15	0.6%
1593.07 ± 0.15	0.6%
1596.61 ± 0.15	1.9%
$_{63}Eu^{155}$	
43.37 ± 0.05	†
58.12 ± 0.10	0.06
60.07 ± 0.05	2.8
86.60 ± 0.05	72.4
105.36 ± 0.05	48.2
$_{63}Eu^{156}$	
88.97 ± 0.05	45%
190.37 ± 0.10	0.5%
199.27 ± 0.07	0.6%
434.33 ± 0.08	0.23%
472.57 ± 0.08	0.17%
490.35 ± 0.08	0.17%
599.43 ± 0.07	2.1%
646.23 ± 0.07	6.3%
709.85 ± 0.07	0.9%
723.44 ± 0.07	5.4%
768.27 ± 0.20	0.09%
797.67 ± 0.10	0.09%
811.73 ± 0.07	9.7%
820.32 ± 0.07	0.16%
836.37 ± 0.07	0.11%
841.10 ± 0.31	0.22%
858.19 ± 0.13	0.27%
866.98 ± 0.07	1.5%
944.06 ± 0.07	2.4%
960.41 ± 0.07	1.6%
969.81 ± 0.07	0.38%
1111.81 ± 0.10	0.55%
1027.53 ± 0.07	0.38%
1040.36 ± 0.10	0.54%
1065.08 ± 0.07	5.3%
1075.92 ± 0.07	2.4%
1079.19 ± 0.07	4.7%
1129.31 ± 0.18	0.17%
1140.51 ± 0.07	0.31%
1153.72 ± 0.07	12.6%
1168.94 ± 0.15	0.31%
1230.66 ± 0.07	8.7%
1242.47 ± 0.07	7.1%
1258.30 ± 0.6	0.21%
1277.50 ± 0.07	3.2%
1366.44 ± 0.07	1.8%
1455.02 ± 0.20	0.14%
1587.06 ± 0.20	0.20%
1675.92 ± 0.50	0.17%
1681.13 ± 0.10	0.42%
1857.24 ± 0.10	0.32%
1876.91 ± 0.10	1.5%
1937.66 ± 0.10	2.0%
1945.92 ± 0.10	0.12%
1965.91 ± 0.10	4.4%
2026.68 ± 0.10	3.8%
2031.57 ± 0.10	0.44%
2097.77 ± 0.10	5.3%
2110.64 ± 0.30	0.16%
2116.46 ± 0.15	0.17%
2181.05 ± 0.10	4.1%
2186.82 ± 0.10	6.1%
2205.57 ± 0.10	1.7%
2211.85 ± 0.10	0.20%
2269.85 ± 0.10	2.4%
$_{63}Eu^{157}$	
40.6	
51.8	
54.5	
64.0	85†
77.0	
131.4	
276.6	
320.5	
330.7	10
361.8	
372.8	
413.1	100
477.3	20
527.0	
623.3	20
687.5	10
729.4	5
$_{63}Eu^{158}$	
80	100†
180	
520	25
610	8

GAMMA ENERGIES AND INTENSITIES OF RADIONUCLIDES (*Continued*)

Gamma energies (KeV)	Gamma intensities
750	8
900	26
950	49
980	20
1110	11
1190	16
1260	5
1330	4
1960	3
2180	2
$_{63}Eu^{159}$	
54	
75	
90	90†
130	
145	68
165	16
220	26
670	100
740	50
880	64
970	10
1050	25
1100	25
1270	8
1350	8
1370	25
1440	25
1500	
1600	
$_{64}Gd^{145}$	
80	9†
1030	10
1750	100
$_{64}Gd^{146}$	
22.9	
77.7	30†
114.7	
115.5	
154.8	48
198.8	
329	
$_{64}Gd^{147}$	
106.6	
111	
166.3	
194.0	
214.8	
229.4	150†
240.5	
260.8	
297.5	
310.0	
328.6	
370.0	
396.1	
485	
559	
609	
617	
625	
632	75†
780	60
940	80
1350	6
$_{64}Gd^{149}$	
149.4 ± 0.5	100†
214.2 ± 0.7	
252.2 ± 0.7	
260.6 ± 0.6	
272.2 ± 0.6	
299.3 ± 0.5	69
346.6 ± 0.5	42
404.2 ± 0.7	
459.8 ± 0.7	<4
478.7 ± 0.7	
496.3 ± 0.6	4
516.5 ± 0.6	10
534.3 ± 0.6	9
645.2 ± 0.6	4
663.3 ± 0.8	
666.2 ± 0.6	
748.4 ± 0.5	19
762.9 ± 0.7	
780.8 ± 0.7	
788.8 ± 0.5	16
812.6 ± 0.6	
876.0 ± 0.6	1.3
933.0 ± 0.6	2.1
938.5 ± 0.5	6
947.8 ± 0.5	
$_{64}Gd^{151}$	
21.6	4.4†
153.7	6
175	2.7
244	4.5
308	0.8
350	0.3
$_{64}Gd^{153}$	
68.23	0.5†
69.68	11
75.43	0.4
83.37	1.0
89.48	0.3
97.430 ± 0.005	132
103.178 ± 0.003	100
172.85	
$_{64}Gd^{159}$	
58	1.9†
80	0.43
225	1.9
290	0.35
306	0.8
334	0.42
348	2.4
363	100
536	
558	0.21
579	0.6
617	0.17
$_{64}Gd^{161}$	
56.6	†
77.6	0.3
102.4	17
105.5	1.0
133.7	
165.3	7
180.6	3.5
258.8	3
273	2
283.8	12
315.3	37
361.0	100
482	3.2
531	2.6
$_{65}Tb^{147}$	
Ann. Rad.	
305	
$_{65}Tb^{148}$	
Ann. Rad.	
780	
1120	
$_{65}Tb^{149m}$	
Ann. Rad.	
$_{65}Tb^{149}$	
165.5	
187.3	
352	
388	
464	
652	
818	
856	
1206	
$_{65}Tb^{150}$	
637	100†
930	35
$_{63}Tb^{151}$	
108.2	
180.1	50†
192.0	
251.8	100
287.2	100
318.5	
380.1	
384.6	
395.2	
416.0	
426.4	
443.8	
479.2	80†
499.8	
512	
588	
604	100†
704	
732	
794	
799	
805	
906	
939	
979	
1010	
1025	
1053	
1090	
1097	
1171	
1250	
1308	
1312	
$_{65}Tb^{152m}$	
Ann. Rad.	
140	25†
230	30
$_{65}Tb^{152}$	
271.2	13†
344.4	100
351.5	
411.2	
432.5	
496.7	
527.1	
586.7	
615.6	
623.1	
676	
679	
704	2.5
753	
766	
779	
929	
931	
975	
991	
1048	
1110	
1138	
1187	
1206	
1263	
1300	
1303	
1319	
1328	
1348	
1352	
1361	
1412	
1519	
1588	
1598	
1670	
1783	
1791	
1910	
1940	
2040	
2100	
2190	
2260	

GAMMA ENERGIES AND INTENSITIES OF RADIONUCLIDES (*Continued*)

Gamma energies (KeV)	Gamma intensities
2370	
2400	
2530	
2580	
2600	
2670	
2710	
2800	
2920	
$_{65}Tb^{153}$	
19.4	
41.6	
51.8	
68.2	≈20
82.8	190
87.6	50
88.4	45
102.3	190
109.7	180
129.2	≈13
139.8	<10
141.9	35
170.5	220
174.4	<50
183.5	25
185.9	<5
193.6	≈4
195.2	28
208.1	<50
212.1	1000
249.6	72
262	15
266	<2
271	<2
275	7
278	<2
299	<2
303	28
315	30
320	≈7
327	≈5
333	<2
340	5
349	<1
355	5
362	4
393	≈3
400	≈4
407	≈3
419	≈3
436	25
442	6
448	4
455	13
467	15
496	10
507	16
526	6
531	7
549	6
566	<2
571	<2
599	<2
629	10
636	<2
653	5
665	10
689	8
740	7
762	<2
816	7
836	29
847	6
853	<5
858	<5
904	35
918	<5
946	32
974	10
993	37
1102	13
1203	2.4
1276	1.9
1364	0.9
$_{65}Tb^{154}$	
123.07 ± 0.2	
225.7 ± 0.3	
231.7 ± 0.3	
247.7 ± 0.2	
346.6 ± 0.2	
381.9 ± 0.4	
415.9 ± 0.3	
444.6 ± 0.4	
517.9 ± 0.3	
540.10 ± 0.25	
557.2 ± 0.2	
598.0 ± 0.4	
642.4 ± 0.6	
649.5 ± 0.3	
676.4 ± 0.4	
692.3 ± 0.3	
704.6 ± 0.5	
714.5 ± 0.5	
722.1 ± 0.5	
756.9 ± 0.3	
815.5 ± 1.0	
873.2 ± 0.2	
892.7 ± 0.3	
925.3 ± 1.0	
996.4 ± 0.2	
1004.8 ± 0.2	
1060.0 ± 1.0	
1148.9 ± 1.0	
1152.4 ± 1.0	
1258.4 ± 1.5	
1274.9 ± 1.0	
1289.0 ± 1.5	
1292.0 ± 1.5	
1410.0 ± 2.0	
1491.0 ± 3.0	
1967.0 ± 2.0	
2186.0 ± 1.5	
$_{65}Tb^{154}$	
123.1	
141.4	
225.7	
247.7	
265.9	
346.6	
426.9	
444.6	
557.2	
602.6	
680.8	
692.3	
815.5	
873.2	
996.4	
1124	
1274.9	
1292.0	
1460	
1994	
2062	
2116	
2186	
$_{65}Tb^{155}$	
45.35 ± 0.10	
58.03 ± 0.10	
60.03 ± 0.10	7†
72.83 ± 0.10	
74.99 ± 0.10	
86.53 ± 0.07	100
99.02 ± 0.10	
101.14 ± 0.10	
105.29 ± 0.07	69
120.53 ± 0.10	
148.63 ± 0.08	8
161.26 ± 0.08 }	23
163.28 ± 0.07 }	
175.20 ± 0.10	
203.05 ± 0.10	
216.04 ± 0.10	
220.69 ± 0.10	
226.94 ± 0.10	
239.53 ± 0.10	
262.30 ± 0.10	
268.56 ± 0.10	
281.03 ± 0.10	
286.93 ± 0.10	
321.74 ± 0.10	
367.32 ± 0.10	10
370.70 ± 0.10	
402.05 ± 0.10	
559.09 ± 0.10	
180.06 ± 0.10	
181.59 ± 0.10	
206.60 ± 0.10	
208.22 ± 0.10	
200.48 ± 0.10	
$_{65}Tb^{156m}$	
88.4 (IT)	
$_{65}Tb^{156}$	
88.97 ± 0.05	
111.9	
115.6	
155.2	
170.8	
199.27 ± 0.07	
262.5	
266.9	
296.3	
356.4	
374.0	
381.4	
422.4	
535.2	
687.5	
782	
844	
867	
927	
944.06 ± 0.07	
951	
963	
1012	
1040.36 ± 0.1	
1043	
1068	
1070	
1157	
1162	
1225	
1338	
1425	
1649	
1849	
1951	
2014	
2090	
2105	
2141	
2268	
2281	
2310	
$_{65}Tb^{158m}$	
0.111	100%
$_{65}Tb^{158}$	
79.5	12%
99	4%
182	10%
218	1%
782	10%
928	0.2%
946	47%
964	22%
1110	2.2%
1190	1.8%
$_{65}Tb^{160}$	
86.788 ± 0.002	
93.919 ± 0.006	
176.490 ± 0.030	
197.035 ± 0.008	5.1
215.646 ± 0.008	3.97
230.628 ± 0.013	0.09
237.638 ± 0.086	
246.489 ± 0.016	0.03
298.582 ± 0.010	26.3
309.557 ± 0.018	0.87
337.324 ± 0.030	0.34
349.935 ± 0.110	
379.449 ± 0.090	
392.514 ± 0.026	5.1
486.075 ± 0.080	3.97
682.349 ± 0.110	0.09
765.194 ± 0.110	

GAMMA ENERGIES AND INTENSITIES OF RADIONUCLIDES (*Continued*)

Gamma energies (KeV)	Gamma intensities
871.9 ± 1.5	0.45
879.333 ± 0.070	31.8
962.085 ± 0.220	12.6
966.099 ± 0.120	27.6
1003.26 ± 0.30	1.14
1068.9 ± 1.5	0.13
1103.22 ± 0.25	0.7
1115.29 ± 0.19	1.70
1178.12 ± 0.19	16.6
1200.21 ± 0.37	2.67
1251.3 ± 0.5	0.11
1271.87 ± 0.25	8.28
1311.90 ± 0.30	3.16
1358.6 ± 0.2	0.02
1460.9 ± 0.2	0.05
1592.7 ± 0.2	0.02
$_{65}Tb^{161}$	
25.55 ± 0.2	
43.8	
57.38 ± 0.18	16† 21%
74.68 ± 0.15	100
77.41 ± 0.15	0.5
86.81 ± 0.15	2.0
103.06 ± 0.10	0.12
105.92 ± 0.10	0.9
131.61 ± 0.10	<1
298.55 ± 0.10	
340	
410	
470	
540	
$_{65}Tb^{162}$	
40	17†
80.82 ± 0.3	8
185.18 ± 0.25	26
260.01 ± 0.15	100
795.80 ± 0.5	0.8
807.53 ± 0.2	44
882.27 ± 0.25	18
888.16 ± 0.2	36
$_{65}Tb^{163m}$	
180	
$_{65}Tb^{163}$	
25	80†
120	
160	5
235	30
330	100
360	1
440	1
510	35
$_{66}Dy^{150}$	
390	
Ann. Rad.	

Gamma energies (KeV)	Gamma intensities
$_{66}Dy^{151}$	
145	
$_{66}Dy^{152}$	
Ann. Rad.	
275.2	
$_{66}Dy^{153}$	
80.8	
82.5	
93	
99.7	
147.8	
149.0	
174.1	
218.9	
240.7	
244.5	
254.5	
274.4	
303.8	
389.8	
415.7	
512.4	
544.7	
$_{66}Dy^{155}$	
Ann. Rad.	
65.43	
90.38	2.3†
115.4	
155.8	
161.4	
181	
184.6	
205.7	
227	100
271.4	
389	
433	
451	
461	
484	
498	
508	
550	
587	
642	
654	
664	
723	
745	
761	
813	
837	
844	
892	
905	
929	
1000	9†
1091	8
1155	
1164	
1250	6
1390	5
1450	6

Gamma energies (KeV)	Gamma intensities
1660	4
$_{66}Dy^{157}$	
60.8	
83.01	0.5†
143.8	0.0016
182.2	2
265.3	
326.2	100
554	
577	
597	
636	
743	
769	
775	
985	
992	
$_{66}Dy^{159}$	
58.0	4%
80.0	
138	
211	4×10^{-5}%
290	1.2×10^{-4}%
348	9×10^{-4}%
$_{66}Dy^{165m}$	
108.16	100†
152	9
361.7	17
514	55
650	1
750	0.5
$_{66}Dy^{165}$	
94.70	370†
279.6	57
361.5	100
479	3
501	0.6
514	2.9
545	17
566	14
575	8
621	10
633	64
695	2
716	64†
995	6
1056	3
1080	10
$_{66}Dy^{166}$	
28.23	†
54.23	
82.45	100
288	0.7
344	2
371.6	8
425.8	10

Gamma energies (KeV)	Gamma intensities
$_{67}Ho^{151}$	
Ann. Rad.	
$_{67}Ho^{151}$	
Ann. Rad.	
$_{67}Ho^{152}$	
Ann. Rad.	
$_{67}Ho^{152}$	
Ann. Rad.	
$_{67}Ho^{153}$	
Ann. Rad.	
355	
$_{67}Ho^{155}$	
Ann. Rad.	
92	
117	
138	
209	
243	
326	
$_{67}Ho^{156}$	
138	100†
266.4	99
366.7	23
685.2	
890	
1200	
1410	
$_{67}Ho^{157}$	
Ann. Rad.	
87	
152	
190	
277	
710	
860	
900	
1200	
$_{67}Ho^{158m}$	
Ann. Rad.	
67.3(IT)	
98.9	
193.7	
218.3	
309	
320	
356	
463	
471	
647	
848	

GAMMA ENERGIES AND INTENSITIES OF RADIONUCLIDES (*Continued*)

Gamma energies (KeV)	Gamma intensities
851	
946	
948	
1450	
1600	
1800	
2060	
2200	
2620	
$_{67}Ho^{158}$	
Ann. Rad. see Ho^{158m}	
$_{67}Ho^{159m}$	
206	
$_{67}Ho^{159}$	
69.51 ± 0.30	
79.79 ± 0.30	
100.55 ± 0.30	
120.94 ± 0.30	
131.86 ± 0.30	
152.23 ± 0.40	
155.88 ± 0.40	
159.43 ± 0.40	
173.02 ± 0.40	
177.51 ± 0.40	
186.09 ± 0.40	
252.96 ± 0.30	
258.74 ± 0.40	
265.91 ± 0.40	
296.59 ± 0.40	
309.51 ± 0.30	
538.20 ± 0.40	
706.54 ± 0.40	
807.05 ± 0.40	
838.36 ± 0.40	
$_{67}Ho^{160m}$	
Ann. Rad.	
60.1 (IT)	Equilibrium with Ho^{160}
86.91 ± 0.30	
107.7 ± 0.7	
196.76 ± 0.20	
297.79 ± 0.20	
$_{67}Ho^{160}$	
405.5 ± 1.0	
645.20 ± 0.20	
728.00 ± 0.20	
753.0 ± 0.7	
765.2 ± 0.7	
826.1 ± 1.0	
871.88 ± 0.40	
879.28 ± 0.20	
962.32 ± 0.25	
966.24 ± 0.25	
1004.4 ± 0.7	
1068.8 ± 0.7	
1131.3 ± 1.0	

Gamma energies (KeV)	Gamma intensities
1155.4 ± 1.0	
1198.9 ± 1.0	
1271.7 ± 1.0	
1285.3 ± 1.0	
1522.6 ± 1.0	
1594.7 ± 1.0	
1613.2 ± 1.0	
1654.5 ± 1.0	
1719	
1784	
1807	
1872	
1927	
1959	
2008	
2069	
2190	
2550	
2619	
2638	
2675	
2728	
2763	
$_{67}Ho^{161m}$	
211.1	
$_{67}Ho^{161}$	
25.7	100†
60.03 ± 0.30	4
75.03 ± 0.40	
77.79 ± 0.20	10
98.1 ± 0.7	1.8
103.00 ± 0.20	1.2
137.9 ± 1.0	
156.90 ± 0.30	1.7
175.08 ± 0.30	1.6
210.8	0.6
234.5	
339.7	
416.0	
760	0.3
$_{67}Ho^{162m}$	
38.63	(IT)
58.88 ± 0.40	
81.15 ± 0.20	
90.28 ± 0.40	
95.46 ± 0.40	
184.70 ± 0.20	
275.2 ± 0.7	
282.48 ± 0.20	
302.8 ± 0.7	
333.8 ± 0.7	
392.6 ± 1.0	
424.3 ± 1.0	
936.72 ± 0.20	
1220.0 ± 0.40	
1319.9 ± 0.8	
1374.4 ± 0.8	
$_{67}Ho^{162}$	
Ann. Rad.	~50%
81.15 ± 0.2	

Gamma energies (KeV)	Gamma intensities
$_{67}Ho^{163m}$	
305	~80%
$_{67}Ho^{164}$	
37.3	3.6†
73.4 ± 0.2	3.3
91.4 ± 0.2	3.5
$_{67}Ho^{166}$	
80.573 ± 0.015	%
184.407 ± 0.015	0.002 ± 0.0005
674.991 ± 0.040	0.020 ± 0.002
705.342 ± 0.040	0.015 ± 0.002
785.949 ± 0.040	0.013 ± 0.002
1379.432 ± 0.065	0.93 ± 0.05
1581.89 ± 0.08	0.181 ± 0.009
1662.44 ± 0.08	0.116 ± 0.006
1749.94 ± 0.10	0.025 ± 0.002
1830.57 ± 0.15	0.008 ± 0.001
$_{67}Ho^{166m}$	
80.573 ± 0.015	12.5 ± 0.6%
94.653 ± 0.030	0.14 ± 0.01%
119.038 ± 0.030	0.18 ± 0.02%
121.161 ± 0.030	0.26 ± 0.03%
135.238 ± 0.035	0.10 ± 0.01%
140.618 ± 0.040	0.04 ± 0.01%
160.064 ± 0.045	0.10 ± 0.01%
161.748 ± 0.045	0.11 ± 0.01%
184.407 ± 0.015	73.2 ± 3.7%
190.711 ± 0.025	0.22 ± 0.02%
214.763 ± 0.045	0.55 ± 0.07%
215.875 ± 0.030	2.6 ± 0.2%
231.282 ± 0.040	0.24 ± 0.02%
233.28 ± 0.09	0.16 ± 0.02%
239.99 ± 0.08	0.05 ± 0.01%
259.716 ± 0.020	1.10 ± 0.06%
280.456 ± 0.020	29.8 ± 1.5%
300.744 ± 0.020	3.75 ± 0.19%
329.12 ± 0.08	0.16 ± 0.02%
339.78 ± 0.08	0.17 ± 0.02%
365.739 ± 0.025	2.52 ± 0.13%
410.941 ± 0.025	11.6 ± 0.6%
451.524 ± 0.025	3.06 ± 0.15%
464.825 ± 0.040	1.52 ± 0.08%
529.813 ± 0.030	10.2 ± 0.5%
570.992 ± 0.030	5.75 ± 0.29%
594.481 ± 0.040	0.70 ± 0.04%
611.522 ± 0.065	1.39 ± 0.08%
639.770 ± 0.060	
644.45 ± 0.10	0.18 ± 0.02%
670.509 ± 0.040	5.77 ± 0.29%
691.211 ± 0.050	1.53 ± 0.08%
711.693 ± 0.040	58.7 ± 2.9%
736.67 ± 0.08	0.40 ± 0.03%
752.265 ± 0.040	13.1 ± 0.7%
778.817 ± 0.040	3.30 ± 0.17%
810.309 ± 0.040	62.7 ± 3.1%
830.560 ± 0.040	10.6 ± 0.6%
875.64 ± 0.05	0.79 ± 0.06%
1010.25 ± 0.10	0.10 ± 0.01%
1120.31 ± 0.07	0.23 ± 0.02%
1146.82 ± 0.07	0.22 ± 0.02%

GAMMA ENERGIES AND INTENSITIES OF RADIONUCLIDES (*Continued*)

Gamma energies (KeV)	Gamma intensities
1241.44 ± 0.06	1.00 ± 0.05%
1282.12 ± 0.07	0.23 ± 0.02%
1400.72 ± 0.08	0.55 ± 0.03%
1427.05 ± 0.08	0.59 ± 0.03%
$_{67}Ho^{167}$	
57.1	
73.7	
79.3	
83.4	
133.4	
136.9	
150.4	
207.78 ± 0.08	
237.7	
241.1	
260.1	
266.5	
271.7	
304.3	
320.7	
322.8	
331.0	
347.7	
386.6	
403.4	
459.3	
531.58 ± 0.1	
$_{67}Ho^{168}$	
79.80 ± 0.2	
184.2 ± 0.2	2†
556.9 ± 0.5	2
741.3 ± 0.2	100
821.1 ± 0.3	92
$_{67}Ho^{169}$	
80	
150	12†
680	10
760	30
840	15
920	5
$_{67}Ho^{170\sigma}$	
430	
$_{68}Er^{158}$	
Ann. Rad.	
67.2	
71.9	
315	
250	
387	
875	
906	
978	
$_{68}Er^{159}$	
205.9 (IT)	
370	
627	
840	
1200	
1400	
1800	
2600	
$_{68}Er^{161}$	
Ann. Rad.	
45.6	
59.5	
84.5	
94.3	
106.1	
112.6	
122.6	
138.7	
144.0	
146.7	
156.5	
172.0	
172.6	
190.6	
207.0	
215.9	
218.2	
244.5	
250.2	
252.5	
265.3	
283.3	
354.0	
372.7	
$_{68}Er^{163}$	
Ann. Rad.	
300	0.01%
430	0.06%
1100	0.04%
$_{68}Er^{167m}$	
207.8	50%
$_{68}Er^{169}$	
8.42	0.1%
$_{68}Er^{171}$	
5.1	
12.4	
86	
111.63	25%
116.69	3%
124	
166.4	0.6%
210.4	0.6%
236	0.5%
277.0	0.65%
295.6	28%
308.2	63%
371.2	0.4%
404	0.03%
418.9	0.09%
544	0.04%
557	0.08%
573	0.07%
606	0.1%
619	0.13%
670	0.3%
675	0.4%
732	0.2%
783	0.31%
796	0.80%
842	0.04%
869	0.06%
882	0.07%
906	0.9%
911	0.2%
962	0.11%
$_{68}Er^{172}$	
29.2	
59.6	
67.9	
128	
164	0.9%
203	0.9%
346	0.6%
384	2.4%
407	40%
446	2.5%
476	1.0%
535	0.3%
610	40%
$_{69}Tm^{161}$	
45.6	
59.5	
84.5	
94.3	
106.1	
112.6	
122.6	
138.7	
144	
146.7	
156.7	
172.0	
172.6	
190.6	
207.0	
215.9	
218.2	
244.5	
250.2	
252.5	
265.3	
283.3	
354.0	
327.7	
$_{69}Tm^{162}$	
Ann. Rad.	
102	20†
236	10
$_{69}Tm^{162}$	
Ann. Rad.	
102	

GAMMA ENERGIES AND INTENSITIES OF RADIONUCLIDES (*Continued*)

Gamma energies (KeV)	Gamma intensities
$_{69}Tm^{163}$	
Ann. Rad	
22.2	
60.2	
69.2	
80.5	
83.9	
85.1	
104.3	
145.3	
164.4	
165.4	
190.1	
239.7	
241.5	
249.2	
275.3	
299.9	
335.4	
345.6	
393.5	
404.2	
471.2	
505.1	
531.0	
549.9	
579.9	
613.0	
655.5	
671.1	
$_{69}Tm^{164}$	
Ann. Rac	
91	
211	
356	
361	
391	
773	
862	
907	
930	
$_{69}Tm^{165}$	
15.45	
27.75	
30.05	
35.2	
47.15	
53.2	
54.45	
59.15	
60.4	
70.55	
77.2	
82.25	
86.9	
88.2	
113.6	
141.4	
150.8	
156.0	
165.5	
175.7	†
181.5	
195.6	
197.2	

Gamma energies (KeV)	Gamma intensities
205.2	
209.9	
218.2 ± 0.2	100 ± 5
221.5	
223.9	
229	
233.2	
234.6	
242.4 ± 0.2	1040 ± 30
248.9	
249.7	
264.4	17 ± 2
275.6	
279.9 ± 0.3	20 ± 2
286.1	
292.3	
295.9	
296.9 ± 0.3	
306.9	4.8 ± 1.5
312.2	14 ± 2
321	
330.5	3.6 ± 1.8
336	
345.7 ± 0.3	
346.6	89 ± 4
356.3 ± 0.3	81 ± 3
363	
365.3	16 ± 3
377	w
384.1	
389.2 ± 0.3	80 ± 4
400.2	5.8 ± 1.7
414	w
420.7	10 ± 3
426.9	w
429.9	w
436	w
442.4	24 ± 3
448.1	57 ± 4
456.0	47 ± 16
459.9 ± 0.3	110 ± 9
471.8	9 ± 2
477.7	12 ± 2
487.0	31 ± 3
502	w
507	w
513	w
513.5	10 ± 3
526.8	32 ± 5
530.9	c
542.2	46 ± 4
558.3	19 ± 5
563.7	67 ± 5
570	w
573.8	14 ± 3
577.3	7 ± 4
589.4	57 ± 5
605.6	
608.1	
622.8	9 ± 3
630	w
663	w
664.6	12 ± 3
677.3	≈6
680.5	≈4
698.6	41 ± 3
734	w
747.4	8 ± 3
791.0	18 ± 4
806.4 ± 0.1	343 ± 12
821.2	w

Gamma energies (KeV)	Gamma intensities
826.9	
837.7	13 ± 2
855	3.9 ± 0.5
892.8	
907.8	
932.4	
952.8	
991	
1042	
1045.7	
1067	
1096	
1131.0	48 ± 2
1184.5	74 ± 4
1232	0.9 ± 0.3
1246.4	
1262	0.3 ± 0.1
1289	3.2 ± 0.5
1333	0.2 ± 0.1
1339	0.4 ± 0.1
1344	0.3 ± 0.1
1365	1.8 ± 0.4
1379.5	8.5 ± 1.0
1426.9	19.5 ± 1.6
$_{69}Tm^{166}$	
Ann. Rad.	
80.573 ± 0.015	
184.39 ± 0.05	12.6
214.76 ± 0.04	4.4
594.48 ± 0.04	4.5
598.8 ± 0.3	2.9
654.50 ± 0.10	0.4
672.28 ± 0.08	7.6
674.91 ± 0.08	2.7
691.28 ± 0.08	8.4
705.83 ± 0.15	12.8
712.83 ± 0.15	0.6
727.98 ± 0.15	0.5
757.84 ± 0.10	2.9
778.83 ± 0.08	23.4
785.91 ± 0.08	12.7
810.33 ± 0.08	1.34
875.63 ± 0.08	4.9
1030.37 ± 0.03	7.9
1078.86 ± 0.15	0.6
1084.78 ± 0.20	0.3
1119.15 ± 0.13	0.2
1152.32 ± 0.05	2.4
1161.91 ± 0.07	1.1
1176.68 ± 0.04	12.4
1216.08 ± 0.09	0.9
1235.41 ± 0.04	2.4
1248.80 ± 0.12	0.2
1263.44 ± 0.07	1.3
1273.53 ± 0.04	19.2
1300.74 ± 0.06	1.8
1313.31 ± 0.28	0.3
1347.12 ± 0.05	1.3
1374.21 ± 0.04	7.4
1432.18 ± 0.12	1.3
1447.84 ± 0.09	0.8
1505.04 ± 0.08	1.1
1622.48 ± 0.20	0.7
1652.80 ± 0.07	1.4
1868.0 ± 0.1	5.3
1907.63 ± 0.16	0.7
2008.41 ± 0.26	0.4
2052.2 ± 0.2	23.2
2079.5 ± 0.2	8.3

Gamma energies (KeV)	Gamma intensities
2092.11 ± 0.06	2.1
2192.29 ± 0.13	0.2
$_{69}Tm^{167}$	
57.1	
207.78 ± 0.08	100†
266.5	
322.8	
531.58 ± 0.1	4.18 ± 0.3
$_{69}Tm^{168}$	
79.80 ± 0.2	9.19
99.0 ± 0.2	0.6
99.3 ± 0.2	3.3
173.5 ± 0.3	0.054
184.2 ± 0.2	17.2
198.2 ± 0.2	52.1
272.9 ± 0.4	0.103
348.5 ± 0.3	0.33
422.2 ± 0.4	0.32
447.4 ± 0.2	24.5
546.7 ± 0.3	2.84
556.9 ± 0.5	0.206
631.6 ± 0.5	8.98
645.7 ± 0.5	1.52
673.4 ± 0.3	0.223
720.3 ± 0.2	12.2
730.6 ± 0.2	5.14
741.3 ± 0.2	12.6
748.2 ± 0.3	0.451
815.9 ± 0.2	50.0
821.1 ± 0.3	11.7
829.9 ± 0.2	6.95
853.5 ± 0.7	0.05
914.9 ± 0.3	3.0
928.9 ± 0.5	0.05
1014.1 ± 0.5	0.06
1167.5 ± 0.5	0.085
1277.3 ± 0.3	1.70
1322.6 ± 0.7	0.03
1351.2 ± 0.7	0.10
1358.0 ± 1.0	0.01
1461.3 ± 0.4	0.42
1592.5 ± 0.4	
$_{69}Tm^{170}$	
84.262 ± 0.005	3.4%
$_{69}Tm^{171}$	
66.7	~2%
$_{69}Tm^{172}$	
78.70 ± 0.08	58.0 ± 4
90.6 ± 0.1	0.32 ± 0.06
112.8 ± 0.2[b]	~0.1
131.8 ± 0.2[a]	~0.1
133.6 ± 0.2[a]	~0.1
142.50 ± 0.10	1.8 ± 0.2
145.0 ± 0.2	~0.2
181.55 ± 0.08	44.0 ± 4
203.6 ± 0.3	~0.1
238.5 ± 0.2[a]	~0.2

Gamma energies (KeV)	Gamma intensities
351.8 ± 0.2[a]	~0.2
374.1 ± 0.3	~0.1
399.63 ± 0.08	2.5 ± 0.5
423.0 ± 0.2	~0.5
429.0 ± 1.0	~0.1
436.07 ± 0.08	4.2 ± 0.4
490.31 ± 0.08	6.7 ± 0.6
528.23 ± 0.10	2.0 ± 0.3
542.0 ± 0.4[b]	~0.1
565.5 ± 0.15	0.7 ± 0.2
858.7 ± 0.3	2.0 ± 0.6
912.10 ± 0.08	25.0 ± 3
964.20 ± 0.10	5.3 ± 0.6
1003.0 ± 0.3	~1.0
1040.1 ± 0.3	2.6 ± 0.6
1076.3 ± 0.1[a]	13.0 ± 2
1093.68 ± 0.08	≡104.0
1116.0 ± 0.3	~1
1119.2 ± 0.3	3.8 ± 1.0
1154.9 ± 0.2[a]	3.3 ± 1.0
1184.0 ± 0.5	~1
1205.7 ± 0.2	3.4 ± 0.8
1289.0 ± 0.15	8.4 ± 1.0
1348.4 ± 0.15	2.6 ± 0.6
1387.27 ± 0.08	95.0 ± 6
1398.17 ± 0.08	11.0 ± 2
1402.6 ± 0.15	2.5 ± 0.6
1440.8 ± 0.4	~0.5
1461.0 ± 0.1[a]	4.0 ± 0.5
1466.07 ± 0.08	76.0 ± 6
1470.47 ± 0.08	27.0 ± 3
1476.8 ± 0.10	4.5 ± 0.6
1491.0 ± 1.0	~0.5
1529.87 ± 0.08	84.0 ± 6
1545.0 ± 1.00	~0.5
1584.2 ± 0.10	10.0 ± 1.5
1592.7 ± 0.10[a]	1.5 ± 0.3
1608.62 ± 0.08	69.0 ± 5
1622.1 ± 0.10	1.0 ± 0.2
$_{69}Tm^{173}$	
66	1.1†
399	89
465	7.6
$_{69}Tm^{174m}$	
76	9†
176	75
273	95
366	104
500	17
630	5
990	100
1270	8
$_{69}Tm^{175}$	
513	
$_{70}Yb^{165}$	
Ann. Rad.	
$_{70}Yb^{166}$	
820	

Gamma energies (KeV)	Gamma intensities
$_{70}Yb^{167}$	
Ann. Rad.	
25.6	
37.1	
63.0	
106	4†
113.3	10
116.5	
131.8	
150.3	
168.9	
176.1	2.3
920	
1040	
1250	
1440	
1520	
$_{70}Yb^{169}$	
8.42	
20.74	†
63.12	164
93.6	6.6
109.77	51
118.17	4.8
130.51	32
177.18	58
197.97	100
261.05	4.6
307.68	31.0
336.0	0.02
$_{70}Yb^{175m}$	
513	
$_{70}Yb^{175}$	
113.81	31†
137.65	2.2
144.85	5.9
251.3	3.8
282.6	62
396.1	100
$_{70}Yb^{177m}$	
104	
228	
$_{70}Yb^{177}$	
121.6	0.23†
138.4	~1
147.4	
150.8	
162.6	
790	0.2
940	1.1
1079	5
1120	0.7
1239	5†
$_{71}Lu^{167}$	
Ann. Rad.	

Gamma energies (KeV)	Gamma intensities
29.6	
56.2	
78.6	
100.1	
102.5	
123.1	
129.1	
151.8	
178.7	
181.9	
188.5	
198.9	
213.0	
222.5	
229.5	
235.6	
239.0	
258.2	
261.5	
278.1	
307.8	
317.2	
361.4	
371.8	
401.2	
601	
$_{71}Lu^{168}$	
Ann. Rad.	
87	7†
900	10
990	13
1440	
1810	
2100	
$_{71}Lu^{169m}$	
29.0	
$_{71}Lu^{169}$	
Ann. Rad.	
24.2	
62.8	
70.9	
75.0	
87.4	
92.0	
104.4	
110.9	
133.5	
144.6	
157	
165.2	
191.4	
198.4	
258.7	
291.6	
369.6	
379.2	
404.6	
456.8	
470.8	
484.0	
489.7	
549.1	
577.1	

Gamma energies (KeV)	Gamma intensities
647.5	
881	
891	
962	
1062	
1076	
1173	
1187	
1207	
1273	
1380	
1394	
$_{71}Lu^{170m}$	
44.6	
48.4	
$_{71}Lu^{170}$	
Ann. Rad.	
84.262 ± 0.005 (IT)	
152.8	
193.5	
221.2	
222.7	
223.8	
228.5	
236.1	
241.7	
251.3	
283.4	
286.8	
301.9	
323.9	
366.4	
369.9	
372.2	
382.8	
384.8	
389.1	
396.2	
410.6	
419.8	
443.6	
455.7	
479.0	
492.3	
497.1	
540.4	
544.6	
572.2	
579.6	
689.1	
841	
857	
940	
955	
987	
989	
1001	
1005	
1030	
1056	
1063	
1103	
1136	
1141	
1148	
1225	
1228	
1231	
1260	
1266	
1283	
1298	
1310	
1325	
1344	
1368	
1398	
1408	
1431	
1453	
1458	
1483	
1515	
1517	
1553	
1569	
1577	
1688	
1816	
1861	
1939	
1956	
2039	
2124	
2359	
2488	
2512	
2655	
2684	
2700	
2740	
2775	
2836	
2872	
2930	
2955	
3023	
$_{71}Lu^{171m}$	
71.2	
$_{71}Lu^{171}$	
Ann. Rad.	
19.3	
27.0	
46.5	
55.7	
66.7	
72.3	
75.9	
85.5	
91.3	
109.2	
122.2	
132.2	
141.3	
154.6	
163.8	
170.6	
195.0	
518	
627	
668	
690	
714	
741	
768	
782	
827	
842	
854	
$_{71}Lu^{172}$	
78.73 ± 0.12	9.6
90.60 ± 0.15	4.5
112.7 ± 0.2	1.20
119.1 ± 1.17	0.050
134.4 ± 0.3	0.058
145.8 ± 1.3	0.10
151.7 ± 1.0	0.034
163.4 ± 1.6	0.093
174.7 ± 1.0	0.12
181.4 ± 0.1	19.3
196.2 ± 0.4	0.090
203.2 ± 0.1	4.60
210.2 ± 0.5	0.10
228.9 ± 0.6	0.35
247.0 ± 0.4	0.62
264.6 ± 0.2	0.65
269.9 ± 0.1	1.63
279.5 ± 0.1	0.98
318.8 ± 0.8	0.12
323.8 ± 0.1	1.30
330.3 ± 0.7	0.52
337.4 ± 1.5	~0.03
352.4 ± 1.5	<0.03
358.2 ± 1.5	~0.10
366.5 ± 0.6	0.30
372.4 ± 0.3	2.50
373.3 ± 1.0	<0.4
377.4 ± 0.1	3.37
399.6 ± 0.2	0.55
410.2 ± 0.1	2.08
415.9 ± 1.5	0.11
422.4 ± 1.5	0.10
427.0 ± 1.5	<0.1
432.5 ± 0.3	1.65
437.1 ± 1.5	~0.25
442.6 ± 1.5	~0.18
482.3 ± 0.5	0.49
486.1 ± 0.5	0.60
490.4 ± 0.1	1.83
512.8 ± 1.5	0.16
524.0 ± 1.5	0.21
528.2 ± 0.1	3.80
536.2 ± 1.2	0.56
540.2 ± 0.2	1.44
551.2 ± 1.5	0.38
566.3 ± 1.5	0.10
577.0 ± 1.5	0.27
584.8 ± 0.8	0.32
594.5 ± 0.7	0.43
607.2 ± 0.7	0.50
623.1 ± 1.5	0.22
631.4 ± 1.5	0.39
681.8 ± 0.6	0.67
697.3 ± 0.2	6.20
709.1 ± 0.6	0.79
723.2 ± 1.1	0.40
810.1 ± 0.1	16.5
816.2 ± 1.4	1.02
900.8 ± 0.1	29.4
910 ± 1.0	~0.68
912.1 ± 0.1	15.0
927.0 ± 1.0	
929.1 ± 0.1	3.10
967.5 ± 1.5	0.18
999.5 ± 1.0	<0.2
1002.8 ± 0.13	5.59
1013.7 ± 1.0	
1020.2 ± 1.0	0.16
1022.4 ± 0.5	1.4
1039.7 ± 1.0	~0.1
1041.2 ± 0.2	0.40
1080.8 ± 0.2	0.93
1093.7 ± 0.1	64.75
1113.2 ± 0.2	1.7
1115.7 ± 1.0	0.4
1126.6 ± 1.0	<0.1
1184.4 ± 0.2	0.3
1212.0 ± 1.0	<0.1
1283.8 ± 0.7	<0.1
1288.3 ± 0.7	0.22
1322.8 ± 1.0	<0.1
1370.3 ± 0.2	0.70
1387.2 ± 0.1	0.88
1397.1 ± 0.4	0.23
1399.1 ± 1.5	0.13
1402.7 ± 0.2	0.58
1440.4 ± 0.2	0.64
1466.2 ± 0.2	0.60
1470.3 ± 1.0	0.70
1477.0 ± 1.0	~0.5
1490.0 ± 1.0	1.10
1532.5 ± 1.0	0.09
1543.0 ± 0.2	0.90
1579.3 ± 1.0	0.14
1584.2 ± 0.1	2.47
1602.8 ± 0.2	0.26
1608.8 ± 0.8	0.095
1622.0 ± 0.1	2.06
1666.7 ± 0.3	0.24
1670.6 ± 0.2	0.51
1724.5 ± 0.2	0.55
1813.0 ± 0.3	0.20
1914.7 ± 0.4	0.63
1920.2 ± 1.5	0.03
1931.9 ± 1.5	0.03
1955.3 ± 1.5	<0.03
1994.4 ± 0.4	0.16
2025.2 ± 0.9	0.059
2083.5 ± 0.3	0.29
2096.1 ± 0.5	0.11
$_{71}Lu^{173}$	
78.70	12.5%
100.63	6.1%
171.36	3.2%
179.33	1.9%
272.05	26%
285.40	0.8%
351.0	1.0%
457	0.3%
558	0.8%
637	0.05%
$_{71}Lu^{174m}$	
44.6 (IT)	
59.1 (IT)	
67.1 (IT)	
76.50	8%

Gamma energies (KeV)	Gamma intensities
176.66	0.5%
272.87	0.7%
449.6	0.07%
992.1	0.08%
$_{71}Lu^{174}$	
76.5	8.1%
1241.76 ± 0.2	9%
1318.24 ± 0.3	0.12%
$_{71}Lu^{176m}$	
88.36	0.0004%
1050	0.0009%
1140	
$_{71}Lu^{176}$	
88.36	16†
203	89
306	100
$_{71}Lu^{177m}$	
71.64	6.8†
105.36	100
112.97	179
*115.83	5.0
117.2	1.8
*121.6	52
128.5	127
136.7	12
145.8	7
*147.1	29
153.3	133
159.7	5.4
*171.8	37
174.4	96
177.0	26
182.0	0.8
*195.5	7
204.1	114
208.36	485
214.4	48
*218.1	27
228.4	287
233.8	45
249.7	47
*268.8	25
281.8	108
283.4	5
291.4	8
292.5	8
296.5	38
299.0	12
305.5	14
313.7	9
*319.0	78
321.4	9
327.7	136
341.6	13
*367.4	23
378.5	222
385.0	24
*413.6	131
418.5	161
426.3	3.4
466.0	19
$_{71}Lu^{177}$	
71.64	2.4†
112.97	100
136.7	0.74
208.36	171
249.7	3.3
321.4	3.4
$_{71}Lu^{178}$	
93.179	
215	
325	
425	
$_{71}Lu^{178}$	
93.179	
215	
325	
425	
$_{71}Lu^{179}$	
90	
213	
$_{72}Hf^{168}$	
Ann. Rad.	
129	
170	
$_{72}Hf^{169}$	
Ann. Rad.	
115	
$_{72}Hf^{170}$	
Ann. Rad.	
44.6	
48.4	
120.2	
164.7	
990	
1280	
1650	
2030	
2360	
2520	
2940	
$_{72}Hf^{171}$	
70.1	
122	
188	
290	
340	
470	
660	
860	
1070	
$_{72}Hf^{172}$	
23.99	
41.86	
44.08	
81.8	
114	
125.8	
$_{72}Hf^{173}$	
4.65	
17.8	
123.6	
134.9	
139.6	
161.8	
171.3	
297	
306	
311	18†
357	
422	
539	
549	
555	
567	
577	
617	
624	
717	
759	
764	
853	
857	
874	
878	
898	
1032	
1037	
1069	
1204	
1209	
1213	
1351	
1485	
1551	
1781	
$_{72}Hf^{175}$	
89.36	40†
113.81	3.6
161.3	0.3
229.6	7.3
318.9	
343.4	1000
433.0	16
$_{72}Hf^{178m_2}$	
88.878 ± 0.015	67
93.181 ± 0.015	18
213.444 ± 0.015	86
216.673 ± 0.015	71
237.40 ± 0.03	10
238.66 ± 0.1[b]	2.5
257.62 ± 0.02	18
277.35 ± 0.05	1.6
296.80 ± 0.04	10
325.562 ± 0.015	100
426.371 ± 0.015	105
454.05 ± 0.04	19
495.01 ± 0.03	82
535.02 ± 0.04	11
574.21 ± 0.03	101
$_{72}Hf^{178m_1}$	
88.878 ± 0.015	67%
93.181 ± 0.015	18%
213.44 ± 0.015	86%
325.562 ± 0.015	96%
426.371 ± 0.015	100%
$_{72}Hf^{179m}$	
161	5%
217	99.5%
325	0.5%
$_{72}Hf^{180m}$	
57.54	29†
93.33	10
215.3	50
332.5	57
444	49
501	8
$_{72}Hf^{181}$	
3.9	
6.3	
133.02	40†
136.25	6
136.86	1.7
345.9	13
476.0	~1.7
482.0	81
615.5	0.2
$_{72}Hf^{182}$	
97.89 ± 0.1	0.08%
114.33 ± 0.02	2.6%
156.09 ± 0.02	7.0%
172.54 ± 0.1	0.2%
270.405 ± 0.01	80%
$_{72}Hf^{183}$	
140	2†
170	0.9
210	2
280	3

Gamma energies (KeV)	Gamma intensities
320	4
400	5
460	58
610	2†
720	6
820	100
900	6
$_{73}Ta^{172}$	
Ann. Rad.	
92	
115	
155	
208	
270	
1100	
1300	
$_{73}Ta^{173}$	
Ann. Rad.	
37.4	
58.0	
69.8	
81.5	
90.2	
115.0	
160.4	
172.1	
180.6	
$_{73}Ta^{174}$	
Ann. Rad.	
90.9	
125	
160	
205	
280	
350	
$_{73}Ta^{175}$	
35.8	
50.5	
70.5	
77.3	
81.5	
100.8	
104.3	
125.9	
126.6	
132.0	
142.0	
162.0	
162.5	
179.1	
185.8	
192.7	
196.4	
207.4	
213.4	
230.8	
266.9	
280.5	
288.9	
294.0	
308.9	
348.5	
361.4	
365.7	
386.0	
393.2	
404	
433	
436	
443	
450	
462	
475	
525	
533	
540	
602	
622	
694	
732	
751	
762	
810	
815	
852	
854	
860	
867	
878	
902	
964	
1000	
1013	
1038	
1121	
1146	
1252	
1376	
1383	
1470	
1639	
$_{73}Ta^{176}$	
88.35	
91.3	
99.6	
103.4	
125.6	
131.1	
146.7	
156.8	
158.2	
175.6	
190.4	
201.9	
240.0	
366	
415	
639	
646	
678	
686	
711	
723	
741	
938	
1025	
1140	
1162	
1192	
1203	
1225	
1255	
1272	
1296	
1343	
1359	
1489	
1504	
1557	
1586	
1618	
1633	
1672	
1681	
1699	
1707	
1723	
1778	
1825	
1864	
1908	
1958	
2045	
2137	
2219	
2247	
2312	
2460	
2467	
2519	
2602	
2622	
2760	
2831	
2925	
$_{73}Ta^{177}$	
51.1	
71.6	0.003%
96.2	
112.97	6%
136.7	
208.38	1.0%
249.7	0.02%
257.3	
298.1	
321.3	0.024%
357.3	0.002%
396.0	
420.9	0.0018%
424.7	0.13%
452.9	
492.5	0.0015%
509	0.0024%
526.9	
549.4	
598.5	0.009%
633.1	0.039%
735.2	0.04%
737.0	0.016%
746.0	0.22%
848.2	
945.4	0.055%
1058.4	0.30%
$_{73}Ta^{178}$	
Ann. Rad.	
93	
203	
214	
970	
1100	
1170	
1180	
1200	
1260	
1330	
1340	
1390	
1430	
1440	
1460	
1480	
$_{73}Ta^{178}$	
88.8	
93.1	
213.6	10†
325.7	11
331.7	3
426.8	10
$_{73}Ta^{179}$	
No γ	
$_{73}Ta^{180m}$	
93.4 ± 0.2	27%
103.6 ± 0.2	3%
200	
$_{73}Ta^{180}$	
400	
$_{73}Ta^{181m}$	
6.3	
$_{73}Ta^{182m}$	
146.785 ± 0.01	75†
171.586 ± 0.01	≡100
184.951 ± 0.01	51
318.401 ± 0.04	14
356.468 ± 0.07	0.6
$_{73}Ta^{182}$	
31.74	
42.71	
65.72	
67.75	8%
84.68	3.34%
100.10	41%
113.67	1.8%
116.41	0.4%
152.441 ± 0.003	57%
156.386 ± 0.003	72%

Gamma energies (KeV)	Gamma intensities
179.394 ± 0.004	33%
222.114 ± 0.004	97%
229.335 ± 0.017	3.6%
264.079 ± 0.009	24%
927.903 ± 0.07	0.6%
1001.654 ± 0.041	2.8%
1113.050 ± 0.08	0.6%
1121.218 ± 0.016	35.6%
1157.357 ± 0.060	1.04%
1188.978 ± 0.019	16.5%
1221.354 ± 0.019	27.9%
1230.982 ± 0.021	11.4%
1257.401 ± 0.036	1.56%
1273.657 ± 0.050	0.67%
1289.126 ± 0.032	1.44%
1342.743 ± 0.010	0.27%
1373.638 ± 0.015	0.24%
1387.2 ± 0.2	0.08%
1410.0 ± 0.2	0.05%
1453.0 ± 0.2	0.03%
$_{73}Ta^{183}$	
40.98	1.8†
46.48	22
52.59	22
82.92	1.4
84.71	5.1
99.08	27
101.93	1.2
102.48	0.51
103.15	0.35
107.93	44
109.73	2.4
120.73	0.23
142.27	1.4
144.12	9.5
160.53	11
161.34	35†
162.32	19
192.64	1.35
203.28	1.5
205.08	3.4
208.81	2.3
209.86	17
244.26	32
245.24	1.2
246.06	100
291.72	14.1
313.00, 313.28	27
353.99	43
365.64	2.0
406.61	1.9
$_{73}Ta^{184}$	
111	
160	
250	
300	
410	
530	38†
660	7
790	34
900	100
950	30
1060	9
1100	1.3
1160	25
1270	10†
1320	1.8
1440	5
$_{73}Ta^{185}$	
75	8†
100	16
110	1
135	5
175	100
245	8
$_{73}Ta^{186}$	
122	25†
200	100
300	25
410	20
510	45
608	45
730	65
940	15
$_{74}W^{175}$	
260	
800	
1300	
1600	
$_{74}W^{176}$	
Ann. Rad.	
33.58	
50.55	
61.29	
84.14	
94.9	
100.2	
$_{74}W^{177}$	
30.5	
70.45	
101.8	
115.7	
142.6	
143.9	
156.0	
186.4	
203.0	
223.3	
259.2	
271.1	
290.0	
308.1	
317.5	
367.6	
377.1	
381.5	
388.9	
417.1	
427.2	
450.8	
502.0	
528.0	
562.0	
568	
615	
630	
639	
673	
712	
722	
760	
787	
797	
827	
858	
877	
1016	
1036	
1067	
1174	
1269	
1295	
$_{74}W^{179m}$	
221.8	
$_{74}W^{179}$	
30.7	
$_{74}W^{180m}$	
0.102	
0.35	
0.5	
0.22	
$_{74}W^{181m}$	
0.37	
$_{74}W^{181}$	
63	35%
136.3	0.11%
152.5	0.2%
$_{74}W^{183m}$	
46	†
52	
105	19
160	3.5
$_{74}W^{185m}$	
60	
131	
175	
$_{74}W^{187}$	
36.3	
49.2	
57.2	†
71.95 ± 0.03	37.3
106.57 ± 0.08	0.09
113.72 ± 0.06	0.33
134.24 ± 0.03	29.4
206.20 ± 0.04	0.53
239.10 ± 0.10	0.28
246.35 ± 0.07	0.38
479.51 ± 0.03	80.47
511.70 ± 0.06	2.53
551.51 ± 0.10	18.98
588.90 ± 0.12	0.65
618.24 ± 0.04	23.25
625.42 ± 0.07	3.93
685.72 ± 0.04	100
745.33 ± 0.10	1.00
772.85 ± 0.07	14.75
864.53 ± 0.10	1.25
879.0 ± 0.5	0.47
$_{74}W^{188}$	
63.58	0.9%
227	0.3%
290.5	0.6%
$_{74}W^{189}$	
94	2†
130	12
178	13
220	3
258	100
360	10
417	96
550	28
860	20
960	17
$_{75}Re^{177}$	
Ann. Rad	
$_{75}Re^{178}$	
Ann. Rad	
$_{75}Re^{179}$	
221.8	
280	
$_{75}Re^{180}$	
Ann. Rad	
110	
880	
$_{75}Re^{180}$	
Ann. Rad.	
$_{75}Re^{181}$	
19.7	
31.1	
38.1	

GAMMA ENERGIES AND INTENSITIES OF RADIONUCLIDES (*Continued*)

Gamma energies (KeV)	Gamma intensities
43.5	
65.0	
71.7	
72.7	
93.7	
102.7	
103.1	
109.9	
110.3	
113.3	
137.2	
144.3	
154.4	
163.9	
164.6	
165.8	
175.2	
177.4	
186.2	
195.0	
197.0	
252.0	
262.6	
276.2	
295.9	
318.6	
331.9	
353.6	
356.1	
360.7	
365.5	
382.3	
389.0	
408.7	
441.8	
475.6	
487.2	
489.0	
515.7	
522.2	
557.7	
587.6	
639.0	
651.2	
661.8	
668.2	
693.9	
738.1	
769.6	
781.4	
804	
805	
823	
840	
854	
880	
883	
908	
954	
989	
994	
1000	
1009	
1076	
1087	
1103	
1384	
1441	
1469	
1538	
$_{75}Re^{182}$	
31.7	
42.0	
42.7	
65.8	
67.8	
84.7	
100.1	
113.7	
116.4	
152.5	
156.4	
179.4	
198.4	
222.0	
229.3	
264.0	
470.0	
514.3	
536.1	
598.5	
649.5	
734.5	
786.6	
810	
836	
895	
901	
928	
960	
1002	
1045	
1122	
1158	
1189	
1222	
1231	
1257	
1274	
1289	
1374	
1387	
1438	
1955	
2014	
2023	
2054	
$_{75}Re^{182}$	
47.05 ± 0.15	
53.04 ± 0.15	
87.31 ± 0.15	
100.07 ± 0.15	
113.62 ± 0.15	
116.45 ± 0.15	
130.75 ± 0.15	
133.72 ± 0.15	
145.39 ± 0.15	
149.00 ± 0.15	
152.37 ± 0.15	
156.34 ± 0.15	
172.83 ± 0.15	
187.47 ± 0.15	
189.62 ± 0.15	
191.33 ± 0.15	
198.29 ± 0.15	
214.24 ± 0.15	
215.83 ± 0.15	
217.48 ± 0.15	
226.16 ± 0.15	
247.44 ± 0.15	
256.44 ± 0.15	
264.07 ± 0.15	
276.30 ± 0.15	
281.45 ± 0.15	
286.53 ± 0.15	
295.70 ± 0.15	
300.15 ± 0.15	
323.36 ± 0.15	
339.06 ± 0.15	
341.93 ± 0.15	
346.13 ± 0.15	
351.07 ± 0.15	
357.11 ± 0.15	
1157.40 ± 0.20	
1181.09 ± 0.20	
1188.78 ± 0.20	
1221.10 ± 0.15	
1230.76 ± 0.15	
1257.18 ± 0.30	
1288.67 ± 0.30	
1293.66 ± 0.30	
1342.33 ± 0.30	
1426.88 ± 0.15	
1438	
$_{75}Re^{183}$	
46.73 ± 0.15	78%
52.77 ± 0.15	25%
67.29 ± 0.15	
69.22 ± 0.15	
82.99 ± 0.15	3.1%
84.79 ± 0.15	12%
99.11 ± 0.15	25%
107.94 ± 0.15	11%
109.74 ± 0.15	11%
115.90 ± 0.15	
117.19 ± 0.15	
118.58 ± 0.15	
126.49 ± 0.15	
125.18 ± 0.15	
144.06 ± 0.15	0.3%
169.01 ± 0.15	
192.54 ± 0.15	0.5%
215.18 ± 0.15	
220.15 ± 0.15	
221.59 ± 0.15	
229.29 ± 0.15	
244.36 ± 0.15	0.8%
245.94 ± 0.15	1.8%
313.11 ± 0.15	0.6%
318.09 ± 0.15	
354.21 ± 0.15	0.7%
$_{75}Re^{184m}$	
83.4	70%
104.7	70%
111.43 ± 0.1	61%
161.27 ± 0.15	10%
216.53 ± 0.15	15%
252.82 ± 0.15	10%
384.21 ± 0.15	~2%
641.9 ± 0.15	1.5%
792.83 ± 0.10	28%
894.68 ± 0.10	23%
903.20 ± 0.10	36%
$_{75}Re^{184}$	
71.11 ± 0.15	
87.50 ± 0.2	
91.42 ± 0.2	
104.80 ± 0.1	
111.43 ± 0.10	59%
117.43 ± 0.10	
118.67 ± 0.20	
124.47 ± 0.20	
126.52 ± 0.20	
161.27 ± 0.15	
169.21 ± 0.20	
170.57 ± 0.20	
178.27 ± 0.20	
180.0 ± 0.3	
216.53 ± 0.15	0.5%
226.72 ± 0.15	
252.82 ± 0.15	4.7%
317.97 ± 0.15	1%
363.97 ± 0.25	
384.21 ± 0.15	
536.61 ± 0.15	
539.24 ± 0.20	0.3%
641.90 ± 0.15	2.3%
769.75 ± 0.15	5%
792.83 ± 0.10	36%
850.78 ± 0.15	
859.19 ± 0.15	
894.68 ± 0.10	19%
903.20 ± 0.10	41%
920.88 ± 0.15	~4%
953.64 ± 0.20	
962.02 ± 0.15	
970.77 ± 0.20	
1005.88 ± 0.20	
1010.11 ± 0.20	
1022.49 ± 0.15	
1109.86 ± 0.15	
1173.72 ± 0.15	
1274.96 ± 0.25	
1384	
$_{75}Re^{186}$	
176	2.1
191	1.2
230	0.6
$_{75}Re^{188m}$	
2.4	
15.9	
63.58	
92.45	
105.96	
$_{75}Re^{188}$	
155.03	1550† 3%
297	1.7
321	1.6
452	7
478	100 0.27%
485	8

Gamma energies (KeV)	Gamma intensities
633	141
641	1.2
672	10
824	4
829	40
846	0.6
881	1.6
932	58
963	1.3
967	1.8†
977	0.7
1019	1.6
1132	9
1150	3.6
1171	1.1
1176	2.1
1191	1.2
1230	0.6
1307	7
1322	1.3
1368	1.1
1457	1.8
1610	11
1732	3.5
1786	2.6
1803	4.1
1866	0.6
1957	1.6
$_{75}Re^{189}$	
30.8	
36.2	
69.5	
95.2	
147.2	
150.1	
185.9	
188.6	
206.4	
211.0	
216.8	
219.4	
245.0	
$_{75}Re^{190}$	
187	10†
392	10
570	10
830	3
$_{75}Re^{192}$	
200	
290	
370	
480	
570	
$_{76}Os^{181}$	
93	
101	
$_{76}Os^{182}$	

Gamma energies (KeV)	Gamma intensities
27.6	†
55.5	
180.2	7
263.2	1.4
509.9	10
$_{76}Os^{183m}$	
67.3	6%
147	
170.7	46%
1.036	6%
1.103	26%
1.1095	22%
$_{76}Os^{183}$	
114.4	
145.4	
151.0	
167.9	
236.2	
259.8	
355.5	
381.8	
477.2	
496.1	
737	
808	
851	
887	
889	
1163	
1440	
$_{76}Os^{185}$	
71.6	1.5%
125.3	2.4%
162.6	1.1%
233.4	1.4%
592.0	1.2%
645.8	81%
718	4.2%
872	6.5%
879	6.5%
$_{76}Os^{189m}$	
30.81	
$_{76}Os^{190m}$	
38.9	<1%
186.7	100%
361.2	100%
502.5	100%
616.5	100%
$_{76}Os^{191m}$	
74.2	
$_{76}Os^{191}$	
49.5	
82.33	

Gamma energies (KeV)	Gamma intensities
96.45	
129.4	
178.9	
186.7	
$_{76}Os^{193}$	
73	1.45†
98	0.04
99	0.02
107	0.37
139	2.93
153	0.02
180	0.09
181	0.18
219	0.16
234	0.033
252	0.134
280	1.00
289	0.19
298	0.16
321	0.76
361	0.21
378	0.08
387	0.75
419	0.133
441	0.065
460	2.32
485	0.113
517	0.02
532	0.05
558	1.10
573	0.01
660	0.002
712	0.011
848	0.003
875	0.0135
$_{76}Os^{194}$	
43.0	10%
78	0.03%
$_{77}Ir^{182}$	
Ann. Rad.	
133	
278	
↓	
4000	
$_{77}Ir^{183}$	
240	
$_{77}Ir^{184}$	
Ann. Rad.	
125	100†
267	200
392	90
830	
960	
1090	
↓	
4300	
$_{77}Ir^{185}$	

Gamma energies (KeV)	Gamma intensities
24.2	
30.4	
33.8	
37.4	
60.0	
90.4	
94.5	
97.4	
100.7	
119.6	
124.9	
126.9	
127.9	
129.4	
153.6	
158.3	
160.7	
185.0	
220.4	
222.3	
223.8	
254.4	
266.5	
300.4	
307.1	
314.4	
321.5	
339.2	
352.4	
377.7	
406.8	
419.1	
431.4	
507.0	
513.7	
539.4	
745.7	
$_{77}Ir^{186}$	
Ann. Rad.	
70.9	
87.19	
102.1	
119.4	
137.15	
143.0	
160.0	
167.1	
220.0	
224.1	
234.5	
252.5	
269.0	
272.8	
276.5	
284.3	
288.8	
293.0	
296.7	
299.4	
302.9	
305.6	
309.6	
311.8	
321.2	
322.6	
326.6	
330.2	
334.0	

GAMMA ENERGIES AND INTENSITIES OF RADIONUCLIDES (*Continued*)

Gamma energies (KeV)	Gamma intensities
342.5	
351.7	
364.9	
387.9	
403.3	
406.6	
412	
420.7	
542.2	
551.4	
558.0	
565.4	
570.3	
584.4	
592	
600	
622.1	
630.3	
636.2	
650.0	
662	
679.5	
684.8	
693.6	
706	
712.6	
729.5	
745	
760.0	
767.3	
773.1	
780.8	
794	
805.5	
841.3	
885	
933.2	
943.6	
960	
1011.0	
1026.5	
1057.1	
1107	
1172	
1187.9	
1314	
1324	
1440	
1467	
1508	
1597	
1647	
1701	
1738	
1743	
1752	
1789	
1801	
1805	
1815	
1869	
1906	
1939	
1953	
1975	
2000	
2260	
2312	
2340	
2360	
2383	
2389	
2397	
2413	
2429	
2498	
2513	
2564	
2578	
2617	
2678	
2733	
2748	
2826	
2854	
2994	
3043	
$_{77}Ir^{186}$	
Ann. Rad.	
137.15	
295	
630	
770	
990	
$_{77}Ir^{187}$	
65.1	
73.7	
85.0	
125.5	
137.9	
162.6	
177.5	
187.1	
313.8	
399.2	
400.9	
426.9	
491.1	
501.5	
576.5	
610.6	
799.4	
912.6	
976.7	
987.0	
1103	
1112	
$_{77}Ir^{188}$	
Ann. Rad.	
85.0	
95.6	
115.7	
123.1	
150.5	
155.0	209
157.0	
158.0	
162.2	
175.4	
189.3	
222.3	
268.3	
279.6	
312.0	
322.9	12
332.4	
346.7	
350.0	1.6
351.7	
352.7	
371.4	
379.6	
385.3	1.4
399.0	
424.5	
447.7	
452.9	
478.0	
478.6	100
514.8	3.2
522.7	
566.7	3
586.2	
594.0	
599.0	
620.6	6
633.1 }	
634.8 }	180
641.5	3.4
652.4	
672.3	9
736.8	
747.6	
757.2	2.0
777.5	†
781.9	
810.5	
821.2	
824.8	
828.6	
829.5	
886.0	
905.9	
940.0	
943.6	
947.4	
983.8	
987.7	13
1013	
1018	6
1053	
1097	8
1143	2
1150	4
1175	6
1193	
1206	†
1210	42
1296	1.5
1304	4
1307	
1323	10
1329	4
1333	4
1337	
1430	
1436	9
1446	
1453	4
1458	7
1463 }	
1466 }	7
1560	4
1575	12
1620	4
1653	10
1689	2
1705	7
1717	25
1721	
1728	
1732	
1737	
1760	
1775	
1784	
1789	
1805	
1812	1
1904	
1931	
1945	18
1968	
2013	3
2052	22
2061	30
2098	45
2100	
2196	10
2217	80†
2294	1.5
2344	
2350	
2494	
2525	
2623	
2700	
$_{77}Ir^{189}$	
25.7	
30.8	
33.3	
36.2	
56.5	
59.1	
59.3	
69.6	
95.2	
97.8	
138.3	
147.1	
149.9	
164.0	
180.5	
185.9	
188.6	
197.4	
206.3	
216.7	
219.3	
233.5	
245.0	
275.8	
$_{77}Ir^{190m_2}$	
38.9	
116.7	
148.7 (IT)	
186.7	75†
206.6	
361.2	100
502.5	115
616.5	103

Gamma energies (KeV)	Gamma intensities
$_{77}Ir^{190m_1}$	
26.3	
$_{77}Ir^{190}$	
97.8	
100.0	
137.8	
182.1	
186.7	
190.5	
196.8	
198.0	
199.3	
207.8	
223.8	
235.3	
251.4	
282.9	
288.2	
294.6	
361.2	
371.1	
380.1	
394.8	
397.3	
402.4	
407.2	
447.5	
462.5	
485.3	
490.8	
502.5	
518.4	
557.8	
569.3	
605.3	
655.9	
690.0	
726.4	
750.7	
768.5	
801	
829.0	
833	
836	
839	
915	
933	
953	
1036	
1079	
1110	
1116	
1131	
1134	
1184	
1200	
1234	
1314	
1324	
1329	
1339	
1355	
1386	
1391	
1397	
1403	
1414	

Gamma energies (KeV)	Gamma intensities
1437	
1464	
1473	
1482	
1517	
1525	
1534	
1543	
1584	
1606	
1677	
1685	
1700	
1711	
$_{77}Ir^{191m}$	
41.8	
47	
82.3	
129.4	
$_{77}Ir^{192m_2}$	
161	
$_{77}Ir^{192m_1}$	
58(IT)	
317	
612	
$_{77}Ir^{192}$	
136.34	2.7†
201.28	5.6
205.79	38
283.4	4
295.9	360
308.43	370
316.5	1000–85%
374.6	6
416.6	≤3
468.05	600
484.6	40
489.1	4
588.6	49
604.38	105
612.43	70
884.6	5
1062	0.5
$_{77}Ir^{193m}$	
80.2	
$_{77}Ir^{194}$	
29.362	6.5†
300.7	0.9
328.54	33
622.1	0.8
645.3	2.7
890.1	0.10
925.3	0.09
938.9	1.5
1000.2	0.08

Gamma energies (KeV)	Gamma intensities
1048.7	0.07
1104.1	0.06
1151.3	1.6
1175.5	0.17
1184.1	0.79
1219.0	0.34
1294	0.26
1342.4	0.11
1431	0.02
1469.4	0.93
1480	<0.05
1488	0.9
1512	0.13
1623	0.24
1671	0.033
1784	0.025
1798	0.06
1805	0.06
1808	0.09
1832	0.009
1926	0.012
2044	0.022
2115	0.009
$_{77}Ir^{195}$	
98.5	100†
129.9	50
330	60
370	40
430	30
660	20
$_{77}Ir^{196m}$	
103.3 ± 0.2	17 ± 2
340.7 ± 0.4	1.6 ± 0.2
355.9 ± 0.2	98 ± 3
393.5 ± 0.2	101 ± 2
420.9 ± 0.3	2.6 ± 0.1
447.1 ± 0.2	98 ± 2
521.4 ± 0.2	100
553.0 ± 0.3	0.66 ± 0.04
566.4 ± 0.4	0.3 ± 0.1
615.9 ± 0.4	0.44 ± 0.05
633.5 ± 0.3	1.15 ± 0.05
647.3 ± 0.2	95 ± 3
673.9 ± 0.2	0.18 ± 0.04
693.9 ± 0.2	4.4 ± 0.3
722.0 ± 0.4	0.67 ± 0.07
727.3 ± 0.2	2.7 ± 0.1
760.6 ± 0.3	0.78 ± 0.05
835.6 ± 0.2	6.6 ± 0.2
849.4 ± 0.3	0.53 ± 0.05
868.1 ± 0.3	0.48 ± 0.04
887.0 ± 0.5	0.11 ± 0.02
893.0 ± 0.5	0.20 ± 0.02
904.6 ± 0.5	0.10 ± 0.02
914.6 ± 0.3	0.30 ± 0.05
926.0 ± 0.5	0.07 ± 0.02
1024.6 ± 0.3	0.26 ± 0.03
1068 ± 2	0.074 ± 0.018
1080.5 ± 0.5	0.12 ± 0.02
1116.7 ± 0.8	0.073 ± 0.022
1281.6 ± 0.5	0.061 ± 0.024
1341.5 ± 0.5	0.29 ± 0.03
1355.8 ± 0.5	0.06 ± 0.01
1394.0 ± 0.5	0.074 ± 0.015
1482.5 ± 0.4	2.4 ± 0.2

Gamma energies (KeV)	Gamma intensities
$_{77}Ir^{196}$	
KX	40 ± 30
332.8 ± 0.3	23 ± 1
355.4 ± 0.3	100
446.6 ± 0.3	24 ± 2
779.4 ± 0.3	55 ± 2
1047.0 ± 0.7	5 ± 1
1228.6 ± 1.2	1.8 ± 0.4
1468.4 ± 0.8	4.4 ± 0.7
1564.2 ± 0.8	4.7 ± 1.0
$_{77}Ir^{197}$	
510	~50%
$_{77}Ir^{198}$	
780	
$_{78}Pt^{174}$	
Ann. Rad.	
$_{78}Pt^{177}$	
Ann. Rad.	
$_{78}Pt^{178}$	
Ann. Rad.	
160	
$_{78}Pt^{180}$	
Ann. Rad.	
$_{78}Pt^{181}$	
Ann. Rad.	
$_{78}Pt^{182}$	
Ann. Rad.	
$_{78}Pt^{183}$	
Ann. Rad.	
$_{78}Pt^{185}$	
35	
63	
1560	
$_{78}Pt^{186}$	
140	
190	
680	
1400	

Gamma energies (KeV)	Gamma intensities
$_{78}Pt^{187}$	
110	
184	
2010	
$_{78}Pt^{188}$	
Ir KX	100†
41.9	0.4
54.8	0.3
96.7	0.2
98.4	0.4
132.8	0.30
140.3	2.2
187.6	19.2
195.1	18.4
280.4	0.22
283.0	0.11
290.6	0.11
381.6	7.0
423.6	4.4
478.3	1.9
$_{78}Pt^{189}$	
71.6	
82.1	
94.2	
113.8	
141.1	
176.3	
181.3	
186.7	
190.8	
203.7	
223.3	
225.6	
243.4	
252.0	
258.2	
300.4	
317.6	
403.8	
542	
544.8	
568.8	
607	
626.9	
644	
720.9	
735	
792	
798	
$_{78}Pt^{191}$	
82.33	
96.45	
129.39	
172.2	10.2%
178.9	3.2%
186.8	
187.8	
219.7	
221.8	
227.4	
268.8	

Gamma energies (KeV)	Gamma intensities
351.2	9.9%
360.0	16.6%
409.3	28.8%
456.5	9.6%
501.4	
539.0	40.5%
569.1	
575.6	
587.9	
624.1	
632.6	
$_{78}Pt^{193m}$	
12.58	
135.5	
$_{78}Pt^{195m}$	
30.8	
98.5	12%
129.4	2.8%
129.9	
$_{78}Pt^{197m}$	
52.9 (IT)	†
130.2	21
279.3	21
346 (IT)	100
$_{78}Pt^{197}$	
77.35	20%†
191.4	5.7%
268	0.006%
279	<0.0001%
$_{78}Pt^{199m}$	
32	~5†
393	100
$_{78}Pt^{199}$	
75	†
179	38
220	w
245	16
315	11
320	22
475 D	45
540	100
715	14
790	10
960	8
$_{79}Au^{186}$	
160	
220	
330	
400	

Gamma energies (KeV)	Gamma intensities
$_{79}Au^{188}$	
Ann. Rad.	
250	
330	
630	
$_{79}Au^{189}$	
39.44	
45.7	
88.6	
166.6	
215.9	
440	
446	
$_{79}Au^{190}$	
Ann. Rad.	
295.7	
301.6	
319	
598	
623	
1400	
1400	
1760	
2080	
2400	
2700	
2940	
3250	
3460	
$_{79}Au^{191}$	
48.3	
91.03	
132.90	
136.10	
158.8	
166.5	
194.1	
206.4	
210.0	
244.4	
253.9	
279.9	
280.5	
284.0	
312.7	
336.3	
368.7	
386.9	
390.4	
399.6	
408.3	
413.8	
421.5	
451.3	
478.8	
487.5	
495.9	
525.9	
586.4	

Gamma energies (KeV)	Gamma intensities
620.3	
674.2	
702.0	
732.5	
$_{79}Au^{192}$	
Ann. Rad.	
45.1	
96.8	
104.7	
137	
157	
167	
205	
283	
295	
308	
316	
415	
437	
466	
588	
612	
783	
1158	
$_{79}Au^{193m}$	
31.9	
220	
290.6 (IT	
$_{79}Au^{193}$	
12.7	
37.4	
44.3	
73.7	
99.8	
112.4	
114.0	
117.9	
119.4	
155.7	0.35%
173.5	2.0%
186.1	9%
187.5	
221.3	
231.8	0.5%
255.6	4.6%
258.4	
268.2	4.4%
290.0	0.2%
303.6	
317.7	1.3%
334.8	
377.6	1.3%
425	0.25%
440	3.1%
477	0.5%
490	0.6%
$_{79}Au^{194}$	
107.2	

Gamma energies (KeV)	Gamma intensities
163.94	
202.9	
216.15	
226.9	
236.3	
285.34	
293.62	
300.71	
328.54	
365.0	
482.8	
528.9	
544.7	
589.4	
593.7	
607.6.	
622.1	
645.3	
668.6	
672	
676.1	
703.6	
736.5	
818.7	
843.9	
847.4	
856.1	
860	
890.1	
894.4	
925.3	
938.9	
948.4	
1002.2	
1007.5	
1030.9	
1038.8	
1048.7	
1081.6	
1104.1	
1119.7	
1141.0	
1150.9	
1156.6	
1175.5	
1183.6	
1187.1	
1195.1	
1208.5	
1219.0	
1267.4	
1292	
1293.9	
1302.6	
1308.6	
1342.4	
1421.8	
1432.3	
1442.0	
1450.5	
1463.7	
1469.4	
1479.8	
1487.9	
1492.7	
1513	
1519	
1547.9	
1563.4	
1592.9	
1596.2	

Gamma energies (KeV)	Gamma intensities
1602.0	
1618	
1622.6	
1633.5	
1671	
1675.9	
1690.2	
1735.8	
1758	
1786.3	
1798.0	
1803	
1806	
1813	
1830.1	
1835.9	
1886.8	
1893.1	
1911.1	
1924.6	
1960	
1970.1	
1983.9	
2044.1	
2085.8	
2114.0	
2164.1	
2216	
2299	
2313	
2357	
2366	
2371	
2399	
2414	
$_{79}Au^{195m}$	
318	
$_{79}Au^{195}$	
30.8	
98.82 ± 0.15	
129.12 ± 0.15	
$_{79}Au^{196}$	
107.13 ± 0.1	1.7†
332.9 ± 0.25	280
355.7 ± 0.3	1000
425.8 ± 0.3	5
520.9 ± 0.5	0.5
758.5 ± 0.4	0.4
960	0.06
1091.3 ± 0.3	2.4
$_{79}Au^{197m}$	
130.2	
279	
$_{79}Au^{198}$	
411.795 ± 0.009	99%
675.883 ± 0.018	1%
1087.678 ± 0.027	0.2%

Gamma energies (KeV)	Gamma intensities
$_{79}Au^{199}$	
49.828 ± 0.004	
158.372 ± 0.004	76%
208.200 ± 0.006	16.6%
$_{79}Au^{200}$	
367.97	30%
1225	8.4%
1594	15.6%
$_{79}Au^{201}$	
530	5%
$_{79}Au^{203}$	
690	
$_{80}Hg^{186}$	
125	
270	
350	
440	
$_{80}Hg^{187}$	
175	
255	
400	
$_{80}Hg^{188}$	
140	
$_{80}Hg^{189}$	
Ann. Rad.	
165	
240	
320	
$_{80}Hg^{190}$	
28.8	
100.8	†
129.8	
142.7	10
155.0	
165.4	
171.5	
220	1
$_{80}Hg^{191}$	
252.6	
274.1	
$_{80}Hg^{192}$	
31.5	
40.9	
47.7	

Gamma energies (KeV)	Gamma intensities
99.4	
104.5	
105.3	
114.5	
120.1	
135.9	
142.5	
146.0	
157.2	
204.6	
245.4	
262.7	
274.8	
279.2	
306.5	
436.7	
$_{80}Hg^{193m}$	
32.2	
38.2	
39.5 (IT)	
101.2 (IT)	
165.9	
218	
220	
257	
281	
291	
299	
342	
345	
361	
364	
382	
383	
408	
462	
488	
500	
510	
535	
537	
574	
601	
624	
675	
701 D	
712	
861	
871	
879	
914	
933	
997	
1113	
1175	
1233	
1243	
1328	
1343	
1489	
1646	
$_{80}Hg^{193}$	
38.2	
186.8	
574	
762	

GAMMA ENERGIES AND INTENSITIES OF RADIONUCLIDES (*Continued*)

Gamma energies (KeV)	Gamma intensities
855	
1040	
1078	
$_{80}Hg^{195m}$	
16.2 (IT)	
37.2 (IT)	
53.3 (IT)	
56.7	
61.4	
123.0 (IT)	
172.6	
200.5	
261.5	
318.5	
368	
386.0	
387.6	
452	
467	
526	
560	
575	
680	
963	
1242	
$_{80}Hg^{195}$	
61.4	
180.2	
200.5	
207.1	
261.5	
600	
780	
821	
842	
930	
1020	
1110	
1172	
$_{80}Hg^{197m}$	
130.2	†
134.0 (IT)	1
165.3 (IT)	0.01
202	0.23
279.3	16
$_{80}Hg^{197}$	
77.34 ± 0.005	34
191.5	1.0†
269	0.06
$_{80}Hg^{199m}$	
158	
368	
$_{80}Hg^{203}$	
279.190 ± 0.007	

Gamma energies (KeV)	Gamma intensities
$_{80}Hg^{205}$	
203	
$_{80}Hg^{206}$	
310	
$_{81}Tl^{191}$	
Ann. Rad.	
$_{81}Tl^{192}$	
Ann. Rad.	
424	
$_{81}Tl^{193m}$	
365	
$_{81}Tl^{193}$	
241	
261	
270	
291	
309	
321	
330	
$_{81}Tl^{194m}$	
97.2	
$_{81}Tl^{194}$	
427	
$_{81}Tl^{195m}$	
99.1	
383	
$_{81}Tl^{195}$	
37.2	
225.8	
242.1	
247.3	
279	
562	
883	
$_{81}Tl^{196m}$	
33.7	
84.4	
120.1	
425.8	
$_{81}Tl^{196}$	
426	
$_{81}Tl^{197m}$	
222	
385	

Gamma energies (KeV)	Gamma intensities
$_{81}Tl^{197}$	
18.18	
133.9	
152.2	
156.3	
173.8	
269.6	
277.5	
308.5	
426.0	
433.1	
444	
451.8	
578.3	
585	
791	
857	
890	
980	
1010	
$_{81}Tl^{198m}$	
23.1 (IT)	
48.7	
259.6 (IT)	
260.9 (IT)	
282.4 (IT)	
411	13
586	10
635	10
$_{81}Tl^{198}$	
194	
227	
284.5	
400.4	
411.8	100†
586.5	
637	
675.7	
1089	3†
1202	24
1440	27
2010	17
2450	6
2780	2
$_{81}Tl^{199}$	
36.83	
49.82	4†
51.9	
158.36	54
195.1	9
208.18	119
247.2	93
284.0	13
333.9	16
403.5	4
455.1	136
492	10
$_{81}Tl^{200}$	
Ann. Rad.	

Gamma energies (KeV)	Gamma intensities
116.5	
134	
181	
227	
251	
276	
289	
309	
368	1000†
388	
545	
555	
579	120†
629	8
661	12
690	
712	8
786	15
829	87
887	5
898	9
939	
1029	
1207	330
1227	58
1265	†
1275	
1341	10
1364	45
1410	18
1479	
1517	42
1569	
1593	
1604	6
1718	7
1745	
1900	
1920	
1966	
1991	
2012	
$_{81}Tl^{201m}$	
230	
331.2	
361.3	
600	
$_{81}Tl^{201}$	
30.60	
32.19	
135.34	7%
165.9	0.3%
167.4	26%
$_{81}Tl^{202}$	
439.4 ± 0.2	100†
509.4 ± 0.3	
520.3 ± 0.3	4
959.7 ± 0.4	0.07
$_{81}Tl^{207m}$	
351	
1000	

GAMMA ENERGIES AND INTENSITIES OF RADIONUCLIDES (*Continued*)

Gamma energies (KeV)	Gamma intensities
$_{81}Tl^{207}$	
897.3	0.16%
$_{81}Tl^{208}$	
233	
252	†
277.33	7
486	0.1
510.72	23
583.169 ± 0.013	86
763.2	2
860	12
1094	0.6
2614.611 ± 0.060	100
$_{81}Tl^{209}$	
120	100%
450	100%
1560	100%
$_{81}Tl^{210}$	
97	4%
296	80%
360	4%
380	3%
795	100%
910	3%
1060	12%
1110	7%
1210	17%
1310	21%
2010	7%
2090	5%
2280	3%
2360	8%
2430	9%
$_{82}Pb^{194}$	
0.204	
$_{82}Pb^{195}$	
99.1	
383	
393	
$_{82}Pb^{196}$	
191.8	
240.3	
253.2	
367	
494	
503	
$_{82}Pb^{197m}$	
84.9 (IT)	
222.4	
234.4 (IT)	
387	
$_{82}Pb^{197}$	
385	
387	
$_{82}Pb^{198}$	
30.8	
116.9	10†
122.6	
173.4	100
259.5	24
290.3	50
365.4	
382.0	
389.5	
397.7	
575.0	
649.0	
743	8
865.3	20
$_{82}Pb^{199m}$	
424	
$_{82}Pb^{199}$	
352.8	19%
367.0	90%
721.0	10%
Ann. Rad.	
$_{82}Pb^{200}$	
32.74	
109.54	
142.30	
147.61	
161.4	
235.61	
257.15	
268.42	
289.1	
289.9	
302.85	
315.30	
450.5	
457.6	
605.3	
$_{82}Pb^{201m}$	
629	
$_{82}Pb^{201}$	
Ann. Rad.	
58.9	
129.8	
285.0	
309.2	
331.2	
345.3	
361.3	
381.4	
395.1	
406.0	
545.9	
584.6	
692.5	
708.6	
753.2	
767.3	
803.5	
820.3	
826.0	
907.4	
945.8	
1069.8	
1098.4	
1115	
1149	
1239	
1309	
$_{82}Pb^{202m}$	
129.3 (IT)	
148.9	
240.3 (IT)	
241.1	
335.6	
389.9	
422.1 (IT)	90%
459.8	10%
490.4	10%
547.2 (IT)	
657.6 (IT)	
787.2 (IT)	
961.4 (IT)	90%
$_{82}Pb^{203m}$	
825.2	
$_{82}Pb^{203}$	
279.190 ± 0.007	99+%
401.28 ± 0.3	4.6%
680.67 ± 0.3	0.8%
$_{82}Pb^{204m}$	
373.4 ± 0.6	
899.10 ± 0.2	
911.66 ± 0.2	
$_{82}Pb^{205m}$	
26.22 ± 0.11	†
284.12 ± 0.10	4.8
310.37 ± 0.07	0.26
703.40 ± 0.03	100
1014.22 ± 0.06	3.3
$_{82}Pb^{207m}$	
569.684 ± 0.014	100%
1063.614 ± 0.04	100%
$_{82}Pb^{210}$	
46.52	4.1%
$_{82}Pb^{211}$	
65.5	2†
84	1
310	0.7
340	1
404.84	100
426.99	53
612	3
702	13
766.3	19
831.8	100
860	0.1
1020	0.5
1076	0.5
1104	3.8
1188	0.4
1265	0.2
$_{82}Pb^{212}$	
115.1	0.7%
176.7	0.2%
238.60 ± 0.007	47%
300.1	3.2%
415.2	0.16%
$_{82}Pb^{214}$	
53.24	
196	
206	
241.92	3.7†
259	
272	
275	
279	
295	†
352.0	36
481	
534	
549	
777	
$_{83}Bi^{200}$	
462	
1027	
$_{83}Bi^{201}$	
629	
$_{83}Bi^{202}$	
422	
961	
$_{83}Bi^{203}$	
Ann. Rad.	
59.99	
126.4	

Gamma energies (KeV)	Gamma intensities
186.5	
263.9	
381.4	
626	
722	
758	0.2†
820.1	240
825.2	100
847.4	
932	
1034	
1184	
1256	
1510	
1523	
1537	
1846 } 1896 }	180
$_{83}Bi^{204}$	
78.5	
80.2	
90.9	
92.2	
100.2	
140.9	
170.2	
175.9	
213.4	
216.0	
219.1	
221.9	
227.5	
240.5	
249.0	
252.9	
289.3	
292.2	
330.5	
368.0	
374.7	
405.0	
412.2	
421.5	
438.5	
473.4	
502	
522	
532	
542	
622	
661	
663	
671	
692	
709	
710	
719	
724	
791	
899	
912	
918	
984	
1203	
1211	
$_{83}Bi^{205}$	
Ann. Rad.	
26.22 ± 0.11	
45.43 ± 0.075	
90.03 ± 0.024	
115.409 ± 0.181	0.2332 ± 0.022
122.466 ± 0.083	0.0465 ± 0.007
129.742 ± 0.127	0.0314 ± 0.006
147.315 ± 0.068	0.0755 ± 0.001
149.505 ± 0.236	0.0216 ± 0.006
157.269 ± 0.219	0.0352 ± 0.010
165.113 ± 0.143	0.0466 ± 0.013
185.293 ± 0.115	0.2806 ± 0.037
205.841 ± 0.273	0.0987 ± 0.017
221.024 ± 0.135	0.0926 ± 0.012
235.997 ± 0.046	0.1664 ± 0.043
260.511 ± 0.052	3.1038 ± 0.338
262.815 ± 0.067	1.1908 ± 0.172
282.261 ± 0.103	1.1461 ± 0.030
284.124 ± 0.103	4.8029 ± 0.545
310.369 ± 0.069	0.2648 ± 0.044
313.052 ± 0.092	0.1771 ± 0.019
349.550 ± 0.045	1.4781 ± 0.083
354.598 ± 0.053	0.0767 ± 0.002
361.601 ± 0.105	0.1787 ± 0.033
488.349 ± 0.183	0.1895 ± 0.055
493.701 ± 0.033	0.9765 ± 0.103
498.840 ± 0.152	0.5467 ± 0.115
511.429 ± 0.062	3.2392 ± 0.091
549.855 ± 0.014	8.8773 ± 0.236
553.723 ± 0.276	0.3873 ± 0.104
561.264 ± 0.203	0.1970 ± 0.057
570.581 ± 0.030	13.2248 ± 0.723
573.867 ± 0.037	1.7325 ± 0.321
576.213 ± 0.144	1.0763 ± 0.073
579.781 ± 0.020	16.7838 ± 0.422
609.782 ± 0.198	0.3556 ± 0.080
626.720 ± 0.024	1.7814 ± 0.062
646.065 ± 0.060	0.1963 ± 0.029
688.595 ± 0.102	0.6343 ± 0.058
703.401 ± 0.028	100.0000 ± 0.000
717.352 ± 0.104	0.7737 ± 0.154
720.752 ± 0.232	0.3716 ± 0.093
723.673 ± 0.118	0.5270 ± 0.093
729.676 ± 0.225	0.2556 ± 0.164
744.881 ± 0.075	2.3298 ± 0.150
759.120 ± 0.049	3.7326 ± 0.203
761.470 ± 0.099	2.1655 ± 0.225
772.096 ± 0.319	0.1131 ± 0.024
780.837 ± 0.081	1.5576 ± 0.270
787.764 ± 0.130	0.3587 ± 0.051
795.692 ± 0.075	0.4267 ± 0.128
800.918 ± 0.141	0.4230 ± 0.064
806.431 ± 0.036	0.5268 ± 0.153
813.813 ± 0.046	1.4921 ± 0.196
828.213 ± 0.126	0.9586 ± 0.107
852.779 ± 0.255	0.2291 ± 0.043
860.103 ± 0.089	1.4043 ± 0.274
871.861 ± 0.089	1.3232 ± 0.132
890.087 ± 0.021	2.0553 ± 0.129
894.649 ± 0.027	2.6969 ± 0.215
901.776 ± 0.144	0.4171 ± 0.067
910.843 ± 0.037	5.2527 ± 0.202
922.027 ± 0.162	0.2010 ± 0.031
931.249 ± 0.686	0.1803 ± 0.004
950.763 ± 0.067	1.2799 ± 0.076
971.488 ± 0.147	0.9149 ± 0.169
978.513 ± 0.350	0.1241 ± 0.027
987.557 ± 0.300	53.3456 ± 1.175
1001.896 ± 0.039	1.7873 ± 0.191
1014.218 ± 0.059	3.2914 ± 0.181
1031.720 ± 0.276	0.1260 ± 0.024
1043.661 ± 0.041	23.8732 ± 1.309
1051.613 ± 0.751	0.1545 ± 0.015
1065.952 ± 0.105	0.3447 ± 0.033
1072.212 ± 0.084	0.9954 ± 0.117
1107.539 ± 0.107	0.3714 ± 0.081
1189.943 ± 0.027	7.3423 ± 0.348
1199.814 ± 0.210	0.6296 ± 0.160
1208.569 ± 0.075	1.6523 ± 0.121
1216.294 ± 0.272	0.2327 ± 0.044
1264.422 ± 0.310	0.8413 ± 0.102
1351.471 ± 0.099	3.5151 ± 0.375
1438.606 ± 0.236	0.3400 ± 0.060
1500.944 ± 0.609	0.9486 ± 0.333
1522.121 ± 1.091	0.6432 ± 0.147
1550.807 ± 0.232	2.5417 ± 0.654
1563.112 ± 0.130	0.7286 ± 0.145
1577.395 ± 0.273	0.5087 ± 0.180
1594.466 ± 2.001	0.3268 ± 0.152
1614.343 ± 0.045	6.3607 ± 0.417
1619.059 ± 0.207	0.7599 ± 0.067
1764.274 ± 0.063	93.0470 ± 3.98
1775.795 ± 0.038	11.0945 ± 0.691
1818.270 ± 0.399	0.1727 ± 0.034
1844.759 ± 0.400	0.0663 ± 0.004
1861.691 ± 0.019	17.4466 ± 0.871
1903.346 ± 0.056	6.7117 ± 0.332
1965.948 ± 0.558	0.0484 ± 0.020
2565.467 ± 0.844	0.1624 ± 0.058
2606.776 ± 0.412	0.0681 ± 0.019
$_{83}Bi^{206}$	
107.2	
123.6	
184.0	21†
202.7	
234.4	
262.8	2.9
313.7	3.0
343.5	26
386.1	
398.1	10
497.1	18
516.3	46
537.6	34
620.7	4.0
632.4	4.5
657.3	5.3
740	3.6
754	2.7
803.1	100
816	
842	
880.8	73
895.0	19
1018.9	8
1099	13
1405	
1596	8
1720	36
1845	
1880	
1903	
$_{83}Bi^{207}$	
569.684 ± 0.014	100†
897.3	0.16
1063.610 ± 0.04	76
1430	0.16
1769.7	8

GAMMA ENERGIES AND INTENSITIES OF RADIONUCLIDES (*Continued*)

Gamma energies (KeV)	Gamma intensities
$_{83}Bi^{208}$	
2614.611 ± 0.06	100%
$_{83}Bi^{210m}$	
262	10†
300	30
340	
610	
$_{83}Bi^{211}$	
350.7	15.9%
$_{83}Bi^{212}$	
39.85	
144.9	
288.2	0.28%
328.0	0.11%
434	
453	
473	
493	
616	
727	7.1%
785	1.1%
893	0.42%
953	0.1%
1074 } 1079 }	0.6%
1513	0.31%
1620	1.8%
1800 } 1809 }	0.1%
$_{83}Bi^{213}$	
320	
437	
$_{83}Bi^{214}$	
63	
191	
450	1%
609.3	47%
666	2%
703	0.8%
721	0.7%
768.7	5.3%
787	1.2%
806	1.5%
821	
825	
874	0.4%
934.8	3.3%
960	0.5%
1050	0.5%
1120	16%
1155	1.8%
1207	0.6%
1238	6.0%
1281	1.7%
1379	4.8%
1390–1400 complex	4%
1416	
1509	2.4%

Gamma energies (KeV)	Gamma intensities
1541	0.8%
1583	0.9%
1600	0.6%
1661	1.2%
1681	0.2%
1728	3.2%
1764	17%
1836	0.3%
1848	2.3%
1877	0.2%
1897	0.3%
2017	0.07%
2090	
2117	1.3%
2162	
2204	6%
2435	2.0%
2700	0.04%
2770	0.04%
2890	0.04%
2990	0.04%
3070	0.04%
$_{84}Po^{206}$	
59.9	
82.9	
107?	
117.6	
140.6	
146?	
170.8	
171.5	
181?	
282.1	
286.5	35†
311.5	
338.4	41†
354.8	
369?	
381?	
453?	
463.5	
469.1	
511.4 } 522.4 }	100†
538?	
555	
580	
646	
669	
678	
807.6	57†
818?	
862	
981	
1008	
1033	84†
1320	
$_{84}Po^{207}$	
100	
139.7?	
149.6	
156.1	
158.0	
177.7	
205.2	
214.4?	

Gamma energies (KeV)	Gamma intensities
222.0	
222.7	
224.0	
249.6	
288.0	
297.1?	
307.5	
330.2	
345.2	
369.3	
380.1?	
390.1?	
402.4	
405.7	
468.5?	
503.3	
531.7	
629.9	
669.5	
688	
698?	
742.7	
771	
892	
911.8	
948	
955	
992.6	
1020	
1149	
1212	
1280	
1318	
1361	
1373	
1586	
1663	
1747	
1763	
1797	
1847	
2061	
2600?	
$_{84}Po^{208}$	
285	≈0.003%
600	≈0.006%
$_{84}Po^{209}$	
260.5 } 262.8 }	≈0.4%
910	≈0.5
$_{84}Po^{210}$	
803	1.2×10^{-3}%
$_{84}Po^{211m}$	
0.90	1.7%
560	≈80
106.0	≈80
$_{84}Po^{211}$	
880	
561	

Gamma energies (KeV)	Gamma intensities
$_{84}Po^{212m}$	
570	≈2%
2610	2.6
$_{84}Po^{214}$	
800	0.014%
$_{84}Po^{215}$	
≈443	
$_{85}At^{206}$	
68	
$_{85}At^{208}$	
180	20†
250	
660	80
120	
$_{85}At^{209}$	
90.8	
195.0	24†
545	66
780	100
$_{85}At^{210}$	
46.5	
116.5	0.68†
202.0	0.17
245.3	79
298.9	0.13
316.9	0.14
402.4	0.78
499.1	0.13
506.9	0.63
518.7	0.13
527.6	1.1
544.8	0.09
584.0	0.32
602?	
615.1	0.31
623.1	0.52
630.9	0.30
639.5	0.23
643.7	0.46
701.1	0.43
726	?
817.8	1.7
853.1	1.4
882	0.25
909	?
930.6	0.96
956.7	1.8
977	0.63
1088	0.28
1180	100
1204	?
1289	0.46
1324	0.36
1436	29
1483	48
1552	0.16
1599	14

Gamma energies (KeV)	Gamma intensities
1648	0.09
1719	0.11
1955	0.38
2002	0.11
2052	0.05
2226	0.05
2239	?
2254	1.4
2273	0.33
2285	?
2306	0.029
2353	0.13
$_{85}At^{211}$	
670	Weak
$_{85}At^{212m}$	
63	
$_{85}At^{212}$	
63	
$_{85}At^{215}$	
≈400	
$_{85}At^{217}$	
268	
455	
595	
$_{86}Rn^{211}$	
γ with EC	
32	
113.9	
168.6	
232	
264	
296	
333	
445	39†
680	100
865	24
946	29
1130	31
1370	52
1800	0.8
γ with α	
68.7	100†
169	6.5
234	6.5
$_{86}Rn^{218}$	
609.4	0.2%
$_{86}Rn^{219}$	
130	
271	9%
371	≈0.002
380	≈0.0005
401	6.5
517	≈0.04
536	≈0.0003
562	≈0.002

Gamma energies (KeV)	Gamma intensities
606	≈0.0026
673	≈0.01
833	≈0.001
888	≈0.001
1053	≈0.0003
$_{86}Rn^{220}$	
542	≈0.07%
$_{86}Rn^{222}$	
510	0.07%
$_{87}Fr^{212}$	
39	
62	
66	
70	
123	
$_{87}Fr^{219}$	
163	
189	
352	
493	
530	
$_{87}Fr^{221}$	
63	9.1%
68	0.4
91?	
98	0.05
99.5	0.12
118.0	0.04
150.0	0.08
171.3	0.08
218.0	12.5
268?	<0.04
282.8	~0.01
303	~0.03
324.1	0.02
359.1	0.04
381.8	0.04
409.1	0.15
$_{87}Fr^{223}$	
49.8	40†
50.2	100
61.5	<0.8
79.8	25
100.3	≈3
134.4	1.6
173.5	0.4
184.8	0.9
205.0	3.2
234.8	10
246	0.13
250.2	0.13
256.2	0.13
286.1	0.05
289.7	≈0.9
300.0	0.17
304.5	0.07
307	0.09
312.7	0.08
318	≈1.7

Gamma energies (KeV)	Gamma intensities
329.8	0.10
334.3	0.05
338.7	0.2
342.5	0.09
369.0	0.32
723	0.15
746.5	0.07
756	0.04
767	0.08
776.7	1.23
784	0.07
793	0.03
798	0.03
804	0.17
813	0.06
822	0.03
826	0.14
835	0.02
841	0.02
847	0.14
860	0.01
864	0.01
876.5	0.13
892	0.01
898	0.05
908	0.04
$_{88}Ra^{220}$	
465	1%
$_{88}Ra^{221}$	
89	3.5%
152	13
176	2
219	0.1
293	0.6
320	0.7
415	0.5
$_{88}Ra^{222}$	
324.6	3.9%
325	8×10^{-3}
473	7×10^{-3}
526	3×10^{-3}
795	3×10^{-3}
846	4×10^{-3}
$_{88}Ra^{223}$	
31.2	
≈68?	≈0.4%
122.2	2
131	≈3
143.1?	
144.2	4.1
148	
154.1	5.5
158.6	1.7
179.6	0.5
269.6	10
288	≈0.2
324	2.3
330	≈5.1
338	2.8
362	≈0.07
371	0.45
441	0.2

Gamma energies (KeV)	Gamma intensities
446	0.4
580	≤0.1
$_{88}Ra^{224}$	
240.98	
290	3.7%
410	9×10^{-3}
650	4×10^{-3}
	6×10^{-3}
$_{88}Ra^{225}$	
40	33%
$_{88}Ra^{226}$	
185.7	
260	100†
420	0.29
450	0.021
610	0.009
	0.033
$_{88}Ra^{227}$	
27.3	
291	
498	4%
	0.6
$_{88}Ra^{228}$	
10.3	
26.3	
$_{89}Ac^{223}$	
82	
96	0.23%
120	0.2
≈170	?<0.05
	?≤0.8
$_{89}Ac^{224}$	
132	
216	100†
	224
$_{89}Ac^{225}$	
36.6	
38.4	
62.8	
75	0.5%
85.5	1.2
87.5	3
90	≤2
96	≤0.5
100	≤1
109	~3
123	0.6
134	0.3
146	~0.1
150	0.2
158	1
187	0.8
196	0.8
216	0.3
241	1
254	0.03
324	0.3
340	Weak

GAMMA ENERGIES AND INTENSITIES OF RADIONUCLIDES (*Continued*)

Gamma energies (KeV)	Gamma intensities
351	0.06
362	0.1
373	Weak
384	0.06
453	0.3
482	0.1
517	0.04
529	0.06
$_{89}Ac^{226}$	
γ with β^-	
72.13	
158.5	100†
230.4	151
γ with EC	
67.8	
185	100†
255	132
$_{89}Ac^{227}$	
γ with β^-	
9.3	
15.2	
24.5	
γ with α	
12.7	
70.2	12†
83 } 87 }	≈27
99.5	20
107	1
121.5	4
134	2
147	6
160	14
172	4
190?	<0.5
$_{89}Ac^{228}$	
57.5	
78.1	
99	
113?	
129.1	5†
155?	
178.2	
184	1.7
209	4.3
232	
270	4.1
282	0.3
321	
327.5	5.3
338	15
410	2.5
443	
463	3.6
473	
484	
510	
555	
573	
619	
752	
769	
783	1.4
796	4
836	2.5
831	8
912.2	23
922	
966	20
1035	
1095	
1464	1.1
1503	2.5
1593	4.5
1642	2.4
$_{89}Ac^{231}$	
185	
280	
390	
710	
$_{90}Th^{224}$	
177	9%
235	0.4
297	0.3
410	0.8
$_{90}Th^{225}$	
151	1†
246	5
322	30
362	5
450	1
490	1
$_{90}Th^{226}$	
111.1	4.8%
131.3	0.34
190.6	0.15
206.5	0.26
197	0.4
242.4	1.2
$_{90}Th^{227}$	
29.9	
31.6	
44.5	1.4%
45.1	
48.3	
50.2	10.5
56.1	<0.2
61.5	0.8
68.8	≈0.1
79.8	≈2.6
94.0	≈3
100.3	<0.3
113.1	≈2
141.6	0.13
154.2	0.1
173.5	≲0.1
184.8	≲0.1
205.0	0.32
206.2	0.5
210.7	1.0
212.0	
218.8	0.08
234.8	
236.0	12.0
247.7	1.4
250.2	0.7
250.4	
256.26	6.4
273.0	0.46
279.8	≲0.05
281.5	0.30
286.1	1.8
289.7	<0.3
296.6	0.40
300.0	2.5
304.5	1.2
312.7	0.83
318	0.2
323	≲0.1
329.8	2.8
334.3	1.2
342.5	0.46
351?	≲0.2
371	0.03
384	0.06
404?	≲0.04
671	0.002
760	0.002
776	0.0015
$_{90}Th^{228}$	
84.5	1.6%
132	0.19
167	0.12
205	0.03
216	0.29
234?	7×10^{-5}
$_{90}Th^{229}$	
17.3	
23.7	
25.3	
32	
42.7	
56.7	
68.9	
75.1	
86.3	
107.2	
124.4	
131.9	
137.0	
143.0	
154.2	
156.5	
179.9	
193.4	
210.7	
217.0	
242.2	
269?	
$_{90}Th^{230}$	
67.8	0.59%
110	1×10^{-4}%
142	0.07%
184	1.4×10^{-2}%
206	$\approx 5 \times 10^{-6}$%
235	$\approx 5 \times 10^{-6}$%
253	1.7×10^{-2}%
257	$\approx 8 \times 10^{-4}$%
$_{90}Th^{231}$	
17.21	
18.07	
25.6	
42.8	0.7
58.54	6
63.79	
68.50	
72.7	3.4
76.04	
81.22	13
82.07	7
84.20	100
89.94	15
99.27	2.1
102.2	7
106.8	0.34
112.6	0.29
115.4?	0.03
116.8	0.38
124.9	1.0
134.0	0.42
135.66	1.3
145.90	0.5
163.10	2.4
164.7	0.13
174.1	0.29
183.4	0.5
188.7	0.05
218.0	0.6
236.1	0.13
249.8	0.017
267.8	0.02
$_{90}Th^{232}$	
59	
$_{90}Th^{233}$	
29.35	690†
57.15	23
74.7	22
86.00	1150
131.1	24
148.2	6
151.5	4
153.6	28
162.5	96
169.1	190
179.0	16
186.8	14
190.54	55
195.0	70
201.6	13
210.1	15
211.3	8
216.6	6.3
226.1	9.6
250.5	2
252.9	5
257.3	29
278.7	3.3

GAMMA ENERGIES AND INTENSITIES OF RADIONUCLIDES (*Continued*)

Gamma energies (KeV)	Gamma intensities
285.5	9
346.6	5
359.9	50
361.4	16
368.0	2
377.2	16
399.0	6
408.8	1.6
412.5	5.7
418.4	5
431.0	9.8
433.0	6
441.0	103
447.7	62
459.2	600
467.5	7.8
473.9	1.5
490.8	74
497.1	9
499.0	91
505.5	2.1
513.6	8.6
517.0	2.9
526.5	27
531.8	1.8
552.1	10
554.9	1.5
562.7	30
573.5	18
595.2	69
599.1	20
609.9	36
642.3	12
663.4	1
669.7	290
677.8	37
681.2	8
698.4	5
703.6	4.5
707.8	5
716.8	24
724.9	37
740.9	13
744.9	2.9
751.6	1
757.6	18
764.3	49
774.0	5.9
782.7	2.6
784.2	2.1
804.8	13
806.4	5.7
811.4	3.3
815.9	12
817.0	6.7
846.8	0.6
849.3	2
871.0	0.9
873.8	3.5
880.7	3.3
890.6	61
898.4	1.4
935.9	21
942.0	3.3
948.0	3.2
960.8	2.9
962.8	0.6
969.1	4.9
978.1	3.2
984.8	6.2
995	0.4

Gamma energies (KeV)	Gamma intensities
1001	0.5
1008	1.2
1012	1.7
1146	1.2
$_{90}Th^{234}$	
29.9	
63.2 }	
63.6 }	85†
69.8	
93.1 }	
93.5 }	100
$_{91}Pa^{227}$	
50	<20†
65	62
67	12
110	20
$_{91}Pa^{228}$	
γ with α	
130	27†
150	34
170	11
200	14
220	10
240	55
280	49
310	100
345	21
γ with EC	
57.5	
77.7	
99	13†
129.1	
137.8	
156	
178.0	
184	
191.3?	55
209	320
224	122
270	320
278	71
282	310
327.5	950
338	740
341.1	250
397	
410	1350
445.4	
462	320
463	2900
469.2	
617	
622	
641	
662	
666	
670	
680	
694	
704	
713	135
732	

Gamma energies (KeV)	Gamma intensities
739.6	
745.6	
756.2	230
773.4	375
782.6	
792.2	≈70
796	≤600
817.4	
831.4	375
835.8	
836.4	730
841.0	170
853.8	
871.0	520
889.0	
905.2	310
912.2	3100
923	930
966.0	2200
970.0	1680
976.8	620
1034.1	
1095	
1123	
1168	
1253	130
1293	83
1423	
1464	155
1489	
1503	
1563	15
1593	620
1624	
1678	
1708	
1744	230
1758	
1838	
1888	415
$_{91}Pa^{232}$	
γ with EC	
42.37	
γ with α	
26?	
40	
51	
52	
64	
65	
71	
$_{91}Pa^{230}$	
γ with Ec	
53	
122	
230	
280	
320	
340	
379	
380	
398	135†
401	100
444	
456	500

Gamma energies (KeV)	Gamma intensities
464	
508	420
520	
572	100
624	
634	
730	210
783	160
901 }	
920 }	1600
954	3500
1012	
1027	240
γ with β^-	
51.7	
$_{91}Pa^{231}$	
18.9	
25.3	
27.3	
29.9	
38.1	
44.0?	
52.6	
57.1	
63.5	
74.0	
77.2	
96.7	
100.7	
102.6	
244	
257	
259.8	
274	
283.1	
299.4	
302.0	
313	
329.2	
342	
356.3	
409	
437	
439	
488	
513	
517	
$_{91}Pa^{232}$	
47.6	
80.2	0.1%
81.2?	0.02
105.4	2.1
109.0	3.0
132.5	0.02
139.2	0.7
150.1	12
174.9	0.025
178	0.020
183.9	1.6
290	0.03
388.0	
422.0	
454.2	
472.8	4
516	3
564	

GAMMA ENERGIES AND INTENSITIES OF RADIONUCLIDES (*Continued*)

Gamma energies (KeV)	Gamma intensities
584	6
645	
676	
687	
693	
712	
757	
820	
865	
868	
895	
971	
$_{91}Pa^{233}$	
17.26	
28.54	
40.35	0.015%
41.65	
57.9	
75.28	0.8
86.59	1.7
95.8	
103.86	0.7
145.4	0.44
271.6	0.29
300.2	6.3
303.1	
311.9	34
320	
340.5	3.9
375.4	0.56
398.6	1.1
415.9	1.5
$_{91}Pa^{234m}$	
43.5	
236	
255	0.05%
746	Weak
765	0.36
790	Weak
811	
1001	0.59
$_{91}Pa^{234}$	
43.3	
99.8	
126.3	≈24†
153.0	9
186.0	≈3
196.7	4
208	? ≤16
223.9	≈2
226.7?	
227.3	12
287	≈10
294.3?	
323	? ≈3
355	? ≈6
370.1	5.3
565	15
694	
727	
791	≤5
804	

Gamma energies (KeV)	Gamma intensities
822	
873	
875	
878	70
920	
922	
941	
976	
1020	≈8
1130	?3
1340	?2
1410	?6
1620	?3
1850	?1
$_{91}Pa^{235}$	
No γ	
$_{91}Pa^{237}$	
90	50†
145	45
205	55
275	20
330	40
405	30
460	100
550	30
590	25
750	50
800	45
870	100
920	100
1040	35
1320	10
1420	15
$_{92}U^{228}$	
152	20†
187	30
246	40
$_{92}U^{230}$	
72.13	0.54%
154.4	0.21
158.8	0.13
230.6	0.18
$_{92}U^{231}$	
18.05	
25.64	12%
58.54	
68.5	
81.3	
82.1	
84.18	7
108.2	
220	1
$_{92}U^{232}$	
57.6	0.21%
129.0	0.082
270.5	3.8×10^{-3}
327.8	3.4×10^{-3}
500	2.2×10^{-5}

Gamma energies (KeV)	Gamma intensities
$_{92}U^{233}$	
29.0	60†
42.4	310
54.6	68
66.0	3.9
71.8	12
97.1	100
117.1	13
118.9	16
120.7	11
124.0	2.7
135.3	10
144.8	9
146.4	26
164.5	27
187.9	8
208.1	9
217.0	16
245.4	14
248.4	6
260	1.2
261.5	1.4
268.1	1.0
274.5	2.0
277.9	6
288.0	4.9
291.2	23
316.8	32
320.2	11
323.1	3.2
328.2	0.1
336.4	2.3
353.9	0.23
365.5	3.2
383.0	0.4
393.3	0.07
$_{92}U^{234}$	
53.3	100†
120.9	34
460	
510	
580	
$_{92}U^{235}$	
110	2.5%
143	11
163	5
180	0.5
185	54
201	0.8
204	5
$_{92}U^{235m}$	
0.075	
$_{92}U^{236}$	
≈50	
$_{92}U^{237}$	
13.81	
26.35	2.4%
33.20	
38.54	
43.42	

Gamma energies (KeV)	Gamma intensities
51.01	0.2
59.54	36
64.83	1.3
114.09	
164.61	2.0
208.00	23
221.80	0.022
234.40	0.021
267.54	0.76
292.7	0.003
332.36	1.3
335.38	0.10
337.7	0.008
368.59	0.050
370.94	0.12
$_{92}U^{238}$	
48	
$_{92}U^{239}$	
43.520 ± 0.042	104.553%
74.660 ± 0.022	1106.989%
83.822 ± 0.307	3.094%
86.716 ± 0.065	1.392%
110.973 ± 0.222	0.442%
117.658 ± 0.030	3.204%
174.009 ± 0.028	0.398%
186.149 ± 0.020	0.729%
187.396 ± 0.036	0.148%
190.345 ± 0.067	0.067%
191.973 ± 0.060	0.062%
196.850 ± 0.100	0.049%
201.186 ± 0.065	0.045%
210.960 ± 0.091	0.072%
220.531 ± 0.017	0.575%
228.141 ± 0.016	
231.70 ± 0.10	0.069%
255.27 ± 0.123	0.063%
258.438 ± 0.046	0.058%
260.792 ± 0.112	0.051%
296.963 ± 0.070	0.035%
301.862 ± 0.074	0.036%
304.123 ± 0.117	0.033%
306.836 ± 0.042	0.103%
312.049 ± 0.026	0.090%
317.182 ± 0.107	0.076%
321.901 ± 0.132	0.027%
326.227 ± 0.036	0.080%
334.255 ± 0.018*	
345.060 ± 0.038	0.045%
357.858 ± 0.036	0.106%
361.814 ± 0.063	0.072%
363.064 ± 0.217	0.020%
373.524 ± 0.016	0.426%
378.055 ± 0.028	0.189%
395.353 ± 0.127	0.022%
399.3 ± 0.2	0.011%
407.645 ± 0.051	0.057%
434.751 ± 0.045	0.075%
445.732 ± 0.085	0.039%
448.141 ± 0.024	0.177%
452.054 ± 0.089	0.039%
462.435 ± 0.124	0.036%
474.487 ± 0.109	0.067%
486.865 ± 0.016	1.18%
492.594 ± 0.045	0.086%
499.091 ± 0.177	0.033%
504.707 ± 0.053	0.089%
506.462 ± 0.109	0.039%
517.851 ± 0.084	0.065%

GAMMA ENERGIES AND INTENSITIES OF RADIONUCLIDES (*Continued*)

Gamma energies (KeV)	Gamma intensities
522.080 ± 0.076	0.053%
532.887 ± 0.116	0.040%
541.221 ± 0.078	0.054%
544.608 ± 0.059	0.089%
548.099 ± 0.126	0.044%
560.511 ± 0.047	0.131%
575.295 ± 0.026	0.269%
577.640 ± 0.222	0.028%
587.726 ± 0.038	0.486%
589.709 ± 0.116	0.137%
602.659 ± 0.076	0.101%
624.016 ± 0.045	0.168%
631.080 ± 0.017	1.518%
662.238 ± 0.016	3.956%
664.178 ± 0.093	0.203%
691.091 ± 0.044	0.127%
695.246 ± 0.083	0.084%
697.394 ± 0.232	0.031%
700.748 ± 0.079	0.062%
703.239 ± 0.077	0.065%
707.379 ± 0.084	0.072%
710.171 ± 0.208	0.029%
714.155 ± 0.073	0.092%
722.865 ± 0.025	0.623%
727.345 ± 0.155	0.051%
730.972 ± 0.026	0.254%
748.108 ± 0.016	2.234%
767.643 ± 0.100	0.097%
772.809 ± 0.175	0.075%
774.742 ± 0.034	0.372%
779.78 ± 0.200	0.010%
788.187 ± 0.056	0.105%
791.322 ± 0.031	0.204%
793.576 ± 0.084	0.058%
812.933 ± 0.017	1.732%
819.232 ± 0.018	3.176%
831.834 ± 0.075	0.079%
840.806 ± 0.250	0.080%
844.091 ± 0.015	3.515%
846.466 ± 0.060	0.880%
866.89 ± 0.300	0.040%
868.06 ± 0.300	0.040%
869.88 ± 0.300	0.040%
874.215 ± 0.070	0.093%
876.029 ± 0.140	0.048%
881.353 ± 0.169	0.027%
884.497 ± 0.031	0.264%
889.506 ± 0.036	0.525%
895.412 ± 0.130	0.035%
913.569 ± 0.076	0.057%
917.30 ± 0.30	0.070%
918.80 ± 0.40	0.080%
920.70 ± 0.40	0.070%
921.80 ± 0.40	0.050%
922.80 ± 0.50	0.030%
928.154 ± 0.043	0.126%
933.057 ± 0.037	0.791%
939.241 ± 0.284	0.014%
943.145 ± 0.180	0.034%
948.713 ± 0.173	0.027%
953.677 ± 0.190	0.030%
959.322 ± 0.305	0.167%
961.235 ± 0.167	0.403%
964.351 ± 0.037	2.156%
974.609 ± 0.079	0.082%
992.169 ± 0.066	0.076%
1000.996 ± 0.300	0.035%
1002.392 ± 0.200	0.035%
1009.744 ± 0.114	0.044%
1012.659 ± 0.117	0.040%
1032.238 ± 0.084	0.076%

Gamma energies (KeV)	Gamma intensities
1040.567 ± 0.199	0.021%
1065.70 ± 0.300	0.010%
1097.070 ± 0.102	0.052%
1122.40 ± 0.300	0.015%
1144.80 ± 0.300	0.015%
$_{92}U^{240}$	
440	
$_{93}Np^{234}$	
43.49	
99.7	
233.6	
234.6	
238.6?	
247.9	
266?	
298?	
450	
483?	
485?	
516	
526	
557	
744	
752	
768	
788	
793	
810	
812	
854	
1003	
1105	
1196	
1240	
1395	
1439	
1531	
1562	
1575	
1606	
$_{93}Np^{235}$	
26	15†
84	9
$_{93}Np^{236}$	
γ with EC	
45.32	
643	
688	
γ with β^-	
44.6	
$_{93}Np^{237}$	
20?	
29.6	14†
55.0?	
56.8	
84.5?	
85.9	14
105.0	
108.4	
133.5	
144.5	0.8

Gamma energies (KeV)	Gamma intensities
160?	
207	
240	
$_{93}Np^{238}$	
44.1	
101.7	
119.8	
221.0	
292.6?	
871	
885	
925	
940	
943	
986	83†
1027 }	100
1030 }	
1034	
1095?	
$_{93}Np^{239}$	
44.65	
49.41	
57.26	
61.48	
67.86	
88.06	
106.13	22.8†
106.47	
166.4	
181.7	0.077
209.726 ± 0.02	3.4
226.4	
228.2 ± 0.1	11.8
254.2	0.12
272.809 ± 0.09	0.089
277.56 ± 0.015	14.1
285.37 ± 0.04	0.82
315.83 ± 0.04	1.5
334.25 ± 0.02	2.1
392.4	0.0016
429.5	0.0037
434.7	0.012
447.6	2.6×10^{-4}
454.2	8.0×10^{-4}
461.9	0.0016
469.8	0.0011
484.3	0.0010
492.3	0.0060
497.8	0.0031
498.7	
504.2	7.7×10^{-4}
$_{93}Np^{240m}$	
42.9	
260	1.9%
304	0.9
557	21
599	13
760	1.3
792?	
816	1.6
820	0.3
844?	
860	
863	

Gamma energies (KeV)	Gamma intensities
898	1.2
940	0.3
945	1.9
1092?	
1490	1.5
1530	1.9
1620	0.7
$_{93}Np^{240}$	
85	
160	
245	
440	
560	
600	
920	
1000	
1160	
$_{94}Pu^{234}$	
0.047	
$_{94}Pu^{236}$	
47	0.031%
110	0.012
165	6.6×10^{-4}
452	1.7×10^{-4}
570	1.0×10^{-4}
640	2.4×10^{-4}
$_{94}Pu^{237m}$	
145	100%
$_{94}Pu^{237}$	
33.2	
43.5	
55.6?	
59.6	
76.4?	
96.0?	
$_{94}Pu^{238}$	
43.49	0.038%
99.8	8×10^{-3}
153.1	1×10^{-3}
203	4×10^{-6}
743	9×10^{-6}
767	3.5×10^{-5}
786	6×10^{-6}
810	$<1 \times 10^{-6}$
$_{94}Pu^{239}$	
38.70 ± 0.03	$(1.63 \pm 0.16)^b$
41.99 ± 0.10	$(2.8 \pm 0.4)^d$
46.19 ± 0.07	$(2.35 \pm 0.30)^c$
51.61 ± 0.03	$(1.35 \pm 0.14)^e$
54.01 ± 0.05	$(1.13 \pm 0.20)^c$
56.82 ± 0.04	$(7.3 \pm 0.8)^c$
59.57	^{241}Am
65.73 ± 0.08	$(2.7 \pm 0.5)^d$
67.70 ± 0.05	$(1.4 \pm 0.3)^c$
68.72 ± 0.06	$(4.2 \pm 0.6)^c$
72.79	Pb $K_{\alpha 2}$
74.96	Pb $K_{\alpha 1}$
77.60 ± 0.05	$(4.1 \pm 0.6)^d$

$a = \times 10^{-2}$, $b = \times 10^{-3}$, $c = \times 10^{-4}$, $d = \times 10^{-5}$, $e = \times 10^{-6}$.

Gamma energies (KeV)		Gamma intensities
78.48 ± 0.05		(1.7 ± 0.2)^c
84.30	Pb $K_{\beta3}$	
84.82	Pb $K_{\beta1}$	
87.34	Pb $K_{\beta2}$	
89.94 ± 0.06		(1.1 ± 0.2)^d
94.64	U $K_{\alpha2}$	
98.46	U $K_{\alpha1}$	
99.46	Pu $K_{\alpha2}$	
102.93 ± 0.08		(2.6 ± 0.4)^c
103.64	Pu $K_{\alpha1}$	
110.38	U $K_{\beta3}$	
111.29	U $K_{\beta1}$	
114.55	U $K_{\beta2}$	
115.32 ± 0.05		(7.4 ± 0.9)^c
116.18 ± 0.05		(7.1 ± 0.8)^c
117.15	Pu $K_{\beta1}$	
119.89 ± 0.08		(2.8 ± 0.4)^d
120.59	Pu $K_{\beta2}$	
122.35 ± 0.12		(3.0 ± 1.5)^e
123.73 ± 0.08		(1.9 ± 0.3)^d
124.41 ± 0.12		(4.8 ± 0.8)^d
125.00 ± 0.12		(4.8 ± 0.8)^d
129.27 ± 0.03		(5.6 ± 0.7)^b
141.66 ± 0.06		(9.8 ± 0.9)^d
144.18 ± 0.08		(2.7 ± 0.4)^d
146.07 ± 0.08		(9.8 ± 1.0)^d
159.6 ± 0.2		(2.0 ± 1.0)^e
160.07 ± 0.13		(5.0 ± 2.0)^e
161.47 ± 0.05		(1.0 ± 0.2)^c
167.95 ± 0.18		(2.0 ± 0.1)^e
171.37 ± 0.08		(9.8 ± 0.9)^d
173.61 ± 0.10		(2.9 ± 0.4)^e
179.20 ± 0.08		(2.9 ± 0.3)^d
184.53 ± 0.15		(2.0 ± 0.1)^e
188.13 ± 0.10		(4.2 ± 0.6)^e
189.34 ± 0.07		(7.4 ± 0.9)^d
193.13 ± 0.12		(5.9 ± 0.8)^e
195.68 ± 0.07		(8.9 ± 1.0)^d
203.55 ± 0.04		(4.8 ± 0.4)^c
209.34 ± 0.08		(1.1 ± 0.3)^d
225.46 ± 0.08		(1.1 ± 0.2)^d
237.6 ± 0.2 238.2 ± 0.2		(1.4 ± 0.2)^d
242.08 ± 0.08		(9.3 ± 0.9)^e
243.38 ± 0.07		(2.0 ± 0.5)^d
244.90 ± 0.09		(4.9 ± 0.8)^e
248.82 ± 0.08		(7.5 ± 0.9)^e
255.38 ± 0.07		(6.6 ± 0.8)^d
258.20 ± 0.10		(3.8 ± 0.6)^e
263.92 ± 0.07		(1.9 ± 0.3)^d
265.68 ± 0.12		(3.0 ± 0.9)^e
297.48 ± 0.07		(4.1 ± 0.6)^d
302.84 ± 0.10		(7.2 ± 0.9)^e
307.71 ± 0.10		(3.9 ± 0.8)^e
311.74 ± 0.06		(2.0 ± 0.2)^d
316.44 ± 0.08		(9.9 ± 1.2)^e
320.83 ± 0.06		(4.6 ± 0.5)^d
323.81 ± 0.06		(4.4 ± 0.5)^d
332.84 ± 0.04		(4.3 ± 0.4)^c
336.14 ± 0.05		(9.4 ± 1.1)^d
341.51 ± 0.06		(6.0 ± 0.8)^d
345.02 ± 0.03		(5.0 ± 0.5)^c
353.60 ± 0.10		(2.7 ± 0.6)^e
361.90 ± 0.08		(1.1 ± 0.2)^d
367.08 ± 0.08		(8.2 ± 0.9)^d
368.57 ± 0.08		(8.1 ± 0.9)^d
371.67 ± 0.10		(3.9 ± 0.6)^e
375.04 ± 0.03		(1.5 ± 0.1)^b
380.19 ± 0.05		(3.0 ± 0.3)^c
382.77 ± 0.05		(2.5 ± 0.3)^c
392.7 ± 0.2		(2.7 ± 0.5)^c

Gamma energies (KeV)	Gamma intensities
393.1 ± 0.2	(2.7 ± 0.5)^c
399.54 ± 0.09	(9.0 ± 1.5)^e
410.92 ± 0.06	(1.4 ± 0.3)^d
413.71 ± 0.03	(1.5 ± 0.1)^b
422.62 ± 0.05	(1.2 ± 0.2)^c
426.79 ± 0.08	(3.3 ± 0.5)^d
430.12 ± 0.12	(4.8 ± 0.8)^e
445.78 ± 0.12	(7.5 ± 1.1)^e
451.49 ± 0.05	(1.8 ± 0.2)^c
457.57 ± 0.10	(1.2 ± 0.4)^e
461.09 ± 0.10	(3.0 ± 0.6)^e
481.58 ± 0.10	(5.1 ± 1.0)^e
558.31 ± 0.12	(2.4 ± 0.5)^e
617.45 ± 0.10	(2.1 ± 0.4)^e
618.6 ± 0.2	(2.1 ± 0.4)^e
633.3 ± 0.2	(1.2 ± 0.3)^e
640.05 ± 0.08	(5.7 ± 0.9)^e
645.80 ± 0.06	(1.3 ± 0.3)^d
651.98 ± 0.10	(5.1 ± 1.0)^e
655.20 ± 0.20	(1.5 ± 0.3)^e
658.82 ± 0.10	(9.6 ± 1.2)^e
703.52 ± 0.10	(4.8 ± 0.9)^e
717.59 ± 0.15	(1.8 ± 0.4)^e
756.10 ± 0.20	(3.9 ± 0.7)^e
769.12 ± 0.15	(9.6 ± 1.2)^e
$_{94}Pu^{240}$	
45.28	
103.6	
160	
642.3	1.4×10^{-5}%
687.8	4×10^{-6}
$_{94}Pu^{241}$	
76.8	0.7%
103.5	4.4
148.5	9.0
160.0	0.3
$_{94}Pu^{243}$	
12.2?	
29.7?	
36.7	
42.2	1%
54	
84	21
96	
134?	
381	0.7
$_{94}Pu^{246}$	
27	13%
44	≈30
80	3
100?	
180.4	10
224.0	25
$_{95}Am^{238}$	
360	
580	30†
950	
980	80
1350	

Gamma energies (KeV)	Gamma intensities
$_{95}Am^{239}$	
γ with α	
48	
γ with EC	
44.70	Weak
49.47	
57.31	
67.91	
181.8	
209.9	
226.5	
228.3	
277.6	
$_{95}Am^{240}$	
42.87	
98.9	
900	23%
1000	77
1400	<10
$_{95}Am^{241}$	
26.36	2.5%
33.21	0.17
43.44	0.07
55.56	
59.57	36
67.5	
99.00	0.024
102.97	0.019
106?	<0.001
114.2	4.3×10^{-4}
118.2	4×10^{-4}
123.2	1.8×10^{-3}
125.3	4.5×10^{-3}
146.6	1.6×10^{-4}
150.1	3.5×10^{-5}
154.4	8×10^{-6}
156.4	1.2×10^{-5}
158.3	9×10^{-6}
161.7	9×10^{-6}
164.6	2.5×10^{-5}
166.2	5×10^{-6}
169.6	8.5×10^{-5}
173.1	5×10^{-6}
175.0	7×10^{-6}
191.9	3×10^{-5}
197.0	1.3×10^{-5}
208.00	5.6×10^{-4}
221.4	3×10^{-5}
242.4	3×10^{-5}
245.0	3×10^{-5}
248.0	5.5×10^{-5}
267.4	3.5×10^{-5}
292.8	2×10^{-5}
294.9	5.5×10^{-5}
310.3	1.5×10^{-5}
312.1	8×10^{-5}
322.5	1.9×10^{-4}
332.4	1.4×10^{-4}
335.4	5.7×10^{-4}
343.3?	1.1×10^{-4}
351.8	9×10^{-5}
368.6	2.8×10^{-4}
371.0	6×10^{-5}
376.8	1.4×10^{-4}
383.6	5.5×10^{-5}
391.0	4×10^{-5}
419.4	6×10^{-5}

$a = \times 10^{-2}$, $b = \times 10^{-3}$, $c = \times 10^{-4}$, $d = \times 10^{-5}$, $e = \times 10^{-6}$.

GAMMA ENERGIES AND INTENSITIES OF RADIONUCLIDES (*Continued*)

Gamma energies (KeV)	Gamma intensities
426.8	3×10^{-5}
445.1	4.5×10^{-5}
454?	$<1.7 \times 10^{-5}$
462	1.4×10^{-5}
511.6	8.5×10^{-5}
525.7	8×10^{-5}
570	1.0×10^{-5}
591.5?	7×10^{-5}
597	1.0×10^{-5}
609.8	2.3×10^{-4}
619.6	8×10^{-5}
641	1.0×10^{-5}
653.1	9×10^{-5}
663.0	5.3×10^{-4}
676?	$<2 \times 10^{-6}$
680	6×10^{-6}
688.7	4×10^{-5}
695	1.0×10^{-5}
710	9×10^{-6}
723.2	4.8×10^{-4}
737	1.2×10^{-5}
757.3	1.0×10^{-4}
767.0	8×10^{-6}
770.7	1×10^{-5}
$_{95}Am^{242m}$	
γ with IT	
48.6	
γ with α	
42	
49.3	41
66.8	4.6
67.9	1.6
69	
73.3?	1.3
86.7	8
92.5	0.9
109.6	5.3
111.1	0.6
121.8	1.3
135.5	2.3
136.1	2.1
152.9	0.3
153.9	1.0
163.4	5.2
194.9	
206.4	
$_{95}Am^{242}$	
γ with β^-	
42.12	
γ with EC	
44.50	
$_{95}Am^{243}$	
31.42	0.03%
43.53	5.3
50.6	0.0027
55.4	0.0094
74.673	61
86.7	0.37
117.8	0.75
142.0	0.13
169	0.0012
195	0.00085
500?	
570?	

Gamma energies (KeV)	Gamma intensities
620	0.0003
650	0.0010
$_{95}Am^{244m}$	
42.9	
980?	<1%
1050	<1
$_{95}Am^{244}$	
42.9	
99.4	5%
154	19
206	
540	≈0.4
746	66
900	25
$_{95}Am^{245}$	
36	
56?	
78?	
140	
153	
240	1.2%
252.3	20
296	0.55
$_{95}Am^{246}$	
171	
228	
238	0.3%
244.6	1.5
245.4	1.5
256	0.2
262	0.14
270.3	1.0
288	0.1
402	0.3
734	1.2
752	1
759	0.8
799.0	28.6
834	2.2
987	1.1
1037	14.4
1063	19.3
1079	31.5
1086	1.7
$_{96}Cm^{239}$	
188	
$_{96}Cm^{241}$	
γ with α	
145	
γ with EC	
470	
600	
$_{96}Cm^{242}$	
44.11	0.039%
101.9	0.0035
157.7	0.0023
210	$\approx 2 \times 10^{-5}$
562	1.8×10^{-4}

Gamma energies (KeV)	Gamma intensities
605	1.4×10^{-4}
≈900	2×10^{-5}
941	3×10^{-5}
$_{96}Cm^{243}$	
106	
210	~5%
228	~6.5
227	11
$_{96}Cm^{244}$	
42.88	0.021%
100	1.5×10^{-3}
150	1.3×10^{-3}
264	1.2×10^{-4}
~570, ~610	2.5×10^{-4}
~830	7×10^{-5}
$_{96}Cm^{245}$	
≈130	5%
173	14
$_{97}Bk^{243}$	
γ with EC	
755	10†
840	30
946	8
γ with α	
42	4†
146.4	8
187.1	41
558	13
$_{97}Bk^{244}$	
γ with EC	
144.5	7†
154.0	3.6
177.0	5.0
187.6	16
217.6	100
233.8	2.9
333.5	10
490	14
745	6
870	5
892	88
922	17
988	4.0
1153	7
1178	4.0
1233	2.8
1252	2.2
1333	0.9
1505	2.0
$_{97}Bk^{245}$	
γ with EC	
252.7	31%
380.5	3.5
384.6	1.5
γ with α	
165.5	
207.4	
474	

Gamma energies (KeV)	Gamma intensities
$_{97}Bk^{246}$	
734	4.6†
800	100
834	8
986	0.5
1037	2.8
1063	5
1079	5
1082	10
1124	8
$_{97}Bk^{247}$	
84	40%
≈270	30
$_{97}Bk^{248}$	
No γ	
$_{97}Bk^{249}$	
307	2.2†
327.2	12
$_{97}Bk^{250}$	
42.2	
98.2	
990	47%
1032	39
$_{98}Cf^{246}$	
42	0.014%
96	0.012
146	0.0035
$_{98}Cf^{247}$	
295	20†
417	13
460	9
$_{98}Cf^{249}$	
43.0	0.03%
55.0	0.20
66.7	0.04
92.6	0.4
241.0	0.3
252.8	3.0
266.7	0.7
295.9	0.17
333.4	16
388.4	72
$_{98}Cf^{250}$	
42.9	17%
$_{98}Cf^{251}$	
177	
224	
$_{98}Cf^{252}$	
43.3	0.014%
100.2	0.01
≈160	0.002

Gamma energies (KeV)	Gamma intensities
$_{99}Es^{252}$	
74	0.3%
154	0.07
198	0.08
228	0.23
278	0.21
~400	1.1
~520	0.15
~570	0.14
$_{99}Es^{253}$	
8.80	
30.83	
41.79	
42.98	
51.96	
55.11	
62.09	
66.85	
73.43	
73.82	
78.6	
93.75	
98.09	
114.06	
121.98	
135.5	
145.4	
368.8	0.7†
381.2	10.4
386.1	
387.2	36
389.18	55
428.94	11
433.2	5
442.2	
448.2	1.5
475.0	
500.2	
≈750	
≈900	
$_{99}Es^{254m}$	
44	
104	0.6†
544	3
583.3	9
648.1	100
688.2	~40
693.1	~80
~990	2.0
$_{99}Es^{254}$	
34.4	
35.5	
42.6	
63	2.0%
69.7	
70.4	
80.8	
85.1	
150	0.02
233?	0.008
249	0.025
264	0.05
278	0.03
285	0.01
304	0.07
316	0.15
342	0.009
348	0.007
377	0.015
385	0.05
$_{100}Fm^{254}$	
41	0.2%
98	0.028
151	0.001
$_{100}Fm^{255}$	
58.6	0.9%
60.0	0.19
72.9	0.034
80.5 } 81.2 }	1.34
85.7	0.009
98.5	0.007
~120	0.03
178	0.007
~340	0.003
~370	0.005
~440	0.003
~490	0.003
$_{100}Fm^{257}$	
62	
78	
103	1%
180	8
242	10

PERMISSIBLE QUARTERLY INTAKES OF RADIONUCLIDES

Karl Z. Morgan

The following table lists the permissible quarterly intakes (oral and inhalation) or radionuclides and critical body organs recommended by NCPR for occupational exposure.

PERMISSIBLE QUARTERLY INTAKES (ORAL AND INHALATION) OF RADIONUCLIDES AND CRITICAL BODY ORGANS RECOMMENDED BY NCRP FOR OCCUPATIONAL EXPOSURE (1)

Radionuclide	Radioactive half-life days	Solubility state	Oral intake: Critical organ	Oral intake: Permissible quarterly intake microcuries	Intake by inhalation: Critical organ	Intake by inhalation: Permissible quarterly intake microcuries
3-H	4.50E 03	Sol.	Body tissue	6.4E 03(2)	Body tissue	3.1E 03(2)
7-BE	5.36E 01	Sol.	GI LLI	3.6E 03	Tot. body	3.5E 03
		Insol.	GI LLI	3.6E 03	Lungs	7.5E 02
14-C	2.00E 06	Sol.	Fat	1.6E 03	Fat	2.2E 03
18-F	7.80E-02	Sol.	GI SI	1.7E 03	GI SI	3.3E 03
		Insol.	GI ULI	1.0E 03	GI ULI	1.6E 03
22-NA	9.50E 02	Sol.	Tot. body	8.0E 01	Tot. body	1.1E 02
		Insol.	GI LLI	6.0E 01	Lungs	5.3E 00
24-NA	6.30E-01	Sol.	GI SI	3.8E 02	GI SI	7.6E 02
		Insol.	GI LLI	5.6E 01	GI LLI	8.9E 01
31-SI	1.10E-01	Sol.	GI S	1.8E 03	GI S	3.5E 03
		Insol.	GI ULI	3.8E 02	GI ULI	6.2E 02
32-P	1.43E 01	Sol.	Bone	3.8E 01	Bone	4.4E 01
		Insol.	GI LLI	4.6E 01	Lungs	4.9E 01
35-S	8.71E 01	Sol.	Testes	1.3E 02	Testes	1.7E 02
		Insol	GI LLI	5.4E 02	Lungs	1.6E 02
36-CL	1.20E 08	Sol.	Tot. body	1.6E 02	Tot. body	2.2E 02
		Insol.	GI LLI	1.2E 02	Lungs	1.4E 01
38-CL	2.60E-02	Sol.	GI S	8.0E 02	GI S	1.6E 03
		Insol.	GI S	8.0E 02	GI S	1.3E 03
42-K	5.20E-01	Sol.	GI S	6.2E 02	GI-S	1.2E 03
		Insol.	GI LLI	4.0E 01	GI LLI	6.7E 01
45-CA	1.64E 02	Sol.	Bone	1.8E 01	Bone	2.0E 01
		Insol.	GI LLI	3.6E 02	Lungs	7.5E 01
47-CA	4.90E 00	Sol.	Bone	1.0E 02	Bone	1.1E 02
		Insol.	GI LLI	6.6E 01	GI LLI	1.1E 02
					Lungs	1.2E 02
46-SC	8.50E 01	Sol.	GI LLI	7.6E 01	Liver	1.5E 02
					GI LLI	1.5E 02
		Insol.	GI LLI	7.6E 01	Lungs	1.4E 01
47-SC	3.43E 00	Sol.	GI LLI	1.8E 02	GI LLI	3.6E 02
		Insol.	GI LLI	1.8E 02	GI LLI	2.9E 02
48-SC	1.83E 00	Sol.	GI LLI	5.4E 01	GI LLI	1.1E 02
		Insol.	GI LLI	5.4E 01	GI LLI	8.7E 01
48-V	1.61E 01	Sol.	GI LLI	5.8E 01	GI LLI	1.2E 02
		Insol.	GI LLI	5.8E 01	Lungs	3.5E 01
51-CR	2.78E 01	Sol.	GI LLI	3.2E 03	GI LLI	6.4E 03
					Tot. body	6.7E 03
		Insol.	GI LLI	3.0E 03	Lungs	1.4E 03
52-MN	5.55E 00	Sol.	GI LLI	6.6E 01	GI LLI	1.3E 02
		Insol.	GI LLI	6.0E 01	Lungs	8.7E 01
					GI LLI	9.5E 01
54-MN	3.00E 02	Sol.	GI LLI	2.6E 02	Liver	2.4E 02
		Insol.	GI LLI	2.4E 02	Lungs	2.2E 01
56-MN	1.10E-01	Sol.	GI LLI	2.4E 02	GI LLI	4.7E 02
		Insol.	GI LLI	2.0E 02	GI LLI	3.3E 02
55-FE	1.10E 03	Sol.	Spleen	1.6E 03	Spleen	5.3E 02
		Insol.	GI LLI	4.6E 03	Lungs	6.4E 02
59-FE	4.51E 01	Sol.	GI LLI	1.2E 02	Spleen	9.3E 01
		Insol.	GI LLI	1.1E 02	Lungs	3.3E 01
57-CO	2.70E 02	Sol.	GI LLI	1.1E 03	GI LLI	2.2E 03
		Insol.	GI LLI	7.6E 02	Lungs	1.0E 02
58M-CO	3.80E-01	Sol.	GI LLI	5.6E 03	GI LLI	1.1E 04
		Insol.	GI LLI	4.0E 03	Lungs	5.5E 03
58-CO	7.20E 01	Sol	GI LLI	2.6E 02	GI LLI	5.3E 02
					Tot. body	6.0E 02
		Insol	GI LLI	1.8E 02	Lungs	3.5E 01

PERMISSIBLE QUARTERLY INTAKES (ORAL AND INHALATION) OF RADIONUCLIDES AND CRITICAL BODY ORGANS RECOMMENDED BY NCRP FOR OCCUPATIONAL EXPOSURE (1) —(Continued)

			Oral intake		Intake by inhalation	
Radionuclide	Radioactive half-life days	Solubility state	Critical organ	Permissible quarterly intake microcuries	Critical organ	Permissible quarterly intake microcuries
60-CO	1.90E 03	Sol	GI LLI	9.8E 01	GI LLI	2.0E 02
					Tot. body	2.2E 02
		Insol.	GI LLI	7.0E 01	Lungs	5.5E 00
59-NI	2.90E 07	Sol.	Bone	4.0E 02	Bone	2.9E 02
		Insol.	GI LLI	4.0E 03	Lungs	4.7E 02
63-NI	2.90E 04	Sol	Bone	5.4E 01	Bone	4.0E 01
		Insol	GI LLI	1.4E 03	Lungs	1.7E 02
65-NI	1.10E-01	Sol.	GI ULI	2.8E 02	GI ULI	5.6E 02
		Insol.	GI ULI	2.0E 02	GI ULI	3.3E 02
64-CU	5.30E-01	Sol.	GI LLI	6.6E 02	GI LLI	1.3E 03
		Insol.	GI LLI	4.2E 02	GI LLI	6.6E 02
65-ZN	2.45E 02	Sol.	Tot. body	2.0E 02	Tot. body	6.6E 01
			Prostate	2.4E 02	Prostate	8.0E 01
			Liver	2.6E 02		
		Insol.	GI LLI	3.6E 02	Lungs	3.6E 01
69M-ZN	5.80E-01	Sol.	GI LLI	1.4E 02	Prostate	2.4E 02
		Insol.	GI LLI	1.2E 02	GI LLI	2.0E 02
69-ZN	3.60E-02	Sol.	GI S	3.6E 03	Prostate	4.4E 03
		Insol.	GI S	3.6E 03	GI S	5.8E 03
72-GA	5.90E-01	Sol.	GI LLI	7.4E 01	GI LLI	1.5E 02
		Insol.	GI LLI	7.4E 01	GI LLI	1.2E 02
71-GE	1.20E 01	Sol.	GI LLI	3.2E 03	GI LLI	6.4E 03
		Insol.	GI LLI	3.2E 03	Lungs	4.0E 03
73-AS	7.60E 01	Sol.	GI LLI	9.6E 02	Tot. body	1.3E 03
		Insol.	GI LLI	9.2E 02	Lungs	2.4E 02
74-AS	1.75E 01	Sol.	GI LLI	1.1E 02	GI LLI	2.2E 02
		Insol.	GI LLI	1.0E 02	Lungs	7.8E 01
76-AS	1.10E 00	Sol.	GI LLI	4.0E 01	GI LLI	8.0E 01
		Insol.	GI LLI	3.8E 01	GI LLI	6.2E 01
77-AS	1.60E 00	Sol.	GI LLI	1.6E 02	GI LLI	3.3E 02
		Insol.	GI LLI	1.6E 02	GI LLI	2.5E 02
75-SE	1.27E 02	Sol.	Kidneys	6.0E 02	Kidneys	7.8E 02
			Tot. body	6.8E 02		
		Insol.	GI LLI	5.4E 02	Lungs	7.6E 01
82-BR	1.50E 00	Sol.	Tot. body	5.2E 02	Tot. body	7.1E 02
			GI SI	5.6E 02		
		Insol.	GI LLI	7.4E 01	GI LLI	1.2E 02
86-RB	1.86E 01	Sol.	Tot. body	1.3E 02	Tot. body	1.8E 02
			Pancreas	1.3E 02	Pancreas	1.8E 02
					Liver	2.5E 02
		Insol.	GI LLI	4.8E 01	Lungs	4.2E 01
87-RB	1.80E 13	Sol.	Pancreas	2.2E 02	Pancreas	3.1E 02
					Tot. body	4.0E 02
					Liver	4.2E 02
		Insol.	GI LLI	3.4E 02	Lungs	4.0E 01
85M-SR	4.90E-02	Sol.	GI SI	1.3E 04	GI SI	2.5E 04
		Insol.	GI SI	1.3E 04	GI SI	2.2E 04
85-SR	6.50E 01	Sol.	Tot. body	1.9E 02	Tot. body	1.4E 02
		Insol.	GI LLI	3.4E 02	Lungs	6.6E 01
89-SR	5.05E 01	Sol.	Bone	2.4E 01	Bone	1.7E 01
		Insol.	GI LLI	5.6E 01	Lungs	2.2E 01
90-SR	1.00E 04	Sol.	Bone	8.0E-01	Bone	7.3E-01
		Insol.	GI LLI	7.0E 01	Lungs	3.5E 00
91-SR	4.00E-01	Sol.	GI LLI	1.4E 02	GI LLI	2.7E 02
		Insol.	GI LLI	9.8E 01	GI LLI	1.6E 02
92-SR	1.10E-01	Sol.	GI ULI	1.4E 02	GI ULI	2.7E 02
		Insol.	GI ULI	1.2E 02	GI ULI	1.8E 02
90-Y	2.68E 00	Sol.	GI LLI	4.0E 01	GI LLI	8.0E 01
		Insol.	GI LLI	4.0E 01	GI LLI	6.4E 01
91M-Y	3.50E-02	Sol.	GI SI	6.8E 03	GI SI	1.4E 04
		Insol.	GI SI	6.8E 03	GI SI	1.1E 04
91-Y	5.80E 01	Sol.	GI LLI	5.2E 01	Bone	2.2E 01
		Insol.	GI LLI	5.2E 01	Lungs	2.0E 01

PERMISSIBLE QUARTERLY INTAKES (ORAL AND INHALATION) OF RADIONUCLIDES AND CRITICAL BODY ORGANS RECOMMENDED BY NCRP FOR OCCUPATIONAL EXPOSURE (1) —(Continued)

Radionuclide	Radioactive half-life days	Solubility state	Oral intake Critical organ	Oral intake Permissible quarterly intake microcuries	Intake by inhalation Critical organ	Intake by inhalation Permissible quarterly intake microcuries
92-Y	1.50E-01	Sol.	GI ULI	1.2E 02	GI ULI	2.4E 02
		Insol.	GI ULI	1.2E 02	GI ULI	1.8E 02
93-Y	4.20E-01	Sol.	GI LLI	5.4E 01	GI LLI	1.1E 02
		Insol.	GI LLI	5.4E 01	GI LLI	8.6E 01
93-ZR	4.00E 08	Sol.	GI LLI	1.6E 03	Bone	8.0E 01
		Insol.	GI LLI	1.6E 03	Lungs	2.0E 02
95-ZR	6.33E 01	Sol.	GI LLI	1.3E 02	Tot. body	8.0E 01
		Insol.	GI LLI	1.3E 02	Lungs	2.0E 01
97-ZR	7.10E-01	Sol.	GI LLI	3.6E 01	GI LLI	7.3E 01
		Insol.	GI LLI	3.6E 01	GI LLI	5.6E 01
93M-NB	3.70E 03	Sol.	GI LLI	8.0E 02	Bone	7.6E 01
		Insol.	GI LLI	8.0E 02	Lungs	1.0E 02
95-NB	3.50E 01	Sol.	GI LLI	1.9E 02	Tot. body	2.9E 02
					GI LLI	3.8E 02
		Insol.	GI LLI	1.9E 02	Lungs	6.2E 01
97-NB	5.10E-02	Sol.	GI ULI	1.9E 03	GI ULI	3.8E 03
		Insol.	GI ULI	1.9E 03	GI ULI	2.9E 03
99-MO	2.79E 00	Sol.	Kidneys	3.6E 02	Kidneys	4.5E 02
			GI LLI	4.8E 02		
		Insol.	GI LLI	7.8E 01	GI LLI	1.3E 02
96M-TC	3.60E-02	Sol.	GI LLI	2.4E 04	GI LLI	4.7E 04
		Insol.	GI LLI	2.0E 04	Lungs	1.8E 04
96-TC	4.30E 00	Sol.	GI LLI	2.0E 02	GI LLI	4.0E 02
		Insol.	GI LLI	9.4E 01	GI LLI	1.5E 02
97M-TC	9.20E 01	Sol.	GI LLI	7.2E 02	GI LLI	1.4E 03
		Insol.	GI LLI	3.4E 02	Lungs	9.5E 01
97-TC	3.70E 06	Sol.	GI LLI	3.4E 03	GI LLI	6.7E 03
					Kidneys	8.0E 03
		Insol.	GI LLI	1.6E 03	Lungs	1.8E 02
99M-TC	2.50E-01	Sol.	GI ULI	1.1E 04	GI ULI	2.4E 04
		Insol.	GI ULI	5.6E 03	GI ULI	8.7E 03
99-TC	7.70E 07	Sol.	GI LLI	6.6E 02	GI LLI	1.3E 03
		Insol.	GI LLI	3.2E 02	Lungs	3.8E 01
97-RU	2.80E 00	Sol.	GI LLI	7.2E 02	GI LLI	1.4E 03
		Insol.	GI LLI	7.0E 02	GI LLI	1.1E 03
103-RU	4.10E 01	Sol.	GI LLI	1.6E 02	GI LLI	3.3E 02
		Insol.	GI LLI	1.6E 02	Lungs	5.3E 01
105-RU	1.90E-01	Sol.	GI ULI	2.2E 02	GI ULI	4.4E 02
		Insol.	GI ULI	2.0E 02	GI ULI	3.3E 02
106-RU	3.65E 02	Sol.	GI LLI	2.4E 01	GI LLI	4.7E 01
		Insol.	GI LLI	2.4E 01	Lungs	3.5E 00
103M-RH	3.80E-02	Sol.	GI S	2.4E 04	GI S	4.7E 04
		Insol.	GI S	2.4E 04	GI S	3.8E 04
105-RH	1.52E 00	Sol.	GI LLI	2.6E 02	GI LLI	5.3E 02
		Insol.	GI LLI	2.0E 02	GI LLI	3.3E 02
103-PD	1.70E 01	Sol.	GI LLI	6.8E 02	Kidneys	8.4E 02
		Insol.	GI LLI	5.4E 02	Lungs	4.7E 02
109-PD	5.70E-01	Sol.	GI LLI	1.8E 02	GI LLI	3.5E 02
		Insol.	GI LLI	1.4E 02	GI LLI	2.2E 02
105-AG	4.00E 01	Sol.	GI LLI	1.9E 02	GI LLI	3.8E 02
		Insol.	GI LLI	1.9E 02	Lungs	5.1E 01
110M-AG	2.70E 02	Sol.	GI LLI	6.0E 01	GI LLI	1.2E 02
		Insol.	GI LLI	6.0E 01	Lungs	6.4E 00
111-AG	7.50E 00	Sol.	GI LLI	8.8E 01	GI LLI	1.8E 02
		Insol.	GI LLI	8.6E 01	GI LLI	1.4E 02
109-CD	4.75E 02	Sol.	GI LLI	3.4E 02	Liver	3.3E 01
					Kidneys	3.5E 01
		Insol.	GI LLI	3.4E 02	Lungs	4.5E 01
115M-CD	4.30E 01	Sol.	GI LLI	5.0E 01	Liver	2.2E 01
		Insol.	GI LLI	5.0E 01	Lungs	2.2E 01
115-CD	2.20E 00	Sol.	GI LLI	6.8E 01	GI LLI	1.4E 02
		Insol.	GI LLI	7.2E 01	GI LLI	1.1E 02

PERMISSIBLE QUARTERLY INTAKES (ORAL AND INHALATION) OF RADIONUCLIDES AND CRITICAL BODY ORGANS RECOMMENDED BY NCRP FOR OCCUPATIONAL EXPOSURE (1) —(Continued)

			Oral intake		Intake by inhalation	
Radionuclide	Radioactive half-life days	Solubility state	Critical organ	Permissible quarterly intake microcuries	Critical organ	Permissible quarterly intake microcuries
113M-IN	7.30E-02	Sol.	GI ULI	2.6E 03	GI ULI	5.3E 03
		Insol.	GI ULI	2.6E 03	GI ULI	4.2E 03
114M-IN	4.90E 01	Sol.	GI LLI	3.4E 01	Kidneys	6.4E 01
					GI LLI	6.7E 01
					Spleen	7.1E 01
		Insol.	GI LLI	3.4E 01	Lungs	1.3E 01
115M-IN	1.90E-01	Sol.	GI ULI	7.4E 02	GI ULI	1.5E 03
		Insol.	GI ULI	7.4E 02	GI ULI	1.2E 03
115-IN	2.20E 17	Sol.	GI LLI	1.8E 02	Kidneys	1.5E 02
		Insol.	GI LLI	1.8E 02	Lungs	2.2E 01
113-SN	1.12E 02	Sol.	GI LLI	1.7E 02	Bone	2.2E 02
		Insol.	GI LLI	1.6E 02	Lungs	3.3E 01
125-SN	9.50E 00	Sol.	GI LLI	3.6E 01	GI LLI	7.3E 01
		Insol.	GI LLI	3.4E 01	Lungs	5.3E 01
					GI LLI	5.5E 01
122-SB	2.80E 00	Sol.	GI LLI	5.8E 01	GI LLI	1.2E 02
		Insol.	GI LLI	5.8E 01	GI LLI	9.1E 01
124-SB	6.00E 01	Sol.	GI LLI	4.6E 01	GI LLI	9.3E 01
		Insol.	GI LLI	4.6E 01	Lungs	1.2E 01
125-SB	8.77E 02	Sol.	GI LLI	2.0E 02	Lungs	3.3E 02
					Tot. body	3.6E 02
					GI LLI	4.0E 02
					Bone	4.5E 02
		Insol.	GI LLI	2.0E 02	Lungs	1.7E 01
125M-TE	5.80E 01	Sol.	Kidneys	3.2E 02	Kidneys	2.2E 02
			GI LLI	3.6E 02		
			Testes	4.4E 02		
		Insol.	GI LLI	2.4E 02	Lungs	8.0E 01
127M-TE	1.05E 02	Sol.	Kidneys	1.2E 02	Kidneys	8.2E 01
					Testes	8.7E 01
		Insol.	GI LLI	1.1E 02	Lungs	2.5E 01
127-TE	3.90E-01	Sol.	GI LLI	5.2E 02	GI LLI	1.0E 03
		Insol.	GI LLI	3.4E 02	GI LLI	5.3E 02
129M-TE	3.30E 01	Sol.	GI LLI	6.6E 01	Kidneys	4.9E 01
					Testes	5.6E 01
		Insol.	GI LLI	4.0E 01	Lungs	2.0E 01
129-TE	5.10E-02	Sol.	GI S	1.6E 03	GI S	3.3E 03
		Insol.	GI ULI	1.6E 03	GI ULI	2.5E 03
131M-TE	1.25E 00	Sol.	GI LLI	1.2E 02	GI LLI	2.4E 02
		Insol.	GI LLI	7.4E 01	GI LLI	1.2E 02
132-TE	3.20E 00	Sol.	GI LLI	6.4E 01	GI LLI	1.3E 02
		Insol.	GI LLI	4.2E 01	GI LLI	6.6E 01
126-I	1.33E 01	Sol.	Thyroid	3.4E 00	Thyroid	4.5E 00
		Insol.	GI LLI	1.8E 02	Lungs	2.0E 02
129-I	6.30E 09	Sol.	Thyroid	7.6E-01	Thyroid	1.0E 00
		Insol.	GI LLI	4.2E 02	Lungs	4.5E 01
131-I	8.00E 00	Sol.	Thyroid	4.0E 00	Thyroid	5.3E 00
		Insol.	GI LLI	1.3E 02	GI LLI	2.0E 02
					Lungs	2.0E 02
132-I	9.70E-02	Sol.	Thyroid	1.1E 02	Thyroid	1.5E 02
		Insol.	GI ULI	3.6E 02	GI ULI	5.8E 02
133-I	8.70E-01	Sol.	Thyroid	1.5E 01	Thyroid	2.0E 01
		Insol.	GI LLI	8.2E 01	GI LLI	1.3E 02
134-I	3.60E-02	Sol.	Thyroid	2.4E 02	Thyroid	3.1E 02
		Insol.	GI S	1.2E 03	GI S	2.0E 03
135-I	2.80E-01	Sol.	Thyroid	4.8E 01	Thyroid	6.4E 01
		Insol.	GI LLI	1.4E 02	GI LLI	2.2E 02
131-CS	1.00E 01	Sol.	Tot. body	4.8E 03	Tot. body	6.6E 03
					Liver	8.0E 03
		Insol.	GI LLI	1.9E 03	Lungs	2.0E 03
134M-CS	1.30E-01	Sol.	GI S	1.1E 04	GI S	2.2E 04
		Insol.	GI ULI	2.2E 03	GI ULI	3.6E 03

PERMISSIBLE QUARTERLY INTAKES (ORAL AND INHALATION) OF RADIONUCLIDES AND CRITICAL BODY ORGANS RECOMMENDED BY NCRP FOR OCCUPATIONAL EXPOSURE (1) —(Continued)

Radionuclide	Radioactive half-life days	Solubility state	Oral intake: Critical organ	Oral intake: Permissible quarterly intake microcuries	Intake by inhalation: Critical organ	Intake by inhalation: Permissible quarterly intake microcuries
134-CS	8.40E 02	Sol.	Tot. body	1.7E 01	Tot. body	2.4E 01
		Insol.	GI LLI	8.0E 01	Lungs	8.0E 00
135-CS	1.10E 09	Sol.	Liver	2.2E 02	Liver	2.9E 02
			Spleen	2.4E 02	Spleen	3.3E 02
			Tot. body	2.6E 02	Tot. body	3.6E 02
		Insol.	GI LLI	4.6E 02	Lungs	5.6E 01
136-CS	1.30E 01	Sol.	Tot. body	1.7E 02	Tot. body	2.4E 02
		Insol.	GI LLI	1.3E 02	Lungs	1.1E 02
137-CS	1.10E 04	Sol.	Tot. body	3.0E 01	Tot. body	4.0E 01
			Liver	3.6E 01		
			Spleen	4.4E 01		
			Muscle	4.8E 01		
		Insol.	GI LLI	8.8E 01	Lungs	9.1E 00
131-BA	1.16E 01	Sol.	GI LLI	3.6E 02	GI LLI	7.3E 02
		Insol.	GI LLI	3.4E 02	Lungs	2.2E 02
140-BA	1.28E 01	Sol.	GI LLI	5.2E 01	Bone	8.0E 01
		Insol.	GI LLI	5.0E 01	Lungs	2.7E 01
140-LA	1.68E 00	Sol.	GI LLI	4.8E 01	GI LLI	9.6E 01
		Insol.	GI LLI	4.8E 01	GI LLI	7.6E 01
141-CE	3.20E 01	Sol.	GI LLI	1.8E 02	Liver	2.7E 02
					GI LLI	3.6E 02
					Bone	3.6E 02
		Insol.	GI LLI	1.8E 02	Lungs	9.8E 01
143-CE	1.33E 00	Sol.	GI LLI	8.0E 01	GI LLI	1.6E 02
		Insol.	GI LLI	8.0E 01	GI LLI	1.3E 02
144-CE	2.90E 02	Sol.	GI LLI	2.4E 01	Bone	6.0E 00
		Insol.	GI LLI	2.4E 01	Lungs	4.0E 00
142-PR	8.00E-01	Sol.	GI LLI	6.0E 01	GI LLI	1.2E 02
		Insol.	GI LLI	6.0E 01	GI LLI	9.8E 01
143-PR	1.37E 01	Sol.	GI LLI	9.8E 01	GI LLI	2.0E 02
		Insol.	GI LLI	9.8E 01	Lungs	1.1E 02
144-ND	7.30E 17	Sol.	Bone	1.3E 02	Bone	5.1E-02
		Insol.	GI LLI	1.6E 02	Lungs	1.8E-01
147-ND	1.13E 01	Sol.	GI LLI	1.2E 02	Liver	2.2E 02
					GI LLI	2.4E 02
		Insol.	GI LLI	1.2E 02	Lungs	1.4E 02
149-ND	8.30E-02	Sol.	GI LLI	5.6E 02	GI LLI	1.1E 03
		Insol.	GI ULI	5.6E 02	GI ULI	9.1E 02
147-PM	9.20E 02	Sol.	GI LLI	4.4E 02	Bone	4.0E 01
		Insol.	GI LLI	4.4E 02	Lungs	6.0E 01
149-PM	2.20E 00	Sol.	GI LLI	8.8E 01	GI LLI	1.8E 02
		Insol.	GI LLI	8.8E 01	GI LLI	1.4E 02
147-SM	4.80E 13	Sol.	Bone	1.1E 02	Bone	4.4E-02
		Insol.	GI LLI	1.4E 02	Lungs	1.6E-01
151-SM	3.70E 04	Sol.	GI LLI	7.4E 02	Bone	4.0E 01
		Insol.	GI LLI	7.4E 02	Lungs	8.7E 01
153-SM	1.96E 00	Sol.	GI LLI	1.5E 02	GI LLI	3.1E 02
		Insol.	GI LLI	1.5E 02	GI LLI	2.5E 02
152M-EU	3.80E-01	Sol.	GI LLI	1.3E 02	GI LLI	2.5E 02
		Insol.	GI LLI	1.3E 02	GI LLI	2.0E 02
152-EU	4.70E 03	Sol.	GI LLI	1.5E 02	Kidneys	7.6E 00
		Insol.	GI LLI	1.5E 02	Lungs	1.1E 01
154-EU	5.80E 03	Sol.	GI LLI	4.4E 01	Kidneys	2.4E 00
					Bone	2.4E 00
		Insol.	GI LLI	4.4E 01	Lungs	4.4E 00
155-EU	6.21E 02	Sol.	GI LLI	4.0E 02	Kidneys	5.8E 01
					Bone	6.2E 01
		Insol.	GI LLI	4.0E 02	Lungs	4.5E 01
153-GD	2.36E 02	Sol.	GI LLI	4.2E 02	Bone	1.4E 02
		Insol.	GI LLI	4.2E 02	Lungs	5.6E 01
159-GD	7.50E-01	Sol.	GI LLI	1.6E 02	GI LLI	3.1E 02
		Insol.	GI LLI	1.6E 02	GI LLI	2.5E 02

PERMISSIBLE QUARTERLY INTAKES (ORAL AND INHALATION) OF RADIONUCLIDES AND CRITICAL BODY ORGANS RECOMMENDED BY NCRP FOR OCCUPATIONAL EXPOSURE (1) —(Continued)

Radionuclide	Radioactive half-life days	Solubility state	Oral intake: Critical organ	Oral intake: Permissible quarterly intake microcuries	Intake by inhalation: Critical organ	Intake by inhalation: Permissible quarterly intake microcuries
160-TB	7.30E 01	Sol.	GI LLI	8.8E 01	Bone	6.2E 01
		Insol.	GI LLI	9.0E 01	Lungs	2.0E 01
165-DY	9.70E-02	Sol.	GI ULI	8.0E 02	GI ULI	1.6E 03
		Insol.	GI ULI	8.0E 02	GI ULI	1.3E 03
166-DY	3.40E 00	Sol.	GI LLI	7.6E 01	GI LLI	1.5E 02
		Insol.	GI LLI	7.6E 01	GI LLI	1.2E 02
166-HO	1.10E 00	Sol.	GI LLI	6.2E 01	GI LLI	1.2E 02
		Insol.	GI LLI	6.2E 01	GI LLI	1.0E 02
169-ER	9.40E 00	Sol.	GI LLI	1.9E 02	GI LLI	3.8E 02
		Insol.	GI LLI	1.9E 02	Lungs	2.4E 02
171-ER	3.10E-01	Sol.	GI ULI	2.2E 02	GI ULI	4.4E 02
		Insol.	GI ULI	2.2E 02	GI ULI	3.6E 02
170-TM	1.27E 02	Sol.	GI LLI	9.2E 01	Bone	2.2E 01
		Insol.	GI LLI	9.2E 01	Lungs	2.2E 01
171-TM	6.94E 02	Sol.	GI LLI	1.0E 03	Bone	6.9E 01
		Insol.	GI LLI	1.0E 03	Lungs	1.4E 02
175-YB	4.10E 00	Sol.	GI LLI	2.2E 02	GI LLI	4.4E 02
		Insol.	GI LLI	2.2E 02	GI LLI	3.6E 02
177-LU	6.80E 00	Sol.	GI LLI	2.0E 02	GI LLI	4.0E 02
		Insol.	GI LLI	2.0E 02	GI LLI	3.3E 02
					Lungs	4.4E 02
181-HF	4.60E 01	Sol.	GI LLI	1.4E 02	Spleen	2.4E 01
		Insol.	GI LLI	1.4E 02	Lungs	4.5E 01
182-TA	1.12E 02	Sol.	GI LLI	8.0E 01	Liver	2.4E 01
		Insol.	GI LLI	8.0E 01	Lungs	1.4E 01
181-W	1.40E 02	Sol.	GI LLI	7.2E 02	GI LLI	1.4E 03
		Insol.	GI LLI	6.4E 02	Lungs	7.8E 01
185-W	7.40E 01	Sol.	GI LLI	2.4E 02	GI LLI	4.7E 02
		Insol.	GI LLI	2.2E 02	Lungs	6.9E 01
187-W	1.00E 00	Sol.	GI LLI	1.4E 02	GI LLI	2.7E 02
		Insol.	GI LLI	1.2E 02	GI LLI	2.0E 02
183-RE	7.30E 01	Sol.	GI LLI	1.1E 03	Tot. body	1.6E 03
		Insol.	GI LLI	5.6E 02	Lungs	9.8E 01
186-RE	3.79E 00	Sol.	GI LLI	1.9E 02	GI LLI	3.8E 02
		Insol.	GI LLI	9.4E 01	GI LLI	1.5E 02
187-RE	1.80E 13	Sol.	GI LLI	5.0E 03	Skin	5.6E 03
			Skin	5.6E 03		
		Insol.	GI LLI	5.0E 03	Lungs	3.1E 02
188-RE	7.10E-01	Sol.	GI LLI	1.3E 02	GI LLI	2.5E 02
		Insol.	GI LLI	6.2E 01	GI LLI	1.0E 02
185-OS	9.50E 01	Sol.	GI LLI	1.5E 02	GI LLI	2.9E 02
		Insol.	GI LLI	1.3E 02	Lungs	2.9E 01
191M-OS	5.80E-01	Sol.	GI LLI	5.0E 03	GI LLI	1.0E 04
		Insol.	GI LLI	4.8E 03	Lungs	5.8E 03
191-OS	1.60E 01	Sol.	GI LLI	3.4E 02	GI LLI	6.7E 02
		Insol.	GI LLI	3.2E 02	Lungs	2.5E 02
193-OS	1.30E 00	Sol.	GI LLI	1.2E 02	GI LLI	2.4E 02
		Insol.	GI LLI	1.1E 02	GI LLI	1.7E 02
190-IR	1.20E 01	Sol.	GI LLI	4.0E 02	GI LLI	8.0E 02
		Insol.	GI LLI	3.6E 02	Lungs	2.5E 02
192-IR	7.45E 01	Sol.	GI LLI	8.0E 01	Kidneys	7.8E 01
		Insol.	GI LLI	7.2E 01	Lungs	1.6E 01
194-IR	7.90E-01	Sol.	GI LLI	6.8E 01	GI LLI	1.4E 02
		Insol.	GI LLI	6.0E 01	GI LLI	9.6E 01
191-PT	3.00E 00	Sol.	GI LLI	2.4E 02	GI LLI	4.7E 02
		Insol.	GI LLI	2.2E 02	GI LLI	3.5E 02
193M-PT	3.40E 00	Sol.	GI LLI	2.2E 03	GI LLI	4.4E 03
		Insol.	GI LLI	2.0E 03	GI LLI	3.3E 03
					Lungs	4.0E 03
193-PT	1.80E 05	Sol.	Kidneys	1.9E 03	Kidneys	6.4E 02
		Insol.	GI LLI	3.0E 03	Lungs	2.0E 02

PERMISSIBLE QUARTERLY INTAKES (ORAL AND INHALATION) OF RADIONUCLIDES AND CRITICAL BODY ORGANS RECOMMENDED BY NCRP FOR OCCUPATIONAL EXPOSURE (1) —(Continued)

Radionuclide	Radioactive half-life days	Solubility state	Oral intake: Critical organ	Oral intake: Permissible quarterly intake microcuries	Intake by inhalation: Critical organ	Intake by inhalation: Permissible quarterly intake microcuries
197M-PT	5.60E-02	Sol.	GI ULI	2.0E 03	GI ULI	4.0E 03
		Insol.	GI ULI	1.9E 03	GI ULI	2.9E 03
197-PT	7.50E-01	Sol.	GI LLI	2.4E 02	GI LLI	4.7E 02
		Insol.	GI LLI	2.2E 02	GI LLI	3.5E 02
196-AU	5.60E 00	Sol.	GI LLI	3.2E 02	GI LLI	6.4E 02
		Insol.	GI LLI	3.0E 02	Lungs	3.8E 02
198-AU	2.70E 00	Sol.	GI LLI	1.0E 02	GI LLI	2.0E 02
		Insol.	GI LLI	9.2E 01	GI LLI	1.5E 02
199-AU	3.15E 00	Sol.	GI LLI	3.4E 02	GI LLI	6.7E 02
		Insol.	GI LLI	3.2E 02	GI LLI	5.1E 02
197M-HG	1.00E 00	Sol.	Kidneys	3.8E 02	Kidneys	4.5E 02
		Insol.	GI LLI	3.4E 02	GI LLI	5.3E 02
197-HG	2.70E 00	Sol.	Kidneys	6.0E 02	Kidneys	7.3E 02
		Insol.	GI LLI	9.8E 02	GI LLI	1.5E 03
203-HG	4.58E 01	Sol.	Kidneys	3.6E 01	Kidneys	4.4E 01
		Insol.	GI LLI	2.2E 02	Lungs	7.8E 01
200-TL	1.13E 00	Sol.	GI LLI	8.8E 02	GI LLI	1.6E 03
		Insol.	GI LLI	4.4E 02	GI LLI	7.1E 02
201-TL	3.00E 00	Sol.	GI LLI	6.2E 02	GI LLI	1.2E 03
		Insol.	GI LLI	3.4E 02	GI LLI	5.5E 02
202-TL	1.20E 01	Sol.	GI LLI	2.4E 02	GI LLI	4.7E 02
		Insol.	GI LLI	1.4E 02	Lungs	1.5E 02
204-TL	1.10E 03	Sol.	GI LLI	2.2E 02	Kidneys	3.8E 02
					GI LLI	4.4E 02
		Insol.	GI LLI	1.2E 02	Lungs	1.6E 01
203-PB	2.17E 00	Sol.	GI LLI	7.8E 02	GI LLI	1.6E 03
		Insol.	GI LLI	7.0E 02	GI LLI	1.1E 03
210-PB	7.10E 03	Sol.	Tot. body	2.4E-01	Kidneys	7.8E-02
			Kidneys	2.8E-01		
		Insol.	GI LLI	3.6E 02	Lungs	1.5E-01
212-PB	4.40E-01	Sol.	GI LLI	3.8E 01	Kidneys	1.1E 01
			Kidneys	4.0E 01		
		Insol.	GI LLI	3.4E 01	Lungs	1.2E 01
206-BI	6.40E 00	Sol.	GI LLI	7.6E 01	Kidneys	1.2E 02
		Insol.	GI LLI	7.6E 01	Lungs	9.1E 01
207-BI	2.90E 03	Sol.	GI LLI	1.3E 02	Kidneys	1.1E 02
		Insol.	GI LLI	1.3E 02	Lungs	8.6E 00
210-BI	5.00E 00	Sol.	GI LLI	8.2E 01	Kidneys	4.0E 00
		Insol.	GI LLI	8.2E 01	Lungs	3.6E 00
212-BI	4.20E-02	Sol.	GI S	7.0E 02	Kidneys	6.0E 01
		Insol.	GI S	7.0E 02	Lungs	1.3E 02
210-PO	1.38E 02	Sol.	Spleen	1.4E 00	Spleen	3.1E-01
					Kidneys	3.1E-01
		Insol.	GI LLI	5.8E 01	Lungs	1.3E-01
211-AT	3.00E-01	Sol.	Thyroid	3.4E 00	Thyroid	4.4E 00
			Ovaries	3.4E 00		
		Insol.	GI ULI	1.5E 02	Lungs	2.2E 01
220-RN(3)	6.40E-04				Lungs	1.8E 02
222-RN(3)	3.82E 00				Lungs	1.8E 01
223-RA	1.17E 01	Sol.	Bone	1.4E 00	Bone	1.1E 00
		Insol.	GI LLI	8.4E 00	Lungs	1.5E-01
224-RA	3.64E 00	Sol.	Bone	4.6E 00	Bone	3.5E 00
		Insol.	GI LLI	1.1E 01	Lungs	4.5E-01
226-RA	5.90E 05	Sol.	Bone	2.4E-02	Bone	1.8E-02
		Insol.	GI LLI	6.4E 01	Lungs	3.3E-02
228-RA	2.40E 03	Sol.	Bone	5.6E-02	Bone	4.2E-02
		Insol.	GI LLI	5.0E 01	Lungs	2.4E-02
227-AC	8.00E 03	Sol.	Bone	3.8E 00	Bone	1.5E-03
		Insol.	GI LLI	6.0E 02	Lungs	1.6E-02
228-AC	2.60E-01	Sol.	GI ULI	1.7E 02	Liver	4.7E 01
					Bone	5.3E 01
		Insol.	GI ULI	1.7E 02	Lungs	1.1E 01

PERMISSIBLE QUARTERLY INTAKES (ORAL AND INHALATION) OF RADIONUCLIDES AND CRITICAL BODY ORGANS RECOMMENDED BY NCRP FOR OCCUPATIONAL EXPOSURE (1) —(Continued)

Radionuclide	Radioactive half-life days	Solubility state	Oral intake Critical organ	Oral intake Permissible quarterly intake microcuries	Intake by inhalation Critical organ	Intake by inhalation Permissible quarterly intake microcuries
227-TH	1.84E 01	Sol.	GI LLI	3.6E 01	Bone	2.2E-01
		Insol.	GI LLI	3.6E 01	Lungs	1.1E-01
228-TH	7.00E 02	Sol.	Bone	1.5E 01	Bone	5.6E-03
		Insol.	GI LLI	2.6E 01	Lungs	3.8E-03
230-TH	2.90E 07	Sol.	Bone	3.4E 00	Bone	1.4E-03
		Insol.	GI LLI	6.4E 01	Lungs	6.4E-03
231-TH	1.07E 00	Sol.	GI LLI	4.6E 02	GI LLI	9.3E 02
		Insol.	GI LLI	4.6E 02	GI LLI	7.5E 02
232-TH	5.10E 12	Sol.	Bone	3.0E 00	Bone	1.2E-03(4)
		Insol.	GI LLI	7.6E 01	Lungs	7.3E-03
234-TH	2.41E 01	Sol.	GI LLI	3.4E 01	Bone	3.6E 01
		Insol.	GI LLI	3.4E 01	Lungs	2.0E 01
NAT-TH(5)		Sol.	Bone	2.6E 00	Bone	1.0E-03(4)
		Insol.	GI LLI	1.9E 01	Lungs	2.5E-03
230-PA	1.77E 01	Sol.	GI LLI	4.8E 02	Bone	1.1E 00
		Insol.	GI LLI	5.0E 02	Lungs	4.9E-01
231-PA	1.30E 07	Sol.	Bone	1.7E 00	Bone	7.1E-04
		Insol.	GI LLI	5.4E 01	Lungs	6.7E-02
233-PA	2.74E 01	Sol.	GI LLI	2.4E 02	Kidneys	3.8E 02
		Insol.	GI LLI	2.4E 02	Lungs	1.1E 02
230-U	2.08E 01	Sol.	Kidneys	4.8E 00	Kidneys	1.8E-01
		Insol.	GI LLI	9.2E 00	Lungs	7.1E-02
232-U	2.70E 04	Sol.	Bone	1.7E 00	Bone	6.4E-02
		Insol.	GI LLI	5.8E 01	Lungs	1.7E-02
233-U	5.90E 07	Sol.	Bone	8.4E 00	Bone	3.3E-01
		Insol.	GI LLI	6.4E 01	Lungs	7.5E-02
234-U	9.10E 07	Sol.	Bone	8.6E 00	Bone	3.5E-01
		Insol.	GI LLI	6.4E 01	Lungs	7.5E-02
235-U(5)	2.60E 11	Sol.	Kidneys	7.4E 00	Kidneys	2.9E-01
					Bone	3.6E-01
		Insol.	GI LLI	5.6E 01	Lungs	8.0E-02
236-U(5)	8.70E 09	Sol.	Bone	9.0E 00	Bone	3.6E-01
		Insol.	GI LLI	6.8E 01	Lungs	7.8E-02
238-U(5)	1.60E 12	Sol.	Kidneys	1.2E 00	Kidneys	4.5E-02
		Insol.	GI LLI	7.0E 01	Lungs	8.6E-02
240-U	5.88E-01	Sol.	GI LLI	6.8E 01	GI LLI	1.4E 02
		Insol.	GI LLI	6.8E 01	GI LLI	1.1E 02
NAT-U(5)		Sol.	Kidneys	1.2E 00	Kidneys	4.5E-02
		Insol.	GI LLI	3.2E 01	Lungs	4.0E-02
237-NP	8.00E 08	Sol.	Bone	6.2E 00	Bone	2.5E-03
		Insol.	GI LLI	7.0E 01	Lungs	7.5E-02
239-NP	2.33E 00	Sol.	GI LLI	2.6E 02	GI LLI	5.3E 02
		Insol.	GI LLI	2.6E 02	GI LLI	4.2E 02
238-PU	3.30E 04	Sol.	Bone	1.0E 01	Bone	1.2E-03
		Insol.	GI LLI	5.6E 01	Lungs	2.2E-02
239-PU	8.90E 06	Sol.	Bone	9.0E 00	Bone	1.1E-03
		Insol.	GI LLI	5.8E 01	Lungs	2.4E-02
240-PU	2.40E 06	Sol.	Bone	9.0E 00	Bone	1.1E-03
		Insol.	GI LLI	5.8E 01	Lungs	2.4E-02
241-PU	4.80E 03	Sol.	Bone	4.6E 02	Bone	5.6E-02
		Insol.	GI LLI	2.6E 03	Lungs	2.4E 01
242-PU	1.40E 08	Sol.	Bone	9.4E 00	Bone	1.1E-03
		Insol.	GI LLI	6.2E 01	Lungs	2.4E-02
243-PU	2.08E-01	Sol.	GI ULI	6.8E 02	GI ULI	1.1E 03
		Insol.	GI ULI	6.8E 02	GI ULI	1.4E 03
244-PU	2.80E 10	Sol.	Bone	8.6E 00	Bone	1.0E-03
		Insol.	GI LLI	2.2E 01	Lungs	2.0E-02
241-AM	1.70E 05	Sol.	Kidneys	7.6E 00	Bone	3.6E-03
					Kidneys	3.8E-03
		Insol.	GI LLI	5.4E 01	Lungs	6.4E-02
242M-AM	5.60E 04	Sol.	Bone	8.8E 00	Bone	3.5E-03

PERMISSIBLE QUARTERLY INTAKES (ORAL AND INHALATION) OF RADIONUCLIDES AND CRITICAL BODY ORGANS RECOMMENDED BY NCRP FOR OCCUPATIONAL EXPOSURE (1) —(Continued)

Radionuclide	Radioactive half-life days	Solubility state	Oral intake: Critical organ	Oral intake: Permissible quarterly intake microcuries	Intake by inhalation: Critical organ	Intake by inhalation: Permissible quarterly intake microcuries
		Insol.	GI LLI	1.8E 02	Lungs	1.6E-01
242-AM	6.77E-01	Sol.	GI LLI	2.6E 02	Liver	2.4E 01
		Insol.	GI LLI	2.6E 02	Lungs	2.9E 01
243-AM	2.90E 06	Sol.	Bone	8.8E 00	Bone	3.5E-03
					Kidneys	3.8E-03
		Insol.	GI LLI	5.6E 01	Lungs	6.7E-02
244-AM	1.81E-02	Sol.	GI SI	9.6E 03	Bone	2.5E 03
					Kidneys	2.7E 03
		Insol.	GI SI	9.6E 03	Lungs	1.5E 04
					GI(SI)	1.5E 04
242-CM	1.62E 02	Sol.	GI LLI	4.8E 01	Liver	7.5E-02
		Insol.	GI LLI	5.0E 01	Lungs	1.0E-01
243-CM	1.30E 04	Sol.	Bone	1.0E 01	Bone	4.0E-03
		Insol.	GI LLI	5.0E 01	Lungs	6.2E-02
244-CM	6.70E 03	Sol.	Bone	1.4E 01	Bone	5.6E-03
		Insol.	GI LLI	5.2E 01	Lungs	6.2E-02
245-CM	7.30E 06	Sol.	Bone	7.0E 00	Bone	2.9E-03
		Insol.	GI LLI	5.6E 01	Lungs	6.7E-02
246-CM	2.40E 06	Sol.	Bone	7.2E 00	Bone	2.9E-03
		Insol.	GI LLI	5.6E 01	Lungs	6.6E-02
247-CM	3.30E 10	Sol.	Bone	7.2E 00	Bone	2.9E-03
		Insol.	GI LLI	4.4E 01	Lungs	6.7E-02
248-CM	1.70E 08	Sol.	Bone	8.8E-01	Bone	3.6E-04
		Insol.	GI LLI	2.6E 00	Lungs	8.2E-03
249-CM	4.40E-02	Sol.	GI S	4.4E 03	Bone	7.8E 03
		Insol.	GI S	4.4E 03	GI S	7.1E 03
249-BK	2.90E 02	Sol.	GI LLI	1.2E 03	Bone	5.8E-01
		Insol.	GI LLI	1.2E 03	Lungs	7.5E 01
250-BK	1.34E-01	Sol.	GI ULI	4.4E 02	Bone	9.1E 01
		Insol.	GI ULI	4.4E 02	GI ULI	7.1E 02
249-CF	1.70E 05	Sol.	Bone	8.2E 00	Bone	9.8E-04
		Insol.	GI LLI	4.8E 01	Lungs	6.2E-02
250-CF	3.70E 03	Sol.	Bone	2.6E 01	Bone	3.1E-03
		Insol.	GI LLI	5.0E 01	Lungs	6.2E-02
251-CF	2.90E 05	Sol.	Bone	8.6E 00	Bone	1.0E-03
		Insol.	GI LLI	5.2E 01	Lungs	6.2E-02
252-CF	8.03E 02	Sol.	GI LLI	1.5E 01	Bone	4.0E-03
		Insol.	GI LLI	1.5E 01	Lungs	2.0E-02
253-CF	1.80E 01	Sol.	GI LLI	2.8E 02	Bone	5.3E-01
		Insol.	GI LLI	2.8E 02	Lungs	4.7E-01
254-CF	5.60E 01	Sol.	GI LLI	2.4E-01	Bone	3.3E-03
		Insol.	GI LLI	2.4E-01	Lungs	3.1E-03
253-ES	2.00E 01	Sol.	GI LLI	4.6E 01	Bone	4.7E-01
		Insol.	GI LLI	4.6E 01	Lungs	3.8E-01
254M-ES	1.60E 00	Sol.	GI LLI	3.8E 01	Bone	3.3E 00
		Insol.	GI LLI	3.8E 01	Lungs	3.6E 00
254-ES	4.80E 02	Sol.	GI LLI	2.8E 01	Bone	1.2E-02
		Insol.	GI LLI	2.8E 01	Lungs	6.7E-02
255-ES	3.00E 01	Sol.	GI LLI	5.6E 01	Bone	3.1E-01
		Insol.	GI LLI	5.6E 01	Lungs	2.5E-01
254-FM	1.35E-01	Sol.	GI ULI	2.4E 02	Bone	4.0E 01
		Insol.	GI ULI	2.4E 02	Lungs	4.4E 01
255-FM	8.96E-01	Sol.	GI LLI	6.6E 01	Bone	1.0E 01
		Insol.	GI LLI	6.6E 01	Lungs	6.7E 00
256-FM	1.11E-01	Sol.	GI ULI	1.8E 00	Bone	1.7E 00
		Insol.	GI ULI	1.8E 00	Lungs	1.1E 00

[1] These quarterly intakes are calculated from values of $(MPC)_a$ and $(MPC)_w$ recommended by the NCRP for occupational exposure (see NBS Handbook 69 and NCRP Report No. 32)* except for uranium, ^{90}Sr, and certain transuranic isotopes given by the ICRP (see ICRP Publications 2 and 6).** Specifically, except where indicated in the footnotes, the quarterly intakes are calculated to two-digit accuracy as $91 \times 2 \times 10^7 \times (MPC)_a$ and $91 \times 2200 \times (MPC)_w$ where 91 ~ days/quarter, 2×10^7 ~ cc of air breathed/day, 2200 ~ cc of water intake/day, and $(MPC)_a$, $(MPC)_w$ are μCi/cc of air and water recommended as limits for occupational exposure for the 168-hour week. These quarterly intakes are calculated for standard adult man and would be expected to deliver a total dose (= dose commitment) to the critical organ of 1/4 the maximum permissible annual dose (MPD) recommended as limiting by the NCRP for

PERMISSIBLE QUARTERLY INTAKES (ORAL AND INHALATION) OF RADIONUCLIDES AND CRITICAL BODY ORGANS RECOMMENDED BY NCRP FOR OCCUPATIONAL EXPOSURE (1) —(Continued)

occupational exposure. These limiting MPD are the following: Bone, skin, thyroid—MPD = 30 rem; blood-forming organs, lenses of the eyes, gonads—MPD = 5 rem/yr; other organs—MPD = 15 rem/yr. These limits are not appropriate for exposure of the population or, specifically, for persons below the age of 18 years. For more detailed information on appropriate exposure limits recommended by the NCRP, see NBS Handbook 69 and NCRP Report No. 32.*

The abbreviations GI, S, SI, ULI, and LLI refer to gastrointestinal tract, stomach, small intestine, upper large intestine, and lower large intestine, respectively.

With few exceptions, the above intakes do not take account of chemical toxicity which should be considered separately.

2 This intake includes absorption through the skin (see ICRP Publication 2, p. 22).**

3 The daughter isotopes of ^{220}Rn and ^{222}Rn are assumed present to the extent they occur in unfiltered air (see NBS Handbook 69, p. 12).* For all other isotopes, the daughter elements are not considered as part of the intake and if present must be considered on the basis of the rules for mixtures.

4 These are provisional values for ^{232}Th and nat-Th (see footnote, p. 83, NBS Handbook 69).*

5 A curie of natural uranium is here considered to correspond to 2.94 g and a curie of natural thorium to 9.0 g of these materials (see NBS Handbook 69, p. 14).* For soluble nat-U, ^{238}U, ^{236}U, and ^{235}U these limits are based on considerations of chemical toxicity (see ICRP Publication 6, p. 13).**

* NBS Handbook 69, *Maximum Permissible Body Burdens and Maximum Permissible Concentrations of Radionuclides in Air and Water for Occupational Exposure*, 1959 (available from the Superintendent of Documents, U.S. Government Printing Office, Washington, D.C. 20402). NCRP Report No. 32, *Radiation Protection in Educational Institutions*, 1966 (NCRP Publications, P.O. Box 4867, Washington, D.C. 20008).

** *Recommendations of the International Commission on Radiological Protection*, ICRP Publication 2 (Pergamon Press, London, 1959) or with the bibliography, *Healthy Phys.* 3, 1–380 (June 1960); ICRP Publication 6 (A Pergamon Press Book, 1964, distributed by The Macmillan Co., New York).

CRYOGENIC PROPERTIES OF GASES

Property and conditions		He	Ne	Ar	Kr	Xe	H_2	CH_4	NH_3	N_2	O_2	F_2
		Gas										
Density	32°F, 1 atm, lb/ft³	0.01114	0.0562	0.1113	0.234	0.368	0.00561	0.0448	0.0481	0.0781	0.0892	0.106
	O°C, 1 atm, kg/m³	0.01784	0.9002	1.783	3.748	5.895	0.0899	0.718	0.770	1.251	1.429	1.698
Boiling point	°F, 1 atm	–452.08	–410.89	–302.3	–242.1	–160.8	–423.2	–263.2	–28.03	–320.4	–297.35	–306.7
	°C, 1 atm	–268.934	–246.048	–185.7	–152.90	–107.1	–252.87	–164.0	–33.35	–195.8	–182.97	–188.14
	°K, 1 atm	4.95	27.10	87.45	120.25	166.05	20.28	109.15	239.80	77.35	90.18	85.01
Melting point	°F, 1 atm	–458.0[a]	–415.6	–308.6	–249.9	–169.4	–434.5	–296.46	–107.9	–345.87	–361.1	–363.3
	°C, 1 atm	–272.2[a]	–248.67	–189.2	–156.6	–111.9	–259.14	–182.48	–77.7	–209.86	–218.4	–219.62
	°K, 1 atm	0.95[a]	24.48	83.95	116.55	161.25	14.01	90.67	195.45	63.29	54.75	53.53
Vapor density at boiling point	lb/ft³	0.999	0.593	0.368	0.518	0.606	0.0830	0.1124	0.0556	0.288	0.279	
	kg/m³	16.002	9.499	5.895	8.289	9.707	1.329	1.8004	0.8906	4.613	4.4756	
Liquid density at boiling point	lb/ft³	7.803	74.91	86.77	149.8	193.5	4.37	26.47	42.58	50.19	71.23	94.4
	kg/m³	125.	1200.	1390.	2400.	3100.	70.0	424.	682.1	804.	1142	1512
Vapor pressure of solid at melting point	lb/in.²		6.25	9.98	10.6	11.8	1.04	1.35	0.87	1.86	0.038	0.002
	kg/m²		323.	516	549.	612.	54.	70.	45.2	96.4	2.0	0.12
	(N/m²) × 10⁴		4.34	6.93	7.36	8.20	0.723	0.938	0.604	1.29	0.0026	0.00014
Heat of vaporization at boiling point	Btu/lb	10.3	37.4	70.0	46.4	41.4	194.4	248.4	588.6	85.7	91.588	73.7
	kcal/kg	5.72	20.8	38.9	25.8	23.0	108	138	327	47.6	50.88	40.9
	(J/kg) × 10³	23.932	87.027	162.76	107.95	96.23	451.9	577.4	1368.2	199.2	212.9	171.1
Heat of fusion at melting point	Btu/lb	1.8	7.2	12.1	7.0	5.9	25.2	26.1	152.1	11.0	5.9	5.8
	kcal/kg	1.0	4.0	6.7	3.9	3.3	14.0	14.5	84.0	6.1	3.27	3.2
	(J/kg) × 10³	4.184	16.74	28.03	16.3	13.8	58.6	60.7	351.5	25.5	13.7	13.4
C_p	59°F, 1 atm, Btu/lb-°F or 15°C, 1 atm, kcal/kg-°C	1.25[b]	0.25[c]	0.125	0.06[c]	0.04[c]	3.39	0.528	0.523	0.248	0.220	0.180
	288.15°K, 1 atm, (J/kg) × 10³	5.23[b]	1.05[c]	0.523	0.251[c]	0.167[c]	14.2	2.21	2.188	1.038	0.9205	0.753
C_p/C_v	15–20°C, 1 atm; 288–293°K, 1 atm	1.66[b]	1.64[c]	1.67	1.68[c]	1.66[c]	1.41	1.31	1.31	1.40	1.40	
Critical temperature	°F	–450.2	–397.7	–188.5	–82.7	61.9	–399.8	–116.5	270.3	–232.8	–181.3	–200.2
	°C	–267.9	–228.7	–122.5	–63.7	16.6	–239.9	–82.5	132.4	–147.1	–118.57	–129.0
	°K	5.25	44.45	150.65	209.45	289.75	33.25	190.65	405.55	126.05	154.58	144.15
Critical pressure	lb/in.² (absolute)	33.2	394.6	705.2	798	855	188.1	672	1639	492.3	731.4	808.3
	kg/cm²	2.33	27.7	49.6	46.1	60.1	13.2	47.2	115.5	34.6	51.4	56.8
	(kg/m²) × 10³	23.3	277	496	561	601	132	472	1155	346	514	568
	(N/m²) × 10⁴	23.1	274.1	489.9	554.4	594	130.7	466.9	1139	342.	508.1	561.6

Note: For conversion factors see p. B-439.

[a] At 26 atmospheres.
[b] At –292°F or –180°C.
[c] Approximate.

CONVERSION FACTORS FOR TABLE OF CRYOGENIC PROPERTIES OF GASES

To convert from	To	Multiply by
lbs ft^{-3}	kg m^{-3}	16.018
lbs $in.^{-2}$	N m^{-2}	6894.8
lbs ft^{-2}	N m^{-2}	47.880
BTU lb^{-1}	J kg^{-1}	2324.4
cal g^{-1}	J kg^{-1}	4184
cal g^{-1} °F	J kg^{-1} °C	4184

VISCOSITY AND THERMAL CONDUCTIVITY OF NITROGEN AT CRYOGENIC TEMPERATURES

The viscosity and thermal conductivity of nitrogen gas for the temperature range 5 K–135 K have been computed from the second Chapman-Enskog approximation. Quantum effects, which become appreciable at the lower temperatures, are included by utilizing collision integrals based on quantum theory. A Lennard-Jones (12-6) potential was assumed. The computations yield viscosities about 20% lower than those predicted for the high end of this temperature range by the method of corresponding states, but the agreement is excellent when the computed values are compared with existing experimental data.

T, °K	η, micropoise	λ, $\frac{\mu cal}{cm\ sec\ °K}$
4.575	4.16639	1.10990
5.49	4.85525	1.29307
6.405	5.54773	1.47715
7.320	6.24722	1.63156
8.235	6.95025	1.85020
9.15	7.65067	2.03666
13.725	10.95231	2.91729
18.3	13.85738	3.69313
22.875	16.55204	4.41068
27.45	19.21670	5.11847
32.025	21.9423	5.84208
36.60	24.7606	6.59077
41.175	27.67445	7.36545
45.75	30.67422	8.16348
54.9	36.8747	9.81383
64.05	43.2469	11.51005
82.35	56.14634	14.94298
91.5	62.54767	16.64610
137.25	92.96598	24.74580

From Pearson, W. E., NASA Technical Note D-7565, National Aeronautics and Space Administration, 1974 (available from Superintendent of Documents, U.S. Government Printing Office, Washington, D.C.).

DEFINITIVE RULES FOR NOMENCLATURE OF ORGANIC CHEMISTRY

IUPAC 1957 Rules.

Section A. Hydrocarbons

Section B. Fundamental Heterocyclic Systems

These rules are taken from 'Definitive Rules for the Nomenclature of Organic Chemistry' which were adopted unanimously by the Commission on Nomenclature and by The Council of the International Union of Pure and Applied Chemistry at Paris 1957, and subsequently published by Butterworths Scientific Publications on behalf of the Union. The extracts are printed here by permission of the Union and of Butterworths Scientific Publications. Future 'tentative' rules will be published in the Bulletin of the Union, and when made 'definitive' in its Journal 'Pure and Applied Chemistry.'

RULES

A. HYDROCARBONS

Acyclic Hydrocarbons

A-1

1.1.—The first four saturated unbranched acyclic hydrocarbons are called methane, ethane, propane and butane. Names of the higher members of this series consist of a numerical prefix and the termination "-ane." Examples of these numerical prefixes are shown in the table below. The generic name of saturated acyclic hydrocarbons (branched or unbranched) is "alkane."

Examples:

n		n		n		n	
1	Methane	12	Dodecane	22	Docosane	32	Dotriacontane
2	Ethane	13	Tridecane	23	Tricosane	33	Tritriacontane
3	Propane	14	Tetradecane	24	Tetracosane	40	Tetracontane
4	Butane	15	Pentadecane	25	Pentacosane	50	Pentacontane
5	Pentane	16	Hexadecane	26	Hexacosane	60	Hexacontane
6	Hexane	17	Heptadecane	27	Heptacosane	70	Heptacontane
7	Heptane	18	Octadecane	28	Octacosane	80	Octacontane
8	Octane	19	Nonadecane	29	Nonacosane	90	Nonacontane
9	Nonane	20	Eicosane	30	Triacontane	100	Hectane
10	Decane	21	Heneicosane	31	Hentriacontane	132	Dotriacontahectane
11	Undecane						

1.2.—Univalent radicals derived from saturated unbranched acyclic hydrocarbons by removal of hydrogen from a terminal carbon atom are named by replacing the ending "-ane" of the name of the hydrocarbon by "-yl." The carbon atom with the free valence is numbered as 1. As a class, these radicals are called normal, or unbranched-chain, alkyls.

Examples:

Pentyl $\overset{5}{C}H_3 - \overset{4}{C}H_2 - \overset{3}{C}H_2 - \overset{2}{C}H_2 - \overset{1}{C}H_2 -$

Undecyl $\overset{11}{C}H_3 - [\overset{10-2}{C}H_2]_9 - \overset{1}{C}H_2 -$

A-2

2.1.—A saturated branched acyclic hydrocarbon is named by prefixing the designations of the side chains to the name of the longest chain present in the formula.

Example:

$$\overset{5}{C}H_3 - \overset{4}{C}H_2 - \overset{3}{C}H - \overset{2}{C}H_2 - \overset{1}{C}H_3$$
$$|$$
$$CH_3$$

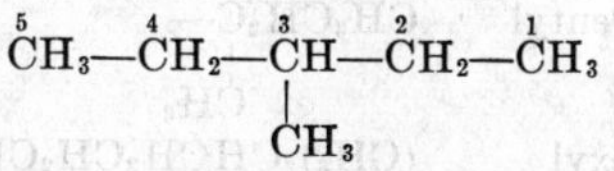

3-Methylpentane

These names are retained for unsubstituted hydrocarbons only

Isobutane	$(CH_3)_2CH—CH_3$
Isopentane	$(CH_3)_2CH—CH_2—CH_3$
Neopentane	$(CH_3)_4C$
Isohexane	$(CH_3)_2CH—CH_2—CH_2—CH_3$

Chemical Abstracts index names for the above compounds are 2-methylpropane, 2-methylbutane, 2,2-dimethylpropane, and 2-methylpentane.

2.2.—The longest chain is numbered from one end to the other by arabic numerals, the direction being so chosen as to give the lowest numbers possible to the side chains. When series of locants containing the same number of terms are compared term by term, that series is "lowest" which contains the lowest number on the occasion of the first difference. This principle is applied irrespective of the nature of the substituents.

Examples:

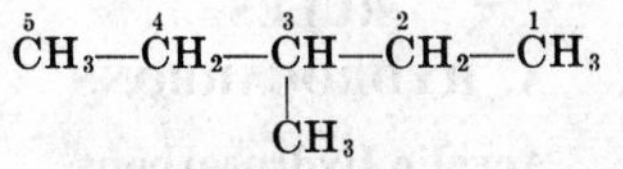

$$\overset{5}{C}H_3—\overset{4}{C}H_2—\overset{3}{C}H—\overset{2}{C}H_2—\overset{1}{C}H_3$$
$$|$$
$$CH_3$$

3-Methylpentane

$$\overset{6}{C}H_3—\overset{5}{C}H—\overset{4}{C}H_2—\overset{3}{C}H—\overset{2}{C}H—\overset{1}{C}H_3$$
$$|\qquad\qquad|\quad\;|$$
$$CH_3\qquad CH_3\; CH_3$$

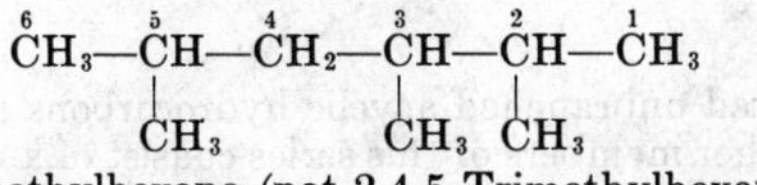

2,3,5-Trimethylhexane (not 2,4,5-Trimethylhexane)

$$\overset{10}{C}H_3—\overset{9}{C}H_2—\overset{8}{C}H—\overset{7}{C}H—\overset{6}{C}H_2—\overset{5}{C}H_2—\overset{4}{C}H_2—\overset{3}{C}H_2—\overset{2}{C}H—\overset{1}{C}H_3$$
$$|\qquad|\qquad\qquad\qquad\qquad\qquad|$$
$$CH_3\; CH_3\qquad\qquad\qquad\qquad CH_3$$

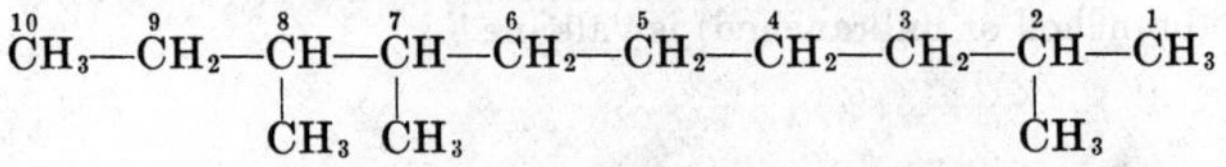

2,7,8-Trimethyldecane (not 3,4,9-Trimethyldecane)

$$\overset{9}{C}H_3—\overset{8}{C}H_2—\overset{7}{C}H_2—\overset{6}{C}H_2—\overset{5}{C}H —\overset{4}{C}H—\overset{3}{C}H_2—\overset{2}{C}H_2—\overset{1}{C}H_3$$
$$|\qquad\;\;|$$
$$CH_3\;\; CH_2—CH_2—CH_3$$

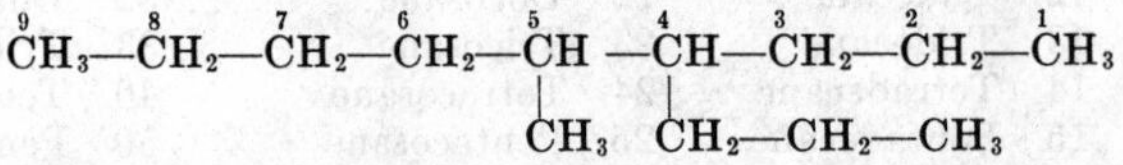

5-Methyl-4-propylnonane (not 5-Methyl-6-propylnonane since 4,5 is lower than 5,6)

2.25.—Univalent branched radicals derived from alkanes are named by prefixing the designation of the side chains to the name of the unbranched alkyl radical possessing the longest possible chain starting from the carbon atom with the free valence, the starting atom being numbered as 1.

Examples:

1-Methylpentyl	$\overset{5}{C}H_3\overset{4}{C}H_2\overset{3}{C}H_2\overset{2}{C}H_2\overset{1}{C}H(CH_3)—$
2-Methylpentyl	$CH_3CH_2CH_2CH(CH_3)CH_2—$
5-Methylhexyl	$(CH_3)_2CHCH_2CH_2CH_2CH_2—$

These names are retained for the unsubstituted radicals only

Isopropyl	$(CH_3)_2CH—$
Isobutyl	$(CH_3)_2CHCH_2—$
sec-Butyl	$CH_3CH_2CH—$
	$\qquad\qquad\; |$
	$\qquad\qquad CH_3$
tert-Butyl	$(CH_3)_3C—$
Isopentyl	$(CH_3)_2CHCH_2CH_2—$
Neopentyl	$(CH_3)_3CCH_2—$
	$\qquad\qquad\quad |$
	$\qquad\qquad\quad CH_3$
tert-Pentyl	$CH_3CH_2C—$
	$\qquad\qquad\; |$
	$\qquad\qquad CH_3$
Isohexyl	$(CH_3)_2CHCH_2CH_2CH_2—$

2.3.—If two or more side chains of different nature are present, they may be cited (a) in order of increasing complexity or (b) in alphabetical order.

(a) The side chains are arranged in order of increasing complexity by applying the following criteria in series until a decision is reached:

(i) The less complex is that containing the smaller total number of carbon atoms.

Examples:

$$CH_3$$
$$|$$
$$CH_3—C— \quad \text{less complex than}$$
$$|$$
$$CH_3$$

$$\overset{5}{C}H_3—\overset{4}{C}H_2—\overset{3}{C}H_2—\overset{2}{C}H_2—\overset{1}{C}H_2—$$

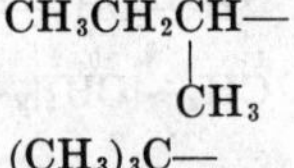

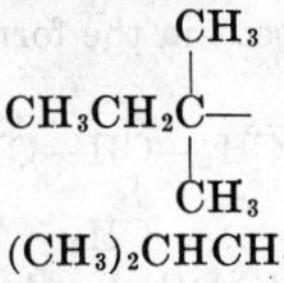

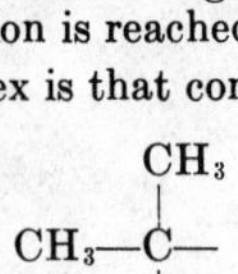

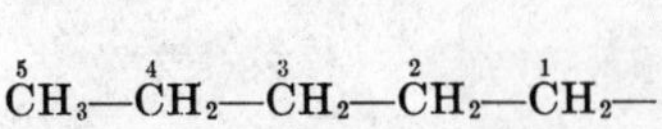

(ii) The less complex is that containing the longer straight chain.

Example:

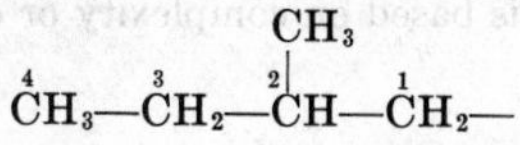

less complex than

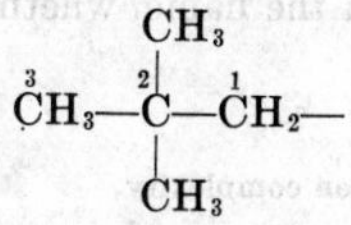

(iii) The less complex is that whose longest substituent has the lower locant.

Example:

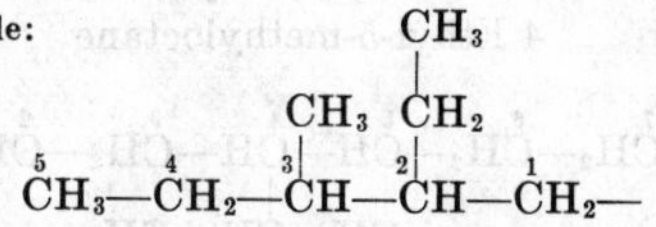

less complex than

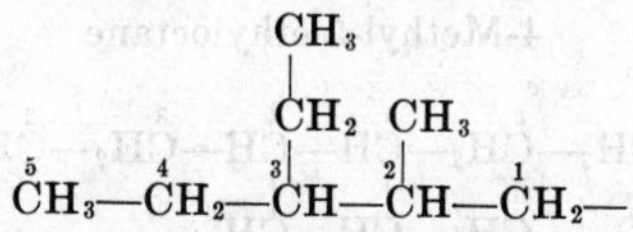

(iv) The less complex is that whose next longest substituent has the lower locant.

Example:

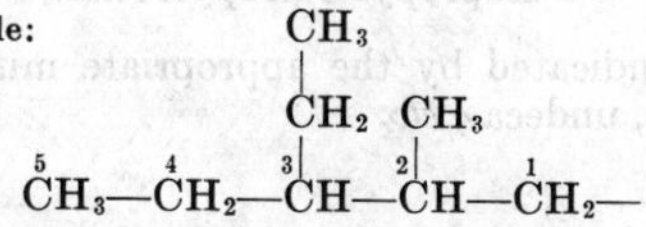

less complex than

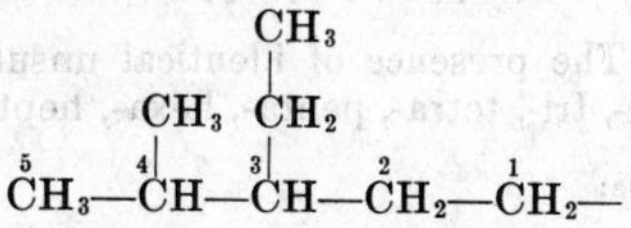

(v) The less complex is that which is the more saturated.

Example:

$\overset{3}{C}H_3\text{—}\overset{2}{C}H_2\text{—}\overset{1}{C}H_2\text{—}$ less complex than $\overset{3}{C}H_3\text{—}\overset{2}{C}H{=}\overset{1}{C}H\text{—}$

(vi) The less complex is that whose multiple linkage has the lower locant.

Example:

$\overset{3}{C}H_3\text{—}\overset{2}{C}H{=}\overset{1}{C}H\text{—}$ less complex than $\overset{3}{C}H_2{=}\overset{2}{C}H\text{—}\overset{1}{C}H_2\text{—}$

(b) The alphabetical order is decided as follows:

(i) The names of simple radicals are first alphabetized and the multiplying prefixes are then inserted.

Example:

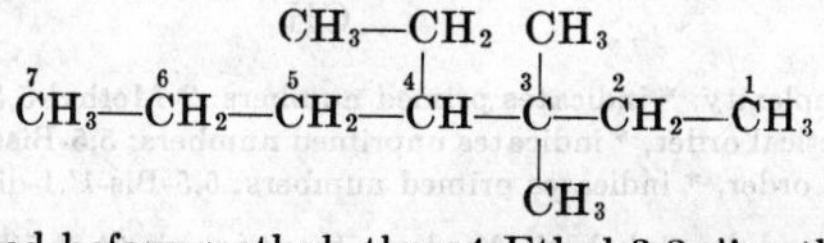

ethyl is cited before methyl, thus 4-Ethyl-3,3-dimethylheptane

(ii) The name of a complex radical is considered to begin with the first letter of its complete name.

Example:

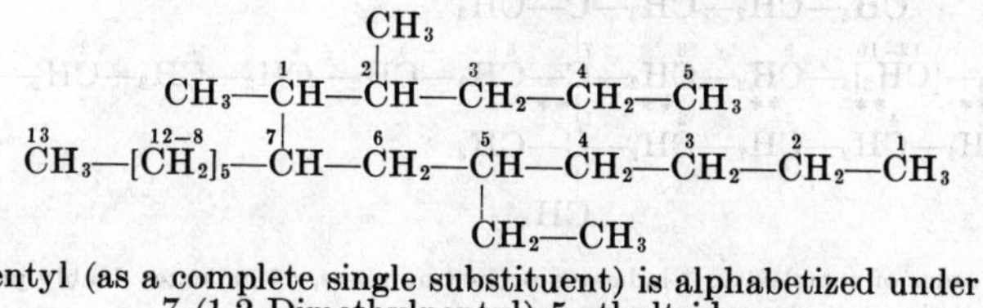

dimethylpentyl (as a complete single substituent) is alphabetized under "d," thus 7-(1,2-Dimethylpentyl)-5-ethyltridecane

(iii) In cases where complex radicals are composed of identical words, priority for citation is given to that radical which contains the lowest locant at the first cited point of difference in the radical.

Example:

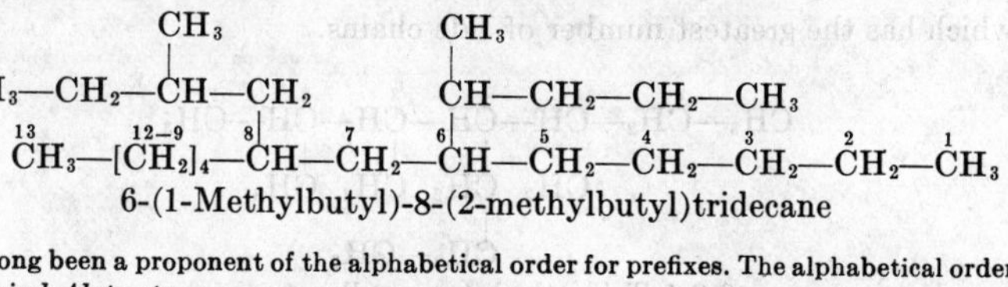

6-(1-Methylbutyl)-8-(2-methylbutyl)tridecane

***Chemical Abstracts* has long been a proponent of the alphabetical order for prefixes. The alphabetical order as given in A-2.3(b) is the order used by *Chemical Abstracts*.**

2.4.—If two or more side chains are in equivalent positions, the one to be assigned the lower number is that cited first in the name, whether the order of citation is based on complexity or on the alphabetical order.

Examples:

(a) Order based on complexity

$$\overset{8}{C}H_3-\overset{7}{C}H_2-\overset{6}{C}H_2-\overset{5}{C}H-\overset{4}{C}H-\overset{3}{C}H_2-\overset{2}{C}H_2-\overset{1}{C}H_3$$
$$CH_3-CH_2\ \ CH_3$$

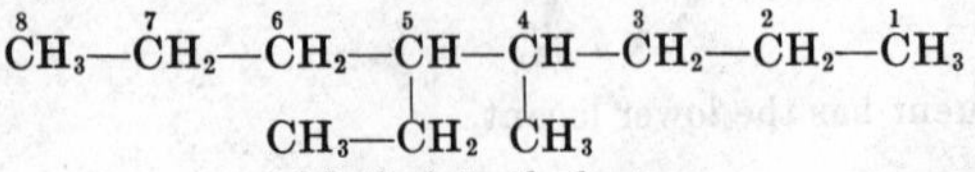

4-Methyl-5 ethyloctane

(b) Alphabetical order

$$\overset{8}{C}H_3-\overset{7}{C}H_2-\overset{6}{C}H_2-\overset{5}{C}H-\overset{4}{C}H-\overset{3}{C}H_2-\overset{2}{C}H_2-\overset{1}{C}H_3$$
$$CH_3\ \ CH_2-CH_3$$

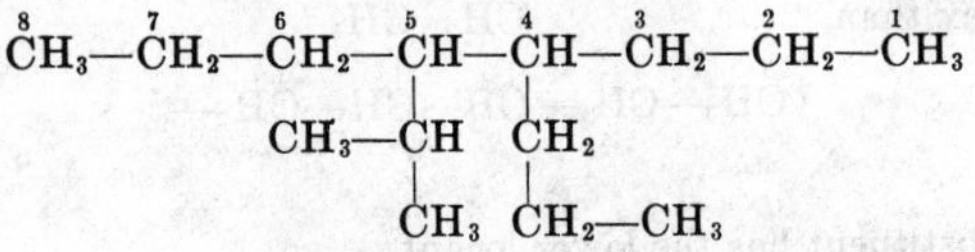

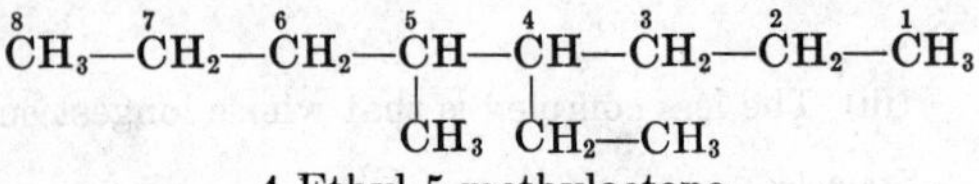

4 Ethyl-5-methyloctane

$$\overset{8}{C}H_3-\overset{7}{C}H_2-\overset{6}{C}H_2-\overset{5}{C}H-\overset{4}{C}H-\overset{3}{C}H_2-\overset{2}{C}H_2-\overset{1}{C}H_3$$
$$CH_3-CH\ \ CH_2$$
$$CH_3\ \ CH_2-CH_3$$

4-Propyl-5-isopropyloctane

$$\overset{8}{C}H_3-\overset{7}{C}H_2-\overset{6}{C}H_2-\overset{5}{C}H-\overset{4}{C}H-\overset{3}{C}H_2-\overset{2}{C}H_2-\overset{1}{C}H_3$$
$$CH_2\ \ CH-CH_3$$
$$CH_3-CH_2\ \ CH_3$$

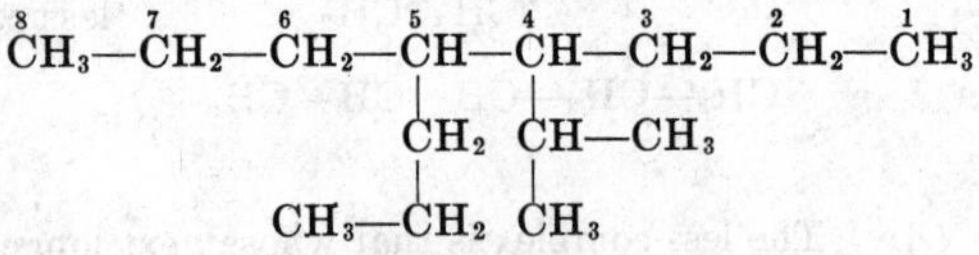

4-Isopropyl-5-propyloctane

2.5.—The presence of identical unsubstituted radicals is indicated by the appropriate multiplying prefix di-, tri-, tetra-, penta-, hexa-, hepta-, octa-, nona-, deca-, undeca-, *etc*.

Example:

$$CH_3$$
$$\overset{5}{C}H_3-\overset{4}{C}H_2-\overset{3}{C}-\overset{2}{C}H_2-\overset{1}{C}H_3$$
$$CH_3$$

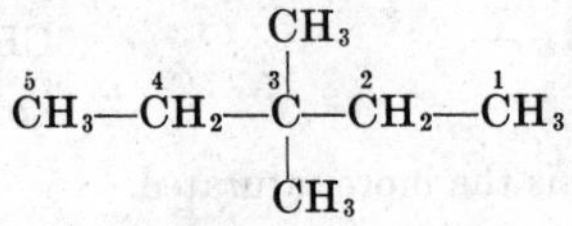

3,3-Dimethylpentane

The presence of identical radicals each substituted in the same way may be indicated by the appropriate multiplying prefix bis-, tris-, tetrakis-, pentakis-, *etc*. The complete expression denoting a side chain may be enclosed in parentheses or the carbon atoms in side chains may be indicated by primed numbers.

Example:

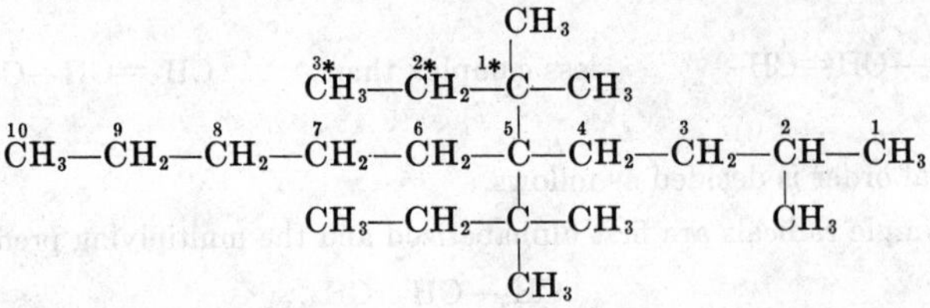

$$CH_3$$
$$\overset{3*}{C}H_3-\overset{2*}{C}H_2-\overset{1*}{C}-CH_3$$
$$\overset{10}{C}H_3-\overset{9}{C}H_2-\overset{8}{C}H_2-\overset{7}{C}H_2-\overset{6}{C}H_2-\overset{5}{C}-\overset{4}{C}H_2-\overset{3}{C}H_2-\overset{2}{C}H-\overset{1}{C}H_3$$
$$CH_3-CH_2-C-CH_3\qquad CH_3$$
$$CH_3$$

(a) Use of primes and order of complexity, * indicates primed numbers: 2-Methyl-5,5-bis-1',1'-dimethylpropyldecane.
(b) Use of parentheses and alphabetical order, * indicates unprimed numbers: 5,5-Bis(1,1-dimethylpropyl)-2-methyldecane.
(c) Use of primes and alphabetical order, * indicates primed numbers: 5,5-Bis-1',1-dimethylpropyl-2-methyldecane.

Chemical Abstracts uses parentheses and the alphabetical order without primes (A-2.5(b)). *Chemical Abstracts* would name the example in (b) 2-Methyl-5,5-di-*tert*-pentyldecane.

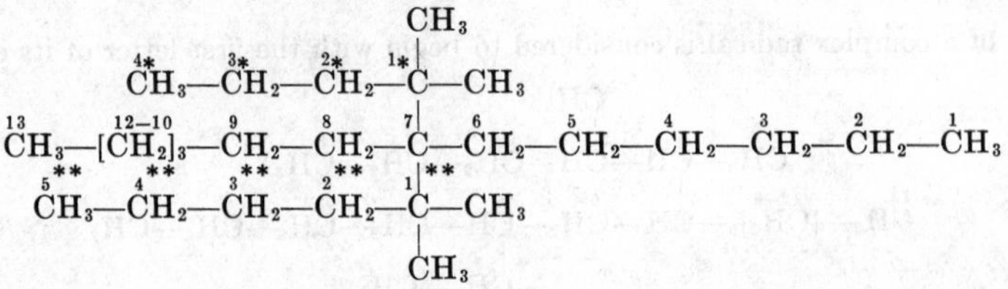

$$CH_3$$
$$\overset{4*}{C}H_3-\overset{3*}{C}H_2-\overset{2*}{C}H_2-\overset{1*}{C}-CH_3$$
$$\overset{13}{C}H_3-\overset{12-10}{[CH_2]_3}-\overset{9}{C}H_2-\overset{8}{C}H_2-\overset{7}{C}-\overset{6}{C}H_2-\overset{5}{C}H_2-\overset{4}{C}H_2-\overset{3}{C}H_2-\overset{2}{C}H_2-\overset{1}{C}H_3$$
$$\overset{5**}{C}H_3-\overset{4**}{C}H_2-\overset{3**}{C}H_2-\overset{2**}{C}H_2-\overset{1**}{C}-CH_3$$
$$CH_3$$

(a) Use of primes and order of complexity, * indicates primed numbers, ** indicates doubly primed numbers: 7-1',1'-Dimethylbutyl-7-1'',1''-dimethylpentyltridecane.
(b) Use of parentheses and alphabetical order, * and ** indicate unprimed numbers: 7-(1,1-Dimethylbutyl)-7-(1,1-dimethylpentyl)-tridecane

2.6.—If chains of equal length are competing for selection as main chain in a saturated branched acyclic hydrocarbon, then the choice goes in series to:

(a) The chain which has the greatest number of side chains.

Example:

$$\overset{7}{C}H_3-\overset{6}{C}H_2-\overset{5}{C}H-\overset{4}{C}H-\overset{3}{C}H-\overset{2}{C}H-\overset{1}{C}H_3$$
$$CH_3\ \ CH_2\ \ CH_3\ \ CH_3$$
$$CH_2-CH_3$$

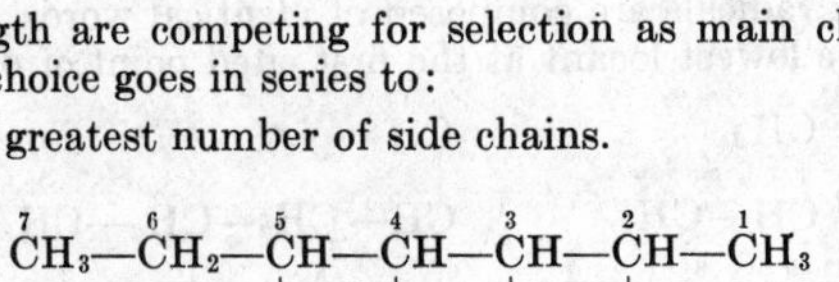

2,3,5-Trimethyl-4-propylheptane

(b) The chain whose side chains have the lowest-numbered locants.

Example:

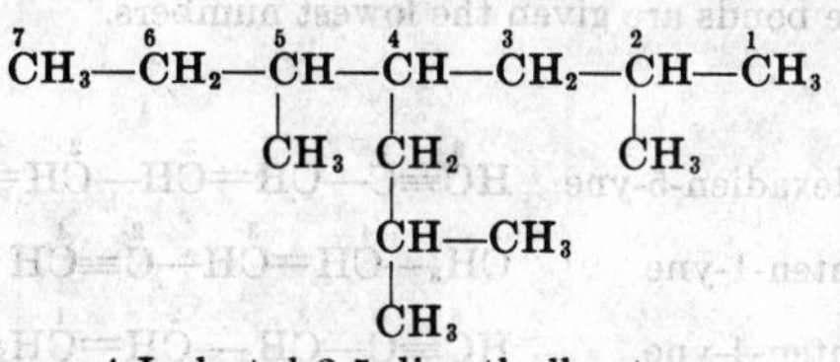

4-Isobutyl-2,5-dimethylheptane

(c) The chain having the greatest number of carbon atoms in the smaller side chains.

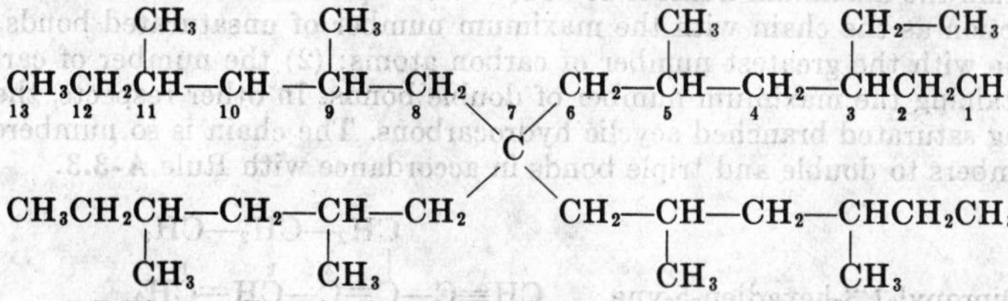

7,7-Bis(2,4-dimethylhexyl)-3-ethyl-5,9,11-trimethyltridecane

(d) The chain having the least branched side chains.

(1) Here the choice lies between two possible main chains of equal length, each containing six side chains in the same positions. Listing in increasing order, the number of carbon atoms in the several side chains of the first choice as shown and of the alternate second choice results as follows:

first choice	1, 1, 1, 2, 8, 8
second choice	1, 1, 1, 1, 8, 9

The expression, "the greatest number of carbon atoms in the smaller side chains," is taken to mean the largest side chain at the first point of difference when the size of the side chains is examined step by step. Thus, the selection in this case is made at the fourth step where 2 is greater than 1.

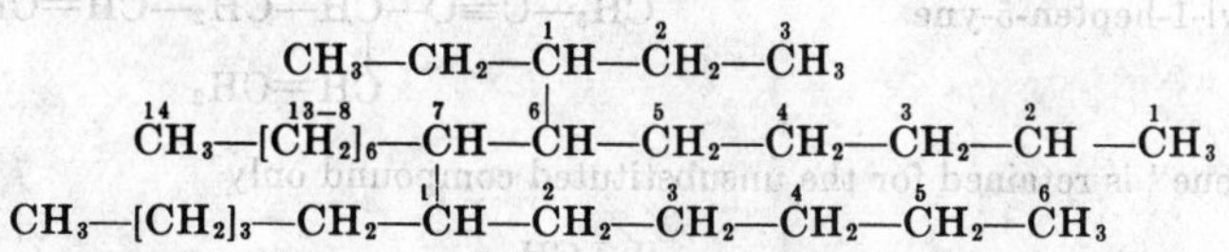

6-(1-Ethylpropyl)-7-(1-pentylhexyl)tetradecane

A-3

3.1.—Unsaturated unbranched acyclic hydrocarbons having one double bond are named by replacing the ending "-ane" of the name of the corresponding saturated hydrocarbon with the ending "-ene." If there are two or more double bonds, the ending will be "-adiene," "-atriene," *etc.* The generic names of these hydrocarbons (branched or unbranched) are "alkene," "alkadiene," "alkatriene," *etc.* The chain is so numbered as to give the lowest possible numbers to the double bonds.

Examples:

2-Hexene	$\overset{6}{C}H_3—\overset{5}{C}H_2—\overset{4}{C}H_2—\overset{3}{C}H{=}\overset{2}{C}H—\overset{1}{C}H_3$
1,4-Hexadiene	$\overset{6}{C}H_3—\overset{5}{C}H{=}\overset{4}{C}H—\overset{3}{C}H_2—\overset{2}{C}H{=}\overset{1}{C}H_2$

These non-systematic names are retained

Ethylene	$CH_2{=}CH_2$	Allene	$CH_2{=}C{=}CH_2$

Chemical Abstracts retains allene for the unsubstituted hydrocarbon only.

3.2.—Unsaturated unbranched acyclic hydrocarbons having one triple bond are named by replacing the ending "-ane" of the name of the corresponding saturated hydrocarbon with the ending "-yne." If there are two or more triple bonds, the ending will be "-adiyne," "-atriyne," *etc.* The generic names of these hydrocarbons (branched or unbranched) are "alkyne," "alkadiyne," "alkatriyne," *etc.* The chain is so numbered as to give the lowest possible numbers to the triple bonds.

The name "acetylene" for $HC{\equiv}CH$ is retained.

3.3.—Unsaturated unbranched acyclic hydrocarbons having both double and triple bonds are named by replacing the ending "-ane" of the name of the corresponding saturated hydrocarbon with the ending "-enyne," "-adienyne," "-atrienyne," "-enediyne," *etc.* Numbers as low as possible are given to double

and triple bonds even though this may at times give "-yne" a lower number than "-ene." When there is a choice in numbering the double bonds are given the lowest numbers.

Examples:

1,3-Hexadien-5-yne	$\overset{6}{H}C\equiv\overset{5}{C}-\overset{4}{C}H=\overset{3}{C}H-\overset{2}{C}H=\overset{1}{C}H_2$
3-Penten-1-yne	$\overset{5}{C}H_3-\overset{4}{C}H=\overset{3}{C}H-\overset{2}{C}\equiv\overset{1}{C}H$
1-Penten-4-yne	$\overset{5}{H}C\equiv\overset{4}{C}-\overset{3}{C}H_2-\overset{2}{C}H=\overset{1}{C}H_2$

3.4.—Unsaturated branched acyclic hydrocarbons are named as derivatives of the unbranched hydrocarbons which contain the maximum number of double and triple bonds. If there are two or more chains competing for selection as the chain with the maximum number of unsaturated bonds, then the choice goes to (1) that one with the greatest number of carbon atoms; (2) the number of carbon atoms being equal, the one containing the maximum number of double bonds. In other respects, the same principles apply as for naming saturated branched acyclic hydrocarbons. The chain is so numbered as to give the lowest possible numbers to double and triple bonds in accordance with Rule **A-3.3.**

Examples:

3,4-Dipropyl-1,3-hexadien-5-yne

$$CH_2-CH_2-CH_3$$
$$CH\equiv\overset{}{\underset{6}{C}}-\overset{3}{\underset{5}{C}}=\overset{2}{\underset{4}{C}}-\overset{1}{C}H=CH_2$$
$$CH_2-CH_2-CH_3$$

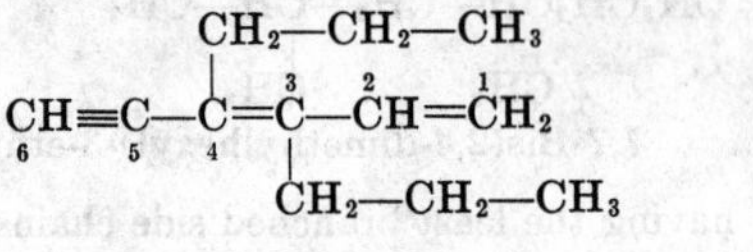

5-Ethynyl-1,3,6-heptatriene

$$\overset{7}{C}H_2=\overset{6}{C}H-\overset{5}{C}H-\overset{4}{C}H=\overset{3}{C}H-\overset{2}{C}H=\overset{1}{C}H_2$$
$$C\equiv CH$$

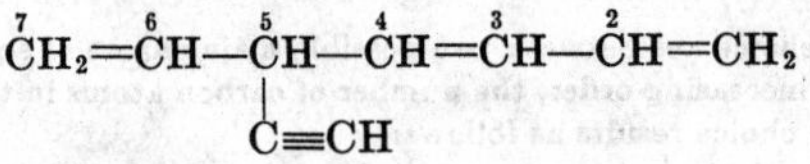

5,5-Dimethyl-1-hexene

$$CH_3$$
$$\overset{6}{C}H_3-\overset{5}{C}-\overset{4}{C}H_2-\overset{3}{C}H_2-\overset{2}{C}H=\overset{1}{C}H_2$$
$$CH_3$$

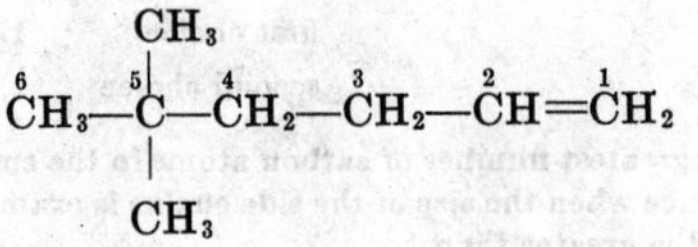

4-Vinyl-1-hepten-5-yne

$$\overset{7}{C}H_3-\overset{6}{C}\equiv\overset{5}{C}-\overset{4}{C}H-\overset{3}{C}H_2-\overset{2}{C}H=\overset{1}{C}H_2$$
$$CH=CH_2$$

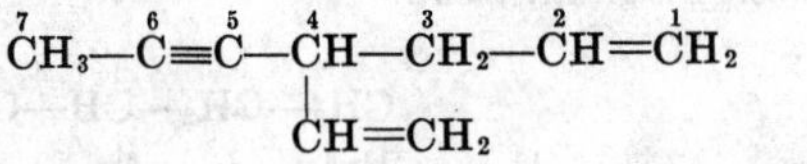

The name "isoprene" is retained for the unsubstituted compound only

$$CH_3$$
$$CH_2=CH-C=CH_2$$

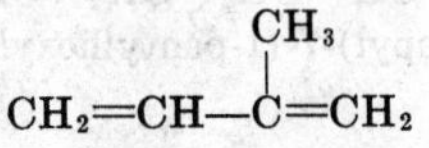

3.5.—The names of univalent radicals derived from unsaturated acyclic hydrocarbons have the endings "-enyl," "-ynyl," "-dienyl," etc., the positions of the double and triple bonds being indicated where necessary. The carbon atom with the free valence is numbered as 1.

Examples:

Ethynyl	$CH\equiv C-$
2-Propynyl	$CH\equiv C-CH_2-$
1-Propenyl	$CH_3-CH=CH-$
2-Butenyl	$CH_2-CH=CH-CH_2-$
1,3-Butadienyl	$CH_2=CH-CH=CH-$
2-Pentenyl	$CH_3-CH_2-CH=CH-CH_2-$
2-Penten-4-ynyl	$CH\equiv C-CH=CH-CH_2-$

Exceptions: These names are retained

Vinyl (for ethenyl)	$CH_2=CH-$
Allyl (for 2-propenyl)	$CH_2=CH-CH_2-$
Isopropenyl	$CH_2=C-$ (for unsubstituted radical only)
(for 1-methylvinyl)	CH_3

3.6.—When there is a choice for the fundamental chain of a radical, that chain is selected which contains (1) the maximum number of double and triple bonds; (2) the largest number of carbon atoms, and (3) the largest number of double bonds.

Examples:

$$\overset{10}{C}H_3-\overset{9}{C}H=\overset{8}{C}H-\overset{7}{C}H=\overset{6}{C}H-\overset{5}{C}H-\overset{4}{C}H=\overset{3}{C}H-\overset{2}{C}\equiv\overset{1}{C}-$$
$$CH_2-CH_2-CH=CH-CH_3$$

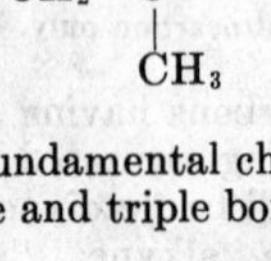

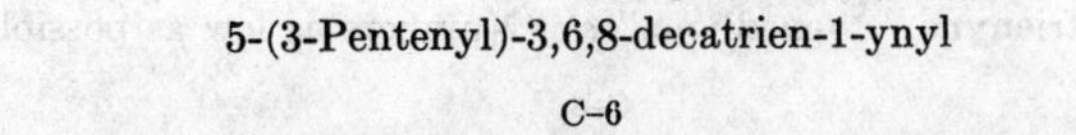

5-(3-Pentenyl)-3,6,8-decatrien-1-ynyl

$$\overset{12}{C}H_3-\overset{11}{C}H_2-\overset{10}{C}\equiv\overset{9}{C}-\overset{8}{C}H=\overset{7}{C}H-\overset{6}{C}H-\overset{5}{C}H=\overset{4}{C}H-\overset{3}{C}H=\overset{2}{C}H-\overset{1}{C}H_2-$$
$$CH=CH-CH=CH-CH_3$$

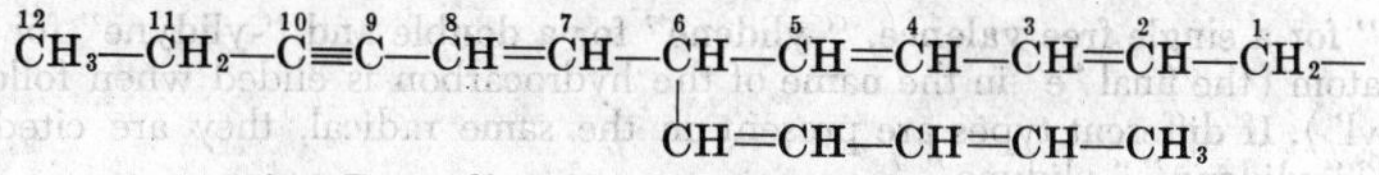

6-(1,3-Pentadienyl)-2,4 7-dodecatrien-9-ynyl

$$\overset{11}{C}H_3-\overset{10}{C}H=\overset{9}{C}H-\overset{8}{C}H=\overset{7}{C}H-\overset{6}{C}H-\overset{5}{C}H=\overset{4}{C}H-\overset{3}{C}H=\overset{2}{C}H-\overset{1}{C}H_2$$
$$CH=CH-C\equiv C-CH_3$$

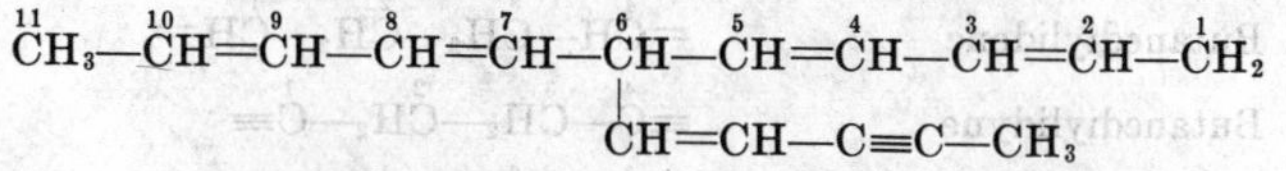

6-(1-Penten-3-ynyl)-2,4,7,9-undecatetraenyl

$$\overset{4}{C}H_3-\overset{3}{C}H=\overset{2}{C}-\overset{1}{C}H_2-$$
$$CH_2-CH_2-CH_2-CH_2-CH_2-CH_2-CH_2-CH_2-CH_3$$

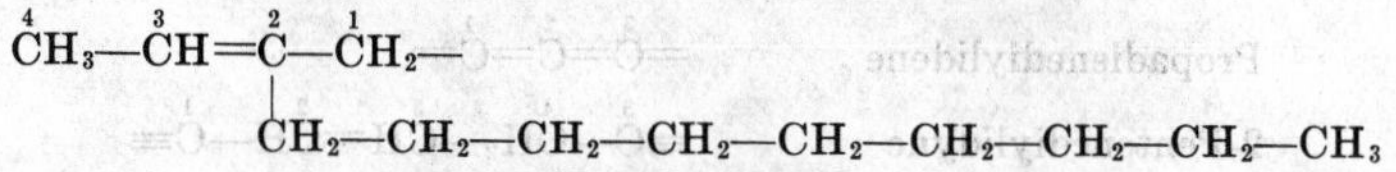

2-Nonyl-2-butenyl

A-4

4.1.—Bivalent and trivalent radicals derived from univalent acyclic hydrocarbon radicals whose authorized names end in "-yl" by removal of one or two hydrogen atoms from the carbon atom with the free valences are named by adding "-idene" or "-idyne," respectively, to the name of the corresponding univalent radical. The carbon atom with the free valence is numbered as 1.

The name "methylene" is retained for the radical $CH_2=$.

Examples:

Methylidyne[1]	$CH\equiv$
Ethylidene	$CH_3-CH=$
Ethylidyne	$CH_3-C\equiv$
Vinylidene	$CH_2=C=$
Isopropylidene[2]	$(CH_3)_2C=$

4.2.—The names of bivalent radicals derived from normal alkanes by removal of a hydrogen atom from each of the two terminal carbon atoms of the chain are ethylene, trimethylene, tetramethylene, *etc.*

Examples:

Pentamethylene	$-CH_2-CH_2-CH_2-CH_2-CH_2-$
Hexamethylene	$-CH_2-CH_2-CH_2-CH_2-CH_2-CH_2-$

Names of the substituted bivalent radicals are derived in accordance with Rules **A-2.2** and **A-2.25.**

Example:

$$Ethylethylene \qquad -\overset{2}{C}H_2-\overset{1}{C}H-$$
$$CH_2-CH_3$$

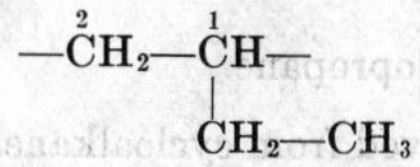

The name "propylene" is retained

$$CH_3-CH-CH_2-$$

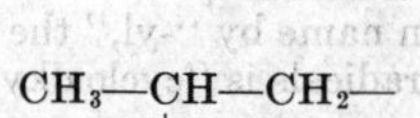

4.3.—Bivalent radicals similarly derived from unbranched alkenes, alkadienes, alkynes, *etc.*, by removing a hydrogen atom from each of the terminals carbon atoms are named by replacing the endings "-ene," "-diene," "-yne," *etc.*, of the hydrocarbon name by "-enylene," "-dienylene," "-ynylene," *etc.*, the positions of the double and triple bonds being indicated where necessary.

Example:

$$Propylene \qquad -\overset{3}{C}H_2-\overset{2}{C}H=\overset{1}{C}H-$$

The name "vinylene" is retained (for ethenylene)

$$-CH=CH-$$

Names of the substituted bivalent radicals are derived in accordance with Rule **A-3.4.**

Examples:

$$4 \ Propyl\text{-}2\text{-}pentenylene \qquad -\overset{5}{C}H_2-\overset{4}{C}H-\overset{3}{C}H=\overset{2}{C}H-\overset{1}{C}H_2-$$
$$CH_2-CH_2-CH_3$$

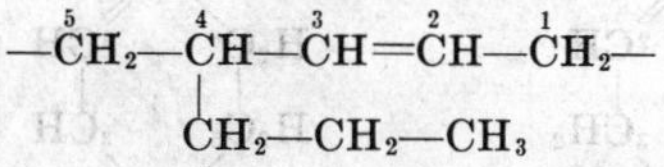

4.4.—Trivalent, quadrivalent and higher valent acyclic hydrocarbon radicals of two or more carbon atoms with the free valences at each end of a chain are named by adding to the hydrocarbon name the

---

(1) The group $=CH-$ may be referred to as the "methine" group.
(2) For unsubstituted radical only.

terminations "-yl" for a single free valence, "-ylidene" for a double and "-ylidyne" for a triple free valence on the same atom (the final "e" in the name of the hydrocarbon is elided when followed by a suffix beginning with "-yl"). If different types are present in the same radical, they are cited and numbered in the order of "-yl" "-ylidene," "-ylidyne."

Examples:

Butanediylidene	$=\overset{4}{C}H-\overset{3}{C}H_2-\overset{2}{C}H_2-\overset{1}{C}H=$
Butanediylidyne	$\equiv\overset{4}{C}-\overset{3}{C}H_2-\overset{2}{C}H_2-\overset{1}{C}\equiv$
1-Propanyl-3-ylidene	$=\overset{3}{C}H-\overset{2}{C}H_2-\overset{1}{C}H_2-$
Propadienediylidene	$=\overset{3}{C}=\overset{2}{C}=\overset{1}{C}=$
2-Pentenediylidyne	$\equiv\overset{5}{C}-\overset{4}{C}H_2-\overset{3}{C}H=\overset{2}{C}H-\overset{1}{C}\equiv$
1-Butanyliden-4-ylidyne	$\equiv\overset{4}{C}-\overset{3}{C}H_2-\overset{2}{C}H_2-\overset{1}{C}H=$

4.5.—Multivalent radicals containing three or more carbon atoms with free valences at each end of a chain and additional free valences at intermediate carbon atoms are named by adding the ending "-triyl," "-tetrayl," "-diylidene," "diylylidene," *etc.*, to the hydrocarbon name.

Examples:

$$-\overset{3}{C}H_2-\overset{2}{C}H-\overset{1}{C}H_2- \qquad\qquad -\overset{3}{C}H_2-\overset{2}{C}-\overset{1}{C}H_2-$$

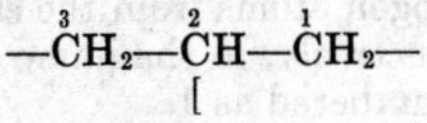

1,2,3-Propanetriyl   1,3-Propanediyl-2-ylidene

Monocyclic Hydrocarbons

A-11

11.1.—The names of saturated monocyclic hydrocarbons (with no side chains) are formed by attaching the prefix "cyclo" to the name of the acyclic saturated unbranched hydrocarbon with the same number of carbon atoms. The generic name of saturated monocyclic hydrocarbons (with or without side chains) is "cycloalkane."

Examples:

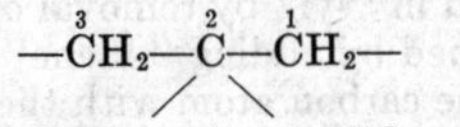

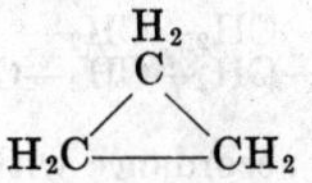

Cyclopropane   Cyclohexane

11.2.—Univalent radicals derived from cycloalkanes (with no side chains) are named by replacing the ending "-ane" of the hydrocarbon name by "-yl," the carbon atom with the free valence being numbered as 1. The generic name of these radicals is "cycloalkyl."

Examples:

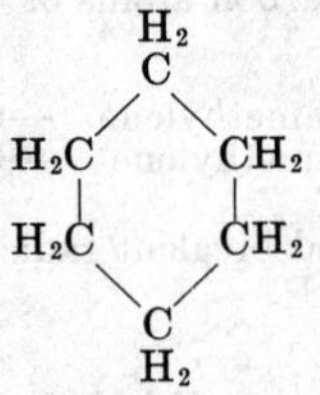

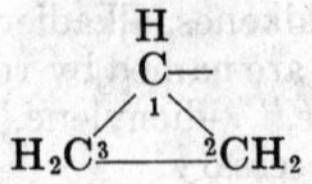

Cyclopropyl   Cyclohexyl

11.3.—The names of unsaturated monocyclic hydrocarbons (with no side chains) are formed by substituting "-ene," "-adiene," "-atriene," "-yne," "-adiyne," *etc.*, for "-ane" in the name of the corresponding cycloalkane. The double and triple bonds are given numbers as low as possible as in Rule **A-3.3.**

Examples:

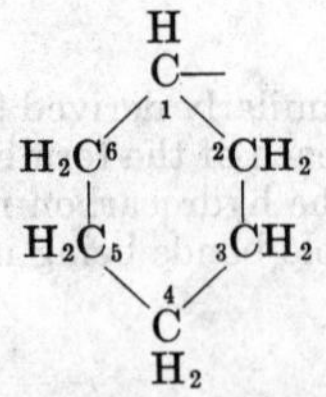

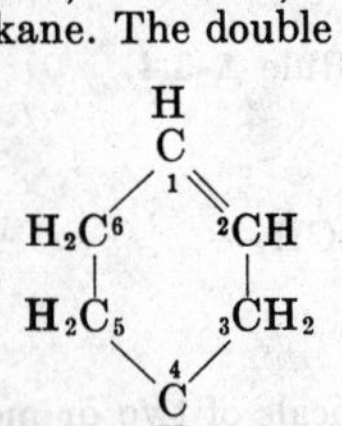

Cyclohexene   1,3-Cyclohexadiene   1-Cyclodecen-4-yne

The names "fulvene" (for methylenecyclopentadiene) and "benzene" are retained.

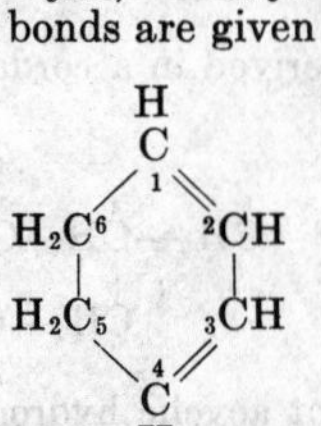

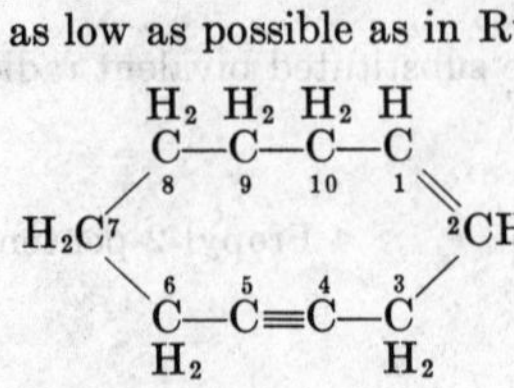

11.4.—The names of univalent radicals derived from unsaturated monocyclic hydrocarbons have the endings "-enyl," "-ynyl," "-dienyl," *etc.*, the positions of the double and triple bonds being indicated according to the principles of Rule **A-3.3**. The carbon atom with the free valence is numbered as 1, except as stated in the rules for terpenes(see Rules **A-72** to **A-75**).

Examples:

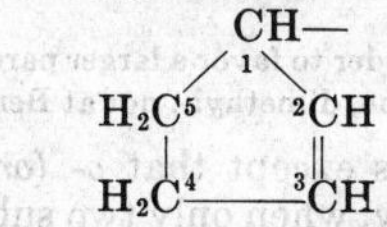

2-Cyclopenten-1-yl

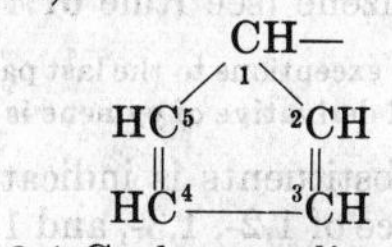

2,4-Cyclopentadien-1-yl

The radical name "phenyl" is retained.

The point of attachment is numbered 1 and the double and triple bonds are given numbers as low as possible. The number 1 for the point of attachment is given in the radical name to emphasize the fact that the point of attachment takes precedence over the double and triple bonds (except with terpenes which have fixed numberings).

11.5.—Names of bivalent radicals derived from saturated or unsaturated monocyclic hydrocarbons by removal of two atoms of hydrogen from the same carbon atom of the ring are obtained by replacing the endings "-ane," "-ene," "-yne," by "-ylidene," "-enylidene" and "-ynylidene," respectively. The carbon atom with the free valences is numbered as 1, except as stated in the rules for terpenes.

Examples:

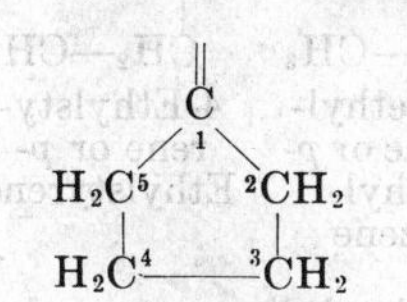

Cyclopentylidene

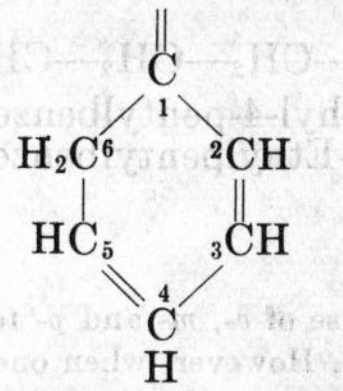

2,4-Cyclohexadien-1-ylidene

11.6.—Bivalent radicals derived from saturated or unsaturated monocyclic hydrocarbons by removing a hydrogen atom from each of two different carbon atoms of the ring are named by replacing the endings "-ane," "-ene," "-diene," "-yne," *etc.*, of the hydrocarbon name by "-ylene," "-enylene," "-dienylene," "-ynylene," *etc.*, the positions of the double and triple bonds and of the points of attachment being indicated. Preference in lowest numbers is given to the carbon atoms having the free valences.

Examples:

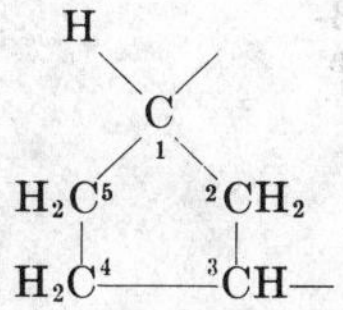

1,3-Cyclopentylene

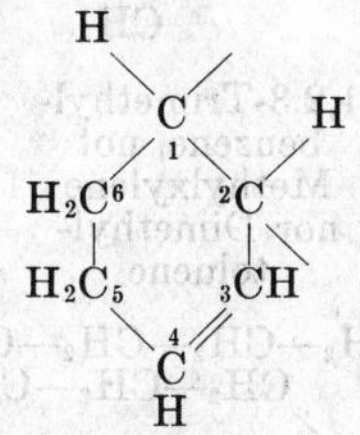

3-Cyclohexen-1,2-ylene

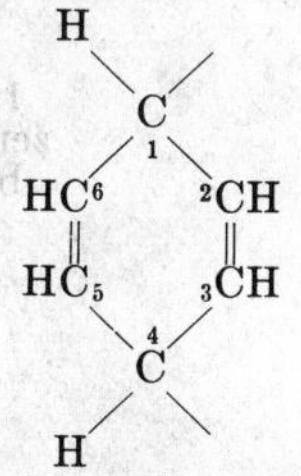

2,5-Cyclohexadien-1,4-ylene

The name "phenylene" is retained

Phenylene (*p*-shown)

A-12

12.1.—These names for substituted monocyclic aromatic hydrocarbons are retained

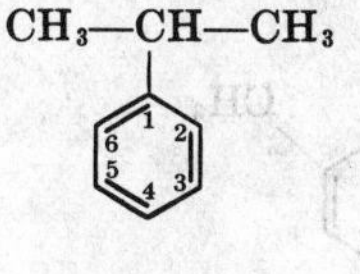

Cumene

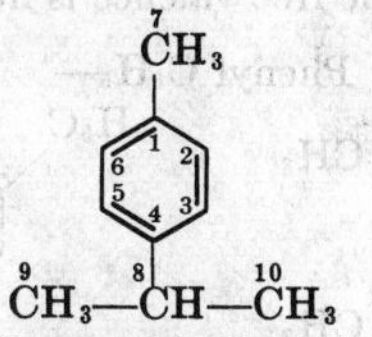

Cymene (*p*-shown)

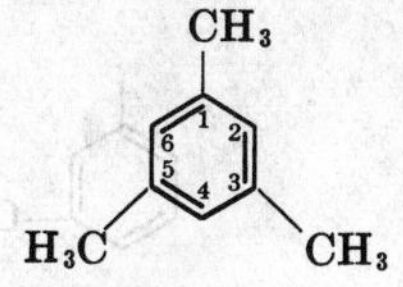

Mesitylene

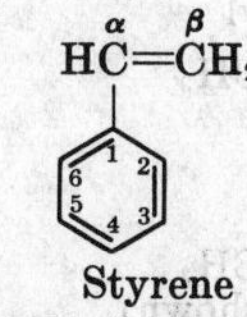

Styrene

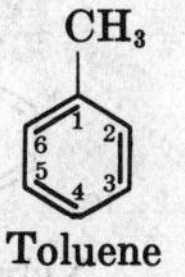

Toluene

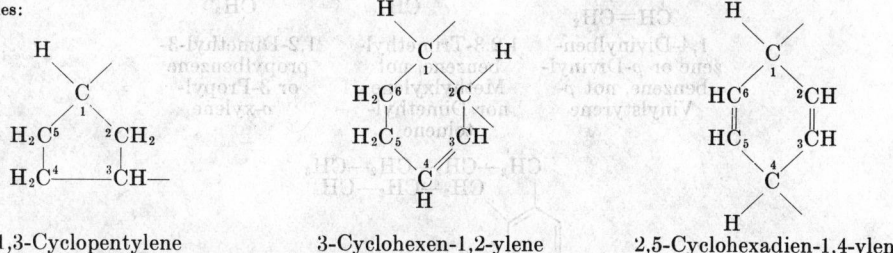

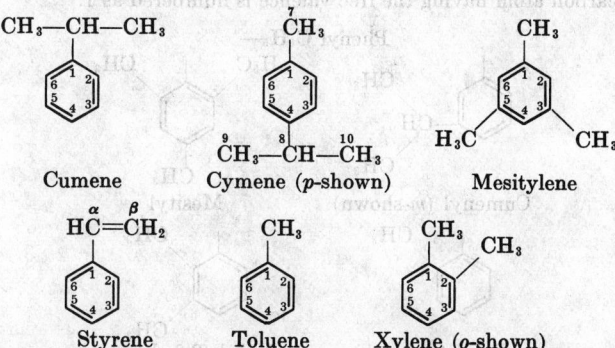

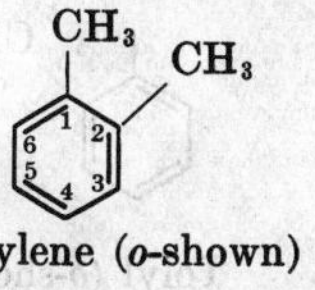

Xylene (*o*-shown)

12.2.—Other substituted monocyclic aromatic hydrocarbons are named as derivatives of benzene or of one of the compounds listed in Part 1 of this rule. However, if the substituent introduced into such a compound, is identical with one already present in that compound, then the substituted compound is named as a derivative of benzene (see Rule **61.4**).

Chemical Abstracts **makes certain exceptions to the last part of this rule in order to favor a larger parent, although still treating like groups alike, *e.g.*, an *ar*-methyl derivative of cymene is indexed at Cumene, dimethyl-, not at Benzene, isopropyldimethyl-.**

12.3.—The position of substituents is indicated by numbers except that *o-* (*ortho-*), *m-* (*meta-*) and *p-* (*para-*) may be used in place of 1,2-, 1,3-, and 1,4-, respectively, when only two substituents are present. The lowest numbers possible are given to substituents, choice between alternatives being governed by Rule **A-2** so far as applicable, except that when names are based on those of compounds listed in Part 1 of this rule the first priority for lowest numbers is given to the substituent(s) already present in those compounds.

Examples:

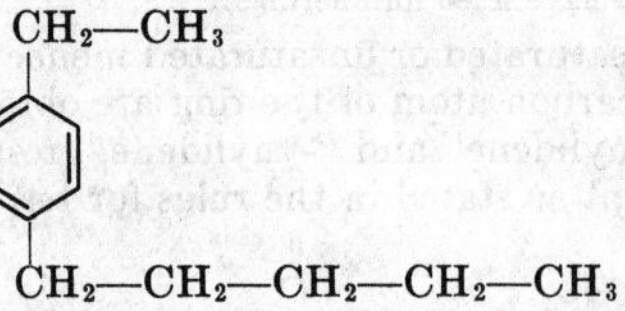

1-Ethyl-4-pentylbenzene or *p*-Ethylpentylbenzene

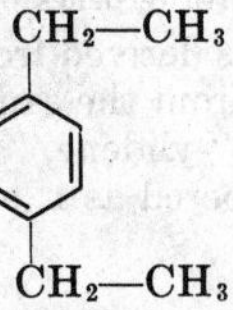

1,4-Diethylbenzene or *p*-Diethylbenzene

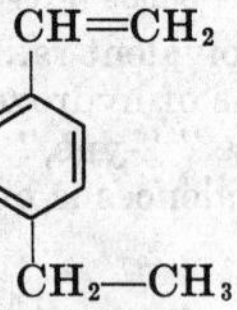

4-Ethylstyrene or *p*-Ethylstyrene

Chemical Abstracts **limits the use of *o-*, *m-* and *p-* to like substituents when both are expressed as prefixes, *e.g.*, *p*-diethylbenzene, 1-ethyl-4-pentylbenzene. However, when one substituent is expressed in the name, *m-*, *o-* or *p-* is used to indicate the position of the second and different substituent, *e.g.*, *o*-ethylcumene.**

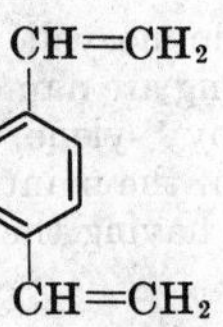

1,4-Divinylbenzene or *p*-Divinylbenzene, not *p*-Vinylstyrene

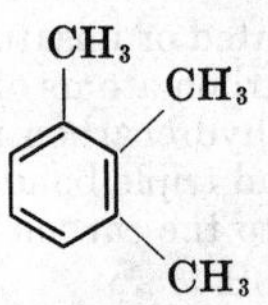

1,2,3-Trimethylbenzene, not Methylxylene nor Dimethyltoluene

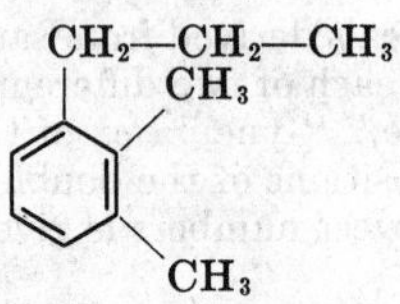

1,2-Dimethyl-3-propylbenzene or 3-Propyl-*o*-xylene

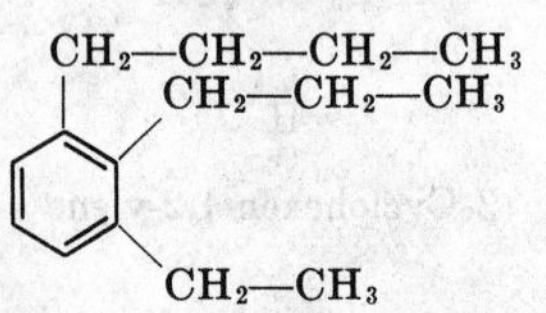

1-Ethyl-2-propyl-3-butylbenzene (Order of complexity)
or 1-Butyl-3-ethyl-2-propylbenzene (Alphabetical order)

12.4.—The generic name of monocyclic and polycyclic aromatic hydrocarbons is "arene."

A-13

13.1.—Univalent radicals derived from monocyclic aromatic hydrocarbons and having the free valence at a ring atom are given the names listed below. Such radicals not listed below are named as substituted phenyl radicals. The carbon atom having the free valence is numbered as 1.

Phenyl C_6H_5—

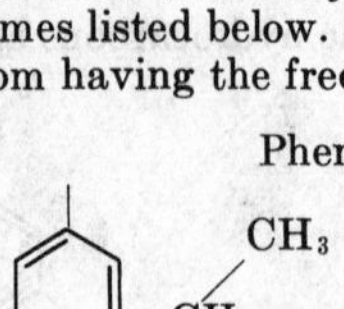

Cumenyl (*m*-shown)

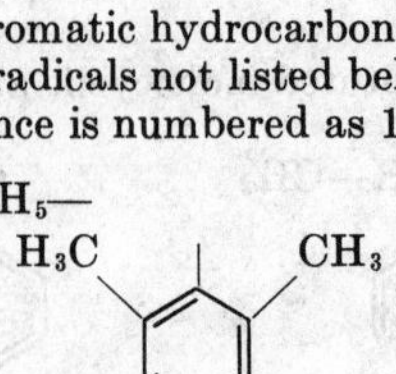

Mesityl

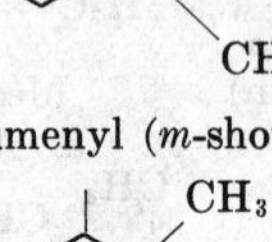

Tolyl (*o*-shown)

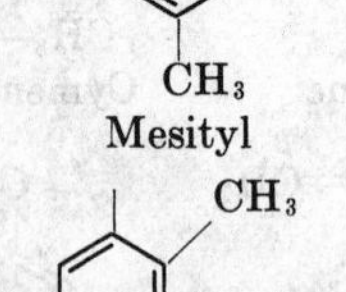

Xylyl (2,3-shown)

13.2.—Since the name phenylene (*o*-, *m*-, or *p*-) is retained for the radical —C_6H_4— (exception to Rule **A-11.6**) bivalent radicals formed from substituted benzene derivatives and having the free valences at ring atoms are named as substituted phenylene radicals. The carbon atoms having the free valences are numbered 1,2-, 1,3-, or 1,4- as appropriate.

13.3.—These trivial names for radicals having a single free valence in the side chain are retained:

Benzyl	$C_6H_5—\overset{\alpha}{C}H_2—$
Cinnamyl	$C_6H_5—\overset{\gamma}{C}H=\overset{\beta}{C}H—\overset{\alpha}{C}H_2—$
Phenethyl	$C_6H_5—\overset{\beta}{C}H_2—\overset{\alpha}{C}H_2—$
Styryl	$C_6H_5—\overset{\beta}{C}H=\overset{\alpha}{C}H—$
Trityl	$(C_6H_5)_3C—$

Chemical Abstracts limits trityl to the unsubstituted radical.

13.4.—Multivalent radicals of aromatic hydrocarbons with free valences in the side chain are named in accordance with Rule **A-4.**

Examples:		
	Benzylidyne	$C_6H_5—C\equiv$
	Cinnamylidene	$C_6H_5—\overset{\gamma}{C}H=\overset{\beta}{C}H—\overset{\alpha}{C}H=$

A-14

14.1.—The generic names of univalent and bivalent aromatic hydrocarbon radicals are "aryl" and "arylene," respectively.

Fused Polycyclic Hydrocarbons

A-21

21.1.—The names of polycyclic hydrocarbons with maximum number of non-cumulative[1] double bonds end in "-ene." The names listed on pp. 11 and 12 are retained.

21.2.—The names of hydrocarbons containing five or more fused benzene rings in a straight linear arrangement are formed from a numerical prefix as specified in Rule **A-1.1** followed by "-acene."

Examples:

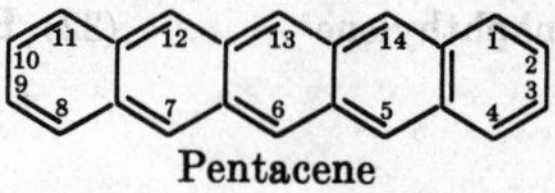

Pentacene

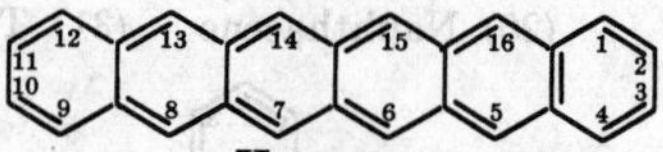

Hexacene

The following list contains the names of polycyclic hydrocarbons which are retained (see Rule A-21.1).

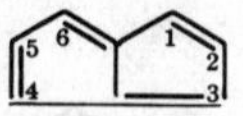

(1) Pentalene

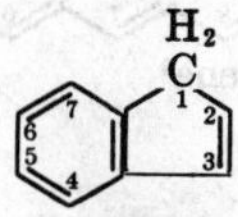

(2) Indene

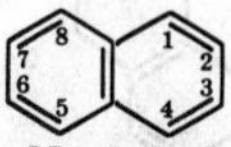

(3) Naphthalene

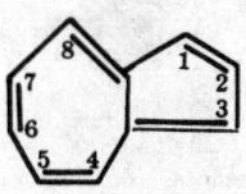

(4) Azulene

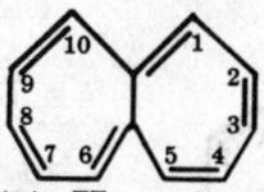

(5) Heptalene

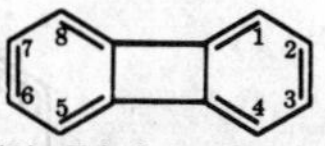

(6) Biphenylene

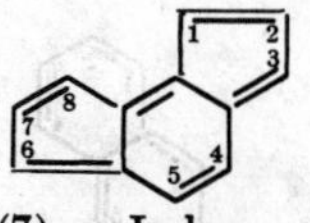

(7) *as*-Indacene

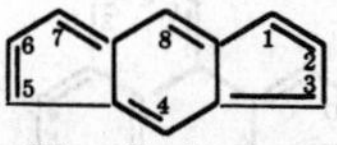

(8) *s*-Indacene

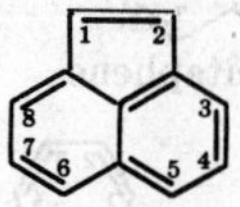

(9) Acenaphthylene

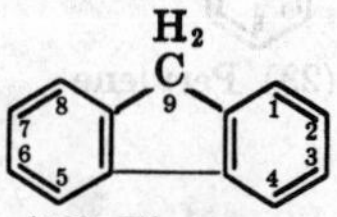

(10) Fluorene

(1) Cumulative double bonds are those present in a chain in which at least three contiguous carbon atoms are joined by double bonds; non-cumulative double bonds comprise every other arrangement of two or more double bonds in a single structure. The generic name "cumulene" is given to compounds containing three or more cumulative double bonds.

$$CH_2=C=C=C=CH_2$$

Cumulative

$$CH_3—CH=CH—CH=CH—CH=CH_2$$

or

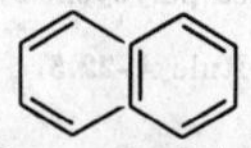

Non-cumulative

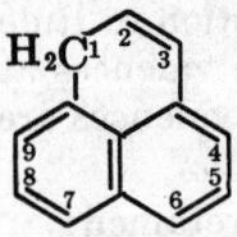

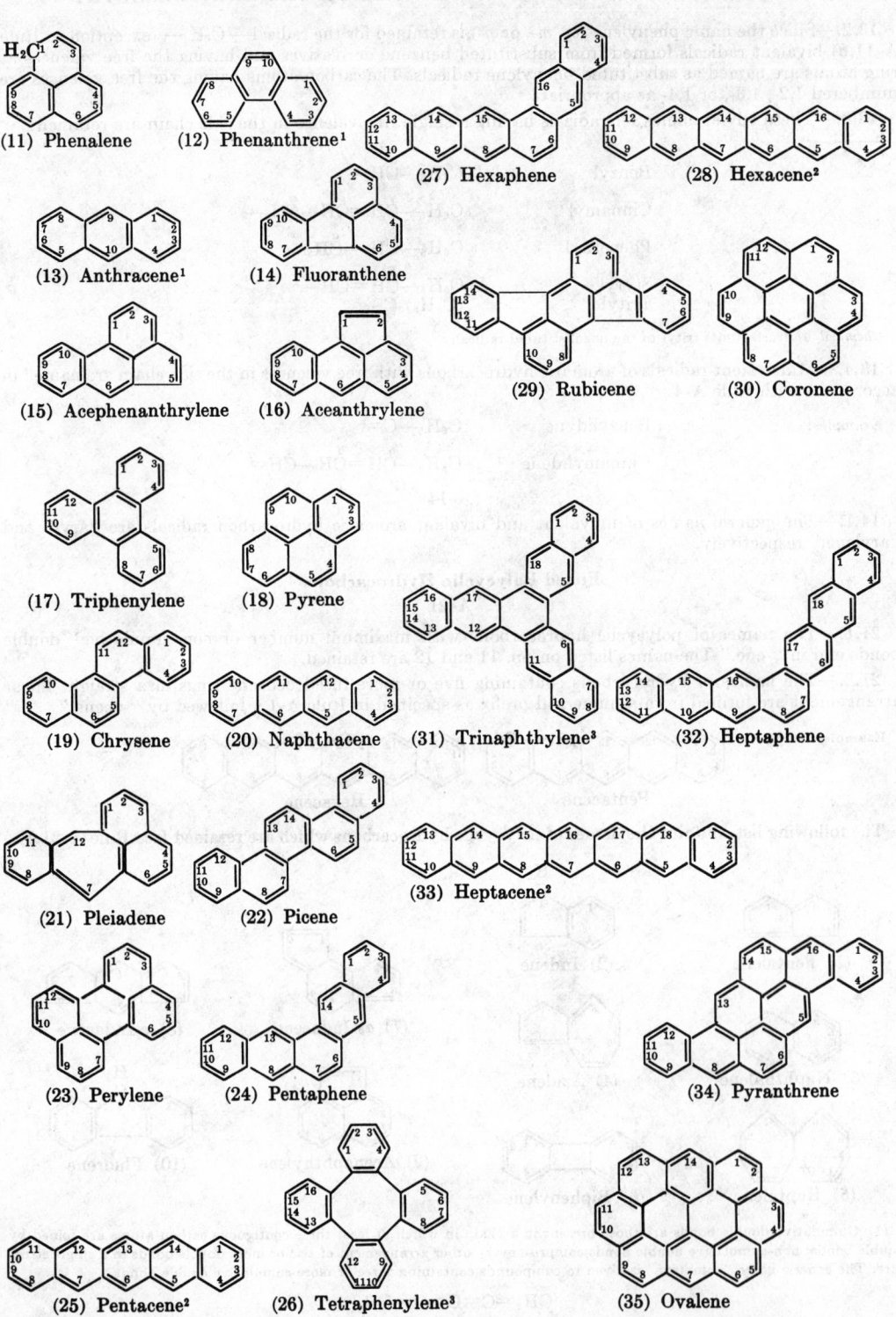

(11) Phenalene

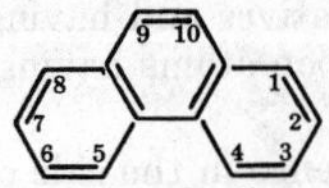

(12) Phenanthrene[1]

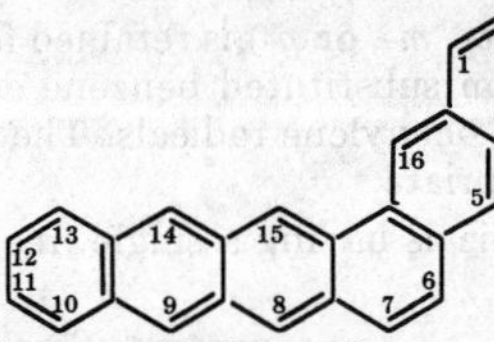

(27) Hexaphene

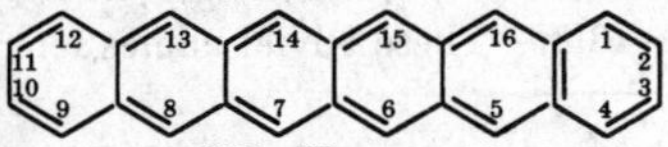

(28) Hexacene[2]

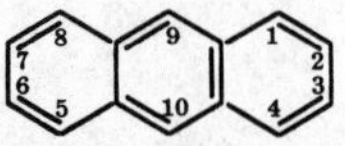

(13) Anthracene[1]

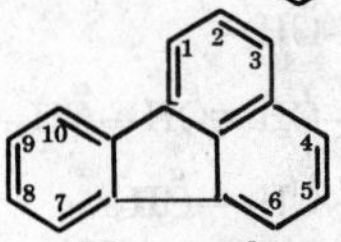

(14) Fluoranthene

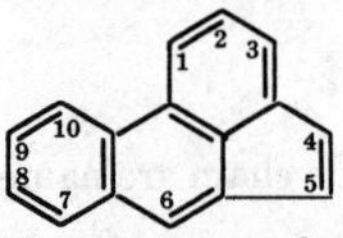

(15) Acephenanthrylene

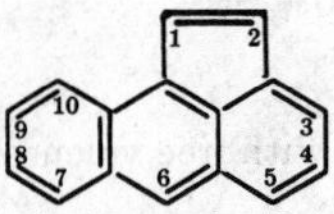

(16) Aceanthrylene

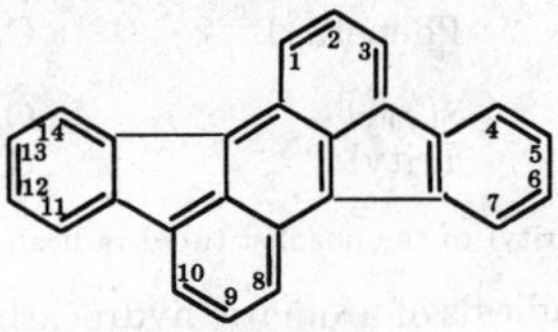

(29) Rubicene

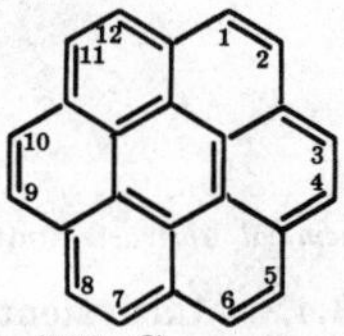

(30) Coronene

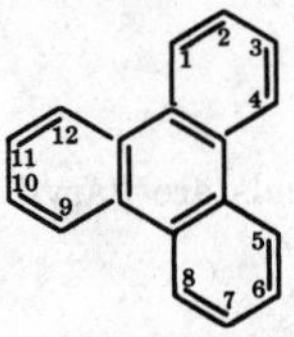

(17) Triphenylene

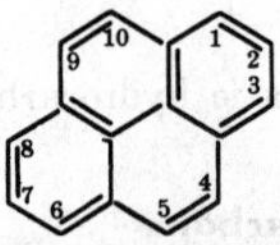

(18) Pyrene

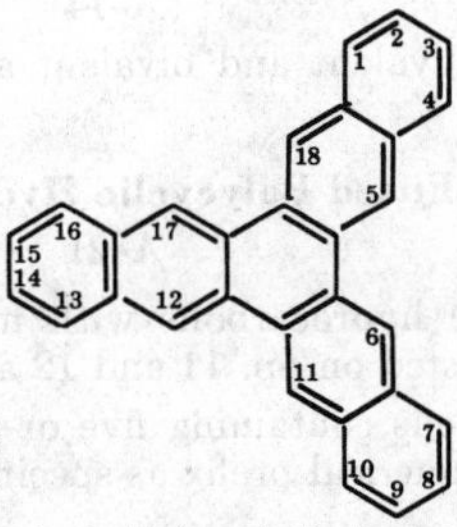

(31) Trinaphthylene[3]

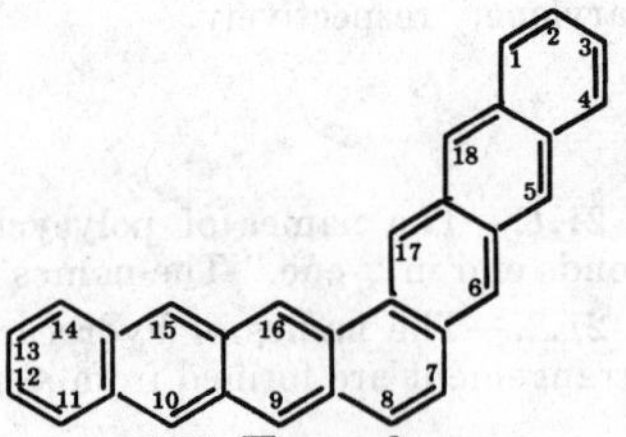

(32) Heptaphene

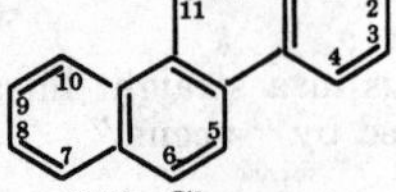

(19) Chrysene

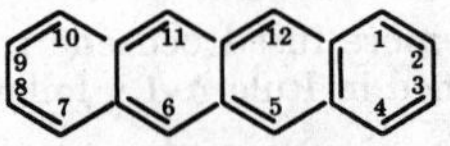

(20) Naphthacene

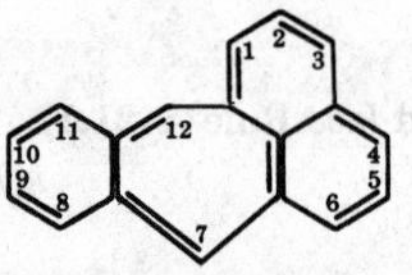

(21) Pleiadene

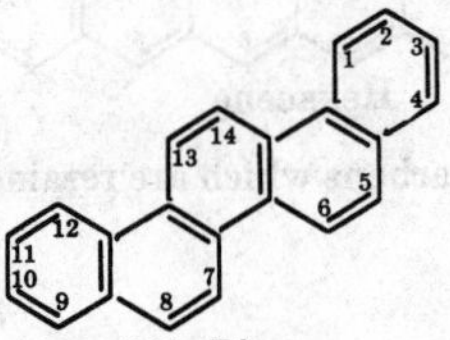

(22) Picene

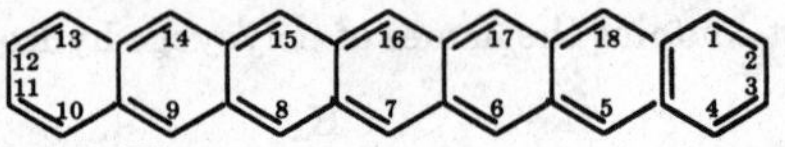

(33) Heptacene[2]

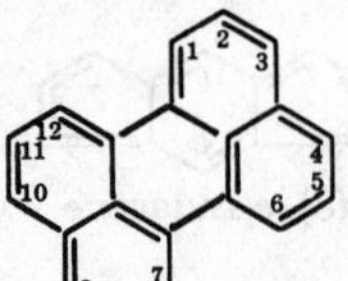

(23) Perylene

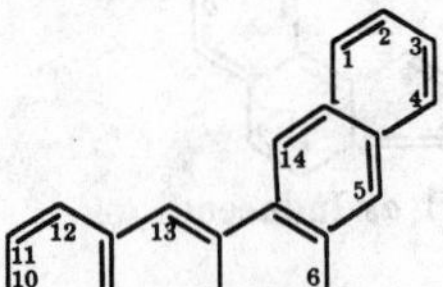

(24) Pentaphene

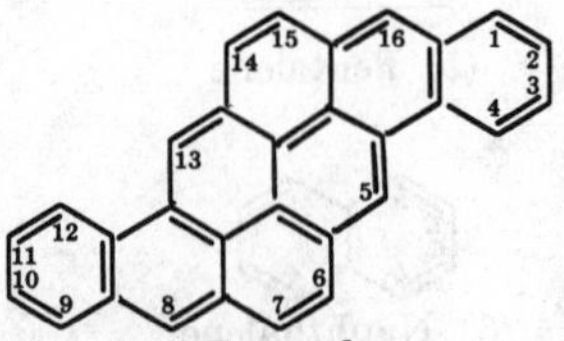

(34) Pyranthrene

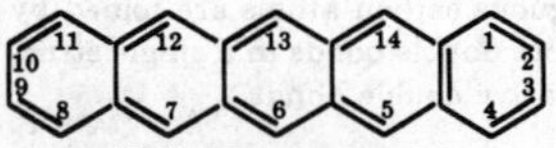

(25) Pentacene[2]

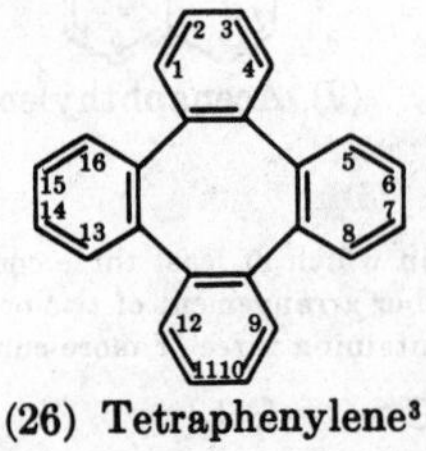

(26) Tetraphenylene[3]

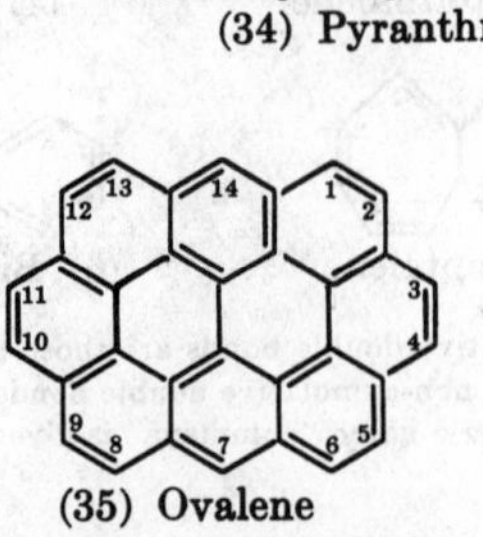

(35) Ovalene

Beginning with Volume 51 Subject Index *Chemical Abstracts* uses biphenylene in place of cyclobutadibenzene and phenalene in place of benzonaphthene. The polycyclic hydrocarbons which have the trivial names listed above may be used as the base components in naming orthofused or ortho- and peri-fused polycyclic systems for which there are no acceptable trivial names.

(1) Denotes exception to systematic numbering. See Rule **A-22.5.**

(2) See Rule **A-21.2.**

(3) For isomer shown only.

21.3.—"Ortho-fused"[4] or "ortho- and peri-fused"[5] polycyclic hydrocarbons with maximum number of noncumulative double bonds which contain at least two rings of five or more members and which have no accepted trivial name such as those of **21.1** of this Rule, are named by prefixing to the name of a component ring or ring system (the base component) designations of the other components. The base component should contain as many rings as possible (provided it has a trivial name), and should occur as far as possible from the beginning of the list of Rule **A-21.1.** The attached components should be as simple as possible.

Example:

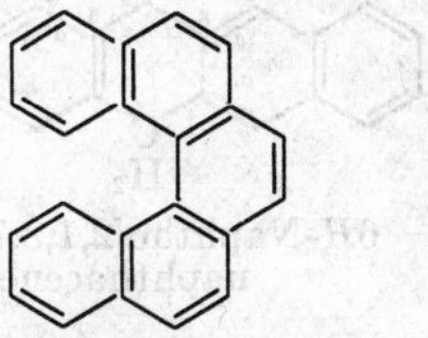

(not Naphthophenanthrene; benzo is "simpler" than naphtho, even though there are two benzo rings and only one naphtho)

Dibenzophenanthrene

(4) Polycyclic compounds in which two rings have two, and only two, atoms in common are said to be "ortho-fused." Such compounds have *n* common faces and 2*n* common atoms (Example I).

(5) Polycyclic compounds in which one ring contains two, and only two, atoms in common with each of two or more rings of a contiguous series of rings are said to be "ortho- and peri-fused." Such compounds have *n* common faces and fewer than 2*n* common atoms (Examples II and III).

Examples:

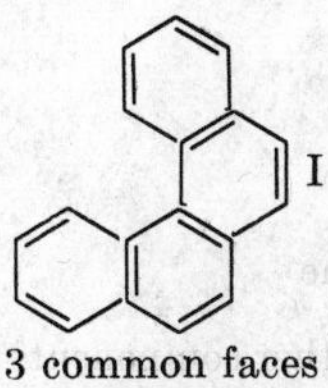

I

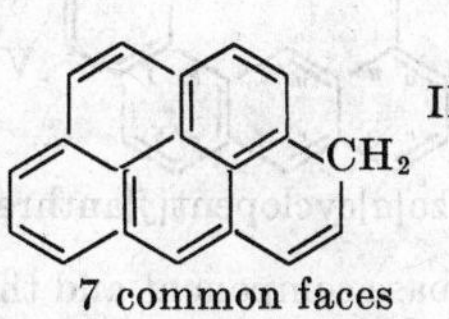

II

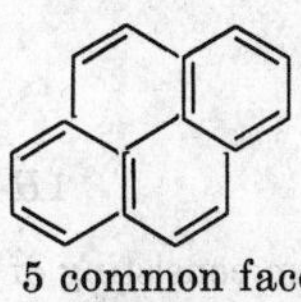

III

3 common faces, 6 common atoms — "Ortho-fused" system

7 common faces, 8 common atoms; 5 common faces, 6 common atoms — "Ortho- and peri-fused" systems

21.4.—The prefixes designating attached components are formed by changing the ending "-ene" of the name of the component hydrocarbon into "-eno," *e.g.*, "pyreno" (from pyrene). When more than one prefix is present, they are arranged in alphabetical order. These common abbreviated prefixes are recognized (see list in **21.1** of this Rule)

Acenaphtho	from	Acenaphthylene
Anthra	from	Anthracene
Benzo	from	Benzene
Naphtho	from	Naphthalene
Perylo	from	Perylene
Phenanthro	from	Phenanthrene

For monocyclic prefixes other than "benzo-," the following names are recognized, each to represent the form with the maximum number of non-cumulative double bonds: cyclopenta, cyclohepta, cycloocta, cyclonona, *etc.* When the base component is a monocyclic system, the ending "-ene" signifies the maximum number of noncumulative double bonds, and does not denote one double bond only.

Examples:

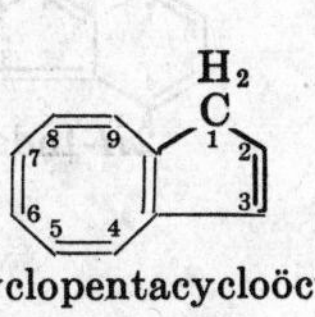

1*H*-Cyclopentacycloöctene

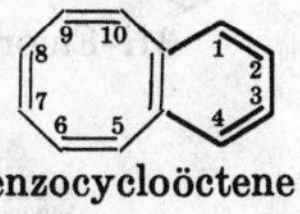

Benzocycloöctene

In forming names of polycyclic systems from the above carbocyclic prefixes the final "o" or "a" of the prefix is elided before another vowel, except in the case of anthra, phenanthro and prefixes ending in -eno from which the "a" and "o" are never elided. Examples: benz[*a*]anthracene, cyclopent[*a*]acenaphthylene, anthra[1,2 − *a*]anthracene, fluoreno[4,3,2-*de*]anthracene.

The ending "-ene" to signify the maximum number of noncumulative double bonds in all fused carbocyclic systems means that the endings -diene, triene, tetraene, *etc.*, are unnecessary when the base component is a monocyclic ring of seven or more carbon atoms. Higher stages of hydrogenation are indicated by the use of the prefixes dihydro-, tetrahydro-, *etc.* (*cf.* Rule **A-23.1**). *Chemical Abstracts* follows Rules **A-21.4** and **A-23.1** instead of Alternate Rule **A-23.5.**

21.5.—Isomers are distinguished by lettering the peripheral sides of the base component *a*, *b*, *c*, *etc.*, beginning with "*a*" for the side "1,2," "*b*" for "2,3" (or in certain cases "2,2a") and lettering every side around the periphery. To the letter as early in the alphabet as possible, denoting the side where fusion occurs, are prefixed, if necessary, the numbers of the positions of attachment of the other component. These numbers are chosen to be as low as is consistent with the numbering of the component, and their order conforms to the direction of lettering of the base component (see Examples II and IV). When two

or more prefixes refer to equivalent positions so that there is a choice of letters, the prefixes are cited in alphabetical order according to Rule **A-21.4** and the location of the first cited prefix is indicated by a letter as early as possible in the alphabet (see Example V). The numbers and letters are enclosed in brackets and placed immediately after the designation of the attached component. This expression merely defines the manner of fusion of the components.

Examples:

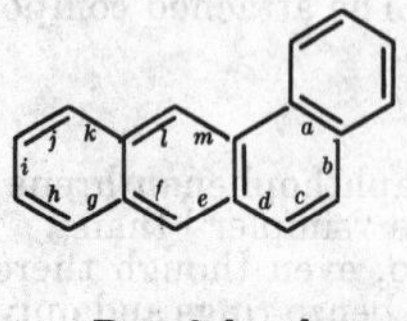
I

Benz[*a*]anthracene

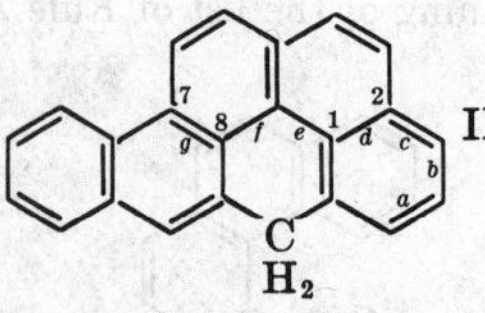
II

6*H*-Naphtho[2,1,8,7-*defg*]-naphthacene

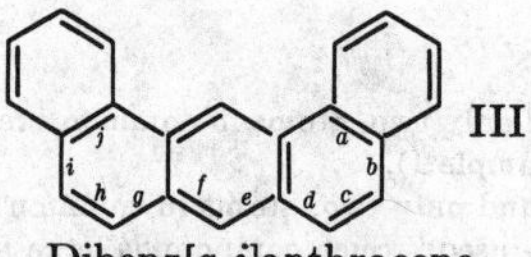
III

Dibenz[*a,j*]anthracene
(not Naphtho[2,1-*b*]phenanthrene)

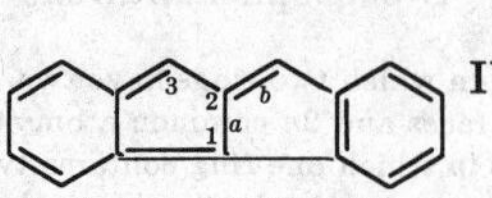
IV

Indeno[1,2-*a*]indene

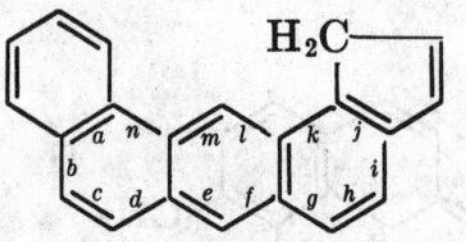
V

1*H*-Benzo[*a*]cyclopent[*j*]anthracene

The completed system consisting of the base component and the other components is then renumbered according to Rule **A-22,** the numbering of the component parts being ignored.

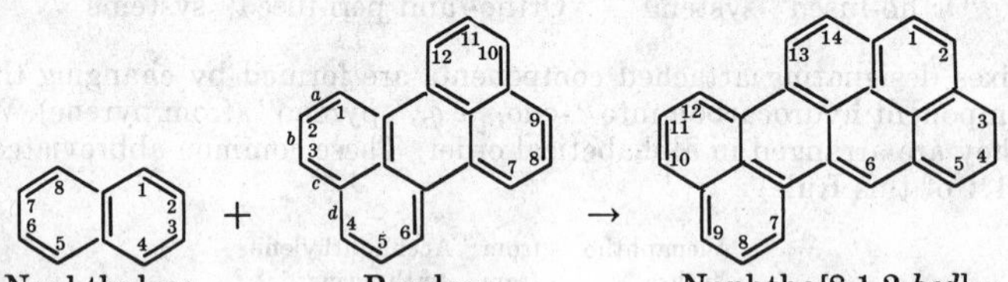

Naphthalene + Perylene → Naphtho[8,1,2-*bcd*]-perylene

21.6.—When a name applies equally to two or more isomeric condensed parent ring systems with the maximum number of non-cumulative double bonds and when the name can be made specific by indicating the position of one or more hydrogen atoms in the structure, this is accomplished by modifying the name with a locant, followed by italic capital *H* for each of these hydrogen atoms. Such symbols ordinarily precede the name. The said atom or atoms are called "indicated hydrogen." The same principle is applied to radicals and compounds derived from these systems.

Examples:

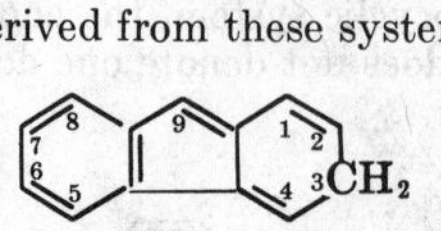

3*H*-Fluorene

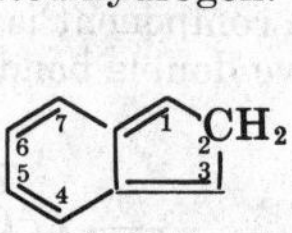

2*H*-Indene

A-22

22.1.—For the purposes of numbering, the individual rings of a polycyclic "ortho-fused" or "ortho- and peri-fused" hydrocarbon system usually are drawn as follows:

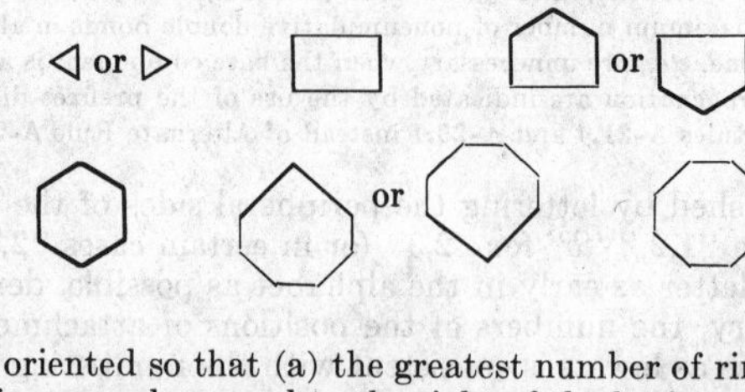

and the polycyclic system is oriented so that (a) the greatest number of rings are in a horizontal row and (b) a maximum number of rings are above and to the right of the horizontal row (upper right quadrant).

If two or more orientations meet these requirements, the one is chosen which has as few rings as possible in the lower left quadrant.

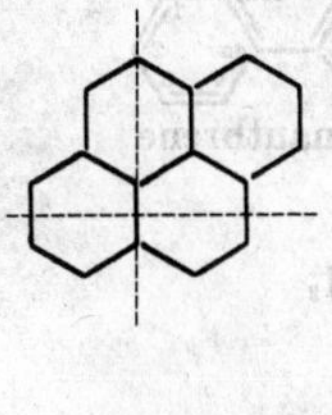

Correct orientation

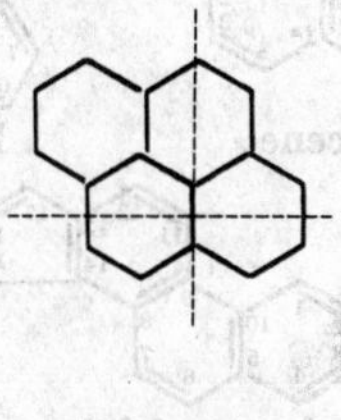

Incorrect orientation

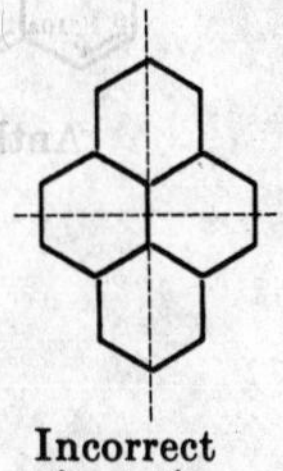

Incorrect orientation

The system thus oriented is numbered in a clockwise direction commencing with the carbon atom not engaged in ring-fusion in the most counterclockwise position of the uppermost ring, or if there is a choice, of the uppermost ring farthest to the right, and omitting atoms common to two or more rings.

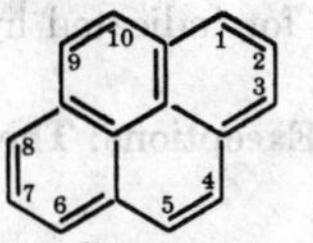

Correct

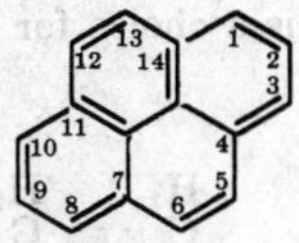

Incorrect

22.2.—Atoms common to two or more rings are designated by adding roman letters "a," "b," "c," *etc.*, to the number of the position immediately preceding. Interior atoms follow the highest number, taking a clockwise sequence wherever there is a choice.

Examples:

[*cf.* Note below]

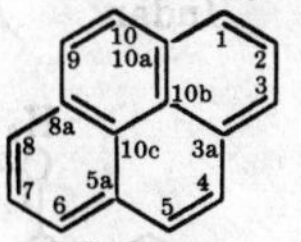

Correct

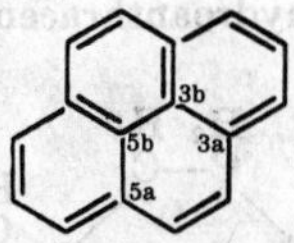

Incorrect

22.3.—When there is a choice, carbon atoms common to two or more rings follow the lowest possible numbers.

Examples:

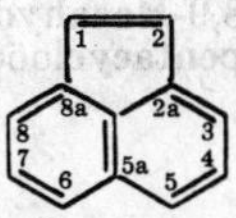

Correct

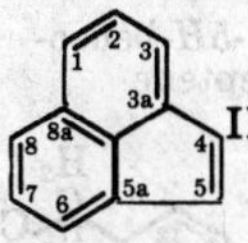

II

Incorrect

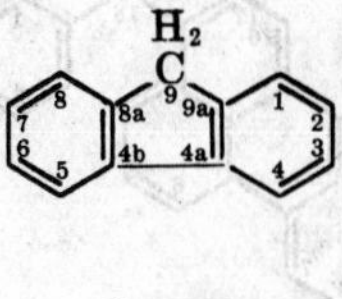

Correct

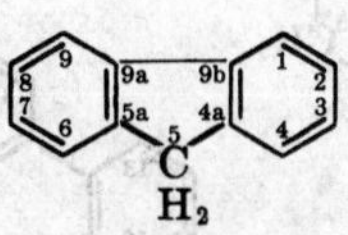

I

Incorrect

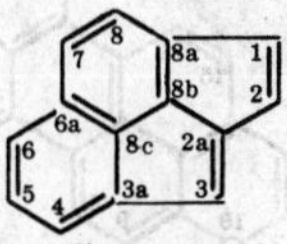

Correct

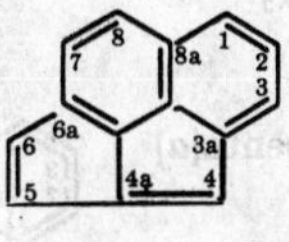

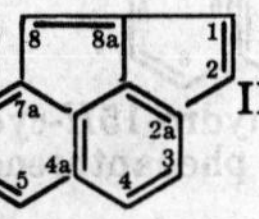

III

Incorrect

Note: I. 4, 4, 8, 9 is lower than 4, 5, 9, 9.
II. 2, 5, 8 is lower than 3, 5, 8.
III. 2, 3, 6, 8 is lower than 3, 4, 6, 8 or 2, 4, 7, 8.

22.4.—When there is a choice, the carbon atoms which carry an indicated hydrogen atom are numbered as low as possible.

Examples:

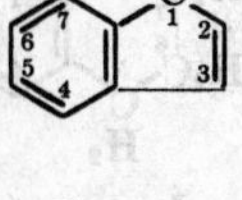

Correct

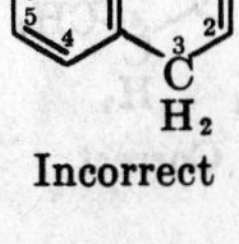

Incorrect

22.5.—The following are recommended exceptions to the above rules on numbering

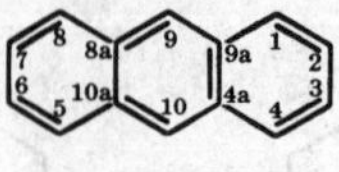

Anthracene

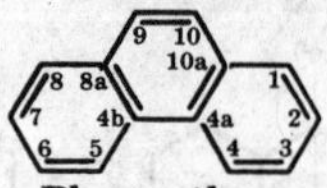

Phenanthrene

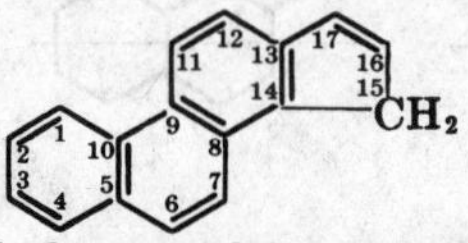

Cyclopenta[*a*]phenanthrene
(15*H*- shown)
See also rules on steroids

A-23

23.1.—The names of "ortho-fused" or "ortho- and peri-fused" polycyclic hydrocarbons with less than maximum number of non-cumulative double bonds are formed from a prefix "dihydro-," "tetrahydro-," *etc.*, followed by the name of the corresponding unreduced hydrocarbon. The prefix "perhydro-" signifies full hydrogenation. When there is a choice for *H* used for indicated hydrogen it is assigned the lowest available number.

Examples:

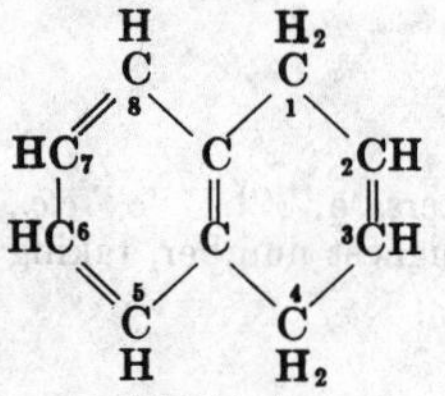

1,4-Dihydro-naphthalene

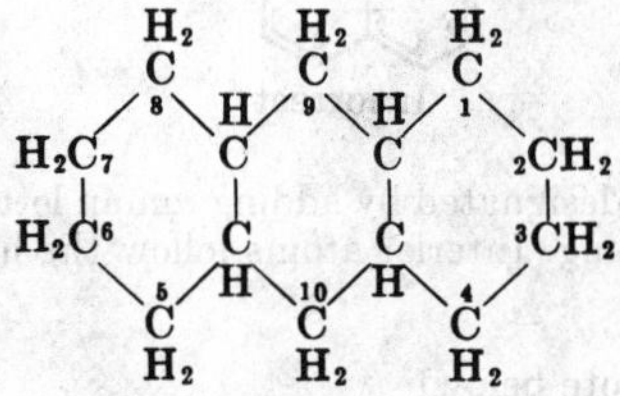

Perhydroanthracene

Exceptions: These names are retained

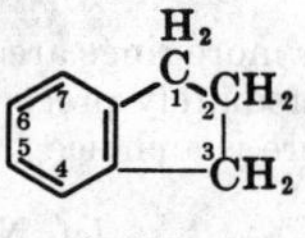

Indan

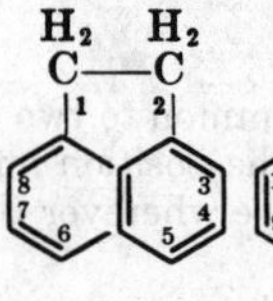

Acenaphthene

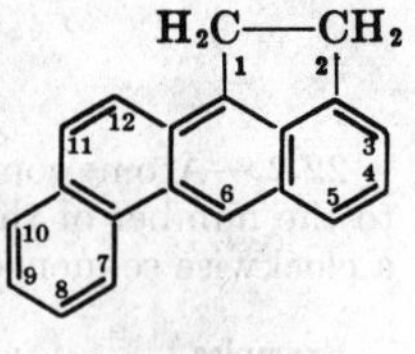

Cholanthrene

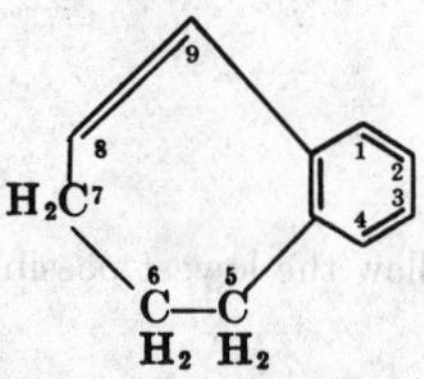

6,7-Dihydro-5*H*-benzo-cycloheptene

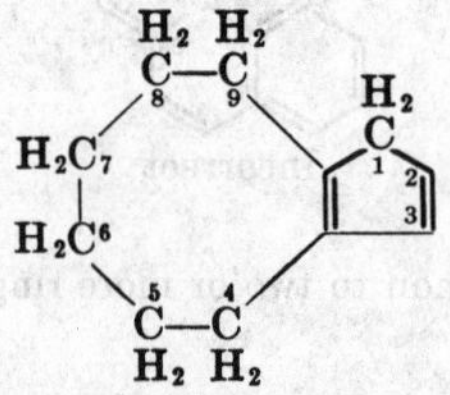

4,5,6,7,8,9-Hexahydro-1*H*-cyclopentacycloöctene

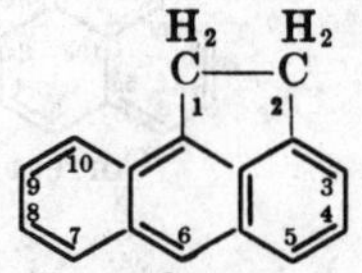

Aceanthrene

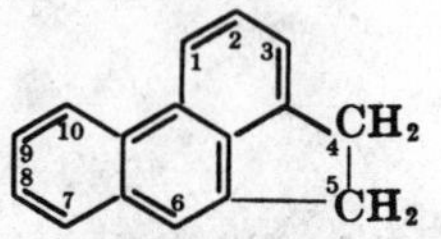

Acephenanthrene

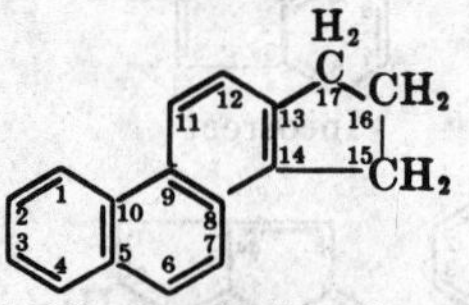

16,17-Dihydro-15*H*-cyclopenta[*a*]phenanthrene

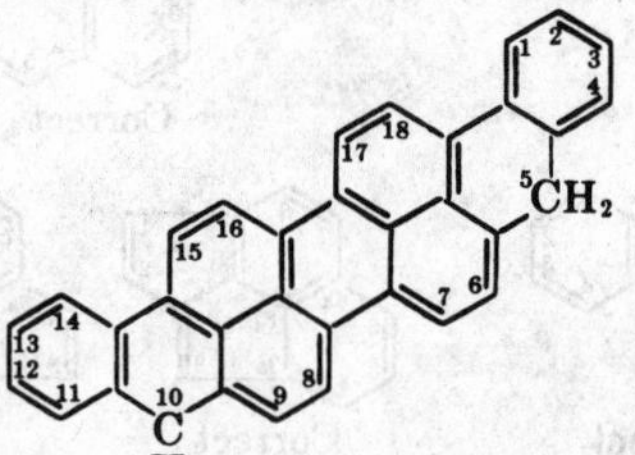

Violanthrene

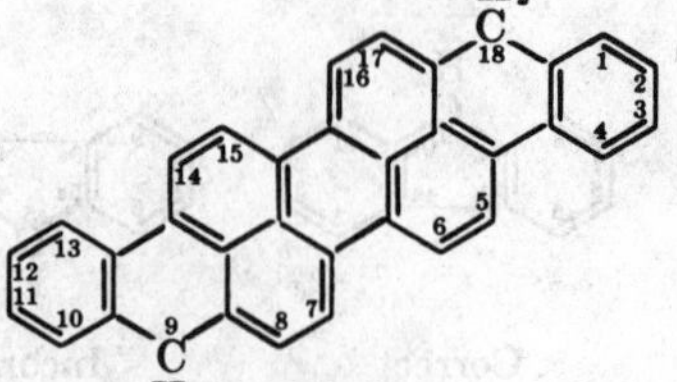

Isoviolanthrene

Chemical Abstracts follows this Rule instead of the alternate Rule 23.5.

23.2.—When there is a choice, the carbon atoms to which hydrogen atoms are added are numbered as low as possible.

Example:

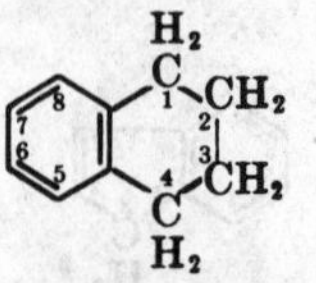

Correct

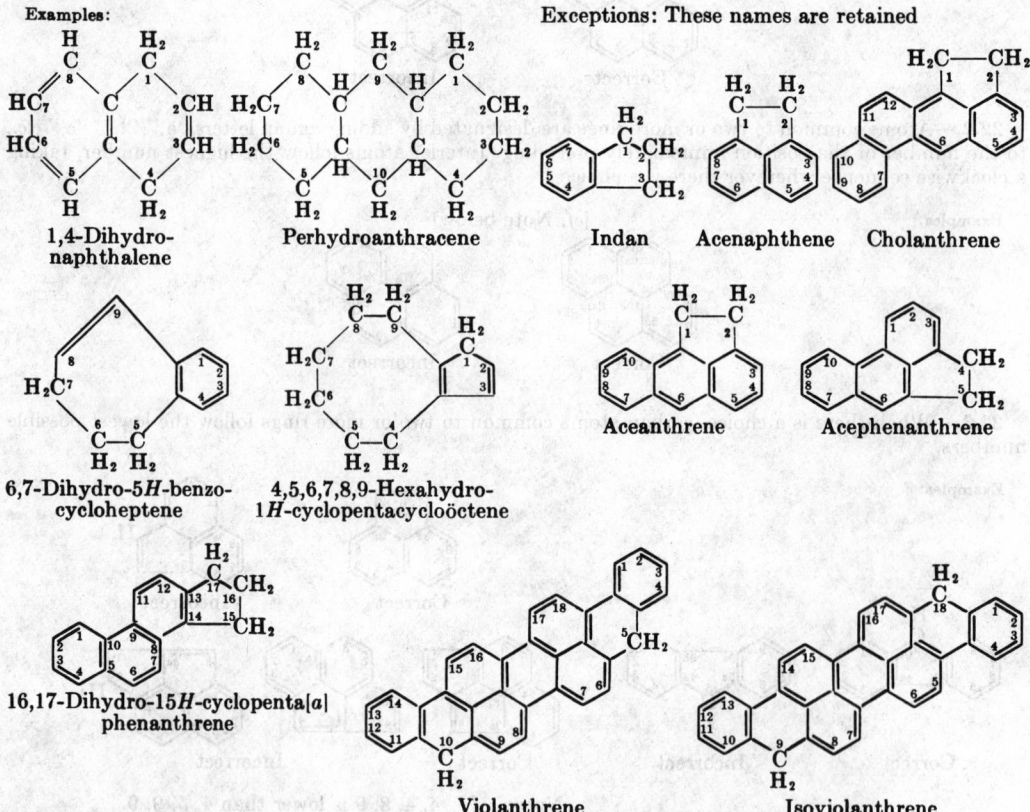

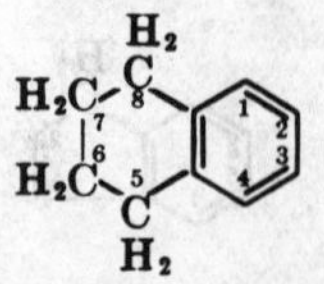

Incorrect

23.3.—Substituted polycyclic hydrocarbons are named according to the same principles as substituted monocyclic hydrocarbons (see Rules **A-12** and **A-61**).

23.5 (Alternate to part of Rule **A-23.1**).—The names of "ortho-fused" polycyclic hydrocarbons which have (a) less than the maximum number of non-cumulative double bonds, (b) at least one terminal unit which is most conveniently named as an unsaturated cycloalkane derivative, and (c) a double bond at the positions where rings are fused together, may be derived by joining the name of the terminal unit to that of the other component by means of a letter "o" with elision of a terminal "e". The abbreviations for fused aromatic systems laid down in Rule **A-21.4** are used, and the exceptions of Rule **A-23.1** apply.

Examples:

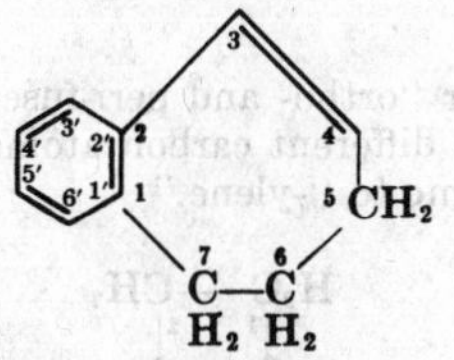

1,2-Benzo-1,3-cycloheptadiene

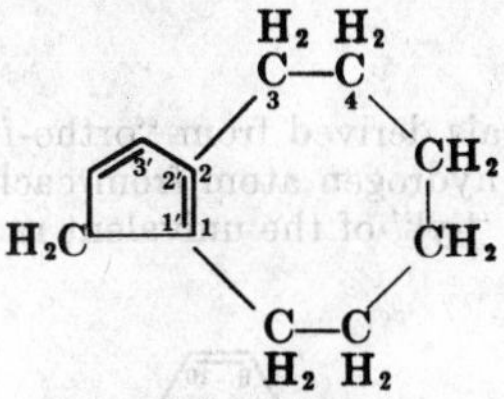

1,2-Cyclopenta-1′,3′-dienocycloöctene

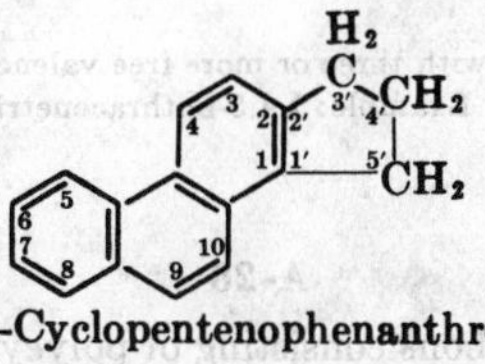

1,2-Cyclopentenophenanthrene

Chemical Abstracts follows Rules **A-21.4** and **A-23.1** instead of this alternate rule.

A-24

24.1.—For radicals derived from polycyclic hydrocarbons, the numbering of the hydrocarbon is retained. The point or points of attachment are given numbers as low as is consistent with the fixed numbering of the hydrocarbon.

24.2.—Univalent radicals derived from "ortho-fused" or "ortho- and peri-fused" polycyclic hydrocarbons with names ending in "-ene" by removal of a hydrogen atom from an aromatic or alicyclic ring are named in principle by changing the ending "-ene" of the names of the hydrocarbons to "-enyl."

Examples:

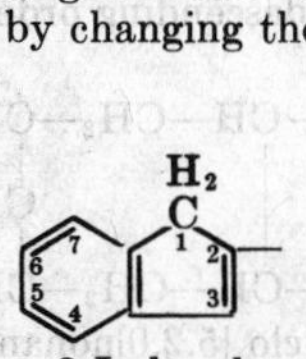

2-Indenyl

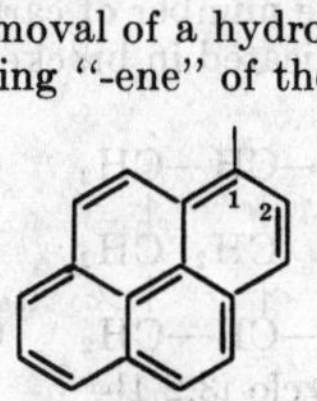

1-Pyrenyl

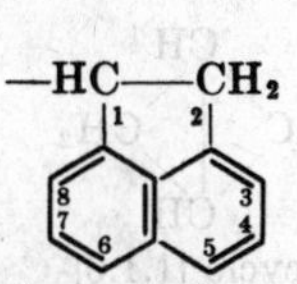

1-Acenaphthenyl

Exceptions:

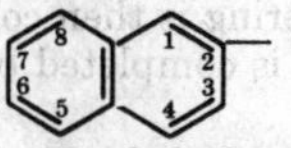

Naphthyl (2-shown)

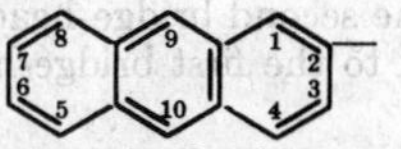

Anthryl (2-shown)

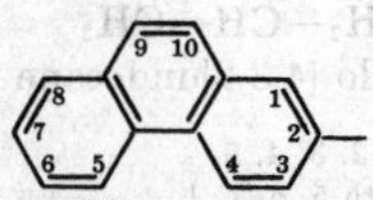

Phenanthryl (2-shown)

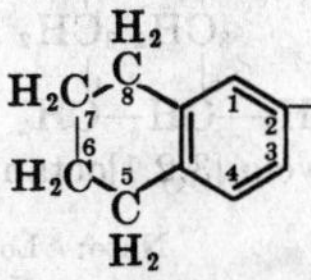

5,6,7,8-Tetrahydro-2-naphthyl

The above exceptions are limited to the simple rings as shown. Radicals from names of fused derivatives of these rings are formed according to the Rule, *e.g.*, benz[*a*]anthracenyl.

24.3.—Bivalent radicals derived from univalent polycyclic hydrocarbon radicals whose names end in "-yl", by removal of one hydrogen atom from the carbon atom with the free valence are named by adding "-idene" to the name of the corresponding univalent radical.

Example:

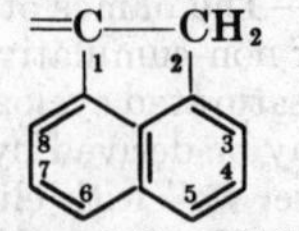

1-Acenaphthenylidene

24.4.—Bivalent radicals derived from "ortho-fused" or "ortho- and peri-fused" polycyclic hydrocarbons by removing of a hydrogen atom from each of two different carbon atoms of the ring are named by changing the ending "-yl" of the univalent radical name to "-ylene."

Examples:

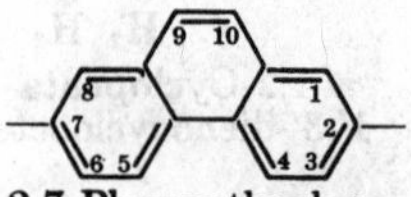

2,7-Phenanthrylene

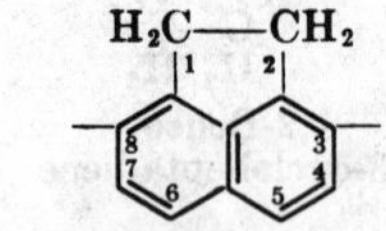

3,8-Acenaphthenylene

Chemical Abstracts names multivalent radicals with three or more free valences derived from polycyclic hydrocarbons by an extension of Rule **B-5.13** to carbocyclic systems. Example: 1,4,5-anthracenetriyl.

A-28

28.1.—Radicals formed from hydrocarbons consisting of polycyclic systems and side chains are named according to the principles of the preceding rules.

Bridged Hydrocarbons

A-31. von Baeyer System

31.1.—Saturated alicyclic hydrocarbon systems consisting of two rings only, having two or more atoms in common, take the name of an open chain hydrocarbon containing the same total number of carbon atoms preceded by the prefix "bicyclo-." The number of carbon atoms in each of the three bridges[1] connecting the two tertiary carbon atoms is indicated in brackets in descending order.

Examples:

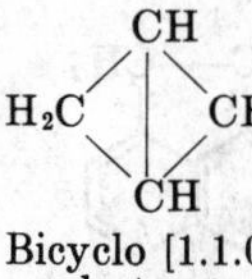

Bicyclo [1.1.0]-butane

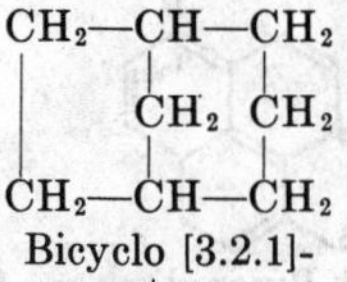

Bicyclo [3.2.1]-octane

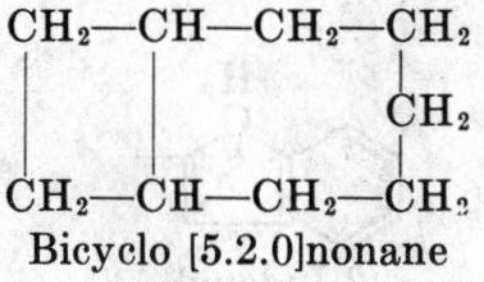

Bicyclo [5.2.0]nonane

31.2.—The system is numbered commencing with one of the bridge-heads, numbering proceeding by the longest possible path to the second bridge-head; numbering is then continued from this atom by the longer unnumbered path back to the first bridge-head and is completed by the shortest path.

Examples:

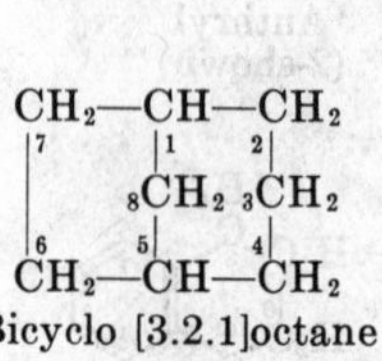

Bicyclo [3.2.1]octane

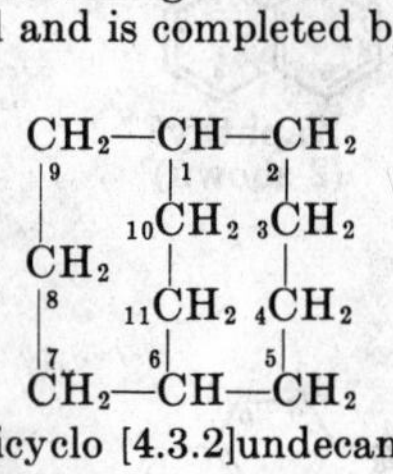

Bicyclo [4.3.2]undecane

Note: Longest path 1, 2, 3, 4, 5
Next longest path 5, 6, 7, 1
Shortest path 1, 8, 5

(1) A bridge is a valence bond or an atom or an unbranched chain of atoms connecting two different parts of a molecule. The two tertiary carbon atoms connected through the bridge are termed "bridge-heads."

31.3.—Unsaturated hydrocarbons are named in accordance with the principles set forth in Rule **A-11.3.** When there is a choice in numbering unsaturation is given the lowest numbers.

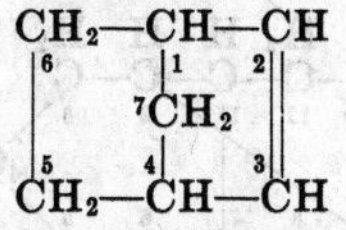

Bicyclo [2.2.1]hept-2-ene

Chemical Abstracts names the above ring system 2-norbornene by Rule **A-74.2**, but follows the above Rule for hydrocarbons which do not have terpene names.

31.4.—Radicals derived from these hydrocarbons are named in accordance with the principles set forth in Rule **A-11.4.** The numbering of the hydrocarbon is retained and the point or points of attachment are given numbers as low as is consistent with the fixed numbering of the hydrocarbon.

Example:

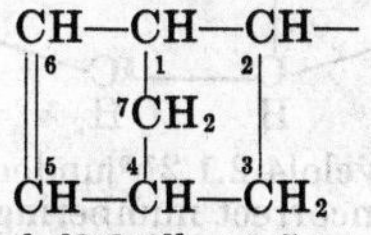

Bicyclo[2.2.1]hept-5-en-2-yl

Chemical Abstracts names the above radical 5-norbornen-2-yl by Rule **A-75.2**, but follows the above Rule for radicals derived from hydrocarbons which do not have terpene names. Radicals derived from the saturated hydrocarbons (except terpenes) are named by replacing the ending "-ane" of the hydrocarbon name by ,"-yl." The point of attachment is given a number as low as is consistent with the fixed numbering of the hydrocarbon.

A-32

32.11.—Cyclic hydrocarbon systems consisting of three or more rings may be named in accordance with the principles stated in Rule **A-31.** The appropriate prefix "tricyclo-," "tetracyclo-," *etc.*, is substituted for "bicyclo-" before the name of the open-chain hydrocarbon containing the same total number of carbon atoms.

32.12.—A polycyclic system is regarded as containing a number of rings equal to the number of scissions required to convert the system into an open-chain compound.

32.13.—The word "cyclo" is followed by brackets containing, in decreasing order, numbers indicating the number of carbon atoms in: the two branches of the main ring, the main bridge, the secondary bridges.

Examples:

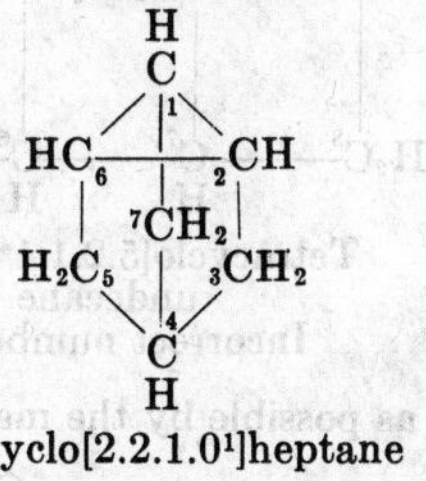

Tricyclo[2.2.1.0[1]]heptane

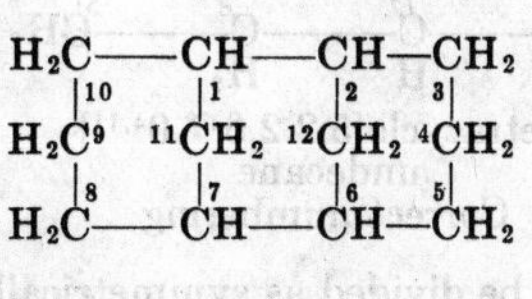

Tricyclo[5.3.1.1[1]]dodecane

(1) For location and numbering of the secondary bridge see Rules **A-32.22. A32.23, A-32.31.**

32.21.—The main ring and the main bridge form a bicyclic system whose numbering is made in compliance with Rule **A-31.**

32.22.—The location of the other or so-called secondary bridges is shown by superscripts following the number indicating the number of carbon atoms in the said bridges.

32.23.—For the purpose of numbering, the secondary bridges are considered in decreasing order. The numbering of any bridge follows from the part already numbered, proceeding from the highest-numbered bridge-head.

Numbering of the secondary bridges by proceeding from the highest-numbered bridge-head of each bridge is an improvement over the former practice of "following the shortest path from the highest previous number."

32.31.—When there is a choice, the following criteria are considered in turn until a decision is made:

(a) The main ring shall contain as many carbon atoms as possible, two of which must serve as bridgeheads for the main bridge.

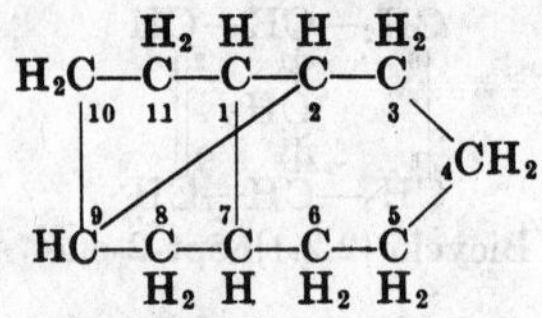

Tricyclo[5.4.0.0^{2,9}]undecane
Correct numbering

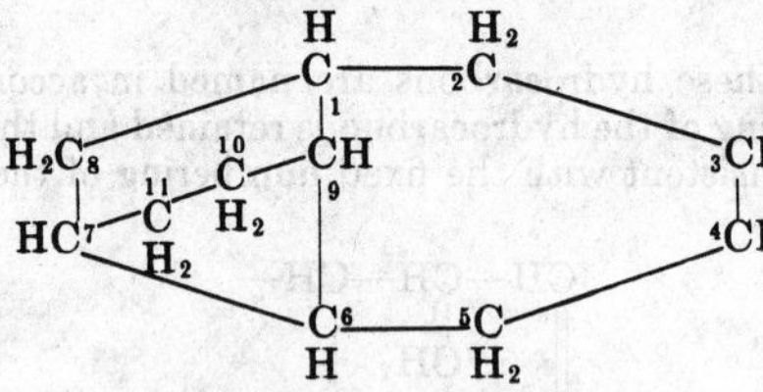

Tricyclo[4.2.1.2^{7,9}]undecane
Incorrect numbering

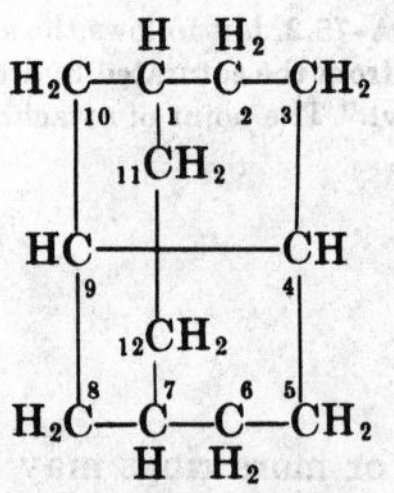

Tricyclo[5.3.2.0^{4,9}]dodecane
Correct numbering

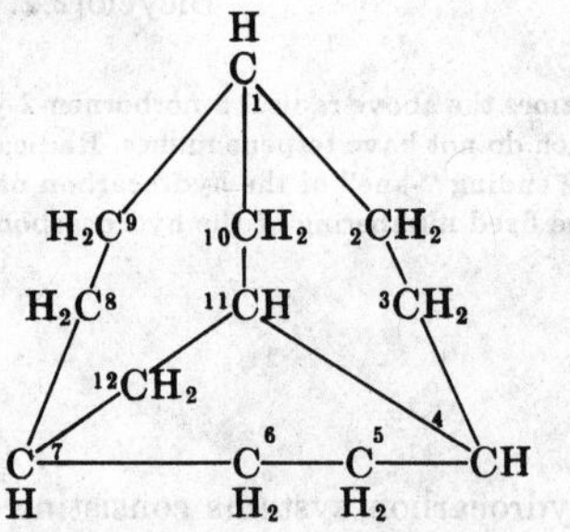

Tricyclo[5.2.3.0^{4,11}]dodecane
Incorrect numbering

(b) The main bridge shall be as large as possible.

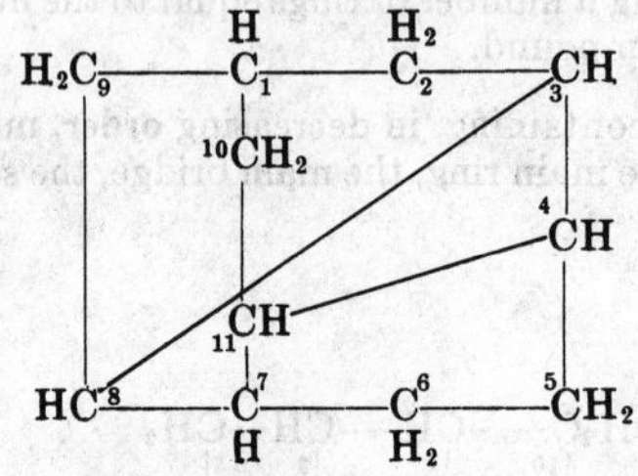

Tetracyclo[5.2.2.0^{3,8}.0^{4,11}]-undecane
Correct numbering

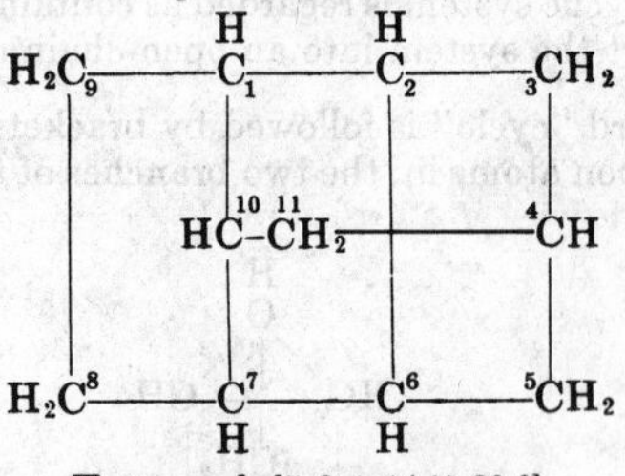

Tetracyclo[5.2.1.1^{4,10}.0^{2,6}]-undecane
Incorrect numbering

(c) The main ring shall be divided as symmetrically as possible by the main bridge.

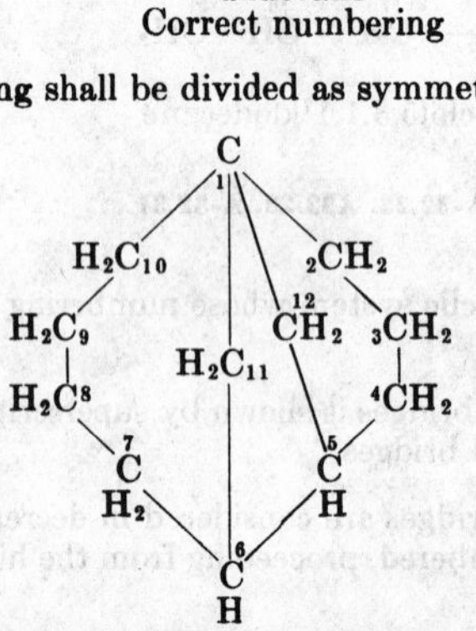

Tricyclo[4.4.1.1^{1,5}]dodecane:
Correct numbering

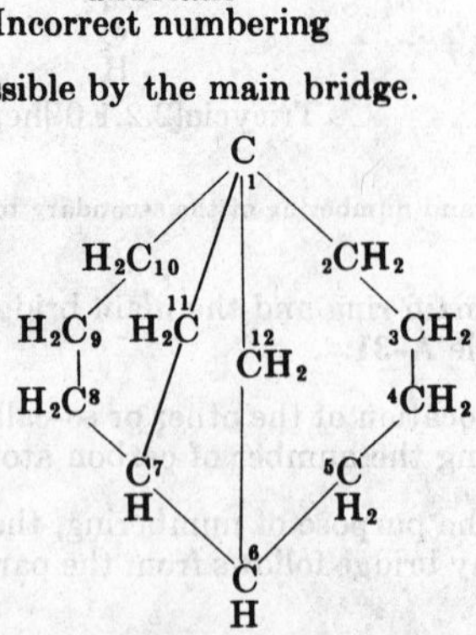

Tricyclo[5.3.1.1^{1,6}]dodecane:
Incorrect numbering

(d) The superscripts locating the other bridges shall be as small as possible (in the sense indicated in Rule **A-2.2**)

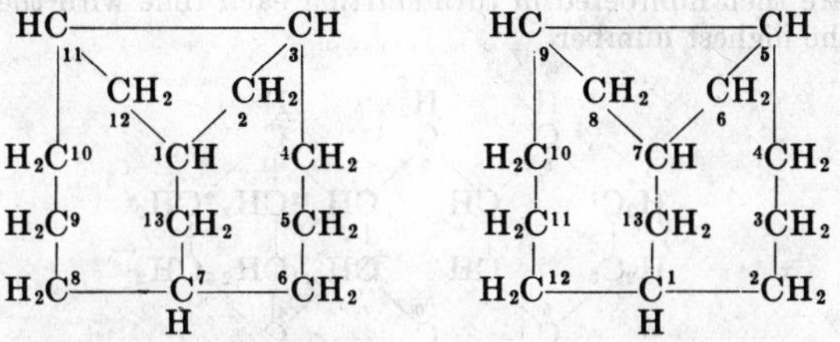

Tricyclo[5.5.1.$0^{3,11}$]tridecane Correct numbering — Tricyclo[5.5.1.$0^{5,9}$]tridecane Incorrect numbering

A-34 (Alternate to Rule A-35)

34.1.—Polycyclic hydrocarbon systems which can be regarded as "ortho-fused" or "ortho- and peri-fused" systems according to Rule **A-21** and which, at the same time, have other bridges,[1] are first named as "ortho-fused" or "ortho- and peri-fused" systems. The other bridges are then indicated by prefixes derived from the corresponding hydrocarbon by replacing the final "-ane," "-ene," *etc.*, by "-ano," "-eno," *etc.*, and their positions are indicated by the points of attachment in the parent compound. If bridges of different types are present, they are cited in alphabetical order.

Examples of bridge names

Butano	$—CH_2—CH_2—CH_2—CH_2—$	Etheno	$—CH{=}CH—$
Benzeno (*o*-, *m*-, *p*-)	$—C_6H_4—$	Methano	$—CH_2—$
Ethano	$—CH_2—CH_2—$	Propano	$—CH_2—CH_2—CH_2—$

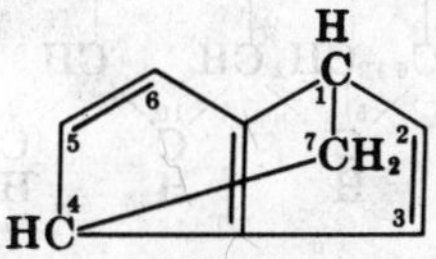

1,4-Dihydro-1,4-methanopentalene

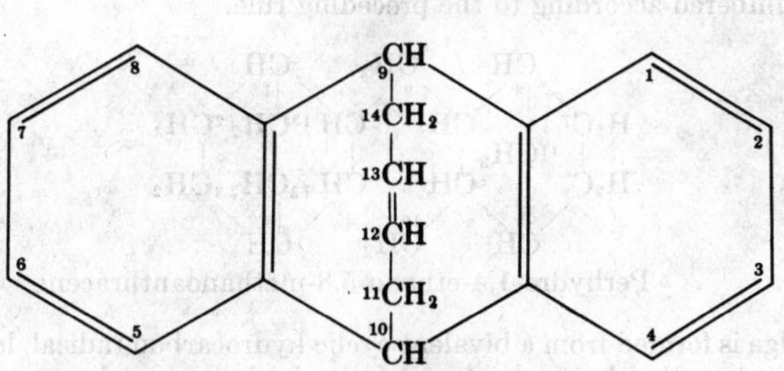

9,10-Dihydro-9,10-(2-buteno)-anthracene

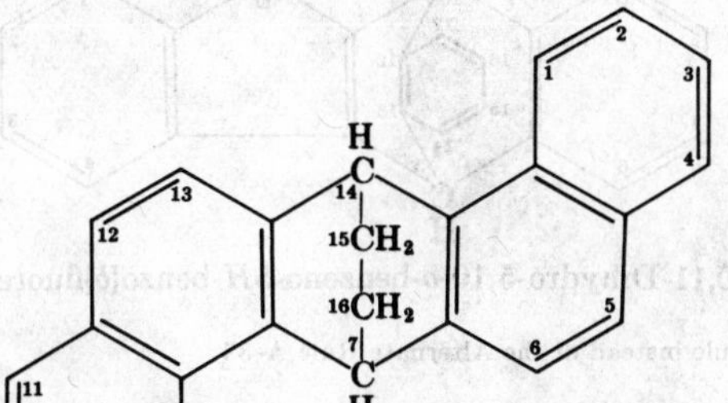

7,14-Dihydro-7,14-ethano-dibenz[*a,h*]anthracene

(1) The term "bridge," when used in connection with an "ortho-fused" or "ortho- and peri-fused" polycyclic system as defined in note to Rule A-31.1, also includes "bivalent cyclic systems."

34.2.—The parent "ortho-fused" "or ortho- and peri-fused" system is numbered as prescribed in Rule **A-22.** Where there is a choice, the position numbers of the bridge-heads should be as low as possible. The remaining bridges are then numbered in turn starting each time with the bridge atom next to the bridge-head possessing the highest number.

Example:

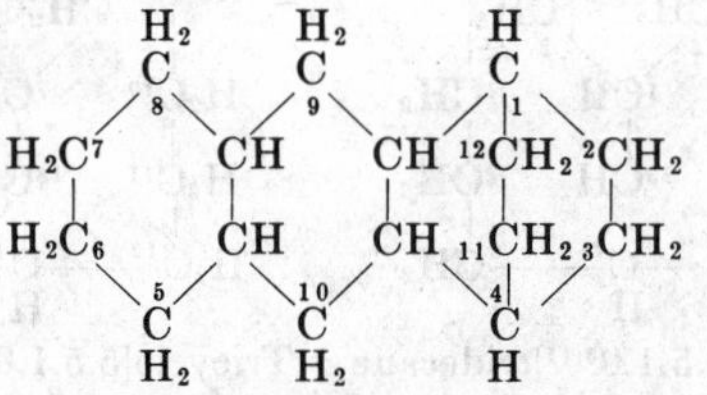

Perhydro-1,4-ethanoanthracene

not

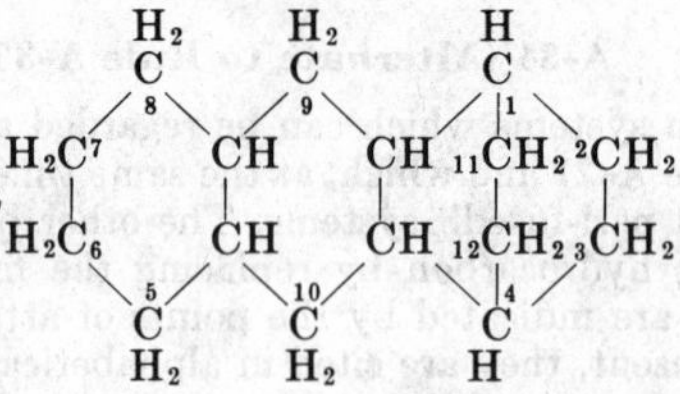

or

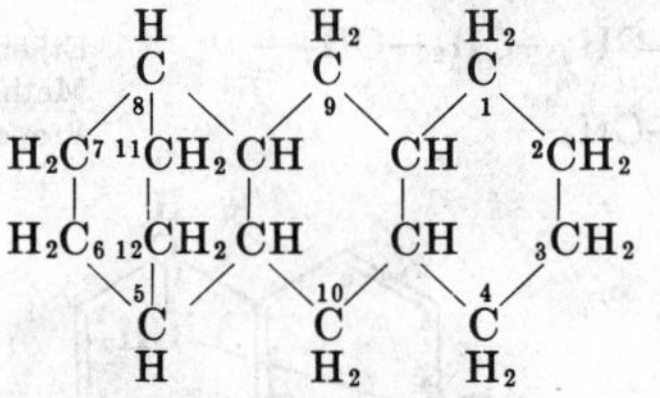

34.3.—When there is a choice of position numbers for the points of attachment for several individual bridges, the lowest numbers are assigned to the bridge-heads in the order of citation of the bridges and the bridge atoms are numbered according to the preceding rule.

Example:

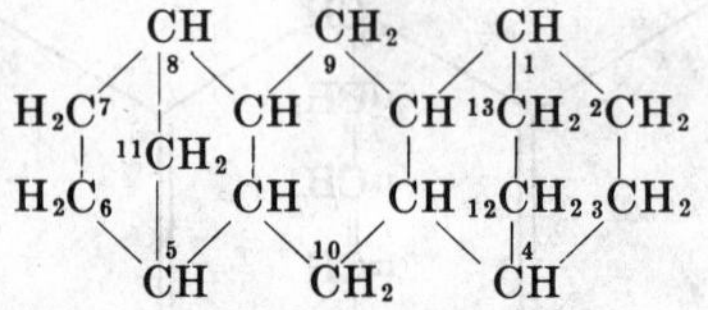

Perhydro-1,4-ethano-5,8-methanoanthracene

34.4.—When the bridge is formed from a bivalent cyclic hydrocarbon radical, low numbers are given to the carbon atoms constituting the shorter bridge and numbering proceeds around the ring.

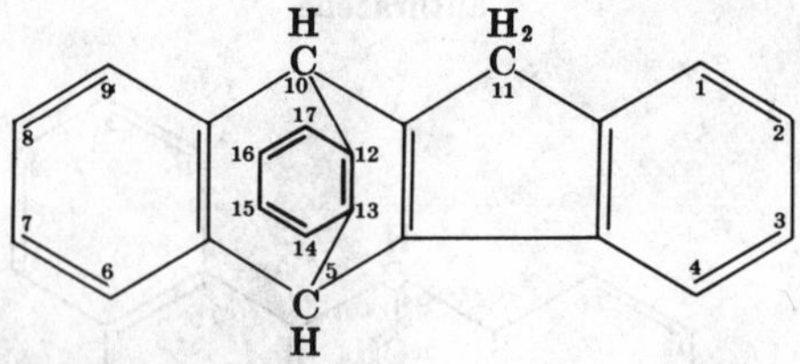

10,11-Dihydro-5,10-*o*-benzeno-5*H*-benzo[*b*]fluorene

Chemical Abstracts follows this rule instead of the Alternate Rule A-35.

A-35 (Alternate to Rule A-34)

35.1.—Polycyclic hydrocarbons which can be regarded as "ortho-fused" or "ortho- and peri-fused" according to Rule **A-21** and which have also other bridges may be named as "ortho-fused" or "ortho- and peri-fused" systems into which bivalent radicals are substituted.

35.2.—Compounds which are named in accordance with **35.1** of this rule are numbered as prescribed in Rule **A-34.** The bridge is numbered independently. Numbering of the bridge is started at the lower-numbered bridge head. Numbers assigned to the bridge atoms are primed.

Examples:

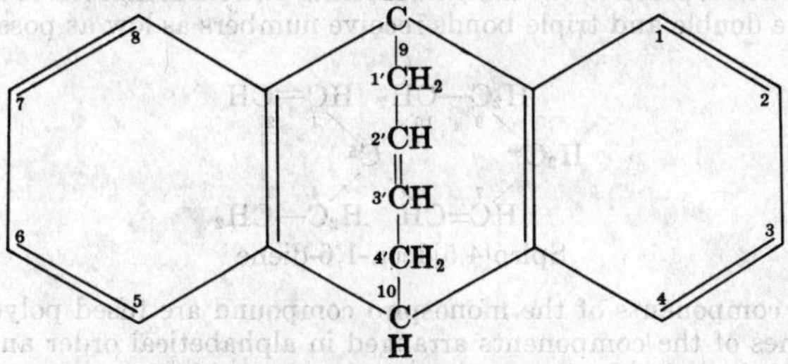

9,10-Dihydro-9,10-(2'-butylene)anthracene

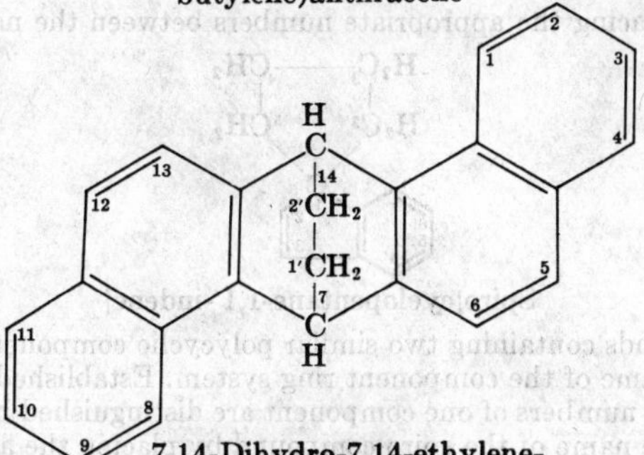

7,14-Dihydro-7,14-ethylene-dibenz[*a*,*h*]anthracene

Chemical Abstracts follows Rule A-34 instead of this Alternate Rule.

Spiro Hydrocarbons

A "spiro union" is one formed by a single atom which is the only common member of two rings. A "free spiro union" is one constituting the only union direct or indirect between two rings.[1] The common atom is designated as the "spiro atom." According to the number of spiro atoms present, the compounds are distinguished as monospiro, dispiro, trispiro compounds, *etc.* The following rules apply to the naming of compounds containing free spiro unions.

(1) An example of a compound where the spiro union is *not* free is:

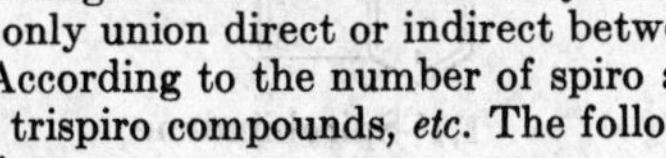

A-41 (Alternate to Rule A-42)

41.1.—Monospiro compounds consisting of only two alicyclic rings as components are named by placing "spiro" before the name of the normal acyclic hydrocarbon of the same total number of carbon atoms. The number of carbon atoms linked to the spiro atom in each ring is indicated in ascending order in brackets placed between the spiro prefix and the hydrocarbon name.

Examples:

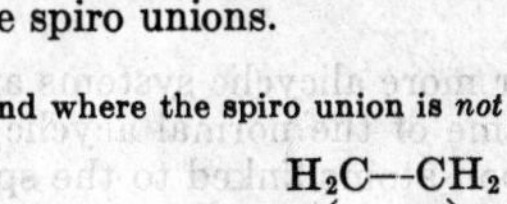

Spiro[3.4]octane

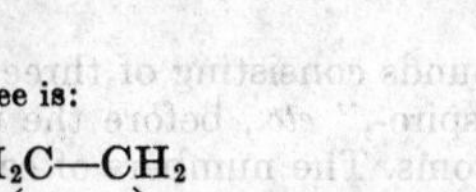

Spiro[3.3]heptane

41.2.—The carbon atoms in monospiro hydrocarbons are numbered consecutively starting with a ring atom next to the spiro atom, first through the smaller ring (if such be present) and then through the spiro atom and around the second ring.

Example:

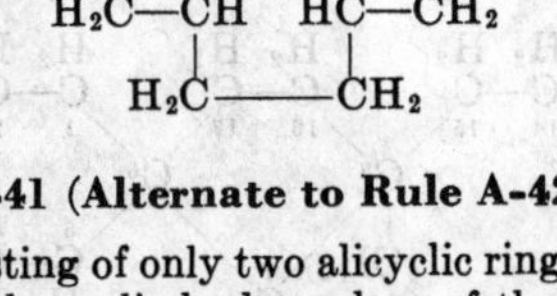

Spiro[4.5]decane

41.3.—When unsaturation is present, the same numbering pattern is maintained, but in such a direction around the rings that the double and triple bonds receive numbers as low as possible in accordance with Rule **A-11.**

Example:

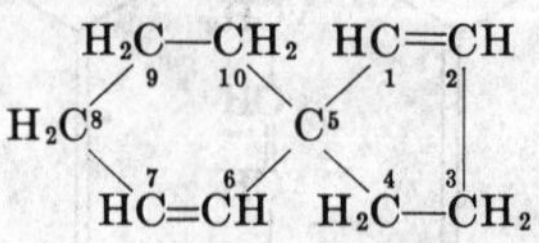

Spiro[4.5]deca-1,6-diene

41.4.—If one or both components of the monospiro compound are fused polycyclic systems, "spiro" is placed before the names of the components arranged in alphabetical order and enclosed in brackets. Established numbering of the individual components is retained. The lowest possible number is given to the spiro atom, and the numbers of the second component are marked with primes. The position of the spiro atom is indicated by placing the appropriate numbers between the names of the two components.

Example:

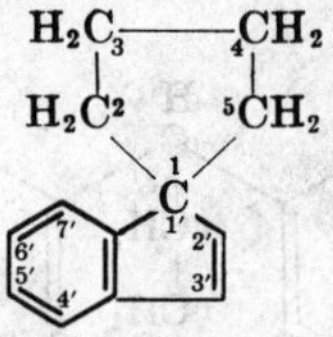

Spiro[cyclopentane-1,1′-indene]

41.5.—Monospiro compounds containing two similar polycyclic components are named by placing the prefix "spirobi" before the name of the component ring system. Established numbering of the polycyclic system is maintained and the numbers of one component are distinguished by primes. The position of the spiro atom is indicated in the name of the spiro compound by placing the appropriate locants before the name.

Example:

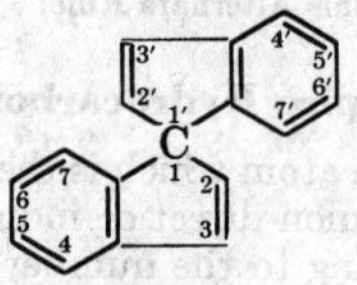

1,1′-Spirobiindene

41.6.—Polyspiro compounds consisting of three or more alicyclic systems are named by placing "dispiro-," "trispiro-," "tetraspiro-," *etc*., before the name of the normal acyclic hydrocarbon of the same total number of carbon atoms. The numbers of carbon atoms linked to the spiro atoms in each ring are indicated in brackets in the same order as the numbering proceeds about the ring. Numbering starts with a ring atom next to a terminal spiro atom and proceeds in such a way as to give the spiro atoms as low numbers as possible.

Example:

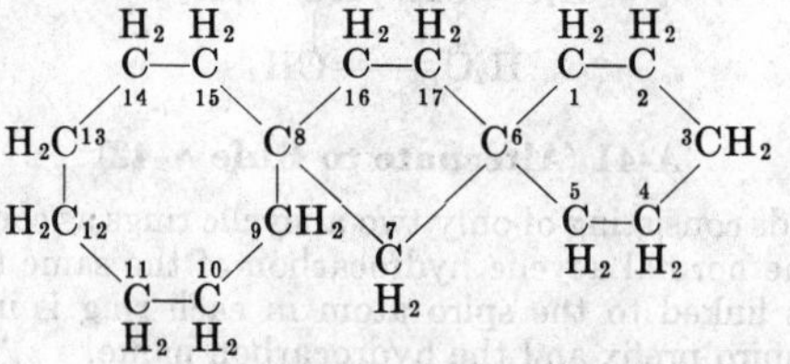

Dispiro[5.1.7.2]heptadecane

41.7.—Polycyclic compounds containing more than one spiro atom and at least one fused polycyclic component are named in accordance with **41.4** of this rule by replacing "spiro" with "dispiro," "trispiro," etc., and choosing the end components by alphabetical order.

Example:

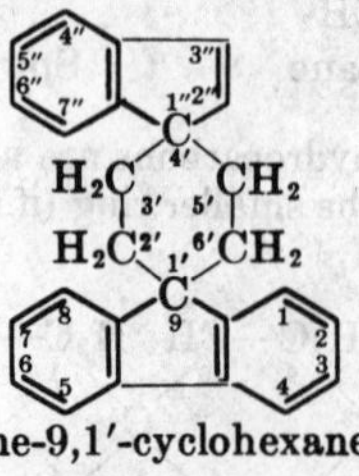

Dispiro[fluorene-9,1′-cyclohexane-4′,1″-indene]

This represents the method of naming spiro compounds which is followed by *Chemical Abstracts* instead of the Alternate Rule A-42.

A-42

42.1 (Alternate to **A-41.1** and **A-41.2**).—When two dissimilar cyclic components are united by a spiro union the name of the larger component is followed by the affix "spiro" which, in turn, is followed by the name of the smaller component. Between the affix "spiro" and the name of each component system is inserted the number denoting the spiro-position in the appropriate ring system, these numbers being as low as permitted by any fixed numbering of the component. The components retain their respective numberings but numbers for the component mentioned second are primed. Numbers 1 may be omitted when a free choice is available for a component.

Examples:

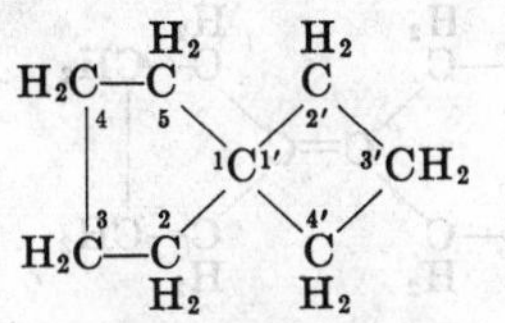

Cyclopentanespiro-cyclobutane

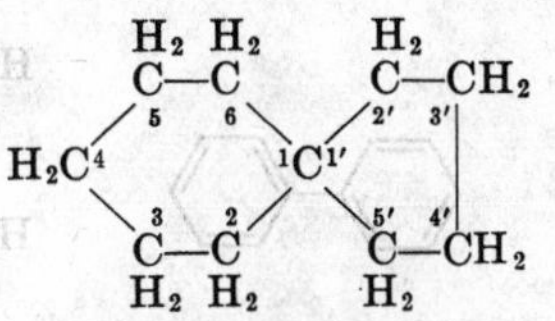

Cyclohexanespirocyclo-pentane

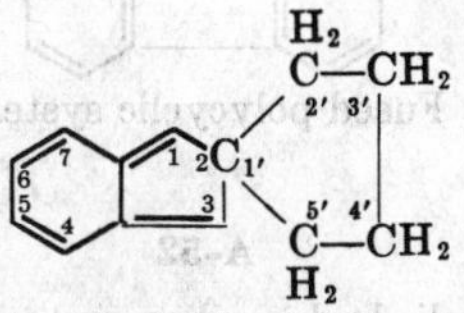

2*H*-Indene-2-spiro-1′-cyclopentane

42.2 (Alternate to **A-41.3**).—Rule **A-41.3** applies also with appropriate different numbering where nomenclature is according to Rule **A-42.1**, but the spiro-junction has priority for lowest numbers over unsaturation.

Example:

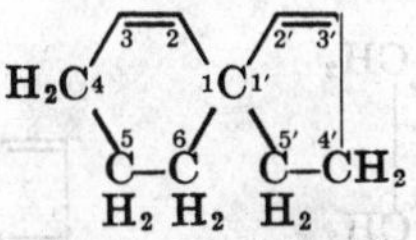

2-Cyclohexenespiro-(2′-cyclopentene)

42.3 (Alternate to **A-41.5**).—The nomenclature of Rule **A-41.5** is applied also to monocyclic components with identical saturation, the spiro-union being numbered 1.

Examples:

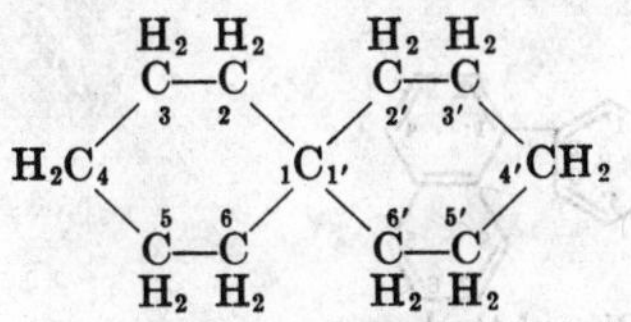

Spirobicyclohexane

but

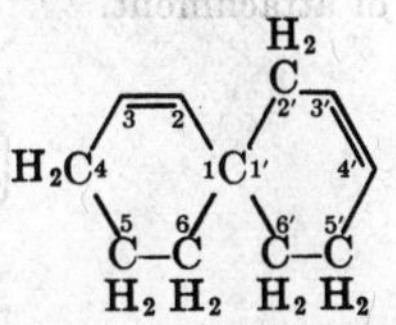

2-Cyclohexenespiro-(3′-cyclohexene)

42.4 (Alternate to **A-41.6** and **A-41.7**).—Polycyclic compounds containing more than one spiro atom are named in accordance with Rule **A-42.1** starting from the senior[1] end-component irrespective of whether the components are simple or fused rings.

Examples:

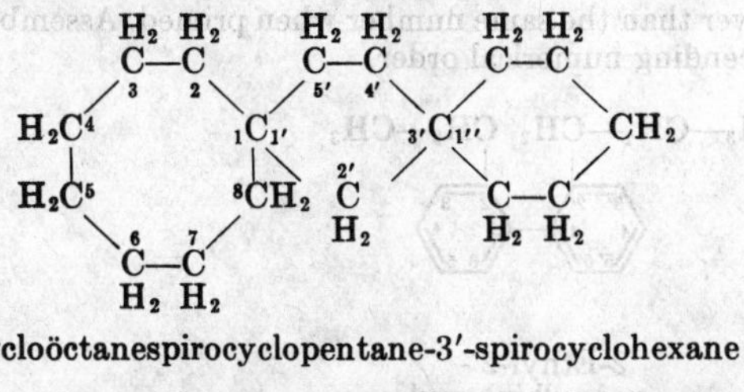

Cycloöctanespirocyclopentane-3′-spirocyclohexane

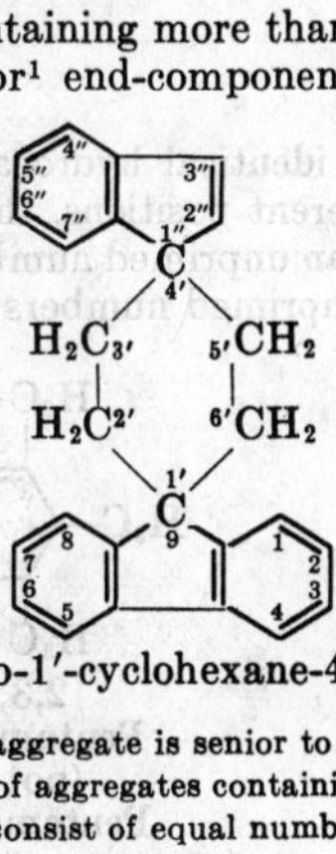

Fluorene-9-spiro-1′-cyclohexane-4′-spiro-1″-indene

(1) "Seniority" in respect to spiro compounds is based on the principles: (i) an aggregate is senior to a monocycle; (ii) of aggregates, the senior is that containing the largest number of individual rings; (iii) of aggregates containing the same number of individual rings, the senior is that containing the largest ring; (iv) if aggregates consist of equal number of equal rings the senior is the first occurring in the alphabetical list of names.

Chemical Abstracts follows Rule A-41 instead of this alternate Rule.

Hydrocarbon Ring Assemblies

A-51

51.1.—Two or more cyclic systems (single rings or fused systems) which are directly joined to each other by double or single bonds are named "ring assemblies" when the number of such direct ring junctions is one less than the number of cyclic systems involved.

Examples:

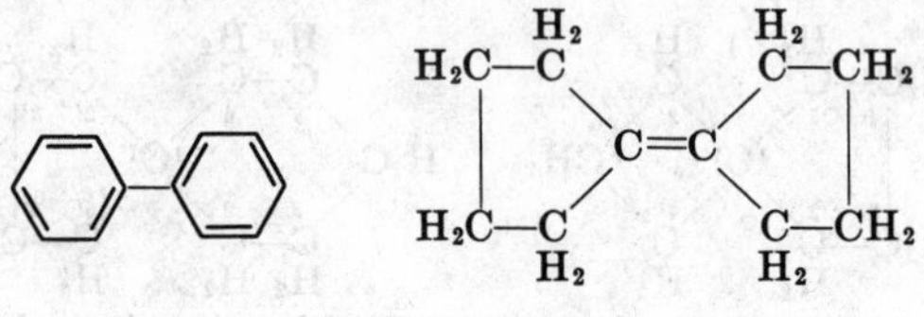

Ring assemblies

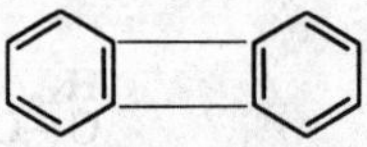

Fused polycyclic system

A-52

52.1.—Assemblies of two identical cyclic hydrocarbon systems are named in either of two ways: (a) by placing the prefix "bi-" before the name of the corresponding radical, or (b) for systems joined by a single bond by placing the prefix "bi-" before the name of the corresponding hydrocarbon. In each case, the numbering of the assembly is that of the corresponding radical or hydrocarbon, one system being assigned unprimed numbers and the other primed numbers. The points of attachment are indicated by placing the appropriate locants before the name.

Examples:

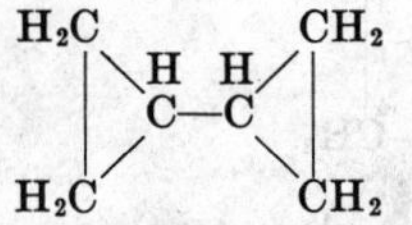

1,1′-Bicyclopropyl
or 1,1′-Bicyclopropane

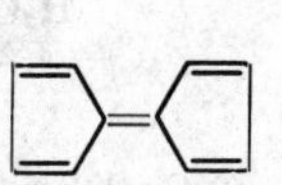

1,1′-Bicyclopentadienylidene

52.2.—If there is a choice in numbering, unprimed numbers are assigned to the system which has the lower-numbered point of attachment.

Example:

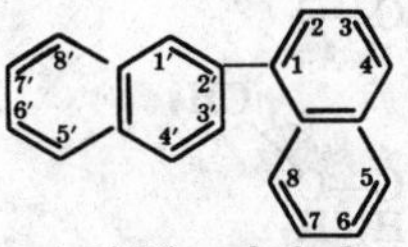

1,2′-Binaphthyl
or 1,2′-Binaphthalene

52.3.—If two identical hydrocarbon systems have the same point of attachment and contain substituents at different positions, the locants of these substituents are assigned according to Rule A-2.2; for this purpose an unprimed number is considered lower than the same number when primed. Assemblies of primed and unprimed numbers are arranged in ascending numerical order.

Examples:

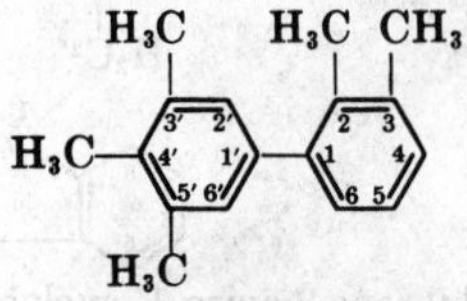

2,3,3′,4′,5′-
Pentamethylbiphenyl
(not 2′,3,3′,4,5-
Pentamethylbiphenyl)

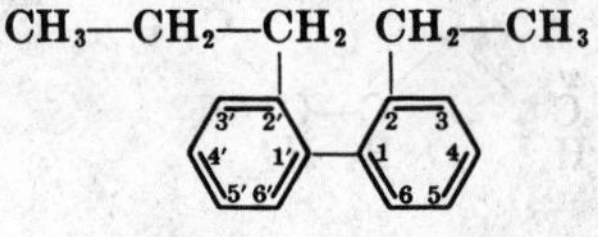

2-Ethyl-2′-
propylbiphenyl

52.4.—The name "biphenyl" is used for the assembly consisting of two benzene rings:

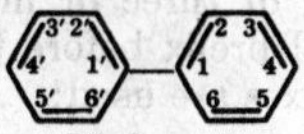

Biphenyl

A-53

53.1.—Other hydrocarbon ring assemblies are named by selecting one ring system as the base component and considering the other system as substituents of the base component. Such substituents are arranged in alphabetical order. The base component is assigned unprimed numbers and the substituents are assigned numbers with primes.

53.2.—The base component is chosen by considering these characteristics in turn until a decision is reached:

(a) The system containing the larger number of rings.

Examples:

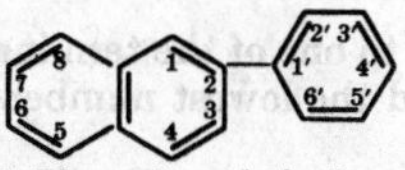

2-Phenylnaphthalene

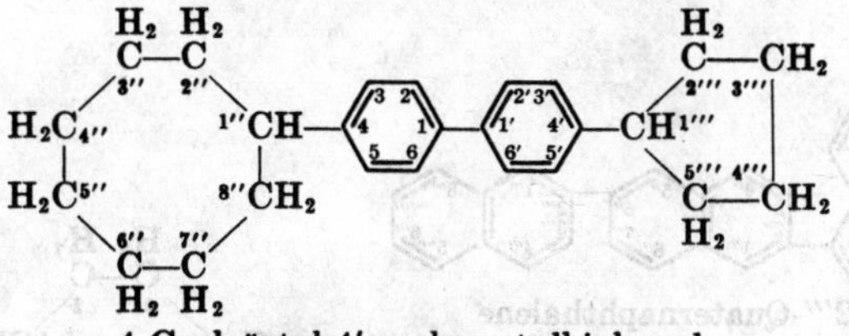

4-Cycloöctyl-4′-cyclopentylbiphenyl

(b) The system containing the larger ring.

Examples:

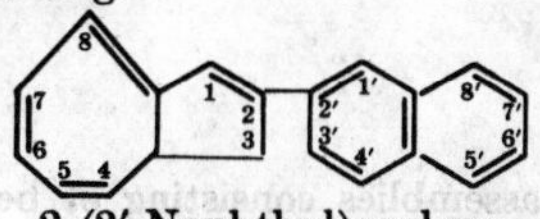

2-(2′-Naphthyl)azulene

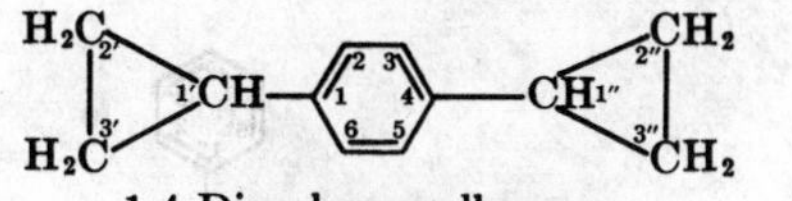

1,4-Dicyclopropylbenzene
or *p*-Dicyclopropylbenzene

(c) The system in the lowest state of hydrogenation (see also **53.3** of this rule).

Example:

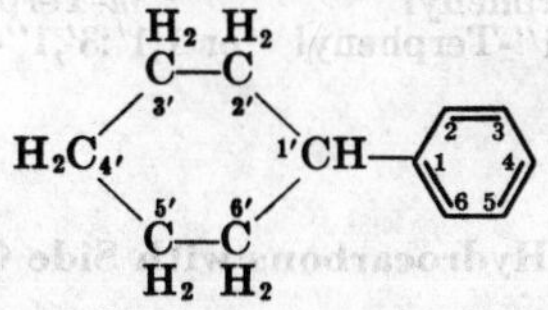

Cyclohexylbenzene

(d) The order of ring systems as set forth in the list of Rule **A-21.1**.

53.3.—Compounds covered by **53.2**(c) of this rule may also be named as hydrogenation products according to Rule **A-23**.

Example:

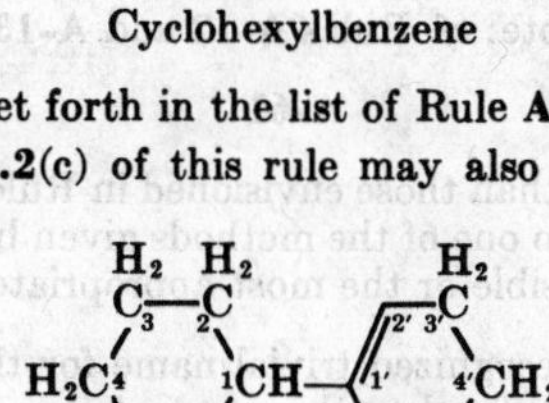

1,2,3,3′,4,4′-Hexahydro-1,1′-binaphthyl

The name of the second example in (a) which contains as a base component the ring assembly "biphenyl" indicates that the word system as used here includes ring assemblies as base components.

Chemical Abstracts does not use primed numbers with the groups which may be considered as substituent groups expressed as prefixes and attached to the name of the base component, Examples: 2-(2,4-dichlorophenyl)naphthalene; 2(2-naphthyl)-azulene.

A-54

54.1.—Unbranched assemblies consisting of three or more identical hydrocarbon ring systems are named by placing an appropriate numerical prefix before the name of the hydrocarbon corresponding to the repetitive unit. These numerical prefixes are used:

3. ter-	5. quinque-	7. septi-	9. novi-
4. quater-	6. sexi-	8. octi-	10. deci-

Example:

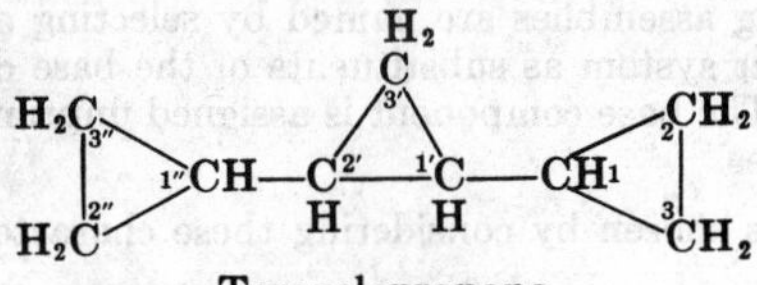

Tercyclopropane

54.2.—Unprimed numbers are assigned to one of the terminal systems, the other systems being primed serially. Points of attachment are assigned the lowest numbers possible.

Examples:

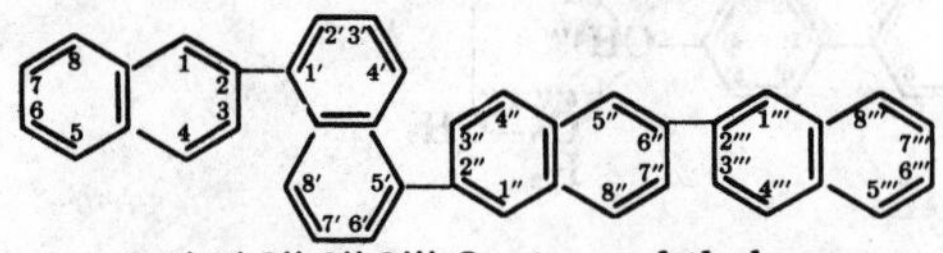

2,1′:5′,2″:6″,2‴-Quaternaphthalene

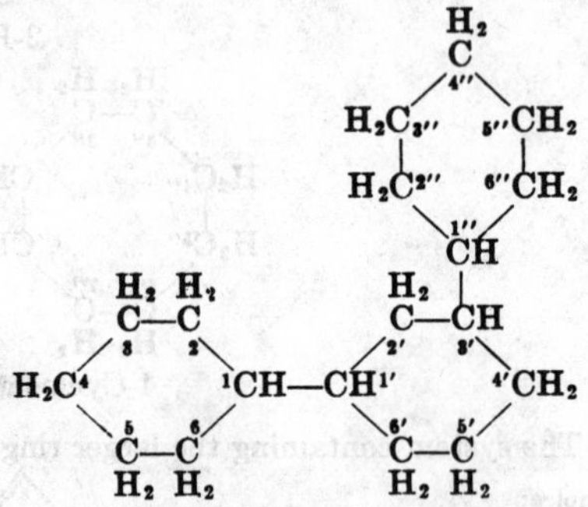

1,1′:3′,1″-Tercyclohexane

54.3.—As exceptions, unbranched assemblies consisting of benzene rings are named by using the appropriate prefix with the radical name "phenyl."

Examples:

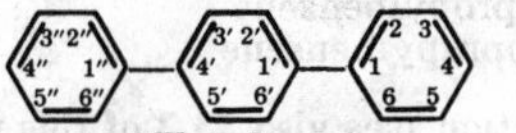

p-Terphenyl or 1,1′:4′,1″-Terphenyl

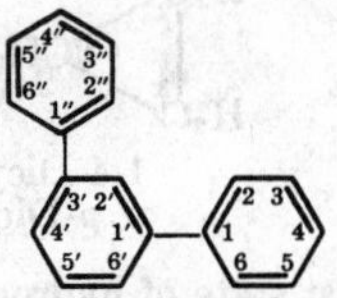

m-Terphenyl or 1,1′:3′,1″-Terphenyl

Cyclic Hydrocarbons with Side Chains

(Note: *cf.* Rules **A-12** and **A-13**)

A-61

61.1.—Hydrocarbons more complex than those envisioned in Rule **A-12,** composed of cyclic nuclei and aliphatic chains, are named according to one of the methods given below. Choice is made so as to provide the name which is the simplest permissible or the most appropriate for the chemical intent.

61.2.—When there is no generally recognized trivial name for the hydrocarbon, then (1) the radical name denoting the aliphatic chain is prefixed to the name of the cyclic hydrocarbon, or (2) the radical name for the cyclic hydrocarbon is prefixed to the name of the aliphatic compound. Choice between these methods is made according to the more appropriate of the following principles: (a) the maximum number of substitutions into a single unit of structure; (b) treatment of a smaller unit of structure as a substituent into a larger.

61.3.—In accordance with the principle (a) of **61.2** of this rule, hydrocarbons containing several chains attached to one cyclic nucleus generally are named as derivatives of the cyclic compound; and compounds containing several side chains and/or cyclic radicals attached to one chain are named as derivatives of the acyclic compound.

Examples:

CH₃

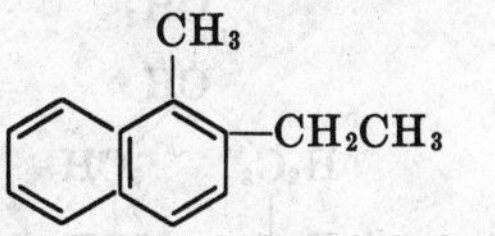

2-Ethyl-1-methylnaphthalene   Diphenylmethane

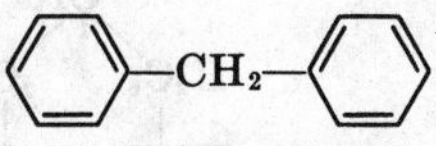

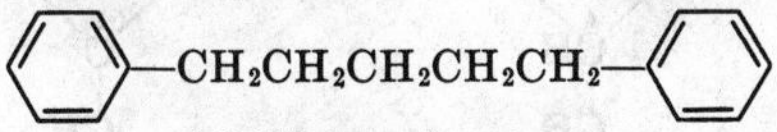

1,5-Diphenylpentane

$$CH_3CH_2CH_2-\overset{CH_3}{\underset{|}{C}H}-\overset{CH_3}{\underset{|}{C}}=CH-$$

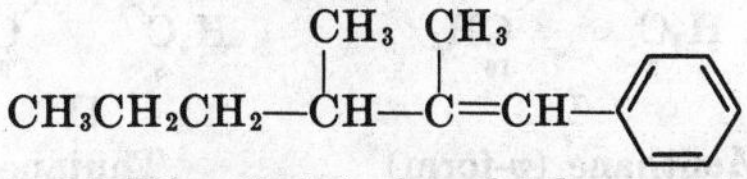

2,3-Dimethyl-1-phenyl-1-hexene

61.4.—In accordance with principle (b) of **61.2** of this rule, a hydrocarbon containing a small cyclic nucleus attached to a long chain is generally named as a derivative of the acyclic hydrocarbon; and a hydrocarbon containing a small group attached to a large cyclic nucleus is generally named as a derivative of the cyclic hydrocarbon.

Examples: $CH_3[CH_2]_{10}CH_2CH_2CH_2CH_2CH_2-$

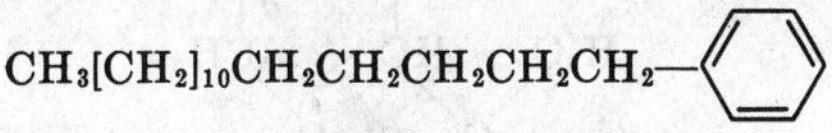

1-Phenylhexadecane

$CH_3[CH_2]_{10}CH_2CH_2CH_2CH=CH-$

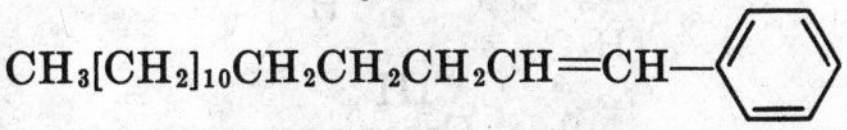

1-Phenyl-1-hexadecene

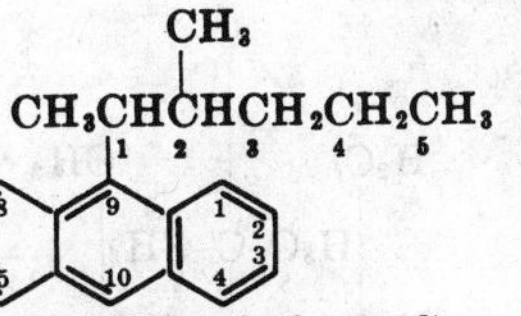

9-(1,2-Dimethylpentyl)-anthracene   7-(3-Phenylpropyl)-benz[a]anthracene

61.5.—Recognized trivial names for composite radicals are used if they lead to simplifications in naming.

Examples:

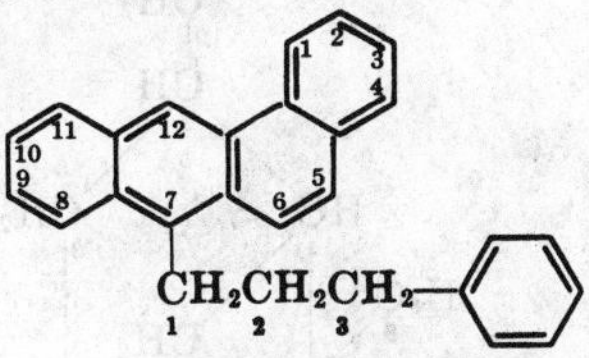

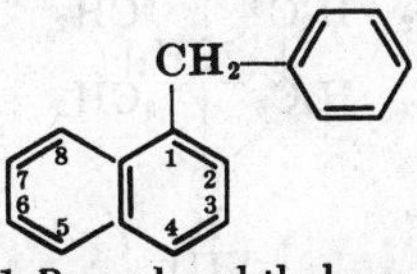

1-Benzylnaphthalene   1,2,4-Tris(3-p-tolylropyl)benzene

Terpene Hydrocarbons

Owing to long-established custom, terpenes are given exceptional treatment in these rules.

A-71. Acyclic Terpenes[1]

71.1.—The acyclic terpene hydrocarbons are named in a manner similar to that used for other unsaturated acyclic hydrocarbons when pure compounds are involved.

Example:

$$\underset{8}{CH_3}-\underset{7}{\overset{CH_3}{\underset{|}{C}}}=\underset{6}{CH}-\underset{5}{CH_2}-\underset{4}{CH_2}-\underset{3}{\overset{CH_2}{\underset{||}{C}}}-\underset{2}{CH}=\underset{1}{CH_2}$$

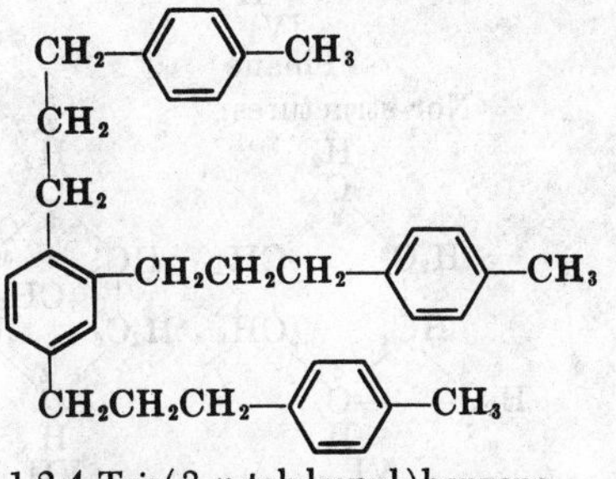

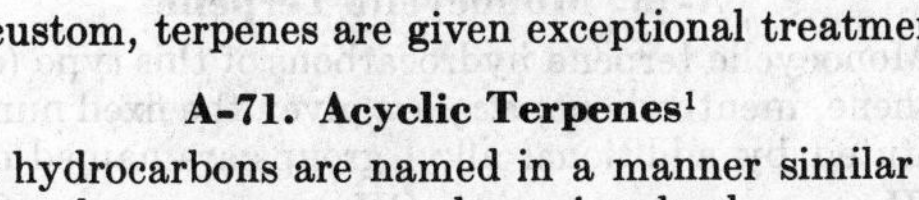

7-Methyl-3-methylene-1,6-octadiene

A-72. Cyclic Terpenes

72.1.—The following structural types with their special names and special systems of numbering are used as the basis for the specialized nomenclature of monocyclic and bicyclic terpene hydrocarbons. The name "bornane" replaces camphane and bornylane; "norbornane" replaces norcamphane and norbornylane.

(1) For a more complete discussion of terpene nomenclature see "Nomenclature for Terpene Hydrocarbons," *Advances in Chemistry Series No. 14* (American Chemical Society).

Fundamental terpene types:

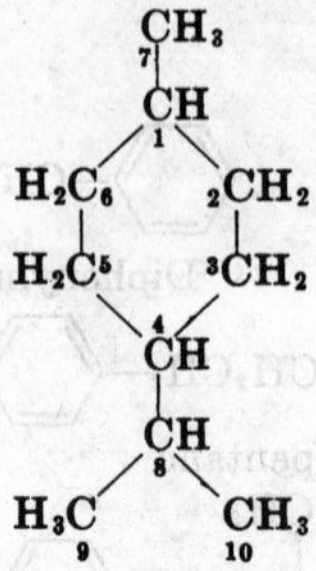

I
Menthane (*p*-form)

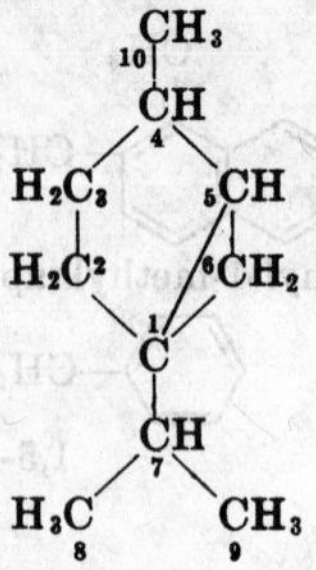

II
Thujane

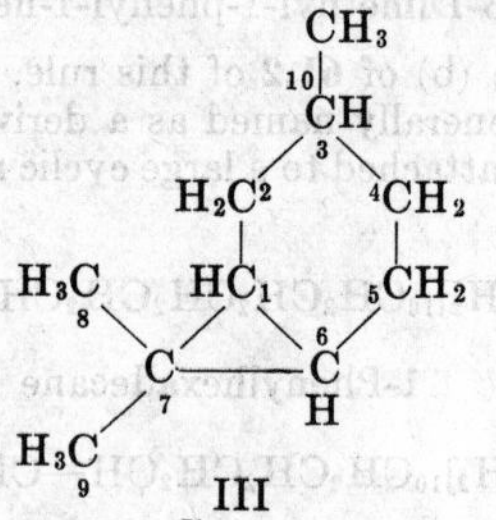

III
Carane

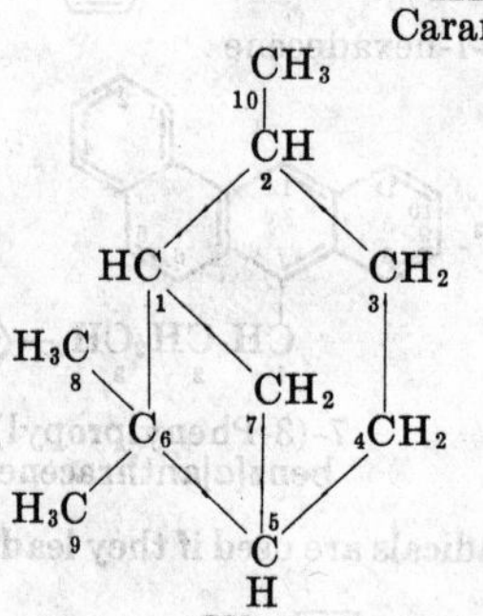

IV
Pinane

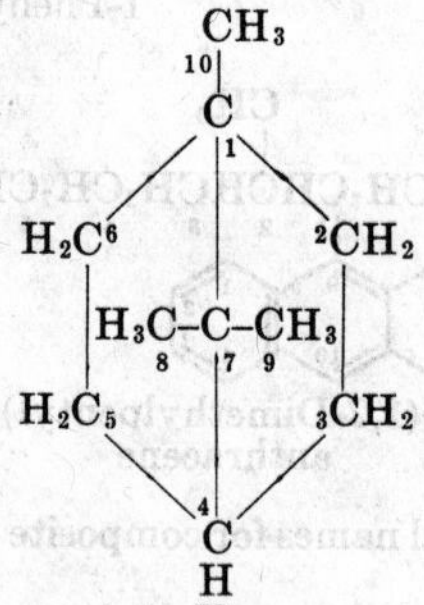

V
Bornane

Nor-structures:

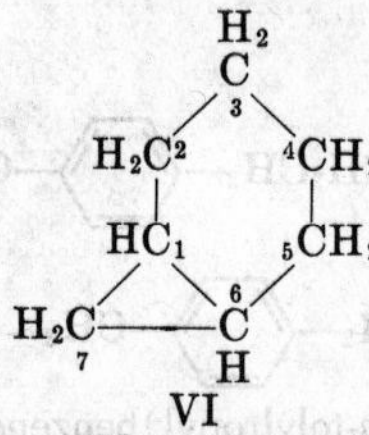

VI
Norcarane

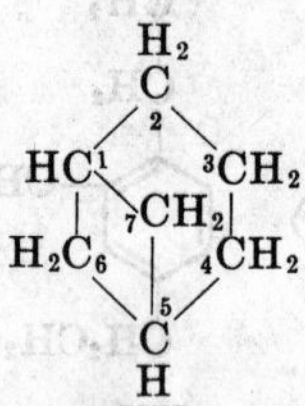

VII
Norpinane

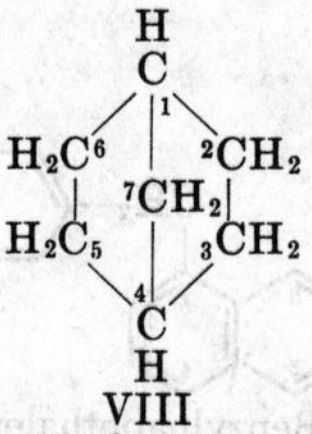

VIII
Norbornane

A-73. Monocyclic Terpenes

73.1.—Menthane Type: Monocyclic terpene hydrocarbons of this type (ortho, meta, and para isomers) are named menthane, menthene, menthadiene, *etc.*, are given the fixed numbering of menthane (Formula I). Such compounds substituted by additional alkyl groups are named in accordance with Rule **A-11.**

Examples:

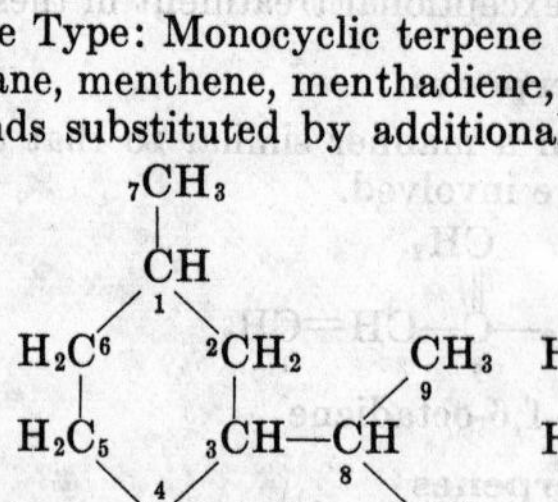

m-Menthane

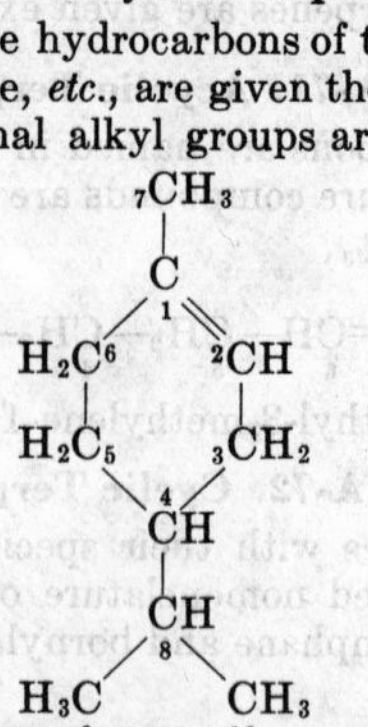

1-*p*-Menthene

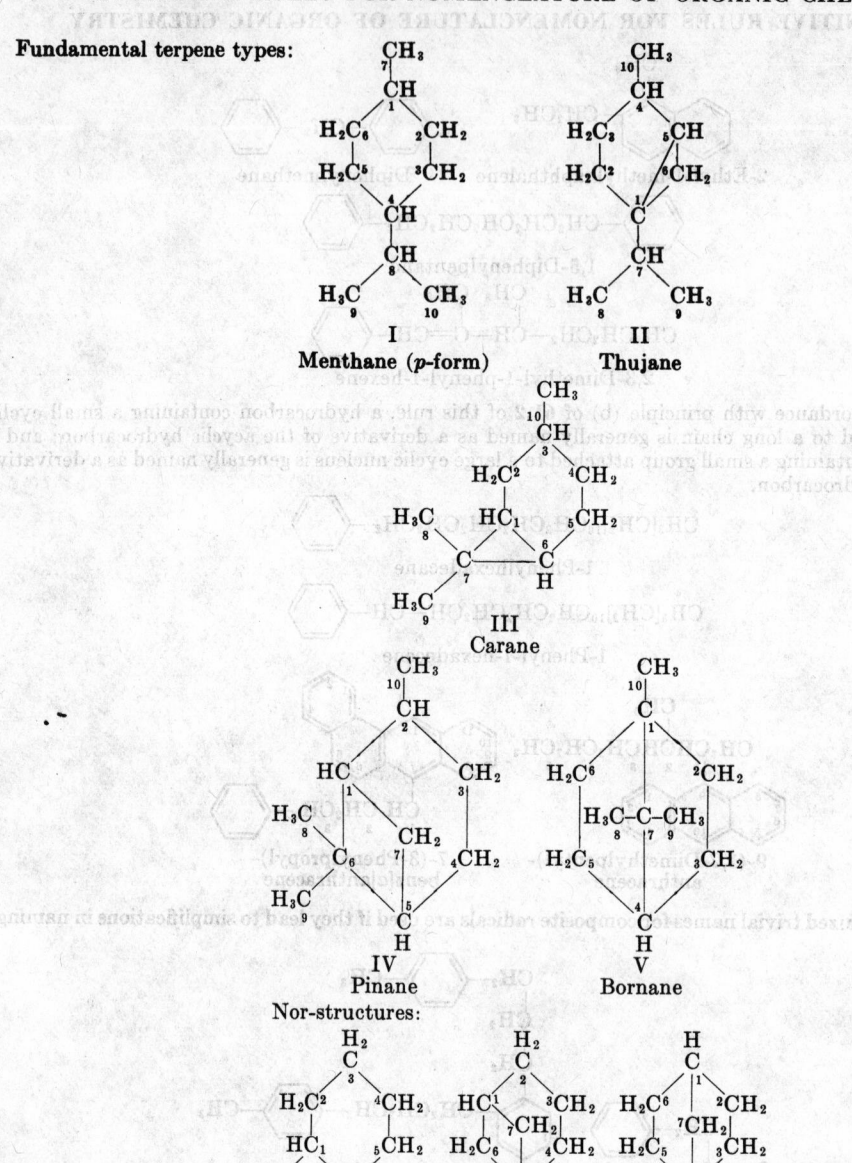

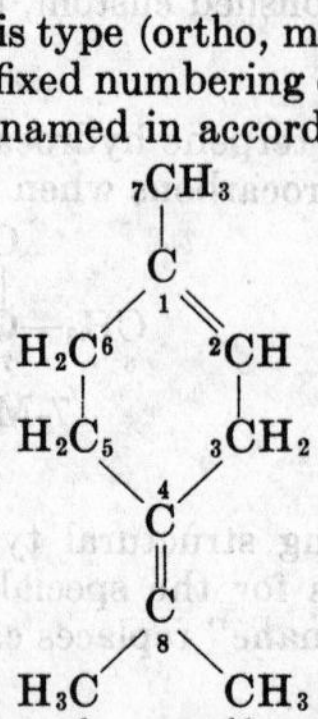

1,4(8)-*p*-Menthadiene

73.2.—Tetramethylcyclohexane Type: Monocyclic terpene hydrocarbons of this type are named systematically as derivatives of cyclohexane, cyclohexene, and cyclohexadiene (see Rule **A-11**).

Examples:

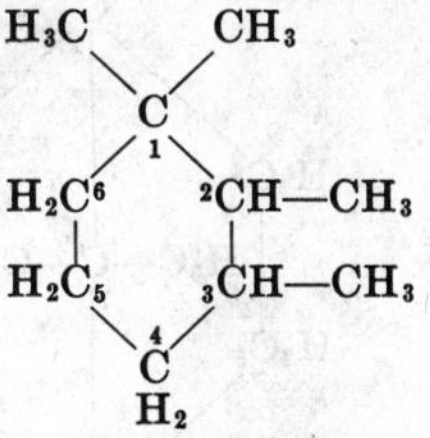

1,1,2,3-Tetramethylcyclohexane

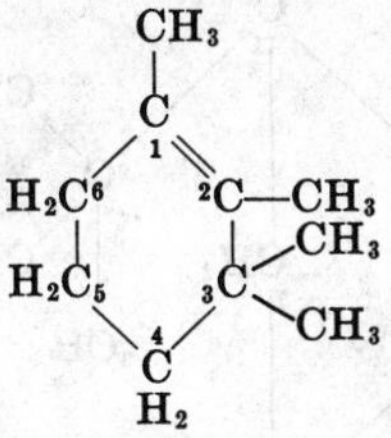

1,2,3,3-Tetramethylcyclohexene

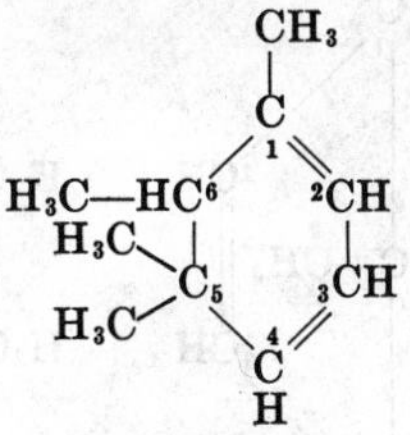

1,5,5,6-Tetramethyl-1,3-cyclohexadiene

A-74. Bicyclic Terpenes

74.1.—Bicyclic terpene hydrocarbons having the skeleton of Formula II or this skeleton and additional side chains except methyl or isopropyl (or methylene if one methylene group is already present) are named as thujane, thujene, thujadiene, *etc.*, and are given the fixed numbering shown for thujane (Formula II). Other hydrocarbons containing the thujane ring-skeleton are named from bicyclo[3.1.0]hexane and are given systematic bicyclo numbering (*cf.* Rule **A-31**).

Example:

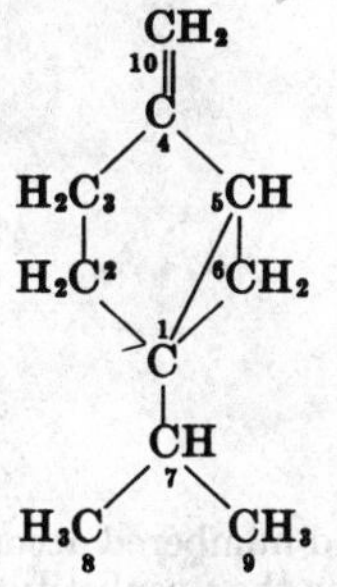

4(10)-Thujene

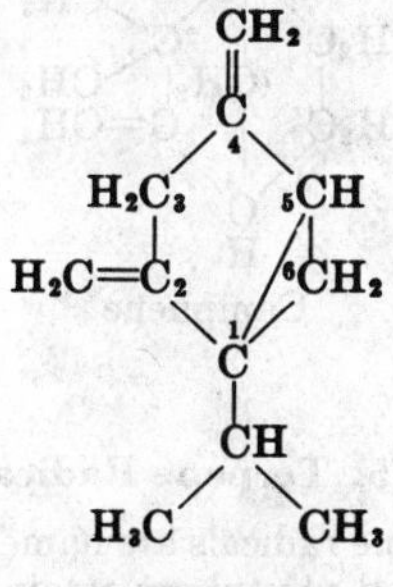

1-Isopropyl-2,4-dimethylenebicyclo-[3.1.0]hexane

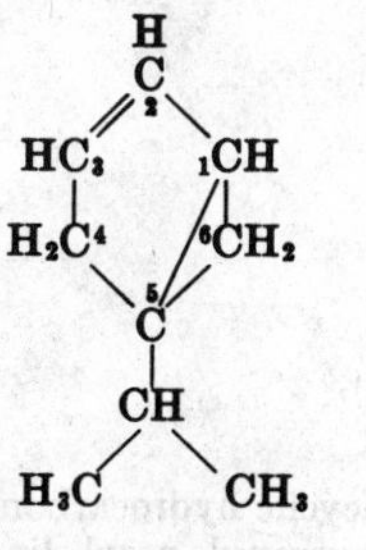

5-Isopropyl-bicyclo[3.1.0]hex-2-ene

74.2.—Bicyclic terpene hydrocarbons having the skeleton of Formula III, IV, or V and additional side chains except methyl (or methylene if one methylene group is already present) are named, respectively, as carane, carene, caradiene, *etc.;* pinane, pinene, pinadiene, *etc.;* bornane, bornene, bornadiene, *etc.* They are given, respectively, the fixed numbering shown for carane (Formula III), pinane (Formula IV), and bornane (Formula V). Other hydrocarbons containing the ring-skeleton of carane, pinane, or bornane are named, respectively, from norcarane (Formula VI), norpinane (Formula VII), or norbornane (Formula VIII). These names are preferred to those from bicyclo [4.1.0]heptane, bicyclo[3.1.1]heptane, or bicyclo [2.2.1]heptane. The nor-names are given systematic bicyclo numbering (*cf.* Rule **A-31**).

Examples:

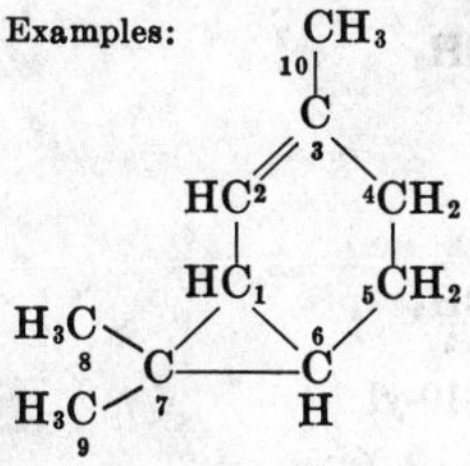

2-Carene

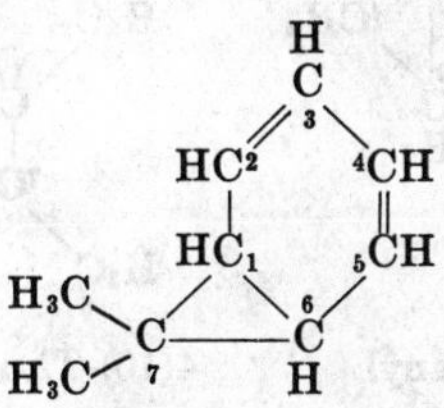

7,7-Dimethyl-2,4-norcaradiene

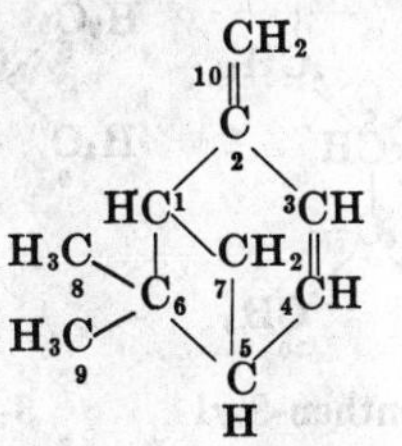

2(10),3-Pinadiene

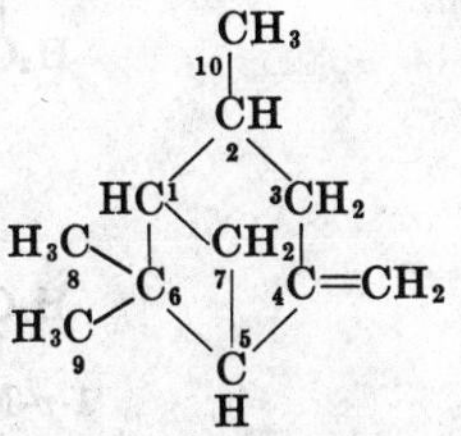

4-Methylenepinane

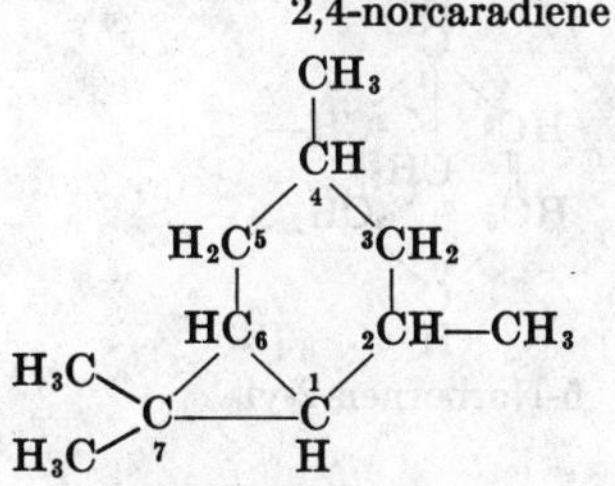

2,4,7,7-Tetramethylnorcarane

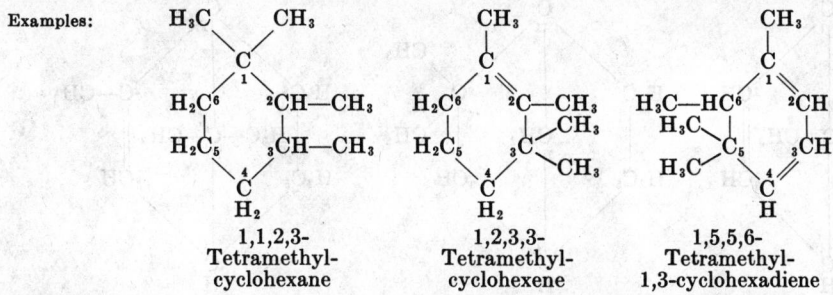

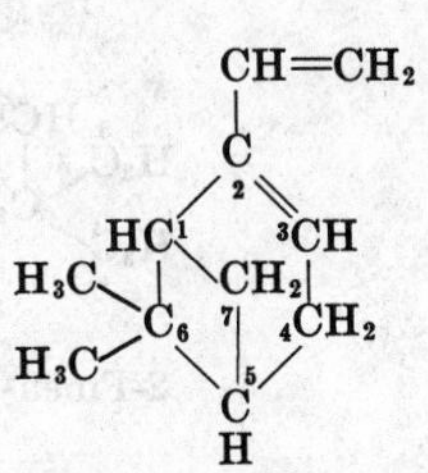

6,6-Dimethyl-2-vinyl-2-norpinene

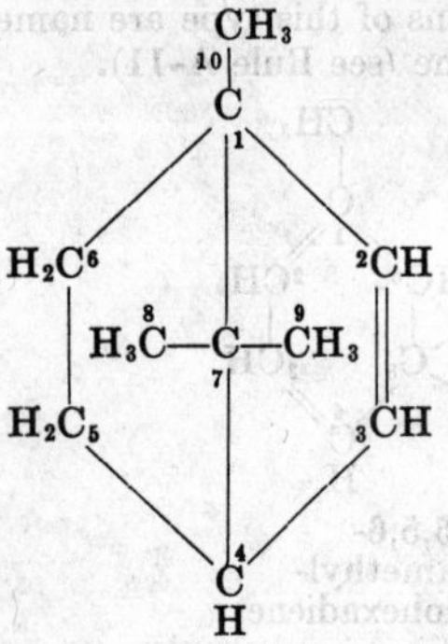

2-Bornene

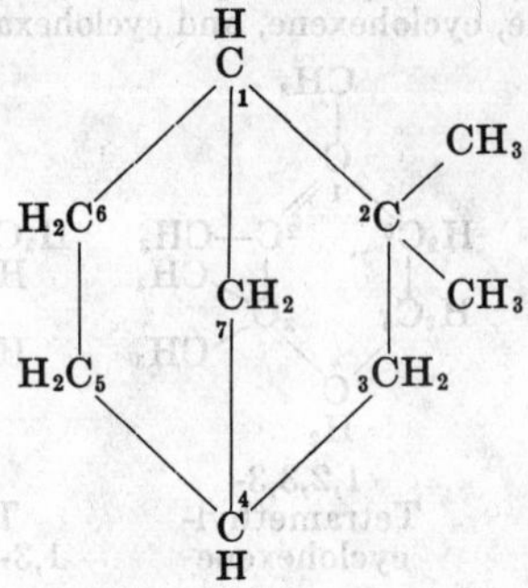

2,2-Dimethylnorbornane

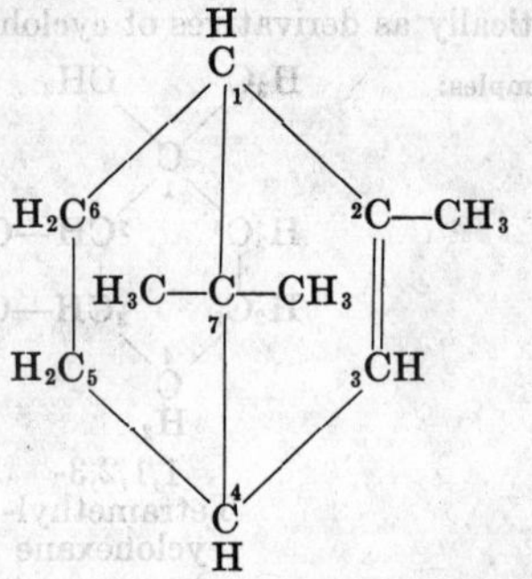

2,7,7-Trimethyl-2-norbornene

74.3.—The name "camphene" is retained for the unsubstituted compound 2,2-dimethyl-3-methylenenorbornane.

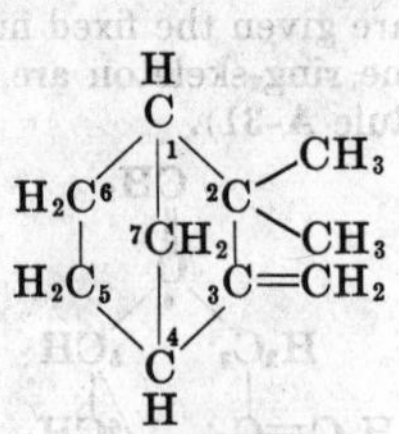

Camphene

A-75. Terpene Radicals

75.1.—Simple acyclic hydrocarbon terpene radicals are named and numbered according to Rule A-3.5. The trivial names geranyl, neryl, linalyl and phytyl are retained for the unsubstituted radicals.

75.2.—Radicals derived from menthane, pinane, thujane, carane, bornane, norcarane, norpinane, and norbornane are named in accordance with the principles set forth in Rules A-1.2 and A-11.4 except that the saturated radicals of pinane are named pinanyl, pinanylene, and pinanylidene. The numbering of the hydrocarbon is retained and the point or points of attachment, whether in the ring or side chain, are given numbers as low as is consistent with the fixed numbering of the hydrocarbon.

Examples:

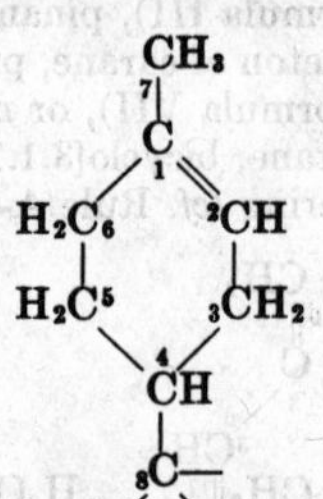

1-*p*-Menthen-8-yl

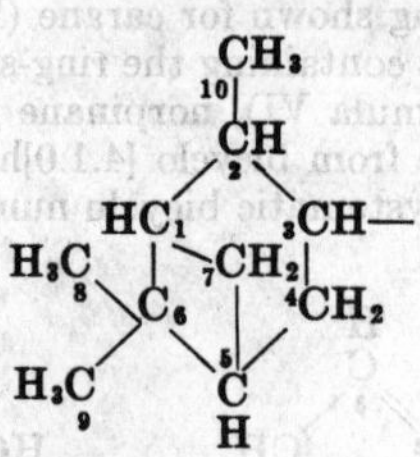

3-Pinanyl

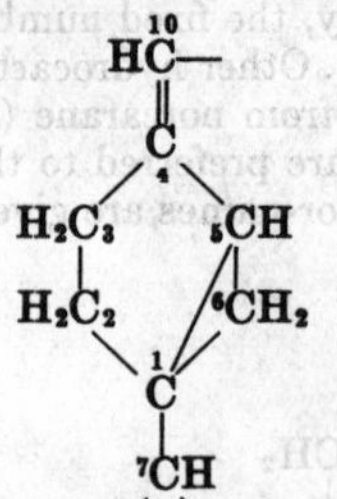

4(10)-Thujen-10-yl

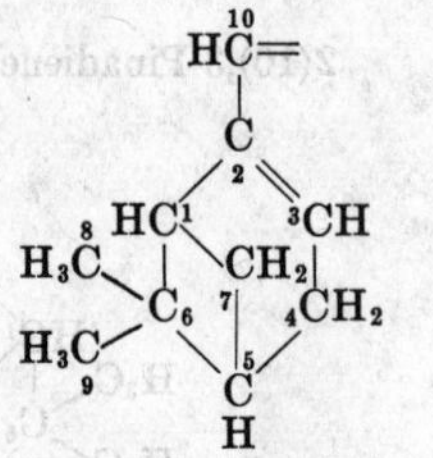

2-Pinen-10-ylidene

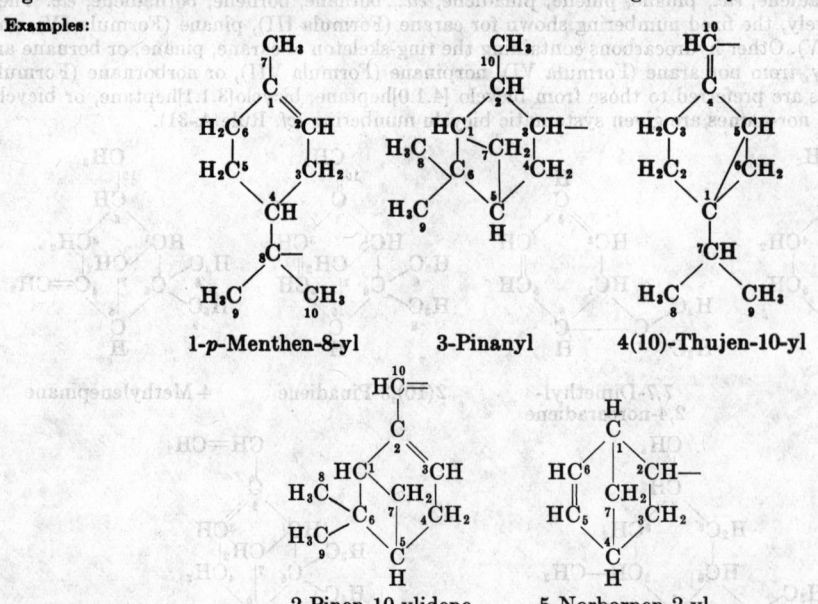

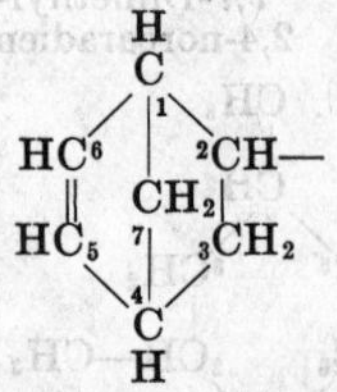

5-Norbornen-2-yl

75.3.—Radicals not named in Rules A-75.1 and A-75.2 are named as described in Rules A-11 and A-31.4.

DEFINITIVE RULES FOR NOMENCLATURE OF ORGANIC CHEMISTRY

B. FUNDAMENTAL HETEROCYCLIC SYSTEMS

B-1. Extension of Hantzsch-Widman System

1.1.—Monocyclic compounds containing one or more hetero atoms in a three- to ten-membered ring are named by combining the appropriate prefix or prefixes from Table I (eliding "a" where necessary) with a stem from Table II. The state of hydrogenation is indicated either in the stem, as shown in Table II, or by the prefixes "dihydro-," "tetrahydro-," *etc.*, according to Rule **B-1.2.**

TABLE I

Element	Valence	Prefix
Oxygen	II	Oxa
Sulfur	II	Thia
Selenium	II	Selena
Tellurium	II	Tellura
Nitrogen	III	Aza
Phosphorus	III	Phospha[a]
Arsenic	III	Arsa[a]
Antimony	III	Stiba[a]
Bismuth	III	Bisma
Silicon	IV	Sila
Germanium	IV	Germa
Tin	IV	Stanna
Lead	IV	Plumba
Mercury	II	Mercura

[a] When immediately followed by "in-" or "-ine," "phospha-" should be replaced by "phosphor-," "arsa-" should be replaced by "arsen-" and "stiba" should be replaced by "antimon-".

"Bora" for boron is used frequently. *Chemical Abstracts* has placed boron with the valence of III between lead and mercury in Table I. This table shows the valence of tin and lead as IV instead of II (a correction of a typographical error).

TABLE II

No. of members in the ring	Rings containing nitrogen: Unsaturation[a]	Rings containing nitrogen: Saturation	Rings containing no nitrogen: Unsaturation[a]	Rings containing no nitrogen: Saturation
3	-irine	-iridine	-irene	irane[e]
4	-ete	-etidine	-ete	-etane
5	-ole	-olidine	-ole	-olane
6	-ine[b]	[c]	-ine[b]	-ane[d]
7	-epine	[c]	-epine	-epane
8	-ocine	[c]	-ocin	-ocane
9	-onine	[c]	-onin	-onane
10	-ecine	[c]	-ecin	-ecane

[a] Corresponding to the maximum number of non-cumulative double bonds, the hetero elements having the normal valences shown in Table I. [b] For phosphorus, arsenic, antimony, see the special provisions of Table I. [c] Expressed by prefixing "perhydro" to the name of the corresponding unsaturated compound. [d] Not applicable to silicon, germanium, tin and lead. In this case, "perhydro-" is prefixed to the name of the corresponding unsaturated compound. [e] The syllables denoting the size of rings containing 3, 4 or 7–10 members are derived as follows: "ir" from *tri-*, "et" from *tetra*, "ep" from *hepta*, "oc" from *octa*, "on" from *nona*, and "ec" from *deca*.

Examples:

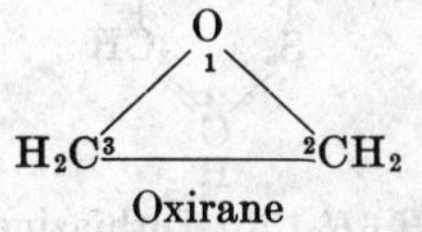

Oxirane

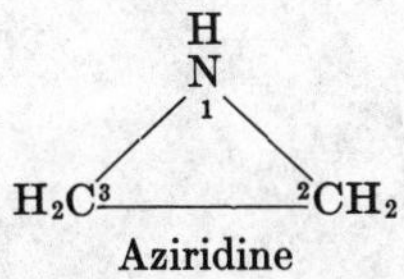

Aziridine

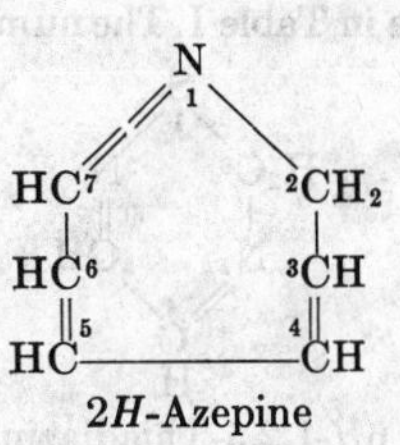

2*H*-Azepine

1.2.—Heterocyclic systems whose unsaturation is less than the one corresponding to the maximum number of noncumulative double bonds are named by using the prefixes "dihydro-," "tetrahydro-," etc. In the case of 4- and 5-membered rings, a special termination is used for the structures containing one double bond, when there can be more than one noncumulative double bond.

No. of members of the partly saturated rings	Rings containing nitrogen	Rings containing no nitrogen
4	-etine	-etene
5	-oline	-olene

Examples:

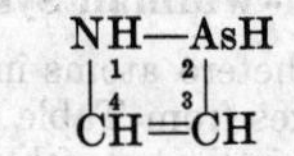

1,2-Azarset-3-ine

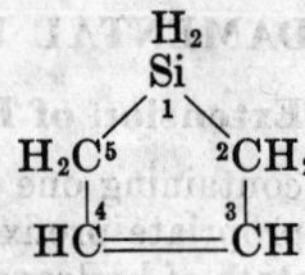

3-Silolene

1.3.—Multiplicity of the same hetero atom is indicated by a prefix "di," "tri-," etc., placed before the appropriate "a" term (Table I).

Example:

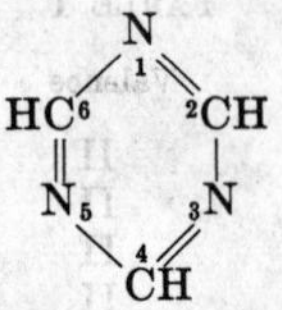

1,3,5-Triazine (or *s*-Triazine)

1.4.—If two or more kinds of "a" terms occur in the same name, their order of citation is by descending group number of the Periodic Table and increasing atomic number in the group as illustrated by the sequence in Table I.

Examples:

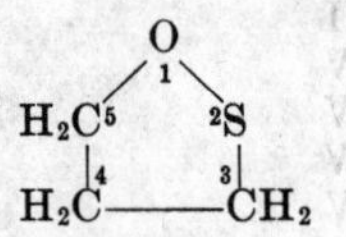

1,2-Oxathiolane

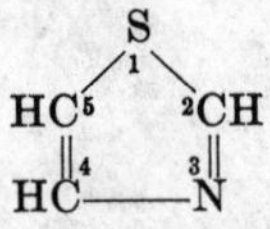

Thiazole

1.51.—The position of a single hetero atom determines the numbering in a monocyclic compound.

Example:

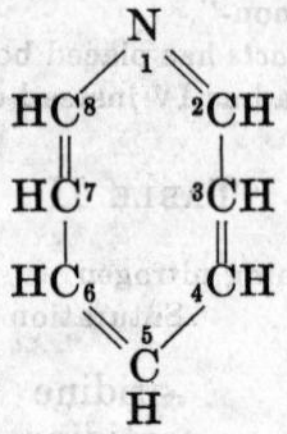

Azocine

1.52.—When the same hetero atom occurs more than once in a ring, the numbering is chosen to give the lowest locants to the hetero atoms.

Example:

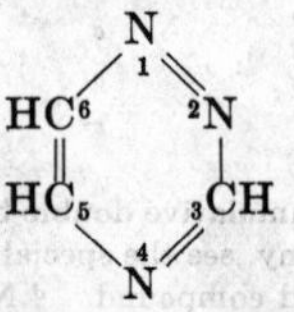

1,2,4-Triazine (or *as*-Triazine)

1.53.—When hetero atoms of different kinds are present, the locant 1 is given to a hetero atom which is as high as possible in Table I. The numbering is then chosen to give the lowest locants to the hetero atoms.

Examples:

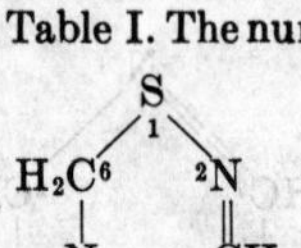

6*H*-1,2,5-Thiadiazine
(not 2,1,4-Thiadiazine)
(not 1,3,6-Thiadiazine)

The numbering must begin with the sulfur atom. This condition eliminates 2,1,4-thiadiazine. Then the nitrogen atoms receive the lowest possible locant, which eliminates 1,3,6-thiadiazine.

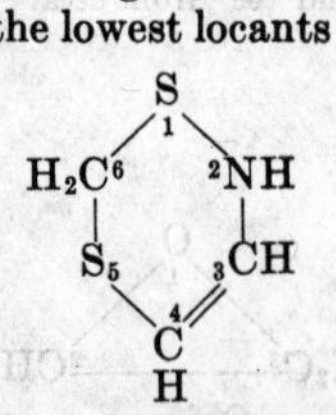

2*H*,6*H*-1,5,2-Dithiazine
(not 1,3,4-Dithiazine)
(not 1,3,6-Dithiazine)
(not 1,5,4-Dithiazine)

The numbering must begin with a sulfur atom. The choice of this atom is determined by the set of locants which can be attributed to the remaining hetero atoms of any kind. As the set 1,2,5 is lower than 1,3,4 or 1,3,6 or 1,5,4 in the usual sense, the name is 1,5,2-dithiazine.

B-2. Trivial and Semi-trivial Names

2.11.—The following trivial and semi-trivial names constitute a partial list of such names which are retained for the compound and as a basis of fusion names. They are arranged in the inverse order of the precedence prescribed in Rule **B-3.** The names of the radicals shown are formed according to Rule **B-5.**

	Parent Compound	Radical Name

(1)
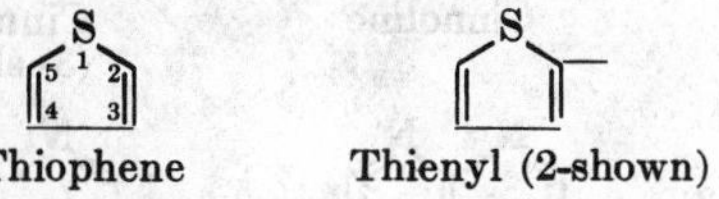
Thiophene — Thienyl (2-shown)

(2)
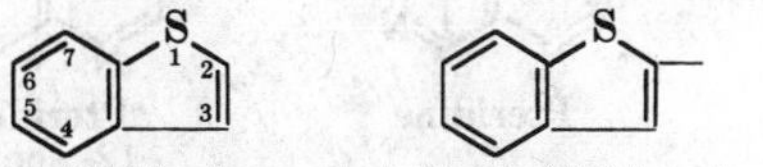
Benzo[*b*]thiophene (replacing thianaphthene) — Benzo[*b*]thienyl (2-shown)

(3)
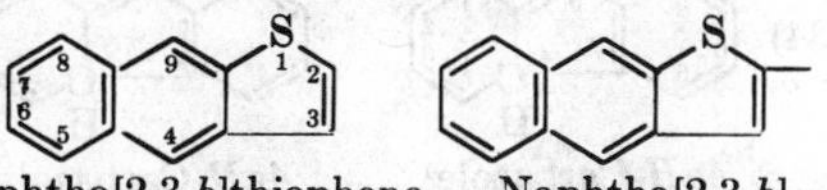
Naphtho[2,3-*b*]thiophene (replacing thiophanthrene) — Naphtho[2,3-*b*]thienyl (2-shown)

(4)
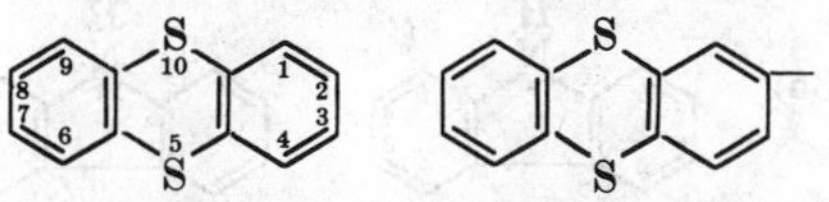
Thianthrene — Thianthrenyl (2-shown)

(5)

Furan — Furyl (3-shown)

(6)

Pyran (2*H*-shown) — Pyranyl (2*H*-Pyran-3-yl shown)

(7)

Isobenzofuran — Isobenzofuranyl (1-shown)

(8)
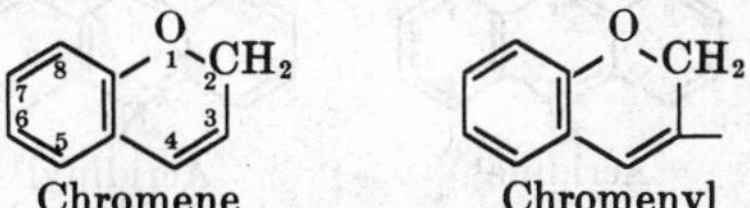
Chromene (2*H*- shown) — Chromenyl (2*H*-Chromen-3-yl shown)

(9)
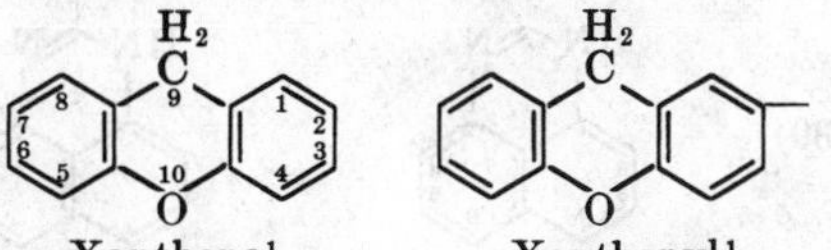
Xanthene[1] — Xanthenyl[1] (2- shown)

(10)
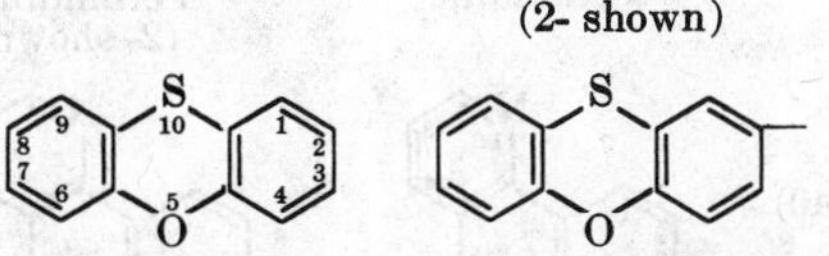
Phenoxathiin — Phenoxathiinyl (2- shown)

(11)

2*H*-Pyrrole — 2*H*-Pyrrolyl (2*H*-Pyrrol-3-yl shown)

	Parent Compound	Radical Name

(12)

Pyrrole — Pyrrolyl (3- shown)

(13)

Imidazole — Imidazolyl (2- shown)

(14)

Pyrazole — Pyrazolyl (1- shown)

(15)

Pyridine — Pyridyl (3- shown)

(16)

Pyrazine — Pyrazinyl

(17)

Pyrimidine — Pyrimidinyl (2-shown)

(18)

Pyridazine — Pyridazinyl (3- shown)

(19)

Indolizine — Indolizinyl (2-shown)

(20)

Isoindole — Isoindolyl (2- shown)

(21)

3*H*-Indole — 3*H*-Indolyl (3*H*-Indol-2-yl shown)

(1) Denotes exceptions to systematic numbering.

DEFINITIVE RULES FOR NOMENCLATURE OF ORGANIC CHEMISTRY

	Parent Compound	Radical Name
(22)	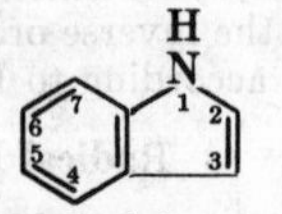 Indole	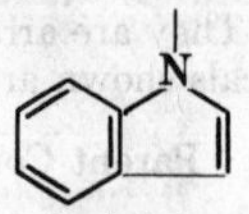 Indolyl (1- shown)
(23)	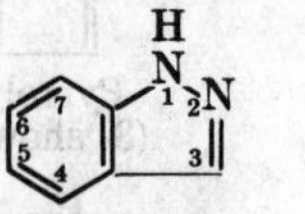 1*H*-Indazole	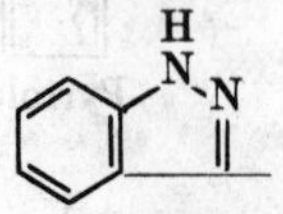 Indazolyl (1*H*-Indazol-3-yl shown)
(24)	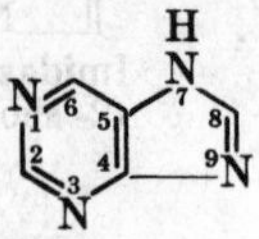 Purine[1]	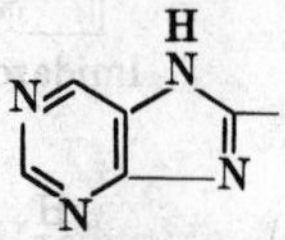 Purinyl[1] (8- shown)
(25)	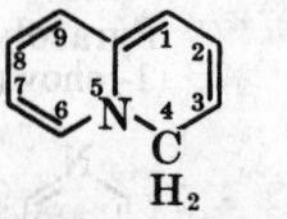 4*H*-Quinolizine	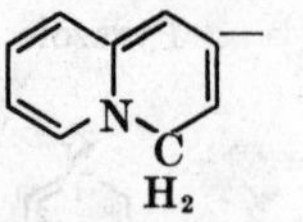 4*H*-Quinolizinyl (4*H*-Quinolizin-2-yl shown)
(26)	 Isoquinoline	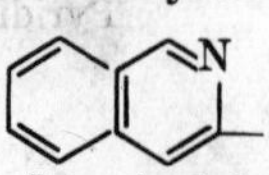 Isoquinolyl (3- shown)
(27)	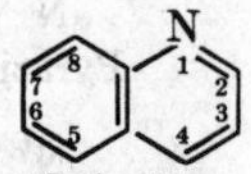 Quinoline	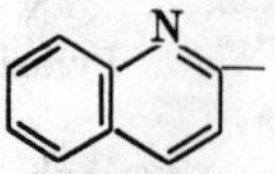 Quinolyl (2- shown)
(28)	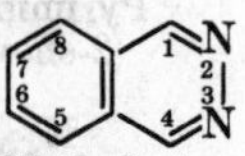 Phthalazine	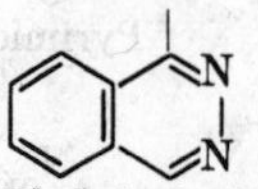 Phthalazinyl (1- shown)
(29)	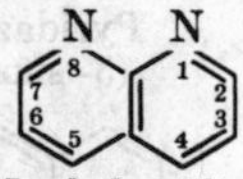 Naphthyridine (1,8- shown)	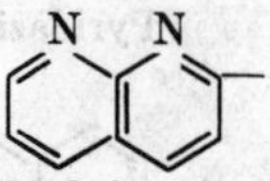 Naphthyridinyl (1,8-Naphthyridin-2-yl shown)
(30)	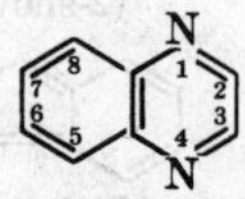 Quinoxaline	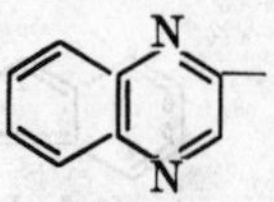 Quinoxalinyl (2- shown)
(31)	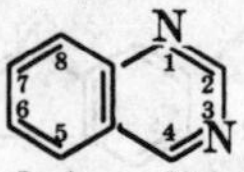 Quinazoline	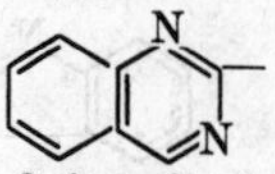 Quinazolinyl (2- shown)
(32)	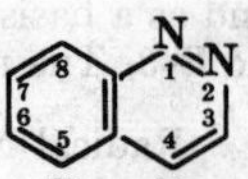 Cinnoline	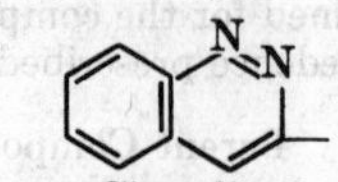 Cinnolinyl (3- shown)
(33)	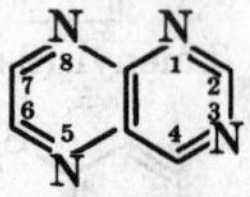 Pteridine	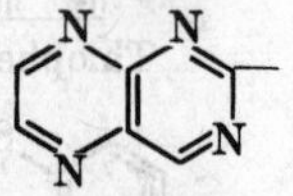 Pteridinyl (2- shown)
(34)	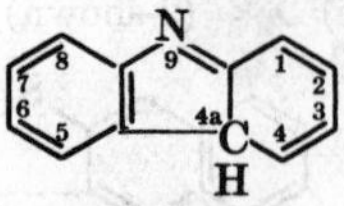 4a*H*-Carbazole[1]	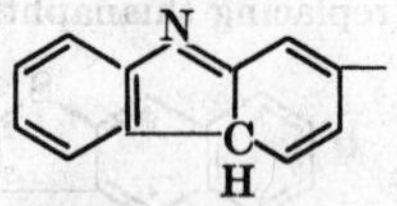 4a*H*-Carbazolyl[1] (4a*H*-Carbazol-2-yl shown)
(35)	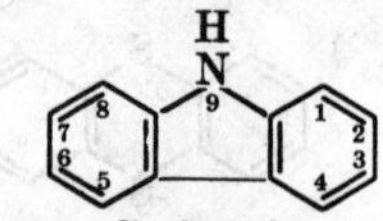 Carbazole[1]	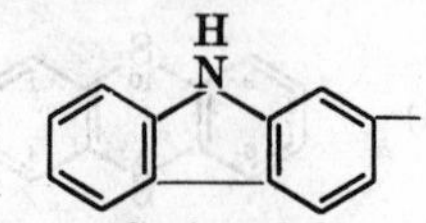 Carbazolyl[1] (2- shown)
(36)	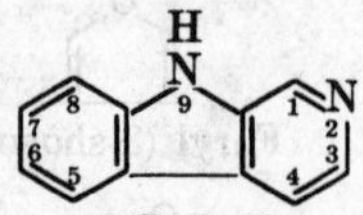 β-Carboline	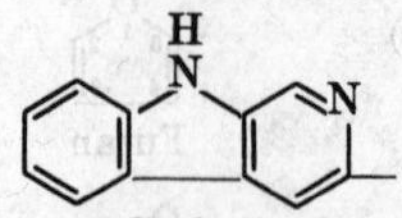 β-Carbolinyl (β-Carbolin-3-yl shown)
(37)	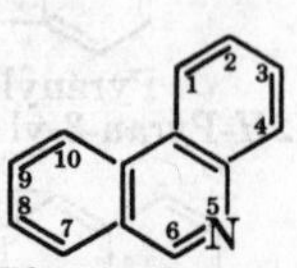 Phenanthridine	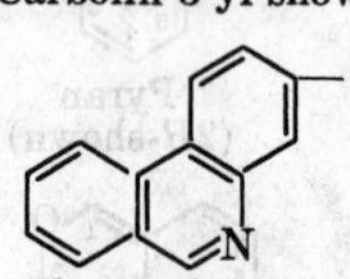 Phenanthridinyl (3- shown)
(38)	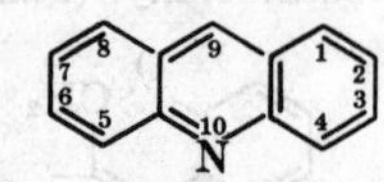 Acridine[1]	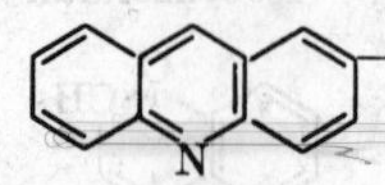 Acridinyl[1] (2- shown)
(39)	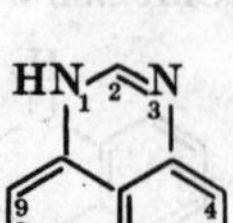 Perimidine	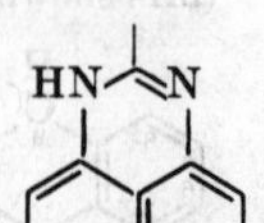 Perimidinyl (2- shown)
(40)	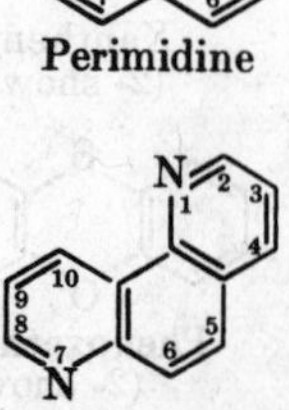 Phenanthroline (1,7- shown)	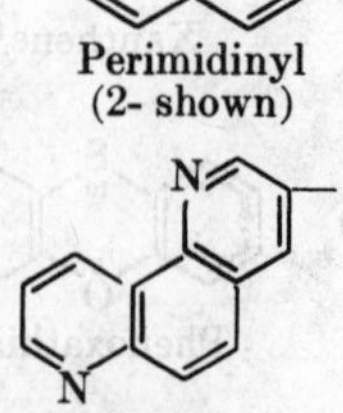 Phenanthrolinyl (1,7-Phenanthrolin-3-yl shown)

	Parent Compound	Radical Name
(41)	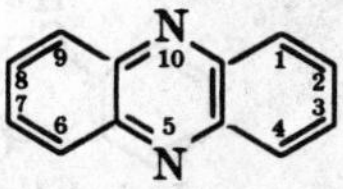 Phenazine	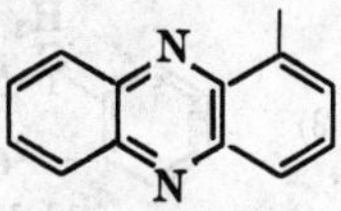 Phenazinyl (1- shown)
(42)	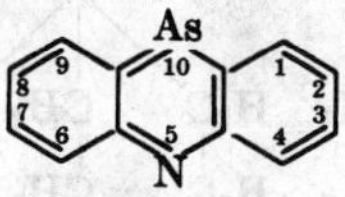 Phenarsazine	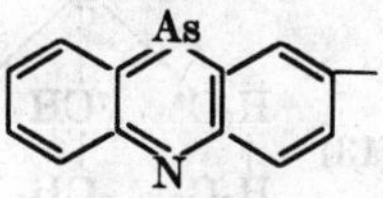 Phenarsazinyl (2- shown)
(43)	 Isothiazole	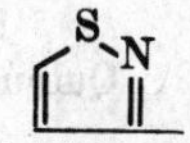 Isothiazolyl (3- shown)
(44)	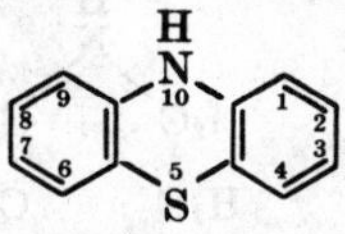 Phenothiazine	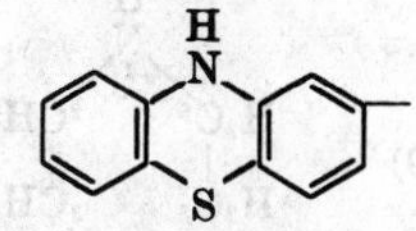 Phenothiazinyl (2- shown)
(45)	 Isoxazole	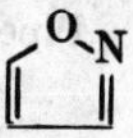 Isoxazolyl (3- shown)
(46)	 Furazan	 Furazanyl (3- shown)
(47)	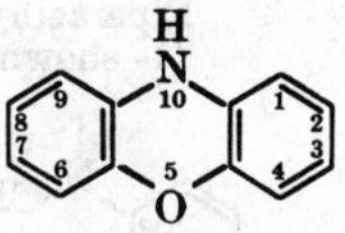 Phenoxazine	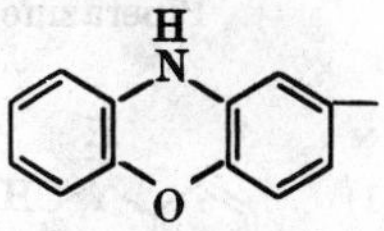 Phenoxazinyl (2- shown)

Chemical Abstracts **uses 2*H*-1-benzopyran instead of chromene, and 9*H*-pyrido[3,4-*b*]indole instead of β-carboline.**

B-2.12.—The following trivial and semi-trivial names are retained but are not recommended for use in fusion names. The names of the radicals shown are formed according to Rule B-5.

	Parent Compound	Radical Name
(1)	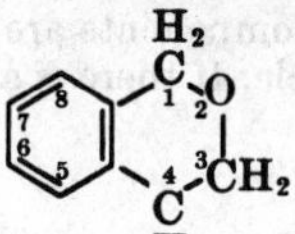 Isochroman	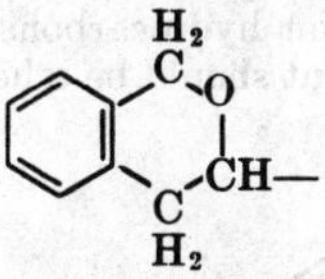 Isochromanyl (3- shown)
(2)	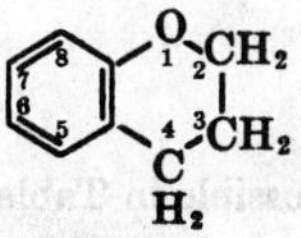 Chroman	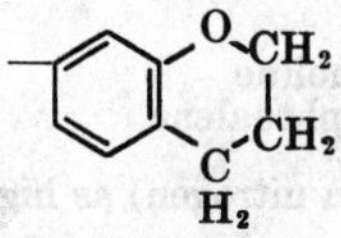 Chromanyl (7- shown)
(3)	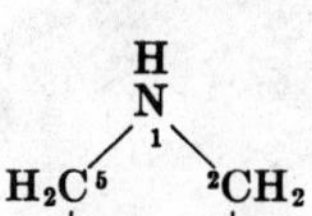 Pyrrolidine	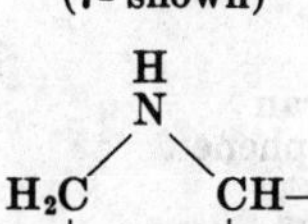 Pyrrolidinyl (2- shown)
(4)	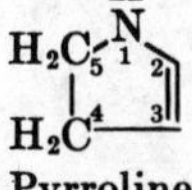 Pyrroline (2- shown[1])	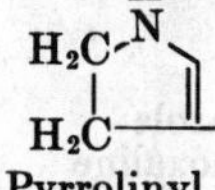 Pyrrolinyl (2-Pyrrolin-3-yl[1] shown)
(5)	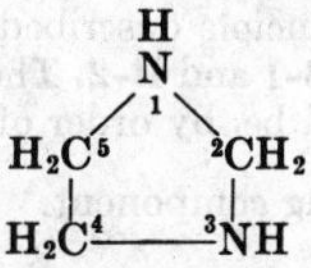 Imidazolidine	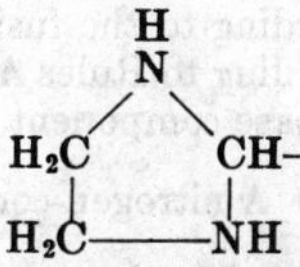 Imidazolidinyl (2- shown)
(6)	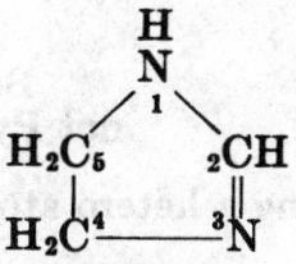 Imidazoline (2- shown[1])	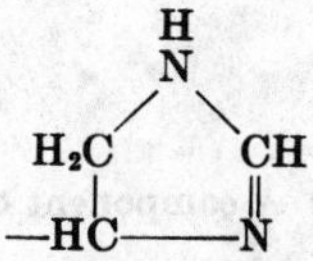 Imidazolinyl (2-Imidazolin-4-yl[1] shown)
(7)	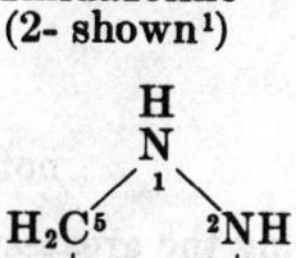 Pyrazolidine	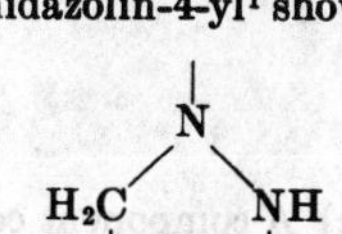 Pyrazolidinyl (1- shown)
(8)	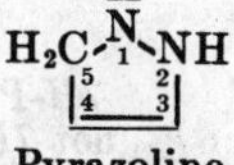 Pyrazoline (3- shown[2])	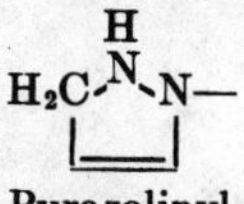 Pyrazolinyl (3-Pyrazolin-2-yl[2] shown)

(1) The "2-" denotes the position of the double bond.
(2) The "3-" denotes the position of the double bond.
(3) For 1-piperidyl use piperidino.

	Parent Compound	Radical Name		Parent Compound	Radical Name
(9)	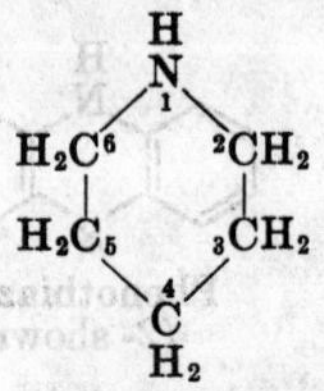 Piperidine	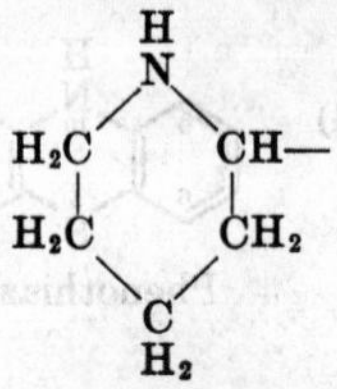 Piperidyl[3] (2- shown)	(12)	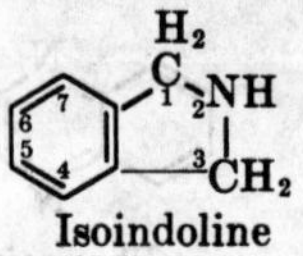 Isoindoline	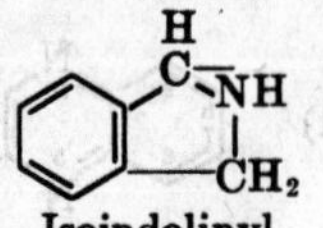 Isoindolinyl (1- shown)
(10)	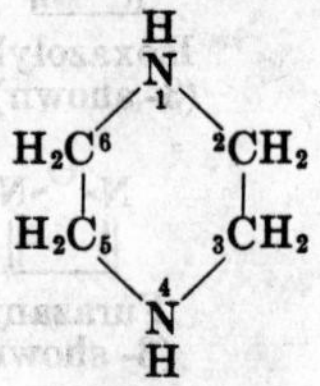 Piperazine	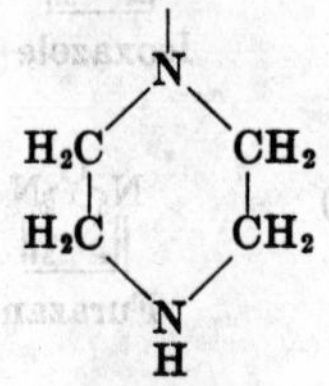 Piperazinyl (1- shown)	(13)	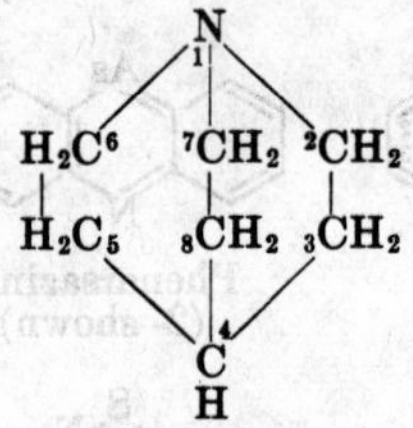 Quinuclidine	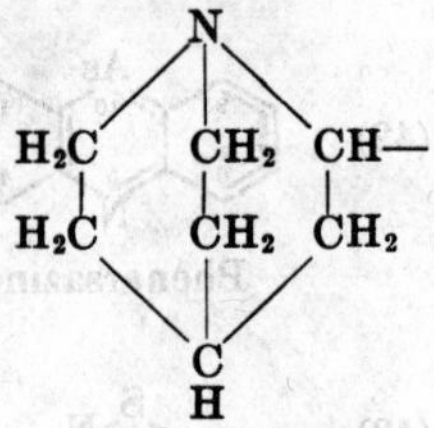 Quinuclidinyl (2- shown)
(11)	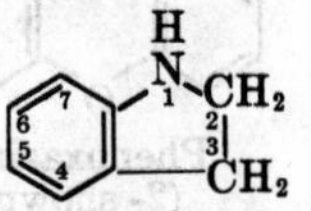 Indoline	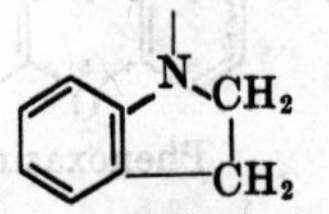 Indolinyl (1- shown)	(14)	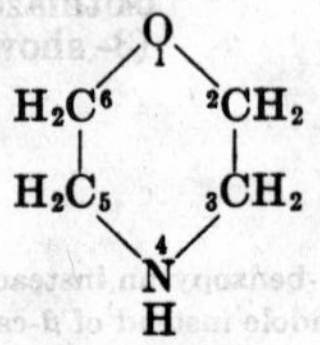 Morpholine	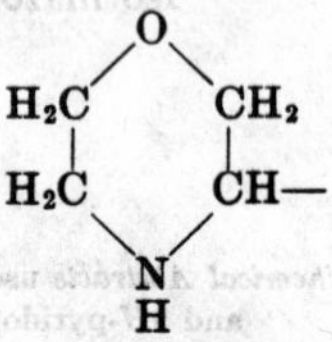 Morpholinyl[1] (3- shown)

B-3. Fused Heterocyclic Systems

3.1.—"Ortho-fused" and "ortho- and peri-fused" ring compounds containing hetero atoms are named according to the fusion principle described in Rule **A-21** for hydrocarbons. The components are named according to Rules **A-21, B-1** and **B-2.** The base component should be a heterocycle. If there is a choice, the base component should be, by order of preference:

(a) A nitrogen-containing component.

Example:

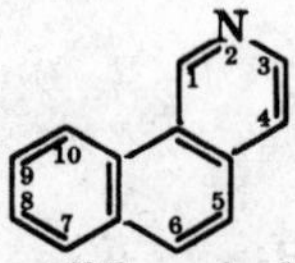

Benzo[*h*]isoquinoline
not Pyrido[3,4-*a*]naphthalene

(b) A component containing a hetero atom (other than nitrogen) as high as possible in Table I.

Example:

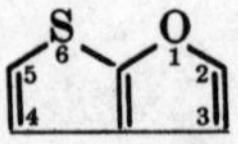

Thieno[2,3-*b*]furan
not Furo[2,3-*b*]thiophene

(c) A component containing the greatest number of rings.

Example:

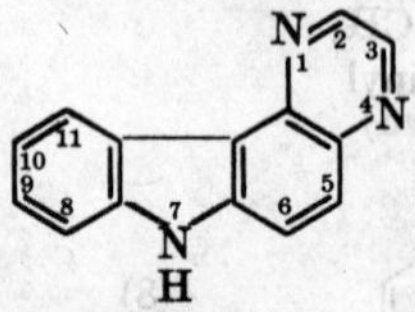

7*H*-Pyrazino[2,3-*c*]carbazole
not 7*H*-Indolo[3,2-*f*]quinoxaline

(d) A component containing the largest possible individual ring.

Example:

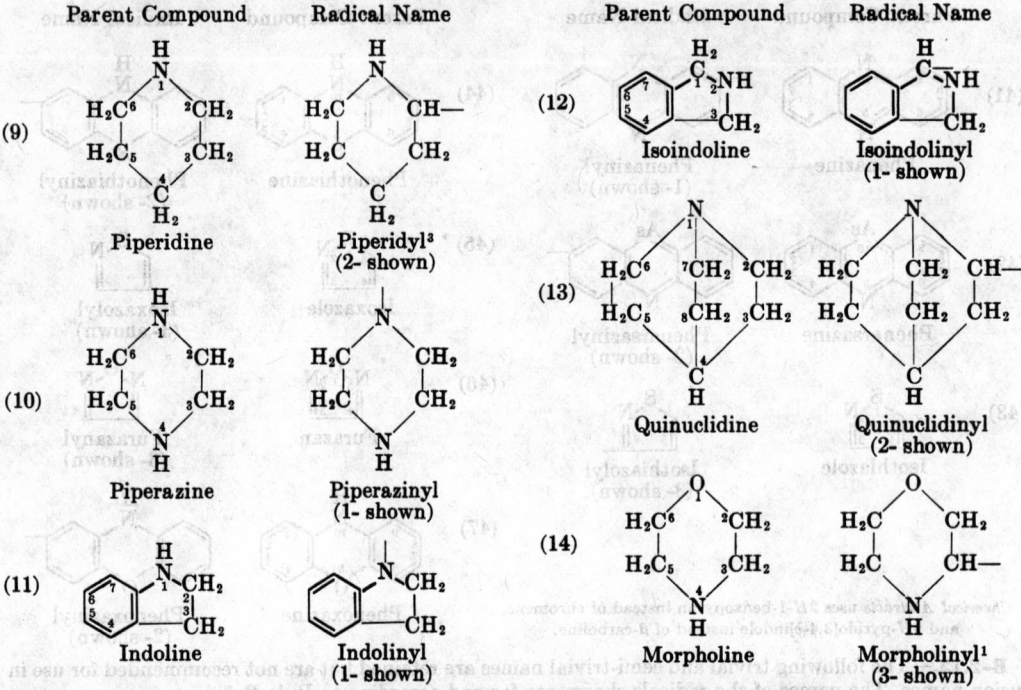

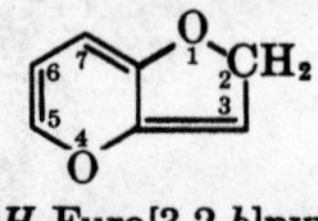

2*H*-Furo[3,2-*b*]pyran
not 2*H*-Pyrano[3,2-*b*]furan

(1) For 4-morpholinyl use morpholino.

(e) A component containing the greatest number of hetero atoms of any kind.

Example:

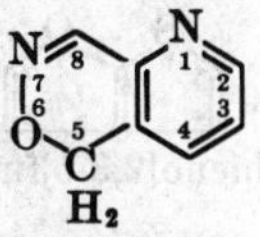

5*H*-Pyrido[2,3-*d*][1,2]oxazine
not [1,2]Oxazino[4,5-*b*]pyridine

(f) A component containing the greatest variety of hetero atoms.

Examples:

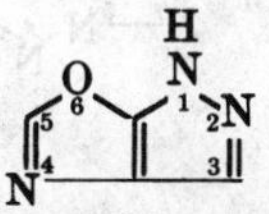

1*H*-Pyrazolo[4,3-*d*]oxazole
not 1*H*-Oxazolo[5,4-*c*]-pyrazole

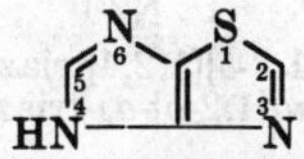

4*H*-Imidazo[4,5-*d*]thiazole
not 4*H*-Thiazolo[4,5-*d*]-imidazole

(g) A component containing the greatest number of hetero atoms first listed in Table I.

Example:

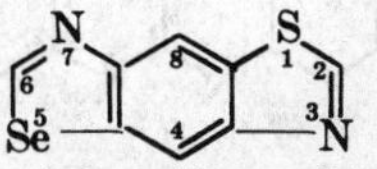

Selenazolo[5,4-*f*]benzothiazole[1]
not Thiazolo[5,4-*f*]benzoselenazole

(1) In this example the hetero atom first listed in Table I is sulfur and the greatest number of sulfur atoms in a ring is one.

(h) If there is a choice between components of the same size containing the same number and kind of hetero atoms choose as the base component that one with the lower numbers for the hetero atoms before fusion.

Example:

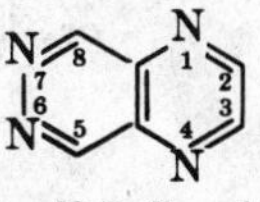

Pyrazino[2,3-*d*]pyridazine

3.2.—If a position of fusion is occupied by a hetero atom, the names of the component rings to be fused are so chosen as both to contain the hetero atom.

Example:

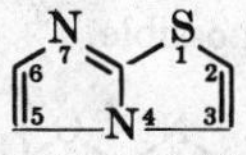

Imidazo[2,1-*b*]thiazole

3.3.—These contracted fusion prefixes may be used: furo, imidazo, isoquino, pyrido, pyrimido, quino and thieno.

Examples:

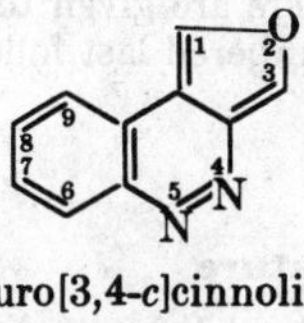

Furo[3,4-*c*]cinnoline

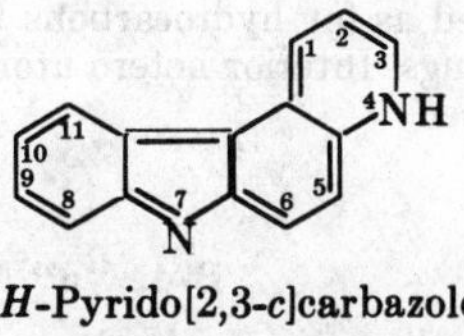

4*H*-Pyrido[2,3-*c*]carbazole

3.4.—In peripheral numbering of the complete fused systems, the ring system is oriented and numbered according to the principles of Rule A-22. When there is a choice of orientations, it is made in the following sequence in order to:

(a) Give low numbers to hetero atoms.

Examples:

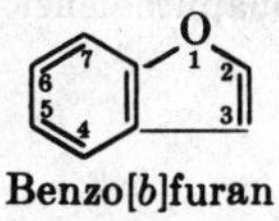

Benzo[*b*]furan

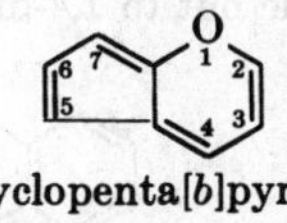

Cyclopenta[*b*]pyran

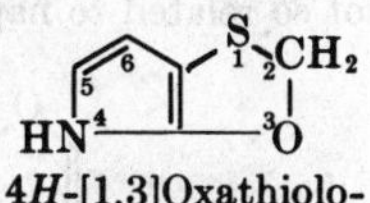

4*H*-[1,3]Oxathiolo-[5,4-*b*]pyrrole (N.B. 1,3,4 lower than 1,3,6)

(b) Give low numbers to hetero atoms in order of Table I.

Example:

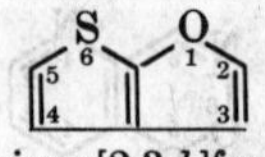

Thieno[2,3-*b*]furan

(c) Allow carbon atoms common to two or more rings to follow the lowest possible numbers (see Rules **A-22.2** and **A-22.3**). [A hetero atom common to the two rings is numbered according to Rule **B-3.4**(e).]

Examples:

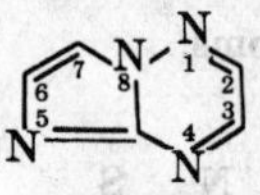 not 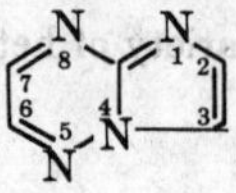 or 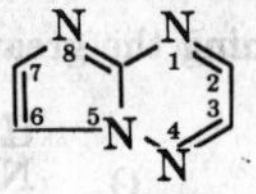

Imidazo[1,2-*b*][1,2,4]triazine
(or Imidazo[1,2-*b*]-*as*-triazine)

In a compound name for a fusion prefix (*i.e.*, when more than one pair of brackets is required), the points of fusion in the compound prefix are indicated by the use of unprimed and primed numbers, the unprimed numbers being assigned to the ring attached directly to the base component, thus

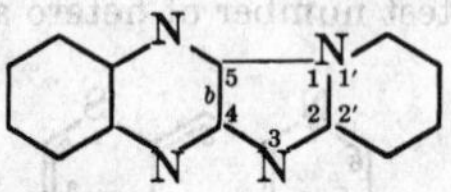

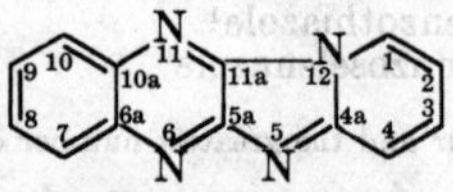 not 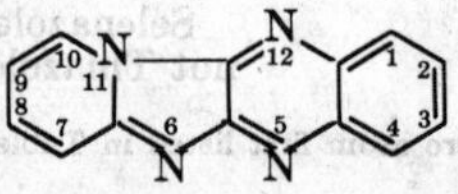

Pyrido[1′,2′:1,2]imidazo-
[4,5-*b*]quinoxaline

or 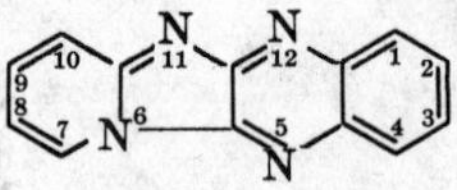

or 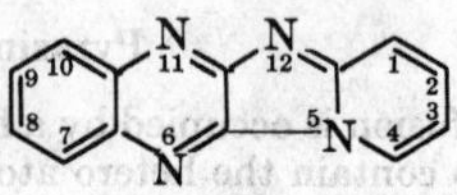

(d) Give hydrogen atoms lowest numbers possible.

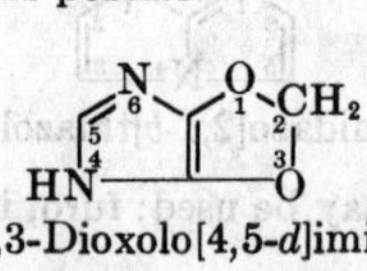

4*H*-1,3-Dioxolo[4,5-*d*]imidazole

(e) The ring is numbered as for hydrocarbons but numbers are given to all hetero atoms even when common to two or more rings. Interior hetero atoms are numbered last following the shortest path from the highest previous number.

B-4. "a" Nomenclature

4.1.—Names of heterocyclic compounds also may be formed by prefixing "a" terms (see Table I of Rule **B-1.1**) to the name of the corresponding homocyclic compound. The letter "a" should not be elided. There are two methods of applying this principle:

4.1(a). Stelzner Method.—In this method, the "a" term name relates to that of the hydrocarbon with the same distribution of bonds in the rings. Thus, I is not so related to benzene but to 1,4-cyclohexadiene, and II is not so related to naphthalene but to 1,4-dihydronaphthalene.

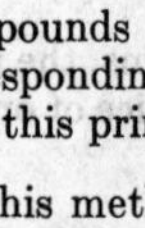

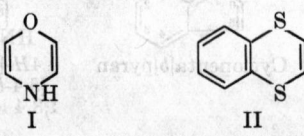

I

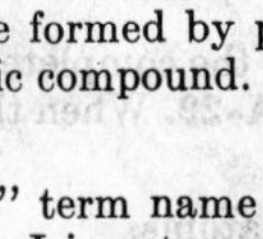

II

4.1(b). Chemical Abstracts Method.—If the corresponding homocyclic compound is partially or completely hydrogenated and if this state of hydrogenation is denoted in its name without the use of hydro prefixes, as indan and cyclohexane, the procedure is the same as in (a). In other cases, positions in the skeleton of the corresponding homocyclic compound which are occupied by hetero atoms are denoted by the "a" terms, and the parent heterocyclic compound is considered to be that which contains the maximum number of conjugated or isolated[1] double bonds; hydrogen is added, as necessary, as hydro prefixes and/or as *H* to the "a" name thus obtained.

Examples:

	Stelzner Method	*Chemical Abstracts* Method
	Sila-2,4-cyclopentadiene	Sila-2,4-cyclopentadiene
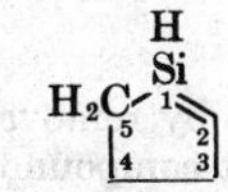	Sila-1,3-cyclopentadiene	Sila-1,3-cyclopentadiene
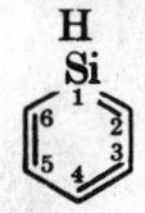	Silabenzene	Silabenzene
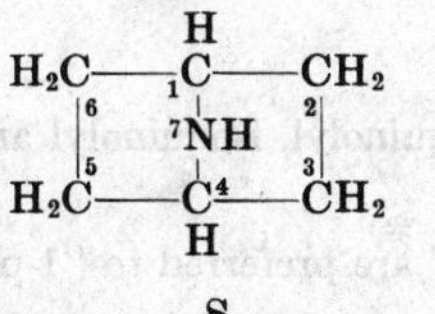	7-Azabicyclo-[2.2.1]heptane	7-Azabicyclo-[2.2.1]heptane
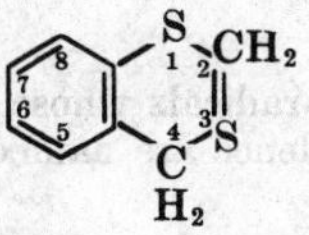	1,3-Dithia-1,2,3,4-tetrahydronaphthalene	4*H*-1,3-Dithianaphthalene
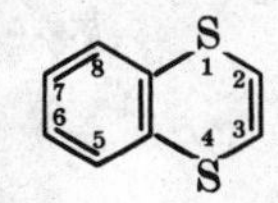	1,4-Dithia-1,4-dihydronaphthalene	1,4-Dithianaphthalene

(1) Isolated double bonds are those which are neither conjugated nor cumulative as in

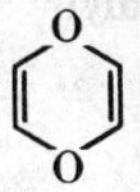

or the B ring of

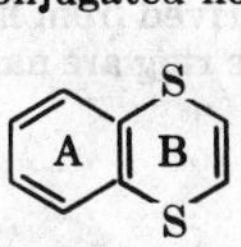

	Stelzner Method	*Chemical Abstracts* Method
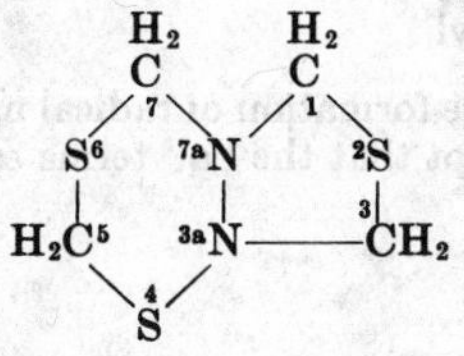	2,4,6-Trithia-3a,7a-diazaperhydroindene	2,4,6-Trithia-3a,7a-diazaindene
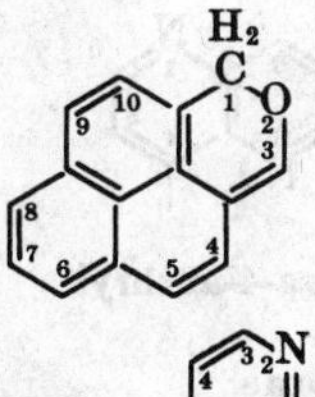	2-Oxa-1,2-dihydropyrene	1*H*-2-Oxapyrene
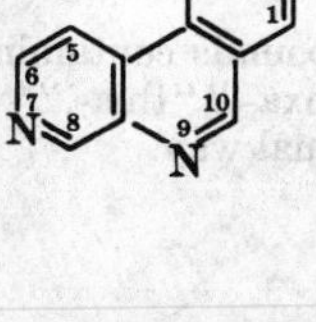	2,7,9-Triazaphenanthrene	2,7,9-Triazaphenanthrene

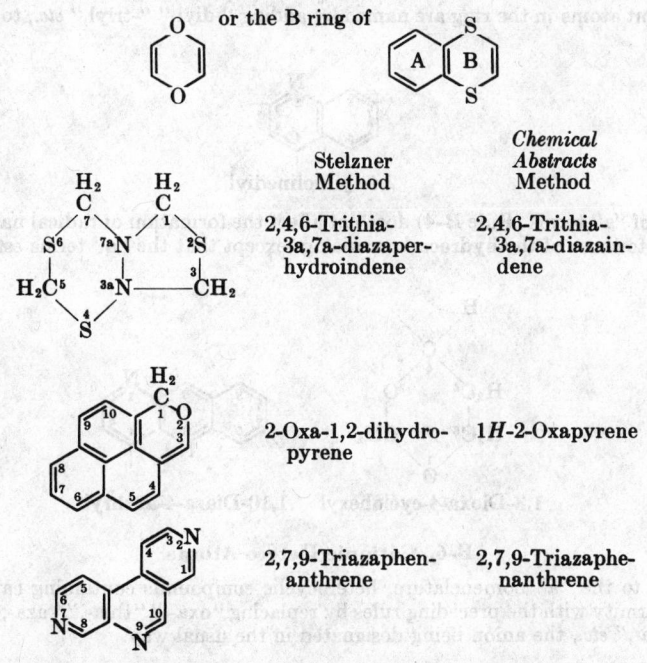

4.2.—In fusion names, the "a" terms precede the completed name of the parent hydrocarbon. If two or more kinds of "a" terms occur in the same name, the procedure described in Rule **B-1.4** applies. Prefixes denoting ordinary substitution procede the "a" terms.

Example:

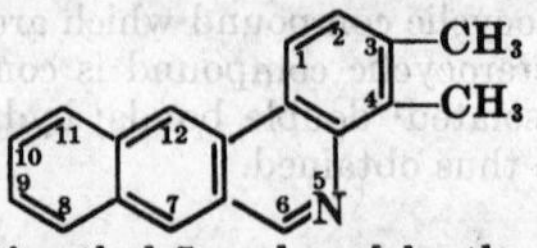

3,4-Dimethyl-5-azabenz[a]anthracene

Chemical Abstracts follows Rule **B-4.1** (b). Since "a" names for heterocyclic systems are based on the name of the carbocyclic system, the numbering of the "a" name must agree with the numbering of the carbocyclic parent. When there is a choice in direction of numbering the ring is numbered so as to give (1) lowest numbers to the hetero atoms, (2) lowest numbers to hetero atoms as high as possible in Table 1 (Rule **B-1.1**).

B-5. Radicals

5.11.—Univalent radicals derived from heterocyclic compounds by removal of hydrogen from a ring are in principle named by adding "yl" to the names of the parent compounds (with elision of final "e" if present).

Examples:

Indolyl from indole
Pyrrolinyl from pyrroline
Triazolyl from triazole
Triazinyl from triazine

(For further examples see Rule **B-2.11**).

These exceptions are retained: furyl, pyridyl, piperidyl, quinolyl, isoquinolyl and thienyl (from thiophene) (see also Rule **B-2.12**).

As exceptions, the names "piperidino" and "morpholino" are preferred to "1-piperidyl" and "4-morpholinyl."

5.12.—Bivalent radicals derived from univalent heterocyclic radicals whose names end in "-yl" by removal of one hydrogen atom from the atom with the free valence are named by adding "-idene" to to the name of the corresponding univalent radical.

Example:

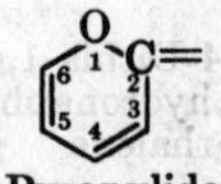

2-Pyranylidene

5.13.—Multivalent radicals derived from heterocyclic compounds by removal of two or more hydrogen atoms from different atoms in the ring are named by adding "-diyl," "-triyl," *etc.*, to the name of the ring system.

Example:

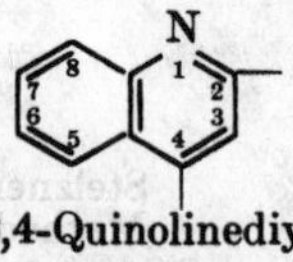

2,4-Quinolinediyl

5.21.—The use of "a" terms (Rule B-4) does not affect the formation of radical names. Such names are strictly analogous to those of the hydrocarbon analogs except that the "a" terms establish numbering in whole or in part.

Examples:

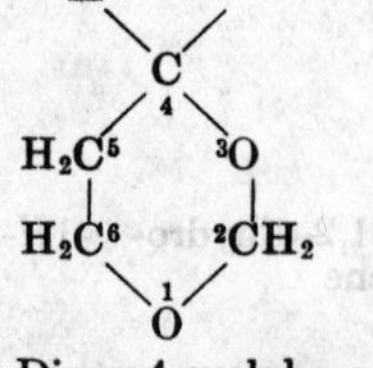

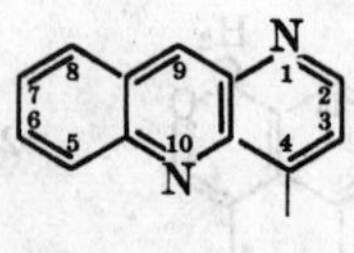

1,3-Dioxa-4-cyclohexyl 1,10-Diaza-4-anthryl

B-6. Cationic Hetero Atoms

6.1.—According to the "a" nomenclature, heterocyclic compounds containing cationic hetero atoms are named in conformity with the preceding rules by replacing "oxa-," "thia-," "aza-," *etc.*, by "oxonia-," "thionia-," "azonia-," *etc.*, the anion being designated in the usual way.

Examples:

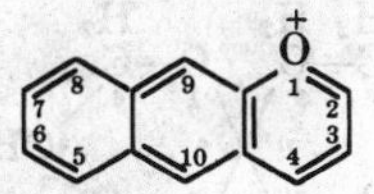
Cl⁻ 1-Oxoniaanthracene chloride

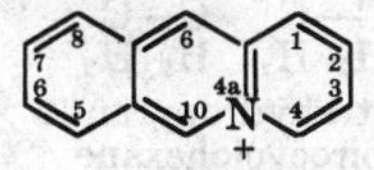
Cl⁻ 4a-Azoniaanthracene chloride

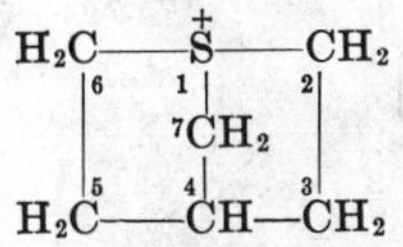
Cl⁻ 1-Thioniabicyclo[2.2.1]heptane chloride

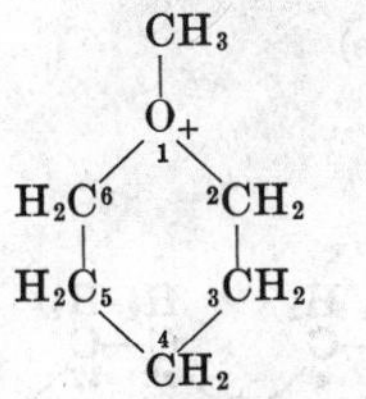
Cl⁻ 1-Methyl-1-oxoniacyclohexane chloride

HETEROCYCLIC SPIRO COMPOUNDS

B-10 (Alternate to B-11)

10.1.—Heterocyclic spiro compounds containing single-ring units only may be named by prefixing "a" terms (see Table I, Rule **B-1.1**) to the names of the spiro hydrocarbons formed according to Rules **A-41.1, A-41.2, A-41.3** and **A-41.6.** The numbering of the spiro hydrocarbon is retained and the hetero atoms in the order of Table I are given as low numbers as are consistent with the fixed numbering of the ring. When there is a choice, hetero atoms are given lower numbers than double bonds.

Examples:

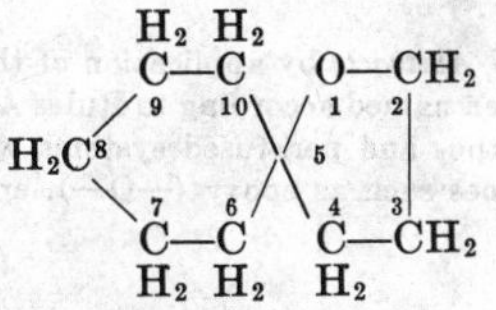
1-Oxaspiro[4.5]decane

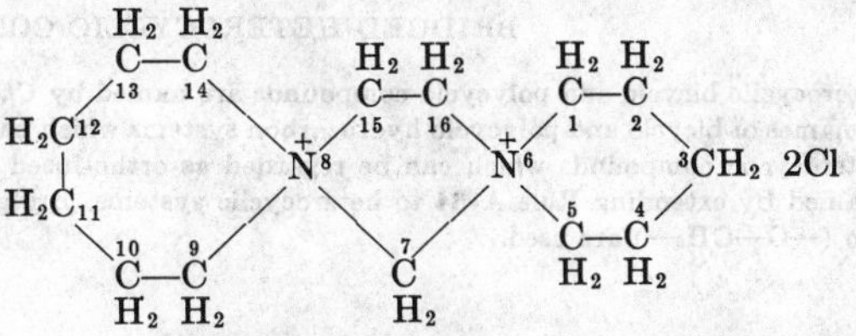
6,8-Diazoniadispiro[5.1.6.2]hexadecane dichloride

10.2.—If at least one component of a mono- or polyspiro compound is a fused polycyclic system, the spiro compound is named according to Rule **A-41.4** or **A-41.7,** giving the spiro atom as low a number as possible consistent with the fixed numberings of the component systems.

Examples:

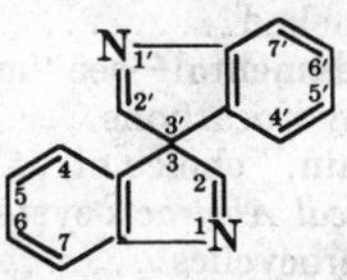
3,3′-Spirobi[3*H*-indole]

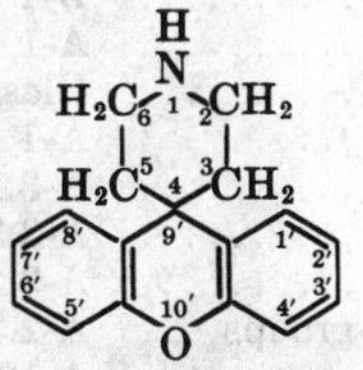
Spiro[piperidine-4,9′-xanthene]

Chemical Abstracts names heterocyclic spiro compounds according to Rules **B-10.1** and **B-10.2.**

B-11 (Alternate to B-10)

11.1.—Heterocyclic spiro compounds are named according to Rule **A-42,** the following criteria being applied where necessary: (a) spiro atoms have numbers as low as consistent with the numbering of the individual component systems; (b) heterocyclic components have priority over homocyclic components of the same size; (c) priority of heterocyclic components is decided according to Rule **B-3.** Parentheses are used where necessary for clarity in complex expressions.

Examples:

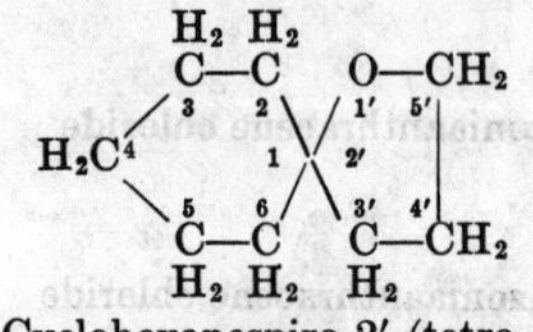

Cyclohexanespiro-2′-(tetrahydrofuran)

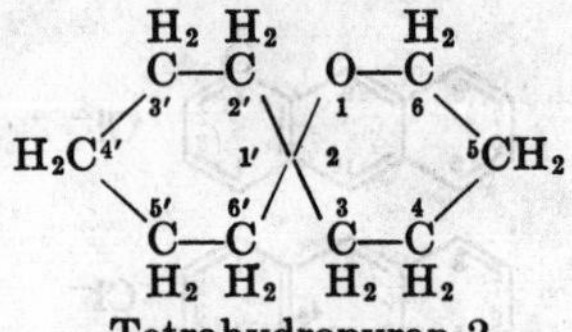

Tetrahydropyran-2-spirocyclohexane

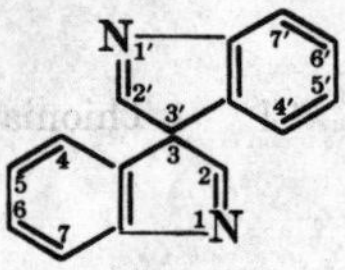

3,3′-Spirobi(3*H*-indole)

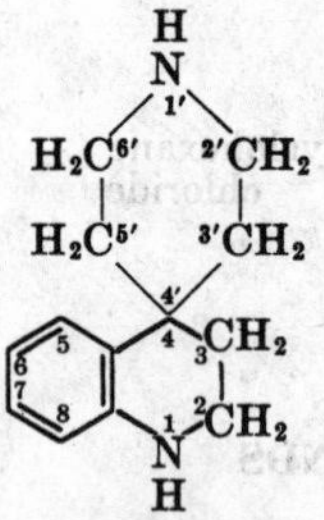

1,2,3,4-Tetrahydroquinoline-4-spiro-4′-piperidine

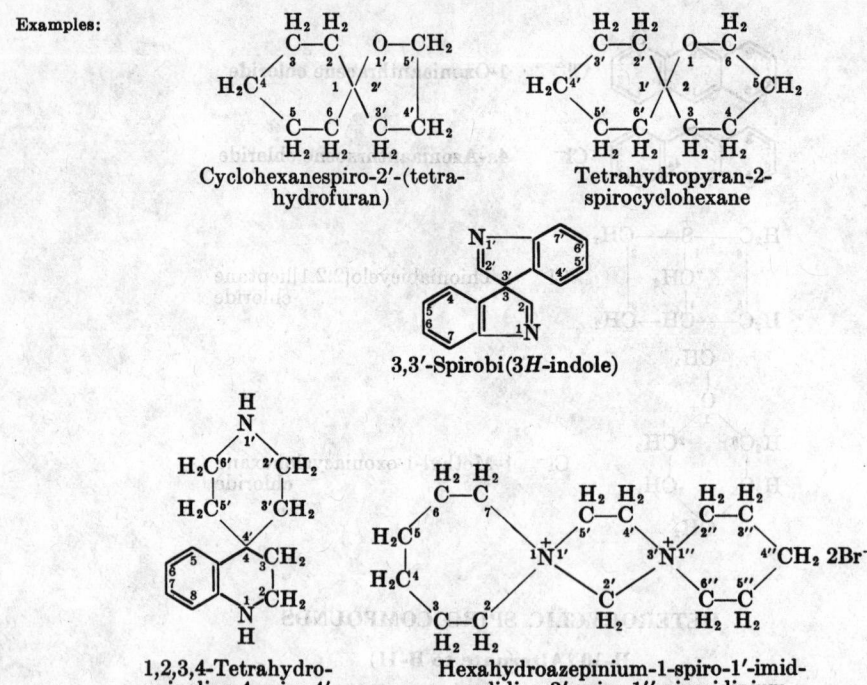

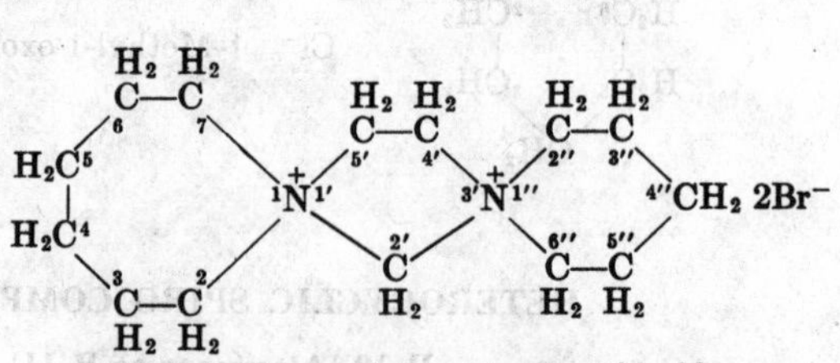

Hexahydroazepinium-1-spiro-1′-imidazolidine-3′-spiro-1″-piperidinium dibromide

Chemical Abstract follows Rule **B-10.**

BRIDGED HETEROCYCLIC COMPOUNDS

Bridged heterocyclic bicyclo and polycyclo compounds are named by *Chemical Abstracts* by application of the "a" nomenclature to the names of bicyclo and polycyclo hydrocarbon systems which have been named according to Rules **A-31** and **A-32.**

Bridged heterocyclic compounds which can be regarded as ortho-fused or ortho- and peri-fused systems with additional bridges are named by extending Rule **A-34** to heterocyclic systems. Bridge names such as epoxy (—O—), epithio (—S—), epoxymethano (—O—CH_2—) are used.

INDEX

ILLUSTRATIVE LIST OF SUBSTITUENT PREFIXES

Groups always expressed as substituents now include isocyano, isocyanató, isothiocyanato, etc., and groups terminating in oxy, thio, sulfinyl, sulfonyl, and their analogs, such as seleno and telluronyl.

Changes in radical names include iodosyl (instead of iodoso) for –IO; and iodyl (instead of iodoxy) for $-IO_2$. Morpholino has been changed to 4-morpholinyl; piperidino to 1-piperidinyl; the unsubstituted diazeno radical, HN=N–, to diazenyl; *p*-phenylene, etc., to 1,4-phenylene, etc.; *v*-phenenyl to 1,2,3-benzenetriyl; benzyl to (phenylmethyl); styryl to (2-phenylethenyl); *p*-tolyl to (4-methylphenyl); 2,4-xylyl to (2,4-dimethylphenyl); and similarly for other radicals containing a benzene ring and one or more acyclic chains. Thenyl becomes (thienylmethyl).

Ring-assembly radicals are now based on the ring assembly names, as, [1,2′-binaphthalen]-8′-yl and [1,1′-biphenyl]-4,4′-diyl.

Acyl radicals in substitutive names are replaced by substituted radicals (exceptions are acetyl, benzoyl, carbonyl, and, when unsubstituted, formyl). Hence propionyl becomes (1-oxopropyl); acetimidoyl becomes (1-iminoethyl); succinoyl becomes (1,4-dioxo-1,4-butanediyl). The three radicals carbonimidoyl, –C(:NH)–; carbonohydrazonoyl, $-C(:NNH_2)-$, and carbonothioyl, –C(:S)–, are introduced for use in multiplicative nomenclature and for cases in which both free valencies are attached to the same atom. In other cases, (iminomethyl) (formerly formimidoyl), (thioxomethyl), etc., are employed.

Replacement ("a") names for acyclic radicals have been introduced under the same restrictions as for heading parents. The free valencies, not the hetero atoms, are preferred for lowest locants, which are always cited; e.g., 4,12-dioxa-7,9-dithiatetradec-1-yl.

Compound and complex radicals are constructed in accordance with revised rules which reflect the new policies for compounds. The most important changes are emphasis on hetero-atom content of the parent radical, abandonment of "like treatment of like things" and the "complexity" principle, elimination of preference for unsaturated acyclic radicals regardless of size, consistent application of the principle of "lowest locants" for substituents on the parent radical, and adoption of more systematic radical names.

abietamido = [[[1,2,3,4,4a,4b,5,6,10,10a-decahydro-1,4a-dimethyl-7-(1-methyl⊃ethyl)-1-phenanthrenyl]carbonyl]⊃amino]- $C_{19}H_{29}CONH$-
acenaphthenyl = (1,2-dihydroacenaph⊃thylenyl) $(C_{12}H_9)$-
1,2-acenaphthenylene = (1,2-dihydro-1,2-acenaphthylenediyl) $-(C_{12}H_8)$-
1-acenaphthenylidene = 1(2*H*)-acena⊃phthylenylidene $(C_{12}H_8)$=
acetamido = (acetylamino) AcNH-
acetenyl = ethynyl HC≡C-
acetimido = (acetylimino) or (1-iminoethyl) AcN= or MeC(=NH)-
acetimidoyl = (1-iminoethyl) MeC(=NH)-
acetoacetamido = [(1,3-dioxobutyl)amino] $MeCOCH_2CONH$-
acetoacetyl = (1,3-dioxobutyl) $MeCOCH_2CO$-
acetohydroximoyl = [1-(hydroxyimino)⊃ethyl] MeC(=NOH)-
acetonyl = (2-oxopropyl) $MeCOCH_2$-
acetonylidene = (2-oxopropylidene) MeCOCH=
acetoxy = (acetyloxy) AcO-
acetyl Ac (MeCO-)
acetylene = 1,2-ethanediylidene =CHCH=
acridanyl = (9,10-dihydroacridinyl) $(C_{13}H_{10}N)$-
acryloyl = (1-oxo-2-propenyl) $H_2C=CHCO$-
acrylyl = (1-oxo-2-propenyl) $H_2C=CHCO$-
adamantyl = tricyclo[3.3.1.1^{3,7}]decyl $(C_{10}H_{15})$-
adamantylene = tricyclo[3.3.1.1^{3,7}]⊃decanediyl $-(C_{10}H_{14})$-
adipaldehydoyl = (1,6-dioxohexyl) $HCO(CH_2)_4CO$-
adipamoyl = (6-amino-1,6-dioxohexyl) $H_2NCO(CH_2)_4CO$-
adipaniloyl = [1,6-dioxo-6-(phenylamino)⊃hexyl] $PhNHCO(CH_2)_4CO$-
adipoyl = (1,6-dioxo-1,6-hexanediyl) $-CO(CH_2)_4CO$-
adipyl = (1,6-dioxo-1,6-hexanediyl) $-CO(CH_2)_4CO$-
alaninamido = [(2-amino-1-oxopropyl)⊃amino] $MeCH(NH_2)CONH$-
alanyl[1] = (2-amino-1-oxopropyl) $MeCH(NH_2)CO$-
β-alanyl[1] = (3-amino-1-oxopropyl) $H_2N(CH_2)_2CO$-
aldo[2] = oxo O =
alloisoleucyl[1] = (2-amino-3-methyl-1-oxopentyl) $EtCHMeCH(NH_2)CO$-
allophanamido = [[[(aminocarbonyl)⊃amino]carbonyl]amino] $H_2N(CONH)_2$-
allophanoyl = [[(aminocarbonyl)amino]⊃carbonyl] $H_2NCONHCO$-
allothreonyl[1] = (2-amino-3-hydroxy-1-oxobutyl) $MeCH(OH)CH(NH_2)CO$-
allyl = 2-propenyl $H_2C=CHCH_2$-
β-allyl = (1-methylethenyl) $H_2C=CMe$-
π-allyl = (η^3-2-propenyl) $(H_2C \cdots CH \cdots CH_2)$-
allylidene = 2-propenylidene $H_2C=CHCH=$
ambrosan-6-yl = [decahydro-3a,8-dimethyl-5-(1-methylethyl)-4-azulenyl] $(C_{15}H_{27})$-
amidino = (aminoiminomethyl) $H_2NC(=NH)$-
amidoxalyl = (aminooxoacetyl) H_2NCOCO-
amino H_2N-
(aminoamidino) = (aminohydrazonomethyl) or (hydrazinoiminomethyl) $H_2NC(=NNH_2)$- or $H_2NNHC(=NH)$-
(aminoiminophosphoranyl) = (*P*-amino⊃phosphinimyl) $H_2NPH(=NH)$-
amoxy = (pentyloxy) $Me(CH_2)_4O$-
amyl = pentyl $Me(CH_2)_4$-
tert-amyl = (1,1-dimethylpropyl) $EtCMe_2$-
amylidene = pentylidene BuCH=
anilino = (phenylamino) PhNH-
anisal = [(methoxyphenyl)methylene] $MeOC_6H_4CH=$
anisidino = [(methoxyphenyl)amino] $MeOC_6H_4NH$-
anisoyl = (methoxybenzoyl) $MeOC_6H_4CO$-
anisyl = (methoxyphenyl) or [(methoxy⊃phenyl)methyl] $MeOC_6H_4$- or $MeOC_6H_4CH_2$-
anisylidene = [(methoxyphenyl)methylene] $MeOC_6H_4CH=$
anthranilamido = [(2-aminobenzoyl)⊃amino] 2-$H_2NC_6H_4CONH$-
anthraniloyl = (2-aminobenzoyl) 2-$H_2NC_6H_4CO$-
anthranoyl = (2-aminobenzoyl) 2-$H_2NC_6H_4CO$-
anthraquinonyl = (9,10-dihydro-9, 10-di⊃oxoanthracenyl) $(C_{14}H_7O_2)$-
anthraquinonylene = (9,10-dihydro-9,10-dioxoanthracenediyl) $-(C_{14}H_6O_2)$-
anthroyl = (anthracenylcarbonyl) $(C_{14}H_9)CO$-
anthryl = anthracenyl $(C_{14}H_9)$-
anthrylene = anthracenediyl $-(C_{14}H_8)$-
antimono = 1,2-distibenediyl -Sb=Sb-
antipyrinyl (antipyryl) = (2,3-dihydro-1,5-dimethyl-3-oxo-2-phenyl-1*H*-pyrazol-4-yl)

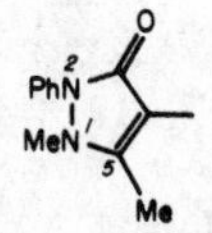

antipyroyl = [(2,3-dihydro-1,5-dimethyl-3-oxo-2-phenyl-1*H*-pyrazol-4-yl)carbonyl]

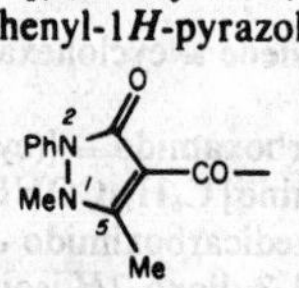

apocamphanyl = (7,7-dimethylbicyclo[2.2.1]heptyl) (C_9H_{15})-
apotrichothecanyl = (decahydro-3a,6,8a,8b-tetramethyl-1*H*-cyclopenta[*b*]benzofuranyl) ($C_{15}H_{25}O$)-
arginyl[1] = [2-amino-5-[(aminoiminomethyl)amino]-1-oxopentyl] $H_2NC(=NH)NH(CH_2)_3CH(NH_2)CO$-
arseno = 1,2-diarsenediyl -As=As-
arsenoso OAs-
arsinico[3,4] HOAs(O)=
arsinimyl $AsH_2(=NH)$-
arsino AsH_2-
arsinothioyl $AsH_2(S)$-
arsinyl $AsH_2(O)$-
arsinylidene AsH(O) =
arso O_2As-
arsono[4] $(HO)_2As(O)$-
arsononitridyl $AsH(\equiv N)$-
arsoranyl AsH_4-
arsoranylidyne $AsH_2\equiv$
arsylene = arsinidine AsH=
arsylidyne = arsinidyne As≡
asaryl = (2,4,5-trimethoxyphenyl) 2,4,5-$(MeO)_3C_6H_2$-
asparaginyl[1] = (2,4-diamino-1,4-dioxobutyl) $H_2NCOCH_2CH(NH_2)$CO-
asparagyl = asparaginyl[1] or (2,4-diamino-1,4-dioxobutyl) $H_2NCOCH_2CH(NH_2)CO$-
aspartoyl[1] = (2-amino-1,4-dioxo-1,4-butanediyl) -$COCH_2CH(NH_2)CO$-
aspartyl = aspartoyl or unspecified aspartyl (see below)
α-aspartyl[1] = (2-amino-3-carboxy-1-oxopropyl) $HO_2CCH_2CH(NH_2)CO$-
β-aspartyl[1] = (3-amino-3-carboxy-1-oxopropyl) $HO_2CCH(NH_2)CH_2CO$-
astato At-
astatoxy = astatyl O_2At-
atisanyl (from atisane) ($C_{20}H_{33}$)-
atropoyl = (1-oxo-2-phenyl-2-propenyl) $H_2C=CPhCO$-
azelaoyl = (1,9-dioxo-1,9-nonanediyl) -$CO(CH_2)_7CO$-
azelaaldehydoyl = (1,9-dioxononyl) $HCO(CH_2)_7CO$-
azi[5] (see also azo) -N=N-
azido N_3-
(azidoformyl) = (azidocarbonyl) N_3CO-
azino =NN=
azo[6] (see also azi) -N=N-
azoxy -N(O)N-
benzal = (phenylmethylene) PhCH=
benzamido = (benzoylamino) BzNH-
benzenesulfenamido = [(phenylthio)amino] PhSNH-
benzenesulfonamido = [(phenylsulfonyl)amino] $PhSO_2NH$-
benzenyl = (phenylmethylidyne) PhC≡
benzhydryl = (diphenylmethyl) Ph_2CH-
benzhydrylidene = (diphenylmethylene) Ph_2C=
benzidino = [(4′-amino[1,1′-biphenyl]-4-yl)amino] 4-$(4-H_2NC_6H_4)C_6H_4NH$-
benziloyl = (hydroxydiphenylacetyl) $Ph_2C(OH)CO$-
(3-benziloylpropyl) = (5-hydroxy-4-oxo-5,5-diphenylpentyl) $Ph_2C(OH)CO(CH_2)_3$-
benzimidazolinyl = (2,3-dihydro-1*H*-benzimidazolyl) ($C_7H_7N_2$)-
2-benzimidazolyl = 1*H*-benzimidazol-2-yl ($C_7H_5N_2$)-
benzimido = (benzoylimino) or (iminophenylmethyl) BzN= or PhC(=NH)-
benzimidoyl = (iminophenylmethyl) PhC(=NH)-
benzofuryl = benzofuranyl (C_8H_5O)-
benzohydroximoyl = [(hydroxyimino)phenylmethyl] PhC(=NOH)-
o-benzoquinon-3-yl = (5,6-dioxo-1,3-cyclohexadien-1-yl) ($C_6H_3O_2$)-
p-benzoquinon-2,5-ylene = (3,6-dioxo-1,4-cyclohexadiene-1,4-diyl) -($C_6H_2O_2$)-
benzoxy = (benzoyloxy) BzO-
benzoyl Bz (PhCO-)
(benzoylacetyl) = (1,3-dioxo-3-phenylpropyl) $PhCOCH_2CO$-
(benzoylformyl) = (oxophenylacetyl) PhCOCO-
benzyl = (phenylmethyl) $PhCH_2$-
benzylidene = (phenylmethylene) PhCH=
benzylidyne = (phenylmethylidyne) PhC≡
(benzyloxy) = (phenylmethoxy) $PhCH_2O$-
(benzylselenyl) = [(phenylmethyl)seleno] $PhCH_2Se$-
bicarbamoyl = (hydrazodicarbonyl) -CONHNHCO-
bicyclo[1.1.0]butylene = bicyclo[1.1.0]butanediyl -(C_4H_4)-
biphenylyl = [1,1′-biphenyl]yl PhC_6H_4-
biphenylene = [1,1′-biphenyl]diyl -($C_{12}H_8$)-
biphenylylene = [1,1′-biphenyl]diyl -($C_{12}H_8$)-
bismuthino BiH_2-
bismuthylene BiH=
bismuthylidyne Bi≡
2-bornyl = (1,7,7-trimethylbicyclo[2.2.1]hept-2-yl) ($C_{10}H_{17}$)-
3-bornylidene = (4,7,7-trimethylbicyclo[2.2.1]hept-2-ylidene) ($C_{10}H_{16}$)=
borono[4] $(HO)_2B$-
boryl BH_2-
borylene BH=
borylidyne B≡
bromo Br-
1,3-butadienediylidene = 1,3-butadiene-1,4-diylidene =C=CHCH=C=
butadiynylene = 1,3-butadiyne-1,4-diyl -C≡CC≡C-
2-butenylene = 2-butene-1,4-diyl -$CH_2CH=CHCH_2$-
butoxy BuO-
sec-butoxy = (1-methylpropoxy) EtCHMeO-
tert-butoxy = (1,1-dimethylethoxy) Me_3CO-
butyl Bu ($Me(CH_2)_3$-)
butyl[β] = (2-methylpropyl) Me_2CHCH_2-
butyl[γ] = (1,1-dimethylethyl) Me_3C-
sec-butyl = (1-methylpropyl) EtCHMe-
tert-butyl = (1,1-dimethylethyl) Me_3C-
1,4-butylene = 1,4-butanediyl -$(CH_2)_4$-
sec-butylidene = (1-methylpropylidene) EtCMe=
(butyloxy) = butoxy BuO-
butynedioyl = (1,4-dioxo-2-butyne-1,4-diyl) -COC≡CCO-
butyryl = (1-oxobutyl) PrCO-

cacodyl = (dimethylarsino) Me_2As-
cadinan-1-yl = [octahydro-4,7-dimethyl-1-(1-methylethyl)-4a(2*H*)-naphthalenyl] ($C_{15}H_{29}$)-
2-camphanyl = (4,7,7-trimethylbicyclo[2.2.1]hept-2-yl) ($C_{10}H_{17}$)-
camphoroyl (from camphoric acid) = [(1,2,2-trimethyl-1,3-cyclopentanediyl)dicarbonyl] -$CO(C_8H_{14})CO$-
5-camphoryl (from camphor) = (4,7,7-trimethyl-5-oxobicyclo[2.2.1]hept-2-yl) ($C_{10}H_{15}O$)-
canavanyl = [2-amino-4-[[(aminoiminomethyl)amino]oxy]-1-oxobutyl] $H_2NC(=NH)NHO(CH_2)_2CH(NH_2)CO$-
caprinoyl = (1-oxodecyl) $Me(CH_2)_8CO$-
caproyl (from caproic acid) = (1-oxohexyl) $Me(CH_2)_4CO$-
capryl (from capric acid) = (1-oxodecyl) $Me(CH_2)_8CO$-
capryloyl (from caprylic acid) = (1-oxooctyl) $Me(CH_2)_6CO$-
caprylyl (from caprylic acid) = (1-oxooctyl) $Me(CH_2)_6CO$-
carbamido = [(aminocarbonyl)amino] H_2NCONH-
carbamoyl = (aminocarbonyl) H_2NCO-
carbamyl = (aminocarbonyl) H_2NCO-
carbanilino = [(phenylamino)carbonyl] PhNHCO-
carbaniloyl = [(phenylamino)carbonyl] PhNHCO-
carbazimidoyl = (hydrazinoiminomethyl) $H_2NNHC(=NH)$-
carbazol-9-yl = 9*H*-carbazol-9-yl ($C_{12}H_8N$)-
carbazoyl = (hydrazinocarbonyl) H_2NNHCO-
carbethoxy = (ethoxycarbonyl) EtO_2C-
carbobenzoxy = [(phenylmethoxy)carbonyl]$PhCH_2O_2C$-
carbonimidoyl[9] -C(=NH)-

ILLUSTRATIVE LIST OF SUBSTITUENT PREFIXES (*continued*)

carbonothioyl[9] -CS-
carbonyl -CO-
(carbonyldioxy) = [carbonylbis(oxy)] $-OCO_2-$
(1-carbonylethyl) = (methyloxoethenyl) O=C=CMe-
(carbonylmethyl) = (oxoethenyl) O=C=CH-
(carbonylmethylene) = (1-oxo-1,2-ethanediyl) or (oxoethenylidene) $-COCH_2-$ or O=C=C=
carboxy[4] HO_2C-
(carboxyformyl) = (carboxycarbonyl) HO_2CCO-
(5-carboxyvaleryl) = (5-carboxy-1-oxo⊃pentyl) $HO_2C(CH_2)_4CO-$
carbyl -C-
carnosyl = (*N*-β-alanylhistidyl)[1] or [2-[(3-amino-1-oxopropyl)amino]-3-(1*H*-imidazol-4-yl)-1-oxopropyl]

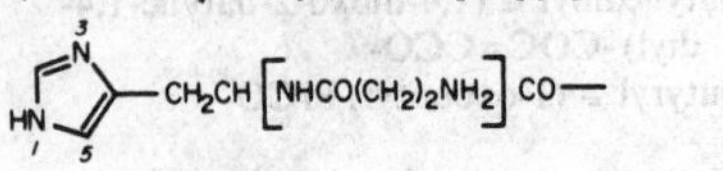

caronaldehydoyl = [(3-formyl-2,2-di⊃methylcyclopropyl)carbonyl]

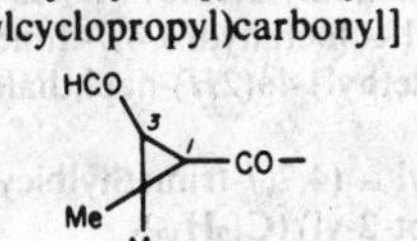

carvacryl = [2-methyl-5-(1-methylethyl)⊃phenyl]

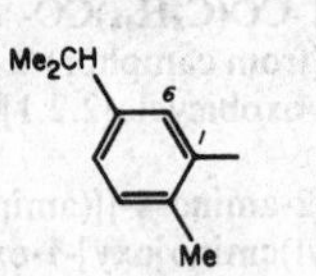

carvomenthyl = [2-methyl-5-(1-methylethyl)⊃cyclohexyl)]

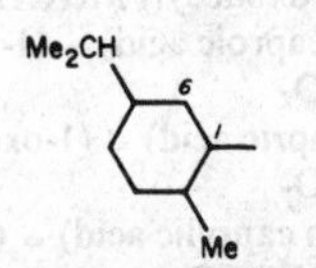

10-caryl = [(7,7-dimethylbicyclo[4.1.0]⊃hept-3-yl)methyl] $(C_9H_{15})CH_2-$
cathyl = [(ethoxycarbonyl)oxy] $EtOCO_2-$
cedranyl = (octahydro-3a,6,8,8-tetra⊃methyl-1*H*-3a,7-methanoazulenyl) $(C_{15}H_{25})-$
cetyl = hexadecyl $Me(CH_2)_{15}-$
chaulmoogroyl (from chaulmoogric acid) = [13-(2-cyclopenten-1-yl)-1-oxotridecyl] $C_5H_7(CH_2)_{12}CO-$
chaulmoogryl (from chaulmoogryl alcohol) = [13-(2-cyclopenten-1-yl)tridecyl] $C_5H_7(CH_2)_{13}-$
chloro Cl-
(chloroformyl) = (chlorocarbonyl) ClCO-
(chloroglyoxyloyl) = (chlorooxoacetyl) ClCOCO-
(chlorooxalyl) = (chlorooxoacetyl) ClCOCO-

chlorosyl OCl-
chloryl O_2Cl-
cholesteryl (from cholesterol) = cholest-5-en-3-yl (from cholestene) $(C_{25}H_{45})-$
choloyl (from cholic acid) = (3,7,12-trihydroxy-24-oxocholan-24-yl) $(HO)_3(C_{23}H_{36})CO-$
chromanyl = (3,4-dihydro-2*H*-1-benzo⊃pyranyl) $(C_9H_9O)-$
cinchoninoyl (from cinchoninic acid) = (4-quinolinylcarbonyl) $(4-C_9H_6N)CO-$
cinnamal = (3-phenyl-2-propenylidene) PhCH=CHCH=
cinnamenyl = (2-phenylethenyl) PhCH=CH-
cinnamoyl = (1-oxo-3-phenyl-2-propenyl) PhCH=CHCO-
cinnamyl = (3-phenyl-2-propenyl) $PhCH=CHCH_2-$
cinnamylidene = (3-phenyl-2-propen⊃ylidene) PhCH=CHCH=
citraconimido = (2,5-dihydro-3-methyl-2,5-dioxo-1*H*-pyrrol-1-yl)

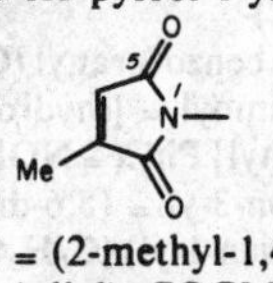

citraconoyl = (2-methyl-1,4-dioxo-2-butene-1,4-diyl) -COCMe=CHCO-
conaninyl (from conanine) $(C_{22}H_{36}N)-$
cresotoyl (from cresotic acid) = (hydroxymethylbenzoyl) $HO(Me)C_6H_3CO-$
cresoxy = (methylphenoxy) MeC_6H_4O-
cresyl = (hydroxymethylphenyl) or (methylphenyl) $HO(Me)C_6H_3-$ or MeC_6H_4-
cresylene = (methylphenylene) $-(MeC_6H_3)-$
crotonoyl = (1-oxo-2-butenyl) MeCH=CHCO-
crotonyl = (1-oxo-2-butenyl) MeCH=CHCO-
crotyl = 2-butenyl $MeCH=CHCH_2-$
cumal = [[4-(1-methylethyl)phenyl]⊃methylene] $4-(Me_2CH)C_6H_4CH=$
cumenyl = [(1-methylethyl)phenyl] $Me_2CHC_6H_4-$
cumidino = [[4-(1-methylethyl)phenyl]⊃amino] $4-(Me_2CH)C_6H_4NH-$
cuminal = [[4-(1-methylethyl)phenyl]⊃methylene] $4-(Me_2CH)C_6H_4CH=$
cuminyl = [[4-(1-methylethyl)phenyl]⊃methyl] $4-(Me_2CH)C_6H_4CH_2-$
cuminylidene = [[4-(1-methylethyl)⊃phenyl]methylene] $4-(Me_2CH)C_6H_4CH=$
cumoyl = [4-(1-methylethyl)benzoyl] $4-(Me_2CH)C_6H_4CO-$
cumyl[13] = [(1-methylethyl)phenyl] $(Me_2CH)C_6H_4-$
α-cumyl = (1-methyl-1-phenylethyl)- $PhCMe_2-$
cyanamido = (cyanoamino) NCNH-

cyanato NCO-
cyano NC-
cyclodisiloxan-2-yl

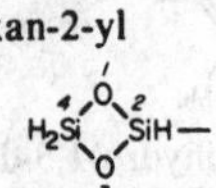

cyclohexadienylene = cyclohexadienediyl $-C_6H_6-$
cyclohexanecarboxamido = [(cyclohexyl⊃carbonyl)amino] $C_6H_{11}CONH-$
1,2-cyclohexanedicarboximido = (octahydro-1,3-dioxo-2*H*-isoindol-2-yl)

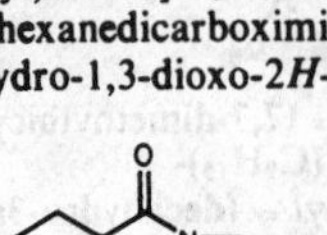

cymyl = [methyl(1-methylethyl)phenyl] $Me(Me_2CH)C_6H_3-$
cysteinyl[1] = (2-amino-3-mercapto-1-oxopropyl) $HSCH_2CH(NH_2)CO-$
cysteyl = (3-sulfoalanyl)[1] or (2-amino-1-oxo-3-sulfopropyl) $HO_3SCH_2CH(NH_2)CO-$
cystyl[1] = [dithiobis(2-amino-1-oxo-3,1-propanediyl)] $-COCH(NH_2)CH_2SSCH_2CH(NH_2)CO-$
decanedioyl = (1,10-dioxo-1,10-decanediyl) $-CO(CH_2)_8CO-$
decanoyl = (1-oxodecyl) $CH_3(CH_2)_8CO-$
decasiloxanylene = 1,19-decasiloxanediyl $-SiH_2(OSiH_2)_8OSiH_2-$
desyl = (2-oxo-1,2-diphenylethyl) PhCOCHPh-
diarsenyl HAs=As-
diarsinetetrayl = 1,2-diarsinediylidene =AsAs=
diarsinyl $H_2AsAsH-$
1,2-diazenediyl = azi or azo -N=N-
diazeno = diazenyl[4] (see also azo) HN=N-
diazo $N_2=$
diazoamino = 1-triazene-1,3-diyl -NHN=N-
diazonio N_2^+-
dibenzothiophene-yl = dibenzothienyl $(C_{12}H_7S)-$
diborane(4)tetrayl = 1,2-diborane(4)⊃diylidene =BB=
1,2-dicarbadodecaboran(12)-1-yl $(C_2H_{11}B_{10})-$
digermanylene = 1,2-digermanediyl $-GeH_2GeH_2-$
digermthianyl = digermathianyl $H_3GeSGeH_2-$
diglycoloyl = [oxybis(1-oxo-2,1-ethanediyl)] $-COCH_2OCH_2CO-$
dimethyliminio $Me_2N^{\pm}$
dioxy[7] (see also epidioxy) -OO-
1,2-diphosphinediyl -PHPH-
diphosphinetetrayl = 1,2-diphosph⊃inediylidene =PP=
diphosphino = diphosphinyl H_2PPH-
diseleno[7] (see also epidiseleno and perseleno) -SeSe-

disilanoxy = (disilanyloxy) H_3SiSiH_2O-
disilanyl H_3SiSiH_2-
disilanylene = 1,2-disilanediyl -SiH_2SiH_2-
disilazanoxy = (disilazanyloxy) $H_3SiNHSiH_2O$-
2-disilazanyl = (disilylamino) $(H_3Si)_2N$-
disiloxanediylidene = 1,3-disiloxanediylidene =SiHOSiH=
disiloxanoxy = (disiloxanyloxy) $H_3SiOSiH_2O$-
disiloxanylene = 1,3-disiloxanediyl -SiH_2OSiH_2-
disilthianoxy = (disilathianyloxy) $H_3SiSSiH_2O$-
distannanylene = 1,2-distannanediyl -SnH_2SnH_2-
distannthianediylidene = 1,3-distannathianediylidene = SnHSSnH=
disulfinyl[6] -S(O)S(O)-
dithio[7] (see also epidithio and perthio) -SS-
(dithiobicarbamoyl) = (hydrazodicarbonothioyl) -CSNHNHCS-
(dithiohydroperoxy) = thiosulfeno[4] HSS-
dodecanoyl = (1-oxododecyl) $Me(CH_2)_{10}CO$-
duryl = (2,3,5,6-tetramethylphenyl) 2,3,5,6-Me_4C_6H-
durylene = (2,3,5,6-tetramethyl-1,4-phenylene) -(2,3,5,6-Me_4C_6)-
enanthoyl = (1-oxoheptyl) $Me(CH_2)_5CO$-
enanthyl = (1-oxoheptyl) $Me(CH_2)_5CO$-
epidoxy[8] (see also dioxy) -OO-
epidiseleno[8] (see also diseleno and perseleno) -SeSe-
epidithio[8] (see also dithio and perthio) -SS-
epioxy = epoxy[8] (see also oxy and oxo) -O-
episeleno[8] (see also seleno and selenoxo) -Se-
epithio[8] (see also thio and thioxo)-S-
epoxy[8] (see also oxy and oxo)-O-
(epoxyethyl) = oxiranyl

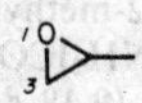

(2,3-epoxypropyl) = (oxiranylmethyl)

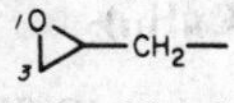

eremophilan-1-yl = [decahydro-4,4a-dimethyl-5-(1-methylethyl)-1-naphthalenyl] $(C_{15}H_{27})$-
ethanediylidene = 1,2-ethanediylidene =CHCH=
ethene = 1,2-ethanediyl -CH_2CH_2-
ethinyl = ethynyl HC≡C-
ethoxalyl = (ethoxyoxoacetyl) EtO_2CCO-
ethoxy EtO-
(ethoxycarbonyl) EtO_2C-
(1-ethoxyformimidoyl) = (ethoxyiminomethyl) EtOCH(=NH)-
(ethoxyphosphinyl) EtOPH(O)-
ethyl Et ($MeCH_2$-)
ethylene = 1,2-ethanediyl -CH_2CH_2-
[ethylenebis(nitrilodimethylene)] = [1,2-ethanediylbis[nitrilobis(methylene)]] (-$CH_2)_2N(CH_2)_2N(CH_2$-$)_2$
(ethylenedioxy) = [1,2-ethanediylbis(oxy)] -$O(CH_2)_2O$-
(ethyloxy) = ethoxy EtO-
(ethylselenyl) = (ethylseleno) EtSe-
(ethylthio) EtS-
eudesman-8-yl = [decahydro-5,8a-dimethyl-2-(1-methylethyl)-2-naphthalenyl] $(C_{15}H_{27})$-
farnesyl (from farnesol) = (3,7,11-trimethyl-2,6,10-dodecatrienyl) $Me_2C=CH(CH_2)_2CMe=CH(CH_2)_2CMe=CHCH_2$-
fenchyl = (1,3,3-trimethylbicyclo[2.2.1]hept-2-yl) $(C_{10}H_{17})$-
ferrocenediyl -$(C_5H_4)Fe(C_5H_4)$-
fluoranyl = (3-oxospiro[isobenzofuran-1(3*H*),9′-[9*H*]xanthen]yl) $(C_{20}H_{11}O_3)$-
fluoren-9-ylidene = 9*H*-fluoren-9-ylidene $(C_{13}H_8)$=
fluoro F-
formamido = (formylamino) HCONH-
1,5-formazandiyl -N=NCH=NNH-
1-formazano $H_2NN=CHN=N$-
5-formazano HN=NCH=NNH-
1,3,5-formazantriyl -N=NC=NNH-
formazanyl HN=NC(=NNH_2)-
formazyl = (1,5-diphenylformazanyl) PhN=NC(=NNHPh)-
formimidoyl = (iminomethyl) CH(=NH)-
(1-formimidoylformimidoyl) = (1,2-diiminoethyl) HC(=NH)C(=NH)-
(formimidoylformyl) = (iminoacetyl) CH(=NH)CO-
(formimidomethyl) = (2-iminoethyl) HC(=NH)CH_2-
formyl[4] HCO-
fucosyl = (6-deoxygalactosyl) $(C_6H_{11}O_4)$-
fumaraniloyl = [1,4-dioxo-4-(phenylamino)-2-butenyl] PhNHCOCH=CHCO-
fumaroyl = (1,4-dioxo-2-butene-1,4-diyl) -COCH=CHCO-
furfural = (2-furanylmethylene)

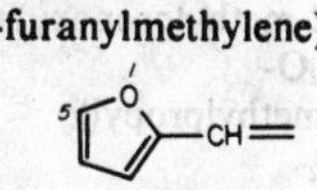

furfuryl = (2-furanylmethyl)

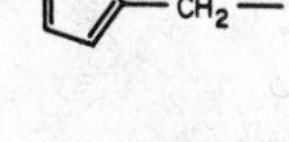

furfurylidene = (2-furanylmethylene)

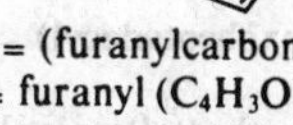

furoyl = (furanylcarbonyl) $(C_4H_3O)CO$-
furyl = furanyl (C_4H_3O)-
galloyl = (3,4,5-trihydroxybenzoyl) 3,4,5-$(HO)_3C_6H_2CO$-
gentisoyl = (2,5-dihydroxybenzoyl) 2,5-$(HO)_2C_6H_3CO$-
geranyl (from geraniol) = (3,7-dimethyl-2,6-octadienyl) $Me_2C=CH(CH_2)_2CMe=CHCH_2$-
germacran-6-yl = [5,9-dimethyl-2-(1-methylethyl)cyclodecyl] $(C_{15}H_{25})$-
germanetetrayl Ge≡
germyl H_3Ge-
germylene H_2Ge=
germylidyne HGe≡
gibbanyl (from gibbane) $(C_{15}H_{23})$-
(glucosyloxy) $(C_6H_{11}O_5)O$-
glutaminyl[1] = (2,5-diamino-1,5-dioxopentyl) $H_2NCO(CH_2)_2CH(NH_2)CO$-
glutamoyl[1] = (2-amino-1,5-dioxo-1,5-pentanediyl) -$CO(CH_2)_2CH(NH_2)CO$-
glutamyl = glutamoyl or unspecified glutamyl (see below)
α-glutamyl[1] = (2-amino-4-carboxybutyl) $HO_2C(CH_2)_2CH(NH_2)CO$-
γ-glutamyl[1] = (4-amino-4-carboxybutyl) $HO_2CCH(NH_2)(CH_2)_2CO$-
glutaryl = (1,5-dioxo-1,5-pentanediyl) -$CO(CH_2)_3CO$-
glyceroyl = (2,3-dihydroxy-1-oxopropyl) $HOCH_2CH(OH)CO$-
glyceryl = 1,2,3-propanetriyl -$CH(CH_2$-$)_2$
glycidyl = (oxiranylmethyl)

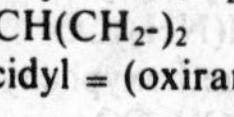

glycinamido = [(aminoacetyl)amino] H_2NCH_2CONH-
glycinimidoyl = (2-amino-1-iminoethyl) $H_2NCH_2C(=NH)$-
glycoloyl = (hydroxyacetyl) $HOCH_2CO$-
glycolyl = (hydroxyacetyl) $HOCH_2CO$-
glycyl[1] = (aminoacetyl) H_2NCH_2CO-
glyoxalinyl = imidazolyl $(C_3H_3N_2)$-
glyoxalyl = imidazolyl $(C_3H_3N_2)$-
glyoxylimidoyl = (1-imino-2-oxoethyl) HCOC(=NH)-
glyoxyloyl = (oxoacetyl) HCOCO-
(glyoxyloylmethyl) = (2,3-dioxopropyl) $HCOCOCH_2$-
glyoxylyl = (oxoacetyl) HCOCO-
guaiacyl = (2-methoxyphenyl) 2-$(MeO)C_6H_4$-
guaian-8-yl = [decahydro-1,4-dimethyl-7-(1-methylethyl)-6-azulenyl] $(C_{15}H_{27})$-
guanidino = [(aminoiminomethyl)amino] $H_2NC(=NH)NH$-
(guanidinoazo) = [3-(aminoiminomethyl)-1-triazenyl] $H_2NC(=NH)NHN=N$-

guanyl = (aminoiminomethyl) $H_2NC(=NH)$-

heptadecanoyl = (1-oxoheptadecyl) $Me(CH_2)_{15}CO$-
heptanamido = [(1-oxoheptyl)amino] $Me(CH_2)_5CONH$-
heptanedioyl = (1,7-dioxo-1,7-heptanediyl) $-CO(CH_2)_5CO$-
heptanoyl = (1-oxoheptyl) $Me(CH_2)_5CO$-
hexadecanoyl = (1-oxohexadecyl) $Me(CH_2)_{14}CO$-
2,4-hexadiynylene = 2,4-hexadiyne-1,6-diyl $-CH_2C\equiv CC\equiv CCH_2$-
hexamethylene = 1,6-hexanediyl $-(CH_2)_6$-
hexanedioyl = (1,6-dioxo-1,6-hexanediyl) $-CO(CH_2)_4CO$-
hexanethioyl = (1-thioxohexyl) $Me(CH_2)_4CS$-
hippuroyl = (*N*-benzoylglycyl)[1] or [(benzoylamino)acetyl] $BzNHCH_2CO$-
hippuryl = (*N*-benzoylglycyl)[1] or [(benzoylamino)acetyl] $BzNHCH_2CO$-
histidyl[1] = [2-amino-3-(1*H*-imidazol-4-yl)-1-oxopropyl]

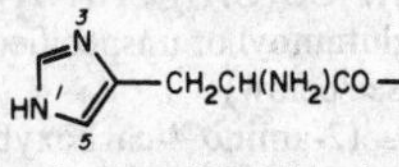

homocysteinyl[1] = (2-amino-4-mercapto-1-oxobutyl) $HS(CH_2)_2CH(NH_2)CO$-
homomyrtenyl = [2-(6,6-dimethylbicyclo[3.1.1]hept-2-en-2-yl)ethyl] $(C_9H_{13})CH_2CH_2$-
homopiperonyl = [2-(1,3-benzodioxol-5-yl)ethyl] $(C_7H_5O_2)CH_2CH_2$-
homoseryl[1] = (2-amino-4-hydroxy-1-oxobutyl) $HO(CH_2)_2CH(NH_2)CO$-
homoveratroyl = [(3,4-dimethoxyphenyl)acetyl] 3,4-$(MeO)_2C_6H_3CH_2CO$-
homoveratryl = [2-(3,4-dimethoxyphenyl)ethyl] 3,4-$(MeO)_2C_6H_3CH_2CH_2$-
hydantoyl = [*N*-(aminocarbonyl)glycyl][1] or [[(aminocarbonyl)amino]acetyl] $H_2NCONHCH_2CO$-
hydnocarpoyl (from hydnocarpic acid) = [11-(2-cyclopenten-1-yl)-1-oxoundecyl] $(C_5H_7)(CH_2)_{10}CO$-
hydnocarpyl (from hydnocarpyl alcohol) = [11-(2-cyclopenten-1-yl)undecyl] $(C_5H_7)(CH_2)_{10}CH_2$-
hydracryloyl = (3-hydroxy-1-oxopropyl) $HO(CH_2)_2CO$-
hydratropoyl = (1-oxo-2-phenylpropyl) PhCHMeCO-
hydrazi[5] (see also hydrazo) -NHNH-
hydrazinediylidene = azino =NN=
hydrazino H_2NNH-
1-hydrazinyl-2-ylidene -NHN =
hydrazo[6] (see also hydrazi) -NHNH-
hydrazono H_2NN=
hydrocinnamoyl = (1-oxo-3-phenylpropyl) $Ph(CH_2)_2CO$-
hydrocinnamyl = (3-phenylpropyl) $Ph(CH_2)_3$-
hydroperoxy[4] HOO-
(hydroperoxyformyl) = (hydroperoxycarbonyl) HOOCO-
hydroxamino = (hydroxyamino) HONH-
hydroximino = (hydroxyimino) HON=
hydroxy[4] HO-
hydroxyl = hydroxy[4] HO-
(hydroxyphosphinyl) HOPH(O)-
hygroyl = (1-methylprolyl)[1] or [(1-methyl-2-pyrrolidinyl)carbonyl]

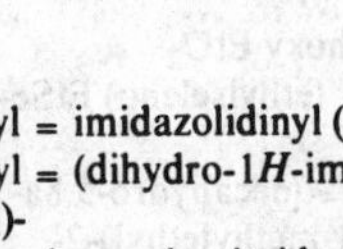

imidazolidyl = imidazolidinyl $(C_3H_7N_2)$-
imidazolinyl = (dihydro-1*H*-imidazolyl) $(C_3H_5N_2)$-
imidocarbonyl = carbonimidoyl[9] $-C(=NH)$-
(imidocarbonylamino) = (carbonimidoylamino) HN=C=N-
imino HN=
(3-iminoacetonyl) = (3-imino-2-oxopropyl) $HN=CHCOCH_2$-
(iminodisulfonyl) = [iminobis(sulfonyl)] $-SO_2NHSO_2$-
(iminonitrilo) = 1-hydrazinyl-2-ylidene -NHN=
(iminophosphoranyl) = phosphinimyl $H_2P(=NH)$-
indanyl = (2,3-dihydro-1*H*-indenyl) (C_9H_9)-
indenyl = 1*H*-indenyl (C_9H_7)-
1-indolinyl = (2,3-dihydro-1*H*-indol-1-yl) (C_8H_8N)-
2-indolinylidene = (1,3-dihydro-2*H*-indol-2-ylidene) (C_8H_7N)=
indyl = 1*H*-indolyl (C_8H_6N)-
iodo I-
iodoso = iodosyl OI-
iodoxy = iodyl O_2I-
isoallyl = 1-propenyl MeCH=CH-
isoamoxy = (3-methylbutoxy) $Me_2CH(CH_2)_2O$-
isoamyl = (3-methylbutyl) $Me_2CH(CH_2)_2$-
sec-isoamyl = (1,2-dimethylpropyl) $Me_2CHCHMe$-
isoamylidene = (3-methylbutylidene) Me_2CHCH_2CH=
isobornyl (from isoborneol) = (1,7,7-trimethylbicyclo[2.2.1]hept-2-yl) $(C_{10}H_{17})$-
isobutenyl = (2-methyl-1-propenyl) $Me_2C=CH$-
isobutoxy = (2-methylpropoxy) Me_2CHCH_2O-
isobutyl = (2-methylpropyl) Me_2CHCH_2-
isobutylidene = (2-methylpropylidene) Me_2CHCH=
isobutyryl = (2-methyl-1-oxopropyl) Me_2CHCO-
isocrotyl = (2-methyl-1-propenyl) $Me_2C=CH$-
isocyanato OCN-
isocyano CN-
isohexyl = (4-methylpentyl) $Me_2CH(CH_2)_3$-
isohexylidene = (4-methylpentylidene) $Me_2CH(CH_2)_2CH$=
2-isoindolinyl = (1,3-dihydro-2*H*-isoindol-2-yl) (C_8H_8N)-
isoleucyl[1] = (2-amino-3-methyl-1-oxopentyl) $EtCHMeCH(NH_2)CO$-
isonicotinoyl = (4-pyridinylcarbonyl)

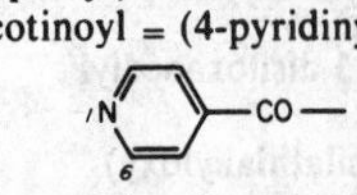

isonipecotoyl = (4-piperidinylcarbonyl)

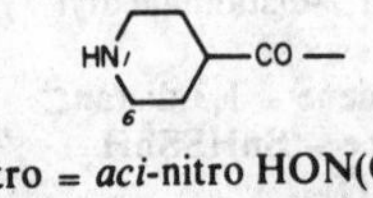

isonitro = *aci*-nitro HON(O)=
isonitroso = (hydroxyimino) HON=
1-isopentenyl = (3-methyl-1-butenyl) $Me_2CHCH=CH$-
isopentyl = (3-methylbutyl) $Me_2CH(CH_2)_2$-
isopentylidene = (3-methylbutylidene) Me_2CHCH_2CH=
isophthalal = (1,3-phenylenedimethylidyne) 1,3-$C_6H_4(CH=)_2$
isophthalaldehydoyl = (3-formylbenzoyl) 3-$(HCO)C_6H_4CO$-
isophthaloyl = (1,3-phenylenedicarbonyl) 1,3-$C_6H_4(CO-)_2$
isophthalylidene = (1,3-phenylenedimethylidyne) 1,3-$C_6H_4(CH=)_2$
isopropenyl = (1-methylethenyl) $H_2C=CMe$-
isopropoxy = (1-methylethoxy) Me_2CHO-
isopropyl = (1-methylethyl) Me_2CH-
isopropylidene = (1-methylethylidene) Me_2C=
(isopropylidenedioxy) = [(1-methylethylidene)bis(oxy)] $-OCMe_2O$-
isosemicarbazido = [(aminohydroxymethylene)hydrazino] $H_2NC(OH)=NNH$-
isothiocyanato SCN-
isothiocyano = isothiocyanato SCN-
isovaleryl = (3-methyl-1-oxobutyl) Me_2CHCH_2CO-
isovalyl[1] = (2-amino-2-methyl-1-oxobutyl) $EtCMe(NH_2)CO$-
isoviolanthrenylene = (9,18-dihydrodinaphtho[1,2,3-*cd*:1′,2′,3′-*lm*]perylenediyl) $-(C_{34}H_{18})$-

kauranyl (from kaurane) $(C_{20}H_{33})$-
kauranylene = kauranediyl (from kaurane) $-(C_{20}H_{32})$-
kaurenyl (from kaurene) $(C_{20}H_{31})$-
keto[10] = oxo O=

labdan-15-yl = [5-(decahydro-2,5,5,8a-

tetramethyl-1-naphthalenyl)-3-methyl⊃ pentyl] $(C_{20}H_{37})$-
lactosyl = (4-*O*-*β*-D-galactopyranosyl-*β*-D-glucopyranosyl) $(C_6H_{11}O_5)O(C_6H_{10}O_5)$-
lactoyl = (2-hydroxy-1-oxopropyl) MeCH(OH)CO-
lanostenylene = lanostenediyl (from lanostane) -$(C_{30}H_{50})$-
lauroyl = (1-oxododecyl) $Me(CH_2)_{10}CO$-
leucyl[1] = (2-amino-4-methyl-1-oxopentyl) $Me_2CHCH_2CH(NH_2)CO$-
levulinoyl = (1,4-dioxopentyl) $MeCO(CH_2)_2CO$-
linalyl (from linalool) = (1-ethenyl-1,5-dimethyl-4-hexenyl) $Me_2C=CH(CH_2)_2CMe(CH=CH_2)$-
linolelaidoyl = (1-oxo-9, 12-octadeca⊃ dienyl) $Me(CH_2)_4CH=CHCH_2CH=CH(CH_2)_7CO$-
linolenoyl = (1-oxo-9,12,15-octadecatrienyl) $EtCH=CHCH_2CH=CHCH_2CH=CH(CH_2)_7CO$-
γ-linolenoyl = (1-oxo-6,9,12-octadeca⊃ trienyl) $Me(CH_2)_4CH=CHCH_2CH=$⊃ $CHCH_2CH=CH(CH_2)_4CO$-
linoleoyl = (1-oxo-9,12-octadecadienyl) $Me(CH_2)_4CH=CHCH_2CH=$⊃ $CH(CH_2)_7CO$-
lupanyl (from lupane) $(C_{30}H_{51})$-
lysyl[1] = (2,6-diamino-1-oxohexyl) $H_2N(CH_2)_4CH(NH_2)CO$-
maleoyl = (1,4-dioxo-2-butene-1,4-diyl) -COCH=CHCO-
malonaldehydoyl = (1,3-dioxopropyl) $HCOCH_2CO$-
malonamoyl = (3-amino-1,3-dioxopropyl) H_2NCOCH_2CO-
malonaniloyl = [1,3-dioxo-3-(phenyl⊃ amino)propyl] $PhNHCOCH_2CO$-
malonimido = (2,4-dioxo-1-azetidinyl)

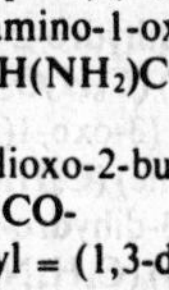

malonyl = (1,3-dioxo-1,3-propanediyl) $-COCH_2CO$-
maloyl = (2-hydroxy-1,4-dioxo-1,4-butanediyl) $-COCH(OH)CH_2CO$-
maltosyl = (4-*O*-*α*-D-glucopyranosyl-*β*-D-glucopyranosyl) $(C_6H_{11}O_5)O(C_6H_{10}O_5)$-
mandeloyl = (hydroxyphenylacetyl) PhCH(OH)CO-
p-menth-2-yl = [2-methyl-5-(1-methyl⊃ ethyl)cyclohexyl] $2,5\text{-}Me(Me_2CH)C_6H_9$-
p-menth-3,5-ylene = [5-methyl-2-(1-methylethyl)-1,3-cyclohexanediyl] $5,2,1,3\text{-}Me(Me_2CH)C_6H_8$
mercapto[4] HS-
mesaconoyl = (2-methyl-1.4-dioxo-2-butene-1,4-diyl) -COCMe=CHCO-
mesidino = [(2,4,6-trimethylphenyl)⊃ amino] $2,4,6\text{-}Me_3C_6H_2NH$-
mesityl = (2,4,6-trimethylphenyl) $2,4,6\text{-}Me_3C_6H_2$-
α-mesityl = [(3,5-dimethylphenyl)methyl] $3,5\text{-}Me_2C_6H_3CH_2$-
mesoxalyl = (1,2,3-trioxo-1,3-propanediyl) -COCOCO-
mesyl = (methylsulfonyl) $MeSO_2$-
metanilamido = [[(3-aminophenyl)sul⊃ fonyl]amino] $3\text{-}H_2NC_6H_4SO_2NH$-
metanilyl = [(3-aminophenyl)sul⊃ fonyl] $3\text{-}H_2NC_6H_4SO_2$-
methacryloyl = (2-methyl-1-oxo-2-pro⊃ penyl) $H_2C=CMeCO$-
methallyl = (2-methyl-2-propenyl) $H_2C=CMeCH_2$-
methanetetrayl C≡
methene = methylene $H_2C=$
methenyl = methylidyne HC≡
methionyl[1] = [methylenebis(sul⊃ fonyl)] or [2-amino-4-(methylthio)-1-oxobutyl] $-SO_2CH_2SO_2$- or $MeS(CH_2)_2CH(NH_2)CO$-
methoxalyl = (methoxyoxoacetyl) MeO_2CCO-
methoxy MeO-
(methoxycarbonyl) MeO_2C-
methyl Me $(H_3C\text{-})$
[(methyldithio)sulfonyl] $MeSSSO_2$-
methylene $H_2C=$
(methylenedioxy) = [methylenebis(oxy)] $-OCH_2O$-
[(methylenedioxy)phenyl] = 1,3-benzodioxol-*ar*-yl $(CH_2O_2)C_6H_3$-
(methylenedisulfonyl) = [methylenebis⊃ (sulfonyl)] $-SO_2CH_2SO_2$-
methylidyne HC≡
methylol = (hydroxymethyl) $HOCH_2$-
(methyloxy) = methoxy MeO-
(1-methyl-2*H*-pyranium-2-yl)

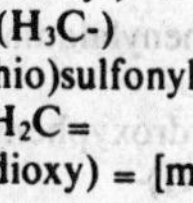

(1-methylpyridinium-2-yl)

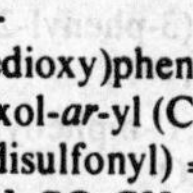

(methylselenyl) = (methylseleno) MeSe-
(methylthio) MeS-
(methyltelluro) MeTe-
(methyltrioxy) MeOOO-
morpholino = 4-morpholinyl

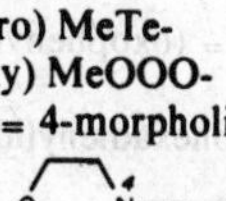

myristoyl = (1-oxotetradecyl) $Me(CH_2)_{12}CO$-
naphthal = (naphthalenylmethylene) $(C_{10}H_7)CH=$
naphthalimido = (1,3-dioxo-1*H*-benz⊃ [*de*]isoquinolin-2(3*H*)-yl) $(C_{12}H_6NO_2)$-
naphthenyl = (naphthalenylmethylidyne) $(C_{10}H_7)C\equiv$
naphthionyl = [(4-amino-1-naphthalenyl)⊃ sulfonyl] $4,1\text{-}H_2NC_{10}H_6\text{-}SO_2$-
naphthobenzyl = (naphthalenylmethyl) $(C_{10}H_7)CH_2$-
naphthothiophene-yl = naphthothienyl $(C_{12}H_7S)$-
naphthoxy = (naphthalenyloxy) $(C_{10}H_7)O$-
naphthoyl = (naphthalenylcarbonyl) $(C_{10}H_7)CO$-
naphthyl = naphthalenyl $(C_{10}H_7)$-
naphthylene = naphthalenediyl -$(C_{10}H_6)$-
naphthylidene = naphthalenylidene $(C_{10}H_6)=$
(naphthylnaphthyl) = [binaphthalen]yl $C_{10}H_7C_{10}H_6$-
nazyl = (naphthalenylmethyl) $(C_{10}H_7)CH_2$-
neopentyl = (2,2-dimethylpropyl) Me_3CCH_2-
neophyl = (2-methyl-2-phenylpropyl) $PhCMe_2CH_2$-
neryl (from nerol) = (3,7-dimethyl-2,6-octadienyl)- $Me_2C=CH(CH_2)_2CMe=CHCH_2$-
nicotinimidoyl = (imino-3-pyridinylmethyl) $(C_5H_4N)C(=NH)$-
nicotinoyl = (3-pyridinylcarbonyl)

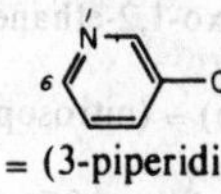

nipecotoyl = (3-piperidinylcarbonyl)

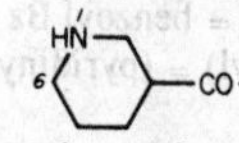

nitramino = (nitroamino) O_2NNH-
aci-nitramino = (*aci*-nitroamino) HON(O)=N-
nitrilo N≡
(nitrilophosphoranyl) = phosphono⊃ nitridyl HP(≡N)-
nitro O_2N-
aci-nitro HON(O)=
nitrosamino = (nitrosoamino) ONNH-
nitrosimino = (nitrosoimino) ONN=
nitroso ON-
(nitrothio) O_2NS-
nonanedioyl = (1,9-dioxo-1,9-nonanediyl) $-CO(CH_2)_7CO$-
nonanoyl = (1-oxononyl) $Me(CH_2)_7CO$-
norbornyl = bicyclo[2.2.1]heptyl (C_7H_{11})-
norbornylene = bicyclo[2.2.1]hept⊃ anediyl -(C_7H_{10})-
norcamphanyl = bicyclo[2.2.1]heptyl (C_7H_{11})-
norcaryl (from norcarane) = bicyclo⊃ [4.1.0]heptyl (C_7H_{11})-
norpinyl (from norpinane) = bicyclo⊃ [3.1.1]heptyl (C_7H_{11})-
norleucyl[1] = (2-amino-1-oxohexyl) $Me(CH_2)_3CH(NH_2)CO$-
norvalyl[1] = (2-amino-1-oxopentyl) $Me(CH_2)_2CH(NH_2)CO$-

octadecanoyl = (1-oxooctadecyl) $Me(CH_2)_{16}CO$-
octanedioyl = (1,8-dioxo-1,8-octanediyl) $-CO(CH_2)_6CO$-
octanoyl = (1-oxooctyl) $Me(CH_2)_6CO$-
tert-octyl = (1,1,3,3-tetramethylbutyl) $Me_3CCH_2CMe_2$-
oenanthyl = (1-oxoheptyl) $Me(CH_2)_5CO$-
oleananyl (from oleanane) $(C_{30}H_{51})$-
oleoyl = (1-oxo-9-octadecenyl) $Me(CH_2)_7CH=CH(CH_2)_7CO$-
ornithyl[1] = (2,5-diamino-1-oxopentyl) $H_2N(CH_2)_3CH(NH_2)CO$-
oxalaldehydoyl = (oxoacetyl) HCOCO-
oxalyl = (1,2-dioxo-1,2-ethanediyl) -COCO-
oxamido = [(aminooxoacetyl)amino] $H_2NCOCONH$-
oxamoyl = (aminooxoacetyl) H_2NCOCO-
oxamyl = (aminooxoacetyl) H_2NCOCO-
oxaniloyl = [oxo(phenylamino)acetyl] PhNHCOCO-
oxazolinyl = (dihydrooxazolyl) (C_3H_4NO)-
oximido = (hydroxyimino) HON=
oxo[5] (see also epoxy and oxy) O=
(oxobornyl) = (trimethyloxobicyclo[2.2.1]heptyl) $(C_{10}H_{15}O)$-
(oxoboryl) OB-
(1-oxoethyl) = acetyl Ac (MeCO-)
(oxoethylene) = (1-oxo-1,2-ethanediyl) $-COCH_2$-
(oxophenylhydrazino) = (nitrosophenylamino) PhN(NO)-
(oxophenylmethyl) = benzoyl Bz (PhCO-)
(oxopyridinylmethyl) = (pyridinylcarbonyl) $(C_5H_4N)CO$-
(2-oxotrimethylene) = (2-oxo-1,3-propanediyl) $-CH_2COCH_2$-
(2-oxovinyl) = (oxoethenyl) OC=CH-
oxy[7] (see also epoxy and oxo) -O-
[oxybis(methylenecarbonylimino)] = [oxybis[(1-oxo-2,1-ethanediyl)imino]] $-NHCOCH_2OCH_2CONH$-

palmitoyl = (1-oxohexadecyl) $Me(CH_2)_{14}CO$-
pantothenoyl = *N*-(2,4-dihydroxy-3,3-dimethyl-1-oxobutyl)-β-alanyl[1] or [3-[(2,4-dihydroxy-3,3-dimethyl-1-oxobutyl)amino]-1-oxopropyl] $HOCH_2CMe_2CH(OH)CONH(CH_2)_2CO$-
pelargonoyl = (1-oxononyl) $Me(CH_2)_7CO$-
pelargonyl = (1-oxononyl) $Me(CH_2)_7CO$-
pentadecanoyl = (1-oxopentadecyl) $Me(CH_2)_{13}CO$-
pentamethylene = 1,5-pentanediyl $-(CH_2)_5$-
3-pentanesulfonamido = [[(1-ethylpropyl)sulfonyl]amino] Et_2CHSO_2NH-
1,3-pentazadieno = 1,3-pentazadienyl $H_2NN=NN=N$-
2-pentenediylidyne = 2-pentene-1,5-diylidyne $\equiv CCH=CHCH_2C\equiv$
tert-pentyl = (1,1-dimethylpropyl) $EtCMe_2$-
pentyl $Me(CH_2)_4$-
pentylidyne BuC≡
perchloryl O_3Cl-
perseleno[5] (see also epidiseleno and diseleno) Se=Se=
perthio[5] (see also epidithio and dithio) S=S=
phenacyl = (2-oxo-2-phenylethyl) $PhCOCH_2$-
phenacylidene = (2-oxo-2-phenylethylidene) PhCOCH=
phenanthrothiophene-yl = phenanthrothienyl $(C_{16}H_9S)$-
phenanthryl = phenanthrenyl $(C_{14}H_9)$-
phenanthrylene = phenanthrenediyl $-(C_{14}H_8)$-
phenenyl = benzenetriyl $(C_6H_3)\equiv$
phenethyl = (2-phenylethyl) $PhCH_2CH_2$-
phenethylidene = (2-phenylethylidene) $PhCH_2CH=$
phenetidino = [(ethoxyphenyl)amino] $(EtO)C_6H_4NH$-
phenetyl = (ethoxyphenyl) $(EtO)C_6H_4$-
phenoxy PhO-
phenyl Ph (C_6H_5-)
phenylalanyl[1] (from phenylalanine) $PhCH_2CH(NH_2)CO$-
(phenylarsinico) = (hydroxyphenylarsinyl) PhAs(O)(OH)-
[(phenylazo)imino] = (3-phenyl-2-triazenylidene) PhN=NN=
(phenylbenzoyl) = ([1,1'-biphenyl]ylcarbonyl) PhC_6H_4CO
(phenyldiazenyl) = (phenylazo) PhN=N-
phenylene $-(C_6H_4)$-
[phenylenebis(azo)] $-N=NC_6H_4N=N$-
[phenylenebis[azo(methylimino)]] = [phenylenebis(1-methyl-2-triazene-3,1-diyl)] $-NMeN=NC_6H_4N=NNMe$-
[phenylenebis(1-oxo-1-ethanyl-2-ylidene)] = [phenylenebis(2-oxo-2-ethanyl-1-ylidene)] $=CHCOC_6H_4COCH=$
(phenylenedimethylene) = [phenylenebis(methylene)] $-CH_2C_6H_4CH_2$-
(phenylenedimethylidyne) $=CHC_6H_4CH=$
(phenylenedioxy) = [phenylenebis(oxy)] $-OC_6H_4O$-
(phenylglyoxyloyl) = (oxophenylacetyl) PhCOCO-
phenylidene = cyclohexadienylidene $(C_6H_6)=$
(phenylimidocarbonyl) = (phenylcarbonimidoyl)[9] PhN=C=
(phenyloxalyl) = (oxophenylacetyl) PhCOCO-
(phenyloxy) = phenoxy PhO-
(phenylphenoxy) = ([1,1'-biphenyl]yloxy) PhC_6H_4O-
(phenylsulfenyl) = (phenylthio) PhS-
(phenylsulfinyl) PhS(O)-
(*S*-phenylsulfonimidoyl) PhS(O)(=NH)-
phorbinyl (from phorbine) $(C_{22}H_{17}N_4)$-
phosphinico[3,4] HOP(O)=
phosphinidene HP=
phosphinidyne P≡
phosphinimyl $H_2P(=NH)$-
phosphino H_2P-
phosphinothioyl $H_2P(S)$-
phosphinothioylidene HP(S)=
phosphinyl $H_2P(O)$-
phosphinylidene HP(O)=
phosphinylidyne P(O)≡
phospho O_2P-
phosphono[4] $(HO)_2P(O)$-
(phosphonoformyl) = (phosphonocarbonyl) $(HO)_2P(O)CO$-
phosphononitridyl HP(≡N)-
phosphoranyl H_4P-
phosphoranylidene $H_3P=$
phosphoranylidyne $H_2P\equiv$
phosphoro = 1,2-diphosphenediyl -P=P-
phosphoroso OP-
phthalal = (1,2-phenylenedimethylidyne) $1,2\text{-}C_6H_4(CH=)_2$
phthalaldehydoyl = (2-formylbenzoyl) $2\text{-}(HCO)C_6H_4CO$-
phthalamoyl = [2-(aminocarbonyl)benzoyl] $2\text{-}(H_2NCO)C_6H_4CO$-
phthalanyl = (1,3-dihydroisobenzofuranyl) (C_8H_7O)-
phthalidyl = (1,3-dihydro-2-oxo-1-isobenzofuranyl) $(C_8H_5O_2)$-
phthalidylidene = (3-oxo-1(3*H*)-isobenzofuranylidene) $(C_8H_4O_2)$-
phthalimido = (1,3-dihydro-1,3-dioxo-2*H*-isoindol-2-yl) $(C_8H_4NO_2)$-
phthalocyaninyl[1] (from phthalocyanine) $(C_{32}H_{17}N_8)$-
phthaloyl = (1,2-phenylenedicarbonyl) $1,2\text{-}C_6H_4(CO\text{-})_2$
phthalylidene = (1,2-phenylenedimethylidyne) $1,2\text{-}C_6H_4(CH=)_2$
phyllocladanyl = kauranyl (from kaurane) $(C_{20}H_{33})$-
phytyl = (3,7,11,15-tetramethyl-2-hexadecenyl) $Me[CHMe(CH_2)_3]_3CMe=CHCH_2$-
picolinoyl = (2-pyridinylcarbonyl)

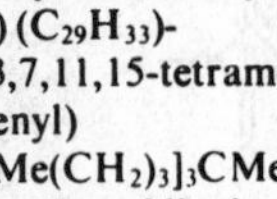

picryl = (2,4,6-trinitrophenyl) $2,4,6\text{-}(O_2N)_3C_6H_2$-
pimeloyl = (1,7-dioxo-1,7-heptanediyl) $-CO(CH_2)_5CO$-
4-pinanyl (from pinane) = (4,6,6-trimethylbicyclo[3.1.1]hept-2-yl) $(C_{10}H_{17})$-
pinanylene = (trimethylbicyclo[3.1.1]heptanediyl) $-(C_{10}H_{16})$-
pipecoloyl = (2-piperidinylcarbonyl)

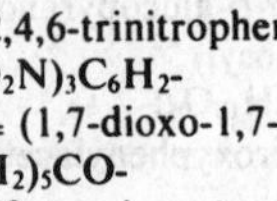

piperidino = 1-piperidinyl

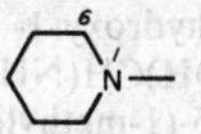

piperidyl = piperidinyl $(C_5H_{10}N)$-
piperidylidene = piperidinylidene (C_5H_9N)=
piperonyl = (1,3-benzodioxol-5-ylmethyl) 3,4-$(CH_2O_2)C_6H_3CH_2$-
piperonylidene = (1,3-benzodioxol-5-ylmethylene) 3,4-$(CH_2O_2)C_6H_3CH$=
piperonyloyl = (1,3-benzodioxol-5-ylcarbonyl) 3,4-$(CH_2O_2)C_6H_3CO$-
pivaloyl = (2,2-dimethyl-1-oxopropyl) Me_3CCO-
pivalyl = (2,2-dimethyl-1-oxopropyl) Me_3CCO-
plumbanetetrayl Pb≡
plumbyl H_3Pb-
plumbylene H_2Pb=
plumbylidyne HPb≡
podocarpan-13-yl = (tetradecahydro-4b,8,8-trimethyl-2-phenanthrenyl) $(C_{17}H_{29})$-
porphinyl (from porphine) $(C_{20}H_{13}N_4)$-
pregna-5,16-dien-21-yl (from pregnadiene) $(C_{21}H_{31})$-
prenyl = (3-methyl-2-butenyl) $Me_2C=CHCH_2$-
prolyl[1] = (2-pyrrolidinylcarbonyl)

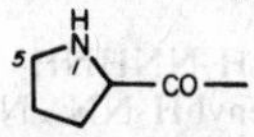

2-propanesulfonamido = [[(1-methylethyl)sulfonyl]amino] Me_2CHSO_2NH-
propargyl = 2-propynyl $HC \equiv CCH_2$-
propenyl = 1-propenyl MeCH=CH=
propenylene = 1-propene-1,3-diyl -CH=$CHCH_2$-
propenylidene = 1-propenylidene MeCH=C=
propioloyl = (1-oxo-2-propynyl) HC≡CCO-
propiolyl = (1-oxo-2-propynyl) HC≡CCO-
propionamido = [(1-oxopropyl)amino] EtCONH-
propionyl = (1-oxopropyl) EtCO-
(propionyldioxy) = [(1-oxopropyl)dioxy] EtC(O)OO-
propionyloxy = (1-oxopropoxy) $EtCO_2$-
propoxy PrO-
propyl Pr $(CH_3CH_2CH_2$-)
sec-propyl = (1-methylethyl) Me_2CH-
propylene = (1-methyl-1,2-ethanediyl) $-CHMeCH_2$-
propylidene EtCH=
propylidyne EtC≡
(propyloxy) = propoxy PrO-
protocatechuoyl = (3,4-dihydroxybenzoyl) 3,4-$(HO)_2C_6H_3CO$-
pseudoallyl = (1-methylethenyl) H_2C=CMe-
pseudocumidino = [(2,4,5-trimethylphenyl)amino] 2,4,5-$Me_3C_6H_2NH$-

as-pseudocumyl = (2,3,5-trimethylphenyl) 2,3,5-$Me_3C_6H_2$-
s-pseudocumyl = (2,4,5-trimethylphenyl) 2,4,5-$Me_3C_6H_2$-
v-pseudocumyl = (2,3,6-trimethylphenyl) 2,3,6-$Me_3C_6H_2$-
pseudoindolyl = 1*H*-indolyl (C_8H_6N)-
pteroyl = [4-[[(2-amino-1,4-dihydro-4-oxo-6-pteridinyl)methyl]amino]benzoyl] $(C_{14}H_{11}N_6O_2)$-
2*H*-pyranio

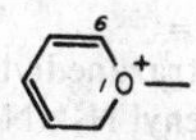

2*H*-pyran-2-ylium-2-yl

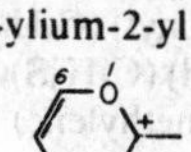

pyrazolidyl = pyrazolidinyl $(C_3H_7N_2)$-
pyrazolinyl = (dihydropyrazolyl) $(C_3H_5N_2)$-
pyridinio

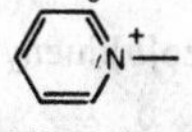

pyridyl = pyridinyl (C_5H_4N)-
pyroglutamoyl = (5-oxoprolyl)[1] or [(5-oxo-2-pyrrolidinyl)carbonyl]

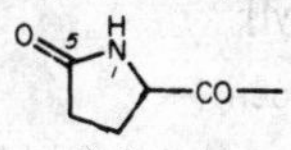

pyromucyl = (2-furanylcarbonyl)

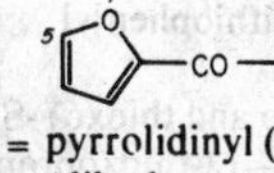

pyrrolidyl = pyrrolidinyl (C_4H_8N)-
pyrrolinyl = (dihydropyrrolyl) (C_4H_6N)-
pyrrol-1-yl = 1*H*-pyrrol-1-yl (C_4H_4N)
pyrroyl = (pyrrolylcarbonyl) $(C_4H_3N)CO$-
pyrryl = pyrrolyl (C_4H_4N)-
pyruvoyl = (1,2-dioxopropyl) MeCOCO-
p-quaterphenylyl = [1,1':4',1'':4'',1'''-quaterphenyl]yl $(C_{24}H_{17})$-
quinaldoyl = (2-quinolinylcarbonyl) $(2-C_9H_6N)CO$-
quinolyl = quinolinyl (C_9H_6N)-
quinonyl = (dioxocyclohexadienyl) $(C_6H_3O_2)$-
quinuclidinyl = 1-azabicyclo[2.2.2]octyl $(C_7H_{12}N)$-
α-resorcyloyl = (3,5-dihydroxybenzoyl) 3,5-$(HO)_2C_6H_3CO$-
β-resorcyloyl = (2,4-dihydroxybenzoyl) 2,4-$(HO)_2C_6H_3CO$-
γ-resorcyloyl = (2,6-dihydroxybenzoyl) 2,6-$(HO)_2C_6H_3CO$-
rhamnosyl = (6-deoxymannopyranosyl)

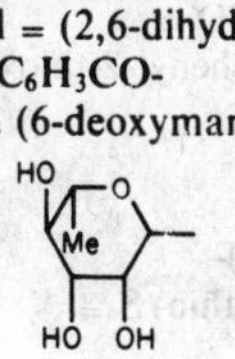

ricinelaidoyl = (12-hydroxy-1-oxo-9-octadecenyl) $Me(CH_2)_5CH(OH)CH_2CH=CH(CH_2)_7CO$-
ricinoleoyl = (12-hydroxy-1-oxo-9-octadecenyl) $Me(CH_2)_5CH(OH)CH_2CH=CH(CH_2)_7CO$-
rosan-6-yl = (2-ethyltetradecahydro-2,4a,8,8-tetramethyl-9-phenanthrenyl) $(C_{20}H_{35})$-
salicyl = [(2-hydroxyphenyl)methyl] 2-$HOC_6H_4CH_2$-
salicylidene = [(2-hydroxyphenyl)methylene] 2-HOC_6H_4CH=
salicyloyl = (2-hydroxybenzoyl) 2-HOC_6H_4CO-
sarcosyl = (*N*-methylglycyl)[1] or [(methylamino)acetyl] $MeNHCH_2CO$-
sebacoyl = (1,10-dioxo-1,10-decanediyl) $-CO(CH_2)_8CO$-
seleneno[4] HOSe-
selenino[4] HOSe(O)-
seleninyl OSe=
seleno[7] (see also episeleno and selenoxo) -Se-
selenocyanato NCSe-
selenono[4] $(HO)_2SeO_2$-
selenonyl O_2Se=
selenophenyl = selenophene-yl (C_4H_3Se)-
selenoxo[5] (see also episeleno and seleno) Se=
selenyl[4] HSe-
semicarbazido = [2-(aminocarbonyl)hydrazino] $H_2NCONHNH$-
semicarbazono = [(aminocarbonyl)hydrazono] $H_2NCONHN$=
senecioyl = (3-methyl-1-oxo-2-butenyl) $Me_2C=CHCO$-
seryl[1] = (2-amino-3-hydroxy-1-oxopropyl) $HOCH_2CH(NH_2)CO$-
siamyl = (1,2-dimethylpropyl) $Me_2CHCHMe$-
silanetetrayl Si≡
siloxy = (silyloxy) H_3SiO-
silyl H_3Si-
silylene H_2Si=
silylidyne HSi≡
sorboyl = (1-oxo-2,4-hexadienyl) MeCH=CHCH=CHCO-
spirostanyl (from spirostane) $(C_{27}H_{43}O_2)$-
stannanetetrayl Sn≡
stannono[4] HOSn(O)-
stannyl H_3Sn-
stannylene H_2Sn=
stannylidyne HSn≡
stearoyl = (1-oxooctadecyl) $Me(CH_2)_{16}CO$-
stibinico[3,4] HOSb(O)=
stibino H_2Sb-
stibo O_2Sb-
stibono[4] $(HO)_2Sb(O)$-
stiboso OSb-

stibyl = stibino H_2Sb-
stibylene $HSb=$
stibylidyne $Sb\equiv$
styrene = (1-phenyl-1,2-ethanediyl) $-CHPhCH_2-$
styrolene = (1-phenyl-1,2-ethanediyl) $-CHPhCH_2-$
styryl = (2-phenylethenyl) $PhCH=CH-$
suberoyl = (1,8-dioxo-1,8-octanediyl) $-CO(CH_2)_6CO-$
succinaldehydoyl = (1,4-dioxobutyl) $HCO(CH_2)_2CO-$
succinamoyl = (4-amino-1,4-dioxobutyl) $H_2NCO(CH_2)_2CO-$
succinamyl = (4-amino-1,4-dioxobutyl) $H_2NCO(CH_2)_2CO-$
succinaniloyl = [1,4-dioxo-4-(phenylamino)butyl] $PhNHCO(CH_2)_2CO-$
succinimido = (2,5-dioxo-1-pyrrolidinyl)

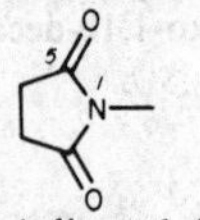

succinyl = (1,4-dioxo-1,4-butanediyl) $-CO(CH_2)_2CO-$
sulfamino = (sulfoamino) $HOSO_2NH-$
sulfamoyl = (aminosulfonyl) H_2NSO_2-
sulfamyl = (aminosulfonyl) H_2NSO_2-
sulfanilamido = [[(4-aminophenyl)sulfonyl]amino] $4\text{-}H_2NC_6H_4SO_2NH-$
sulfanilyl = [(4-aminophenyl)sulfonyl] $4\text{-}H_2NC_6H_4SO_2-$
sulfeno[4] $HOS-$
sulfhydryl = mercapto[4] $HS-$
sulfino[4] $HOS(O)-$
sulfinyl $OS=$
sulfo[4] HO_3S-
sulfonyl $-SO_2-$
sulfuryl = sulfonyl $-SO_2-$

tartaroyl = (2,3-dihydroxy-1,4-dioxo-1,4-butanediyl) $-COCH(OH)CH(OH)CO-$
tartronoyl = (2-hydroxy-1,3-dioxo-1,3-propanediyl) $-COCH(OH)CO-$
tauryl = [(2-aminoethyl)sulfonyl] $H_2N(CH_2)_2SO_2-$
telluro[6] (see also telluroxo) $-Te-$
telluroxo[5] (see also telluro) $Te=$
telluryl[4] $HTe-$
terephthalal = (1,4-phenylenedimethylidyne) $1,4\text{-}C_6H_4(CH=)_2$
terephthalaldehydoyl = (4-formylbenzoyl) $4\text{-}HCOC_6H_4CO-$
terephthalamoyl = [4-(aminocarbonyl)benzoyl] $4\text{-}H_2NC_6H_4CO-$
terephthalaniloyl = [4-[(phenylamino)carbonyl]benzoyl] $4\text{-}(PhNHCO)C_6H_4CO-$
terephthaloyl = (1,4-phenylenedicarbonyl) $1,4\text{-}C_6H_4(CO-)_2$
terephthalylidene = (1,4-phenylenedimethylidyne) $1,4\text{-}C_6H_4(CH=)_2$
m-terphenylyl = [1,1':3',1''-terphenyl]yl $(C_{18}H_{13})-$
terphenylylene = [terphenyl]diyl $-(C_{18}H_{12})-$
tetradecanoyl = (1-oxotetradecyl) $Me(CH_2)_{12}CO-$
tetramethylene = 1,4-butanediyl $-(CH_2)_4-$
1,4-tetraphosphinediyl $-(PH)_4-$
tetrasiloxanylene = 1,7-tetrasiloxanediyl $-SiH_2(OSiH_2)_2OSiH_2-$
tetrathio[11] $-SSSS-$
tetrazanediylidene = 1,4-tetrazanediylidene $=N(NH)_2N=$
tetrazanylene = 1,4-tetrazanediyl $-(NH)_4-$
1-tetrazeno = 1-tetrazenyl $H_2NNHN=N-$
thenoyl = (thienylcarbonyl) $(C_4H_3S)CO-$
thenyl = (thienylmethyl) $(C_4H_3S)CH_2-$
thenylidene = (thienylmethylene) $(C_4H_3S)CH=$
(thenyloxy) = (thienylmethoxy) $(C_4H_3S)CH_2O-$
thexyl = 1,1,2-trimethylpropyl Me_2CHCMe_2-
thianaphthenyl = benzo[*b*]thienyl $(C_8H_5S)-$
thiazolidyl = thiazolidinyl $(C_3H_6NS)-$
thiazolinyl = (dihydrothiazolyl) $(C_3H_4NS)-$
[(5-thiazolylcarbonyl)methyl] = [2-oxo-2-(5-thiazolyl)ethyl]

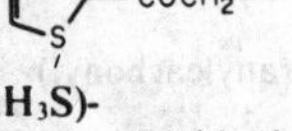

thienyl $(C_4H_3S)-$
(thienylthienyl) = [bithiophen]yl $(C_4H_3S)(C_4H_2S)-$
thio[7] (see also epithio and thioxo) $-S-$
(thioacetonylidene) = (2-thioxopropylidene) $MeCSCH=$
(thiobenzoyl) = (phenylthioxomethyl) $PhCS-$
(thiocarbamoyl) = (aminothioxomethyl) H_2NCS-
thiocarbamyl = (aminothioxomethyl) H_2NCS-
(thiocarbonyl) = carbonothioyl[9] $-CS-$
(thiocarboxy)[12] $HOSC-$
thiocyanato $NCS-$
thiocyano = thiocyanato $NCS-$
(thioformyl) = (thioxomethyl) $HCS-$
(thiohexanoyl) = (1-thioxohexyl) $Me(CH_2)_4CS-$
thiohydroperoxy = sulfeno[4] $HOS-$
(thiohydroxy) = mercapto[4] $HS-$
thiomorpholino = 4-thiomorpholinyl

(thionitroso) $SN-$
thionyl = sulfinyl $-SO-$
thiophenacyl = (2-phenyl-2-thioxoethyl) $PhCSCH_2-$
thioseleneno[4] $HSSe-$
thiosulfeno[4] $HSS-$
(thiosulfo)[4] $(HO_2S_2)-$
thioxo[5] (see also epithio) $S=$
(thioxoarsino) $SAs-$
thiuram = (aminothioxomethyl) H_2NCS-
threonyl[1] (2-amino-3-hydroxy-1-oxobutyl) $MeCH(OH)CH(NH_2)CO-$
4-thujyl = [2-methyl-5-(1-methylethyl)bicyclo[3.1.0]hex-2-yl] $(C_{10}H_{17})-$
thymyl (from thymol) = [5-methyl-2-(1-methylethyl)phenyl]

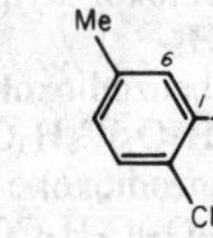

thyronyl = [*O*-(4-hydroxyphenyl)tyrosyl][1] or [2-amino-3-[4-(4-hydroxyphenoxy)phenyl]-1-oxopropyl] $[4\text{-}(4\text{-}HOC_6H_4O)C_6H_4]CH_2CH(NH_2)CO-$
toloxy = (methylphenoxy) MeC_6H_4O-
p-toluenesulfonamido = [[(4-methylphenyl)sulfonyl]amino] $4\text{-}MeC_6H_4SO_2NH-$
toluidino = [(methylphenyl)amino] MeC_6H_4NH-
toluoyl = (methylbenzoyl) MeC_6H_4CO-
toluyl = (methylbenzoyl) MeC_6H_4CO-
tolyl = (methylphenyl) MeC_6H_4-
α-tolyl = (phenylmethyl) $PhCH_2-$
tolylene = (methylphenylene) $-(MeC_6H_3)-$
α-tolylene = (phenylmethylene) $PhCH=$
tosyl = [(4-methylphenyl)sulfonyl] $4\text{-}MeC_6H_4SO_2-$
triazano = triazanyl $H_2NNHNH-$
1-triazeno = 1-triazenyl $H_2NN=N-$
s-triazin-2-yl = 1,3,5-triazin-2-yl

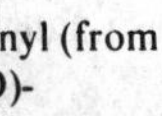

trichothecanyl (from trichothecane) $(C_{15}H_{25}O)-$
tridecanoyl (1-oxotridecyl) $Me(CH_2)_{11}CO-$
(trimethylammonio) Me_3N^+-
(trimethylarsonio) Me_3As^+-
trimethylene = 1,3-propanediyl $-(CH_2)_3-$
(1,3,3-trimethyl-2-norbornyl) = (1,3,3-trimethylbicyclo[2.2.1]hept-2-yl) $(C_{10}H_{17})-$
(trimethylphosphonio) Me_3P^+-
triseleno[11] $-SeSeSe-$
trisilanylene = 1,3-trisilanediyl $-(SiH_2)_3-$
trisiloxane-1,3,5-triyl = 1,3,5-trisiloxanetriyl $-SiH(OSiH_2-)_2$
trithio[11] $-SSS-$
trityl = (triphenylmethyl) Ph_3C-
tropanyl = 8-azabicyclo[3.2.1]octyl $(C_8H_{14}N)-$
tropoyl = (3-hydroxy-1-oxo-2-phenylpropyl) $HOCH_2CHPhCO-$
tryptophyl[1] = [2-amino-3-(1*H*-indol-3-yl)-1-oxopropyl] $(C_8H_6N)CH_2CH(NH_2)CO-$
tyrosyl[1] = [2-amino-3-(4-hydroxyphenyl)-1-oxopropyl] $4\text{-}HOC_6H_4CH_2CH(NH_2)CO-$

undecanoyl = (1-oxoundecyl) $Me(CH_2)_9CO$-
uramino = [(aminocarbonyl)amino] H_2NCONH-
ureido = [(aminocarbonyl)amino] H_2NCONH-
ureylene = (carbonyldiimino) -NHCONH-
(ureylenediureylene) = [carbonylbis(hydrazocarbonylimino)] -NHCONHNHCONHNHCONH-
ursanyl (from ursane) ($C_{30}H_{51}$)-

valeryl = (1-oxopentyl) BuCO-
valyl[1] = (2-amino-3-methyl-1-oxobutyl) $Me_2CHCH(NH_2)CO$-
vanillal = [(4-hydroxy-3-methoxyphenyl)methylene] $4,3\text{-}HO(MeO)C_6H_3CH=$
vanilloyl = (4-hydroxy-3-methoxybenzoyl) $4,3\text{-}HO(MeO)C_6H_3CO$-
vanillyl = [(4-hydroxy-3-methoxyphenyl)methyl] $4,3\text{-}HO(MeO)C_6H_3CH_2$-
vanillylidene = [(4-hydroxy-3-methoxyphenyl)methylene] $4,3\text{-}HO(MeO)C_6H_3CH=$
vanilmandeloyl = [hydroxy(4-hydroxy-3-methoxyphenyl)acetyl] $4,3\text{-}HO(MeO)C_6H_3CH(OH)CO$-
veratral = [(3,4-dimethoxyphenyl)methylene] $3,4\text{-}(MeO)_2C_6H_3CH=$
veratroyl = (3,4-dimethoxybenzoyl) $3,4\text{-}(MeO)_2C_6H_3CO$-
o-veratroyl = (2,3-dimethoxybenzoyl) $2,3\text{-}(MeO)_2C_6H_3CO$-
veratryl = [(3,4-dimethoxyphenyl)methyl] $3,4\text{-}(MeO)_2C_6H_3CH_2$-
o-veratryl = [(2,3-dimethoxyphenyl)methyl] $2,3\text{-}(MeO)_2C_6H_3CH_2$-
veratrylidene = [(3,4-dimethoxyphenyl)methylene] $3,4\text{-}(MeO)_2C_6H_3CH=$
vinyl = ethenyl $H_2C=CH$-
vinylene = 1,2-ethenediyl -CH=CH-
vinylidene = ethenylidene $H_2C=C=$

xanthen-9-yl = 9*H*-xanthen-9-yl ($C_{13}H_9O$)-
xanth-9-yl = 9*H*-xanthen-9-yl ($C_{13}H_9O$)-
xenyl = [1,1′-biphenyl]yl ($C_6H_5C_6H_4$)-
xylidino = [(dimethylphenyl)amino] $Me_2C_6H_3NH$-
xyloyl = (dimethylbenzoyl) $Me_2C_6H_3CO$-
xylyl = (dimethylphenyl) $Me_2C_6H_3$-
xylylene = [phenylenebis(methylene)] $-CH_2C_6H_4CH_2$-

[1] This prefix is used in peptide nomenclature.
[2] This prefix may be used in a generic sense, e.g., aldoxime.
[3] This prefix is used only as a multiplying radical.
[4] This prefix is used only when unsubstituted.
[5] This prefix is used when both free valencies are attached to the same atom.
[6] This prefix is used when the free valencies are attached to different atoms which are usually not otherwise connected.
[7] This prefix is used when the free valencies are attached to different atoms which are not otherwise connected.
[8] This prefix is used when the free valencies are attached to different atoms which are otherwise connected.
[9] This prefix is used as a multiplying radical or when both free valencies are attached to the same atom.
[10] This prefix may be used in a generic sense, e.g., ketoxime.
[11] This prefix is used to denote a series of chalcogen atoms in a chain or an indefinite structure.
[12] This prefix is not used when the hydrogen atom has been substituted by another atom or group if a definite structure can be determined.
[13] The prefix "cumyl" has been used in the recent literature to mean α-cumyl.

From *Chem. Abstracts,* Vol. 76, p. 1251. With permission of the American Chemical Society.

ORGANIC RING COMPOUNDS

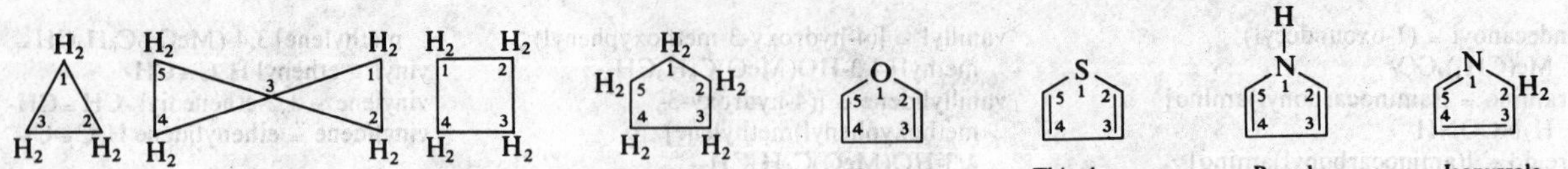

Cyclopropane | Spiropentane | Cyclobutane | Cyclopentane | Furan | Thiophene (Thiofuran) | Pyrrole (Azole) | Isopyrrole (Isoazole) (α-Pyrrolenine)

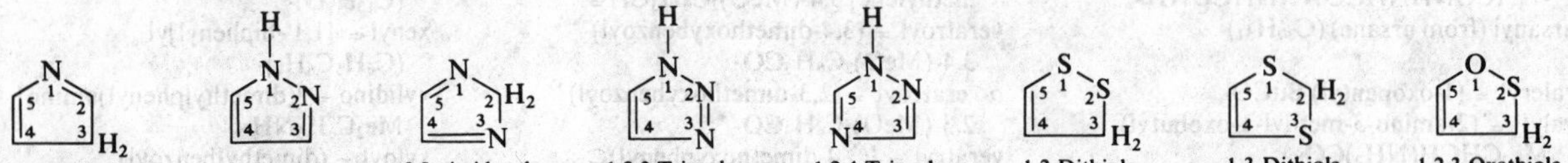

3-Isopyrrole (Isoazole) (β-Pyrrolenine) | Pyrazole (1,2-Diazole) | 2-Isoimidazole (1,3-Isodiazole) | 1,2,3-Triazole | 1,2,4-Triazole | 1,2-Dithiole | 1,3-Dithiole | 1,2,3-Oxathiole

Isoxazole (Furo[α]monazole) | Oxazole (Furo[β]monazole) | Thiazole | Isothiazole | 1,2,3-Oxadiazole | 1,2,4-Oxadiazole (Azoxime) | 1,2,5-Oxadiazole (Furazan) | 1,3,4-Oxadiazole

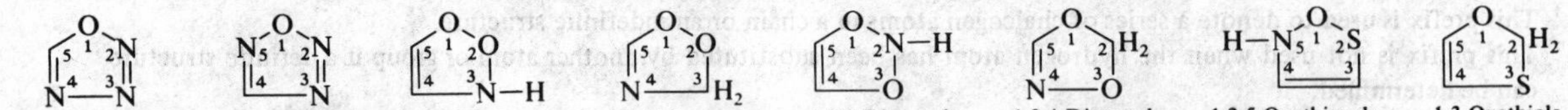

1,2,3,4-Oxatriazole | 1,2,3,5-Oxatriazole | 1,2,3-Dioxazole | 1,2,4-Dioxazole | 1,3,2-Dioxazole | 1,3,4-Dioxazole | 1,2,5-Oxathiazole | 1,3-Oxathiole

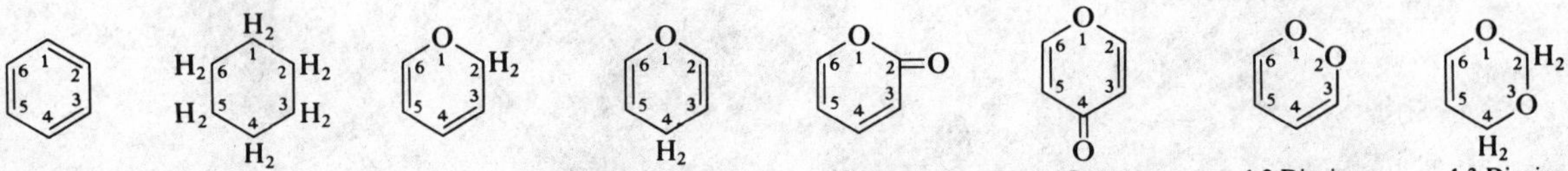

Benzene | Cyclohexane | 1,2-Pyran | 1,4-Pyran | 1,2-Pyrone (α-Pyrone) | 1,4-Pyrone (γ-Pyrone) | 1,2-Dioxin | 1,3-Dioxin, Dihydro form

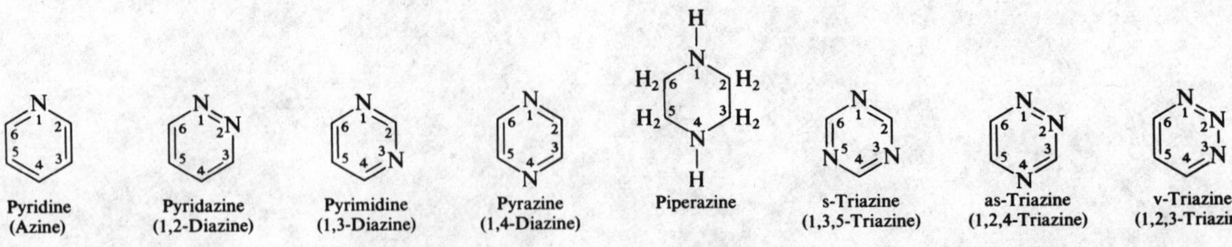

Pyridine (Azine) | Pyridazine (1,2-Diazine) | Pyrimidine (1,3-Diazine) | Pyrazine (1,4-Diazine) | Piperazine | s-Triazine (1,3,5-Triazine) | as-Triazine (1,2,4-Triazine) | v-Triazine (1,2,3-Triazine)

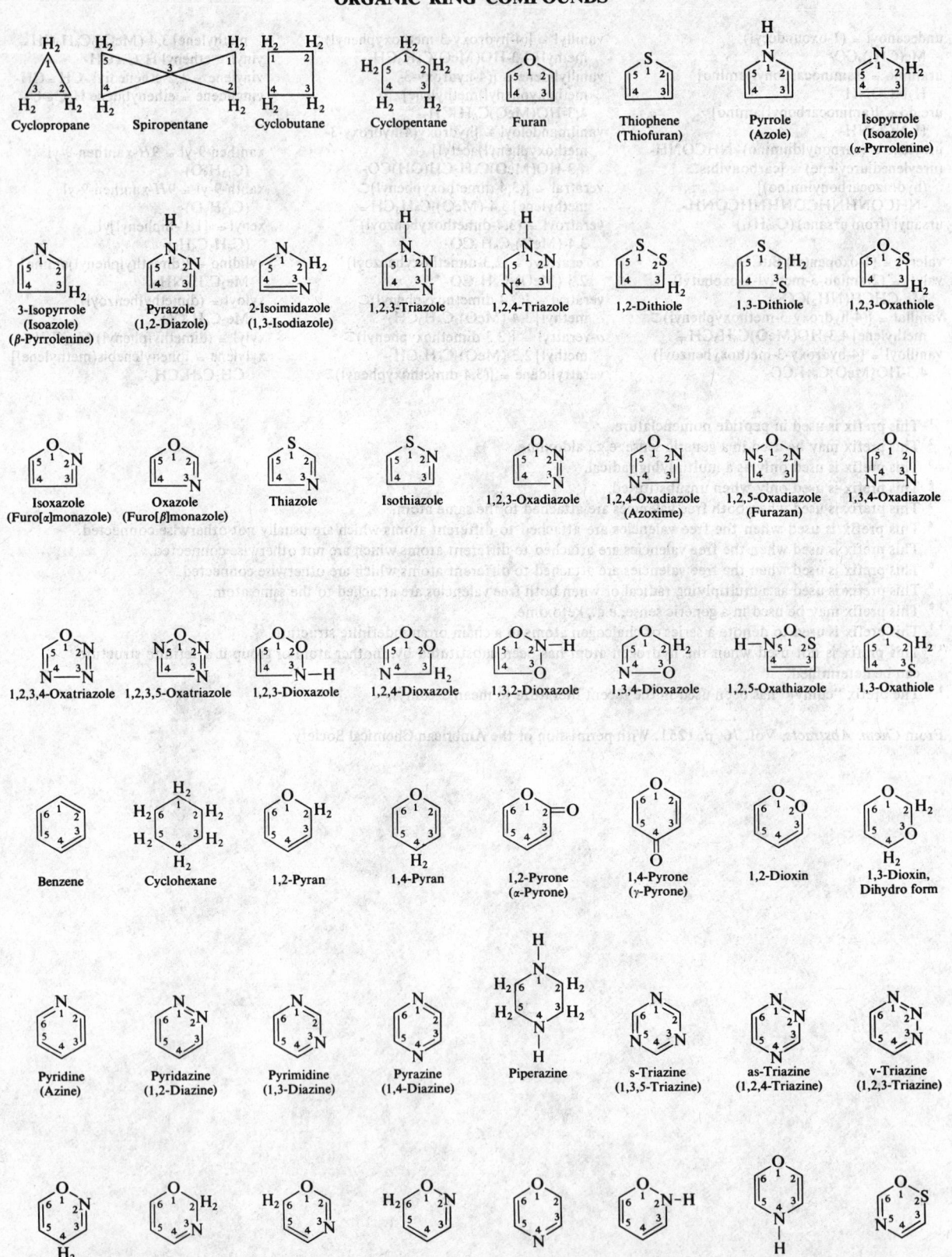

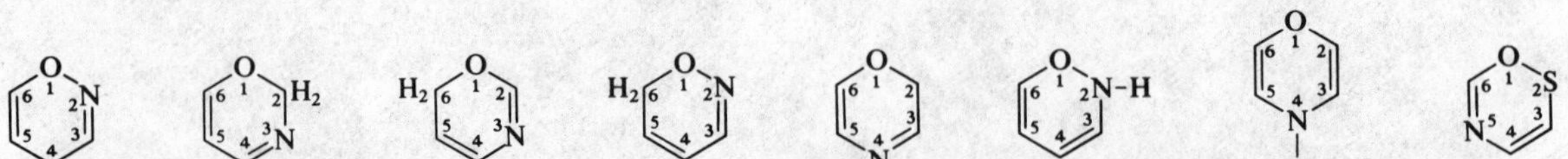

1,2,4-Oxazine | 1,3,2-Oxazine | 1,3,6-Oxazine (Pentoxazole) | 1,2,6-Oxazine | 1,4-Oxazine | o-Isoxazine | p-Isoxazine | 1,2,5-Oxathiazine

ORGANIC RING COMPOUNDS (Continued)

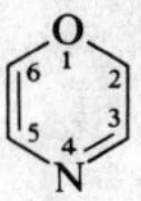
1,4-Oxazine

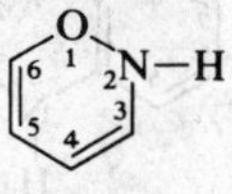
o-Isoxazine

p-Isoxazine

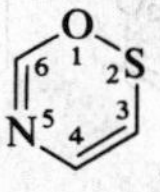
1,2,5-Oxathiazine

1,2,6-Oxathiazine

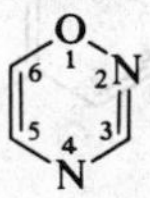
1,4,2-Oxadiazine

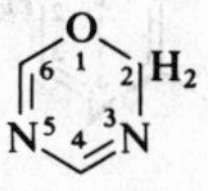
1,3,5,2-Oxadiazine

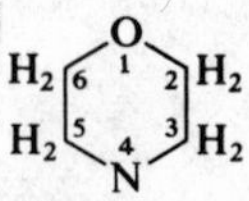
Morpholine (Tetrahydro-p-isoxazine)

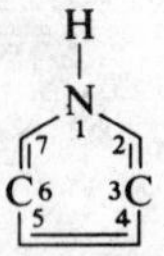
Azepine

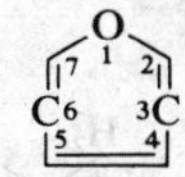
Oxepin

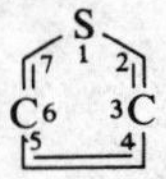
Thiepin

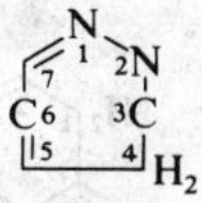
1,2,4-Diazepine

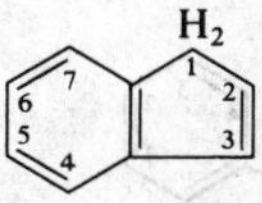
Indene

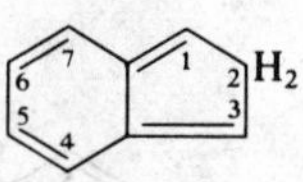
Isoindene

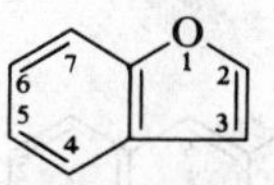
Benzofuran (Coumarone)

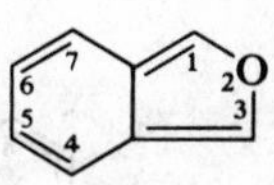
Isobenzofuran

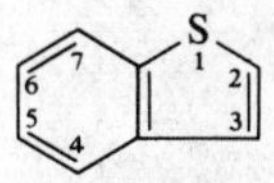
Thionaphthene (Benzothiofuran)

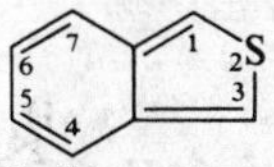
Isothionaphthene (Isobenzothiofuran)

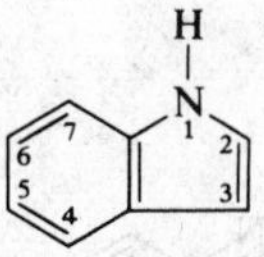
Indole

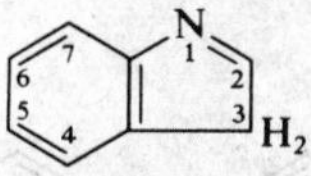
Indolenine (3-Pseudoindole)

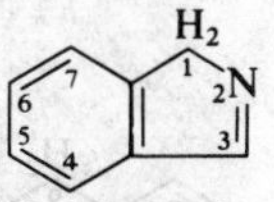
2-Isobenzazole (Pseudoisoindole)

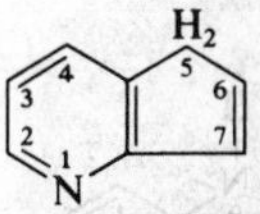
1,5-Pyrindine (4-Pyrindine)

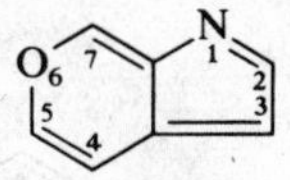
Pyrano[3,4-b]-pyrrole

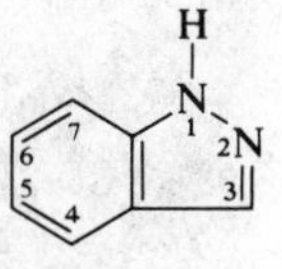
Isoindazole (Benzpyrazole)

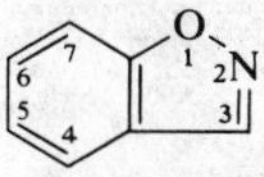
Indoxazine (Benzisoxazole)

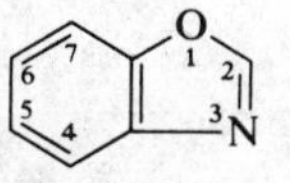
Benzoxazole

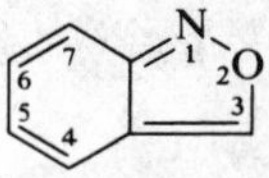
Anthranil

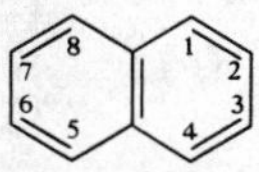
Naphthalene

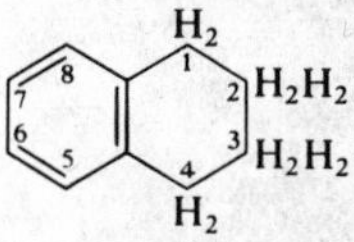
Tetralin

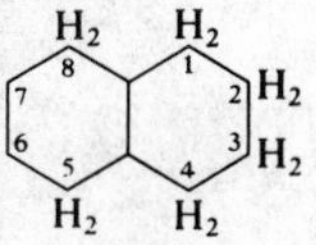
Decalin (Bicyclo[4,4,0]-decane)

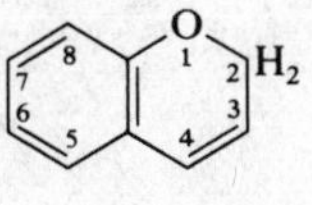
1,2-Benzopyran (1,2-Chromene)

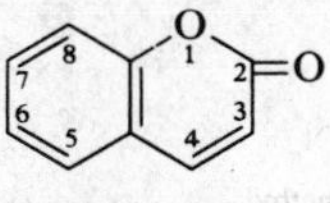
Coumarin (1,2-Benzopyrone)

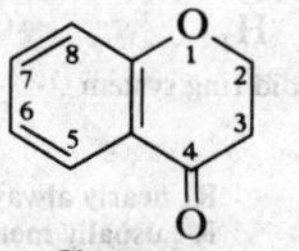
Chromone (1,4-Benzopyrone)

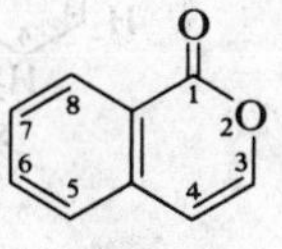
Isocoumarin (2,1-Benzopyrone)

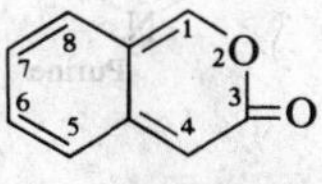
2,3-Benzopyrone

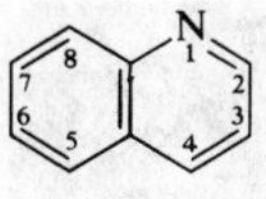
Quinoline (1-Benzazine)

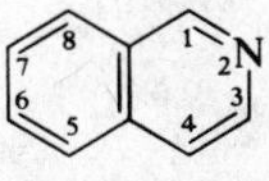
Isoquinoline (2-Benzazine)

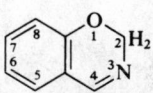

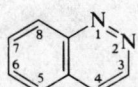

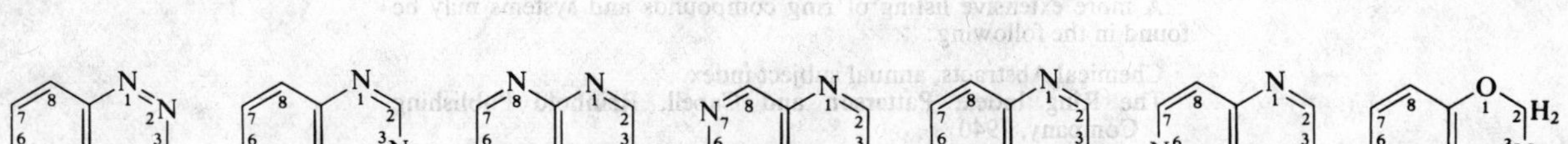
Cinnoline (1,2-Benzodiazine) — Quinazoline (1,3-Benzodiazine) — Naphthyridine — Pyrido[3,4-b]-pyridine — Pyrido[3,2-b]-pyridine — Pyrido[4,3-b]-pyridine — 1,3,2-Benzoxazine

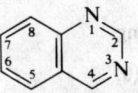

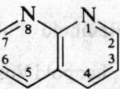

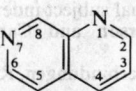

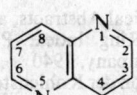

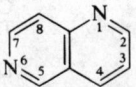

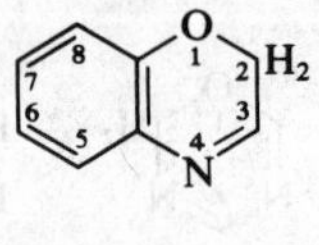
1,4,2-Benzoxazine

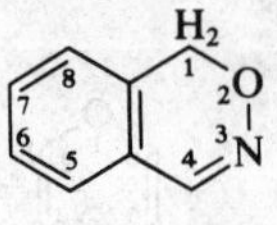
2,3,1-Benzoxazine

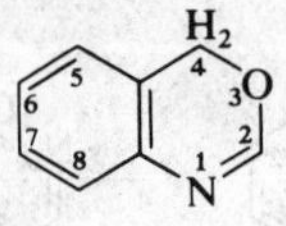
3,1,4-Benzoxazine

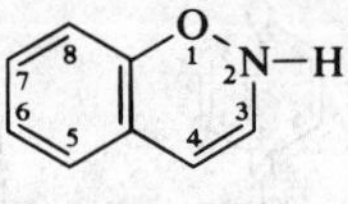
1,2-Benzisoxazine

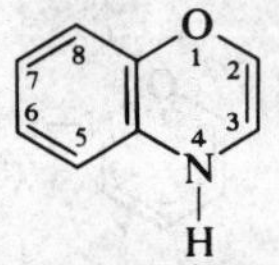
1,4-Benzisoxazine

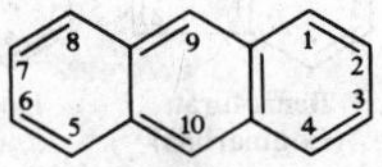
Anthracene

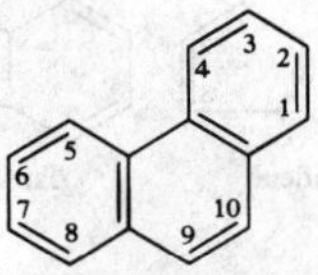
Phenanthrene

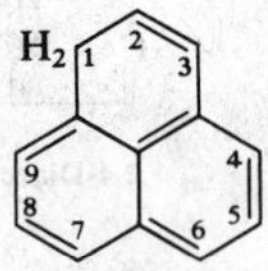
Benzonaphthene
(Phenalene)

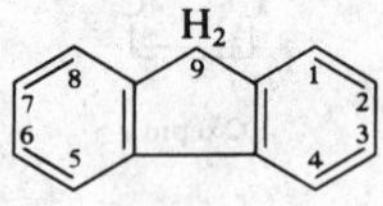
Fluorene

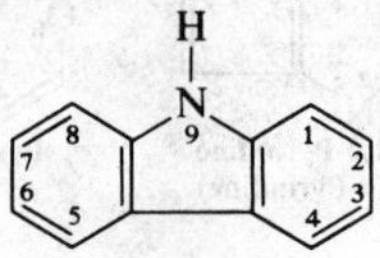
Carbazole

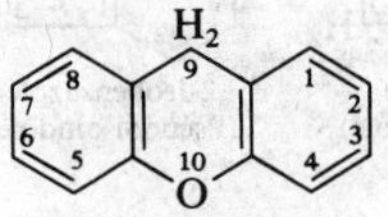
Xanthene

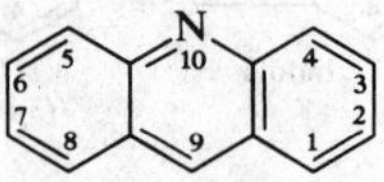
Acridine

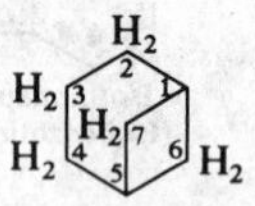
Norpinane

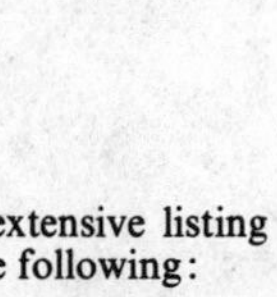
Purine

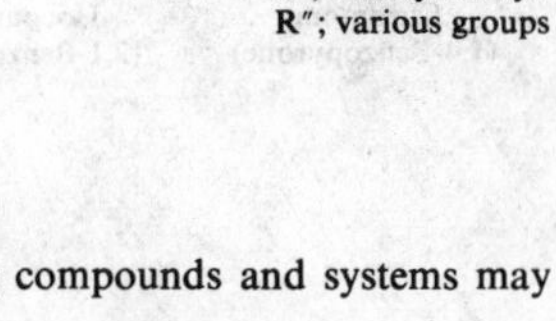
Steroid ring system

R; nearly always methyl
R′; usually methyl
R″; various groups

A more extensive listing of ring compounds and systems may be found in the following:

Chemical Abstracts, annual subject index.
The Ring Index, Patterson and Capell, Reinhold Publishing Company, 1940.
Lexikon der Kohlenstoffverbindungen, Richter, Leopold Voss, 1910.

The numbering system for the compounds listed above is that used in The Ring Index.

TABLE OF PHYSICAL PROPERTIES OF ORGANIC COMPOUNDS
EXPLANATION OF TABLE

The section of this book dealing with Physical Constants of Organic Compounds was prepared by Saul Patai, Ph.D. and J. Zabicky, Ph.D. Enlarged and revised by Zvi Rappoport, Ph.D., 1969

KEY TO THE EXPLANATION
(numbers refer to paragraphs)

TABLE OF PHYSICAL PROPERTIES OF ORGANIC COMPOUNDS

EXPLANATION OF TABLE

1. The table of Physical Constants of Organic Compounds is a compilation of data on some 13600 organic compounds that include those of wide application in teaching, industry, medicine and research.

CONTENTS OF TABLE

2. **Number of entry**. Each entry has a number preceded by a letter corresponding to the first letter of the name of the parent compound (see paragraphs 17–26, 31, 32). Cross references are left unnumbered. The number of entry is used in the formula, melting point and boiling point indexes (see paragraphs 13–15).

3. **Name of compound.** See rules for naming of compounds (paragraphs 16 ff.).

4. **Synonyms and formula.**

a) *Synonyms*. For the sake of brevity, the number of synonyms in an entry is limited. Additional synonyms may readily be formed by derivation from the names given to simpler or less substituted compounds; *e.g.*: **Benzoic acid, 4-nitro,** ethyl ester is given no synonyms but they may be derived from those appearing under **Benzoic acid,** for example ethyl *p*-nitrobenzene-carboxylate.

b) *Formula*. Each entry has either a structural or an empirical formula. For compounds having complex structural formula both are given.

i) Structural formulas—simple ones are given in this column, complex ones at the end of the Table.

ii) Empirical formulas are given if the compound is cyclic and appears among a series of compounds derived from the same cyclic parent compound; e.g.:

u134 **Uric acid, 7-methyl-** $C_6H_6N_4O_3$. *See* u131 means that in entry u131 appears the numbered skeleton of **Uric acid,** and the structural formula of u134 may be derived from that of u131.

Structural formulas in the Table give generally no configurational details, and when they do, they seldom give absolute configurations. These details are to be looked for either in the original literature or in specialized textbooks.

5. **Molecular weight.** Computed according to International Atomic Weight values of 1961.

6. **Color, crystalline form, specific rotation and max(log ε).**

a) *Color*: Solid compounds are to be considered as crystalline if not otherwise stated. Data on solvents used and solvents of crystallization are given in brackets. Some examples will explain the usage of the Table (see also list of abbreviations):

ABBREVIATION	MEANING
pl	plates
pl(al)	plates obtained from alcohol as the solvent
pl(al + $1\frac{1}{2}$)	plates obtained from alcohol crystallizing with $1\frac{1}{2}$ molecules of the **same solvent** per molecule of compound
pl(al + ?)	as above but the amount of solvent of crystallization per molecule of compound is undefined
pl(aq ace + 2w)	here the solvent of crystallization is one of the components of the solvent, *i.e.*, two molecules of water
(al)	compound crystallizes from alcohol but crystalline form is not reported.

If no melting point is given in the Table the compound is a liquid at room temperature, unless otherwise stated.

b) *Crystalline form*: No special remark is made if the compound is colorless or no data on its color are available; otherwise an abbreviation appears before the crystalline form. (See table of abbreviations).

c) *Specific rotation* appears according to the common usage, *e.g.*:

$$[\alpha]_D^{25} - 25.8(w, c = 4)$$

d) λ_{max}(log ε): The main maxima in the ultraviolet spectra, in the solvent shown in the superscript (see list of abbreviations) are given with the logarithm of the molar absorption coefficient ε. Solvent is missing when it was not given in the original reference, log ε is missing when it was not determined. Mostly the long wave absorption maxima is only given, but for important compounds (especially aromatic ones) some of the lower wavelength maxima are also given.

Bracketed abbreviations indicate solvent and concentration, *i.e.*, in the above example, solvent water and concentration 4%.

EXPLANATION OF TABLE (Continued)

7. **Melting point.** Remarks on this physical property appear as abbreviations after the m.p. Melting points of questionable accuracy are given in brackets.

8. **Boiling point.** The pressure at which this physical property was determined appears as a superscript. If no superscript is given, pressure is about one atmosphere. Remarks on this physical property appear as abbreviations. Boiling points of questionable accuracy are given in brackets.

9. **Density** is relative to water, otherwise it has the dimensions g/ml. A superscript indicates the temperature of the liquid and a subscript indicates the temperature of water to which the density is referred.

10. **Refractive index** is reported for the D line of the spectrum of sodium (n_D). The temperature of determination of this physical property appears as a superscript.

11. **Solubility.** Data are given for the most common solvents. As numerical data on this property scarcely appear in the literature, and when they appear their degree of approximation is not great, it was preferred to rely on the intuitive meaning that the following scale has for the chemist (corresponding abbreviations in brackets): insoluble (i), slightly soluble (δ), soluble (s), very soluble (v), miscible (∞), decomposes (d). If no special remark is made about the temperature, the reference is to room temperature; otherwise, a superscript appears.

12. **Literature references** are given to general sources of data such as *Beilstein's Handbook, Elsevier's Encyclopaedia,* etc. The superscript after the boldtype number indicates the number of the supplement (Ergenzungswerk) of *Beilstein's Handbook;* e.g.: **B7**^{1}, 12 means page 12 of the first supplement to volume 7 of *Beilstein.*

INDEXES

Three indexes are found at the end of the Table:

13. **Empirical formula index** is arranged according to increasing C, H, and remaining elements in alphabetical order. Hydrates are entered under the formula of the *anhydrous* compound. Salts, complexes, etc., are found under their total formulae.

14. **Melting point index** is arranged according to increasing melting or freezing points. Only the lower temperature of the melting range is given. Remarks to the melting point in the main table, such as decomposition, sublimation, etc., are omitted. Melting points are rounded off to the nearest unit.

15. **Boiling point index** is arranged according to increasing boiling point of compounds of which the boiling point is given at a pressure near to 1 atm. (*i.e.,* 700–780 mm. Hg). Only the lower temperature of the boiling range is given. No special entries are made for remarks on the boiling point contained in the main table. Boiling points are rounded off to the nearest unit.

RULES FOR THE NAMING OF COMPOUNDS*

16. Naming of compounds is based mainly on *International Union of Chemistry* (*IUC*) and *Chemical Abstracts* usages, with variations adapted to the scope and range of the Table, in order to enable the user to find compounds even when their trivial names are not known.

17. A suitable systematic nomenclature is used in order to bring as near as possible compounds that are isomeric derivatives of the same parent compound, *e.g.,* physical properties appear under the entries **Benzene, 1,2-dihydroxy-, Benzene, 1,3-dihydroxy-,** and **Benzene, 1,4-dihydroxy-;** under **Hydroquinone, Pyrocatechol** and **Resorcinol** a cross reference directs the user towards the systematic names. No special entry or cross-reference is made for *e.g.,* **Bromoresorcinol;** this will be found under **Benzene, bromo(dihydroxy)-,** with the appropriate numbering before the names of the substituents. Another example: **4-Chloroquinaldine;** the trivial name of the parent compound is **Quinaldine,** it appears as: **Quinaldine** *see* **Quinoline, 2-methyl-;** accordingly the properties of **4-Chloroquinaldine** will appear under **Quinoline, 4-chloro-2-methyl-.**

18. Physical constants of a compound having widely used trivial and systematic names will appear under one of them only. The remaining names will appear with a cross-reference to the assigned name, *e.g.:*

Resorcinol *see* **Benzene, 1,3-dihydroxy-**

Propanoic acid, 2-amino- *see* **Alanine**

Very complicated compounds such as some natural products, dyes, etc. appear almost exclusively under their trivial names.

---

* The rules for nomenclature followed by the Table were adapted to the present, relatively limited scope of some thousands of compounds. For other use the IUPAC 1957 rules, which appear in this Handbook, may be consulted.

EXPLANATION OF TABLE (Continued)

19. Rules adopted by the Table for the naming of compounds according to their functional groups appear in paragraphs 27 ff.

20. The naming of a compound has three phases:

a) Determination of the parent compound.
b) Determination and numbering of the substituents in the skeleton of the parent compound.
c) Determination of the derivatives of the principal functions of the compound.

e.g.:

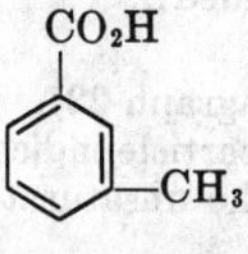

Phase a) **Benzoic acid**
Phase b) **3-methyl-**

Accordingly it appears as **Benzoic acid, 3-methyl-**

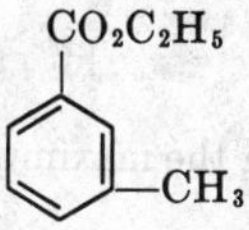

Phase a) **Benzoic acid**
Phase b) **3-methyl-**
Phase c) ethyl ester

therefore it appears as **Benzoic acid, 3-methyl-**, ethyl ester.

21. The parent compound has at most one kind of principal function, (*e.g.:* acid) but this function may be multiple, (*e.g.:* diacid, triacid, etc.); examples:

$CH_3COCH_2CH_2CO_2H$ appears as **Pentanoic acid, 4-oxo-**
not as: **Pentanon-4-on-1-oic acid**

$HO_2CCH_2CH(CH_3)COCH_2CO_2H$ appears as **Hexanedioic acid, 3-methyl-4-oxo-**

Unsaturation is not considered as a function and, when possible, it is included in the name of the parent compound; *e.g.:*

$CH_2{:}CHCH_2CO_2H$ appears as **3-Butenoic acid.**

Exceptions to this rule are some cyclic ketones that may be designated as ring systems (see paragraph 83) and some natural products, such as **Pregnenolone** (the functions alcohol and ketone appear in the parent compound).

22. Entries in the Table are made first according to alphabetical order of parent compounds (paragraph 20, phase a), and then according to alphabetical order of the substituents in the skeleton of the parent compound (paragraph 20, phase b). It should be emphasized that numbers or single letters that are used in the designation of the compound are not considered for the alphabetisation. If the designation of a compound includes phase c of paragraph 20 this compound appears immediately after the compound that is designated without phase c and has the same substituents bearing the same numbers; *e.g.:*

Benzoic acid
Benzoic acid, 2-chloro-
Benzoic acid, 2-chloro-, ethyl ester
Benzoic acid, 3-chloro-
Benzoic acid, 3-chloro-, allyl ester

23. Prefixes indicating multiplicity of substituents as **di-, tri-, bis-, tris**, etc. and the prefix **iso-**, as in **dimethyl-, tripropyl-, bis**(**nitrophenyl**)**-, tris**(**aminomethyl**)**-, isopropyl-**, etc., are indexed under **d, t, b, t, i** respectively. The prefixes *sec-* and *tert-* are regarded as numbers, and therefore, they are not considered for the alphabetization.

24. If the designation of a compound is difficult, the use of the empirical formula index is suggested (paragraph 13).

25. When a compound is designated by the name of its principal functional group that alone is not a compound, as is the case with some amines, ketones, azo-compounds, etc., after the name of the functions the skeleton is designated by alphabetical order, and then the substituents on each of the groups attached to the principal function; *e.g.:*

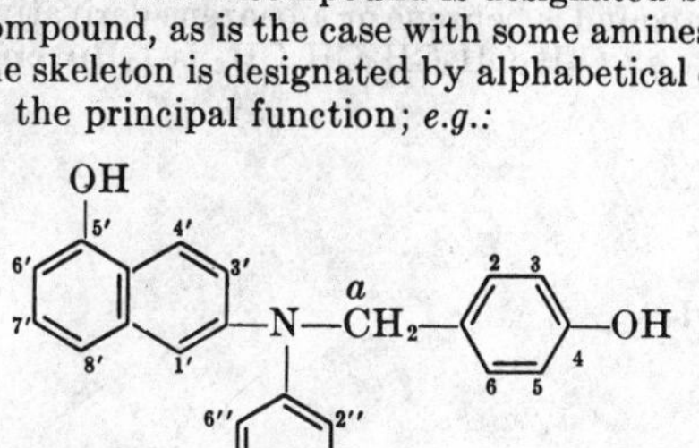

Amine, benzyl 2-naphthyl phenyl, 4,5′-dihydroxy-4″-nitro-

EXPLANATION OF TABLE (Continued)

It is to be noted that substituents of benzyl (the first radical in the name) are unprimed; substituents of naphthyl (the second radical in the name) are primed; substituents of phenyl (the third radical in the name) are doubly primed.

26. The principal function of the parent compound has a number as low as possible.

HYDROCARBONS*

27. **Aliphatic hydrocarbons.** IUC rules 4–10 are followed, but some remarks are added:

a) For C_{11} hendecane (and not undecane) is used (see IUC rule 5).

b) From the possibilities in IUC rule 7 the alphabetical order is used always (see paragraph 22).

c) Regarding IUC rules 8 and 9: numbers before the name of the hydrocarbon or the particle indicating unsaturation indicate the carbon with the lower number of the two connected by the unsaturated link, *e.g.:*

$CH_3CH_2CH_2CH:CH_2$	**1-Pentene**
$CH_2:C:CHCH_3$	**1,2-Butadiene**
$CH_3C\vdots CCH:CHCH:CHCH_3$	**2,4-Octadien-6-yne**

d) Regarding IUC rule 10: in an unsaturated hydrocarbon the longest chain containing the maximum number of double and triple bonds will be considered; *e.g.:*

$CH_2:C—CH:CH_2$
  |
  CH_2CH_3

appears as **1-3-Butadiene, 2-ethyl-**
not as **1-Pentene, 3-methylene-**
not as **1-Butene, 2-ethenyl-**

Preference is given to the double bond over the triple bond in the fundamental chain; *e.g.:*

$CH_2:CHCHCH:CH_2$
  |
  $C\vdots CH$

Appears as **1,4-Pentadiene, 3-ethenyl-**
not as **1-Penten-4-yne, 3-ethenyl-**

28. **Cyclic hydrocarbons:**

a) *Saturated monocyclic* hydrocarbons follow IUC rule 11.

b) *Unsaturated cyclic* hydrocarbons follow IUC rule 12 with the following comments: the names **tetralin** for 1,2,3, 4-tetrahydronaphthalene, **decalin** for decahydronaphthalene, **indan** for 2,3-dihydroindene, and special names for steroidal skeletons are retained.

c) *Aromatic parent hydrocarbons.* The names **benzene, naphthalene, anthracene** are used. For linear arrangements of four or more fused benzene rings names are formed by a numeric prefix denoting the number of fused benzene rings and the suffix -cene: **tetracene, pentacene,** etc. The usual trivial names are given to other arrangements: **phenanthrene, chrysene, indene, fluorene, benzanthrene, dibenzophenanthrene,** etc.

d) *Bridged hydrocarbon ring systems* appear under the headings: **bicyclo-** or **tricyclo-** according to Baeyer's system; *e.g.:*

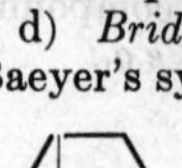

Bicyclo[3.1.0]hexane

However in most cases compounds having such structures appear only under their trivial names.

e) *Spiro, dispiro, etc.* compounds appear under these headings; *e.g.:*

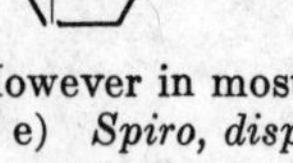

Spiro[4,5]decane

f) *Cyclic hydrocarbons with side chains.* IUC rules 49aI and 49aII are followed with some variations: the *Chemical Abstracts* usage is followed in conjunction with IUC rule 49aI: the parent compound is (when possible) the one containing the higher number of carbon atoms; *e.g.:* **Benzene, pentyl-,** *not* Pentane, 1-phenyl-; **Docosane, 1(1-naphthyl)-,** *not* Naphthalene, 1-docosyl-. The form **Hexane, phenyl-** is preferred. When the side chain is unsaturated and the cyclic compound is benzene or a benzene derivative, preference is given to the unsaturated aliphatic hydrocarbon; *e.g.:* $C_6H_5CH_2CH_2CH:CH_2$ is **1-Butene, 4-phenyl-;** $C_6H_5C\vdots CCH_2CH_3$ is **1-Butyne, 1-phenyl-.**

The following are exceptions:

$C_6H_5C\vdots CH$	**Benzene, ethynyl-**
$C_6H_5CH_2CH:CH_2$	**Benzene, allyl-**
$C_6H_5CH:CHCH_3$	**Benzene, propenyl-**
$C_6H_5CH:CH_2$	**Styrene**
$C_6H_5CH:CHC_6H_5$	**Stilbene**

---

* In order to avoid unnecessary repetition, the 1930 Liége rules for nomenclature of the International Union of Chemistry are referred to only by number, *e.g.* IUC rule 1. These rules may be found in *J. Am. Chem. Soc.* 55, 3905–25 (1933); the IUPAC 1957 rules appeared after preparation of this table.

For alkyl derivatives of benzene the only trivial name retained is **toluene** for $C_6H_5CH_3$. Toluene derivatives appear under the heading **Toluene** if no other carbon atom is attached either to the methyl group or to the ring; otherwise the name will appear under **Benzene** or elsewhere as suitable; *e.g.:*

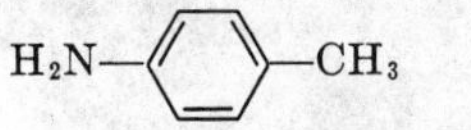

Toluene, 4-amino-

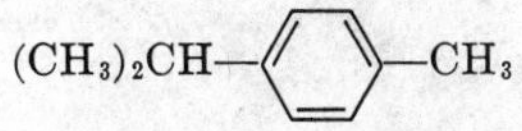

Benzene, 1-isopropyl-4-methyl-

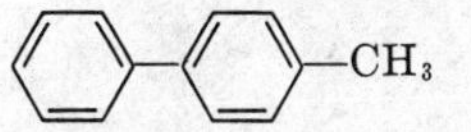

Biphenyl, 4-methyl-

Substitution in the methyl group of toluene will be labelled *a* and the word benzyl will not appear as a parent compound (only as a substituent); *e.g.:*

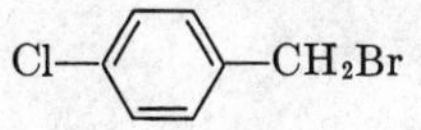

Toluene, *a*-bromo-4-chloro-

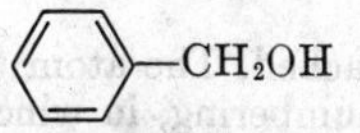

Toluene, *a*-hydroxy-
(cross-reference under **Benzyl alcohol**).

Regarding IUC rule 49aII the trivial name **Stilbene** is used instead of **Ethene, 1,2-diphenyl-.**

29. **Heterocyclic systems.** Trivial names are used, or when these are lacking systematic names are formed by using the homocyclic system name and the particles **oxa-, aza-, thia-,** etc.
The following trivial names are used for simple heterocyclic compounds: **furan, pyran, pyrrole, pyrazole, imidazole, piperidine, pyridine, pyrazine, pyrimidine, pyrazidine, thiophene,** etc. Fused heterocyclic systems follow the same rule; trivial names such as acridine, indole, quinoline, isoquinoline, quinazoline, quinoxaline, xanthene, etc., or juxtaposed names such as **1,2-benzopyran,** etc. are used. All names of fused ring systems beginning with **benzo-, dibenzo-,** etc. appear under **b, d,** etc. respectively.

30. **Radicals.** Normal alkyl radicals are named according to the Geneva designation. The following radical names are also used: **phenyl, benzyl, tolyl, styryl, allyl, isopropyl, isobutyl, *sec*-butyl, *tert*-butyl, ethenyl, ethynyl.** Larger non-normal alkyl radicals are designated by forming the longest chain containing the radical linkage and numbering the carbon on which the radical links itself to other groups according to the nearest end of the longest chain; *e.g.:*

$$\overset{8}{C}H_3\overset{7}{C}H_2\overset{6}{C}H\overset{5}{C}H_2\overset{4}{C}H\overset{3}{C}H_2\overset{2}{C}H\overset{1}{C}H_3 \quad (\text{CH}_3 \text{ on C-6 and C-2; linkage at C-4})$$

(2,6-dimethyl-4-octyl)-

The following IUC rules are also used: 54–62, 64, 65, 67.

31. **Substituted Hydrocarbons.** The following substituents never appear in a parent compound: **arsono-**(—AsO), **arso-**(—AsO_2), **boryl-** (—BO), **bromo-** (—Br), **chloro-** (—Cl), **fluoro-** (—F), **iodo-** (—I), **iodoso-** (—IO), **iodoxy-** (—IO_2), **nitro-** (—NO_2), **nitroso-** (—NO), **phospho-** (—PO_2), **phosphoroso-** (—PO), **stibo-** (—SbO_2), **triazo-** (—N_3). These are always treated as substituents (paragraph 20, phase b).

COMPOUNDS CONTAINING FUNCTIONAL GROUPS

32. **Order of precedence of functions.** When a compound bears one or more functions the parent compound is designated according to the following order of precedence: onium compound, acid (and derivatives of the acidic function), aldehyde, ketone, alcohol, amine, ether, sulfide, sulfone, sulfoxide, etc.

ONIUM COMPOUNDS

33. Onium compounds appear mainly under their trivial names.

34. **Ammonium compounds.** When four organic radicals are attached to the N atom, the compound appears under **Ammonium,** otherwise it appears as the corresponding salt of an amine; *e.g.:*

$C_6H_5\overset{+}{N}(CH_3)_3\overset{-}{I}$ — **Ammonium, phenyl trimethyl,** iodide

$(C_2H_5)_3\overset{+}{N}H\ \overset{-}{Cl}$ — **Amine, triethyl, hydrochloride**

$\overset{-}{Cl}\ \overset{+}{N}H_3CH_2CO_2H$ — **Glycine,** hydrochloride

35. **Betaines.** The word betaine appears after the name of the corresponding acid ("betaine" alone is the name of a compound); *e.g.:*

$C_6H_5(CH_3)_2\overset{+}{N}CH_2CO\overset{-}{O}$ — **Glycine, *N*,*N*-dimethyl-*N*-phenyl-,** betaine

ACIDS AND ACID DERIVATIVES

36. **Aliphatic monocarboxylic acids.** The longest chain containing the carboxylic group is taken as the parent compound. The carbon of the carboxyl group is numbered 1. The Geneva nomenclature is followed. (IUC rule 29); *e.g.:*

$\overset{6}{C}H_3\overset{5}{C}H_2\overset{4}{C}H_2\overset{3}{C}H\overset{2}{C}H_2\overset{1}{C}O_2H$ (with $CH{:}CH_2$ on C-3) — **Hexanoic acid, 3-ethenyl-**

$\overset{5}{C}H_3\overset{4}{C}H_2\overset{3}{C}H{:}\overset{2}{C}\overset{1}{C}O_2H$ (with CH_3 on C-2) — **2-Pentenoic acid, 2-methyl-**

In exception to this rule, the following trivial names are retained:

HCO_2H	**Formic acid**
CH_3CO_2H	**Acetic acid**
$C_6H_5CH{:}CHCO_2H$	**Cinnamic acid**

37. **Cyclic acids.** After the name of the ring the suffix **-carboxylic acid** is attached. The atom to which the carboxyl is attached is numbered 1 unless the ring system has a fixed numbering, in which case the carboxylic group takes a number as low as possible; *e.g.:*

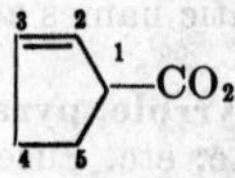

2-Cyclopentenecarboxylic acid

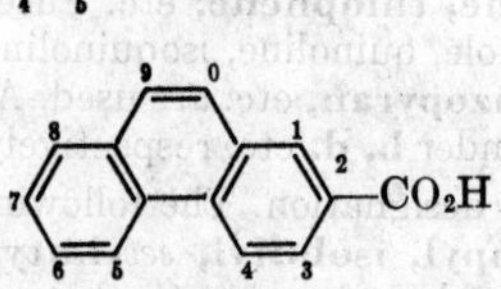

2-Phenanthrenecarboxylic acid

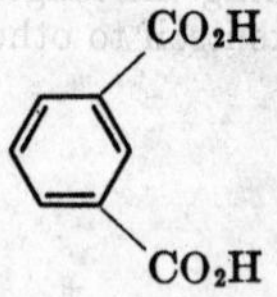

1,3-Benzenedicarboxylic acid
(cross reference under **Isophthalic acid**)

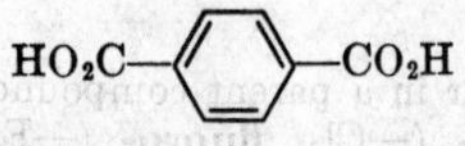

1,4-Benzenedicarboxylic acid
(cross reference under **Terephthalic acid**)

Linear acids are given preference over cyclic acids; *e.g.:*

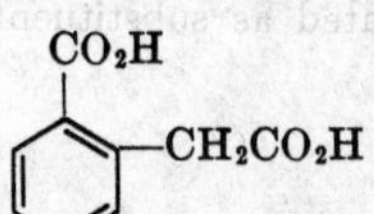

Acetic acid, (2-carboxyphenyl)-
(cross reference under **Homophthalic acid**)

Exceptions: **benzoic acid, 1-naphthoic acid, 2-naphthoic acid, phthalic acid.**

38. **Aliphatic, dicarboxylic acids** are found under the Geneva name of the chain containing both carboxylic groups; *e.g.:*

$HO_2\overset{5}{C}\overset{4}{C}H_2\overset{3}{C}{:}\overset{2}{C}H\overset{1}{C}O_2H$ (with $CH_2(CH_2)_6CH_3$ on C-3) — **2-Pentenedioic acid, 3-octyl-**

Trivial names are used for the following acids: **oxalic, malonic, succinic, maleic, fumaric, tartaric, citric.**

39. **Aliphatic polycarboxylic acids** are designated using the **-carboxylic acid** suffix; *e.g.:*

$HO_2CCH_2CH_2CHCH_2CO_2H$ (with CO_2H on the CH) — **1,2,4-Butanetricarboxylic acid**

40. **Thioacids.** After the name of the acid one of the following particles will appear: **thio-, dithio-, thiolo-, thiono-.** The last two appear only when it is certain that the ketonic or the hydroxylic oxygen is substituted by a sulfur atom.

41. Imidic, hydroxamic, nitrolic and nitrosolic acids. Only simple compounds of this type appear in the Table. The various cases arising are exemplified as follows:

$CH_3CH_2CH_2C(:NH)OH$	**Butanimidic acid**
$C_6H_5C(:NH)OH$	**Benzimidic acid**
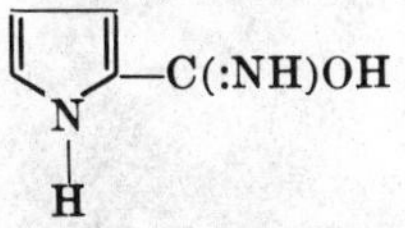	**2-Pyrrolecarboxymidic acid**
$CH_3CH_2C(:NOH)OH$	**Propanehydroxamic acid**
$C_6H_5C(:NOH)OH$	**Benzohydroxamic acid**
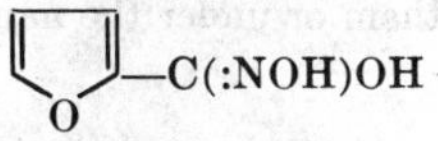	**2-Furancarbohydroxamic acid**
$CH_3C(:NOH)NO_2$	**Acetonitrolic acid**
$CH_3C(:NOH)NO$	**Acetonitrosolic acid**

42. Sulfonic and sulfinic acids. According to IUC rule 47, with the following comment: onium and carboxylic acids are functions that precede the sulfonic and sulfinic acids, and so when the former functions are present in the compound the particles **sulfo-** and **sulfino-** will be used; *e.g.:*

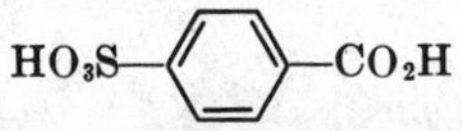

Benzoic acid, 4-sulfo-

43. Sulfenic acids are found under the corresponding hydrocarbon structure followed by the suffix **sulfenic acid;** *e.g.:*

$C_6H_5—S—OC_2H_5$ **Benzenesulfenic acid,** ethyl ester.

44. Aldehydic and ketonic acids. When the carbonyl group is present in the principal chain the particle **oxo-** is used; *e.g.:*

$CH_3C(=O)CH_2COOH$	**Butanoic acid, 3-oxo-** (cross reference under **Acetoacetic acid**)
$CH_3CH(CHO)CH_2CH_2CO_2H$	**Pentanoic acid, 4-methyl-5-oxo-**

If the aldehyde group is not in the principal chain the particles **formyl-** or **oxo-** are used; *e.g.:*

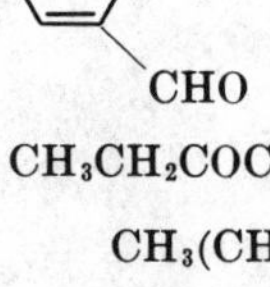

Benzoic acid, 3-formyl-

$CH_3CH_2COCH_2CHCO_2H$ with $CH_3(CH_2)_3CH_2$ on the CH — **Dodecanoic acid, 2(2-oxobutyl)-**

See however paragraphs 83, 84.

45. Hydroxyacids. The hydroxy group is considered as a substituent; *e.g.:*

$CH_3CHOHCO_2H$ **Propanoic acid, 2-hydroxy-** (cross reference under **Lactic acid**).

Exceptions: **tartaric, citric,** and acids derived from carbohydrates such as gluconic, mannonic, etc.

46. Aminoacids. Most of them appear under their trivial names with cross references under their systematic names.

47. Orthoacids appear under the heading **Ortho-;** *e.g.:*

$CH_3C(OCH_3)_3$	**Orthoacetic acid,** trimethyl ester.
$C_6H_5C(OC_2H_5)_3$	**Orthobenzoic acid,** triethyl ester.

48. **Acids attached by a non-carbon link.** The following examples explain the usage of the Table where the same acid appears attached twice to a non-carbon atom:

$NH(CH_2CO_2H)_2$ **Acetic acid, iminodi-**

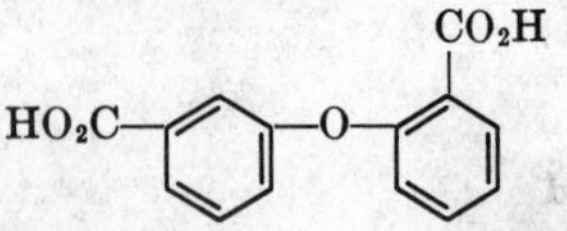

Benzoic acid, 2,3′-oxydi-

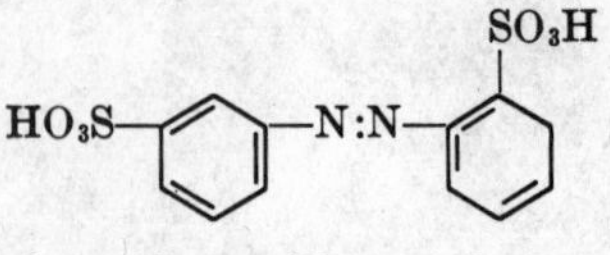

Benzenesulfonic acid, 2,3′-azodi-

If the acids are different the name of the compound may appear under one of them or under the name of the linking functions.

49. After the name of the parent acid the particle **per-** is used; *e.g.:*

CH_3CO_3H **Acetic acid, per-**

50. **Organic acids other than carboxylic, sulfonic, sulfinic and sulfenic.** See IUC rule 34.

51. **Inorganic acids.** Their derivatives may appear either under the name of the acid or under the name of the organic base; *e.g.:*

$SO_2(OCH_3)_2$ **Sulfuric acid,** dimethyl ester
$C_6H_5NH_2.HNO_3$ **Aniline** nitrate.

52. **Isocyanides.** The prefix **isocyano-** is used for the substituent —NC; *e.g.:*

C_6H_5—NC **Benzene, isocyano-**

53. **Anhydrides.**

a) *Symmetrical anhydrides of monobasic acids.* After the name of the acid comes the word anhydride; *e.g.:*

$(CH_3CHBrCO)_2O$ **Propanoic acid, 2-bromo-,** anhydride.

b) *Mixed anhydrides:* Both possible names appear (one of them as a cross-reference); *e.g.:*

$CH_3CHClCOOCOCH_2CH_2Cl$ **Propanoic acid, 2-chloro-,** anhydride with 3-chloropropanoic acid; and also:
Propanoic acid, 3-chloro-, anhydride with 2-chloropropanoic acid.

c) *Cyclic anhydrides of polyacids* are named by placing the word anhydride after the name of the acid; *e.g.:*

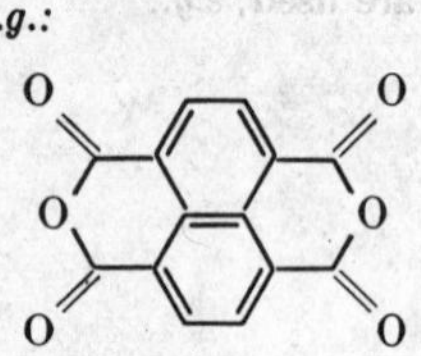

1,4,5,8-Naphthalenetetracarboxylic acid, 1:8,4:5-dianhydride

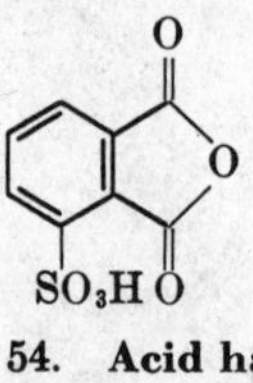

Phthalic acid, 3-sulfo-, anhydride.

54. **Acid halides and azides.** The name of the substituent of the hydroxyl group appears after the name of the corresponding acid; *e.g.:*

$CH_3CHClCOCl$ **Propanoic acid, 2-chloro-,** chloride

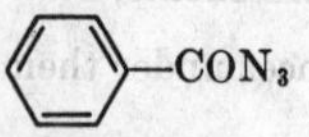

Benzoic acid, azide

55. **Amides.**

a) *Primary amides* appear under the name of the corresponding acid followed by the word amide; *e.g.:*

CH_3CONH_2 **Acetic acid,** amide
(cross reference under Acetamide)

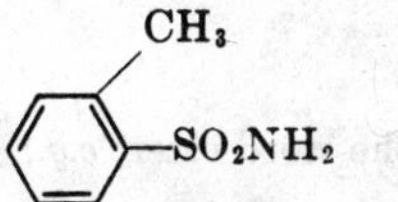

2-Toluenesulfonic acid, amide

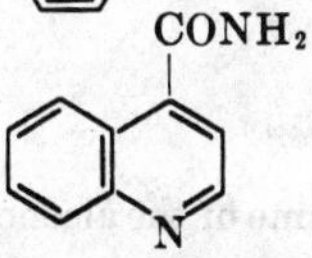

4-Quinolinecarboxylic acid, amide

Nevertheless, when in forming the name of the corresponding acid the particle **carboxy-** is used, and this group turns to an amide group, the prefix **carbamyl-** is used; *e.g.:*

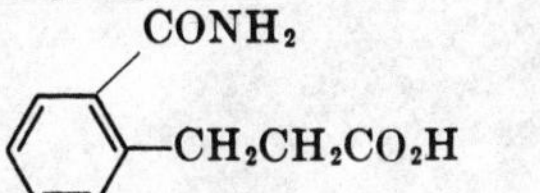

Propanoic acid, 3(2-carbamylphenyl)- (see paragraph 37),

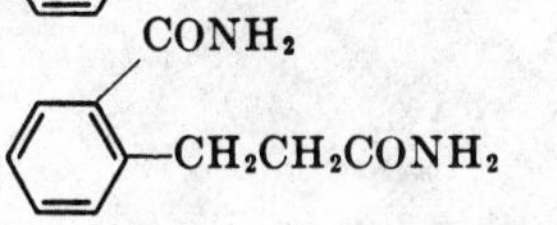

Propanoic acid, 3(2-carbamylphenyl)-, amide

b) *N-Alkyl, N-Aryl and other N-derivatives of amides.* Substituents in the N atom appear in alphabetical order after the word amide; *e.g.:*

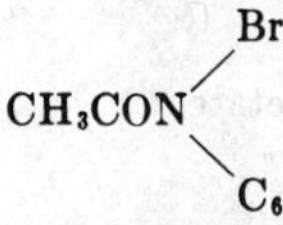

Acetic acid, amide, *N*-bromo-*N*-phenyl-

Compounds with heterocyclic radicals attached to the *N*-atom of formamide, acetamide and benzamide are named with the heterocycle as parent compound using the particles **formamido-, acetamido-** and **benzamido;** *e.g.:*

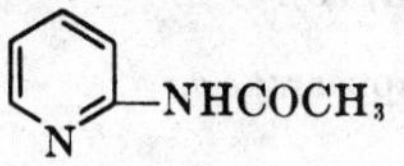

Pyridine, 2-acetamido-

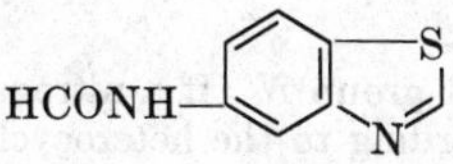

Benzothiazole, 5-formamido-

Derivatives of other amides follow the general rule stated above, *e.g.:*

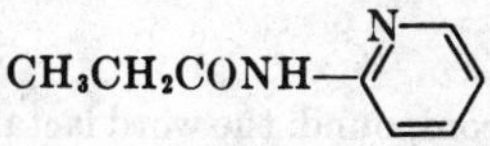

Propanoic acid, amide, *N*(2-pyridyl)-

Amides in which the amide nitrogen is part of a heterocyclic system are named as acyl derivatives of that heterocycle:

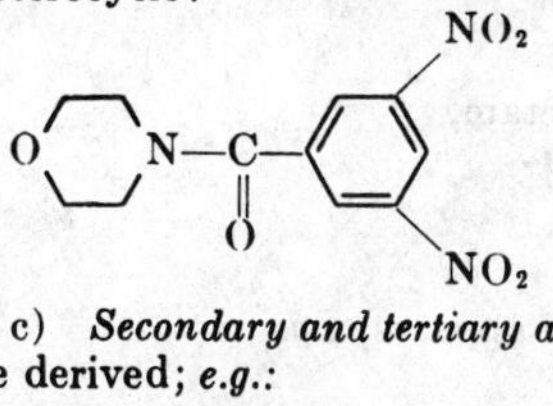

Morpholine, 4(3,5-dinitrobenzoyl)-

c) *Secondary and tertiary amides* are named according to the largest carboxylic acid from which it can be derived; *e.g.:*

$CH_3CONHCOCH_3$ — **Acetic acid,** amide, *N*-acetyl- (cross reference under **Diacetamide**)

$C_6H_5CONHCOCH_3$ — **Benzoic acid,** amide, *N*-acetyl-

$(C_6H_5CO)_3N$ — **Benzoic acid,** amide, *N,N*-dibenzoyl- (cross reference under **Tribenzamide**)

A mixed amide of a carboxylic and a sulfonic acid is named as a derivative of the amide of the carboxylic acid *e.g.:*

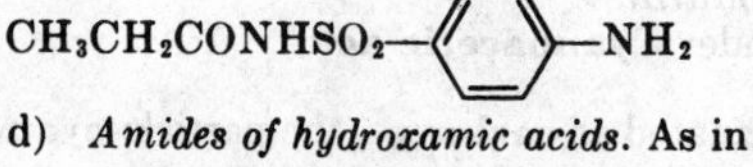

Propanoic acid, amide, *N*(4-aminobenzenesulfonyl)-

d) *Amides of hydroxamic acids.* As in paragraph 52a.

56. **Polypeptides.** A small number of polypeptides appear in the Table. They are found under their trivial name, or under the name of the amino acid that retains its carboxylic group and shares its amino group; *e.g.:*

$CH_3CH(NH_2)CONHCH_2CONHCH_2CO_2H$

Glycine, *N*(*N*-alanylglycyl)-

57. **Hydrazides of acids.** See paragraph 110.

58. **Esters.** After the name of the corresponding acid comes the radical and the word ester; *e.g.:*

$C_6H_5COOC_6H_5$	**Benzoic acid,** phenyl ester
$HOSO_2OCH_3$	**Sulfuric acid,** monomethyl ester

Nevertheless, esters of complicated and polyvalent alcohols may appear under the name of the alcohol; *e.g.:*

$CH_2O_2C(CH_2)_{15}CH_3$
|
$CHO_2C(CH_2)_{15}CH_3$ — **Glycerol,** triheptadecenoate
|
$CH_2O_2C(CH_2)_{15}CH_3$

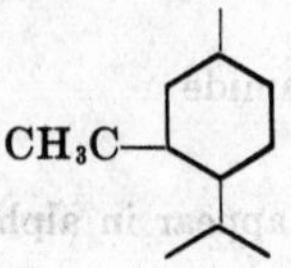

Menthol, acetate

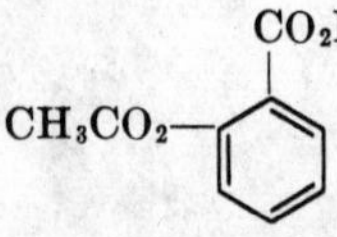

Aspirin, *see* **Benzoic acid, 2-hydroxy-,** acetate

59. **Orthoesters.** See paragraph 47.

60. **Amidoximes** are treated as amides of hydroxamic acid (see paragraphs 41, 55).

61. **Amidines.** After the name of the corresponding acid the word **amidine** appears; *e.g.:*

$CH_3(CH_2)_6C(:NH)NH_2$	**Octanoic acid,** amidine

The nitrogen atom of the $—NH_2$ group is designated *N* and that of the $=NH$ group *N′*. If a nitrogen atom of the amidine function is a part of a ring the compound is named according to the heterocyclic system.

62. **Betaines.** See paragraph 35.

63. **Lactams and lactones.** If there is no name for the ring system of the compound, the word lactam or lactone appears after the name of the corresponding aminoacid or hydroxyacid; *e.g.:*

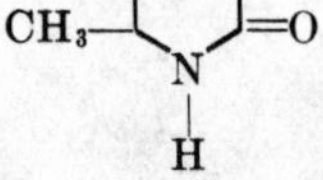

Hexanoic acid, 5-amino-, lactam, *see* **2-Piperidone, 6-methyl-**

64. **Nitriles.** After the name of the corresponding acid the word nitrile appears; *e.g.:*

$CH_3CH(NH_2)CN$	**Alanine,** nitrile

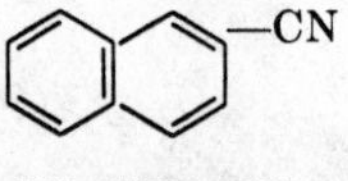

2-Naphthoic acid, nitrile

$CH_3(CH_2)_5CN$	**Heptanoic acid,** nitrile
$NCCH_2CO_2H$	**Malonic acid,** mononitrile (cross-reference under **Cyanoacetic acid**)

If the group —CN comes instead of a $—CO_2H$ group that is designated as **carboxy-,** the particle **cyano-** is used; *e.g.:*

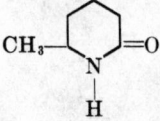

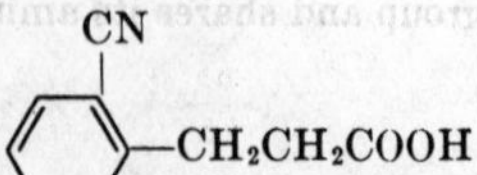

Propanoic acid, 3(2-cyanophenyl)-

65. **Imides.** The word imide appears after the name of the corresponding dicarboxylic acid:

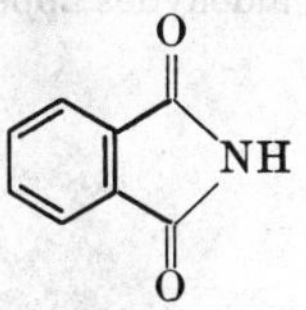

Phthalic acid, imide
(cross-reference under **Phthalimide**)

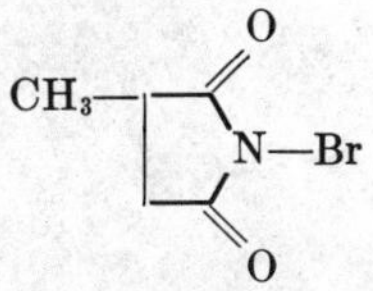

Succinic acid, 2-methyl-, imide, *N*-bromo-

For imides of monobasic acids see paragraph 55c.

66. **Cyanates, isocyanates, thiocyanates and isothiocyanates,** appear under the headings **Cyanic acid, Ioscyanic acid, Thiocyanic acid** and **Isothiocyanic acid** respectively, or, when the radical attached to these groups is complicated, the compound appears under the name of the corresponding alcohol (these rules are identical to those given for esters, amides, etc. in paragraphs 58, ff.).

ALDEHYDES

67. Geneva names for saturated and unsaturated aliphatic mono- and di-aldehydes are formed by dropping the final **e** of the name of the hydrocarbon and adding -al or (without dropping the **e**) **-dial.** The largest chain having the aldehydic group (or groups) as terminals is chosen.

Examples:

$CH_3(CH_2)_9CHO$	**Hendecanal**
$\overset{5}{C}H_3\overset{4}{C}H_2\overset{3}{C}H{:}\overset{2}{C}H\overset{1}{C}HO$	**2-Pentenal**
$OCHCH_2CH{:}CHCH{:}CHCH_2CHO$	**3,5-Octadienedial**

Some trivial names are used whenever the trivial name of the corresponding acid is used, *e.g.*: **formaldehyde, acetaldehyde, benzaldehyde, 1- and 2-naphthaldehyde, cinnamaldehyde,** etc.

When the **-carboxylic acid** nomenclature is used for the corresponding acid (see paragraphs 37, 39), the name of the aldehyde is formed by the use of the suffix **-carboxaldehyde;** *e.g.:*

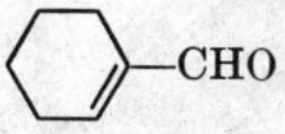

1-Cyclohexenecarboxaldehyde

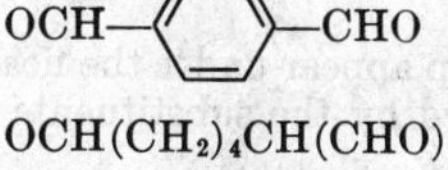

1,4-Benzenedicarboxaldehyde

$OCH(CH_2)_4CH(CHO)_2$ **1,1,5-Pentanetricarboxaldehyde**

When a function of higher rank (see paragraph 32) appears in the molecule the —CHO group is designated as **oxo-** or as **formyl-** as the case demands; *e.g.:*

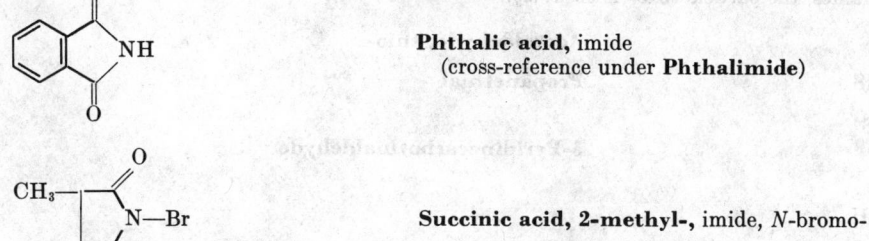

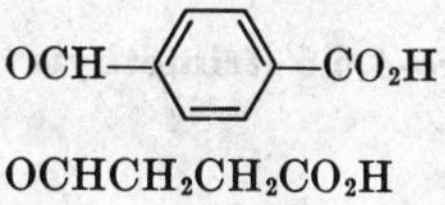

Benzoic acid, 4-formyl-

$OCHCH_2CH_2CO_2H$ **Butanoic acid, 4-oxo-**

$CH_3CH_2CH_2CH(CHO)CH_2CO_2H$ **Hexanoic acid, 3-formyl-**

68. **Acetals, hemiacetals and hydrates of aldehydes or ketones** appear in the Table immediately after the corresponding aldehyde or ketone. The following examples illustrate the nomenclature:

$CH_3CH_2CH(OC_2H_5)$	**Propanal,** diethyl acetal
$CH_3CH_2CH(OCH_3)OH$	**Propanal,** methyl hemiacetal
$Cl_3CCH(OH)_2$	**Acetaldehyde, trichloro-,** hydrate

69. **Thioaldehydes** are named by using the particles **-thial, -carbothialdehyde, thioformyl-** or **thioxo-** corresponding to the oxygen analogs (see paragraph 67). If the corresponding aldehydes appear under trivial names, the particle **thio-** is used; *e.g.:*

CH_3CHS	**Acetaldehyde, thio-**
CH_3CH_2CHS	**Propanethial**
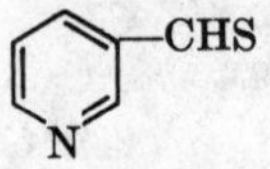	**3-Pyridinecarbothialdehyde**
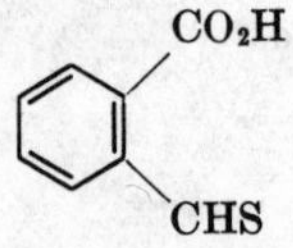	**Benzoic acid, 2-thioformyl-**
$SCHCH_2CO_2H$	**Propanoic acid, 3-thioxo-**

70. **Azines.** See paragraph 109.

71. **Imines.** See paragraph 99.

72. **Oximes.** The word oxime appears after the name of the corresponding aldehyde (or ketone).

KETONES

73. Ketones with no ring attached directly to the carbonyl group are named according to IUC rule 27.

74. Ketones with one aromatic ring attached to the carbonyl group are named according to IUC rule 27; *e.g.:*

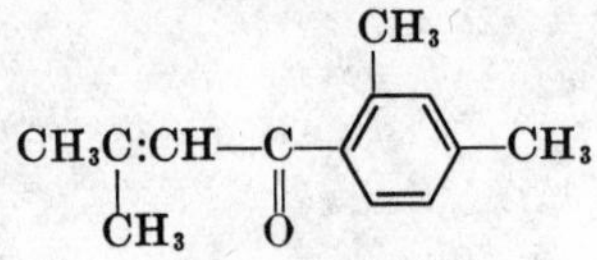	**2-Buten-1-one, 1(2,4-dimethylphenyl)-3-methyl-**
$CH_3CH_2COC_6H_5$	**1-Propanone, 1-phenyl-** (cross-reference under **Propiophenone**)

The following are exceptions:

$C_6H_5COCH_3$	**Acetophenone**
$C_6H_5CH:CHCOC_6H_5$	**Chalcone**
$C_6H_5COC_6H_5$	**Benzophenone.**

75. Monoketones with two rings attached to both sides of the carbonyl group appear under the heading **Ketone,** followed by the two rings in alphabetical order and then followed by the substituents in the rings (see paragraph 25); *e.g.:*

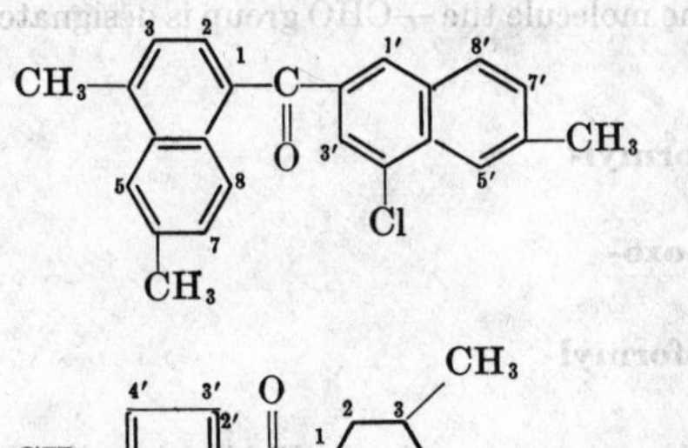	**Ketone, 1,2′-dinaphthyl, 4′-chloro-4,6,6′-trimethyl-**
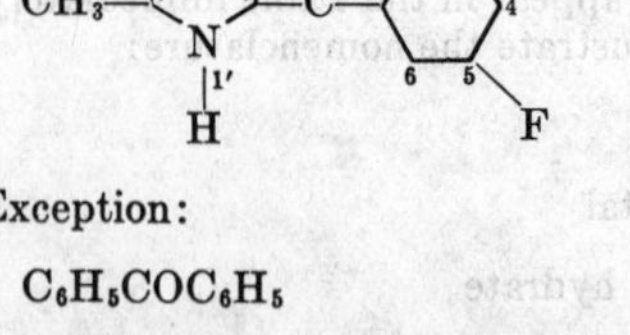	**Ketone, cyclohexyl 2-pyrryl, 3,5′-dimethyl-5-fluoro-**

Exception:

$C_6H_5COC_6H_5$	**Benzophenone**

76. Acetyl derivatives of cyclic compounds other than acetophenone are named according to the name of the cyclic compound:

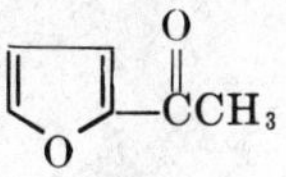

Furan, 2-acetyl-

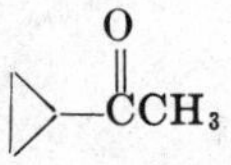

Cyclopropane, acetyl-

77. **Acetals, hemiacetals and hydrates of ketones.** See paragraph 68.

78. **Thiones (thioketones)** follow the Geneva system and IUC rule 27. When the corresponding ketone was given a trivial name the thione has the same name followed by the particle **thio-;** *e.g.:*

$C_6H_5CSCH_3$ **Acetophenone, thio-**

The particle **thioacetyl-** is used to designate the group CH_3CS— and its use is similar to that of the particle **acetyl-** (see paragraph 76). The heading corresponding to **Ketone** is **Thione;** and it is used as described in paragraph 75; *e.g.:*

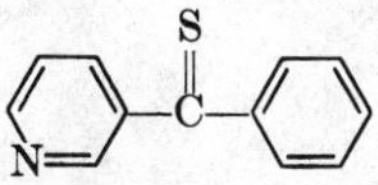

Thione, cyclohexyl 3-pyridyl

The particle corresponding to **oxo-** is **thioxo-;** *e.g.:*

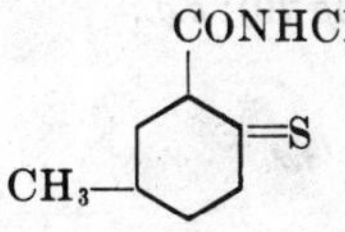

Cyclohexanecarboxylic acid, 5-methyl-2-thioxo-, amide, *N*-methyl-

79. **Ketenes.** According to IUC rule 28.

80. **Azines.** See paragraph 109.

81. **Imines.** See paragraph 99.

82. **Oximes.** See paragraph 72.

83. **Cyclic ketones and thiones.** When a cyclic system whose leading function is ketone or thione originally had two hydrogens in the site where O= or S= are attached, the compound gets the name of the parent compound with the ending **-one** or **-thione** *e.g.:*

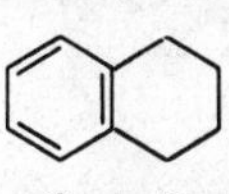

Tetralin

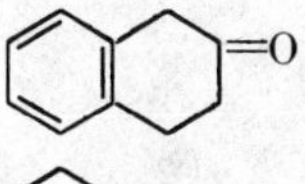

2-Tetralone

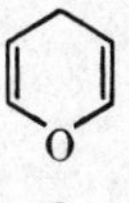

Pyran

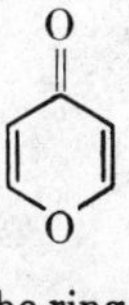

4-Pyrone

The ring system so obtained (including oxo- or thioxo- functions) may be used as a new ring system for the naming of compounds containing higher order functions (see paragraph 21); *e.g.:*

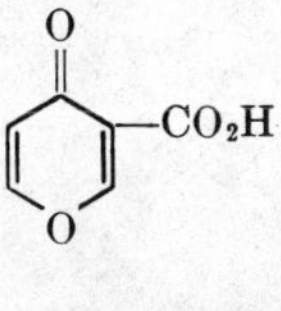

4-Pyrone-3-carboxylic acid

EXPLANATION OF TABLE (Continued)

If the compound originally had only one hydrogen at the site of the ketonic function, the particles **oxo-** or **thioxo-** together with **dihydro-**, **tetrahydro-**, etc. are used; *e.g.:*

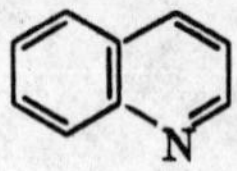

Quinoline

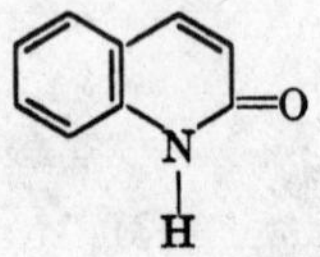

Quinoline, 1,2-dihydro-2-oxo-

84. Quinones. IUC rule 46 is followed. The quinone ring system may be used as a basis for the naming of compounds containing higher order functions.

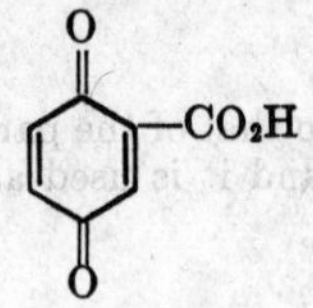

1,4-Benzoquinone-2-carboxylic acid

ALCOHOLS

85. Aliphatic alcohols appear under their Geneva names (IUC rule 20).

86. Monocyclic aliphatic alcohols appear under their Geneva names (IUC rule 20); *e.g.:*

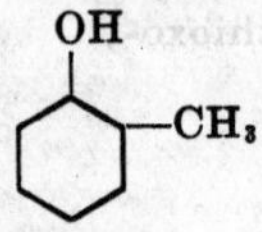

Cyclohexanol, 2-methyl-

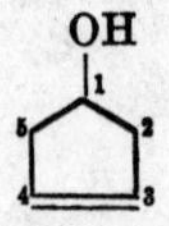

3-Cyclopenten-1-ol

87. Heterocyclic and aromatic alcohols (phenols). The name **Phenol** is used for C_6H_5OH and ring substituted derivatives with **no carbon** attached to the ring and no other hydroxyl group or higher order of precedence functions; *e.g.:*

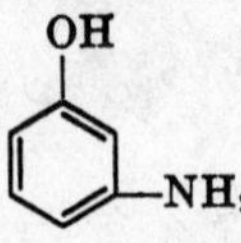

Phenol, 3-amino-

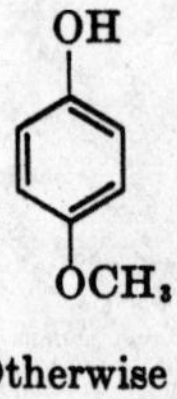

Phenol, 4-methoxy-

Otherwise the compound appears under benzene, toluene or elsewhere; *e.g.:*

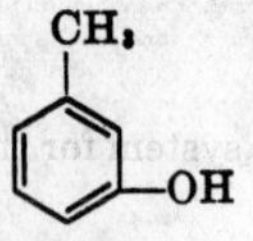

Toluene, 3-hydroxy-

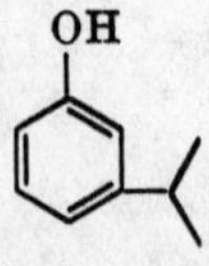

Benzene, 1-hydroxy-3-isopropyl-

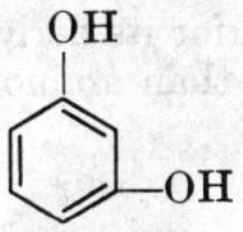

Benzene, 1,3-dihydroxy-
(cross-reference under **Resorcinol**)

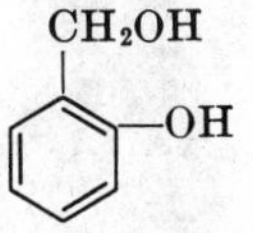

Toluene, *a*,2-dihydroxy-

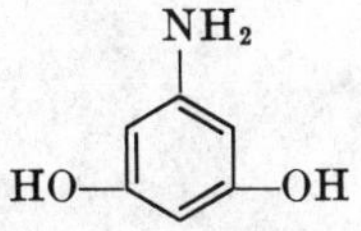

Benzene, 1-amino-, 3,5-dihydroxy-

Alcohols precede phenols in order; *e.g.:*

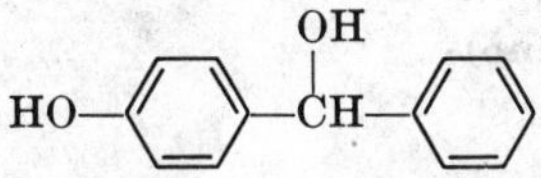

Methanol, (4-hydroxyphenyl)phenyl-

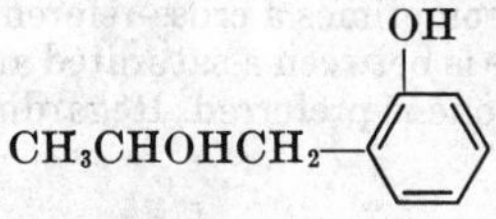

2-Propanol, 1(2-hydroxyphenyl)-

Exception: **Toluene, *a*-hydroxy-**, *not* **Methanol, phenyl-**
Other aromatic and heterocyclic phenols or alcohols will be named considering the hydroxyl group as an ordinary substituent; *e.g.:*

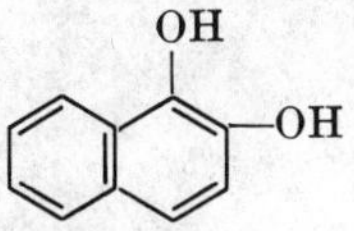

Naphthalene, 1,2-dihydroxy-

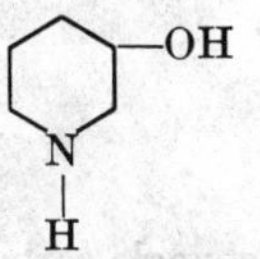

Piperidine, 3-hydroxy-

88. **Polyalcohols** such as **glycerol, erythrol,** etc. appear under their trivial names. Exception: **1,2-ethanediol.**

89. **Sugars** appear under their trivial names.

90. **Esters of complicated alcohols.** See paragraph 58.

91. **Hemiacetals and hydrates of aldehydes and ketones.** See paragraph 68.

92. The **carbinol** nomenclature for alcohols **is not used.**

93. **Thiols.** Aliphatic thiols follows IUC rule 22. Aromatic and heterocyclic thiols and molecules containing higher order functions together with the SH group will be named using the particle **mercapto.** Thiol is lower than alcohol but higher than amine in the order of precedence; *e.g.:*

CH_3CH_2SH **Ethanethiol**

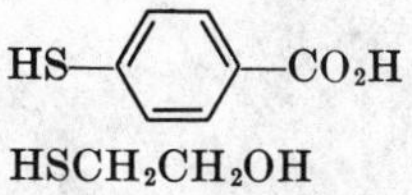

Benzoic acid, 4-mercapto-

$HSCH_2CH_2OH$ **Ethanol, 2-mercapto-**

94. **Hydroperoxides.** See paragraph 112.

AMINES AND IMINES

95. The amine function is generally considered as an ordinary substituent (like **chloro-, nitro-,** etc.) *e.g.:*

$CH_3CH_2NH_2$ **Ethane, amino-**

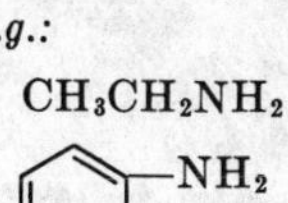

Pyridine, 3-amino-

96. The name **aniline** is used for the compound $C_6H_5NH_2$ and as a parent compound for its derivatives unless a carbon atom is attached to the benzene ring or a function of rank higher than amino is present in the molecule; *e.g.:*

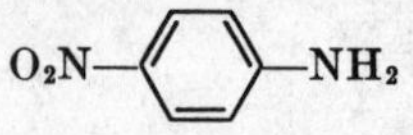

Aniline, 4-nitro-

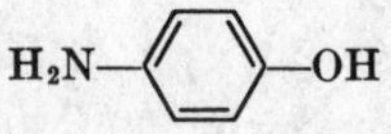

Phenol, 4-amino-

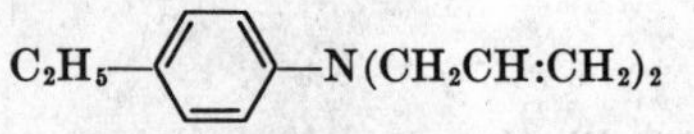

Benzene, 1-diallylamino-4-ethyl-

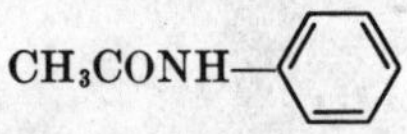

Acetic acid, amide, *N*-phenyl-
(cross reference under **Acetanilide**)

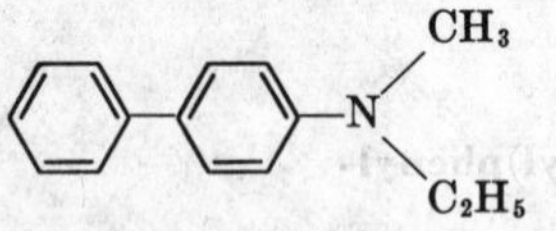

Biphenyl, 4(ethylmethylamino)-

97. A list of names beginning with the word *Amine* appears in the Table; sometimes a cross-referen is given to the preferred name based on the largest radical present. If the choice is between a saturated aı an unsaturated radical of the same number of carbon atoms, the unsaturated one is preferred. Regardiı this nomenclature see paragraph 25.

$NH(CH_2C_6H_5)_2$ — **Amine, dibenzyl-**

C_2H_5—N—CH_3 (with C_6H_5 on N) — **Amine, ethyl methyl phenyl,** *see* **Aniline, *N*-ethyl-*N*-methyl-**

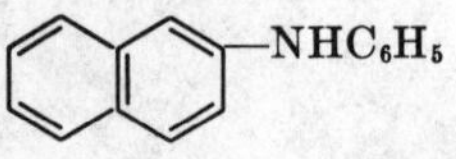

Naphthalene, 2(phenylamino)-, *see* **Amine, 2-naphthyl phenyl**

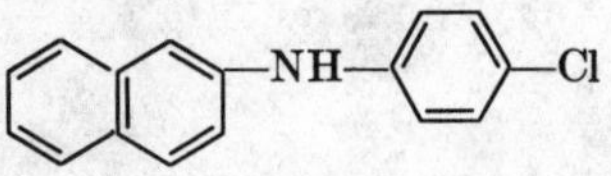

Amine, 2-naphthyl phenyl, 1,4′-dichloro-
(see paragraph 25)

98. **Hydroxylamine derivatives.** See paragraph 111.

99. **Imines.** The particle **imino-** is used for the group NH= when it is attached to a single carbı atom; *e.g.:*

$CH_3CH{=}NH$ — **Ethane, imino-**

In some instances when the *imine* group comes in place of the oxygen of a keto-group the word imiı appears after the name of the ketone; *e.g.:*

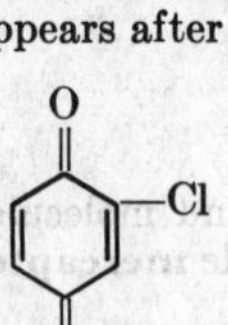

1,4-Benzoquinone, 2-chloro, 4-imine

(This is similar to the usage in the case of oximes, see paragraph 72.)

ETHERS

100. Simple and complex monoethers are found under the name **Ether** followed by the two radicı (without substituents) attached to the oxygen in alphabetical order and finally followed by the su stituents in the skeleton of the ether (see paragraph 25); *e.g.:*

CH_3OCH_3	**Ether, dimethyl**
$BrCH_2OCH_2Br$	**Ether, dimethyl, 1,1′-dibromo-**
$(C_6H_5)_2CHOCH_3$	**Ether, (diphenylmethyl) methyl**
$C_6H_5CH_2OC_6H_5$	**Ether, benzyl phenyl**
$HC{:}COC_6H_5$	**Ether, ethynyl phenyl**

EXPLANATION OF TABLE (Continued)

101. Alkoxy derivatives of benzene (linear, saturated and with less than 6 carbons) and alkoxy derivatives of polycyclic or heterocyclic hydrocarbons appear under the cyclic hydrocarbon as a parent compound; *e.g.:*

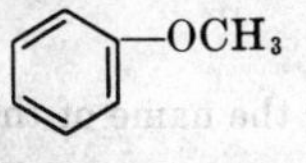

Benzene, methoxy
(cross-reference under **Anisole**)

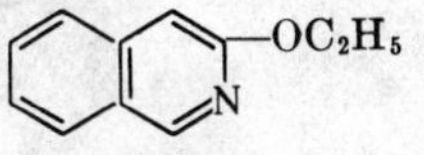

Isoquinoline, 3-ethoxy-

102. **Epoxy compounds.** Follow IUC rule 24, but cyclic ethers such as **furan, dioxan, dioxolan,** etc. are designated by their heterocyclic system name.

103. **Polyethers.** Some appear under their trivial names. Ethers of polyalcohols sometimes appear after the name of the corresponding alcohol.

104. **Peroxides.** See paragraph 112.

105. **Thioethers.** See paragraph 114.

106. **Orthoesters.** See paragraph 47.

MISCELLANEOUS FUNCTIONS

107. **Azo and azoxy compounds. Azo** and **Azoxy** are the headings for these compounds. After the word **azo** or **azoxy** comes the names of the hydrocarbon (if symmetrical) or hydrocarbons (if unsymmetrical) attached to the azo or azoxy group in alphabetical order; places of substitution are given with unprimed numbers for the first hydrocarbon and primed numbers for the second hydrocarbon (see paragraph 25); *e.g.:*

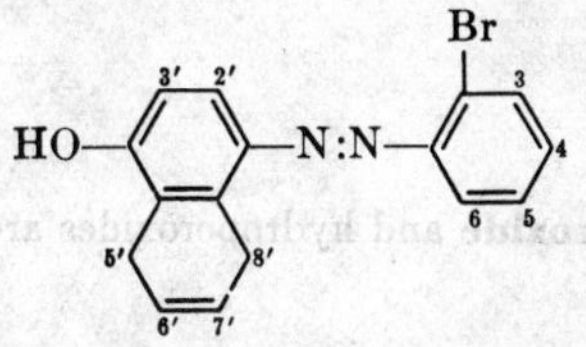

Azo, benzene 1-naphthalene, 2-bromo$_2$, 4′-hydroxy-

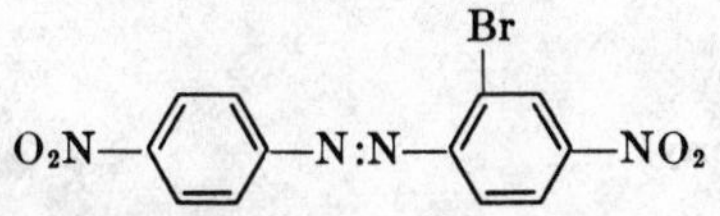

Azobenzene, 2-bromo-4,4′-dinitro-

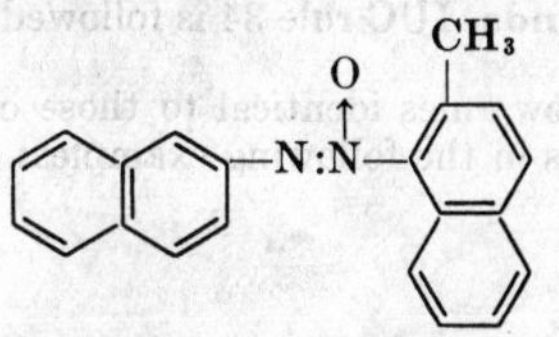

1,2′-Azoxynaphthalene, 4′-methoxy-2-methyl-

In the case of **azoxy** no attempt is made to distinguish between the nitrogen atoms.
Some of the azo compounds are cross-referred to other parent compounds (see paragraph 48).
Some multiple azo compounds are named by derivation from the most highly substituted nucleus; *e.g.:*

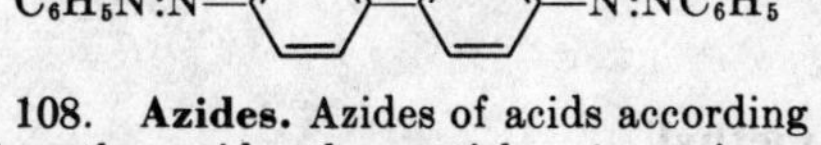

Biphenyl, 4,4′-bis(phenylazo)-

108. **Azides.** Azides of acids according to paragraph 54.
For other azides the particle **triazo-** is used (see paragraph 31); *e.g.:*

$CH_3CH_2N_3$ **Ethane, triazo-**

109. **Azines.** Appear under the name of the corresponding aldehyde or ketone if they are symmetrical; if not, under two headings with cross references; *e.g.:*

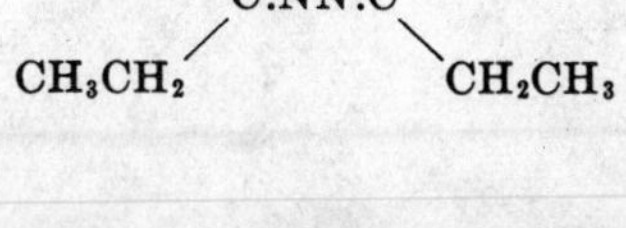

2-Butanone, azine

CH₃

$\qquad$ C:NN:CHCH₃

CH₃

2-Propanone, azine with acetaldehyde

see **Acetaldehyde**, azine with 2-propanone

110. Hydrazine derivatives. Monoacyl derivatives of hydrazine are entered under the name of the acid followed by the word hydrazide; *e.g.:*

CH₃CONHNH₂ $\qquad$ **Acetic acid**, hydrazine

Diacyl hydrazines appear under **Hydrazine**, *e.g.:*

CH₃CONHNHCOC₆H₅
$\quad$ 1 $\qquad$ 2

Hydrazine, 1-acetyl-2-benzoyl-

When a function of higher order of precedence (amino or higher) is also present, it may be convenient to use the particle **hydrazino-**; *e.g.:*

NH₂NHCH₂CH₂CN $\qquad$ **Propanoic acid**, 3-hydrazino, nitrile

NH₂NHCONH₂
1 $\quad$ 2 $\quad$ 3 $\quad$ 4

is **Semicarbazide**, with derivatives under this name, but semicarbazones appear under the corresponding ketone or aldehyde.

NH₂CONHNHCONH₂
1 $\quad$ 2 $\quad$ 3 $\quad$ 4 $\quad$ 5 $\quad$ 6

is **Biurea**, with all its derivatives appearing under this name.

111. Hydroxylamine derivatives. Hydroxylamino- is used for HONH—, and **aminooxy** for NH₂O—. Names of *N*-acyl derivatives of hydroxylamine are derived from that of the corresponding amide; *e.g.:*

CH₃CONHOH $\qquad$ **Acetic acid**, amide, *N*-hydroxy-

Oximes are explained in paragraph 72.

112. Peroxides and hydroperoxides. Peroxides are listed under **Peroxide** and hydroperoxides are listed under **Hydroperoxide**; *e.g.:*

C₂H₅OOH	**Hydroperoxide, ethyl**
C₆H₅CH₂OOCH₂C₆H₅	**Peroxide, dibenzyl**
C₆H₅OOCH₃	**Peroxide, methyl phenyl**

113. Miscellaneous Phosphorus, Arsenic and Antimony Compounds. IUC rule 34 is followed.

114. Sulfides, disulfides, trisulfides sulfoxides and sulfones follow rules identical to those of **ethers** see (paragraphs 100, 101), using an appropriate word or particle, as in the following examples:

CH₃OCH₃	**Ether,**	**dimethyl**
CH₃SCH₃	**Sulfide,**	,,
CH₃SSCH₃	**Disulfide,**	,,
CH₃SSSCH₃	**Trisulfide,**	,,
CH₃S(O)CH₃	**Sulfoxide,**	,,
CH₃S(O₂)CH₃	**Sulfone,**	,,
CH₃OC₂H₅	**Ether,**	**ethyl methyl**
CH₃SC₂H₅	**Sulfide,**	,, ,,
CH₃SSC₂H₅	**Disulfide,**	,, ,,
CH₃SSSC₂H₅	**Trisulfide,**	,, ,,
CH₃S(O)C₂H₅	**Sulfoxide,**	,, ,,
CH₃S(O₂)C₂H₅	**Sulfone,**	,, ,,
C₆H₅OCH₃	**Benzene, methoxy-**	

EXPLANATION OF TABLE (Continued)

$C_6H_5SCH_3$	**Benzene, methylthio-**
$C_6H_5SSCH_3$	,, **methyldithio**
$C_6H_5SSSCH_3$	,, **methyltrithio-**
$C_6H_5S(O)CH_3$	,, **methylsulfinyl-**
$C_6H_5S(O_2)CH_3$	,, **methylsulfonyl-**

115. **Urea derivatives.** According to IUC rule 36.

Cyclic urea derivatives are named according to their ring system or trivial names:

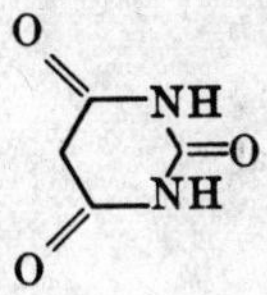

Barbituric acid

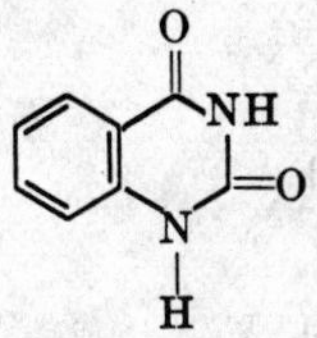

Quinazoline, 2,4-dioxo-1,2,3,4-tetrahydro-

The ureido function follows in order of precedence after the onium and the acid functions. The particle **ureido-** is used for the group NH_2CONH— and **thioureido-** for NH_2CSNH—; *e.g.:*

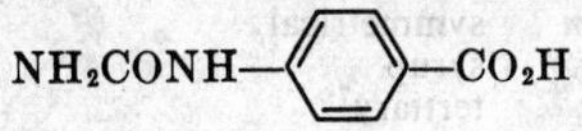

Benzoic acid, 4-ureido-

SYMBOLS AND ABBREVIATIONS

[α]	specific rotation
δ	slightly
>	above, more than
<	below, less than
∞	soluble in all proportions
*	name approved by the International Union of Chemists (I.U.C.)[1]
Ω	IR, or UV, or NMR spectra referenced
?	unknown
aa	acetic acid
abs	absolute
ac	acid
Ac	acetyl
ace	acetone
al	alcohol[2]
alk	alkali
Am	J. Am. Chem. Soc.
Am	amyl (pentyl)
amor	amorphous
anh	anhydrous
aqu	aqueous
as	asymmetric
atm	atmospheres
b	boiling
B	Beilstein
Ber	Chem. Ber.
bipym	bipyramidal
bk	black[3]
bl	blue[3]
br	brown[3]
bt	bright
Bu	butyl
bz	Benzene
C	Chem. Abs.
c	percentage concentration
ca	about (circa)
chl	chloroform
co	columns
col	colorless
con	concentrated
cor	corrected
cr	crystals
cy	cyclohexane
d	decomposes
D	line in the spectrum of sodium (subscript)
D, d	dextro[4]
δd	slight decomposition
dil	diluted
diox	dioxane
distb	distillable
dk	dark
Dl, dl	racemic[4]
dlq	deliquescent
DMF	dimethyl formamide
E	Elsevier's
eff	efflorescent
Et	ethyl
eth	ether[5]
exp	explodes
extrap	extrapolated
fl	flakes
flr	fluorescent
fr	freezes
fr. p.	freezing point
fum	fuming
gel	gelatinous
gl	glacial
gold	golden
gr	green[3]
gran	granular
gy	gray[3]
h	hot
H	Helv. Chim. Acta
hex	hexagonal
hp	heptane
htng	heating
hx	hexane
hyd	hydrate
hyg	hygroscopic
i	insoluble
i-	iso-
ign	ignites
in	inactive
inflam	inflammable
infus	infusible
irid	iridescent
iso	isooctane
J	J. Chem. Soc.
JOC	J. Org. Chem.
L, l	levo[4]
la	large
lf	leaf
lig	ligroin
liq	liquid
lo	long
lt	light
m	melting
m-	meta-
M	molar (concentration)
M	Merck Index, 7th Edition
mcl	monoclinic
Me	methyl
met	metallic
micr	microscopic
min	mineral
mod	modification
mut	mutarotatory
n	normal chain, refractive index
N	normal (concentration)
N	nitrogen[6]
nd	needles
o-	ortho-
oct	octahedral
og	orange[3]
oos	ordinary organic solvents
or	or
ord	ordinary
org	organic
orh	orthorhombic
os	organic solvents
p-	para-
pa	pale
par	partial
peth	petroleum ether
pk	pink[3]
Ph	phenyl
pl	plates
pr	prisms
Pr	propyl
Prak	J. Prak. Chem.
purp	purple[3]
pw	powder
Py	pyrimidene
pym	pyramids
rac	racemic
rect	rectangular
red	red
res	resinous
rh	rhombic
rhd	rhombohedral
s	soluble
s	secondary[7]
sc	scales
sec	secondary[7]
sf	softens
sh	shoulder
silv	silvery
sl	slightly (δ)
so	solid
sol	solution
solv	solvent
sph	sphenoidal
st	stable
sub	sublimes
suc	supercooled
sulf	sulfuric acid
sym	symmetrical
syr	syrup
t	tertiary[7]
ta	tablets
tcl	triclinic
tert	tertiary[7]
Tet	Tetrahedron
tetr	tetragonal
THF	tetrahydrofuran
to	toluene
tr	transparent
trg	trigonal
undil	undiluted
uns	unsymmetrical
unst	unstable
v	very
vac	vacuum
var	variable
vap	vapor
vic	vicinal
visc	viscous
volat	volatile or volatilises
vt	violet[3]
w	water
wh	white[3]
wr	warm
wx	waxy
ye	yellow[3]
xyl	xylene

1 For I.U.C. rules of nomenclature see General Index.
2 Generally means ethyl alcohol.
3 The abbreviation of a color ending in "sh" is to be read as ending with the suffix "-ish," *e.g.*, grsh means greenish.
4 *D, L* generally mean configuration and *d, l* generally mean optical rotation, but there are many examples in the chemical literature for which the meaning of these symbols is ambiguous and/or interchangeable.
5 Generally means diethyl ether.
6 *N* indicates a position in the molecule.
7 *s* and *sec*, or *t* and *tert*, are used as convenient.

No.	Name	Synonyms and Formula	Mol. wt.	Color, crystalline form, specific rotation and λ_{max} (log ε)	m.p. °C	b.p. °C	Density	n_D	Solubility w	al	eth	ace	bz	other solvents	Ref.
	Abietic acid														
Ω a1	**Abietic acid**	Sylvic acid. $C_{20}H_{30}O_2$.	302.46	mcl pl (al-w) $[\alpha]_D^{25}$−116 (al, c = 1) λ^{al} 234 (2.50), 241 (4.37)	173–4	250^{9}	...	...	i	s	v	s	v	CS_2 s MeOH s	**B9**[2], 424
Ω a2	—,methyl ester	$C_{21}H_{32}O_2$. *See* a1	316.49	...	...	$225–6^{16}$	1.049_4^{20}	1.5344	i	s	...	...	...	aa s	**B9**[2], 430
—	**Acacetin**	*see* **Flavone, 5,7-dihydroxy-4′-methoxy-**													
a3	**Acenaphth-anthracene**	Naphtho-2′,3′:4,5-acenaphthene.	254.34	pa ye lf (lig) λ^{al} 227 (4.86), 289.5 (3.80), 321 (3.19)	192.5–3.5	...	...	...	i	...	...	...	s	con sulf s aa s lig s^h	**E14s**, 565
Ω a4	**Acenaphthene**	Naphthyleneethylene.	154.21	nd (al) λ^{al} 227.5 (4.86), 289 (3.80), 300 (3.60), 321 (3.19)	96.2	279^{760}	1.0242_4^{99} 1.225_4^{0}	1.6048^{95}	i	δ s^h	...	...	v	aa s	**B5**[2], 495
a5	—,1-amino-	$C_{12}H_{11}N$. *See* a4	169.23	cr (peth)	135	sub	...	...	δ	v	...	s	...	os s	**B12**[2], 764
a6	—,3-amino-	$C_{12}H_{11}N$. *See* a4	169.23	pl (al), nd (peth) pk in air	81.5	...	...	...	...	s	...	...	...	chl s (red sol)	**E13**, 148
a7	—,4-amino-	$C_{12}H_{11}N$. *See* a4	169.23	nd (al, w)	87	...	...	...	s^h	...	...	s	...	os v lig s^h	**B12**[2], 764
a8	—,5-amino-	$C_{12}H_{11}N$. *See* a4	169.23	nd (lig) red in air	108	...	...	...	...	s	...	...	...	...	**E13**, 149
Ω a9	—,5-bromo-	$C_{12}H_9Br$. *See* a4	233.12	pl (al) λ^{al} 250 (3.3), 290 (4.1), 300 (4.2), 313 (4.0)	52	335^{760}	1.4392_4^{52}	1.6565^{54}	i	δ s^h	...	...	...	...	**B5**[1], 276
a10	—,5-chloro-	$C_{12}H_9Cl$. *See* a4	188.66	pl or nd (al)	70.5	319.2^{770} 163^{13}	1.1954_4^{70}	1.6288^{7}	i	δ s^h	...	...	...	...	**B5**[1], 276
a11	—,5-iodo-	$C_{12}H_9I$. *See* a4	280.11	nd (al)	65	...	1.6738_4^{62}	1.6909^{65}	i	s	...	...	s	aa s	**E13**, 145
a12	—,3-nitro-	$C_{12}H_9NO_2$. *See* a4	199.21	gr-ye nd (aa)	151.5	...	...	...	i	...	...	...	...	sulf s (gy-bl → red) con alk s (red) aa s^h	**E13**, 147
—	—,1-oxo-	*See* **1-Acenaphthenone**													
a13	**5-Acenaphthene-carboxylic acid**	5-Acenaphthoic acid.	198.22	nd (bz or lig) λ^{al} 236.5 (4.45), 311 (3.9)	220–1	...	...	...	...	...	...	...	δ s^h	lig s^h	**B9**[1], 280
Ω a14	**Acenaphthene-quinone**	Acenaphthaquinone.	182.18	ye nd (aa) λ^{al} 225 (4.45), 327 (3.86), 338 (3.86)	261	...	...	...	i	δ	...	...	δ s^h	lig s^h aa δ, s^h	**E13**, 169
a15	**3-Acenaphthene sulfonic acid**	...	234.28	hyg nd (bz)	87–9	...	...	...	s	...	...	...	...	...	**E13**, 181
a16	**1-Acenaphthenone**	1-Oxoacenaphthene.	168.21	nd (al)	121	...	...	...	s^h	s^h	...	...	v	chl s sulf s (ye-gr) NaOH s (vt)	**E13**, 164
Ω a17	**Acenaphthylene**	...	152.21	pr (eth), pl (al) λ^{cy} 276 (3.48), 311 (3.88), 323 (3.98), 335 (3.66), 340 (3.67)	92–3	265–75 par d 156–60^{28}	$\sim1.194^{25}$	...	i	v	v	...	v	...	**B5**[2], 530
Ω a18	**Acetaldehyde**	Acetic aldehyde. Ethanal*. CH_3CHO	44.05	λ^{cy} 290 (1.23), λ^{gas} 178 (3.48), 181 (3.60), 181.5 (4.05)	−121	20.8^{760}	0.7834_4^{18}	1.3316^{20}	∞^h	∞	∞	...	∞	...	**B1**[3], 2617
a19	—,bis(2-chloro-ethyl) acetal	1,1-Bis(2-chloroethoxy)-ethane*. $CH_3CH(OCH_2CH_2Cl)_2$	187.07	...	...	194–6d 106–7^{14}	1.1737_4^{20}	1.45266^{20}	d^h	δ	δ	...	...	...	**B1**[3], 2644
a20	—,diacetate	Ethylidene diacetate. $CH_3CH(O_2CCH_3)_2$	146.14	...	18.9	169^{760} 65–7^{10}	1.3985^{25}	1.070^{25}	...	s	s	...	...	...	**B2**[2], 167
Ω a21	—,diethyl acetal	Acetal. 1,1-Diethoxyethane* Ethylidene diethyl ether. $CH_3CH(OC_2H_5)_2$	118.18	volat λ^{w} 270 (0.9)	...	103.2^{760} 21^{22}	0.8314_4^{20}	1.3834^{20}	s	∞	∞	v	...	chl s	**B1**[3], 2641
Ω a22	—,dimethyl acetal	1,1-Dimethoxyethane*. Ethylidene dimethyl ether. $CH_3CH(OCH_3)_2$	90.12	...	−113.2	64.5	0.85015_4^{20}	1.3668^{20}	s	s	s	v	...	chl s	**B1**[2], 671
Ω a23	—,2,4-dinitro-phenylhydrazone (stable form)	...	224.19	ye sc (al) λ^{al} 355 (4.3)	168.5 (164)	...	...	...	i	s^h	s	s	v	chl v	**B15**[1], 490
a24	—,—(unstable form)	$C_8H_8N_4O_4$. *See* a23	224.19	og-red	146 (157)	...	...	...	i	...	...	...	...	...	**B15**[1], 490
Ω a25	—,oxime	Acetaldoxime. $CH_3CH{:}NOH$	59.07	nd λ^{al} < 220	47	115^{760}	0.9656_4^{20}	1.42567^{20}	s^h	∞	∞	...	...	...	**B1**[3], 675
a26	—,phenylhydra-zone	*N*-Ethylidene-*N*′-phenyl-hydrazine. $CH_3CH{:}NNHC_6H_5$	134.18	(i) nd or ta (ii) nd λ^{al} 268 (4.25)	(i) 98–101 (ii) 57	133–6^{21}	...	...	...	s	...	...	...	lig δ^h	**B15**[2], 54

For explanations, symbols and abbreviations see beginning of table. For structural formulas see end of table.

No.	Name	Synonyms and Formula	Mol. wt.	Color, crystalline form, specific rotation and λ_{max} (log ε)	m.p. °C	b.p. °C	Density	n_D	Solubility w	al	eth	ace	bz	other solvents	Ref.
	Acetaldehyde														
a27	—,semicarbazone..	$CH_3CH:NNHCONH_2$	101.11	nd (w or al) λ^{al} 262 (2.38), 266 (2.25)	163	....	1.0300^{0}_{4}	....	δ s^h	s	...	...	...	....	**B3**^{3}, 48
a28	—,**amino-**, diethyl acetal	Acetalylamine. Aminoacetal. β,β-Diethoxyethylamine. $H_2NCH_2CH(OC_2H_5)_2$	133.19	....	....	163	0.9159^{25}	1.4170^{20}	v	v	v	...	...	chl v	**B4**^{2}, 758
—	—,**ammonia**.......	*see* **Ethanol, 1-amino-**													
a29	—,**bromo-**, diethyl acetal	2-Bromoacetal. $BrCH_2CH(OC_2H_5)_2$	197.08	....	....	180^{760} (170) d 66^{18}	1.280^{20}_{4}	1.4376^{20}	...	s^h	s	...	...	....	**B1**^{2}, 682
a30	—,**chloro-**	Chloroethanal. $ClCH_2CHO$	78.50	λ^{cy} 279.5 (1.54)	....	85–5.5^{748}	....	....	...	...	s	...	...	....	**B1**^{2}, 675
Ω a31	—,—,diethyl acetal	Chloroacetal. $ClCH_2CH(OC_2H_5)_2$	152.62	....	....	157.4^{760} 71–2^{35}	1.0180^{20}_{4}	1.4170^{20}	δ^h	∞	∞	...	...	dil sulf d	**B1**^{2}, 676
Ω a32	—,—,dimethyl acetal	$ClCH_2CH(OCH_3)_2$	124.57	....	....	127–8^{760}	1.068^{20}_{4}	1.4150^{20}	...	s	s	...	s	....	**B1**^{3}, 2660
Ω a33	—,**dichloro-**.......	Cl_2CHCHO	112.94	λ^{cy} 301.5 (1.59)	....	90–1^{760}	....	....	...	s	...	...	...	....	**B1**^{2}, 676
a35	—,—,diethyl acetal	Dichloroacetal. $Cl_2CHCH(OC_2H_5)_2$	187.07	....	....	183–4 67–71^{12}	1.1383^{14}	....	...	s	s	...	...	....	**B1**^{2}, 677
a36	—,—,ethyl hemiacetal	$Cl_2CHCH(OH)OC_2H_5$	159.02	....	....	109.5–111 63–5^{56}	1.314^{26}	1.4360^{20}	δ	∞	...	...	∞	lig δ	**B1**^{2}, 677
a37	—,—,hydrate	$Cl_2CHCH(OH)_2$	130.96	cr (bz), ta	56–7 (43)	97 (118)	....	....	v	s	v	...	...	CS_2 s	**B1**^{1}, 614
a38	—,**dimethyl-amino-**, diethyl acetal	Dimethylaminoacetal. $(CH_3)_2NCH_2CH(OC_2H_5)_2$	161.25	ye	....	170–1 (165)	0.885^{7}	1.4129^{20}	v	s	s	s	...	os s	**B4**, 308
Ω a39	—,**diphenyl-**.......	2,2-Diphenylethanal*. $(C_6H_5)_2CHCHO$	196.25	λ^{al} 255 (3.34), 300 sh (2.41)	....	315 δd 157.5^{7}	1.1061^{21}_{4}	1.5920^{21}	i	v	v	...	v	....	**B7**^{2}, 370
a40	—,**ethoxy-**........	$C_2H_5OCH_2CHO$	28.12	....	....	71–3	0.942^{20}_{4}	1.3956^{20}	s	s	...	s	...	....	**B1**, 818
a41	—,**ethylmethyl-amino-**, diethyl acetal	$CH_3(C_2H_5)NCH_2CH(OC_2H_5)_2$	175.27	....	....	179–80	....	....	i	s	s	s	...	os v	**B4**, 309
a42	—,**hydroxy-**.......	Glycolaldehyde. Glycolic aldehyde. $HOCH_2CHO$	60.05	pl	97	....	1.366^{100}	1.4772^{19}	v	v	δ	...	...	....	**B1**^{2}, 863
a43	—,—,diethyl acetal	2,2-Diethoxyethanol*. $HOCH_2CH(OC_2H_5)_2$	134.18	....	....	167 57–8^{8}	0.888^{24}_{4}	1.4073^{20}	...	s	s	...	...	....	**B1**^{2}, 864
a44	—,**methoxy-**	CH_3OCH_2CHO..........	74.08	....	....	92.3^{770}	1.005^{25}_{4}	1.3950^{20}	s	s	s	s	...	....	**B1**^{3}, 3180
a45	—,**2-oxo-2-phenyl-**, hydrate	Benzoyl formaldehyde hydrate. Phenyl glyoxal hydrate. $C_6H_5COCH(OH)_2$	152.15	nd (w, chl, al-lig, eth-lig)	....	93–4	....	....	δ	s	s	...	...	chl, CS_2 s	
a46	—,—,dioxime.....	Phenyl glyoxime. $C_6H_5C(:NOH)CH:NOH$	164.18	nd (chl) λ^{al} 231 (4.15)	180	....	....	....	v^h	s	s	...	...	chl δ	**B7**^{2}, 602
a47	—,—,1-oxime.....	Isonitrosoacetophenone. $C_6H_5COCH:NOH$	149.15	mcl pr or lf (chl, w)	129	....	....	....	δ s^h	...	...	...	...	chl s^h	**B7**^{2}, 600
Ω a48	—,**phenyl-**	α-Tolualdehyde. α-Toluic aldehyde. $C_6H_5CH_2CHO$	120.16	(w) λ^{al} 252 (291), 288 sh (2.0), 291 sh (2.33)	33–4	195 88^{18}	1.0272^{20}_{4}	1.5255^{20}	δ	∞	∞	s	...	....	**B7**^{1}, 292
a49	—,**tribromo-**	Bromal. Br_3CCHO	280.76	....	....	174^{760} 61^{9}	2.6650^{25}_{4}	1.5939^{20}	d	s	s	s	...	....	**B1**^{2}, 683
a50	—,—,ethyl hemiacetal	Bromal monoethyl acetal. $Br_3CCH(OH)OC_2H_5$	326.83	nd	44	d	....	....	δ	v	v	...	...	....	**B1**^{2}, 684
a51	—,—,hydrate	$Br_3CCH(OH)_2$...........	298.77	mcl pr (w + 1)	53.5	d	2.5662^{40}_{4}	....	s	s	s	...	...	chl s	**B1**^{2}, 683
Ω a52	—,**trichloro-**	Chloral. Cl_3CCHO	147.39	λ^{cy} 290 (1.58)	−57.5	97.75^{760}	1.5121^{20}_{4}	1.45572^{20}	v^h	s^h	s^h	...	...	....	**B1**^{2}, 677
a53	—,—,butyl hemiacetal	Chloral *n*-butyl alcoholate. $Cl_3CCH(OH)OC_4H_9^n$	221.51	nd	49–50	129–30^{743}	....	....	...	v^h	s	...	...	....	**B1**^{3}, 267
a54	—,—,diethyl acetal	Trichloroacetal. $Cl_3CCH(OC_2H_5)_2$	221.51	....	....	205^{760} 84–5^{10}	1.266^{25}_{4}	1.4586^{25}	δ	∞	∞	...	...	glycerol ∞	**B1**, 621
a55	—,—,ethyl hemiacetal	Chloral alcoholate. $Cl_3CCH(OH)OC_2H_5$	193.46	nd (al)	56–7 (50)	115–6^{760}	1.143^{40}	....	s^h	s^h	s^h	...	...	....	**B1**^{2}, 681
a56	—,—,hydrate	Trichloroethylidene glycol. $Cl_3CCH(OH)_2$	165.40	mcl pl (w)	57	96.3^{764} d	1.9081^{20}_{4}	....	v	v	v	v	v	chl, Py, to v	**B1**^{2}, 680
a57	—,—,β,β,β-trichloro-*tert*-butyl hemiacetal	Chloralacetone chloroform. $Cl_3CCH(OH)OC(CCl_3)(CH_3)_2$	324.86	nd (bz)	65	sub	....	....	δ s^h	s	s	...	s^h	....	**B1**, 622
—	**Acetaldoxime**	*see* **Acetaldehyde,** oxime													
—	**Acetamide**........	*see* **Acetic acid,** amide													
—	—,**α-cyano-**.......	*see* **Malonic acid,** monoamide mononitrile													
—	—,**diisobutyl-**	*see* **Pentanoic acid, 2-isobutyl-4-methyl-,** amide													
—	**Acetamidine**	*see* **Acetic acid,** amidine													
—	**Acetanilide**	*see* **Acetic acid,** amide, *N*-phenyl-													
—	—,**thio-**	*see* **Acetic acid, thiono-,** amide, *N*-phenyl-													
Ω a58	**Acetic acid**........	Ethanoic acid*. CH_3CO_2H ..	60.05	rh (hyg) λ^{al} 208 (1.5)	16.604	117.9^{760} 17^{10}	1.0492^{20}_{4}	1.3716^{20}	∞	∞	∞	∞	∞	os s CS_2 s	**B2**^{2}, 91 **B5**^{2}, 761
Ω a59	—,allyl ester	Allyl acetate. $CH_3CO_2CH_2CH:CH_2$	100.13	....	....	103.5^{760}	0.9276^{20}_{4}	1.4049^{20}	δ	∞	∞	s	...	....	**B2**^{2}, 150

For explanations, symbols and abbreviations see beginning of table. For structural formulas see end of table.

No.	Name	Synonyms and Formula	Mol. wt.	Color, crystalline form, specific rotation and λ_{max} (log ε)	m.p. °C	b.p. °C	Density	n_D	Solubility w	al	eth	ace	bz	other solvents	Ref.
	Acetic acid														
Ω a60	—,amide	Acetamide. Ethanamide*. CH_3CONH_2	59.07	trg mcl (al-eth) λ^{MeOH} 205 (2.21)	82.3	221.2^{760} 120^{20}	0.9986^{85}_4 1.1590^{20}_4	1.4278^{78}	s	v	i	...	δ s^h	chl s Py s	$B2^2$, 177
a61	—,—(labile form)	CH_3CONH_2	59.07	rh	48.5	221.2^{760} 99^7	...	1.4158^{110}	v	v	...	...	...	...	B2, 175
Ω a62	—,—,N-acetyl-	N-Acetylacetamide. Diacetamide. $(CH_3CO)_2NH$	101.11	nd (eth) λ^{eth} 260 (1.95)	79	223.5^{760} 113^{13}	...	...	s	s	s	...	...	lig s	$B2^2$, 180
a63	—,—,N-acetyl-N-(4-chlorophenyl)-	p-Chlorodiacetanilide.	211.65	cr (al, eth) λ^{al} 251 (4.23)	66–7	...	...	...	...	v	v	...	v	...	B12, 612
a64	—,—,N-acetyl-N(4-ethoxyphenyl)-	p-Ethoxydiacetanilide.	221.26	nd (lig)	53.5–4	182^{12}	...	...	δ	v	δ	...	δ	aa v lig δ^h	B13, 468
a65	—,—,N-acetyl-N-ethyl-	Diacetylethylamine. $(CH_3CO)_2NC_2H_5$	129.16	...	...	$195–9^{760}$	1.0092^{20}	1.4513^{20}	i	s	...	...	...	...	$B4^1$, 352
a66	—,—,N-acetyl-N-methyl-	Diacetylmethylamine. $(CH_3CO)_2NCH_3$	115.13	...	−25	194.5^{741} 114.5^{61}	1.0663^{25}_4	1.4502^{25}	∞	...	i	...	...	...	$B4^1$, 329
a67	—,—,N-acetyl-N-phenyl-	Diacetanilide. $(CH_3CO)_2NC_6H_5$	177.21	ta (lig) λ^{chl} 242 (3.01)	37–8	200^{100} 142^{41}	...	...	δ	s	...	...	s	to s	$B12^2$, 145
a68	—,—,N(4-acetylphenyl)-	p-Acetylacetanilide. 4-Acetylaminoacetophenone.	177.21	nd (w) λ^{al} 283 (4.53)	166–7	...	...	...	δ v^h	v	δ	...	δ	lig s to s	$B14^2$, 33
Ω a69	—,—,N(2-aminoethyl)-	N-Acetylethylenediamine. $CH_3CONHCH_2CH_2NH_2$	102.14	wh (bz)	51 (cor)	128^3	...	...	s	s	i	...	s^h	...	Am 63, 852
a70	—,—,N(2-aminophenyl)-	o-Aminoacetanilide.	150.18	pl (bz) λ^{al} 266 (3.45)	132	...	...	...	s	s	δ	...	s^h	...	$B13^2$, 15
a71	—,—,N(3-aminophenyl)-	m-Aminoacetanilide.	150.18	nd or pl (bz) λ^{al} 271 (3.32)	87–9	...	...	...	v	v	δ	v	δ	...	$B13^2$, 27
Ω a72	—,—,—,hydrochloride	$C_8H_{10}N_2O.HCl$. See a71	186.64	redsh pl (al)	248–51	...	...	...	v	s	i	...	i	...	$B13^2$, 27
Ω a73	—,—,N(4-aminophenyl)-	p-Aminoacetanilide.	150.18	nd (w) λ^{al} 260 (4.26), 312 sh (3.15)	165–8 (162)	267^{760}	...	...	s v^h	v	v	...	...	...	$B13^2$, 50
a74	—,—,N-benzyl-	N-Acetylbenzylamine. $CH_3CONHCH_2C_6H_5$	149.19	lf (eth) λ^{al} 258 (2.34)	61	$>300^{760}$ 157^2	...	...	...	v	v	...	...	...	$B12^2$, 588
a75	—,—,N-bromo-, hydrate	Acetobromamide. $CH_3CONHBr.H_2O$	155.99	bt ye pl (w + 1)	70–80 (hyd) 108–9 (anh)	$39–40^{13}$	...	...	s^h	s	s	...	...	chl δ	B2, 181
a76	—,—,N-bromo-N-phenyl-	N-Bromoacetanilide. $CH_3CONBrC_6H_5$	214.08	ye pl (peth)	88	...	...	...	...	...	...	...	...	chl v lig δ	$B12^2$, 296
a77	—,—,N(2-bromophenyl)-	o-Bromoacetanilide.	214.08	nd (al) λ^{al} 234 (3.88)	99	...	...	...	i	s	s	...	...	...	$B12^2$, 342
a78	—,—,N(3-bromophenyl)-	m-Bromoacetanilide.	214.08	nd (aq al) λ^{al} 246 (4.15), 278 (3.04), 286 (2.93)	87.5	...	...	...	...	v	v	...	...	...	B12, 634
Ω a79	—,—,N(4-bromophenyl)-	p-Bromoacetanilide.	214.08	nd (60 % al) mcl pr λ^{al} 252 (4.27)	168	...	1.717	...	i^h δ^h	s	δ	...	δ	chl s	$B12^2$, 348
Ω a80	—,—,N-butyl-	$CH_3CONH(CH_2)_3CH_3$	115.18	...	...	229	...	1.4388^{25}	...	...	...	...	...	...	$B4^2$, 634
Ω a81	—,—,N-butyl-N-phenyl-	N-Butylacetanilide. $CH_3CON(C_6H_5)(CH_2)_3CH_3$	191.28	$\lambda^{95\%al}$ <220	24.5	281^{760} 141^{10}	0.9912^{20}_4	1.5146^{20}	...	...	...	...	...	...	B12, 247
a82	—,—,N(4-butylphenyl)-	p-Butylacetanilide.	191.28	wh pl (al)	105	...	...	...	i	s	...	...	...	...	$B12^2$, 634
a83	—,—,N(2-chloro-4-nitrophenyl)-	2-Chloro-4-nitroacetanilide.	214.62	pr (al)	139–40	...	...	...	i	δ	...	...	...	...	B12, 733
a84	—,—,N(2-chloro-5-nitrophenyl)-	2-Chloro-5-nitroacetanilide.	214.62	nd (al)	156	...	...	...	i	δ	...	...	...	...	B12, 732
Ω a85	—,—,N(3-chloro-4-nitrophenyl)-	3-Chloro-4-nitroacetanilide.	214.62	pa ye nd (al)	145	...	...	...	i	δ	...	...	δ	...	$B12^2$, 399
a86	—,—,N(4-chloro-2-nitrophenyl)-	4-Chloro-2-nitroacetanilide.	214.62	ye nd (al)	104	...	...	...	i	δ	...	...	...	...	$B12^2$, 397
Ω a87	—,—,N(4-chloro-3-nitrophenyl)-	4-Chloro-3-nitroacetanilide.	214.62	ye nd (al)	150	...	...	...	i	s^h	...	...	...	...	$B12^2$, 732
a88	—,—,N-chloro-N-phenyl-	N-Chloroacetanilide. $CH_3CON(Cl)C_6H_5$	169.61	nd (dil aa), pl (peth-chl)	91	...	...	...	δ	...	...	...	...	...	$B12^2$, 295
Ω a89	—,—,N(2-chlorophenyl)-	o-Chloroacetanilide.	169.61	nd (dil aa) λ^{al} 240 (4.02)	87–8 (sub 50–60)	...	...	...	i	s	v	...	s v^h	...	$B12^2$, 317
Ω a90	—,—,N(3-chlorophenyl)-	m-Chloroacetanilide.	169.61	nd (50 % aa) λ^{al} 278 (3.04), 286 (2.95)	79	...	...	...	δ	v	v	...	v	CS_2 v	$B12^2$, 321
a91	—,—,N(4-chlorophenyl)-	p-Chloroacetanilide.	169.61	nd (aq aa), ta (al, ace), cr (w) λ^{al} 249 (4.25)	179	...	1.385^{22}_4	...	i s^h	s v^h	v	...	...	CCl_4 δ	$B12^2$, 327
a92	—,—,N(2-chloro-3-tolyl)-	3-Acetamino-2-chlorotoluene.	183.64	nd (al)	133–4	...	...	...	i	s	...	s	...	lig δ	B12, 871
a93	—,—,N(3-chloro-2-tolyl	2-Acetamino-6-chlorotoluene.	183.64	nd (dil al)	157–9 (136)	...	...	...	δ^h	s	...	...	s	...	B12, 836

For explanations, symbols and abbreviations see beginning of table. For structural formulas see end of table.

No.	Name	Synonyms and Formula	Mol. wt.	Color, crystalline form, specific rotation and λ_{max} (log ε)	m.p. °C	b.p. °C	Density	n_D	Solubility w	al	eth	ace	bz	other solvents	Ref.
	Acetic acid														
a94	—,—,*N*(3-chloro-4-tolyl)-	4-Acetamino-2-chloro-toluene.	183.64	tcl cr	105	...	...	...	i	s	...	...	...	...	**B12**, 989
a95	—,—,*N*(4-chloro-2-tolyl)-	2-Acetamino-5-chloro-toluene.	183.64	lf (al)	140	...	...	...	i	s^h	...	...	...	...	**B12**, 836
a96	—,—,*N*(4-chloro-3-tolyl)-	5-Acetamino-2-chloro-toluene.	183.64	lf (al)	91.2–.7	...	...	...	i	s	...	s	...	...	**B12**, 871
a97	—,—,*N*(5-chloro-2-tolyl)-	2-Acetamino-4-chloro-toluene.	183.64	nd (w)	139–40	...	...	...	s^h	s	s	...	...	...	**B12**^{1}, 389
a98	—,—,*N*(5-chloro-3-tolyl)-	5-Acetamino-3-chloro-toluene.	183.64	nd (al)	151	...	...	...	...	s	...	...	...	...	**B12**, 871
a99	—,—,*N*(4-cyclo-hexylphenyl)-	4-Cyclohexylacetanilide.	217.31	nd (peth)	129	...	...	...	δ^h	v	v	...	...	chl v	**B12**^{2}, 667
a100	—,—,*N*,*N*-diacetyl-	Triacetamide. $(CH_3CO)_3N$	143.14	nd (eth)	79	...	...	...	...	s	...	...	...	...	**B2**^{1}, 82
a101	—,—,*N*,*N*-diethyl-.	$CH_3CON(C_2H_5)_2$	115.18	λ^w *ca.* 200 (3.96)	...	185–6 91^{30}	$0.9130_4^{17.4}$	$1.4374_4^{17.4}$	s	s	∞	∞	∞	...	**B4**^{2}, 602
a102	—,—,*N*,*N*-dimethyl-	$CH_3CON(CH_3)_2$	87.12	λ^{hp} 218 (3.00)	−20	165^{758} 84^{32}	0.9366_4^{25}	1.4380^{20}	∞	∞	∞	∞	∞	chl ∞	**B4**^{2}, 564
a103	—,—,*N*(2,4-di-methyl-6-nitro-phenyl)-	4-Acetamino-5-nitro-*m*-xylene.	208.22	yesh nd (w)	172–3	...	...	...	s^h	s	...	...	s	...	**B12**^{2}, 612
a104	—,—,*N*(2,4-di-methylphenyl)-	2,4-Acetoxylide.	163.22	nd (al)	128–9	170^{10}	...	...	δ	v	...	...	...	...	**B12**^{2}, 608
Ω a105	—,—,*N*(2,3-dinitrophenyl)-	2,3-Dinitroacetanilide.	225.16	nd (al)	187	...	...	...	...	δ	δ	...	δ	chl δ	**B12**^{2}, 405
Ω a106	—,—,*N*(2,4-dinitrophenyl)-	2,4-Dinitroacetanilide.	225.16	ye nd (al or bz)	125–6 (121)	...	...	...	i	v^h	s	...	s^h	...	**B12**^{2}, 410
a107	—,—,*N*(2,5-dinitrophenyl)-	2,5-Dinitroacetanilide.	225.16	nd (al)	121	...	...	...	i	s v^h	...	...	...	...	**B12**, 758
a108	—,—,*N*(2,6-dinitrophenyl)-	2,6-Dinitroacetanilide.	225.16	nd (aa)	197	...	...	...	i	s	...	...	...	aa s	**B12**, 758
a109	—,—,*N*(3,4-dinitrophenyl)-	3,4-Dinitroacetanilide.	225.16	ye cr (al), nd (w)	144–5	...	...	...	s^h	v^h	...	...	...	...	**B12**^{2}, 414
a110	—,—,*N*(3,5-dinitrophenyl)-	3,5-Dinitroacetanilide.	225.16	ye nd (dil aa, w)	191	...	...	...	δ	v^h	i	...	...	aa v^h	**B12**, 759
a111	—,—,*N*,*N*-diphenyl-	*N*-Acetyldiphenylamine. *N*-Phenylacetanilide. $CH_3CON(C_6H_5)_2$	211.27	rh or nd (w or lig) λ^{al} 237 (4.29)	103	sub	...	...	δ	s	δ	...	...	...	**B12**, 247
a112	—,—,*N*,*N*-dipropyl-	$CH_3CON(CH_2CH_2CH_3)_2$	143.23	...	...	209–210 101^{16}	$0.8992_4^{17.4}$	$1.4419_4^{17.4}$	...	s	...	...	...	...	**B4**, 142
a113	—,—,*N*(2-ethoxy-4-nitrophenyl)-	2-Acetamino-5-nitro-phenetole.	224.22	nd (al)	202	...	...	...	i	s	...	...	...	...	**B13**^{2}, 194
a114	—,—,*N*(2-ethoxy-5-nitrophenyl)-	2-Acetamino-4-nitro-phenetole.	224.22	ye nd (al)	199	...	...	...	...	v^h	...	v	...	chl v	**B13**^{2}, 193
a115	—,—,*N*(4-ethoxy-2-nitrophenyl)-	4-Acetamino-3-nitro-phenetole.	224.22	ye nd (w)	104	...	...	...	...	v	v	v	...	chl v	**B13**^{2}, 287
a116	—,—,*N*(4-ethoxy-3-nitrophenyl)-	4-Acetamino-2-nitro-phenetole.	224.22	nd (dil al)	123	...	...	...	i	v	δ	v	v	chl v lig δ aa s	**B13**^{2}, 285
a117	—,—,*N*(5-ethoxy-2-nitrophenyl)-	3-Acetamino-4-nitro-phenetole.	224.22	nd (al)	95	...	...	...	i	v	s	v	s	chl v lig s^h	**B13**^{1}, 136
a118	—,—,*N*(2-ethoxy-phenyl)-	*o*-Acetophenetidide.	179.22	lf (dil al)	79	>250	...	...	i	s	s	...	...	os s	**B13**^{2}, 172
a119	—,—,*N*(3-ethoxy-phenyl)-	*m*-Acetophenetidide.	179.22	gy pl (w)	97–9	...	...	...	δ	s	s	...	...	...	**B13**^{1}, 133
a120	—,—,*N*(4-ethoxy-phenyl)-	*p*-Acetophenetidide. Phenacetin.	179.22	mcl pr	137–8	d	...	1.571	δ	s	δ	s	δ	Py v chl s	**B13**^{2}, 244
Ω a121	—,—,*N*-ethyl-.	Acetoethylamine. $CH_3CONHC_2H_5$	87.12	...	...	205 $104–5^{18}$	$0.942_4^{4.5}$	...	∞	∞	...	...	...	aa s	**B4**, 109
a122	—,—,*N*-ethyl-*N*(3-nitrophenyl)-	*N*-Ethyl-3-nitroacetanilide.	208.22	pa ye nd (dil al)	88–9	...	...	...	...	v	...	...	s	...	**B12**, 704
a123	—,—,*N*-ethyl-*N*-(4-nitrophenyl)-	*N*-Ethyl-4-nitroacetanilide.	208.22	lf or pr (dil al)	118–9	...	...	...	δ	v	δ	...	v	lig, CS_2 i	**B12**, 720
Ω a124	—,—,*N*-ethyl-*N*-phenyl-	*N*-Ethylacetanilide. $CH_3CON(C_2H_5)C_6H_5$	163.22	(w), rh (eth) λ^{hp} 238 (3.74)	55	258^{731}	...	...	s	...	s	...	...	...	**B12**^{2}, 143
—	—,—,*N*(formyl-phenyl)-	*see* **Benzaldehyde, acetamido-**													
a125	—,—,*N*(2-hydroxy-ethyl)-	*N*-Acetylethanolamine. $CH_3CONHCH_2CH_2OH$	103.12	nd (ace)	63–5	$166–7^8$	1.1079_4^{25}	1.4674^{20}	∞	...	...	s^h	δ	lig δ	**B4**^{1}, 430
a126	—,—,*N*(2-hydroxy-1-naphthyl)-	1-Acetamino-2-naphthol.	201.23	lf (w, dil al, MeOH)	235 d	sub	...	...	...	s	s	s	s	NaOH v CS_2 s	**B13**^{2}, 414
a127	—,—,*N*(4-hydroxy-3-nitrophenyl)-*N*-methyl-	...	210.19	(al)	161–2	...	...	...	...	s	...	...	...	...	**B13**^{1}, 186
Ω a128	—,—,*N*(2-hydroxy-phenyl)-	2-Acetaminophenol. *o*-Hydroxyacetanilide.	151.17	pl (dil al)	209	...	...	...	δ v^h	v	v	...	v	alk v	**B13**^{2}, 171
Ω a129	—,—,*N*(3-hydroxy-phenyl)-	3-Acetaminophenol. *m*-Hydroxyacetanilide.	151.17	nd (w)	148–9	...	...	...	v	v	δ	...	δ	chl δ	**B13**^{2}, 213
Ω a130	—,—,*N*(4-hydroxy-phenyl)-	4-Acetaminophenol. *p*-Hydroxyacetanilide.	151.17	mcl pr (w, al)	169–71	...	1.293^{21}	...	i s^h	v	...	...	...	...	**B13**^{2}, 243
a131	—,—,*N*(2-hydroxy-phenyl)-*N*-methyl-	2-(Acetylmethylamino)-phenol.	165.19	nd (bz-peth)	151	...	...	...	δ	v	v	...	v	peth i	**B13**^{2}, 173

For explanations, symbols and abbreviations see beginning of table. For structural formulas see end of table.

No.	Name	Synonyms and Formula	Mol. wt.	Color, crystalline form, specific rotation and λ_{max} (log ε)	m.p. °C	b.p. °C	Density	n_D	Solubility w	al	eth	ace	bz	other solvents	Ref.
	Acetic acid														
a132	—,—,N(4-hydroxyphenyl)-N-methyl-	4-(Acetylmethylamino)-phenol.	165.19	sc (w)	245	....	....	....	δ	v	s	...	...	....	**B13[1], 162**
a133	—,—,N(4-iodophenyl)-	p-Iodoacetanilide.	261.07	ta (w), pr (w, al) λ^{al} 254 (4.36)	184.5	....	....	....	δ v^h	v	v	...	v	....	**B12[2], 362**
a134	—,—,N-isopropyl-N-phenyl-	N-Isopropylacetanilide $CH_3CON(C_6H_5)CH(CH_3)_2$	177.22	lf (lig)	39	262–3^{712}	....	....	...	...	...	...	...	lig s^h	**B12, 246**
a135	—,—,N(2-methoxy-3-nitrophenyl)-	3-Nitro-o-acetanisidide.....	210.19	pa ye pr (dil aa, MeOH)	103–4	....	....	....	...	s	...	...	...	MeOH s^h aa s^h	**B13[2], 195**
a136	—,—,N(2-methoxy-4-nitrophenyl)-	4-Nitro-o-acetanisidide.....	210.19	pa ye cr (AcOEt)	153–4	....	....	....	...	s	...	...	s^h	lig i AcOEt s^h	**B13[2], 194**
a137	—,—,N(2-methoxy-5-nitrophenyl)-	5-Nitro-o-acetanisidide.....	210.19	nd (w)	178	....	....	....	δ s^h	...	...	...	δ	....	**B13[2], 193**
a138	—,—,N(2-methoxy-6-nitrophenyl)-	6-Nitro-o-acetanisidide.....	210.19	pa ye nd (dil al)	158–9	....	....	....	...	s^h	...	...	s	....	**B13[2], 192**
a139	—,—,N(3-methoxy-2-nitrophenyl)-	2-Nitro-m-acetanisidide	210.19	br amor (al)	265 sub	....	....	....	δ	s	...	...	s	lig i	**B13[1], 136**
a140	—,—,N(3-methoxy-4-nitrophenyl)-	4-Nitro-m-acetanisidide	210.19	ye nd (w)	165	....	....	....	δ	v	...	...	...	aa v lig i	**B13[1], 137**
a141	—,—,N(3-methoxy-5-nitrophenyl)-	5-Nitro-m-acetanisidide	210.19	nd (aa)	201	....	....	....	i	s	δ	...	s	aa s AcOEt s chl, lig δ	**B13[2], 216**
Ω a142	—,—,N(4-methoxy-2-nitrophenyl)-	2-Nitro-p-acetanisidide.....	210.19	ye nd (al)	116.5–7	....	....	....	s^h	s	s	...	s	aa s	**B13[2], 287**
a143	—,—,N(4-methoxy-3-nitrophenyl)-	3-Nitro-p-acetanisidide.....	210.19	og-ye nd (w or dil al)	153	....	....	....	s	s^h	...	...	δ	lig δ	**B13[1], 186**
a144	—,—,N(5-methoxy-2-nitrophenyl)-	6-Nitro-m-acetanisidide.....	210.19	wh nd (al)	125	....	....	....	δ	s^h	...	...	...	lig δ	**B13[1], 136**
Ω a145	—,—,N(2-methoxyphenyl)-	o-Acetanisidide	165.19	nd (w) λ^{al} 244 (4.02), 280 (3.66)	87–8	303–5	....	....	v^h	v	s	s	...	os s aa s	**B13[2], 172**
a146	—,—,N(3-methoxyphenyl)-	m-Acetanisidide........	165.19	nd or pl (w) λ^{al} 245 (4.07), 280 (3.49)	81	....	....	....	v^h	v	s	s	...	os s	**B13[1], 133**
Ω a147	—,—,N(4-methoxyphenyl)-	p-Acetanisidide. Methacetin.	165.19	pl (w) λ^{al} 249 (4.17), 280 sh (3.34)	130–2	....	....	....	δ	s	δ	s	...	chl s	**B13[2], 243**
Ω a148	—,—,N-methyl-...	$CH_3CONHCH_3$.........	73.10	nd λ^{w} < 210	28	204–6^{760} 95^{14}	0.9571$^{25}_{4}$	1.4301^{20}	v	v	v	v	v	chl v lig i	**B4[2], 563**
Ω a150	—,—,N-methyl-N-(1-naphthyl)-	$CH_3CON(CH_3)C_{10}H_7^{\alpha}$......	199.26	nd (lig)	95–7	....	....	....	δ	s	s	...	...	lig δ	**B12[2], 741**
a151	—,—,N-methyl-N-(4-nitrophenyl)-	....	194.19	pl (w) λ^{al} 288 (4.04)	153	....	....	....	δ s^h	s	s	...	...	....	**B12[1], 352**
Ω a152	—,—,N-methyl-N-phenyl-	Exalgin. $CH_3CON(CH_3)C_6H_5$	149.19	nd (eth), pr (al), lf (lig), λ^{al} 226 (3.75)	102–4	253^{712}	1.0036$^{105}_{4}$	1.576	s	s	s^h	...	...	lig s^h	**B12[2], 142**
a153	—,—,N-methyl-N-(2-tolyl)-	N-Methyl-o-acetotoluidide..	163.22	λ^{al} 263 (2.5), 270 (2.5)	55–6	260	....	....	...	s	...	...	...	....	**B12, 793**
a154	—,—,N-methyl-N(3-tolyl)-	N-Methyl-m-acetotoluidide .	163.22	cr	66	....	....	....	...	s	...	...	...	....	**B12, 861**
a155	—,—,N-methyl-N(4-tolyl)-	N-Methyl-p-acetotoluidide..	163.22	lf (eth-al)	83	283^{760}	....	....	...	s	s	...	...	lig s	**B12[2], 502**
Ω a156	—,—,N(1-naphthyl)-	Aceto-1-naphthalide. $CH_3CONHC_{10}H_7^{\alpha}$	185.23	cr (al)	160	....	....	....	s^h	s^h	δ	...	...	....	**B12, 1230**
Ω a157	—,—,N(2-naphthyl)-	Aceto-2-naphthalide. $CH_3CONHC_{10}H_7^{\beta}$	185.23	lf (al, w)	134	....	....	....	s^h	s^h	...	...	...	....	**B12, 1284**
a158	—,—,N(1-nitro-2-naphthyl)-	....	230.23	ye nd (al) λ^{cy} 340 (3.63)	126 (123.5)	....	....	....	δ	s	s	...	s	aa s lig i	**B12, 1313**
Ω a159	—,—,N(2-nitrophenyl)-	o-Nitroacetanilide........	180.18	ye pr (lig), lf (dil al) λ^{al} 233 (4.23), 270 (3.62), 340 (3.40)	94	100$^{0.1}$	1.419^{15}	....	s^h	s	v	...	s v^h	chl s lig s^h	**B12[2], 371**
Ω a160	—,—,N(3-nitrophenyl)-	m-Nitroacetanilide........	180.18	wh lf (al) λ^{al} 242 (4.35), 270 sh (3.75), 326 (3.18)	154–6	100$^{0.0074}$	....	....	s^h	s	i	...	...	alk s chl s Ph NO_2 s	**B12[2], 380**
Ω a161	—,—,N(4-nitrophenyl)-	p-Nitroacetanilide........	180.18	ye pr (w) λ^{al} 222 (4.12), 316 (4.06)	216 (207)	100$^{0.0084}$	....	....	δ v^h	s	δ	...	...	aa δ chl, lig δ alk s (og)	**B12[2], 389**
Ω a162	—,—,N(2-nitro-4-tolyl)-	2-Nitro-p-acetotoluidide.....	194.19	dk ye nd (peth) (metast) λ^{cy} 368 (3.62)	96	(metast 93.5)	....	....	δ	δ	...	...	...	peth δ	**B12[2], 536**
a163	—,—,N(4-octylphenyl)-	p-Octylacetanilide........	247.39	lf or pl (al)	94	....	....	....	i	v^h	v	...	...	....	**B12, 1185**
Ω a164	—,—,N-phenyl-...	Acetanilide. Antifebrin. $CH_3CONHC_6H_5$	135.17	rh or pl (w) λ^{al} 242 (4.16)	114.3 (115–6)	304^{760}	1.2190^{15}	....	δ s^h	v	s	v	s	chl, CCl_4, MeOH v to s, v^h	**B12[2], 137**
a165	—,—,N-phenyl-N-propyl-	$CH_3CON(C_6H_5)CH_2CH_2CH_3$	177.25	mcl lf (eth, lig)	49 (56)	266^{712}	....	....	i	v	v	...	...	....	**B12, 246**

For explanations, symbols and abbreviations see beginning of table. For structural formulas see end of table.

PHYSICAL CONSTANTS OF ORGANIC COMPOUNDS (Continued)

No.	Name	Synonyms and Formula	Mol. wt.	Color. crystalline form. specific rotation and λ_{max} (log ε)	m.p. °C	b.p. °C	Density	n_D	Solubility w	al	eth	ace	bz	other solvents	Ref.
	Acetic acid														
Ω a166	—,—,*N*(2-tolyl)-	*o*-Acetotoluidide	149.19	nd (al) λ^{al} 230 (3.80), 280 sh (2.32)	110	296	1.168^{15}	...	δ	s v^h	s	s v^h	δ s^h	chl s aa, MeOH s	**B12²**, 439
Ω a167	—,—,*N*(3-tolyl)-	*m*-Acetotoluidide	149.19	nd (w) λ^{al} 245 (4.15), 282 sh (2.81)	65.5	303 $182–3^{14}$	1.141^{15}	...	δ	v	v	...	...	...	**B12²**, 468
Ω a168	—,—,*N*(4-tolyl)-	*p*-Acetotoluidide	149.19	mcl cr or nd (dil al) λ^{al} 245 (4.20), 286 sh (2.79)	148.5 (153)	307 sub	1.212^{15}	...	δ s^h	s	s	...	δ	aa, chl, MeOH s lig δ	**B12²**, 501
a169	—,amidine	Acetamidine. Ethanamidine.* $CH_3C(:NH)NH_2$	58.08	$\lambda^{w,pH=13}$219 (3.04)	−35	...	...	...	d^h δ	s	...	...	...	ac s	**B2²**, 183
Ω a170	—,—,hydrochloride	$CH_3C(:NH)NH_2 . HCl$	94.55	nd or pr (al)	177–8 (167)	...	...	...	s	...	...	...	i	...	**B2²**, 184
a171	—,—,*N*,*N′*-diphenyl-	$CH_3C(:NC_6H_5)NHC_6H_5$	210.28	nd (al)	131–2	...	...	...	δ v^h	...	v	...	...	ac s	**B12²**, 144
Ω a172	—,anhydride	Acetic anhydride. Ethanoic anhydride*. $(CH_3CO)_2O$	102.09	λ^{iso} 224.5 (1.68)	−73.1	139.55 44^{15}	1.0820_4^{20}	1.39006^{20}	v	s	∞	...	s	chl s	**B2²**, 170
Ω a173	—,benzyl ester	$CH_3CO_2CH_2C_6H_5$	150.18	...	−51.5	215.5^{760} $93–4^{10}$	1.0550_4^{20}	1.5232^{20}	δ	∞	s	s	...	...	**B6²**, 415
Ω a174	—,bromide	Acetyl bromide. Ethanoyl bromide*. CH_3COBr	122.96	ye in air	−96	76^{760} (81)	1.6625_4^{16}	1.45376^{16}	d	d	∞	s	∞	chl ∞	**B2²**, 176
a175	—,2-bromoethyl ester	$CH_3CO_2CH_2CH_2Br$	167.01	...	−13.8	162–3	1.514_4^{20}	1.457^{23}	v	∞	∞	...	...	chl v	**B2¹**, 57
a176	—,bromomethyl ester	$CH_3CO_2CH_2Br$	152.98	...	...	$130–3^{750}$	1.6350_4^{20}	1.4520^{20}	i	s	v	...	...	chl v	**B2²**, 166
a177	—,2-buten-1-yl ester	Crotonyl acetate. Crotyl acetate. $CH_3CO_2CH_2CH:CHCH_3$	114.15	...	...	132^{761}	0.9192_4^{20}	1.4181^{20}	δ	s	s	s	...	...	**B2³**, 283
Ω a178	—,butyl ester	$CH_3CO_2(CH_2)_3CH_3$	116.16	...	−77.9	126.5 (125)	0.8825_4^{20}	1.3941^{20}	δ	∞	∞	s	...	...	**B2²**, 140
a179	—,*sec*-butyl ester (*d*)	$CH_3CO_2CH(CH_3)CH_2CH_3$	116.16	$[\alpha]_D^{20}$ +25.43	...	112^{760}	0.8758_4^{16}	1.3877^{20}	i	s	s	s	...	...	**B2¹**, 59
Ω a180	—,—(*dl*)	$CH_3CO_2CH(CH_3)CH_2CH_3$	116.16	...	...	112.2^{760}	0.8716_4^{20}	1.3888^{20}	i	s	s	s	...	...	**B2²**, 141
a181	—,—(*l*)	$CH_3CO_2CH(CH_3)CH_2CH_3$	116.16	$[\alpha]_{546}^{19}$ − 20.2 (c = 5)	...	$116–7^{760}$	0.8730_4^{19}	1.3899^{18}	i	s	s	s	...	...	**B2³**, 241
Ω a182	—,*tert*-butyl ester	$CH_3CO_2C(CH_3)_3$	116.16	...	...	$97–8^{760}$	0.8665_4^{20}	1.3853^{20}	...	s	s	...	...	aa s	**B2²**, 142
Ω a183	—,chloride	Acetyl chloride. Ethanoyl chloride*. CH_3COCl	78.50	fum λ^{peth} 220 (2.01)	−112	50.9^{760}	1.1051_4^{20}	1.38976^{20}	d	d	∞	∞	∞	chl ∞	**B2²**, 175
a184	—,1-chloroethyl ester	$CH_3CO_2CHClCH_3$	122.55	...	...	121.5^{746} δd	1.110_4^{20}	1.409^{20}	d	...	s	...	...	...	**B2¹**, 71
Ω a185	—,2-chloroethyl ester	$CH_3CO_2CH_2CH_2Cl$	122.55	...	...	145 50^{18}	1.178_4^{20}	1.4234^{20}	i	∞	∞	...	...	...	**B2²**, 136
a186	—,2-chloroisopropyl ester	$CH_3CO_2CH(CH_3)CH_2Cl$	136.58	...	...	149–50	1.0788^{20}	1.4223^{20}	i	s	s	...	...	...	**B2**, 130
a187	—,chloromethyl ester	$CH_3CO_2CH_2Cl$	108.53	...	...	$115–6^{757}$	1.194_4^{20}	1.409^{20}	i	s	s	...	...	con sulf s	**B2²**, 166
Ω a188	—,3-chloropropyl ester	$CH_3CO_2CH_2CH_2CH_2Cl$	136.58	...	...	$163–5^{747}$ $62–3^{10}$	1.250^{19}	1.431^{20}	i	s	s	s	...	...	**B2²**, 139
Ω a189	—,cyclohexyl ester	...	142.20	...	...	173 63^{12}	0.9698_4^{20}	1.4401^{20}	i	∞	∞	...	...	...	**B6²**, 10
Ω a190	—,decyl ester	$CH_3CO_2(CH_2)_9CH_3$	220.33	...	−15.05	244 $125.8^{15.5}$	0.8671_4^{20}	1.4273^{20}	i	s	s	...	s	aa s	**B2**, 135
a191	—,1,3-dichloroisopropyl ester	$CH_3CO_2CH(CH_2Cl)_2$	171.03	...	...	202–8 81^{15}	1.281^{20}	1.4542^{20}	...	s	s	...	...	chl s	**B2²**, 140
a193	—,1,1-dimethylbutyl ester	$CH_3CO_2C(CH_3)_2CH_2CH_2CH_3$	144.22	...	...	$152–3^{755}$ 34.5^{10}	0.8798_4^{18}	1.4068^{18}	i	s	s	...	...	...	**B2²**, 145
—	—, 1,1-dimethylpropyl ester	*See* **Acetic acid,** 2-methyl-2-butyl ester													
a195	—,2,4-dinitrophenyl ester	...	226.15	cr (MeOH)	72–3	...	...	...	i	...	...	...	...	...	**B6¹**, 127
a196	—,3,5-dinitrophenyl ester	...	226.15	cr (bz-peth)	126–7	...	...	...	i	...	...	...	v	aa v^h	**B6**, 258
Ω a197	—,ethenyl ester	Vinyl acetate. $CH_3CO_2CH:CH_2$	86.09	wh (unst), bl-gr (st) λ^{hx} 258 sh (−0.3)	−93.2	72.2– 2.3^{760}	0.9317_4^{20}	1.3959^{20}	i s^h	∞	s	s	s	os s chl s CCl_4 s	**B2²**, 147
Ω a198	—,2-ethoxyethyl ester	Cellosolve acetate. $CH_3CO_2CH_2CH_2OCH_2CH_3$	132.16	...	−61.7	156.4 49^{12}	0.9749_4^{20}	1.4058^{20}	v	∞	∞	s	...	os v	**B2²**, 155
Ω a199	—,ethyl ester	Ethyl acetate. $CH_3CO_2CH_2CH_3$	88.12	λ^{MeOH} 209 (1.86)	−83.578	77.06^{760}	0.9003_4^{20}	1.3723^{20}	s	∞	∞	v	v	chl ∞, os v	**B2²**, 129
Ω a200	—,2-ethylbutyl ester	$CH_3CO_2CH_2CH(C_2H_5)_2$	144.22	...	< −100	$162–3^{760}$ 63^{20}	0.8790_4^{20}	1.4109^{20}	i	s	s	...	...	...	**B2³**, 257
Ω a201	—,2-ethylhexyl ester	$CH_3CO_2CH_2CH(C_2H_5)C_4H_9^n$	172.27	...	−93	199 95^{25}	0.8734_{20}^{20}	1.4204^{20}	i	s	s	...	...	...	**B2³**, 261
a202	—,3-ethyl-3-pentyl ester	$CH_3CO_2C(C_2H_5)_3$	158.24	...	...	160–3	...	...	i	s	...	...	...	...	**B2**, 134
Ω a203	—,fluoride	Acetyl fluoride. Ethanoyl fluoride*. CH_3COF	62.04	fum	< −60	20.8^{760}	1.002_4^{15}	...	d	∞^d	∞	...	s	CS_2 δ chl s	**B2²**, 175
Ω a204	—,furfuryl ester	...	140.15	...	...	177	1.1175_4^{20}	1.4627^{20}	i	s	s	...	...	...	**B17²**, 115

For explanations, symbols and abbreviations see beginning of table. For structural formulas see end of table.

No.	Name	Synonyms and Formula	Mol. wt.	Color, crystalline form, specific rotation and λ_{max} (log ε)	m.p. °C	b.p. °C	Density	n_D	Solubility						Ref.
									w	al	eth	ace	bz	other solvents	
	Acetic acid														
Ω a205	—,heptyl ester	$CH_3CO_2(CH_2)_6CH_3$	158.24	...	−50.2	192.4^{760} 96^{28}	0.8750$^{15}_{4}$	1.4150^{20}	i	s	s	...	...	...	**B2**, 134
a206	—,4-heptyl ester	$CH_3CO_2CH(C_3H_7)_2$	158.24	...	...	170–2^{760} 69–70^{17}	0.8742$^{0}_{0}$	1.4105^{19}	i	...	s	...	...	...	**B2**, 134
Ω a207	—,hexadecyl ester	Cetyl acetate. $CH_3CO_2(CH_2)_{15}CH_3$	284.49	α: nd, β: wax	α: 18.5 β: 24.2	220–5^{205} 180–1^{6}	0.8574$^{25}_{4}$	1.4438^{20}	i	δ	...	...	...	...	**B2**^{2}, 146
Ω a208	—,hexyl ester	$CH_3CO_2(CH_2)_5CH_3$	144.22	...	−80.9	171.5^{760} 61.5^{12}	0.8779$^{15}_{4}$	1.4092^{20}	i	v	v	...	...	...	**B2**^{2}, 145
a209	—,2-hexyl ester (*d*)	$CH_3CO_2CH(CH_3)C_4H_9^n$	144.22	...	...	158^{760} 57^{20}	0.8658$^{18}_{4}$	1.4014^{25}	i	v	v	...	...	...	**B2**^{2}, 145
a210	—,—(*dl*)	$CH_3CO_2CH(CH_3)C_4H_9^n$	144.22	...	...	157–8^{760}	0.865$^{15}_{15}$	1.4014^{25}	i	v	v	...	...	...	**B2**^{3}, 256
a211	—,3-hexyl ester (*d*)	$CH_3CO_2CH(C_2H_5)CH_2CH_2CH_3$	144.22	$[\alpha]^{20}_{D}$ + 0.55	...	149–51	0.8672$^{20}_{4}$	1.4037^{20}	i	v	v	...	...	...	**B2**^{1}, 61
Ω a212	—,2-hydroxyethyl ester	Glycol monoacetate. $CH_3CO_2CH_2CH_2OH$	104.12	...	...	187–9	1.108^{15}	...	∞	∞	∞	...	...	...	**B2**^{2}, 154
Ω a213	—,3-hydroxyphenyl ester	Resorcinol monoacetate. Euresol.	152.16	ye	...	283d	...	...	i	s	...	...	...	...	**B6**^{2}, 817
a214	—,iodide	Acetyl iodide. Ethanoyl iodide*. CH_3COI	169.95	turn br (separate I_2) fum	...	108 36^{50}	2.0674$^{20}_{4}$	1.5491^{20}	d	d	s	...	...	...	**B2**^{2}, 177
Ω a215	—,isobutyl ester	$CH_3CO_2CH_2CH(CH_3)_2$	116.16	...	−98.58	117.2^{760}	0.8712$^{20}_{4}$	1.3902^{20}	δ	∞	∞	s	...	...	**B2**^{2}, 142
Ω a216	—,isopropenyl ester	$CH_3CO_2C(CH_3){:}CH_2$	100.12	...	−92.9	92–4^{732}	0.9090^{20}	1.4033^{20}	δ	s	v	s	...	...	**B2**^{3}, 278
Ω a217	—,isopropyl ester	$CH_3CO_2CH(CH_3)_2$	102.13	...	−73.4	90^{760}	0.8718$^{20}_{4}$	1.3773^{20}	s	s	∞	s	...	...	**B2**^{2}, 139
Ω a218	—,menthyl ester (*l*)	...	198.31	$[\alpha]^{20}_{D}$ − 79.42	...	109^{10}	0.9185$^{20}_{4}$	1.4469^{20}	...	...	...	...	...	...	**B6**^{2}, 42, 50
Ω a219	—,2-methoxyethyl ester	Methyl cellosolve acetate. $CH_3CO_2CH_2CH_2OCH_3$	118.13	...	...	144.5–5 40–1^{12}	1.0090$^{19}_{19}$	1.4002^{20}	s	s	s	...	...	...	**B2**^{2}, 154
a220	—,2-methoxyphenyl ester	...	166.18	λ^{al} 207 (3.3), 277 (3.3)	31–2	134–5^{18}	...	...	i	s	s	...	...	...	**B6**^{1}, 416
Ω a221	—,methyl ester	Methyl acetate. $CH_3CO_2CH_3$	74.08	...	−98.1	57	0.9330$^{20}_{4}$	1.3593^{20}	v	∞	∞	v	s	chl v	**B2**^{2}, 125
a222	—,2-methyl-2-butyl ester	*tert*-Amyl acetate. $CH_3CO_2C(CH_3)_2CH_2CH_3$	130.19	...	...	124–4.5	0.8740^{20}	1.4010^{20}	δ	s	s	s	...	diox s aa s	**B2**^{2}, 143
Ω a223	—,3-methylbutyl ester	Isoamyl acetate. $CH_3CO_2CH_2CH_2CH(CH_3)_2$	130.19	...	−78.5	142	0.8670$^{20}_{4}$	1.4003^{20}	δ	∞	∞	s	...	AmOH s	**B2**^{2}, 144
a224	—,methylene diester	Methanediol diacetate. $(CH_3CO_2)_2CH_2$	132.13	...	−23	164–5	1.136$^{20}_{4}$	1.4025^{24}	δ	∞	∞	...	...	os v	**B2**^{2}, 163
a225	—,2-methyl-3-heptyl ester	$CH_3CO_2CH(C_4H_9^n)CH(CH_3)_2$	172.27	...	...	172	0.875^{20}	1.4166^{20}	i	s	...	...	...	...	**B2**, 134
a226	—,3-methyl-2-heptyl ester	$CH_3CO_2CH(CH_3)CH(CH_3)C_4H_9^n$	172.27	...	...	185	0.8545$^{21}_{4}$	1.418^{21}	i	s	...	...	...	...	**B2**^{2}, 146
a227	—,6-methyl-2-heptyl ester	$CH_3CO_2CH(CH_3)(CH_2)_3CH(CH_3)_2$	172.27	...	...	187–8^{768}	0.8494^{20}	1.4137^{20}	i	s	...	...	...	...	**B2**, 135
a228	—,6-methyl-3-heptyl ester	$CH_3CO_2CH(C_2H_5)(CH_2)_2CH(CH_3)_2$	172.27	...	...	184–5	0.8554^{20}	1.41602^{20}	i	s	...	...	...	...	**B2**, 146
a229	—,2-methyl-1-naphthyl ester	1-Acetoxy-2-methylnaphthalene.	200.24	nd (eth-peth)	81–2	...	...	...	i	...	s	...	...	...	**B6**^{3}, 3027
a230	—,4-methyl-1-naphthyl ester	1-Acetoxy-4-methylnaphthalene.	200.24	nd (eth-peth)	86–8	...	...	...	i	...	...	...	...	...	**C64**, 19519
a231	—,6-methyl-1-naphthyl ester	5-Acetoxy-2-methylnaphthalene.	200.24	...	...	124^{2}	...	...	i	...	...	...	...	...	**B6**^{3}, 3028
a232	—,7-methyl-1-naphthyl ester	1-Acetoxy-7-methylnaphthalene.	200.24	...	39–41	188^{15}	...	...	i	...	...	...	...	...	**C25**, 1515
a233	—,2-methyl-3-pentyl ester	$CH_3CO_2CH(C_3H_7^i)C_2H_5$	144.22	...	...	148.5^{747}	0.8688^{20}	...	i	s	s	...	...	...	**B2**, 133
a234	—,3-methyl-3-pentyl ester	$CH_3CO_2C(C_2H_5)_2CH_3$	144.22	...	...	148	0.8834$^{20}_{20}$	1.4109^{18}	i	s	s	...	...	...	**B2**^{2}, 145
a237	—,4-methyl-2-pentyl ester	$CH_3CO_2CH(C_4H_9^i)CH_3$	144.22	...	...	147–8	0.8805^{0}	1.3980^{20}	i	s	s	...	...	...	**B2**, 133
a238	—,2-methylpentyl ester	$CH_3CO_2CH_2CH(CH_3)(CH_2)_2CH_3$	144.22	...	...	162.2^{746}	0.8717$^{25}_{25}$	...	i	s	s	...	...	...	**B2**, 133
a239	—,1-naphthyl ester	α-Naphthyl acetate	186.21	nd or pl (al) $\lambda^{95\% al}$ 232 (4.5), 282 (3.65)	49	...	...	...	i	s	s	...	...	...	**B6**^{2}, 580
Ω a240	—,2-naphthyl ester	β-Naphthyl acetate	186.21	nd (al) $\lambda^{95\% al}$ 228 (4.8), 277 (3.7)	70	...	...	...	i	s	s	...	...	chl s	**B6**^{2}, 600
Ω a241	—,nitrile	Acetonitrile. Cyanomethane. Ethanenitrile*. Methyl cyanide. CH_3CN	41.05	λ^{undil} 274 (−2.7)	−45.72	81.6	0.7857^{20}	1.34423^{20}	∞	∞	∞	∞	∞	CCl_4 ∞	**B2**^{2}, 181
Ω a242	—,nitrilotri-	Triglycolamic acid. Trimethylamine-α,α′,α″-tricarboxylic acid. $N(CH_2CO_2H)_3$	191.14	pr	242 d (258–9)	...	...	...	δ	s^h	...	...	...	...	**B4**^{2}, 801
a243	—,2-nitrophenyl ester	...	181.15	nd or pr (lig) $\lambda^{95\% al}$ 256 (3.75)	40–1	253d	...	...	s	v	v	v	v	lig δ, s^h	**B6**^{2}, 210
a244	—,3-nitrophenyl ester	...	181.15	nd (peth) $\lambda^{95\% al}$ 257 (3.9)	55–6	...	...	...	s	s	...	...	...	lig s	**B6**^{2}, 214
Ω a245	—,4-nitrophenyl ester	...	181.15	lf (dil al) λ^{al} 270 (4.00)	81–2	...	...	...	v^h	s	...	...	v	os s lig s	**B6**^{2}, 223

For explanations, symbols and abbreviations see beginning of table. For structural formulas see end of table.

No.	Name	Synonyms and Formula	Mol. wt.	Color, crystalline form, specific rotation and λ_{max} (log ε)	m.p. °C	b.p. °C	Density	n_D	Solubility w	al	eth	ace	bz	other solvents	Ref.
	Acetic acid														
a246	—,octadecyl ester	$CH_3CO_2(CH_2)_{17}CH_3$	312.54	...	34.5	$222–3^{15}$ $166–8^{1}$	...	...	i	s	...	...	...	...	**B2^2, 147**
Ω a247	—,octyl ester	$CH_3CO_2(CH_2)_7CH_3$	172.27	...	−38.5	210 $112–3^{30}$	0.8705_4^{20}	1.4190^{20}	i	s	s	...	...	os s	**B2^2, 146**
a248	—,2-octyl ester (*d*)	$CH_3CO_2CH(CH_3)(CH_2)_5CH_3$	172.27	$[\alpha]_D^{20}$ + 7.04 $[\alpha]_D^{17}$ + 5.64 (undil)	...	196^{760} 84^{15}	0.8606_4^{19}	1.4141^{20}	i	s	s	...	s	CS_2, Py s	**B2^2, 146**
a249	—,—(*dl*)	$CH_3CO_2CH(CH_3)(CH_2)_5CH_3$	172.27	...	...	194.5^{744}	0.8626_4^{14}	1.4146^{20}	i	s	s	...	...	...	**B2^2, 146**
a250	—,—(*l*)	$CH_3CO_2CH(CH_3)(CH_2)_5CH_3$	172.27	$[\alpha]_D^{14.5}$ − 6.0	...	196^{760} $89–90^{17}$	0.8570_4^{20}	1.4140^{20}	i	s	s	...	...	...	**B2^2, 146**
a251	—,3-octyl ester (*l*)	$CH_3CO_2CH(C_2H_5)(CH_2)_4CH_3$	172.27	$[\alpha]_D^{20}$ − 4.30	...	191–1.5^{760} 56.5^{3}	0.8641_4^{20}	1.4152^{20}	i	s	s	...	...	CS_2 s	**B2^1, 62**
a252	—,2-oxopropyl ester	Acetonyl acetate. Acetoxyacetone. $CH_3CO_2CH_2COCH_3$	116.13	...	...	$170–1^{755}$ 65^{11}	1.0757_4^{20}	1.4141^{20}	v	v	v	...	...	...	**B2^2, 168**
Ω a253	—,pentyl ester	Amyl acetate. $CH_3CO_2(CH_2)_4CH_3$	130.19	...	−70.8	149.25^{760}	0.8756_4^{20}	1.4023^{20}	δ	∞	∞	...	...	...	**B2^2, 143**
a253[1]	—,2-pentyl ester (*d*)	*d*-2-Amyl acetate. $CH_3CO_2CH(CH_3)CH_2CH_2CH_3$	130.19	$[\alpha]_D^{20}$ + 17.16 (undil) $[\alpha]_D^{20}$ + 15.41 (al)	...	130–1	0.8692_4^{18}	1.3960^{20}	i	s	s	s	...	...	**B2^1, 60**
a253[2]	—,—(*dl*)	$CH_3CO_2CH(CH_3)CH_2CH_2CH_3$	130.19	...	...	134	0.8692_4^{18}	1.3960^{20}	i	s	s	...	...	...	**B2^2, 143**
a253[3]	—,—(*l*)	$CH_3CO_2CH(CH_3)CH_2CH_2CH_3$	130.19	$[\alpha]_D^{20}$ + 3.30	...	142	0.8803^{13}	1.4012^{20}	i	s	s	...	...	...	**B2, 131**
a254	—,3-pentyl ester	$CH_3CO_2CH(C_2H_5)_2$	130.19	...	...	132^{741}	0.8712_4^{20}	1.4005^{20}	i	s	s	...	...	...	**B2^2, 143**
Ω a255	—,phenacyl ester	ω-Acetoxyacetophenone. $CH_3CO_2CH_2COC_6H_5$	178.19	rh pl	49–9.5	270	1.1169_4^{65}	1.5036^{65}	i	v	v	...	δ	chl v lig δ	**B8^2, 89**
Ω a256	—,phenyl ester	*O*-Acetylphenol. $CH_3CO_2C_6H_5$	136.16	λ^{al} 265 (2.7)	...	195.7 $75–6^{8}$	1.0780_4^{20}	1.5033^{20}	δ	∞	∞	...	...	chl ∞	**B6^2, 153**
a257	—,2-phenylethyl ester	β-Phenethyl acetate. $CH_3CO_2CH_2CH_2C_6H_5$	164.21	...	−31.1	232.6^{760} 153^{76}	1.0883_4^{20}	1.5171^{20}	i	s	s	...	...	...	**B6^2, 451**
a258	—,piperazinium salt	$2CH_3CO_2H.C_4H_{10}N_2$	206.24	cr (n-BuOH)	208.5–9	...	...	...	s	s	i	...	...	...	**Am 56, 1759**
Ω a259	—,propyl ester	$CH_3CO_2CH_2CH_2CH_3$	102.13	...	−95	101.6^{760}	0.8878_4^{20}	1.3842^{20}	δ	∞	∞	...	...	...	**B2^2, 137**
Ω a260	—,tetrahydrofurfuryl ester	...	144.17	...	...	$204–7^{756}$	1.0624_4^{25}	1.4350^{25}	∞	∞	∞	...	...	chl ∞	**B17^2, 107**
a261	—,2-tolyl ester	*o*-Cresyl acetate	150.18	...	...	208 89^{10}	1.0533^{15}	1.5002^{20}	δ	v	v	...	...	...	**B6^2, 330**
Ω a262	—,3-tolyl ester	*m*-Cresyl acetate	150.18	...	12	212	1.043_4^{26}	1.4978^{26}	i	∞	∞	...	∞	chl ∞ lig ∞	**B6^2, 352**
Ω a263	—,4-tolyl ester	*p*-Cresyl acetate	150.18	...	...	212.5 $107–9^{5}$	1.0512_4^{17}	1.5163^{23}	δ	s	s	...	...	chl s	**B6^2, 378**
a264	—,1,2,2-trimethylpropyl ester (*d*)	$CH_3CO_2CH(CH_3)C(CH_3)_2CH_3$	144.22	$[\alpha]_D^{25}$ + 16.2	...	141^{756}	0.856_4^{25}	1.4001^{25}	i	...	s	...	...	...	**B2^3, 258**
a265	—,—(*dl*)	$CH_3CO_2CH(CH_3)C(CH_3)_2CH_3$	144.22	...	...	143^{757}	...	1.402^{23}	i	...	s	...	...	...	**B2, 133**
—	—,vinyl ester	*See* **Acetic acid,** ethenyl ester													
a266	—,**4-acetylamino-phenyoxy-**, amide	...	208.22	nd (w)	208	...	...	...	δ v^h	v	i	...	i	...	**B13, 465**
—	—,**amino-**	*See* **Glycine**													
a267	—,—,**phenyl-(*dl*)-**	α-Amino-α-toluic acid. Phenyl glycine. $C_6H_5CH(NH_2)CO_2H$	151.17	rh lf	290 (258)	sub slowly	...	...	i	δ	...	...	...	os δ	**B14^2, 282**
a268	—,—,—,nitrile (*dl*)	α-Cyanobenzylamine. $C_6H_5CH(NH_2)CN$	132.18	hyg lf (lig)	55	...	...	...	...	...	...	...	...	lig δ	**B14^2, 285**
a269	—,**2-aminophenyl,** nitrile	2-Aminobenzyl cyanide	132.18	lf (dil al) λ^{al} 240 (4.0), 288 (3.4)	72	...	...	...	i	s	δ	...	v	...	**B14^2, 279**
a270	—,**4-aminophenyl-**	*p*-Amino-α-toluic acid	151.17	pl (w)	199–200d	...	...	...	i s^h	δ	...	...	...	...	**B14^2, 280**
a271	—,—,nitrile	...	132.18	lf (w) λ^{al} 295 (2.21)	46	312 177^{11}	...	...	δ^h	s	...	...	...	CS_2 δ^h os s	**B14^2, 281**
—	—,**(2-aminophenyl)-oxo-**	*see* **Isatic acid**													
a272	—,**arsono-**	$H_2O_3AsCH_2CO_2H$	183.98	pl (w or aa)	152d	...	2.425_4^{20}	...	v	i	i	...	i	AcOEt i	**B4^2, 999**
—	—,**benzoyl-**	*see* **Propanoic acid, 3-oxo-3-phenyl-**													
a273	—,**bis(2,4-dinitrophenyl)-**, ethyl ester	...	420.30	lf (al), nd (bz or aa)	154	...	...	...	i	i	δ	...	δ	chl, CS_2 δ	**B9, 675**
a274	—,—,methyl ester	...	406.27	lf (chl-MeOH)	159	...	...	...	i	...	...	...	...	MeOH i chl v	**B9, 675**
Ω a275	—,**bromo-**	$BrCH_2CO_2H$	138.95	hex or rh	50	208^{760} 127.2–7.6^{30}	1.9335_4^{50}	1.4804^{50}	∞	∞	∞	s	s	...	**B2^2, 201**
a276	—,—,amide	$BrCH_2CONH_2$	137.97	nd (al or bz)	91	...	...	...	v	δ	i	...	v^h	...	**B2^1, 97**
Ω a277	—,—,bromide	$BrCH_2COBr$	201.86	...	...	150	2.317_{22}^{22}	1.5449^{20}	d	d	...	s	...	...	**B2^2, 215**

For explanations, symbols and abbreviations see beginning of table. For structural formulas see end of table.

PHYSICAL CONSTANTS OF ORGANIC COMPOUNDS (Continued)

No.	Name	Synonyms and Formula	Mol. wt.	Color, crystalline form, specific rotation and λ_{max} (log ε)	m.p. °C	b.p. °C	Density	n_D	Solubility w	al	eth	ace	bz	other solvents	Ref.
	Acetic acid														
a278	—,—,*tert*-butyl ester	$BrCH_2CO_2C(CH_3)_3$	195.06			$73–4^{25}$ 50^{14}		1.4430^{30}	i	s	s	...	...	os v	**B2**[1], 96
Ω a279	—,—,ethyl ester	$BrCH_2CO_2CH_2CH_3$	167.01			168–9 $58–9^{15}$	1.5059^{20}_{20}	1.4489^{20}	i	∞	∞	s	...	os v	**B2**[2], 202
a280	—,—,isobutyl ester	$BrCH_2CO_2CH_2CH(CH_3)_2$	195.06			188^{752} 74.5^{10}	1.3269^{20}_{4}		i	v	v	s	...	os v	**B2**[1], 96
Ω a281	—,—,methyl ester	$BrCH_2CO_2CH_3$	152.98			144d 64^{33}			i	s	s	s	...	os v	**B2**[2], 202
a282	—,—,nitrile	$BrCH_2CN$	119.95	pa ye		$150–1^{752}$ 46^{13}			...	δ	s	...	...		**B2**, 216
a283	—,—,phenyl ester	$BrCH_2CO_2C_6H_5$	215.05	pl (al)	32	140^{20}			...	s	v	...	...		**B6**[1], 87
a284	—,—,propyl ester	$BrCH_2CO_2CH_2CH_2CH_3$	181.04			176^{762}	1.4099^{20}_{4}	1.4518^{20}	i	s	s	s	...	os s	**B2**, 215
a285	—,**bromochloro-**	$BrClCHCO_2H$	173.40		38 (32)	215 δ d $103–4^{11}$	1.9848^{31}_{4}	1.5014^{31}	v	v	v	v	...		**B2**[2], 204
a286	—,—,ethyl ester	$BrClCHCO_2CH_2CH_3$	201.45			174d	1.5890^{22}_{4}	1.4659^{24}	...	s	s	...	...		**B2**[2], 204
a287	—,**bromodifluoro-**	F_2BrCCO_2H	174.94	lf (chl)	40	145–60 87^{82}			v	s	...	...	...	chl s^h	**B2**, 217
a288	—,**bromo-(diphenyl)-**	$(C_6H_5)_2CBrCO_2H$	291.15	(chl-peth)	133–4				...	...	...	...	...	to v^h, chl s MeOH δ	**B9**[2], 471
a289	—,—,bromide	$(C_6H_5)_2CBrCOBr$	354.05	nd (lig)	65–6				d	v d^h	v	...	v	chl v	**B9**[1], 283
a290	—,**bromofluoro-**	$BrFCHCO_2H$	156.95		48	183 103^{20}			s	v	...	...	...	chl v	**B2**, 216
a291	—,—,amide	$BrFCHCONH_2$	155.96	nd (CCl_4)	44				v	v	v	...	...	CCl_4 s^h	**B2**, 217
a292	—,**(2-bromo-phenoxy)-**		231.05	nd (dil al), cr (w) $\lambda^{0.1\,N\,NaOH}$ 273.5 (3.31)	142.5				...	v	v	...	...	chl v AcOEt v	**B6**, 198
a293	—,**(4-bromo-phenoxy)-**		231.05	pr (al), cr (w) $\lambda^{0.1\,N\,NaOH}$ 279 (3.14)	161.4–1.8				δ	v	v	...	...	CS_2 δ	**B6**, 200
a294	—,**bromo-(phenyl)-**, nitrile	$C_6H_5CHBrCN$	196.05	yesh cr (dil al) λ^{hx} 235 (3.8)	29	242d $132–4^{12}$	1.539^{29}_{4}		i	v	v	v	v	chl v	**B9**[2], 311
a295	—,**(2-bromo-phenyl)-**	*o*-Bromotoluic acid	215.05	cr (aa)	105–6				...	v	v	...	...	CS_2 v aa v^h lig δ, v^h	**B9**, 450
a296	—,—,nitrile	2-Bromobenzyl cyanide.	196.05	λ^{al} <235	1	$145–7^{14}$			i	v	...	...	...		**B9**[1], 181
a297	—,**(3-bromo-phenyl)-**	*m*-Bromotoluic acid.	215.05	nd (w)	100–1				δ v^h	...	...	...	...		**B9**[1], 181
a298	—,**(4-bromo-phenyl)-**		215.05	nd (w)	116	sub			δ s^h	v	v	...	...	CS_2 v	**B9**[2], 309
a299	—,—,nitrile	4-Bromobenzyl cyanide.	196.05	pa ye cr (al) λ^{al} 266 (2.37)	47				i	s	...	...	v	CS_2 v	**B9**[1], 181
a300	—,**(4-bromo-phenyl)-α-hydroxy-(*dl*)**	*p*-Bromomandelic acid.	231.05	nd (bz) λ^{al} 259 (2.40), 265 (2.42)	118–20				v	v	v	...	v	chl v^h	**B10**[2], 125
a301	—,**(2-carboxy-phenoxy)-**	*o*-(Carboxymethoxy) benzoic acid. Salicylacetic acid.	196.17	nd (w)	190–2 (151)				δ s^h	s	s	s	s^h	aa s chl δ lig δ	**B10**[1], 31
a302	—,**(3-carboxy-phenoxy)-**	*m*-(Carboxymethoxy) benzoic acid.	196.17	cr	206–7				...	...	...	...	...		**B10**[1], 65
a303	—,**(4-carboxy-phenoxy)-**	*p*-(Carboxymethoxy) benzoic acid.	196.17	nd (ace or w)	280–2				δ	v	v	v	s	chl v	**B10**[2], 94
a304	—,**(2-carboxy-phenyl)-**	*o*-Carboxy-α-toluic acid. Homophthalic acid.	180.17	cr (w, eth)	185–7				s^h	s	δ	...	i	chl i	**B9**[2], 617
a305	—,—,nitrile	*o*-Carboxyphenylacetonitrile.	161.16	cr (aa), nd (w or bz)	116d (126)				δ s^h	v	v	...	v	CCl_4 v	**B9**[2], 618
a306	—,**(3-carboxy-phenyl)-**	*m*-Carboxy-α-toluic acid. Homoisophthalic acid.	180.17	nd or pl (w)	184–5				δ s^h	s	s	...	δ	chl δ	**B9**, 860
a307	—,**(4-carboxy-phenyl)-**	*p*-Carboxy-α-toluic acid. Homoterephthalic acid.	180.17	cr (dil al)	239–41				δ	s	s	...	s		**B9**, 861
a308	—,**(2-carboxy-phenyl)-2-oxo-**	*o*-Carboxyphenylglyoxylic acid. Phthalonic acid.	194.15	cr (w + 2)	146(anh)				...	s	s	s	δ	chl δ	**B10**[2], 604
Ω a309	—,**chloro-(α)**	$ClCH_2CO_2H$	94.50	mcl pr λ^{al} 218 sh(1.7)	63	187.85 104^{20}	1.4043^{40}_{4}	1.4351^{55}	v	s	s	...	s	chl, CS_2 s	**B2**[2], 187
a310	—,—(β)	$ClCH_2CO_2H$	94.50	mcl pr	56.2	187.85 104^{20}	1.4043^{40}_{4}	1.4351^{55}	v	s	s	...	s	chl, CS_2 s	**B2**[2], 187
a311	—,—(γ)	$ClCH_2CO_2H$	94.50	mcl	52.5	187.85 104^{20}	1.3978^{65} 1.4043^{40}_{4}	1.4351^{55}	v	s	s	...	s	chl, CS_2 s	**B2**[2], 187
Ω a312	—,—,amide	$ClCH_2CONH_2$	93.51	mcl pr λ^{MeOH} 442 sh(2.1)	121 (118)	$224–5^{743}$			s	v	δ	...	...		**B2**, 199
a313	—,—,—,*N*-allyl-*N*-phenyl-	$ClCH_2CON(CH_2CH{:}CH_2)C_6H_5$	217.74			$119^{0.5}$		1.5079^{25}	...	...	...	...	...		**Am 78**, 2556
a314	—,—,—,*N*,*N*-bis(2-chloroallyl)-	$ClCH_2CON(CH_2CCl{:}CH_2)_2$	242.53			$161–3^{12}$ $130^{2.1}$		1.5220^{25}	...	...	...	...	...		**Am 78**, 2556
a315	—,—,—,*N*,*N*-bis(3-chloroallyl)-	$ClCH_2CON(CH_2CH{:}CHCl)_2$	242.53			$140–5^{1}$		1.5220^{25}	...	...	...	...	...		**Am 78**, 2556
a316	—,—,—,*N*,*N*-bis(2-chloropropyl)-	$ClCH_2CON(CH_2CHClCH_3)_2$	246.57			$134^{0.7}$		1.5018^{25}	...	...	...	...	...		**Am 78**, 2556
a318	—,—,—,*N*,*N*-bis(2-ethylhexyl)-	$ClCH_2CON[CH_2CH(C_2H_5)C_4H_9^n]_2$	317.95			$154^{0.8}$		1.4622^{25}	...	...	...	...	...		**Am 78**, 2556

For explanations, symbols and abbreviations see beginning of table. For structural formulas see end of table.

No.	Name	Synonyms and Formula	Mol. wt.	Color, crystalline form, specific rotation and λ_{max} (log ε)	m.p. °C	b.p. °C	Density	n_D	Solubility						Ref.
									w	al	eth	ace	bz	other solvents	
	Acetic acid														
a319	—,—,—,*N*,*N*-bis(2-methylallyl)-	$ClCH_2CON(CH_2C(CH_3):CH_2)_2$	201.70			$133–5^{20}$ $110^{2.2}$		1.4882^{2}							Am 78, 2556
a320	—,—,—,*N*,*N*-bis(3-methylbutyl)-	$ClCH_2CON[CH_2CH_2CH(CH_3)_2]_2$	233.79			$109^{0.6}$		1.4625^{25}							Am 78, 2556
a321	—,—,—,*N*-butyl-	$ClCH_2CONHC_4H_9^n$	149.62			110^{7}		1.4665^{25}							Am 78, 2556
a322	—,—,—,*N*-*sec*-butyl-	$ClCH_2CONHC_4H_9^s$	149.62	cr (peth)	45–5.5	$68^{0.7}$								peth s^h	Am 78, 2556
a323	—,—,—,*N*-*tert*-butyl-	$ClCH_2CONHC_4H_9^t$	149.62	cr (peth)	84									peth s^h	Am 78, 2556
a324	—,—,—,*N*-butyl-*N*-ethyl-	$ClCH_2CON(C_4H_9^n)C_2H_5$	177.68			$90^{1.5}$ (115^{1})		1.4665^{25}							Am 78, 2556
a325	—,—,—,*N*-butyl-*N*-isopropyl-	$ClCH_2CON(C_4H_9^n)C_3H_7^i$	191.70			$82^{0.5}$		1.4655^{25}							Am 78, 2556
a326	—,—,—,*N*(2-chloroallyl)-	$ClCH_2CONHCH_2CCl:CH_2$	168.02			$101^{1.4}$		1.5078^{25}							Am 78, 2556
a327	—,—,—,*N*(3-chloroallyl)-	$ClCH_2CONHCH_2CH:CHCl$	168.02	cr (peth)	52–3.5	$112^{0.5}$								peth s^h	Am 78, 2556
a328	—,—,—,*N*(2-chloroallyl)-*N*-phenyl-	$ClCH_2CON(CH:CClCH_2)C_6H_5$	243.12			$138^{0.7}$		1.5602^{25}							Am 78, 2556
a329	—,—,—,*N*(2-chloro-4-nitrophenyl)-		263.08	cr (peth)	118–9										Am 78, 2556
a330	—,—,—,*N*(4-chlorophenyl)-*N*-ethyl-		323.11	cr (peth)	70–1									peth s	Am 78, 2556
a331	—,—,—,*N*(2-chloropropyl)-	$ClCH_2CONHCH_2CHClCH_3$	170.04			$88^{1.5}$		1.4942^{25}							Am 78, 2556
a332	—,—,—,*N*(3-chloropropyl)-	$ClCH_2CONHCH_2CH_2CH_2Cl$	170.04		36–7										Am 78, 2556
a333	—,—,—,*N*,*N*-diallyl-	$ClCH_2CON(CH_2CH:CH_2)_2$	173.64			$92^{0.7}$		1.4932^{25}							Am 78, 2556
a334	—,—,—,*N*,*N*-dibenzyl-	$ClCH_2CON(CH_2C_6H_5)_2$	273.77			$190^{1.8}$		1.5837^{25}							Am 78, 2556
a335	—,—,—,*N*,*N*-di-*sec*-butyl-	$ClCH_2CON(C_4H_9^s)_2$	205.73			$92^{0.7}$		1.4681^{25}							Am 78, 2556
a336	—,—,—,*N*(2,3-dichloroallyl)-	$ClCH_2CONHCH_2CCl:CHCl$	202.47			$126–31^{1.8}$		1.5311^{25}							Am 78, 2556
a337	—,—,—,*N*(2,4-dichlorobenzyl)-		252.54	cr (bz)	96.5–7.5								s^h		Am 78, 2556
a338	—,—,—,*N*(3,4-dichlorobenzyl)-		252.54	cr (dil al)	105–6					s					Am 78, 2556
a339	—,—,—,*N*(2,4-dichlorophenyl)-		238.50	cr (dil al)	101–2					s				peth s^h	Am 78, 2556
a340	—,—,—,*N*(2,5-dichlorophenyl)-		238.50	cr (dil al)	116–7					s				peth s^h	Am 78, 2556
a342	—,—,—,*N*(2,3-dichloropropyl)-	$ClCH_2CONHCH_2CHClCH_2Cl$	204.50	cr (peth)	65–6									peth s^h	Am 78, 2556
a343	—,—,—,*N*,*N*-diethyl-	$ClCH_2CON(C_2H_5)_2$	149.62			$190–5^{25}$			v						B4², 602
a344	—,—,—,*N*,*N*-dihexyl-	$ClCH_2CON[(CH_2)_5CH_3]_2$	261.84	cr (peth)	114.5–5.5										Am 78, 2556
a345	—,—,—,*N*,*N*-diisobutyl-	$ClCH_2CON(C_4H_9^i)_2$	205.73			99^{2}		1.4642^{25}							Am 78, 2556
a346	—,—,—,*N*,*N*-diisopropyl-	$ClCH_2CON(C_3H_7^i)_2$	177.68	cr (peth)	48.5–9.5	$86^{2.7}$		1.4619^{25}							Am 78, 2556
a347	—,—,—,*N*(2,4-dinitrophenyl)-		259.61	nd (al)	114–4.5										B9², 315
a348	—,—,—,*N*,*N*-dipentyl-	$ClCH_2CON(C_5H_{11}^n)_2$	233.79			126^{1}		1.4651^{25}							Am 78, 2556
a349	—,—,—,*N*,*N*-dipropyl-	$ClCH_2CON(C_3H_7^n)_2$	177.68			120^{8} $90–2^{0.8}$		1.4670^{20}							Am 78, 2556
a350	—,—,—,*N*-ethyl-*N*-hexyl-	$ClCH_2CON[(CH_2)_5CH_3]C_2H_5$	205.73			$120^{1.1}$		1.4978^{25}							Am 78, 2556
a351	—,—,—,*N*-furfuryl-		173.61	cr (peth)	58–8.5									peth s^h	Am 78, 2556
a352	—,—,—,*N*-hexyl-	$ClCH_2CONH(CH_2)_5CH_3$	177.68	cr (peth)	108.5–9.5	$95–105^{0.2}$								peth s^h	Am 78, 2556
a353	—,—,—,*N*-hexyl-*N*-methyl-	$ClCH_2CON[(CH_2)_5CH_3]CH_3$	191.70			$134^{3.8}$		1.5005^{25}							Am 78, 2556
a354	—,—,—,*N*(3-isopropoxypropyl)-	$ClCH_2CONHCH_2CH_2CH_2OCH(CH_3)_2$	193.68			$92^{0.5}$		1.4625^{25}							Am 78, 2556
a355	—,—,—,*N*-isopropyl-	$ClCH_2CONHCH(CH_3)_2$	136.60	cr (peth)	62–2.5									peth s^h	Am 78, 2556
a356	—,—,—,*N*(3-methoxypropyl)-	$ClCH_2CONHCH_2CH_2CH_2OCH_3$	165.62		30	$88^{0.5}$		1.4712^{25}							Am 78, 2556
a357	—,—,—,*N*(2-methylallyl)-	$ClCH_2CONHCH_2C(CH_3):CH_2$	147.61			96^{1}		1.4860^{25}							Am 78, 2556
a358	—,—,—,*N*(3-methylbutyl)-	$ClCH_2CONHCH_2CH_2CH(CH_3)_2$	163.65		−15	$134–5^{13}$									B4², 647

For explanations, symbols and abbreviations see beginning of table. For structural formulas see end of table.

No.	Name	Synonyms and Formula	Mol. wt.	Color, crystalline form, specific rotation and λ_{max} (log ε)	m.p. °C	b.p. °C	Density	n_D	Solubility w	al	eth	ace	bz	other solvents	Ref.
	Acetic acid														
a359	—,—,—,*N*-pentyl-	$ClCH_2CONH(CH_2)_4CH_3$	163.65	...	...	$82^{0.5}$	...	1.4665^{25}	...	...	...	...	...	...	**Am 78,** 2556
a360	—,—,—,*N*-propyl-	$ClCH_2CONHCH_2CH_2CH_3$	135.59	...	62–2.5	$105–6^{10.5}$	...	...	s	s	s	s	...	chl s peth s[h]	**B4**[1], 365
a361	—,—,—,*N*-tetradecyl-	$ClCH_2CONH(CH_2)_{13}CH_3$	289.89	cr (peth)	64–5	...	...	...	...	...	...	...	...	...	**Am 78,** 2556
a362	—,—,—,*N*-tetrahydrofurfuryl-	...	177.63	cr (peth)	62.5–3.5	...	...	...	...	...	...	...	...	peth s[h]	**Am 78,** 2556
a363	—,—,anhydride	$(ClCH_2CO)_2O$	170.98	pr (bz)	46	203^{760}	1.5497^{20}	...	d	d[h]	δ	...	...	chl δ	**B2**[2], 193
a364	—,—,benzyl ester	$ClCH_2CO_2CH_2C_6H_5$	184.62	...	...	147.5^{9} $84–6^{0.4}$	1.2223_4^4	1.5426^{18}	...	s	s	...	...	chl s	**B6,** 435
Ω a365	—,—,butyl ester	$ClCH_2CO_2(CH_2)_3CH_3$	150.61	...	...	183 94^{38}	1.0704_4^{20}	1.4297^{20}	i	s	s	...	...	...	**B2**[2], 192
a366	—,—,*sec*-butyl ester	$ClCH_2CO_2CH(CH_3)CH_2CH_3$	150.61	...	...	$163–4^{760}$	1.062_{20}^{20}	1.4251^{19}	i	s	s	...	...	...	**B2**[3], 443
Ω a367	—,—,chloride	Chloroacetyl chloride. $ClCH_2COCl$	112.94	...	...	107^{760} (108–10)	1.4202_4^{20}	1.4541^{20}	d	d	∞	s	...	...	**B2**[2], 193
a368	—,—,2-chloroethyl ester	$ClCH_2CO_2CH_2CH_2Cl$	157.00	...	...	202^{760} 89^{10}	1.3600_4^{25}	1.4619^{25}	d[h]	...	s	...	...	...	**B2,** 198
Ω a369	—,—,ethyl ester	$ClCH_2CO_2CH_2CH_3$	122.55	...	−26	144^{740} 52^{20}	1.1585_4^{20}	1.4215^{20}	i	∞	∞	∞	s	...	**B2**[2], 191
a370	—,—,2-hydroxyethyl ester	$ClCH_2CO_2CH_2CH_2OH$	138.55	...	...	240d $86^{1.6}$	1.3300_4^{20}	1.4609^{20}	∞	∞	...	...	...	...	**C27,** 4216
a371	—,—,isobutyl ester	$ClCH_2CO_2CH_2CH(CH_3)_2$	150.61	...	...	170	1.0612_4^{20}	1.4255^{20}	s	...	s	s	...	os s	**B2**[1], 89
a372	—,—,isopropyl ester	$ClCH_2CO_2CH(CH_3)_2$	136.58	...	...	$150.4–1.6^{760}$	1.0888_4^{20}	1.4192^{20}	i	s	s	...	...	...	**B2**[2], 192
a373	—,—,2-methoxyethyl ester	Methyl cellosolve chloroacetate. $ClCH_2CO_2CH_2CH_2OCH_3$	152.58	...	...	$85–6^{9}$ $60^{1.3}$	1.2015_4^{20}	1.4382^{20}	d	...	v	...	...	...	**C27,** 4216
Ω a374	—,—,methyl ester	$ClCH_2CO_2CH_3$	108.53	...	−32.12	129.8^{760} 29^{10}	1.2337_4^{20}	1.4218^{20}	δ	∞	∞	∞	∞	...	**B2**[2], 191
Ω a375	—,—,nitrile	Chloromethyl cyanide. $ClCH_2CN$	75.50	...	...	126–7 $30–2^{15}$	1.1930^{20}	1.4202^{25}	...	...	s	...	...	...	**B2**[2], 194
a376	—,—,phenyl ester	$ClCH_2CO_2C_6H_5$	170.60	nd or pl (al)	44–5	230–5 114^{8}	1.2202_4^{44}	1.5146^{44}	i	v	v	...	...	...	**B6**[2], 154
a377	—,—,4-phenylphenacyl ester	...	288.73	cr	116	...	...	...	i	...	...	...	...	...	**C24,** 503
a378	—,—,piperazinium salt	$2ClCH_2CO_2H.C_4H_{10}N_2$	275.13	cr (al)	145–6	...	...	...	s	s[h]	i	...	...	...	**Am 56,** 1759
a379	—,—,propyl ester	$ClCH_2CO_2CH_2CH_2CH_3$	136.58	...	...	161^{764}	1.1033_4^{20}	1.4261^{20}	i	...	s	...	...	...	**B2**[1], 89
a380	—,—,4-tolyl ester	...	184.62	pl	32	162^{45}	...	...	i	s	s	...	...	os s	**B6**[2], 378
a381	**—,chloro-(diphenyl)-**	$(C_6H_5)_2CClCO_2H$	246.70	pl (bz-lig)	118–9d	...	...	...	...	s	s	s	s	aa v[h] lig v[h]	**B9**[2], 471
a382	—,—,amide	$(C_6H_5)_2CClCONH_2$	245.71	cr (to)	115	...	...	...	...	s	s	...	s	chl v	**B9**[1], 283
a383	—,—,chloride	$(C_6H_5)_2CClCOCl$	265.14	cr (lig)	50–1	180^{14}	...	...	d	d	δ	...	...	...	**B9**[2], 471
a384	—,—,ethyl ester	$(C_6H_5)_2CClCO_2C_2H_5$	274.75	pl (chl), cr (al)	43–4	185^{14} $108^{0.005}$	...	...	...	s	s	...	...	chl s[h]	**B9**[1], 282
a385	**—,(4-chloro-2-methylphenoxy)-**	...	200.62	pl (bz or to)	120	...	...	...	δ[h]	v	v	...	s	CCl_4 s	**B6**[3], 1265
a386	**—,(6-chloro-2-methyl-5-pyrimidyl)-, ethyl ester**	...	214.65	...	35–6	$108–12^{11}$	...	...	i	...	s	...	s	...	**Am 59,** 1714
Ω a387	**—,(2-chlorophenoxy)-**	...	186.60	nd (w or al) $\lambda^{w,0.1\,N\,HCl}$ 272 (3.20), 279 (3.15)	148–9	...	...	...	s[h]	s[h]	...	...	...	...	**B6**[2], 172
Ω a388	**—,(3-chlorophenoxy)-**	...	186.60	cr (w) $\lambda^{w,0.1\,N\,HCl}$ 272 (3.18), 279.5 (3.14)	109.7–10.2	...	...	...	i	...	...	...	...	...	**B6**[3], 683
Ω a389	**—,(4-chlorophenoxy)-**	...	186.60	pr or nd (w) $\lambda^{w,0.1\,N\,HCl}$ 278 (3.11), 284 sh(3.01)	156–7	...	...	...	δ	...	...	...	...	...	**B6**[2], 177
a390	**—,chloro-(phenyl)-(*d*)**	*d*-α-Chloro-α-toluic acid. $C_6H_5CHClCO_2H$	170.60	cr (peth), $[\alpha]_D^{20}$ +191.9(bz)	60–1	...	...	...	δ	s	v	...	s	lig s[h] chl s	**B9**[2], 306
Ω a391	—,—(*dl*)	$C_6H_5CHClCO_2H$	170.60	lf (peth)	78	...	...	...	δ	v	v	...	...	lig δ, s[h]	**B9**[2], 307
a392	—,—(*l*)	$C_6H_5CHClCO_2H$	170.60	nd (peth), $[\alpha]_D^{18}$ −191.3 (bz)	61	...	...	...	δ	v	v	...	v	chl v lig s[h]	**B9**[2], 307
a393	**—,(2-chlorophenyl)-**	*o*-Chloro-α-toluic acid.	170.60	nd (w)	96	...	...	...	δ	v[h]	...	...	...	...	**B9**[1], 178
a394	—,—,nitrile	2-Chlorobenzyl cyanide.	151.60	grsh-ye nd λ^{al} 266 (2.33), 273 (2.22)	24	251^{760} 170^{120}	1.1737_4^{18}	1.15341^{18}	...	...	...	...	...	...	**B9,** 448
Ω a395	**—,(3-chlorophenyl)-**	*m*-Chloro-α-toluic acid.	170.60	nd (hp), pl (aq al)	77.5–8.5 (74)	...	...	...	δ	δ	∞	...	δ	...	**B9**[2], 306
a396	**—,(4-chlorophenyl)-**	*p*-Chloro-α-toluic acid.	170.60	nd (w)	105–6	...	...	...	s	s	s	...	s	...	**B9**[2], 306
a397	**—,(4-chlorophenyl)hydroxy-**	*p*-Chloromandelic acid.	186.60	nd (bz)	119–20	...	...	...	s	v	...	...	s[h]	...	**B10**[1], 92
—	**—,cyano-**	*see* **Malonic acid,** mononitrile													

For explanations, symbols and abbreviations see beginning of table. For structural formulas see end of table.

No.	Name	Synonyms and Formula	Mol. wt.	Color, crystalline form, specific rotation and λ_{max} (log ε)	m.p. °C	b.p. °C	Density	n_D	Solubility						Ref.
									w	al	eth	ace	bz	other solvents	
	Acetic acid														
a398	—,**(3-cyano-phenyl)-**, nitrile	3-Cyanobenzyl cyanide. Homoisophthalonitrile.	142.16	nd (w)	84				s^h	s	s		s	chl s	**B9**, 860
a399	—,**1-cyclohexenyl-**, nitrile		121.18	λ^{al} 217 (4.14)		144^{90} 105^{22}	0.9473_4^{21}	1.4787^{21}	i	s	s				**B9**^{2}, 32
a400	—,**cyclohexyl-**		142.20	nd (HCO_2H)	33	244–6 135^{13}	1.0423_4^{18}	1.4775^{20}	δ		s	s		os s	**B9**^{2}, 9
a401	—,**cyclohexylidene,** nitrile		121.18			$107–8^{22}$ 90^{10}	0.9483_4^{15}	1.4832^{25}		s	s				**B9**^{2}, 34
Ω a402	—,**diazo,** ethyl ester	Diazoacetic ester. Ethyl diazoacetate. $N_2CHCO_2CH_2CH_3$	114.10	ye rh λ^{al} 247 (4.13)	−22	140–1^{720}δd 44.5–6.5^{13}	1.0852_4^{18}	1.4605^{20}	δ	∞	∞		∞	lig ∞	**B3**^{2}, 390
Ω a403	—,**dibromo-**	Br_2CHCO_2H	217.86	dlq cr	48	232–4d 195^{250}			v	v	s				**B2**^{2}, 205
a404	—,—,amide,*N,N*-dimethyl-	$Br_2CHCON(CH_3)_2$	244.93	pr (w or eth)	79–80	128^{16}					δ				**B4**, 59
a405	—,—,ethyl ester	$Br_2CHCO_2CH_2CH_3$	245.91			194^{760} 121^{74}	1.9025_{20}^{20}	1.5017^{13}	i	∞	∞				**B2**^{1}, 97
a406	—,—,methyl ester	$Br_2CHCO_2CH_3$	231.88			181.5–3.5				s	s				**B2**, 219
Ω a407	—,**dichloro-**	Cl_2CHCO_2H	128.94		13.5	194^{760} 91–2^{12}	1.5634_4^{20}	1.4658^{20}	∞	∞	∞	s			**B2**^{2}, 194
Ω a408	—,—,amide	$Cl_2CHCONH_2$	127.96	mcl pr (w)	99.4	233–4^{745}	sub		s^h	s	s				**B2**^{2}, 196
a409	—,—,anhydride	$(Cl_2CHCO)_2O$	239.87			214–6d 100–2^{16}	1.574^{24}		d	d					**B2**, 204
a410	—,—,butyl ester	$Cl_2CHCO_2(CH_2)_3CH_3$	185.05			193.4–4.6 102^{37}	1.1820_4^{20}	1.4420^{20}	i	s	s			os s	**B2**, 204
Ω a411	—,—,chloride	$Cl_2CHCOCl$	147.39			108–10^{760}	1.5315_4^{16}	1.4591^{20}	d	d	∞				**B2**^{2}, 193
Ω a412	—,—,ethyl ester	$Cl_2CHCO_2CH_2CH_3$	157.00			155.5^{764} (158) 56^{10}	1.2827_4^{20}	1.4386^{20}	δ	∞	∞	s			**B2**^{2}, 196
a413	—,—,(2-hydroxy-ethyl) ester	$Cl_2CHCO_2CH_2CH_2OH$	173.00			81–$2^{0.5}$	1.438_4^{20}	1.4735^{20}	δ	s					**B2**^{3}, 460
a414	—,—,isopropyl ester	$Cl_2CHCO_2CH(CH_3)_2$	171.03			163.8–4.8	1.2053_4^{20}	1.4328^{20}		s	s				**B2**^{3}, 459
Ω a415	—,—,methyl ester	$Cl_2CHCO_2CH_3$	142.97		−51.91	142.8^{760} 38^{10}	1.3774_4^{20}	1.4429^{20}	i	s					**B2**^{2}, 196
a416	—,—,nitrile	Cl_2CHCN	109.94			112–3	1.369^{20}	1.4391^{25}						MeOH s	**B2**^{1}, 92
a417	—,—,propyl ester	$Cl_2CHCO_2CH_2CH_2CH_3$	171.03			176^{760}	1.2240^{20}	1.4398^{20}	i	s	s				**B2**^{2}, 196
Ω a418	—,**(2,4-dichloro-phenoxy)-**		221.04	cr (bz) $\lambda^{w,0.1\,N\,HCl}$ 282 (3.26), 289 sh(3.20)	140–1	$160^{0.4}$			i	s			δ		**Am 63**, 1768
a419	—,**diethoxy-,** ethyl ester	$(C_2H_5O)_2CHCO_2CH_2CH_3$	176.22			199 83–5^{13}	0.994^{18}	1.4089^{25}		v	v				**B3**^{2}, 389
Ω a420	—,**difluoro-**	F_2CHCO_2H	96.03		−0.35	134.2^{766} 67–70^{20}	1.5255^{20}	1.3420^{20}	∞	∞	∞	∞	∞		**B2**^{2}, 185
Ω a421	—,**(2,5-dihydroxy-phenyl)-**	Homogentisic acid	168.16	pr (w+1), lf (al-chl) λ^w 290 (3.58)	152–4				v	v	v		i	chl i	**B10**^{2}, 267
a422	—,**diiodo-**	I_2CHCO_2H	311.85	lt ye cr or wh nd (bz)	110 (96)				s	s^h	s^h		s^h		**B2**^{1}, 99
Ω a423	—,**(3,4-dimeth-oxyphenyl)-**	Homoveratric acid	196.21	nd (w+1), cr (bz-peth) λ 279 (3.45)	80–2 (hyd), 98–9 (anh)				s	v	v				**B10**^{2}, 268
a424	—,—,amide		195.22	cr (w)	145–7 (142)					s	s				**B10**^{1}, 198
a425	—,**(2,4-dimethyl-phenyl) hydroxy-**	2,4-Dimethylmandelic acid	180.21	rh (w), lf (to or peth-chl), nd (bz)	119 (115)				v^h	v	v		s^h	chl v	**B10**, 275
a426	—,**(2,5-dimethyl-phenyl) hydroxy-**	2,5-Dimethylmandelic acid	180.21	nd or pr (bz)	116.5–7				δ	v	v			chl v	**B10**^{2}, 168
a427	—,**(3,4-dimethyl-phenyl) hydroxy-**	3,4-Dimethylmandelic acid	180.21	lf (bz)	135				s	s^h			v^h		**B10**^{2}, 167
a428	—,**(2,4-dinitro-phenoxy)-**		242.15	pa ye pr (w)	147–8				δ	s			i	alk s	**B6**^{2}, 244
a429	—,**(3,5-dinitro-phenoxy)-**		242.15	pa br cr pw	207				δ	s				alk s	**B6**, 259
Ω a430	—,**(2,4-dinitro-phenyl)-**		226.15	nd (w)	179–80d	d			δ^h	s	s				**B9**^{2}, 315
a431	—,—,chloride		244.59	ye lf (CS_2)	77				d		v		v	chl v	**B9**^{1}, 185
a432	—,—,ethyl ester		254.20	nd (w)	37				δ	s	s			os v	**B9**, 459
a433	—,**(2,6-dinitro-phenyl)-**		226.15	ye lf (aa)	201–2d				δ	s^h				aa δ	**B9**^{1}, 185
Ω a434	—,**diphenyl-**	$(C_6H_5)_2CHCO_2H$	212.25	nd (w), lf (al) λ^{al} 260 (2.6)	148	195^{25} (sub)	1.258_{15}^{15}		δ v^h	v^h	s			chl s, v^h	**B9**^{2}, 466
Ω a435	—,—,amide	$(C_6H_5)_2CHCONH_2$	211.25	pl (al)	167.5–8.5					s^h					**B9**^{2}, 468
a436	—,—,anhydride	$[(C_6H_5)_2CHCO]_2O$	406.49	nd (eth)	98	220–5^{15}			d^h	d^h	δ		v	chl v lig δ	**B9**^{1}, 281
Ω a437	—,—,chloride	$(C_6H_5)_2CHCOCl$	230.70	pl (lig)	56–7	178^{15}			d	d				lig s^h	**B9**^{1}, 281

For explanations, symbols and abbreviations see beginning of table. For structural formulas see end of table.

No.	Name	Synonyms and Formula	Mol. wt.	Color, crystalline form, specific rotation and λ_{max} (log ε)	m.p. °C	b.p. °C	Density	n_D	Solubility w	al	eth	ace	bz	other solvents	Ref.
	Acetic acid														
a438	—,—,ethyl ester	$(C_6H_5)_2CHCO_2CH_2CH_3$	240.31	nd (al), rh (AcOEt)	59	195^{25}	...	...	i	v	v	...	...	CS_2 v	**B9**^{1}, 281
a439	—,—,methyl ester	$(C_6H_5)_2CHCO_2CH_3$	226.28	mcl pl (AcOEt), lf (dil al)	60	...	...	...	i	s	s	...	...	AcOEt s^h	**B9**^{2}, 467
Ω a440	—,—,nitrile	$(C_6H_5)_2CHCN$	193.25	pr (eth, peth), lf (dil al)	72–3 (75–6)	$181–4^{12}$	...	...	...	s^h	v	...	...	lig δ, s^h	**B9**^{2}, 469
Ω a441	**—,diphenyl-(hydroxy)-**	Benzilic acid. $(C_6H_5)_2C(OH)CO_2H$	228.25	mcl nd (w)	151	d 180	...	...	δ v^h	v	v	...	...	con sulf s (red)	**B10**^{2}, 223
a442	—,—,ethyl ester	Ethyl benzilate. $(C_6H_5)_2C(OH)CO_2CH_2CH_3$	256.31	pr or nd	34	201^{21}	...	1.5620^{20}	i	s	s	...	...	con sulf s	**B10**^{2}, 225
a443	—,—,methyl ester	Methyl benzilate. $(C_6H_5)_2C(OH)CO_2CH_3$	242.28	mcl or tcl cr (al)	75	187^{13}	...	...	i	s^h	s	...	...	con sulf s, aa s	**B10**^{2}, 225
—	**—,diureido-**	*see* **Allantoic acid**													
Ω a444	**—,ethoxy-**	Glycolic acid ethyl ether. $C_2H_5OCH_2CO_2H$	104.12	...	...	$206–7^{760}$ 111^{25}	1.1021_4^{20}	1.4194^{20}	v^h	v^h	v^h	...	...	...	**B3**^{2}, 170
a445	—,—,chloride	$C_2H_5OCH_2COCl$	122.55	...	...	123–4 $49–50^{37}$	1.1170_4^{20}	1.4204^{20}	d	d	s	s	...	...	**B3**^{2}, 173
Ω a446	—,—,ethyl ester	$C_2H_5OCH_2CO_2CH_2CH_3$	132.16	λ^{al} 214 (1.92)	...	158^{760} (152^{760}) 52^{12}	0.9701_4^{20} 0.9945^{50}	1.4029^{20}	...	s	s	s	...	...	**B3**^{2}, 172
a447	—,—,*l*-menthyl ester	...	242.36	$[\alpha]_D^{20}$ −66.35	...	155^{20}	0.9545_4^{20}	...	δ	s	s	...	...	chl s	**B6**^{2}, 47
a448	—,—,methyl ester	$C_2H_5OCH_2CO_2CH_3$	118.13	...	...	$147–8^{734}$	1.0112^{15}	...	...	s	s	s	...	...	**B3**^{1}, 91
a449	—,—,piperazinum salt	$2C_2H_5OCH_2CO_2H.C_4H_{10}N_2$	294.35	...	120–1 (cor)	...	...	...	s	s	δ	...	...	...	**Am 70,** 2759
a450	**—,(2-ethoxyphenyl)-**	...	180.21	nd (lig), cr (w)	103–4	...	...	...	δ	δ	...	...	...	lig s^h	**B10**^{1}, 82
Ω a451	**—,(ethylthio)-**	$C_2H_5SCH_2CO_2H$	120.18	...	−8.5	164^{83} 109^5	1.1497_4^{20}	...	v	v	v	...	...	...	**B3**^{1}, 95
a452	**—,9-fluorenyl-**	...	224.26	mcl nd (al)	138–9	$218–20^{11}$	...	...	...	...	...	...	...	...	**E13,** 95
Ω a453	**—,fluoro-**	FCH_2CO_2H	78.04	nd	35.2	165^{760}	1.3693^{36}	...	s^h	s^h	...	...	...	...	**B2**^{2}, 185
Ω a456	**—,2-furyl-**	2-Furanacetic acid.	126.11	lf (bz, peth)	68–9	$102–4^{0.4}$	...	...	s	...	...	...	s^h	MeOH s, peth s^h	**Am 62,** 1512
a457	—,—,nitrile	Furfuryl cyanide.	107.11	...	...	$78–80^{20}$	1.0854_4^{25}	1.4693^{20}	δ	v	v	...	...	...	**C49,** 1695
—	**—,guanidino-**	*see* **Glycocyamine**													
Ω a458	**—,hydroxy-**	Glycolic acid. $HOCH_2CO_2H$	76.05	rh, nd (w), lf (eth)	80	d	...	...	s	s	s	...	...	...	**B3**^{2}, 167
a459	—,—,acetate	Acetoxyacetic acid. $CH_3CO_2CH_2CO_2H$	118.09	nd (bz)	67–8	$144–5^{12}$ $113–4^{4.6}$	...	...	v	v	v	v	i	chl v	**B3**^{2}, 171
a460	—,—,—,ethyl ester	Ethyl acetoxyacetate. Ethyl acetylglycolate. $CH_3CO_2CH_2CO_2CH_2CH_3$	146.14	...	...	179	1.0880_4^{20}	1.4112^{20}	δ	s	s	...	...	aa s	**B3**^{2}, 172
a461	—,—,amide	Glycolamide. Glycolic amide. $HOCH_2CONH_2$	75.07	lf (al), rh (AcOEt)	120	...	1.415^{13}	...	v	δ	δ	...	...	...	**B3**^{2}, 173
a462	—,—,anhydride	Glycolic anhydride. $(HOCH_2CO)_2O$	134.09	pw	128–30	d	...	...	i	i	i	...	...	...	**B3**^{1}, 92
a463	—,—,ethyl ester	Ethyl glycolate. $HOCH_2CO_2CH_2CH_3$	104.11	...	...	160 69^{25}	1.0826_4^{23}	1.4180^{20}	...	v	v	...	...	...	**B3**^{2}, 171
Ω a464	—,—,methyl ester	Methyl glycolate. $HOCH_2CO_2CH_3$	90.08	...	...	151.1	1.1677_4^{18}	...	s	∞	∞	...	...	...	**B3**^{2}, 171
Ω a465	—,—,nitrile	Formaldehyde cyanohydrin. Glycolonitrile. $HOCH_2CN$	57.05	...	< −72	183 δd 119^{24}	...	1.4117^{19}	v	v	v	...	i	chl i	**B3**^{2}, 174
a466	**—,hydroxy(4-iodophenyl)-,(*dl*)**	*p*-Iodomandelic acid.	278.05	...	135	...	...	...	v	v	v	...	...	lig δ	**B10,** 210
a467	**—,hydroxy-(4-isopropylphenyl)-, (*d*)**	*d-p*-Isopropylmandelic acid	194.23	lf (w), $[\alpha]_D^{17}$ +134.9 (abs al, c = 4.06)	153–4	...	...	...	δ s^h	s	v	...	...	...	**B10,** 279
a468	—,—(*dl*)	$C_{11}H_{14}O_3$. *See* a467	194.23	nd (w)	159.2–60	...	...	...	δ	s	v	...	...	...	**B10**^{1}, 120
a469	—,—(*l*)	$C_{11}H_{14}O_3$. *See* a467	194.23	ta (20% al), $[\alpha]_D^{17}$ −135 (abs al, c = 4.09)	153–4	...	...	...	δ s^h	v	v	...	...	...	**B10,** 279
a470	**—,hydroxy(phenyl)-, (*D*, −)**	*l*-Mandelic acid. *l*-Phenylglycolic acid. $C_6H_5CH(OH)CO_2H$	152.16	ta, $[\alpha]_D^{20}$ −158.0 (w, c = 2.5) −187.44 (aa)	133–5	...	1.341	...	s	s	v	...	...	aa s, chl s	**B10**^{2}, 114
Ω a471	—,—(*DL*)	*dl*-Mandelic acid. *dl*-Phenylglycolic acid. $C_6H_5CH(OH)CO_2H$	152.16	pl (w), rh (bz) λ^{al} 252 (3.1), 258 (2.0), 265 (1.9)	121–3 (119)	d	1.300_4^{20}	...	s	v	v	...	...	...	**B10**^{2}, 118
a472	—,—,—,(solvate)	$C_6H_5CH(OH)CO_2H$	152.16	mcl (bz + 1)	32.6–.7	...	...	...	...	...	...	...	...	...	**B10**^{2}, 118
a473	—,—(*L*, +)	*d*-Mandelic acid. *d*-Phenylglycolic acid. $C_6H_5CH(OH)CO_2H$	152.16	pl (w), $[\alpha]_D^{20}$ +156.6 (w, c = 2.9)	134–5	...	...	...	s v^h	s	v	...	...	chl s	**B10**^{2}, 114
a474	—,—,acetate (*D*)	*l-O*-Acetylmandelic acid. $C_6H_5CH(O_2CCH_3)CO_2H$	194.19	nd (w + 1), $[\alpha]_D^{20}$ −156.4 (ace)	96.5–8 (anh)	...	...	...	δ s^h	v	v	v	s	chl v, CCl_4 δ, lig δ	**B10**^{1}, 85

For explanations, symbols and abbreviations see beginning of table. For structural formulas see end of table.

No.	Name	Synonyms and Formula	Mol. wt.	Color, crystalline form, specific rotation and λ_{max} (log ε)	m.p. °C	b.p. °C	Density	n_D	Solubility: w	al	eth	ace	bz	other solvents	Ref.
	Acetic acid														
Ω a475	—,—,—(*DL*)	$C_6H_5CH(O_2CCH_3)CO_2H$	194.19	cr (bz), amor (chl-peth), cr (w + 1), nd (w + ½)	38–9 (+1 w), 79–80 (anh)	...	...	...	δ s^h	v	v	...	v	chl v	**B10**[2], 120
Ω a476	—,—,acetate chloride (*dl*)	*dl-O*-Acetylmandelyl chloride. $C_6H_5CH(O_2CCH_3)COCl$	212.64	...	...	$150–5^{33}$ 129^{10}	...	...	...	...	v	...	v	chl v lig v	**B10**[1], 89
a477	—,—,amide (*dl*)	α-Hydroxy-α-toluamide. *dl*-Mandelamide. $C_6H_5CH(OH)CONH_2$	151.17	pl (bz or al)	134–5	...	...	...	δ	v^h	δ	...	s^h	...	**B10**[2], 122
a478	—,—,—,*N*-ethyl-(*D*)	*l-N*-Ethylmandelamide. $C_6H_5CH(OH)CONHC_2H_5$	179.22	pl (chl-peth), $[\alpha]_D^{16}$ − 34.4 (al, c = 3.7) $[\alpha]_D^{18}$ − 103.6 (ace)	65.5–6.5	...	...	...	v	s	s	s	s	chl s	**B10**[2], 117
a479	—,—,—,—(*DL*)	$C_6H_5CH(OH)CONHC_2H_5$	179.22	pl (bz-peth)	53–4	...	...	...	s	s	s	s	s	peth δ	**B10**[1], 89
a480	—,—,ethyl ester (*D*, −)	*l*-Ethyl mandelate. $C_6H_5CH(OH)CO_2CH_2CH_3$	180.21	(peth), $[\alpha]_D^{20}$ −128.4 (chl, c = 6.67), $[\alpha]_D^{14}$ −200.2 (CS_2, c = 2)	35	150^{21}	1.1270_4^{20}	...	...	s	s	...	...	CS_2 s lig s^h	**B10**[2], 116
Ω a481	—,—,—(*DL*)	$C_6H_5CH(OH)CO_2CH_2CH_3$	180.21	nd (peth)	37	253–5 141^{15}	...	...	...	s	s	...	...	CS_2 s lig s	**B10**[2], 121
a482	—,—,—(*L*, +)	$C_6H_5CH(OH)CO_2CH_2CH_3$	180.21	(peth), $[\alpha]_D^{20}$ +205.1 (CS_2, c = 0.7)	33	...	1.1270_4^{20}	...	...	s	s	...	...	CS_2 s lig s^h	**B10**[2], 114
a483	—,—,methyl-ester (*D*, −)	*l*-Methyl mandelate. $C_6H_5CH(OH)CO_2CH_3$	166.18	cr (peth) $[\alpha]_D^{20}$ −131.5 (w) $[\alpha]_D^{18}$ − 214 (CS_2)	55	160^{32} 135^{12}	1.1756^{20}	...	s	s	...	s	s	chl s peth s^h	**B10**[2], 115
Ω a484	—,—,—(*DL*)	$C_6H_5CH(OH)CO_2CH_3$	166.18	pl (bz-lig)	58	250 δd 144^{20}	1.1756^{20}	...	δ	s	...	...	s^h	chl s peth s^h	**B10**[2], 120
a485	—,—,— (*L*, +)	$C_6H_5CH(OH)CO_2CH_3$	166.18	cr (peth) $[\alpha]_D^{20}$ + 133.6 (w) $[\alpha]_D^{20}$ + 252 (CS_2)	55.5	160^{32}	1.1756^{20}	...	s	s	...	s	s	chl s peth s^h	**B10**[2], 114
a486	—,—,nitrile (*D*, +)	*D*(+)-Benzaldehyde cyanohydrin. *D*(+)-Mandelonitrile. $C_6H_5CH(OH)CN$	133.15	nd, $(\alpha)_{5461}^{25}$ +46.9 (bz)	28.5–9.5	...	...	...	i	s	s	...	...	...	**B10**[2], 117
a487	—,—,—(*DL*)	$C_6H_5CH(OH)CN$	133.15	ye pr	22	170d	1.1165_4^{20}	1.5201^{20}	i	s	...	...	...	...	**B10**[2], 123
a488	—,—,—(*L*, −)	$C_6H_5CH(OH)CN$	133.15	$\lambda^{3:1\ hx-eth}$ 250 (2.05), 255 (2.12), 263 (2.10)	...	...	...	...	i	s	s	...	...	...	**B10**[1], 86
a489	—,**(3-hydroxy-4-methoxy-phenyl)-**	Homoisovanillic acid	182.17	...	130–1	...	...	...	s	s	s	...	s^h	...	**B10**[2], 268
a490	—,**(2-hydroxy-5-nitrophenyl)-**	...	197.15	nd	160–2	...	...	...	v	s	s	...	δ	chl δ	**B10**, 189
a491	—,—,ethyl ester	...	225.20	pr or pl(al)	154–5	...	...	...	...	s^h	δ	...	s	chl s lig δ	**B10**, 189
Ω a492	—,**(3-hydroxy-phenoxy)-**	Resorcinol-*O*-acetic acid	168.16	nd or pr (w, to)	158–9	...	...	...	s^h	s	...	...	...	to s^h	**B6**, 817
a493	—,—,ethyl ester	...	196.21	pr (w, bz)	55	274 δd $170–3^{11}$	...	...	i s^h	...	...	...	s^h	lig δ	**B6**, 817
a494	—,**(4-hydroxy-phenoxy)-**	Hydroquinone-*O*-acetic acid	168.16	nd (to), pr (w + ⅓)	154 (hyd)	...	...	...	s^h	...	...	...	...	to s^h	**B6**, 847
Ω a495	—,**(2-hydroxy-phenyl)-**	*o*-Hydroxy-α-toluic acid	152.16	nd (eth), pr (chl) $\lambda^{0.1N\ HCl}$ 271 (3.3), $\lambda^{0.1N\ NaOH}$ 293 (3.6)	147–9	240–3d	...	...	δ	...	s	...	...	chl δ	**B10**[2], 112
a496	—,—,amide	...	151.17	lf (al-chl)	118	...	...	...	...	s^h	...	...	...	chl s^h	**B10**, 188
a497	—,—,hydrazide	...	166.18	lf (chl or bz), nd (al)	154	...	...	...	...	s^h	...	...	s	chl s^h	**B10**[2], 112
a498	—,—,nitrile	*o*-Hydroxybenzyl cyanide	133.15	nd (bz-lig)	117–9	...	...	...	s^h	...	s	s	s	os v lig i	**B10**, 188
Ω a499	—,**(3-hydroxy-phenyl)-**	*m*-Hydroxy-α-toluic acid	152.16	nd (bz-lig)	131–4	190^{11}	...	...	v	v	v	...	s	lig δ	**B10**[2], 112
a500	—,—,nitrile	*m*-Hydroxybenzyl cyanide	133.15	pl (w) λ^{alk} 245 (4.01), 294 (3.46)	52–3	...	...	...	s v^h	s	s	...	...	...	**B10**, 189
a501	—,**(4-hydroxy-phenyl)-**	*p*-Hydroxy-α-toluic acid	152.16	nd (w) λ^{al} 225 (3.9), 278 (3.2)	149–51	sub	...	...	δ v^h	v	v	...	...	...	**B10**[2], 112
a502	—,—,nitrile	*p*-Hydroxybenzyl cyanide	133.15	pl (w), mcl pr	69–70	330.5^{756} 210^{10}	...	...	δ	v	v	...	...	...	**B10**[2], 113
Ω a503	—,**iminodi-**	Diglycolamidic acid. $HN(CH_2CO_2H)_2$	133.11	rh pr	247.5 (225d)	...	...	...	δ	i	i	...	...	...	**B4**[2], 800

For explanations, symbols and abbreviations see beginning of table. For structural formulas see end of table.

PHYSICAL CONSTANTS OF ORGANIC COMPOUNDS (Continued)

No.	Name	Synonyms and Formula	Mol. wt.	Color, crystalline form, specific rotation and λ_{max} (log ε)	m.p. °C	b.p. °C	Density	n_D	Solubility w	al	eth	ace	bz	other solvents	Ref.
	Acetic acid														
Ω a504	—,—,dinitrile	α,α′-Dicyanodimethylamine. Iminodiacetonitrile. $HN(CH_2CN)_2$	95.11	lf(al), cr(w)	78	...	...	...	s	s^h	δ	...	δ	chl δ	**B4**[1], 481
a505	—,**(3-indolyl)-**	Heteroauxin.	175.19	lf(bz), pl(chl) λ^w 282 (3.70)	165–6	...	...	...	i	v	s	s	s^h	chl δ, s^h	**B22**[2], 50
Ω a506	—,**iodo-**	ICH_2CO_2H	185.95	pl (w, peth) λ^{al} 224 (2.71), 273 (2.66)	83	d	...	...	s^h	s^h	δ	...	...	peth s^h	**B2**[2], 206
Ω a507	—,—,amide	ICH_2CONH_2	184.97	cr (w)	95	...	...	...	s^h	...	...	...	...	...	**B2**[1], 99
Ω a508	—,—,ethyl ester	$ICH_2CO_2CH_2CH_3$	214.00	oil	...	178–80^{760} 73^{16}	1.8173_4^{13}	1.5079^{13}	...	s	s	...	...	...	**B2**[2], 206
Ω a509	—,**(2-isopropyl-5-methylphenoxy)-**	Thymoxyacetic acid.	208.24	nd (dil al), cr (bz)	149–50	...	...	...	δ s^h	v	v	...	s^h	...	**B6**[2], 499
a510	—,**(4-isopropyl-phenyl)-**, amide	...	177.25	pl (bz)	170	...	...	...	i	s	δ	...	s^h	...	**B9**, 561
a512	—,**Isothiocyanato-** ethyl ester	Carbethoxymethyl isothiocyanate. $SCNCH_2CO_2CH_2CH_3$	145.18	red in air	...	215 δd 110^{10}	1.1649_4^{18}	...	...	...	s	...	s	...	**B4**[1], 480
Ω a513	—,**menthoxy-(*l*)**	...	214.31	cr (eth), $[\alpha]_D^{20}$ −92.93 (MeOH)	53–5	171^{11}	...	...	...	v	s	v	...	os v	**B6**[1], 25
a514	—,—,chloride(*l*)	...	232.75	$[\alpha]_D^{13}$ −84.8 (chl)	...	128–31^{11}	...	...	d^h	d^h	s	...	...	...	**B6**[1], 25
Ω a515	—,**mercapto-**	Thioglycolic acid. $HSCH_2CO_2H$	92.12	$\lambda^{w, pH=10}$ 240 (3.26)	−16.5	120^{20} 96^{5}	1.3253^{20}	1.5030^{20}	∞	∞	∞	...	...	...	**B3**[2], 175
a516	—,—,acetate	Acetylthioglycolic acid. $CH_3COSCH_2CO_2H$	134.16	ye λ^{MeOH} 231 (3.53)	...	158–9^{17} 115–8$^{2.5}$	...	...	s	...	...	...	...	...	**B3**[1], 96
Ω a516[1]	—,—,amide,*N*(2-naphthyl)-	Thionalide. $C_{10}H_7^{\beta}NHCOCH_2SH$	217.29	nd	111–2	...	...	...	i	v	...	...	...	os v	**C29**, 3330
Ω a517	—,—,—,*N*-phenyl-	Thioglycolic acid anilide. $HSCH_2CONHC_6H_5$	167.23	nd (w or al)	110.5–11	...	...	...	s^h	v	v	...	δ	lig δ	**B12**[1], 265
Ω a518	—,—,ethyl ester	$HSCH_2CO_2CH_2CH_3$	120.18	...	...	156–8 55^{17}	1.0964^{15}	1.4582^{20}	...	s	...	...	...	...	**B3**[2], 180
a519	—,**(2-mercapto-phenyl)-**	...	168.23	pl (w, bz-lig)	96–7	...	...	...	δ s^h	v	v	...	s	lig δ	**B10**[1], 82
Ω a520	—,**methoxy-**	Methylglycolic acid. $CH_3OCH_2CO_2H$	90.08	hyg	...	203–4 96^{13}	1.1768_4^{20}	1.4168^{20}	s	s	s	...	...	...	**B3**[2], 170
a521	—,—,amide,*N*(4-ethoxyphenyl)-	*p*-Phenetitide methoxyacetic acid.	209.25	nd (dil al)	99–100	...	...	...	δ^h	v	v	...	...	...	**B13**, 489
Ω a522	—,—,chloride	CH_3OCH_2COCl	108.53	...	...	99 46–9^{62}	1.1871_4^{20}	1.4196^{20}	d	d	s	s	...	chl v	**C55**, 18727
Ω a523	—,—,ethyl ester	$CH_3OCH_2CO_2CH_2CH_3$	118.13	...	...	142^{758} 44–5.5^{9}	1.0118^{15}	1.4050^{20}	δ	v	v	...	...	os s	**B3**[2], 172
Ω a524	—,—,methyl ester	$CH_3OCH_2CO_2CH_3$	104.12	...	...	131^{763} 57^{50}	1.0511_4^{20}	1.3962^{20}	δ	v	v	s	...	os s	**B3**[2], 171
Ω a525	—,—,nitrile	CH_3OCH_2CN	71.08	...	...	118.1	0.9492_4^{20}	1.3831^{20}	δ	s	s	s	...	ac, alk, os s	**B3**[2], 174
a526	—,—,piperazinium salt	$2CH_3OCH_2CO_2H.C_4H_{10}N_2$	266.30	cr (al)	155.7–6 (cor)	...	...	...	s	s	i	...	...	...	**Am 70**, 2759
a527	—,**methoxy-(phenyl)-(*D*, −)**	$C_6H_5CH(OCH_3)CO_2H$	166.18	nd (peth), $[\alpha]_D^{13}$ −150.0 (al), −168.5 (w)^2	63–4	...	...	...	s	s	...	s	...	lig s^h	**B10**[1], 85
a528	—,—(*dl*)	$C_6H_5CH(OCH_3)CO_2H$	166.18	pl (lig)	71–2	...	...	...	δ	v	v	...	...	lig δ, s^h	**B10**[1], 87
a529	—,**(4-methoxy-2-nitrophenyl)-**	...	211.18	ye nd (50 % al)	157–8 d	...	...	...	...	s	...	...	...	aa s^h	**B10**[2], 113
a530	—,**(2-methoxy-phenoxy)-**	Pyrocatechol methyl ether *O*-acetic acid.	182.18	nd (w)	123–5 (129)	...	...	...	v^h	v	v	...	s	aa s lig i	**B6**[2], 784
a531	—,**(3-methoxy-phenoxy)-**	Resorcinol methyl ether *O*-acetic acid.	182.18	nd (w)	118	...	...	...	δ v^h	...	...	...	...	...	**B6**, 817
a532	—,**(2-methoxy-phenyl)-**	*o*-Methoxy-α-toluic acid.	166.18	nd (w)	123	100–1^{2}	...	...	s^h	v	v	v	v	to v^h chl v aa v lig δ	**B10**, 188
a533	—,—,nitrile	*o*-Methoxybenzyl cyanide. *o*-Methoxy-α-tolunitrile.	147.18	pr(bz-lig)	68.5	141–3^{11} 114^{3}	...	...	...	...	...	...	s^h	...	**B10**[2], 112
Ω a534	—,**(4-methoxy-phenyl)-**	Homoanisic acid. *p*-Methoxy-α-toluic acid.	166.18	pl (w)	86–8.5	138–40$^{2-3}$	...	...	i s^h	v	s	...	s^h	lig δ	**B10**[2], 113
Ω a535	—,—,nitrile	*p*-Methoxybenzyl cyanide. *p*-Methoxy-α-tolunitrile.	147.18	...	...	286–7 152^{16}	1.0845_4^{20}	1.5309^{20}	...	s	s	...	...	...	**B10**[2], 113
—	—,**methylguanido-**	*see* **Creatine**													
a537	—,**(2-methyl-3-indolyl)-**	...	189.22	(ace)	195–200 (196–7d)	...	...	...	δ^h	s^h	s	s^h	...	chl δ^h	**B22**, 69
Ω a538	—,**(1-naphthoxy)-**	$C_{10}H_7^{\alpha}OCH_2CO_2H$	202.21	pr	190	...	...	...	δ	v	v	...	...	...	**B6**[2], 580
Ω a539	—,**(2-naphthoxy)-**	$C_{10}H_7^{\beta}OCH_2CO_2H$	202.21	pr (w) λ^{chl} 272 (3.69), 311 (3.14), 325.5 (3.25)	156	...	...	...	s^h	s	s	...	...	aa s	**B6**[2], 602
Ω a540	—,**(1-naphthyl)-**	α-Naphthaleneacetic acid. $C_{10}H_7^{\alpha}CH_2CO_2H$	186.21	wh nd (w)	133	d	...	...	δ s^h	δ s^h	v	v	s	chl v aa s	**B9**[2], 456

For explanations, symbols and abbreviations see beginning of table. For structural formulas see end of table.

No.	Name	Synonyms and Formula	Mol. wt.	Color, crystalline form, specific rotation and λ_{max} (log ε)	m.p. °C	b.p. °C	Density	n_D	Solubility w	al	eth	ace	bz	other solvents	Ref.
	Acetic acid														
a541	—,—,amide	$C_{10}H_7^{\alpha}CH_2CONH_2$	185.23	nd (w, al)	180–1 sub	...	...	...	i s^h	...	s	...	s	CS_2 s aa s	**E12B,** 3267
Ω a542	—,—,nitrile	$C_{10}H_7^{\alpha}CH_2CN$	167.21	wx so	32–3	162–4[12]	...	1.6192^{20}	...	s	...	...	...	...	**E12B,** 3268
a543	—,(2-naphthyl)-, amide	$C_{10}H_7^{\beta}CH_2CONH_2$	185.23	lf (w)	202–4.6d	...	...	...	δ^h	s	s	...	...	...	**E12B,** 3275
a544	—,—,nitrile	$C_{10}H_7^{\beta}CH_2CN$	167.21	nd or lf (dil al or tetralin)	85.6–6.2	145–50[2] 202–5[28]	...	...	δ^h	δ	v	...	v	chl v, CS_2, AcOEt v peth i	**E12B,** 3275
a545	**—,nitro-**	$O_2NCH_2CO_2H$	105.05	nd (chl)	92–3d (89)	...	...	...	d	v	v	..	v^h	chl v^h peth i	**B2**[2], 207
a546	—,—,ethyl ester	$NO_2CH_2CO_2CH_2CH_3$	133.11	λ^{MeOH} 270 (2.2)	...	105–7[25] 83[6]	1.1953_4^{20}	1.4250^{20}	δ	∞	v	...	...	dil alk s	**B2**[2], 207
a547	**—,(2-nitro-phenoxy)-**	...	197.15	pr (w)	158.2–8.5	...	...	...	δ	s	s	...	...	...	**B6**[2], 211
a548	**—,nitro(phenyl)-,** nitrile	α-Nitrobenzyl cyanide. $C_6H_5CH(NO_2)CN$	162.15	...	39–40	...	...	...	d	s	...	...	...	...	**B9**[2], 313
Ω a549	**—,(2-nitro-phenyl)-**	...	181.15	nd (w), pl (dil al) λ^{al} 259 (3.75)	141–2	...	...	...	s^h	s	...	...	...	...	**B9**[2], 311
a550	—,—,nitrile	*o*-Nitrobenzyl cyanide	162.15	nd (al-w), pr (aa or al)	84	178[12] 137–9[1]	...	...	s^h	s	s	s	s	chl s alk s aa s^h	**B9**[2], 311
a551	**—,(3-nitro-phenyl)-**	...	181.15	nd (w)	122	...	...	...	s^h	s	...	...	...	...	**B9**[2], 311
a552	—,—,nitrile	*m*-Nitrobenzyl cyanide.	162.15	(eth-lig)	63	180[1.5]	...	...	v	s	v	...	v	chl v lig δ	**B9**[2], 312
Ω a553	**—,(4-nitro-phenyl)-**	...	181.15	pa ye nd (w) λ^{al} 272 (3.98)	153–5	...	...	...	δ s^h	s	s	...	s	...	**B9**[2], 312
Ω a554	—,—,nitrile	*p*-Nitrobenzyl cyanide.	162.15	pl λ^{al} 262 (4.81)	116–7	195–7[12]	...	...	δ	s	s	...	s	chl s	**B9**[2], 313
a555	—,—,piperazinium salt	$2(C_8H_7NO_4).C_4H_{10}N_2$. *See* a553	448.44	cr (al)	205.5–9.5 (cor)	...	...	...	s^h	s^h	i	...	...	...	Am 70, 2759
a556	**—,(2-nitro-phenyl)oxo-**	*o*-Nitrobenzoylformic acid. *o*-Nitrophenylglyoxylic acid.	195.13	pr (w + 1)	123 (156–7d) (anh) 76–7 (hyd)	...	...	...	∞^h	s	δ	v	δ	aa v lig δ to δ	**B10**[1], 315
a557	**—,oxo-,**	Glyoxylic acid. Aldehydo-formic acid. $OHCCO_2H$	74.04	rh pr (w + ½) $\lambda^{w,pH=7.1}$ 565 (0.1), 576 (0.9), 580 (1.0), 584 (0.9)	70–5 (+ ½w) 98 (anh)	...	...	...	v	δ	δ	...	δ	...	**B3**[2], 385
Ω a558	**—,oxo(phenyl)-**	Benzoylformic acid. Phenyl-glyoxylic acid. $C_6H_5COCO_2H$	150.14	pr (CCl_4)	66	147–51[12]	...	...	v	s	s	...	...	CS_2 i	**B10**[2], 454
Ω a559	—,—,methyl ester	Methyl benzoylformate. $C_6H_5COCO_2CH_3$	164.16	ye	...	246–8 137[14]	...	1.5268^{20}	...	...	...	...	...	...	**B10**[2], 455
Ω a560	—,—,nitrile	Benzoyl cyanide. C_6H_5COCN	131.14	ta	32–3	206–8 99[19]	...	...	i	v	v	...	...	...	**B10**[2], 457
a561	—,—,nitrile oxime	Phenylglyoxylonitrile oxime. $C_6H_5C(:NOH)CN$	146.15	lf(w), nd	127–8	...	...	...	s^h	v	s	...	v	...	**B10**[2], 457
a562	**—,oxo(2-thienyl)-**	2-Thienylglyoxylic acid.	156.16	vlt cr (w + 1)	91.5 (anh) 58–9 (hyd)	...	...	...	v	...	v	...	...	...	**B18,** 407
a563	**—,oxydi-**	Diglycolic acid. Oxydiethanoic acid. $O(CH_2CO_2H)_2.H_2O$	152.11	mcl pr (w + 1)	148	d	...	...	s	s	s	...	...	...	**B3**[1], 90
a564	—,—,dichloride	Diglycolyl dichloride. $O(CH_2COCl)_2$	170.99	...	...	116[15]	...	...	d	d	...	...	...	chl s	**B3,** 240
a566	**—,per-**	Acetyl hydroperoxide. Peracetic acid. CH_3CO_2OH	76.05	λ^w <240 (>1.4)	0.1	105 (exp 110)	1.226_4^{15}	1.3974^{20}	v	s	v	...	...	sulf v	**B2**[3], 379
Ω a567	**—,phenoxy-**	Glycolic acid phenyl ether. $C_6H_5OCH_2CO_2H$	152.16	nd or pl (w) λ^{hx} 270 (3.10), 276 (3.06)	98–9	285 d	...	...	s	v	v	...	v	CS_2 v aa v	**B6**[2], 157
a568	—,—,amide	$C_6H_5OCH_2CONH_2$	151.17	nd (w, al)	101.5	...	...	...	i δ^h	v^h	...	...	...	...	**B6,** 162
a569	—,—,anhydride	$(C_6H_5OCH_2CO)_2O$	286.29	lf (eth)	67–9	...	...	...	i	δ	δ	...	v	lig δ	**B6,** 162
Ω a570	—,—,chloride	$C_6H_5OCH_2COCl$	170.60	...	...	225–6 111[13]	...	...	d	d	s	...	...	...	**B6**[2], 158
a571	—,—,ethyl ester	$C_6H_5OCH_2CO_2CH_2CH_3$	180.21	...	...	250–1 136[19]	1.104^{18}	...	i	s	s	...	...	...	**B6**[2], 157
a572	—,—,methyl ester	$C_6H_5OCH_2CO_2CH_3$	166.18	...	...	245 130[14]	1.1493_4^{20}	1.5155^{20}	i	∞	∞	...	...	CS_2 ∞	**B6,** 162
a573	—,—,nitrile	$C_6H_5OCH_2CN$	133.15	...	...	239–40[760] 128[17]	1.0991_4^{20}	1.5246^{20}	...	s	s	...	...	...	**B6**[2], 158

For explanations, symbols and abbreviations see beginning of table. For structural formulas see end of table.

No.	Name	Synonyms and Formula	Mol. wt.	Color, crystalline form, specific rotation and λ_{max} (log ε)	m.p. °C	b.p. °C	Density	n_D	Solubility w	al	eth	ace	bz	other solvents	Ref.
	Acetic Acid														
Ω a574	—,**phenyl-**	Phenylethanoic acid*. α-Toluic acid. $C_6H_5CH_2CO_2H$	136.16	lf, pl (peth) λ^{al} 247.5 (2.05), 258.5 (2.26), 267.5 sh (1.85)	77	265.5 144.5^{12}	1.091^{77}_{4} 1.228^{4}	...	δ	v	v	s	...	CS_2 v lig i	**B9**², 294
Ω a575	—,—,amide	α-Toluamide. $C_6H_5CH_2CONH_2$	135.17	pl or lf (w) λ 242 (2.04), 258 (2.33), 264 (2.29), 268 (2.12)	157	...	...	...	δ s^h	s	δ	...	δ s^h	...	**B9**¹, 175
a576	—,—,—,*N*-methyl-	$C_6H_5CH_2CONHCH_3$	149.19	cr (bz) λ^{MeOH} 258 (2.3)	58	...	...	...	δ	v	v	...	δ	chl v	**B9**², 300
a577	—,—,—,*N*-phenyl-	α-Phenylacetanilide. $C_6H_5CH_2CONHC_6H_5$	211.27	pr (al) λ^{al} 244 (4.26)	117–8	...	...	...	i	s	v	...	...	...	**B12**, 275
a578	—,—,anhydride	α-Toluic anhydride. $(C_6H_5CH_2CO)_2O$	254.29	pr or nd (eth)	71–2	$195–8^{12}$	...	...	...	s^h	s	...	...	chl s	**B9**², 299
Ω a579	—,—,chloride	Phenacetyl chloride. α-Toluyl chloride. $C_6H_5CH_2COCl$	154.60	...	...	170^{250} $104–5^{24}$	1.16817^{20}_{4}	1.5325^{20}	d	d	v	...	...	...	**B9**², 300
Ω a580	—,—,ethyl ester	$C_6H_5CH_2CO_2CH_2CH_3$	164.21	λ^{al} 220 (3.26)	...	227^{760} $120–1^{20}$	1.0333^{20}_{4}	1.4980^{20}	i	∞	∞	...	...	...	**B9**², 297
Ω a581	—,—,isobutyl ester.	Eglantine. $C_6H_5CH_2CO_2CH_2CH(CH_3)_2$	192.24	...	...	247 $123–5^{14}$	0.999^{18}	...	i	s	s	...	...	...	**B9**², 298
Ω a582	—,—,methyl ester	$C_6H_5CH_2CO_2CH_3$	150.18	...	...	218^{760} $131–2^{50}$	1.0633^{16}_{16}	1.5075^{20}	i	∞	∞	s	...	...	**B9**², 297
Ω a583	—,—,nitrile	Benzyl cyanide. α-Tolunitrile. $C_6H_5CH_2CN$	117.15	λ^{al} 258 (2.25), 264 (2.15)	−23.8	234 107^{12}	1.0157^{20}_{4}	1.5230^{20}	i	∞	∞	s	...	...	**B9**², 302
Ω a584	—,—,2-phenylethyl ester	$C_6H_5CH_2CO_2CH_2CH_2C_6H_5$	240.31	...	26.5	$177–8^{4–5}$	1.080^{25}_{25}	...	i	s	...	...	...	...	**B9**², 299
a585	—,—,piperazinium salt	$2C_6H_5CH_2CO_2H.\ C_4H_{10}N_2$	358.44	nd (al)	146.5–7.5	...	...	...	s^h	s^h	i	...	...	...	**Am 56**, 150
a586	—,**3-pyrenyl-**	...	260.30	nd or pl (PhCl)	220d	...	...	...	...	...	...	...	...	dil alk s con sulf s PhCl s	**E14s**, 441
a587	—,**sulfo-**	Sulfoethanoic acid*. $HO_3SCH_2CO_2H$	140.12	hyg ta (w + 1)	84–6	245^{760}d	...	...	s	s	i	s	...	chl i	**B4**², 531
a588	—,**2-thienyl-**	2-Thiopheneacetic acid.	142.18	cr (w)	76 (62)	...	...	...	s^h	s	s	...	...	...	**B18**, 293
Ω a589	—,**thiolo-**	Ethanethiolic acid. Thioacetic acid. CH_3COSH	76.12	ye λ^{MeOH} 240 sh (2.10)	< −17	87^{760} $26–7^{35}$	1.064^{20}_{4}	1.4648^{20}	s v^h	v	∞	v	...	...	**B2**², 208
Ω a591	—,—,ethyl ester	$CH_3COSCH_2CH_3$	104.18	...	...	116.4^{760}	0.9792^{20}_{4}	1.4583^{21}	i	v	v	...	...	...	**B2**², 209
a592	—,**thiolo-thiono-**	Dithioacetic acid. Methyl carbithionic acid. CH_3CS_2H	92.18	ye-red oil	...	66^{85} 37^{15}	1.24^{20}	...	s	v	v	v	v	chl v aa v	**B2**², 212
Ω a593	—,**thiono-**,amide	Thioacetamide. CH_3CSNH_2	75.13	cr (al), pl (eth) λ^{al} 210 (3.66), 260 (4.08)	115–6	...	...	...	v	v	δ	...	δ	lig δ	**B2**², 210
a594	—,—,—,*N*-phenyl-	Thioacetanilide. $CH_3CSNHC_6H_5$	151.23	nd (w)	74.5–6	d	...	...	i	...	i	...	...	alk s	**B12**², 142
a595	—,**2-tolyl-**	*o*-Methyl-α-toluic acid.	150.18	nd (w)	88–90	...	...	...	s^h	...	...	...	...	...	**B9**¹, 207
Ω a597	—,—,nitrile	*o*-Methyl benzyl cyanide. *o*-Methyl-α-tolunitrile.	131.18	λ^{al} 262 (2.43)	...	244^{760}	1.0156^{22}	1.5252^{20}	i	s	s	...	s	...	**B9**², 349
a598	—,**3-tolyl-**	*m*-Methyl-α-toluic acid.	150.18	nd (w)	62	$120–3^{26}$	...	...	s^h	...	...	...	...	...	**B9**¹, 208
a599	—,—,nitrile	*m*-Methyl benzyl cyanide. *m*-Methyl-α-tolunitrile.	131.18	...	...	$245–7^{745}$ δd (241) 133^{15}	1.0022^{22}	1.5233^{20}	i	s	s	...	s	...	**B9**², 349
Ω a600	—,**4-tolyl-**	*p*-Methyl-α-toluic acid.	150.18	nd or pl (al or w)	91–3	265–7 (sub)	...	...	δ v^h	s	s	...	s	chl s	**B9**², 349
Ω a601	—,—,nitrile	*p*-Methyl benzyl cyanide. *p*-Methyl-α-tolunitrile.	131.18	λ^{al} 267 (2.49)	18	242–3 122^{13}	0.9922^{22}	1.5167^{20}	i	s	s	...	s	...	**B9**², 349
Ω a602	—,**2-tolyloxy-**	*o*-Cresoxyacetic acid.	166.18	lf (w) λ^{al} 270 (3.17), 278 (3.11)	156.8–7.4	...	...	...	s^h	s	...	...	δ	os δ con sulf s	**B6**², 331
Ω a603	—,**3-tolyloxy-**	*m*-Cresoxyacetic acid.	166.18	nd (w)	103–4	...	...	...	s^h	s	...	...	s	...	**B6**², 353
a604	—,**4-tolyloxy-**	*p*-Cresoxyacetic acid.	166.18	nd (w)	136	...	...	...	s^h	s	...	...	s	alk s	**B6**², 380
Ω a605	—,**tribromo-**	Br_3CCO_2H	296.76	mcl	135 (133)	245^{760}	...	...	s	s	s	...	...	...	**B2**², 205
a606	—,—,amide	Br_3CCONH_2	295.77	mcl pr (al)	121–2	sub	...	...	v	v^h	v	...	δ	chl δ	**B2**², 206
a607	—,—,bromide	Br_3CCOBr	359.66	...	...	210–5 86^{11}	...	...	d	d	s	...	s	chl s	**B2**², 206
Ω a608	—,—,ethyl ester	$Br_3CCO_2CH_2CH_3$	324.81	...	...	225 148^{73}	2.2300^{20}_{20}	1.5438^{13}	...	s	s	...	...	...	**B2**², 205
a609	—,**(2,4,6-tribromo-phenoxy)-**	...	388.85	nd (dil al)	200	...	...	...	i	v	v	...	...	CS_2 i lig i	**B6**, 205
Ω a610	—,**trichloro-**	Cl_3CCO_2H	163.39	dlq cr	α 58 β49.6	197.55^{760} $141–2^{25}$	1.62^{25}_{4} 1.6218^{64}_{1} 1.6237^{70}_{70}	1.4603^{61}	v	s	s	...	...	...	**B2**², 196
a611	—,—,amide	Cl_3CCONH_2	162.40	mcl pr (w)	142	$238–9^{746}$	...	...	δ	v	v	...	...	...	**B2**¹, 94
a612	—,—,—,*N,N*-diethyl-	$Cl_3CCON(C_2H_5)_2$	218.52	pr	27	109^{9}	...	1.4900^{24}	δ	...	...	...	...	...	**B4**, 110
a613	—,—,—,*N,N*-dimethyl-	$Cl_3CCON(CH_3)_2$	190.46	...	12	230–3 δd 84^{4}	1.390^{20}	1.5017^{25}	δ	...	...	...	s	chl s	**B4**, 59

For explanations, symbols and abbreviations see beginning of table. For structural formulas see end of table.

PHYSICAL CONSTANTS OF ORGANIC COMPOUNDS (Continued)

No.	Name	Synonyms and Formula	Mol. wt.	Color, crystalline form, specific rotation and λ_{max} (log ε)	m.p. °C	b.p. °C	Density	n_D	Solubility: w	al	eth	ace	bz	other solvents	Ref.
	Acetic acid														
a614	—,—,—,N-phenyl-	α-Trichloroacetanilide. $Cl_3CCONHC_6H_5$	238.50	lf (dil al)	95–7	$168–70^{760}$	...	...	i δ^h	s	...	...	...	...	**B12²**, 142
a615	—,—,anhydride	$(Cl_3CCO)_2O$	308.76	...	...	222–4d $98–100^{11}$	1.6908^{20}	...	d	d	s	...	...	aa s	**B2²**, 200
a616	—,—,bromide	Cl_3CCOBr	226.29	...	...	143	1.900_{15}^{15}	...	d	d	s	...	s	...	**B2**, 211
a617	—,—,butyl ester	$Cl_3CCO_2(CH_2)_3CH_3$	219.50	...	...	203–5 $97–9^{19}$	1.2778_4^{20}	1.4525^{25}	i	s	s	s	s	...	**B2³**, 471
a618	—,—,sec-butyl ester	$Cl_3CCO_2CH(CH_3)CH_2CH_3$	219.50	...	...	$93–4^{24}$	1.2636_4^{20}	1.4483^{20}	i	s	s	...	s	...	**B2³**, 472
a619	—,—,tert-butyl ester	$Cl_3CCO_2C(CH_3)_3$	219.50	cr (pentane, MeOH)	25.5	$54–5^{7}$ 37^{1}	1.2363_4^{25}	1.4398^{25}	i	v	s	...	...	MeOH v	**B2³**, 473
a620	—,—,chloride	Cl_3CCOCl	181.83	...	...	118	1.6202_4^{20}	1.4695^{20}	d	d	∞	...	...	...	**B2²**, 200
a621	—,—,2-chloroethyl ester	$Cl_3CCO_2CH_2CH_2Cl$	225.89	...	...	217^{766} 100^{14}	1.5357_4^{20}	1.4813^{20}	d	d^h	s	...	...	...	**B2**, 209
a622	—,—,2-hydroxy-ethyl ester	$Cl_3CCO_2CH_2CH_2OH$	207.44	...	...	$130–4^{12}$	1.532_4^{20}	1.4775^{20}	δ	v	...	...	...	CCl_4 v	**B2³**, 474
Ω a623	—,—,ethyl ester	$Cl_3CCO_2CH_2CH_3$	191.44	...	...	167.5–8 62^{12}	1.3836_4^{20}	1.4505^{20}	i	s	s	...	s	...	**B2²**, 200
a624	—,—,isobutyl ester	$Cl_3CCO_2CH_2CH(CH_3)_2$	219.50	...	...	$187–9^{760}$ $93–4^{20}$	1.2636_4^{20}	1.4483^{20}	i	s	s	...	s	...	**B2**, 209
Ω a625	—,—,isopropyl ester	$Cl_3CCO_2CH(CH_3)_2$	205.47	...	...	173.5^{747} $65.5–7^{15}$	1.3034_4^{20}	1.4428^{20}	i	s	s	...	s	...	**B2³**, 471
a626	—,—,2-methoxy-ethyl ester	$Cl_3CCO_2CH_2CH_2OCH_3$	221.47	...	14.6–4.8	$98.0–9.5^{17}$	1.3826_4^{20}	1.4563^{20}	i	s	s	...	s	...	**B2³**, 474
a627	—,—,methyl ester	$Cl_3CCO_2CH_3$	177.42	...	−17.5	153.8 44.5^{12}	1.4874_4^{20}	1.4572^{20}	i	v	v	...	...	...	**B2²**, 199
a628	—,—,2(2-methyl-butyl) ester	$Cl_3CCO_2C(CH_3)_2CH_2CH_3$	233.52	...	...	191^{756} 105^{30}	1.2505_{20}^{20}	...	i	s	s	...	...	...	**B2³**, 473
a629	—,—,3-methylbutyl ester	$Cl_3CCO_2CH_2CH_2CH(CH_3)_2$	233.52	...		217^{760} $92–5^{11}$	1.2314_4^{20}	1.4521^{20}	i	s	s	...	...	...	**B2²**, 200
a630	—,—,nitrile	Cl_3CCN	144.39	...	−42	84.6^{741}	1.4403_4^{25}	1.4409^{20}	i	...	...	...	...	...	**B2**, 212
a631	—,—,pentyl ester	$Cl_3CCO_2(CH_2)_4CH_3$	233.52	...	...	220.3–2.3 118^{30}	1.2475_{20}^{20}	...	i	s	s	...	...	...	**B2³**, 473
a632	—,—,piperazinium salt	$2Cl_3CCO_2H.C_4H_{10}N_2$	412.91	cr (al)	121–1.5	...	...	...	s	s^h	i	...	...	...	**Am 56**, 1759
a633	—,—,propyl ester	$Cl_3CCO_2CH_2CH_2CH_3$	205.47	...	...	187 69^{10}	1.3221_4^{20}	1.4501^{20}	i	s	s	...	...	...	**B2³**, 471
a634	—,—,trichloro-methyl ester	$Cl_3CCO_2CCl_3$	280.75	...	34	191–2 $73–4^{10}$	1.6733_4^{35}	...	...	...	s	...	s	chl s lig s	**B3**, 17
Ω a635	—,(2,4,5-trichloro-phenoxy)-	...	255.49	cr (bz) $\lambda^{0.1N\ HCl}$ 287.5 (3.35), 295 (3.32)	157–8 (153)	...	...	...	i	s	...	v^h	...	...	**Am 63**, 1768
Ω a636	—,(2,4,6-trichloro-phenoxy)-	...	255.49	cr (al) $\lambda^{0.1N\ HCl}$ 278.5 (2.82), 286 (2.82)	177	...	...	...	δ^h	s^h	...	...	δ	lig δ	**B6**, 192
a637	—,trifluoro-	F_3CCO_2H	114.02	...	−15.25	72.4	1.5351^{0}	...	s	s	s	s	...	...	**B2²**, 186
a638	—,—,amide,N-phenyl-	Trifluoroacetanilide. $F_3CCONHC_6H_5$	189.14	(60 % al) λ^{al} 245 (4.04)	87.6	...	...	...	...	v	...	...	...	...	**B12²**, 141
a639	—,—,anhydride	$(F_3CCO)_2O$	210.03	...	−65	39.5–40.1	1.490_4^{25}	1.269^{25}	d	d	s	...	...	aa s	B2², 186
Ω a640	—,—,nitrile	F_3CCN	95.03	...	...	-63.9^{760}	...	...	...	...	...	...	...	...	**B2³**, 428
a641	—,triiodo-	I_3CCO_2H	437.74	ye lf	150d	...	...	...	s	s	s	...	...	...	**B2**, 225
a642	—,trinitro-,nitrile	$(NO_2)_3CCN$	176.05	wx	41.5	220 exp	...	...	d	d	s	...	...	...	**B2**, 229
Ω a643	—,triphenyl-	$(C_6H_5)_3CCO_2H$	288.35	mcl pr (al), lf (aa)	271 (267–71)	...	...	...	i	s	δ	...	δ	MeOH s, aa, lig s CS_2 chl δ	**B9²**, 500
—	—,ureido-	*see* **Hydantoic acid**													
—	**Acetic anhydride**	*see* **Acetic acid,** anhydride													
—	**Acetoacetic acid**	*see* **Butanoic acid, 3-oxo-***													
—	**Acetoin**	*see* **2-Butanone, 3-hydroxy-***													
—	**Acetol**	*see* **2-Propanone, 1-hydroxy-***													
—	**Acetonaphthone**	*see* **Naphthalene, acetyl-**													
—	**Acetone**	*see* **2-Propanone***													
—	**Acetonedicar-boxylic acid**	*see* **Pentanedioic acid, 3-oxo-***													
—	**Acetonic acid**	*see* **Propanoic acid, 2-hydroxy-2-methyl-***													
—	**Acetonitrile**	*see* **Acetic acid,** nitrile													
a644	**Acetonitrolic acid**	Ethylnitrolic acid. $CH_3C(:NOH)NO_2$	104.07	ye rh (w, al, eth)	87–8d	...	...	...	s	s	s	s	...	oos s	**B2²**, 185
—	**Acetonylacetone, dioxime**	*see* **2,5-Hexanedione, dioxime***													
Ω a645	**Acetophenone**	Acetylbenzene. Methyl phenyl ketone	120.16	mcl pr or pl λ^{al} 243 (4.12)	20.5	202.0^{760} 79^{10}	1.0281_4^{20}	1.53718^{20}	i	s	s	s	s	chl s con sulf s (og)	**B7²**, 208
Ω a646	—,oxime	$C_6H_5C(:NOH)CH_3$	135.17	nd (w) λ^{al} 246 (4.0)	60	245 119^{20}	...	...	δ s^h	v	v	v	v	chl v lig v	**B7²**, 216

For explanations, symbols and abbreviations see beginning of table. For structural formulas see end of table.

No.	Name	Synonyms and Formula	Mol. wt.	Color, crystalline form, specific rotation and λ_{max} (log ε)	m.p. °C	b.p. °C	Density	n_D	Solubility: w	al	eth	ace	bz	other solvents	Ref.
	Acetophenone														
—	**—,4-acetylamino-**	*see* **Acetic acid,** amide, *N*(4–acetylphenyl)-													
Ω a647	**—,2-amino-**	*o*-Acetylaniline C_8H_9NO. *See* a645	135.17	ye cr λ^{al} 226 (4.32), 254 (3.75), 359 (3.64)	20	250–2 δd 135^{17}		1.6160^{20}	i		s				**B14²**, 28
Ω a648	**—,3-amino-**	*m*-Acetylaniline. C_8H_9NO. *See* a645	135.17	pa ye pl (al), lf (eth) λ^{al} 231 (4.36), 255 sh (3.85), 338 (3.28)	98–9	289–90			δ	s^h					**B14²**, 30
Ω a649	**—,4-amino-**	*p*-Acetylaniline. C_8H_9NO. *See* a645	135.17	ye mcl pr (al) λ^{al} 316 (4.30)	106	293–5 195–200^{15}			δ s^h	s	s		δ	chl δ lig δ	**B14²**, 30
Ω a650	**—,4-amino-α-chloro-**	*p*-Aminophenacyl chloride. C_8H_8ClNO. *See* a645	169.62	ye pl	148										**B14¹**, 367
a651	**—,4-amino-3-chloro-**	C_8H_8ClNO. *See* a645	169.62	pr (chl-peth)	92										**B14**, 49
Ω a652	**—,3-amino-4-methoxy-**	$C_9H_{11}NO_2$. *See* a645	165.19	pr (al)	102					s	s		v		**B14²**, 141
—	**—,benzal-**	*see* **Chalcone**													
—	**—,benzylidene**	*see* **Chalcone**													
Ω a653	**—,α-bromo-**	Phenacyl bromide. $C_6H_5COCH_2Br$	199.05	nd (al), rh pr (dil al), pl (peth) λ^{al} 251	50–1	135^{18}	1.647_4^{20}		i	s v^h	v		v	chl v peth s^h	**B7²**, 220
Ω a654	**—,2-bromo-**	C_8H_7BrO. *See* a645	199.05	ye λ^{al} 236 (3.69), 282 (2.89)		131–5^{20} 112^{10}		1.5678^{20}							**B7²**, 220
Ω a655	**—,4-bromo-**	C_8H_7BrO. *See* a645	199.05	lf (al) λ^{al} 253 (4.20)	50–1	255.5^{736} 130^{11}	1.647		i	s	s		s	CS_2, aa, lig s	**B7²**, 220
a656	**—,α-bromo-3-chloro-**	*m*-Chlorophenacyl bromide. C_8H_6BrClO. *See* a645	233.50	nd	39.5–40					v					**Am 83**, 4277
Ω a657	**—,α-bromo-4-chloro-**	*p*-Chlorophenacyl bromide. C_8H_6BrClO. *See* a645	233.50	nd	96–7										**B7**, 285
a658	**—,4-bromo-α-chloro-**	*p*-Bromophenacyl chloride. C_8H_6BrClO. *See* a645	233.50	nd (al)	116–7					v^h					**B7**, 285
a659	**—,α-bromo-4-methyl-**	*p*-Methylphenacyl bromide. C_9H_9BrO. *See* a645	213.08	nd or lf (al)	51	155–9^{14}			d	s	s				**B7²**, 239
a660	**—,4-*tert*-butyl-2,6-dimethyl-3,5-dinitro-**	Musk ketone. $C_{14}H_{18}N_2O_5$. *See* a645	294.31	ye	134.5–6.5				i	δ					**C52**, 8988
Ω a661	**—,α-chloro-**	Phenacyl chloride. Chloromethyl phenyl ketone. $C_6H_5COCH_2Cl$	154.60	pl (dil al), rh, lf (peth), λ^{hx} 246 (3.92), 280 (3.00), 291 (2.86), 329 (1.83)	56–5	247^{760} 139–41^{14}	1.324_4^{15}		i	v	v	s	v	CS_2 v peth s^h	**B7²**, 219
a662	**—,2-chloro-**	*o*-Chlorophenyl methyl ketone. C_8H_7ClO. *See* a645	154.60	λ^{al} 238 (3.74), 281 (2.88)		227–8^{738} 113^{18}	1.2016_4^{17}	1.685^{25}	δ		s				**B7²**, 218
a663	**—,3-chloro-**	*m*-Chlorophenyl methyl ketone. C_8H_7ClO. *See* a645	154.60	λ^{al} 240 (4.00), 286 (3.02)		241–5^{744} (227–9) 127–31^{30}	1.2130_4^0	1.5494^{20}		s	s	s			**B7²**, 218
Ω a664	**—,4-chloro-**	*p*-Chlorophenyl methyl ketone. C_8H_7ClO. *See* a645	154.60	λ^{al} 249 (4.23), 272 (2.95), 284 (2.70)	20	273^{760} 106^{10}	1.1922_4^{20}	1.5550^{20}	i	∞	∞				**B7²**, 219
a665	**—,α-chloro-2,4-dimethyl-**	2,4-Dimethylphenacyl chloride. $C_{10}H_{11}ClO$. *See* a645	182.65	nd	62					v	v		v		**B7²**, 249
a666	**—,α-chloro-α-isonitroso-**	Benzoylformyl chloridoxime. $C_6H_5COCCl(:NOH)$	183.60	lf (bz), pr (chl)	133				i	v	v		v	chl v CCl_4 δ	**B10²**, 460
a667	**—,α-chloro-4-methyl-**	*p*-Methylphenacyl chloride. C_9H_9ClO. *See* a645	168.63	nd (al)	57–8	260–3 δd 113^4			d	v	v				**B7¹**, 165
Ω a668	**—,α,4-dibromo-**	*p*-Bromophenacyl bromide. $C_8H_6Br_2O$. *See* a645	277.96	nd (al)	110–2				i	s^h	s				**B7²**, 222
Ω a669	**—,α,α-dichloro-**	Phenacylidene chloride. $C_6H_5COCHCl_2$	189.04	amor	20–1.5	249 143^{25}	1.340^{16}	1.5686^{20}		s			s		**B7²**, 220
a670	**—,α,4-dichloro-**	*p*-Chlorophenacyl chloride. $C_8H_6Cl_2O$. *See* a645	189.04	nd (al)	101–2	270				s v^h			s	MeOH s	**B7¹**, 152
Ω a671	**—,2,4-dichloro-**	$C_8H_6Cl_2O$. *See* a645	189.04	$\lambda^{dil\ MeOH}$ 252 (3.95) 290 sh (3.04)	33–4	245–7 140–50^{15}		1.5640^{20}	i						**B7²**, 219
Ω a672	**—,3,4-dichloro-**	$C_8H_6Cl_2O$. *See* a645	189.04	nd (peth) $\lambda^{dil\ MeOH}$ 254 (4.16), 290 sh (3.18)	76	135^{12}			i					lig s^h	**B7²**, 219

For explanations, symbols and abbreviations see beginning of table. For structural formulas see end of table.

No.	Name	Synonyms and Formula	Mol. wt.	Color, crystalline form, specific rotation and λ_{max} (log ε)	m.p. °C	b.p. °C	Density	n_D	Solubility w	al	eth	ace	bz	other solvents	Ref.
	Acetophenone														
a673	—,2,3-dihydroxy-	3-Acetylcatechol. $C_8H_8O_3$. *See* a645	152.16	ye pr (bz-lig)	97–8	...	...	...	...	s	...	...	...	alk s	**B8**[1], 613
Ω a674	—,2,4-dihydroxy-	Resacetophenone. $C_8H_8O_3$. *See* a645	152.16	nd or lf λ^{iso} 232 (3.84), 270 (4.17), 316 (3.84)	147	...	1.1800^{141}	...	i	s^h	δ	...	δ	chl i, Py s aa s	**B8**[2], 294
Ω a675	—,2,5-dihydroxy-	2-Acetylhydroquinone. Quinacetophenone. $C_8H_8O_3$	152.16	ye-gr nd (dil al or w) λ^{al} 230 (4.48), 257 (4.04), 370 (3.95)	204–5	...	...	...	δ s^h	s	δ	...	δ	...	**B8**[2], 297
a676	—,3,4-dihydroxy-	Acetylpyrocatechol. $C_8H_8O_3$. *See* a645	152.16	nd (w or chl) λ^{al} 230 (4.30), 276 (4.00), 307 (3.90)	115–6	...	...	...	s^h	...	...	...	...	chl s^h	**B8**[2], 298
a677	—,3,5-dihydroxy-	$C_8H_8O_3$. *See* a645	152.16	cr (w) λ^{al} 218 (4.30), 265 (3.90), 321 (3.60)	147–8	...	...	...	s v^h	v	v	v	δ	...	**B8**[2], 301
—	—,3,4-dihydroxy-β-methylamino-	*see* **Adrenalone**													
a677[1]	—,3,4-dimethoxy-	Acetoveratrone. $C_{10}H_{12}O_3$. *See* a645	180.21	pr (dil al) λ^{al} 228 (4.21), 272.5 (4.02), 303.5 (3.88)	51	$286–8^{760}$ $160–2^{10–15}$	...	...	s^h	s^h	s^h	...	s^h	chl s^h	**B8**[2], 298
a677[2]	—,3,5-dimethoxy-4-hydroxy-	Acetosyringone. $C_{10}H_{12}O_4$. *See* a645	196.21	nd (w), pr (peth) λ^{al} 302 (4.04)	122–3	...	...	...	s^h	s	s	s	...	aa s lig i	**B8**[2], 244
Ω a678	—,2,4-dimethyl-	$C_{10}H_{12}O$. *See* a645	148.21	λ^{al} 251 (4.15), 282 (3.23), 291 (3.11)	...	228^{760} 110^{13}	1.0121^{15}	1.5340^{20}	i	s	...	...	...	...	**B7**[2], 248
a679	—,2,5-dimethyl-	$C_{10}H_{12}O$. *See* a645	148.21	...	...	$232–3^{760}$ 107^{13}	0.9963^{19}_{4}	1.5291^{20}	i	v	v	...	v	CS_2 v	**B7**[2], 248
a680	—,3,4-dimethyl-	$C_{10}H_{12}O$. *See* a645	148.21	λ^{hp} 249 (4.18), 282 (3.11), 294 (2.90)	...	$246–7$ 213^{310}	1.0090^{14}_{4}	1.5413^{15}	i	v	v	...	v	chl v aa v	**B7**[2], 248
a681	—,3-dimethyl-amino-	$C_{10}H_{13}NO$. *See* a645	163.22	λ^{al} 242 (4.60), 360.5 (3.54)	43	148^{13}	...	...	...	...	v	...	...	...	**B14**, 45
a682	—,4-dimethyl-amino-	$C_{10}H_{13}NO$. *See* a645	163.22	nd (w, peth) λ^{al} 239 (3.82), 334 (4.43)	105.5	$172–5^{11}$	...	...	v^h	...	v	...	...	lig v^h	**B14**[2], 32
a683	—,2-ethoxy-	*o*-Acetylphenetole. $C_{10}H_{12}O_2$. *See* a645	164.21	pr (dil al), pl (lig)	43	243–4	1.0036^{78}	...	δ	v	v	...	...	lig s^h	**B8**, 85
a684	—,4-ethoxy-	*p*-Acetylphenetole. $C_{10}H_{12}O_2$. *See* a645	164.21	pl (eth) λ^{al} 270 (4.26), 279 (4.23)	39	...	...	...	...	v	v^h	...	...	...	**B8**[2], 85
Ω a685	—,α-hydroxy-	Benzoyl carbinol. Phenacyl alcohol. $C_6H_5COCH_2OH$	136.16	hex pl (al or eth), pl (w or dil al, +w), pr (lig)	89.5–90.5 (anh) 73–4 (hyd)	$124–6^{12}$ (sub 56^1)	1.0963^{99}_{4}	...	s^h	s	s	...	...	chl s lig δ	**B8**[2], 88
a686	—,—,acetate	Phenacyl acetate. $C_6H_5COCH_2O_2CCH_3$	178.19	rh pl (eth, lig or peth)	49–9.5	270 $150–2^{10}$	1.1169^{65}_{4}	...	i	v	v	...	s	chl v lig s	**B8**[2], 89
Ω a687	—,2-hydroxy-	*o*-Acetylphenol. $C_8H_8O_2$. *See* a645	136.16	λ^{iso} 253 (4.00), 325 (3.70)	4–6	218^{760} 106^{17}	1.1307^{20}_{4}	1.5584^{20}	δ	∞	∞	...	...	aa ∞	**B8**[2], 81
Ω a688	—,3-hydroxy-	*m*-Acetylphenol. $C_8H_8O_2$. *See* a645	136.16	nd or lf λ^{al} 218 (4.30), 252.5 (3.95), 311 (3.60)	96	296^{756} 153^5	1.0992^{109}	1.5348^{109}	δ v^h	v	v	...	v	chl v lig i	**B8**[2], 84
Ω a689	—,4-hydroxy-	*p*-Acetylphenol. $C_8H_8O_2$. *See* a645	136.16	nd (eth, dil al) λ^{al} 279 (4.15), 315 sh (2.2)	109–10	$147–8^3$	1.1090^{109}	1.5577^{109}_{α}	δ s^h	v	v	...	...	...	**B8**[2], 84
a690	—,α-hydroxy-4-methoxy-	Anisoyl carbinol. $C_9H_{10}O_3$. *See* a645	166.18	pl (dil al)	104	...	...	...	...	s	...	...	s	...	**B8**[1], 618
a691	—,2-hydroxy-3-methoxy-	*o*-Acetovanillon. $C_9H_{10}O_3$. *See* a645	166.18	pa ye nd (peth or eth-peth)	53–4	...	...	...	...	...	s	...	s	...	**B8**[2], 293
Ω a692	—,2-hydroxy-4-methoxy-	Peonol. $C_9H_{10}O_3$. *See* a645	166.18	nd (al) λ^{al} 230 (4.0), 273.5 (4.20), 314 (3.90)	52–3	158^{20}	1.3102^{81}	1.5432^{81}	δ	s	s	...	s	chl s CS_2 s	**B8**[2], 294
Ω a693	—,2-hydroxy-5-methoxy-	$C_9H_{10}O_3$. *See* a645	166.18	pa ye pr (dil al) λ^{al} 227.5 (4.60), 255.5 (3.95), 357.5 (3.78)	52	...	...	...	...	s^h	...	...	s	...	**B8**, 271
a694	—,3-hydroxy-4-methoxy-	Isoacetovanillon. $C_9H_{10}O_3$. *See* a645	166.18	cr (eth-lig) or cr (w+1) λ^{al} 231 (4.60), 274 (4.20), 310 (4.04)	67–8 (+1w) 91 (anh)	...	...	...	s^h	...	s^h	...	...	...	**B8**[2], 298

For explanations, symbols and abbreviations see beginning of table. **For structural formulas see end of table.**

No.	Name	Synonyms and Formula	Mol. wt.	Color, crystalline form, specific rotation and λ_{max} (log ε)	m.p. °C	b.p. °C	Density	n_D	Solubility						Ref.
									w	al	eth	ace	bz	other solvents	
	Acetophenone														
a695	**—,4-hydroxy-α-methoxy-**	$C_9H_{10}O_3$. *See* a645	166.18	nd (bz), pr (w + ¼)	130–1	...	...	...	s^h	v	s	v	s^h	lig δ	**B8²**, 302
a696	**—,4-hydroxy-2-methoxy-**	Isopeonol. Resacetophenone 2-methyl ether. $C_9H_{10}O_3$. *See* a645	166.18	nd (w) λ^{al} 230.5 (4.00), 278 (4.3), 309 (4.0)	138	...	...	...	s^h	...	...	...	s	lig s	**B8¹**, 618
Ω a697	**—,4-hydroxy-3-methoxy-**	Acetovanillon. Apocynin. $C_9H_{10}O_3$. *See* a645	166.18	pr (w) λ^{al} 229 (4.6) 275 (4.30), 308 (4.00)	115	295–300 233–5^{15-20}	...	...	δ v^h	s	v	s	s	chl v lig δ	**B8²**, 298
a698	**—,5-isopropyl-2-methyl-**	2-Acetyl-*p*-cymene. Carvacryl methyl ketone. $C_{12}H_{16}O$. *See* a645	176.25	...	< −20	249–50 (244) 124–5^{12}	0.956_4^{20}	1.5181^{20}	...	...	...	...	...	...	**B7²**, 262
Ω a699	**—,2-methoxy-**	*o*-Acetylanisole. $C_9H_{10}O_2$. *See* a645	150.17	ye λ^{al} 211 (4.32), 247 (3.92), 306 (3.58)	...	245	1.0897_4^{20}	1.5393^{20}	δ	s	...	s	...	...	**B3**, 232
Ω a700	**—,3-methoxy-**	*m*-Acetylanisole. $C_9H_{10}O_2$. *See* a645	150.17	λ^{al} 217 (4.35), 249 (3.94), 306 (3.39)	95–6	240 125–6^{12}	1.0343^{19}	1.5410^{20}	s	s	...	s	...	...	**B8¹**, 535
Ω a701	**—,4-methoxy-**	*p*-Acetylanisole. $C_9H_{10}O_2$. *See* a645	150.17	pl (eth) λ^{al} 217 (4.07), 271.5 (4.22)	38–9	258 138–9^{15}	1.0818_4^{41}	1.53349^{20} 1.547^{41}	δ	s	s	s	...	...	**B8²**, 84
Ω a702	**—,4-methoxy-3-nitro-**	4-Acetyl-2-nitroanisole $C_9H_9NO_4$. *See* a645	195.17	nd (al)	99.5	...	...	...	...	s^h	s	s	s	...	**B8¹**, 538
Ω a703	**—,2-methyl-**	*o*-Acetyltoluene. $C_9H_{10}O$. *See* a645	134.17	λ^{al} 242 (3.92)	...	214^{761} 89–92^{10}	1.026_4^{20}	1.5276^{20}	...	...	...	...	...	...	**B7²**, 237
Ω a704	**—,3-methyl-**	*m*-Acetyltoluene. $C_9H_{10}O$. *See* a645	134.17	λ^{al} 244.5 (4.05), 285 (3.11), 291 sh (3.08)	...	220^{760} 109^{12}	1.0070_4^{20}	1.5270^{20}	...	s	s	s	...	...	**B7²**, 237
Ω a705	**—,4-methyl-**	*p*-Acetyltoluene. $C_9H_{10}O$. *See* a645	134.17	nd λ^{al} 252 (4.18)	28	226^{760} 113^{11}	1.0051_4^{20}	1.5335^{20}	i	s	s	...	s	chl s	**B7¹**, 238
Ω a706	**—,2-nitro-**	$C_8H_7NO_3$. *See* a645	165.15	(al) λ^{al} 257 (3.78)	28–9	158^{16}	...	1.5468^{20}	i	v	v	...	...	chl v	**B7²**, 222
Ω a707	**—,3-nitro-**	$C_8H_7NO_3$. *See* a645	165.15	nd (al) λ^{al} 226 (4.35), 260 sh (3.81), 300 sh (2.90)	81	202 167^{18}	...	...	v^h	δ	v	...	...	...	**B7²**, 223
Ω a708	**—,4-nitro-**	$C_8H_7NO_3$. *See* a645	165.15	yesh pr (al) λ^{al} 261 (4.15), 298 sh (3.34), 312 sh (3.08)	80–2	...	...	...	...	s	v	...	...	...	**B7²**, 223
a709	**—,α,α,α-trichloro-**	$C_6H_5COCCl_3$	223.50	...	...	256–7 145^{25}	1.425^{16}	...	i	s	v	...	...	...	**B7¹**, 152
Ω a710	**—,2,3,4-tri-hydroxy-**	4-Acetylpyrogallol. Gallacetophenone. $C_8H_8O_4$. *See* a645	168.15	pa ye nd or lf (w) λ^{MeOH} 237 (3.93), 296 (4.10)	173	...	...	...	s^h	v	s	v	δ	chl δ aa v lig v	**B8²**, 493
Ω a711	**—,2,4,6-tri-hydroxy-**	2-Acetylphloroglucinol. Phloroacetophenone. $C_8H_8O_4$. *See* a645	168.15	ye (in alk), nd (w + 1)	222–4 (anh)	...	...	...	δ s^h	v	v	v	$δ^h$	chl δ aa v	**B8²**, 442
Ω a712	**—,2,4,5-tri-methyl-**	5-Acetylpseudocumene. $C_{11}H_{14}O$. *See* a645	162.22	...	10–1	246–7 137–8^{20}	1.0039_4^{15}	1.541^{15}	i	v	v	...	v	CS_2 v aa v	**B7²**, 256
Ω a713	**—,2,4,6-tri-methyl-**	Acetylmesitylene. $C_{11}H_{14}O$. *See* a645	162.22	λ^{al} 245 sh (3.40)	...	240.5^{735} 120^{12}	0.9754_4^{20}	1.5175^{20}	i	s	s	s	s	...	**B7²**, 257
a714	**—,α,α,α-triphenyl- (α form)**	α-Benzopinacolone. $C_6H_5COC(C_6H_5)_3$	348.45	nd	206–7	...	...	...	...	i	...	...	s	chl, CS_2 s aa i	**C55**, 22234
a715	**—,—(β form)**	β-Benzopinacolone. $C_6H_5COC(C_6H_5)_3$	348.45	nd (al)	182	...	...	...	i	δ s^h	s	...	s	chl, CS_2 s aa s^h	**B7²**, 513
—	**Acetopyruvic acid**	*see* **Pentanoic acid, 2,4-dioxo-***													
—	**Acetosyringone**	*see* **Acetophenone, 3,5-dimethoxy-4-hydroxy-**													
—	**Acetovanillon**	*see* **Acetophenone, 4-hydroxy-3-methoxy-**													
—	**Acetoveratrone**	*see* **Acetophenone, 3,4-dimethoxy-**													
—	**Acetoxime**	*see* **2-Propanone**, oxime*													
—	**Aceturic acid**	*see* **Glycine**, *N*-**acetyl-**													
—	**Acetylacetone**	*see* **2,4-Pentanedione***													
—	**Acetylbenzoyl**	*see* **1,2-Propanedione, 1-phenyl-***													
—	**Acetyl bromide**	*see* **Acetic acid**, bromide													
—	**Acetyl chloride**	*see* **Acetic acid**, chloride													
—	**Acetylene**	*see* **Ethyne***													
—	**Acetylenedi-carboxylic acid**	*see* **2-Butynedioic acid***													
—	**Acetyl fluoride**	*see* **Acetic acid**, fluoride													
—	**Acetyl iodide**	*see* **Acetic acid**, iodide													

For explanations, symbols and abbreviations see beginning of table. For structural formulas see end of table.

No.	Name	Synonyms and Formula	Mol. wt.	Color, crystalline form, specific rotation and λ_{max} (log ε)	m.p. °C	b.p. °C	Density	n_D	Solubility w	al	eth	ace	bz	other solvents	Ref.
	Achroodextrin														
a718	**Achroodextrin**	$C_{36}H_{62}O_{31}$	990.88	amor					s	i	...	...	...	...	**C25,** 3018
a719	**Aconic acid**		128.08	lf (eth), rh (w, al) λ^{al} 216.5 (4.6), 270 sh (2.1)	164	d			v	s	...	...	...	MeOH s	**B18,** 395
a720	**Aconine**	$C_{25}H_{41}NO_9$.	499.61	amor, [α] +23 $\lambda^{1\,N\,HCl}$ < 220	132				s	s	δ	...	...	chl s lig δ	**C19,** 291
—	**Aconitic acid**	*see* **1,2,3-Propanetricarboxylic acid***													
Ω a721	**Aconitine**	$C_{34}H_{47}NO_{11}$	645.76	rh lf, $[\alpha]^{20}_D$ +19 (chl, c = 0.95) λ^{chl} 230 (4.15), 273 (3.03), 280 (2.95)	204				i	s	δ	...	s	chl s peth i	**C51,** 446
a723	—,hydrobromide	$C_{34}H_{47}NO_{11}$.HBr. $1\frac{1}{2}$ H_2O. *See* a721	753.71	yesh pr (w), $[\alpha]_D$ −31 (w, c = 2.3)	209–10				s[h]	s	s	...	...	...	**C51,** 446
a724	—,hydrochloride(*l*)	$C_{34}H_{47}NO_{11}$. HCl. $3\frac{1}{2}$ H_2O. *See* a721	745.27	(w + $3\frac{1}{2}$) $[\alpha]_D$ − 30.5 (w)	170–2				s	s	s	...	...	...	**M,** 15
a725	—,nitrate(*l*)	$C_{34}H_{47}NO_{11}$. HNO_3. *See* a721	708.77	$[\alpha]^{20}_D$ − 35 (2 % aq sol)	ca. 200d				s	s		...	...	...	**M,** 15
Ω a727	**Acridine**	2,3,5,6-Dibenzopyridine.	179.22	rh nd or pr (al), mcl, orh (2 forms) λ^{al} 240 (5.2), 340 (3.8)	111 (sub 100)	345–6 (> 360)	1.005^{20}_4		δ[h]	v	v	...	v	CS_2 v	**B20**[2], 300
a728	—,**2-amino-**	$C_{13}H_{10}N_2$. *See* a727	194.24	ye nd (w or al + 1) λ^{al} 240 (4.7), 260 (4.9), 358 (4.0)	213–4				s[h]	v	s	...	...	dil ac s	**B22,** 462
a729	—,**3-amino-**	$C_{13}H_{10}N_2$. *See* a727	194.24	ye-og nd (dil al or bz-lig) $\lambda^{2:1\,al-eth}$ 240 (4.5), 280 (1.8), 355 (4.1), 440 (5.0), 460 (4.9)	224 (cor)				s	...	v	s	s	dil ac v	**B22**[2], 376
a730	—,**4-amino-**	$C_{13}H_{10}N_2$. *See* a727	194.24	red-br pr (peth), og nd (MeOH) λ^{al-eth} 212 (4.4), 243 (4.6), 287 (4.6), 350 (3.1), 460 (3.5)	108	183–4			...	v	s	s	s	MeOH s[h] con sulf s dil HCl s lig δ	**B22**[2], 376
a731	—,—hydrochloride	$C_{13}H_{10}N_2$. HCl. *See* a727	230.70	ye nd (dil al)	234d				v	s	...	...	...	...	**B22**[2], 376
Ω a732	—,**9-amino-**	$C_{13}H_{10}N_2$. *See* a727	194.24	ye nd (ace or al) λ^{al} 260 (5.4), 410 (3.9)	241 (cor) (237)				...	s[h]	...	s[h]	...	dil HCl v	**B21**[2], 280
a733	—,**2-amino-5(4-aminophenyl)-**	Chrysaniline. $C_{19}H_{15}N_3$. $2H_2O$. *See* a727	321.38	ye nd (95 % al + 2w)	260–7				i	δ	...	...	...	...	**B22**[2], 403
a734	—,**9-chloro-**	$C_{13}H_8ClN$. *See* a727	213.68	nd (al)	122	sub			δ	v	...	...	...	...	**B20**[2], 301
a735	—,**3,6-diamino-**	$C_{13}H_{11}N_3$. *See* a727	209.25	ye nd (al or w) λ^{al} 265 (4.8), 290 sh (4.4), 460 (4.7)	284–6				s[h]	v	δ	...	δ	con sulf s	**B22**[2], 397
a736	—,**6,9-diamino-2-ethoxy-**	Rivanol. $C_{15}H_{15}N_3O$. *See* a727	253.31	ye nd	123–4d				...	...	...	...	...	...	**B22**[2], 458
Ω a737	—,**9,10-dihydro-**	Acridan. Carbazine. *ms*-Dihydroacridine. $C_{13}H_{11}N$. *See* a727	181.24	pl or pr (al) λ^{al} 250 sh, 290 (4.14)	169–71	d300 sub			i	v[h]	s	s	...	...	**B20**[2], 291
Ω a738	—,**9,10-dihydro-9-oxo-**	*ms*-Acridone. $C_{13}H_9NO$. *See* a727	195.22	ye lf (al) λ^{al} 379 (3.90), 398 (3.95)	354	> 354 (no d)			i	δ v[h]	i	...	i	alk, aa s[h] con sulf s chl i	**B21**[2], 280
a739	—,**2-methyl-**	$C_{14}H_{11}N$. *See* a727	193.25	ye nd (dil al)	134				δ	s	s	...	s	con sulf s lig δ	**B20**[1], 173
a740	—,**9-phenyl-**	$C_{19}H_{13}N$. *See* a727	255.32	lf, ye nd (al) λ^{chl} 260 (4.49), 360 (4.01)	184	403–4 sub			i	δ	s	...	v	con sulf v	**B20**[2], 332
—	*ms*-**Acridone**	*see* **Acridine, 9, 10-dihydro-9-oxo-**													
—	**Acrolein**	*see* **Propenal***													
—	*β*-**Acrose**	*see* **Sorbose** (*L*)													
—	**Acrylamide**	*see* **Propenoic acid, amide***													

For explanations, symbols and abbreviations see beginning of table. For **structural formulas see end of table.**

No.	Name	Synonyms and Formula	Mol. wt.	Color, crystalline form, specific rotation and λ_{max} (log ε)	m.p. °C	b.p. °C	Density	n_D	Solubility w	al	eth	ace	bz	other solvents	Ref.
	Acrylic acid														
—	**Acrylic acid**	see **Propenoic acid***													
—	**Acrylyl chloride**	see **Propenoic acid,** chloride													
a741	**Actidione** (*l*)	Cycloheximide.	281.36	pl (AmOH, 30% MeOH) $[\alpha]_D^{25}$ −33 (chl, c = 1)	144–5	...	...	...	δ	v	δ	...	...	os v lig i	**Am 70,** 1223
—	**Adabine**	see **Urea, 1(2-bromo-2-ethylbutanoyl)-**													
Ω a742	**Adamantane**	Tricyclo [3,3,1,$1^{3,7}$] decane.	136.24	nd (sub)	268 (sealed tube)	sub	1.07	1.568	...	...	...	...	s	...	**E13,** 1058
—	**Adenine**	see **Purine, 6-amino-**													
Ω a743	**Adenosine**	Adenine-9-*d*-ribofuranoside. 6-Amino-g-ribofuranoside purine.	267.25	nd (w + 1½), $[\alpha]_D^{20}$ −60.0 (w, c = 1), −43.5 (10% HCl, c = 2) λ^{al} 260 (4.23)	235–6 (anh)	...	...	...	s v^h	i	...	...	...	...	**B31,** 27
Ω a744	**5-Adenylic acid**	Adenosine-5-phosphate. Muscle adenylic acid. Synadenylic acid. $C_{10}H_{14}N_5O_7P$.	347.23	pw, nd (w, dil al) cr (+w, w–ace) $[\alpha]_D^{50}$ −41.78 (w) $\lambda^{w,pH=1.1}$ 255 (4.16)	195d 208d (sealed tube)	...	...	...	v^h	s	...	...	...	10% HCl s	**B31,** 27
—	**Adipamic acid**	see **Hexanedioic acid, monoamide***													
—	**Adipic acid**	see **Hexanedioic acid***													
—	**Adipoin**	see **Cyclohexanone,2-hydroxy-**													
Ω a745	**Adonitol**	Adonite. Ribitol.	152.15	pr (w), nd (al)	104	...	...	...	s	s^h	i	...	...	lig i	**B1**², 604
a746	**Adrenaline**(*d*)	*d*-Epinephrine. *d*-1-[3,4-Dihydroxyphenyl]-2-methylaminoethanol.	183.21	$[\alpha]_D^{20}$ + 50.5 (HCl) λ^w 290 (2.90)	215d	...	...	...	δ	i	...	...	...	aa s	**B13**², 525
a747	—(*l*)	$C_9H_{13}NO_3$. *See* a746	183.21	br (in air), pw $[\alpha]_D^{20}$ −53 (aq HCl)	211–2 (215 rapid htng)	...	...	...	δ	i	...	...	...	aa s min ac s	**B13**², 523
a748	**Adrenalone**	...	181.19	nd λ^{al} 232 (4.08), 279 (3.93), 312 (3.86)	235–6d	...	...	...	δ	δ	δ	...	...	...	**B14**², 157
a749	**Adrenochrome** (*dl*)	...	179.18	red-br rods (MeOH-HCO_2H, +½w) $[\alpha]_D^{18}$ −106 $\lambda^{w,pH=6.5}$ 485 (3.3)	125d (anh)	...	...	...	v	v	i	...	i	...	**C31,** 5397
a750	**Adrenosterone**	Δ^4-Androstene-3,11,17-trione. $C_{19}H_{24}O_3$.	300.40	nd (al), lf (eth) $[\alpha]_D^{20}$ +262 (abs al, c = 1) $[\alpha]_D^{25}$ +281 (ace) λ 235	222 (cor)	sub (vac)	...	...	δ	s	s	s	...	chl s	**E14s,** 2979
—	**Adurol**	see **Benzene, 2-bromo-1,4-dihydroxy-***													
—	**Agaric acid**	see **1,2,3-Nonadecanetricarboxylic acid, 2-hydroxy-***													
a751	**Ajmalicine**	Py-tetrahydroserpentine. δ-Yohimbine. $C_{21}H_{24}N_2O_3$.	352.44	pr (MeOH) $[\alpha]_D^{24}$ −60 (chl, c = 0.5) λ^{al} 225 (4.68), 280 (3.91)	258d 261–3(vac)	...	...	...	i	...	...	...	...	MeOH s	**Am 76,** 1332
Ω a752	**Ajmaline**	$C_{20}H_{26}N_2O_2$.	326.44	pl (+3.5 w, aq AcOEt) $[\alpha]_D^{20}$ +144 (chl) λ^{al} 249 (3.9), 290 (3.4)	158–60 (hyd) 250–7 (anh)	...	...	...	i	s	δ	...	δ	chl s AcOEt s	**J1954,** 1242
a753	**α-Alanine** (*D*)	*l*-α-Aminopropionic acid. $CH_3CH(NH_2)CO_2H$	89.10	nd (w, al), $[\alpha]_D^{25}$ −13.6 (6N HCl, c = 1)	314d (297d)	sub	...	...	s v^h	δ	i	...	...	...	**B4**², 812
Ω a754	—(*DL*)	$CH_3CH(NH_2)CO_2H$	89.10	orh pr or nd (w) $\lambda^{w,pH=10}$ 205 (0.7)	295–6d (289d)	sub 258	1.424	...	s v^h	δ	i	i	...	Py δ	**B4**², 814
a755	—(*L*)	$CH_3CH(NH_2)CO_2H$	89.10	rh (w), $[\alpha]_D^{25}$ +2.8 (w, c = 6), +9.55 (HCl)	314d (297d)	sub 160–5	1.432^{23}	...	v	δ	i	i	...	...	**B4**², 809
Ω a756	—,ethyl ester hydrochloride (*dl*)	$CH_3CH(NH_2)CO_2CH_2CH_3 \cdot HCl$	153.61	pr or hyg nd (al)	87–8 (129)	d	...	...	v	v	v	...	...	...	**B4**², 819

For explanations, symbols and abbreviations see beginning of table. For structural formulas see end of table.

No.	Name	Synonyms and Formula	Mol. wt.	Color, crystalline form, specific rotation and λ_{max} (log ε)	m.p. °C	b.p. °C	Density	n_D	Solubility						Ref.
									w	al	eth	ace	bz	other solvents	
	β-Alanine														
Ω a756[1]	**β-Alanine**	β-Aminopropionic acid. $H_2NCH_2CH_2CO_2H$	89.10	nd, rh pr (al)	207d (199d)	...	1.437^{19}	...	s	δ	i	i	...	...	**B4**^{2}, 827
a757	**α-Alanine, *N*-alanyl-(*l*)**	$H_2NCH(CH_3)CONHCH(CH_3)CO_2H$	160.17	lf, $[\alpha]^{20}_D$ −21.6 (w) $\lambda^{w,pH=1}$ 270 sh (3.1)	298	...	...	...	v	δ	i	...	...	...	**B4**^{2}, 813
—	**—,3-anthraniloyl-(*l*)**	*see* **Kynurenine (*l*)**													
a758	**—,*N*-benzoyl-(*d*)**	$CH_3CH(NHCOC_6H_5)CO_2H$	193.21	pl (w), $[\alpha]^{20}_D$ −2.4 (w, c = 1), +37.3 (alk)	152–4	...	...	...	...	...	...	...	...	...	**B9**^{2}, 179
Ω a759	**—,—(*dl*)**	$CH_3CH(NHCOC_6H_5)CO_2H$	193.21	pl or pr, lf (eth) $\lambda^{w,pH=7}$ 248 (4.1), 287 (3.1)	165–6	d	...	...	s	s	δ	...	...	...	**B9**^{2}, 179
a760	**—,—(*l*)**	$CH_3CH(NHCOC_6H_5)CO_2H$	193.21	$[\alpha]^{20}_D$ +2.4 (w, c = 1) −37.3 (alk)	151	...	...	...	δ	...	...	...	...	...	**B9**^{2}, 179
a761	**—,*N*(carboxy-methyl)-(*dl*)**	$HO_2CCH_2NHCH(CH_3)CO_2H$.	147.13	cr (dil al or w)	222–3	...	...	...	s	i	i	...	...	...	**B4**^{1}, 497
a762	**—,*N*(4-chloro-phenyl)-**, nitrile	...	180.64	lf (eth-peth)	114.5	...	...	...	...	s	s	...	v^h	chl v lig i	**B12**, 617
a763	**—,*N*,*N*-diethyl-**, nitrile	$(C_2H_5)_2NCH(CH_3)CN$	126.20	...	...	81^{27} 55^{11}	0.857^{16}_4	...	s	s	s	...	...	...	**C35**, 1404
Ω a764	**—,3(3,4-dihy-droxyphenyl)-(*L*)**	*l*-Dopa	197.19	pl (dil al), pr or nd (w + SO_2) $[\alpha]^{15}_D$ −39.5 (w, p = 1.3) −12.7(4 % HCl, c = 1)	285.5d	...	...	...	s	i	i	i	i	MeOH s alk s aa i	**B14**^{2}, 399
a765	**—,*N*-fumaryl-(*DL*)**	Fumaroalanide. $HO_2CCH{:}CHCONHCH(CH_3)CO_2H$	187.15	nd	229d	...	...	...	δ s^h	s	δ	...	...	$NaHCO_3$ s lig i	**C37**, 2030
a766	**—,*N*-methyl-(*dl*)**	$CH_3CH(NHCH_3)CO_2H$	103.12	rh pr (abs al)	sinters 280	sub 292	...	...	δ	i s^h	...	...	...	...	**B4**^{2}, 821
a767	**β-Alanine, *N*-methyl-**	$CH_3NHCH_2CH_2CO_2H$	103.12	pl (al, +1w)	99–100 (+1w) 146 (anh)	...	...	...	v	v^h	...	δ	...	chl δ	**B4**^{2}, 828
a768	**—,—,ethyl ester**	$CH_3NHCH_2CH_2CO_2CH_2CH_3$	131.18	...	...	80^{21}	1.0082^{20}_{20}	1.4443^{20}	...	s	s	...	...	dil HCl s	**B4**^{2}, 828
a769	**—,—,nitrile**	β-Methylaminopropionitrile*. $CH_3NHCH_2CH_2CN$	84.13	...	...	$101–4^{49}$ 74^{16}	0.8992^{20}_4	1.4320^{20}	s	...	...	s	s	MeOH, chl s	**B4**^{3}, 1264
—	**α-Alanine,3-phenyl-**	*see* **Phenylalanine**													
a770	**—,*N*-phenyl-**, nitrile	α-Cyanoethyl aniline. $C_6H_5NHCH(CH_3)CN$	146.19	nd (bz or dil al), lf (dil al, bz, eth)	92	...	...	...	i δ^h	v	v	...	v	chl v	**B12**^{1}, 266
—	**Alantolactone**	*see* **Helenin**													
—	**Aldehydin**	*see* **Pyridine, 5-ethyl-2-methyl-**													
—	**Aldol**	*see* **Butanal, 3-hydroxy-***													
Ω a772	**Aldrin**	Octalene.	364.93	...	104–4.5	...	...	...	i	s	s	s	s	oos s	**C44**, 2690
—	**Aleuritic acid**	*see* **Hexadecanoic acid, 9,10,16-trihydroxy-***													
—	**Alizarin**	*see* **9,10-Anthraquinone, 1,2-dihydroxy-***													
—	**Alizarin cyanine R**	*see* **9,10-Anthraquinone, 1,2,4,5,8-pentahydroxy-***													
—	**Alkanin (*d*)**	*see* **Shikonine**													
Ω a773	**—(*l*)**	$C_{16}H_{16}O_5$	288.30	red br pr (bz, sub), nd (eth-al) $[\alpha]^{20}_{Cd}$ −157 (bz)	149	sub $140^{0.001}$	...	...	i	δ	δ	...	...	alk s chl δ aa s	**C30**, 1050
a774	**Allantoic acid**	Dicarbamidoacetic acid. Diureidoacetic acid. $(H_2NCONH)_2CHCO_2H$	176.14	nd, lf (MeOH)	173d	...	...	...	δ	s	...	...	...	os δ dil ac δ	**B3**^{2}, 388
Ω a775	**Allantoin**	Glyoxyldiureide. 5-Ureidohydantoin.	158.12	mcl pl or pr (w)	238–40	...	...	...	δ s^h	s^h	i	...	...	MeOH i NaOH s	**B25**^{2}, 379
a776	**Allanturic acid**	Glyoxalylurea. 5-Hydroxy-2,4-imidazoledione.	116.08	amor hyg pw	turns br at 180	...	...	...	δ	i	...	...	...	hot alk d os i	**B25**^{2}, 388
—	**Allene**	*see* **Propadiene***													
a777	**Allicin**	*S*-Oxodiallyl disulfide.	162.27	...	...	d	1.112^{20}	1.561^{20}	δ	∞	∞	...	∞	...	**Am 66**, 1950
a778	**Allitol(*D*)**	*D*-Allodulcitol.	182.18	...	150–1	...	...	...	s	...	i	...	i	...	**Am 68**, 1443
—	**Allocholesterol**	*see* **Coprostenol**													
—	**Allocinnamic acid**	*see* **Cinnamic acid (*cis*)**													

For explanations, symbols and abbreviations see beginning of table. For structural formulas see end of table.

No.	Name	Synonyms and Formula	Mol. wt.	Color, crystalline form, specific rotation and λ_{max} (log ε)	m.p. °C	b.p. °C	Density	n_D	Solubility w	al	eth	ace	bz	other solvents	Ref.
	Alloisoleucine														
Ω a779	**Alloisoleucine(*l*)**	$CH_3CH_2CH(CH_3)CH(NH_2)CO_2H$	131.18	lf (w), $[\alpha]_D^{20}$ −14.2 (w, c = 2)	280–1d				s	δ					**B4,** 457
a780	**Allomucic acid**	Tetrahydroxyadipic acid. $HO_2C(CHOH)_4CO_2H$	210.14	nd or pr (w)	198–200 (176d)				s^h	δ					**B3**[2], 376
a781	**Allonic acid,** γ-lactone(*D*)		178.14	pr(al) $[\alpha]_D^{20}$ −6.8(w, c = 11)	120				s	δ s^h					**B18**[1], 406
—	**Allocimene A**	*see* **2,4,6-octatriene, 2,6-dimethyl-***													
—	**Allophanamide**	*see* **Biuret**													
—	**Allophanic acid**	*see* **1-Ureacarboxylic acid**													
a782	**Allose** (*D*)		180.16	cr (w) $[\alpha]_D$ +0.58 → +14.41 (w, c = 5) (mut)	128–8.5				s	i					**B31,** 82
a783	—(*L*)	$C_6H_{12}O_6$. *See* a782	180.16	pr (dil al) $[\alpha]_D^{20}$ −0.58(al), −1.90 → −13.88 (w, c = 5) (mut)	128–9				s	δ					**Am 55,** 2167
a784	**Alloxan**	Mesoxalylurea. Pyrimidinetetrone.	142.07	wh tcl, rh pr (w + 4), orh cr (aa, ace, sub) $\lambda^{w, pH=4}$ 265 sh (1.9), $\lambda^{w, pH=6.8}$ 243 (3.40)	256d (anh) (turn pk at 230)	sub (vac)			v	s		s	s	aa s	**B24**[2], 301
a785	**Alloxanic acid**		160.09	tcl pr (eth)	162–3d				s	s	δ				**B25**[2], 266
a786	**Alloxantin**		286.16	rh pr (w + 2) $\lambda^{0.03N HCl}$ 270 (3.57)	253–5d (ye at 225)				δ s^h	δ	δ				**B26**[2], 335
—	**Allylacetic acid**	*see* **4-Pentenoic acid***													
—	**Allyl alcohol**	*see* **2-Propenol***													
—	**Allyl amine**	*see* **Propene, 3-amino-***													
—	**Allyl bromide**	*see* **Propene, 3-bromo-***													
—	**Allyl cellosolve**	*see* **Ethanol, 2-allyloxy-**													
—	**Allyl chloride**	*see* **Propene, 3-chloro-***													
—	**Allyl cyanide**	*see* **3-Butenoic acid,** nitrile													
—	**Allyl fluoride**	*see* **Propene, 3-fluoro-***													
—	**Allyl iodide**	*see* **Propene, 3-iodo-***													
—	**Allyl mustard oil**	*see* **Isothiocyanic acid,** allyl ester													
—	**Aloe-emodine**	*see* **9,10-Anthraquinone, 1,8-dihydroxy-3-hydroxy-methyl-**													
a787	**Aloetic acid**		450.24	og-ye nd (aa), ye cr (w + 1)	285d (>285 exp)				δ s^h					NH_4OH v alk s (red) acs[h]	**E13,** 575
Ω a788	**Aloin**	Barbaloin. $C_{21}H_{22}O_9$	386.36	ye nd (al) $[\alpha]_D$ −8.3 (dil al), −10.4 (AcOEt)	148.5 (150)				s	s	δ	s	δ	KOH s chl i	**J1956,** 3141
—	**Alphol**	*see* **Benzoic acid,** 2-hydroxy-, 1-naphthyl ester													
a789	**Alstonine**	$C_{21}H_{20}N_2O_3$	348.41	ye nd (ace) λ^{al} 250 (4.5), 308 (4.3), 375 (3.7)	205–10d (>300d)										**B27**[2], 824
a790	**Altrose(*d*)**		180.16	pr (MeOH-al) $[\alpha]_D^{20}$ +11.7 → 33.1 (w, c = 7.6) (mut)	103–5				s	i				MeOH s^h	**B31,** 83
a791	—(*l*,β)	$C_6H_{12}O_6$. *See* a790	180.16	pr (al, aa) −28.55 → −32.30 (w) (mut)	107–9.5				s	i s^h					**Am 56,** 1153
a792	**Amalic acid**	Tetramethylalloxantin. $C_{12}H_{14}N_4O_8$.	324.25	cr (w)	245d				δ^h	i				KOH s	**B26**[2], 336
a793	**Amarine**	$C_{21}H_{18}N_2$	298.39	pr (eth, bz-lig)	136	d198			i	s	s		s	lig i	**B23**[2], 274
a794	—,hydrate	$C_{21}H_{18}N_2.\frac{1}{2}H_2O$	307.40	pr (aq al + ½w)	106				s						**B23**[2], 274
Ω a795	**Amaron**	Benzoin imide. Ditolan azotide. Tetraphenyl-pyrazine.	384.49	tcl nd or pr (ace, al), nd (aa)	246	sub			i	δ^h	δ	s^h	s^h	chl, CS_2 s aa s^h con sulf s (red)	**B23**[2], 304
a796	**Amine, allyl methyl**	$CH_2:CHCH_2NHCH_3$	71.12			65		1.4065^{20}	∞	∞	v	v			**B4,** 206

For explanations, symbols and abbreviations see beginning of table. For structural formulas see end of table.

No.	Name	Synonyms and Formula	Mol. wt.	Color, crystalline form, specific rotation and λ_{max} (log ε)	m.p. °C	b.p. °C	Density	n_D	Solubility w	al	eth	ace	bz	other solvents	Ref.
	Amine														
a797	**—,benzyl diphenyl**	$C_6H_5CH_2N(C_6H_5)_2$	259.36	nd (al)	95	....	....	....	δ	δ v^h	v	v	v	aa δ lig v	**B12²**, 551
Ω a798	**—,benzyl ethyl phenyl**	*N*-Benzyl-*N*-ethylaniline. $C_6H_5CH_2N(C_2H_5)C_6H_5$	211.31	pa ye	34–6	$285–6^{710}$ δd 186^{22}	1.034_4^{19}	1.5930^{20}	i	s	s	...	...	....	**B12²**, 550
a799	**—,benzyl ethyl 2-tolyl**	*N*-Benzyl-*N*-ethyl-*o*-toluidine	225.34	ye	....	230^{20-5}	....	....	i	...	...	...	...	....	**B12,** 1033
a800	**—,benzyl methyl 2-tolyl**	*N*-Benzyl-*N*-methyl-*o*-toluidine.	211.31	ye	....	167^{13}	....	....	i	s	...	...	...	os s	**B12,** 1033
a801	**—,benzyl *N*-nitroso phenyl**	$C_6H_5CH_2N(NO)C_6H_5$	212.25	ye nd (al)	58	....	....	....	i	s	s	...	...	chl s lig s	**B12²**, 572
Ω a802	**—,benzyl phenyl**	*N*-Benzylaniline. $C_6H_5CH_2NHC_6H_5$	183.26	pr (MeOH) λ^{al} 250 (4.1), 294 (3.3)	37–8	306–7 (321^{775}) 171.5^{10}	1.0698_{15}^{15} 1.0298_4^{65}	1.6118^{25}	i	s	s	...	...	MeOH s^h	**B12²**, 548
a803	**—,benzyl 2-tolyl**	*N*-Benzyl-*o*-toluidine	197.28	cr (al, eth)	60	$300–5^{760}$ 176^{10}	1.0142_4^{65}	1.5861^{65}	i	s	...	s	...	os s chl s	**B12²**, 551
a804	**—,benzyl 3-tolyl-**	*N*-Benzyl-*m*-toluidine	197.28	pa ye oil	....	312^{760} 180^{10}	1.0083_4^{65}	1.5845^{65}	i	δ	v	s	v	chl v	**B12²**, 552
a805	**—,benzyl 4-tolyl**	*N*-Benzyl-*p*-toluidine	197.28	ye lf	19–20	319^{765} 181^{10}	1.0064_4^{65}	1.5832^{65}	i	v	v	...	v	chl v	**B12²**, 553
a806	**—,benzylidene ethyl**	$C_6H_5CH{:}NC_2H_5$	133.20	λ^{al} 245 (4.14), 278 (3.1), 290 sh (2.9)	....	195^{749} 117^{12}	0.9370_4^{20}	1.5365^{26}	i	s	s	s	...	....	**B7²**, 163
a807	**—,benzylidene methyl**	$C_6H_5CH{:}NCH_3$	119.17	λ^{al} 245 (4.18)	....	185^{760} $90–1^{30}$	0.9672_2^{14}	1.5526^{20}	...	s	s	s	...	....	**B7²**, 162
a808	**—,bis(dimethyl-phosphino)**	Amino-bis(dimethyl phosphine). $(CH_3)_2PNHP(CH_3)_2$	137.10	....	39.5	d $33.5^{5.4}$ sub	....	....	d	d	...	...	...	os ∞	**Am 75,** 3869
—	**—,*tert*-butyl**	*see* **Propane, 2-amino-2-methyl-***													
a809	**—,butyl diethyl, 2′2″-dihydroxy-**	Bis(2-hydroxyethyl)butyl-amine. $CH_3(CH_2)_3N(CH_2CH_2OH)_2$	161.25	....	....	$273–5^{741}$ 176–80^{35}	0.9692_4^{20}	1.4625^{20}	∞	∞	∞	s	∞	lig i	**B4,** 285
a810	**—,butyl dimethyl**	$CH_3(CH_2)_3N(CH_3)_2$	101.19	....	....	95^{761}	0.7206_4^{20}	1.3970^{20}	∞	∞	∞	∞	∞	....	**B4²**, 632
a811	**—,butyl ethyl**	$CH_3(CH_2)_3NHCH_2CH_3$	101.19	....	....	108–9	0.7398_4^{20}	1.4040^{20}	...	∞	∞	∞	∞	os s	**B4,** 157
a812	**—,*sec*-butyl ethyl(*d*)**	$CH_3CH_2CH(CH_3)NHCH_2CH_3$	101.19	$[\alpha]_D^{15}$ +18	....	98	0.7396_4^{15}	1.4043^{15}	...	∞	∞	∞	∞	os s	**B4³**, 307
a813	**—,—(*dl*)**	$CH_3CH_2CH(CH_3)NHCH_2CH_3$	101.19	....	−104.5	$97–8^{741}$	0.7358_4^{20}	....	...	∞	∞	∞	∞	os s	**B4²**, 636
a814	**—,butyl ethyl, 2,2′-dihydroxy-**	$HOCH_2CH_2NHCH_2CH(OH)CH_2CH_3$	133.19	yesh	....	137^9	1.0310_4^{20}	1.4690^{20}	s	s	...	s	...	lig i	**C44,** 7853
a815	**—,—,3,2′-dihydroxy-**	$HOCH_2CH_2NHCH_2CH_2CH(OH)CH_3$	133.19	yesh	....	$107–9^1$	1.0331_4^{20}	1.4718^{20}	s	...	...	s	...	lig i	**C59,** 6392
—	**—,cyano diphenyl**	*see* **Cyanamide, diphenyl-**													
—	**—,cyclohexyl phenyl**	*see* **Aniline, *N*-cyclohexyl-**													
Ω a816	**—,diallyl**	$NH(CH_2CH{:}CH_2)_2$	97.16	$\lambda^{aa-HNO_3\ salt}$ 296 (0.91)	....	111^{760}	....	1.4387^{20}	...	s	s	...	...	....	**B4²**, 663
Ω a817	**—,dibenzyl**	$NH(CH_2C_6H_5)_2$	197.28	$\lambda^{aa-HNO_3\ salt}$ 296 (0.94)	−26	300d 270^{250}	1.0256_4^{22}	1.5731^{20}	i	v	v	...	...	....	**B12²**, 553
a818	**—,dibenzyl ethyl**	$CH_3CH_2N(CH_2C_6H_5)_2$	225.34	....	....	306 $131^{<1}$	....	....	i	v	v	...	...	ac s	**B12²**, 553
Ω a819	**—,dibenzyl phenyl**	$C_6H_5N(CH_2C_6H_5)_2$	273.38	nd or pr (al)	71–2	>300d 226^{10}	1.0444_4^{80}	1.6065^{80}	i	δ s^h	s	...	s	aa δ, v^h	**B12²**, 554
Ω a820	**—,dibutyl**	$NH(CH_2CH_2CH_2CH_3)_2$	129.25	$\lambda^{al-HNO_3\ salt}$ 301 (0.83)	−60 to −59	159^{761} 48^{13}	0.7670_4^{20}	1.4177^{20}	s	v	v	s	s	....	**B4²**, 633
Ω a821	**—,di-*sec*-butyl**	$[CH_3CH_2CH(CH_3)]_2NH$	129.25	....	....	135^{765}	0.7534_4^{20}	1.4162^{20}	v	s	...	...	...	os s	**B4²**, 636
a822	**—,—,3,3′-di-hydroxy-**	$[CH_3CH(OH)CH(CH_3)]_2NH$	161.25	yesh	....	$112–5^3$	0.9775_4^{20}	1.4162^{20}	v	v	δ	v	...	lig δ	**C49,** 11689
a823	**—,dibutyl,3,3′-dimethyl-**	Diisoamyl amine. $[(CH_3)_2CHCH_2CH_2]_2NH$	157.30	....	−44	188	0.7672_4^{21}	1.4235^{20}	i	s	∞	...	...	....	**B4²**, 646
Ω a825	**—,dicyclohexyl**	....	181.33	....	−0.10	255.8^{760} δd 113.5^9	0.9123_4^{20}	1.4842^{20}	δ^h	s	s	...	s	....	**B12²**, 7
a826	**—,didodecyl**	$[CH_3(CH_2)_{10}CH_2]_2NH$	353.68	....	α 55–6	....	....	....	...	v	v	δ	v	chl v diox s	**Am 77,** 485
Ω a827	**—,diethyl**	$(CH_3CH_2)_2NH$	73.14	λ^{gas} 194.2 (3.47), 222 sh (2.47)	−48 fp −50	56.3^{760}	0.7056_4^{20}	1.3864^{20}	v	∞	s	...	...	....	**B4²**, 590
Ω a828	**—,—,hydrochloride**	$(CH_3CH_2)_2NH.HCl$	109.60	lf (al-eth)	227–30	320–30	1.04775_4^{21}	....	v	v	δ	δ	...	chl s	**B4²**, 592
a829	**—,—,2-amino-2′-hydroxy-**	*N*(2-Hydroxyethyl)ethylene-diamine. $H_2NCH_2CH_2NHCH_2CH_2OH$	104.15	....	....	238–40 123^{10}	1.0254_4^{25}	1.4863^{20}	∞	∞	...	s	δ	lig δ	**B4,** 286
—	**—,—,2,2′-diamino-**	*see* **Diethylenetriamine**													
Ω a830	**—,—,2,2′-dihydroxy-**	Diethanolamine. Diethylol-amine. $NH(CH_2CH_2OH)_2$	105.14	pr	28	271^{760} $154–5^{10}$	1.09664_4^{20}	1.4776^{20}	v	v	δ	...	δ	....	**B4²**, 729
Ω a831	**—,—,1,1′-diphenyl-**	$NH[CH(CH_3)C_6H_5]_2$	225.34	ye	....	295–8 190^{10}	1.018^{13}	1.573	δ	...	...	...	...	....	**B12²**, 589

For explanations, symbols and abbreviations see beginning of table. For structural formulas see end of table.

No.	Name	Synonyms and Formula	Mol. wt.	Color, crystalline form, specific rotation and λ_{max} (log ε)	m.p. °C	b.p. °C	Density	n_D	Solubility w	al	eth	ace	bz	other solvents	Ref.
	Amine														
Ω a832	—,—,**2,2′-diphenyl-**	$NH(CH_2CH_2C_6H_5)_2$	225.34	...	28–30	$335–7^{603}$ 190^{15}	...	1.5550^{25}	i	v	v	...	...	ac s	**B12²**, 593
a833	—,**diethyl methyl**	$(CH_3CH_2)_2NCH_3$	87.17	...	...	66^{760}	0.703_4^{23}	1.3879^{25}	v	v	v	...	...	...	**B4²**, 593
a834	—,—,**2,2′-dichloro-**, hydrochloride	$(ClCH_2CH_2)_2NCH_3.HCl$	192.52	hyg nd	111–2	...	...	...	v	s	...	...	...	...	**Am 81**, 5167
—	—,—,**2,2-diethoxy-**	*see* **Acetaldehyde, ethylmethylamino-**, diethyl acetal													
a835	—,—,**2,2′-dihydroxy-**	$(HOCH_2CH_2)_2NCH_3$	119.17	...	...	$246–8^{747}$ $123–5^{4}$	1.0377^{20}	1.4642^{20}	∞	∞	δ	...	...	...	**B4²**, 729
Ω a836	—,—,**1‴-methoxy-**	$CH_3OCH_2N(CH_2CH_3)_2$	117.19	...	...	116.5–7^{763} 53^{68}	...	...	s	s	s	...	...	HCl s	**B4²**, 598
a837	—,**diethyl *N*-nitro**	$(CH_3CH_2)_2NNO_2$	118.14	...	...	206.5^{757} 93^{16}	1.057^{15}	...	δ	∞	∞	...	...	...	**B4**, 130
a838	—,**diethyl *N*-nitroso**	$(CH_3CH_2)_2NNO$	102.14	ye λ^{al} 233 (3.7), 350 (1.95)	...	176.9^{760}	0.9422_4^{20}	1.4386^{20}	s	s	s	...	...	...	**B4²**, 617
a839	—,**diethyl phenyl, 2,2′-dihydroxy-**	*N*-Phenyldiethanolamine. $(HOCH_2CH_2)_2NC_6H_5$	181.24	pl (al)	58	228^{15}	...	...	δ	v	v	v	v	...	**B12²**, 109
a840	—,**difurfuryl**	...	177.21	...	...	$135–42^{15}$ $102–3^{1}$	1.1045_4^{20}	1.5168^{20}	i	...	s	...	...	...	**B18²**, 418
Ω a841	—,**diheptyl**	$[CH_3(CH_2)_6]_2NH$	213.41	nd	30	271^{750} $134–6^{9}$	...	...	δ	s	v	...	...	...	**B4²**, 652
Ω a843	—,**dihexyl**	$[CH_3(CH_2)_5]_2NH$	185.36	...	...	192–5 $112–4^{12}$	...	1.4339^{20}	...	s	s	...	...	...	**B4²**, 650
a844	—,**di-2-hydrindyl,1,1′-dihydroxy-(*d*)**	...	281.36	nd (al) $[\alpha]_D^{18}$ +83.2 (dil HCl, c = 0.3)	223	...	...	...	...	s^h	...	...	...	os δ	**B13¹**, 265
a845	—,—(*dl*)	$C_{18}H_{19}NO_2$. *See* a844	281.36	nd or pl (dil al)	205	...	...	...	δ	v	v	...	v	...	**B13¹**, 267
a846	—,—(*l*)	$C_{18}H_{19}NO_2$. *See* a844	281.36	nd (al), $[\alpha]_D^{18}$ −83.3 (dil HCl, c = 0.3)	223	...	...	...	δ	s^h	v	...	...	os δ	**B13¹**, 266
Ω a847	—,**diisobutyl**	$[(CH_3)_2CHCH_2]_2NH$	129.25	...	−73.5 (−77)	139–40^{760}	...	1.4090^{20}	δ	s	s	s	s	...	**B4²**, 638
Ω a848	—,**diisopropyl**	$[(CH_3)_2CH]_2NH$	101.19	λ^{al} 301 (0.82)	−61	84^{760}	0.7169_4^{20}	1.3924^{20}	δ	s	s	s	s	...	**B4²**, 630
a849	—,**diisopropyl *N*-nitroso**	Nitrous diisopropylamide. $[(CH_3)_2CH]_2NNO$	130.19	cr (eth, w)	48	194.5 $76–81^{14}$	0.9422_4^{20}	...	δ s^h	s	s	...	s	...	**B4**, 156
Ω a850	—,**dimethyl**	$(CH_3)_2NH$	45.09	λ^{gas} 190.5 (3.51), 222 (2.0)	−93	7.4	0.6804_4^{0}	1.350^{17}	v	s	s	...	...	...	**B4²**, 550
Ω a851	—,—,hydrochloride	Dimethylammonium chloride. $(CH_3)_2NH.HCl$	81.56	rh, nd (al)	171	...	...	...	v	s	i	...	...	chl v	**B4²**, 552
a852	—,—,**hexafluoro-**	Bis(trifluoromethyl)amine. $(CF_3)_2NH$	153.03	...	−130	-6.7^{760}	$1.561_4^{-14.4}$	...	d	...	...	...	...	...	**C35**, 434
a853	—,—,**perfluoro-**	$(CF_3)_2NF$	171.02	...	...	−37	...	...	i	...	...	...	...	...	**J1949**, 3080
a854	—,**dimethyl isobutyl**	$(CH_3)_2CHCH_2N(CH_3)_2$	101.19	...	...	$80–1^{760}$	0.7097_4^{20}	1.3907^{20}	s	...	...	...	...	ac v	**B4²**, 638
a855	—,**dimethyl *N*-nitro**	Dimethylnitramine. Nitric dimethylamide. $(CH_3)_2NNO_2$	90.08	nd (eth)	58	187^{759}	1.1090_4^{72}	1.4462^{72}	s	s	s	s	s	...	**B4¹**, 342
Ω a856	—,**dimethyl *N*-nitroso**	Dimethylnitrosamine. Nitrous dimethylamide. $(CH_3)_2NNO$	74.08	ye λ^{al} 231 (3.85), 346 (2.00)	...	154^{760}	1.0059_4^{20}	1.4358^{20}	s	s	s	...	...	...	**B4²**, 585
a857	—,**dimethyl pentyl**	$CH_3(CH_2)_4N(CH_3)_2$	115.22	...	...	123	0.743_4^{20}	1.4083^{20}	δ^h	...	∞	...	...	...	**B4²**, 642
a858	—,**di-2-naphthyl**	$(C_{10}H_7)_2NH$	269.35	lf (bz)	172.2	471	...	...	i	δ^h	s	...	δ s^h	aa s^h	**B12²**, 717
a859	—,**dioctadecyl**	$[CH_3(CH_2)_{16}CH_2]_2NH$	522.01	...	73–4	...	...	...	i	i	δ	i	δ	chl v	**C35**, 3963
Ω a860	—,**dioctyl**	$[CH_3(CH_2)_6CH_2]_2NH$	241.47	nd	35.6	297–8 175^{14}	0.7968_4^{26}	1.4415^{26}	i	v	v	...	...	...	**B4²**, 655
a860¹	—,**di-2-octyl**	$[CH_3(CH_2)_5CH(CH_3)]_2NH$	241.47	...	...	281.5^{739}	0.7948_4^{20}	...	i	v	v	...	...	...	**B4**, 197
Ω a861	—,**dipentyl**	$[CH_3(CH_2)_4]_2NH$	157.30	...	...	$202–3^{760}$ $91–3^{14}$	0.7771_4^{20}	1.4272^{20}	δ	v	∞	s	...	...	**B4²**, 642
a862	—,—,**2,2′-dihydroxy-2,2′-dimethyl-**	$[CH_3CH_2CH_2C(CH_3)(OH)CH_2]_2NH$	217.36	...	...	$165–70^{15}$	0.9264_4^{20}	1.4584^{20}	v	v	s	...	...	lig i	**C53**, 16829
Ω a863	—,**diphenyl**	*N*-Phenylaniline	169.23	mcl lf (dil al) λ^{al} 208 (4.33), 286 (4.29)	54–5	302 179^{22}	1.160_{20}^{22}	...	i	v	s	v	v	Py, CCl_4 v AcOEt v aa s lig s	**B12²**, 101
a864	—,—,hydrobromide	$(C_6H_5)_2NH.HBr$	250.15	pl (dil al)	230d	...	...	...	s	δ	i	...	i	...	**B12**, 180
Ω a865	—,—,**2-amino-**	*N*-Phenyl-*o*-phenylenediamine. $C_{12}H_{12}N_2$. *See* a863	184.24	nd (w) $\lambda^{95\% al}$ 237 (3.95), 280 (3.85)	80–1	312.5^{744}	...	...	δ	...	...	s	s	chl s lig δ	**B13²**, 13

For explanations, symbols and abbreviations see beginning of table. For structural formulas see end of table.

No.	Name	Synonyms and Formula	Mol. wt.	Color, crystalline form, specific rotation and λ_{max} (log ε)	m.p. °C	b.p. °C	Density	n_D	Solubility w	al	eth	ace	bz	other solvents	Ref.
	Amine														
a866	**—,—,4-amino-**	*N*-Phenyl-p-phenylene-diamine. black base *P*. $C_{12}H_{12}N_2$. *See* a863	184.24	nd (al), cr (lig) $\lambda^{95\% al}$ 281 (4.18)	66 (al), 75 (lig)	354 $155^{0.026}$	...	...	δ	v^h	s	...	...	lig s^h	**B13**, 76
a867	**—,—,4,4′-bis(di-methylamino)-**	Leuco base of Bindschedler green. $C_{16}H_{21}N_3$. *See* a863	255.37	tetr pl (CS_2)	119 (116)	...	...	...	δ s^h	δ	s	...	...	lig δ CS_2 s^h	**B13**[2], 56
a868	**—,—,4,4′-di-amino-**	*p,p′*-Iminodianiline. $C_{12}H_{13}N_3$. *See* a863	199.26	lf (w)	158	d	...	...	δ s^h	s	s	...	...	...	**B13**[2], 55
a869	**—,—,2,2′-dinitro-**	$C_{12}H_9N_3O_4$. *See* a863	259.22	lf (al, ace MeOH), ye cr (al, aa) λ^{al} 255 (4.18), 264 (4.22), 420 (3.97)	169	...	...	...	...	v^h	...	s	...	MeOH s^h aa v^h	**B12**[1], 341
Ω a870	**—,—,2,4-dinitro-**	$C_{12}H_9N_3O_4$. *See* a863	259.22	ye red nd (al) λ^{al} 231.5 (4.11), 350.5 (4.23)	157	...	...	...	i	s^h	δ	s	δ s^h	chl s Py s	**B12**[2], 407
Ω a871	**—,—,2,4′-dinitro-**	$C_{12}H_9N_3O_4$. *See* a863	259.22	red nd (aa)	222–3 (219)	...	...	...	i	δ	...	δ	s^h	to s chl s	**B12**[2], 387
a872	**—,—,2,6-dinitro-**	$C_{12}H_9N_3O_4$. *See* a863	259.22	og lf (al, aa)	107–8	...	...	...	i	s^h	...	...	...	aa s^h	**B12**[2], 413
a873	**—,—,3,4′-dinitro-**	$C_{12}H_9N_3O_4$. *See* a863	259.22	pa ye (chl, aq Py)	217 (softens 205)	...	...	...	...	δ	...	v	...	chl s^h con ac s	**B12**[2], 387
a874	**—,—,4,4′-dinitro-**	$C_{12}H_9N_3O_4$. *See* a863	259.22	ye nd (al) λ^{al} 232 (4.06), 402 (4.58)	216–6.5	...	...	...	i	δ s^h	...	s	δ	to i aa s	**B12**[2], 387
a875	**—,—,2,4-dinitro-4′-hydroxy-**	$C_{12}H_9N_3O_5$. *See* a863	275.22	red lf	195–6	...	...	...	...	...	...	...	...	alk s	**B13**[1], 150
a876	**—,—,2,4-dinitro-5-hydroxy-**	$C_{12}H_9N_3O_5$. *See* a863	275.22	dk ye nd (al)	139	...	...	...	...	s^h	...	...	...	...	**B13**[1], 138
a877	**—,—,2,6-dinitro-2′-hydroxy-**	$C_{12}H_9N_3O_5$. *See* a863	275.22	red vt nd (al)	191	...	...	...	...	s^h	v^h	...	v^h	aa v^h	**B13**, 365
a878	**—,—,2,6-dinitro-3-hydroxy-**	$C_{12}H_9N_3O_5$. *See* a863	275.22	ye br nd (MeOH)	124–5	...	...	...	...	s^h	...	...	...	MeOH s^h	**B13**[2], 216
Ω a880	**—,—,2,2′,4,4′,6,6′-hexanitro-**	Dipicrylamine. $C_{12}H_5N_7O_{12}$. *See* a863	439.21	pa ye pr (aa), cr (93% HNO_3) λ^{eth} 250 (4.10), 372 (4.18)	244d	...	...	...	i	i	δ	δ	i	aq alk s Py v to CCl_4 i	**B12**[2], 422 **B12**[2], 422
a881	**—,—,2-hydroxy-**	2-Anilinophenol. $C_{12}H_{11}NO$. *See* a863	185.23	pr (w) λ^{al} 274 (4.2), 298 (4.0)	69–70	$180–9^{20}$	...	...	δ^h	s	s	...	δ	aa s	**B13**, 365
Ω a882	**—,—,3-hydroxy-**	3-Anilinophenol. $C_{12}H_{11}NO$. *See* a863	185.23	lf (w)	81.5–2	340	...	...	δ s^h	v	v	v	s	dil ac s lig δ	**B13**[2], 213
a883	**—,—,4-hydroxy-**	4-Anilinophenol. $C_{12}H_{11}NO$. *See* a863	185.23	lf (w) λ^{al} 288 (4.1)	73	330 $215–6^{12}$	...	...	δ	v	v	...	v^h	chl v lig δ ac, alk s	**B13**[2], 231
Ω a884	**—,—,2-nitro-**	$C_{12}H_{10}N_2O_2$. *See* a863	214.23	og pl (dil al), rh bipym λ^{al} 220 (4.13), 259 (4.15), 428 (3.82)	75.5	...	...	...	i	s	...	...	...	...	**B12**[2], 369
a885	**—,—,3-nitro-**	$C_{12}H_{10}N_2O_2$. *See* a863	214.23	red nd or pl (dil al) λ^{al} 265 (4.26), 282 (4.21), 367 (3.17)	114	...	...	...	...	v	v	...	v	lig δ	**B12**[2], 377
Ω a886	**—,—,4-nitro-**	$C_{12}H_{10}N_2O_2$. *See* a863	214.23	ye nd, tab (CCl_4) λ^{al} 228 (3.82), 258 (3.98), 391 (4.29)	133.5–4.5	211^{30}	...	...	i	v	...	...	...	aa v con sulf s (vt)	**B12**[2], 386
Ω a887	**—,—,4-nitroso-**	$C_{12}H_{10}N_2O$. *See* a863	198.23	blsh bk (al), ye pw, pl (bz) λ^{al-HCl} 259 (3.62), 406 (4.31)	143	...	...	...	δ	v^h	v	...	v	chl v lig δ	**B12**[2], 122
—	**—,—,4-sulfo-**	*see* **Benzenesulfonic acid, 4-phenylamino-***													
a888	**—,—,2,2′,4,4′-tetrabromo-**	$C_{12}H_7Br_4N$. *See* a863	484.83	nd (chl or bz)	187.5 (182)	...	...	...	i	δ^h	...	...	s^h	chl s^h	**B12**[2], 356
a889	**—,—,2,2′,4,4′-tetranitro-**	$C_{12}H_7N_5O_8$. *See* a863	349.22	ye nd or cr (aa), red-br pl (al) λ^{al} 219.5 (4.35), 358 (4.24), 401.5 (4.36)	201–1.5	...	...	...	i^h	δ^h	δ	δ s^h	δ^h	chl, AcOEt δ^h aa s^h Py s	**B12**[2], 409
a890	**—,diphenyl ethyl**	$(C_6H_5)_2NCH_2CH_3$	197.28	...	...	295.5–6.5[760] 148^{11}	1.0396^{20}_{20}	1.6095^{20}	i	s	s	...	...	...	**B12**[2], 105
a891	**—,diphenyl methyl**	$(C_6H_5)_2NCH_3$	183.26	λ^{al} 245 (3.9), 290 (4.0)	−7.55	293–4 145^{10}	1.0476^{20}_4	1.6193^{20}	i	δ	...	...	...	MeOH δ	**B12**[2], 105

For explanations, symbols and abbreviations see beginning of table. For structural formulas see end of table.

No.	Name	Synonyms and Formula	Mol. wt.	Color, crystalline form, specific rotation and λ_{max} (log ε)	m.p. °C	b.p. °C	Density	n_D	Solubility w	al	eth	ace	bz	other solvents	Ref.
	Amine														
Ω a892	**—,diphenyl N-nitroso**	Nitrous diphenylamide. $(C_6H_5)_2NNO$	198.23	ye pl (lig) λ^{al} 290 (3.88)	66.5	...	...	...	...	δ v^h	...	...	s^h	...	**B12**², 310
—	**—,dipicryl**	*see* **Amine, diphenyl, 2,2′, 4,4′,6,6′-hexanitro-**													
Ω a893	**—,dipropyl**	$(CH_3CH_2CH_2)_2NH$	101.19	...	−39.6 (−63)	109.4–10.4	0.7400_4^{20}	1.4050^{20}	s	s	∞	v	v	...	**B4**², 622
a894	**—,—,3,3′-diallyloxy-2,2′-dihydroxy-**	$[CH_2{:}CHCH_2OCH_2CH(OH)CH_2]_2NH$	245.32	...	73	$180–5^{2–3}$	...	...	s	s	...	s	...	lig i	**C52,** 15129
a895	**—,—,2,2′-dihydroxy-**	Diisopropanolamine. $[CH_3CH(OH)CH_2]_2NH$	133.19	cr	44.5–5.5	249–50^{745} 151^{23}	0.989_4^{20}	...	s	s	δ	...	...	...	**B4**², 737
a896	**—,dipropyl ethyl, 2,2′,2″-trihydroxy-**	$[CH_3CH(OH)CH_2]_2NCH_2CH_2OH$	177.25	...	...	$155–6^1$	1.0458_4^{20}	1.4708^{20}	v	v	...	v	...	lig δ	**C48,** 7628
a897	**—,dipropyl isobutyl, perfluoro-**	$(CF_3)_2CFCF_2N(CF_2CF_2CF_3)_2$	571.08	...	...	$146–8^{760}$	1.84_4^{25}	1.283^{25}	...	s	s	...	...	...	**J1951,** 102
a898	**—,dipropyl methyl, 3,3′-dihydroxy-**	$(HOCH_2CH_2CH_2)_2NCH_3$	147.22	hyg	...	$164–5^{13}$	...	...	∞	∞	δ	...	v	...	**B4**², 735
a899	**—,dipropyl N-nitroso**	Dipropyl nitrosamine. Nitrous dipropylamide. $(CH_3CH_2CH_2)_2NNO$	130.19	golden λ^{al} 233 (3.85), 350 (1.95)	...	206^{760} 89^{13}	0.9163_4^{20}	1.4437^{20}	δ	∞	∞	...	...	...	**B4**², 628
a900	**—,di-2-tolyl**	...	197.28	bl flr, wh cr $\lambda^{95\%al}$ 240 (3.84)	52–3	312^{727} 192^{23}	...	...	δ	...	...	...	...	...	**B12**², 437
a901	**—,di-3-tolyl**	...	197.28	pa ye (peth) $\lambda^{95\%al}$ 259 (3.90)	53	319–20	...	...	δ	s	s	...	...	peth s^h	**B12**², 467
a902	**—,di-4-tolyl**	...	197.28	nd (peth)	79	330.5^{760}	...	...	δ	...	s	...	...	peth s^h	**B12**², 494
a903	**—,3,4-ditolyl, hydrochloride**	...	233.74	cr $\lambda^{95\%al}$ 255 (4.12)	202–3	...	...	...	s	v	i	v	v	chl v lig i	**B12**¹, 414
a904	**—,ethyl methyl**	$CH_3NHCH_2CH_3$	59.11	...	...	36–7	...	...	v	v	s	s	...	os s	**B4**², 589
a905	**—,—,hydrochloride**	$CH_3NHCH_2CH_3.HCl$	95.57	pl (al-eth)	126–30	...	...	...	v	v	i	...	...	chl s	**B4**², 589
a906	**—,ethyl N-nitroso-phenyl**	*N*-Ethylphenylnitrosamine. $C_6H_5N(NO)CH_2CH_3$	150.18	yesh λ^{al} 271.5 (3.83)	...	119.5–20^{15}	1.0874_4^{20}	...	i	...	...	...	...	aa s	**B12**², 310
Ω a907	**—,ethyl propyl, 2,2′-dihydroxy-**	$HOCH_2CH_2NHCH_2CH(OH)CH_3$	119.17	yesh	...	$120–3^1$	1.0455_4^{20}	1.4695^{20}	s	s	...	s	...	...	**C40,** 4732
a908	**—,ethyl 2-tolyl, 2-hydroxy-**	...	151.21	...	...	285–6 149^4	1.0794_4^{20}	1.5675^{20}	...	s	s	...	...	chl s	**B12**², 437
a909	**—,ethyl 4-tolyl, 2-hydroxy-**	...	151.21	pl (eth-lig)	42–3	286–8 $153–5^4$	...	...	δ d^h	v	v	...	v	chl v ac v	**B12**², 495
a910	**—,methyl propyl**	$CH_3NHCH_2CH_2CH_3$	73.14	...	...	$61–2.5^{760}$	0.7204^{17}	1.3858^{25}	s	s	...	s	...	...	**B4**², 621
Ω a911	**—,1-naphthyl phenyl**	$C_{10}H_7^{\alpha}NHC_6H_5$	219.29	lf (lig), pr or nd (al) λ^{MeOH} 217 (4.78), 252.5 (4.25), 339 (3.94)	62	335^{528} 226^8	...	...	δ	s	s	...	s	chl s aa s	**B12**², 682
Ω a912	**—,2-naphthyl phenyl**	$C_{10}H_7^{\beta}NHC_6H_5$	219.29	nd (MeOH) λ^{al} 220 (4.7), 275 (4.4), 312 (4.3)	108	395–9.5 237^{13}	...	...	i	s	s	...	s	aa s (bl flr)	**B12,** 1275
a913	**—,1-naphthyl propyl**	$C_{10}H_7^{\alpha}NHCH_2CH_2CH_3$	185.27	ye	...	$316–8^{771}$	...	...	i	δ	δ	...	δ	aa δ	**B12,** 1224
a914	**—,1-naphthyl 2-tolyl**	...	233.32	nd (lig)	94–5	$198–202^9$	...	...	i	s	s	...	s	lig s^h	**B12,** 1255
a915	**—,1-naphthyl 4-tolyl**	...	233.32	pr (al)	79	360^{528} 236^{15}	...	...	δ	δ s^h	s	...	s	con sulf s peth s^h	**B12,** 1278
a916	**—,2-naphthyl 2-tolyl**	Yellow OB	233.32	lf (lig)	95–6 (105)	400–5 $235–7^{14}$	...	...	...	s	s	s	s	chl v lig v	**B12,** 1277
a917	**—,2-naphthyl 4-tolyl**	...	233.32	red lf (al)	103–3.5	...	...	...	...	δ s^h	s	...	s	PrOH s lig δ	**B12,** 1277
Ω a918	**—,triallyl**	$(CH_2{:}CHCH_2)_3N$	137.23	...	...	155–6 (cor)	0.809_4^{20}	1.4502^{20}	...	s	s	s	s	ac s	**B4,** 208
Ω a919	**—,tribenzyl**	$(C_6H_5CH_2)_3N$	287.41	pl (eth), mcl (al) λ^{al} 247.5 sh (2.9), 252.5 (2.9)	91–2	380–90 230^{13}	0.9912_4^{95}	...	δ	δ s^h	s	...	...	...	**B12**², 555
Ω a920	**—,tributyl**	$(CH_3CH_2CH_2CH_2)_3N$	185.36	hyg λ^{chl} 242 (3.0)	...	213^{760} $91–2^9$	0.7771_0^{20}	1.4297^{20}	δ	v	v	s	s	...	**B4**², 633
Ω a921	**—,—,perfluoro-**	$(CF_3CF_2CF_2CF_2)_3N$	671.10	...	...	179^{760}	1.873_4^{25}	1.291^{25}	...	...	...	s	...	...	**J1951,** 102
a922	**—,—,2,2′,2″-trimethyl-**	Tri-*active*-amylamine. $[CH_3CH_2CH(CH_3)CH_2]_3N$	227.44	...	...	230–7 94^4	0.7964^{13}	1.43305^{20}	i	s	v	v	v	chl v	**B4**², 642
Ω a923	**—,—,3,3′,3″-trimethyl-**	Triisoamylamine. $[(CH_3)_2CHCH_2CH_2]_3N$	227.44	...	...	235^{760} 94^4	0.7848_4^{20}	1.4331^{20}	i	v	∞	...	∞	CCl_4 ∞	**B4**², 646
a924	**—,tridodecyl**	$[CH_3(CH_2)_{10}CH_2]_3N$	522.01	...	fr 15.7	$220–8^{0.03}$	...	1.4567^{25}	...	δ	∞	δ	∞	chl ∞	**Am 74,** 4287

For explanations, symbols and abbreviations see beginning of table. For structural formulas see end of table.

No.	Name	Synonyms and Formula	Mol. wt.	Color, crystalline form, specific rotation and λ_{max} (log ε)	m.p. °C	b.p. °C	Density	n_D	Solubility w	al	eth	ace	bz	other solvents	Ref.
	Amine														
Ω a925	—,**triethyl**	$(CH_3CH_2)_3N$	101.19	λ^{hp} 196 (3.70)	−114.7	89.3^{760}	0.7275_4^{20}	1.4010^{20}	s δ^h	s	s	v	v	chl v	**B4**^{2}, 593
Ω a926	—,—,hydrochloride	$(CH_3CH_2)_3N.HCl$	137.65	hex (al)	260d	sub 245	1.0689_4^{21}	...	v	s	i	...	δ	chl v	**B4**^{1}, 348
a927	—,—,hydroiodide	$(CH_3CH_2)_3N.HI$	229.11	pr (95 % al)	181		1.924	...	v	s	i	...	...	chl s	**B4,** 101
Ω a928	—,—,**2-bromo-,** hydrobromide	$BrCH_2CH_2N(CH_2CH_3)_2.HBr$.	261.01	nd (al-eth)	209	...	...	...	v	δ v^h	...	...	...	...	**B4**^{2}, 619
Ω a929	—,—,**2,2-diethoxy-**	Diethylaminoacetal. $(CH_3CH_2)_2NCH_2CH(OCH_2CH_3)_2$	189.30	...	...	194–5	0.863^{16}	1.4189^{20}	s	s	s	...	...	os s	**B4,** 309
a930	—,—,**2,2′-dihydroxy-**	2,2′-Ethyliminodiethanol. $(HOCH_2CH_2)_2NC_2H_5$	133.19	ye	...	$246–8^{760}$ 118^3	1.0135_4^{20}	1.4663^{20}	v	v	δ	...	...	...	**B4,** 284
Ω a931	—,—,**perfluoro-**	$(CF_3CF_2)_3N$	371.05	...	...	70.3^{760}	1.736_4^{25}	1.262^{25}	...	...	...	...	...	...	**J1951,** 102
a932	—,—,**2,2′,2″-trichloro-**	Nitrogen mustard gas. $(ClCH_2CH_2)_3N$	204.53	pa ye	−4	$143–4^{15}$	...	...	δ	∞	∞	...	∞	HCl s	**J1935,** 1217
a933	—,—,**2,2′,2″-trihydroxy-**	Triethanolamine. $(HOCH_2CH_2)_3N$	149.19	hyg cr	21.2	277^{150} $206–7^{15}$	1.1242_4^{20}	1.4852^{20}	∞	∞	δ	...	δ	chl s lig δ	**B4**^{2}, 729
a934	—,—,—,hydrochloride	$(HOCH_2CH_2)_3N.HCl$	185.65	cr (al)	179–80	...	...	...	δ	δ s^h	...	...	...	...	**B4,** 285
a935	—,**trifurfuryl**	...	257.29	...	fr in liq air	$136–8^1$	...	...	i	...	s	...	...	...	**B18**^{2}, 418
a936	—,**triheptyl**	$[CH_3(CH_2)_6]_3N$	311.60	...	...	330^{762} $151–4^1$	...	...	i	v	v	...	...	ac s	**B4,** 193
a937	—,**trihexyl**	$[CH_3(CH_2)_5]_3N$	269.52	...	...	263–4 119^{15}	...	...	i	v	v	...	...	ac s	**B4**^{2}, 650
a938	—,**triisobutyl**	$[(CH_3)_2CHCH_2]_3N$	185.36	...	−21.8	191.5^{760} 84^{15}	0.7684_4^{20}	1.4252^{17}	i	v	∞	...	...	...	**B4**^{2}, 639
Ω a939	—,**trimethyl**	$(CH_3)_3N$	59.11	λ^{gas} 161 (3.4), 190.5 (3.59), 227 (2.95)	−117.2	2.87^{760}	0.6709_4^0 0.6356_4^{20}	1.3631^0	v	s	s	...	s	chl, to v	**B4**^{2}, 553
Ω a940	—,—,hydrochloride	$(CH_3)_3N.HCl$	95.57	mcl dlq nd (al)	277–8d	sub at 200	...	...	s	s	i	...	...	chl s	**B4**^{2}, 555
a941	—,—,oxide	$(CH_3)_3NO$	75.11	hyg nd (w + 2)	255–7 (hyd) 96 (anh)	...	...	...	s	s	i	...	i	chl s^h	**B4**^{2}, 556
a942	—,—,oxide hydrochloride	$(CH_3)_3NO.HCl$	111.57	nd (al)	218–20d	...	...	...	v	s^h	...	...	...	MeOH s^h	**B4**^{2}, 556
a943	—,—,oxide hydroiodide	$(CH_3)_3NO.HI$	203.02	pr (al)	130d	...	...	...	v	v	i	...	...	...	**B4,** 50
a944	—,—,**perfluoro-**	$(CF_3)_3N$	221.03	...	...	−7 to −6	...	...	...	...	...	...	...	...	**J1951,** 102
—	—,—,**1,1′,1″-tricarboxy-**	*see* **Acetic acid,** nitrilotri-													
a945	—,**trioctadecyl**	$[CH_3(CH_2)_{16}CH_2]_3N$	774.50	...	54.6	...	...	...	...	i	v	...	v	chl s	**C34,** 6221
Ω a946	—,**tripentyl**	$[CH_3(CH_2)_4]_3N$	227.44	...	...	240–5 130^{14}	0.7907_4^{20}	1.43665^{20}	i	s	s	...	...	ac s	**B4**^{2}, 642
Ω a947	—,**triphenyl**	$(C_6H_5)_3N$	245.33	mcl (MeOH, AcOEt, bz) λ^{al} 297 (4.30)	127	365	0.774_0^0	1.353^{16}	i	δ s^h	s	...	s	MeOH s^h	**B12**^{2}, 106
Ω a948	—,**tripropyl**	$(CH_3CH_2CH_2)_3N$	143.28	...	−93.5	156^{760}	0.7558_4^{20}	1.4181^{20}	δ	∞	s	...	...	...	**B4**^{2}, 623
Ω a949	—,—,**perfluoro-**	$(CF_3CF_2CF_2)_3N$	521.07	...	...	130^{760}	1.822_4^4	1.279^{25}	...	...	...	...	...	...	**J1951,** 102
a950	—,—,**2,2′,2″-trihydroxy-**	Triisopropanolamine. $[CH_3CH(OH)CH_2]_3N$	227.31	...	45	$170–80^{10}$	1.0^{20}	...	s	s	...	...	...	...	**Am 73,** 3635
a951	—,**3,3′,4″-tritolyl**	...	287.41	nd (al)	89–90	...	...	...	i	s	...	...	...	os v con sulf s	**B12**^{1}, 415
a952	—,**tri(4-tolyl)**	...	287.41	cr (aa)	117	...	...	...	...	δ	s	s	v	chl v sulf s lig δ	**B12**^{2}, 494
—	**Aminomethanamidine**	*see* **Guanidine**													
—	**Aminonaphthosulfonic acid**	*see* **1-Naphthalenesulfonic acid, 4-amino-5-hydroxy-***													
Ω a953	**Ammelide**	6-Amino-*s*-triazine-2,4-diol* Cyanuromonoamide.	128.09	mcl pr (w) $\lambda^{w, pH=4}$ 222 (4.15), $\lambda^{w, pH=12}$ 226 (3.99)	d	...	...	...	δ^h	i	i	...	i	dil sulf δ ac, alk s aa δ oos i	**B26**^{2}, 132
Ω a954	**Ammeline**	4,6-Diamino-*s*-triazin-2-ol.* Cyanurodiamide.	127.11	nd aq (Na_2CO_3) $\lambda^{w, pH=7}$ 230 (3.88)	d	...	...	...	i	i	i	...	i	ac s alk s aa i	**B26**^{2}, 132
Ω a955	**Ammonium, benzyl trimethyl,** bromide	$C_6H_5CH_2N(CH_3)_3Br$	230.16	pl (al-lig), (w)	235	...	...	...	s^h	s^h	...	...	...	...	**B12**^{2}, 546
Ω a956	—,—,chloride	$C_6H_5CH_2N(CH_3)_3Cl$	185.70	(ace)	243	...	...	...	v	...	...	s^h	...	...	**B12**^{1}, 448
Ω a957	—,—,iodide	$C_6H_5CH_2N(CH_3)_3I$	277.15	(al)	180	...	...	...	...	s	...	...	...	...	**B12**^{1}, 448
a958	—,—,nitrate	$C_6H_5CH_2N(CH_3)_3NO_3$	212.25	...	151–60	...	...	...	s	s	...	...	...	...	**B12**^{2}, 546
a959	—,**dodecyl trimethyl,** chloride	$CH_3(CH_2)_{11}N(CH_3)_3Cl$	263.90	...	246d	...	...	...	v	v	i	v	i	MeOH v chl v	**C55,** 1442

For explanations, symbols and abbreviations see beginning of table. For structural formulas see end of table.

No.	Name	Synonyms and Formula	Mol. wt.	Color, crystalline form, specific rotation and λ_{max} (log ε)	m.p. °C	b.p. °C	Density	n_D	Solubility w	al	eth	ace	bz	other solvents	Ref.
	Ammonium														
Ω a960	—,**phenyl trimethyl, bromide**	$C_6H_5N(CH_3)_3Br$	216.13	hyg pr (al or al-eth)	213–4				v	s^h	i	...	...	aa s^h	**B12²**, 88
a961	—,—,chloride	$C_6H_5N(CH_3)_3Cl$	171.67	hyg nd (al-eth) λ^{al} 255 (2.3)	d				s	v	...	...	...	chl δ	**B12²**, 88
a962	—,—,iodide	$C_6H_5N(CH_3)_3I$	263.12	lf (al)	224 (216)				v	s	...	δ	...	chl i, δ^h aa s	**B12²**, 89
a963	—,**propyl trimethyl, 2-acetoxy-chloride** (*D*)	*O*-Acetyl-β-methylcholine chloride. $CH_3CH(O_2CCH_3)CH_2N(CH_3)_3Cl$	195.69	hyg, $[\alpha]_D$ +41.9	172–3	d			s	s	i	...	...	...	**C30**, 4541
a964	—,—,—,—(*L*)	$CH_3CH(O_2CCH_3)CH_2N(CH_3)_3Cl$	195.69	$[\alpha]_D$ −41.3	172–3	d			s	...	...	...	...	...	**C32**, 5583
a965	—,—,**2-hydroxy-, chloride**	β-Methylcholine chloride. $CH_3CH(OH)CH_2N(CH_3)_3Cl$	153.65	pr (BuOH)	165	d			s	s	i	...	...	...	**C41**, 989
—	—,**purpurate**	*see* **Murexide**													
Ω a966	—,**tetrabutyl, iodide**	$[CH_3(CH_2)_3]_4NI$	369.38	lf (w or bz)	148				δ s^h	v	...	...	...	os s MeOH v	**B4**, 157
Ω a967	—,**tetraethyl, bromide**	$(CH_3CH_2)_4NBr$	210.17	dlq (al)			1.3970_4^{20}		v	v	...	...	...	chl v MeOH v	**B4²**, 597
a968	—,—,hydroxide tetrahydrate	$(CH_3CH_2)_4NOH.4H_2O$	219.32	nd (w + 4)	49–50	d (vac)			v	s	...	...	...	...	**B4²**, 596
Ω a969	—,**tetramethyl, bromide**	$(CH_3)_4NBr$	154.06	dlq ditetr bipym	230d	(360 sub in vac)	1.56		v	δ	i	...	i	chl i MeOH s	**B4²**, 558
Ω a970	—,—,chloride	$(CH_3)_4NCl$	109.60	dlq ditetr bipym (dil al)	420 (sealed tube) (d > 320)		1.169_4^{20}		s	δ	i	...	i	chl i MeOH v	**B4²**, 557
a972	—,—,hydroxide, monohydrate	$(CH_3)_4NOH.H_2O$	109.17		130–5d without melting				s	...	i	...	i	...	**B4**, 50
a973	—,—,hydroxide, trihydrate	$(CH_3)_4NOH.3H_2O$	145.20		60				s	...	i	...	i	...	**B4**, 50
a974	—,—,hydroxide, pentahydrate	$(CH_3)_4NOH.5H_2O$	181.23	hyg nd	62–3	d			v ∞^h	v	i	...	...	...	**B4²**, 557
Ω a975	—,—,iodide	$(CH_3)_4NI$	201.05	pr (w)	> 230d (> 355)		1.829		δ	δ	i	δ	...	chl i alk δ liq SO_2 s MeOH δ	**B4²**, 558
a976	—,**tetrapropyl, bromide**	$(CH_3CH_2CH_2)_4NBr$	266.27		252 (cor)				s	...	...	...	...	chl s	**J1929**, 1291
Ω a977	—,—,iodide	$(CH_3CH_2CH_2)_4NI$	313.27	rh bipym	280d		1.3138_4^{25}		v	s	δ	...	...	chl v aa s MeOH v	**B4²**, 624
a978	**Amphetamine** (*d*)	Dexedrine. β-Phenyl isopropylamine. $C_6H_5CH_2CH(NH_2)CH_3$	135.21	$[\alpha]_D^{15}$ +37.6 (bz, c = 8)		203–4 80^{10}	0.949_4^{15}	1.4704^{20}	δ	s	s	...	...	...	**B12²**, 621
a979	—(*dl*)	Benzedrine. $C_6H_5CH_2CH(NH_2)CH_3$	135.21			203^{760} 97^{20}	0.9306_4^{25}	1.518^{26}	δ	s	δ	...	...	chl s	**C49**, 1537
a980	—,sulfate (*d*)	Dexedrine sulfate. $2(C_6H_5CH_2CH(NH_2)CH_3).H_2SO_4$	368.50	$[\alpha]_D^{20}$ +22 (w, c = 8)	d > 300		1.15_4^{25}		v	δ	i	...	...	...	**C50**, 4210
a981	—,—(*dl*)	Benzedrine sulfate. $2(C_6H_5CH_2CH(NH_2)CH_3).H_2SO_4$	368.50		280–1 (d > 300)		1.15_4^{25}		v	δ	i	...	...	...	**C49**, 1537
Ω a982	**Amygdalin**	Amigdaloside. Mandelonitrile β-gentiobioside. $C_{20}H_{27}NO_{11}$.	457.44	$[\alpha]_D^{20}$ −42 (w, c = 1)	223–6 (216)				v^h	δ s^h	i	...	...	chl i	**B31**, 401
a983	—,trihydrate	$C_{20}H_{27}NO_{11}.3H_2O$. *See* a982	511.48	rh (w + 3), $[\alpha]_D^{20}$ −40 (w, c = 1)	200 (remelts 125–30)				s ∞^h	δ s^h	i	...	...	...	**B31**, 400
—	**Amylamine**	*see* **Pentane, 1-amino-***													
—	**Amyl bromide**	*see* **Pentane, 1-bromo-***													
—	**Amyl chloride**	*see* **Pentane, 1-chloro-***													
—	**Amyl iodide**	*see* **Pentane, 1-iodo-***													
a984	**α-Amyrin**	α-Amyrenol. $C_{30}H_{50}O$.	426.74	nd (al), $[\alpha]_D^{17}$ +91.6 (bz, c = 1.3) (+83.5 (chl) λ^{al} 207 (3.82)	186	$243^{0.7}$			...	s^h	s	...	s	chl, aa s peth δ	**E14s**, 1067
a985	**β-Amyrin**	β-Amyrenol. $C_{30}H_{50}O$.	426.74	nd (lig or al), $[\alpha]_D^{19}$ +99.8 (bz, c = 1.3), +88.4 (chl) λ^{al} 207 (3.82)	197–7.5	$260^{0.8}$			i	δ	s	...	s	chl s aa s lig δ	**E14s**, 946
—	**Amytal**	*see* **Barbituric acid, 5-ethyl-5(3-methylbutyl)-**													
a986	**Anabasine** (*l*)	Neonicotine. *l*-2(3-Pyridyl)-piperidine. $C_{10}H_{14}N_2$.	162.24	darkens in air, $[\alpha]_D^{20}$ −83.1 λ^{al} 261 (3.46)	9	276^{760} 105^2	1.0455_{20}^{20}	1.5430^{20}	∞	s	s	...	s	...	**B23²**, 113
a988	**Anagyrine**	Monolupine. Rhombinine. $C_{15}H_{20}N_2O$.	244.34	pa ye glass, $[\alpha]_D^{25}$ −168 (al, c = 4.8)		260–70^{12} 210–5^4			s δ^h	v	s	...	s	chl v dil HCl v lig i	**B24²**, 84
a989	**Analgen**	5-Benzamido-8-ethoxyquinoline. Benzanalgen. Chinalgen. Labordin. Quinolagen.	292.34	ye nd (al)	206				δ	δ s^h	δ	...	...	ac s	**B22**, 503

For explanations, symbols and abbreviations see beginning of table. For **structural formulas** see end of table.

No.	Name	Synonyms and Formula	Mol. wt.	Color, crystalline form, specific rotation and λ_{max} (log ε)	m.p. °C	b.p. °C	Density	n_D	Solubility w	al	eth	ace	bz	other solvents	Ref.
	Anatabine														
a990	**Anatabine (*l*)**	2(3-Pyridyl)-1,2,3,6-tetrahydropyridine. $C_{10}H_{12}N_2$.	160.22	$[\alpha]_D^{17}$ −177.8 λ^{al} 262 (3.53)	...	$145–6^{10}$	1.091_4^{19}	1.5676^{20}	∞	s	s	...	s	...	**C31,** 3055
a991	**Androstane**	Etioallocholane.	260.47	lf (ace-MeOH or chl-ace) $[\alpha]_D^{16}$ +2(chl, c = 1.2)	50–2	$75–80^{0.01}$ (sub $60^{0.003}$)	...	...	i	s	s	s	...	MeOH s chl s lig s	**E14s,** 1396
Ω a992	**Androsterone**	3α-Hydroxy-17-ketoandrostane. $C_{19}H_{30}O_2$.	290.45	lf or nd (al, ace) $[\alpha]_D^{20}$ +94.6 (abs al, c = 0.7) $\lambda^{con\ sulf}$ 305 (3.70), 412 (3.91), 485 (3.10)	185–5.5 (cor)	...	...	...	δ	s	s	s	s	os s	**C46,** 8136
a993	**Anemonin**	Anemone camphor. $C_{10}H_8O_4$.	192.18	rh pl (chl), nd (al or bz)	158	...	...	...	δ s^h	δ s^h	i	...	...	chl s alk s lig δ	**B19²**, 182
—	**Anethole**	*see* **Benzene, 1-methoxy-4-propenyl-***													
—	**Angelic acid**	*see* **2-Butenoic acid, 2-methyl-(*cis*)***													
a994	**Anhalamine**	$C_{11}H_{15}NO_3$	209.25	nd (al)	187–8	...	...	...	s^h	v^h i	δ	v	δ	chl δ^h peth δ	**B21²**, 157
a995	**Anhalonidine**	$C_{12}H_{17}NO_3$	223.28	oct (bz, eth)	160–1	...	...	...	s	s	δ	...	s^h	chl s peth δ	**B21²**, 157
a996	**Anhalonine (*d*)**	$C_{12}H_{15}NO_3$	221.26	$[\alpha]_D^{25}$ +56.7 (chl)	84.5–5.5	$140^{0.02}$	...	...	s	s	v	...	...	chl s peth s	**B27²**, 542
a997	—(*dl*)	$C_{12}H_{15}NO_3$. *See* a996	221.26	rh nd (peth)	85–5.5	...	...	...	s	v	v	...	v	chl v peth s	**B27²**, 542
a998	—(*l*)	$C_{12}H_{15}NO_3$. *See* a996	221.26	nd (peth) $[\alpha]_D^{25}$ −56.3 (chl, c = 4)	86	$140^{0.02}$	...	...	s	v	v	...	...	chl s peth s	**B27²**, 542
a999	—,hydrochloride (*l*)	$C_{12}H_{15}NO_3 \cdot HCl$. *See* a996	257.72	rh pr, $[\alpha]_D^{25}$ −40.5(al)	254–5d	...	...	...	δ v^h	s	δ	...	...	chl δ	**B27²**, 542
a1000	**Anhydroecgonine (*dl*)**	Ecgonidine. $C_9H_{13}NO_2$.	167.20	cr (MeOH, MeOH-eth), cr (w + 1)	226–30d	...	...	...	v	δ	δ	...	δ	chl δ lig δ	**B22²**, 26
a1001	—,hydrochloride	Ecgonidine hydrochloride. $C_9H_{13}NO_2 \cdot HCl$. *See* a1000	203.66	rh nd (al + 1w), ta (w) $[\alpha]_D^{15}$ −62.7(w)	240–1	...	...	...	s^h	s	s	...	...	...	**B22,** 31
Ω a1002	**Aniline**	Aminobenzene. Phenylamine.	93.13	λ^{al} 234.5 (3.90), 284.5 (3.23)	−6.3	184.13^{760} 68.3^{10}	1.02173_4^{20}	1.5863^{20}	s ∞^h	∞	∞	∞	∞	CCl_4 s lig s	**B12²**, 44
Ω a1003	—,benzenesulfonate	$C_6H_5NH_2 \cdot C_6H_5SO_3H$	251.31	nd	240 δd (218)	...	...	...	v	v	i	...	i	CS_2 i	**B12²**, 75
Ω a1004	—,hydrochloride	$C_6H_5NH_2 \cdot HCl$	129.60	lf or nd λ^{HCl} 243 (2.0), 248 (2.1), 254 (2.18), 260 (2.08)	198	245^{760}	1.2215^4	...	v	v	i	...	...	chl i	**B12²**, 65
a1005	—,nitrate	$C_6H_5NH_2 \cdot HNO_3$	156.15	rh	d 190	...	1.356^4	...	v	v	v	...	...	...	**B12²**, 66
a1006	—,oxalate (acid)	$C_6H_5NH_2 \cdot H_2C_2O_4$	183.17	ta (aq al)	150–1	...	...	...	v	v	...	v	...	...	**B12²**, 73
a1007	—, (neutral)	$2(C_6H_5NH_2) \cdot H_2C_2O_4$	276.30	tcl pr (w)	174–5d	...	...	...	v	δ	i	δ	...	...	**B12²**, 73
Ω a1008	—,phosphate	$C_6H_5NH_2 \cdot H_3PO_4$	191.13	nd (al), lf (w-al)	180	...	...	...	v	v	i	...	...	...	**B12,** 117
a1009	—,picrate		322.24	ye or red mcl pr (w), dk gr	181	...	1.558	...	δ	s	...	...	i	...	**B12²**, 72
a1010	—,sulphate(acid)	$C_6H_5NH_2 \cdot H_2SO_4 \cdot \frac{1}{2}H_2O$	200.21	lf(w + ½) $\lambda^{1N\ H_2SO_4}$ 244 (2.08), 256 (2.12), 259 (2.02)	162	...	...	...	...	...	...	...	...	...	**B12²**, 66
Ω a1011	—,—(neutral)	$2(C_6H_5NH_2) \cdot H_2SO_4$	284.34	lf (al)	d	...	1.377^4	...	s	δ	i	...	...	...	**B12²**, 66
a1012	—,**3-acetamido-**	*N*-Acetyl-*m*-phenylenediamine. $C_8H_{10}N_2O$. *See* a1002	150.18	lf(ace-bz), nd or pl (bz)	87–9	d at 100	...	...	v	v	s	v	i	...	**B13²**, 27
a1013	—,**4-acetamido-, sulfate**	*N*-Acetyl-*p*-phenylenediamine sulfate. $C_8H_{10}N_2O \cdot H_2SO_4$.	248.26	nd (eth-al)	285d	...	...	...	v	v	i	...	...	...	**B13,** 95
Ω a1014	—,***N*-allyl-**	$CH_2{:}CHCH_2NHC_6H_5$	133.20	ye	...	$217–9^{736}$ $105–8^{12}$	0.982_4^{25}	1.5630^{20}	δ	s	∞	s	...	...	**B12²**, 96
—	—,***N*-benzyl-**	*see* **Amine, benzyl phenyl**													
Ω a1015	—,***N*-benzylidene-**	Benzalaniline. $C_6H_5CH{:}NC_6H_5$	181.24	pa ye nd (CS_2), pl (dil al) λ^{al} 238 (3.98), 262 (4.24), 306 sh (3.95)	54 (51–2)	310	1.038_4^{55}	1.600^{99}	i	s	s	...	...	liq SO_2, NH_3 s	**B12²**, 113
Ω a1016	—,**2-bromo-**	C_6H_6BrN. *See* a1002	172.03	λ^{cy} 235 (3.93), 289 (3.48)	32 (fr 28.7)	229^{760} 110.5^{19}	1.578_4^{20}	1.6113_4^{20}	i	s	s	...	...	...	**B12²**, 341
Ω a1017	—,**3-bromo-**	C_6H_6BrN. *See* a1002	172.03	λ^{cy} 237.5 (3.89), 290 (3.38)	18.5 (fr 16.7)	251^{760} 130^{12}	$1.5793_4^{20.4}$	$1.6260^{20.4}$	δ	s	s	...	...	...	**B12²**, 342

For explanations, symbols and abbreviations see beginning of table. For structural formulas see end of table.

No.	Name	Synonyms and Formula	Mol. wt.	Color, crystalline form, specific rotation and λ_{max} (log ε)	m.p. °C	b.p. °C	Density	n_D	Solubility w	al	eth	ace	bz	other solvents	Ref.
	Aniline														
Ω a1018	**—,4-bromo-**	C_6H_6BrN. *See* a1002	172.03	rh bipym, nd (60 % al) λ^{al} 245 (4.12), 296.5 (3.20)	66.4	d	1.4970_4^{100}		i	s	s				**B12**^{2}, 344
a1019	**—,2-bromo-4,6-dichloro-**	$C_6H_4BrCl_2N$. *See* a1002	240.92	nd (al)	83.5	273^{760}				v			v	chl v	**B12**^{2}, 355
a1020	**—,4-bromo-2,3-dichloro-**	$C_6H_4BrCl_2N$. *See* a1002	240.92	nd	77.5					v	v		v		**B12**, 653
a1021	**—,4-bromo-2,5-dichloro-**	$C_6H_4BrCl_2N$. *See* a1002	240.92	nd	91					v	v				**B12**, 654
Ω a1022	**—,4-bromo-3,5-dichloro-**	$C_6H_4BrCl_2N$. *See* a1002	240.92	nd (w + al)	129				s^h	v			v	chl v	**B12**, 654
a1023	**—,3-bromo-*N,N*-diethyl-**	$C_{10}H_{14}BrN$. *See* a1002	228.14			142^{10}				s		s			**B12**^{1}, 315
Ω a1024	**—,4-bromo-*N,N*-diethyl-**	$C_{10}H_{14}BrN$. *See* a1002	228.14	nd or pr	33	270			i	v	v				**B12**^{2}, 347
Ω a1025	**—,2-bromo-*N,N*-dimethyl-**	$C_8H_{10}BrN$. *See* a1002	200.09	λ^{hp} 188 (4.6), 216 (4.2), 255 (3.7)		107–8^{14}	1.3880_{25}^{25}	1.5748^{25}		v					**B12**^{1}, 313
a1026	**—,3-bromo-*N,N*-dimethyl-**	$C_8H_{10}BrN$. *See* a1002	200.09	λ^{cy} 254 (4.11), 303 (3.43)	11	259 (cor) 126^{14}				v				aa v con ac s	**B12**^{2}, 343
Ω a1027	**—,4-bromo-*N,N*-dimethyl-**	$C_8H_{10}BrN$. *See* a1002	200.09	lf (al) λ^{cy} 260 (4.29), 310 (3.34)	55 (58)	264^{760}			i	s	v			os s	**B12**^{2}, 345
a1028	**—,2-bromo-3,5-dinitro-**	$C_6H_4BrN_3O_4$. *See* a1002	262.03	ye lf (al)	181					δ v^h				HNO_3 s	**B12**, 762
a1029	**—,2-bromo-4,5-dinitro-**	$C_6H_4BrN_3O_4$. *See* a1002	262.03	pa ye (al)	186					v^h					**B12**, 762
Ω a1030	**—,2-bromo-4,6-dinitro-**	$C_6H_4BrN_3O_4$. *See* a1002	262.03	ye nd (aa or al)	153–4	sub				v^h		v^h		aa s^h	**B12**^{2}, 418
a1030^1	**—,4-bromo-2,5-dinitro-**	$C_6H_4BrN_3O_4$. *See* a1002	262.03	ye cr (al)	186					v^h					**B12**, 761
a1031	**—,4-bromo-2,6-dinitro-**	$C_6H_4BrN_3O_4$. *See* a1002	262.03	og-red pl (abs al)	163 (cor) (160)					v^h					**B12**^{2}, 418
a1032	**—,5-bromo-2,4-dinitro-**	$C_6H_4BrN_3O_4$. *See* a1002	262.03	pa ye nd (dil al)	178.4					v	v		v	chl v	**B12**^{2}, 417
a1033	**—,6-bromo-2,3-dinitro-**	$C_6H_4BrN_3O_4$. *See* a1002	262.03	dk red cr (al)	158					v^h					**B12**, 760
Ω a1034	**—,2-bromo-4-nitro-**	$C_6H_5BrN_2O_2$. *See* a1002	217.03	ye nd (al)	104.5					s				aa v	**B12**^{2}, 403
Ω a1035	**—,2-bromo-5-nitro-**	$C_6H_5BrN_2O_2$. *See* a1002	217.03	pa ye nd (al)	141					s	v		v	chl v lig δ	**B12**^{2}, 402
a1036	**—,3-bromo-4-nitro-**	$C_6H_5BrN_2O_2$. *See* a1002	217.03	ye nd (al)	175–6					v			i	chl, lig i	**B12**^{2}, 403
a1037	**—,4-bromo-2-nitro-**	$C_6H_5BrN_2O_2$. *See* a1002	217.03	og-ye nd (w)	111.5	sub			δ s^h	v					**B12**^{2}, 401
a1038	**—,4-bromo-3-nitro-**	$C_6H_5BrN_2O_2$. *See* a1002	217.03	nd (al)	132					s	s			chl, aa s	**B12**^{2}, 402
a1039	**—,5-bromo-2-nitro-**	$C_6H_5BrN_2O_2$. *See* a1002	217.03	red-ye nd (dil al)	151–2					v				aa v	**B12**^{2}, 402
a1040	**—,6-bromo-2-nitro-**	$C_6H_5BrN_2O_2$. *See* a1002	217.03	og or ye nd (dil al)	74–5		1.988			s					**B12**^{2}, 402
Ω a1041	**—,*N*-butyl-**	*N*-Phenylbutylamine. $C_6H_5NHCH_2CH_2CH_2CH_3$	149.24	λ^{iso} 245 (4.16), 295 (3.36)	−14.4	241.6^{760} 118–20^{15}	0.93226_4^{20}	1.53412^{20}	δ	v	s				**B12**^{2}, 95
Ω a1042	**—,*N-sec*-butyl-**	$C_6H_5NHCH(CH_3)CH_2CH_3$	149.24			225^{759} 112–4^{22}		1.5333^{20}						os s	**B12**^{2}, 96
a1043	**—,2-chloro-(α)**	C_6H_6ClN. *See* a1002	127.57	λ^{al} 237 (3.8), 291 (3.3)	−14	208.84	1.21253_4^{20}	1.58951^{20}	i	∞	s	s		os s	**B12**^{2}, 314
a1044	**—,—(β)**	C_6H_6ClN. *See* a1002	127.57	λ^{al} 236 (3.8), 291 (3.3)	−1.9	208.84^{760} 84.6^{10}	1.21266_4^{20}	1.5889^{20}	i	∞	s	s	∞	CCl_4 ∞ os s	**B12**^{2}, 314
a1045	**—,—,hydrochloride**	$C_6H_6ClN.HCl$. *See* a1002	164.04	pl (w, aq al)	235		1.505^{18}		s	δ					**B12**^{2}, 315
Ω a1046	**—,3-chloro-**	C_6H_6ClN. *See* a1002	127.57	λ^{al} 240 (3.8), 292 (3.3)	−10.3	229.92^{760} 101.2^{10}	1.21606_4^{20}	1.59414^{20}	i	∞	∞	∞	∞	CCl_4 ∞ os s	**B12**^{2}, 319
Ω a1047	**—,—,hydrochloride**	$C_6H_6ClN.HCl$. *See* a1002	164.04	pl	221.5–2.5				v	v					**B12**^{2}, 320
Ω a1048	**—,4-chloro-**	C_6H_6ClN. *See* a1002	127.57	rh pr λ^{al} 242 (3.9), 295 (3.2)	72.5	232^{760}	1.429_4^{19}	1.5546^{87}	s^h	s	s			os s	**B12**^{2}, 322
Ω a1049	**—,4-chloro-*N,N*-diethyl-**	$C_{10}H_{14}ClN$. *See* a1002	183.68	nd (al)	45.5–6.5	251–3 95–$6^{1.5}$				s^h					**B12**^{2}, 324
a1050	**—,4-chloro-*N,N*-dimethyl-**	$C_8H_{10}ClN$. *See* a1002	155.63	nd (al) λ^{cy} 257 (4.17), 308 (3.30)	35.5	231 (233–6)				s^h					**B12**^{1}, 304
a1051	**—,4-chloro-*N,N*-dimethyl-3-nitro-**	$C_8H_9ClN_2O_2$. *See* a1002	200.63	ye nd (dil al)	81.5–2.5					s				lig s	**B12**, 732
a1052	**—,3-chloro-2,6-dinitro-**	$C_6H_4ClN_3O_4$. *See* a1002	217.57	og-ye nd (al)	112					s^h					**B12**^{2}, 416
Ω a1053	**—,4-chloro-2,6-dinitro-**	$C_6H_4ClN_3O_4$. *See* a1002	217.57	og-ye nd (al)	147 (cor)					s^h					**B12**^{2}, 416

For explanations, symbols and abbreviations see beginning of table. For structural formulas see end of table.

No.	Name	Synonyms and Formula	Mol. wt.	Color, crystalline form, specific rotation and λ_{max} (log ε)	m.p. °C	b.p. °C	Density	n_D	Solubility w	al	eth	ace	bz	other solvents	Ref.
	Aniline														
a1054	—,5-chloro-2-ethoxy-	5-Chloro-*o*-phenetidine. $C_8H_{10}ClNO$. *See* a1002	171.63	nd (dil al)	42	...	...	...	δ	s	...	...	s	chl s	**B13**, 383
a1055	—,2-chloro-*N*-ethyl-	$C_8H_{10}ClN$. *See* a1002 ...	155.63	...	...	219^{726}	1.104_4^{25}	...	...	...	...	...	...	...	**B12²**, 316
a1056	—,4-chloro-2-methoxy-	4-Chloro-*o*-anisidine. C_7H_8ClNO. *See* a1002	157.61	nd or pr (dil al)	52	260^{760}	...	...	...	s	s	...	s	...	**B13²**, 184
Ω a1057	—,5-chloro-2-methoxy-	5-Chloro-*o*-anisidine. C_7H_8ClNO. *See* a1002	157.61	nd (dil al)	84	...	...	...	...	s	...	...	...	lig δ	**B13²**, 183
Ω a1058	—,2-chloro-4-nitro-	$C_6H_5ClN_2O_2$. *See* a1002 ...	172.57	ye nd (lig-CS_2, w, 20 % aa)	108	...	...	...	δ	v	v	...	...	CS_2 v lig i	**B12²**, 398
Ω a1059	—,3-chloro-4-nitro-	$C_6H_5ClN_2O_2$. *See* a1002 ...	172.57	ye lf (bz) λ^{al} 233 (3.79), 374 (4.13)	156–7	...	...	...	...	s	s	...	...	aa s lig, CS_2 i	**B12²**, 398
Ω a1060	—,4-chloro-2-nitro-	$C_6H_5ClN_2O_2$. *See* a1002 ...	172.57	dk og-ye pr (dil al), nd (lig, w) λ^{aa} 410 (3.7)	116–7	...	...	...	...	v	v	...	...	aa v lig δ	**B12²**, 396
Ω a1061	—,4-chloro-3-nitro-	$C_6H_5ClN_2O_2$. *See* a1002 ...	172.57	ye nd or pr (w), nd (peth)	103	...	...	...	s^h	v	s	...	...	chl s lig δ	**B12²**, 397
a1062	—,5-chloro-2-nitro-	$C_6H_5ClN_2O_2$. *See* a1002 ...	172.57	gold-ye nd (CS_2), ye lf (al, bz)	126.5	sub	...	...	...	s	s	...	...	lig CS_2 δ dil aa s	**B12²**, 397
a1063	—,5-chloro-3-nitro-	$C_6H_5ClN_2O_2$. *See* a1002 ...	172.57	og-ye nd (al)	133–4	...	...	...	...	s^h	s	...	...	aa s	**B12**, 732
a1064	—,6-chloro-2-nitro-	$C_6H_5ClN_2O_2$. *See* a1002 ...	172.57	ye nd (dil al) λ^{cy} 231.5 (4.22), 275 (3.74), 382 (3.76)	76	...	...	...	...	s	...	...	...	50 % al s	**B12²**, 397
a1065	—,6-chloro-3-nitro-	$C_6H_5ClN_2O_2$. *See* a1002 ...	172.57	ye nd (lig)	121	...	...	...	...	s	s	s	...	aa v CS_2, lig δ	**B12²**, 398
a1066	—,*N*-cyclohexyl- ..	Cyclohexylphenylamine.	175.28	mcl pr	16	279^{764} $134–5^{6}$	1.0155_4^{20}	1.5610^{20}	i	s	s	...	s	...	**B12²**, 98
a1067	—,2,3-dibromo- ...	$C_6H_5Br_2N$. *See* a1002 ...	250.93	pl (dil al)	43	...	...	...	δ	s	s	...	...	aa v AcOEt s	**B12**, 655
Ω a1068	—,2,4-dibromo- ...	$C_6H_5Br_2N$. *See* a1002 ...	250.93	rh bipym (chl), nd or lf (dil al)	79.5–80.5	156^{24}	2.260^{20}	...	...	s	s	...	...	aa s chl s^h	**B12²**, 356
Ω a1069	—,2,5-dibromo- ...	$C_6H_5Br_2N$. *See* a1002 ...	250.93	pr (al)	53–5	...	...	...	...	v	s	...	...	...	**B12²**, 357
Ω a1070	—,2,6-dibromo- ...	$C_6H_5Br_2N$. *See* a1002 ...	250.93	nd (al) λ^{w} 293 (3.45)	87–8 (83–4)	262–4	...	...	...	v	v	...	v	chl v	**B12²**, 357
a1071	—,3,4-dibromo- ...	$C_6H_5Br_2N$. *See* a1002 ...	250.93	lf (dil al)	81	100 sub	...	...	...	s^h	...	...	...	...	**B12¹**, 329
a1072	—,3,5-dibromo- ...	$C_6H_5Br_2N$. *See* a1002 ...	250.93	nd (dil al) λ^{al} 212 (4.5), 250 (4.0), 305 (3.5)	57	...	...	...	...	v	v	...	v	...	**B12²**, 357
a1073	—,2,4-dibromo-6-nitro-	$C_6H_4Br_2N_2O_2$. *See* a1002 ...	295.93	ye cr	128	...	...	...	...	...	...	...	...	...	**B12²**, 403
Ω a1074	—,2,6-dibromo-4-nitro-	$C_6H_4Br_2N_2O_2$. *See* a1002 ...	295.93	ye nd (al or aa)	207	...	...	...	δ	...	...	...	...	aa s	**B12²**, 404
Ω a1075	—,*N*,*N*-dibutyl- ...	$C_6H_5N(CH_2CH_2CH_2CH_3)_2$	205.34	λ^{iso} 259 (4.25), 304 (3.41)	−32.20	274.8^{760} 138.8^{10}	0.9037_4^{20}	1.5186^{20}	i	∞	∞	v	v	ac s	**B12²**, 95
Ω a1076	—,2,3-dichloro- ...	$C_6H_5Cl_2N$. *See* a1002 ...	162.02	nd (lig)	24	252	...	...	...	s	v	s	δ	lig δ	**B12**, 621
a1077	—,2,4-dichloro- ...	$C_6H_5Cl_2N$. *See* a1002 ...	162.02	pr (ace), nd (dil al), (lig)	63–4	245^{759}	1.567^{20}	...	δ	s	s	...	...	...	**B12²**, 333
Ω a1078	—,2,5-dichloro- ...	$C_6H_5Cl_2N$. *See* a1002 ...	162.02	nd (lig) λ^{w} 293 (2.45)	50	251	...	...	δ	s	s	...	s	CS_2 s	**B12²**, 336
Ω a1079	—,3,4-dichloro- ...	$C_6H_5Cl_2N$. *See* a1002 ...	162.02	nd (lig)	72	272 145^{15}	...	...	...	s	s	...	δ	...	**B12²**, 337
Ω a1080	—,3,5-dichloro- ...	$C_6H_5Cl_2N$. *See* a1002 ...	162.02	nd (lig, dil al)	51–3	260^{741}	...	...	i	s	s	...	...	lig s^h	**B12²**, 337
a1081	—,2,6-dichloro-4-ethoxy-	4-Amino-3,5-dichloro-phenetole. $C_8H_9Cl_2NO$. *See* a1002	206.07	nd (dil al)	46	275	...	...	δ	v	v	...	v	chl v	**B13²**, 276
a1082	—,3,5-dichloro-4-ethoxy-	4-Amino-2,6-dichloro-phenetole. $C_8H_9Cl_2NO$. *See* a1002	206.07	nd (peth)	105–7	...	...	...	...	v	v	...	v	aa δ	**B13²**, 275
a1083	—,*N*-dichloro-methylene-	Phenyliminophosgene. Phenylisocyanide dichloride. $C_6H_5N{:}CCl_2$	174.04	oil	19.5	210^{760} 96^{19}	1.285^{15}	...	...	...	...	...	...	...	**B12²**, 245
Ω a1084	—,2,6-dichloro-4-nitro-	Dichloran. $C_6H_4Cl_2N_2O_2$. *See* a1002	207.02	ye nd (al, aa) λ^{hp} 240 (3.90), 322 (4.16)	191	...	...	...	...	s	...	...	...	ac s	**B12²**, 400
Ω a1085	—,*N*,*N*-diethyl- ...	Diethylphenylamine. $C_6H_5N(C_2H_5)_2$	149.24	ye oil λ^{iso} 259 (4.22), 303 (3.38)	−38.8	216.27 92^{10}	0.93507_4^{20}	1.5409^{20}	δ	s	v	s	...	chl v	**B12²**, 92
a1086	—,*N*,*N*-diethyl-2-ethoxy-	*N*,*N*-Diethyl-*o*-phenetidine. $C_{12}H_{19}NO$. *See* a1002	193.29	...	...	231–3	...	...	i	∞	∞	...	∞	chl, CS_2 ∞	**B13**, 365
a1087	—,*N*,*N*-diethyl-3-ethoxy-	*N*,*N*-Diethyl-*m*-phenetidine. $C_{12}H_{19}NO$. *See* a1002	193.29	λ^{al} 236 (3.97), 400 (4.33)	...	268–70 (288) 145^{14}	...	1.5325^{25}	...	s	...	...	s	aa s	**B13¹**, 131

For explanations, symbols and abbreviations see beginning of table. For structural formulas see end of table.

No.	Name	Synonyms and Formula	Mol. wt.	Color, crystalline form, specific rotation and λ_{max} (log ε)	m.p. °C	b.p. °C	Density	n_D	Solubility w	al	eth	ace	bz	other solvents	Ref.
	Aniline														
a1088	**—,*N,N*-diethyl-3-nitro-**	$C_{10}H_{14}N_2O_2$. *See* a1002	194.24	ye	....	288–90	....	....	i	...	...	...	...	....	**B12**[1], 346
a1089	**—,*N,N*-diethyl-4-nitro-**	$C_{10}H_{14}N_2O_2$. *See* a1002	194.24	ye nd (lig), pl (al) λ^{al} 236 (3.93), 316 (3.26), 393.5 (4.34)	77–8	....	1.225	....	...	s^h	...	...	...	lig δ	**B12**[1], 351
Ω a1090	**—,*N,N*-diethyl-4-nitroso-**	$C_{10}H_{14}N_2O$. *See* a1002	178.24	gr mcl pr (eth), gr lf (ace) λ^{al} 236.5 (3.50), 275 (3.73), 425 (4.51)	87–8	....	1.24_4^{15}	....	δ	s	s	s^h	...	....	**B12**[2], 365
a1091	**—,2,4-diiodo-**	$C_6H_5I_2N$. *See* a1002	344.92	br nd or rh cr (al)	95–6	....	2.748	....	δ s^h	s v^h	v	v	s	chl, CS_2 v lig s	**B12**, 675
a1092	**—,2,5-diiodo-**	$C_6H_5I_2N$. *See* a1002	344.92	nd (al)	88–9	....	....	....	...	s	v	s	...	os s	**B12**, 675
a1093	**—,2,6-diiodo-**	$C_6H_5I_2N$. *See* a1002	344.92	nd (al)	122	....	....	....	...	s	s	s	s	os s	**B12**, 675
a1094	**—,3,4-diiodo-**	$C_6H_5I_2N$. *See* a1002	344.92	pa ye lf or pr (bz-peth)	74.5	....	....	....	...	s	s	...	s	peth δ	**B12**, 675
a1095	**—,3,5-diiodo-**	$C_6H_5I_2N$. *See* a1002	344.92	nd (al)	110 (107)	....	....	....	...	s	s	...	...	chl s	**B12**[1], 377
a1096	**—,2,4-diiodo-3-nitro-**	$C_6H_4I_2N_2O_2$. *See* a1002	389.92	pa ye mcl pl	125	....	....	....	...	v	...	...	...	os v lig δ	**B12**, 747
a1097	**—,2,6-diiodo-3-nitro-**	$C_6H_4I_2N_2O_2$. *See* a1002	389.92	ye nd (dil al)	145.5 (149)	....	....	....	...	v	...	...	...	os s^h AcOEt δ	**B12**, 747
Ω a1098	**—,2,6-diiodo-4-nitro-**	$C_6H_4I_2N_2O_2$. *See* a1002	389.92	pa ye lf or nd (bz)	245 (248)	....	....	....	...	δ^h	...	...	s^h	....	**B12**[2], 405
a1099	**—,4,6-diiodo-2-nitro-**	$C_6H_4I_2N_2O_2$. *See* a1002	389.92	ye nd (ace)	154	....	....	....	δ	δ	s	s^h	s	chl s, aa s	**B12**[1], 361
a1100	**—,4,6-diiodo-3-nitro-**	$C_6H_4I_2N_2O_2$. *See* a1002	389.92	pa ye nd (eth-al)	149	....	....	....	...	δ^h	s	...	...	....	**B12**, 747
a1101	**—,2,3-dimethoxy-**	3-Aminoveratrole. $C_8H_{11}NO_2$. *See* a1002	153.18	....	....	137^{15}	....	....	s	...	...	...	...	....	**B13**[2], 464
Ω a1102	**—,2,4-dimethoxy-**	4-Aminoresorcinol dimethyl ether. $C_8H_{11}NO_2$. *See* a1002	153.18	pl (lig)	33.5 (cor) (39–40)	....	....	....	δ	s	s	...	s	lig s^h	**B13**[2], 470
a1103	**—,2,6-dimethoxy-**	2-Aminoresorcinol dimethyl ether. $C_8H_{11}NO_2$. *See* a1002	153.18	pl (al), lf (peth)	75	146^{23}	....	....	δ	v	v	...	v	aa v lig s	**B13**[2], 468
Ω a1104	**—,3,4-dimethoxy-**	4-Aminoveratrole. $C_8H_{11}NO_2$. *See* a1002	153.18	lf (eth)	87–8 (90)	....	....	....	...	...	s^h	...	...	....	**B13**[2], 465
Ω a1105	**—,*N,N*-dimethyl-**	$C_6H_5N(CH_3)_2$	121.18	pa ye λ^{al} 235 (4.1), 251 (4.11), 293 (3.2)	2.45 (fp 1.96)	194.15^{760} 77^{13}	0.9557_4^{20}	1.5582^{20}	δ	s	s	s	s	chl v os s	**B12**[2], 82
a1106	**—,—,hydrochloride**	$C_6H_5N(CH_3)_2$. HCl	157.65	hyg pl (w), (bz)	85–95	....	1.1156_4^{19}	....	s	s	i	...	δ	chl s	**B12**[2], 86
a1107	**—,—,*N,N*-dimethyl-, *N*-oxide**	$C_6H_5N(:O)(CH_3)_2$	137.18	pr λ^{al} 248 (2.4), 282 sh (0.8)	152–3	....	....	....	v	v	δ	...	...	chl v	**B12**[2], 88
a1108	**—,*N,N*-dimethyl-2-nitro-**	$C_8H_{10}N_2O_2$. *See* a1002	166.18	ye-og λ^{al} 246 (4.1), 273 sh (3.4), 429 (3.3)	fp-20	146^{20}	1.1794_4^{20}	1.6102^{20}	s	v^h	s	...	...	chl v	**B12**[2], 369
Ω a1109	**—,*N,N*-dimethyl-3-nitro-**	$C_8H_{10}N_2O_2$. *See* a1002	166.18	og-ye or red mcl pr (eth or eth-al) λ^{al} 249 (4.3), 405 (3.2)	60–1	$280–5^{760}$	1.313^{17}	....	i	s	s	...	...	....	**B12**[2], 377
Ω a1110	**—,*N,N*-dimethyl-4-nitro-**	$C_8H_{10}N_2O_2$. *See* a1002	166.18	ye nd (al) λ^{al} 232 (3.96), 314 (3.27), 392 (4.30)	164.5	....	....	....	i	s^h	s	...	...	aa s^h	**B12**[2], 386
Ω a1111	**—,*N,N*-dimethyl-4-nitroso-**	Accelerene. $C_8H_{10}N_2O$. *See* a1002	150.18	gr pl (eth) λ^{al} 234 (3.67), 273 (3.82), 305 (3.18), 314 (3.14)	92.5–3.5	....	1.145^{20}	....	δ	s	s	...	...	$HCONH_2$ s	**B12**[2], 364
a1112	**—,4-dimethyl-amino-*N*(2,4,6-trinitrobenzylidene)-**	....	359.30	bk gr lf ($PhCO_2Et$), nd($PhNO_2$ +1)	268 exp	....	....	....	i	...	...	δ	δ	chl, AcOEt, aa δ	**B13**, 85
Ω a1113	**—,2,4-dinitro-**	$C_6H_5N_3O_4$. *See* a1002	183.12	ye nd (dil ace), gr-ye ta (al) λ^{aa} 257 (3.79), 335 (2.94)	180 (188)	....	1.615^{14}	....	i δ^h	δ	...	...	...	HCl s^h	**B12**[2], 405
Ω a1114	**—,2,6-dinitro-**	$C_6H_5N_3O_4$. *See* a1002	183.12	gold lf (50 % aa), ye nd (al) λ^{cy} 223 (4.43), 249 (4.15), 307 (2.98), 408 (3.99)	141–2	....	....	....	i	δ	s	...	s^h	lig i	**B12**[2], 413
Ω a1115	**—,*N,N*-dipropyl-**	$C_6H_5N(CH_2CH_2CH_3)_2$	177.29	lt ye	....	245^{760} 127^{10}	0.9104^{20}	1.5271^{20}	i	s	s	s	s	....	**B12**[2], 95
Ω a1116	**—,2-ethoxy-**	*o*-Aminophenetole. *o*-Phenetidine. $C_8H_{11}NO$. *See* a1002	137.18	....	< −21	232.5^{760} $127–8^{14}$	....	1.5560^{20}	δ	s	s	...	...	....	**B13**[2], 166

For explanations, symbols and abbreviations see beginning of table. For **structural formulas** see end of table.

No.	Name	Synonyms and Formula	Mol. wt.	Color, crystalline form, specific rotation and λ_{max} (log ε)	m.p. °C	b.p. °C	Density	n_D	Solubility w	al	eth	ace	bz	other solvents	Ref.
	Aniline														
a1117	**—,3-ethoxy-**	*m*-Aminophenetole. *m*-Phenetidine. $C_8H_{11}NO$. *See* a1002	137.18	...	...	248^{760} $127–8^{11}$	...	...	δ	s	s	...	...	...	**B13²**, 211
Ω a1118	**—,4-ethoxy-**	*p*-Aminophenetole. *p*-Phenetidine. $C_8H_{11}NO$. *See* a1002	137.18	...	2.4	254^{760} 125^{12}	1.0652_4^{16}	1.5528^{20}	δ	s	s	...	...	...	**B13²**, 224
a1119	**—,4-ethoxy-*N*-(2-hydroxybenzylidene)-**	*N*-Salicylidene-*p*-phenetidine. Malakin.	241.29	pa ye or grsh pl or nd (al)	94 (92)	...	...	...	i	s^h	s	...	s	...	**B13²**, 241
a1120	**—,2-ethoxy-4-nitro-**	4-Nitro-*o*-phenetidine. $C_8H_{10}N_2O_3$. *See* a1002	182.18	ye nd (dil al)	91	...	...	...	δ	s	s	s	...	lig δ	**B13**, 390
a1121	**—,2-ethoxy-5-nitro-**	5-Nitro-*o*-phenetidine. $C_8H_{10}N_2O_3$. *See* a1002	182.18	ye nd (dil al)	96–7	$205–6^{14}$	...	...	δ	s	s	...	...	...	**B13²**, 192
a1122	**—,2-ethoxy-6-nitro-**	6-Nitro-*o*-phenetidine. $C_8H_{10}N_2O_3$. *See* a1002	182.18	ye or og cr (w)	60	...	...	...	s^h	...	...	...	...	...	**B13**, 388
a1123	**—,3-ethoxy-4-nitro-**	4-Nitro-*m*-phenetidine. $C_8H_{10}N_2O_3$. *See* a1002	182.18	nd (dil al)	122–3	...	...	...	δ	s	s	s	s	os s lig i	**B13¹**, 137
a1124	**—,4-ethoxy-2-nitro-**	2-Nitro-*p*-phenetidine. $C_8H_{10}N_2O_3$. *See* a1002	182.18	red pr (al)	113	...	...	...	...	δ s^h	v	...	...	chl v	**B13²**, 286
a1125	**—,4-ethoxy-3-nitro-**	3-Nitro-*p*-phenetidine. $C_8H_{10}N_2O_3$. *See* a1002	182.18	og-ye nd (bz, to-lig, dil al)	41	...	...	...	$δ^h$	...	...	s	s	os v lig i	**B13²**, 284
a1126	**—,5-ethoxy-2-nitro-**	6-Nitro-*m*-phenetidine. $C_8H_{10}N_2O_3$. *See* a1002	182.18	ye nd (dil al)	105–6	...	...	...	i	s^h	...	s	s	os s	**B13¹**, 136
a1127	**—,5-ethoxy-3-nitro-**	5-Nitro-*m*-phenetidine. $C_8H_{10}N_2O_3$. *See* a1002	182.18	ye, og-red nd (al)	115	...	...	...	δ v^h	v	...	s	s	os s	**B13**, 422
Ω a1128	**—,*N*-ethyl-**	Ethyl phenylamine. $C_6H_5NHC_2H_5$	121.18	λ^{iso} 245 (4.12), 294 (3.35)	−63.5	204.72^{760} $97.5–8^{18}$	0.9625_4^{20}	1.5559^{20}	i	∞	∞	v	v	os s	**B12²**, 90
Ω a1129	**—,—,hydrochloride**	$C_6H_5NHC_2H_5 \cdot HCl$	157.65	nd	178.5 (176)	...	1.0085_4^{182}	...	s	s	i	...	...	chl s	**B12²**, 91
a1130	**—,*N*-ethyl-*N*-methyl-**	$C_6H_5N(CH_3)C_2H_5$	135.21	...	...	$203–5^{760}$ $93–5^{12}$	0.9193_4^{55}	...	i	∞	∞	...	...	...	**B12²**, 91
a1131	**—,*N*-ethyl-*N*-nitroso-**	$C_6H_5N(NO)C_2H_5$	150.18	yesh λ^{al} 272 (3.80)	...	130^{20}	1.0874_4^{20}	1.55977^{20}	i	...	...	...	...	aa s	**B12**, 580
Ω a1132	**—,2-fluoro-**	C_6H_6FN. *See* a1002	111.12	pa ye λ^{cy} 234 (4.00), 283 (3.26)	−28.5	174.5–6^{757} 58^{11}	1.1513^{21}	1.5421^{20}	i	s	s	...	...	...	**B12²**, 314
Ω a1133	**—,3-fluoro-**	C_6H_6FN. *See* a1002	111.12	pa ye	...	$187–9^{760}$ 82.3^{18}	1.1561^{19}	1.5436^{20}	δ	s	s	...	...	...	**B12²**, 314
Ω a1134	**—,4-fluoro-**	C_6H_6FN. *See* a1002	111.12	pa ye λ^{cy} 230 (3.85), 293 (3.34), 296 (3.34), 300 sh (3.32)	−0.8	180.5–2.5^{757} 85^{19}	1.1725_4^{20}	1.5195_4^{20}	δ	s	s	...	...	...	**B12²**, 314
Ω a1135	**—,*N*-formyl-*N*-methyl-**	*N*-Methylformanilide. $C_6H_5N(CHO)CH_3$	135.17	λ^{hp} 240 (4.00)	14–5	243^{760} 131^{22}	1.0948_4^{20}	1.5589^{20}	δ	s	...	s	...	...	**B12²**, 136
—	**—,*N*(2-hydroxyethyl)-*N*-methyl-**	*see* **Ethanol, 2[methyl-(phenyl)amino]-***													
—	**—,p,p′-iminodi-**	*see* **Amine, diphenyl, diamino-**													
Ω a1136	**—,2-iodo-**	C_6H_6IN. *See* a1002	219.03	nd (dil al) λ^{cy} 237 (3.92), 292 (3.51)	60–1	...	...	...	δ	v	v	s	...	os s	**B12²**, 360
Ω a1137	**—,3-iodo-**	C_6H_6IN. *See* a1002	219.03	lf or nd λ^{cy} 232 sh (4.02), 239 (3.94), 292 (3.43), 299 (3.34)	33	280 $145–6^{15}$	...	1.6811^{20}	i	s	...	...	...	chl s	**B12²**, 360
Ω a1138	**—,4-iodo-**	C_6H_6IN. *See* a1002	219.03	nd (w) λ^{al} 250 (4.27)	67–8	...	...	...	δ	s	s	...	...	os s peth δ	**B12²**, 361
a1139	**—,*N*-isobutyl-**	$C_6H_5NHCH_2CH(CH_3)_2$	149.24	$\lambda^{dil\ al-HCl}$ 248 (2.29), 254 (2.34), 260 (2.20)	...	$231–2^{760}$ (225–7) 109–10^{13}	0.9400_4^{18}	1.5328^{20}	i	...	v	...	v	...	**B12²**, 96
Ω a1140	**—,*N*-isopropyl-**	$C_6H_5NHCH(CH_3)_2$	135.21	$\lambda^{dil\ al-HCl}$ 248 (2.19), 254 (2.27), 260 (2.13)	...	$203–4^{760}$ $72–5^{14}$	...	1.5380^{20}	...	s	s	s	s	...	**B12²**, 95
Ω a1141	**—,2-mercapto-**	2-Aminothiophenol. C_6H_7NS. *See* a1002	125.19	nd	26	234 $125–7^{6}$	...	1.4606^{20}	...	s	...	...	...	...	**B13²**, 198
a1142	**—,3-mercapto-**	3-Aminothiophenol. C_6H_7NS. *See* a1002	125.19	...	...	$180–90^{16}$	...	...	...	v	i	...	...	aa s peth i	**B13¹**, 140
a1143	**—,—,hydrochloride**	$C_6H_7NS \cdot HCl$. *See* a1002	161.66	...	232	sub	...	...	v	v	i	...	...	...	**B13**, 425
a1144	**—,4-mercapto-**	4-Aminothiophenol. C_6H_7NS. *See* a1002	125.19	cr	46	$140–5^{15}$	...	...	s	s	...	...	...	...	**B13²**, 296
a1145	**—,2-methoxy-**	*o*-Anisidine. C_7H_9NO. *See* a1002	123.16	λ^{al} 236 (3.87), 286 (3.47)	6.22	224^{760} 90^{4}	1.0923_4^{20}	1.5713^{20}	δ	s	s	s	s	...	**B13²**, 165
a1146	**—,3-methoxy-**	*m*-Anisidine. C_7H_9NO. *See* a1002	123.16	λ^{al} 236 (3.89), 286 (3.40)	−1–1	251 (cor)	1.096_4^{20}	1.5794^{20}	δ	s	s	s	s	...	**B13²**, 211

For explanations, symbols and abbreviations see beginning of table. For structural formulas see end of table.

No.	Name	Synonyms and Formula	Mol. wt.	Color, crystalline form, specific rotation and λ_{max} (log ε)	m.p. °C	b.p. °C	Density	n_D	Solubility w	al	eth	ace	bz	other solvents	Ref.
	Aniline														
a1147	**—,4-methoxy-**.....	*p*-Anisidine. C_7H_9NO. *See* a1002	123.16	ta (w), rh pl λ^{al} 235 (3.97), 300 (3.35)	57.2	243 (cor) 115^{13}	1.071_4^{57}	1.5559^{67}	s	v	v	s	s	....	**B13**^{2}, 223
a1148	**—,—,hydrochloride**	$C_7H_9NO.HCl$. *See* a1002...	159.62	lf, nd	236	....	....	....	s	s	...	...	...	....	**B13**^{2}, 223
a1149	**—,2-methoxy-4-nitro-**	4-Nitro-*o*-anisidine. $C_7H_8N_2O_3$. *See* a1002	168.16	pa ye nd (dil al)	139–40	....	1.2112^{156}	....	...	v	...	v	...	....	**B13**^{2}, 194
a1150	**—,2-methoxy-5-nitro-**	5-Nitro-*o*-anisidine. $C_7H_8N_2O_3$. *See* a1002	168.16	og-red nd (al, eth, w) $\lambda^{w,pH=7}$ 219 (3.90), 257 (3.78), 310 sh (3.38), 400 (4.09)	118	....	1.2068^{156}	....	s^h	v	s^h	v	v^h	AcOEt, aa v lig δ	**B13**^{2}, 192
a1151	**—,2-methoxy-6-nitro-**	6-Nitro-*o*-anisidine. $C_7H_8N_2O_3$. *See* a1002	168.16	ye, pa red nd (al) λ^{cy} 236 (4.15), 284 (3.70), 393 (3.66)	76	....	....	....	...	s^h	...	s	...	....	**B13**^{2}, 191
a1152	**—,3-methoxy-2-nitro-**	2-Nitro-*m*-anisidine. $C_7H_8N_2O_3$. *See* a1002	168.16	ye nd (bz) λ^{cy} 233 (4.02), 290 (3.37), 355.5 (3.37)	124–4.5 (143)	....	....	....	...	s	...	s	...	....	**B13**^{1}, 136
a1153	**—,3-methoxy-4-nitro-**	4-Nitro-*m*-anisidine. $C_7H_8N_2O_3$. *See* a1002	168.16	ye nd (al)	169	sub	....	....	...	s	...	s	...	aa s	**B13**^{1}, 136
a1154	**—,3-methoxy-5-nitro-**	5-Nitro-*m*-anisidine. $C_7H_8N_2O_3$. *See* a1002	168.16	og cr (w)	120	....	1.2034^{156}	....	s^h	s	...	s	s	lig i	**B13**^{2}, 216
a1155	**—,4-methoxy-2-nitro-**	2-Nitro-*p*-anisidine. $C_7H_8N_2O_3$. *See* a1002	168.16	dk red pr (w or al)	129	....	....	....	s	s	s	s	δ	....	**B13**^{2}, 286
a1156	**—,4-methoxy-3-nitro-**	3-Nitro-*p*-anisidine. $C_7H_8N_2O_3$. *See* a1002	168.16	red (eth), og pr or pl (eth-lig) $\lambda^{w,pH=11}$ 230.5 (4.30), 259 sh (3.76), 285 (3.72), 448 (3.67)	57–7.5	....	....	....	s^h	s	s	s	s	to δ	**B13**^{2}, 284
Ω a1157	**—,*N*-methyl-**......	Methyl phenylamine. $C_6H_5NHCH_3$	107.16	λ^{al} 240 (4.0), 294 (3.18)	−57	196.25^{760} 86^{15}	0.98912_4^{20}	1.5684^{20}	i	s	s	...	...	chl s	**B12**^{2}, 79
Ω a1158	**—,—,hydrochloride**	$C_6H_5NHCH_3.HCl$	143.62	nd(chl-eth)	122.5–3 (cor)	....	1.0660_4^{131}	....	v	s	i	...	i	chl v	**B12**^{2}, 81
a1159	**—,*N*-3-methyl-butyl-**	*N*-Isoamylaniline. $C_6H_5NHCH_2CH_2CH(CH_3)_2$	163.26	....	....	254.5^{760} $126–7^{14}$	0.8912_4^{55}	1.5305^{20}	i	∞	∞	...	...	....	**B12**^{2}, 96
a1160	**—,*N*-methyl-2-nitro-**	$C_7H_8N_2O_2$. *See* a1002......	152.16	red or og nd (peth) λ^{al} 279 (3.5), 429 (3.7)	38	d	....	....	δ^h	s	s	s	s	os, con ac s lig δ	**B12**^{2}, 369
a1161	**—,*N*-methyl-3-nitro-**	$C_7H_8N_2O_2$. *See* a1002......	152.16	red-ye nd or pr (al), cr (lig)	68	....	....	....	s^h	s	s	...	s	lig s^h	**B12**^{2}, 377
Ω a1162	**—,*N*-methyl-4-nitro-**	$C_7H_8N_2O_2$. *See* a1002......	152.16	br-ye pr (al), cr (eth) λ^{al} 315 (3.2), 387 (4.28)	152	d	1.201_4^{155}	....	i	s	δ	...	s	lig δ	**B12**^{2}, 385
a1163	**—,*N*-methyl-*N*-nitroso-**	Methylnitroso phenylamine. $C_6H_5N(NO)CH_3$	136.17	ye	14.7	225 d 121^{13}	1.1240_4^{20}	1.57688^{20}	i	s	s	...	...	....	**B12**^{2}, 309
Ω a1164	**—,*N*-methyl-4-nitroso-**....	$C_7H_8N_2O$. *See* a1002.......	136.17	bl pl (bz)	118	....	....	....	δ	s	s	...	δ s^h	chl s lig δ	**B7**^{2}, 575
Ω a1165	**—,*N*-methyl-*N*,2,4,6-tetranitro-**	Tetryl.	287.15	ye pr (al) λ^{al} 225 (4.4), 300 (3.4)	131–2	exp 187	1.57^{19}	....	i	δ s^h	δ	s	s	Py s chl δ CS_2 i	**B12**, 770
Ω a1166	**—,4-methylthio-**...	4-Aminothioanisole. C_7H_9NS. *See* a1002	139.22	λ^{al} 240 (3.91), 264 (4.14), 295 (3.21)	....	272–3 140^{15}	1.1379_4^{20}	1.6395^{20}	...	s	s	s	s	....	**B13**^{2}, 297
a1167	**—,*N*-nitro-**.......	Diazobenzolic acid. Nitranilide. Phenyl-nitramine. $C_6H_5NHNO_2$	138.13	lf (peth) λ^{al} 278(2.71)	43	exp on htng	....	....	v	s	s	s	s	peth δ	**B16**^{2}, 343
Ω a1168	**—,2-nitro-**.......	$C_6H_6N_2O_2$. *See* a1002......	138.13	gold-ye pl or nd (w) λ^{al} 231 (4.23), 276 (3.70), 402 (3.74)	71.5 (cor) (74–6)	284 $165–6^{28}$	1.442^{15}	....	δ s^h	s v^h	v	v	v	chl v	**B12**^{2}, 367
Ω a1169	**—,3-nitro-**.......	$C_6H_6N_2O_2$. *See* a1002......	138.13	ye nd, rh bipym (w) λ^{al} 236 (4.21), 275 sh (3.64), 374 (3.17)	114	305–7d $100^{0.16}$	1.1747_4^{160} 1.430_4	....	δ s^h	s	s	s v^h	δ s^h	MeOH v chl s	**B12**^{2}, 374
Ω a1170	**—,4-nitro-**.......	$C_6H_6N_2O_2$. *See* a1002......	138.13	pa ye mcl nd (w) λ^{al} 227 sh (3.86), 375 (4.19)	148.5–9.5	331.73 $106^{0.03}$	1.424_4^{20}	....	i δ^h	s	s	s v^h	δ	MeOH v to, chl s	**B12**^{2}, 383
a1171	**—,5-nitro-2-propoxy-**	$C_9H_{12}N_2O_3$. *See* a1002.....	196.21	og (PrOH-peth)	49–9.5	....	....	....	i	s	...	...	...	lig i	**C45**, 7544

For explanations, symbols and abbreviations see beginning of table. For structural formulas see end of table.

No.	Name	Synonyms and Formula	Mol. wt.	Color. crystalline form. specific rotation and λ_{max} (log ε)	m.p. °C	b.p. °C	Density	n_D	Solubility w	al	eth	ace	bz	other solvents	Ref.
	Aniline														
a1172	—,**4-nitroso-**	*p*-Benzoquinoneimine oxime. $C_6H_6N_2O$. *See* a1002	122.13	bl nd (bz) λ^{MeOH} 270 (3.6), 415 (4.5)	173–4	d on htng	...	...	s	s	...	...	δ	...	**B7**[2], 575
a1173	—,**2,3,4,5,6-pentabromo-**	$C_6H_2Br_5N$. *See* a1002	487.66	nd (al-to)	265–6	...	...	...	...	s	...	...	δ	to δ	**B12**, 669
a1174	—,**2,3,4,5,6-pentachloro-**	$C_6H_2Cl_5N$. *See* a1002	265.35	nd (al)	232	...	...	...	...	s	s	...	...	lig s	**B12**[2], 341
Ω a1175	—,**2-phenoxy-**	2-Aminodiphenyl ether. $C_{12}H_{11}NO$. *See* a1002	185.23	cr (lig) λ^{al} 233 sh (4.0), 270 (3.44), 277 (3.47), 289 (3.43)	44–5	$307–8^{728}$ $172–3^{14}$	...	...	...	s	s	s	s	os v	**B13**[2], 167
Ω a1176	—,**3-phenoxy-**	3-Aminodiphenyl ether. $C_{12}H_{11}NO$. *See* a1002	185.23	pr (lig) λ^{al} 220 sh (4.30), 272 (3.32), 278 (3.37), 288 (3.40)	37	315 $190–1^{14}$	...	...	...	s	s	s	s	os s lig δ	**B13**, 404
Ω a1177	—,**4-phenoxy-**	4-Aminodiphenyl ether. $C_{12}H_{11}NO$. *See* a1002	185.23	nd (w), cr (dil al) λ^{al} 241.5 (4.11), 270 sh (3.42), 278 (3.33), 298 (3.27)	85–6	$315–20^{30}$ $187–9^{14}$	...	...	s^h	v	v	...	...	lig δ	**B13**[2], 227
—	—,***N*-phenyl-**	*see* **Amine, diphenyl**													
a1178	—,***N*-propyl-**	Phenylpropylamine. $C_6H_5NHCH_2CH_2CH_3$	135.21	...	...	222 (cor) 100^{11}	0.9443_4^{20}	1.5428^{20}	i	v	s	...	...	lig s	**B12**[2], 94
a1179	—,**2,3,4,5-tetrachloro-**	$C_6H_3Cl_4N$. *See* a1002	230.91	nd (al)	118–20	...	...	...	...	s	v	...	v	aa v	**B12**[2], 340
a1180	—,**2,3,5,6-tetrachloro-**	$C_6H_3Cl_4N$. *See* a1002	230.91	nd (lig, al)	108	...	...	...	i	s	v	...	...	CS_2, lig v	**B12**[2], 340
a1181	—,**2,4,5-tribromo-**	$C_6H_4Br_3N$. *See* a1002	329.83	nd (al) λ^{al} 268 (2.2), 274 (2.3), 284 (2.05)	80–1 (85)	...	...	...	...	v	v	...	v	dil ac s	**B12**, 662
Ω a1182	—,**2,4,6-tribromo-**	$C_6H_4Br_3N$. *See* a1002	329.83	nd (al, bz) λ^{al} 249 (4.1), 315 (3.4)	122	300	2.35_{20}^{20}	...	i	δ s^h	s	...	...	chl s	**B12**[2], 358
a1183	—,—,hydrobromide	$C_6H_4Br_3N$. HBr. *See* a1002	410.75	nd	195–6d	sub	...	...	d	i	i	...	i	lig i	**B12**[1], 330
a1184	—,**3,4,5-tribromo-**	$C_6H_4Br_3N$. *See* a1002	329.83	nd (al)	123	...	...	...	i	s	s	...	...	...	**B12**, 668
a1185	—,**2,3,4-trichloro-**	$C_6H_4Cl_3N$. *See* a1002	196.46	nd (lig)	73 (68)	292^{774}	...	...	...	v	...	...	...	lig s	**B12**, 626
Ω a1186	—,**2,4,5-trichloro-**	$C_6H_4Cl_3N$. *See* a1002	196.46	nd (lig or 50 % al) λ^{al} 248 (4.15), 313 (4.06)	96.5	*ca.* 270	...	...	...	s	s	...	...	CS_2 v lig δ	**B12**[2], 338
a1187	—,**2,4,6-trichloro-**	*sym*-Trichloroaniline. $C_6H_4Cl_3N$. *See* a1002	196.46	cr (al), nd (lig or peth) λ^{al} 244 (3.9), 310 (3.4)	78.5	262^{746}	...	...	i	s	s	...	...	CS_2 v lig s	**B12**[2], 339
a1188	—,**2,3,5-triiodo-**	$C_6H_4I_3N$. *See* a1002	470.82	nd	116	...	...	...	...	s	δ	...	s	aa s lig δ	**B12**, 676
a1189	—,**2,3,6-triiodo-**	$C_6H_4I_3N$. *See* a1002	470.82	nd (al or al-eth)	116.8	...	...	...	...	s^h	i	...	...	...	**B12**, 676
a1190	—,**2,4,6-triiodo-**	$C_6H_4I_3N$. *See* a1002	470.82	ye nd or pl (al), pr (aa) λ^{MeOH} 229 (4.50), 255 sh (4.03), 315 (3.63)	185.5	...	...	...	...	δ s^h	...	...	...	AcOEt v CS_2, aa s	**B12**[2], 364
a1191	—,**3,4,5-triiodo-**	$C_6H_4I_3N$. *See* a1002	470.82	nd (al-ace)	174.5 d	...	...	...	...	s	s	v	s	chl, peth v	**B12**, 676
a1192	—,**2,4,6-trinitro-**	Picramide. TNA. $C_6H_4N_4O_6$. *See* a1002	228.12	dk ye pr (aa) λ^{aa} 324 (4.0), 408 (3.9)	192–5	exp	1.762^{14}	...	i	δ	δ	s	s	AcOEt s aa s^h chl δ	**B12**[2], 421
—	**Anisaldehyde**	*see* **Benzaldehyde, 4-methoxy-**													
—	**Anisic acid**	*see* **Benzoic acid, 4-methoxy-**													
—	**Anisidine**	*see* **Aniline, methoxy-**													
—	**Anisoin**	*see* **Benzoin, 4,4'-dimethoxy-**													
—	**Anisole**	*see* **Benzene, methoxy-***													
—	**Anol**	*see* **Benzene, 1-hydroxy-4(1-propenyl)-***													

For explanations, symbols and abbreviations see beginning of table. For structural formulas see end of table.

No.	Name	Synonyms and Formula	Mol. wt.	Color, crystalline form, specific rotation and λ_{max} (log ε)	m.p. °C	b.p. °C	Density	n_D	Solubility: w	al	eth	ace	bz	other solvents	Ref.
	Anthracene														
Ω a1193	**Anthracene**		178.24	ta or mcl pr (al) λ^{al} 218 (4.04), 252 (3.23), 295 (2.78), 310 (3.11), 324 (3.45), 340 (3.73), 357 (3.89) 376 (3.87)	216.2–.4	340 (cor) 226.5^{53} sub	1.283_4^{25}	...	i	δ^h	δ	δ s^h	δ s^h	to δ, s^h chl, CCl_4, CS_2 δ	**B5²**, 569
a1194	**—,1-acetyl-**	$C_{16}H_{12}O$. *See* a1193	220.27	pa ye (al) λ^{al} 240 (4.7), 251 (4.7), 375 (3.7)	107.5–9.0	...	...	...	...	s^h	...	...	...	...	**B7²**, 450
a1195	**—,2-acetyl-**	$C_{16}H_{12}O$. *See* a1193	220.27	ye (al), cr (AcOEt-peth) λ^{al} 240 (4.4), 260 (4.8), 350 (3.7)	190–2 (186)	...	...	...	...	s^h	...	...	...	...	**B7²**, 450
a1196	**—,9-acetyl-**	$C_{16}H_{12}O$. *See* a1193	220.27	pa ye (al) λ^{cy} 255 (5.2), 347 (3.7), 364 (3.9), 382 (3.9)	76	...	...	...	...	s^h	...	...	...	aa s	**B7²**, 450
a1197	**—,1-amino-**	1-Anthrylamine. $C_{14}H_{11}N$. *See* a1193	193.25	gold-ye nd (al) λ^{al} 240 (4.9), 259 (4.8), 369 (3.6)	130 (119)	...	...	...	i	v	...	...	...	con HCl i	**B12²**, 785
a1198	**—,2-amino-**	2-Anthrylamine. $C_{14}H_{11}N$. *See* a1193	193.25	ye lf (al) λ^{eth} 244 (4.3), 263 (4.6), 408 (3.6)	238–41	sub	...	...	i	s	...	...	...	os δ con sulf i	**B12²**, 785
a1199	**—,9-amino-**	9-Anthrylamine. $C_{14}H_{11}N$. *See* a1193	193.25	ye lf (dil al), br (bz) λ^{chl} 290 (5.0), 390 (3.8), 410 (3.8)	145–50	...	...	...	...	s	s	...	s	chl s	**B7²**, 416
a1200	**—,9-benzoyl-**	9-Anthrophenone. $C_{21}H_{14}O$. *See* a1193	282.35	ye nd (bz, aa, AcOEt)	148	...	...	...	...	...	...	s	s^h	CCl_4, CS_2, Ac_2O s aa s^h	**B13**, 321
a1201	**—,1-chloro-**	$C_{14}H_9Cl$. *See* a1193	212.68	lf (aa) λ^{chl} 257 (4.02), 365 (3.80), 384 (3.76)	83.5	...	1.1707_4^{100}	1.6959^{100}	i	s	s	...	s	CCl_4 s	**E13**, 235
a1202	**—,2-chloro-**	$C_{14}H_9Cl$. *See* a1193	212.68	nd or lf λ^{chl} 255.5 (5.18), 342 (3.57), 380 (3.69)	215 (223)	...	...	...	i	...	...	...	...	CCl_4 s	**E13**, 236
a1203	**—,9-chloro-**	$C_{14}H_9Cl$. *See* a1193	212.68	gold-ye nd (al) λ^{al} 255 (5.5), 340 (3.4), 392 (3.9)	106	...	...	...	i	s^h	s	...	s	CCl_4, CS_2, con sulf s (gr)	**E13**, 236
a1204	**—,10-chloro-9,10-dihydro-9-nitro-**	$C_{14}H_{11}ClNO_2$. *See* a1193	259.70	nd (bz)	163	...	...	...	...	δ	...	...	s^h	chl δ	**B5²**, 549
a1207	**—,9,10-diamino-**	$C_{14}H_{12}N_2$. *See* a1193	208.27	red cr	196	...	...	...	i	...	...	...	...	...	**C35**, 3998
Ω a1208	**—,9,10-dibromo-**	$C_{14}H_8Br_2$. *See* a1193	336.05	ye nd (to or xyl) λ^{hx} 250 (5.32), 265 (5.32)	226	sub	...	...	i	δ	δ	...	δ s^h	chl s to s^h	**B5²**, 577
a1209	**—,9,10-dichloro-**	$C_{14}H_8Cl_2$. *See* a1193	247.14	ye nd (MeCOEt or CCl_4) λ^{al} 252 (5.0), 259 (5.3), 325 (3.1), 341 (3.5), 379 (4.0), 401 (4.1)	212–2.5	...	...	...	...	δ	δ	...	s	...	**B5²**, 575
Ω a1210	**—,9,10-dihydro-**	$C_{14}H_{12}$. *See* a1193	180.25	ta or pr λ^{al} 260 (4.0), 270 (4.0)	111	305 $165–70^{13}$ sub	0.8976_4^{11}	...	i	s	s	...	s	...	**B5²**, 545
a1211	**—,9,10-dihydro-9-ethyl-**	$C_{16}H_{16}$. *See* a1193	208.31	...	...	320–3 δd (cor)	1.049_{18}^{18}	...	i	∞	∞	...	∞	aa s	**B5**, 649
a1212	**—,9,10-dihydro-9-hydroxy-**	Hydroanthrol. $C_{14}H_{12}O$. *See* a1193	196.25	nd (peth)	76	...	...	...	s^h	s	s	...	s	CS_2, lig s	**B6²**, 660
a1213	**—,9,10-dihydro-10-nitro-9-oxo-**	10-Nitroanthrone. $C_{14}H_9NO_3$. *See* a1193	239.23	cr (bz-lig), nd (CS_2)	140 (148d)	...	...	...	...	s	...	...	v^h	alk s CS_2 s^h	**B7²**, 240
Ω a1214	**—,9,10-dihydro-9-oxo-**	Anthrone. $C_{14}H_{10}O$. *See* a1193	194.24	nd (bz-lig, aa) λ^{al} 253 (4.35)	155 (165–70)	...	...	...	...	...	...	s	s^h	con sulf s dil alk s^h	**B7²**, 414

For explanations, symbols and abbreviations see beginning of table. For structural formulas see end of table.

No.	Name	Synonyms and Formula	Mol. wt.	Color, crystalline form, specific rotation and λ_{max} (log ε)	m.p. °C	b.p. °C	Density	n_D	Solubility w	al	eth	ace	bz	other solvents	Ref.
	Anthracene														
a1215	**—,1,2-dihydroxy-**	1,2-Anthradiol. $C_{14}H_{10}O_2$. *See* a1193	210.24	pa gr lf	160–2	...	...	...	...	v	v	...	...	aa v alk s (dk)	**B6²**, 998
a1216	**—,1,5-dihydroxy-**	1,5-Anthradiol. Rufol. $C_{14}H_{10}O_2$. *See* a1193	210.24	ye nd	265d	...	...	...	...	s	s	...	s	alk, aa s	**B6**, 1032
a1217	**—,1,8-dihydroxy-**	1,8-Anthradiol. Chrysazol. $C_{14}H_{10}O_2$. *See* a1193	210.24	ye nd (dil al), lf (al-aa)	225d	...	...	...	i	s	s	...	s	alk, AcOEt s	**B6**, 1033
a1218	**—,2,6-dihydroxy-**	2,6-Anthradiol. Flavol. $C_{14}H_{10}O_2$. *See* a1193	210.24	pa ye lf (al)	295–300d (dk at 270)	...	...	...	i	v	v	...	...	aa s	**B6²**, 999
a1219	**—,9,10-dihydroxy-**	9,10-Anthradiol. Anthraquinol. $C_{14}H_{10}O_2$. *See* a1193	210.24	br or ye nd	180	...	...	...	...	s	s	...	δ	chl δ	**B8²**, 214
Ω a1220	**—,1,3-dimethyl-**	$C_{16}H_{14}$. *See* a1193	206.29	pa bl flr lf (eth), cr (al)	83	$140–5^{2}$	...	...	i	v	v	...	...	...	**B5²**, 592
Ω a1221	**—,2,3-dimethyl-**	$C_{16}H_{14}$. *See* a1193	206.29	bl gr flr lf (bz)	252	...	...	...	i	s	...	...	v	...	**B5²**, 592
a1222	**—,9-ethyl-**	$C_{16}H_{14}$. *See* a1193	206.29	bl flr lf (al or MeOH)	59	...	1.0413^{99}_{4}	1.6762^{99}	i	s	s	...	...	...	**B5²**, 591
a1223	**—,1,2,3,4,5,6-hexahydro-**	$C_{14}H_{16}$. *See* a1193	184.28	lf (MeOH), cr (al)	67 (70)	160^{15}	...	...	...	δ	...	...	v	...	**B5²**, 472
Ω a1224	**—,1-hydroxy-**	1-Anthrol. $C_{14}H_{10}O$. *See* a1193	194.24	cr (bz), br nd or lf (al or aa) λ^{DMF} 395 (3.60)	158 (150–3)	224^{13}	...	...	i	v	v	...	...	NaOH, os s	**B6²**, 669
a1225	**—,2-hydroxy-**	2-Anthrol. $C_{14}H_{10}O$. *See* a1193	194.24	ye (bz), br lf or nd (dil al)	255 (200d)	...	...	...	i	v	v	v	s^h	KOH s	**B6²**, 669
a1226	**—,9-hydroxy-**	9-Anthrol. Anthranol. $C_{14}H_{10}O$. *See* a1193	194.24	ye red lf (dil al), pa ye nd (aa)	160–4 (120 rapid htng)	...	...	...	i	s	...	...	s^h	os v aq alk s	**B7²**, 414
a1227	**—,1-hydroxy-9, 10-dihydro-**	$C_{14}H_{12}O$. *See* a1193	196.25	sl grsh flr lf or nd (bz, peth)	94	...	...	...	...	s	s	...	...	aa s	**B6²**, 660
a1228	**—,2-hydroxy-9, 10-dihydro-**	$C_{14}H_{12}O$. *See* a1193	196.25	(bz-peth), lf (dil al)	129	...	...	...	δ	...	...	...	...	os v	**B6²**, 660
a1229	**—,1-methyl-**	$C_{15}H_{12}$. *See* a1193	192.26	bl nd (MeOH), lf (al) λ^{iso} 253 (5.22), 343 (3.74), 359 (3.93), 377 (3.92)	85–6	199–200	1.0471^{99}_{4}	1.6802^{99}	i	s	s	...	s	chl, sulf s al (bl flr)	**B5²**, 585
Ω a1230	**—,2-methyl-**	$C_{15}H_{12}$. *See* a1193	192.26	gr-bl flr lf (sub)	209–9.5	sub	1.81^{0}_{4}	...	i	δ	δ	i	s	chl, CS_2 s MeOH, aa δ	**B5²**, 586
Ω a1231	**—,9-methyl-**	$C_{15}H_{12}$. *See* a1193	192.26	yesh nd (dil al), pr (bz, al, MeOH) λ^{al} 252 (5.18), 256 (5.33), 318 (3.04), 331 (3.44), 348 (3.76), 366 (3.96), 386 (3.95)	81.5	$196–7^{12}$	1.065^{99}_{4}	1.6959^{99}	...	s	s	s	s	os v	**B5²**, 587
Ω a1232	**—,9-nitro-**	$C_{14}H_9NO_2$. *See* a1193	223.23	ye nd (al), pr (aa or xyl) λ^{al} 331 (3.42), 347 (3.61), 363 (3.66), 382 (3.58)	146	>360 ca. 275^{17}	...	...	i	δ s^h	...	...	v	CS_2 v aa δ, s^h	**B6²**, 1245
Ω a1233	**—,1,2,3,4,5,6,7,8-octahydro-**	Octhracene, $C_{14}H_{18}$. *See* a1193	186.30	pl (al) λ^{al} 270 (3.0), 280 (3.2), 285 (3.29)	78 (74)	$293–5^{760}$ 167^{12}	1.131^{0}_{4} 0.9703^{80}_{4}	1.5372^{80}	i	s v^h	...	...	v	aa s, v^h	**B5²**, 422
a1234	**—,9-phenyl-**	$C_{20}H_{14}$. *See* a1193	254.34	bl flr in sol, lf (al), (aa or al) λ^{al} 255.5 (5.16), 346.5 (3.90), 384 (4.08)	156	417	...	...	i	s^h	s^h	...	s^h	CS_2, chl s^h	**B5²**, 639
a1235	**—,1,2,9-trihydroxy-**	$C_{14}H_{10}O_3$. *See* a1193	226.24	og ye lf	149–51	...	...	...	...	...	...	...	...	con sulf, alk s	**B8²**, 371
a1236	**—,1,2,10-trihydroxy-**	$C_{14}H_{10}O_3$. *See* a1193	226.24	ye lf, nd (al-w)	208	...	...	...	δ	v	v	v	s	aa v	**B8²**, 372
a1237	**—,1,4,9-trihydroxy-**	$C_{14}H_{10}O_3$. *See* a1193	226.24	og-red nd (al)	156	...	...	...	...	s^h	...	...	...	...	**B8**, 330
a1238	**—,1,5,9-trihydroxy-**	$C_{14}H_{10}O_3$. *See* a1193	226.24	gold lf (al)	200d without melting	...	...	...	...	s^h	...	...	...	...	**B8**, 330
a1239	**—,1,8,9-trihydroxy-**	Anthralin. Cignolin. $C_{14}H_{10}O_3$. *See* a1193	226.24	ye pl or nd (lig)	178–80	...	...	...	i	s	δ	s	s	chl, Py v dil NaOH s(ye) lig s^h	**E13**, 372

For explanations, symbols and abbreviations see beginning of table. **For structural formulas see end of table.**

No.	Name	Synonyms and Formula	Mol. wt.	Color, crystalline form, specific rotation and λ_{max} (log ε)	m.p. °C	b.p. °C	Density	n_D	Solubility w	al	eth	ace	bz	other solvents	Ref.
	Anthracene														
a1240	**—,1,9,10-trihydroxy-(enol form)**	$C_{14}H_{10}O_3$. *See* a1193	226.24	gr nd (eth)	204–6	...	...	...	...	s	s	...	...	...	**B8**[2], 372
a1241	**—,1,9,10-trihydroxy-(keto form)**	...	226.24	ye nd (lig)	135–7	...	...	...	...	s	...	...	...	os δ	**B8**[2], 372
a1242	**—,2,3,9-trihydroxy-**	$C_{14}H_{10}O_3$. *See* a1193	226.24	ye br nd (al)	288–9	...	...	...	...	s	s	s	...	aa s	**B8**[2], 372
a1243	**9-Anthracenecarboxaldehyde**	9-Anthraldehyde.	206.25	og nd (dil aa) λ^{al} 234 (4.5), 262 (4.7), 372 (3.8), 404 (3.9)	104–5	...	...	...	i	...	...	...	s	aa s	**E13**, 320
a1244	**1-Anthracenecarboxylic acid**	1-Anthroic acid.	222.25	ye nd (aa), ye pr (al or AcOEt), bt ye nd (sub) λ^{al} 250 (5.0), 355 (3.7), 375 (3.7)	251–2	sub	...	...	i	s^h	s	...	δ	alk s, aa s^h, chl δ	**B9**[2], 493
a1245	**2-Anthracenecarboxylic acid**	2-Anthroic acid.	222.25	ye lf (al), nd, lf (sub)	281	sub	...	...	i	s^h	δ	...	δ	aa s, alk s (bl flr), chl δ	**B9**[2], 494
Ω a1246	**9-Anthracenecarboxylic acid**	9-Anthroic acid.	222.25	pa ye nd (bz, al) λ^{al} 254 (5.17), 344 (3.73), 362 (3.89), 381 (3.83)	217d	sub	...	...	i δ^h	s	...	...	...	...	**B9**[2], 494
—	**Anthrachrysone**	*see* **9,10-Anthraquinone, 1,3,5,7-tetrahydroxy-***													
—	**Anthraflavin**	*see* **9,10-Anthraquinone, 2,6-dihydroxy-***													
—	**Anthragallol**	*see* **9,10-Anthraquinone, 1,2,3-trihydroxy-***													
—	**Anthralin**	*see* **Anthracene, 1,8,9-trihydroxy-***													
a1247	**Anthranil**	3,4-Benzoisoxazol.	119.12	λ 304 (3.5)	< −18	215^{760} δd, 99^{13}	1.8127^{20}_{4}	1.5845^{20}	δ^h	s	...	s	...	os, ac s	**B27**[2], 17
—	**Anthranilaldehyde**	*see* **Benzaldehyde, 2-amino-**													
—	**Anthranilic acid**	*see* **Benzoic acid, 2-amino-**													
—	**Anthranol**	*see* **Anthracene, hydroxy-***													
—	**Anthrapurpurin**	*see* **9,10-Anthraquinone, 1,2,7-trihydroxy-***													
Ω a1248	**9,10-Anthraquinone***	...	208.23	ye rh nd (al, bz) λ^{al} 252 (4.7), 278 (4.1), 330 (3.7)	286 (sub)	379.8^{760}	1.438^4	...	i	δ	i	...	δ s^h	con sulf s, chl, CCl_4 δ, to δ, s^h	**B7**[2], 709
Ω a1249	**—,1-amino-***	$C_{14}H_9NO_2$. *See* a1248	223.23	red nd (al), (gl aa) λ^{al} 234 (4.53), 306 (3.82), 497 (2.83)	253–4	sub	...	...	i	δ	s	s	s	con HCl s, chl aa s, $phNO_2$ s^h	**B22**[2], 618
Ω a1250	**—,2-amino-***	$C_{14}H_9NO_2$. *See* a1248	223.23	red nd (al, aa, sub) λ^{MeOH} 242 (4.49), 298 (4.37), 327 (3.95), 440 (3.65)	303–6	sub	...	...	i	δ	i	s	s	chl s	**B14**[2], 107
a1251	**—,1-amino-2-benzoyl-**	$C_{21}H_{13}NO_3$. *See* a1248	327.34	red nd (aa)	190	...	...	...	...	δ	...	...	...	aa s^h	**B14**[1], 482
a1252	**—,2-amino-3-benzoyl-**	$C_{21}H_{13}NO_3$. *See* a1248	327.34	ye pl (Py)	331	...	...	...	...	δ	...	...	δ	Py v, aa δ	**B14**[1], 482
a1253	**—,1-amino-2-bromo-***	$C_{14}H_8BrNO_2$. *See* a1248	302.15	ye-red nd (aa), nd (xyl) λ^{al} 308 (3.64), 469 (3.88)	182	...	...	...	i	δ	δ	δ	δ	Py v, aa, to s, lig δ	**B14**[1], 446
a1254	**—,1-amino-3-bromo-***	$C_{14}H_8BrNO_2$. *See* a1248	302.15	red nd (to) λ^{al} 305 sh (3.87), 471 (3.84)	243	...	...	...	i	i	i	i	δ	Py, $PhNO_2$ s, to δ	**B14**[1], 446
a1255	**—,3-amino-1,2-dihydroxy-***	3-Aminoalizarin. $C_{14}H_9NO_4$. *See* a1248	255.23	dk red pr (aa)	> 300	sub δd	...	...	...	δ	...	...	...	aq NH_3 s, aq HCl δ	**E13**, 565
a1256	**—,4-amino-1,2-dihydroxy-***	4-Aminoalizarin. $C_{14}H_9NO_4$. *See* a1248	255.23	gr-bl nd (al)	d	...	...	...	...	s^h	...	...	...	alk s (crimson), $PhNO_2$ s^h	**E13**, 565
a1257	**—,2-amino-1 hydroxy-***	2-Aminoerythroxyanthraquinone. $C_{14}H_9NO_3$. *See* a1248	239.23	red br nd (al)	226–7	sub	...	...	i	s	s	...	s	Py v, con sulf s, aq NH_3 δ	**B14**[2], 167

For explanations, symbols and abbreviations see beginning of table. For structural formulas see end of table.

No.	Name	Synonyms and Formula	Mol. wt.	Color, crystalline form, specific rotation and λ_{max} (log ε)	m.p. °C	b.p. °C	Density	n_D	Solubility w	al	eth	ace	bz	other solvents	Ref.
	Anthraquinone														
a1258	**—,1-bromo-***	$C_{14}H_7BrO_2$. *See* a1248	287.12	ye nd (bz), cr (xy, phNO$_2$)	188	sub	...	...	i	s	...	...	s	PhNO$_2$, con sulf s	**B7**[2], 717
a1259	**—,2-bromo-***	$C_{14}H_7BrO_2$. *See* a1248	287.12	ye nd (to), (gl aa)	204.5	sub	...	...	i	δ	...	...	s^h	AmOH s to, aa s^h	**B7**[2], 717
a1260	**—,3-bromo-1,2-dihydroxy-***	3-Bromoalizarin. $C_{14}H_7BrO_4$. *See* a1248	319.12	br-red nd (to)	260–1	sub	...	...	s	δ	...	...	δ	Py, dil alk s (bl-vt) con alk s (red) con sulf s (red) aa δ	**E13**, 555
Ω a1261	**—,1-bromo-4-methylamino-***	$C_{15}H_{10}BrNO_2$. *See* a1248	316.16	br-red nd (Py)	194		...	...	i	δ	...	...	...	Py, ac s	**B14**[1], 447
a1262	**—,2-bromo-1-methylamino-***	$C_{15}H_{10}BrNO_2$. *See* a1248	316.16	br nd (aa)	170–2		...	...	i	δ	...	...	...	Py, aa s^h	**B14**[1], 446
Ω a1263	**—,1-chloro-***	$C_{14}H_7ClO_2$. *See* a1248	242.66	ye nd (to or al) λ^{MeOH} 253 (4.63), 266 (4.15), 333 (3.70), 415 (2.00)	162	sub	...	...	i	δ^h	∞	...	s^h	PhNO$_2$ aa s AmOH s^h	**B7**[2], 714
Ω a1264	**—,2-chloro-***	$C_{14}H_7ClO_2$. *See* a1248	242.66	pa ye nd (aa or al) λ^{MeOH} 256 (4.69), 265 (4.36), 274 (4.23), 325 (3.59)	211	sub	...	...	i	δ	i	...	δ s^h	to v^h PhNO$_2$, con sulf s	**B7**[2], 714
a1265	**—,1,2-diamino-***	$C_{14}H_{10}N_2O_2$. *See* a1248	238.25	vt nd (PhNO$_2$) λ^{sulf} 262 (4.72), 360 (3.82)	303–4		...	...	...	δ	δ	...	...	Py, con sulf s con HCl i dil HCl, chl, xy δ	**B14**[2], 112
a1266	**—,1,3-diamino-***	$C_{14}H_{10}N_2O_2$. *See* a1248	238.25	red (PhNO$_2$)	290		...	...	...	δ	i	...	...	Py v con sulf s (gr) PhNO$_2$ δ, s^h	**B14**[2], 112
Ω a1267	**—,1,4-diamino-***	$C_{14}H_{10}N_2O_2$. *See* a1248	238.25	dk vt nd (Py), vt cr (al) λ^{al} 249 (4.45), 288 (4.09), 300 (4.08), 522 (2.83), 551 (3.04), 592 (3.04)	268		...	...	δ v^h	s	...	...	s	PhNO$_2$, Py v aa s	**B14**[2], 113
Ω a1268	**—,1,5-diamino-***	$C_{14}H_{10}N_2O_2$. *See* a1248	238.25	dk red nd (al, aa) $\lambda^{oClC_6H_4OH}$ 500 (4.08)	319 (cor)	sub	...	...	i	δ	δ	δ	δ	con sulf s PhNO$_2$ s^h chl δ	**B14**[2], 116
a1269	**—,1,6-diamino-***	$C_{14}H_{10}N_2O_2$. *See* a1248	238.25	red nd (aa), lf (MeOPh)	297		...	...	i	δ	...	...	...	PhNO$_2$ aa s^h	**B14**[2], 119
a1270	**—,1,7-diamino-***	$C_{14}H_{10}N_2O_2$. *See* a1248	238.25	red nd (PhNO$_2$)	290		...	...	i	δ	δ	...	...	PhNO$_2$ s^h, os δ	**B14**[1], 470
a1271	**—,1,8-diamino-***	$C_{14}H_{10}N_2O_2$. *See* a1248	238.25	red (al or aa, Py, PhNO$_2$) λ^{diox} 250 (4.4), 435 (4.0), 458 (4.1), 490 (4.0)	265		...	...	i	s	δ	...	...	Py, PhNO$_2$ s aa δ	**B14**[2], 119
a1272	**—,2,3-diamino-***	$C_{14}H_{10}N_2O_2$. *See* a1248	238.25	red (PhNO$_2$)	353		...	...	...	...	δ	...	...	sulf, Py, PhNO$_2$ s chl, xyl δ	**B14**[2], 120
Ω a1273	**—,2,6-diamino-***	$C_{14}H_{10}N_2O_2$. *See* a1248	238.25	red-br pr (aq Py) λ^{diox} 281 (4.0), 333 (4.1), 384 (3.7)	> 325 (320d)		...	...	δ^h	s^h	...	...	...	con sulf, chl, xy s Py s^h	**B14**[2], 120
a1274	**—,2,7-diamino-***	$C_{14}H_{10}N_2O_2$. *See* a1248	238.25	og-ye nd (al or PhNO$_2$), dk red nd (sub)	> 330	sub	...	...	i	δ	δ	...	...	dil sulf, con ac s	**B14**[1], 473
a1275	**—,2,3-dibromo-***	$C_{14}H_6Br_2O_2$. *See* a1248	366.02	ye nd (to)	283	sub	...	...	...	δ	...	...	s	chl, con sulf s	**B7**[2], 718
a1276	**—,2,7-dibromo-***	$C_{14}H_6Br_2O_2$. *See* a1248	366.02	lt ye lf (MeOPh), ye pl (sub)	248	sub	...	...	i	δ^h	...	...	s	con sulf s aa s^h	**B7**[2], 718
a1277	**—,1,3-dichloro-***	$C_{14}H_6Cl_2O_2$. *See* a1248	277.11	ye nd (aa)	209–10		...	...	i	δ^h	δ^h	δ^h	...	PhNO$_2$ s aa s^h	**B7**[1], 411
a1278	**—,1,4-dichloro-***	$C_{14}H_6Cl_2O_2$. *See* a1248	277.11	og-ye nd (aa) λ^{cy} 213 (4.45), 256 (4.6), 357 (4.6)	187.5–8		...	...	i	δ	δ	...	δ s^h	PhNO$_2$, Py v aa s^h	**B7**[2], 715

For explanations, symbols and abbreviations see beginning of table. For structural formulas see end of table.

No.	Name	Synonyms and Formula	Mol. wt.	Color, crystalline form, specific rotation and λ_{max} (log ε)	m.p. °C	b.p. °C	Density	n_D	Solubility w	al	eth	ace	bz	other solvents	Ref.
	Anthraquinone														
Ω a1279	**—,1,5-dichloro-***	$C_{14}H_6Cl_2O_2$. *See* a1248	277.11	yesh (to), ye nd (aa) λ^{al} 254 (5.24), 344 (4.43)	252	...	...	...	i	δ	...	δ	δ	$PhNO_2$, con sulf s aa s^h	**B7**[2], 715
a1280	**—,1,6-dichloro-***	$C_{14}H_6Cl_2O_2$. *See* a1248	277.11	pa ye nd (aa)	203–4	...	...	...	i	δ	δ	v	...	$PhNO_2$, to, anisole v	**B7**[2], 715
a1281	**—,1,7-dichloro-***	$C_{14}H_6Cl_2O_2$. *See* a1248	277.11	ye nd (aa) λ^{cy} 213 (4.42), 263 (4.7), 270 (4.3), 282 (4.3)	213–4	...	...	...	i	δ	...	...	...	$PhNO_2$, aa s^h	**B7**[2], 716
Ω a1282	**—,1,8-dichloro-***	$C_{14}H_6Cl_2O_2$. *See* a1248	277.11	ye nd (aa)	202–3	...	...	...	i	δ	...	...	s^h	to, phNO_2 s^h	**B7**[2], 716
a1283	**—,2,3-dichloro-***	$C_{14}H_6Cl_2O_2$. *See* a1248	277.11	ye nd (aa)	271	...	...	...	i	δ	...	...	s^h	aa δ, s^h	**B7**[2], 716
a1284	**—,2,6-dichloro-***	$C_{14}H_6Cl_2O_2$. *See* a1248	277.11	ye nd (aa or al), pa ye lf (PhCl)	291–1.5	...	...	...	i	s^h	...	...	v^h	aa s^h con sulf s (ye)	**B7**[2], 716
a1285	**—,2,7-dichloro-***	$C_{14}H_6Cl_2O_2$. *See* a1248	277.11	yesh nd (MeOPh)	212 (231)	...	...	...	i	...	s	...	...	MeOPh s^h	**B7**[2], 717
a1286	**—,1,2-dihydroxy-***	Alizarin. $C_{14}H_8O_4$. *See* a1248	240.23	og or red tcl nd or pr. (al, sub) λ^{al} 248 (4.4), 435 (3.8)	289–90 (cor)	430 sub	...	...	δ	s	s	s	s	Py ∞ CS_2 s MeOH s^h chl i	**B8**[2], 487
a1287	**—,1,3-dihydroxy-***	Purpuroxanthin. Xanthopurpurin. $C_{14}H_8O_4$. *See* a1248	240.23	ye-red nd (sub), pr (aa + 2) λ^{al} 246 (4.43), 284 (4.36), 415 (3.79)	268–70	...	...	...	i	s	...	v	s	$PhNO_2$ s alk s(red) aa s^h	**E13**, 526
Ω a1288	**—,1,4-dihydroxy-***	Quinizarin. $C_{14}H_8O_4$. *See* a1248	240.23	ye red lf (eth), dk red nd (al), red cr (to, aa) λ^{DMF} 330 sh (3.5), 470 (3.9), 520 sh (3.7)	200–2 (aa) 194 (to)	...	...	...	s^h	s^h	s^h	...	s^h	KOH, sulf s	**B9**[2], 889
a1289	**—,1,5-dihydroxy-***	Anthrarufin. $C_{14}H_8O_4$. *See* a1248	240.23	pa ye pl (gl aa, sub) λ^{al} 225 (4.57), 253 (4.24), 287 (3.98), 418 (3.98), 437 (3.98)	280	sub	...	...	i	δ	δ	δ	s	con sulf s CS_2 δ	**B8**[2], 496
a1290	**—,1,6-dihydroxy-***	$C_{14}H_8O_4$. *See* a1248	240.23	og-ye nd (gl aa)	276	...	...	...	...	δ s^h	...	s	δ s^h	dil alk, $PhNO_2$ s	**B8**[1], 721
a1291	**—,1,7-dihydroxy-***	$C_{14}H_8O_4$. *See* a1248	240.23	ye nd (sub)	292–3	sub	...	...	i	s	s	...	s	chl, CS_2, aa s	**B8**[1], 721
Ω a1292	**—,1,8-dihydroxy-***	Chrysazin. Istizin. $C_{14}H_8O_4$. *See* a1248	240.23	red or redsh-ye nd or lf (al) λ^{al} 222 (4.6), 253 (4.4), 284 (4.1), 430 (4.1)	193	sub	...	...	i	s	s	s	...	alk, aa, chl s $PhNO_2$ δ	**B8**[2], 500
a1293	**—,2,3-dihydroxy-***	Hystazarin. Hystazin $C_{14}H_8O_4$. *See* a1248	240.23	ye-br nd (aa), ye nd (sub)	> 330	sub	...	...	i	δ	δ	δ	i	sulf, NH_3, NaOH s	**B8**[2], 504
Ω a1294	**—,2,6-dihydroxy-***	Anthraflavin. Anthraflavic acid. $C_{14}H_8O_4$. *See* a1248	240.23	ye nd (al) λ^{al} 274 (4.53), 301.5 (4.28), 349 (3.90)	360d	...	...	...	δ	δ	i	...	i	con sulf, alk s chl i	**B8**[2], 504
Ω a1295	**—,2,7-dihydroxy-***	Isoanthraflavin. Isoanthraflavic acid. $C_{14}H_8O_4$. *See* a1248	240.23	ye nd (+1w, dil al), nd (sub)	350–5	sub	...	...	i	s	δ	...	δ	aa s chl δ	**B9**[2], 890
a1296	**—,1,8-dihydroxy-3-hydroxymethyl-***	Aloe-emodin. Rhabarberone. $C_{15}H_{10}O_5$. *See* a1248	270.25	og-ye nd (to, al) λ^{MeOH} 220 (3.7), 255 (3.3), 283 (3.0), 430 (3.0)	223–4	sub	...	...	...	v^h	s	...	s	NH_4OH, sulf, Na_2CO_3 s (red)	**E13**, 571
—	**—,1,8-dihydroxy-3-hydroxy-methyl-2,4,5,7-tetranitro-***	*see* **Aloetic acid**													
a1296[1]	**—,1,2-dihydroxy-3-iodo-***	β-Iodoalizarin. $C_{14}H_7IO_4$. *See* a1248	366.11	og-red nd (xyl)	228.6–9.7	...	...	...	s	...	...	...	...	sulf s(og-red)	**E13**, 555
a1297	**—,1,2-dihydroxy-3-methyl-***	β-Methylalizarin. $C_{15}H_{10}O_4$. *See* a1248	254.25	og nd	245	sub	...	...	...	s	s	s	...	...	**B8**[2], 510
Ω a1298	**—,1,8-dihydroxy-3-methyl-***	Chrysophanic acid, Chrysophanol. $C_{15}H_{10}O_4$. *See* a1248	254.25	ye hex or mcl nd (sub)	196	sub	0.92	...	δ	δ s^h	δ	s	s^h	chl. aa s lig δ	**B8**[2], 510
a1299	**—,1,2-dihydroxy-3-nitro-***	Alizarin orange. β-Nitroalizarin. $C_{14}H_7NO_6$. *See* a1248	285.22	og-ye nd (bz), ye pl (gl aa, al)	244d	sub d	...	...	δ s^h	s	...	...	s	sulf, chl, aa s	**B8**[2], 491

For explanations, symbols and abbreviations see beginning of table. For structural formulas see end of table.

No.	Name	Synonyms and Formula	Mol. wt.	Color, crystalline form, specific rotation and λ_{max} (log ε)	m.p. °C	b.p. °C	Density	n_D	Solubility w	al	eth	ace	bz	other solvents	Ref.
	Anthraquinone														
a1300	**—,1,2-dihydroxy-4-nitro-***	4-Nitroalizarin. $C_{14}H_7NO_6$. *See* a1248	285.22	gold-ye nd (aa or al)	289d	sub d	....	....	δ	s	...	...	s	sulf, chl, aa s	**B8**^{2}, 491
a1301	**—,1,8-dihydroxy-2,4,5,7-tetrabromo-***	2,4,5,7-Tetrabromochrysazin. $C_{14}H_4Br_4O_4$. *See* a1248	555.82	og-ye nd (bz)	312	....	....	....	...	...	...	...	δ	dil KOH s aa δ	**B8**^{1}, 722
a1302	**—,1,4-dihydroxy-5,6,7,8-tetrachloro-***	5,6,7,8-Tetrachloroquinizarine. $C_{14}H_4Cl_4O_4$. *See* a1248	378.00	red pl (aa)	270	....	....	....	...	δ	δ	...	s	NaOH s aa s^h	**B8**^{1}, 716
a1303	**—,1,8-dihydroxy-2,4,5,7-tetranitro-***	Chrysammic acid. Chrysamminic acid. $C_{14}H_4N_4O_{12}$. *See* a1248	420.21	ye pl or lf	exp	d	....	....	δ	s	s	...	...	ac s	**B8**^{1}, 723
a1304	**—,1,2-dimethyl-***	$C_{16}H_{12}O_2$. *See* a1248	236.27	nd (ace or aa)	156	....	....	....	i	s	s	s^h	s	aa s^h	**B7**^{2}, 743
a1305	**—,1,3-dimethyl-***	$C_{16}H_{12}O_2$. *See* a1248	236.27	nd (aa)	162	....	....	....	i	δ	...	...	δ	aa s^h	**B7**^{2}, 743
a1306	**—,1,4-dimethyl-***	$C_{16}H_{12}O_2$. *See* a1248	236.27	ye nd (al, sub)	140–1	sub	....	....	i	δ s^h	...	...	s	xyl, aa s	**B7**^{2}, 743
Ω a1307	**—,2,3-dimethyl-***	$C_{16}H_{12}O_2$. *See* a1248	236.27	ye nd (al, to or xyl), cr (aa)	210	sub	....	....	...	s^h	...	...	s^h	xyl s^h, aa s^h	**B7**^{2}, 744
a1308	**—,2,6-dimethyl-***	$C_{16}H_{12}O_2$. *See* a1248	236.27	ye nd (aa or al)	242	sub	....	....	...	δ^h	...	...	...	to, $PhNO_2$ s aa s^h	**B7**^{2}, 744
a1309	**—,2,7-dimethyl-***	$C_{16}H_{12}O_2$. *See* a1248	236.27	yesh nd (al)	170	....	....	....	i	s^h	...	...	...	....	**B7**^{2}, 744
a1310	**—,1,3-dinitro-***	$C_{14}H_6N_2O_6$. *See* a1248	298.21	ye nd (HNO_3)	246–50 (240)	....	....	....	i	...	...	...	...	HNO_3 s	**E13**, 436
a1311	**—,1,5-dinitro-***	$C_{14}H_6N_2O_6$. *See* a1248	298.21	pa ye nd (xyl or $PhNO_2$), ye cr (sub) λ^{diox} 258 (4.6), 322 (3.7), 400 sh (2.2)	422 (385)	sub	....	....	i	δ^h	δ^h	...	δ^h	$PhNO_2$ v con sulf δ, aa δ	**B7**^{2}, 721
a1312	**—,2-ethyl-1-nitro-***	$C_{16}H_{11}NO_4$. *See* a1248	281.27	yesh br (aa)	226	....	....	....	...	...	...	...	...	aa δ	**B7**^{2}, 743
a1313	**—,1,2,3,5,6,7-hexahydroxy-***	Rufigallol. Rufigallic acid. $C_{14}H_8O_8$. *See* a1248	304.23	red rh, red-ye nd (sub) λ^{al} 222 (4.18), 258 (3.85), 295 (4.64), 349 (3.92), 438 (3.92)	....	sub d	....	....	i	δ s^h	δ s^h	s	...	alk s	**E13**, 613
Ω a1314	**—,1-hydroxy-***	Erythroxyanthraquinone. $C_{14}H_8O_3$. *See* a1248	224.23	red-og nd (al, sub) λ^{MeOH} 252 (4.46), 266 (4.15), 327 (3.52), 402 (3.74)	194–5	sub	....	....	i	s	s	...	s	liq NH_3 δ	**B8**^{2}, 388
Ω a1315	**—,2-hydroxy-***	$C_{14}H_8O_3$. *See* a1248	224.23	ye pl or nd (al or aa), ye lf (sub) λ^{al} 250 (4.4), 277 (4.3), 340 (3.5), 406 (3.7)	306	sub	....	....	i	s	s	...	...	aq NH_3, KOH s	**B8**^{2}, 393
Ω a1316	**—,2-methyl-***	$C_{15}H_{10}O_2$. *See* a1248	222.25	yesh nd (al, aa or sub)	182–3 (177–9)	sub	....	....	i	v	δ	...	s	con sulf v aa s	**B7**^{2}, 733
Ω a1317	**—,1-methylamino-***	$C_{15}H_{11}NO_2$. *See* a1248	237.26	ye-red nd	170	....	....	....	...	s	...	...	s	chl, aa s	**B14**^{2}, 100
a1318	**—,2-methylamino-***	$C_{15}H_{11}NO_2$. *See* a1248	237.26	red nd (aa)	226–7	....	....	....	...	s	δ	...	...	aa, to v chl s	**B14**^{2}, 108
a1319	**—,2-methyl-1-nitro-***	$C_{15}H_9NO_4$. *See* a1248	267.24	pa ye nd (aa)	270–1	....	....	....	i	i^h	δ^h	...	δ^h	$PhNO_2$ s chl δ aa δ^h	**E13**, 433
a1320	**—,6-methyl-1,2,5-trihydroxy-***	Morindone. $C_{15}H_{10}O_5$. *See* a1248	270.25	og-red nd (to)	282	....	....	....	i	v	v	...	v	Py v sulf s (bl-vt)	**E13**, 588
a1321	**—,6-methyl-1,3,8-trihydroxy-***	Emodin. Frangula-emodine. $C_{15}H_{10}O_5$. *See* a1248	270.25	og-red mcl nd (aa), cr (dil aa + 1w) og nd (dil Py) λ^{MeOH} 220 (4.5), 245 (4.2), 290 (4.3), 440 (3.95)	256–7	sub	....	....	i	s	δ	...	i	alk s aa δ chl, CCl_4, CS_2 i	**B8**^{2}, 563
a1322	**—,1-nitro-***	$C_{14}H_7NO_4$. *See* a1248	253.22	yesh pr (ace), nd (aa) λ^{MeOH} 255 (4.57), 325 (3.63)	232.5–3.5	$270–1^{7}$	....	....	i	δ	δ	s	s	chl, aa s	**B7**^{2}, 719
a1323	**—,2-nitro-***	$C_{14}H_7NO_4$. *See* a1248	253.22	ye nd (aa or al) λ^{MeOH} 258 (4.60), 323 (3.72), 420 (1.30)	184.5–5	sub	....	....	i	δ	δ	s	...	sulf, chl s aa s^h lig δ	**B7**^{2}, 720
a1324	**—,1,2,4,5,8-pentahydroxy-***	Alizarin cyanine R. Alizarin pentacyanine. $C_{14}H_8O_7$. *See* a1248	288.24	br lf ($PhNO_2$)	d	sub	....	....	...	...	...	...	...	alk, sulf s (bl with red flr)	**E13**, 613

For explanations, symbols and abbreviations see beginning of table. For structural formulas see end of table.

No.	Name	Synonyms and Formula	Mol. wt.	Color, crystalline form, specific rotation and λ_{max} (log ε)	m.p. °C	b.p. °C	Density	n_D	Solubility: w	al	eth	ace	bz	other solvents	Ref.
	Anthraquinone														
a1325	**—,1,2,4,6-tetra-hydroxy-***	Hydroxyflavopurpurin. $C_{14}H_8O_6$. *See* a1248	272.23	dk red nd (sub)	......	......	......	......	...	v	...	...	...	Py v aa δ	**B8**[2], 582
a1326	**—,1,2,4,7-tetra-hydroxy-***	4-Hydroxyanthrapurpurin. $C_{14}H_8O_6$. *See* a1248	272.23	red-ye (al, Py or aa)	......	......	......	......	...	s	...	...	...	con sulf, NH_3 s	**B8**[2], 582
a1327	**—,1,2,5,6-tetra-hydroxy-***	Rufiopin. $C_{14}H_8O_6$. *See* a1248	272.23	og-red nd (Py)	340	sub	......	......	s^h	s	δ	...	δ	aa s chl δ	**B8**[2], 583
a1328	**—,1,2,5,8-tetra-hydroxy-***	Quinalizarin. $C_{14}H_8O_6$. *See* a1248	272.23	red nd ($PhNO_2$)	>275	sub	......	......	i	δ	δ	δ	δ	alk, os, aa δ	**B8**[2], 584
a1329	**—,1,2,6,7-tetra-hydroxy-***	$C_{14}H_8O_6$. *See* a1248........	272.23	og nd ($PhNO_2$)	>330	......	......	......	...	δ	...	...	δ^h	xyl δ^h $PhNO_2$ δ, s^h	**B8**[2], 584
a1330	**—,1,2,7,8-tetra-hydroxy-***	$C_{14}H_8O_6$. *See* a1248........	272.23	red nd or pr (aa)	318d	......	......	......	...	δ	...	...	δ^h	sulf, Py, NaOH s, $PhNO_2$, aa δ	**B8**[2], 585
a1331	**—,1,3,5,7-tetra-hydroxy-***	Anthrachrysone. $C_{14}H_8O_6$. *See* a1248	272.23	yesh nd (al + 2w)	150–60d (+2w) >360 (anh)	sub	......	......	i	s	δ	s	s	CS_2 i chl, lig δ aa s	**B8**[2], 585
a1332	**—,1,4,5,8-tetra-hydroxy-***	$C_{14}H_8O_6$. *See* a1248........	272.23	gr nd (aa), br nd (bz-lig) λ^{peth} 488 (4.1), 523 (4.4), 548 (4.4), 562 (4.5)	>300	sub	......	......	i	s	...	...	...	alk s Ac_2O δ	**B8**[2], 586
Ω a1333	**—,1,2,3-tri-hydroxy-***	Anthragallic acid. Anthragallol. $C_{14}H_8O_5$. *See* a1248	256.23	ye nd (dil al), br (aa), og nd (sub) λ^{al} 245 (4.30), 287 (4.49), 414 (3.81)	313	sub at 290	......	......	δ	s	s	...	...	CS_2, aa s sulf, alk δ	**B8**[2], 549
a1334	**—,1,2,4-tri-hydroxy-***	Purpurin. $C_{14}H_8O_5$. *See* a1248	256.23	og-red, dk red or og-ye nd (al)	259	sub	......	......	δ^h	v^h	s	...	v^h	aa v^h	**B8**, 509
a1335	**—,1,2,5-tri-hydroxy-***	2-Hydroxyanthrarufin. $C_{14}H_8O_5$. *See* a1248	256.23	red nd (gl aa, sub)	278 (274)	......	......	......	i	...	s	...	...	....	**B8**[2], 554
Ω a1336	**—,1,2,6-tri-hydroxy-***	Flavopurpurin. $C_{14}H_8O_5$. *See* a1248	256.23	ye nd (al)	>330 (sub >160)	459 par d	......	......	s^h	s	δ	...	s	alk (vt)	**B8**[2], 555
Ω a1337	**—,1,2,7-tri-hydroxy-***	Anthrapurpurin. $C_{14}H_8O_5$. *See* a1248	256.23	og nd (al)	374 (sub 170)	462^{760} par d	......	......	δ^h	s^h	δ	...	i	chl, aa s^h	**B8**[2], 555
a1338	**—,1,2,8-tri-hydroxy-***	2-Hydroxychrysazin. $C_{14}H_8O_5$. *See* a1248	256.23	red nd (aa, sub)	239–40	sub	......	......	i	δ	...	...	...	....	**B8**[2], 557
a1339	**—,1,3,8-tri-hydroxy-***	$C_{14}H_8O_5$. *See* a1248	256.23	bt red-br nd (bz), gold-ye pl (AcOEt) λ^{al} 256 (4.2), 292 (4.1), 450 (3.9)	287–8	......	......	......	δ	...	...	...	...	con sulf (red-og) con alk s (red)	**B8**[2], 557
a1340	**—,1,4,5-tri-hydroxy-***	5-Hydroxyquinizarin. $C_{14}H_8O_5$. *See* a1248	256.23	red-br nd or lf ($PhNO_2$), dk red nd (Py) λ^{cy} 480 (4.08), 495 (4.13), 516 (4.04), 529 (4.04)	271	......	......	......	δ	δ	...	...	...	aa s^h	**B8**[2], 557
a1341	**—,1,4,6-tri-hydroxy-***	6-Hydroxyquinizarin. $C_{14}H_8O_5$. *See* a1248	256.23	bt red nd (al), br pw	>300 (256)	......	......	......	δ	s	...	...	...	alk, Py s chl δ	**B8**[2], 558
Ω a1342	**9,10-Anthra-quinone-2-carboxylic acid**	....	252.23	ye nd (aa)	290–2	sub	......	......	...	δ	i	s	i	aa δ	**B10**[2], 581
a1342[1]	**—,1,3-dihydroxy-**..	Munjistin. $C_{15}H_8O_6$. *See* a1342	284.24	ye nd (al-w + 1w), lf (aa)	230–1	sub	......	......	δ s^h	v^h	s	...	s	chl, sulf, alk s aa s	**B10**[2], 761
a1343	**—,1,4-dihydroxy-**..	$C_{15}H_8O_6$. *See* a1342........	284.24	ye br or red nd ($PhNO_2$)	249–50	......	......	......	...	δ	δ	s	δ	to, alk s aa s^h	**B10**[1], 509
a1344	**—,4,5-dihydroxy-**..	Cassic acid. Rhein. $C_{15}H_8O_6$. *See* a1342	284.24	ye or og nd (MeOH, Py) $\lambda^{95\%al}$ 230 (4.62), 255 (4.32), 440 (4.04)	321–1.5	sub	......	......	δ^h	δ	δ	δ	δ	Py v, sulf, alk s chl δ	**B10**[1], 510
a1345	**—,4,5-dihydroxy-7-methoxy-**	Emodic acid monomethyl ether. Parietinic acid. $C_{16}H_{10}O_7$. *See* a1342	314.26	red-br or ye nd (sub)	300	sub	......	......	...	...	...	...	...	os δ	**B10**[2], 767
a1346	**—,1-nitro-**.......	$C_{15}H_7NO_6$. *See* a1342......	297.23	nd (al or aa)	288d	......	......	......	...	s^h	i	s^h	...	aa s^h lig i	**B10**[2], 586
a1347	—,—,chloride.....	$C_{15}H_6ClNO_5$. *See* a1342....	315.67	....	243–4	......	......	......	d^h	d^h	...	...	...	....	**E13**, 665
a1348	**—,5-nitro-**......	$C_{15}H_7NO_6$. *See* a1342.....	297.23	yesh nd (aa)	......	......	......	......	...	...	...	...	...	sulf, aa s	**E13**, 665

For explanations, symbols and abbreviations see beginning of table. **For structural formulas see end of table.**

No.	Name	Synonyms and Formula	Mol. wt.	Color, crystalline form, specific rotation and λ_{max} (log ε)	m.p. °C	b.p. °C	Density	n_D	Solubility w	al	eth	ace	bz	other solvents	Ref.
	9,10-Anthraquinone														
a1349	**9,10-Anthraquinone-1,5-disulfonic acid**	...	368.35	ye nd (HCl +4w), pl (dil aa +4w) $\lambda^{NH_3 salt}$ 255 (4.5)	310–1d	...	...	...	v	s	...	δ	δ	sulf, aa s chl δ	**B11**², 195
a1350	**9,10-Anthraquinone-1,6-disulfonic acid**	...	368.35	ye nd (HCl+5w), gold pr (dil aa + 5w)	215–7d	...	...	...	v	...	...	...	...	aa s	**B11**², 196
a1351	**9,10-Anthraquinone-1,7-disulfonic acid**	...	368.35	ye hyg pw (dil aa +4w)	d at 120	...	...	...	v	v	...	...	...	aa s	**B11**², 197
a1352	**9,10-Anthraquinone-1,8-disulfonic acid**	...	368.35	ye nd(+5w)	293–4d	...	...	...	v	s	...	...	...	...	**B11**², 197
Ω a1353	**9,10-Anthraquinone-1-sulfonic acid**	α-Sulfoanthraquinone	288.29	lf (aa), ye lf (con HCl +3w) λ^w 257 (4.61), 328 (3.63)	214 (cor) (anh) 218 (hyd)	...	...	...	v	s	...	...	...	aa v^h	**B11**², 192
Ω a1354	**9,10-Anthraquinone-2-sulfonic acid**	β-Sulfoanthraquinone	288.29	ye lf (+3w)	...	...	...	...	v	s	i	...	...	...	**B11**², 193
a1355	—,amide	$C_{14}H_9NO_4S$. *See* a1354	287.30	ye nd (aa)	261	...	...	...	...	δ	...	...	...	CS_2, to, chl δ	**B11**, 339
a1356	**9,10-Anthraquinone-1-sulfonic acid, 5-chloro-**	$C_{14}H_7ClO_5S$. *See* a1353	322.73	ye rh pr (HCl or aa +4w)	236–7	...	...	...	v	...	...	...	...	sulf s (ye)	**E13**, 707
a1357	**9,10-Anthraquinone-2-sulfonic acid, 5-nitro-**	$C_{14}H_7NO_7S$. *See* a1354	333.28	yesh pl (dil HNO_3)	255d	...	...	...	δ v^h	...	...	...	...	dil HNO_3 s^h	**B11**², 195
—	**Anthrarufin**	*see* **9,10-Anthraquinone, 1,5-dihydroxy-***													
—	**Anthroic acid**	*see* **Anthracenecarboxylic acid**													
—	**9-Anthrone**	*see* **Anthracene,9,10-dihydro-9-oxo-**													
a1358	**Antimalarine**	Plasmocid.	287.41	...	...	$182^{1.0}$	1.0569_4^{24}	1.5855^{24}	...	...	δ	...	...	dil HCl s	**C40**, 3563
a1359	**Antipyrine** (α-form)	Analgesine. 2,3-Dimethyl-1-phenyl-5-pyrazolone. Phenazone.	188.23	mcl lf or sc (w, bz or eth) λ^{al} 223 (3.91), 247 (3.97), 273 (3.98)	114	319^{741} $211–2^{10}$	1.0747_4^{130}	1.5697	v	v	δ	...	s	chl v Py s lig, liq SO_2 δ	**B24**², 11
a1360	—(β-form)	$C_{11}H_{12}N_2O$. *See* a1359	188.23	cr (unst)	109 (β→α at 18)	...	...	...	...	...	δ	...	s	...	**B24**², 11
a1361	—,2-hydroxybenzoate	Salazolon. Salipyrazolone. Salipyrine. $C_{11}H_{12}N_2O.C_7H_6O_3$. *See* a1359	326.35	pw	92	...	...	...	δ s^h	v^h	δ	...	...	chl v^h	**B24**¹, 197
a1362	—,*m*-amino-	$C_{11}H_{13}N_3O$. *See* a1359	203.25	redsh in air (bz)	148	...	...	...	s	s	i	...	...	chl s	**B24**¹, 210
a1363	—,*o*-amino-	$C_{11}H_{13}N_3O$. *See* a1359	203.25	nd (AcOEt-eth)	165	...	...	...	s	s	...	...	...	...	**B24**¹, 210
a1364	—,*p*-bromo-	$C_{11}H_{11}BrN_2O$. *See* a1359	267.13	nd (w)	122	300^9	...	...	s s^h	v	s^h	...	...	chl, to v	**B24**, 33
a1365	—,*p*-dimethylamino-	$C_{13}H_{17}N_3O$. *See* a1359	231.30	pr or pl (lig or AcOEt) λ^{al} 255 (4.04)	134–5	...	...	...	s	s	δ	...	s	lig δ	**B24**, 46
—	**Antipyrine chloral**	*see* **Hypnal**													
a1366	**Aphanin**	$C_{40}H_{54}O$	550.88	bl-bk lf (bz-MeOH) λ^{chl} 474, 504	178	...	...	...	i	δ	δ	...	δ	CS_2, chl v	**C32**, 3454
a1367	**Aphylline**	$C_{15}H_{24}N_2O$	248.37	$[\alpha]_D^{20}$ +10.3 (MeOH, c = 20)	52–7	200^4	...	...	...	s	s	s	s	dil HCl v os s	**C26**, 2742
—	**Apigenin**	*see* **Flavone,4′,5-,7-trihydroxy-**													
a1368	**Apiol**	2,5-Dimethoxysafrole. Parsley camphor.	222.24	nd	29.5	294 179^{33}	1.015_4^{20}	1.5360^{20}	i	s	s	s	s	chl, lig s	**B19**², 98
a1369	**Apoatropine**	Atropamine. $C_{17}H_{21}NO_2$	271.36	pr (chl)	62	...	...	...	δ	v	v	v	v	chl, CS_2 v lig δ	**B21**², 18
Ω a1370	—,hydrochloride	$C_{17}H_{21}NO_2.HCl$. *See* a1369	307.82	lf (w)	239	...	...	...	δ s^h	δ	i	δ	...	...	**B21**¹, 197
a1371	**Apocinchonidine**	$C_{19}H_{22}N_2O$. *See* a1372	294.40	lf (al), $[\alpha]_D^{20}$ −139.3 (chl-al, c = 2)	252	...	...	...	i	v	δ	...	...	chl v lig δ	**B23**¹, 131
a1372	**Apocinchonine**	Allocinchonine. $C_{19}H_{22}N_2O$.	294.40	pr (al), $[\alpha]_D^{20}$ +167.4 (abs al, c = 3)	219	...	...	...	i	s v^h	δ	...	δ	chl δ, lig δ	**B23**¹, 131

For explanations, symbols and abbreviations see beginning of table. For structural formulas see end of table.

No.	Name	Synonyms and Formula	Mol. wt.	Color, crystalline form, specific rotation and λ_{max} (log ε)	m.p. °C	b.p. °C	Density	n_D	Solubility w	al	eth	ace	bz	other solvents	Ref.
	Apocinchonine														
a1373	—,hydrochloride	$C_{19}H_{22}N_2O.HCl.2H_2O$. *See* a1372	366.89	nd (+2w) $[\alpha]_D^{16}$ +139 (w, c = 0.006)	...	...	...	...	s	...	i	...	i	...	**B23**, 419
a1374	**Apocodeine**	Apomorphine 3-methyl ether. Pseudoapocodeine. $C_{18}H_{19}NO_2$.	281.36	pr (MeOH) $[\alpha]_D^{23}$ −90 (abs al, c = 0.449) λ 280 (4.05), 313 (3.50)	122.5–4.5 (anh)	...	...	...	...	δ	s	s	s	lig s	**B21**, 188
a1375	—,(ethanol solvate)	$C_{18}H_{19}NO_2.C_2H_5OH$. *See* a1374	327.43	lf(al + 1)	104.5–6.5	...	...	...	...	s^h	...	...	...	...	**B21**, 188
a1376	—,(methanol solvate)	$C_{18}H_{19}NO_2.CH_3OH$. *See* a1374	313.40	nd(MeOH + 1)	85	...	...	...	...	...	...	...	...	MeOH s^h	**B21**, 188
—	**Apocupreine**	*see* **Apoquinine**													
a1377	**Apocyclene**	...	122.21	cr (al)	42.5–43	$138–9^{764}$	0.8710_4^{40}	1.4514^{40}	i	s	s	...	s	aa s	**E13**, 1042
—	**Apocynin**	*see* **Acetophenone,4-hydroxy-3-methoxy-**													
—	**Apofencho-camphoric acid**	*see* **1,3-Cyclopentane dicarboxylic acid, 4,4-dimethyl-***													
a1378	**Apomorphine**	$C_{17}H_{17}NO_2$	267.33	hex pl (chl-peth), rods (eth + 1), gr in air λ^{cy} 280 (384), 288 (3.81)	195d	...	...	...	δ	s	s	s	s	chl, alk s HCl, lig δ	**B21**[2], 142
Ω a1379	—,hydrochloride	$C_{17}H_{17}NO_2.HCl.\frac{1}{2}H_2O$ *See* a1378	312.80	gr in air, mcl pr $[\alpha]_D^{25}$ −48 (w, c = 1.2) λ^w 274 (4.22), 306 sh (3.52)	200–10d	...	...	...	δ s^h	δ	δ	...	...	chl δ	**B21**[2], 143
—	**Aponal**	*see* **Carbamic acid,** 2-methyl-2-butyl ester.													
a1380	**Apoquinine** (α-form)	Apocupreine. $C_{19}H_{22}N_2O_2$.	310.40	pr (eth), $[\alpha]_D^{20}$ −214.8 (al)	184d	...	...	...	i s^h	s	δ	δ	s	KOH s chl δ	**B23**[2], 412
a1381	—(β-form)	$C_{19}H_{22}N_2O_2.2H_2O$. *See* a1380	346.43	$[\alpha]_D^{20}$ −194 (al)	190d	...	...	...	i	s	...	δ	...	...	**B23**[2], 412
a1382	**Aposafranone**	Benzeneindone. 10-Phenyl-2-phenazinone.	272.31	br, gr (red in sol), nd (al)	248–9	...	...	...	δ^h	s	...	...	s	dil ac s	**B23**[2], 364
a1383	**Arabinose** (*D*) (α-form)	...	150.13	cr (MeOH), $[\alpha]_D^{20}$ +105	155.5–6.5 (cor)	...	1.585	...	v	δ	i	i	...	MeOH i	**B31**, 32
a1384	—(*D*) (β-form)	$C_5H_{10}O_5$. *See* a1383	150.13	(aq MeOH) $[\alpha]_D^{20}$ −175 → −108 (mut)	155.5–6.5	...	1.625	...	v	δ	i	i	...	MeOH i	**B31**, 32
a1385	—(*DL*)	Pectinose. $C_5H_{10}O_5$. *See* a1383	150.13	pr, nd (al)	164.5 (cor)	...	1.585_4^{20}	...	v	δ	i	...	i	...	**B31**, 46
a1386	—(*L*) (α-form)	...	150.13	orh cr (pr), cr (dil al)	159–60	...	1.585_4^{20}	...	v	δ	i	...	...	...	**B31**, 34
a1387	—(*L*) (β-form)	$C_5H_{10}O_5$. *See* a1386	150.13	orh, $[\alpha]_D^{20}$ +190.6 → +104.5 (c = 3) (mut)	159–60	...	1.625_4^{20}	...	v	δ	i	...	...	...	**B31**, 34
a1388	—, diphenyl-hydrazone (*D*)	...	316.36	orh pr (dil al)	207	...	...	...	δ	δ	...	...	i	...	**B31**, 33
a1389	—,—(*DL*)	$C_{17}H_{20}N_2O_4$. *See* a1388	316.36	nd (aq Py)	206	...	...	...	i δ^h	i δ^h	i	...	...	Py v CS_2 i	**B31**, 47
a1390	—,—(*L*)	...	316.36	orh pr, nd (dil al), $[\alpha]_D^{20}$ +18.5 (Py)	204–5	...	...	...	δ	s^h	...	...	...	...	**B31**, 44
Ω a1391	**Arabitol** (*D*)	Arabite.*D*-Lyxitol. 1,2,3,4,5-Pentanepentol	152.15	pr (dil al) $[\alpha]_D^{20}$ +11.8 (borax sol, c = 9.5)	103–4	...	...	...	v	δ	i	...	...	...	**B1**, 531
a1392	—(*DL*)	$C_5H_{12}O_5$. *See* a1391	152.15	pr (90 % al), cr (al-ace)	106	...	...	...	v	δ	i	...	...	...	**B1**, 531
Ω a1393	—(*L*)	$C_5H_{12}O_5$. *See* a1391	152.15	$[\alpha]_D$ −5.4 (borax sol)	102–3	...	...	...	v	δ	i	...	...	...	**B1**[2], 604
a1394	**Arabonic acid** (*D*)	...	166.13	(dil aa), $[\alpha]_D^{25}$ +10.5 (c = 6.0)	114–16	...	...	...	v	s	i	...	...	aa δ	**B3**[2], 303
a1395	—(*DL*)	$C_5H_{10}O_6$. See a1394	166.13	...	...	...	...	...	s	...	i	...	i	...	**B2**[2], 982
a1396	—(*L*)	$C_5H_{10}O_6$. See a1394	166.13	cr (al or MeOH) $[\alpha]_D^{20}$ −9.6 → −41.7 (21 days) (w, c = 2.5) (mut)	118–9	...	...	...	v	...	i	...	i	...	**B3**[2], 303
—	**Arachidic acid**	*see* **Eicosanoic acid***													
—	**Arachidonic acid**	*see* **6,10,14,18-Eicosatetrae-noic acid***													

For explanations, symbols and abbreviations see beginning of table. For structural formulas see end of table.

No.	Name	Synonyms and Formula	Mol. wt.	Color, crystalline form, specific rotation and λ_{max} (log ε)	m.p. °C	b.p. °C	Density	n_D	Solubility w	al	eth	ace	bz	other solvents	Ref.
	Aramite														
a1397	**Aramite**	Niagaramite. $C_{15}H_{23}ClO_4S$	334.87	...	−37.1	195^{2} $175^{00.1}$	1.145^{20}_{20}	1.5100^{20}	i	s	v	v	v	os v	**M**, 97
Ω a1398	**Arbutin**	Hydroquinone β-d-glucopyranoside. Uvasol. $C_{12}H_{16}O_7$.	272.24	nd (w + 1) $[\alpha]^{25}_{D}$ −64 (w, c = 3) λ^{al} 290 (1.55)	199.5–200(st) (anh) 165 (unst)	...	...	...	v	s	δ	...	i	chl, CS_2 i	**B31**, 210
a1399	—, hydrate	$C_{12}H_{16}O_7.H_2O$. *See* a1398	290.27	$[\alpha]^{17}_{D}$ −60.3 (w, p = 5)	142	...	...	...	...	s	...	...	i	...	**B31**, 210
a1400	**Arecaidine**	Arecaine. N-Methylguvacine. 1-Methyl-1,2,5,6-tetrahydronicotinic acid. $C_7H_{11}NO_2$.	141.16	pl (dil al), ta (dil al + 1 w)	232(anh)	...	...	...	v	i	i	...	i	dil al v chl i	**B22**^{2}, 12
a1401	**Arecoline**	Arecaidine methyl ester. Methyl 1-methyl-1,2,5,6-tetrahydro nicotinate. $C_8H_{13}NO_2$.	155.20	λ^{al}214(4.03)	...	209(220) 94^{17}	1.0504^{20}_{20}	1.4860^{20}	∞	∞	∞	...	...	chl s	**B22**^{2}, 12
Ω a1402	—, hydrobromide	$C_8H_{13}NO_2.HBr$. *See* a1401	236.12	mcl pr (al)	172(177.4 −7.9)	...	...	...	v	s v^h	δ	...	...	chl δ	**B22**^{2}, 12
Ω a1403	—, hydrochloride	$C_8H_{13}NO_2.HCl$. *See* a1401	191.66	nd (al)	157–8	...	...	...	s	s	δ	...	...	chl δ	**B22**^{1}, 490
Ω a1404	**Arginine** (*DL*)	*DL*-2-Amino-5-guanidopentanoic acid*. $H_2NC(:NH)NH(CH_2)_3CH(NH_2)CO_2H$	174.20	λ^{w}205 (3.28)	217–8	...	...	...	i	i	i	...	i	...	**B4**^{2}, 850
Ω a1405	—(*L*)	$H_2NC(:NH)NH(CH_2)_3CH(NH_2)CO_2H$	174.20	pr (+2w), anh mcl pl (60 % al), $[\alpha]^{20}_{D}$ + 12.5 (w, c = 3.5) λ^{w} 205 (3.28)	244d (223d) (anh) 105d (hyd)	...	...	...	s	δ	i	...	...	...	**B4**^{2}, 845
a1406	—, diflavianate (*l*)	$C_6H_{14}N_4O_2.(C_{10}H_6N_2O_8S)_2$. *See* a1405	802.67	ye nd	202d	...	...	...	d	d	...	...	...	ac i	**B4**^{2}, 848
a1407	—, dipicrate (*dl*)	$C_6H_{14}N_4O_2.2C_6H_3N_3O_7$. *See* a1404	632.42	...	196d	...	...	...	...	...	...	...	...	...	**B4**^{2}, 850
a1408	—, —(*l*)	$C_6H_{14}N_4O_2.2C_6H_3N_3O_7$. *See* a1405	632.42	...	200d	...	...	...	...	i	...	...	...	...	**B4**^{2}, 848
a1409	—, flavianate (*l*)	$C_6H_{14}N_4O_2.C_{10}H_6N_2O_8S$. *See* a1405	488.44	ye-og lf	258–60d	...	...	...	i	i	i	...	...	...	**B4**^{2}, 847
a1410	—, picrate (*d*)	$C_6H_{14}N_4O_2.C_6H_3N_3O_7.2H_2O$. *See* a1404	439.34	nd (w)	217–8d	...	...	...	δ	...	...	...	...	...	**B6**, 287
a1411	—, —(*dl*)	$C_6H_{14}N_4O_2.C_6H_3N_3O_7$. *See* a1404	403.31	pr (w)	200(223d)	...	...	...	i	i	i	...	...	...	**B4**^{2}, 851
a1412	—, —(*l*)	$C_6H_{14}N_4O_2.C_6H_3N_3O_7.2H_2O$. *See* a1405	439.34	ye nd (w)	217–8d	...	...	...	δ	i	i	...	...	...	**B4**^{2}, 848
a1413	—, **benzylidene** (*l*)	$C_{13}H_{18}N_4O_2$. *See* a1405	262.32	lf(w)	204–5	...	...	...	s^h	...	...	...	...	MeOH, ac s alk i	**B4**^{2}, 848
—	**Arsanilic acid**	*see* **Benzenearsonic acid, 4-amino-***													
a1414	**Arsenic acid,** triethyl ester*	Ethyl arsenate*. $(C_2H_5O)_3AsO$	226.11	...	...	235–8 $118–20^{15.5}$	1.3023^{20}_{0}	1.4343^{20}	d	...	...	...	...	...	**B1**^{2}, 332
a1415	**Arsenous acid,** triethyl ester*.	Ethyl arsenite.* $(C_2H_5O)_3As$	210.11	...	...	165–6 $66.5–7^{12}$	1.2239^{20}_{4}	1.4369^{13}	d	...	...	...	...	...	**B1**^{2}, 332
a1416	**Arsine, bis(pentafluoroethyl)-iodo-***	$(CF_3CF_2)_2AsI$	439.86	...	...	120^{760}	...	...	...	...	...	...	...	...	**C49**, 14635
a1416^{1}	—, **bis(trifluoromethyl)-***	$(CF_3)_2AsH$	213.94	...	...	19^{760} -25^{108}	...	...	...	...	...	...	...	...	**J1953**, 1552
a1417	—, **bis(trifluoromethyl)bromo-***	$(CF_3)_2AsBr$	292.84	...	...	59.5^{745}	...	1.398^{20}	i	...	...	...	...	...	**J1953**, 1552
a1418	—, **bis(trifluoromethyl) chloro***	$(CF_3)_2AsCl$	248.39	...	...	46^{760}	...	1.351^{19}	...	...	...	...	...	...	**J1953**, 1552
a1419	—, **bis(trifluoromethyl) fluoro-***	$(CF_3)_2AsF$	231.93	...	...	25^{760}	...	...	...	...	...	...	...	...	**J1953**, 1552
a1420	—, **bis(trifluoromethyl)iodo-***	$(CF_3)_2AsI$	339.84	ye oil λ^{eth} 287 (3.08)	...	92^{760}	...	1.425^{25}	i δ^hd	...	s	...	...	...	**J1953**, 1552
a1421	—, **bis(trifluoromethyl)methyl-***	$(CF_3)_2AsCH_3$	227.97	...	...	52^{760}	...	...	...	...	...	...	...	...	**J1953**, 1552
a1422	—, **chloro-(diphenyl)-***	$(C_6H_5)_2AsCl$	264.59	rh pl (peth)	44	333^{760} 193^{20}	1.48204^{16}_{4}	1.6332^{56}	i	v	s	s	s	...	**B16**^{1}, 437
a1423	—, **dibromo(trifluoromethyl)-***	CF_3AsBr_2	303.75	...	...	118^{745}	...	1.528^{20}	...	...	...	...	...	...	**J1953**, 1552
a1424	—, **dichloro-(methyl)-***	CH_3AsCl_2	160.86	...	−42.5	133 37^{25}	1.8358^{20}	1.5677^{15}	δ	v	s	...	...	os s	**B4**^{2}, 979
a1425	—, **dichloro-(phenyl)-***	$C_6H_5AsCl_2$	222.93	...	...	254.4–7.6 131^{14}	1.6516^{19}_{4}	1.6386^{15}	i	v	s	s	s	...	**B16**^{2}, 411
a1426	—, **dichloro(trifluoromethyl)-***	CF_3AsCl_2	214.83	...	...	71	...	1.431^{20}	i	s	...	s	...	...	**J1953**, 1552
a1427	—, **diethyl-***	$(C_2H_5)_2AsH$	134.05	...	...	105(97)	1.1338^{24}_{4}	1.4709	δ	s	v	s	v	...	**B4**^{2}, 980
a1428	—, **(difluoro)-ethenyl, 2-chloro-***	$ClCH:CHAsF_2$	174.41	...	...	>140d 43.5^{15}	...	...	d	...	...	...	...	...	**B4**^{3}, 1809

For explanations, symbols and abbreviations see beginning of table. For **structural formulas see end of table.**

No.	Name	Synonyms and Formula	Mol. wt.	Color, crystalline form, specific rotation and λ_{max} (log ε)	m.p. °C	b.p. °C	Density	n_D	Solubility w	al	eth	ace	bz	other solvents	Ref.
	Arsine														
a1429	—, **difluoro-(ethyl)-***	$C_2H_5AsF_2$	141.98	fum in air	−38.7	94.3 74^{100}	1.708^{17}		d	...	...	...	...		**B4**³, 1799
a1430	—, **difluoro-(methyl)-***	CH_3AsF_2	127.95	fum in air	−29.7	76.5	1.924^{18}		d	...	...	...	...		**B4**³, 1796
a1431	—, **difluoro-(phenyl)-***	$C_6H_5AsF_2$	190.03	wax	42	110^{48}			d	...	...	...	...		**J1946,** 1123
a1432	—, **diiodo(tri-fluoromethyl)-***	CF_3AsI_2	397.74	λ^{eth} 230 (3.74), 267 (3.60)		183 δd 100^{48}			d	...	...	...	...		**J1953,** 1552
a1433	—, **dimethyl-***	Cacodyl hydride. $(CH_3)_2AsH$	106.00	ign in air		36^{760}	1.213^{29}_{29}		δ	∞	∞	∞	∞	chl, CS_2, aa ∞	**B4**², 978
a1434	—, **dimethyl(tri-fluoromethyl)-***	$(CH_3)_2AsCF_3$	174.00			58			...	...	...	...	...		**J1953,** 1552
a1435	—, **diphenyl-***	$(C_6H_5)_2AsH$	230.14	oil		174^{25}	1.30^{25}_{25}		i	s	s	...	...		**B16**², 406
a1436	—, **ethyl-***	$C_2H_5AsH_2$	106.00			36^{760}	1.217^{22}_{22}		i	s	s	...	...		**B4**², 980
a1437	—, **methyl-***	CH_3AsH_2	91.97		−143	2^{760}			δ	∞	∞	...	...	CS_2 ∞	**B4**², 978
a1438	—, **triethyl-***	$(C_2H_5)_3As$	162.11			$138–9^{760}$	1.150^{20}_{4}	1.467^{20}	i	∞	∞	...	...		**B4**², 980
a1439	—, **(trifluoro-methyl)-***	CF_3AsH_2	145.94			-11.6^{781}			...	...	...	...	...		**J1953,** 1552
a1440	—, **trimethyl-***	$(CH_3)_3As$	120.03		−87.3	52^{760}	1.144^{15}		δ	s	s	...	s		**B4**², 978
a1441	—, **triphenyl-***	$(C_6H_5)_3As$	306.24	tcl pr (bz, eth-peth) λ^{al} 248 (4.12)	60.5–1.5	360 (under CO_2)	1.2634^{48}_{4}	1.6888^{21}	i	δ	v	...	v		**B16**², 407
a1442	—, **tris(penta-fluoroethyl)-***	$(CF_3CF_2)_3As$	431.96			96–3			...	...	...	...	...		**C49,** 14635
a1443	—, **tris(trifluoro-methyl)-***	$(CF_3)_3As$	281.94			33.3			i	...	s	...	...		**J1952,** 2198
a1444	**Arsine oxide, 4-methoxy-phenyl-**		198.05	cr (chl-eth or bz-peth)	114–6 (anh)				...	...	...	...	s	chl s	**B16**², 448
a1445	—, **methyl-***	Methylarsinic acid anhydride. CH_3AsO	105.96	pr (CS_2), (al)	95	275d			i	s	δ	...	s	chl s	**B4**², 992
a1446	—, **phenyl-**	C_6H_5AsO	168.03	cr (bz-eth or chl-eth) λ^{al} 216 (3.97), 234 (3.72)	144–6 (127–30)				i	δ v^h	i	...	v^h	chl v	**B16**², 443
a1447	—, **2-tolyl-**		182.06	pw	145–6				v^h	...	i	...	...	alk δ	**B16,** 861
a1448	—, **4-tolyl-**		182.06	pw	156				v^h	...	i	...	...	alk δ	**B16,** 861
a1449	**Arsinic acid, bis(trifluoro-methyl)-***	Perfluorocacodylic acid, $(CF_3)_2AsO_2H$	245.94	nd (sub), rh(w)	150d	10^{-3} sub			s	...	...	...	...	chl δ^h	**J1953,** 1552
a1450	—, **diphenyl-***	$(C_6H_5)_2AsO_2H$	262.14	nd, pr $\lambda^{0.1\,NHCl}$ 220 (4.26), 263 (3.14)	178				δ s^h	s	δ	...	δ	chl s	**B16**², 443
a1451	**Arsonium, tetra-phenyl-, bromide**	$(C_6H_5)_4AsBr$	463.26		281–4				δ	s	...	δ	...	MeOH s	**J1940,** 1192
a1452	—, —, chloride	$(C_6H_5)_4AsCl$	418.80	λ^{diox} 259 (3.44), 265 (3.48), 272 (3.45)	264–5				v	s	...	δ	...	MeOH s	**Am 55,** 3056
a1453	**Artemisic acid**	$C_{15}H_{16}O_3$	244.29	nd (dil aa) $[\alpha]^{18.5}_D$ +70.4 (al)	135–6				...	v	v	...	...	aa, lig v	**E12B,** 3469
—	**Arterenol (*l*)**	*see* **Noradrenalin (*l*)**													
—	**Artificial musk**	*see* **Benzene,1-*tert*-butyl-3-methyl-2,4,6-trinitro-***													
—	**Asaron**	*see* **Benzene,1-propenyl-2,4,5-trimethoxy-***													
—	**Asaronic acid**	*see* **Benzoic acid, 2,4,5-tri-methoxy-***													
Ω a1454	**Ascaridole**	$C_{10}H_{16}O_2$	168.24	unst, $[\alpha]_D$ −4.14 λ^{al} 238 (3.56), 241 (3.56)	3.3	exp^{760} 115^{15} $39–40^{0.2}$	1.0103^{20}_{4}	1.4769^{20}	i d^h	s	...	s	s	os, to, peth s	**B19**², 18
a1455	**Ascorbic acid (*D*)**	$C_6H_8O_6$	176.14	pl or mcl nd $[\alpha]^{18}_D$ −48 (MeOH, c = 1) $[\alpha]^{20}_D$ −23.8 (w, c = 3) λ^{al} 254 (3.9)	192d (189d)				v	s	i	...	i	chl i	**C54,** 3229
a1456	—(*DL*)	$C_6H_8O_6$. *See* a1455	176.14	λ^{al} 245 (3.9)	168–9				v	s	i	...	i	chl i	**Am 74,** 5162
Ω a1457	—(*L*)	Antiscorbutin. Vitamin C. $C_6H_8O_6$. *See* a1455	176.14	pl or mcl nd $[\alpha]^{20}_D$ +24 (w, c = 1) $[\alpha]^{25}_D$ +48 (MeOH, c = 1) λ^w 244 (3.98)	192d (189d)		1.65		v	s	i	...	i	chl, peth i	**Am 74,** 5162
a1458	—, **6-desoxy-(*L*)**		160.14	pr (AcOEt) $[\alpha]^{22}_D$ +36.7 (0.1N HCl, c = 1)	168	sub $160^{0.001}$			v	s	i	s	i	AcOEt δ	**H21,** 273

For explanations, symbols and abbreviations see beginning of table. For structural formulas see end of table.

No.	Name	Synonyms and Formula	Mol. wt.	Color, crystalline form, specific rotation and λ_{max} (log ε)	m.p. °C	b.p. °C	Density	n_D	Solubility w	al	eth	ace	bz	other solvents	Ref.
	β-Asparagine														
Ω a1459	**β-Asparagine** (*D*)	$H_2NCOCH_2CH(NH_2)CO_2H.H_2O$	150.14	$[\alpha]^{20}_D$ +5.41 (w, c = 1.3)	234.5 (anh) 215 (hyd)	....	1.543^{15}_4	....	δ s^h	i	i	...	i	MeOH i	**B4**[1], 531
a1460	—(*DL*)	$C_4H_8N_2O_3.H_2O$. *See* a1459	150.14	tcl cr (w+1), pr(w-al) λ^w 220 sh (2.0)	182–3	213–5d	1.4540^{15}_4	....	δ s^h	i	i	...	i	MeOH i	**B4**[2], 900
Ω a1461	—(*L*)	$C_4H_8N_2O_3$. *See* a1459	132.13	rh (w+1) $[\alpha]^{20}_D$ −5.42 (w, c = 1.3) $\lambda^{w, pH=10.6}$ 217sh (2.5)	236 (anh)	....	1.543^{15}_4	....	s v^h	i	i	...	..	MeOH i	**B4**[2], 896
a1462	**Aspartic acid** (*D*)	Aminosuccinic acid. Asparaginic acid. $HO_2CCH_2CH(NH_2)CO_2H$	133.11	$[\alpha]^{20}_D$ −25.5 (HCl)	269–71	....	1.6613^{13}_{13}	....	s^h	i	i	...	i	dil HCl s	**B4**[1], 531
Ω a1463	—(*DL*)	Asparacemic acid. $C_4H_7NO_4$. *See* a1462	133.11	mcl pr (w)	338–9 (275 sealed tube)	....	1.6632^{13}_{13}	....	δ s^h	i	i	...	i	Py i	**B4**[2], 900
a1464	—(*L*)	$C_4H_7NO_4$. *See* a1462	133.11	rh lf (w) $[\alpha]^{20}_D$ +4.36 (w) $[\alpha]^{24}_D$ +24.6 (6N HCl, c = 2)	324d (rapid htng) 270–1 (sealed tube)	....	1.6613^{13}_{13}	....	δ s^h	i	i	...	i	dil HCl s Py i	**B4**[2], 892 **B14**[2], 653
Ω a1465	—, *N*-**benzoyl-(*L*)**	$HO_2CCH_2CH(NHCOC_6H_5)CO_2H$	237.21	nd or lf $[\alpha]^{20}_D$ +37 (0.76 N NaOH)	171–3	....	....	....	δ^h	δ	i	...	..	chl i	**B9**[2], 185
a1466	**Aspidospermine**	Vallesine. $C_{22}H_{30}N_2O_2$	354.50	nd or pr(al), nd(peth) $[\alpha]^{15}_D$ −100.2 (al) λ^{MeOH} 220 (4.52), 256 (4.04)	212 sub 180	220^2	....	....	δ	s	δ	...	s	chl s	**J1940**, 1051
—	**Aspirin**	*see* **Benzoic acid, 2-hydroxy-, acetate**													
—	**Atabrin**	*see* **Quinacrine**, dihydrochloride (*dl*)													
a1467	**Atisine**	Anthorine. $C_{22}H_{33}NO_2$	343.51	rh bipym	57–60	....	....	....	δ	s	s	...	...	chl MeOH-KOH s	**Am 78**, 4139
a1468	—, hydrochloride	$C_{22}H_{33}NO_2$. HCl. *See* a1467	377.96	nd (dil al) $[\alpha]_D$ +28 (w, c = 1.1) λ^{al} < 220	340 (311)	....	....	....	v	v	i	...	...	....	**C36**, 4826
—	**Atophan**	*see* **4-Quinolinecarboxylic acid, 2-phenyl-**													
—	**Atoquinol**	*see* **4-Quinolinecarboxylic acid, 2-phenyl-**, allyl ester													
—	**Atranol**	*see* **Benzaldehyde, 2,6-dihydroxy-4-methyl-**													
—	**Atrolactic acid**	*see* **Propanoic acid, 2-hydroxy-2-phenyl-***													
—	**Atropamine**	*see* **Apoatropine**													
—	**Atropic acid**	*see* **Propenoic acid, 2-phenyl-***													
Ω a1469	**Atropine**	*dl*-Daturine. *dl*-Hyoscyamine. Troyl tropate. $C_{17}H_{23}NO_3$.	289.36	rh nd (dil al), orh pr (ace) λ^{MeOH} 252 (2.22), 258 (2.29), 262 (2.21)	118–9	sub (vac) 93–110	....	....	δ^h	v	s	...	s	chl v dil ac s lig i	**B21**[2], 19
a1470	—, hydrochloride	$C_{17}H_{23}NO_3$. *See* a1469	325.84	nd (al)	165	....	....	....	s	s	...	...	...	....	**B21**, 30
a1471	—, pentanoate	$C_{17}H_{23}NO_3 . C_5H_{10}O_2 . \frac{1}{2}H_2O$. *See* a1469.	400.52	cr	42	....	....	....	∞	∞	δ	...	...	....	**M**, 111
a1472	—, sulfate	$2(C_{17}H_{23}NO_3). H_2SO_4 . H_2O$. *See* a1469.	694.82	nd (al-eth or al-ace) λ^w 252 (2.5), 259 (2.6), 264 (2.4)	194 (anh)	sub	....	....	v	v	i	...	...	chl δ	**B21**[2], 20
—	**Auligen**	*see* **Xantogen,** diethyl-													
a1473	**Auramine** (base)	Bis (*p*-dimethylaminophenyl) methyleneimine. Apyonin.	267.38	ye or col pl (al) λ^{aa} 270 (3.48), 372 (3.54), 440.5 (3.74)	136	....	....	....	i	s	δ	...	...	....	**B14**[2], 58

For explanations, symbols and abbreviations see beginning of table. For structural formulas see end of table.

No.	Name	Synonyms and Formula	Mol. wt.	Color, crystalline form, specific rotation and λ_{max} (log ε)	m.p. °C	b.p. °C	Density	n_D	Solubility w	al	eth	ace	bz	other solvents	Ref.
	Auramine														
a1474	—(dye)	Auramine hydrochloride. Auramine O. $C_{17}H_{21}N_3$. HCl. H_2O. *See* a1473	321.85	ye nd (w) λ^{al} 365 (4.30), 425 (4.79)	267				δ	s	δ			chl v	**B14**^{2}, 58
a1475	—, *N*-methyl-	$C_{18}H_{23}N_3$. *See* a1473	281.41	ye cr (al)	133				i	s		s	δ	aa s	**B14**, 93
a1476	**Aureomycin**	Biomycine. Chlorotetracycline. CTC. Duomycine. $C_{22}H_{23}ClN_2O_8$.	478.89	gold ye, ye flr $[\alpha]_D^{23}$ −275 (MeOH) λ^{al} 380 (4.0)	168–9				i	δ	i	δ	δ	Cellosolve, diox s peth i	**C47**, 7169
Ω a1477	—, hydrochloride	$C_{22}H_{23}ClN_2O_8$. HCl. *See* a1476	515.35	ye orh $[\alpha]_D^{20}$ −106.5 (dil HCl) $\lambda^{0.01\ N HCl}$ 230 (4.2), 270 (4.2), 345 sh (4.9), 370 (4.0)	216d				δ						**C50**, 9696
Ω a1478	**Aurin**	Corallin. Pararosolic acid. Rosolic acid.	290.32	dk red lf or rh $\lambda^{10\%\ Na_2CO_3}$ 279 (4.2), 525 (4.85)	d 308–10				i	s	δ		i	alk, aa s chl δ	**B8**^{2}, 417
a1479	**Auxin A**	Auxenetriolic acid. $C_{18}H_{32}O_5$.	328.45	hex (al-lig) $[\alpha]_D^{20}$ −3.19 (al)	196		1.292_4^{19}		δ	s	δ		i	MeOH, AcOEt s peth i	**C45**, 2416
a1480	**Auxin B**	Auxenolonic acid. $C_{18}H_{30}O_4$.	310.44	cr (al-lig) $[\alpha]_D^{20}$ −2.8 (al)	183		1.269_4^{20}			s	s				**C45**, 2416
a1481	**1-Azacyclooctane, 2-methyl-**	α-Methyl heptamethyleneimine	127.23	nd (ace)	156–7	162–3^{746}	0.853^{20}	1.4620^{21}							**B20**^{1}, 30
—	**Azelaic acid**	*see* **Nonanedioic acid***													
a1482	**Azetidine**	Trimethyleneimine.	57.09			63^{748}	0.8436^{20}	1.4287^{25}	∞	∞	v	v	v		**B20**, 2
—	**Azibenzyl**	*see* **Ketone, benzyl phenyl, α-diazo-**													
a1483	**Aziridine**	Azirane. Dihydroazirine. Ethyleneimine.	43.07			56^{756}	0.8321_4^{20}		∞	s	v	v	v	os ∞	**B20**^{2}, 3
a1484	**Azo, benzene ethane**	Ethylphenyl diimide. $C_6H_5N{:}NC_2H_5$	134.18	bt ye oil		175–85^{760} 82.5–3^{20}	0.9628_4^{22}	α: 1.5313 β: 1.5579	i	s	s		s	dil ac δ	**B16**^{2}, 3
a1485	—, **benzene methane**	Methylphenyl diimide. $C_6H_5N{:}NCH_3$	120.17	ye oil λ^{al} 287 (3.9), 396 (2.2)		150^{760}d 60^{15}				s	s				**B16**, 7
a1485^1	—, **benzene 1-naphthalene**		231.30	dk red lf (al) λ^{al} 219 (4.58), 266 (4.03), 273 (4.03), 290 (3.94), 372 (4.10)	70					s	s		s	lig s	**B16**, 78
a1485^2	—, **benzene 2-naphthalene**		231.30	ye (al) λ^{al} 219 (4.48), 265 (4.13), 277 (4.13), 287 (4.13), 328 (4.28)	131					s	s		s	lig s	**B16**^{1}, 231
a1486	—, **benzene 1-naphthalene, 2′-amino-**	Yellow AB. $C_{16}H_{13}N_3$. *See* a1485^1	247.30	red pl (al) λ^{al} 340 (3.9), 450 (4.2)	102–4				i	v				aa v con sulf s (bl)	**B16**^{2}, 193 **B25**^{2}, 545
a1487	—, —, **4′-amino-**	Naphthyl red. $C_{16}H_{13}N_3$. *See* a1485^1	247.30	red lf (dil al), nd (dil al) λ^{al} 285 (4.2), 330 (3.4), 350 (3.5), 455 (4.3)	125–5.5					s	s		s		**B16**^{2}, 186
a1488	—, —, —, hydrochloride	$C_{16}H_{13}N_3$. HCl. *See* a1485^1	283.76	gr nd, pr (al or aa)	205–6				δ	s^h				aa s^h	**B16**^{1}, 324
a1489	—, —, **2,3-dimethyl-2′-hydroxy-**	$C_{18}H_{16}N_2O$. *See* a1485^1	276.34	pa ye amor (al-bz)	125–30 (189)				i	s^h			δ s^h	to, CCl_4 s	**B16**^{2}, 71
a1490	—, —, **2,4-dimethyl-2′-hydroxy-**	$C_{18}H_{16}N_2O$. *See* a1485^1	276.34	red nd (al)	166				i	s	s				**B16**^{2}, 72
a1491	—, —, **2,5-dimethyl-2′-hydroxy-**	$C_{18}H_{16}N_2O$. *See* a1485^1	276.34	nd (al)	153				i	δ s^h			δ		**B16**, 168
a1492	—, —, **3,4-dimethyl-2′-hydroxy-**	$C_{18}H_{16}N_2O$. *See* a1485^1	276.34	red nd (al)	146				i	δ s^h	δ		v	chl v	**B16**^{1}, 260
a1493	—, **benzene 2-naphthalene, 2,4-dimethyl-1′-hydroxy-**	$C_{18}H_{16}N_2O$. *See* a1485^2	276.34	gold-red lf or nd (al-chl)	186					δ^h			s	chl, CS_2 s lig, alk i	**B16**^{1}, 249

For explanations, symbols and abbreviations see beginning of table. For structural formulas see end of table.

No.	Name	Synonyms and Formula	Mol. wt.	Color, crystalline form, specific rotation and λ_{max} (log ε)	m.p. °C	b.p. °C	Density	n_D	Solubility w	al	eth	ace	bz	other solvents	Ref.
	Azo														
a1494	**—,—,4-dimethyl-amino-**	$C_{18}H_{17}N_3$. *See* 1485²	275.36	ye-br (bz-lig) $\lambda^{50\%al-1.2N\,HCl}$ 325 (4.15), 545 (4.54)	174	......	......	......	...	...	...	...	s^h	....	**B16**, 321
a1495	**—,benzene 1-naphthalene 4,4′-dinitro-2-hydroxy-**	$C_{16}H_{10}N_4O_5$. *See* a1485¹ ...	338.28	cr (gl aa) purp bl sol in al-NaOH	231	......	......	......	...	...	...	...	...	aa s^h os δ aq alk i	**B16¹**, 268
a1496	**—,—,4,8′-dinitro-4′-hydroxy-**	$C_{16}H_{10}N_4O_5$. *See* a1485¹ ...	338.28	red ($PhNO_2$)	252–60d	......	......	......	...	...	i	...	...	alk s $PnNO_2$ s^h to lig i	**B16**, 161
a1497	**—,benzene 2-naphthalene, 4,4′-dinitro-1′-hydroxy-**	$C_{16}H_{10}N_4O_5$. *See* a1485² ...	338.28	red nd ($PhNO_2$ -aa)	255d	......	......	......	...	δ	...	δ	...	$PhNO_2$ s aa δ	**B16²**, 67
a1498	**—,—,4,5′-dinitro-1′-hydroxy-**	$C_{16}H_{10}N_4O_5$. *See* a1485² ...	338.28	cr (bz)	210	235d	......	......	...	s	s	...	s^h	$PhNO_2$ y chl s	**B16**, 154
a1499	**—,benzene 1-naphthalene, 2′-hydroxy-**	Sudan Yellow. $C_{16}H_{12}N_2O$. *See* a1485¹	248.29	red-gold lf or nd (al) λ^{al} 315 (3.9), 422 (4.1), 480 (4.3)	133–4	......	......	......	...	s^h	s	...	s	lig, CS_2, con HCl s	**B16²**, 70
a1500	**—,—,4′-hydroxy-**	$C_{16}H_{12}N_2O$. *See* a1485¹	248.29	vt-br lf (bz) λ^{al} 410 (4.2), 470 (4.0)	205–6d	......	......	......	i	s^h	...	...	s^h	dil NaOH, con sulf v aa s^h	**B16²**, 67
a1501	**—,benzene 2-naphthalene, 1′-hydroxy-**	$C_{16}H_{12}N_2O$. *See* a1485²	248.29	red nd (al) λ^{al} 294 (4.12), 356 (3.92), 490 (4.12)	138	sub	......	......	i	v^h	...	...	...	aa s con alk i	**B16¹**, 248
a1502	**—,benzene 1-naphthalene 2′-hydroxy-2-nitro-**	$C_{16}H_{11}N_3O_3$. *See* a1485¹ ...	293.29	og-red nd (gl aa) λ^{chl} 410 (3.9), 480 (4.3), 500 (4.4)	294	......	......	......	...	δ	...	...	...	al-NaOH, aa s^h	**B16²**, 70
a1503	**—,—,2′-hydroxy-4-nitro-**	Para red. Paranitraniline red. $C_{16}H_{11}N_3O_3$. *See* a1485¹	293.29	br-og pl (to or bz)	257 (>300)	......	......	......	i	s	...	s	...	....	**B16²**, 70
a1504	**—,—,2′-hydroxy-4′-nitro-**	$C_{16}H_{11}N_3O_3$. *See* a1485¹ ...	293.29	dk brsh-red nd (bz)	206–7	......	......	......	...	i	...	...	s^h	dil NaOH i	**B16¹**, 267
a1506	**—,—,4′-hydroxy-2′-nitro-**	$C_{16}H_{11}N_3O_3$. *See* a1485¹ ...	293.29	og-red nd (bz) λ^{al} 467	164	......	......	......	...	...	...	...	s^h	aa s dil KOH s^h	**E12B**, 1915
a1507	**—,—,4′-hydroxy-3-nitro-**	$C_{16}H_{11}N_3O_3$. *See* a1485¹ ...	293.29	dk red nd (al) λ^{al} 447	242–3d	......	......	......	...	s	...	...	...	NaOH s	**E12B**, 1762
a1508	**—,—,4′-hydroxy-4-nitro-**	$C_{16}H_{11}N_3O_3$. *See* a1485¹ ...	293.29	red nd	277–9d	......	......	......	i	s	δ	...	s	aa, chl s $PhNO_2$ s^h	**B16¹**, 237
a1509	**—,benzene 2-naphthalene, 1′-hydroxy-4′-nitro-**	$C_{16}H_{11}N_3O_3$. *See* a1485² ...	293.29	dk red nd (gl aa)	182–2	......	......	......	...	δ^h	...	...	s^h	aa s^h	**E12B**, 1914
a1510	**—,1-naphthalene 2-pyridine, 2-hydroxy-**	....	249.28	og-red nd (al)	137	......	......	......	i	s	s	...	...	dil ac s, dil alk s^h	**B22¹**, 694
a1511	**—,—,4-hydroxy-**	....	249.28	gr-red nd	......	......	......	......	i	...	...	...	...	dil ac, dil alk, con sulf s	**B22¹**, 694
a1512	**Azobenzene** (*cis*) ...	Azobenzide. Benzene-azobenzene. Diphenyldiimide.	182.23	og-red pl (peth) λ^{al} 243 (4.02), 281 (3.72), 433 (3.18)	71	......	......	......	δ	s	s	...	s	aa, lig s	**B16²**, 4
a1513	—(*trans*).........	....	182.23	og red mcl lf (al) λ^{al} 226 (3.99), 320 (4.28), 350 (2.63)	68.5	293^{760}	1.203_4^{20}	1.6266^{78}	δ	s	s	...	s	Py v MeOH, chl, aa s lig δ	**B16²**, 4
a1514	**—,4-acetamido-** ...	$C_{14}H_{13}N_3O$. *See* a1513	239.28	gold-ye nd (al) λ^{chl} 347 (4.37)	144–6	......	......	......	δ	...	δ	...	δ	....	**B16²**, 155
a1515	**—,2-acetamido-4′,5-dimethyl-**	$C_{16}H_{17}N_3O$. *See* a1513	267.33	ye nd (al-aa)	157	......	......	......	i	s	s	...	...	chl s	**B16²**, 182
a1516	**—,4-acetamido-2′,3-dimethyl-**	$C_{16}H_{17}N_3O$. *See* a1513.....	267.33	red nd (al)	186–7	......	......	......	i	δ	s	...	...	chl s	**B16²**, 179
a1517	**—,4-acetoxy-2′-methyl-**	$C_{15}H_{14}N_2O_2$. *See* a1513.....	254.29	red-ye lf (al)	68	......	......	......	i	s^h	...	...	...	....	**B16**, 105
a1518	**—,4-acetoxy-3-methyl-**	$C_{15}H_{14}N_2O_2$. *See* a1513	254.29	ye pl (dil al)	81–2	......	......	......	...	s	s	...	s	chl s	**B16**, 130
a1519	**—,4-acetoxy-4′-methyl-**	$C_{15}H_{14}N_2O_2$. *See* a1513 ...	254.29	og nd (al, bz)	98	......	......	......	...	s^h	...	...	...	....	**B16²**, 42

For explanations, symbols and abbreviations see beginning of table. For structural formulas see end of table.

No.	Name	Synonyms and Formula	Mol. wt.	Color, crystalline form, specific rotation and λ_{max} (log ε)	m.p. °C	b.p. °C	Density	n_D	Solubility w	al	eth	ace	bz	other solvents	Ref.
	Azobenzene														
a1520	**—,2-amino-**	2-Benzeneazoaniline. $C_{12}H_{11}N_3$. *See* a1513	197.24	red nd or pr (al) λ^{al} 222 (4.17), 258 sh (3.95), 313 (4.17), 424 (3.91)	59	...	...	...	i	δ	v	v	v	os v	**B16**[2], 147
a1521	**—,3-amino-**	3-Benzeneazoaniline. $C_{12}H_{11}N_3$. *See* a1513	197.24	(i) og-ye nd (peth) (ii) br-red cr λ^{al} 301 (4.2), 461 sh (3.3)	(i) 69–70 (ii) 90–1	...	...	...	i	s	s	s	s	chl s	**B16**, 304
a1522	**—,4-amino-**	4-Benzeneazoaniline. $C_{12}H_{11}N_3$. *See* a1513	197.24	og mcl nd (al) λ^{al} 251 (4.0), 384 (4.4)	127	>360	...	...	δ^h	s^h	s	...	s^h	chl s lig δ	**B16**[2], 149
a1523	**—,—,hydrobromide**	$C_{12}H_{11}N_3$. HBr. *See* a1513	278.04	bk-vt nd (dil al)	206–7	...	...	...	...	δ	δ	...	...	...	**B16**, 307
a1524	**—,—,hydrochloride**	$C_{12}H_{11}N_3$. HCl. *See* a1513	233.70	bl-vt or pa red nd or pw (w or HCl)	240	...	...	...	s^h	s^h	...	...	...	HCl s^h	**B16**[2], 149
a1525	**—,4-amino-2,2′-dimethyl-**	*o*-Tolueneazo-*m*-toluidine. $C_{14}H_{15}N_3$. *See* a1513	225.30	ye nd (lig)	116–7	...	...	...	δ	s	...	...	...	lig v^h	**B16**[2], 180
a1526	**—,4-amino-2,3′-dimethyl-**	*m*-Tolueneazo-*m*-toluidine. $C_{14}H_{15}N_3$. *See* a1513	225.30	ye-gold nd (al), ye-br nd (lig)	80	...	...	...	δ	s	...	...	...	lig s^h	**B16**[2], 181
a1527	**—,4-amino-2′,3-dimethyl-**	*o*-Tolueneazo-*o*-toluidine. $C_{14}H_{15}N_3$. *See* a1513	225.30	ye lf (al)	101.5–3	...	...	...	i	s	s	...	...	chl s	**B16**[2], 178
a1528	**—,4-amino-2,4′-dimethyl-**	*p*-Tolueneazo-*m*-toluidine. $C_{14}H_{15}N_3$. *See* a1513	225.30	ye pl (al), gold-ye (lig)	127	...	...	...	...	s^h	...	...	...	lig s^h	**B16**, 348
a1529	**—,4-amino-3,3′-dimethyl-**	*m*-Tolueneazo-*o*-toluidine. $C_{14}H_{15}N_3$. *See* a1513	225.30	ye br lf or nd (lig)	124	...	...	...	...	s^h	...	...	...	lig s^h	**B16**, 345
a1530	**—,4-amino-3,4′-dimethyl-**	*p*-Tolueneazo-*o*-toluidine. $C_{14}H_{15}N_3$. *See* a 1513	225.30	og-ye nd (lig), ye pl (al)	128	...	...	...	i	δ s^h	...	...	...	lig δ, s^h	**B16**[1], 322
a1531	**—,4-benzoxy-4′-methyl-**	4-*p*-Tolueneazophenol benzyl ether. $C_{20}H_{18}N_2O$. *See* a1513	302.38	pa ye lf (lig)	128	...	...	...	...	s	v	...	v	lig s^h	**B16**, 107
a1532	**—,3,3′-bis(dimethylamino)-**	3,3′-Azo-bis(*N,N*-dimethylaniline). $C_{16}H_{20}N_4$. *See* a1513	268.37	red nd (al)	118	...	...	...	i	s	...	...	s	lig δ	**B16**, 305
a1533	**—,4,4′-bis(dimethylamino)-**	4,4′-Azo-bis(*N,N*-dimethylaniline). $C_{16}H_{20}N_4$. *See* a1513	268.37	og nd (bz) $\lambda^{95\%sulf-al}$ 405 (4.52), 485 sh (3.80)	273 (266)	sub	...	...	...	δ	s	...	v	chl v	**B16**[2], 174
a1534	**—,4-bromo-**	$C_{12}H_9BrN_2$. *See* a1513	261.13	og nd (lig) $\lambda^{15\%sulf}$ 328 (4.3)	90–1	sub	...	...	i	s	s	s	...	lig s	**B16**[2], 14
a1535	**—,4-diacetamido-2,3′-dimethyl-**	$C_{18}H_{19}N_3O_2$. *See* a1513	309.37	(i) red nd (lig), (ii) cr	(i) 65 (ii) 75	...	...	...	...	s	s	s	...	os v	**B16**[1], 322
a1536	**—,2,2′-diamino-**	*o,o′*-Azodianiline. $C_{12}H_{12}N_4$. *See* a1513	212.26	red pl (al or bz)	134	...	...	...	δ	δ s^h	s	s	δ	...	**B16**[2], 148
a1537	**—,2,4-diamino-**	Chrysoidine. $C_{12}H_{12}N_4$. *See* a1513	212.26	pa ye nd (w)	117.5	...	...	...	δ^h	s	s	...	s	chl s	**B16**[2], 203
a1538	**—,4,4′-diamino-**	*p,p′*-Azodianiline. $C_{12}H_{12}N_4$. *See* a1513	212.26	gold-ye nd (al), og-ye pr (al or bz) λ^{al} 305 sh (3.7), 399 (4.5), 430 sh (4.4)	250–1	...	...	...	δ	s	...	...	v	chl v lig δ	**B16**[2], 174
a1539	**—,2,2′-diethoxy-**	*o*-Azophenetole. $C_{16}H_{18}N_2O_2$. *See* a1513	270.34	red pr (al) λ^{bz} 317 (4.0), 370 (4.1)	131	240d	...	...	i	s	s	...	...	HCl s	**B16**, 92
a1540	**—,4,4′-diethoxy-**	*p*-Azophenetole. $C_{16}H_{18}N_2O_2$. *See* a1513	270.34	ye-lf (al) λ^{bz} 362 (4.4), 428 sh (3.3)	162 (159)	d	...	...	i	δ v^h	s	...	s	chl s aa v^h	**B16**[2], 44
a1541	**—,2,2′-dihydroxy-**	*o*-Azophenol. $C_{12}H_{10}N_2O_2$. *See* a1513	214.23	gold-ye lf (bz), nd (al) λ^{al} 259 (3.80)	172	sub	...	...	i	δ s^h	v	...	δ s^h	con alk s	**B16**[2], 33
a1541[1]	**—,2,4-dihydroxy-**	Benzeneazoresorcinol. Sudan G. $C_{12}H_{10}N_2O_2$. *See* a1513	214.23	dk red nd (dil al)	161 (½w) 170 (anh)	...	...	...	i	v	v	...	v	chl, aa v	**B16**[2], 80
a1542	**—,3,3′-dihydroxy-**	*m*-Azophenol. $C_{12}H_{10}N_2O_2$. *See* a1513	214.23	ye lf (dil al)	207	...	...	...	δ	s^h	v	v	δ	alk s lig δ	**B16**[2], 37
a1543	**—,4,4′-dihydroxy-**	*p*-Azophenol. $C_{12}H_{10}N_2O_2.H_2O$. *See* a1513	232.24	og-ye pl (dil al + 1w), gr (st), red (metast), ta (eth) λ^{al} 355 (4.3)	216–6.5 (anh)	...	...	...	δ	s	s	s	δ	aa, lig δ	**B16**[2], 43
a1544	**—,2,4-dihydroxy-4′-nitro-**	4-(*p*-Nitrobenzeneazo)-resorcinol. $C_{12}H_9N_3O_4$. *See* a1513	259.22	red pw (al or MeOH)	200	...	...	...	i	δ^h	...	...	δ	aa, to δ	**B16**[2], 81
a1545	**—,2,2′-dimethyl-**	*o*-Azotoluene. $C_{14}H_{14}N_2$. *See* a1513	210.28	dk red mcl pr (eth) λ^{al} 235 (3.98), 332 (4.24)	55–6	...	1.0215^{65}_{4}	1.6180^{65}	i	s	v	...	v	CS_2, chl s	**B16**[2], 19

For explanations, symbols and abbreviations see beginning of table. For structural formulas see end of table.

No.	Name	Synonyms and Formula	Mol. wt.	Color, crystalline form, specific rotation and λ_{max} (log ε)	m.p. °C	b.p. °C	Density	n_D	Solubility w	al	eth	ace	bz	other solvents	Ref.
	Azobenzene														
a1546	—,3,3′-dimethyl- (*cis*)	*m*-Azotoluene. $C_{14}H_{14}N_2$. *See* a1512	210.28	red (peth) λ^{chl} 329 (3.84), 424 (3.24)	46	...	...	...	...	...	...	...	...	...	**J1953** 2143
a1547	—,—(*trans*)	$C_{14}H_{14}N_2$. *See* a1513	210.28	og-red orh λ^{chl} 331 (4.18), 447 (2.79)	54–4.5	...	1.0123^{66}	1.6152^{66}	...	v	v	...	v	...	**B16**[2], 20
a1548	—,4,4′-dimethyl- (*cis*)	*p*-Azotoluene. $C_{14}H_{14}N_2$. *See* a1512	210.28	dk red λ^{al} 247.5 (4.04), 296 (3.80), 438 (3.31)	104	...	...	...	i	...	...	...	s	lig s^h	**C48,** 2475
a1549	—,—(*trans*)	$C_{14}H_{14}N_2$. *See* a1513	210.28	og-ye nd (al) λ^{al} 233 (4.19), 336 (4.41), 438 (2.87)	144	...	...	...	i	δ	s	...	s	lig s dil ac i	**B16**[2], 21
a1549[1]	—,4-dimethyl-amino-	Butter yellow. Methyl yellow. $C_{14}H_{15}N_3$. *See* a1513	225.30	ye lf (al) λ^{al} 410 (4.44)	117	d	...	...	i	v	s	...	...	Py v ac s lig δ^h	**B16**[2], 151
a1550	—,2,2′-dimethyl-4-ethoxy-	$C_{16}H_{18}N_2O$. *See* a1513	254.34	dk red nd (al)	64	...	...	...	...	s	s	s	s	lig s	**B16,** 134
a1551	—,2,3′-dimethyl-4-ethoxy-	$C_{16}H_{18}N_2O$. *See* a1513	254.34	red pr (al)	73	...	...	...	...	v	s	s	s	os v	**B16,** 135
a1552	—,2′,3-dimethyl-4-ethoxy-	$C_{16}H_{18}N_2O$. *See* a1513	254.34	red cr (lig)	35–7	...	...	...	...	s	δ	...	s	lig s^h	**B16,** 131
a1553	—,2,4′-dimethyl-4-ethoxy-	$C_{16}H_{18}N_2O$. *See* a1513	254.34	og-red pl (al)	64	...	...	...	...	v	v	...	v	lig v	**B16,** 135
a1554	—,3,3′-dimethyl-4-ethoxy-	$C_{16}H_{18}N_2O$. *See* a1513	254.34	red-ye pl (al)	46–7	...	...	...	...	s	s	s	s	lig s	**B16,** 131
a1555	—,3,4′-dimethyl-4-ethoxy-	$C_{16}H_{18}N_2O$. *See* a1513	254.34	og-ye nd (al)	73–4	251^{42}	...	...	...	s	s	s	s	os v	**B16,** 131
a1556	—, 4′,5-dimethyl-2-ethoxy-	$C_{16}H_{18}N_2O$. *See* a1513	254.34	pa red nd (abs al)	43	$253–5^{63}$	...	...	...	s^h	...	...	...	...	**B16,** 141
a1557	—, 2,2′-dimethyl-4-hydroxy-	4-*o*-Tolueneazo-*m*-cresol. $C_{14}H_{14}N_2O$. *See* a1513	226.28	og-red pl (bz), red cr (w + 1)	113(anh) 83(hyd)	...	...	...	...	v	v	...	v	...	**B16**[2], 61
a1558	—, 2,3-dimethyl-4-hydroxy-	4-Benzeneazo-*o*-3-xylenol. $C_{14}H_{14}N_2O$. *See* a1513	226.28	red pr (al, lig)	132	...	...	...	i	s	s	s	s	oos s lig s^h	**B16**[2], 60
a1559	—, 2,3′-dimethyl-4-hydroxy-	4-*m*-Tolueneazo-*m*-cresol. $C_{14}H_{14}N_2O$. *See* a1513	226.28	og-ye pl (bz)	106–7	...	...	...	...	δ	v	...	v	lig v	**B16**[2], 61
a1560	—, 2,4-dimethyl-5-hydroxy-	5-Benzeneazo-2,4-xylenol. $C_{14}H_{14}N_2O$. *See* a1513	226.28	og-ye nd (lig-peth)	114	...	...	...	...	v	s	v	v	MeOH, aa peth v	**B16,** 146
a1561	—, 2,4′-dimethyl-4-hydroxy-	4-*p*-Tolueneazo-*m*-cresol. $C_{14}H_{14}N_2O$. *See* a1513	226.28	og-ye pr (bz)	135	...	...	...	...	v	v	...	v	lig δ	**B16**[2], 62
a1562	—, 3,3′-dimethyl-4-hydroxy-	4-*m*-Tolueneazo-*o*-cresol. $C_{14}H_{14}N_2O$. *See* a1513	226.28	gold-ye nd (bz)	115	...	...	...	...	v	v	...	v	...	**B16**[2], 60
a1563	—, 3,4′-dimethyl-4-hydroxy-	4-*p*-Tolueneazo-*o*-cresol. $C_{14}H_{14}N_2O$. *See* a1513	226.28	og nd (bz)	163	...	...	...	...	s	s	...	s	os v	**B16**[2], 60
a1564	—,3,5-dimethyl-2-hydroxy-	6-Benzeneazo-*m*-4-xylenol. $C_{14}H_{14}N_2O$. *See* a1513	226.28	dk red nd (al, lig) λ^{al} 332 (4.38), 395 (3.93)	90	...	...	...	...	v	v	...	v	aa, lig v	**B16,** 145
a1565	—, 3,5-dimethyl-4-hydroxy-	4-Benzeneazo-2,6-xylenol. $C_{14}H_{14}N_2O$. *See* a1513	226.28	ye nd, pr (lig) λ^{al} 351 (4.45)	95–6	...	...	...	...	s	s	...	s	os v aa s lig s^h	**B16,** 145
a1566	—, 4,4′-dimethyl-2-hydroxy-	6-*p*-Tolueneazo-*m*-cresol. $C_{14}H_{14}N_2O$. *See* a1513	226.28	og-red cr (lig)	150–1	...	...	...	...	s	s	...	s	lig v^h alk δ	**B16**[2], 61
a1567	—, 4′,5-dimethyl-2-hydroxy-	2-*p*-Tolueneazo-*p*-cresol. $C_{14}H_{14}N_2O$. *See* a1513	226.28	red cr, ye pl (to)	112–3	...	...	...	...	δ v^h	v	...	v	chl v to s	**B16**[2], 63
a1569	—, 2-ethoxy-	*o*-Benzeneazophenetole. $C_{14}H_{14}N_2O$. *See* a1513	226.28	red pl or mcl pr (peth)	44	...	...	...	...	v	v	v	v	os v peth v^h	**B16,** 91
a1570	—, 3-ethoxy-	*m*-Benzeneazophenetole. $C_{14}H_{14}N_2O$. *See* a1513	226.28	pl (peth)	63.5–4	200^{22}	...	...	...	v	v	v	v	os v	**B16,** 95
a1571	—, 4-ethoxy-	*p*-Benzeneazophenetole. $C_{14}H_{14}N_2O$. *See* a1513	226.28	og nd (60–70 % al) $\lambda^{15\ \%\ al}$ 225 sh (4.2), 348 (4.3), 415 sh (2.9)	85	339–40	1.0400_4^{100}	1.6419_{5875}^{100}	...	v	v	s	v	os v	**B16**[2], 40
a1572	—, 4-ethoxy-2-methyl-	$C_{15}H_{16}N_2O$. *See* a1513	240.31	og-red nd (dil al)	51.5–2.0	...	...	...	δ	v	v	...	...	lig v	**B16,** 134
a1573	—, 4-ethoxy-2′-methyl-	$C_{15}H_{16}N_2O$. *See* a1513	240.31	og pl (al)	53	...	...	...	δ	s	s	...	s	chl s	**B16,** 105
a1574	—, 4-ethoxy-3-methyl-	$C_{15}H_{16}N_2O$. *See* a1513	240.31	og nd or mcl pr (al)	60	...	...	...	...	s	s	...	s	...	**B16,** 130
a1575	—, 4-ethoxy-3′-methyl-	$C_{15}H_{16}N_2O$. *See* a1513	240.31	og-red pr (al)	65	...	...	...	...	s	v	...	v	...	**B16,** 106
a1576	—, 4-ethoxy-4′-methyl-	$C_{15}H_{16}N_2O$. *See* a1513	240.31	red lf (al)	121–2	251^{47}	...	...	...	s v^h	...	...	s	chl v	**B16,** 107
a1577	—, 2-hydroxy-	*o*-Benzeneazophenol. $C_{12}H_{10}N_2O$. *See* a1513	198.23	og-red nd (eth) λ^{al} 323 (4.26)	82.5–3	...	...	...	δ	s	s	s	s	os, dil alk s (og-red)	**B16**[2], 32
a1578	—, —, benzoate	$C_{19}H_{14}N_2O_2$. *See* a1513	302.34	og-red nd (lig) λ^{eth} 260 (> 3.0), 380 (1.8)	93	...	...	...	...	s^h	...	...	s^h	lig s^h	**B16,** 91

For explanations, symbols and abbreviations see beginning of table. For structural formulas see end of table.

No.	Name	Synonyms and Formula	Mol. wt.	Color, crystalline form, specific rotation and λ_{max} (log ε)	m.p. °C	b.p. °C	Density	n_D	Solubility w	al	eth	ace	bz	other solvents	Ref.
	Azobenzene														
a1579	**—, 3-hydroxy-**	*m*-Benzeneazophenol. $C_{12}H_{10}N_2O$. *See* a1513	198.23	ye nd (w, lig), pr (bz) λ^{eth} 248 (2.8), 380 (1.8)	116.5–7	...	...	...	i	s	s	s	s	dil alk s lig δ	**B16**, 94
a1580	—, —, acetate	$C_{14}H_{12}N_2O_2$. *See* a1513	240.27	og pl (peth)	67.5	...	...	...	...	...	...	...	...	peth s^h	**B16**, 95
a1581	—, —, benzoate	$C_{19}H_{14}N_2O_2$. *See* a1513	302.34	og-red pl (peth)	92	...	...	...	...	...	...	...	...	peth s^h	**B16**, 95
a1582	**—, 4-hydroxy-**	*p*-Benzeneazophenol. $C_{12}H_{10}N_2O$. *See* a1513	198.23	ye lf (bz), og pr (al) λ^{al} 236.5 (4.0)	155–7	220–30[20] δd	...	...	i	v	v	...	s	con sulf, dil alk s (ye)	**B16**[2], 38
a1583	—, —, acetate	$C_{14}H_{12}N_2O_2$. *See* a1513	240.27	og lf (al), nd (al, lig) λ^{al} 325 (4.34), 440 (2.80)	89	> 360d	...	...	...	s^h	...	...	...	lig s^h	**B16**[1], 236
a1584	—, —, benzoate	$C_{19}H_{14}N_2O_2$. *See* a1513	302.34	og-ye nd (al, lig), ye-red pr (eth, al), ye lf (al)	138	...	...	...	...	δ s^h	δ s^h	...	...	to v lig s^h	**B16**[2], 41
a1585	**—, 4-hydroxy-2′-methoxy-2-methyl-**	$C_{14}H_{14}N_2O_2$. *See* a1513	242.28	og	161	...	...	...	δ	v	s	...	...	...	**B16**, 135
a1586	**—, 4-hydroxy-2′-methoxy-3-methyl-**	$C_{14}H_{14}N_2O_2$. *See* a1513	242.28	og red pl (dil al)	68	...	...	...	...	v	v	...	v	...	**B16**, 131
a1587	**—, 2-hydroxy-4-methyl-**	6-Benzeneazo-*m*-cresol. $C_{13}H_{12}N_2O$. *See* a1513	212.25	red pl (lig)	122	...	...	...	...	s	...	s	s	chl, lig s alk δ	**B16**[1], 241
a1588	—, —, benzoate	$C_{20}H_{16}N_2O_2$. *See* a1513	316.36	og-ye nd (lig)	98	...	...	...	...	δ	...	...	v	MeOH δ	**B16**[1], 241
a1589	**—, 2-hydroxy-5-methyl-**	2-Benzeneazo-*p*-cresol. $C_{13}H_{12}N_2O$. *See* a1513	212.25	og-ye lf (bz, bz-lig), gold lf (w-al), λ^{bz} 330 (4.49), 410 (4.28)	108–9	sub	...	...	...	s	s	...	s	chl s	**B16**[2], 62
a1590	**—, 4-hydroxy-2-methyl-**	4-Benzeneazo-*m*-cresol. $C_{13}H_{12}N_2O$. *See* a1513	212.25	ye nd (lig) λ^{al} 352 (4.2)	109	...	...	...	...	s	s	...	s	chl s	**B16**[2], 61
a1591	**—, 4-hydroxy-2′-methyl-**	*o*-Tolueneazo-*p*-phenol. $C_{13}H_{12}N_2O$. *See* a1513	212.25	red pl or lf (bz-lig), og-ye nd (bz)	107–8	...	...	...	...	s	s	...	s	chl s lig δ	**B16**[2], 41
a1592	**—, 4-hydroxy-3-methyl-**	4-Benzeneazo-*o*-cresol. $C_{13}H_{12}N_2O$. *See* a1513	212.25	gold-ye lf or nd (al) $\lambda^{DMF-alk}$ 489 (4.62)	128–30	...	...	...	i δ^h	s	s	...	s	chl, lig s	**B16**[2], 59
a1593	—, —, benzoate	$C_{20}H_{16}N_2O_2$. *See* a1513	316.36	ye nd (al)	110–1	...	...	...	...	δ v^h	v	v	...	chl v	**B16**[2], 130
a1594	**—, 4-hydroxy-3′-methyl-**	*m*-Tolueneazo-*p*-phenol. $C_{13}H_{12}N_2O$. *See* a1513	212.25	ye pl (al), dk ye pr (bz-lig)	144–5	...	...	...	...	s^h	...	...	s	lig δ	**B16**[2], 41
a1595	**—, 4-hydroxy-4′-methyl-**	*p*-Tolueneazo-*p*-phenol. $C_{13}H_{12}N_2O$. *See* a1513	212.25	og-red mcl, ye (Na_2CO_3 sol or bz-lig), og-ye (bz-lig) $\lambda^{10\% Sulf}$ 240 (4.8), 360 (4.32), 470 (4.04)	152	...	...	...	δ s^h	v	v	...	v	alk s	**B16**[2], 42
a1596	—, —benzoate	$C_{20}H_{16}N_2O_2$. *See* a1513	316.36	og-red pr (bz), redsh-ye nd (lig)	178	...	...	...	...	...	v	v	s^h	lig s^h	**B16**[1], 237
a1597	**—, 2-methoxy-**	*o*-Benzeneazoanisole. $C_{13}H_{12}N_2O$. *See* a1513	212.25	og-red nd (dil al) λ^{al} 317 (4.2), 353 (4.2), 430 (3.0)	41	195–7[14]	1.0728_4^{100}	...	i	s^h	...	s	...	os v	**B16**[2], 33
a1598	**—, 3-methoxy-**	*m*-Benzeneazoanisole. $C_{13}H_{12}N_2O$. *See* a1513	212.25	og-red pl (MeOH) λ 312.5 (4.34), 444.5 (2.75)	32.5–3.5	193–3.5[15]	...	...	i	s^h	...	s	...	os v MeOH s^h	**B16**, 95
Ω a1599	**—, 4-methoxy-**	*p*-Benzeneazoanisole. $C_{13}H_{12}N_2O$. *See* a1513	212.25	og-red pl, lf (al), (peth) λ^{al} 342 (4.3), 432 (3.2)	56(64)	340	1.12[75]	...	i	s^h	s	s	...	os v	**B16**[2], 40
a1600	**—, 4-methoxy-4′-methyl-**	$C_{14}H_{14}N_2O$. *See* a1513	226.28	og-ye pr (al)	110–1	...	...	...	...	s^h	s	s	s	os s	**B16**[1], 236
Ω a1601	**—, 4-nitro-**	$C_{12}H_9N_3O_2$. *See* a1513	227.23	lf, nd (al, lig) λ^{al} 333 (4.4), 446 (3.0)	135(155)	...	...	...	...	s^h	...	s	s	chl, aa, lig s	**B16**[2], 17
a1602	**—, 4-phenyl-amino-**	4-Anilineazobenzene. $C_{18}H_{15}N_3$. *See* a1513	273.34	ye pl or pr λ^{al} 260 (4.10), 380 (4.19), 450 sh (4.80)	82	...	...	...	i	v	v	...	...	lig v	**B16**, 315
a1603	**—, 2,4,3′-triamino-**	Bismark's brown. $C_{12}H_{13}N_5$. *See* a1513	227.27	og lf (w), red (bz)	143–5	...	...	...	i	v	v	...	...	...	**B16**, 386
a1604	**2-Azobenzene-carboxylic acid**	*o*-Benzeneazobenzoic acid.	226.24	og-red nd or pl (al)	97–8(95)	...	...	...	...	v	v	s	v	aa v	**B16**[2], 97

For explanations, symbols and abbreviations see beginning of table. For structural formulas see end of table.

No.	Name	Synonyms and Formula	Mol. wt.	Color, crystalline form, specific rotation and λ_{max} (log ε)	m.p. °C	b.p. °C	Density	n_D	Solubility w	al	eth	ace	bz	other solvents	Ref.
	3-Azobenzenecarboxylic acid														
a1605	**3-Azobenzene-carboxylic acid**	*m*-Benzeneazobenzoic acid.	226.24	og-red pl (w), red lf (al)	170–1	...	...	...	δ^h	s	v	...	v	chl, AcOEt s	**B16**, 229
a1606	**4-Azobenzene-carboxylic acid**	*p*-Benzeneazobenzoic acid.	226.24	red pl or lf (al) λ^{PrOH} 323 (4.42)	248.5–9.5 (241)	...	...	...	...	s^h	s	s	s^h	chl s aa s^h lig, CS_2 δ	**B16**[2], 98
—	**2-Azobenzene-carboxylic acid, 4′-dimethyl-amino-**	*see* **Methyl red**													
a1607	**3-Azobenzene-carboxylic acid, 4-hydroxy-**	5-Benzeneazosalicylic acid. $C_{13}H_{10}N_2O_3$. *See* a1605	242.24	nd (bz) λ^{al} 350 (4.4)	220d	...	...	...	...	s	s	s	δ	chl δ CS_2 δ^h	**B16**[2], 100
a1608	—,4-hydroxy-2′-nitro-	5-*o*-Nitrobenzeneazosalicylic acid. $C_{13}H_9N_3O_5$. *See* a1605	287.23	br-red cr (al)	215–7	...	...	...	δ	s^h	...	...	...	aa s^h chl δ	**B16**[2], 101
a1609	—,4-hydroxy-3′-nitro-	5-*m*-Nitrobenzeneazosalicylic acid. $C_{13}H_9N_3O_5$. *See* a1605	287.23	red-br nd (al)	237	...	...	...	i	v	v	...	v	chl, aa v	**B16**[2], 101
a1610	—,4-hydroxy-4′-nitro-	5-*p*-Nitrobenzeneazosalicylic acid. $C_{13}H_9N_3O_5$. *See* a1605	287.23	og-br nd (dil aa)	257	...	...	...	i	s	...	...	...	aa s to s^h	**B16**[2], 101
a1611	—,3′-sulfo-	$C_{13}H_{10}N_2O_5S$. *See* a1605	306.30	ye lf (al)	d	...	...	...	v	v	i	...	...	...	**B16**, 268
a1612	**2,2′-Azobenzene-dicarboxylic acid**	*o*-Azobenzoic acid.	270.25	dk ye nd (al)	245–5.5 (cor)	...	...	...	δ^h	s^h	v	...	i	...	**B16**[1], 287
a1613	**3,3′-Azobenzene-dicarboxylic acid**	*m*-Azobenzoic acid.	270.25	ye nd (aa)	340d	...	...	...	i	δ^h	i	...	...	...	**B16**, 233
a1614	**4,4′-Azobenzene-dicarboxylic acid**	*p*-Azobenzoic acid.	270.25	dk ye nd (al), red nd (aa) $\lambda^{0.1N NaOH}$ 225 (4.08), 331 (4.41), 430 (3.11)	d330	...	...	...	i	i	i	...	...	aa δ^h os i	**B16**, 236
a1614[1]	**3,3′-Azobenzene-disulfonic acid**	...	342.35	ye lf (w + 5)	...	...	...	...	s	s	s	...	...	...	**B16**[2], 114
a1615	—,diamide	$C_{12}H_{12}N_4O_4S_2$. *See* a1614[1]	340.38	ye nd (al)	305	...	...	...	...	s^h	...	...	...	os δ	**B16**, 268
a1616	—,dichloride	$C_{12}H_8Cl_2N_2O_4S_2$. *See* a1614[1]	379.25	red nd (eth)	166–7	...	...	...	d	d^h	s^h	...	δ	...	**B16**, 268
a1617	—,diethyl ester	$C_{16}H_{18}N_2O_6S_2$. *See* a1614[1]	398.46	gold-ye nd (eth)	100	...	...	...	δ	s	s	...	...	...	**B16**, 268
a1617[1]	**3,4′-Azobenzene-disulfonic acid**	...	342.35	syr	...	...	...	...	...	...	...	...	...	...	**B16**, 279
a1618	—,diamide	$C_{12}H_{12}N_4O_4S_2$. *See* a1617[1]	340.38	ye nd (al)	172 (288)	...	...	...	δ	s^h	...	...	...	...	**B16**, 279
a1619	—,dichloride	$C_{12}H_8Cl_2N_2O_4S_2$. *See* a1617[1]	379.25	red nd (eth, CS_2)	123–5	...	...	...	d^h	d^h	s^h	...	...	CS_2 s^h	**B16**, 279
a1619[1]	**4,4′-Azobenzene-disulfonic acid**	...	342.35	red nd (w + 2,3 or 5)	169d (anh)	...	...	...	s	...	...	...	...	...	**B16**[2], 119
a1620	—,diamide	$C_{12}H_{12}N_4O_4S_2$. *See* a1619[1]	340.38	og nd (al) $\lambda^{50\% al}$ 232 (4.0), 337 (4.3)	> 250d	...	...	...	δ	δ^h	i	...	i	...	**B16**, 280
a1621	—,dichloride	$C_{12}H_8Cl_2N_2O_4S_2$. *See* a1619[1]	379.25	br-red nd (eth, bz)	222	...	...	...	d^h	d^h	δ s^h	...	v	chl v	**B16**, 280
a1622	**3,3′-Azobenzene-disulfonic acid, 4,4′-dimethyl-,** dichloride	$C_{14}H_{12}Cl_2N_2O_4S_2$. *See* a1614	407.30	dk red (bz)	194	...	...	...	d^h	d^h	...	...	s^h	...	**B16**, 283
a1623	—,6,6′-dimethyl-, dichloride	$C_{14}H_{12}Cl_2N_2O_4S_2$. *See* a1614	407.30	red pr (bz + 2)	220	...	...	...	d^h	d^h	δ	...	s^h	...	**B16**, 285
a1624	**4,4′-Azobenzene-disulfonic acid, 2,2′-dimethyl-,** diamide	$C_{14}H_{16}N_4O_4S_2$. *See* a1619[1]	368.44	pl(NH_3)	> 250	...	...	...	δ	...	...	...	...	NH_3 v	**B16**, 284
a1625	—,—,dichloride	$C_{14}H_{12}Cl_2N_2O_4S_2$. *See* a1619[1]	407.30	dk red nd (bz)	218	...	...	...	d^h	d^h	s	...	v^h	...	**B16**, 284
a1626	**3,3′-Azobenzene-disulfonic acid, 2,2′,4,4′,6,6′-hexabromo-,** dichloride	$C_{12}H_2Br_6Cl_2N_2O_4S_2$. *See* a1614[1]	852.65	vt pl (bz)	222–4	...	...	...	d	...	δ	...	δ s^h	...	**B16**, 269
a1627	—,4,4′,6,6′-tetrabromo-, dichloride	$C_{12}H_4Br_4Cl_2N_2O_4S_2$. *See* a1614[1]	694.85	red nd (bz)	232–3	...	...	...	d	d^h	δ	...	s^h	...	**B16**, 269
a1628	**4,4′-Azobenzene-disulfonic acid, 2,2′,6,6′-tetrabromo-,**	$C_{12}H_6Br_4N_2O_6S_2$. *See* a1619[1]	657.96	ye-br lf (bz)	258–62	...	...	...	d^h	d^h	δ	...	v	...	**B16**, 281
a1629	**4-Azobenzene-sulfonic acid**	4-Phenylazobenzenesulfonic acid*.	262.29	og-red lf (w + 3) λ^{al} 232 (4.07), 320 (4.32), 439 (2.90)	127(hyd)	...	...	...	δ v^h	δ	δ	...	...	...	**B16**[2], 115
a1630	—,chloride	$C_{12}H_9ClN_2O_2S$. *See* a1629	280.74	og-red nd or pl (bz) λ^{hp} 230, 320	82	...	...	...	i	s	...	...	s^h	...	**B16**, 272

For explanations, symbols and abbreviations see beginning of table. For structural formulas see end of table.

C-136

No.	Name	Synonyms and Formula	Mol. wt.	Color, crystalline form, specific rotation and λ_{max} (log ε)	m.p. °C	b.p. °C	Density	n_D	Solubility w	al	eth	ace	bz	other solvents	Ref.
	4-Azobenzenesulfonic acid														
a1631	—,4′-chloro-	$C_{12}H_9ClN_2O_3S$. *See* a1629	296.74	br nd (w) λ^{al} 295 (4.4), 415 (3.1)	148	...	...	...	v	v	...	...	...	...	**B16**, 271
a1632	—,—,amide	$C_{12}H_{10}ClN_3O_2S$. *See* a1629	295.75	ye-br pr (al)	211	...	...	...	...	δ s^h	δ	...	...	...	**B16**, 272
a1633	—,—,chloride	$C_{12}H_8Cl_2N_2O_2S$. *See* a1629	315.19	red pr (eth)	130	...	...	...	d^h	s	s	...	...	...	**B16**, 272
—	**2-Azobenzene-sulfonic acid, 4′-dimethylamino-**, sodium salt	*see* **Methyl orange**													
a1634	**3-Azobenzene-sulfonic acid, 4′-formyl-3′-hydroxy-**	...	306.30	red lf	>270	...	...	...	v	v	i	...	...	HCl δ	**B16**, 267
a1635	**4-Azobenzene-sulfonic acid, 4′-hydroxy-**	$C_{12}H_{10}N_2O_4S$. *See* a1629	278.29	ye red pr (dil al), amor (w) $\lambda^{w, pH=2.9}$ 347 (4.36)	200d	...	...	...	s	v^h	δ	...	...	...	**B16**[2], 115
a1636	**3-Azobenzene-sulfonic acid, 4′-hydroxy-3′-nitro-**	...	323.29	gold-ye pr (dil (HCl + w)	116(+w), 235d (anh)	...	...	...	v	v	...	...	...	...	**B16**, 267
a1637	**4-Azobenzene-sulfonic acid, 4′-methyl-**, chloride	$C_{13}H_{11}ClN_2O_2S$. *See* a1629	294.76	red pr (bz)	130–2	...	...	...	d^h	d^h	v	...	s^h	chl v	**B16**, 272
—	**Azobenzoic acid**	*see* **Azobenzenedicarboxylic acid**													
a1638	***p,p′*-Azobiphenyl**	Di-*p*-xenyldiimide.	334.42	og-red pl (bz) λ 334.5 (4.48) 450.5 (2.95)	256(250)	...	...	...	i	i	s	...	...	con sulf s aa i	**B16**[2], 27
a1639	**Azodicarboxylic acid, diamide**	Azobisformamide. Azodicarbonamide. $H_2NCON:NCONH_2$	116.08	og nd (w)	225 (180–4)	...	...	...	s^h	i	δ	...	...	os i	**B3**[2], 99
a1640	—,diethyl ester	$C_2H_5O_2CN:NCO_2C_2H_5$	174.16	og-ye	...	106^{13}	$1.1104_4^{19.2}$	1.4199^{19}	...	s	s	s	...	os s	**B3**[2], 98
a1641	**Azomethane** (*trans*)	Dimethyldiazene. Dimethyldiimide. $CH_3N:NCH_3$	58.08	col or pa ye gas λ^w 345(1.3)	−78	1.5	0.744_{15}^0	...	...	...	...	...	...	C_7H_{16}, CCl_4 s	**B4**[3], 1747
a1642	—,hexafluoro-	$F_3CN:NCF_3$	166.03	pa grsh gas	−133	−31.6	...	...	δ	...	...	...	...	...	**C35**, 433
Ω a1643	**1,1′-Azo-naphthalene**	Di-α-naphthyldiimide.	282.35	red nd (gl aa) λ^{al} 214 (4.87), 266 (4.26), 400 (4.21)	190	sub	...	...	i	δ	...	s	v	con sulf, AmOH s aa s^h	**B16**[2], 26
a1644	**1,2′-Azo-naphthalene**	α,β-Naphthyldiimide.	282.35	red nd, br lf (aa) λ^{al} 216 (4.76), 264 (4.35), 310 (402), 381 (4.24)	136 (144–5)	...	...	...	i	s	...	...	s	aa, con sulf s (vt)	**E12B**, 936
Ω a1645	**2,2′-Azo-naphthalene**	Di-β-naphthyldiimide	282.35	red lf (bz), og-ye nd (chl, al) λ^{al} 214 (4.57), 262 (4.39), 290 (4.18), 335 (4.35)	208	sub	...	...	i	δ	δ	...	s	con sulf s (red) chl s^h	**E12B**, 936
a1646	**1,1′-Azo-naphthalene, 4-amino-**	$C_{20}H_{15}N_3$. *See* a1643	297.36	red-br nd $\lambda^{al-0.1NHCl}$ 355, 530	183(cor)	...	...	...	i	δ	δ	...	δ	...	**B16**[1], 325
—	**Azophenol**	*see* **Azobenzene, dihydroxy-**													
—	**Azotoluene**	*see* **Azobenzene, dimethyl-**													
a1647	**Azoxybenzene** (*cis*)	...	198.23	nd λ^{al} 231 (3.92), 260 (3.85), 323 (4.16)	87	...	1.166_4^{20}	1.633^{20}	...	...	...	...	...	...	**B16**, 624
a1648	—,(*trans*)	...	198.23	bt ye nd λ^{al} 322 (4.19)	36	d	1.1590_4^{26}	1.652^{20}	i	s	s	...	...	lig v	**B16**[2], 313
a1649	—,4-bromo-	$C_{12}H_7BrN_2O$. *See* a1648	277.13	dk ye pl $\lambda^{dil\ sulf}$ 250 (3.87), 355 (4.40)	93.5–4.5	...	1.4138_4^{100}	...	i	s	...	...	s^h	MeOH s^h	**B16**[2], 315
a1650	—,2,2′-dimethoxy-	*o*-Azoxyanisole. $C_{14}H_{14}N_2O_3$. *See* a1648	258.28	ye-og cr (al), og pr (MeOH) λ^{al} 227 (3.59), 304 (3.41), 330 (3.37), 342 (3.37)	81	...	...	...	i	s	s	s	s	MeOH s^h, chl v	**B16**, 635
a1651	—,3,3′-dimethoxy-	*m*-Azoxyanisole. $C_{14}H_{14}N_2O_3$. *See* a1648	258.28	ye cr (al) $\lambda^{50\%al}$ 235 (4.0), 267 (3.8), 324 (4.1)	51–2	...	...	...	...	s^h	s^h	...	...	...	**B16**[2], 325

For explanations, symbols and abbreviations see beginning of table. For structural formulas see end of table.

No.	Name	Synonyms and Formula	Mol. wt.	Color, crystalline form, specific rotation and λ_{max} (log ε)	m.p. °C	b.p. °C	Density	n_D	Solubility w	al	eth	ace	bz	other solvents	Ref.
	Azoxybenzene														
Ω a1652	**—,4,4′-dimethoxy-**	*p*-Azoxyanisole. $C_{14}H_{14}N_2O_3$. *See* a1648	258.28	ye mcl nd (al) λ^{al} 242 (4.06), 356 (4.41)	119–21	1.171_4^{115}	....	....	...	s^h	...	s	s	....	**B16²**, 326
a1653	**2,2′-Azoxybenzenedicarboxylic acid**	*o*-Azoxybenzoic acid.	286.25	ye pl or pr (al)	254.5	....	....	....	δ	s^h	δ	s	δ	Py v aa s chl, lig i	**B16²**, 335
a1654	**3,3′-Azoxybenzenedicarboxylic acid**	*m*-Azoxybenzoic acid.	286.25	pa ye nd or lf (gl aa)	320d	....	....	....	i	...	δ	...	...	....	**B16¹**, 388
Ω a1655	**4,4′-Azoxybenzenedicarboxylic acid**	*p*-Azoxybenzoic acid.	286.25	ye amor pw $\lambda^{1N\ NaOH}$ 262 (4.06), 335 (4.24)	360d (398, 242)	....	....	....	i	i	i	...	...	Py s oos i	**B16²**, 336
—	**Azoxybenzoic acid**	*see* **Azoxybenzenedicarboxylic acid**													
a1656	**1,1′-Azoxynaphthalene**	$C_{10}H_7^{\alpha}N{:}NC_{10}H_7^{\alpha}$ (↓O)	298.35	ye or red orh pl (al) λ^{al} 218 (4.93), 262 (4.11), 288 (3.91), 364 (4.06)	127	....	....	....	i	s^h	δ	...	...	con sulf s (vt)	**E12B**, 1038
a1657	**2,2′-Azoxynaphthalene**	$C_{10}H_7^{\beta}N{:}NC_{10}H_7^{\beta}$ (↓O)	298.35	ye or red nd (al, eth, gl aa) λ^{al} 216 (4.71), 263 (4.41), 289 (4.24), 346 (4.41)	167–8	....	....	....	i	δ	δ	...	s	chl s aa s^h MeOH δ	**E12B**, 1039
Ω a1658	**Azulene**	....	128.19	bl or grsh-bk lf (al) λ^{hx} 274 (4.79), 340 (3.67), 579 (2.51), 632 (2.51), 659 (2.18), 665 sh (2.15)	99–100.5	270d 115–35¹⁰	....	....	i	s^h	s	s	...	ac, os s	**E12A**, 421
a1659	**—,1,3,5-trinitro-, benzene addition compound**	$C_{16}H_{11}N_3O_6$. *See* a1658	341.28	br nd	166.5–7.5	....	....	....	...	s^h	...	...	...	....	**H20**, 224
a1660	**—,1,4-dimethyl-7-isopropyl-**	Guaiazulene. $C_{15}H_{18}$. *See* a1658	198.31	bl-vt pl (al) λ^{al} 254 (4.4), 287 (4.7), 350 (3.7), 370 (3.6), 610 (2.6), 625 (2.6), 665 (2.5)	31.5	167–8¹²	0.973_0^{20}	....	...	s^h	s	...	...	AcOEt s	**E12A**, 424
a1661	**—,1,5-dimethyl-8-isopropyl-**	Chamazulene. $C_{15}H_{18}$. *See* a1658	198.31	bl liq λ^{lig} 286 (4.59), 305 sh (3.98), 350 (3.63), 368 (3.50)	....	161¹²	0.9883_4^{20}	....	i	s	s	...	...	AcOEt, peth s	**B5²**, 474
a1662	**—,4,8-dimethyl-2-isopropyl-**	Elemazulene. Vetivazulene. $C_{15}H_{18}$. *See* a1658	198.31	red-vt nd (al) $\lambda^{50\%sulf}$ 228 (4.31), 272 (4.44), 374 (4.25)	31.5–2.0	140–60²	....	....	...	s^h	s	...	...	....	**E12A**, 427
a1663	**—,Δ⁴-octahydro-** (*trans*)	*trans*-Bicyclo-[5,3,0]-Δ² decene. $C_{10}H_{16}$. *See* a1658	136.24	....	....	68–9¹²	$0\ 8996_4^{21}$	1.4419²⁰	i	s	...	...	...	aa s	**E12A**, 421

For explanations, symbols and abbreviations see beginning of table. For structural formulas see end of table.

No.	Name	Synonyms and Formula	Mol. wt.	Color, crystalline form, specific rotation and λ_{max} (log ε)	m.p. °C	b.p. °C	Density	n_D	Solubility w	al	eth	ace	bz	other solvents	Ref.
	Baeyer's acid														
—	**Baeyer's acid**	*see* **2-Naphthalenesulfonic acid, 7-amino-**													
—	**Baicalin**	*see* **Flavone, 5, 6, 7-tri-hydroxy-**													
—	**B.A.L.**	*see* **1-Propanol, 2,3-dimercapto-**													
b1	**Bandrowski's base**	1,4-Benzoquinonebis(2,5-diaminoanil)	318.38	dk red or br lf	238	...	...	...	δ	...	...	...	...	os δ dil HCl s	**B13**[2], 146
—	**Banthine bromide**	*see* **Methantheline bromide**													
—	**Barbital**	*see* **Barbituric acid, 5,5-diethyl-**													
Ω b2	**Barbituric acid**	Pyrimidinetrione.	128.09	rh pr (w+2) λ^{al} 269 (3.74), 311 (4.23)	248	260d	...	...	s^h δ	δ	s	...	...	...	**B24**[2], 267
b3	**—,5-allyl-5-butyl-**	$C_{11}H_{16}N_2O_3$. *See* b2	224.26	(w, dil al)	128	...	...	...	s^h	s^h	...	...	...	...	**B24**[2], 292
b4	**—,5-allyl-5(2-cyclopenten-1-yl)-**	Cyclopal.	234.26	(w, dil al)	139–40	...	...	...	δ s^h	v	...	...	...	...	**C24,** 5308
b5	**—,5-allyl-5-isobutyl-**	Sandoptal. $C_{11}H_{16}N_2O_3$. *See* b2	224.26	(w, dil al)	138	...	...	...	δ v^{wr}	s	s	s	...	chl s lig i	**B24**[2], 292
b6	**—,5-allyl-5-isopropyl-1-methyl-**	Narconumal. $C_{11}H_{16}N_2O_3$ *See* b2	224.26	(w, dil al)	56–7	176–8[12]	...	...	...	s	s	s	s	...	**C32,** 1052
b7	**—,5-allyl-5-phenyl-**	Alphenal. Prophenal. $C_{13}H_{12}N_2O_3$. *See* b2	244.25	...	156–7.5	...	...	...	δ	v	v	...	δ	lig i chl s	**C26,** 4828
Ω b8	**—,5-amino-**	Uramil. $C_4H_5N_3O_3$. *See* b2	143.10	nd or pl (w)	>400	...	...	...	s^h i	...	i	...	i	chl, con sulf s	**B25**[1], 704
Ω b9	**—,5-benzylidene-**		216.21	pr (aa) λ^{al} 331	256	...	...	...	i	i	i	s^h	...	...	**B24**[2], 299
b10	**—,5(2-bromoallyl)-5-isopropyl-**	Propallylonal. Nostal. $C_{10}H_{13}BrN_2O_3$. *See* b2	289.14	(dil aa, dil al)	181	...	...	...	δ	v	δ	v	δ	chl δ aa v	**B24**[2], 291
Ω b11	**—,5(1-cyclo-hexenyl)-5-ethyl-**	Cyclobarbital. Cyclobarbitone.	236.27	lf (w)	172–4	...	...	...	i s^h	v	s	...	...	dil alk s aa δ	**B24**[2], 294
b12	**—,5,5-diallyl-**	Allobarbitone. Curral. Dial. $C_{10}H_{12}N_2O_3$. *See* b2	208.22	pl (w or 50 % al)	174	...	...	...	δ	s	s	...	s	...	**B24**[2], 293
b13	**—,5,5-dibromo-**	Dibromin. $C_4H_2Br_2N_2O_3$. *See* b2	285.89	pl (MeOH-bz), lf (dil HNO_3)	235	d	...	...	δ s^h	v	s	...	...	sulf s	**B24**[2], 272
Ω b14	**—,5,5-diethyl-**	Barbital. Barbitone. Veronal. $C_8H_{12}N_2O_3$. *See* b2	184.20	(i) trig (w) (ii) mcl pr (iii) mcl nd (iv) tcl $\lambda^{w,pH=1}$ 239 (3.0)	(i) 190 (ii) 183 (iii) 181 (iv) 176	...	1.220	...	δ s^h	s^h	s	s	...	chl, lig, peth, aa s	**B24**[2], 279
b15	**—,5,5-diethyl-1-methyl-**	Metharbital. $C_9H_{14}N_2O_3$. *See* b2	198.22	nd $\lambda^{w,p0=1-6}$ 222.5 (3.90) $\lambda^{w,pH=11-14}$ 245 (3.94)	150–1	...	...	...	s^h	...	...	...	...	...	**B24**[2], 281
b16	**—,5,5-dipropyl-**	Proponal. Propytal. $C_{10}H_{16}N_2O_3$. *See* b2	212.25	pl (w)	146 (166)	...	...	...	i δ^h	v	v	...	v	dil alk s chl v	**B24**[2], 286
b17	**—,5-ethyl-5-hexyl-**	$C_{12}H_{20}N_2O_3$. *See* b2	240.31	nd (w)	112–3	...	...	...	δ	s	s	...	s	...	**B24**[2], 288
b18	**—,5-ethyl-5-isopropyl-**	Ipral. Probarbital. $C_9H_{14}N_2O_3$. *See* b2	198.22	nd (w)	203	...	...	...	δ	i	s	...	...	...	**B24**[2], 284
b19	**—,5-ethyl-5(2-methylallyl)-2-thio-**	...	226.28	...	160–1	...	...	...	i	...	...	...	...	dil alk s	**C33,** 5599
b20	**—,5-ethyl-5(3-methylbutyl)-**	Amobarbital. Amytal. $C_{11}H_{18}N_2O_3$. *See* b2	226.28	lf (w, dil al) $\lambda^{w,pH=10.5}$ 242	156–8	...	...	...	δ	v	v	...	∞	chl s peth i	**B24**[2], 287
Ω b21	**—,5-ethyl-1-methyl-5-phenyl-**	Mephobarbital. Prominal. $C_{13}H_{14}N_2O_3$. *See* b2	246.27	wh cr (w)	176	...	...	...	δ s^h	v	δ	...	...	chl s	**C28,** 781
b22	**—,5-ethyl-5-pentyl-**	$C_{11}H_{18}N_2O_3$. *See* b2	226.28	cr (dil al)	135–6	...	...	...	δ s^h	v	v	...	...	chl s	**B24**[2], 286
b23	**—,5-ethyl-5(2-pentyl)-**	Nembutal. Pentobarbital. $C_{11}H_{18}N_2O_3$. *See* b2	226.28	nd (w) $\lambda^{w,pH=9.65}$ 240 (4.11) $\lambda^{w,pH=14}$ 256 (3.88)	130	...	...	...	δ	s	s	...	...	...	**B24**[2], 287
b24	**—,5-ethyl-5-phenyl-**	Luminal. Phenobarbital. $C_{12}H_{12}N_2O_3$. *See* b2	232.24	pl (w) $\lambda^{w,pH=7.3}$ 240 (3.6) $\lambda^{w,pH=13}$ 255 (3.9)	(i) 174 (st) (ii) 156–7 (unst) (iii) 166–7 (unst)	...	...	...	i s^h	s	s	...	i	...	**B24**[2], 297
Ω b25	**—,5-ethyl-5(1-piperidyl)-**	Eldoral.	239.28	wh cr (dil al)	215	...	...	...	δ s^h	s	s	s	i	chl, alk, ac s	**C36,** 3160
Ω b26	**—,5(2-furfuryl-idene)-2-thio-**	...	222.22	ye pl	>280d	...	...	...	i	i	...	...	i	aa i py s	**B27,** 607
b27	**—,5-hydroxy-**	Dialuric acid. $C_4H_4N_2O_4$. *See* b2	144.09	pr or pl nd (+1w, w) $\lambda^{w,pH=7.4}$ 275 (4.22)	224 (anh) 214–5 (+w)	...	...	...	δ v^h	s	...	s	i	aa, py s	**B25**[2], 61

For explanations, symbols and abbreviations see beginning of table. For structural formulas see end of table.

No.	Name	Synonyms and Formula	Mol. wt.	Color, crystalline form, specific rotation and λ_{max} (log ε)	m.p. °C	b.p. °C	Density	n_D	Solubility w	al	eth	ace	bz	other solvents	Ref.
	Barbituric acid														
b28	—,**5-methyl-5-phenyl-**	Rutonal. $C_{11}H_{10}N_2O_3$. *See* b2	218.21	cr	220	...	...	...	i	s	s	...	...	alk s	**B24**^{2}, 296
b29	—,**5-nitro-**	Dilituric acid. $C_4H_3N_3O_5$. *See* b2	173.09	pr, lf(w + 3) λ^{w} 314 (4.0) $\lambda^{w,pH=13}$ 334 (4.1)	180–1 176 (anh)	...	...	...	s^h	s	i	...	...	...	**B24**^{2}, 273
b30	—,**5(2-phenylethyl)-**	$C_{12}H_{12}N_2O_3$. *See* b2	232.24	cr	212–3	...	...	...	δ	v	...	...	...	...	**B24**^{2}, 296
Ω b31	—,**2-thio-**	...	144.15	pl (w) λ^{al} 235 (3.74), 285 (4.26)	235 d	...	...	...	δ s^h	s	...	...	...	dil alk, dil HCl s	**B24**^{2}, 275
—	**Batyl alcohol**	*see* **Glycerol,** 1-octadecyl ether													
b32	**Bebeerine**(*d*)	Chondrodendrin. $C_{36}H_{38}N_2O_6$	594.72	cr (bz, eth, chl-MeOH), $[\alpha]_D^{25}$ + 297 (al). $[\alpha]_D^{20}$ + 332 (Py) λ^{MeOH} 282 (3.96)	221–1.5, 161 (+bz)	...	...	...	...	s	s	v	...	chl v MeOH s	**B27**^{2}, 896
b33	—(*l*)	Curine. $C_{36}H_{38}N_2O_6$.	594.72	pr, nd (chl-MeOH), $[\alpha]_D^{20}$ − 328 (Py) $[\alpha]_D^{28}$ − 298(al)	221, 165–7 (+bz), 159–63 (+al)	...	...	...	δ	δ s^h	δ	s	s	chl, Py, ac s	**B27**^{2}, 894
—	**Behenic acid**	*see* **Docosanoic acid**													
—	**Behenolic acid**	*see* **13-Docosynoic acid**													
Ω b35	**Benzaldehyde**	Benzenecarbonal*. Benzenecarboxaldehyde.	106.13	λ^{cy} 241 (4.15), 247 sh (4.06), 277.5 (3.08), 287 sh (3.00)	−26 (fr −56.9 (−55.6)	178.1^{760} 62^{10}	1.0415_4^{15}	1.5463^{20}	δ	∞	∞	v	v	liq NH_3 s lig v	**B7**^{2}, 145
Ω b36	—,**azine**	Dibenzalhydrazine. $C_6H_5CH{:}NN{:}CHC_6H_5$	208.27	ye pr (al) λ^{cy} 300 (4.6)	93	...	...	...	i	s^h	s	s	s	chl s	**B7**^{2}, 171
b37	—,**diacetate**	Benzylidene diacetate. $C_6H_5CH(O_2CCH_3)_2$	208.22	pl (eth)	46	220^{760} 154^{20}	1.11^{20}	...	i	v	v	...	s	...	**B7**^{2}, 161
b38	—,**hydrazone**	$C_6H_5CH{:}NNH_2$	120.15	lf λ^{al} 267 (4.2)	16	140^{14}	...	...	d	s	...	...	...	...	**B7**^{2}, 171
b39	—,**imine,** *N***-ethyl-**	*N*-Benzalethylamine. $C_6H_5CH{:}NC_2H_5$	133.20	λ^{al} 246 (4.2), 278 sh (3.1), 289 (2.9)	...	195^{749}	0.937_4^{20}	1.5378^{25}	i	s	s	...	...	...	**Am 80,** 4320
b40	—,—,*N***-methyl-**	*N*-Benzalmethylimine. $C_6H_5CH{:}NCH_3$	119.17	λ^{al} 245 (4.18)	...	185^{760} $90–1^{30}$	0.9672_4^{15}	1.5540^{20}	δ	...	...	...	s	...	**B7**^{2}, 162
b41	—,—,*N***(4-tolyl)-**	*N*-Benzal-*p*-toluidine.	195.27	ye cr	35	318^{755} 178^{11}	...	...	i	...	...	s	...	...	**B12**^{2}, 496
b42	—,**oxime**(*anti*)	*anti*-Benzaldoxime. $C_6H_5CH{:}NOH$	121.14	nd (eth) λ^{al} 247 (4.2), 288 sh (2.8)	130	...	1.145_4^{20}	...	s^h	v	v	...	...	...	**B7**^{2}, 169
Ω b43	—,—(*syn*)	*syn*-Benzaldoxime. $C_6H_5CH{:}NOH$	121.14	pr λ^{al} 255 (4.2), 282 sh (3.2), 292 (3.0)	36–7	200 $118–9^{10}$	1.1111_4^{20}	1.5908^{20}	δ	s	v	...	v	...	**B7**^{2}, 167
Ω b44	—,**phenylhydrazone**	$C_6H_5CH{:}NNHC_6H_5$	196.26	nd (lig), pr λ^{al} 237 (4.17), 302 (4.06), 345 (4.39)	156	...	...	...	...	δ s^h	δ	s^h	s^h	liq NH_3 s	**B15**^{2}, 57
b45	—,**2-acetamido-**	$C_9H_9NO_2$. *See* b35	163.18	nd (w) λ^{MeOH} 292 (3.4)	70–1	...	...	...	s^h	...	s	s	...	os s	**B14,** 26
b46	—,**3-acetamido-**	$C_9H_9NO_2$. *See* b35	163.18	pl (bz)	84	...	...	...	δ	v	s	...	s^h	...	**B14,** 29
b47	—,**4-acetamido-**	$C_9H_9NO_2$. *See* b35	163.18	pr (w)	156	...	...	...	s	δ	...	...	s	...	**B14,** 38
Ω b48	—,**2-amino-**	Anthranilaldehyde. C_7H_7NO. *See* b35	121.14	silv lf λ^{hx} 255 (3.75), 262 sh (3.54), 352 (3.66)	39–40	...	...	...	δ	v	v	...	s	chl s lig i	**B14,** 21
b49	—,—,**oxime**	...	136.17	nd (bz)	135–6	...	...	...	δ	v	v	...	δ	CS_2 v aa s lig δ	**B14,** 24
Ω b50	—,**3-amino-**	C_7H_7NO. *See* b35	121.14	nd (AcOEt) λ^{hx} 245 (3.95), 256 (3.87)	28–30	...	...	...	...	...	s	...	...	ac s	**B14**^{2}, 21
b51	—,—,**oxime**	...	136.17	nd (bz)	88 (195)	...	...	...	...	v	v	...	s	lig δ	**B14,** 28
b52	—,**4-amino-**	C_7H_7NO. *See* b35	121.14	pl (w) λ^{hx} 291 (4.21), 296 (4.21)	71–2	...	...	...	s	s	s	...	...	ac s	**B14**^{2}, 22
b53	—,—,**oxime**	$C_7H_8N_2O$. *See* b42	136.17	ye cr (w)	124	...	...	...	...	v	v	...	...	ac s	**B14,** 31
Ω b54	—,**2-bromo-**	C_7H_5BrO. *See* b35	185.03	λ^{hx} 245.5 (4.10), 252 (4.00), 292 (3.32), 301 (3.24)	21–2	230 118^{12}	...	1.5928^{20}	i	v	...	...	v	...	**B7**^{2}, 181
b55	—,—,**diacetate**	...	287.12		84–6	...	...	...	...	s^h	s	...	...	...	**B7**^{2}, 181
Ω b56	—,**3-bromo-**	C_7H_5BrO. *See* b35	185.03	λ^{hx} 242 (4.06)	...	$233–6^{760}$	...	1.5935^{20}	i	v	v	...	...	...	**B7**^{2}, 182
Ω b57	—,**4-bromo-**	C_7H_5BrO. *See* b35	185.03	lf (dil al) λ^{hx} 257.5 (4.27)	67	$66–8^{2}$	...	...	i	v	...	...	v	...	**B7**^{2}, 182

For explanations, symbols and abbreviations see beginning of table. For structural formulas see end of table.

No.	Name	Synonyms and Formula	Mol. wt.	Color, crystalline form, specific rotation and λ_{max} (log ε)	m.p. °C	b.p. °C	Density	n_D	Solubility w	al	eth	ace	bz	other solvents	Ref.
	Benzaldehyde														
b58	—,—,diacetate	....	287.12	ye	94–5	....	....	....	...	s^h	s	...	...	....	**B7**[2], 182
b59	—,**2-bromo-4-hydroxy-**	$C_7H_5BrO_2$. *See* b35	201.03	pa ye nd (w)	159.5	....	....	....	δ s^h	...	...	...	...	....	**B8**[2], 74
b60	—,**2-bromo-5-hydroxy-**	$C_7H_5BrO_2$. *See* b35	201.03	nd (w)	135	....	....	....	i	v	v	s	v	chl s	**B8**[2], 57
b61	—,**3-bromo-2-hydroxy-**	$C_7H_5BrO_2$. *See* b35	201.03	nd (dil al)	49	....	....	....	δ	s	s	s	s	alk, os s	**B8**, 54
b62	—,**3-bromo-4-hydroxy-**	$C_7H_5BrO_2$. *See* b35	201.03	lf (w)	124	....	....	....	δ	v	v	v	s	chl, aa s lig δ	**B8**, 82
b63	—,**4-bromo-2-hydroxy-**	$C_7H_5BrO_2$. *See* b35	201.03	nd (dil al)	52	....	....	....	δ	s	v	s	s	os s	**B8**, 54
b64	—,**4-bromo-3-hydroxy-**	$C_7H_5BrO_2$. *See* b35	201.03	....	131.5	....	....	....	...	...	...	s	s	chl v	**B8**[2], 56
Ω b65	—,**5-bromo-2-hydroxy-**	$C_7H_5BrO_2$. *See* b35	201.03	nd (al), lf (eth) λ^{al} 240 (3.7), 278 (2.0), 337 (3.5)	105–6	....	....	....	i	s	s	...	...	....	**B8**[2], 45
b66	—,**5-bromo-4-hydroxy-3-methoxy-**	5-Bromovanillin. $C_8H_7BrO_3$. *See* b35	231.05	pl (aa), nd, pl (al) λ^{al} 213.5 (4.36), 234 (4.29), 291 (4.10)	164–6	....	....	....	i	s^h	δ	...	δ	....	**B8**[2], 286
Ω b67	—,**2-chloro-**	C_7H_5ClO. *See* b35	140.57	nd λ^{hx} 246 (4.04), 252 (3.93), 292 (3.24), 300 (3.15), 302 (3.15)	12.39	211.9^{760} 84.3^{10}	1.2483^{20}_{4}	1.5662^{20}	δ	s	s	s	s	....	**B7**[2], 177
Ω b68	—,**3-chloro-**	C_7H_5ClO. *See* b35	140.57	pr λ^{hx} 241.5 (4.06), 248 (4.00), 288 (3.15), 298 (3.08)	17–8	$213–4^{760}$ 55^{1}	1.2410^{20}_{4}	1.5650^{20}	δ	s	s	s	s	....	**B7**[2], 178
Ω b69	—,**4-chloro-**	C_7H_5ClO. *See* b35	140.57	pl λ^{hx} 253 (4.26), 259 (4.19), 276 (3.18), 286 (3.00)	47.5	$213–4^{760}$ $72–5^{3}$	1.196^{61}_{4}	1.5552^{61}	s	v	v	s	v	CS_2, aa s	**B7**[2], 178
b70	—,**2-chloro-3-hydroxy-**	$C_7H_5ClO_2$. *See* b35	156.57	cr (aq aa)	139.5	....	....	....	...	v	...	...	...	aa v chl δ	**B8**[2], 54
b71	—,**2-chloro-4-hydroxy-**	$C_7H_5ClO_2$. *See* b35	156.57	nd (w or aa)	147–8	....	....	....	s^h	s	s	...	...	aa v^h	**B8**, 81
b72	—,**2-chloro-5-hydroxy-**	$C_7H_5ClO_2$. *See* b35	156.57	nd (w or aa)	110.5–11.5	....	....	....	...	s	...	...	...	aa v	**B8**[1], 526
b73	—,**3-chloro-2-hydroxy-**	$C_7H_5ClO_2$. *See* b35	156.57	nd (MeOH)	55	....	....	....	...	s	s	s	s	os v	**B8**[1], 523
b74	—,**3-chloro-4-hydroxy-**	$C_7H_5ClO_2$. *See* b35	156.57	nd (w)	139	$149–50^{14}$	....	....	δ s^h	v	v	...	...	chl δ	**B8**, 81
b75	—,**4-chloro-2-hydroxy-**	$C_7H_5ClO_2$. *See* b35	156.57	nd (al or aq aa)	52.5	....	....	....	s	...	s	s	s	os v	**J1927**, 1740
b76	—,**4-chloro-3-hydroxy-**	$C_7H_5ClO_2$. *See* b35	156.57	nd (al)	121	....	....	....	s	s	s	s	s	os s	**B8**[2]; 55
b77	—,**5-chloro-2-hydroxy-**	$C_7H_5ClO_2$. *See* b35	156.57	pl (al) λ^{al} 338 (3.57)	99–100	105^{12}	....	....	i	v	s	...	...	alk s	**B8**[2], 45
b78	—,—.oxime	$C_7H_6ClNO_2$. *See* b42	171.58	nd (w)	128	....	....	....	...	s^h	...	...	...	....	**B8**, 53
b79	—,**5-chloro-4-hydroxy-3-methoxy-**	5-Chlorovanillin. $C_8H_7ClO_3$. *See* b35	186.60	tetr λ^{diox} 280 (4.1), 305 sh (3.8)	165	....	....	....	i	s^h	...	...	...	aa s^h	**B8**[2], 286
b80	—,**2,3-dichloro-**	$C_7H_4Cl_2O$. *See* b35	175.02	cr (dil al)	65–7	....	....	....	...	v	v	...	...	....	**Am 68**, 861
Ω b81	—,**2,4-dichloro-**	$C_7H_4Cl_2O$. *See* b35	175.02	pr $\lambda^{aq\ MeOH}$ 218 (4.23), 263 (4.14), 300 sh (3.30)	72	....	....	....	i	s	s	...	s	aa s	**B7**[2], 180
b82	—,**2,5-dichloro-**	$C_7H_4Cl_2O$. *See* b35	175.02	nd (al)	58	231–3	....	....	i	v	v	...	v	chl v CS_2 s	**B7**, 237
b83	—,**2,6-dichloro-**	$C_7H_4Cl_2O$. *See* b35	175.02	nd (lig)	71	....	....	....	i	s^h	s^h	...	...	lig s^h	**B7**, 237
Ω b84	—,**3,4-dichloro-**	$C_7H_4Cl_2O$. *See* b35	175.02	....	44	247–8	....	....	i	s	s	...	...	....	**B7**, 238
b85	—,**3,5-dichloro-**	$C_7H_4Cl_2O$. *See* b35	175.02	nd or lf (dil al)	65	$235–40^{748}$	....	....	δ^h	s	s	v	v	os v	**C28**, 754
b86	—,**2,4-dichloro-3-hydroxy-**	$C_7H_4Cl_2O_2$. *See* b35	191.02	cr (aa)	141	....	....	....	...	...	...	...	...	lig δ, v^h	**B8**[2], 55
b87	—,**2,6-dichloro-3-hydroxy-**	$C_7H_4Cl_2O_2$. *See* b35	191.02	cr (w)	142	....	....	....	δ^h	...	...	...	...	aa s	**B8**[2], 56
b88	—,**3,5-dichloro-2-hydroxy-**	$C_7H_4Cl_2O_2$. *See* b35	191.02	ye rh (aa)	95	....	....	....	i	...	...	...	...	....	**B8**, 54
b89	—,—,oxime	$C_7H_5Cl_2NO_2$. *See* b42	206.03	nd (dil al)	195–6	....	....	....	i	v	v	...	v	....	**B8**, 54
b90	—,**3,5-dichloro-4-hydroxy-**	$C_7H_4Cl_2O_2$. *See* b35	191.02	nd (chl, dil al)	158–9	....	....	....	...	v	v	...	δ	aa s chl, lig δ	**B8**, 81
b91	—,**4,6-dichloro-3-hydroxy-**	$C_7H_4Cl_2O_2$. *See* b35	191.02	nd	130	....	....	....	v^h	s	s	...	v	lig i	**B8**[1], 526
b92	—,**2,3-diethoxy-**	$C_{11}H_{14}O_3$. *See* b35	194.23	....	....	169^{37}	....	....	...	s	s	...	...	....	**B8**[2], 268

For explanations, symbols and abbreviations see beginning of table. For structural formulas see end of table.

No.	Name	Synonyms and Formula	Mol. wt.	Color, crystalline form, specific rotation and λ_{max} (log ε)	m.p. °C	b.p. °C	Density	n_D	Solubility w	al	eth	ace	bz	other solvents	Ref.
	Benzaldehyde														
b93	—,3,4-diethoxy-	Protocatechualdehyde diethyl ether. $C_{11}H_{14}O_3$. *See* b35	194.23	...	...	278–80	...	...	...	s	...	...	...	...	**B8**, 256
Ω b94	—,4-diethyl-amino-	$C_{11}H_{15}NO$. *See* b35	177.25	ye nd (w) λ^{al} 243 (3.78), 348 (4.52)	41	174^{7}	...	...	v	s	s	...	s	...	**B14²**, 25
Ω b95	—,2,4-dihydroxy-	β-Resorcylaldehyde. $C_7H_6O_3$. *See* b35	138.12	nd (eth-lig) λ^{eth} 232 (3.93), 280 (4.14), 310 (3.82)	201–2	$220–8^{22}$	...	...	s	v	v	...	δ	chl v aa s	**B8²**, 272
b96	—,—,oxime	$C_7H_7NO_3$. *See* b42	153.14	...	192	...	...	...	s	v	v	...	...	...	**B8²**, 274
Ω b97	—,2,5-dihydroxy-	Gentisic aldehyde. $C_7H_6O_3$. *See* b35	138.12	ye nd (bz)	99	...	...	...	v	v	v	...	s^h	chl v lig δ	**B8²**, 276
Ω b98	—,3,4-dihydroxy-	Protocatechualdehyde. $C_7H_6O_3$. *See* b35	138.12	lf (w, to) λ^{al} 250 (2.55), 300 (2.05)	153–4	...	...	...	s v^h	v	v	...	...	...	**B8²**, 277
b99	—,—,oxime	$C_7H_7NO_3$. *See* b42	153.14	nd (w)	157d	...	...	...	v	v	δ	...	δ	chl δ	**B8²**, 285
b100	—,2,6-dihydroxy-4-methyl-	Atranol. $C_8H_8O_3$. *See* b35	152.16	ye nd (w)	124 118–9 (+½w)	...	...	...	v^h	v	v	...	s^h	chl v alk s lig s^h	**B8²**, 304
b101	—,3,5-diiodo-4-hydroxy-	$C_7H_4I_2O_2$. *See* b35	373.91	nd (al)	206.5	...	...	...	δ	s^h	...	s	...	aa s	**B8²**, 48
Ω b102	—,2,4-dimethoxy-	$C_9H_{10}O_3$. *See* b35	166.18	nd (al or lig)	72	165^{10}	...	...	i	s	s	...	s	lig s	**B8²**, 273
Ω b103	—,3,4-dimethoxy-	Veratraldehyde. $C_9H_{10}O_3$. *See* b35	166.18	nd (eth, lig, tol) λ^{w} 230 (4.19), 278 (4.05), 308 (3.97)	44 (58)	258 $172–5^{18}$	...	...	δ	v	v	...	...	...	**B8²**, 282
b104	—,2,4-dimethoxy-6-hydroxy-	$C_9H_{10}O_4$. *See* b35	182.18	nd, pl (dil al) λ^{al} 215 (4.2), 294 (4.31), 330 sh (3.58)	71	$190–5^{25}$ 165^{10}	...	...	i	v	v	...	v	chl, aa v	**B8²**, 436
b105	—,2,6-dimethoxy-4-hydroxy-	$C_9H_{10}O_4$. *See* b35	182.18	nd or pl (MeOH), pr (bz)	70–1	...	...	...	...	...	...	...	...	...	**B8²**, 436
b106	—,3,4-dimethoxy-5-hydroxy-	$C_9H_{10}O_4$. *See* b35	182.18	...	62–3	$177–80^{12}$	...	...	...	v	...	...	v	aa v lig v^h	**B8²**, 437
b106[1]	—,3,5-dimethoxy-4-hydroxy-	Syringaldehyde. $C_9H_{10}O_4$. *See* b35	182.18	br nd (lig) λ^{al} 215 (4.31), 231 (4.25), 308.5 (4.18)	113	$192–3^{14}$	...	...	δ	v	v	...	v^h	chl, aa v lig δ	**B8²**, 437
b107	—,2,4-dimethyl-	$C_9H_{10}O$. *See* b35	134.18	...	−9 – −8	$218–99^{10}$ 99^{10}	...	...	...	s	s	s	s	...	**B7²**, 240
b108	—,2,5-dimethyl-	$C_9H_{10}O$. *See* b35	134.18	...	...	220^{738} 100^{10}	...	...	...	v	s	s	s	os s	**B7¹**, 166
b109	—,3,4-dimethyl-	$C_9H_{10}O$. *See* b35	134.18	...	...	$223–5^{760}$	...	...	...	s	s	s	...	...	**B7²**, 240
b110	—,3,5-dimethyl-	$C_9H_{10}O$. *See* b35	134.18	...	9	220–2	...	...	...	s	s	s	s	...	**B7²**, 240
Ω b111	—,4-dimethyl-amino-	$C_9H_{11}NO$. *See* b35	149.19	lf (w) λ^{al} 240 (3.7), 340 (4.5)	74	$176–7^{17}$	...	...	δ	s	s	s	s	os s	**B14²**, 23
b112	—,2,4-dinitro-	$C_7H_4N_2O_5$. *See* b35	196.12	pa ye pr (al), pl (bz) λ^{cy} 230 (1.00), 300 (0.07)	72	190–$210^{10–20}$	...	...	δ	s	s	...	s	lig δ	**B7²**, 205
b113	—,2,6-dinitro-	$C_7H_4N_2O_5$. *See* b35	196.12	lf (dil aa)	123	...	...	...	s^h	s	s	...	s	chl, aa s CS_2 δ	**B7²**, 206
b114	—,2-ethoxy-	$C_9H_{10}O_2$. *See* b35	150.18	...	20–2	247–9 $143–7^{25}$	...	...	...	∞	∞	...	...	...	**B8**, 43
b115	—,3-ethoxy-	$C_9H_{10}O_2$. *See* b35	150.18	...	...	245.5^{760}	1.0768^{20}_{4}	1.5408^{20}	...	s	s	...	s	...	**B8**, 60
b116	—,4-ethoxy-	$C_9H_{10}O_2$. *See* b35	150.18	...	13–4	249 140^{20}	1.08^{21}_{21}	...	...	v	v	...	v	...	**B8**, 73
b117	—,3-ethoxy-2-hydroxy-	$C_9H_{10}O_3$. *See* b35	166.18	...	64–5	$263–4^{740}$	...	...	δ	...	...	...	...	os s	**B8²**, 267
Ω b118	—,3-ethoxy-4-hydroxy-	Bourbonal. Ethylvanillin. $C_9H_{10}O_3$. *See* b35	166.18	pl or sc (w)	77–8	...	...	...	δ	s^h	s	...	s	chl s	**B8²**, 282
b119	—,4-ethoxy-2-hydroxy-	$C_9H_{10}O_3$. *See* b35	166.18	nd (dil al)	35	...	...	...	...	v	...	...	...	...	**B8**, 242
b120	—,5-ethoxy-2-hydroxy-	$C_9H_{10}O_3$. *See* b35	166.18	ye pr	51.5	230	...	...	δ	v	v	...	...	chl v	**B8**, 245
Ω b121	—,4-ethoxy-3-methoxy-	$C_{10}H_{12}O_3$. *See* b35	180.21	mcl pr	64–5 (73–4)	sub	...	...	$δ^h$	s	s	...	s	chl, aa s	**B8²**, 283
Ω b122	—,2-hydroxy-	Salicylaldehyde. $C_7H_6O_2$. *See* b35	122.13	λ^{hx} 257 (4.07), 324 (3.56)	−7	197^{760} 93^{25}	1.1674^{20}_{4}	1.5740^{20}	δ	∞	∞	v	v	...	**B8**, 31
Ω b123	—,—,azine	Salazine. $C_{14}H_{12}N_2O_2$. *See* b36	240.27	ye nd or lf (al) λ^{al} 295 (4.48)	214	...	...	...	i	s	...	...	v	chl s alk v	**B8²**, 43
Ω b124	—,—,oxime	Salicylaldoxime. $C_7H_7NO_2$. *See* b42	137.14	pr (bz-peth) λ^{al} 265 (3.8), 305 (3.6)	63	...	...	...	δ	v	v	...	v	lig i	**B8²**, 42
Ω b125	—,3-hydroxy-	$C_7H_6O_2$. *See* b35	122.13	nd (w) λ^{hx} 250 (3.98), 302 (3.46), 307 (3.45)	108	240^{760} 161^{20}	...	...	δ	s	s	s	s	lig i	**B8²**, 52

For explanations, symbols and abbreviations see beginning of table. For structural formulas see end of table.

No.	Name	Synonyms and Formula	Mol. wt.	Color, crystalline form, specific rotation and λ_{max} (log ε)	m.p. °C	b.p. °C	Density	n_D	Solubility: w	al	eth	ace	bz	other solvents	Ref.
	Benzaldehyde														
b126	—,—,azine	$C_{14}H_{12}N_2O_2$. *See* b36	240.27	ye pl (al) λ^{al} 306 (4.45), 330 (4.33)	162	...	...	...	i	δ	...	...	δ	chl δ	**B8**^{1}, 526
b127	—,—,oxime	$C_7H_7NO_2$. *See* b42	137.14	(bz)	90	...	...	...	v	v	v	...	i	chl δ lig i	**B8**^{1}, 525
Ω b128	—,**4-hydroxy-**	$C_7H_6O_2$. *See* b35	122.13	nd (w) λ^{hx} 281 (4.08), 288 (3.78)	117	...	1.129$_4^{130}$	1.5705$_\alpha^{130}$	δ s^h	v	v	...	s	...	**B8**^{2}, 63
b129	—,—,azine	$C_{14}H_{12}N_2O_2$. *See* b36	240.27	ye (al) λ^{al} 335 (4.64)	239–40 (268d) 273 (+w)	...	...	...	δ^h	v	...	...	v	CCl_4, lig δ	**B8**^{1}, 531
b131	—,4-hydroxy-2-iodo-3-methoxy-	2-Iodovanillin. $C_8H_7IO_3$. *See* b35	278.05	...	155–6	...	...	...	i	s	δ	...	...	...	**Am 57**, 2500
b132	—,4-hydroxy-5-iodo-3-methoxy-	5-Iodovanillin. $C_8H_7IO_3$. *See* b35	278.05	pa ye	180	...	...	...	i	δ	δ	...	...	...	**B8**^{2}, 289
Ω b133	—,**2-hydroxy-3-methoxy-**	*o*-Vanillin. $C_8H_8O_3$. *See* b35	152.16	lt ye-lt gr nd (w, lig) λ^{diox} 265 (4.0), 340 (3.5)	44–5	265–6^{760} 128^{10}	...	...	δ	v	v	...	...	CCl_4 v lig s	**B8**^{2}, 267
b134	—,2-hydroxy-4-methoxy-	4-Methoxysalicylaldehyde. $C_8H_8O_3$. *See* b35	152.16	nd (w), cr (al)	40–2	...	...	...	δ s^h	s v^h	v	...	v	lig v	**B8**^{2}, 272
b135	—,2-hydroxy-5-methoxy-	5-Methoxysalicylaldehyde. $C_8H_8O_3$. *See* b35	152.16	ye (w)	4	247–8^{760}	...	...	δ	v	v	...	...	...	**B8**^{2}, 276
Ω b136	—,**3-hydroxy-4-methoxy-**	Isovanillin. $C_8H_8O_3$. *See* b35	152.16	pl (w) $\lambda^{0.01 N NaOH}$ 292 (3.92) 360 (3.78)	116–7	179^{15}	1.196	...	δ s^h	s	s	...	s^h	chl v aa s	**B8**^{2}, 282
b137	—,3-hydroxy-5-methoxy-	$C_8H_8O_3$. *See* b35	152.16	...	130–1	110–20 sub	...	...	...	v	...	...	v	aa v lig δ	**B8**^{2}, 291
b138	—,4-hydroxy-2-methoxy-	$C_8H_8O_3$. *See* b35	152.16	lf (bz), nd (w)	153	...	...	...	δ	s^h	s	...	δ	chl s lig δ	**B8**^{2}, 272
Ω b139	—,**4-hydroxy-3-methoxy-**	Vanillin. $C_8H_8O_3$. *See* b35	152.16	(i) nd (w, lig) (ii) tetr (w, lig) λ^{al} 279 (4.01), 309 (4.02)	(i) 77–9 (ii) 81–2	285 170^{15}	1.056	...	δ v^h	v	v	v	s^h	CS_2, chl v lig s^h	**B8**^{2}, 278
b140	—,—,acetate	Vanillin acetate.	194.19	nd (eth) λ^{al} 257 (4.05), 305 (3.63)	102–3 (78–9)	...	...	...	δ	v	v	...	...	...	**B8**^{2}, 283
b141	—,2-hydroxy-3-nitro-	3-Nitrosalicylaldehyde. $C_7H_5NO_4$. *See* b35	167.12	nd (aa)	109–10	...	...	...	...	s	...	...	s	...	**B8**^{2}, 48
b142	—,2-hydroxy-5-nitro-	5-Nitrosalicylaldehyde. $C_7H_5NO_4$. *See* b35	167.12	(dil aa)	126	...	...	...	...	s	s	...	...	...	**B8**^{2}, 48
b143	—,—,oxime	$C_7H_6N_2O_4$. *See* b42	182.14	...	225	...	...	...	...	...	...	...	...	...	**B8**^{2}, 48
b144	—,3-hydroxy-2-nitro-	$C_7H_5NO_4$. *See* b35	167.12	nd or pl (bz, lig)	152	...	...	...	...	s	...	...	s^h	lig δ	**B8**^{2}, 58
b145	—,—,oxime	$C_7H_6N_2O_4$. *See* b42	182.14	pa ye nd	172.5	...	...	...	...	s	s	...	...	...	**B8**^{2}, 58
b146	—,3-hydroxy-4-nitro-	$C_7H_5NO_4$. *See* b35	167.12	ye lf	128	...	...	...	δ	v	v	...	v	chl s lig δ	**B8**^{2}, 58
b147	—,—,oxime	$C_7H_6N_2O_4$. *See* b42	182.14	ye nd (chl)	164	...	...	...	...	v	...	...	δ	chl s	**B8**^{2}, 58
b148	—,4-hydroxy-2-nitro-	$C_7H_5NO_4$. *See* b35	167.12	ye nd	67	...	...	...	δ	v	v	...	v	...	**B8**, 83
b149	—,4-hydroxy-3-nitro-	$C_7H_5NO_4$. *See* b35	167.12	dk ye nd (al, w)	144.5	...	...	...	s^h	s v^h	δ	...	...	chl δ	**B8**^{2}, 77
b150	—,—,oxime	$C_7H_6N_2O_4$. *See* b42	182.14	pr or nd (al-chl)	169	...	...	...	...	s	s	...	δ	aa s chl, lig δ	**B8**^{2}, 77
Ω b151	—,**4-isopropyl-**	Cumaldehyde. Cuminal. $C_{10}H_{12}O$. *See* b35	148.21	λ^{iso} 251.5 (4.24), 277.5 (3.19)	...	235–6 103–4^{10}	0.9755$_4^{20}$	1.5301^{20}	i	s	s	...	...	...	**B7**^{2}, 247
Ω b152	—,**2-methoxy-**	*o*-Anisaldehyde. $C_8H_8O_2$. *See* b35	136.16	pr λ^{hx} 253 (3.93), 306 (3.66), 314 (3.62)	37–8	243–4^{760} 124–5^{18}	1.1326$_4^{20}$	1.5600^{20}	i	s	v	v	s	chl v	**B8**^{2}, 40
b153	—,3-methoxy-	*m*-Anisaldehyde. $C_8H_8O_2$. *See* b35	136.16	λ^{hx} 251 (3.81), 304 (3.48), 313 (3.43)	...	230 62^1	1.1187$_4^{20}$	1.5530^{20}	i	s	s	v	s	chl v	**B8**^{2}, 53
Ω b154	—,**4-methoxy-**	*p*-Anisaldehyde. $C_8H_8O_2$. *See* b35	136.16	λ^{hx} 272 (4.20), 278 (4.10), 286 (3.74)	0	249.5^{760} 83^2	1.1191$_4^{15}$	1.5730^{20}	i	∞	∞	v	s	chl v	**B8**^{2}, 64
b155	—,—,oxime	$C_8H_9NO_2$. *See* b42	151.17	nd (bz) λ^{al} 265 (4.24)	64–5 (*anti*) 133 (*syn*)	...	...	...	...	...	...	...	δ	...	**B8**^{2}, 69
b156	—,2-methoxy-3-nitro-	$C_8H_7NO_4$. *See* b35	181.15	ye pr (dil al), nd (bz)	89–90	...	...	...	i	s	s	...	...	...	**B8**, 56
Ω b157	—,**2-methyl-**	*o*-Tolualdehyde. C_8H_8O. *See* b35	120.16	λ^{hx} 251 (4.11), 291 (3.23)	...	200 94^{10}	1.0386$_4^{19}$	1.5481^{20}	δ^h	s	s	v	s	chl s	**B7**, 295
b158	—,3-methyl-	*m*-Tolualdehyde. C_8H_8O. *See* b35	120.16	λ^{hx} 251 (4.08) 280 (4.08), 290 (2.90)	...	199 93–4^{17}	1.0189$_4^{21}$	1.5413^{21}	δ	∞	∞	v	s	chl s	**B7**, 296
Ω b159	—,**4-methyl-**	*p*-Tolualdehyde. C_8H_8O. *See* b35	120.16	λ^{hx} 251 (4.18), 257 (4.10), 279 (3.08), 284 (3.00)	...	204–5^{760} 106^{10}	1.0194$_4^{17}$	1.5454^{20}	δ	∞	∞	∞	...	chl v	**B7**, 297

For explanations, symbols and abbreviations see beginning of table. For structural formulas see end of table.

No.	Name	Synonyms and Formula	Mol. wt.	Color, crystalline form, specific rotation and λ_{max} (log ε)	m.p. °C	b.p. °C	Density	n_D	Solubility w	al	eth	ace	bz	other solvents	Ref.
	Benzaldehyde														
Ω b160	**—,3,4-methylene-dioxy-**	Piperonal.	150.14	wh-ye (w) λ^{al} 275 (3.8), 315 (4.0)	37	263^{760} 140^{15}	...	...	δ s^h	v	∞	s	...	...	**B19²**, 141
b161	—,—,oxime	...	165.15	(MeOH) λ^{al} 265 (4.1), 310 (3.9)	146	...	...	...	...	...	s	...	...	...	**B19²**, 147
Ω b162	**—,2-nitro-**	$C_7H_5NO_3$. *See* b35	151.12	ye nd (w) λ^{hx} 222 (4.18), 270 (3.56), 285 (3.23)	43.5–4	153^{23}	1.2844^{50}_{4}	...	δ	v	v	v	v	os v	**B7²**, 185
b163	—,—,diacetate	...	253.21	pr (lig)	90	...	...	...	...	s^h	s	s	s	os s	**B7²**, 187
b164	—,—,dimethyl acetal	...	197.19	gr-ye	...	$274–6^{762}$ $138–9^{11}$	...	...	...	...	...	...	s	...	**B7¹**, 137
Ω b165	**—,3-nitro-**	$C_7H_5NO_3$. *See* b35	151.12	lt ye nd (w) λ^{hx} 225 (4.41), 258 (2.85), 287 (3.00)	58	164^{23}	1.2792^{20}_{4}	...	δ	s	s	v	v	os s	**B7²**, 190
b166	—,—,diacetate	...	253.21	pr or nd (al)	72	...	1.393	...	...	s	s	s	s	os, aa s	**B7²**, 192
b167	—,—,dimethyl acetal	...	197.19	...	...	$162–4^{19}$	1.209^{15}	...	...	...	...	...	s	...	**B7**, 253
Ω b168	**—,4-nitro-**	$C_7H_5NO_3$. *See* b35	151.12	lf, pr (w) λ^{al} 259 (4.14), 284 (3.53), 295 (3.32), 305 (3.08)	106	sub	1.496	...	δ	v	δ	...	s	aa s lig δ	**B7²**, 196
b169	—,—,diacetate	...	253.21	pr (al)	127	...	...	...	...	s	s	s	s	os s	**B7²**, 198
b170	—,—,dimethyl acetal	...	197.19	...	23–5	$294–6^{774}$	...	...	...	...	...	...	s	...	**B7²**, 198
b171	**—,pentachloro-**	C_7HCl_5O. *See* b35	278.35	nd (bz, al) λ^{iso} 262 (3.8), 290 (2.9)	202.5	...	...	...	...	δ	v	v	v	CS_2 v	**B7¹**, 134
b172	**—,2,3,4,5-tetra-chloro-**	$C_7H_2Cl_4O$. *See* b35	243.91	...	106	...	...	...	...	...	...	...	...	...	**B7²**, 181
b173	**—,2,3,4-trichloro-**	$C_7H_3Cl_3O$. *See* b35	209.46	nd (al)	91	...	...	...	...	δ v^h	...	...	...	...	**B7**, 238
b174	**—,2,3,5-trichloro-**	$C_7H_3Cl_3O$. *See* b35	209.46	nd (al)	56	...	...	...	...	v	v	v	v	os s	**B7²**, 180
b175	**—,2,3,6-trichloro-**	$C_7H_3Cl_3O$. *See* b35	209.46	nd (lig)	86–7	...	...	...	...	...	v	v	v	lig δ, v^h	**B7**, 238
b176	**—,2,4,5-trichloro-**	$C_7H_3Cl_3O$. *See* b35	209.46	nd (al)	112–3	...	...	...	δ	v	v	v	v	chl, CS_2 v	**B7**, 238
b177	**—,2,4,6-trichloro-**	$C_7H_3Cl_3O$. *See* b35	209.46	λ^{al} 260 (3.7), 304 (2.8), 346 (1.5)	58–9	...	...	...	...	...	v	...	...	lig δ, v^h	**B7**, 238
b178	**—,3,4,5-trichloro-**	$C_7H_3Cl_3O$. *See* b35	209.46	nd (al)	90–1	...	...	...	$δ^h$	v^h	v	v	v	chl v peth s	**B7²**, 180
b179	**—,2,4,6-tri-methyl-**	Mesitylaldehyde. $C_{10}H_{12}O$. *See* b35	148.21	λ^{hx} 264 (4.16), 300 (3.32)	14	237–40 192^{50}	...	...	i	s	s	s	s	...	**B7²**, 250
b180	**—,2,4,6-trinitro-**	$C_7H_3N_3O_7$. *See* b35	241.12	dk gr pl (bz)	119	...	...	...	i	s	s	s	s	chl, aa s	**B7²**, 207
—	**Benzaldoxime**	*see* **Benzaldehyde, oxime**													
—	**Benzamarone**	*see* **1,5-Pentanedione, 1,2,3,4,5-pentaphenyl-**													
—	**Benzamide**	*see* **Benzoic acid,** amide													
—	**Benzamidine**	*see* **Benzoic acid,** amidine													
—	**Benzanilide**	*see* **Benzoic acid,** amide, *N*-phenyl-													
Ω b181	**1,2-Benzanthra-cene**	2,3-Benzophenanthrene. Naphthanthracene. Tetraphene. $C_{18}H_{12}$	228.30	ye br flr pl or lf (al-aa) λ^{bz} 281 (4.85), 292 (4.98), 301 (4.04), 317 (3.64), 331 (3.79), 344 (3.84), 376 (2.80), 386 (2.96)	162 (167)	435^{760} sub	...	...	i	$δ^h$	s	s	v	os s aa δ	**E14**, 329
—	**2,3-Benzanthra-cene**	*see* **Naphthacene**													
b182	**1,2-Benzanthra-cene, 9,10-di-hydro-**	$C_{18}H_{14}$. *See* b181	230.31	lf (bz, al, aa)	112	...	...	...	i	δ	...	...	δ	aa δ	**E14**, 330
b183	**—,9,10-dimethyl-**	$C_{20}H_{16}$. *See* b181	256.35	pa ye pl (al, aa) λ^{al} 296.5 (4.90), 345 (3.83), 364 (3.54), 384 (3.83)	122–3	...	...	...	i	δ	...	s	v	CS_2, to s	**E14s**, 139
b184	**—,4′-hydroxy-**	$C_{18}H_{12}O$. *See* b181	244.30	pa ye or og nd (bz) λ^{al} 278 (4.7), 288 (4.7), 315 (4.3)	225	...	...	...	...	...	...	...	s	Py, to s aa v	**E14s**, 171

For explanations, symbols and abbreviations see beginning of table. For structural formulas see end of table.

No.	Name	Synonyms and Formula	Mol. wt.	Color, crystalline form, specific rotation and λ_{max} (log ε)	m.p. °C	b.p. °C	Density	n_D	Solubility w	al	eth	ace	bz	other solvents	Ref.
	1,2-Benzanthracene														
Ω b185	—,3-methyl-	$C_{19}H_{14}$. *See* b181	242.32	pl (bz-al), nd (bz-lig) λ^{al} 270.5 (4.68), 291 (4.96), 340.5 (3.86), 377 (2.76)	156–7	$160^{0.1}$ sub	...	...	i	s	...	s	s	xyl, aa, Py, peth s	**E14s,** 116
Ω b186	—,4-methyl-	$C_{19}H_{14}$. *See* b181	242.32	nd (al) λ^{hx} 252 (4.51), 257 (4.51), 280 (4.85), 288 (4.91), 300 (4.27)	126.2–7.2	...	...	...	i	s	s	...	...	xyl s	**E14s,** 117
Ω b187	—,5-methyl-	$C_{19}H_{14}$. *See* b181	242.32	ye mcl pl (bz-al) λ^{al} 261 (4.47), 269 (4.66), 278 (4.85), 285 (4.81), 315 (3.70), 328 (3.80)	160–0.6	...	...	...	i	s	s	...	s	xyl s	**E14s,** 118
Ω b188	—,6-methyl-	$C_{19}H_{14}$. *See* b181	242.32	pl (al) λ^{al} 257 (4.58), 269.5 (4.61), 279 (4.85), 289 (4.92), 330 (3.83), 360 (3.60), 385 (2.48)	150.5–1.5	...	...	...	i	s^h	s	...	...	chl s xyl, CS_2 s^h	**E14s,** 119
b189	—,7-methyl-	$C_{19}H_{14}$. *See* b181	242.32	ye pl (al) λ^{hx} 258 (4.55), 272 (4.58), 282 (4.87), 293 (4.98), 304 (4.10)	183–3.6	...	...	...	i	s^h	...	...	...	aa s^h	**E14s,** 119
Ω b190	—,8-methyl-	$C_{19}H_{14}$. *See* b181	242.32	pl or nd (al, aa)	118	...	...	...	i	s	...	...	...	aa s	**E14s,** 120
Ω b191	—,9-methyl-	$C_{19}H_{14}$. *See* b181	242.32	pa ye nd or pl (aa, MeOH) λ^{al} 223 (4.5), 260 (4.6), 270 (4.6), 280 (4.8), 290 (4.9), 353 (3.8), 370 (3.7), 390 (3.0)	138	...	...	...	i	s	...	...	...	CS_2, aa s	**E14s,** 121
Ω b192	—,10-methyl-	$C_{19}H_{14}$. *See* b181	242.32	λ^{al} 232 (4.5), 258 (4.6), 280 (4.9), 298 (5.0), 353 (3.9), 370 (3.8), 388 (2.9)	141	...	...	...	i	s	s	s	...	CS_2, CCl_4, aa s	**E14s,** 122
b193	**1,2-Benzanthracene-10-carboxaldehyde**		256.31	ye pr or nd λ^{al} 292 (4.61), 302 (4.65), 383 (3.91)	148	...	...	...	...	s	s	s	s	MeOH s	**E14s,** 186
b194	**1,2-Benz-3,4-anthraquinone**	2,3-Benzo-9,10-phenanthraquinone. Tetraphene-5,6-quinone.	258.28	red nd (to) λ^{al} 234 (4.40), 256 (4.46), 400 (3.78)	262–3d	...	...	...	...	δ	...	...	δ	sulf s to, aa δ	**B7**[2], 759
b195	**1,2-Benz-9,10-anthraquinone**	...	258.28	ye pr (aa) λ^{al} 224 (4.4), 255 (4.5), 266 (4.7), 277 (4.7), 288 (4.9), 300 (4.0), 325 (3.7), 340 (3.7), 372 (2.7), 384 (2.8)	169	sub	...	...	...	δ	δ	s	v	chl v aa, sulf s lig δ	**B7**[2], 760
b196	**Benzanthrene**	...	216.29	lf (al) λ^{al} 227 (4.67), 247 (4.35), 329 (4.26), 344 (4.17)	84	...	...	...	...	s	...	...	...	sulf s (red) aa, xyl s	**E14s,** 343
Ω b197	**1,9-Benzanthr-10-one**	...	230.27	ye nd (xyl or al) λ^{al} 394 (4.00)	170	...	...	...	i	...	...	...	...	sulf s^h → og-red	**B7**[2], 468
b198	—,*bz*-1-bromo-	$C_{17}H_9BrO$. *See* b197	309.17	ye nd (aa)	178 (170)	...	...	...	...	s	...	...	...	MeOH, sulf s	**E14s,** 355
b199	—,*bz*-1-chloro-	$C_{17}H_9ClO$. *See* b197	264.71	ye nd (aa)	182–3	...	...	...	...	...	...	...	δ	PhCl, aa δ	**E14s,** 354

For explanations, symbols and abbreviations see beginning of table. For structural formulas see end of table.

No.	Name	Synonyms and Formula	Mol. wt.	Color, crystalline form, specific rotation and λ_{max} (log ε)	m.p. °C	b.p. °C	Density	n_D	Solubility w	al	eth	ace	bz	other solvents	Ref.
	1,9-Benzanthr-10-one														
b200	**—,*bz*-2-hydroxy-**	$C_{17}H_{10}O_2$. *See* b197	246.27	ye nd ($PhCH_2OH$)	294	...	...	...	...	...	...	...	...	sulf, alk s os δ	**E14s**, 373
b201	**—,4-hydroxy-**	$C_{17}H_{10}O_2$. *See* b197	246.27	ye nd (aa)	178–9	...	...	...	...	...	s	...	v^h	sulf s	**E14s**, 371
—	**Benzaurine**	*see* **Methanol, bis (4-hydroxyphenyl)-phenyl-**													
—	**Benzedrine**	*see* **Amphetamine (*dl*)**													
Ω b202	**Benzene***	Benzol. Phene*	78.12	rh pr λ^{al} 243 (2.2), 249 (2.3), 256 (2.4), 261 (2.2)	5.5 (3.3)	80.1^{760}	0.87865_4^{20}	1.5011^{20}	δ	∞	∞	∞	...	chl, aa ∞	**B5**^{2}, 119
—	**—,acetyl-**	*see* **Acetophenone**													
Ω b203	**—,allyl-**	3-Phenylpropene*. C_9H_{10}. *See* b202	118.18	λ^{al} 210 (3.9), 260 (2.3), 270 (2.3)	−40	156^{760} 47^{13}	0.8920_4^{20}	1.5131^{20}	i	s	s	...	s	chl, CCl_4s	**B5**^{2}, 373
b204	**—,1-allyl-4-bromo-**	C_9H_9Br. *See* b202	197.08	...	...	$222–3^{730}$ 96^{12}	1.324_4^{15}	1.559^{15}	i	...	s	...	s	chl s	**B5**^{2}, 373
b205	**—,2-allyl-4-chloro-1-hydroxy-**	C_9H_9ClO. *See* b202	168.63	pr (lig)	48	$130–2^{15}$	1.171^{15}	...	...	s	...	...	s	chl, alk s	**B6**^{1}, 282
Ω b206	**—,1-allyl-3,4-dimethoxy-**	$C_{11}H_{14}O_2$. *See* b202	178.23	cr (hx) λ^{al} 230 (3.8), 280 (3.4)	−4	254.7^{760} 104^2	1.0396_4^{20}	1.5340^{20}	i	s	s	...	...	...	**B6**^{3}, 5024
b207	**—,1-allyl-2-hydroxy-**	$C_9H_{10}O$. *See* b202	134.18	λ^{al} 260 sh (3.1), 275 (3.4)	−6	220^{760} $93–4^8$	1.0255_{15}^{15}	1.5181^{20}	...	...	v	...	...	...	**B6**, 282
b208	**—,1-allyl-4-hydroxy-**	Chavicol. $C_9H_{10}O$. *See* b202	134.18	...	16	$235–6^{760}$ 120^{12}	1.033_4^{18}	1.5448^{20}	s^h	∞	∞	...	...	chl, peth ∞	**B6**^{2}, 529
Ω b209	**—,5-allyl-1-hydroxy-2-methoxy-**	Betel phenol. Chavibetol. $C_{10}H_{12}O_2$. *See* b202	164.21	...	8.5	$253–4^{760}$ 111^6	1.0613_4^{25}	1.5413^{20}	...	s	s	...	...	...	**B6**^{2}, 923
—	**—,amino-***	*see* **Aniline**													
b210	**—,1-amino-2-butyl-**	*o*-Butylaniline. $C_{10}H_{15}N$. *See* b202	149.24	ye oil	...	$122–5^{12}$	0.953_4^{20}	...	...	s	...	s	s	os, ac s	**B12**^{2}, 633
b211	**—,1-amino-4-butyl-***	*p*-Butylaniline. $C_{10}H_{15}N$. *See* b202	149.24	pa ye		261^{760} $133–4^{14}$	0.945_4^{20}	...	i	s	...	s	s	os. ac s	**B12**^{2}. 633
b212	**—,1-amino-4-*sec*-butyl-(*dl*)**	$C_{10}H_{15}N$. *See* b202	149.24	...	...	238^{762} 118^{15}	0.949_4^{15}	1.5360^{20}	...	...	v	...	v	...	**B12**^{2}, 635
b213	**—,1-amino-2-*tert*-butyl-**	$C_{10}H_{15}N$. *See* b202	149.24	λ^{iso} 237 (3.89), 286 (3.35)	...	233–5	0.977^{15}	1.5453^{20}	...	v	v	...	v	ac s	**B12**^{2}, 636
b214	**—,1-amino-3-*tert*-butyl-**	$C_{10}H_{15}N$. *See* b202	149.24	...	...	229^{708}	...	...	...	v	v	...	v	...	**B12**^{2}, 637
Ω b215	**—,1-amino-4-*tert*-butyl-**	$C_{10}H_{15}N$. *See* b202	149.24	ye-red (peth) λ^{iso} 237 (4.03), 292 (3.25)	17	240^{740} (228^{762})	0.9525_4^{15}	1.5380^{20}	δ	∞	∞	...	v	...	**B12**^{3}, 637
b216	**—,1-amino-4-cyclohexyl-***	...	175.28	pl (lig)	55	...	...	...	...	s	...	...	s	...	**B12**^{2}, 667
b217	**—,2-amino-3,5-dibromo-1-ethoxy-***	$C_8H_9Br_2NO$. *See* b202	294.99	pr (dil al)	52	...	...	...	...	s	...	...	...	lig s^h	**B13**, 387
b218	**—,5-amino-1,3-dibromo-2-methoxy-***	$C_7H_7Br_2NO$. *See* b202	280.96	pl (lig) λ^w 302.5 (3.69)	66	...	...	...	...	s	s	s	s	chl, aa s	**B13**, 517
b219	**—,1-amino-2-diethylamino-***	$C_{10}H_{16}N_2$. *See* b202	164.25	oil	...	312^{744} 127^{25}	...	...	δ	s	...	s	s	os s	**B13**^{1}, 16
b220	**—,1-amino-3-diethylamino-***	$C_{10}H_{16}N_2$. *See* b202	164.25	...	...	$276–8^{760}$ 117^4	...	...	δ	s	...	...	...	...	**B13**^{2}, 26
b221	**—,1-amino-4-diethylamino-***	$C_{10}H_{16}N_2$. *See* b202	164.25	ye oil	...	260–2 $139–40^{10}$	...	...	...	...	...	...	s	...	**B13**^{2}, 40
b222	**—,1-amino-3,5-dihydroxy-***	Phloramin. $C_6H_7NO_2$. *See* b202	125.13	nd	146–52	...	...	...	δ	s	i	...	...	...	**B13**, 787
b223	**—,4-amino-1,2-dihydroxy-**	$C_6H_7NO_2$. *See* b202	125.13	br-vt nd (al-bz)	124–5d	...	...	...	s	s	v	δ	δ	...	**B13**^{2}, 464
Ω b224	**—,1-amino-2,3-dimethyl-***	2,3-Xylidine. $C_8H_{11}N$. *See* b202	121.18	λ^{iso} 236 (3.88), 286 (3.29)	< −15	$221–2^{760}$ 106^{15}	0.9931^{20}	1.5684^{20}	δ	v	v	...	...	...	**B12**^{2}, 601
Ω b225	**—,1-amino-2,4-dimethyl-***	2,4-Xylidine. $C_8H_{11}N$. *See* b202	121.18	λ^{al} 290 (3.2)	−14.3	214^{760} 91^{10}	0.9723_4^{20}	1.5569^{20}	δ	s	s	...	s	...	**B12**^{2}, 602
Ω b226	**—,1-amino-3,5-dimethyl-***	3.5-Xylidine. $C_8H_{11}N$. *See* b202	121.18	λ^{iso} 239 (3.85). 289 (3.24)	9.8	220–1 $99–100^{20}$	0.9706_4^{20}	1.5581^{20}	δ	...	s	...	...	...	**B12**^{2}, 613
Ω b227	**—,2-amino-1,3-dimethyl-***	2,6-Xylidine. $C_8H_{11}N$. *See* b202	121.18	λ^{iso} 233 (3.92), 284 (3.37)	11.2	214^{739}	0.9842^{20}	1.5610^{20}	i	s	s	...	...	...	**B12**^{2}, 604
Ω b228	**—,2-amino-1,4-dimethyl-***	2,5-Xylidine. $C_8H_{11}N$. *See* b202	121.18	ye lf (lig) λ^{iso} 236 (3.88), 288 (3.39)	15.5	214^{760} $97–101^{10}$	0.9790_4^{21}	1.5591^{21}	δ	...	s	...	...	...	**B12**^{2}, 614
Ω b229	**—,4-amino-1,2-dimethyl-***	3,4-Xylidine. $C_8H_{11}N$. *See* b202	121.18	pl or pr (lig) λ^{al} 240 (3.97)	51	226^{760}	1.076^{18}	...	δ	...	s	...	...	lig v	**B12**^{2}, 602
b230	**—,1-amino-2-dimethylamino-**	$C_8H_{12}N_2$. *See* b202	136.20	oil λ^{al} 240 sh (3.81), 293 (3.45)	...	218^{751} 117^{25}	0.995^{25}	...	δ	v	v	v	v	os v	**B13**, 15
b231	**—,1-amino-3-dimethylamino-**	$C_8H_{12}N_2$. *See* b202	136.20	λ^{al} 245 sh (3.9), 298 (3.4)	< −20	$268–70^{740}$ 138^{10}	0.995^{25}	...	δ	v	v	...	...	...	**B13**, 38

For explanations, symbols and abbreviations see beginning of table. For structural formulas see end of table.

No.	Name	Synonyms and Formula	Mol. wt.	Color, crystalline form, specific rotation and λ_{max} (log ε)	m.p. °C	b.p. °C	Density	n_D	Solubility						Ref.
									w	al	eth	ace	bz	other solvents	
	Benzene														
b232	**—,1-amino-4-dimethylamino-**	$C_8H_{12}N_2$. *See* b202	136.20	nd (bz) λ^{eth} 284 (4.24), 328 (3.46)	53 (40.5)	262, 158^{11}	1.036_4^{20}	...	s	v	v	...	v	chl s lig δ	**B13**, 72
b233	**—,1-amino-2,3-dimethyl-4-nitro-***	$C_8H_{10}N_2O_2$. *See* b202	166.18	ye pr (al) λ^{al} 382 (3.99)	114	...	...	...	...	v^h	...	...	...	...	**B12**, 1103
b234	**—,1-amino-2,3-dimethyl-5-nitro-***	$C_8H_{10}N_2O_2$. See b202	166.18	pa ye nd (al)	111–2	...	...	...	...	v^h	...	...	...	...	**B12**[1], 479
b235	**—,1-amino-2,4-dimethyl-3-nitro-***	$C_8H_{10}N_2O_2$. *See* b202	166.18	ye nd	81–2 (78)	...	...	...	s^h	s	...	...	...	lig s	**B12**[2], 612
Ω b236	**—,1-amino-2,4-dimethyl-5-nitro-***	$C_8H_{10}N_2O_2$. *See* b202	166.18	og ye nd (al)	123	...	...	...	...	v^h	...	...	...	...	**B12**[2], 612
b237	**—,1-amino-3,4-dimethyl-2-nitro-***	$C_8H_{10}N_2O_2$. *See* b202	166.18	red pr (al)	65–6	...	...	...	i	v^h	...	...	...	...	**B12**[1], 481
b238	**—,1-amino-3,5-dimethyl-2-nitro-***	$C_8H_{10}N_2O_2$. *See* b202	166.18	ye nd (lig)	56	...	...	...	...	...	...	...	...	...	**B12**[2], 613
b239	**—,1-amino-4,5-dimethyl-2-nitro-***	$C_8H_{10}N_2O_2$. *See* b202	166.18	red-br pr (al)	140	...	...	...	...	...	δ	s	s	os v lig i	**B12**, 1106
b240	**—,2-amino-1,3-dimethyl-4-nitro-***	$C_8H_{10}N_2O_2$. *See* b202	166.18	ye nd (dil al) λ^{al} 380 (4.16)	81–2	...	...	...	...	s	...	...	...	...	**B12**[2], 605
b241	**—,2-amino-1,3-dimethyl-5-nitro-***	$C_8H_{10}N_2O_2$. *See* b202	166.18	og-ye nd	158	...	...	...	...	s	...	...	...	...	**B12**[2], 606
b242	**—,2-amino-1,5-dimethyl-3-nitro-***	$C_8H_{10}N_2O_2$. *See* b202	166.18	og-red nd or pl (lig)	76	...	...	...	...	s	...	...	...	...	**B12**[2], 612
b243	**—,2-amino-3,4-dimethyl-1-nitro-***	$C_8H_{10}N_2O_2$. *See* b202	166.18	red pl (al)	118–9	...	...	...	...	v^h	...	...	...	con HCl v	**B12**, 1102
b244	**—,5-amino-1,2-dimethyl-3-nitro-***	$C_8H_{10}N_2O_2$. *See* b202	166.18	og lf (al)	74–5	...	...	...	...	v^h	...	...	...	...	**B12**, 1106
b245	**—,5-amino-1,3-dimethyl-2-nitro-***	$C_8H_{10}N_2O_2$. *See* b202	166.18	og pr (bz), lf (lig) λ^{al} 240 (3.87), 301 (3.48), 385 (3.68)	133	...	...	...	s^h	v	...	...	s	lig δ	**B12**[2], 613
b245[1]	**—,(1-amino-ethyl)-(*d*)***	$C_6H_5CH(CH_3)NH_2$	121.18	$[\alpha]_D^{25}$ +39.2 (MeOH)	...	187^{760}	0.9651^{15}	...	s	v	v	...	...	...	**B12**[2], 586
Ω b245[2]	**—,—(*dl*)***	$C_6H_5CH(CH_3)NH_2$	121.18	...	...	187^{760}, 87^{24}	0.9395^{15}	1.5238^{25}	s	∞	∞	...	...	...	**B12**[2], 586
b245[3]	**—,—(*l*)***	$C_6H_5CH(CH_3)NH_2$	121.18	$[\alpha]_D^{18}$ −38 (MeOH)	...	184–6, 77^{16}	0.9520_4^{20}	...	s	v	v	...	...	...	**B12**[2], 586
Ω b246	**—,1-amino-2-ethyl-***	*o*-Ethylaniline. $C_8H_{11}N$. *See* b202	121.18	λ^{iso} 234 (3.90), 286 (3.37)	−43	$209–10^{760}$	0.983_4^{22}	1.5584^{22}	δ	v	v	...	...	...	**B12**[2], 584
b247	**—,1-amino-3-ethyl-***	*m*-Ethylaniline. $C_8H_{11}N$. *See* b202	121.18	...	−64	$214–5^{764}$, $93–5^{6}$	0.9896^{0}	...	s^h	v	v	...	...	...	**B12**[1], 468
Ω b248	**—,1-amino-4-ethyl-***	*p*-Ethylaniline. $C_8H_{11}N$. *See* b202	121.18	...	−4.87	217.8^{760}, 92.3^{10}	0.9679_4^{20}	1.5554^{20}	δ	v	v	...	...	...	**B12**[2], 584
Ω b248[1]	**—,(2-amino-ethyl)-***	*β*-Phenethylamine. $C_6H_5CH_2CH_2NH_2$	121.18	...	...	197–8	0.9580_4^{24}	1.5290^{25}	s	v	v	...	...	...	**B12**[12], 591
b249	**—,—,hydro-chloride***	$C_6H_5CH_2CH_2NH_2.HCl$	157.65	pl or lf (al)	218–9	...	...	...	v	v	i	...	...	...	**B12**[2], 592
b250	**—,4(2-amino-ethyl)-1,3-dihydroxy-, hydrochloride***	$C_8H_{11}NO_2.HCl$. *See* b202	189.64	nd (w)	237d	...	...	...	v	v	δ	...	...	...	**B13**[2], 486
b251	**—,1(2-amino-ethyl)-2-hydroxy-***	$C_8H_{11}NO$. *See* b202	137.18	rh (al-eth)	152–3	...	...	...	...	...	...	...	...	...	**B13**, 624
b252	**—,1(2-amino-ethyl)-3-hydroxy-, hydrochloride***	$C_8H_{11}NO.HCl$. *See* b202	173.65	cr (al-eth)	145	...	...	...	...	...	...	...	...	...	**B13**[1], 233
—	**—,1(2-amino-ethyl)-4-hydroxy-***	*see* **Tyramine**													
b253	**—,5-amino-2-hydroxy-4-isopropyl-1-methyl-**	4-Aminocarvacrol. $C_{10}H_{15}NO$. *See* b202	165.24	cr (MeOH)	134	...	...	...	...	...	...	...	...	MeOH s^h	**B13**[2], 392
b254	**—,6-amino-3-hydroxy-4-isopropyl-1-methyl-**	4-Aminothymol. $C_{10}H_{15}NO$. *See* b202	165.24	nd, sc (bz)	178–9	...	...	...	...	s	δ	...	v	os s	**B13**[2], 393

For explanations, symbols and abbreviations see beginning of table. For structural formulas see end of table.

No.	Name	Synonyms and Formula	Mol. wt.	Color, crystalline form, specific rotation and λ_{max} (log ε)	m.p. °C	b.p. °C	Density	n_D	Solubility w	al	eth	ace	bz	other solvents	Ref.
	Benzene														
b255	**—,1-amino-4-isobutyl-(*dl*)**	$C_{10}H_{15}N$. *See* b202	149.24	pa ye	...	238^{762} 112^{11}	0.949_4^{15}	...	i	∞	...	∞	v	os ∞	**B12**², 635
Ω b256	**—,1-amino-2-isopropyl-**	$C_9H_{13}N$. *See* b202	135.21	λ^{iso} 234 (3.89), 286 (3.36)	...	$220–1^{745}$ 95^{13}	0.9760_4^{12}	...	i	...	s	...	s	...	**B12**², 624
b257	**—,1-amino-4-isopropyl-**	Cumidine. $C_9H_{13}N$. *See* b202	135.21	λ^{al} 234.8 (3.99), 288.2 (3.23)	−63	225^{760}	0.953_4^{20}	1.5415^{20}	i v^h	s^h	s	...	s	...	**B12**², 625
b258	**—,2-amino-1-isopropyl-4-methyl-**	Thymylamine. $C_{10}H_{15}N$. *See* b202	149.24	oil λ^{al} 238 (3.81), 286.5 (3.27)	...	$238–42^{760}$	...	...	δ	∞	∞	...	...	...	**B12**², 642
b259	**—,2-amino-4-isopropyl-1-methyl-**	Carvacrylamine. $C_{10}H_{15}N$. *See* b202	149.24	...	−16	241^{760} 118^{12}	0.9942_4^{20}	1.5387^{20}	δ	s	s	...	...	...	**B12**², 638
Ω b260	**—,1-amino-2-mercapto-***	*o*-Aminothiophenol. C_6H_7NS. *See* b202	125.19	nd λ 360 (3.5)	26	234 $125–7^{6}$	...	...	...	...	s	...	...	...	**B13**, 397
b261	**—,1-amino-3-mercapto-***	*m*-Aminothiophenol. C_6H_7NS. *See* b202	125.19	pa ye oil	...	$180–90^{16}$	...	...	i	s	s	...	...	aa s peth i	**B13**, 425
Ω b262	**—,1-amino-4-mercapto-***	*p*-Aminothiophenol. C_6H_7NS. *See* b202	125.19	wh gran	46	$140–5^{16}$	...	...	s	...	...	...	...	...	**B13**, 533
b263	**—,1-amino-2-methoxy-3-nitro-**	$C_7H_8N_2O_3$. *See* b202	168.16	pa ye nd (lig)	67	...	...	...	...	s	...	...	...	lig s^h	**B13**², 195
b264	**—,2-amino-4-methoxy-1-nitro-***	$C_7H_8N_2O_3$. *See* b202	168.16	br nd	131	sub	...	...	...	s	...	...	...	...	**B13**², 215
b265	**—,1-amino-4-methylamino-***	$C_7H_{10}N_2$. *See* b202	122.17	lf (eth-peth) λ^{eth} 253 (4.16), 329 (3.48)	36	257–9.5 152^{20}	...	...	v	v	v	v	v	os s	**B13**, 71
b266	**—,1(amino-methyl)-4-isopropyl-***	Cuminylamine. $C_{10}H_{15}N$. *See* b202	149.24	oil	...	$225–7^{724}$ $111–2^{11}$	...	...	i	v	v	...	...	...	**B12**, 1172
b267	**—,1(amino-methyl)-2,4,5-trimethyl-**	$C_{10}H_{15}N$. *See* b202	149.24	nd (dil al)	65	...	...	...	i δ^h	s^h	...	...	...	chl s	**B12**, 1177
b268	**—,5(amino-methyl)-1,2,3-trimethyl-**	$C_{10}H_{15}N$. *See* b202	149.24	lf (w)	123	...	...	...	s^h	...	...	...	...	...	**B12**, 1176
b269	**—,1-amino-4-octyl-***	1(4-Aminophenyl)octane*. $C_{14}H_{23}N$. *See* b202	205.35	...	20	$310–1^{760}$ $170–2^{17}$	...	...	i	...	∞	...	...	...	**B12**, 1185
b270	**—,1-amino-2,3,4,5,6-penta-methyl-***	$C_{11}H_{17}N$. *See* b202	163.26	mcl pr (al)	152–3	277–8	...	...	i	s	s	...	...	...	**B12**², 646
b271	**—,1-amino-2-propyl-***	*o*-Propylaniline. $C_9H_{13}N$. *See* b202	135.21	λ^{cy} 287.5 (3.34)	...	226^{758} 116^{15}	0.9602_4^{30}	1.5427^{30}	i	s	s	...	...	...	**B12**², 620
b272	**—,1-amino-4-propyl-***	*p*-Propylaniline. $C_9H_{13}N$. *See* b202	135.21	...	...	$224–6^{760}$	...	...	δ	...	...	...	...	...	**B12**, 1143
b273	**—,1(2-amino-propyl)-4-hydroxy-(*l*)***	*L*-Paredrine. $C_9H_{13}NO$. *See* b202	151.21	$[\alpha]_D^{17}$ − 52 (al)	111	...	...	...	...	s^h	s	...	...	chl s	**H34**, 2202
b274	**—,1-amino-2,3,4,5-tetramethyl-***	Prehnidine. $C_{10}H_{15}N$. *See* b202	149.24	lf (w)	70	259–60	...	...	s^h	v	v	...	...	lig v	**B12**, 1175
b275	**—,1-amino-2,3,4,6-tetramethyl-***	Isoduridine. $C_{10}H_{15}N$. *See* b202	149.24	...	23–4	255	0.978^{24}	...	...	s	...	...	...	...	**B12**¹, 506
Ω b276	**—,1-amino-2,4,5-trimethyl-***	Pseudocumidine. $C_9H_{13}N$. *See* b202	135.21	nd (w)	68	234–5	0.957	...	v	s	...	...	...	...	**B12**², 629
Ω b277	**—,2-amino-1,3,5-trimethyl-***	Mesidine. $C_9H_{13}N$. *See* b202	135.21	λ^{iso} 237 (3.94), 289 (3.39)	−5	232–3	0.9633	1.5495^{20}	...	...	...	...	...	...	**B12**², 631
—	**—,azido** ...	*see* **Benzene, triazo-***													
Ω b278	**—,1-benzyl-4-ethyl-**	*p*-Ethyldiphenylmethane. $C_{15}H_{16}$. *See* b202	196.30	λ^{iso} 260 (2.77), 263 (2.75), 266 (2.80), 269 (2.75), 274 (2.69)	−24	297^{760} $85^{0.2}$	0.9777^{20}	1.5616^{20}	δ	s	s	...	...	chl s	**B5**, 614
b279	**—,1-benzyl-2-hydroxy-**	2-Benzylphenol. $C_{13}H_{12}O$. *See* b202	184.24	cr (peth)	(i) 21 (metast) (ii) 51.5–3 (st)	312 $159–62^{12}$	...	...	...	...	...	...	...	...	**B6**, 675
b280	**—,1-benzyl-4-hydroxy-**	4-Benzylphenol. $C_{13}H_{12}O$. *See* b202	184.24	nd (al)	84	320–2 198–200^{10}	...	...	s^h	s	s	...	s	chl, aa s	**B6**, 678
Ω b281	**—,1,2-bis(bromo-methyl)-***	*o*-Xylylene bromide. $C_8H_8Br_2$. *See* b202	263.98	rh (chl)	95	$128–30^{4.5}$	1.988^{0}	...	i	s	s	...	...	chl, lig, peth s	**B5**², 285
Ω b282	**—,1,3-bis(bromo-methyl)-***	*m*-Xylylene bromide. $C_8H_8Br_2$. *See* b202	263.98	nd (chl), pr (ace)	77	$135–40^{20}$	1.959^{0}	...	i	s	s	...	...	chl, lig s	**B5**², 294
Ω b283	**—,1,4-bis(bromo-methyl)-***	*p*-Xylylene bromide. $C_8H_8Br_2$. *See* b202	263.98	mcl pr (al), cr (chl, bz)	145–7	245 $155–8^{14}$	2.012^{0}	...	i	v	δ	...	s	chl v	**B5**, 385
b284	**—,1,3-bis(bromo-methyl)-5-methyl-***	$C_9H_{10}Br_2$. *See* b202	278.00	pr	66	...	...	...	i	δ v^h	s	...	s	lig s	**B5**², 315
b285	**—,1,3-bis(car-boxymethoxy)-***	$C_{10}H_{10}O_6$. *See* b202	226.19	nd (aa, w)	195	...	...	...	s^h	...	...	...	...	aa s^h	**B6**², 817

For explanations, symbols and abbreviations see beginning of table. For structural formulas see end of table.

No.	Name	Synonyms and Formula	Mol. wt.	Color, crystalline form, specific rotation and λ_{max} (log ε)	m.p. °C	b.p. °C	Density	n_D	Solubility						Ref.
									w	al	eth	ace	bz	other solvents	
	Benzene														
b286	—,—.diethyl ester*	$C_{14}H_{18}O_6$. *See* b202	282.30	nd (eth)	42	228^{32}	...	...	i	...	...	...	...	lig δ	**B6**, 818
Ω b287	**—,1,2-bis(chloro-methyl)-***	*o*-Xylylene chloride. $C_8H_8Cl_2$. *See* b202	175.07	mcl (lig)	55	239–41 $130–5^{19}$	1.393^0	...	i	v	v	...	...	chl, lig v	**B5**^{2}, 283
Ω b288	**—,1,3-bis(chloro-methyl)-***	*m*-Xylylene chloride. $C_8H_8Cl_2$. *See* b202	175.07	...	34.2	250–5	1.302^{20}	...	i	v	v	...	...	...	**B5**^{2}. 291
Ω b289	**—,1,4-bis(chloro-methyl)-***	*p*-Xylylene chloride. $C_8H_8Cl_2$. *See* b202	175.07	pl (al) λ^{al} 268 (2.30)	100	240–5d 135^{16}	1.417^0	...	i	v^h	v	v	...	chl v aa δ	**B5**^{2}, 300
b290	**—,1,2-bis(cyano-methyl)-***	*o*-Xylylene dicyanide. $C_{10}H_8N_2$. *See* b202	156.20	(i) nd or pr (MeOH) (ii) pr (al)	(i) 18 (unst) (ii) 60 (st)	...	...	...	i	s^h	s	...	...	lig i	**B9**^{2}, 623
b291	**—,1,3-bis(cyano-methyl)-***	*m*-Xylylene dicyanide. $C_{10}H_8N_2$. *See* b202	156.20	nd (w), pr (eth)	30	$305–10^{300}$ $170^{20–30}$	...	...	i	s^h	s	...	s	chl s lig i	**B9**^{2}, 624
b292	**—,1,4-bis(cyano-methyl)-***	*p*-Xylylene dicyanide. $C_{10}H_8N_2$. *See* b202	156.20	pr (al), nd (w)	98	...	...	...	i s^h	s^h	s	...	...	chl s	**B9**^{2}, 624
Ω b293	**—,1,2-bis(di-bromomethyl)-***	$C_8H_6Br_4$. *See* b202	421.77	mcl	116–7	...	...	...	δ s^h	...	...	...	...	chl v lig i	**B5**, 367
b294	**—,1,3-bis(di-bromomethyl)-***	$C_8H_6Br_4$. *See* b202	421.77	nd (al), pl (chl)	107	...	...	...	i	s^h	...	...	v	chl v lig, aa s	**B5**, 375
b295	**—,1,4-bis(di-bromomethyl)-***	$C_8H_6Br_4$. *See* b202	421.77	mcl pr (chl)	172	...	...	...	δ^h	δ	δ	...	v	chl δ, s^h	**B5**, 386
b296	**—,1,2-bis(diethyl-amino)-***	$C_{14}H_{24}N_2$. *See* b202	220.36	...	...	119^{10}	0.9267_4^{13}	1.5213^{13}	...	s	s	...	...	...	**B13**^{2}, 12
b297	**—,1,3-bis(diethyl-amino)-***	$C_{14}H_{24}N_2$. *See* b202	220.36	...	...	148^9	0.9522_4^{12}	1.5537^{12}	...	s	s	...	...	...	**B13**^{2}, 26
b298	**—,1,4-bis(diethyl-amino)-***	$C_{14}H_{24}N_2$. *See* b202	220.36	mcl pr, pl (al-w)	52	280	...	...	...	v	v	...	v	chl, lig v	**B13**^{2}, 40
b299	**—,1,4-bis-(dimethyl-amino)-***	$C_{10}H_{16}N_2$. *See* b202	164.25	lf (dil al or lig) λ^{cy} 228 sh (3.67), 264 (4.21), 304 sh (3.27)	51	260	...	...	δ s^h	v	v	...	v	chl v lig s	**B13**^{2}, 40
b301	**—,1,3-bis(1,1-dimethyl-propyl)-2-hydroxy-5-methyl-**	$C_{17}H_{28}O$. *See* b202	248.41	...	...	283^{760} 165^{20}	0.931_4^{25}	1.4950^{20}	...	...	...	...	...	alk i	**C38**, 4170
b302	**—,1,2-bis(hy-droxymethyl)-***	Phthalyl alcohol. *o*-Xylylene glycol. $C_8H_{10}O_2$. *See* b202	138.17	pl (eth, peth) λ^{al} 261 (2.3), 267 (2.3)	65–6.5	...	...	...	s	s	v	...	δ	...	**B6**, 910
b303	**—,1,3-bis(hy-droxymethyl)-***	Isophthalyl alcohol *m*-Xylylene glycol. $C_8H_{10}O_2$. *See* b202	138.17	nd (bz)	57 (46–7)	$154–9^{13}$	1.1359^{53}	...	v	s	s	...	...	AcOEt v	**B6**^{1}, 446
b304	**—,1,4-bis(hy-droxymethyl)-***	Terephthalyl alcohol *p*-Xylylene glycol. $C_8H_{10}O_2$. *See* b202	138.17	nd (w)	115–6	...	...	...	v	v	v	s	...	...	**B6**^{2}, 891
b304^1	**—,1,2-bis(meth-ylamino)-4-chloro-***	$C_8H_{11}ClN_2$. *See* b202	170.64	pr (lig)	61	...	...	...	s	...	...	...	...	...	**B13**, 25
b305	**—,1,3-bis(3-methylbutoxy)-***	Resorcinol diisoamyl ether. $C_{16}H_{26}O_2$. *See* b202	250.39	...	47	...	...	...	i	...	...	...	...	...	**B6**, 815
b306	**—,1,3-bis(phen-ylazo)-2,4-di-hydroxy-***	$C_{18}H_{14}N_4O_2$. *See* b202	318.34	red nd (chl-al)	222	...	...	...	...	δ	...	...	...	con ac, con alk s, chl v^h	**B16**, 185
b307	**—,1,5-bis(phen-ylazo)-2,4-di-hydroxy-***	$C_{18}H_{14}N_4O_2$. *See* b202	318.34	br-red nd	217	...	...	...	i	δ	δ	...	...	sulf, chl s	**B16**, 186
b308	**—,1,3-bis(4-tolyl-amino)-***	$C_{20}H_{20}N_2$. *See* b202	288.40	nd (al)	138–9	...	...	...	i	δ	δ	...	...	...	**B13**, 42
Ω b309	**—,bromo-***	Phenyl bromide. C_6H_5Br. *See* b202	157.02	λ^{al} 213.5 (3.86)	− 30.82	156^{760} 43^{18}	1.4950_4^{20}	1.5597^{20}	i	v	v	...	v	CCl_4 s	**B5**^{2}, 158
Ω b310	**—,(4-bromo-butoxy)-***	$C_{10}H_{13}BrO$. *See* b202	229.12	cr (al)	41	$153–6^{18}$	...	...	...	δ s^h	...	...	...	...	**B6**^{2}, 146
Ω b311	**—,1-bromo-4-*tert*-butyl-**	$C_{10}H_{13}Br$. *See* b202	213.13	...	19	$231–2^{760}$ 103^{10}	1.2286_4^{20}	1.5436^{20}	i	...	s	...	s	chl s	**B5**^{2}, 320
Ω b312	**—,1-bromo-2-chloro-***	C_6H_4BrCl. *See* b202	191.46	...	− 12.3	204^{765}	1.6382_4^{25}	1.5809^{20}	i	...	...	...	v	...	**B5**^{2}, 161
Ω b313	**—,1-bromo-3-chloro-***	C_6H_4BrCl. *See* b202	191.46	...	− 21.5	196	1.6302_4^{20}	1.5771^{20}	i	v	v	...	...	...	**B5**^{2}, 161
Ω b314	**—,1-bromo-4-chloro-***	C_6H_4BrCl. *See* b202	191.46	nd or pl (al, eth)	68	196^{756}	1.576_4^{71}	1.5531^{70}	i	δ s^h	s	...	s	chl s	**B5**^{2}, 162
b315	**—,1-bromo-4-cyclohexyl-***	$C_{12}H_{15}Br$. *See* b202	239.16	...	...	160^{23}	1.283_4^{25}	1.5584^{20}	i	...	...	...	s	chl s	**B5**^{2}, 396
b316	**—,1-bromo-2,3-dichloro-***	$C_6H_3BrCl_2$. *See* b202	225.91	pl or lf (al)	60	243^{705}	...	...	i	δ v^h	s	...	v	...	**B5**, 209
b318	**—,1-bromo-3,5-dichloro-***	$C_6H_3BrCl_2$. *See* b202	225.91	pr (al)	82–4	232^{757}	...	...	i	s^h	s	...	v	chl s	**B5**, 210
b319	**—,2-bromo-1,3-dichloro-***	$C_6H_3BrCl_2$. *See* b202	225.91	pl (al)	65	242^{765}	...	...	i	...	v	...	v	chl v	**B5**, 210
b320	**—,2-bromo-1,4-dichloro-***	$C_6H_3BrCl_2$. *See* b202	225.91	pr or nd (al)	35	235^{731} 119^{26}	...	...	i	s v^h	v	...	v	chl, lig v	**B5**, 210

For explanations, symbols and abbreviations see beginning of table. For structural formulas see end of table.

No.	Name	Synonyms and Formula	Mol. wt.	Color, crystalline form, specific rotation and λ_{max} (log ε)	m.p. °C	b.p. °C	Density	n_D	Solubility: w	al	eth	ace	bz	other solvents	Ref.
	Benzene														
b321	—,4-bromo-1,2-dichloro-*	$C_6H_3BrCl_2$. *See* b202	225.91	pr	25	237^{757} 124^{33}	......	......	i	δ v^h	v	...	v	chl v	**B5**, 210
b322	—,1-bromo-2,4-dihydroxy-*	$C_6H_5BrO_2$. *See* b202	189.02	......	103	$138–41^{12}$	......	......	v	δ	v	...	δ	CS_2, chl δ	**B6**^{2}, 819
b323	—,1-bromo-3,5-dihydroxy-*	$C_6H_5BrO_2$. *See* b202	189.02	nd (bz), pr(w + 1)	79(+ w), 87	......	......	......	v^h	...	s	...	v^h s	......	**B6**^{2}, 820
b324	—,2-bromo-1,3-dihydroxy-*	$C_6H_5BrO_2$. *See* b202	189.02	nd (chl), cr (peth)	102–3	......	......	......	s	s	s	...	...	chl s	**B6**^{2}, 819
Ω b325	—,2-bromo-1,4-dihydroxy-*	Adurol. $C_6H_5BrO_2$. *See* b202	189.02	lf (lig) cr (chl)	110–1	sub	......	......	v	v	v	...	v	chl. lig δ aa s	**B6**^{2}, 846
b326	—,4-bromo-1,2-dihydroxy-*	4-Bromocatechol. $C_6H_5BrO_2$. *See* b202	189.02	pr or nd (chl)	87	......	......	......	v	v	v	...	v^h	chl, peth δ	**B6**^{2}, 781
b327	—,1-bromo-2,3-dimethyl-*	C_8H_9Br. *See* b202	185.07	......	......	214 83^{11}	1.365_4^{20}	......	i	...	v	s	s	......	**B5**^{1}, 180
b328	—,1-bromo-2,4-dimethyl-*	C_8H_9Br. *See* b202	185.07	λ^{hx} 359 (2.83), 367 (2.84), 379 (2.76)	0	205 84^{13}	......	1.5501^{20}	i	v	v	...	s	......	**B5**^{2}, 293
Ω b329	—,2-bromo-1,4-dimethyl-*	C_8H_9Br. *See* b202	185.07	lf or pl λ^{hx} 359.5 (2.88), 369 (2.83), 377.5 (2.72)	9	199–200 $88–9^{13}$	1.3582^{18}	1.5514^{18}	i	v	...	...	s	......	**B5**^{2}, 301
Ω b330	—,4-bromo-1,2-dimethyl-*	C_8H_9Br. *See* b202	185.07	λ^{hx} 359 (2.75) 370 (2.78), 379.5 (2.63)	– 0.2	214.5	1.3708_4^{20}	1.5530^{20}	i	v	v	...	...	......	**B5**^{2}, 285
b331	—,1-bromo-2,3-dinitro-*	$C_6H_3BrN_2O_4$. *See* b202	247.01	pa ye pl (al)	101–2	320	......	......	i	v^h	δ	...	...	......	**B5**^{1}, 138
b332	—,1-bromo-2,4-dinitro-*	$C_6H_3BrN_2O_4$. *See* b202	247.01	ye nd (al) $\lambda^{w, pH = 7.4}$ 265 (4.01)	75	......	......	......	d^h	v^h	...	...	...	......	**B5**^{1}, 138
b333	—,2-bromo-1,3-dinitro-*	$C_6H_3BrN_2O_4$. *See* b202	247.01	ye pr (al)	107	......	......	......	d^h	δ v^h	...	...	...	......	**B5**^{1}, 138
b334	—,2-bromo-1,4-dinitro-*	$C_6H_3BrN_2O_4$. *See* b202	247.01	nd (al), pr (al-eth)	70	......	......	......	...	δ v^h	v	...	...	......	**B5**^{1}, 139
b335	—,4-bromo-1,2-dinitro-*	$C_6H_3BrN_2O_4$. *See* b202	247.01	nd (al), pl (al-eth)	34–5 (unst), 59–60 (st)	......	......	......	...	δ v^h	v	...	...	MeOH δ peth i	**B5**^{1}, 138
b336	—,1-bromo-2-ethoxy-*	C_8H_9BrO. *See* b202	201.07	......	......	222–6	......	......	i	v	v	...	...	......	**B6**, 197
b337	—,1-bromo-3-ethoxy-*	C_8H_9BrO. *See* b202	201.07	......	......	222 (228–31)	......	......	i	v	v	...	...	......	**B6**^{2}, 184
Ω b338	—,1-bromo-4-ethoxy-*	C_8H_9BrO. *See* b202	201.07	......	4	230–2 109^{17}	1.4071_4^{25}	1.5517^{20}	i	v	v	...	...	......	**B6**^{2}, 185
b338[1]	—,(1-bromo-ethyl)-(*d*)*	C_8H_9Br. *See* b202	185.07	$[\alpha]_D^{14}$ + 1.5	......	203 $86–8^{15}$	1.3108^{23}	1.5612^{20}	i	s	s	...	...	......	**B5**^{2}, 278
Ω b339	—,—,(*dl*)*	C_8H_9Br. *See* b202	185.07	......	......	$202–3^{760}$ 85^{13}	1.3605_4^{20}	1.5612^{20}	d^h	s	s	...	s	......	**B5**^{2}, 278
Ω b340	—,(2-bromo-ethyl)-*	C_8H_9Br. *See* b202	185.07	......	......	$217–8^{734}$ 92^{11}	1.3587_4^{20}	1.5572^{20}	i	...	s	...	s	......	**B5**^{2}, 278
Ω b341	—,1-bromo-4-fluoro-*	C_6H_4BrF. *See* b202	175.01	λ^{cy} 215 (3.85) 265 (2.90), 271 (3.02), 278 (2.95)	–8 (fr – 17.4)	152^{764}	1.4946_4^{20}	1.5604^{20}	i	s	s	...	...	......	**B5**^{2}, 161
Ω b342	—,1-bromo-2-iodo-*	C_6H_4BrI. *See* b202	282.91	......	9–10	257^{754} $120–1^{15}$	2.2571_4^{25}	1.6618^{25}	i	δ	...	s	...	aa δ	**B5**^{2}, 167
b343	—,1-bromo-3-iodo-*	C_6H_4BrI. *See* b202	282.91	λ^{peth} 235 (4.06)	– 9.3	252^{754} 120^{18}	......	......	i	δ	...	...	...	aa δ	**B5**^{2}, 168
Ω b344	—,1-bromo-4-iodo-*	C_6H_4BrI. *See* b202	282.91	pr or pl (eth-al) λ^{al} 237.5 (4.24)	92	252^{754}	......	......	i	δ	s	...	...	......	**B5**^{2}, 168
Ω b345	—,1-bromo-4-isopropyl-	$C_9H_{11}Br$. *See* b202	199.10	......	– 22.54 (– 17.3)	218.7^{760} $97–8^5$	1.3145_4^{20}	1.5569^{20}	i	...	s	...	s	chl s	**B5**^{2}, 307
b346	—,(β-bromo-isopropyl)-	$C_9H_{11}Br$. *See* b202	199.10	$[\alpha]_D^{20}(d)$ + 15.6, (l) – 15.6	......	188–9 $106–8^{18}$	1.3155^{20}	1.5548^{28}	i	...	...	...	s	chl s	**B5**^{1}, 191
b347	—,2-bromo-4-isopropyl-1-methyl-	$C_{10}H_{13}Br$. *See* b202	213.13	......	......	234.3^{760} 99^9	1.2689_4^{18}	1.5360^{20}	i	v	s	...	...	chl s	**B5**^{1}, 205
Ω b348	—,1-bromo-4-mercapto-*	C_6H_5BrS. *See* b202	189.08	lf (al)	75	230–1	......	......	δ^h	δ v^h	v	...	...	sulf, CCl_4, chl v	**B6**^{2}, 300
b349	—,1-bromo-2-methoxy-*	C_7H_7BrO. *See* b202	187.04	λ^{cy} 218 sh (3.92), 266 sh (3.26), 273.5 (3.44), 280.5 (3.45)	2.5	216 (223) 94^{10}	1.5018_4^{20}	1.5727^{20}	i	v	v	...	...	......	**B6**^{2}, 183
b350	—,1-bromo-3-methoxy-*	C_7H_7BrO. *See* b202	187.04	λ^{cy} 218 sh (3.93), 267 sh (3.16), 272.5 (3.37), 279.5 (3.36)	......	$210–1^{752}$ 105^{16}	......	1.5635^{20}	i	s	s	...	s	CS_2 s	**B6**^{2}, 184

For explanations, symbols and abbreviations see beginning of table. For structural formulas see end of table.

No.	Name	Synonyms and Formula	Mol. wt.	Color, crystalline form, specific rotation and λ_{max} (log ε)	m.p. °C	b.p. °C	Density	n_D	Solubility w	al	eth	ace	bz	other solvents	Ref.
	Benzene														
Ω b351	**—,1-bromo-4-methoxy-***	C_7H_7BrO. *See* b202	187.04	λ^{cy} 226 (4.10), 274 sh (3.13), 279 (3.24), 287 (3.19)	13–4	215 100^{16}	1.4564_4^{20}	1.5642^{20}	δ	v	v	...	...	chl v	**B6**^{2}, 185
b352	**—,1(bromo-methyl)-3,5-dimethyl-***	Mesityl bromide. $C_9H_{11}Br$. *See* b202	199.10	nd (eth)	40	229–31^{740} δd 118^{22}	...	...	...	v	v	...	s	chl v	**B5**^{1}, 200
Ω b353	**—,1(bromo-methyl)-2-methyl-***	*o*-Xylyl bromide. C_8H_9Br. *See* b202	185.07	pr	21	216–7^{742} 108^{16}	1.3811^{23}	1.5730^{20}	i	s	s	s	s	os s	**B5**^{2}, 285
Ω b354	**—,1(bromo-methyl)-3-methyl-***	*m*-Xylyl bromide. C_8H_9Br. *See* b202	185.07	...	...	212–3 δd 105^{13}	1.3711^{23}	1.5660^{20}	i	v	v	...	...	os s	**B5**^{2}, 293
Ω b355	**—,1(bromo-methyl)-4-methyl-***	*p*-Xylyl bromide. C_8H_9Br. *See* b202	185.07	nd (al) λ^{hx} 236 (3.7)	35	218–20^{740}	1.324	...	i	s	v^h	...	...	chl v^h	**B5**^{2}, 301
Ω b356	**—,1-bromo-2-nitro-***	$C_6H_4BrNO_2$ *See* b202	202.01	pa ye (al) λ^{hx} 293 (3.12)	43	258^{756}	1.6245_4^{80}	...	i	v	s	s	s	...	**B5**^{2}, 188
Ω b357	**—,1-bromo-3-nitro-***	$C_6H_4BrNO_2$. *See* b202	202.01	rh λ^{hx} 255 (3.85), 293 (3.11), 302 (3.12)	17 (unst) 56 (st)	265^{760}	1.7036_4^{20}	1.5979^{20}	δ	s	s	...	s	...	**B5**^{2}, 188
Ω b358	**—,1-bromo-4-nitro-***	$C_6H_4BrNO_2$. *See* b202	202.01	rh or mcl pr (al) λ^{hx} 269.2 (4.6)	127	256^{760}	1.948	...	i	s	s	...	s	...	**B5**^{2}, 189
b359	**—,1-bromo-2-nitroso-***	C_6H_4BrNO. *See* b202	186.01	nd	98	...	...	...	...	...	...	...	s	chl s	**B5**, 232
b360	**—,1-bromo-4-nitroso***	C_6H_4BrNO. *See* b202	186.01	nd (al) λ^{al} 227 (3.80), 293 (4.00), 315 (4.08), 750 (1.62)	95	...	...	...	...	s^h	...	...	s	chl v lig s^h	**B5**^{2}, 171
b361	**—,(3-bromo-1-propenyl)-***	Cinnamyl bromide. C_9H_9Br. *See* b202	197.08	nd (al, eth)	34	130^{10}	1.3428^{30}	1.610–1.613^{20}	i	s^h	...	...	...	hx s	**B5**^{2}, 372
b362	**—,(3-bromo-propoxy)-***	γ-Phenoxypropyl bromide. $C_9H_{11}BrO$. *See* b202	215.10	...	7–8	127^{18}	1.365_{16}^{16}	...	...	...	v	...	...	...	**B6**^{2}, 145
b363	**—,(1-bromo-propyl)-(*dl*)***	$C_9H_{11}Br$. *See* b202	199.10	$d[\alpha]_D$ +5.9 (al, c = 1) $l[\alpha]_D^{20}$ −5.7 (eth, c = 10)	...	105^{17}	1.3098_4^{19}	1.5517^{19}	i d^h	...	v	...	s	...	**B5**^{2}, 305
b364	**—,(2-bromo-propyl)-***	$C_9H_{11}Br$. *See* b202	199.10	...	...	107–9^{16}	1.2908^{16}	1.5450^{20}	i d^h	...	...	...	...	chl s	**B5**^{1}, 190
b365	**—,(3-bromo-propyl)-***	$C_9H_{11}Br$. *See* b202	199.10	λ^{al} 242.5 (2.13), 248 (2.23), 253.5 (2.33), 259 (2.38), 264.5 (2.27), 268 (2.24)	...	110^{12}	1.3106_4^{25}	1.5440^{25}	i	...	v	...	...	...	**B5**^{2}, 305
b366	**—,1-bromo-4-triazo-***	$C_6H_4BrN_3$. *See* b202	198.03	pl λ^{al} 253 (4.0), 291 (3.2)	20	105^{10}	...	...	i	δ	v	...	v	...	**B5**^{2}, 208
b367	**—,1-bromo-2,3,5-trimethyl-***	$C_9H_{11}Br$. *See* b202	199.10	...	< −15	238^{760} (223) 117^{17}	...	1.5516^{26}	i	...	...	...	s	sulf s	**B5**^{1}, 196
Ω b368	**—,1-bromo-2,4,5-trimethyl-***	$C_9H_{11}Br$. *See* b202	199.10	nd (al)	73	233–5	...	...	i	v^h	...	...	...	...	**B5**^{1}, 196
b369	**—,2-bromo-1,3,5-trimethyl-***	Bromomesitylene. $C_9H_{11}Br$. *See* b202	199.10	λ^{al} 269 (2.36)	−1	225 117^{25}	1.3191^{10}	1.5510^{20}	i	...	v	...	s	...	**B5**^{2}, 315
Ω b370	**—,butoxy-***	Butyl phenyl ether. $C_{10}H_{14}O$. *See* b202	150.22	...	−19.4	210^{760} 95^{17}	0.9351_4^{20}	1.4969^{20}	...	...	s	s	...	...	**B6**^{2}, 145
b371	**—,*sec*-butoxy-**	*sec*-Butyl phenyl ether. $C_{10}H_{14}O$. *See* b202	150.22	...	...	194.5^{760} 70–2^5	0.9415_4^{20}	1.4926^{25}	i	...	s	...	...	...	**B6**^{2}, 146
Ω b372	**—,butyl-***	$C_{10}H_{14}$. *See* b202	134.22	λ^{iso} 248 (2.1), 253 (2.2), 259 (2.3), 262 (2.3), 265 (2.2), 268 (2.2)	−88	183^{760} 62.2^{10}	0.8601_4^{20}	1.4898^{20}	i	∞	∞	∞	∞	peth, CCl_4 ∞	**B5**^{2}, 317
b373	**—,*sec*-butyl-(*d*)**	$C_{10}H_{14}$. *See* b202	134.22	$[\alpha]_D^{20}$ +27.3 (undil)	−75	173^{760} 63^{15}	0.8621_4^{20}	1.4895^{20}	i	∞	∞	∞	∞	peth, CCl_4 ∞	**B5**^{2}, 319
Ω b374	**—,—(*dl*)**	$C_{10}H_{14}$. *See* b202	134.22	λ^{iso} 248 (2.0), 253 (2.2), 258 (2.3), 261 (2.3), 264 (2.2), 267 (2.2)	−75.47	173^{760} 53.6^{10}	0.8621_4^{20}	1.4902^{20}	i	∞	∞	∞	∞	peth, CCl_4 ∞	**B5**^{2}, 319
b375	**—,—(*l*)**	$C_{10}H_{14}$. *See* b202	134.22	$[\alpha]_D^{23}$ −17.9 (al, c = 5)	...	61^{18}	0.868_4^{24}	1.4891^{20}	i	∞	∞	∞	∞	peth, CCl_4 ∞	**B5**^{3}, 931

For explanations, symbols and abbreviations see beginning of table. For structural formulas see end of table.

No.	Name	Synonyms and Formula	Mol. wt.	Color, crystalline form, specific rotation and λ_{max} (log ε)	m.p. °C	b.p. °C	Density	n_D	Solubility w	al	eth	ace	bz	other solvents	Ref.
	Benzene														
Ω b376	—,*tert*-butyl-	$C_{10}H_{14}$. *See* b202	134.22	λ^{hx} 253 (2.2), 257 (2.3), 264 sh (2.2), 267 (2.1)	−57.85	169^{760} 50.7^{10}	0.8665^{20}_{4}	1.4927^{20}	i	v	v	∞	∞	peth, CCl_4 ∞	**B5**², 320
b377	—,1-butyl-2,4-dihydroxy-*	$C_{10}H_{14}O_2$. *See* b202	166.22	...	47–8	$164–6^{6–7}$	...	...	i	δ	...	...	δ	chl δ	**B6**², 898
b378	—,1-*tert*-butyl-2,3-dimethyl-4-hydroxy-	$C_{12}H_{18}O$. *See* b202	178.28	...	...	259^{760} 145^{20}	...	...	...	...	...	...	...	...	**B6**³, 2019
b379	—,1-*tert*-butyl-2,5-dimethyl-4-hydroxy-	$C_{12}H_{18}O$. *See* b202	178.28	...	71.2	264^{760} 136^{10}	0.939^{80}_{4} 1.001^{27}_{4}	1.5311^{20}	...	...	...	...	...	alk s	**B6**³, 2019
b380	—,1-*tert*-butyl-3,5-dimethyl-2-hydroxy-	$C_{12}H_{18}O$. *See* b202	178.28	...	22.3	249^{760} 115^{10}	0.917^{80}_{4}	1.5183^{20}	...	...	...	...	...	alk i	**B6**³, 2020
b381	—,1-*tert*-butyl-4,5-dimethyl-2-hydroxy-	$C_{12}H_{18}O$. *See* b202	178.28	...	46.0	145^{20} 127^{10}	0.920^{80}_{4} 0.973^{27}_{4}	1.5222^{20}	...	...	...	...	...	alk i	**B6**³, 2019
b382	—,5-*tert*-butyl-1,3-dimethyl-2-hydroxy-	$C_{12}H_{18}O$. *See* b202	178.28	pr (peth)	82.4	248^{760} 119^{10}	0.916^{80}_{4} 0.959^{27}_{4}	...	...	...	...	...	...	alk s	**B6**³, 2019
b383	—,1-*tert*-butyl-3,5-dimethyl-2,4,6-trinitro-	Musk xylene. $C_{12}H_{15}N_3O_6$. *See* b202	297.27	pl, nd (al) λ^{al} <220	110 (unst) (114)	...	...	...	i	δ v^h	s	...	...	...	**B5**², 340
b384	—,2-*tert*-butyl-3,5-dinitro-1-iso-propyl-4-methyl-	Moskene. $C_{14}H_{20}N_2O_4$. *See* b202	280.33	pa ye cr	132–3	...	...	...	i	δ	...	...	...	...	**C49**, 3044
Ω b385	—,5-*tert*-butyl-1,3-dinitro-4-methoxy-2-methyl-	Musk ambrette. $C_{12}H_{16}N_2O_5$. *See* b202	268.25	pa ye lf (al)	84–6	185^{16}	...	...	i	δ	s	...	...	...	**B6**³, 1984
b386	—,2-*tert*-butyl-1,3-dinitro-4,5,6-trimethyl-	Musk tebetene. $C_{13}H_{18}N_2O_4$. *See* b202	266.30	pa ye pr (al)	135–6.5	...	...	...	i	δ	s	...	...	...	**B5**³, 1055
b387	—,2-*tert*-butyl-4-ethyl-1-hydroxy-	$C_{12}H_{18}O$. *See* b202	178.28	...	23	250^{760} 123^{10}	...	...	...	s	...	...	...	alk i	**B6**², 2012
b388	—,4-*tert*-butyl-2-ethyl-1-hydroxy-	$C_{12}H_{18}O$. *See* b202	178.28	...	...	257^{760} 141^{20}	...	...	...	...	...	...	...	...	**B6**³, 2012
Ω b389	—,1-butyl-2-hydroxy-*	*o*-Butylphenol. $C_{10}H_{14}O$. *See* b202	150.22	...	...	235^{760} 106.5^{10}	0.975^{20}_{4}	$1.5180^{25.5}$	i	s	s	...	...	alk s	**B6**³, 1843
Ω b390	—,1-butyl-3-hydroxy-*	*m*-Butylphenol. $C_{10}H_{14}O$. *See* b202	150.22	...	...	248^{760} 123^{10}	0.974^{20}_{4}	...	i	s	s	...	...	alk s	**B6**², 485
Ω b391	—,1-butyl-4-hydroxy-*	*p*-Butylphenol. $C_{10}H_{14}O$. *See* b202	150.22	...	22	248^{760} $138–9^{18}$	0.978^{20}_{4}	$1.5165^{25.5}$	i	s	s	...	...	alk s	**B6**², 485
Ω b392	—,1-*sec*-butyl-4-hydroxy-	*p-sec*-Butylphenol. $C_{10}H_{14}O$. *See* b202	150.22	nd, $[\alpha]^{20}_{D}$ +13.3 (*m*-xyl)	61–2	$240–2^{760}$	0.9883^{20}_{20}	1.5182^{21}	i	s	v	...	...	alk s	**B6**², 487
Ω b393	—,1-*tert*-butyl-2-hydroxy-	*o-t*-Butylphenol. $C_{10}H_{14}O$. *See* b202	150.22	λ^{al} 216 (3.71), 271 (3.36), 277 (3.31)	...	221^{760} 99^{10}	0.9783^{20}_{4}	1.5160^{20}	...	s	v	...	...	alk s	**B6**², 489
Ω b394	—,1-*tert*-butyl-3-hydroxy-	*m-t*-Butylphenol. $C_{10}H_{14}O$. *See* b202	150.22	nd (peth)	41–2	240^{760} 132.5^{20}	...	...	...	s	v	...	...	alk s	**B6**³, 1862
Ω b395	—,1-*tert*-butyl-4-hydroxy-	*p-t*-Butylphenol. $C_{10}H_{14}O$. *See* b202	150.22	nd (lig) λ^{al} 225 (3.9), 278 (3.4), 285 sh (3.2)	101	239.5^{760} 114^{10}	0.908^{80}_{4}	1.4787^{114}	s^h	s	s	...	...	alk s	**B6**², 489
Ω b396	—,1-*tert*-butyl-2-hydroxy-4-methyl-	$C_{11}H_{16}O$. *See* b202	164.25	...	46–7 37 (+ ⅓w)	224^{760} 127^{11}	0.922^{80}_{4}	1.5250^{20}	i	s	s	s	...	os s	**B6**², 507
Ω b397	—,2-*tert*-butyl-1-hydroxy-4-methyl-	$C_{11}H_{16}O$. *See* b202	164.25	nd (peth) λ^{al} 281 (3.22)	55 (44)	237^{760} 111^{10}	0.9247^{75}_{4}	1.4969^{75}	i	...	s	s	s	os s	**B6**³, 1978
Ω b398	—,4-*tert*-butyl-1-hydroxy-2-methyl-	$C_{11}H_{16}O$. *See* b202	164.25	yesh	27–8	$235–7^{740}$ 132^{20}	0.965^{20}_{4}	1.5230^{20}	i	...	s	s	s	os s	**B6**², 507
b399	—,1-butyl-2-methyl-*	$C_{11}H_{16}$. *See* b202	148.25	...	...	208^{760} 81^{10}	0.8710^{20}_{4}	1.4960^{20}	i	δ	s	s	v	...	**B5**, 437
b400	—,1-butyl-3-methyl-*	$C_{11}H_{16}$. *See* b202	148.25	...	...	205^{760} 79^{10}	0.8590^{20}_{4}	1.4910^{20}	i	δ	s	s	v	...	**B5**, 437
b401	—,1-butyl-4-methyl-*	$C_{11}H_{16}$. *See* b202	148.25	...	−85	207^{760} 80.6^{10}	0.8586^{20}_{4}	1.4916^{20}	i	δ	s	s	s	...	**B5**, 437
b402	—,1-*sec*-butyl-4-methyl-	$C_{11}H_{16}$. *See* b202	148.25	...	...	197^{760} 73^{10}	0.8640^{19}	1.493^{20}	i	...	v	...	s	chl v	**B5**², 333
Ω b403	—,1-*tert*-butyl-4-methyl-	$C_{11}H_{16}$. *See* b202	148.25	...	−52	193^{760} 70^{10}	0.8612^{20}_{4}	1.4918^{20}	i	δ	v	s	s	chl v	**B5**¹, 210
b404	—,1-*tert*-butyl-3-methyl-2,4,6-trinitro-	Artificial musk. $C_{11}H_{13}N_3O_6$. *See* b202	283.24	ye nd (al)	96–7	...	...	...	i	v	v	...	s	chl s	**B5**, 438
b405	—,2-*tert*-butyl-4-methyl-1,3,5-trinitro-	Musk baur. $C_{11}H_{13}N_3O_6$. *See* b202	283.24	ye nd (al)	97	...	...	...	i	s	s	...	s	chl, peth s	**B5**, 438

For explanations, symbols and abbreviations see beginning of table. For structural formulas see end of table.

No.	Name	Synonyms and Formula	Mol. wt.	Color, crystalline form, specific rotation and λ_{max} (log ε)	m.p. °C	b.p. °C	Density	n_D	Solubility: w	al	eth	ace	bz	other solvents	Ref.
	Benzene														
	—,(1-butynyl)-* ...	*see* **Butyne, phenyl-**													
Ω b407	—,chloro-*	Phenyl chloride C_6H_5Cl. *See* b202	112.56	λ^{al} 245 (1.95), 251 (2.34), 258 (2.13), 264 (2.45), 272 (2.32)	−45.6	132^{760} 22^{10}	1.1058^{20}_{4}	1.5241^{20}	i	∞	∞	...	v	chl, CCl_4, CS_2 v	**B5**[2], 148
b408	—,2-chloro-1,4-diacetamido-*	$C_{10}H_{11}ClN_2O_2$. *See* b202 ...	226.67	nd	196–7	....	....	....	...	s	...	...	δ	....	**B13**, 118
b409	—,1-chloro-2,4-diamino-*	$C_6H_7ClN_2$. *See* b202	142.59	pl or nd	91	....	....	....	δ	s	...	...	...	lig i	**B13**[2], 29
b410	—,1-chloro-3,5-diamino-*	$C_6H_7ClN_2$. *See* b202	142.59	rh pr (al), nd (to-al)	105–6	....	....	....	s^h	s	s	s	δ	chl s lig i	**B13**[2], 29
b411	—,2-chloro-1,3-diamino-*	$C_6H_7ClN_2$. *See* b202	142.59	....	85–6	....	....	....	...	...	s	...	...	....	**B13**[1], 15
b412	—,2-chloro-1,4-diamino-*	$C_6H_7ClN_2$. *See* b202	142.59	nd (bz-lig)	64	....	....	....	s	...	...	...	s^h	lig s^h	**B13**[2], 58
Ω b413	—,4-chloro-1,2-diamino-*	$C_6H_7ClN_2$. *See* b202	142.59	pl (bz-lig), lf (w)	76	....	....	....	δ	v	v	...	s	lig s	**B13**, 25
b414	—,1-chloro-2,3-dihydroxy-*	$C_6H_5ClO_2$. *See* b202	144.56	cr (lig)	46–8	$110–1^{11}$	....	....	...	...	...	...	...	lig v	**B6**[1], 388
Ω b415	—,1-chloro-2,4-dihydroxy-*	$C_6H_5ClO_2$. *See* b202	144.56	(i) cr (al) (ii) cr (bz) λ^{w} 215 sh (3.78), 282.5 (3.45)	(i) 89 (ii) 105	259 147^{18}	....	....	v	v	v	v	v	CS_2 v	**B6**[2], 818
b416	—,1-chloro-3,5-dihydroxy-*	$C_6H_5ClO_2$. *See* b202	144.56	hyg nd (sub) λ^{al} 222 sh (3.9), 280 (3.3), 285 (3.3)	117	sub	....	....	s	...	s	s	...	os s peth i	**B6**[2], 819
Ω b417	—,2-chloro-1,3-dihydroxy-*	$C_6H_5ClO_2$. *See* b202	144.56	....	97–8	....	....	....	...	...	...	...	...	....	**C40**, 565
Ω b418	—,2-chloro-1,4-dihydroxy-*	$C_6H_5ClO_2$. *See* b202	144.56	red lf (chl), nd (bz)	108	263	....	....	v	s	s	...	v^h	chl s^h	**B6**[2], 844
b419	—,4-chloro-1,2-dihydroxy-*	$C_6H_5ClO_2$. *See* b202	144.56	lf (bz-peth)	90–1	$139^{10.5}$	....	....	v	v	v	v	...	aa v lig δ	**B6**[2], 787
b420	—,1-chloro-2,4-dimethyl-*	C_8H_9Cl. *See* b202	140.61	λ^{hx} 358 (2.94), 369.5 (2.92), 374.5 (2.87)	....	$187–8^{760}$ 89^{24}	1.0598^{20}_{20}	1.5230^{25}	i	...	...	s	v	....	**B5**[2], 291
b421	—,1-chloro-3,5-dimethyl-*	C_8H_9Cl. *See* b202	140.61	λ^{hx} 364.5 (2.41), 374.5 (2.48), 386 (2.40)	....	$187–8^{760}$ 66^{12}	....	....	i	...	...	s	s	....	**B5**, 373
Ω b422	—,2-chloro-1,4-dimethyl-*	C_8H_9Cl. *See* b202	140.61	....	1.6	187 (192)	1.0589^{15}_{4}	....	i	...	...	s	v	....	**B5**[1], 186
Ω b423	—,4-chloro-1,2-dimethyl-*	C_8H_9Cl. *See* b202	140.61	λ^{hx} 359 (2.67), 370 (2.70), 378.5 (2.57)	−6	194^{755}	1.0692^{15}_{15}	....	i	...	...	s	v	....	**B5**, 363
b424	—,1-chloro-2,3-dimethyl-4-hydroxy-*	C_8H_9ClO. *See* b202	156.61	nd (peth)	85	....	....	....	...	s	s	s	...	os s peth δ	**B6**, 454
b425	—,1-chloro-2,3-dimethyl-5-hydroxy-*	C_8H_9ClO. *See* b202	156.61	nd (peth)	98	....	....	....	...	s	...	s	...	chl v peth δ	**B6**[2], 456
b426	—,1-chloro-2,4-dimethyl-5-hydroxy-*	C_8H_9ClO. *See* b202	156.61	nd (lig or w)	90–1	....	....	....	s^h	s	s	s	...	os s	**B6**[1], 242
b427	—,1-chloro-2,5-dimethyl-4-hydroxy-*	C_8H_9ClO. *See* b202	156.61	silv-gy nd (lig)	74–5	....	....	....	v	s	...	...	s	CS_2 v MeOH, aa, peth s lig δ	**B6**[2], 467
b428	—,1-chloro-3,4-dimethyl-2-hydroxy-*	C_8H_9ClO. *See* b202	156.61	....	....	$221–3^{760}$ 100^{17}	....	....	s	s	...	s	s	os s	**B6**[1], 241
b429	—,1-chloro-4,5-dimethyl-2-hydroxy-*	C_8H_9ClO. *See* b202	156.61	nd (peth)	72	....	....	....	v	s	...	...	s	peth, CS_2 v aa s	**B6**[2], 456
b430	—,2-chloro-1,3-dimethyl-5-hydroxy-*	C_8H_9ClO. *See* b202	156.61	λ^{w} 260 (3.12), 286.5 (3.14)	115–6	246	....	....	δ	s	s	...	δ	peth δ	**B6**[2], 463
b431	—,2-chloro-1,5-dimethyl-3-hydroxy-*	C_8H_9ClO. *See* b202	156.61	nd	49–50	....	....	....	...	...	s	...	...	....	**B6**[2], 464
b432	—,2-chloro-3,4-dimethyl-1-hydroxy-*	C_8H_9ClO. *See* b202	156.61	cr (peth)	27	187–9	....	1.5538^{20} (1.5300^{20})	...	...	...	...	...	peth s^h	**B6**[2], 456
b433	—,5-chloro-1,3-dimethyl-2-hydroxy-*	C_8H_9ClO. *See* b202	156.61	nd (w)	83	....	....	....	δ	s	...	...	s	aa s peth δ	**B6**[3], 1738
b434	—,1-chloro-2,3-dinitro-*	$C_6H_3ClN_2O_4$. *See* b202....	202.56	pr (al), cr (MeOH)	78	....	....	....	i	δ	s	...	...	....	**B5**[2], 196

For explanations, symbols and abbreviations see beginning of table. For structural formulas see end of table.

No.	Name	Synonyms and Formula	Mol. wt.	Color, crystalline form, specific rotation and λ_{max} (log ε)	m.p. °C	b.p. °C	Density	n_D	Solubility w	al	eth	ace	bz	other solvents	Ref.
	Benzene														
Ω b435	**—,1-chloro-2,4-dinitro-***	$C_6H_3ClN_2O_4$. *See* b202.....	202.56	α: ye rh (eth) β: ye rh (eth), nd (al)	α: 53 (st) β: 43 (unst) γ: 27 (unst)	315^{762} $158–60^{1}$	1.4982^{75}_{4} α: 1.697^{22} β 1.6867^{16}	1.5857^{60}	i $δ^h$	δ s^h	s	...	s	CS_2 s	**B5**[2], 196
b436	**—,1-chloro-3,5-dinitro-***	$C_6H_3ClN_2O_4$. *See* b202.....	202.56	nd (al, MeOH, peth)	59	....	....	....	i	s	v	...	...	....	**B5**, 264
Ω b437	**—,2-chloro-1,3-dinitro-***	$C_6H_3ClN_2O_4$. *See* b202.....	202.56	ye nd (al, aa)	88	315	1.6867^{16}	....	i	s	s	...	...	to s	**B5**[1], 137
b438	**—,4-chloro-1,2-dinitro-***	$C_6H_3ClN_2O_4$. *See* b202.....	202.56	α: cr (al) β: pr (lig) γ: mcl or rh nd (eth or lig)	α: 36 β: 37 γ: 40–1	315d 160^{4}	1.6867^{20}_{4}	....	i	δ s^h	v	...	v	CS_2 s	**B5**[2], 196
Ω b439	**—,1-chloro-2-ethoxy-***	C_8H_9ClO. *See* b202	156.61	....	....	210 $97–8^{15}$	1.1288^{25}_{4}	1.5284^{25}	i	s	s		s	....	**B6**[2], 171
b440	**—,1-chloro-3-ethoxy-***	C_8H_9ClO. *See* b202	156.61	....	....	$204–5^{717}$	1.1712^{20}_{4}	....	i	v	v		v	aa v	**B6**, 185
Ω b441	**—,1-chloro-4-ethoxy-***	C_8H_9ClO. *See* b202	156.61	....	21	212–4 98^{17}	1.1254^{20}_{4}	1.5252^{20}	i	s	s		v	aa s	**B6**[2], 176
Ω b442	**—,(2-chloro-ethoxy)-***	C_8H_9ClO. *See* b202	156.61	pr	28	$217–20^{760}$ $100–2^{12}$	....	....	i	v	v	v	v	lig v	**B6**[2], 144
b443	**—,(1-chloro-ethyl)-(*d*)***	*d*-α-Phenethyl chloride. C_8H_9Cl. *See* b202	140.61	$[α]^{20}_{D}$ + 50.6 (undil)	....	85^{20}	1.0631^{20}_{4}	1.5250^{25}	d^h	s	v	...	s	....	**B5**[2], 277
b444	**—,—(*dl*)***	α-Phenethyl chloride. C_8H_9Cl. *See* b202	140.61	$λ^{cy}$ 218 (3.73), 260 (2.3)	....	$81–2^{17}$	1.0620^{20}_{4}	1.5276^{20}	d^h	s	v	...	s	....	**B5**[2], 277
b445	**—,—(*l*)***	*l*-α-Phenethyl chloride. C_8H_9Cl. *See* b202	140.61	$[α]^{20}_{D}$–30.1 (undil)	....	85^{20}	1.0632^{20}_{4}	....	d^h	s	v	...	s	....	**B5**[1], 177
Ω b446	**—,(2-chloro-ethyl)-***	β-Phenethyl chloride. C_8H_9Cl. *See* b202	140.61	$λ^{al}$ 209 (3.87)	....	$197–8^{760}$ 92^{20}	1.069^{25}_{4}	1.5276^{20}	i	s	s	s	s	CS_2, lig s	**B5**[2], 277
b447	**—,1-chloro-4-ethyl-***	C_8H_9Cl. *See* b202.........	140.61	....	–62.6	184.4^{760} 63.5^{10}	1.0455^{20}_{4}	1.5175^{20}	i	∞	∞	∞	v	peth, CCl_4 ∞ aa s	**B5**[1], 177
b448	**—,(2-chloro-ethylthio)-***	C_8H_9ClS. *See* b202.........	172.68	....	....	$117–8^{11}$	1.1769^{25}_{4}	1.5828^{20}	...	...	s	...	...	chl s	**B6**[2], 287
Ω b449	**—,1-chloro-2-fluoro-***	C_6H_4ClF. *See* b202	130.55	$λ^{hx}$ 210 (4.00), 214 (4.00), 260 (3.00), 266 (3.15), 272 (3.18)	–43	137.6^{760}	1.2233^{30}_{4}	1.4968^{30}	i	...	...	s	s		**B5**[2], 153
Ω b450	**—,1-chloro-4-fluoro-***	C_6H_4ClF. *See* b202	130.55	$λ^{hx}$ 209 (3.74), 215 (3.74), 263 (2.89), 271 (3.04) 277 (3.06)	–26.85	130^{757}	1.226^{20}_{4}	1.4990^{15}	i	s	s	...	s	....	**B5**[2], 153
Ω b451	**—,1-chloro-4-hydroxy-5-isopropyl-2-methyl-***	6-Chlorothymol. $C_{10}H_{13}ClO$. *See* b202......	184.67	nd or pl (lig)	59–61	259–63	....	....	v	s	s	...	s	alk, peth s	**B6**[3], 499
b452	**—,1-chloro-3-hydroxylamino-***	C_6H_6ClNO. *See* b202.......	143.57	pl (bz)	49	....	....	....	s	...	...	...	s	peth s	**B15**[2], 8
b453	**—,1-chloro-4-hydroxylamino-***	C_6H_6ClNO. *See* b202.......	143.57	lf (dil al)	87–8	....	....	....	δ s^h	s	...	...	v	chl v peth δ	**B15**[2], 9
Ω b454	**—,1-chloro-2-iodo-***	C_6H_4ClI. *See* b202	238.46	$λ^{cy}$ 231 (4.01), 238 (4.01)	....	$234–5^{760}$ 110^{16}	1.9515^{25}	1.6331^{25}	i	...	...	s	...	....	**B5**[1], 119
b455	**—,1-chloro-3-iodo-***	C_6H_4ClI. *See* b202	238.46	$λ^{cy}$ 231 (4.01), 238 (4.01)	....	230	....	....	i	...	...	s	...	....	**B5**[2], 167
Ω b456	**—,1-chloro-4-iodo-***	C_6H_4ClI. *See* b202	238.46	lf (ace), (al) $λ^{cy}$ 236.3 (4.20)	57	227 108^{13}	1.886^{57}_{4}	....	i	s	...	...	...	$PhNO_2$ s	**B5**[2], 167
b457	**—,1-chloro-4-isopropyl-***	$C_9H_{11}Cl$. *See* b202	154.64	....	–12.27	198.3^{760} 74^{10}	1.0208^{20}_{4}	1.5117^{20}	i	∞	∞	∞	v	peth, CCl_4 ∞	**B5**[2], 307
Ω b458	**—,1-chloro-2-mercapto-***	*o*-Chlorobenzenethiol. C_6H_5ClS. *See* b202	144.62	....	....	205–6 117^{15}	1.2752^{20}	....	δ	δ	...	...	...	....	**B6**[3], 1032
Ω b459	**—,1-chloro-3-mercapto-***	*m*-Chlorobenzenethiol. C_6H_5ClS. *See* b202	144.62	....	....	205–7	1.2637^{13}	....	i	s	s			peth, chl s	**B6**[3], 1034
Ω b460	**—,1-chloro-4-mercapto-***	*p*-Chlorobenzenethiol. C_6H_5ClS. *See* b202	144.62	pr or pl (al)	61 (54)	205–7	....	....	i	v^h	v	...	v	....	**B6**[2], 297
Ω b461	**—,1-chloro-2-methoxy-***	C_7H_7ClO. *See* b202	142.59	$λ^{cy}$ 222 (3.87), 267 sh (3.20), 272.5 (3.38), 280 (3.39)	–26.6	198.5–9.5^{760} $90–1^{16}$	1.1911^{20}_{4}	1.5480^{20}	i	s	s	...	...	....	**B6**[2], 171
b462	**—,1-chloro-3-methoxy-***	C_7H_7ClO. *See* b202	142.59	$λ^{cy}$ 219 (3.86), 272 (3.33), 279 (3.33)	....	193–4 70^{9}	1.1759^{12}_{4}	1.5365^{20}	i	s	s	...	...	....	**B6**[1], 100
Ω b463	**—,1-chloro-4-methoxy-***	C_7H_7ClO. *See* b202	142.59	$λ^{cy}$ 227 (4.07), 279 (3.28), 287 (3.27)	< –18	197.5^{760} 75^{10}	1.201^{20}_{4}	1.5390^{20}	i	v	v	...	...	chl v	**B6**[2], 175
b464	**—,1-chloro-3-methoxy-2-nitro-***	$C_7H_6ClNO_3$. *See* b202......	187.58	....	56	....	....	....	...	s^h	s	...	s	chl s	**B6**[2], 226

For explanations, symbols and abbreviations see beginning of table. For structural formulas see end of table.

No.	Name	Synonyms and Formula	Mol. wt.	Color, crystalline form, specific rotation and λ_{max} (log ε)	m.p. °C	b.p. °C	Density	n_D	Solubility w	al	eth	ace	bz	other solvents	Ref.
	Benzene														
Ω b465	—,4-chloro-1-methoxy-2-nitro-*	$C_7H_6ClNO_3$. *See* b202	187.58	ye nd or pr (al) $\lambda^{aq\ MeOH}$ 220.5 (4.28), 265 (3.59), 349 (3.43)	98	....	....	....	...	s^h	δ	...	...	MeOH s lig δ	**B6**^{2}, 226
b466	—,1(chloromethyl)-2,4-dimethyl-*	$C_9H_{11}Cl$. *See* b202	154.64	....	....	215–6 86–7^{12}	....	....	i d^h	v	v	...	v	....	**B5**^{2}, 313
b467	—,1(chloromethyl)-4-ethyl-*	$C_9H_{11}Cl$. *See* b202	154.64	....	....	95–6^{15}	....	1.5290^{25}	d^h	s	...	...	s	chl s	**B5**^{2}, 310
b468	—,1(chloromethyl)-2-methyl-*	*o*-Xylyl chloride. C_8H_9Cl. *See* b202	140.61	....	....	197–9 80^{12}	....	1.5410^{20}	i	∞	∞	...	...	....	**B5**^{2}, 283
b469	—,1(chloromethyl)-3-methyl-*	*m*-Xylyl chloride. C_8H_9Cl. *See* b202	140.61	....	....	195–6 101–2^{30}	1.064^{20}	1.5345^{20}	i	s	s	...	...	....	**B5**^{2}, 291
Ω b470	—,1(chloromethyl)-4-methyl-*	*p*-Xylyl chloride. C_8H_9Cl. *See* b202	140.61	....	....	200–2 81^{15}	1.0512$_4^{20}$	1.5380	i	s	∞	...	...	....	**B5**^{2}, 299
b471	—,(chloro-*tert*-butyl)-	$C_{10}H_{13}Cl$. *See* b202	168.67	....	....	222^{741} 104–5^{18}	1.047$_4^{20}$	1.5247$_4^{20}$	i	∞	∞	∞	∞	CCl_4, lig ∞	**B5**^{2}, 320
Ω b472	—,1-chloro-2-nitro-*	$C_6H_4ClNO_2$. *See* b202	157.56	mcl nd λ^{MeOH} 252 (3.51)	33.5–35	246^{760} 119^8	1.368^{242}	....	i	s	s	v	s	to, Py v CCl_4 s^h MeOH s	**B5**^{2}, 180
Ω b473	—,1-chloro-3-nitro-*	$C_6H_4ClNO_2$. *See* b202	157.56	pa ye rh pr (al) λ^{MeOH} 257.5 (3.89)	24 (unst) 46 (st)	235–6^{760}	1.343$_4^{50}$	1.5374$^{80}_{\alpha}$	i	s v^h	s	...	s	CS_2, chl, aa s	**B5**^{2}, 182
Ω b474	—,1-chloro-4-nitro-*	$C_6H_4ClNO_2$. *See* b202	157.56	mcl pr λ^{MeOH} 270.5 (4.03)	83.6	242^{760} 113^8	1.2979$^{90.5}$	1.5376$^{100}_{\alpha}$	i	δ v^h	s	...	...	CS_2 s	**B5**^{2}, 183
b475	—,1-chloro-2-nitroso-*	C_6H_4ClNO. *See* b202	141.56	nd (al)	65.5–6.5	....	....	....	...	v	v	...	v	chl v peth v^h	**B5**^{2}, 171
b476	—,1-chloro-3-nitroso-*	C_6H_4ClNO. *See* b202	141.56	nd (bz)	72 (77)	....	....	....	...	v	v	v	v	chl v	**B5**^{2}, 171
b477	—,1-chloro-4-nitroso-*	C_6H_4ClNO. *See* b202	141.56	(aa or al) λ^{al} 226 (3.87), 288 (4.03), 313 (4.08), 750 (1.66)	92–3	....	....	....	...	s	...	...	...	chl v aa s^h	**B5**^{2}, 171
b478	—,(3-chloropropoxy)-*	$C_9H_{11}ClO$. *See* b202	170.64	....	12	245–55 139^{25}	1.1167^{20}	1.5235^{25}	...	...	v	s	...	....	**B6**, 142
b479	—,(3-chloropropylthio)-*	$C_9H_{11}ClS$. *See* b202	186.71	....	....	116–7^4	1.1536$_4^{20}$	1.5752^{20}	...	...	...	s	...	Py s	**B6**^{2}, 288
b480	—,1-chloro-4-triazo-*	$C_6H_4ClN_3$. *See* b202	153.57	λ^{al} 267 (3.82), 279 (3.85)	20	96^{20}	....	....	i	...	s	...	...	....	**B5**^{3}, 648
b481	—,2-chloro-1,3,5-trimethyl-*	$C_9H_{11}Cl$. *See* b202	154.64	λ^{al} 268 (2.35)	< −20	204–6 104^{25}	1.0337^{30}	1.5212^{30}	i	v	v	...	...	....	**B5**^{2}, 315
b482	—,1-chloro-2,4,5-trinitro-	$C_6H_2ClN_3O_6$. *See* b202	247.55	ye lf (al)	116	....	....	....	i	v^h	...	...	s	aa s	**B5**^{2}, 205
Ω b483	—,2-chloro-1,3,5-trinitro-*	Picryl chloride. $C_6H_2ClN_3O_6$. *See* b202	247.55	wh nd or pl (chl, al-lig) λ^{al} 340 (2.8)	83	....	1.797^{20}	....	i	s^h	δ	v	s v^h	to, Py v chl v^h peth δ	**B5**^{2}, 205
Ω b484	—,cyclohexyl-*	Phenylcyclohexane*. $C_{12}H_{16}$. *See* b202	160.26	pl λ^{iso} 254.5 (2.23), 259 (2.33), 264 (2.21), 267.5 (2.25)	7–8	235–6 127–8^{30}	0.9502$_4^{20}$	1.5329^{20}	i	v	s	...	...	....	**B5**, 503
Ω b485	—,4-cyclohexyl-1,3-dihydroxy-*	$C_{12}H_{16}O_2$. *See* b202	192.24	nd (bz, chl), cr (bz-peth)	127–8	....	....	....	δ	δ	...	...	...	....	**B6**^{2}, 929
Ω b486	—,1-cyclohexyl-2-hydroxy-*	$C_{12}H_{16}O$. *See* b202	176.24	nd (lig)	56–7	283^{760} 147^{17}	....	....	i	s	...	...	...	aa s lig s^h	**B6**^{2}, 548
Ω b487	—,1-cyclohexyl-4-hydroxy-*	$C_{12}H_{16}O$. *See* b202	176.24	nd (bz)	133	293.5–5.5^{752} 132–5^4	....	....	i	v	v	...	s	CCl_4 v^h lig δ	**B6**^{2}, 548
b488	—,cyclopentyl-*	Phenylcyclopentane*. $C_{11}H_{14}$. *See* b202	146.24	....	....	219^{760} 102^{18}	0.9462$_4^{20}$	1.5280^{20}	i	...	s	...	...	....	**B5**^{2}, 393
b489	—,1,2-diacetamido-*	$C_{10}H_{12}N_2O_2$. *See* b202	192.22	nd (w) λ^{al} 278 sh (3.0)	185–6	....	....	....	s^h	s	δ	s	δ	chl, aa s	**B13**, 20
Ω b490	—,1,4-diacetyl-	$C_{10}H_{10}O_2$. *See* b202	162.19	pr (al, eth)	114	sub	....	....	...	v^h	...	...	...	....	**B7**^{2}, 624
b491	—,1,5-diacetyl-2,4-dihydroxy-	Resodiacetophenone. $C_{10}H_{10}O_4$. *See* b202	194.19	nd (al) λ^{al} 252 (4.66), 275 (4.04), 318 (3.91)	185	....	....	....	i	s^h	s	s	s^h	os s^h	**B8**^{2}, 456
Ω b492	—,1,2-diamino-*	*o*-Phenylenediamine. $C_6H_8N_2$. *See* b202	108.15	brsh ye lf (w), pl (chl) λ^{cy} 235.5 (3.82), 289 (3.54)	102–3	256–8^{760}	....	....	s^h	v	s	...	s	chl s	**B13**, 6

For explanations, symbols and abbreviations see beginning of table. For structural formulas see end of table.

No.	Name	Synonyms and Formula	Mol. wt.	Color, crystalline form, specific rotation and λ_{max} (log ε)	m.p. °C	b.p. °C	Density	n_D	Solubility w	al	eth	ace	bz	other solvents	Ref.
	Benzene														
Ω b493	**—,1,3-diamino-***	*m*-Phenylenediamine. $C_6H_8N_2$. *See* b202	108.15	rh (al) λ^{cy} 240 sh (3.85), 293 (3.42)	63–4	$282–4^{760}$	1.0696_4^{58} 1.1421_{10}^{10}	1.6339^{58}	v	s	s	...	s v^h	...	**B13**[1], 10
Ω b494	**—,1,4-diamino-***	*p*-Phenylenediamine. $C_6H_8N_2$. *See* b202	108.15	wh pl (bz, eth) λ^{cy} 246 (3.93), 315 (3.30)	140	267	...	...	s^h δ	s^h	s	...	s^h	chl s	**B13,** 61
b495	**—,1,2-diamino-3,5-dichloro-***	$C_6H_6Cl_2N_2$. *See* b202	177.03	nd (al)	60.5	...	...	...	...	s^h	...	...	...	...	**B13,** 27
b496	**—,1,3-diamino-2,5-dichloro-***	$C_6H_6Cl_2N_2$. *See* b202	177.03	nd (w)	100	...	...	...	v^h	s	...	...	v	...	**B13**[2], 29
b497	**—,1,4-diamino-2,5-dichloro-*-**	$C_6H_6Cl_2N_2$. *See* b202	177.03	pr (w)	170	...	...	...	δ	δ	...	...	...	...	**B13,** 118
b498	**—,1,5-diamino-2,4-dichloro-***	$C_6H_6Cl_2N_2$. *See* b202	177.03	nd (dil al)	136–7	...	...	...	...	s	...	...	...	...	**B13,** 54
b499	**—,2,3-diamino-1,4-dichloro-***	$C_6H_6Cl_2N_2$. *See* b202	177.03	nd (50% al)	100	...	...	...	s	s	s	v	...	os v peth δ	**B13**[2], 20
b500	**—,2,5-diamino-1,3-dichloro-***	$C_6H_6Cl_2N_2$. *See* b202	177.03	nd, pr (dil al)	124–5.5	...	...	...	...	s	s	s	s	os s	**B13**[1], 37
b501	**—,1,2-diamino-4,5-dimethoxy-***	$C_8H_{12}N_2O_2$. *See* b202	168.20	bl pr	131–2	...	...	...	s	s	i	...	...	to δ	**B13,** 782
b502	**—,1,2-diamino-4-methoxy-***	$C_7H_{10}N_2O$. *See* b202	138.17	gr pl	50–2 (130–1.5)	$167–70^{11}$	...	...	...	...	∞	...	...	...	**Am 69,** 586
Ω b503	**—,2,4-diamino-1-methoxy-***	$C_7H_{10}N_2O$. *See* b202	138.17	nd (eth)	67–8	...	...	...	...	s	s^h	...	...	...	**B13**[2], 308
b504	**—,1,4-diamino-2-methoxy-5-methyl-***	$C_8H_{12}N_2O$. *See* b202	152.20	...	166	...	...	...	δ	s	s	...	...	...	**B13**[2], 349
b505	**—,2,3-diamino-1-methoxy-4-methyl-***	$C_8H_{12}N_2O$. *See* b202	152.20	pr (eth-bz)	75–6	...	...	...	δ	v	v	...	δ	...	**B13**[2], 349
b506	**—,1,2-diamino-4-methyl-***	$C_7H_{10}N_2$. *See* b202	122.17	pl (lig)	89–90	265 92^1	...	...	s	...	...	...	...	...	**B13,** 148
Ω b507	**—,1,2-diamino-3-nitro-***	$C_6H_7N_3O_2$. *See* b202	153.14	dk red nd (dil al)	158–9	...	...	...	δ	...	...	...	...	ac s	**B13**[1], 10
Ω b508	**—,1,2-diamino-4-nitro-***	$C_6H_7N_3O_2$. *See* b202	153.14	dk red nd $\lambda^{20\%sulf}$ 380 (4.08)	199–200	...	...	...	...	...	...	...	...	ac s	**B13,** 29
b509	**—,1,4-diamino-2,3,5,6-tetramethyl-***	Diaminodurene. $C_{10}H_{16}N_2$. *See* b202	164.25	nd (w) λ^{cy} 288 sh (3.67), 264 (4.21), 304 sh (3.27)	149	...	...	...	s^h	v	s	...	...	chl v	**B13**[2], 76
b510	**—,1,2-dibenzoxy-**	Pyrocatechol dibenzyl ether. $C_{20}H_{18}O_2$. *See* b202	290.37	yesh nd or pr (al)	63–4	...	...	...	i	s^h	s	...	...	peth s	**B6,** 772
b511	**—,1,4-dibenzoxy-**	Hydroquinone dibenzyl ether. $C_{20}H_{18}O_2$. *See* b202	290.37	pl (al)	130	...	...	...	i	s^h	s	...	...	aa s	**B6,** 845
Ω b512	**—,1,2-dibromo-***	$C_6H_4Br_2$. *See* b202	235.92	λ^{hx} 257 (2.40), 263 (2.40) 270 (2.46), 278 (2.39)	7.1	225^{760} 92^{10}	1.9843_4^{20}	1.6155^{20}	i	s	∞	∞	∞	peth, CCl_4 ∞	**B5**[2], 162
Ω b513	**—,1,3-dibromo-***	$C_6H_4Br_2$. *See* b202	235.92	λ^{hx} 257 (2.33), 264 (2.62), 270 (2.77), 278 (2.63)	−7	218^{760} 66^5	1.9523_4^{20}	1.6083^{17}	i	s	∞	...	...	...	**B5**[2], 162
Ω b514	**—,1,4-dibromo-***	$C_6H_4Br_2$. *See* b202	235.92	pl λ^{al} 228 (4.19), 265.5 (2.46), 273 (2.48), 282 (2.29)	87.33	218–9	1.8322_4^{100} 2.261^{17}	1.5742	i	s v^h	v	v	s v^h	CS_2 v to s, v^h	**B5**[2], 163
b515	**—,1,5-dibromo-2,4-dimethyl-***	$C_8H_8Br_2$. *See* b202	263.98	pl (al)	68	255–6	...	...	i	s^h	...	...	...	...	**B5**[2], 294
b516	**—,1,2-dibromo-3-nitro-***	$C_6H_3Br_2NO_2$. *See* b202	280.91	mcl pr	85	...	...	...	i	...	s	v	...	chl v aa s^h	**B5**[2], 190
b517	**—,1,2-dibromo-4-nitro-***	$C_6H_3Br_2NO_2$. *See* b202	280.91	nd (al, aa), mcl pr	58–9	296 180^{20}	2.354^8	1.9835^{111}	i	s	...	...	s	aa s	**B5,** 249
b518	**—,1,3-dibromo-2-nitro-***	$C_6H_3Br_2NO_2$. *See* b202	280.91	nd (al), mcl pr (al)	84	sub	1.9211^{111} 2.211^8	...	i	s^h	...	s	s	...	**B5,** 250
b519	**—,1,3-dibromo-5-nitro-***	$C_6H_3Br_2NO_2$. *See* b202	280.91	pl or pr (eth), lf or nd (al)	106	...	1.9341^{111} 2.363^8	...	i	s^h	s^h	...	s	...	**B5,** 250
Ω b520	**—,1,4-dibromo-2-nitro-***	$C_6H_3Br_2NO_2$. *See* b202	280.91	yesh pl (ace) λ^{al} 233.5 (4.24), 309 (3.18)	85–6	...	1.9416^{111} 2.368^8	...	i	...	...	s^h	s		**B5,** 250
b521	**—,2,4-dibromo-1-nitro-***	$C_6H_3Br_2NO_2$. *See* b202	280.91	ye pl or pr (al)	62	...	1.9581^{111} 2.356^8	...	i	s^h	...	s	s	chl s	**B5,** 250
b522	**—,2,3-dibromo-1,4,5-trimethyl-***	$C_9H_{10}Br_2$. *See* b202	278.00	nd (al)	63–4	293–4	...	...	i	v	v	...	v	chl v	**B5,** 403
Ω b523	**—,2,4-dibromo-1,3,5-trimethyl-***	$C_9H_{10}Br_2$. *See* b202	278.00	nd (al)	65.5	285 (278–9)	...	...	i	δ v^h	...	...	s	...	**B5,** 408
b524	**—,1,2-dibutoxy-***	Pyrocatechol dibutyl ether. $C_{14}H_{22}O_2$. *See* b202	222.33	ye	...	241^{765} $135–8^{12}$	...	...	...	...	...	...	...	...	**C24,** 3500

For explanations, symbols and abbreviations see beginning of table. For structural formulas see end of table.

No.	Name	Synonyms and Formula	Mol. wt.	Color, crystalline form, specific rotation and λ_{max} (log ε)	m.p. °C	b.p. °C	Density	n_D	Solubility w	al	eth	ace	bz	other solvents	Ref.
	Benzene														
Ω b525	—,1,4-di-*tert*-butyl-	$C_{14}H_{22}$. *See* b202	190.33	nd (MeOH) λ^{iso} 217 (4.0), 250 (2.2), 256 (2.4), 263 (2.5), 270 (2.5)	80–1	237^{743} 109^{15}			i	s	s				**B5**[2], 344
b526	—,1,3-di-*tert*-butyl-5-ethyl-2-hydroxy-	$C_{16}H_{26}O$. *See* b202	234.39	λ^{iso} 277.5 (3.34), 283 (3.34), 307 (1.55)	44	272^{760} 140^{10}								alk i	**Am 67,** 304
Ω b527	—,1,4-di-*tert*-butyl-2,5-dihydroxy-	$C_{14}H_{22}O_2$. *See* b202	222.33	cr (aq aa) λ^{diox} 229 (3.5), 294 (3.5)	213–4										**Am 64,** 937
Ω b529	—,1,3-di-*tert*-butyl-2-hydroxy-	$C_{14}H_{22}O$. *See* b202	206.33	pr (al) λ^{al} 268 (3.19), 274 (3.18)	39	133^{20}		1.5001^{25}		s^h				alk i	**Am 73,** 3179
Ω b530	—,2,4-di-*tert*-butyl-1-hydroxy-	$C_{14}H_{22}O$. *See* b202	206.33	λ^{iso} 226 (3.7), 268 (3.2), 274 (3.3), 276 (3.3), 283 (3.3)	56.5	263.5^{760} 146^{20}		1.5080^{20}						alk i	**Am 60,** 2496
b531	—,1,3-di-*tert*-butyl-2-hydroxy-5-methyl-	Ionol. $C_{15}H_{24}O$. *See* b202	220.36	λ^{iso} 227 (3.75), 277 (3.34), 283 (3.34)	71	265^{760} 136^{10}	0.8937^{75}_{4}	1.4859^{75}	i	s		s	s	peth s alk i	**Am 69,** 1624
b532	—,1,5-di-*tert*-butyl-2-hydroxy-3-methyl-	$C_{15}H_{24}O$. *See* b202	220.36	λ^{cy} 275 (3.41), 280 sh (3.38)	51 (52.5)	269^{760} 138.5^{10}	0.891^{80}_{4}							alk i	**Am 67,** 305
b533	—,1,5-di-*tert*-butyl-2-hydroxy-4-methyl-	$C_{15}H_{24}O$. *See* b202	220.36		62.1	282^{760} 167^{20}	0.912^{80}_{4}		i	s	s	s	s	CCl_4 s alk i	**C38,** 4170
b534	—,1,3-di-*tert*-butyl-2-hydroxy-5(2-methyl-2-butyl)-	$C_{19}H_{32}O$. *See* b202	276.47		47	$135–8^{6}$									**Am 67,** 304
Ω b535	—,1,2-dichloro-*	$C_6H_4Cl_2$. *See* b202	147.01	λ^{al} 250 (1.98), 256 (2.13), 263 (2.36), 270 (2.44), 277 (2.37)	−17.0	180.5^{760} 86^{18}	1.3048^{20}_{4}	1.5515^{20}	i	s	s	∞	∞	lig, CCl_4 ∞	**B5**[2], 153
Ω b536	—,1,3-dichloro-*	$C_6H_4Cl_2$. *See* b202	147.01	λ^{al} 250 (1.90), 256 (2.15), 263 (2.40), 270 (2.52), 278 (2.43)	−24.7	173^{760} 53^{10}	1.2884^{20}_{4}	1.5459^{20}	i	s	s	∞	s	lig, CCl_4 ∞	**B5**[2], 154
Ω b537	—,1,4-dichloro-*	$C_6H_4Cl_2$. *See* b202	147.01	mcl pr, lf (ace) λ^{al} 258 (2.24), 266 (2.46), 273 (2.60), 280 (2.51)	53.1	174^{760} 55^{10}	1.2475^{20}_{4}	1.5285^{20}	i	∞	s	∞	s	CS_2, chl s	**B5**[2], 154
b538	—,1,2-dichloro-4,5-dihydroxy-*	$C_6H_4Cl_2O_2$. *See* b202	179.00	pr (chl-CS_2), nd (bz-peth) $\lambda^{w, pH=1}$ 258 (3.45), 316 (4.15), 325 (4.14),	116–7				v	v			v		**B6**[1], 389
b539	—,1,3-dichloro-2,5-dihydroxy-*	$C_6H_4Cl_2O_2$. *See* b202	179.00	nd or lf (w, bz)	164				s^h	s		v	s^h		**B6**[2], 845
b540	—,1,4-dichloro-2,5-dihydroxy-*	$C_6H_4Cl_2O_2$. *See* b202	179.00	nd or pr (w, ace, bz) λ^{al} 305 (3.7)	172.5				s	v	v	v			**B6**[2], 845
b541	—,1,5-dichloro-2,3-dihydroxy-*	$C_6H_4Cl_2O_2$. *See* b202	179.00	pr	83–4				δ v^h	s		v			**B6,** 783
Ω b542	—,1,5-dichloro-2,4-dihydroxy-*	$C_6H_4Cl_2O_2$. *See* b202	179.00		113	254 sub			v	v	v	v		lig δ	**B6**[2], 819
b543	—,2,3-dichloro-1,4-dihydroxy-*	$C_6H_4Cl_2O_2$. *See* b202	179.00	cr (sub) nd (w + 2)	146–8 (anh)					v				lig i	**B6**[2], 845
b544	—,1,2-dichloro-4,5-dimethyl-3-hydroxy-*	$C_8H_8Cl_2O$. *See* b202	191.07	cr (peth)	90						v	v	s	chl v peth δ	**B6**[2], 454
b545	—,1,3-dichloro-4,5-dimethyl-2-hydroxy-*	$C_8H_8Cl_2O$. *See* b202	191.07	cr (peth)	52						s			peth δ	**B6**[2], 456
b546	—,1,4-dichloro-2,3-dimethyl-5-hydroxy-*	$C_8H_8Cl_2O$. *See* b202	191.07	nd (peth)	84						s			peth δ, s^h	**B6**[2], 456
b547	—,2,4-dichloro-1,3-dimethyl-5-hydroxy-*	$C_8H_8Cl_2O$. *See* b202	191.07		83						s			lig δ, s^h	**B6**[2], 464
b548	—,2,4-dichloro-1,5-dimethyl-3-hydroxy-*	$C_8H_8Cl_2O$. *See* b202	191.07	cr (peth)	87–8	$105–10^{1}$					s			chl v	**Am 75,** 6311

For explanations, symbols and abbreviations see beginning of table. For structural formulas see end of table.

No.	Name	Synonyms and Formula	Mol. wt.	Color, crystalline form, specific rotation and λ_{max} (log ε)	m.p. °C	b.p. °C	Density	n_D	Solubility w	al	eth	ace	bz	other solvents	Ref.
	Benzene														
b549	**—,3,4-dichloro-1,2-dimethyl-5-hydroxy-***	$C_8H_8Cl_2O$. *See* b202	191.07	nd (peth)	102.5				...	...	s	s	s	os s	**B6**^{2}, 456
b550	**—,2,4-dichloro-1-ethoxy-***	$C_8H_8Cl_2O$. *See* b202	191.07			237			i	s	s	...	s		**B6**, 189
b551	**—,(1,2-dichloro-ethyl)-***	Styrene dichloride. $C_8H_8Cl_2$. *See* b202	175.07			$233–4^{759}$ 115^{15}	1.240_4^{15}	1.5544^{15}	i	...	...	s	s		**B5**^{2}, 278
b552	**—,1,4-dichloro-2-iodo-***	$C_6H_3Cl_2I$. *See* b202	272.90	pl (al)	21	255–6 134^{12}			i	s	s	...	s	chl s	**B5**, 221
b553	**—,2,4-dichloro-1-methoxy-***	$C_7H_6Cl_2O$. *See* b202	177.03	pr	28–9	233^{740} 125^{10}			i	s	s	...	...		**B6**^{1}, 103
Ω b554	**—,1,2-dichloro-3-nitro-***	$C_6H_3Cl_2NO_2$. *See* b202	192.00	mcl nd (peth, aa) $\lambda^{aq\ MeOH}$ 211 (4.15), 276 (3.56) 320 sh (3.30)	61–2	257–8	1.721^{14}		i	s	s	s	s	peth, aa s os v	**B5**^{2}, 185
Ω b555	**—,1,2-dichloro-4-nitro-***	$C_6H_3Cl_2NO_2$. *See* b202	192.00	(i) nd (al, CCl_4) (ii) liq	43	255–6 189^{100}	1.4558_4^{75}		i	s^h	s	...	...		**B5**^{2}, 186
b556	**—,1,3-dichloro-2-nitro-***	$C_6H_3Cl_2NO_2$. *See* b202	192.00	nd or pr (al, CS_2)	72.5	130^{8}	1.603^{17} 1.4094^{80}		i	s^h	s	...	...		**B5**^{2}, 186
b557	**—,1,3-dichloro-5-nitro-***	$C_6H_3Cl_2NO_2$. *See* b202	192.00	mcl pr or lf (aa, al) λ^{hx} 257 (3.79), 307 (3.08)	65.4		1.692^{14} 1.4000^{100}		i	s^h	s	...	...		**B5**^{2}, 186
Ω b558	**—,1,4-dichloro-2-nitro-***	$C_6H_3Cl_2NO_2$. *See* b202	192.00	pl or pr (al), pl (AcOEt) $\lambda^{aq\ MeOH}$ 220 (4.28), 258 (3.52), 314 (3.18)	56	267	1.669^{22} 1.4390_4^{75}		i	s^h	s	...	s	chl, CS_2 s	**B5**, 245
Ω b559	**—,2,4-dichloro-1-nitro-***	$C_6H_3Cl_2NO_2$. *See* b202	192.00	nd (al)	34	258.5 $100–1^{4}$	1.4390^{80}	$1.5512_α^{78}$	i	s	s	...	...		**B5**^{2}, 185
b560	**—,1-dichloro-phosphino-4-isopropyl-***	$C_9H_{11}Cl_2P$. *See* b202	221.07			268–70 $129–30^{10}$	1.1917^{25}	1.5677^{25}	d^h	d	...	s			**B16**, 773
b561	**—,2,4-dichloro-1-triazo-***	$C_6H_3Cl_2N_3$. *See* b202	188.02	ye nd (al), pr (ace or bz)	54				i	s	s	...	s	peth, chl s	**B5**^{2}, 208
Ω b562	**—,1,2-diethoxy-***	Catechol diethyl ether. $C_{10}H_{14}O_2$. *See* b202	166.22	pr (peth, dil al) λ^{al} 224.5 (3.82), 275 (3.28)	43–5	219–9.5	1.0075_4^{25}	1.5083^{25}	...	s	v	...			**B6**^{2}, 780
b563	**—,1,3-diethoxy-***	Resorcinol diethyl ether. $C_{10}H_{14}O_2$. *See* b202	166.22	pr λ^{hx} 269 (3.35), 274 (3.45), 280 (3.41)	12.4	235^{756}			i	s	s	...			**B6**^{2}, 814
Ω b564	**—,1,4-diethoxy-***	Hydroquinone diethyl ether. $C_{10}H_{14}O_2$. *See* b202	166.22	pl (dil al) λ^{hx} 292 (3.48)	72	246			...	v	s	...	s	chl s	**B6**^{2}, 840
Ω b565	**—,1,2-diethyl-***	$C_{10}H_{14}$. *See* b202	134.22	λ^{iso} 257 (2.28), 264 (2.38), 272 (2.32)	−31.2	183.4^{760} 63^{10}	0.8800^{20}	1.5035^{20}	i	∞	∞	∞	∞	CCl_4, lig ∞	**B5**^{2}, 327
Ω b566	**—,1,3-diethyl-***	$C_{10}H_{14}$. *See* b202	134.22	λ^{iso} 258 (2.28), 264 (2.39), 268 (2.26), 272 (2.32)	−83.89	181^{760} 60^{10}	0.8602_4^{20}	1.4955^{20}	i	∞	∞	∞	∞	CCl_4, lig ∞	**B5**^{2}, 327
Ω b567	**—,1,4-diethyl-***	$C_{10}H_{14}$. *See* b202	134.22	λ^{iso} 259 (2.47), 265 (2.60), 267 (2.58), 273 (2.62)	−42.85	183.8^{760} 63^{10}	0.8620_4^{20}	1.4967^{20}	i	∞	∞	∞	∞	CCl_4, lig ∞	**B5**^{2}, 327
Ω b568	**—,1,3-diethyl-5-methyl-***	$C_{11}H_{16}$. *See* b202	148.25		−74.12	205^{760} 79^{10}	0.8748_4^{20}	1.5027^{20}	i	∞	∞	∞	∞	CCl_4, lig ∞	**B5**, 441
Ω b569	**—,1,2-dihydroxy-***	Catechol. Pyrocatechol. $C_6H_6O_2$. *See* b202	110.11	cr λ^{cy} 214 (3.83), 270 sh (3.34), 274 (3.40), 280 (3.35)	105	245^{750}	1.1493^{21}	1.604	s	s	s	v	s^h δ	alk. chl, CCl_4 s	**B6**^{3}, 4187
b570	—,—,carbonate		136.11	pr (ace or al)	120	225–30			...	s^h	...	...	s^h		**B19**^{1}, 660
Ω b571	—,—,diacetate	$C_{10}H_{10}O_4$. *See* b202	194.19	nd (al)	64–5	$142–3^{9}$			i	v	v	...	...	chl v peth s	**B6**^{2}, 784
b572	—,—,dibenzoate	$C_{20}H_{14}O_4$. *See* b202	318.33	lf (eth-al)	86				i	s	s	...	s		**B9**^{2}, 112
b573	—,—,monoacetate	$C_8H_8O_3$. *See* b202	152.16	pl	57–8	$189–91^{102}$ 148^{25}			v	v	...	v	s^h	peth s	**B6**^{2}, 783
—	—,—,mono-benzoate	*see* **Benzoic acid,** 2-hydroxyphenyl ester													
Ω b574	**—,1,3-dihydroxy-***	Resorcinol. $C_6H_6O_2$. *See* b202	110.11	nd (bz) pl (w) λ^{al} 220 (3.79), 276 (3.33)	(i) 111 (st) (ii) 108.5 (unst)	276.5^{760} 178^{16}	1.2717		s	s	s	...	δ	CCl_4 ∞ aa v chl δ, s^h	**B6**^{2}, 802
Ω b575	—,—,diacetate	$C_{10}H_{10}O_4$. *See* b202	194.19			278 $153–4^{12}$			i	v	v	...			**B6**^{2}, 817
Ω b576	—,—,dibenzoate	$C_{20}H_{14}O_4$. *See* b202	318.33	pl (dil al)	117				...	s	s	...			**B9**^{2}, 113

For explanations, symbols and abbreviations see beginning of table. For structural formulas see end of table.

No.	Name	Synonyms and Formula	Mol. wt.	Color, crystalline form, specific rotation and λ_{max} (log ε)	m.p. °C	b.p. °C	Density	n_D	Solubility w	al	eth	ace	bz	other solvents	Ref.
	Benzene														
Ω b577	**—,1,4-dihydroxy-***	Hydroquinone. Quinol. $C_6H_6O_2$. *See* b202	110.11	mcl pr (sub), nd (w), pr (MeOH) λ^{w} 288 (3.36)	173–4	285^{730}	1.328^{15}		s v^h	v	s	v	i	CCl_4 ∞	**B6²**, 832
Ω b578	**—,—,diacetate**	$C_{10}H_{10}O_4$. *See* b202	194.19	pl (w, al) λ^{al} 265 (2.7), 270 sh (2.6)	123–4		0.8731^{25}_{4}		s^h	v^h	v			lig, chl v aa v^h	**B6²**, 843
b579	**—,—,dibenzoate**	$C_{20}H_{14}O_4$. *See* b202	318.33	mcl nd (al or to)	204				i	s^h	i			to s^h	**B9²**, 114
b580	**—,1,2-dihydroxy-3,5-dimethyl-***	$C_8H_{10}O_2$. *See* b202	138.17	pr (w), nd (peth-bz) λ^{al} 282 (3.30)	73–4				v	v	v				**B6**, 911
b581	**—,1,2-dihydroxy-4,5-dimethyl-***	$C_8H_{10}O_2$. *See* b202	138.17	mcl pr or nd (peth) λ^{al} 286.5 (3.50)	87–8	sub			v	v	v		δ	sulf s peth δ^h	**B6¹**, 444
b582	**—,1,3-dihydroxy-2,4-dimethyl-***	$C_8H_{10}O_2$. *See* b202	138.17	nd (peth) λ^{al} 279 (3.19)	149–50	sub			v	v	v				**B6³**, 4588
b583	**—,1,3-dihydroxy-2,5-dimethyl-***	$C_8H_{10}O_2$. *See* b202	138.17	nd (bz), pr (w) λ^{MeOH} 272 (3.88), 281 (3.85)	163	277–80			s^h	s	s				**B6²**, 891
b584	**—,1,4-dihydroxy-2,3-dimethyl-***	$C_8H_{10}O_2$. *See* b202	138.17	cr (w), pr (MeCN) λ^{diox} 225 (3.8), 300 (3.6)	224–5 δd				s	s	s				**B6**, 908
b585	**—,1,4-dihydroxy-2,5-dimethyl-***	$C_8H_{10}O_2$. *See* b202	138.17	lf (w, al, MeOH, bz)	217 sub				s^h	s	s		δ	CS_2 δ chl s	**B6**, 915
b586	**—,1,5-dihydroxy-2,4-dimethyl-***	$C_8H_{10}O_2$. *See* b202	138.17	mcl pr, (w+1), nd (sub) λ^{al} 286 (3.53)	125	276–9 sub			s	s	s				**B6²**, 889
b587	**—,1,5-dihydroxy-3,4-dimethyl-***	$C_8H_{10}O_2$. *See* b202	138.17	nd (bz), pr (w+1) λ^{al} 282 (3.35)	136–7 115–7 (+1w)	sub			s	v	s		δ	aa s, peth, CS_2, chl δ	**B6**, 908
b588	**—,2,5-dihydroxy-1,3-dimethyl-***	$C_8H_{10}O_2$. *See* b202	138.17	nd (xyl), cr(w)	153–4				s	s	s				**B6³**, 4588
b589	**—,1,3-dihydroxy-2,4-dinitro-***	$C_6H_4N_2O_6$. *See* b202	200.11	ye lf (al) λ^{al} 400 (4.30)	147–8				δ	δ					**B6²**, 823
b590	**—,1,3-dihydroxy-2,4-dinitroso-***	$C_6H_4N_2O_4$. *See* b202	168.11	ye rh pl (50% al) lf (+1w, aq MeOH) λ^{al} 253 (4.01), 292.5 (3.96)	168				s	s					**B7²**, 851
b591	**—,1,3-dihydroxy-2,4-di(2-tolylazo)-**	$C_{20}H_{18}N_4O_2$. *See* b202	346.39	red nd (chl-al)	212					δ			s	chl s	**B16**, 186
b592	**—,1,3-dihydroxy-2,4-di(4-tolylazo)-**	$C_{20}H_{18}N_4O_2$. *See* b202	346.39	nd (chl-al)	230.5				δ					os s	**B16**, 186
b593	**—,2,4-dihydroxy-1,5-di(4-tolylazo)-**	$C_{20}H_{18}N_4O_2$. *See* b202	346.39	ye nd (chl-al)	255–6				s	δ				chl v^h	**B16**, 187
Ω b594	**—,2,4-dihydroxy-1-ethyl-***	$C_8H_{10}O_2$. *See* b202	138.17	pr (chl or bz) λ^{al} 281 (3.46)	98–9	131^{15}			δ	δ	δ				**B6²**, 885
Ω b595	**—,2,4-dihydroxy-1-hexyl-***	$C_{12}H_{18}O_2$. *See* b202	194.28	pa ye nd (lig) λ^{w} 280 (3.42)	68–9	333 178^{8}			i	s	s	s	s	chl s lig δ	**B6²**, 904
b596	**—,1,2-dihydroxy-4-iodo-***	$C_6H_5IO_2$. *See* b202	236.01	lf (CCl_4)	92	sub			δ	s	s	s	s	chl, peth δ	**B6³**, 4262
b597	**—,1,3-dihydroxy-5-iodo-***	$C_6H_5IO_2$. *See* b202	236.01	nd (+lw, bz), nd (sub)	92.3 (105–13)	sub			i	v				peth i	**B6²**, 821
b598	**—,1,4-dihydroxy-2-iodo-***	$C_6H_5IO_2$. *See* b202	236.01		115–6										**B6³**, 4440
b599	**—,2,4-dihydroxy-1-iodo-***	$C_6H_5IO_2$. *See* b202	236.01	pr, nd (+lw, w)	67				δ	s	s	s		chl s	**B6²**, 821
b600	**—,2,4-dihydroxy-1-isobutyl-**	$C_{10}H_{14}O_2$. *See* b202	166.22		62–3.5	$166–8^{6–7}$			δ	s	s				**B6²**, 899
Ω b601	**—,1,2-dihydroxy-4-isopropyl-**	$C_9H_{12}O_2$. *See* b202	152.20	lf (lig)	78	270–2 168^{26}								peth s^h	**B6**, 929
b602	**—,1,4-dihydroxy-2-isopropyl-**	$C_9H_{12}O_2$. *See* b202	152.20	nd (w) λ^{sulf} 289 (3.53)	130–1									lig s^h	**B6**, 929
b603	**—,2,4-dihydroxy-1-isopropyl-**	$C_9H_{12}O_2$. *See* b202	152.20	cr (aq aa)	105	265–81 $114^{0.2}$			δ	s	s				**B6²**, 896
b604	**—,1,4-dihydroxy-2-isopropyl-5-methyl-**	Thymohydroquinone. $C_{10}H_{14}O_2$. *See* b202	166.22	pr (dil al) λ^{al} 293 (3.59)	143	290 sub			δ s^h	s	s		i	peth i	**B6²**, 901
Ω b605	**—,2,3-dihydroxy-1-isopropyl-4-methyl-**	$C_{10}H_{14}O_2$. *See* b202	166.22		48	270			δ		s	s		os s peth i	**B6²**, 900
Ω b606	**—,1,2-dihydroxy-3-methoxy-***	$C_7H_8O_3$. *See* b202	140.15	nd	43–4	129^{10}									**B6²**, 1065
b607	**—,1,3-dihydroxy-2-methoxy-***	$C_7H_8O_3$. *See* b202	140.15	cr (bz)	85–7	$154–5^{24}$							s^h δ		**B6**, 1081

For explanations, symbols and abbreviations see beginning of table. For structural formulas see end of table.

No.	Name	Synonyms and Formula	Mol. wt.	Color, crystalline form, specific rotation and λ_{max} (log ε)	m.p. °C	b.p. °C	Density	n_D	Solubility w	al	eth	ace	bz	other solvents	Ref.
	Benzene														
b608	**—,1,3-dihydroxy-5-methoxy-***	$C_7H_8O_3$. *See* b202	140.15	pl (bz) λ^{al} <215 (>4.15), 265 (2.88)	78–81	213^{16}	...	...	δ	v	v	...	δ	...	**B6**, 1101
Ω b609	**—,1,2-dihydroxy-3-methyl-***	$C_7H_8O_2$. *See* b202	124.15	lf (bz)	68 (47)	241 127^{12}	...	...	s	s	...	...	s	chl s	**B6**², 858
Ω b610	**—,1,3-dihydroxy-2-methyl-***	$C_7H_8O_2$. *See* b202	124.15	pr (bz) λ^{cy} 274 (3.02)	119–21	264 168^{16}	...	...	s	s	s	...	s	...	**B6**, 878
b611	**—,2,4-dihydroxy-1(3-methylbutyl)-***	$C_{11}H_{16}O_2$. *See* b202	180.25		68–70	$177–8^{6}$	...	...	δ	s	s	...	s	...	**B6**², 903
b612	**—,2,4-dihydroxy-1(4-methylpentyl)-***	$C_{12}H_{18}O_2$. *See* b202	194.28	lf (peth)	71	190^{12}	...	1.5292^{25}	δ	s	s	...	...	...	**B6**, 905
Ω b613	**—,1,3-dihydroxy-2-nitro-***	$C_6H_5NO_4$. *See* b202	155.11	og-red pr (al) λ^{w} 345 (4.15)	87–8 (85)	...	...	...	...	s^h	...	...	...	...	**B6**², 822
b614	**—,1,4-dihydroxy-2-nitro-***	$C_6H_5NO_4$. *See* b202	155.11	og-red rh (w) $\lambda^{1N\ HCl}$ 280 (3.74), 395 (3.45), $\lambda^{1N\ NaOH}$ 238 (4.10), 540 (3.59)	133–4	...	...	...	s^h	v	v	...	s	lig δ	**B6**³, 4442
b615	**—,2,4-dihydroxy-1-nitroso-***	2-Hydroxy-1,4-benzoquinone oxime. $C_6H_5NO_3$. *See* b202	139.11	ye nd (+w), br nd (chl)	150d	...	...	...	s	v	s	v	i	chl s CS_2 i	**B8**, 235
b616	**—,1,3-dihydroxy-4-pentyl-***	$C_{11}H_{16}O_2$. *See* b202	180.25	...	72–3 (43–4)	168– 70^{6-7}	...	...	i	s	s	...	s	...	**B6**², 902
b617	**—,1,3-dihydroxy-5-pentyl-***	Olivetol. $C_{11}H_{16}O_2$. *See* b202	180.25	nd (+w), pr(bz-lig)	49 (anh) 41 (+w)	164^{5}	...	...	v	s	s	s	s	...	**J1945**, 311
b618	**—,1,2-dihydroxy-4-propyl-***	$C_9H_{12}O_2$. *See* b202	152.20	pr (w, bz)	60	152^{13} $111–2^{0.2}$	1.100^{18}_{4}	1.4440^{18}	δ	s	s	s	s^h	...	**B6**², 892
b619	**—,1,3-dihydroxy-5-propyl-***	Divarinol. $C_9H_{12}O_2$. *See* b202	152.20	lf (+1w, w) lf (bz)	83–4 (anh) 51 (hyd)	$148–9^{3}$	...	...	s^h δ	s	s	s	s	aa s	**B6**², 893
Ω b620	**—,2,4-dihydroxy-1-propyl-***	$C_9H_{12}O_2$. *See* b202	152.20	pr (bz), nd (peth)	82–3	172^{14}	...	...	s	s	s	...	...	...	**B6**³, 4611
Ω b621	**—,1,2-dihydroxy-3,4,5,6-tetrabromo-***	$C_6H_2Br_4O_2$. *See* b202	425.72	nd (bz, al) lf (bz-peth)	192–3	...	...	...	...	s	...	δ	s	...	**B6**², 788
b622	**—,1,4-dihydroxy-2,3,5,6-tetrabromo-***	$C_6H_2Br_4O_2$. *See* b202	425.72	mcl pr (al-eth)	244	...	3.023^{21}	...	i	v	v	...	...	aa s	**B6**², 848
b623	**—,1,2-dihydroxy-3,4,5,6-tetrachloro-***	$C_6H_2Cl_4O_2$. *See* b202	247.89	cr (dil al, bz), cr (+1w, aq aa), cr (+3w, aq aa) λ^{iso} 287 (3.28)	110 (anh) 194–5 (+1w) 94 (+3w)	...	...	...	...	...	...	...	δ	...	**B6**², 787
b625	**—,1,3-dihydroxy-2,4,5,6-tetrachloro-***	$C_6H_2Cl_4O_2$. *See* b202	247.89	nd (w)	141	...	...	...	v^h s	v	v	...	v	aa v	**B6**², 819
Ω b626	**—,1,4-dihydroxy-2,3,5,6-tetrachloro-***	$C_6H_2Cl_4O_2$. *See* b202	247.89	nd (aa) λ^{chl} 305 (3.7)	232	sub	...	...	i	v	v	...	i	aa δ CS CCl_4 i	**B6**², 846
b627	**—,1,4-dihydroxy-2,3,5,6-tetraiodo-***	$C_6H_2I_4O_2$. *See* b202	613.70	(aa)	258	...	...	...	...	δ	s	...	δ	chl s alk s^h aa δ	**B6**¹, 417
b628	**—,1,4-dihydroxy-2,3,5,6-tetramethyl-***	Durohydroquinone. $C_{10}H_{14}O_2$. *See* b202	166.22	nd (al) λ^{w} 283 (3.33)	233	...	...	...	...	s^h	δ	...	...	...	**Am 48**, 1422
b629	**—,1,3-dihydroxy-2,4,6-tribromo-***	$C_6H_3Br_3O_2$. *See* b202	346.82	nd (w)	112	...	...	...	δ	s	s	...	...	...	**B6**, 822
b630	**—,1,4-dihydroxy-2,3,5-tribromo-***	$C_6H_3Br_3O_2$. *See* b202	346.82	nd (chl)	136–7	...	...	...	v^h	v	v	...	v	aa v lig s	**B6**², 848
b631	**—,1,2-dihydroxy-3,4,5-trichloro-***	$C_6H_3Cl_3O_2$. *See* b202	213.46	(i) pr (+1w, aa) (ii) pr (+½w, bz)	(i) 115 (ii) 134–5	...	...	...	i	s	s	...	...	aa s	**B6**¹, 389
b632	**—,2,4-dihydroxy-1,3,5-trichloro-***	$C_6H_3Cl_3O_2$. *See* b202	213.46	nd (w)	83	...	...	...	δ	v	v	...	...	...	**B6**, 820
b633	**—,1,3-dihydroxy-2,4,5-trimethyl-***	$C_9H_{12}O_2$. *See* b202	152.20	cr	156	...	...	...	...	...	...	...	...	...	**B6**, 931
b634	**—,1,3-dihydroxy-4,5,6-trimethyl-***	$C_9H_{12}O_2$. *See* b202	152.20	nd or lf	163–4	...	...	...	δ	v	v	v	s	chl s peth i	**B6**, 930
b635	**—,1,4-dihydroxy-2,3,5-trimethyl-***	$C_9H_{12}O_2$. *See* b202	152.20	nd (w)	168–70d	...	...	...	v^h δ	v	v	...	v	...	**B6**, 930
b636	**—,2,4-dihydroxy-1,3,5-trimethyl-***	Mesorcinol. $C_9H_{12}O_2$. *See* b202	152.20	pl (al), lf	149–50	275	...	...	δ	v	v	...	s	...	**B6**, 939

For explanations, symbols and abbreviations see beginning of table. For structural formulas see end of table.

No.	Name	Synonyms and Formula	Mol. wt.	Color, crystalline form, specific rotation and λ_{max} (log ε)	m.p. °C	b.p. °C	Density	n_D	Solubility w	al	eth	ace	bz	other solvents	Ref.
	Benzene														
Ω b637	**—,2,4-dihydroxy-1,3,5-trinitro-***	Styphnic acid. $C_6H_3N_3O_8$	245.11	hex (al), ye cr (aa) λ^{al} 400 (4.20)	179–80	sub	...	...	δ	s	s	...	...	...	**B6²**, 825
b638	**—,1,2-diiodo-***	$C_6H_4I_2$. *See* b202	329.91	pl or pr (lig)	27	286^{750} 109^{3}	2.54^{20}	1.7179^{20}	δ	δ	v	...	...	...	**B5²**, 168
b639	**—,1,3-diiodo-***	$C_6H_4I_2$. *See* b202	329.91	rh pl or pr (eth-al)	40.4 (36.5)	285^{760}	2.47^{25}	...	i	s	s	...	...	chl s	**B5²**, 168
Ω b640	**—,1,4-diiodo-***	$C_6H_4I_2$. *See* b202	329.91	rh lf (al) λ^{al} 241.6 (4.27)	131–2	285 sub	...	...	i	s	v	...	...	...	**B5²**, 168
b641	**—,1,2-diisopropyl-**	$C_{12}H_{18}$. *See* b202	162.28	λ^{iso} 257 (2.28), 263 (2.39), 267.5 (2.25), 271 (2.29)	−56.7	204^{760} 115.4^{50}	0.8701^{20}_{4}	1.4960^{20}	i	∞	∞	∞	∞	CCl_4 ∞	**B5**, 447
Ω b642	**—,1,3-diisopropyl-**	$C_{12}H_{18}$. *See* b202	162.28	λ^{iso} 257 (2.26), 263 (2.37), 267 (2.23), 270 (2.28)	−63.1	203.2^{760} 75^{9}	0.8559^{20}_{4}	1.4883^{20}	i	∞	∞	∞	∞	CCl_4 ∞	**B5**, 447
Ω b643	**—,1,4-diisopropyl-**	$C_{12}H_{18}$. *See* b202	162.28	λ^{iso} 259 (2.36), 264 (2.54), 272 (2.58)	−17.1	210.3^{760} 120^{50}	0.8568^{20}_{4}	1.4898^{20}	i	∞	∞	∞	∞	CCl_4 ∞	**B5²**, 339
b644	**—,1,2-dimercapto-***	Dithiocatechol. $C_6H_6S_2$. *See* b202	142.25	λ 230 (4.3), 285 sh (3.2)	28–9	238–9 120^{17}	...	...	...	v	v	...	v	AcOEt s	**B6²**, 799
b645	**—,1,3-dimercapto-***	Dithioresorcinol. $C_6H_6S_2$. *See* b202	142.25	lf (dil al)	27	245 123^{17}	...	...	δ	v	v	v	v	os, alk v	**B6²**, 829
b646	**—,1,4-dimercapto-***	Dithioquinol. $C_6H_6S_2$. *See* b202	142.25	lf (al)	98	...	...	...	...	v	...	...	v	aa v lig δ	**B6²**, 854
Ω b647	**—,1,2-dimethoxy-***	Veratrole. $C_8H_{10}O_2$. *See* b202	138.17	cr (lig) λ^{cy} 225 (3.85), 271 (3.39), 275 (3.41), 280 (3.28)	22.5	206^{759} 90^{10}	1.0842^{25}_{25}	1.5827^{21}_{α}	δ	s	s	...	...	...	**B6²**, 779
Ω b648	**—,1,3-dimethoxy-***	Resorcinol dimethyl ether. $C_8H_{10}O_2$. *See* b202	138.17	λ^{cy} 220 (3.86), 267 sh (3.23), 271.5 (3.34), 277.5 (3.33)	−52	217–8	1.0552^{25}_{25} 1.0705^{4}_{4}	1.5231^{20}	δ	s	s	...	s	sulf s	**B6²**, 813
Ω b649	**—,1,4-dimethoxy-***	Hydroquinone dimethyl ether. $C_8H_{10}O_2$. *See* b202	138.17	lf (w) λ^{cy} 226 (3.99), 280 sh (3.40), 287 (3.51), 290 (3.51), 297 (3.41)	58–60	212.6 109^{20}	1.0526^{55}_{55}	...	δ	s	v	...	v	...	**B6²**, 839
b650	**—,1,2-dimethoxy-4,5-dinitro-***	$C_8H_8N_2O_6$. *See* b202	228.17	ye nd (al)	130–2	...	1.3164^{140}_{4}	...	i	δ	...	...	...	MeOH δ	**B6²**, 794
b651	**—,1,2-dimethoxy-4-iodo-***	$C_8H_9IO_2$. *See* b202	264.06	nd (dil MeOH)	35	$163–4^{26}$ $90^{0.2}$	...	...	...	s	...	...	s	MeOH s^h	**B6¹**, 390
b652	**—,1,4-dimethoxy-2-isopropyl-**	$C_{11}H_{16}O_2$. *See* b202	180.25	...	...	$114–6^{15}$	1.0129^{17}_{4}	1.5105^{17}	...	...	s	...	s	...	**B6**, 929
b653	**—,1,2-dimethoxy-3-nitro-***	$C_8H_9NO_4$. *See* b202	183.17	nd (al)	64–5	...	1.1404^{133}_{4}	...	i	v	v	...	v	aa s lig i	**B6²**, 790
Ω b654	**—,1,2-dimethoxy-4-nitro-***	$C_8H_9NO_4$. *See* b202	183.17	ye nd (al-w) λ^{al} 322.5 (3.90)	98	230^{15-20}	1.1888^{133}_{4}	...	i	v	v	...	...	chl s lig δ	**B6²**, 790
b655	**—,1,3-dimethoxy-2-nitro-***	$C_8H_9NO_4$. *See* b202	183.17	ye nd (al or aa)	131	...	...	...	i	s^h	δ	...	s	chl s aa s^h lig i	**B6¹**, 404
b656	**—,1,3-dimethoxy-5-nitro-***	$C_8H_9NO_4$. *See* b202	183.17	pa ye nd (AcOEt)	89	...	1.1693^{132}_{4}	...	i	s	δ	...	s	os s lig i	**B6²**, 822
Ω b657	**—,1,4-dimethoxy-2-nitro-***	$C_8H_9NO_4$. *See* b202	183.17	gold-ye nd (dil al) $\lambda^{aq\ MeOH}$ 273 (3.52), 372 (3.42)	72–3	169^{13} sub	1.1666^{132}_{4}	...	i	s^h	...	...	s	chl, sulf s	**B6²**, 849
Ω b658	**—,2,4-dimethoxy-1-nitro-***	$C_8H_9NO_4$. *See* b202	183.17	nd (al) λ^{al} 322.5 (3.90)	76–7	...	1.1876^{132}_{4}	...	i	s	...	...	...	os s lig i	**B6²**, 822
b659	**—,1,2-dimethoxy-4-propenyl-(*cis*)***	Isoeugenol methyl ether. 4-Propenylveratrol. $C_{11}H_{14}O_2$. *See* b202	178.23	λ^{al} 256 (4.16), 289 (3.58)	...	270.5^{760} $138–40^{12}$	1.0521^{20}_{4}	1.5616^{20}	i	...	...	s	s	...	**B6**, 956
Ω b660	**—,1,2-dimethyl-***	*o*-Xylene. C_8H_{10}. *See* b202	106.17	λ^{cy} 265 (2.3), 271 (2.2)	−25.18	144.4^{760} 32^{10}	0.8802^{20}_{4}	1.5055^{20}	i	∞	∞	∞	∞	peth, CCl_4 ∞	**B5²**, 281
Ω b661	**—,1,3-dimethyl-***	*m*-Xylene. C_8H_{10}. *See* b202	106.17	λ^{cy} 269 (2.3), 274 (2.3)	−47.87	139.1^{760} 28.1^{10}	0.8642^{20}_{4}	1.4972^{20}	i	∞	∞	∞	∞	os, peth ∞	**B5²**, 287
Ω b662	**—,1,4-dimethyl-***	*p*-Xylene. C_8H_{10}. *See* b202	106.17	mcl pr (al) λ^{al} 267 (2.7), 275 (2.7)	13.26	138.35^{760} 27.2^{10}	0.8611^{20}_{4}	1.4958^{20}	i	∞	∞	∞	∞	os, peth ∞	**B5²**, 296
b663	**—,1,2-dimethyl-3,4-dinitro-***	3,4-Dinitro-*o*-xylene. $C_8H_8N_2O_4$. *See* b202	196.17	nd (al)	82	...	...	...	...	δ s^h	s	...	s	CS_2, CCl_4 os s^h	**B5²**, 287
b664	**—,1,2-dimethyl-3,5-dinitro-***	3,5-Dinitro-*o*-xylene. $C_8H_8N_2O_4$. *See* b202	196.17	ye nd (al or peth)	77	...	...	...	...	s	...	s	s	chl s	**B5¹**, 181

For explanations, symbols and abbreviations see beginning of table. For structural formulas see end of table.

No.	Name	Synonyms and Formula	Mol. wt.	Color, crystalline form, specific rotation and λ_{max} (log ε)	m.p. °C	b.p. °C	Density	n_D	Solubility w	al	eth	ace	bz	other solvents	Ref.
	Benzene														
b665	**—,1,2-dimethyl-4,5-dinitro-***	4,5-Dinitro-*o*-xylene. $C_8H_8N_2O_4$. *See* b202	196.17	nd (al, bz or aa)	118	....	....	....	δ^h	s^h	s	s	s	chl CS_2 s peth δ os δ	**B5**[1], 181
b666	**—,1,3-dimethyl-2,5-dinitro-***	2,5-Dinitro-*m*-xylene. $C_8H_8N_2O_4$. *See* b202	196.17	ye (al)	101	....	....	....	...	s	s	s	s	chl, CS_2 s	**B5**[2], 295
b667	**—,1,4-dimethyl-2,3-dinitro-***	2,3-Dinitro-*p*-xylene. $C_8H_8N_2O_4$. *See* b202	196.17	mcl pr (al) λ^{hx} 295 (3.16)	93	....	....	....	i	v^h	s	s	s	chl s	**B5**[2], 302
b668	**—,1,4-dimethyl-2,5-dinitro-***	2,5-Dinitro-*p*-xylene. $C_8H_8N_2O_4$. *See* b202	196.17	ye nd (al)	147–8 (142)	....	....	....	i	s^h	δ	s	s	chl s	**B5**[1], 188
b669	**—,2,3-dimethyl-1,4-dinitro-***	3,6-Dinitro-*o*-xylene. $C_8H_8N_2O_4$. *See* b202	196.17	nd (al)	89–90	....	....	....	i	s	s	s	s	peth, chl s	**B5**[1], 181
b670	**—,2,5-dimethyl-1,3-dinitro-***	2,6-Dinitro-*p*-xylene. $C_8N_8N_2O_4$. *See* b202	196.17	nd (al) λ^{hx} 295 (3.18)	123–4	....	....	....	i	s	s^h	s	s	chl s	**B5**[2], 302
b671	**—,1,2-dimethyl-4-ethyl-***	$C_{10}H_{14}$. *See* b202	134.22	λ^{iso} 266 (2.6), 275 (2.7)	−67.0	189.75^{760} 67.8^{10}	0.8745^{20}_4	1.5031^{20}	i	∞	∞	∞	∞	peth, CCl_4 s	**B5**[2], 328
b672	**—,1,3-dimethyl-5-ethyl-***	$C_{10}H_{14}$. *See* b202	134.22	λ^{hx} 200 (4.02)	−84.33	183.75^{760} 63.0^{10}	0.8648^{20}_4	1.4981^{20}	i	∞	∞	∞	∞	peth, CCl_4 s	**B5**[2], 328
b673	**—,1,4-dimethyl-2-ethyl-***	$C_{10}H_{14}$. *See* b202	134.22	λ^{iso} 267 (2.67), 276 (2.70)	−53.68	186.91^{760} 64.9^{10}	0.8772^{20}_4	1.5043^{20}	i	∞	∞	∞	∞	peth, CCl_4 s	**B5**[2], 328
b674	**—,2,4-dimethyl-1-ethyl-***	$C_{10}H_{14}$. *See* b202	134.22	λ^{iso} 267 (2.7), 276 (2.7)	−62.9	188.4^{760} 66.5^{10}	0.8763^{20}_4	1.5038^{20}	i	∞	∞	∞	∞	peth, CCl_4 s	**B5**[2], 328
Ω b675	**—,1,2-dimethyl-3-hydroxy-***	2,3-Dimethylphenol. *o*-3-Xylenol. $C_8H_{10}O$. *See* b202	122.17	nd (w or dil al) λ^{hx} 273 (3.15), 278 (3.19)	75	218^{760} 95.4^{10}	....	1.5420^{20}	δ	s	s	...	...	....	**B6**[2], 453
Ω b676	**—,1,2-dimethyl-4-hydroxy-***	3,4-Dimethylphenol. $C_8H_{10}O$. *See* b202	122.17	nd (w) λ^{hx} 277.5 (3.28), 284 (3.24)	66–8	225^{760} 106.8^{10}	0.9830^{20}_4	....	δ	s	∞	...	...	....	**B6**[2], 455
b677	—,—,acetate	$C_{10}H_{12}O_2$. *See* b202	164.21	....	22	235 140^{80}	....	....	i	s	s	...	s	....	**B6**[2], 455
Ω b678	**—,1,3-dimethyl-2-hydroxy-***	2,6-Dimethylphenol. $C_8H_{10}O$. *See* b202	122.17	lf or nd (al) λ^{al} 271 (3.19), 275.5 (3.16)	49.0	212^{760} 91.2^{10}	....	....	s	s	s	...	...	....	**B6**[2], 457
Ω b679	**—,1,3-dimethyl-5-hydroxy-***	3,5-Dimethylphenol. $C_8H_{10}O$. *See* b202	122.17	nd (w or peth) λ^{cy} 281.5 (3.21)	68	219.5 sub 102.3^{10}	0.9680^{20}_4	....	s	s	...	...	...	....	**B6**[2], 462
Ω b680	**—,1,4-dimethyl-2-hydroxy-***	2,5-Dimethylphenol. $C_8H_{10}O$. *See* b202	122.17	nd (w), pr (al-eth) λ^{cy} 276 (3.28)	75	211.5^{762}	....	....	s	s	v	...	...	....	**B6**[2], 466
b681	—,—,acetate	$C_{10}H_{12}O_2$. *See* b202	164.21	....	< −20	237^{768}	1.0624^{15}	....	i	s	s	...	...	....	**B6**, 495
Ω b682	**—,2,4-dimethyl-1-hydroxy-***	2,4-Dimethylphenol. $C_8H_{10}O$. *See* b202	122.17	nd (w) λ^{cy} 280 (3.31)	27–8	210^{760} 89.3^{10}	0.9650^{20}_4	1.5420^{14}	δ	∞	∞	...	...	....	**B6**[2], 458
b683	—,—,acetate	$C_{10}H_{12}O_2$. *See* b202	164.21	....	....	226 108^{13}	$1.0298^{15.5}_4$	1.4990^{15}	i	s	s	...	...	....	**B6**, 487
b684	**—,1,5-dimethyl-2-hydroxy-3-(hydroxymethyl)-***	$C_9H_{12}O_2$. *See* b202	152.20	nd (bz-peth)	57–8	....	....	....	...	s	s	...	s^h	peth s^h	**B6**, 939
b685	**—,1,2-dimethyl-3-hydroxy-5-nitro-***	$C_8H_9NO_3$. *See* b202	167.17	og-ye nd (bz or w)	109 (120–1)	....	....	....	s^h	v	...	v	s^h	aa v chl s^h peth i	**B6**[2], 455
b686	**—,1,2-dimethyl-4-hydroxy-5-nitro-***	$C_8H_9NO_3$. *See* b202	167.17	ye rh (al) λ^{hx} 283 (3.97), 357 (3.66)	86.8–7.5	....	....	....	...	v	v	...	v	chl v peth δ	**B6**, 484
b687	**—,1,3-dimethyl-2-hydroxy-4-nitro-***	$C_8H_9NO_3$. *See* b202	167.17	lf pr (bz), nd (lig)	99–100	....	....	....	...	v	...	...	s^h	chl v lig s^h	**B6**, 485
Ω b688	**—,1,3-dimethyl-2-hydroxy-5-nitro-***	$C_8H_9NO_3$. *See* b202	167.17	pr (MeOH) λ^{al} 322.5 (4.05)	171	....	....	....	...	s	...	s	δ	chl v MeOH s^h lig δ	**B6**, 486
b689	**—,1,4-dimethyl-2-hydroxy-3-nitro-***	$C_8H_9NO_3$. *See* b202	167.17	nd (peth) λ^{hx} 285 (3.77), 358 (3.38)	34–5	236d 150^{15}	....	....	...	v	s	s	s	os v	**B6**[1], 246
b690	**—,1,4-dimethyl-2-hydroxy-5-nitro-***	$C_8H_9NO_3$. *See* b202	167.17	pa ye nd (dil al) λ^{hx} 234 (3.77), 310 (3.83)	122–3	....	....	....	s^h	s	s	s	...	....	**B6**[1], 246
b691	**—,1,5-dimethyl-2-hydroxy-3-nitro-***	$C_8H_9NO_3$. *See* b202	167.17	ye nd (al) λ^{hx} 282.5 (3.92), 366 (3.58)	73	....	....	....	s^h	...	...	...	...	....	**B6**[2], 461
Ω b692	**—,1,5-dimethyl-3-hydroxy-2-nitro-***	$C_8H_9NO_3$. *See* b202	167.17	ye nd (lig or dil MeOH) λ^{hx} 287 (3.96), 352 (3.57)	66–6.5	....	....	....	...	v	s	s	s	os s lig s^h peth i	**B6**[1], 244
b693	**—,2,5-dimethyl-1-hydroxy-3-nitro-***	$C_8H_9NO_3$. *See* b202	167.17	ye lf (peth)	91	....	....	....	δ	v	v	...	...	peth s^h	**B6**, 497
b694	**—,1,2-dimethyl-4-hydroxy-3,5,6-tribromo-***	$C_8H_7Br_3O$. *See* b202	358.87	nd (al)	173–4	....	....	....	i	s	...	...	...	....	**B6**[1], 240
b695	**—,1,3-dimethyl-5-hydroxy-2,4,6-tribromo-***	$C_8H_7Br_3O$. *See* b202	358.87	nd (al)	166	....	....	....	s	s	...	...	...	....	**B6**[2], 465

For explanations, symbols and abbreviations see beginning of table. For structural formulas see end of table.

No.	Name	Synonyms and Formula	Mol. wt.	Color, crystalline form, specific rotation and λ_{max} (log ε)	m.p. °C	b.p. °C	Density	n_D	Solubility w	al	eth	ace	bz	other solvents	Ref.
	Benzene														
b696	—,1,2-dimethyl-3-hydroxy-4,5,6-trichloro-*	$C_8H_7Cl_3O$. *See* b202	225.50	nd (dil al or peth)	180–1	...	...	...	...	v	...	...	...	aa s^h peth δ	**B6**^{2}, 454
b697	—,1,2-dimethyl-4-hydroxy-3,5,6-trichloro-*	$C_8H_7Cl_3O$. *See* b202	225.50	nd (peth)	182.5	...	...	...	...	v	...	...	...	peth v^h aa s^h	**B6**^{2}, 456
b698	—,1,3-dimethyl-4-hydroxy-2,5,6-trichloro-*	$C_8H_7Cl_3O$. *See* b202	225.50	pa ye nd	174	...	...	...	...	...	s	...	...	...	**B6**^{2}, 460
Ω b699	—,1,3-dimethyl-5-hydroxy-2,4,6-trichloro-*	$C_8H_7Cl_3O$. *See* b202	225.50	ye nd (peth) λ^{cy} 290 (3.2)	117–8	...	...	...	...	...	...	...	...	peth v^h	**B6**^{3}, 1761
b700	—,1,4-dimethyl-2-hydroxy-3,5,6-trichloro-*	$C_8H_7Cl_3O$. *See* b202	225.50	pa gr nd (al, aq MeOH)	175–6	...	...	...	i	s	v	...	v	chl v	**B6**^{2}, 467
b701	—,1,2-dimethyl-3-iodo-*	C_8H_9I. *See* b202	232.07	...	...	228–230.6 125–6^{15}	1.6395^{20}_4	1.6074^{20}	...	...	...	s	...	...	**B5**^{2}, 286
b702	—,1,2-dimethyl-4-iodo-*	C_8H_9I. *See* b202	232.07	...	...	231–2.4 111^{11}	1.6334^{18}_4	1.6049^{18}	...	...	...	s	...	...	**B5**^{2}, 286
b703	—,1,3-dimethyl-2-iodo-*	C_8H_9I. *See* b202	232.07	oil	11.2	229.8–30.8 102.3^{14}	1.6158^{20}_4	1.6035^{20}	i	...	...	s	s	...	**B5**, 375
b704	—,1,3-dimethyl-5-iodo-*	C_8H_9I. *See* b202	232.07	oil	...	230.2–31.8 117^{27}	$1.6085^{18.5}_4$	$1.5967^{18.5}$	i	...	...	s	...	os s	**B5**^{2}, 294
b705	—,1,4-dimethyl-2-iodo-*	C_8H_9I. *See* b202	232.07	...	...	227–8δd 106–8^{13}	1.6168^{17}_4	1.5992^{17}	i	...	...	s	s	...	**B5**^{2}, 302
Ω b706	—,2,4-dimethyl-1-iodo-*	C_8H_9I. *See* b202	232.07	...	...	230.7–31.7δd 111^{14}	1.6282^{16}_4	1.6008^{16}	i	...	...	s	s	...	**B5**^{2}, 294
Ω b707	—,1,2-dimethyl-3-methoxy-*	$C_9H_{12}O$. *See* b202	136.20	λ^{cy} 270.5 (3.1), 274 (3.2), 278.5 (3.2)	29	199 85^{18}	0.9596^{40}	1.5120^{40}	i	v	v	s	v	...	**B6**^{2}, 453
Ω b708	—,1,2-dimethyl-4-methoxy-*	$C_9H_{12}O$. *See* b202	136.20	λ^{cy} 278.5 (3.3), 284.7 (3.3)	...	204–5 96–7^{17}	0.9744^{14}_4	1.5198^{14}	i	s	s	...	s	...	**B6**^{2}, 455
Ω b709	—,1,3-dimethyl-2-methoxy-*	$C_9H_{12}O$. *See* b202	136.20	λ^{al} 262 (2.50), 266 (2.49), 270 (2.42)	...	182–3	0.9619^{14}_4	1.5053^{14}	i	s	s	...	s	...	**B6**^{2}, 457
Ω b710	—,1,3-dimethyl-5-methoxy-*	$C_9H_{12}O$. *See* b202	136.20	λ^{cy} 220 (3.86), 264 sh (2.95), 271 (3.16), 278 (3.23)	...	194.5 89^{15}	0.9627^{15}_4	1.5110^{20}	i	s	s	...	s	CS_2, aa s	**B6**^{2}, 462
Ω b711	—,1,4-dimethyl-2-methoxy-*	$C_9H_{12}O$. *See* b202	136.20	λ^{cy} 219 (3.86), 270 sh (3.28), 273 (3.32), 278.5 (3.33)	...	194^{772}	0.9693^{13}_4	1.5182^{13}	i	s	s	...	s	CS_2, peth s	**B6**^{2}, 466
Ω b712	—,2,4-dimethyl-1-methoxy-*	$C_9H_{12}O$. *See* b202	136.20	...	...	192 83–4^{15}	0.9740^{16}_4	1.5190^{16}	i	s	s	...	s	CS_2 s	**B6**^{2}, 459
Ω b713	—,1,2-dimethyl-3-nitro-*	$C_8H_9NO_2$. *See* b202	151.17	nd (al) λ^{iso} 251 (3.62)	15	240^{760} 131^{20}	1.1402^{20}	1.5441^{20}	i	s	...	...	...	...	**B5**^{2}, 286
Ω b714	—,1,2-dimethyl-4-nitro-	$C_8H_9NO_2$. *See* b202	151.17	ye pr (al) λ^{al} 278 (3.96)	30–1	254^{750} (231^{760}) 143^{21}	...	...	i	∞^h	...	...	...	...	**B5**^{2}, 286
Ω b715	—,1,3-dimethyl-2-nitro-*	$C_8H_9NO_2$. *See* b202	151.17	...	13	222^{760} 84.6$^{0.05}$	1.112^{15}	1.5202^{20}	i	v	...	...	...	...	**B5**^{2}, 294
b716	—,1,3-dimethyl-5-nitro-*	$C_8H_9NO_2$. *See* b202	151.17	nd (al) λ^{al} 271 (3.87)	75	273^{739}	...	...	i	v	v	...	...	...	**B5**^{2}, 295
Ω b717	—,1,4-dimethyl-2-nitro-*	$C_8H_9NO_2$. *See* b202	151.17	pa ye	...	239.5–41^{760} 64–5$^{0.35}$	1.132^{15}	1.5415^{20}	i	s	...	...	...	...	**B5**^{2}, 302
Ω b718	—,2,4-dimethyl-1-nitro-*	$C_8H_9NO_2$. *See* b202	151.17	...	9	245^{744} 122^{18}	$1.126^{17.5}$	...	i	s	s	...	...	...	**B5**^{2}, 294
b721	—,1,2-dimethyl-3,4,5,6-tetrabromo-*	$C_8H_6Br_4$. *See* b202	421.77	nd (bz)	262	374–5	...	...	i	δ^h	...	...	s^h	...	**B5**^{2}, 285
b722	—,1,3-dimethyl-2,4,5,6-tetrabromo-*	$C_8H_6Br_4$. *See* b202	421.77	nd (xyl or al)	248 (252)	...	...	...	i	i	...	...	v	xyl s^h	**B5**^{2}, 294
b723	—,1,3-dimethyl-2,4,6-trinitro-*	$C_8H_7N_3O_6$. *See* b202	241.16	pa ye pr or lf (al-bz) λ^{Py} 413 (4.22), 600 (4.00)	184	...	1.604^{19}	...	i	δ s^h	δ	s^h	v	Py v, chl v^h AcOEt s HNO_3 δ^h	**B5**^{2}, 295
b724	—,1,4-dimethyl-2,3,5-trinitro-*	$C_8H_7N_3O_6$. *See* b202	241.16	mcl nd (al), lf (al-bz)	139–40	410 exp	1.59^{19}	...	i	s	...	...	...	...	**B5**^{2}, 303
Ω b725	—,1,2-dinitro-*	$C_6H_4N_2O_4$. *See* b202	168.11	nd (bz), pl (al)	118.5	319^{773} 194^{30}	1.3119^{120}_4 1.565^{17}	...	i	s v^h	...	...	s	chl, MeOH, AcOEt, to s	**B5**^{2}, 193

For explanations, symbols and abbreviations see beginning of table. For structural formulas see end of table.

No.	Name	Synonyms and Formula	Mol. wt.	Color, crystalline form, specific rotation and λ_{max} (log ε)	m.p. °C	b.p. °C	Density	n_D	Solubility w	al	eth	ace	bz	other solvents	Ref.
	Benzene														
Ω b726	**—,1,3-dinitro-***	$C_6H_4N_2O_4$. *See* b202	168.11	rh pl (al) λ^{al} 242 (4.21), 305 (3.04)	90.02	291^{756} 167^{14}	1.575_4^{18}	....	δ^h	v^h	s	v	v^h	Py v to, chl, MeOH s	**B5²**, 193
Ω b727	**—,1,4-dinitro-***	$C_6H_4N_2O_4$. *See* b202	168.11	nd (al) λ^{al} 266 (4.16)	174	299^{777} 183^{34} sub	1.625_4^{18}		i	δ	...	s	s	aa, to, AcOEt s chl δ	**B5²**, 195
b728	**—,1,3-dinitro-2-ethoxy-***	$C_8H_8N_2O_5$. *See* b202	212.17	nd (eth)	59.5–60.5	$137–9^3$	....	....	i	δ	s	...	...	....	**B6¹**, 127
b729	**—,1,3-dinitro-5-ethoxy-***	$C_8H_8N_2O_5$. *See* b202	212.17	nd (al)	97.5	....	....	....	δ	δ	s	...	...	....	**B6**, 258
b730	**—,1,4-dinitro-2-ethoxy-***	$C_8H_8N_2O_5$. *See* b202	212.17	lf (al)	96–8 (85)	....	....	....	i	δ	δ	...	...	....	**B6²**, 245
Ω b731	**—,2,4-dinitro-1-ethoxy-***	$C_8H_8N_2O_5$. *See* b202	212.17	nd or lf (al) λ^{cy} 283	86–7	....	....	....	i	δ s^h	...	s	δ	....	**B6³**, 858
Ω b732	**—,2,4-dinitro-1-fluoro-***	$C_6H_3FN_2O_4$. *See* b202	186.10	pa ye (al) λ^{iso} 260 (3.8), 295 sh (2.9)	25.8	296 178^{25}	1.4718^{84}	....	...	s^h	...	...	...	....	**B5³**, 634
b733	**—,1,3-dinitro-5-isopropyl-4-hydroxy-6-methyl-**	2,6-Dinitrothymol. $C_{10}H_{12}N_2O_5$. *See* b202	240.22	pr (peth)	55.5	....	....	....	i	s	s	...	δ	lig s^h	**C37**, 6254
b734	**—,1,2-dinitro-3-methoxy-***	$C_7H_6N_2O_5$. *See* b202	198.14	nd (al), pl (to)	119	....	1.2290^{137} 1.524^{20}	....	i	s^h	...	...	...	lig s^h	**B6²**, 239
b735	**—,1,2-dinitro-4-methoxy-***	$C_7H_6N_2O_5$. *See* b202	198.14	ye nd (dil al)	71	....	1.3332^{110}	....	...	s	...	...	...	CS_2, MeOH s	**B6¹**, 127
b736	**—,1,3-dinitro-2-methoxy-***	$C_7H_6N_2O_5$. *See* b202	198.14	nd (al, MeOH)	118	....	1.3000^{128} 1.319^{20}	....	...	s	...	...	...	CS_2, MeOH s	**B6²**, 245
b737	**—,1,3-dinitro-5-methoxy-***	$C_7H_6N_2O_5$. *See* b202	198.14	nd (al) λ^{w} 254 (4.12), 340 (3.45)	105.5	....	1.558^{12}	....	δ^h	δ	...	s	s	CS_2, MeOH s	**B6³**, 869
b738	**—,1,4-dinitro-2-methoxy-***	$C_7H_6N_2O_5$. *See* b202	198.14	nd (bz-lig) $\lambda^{90\% al}$ 220 (4.1), 310 (4.02)	97	>360 $136–8^2$	1.476^{18}	....	δ	s^h	...	s	s	CS_2, MeOH s lig δ^h	**B6¹**, 127
Ω b739	**—,2,4-dinitro-1-methoxy-***	$C_7H_6N_2O_5$. *See* b202	198.14	nd (al or w)	(i) 94.5–5.5 (st) (ii) 87–8 (unst)	$206–7^{12}$ sub	1.3364^{131} 1.546^{15}	....	δ^h	s^h	s	s	s	to, Py v CS_2 s	**B6²**, 241
Ω b740	**—,1,4-dinitro-2,3,5,6-tetramethyl-***	Dinitrodurene. $C_{10}H_{12}N_2O_4$. *See* b202	224.22	pr (al)	211–2	sub	....	....	i	δ^h	v	...	s	....	**B5²**, 330
b741	**—,2,4-dinitro-1,3,5-trimethyl-***	$C_9H_{10}N_2O_4$. *See* b202	210.19	rh (al) λ^{cy} 238 sh (3.59)	86	418 exp	....	....	i	s^h	...	...	...	....	**B5²**, 316
—	**—,diphenyl-***	*see* **Terphenyl**													
Ω b742	**—,1,2-dipropoxy-***	Catechol dipropyl ether. $C_{12}H_{18}O_2$. *See* b202	194.28	....	....	234–7 $117–20^{12}$	0.9554_4^{33}	1.4950^{27}	...	...	...	...	...	....	**C45**, 1974
b743	**—,1,3-dipropoxy-***	Resorcinol dipropyl ether. $C_{12}H_{18}O_2$. *See* b202	194.28	....	....	251 $127–8^{12}$	1.035_{21}^{20}	1.5138^{83}	...	...	...	...	...	....	**B6²**, 815
Ω b744	**—,1,3-dipropyl-2-hydroxy-***	$C_{12}H_{18}O$. *See* b202	178.28	....	28	256 $114–6^5$	....	....	s	...	s	...	...	....	**B6²**, 511
Ω b745	**—,(1,2-epoxy-ethyl)-***	Styrene oxide. C_8H_8O. *See* b202	120.16	λ^{al} 250 sh (2.2), 254 (2.24), 260 (2.28), 265 sh (2.10)	....	194.1 $84–5^{15}$	1.0523_4^{16}	1.5342^{20}	i	s	s	...	...	....	**B17**, 49
—	**—,ethenyl-***	*see* **Styrene**													
Ω b746	**—,ethoxy-***	Phenetole. $C_8H_{10}O$. *See* b202	122.17	λ^{hx} 271 (3.29)	−29.5	170^{760} 60^9	0.9666_4^{20}	1.5076^{20}	i	s	s	...	...	....	**B6**, 140
b747	**—,1-ethoxy-4-ethyl-***	$C_{10}H_{14}O$. *See* b202	150.22	....	....	211 $92–3^{12}$	0.9385_4^{17}	....	i	s	...	s	s	....	**B6¹**, 234
Ω b748	**—,1-ethoxy-2-fluoro-***	C_8H_9FO. *See* b202	140.16	....	−16.7	171.4 64^{11}	1.0874^{17}	1.4932^{17}	i	...	...	s	s	....	**B6²**, 169
b749	**—,1-ethoxy-3-fluoro-***	C_8H_9FO. *See* b202	140.16	....	−27.5	171.4^{755} 65.2^{15}	1.0716_4^{16}	1.4847^{17}	i	...	...	...	s	....	**B6²**, 170
Ω b750	**—,1-ethoxy-4-fluoro-***	C_8H_9FO. *See* b202	140.16	....	−8.5	173^{766} 54^7	1.0715^{18}	1.4826^{18}	i	...	...	...	s	chl s	**B6²**, 170
b751	**—,1-ethoxy-2-iodo-***	C_8H_9IO. *See* b202	248.07	....	....	245^{736} $121–31^{18}$	....	....	i	s	s	s	s	os s	**B6**, 207
b752	**—,1-ethoxy-4-iodo-***	C_8H_9IO. *See* b202	248.07	cr (dil MeOH)	29	$249–50^{729}$	....	....	i	s	v	...	s	chl v	**B6**, 208
b753	**—,1-ethoxy-3-mercapto-***	$C_8H_{10}OS$. *See* b202	154.23	....	....	238–9	....	....	...	s	s	s	...	CS_2, alk s	**B6**, 833
b754	**—,1-ethoxy-4-mercapto-***	$C_8H_{10}OS$. *See* b202	154.23	....	1.6	238	....	....	...	s	s	s	s	CS_2 s	**B6²**, 852
b755	**—,1-ethoxy-2-nitro-***	$C_8H_9NO_3$. *See* b202	167.17	br ye	2.1	267 149^{16}	1.1903^{15}	1.5425^{20}	i	v	v	...	...	....	**B6²**, 210
b756	**—,1-ethoxy-3-nitro-***	$C_8H_9NO_3$. *See* b202	167.17	br ye	36	284 (264δd), 169^{70}d	....	....	i	s	s	...	...	....	**B6²**, 214
Ω b757	**—,1-ethoxy-4-nitro-***	$C_8H_9NO_3$. *See* b202	167.17	pr (al-w or eth) λ^{hx} 295 (4.12)	60 (57–8)	283 168^{15}	1.1176_4^{100}	....	δ	v^h δ	v	∞	∞	peth s^h	**B6²**, 221

For explanations, symbols and abbreviations see beginning of table. For structural formulas see end of table.

No.	Name	Synonyms and Formula	Mol. wt.	Color, crystalline form, specific rotation and λ_{max} (log ε)	m.p. °C	b.p. °C	Density	n_D	Solubility w	al	eth	ace	bz	other solvents	Ref.
	Benzene														
Ω b758	—,ethyl-*	C_8H_{10}. *See* b202	106.17	λ^{al} 206 (4.5), 259 (2.2)	−94.97	136.2^{760} 25.8^{10}	0.8670^{20}_{4}	1.4959^{20}	i	∞	∞	...	...	....	**B5**[2], 274
Ω b759	—,1-ethyl-2-hydroxy-*	2-Ethylphenol. Phlorol. $C_8H_{10}O$. *See* b202	122.17	λ^{cy} 270 (3.30), 276 (3.28)	< −18	207^{760} 84.1^{10}	1.0371^{0}	1.5367^{20}	δ	∞	∞	s	s	....	**B6**, 442
Ω b760	—,1-ethyl-3-hydroxy-*	3-Ethylphenol. $C_8H_{10}O$. *See* b202	122.17	λ^{cy} 273 (3.25)	−4	214^{760} 99.3^{10}	1.0283^{20}_{4}	....	δ	v	v	...	...	....	**B6**, 471
Ω b761	—,1-ethyl-4-hydroxy-*	4-Ethylphenol. $C_8H_{10}O$. *See* b202	122.17	nd λ^{cy} 273 (3.26), 276 (3.34), 282 (3.30)	47–8	219^{760} 99.5^{10}	....	1.5239^{25}	δ	v	v	s	v	CS_2 v	**B6**, 472
b762	—,1-ethyl-2-iodo-*	C_8H_9I. *See* b202	232.07	....	....	226	1.6189^{16}_{4}	1.5941^{22}	i	...	...	s	s	os s	**B5**[1], 177
b763	—,1-ethyl-4-iodo-*	C_8H_9I. *See* b202	232.07	....	−17	209	1.6095^{16}_{4}	1.5909^{22}	i	...	...	s	s	os s	**B5**[1], 178
b764	—,1-ethyl-4-isobutyl-	$C_{12}H_{18}$. *See* b202	162.28	....	....	211	....	....	i	...	s	s	...	....	**C46**, 11144
b765	—,1-ethyl-3-isopropyl-	$C_{11}H_{16}$. *See* b202	148.25	....	< −20	192.0^{760} 69.2^{10}	0.859^{20}_{4}	1.4921^{20}	i	s	s	...	v	....	**B5**, 440
b766	—,1-ethyl-4-isopropyl-	$C_{11}H_{16}$. *See* b202	148.25	....	< −20	136.6^{760} 73^{10}	0.8585^{20}_{4}	1.4923^{20}	i	...	s	...	v	....	**B5**[2], 334
b767	—,1-ethyl-2-methoxy-*	$C_9H_{12}O$. *See* b202	136.20	....	....	$186–8^{755}$ 80^{14}	0.9636^{19}_{4}	1.5142^{20}	i	...	v	...	s	....	**B6**, 471
Ω b768	—,1-ethyl-3-methoxy-*	$C_9H_{12}O$. *See* b202	136.20	....	....	$196–7^{758}$ 74^{10}	0.9575^{18}_{4}	1.5102	i	...	v	...	s	....	**B6**, 472
b769	—,1-ethyl-4-methoxy-*	$C_9H_{12}O$. *See* b202	136.20	....	....	195–6 $83–4^{16}$	0.9624^{15}_{4}	1.5120^{20}	i	...	v	...	s	....	**B6**, 472
Ω b770	—,1-ethyl-2-methyl-*	C_9H_{12}. *See* b202	120.20	λ^{iso} 263.5 (2.40), 271 (2.32)	−80.83	165.2^{760} 62.3^{20}	0.8807^{20}_{4}	1.5046^{20}	i	∞	∞	∞	∞	peth, CCl_4 ∞	**B5**, 396
b771	—,1-ethyl-3-methyl-*	C_9H_{12}. *See* b202	120.20	λ^{iso} 268.5 (2.31), 272 (2.40)	−95.55	161.3^{760} 45.6^{10}	0.8645^{20}_{4}	1.4966^{20}	i	v	v	∞	∞	peth, CCl_4 ∞	**B5**[2], 309
b772	—,1-ethyl-4-methyl-*	C_9H_{12}. *See* b202	120.20	λ^{iso} 264 (2.53), 268 (2.66), 272 (2.33)	−62.35	162^{760} 45.6^{10}	0.8614^{20}_{4}	1.4959^{20}	i	v	v	∞	∞	peth, CCl_4 ∞	**B5**[2], 310
Ω b773	—,1-ethyl-2-nitro-*	$C_8H_9NO_2$. *See* b202	151.17	λ^{iso} 250 (3.72)	−23	228 116^{22}	1.1139^{20}_{4}	1.5354^{20}	i	v	v	s	...	....	**B5**[2], 279
b774	—,1-ethyl-3-nitro-*	$C_8H_9NO_2$. *See* b202	151.17	....	....	242–3	1.1345^{0}	....	i	v	v	s	...	....	**B5**, 358
Ω b775	—,1-ethyl-4-nitro-*	$C_8H_9NO_2$. *See* b202	151.17	λ^{gas} 251 (3.97)	−12.3 (−32)	245–6 $134–6^{23}$	1.1192^{20}_{4}	1.5455^{20}	i	v	v	s	...	....	**B5**[2], 279
b776	—,1-ethyl-4-propyl-*	$C_{11}H_{16}$. *See* b202	148.25	....	....	205^{760}	0.8594^{20}_{4}	1.4921^{20}	i	v	s	s	∞	....	**B5**, 439
b777	—,(ethylthio-)*	Ethyl phenyl sulfide. $C_8H_{10}S$. *See* b202	138.23	λ^{hx} 257 (3.89)	....	205^{760} 84^{10}	1.0211^{20}_{4}	1.5670^{20}	...	s	...	...	...	....	**B6**[2], 287
Ω b778	—,ethynyl-*	Phenylacetylene. C_8H_6. *See* b202	102.14	λ^{al} 235 (4.18), 244 (4.13)	−44.8	142–4 44^{18}	0.9281^{20}_{4}	1.5485^{20}	i	∞	∞	s	...	....	**B5**[2], 406
Ω b779	—,fluoro-*	C_6H_5F. *See* b202	96.11	λ^{al} 254 (3.00), 259 (3.11), 266 (3.08)	−41.2 fr −39.2	85.1^{760}	1.0225^{20}_{4}	1.4684^{40}	i	∞	∞	∞	∞	CCl_4, lig ∞	**B5**[2], 147
Ω b780	—,1-fluoro-4-iodo-*	C_6H_4FI. *See* b202	222.00	λ^{cy} 266 (3.27), 271 sh (3.20), 280 sh (3.00)	(i) −27.2 (ii) −18	182–4 $67–9^{11}$	1.9523^{15}	1.5270^{22}	i	s	s	s	...	....	**B5**[2], 167
Ω b781	—,1-fluoro-2-methoxy-*	C_7H_7FO. *See* b202	126.13	λ^{cy} 268 (3.23), 293 (3.16)	−39	154–5 59.2^{12}	1.5489^{17}	$1.4969^{17.5}$	i	...	s	...	...	....	**B6**[2], 169
Ω b782	—,1-fluoro-4-methoxy-*	C_7H_7FO. *See* b202	126.13	λ^{cy} 275 (3.44), 278.5 (3.51), 284.5 (3.44)	−45	154^{760}	1.178^{18}_{4}	1.4886^{18}	...	...	s	...	...	....	**B6**[1], 98
Ω b783	—,1-fluoro-2-nitro-*	$C_6H_4FNO_2$. *See* b202	141.10	ye	−5.9	214.6d $86–7^{11}$	1.3285^{18}_{4}	1.5489^{17}	i	s	s	...	...	....	**B5**[2], 180
Ω b784	—,1-fluoro-3-nitro-*	$C_6H_4FNO_2$. *See* b202	141.10	ye λ^{al} 254 (3.8), 290 sh (3.3)	(i) 41 (st) (ii) 3.1 (unst) (iii) 3.6 (unst)	198–200^{760} 86^{19}	1.3254^{19}_{4}	1.5262^{15}	i	s	s	...	...	....	**B5**[2], 180
Ω b785	—,1-fluoro-4-nitro-*	$C_6H_4FNO_2$. *See* b202	141.10	ye nd λ^{al} 212 (4.0), 263 (3.9)	(i) 27 (st) (ii) 21.5 (unst)	206–7 87^{14}	1.3300^{20}_{4}	1.5316^{20}	i	s	s	...	...	....	**B5**[2], 180
b786	—,hexabromo-*	C_6Br_6. *See* b202	551.52	mcl nd (bz)	327 (306)	....	....	....	i	δ	δ	...	s^h	chl, peth s aa s^h	**B5**[2], 164
Ω b787	—,hexachloro-*	C_6Cl_6. *See* b202	284.79	nd (bz-al) λ^{iso} 291 (2.32), 301 (2.25)	230	322 sub	$1.5691^{23.6}$	....	i	δ	s	...	v^h	chl s	**B5**[2], 157
Ω b788	—,hexaethyl-*	$C_{18}H_{30}$. *See* b202	246.44	mcl pr (al or bz) λ^{iso} 272.5 (2.36), 276 sh (2.33), 281 (2.25)	129	298	0.8305^{130}	1.4736^{130}	i	s^h	v	...	v	sulf s	**B5**[2], 358
Ω b789	—,hexahydroxy-*	Hexaphenol-. $C_6H_6O_6$. *See* b202	174.11	nd (w)	>300	....	....	....	δ	δ	δ	...	δ	....	**B5**[2], 116
b790	—,hexaiodo-*	C_6I_6. *See* b202	833.49	red-br nd or mcl pr (bz)	350d	....	....	....	i	i	i	...	...	$PhNO_2$ s^h	**B5**[2], 169

For explanations, symbols and abbreviations see beginning of table. For structural formulas see end of table.

No.	Name	Synonyms and Formula	Mol. wt.	Color, crystalline form, specific rotation and λ_{max} (log ε)	m.p. °C	b.p. °C	Density	n_D	Solubility w	al	eth	ace	bz	other solvents	Ref.
	Benzene														
Ω b791	**—,hexamethyl-***	Mellitene. $C_{12}H_{18}$. *See* b202	162.28	rh pr or nd (al) λ^{al} 275 (2.26)	166–7	265	1.0630^{25}	...	i	v^h s	s	s	v^h s	aa s	**B5**^{2}, 34
—	**—,hydroxy-***	*see* **Phenol***													
b792	**—,1-hydroxy-2-hydroxymethyl-4-methyl-***	Homosaligenine. $C_8H_{10}O_2$. *See* b202	138.17	lf (w or chl), pr (al), lf (w)	106–7	...	...	...	v^h s	v	v	...	...	...	**B6**^{2}, 889
b793	**—,1-hydroxy-4-isobutyl-**	$C_{10}H_{14}O$. *See* b202	150.22	...	51–2	235–9 $125–6^{11}$	0.9796^{20}_{20}	1.5319^{25}	δ	s	s	...	s	...	**B6**^{3}, 1859
Ω b794	**—,1-hydroxy-2-isopropyl-**	*o*-Cumenol. $C_9H_{12}O$. *See* b202	136.20	λ^{al} 271 (3.55), 276 sh (3.30)	15–6	$213–4^{760}$	1.012^{20}_4	1.5315^{20}	δ	s	s	...	s	...	**B6**, 504
b795	**—,1-hydroxy-3-isopropyl-**	$C_9H_{12}O$. *See* b202	136.20	...	26	228	...	1.5261^{20}	δ	...	s	...	...	...	**B6**, 505
Ω b796	**—,1-hydroxy-4-isopropyl-**	$C_9H_{12}O$. *See* b202	136.20	nd (peth) λ^{cy} 272 sh (3.23), 275 (3.32), 281 (3.28)	62–3	228–30^{745} $109–10^{10}$	0.990^{20}	1.5228^{20}	δ	s	...	...	...	...	**B6**, 505
Ω b797	**—,2-hydroxy-1-isopropyl-4-methyl-**	Thymol. $C_{10}H_{14}O$. *See* b202	150.22	pl (aa or ace) λ^{cy} 276 (3.34), 282 (3.34)	52	233 $92^{2–8}$	0.925^{80}_4	1.5227^{20}	i	v	v	...	...	AcOEt, chl, aa v	**B6**^{2}, 494
b798	—,—,acetate	$C_{12}H_{16}O_2$. *See* b202	192.24	...	...	245 131^{21}	1.009^0	...	i	∞	∞	...	∞	chl ∞	**B6**^{2}, 499
b799	—,—,carbonate	Thymotal. Tyratol. $C_{21}H_{26}O_3$. *See* b202	326.44	nd or pr	49	...	...	...	i	v^h	s	...	...	chl s	**B6**^{2}, 499
Ω b800	**—,2-hydroxy-4-isopropyl-1-methyl-**	Carvacrol. $C_{10}H_{14}O$. *See* b202	150.22	nd λ^{al} 277.5 (3.26)	1	237.7 $101–2.5^{10}$	0.9772^{20}_4	1.5230^{20}	δ	s	s	v	...	alk s	**B6**^{2}, 492
b801	—,—,acetate	$C_{12}H_{16}O_2$. *See* b202	192.24	...	...	245–8	0.9896^{25}	1.4913^{28}	...	s	s	...	...	...	**B6**^{2}, 494
b802	**—,1-hydroxy-2-isopropyl-5-methyl-4-nitroso-**	$C_{10}H_{13}NO_2$. *See* b202	179.22	yesh red	175	...	...	...	i	v	v	...	...	chl v	**B7**^{2}, 596
b803	**—,1-hydroxy-5-isopropyl-2-methyl-3-nitroso-**	$C_{10}H_{13}NO_2$. *See* b202	179.22	yesh pr (bz), nd (dil al)	153	...	...	...	i	v	v	...	v	chl v	**B7**^{2}, 596
b804	**—,hydroxylamino-***	Phenylhydroxylamine. C_6H_7NO. *See* b202	109.13	nd (w, bz or peth) λ^{al} 236 (3.8), 279 (3.0)	83–4	...	...	...	s^h	v	v	...	v^h	chl v lig i	**B15**^{2}, 4
b805	—,—,*N*-nitroso-	$C_6H_6N_2O_2$. *See* b202	138.13	nd (lig)	59	...	...	...	i	v	v	...	...	os v	**B16**^{2}, 344
b806	**—,1-hydroxy-2-methoxy-4-propenyl-(*cis*)***	Isoeugenol (*cis*). $C_{10}H_{12}O_2$. *See* b202	164.21	λ^{al} 258 (4.14), 292 (3.58)	...	$134–5^{13}$ $80–1^{0.5}$	1.0837^{20}_4	1.5726^{20}	δ	s	s	...	...	...	**B6**, 955
b807	—,—,acetate	$C_{12}H_{14}O_3$. *See* b202	206.24	...	...	$160–2^{13}$	1.0947^{19}_4	1.5418^{20}	i	...	s	...	...	...	**B6**^{3}, 5007
Ω b808	—,—(*trans*)*	Isoeugenol (*trans*). $C_{10}H_{12}O_2$. *See* b202	164.21	λ^{al} 260 (4.22), 267 (4.18), 300 sh (3.70)	33–4	$141–2^{13}$	1.0852^{20}_4	1.5784^{20}	δ	s	s	...	...	...	**B6**, 955
b809	—,—,acetate	$C_{12}H_{14}O_3$. *See* b202	206.24	nd (al or bz-lig)	80–1	282–3	1.0251^{100}_4	1.5052^{100}_4	i	...	s	...	...	...	**B6**^{2}, 919
b809^1	**—,1-hydroxy-4(2-methyl-2-butyl)-***	4-*tert*-Amylphenol. $C_{11}H_{16}O$. *See* b202	164.25	nd	94–6	262.5^{760} (265–7) 138^{15}	...	...	...	...	...	...	...	...	**B6**, 548
b810	**—,1-hydroxy-2-methyl-4(3-methyl-3-pentyl)-***	$C_{13}H_{20}O$. *See* b202	192.31	...	35	$145–6^{13}$	...	1.5200^{25}	i	...	...	s	...	os s	**C48**, 1328
b811	**—,1-hydroxy-4(3-methyl-1-butyl)-***	$C_{11}H_{16}O$. *See* b202	164.25	nd (w)	93	255 126^{14}	0.9579^{23}_{20}	1.5050^{27}	i	v	v	...	...	...	**B6**^{3}, 1960
b812	**—,1-hydroxymethyl-4-isopropyl-**	Cumic alcohol. $C_{10}H_{14}O$. *See* b202	150.22	...	28	249	0.9818^{18}_4	1.5210^{20}	δ	∞	∞	...	...	...	**B6**, 543
b813	**—,1-hydroxy-2-methyl-3,4,5-trichloro-***	$C_7H_5Cl_3O$. *See* b202	211.48	nd (lig)	77	...	...	...	δ	...	...	s	...	os v alk s lig s^h	**B6**^{1}, 175
b814	**—,1-hydroxy-4-methyl-2,3,5-trichloro-***	$C_7H_5Cl_3O$. *See* b202	211.48	nd (aa or lig)	66–7	...	...	...	δ	...	...	s	...	os v alk s	**B6**, 404
b815	**—,2-hydroxy-4-methyl-1,3,5-trichloro-***	$C_7H_5Cl_3O$. *See* b202	211.48	nd (w), pl (peth)	47	265 $162–3^{28}$	...	...	s^h δ	s	s	...	δ	chl s aa s^h peth δ	**B6**^{2}, 356
Ω b816	**—,hydroxy-(pentamethyl)-***	$C_{11}H_{16}O$. *See* b202	164.25	nd (al, peth or ace) λ^{al} 280.5 (3.20)	128	267	...	...	i	s	...	...	...	...	**B6**^{1}, 270
b817	**—,1-hydroxy-4-pentyl-***	$C_{11}H_{16}O$. *See* b202	164.25	...	23	250.5^{760} $119–20^3$	0.960^{20}_4	1.5272^{25}	i	s	s	...	...	peth δ	**B6**^{3}, 1950
b818	**—,1-hydroxy-4(1-propenyl)-***	Anol. $C_9H_{10}O$. *See* b202	134.18	pl (w), lf λ^{al} 259 (4.27), 289 (3.45)	93–4	250d $140–5^{15}$	...	...	δ^h	s	s	...	...	os s	**B6**, 566
b819	**—,1-hydroxy-2-propyl-***	$C_9H_{12}O$. *See* b202	136.20	...	...	220^{760} 106.7^{10}	1.015^{20}_4	...	δ	s	s	...	...	...	**B6**^{2}, 469

For explanations, symbols and abbreviations see beginning of table. For structural formulas see end of table.

No.	Name	Synonyms and Formula	Mol. wt.	Color, crystalline form, specific rotation and λ_{max} (log ε)	m.p. °C	b.p. °C	Density	n_D	Solubility w	al	eth	ace	bz	other solvents	Ref.
	Benzene														
Ω b820	**—,1-hydroxy-3-propyl-***	$C_9H_{12}O$. *See* b202	136.20	....	26	228^{760} 111.2^{10}	0.987^{20}	1.5223^{20}	δ	s	s	...	...	....	**B6**, 499
b821	**—,1-hydroxy-4-propyl-***	$C_9H_{12}O$. *See* b202	136.20	....	22	232.6^{760} 111.7^{10}	1.009^{20}_4	1.5379^{25}	δ	s	...	...	...	....	**B6²**, 469
b822	**—,1-hydroxy-2,3,4,6-tetramethyl-***	Isodurenol. $C_{10}H_{14}O$. *See* b202	150.22	cr (peth) λ^{al} 280.2 (3.23)	80–1	230–50	....	....	...	s	...	...	...	....	**B6**, 546
Ω b823	**—,1-hydroxy-2,3,5,6-tetramethyl-***	Durenol. $C_{10}H_{14}O$. *See* b202.	150.22	nd (lig), pr (al) λ^{al} 276 (3.08)	118–9	247^{760}	....	....	...	...	...	...	...	peth, aa s	**B6³**, 1919
Ω b824	**—,2-hydroxy-1,3,5-tri-*tert*-butyl-**	$C_{18}H_{30}O$. *See* b202	262.44	cr (al, peth) λ^{al} 273 (3.3), 281 sh (3.2)	131	278^{760} 130^{15}	0.864^{27}_4	....	i	s^h	...	s	...	os s alk i	**B6³**, 2094
Ω b825	**—,1-hydroxy-2,4,5-trimethyl-***	Pseudocumenol. $C_9H_{12}O$. *See* b202	136.20	nd (lig)	72	232^{760}	....	....	i	v	v	...	...	....	**B6**, 509
b826	**—,—,acetate.**	$C_{11}H_{14}O_2$. *See* b202	178.23	nd (peth)	34–4.5	245–6	....	....	...	s	s	...	...	peth s^h	**B6²**, 482
Ω b827	**—,2-hydroxy-1,3,5-trimethyl-***	Mesitol. $C_9H_{12}O$. *See* b202	136.20	nd (peth or MeOH) λ^{cy} 284.5 (3.28)	72	221 sub	....	....	δ	v	v	...	...	alk s	**B6²**, 483
Ω b828	**—,iodo-***	Phenyl iodide. C_6H_5I. *See* b202	204.01	λ^{al} 227.5 (4.08)	−31.27	188.3^{760} 75^{10}	1.8308^{20}_4	1.6200^{20}	i	s	∞	∞	∞	CCl_4, lig ∞ chl s	**B5²**, 165
b829	**—,(2-iodoethoxy)-***	C_8H_9IO. *See* b202	248.07	cr (dil al)	31–2	....	....	....	...	...	...	...	...	os s	**B6¹**, 81
Ω b830	**—,1-iodo-2-methoxy-***	C_7H_7IO. *See* b202	234.04	λ^{cy} 269 (3.38), 275.5 (3.53), 283 (3.51)	....	$239–40^{730}$ $91–2^2$	1.8^{20}	....	...	v	v	v	v	chl, lig v	**B6²**, 198
Ω b831	**—,1-iodo-3-methoxy-***	C_7H_7IO. *See* b202	234.04	λ^{cy} 274.5 (3.43), 282 (3.41)	....	244–5 123^{14}	....	....	...	v	v	...	...	....	**B6¹**, 109
Ω b832	**—,1-iodo-4-methoxy-***	C_7H_7IO. *See* b202	234.04	lf (al), nd (MeOH) λ^{cy} 279 (3.20), 288 (3.08)	53	237^{726} 139^{25}	....	....	...	s^h	s	...	...	....	**B6³**, 774
b833	**—,1-iodomethyl-2-methyl-***	C_8H_9I. *See* b202	232.07	nd (eth), (peth)	33–4	....	....	....	...	...	s^h	...	...	peth s^h	**B5²**, 286
b834	**—,1-iodomethyl-4-methyl-***	C_8H_9I. *See* b202	232.07	nd (eth), (peth)	46–7	....	....	....	i	...	s^h	...	s	peth s^h	**B5²**, 302
Ω b835	**—,1-iodo-2-nitro-***	$C_6H_4INO_2$. *See* b202	249.01	ye rh nd (al) λ^{al} 260 sh (3.5), 310 (3.1)	54	$288–9^{729}$ 162^{18}	1.9186^{75}	....	i	s	s	...	...	....	**B5²**, 190
Ω b836	**—,1-iodo-3-nitro-***	$C_6H_4INO_2$. *See* b202	249.01	mcl pr λ^{al} 262 (3.81), 308 (3.00)	(i) 38.5 (st) (ii) 9.9 (unst)	280 153^{14}	1.9477^{50}_4	....	i	s	s	...	...	....	**B5²**, 191
Ω b837	**—,1-iodo-4-nitro-***	$C_6H_4INO_2$. *See* b202	249.01	ye nd (al) λ^{al} 223 (3.9), 294 (4.0)	174	289^{772}	....	....	i	s	...	...	...	aa s	**B5²**, 191
Ω b838	**—,iodoso-***	C_6H_5IO. *See* b202	220.01	ye pw	210 exp	....	....	....	s^h	s^h	i	i	i	peth i	**B5²**, 166
b839	**—,1-iodo-2-triazo-***	$C_6H_4IN_3$. *See* b202	245.03	ye	....	$90–10^{0.9}$	1.8893^{25}_4	1.6631^{25}	...	s	...	s	...	os s	**B5¹**, 142
Ω b840	**—,iodoxy-***	$C_6H_5IO_2$. *See* b202	236.01	nd (w)	236–7 exp	....	....	....	s^h	i	...	i	i	aa s^h chl i	**B5²**, 167
b841	**—,isobutoxy-***	Isobutyl phenyl ether. $C_{10}H_{14}O$. *See* b202	150.22	....	....	196	0.9240^{24}_{15}	1.4932^{24}	...	...	...	s	s	peth δ	**B6²**, 146
b842	**—,1-isobutoxy-2-nitro-***	$C_{10}H_{13}NO_3$. *See* b202	195.22	ye oil	....	275–80	1.1361^{20}	....	...	...	s	...	...	....	**B6**, 218
Ω b843	**—,isobutyl-***	$C_{10}H_{14}$. *See* b202	134.22	λ^{iso} 248 (2.1), 259 (2.3), 265 (2.2), 268 (2.2)	−51.5	172.8^{760} 53.1^{10}	0.8532^{20}_4	1.4866^{20}	i	∞	∞	∞	∞	peth, CCl_4 ∞	**B5**, 319
Ω b844	**-,isocyanato-**	Phenyl isocyanate. C_7H_5NO. *See* b202	119.12	λ^{hx} 226 (4.04), 256 (2.59), 270 (2.76), 277 (2.67)	....	$162–3^{751}$ 55^{13}	$1.0956^{19.6}_4$	$1.5368^{19.6}_4$	d	d	s	s	...	....	**J1934**, 2011
b845	**—,isopropenyl-**	C_9H_{10}. *See* b202	118.18	λ^{peth} 242.5 (4.04), 277 (2.44)	−23.21	165.4^{760} 48.5^{10}	0.9106^{20}_4	1.5386^{20}	i	s	s	∞	∞	peth, CCl_4 ∞	**B5²**, 374
Ω b846	**—,isopropoxy-**	$C_9H_{12}O$. *See* b202	136.20	λ^{hx} 273 (3.28)	−33.05	176.8^{760}	0.9408^{20}_4	1.4975^{20}	s	s	...	s	s	....	**B6²**, 145
Ω b847	**—,isopropyl-**	Cumene. C_9H_{12}. *See* b202	120.20	λ^{hx} 254 (2.3), 251 (2.4), 267 (2.3)	−96	152.4^{760} 38.2^{10}	0.8618^{20}_4	1.4915^{20}	i	∞	∞	∞	∞	CCl_4, peth ∞	**B5²**, 306
b848	**—,1-isopropyl-2-methyl-**	*o*-Cymene. $C_{10}H_{14}$. *See* b202	134.22	λ^{iso} 257 (2.2), 263 (2.4), 267 (2.2), 270 (2.3)	−71.54	178.15^{760} 57.3^{10}	0.8766^{20}_4	1.5006^{20}	i	∞	∞	∞	∞	peth, CCl_4 ∞	**B5¹**, 322
b849	**—,1-isopropyl-3-methyl-**	*m*-Cymene. $C_{10}H_{14}$. *See* b202	134.22	λ^{iso} 257 (2.2), 264 (2.4), 268 (2.2), 271 (2.3)	−63.75	175.14^{760} 55.0^{10}	0.8610^{20}_4	1.4930^{20}	i	∞	∞	∞	∞	CCl_4, peth ∞ chl s	**B5²**, 322

For explanations, symbols and abbreviations see beginning of table. For structural formulas see end of table.

No.	Name	Synonyms and Formula	Mol. wt.	Color, crystalline form, specific rotation and λ_{max} (log ε)	m.p. °C	b.p. °C	Density	n_D	Solubility w	al	eth	ace	bz	other solvents	Ref.
	Benzene														
Ω b850	**—,1-isopropyl-4-methyl-**	*p*-Cymene. $C_{10}H_{14}$. *See* b202	134.22	λ^{hx} 261 (2.4), 267 (2.6), 274 (2.6)	−67.94	177.1^{760} 56.3^{10}	0.8573^{20}_{4}	1.4909^{20}	i	∞	∞	∞	∞	CCl_4, peth ∞ chl s	**B5**^{2}, 322
b851	**—,4-isopropyl-1-methyl-2-nitro-**	$C_{10}H_{13}NO_2$. *See* b202	179.22	...	...	126^{10}	1.0744^{20}_{4}	1.5301^{20}	i	v	v	...	...	...	**B5**^{2}, 326
b852	**—,1-isopropyl-2-nitro-**	2-Nitrocumene. $C_9H_{11}NO_2$. *See* b202	165.19	pa ye λ^{iso} 247 (3.62)	...	103^{9}	1.101^{12}	1.5259^{20}	i	...	...	s	s	aa s	**B5**^{2}, 307
b853	**—,1-isopropyl-4-nitro-**	$C_9H_{11}NO_2$. *See* b202	165.19	pa ye oil λ^{gas} 251.3 (3.96)	...	122^{9}	1.0830^{20}_{4}	1.5367^{20}	i	...	...	s	s	lig s	**B5**^{2}, 308
Ω b854	**—,mercapto-***	Thiophenol. Benzenethiol. C_6H_6S *See* b202	110.18	λ^{al} 237 (3.85)	−14.8	168.7^{760} 46.4^{10}	1.0766^{20}_{4}	1.5893^{20}	i	s	s	...	s	...	**B6**^{2}, 284
Ω b855	**—,1-mercapto-4-methoxy-***	4-Mercaptoanisole. C_7H_8OS. *See* b202	140.22	...	...	227–9 $89–90^{5}$	1.1313^{25}_{4}	1.5801^{25}	...	s	s	...	s	...	**B6**^{2}, 852
b856	**—,1-mercapto-2-nitro-***	$C_6H_5NO_2S$. *See* b202	155.18	ye nd (peth or CCl_4) λ^{al} 270.5 (3.76), 354 (3.57)	58.5 (61)	...	...	...	...	s	s	...	s	alk s	**B6**^{2}, 303
b857	**—,1-mercapto-4-nitro-***	$C_6H_5NO_2S$. *See* b202	155.18	(eth, chl, ace) λ^{al} 317.5 (4.38)	79	...	...	...	s^h	s	v	v	...	alk s lig, aa δ	**B6**^{2}, 309
b858	**—,2-mercapto-1,3,5-trinitro-***	Thiopicric acid. $C_6H_3N_3O_6S$. *See* b202	245.17	ye nd	114	...	...	...	v d^h	v d^h	v	v	v	chl v peth i	**B6**^{2}, 316
Ω b859	**—,methoxy-***	Anisole. C_7H_8O. *See* b202	108.15	...	−37.5	155	0.9961^{20}_{4}	1.5179^{20}	i	s	s	v	v	...	**B6**^{2}, 139
b860	**—,1-methoxy-4-(3-methyl-butyl)-***	*p*-Isopentylanisole. $C_{12}H_{13}O$. *See* b202	178.28	...	...	121^{14}	...	...	i	...	...	...	s	...	**B6**^{1}, 296
Ω b861	**—,1-methoxy-2-nitro-***	$C_7H_7NO_3$. *See* b202	153.14	λ^{cy} 249 (3.53), 304 (3.40)	10.45	276.8^{757} 144^{4}	1.2540^{20}_{4}	1.5161^{20}	i	∞	∞	...	...	...	**B6**^{2}, 209
Ω b862	**—,1-methoxy-3-nitro-***	$C_7H_7NO_3$. *See* b202	153.14	nd (al), pl (bz-lig) λ^{cy} 260 (3.79), 313 (3.38)	38–9	258	1.373^{18}	...	i	s	v	...	...	...	**B6**^{2}, 214
Ω b863	**—,1-methoxy-4-nitro-***	$C_7H_7NO_3$. *See* b202	153.14	pr (al), nd (dil al) λ^{al} 306 (4.04)	54	274 (258–60)	1.2192^{60}_{4}	1.5070^{60}	i	v	v	...	...	peth δ	**B6**^{2}, 220
Ω b864	**—,1-methoxy-4-propenyl-*** (*trans*)	Anethole. $C_{10}H_{12}O$. *See* b202	148.21	lf (al) λ^{iso} 258 (4.25)	21.35	234.5^{763} 115^{12}	0.9882^{20}_{4}	1.5615^{20}	δ	∞	∞	s	v	os ∞	**B6**^{2}, 523
b865	**—,1-methoxy-2,3,4,6-tetra-bromo-***	$C_7H_4Br_4O$. *See* b202	423.75	nd (dil al or aa)	113–4 (104)	340	...	...	i	s	...	...	...	...	**B6**^{2}, 196
b866	**—,1-methoxy-2,3,4,5-tetra-chloro-***	$C_7H_4Cl_4O$. *See* b202	245.92	cr (MeOH)	83	...	...	...	i	v	...	...	v	MeOH v	**B6**^{2}, 182
b867	**—,1-methoxy-2,3,5,6-tetra-chloro-***	$C_7H_4Cl_4O$. *See* b202	245.92	nd (al)	89–90	...	...	...	i	s	v	...	v	MeOH s^h	**B6**^{2}, 182
b868	**—,2-methoxy-1,3,4,5-tetra-chloro-***	$C_7H_4Cl_4O$. *See* b202	245.92	nd (MeOH), pr (al)	64–5	...	...	...	i	s^h	v	...	v^h	MeOH s^h	**B6**^{2}, 182
b869	**—,1-methoxy-2,3,4-tribromo-***	$C_7H_5Br_3O$. *See* b202	344.84	nd (al)	106	...	...	...	i	v	...	v	v	os v chl s	**B6**^{2}, 192
b870	**—,1-methoxy-2,3,5-tribromo-***	$C_7H_5Br_3O$. *See* b202	344.84	pr (dil al)	82	305–12	...	...	i	v	...	v	v	os v	**B6**^{2}, 192
b871	**—,1-methoxy-2,4,5-tribromo-***	$C_7H_5Br_3O$. *See* b202	344.84	nd (al)	105	$306–9^{775}$	...	...	i	s	...	s	s	os s	**B6**^{2}, 192
Ω b872	**—,2-methoxy-1,3,5-tribromo-***	$C_7H_5Br_3O$. *See* b202	344.84	nd (al) λ^{al} 282 (3.00), 289.5 (3.00)	88	297–9	2.491	...	δ	δ	...	v	v	os s	**B6**^{2}, 193
b873	**—,5-methoxy-1,2,3-tribromo-***	$C_7H_5Br_3O$. *See* b202	344.84	cr (al)	91–4	300–10	...	...	i	s	...	v	v	os s	**B6**^{2}, 195
b874	**—,1-methoxy-2,3,5-trichloro-***	$C_7H_5Cl_3O$. *See* b202	211.48	nd (al)	84	...	...	...	i	s	...	v	...	os s	**B6**^{2}, 180
b875	**—,1-methoxy-2,4,5-trichloro-***	$C_7H_5Cl_3O$. *See* b202	211.48	nd (dil al)	77.5	$252–5^{742}$	...	...	...	v	...	v	...	os s	**B6**^{2}, 180
b876	**—,2-methoxy-1,3,4-trichloro-***	$C_7H_5Cl_3O$. *See* b202	211.48	pr (al)	45	$227–9^{756}$	...	...	...	s	...	v	...	os s	**B6**^{2}, 180
Ω b877	**—,2-methoxy-1,3,5-trichloro-***	$C_7H_5Cl_3O$. *See* b202	211.48	mcl nd (al) λ^{al} 280.5 (2.94), 287.5 (2.94)	61–2 (65)	240^{738}	1.640	...	...	s	...	v	s	os s	**B6**^{2}, 181
b878	**—,2-methoxy-1,3,5-triiodo-***	$C_7H_5I_3O$. *See* b202	485.83	lf (bz), nd (eth or al) λ^{al} 277 sh (2.99)	98–9	...	...	...	i	v	...	v	...	os s	**B6**^{2}, 204
b879	**—,1-methoxy-2,3,4-trinitro-***	$C_7H_5N_3O_7$. *See* b202	243.13	pa ye lf (al)	155	exp	...	...	...	s^h δ	...	v	δ	aa v lig i	**B6**^{1}, 129
b880	**—,1-methoxy-2,3,5-trinitro-***	$C_7H_5N_3O_7$. *See* b202	243.13	ye nd (al)	106.8	...	1.618^{15}	...	s^h	v	...	s	v	...	**B6**^{2}, 253

For explanations, symbols and abbreviations see beginning of table. For structural formulas see end of table.

No.	Name	Synonyms and Formula	Mol. wt.	Color, crystalline form, specific rotation and λ_{max} (log ε)	m.p. °C	b.p. °C	Density	n_D	Solubility w	al	eth	ace	bz	other solvents	Ref.
	Benzene														
b881	**—,1-methoxy-2,4,5-trinitro-***	$C_7H_5N_3O_7$. *See* b202	243.13	ye (al)	106–7	....	....	....	v^h	...	s	...	s	chl, aa s lig i	**B6²**, 253
Ω b882	**—,2-methoxy-1,3,5-trinitro-***	$C_7H_5N_3O_7$. *See* b202	243.13	(i) nd (dil MeOH) (ii) 56–7 (MeOH) $\lambda^{al-KOEt}$ 410 (4.4), 490 (4.2)	(i) 68.5–9.0 (ii) 56–7	....	1.4947^{80}	....	i	v^h	s	...	v	chl v	**B6²**, 280
—	**—,methyl-***	*see* **Toluene**													
b883	**—,(3-methylbutoxy)-***	$C_{11}H_{16}O$. *See* b202	164.25	....	....	225^{760} (216^{718})	0.9198_4^{22}	1.4872^{20}	s	...	...	...	...	....	**B6²**, 146
Ω b884	**—,(2-methyl-2-butyl)-**	$C_{11}H_{16}$. *See* b202	148.25	....	....	192.4^{760} $71–2^{12}$	0.8587_4^{20}	1.4934^{20}	i	∞	∞	...	...	....	**B5²**, 338
b885	**—,(3-methylbutyl)-***	$C_{11}H_{16}$. *See* b202	148.25	....	....	199^{760}	0.8558_4^{20}	1.4853^{20}	i	s	s	...	v	....	**B5²**, 332
b886	**—,(4-methylpentyl)-***	$C_{12}H_{18}$. *See* b202	162.28	....	....	$214–5^{740}$	0.8568^{16}	....	i	s	s	...	v	....	**B5**, 444
b887	**—,1-methyl-2-propyl-***	$C_{10}H_{14}$. *See* b202	134.22	λ^{iso} 263 (2.2), 271 (2.1)	−60.2	185^{760} 63.5^{10}	0.8744^{20}	1.4998^{20}	...	...	...	...	...	....	**B5²**, 321
b888	**—,1-methyl-3-propyl-***	$C_{10}H_{14}$. *See* b202	134.22	λ^{iso} 259 (2.3), 265 (2.4), 268 (2.3), 272 (2.4)	....	182^{760} 61.4^{10}	0.8610^{20}	1.4936^{20}	...	...	...	...	...	....	**B5²**, 321
b889	**—,1-methyl-4-propyl-***	$C_{10}H_{14}$. *See* b202	134.22	λ^{iso} 259 (2.5), 268 (2.7), 274 (2.7)	−63.6	183^{760} 61.9^{10}	0.8584^{20}	1.4919^{20}	i	s	s	...	...	....	**B5²**, 321
b890	**—,methylthio-***	Thioanisole. Methyl phenyl sulfide. C_7H_8S. *See* b202	124.22	λ^{al} 254 (3.98)	....	193^{760} 74^{10}	1.0579_4^{20}	1.5868^{20}	i	s	...	v	...	os s	**B6²**, 287
Ω b891	**—,nitro-***	$C_6H_5NO_2$. *See* b202	123.11	λ^{al} 260 (3.91)	5.7	210.8^{760}	1.2037_4^{20}	1.5562^{20}	δ	v	v	v	v	....	**B5²**, 171
b892	**—,(α-nitroisopropyl)-**	$C_9H_{11}NO_2$. *See* b202	165.19	....	....	224d $125–7^{15}$	1.1025_0^{20}	1.5209^{20}	i	...	...	s	s	....	**B5²**, 308
Ω b893	**—,nitro(pentachloro)*-**	Brassicol. Tritisan. $C_6Cl_5NO_2$. *See* b202	295.34	cr (al)	144	....	1.718_4^{25}	....	i	δ	...	...	s	chl s	**C50**, 3196
Ω b894	**—,nitroso-***	C_6H_5NO. *See* b202	107.11	rh or mcl (al-eth) λ^{chl} 750 (1.65)	68–9	$57–9^{18}$	....	....	i	s	s	...	s	lig v	**B5²**, 169
Ω b895	**—,1-nitro-2,3,5,6-tetrachloro-***	$C_6HCl_4NO_2$. *See* b202	260.89	....	99–100	....	1.744_4^{25}	....	i	s	...	...	s	chl s	**C29**, 6884
Ω b896	**—,1-nitro-2-triazo-***	$C_6H_4N_4O_2$. *See* b202	164.12	ye nd (bz-al) pr (al) λ^{al} 241 (4.2), 264 (3.8), 322 (3.5)	53–5	....	....	....	...	v	...	...	v	aa v chl s	**B5**, 278
Ω b897	**—,1-nitro-3-triazo-***	$C_6H_4N_4O_2$. *See* b202	164.12	wh nd (dil al or peth) λ^{al} 245 (4.3), 318 (3.3)	56	....	....	....	i	v	v	...	v	CS_2 v	**B5**, 278
Ω b898	**—,1-nitro-4-triazo-***	$C_6H_4N_4O_2$. *See* b202	164.12	wh pl (dil al) λ^{al} 306 (4.02), 400 (2.4)	75	....	....	....	i	v^h	v^h	...	v^h	CS_2, aa v^h	**B5**, 278
b899	**—,1-nitro-2,3,4-tribromo-***	$C_6H_2Br_3NO_2$. *See* b202	359.82	cr (al)	85.4	....	....	....	i	s	v	...	v	AcOEt v	**B5**, 251
b900	**—,1-nitro-2,3,5-tribromo-***	$C_6H_2Br_3NO_2$. *See* b202	359.82	nd	119.5 (81)	....	....	....	i	...	v	...	v	....	**B5**, 251
b901	**—,1-nitro-2,4,5-tribromo-***	$C_6H_2Br_3NO_2$. *See* b202	359.82	nd (al)	95–5.5	sub	....	....	i	v^h	v	...	...	CS_2 v	**B5**, 251
b902	**—,2-nitro-1,3,4-tribromo-***	$C_6H_2Br_3NO_2$. *See* b202	359.82	pl or pr (eth-al)	185 sub	....	....	....	i	s^h	v	...	v	....	**B5**, 251
b903	**—,2-nitro-1,3,5-tribromo-***	$C_6H_2Br_3NO_2$. *See* b202	359.82	mcl pr (chl)	125	177^{11}	....	....	i	δ^h	s	...	...	chl v aa s^h	**B5²**, 190
b904	**—,5-nitro-1,2,3-tribromo-***	$C_6H_2Br_3NO_2$. *See* b202	359.82	tcl (eth-al)	112	sub	2.645	....	i	...	v	...	v	....	**B5¹**, 133
Ω b905	**—,1-nitro-2,3,4-trichloro-***	$C_6H_2Cl_3NO_2$. *See* b202	226.45	nd (al)	55.5	....	....	....	i	δ	...	...	...	CS_2 s	**B5²**, 186
Ω b906	**—,1-nitro-2,4,5-trichloro-***	$C_6H_2Cl_3NO_2$. *See* b202	226.45	pr (al or CS_2), nd (al)	57–8	288	1.790^{22}	....	i	s^h δ	s	...	s	CS_2 s	**B5²**, 187
b907	**—,2-nitro-1,3,4-trichloro-***	$C_6H_2Cl_3NO_2$. *See* b202	226.45	nd (al)	89	....	....	....	i	s	...	...	...	lig δ	**B5²**, 187
b908	**—,2-nitro-1,3,5-trichloro-***	$C_6H_2Cl_3NO_2$. *See* b202	226.45	nd (al) λ^{al} 345 (2.3)	71	....	....	....	i	s^h δ	...	...	...	CS_2, lig v	**B5²**, 187
b909	**—,5-nitro-1,2,3-trichloro-***	$C_6H_2Cl_3NO_2$. *See* b202	226.45	pa ye tcl (al)	72.5	....	1.807	....	i	s^h	...	...	...	....	**B5²**, 187
b910	**—,1-nitro-2,3,5-triiodo-***	$C_6H_2I_3NO_2$. *See* b202	500.80	ye pr	124	....	....	....	i	s^h	...	s	...	os s	**B5**, 256
b911	**—,1-nitro-2,4,5-triiodo-***	$C_6H_2I_3NO_2$. *See* b202	500.80	pa ye nd (CS_2)	178	....	....	....	i	δ	δ	δ	...	CS_2 s^h chl δ	**B5**, 256
b912	**—,2-nitro-1,3,4-triiodo-***	$C_6H_2I_3NO_2$. *See* b202	500.80	nd (aa), pr (CS_2)	137	....	....	....	i	...	...	...	...	CS_2, aa s^h	**B5**, 256

For explanations, symbols and abbreviations see beginning of table. For structural formulas see end of table.

No.	Name	Synonyms and Formula	Mol. wt.	Color, crystalline form, specific rotation and λ_{max} (log ε)	m.p. °C	b.p. °C	Density	n_D	Solubility w	al	eth	ace	bz	other solvents	Ref.
	Benzene														
b913	**—,5-nitro-1,2,3-triiodo-***	$C_6H_2I_3NO_2$. *See* b202	500.80	ye pr (chl), nd (al)	167	...	3.256	...	i	δ	v	...	v	aa v^h chl s	**B5**[2], 192
b914	**—,1-nitro-2,3,5-trimethyl-***	$C_9H_{11}NO_2$. *See* b202	165.19	pr (al)	20	139–40[7]	...	...	...	s	...	...	...	...	**B5**, 404
b915	**—,1-nitro-2,4,5-trimethyl-***	$C_9H_{11}NO_2$. *See* b202	165.19	ye nd (al)	71	265	...	...	...	s	...	...	...	peth s	**B5**[2], 313
Ω b916	**—,2-nitro-1,3,5-trimethyl-***	$C_9H_{11}NO_2$. *See* b202	165.19	rh pr (al) λ^{iso} 250 (3.34)	44	255^{760}	1.51	...	...	v^h	...	...	...	...	**B5**[2], 316
b917	**—,pentaamino-***	$C_6H_{11}N_5$. *See* b202	153.19	dk br nd	228d	...	...	...	s	i	i	...	...	os i	**B13**[2], 155
b918	**—,pentabromo-***	C_6HBr_5. *See* b202	472.62	wh nd (aa or al)	160–1	sub	...	...	...	δ s^h	δ	...	s	chl s lig, aa δ	**B5**[1], 117
b919	**—,pentachloro-***	C_6HCl_5. *See* b202	250.34	nd (al) λ^{al} 288 (2.56), 298 (2.54)	86	277	$1.8342^{16.5}$	...	i	s^h i	δ	...	δ	CS_2, chl δ	**B5**[2], 157
b920	**—,pentaethyl-***	$C_{16}H_{26}$. *See* b202	218.39	...	< −20	277	0.8985^{19}_{19}	1.5127^{20}	i	...	...	...	...	...	**B5**[2], 358
b921	**—,pentaiodo-***	C_6HI_5. *See* b202	707.60	nd (al)	172	sub	...	...	i	δ	δ	...	s	chl, aa s	**B5**, 229
Ω b922	**—,pentamethyl-***	$C_{11}H_{16}$. *See* b202	148.25	pr (al) λ^{hx} 270 (1.3)	54.3	232^{760} 100^{10}	0.917^{20}_4	1.527^{20}	i	v	...	...	v	...	**B5**[2], 336
b923	**—,pentoxy-***	$C_{11}H_{16}O$. *See* b202	164.25	...	...	230^{756} 111^{17}	0.9270^{20}_4	1.4947^{20}	i	s	...	s	...	...	**B6**[1], 82
Ω b924	**—,pentyl-***	$C_{11}H_{16}$. *See* b202	148.25	...	−75	205.4^{760} 80.6^{10}	0.8585^{20}_4	1.4878^{20}	i	∞	∞	∞	∞	CCl_4, peth ∞	**B5**[2], 331
b924[1]	**—,(2-pentyl)-***	$C_{11}H_{16}$. *See* b202	148.25	λ^{al} 247.5 (2.06), 255 (2.20), 258 (2.31), 264 (2.20), 267.5 (2.21)	...	$198–9^{757}$	0.8594^{21}_4	1.4875^{21}	i	s	s	...	...	...	**B5**, 434
Ω b925	**—,propenyl-(*cis*)***	C_9H_{10}. *See* b202	118.18	...	−60.5	69^{28}	0.9088^{20}_4	1.5420^{20}	i	∞	∞	∞	∞	peth, CCl_4 ∞	**B5**[2], 371
b925[1]	**—,—(*trans*)***	C_9H_{10}. *See* b202	118.18	λ^{al} 249.5 (4.22)	−27.1	175–6	0.9019^{25}	1.5508^{20}	i	∞	∞	∞	∞	...	**B5**[2], 371
Ω b926	**—,1-propenyl-2,4,5-trimethoxy-*(*α-form*)**	α-Asaron. $C_{12}H_{16}O_3$. *See* b202	208.24	mcl nd (w)	67 (62)	296^{760} $167–8^{12}$	1.165^{20}_4	1.5683^{20}	s^h	s	s	...	...	peth, chl, CCl_4 s	**B6**[2], 1092
Ω b927	**—,propoxy-***	Phenyl propyl ether. $C_9H_{12}O$. *See* b202	136.20	...	−27.09	189.9^{760}	0.9474^{20}_4	1.5014^{20}	...	s	s	...	...	...	**B6**[2], 145
Ω b928	**—,propyl-***	C_9H_{12}. *See* b202	120.20	λ^{iso} 249 (2.07), 261.5 (2.31), 264.5 (2.19), 268 (2.20)	−99.5	159.2^{760} 43.3^{10}	0.8620^{20}_4	1.4920^{20}	i	∞	∞	∞	∞	peth, CCl_4 ∞	**B5**[2], 303
b929	**—,1,2,3,5-tetrabromo-***	$C_6H_2Br_4$. *See* b202	393.72	nd (al)	99–100	329	...	...	i	s^h	s	...	s	CS_2 s	**B5**[1], 117
b930	**—,1,2,4,5-tetrabromo-***	$C_6H_2Br_4$. *See* b202	393.72	mcl pr (CS_2)	182	...	3.072^{20}	...	i	δ	v	...	...	...	**B5**[2], 164
Ω b931	**—,1,2,3,4-tetrachloro-***	$C_6H_2Cl_4$. *See* b202	215.90	nd (al) λ^{al} 274 (2.42), 280 (2.52), 291 (2.46)	47.5	254^{760}	...	...	i	s^h δ	v	...	...	CS_2, aa, lig v	**B5**[2], 156
b932	**—,1,2,3,5-tetrachloro-***	$C_6H_2Cl_4$. *See* b202	215.90	nd (al) λ^{al} 274 (2.32), 281 (2.50), 291 (2.44)	54.5	246	...	...	s^h	δ	s	...	s	CS_2, lig v	**B5**[2], 157
Ω b933	**—,1,2,4,5-tetrachloro-***	$C_6H_2Cl_4$. *See* b202	215.90	nd, mcl pr (eth, al or bz) λ^{al} 276 (2.81), 285 (3.02), 294 (3.02)	139.5–40.5	243–6 (240)	1.858^{22}	...	i	δ^h	s	...	s	chl, CS_2 s	**B5**[2], 157
b934	**—,1,2,3,4-tetraethyl-***	$C_{14}H_{22}$. *See* b202	190.33	...	11.6	251^{734} 121.7^{14}	0.8875^{20}_4	1.5125^{20}	i	v	s	...	...	...	**B5**[2], 344
b935	**—,1,2,4,5-tetraethyl-***	$C_{14}H_{22}$. *See* b202	190.33	...	10	250	0.8788^{20}_4	1.5054^{20}	i	v	v	...	...	...	**B5**, 455
b936	**—,1,2,3,5-tetrahydroxy-***	$C_6H_6O_4$. *See* b202	142.11	nd (w), lf (eth AcOEt)	−165–7	...	...	...	v	v	...	v	i	AcOEt s chl, aa i	**B6**[1], 570
b937	**—,1,2,4,5-tetrahydroxy-***	$C_6H_6O_4$. *See* b202	142.11	lf (w or aa)	232–3	...	...	...	v	v	v	...	...	aa δ	**B6**[2], 1120
b938	**—,1,2,3,4-tetraiodo-***	$C_6H_2I_4$. *See* b202	581.70	pr (CS_2 or eth-aa)	136 (114)	sub	...	...	...	v	v	...	...	chl v	**B5**, 229
b939	**—,1,2,3,5-tetraiodo-***	$C_6H_2I_4$. *See* b202	581.70	pr (eth or aa)	148	sub	...	...	i	δ	δ	...	...	aa v^h chl δ	**B5**, 229
b940	**—,1,2,4,5-tetraiodo-***	$C_6H_2I_4$. *See* b202	581.70	pr (bz), nd (eth)	254 (165)	sub (vac)	...	...	i	δ	δ	...	...	CS_2 v aa s	**B5**, 229
Ω b941	**—,1,2,3,4-tetramethyl-***	Prehnitene. $C_{10}H_{14}$. *See* b202	134.22	cr (peth) λ^{al} 268 (2.50), 272.5 (2.41), 276.5 (2.39)	−6.25	205^{760} 79.4^{10}	0.9052^{20}_4	1.5203^{20}	i	∞	∞	∞	∞	CCl_4, peth ∞	**B5**[2], 329
Ω b942	**—,1,2,3,5-tetramethyl-***	Isodurene. $C_{10}H_{14}$. *See* b202.	134.22	λ^{al} 269 (2.7), 275 (2.7), 278 (2.3)	−23.68	198^{760} 74.4^{10}	0.8903^{20}_4	1.5130^{20}	i	∞	∞	∞	∞	CCl_4, peth ∞	**B5**[3], 976

For explanations, symbols and abbreviations see beginning of table. For structural formulas see end of table.

PHYSICAL CONSTANTS OF ORGANIC COMPOUNDS (Continued)

No.	Name	Synonyms and Formula	Mol. wt.	Color, crystalline form, specific rotation and λ_{max} (log ε)	m.p. °C	b.p. °C	Density	n_D	Solubility w	al	eth	ace	bz	other solvents	Ref.
	Benzene														
Ω b943	**—,1,2,4,5-tetra-methyl-***	Durene. $C_{10}H_{14}$. *See* b202 ...	134.22	λ^{al} 268 (2.8), 272 (2.8), 278 (2.84)	79.24	196.8^{760} 73.5^{10}	0.8380^{81}_{4} 0.8875^{20}_{4}	1.4790^{81} 1.5116^{20}	i	∞	∞	∞	∞	CCl_4, peth ∞	**B5**², 329
Ω b944	**—,1,3,5-triacetyl-**	$C_{12}H_{12}O_3$. *See* b202 ...	204.23	nd (w, aa or al) λ^{al} 226 (4.6), 294 (2.7)	163		...	...	s^h	s^h	δ	...	...	aa v	**B7**², 831
b945	**—,1,2,3-triamino-***	$C_6H_9N_3$. *See* b202 ...	123.16	cr (dil HCl)	103	336	...	...	v	v	v	...	...	...	**B13**², 145
b946	**—,1,2,4-triamino-***	$C_6H_9N_3$. *See* b202 ...	123.16	pl or lf (chl)	95–8 (99)	340	...	...	v	v	δ	...	...	chl s	**B13**², 146
b947	**—,triazo-***	Azidobenzene. Phenyl azide. $C_6H_5N_3$. *See* b202	119.13	pa ye oil λ^{cy} 248 (3.9), 256 (3.1), 278 (3.3)	−27.5	70^{11}	1.0880^{20}_{20}	1.5589^{25}	i	δ	δ	...	...	...	**B5**², 207
b948	**—,1,2,3-tri-bromo-***	$C_6H_3Br_3$. *See* b202 ...	314.82	pl (al)	87.8	...	2.658	...	i	δ^h	v	...	...	...	**B5**¹, 117
b949	**—,1,2,4-tri-bromo-***	$C_6H_3Br_3$. *See* b202 ...	314.82	nd (al or eth)	44–5	275	...	...	i	s	v	v	δ	CCl_4, CS_2 v	**B5**², 164
Ω b950	**—,1,3,5-tri-bromo-***	$C_6H_3Br_3$. *See* b202 ...	314.82	nd or pr (al) λ^{al} 267 (2.3), 277 (2.4), 289 (2.1)	121.5–2.5 (124, sub)	271^{765}	...	...	i	δ^h	s	...	s	chl s	**B5**², 164
b951	**—,2,4,6-tri-bromo-1,3,5-trimethyl-***	$C_9H_9Br_3$. *See* b202 ...	356.90	tcl nd (al or bz)	227–8	...	...	...	i	i	...	...	s	...	**B5**², 315
Ω b952	**—,1,2,3-tri-chloro-***	$C_6H_3Cl_3$. *See* b202 ...	181.45	pl (al) λ^{al} 265 (2.10), 273 (2.20), 280 (2.10)	53–4	218–9	...	...	i	δ	v	...	v	CS_2 v	**B5**², 156
Ω b953	**—,1,2,4-tri-chloro-***	$C_6H_3Cl_3$. *See* b202 ...	181.45	rh λ^{al} 270 (2.54), 278 (2.75), 287 (2.72)	16.95	213.5^{760} 84.8^{10}	1.4542^{20}_{4}	1.5717^{20}	i	δ	v	...	...	...	**B5**², 156
Ω b954	**—,1,3,5-tri-chloro-***	$C_6H_3Cl_3$. *See* b202 ...	181.45	nd λ^{al} 266 (2.26), 273 (2.37), 281 (2.23)	63–4	208^{763}	...	...	i	δ	v	...	v	CS_2, lig v	**B5**², 156
b955	**—,1,2,3-trichloro-4,5,6-trihydroxy-***	$C_6H_3Cl_3O_3$. *See* b202 ...	229.45	nd (al or bz) nd (+3w, w)	185		...	...	v^h	v	v	...	δ	chl, CCl_4, CS_2 δ	**B6**, 1084
b956	**—,1,2,4-trichloro-3,5,6-trihydroxy-***	$C_6H_3Cl_3O_3$. *See* b202 ...	229.45	nd (bz or aa)	160		...	...	δ	v	v	...	...	peth δ	**B6**², 1072
b957	**—,1,3,5-trichloro-2,4,6-trihydroxy-***	$C_6H_3Cl_3O_3$. *See* b202 ...	229.45	cr (al)	136	sub	...	...	i	s	...	...	i	...	**B6**, 1104
b958	**—,1,3,5-triethoxy-***	Phloroglucinol triethyl ether. $C_{12}H_{18}O_3$. *See* b202	210.28	cr (al, dil, al)	43.5	175^{24}	...	...	i	v	v	...	...	...	**B6**², 1079
b959	**—,1,2,4-triethyl-***	$C_{12}H_{18}$. *See* b202 ...	162.28	...	...	217.5^{755} 99^{15}	0.8738^{20}_{4}	1.5024^{20}	i	s	s	...	...	...	**B5**², 340
Ω b960	**—,1,3,5-triethyl-***	$C_{12}H_{18}$. *See* b202 ...	162.28	λ^{iso} 217.5 (3.94), 258 (2.18), 264 (2.31), 271 (2.23)	−66.5	216^{760}	0.8631^{20}_{4}	1.4969^{20}	i	v	v	...	...	...	**B5**², 340
Ω b961	**—,1,2,3-tri-hydroxy-***	Pyrogallol. $C_6H_6O_3$. *See* b202	126.11	lf or nd (bz) λ^{al} 266 (2.92)	133–4	309^{760} 171^{12}	1.453^{4}_{4}	1.561^{134}	v	v	v	...	i	NH_3 v	**B6**², 1059
b962	**—,—,triacetate** ...	$C_{12}H_{12}O_6$. *See* b202 ...	252.23	pr (al)	165		...	...	i	s	...	...	...	...	**B6**², 1066
b963	**—,1,2,4-tri-hydroxy-***	Hydroxyquinol. $C_6H_6O_3$. *See* b202	126.11	pl (eth), lf or pl (w) $\lambda^{w, pH=5.4}$ 288 (3.44)	140.5–1.0		...	...	v	v	v	...	i	CS_2, chl i	**B6**², 1071
b964	**—,—,triacetate** ...	$C_{12}H_{12}O_6$. *See* b202 ...	252.23	nd (MeOH)	97–8	>300	...	...	...	s^h	...	...	...	MeOH s^h	**B6**², 1072
Ω b965	**—,1,3,5-tri-hydroxy-***	Phloroglucinol. $C_6H_6O_3$. *See* b202	126.11	lf or pl (w + 2) λ^{al} 269 (2.70)	117 (hyd) 218–9 (anh)	sub	1.46	...	δ	v	v	...	v	Py v	**B6**, 1075
b966	**—,—,diacetate** ...	$C_{10}H_{10}O_5$. *See* b202 ...	210.19	pl (w)	104		...	...	s^h	s	s	s	...	lig δ	**B6**¹, 547
b967	**—,—,triacetate** ...	$C_{12}H_{12}O_6$. *See* b202 ...	252.23	pr (w), nd (dil. al) λ^{MeOH} 261.5 (2.56)	105–6		...	...	...	s^h	...	...	...	...	**B6**², 1079
b968	**—,1,2,3-triiodo-***	$C_6H_3I_3$. *See* b202 ...	455.80	nd (al), pr (bz)	116	sub	...	...	i	s	s	...	...	chl s	**B5**¹, 122
b969	**—,1,2,4-triiodo-***	$C_6H_3I_3$. *See* b202 ...	455.80	nd (al)	91.5	sub	...	...	i	s^h	s	...	...	chl s	**B5**¹, 122
b970	**—,1,3,5-triiodo-***	$C_6H_3I_3$. *See* b202 ...	455.80	nd (aa) λ^{MeOH} 232 (4.56)	184.2	sub	...	...	i	δ	δ	...	δ	aa s^h chl δ	**B5**¹, 122
Ω b971	**—,1,2,3-tri-methoxy-***	Pyrogallol trimethyl ether. $C_9H_{12}O_3$. *See* b202	168.20	rh nd (al) λ^{MeOH} 267 (2.82)	48–9	235^{760} 140^{12}	1.1118^{45}_{45}	...	i	s	s	...	s	...	**B6**², 1066
Ω b972	**—,1,3,5-tri-methoxy-***	Phloroglucinol trimethyl ether. $C_9H_{12}O_3$. *See* b202	168.20	pr (al), lf (peth) λ^{MeOH} 265 (2.68)	54–5	255.5^{760}	...	...	i	s	s	...	s	...	**B6**², 1078
b973	**—,1,2,3-tri-methyl-***	Hemimellitene. C_9H_{12}. *See* b202	120.20	λ^{iso} 195 (4.74), 204–220 (4.00)	−25.37	176.1^{760} 56.7^{10}	0.8944^{20}_{4}	1.5139^{20}	i	∞	∞	∞	∞	CCl_4, peth ∞	**B5**², 311

For explanations, symbols and abbreviations see beginning of table. For structural formulas see end of table.

No.	Name	Synonyms and Formula	Mol. wt.	Color, crystalline form, specific rotation and λ_{max} (log ε)	m.p. °C	b.p. °C	Density	n_D	Solubility w	al	eth	ace	bz	other solvents	Ref.
	Benzene														
Ω b974	—,1,2,4-tri-methyl-*	Pseudocumene. C_9H_{12}. *See* b202	120.20	λ^{iso} 196 (4.71), 204–220 (3.91)	−43.8	169.35^{760} 51.6^{10}	0.8758_4^{20}	1.5048^{20}	i	∞	∞	∞	∞	CCl_4, peth ∞	**B5²**, 312
Ω b975	—,1,3,5-tri-methyl-*	Mesitylene. C_9H_{12}. *See* b202.	120.20	λ^{al} 258 (2.2), 263 (2.2), 267 (2.2), 273 (2.3)	−44.7	164.7^{760} 48.7^{10}	0.8652_4^{20}	1.4994^{20}	i	∞	∞	∞	∞	CCl_4, peth ∞	**B5²**, 313
b976	—,1,2,3-trimethyl-4,5,6-trinitro-*	$C_9H_9N_3O_6$. *See* b202	255.19	pr (al)	209	...	...	...	i	s	...	s	s^h	...	**B5**, 400
b977	—,1,2,4-trimethyl-3,5,6-trinitro-*	$C_9H_9N_3O_6$. *See* b202	255.19	rh pr (al)	185	...	...	...	i	δ	...	...	v^h	to v^h	**B5²**, 316
b978	—,1,3,5-trimethyl-2,4,6-trinitro-*	Trinitromesitylene. $C_9H_9N_3O_6$. *See* b202	255.19	tcl nd (al), pr (ace) λ^{al} <220	238.2	415 exp	...	...	i	δ^h	δ^h	s	s	...	**B5³**, 923
b979	—,1,2,3-trinitro-*	$C_6H_3N_3O_6$. *See* b202	213.11	ye nd or pr (MeOH)	127.5	...	...	...	i	s^h	...	...	...	...	**B5²**, 203
b980	—,1,2,4-trinitro-*	$C_6H_3N_3O_6$. *See* b202	213.11	lf (eth), pa ye pr (al) $\lambda^{5\%al}$ 235 (4.19)	61–2	...	...	...	δ^h	s	s	s	v	chl v MeOH s	**B5²**, 203
Ω b981	—,1,3,5-trinitro-*	$C_6H_3N_3O_6$. *See* b202	213.11	rh pl (bz), lf (w) λ^{al} 222 (4.5) 330 (2.5)	(i) 121–2 (ii) 61	315^{760} 175^2	1.4775_4^{152}	...	δ^h	δ	δ	v	s	chl, Py s to v^h, CS_2 δ	**B5²**, 203
Ω b982	—,1,3,5-tri-phenyl-*	$C_{24}H_{18}$. *See* b202	306.41	rh nd (al or aa) λ^{hx} 253 (4.76)	176 (cor)	459^{717}	$1.199_4^{30.3}$	...	i	s	s	...	v	CS_2 s aa s^h	**B5²**, 670
b983	—,1,3,5-tris-(phenylamino)-*	*sym*-Trianilinobenzene. $C_{24}H_{21}N_3$. *See* b202	351.46	nd (al)	193	...	...	...	i	δ	v	...	δ	sulf s (vt-red)	**B13²**, 147
b984	—,1,3,5-tris-(4-tolylamino)-	$C_{27}H_{27}N_3$. *See* b202	393.54	nd (al)	186–7	...	...	...	...	δ	...	...	...	sulf s (bl-gr)	**B13**, 299
Ω b985	**Benzenearsonic acid***	Phenylarsonic acid.	202.02	cr (w) λ^{al} 256 (4.0), 262 (4.1), 270 (4.1)	158–62d	...	...	...	s	v	...	...	...	...	**B16²**, 457
Ω b986	—,2-amino-*	$C_6H_8AsNO_3$. *See* b985	217.07	nd (al-eth)	153–4	...	...	...	v	v	δ	...	...	alk, aa s	**B16²**, 483
b987	—,3-amino-*	$C_6H_8AsNO_3$. *See* b985	217.07	pr (w)	215	...	...	...	s^h	δ	δ	...	δ	alk s os δ	**B16²**, 489
Ω b988	—,4-amino-*	Arsanilic acid. $C_6H_8AsNO_3$. *See* b985	217.07	mcl nd (w or al)	232	...	1.9571^{20}	...	s^h	s^h δ	s	i	i	aa δ chl i	**B16²**, 491
b989	—,3-amino-4-hydroxy-*	$C_6H_8AsNO_4$. *See* b985	233.07	pr (w + ½)	290d (darkens 170)	...	...	...	i	δ	...	...	...	os δ	**B16²**, 521
b990	—,2-bromo-*	$C_6H_6AsBrO_3$. *See* b985	280.94	lf (w), pr (dil al)	201d 173–5 (+1w)	...	...	...	s^h	v	...	...	...	...	**B16²**, 458
b991	—,2-chloro-*	$C_6H_6AsClO_3$. *See* b985	236.49	nd (w or dil al)	186–7	...	...	...	s^h	s	i	...	i	MeOH s chl i	**B16²**, 457
Ω b992	—,3-chloro-*	$C_6H_6AsClO_3$. *See* b985	236.49	cr (w)	175	...	...	...	s	s	...	...	...	MeOH s	**B16²**, 457
Ω b993	—,4-chloro-*	$C_6H_6AsClO_3$. *See* b985	236.49	wh nd (al)	383–5d	...	...	...	...	s	s^h	...	...	alk v	**B16²**, 457
b994	—,2-hydroxy-*	$C_6H_7AsO_4$. *See* b985	218.04	nd (w)	190–1	...	...	...	v^h	v	i	v^h	...	aa v^h chl δ	**B16²**, 464
b995	—,3-hydroxy-*	$C_6H_7AsO_4$. *See* b985	218.04	cr (w)	162–73d	...	...	...	v	v	...	δ^h	i	aa v^h chl i	**B16¹**, 454
Ω b996	—,4-hydroxy-*	$C_6H_7AsO_4$. *See* b985	218.04	nd (ace)	170–4	...	...	...	v	v	δ	...	...	aa δ	**B16²**, 466
b997	—,2-hydroxy-3-nitro-*	$C_6H_6AsNO_6$. *See* b985	263.04	pr (w) cr (w)	252–4	...	...	...	s^h	s^h	...	...	...	...	**B16²**, 465
b998	—,2-hydroxy-5-nitro-*	$C_6H_6AsNO_6$. *See* b985	263.04	pa ye (w)	250d	...	...	...	v^h	v^h	i	s	...	...	**B16²**, 465
b999	—,3-hydroxy-2-nitro-*	$C_6H_6AsNO_6$. *See* b985	263.04	ye pl (al)	208d	...	...	...	...	s^h	...	...	...	...	**B16²**, 466
b1000	—,4-hydroxy-2-nitro-*	$C_6H_6AsNO_6$. *See* b985	263.04	ye nd (w)	228	...	...	...	v^h	s	...	s	...	...	**B16²**, 469
—	—,methyl-*	*see* **Toluenearsonic acid**													
Ω b1001	—,2-nitro-*	$C_6H_6AsNO_5$. *See* b985	246.03	lt ye nd (w) λ^{al} 262 (3.76)	235d	...	...	...	δ s^h	δ s^h	...	δ	...	aa s	**B16¹**, 449
b1002	—,3-nitro-*	$C_6H_6AsNO_5$. *See* b985	246.03	lt ye lf (w) λ^{al} 252.5 (3.89)	200 (182)	...	...	...	δ	...	i	...	δ	chl δ	**B16²**, 458
Ω b1003	—,4-nitro-*	$C_6H_6AsNO_5$. *See* b985	246.03	lf or nd (w)	>310 (298–300)	...	...	...	δ s^h	δ s^h	...	...	...	...	**B16¹**, 450
b1004	—,4-nitroso-*	$C_6H_6AsNO_4$. *See* b985	230.03	ye nd	180d	...	...	...	δ s^h	δ	i	...	...	aa v chl i	**B16¹**, 449
Ω b1005	—,4-ureido	Carbarsone. $C_7H_9AsNO_4$. *See* b985	260.08	nd (w)	174	...	...	...	δ v^h	δ	i	...	...	alk s chl i	**B16²**, 497
Ω b1006	**Benzeneboronic acid***	Phenylboric acid.	121.93	nd (w) λ^{hp} 225 sh (3.95), 235 (3.81), 267.5 (2.88), 272.5 (2.89), 280 (2.78)	218–20	...	...	...	δ v^h	s	s	...	s	...	**B16²**, 638
Ω b1007	—,4-bromo-*	$C_6H_6BBrO_2$. *See* b1006	200.83	cr (w)	266–72	...	...	...	δ	...	s	...	...	...	**B16²**, 638
b1008	—,2-chloro-*	$C_6H_6BClO_2$. *See* b1006	156.38	cr (w)	149	...	...	...	δ	s	s	...	s	lig s	**Am 56**, 975

For explanations, symbols and abbreviations see beginning of table. For structural formulas see end of table.

No.	Name	Synonyms and Formula	Mol. wt.	Color, crystalline form, specific rotation and λ_{max} (log ε)	m.p. °C	b.p. °C	Density	n_D	Solubility w	al	eth	ace	bz	other solvents	Ref.
	Benzeneboronic acid														
b1009	—,**4-chloro-***	$C_6H_6BClO_2$. *See* b1006	156.38	nd (w)	306–7 (270–5)	...	...	...	δ	...	s	...	...	...	**B16**^{2}, 638
—	**Benzenecarboxylic acid***	*see* **Benzoic acid**													
b1010	**Benzene-diazonium chloride***	Diazobenzene chloride. $C_6H_5N(:N)Cl$	140.57	nd (al) λ^w 263 (4.05)	exp	...	...	...	s	s	δ	s	i	aa v chl i	**B16**^{1}, 352
b1011	**Benzene-diazonium cyanide***	$C_6H_5N(:N)CN$	131.14	ye pr (w)	69	...	...	...	δ	...	...	...	...	...	**B16**, 432
b1012	**Benzene-diazonium nitrate***	$C_6H_5N(:N)NO_3$	167.13	nd (al-eth)	90 exp	...	...	...	s	δ	i	...	i	chl i	**B16**^{2}, 268
b1013	**Benzene-diazonium tribromide***	$C_6H_5N(:N)Br_3$	344.85	pl (al)	63.5	...	...	...	i	δ	i	...	...	...	**B16**, 431
—	**1,2-Benzenedi-carboxaldehyde**	*see* **Phthalaldehyde**													
Ω b1014	**1,3-Benzenedi-carboxaldehyde**	Isophthalaldehyde.	134.14	nd (dil al) λ^{MeOH} 225 (4.15), 240 (4.11), 290 (3.18), 330 (2.36)	89–90	$245–8^{771}$	...	...	δ	v	δ	s	s	os s lig δ	**B7**^{2}, 606
Ω b1015	**1,4-Benzenedi-carboxaldehyde**	Terephthalaldehyde.	134.14	nd (w)	116	247^{771}	...	...	δ s^h	v	s	...	...	alk s	**B7**, 675
—	**1,2-Benzenedi-carboxylic acid***	*see* **Phthalic acid**													
Ω b1016	**1,3-Benzenedi-carboxylic acid***	Isophthalic acid.	166.14	nd (w or al) λ^{al} 208 (4.60), 225 (4.07), 280 (2.93), 288 (2.92)	348	sub	...	...	δ	s	i	...	i	aa s lig i	**B9**^{2}, 608
b1016^1	—,diamide	...	164.18	pl (w)	280	...	...	...	δ^h	δ^h	...	...	i	os i	**B9**, 834
b1017	—,—,*N,N,N′,N′*-tetraethyl-	$C_{16}H_{24}N_2O_2$. *See* b1016^1	276.38	cr (eth-bz)	85	242^{12}	...	...	i	v	δ	v	v	lig i	**B9**^{2}, 609
Ω b1018	—,dichloride	$C_8H_4Cl_2O_2$. *See* b1016	203.03	pr (eth)	43–4	276 136^{11}	1.3880_4^{17}	1.570^{47}	δ	δ	s	...	...	...	**B9**^{2}, 609
Ω b1019	—,diethyl ester*	$C_{12}H_{14}O_4$. *See* b1016	222.24	λ^{al} 280, 289	11.5	302^{760} $170^{2.4}$	1.1239_4^{17}	1.508^{18}	i	...	...	...	...	...	**B9**^{1}, 372
Ω b1020	—,dimethyl ester*	$C_{10}H_{10}O_4$. *See* b1016	194.19	nd (dil al) λ^{al} 222 sh (3.11)	67.8–8.3 (cor)	282^{760} 124^{12}	1.194_4^{20}	1.5168^{20}	δ	...	...	...	...	...	**B9**^{2}, 609
Ω b1021	—,dinitrile	1,3-Dicyanobenzene	128.14	nd (al) λ^{MeOH} 342 (2.68), 355 (2.65)	162	...	...	...	δ^h	v^h	s	...	s	chl s peth i	**B9**, 836
b1022	—,mononitrile	3-Cyanobenzoic acid.	147.13	nd (w)	217	sub δd	...	...	v^h	v	v	...	...	...	**B9**^{2}, 609
b1023	—,piperazinium salt	$C_8H_6O_4 . C_4H_{10}N_2$. *See* b1016	252.27	...	252d	...	...	...	s^h	s^h	s^h	...	...	...	**Am 70,** 2758
Ω b1024	**1,4-Benzenedi-carboxylic acid***	Terephthalic acid.	166.14	nd (sub) λ^{diox} 242 (4.21), 286 (3.23)	>300 sub without melting	sub	...	...	i	i s^h	i	...	...	chl, aa i	**B9**, 841
b1025	—,diamide	...	164.18	nd (w), pl (aa)	>250	...	...	...	δ^h	i	i	...	i	chl i	**B9**^{1}, 376
b1026	—,—,*N,N,N′,N′*-tetraethyl-	$C_{16}H_{24}N_2O_2$. *See* b1025	276.38	cr (eth-al)	127	...	...	...	i	v	δ	...	s	lig i	**B9**^{2}, 613
Ω b1027	—,dichloride	...	203.03	nd or pl (lig)	83–4	$125–7^9$	...	...	d	d	s	...	...	...	**B9**, 844
Ω b1028	—,diethyl ester*	$C_{12}H_{14}O_4$. *See* b1024	222.24	mcl pr (al, peth) λ^{al} 227 (3.9), 275 (3.1)	44	302^{760} 142^2	1.1098_{45}^{45}	...	i	v	s	...	...	...	**B9**, 844
Ω b1029	—,dimethyl ester*	$C_{10}H_{10}O_4$. *See* b1024	194.19	nd (eth) λ^{al} 241.6 (3.32)	141.0–1.8 (cor)	sub	...	...	δ^h	δ v^h	s	...	...	MeOH δ	**B9**, 843
b1030	—,dinitrile	1,4-Dicyanobenzene.	128.14	nd (w, MeOH) λ^{MeOH} 348 (3.25), 360 (3.21), 405 (4.35), 425 (4.30)	222	sub	...	...	i	δ s^h	δ^h	...	s^h	aa v^h	**B9**^{2}, 613
b1031	—,monoamide	Terephthalamic acid.	165.15	...	>300	sub 250	...	...	i	i	i	...	i	chl, os i	**B9**, 845
Ω b1032	—,mononitrile	4-Cyanobenzoic acid.	147.14	pl or lf (w)	219	...	...	...	s^h	s	s	...	...	aa s^h	**B9**^{2}, 613
b1033	—,piperazinium salt	$C_8H_6O_4 . C_4H_{10}N_2$. *See* b1024	252.27	...	>350d	...	...	...	δ^h	s^h	δ^h	...	...	...	**Am 70,** 2758
b1034	**1,3-Benzenedicar-boxylic acid, 2-amino-***	$C_8H_7NO_4$. *See* b1016	181.15	pl (al), nd (aa)	>260	sub 267–9	...	...	δ	s	s	...	...	...	**B14**^{2}, 337
b1035	—,—,dimethyl ester*	$C_{10}H_{11}NO_4$. *See* b1016	209.20	nd (al)	103–4	...	...	...	...	δ	...	...	...	...	**B14**^{2}, 337
b1036	—,**4-amino-***	$C_8H_7NO_4$. *See* b1016	181.15	nd (w)	336–7	...	...	...	δ	v^h	s	s	...	aa v^h	**B14**^{1}, 633

For explanations, symbols and abbreviations see beginning of table. For structural formulas see end of table.

No.	Name	Synonyms and Formula	Mol. wt.	Color, crystalline form, specific rotation and λ_{max} (log ε)	m.p. °C	b.p. °C	Density	n_D	Solubility w	al	eth	ace	bz	other solvents	Ref.
	1,3-Benzenedicarboxylic acid														
b1037	—,—,dimethyl ester*	$C_{10}H_{11}NO_4$. *See* b1016	209.20	nd (al)	131.5	...	...	...	δ	δ	...	...	...	MeOH δ	**B14**², 337
Ω b1038	—,5-amino-*	$C_8H_7NO_4$. *See* b1016	181.15	pr (al), pl (w)	>360	sub	...	...	i	δ	...	...	...	...	**B14**¹, 636
Ω b1039	—,—,dimethyl ester*	$C_{10}H_{11}NO_4$. *See* b1016	209.20	lf or pl (MeOH)	176	...	...	...	i	...	s	...	...	...	**B14**, 556
b1040	**1,4-Benzenedicarboxylic acid, 2-amino-***	$C_8H_7NO_4$. *See* b1024	181.15	ye cr (w)	324–5d (cor)	...	...	...	δ	δ	δ	i	i	peth. chl, aa i	**B14**¹, 637
b1041	—,—,dimethyl ester*	$C_{10}H_{11}NO_4$. *See* b1024	209.20	nd (bz), cr (al)	134	...	...	...	...	s^h	s^h	s	...	chl s	**B14**, 559
b1042	—,2-benzoyl-	Benzophenone-2,5-dicarboxylic acid. $C_{15}H_{10}O_5$. *See* b1024	270.25	nd	291–2	...	...	...	i	s	s	...	i	to i	**B10**, 881
Ω b1043	**1,3-Benzenedicarboxylic acid, 4-bromo-***	$C_8H_5BrO_4$. *See* b1016	245.04	nd (al)	287 (283)	...	...	...	i	v^h	...	...	i	...	**B9**, 838
Ω b1044	**1,4-Benzenedicarboxylic acid, 2-bromo-***	$C_8H_5BrO_4$. *See* b1024	245.04	nd (w or al)	299	...	...	...	δ v^h	v	i	...	i	...	**B9**², 614
b1045	**1,3-Benzenedicarboxylic acid, 4-chloro-***	$C_8H_5ClO_4$. *See* b1016	200.58	nd (w)	295	...	...	...	δ s^h	v	δ	...	i	chl i	**B9**¹, 372
b1046	—,5-chloro-*	$C_8H_5ClO_4$. *See* b1016	200.58	nd (w + ½)	278 (anh)	...	...	...	δ v^h	s	...	...	i	chl i	**B9**, 838
Ω b1047	**1,4-Benzenedicarboxylic acid, 2-chloro-***	$C_8H_5ClO_4$. *See* b1024	200.58	cr (w)	320	...	...	...	δ^h	v	v	...	...	...	**B9**, 847
b1048	**1,3-Benzenedicarboxylic acid, 4,6-dichloro-***	$C_8H_4Cl_2O_4$. *See* b1016	235.03	nd (w or dil al)	280	...	...	...	i	v	v	...	...	chl v	**B9**, 838
Ω b1049	**1,4-Benzenedicarboxylic acid, 2,5-dichloro-***	$C_8H_4Cl_2O_4$. *See* b1024	235.03	nd (w)	306	sub	...	...	s^h δ	s	v	...	...	...	**B9**, 847
b1050	—,2,5-dihydroxy-*	$C_8H_6O_6$. *See* b1024	198.13	ye cr (al, w)	d	...	...	...	s^h	s^h	s^h	...	...	...	**B10**, 554
b1051	**1,3-Benzenedicarboxylic acid, 4,5-dimethoxy-***	Isohemipinic acid. $C_{10}H_{10}O_6$. *See* b1016	226.19	nd (w)	245–6	...	...	...	s^h δ	s	s	...	...	...	**B10**, 553
b1052	—,4,6-dimethyl-*	α-Cumidic acid. $C_{10}H_{10}O_4$. *See* b1016	194.19	nd (w), pr (al-bz), lf (sub)	266	sub	...	...	i δ^h	s	...	...	...	...	**B9**², 627
b1053	—,2-hydroxy-*	$C_8H_6O_5$. *See* b1016	182.13	nd (w + 1) λ^{al} 213 (4.51), 319 (3.70)	244.5 (239) (hyd) 250 (anh)	...	...	...	δ^h	v	v	...	...	chl s	**B10**², 352
Ω b1054	—,4-hydroxy-*	$C_8H_6O_5$. *See* b1016	182.13	nd (w), lf (dil al) λ^{al} 218 (4.45), 255 (4.05), 301 (3.50)	310	...	...	...	i δ^h	v	v	...	...	aa s^h chl i	**B10**², 352
b1055	—,5-hydroxy-*	$C_8H_6O_5$. *See* b1016	182.13	nd (w + 2), cr (aq al)	293	sub	...	...	i v^h	v	v	...	s	...	**B10**¹, 257
Ω b1056	—,5-methyl-*	Uvitic acid. $C_9H_8O_4$. *See* b1016	180.17	nd (w)	298	...	...	...	δ^h i	s	s	s	...	chl, peth δ	**B9**¹, 380
b1057	—,2-nitro-, dimethyl ester*	$C_{10}H_9NO_6$. *See* b1016	239.19	nd (w or al)	135	...	...	...	δ	s	s	...	...	MeOH δ peth i	**B9**¹, 373
Ω b1058	—,5-nitro-*	$C_8H_5NO_6$. *See* b1016	211.13	gr lf (+1½w)	260–1 (anh)	...	...	...	i v^h	v	v	...	...	...	**B9**¹, 373
b1059	—,—,diethyl ester*	$C_{12}H_{13}NO_6$. *See* b1016	267.24	nd (al)	83.5	...	...	...	δ	δ v^h	...	...	...	...	**B9**, 840
Ω b1060	—,—,dimethyl ester*	$C_{10}H_9NO_6$. *See* b1016	239.19	nd (dil al)	123	...	...	...	i	s	s	...	...	...	**B9**², 611
Ω b1061	**1,4-Benzenedicarboxylic acid, 2-nitro-***	$C_8H_5NO_6$. *See* b1024	211.13	nd (w)	270–5	...	...	...	s^h	v^h	...	...	...	...	**B9**, 851
b1062	**1,3-Benzenedicarboxylic acid, tetrabromo-***	$C_8H_2Br_4O_4$. *See* b1016	481.74	nd (w)	288–92	...	...	...	s^h	...	...	...	i	...	**B9**, 839
b1063	**1,4-Benzenedicarboxylic acid, tetrabromo-***	$C_8H_2Br_4O_4$. *See* b1024	481.74	nd (w)	266 → anhydride	...	...	...	i s^h	δ	δ	...	δ	aa δ	**B9**, 850
b1064	**1,3-Benzenedisulfonic acid, 4-amino-***	...	253.25	nd (w + 2)	120d	...	...	...	s	s	i	...	...	alk s	**B14**², 470
b1065	—,4-hydroxy-*	...	254.24	nd (w)	>100d	...	...	...	v	v	i	...	...	...	**B11**², 139
Ω b1066	**Benzenehexacarboxylic acid***	Mellitic acid.	342.18	nd (al) λ^{w} 236 sh (4.3), 296 (3.1)	286–8d	...	...	...	v	s	...	...	...	sulf s^h	**B9**², 741
Ω b1067	**Benzenepentacarboxylic acid***	...	298.17	nd (w + 5) λ^{w} 393 (3.2)	228–30 (+5w) 238 (anh)	...	...	...	v^h	s	δ	...	i	AcOEt δ peth i	**B9**², 737

For explanations, symbols and abbreviations see beginning of table. For structural formulas see end of table.

No.	Name	Synonyms and Formula	Mol. wt.	Color, crystalline form, specific rotation and λ_{max} (log ε)	m.p. °C	b.p. °C	Density	n_D	Solubility w	al	eth	ace	bz	other solvents	Ref.
	Benzenephosphinic acid														
Ω b1068	**Benzenephosphinic acid***	$C_6H_5P(OH)_2$	142.10	pl (aq MeOH) λ^{al} 216 (3.85), 264.5 (2.79)	82–3	...	...	...	s v^h	v	δ	...	...	...	**B16**, 791
b1069	—,diethyl ester*	$C_6H_5P(OC_2H_5)_2$	198.20	...	...	235	1.032^{16}	...	i	...	...	...	...	...	**B16**, 791
b1070	**Benzenephosphonic acid***	$C_6H_5PO_3H_2$	158.10	lf (w) λ^{al} 258 (2.59), 263.5 (2.72), 270 (2.64)	160	...	...	...	v	s	s	...	i	...	**B16^2**, 390
b1071	—,dichloride	$C_6H_5POCl_2$	194.99	...	...	258 $137–8^{15}$	1.197^{25}	1.5581^{25}	d	...	...	...	...	...	**B16**, 804
b1072	—,tetrachloride	$C_6H_5PCl_4$	249.89	...	73	...	...	...	d	d^h	...	...	...	...	**B16**, 804
b1073	**Benzenephosphothionic acid,** ethyl 4-nitrophenyl ester*	EPN.	323.31	...	36	...	1.27^{25}_4	1.5978^{30}	i	s	s	...	s	...	**C46**, 2742
b1074	**Benzeneseleninic acid***	$C_6H_5SeO_2H$	189.07	pl (w) λ^{al} 220 (4.04), 260 (3.12)	124–5	...	1.652^{123}_4	...	δ	...	...	...	i	alk v	**B11^2**, 241
b1075	**Benzeneselenonic acid***	$C_6H_5SeO_3H$	205.07	nd	142	d	...	...	v	v	i	...	i	...	**B11^1**, 111
b1076	**Benzenesiliconic acid***	$C_6H_5SiO_2H$	138.20	glassy so (eth)	92	...	...	...	i	s	s	...	...	...	**B16**, 911
b1077	**Benzenestibonic acid***	$C_6H_5SbO_3H_2$	248.87	nd (aa)	139	...	...	...	i	δ	...	δ	...	C_2Cl_4 s	**B16^2**, 584
b1078	**Benzenesulfenic acid, chloride**	Benzenesulfenyl chloride.	144.62	red liq	...	$73–5^9$ 58^3	...	...	...	...	...	...	s	...	**B6^2**, 295
Ω b1079	—,**2,4-dinitro-**, chloride	$C_6H_3ClN_2O_4S$. *See* b1078	234.62	ye pr (bz-peth or CCl_4) $\lambda^{CH_2ClCH_2Cl}$ 260 (3.60)	99 (95–6)	...	...	...	...	...	...	...	v	chl, aa v peth δ	**B6^2**, 316
Ω b1080	—,**2-nitro-**, chloride	$C_6H_4ClNO_2S$. *See* b1078	189.62	ye nd (bz)	75	...	...	...	...	...	v	...	v	chl v	**B6^2**, 308
b1081	—,**4-nitro-**, chloride	$C_6H_4ClNO_2S$. *See* b1078	189.62	ye lf (peth)	52	$125^{0.1}$	...	...	d	d	...	...	v	aa d	**B6^1**, 160
b1082	**Benzenesulfinic acid**	...	142.18	pr (w) λ^{al} 218 (3.99), 245 (3.40)	84	d at 100	...	...	s^h δ	s	s	...	s	peth i	**B11^2**, 3
Ω b1083	—,chloride	C_6H_5SOCl	160.62	pl (peth)	38	$71–2^{1.5}$	1.3469^{25}	1.3470^{25}	d	d	s	...	...	chl s^h	**B11^2**, 4
b1084	—,ethyl ester*	$C_6H_5SO_2C_2H_5$	170.23	liq	...	d	...	...	i	∞	∞	...	∞	...	**B11**, 6
b1085	—,**3-acetamido-**	$C_8H_9NO_3S$. *See* b1082	199.23	cr (w)	145	...	...	...	s^h	...	...	...	...	...	**B14^2**, 427
Ω b1086	—,**4-acetamido-**	$C_8H_9NO_3S$. *See* b1082	199.23	cr (w)	160 (153)	...	...	...	s^h	...	...	...	...	...	**B14^2**, 427
b1087	—,**4-bromo-***	$C_6H_5BrO_2S$. *See* b1082	221.08	nd (w)	114	...	...	...	d^h	v	v	...	...	...	**B11^2**, 5
b1088	—,**4-chloro-***	$C_6H_5ClO_2S$. *See* b1082	176.62	lf or nd (w)	99 (93)	...	...	...	s^h	v	v	...	...	...	**B11^2**, 4
	—,**methyl-***	*see* **Toluenesulfinic acid**													
b1089	—,**3-nitro-***	$C_6H_5NO_4S$. *See* b1082	187.18	nd	98	...	...	...	s	s	s	s	i	...	**B11^2**, 5
b1090	—,**4-nitro-***	$C_6H_5NO_4S$. *See* b1082	187.18	pr or nd (w) λ^w 264 (4.05)	159	...	...	...	δ	v	s	...	...	aa v	**B11^2**, 5
Ω b1091	**Benzenesulfonic acid***	...	158.18	nd (bz), λ^{al} 213 (3.89), 253 (2.27), 259 (2.41), 263 (2.46), 270 (2.38)	65–6 (anh)	...	...	...	v	v	i	...	δ	aa s CS_2 i	**B11^2**, 18
Ω b1092	—,hydrate*	$C_6H_5SO_3H$. $1.5H_2O$	176.20	pl (w)	45–6	...	...	...	s	s	i	...	δ	CS_2 i	**B11^2**, 18
Ω b1093	—,amide	Benzenesulfonamide.	157.19	lf, nd (w) λ^{al} 219 (3.95), 259 (2.50), 264 (2.70), 271 (2.50)	156	...	...	...	δ	s^h	s	...	...	...	**B11^2**, 24
Ω b1094	—,—,*N,N*-dichloro-	Dichloramine B. $C_6H_5SO_2NCl_2$	226.08	ye mcl or pl	76	...	...	...	...	s	...	...	...	...	**B11^2**, 28
Ω b1095	—,—,*N*-hydroxy-	$C_6H_5SO_2NHOH$	173.19	pl (w), rh	126d	...	...	...	v^h	v	v	v	δ	aa v chl δ	**B11^1**, 14
b1096	—,—,*N*-phenyl-	Benzenesulfonanilide. $C_6H_5SO_2NHC_6H_5$	233.29	tetr pr (al)	110	...	...	...	δ	v	v	...	...	...	**B12^2**, 298
Ω b1097	—,chloride	$C_6H_5SO_2Cl$	176.62	λ^{al} 210 (4.08), 270 (3.10)	14.5	251–2d 120^{10}	1.3842^{15}_{15}	...	i	v	s	...	...	...	**B11^2**, 23
Ω b1098	—,ethyl ester*	$C_6H_5SO_3C_2H_5$	186.23	...	...	156^{15}	1.2167^{20}_4	1.5081^{20}	δ d^h	s	v	...	...	chl v	**B11^2**, 20
Ω b1099	—,fluoride	$C_6H_5SO_2F$	160.17	λ^{peth} 267 (3.1), 273 (2.9)	...	203–4 $90–1^4$	1.3286^{20}_4	1.4932^{18}	...	s	s	...	...	...	**B11^2**, 23
b1100	—,isopropyl ester	$C_6H_5SO_3C_3H^i_7$	200.26	...	...	...	...	1.5003^{20}	δ	v	v	...	...	...	**B11^2**, 20
Ω b1101	—,methyl ester*	$C_6H_5SO_3CH_3$	172.21	λ^{al} 213 (3.90), 320 (2.6)	...	154^{20}	1.2730^{17}_4	1.5151^{20}	δ	v	v	...	...	chl v	**B11^2**, 20
b1102	—,propyl ester*	$C_6H_5SO_3C_3H^n_7$	200.26	λ^{al} 217.5 (3.93), 259 (2.83), 265 (3.00), 272 (2.93)	...	100d $162–3^{15}$	1.1804^{17}_4	1.5035^{25}	δ	s	v	...	...	chl v	**B11^2**, 20
Ω b1103	—,sodium salt*	$C_6H_5SO_3Na$	180.16	nd or lf (w + 1) λ 257 (2.5), 263 (2.6), 269 (2.5)	450d	...	...	...	s v^h	$δ^h$	...	...	...	...	**B11^2**, 18

For explanations, symbols and abbreviations see beginning of table. For structural formulas see end of table.

No.	Name	Synonyms and Formula	Mol. wt.	Color, crystalline form, specific rotation and λ_{max} (log ε)	m.p. °C	b.p. °C	Density	n_D	Solubility w	al	eth	ace	bz	other solvents	Ref.
	Benzenesulfonic acid														
b1104	**—,3-acetamido-,** amide	$C_8H_{10}N_2O_3S$. *See* b1093	214.25	(aa) $\lambda^{95\%\,al}$ 243 (4.1), 280 (3.1)	216–9							s		aa s	**B14**, 718
b1105	—,—,chloride	$C_8H_8ClNO_3S$. *See* b1091	233.69	nd (bz-peth)	88				d	s	s		δ	aa s	**B14**^{2}, 435
b1106	**—,4-acetamido-,** amide	$C_8H_{10}N_2O_3S$. *See* b1093	214.25	nd (aa) $\lambda^{95\%\,al}$ 255 (4.3)	219–20				s^h	s^h		s^h			**B14**, 702
b1107	—,—,chloride	$C_8H_8ClNO_3S$. *See* b1091	233.69	nd (bz), pr (bz-chl)	149				d	v	v		s^h	chl s^h	**B14**, 439
Ω b1108	**—,2-amino-***	Orthanilic acid $C_6H_7NO_3S$. *See* b1091	173.19	pr (+½w) λ^w 235 (3.8), 294 (3.2)	>320d				δ	i s^h	i				**B14**^{2}, 429
b1109	—,—,amide	$C_6H_8N_2O_2S$. *See* b1093	172.22	mcl pl, nd or pr (w) λ^{al} 206 (4.42), 245.5 (3.93), 306 (3.54)	152.5–3.5				δ v^h	v		v	i	aa s	**B14**^{2}, 429
Ω b1110	**—,3-amino-***	Metanilic acid. $C_6H_7NO_3S$. *See* b1091	173.19	nd, pr (w+½) λ^w 237 (3.2), 262 (2.5), 269 (2.6), 289 (2.6)	d				δ	δ	i				**B14**^{2}, 434
b1111	—,—,amide	$C_6H_8N_2O_2S$. *See* b1093	172.22	lf or nd (w) λ^w 209.5 (4.48), 238.5 (3.86), 297 (3.36)	142				δ s^h	v					**B14**, 690
Ω b1112	**—,4-amino-***	Sulfanilic acid $C_6H_7NO_3S$. *See* b1091	173.19	rh pl or mcl (w+2)	288		1.485$^{25}_{4}$		s^h δ	i	i				**B14**^{2}, 436
Ω b1113	—,—,amide	Sulfanilamide. $C_6H_8N_2O_2S$. *See* b1093	172.22	lf (aq al) λ^{al} 258.5 (4.28)	165–6		1.08		s^h	s	s	s		MeOH s chl, peth δ	**B14**, 698
b1114	—,—,—,*N*-acetyl-	$C_8H_{10}N_2O_3S$. *See* b1093	214.25	pr (w) λ^{al} 257 (4.35)	182–4				δ	s	i	v		alk v	**C50**, 15543
	—,—,—,*N*(4,6-dimethyl-2-pyrimidyl)-	*see* **Sulfamethazine**													
	—,—,—,*N*-guanido-	*see* **Sulfaguanidine**													
	—,—,—,*N*(4-methyl-2-pyrimidyl)-	*see* **Sulfamerazine**													
	—,—,—,*N*(2-pyridyl)-	*see* **Sulfapyridine**													
	—,—,—,*N*(2-quinoxalinyl)-	*see* **Sulfaquinoxaline**													
	—,—,—,*N*(2-thiazolyl)-	*see* **Sulfathiazole**													
b1115	**—,2-amino-4-chloro-***	$C_6H_6ClNO_3S$. *See* b1091	207.64	nd or pl (w)	310–30d				δ s^h	δ				aa δ	**B14**^{2}, 430
b1116	**—,2-amino-5-chloro-***	$C_6H_6ClNO_3S$. *See* b1091	207.64	nd (w)	280d				δ s^h						**B14**^{2}, 430
b1117	**—,2-amino-3,4-dimethyl-***	$C_8H_{11}NO_3S$. *See* b1091	201.25	pr (w+1)	300d				δ						**B14**^{1}, 731
b1118	**—,2-amino-3,5-dimethyl-***	$C_8H_{11}NO_3S$. *See* b1091	201.25	pr or pl (w)	d				i	i	i		i		**B14**^{2}, 452
b1119	**—,2-amino-3,6-dimethyl-***	$C_8H_{11}NO_3S$. *See* b1091	201.25	amor	260				δ		i		i		**B14**^{1}, 732
b1120	**—,2-amino-4,5-dimethyl-***	$C_8H_{11}NO_3S$. *See* b1091	201.25	pl	>300				δ^h						**B14**^{1}, 731
b1121	**—,3-amino-2,5-dimethyl-***	$C_8H_{11}NO_3S$. *See* b1091	201.25	nd (w+1)					δ						**B14**^{1}, 732
b1122	**—,3-amino-4,5-dimethyl-***	$C_8H_{11}NO_3S$. *See* b1091	201.25	red nd (w+1)	315				δ						**B14**^{1}, 731
b1123	**—,4-amino-2,3-dimethyl-***	$C_8H_{11}NO_3S$. *See* b1091	201.25	nd	305										**B14**^{1}, 731
b1124	**—,4-amino-2,5-dimethyl-***	$C_8H_{11}NO_3S$. *See* b1091	201.25	pl or nd	>300				s						**B14**^{1}, 732
b1125	**—,5-amino-2,3-dimethyl-***	$C_8H_{11}NO_3S$. *See* b1091	201.25	pl (w+2)	294d				s	δ			δ		**B14**^{1}, 731
b1126	**—,5-amino-2,4-dimethyl-***	$C_8H_{11}NO_3S$. *See* b1091	201.25	pr or nd	290d				δ	i			i	chl i	**B14**, 734
b1127	**—,3-amino-4-hydroxy-***	$C_6H_7NO_4S$. *See* b1091	189.19	rh (w+½)	>300 (155–6)				δ	i	i				**B14**, 814
b1128	**—,3-amino-4-hydroxy-5-nitro-***	$C_6H_6N_2O_6S$. *See* b1091	234.19	pr (w)					s^h	δ					**B14**^{1}, 748
b1129	**—,4(2-amino-1-naphthylazo)-***		327.37	dk gr nd (w)					s^h δ	v^h	i				**B16**^{2}, 199
b1130	**—,4(4-amino-1-naphthylazo)-***		327.37	dk gr cr (w)					δ v^h	δ					**B16**^{2}, 191

For explanations, symbols and abbreviations see beginning of table. For structural formulas see end of table.

No.	Name	Synonyms and Formula	Mol. wt.	Color, crystalline form, specific rotation and λ_{max} (log ε)	m.p. °C	b.p. °C	Density	n_D	Solubility w	al	eth	ace	bz	other solvents	Ref.
	Benzenesulfonic acid														
b1131	—,**4(4-aminophenylsulfonamido)-,amide**		327.38	$\lambda^{w,pH=2}$ 247.5 (4.14), $\lambda^{w,pH=11}$ 270.5 (4.36)	137	...	...	...	δ s^h	s	s	s	...	chl i peth i	**C51,** 2780
b1132	—,**2-bromo-***	$C_6H_5BrO_3S$. *See* b1091	237.08	nd	...	...	...	...	v	v	...	...	...	...	**B11,** 56
b1133	—,—,amide	$C_6H_6BrNO_2S$. *See* b1093	236.09	nd (w), pr (al)	186	...	...	...	δ	δ	...	...	...	...	**B11,** 56
b1134	—,—,chloride	$C_6H_4BrClO_2S$. *See* b1091	255.52	pr (eth)	51	...	...	...	d^h	d^h	δ	...	...	...	**B11,** 56
b1135	—,**3-bromo-**, amide	$C_6H_6BrNO_2S$. *See* b1093	236.09	nd or lf (w), pr (al)	154	...	...	...	δ s^h	δ s^h	...	...	...	...	**B11,** 57
b1136	—,**4-bromo-***	$C_6H_5BrO_3S$. *See* b1091	237.08	nd (al)	102–3	155^{25}	...	...	v	s	...	...	...	...	**B11,** 57
b1137	—,—,amide	$C_6H_6BrNO_2S$. *See* b1093	236.09	nd (w or dil al)	166	...	...	...	δ s^h	s	...	...	...	...	**B11,** 57
b1138	—,—,bromide	$C_6H_4Br_2O_2S$. *See* b1091	299.98	cr (eth or bz)	77	...	...	...	d^h	d^h	...	...	s	...	**B11**[1], 16
Ω b1139	—,—,chloride	$C_6H_4BrClO_2S$. *See* b1091	255.52	tcl or mcl pr (eth)	76	153^{15}	...	...	i	d	v	...	...	...	**B11,** 57
b1140	—,**2-chloro-**, amide	$C_6H_6ClNO_2S$. *See* b1093	191.64	lf (al)	188	...	...	...	...	s	...	...	...	...	**B11,** 54
b1141	—,**3-chloro-**, amide	$C_6H_6ClNO_2S$. *See* b1093	191.64	lf (dil al)	148	...	...	...	v^h	v	v	...	...	...	**B11,** 54
b1142	—,**4-chloro-***	$C_6H_5ClO_3S$. *See* b1091	192.62	nd (w + 1)	67	$147–8^{25}$	...	...	s	s	i	...	i	...	**B11,** 54
Ω b1143	—,—,amide	$C_6H_6ClNO_2S$. *See* b1093	191.64	nd (w)	145–6.5	...	...	...	v^h	v^h	v^h	...	...	...	**B11,** 55
Ω b1144	—,—,chloride	$C_6H_4Cl_2O_2S$. *See* b1091	211.07	pr or pl (eth)	55	141^{15}	...	...	d^h	d^h	v	...	v	...	**B11,** 55
Ω b1145	—,—,4-chlorophenyl ester*	Chlorfenson. Ovotran. $C_{12}H_8Cl_2O_3S$. *See* b1091	303.18	λ^{al} 230 (4.27), 266 (3.05), 269 (3.04), 277 (2.91)	86.5	...	...	...	i	δ	...	s	...	...	**C51,** 1264
b1146	—,**3,4-diamino-***	$C_6H_8N_2O_3S$. *See* b1091	188.22	nd	d	...	...	...	v^h	...	...	...	i	...	**B14,** 717
b1147	—,**3,5-diamino-***	$C_6H_8N_2O_3S$. *See* b1091	188.22	...	d	...	...	...	δ v^h	δ	i	...	...	...	**B14**[2], 446
b1148	—,**2,5-dichloro-**, dihydrate*	$C_6H_4Cl_2SO_3.2H_2O$. *See* b1091	263.11	nd (w + 2)	< 100	...	...	...	v	s	δ	...	...	...	**B11**[2], 30
b1149	—,—,chloride	$C_6H_3Cl_3SO_2$. *See* b1091	245.51	mcl pr (bz)	38	...	...	...	d^h	d^h	...	...	...	...	**B11**[2], 30
b1150	—,**3,4-dichloro-**, dihydrate*	$C_6H_4Cl_2SO_3.2H_2O$. *See* b1091	263.11	nd	71–2	...	...	...	v	v	v	...	...	...	**B11,** 55
b1151	—,—,chloride	$C_6H_3Cl_3SO_2$. *See* b1091	245.51	mcl pr	fp 22.4	...	...	...	d^h	d^h	...	...	...	...	**B11**[1], 16
b1152	—,**3,5-diiodo-4-hydroxy-**, trihydrate*	Sozoiodolic acid. $C_6H_4I_2SO_4.3H_2O$. *See* b1091	480.02	nd (eth), pr (w + 3)	120 (hyd) 190d (anh)	...	...	...	v	v	v	...	...	...	**B11**[2], 137
b1153	—,**2,3-dimethyl-**, chloride	$C_8H_9ClO_2S$. *See* b1091	204.68	pr (peth)	47	...	...	...	d	d^h	...	...	...	peth v^h	**B11,** 120
b1154	—,**3,4-dimethyl-***	$C_8H_{10}O_3S$. *See* b1091	186.23	pl or pr (chl + 2w)	63–4	...	...	...	v	...	...	...	...	...	**B11**[2], 78
b1155	—,—,chloride	$C_8H_9ClO_2S$. *See* b1091	204.68	pr (eth)	51–2	...	...	...	d	d^h	v	...	...	...	**B11,** 121
b1156	—,**3,5-dimethyl-**, chloride	$C_8H_9ClO_2S$. *See* b1091	204.68	nd (peth or bz)	94	...	...	...	d	d^h	v	...	...	...	**B11,** 126
Ω b1157	—,**2,4-dinitro-***	$C_6H_4N_2O_7S$. *See* b1091	248.17	nd (w + 3)	130 (anh) 108 (hyd)	...	...	...	v	v	δ	...	i	peth i	**B11**[2], 36
b1158	—,**3,5-dinitro-***	$C_6H_4N_2O_7S$. *See* b1091	248.17	ye red (dil al)	235	...	...	...	i s^h	s^h	δ	...	δ	aa δ lig δ	**B11,** 79
b1159	—,**3-ethylamino-***	$C_8H_{11}NO_3S$. *See* b1091	201.25	nd (w)	294d	...	...	...	δ	...	...	...	...	...	**B14**[2], 435
b1160	—,**4-ethylamino-***	$C_8H_{11}NO_3S$. *See* b1091	201.25	pl (w)	258d	...	...	...	v	...	...	...	...	...	**B14**[1], 721
b1161	—,**4-fluoro-**, amide	$C_6H_6FNO_2S$. *See* b1093	175.18	pl or nd (w, al)	126	...	...	...	δ	v	v	v	δ	...	**B11,** 54
Ω b1162	—,—,chloride	$C_6H_4ClFO_2S$. *See* b1091	194.61	pl or nd	36	$95–6^{2.2}$	...	...	d^h	d^h	v	...	v	chl v	**B11,** 53
Ω b1163	—,**2-formyl-**, chloride	2-Benzaldehyde sulfonyl chloride. $C_7H_5ClO_3S$. *See* b1091	204.55	cr (bz)	114	...	...	...	d	d	s	...	...	...	**B11**[2], 78
Ω b1164	—,**4-hydrazino-***	Phenylhydrazine-4-sulfonic acid. $C_6H_8N_2O_3S$. *See* b1091	188.22	nd or lf (w)	286	...	...	...	δ	δ	...	...	...	...	**B15**[2], 305
b1165	—,**2-hydroxy-***	*o*-Phenolsulfonic acid. $C_6H_6O_4S$. *See* b1091	174.18	cr (w + ½)	50d; 145d (+ w)	...	...	...	v	v	...	...	...	...	**B11**[2], 131
b1166	—,**4-hydroxy-***	*p*-Phenolsulfonic acid. $C_6H_6O_4S$. *See* b1091	174.18	nd	...	...	...	...	s	s	...	...	...	...	**B11**[2], 134
b1167	—,—,amide	$C_6H_7NO_3S$. *See* b1093	173.19	cr (al or w)	176–7	...	...	...	s^h	s^h	...	...	...	...	**B11,** 243
b1168	—,**4-hydroxy-5-isopropyl-2-methyl-**	α-Thymolsulfonic acid. $C_{10}H_{14}O_4S$. *See* b1091	230.29	pl (+ 1w)	91–2	...	...	...	v	...	...	...	...	...	**B11**[2], 151
b1169	—,**4-hydroxy-3-nitro-***	$C_6H_5NO_6S$. *See* b1091	219.17	ye pl (AcOEt-bz) nd (w + 3)	141–2 (anh) 51.5 (hyd)	...	...	...	s	v	...	...	...	AcOEt v chl v^h	**B11,** 245
b1170	—,**2-isopropyl-5-methyl-**	$C_{10}H_{14}O_3S$. *See* b1091	214.29	...	130–1	...	...	...	v	s	i	...	...	...	**B11**[2], 84
b1171	—,**4-isopropyl-2-methyl-**	$C_{10}H_{14}O_3S$. *See* b1091	214.29	pl or pr	88–90	...	...	...	v	δ	δ	...	δ	chl δ	**B11,** 139
b1172	—,**5-isopropyl-2-methyl-**	$C_{10}H_{14}O_3S$. *See* b1091	214.29	pl (+ 2w), mcl pr	220, 78–9 (+ 2w)	...	...	...	v	...	...	...	...	...	**B11**[2], 83

For explanations, symbols and abbreviations see beginning of table. For structural formulas see end of table.

No.	Name	Synonyms and Formula	Mol. wt.	Color, crystalline form, specific rotation and λ_{max} (log ε)	m.p. °C	b.p. °C	Density	n_D	Solubility w	al	eth	ace	bz	other solvents	Ref.
	Benzenesulfonic acid														
b1173	—,**2-methoxy-, chloride**	$C_7H_7ClO_3S$. *See* b1091	206.65	nd (peth)	56	126–9[0.3]	...	...	d	d	s	...	...	peth s^h	**B11**, 235
Ωb1174	—,**4-methoxy-, chloride**	$C_7H_7ClO_3S$. *See* b1091	206.65	nd or pr (bz)	42–3	...	...	...	d	s d^h	s	...	s^h	...	**B11**[2], 136
—	—,**methyl-***	*see* **Toluenesulfonic acid**													
b1175	—,**2-nitro-***	$C_6H_5NO_5S$. *See* b1091	203.17	...	85 (70)	...	...	...	v	s	i	...	...	...	**B11**[2], 31
b1176	—,—,chloride	$C_6H_4ClNO_4S$. *See* b1091	221.62	pr (lig, eth-peth) $\lambda^{95\%\,al}$ 280 sh (3.2)	68.5–9.5	...	...	...	d^h	d^h	s	...	...	peth δ	**B11**[2], 32
Ωb1177	—,**3-nitro-***	$C_6H_5NO_5S$. *See* b1091	203.17	pl	48	...	...	...	v^h	s	i	...	i	...	**B11**, 68
Ωb1178	—,—,chloride	$C_6H_4ClNO_4S$. *See* b1091	221.62	mcl pr (eth), nd (lig) $\lambda^{95\%\,al}$ 231–41 sh (4.0)	64	...	...	...	i d^h	s d^h	...	...	...	...	**B11**[2], 33
Ωb1179	—,**4-nitro-***	$C_6H_5NO_5S$. *See* b1091	203.17	(+2w) λ^w 271 (4.01)	95 (109–11)	...	...	...	v	...	...	...	...	...	**B11**[2], 33
Ωb1180	—,—,chloride	$C_6H_4ClNO_4S$. *See* b1091	221.62	mcl nd (peth, lig)	79.5–80.5	...	...	...	d^h	d^h	...	...	...	peth s^h	**B11**[1], 21
b1180[1]	—,**4-phenyl-amino-**	4-Sulfodiphenylamine. $C_{12}H_{11}NO_3S$. *See* b1091	249.29	pl (al-eth)	206.5	...	...	...	v	v	i	...	...	...	**B14**[1], 721
b1181	—,**4-ureido,** amide	$C_7H_9N_3O_3S$. *See* b1091	215.23	nd (al)	206–7	...	...	...	s^h	s^h	...	...	...	...	**C51**, 6692
Ωb1182	**1,2,3,4-Benzene-tetracarboxylic acid***	Prehnitic acid	254.16	pr (+6w) λ^w 257 sh (3.7), 294 (3.2)	241d	...	...	...	s	...	...	s	...	...	**B9**[2], 730
Ωb1183	—,tetramethyl ester*	$C_{14}H_{14}O_8$. *See* b1182	310.26	nd (MeOH or w) λ^{al} 240 sh (3.91), 288 (3.11)	133–5	...	...	...	i δ^h	v	δ	...	v	MeOH δ	**B9**[2], 730
Ωb1184	**1,2,3,5-Benzene-tetracarboxylic acid, tetramethyl ester***	Tetramethyl mellophanate.	310.26	nd (MeOH, al)	111	...	...	...	i	s^h	...	...	...	...	**B9**[1], 435
Ωb1185	**1,2,4,5-Benzene-tetracarboxylic acid***	Pyromellitic acid	254.16	tcl pl (w+2) λ^w 252 sh (3.9), 297 (3.4)	276 (281–4.5) (anh) 242 (+2w)	...	...	...	δ	s	...	...	...	...	**B9**, 998
b1186	—,tetraethyl ester*	$C_{18}H_{22}O_8$. *See* b1185	366.37	nd (al)	54	sub	...	...	i	s^h	...	...	...	...	**B9**, 998
Ωb1187	—,tetramethyl ester*	$C_{14}H_{14}O_8$. *See* b1185	310.26	lf (al, MeOH) λ^{al} 240 sh (3.96), 291 (3.37)	143–4	...	...	...	i	δ^h	...	...	...	...	**B9**, 998
—	**Benzenethiol***	*see* **Benzene, mercapto-***													
Ωb1188	**1,2,3-Benzenetri-carboxylic acid***	Hemimellitic acid	210.15	tcl pl (+2w), nd (w) λ^w 284 (2.9)	197d (anh) 223–4 (hyd)	...	1.546[20]	...	v^h	...	δ	...	...	con HCl i	**B9**[2], 712
b1190	**1,2,4-Benzenetri-carboxylic acid***	Trimellitic acid	210.15	nd (w), cr(aa or al) λ^w 248 sh (3.9), 296 (3.2)	238d	...	...	...	v	v	v	δ	i	CCl_4 i chl i CS_2 i aa s^h	**B9**[2], 712
Ωb1191	**1,3,5-Benzenetri-carboxylic acid***	Trimesic acid	210.15	pr or nd (w+1) λ^w 288 (2.8), 297 sh (2.6)	380 (anh)	...	...	...	δ	v	s	...	...	...	**B9**[2], 712
b1192	—,triethyl ester*	$C_{15}H_{18}O_6$. *See* b1191	294.31	pr or nd (al)	133.5–4.5	...	...	...	i	s^h	v	...	s	CS_2 v aa s	**B9**[2], 712
Ωb1193	—,trimethyl ester*	$C_{12}H_{12}O_6$. *See* b1191	252.23	nd (dil al) λ^{al} 216 (4.61)	144	...	...	...	i	s	...	...	...	...	**B9**[1], 430
b1194	—,**2-chloro-***	$C_9H_5ClO_6$. *See* b1191	244.59	nd or pl (w+1)	285 (anh) 278 (hyd)	sub	...	...	s^h	v	v	...	...	chl i	**B9**, 980
b1195	**1,2,4-Benzenetri-carboxylic acid, 6-hydroxy-***	$C_9H_6O_7$. *See* b1190	226.14	pr (w+2)	278–81d	...	...	...	s	s	...	...	...	...	**B10**, 580
b1196	**1,3,5-Benzenetri-carboxylic acid, 2-hydroxy-***	$C_9H_6O_7$. *See* b1191	226.14	pr (w+1), nd (w+2)	306 (anh)	...	...	...	s^h	v^h	δ	...	i	chl i lig i	**B10**, 580
b1197	**1,3,5-Benzenetri-sulfonic acid***	...	318.30	cr (w+3)	d >100	...	...	...	s	s	i	...	i	...	**B11**[2], 128
—	**Benzestrol**	*see* **Hexane,2,4-bis(4-hydroxyphenyl)-3-ethyl-***													
—	**Benzhydrol**	*see* **Methanol, diphenyl-***													
—	**Benzidine**	*see* **Biphenyl, 4,4′-diamino-**													
—	**Benzidinesulfone**	*see* **Dibenzothiophene, 2,7-diamino-, 9,9-dioxide**													
Ωb1198	**Benzil**	Dibenzoyl. 1,2-Diphenyl-1,2-ethandione. Diphenyl-glyoxal.* $C_6H_5COCOC_6H_5$	210.23	ye pr (al) λ^{al} 260 (4.3), 330 sh (2.4), 385 (1.9)	95–6	346–8d 188[12]	1.084_4^{102}	...	i	v	v	s	v	...	**B7**[2], 764

For explanations, symbols and abbreviations see beginning of table. For structural formulas see end of table.

No.	Name	Synonyms and Formula	Mol. wt.	Color, crystalline form, specific rotation and λ_{max} (log ε)	m.p. °C	b.p. °C	Density	n_D	Solubility w	al	eth	ace	bz	other solvents	Ref.
	Benzil														
Ω b1199	—,dioxime(*anti*)	α-Benzil dioxime. Diphenylglyoxime.	240.27	lf (al or ace) λ^{al} 226 (4.20), 240 (4.20)	238d	...	...	...	i	i	i	...	...	alk s aa i	**B7²**, 680
b1200	—,—(*syn*)	β-Benzil dioxime	240.27	nd (al) λ^{al} 252 (4.41)	207d	...	...	...	δ^h	s	s	...	...	NH_3 s alk s aa s	**B7²**, 681
b1201	—,—(*amphi*)	γ-Benzil dioxime	240.27	nd (al + 1), (aa)	164–6 100 (+al)	...	...	...	i	s	...	s	s	peth δ, lig i alk s CCl_4 δ	**B7²**, 681
b1202	—,monooxime(α)	...	225.25	lf (dil al, bz, MeOH) λ^{al} 250 sh (4.01)	137–8	200d	...	...	δ	v	v	...	δ	chl s aa v lig δ	**B7²**, 678
b1203	—,—(β)	...	225.25	pr or nd (bz + ½) λ^{al} 250 sh (4.01)	70 (+½bz) 113–4	...	...	...	δ	v	v	s	s	lig δ, chl s	**B7²**, 679
b1204	—,osazone(*anti*)	...	390.49	ye nd (al) λ^{al} 239 (4.6), 295 (4.3), 342 (4.6)	230–2	...	...	...	i	δ	...	δ	v^h	chl v^h	**B15²**, 72
b1205	—,—(*syn*)	...	390.49	ye nd (bz-al)	210	...	...	...	δ	s	...	δ	s	aa δ	**B15²**, 72
—	**Benzilic acid**	*see* **Acetic acid, diphenyl (hydroxy)-**													
Ω b1206	**Benzimidazole**	Benzoglyoxaline. 1,3-Benzodiazole.	118.14	rh bipym pl (w) λ^{w} 245 (3.72), 271 (3.79), 278 (3.78)	170.5 (173)	$>360^{760}$	...	...	δ v^h	v	δ	...	i	dil alk s dil HCl v lig i	**B23²**, 151
b1207	—,**2-amino-**	$C_7H_7N_3$. *See* b1206	133.15	pl (w) λ^{al} 244 (3.83), 283 (3.89)	224	...	...	...	s	s	δ	s	δ	...	**B24¹**, 240
b1208	—,**2-hydroxy-**	*o*-Phenylene urea. $C_7H_6N_2O$. *See* b1206	134.14	lf (w or al) λ^{al} 225.5 (3.86), 280 (3.88)	318d	...	...	...	δ	v	δ	s	δ	alk δ con HCl s	**B24²**, 62
Ω b1209	—,**5,6-dimethyl-**	$C_9H_{10}N_2$. *See* b1206	146.19	(eth) $\lambda^{w, pH=8}$ 246 (3.59), 279 (3.66), 286 (3.67)	205–6	140^3 sub	...	...	s	s	s	...	...	chl s	**C50**, 1087
b1210	—,**2-mercapto-**	$C_7H_6N_2S$. *See* b1206	150.20	pl (dil al, aq NH_3)	298 (304)	...	...	...	δ	v	...	...	...	...	**B24²**, 65
b1211	—,**1-methyl-**	$C_8H_8N_2$. *See* b1206	132.18	nd (peth), pl (al) λ^{al} 254 (3.78), 274 (3.70), 281 (3.70)	66	286^{756}	1.1254^{20}_{4}	1.6013^7	...	s^h	...	...	...	peth s^h	**B23²**, 152
Ω b1212	—,**2-methyl-**	$C_8H_8N_2$. *See* b1206	132.18	pr or nd (w) λ^{al} 274 (3.81), 288 (3.86)	176–7	...	...	...	s	δ	δ	...	i	...	**B23²**, 160
b1213	—,**4-methyl-**	$C_8H_8N_2$. *See* b1206	132.18	pl (w), nd (bz) λ^{al} 247 (3.85), 271 (3.60), 279 (3.63)	145	...	...	...	s^h	s	δ	...	...	...	**B23¹**, 38
Ω b1214	—,**5-methyl-**	$C_8H_8N_2$. *See* b1206	132.18	(w) λ^{al} 245 (3.74), 277 (3.71), 283 (3.69)	114	...	...	...	s^h	...	...	...	...	...	**B23¹**, 38
b1215	—,**2-methyl-1(2-naphthyl)-5-nitro-**	$C_{18}H_{13}N_3O_2$. *See* b1206	303.32	nd (al)	162	...	...	...	s	s	s	...	v	aa s	**B23**, 150
b1216	—,**2-methyl-5-nitro-1-phenyl-**	$C_{14}H_{11}N_3O_2$. *See* b1206	253.26	nd (al)	170–1	...	...	...	...	s^h	...	...	...	...	**B23²**, 161
Ω b1217	—,**6-nitro-**	$C_7H_5N_3O_2$. *See* b1206	163.14	nd (w) λ^{al} 235 (4.30), 302 (3.97)	209–10 (205)	...	...	...	i	v	i	...	i	chl i, ac s	**B23²**, 154
Ω b1218	—,**2-phenyl-**	$C_{13}H_{10}N_2$. *See* b1206	194.24	pl (aa, al-w), nd (bz, w) λ^{al} 240 (4.7), 301 (4.35)	293 (300)	...	...	...	δ	s	...	...	δ	chl δ aa s	**B23²**, 238
b1219	**4,5-Benzindane**	1,2-Cyclopentanonaphthalene.	168.24	oil λ 226 (4.89), 279 (3.78), 320 (3.17)	...	$294–5^{757}$ 170^5	1.066^{20}_{4}	1.6290^{20}	i	...	...	...	...	aa s	**E13**, 13
b1220	**1,2-Benzisothiazole, 5-hydroxy-3-phenyl-**	...	227.29	nd (dil al or aa)	159–60	...	...	...	...	v	...	v	δ	alk v aa v lig i	**B27²**, 87
b1221	**Benzisoxazole, 3-methyl-**	...	133.15	...	...	$108–10^{16}$	...	...	i	...	s	...	...	lig δ	**C45**, 2472

For explanations, symbols and abbreviations see beginning of table. For structural formulas see end of table.

No.	Name	Synonyms and Formula	Mol. wt.	Color, crystalline form, specific rotation and λ_{max} (log ε)	m.p. °C	b.p. °C	Density	n_D	Solubility w	al	eth	ace	bz	other solvents	Ref.
	1,2-Benzocarbazole														
b1222	**1,2-Benzo-carbazole,3,4-dihydro-**	...	219.29	(lig or MeOH) λ^{al} 249 (4.36), 320 (4.25), 329 (4.36)	163–4	...	...	...	...	...	...	...	...	MeOH s^h lig s^h	**B20^2**, 315
b1223	**1,2-Benzo-1-cyclooctene-3-one**	...	174.25	λ^{al} 249.6 (3.85)	...	$87–8^{0.001}$	...	1.5577^{25}	...	...	...	...	...	...	**Am 76**, 5462
—	**1,2-Benzodiazole**...	*see* **Indazole**													
b1224	**5,6-Benzoflavone**...	β-Naphthoflavone ...	272.31	nd (al) λ 274, 315	167–8	...	...	...	...	δ^h	s	...	v^h	con sulf s (bl)	**B17^2**, 414
Ω b1225	**7,8-Benzoflavone**...	α-Naphthoflavone ...	272.31	ye pl (al), lf or nd (dil al) λ^{al} 280 (4.48), 348 (3.74)	157–9 (167)	...	...	...	...	δ	...	...	...	sulf s (ye)	**B17^2**, 414
b1226	**3,4-Benzofluor-anthene**	...	252.32	nd (bz) λ^{al} 256 (4.54), 301 (4.54), 338 (3.96), 369 (3.84)	168	...	...	...	i	...	...	...	δ	...	**E14s**, 569
b1227	**10,11-Benzo-fluoranthene**	...	252.32	ye pl (al), nd (aa) λ^{al} 242 (4.60), 318 (4.48), 332 (4.04), 383 (4.00)	166	...	...	...	i	δ	...	...	...	aa δ	**E14s**, 570
b1228	**11,12-Benzofluor-anthene**	...	252.32	pa ye nd (bz) λ^{al} 240 (4.75), 308 (4.75), 400 (4.16)	217	480^{760}	...	...	i	s	...	...	s	aa s	**E14s**, 573
Ω b1229	**1,2-Benzofluorene**..	Chrysofluorene ...	216.29	pl (ace or aa) λ^{al} 263 (4.86), 296 (4.20), 316 (4.14)	189–90	413^{760} (398^{758})	...	...	...	δ	s	...	s^h	chl s	**E14**, 299
b1230	—,**9-phenyl-** ...	$C_{23}H_{16}$. *See* b1229 ...	292.39	nd (aa)	195.5	...	...	...	...	δ^h	v	...	v	sulf s	**E14**, 299
Ω b1231	**Benzofuran** ...	Coumaron.	118.14	λ^{al} 245 (4.0), 278 (3.3), 285 (3.3)	< −18	174^{760} $62–3^{15}$	1.0913^{23}	1.5611^{17}	i	s	s	...	...	...	**B17^2**, 57
b1232	—,**2(chloro-methyl)-2,3-dihydro-**	C_9H_9ClO. *See* b1231 ...	168.63	...	41–2	$118–9^{11}$	$1.2196_{16}^{7.5}$	$1.5620^{7.5}$	i	...	s	s	...	...	**C40**, 3755
—	—,**2,3-dihydro-**...	*see* **Coumaran**													
Ω b1233	—,**2-methyl-** ...	C_9H_8O. *See* b1231 ...	132.17	λ^{al} 250 (4.1), 275 (3.7), 285 (3.7), 310 (2.5)	...	197.3–$.8^{760}$ $93–4^{20}$	1.0540_4^{20}	1.5495^{22}	i	s	s	...	...	...	**B17^2**, 59
Ω b1234	—,**3-methyl-** ...	C_9H_8O. *See* b1231 ...	132.17	λ^{al} 248 (3.97), 276 (3.40), 283 (3.40)	...	$196–7^{742}$ 86^{20}	1.0540_4^{23}	1.5536^{16}	i	s	s	...	...	...	**B17^1**, 26
Ω b1235	—,**5-methyl-** ...	C_9H_8O. *See* b1231 ...	132.17	λ^{al} 245 (4.01), 288 (3.47)	...	197–9 83.5^{17}	1.0603^{19}	1.5570^{19}	i	s	s	...	...	sulf s	**B17^1**, 27
Ω b1236	—,**7-methyl-** ...	C_9H_8O. *See* b1231 ...	132.17	λ^{al} 244 (4.03), 282 (3.25)	...	190–1	1.0490^{19}	1.5525^{17}	i	s	s	...	...	sulf s	**B17**, 61
Ω b1237	**Benzohydroxamic acid**	...	137.14	rh ta, lf (eth) λ^{al} 270 (3.6)	131–2	exp	...	...	s^h	s	δ	...	δ	...	**B9^2**, 213
—	—,amide ...	*see* **Benzoic acid,** amidoxime.													
b1238	—,**2-hydroxy-** ...	Salicylhydroxamic acid. $C_7H_7NO_3$. *See* b1237	153.14	nd (aa)	168	...	...	...	δ	v	v	...	...	aa s^h	**B10^2**, 60
Ω b1239	**Benzoic acid** ...	Benzenecarboxylic acid.	122.13	mcl lf or nd λ^{al} 227 (4.06)	122.4	249^{760} 133^{10} (100 sub)	1.2659_4^{15} 1.0749^{130}	1.504^{132}	δ s^h	v	v	s	v^h s	CCl_4 s lig δ to, chl, MeOH s	**B9^2**, 72
b1240	—,4-acetamido-phenyl ester	*p*-Benzoyloxyacetanilide. $C_{15}H_{13}NO_3$. *See* b1239	255.28	nd (al)	171	...	...	...	δ^h	v	...	...	v	aa v lig i	**Am 70**, 2705
b1241	—,2-acetylphenyl ester	*o*-Benzyloxyacetophenone. $C_{15}H_{12}O_3$. *See* b1239	240.26	nd (al)	88	...	...	...	i	v	v	...	...	aa v	**B9^2**, 133
b1242	—,4-acetylphenyl ester	*p*-Benzyloxyacetophenone. $C_{15}H_{12}O_3$. *See* b1239	240.26	nd (al, aq MeOH)	135–6	...	...	...	i	s^h	s	...	s	aa v	**B9^2**, 133
Ω b1243	—,allyl ester ...	$C_6H_5CO_2CH_2CH{:}CH_2$ ...	162.19	ye	...	242 (228)	1.0578_{15}^{15}	1.5178^{20}	i	s	s	s	...	MeOH s	**B9^2**, 93
Ω b1244	—,amide ...	Benzamide ...	121.14	mcl pr or pl (w) λ^{al} 225 (4.0), 270 sh (2.8)	132.5–3.5	290	1.0792_4^{130} 1.341^4	...	δ v^h	v	δ	...	δ v^h	CCl_4 v CS_2 v	**B9^2**, 163
b1245	—,—,*N*(3-chloro-2-methylphenyl)-	$C_{14}H_{12}ClNO$. *See* b1244 ...	245.71	nd (al)	173	...	...	...	i	s	s	v	δ	aa δ	**B12^2**, 455
b1246	—,—,*N*,*N*-diphenyl-	*N*-Benzoyldiphenylamine. $C_6H_5CON(C_6H_5)_2$	273.34	rh pr (al), nd	180	...	...	...	δ	δ s^h	δ	...	...	...	**B12^2**, 155
Ω b1247	—,—,*N*(1-naph-thyl)-	$C_6H_5CONHC_{10}H_7^{\alpha}$ ...	247.30	nd (al or aa)	161–2	...	...	...	...	δ	...	...	s	aa s	**B12^2**, 525

For explanations, symbols and abbreviations see beginning of table. For structural formulas see end of table.

No.	Name	Synonyms and Formula	Mol. wt.	Color, crystalline form, specific rotation and λ_{max} (log ε)	m.p. °C	b.p. °C	Density	n_D	Solubility w	al	eth	ace	bz	other solvents	Ref.
	Benzoic acid														
b1248	—,—,*N*(2-nitrophenyl)-	$C_{13}H_{10}N_2O_3$. *See* b1244	242.24	gold-ye nd (al) λ^{al} 246 (4.3), 354 (3.5)	98				δ	s	v				**B12**[1], 342
•b1249	—,—,*N*(3-nitrophenyl)-	$C_{13}H_{10}N_2O_3$. *See* b1244	242.24	pl ($C_5H_{11}OH$) λ^{al} 262 (4.3), 319 sh (3.2)	157				i	δ	v			chl v	**B12**[2], 380
b1250	—,—,*N*(4-nitrophenyl)-	$C_{13}H_{10}N_2O_3$. *See* b1244	242.24	ye nd (AcOEt) λ^{al} 321 (4.3)	199				i	δ^h				chl i	**B12**[2], 391
Ω b1251	—,—,*N*-phenyl-	Benzanilide. $C_6H_5CONHC_6H_5$	197.24	lf (al) λ^{al} 222 (4.19), 267 (4.20)	163	$117–9^{10}$ sub	1.315		i	i δ v^h	δ			aa δ	**B12**[2], 152
b1252	—,—,*N*(2-tolyl)-	$C_{14}H_{13}NO$. *See* b1244	211.27	rh nd (AcOEt-ace)	145–6		1.205^{15}		δ^h	s	s				**B12**[2], 441
b1253	—,—,*N*(3-tolyl)-	$C_{14}H_{13}NO$. *See* b1244	211.27	mcl pr (dil al)	125		1.170^{15}			s					**B12**[1], 400
b1254	—,—,*N*(4-tolyl)-	$C_{14}H_{13}NO$. *See* b1244	211.27	rh nd (al)	158	232	1.202^{15}		i	s	s				**B12**[2], 505
b1255	—,amidine	Benzamidine. $C_6H_5C(:NH)NH_2$	120.17	lf (al) λ^{w} 229 (3.96), 268 (2.91)	80				s	v	δ				**B9**[2], 199
Ω b1256	—,—,hydrochloride	$C_6H_5C(:NH)NH_2.HCl$	156.62	rh pr (w + 2)	169, 72 (+2w)				s^h	s^h					**B9**[2], 199
b1257	—,—,*N,N*-diphenyl-	$C_6H_5C(:NH)N(C_6H_5)_2$	272.35	rh pl (eth) $\lambda^{al, 1N\ HCl}$ 225 (4.24), 270 (3.8)	112					s	s		s		**B12**[2], 155
b1258	—,—,—,hydrochloride	$C_6H_5C(:NH)N(C_6H_5)_2.HCl$.	308.82	mcl pr or nd (al) λ^{al} 225 (4.24), 270 (3.8)	223d				v	v					**B12**, 270
b1259	—,—,*N,N*-diphenyl-	$C_6H_5C(:NC_6H_5)NHC_6H_5$	272.35	nd (bz or al) pr (al) $\lambda^{al, 1N\ HCl}$ 273 (4.11)	146–7				δ	v	v				**B12**[2], 157
b1260	—,—,*N*(1-naphthyl)-	$C_6H_5C(:NH)NHC_{10}H_7^{\alpha}$	246.32	pl (al)	141				i	s	s				**B12**, 1233
b1261	—,amidoxime	Benzamidoxime. $C_6H_5C(NH_2):NOH$	136.17	mcl pr (w) λ^{w} 250 (3.73)	80				δ	s	v		s	chl s lig i	**B9**[2], 214
Ω b1262	—,anhydride	Benzoic anhydride. $(C_6H_5CO)_2O$	226.24	pr (eth) λ^{peth} 238 (4.52), 277 (3.41)	42–3	360^{760}	1.1989_4^{15}	1.5767^{15}	i	s	s			lig i	**B9**[2], 147
b1263	—,azide	Benzoyl azide. $C_6H_5CON_3$	147.14	pl (ace)	32	exp			i	s	s				**B9**[2], 219
b1264	—,benzoylmethyl ester	ω-Benzoyloxyacetophenone. $C_6H_5CO_2CH_2COC_6H_5$	240.26	pl (dil al)	118.5					s^h	v		v	chl v	**B9**[2], 133
Ω b1265	—,benzyl ester	$C_6H_5CO_2CH_2C_6H_5$	212.25	nd or lf λ^{al} 229 (4.2), 267 (3.0), 272 (3.0), 280 (2.9)	21	$323–4^{760}$ (cor) $170–1^{11}$	1.1121_4^{25}	1.5680^{20}	i	s	s	s	s	MeOH s chl s peth s	**B9**[2], 100
b1266	—,bromide	Benzoyl bromide. C_6H_5COBr	185.03		−24	218–9 48–$50^{0.05}$	1.570^{15}	1.5868^{25}	d	d	∞				**B9**[2], 162
Ω b1267	—,butyl ester	$C_6H_5CO_2C_4H_9^n$	178.23		−22.4	250.3^{760}	1.000^{20}	1.4940^{25}	i	∞	∞	s			**B9**[2], 90
Ω b1268	—,chloride	Benzoyl chloride. C_6H_5COCl	140.57		0	197.2^{760} 71^9	1.2120_4^{20}	1.5537^{20}	d	d	∞		s	CS_2 s	**B9**[2], 159
b1269	—,1-chloroethyl ester	$C_6H_5CO_2CHClCH_3$	184.62			134^{30}	1.172^{20}		i	s	s				**B9**[2], 127
b1270	—,2-chloroethyl ester	$C_6H_5CO_2CH_2CH_2Cl$	184.62			254^{729} $118–20^2$			i	v	v				**B9**[2], 90
b1271	—,cyclohexyl ester	$C_{13}H_{16}O_2$. *See* b1239	204.27		< −10	285	1.0429^{20}	1.5200^{25}	i	s	s				**B9**[1], 65
b1272	—,2-ethoxyethyl ester	Ethyl cellosolve benzoate. $C_6H_5CO_2(CH_2)_2OC_2H_5$	194.23	λ^{cy} 228 (4.08), 273 (2.94), 280 (2.86)		$260–1^{739}$	1.0585_{25}^{25}	1.4969^{25}	i	v	v	v	s	os v	**Am 54**, 4370
Ω b1273	—,ethyl ester	$C_6H_5CO_2C_2H_5$	150.18		−34.6	213^{760} 87^{10}	1.0468_4^{20}	1.5007^{20}	i	s	∞	s	s	chl s peth s	**B9**[2], 88
Ω b1274	—,fluoride	Benzoyl fluoride. C_6H_5COF	124.12			154–5	> 1		d^h	v	v				**B9**[1], 94
b1275	—,hexyl ester	$C_6H_5CO_2(CH_2)_5CH_3$	206.29	λ^{al} 228 (4.05), 273 (2.94), 280 (2.85)		272^{770} $139–40^8$			i	s		s			**B9**[2], 93
Ω b1276	—,hydrazide	Benzoyl hydrazine. $C_6H_5CONHNH_2$	136.17	pl (w) $\lambda^{0.1N\ NaOH}$ 265 (3.72), 280 (2.16)	113–7	267d			s	s	δ	δ		chl δ	**B9**[2], 214
b1277	—,2-hydroxyethyl ester	1,2-Ethanediol monobenzoate. $C_6H_5CO_2CH_2CH_2OH$	166.18		45	$150–1^{10}$			i	s					**B9**[2], 108
Ω b1278	—,2-hydroxyphenyl ester	Pyrocatechol monobenzoate. $C_{13}H_{10}O_3$. *See* b1239	214.22	nd (w)	130–1				s^h	s					**B9**, 130
Ω b1279	—,3-hydroxyphenyl ester	Resorcinol monobenzoate. $C_{13}H_{10}O_3$. *See* b1239	214.22	pr (dil al) λPr^iOH 320 (1.38)	135–6				i					chl v aa v	**B9**[2], 113

For explanations, symbols and abbreviations see beginning of table. For structural formulas see end of table.

No.	Name	Synonyms and Formula	Mol. wt.	Color, crystalline form, specific rotation and λ_{max} (log ε)	m.p. °C	b.p. °C	Density	n_D	Solubility w	al	eth	ace	bz	other solvents	Ref.
	Benzoic acid														
b1280	—,4-hydroxyphenyl ester	Hydroquinone monobenzoate. $C_{13}H_{10}O_3$. *See* b1239	214.22	nd (al)	163–4d	...	...	...	i	s	v	...	v	chl v	**B9**[2], 113
b1281	—,iodide	Benzoyl iodide. C_6H_5COI	232.02	nd	3	128^{20}	1.748^{18}_{18}	1.137^{20}	d	s	∞	...	...	...	**B9**[2], 162
Ω b1282	—,isobutyl ester	$C_6H_5CO_2C_4H_9^i$	178.23	...	...	242^{760}	0.9990^{20}_{4}	...	i	∞	∞	s	...	...	**B9**[2], 91
Ω b1283	—,isopropyl ester	$C_6H_5CO_2C_3H_7^i$	164.21	λ^{al} 228 (4.01), 273 (2.89), 280 (2.80)	...	218^{760}	1.0172^{15}_{15}	1.4890^{25}	i	s	s	s	...	...	**B9**[2], 90
b1284	—,2-methoxyethyl ester	$C_6H_5CO_2(CH_2)_2OCH_3$	180.21	...	...	$254–6^{760}$	1.0891^{25}_{25}	1.5040^{25}	δ	s	s	s	s	...	**B9**, 129
b1285	—,2-methoxy-phenyl ester	Guaiacyl benzoate. $C_{14}H_{12}O_3$. *See* b1239	228.25	(al) λ^{al} 224 (4.2), 271 (3.6),	58	...	...	...	δ^h	v^h	v	s	...	chl v aa δ	**B9**[2], 112
Ω b1286	—,methyl ester	$C_6H_5CO_2CH_3$	136.16	λ^{al} 227 (4.09)	−12.3	199.6^{760} $96–8^{24}$	1.0888^{20}_{4}	1.5164^{20}	i	s	∞	...	...	MeOH s	**B9**[2], 87
Ω b1287	—,3-methylbutyl ester	Isoamyl benzoate. $C_6H_5CO_2(CH_2)_2CH(CH_3)_2$	192.24	λ^{al} 228 (4.09), 273 (2.96), 280 (2.88)	...	262.3^{760} 133^{14}	1.0040^{20}_{4}	1.4950^{20}	i	s	∞	...	...	...	**B9**[2], 92
b1288	—,methylene diester	Methanediol dibenzoate. $(C_6H_5CO_2)_2CH_2$	256.26	nd or pr (eth)	99	225d	1.275^{22}	...	i	δ	s	s	s	MeOH v^h CCl_4 v peth δ	**B9**[2], 127
b1289	—,1-naphthyl ester	$C_6H_5CO_2C_{10}H_7^{\alpha}$	248.29	pl or pr (al-eth)	56	...	...	...	i	s^h	v	...	...	...	**B9**, 125
Ω b1290	—,2-naphthyl ester	$C_6H_5CO_2C_{10}H_7^{\beta}$	248.29	nd or pr (al)	107	...	...	...	i	s^h	δ	...	...	aa δ v^h	**B9**, 125
Ω b1291	—,nitrile	Phenyl cyanide. Benzonitrile.	103.13	λ^{al} 222 (4.11), 230 (4.04)	−13	190.7 69^{10}	1.0102^{15}_{15}	1.5289^{20}	δ^h	∞	∞	v	v	...	**B9**[2], 196
b1292	—,2-nitrobenzyl ester	$C_{14}H_{11}NO_4$. *See* b1239	257.25	nd (dil al)	101–2	...	...	...	i	s^h	s	...	s	aa s lig δ	**B9**, 121
b1293	—,3-nitrobenzyl ester	$C_{14}H_{11}NO_4$. *See* b1239	257.25	...	71–2	...	...	...	i	s	s	...	...	...	**C25**, 92
b1294	—,4-nitrobenzyl ester	$C_{14}H_{11}NO_4$. *See* b1239	257.25	...	94–5	...	...	...	i	s	s	...	...	...	**B9**[1], 68
Ω b1295	—,phenyl ester	$C_6H_5CO_2C_6H_5$	198.22	mcl pr (eth-al) λ^{al} 265 sh (3.3)	71	314^{760} (299)	1.235^{20}_{4}	...	i	s^h	s^h	...	...	...	**B9**[2], 96
b1296	—,1-phenylethyl ester	$C_6H_5CO_2CH(CH_3)C_6H_5$	226.28	...	...	189^{21} 123.6^{2}	1.1108^{18}	1.5588^{21}	i	s	...	...	...	...	**B9**, 121
b1297	—,2-phenyl-hydrazide	1-Benzoyl-2-phenylhydrazine. $C_6H_5CONHNHC_6H_5$	212.25	pr (al), nd (w), lf (dil al) λ^{al} 279 (3.5)	168	314	...	...	δ	s^h	δ	...	s	chl s	**B15**[2], 97
Ω b1298	—,4-phenyl-phenacyl ester	...	316.34	...	167	...	...	...	...	...	...	...	...	...	**C24**, 5031
Ω b1299	—,propyl ester	$C_6H_5CO_2C_3H_7^n$	164.21	...	−51.6	231^{760}	1.0230^{20}_{4}	1.5000^{20}	i	∞	∞	...	...	...	**B9**[2], 90
b1300	—,tetrahydro-furfuryl ester	...	206.24	...	...	$300–2^{750}$ $138–40^{2}$	1.137^{20}_{0}	...	i	∞	∞	...	...	chl ∞	**B17**[2], 107
b1301	—,2-tolyl ester	$C_{14}H_{12}O_2$. *See* b1239	212.25	λ 273 (3.3)	...	$307–8^{728}$ $154–6^{8.5}$	1.114^{19}	...	i	s	v	...	...	...	**B9**[2], 98
b1302	—,3-tolyl ester	$C_{14}H_{12}O_2$. *See* b1239	212.25	λ 276 sh (3.2)	55–6	314 $168–70^{8}$	...	...	i	s	s	...	...	CS_2 s	**B9**[2], 99
Ω b1303	—,4-tolyl ester	$C_{14}H_{12}O_2$. *See* b1239	212.25	pl (eth-al) λ 275 sh (3.4)	71.5	316	...	...	i	s	s	...	...	CS_2 s	**B9**[2], 99
b1304	—,**2-acetamido-**	$C_9H_9NO_3$. *See* b1239	179.18	nd (aa) λ^{al} 221 (4.41), 252 (4.13), 305 (3.67)	185	...	...	...	δ v^h	s^h	v	v	v	aa v^h	**B14**[2], 219
b1305	—,**3-acetamido-**	$C_9H_9NO_3$. *See* b1239	179.18	nd (al) λ^{al} 297 (3.2)	248–50	...	...	...	i δ^h	v^h	δ	...	...	...	**B14**[2], 241
Ω b1306	—,**4-acetamido-**	$C_9H_9NO_3$. *See* b1239	179.18	nd (aa) λ^{al} 268 (4.3)	256.5 (250d)	...	...	...	i	s	δ	...	...	...	**B14**[2], 264
b1307	—,**2-acetamido-4-ethoxy-**	$C_{11}H_{13}NO_4$. *See* b1239	223.23	nd (al or MeOH)	199d	...	...	...	δ^h	v^h	δ^h	...	δ^h	chl δ^h MeOH v^h	**B14**[1], 657
b1308	—,**5-acetamido-2-ethoxy-**	$C_{11}H_{13}NO_4$. *See* b1239	223.23	nd (w)	190	...	...	...	δ^h	s	...	...	...	...	**B14**, 583
b1309	—,**2-acetyl-**	$C_9H_8O_3$. *See* b1239	164.17	nd (w), pr (bz) λ^{al} 228 (3.9)	114–5	$110–2^{2}$	...	...	s^h	∞	...	...	...	...	**B10**[2], 479
b1310	—,**4-acetyl-**	$C_9H_8O_3$. *See* b1239	164.17	nd (w)	210	sub	...	...	s^h	δ	δ	...	...	chl δ lig i	**B10**[2], 480
b1311	—,—,methyl ester	$C_{10}H_{10}O_3$. *See* b1239	178.19	nd (w)	95	$140–5^{4}$, sub	...	...	s^h	...	...	...	...	...	**B10**, 695
b1312	—,**5-allyl-2-hydroxy-3-methoxy-**	Eugenic acid. $C_{11}H_{12}O_4$. *See* b1239	208.22	pl (w + 1)	85–8 (+1w) 127 (anh)	...	...	...	i s^h	s	s	...	...	...	**B10**[1], 215
Ω b1314	—,**2-amino-**	Anthranilic acid. $C_7H_7NO_2$. *See* b1239	137.14	lf (al) λ^{al} 247 (3.83), 332 (3.65)	146–7	sub	1.412^{20}	...	s v^h	s v^h	s	...	δ s^h	chl v^h Py v	**B14**[2], 205
Ω b1315	—,—,amide	$C_7H_8N_2O$. *See* b1244	136.17	lf (chl or w) λ^{al} 250 (3.8), 335 (3.6)	110–11.5	300	...	...	s^h	s	δ	...	δ	AcOEt v	**B14**[2], 210
Ω b1316	—,—,butyl ester	$C_{11}H_{15}NO_2$. *See* b1239	193.25	...	<0	182^{760}	...	...	...	...	...	...	...	...	**B14**[2], 209
Ω b1317	—,—,ethyl ester	$C_9H_{11}NO_2$. *See* b1239	165.19	λ^{al} 225 (4.3), 250 (3.85), 338 (3.7)	13	268^{760} $145–7^{15}$	1.1174^{20}_{4}	1.5646^{20}	i	s	s	...	...	...	**B14**[2], 209

For explanations, symbols and abbreviations see beginning of table. For structural formulas see end of table.

No.	Name	Synonyms and Formula	Mol. wt.	Color, crystalline form, specific rotation and λ_{max} (log ε)	m.p. °C	b.p. °C	Density	n_D	Solubility w	al	eth	ace	bz	other solvents	Ref.
	Benzoic acid														
Ω b1318	—,—,isobutyl ester	$C_{11}H_{15}NO_2$. *See* b1239	193.25	...	...	$156\text{–}7^{13.5}$	...	...	...	...	...	...	...	...	**B14**^{2}, 209
Ω b1319	—,—,methyl ester	$C_8H_9NO_2$. *See* b1239	151.17	λ^{al} 248 (3.7), 341 (3.7)	24–5	256^{760} 135.5^{15}	1.1682_4^{19}	1.5810^{25}	δ	v	v	...	...	...	**B14**^{2}, 208
b1320	—,—,nitrile	*o*-Cyanoaniline. $C_7H_6N_2$. *See* b1291	118.14	ye pr (CS_2), nd (peth) λ^{al} 245 (3.8), 325 (3.7)	51	263^{751}	...	...	δ	v	v	v	v	Py v chl v peth i CS_2 v	**B14**^{2}, 210
b1321	—,—,phenyl ester	$C_{13}H_{11}NO_2$. *See* b1239	213.24	nd (al)	70	...	...	...	δ	v	v	...	...	...	**B14**^{2}, 210
b1322	—,—,propyl ester	$C_{10}H_{13}NO_2$. *See* b1239	179.22	...	...	270^{760}	...	...	...	...	...	...	...	...	**B14**^{2}, 209
Ω b1323	**—,3-amino-**	$C_7H_7NO_2$. *See* b1239	137.14	ye nd (w) λ^{al} 241 sh (3.85), 319 (3.32)	174 (180)	sub δd	1.5105^4	...	δ s^h	δ s^h	s	v	i	chl δ, v^h MeOH v	**B14**^{2}, 237
b1324	—,—,amide	$C_7H_8N_2O$. *See* b1244	136.17	ye mcl nd (+1w), nd (bz) λ^{al} 240 sh (3.8), 310 (3.3)	79–80 (hyd) 113–14 (anh)	...	...	...	s	s	s	...	δ	chl δ	**B14**^{1}, 559
Ω b1326	—,—,ethyl ester	$C_9H_{11}NO_2$. *See* b1239	165.19	...	...	294 $160\text{–}1^5$	1.1248^{22}	1.5600^{22}	δ	v	v	...	...	...	**B14**^{2}, 239
b1327	—,—,methyl ester	$C_8H_9NO_2$. *See* b1239	151.17	lf λ^{al} 250 sh (3.7), 323 (3.3)	39 (53–4)	$152\text{–}3^{11}$	1.232^{20}	...	...	v	v	...	v	chl v lig s peth δ	**B14**^{2}, 238
b1328	—,—,nitrile	*m*-Cyanoaniline. $C_7H_6N_2$. *See* b1291	118.14	nd (dil al or CCl_4) λ^{al} 250 (3.8), 320 (3.4)	53–4	$288\text{–}90^{760}$	...	...	δ s^h	v	v	v	...	CS_2 s^h chl v	**B14**^{2}, 240
Ω b1329	**—,4-amino-**	$C_7H_7NO_2$. *See* b1239	137.14	mcl pr (w) λ^{al} 220 (3.95), 289 (4.27)	188–9	...	1.374_4^{25}	...	s^h	s v^h	s	...	i	chl i, v^h	**B14**^{2}, 246
b1330	—,—,amide	$C_7H_8N_2O$. *See* b1244	136.17	ye cr (+ ¼w) λ^{al} 285 (4.3)	183	...	...	...	δ	s	s	...	...	...	**B14**, 425
b1331	—,—,—,*N*(2-diethylamino-ethyl)-, hydro-chloride	Procaine amide hydro-chloride. $C_{13}H_{21}N_3O$. HCl. *See* b1244	271.79	cr λ^w 278	165–9 (177–9)	...	...	...	v	s	i	...	i	chl δ	**C54**, 428
b1332	—,—,—,*N*-phenyl-	$C_{13}H_{12}N_2O$. *See* b1244	212.25	λ^{al} 295 (4.1)	135–6	...	...	...	...	s	...	...	...	...	**B14**, 425
Ω b1333	—,—,butyl ester	Butesin. $C_{11}H_{15}NO_2$. *See* b1239	193.25	(al or bz)	58	$173\text{–}4^8$	...	...	i	s	s	...	s	chl s	**B14**^{2}, 249
b1334	—,—,—,picrate	$C_{28}H_{33}N_5O_{11}$. *See* b1239	615.60	ye pw	109–10	...	...	...	δ	s	s	s	s	chl s	**B14**^{2}, 249
Ω b1335	—,2-diethyl-aminoethyl ester	Novocaine. Procaine. $C_{13}H_{20}N_2O_2$. *See* b1239	236.32	nd (w +2), pl (lig or eth) λ^w 221 (3.90), 290 (4.23)	51 (+2w) 61	...	...	...	δ	s	s	...	s	chl s	**C54**, 428
Ω b1336	—,—,—,hydro-chloride	Novocaine hydrochloride Procaine hydrochloride. $C_{13}H_{20}N_2O_2$. HCl. *See* b1239	272.78	nd (al), mcl or tcl pl (w)	156	...	0.707^{17}	...	v	s	i	...	...	chl δ	**B14**^{2}, 251
Ω b1337	—,—,ethyl ester	Benzocaine. $C_9H_{11}NO_2$. *See* b1239	165.19	nd (w), rh (eth) λ^{al} 292 (4.4)	92	310	...	...	i	v	v	...	...	chl s ac s	**B14**^{2}, 248
Ω b1338	—,—,methyl ester	$C_8H_9NO_2$. *See* b1239	151.17	lf or nd (aq MeOH) λ^{al} 294 (4.3)	114	...	...	...	...	...	...	...	...	...	**B14**^{2}, 247
b1339	—,—,nitrile	$C_7H_6N_2$. *See* b1291	118.14	pr or pl (w) λ^{al} 275 (4.3)	86	...	...	...	δ v^h	v	v	v	v	chl v, aa v lig δ CS_2 δ CCl_4 δ	**B14**^{1}, 570
b1340	—,—,phenyl ester	$C_{13}H_{11}NO_2$. *See* b1239	213.24	nd (al)	173	...	...	...	...	v	v	...	...	...	**B14**^{1}, 568
b1341	—,—,propyl ester	Propaesin. $C_{10}H_{13}NO_2$. *See* b1239	179.22	pr	75	...	...	...	δ	v	v	...	v	chl v	**B14**^{2}, 248
b1342	—,—,2,2,2-trichlo-roethyl ester	$C_9H_8Cl_3NO_2$. *See* b1239	268.54	nd (peth)	87	...	...	...	i	...	v	...	...	...	**Am 77**, 1575
b1343	**—,2-amino-3,4-dichloro-**	$C_7H_5Cl_2NO_2$. *See* b1239	206.03	nd (aa)	237–8 (242)	...	...	...	δ	v	v	...	δ	chl s aa δ, s^n	**B14**, 367
Ω b1344	**—,2-amino-3,5-dichloro-**	$C_7H_5Cl_2NO_2$. *See* b1239	206.03	nd or lf (al)	231–2d	...	...	...	i	v	v	v	s	os v	**B14**^{1}, 549
b1345	**—,2-amino-3,6-dichloro-**	$C_7H_5Cl_2NO_2$. *See* b1239	206.03	nd (w or aa)	155	sub	...	...	v^h	v	v	v	v^h	aa s^h	**B14**, 367
b1346	**—,2-amino-4,5-dichloro-**	$C_7H_5Cl_2NO_2$. *See* b1239	206.03	nd (aa)	213–4	...	...	...	δ	s	s	...	...	aa s	**B14**^{1}, 549
b1347	**—,4-amino-3,5-dichloro-**	$C_7H_5Cl_2NO_2$. *See* b1239	206.03	(al) $\lambda^{w,pH=1}$ 220 (4.34), 280 (4.17)	291	...	...	...	i	δ s^h	δ	...	δ	chl δ aa δ peth δ	**B14**^{2}, 271
b1348	**—,6-amino-2,3-dichloro-**	$C_7H_5Cl_2NO_2$. *See* b1239	206.03	nd (MeOH)	176–7d	...	...	...	s^h	v	v	...	...	MeOH v aa v	**B14**, 368
Ω b1349	**—,2-amino-3,5-diiodo-**	$C_7H_5I_2NO_2$. *See* b1239	388.93	pr (al)	232–3	...	...	...	i^h	δ	v	...	δ	...	**B14**^{2}, 233
b1350	—,—,ethyl ester	$C_9H_9I_2NO_2$. *See* b1239	416.99	pr (al)	101	...	...	...	...	δ s^h	...	...	...	...	**B14**^{1}, 555
b1351	**—,2-amino-4,5-diiodo-**	$C_7H_5I_2NO_2$. *See* b1239	388.93	cr (dil NH_3)	210–20d	...	...	...	i^h	s	s	...	s^h	...	**B14**^{1}, 555
b1352	—,—,ethyl ester	$C_9H_9I_2NO_2$. *See* b1239	416.99	pr (al)	137	...	...	...	δ^h	s^h	s	...	s	...	**B14**^{1}, 555

For explanations, symbols and abbreviations see beginning of table. For structural formulas see end of table.

No.	Name	Synonyms and Formula	Mol. wt.	Color, crystalline form, specific rotation and λ_{max} (log ε)	m.p. °C	b.p. °C	Density	n_D	Solubility w	al	eth	ace	bz	other solvents	Ref.
	Benzoic acid														
Ω b1353	—,4-amino-3,5-diiodo-	$C_7H_5I_2NO_2$. *See* b1239	388.93	nd (aa-NH_3)	>350	...	...	...	i	i	...	...	...	$PhNO_2$ δ AcOEt δ aa i	**B14**, 439
b1354	—,—,ethyl ester	$C_9H_9I_2NO_2$. *See* b1239	406.99	nd (al)	148	...	...	...	...	δ	...	...	...	...	**B14**, 439
Ω b1355	—,2-amino-3-hydroxy-	$C_7H_7NO_3$. *See* b1239	153.14	lf (w) $\lambda^{0.5N\ NaOH}$ 230 (4.27), 331 (3.63)	164	...	...	...	s^h δ	s	s	...	...	chl s	**B14**[2], 355
b1356	—,2-amino-4-hydroxy-	$C_7H_7NO_3$. *See* b1239	153.14	nd (w) λ^{al} 261 (4.04), 320 (3.78)	148d	...	...	...	s v^h	s	s	s	δ	chl, to δ	**B14**[2], 359
b1357	—,2-amino-5-hydroxy-	$C_7H_7NO_3$. *See* b1239	153.14	vt pr (w) λ^{al} 250 sh (3.7), 360 (3.5)	252d (darkens 235)	...	...	...	s^h	s	s	s	s	os s	**B14**[2], 357
b1358	—,3-amino-2-hydroxy-	$C_7H_7NO_3$. *See* b1239	153.14	...	235d	...	...	...	i	i	i	...	...	...	**B14**[2], 350
Ω b1359	—,3-amino-4-hydroxy-	$C_7H_7NO_3$. *See* b1239	153.14	pr (w + 1) λ^w 253 (3.84), 297 (3.54)	210 (anh)	...	...	...	s^h	δ^h	i	...	i	chl i aa s^h	**B14**, 593
Ω b1360	—,—,methyl ester	$C_8H_9NO_3$. *See* b1239	167.17	(i) nd (bz or aa) (st) (ii) nd (chl) (unst)	(i) 143 (ii) 111	...	...	...	i s^h	v	s	...	δ	ac s alk s	**B14**[2], 360
Ω b1361	—,4-amino-2-hydroxy-	$C_7H_7NO_3$. *See* b1239	153.14	nd, pl (al-eth) λ^{al} 237 (3.89), 279 (4.09), 305 (4.12)	150–1d (240)	...	...	...	s	s	s	s	i	peth, chl i	**B14**, 579
b1362	—,4-amino-3-hydroxy-	$C_7H_7NO_3$. *See* b1239	153.14	pl (dil al)	216	...	...	...	δ	v	...	v	...	...	**B14**[2], 356
b1363	—,—,methyl ester	Orthoform. $C_8H_9NO_3$. *See* b1239	167.17	pl (bz, w)	120–1 (142)	...	...	...	δ	v	v	s	s	aa v lig i	**B14**[2], 356
Ω b1364	—,5-amino-2-hydroxy-	$C_7H_7NO_3$. *See* b1239	153.14	nd (w)	283	...	...	...	δ^h	i	...	...	...	...	**B14**[2], 352
b1365	—,2-amino-4-iodo-	$C_7H_6INO_2$. *See* b1239	263.04	pr (dil al)	208d	...	...	...	δ	v	v	...	δ	...	**B14**[1], 554
Ω b1366	—,2-amino-5-iodo-	$C_7H_6INO_2$. *See* b1239	263.04	nd or pr (dil al)	210.5d	...	...	...	δ	v	v	v	s	os v peth δ	**B14**[2], 233
Ω b1367	—,2-amino-3-methyl-	$C_8H_9NO_2$. *See* b1239	151.17	nd (al), pr (w)	172	...	...	...	δ	v	v	...	...	...	**B14**[2], 290
b1368	—,2-amino-4-methyl-, nitrile	$C_8H_8N_2$. *See* b1291	132.18	lf (dil al)	94	...	...	...	i	s	...	s	s	chl s	**B14**[2], 291
b1369	—,2-amino-5-methyl-	$C_8H_9NO_2$. *See* b1239	151.17	lf (al), nd (w)	175	...	...	...	δ	s	s	...	...	...	**B14**[2], 291
b1370	—,—,nitrile	$C_8H_8N_2$. *See* b1291	132.18	cr (dil al)	63	...	...	...	s^h	s	s	s	s	os s lig δ	**B14**, 482
b1371	—,2-amino-6-methyl-	$C_8H_9NO_2$. *See* b1239	151.17	nd (MeOH)	125–6d	...	...	...	...	...	...	...	...	MeOH s^h	**B14**[2], 290
b1372	—,—,nitrile	$C_8H_8N_2$. *See* b1291	132.18	ye pr (bz), (w)	128	...	...	...	δ s^h	...	...	...	δ	...	**B14**[2], 290
b1373	—,3-amino-2-methyl-, nitrile	$C_8H_8N_2$. *See* b1291	132.18	nd (w)	95.5	...	...	...	δ	...	...	...	...	...	**B14**, 477
b1374	—,3-amino-4-methyl-	$C_8H_9NO_2$. *See* b1239	151.17	nd (al)	164–6	...	...	...	s	s	...	...	...	...	**B14**, 487
b1375	—,—,nitrile	$C_8H_8N_2$. *See* b1291	132.18	pr (al)	81–2	...	...	...	δ	s	s	v	v	os v	**B14**, 487
b1376	—,3-amino-5-methyl-, nitrile	$C_8H_8N_2$. *See* b1291	132.18	nd (lig)	75	...	...	...	...	s	...	...	...	lig δ	**B14**[1], 600
b1377	—,4-amino-2-methyl-	$C_8H_9NO_2$. *See* b1239	151.17	nd (a) $\lambda^{w, pH = 3.6}$ 223 (3.98), 288 (4.08)	165d (153)	sub	...	...	...	s^h	...	...	...	...	**B14**, 477
b1378	—,—,nitrile	$C_8H_8N_2$. *See* b1291	132.18	rh cr (al)	90	...	...	...	...	s^h	...	...	...	...	**B14**[1], 598
Ω b1379	—,4-amino-3-methyl-	$C_8H_9NO_2$. *See* b1239	151.17	nd (w)	170	...	...	...	v^h	...	...	...	...	...	**B14**[2], 290
b1380	—,—,nitrile	$C_8H_8N_2$. *See* b1291	132.18	nd (w) λ^w 240 (3.89), 320 (3.58)	95	...	...	...	s^h	...	...	...	...	...	**B14**[1], 600
b1381	—,5-amino-2-methyl-	$C_8H_9NO_2$. *See* b1239	151.17	pr (w)	196	...	...	...	s^h	v^h	...	...	...	...	**B14**[2], 290
b1382	—,—,nitrile	$C_8H_8N_2$. *See* b1291	132.17	nd (peth)	88	100–10[22]	...	...	δ^h	s	...	...	...	peth δ^h	**B14**[2], 290
b1383	—,4-amino-methyl-, nitrile, hydrochloride.	$C_8H_8N_2 \cdot HCl$. *See* b1291	168.63	pl (al)	274	...	...	...	...	δ	...	...	...	...	**B14**, 488
b1384	—,2-amino-3-nitro-	$C_7H_6N_2O_4$. *See* b1239	182.14	ye nd (w)	208–9	...	1.558^{15}	...	i	v	v	...	δ	chl δ	**B14**[2], 233
Ω b1385	—,2-amino-4-nitro-	$C_7H_6N_2O_4$. *See* b1239	182.14	og pr (dil al)	269	...	...	...	i δ^h	v	v	v	...	xyl s	**B14**[2], 234
b1386	—,2-amino-5-nitro-	$C_7H_6N_2O_4$. *See* b1239	182.14	lf (al), ye nd (w, dil al)	268–70 (280)	...	...	...	i s^h	s	s	...	i	chl i, xyl i	**B14**[2], 234
b1387	—,2-amino-6-nitro-	$C_7H_6N_2O_4$. *See* b1239	182.14	ye nd or lf (w)	184	...	...	...	v^h	v	v	v	δ	chl δ aa v	**B14**[1], 557
b1388	—,3-amino-2-nitro-	$C_7H_6N_2O_4$. *See* b1239	182.14	ye nd (w, dil al)	156–7	...	...	...	δ s^h	s	s	v	δ	aa s^h lig i	**B14**, 414

For explanations, symbols and abbreviations see beginning of table. For structural formulas see end of table.

No.	Name	Synonyms and Formula	Mol. wt.	Color, crystalline form, specific rotation and λ_{max} (log ε)	m.p. °C	b.p. °C	Density	n_D	Solubility w	al	eth	ace	bz	other solvents	Ref.
	Benzoic acid														
b1389	—,3-amino-4-nitro-	$C_7H_6N_2O_4$. *See* b1239	182.14	red pl or nd (al)	298d	...	...	...	δ	s	s	v	...	aa s^h	**B14**, 415
Ω b1390	—,3-amino-5-nitro-	$C_7H_6N_2O_4$. *See* b1239	182.14	ye pr (w)	209–10	...	...	...	δ	s^h	δ	...	δ	CS_2 δ aa v^h	**B14**, 415
b1391	—,4-amino-2-nitro-	$C_7H_6N_2O_4$. *See* b1239	182.14	red nd (w), pr (dil aa)	239.5d	...	...	...	s^h δ	s	δ	...	δ	aa s	**B14**^{1}, 583
b1392	—,4-amino-3-nitro-	$C_7H_6N_2O_4$. *See* b1239	182.14	red-ye nd (al)	284d	...	...	...	i	δ^h	δ	v	...	aa s	**B14**^{1}, 583
b1393	—,5-amino-2-nitro-	$C_7H_6N_2O_4$. *See* b1239	182.14	ye nd or pr (w)	235d	...	...	...	δ^h	s	δ	v	...	aa s	**B14**^{2}, 245
Ω b1394	—,3-amino-2,4,6-tribromo-	$C_7H_4Br_3NO_2$. *See* b1239	373.84	nd (w)	171.5–3	...	...	...	s^h δ	v	...	...	...	...	**B14**, 413
b1395	—,2-benzamido-	$C_{14}H_{11}NO_3$. *See* b1239	241.25	nd (al or bz)	181	...	...	...	i	v	v	...	...	...	**B14**^{2}, 221
b1396	—,3-benzamido-	$C_{14}H_{11}NO_3$. *See* b1239	241.25	red pr (al)	252–3	...	1.5105^4_4	...	δ	s	δ	...	...	...	**B14**^{1}, 562
b1397	—,4-benzamido-	$C_{14}H_{11}NO_3$. *See* b1239	241.25	nd (al)	278	...	...	...	δ^h	s	s	...	...	aa s	**B14**^{1}, 577
Ω b1398	—,2-benzoyl-	2-Benzophenonecarboxylic acid. $C_{14}H_{10}O_3$. *See* b1239	226.24	tcl nd (w + 1) $\lambda^{H_3PO_4-P_2O_5}$ 234 (4.05), 278 (4.08)	127–9 (anh) 93.4 (+w)	...	...	...	...	v	v	...	s^h	...	**B10**^{2}, 517
b1399	—,—,amide	$C_{14}H_{11}NO_2$. *See* b1244	225.25	nd (to)	165 (cor)	...	...	...	i	v	...	...	v	to s^h	**B10**, 749
Ω b1400	—,—,ethyl ester	$C_{16}H_{14}O_3$. *See* b1239	254.29	rh pl (dil al)	58	...	1.221^{64}_4	1.560^{64}	i	v	v	...	...	sulf s	**B10**^{2}, 517
Ω b1401	—,—,methyl ester	$C_{15}H_{12}O_3$. *See* b1239	240.26	pl or mcl pr (dil al) λ^{MeOH} 247 (4.22)	52	$350–2^{760}$	1.1903^{19}_4	1.591^{20}	i	v	v	...	...	sulf s	**B10**^{2}, 517
b1402	—,3-benzoyl-	3-Benzophenonecarboxylic acid. $C_{14}H_{10}O_3$. *See* b1239	226.24	nd (w), fl (dil al)	161–2	sub	...	...	δ s^h	s	s	...	δ	to δ	**B10**^{2}, 521
Ω b1403	—,4-benzoyl-	4-Benzophenonecarboxylic acid. $C_{14}H_{10}O_3$. *See* b1239	226.24	nd (dil aa), pl (al), mcl lf (w)	198–200 (226–7)	sub	...	...	δ	s	s	...	δ	chl δ aa s	**B10**^{2}, 521
b1404	—,2-benzoyl-4-chloro-	$C_{14}H_9ClO_3$. *See* b1239	260.68	cr (xyl)	180.5	...	...	...	...	s	s	...	...	chl s	**B10**^{1}, 356
b1405	—,2-benzoyl-3,4,5,6-tetra-chloro-, methyl ester	$C_{15}H_8Cl_4O_3$. *See* b1239	378.05	nd (MeOH)	92	...	...	...	i	s^h	...	...	...	lig, CS_2 δ MeOH s^h	**B10**, 750
Ω b1406	—,2-benzyl-	$C_{14}H_{12}O_2$. *See* b1239	212.25	nd (dil al)	118	sub	...	...	δ	s	s	...	s	chl s	**B9**^{2}, 471
Ω b1407	—,3-benzyl-	$C_{14}H_{12}O_2$. *See* b1239	212.25	nd (w), lf (dil al)	108	sub	...	...	δ	v	v	...	...	chl v	**B9**, 676
b1408	—,4-benzyl-	$C_{14}H_{12}O_2$. *See* b1239	212.25	nd (w), lf (dil al)	157–8	sub	...	...	δ	s	s	...	s	chl s	**B9**^{1}, 284
Ω b1409	—,4-(benzyl-sulfonamido)-	Caronamide.	291.33	...	229–30	...	...	...	δ	s	...	...	δ	chl δ aa s	**C45**, 3418
Ω b1410	—,2-bromo-	$C_7H_5BrO_2$. *See* b1239	201.03	mcl pr (w), nd λ^{al} 225 sh (3.92), 280 (2.92)	150	sub	1.929^{25}_4	...	δ s^h	s	s	s	...	chl s	**B9**^{2}, 230
Ω b1411	—,—,amide	C_7H_6BrNO. *See* b1244	200.04	nd (w)	160.5–1.5	sub	...	...	δ	s	δ	...	...	...	**B9**, 348
Ω b1412	—,—,chloride	C_7H_4BrClO. *See* b1239	219.47	nd	11	245^{760} 118^{10}	...	1.5965^{20}	d	d^h	...	...	...	...	**B9**^{2}, 231
b1413	—,—,ethyl ester	$C_9H_9BrO_2$. *See* b1239	229.08	...	...	254–5 135^{15}	1.4438^{15}_4	1.5455^{15}	δ	s	s	s	s	os s	**B9**, 348
b1414	—,—,methyl ester	$C_8H_7BrO_2$. *See* b1239	215.05	...	...	244^{760} 122^{17}	...	...	i	s	...	...	...	...	**B9**, 348
Ω b1415	—,—,nitrile	C_7H_4BrN. *See* b1291	182.03	nd (w)	55.5	$251–3^{754}$	...	...	s^h	v	...	...	...	...	**B9**^{2}, 232
b1416	—,—,piperazinium salt	$C_{18}H_{20}Br_2N_2O_4$. *See* b1239	488.19	...	227–30d (cor)	...	...	...	δ^h	s^h	i	...	...	...	**Am 70**, 2758
Ω b1417	—,3-bromo-	$C_7H_5BrO_2$. *See* b1239	201.03	mcl nd (dil al) λ^{al} 224 sh (3.80), 262 (2.75), 282 (2.88)	155	> 280	1.845^{20}	...	i	s	s	...	...	...	**B9**^{2}, 232
b1418	—,—,amide	C_7H_6BrNO. *See* b1244	200.04	lf (w or al)	155.3	sub	...	...	δ s^h	v	...	...	...	...	**B9**^{2}, 232
b1419	—,—,ethyl ester	$C_9H_9BrO_2$. *See* b1239	229.08	...	...	261^{760} (255) 133^{15}	1.4308^{19}_4	1.5430^{19}	...	s	s	v	s	os s	**B9**^{2}, 232
b1420	—,—,methyl ester	$C_8H_7BrO_2$. *See* b1239	215.05	pl	32	122.5^{15}	...	...	δ	s	s	...	...	...	**B9**^{1}, 143
Ω b1421	—,—,nitrile	C_7H_4BrN. *See* b1291	182.03	(al) λ^{al} 229 (3.98), 288 (2.97)	39–40	225^{760} $112–4^{14}$	...	...	...	v	v	...	...	...	**B9**^{2}, 233
b1422	—,—,piperazinium salt	$C_{18}H_{20}Br_2N_2O_4$. *See* b1239	488.19	...	169–71	...	...	...	δ^h	s^h	i	...	...	...	**Am 70**, 2758
Ω b1423	—,4-bromo-	$C_7H_5BrO_2$. *See* b1239	201.03	nd (eth), lf (w), mcl pr λ^{al} 240 (4.10)	254.5	...	1.894^{20}	...	δ s^h	s	s	...	...	...	**B9**^{2}, 233
Ω b1424	—,—,amide	C_7H_6BrNO. *See* b1244	200.04	nd or pl (w)	192–2.5	...	...	...	s^h	v	δ	...	i	CS_2 i lig i chl, aa s	**B9**, 353
Ω b1425	—,—,chloride	C_7H_4BrClO. *See* b1239	219.47	nd (peth) λ^{cy} 264 (4.33)	42	$245–7^{760}$ δd $123–6^{15}$	...	...	d	v	v	...	v	lig v	**B9**^{2}, 235
b1426	—,—,ethyl ester	$C_9H_9BrO_2$. *See* b1239	229.08	λ^{al} 244 (4.24)	...	262^{737} 129^{15}	1.4332^{17}_4	1.5438^{17}	δ	s	s	s	s	os s	**B9**^{2}, 234

For explanations, symbols and abbreviations see beginning of table. For structural **formulas** see end of table.

No.	Name	Synonyms and Formula	Mol. wt.	Color, crystalline form, specific rotation and λ_{max} (log ε)	m.p. °C	b.p. °C	Density	n_D	Solubility w	al	eth	ace	bz	other solvents	Ref.
	Benzoic acid														
b1427	—,—,methyl ester	$C_8H_7BrO_2$. *See* b1239	215.05	lf (dil al), nd (eth) λ^{al} 245 (4.26)	81	...	1.689	...	...	s	s	s	v	chl v peth s	**B9**[2], 234
Ω b1428	—,—,nitrile	C_7H_4BrN. *See* b1291	182.03	nd (w or al) λ^{al} 241 (4.30), 263 (2.87), 282 (2.68)	114–4.5	235–7[760]	...	...	s^h	s	s	...	...	...	**B9**[2], 236
b1429	—,—,piperazinium salt	$C_{18}H_{20}Br_2N_2O_4$. *See* b1239	488.19	...	224–6	...	...	...	δ^h	s^h	i	...	...	...	**Am 70,** 2758
b1430	—,2-bromo-3-chloro-	$C_7H_4BrClO_2$. *See* b1239	235.47	cr (bz)	143–4	...	...	...	i	...	...	...	v^h	...	**B9**, 355
b1431	—,2-bromo-5-chloro-	$C_7H_4BrClO_2$. *See* b1239	235.47	cr (bz)	153	...	...	...	i	s	...	...	v^h	...	**B9**, 355
b1432	—,3-bromo-2-chloro-	$C_7H_4BrClO_2$. *See* b1239	235.47	cr (al)	165	...	...	...	i	v^h s	...	...	...	...	**B9**, 356
b1433	—,3-bromo-4-chloro-	$C_7H_4BrClO_2$. *See* b1239	235.47	pl (dil aa), cr (al)	215–6	...	...	...	i	s	...	...	...	aa v	**B9**[2], 236
b1434	—,4-bromo-3-chloro-	$C_7H_4BrClO_2$. *See* b1239	235.47	pl (dil aa), cr (al)	218	...	...	...	δ	v	...	...	...	aa v^h	**B9**[2], 236
b1435	—,5-bromo-2-chloro-	$C_7H_4BrClO_2$. *See* b1239	235.47	nd (w), cr (aa)	155–6	...	...	...	δ	v	...	...	...	aa δ, v^h	**B9**, 356
b1436	—,3-bromo-2,4-dihydroxy-	$C_7H_5BrO_4$. *See* b1239	233.03	br or ye nd (w)	202	...	...	...	s^h	...	...	...	...	aa s	**B10**[2], 254
b1437	—,5-bromo-2,3-dihydroxy-	$C_7H_5BrO_4$. *See* b1239	233.03	pr (w + 1), nd (w)	187 (pr), 215 (nd)	...	...	...	δ s^h	v	v	...	δ	aa v	**B10**[1], 175
b1438	—,5-bromo-2,4-dihydroxy-	$C_7H_5BrO_4$. *See* b1239	233.03	micr pr (w + 1)	212	...	...	...	δ s^h	v	v	...	...	...	**B10**[2], 254
b1439	—,5-bromo-3,4-dihydroxy-	$C_7H_5BrO_4$. *See* b1239	233.03	nd (w, aa, dil al)	230	...	...	...	s^h	...	...	...	...	aa s^h	**B10**[1], 192
b1440	—,2-bromo-3,5-dinitro-	$C_7H_3BrN_2O_6$. *See* b1239	291.02	ye nd (w)	213	...	...	...	δ	v	...	...	s	aa v lig s	**B9**[2], 284
b1441	—,2-bromo-3-hydroxy-	$C_7H_5BrO_3$. *See* b1239	217.03	nd (w)	160–1	...	...	...	δ v^h	...	s	...	...	...	**B10**[2], 83
b1442	—,2-bromo-4-hydroxy-	$C_7H_5BrO_3$. *See* b1239	217.03	nd (w)	151	...	...	...	δ v^h	...	s	...	...	...	**B10**[2], 103
b1443	—,2-bromo-5-hydroxy-	$C_7H_5BrO_3$. *See* b1239	217.03	(w)	185d	...	...	...	δ s^h	...	s	...	...	...	**B10**[2], 84
b1444	—,3-bromo-2-hydroxy-	$C_7H_5BrO_3$. *See* b1239	217.03	nd (dil al)	184.5	...	...	...	δ^h	v	v	v	δ	MeOH v, chl δ	**B10**[2], 63
b1445	—,3-bromo-4-hydroxy-	$C_7H_5BrO_3$. *See* b1239	217.03	nd or pr (+w) (w)	177	...	...	...	δ v^h	v	v	...	...	aa v	**B10**[2], 103
b1446	—,4-bromo-2-hydroxy-	$C_7H_5BrO_3$. *See* b1239	217.03	pl, nd (w)	214 (165)	...	...	...	δ v^h	v	...	...	...	...	**B10**[2], 63
b1447	—,4-bromo-3-hydroxy-	$C_7H_5BrO_3$. *See* b1239	217.03	nd (w)	227	...	...	...	δ s^h	...	...	...	...	...	**B10**[2], 63
Ω b1448	—,5-bromo-2-hydroxy-	$C_7H_5BrO_3$. *See* b1239	217.03	nd (w or dil al) λ^{w} 232 sh (3.92), 308 (3.52)	168–9	sub > 100	...	...	v^h δ	v	v	...	...	...	**B10**[2], 63
b1449	—,2-bromo-3-nitro-	$C_7H_4BrNO_4$. *See* b1239	246.02	(dil al)	191	...	...	...	δ s^h	v	...	...	...	...	**B9**[2], 277
b1450	—,2-bromo-4-nitro-	$C_7H_4BrNO_4$. *See* b1239	246.02	nd (w or dil al)	166–7	sub > 155	...	...	δ s^h	v	v	...	...	...	**B9**[2], 277
b1451	—,2-bromo-5-nitro-	$C_7H_4BrNO_4$. *See* b1239	246.02	nd (w)	180–1	sub	...	...	δ s^h	v	v	...	...	chl v	**B9**[2], 277
b1452	—,3-bromo-2-nitro-	$C_7H_4BrNO_4$. *See* b1239	246.02	(eth)	250	...	...	...	δ	...	v^h	...	v^h	...	**B9**[2], 276
b1453	—,3-bromo-4-nitro-	$C_7H_4BrNO_4$. *See* b1239	246.02	nd (dil al)	197	...	...	...	δ	v	v	...	...	chl v	**B9**, 408
b1454	—,3-bromo-5-nitro-	$C_7H_4BrNO_4$. *See* b1239	246.02	nd (w, bz or eth), pl (al)	159–60	...	...	...	δ s^h	v	v	...	s	CS_2 s aa v chl s	**B9**[2], 277
b1455	—,4-bromo-2-nitro-	$C_7H_4BrNO_4$. *See* b1239	246.02	nd (w)	163	...	...	...	s^h	v	v	...	v	chl v	**B9**[2], 276
b1456	—,4-bromo-3-nitro-	$C_7H_4BrNO_4$. *See* b1239	246.02	nd (dil aa)	203–4	sub	...	...	i s^h	v	...	...	...	...	**B9**[2], 277
b1457	—,5-bromo-2-nitro-	$C_7H_4BrNO_4$. *See* b1239	246.02	cr (w, al, bz, to)	140	...	1.920[18]	...	s^h	s	...	...	v^h	...	**B9**, 406
b1458	—,2-*tert*-butyl-	$C_{11}H_{14}O_2$. *See* b1239	178.23	pl (dil al)	80.3–.8	...	...	...	i	v	...	...	...	...	**B9**[2], 365
b1459	—,3-*tert*-butyl-	$C_{11}H_{14}O_2$. *See* b1239	178.23	nd (peth)	127–7.5	...	...	...	...	v	...	...	...	peth v^h	**B9**, 560
Ω b1460	—,4-*tert*-butyl-	$C_{11}H_{14}O_2$. *See* b1239	178.23	nd (dil al) $\lambda^{al-0.01\,N\,HCl}$ 237.5 (4.21)	164.5–5.5	...	...	...	i	v	...	...	v	...	**B9**[2], 365
b1461	—,4-butylamino-(2-dimethylamino-ethyl) ester, hydrochloride	Tetracaine hydrochloride. $C_{15}H_{25}ClN_2O_2$. *See* b1239	300.83	...	147–50	...	...	...	v	s	i	...	i	...	**C54**, 429
—	—,(carboxy-methoxy)-	*see* **Acetic acid, (carboxyphenoxy)-**													
b1462	—,4(4-carboxy-benzoyl)-	*p,p'*-Benzophenonedi-carboxylic acid. $C_{15}H_{10}O_5$. *See* b1239	270.25	nd (al)	> 360	sub	...	...	i	δ	δ	δ	δ	aa s	**B10**[2], 618

For explanations, symbols and abbreviations see beginning of table. For structural formulas see end of table.

No.	Name	Synonyms and Formula	Mol. wt.	Color, crystalline form, specific rotation and λ_{max} (log ε)	m.p. °C	b.p. °C	Density	n_D	Solubility w	al	eth	ace	bz	other solvents	Ref.
	Benzoic acid														
Ω b1465	—.2-chloro-	$C_7H_5ClO_2$. *See* b1239	156.57	mcl pr (w) λ^{al} 229 sh (3.71), 278 (2.88)	142	sub	1.544^{20}	...	s^h	v	v	v	s	CS_2, lig δ peth s^h AcOEt v	$B9^2$, 221
Ω b1466	—.—.amide	C_7H_6ClNO. *See* b1244	155.59	rh nd (w)	142.4	...	...	...	s^h	s	s	...	...	...	B9, 336
b1467	—.—.anhydride	$C_{14}H_8Cl_2O_3$. *See* b1239	295.14	nd (al)	79.6	...	...	...	...	v	...	...	v	chl v lig δ	$B9^2$, 223
Ω b1468	—.—.chloride	$C_7H_4Cl_2O$. *See* b1239	175.02	λ^{cy} 242 (3.95)	−4	238^{760} 110^{15}	...	1.5726^{20}	d	d	...	...	...	...	$B9^2$, 223
b1469	—.—.ethyl ester	$C_9H_9ClO_2$. *See* b1239	184.62	...	...	243^{760} $122–5^{15}$	1.1942^{15}_4	1.5247^{15}	i	s	v	...	...	...	B9, 336
b1470	—.—.methyl ester	$C_8H_7ClO_2$. *See* b1239	170.62	...	...	$234–5^{762}$	...	...	...	s	...	...	...	...	$B9^2$, 222
Ω b1471	—.—.nitrile	C_7H_4ClN. *See* b1291	137.57	nd λ^{cy} 230 (4.00), 272 (2.89), 288 (3.17)	43–6	232	...	...	δ^h	s	s	...	...	...	$B9^2$, 223
b1472	—.—.piperazinium salt	$C_{18}H_{20}Cl_2N_2O_4$. *See* b1239	399.28	...	217–8d	...	...	...	s	s	s	...	...	...	Am 70, 2758
Ω b1473	—.3-chloro-	$C_7H_5ClO_2$. *See* b1239	156.57	pr (w) λ^{al} 230 (3.92), 284 (2.98)	158 (155)	sub	1.496^{25}_4	...	v^h	s	v	...	δ s^h	CS_2 δ lig δ CCl_4 δ	$B9^2$, 223
b1474	—.—.amide	C_7H_6ClNO. *See* b1244	155.59	nd $\lambda^{60\% sulf}$ 248 (3.98)	135.5–7.0	...	...	...	δ s^h	s	s	...	...	...	B9, 338
b1475	—.—.anhydride	$C_{14}H_8Cl_2O_3$. *See* b1239	295.14	nd (al or peth), lf (bz)	95.5	...	...	...	...	v	...	...	v	peth s^h chl v lig δ	$B9^2$, 224
Ω b1476	—.—.chloride	$C_7H_4Cl_2O$. *See* b1239	175.02	λ^{cy} 244 (4.08), 254 (4.06)	...	225 $103–4^{14}$	...	1.5677^{20}	d	d	...	...	...	...	$B9^2$, 224
b1477	—.—.ethyl ester	$C_9H_9ClO_2$. *See* b1239	184.62	...	...	245 121^{20}	1.1859^{15}_4	1.5223^{15}	i	s	s	...	...	...	$B9^1$, 139
b1478	—.—.methyl ester	$C_8H_7ClO_2$. *See* b1239	170.60	...	21	231^{763} 114^{18}	...	...	...	s	...	...	...	...	B9, 338
b1479	—.—.nitrile	C_7H_4ClN. *See* b1291	137.57	nd λ^{cy} 228 (4.02), 271 (2.80), 287 (3.06)	40–2	$99–100^{15}$	...	...	i	s	s	...	...	...	$B9^2$, 225
Ω b1480	—.4-chloro-	$C_7H_5ClO_2$. *See* b1239	156.57	tcl pr (al-eth) λ^{al} 234 (4.18)	243	...	...	...	i s^h	v	δ	δ	i	lig i CCl_4, CS_2 i	$B9^2$, 225
Ω b1481	—.—.amide	C_7H_6ClNO. *See* b1244	155.59	nd (eth)	179–80	...	...	...	s^h	v	v	...	...	...	$B9^2$, 227
b1482	—.—.anhydride	$C_{14}H_8Cl_2O_3$. *See* b1239	295.14	nd or lf (bz)	193–4	...	...	...	...	δ	i	...	δ	CS_2 i chl δ peth i	$B9^2$, 227
Ω b1483	—.—.chloride	$C_7H_4Cl_2O$. *See* b1239	175.02	...	16	222^{760} 111^{18}	1.3770^{20}_4	1.5756^{20}	...	...	...	...	...	...	$B9^2$, 227
Ω b1484	—.—.ethyl ester	$C_9H_9ClO_2$. *See* b1239	184.62	...	...	237–8 122^{15}	...	...	...	s	...	...	...	...	$B9^1$, 140
b1485	—.—.methyl ester	$C_8H_7ClO_2$. *See* b1239	170.60	nd or mcl pr	44	...	1.382^{20}	...	...	s	...	...	...	...	$B9^2$, 226
Ω b1486	—.—.nitrile	C_7H_4ClN. *See* b1291	137.57	nd (al) λ^{cy} 243 (4.22), 270 (2.66), 281 sh (2.18)	94–6	223^{750} 95^5	...	...	δ^h	s	s	...	s	chl s lig δ	$B9^2$, 288
b1487	—.—.piperazinium salt	$C_{18}H_{20}Cl_2N_2O_4$. *See* b1239	399.28	...	219–20d	...	...	...	s	s	δ	...	...	...	Am 70, 2758
Ω b1488	—.2(4-chloro-benzoyl)-	$C_{14}H_9ClO_3$. *See* b1239	260.68	cr (bz, aa) λ^{MeOH} 254.5 (4.25)	150 (148)	...	...	...	δ^h	s	s	...	s^h	...	$B10^2$, 518
b1489	—.2-chloro-3-hydroxy-	$C_7H_5ClO_3$. *See* b1239	172.57	lf (w or bz)	157.5–8.5	...	...	...	δ v^h	...	...	...	v^h	...	B10, 142
b1490	—.2-chloro-4-hydroxy-	$C_7H_5ClO_3$. *See* b1239	172.57	nd (w)	159	...	...	...	δ v^h	...	...	s	...	...	$B10^2$, 102
b1491	—.2-chloro-5-hydroxy-	$C_7H_5ClO_3$. *See* b1239	172.57	(w)	178–9	...	...	...	δ	v^h	...	s	...	...	B10, 143
b1492	—.2-chloro-6-hydroxy-	$C_7H_5ClO_3$. *See* b1239	172.57	nd (w)	166	...	...	...	s	s	s	s	s	os s	B10, 104
b1493	—.3-chloro-2-hydroxy-	$C_7H_5ClO_3$. *See* b1239	172.57	nd (w)	180 (182)	...	...	...	δ	s	...	...	...	chl. aa s	$B10^2$, 61
b1494	—.—.ethyl ester	$C_9H_9ClO_3$. *See* b1239	200.62	nd	21	269–70 147^{12}	...	...	...	s	...	...	...	...	B10, 101
b1495	—.—.methyl ester	$C_8H_7ClO_3$. *See* b1239	186.60	nd (MeOH or al)	38	260 δd	...	...	...	...	...	...	...	...	B10, 101
Ω b1496	—.3-chloro-4-hydroxy-	$C_7H_5ClO_3$. *See* b1239	172.57	nd (w)	170–2	sub	...	...	δ v^h	v	v	v	δ	chl δ lig δ	B10, 175
b1497	—.4-chloro-2-hydroxy-	$C_7H_5ClO_3$. *See* b1239	172.57	nd (w)	207	sub δd	...	...	δ	s	...	...	s	chl s	$B10^2$, 61
b1498	—.4-chloro-3-hydroxy-	$C_7H_5ClO_3$. *See* b1239	172.57	nd (w)	219.5–20.5	...	...	...	δ v^h	...	...	...	...	...	$B10^2$, 83
Ω b1499	—.5-chloro-2-hydroxy-	$C_7H_5ClO_3$. *See* b1239	172.57	nd (w or al) λ^w 230 (3.90), 314 (3.53)	173–4 (168)	...	...	...	s	v	s	...	v	chl v, aa s	$B10^2$, 62
b1500	—.—.ethyl ester	$C_9H_9ClO_3$. *See* b1239	200.62	nd (al)	25	...	...	...	...	s^h	...	...	...	...	B10, 103
b1501	—.—.methyl ester	$C_8H_7ClO_3$. *See* b1239	186.60	nd (al)	50	249d	...	...	...	v	...	...	...	...	B10, 103

For explanations, symbols and abbreviations see beginning of table. For structural formulas see end of table.

No.	Name	Synonyms and Formula	Mol. wt.	Color, crystalline form, specific rotation and λ_{max} (log ε)	m.p. °C	b.p. °C	Density	n_D	Solubility w	al	eth	ace	bz	other solvents	Ref.
	Benzoic acid														
b1502	—.5-chloro-2-methoxy-	$C_8H_7ClO_3$. *See* b1239	186.60	nd (w)	81–2				v	v		v			**B10**, 103
b1503	—.2-chloro-4-methyl-	$C_8H_7ClO_2$. *See* b1239	170.60	nd (al)	155–6				δ v^h	v	v		v^h	chl v	**B9**, 497
b1504	—.2-chloro-5-methyl-	$C_8H_7ClO_2$. *See* b1239	170.60	nd (w or al)	167				v^h	v^h					**B9**, 479
b1505	—.2-chloro-6-methyl-	$C_8H_7ClO_2$. *See* b1239	170.60	nd (w)	102				δ^h						**B9^2**, 320
b1506	—.3-chloro-2-methyl-	$C_8H_7ClO_2$. *See* b1239	170.60	nd (al)	159 (154)					v	s				**B9**, 467
b1507	—.3-chloro-4-methyl-	$C_8H_7ClO_2$. *See* b1239	170.60	nd or lf (dil al)	200–2				δ^h	v					**B9**, 498
b1508	—.3-chloro-5-methyl-	$C_8H_7ClO_2$. *See* b1239	170.60	nd (dil al)	178				δ	v					**B9**, 479
b1509	—.4-chloro-2-methyl-	$C_8H_7ClO_2$. *See* b1239	170.60	nd (w, al, dil aa, bz)	173				v^h	v^h			δ v^h	aa v	**B9**, 468
b1510	—.4-chloro-3-methyl-	$C_8H_7ClO_2$. *See* b1239	170.60	nd (w)	209–10				i δ^h						**B9**, 478
b1511	—.5-chloro-2-methyl-	$C_8H_7ClO_2$. *See* b1239	170.60	nd (al)	168.5–9.5					v^h					**B9^2**, 320
b1512	—.3-chloro-methyl-, nitrile	C_8H_6ClN. *See* b1291	151.60	pr (al)	67	258–60^{760}									**B9**, 479
b1513	—.4-chloro-methyl-, nitrile	C_8H_6ClN. *See* b1291	151.60	pr (al)	79.5	263^{756}									**B9**, 498
Ω b1514	—.2-chloro-4-nitro-	$C_7H_4ClNO_4$. *See* b1239	201.57	nd (w)	140–2				s^h	s	s		s^h		**B9^2**, 276
Ω b1515	—.2-chloro-5-nitro-	$C_7H_4ClNO_4$. *See* b1239	201.57	nd or pr (w)	165 (168)		1.608^{18}		δ	s	s		s		**B9^2**, 275
b1516	—.3-chloro-2-nitro-	$C_7H_4ClNO_4$. *See* b1239	201.57	nd or pl (w)	237–9		1.566^{18}		δ^h	s	s		i		**B9**, 400
b1517	—.3-chloro-5-nitro-	$C_7H_4ClNO_4$. *See* b1239	201.57	nd (w)	147				δ	s	s			aa s	**B9**, 403
Ω b1518	—.4-chloro-2-nitro-	$C_7H_4ClNO_4$. *See* b1239	201.57	pl (bz-lig), pr (bz), nd (w)	142–3				s^h	s	s		s^h		**B9^2**, 274
Ω b1519	—.—.nitrile	$C_7H_3ClN_2O_2$. *See* b1291	182.57	nd (w)	100–1				v^h	s	s				**B9^2**, 275
Ω b1520	—.4-chloro-3-nitro-	$C_7H_4ClNO_4$. *See* b1239	201.57	nd or pl (w)	181–2		1.645^{18}		i s^h	δ					**B9^2**, 275
Ω b1521	—.—.ethyl ester	$C_9H_8ClNO_4$. *See* b1239	229.63	ye nd (al)	59					v			s	aa s	**B9**, 402
b1522	—.—.methyl ester	$C_8H_6ClNO_4$. *See* b1239	215.59	nd (MeOH)	83		1.522^{18}			s					**B9**, 402
b1523	—.5-chloro-2-nitro-, methyl ester	$C_8H_6ClNO_4$. *See* b1239	215.59	pl (MeOH)	48.5		1.453^{18}							MeOH v	**B9**, 401
b1524	—.2,3-diamino-	$C_7H_8N_2O_2$. *See* b1239	152.16	nd (dil al)	190–1d	d			δ	v				aa v	**B14^1**, 585
b1525	—.2,4-diamino-	$C_7H_8N_2O_2$. *See* b1239	152.16		140	>200d			s^h	v				aa v	**B14**, 448
b1526	—.2,5-diamino-	$C_7H_8N_2O_2$. *See* b1239	152.16	br pr (w)	darkens 200				δ	δ	δ				**B14**, 448
Ω b1527	—.3,4-diamino-	$C_7H_8N_2O_2$. *See* b1239	152.16	lf (w)	215–8d				δ s^h						**B14^1**, 586
Ω b1528	—.3,5-diamino-	$C_7H_8N_2O_2$. *See* b1239	152.16	nd (+1w)	240 (rapid htng)				δ s^h	s	v				**B14**, 453
b1529	—.3-diazo-2-hydroxy-	$C_7H_4N_2O_3$. *See* b1239	164.12	ye nd (ace)	155d							s			**B16**, 553
b1530	—.2,3-dibromo-	$C_7H_4Br_2O_2$. *See* b1239	279.93	nd (w)	149–50				δ^h					lig δ s^h	**B9^1**, 146
b1531	—.2,4-dibromo-	$C_7H_4Br_2O_2$. *See* b1239	279.93	lf (w)	174	sub			δ	s	s				**B9^2**, 237
b1532	—.2,5-dibromo-	$C_7H_4Br_2O_2$. *See* b1239	279.93	nd (al or w)	157	sub			s^h	s	s			chl s aa s	**B9^2**, 237
b1533	—.2,6-dibromo-	$C_7H_4Br_2O_2$. *See* b1239	279.93	nd (w), (lig) λ^{al} 262 (2.97), 272 (2.94)	150–1	*ca* 335δd 209–10^{16}			s^h	v	v	s		peth δ chl v	**B9^2**, 237
Ω b1534	—.3,4-dibromo-	$C_7H_4Br_2O_2$. *See* b1239	279.93	nd (w), pl (al)	234–5				s^h	s	s			MeOH s	**B9^2**, 237
b1535	—.3,5-dibromo-2,4-dihydroxy-6-methyl-, ethyl ester	$C_{10}H_{10}Br_2O_4$. *See* b1239	354.01	pr (al)	144					δ v^h	s				**B10^2**, 275
b1536	—.2,6-dibromo-3,4,5-trihydroxy-	Gallobromol. $C_7H_4Br_2O_5$. *See* b1239	327.93	nd, pr or lf (w+1)	139d (hyd)				v	v	v			chl i	**B10^2**, 347
b1537	—.2,3-dichloro-	$C_7H_4Cl_2O_2$. *See* b1239	191.02	nd (w)	168.3 (164)				s	s	s				**B9^2**, 228
Ω b1538	—.2,4-dichloro-	$C_7H_4Cl_2O_2$. *See* b1239	191.02	nd (w or bz) λ^{MeOH} 282 (2.76)	164.2	sub			s^h	s	s		s	chl s	**B9^2**, 228
Ω b1539	—.—.chloride	$C_7H_3Cl_3O$. *See* b1239	209.47	liq	15–8	150^{34} 111$^{7.5}$		1.5895^{20}	d^h	d^h					**B9^2**, 229
Ω b1540	—.2,5-dichloro-	$C_7H_4Cl_2O_2$. *See* b1239	191.02	nd (w)	154.4	301^{760}			δ s^h	s	s				**B9^2**, 229
Ω b1541	—.2,6-dichloro-	$C_7H_4Cl_2O_2$. *See* b1239	191.02	nd (al), pr (w) λ^{MeOH} 273 (2.56)	144 (139)	sub			s^h	s	s		s		**B9^2**, 229
Ω b1542	—.3,4-dichloro-	$C_7H_4Cl_2O_2$. *See* b1239	191.02	nd (w, al, bz) λ^{MeOH} 282 (2.93), 291 (2.96)	208–9				s^h	v	s				**B9^1**, 141

For explanations, symbols and abbreviations see beginning of table. For structural formulas see end of table.

No.	Name	Synonyms and Formula	Mol. wt.	Color, crystalline form, specific rotation and λ_{max} (log ε)	m.p. °C	b.p. °C	Density	n_D	Solubility w	al	eth	ace	bz	other solvents	Ref.
	Benzoic acid														
Ω b1543	—,—,chloride....	$C_7H_3Cl_3O$. *See* b1239	209.47	liq	24–6	242^{760} 160^{42}	...	...	d^h	d^h	...		...	...	**B9**, 344
b1544	—,3,5-dichloro-....	$C_7H_4Cl_2O_2$. *See* b1239	191.02	nd (al or w)	188	sub	...	...	δ	s	s		...	lig δ	**B9**[2], 229
b1545	—,3,5-dichloro-2-hydroxy-	$C_7H_4Cl_2O_3$. *See* b1239	207.01	nd (dil al), rh pr	220–1 (223)	sub δd	...	...	δ^h	v	s		...	...	**B10**[1], 48
Ω b1546	—,3,5-dichloro-4-hydroxy-	$C_7H_4Cl_2O_3$. *See* b1239	207.01	nd (dil al or dil aa)	269	sub δd	...	...	δ	v	v	...	...	...	**B10**[1], 78
b1547	—,2(*N,N*-dichloro-sulfamyl)-	...	270.09	yesh gr pl (chl)	146–8 exp	...	...	...	s^h δ^h	...	...	...	...	chl δ^h	**B11**, 377
b1548	—,4(*N,N*-dichloro-sulfamyl)-	Halazone.	270.09	pr (aa)	213	...	...	...	δ^h	...	...	...	...	chl δ^h aa v^h peth i	**B11**[2], 220
b1549	—,2,3-dihydroxy-..	*o*-Pyrocatechuic acid. $C_7H_6O_4$. *See* b1239	154.12	pr or nd (w + 1)	204 (anh)	...	1.542^{20}_4	...	s	s	s	...	...	...	**B10**[2], 248
Ω b1550	—,2,4-dihydroxy-..	*β*-Resorcylic acid. $C_7H_6O_4$. *See* b1239	154.12	cr + w(w) λ^w 248 (4.05), 292 (3.72), 227 (2.19)	...	...	...	...	s^h	s	s	...	s	CS_2 i	**B10**[2], 251
Ω b1551	—,2,5-dihydroxy-..	Gentisic acid. $C_7H_6O_4$. *See* b1239	154.12	nd or pr (w) λ^w 320 (3.63)	205	...	...	...	v	v	v	...	i	chl i CS_2 i	**B10**[2], 257
Ω b1552	—,2,6-dihydroxy-..	*γ*-Resorcylic acid. $C_7H_6O_4$. *See* b1239	154.12	nd (+ w) λ^{al} 250 (3.80), 277 (3.29), 311 (3.54)	167d (150–70)	...	...	...	s^h	s	s	...	...	chl i	**B10**[2], 259
Ω b1553	—,3,4-dihydroxy-..	Protocatechuic acid. $C_7H_6O_4$. *See* b1239	154.12	mcl nd (w + 1) $\lambda^{1N\ HCl}$ 259 (3.94), 293 (3.64)	200–2d	...	1.524^4	...	v^h δ	v	s	...	i	...	**B10**, 389
Ω b1554	—,3,5-dihydroxy-..	*α*-Resorcylic acid. $C_7H_6O_4$. *See* b1239	154.12	pr or nd (+ 1½w) λ^w 208 (4.49), 248 (4.03), 292 (3.69)	238–40, 232–3 (+ w)	...	...	...	δ v^h	v	v	...	...	...	**B10**[2], 266
Ω b1555	—,2,4-dihydroxy-6-methyl-	*o*-Orsellinic acid. $C_8H_8O_4$. *See* b1239	168.16	nd (dil aa + lw) λ^{al} 260 (4.1), 300 (3.6)	176d	...	...	...	...	s	s	...	...	...	**B10**[2], 272
b1556	—,—,ethyl ester ...	$C_{10}H_{12}O_4$. *See* b1239.......	196.21	lf (aa), pr (al)	132	sub	...	...	δ^h	v	v	...	δ	chl δ lig δ	**B10**[2], 274
b1557	—,2,4-dihydroxy-6-pentyl-	Olivetol carboxylic acid. $C_{12}H_{16}O_4$. *See* b1239	224.24	wh nd	147	...	...	...	...	s	s	...	...	...	**C36**, 7068
b1558	—,2-[(2,4-di-hydroxyphenyl)-phenylmethyl]-	...	320.35	aa, cth, w)	186.9–7.4	...	...	...	s^h	s	s	...	...	aa s^h	**B10**, 455
Ω b1559	—,3,5-diiodo-2-hydroxy-	$C_7H_4I_2O_3$. *See* b1239.......	389.92	nd (al or aa)	235–6	...	...	...	δ	v	v	...	i	chl i	**B10**[2], 65
b1560	—,—,ethyl ester ...	$C_9H_8I_2O_3$. *See* b1239.......	417.98	lf (al)	133	...	...	...	i	s	s	...	s	...	**B10**, 114
b1561	—,3,5-diiodo-4-hydroxy-	$C_7H_4I_2O_3$. *See* b1239.......	389.92	nd (dil al)	237	d260	...	...	i	v	v	...	δ	chl δ lig δ	**B10**[2], 105
b1562	—,—,ethyl ester ...	$C_9H_8I_2O_3$. *See* b1239.......	417.98	nd (dil al)	123	...	...	...	i	s	...	...	...	os s	**B10**[2], 105
b1563	—,2,4-di-methoxy-, nitrile	$C_9H_9NO_2$. *See* b1291.......	163.18	cr (al, lig)	96	...	...	...	...	...	...	...	δ	to δ aa v $PhNO_2$ s^h	**B10**[2], 253
b1564	—,2,5-di-methoxy-, amide	$C_9H_{11}NO_3$. *See* 1244.......	181.19	lf (bz-peth), nd (w)	141–2	...	...	...	δ v^h	v	v^h	v	v	chl v	**B10**[1], 184
b1565	—,—,nitrile........	$C_9H_9NO_2$. *See* b1291.......	163.18	nd (al)	82	...	...	...	...	s^h	...	...	v	chl v lig δ	**B10**[2], 258
Ω b1566	—,2,6-di-methoxy-, nitrile	$C_9H_9NO_2$. *See* b1291.......	163.18	nd or pl	118–20	310^{760}	...	...	δ	v^h	δ	s	v	CS_2 δ lig δ chl v	**B10**[2], 260
Ω b1567	—,3,4-dimethoxy-..	Veratric acid. $C_9H_{10}O_4$. *See* b1239	182.18	nd (w or aa), rh (sub) λ^{al} 258 (4.1), 291 (3.7)	181–2 (anh)	sub	...	...	δ^h i	v	v	...	...	...	**B10**[2], 259
Ω b1568	—,—,amide	Veratramide. $C_9H_{11}NO_3$. *See* b1244	181.19	cr (w)	164	...	...	...	v^h	...	s	...	s	...	**B10**[2], 264
Ω b1569	—,—,nitrile.......	Veratronitrile. $C_9H_9NO_2$. *See* b1291	163.18	nd (w)	67–8	...	...	...	s^h	s^h	...	...	s	...	**B10**[2], 264
b1570	—,3,5-di-methoxy-, amide	$C_9H_{11}NO_3$. *See* b1244.......	181.19	nd (bz) λ^{MeOH} 247 (3.68), 299 (3.37)	148–9	...	...	...	s^h	v	v	...	v^h	lig v^h	**B10**[2], 267
b1571	—,2,3-dimeth-oxy-4-hydroxy-	$C_9H_{10}O_5$. *See* b1239	198.18	pl or lf (w or al)	154–5	...	...	...	δ s^h	s	...	v	...	chl v lig i MeOH v	**B10**[2], 332
b1572	—,2,3-dimeth-oxy-5-hydroxy-	$C_9H_{10}O_5$. *See* b1239	198.18	pl or lf (w)	186–8	...	...	...	s^h	s^h	...	...	...	...	**B10**[2], 333
b1573	—,2,4-dimeth-oxy-6-hydroxy-	$C_9H_{10}O_5$. *See* b1239	198.18	nd (eth-bz)	152–4d	...	...	...	s^h	s	s	v	...	...	**B10**[2], 334
b1574	—,2,6-dimeth-oxy-4-hydroxy-	$C_9H_{10}O_5$. *See* b1239	198.18	pl (w)	175	...	...	...	s^h	δ	δ	v	...	Py v	**B10**[1], 235
b1575	—,3,4-dimeth-oxy-2-hydroxy-	$C_9H_{10}O_5$. *See* b1239	198.18	nd (w)	169–72	...	...	...	s^h	...	...	...	...	chl s	**B10**, 465
b1576	—,3,4-dimeth-oxy-5-hydroxy-	$C_9H_{10}O_5$. *See* b1239	198.18	nd (aa or w)	197–8	...	...	...	s	s	...	...	...	aa s	**B10**[2], 340

For explanations, symbols and abbreviations see beginning of table. For structural formulas see end of table.

No.	Name	Synonyms and Formula	Mol. wt.	Color, crystalline form, specific rotation and λ_{max} (log ε)	m.p. °C	b.p. °C	Density	n_D	Solubility w	al	eth	ace	bz	other solvents	Ref.
	Benzoic acid														
Ω b1577	—,**3,5-dimeth-oxy-4-hydroxy-**	Syringic acid. $C_9H_{10}O_5$. *See* b1239	198.18	nd (w) λ^{al} 228 (4.0), 275 (4.0)	204–5	...	...	...	δ	s	v	v	...	chl v	**B10**, 480
b1578	—,**4,5-dimeth-oxy-2-hydroxy-**	$C_9H_{10}O_5$. *See* b1239	198.18	nd (w) λ^{al} 222 (4.30), 257 (4.03), 318 (3.87)	202d	...	...	...	δ s^h	s	...	...	...	...	**B10**[1], 234
b1579	—,—,methyl ester	$C_{10}H_{12}O_5$. *See* b1239	212.21	nd (w)	95	...	...	...	δ	...	...	...	...	...	**B10**[1], 234
b1580	—,**2,3-dimethyl-**	Hemellitic acid. 2,3-Xylylic acid. $C_9H_{10}O_2$. *See* b1239	150.18	pr (al)	144	...	...	...	δ^h i	s^h	s	...	...	...	**B9**[1], 209
b1581	—,**2,4-dimethyl-**	2,4-Xylylic acid. $C_9H_{10}O_2$. *See* b1239	150.18	mcl or tcl nd (w) λ^{al} 243 (4.1), 282 (3.1), 290 sh (2.8)	127 (anh) 90 (hyd)	267^{727} sub	...	...	δ^h	s^h	...	s	s	chl. to s aa s^h	**B9**[2], 350
b1582	—,**2,5-dimethyl-**	Isoxylylic acid. 2,5-Xylylic acid. $C_9H_{10}O_2$. *See* b1239	150.18	nd (al) λ^{al} 232 (2.61), 269 (2.81), 286 (2.91)	132	268^{760} sub	1.069_4^{21}	...	i	s	s	s	s	...	**B9**[1], 210
b1583	—,**2,6-dimethyl-**	2,6-Xylylic acid. $C_9H_{10}O_2$. *See* b1239	150.18	nd (lig) λ^{al} 270 (2.86)	116	274.5	...	...	δ	s	s	...	...	lig δ, s^h	**B9**[2], 350
Ω b1584	—,**3,4-dimethyl-**	Paraxylylic acid. 3,4-Xylylic acid. $C_9H_{10}O_2$. *See* b1239	150.18	pr (al) $\lambda^{w, pH=3.6}$ 242 (3.95)	166	sub	...	...	i δ^h	s	s	...	s	...	**B9**[2], 353
Ω b1585	—,**3,5-dimethyl-**	Mesitylenic acid. 3,5-Xylylic acid. $C_9H_{10}O_2$. *See* b1239	150.18	nd (w or al)	170–1	sub	...	...	δ^h	v	v	...	...	...	**B9**[2], 354
b1586	—,**2(dimethyl-amino)-**	*N,N*-Dimethylanthranilic acid. $C_9H_{11}NO_2$. *See* b1239	165.19	pr, nd (eth)	72	sub δd	...	...	v	v	s^h	...	...	...	**B14**[2], 213
Ω b1587	—,**3(dimethyl-amino)-**	$C_9H_{11}NO_2$. *See* b1239	165.19	nd (w) λ^{al} 263 (4.0), 338 (3.3)	151	...	...	...	v^h δ	s	s	...	...	...	**B14**, 392
b1588	—,**4(dimethyl-amino)-**	$C_9H_{11}NO_2$. *See* b1239	165.19	nd (al) λ^{al} 227 (3.87), 308 (4.41)	242.5–3.5 (244–6d)	...	...	...	...	s	δ	...	...	...	**B14**[2], 259
Ω b1589	—,**2,4-dinitro-**	$C_7H_4N_2O_6$. *See* b1239	212.12	nd (w)	183	...	...	...	δ s^h	δ s^h	...	...	δ v^h	...	**B9**[2], 279
Ω b1590	—,**2,5-dinitro-**	$C_7H_4N_2O_6$. *See* b1239	212.12	pr (w) λ^{aqNH_3} 515 (0.95)	177	...	...	...	δ s^h	s	s	...	δ	...	**B9**[2], 279
b1591	—,**2,6-dinitro-**	$C_7H_4N_2O_6$. *See* b1239	212.12	nd (w)	202–3	...	...	...	δ s^h	s	s	...	...	...	**B9**[2], 279
Ω b1592	—,**3,4-dinitro-**	$C_7H_4N_2O_6$. *See* b1239	212.12	nd (w) λ^{al} 570 (0.90)	165	...	...	...	δ s^h	v	v	...	...	...	**B9**[1], 167
Ω b1593	—,**3,5-dinitro-**	$C_7H_4N_2O_6$. *See* b1239	212.12	mcl pr (al)	205	...	...	...	δ s^h	s	δ	...	i	aa s	**B9**[2], 279
Ω b1594	—,—,benzyl ester	$C_{14}H_{10}N_2O_6$. *See* b1239	302.25	nd (lig)	112	...	...	...	...	...	...	...	...	...	**J1959**, 3183
Ω b1595	—,—,butyl ester	$C_{11}H_{12}N_2O_6$. *See* b1239	268.23	mcl nd (al)	62.5	...	...	1.488	...	s^h	...	...	...	...	**B9**[2], 280
Ω b1596	—,—,chloride	$C_7H_3ClN_2O_5$. *See* b1239	230.57	ye nd (bz)	74	196^{12}	...	...	d	d	s	...	...	...	**B9**[2], 283
Ω b1597	—,—,ethyl ester	$C_9H_8N_2O_6$. *See* b1239	240.18	nd (al)	92.9	...	1.295^{111}	1.560	...	s^h	...	...	...	...	**B9**, 414
Ω b1598	—,—,furfuryl ester	...	292.22	(bz-Py)	78–81	...	...	...	...	...	...	...	...	...	**B17**[2], 115
Ω b1599	—,—,isobutyl ester	$C_{11}H_{12}N_2O_6$. *See* b1239	268.23	mcl pl or nd (al)	87–8	...	...	...	s^h	s^h	...	...	...	...	**B9**[2], 281
Ω b1600	—,—,isopropyl ester	$C_{10}H_{10}N_2O_6$. *See* b1239	254.20	nd (al)	122	...	...	...	...	s^h	...	...	...	...	**B9**[2], 280
Ω b1601	—,—,methyl ester	$C_8H_6N_2O_6$. *See* b1239	226.15	nd (w)	112	...	...	...	s^h	s^h	...	...	...	...	**B9**, 414
Ω b1602	—,—,pentyl ester	$C_{12}H_{14}N_2O_6$. *See* b1239	282.26	...	46.4	...	...	...	...	s^h	...	...	...	...	**B9**[2], 281
b1603	—,—,phenyl ester	$C_{13}H_8N_2O_6$. *See* b1239	288.23	rods (al)	145–6	...	...	...	i	s^h	δ	...	v	...	**B9**, 119
Ω b1604	—,—,propyl ester	$C_{10}H_{10}N_2O_6$. *See* b1239	254.20	mcl pl (al)	73	...	...	...	...	s^h	...	...	...	...	**B9**[2], 280
Ω b1605	—,—,tetrahydro-furfuryl ester	...	296.24	nd (al)	83–4	...	...	...	i	s^h	...	...	...	...	**Am 62**, 3516
b1606	—,**2,4-dinitro-3-hydroxy-**	$C_7H_4N_2O_7$. *See* b1239	228.12	(w)	204	...	...	...	δ v^h	s	s	...	...	...	**B10**[2], 85
Ω b1607	—,**3,5-dinitro-2-hydroxy-**	$C_7H_4N_2O_7$. *See* b1239	228.12	ye nd or pl (+1w) λ >450	182 (anh) 174 (hyd)	...	...	...	s	s	s	...	s	...	**B10**, 122
b1608	—,**3,5-dinitro-4-hydroxy-**	$C_7H_4N_2O_7$. *See* b1239	228.12	ye lf (al)	248–9	...	...	...	δ s^h	v^h	v	...	...	...	**B10**[2], 108
b1609	—,**2-ethoxy-**	$C_9H_{10}O_3$. *See* b1239	166.18	...	20.7	$211–2^{35}$	...	...	δ s^h	δ	...	...	...	...	**B10**[2], 40
Ω b1610	—,—,amide	$C_9H_{11}NO_2$. *See* b1244	165.19	nd (w, al or AcOEt-peth)	132–4	...	...	...	δ s^h	v^h	v^h	...	...	...	**B10**[2], 58
b1611	—,—,ethyl ester	$C_{11}H_{14}O_3$. *See* b1239	194.23	...	...	251^{760} $180–5^{113}$	1.005^{20}	...	δ	v	s	...	...	...	**B10**, 74
b1612	—,—,nitrile	C_9H_9NO. *See* b1291	147.18		5	260.7 153^{15}	...	...	...	v	v	...	...	lig v	**B10**, 97
b1613	—,**3-ethoxy-**	$C_9H_{10}O_3$. *See* b1239	166.18	nd (w or sub)	137	sub	...	...	s	s	s	...	s	...	**B10**[1], 64
b1614	—,—,amide	$C_9H_{11}NO_2$. *See* b1244	165.19	nd (w)	139–9.5	...	...	...	...	v	s	v	...	peth δ chl v lig δ	**B10**, 141
b1615	—,—,ethyl ester	$C_{11}H_{14}O_3$. *See* b1239	194.23	...	...	264 $172–3^{50}$	1.0725_{20}^{20}	...	i δ^h	v	v	...	...	...	**B10**, 139

For explanations, symbols and abbreviations see beginning of table. For structural formulas see end of table.

No.	Name	Synonyms and Formula	Mol. wt.	Color, crystalline form, specific rotation and λ_{max} (log ε)	m.p. °C	b.p. °C	Density	n_D	Solubility w	al	eth	ace	bz	other solvents	Ref.
	Benzoic acid														
Ω b1616	—,**4-ethoxy-**	$C_9H_{10}O_3$. *See* b1239	166.18	nd (w)	198.5	...	...	...	δ^h	s	s	...	s	...	**B10², 92**
b1617	—,—,amide	$C_9H_{11}NO_2$. *See* b1244	165.19	pr (dil al)	206	...	...	...	δ	v	...	...	...	...	**B10², 101**
Ω b1618	—,—,ethyl ester	$C_{11}H_{14}O_3$. *See* b1239	194.23	...	...	275 148–9^{14}	1.076^{12}	...	i	v	v	...	...	...	**B10², 98**
b1619	—,—,nitrile	C_9H_9NO. *See* b1291	147.18	nd (lig)	61–2	258	...	...	δ	v	v	...	...	lig v^h	**B10², 101**
b1620	—,—,piperazinium salt	$C_{22}H_{30}N_2O_6$. *See* b1239	418.50	...	176–7d	...	...	...	δ	s^h	i	...	...	...	**Am 70, 2758**
b1621	—,**4-ethoxy-2-hydroxy-**	$C_9H_{10}O_4$. *See* b1239	182.18	nd (w or bz)	154				δ s^h	v	v	...	v	...	**B10, 379**
b1622	—,**2-ethyl-**	$C_9H_{10}O_2$. *See* b1239	150.18	nd (w)	68	259^{760}	1.0413_4^{100}	1.5099^{100}	δ	v	s	...	...	lig δ	**B9², 349**
b1623	—,**3-ethyl-**	$C_9H_{10}O_2$. *See* b1239	150.18	nd (w or dil al)	47	...	1.042^{100}	1.5345^{100}	i	s	v	...	...	...	**B9¹, 208**
b1624	—,**4-ethyl-**	$C_9H_{10}O_2$. *See* b1239	150.18	pr (al), pl or lf (w)	113.5	...	...	...	δ v^h	s	...	...	s	chl s	**B9², 349**
b1625	—,**2(ethylamino)-**	$C_9H_{11}NO_2$. *See* b1239	165.19	pr or nd (dil al)	154	...	...	...	...	s	s	s	...	...	**B14², 213**
b1626	—,**3(ethylamino)-**	$C_9H_{11}NO_2$. *See* b1239	165.19	nd or pr (dil al)	112	sub	...	...	δ^h	v	v	s	...	...	**B14, 393**
b1627	—,**4(ethylamino)-**	$C_9H_{11}NO_2$. *See* b1239	165.19	cr (bz)	177–8	...	...	...	...	s	s	s	s	os s	**B14¹, 572**
Ω b1628	—,**2-fluoro-**	$C_7H_5FO_2$. *See* b1239	140.12	nd (w) λ^{al} 223 (4.98), 273 (3.21), 280 (3.11)	126.5	...	1.460_4^{25}	...	δ v^h	v	v	...	i	chl s CS_2 i	**B9², 220**
Ω b1629	—,**3-fluoro-**	$C_7H_5FO_2$. *See* b1239	140.12	lf (w) λ^{al} 225 (4.0), 274 (3.23), 281 (3.15)	124	...	1.474_4^{25}	...	δ	...	s	...	...	...	**B9², 220**
Ω b1630	—,**4-fluoro-**	$C_7H_5FO_2$. *See* b1239	140.12	pr (w) λ^{al} 228 (4.04), 262 (2.70), 268 (2.48)	185		1.479_4^{25}	...	δ s^h	s	s	...	...	...	**B9², 220**
Ω b1631	—,—,nitrile	C_7H_4FN. *See* b1291	121.12	nd (peth) λ^{iso} 228 (4.0), 258 (2.4), 272 (2.4)	34.8	188.8^{750}	1.1070^{55}	1.4925^{55}	...	...	...	...	...	peth s^h	**B9², 221**
b1632	—,**2-formamido-**	$C_8H_7NO_3$. *See* b1291	165.15	nd (w + ½)	169	...	...	...	δ	s	s	...	δ	...	**B14², 219**
Ω b1633	—,**2-formyl-**	Phthalaldehydic acid. $C_8H_6O_3$. *See* b1239	150.14	lf (+w) λ^{hx} 224 (3.90), 276 (2.93), 286 (2.60)	98–9 (hyd) 240–50 (anh)	...	1.404	...	s	v	v	...	...	...	**B10, 666**
b1634	—,**3-formyl-**	Isophthalaldehydic acid. $C_8H_6O_3$. *See* b1239	150.14	nd (w)	175	...	...	...	s^h	v	v	...	...	...	**B10², 465**
b1635	—,—,nitrile	3-Cyanobenzaldehyde. C_8H_5NO. *See* b1291	131.14	nd (eth)	79–81	210	...	...	v^h	v	v	...	...	chl v	**B10, 671**
Ω b1636	—,**4-formyl-**	Terephthalaldehydic acid. $C_8H_6O_3$. *See* b1239	150.14	nd (w) λ^{hx} 249 (4.24), 279 (3.22), 298 (3.20)	256	sub	...	...	δ^h	v	s	...	...	chl s	**B10, 671**
Ω b1637	—,—,nitrile	4-Cyanobenzaldehyde. C_8H_5NO. *See* b1291	131.14	nd (w), pr (eth or dil al)	101–2	133^{12}	...	...	s^h	v	v	...	...	chl v	**B10², 465**
b1638	—,**3-formyl-2-hydroxy-**	$C_8H_6O_4$. *See* b1239	166.14	nd (w + 1)	179	...	...	...	δ s^h	s	...	...	...	...	**B10², 675**
b1639	—,**3-formyl-4-hydroxy-**	$C_8H_6O_4$. *See* b1239	166.14	pr (w)	244	sub	...	...	δ s^h	s	s	...	...	chl δ alk s	**B10², 675**
b1640	—,**4-formyl-3-hydroxy-**	$C_8H_6O_4$. *See* b1239	166.14	nd (w)	234	sub	...	...	δ^h	s	s	...	...	alk s	**B10, 954**
b1641	—,**5-formyl-2-hydroxy-**	$C_8H_6O_4$. *See* b1239	166.14	nd (w)	250d	...	...	...	δ^h	s^h	s	...	...	chl i	**B10², 675**
Ω b1642	—,**2-hydrazino-**	$C_7H_8N_2O_2$. *See* b1239	152.16	nd (w) λ^{al} 280 (3.8), 343 (3.6)	250–1	...	...	...	s^h	δ s^h	δ	...	...	...	**B15², 295**
b1643	—,—,hydrochloride	$C_7H_8N_2O_2$.HCl. *See* b1239	188.62	nd (w)	194–5d	...	...	...	v^h	...	...	i			**B15², 295**
b1644	—,**3-hydrazino-**	$C_7H_8N_2O_2$. *See* b1239	152.16	pa ye lf (w) λ^{al} 276 (4.1), 330 (3.2)	186d	...	...	...	δ^h	δ	i				**B15¹, 205**
Ω b1645	—,**4-hydrazino-**	$C_7H_8N_2O_2$. *See* b1239	152.16	ye nd or pl (w) λ^{al} 322 (4.2)	114 (220–5d)	...	...	...	δ s^h	...	i	...	...	...	**B15², 297**
b1646	—,**2,2′-hydrazodi-**	...	272.26	pl or pr (al)	205	...	...	...	i	s^h	...	...	...	...	**B15², 296**
b1647	—,**3,3′-hydrazodi-**	...	272.26	nd (al)	206d	...	...	...	...	s	s	s	s	os s	**B15, 629**
b1648	—,**4,4′-hydrazodi-**	...	272.26	nd (al), cr (aa + 1)	286d	...	...	...	i	δ^h	...	...	...	aa s^h	**B15, 632**
Ω b1649	—,**2-hydroxy-**	Salicylic acid. $C_7H_6O_3$. *See* b1239	138.12	nd (w), mcl pr (al) λ^{al} 207 (4.46), 236 (3.85), 303 (3.57)	159	211^{20} sub	1.443_4^{20}	1.565	δ v^h	v	v	v	δ s^h	chl, CCl_4, to δ MeOH v	**B10², 25**
b1650	—,—,4-acetamido-phenyl ester	Salophene. $C_{15}H_{13}NO_4$. *See* b1239	271.28	pl (w), lf (al)	187–8	...	...	...	i δ^h	s	s	...	s	peth i	**B13², 247**
Ω b1651	—,—,acetate	Aspirin. $C_9H_8O_4$. *See* b1239	180.17	nd (w), mcl ta (w) λ^{eth} 275 (3.11)	135 (rapid htng)	...			s^h d^h	v	s	...	δ	chl s	**B10², 41**
Ω b1652	—,—,amide	$C_7H_7NO_2$. *See* b1244. Salicylanilide.	137.14	ye lf (dil al) λ^{al} 302 (3.62)	142	181.5^{14}	1.175_4^{140}	...	δ	s	δ	...	...	...	**B10, 87**

For explanations, symbols and abbreviations see beginning of table. For structural formulas see end of table.

No.	Name	Synonyms and Formula	Mol. wt.	Color, crystalline form, specific rotation and λ_{max} (log ε)	m.p. °C	b.p. °C	Density	n_D	Solubility						Ref.
									w	al	eth	ace	bz	other solvents	
	Benzoic acid														
b1653	—,—,—.*N*-phenyl-	$C_{13}H_{11}NO_2$. *See* b1244 ..... Salicylanilide.	213.24	pr (w or al), lf (w) λ^{chl} 265 sh (4.13), 295 (2.16)	136–7	....	....	....	s^h	δ	δ	...	δ	chl δ	**B12**, 500
b1654	—,—,—.*N*(2-tolyl)-	$C_{14}H_{13}NO_2$. *See* b1244	227.27	nd (al)	144	....	....	....	...	s^h	...	...	...	....	**B12²**, 450
b1655	—,—,anhydride ...	$C_{14}H_{10}O_5$. *See* b1239	258.23	amor	255d	....	....	....	i	v	v	...	...	....	**C37**, 3073
Ω b1656	—,—,benzyl ester ..	$C_{14}H_{12}O_3$. *See* b1239	228.25	λ^{al} 240 (4.1), 308 (3.7)	....	170^{5} 320^{760}	1.1799_{4}^{20}	1.5805^{20}	δ	s	s	...	...	....	**B10²**, 51
b1657	—,—,—,acetate ...	$C_{16}H_{14}O_4$. *See* b1239	270.29	....	26	$197–200^{7}$	....	....	...	v	v	v	v	os v lig δ peth δ	**B10²**, 52
Ω b1658	—,—,butyl ester ...	$C_{11}H_{14}O_3$. *See* b1239	194.23	....	−5.9	270–2 (260) $136–8^{10}$	1.0728_{4}^{20}	1.5115^{20}	...	...	...	...	...	....	**B10²**, 49
b1659	—,—,(2-carboxyphenyl) ether	Disalicylic acid. $C_{14}H_{10}O_5$. *See* b1239	258.23	cr ($Cl_2CHCHCl_2$)	230d	....	....	....	i	v^h	s	...	i	aa, MeOH s chl i	**B10²**, 54
b1660	—,—,chloride	$C_7H_5ClO_2$. *See* b1239	156.57	....	19–9.5	92^{15}	1.3112^{20}	1.5812^{20}	d	...	s	...	...	....	**B10¹**, 43
b1661	—,—,2,4-dinitrobenzyl ester	$C_{14}H_{10}N_2O_7$. *See* b1239	318.25	ye pl (aa)	168	....	....	....	i	δ	i	...	i	AcOEt v^h aa v^h	**B10²**, 52
Ω b1662	—,—,ethyl ester ...	$C_9H_{10}O_3$. *See* b1239	166.18	....	2–3	234 132.8^{37}	1.1326_{4}^{20}	1.5296^{20}	i	∞	v	...	...	....	**B10²**, 47
b1663	—,—,—,acetate ...	Ethyl aspirin. $C_{11}H_{12}O_4$. *See* b1239	208.22	....	....	272^{760} $148–50^{15}$	1.1566^{15}	....	i	s	v	v	v	os s	**B10²**, 48
Ω b1664	—,—,hydrazide ...	$C_7H_8N_2O_2$. *See* b1239	152.16	pl (al), pr (w) λ^{al} 295 (3.7)	147	....	....	....	i s^h	v	δ	...	s	....	**B10²**, 61
Ω b1665	—,—,2-hydroxyethyl ester	$C_9H_{10}O_4$. *See* b1239	182.18	....	37	173^{15}	1.2537_{15}^{15}	....	δ	v	v	...	v	chl v	**B10²**, 53
Ω b1666	—,—,isobutyl ester	$C_{11}H_{14}O_3$. *See* b1239	194.23	....	5.9	260–2 $136–8^{10}$	1.0639_{4}^{20}	1.5087^{20}	i	s	s	...	...	....	**B10**, 76
b1667	—,—,isopropyl ester	$C_{10}H_{12}O_3$. *See* b1239	180.21	....	....	240–2 118^{17}	1.0729_{4}^{20}	1.5065^{20}	i	∞	∞	...	...	....	**C50**, 15531
Ω b1668	—,—,menthyl ester	$C_{17}H_{24}O_3$. *See* b1239	276.38	λ^{al} 318 (2.0)	....	190^{15} $143^{0.6}$	1.0467^{20}	1.5198^{26}	i	∞	v	v	∞	os v	**B10²**, 50
b1669	—,—,methoxymethyl ester	Mesotan. $C_9H_{10}O_4$. *See* b1239	182.18	....	....	162^{42}	1.2^{15}	....	δ	∞	∞	∞	∞	chl ∞	**B10²**, 54
b1670	—,—,2-methoxyphenyl ester	$C_{14}H_{12}O_4$. *See* b1239	244.25	....	70	....	....	....	i	s	s	...	...	chl s	**B10**, 81
Ω b1671	—,—,methyl ester ...	$C_8H_8O_3$. *See* b1239	152.16	λ^{al} 238 (3.97), 306 (3.64)	−8 to −7	223.3^{760}	1.1738_{4}^{20}	1.5369^{20}	δ	v	v	...	...	....	**B10²**, 44
b1672	—,—,—,benzoate .	$C_{15}H_{12}O_4$. *See* b1239	256.26	pr (al or eth) λ 277 (3.31)	92	385d $270–80^{120}$	....	....	i δ^h	s	s	...	s	chl s	**B10²**, 47
Ω b1673	—,—,3-methylbutyl ester	Isoamyl salicylate. $C_{12}H_{16}O_3$. *See* b1239	208.24	λ^{al} 310 (3.6)	....	$276–7^{743}$ $151–2^{15}$	1.0535_{4}^{20}	1.5080^{20}	i	v	s	...	...	chl s	**B10²**, 49
b1674	—,—,1-naphthyl ester	Alphol. $C_{17}H_{12}O_3$. *See* b1239	264.28	....	83	....	....	....	...	δ	s	...	...	....	**E12B**, 1183
b1675	—,—,2-naphthyl ester	Betol. $C_{17}H_{12}O_3$. *See* b1239 .	264.28	cr (al)	95.5	....	1.11^{116}	....	i	δ s^h	s	...	s	....	**B10²**, 53
b1676	—,—,nitrile	2-Cyanophenol. C_7H_5NO. *See* b1291	119.12	pr (bz) λ^{w} 231 (3.95), 294 (3.61)	98	149^{14}	1.1052_{4}^{100}	1.5372_{α}^{100}	δ	v	v	...	v	chl v	**B10²**, 60
b1677	—,—,4-nitrobenzyl ester	$C_{14}H_{11}NO_5$. *See* b1239	273.25	cr (dil al)	97–8	....	....	....	δ	v	...	v	...	....	**B10²**, 52
b1678	—,—,pentyl ester ..	$C_{12}H_{16}O_3$. *See* b1239	208.24	λ^{al} 300 (3.9)	....	265^{760}	1.065_{15}^{15}	1.506^{20}	δ	∞	∞	...	...	....	**B10²**, 49
Ω b1679	—,—,phenyl ester ..	Salol. $C_{13}H_{10}O_3$. *See* b1239 .	214.22	pl (MeOH) λ^{al} 300 (3.7)	43	173^{12}	1.2614_{4}^{30}	....	i	v	s	v	v	aa, to s CCl_4, Py v	**B10**, 76
b1680	—,—,propyl ester ..	$C_{10}H_{12}O_3$. *See* b1239	180.21	....	96–8	239^{760} (250)	1.0979_{4}^{20}	1.5161^{20}	δ	∞	∞	...	...	....	**B10**, 75
Ω b1681	**—,3-hydroxy-**	$C_7H_6O_3$. *See* b1239	138.12	nd (w), pl or pr (al) λ^{al} 234 (3.78), 301 (3.36)	201.5–3	....	....	....	δ v^h	s^h	s	s	i	MeOH s	**B10²**, 79
Ω b1682	—,—,amide	$C_7H_7NO_2$. *See* b1244	137.14	pl (w) λ^{al} 250 sh (3.7), 323 (3.3)	170.5	....	....	....	δ s^h	s	s	...	i	chl i CS_2 i	**B10²**, 82
Ω b1683	—,—,—;*N*-phenyl-	$C_{13}H_{11}NO_2$. *See* b1244	213.24	nd or pl (w, dil al)	156	....	....	....	δ	s	δ	...	δ	chl i	**B12¹**, 269
b1684	—,—,chloride	$C_7H_5ClO_2$. *See* b1239	156.57	....	< −15	$110–3^{0.5}$	....	....	d	d	...	...	...	chl s	**B10¹**, 66
Ω b1685	—,—,ethyl ester ...	$C_9H_{10}O_3$. *See* b1239	166.18	pl (bz)	73.8	295 211^{65}	....	....	δ	s	s	...	...	....	**B10²**, 81
Ω b1686	—,—,methyl ester .	$C_8H_8O_3$. *See* b1239	152.16	nd (bz-peth)	71.5 (70)	280^{709} 178^{17}	....	....	...	s	...	...	s^h	peth s^h	**B10²**, 81
Ω b1687	—,—,nitrile	3-Cyanophenol. C_7H_5NO. *See* b1291	119.12	pr (al or eth), lf (w)	83–4	....	....	....	v^h	v	v	...	v	chl v	**B10**, 141
Ω b1688	**—,4-hydroxy-**	$C_7H_6O_3$. *See* b1239	138.12	pr or pl (w, al, xyl-al), cr (dil al or ace-MeOH, + 1w) λ^{al} 293.5 (4.18)	214.5–5.5 (anh)	....	....	....	δ v^h	v^h	s	s	δ	CS_2 i	**B10²**, 88

For explanations, symbols and abbreviations see beginning of table. For structural formulas see end of table.

No.	Name	Synonyms and Formula	Mol. wt.	Color, crystalline form, specific rotation and λ_{max} (log ε)	m.p. °C	b.p. °C	Density	n_D	Solubility w	al	eth	ace	bz	other solvents	Ref.
	Benzoic acid														
b1689	—,—,amide	$C_7H_7NO_2$. *See* b1244	137.14	nd (w + 1) λ^{al} 220 (3.92), 288 (4.24)	162 (hyd)	...	...	...	δ s^h	s	s	...	...	CS_2, chl	**B10²**, 100
b1690	—,—,—,*N*-phenyl-	$C_{13}H_{11}NO_2$. *See* b1244	213.24	pl or nd (w)	201–2 (198)	...	...	...	δ	v	δ	...	i	chl, CS_2 δ	**B12²**, 257
b1691	—,—,butyl ester	$C_{11}H_{14}O_3$. *See* b1239	194.23	...	68–9	...	...	...	δ	s	...	...	...	...	**C51**, 9129
b1692	—,—,ethyl ester	$C_9H_{10}O_3$. *See* b1239	166.18	cr (dil al)	116–8	297–8	...	...	δ	v	v	...	...	CS_2 i lig δ chl, peth δ	**B10²**, 96
b1693	—,—,methyl ester	$C_8H_8O_3$. *See* b1239	152.16	nd (dil al) λ^{al} 258 (4.22)	131	270–80d	...	...	δ	v	v	v	...	...	**B10²**, 95
b1694	—,—,nitrile	4-Cyanophenol. C_7H_5NO. *See* b1291	119.12	lf (w) λ^{al} 247 (4.3)	113	...	...	...	δ s^h	v	v	...	...	chl v	**B10²**, 101
b1695	—,—,propyl ester	$C_{10}H_{12}O_3$. *See* b1239	180.21	pr (eth) λ^{al} 257 (3.21)	96.2–98	...	1.0630^{102}_{4}	1.5050^{102}	i s^h	s^h	s	...	...	...	**B10²**, 97
b1696	—,4(α-hydroxybenzyl)-	$C_{14}H_{12}O_3$. *See* b1239	228.25	nd (w)	164–5	...	...	...	s^h	v	v	...	...	chl δ to δ	**B10**, 346
b1697	—,2-hydroxy-5-bromo-, acetate	$C_9H_7BrO_4$. *See* b1239	259.06	nd (al)	60	...	...	...	i	v	s	...	...	...	**B10²**, 64
b1699	—,2-hydroxy-(dithio)-	Dithiosalicylic acid. *o*-Hydroxydithiobenzoic acid.	170.25	og-ye nd (peth)	48–50	...	...	...	s	s	s	...	s	MeOH s	**B10²**, 78
b1700	—,2-hydroxy-3-iodo-	$C_7H_5IO_3$. *See* b1239	264.02	nd (w)	199	...	...	...	δ s^h	v	v	...	i	chl i	**B10²**, 64
b1701	—,2-hydroxy-4-iodo-	$C_7H_5IO_3$. *See* b1239	264.02	pl or nd (al)	230d	...	...	...	δ	s v^h	s	...	...	chl i	**B10²**, 65
b1702	—,2-hydroxy-5-iodo-	$C_7H_5IO_3$. *See* b1239	264.02	nd (w)	197	...	...	...	δ s^h	v	s	...	i	chl i	**B10²**, 65
b1703	—,3-hydroxy-2-iodo-	$C_7H_5IO_3$. *See* b1239	264.02	nd (chl)	158–9	...	...	...	δ s^h	s	s	...	...	chl s^h	**B10²**, 84
b1704	—,3-hydroxy-4-iodo	$C_7H_5IO_3$. *See* b1239	264.02	nd (w)	226–8d	...	...	...	δ s^h	s	s	...	...	...	**B10²**, 84
b1705	—,4-hydroxy-2-iodo-	$C_7H_5IO_3$. *See* b1239	264.02	nd (w)	215d	...	...	...	s^h	s	s	...	δ	...	**B10²**, 104
b1706	—,4-hydroxy-3-iodo-	$C_7H_5IO_3$. *See* b1239	264.02	nd (w + ½)	173.5–4.5	sub	...	...	s^h	v	v	...	$δ^h$	chl δ aa s lig i	**B10²**, 104
b1707	—,5-hydroxy-2-iodo-	$C_7H_5IO_3$. *See* b1239	264.02	nd (w)	198	sub 160	...	...	s^h	s	s	...	...	...	**B10²**, 84
b1708	—,2-hydroxy-3-isopropyl-6-methyl-	*o*-Thymotinic acid. $C_{11}H_{14}O_3$. *See* b1239	194.23	nd (w, bz, lig)	127	sub	...	...	i $δ^h$	s	s	...	s	lig s^h chl, aa s	**B10²**, 170
b1709	—,2-hydroxy-5-isopropyl-3-methyl-	$C_{11}H_{14}O_3$. *See* b1239	194.23	nd (w)	147	...	...	...	i	v	...	...	...	...	**B10**, 282
b1710	—,2-hydroxy-5-isopropyl-4-methyl-	$C_{11}H_{14}O_3$. *See* b1239	194.23	nd (bz)	189–90	...	...	...	...	...	...	...	...	...	**B10²**, 171
b1711	—,2-hydroxy-6-isopropyl-3-methyl-	*o*-Carvacrotinic acid. $C_{11}H_{14}O_3$. *See* b1239	194.23	nd (w)	136	sub	...	...	δ	v	v	...	...	...	**B10**, 282
b1712	—,4-hydroxy-5-isopropyl-2-methyl-	*p*-Thymotinic acid. $C_{11}H_{14}O_3$. *See* b1239	194.23	pl (dil al)	157	...	...	...	$δ^h$	v	v	...	v	chl v	**B10²**, 171
b1713	—,3-hydroxy-4-methoxy-	Isovanillic acid. $C_8H_8O_4$. *See* b1239	168.16	nd, pr, pl (w) λ^{al} 256 (4.00), 293.5 (4.71)	255–7	sub	...	...	δ s^h	v	v	...	...	...	**B10²**, 261
Ω b1714	—,4-hydroxy-3-methoxy-	Vanillic acid. $C_8H_8O_4$. *See* b1239	168.16	nd (w) λ^{al} 263 (4.00), 292 (3.72)	213–5	sub	...	...	δ	v	s	...	...	...	**B10²**, 261
Ω b1715	—,—,ethyl ester	$C_{10}H_{12}O_4$. *See* b1239	196.21	nd (dil al) λ^{diox} 259 (4.05), 290 (3.72)	44	291–3	...	...	i	v	v	...	...	...	**B10**, 397
Ω b1716	—,—,methyl ester	$C_9H_{10}O_4$. *See* b1239	182.18	nd (dil al)	64	285–7 118^{2}	...	...	...	s^h	...	...	...	peth s^h	**B10¹**, 189
b1717	—,2-hydroxy-4-methoxy-6-methyl-	Everninic acid. $C_9H_{10}O_4$. *See* b1239	182.18	nd (peth or w)	171–2 (157)	...	...	...	δ s^h	v	δ	v	δ	aa v lig δ AcOEt s	**B10²**, 273
b1718	—,—,methyl ester	Sparassol. $C_{10}H_{12}O_4$. *See* b1239	196.21	pr (w), lf (MeOH)	67–8	...	...	...	$δ^h$	s	v	v	...	MeOH s peth s chl v	**B10²**, 273
Ω b1719	—,2-hydroxy-3-methyl-	2,3-Cresotic acid. $C_8H_8O_3$. *See* b1239	152.16	nd (w or dil al)	169–70	...	...	...	δ s^h	s	s	...	s^h	chl s	**B10²**, 131
b1720	—,—,chloride	$C_8H_7ClO_2$. *See* b1239	170.60		27–8	$87–9^{16}$	...	...	d	d^h	...	...	...	...	**B10²**, 132
Ω b1721	—,2-hydroxy-4-methyl-	2,4-Cresotic acid. $C_8H_8O_3$. *See* b1239	152.16	nd (w), pr (al), pl (chl) λ^{MeOH} 238 (3.99), 297 (3.64)	177.8	sub	...	...	δ s^h	s	v	...	s^h	chl s	**B10²**, 137

For explanations, symbols and abbreviations see beginning of table. For structural formulas see end of table.

No.	Name	Synonyms and Formula	Mol. wt.	Color, crystalline form, specific rotation and λ_{max} (log ε)	m.p. °C	b.p. °C	Density	n_D	Solubility w	al	eth	ace	bz	other solvents	Ref.
	Benzoic acid														
b1722	—,**2-hydroxy-5-methyl-**	2,5-Cresotic acid. $C_8H_8O_3$. *See* b1239	152.16	nd (w or peth)	153	...	...	...	δ s^h	s	s	...	s	chl s, peth s^h CS_4 i	**B10**^{2}, 134
b1723	—,**2-hydroxy-6-methyl-**	2,6-Cresotic acid. $C_8H_8O_3$. *See* b1239	152.16	nd (chl) λ^{MeOH} 240 (3.75), 307 (3.52)	173 (184)	...	...	...	δ s^h	v	v	...	...	chl s	**B10**^{2}, 128
b1724	—,**3-hydroxy-2-methyl-**	2,3-Cresotic acid. $C_8H_8O_3$. *See* b1239	152.16	nd (w or dil al)	145–6	...	...	...	δ s^h	v	v	...	...	...	**B10**, 214
b1725	—,**3-hydroxy-4-methyl-**	3,4-Cresotic acid. $C_8H_8O_3$. *See* b1239	152.16	nd or pr (w) λ^{al} 206 (4.67), 244 (4.08), 298 (3.65)	208.5	sub	...	...	δ s^h	s	s	...	δ	peth δ chl i	**B10**^{2}, 140
b1726	—,**3-hydroxy-5-methyl-**	3,5-Cresotic acid. $C_8H_8O_3$. *See* b1239	152.16	nd (w)	210	sub	...	...	s^h	v	v	...	...	chl δ	**B10**, 227
b1727	—,**4-hydroxy-2-methyl-**	4.2-Cresotic acid. $C_8H_8O_3$. *See* b1239	152.16	nd (w + ½) λ^{w} 242 (3.91), 275 sh (3.5)	177–8	236–7 sub	...	...	δ s^h	s s	s	...	...	chl i	**B10**^{2}, 127
b1728	—,**4-hydroxy-3-methyl-**	4.3-Cresotic acid. $C_8H_8O_3$. *See* b1239	152.16	nd (w + ½) λ^{w} 250 (4.10), 280 sh (3.6)	174–5 (anh)	...	...	...	s^h δ	s	s	...	...	chl δ^h	**B10**^{2}, 133
b1729	—,**5-hydroxy-2-methyl-**	3.6-Cresotic acid. $C_8H_8O_3$. *See* b1239	152.16	nd or pr (w)	185	...	...	...	δ s^h	v	v	...	...	chl δ	**B10**, 215
b1730	—,**2-(hydroxy-methyl)-**	2-Carboxybenzyl alcohol. $C_8H_8O_3$. *See* b1239	152.16	nd (w)	128 (120)	...	...	...	s	v	v	...	...	...	**B10**, 218
—	—,—,lactone	*see* **Phthalide**													
b1731	—,**2-hydroxy-3-nitro-**	$C_7H_5NO_5$. *See* b1239	183.12	yesh nd (aa, w + 1) $\lambda^{w, pH = 3.7}$ 357 (3.64)	148–9 (anh) 128–9 (+w)	...	...	...	δ	v	v	s	s	chl s	**B10**^{2}, 66
b1732	—,**2-hydroxy-4-nitro-**	$C_7H_5NO_5$. *See* b1239	183.12	ye nd (w or dil al)	235	...	...	...	δ s^h	v	...	s	δ	chl s aa s lig i	**B10**^{2}, 66
b1733	—,**2-hydroxy-5-nitro-**	$C_7H_5NO_5$. *See* b1239	183.12	nd (w)	229–30	...	1.650^{20}	...	δ s^h	v	v	v	v	os v	**B10**^{2}, 67
b1734	—,**2-hydroxy-6-nitro-**	$C_7H_5NO_5$. *See* b1239	183.12	pa ye nd	...	...	...	...	v	v	...	...	...	...	**J1950**, 2049
b1736	—,**3-hydroxy-2-nitro-**	$C_7H_5NO_5$. *See* b1239	183.12	pl or pr (w + 1)	180.5–1.5 (178)	...	...	...	s	v	v	...	...	...	**B10**^{2}, 84
Ω b1737	—,**3-hydroxy-4-nitro-**	$C_7H_5NO_5$. *See* b1239	183.12	ye lf (w)	235	...	...	...	δ	s	s	...	...	...	**B10**^{2}, 85
b1738	—,**3-hydroxy-5-nitro-**	$C_7H_5NO_5$. *See* b1239	183.12	ye lf or pl (25 % HCl)	167	...	...	...	s^h	s	s	...	...	...	**B10**^{2}, 85
Ω b1739	—,**4-hydroxy-3-nitro-**	$C_7H_5NO_5$. *See* b1239	183.12	nd or lf (w)	186–7	...	...	...	s^h	v	v	...	...	...	**B10**^{2}, 106
b1740	—,**5-hydroxy-2-nitro-**	$C_7H_5NO_5$. *See* b1239	183.12	ye nd or pr (w + 1)	172 (165)	...	...	...	v	v	v	...	...	...	**B10**^{2}, 85
b1741	—,**2-hydroxy-3-sulfo-**	$C_7H_6O_6S$. *See* b1239	218.19	pr (w + 2)	120	...	...	...	v	v	s	...	...	aa s	**B11**^{2}, 231
b1742	—,**2-hydroxy-5-sulfo-**	$C_7H_6O_6S$. *See* b1239	218.19	hyg nd (w + 2)	120 (anh) 198–9d (+2w)	...	...	...	∞	∞	v	...	...	...	**B11**^{2}, 232
b1743	—,**3-hydroxy-4-sulfo-**	$C_7H_6O_6S$. *See* b1239	218.19	ye-gr nd (w + 2½)	208 (213)	...	...	...	v	v	i	...	...	...	**B11**, 413
b1744	—,**3-hydroxy-5-sulfo-**	$C_7H_6O_6S$. *See* b1239	218.19	nd (w + 1)	120d	...	...	...	s	v	v	...	...	...	**B11**, 413
b1745	—,**4-hydroxy-3-sulfo-**	$C_7H_6O_6S$. *See* b1239	218.19	nd or lf (w)	d	...	...	...	v	s	i	...	...	...	**B11**^{2}, 235
Ω b1746	—,**2-iodo-**	$C_7H_5IO_2$. *See* b1239	248.02	nd (w) λ^{al} 233 sh (3.85), 285 (3.16)	163	233 exp	2.249^{25}_4	...	δ	v	v	...	...	...	**B9**^{2}, 239
Ω b1747	—,—,methyl ester .	$C_8H_7IO_2$. *See* b1239	262.05	...	...	277–8^{729} 146^{16}	...	1.6052^{20}	...	s^h	...	...	...	...	**B9**, 364
Ω b1748	—,**3-iodo-**	$C_7H_5IO_2$. *See* b1239	248.02	mcl pr (ace) λ^{al} 217 (4.42), 284 (2.98), 290 sh (2.95)	187–8	sub	...	...	δ	v	δ	...	...	...	**B9**^{2}, 240
Ω b1749	—,—,methyl ester .	$C_8H_7IO_2$. *See* b1239	262.05	nd (dil al)	54–5	276–7^{739} 149–50^{18}	...	...	i	s	v	v	...	os v^h lig i	**B9**, 365
Ω b1750	—,**4-iodo-**	$C_7H_5IO_2$. *See* b1239	248.02	mcl pr (dil al), lf (sub) λ^{al} 252 (4.26), 282 sh (3.23)	(i) 270 (st) (ii) 228–9 (unst)	sub	2.184^{20}	...	δ s^h	δ	δ	...	...	...	**B9**^{2}, 240
Ω b1751	—,—,methyl ester .	$C_8H_7IO_2$. *See* b1239	262.05	nd (eth-al)	114	sub	...	...	...	s^h	s^h	...	...	...	**B9**, 367
b1752	—,**2-iodo-3-nitro-** ..	$C_7H_4INO_4$. *See* b1239	293.02	pr (w)	206	...	...	...	s^h	s	s	...	...	...	**B9**^{2}, 278
b1753	—,**2-iodo-4-nitro-** ..	$C_7H_4INO_4$. *See* b1239	293.02	pa ye pr (w)	146–7 (143)	...	...	...	δ s^h	s	s	...	δ	CCl_4 δ lig δ	**B9**^{1}, 166
b1754	—,**4-iodo-2-nitro-** ..	$C_7H_4INO_4$. *See* b1239	293.02	ye lf or pr (dil al)	192–3	...	...	...	δ^h	v	v	...	s^h	...	**B9**^{2}, 278

For explanations, symbols and abbreviations see beginning of table. For structural formulas see end of table.

No.	Name	Synonyms and Formula	Mol. wt.	Color, crystalline form, specific rotation and λ_{max} (log ε)	m.p. °C	b.p. °C	Density	n_D	Solubility w	al	eth	ace	bz	other solvents	Ref.
	Benzoic acid														
b1755	—,4-iodo-3-nitro-	$C_7H_4INO_4$. *See* b1239	293.02	ye pr (al)	213				δ	s					**B9**[2], 278
b1756	—,5-iodo-3-nitro-	$C_7H_4INO_4$. *See* b1239	293.02	nd (al), pr (peth)	167				s^h	v					**B9**[2], 278
Ω b1757	—,2-iodoso-	$C_7H_5IO_3$. *See* b1239	264.02	lf (w)	223–5d				δ s^h	δ s^h	δ				**B9**[2], 239
b1758	—,3-iodoso-	$C_7H_5IO_3$. *See* b1239	264.02	ye amor	175–80				d^h	δ	δ		i	chl i	**B9**, 365
b1759	—,4-iodoso-	$C_7H_5IO_3$. *See* b1239	264.02	amor	212d				d^h	i			i	chl i	**B9**, 366
b1760	—,2-isopropyl-	*o*-Cuminic acid. $C_{10}H_{12}O_2$. *See* b1239	164.21	pr (w or peth)	64	$160–1^{25}$			δ s^h	v	v		s	peth s	**B9**, 546
Ω b1761	—,4-isopropyl-	Cumic acid. *p*-Cuminic acid. $C_{10}H_{12}O_2$. *See* b1239	164.21	tcl pl (al)	117–8	sub	1.162^4		δ s^h	v	v			peth s	**B9**, 546
Ω b1762	—,2-mercapto-	Thiosalicylic acid. $C_7H_6O_2S$. *See* b1239	154.19	lf or nd (al, w, aa) λ^{al} 220 (4.18), 239 sh (3.81)	168–9 (165)	sub			s^h	s	s			aa s lig δ	**B10**[2], 70
Ω b1763	—,2-methoxy-	*o*-Anisic acid. $C_8H_8O_3$. *See* b1239	152.16	pl (w), fl (al) λ^{al} 226 (3.82), 280 (3.40)	101	200			δ v^h	v	v		s	chl v CCl_4 s	**B10**[2], 39
Ω b1764	—,—,amide	$C_8H_9NO_2$. *See* b1244	151.17	nd (bz), pr (eth), pl (w)	129				δ		v		v^h	sulf s (ye)	**B10**[2], 58
Ω b1765	—,—,chloride	$C_8H_7ClO_2$. *See* 1239	170.60			254^{760} 128^{11}			d	d^h					**B10**[2], 55
b1766	—,—,ethyl ester	$C_{10}H_{12}O_3$. *See* b1239	180.21			261^{760} $135–6^{12}$	1.1124^{20}_4	1.5224^{20}	i	s	s				**B10**[2], 48
b1767	—,—,methyl ester	$C_9H_{10}O_3$. *See* b1239	166.18	λ^{al} 234 (3.82), 294 (3.54)		245^{760} 127^{11}	1.1571^{19}_4	$1.534^{19.5}$	i	s					**B10**[2], 46
b1768	—,—,nitrile	C_8H_7NO. *See* b1291	133.15	λ^{al} 232 (3.95), 294 (3.65)	24.5	255–6 146^{20}	$1.1063^{20.5}_4$			s	v				**B10**[2], 60
b1769	—,—,piperazinium salt	$C_{20}H_{26}N_2O_6$. *See* b1239	390.45		190.4–1.4				s	s	i				**Am 70**, 2758
Ω b1770	—,3-methoxy-	*m*-Anisic acid. $C_8H_8O_3$. *See* b1239	152.16	nd (w)	110	$170–2^{10}$			δ s^h	s	s		s	CCl_4 δ chl v	**B10**[2], 80
b1771	—,—,chloride	$C_8H_7ClO_2$. *See* b1239	170.60	λ^{al} 230 (3.83), 293 (3.39)		$243–4^{760}$ $123–5^{15}$			d	d^h	s	s	v^h		**B10**, 140
b1772	—,—,ethyl ester	$C_{10}H_{12}O_3$. *See* b1239	180.21			$260–1^{760}$ 110^5	1.0993^{20}_4	1.5161^{20}	i	s	s				**B10**[2], 81
b1773	—,—,methyl ester	$C_9H_{10}O_3$. *See* b1239	166.18			252 (238) $121–4^{10}$	1.1310^{20}_4	1.5224^{20}	i	s					**B10**[2], 81
b1774	—,—,piperazinium salt	$C_{20}H_{26}N_2O_6$. *See* b1239	390.45		137–8.5d				s^h	s	$δ^h$				**Am 70**, 2758
Ω b1775	—,4-methoxy-	Anisic acid. $C_8H_8O_3$. *See* b1239	152.16	pr or nd (w) λ^{al} 249 (4.14)	185	275–80			i s^h	v	v			chl s MeOH v	**B10**[2], 91
b1776	—,—,amide	$C_8H_9NO_2$. *See* b1244	151.17	nd or ta (w)	166.5–7.5	295^{760}			s	v	δ				**B10**[2], 100
b1777	—,—,butyl ester	$C_{12}H_{16}O_3$. *See* b1239	208.24			183^{40}	$1.054^{16.5}$	$1.5141^{18.5}$							**B10**[2], 97
Ω b1778	—,—,chloride	$C_8H_7ClO_2$. *See* b1239	170.60	nd	24–5	262–3 91^1	1.261^{20}_4	1.580^{20}	d	d^h	s	s	v^h		**B10**, 77
Ω b1779	—,—,ethyl ester	$C_{10}H_{12}O_3$. *See* b1239	180.21	λ^{al} 254 (4.3)	7–8	269–70^{760} $136–7^{13}$	1.1038^{20}_4	1.5254^{20}	i	s	s				**B10**[2], 96
Ω b1780	—,—,methyl ester	$C_9H_{10}O_3$. *See* b1239	166.18	fl (al or eth)	49	256^{760} 160^{20}			i	s	s				**B10**[2], 95
Ω b1781	—,—,nitrile	C_8H_7NO. *See* b1291	133.15	nd (w), lf (al) λ^{al} 248 (4.29), 273 (3.33), 283 (2.92)	61–2	256–7 (240) $106–8^6$			i s^h	v^h	v		s		**B10**[2], 101
b1782	—,—,piperazinium salt	$C_{20}H_{26}N_2O_6$. *See* b1239	390.45		172–4				δ	s^h	i				**Am 56**, 150
b1783	—,5-methoxy-2(4-methoxyphenoxy)-	$C_{15}H_{14}O_5$. *See* b1239	274.28	cr (bz-lig)	95								s		**B10**[1], 181
b1784	—,3-methoxy-2(methylamino)-, methyl ester	Damascenine. $C_{10}H_{13}NO_3$. *See* b1239	195.22	pr (al)	27–9	270^{750} δd $147–8^{10}$			i $δ^h$	v	v		s	CS_2 v lig v chl v	**B14**[1], 654
Ω b1785	—,2-methyl-	*o*-Toluic acid. $C_8H_8O_2$. *See* b1239	136.16	pr or nd (w) λ^{cy} 231.5 (4.02)	107–8	$258–9^{751}$	1.062^{115}	1.512^{115}	i s^h	v	v			chl s	**B9**, 462
Ω b1786	—,—,amide	C_8H_9NO. *See* b1244	135.17	pl (w) or nd (w) $\lambda^{95\%al}$ 270 (2.85)	147 (143)				δ v^h	v	δ		δ		**B9**, 465
b1786[1]	—,—,anhydride	$C_{18}H_{14}O_3$. *See* b1239	254.29		38–9	$>325^{760}$ 11^{220}			d	d	v				**B9**[2], 319
b1787	—,—,chloride	C_8H_7ClO. *See* b1239	154.60			213–14^{760} $88–90^{12}$		1.5549^{20}	d	d^h	v				**B9**[2], 319
b1788	—,—,ethyl ester	$C_{10}H_{12}O_2$. *See* b1239	164.21	λ 279 (3.02)	< −10	227 102.5^{13}	1.0325^{21}_4	1.507^{22}	i	∞	∞				**B9**, 463
b1789	—,—,methyl ester	$C_9H_{10}O_2$. *See* b1239	150.18		< −50	215^{760} 97^{15}	1.068^{20}_4		i	∞	∞				**B9**, 463

For explanations, symbols and abbreviations see beginning of table. For structural formulas see end of table.

No.	Name	Synonyms and Formula	Mol. wt.	Color, crystalline form, specific rotation and λ_{max} (log ε)	m.p. °C	b.p. °C	Density	n_D	Solubility w	al	eth	ace	bz	other solvents	Ref.
	Benzoic acid														
Ω b1790	—,—,nitrile	C_8H_7N. *See* b1291	117.15	λ^{al} 228 (4.02), 276 (3.15), 284 (3.15)	−14 to −13	205^{760} 90^{15}	0.9955_4^{20}	1.5279^{20}	i	∞	∞	...	...	...	**B9**^{2}, 319
b1791	—,—,4-phenylphenacyl ester	...	330.39	cr (dil al)	94.5	...	...	...	...	...	...	...	...	...	**Am 52**, 3715
Ω b1792	—,**3-methyl-**	*m*-Toluic acid. $C_8H_8O_2$. *See* b1239	136.16	pr (w or al) λ^{cy} 232.5 (4.03)	111–3	263 sub	1.054^{112}	1.509	δ	v	v	...	...	...	**B9**, 475
Ω b1793	—,—,amide	C_8H_9NO. *See* b1244	135.17	nd (eth) $\lambda^{95\%\,al}$ 230 sh (4.05), 276 (2.95)	97	...	...	...	δ	s	δ	...	δ	...	**B9**, 477
b1794	—,—,anhydride	$C_{16}H_{14}O_3$. *See* b1239	254.29	cr (peth)	71	230^{17}	...	...	...	s	v	s	v	chl v	**B9**^{2}, 324
b1795	—,—,chloride	C_8H_7ClO. *See* b1239	154.60	...	−23	219–20^{766} 105^{20}	1.173_4^{20}	1.5485^{20}	d	d^h	s	...	...	...	**B9**^{2}, 324
b1796	—,—,ethyl ester	$C_{10}H_{12}O_2$. *See* b1239	164.21	...	...	234 (227) 103–5^{10}	1.0265_4^{21}	1.505^{22}	i	∞	∞	...	...	...	**B9**, 476
Ω b1797	—,—,methyl ester	$C_9H_{10}O_2$. *See* b1239	150.18	λ 348 (3.04), 360 (3.06)	...	221^{758} (215)	1.061_4^{20}	...	i	s	...	...	...	...	**B9**^{2}, 324
Ω b1798	—,—,nitrile	C_8H_7N. *See* b1291	117.15	λ^{al} 228 (4.03), 276 (3.08), 284 (3.09)	−23	213^{759} 84.5^{10}	1.0316_4^{20}	1.5252^{20}	i δ^h	∞	∞	...	...	...	**B9**^{2}, 325
b1799	—,—,4-phenylphenacyl ester	...	303.39	cr (dil al)	136.5	...	...	...	...	...	...	...	...	...	**Am 52**, 3715
Ω b1800	—,**4-methyl-**	*p*-Toluic acid. $C_8H_8O_2$. *See* b1239	136.16	nd (w) λ^{cy} 239 (4.18)	182	275 (cor) sub	...	...	i δ^h	v	v	...	...	MeOH v	**B9**, 483
Ω b1801	—,—,amide	C_8H_9NO. *See* b1244	135.17	nd or pl (w) $\lambda^{95\%\,al}$ 269 (4.3)	160 (158)	...	...	...	δ v^h	v	v	...	δ	chl δ	**B9**, 486
b1802	—,—,anhydride	$C_{16}H_{14}O_3$. *See* b1239	254.29	pl (MeOH), nd (al) λ^{peth} 242 (4.59)	95 (98)	...	...	...	...	...	s	s	v	chl s	**B9**^{2}, 329
Ω b1803	—,—,chloride	C_8H_7ClO. *See* b1239	154.60	...	−2 (−4)	225–7^{760} 102^{15}	1.1686_4^{20}	1.5547^{20}	d	d^h	...	...	...	...	**B9**^{2}, 329
Ω b1804	—,—,ethyl ester	$C_{10}H_{12}O_2$. *See* b1239	164.21	λ^w 240 (4.08)	...	235.7^{760} 110^{12}	1.0269_4^{18}	1.5089^{18}	i	∞	∞	...	...	...	**B9**, 484
Ω b1805	—,—,methyl ester	$C_9H_{10}O_2$. *See* b1239	150.18	cr (aq MeOH or peth) λ^{undil} 354 (2.75)	33.2	222.5^{760}	...	...	i	v	v	...	...	...	**B9**, 484
Ω b1806	—,—,nitrile	C_8H_7N. *See* b1291	117.15	nd (al) λ^{al} 232 (4.20), 267 (2.82), 279 (2.66)	29.5	217.6^{760} 91^{11}	0.9805_{30}^{30}	...	i	v	v	...	...	...	**B9**^{2}, 330
b1807	—,—,4-phenylphenacyl ester	...	330.39	...	165	...	...	...	...	...	...	...	...	...	**Am 52**, 3715
b1808	—,—,piperazinium salt	$C_{20}H_{26}N_2O_4$. *See* b1239	358.44	...	203.0–.3	...	...	...	δ	s	δ^h	...	...	...	**Am 70**, 2758
b1809	—,**2(methylamino)-**	$C_8H_9NO_2$. *See* b1239	151.17	pl (al or lig)	179	$80^{0.01}$	...	...	δ	v	v	...	v	chl v lig s	**B14**^{2}, 212
Ω b1810	—,—,ethyl ester	$C_{10}H_{13}NO_2$. *See* b1239	179.22	...	39	266^{760} 141–3^{15}	...	...	i	...	v	...	...	...	**B14**^{2}, 213
b1811	—,—,methyl ester	$C_9H_{11}NO_2$. *See* b1239	165.09	cr (peth)	18.5–9.5	255^{760} 130–1^{13}	1.120^{15}	1.5839^{12}	i	s	s	...	...	...	**B14**^{2}, 212
b1812	—,**3(methylamino)-**	$C_8H_9NO_2$. *See* b1239	151.16	pl (peth)	127	...	...	...	i δ^h	v	i	v	v	chl v lig i	**B14**, 391
b1813	—,—,methyl ester	$C_9H_{11}NO_2$. *See* b1239	165.19	cr (al)	72	...	...	...	i	s	v	...	...	...	**B14**, 392
Ω b1814	—,**4(methylamino)-**	$C_8H_9NO_2$. *See* b1239	151.16	nd (bz, w or dil al)	168 (159)	...	...	...	s^h	v	v	...	s^h	AcOEt s	**B14**^{2}, 259
b1815	—,—,methyl ester	$C_9H_{11}NO_2$. *See* b1239	165.19	pl (dil al or lig)	95.5	...	...	...	i	s	s	...	...	...	**B14**^{1}, 571
b1816	—,**3(3-methylbutoxy)-**	*m*-Isoamyloxybenzoic acid. $C_{12}H_{16}O_3$. *See* b1239	208.24	cr (al)	74–5	...	...	...	...	δ s^h	...	...	...	...	**B10**^{1}, 64
Ω b1817	—,**4(3-methylbutoxy)-**	*p*-Isoamyloxybenzoic acid. $C_{12}H_{16}O_3$. *See* b1239	208.24	nd	141–2	...	...	...	...	...	...	...	...	...	**B10**^{1}, 70
b1818	—,—,chloride	$C_{12}H_{15}ClO_2$. *See* b1239	226.71	...	...	180–2^{12}	...	...	d	d^h	...	...	...	...	**B10**^{1}, 77
Ω b1819	—,**3,4-methylenedioxy-**	Piperonylic acid. $C_8H_6O_4$. *See* b1239	166.14	nd (al), pr (sub)	229–31	sub	...	...	i	δ	δ	...	...	chl i	**B19**^{2}, 292
b1820	—,—,chloride	$C_8H_5ClO_3$. *See* b1239	184.58	cr	80	155^{25}	...	...	...	...	...	...	...	...	**B19**, 270
b1821	—,—,ethyl ester	$C_{10}H_{10}O_4$. *See* b1239	194.19	pr	18.5	285.5–6.5 164–5^{11}	...	...	i	v	v	...	...	peth v	**B19**^{2}, 293
b1822	—,—,methyl ester	$C_9H_8O_4$. *See* b1239	180.17	nd or lf (peth)	53	273–4^{760} δd	...	...	...	v	v	...	...	peth v, MeOH s CCl_4 v	**B19**^{2}, 293
b1823	—,**2-methyl-4-nitro-**, nitrile (*colorless*)	$C_8H_6N_2O_2$. *See* b1291	162.15	lf (sub)	100 (105)	sub	...	...	δ^h	δ	...	v	v	chl v	**B9**^{1}, 188
b1824	—,—,—,(*yellow*)	$C_8H_6N_2O_2$. *See* b1291	162.15	ye nd (al)	113–5	...	...	...	δ^h	δ	...	v	v	chl v	**B9**^{1}, 188
b1825	—,**2-methyl-5-nitro-**, nitrile	$C_8H_6N_2O_2$. *See* b1291	162.15	nd (95 % al)	105	174–5^{18}	...	...	s^h	v^h	v	v	v	chl v	**B9**^{2}, 323

For explanations, symbols and abbreviations see beginning of table. For structural formulas see end of table.

No.	Name	Synonyms and Formula	Mol. wt.	Color, crystalline form, specific rotation and λ_{max} (log ε)	m.p. °C	b.p. °C	Density	n_D	Solubility: w	al	eth	ace	bz	other solvents	Ref.
	Benzoic acid														
b1826	—,**2-methyl-6-nitro-**, nitrile	$C_8H_6N_2O_2$. *See* b1291	162.15	pl (bz)	109–10	....	....	....	...	s	...	s	s	os s	**B9²**, 323
b1827	—,**3-methyl-2-nitro-**, nitrile	$C_8H_6N_2O_2$. *See* b1291	162.15	nd (al)	84	....	....	....	...	s^h	...	s	s	chl s	**B9¹**, 191
b1828	—,**3-methyl-4-nitro-**, nitrile	$C_8H_6N_2O_2$. *See* b1291	162.15	pr (al), nd λ^w 225 (4.08), 280 (3.83)	80	....	....	....	i	s^h	...	...	...	....	**B9¹**, 192
b1829	—,**3-methyl-5-nitro-**, nitrile	$C_8H_6N_2O_2$. *See* b1291	162.15	nd (lig)	104–5	....	....	....	...	s	...	...	...	lig s^h	**B9¹**, 192
b1830	—,**4-methyl-2-nitro-**, nitrile	$C_8H_6N_2O_2$. *See* b1291	162.15	nd (w)	101	....	....	....	s^h	v	...	...	v	chl v lig δ	**B9²**, 334
b1831	—,**4-methyl-3-nitro-**, nitrile	$C_8H_6N_2O_2$. *See* b1291	162.15	pa ye nd (w)	107–8	171^{12}	....	....	s	s	v	v	v	chl v	**B9²**, 334
b1832	—,**5-methyl-2-nitro-**, nitrile	$C_8H_6N_2O_2$. *See* b1291	162.15	nd (al)	93–4	....	....	....	v^h	v	...	...	v	aa v	**B9¹**, 192
b1834	—,**2(1-naphthoyl)-**	$C_{18}H_{12}O_3$. *See* b1239	276.30	pr (al-w)	176.4	....	....	....	δ^h i	v	...	v	v	chl s sulf s	**E12B**, 3614
b1835	—,**2(2-naphthoyl)-**	$C_{18}H_{12}O_3$. *See* b1239	276.30	nd (to)	168	....	....	....	δ	v	v	v	s	os v aa v	**E12B**, 3621
Ω b1836	—,**2-nitro-**	$C_7H_5NO_4$. *See* b1239	167.12	tcl nd (w) λ^{al} 250 sh (3.54)	147–8	....	1.575_4^{20}	....	s^h	v	s v^h	v	δ	chl δ, s^h lig δ MeOH v	**B9²**, 242
b1837	—,—,amide	$C_7H_6N_2O_3$. *See* b1244	166.14	nd (dil al) λ^{al} 250 (3.7)	176.6	317	....	....	s^h	s	s	...	...	....	**B9²**, 246
b1838	—,—,—,*N*-phenyl-	$C_{13}H_{10}N_2O_3$. *See* b1244	242.24	nd (al or bz) λ^{al} 246 (4.3), 354 (3.5)	155	....	....	....	i	v	δ	...	s	chl s lig δ	**B12²**, 153
b1839	—,—,azide	$C_7H_4N_4O_3$. *See* b1239	192.14	ye pr (eth)	37.5	....	....	....	...	...	v	...	v	chl v	**B9**, 376
Ω b1840	—,—,ethyl ester	$C_9H_9NO_4$. *See* b1239	195.18	tcl (dil al)	30	275 173^{18}	....	....	i	s	s	...	...	....	**B9²**, 245
Ω b1841	—,—,hydrazide	$C_7H_7N_3O_3$. *See* b1239	181.15	ye-br pr (w) λ^{al} 245 sh (3.6), 350 sh (2.3)	123	....	....	....	s	s	i	...	i	chl i	**B9²**, 246
Ω b1842	—,—,methyl ester	$C_8H_7NO_4$. *See* b1239	181.15	....	−13	275 176^{21}	1.2855^{20}	....	i	s	s	...	s	MeOH s chl s lig i	**B9²**, 244
Ω b1843	—,—,nitrile	$C_7H_4N_2O_2$. *See* b1291	148.12	nd (w or aa) λ^{al} 279 (3.78), 296 sh (3.35)	111	sub	....	....	δ s^h	s	s	v	s	CCl_4, aa s CS_2, peth δ chl v	**B9²**, 246
Ω b1844	—,**3-nitro-**	$C_7H_5NO_4$. *See* b1239	167.12	mcl pr (w) λ^{al} 215 (4.35), 255 (3.85)	140–2	....	1.494_4^{20}	....	δ	v	v	v	δ	chl s MeOH v peth δ	**B9²**, 247
b1845	—,—,amide	$C_7H_6N_2O_3$. *See* b1244	166.14	ye mcl nd (w) λ^{al} 260 (3.8)	142.7	310–5	....	....	s^h	s	s	...	...	....	**B9²**, 252
b1846	—,—,—,*N*-phenyl-	$C_{13}H_{10}N_2O_3$. *See* b1244	242.24	lf (w or al) λ^{al} 262 (4.3) 319 sh (3.2)	153–4	sub	....	....	i	s	s		s	....	**B12²**, 153
b1847	—,—,azide	$C_7H_4N_4O_3$. *See* b1239	192.14	pl (dil al) λ^{al} 230 (4.3), 262 sh (3.9), 330 sh (2.4)	68		....	....	i	v	v	...	v	aa v	**B9**, 338
Ω b1848	—,—,chloride	$C_7H_4ClNO_3$. *See* b1239	185.57	ye cr λ^{cy} 232 (4.40)	35	$275–8^{760}$ $154–5^{18}$	....	....	d	d	v	...	...	....	**B9²**, 252
Ω b1849	—,—,ethyl ester	$C_9H_9NO_4$. *See* b1239	195.18	mcl pr	47	296–8 156^{10}	....	....	i	v	v	...	...	....	**B9²**, 248
Ω b1850	—,—,hydrazide	$C_7H_7N_3O_3$. *See* b1239	181.15	nd (w) λ^{al} 255 (4.0), 350 sh (2.2)	153–4	....	....	....	δ v^h	δ v^h	i	...	i	chl i	**B9²**, 256
Ω b1851	—,—,methyl ester	$C_8H_7NO_4$. *See* b1239	181.15	nd λ^{al} 219 (4.37) 251 (3.86), 284 sh (3.10)	78	....	....	....	i	δ	δ	...	...	MeOH δ	**B9²**, 248
Ω b1852	—,—,nitrile	$C_7H_4N_2O_2$. *See* b1291	148.12	nd (w) λ^{al} 255 (3.8), 335 sh (2.1)	118	sub	....	....	s^h	s	v	v	s^h	peth i aa v	**B9²**, 254
Ω b1853	—,**4-nitro-**	$C_7H_5NO_4$. *See* b1244	167.12	mcl lf (w) λ^{al} 258 (4.08), 294 sh (3.40)	242	sub	1.610^{20}	....	i s^h	s	s	δ s^h	δ	chl s lig i CS_2 δ MeOH v	**B9²**, 256
Ω b1854	—,—,amide	$C_7H_6N_2O_3$. *See* b1244	166.14	nd (w) λ^w 265 (4.06)	201.4	....	....	....	i	s	s	...	...	....	**B9²**, 271
b1855	—,—,—,*N*-phenyl-	$C_{13}H_{10}N_2O_3$. *See* b1244	242.24	lf (eth) λ^{al} 321 (4.3)	211	....	....	....	i	s	s	...	...	....	**B12²**, 153
b1856	—,—,butyl ester	$C_{11}H_{13}NO_4$. *See* b1239	223.23	nd	35.3	160^8	....	....	...	s^h	v	...	v	....	**B9²**, 259
Ω b1857	—,—,chloride	$C_7H_4ClNO_3$. *See* b1239	185.57	ye nd (lig)	75	$202–5^{105}$ $150–2^{15}$	....	....	d	d	s	...	...	....	**B9²**, 270
Ω b1858	—,—,ethyl ester	$C_9H_9NO_4$. *See* b1239	195.18	tcl lf (al) λ^{al} 259 (4.00)	57	186.3^{760}	....	....	i	s	s	...	...	....	**B9²**, 258

For explanations, symbols and abbreviations see beginning of table. For structural formulas see end of table.

No.	Name	Synonyms and Formula	Mol. wt.	Color, crystalline form, specific rotation and λ_{max} (log ε)	m.p. °C	b.p. °C	Density	n_D	Solubility w	al	eth	ace	bz	other solvents	Ref.
	Benzoic acid														
Ω b1859	—,—,hydrazide ...	$C_7H_7N_3O_3$. *See* b1239......	181.15	yesh nd (w) λ^{al} 260 (4.0), 375 (2.5)	214	....	....	....	δ^h	δ^h	i	...	i	chl i	**B9**^{2}, 274
Ω b1860	—,—,methyl ester .	$C_8H_7NO_4$. *See* b1239.......	181.15	ye mcl lf λ^{al} 259 (4.12)	96	....	....	....	i	s	s	...	...	chl s	**B9**^{2}, 258
Ω b1861	—,—,nitrile.......	$C_7H_4N_2O_2$. *See* b1291......	148.12	lf (al), nd (bz) λ^{al} 258 (4.07), 303 (3.00)	149	sub	....	....	δ s^h	δ s^h	δ ...	... ...	... ...	chl s aa s	**B9**^{2}, 273
b1862	—,—,2,2,2-tri-chloroethyl ester	$C_9H_7Cl_3O_2$. *See* b1239	253.51	pr (al)	71	106–7^1	1.352_4^{25}	1.5343^{26}	...	...	...	...	...	....	**Am 70,** 3370
b1863	—,**2-nitroso-**......	$C_7H_5NO_3$. *See* b1239.......	151.12	cr (al or aa)	210d (214d)	....	....	....	...	s^h δ	δ	...	δ	aa, δ, s^h	**B9**^{2}, 241
b1864	—,**3-nitroso-**......	$C_7H_5NO_3$. *See* b1239.......	151.12	cr	230d	....	....	....	...	s	δ	...	δ	....	**B9,** 369
b1865	—,**4-nitroso-**......	$C_7N_5NO_3$. *See* b1239.......	151.12	ye pw	> 350	....	....	....	...	s^h	δ	...	δ	aa δ	**B9**^{2}, 242
b1866	—,**4-octyl-**........	$C_{15}H_{22}O_2$. *See* b1239.......	234.34	lf (al)	139 (100)	....	....	....	...	s^h	...	...	...	....	**B9,** 571
Ω b1867	—,**pentachloro-**....	$C_7HCl_5O_2$. *See* b1239	294.35	nd or pl (bz or dil aa)	208	sub (vac) δd	....	....	...	v	...	...	...	to v aa v	**B9**^{1}, 142
b1868	—,—,chloride.....	C_7Cl_6O. *See* b1239.........	312.80	pl (al)	87	....	....	....	...	v^h	...	...	...	MeOH v^h	**C24,** 108
b1869	—,**pentamethyl-** ...	$C_{12}H_{16}O_2$. *See* b1239.......	192.24	nd (w), lf or nd (dil al) λ^{sulf} 286 (4.6), 344 (3.7)	210.5	sub	....	....	i	v^h	...	...	...	....	**B9,** 569
b1870	—,**per-**...........	Perbenzoic acid. C_6H_5COOOH	138.12	mcl pl (peth)	41–3	97–110$^{13-5}$ sub	....	....	i δ^h	v	v	s	s	chl s peth δ	**B9**^{2}, 157
Ω b1871	—,**2-phenoxy-**.....	$C_{13}H_{10}O_3$. *See* b1239.......	214.22	lf (dil al) $\lambda^{diox-cy}$ 222 (4.14), 285 (3.49)	113–4	355d^{760}	....	....	i δ^h	v	v	...	...	chl s	**B10**^{2}, 40
—	—,**phenyl-**........	*See* **Biphenylcarboxylic acid** .													
b1872	—,**2(phenyl-amino)-**	$C_{13}H_{11}NO_2$. *See* b1239	213.24	lf, nd or pr (al)	184d	....	....	....	i δ^h	v^h	δ	...	δ^h	....	**B14**^{2}, 213
b1873	—,**2-phosphono-** ...	$C_7H_7O_5P$. *See* b1239	202.11	nd (w or al)	> 300	....	....	....	v	δ	...	...	...	....	**B16,** 820
b1874	—,**2-propyl-**.......	$C_{10}H_{12}O_2$. *See* b1239.......	164.21	lf (dil al)	58	272^{739} 164–5^{20}	....	....	s	v	v	...	...	....	**B9**^{1}, 213
Ω b1875	—,**4-propyl-**.......	$C_{10}H_{12}O_2$. *See* b1239.......	164.21	pr or lf (w)	141	....	....	....	v^h	v	v	...	v	CS_2 v lig s chl v	**B9,** 545
b1876	—,**3(1-semicarb-azido)-**, amide	Cryogenine. $C_8H_{10}N_4O_2$. *See* b1244	194.20	cr	172	....	....	....	i	s	s	s	...	chl s	**B15**^{2}, 297
—	—,***x*-sulfamido-**....	*see* **Benzoic acid, *x*-sulfo-, *x*-amide** (*x* = 2, 3, 4)													
b1877	—,**2-sulfo-**........	*o*-Carboxybenzenesulfonic acid. $C_7H_6O_5S$. *See* b1239	202.19	nd (w + 3)	141 (anh) 70 (+3w)	....	....	....	v	v	i	...	...	....	**B11**^{2}, 215
b1878	—,—,1-amide.....	$C_7H_7NO_4S$. *See* b1244.......	201.20	pr(w + 1)	193–4 (anh)	....	....	....	v	v	...	...	...	....	**B11**^{2}, 215
Ω b1879	—,—,2-amide.....	*o*-Sulfamylbenzoic acid. $C_7H_7NO_4S$. *See* b1239	201.20	pl or nd (w)	165–7 (rapid htng) 153–5 (slow htng)	....	....	....	v	v	v	...	...	....	**B11**^{2}, 216
b1880	—,—,anhydride (endo)	....	184.17	pl or pr (bz) ...	129.5	184–6^{18}	....	....	i s^h	...	s	...	s	chl s	**B19**^{2}, 137
—	—,—,imide.......	*see* **Saccharine**													
b1881	—,—,nitrile.......	$C_7H_5NO_3S$. *See* b1291......	183.19	nd (w)	279.5	....	....	....	s^h	s	δ	...	...	chl δ	**B11**^{2}, 372
Ω b1882	—,**3-sulfo-**........	*m*-Carboxybenzenesulfonic acid. $C_7H_6O_5S$. *See* b1239	202.19	cr (w + 2)	141 (anh) 98 (hyd)	....	....	....	v	s	v	...	i	chl, peth i	**B11**^{2}, 217
b1883	—,—,3-amide.....	*m*-Sulfamylbenzoic acid. $C_7H_7NO_4S$. *See* b1239	201.20	pl (w)	246 (238)	....	....	....	s^h δ	s	δ	...	...	....	**B11**^{2}, 218
b1884	—,—,3-chloride ...	$C_7H_5ClO_4S$. *See* b1239......	220.63	pr (bz)	133–4	....	....	....	d^h	d^h	s	...	s^h	....	**B11**^{2}, 218
b1885	—,—,dichloride ...	$C_7H_4Cl_2O_3S$. *See* b1239.....	239.08	....	20.4	153–4^7	....	....	...	...	...	...	...	....	**B11,** 386
b1886	—,**4-sulfo-**........	*p*-Carboxybenzene sulfonic acid. $C_7H_6O_5S$. *See* b1239	202.19	nd (w + 3)	259–60 (anh) 94 (hyd)	....	....	....	v	v	s	...	...	....	**B11**^{2}, 218
Ω b1887	—,—,4-amide.....	*p*-Sulfamylbenzoic acid. $C_7H_7NO_4S$. *See* b1239	201.20	pr or fl (w)	290–2d	....	....	....	i δ^h	v	δ	...	i	....	**B11**^{2}, 219
b1888	—,—,4-chloride ...	$C_7H_5ClO_4S$. *See* b1239	220.63	nd (ace)	237–8d	....	....	....	d^h	d^h	...	s	...	to s	**B11**^{2}, 219
Ω b1889	—,**2,3,4,5-tetra-chloro-**	$C_7H_2Cl_4O_2$. *See* b1239	259.91	nd (al), cr (ace-w)	194–5	....	....	....	δ	v	v	...	...	....	**B9,** 346
b1890	—,**2,3,4,5-tetra-hydroxy-**	$C_7H_6O_6$. *See* b1239	186.12	pr	84–5	....	....	....	v	...	...	...	...	....	**C50,** 14644
—	—,**2(3,4,5,6-tetra-hydroxyxanthyl)-**	*see* **Gallin**													
b1891	—,**thiolo-**.........	$C_6H_5C(:O)SH$	138.19	ye pl (aa)	24	85–7^{10}	....	1.6040_4^{20}	i	v	v	s	s	CS_2 v os s	**B9**^{2}, 286
b1892	—,**thiono-**, amide, *N*-phenyl-	$C_6H_5CSNHC_6H_5$..........	213.30	ye pl or pr (al)	102	d	....	....	i	v	s	...	s^h	chl s lig δ	**B12**^{2}, 154

For explanations, symbols and abbreviations see beginning of table. For structural formulas see end of table.

No.	Name	Synonyms and Formula	Mol. wt.	Color, crystalline form, specific rotation and λ_{max} (log ε)	m.p. °C	b.p. °C	Density	n_D	Solubility w	al	eth	ace	bz	other solvents	Ref.
	Benzoic acid														
b1893	**—,2(2-toluyl)**	$C_{15}H_{12}O_3$. *See* b1239	240.26	nd (w + 1), cr (bz)	130–2 (anh) 84(+ w)	…	…	…	s	s^h	s	…	…	MeOH s	**B10²**, 524
b1894	**—,2(3-toluyl)-**	$C_{15}H_{12}O_3$. *See* b1239	240.26	nd (w + 1), nd (bz)	162.2–2.4	…	…	…	s^h	…	v	v	…	…	**B10²**, 524
b1895	**—,2(4-toluyl)-**	$C_{15}H_{12}O_3$. *See* b1239	240.26	pr (+ 1w, al–to) nd (al)	146	…	…	…	δ^h	v	v	v	v	to v^h	**B10²**, 525
b1896	**—,—,methyl ester**	$C_{16}H_{14}O_3$. *See* b1239	254.29	pl (MeOH)	66 (61)	…	…	…	δ^h	v	…	…	v	…	**B10**, 759
b1897	**—,4(2-toluyl)-**	$C_{15}H_{12}O_3$. *See* b1239	240.26	…	177	…	…	…	…	…	…	…	…	xyl s^h	**B10²**, 527
b1898	**—,4(4-toluyl)-**	$C_{15}H_{12}O_3$. *See* b1239	240.26	nd (MeOH or ace	228	…	…	…	δ^h	v	…	v	δ	chl δ	**B10²**, 528
b1899	**—,2,3,5-triamino-**	$C_7H_9N_3O_2$. *See* b1239	167.13	cr (w)	d	…	…	…	v^h	δ^h	i	…	…	…	**B14**, 455
b1900	**—,3,4,5-triamino-**	$C_7H_9N_3O_2$. *See* b1239	167.13	nd (w + ½)	d	…	…	…	s^h δ	i	i	…	…	…	**B14**, 455
b1901	**—,2,3,4-tribromo-**	$C_7H_3Br_3O_2$. *See* b1239	358.83	nd (bz)	197–8	…	…	…	i	s	s	…	s^h	…	**B9¹**, 147
b1902	**—,2,3,5-tribromo-**	$C_7H_3Br_3O_2$. *See* b1239	358.83	nd (al)	193–4	…	…	…	i	v^h	v	v	δ	lig i, peth δ	**B9¹**, 147
b1903	**—,2,4,5-tribromo-**	$C_7H_3Br_3O_2$. *See* b1239	358.83	nd (al or bz)	195–6	…	…	…	i	s^h	s	…	…	lig s^h	**B9¹**, 147
b1904	**—,2,4,6-tribromo-**	$C_7H_3Br_3O_2$. *See* b1239	358.83	pr (w)	198	…	…	…	δ	s	s	…	s	…	**B9²**, 238
b1905	**—,3,4,5-tribromo-**	$C_7H_3Br_3O_2$. *See* b1239	358.83	nd (bz or al)	240	…	…	…	i	s	s	…	…	aa s^h	**B9¹**, 148
Ω b1906	**—,2,3,4-trichloro-**	$C_7H_3Cl_3O_2$. *See* b1239	225.46	nd (w) λ^{MeOH} 289 (2.79)	187–8	…	…	…	…	…	…	…	…	…	**B9**, 345
Ω b1907	**—,2,3,5-trichloro-**	$C_7H_3Cl_3O_2$. *See* b1239	225.46	nd (w) λ^{MeOH} 295 (3.01)	163	…	…	…	i	s	s	s	s	os s	**B9**, 345
b1908	**—,2,3,6-trichloro-**	$C_7H_3Cl_3O_2$. *See* b1239	225.46	λ^{MeOH} 283 (2.74)	124–5	…	…	…	δ v^h	…	s	…	…	…	**B9**, 345
Ω b1909	**—,2,4,5-trichloro-**	$C_7H_3Cl_3O_2$. *See* b1239	225.46	nd (w or sub) λ^{MeOH} 292 (2.97)	168	sub	…	…	s^h i	s	s	…	…	…	**B9¹**, 141
b1910	**—,2,4,6-trichloro-**	$C_7H_3Cl_3O_2$. *See* b1239	225.46	nd (w) λ^{MeOH} 274 (2.43)	164	…	…	…	δ^h	v^h	v	…	…	chl v	**B9**, 345
b1911	**—,3,4,5-trichloro-**	$C_7H_3Cl_3O_2$. *See* b1239	225.46	nd (dil al) λ^{MeOH} 288 (2.86), 274 (2.43)	210	…	…	…	δ^h	s	s	s	s	CS_2 δ	**B9²**, 230
Ω b1912	**—,2,3,4-trihydroxy-**	$C_7H_6O_5$. *See* b1239	170.12	nd (+ w)	207–8 d (215–20)	sub	…	…	δ	s	s	s	i	CS_2 i AcOEt s	**B10²**, 331
b1913	**—,—,ethyl ester**	$C_9H_{10}O_5$. *See* b1239	198.18	cr (w + 1)	102 (anh) 86(+ w)	…	…	…	s^h i	v	v	…	…	…	**B10**, 467
b1914	**—,—,methyl ester**	$C_8H_8O_5$. *See* b1239	184.16	nd (w + 2½)	151–2	…	…	…	s^h	s	…	…	…	…	**B10**, 466
b1915	**—,2,4,5-trihydroxy-**	$C_7H_6O_5$. *See* b1239	170.12	nd (w + ½)	217–8d	…	…	…	v^h	v	…	…	…	…	**B10²**, 334
Ω b1916	**—,2,4,6-trihydroxy-**	$C_7H_6O_5$. *See* b1239	170.12	(w + 1) λ^{MeOH} 262.5 (4.12), 297.5 (3.44)	100d	…	…	…	s^h δ	s	v	…	i	…	**B10²**, 334
b1917	**—,—,ethyl ester**	$C_9H_{10}O_5$. *See* b1239	198.18	pr or nd (w + 1), pr (lig)	129	…	…	…	s^h	v	v	…	v^h	lig δ	**B10¹**, 236
b1918	**—,—,methyl ester**	$C_8H_8O_5$. *See* b1239	184.16	cr (dil al)	174–6	…	…	…	δ	v	v	…	δ	CCl_4 δ	**B10**, 469
b1919	**—,3,4,5-trihydroxy-**	Gallic acid. $C_7H_6O_5$. *See* b1239	170.12	pr (w + 1) λ^{al} 272.5 (4.06)	253d	…	1.694_4^6	…	v^h δ	v	δ	s	i	chl i	**B10²**, 335
b1920	**—,—,amide**	Gallamide. $C_7H_7NO_4$. *See* b1244	169.14	lf (w + 1½)	244–5d (anh)	…	…	…	s^h δ	s	…	…	…	chl s aa s	**B10²**, 346
b1921	**—,—,—,*N*-phenyl-**	Gallanilide. $C_{13}H_{11}NO_4$. *See* b1244	245.24	lf (+ 2w, dil al)	207	…	…	…	v^h i	s	…	…	i	chl i aa s	**B12²**, 263
Ω b1922	**—,—,ethyl ester**	$C_9H_{10}O_5$. *See* b1239	198.18	mcl pr (w + 2½), nd (chl) λ^{MeOH} 275 (4.00)	160–2 (anh) 158 (hyd)	…	…	…	s^h δ	s	s	…	…	chl δ^h AcOEt s	**B10²**, 343
Ω b1923	**—,—,isopropyl ester**	$C_{10}H_{12}O_5$. *See* b1239	212.21	…	123–4.5	…	…	…	s	s	s	…	i	chl s^h	**B10²**, 343
Ω b1924	**—,—,methyl ester**	Gallicin. $C_8H_8O_5$. *See* b1239	184.16	mcl pr (MeOH) λ^{al} 217 (4.72), 275 (4.32)	202	…	…	…	δ	v	…	…	…	MeOH v^h	**B10²**, 342
Ω b1925	**—,—,propyl ester**	$C_{10}H_{12}O_5$. *See* b1239	212.21	nd (w) λ^{al} 275	150	…	…	…	δ s^h	…	…	…	…	…	**B10²**, 343
Ω b1926	**—,2,3,5-triiodo-**	$C_7H_3I_3O_2$. *See* b1239	499.81	pr (al)	224–6	…	…	…	i	v^h	v	…	δ^h	…	**B9¹**, 150
b1927	**—,2,4,5-triiodo-**	$C_7H_3I_3O_2$. *See* b1239	499.81	nd (al)	248	…	…	…	i	s^h	s	…	i	…	**B9¹**, 150
Ω b1928	**—,3,4,5-triiodo-**	$C_7H_3I_3O_2$. *See* b1239	499.81	pr (al)	292–3	…	…	…	i	v	…	…	…	…	**B9**, 367
b1930	**—,2,3,4-trimethoxy-**	$C_{10}H_{12}O_5$. *See* b1239	212.21	cr (w or peth)	100	…	…	…	s	s	s	…	…	…	**B10²**, 332
b1931	**—,2,4,5-trimethoxy-**	Asaronic acid. $C_{10}H_{12}O_5$. *See* b1239	212.21	nd (al or bz-peth)	144	300^{760}	…	…	s^h i	s	…	…	s	peth s	**B10²**, 334
Ω b1932	**—,3,4,5-trimethoxy-**	$C_{10}H_{12}O_5$. *See* b1239	212.21	mcl nd (w) λ^{al} 215 (4.2), 260 (3.7), 300 sh (3.2)	171–2	$225–7^{10}$	…	…	δ	v	v	…	…	chl v	**B10²**, 340
b1933	**—,2,3,4-trimethyl-**	Prehnitylic acid. $C_{10}H_{12}O_2$. *See* b1239	164.21	pr (al) λ^{al} 237 (3.78), 287 (3.01)	167.5	…	…	…	s^h	s^h	s^h	…	…	…	**B9**, 552
b1934	**—,2,3,5-trimethyl-**	γ-Isodurylic acid. $C_{10}H_{12}O_2$. *See* b1239	164.21	pl (lig) λ^{al} 239 (3.69), 288 (3.15)	127	…	…	…	…	s	…	…	…	…	**B9**, 552
b1935	**—,2,3,6-trimethyl-**	$C_{10}H_{12}O_2$. *See* b1239	164.21	nd (w or peth) λ^{al} 223 (3.99), 279 (2.90)	110–1	…	…	…	s	s	s	…	…	…	**B9**, 552

For explanations, symbols and abbreviations see beginning of table. For structural formulas see end of table.

No.	Name	Synonyms and Formula	Mol. wt.	Color, crystalline form, specific rotation and λ_{max} (log ε)	m.p. °C	b.p. °C	Density	n_D	Solubility w	al	eth	ace	bz	other solvents	Ref.
	Benzoic acid														
b1936	—,**2,4,5-trimethyl-**	Durylic acid. $C_{10}H_{12}O_2$. *See* b1239	164.21	nd (bz) λ^{al} 237 (3.87), 286 (3.21)	152–3	....	....	....	δ^h i	v	v	...	δ	....	**B9**², 361
Ω b1937	—,**2,4,6-tri-methyl-**	β-Isodurylic acid. Mesitoic acid. $C_{10}H_{12}O_2$. *See* b1239	164.21	pr (lig) λ^{al} 235 sh (3.51), 270 sh (2.62)	155	....	....	....	δ	s	s	s	...	chl s	**B9**, 553
b1938	—,—,chloride.....	Mesitoyl chloride. $C_{10}H_{11}ClO$. *See* b1239	182.65	....	....	143–6[60]	....	....	d	...	...	s	...	....	**B9**¹, 360
b1939	—,**3,4,5-tri-methyl-**	α-Isodurylic acid. $C_{10}H_{12}O_2$. *See* b1239	164.21	nd (w) λ^{al} 238.5 (4.01)	215–6	....	....	....	δ^h i	s	s	...	...	....	**B9**, 554
b1940	—,**2,3,6-trinitro-**..	$C_7H_3N_3O_8$. *See* b1239......	257.12	wh nd (w + 2)	160d, 55 (+2w)	....	....	....	δ^h	s	...	...	...	....	**B9**¹, 168
b1941	—,**2,4,5-trinitro-**...	$C_7H_3N_3O_8$. *See* b1239......	257.12	ye lf or pl (w)	194.5d	....	....	....	s^h	v	v	...	s	peth δ	**B9**¹, 168
Ω b1942	—,**2,4,6-trinitro-**...	$C_7H_3N_3O_8$. *See* b1239......	257.12	rh (w) λ^{Et_2NH} 475 (2.90)	228.7d	....	....	....	δ	v	s	s	δ	MeOH v	**B9**², 285
b1943	—,**3,4,5-trinitro-**...	$C_7H_3N_3O_8$. *See* b1239......	257.12	ye nd (eth + 1)	168d	....	....	....	d^h	...	s	...	...	....	**B9**¹, 168
b1944	**Benzoin** (*d*)	Benzoylphenylcarbinol. α-Hydroxybenzyl phenyl ketone.	212.25	nd, $[\alpha]_D^{15}$ +92.8 (Py. c = 1)	133–4	....	....	....	...	v^h	...	v	...	MeOH v^h Py s	**B8**², 193
Ω b1945	—(*dl*)............	$C_{14}H_{12}O_2$. *See* b1944.......	212.25	pr (al) λ^{al} 248 (4.1)	137	344[768] 194[12]	1.310_4^{20}	....	δ^h	s^h	δ	...	...	aa v^h chl s	**B8**², 193
b1946	—(*l*)............	$C_{14}H_{12}O_2$. *See* b1944.......	212.25	nd (MeOH), $[\alpha]_D^{12}$ −117.5 (ace, c = 1.25)	133–4	....	....	....	...	v^h	...	v	...	MeOH s^h	**B8**, 167
Ω b1947	—,acetate (*dl*).....	$C_6H_5CH(OCOCH_3)COC_6H_5$	254.29	pr or pl (eth)	83	....	....	....	...	v	v	...	...	....	**B8**², 196
b1948	—,ethyl ether (*dl*)..	$C_6H_5CH(OC_2H_5)COC_6H_5$..	240.31	nd (lig)	62	194–5[20]	1.1016_4^{17}	1.5727[17]	...	v	v	...	v	lig s^h	**B8**², 195
b1949	—,hydrazone (*dl*)..	$C_6H_5CH(OH)C(:NNH_2)C_6H_5$	226.28	pr (al)	75	....	....	....	i	v^h	...	...	...	....	**B8**, 176
b1950	—,methyl ether (*dl*)	$C_6H_5CH(OCH_3)COC_6H_5$...	226.28	nd (lig)	49–50	188–9[15]	1.1278_4^{14}	....	...	v	v	...	v	lig v^h	**B8**², 194
b1951	—,oxime (*l*).......	$C_6H_5CH(OH)C(:NOH)C_6H_5$	227.27	amor or pr (bz), $[\alpha]_D^{24}$ −3.2 (chl, c = 0.85)	163.5–4.5	....	....	....	i	s	s	s	δ^h	....	**B8**, 167
Ω b1952	—,—,(*dl, anti*).....	$C_{14}H_{13}NO_2$. *See* b1951	227.27	pr (bz) λ^{al} 290 (3.0), 301 (2.0), 311 (1.0)	151–2	....	....	....	i	s	s	s	...	....	**B8**², 196
b1953	—,—,(*dl, syn*)	$C_{14}H_{13}NO_2$. *See* b1951	227.27	pr (eth)	99	....	....	....	i	s	s	s	...	....	**B8**², 196
b1954	—,succinate	$[C_6H_5CH(COC_6H_5)O_2CCH_2-]_2$	506.56	lf (al)	129	....	....	....	i	s	s	...	...	CS_2 s	**B8**, 175
Ω b1955	—,**4,4′-dimethoxy-**.	Anisoin. $C_{16}H_{16}O_4$. *See* b1944	272.30	pr (dil al) λ^{al} 220 (4.38), 277 (4.33)	113	....	....	....	δ^h	v^h δ	δ	s	...	....	**B8**², 470
—	**Benzonitrile**.......	*see* **Benzoic acid,** nitrile													
—	**1,2-Benzophe-nanthrene**	*see* **Chrysene**													
Ω b1956	**3,4-Benzophe-nanthrene**	Benzo(c)phenanthrene	228.30	nd or pl (lig or al) λ^{al} 232 (4.5), 280 (4.9), 302 (4.0), 325 (3.6), 355 (2.5), 372 (2.4)	68	....	....	....	i	δ	...	...	...	lig δ	**E14**, 356
Ω b1957	**9,10-Benzophe-nanthrene**	Triphenylene.	228.30	nd (al, chl, bz) λ^{al} 248.5 (4.94), 257 (5.18), 273 (4.28), 284 (4.23)	199	425	....	....	i	s	...	...	v	chl v aa s	**E14**, 357
b1958	**1,2-Benzophen-azine**	Benzo[α]phenazine. *ang*-Naphthaphenazine.	230.26	pr (al), ye nd (bz) λ^{MeOH} 224 (4.63), 277.5 (4.72), 380 (4.06), 400 (4.08)	142.5	>360	....	....	i	s	s	s	δ	aa s	**B23**², 259
Ω b1959	**Benzophenone**....	Diphenyl ketone. Benzoylbenzene	182.21	(α) rh pr (al or eth); (β) mcl pr λ^{al} 253 (4.24); λ^{hx} 346 (2.1), 361 sh (2.0)	(α) 48.1 (β) 26	305.9[760]	(α) 1.146[20] (β) 1.1076	(α) 1.6077[19] (β) 1.6059[23]	i	v	v	v	s	aa v MeOH s CS_2 v chl v	**B7**², 349
b1960	—,imine	$(C_6H_5)_2C:NH$	181.24	....	....	282 158[12]	1.0847_4^{19}	1.6191[19]	d	...	s	...	...	....	**B7**², 355
Ω b1961	—,oxime.........	$(C_6H_5)_2C:NOH$..........	197.24	nd (al) λ^{al} 231 (4.14), 251 (4.03)	144	....	....	....	i	v^h	s	v	...	MeOH v^h chl s	**B7**², 355
b1962	—,phenyl-hydrazone	$(C_6H_5)_2C:NNHC_6H_5$	272.35	pr or nd (al) λ^{al} 345 (4.26)	137	....	....	....	i	δ	s	...	s	aa s	**B15**², 63

For explanations, symbols and abbreviations see beginning of table. For structural formulas see end of table.

No.	Name	Synonyms and Formula	Mol. wt.	Color, crystalline form, specific rotation and λ_{max} (log ε)	m.p. °C	b.p. °C	Density	n_D	Solubility w	al	eth	ace	bz	other solvents	Ref.
	Benzophenone														
b1963	—,2-amino-	$C_{13}H_{11}NO$. *See* b1959	197.24	pa ye lf or pr (al) λ^{al} 235 (4.3), 385 (3.8)	110–1 (105–6)	....	....	....	i	s^h	s	...	...	....	**B14^2, 51**
b1964	—,3-amino-	$C_{13}H_{11}NO$. *See* b1959	197.24	ye nd (w) λ^{al} 245 (4.3), 335 (3.3)	87	....	....	....	δ	s	s	...	...	....	**B14^1, 388**
b1965	—,4-amino-	$C_{13}H_{11}NO$. *See* b1959	197.24	lf (dil al) λ^{al} 245 (4.2), 335 (4.3)	124	....	....	....	s^h δ	s	s	...	...	aa s	**B14^2, 54**
b1966	—,2-amino-4′-methyl-	$C_{14}H_{13}NO$. *See* b1959	211.27	ye pr or pl (al)	96	....	....	....	...	v	v	...	v	....	**B14^2, 63**
b1967	—,2-amino-5-methyl-	$C_{14}H_{13}NO$. *See* b1959	211.27	ye nd or pl (al)	66	....	....	....	...	v	v	v	...	chl v lig v aa v	**B14, 106**
b1968	—,3-amino-4-methyl	$C_{14}H_{13}NO$. *See* b1959	211.27	pa ye nd (MeOH)	109	....	....	....	δ^h	v	v	v	v	MeOH s^h CS_2 s^h chl v	**B14^2, 63**
b1969	—,3-amino-4′-methyl-	$C_{14}H_{13}NO$. *See* b1959	211.27	pr (al)	111	....	....	....	...	v	v	...	...	....	**B14, 107**
b1970	—,4-amino-3-methyl-	$C_{14}H_{13}NO$. *See* b1959	211.27	ye pr (w)	112	....	....	....	δ s^h	v	v	...	...	....	**B14, 105**
b1971	—,4-amino-4′-methyl-	$C_{14}H_{13}NO$. *See* b1959	211.27	nd (bz)	186–7	....	....	....	...	v	v	...	s^h	chl v lig δ CS_2 v	**B14, 107**
b1973	—,4,4′-bis-(diethylamino)-	$C_{21}H_{28}N_2O$. *See* b1959	324.47	lf (al)	95–6	....	....	....	...	δ v^h	...	...	...	....	**B14^2, 59**
b1974	—,4,4′-bis(di-methylamino)-	Michler's ketone. $C_{17}H_{20}N_2O$. *See* b1959	268.36	lf (al), nd (bz) λ^{al} 370 (4.28), 390 sh (4.36)	179	>360d	....	....	i	δ	i	...	v^h	Py v	**B14^2, 57**
b1975	—,4,4′-bis(di-methylamino), thio-	$C_{17}H_{20}N_2S$. *See* b1959	284.43	pl	204	....	....	....	i	i	δ	...	s	chl s lig i aa s	**B14^2, 60**
b1976	—,2-bromo-	$C_{13}H_9BrO$. *See* b1959	261.13	pr (al), nd (lig)	42	345^{759}	1.517^{14}	....	i	s^h	...	s	...	lig s^h	**B7^2, 360**
b1977	—,3-bromo-	$C_{13}H_9BrO$. *See* b1959	261.13	nd (al)	81.5 (77)	$185–7^5$	....	....	i	s^h	...	...	...	....	**B7^2, 360**
b1978	—,4-bromo-	$C_{13}H_9BrO$. *See* b1959	261.13	lf (al)	82.5	350^{757}	....	....	i	δ^h	δ	...	δ	peth δ	**B7^2, 360**
b1979	—,2-chloro-	$C_{13}H_9ClO$. *See* b1959	216.67	pl (chl-lig)	52–6	330 $185–8^{13}$	....	....	...	...	...	...	...	....	**B7^1, 227**
b1980	—,3-chloro-	$C_{13}H_9ClO$. *See* b1959	216.67	nd	82–3	....	....	....	...	δ	...	...	v	....	**B7^1, 227**
b1981	—,4-chloro-	$C_{13}H_9ClO$. *See* b1959	216.67	nd (al) λ^{al} 260 (4.32)	77–8	332^{771}	....	....	...	s^h	s^h	s	...	....	**B7^2, 359**
b1982	—,2-chloro-3,5-dinitro-	$C_{13}H_7ClN_2O_5$. *See* b1959	306.67	ye nd (aa)	149	....	....	....	...	δ	...	...	...	chl s aa s^h lig i	**B7, 428**
b1983	—,2,2′-diamino-	$C_{13}H_{12}N_2O$. *See* b1959	212.25	lf (dil al), pr (bz)	134–5	....	....	....	i	s	s	...	...	....	**B14, 87**
b1984	—,3,3′-diamino-	$C_{13}H_{12}N_2O$. *See* b1959	212.25	nd (al)	173–4	285^{11}	....	....	i s^h	s	s	...	...	....	**B14^1, 390**
b1985	—,4,4′-diamino-	$C_{13}H_{12}N_2O$. *See* b1959	212.25	nd (al)	244–5	....	....	....	d^h	s	s	...	...	....	**B14^2, 56**
b1986	—,4,4′-dibromo-	$C_{13}H_8Br_2O$. *See* b1959	340.04	pl (al)	177	395	....	....	i	s^h	δ	s	s	chl s CS_2 s	**B7, 423**
b1987	—,2,4′-dichloro-	$C_{13}H_8Cl_2O$. *See* b1959	251.13	pr (al) λ^{al} 262 (4.36)	67	$214–5^{22}$	1.393^{14}	....	...	s^h	...	...	...	....	**B7, 420**
Ω b1988	—,4,4′-dichloro-	$C_{13}H_8Cl_2O$. *See* b1959	251.13	pl (al) λ^{al} 265 (4.39)	147–8	353^{757}	....	....	i	s^h	v	s	...	chl v aa v	**B7, 420**
b1989	—,2,2′-dihydroxy-	$C_{13}H_{10}O_3$. *See* b1959	214.22	lf or pr (lig) λ^{al} 260 (4.07), 336 (4.76)	59.5	330–40	....	....	i	s	s	...	...	chl s	**B8^2, 354**
b1990	—,2,3′-dihydroxy-	$C_{13}H_{10}O_3$. *See* b1959	214.22	nd (w)	126	....	....	....	...	s	s	...	...	....	**B8, 315**
Ω b1991	—,2,4-dihydroxy-	$C_{13}H_{10}O_3$. *See* b1959	214.22	nd (w) λ^{MeOH} 242 (3.94), 290 (3.96), 338 (4.12)	144 (146)	....	....	....	i	s	v	...	δ	aa v	**B8^2, 352**
b1992	—,2,4′-dihydroxy-	$C_{13}H_{10}O_3$. *See* b1959	214.22	pl (w)	150–1	....	....	....	δ^h	s^h	s	...	s^h	....	**B8^2, 354**
b1993	—,2,5-dihydroxy-	$C_{13}H_{10}O_3$. *See* b1959	214.22	ye nd (dil al)	125–6	....	....	....	s^h	s	s	...	s	....	**B8^2, 353**
b1994	—,3,3′-dihydroxy-	$C_{13}H_{10}O_3$. *See* b1959	214.22	nd (w)	170	....	....	....	s^h	s	...	...	...	....	**B8, 316**
b1995	—,3,4′-dihydroxy-	$C_{13}H_{10}O_3$. *See* b1959	214.22	nd (w)	206	....	....	....	s^h	s	s	...	...	....	**B8, 316**
b1996	—,4,4′-dihydroxy-	$C_{13}H_{10}O_3$. *See* b1959	214.22	nd (lig), cr (w) λ^{MeOH} 295 (4.28)	210	1.133^{131}	....	....	δ s^h	s	s	s	i	CS_2, chl i, MeOH s	**B8^2, 354**
b1997	—,2,4-dihydroxy-6-methoxy-	Isocotoin. $C_{14}H_{12}O_4$. *See* b1959	244.25	ye nd (lig)	162	....	....	....	s	...	...	s	...	lig δ	**B8^2, 467**
b1998	—,2,6-dihydroxy-4-methoxy-	Cotoin. $C_{14}H_{12}O_4$. *See* b1959	244.25	yesh pr (chl) lf or nd (w)	130–1	....	....	....	δ^h	s	s	s	s^h	chl s CS_2 s	**B8^2, 467**
b1999	—,4,4′-diiodo-	$C_{13}H_8I_2O$. *See* b1959	434.03	pl (to) nd (bz)	238.5	281^{12}	....	....	i	δ	...	...	...	to s^h	**B7, 425**
b2000	—,2,4-dimethoxy-	$C_{15}H_{14}O_3$. *See* b1959	242.28	pr (dil al) λ^{MeOH} 245 (4.18), 280 (3.86), 310 (3.78)	87–8	218^{10}	....	....	i	v	δ	...	...	chl v lig δ	**B8^2, 353**
Ω b2001	—,2,4′-dimethoxy-	$C_{15}H_{14}O_3$. *See* b1959	242.28	nd (al)	100	....	....	....	i	v^h	v	...	v	aa v	**B8^1, 640**
b2002	—,2,5-dimethoxy-	$C_{15}H_{14}O_3$. *See* b1959	242.28	(lig)	51	225^{38}	....	....	i	δ	δ	...	...	lig δ	**B8^2, 354**
b2003	—,3,4-dimethoxy-	$C_{15}H_{14}O_3$. *See* b1959	242.28	nd or pl (al)	103–4	....	....	....	i	v	...	...	...	....	**B8^2, 354**
b2004	—,3,4′-dimethoxy-	$C_{15}H_{14}O_3$. *See* b1959	242.28	pr (al)	58–9	....	....	....	i	v^h s	...	...	...	....	**B8^2, 354**

For explanations, symbols and abbreviations see beginning of table. For structural formulas see end of table.

No.	Name	Synonyms and Formula	Mol. wt.	Color, crystalline form, specific rotation and λ_{max} (log ε)	m.p. °C	b.p. °C	Density	n_D	Solubility w	al	eth	ace	bz	other solvents	Ref.
	Benzophenone														
Ω b2005	**—,4,4′-dimethoxy-**	$C_{15}H_{14}O_3$. *See* b1959	242.28	nd (al) λ^{al} 220 (4.25), 293 (4.37)	146 (144)	...	...	...	i	v^h	s	s	v	chl v	**B8**, 317
Ω b2006	**—,4,4′-dimethyl-**	$C_{15}H_{14}O$. *See* b1959	210.28	rh (al) λ^{al} 265 (4.32), 328 (2.5)	95	333^{725}	...	...	i	v^h	v	s	...	chl v CS_2 v	**B7**[2], 387
b2007	**—,3(dimethylamino)-**	$C_{15}H_{15}NO$. *See* b1959	225.29	pa ye pl (al) λ^{al} 260 (4.4), 370 (3.1)	47	216^{15}	...	...	...	s^h	...	...	...	...	**B14**[1], 388
b2008	**—,4(dimethylamino)-**	$C_{15}H_{15}NO$. *See* b1959	225.29	ye lf (al), nd (peth) λ^{al} 245 (4.2), 355 (4.4)	92–3	...	...	...	i	δ v^h	v	...	...	peth s^h chl s	**B14**[2], 54
b2009	**—,2,2′-dinitro-**	$C_{13}H_8N_2O_5$. *See* b1959	272.21	nd (to or aa)	188–9	...	...	...	i	...	...	...	...	to s^h aa s^h	**B7**, 427
Ω b2010	**—,2-hydroxy-**	$C_{13}H_{10}O_2$. *See* b1959	198.22	pl (dil al) λ^{al} 260 (4.1), 335 (3.7)	39	250^{560}	...	...	i	v	v	...	v	peth δ aa v	**B8**, 155
b2011	**—,3-hydroxy-**	$C_{13}H_{10}O_2$. *See* b1959	198.22	lf or pl (al)	116	...	...	...	...	v	v	...	...	...	**B8**, 157
Ω b2012	**—,4-hydroxy-**	$C_{13}H_{10}O_2$. *See* b1959	198.22	nd (al) pr (dil al) λ^{al} 290 (4.1), 345 (2.7)	135 (st), 122 (unst)	...	...	...	s^h δ	v	v	...	...	aa v	**B8**[2], 184
b2013	**—,2-methoxy-**	$C_{14}H_{12}O_2$. *See* b1959	212.25	λ^{cy} 245 (4.0), 285 (3.4), 315 (3.0), 360 (1.9)	41	$194–6^{18}$	...	...	i	s	...	...	s	aa s	**B8**[2], 182
b2014	**—,3-methoxy-**	$C_{14}H_{12}O_2$. *See* b1959	212.25	λ^{al} 255 (4.0), 315 (3.3)	44 (37)	$342–3^{730}$ 201^{17}	...	...	i	v	...	...	v	aa v	**B8**[1], 569
b2015	**—,4-methoxy-**	$C_{14}H_{12}O_2$. *See* b1959	212.25	pr (eth) λ^{al} 224 (4.07), 252 (3.96), 290 (4.21)	61–2	$354–5^{729}$ 168^{12}	...	...	i	v	v	s	s	chl s aa s	**B8**[1], 569
Ω b2016	**—,2-methyl-**	$C_{14}H_{12}O$. *See* b1959	196.25	λ^{al} 252 (4.2), 282 sh (3.6), 332 sh (2.2)	< −18	309.5^{762} 128^{12}	...	...	i	v	...	...	...	os s	**B7**[2], 371
Ω b2017	**—,3-methyl-**	$C_{14}H_{12}O$. *See* b1959	196.25	λ^{al} 256 (4.2), 297 sh (3.3), 330 sh (2.2)	oil	$314–5^{725}$ 170^8	$1.088^{17.5}$	...	i	s	s	...	s	chl s aa s	**B7**[2], 372
Ω b2018	**—,4-methyl-**	$C_{14}H_{12}O$. *See* b1959	196.25	mcl pr λ^{al} 260 (4.2), 335 (2.2)	59–60 (st) 55 (unst)	$327–8^{760}$	...	...	i	$δ^h$	s	...	s	lig δ chl s	**B7**[2], 372
b2019	—,—,imine	$C_{14}H_{13}N$. *See* b1959	195.27	...	37	147^5	1.0617^{20}_4	1.6097^{20}	i	...	...	...	...	...	**B7**[1], 235
b2020	—,—,diphenylacetal	$C_{26}H_{22}O_2$. *See* b1959	366.47	(dil al, eth – peth)	134	...	...	...	i	i	...	...	...	os s peth i	**B7**[2], 372
b2021	**—,4-methyl-2-nitro-**	$C_{14}H_{11}NO_3$. *See* b1959	241.25	nd or pl (al)	126–7	sub	...	...	i	s^h	...	...	s	chl s aa s^h	**B7**, 442
b2022	**—,4-methyl-2′-nitro-**	$C_{14}H_{11}NO_3$. *See* b1959	241.25	pr (aa or al)	155	...	...	...	i	$δ^h$	δ	...	s^h	chl s aa s^h lig δ	**B7**, 442
b2023	**—,4-methyl-3-nitro-**	$C_{14}H_{11}NO_3$. *See* b1959	241.25	pa ye pl (al or aa)	130–2	...	...	...	δ	s	s	v	v	CS_2 v aa v chl v	**B7**[2], 374
Ω b2024	**—,4-methyl-3′-nitro-**	$C_{14}H_{11}NO_3$. *See* b1959	241.25	lf (al)	111	...	...	...	i	δ	s	...	s	chl s	**B7**[2], 375
b2025	**—,4-methyl-4′-nitro-**	$C_{14}H_{11}NO_3$. *See* b1959	241.25	nd (al)	122–4	sub	...	...	i	s	s	...	s	chl s CS_2 s aa s	**B7**[2], 375
b2026	—,—,oxime	$C_{14}H_{12}N_2O_3$. *See* b1959	256.26	nd (eth-lig)	145	...	...	...	...	s	s	...	s	lig δ	**B7**, 443
b2027	**—,2-nitro-**	$C_{13}H_9NO_3$. *See* b1959	227.22	mcl (al) λ^{al} 258 (4.3)	105	...	...	...	...	s	...	...	...	...	**B7**[2], 362
b2028	**—,3-nitro-**	$C_{13}H_9NO_3$. *See* b1959	227.22	ye nd (al) λ^{al} 232 (4.32), 248 (4.28)	95	234^{18}	...	...	i	s^h	...	...	...	...	**B7**[2], 362
Ω b2029	**—,4-nitro-**	$C_{13}H_9NO_3$. *See* b1959	227.22	nd or lf (al) λ^{al} 266 (4.30)	138	...	1.406^0_4	...	δ	δ s^h	...	...	s	CS_2 δ lig δ	**B7**[2], 362
b2030	**—,2,2′,3,4-tetrahydroxy-**	$C_{13}H_{10}O_5$. *See* b1959	246.22	ye lf or pl (w + 1)	149 102 (+w)	...	...	...	s^h	v	v	...	δ	aa v lig δ	**B8**[2], 539
Ω b2031	**—,2,2′,4,4′-tetrahydroxy-**	$C_{13}H_{10}O_5$. *See* b1959	246.22	ye nd (w + 1½) λ^{MeOH} 287 (4.01), 348 (4.17)	196–8	...	...	...	s^h	v	v	v	s^h	chl s^h aa s MeOH v	**B8**[2], 540
b2032	**—,2,2′,4,6′-tetrahydroxy-**	Isoeuxanthonic acid. $C_{13}H_{10}O_5$. *See* b1959	246.22	(w + 1)	200d	...	...	...	v^h δ	s	s	...	...	...	**B8**, 496
b2033	**—,2,2′,5,6′-tetrahydroxy-**	Euxanthoic acid. $C_{13}H_{10}O_5$ *See* b1959	246.22	ye nd (w)	200–2d	...	...	...	s v^h	s	...	...	...	...	**B8**[2], 541
b2034	**—,2,3′,4,4′-tetrahydroxy-**	$C_{13}H_{10}O_5$. *See* b1959	246.22	nd (w + 2)	202 (anh)	...	...	...	s^h δ	v	v	v	δ	aa v lig i	**B8**[2], 541
b2035	**—,2,3′,4,6-tetrahydroxy-**	$C_{13}H_{10}O_5$. *See* b1959	246.22	pa ye lf (w)	246d	...	...	...	v^h	v	δ	...	...	...	**B8**[2], 540
b2036	**—,2,4,4′,6-tetrahydroxy-**	$C_{13}H_{10}O_5$. *See* b1959	246.22	pr or nd (w + 2)	210	...	...	...	s^h	s	s	...	...	...	**B8**[2], 540

For explanations, symbols and abbreviations see beginning of table. For structural formulas see end of table.

No.	Name	Synonyms and Formula	Mol. wt.	Color, crystalline form, specific rotation and λ_{max} (log ε)	m.p. °C	b.p. °C	Density	n_D	Solubility w	al	eth	ace	bz	other solvents	Ref.
	Benzophenone														
b2037	—,3,3′,4,4′-tetra-hydroxy-	$C_{13}H_{10}O_5$. *See* b1959	246.22	(w) λ^{al} 236 (4.22), 280 sh (4.00), 323 (4.19)	227–8	......	......	......	s^h	s	...	...	s	peth δ	**B8**[2], 541
b2038	—,2,2′,4,4′-tetra-methoxy-	$C_{17}H_{18}O_5$. *See* b1959	302.33	ye lf (dil al) λ^{MeOH} 278 (4.45), 312 (4.42)	130	......	......	......	δ	v	...	...	...	......	**B8**[2], 540
b2039	—,2,2′,5,5′-tetra-methoxy-	$C_{17}H_{18}O_5$. *See* b1959	302.33	ye (aa or al)	109	......	......	......	...	v^h s	s	...	v	chl v aa s, v^h	**B8**, 497
b2040	—,2,2′,6,6′-tetra-methoxy-	$C_{17}H_{18}O_5$. *See* b1959	302.33	pl (bz)	204	......	......	......	...	δ^h	δ	...	δ	chl v aa δ	**B8**[1], 735
b2041	—,2,3′,4,4′-tetra-methoxy-	$C_{17}H_{18}O_5$. *See* b1959	302.33	nd, lf or pr (al)	126 (107)	......	......	......	...	s	...	...	...	con sulf →ye	**B8**[1], 735
b2042	—,2,3′,4,5′-tetra-methoxy-	$C_{17}H_{18}O_5$. *See* b1959	302.33	nd (bz-peth)	73–4	......	......	......	...	v	v	...	v^h	chl v aa δ^h peth i	**B8**[1], 735
b2043	—,2,3′,4′,5-tetra-methoxy-	$C_{17}H_{18}O_5$. *See* b1959	302.33	pr (dil al)	101–2	......	......	......	...	s	s	...	...	con sulf →og	**B8**[2], 541
b2044	—,2,3,4,6-tetra-methoxy-	$C_{17}H_{18}O_5$. *See* b1959	302.33	nd (lig)	125–6	......	......	......	i	v	...	v	v	lig s^h	**B8**[1], 734
b2045	—,2,4,4′,5-tetra-methoxy-	$C_{17}H_{18}O_5$. *See* b1959	302.33	ye pw (al)	122–4	......	......	......	...	s	...	v	v	chl v	**B8**[1], 734
b2046	—,2,4,4′,6-tetra-methoxy-	$C_{17}H_{18}O_5$. *See* b1959	302.33	pr (al)	146	......	......	......	...	s	s	...	...	......	**B8**, 496
b2047	—,3,3′,4,4′-tetra-methoxy-	Veratrophenone. $C_{17}H_{18}O_5$. *See* b1959	302.33	pr (al) λ^{al} 234 (4.40), 283.5 (4.11), 316 (4.24)	145	......	......	......	...	v^h s	...	...	s	peth δ	**B8**[2], 541
b2048	—,3,3′,4,5′-tetra-methoxy-	$C_{17}H_{18}O_5$. *See* b1959	302.33	nd (bz)	114–5	......	......	......	...	v	v	...	v^h	chl v lig i aa δ	**B8**[1], 735
b2049	—,thio-	$C_6H_5CSC_6H_5$	198.29	bl nd (peth) λ^{al} 595 (2.22)	53–4	174^{14}	......	......	...	δ	δ	...	v	chl v peth δ os v	**B7**[2], 365
b2050	—,2,2′,6-tri-hydroxy-	$C_{13}H_{10}O_4$. *See* b1959	230.22	ye nd (dil al)	133–4	......	......	......	i s^h	s	s	...	s	......	**B8**[2], 468
b2051	—,2,3,4-tri-hydroxy	Callobenzophenone. $C_{13}H_{10}H_4$. *See* b1959	230.22	ye nd (dil al) λ^{MeOH} 227 (4.24), 272 (3.38)	140–1	......	......	......	s^h δ	s	s	s	δ	aa s	**B8**[2], 466
b2052	—,2,4,4′-tri-hydroxy-	$C_{13}H_{10}O_4$. *See* b1959	230.22	ye nd (w +2)	200–1	......	......	......	s^h	s	...	...	...	......	**B8**[2], 468
b2053	—,2,4,6-tri-hydroxy-	$C_{13}H_{10}O_4$. *See* b1959	230.22	ye nd (w + 1, dil al)	165	......	......	......	v^h δ	v	v	...	δ	chl δ lig i	**B8**[2], 466
b2054	—,3,4,5-tri-hydroxy-	$C_{13}H_{10}O_4$. *See* b1959	230.22	ye pl (+1w), col (chl)	177–8 (anh)	......	......	......	v^h δ	v	v	v	δ	peth i	**B8**, 422
b2055	—,2,3,4-tri-methoxy-	$C_{16}H_{16}O_4$. *See* b1959	272.30	pr (dil al)	55	......	......	......	...	s	v	...	...	......	**B8**, 418
b2056	—,2,4,4′-tri methoxy-	$C_{16}H_{16}O_4$. *See* b1959	272.30	nd (al)	73–4	......	......	......	...	s^h	v	...	v	aa v lig i	**B8**[1], 702
b2057	—,2,4,5-tri-methoxy-	$C_{16}H_{16}O_4$. *See* b1959	272.30	ye nd (w)	97	......	......	......	δ^h i	v	...	v	v	......	**B8**[1], 701
b2058	—,2,4,6-tri-methoxy-	$C_{16}H_{16}O_4$. *See* b1959	272.30	mcl pr or rh pl (al)	115	......	......	......	δ^h i	v^h	v	...	...	peth i aa δ, v^h chl v	**B8**, 420
b2059	—,3,3′,4-tri-methoxy-	$C_{16}H_{16}O_4$. *See* b1959	272.30	nd (MeOH)	83–4	......	......	......	...	s	...	v	v	MeOH s^h	**B8**[2], 468
b2060	—,3,4,4′-tri-methoxy-	$C_{16}H_{16}O_4$. *See* b1959	272.30	nd (al)	98–9	......	......	......	δ^h i	s^h	...	v	v	......	**B8**, 422
b2061	—,3,4′,5-tri-methoxy-	$C_{16}H_{16}O_4$. *See* b1959	272.30	nd (bz)	97–8	......	......	......	...	v	v	...	v	chl v lig i aa δ	**B8**[1], 702
—	Benzophenone-carboxylic acid	*see* **Benzoic acid, benzoyl-**													
—	Benzopinacol	*see* **1,2-Ethanediol, tetraphenyl-**													
—	α-Benzopina-colone	*see* **Acetophenone, α,α,α-triphenyl-**													
—	1,4-Benzopyran, 2,3-dihydro-	*see* **Chroman**													
—	1,2-Benzopyrazole	*see* **Indazole**													
Ω b2062	**3,4-Benzopyrene**	......	252.32	pa ye mcl nd (bz-MeOH) λ^{al} 225 (4.44), 265.5 (4.66), 274 (4.50), 284 (4.66), 296.5 (4.76), 347 (4.10), 364 (4.36), 384.5 (4.44), 403 (3.60)	176.5–7.5	......	......	......	i	...	...	...	...	con sulf s → og-red	**E14**, 457
b2063	—,5-amino-	$C_{20}H_{13}N$. *See* b2062	267.33	ye pl (bz-lig)	239–41	......	......	......	δ	s	δ	v	v	MeOH δ chl v	**E14s**, 697

For explanations, symbols and abbreviations see beginning of table. For structural formulas see end of table.

No.	Name	Synonyms and Formula	Mol. wt.	Color, crystalline form, specific rotation and λ_{max} (log ε)	m.p. °C	b.p. °C	Density	n_D	Solubility w	al	eth	ace	bz	other solvents	Ref.
	3,4-Benzopyrene														
b2064	—,**5-hydroxy-**	$C_{20}H_{12}O$. *See* b2062	268.32	nd (eth-lig)	207–9	...	...	...	...	v	...	...	v	alk s^h aa s	**E14s**, 701
b2065	—,**8-hydroxy-**	$C_{20}H_{12}O$. *See* b2062	268.32	ye nd (bz-peth)	226–7d	...	...	...	...	s	s	s	s	...	**E14s**, 704
—	**1,2-Benzopyrone**	*see* **Isocoumarin**													
—	**2,1-Benzopyrone**	*see* **Coumarin**													
—	**1,4-Benzopyrone**	*see* **Chromone**													
—	**Benzo[α]pyrrole**	*see* **Indole**													
Ω b2066	**5,6-Benzo-quinoline**	Benzo[f]quinoline. β-Naphthoquinoline	179.22	lf (peth or w) λ^{al} 316 (3.34), 323 (3.26), 330 (3.62), 338 (3.28), 346 (3.71)	94	350^{721} $202–5^{8}$	...	...	δ^h	v	v	s	v	...	**B20**[2], 302
Ω b2067	**7,8-Benzo-quinoline**	Benzo[h]quinoline. α-Naphthoquinoline	179.22	lf (eth), pl (peth) λ^{al} 316 (3.18), 323 (3.08), 330 (3.45), 346 (3.51)	52	338^{719} 233^{47}	...	...	δ	s	s	s	s	os s	**B20**[2], 302
b2068	—,**2-methyl-**	$C_{14}H_{11}N$. *See* b2067	193.25	...	...	324–6	1.1464_4^{20}	1.6738^{20}	i	s	...	...	...	...	**B20**, 471
b2068[1]	**1,2-Benzoquinone**	*o*-Quinone	108.10	red pl or pr λ^{eth} 385 (3.20), 580 (1.47)	60–70d	...	...	...	d	...	s	s	s	peth i	**B7**, 600
Ω b2069	**1,4-Benzo-quinone***	*p*-Quinone	108.10	ye mcl pr (w) λ^{al} 242 (4.26), 285 sh (2.6), 434 (1.26), 454 (1.22)	115.7	sub	1.318_4^{20}	...	s^h δ	s	s	...	...	peth δ lig s^h	**B7**, 609
b2070	—,diimine	...	106.13	ye nd	124	...	...	...	...	...	...	...	s^h	chl s	**B7**[2], 574
Ω b2071	—,—,*N,N*-di-chloro-	$C_6H_4Cl_2N_2$. *See* b2070	175.02	nd (w)	126d	...	...	...	s^h	v^h	v	...	v^h	...	**B7**, 621
Ω b2072	—,dioxime	...	138.13	pa ye nd (w)	240d	...	...	...	i s^h	...	...	...	...	...	**B7**, 627
b2073	—,monoimine, *N*-chloro-	...	141.56	ye nd (peth) λ^{al} 232 (4.29)	85	...	...	...	v	δ^h	δ	...	...	chl δ^h	**B7**, 619
—	—,monooxime	*see* **Phenol, 4-nitroso-***													
b2074	—,**2-bromo-6-methyl-***	$C_7H_5BrO_2$. *See* b2069	201.03	ye nd (al), pr (eth or lig)	95	sub	...	...	δ	v	v	...	...	chl v	**B7**[2], 591
b2075	—,**5-bromo-2-methyl-***	$C_7H_5BrO_2$. *See* b2069	201.03	ye lf (lig)	106	...	...	...	...	v	...	...	...	lig δ, v^h	**B7**[2], 591
b2076	**1,2-Benzo-quinone, 3-chloro-***	$C_6H_3ClO_2$. *See* b2068[1]	142.54	pa ye-red pr (hx)	68d	...	...	...	...	s	s	...	...	...	**B7**[1], 338
b2077	—,**4-chloro-***	$C_6H_3ClO_2$. *See* b2068[1]	142.54	pa ye-red nd (hx)	78	...	...	...	d	...	s	...	...	lig s^h	**B7**[1], 338
Ω b2078	**1,4-Benzo-quinone, 2-chloro-***	$C_6H_3ClO_2$. *See* b2069	142.54	ye-red rh (hx) λ^{al} 248 (4.1), 321 (2.8), 424 (1.6)	57	...	...	...	v	v	v	...	...	chl v	**B7**[2], 579
b2079	—,—,oxime	3-Chloro-4-nitrosophenol. $C_6H_4NClO_2$. *See* b2069	157.56	gr-ye nd (bz-aa) λ^{eth} 291 (4.1), 403 (3.4)	184d	...	...	...	δ	s	s	...	...	...	**B7**[2], 579
b2080	—,**2-chloro-3-methyl-***	$C_7H_5ClO_2$. *See* b2069	156.57	cr (lig)	55	...	...	...	δ	v	...	...	...	chl v lig v	**B7**[1], 353
b2081	—,**2-chloro-5-methyl-***	$C_7H_5ClO_2$. *See* b2069	156.57	ye nd (w or al)	105	...	...	...	s^h	v	v	...	...	chl v	**B7**[2], 590
b2082	—,**2-chloro-6-methyl-***	$C_7H_5ClO_2$. *See* b2069	156.57	ye nd (w)	90	...	...	...	δ^h	v	v	...	...	chl v	**B7**[1], 353
Ω b2083	—,**2,6-dibromo-**, 1-imine, *N*-chloro-	$C_6H_2Br_2ClNO$. *See* b2073	299.36	ye pr (aa, al)	85–6	...	...	...	i	...	...	...	...	...	**B7**[2], 584
b2084	—,—,4-imine, *N*-chloro-	$C_6H_2Br_2ClNO$. *See* b2073	299.36	ye pr (al or aa)	83	...	...	...	...	s^h	...	...	...	...	**B7**[2], 584
b2085	—,**3,5-dibromo-2,6-dimethyl-***	$C_8H_6Br_2O_2$. *See* b2069	293.95	ye lf (al)	176	sub	...	...	i	s	...	...	...	...	**B7**[2], 593
b2086	—,**3,5-dibromo-2-methyl-***	$C_7H_4Br_2O_2$. *See* b2069	279.93	ye	117	...	...	...	...	v	v	...	...	chl v	**B7**[2], 591
Ω b2087	—,**2,5-di-*tert*-butyl-**	$C_{14}H_{20}O_2$. *See* b2069	220.32	ye (al) λ^{CCl_4} 254 (4.3), 310 (2.48), 460 (1.4)	152.5	...	...	...	i	s^h	s	...	s	aa s	**B7**, 670
b2088	**1,2-Benzo-quinone, 4,5-dichloro-***	$C_6H_2Cl_2O_2$. *See* b2068[1]	176.99	yesh red pr or pl	94d (unst)	...	...	...	...	...	s	...	s	...	**B7**[1], 338
b2089	**1,4-Benzo-quinone, 2,3-dichloro-***	$C_6H_2Cl_2O_2$. *See* b2069	176.99	ye lf	100–1 (96)	...	...	...	...	...	s	...	s	...	**B7**[2], 580
Ω b2090	—,**2,5-dichloro-***	$C_6H_2Cl_2O_2$. *See* b2069	176.99	pa ye mcl pr (al) λ^{hx} 270 (4.34), 327 (2.45)	161–2	...	...	...	i	s^h δ	s	...	...	chl s	**B7**[2], 580

For explanations, symbols and abbreviations see beginning of table. For structural formulas see end of table.

No.	Name	Synonyms and Formula	Mol. wt.	Color, crystalline form, specific rotation and λ_{max} (log ε)	m.p. °C	b.p. °C	Density	n_D	Solubility w	al	eth	ace	bz	other solvents	Ref.
	1,4-Benzoquinone														
Ω b2091	—,**2,6-dichloro-***	$C_6H_2Cl_2O_2$. *See* b2069	176.99	ye rh (lig, bz) λ^{hx} 269 (4.30), 327 (2.79)	120–1	...	...	...	δ^h	δ	...	...	...	chl s	**B7**, 633
Ω b2092	—,—,1-imine, *N*-chloro-	$C_6H_2Cl_3NO$. *See* b2073	210.45	ye nd (al)	67–8	...	...	...	δ	s^h	v	δ	δ	chl v peth:	**B7**^{2}, 581
Ω b2093	—,**2,5-dichloro-3,6-dihydroxy-***	Chloranilic acid. $C_6H_2Cl_2O_4$. *See* b2069	208.99	red lf (w + 2) $\lambda^{50\%\ al}$ 325 (4.1)	283–4	...	...	...	s^h	...	...	...	...	...	**B8**^{2}, 433
Ω b2094	—,**2,5-dihydroxy-***	$C_6H_4O_4$. *See* b2069	140.10	dk ye nd (AcOEt) λ^{al} 285 (4.3), 400 (2.6)	211	215 sub δd	...	...	δ	s^h	i	δ	...	AcOEt δ aa s	**B8**^{2}, 432
b2095	—,**2,5-dihydroxy-3-undecyl-***	Embelin. $C_{17}H_{26}O_4$. *See* b2069	308.42	og-red pl (al-bz)	143	sub	...	...	i	...	...	...	...	os s^h lig δ	**B8**^{2}, 452
Ω b2096	—,**2,5-dihydroxy-3-methoxy-6-methyl-***	Spinulosin. $C_8H_8O_5$. *See* b2069	184.16	red-bl	202–3	120^1 sub	...	...	δ s^h	...	...	...	...	alk s	**C32**, 5027
Ω b2097	—,**2,6-di-methoxy-***	$C_8H_8O_4$. *See* b2069	168.16	ye mcl pr (aa) λ^{al} 287 (4.22), 375 (2.86)	256	sub	...	...	δ^h	δ	δ	...	...	alk v aa v^h	**B8**^{2}, 433
Ω b2098	—,**2,3-dimethyl-***	o-Xyloquinone. $C_8H_8O_2$. *See* b2069	136.16	ye nd λ^{al} 248 (4.22), 254 sh (4.2), 334 (2.60), 432 (1.48)	55	sub	...	...	δ	s	s	...	...	...	**B7**^{2}, 593
Ω b2099	—,**2,6-dimethyl-***	$C_8H_8O_2$. *See* b2069	136.16	ye nd λ^{al} 253 (4.29), 317 (2.54), 426 (1.32)	72–3	sub	1.0479^{78}_4	...	...	...	...	...	...	...	**B7**^{2}, 593
Ω b2100	—,**2,5-diphenyl-***	$C_{18}H_{12}O_2$. *See* b2069	260.30	og-ye pl (bz or aa) λ^{diox} 240 (4.37), 388 (3.94)	214	...	...	...	i	δ	...	...	s^h	aa s^h	**B7**^{2}, 757
b2101	—,**3-hydroxy-2-methoxy-5-methyl-**	Fumigatin. $C_8H_8O_4$. *See* b2069	168.16	br nd or pl (peth)	116	...	...	...	δ	v	v	v	v	chl v peth δ	**C38**, 1218
b2102	**1,2-Benzo-quinone, 5-iso-propyl-4-methyl-**, 1-oxime	$C_{10}H_{13}NO_2$. *See* b2068[1]	179.22	nd (bz-chl)	165–7d	...	...	...	...	...	...	...	s^h	chl s^h	**B7**^{2}, 595
b2103	**1,4-Benzo-quinone, 2-iso-propyl-5-methyl-**, dioxime	Thymoquinone dioxime. $C_{10}H_{14}N_2O_2$. *See* b2072	194.24	pr (al)	235d darkens at 200	...	...	...	s	v^h δ	s	...	δ	chl δ aa δ, v^h	**B7**, 665
b2104	**1,2-Benzo-quinone, 3-methoxy-***	$C_7H_6O_3$. *See* b2068[1]	138.12	br-red pl, pr, nd λ^{al} 375 (3.21)	115–20	...	...	...	s	s	s	...	v	chl v	**B8**^{2}, 264
b2105	—,**4-methoxy-**, 1-oxime	$C_7H_7NO_3$. *See* b2068[1]	153.14	ye pr	158–9	...	...	...	δ	v	...	...	v	aa v	**B8**^{2}, 264
b2106	**1,4-Benzo-quinone, 2-methoxy-***	$C_7H_6O_3$. *See* b2069	138.12	ye nd (w) λ^{hx} 248 (4.02), 343 (3.04)	145	sub 80–90	...	...	s^h	v	...	...	...	lig δ	**B8**^{2}, 265
b2107	—,**2-methyl-6-propyl-**, 4-oxime	$C_{10}H_{13}NO_2$. *See* b2069	179.22	br nd (lig)	93–4	...	...	...	δ	v	...	...	v	aa v lig s^h	**B7**^{2}, 595
b2108	—,**2-methyl-***	Toluquinone. $C_7H_6O_2$. *See* b2069	122.13	ye pl or nd λ^{al} 264 (4.14), 312 (2.77), 429 (1.28)	69	sub	$1.08^{75.5}_4$	...	δ s^h	s	s	...	...	...	**B7**^{2}, 588
b2109	—,**2-methyl-3,5,6-tribromo-***	$C_7H_3Br_3O_2$. *See* b2069	358.83	ye pl (al)	235–6	...	...	...	i	δ	s	...	s	...	**B7**^{2}, 592
b2110	—,**phenyl-***	$C_{12}H_8O_2$. *See* b2069	184.21	ye lf (peth, al) λ^{CCl_4} 250 (4.1), 280 sh (3.4), 369 (3.46), 460 sh (2.0)	114	...	...	...	δ^h	s^h	...	...	s	peth s^h chl v	**B7**, 740
Ω b2111	—,**tetrachloro-***	Chloranil. $C_6Cl_4O_2$. *See* b2069	245.88	ye mcl pr (bz) ye lf (aa) λ^{cy} 290 (4.34)	290 sealed tube	sub	...	...	i	δ^h	s	...	...	CS_2 δ chl δ lig i	**B7**^{2}, 581
Ω b2112	—,**tetrahydroxy-***	$C_6H_4O_6$. *See* b2069	172.10	bl-bk cr $\lambda^{ac-MeOH}$ 310 (4.44)	...	...	...	...	v^h δ	v	δ	...	...	...	**B8**^{2}, 572
Ω b2113	—,**tetramethyl-***	Duroquinone. $C_{10}H_{12}O_2$. *See* b2069	164.21	ye nd (al or lig) λ^{al} 259 (4.25), 267 (4.26), 338 (2.38), 430 sh (1.44)	111–2	...	...	...	i	s	s	s	s	sulf s aa s lig δ, s^h	**B7**^{2}, 597
b2114	—,**trichloro-***	$C_6HCl_3O_2$. *See* b2069	211.43	ye pl (al) λ^{al} 278 (4.1), 362 (2.6)	169–70	...	...	...	i	s^h	v	...	...	...	**B7**^{2}, 581

For explanations, symbols and abbreviations see beginning of table. For structural formulas see end of table.

No.	Name	Synonyms and Formula	Mol. wt.	Color, crystalline form, specific rotation and λ_{max} (log ε)	m.p. °C	b.p. °C	Density	n_D	Solubility w	al	eth	ace	bz	other solvents	Ref.
	α-Benzosuberone														
b2115	**α-Benzosuberone**	1-Benzosuberanone.	160.22	λ^{al} 248 (3.92), 287 (3.24), 316 sh (2.27)		$124–5^{7}$ 108^{1}	1.0780_4^{20}	1.5698^{20}		s					**C50,** 4850
—	**Benzotetronic acid**	*see* **Coumarin, 4-hydroxy-**													
Ω b2116	**Benzothiazole**		135.19	λ^{al} 250 (3.74), 284 (3.22), 296 (3.15)	2	231^{760} (228) 131^{34}	1.2460_4^{20}	1.6379^{20}	δ	v	v	s		CS_2 v	**B27²**, 17
Ω b2117	**—,2,-amino-**	$C_7H_6N_2S$. *See* b2116	150.20	pl (w) λ^{al} 222 (4.52), 264 (4.09)	132				δ	s	s			chl s con HCl s	**B27²**, 225
b2118	**—,6-amino-**	$C_7H_6N_2S$. *See* b2116	150.20	pr (w)	87				i	s	i				**B27**, 366
Ω b2119	**—,2-amino-6-ethoxy-**	$C_9H_{10}N_2OS$. *See* b2116	194.26	nd (al) 225 (4.5), 270 (4.1)	163–4					s^h				MeOH s^h	**B27²**, 335
b2120	**—,5-amino-2-mercapto-**	$C_7H_6N_2S_2$. *See* b2116	182.27	nd ($PhNH_2$)	216									$PhNH_2$ s^h	**B27²**, 475
b2121	**—,6-amino-2-mercapto-**	$C_7H_6N_2S_2$. *See* b2116	182.27		263										**B27²**, 475
b2122	**—,2-amino-4-methyl-**	$C_8H_8N_2S$. *See* b2116	164.22	nd (w), pl (al)	145				δ	δ					**B27²**, 237
b2123	**—,2-amino-5-methyl-**	$C_8H_8N_2S$. *See* b2116	164.24	pl (dil al)	171–2 (145)					s					**B27²**, 240
Ω b2124	**—,2-amino-6-methyl-**	$C_8H_8N_2S$. *See* b2116	164.24	nd (w), pr (dil al)	142				δ^h s^h	s					**B27²**, 241
Ω b2125	**—,2-chloro-**	C_7H_4ClNS. *See* b2116	169.63		24	248, $135–6^{28}$	1.3715_4^{19}	1.6338^{19}		s	s	s			**B27²**, 18
b2126	**—,5-chloro-2-methyl-**	C_8H_6ClNS. *See* b2116	183.66	pl (eth)	69	$70–90^{0.5}$ sub				s				peth s	**B27²**, 22
b2127	**—,6-dimethyl-amino-2-mercapto-**	$C_9H_{10}N_2S_2$. *See* b2116	210.32	ye nd (bz)	230d						δ	s			**B27²**, 475
b2128	**—,2(2,4-dinitro-phenylthio)-**	$C_{13}H_7N_3O_4S_2$. *See* b2116	333.35	ye	162		1.24_4^{20}		i	δ s^h	δ				**C48,** 1443
b2129	**—,2-hydroxy-**	C_7H_5NOS. *See* b2116	151.19	pr (dil al), nd	138	360			i	v	v				**B27²**, 225
b2130	**—,6-hydroxy-2-phenyl-**	$C_{13}H_9NOS$. *See* b2116	227.29	wh nd (dil al)	227–7.5 (cor)				i	s	i	s	s	aa s lig i	**B27²**, 88
Ω b2131	**—,2(2-hydroxy-phenyl)-**	$C_{13}H_9NOS$. *See* b2116	227.29	nd or lf (al) λ^{al} 258 (3.85), 288 (4.12), 333 (4.20)	132–3	$175–93^{3}$				s^h					**B27²**, 91
Ω b2132	**—,2-mercapto-**	$C_7H_5NS_2$. *See* b2116	167.25	nd (al or dil MeOH) λ^{MeOH} 235 (4.12), 282 sh (3.34), 320 (4.43)	180.2–1.7		1.42_4^{20}		i	s^h	δ		δ	aa δ, s^h	**B27²**, 233
b2133	**—,—,benzoate**	$C_{14}H_9NOS_2$. *See* b2116	271.36	ye	132				i	s^h δ	δ				**C51,** 3491
b2134	**—,2-mercapto-4-methyl-**	$C_8H_7NS_2$. *See* b2116	181.28	nd (dil aa)	186									aa s^h	**B27²**, 240
b2135	**—,2-mercapto-5-methyl-**	$C_8H_7NS_2$. *See* b2116	181.28	nd (to)	171–3				i	s		s	s^h	to s^h	**B27²**, 241
b2136	**—,2-mercapto-6-methyl-**	$C_8H_7NS_2$. *See* b2116	181.28		181				i	s		s	s^h	to s^h	**B27²**, 242
b2137	**—,2-mercapto-7-methyl-**	$C_8H_7NS_2$. *See* b2116	181.28		184				i			s	s^h	to s	**B27²**, 242
b2138	**—,2-mercapto-6-nitro-**	$C_7H_4N_2O_2S_2$. *See* b2116	212.25	ye nd (aa)	255–7				i			s		aa s^h	**B27²**, 234
Ω b2139	**—,2-methyl-**	C_8H_7NS. *See* b2116	149.22	λ^{al} 218 (4.32), 252 (3.85), 283 (3.40)	14	238 $150–1^{15}$	1.1763_4^{19}	1.6092^{19}	i	s					**B27²**, 21
Ω b2140	**—,2(methylthio)-**	$C_8H_7NS_2$. *See* b2116	181.28	pr (dil al) λ^{MeOH} 278 (4.1), 288 (4.0), 299 (3.9)	52					s^h					**B27²**, 71
Ω b2141	**—,2-phenyl-**	$C_{13}H_9NS$. *See* b2116	211.29	nd (dil al) λ^{al} 225 (4.30), 297 (4.20)	114	>360			i	s^h	s			CS_2 s	**B27²**, 37
b2142	**—,2-phenylamino**	$C_{13}H_{10}N_2S$. *See* b2116	226.30	nd (al) λ^{al} 222 (4.29), 302 (4.39)	161								δ	MeOH δ AcOEt s	**B27²**, 226
b2143	**Benzothiazoline, 3-methyl-2-imino-**		164.24	pl (w), nd (al) λ^{al} 252 (3.94), 278 (3.77), 286 (3.84)	128				δ v^h	v	v			chl v	**B27²**, 228
b2144	**—,3-methyl-2-thioxo-**		181.28	nd (al), pr (aa) λ^{MeOH} 241 (4.12), 324 (4.41)	90	335^{757}			i	s^h δ	δ		v	CS_2 δ aa δ, s^h chl v	**B27²**, 233

For explanations, symbols and abbreviations see beginning of table. For structural formulas see end of table.

No.	Name	Synonyms and Formula	Mol. wt.	Color, crystalline form, specific rotation and λ_{max} (log ε)	m.p. °C	b.p. °C	Density	n_D	Solubility w	al	eth	ace	bz	other solvents	Ref.
	Benzothiophene														
b2145	**Benzothiophene**	Thionaphthene	134.20	lf λ^{al} 227 (4.45), 257 (3.74), 288 (3.31)	32	221^{760} $103\text{–}5^{20}$	1.1484_4^{32}	1.6374^{37}	i	v	s	s	s	os s	**B17**², 58
b2146	**—,3-hydroxy-**	C_8H_6OS. *See* b2145	150.20	nd (w)	71	...	...	...	δ s^h	v	v	v	v	alk s peth δ	**B17**², 128
b2147	**—,4-hydroxy-**	C_8H_6OS. *See* b2145	150.20	nd (sub), cr (peth)	78–9	sub	...	...	δ	s	...	...	...	alk s, peth s^h	**B17**, 121
b2148	**2,3-Benzothiophenequinone**	Thioisatin.	164.18	gold-ye pr (al)	121	247^{760}	...	...	i	s	...	...	s	aa s	**B17**, 467
Ω b2149	**1,2,3-Benzotriazole**	Azimidobenzene.	119.13	nd (chl or bz)	100	204^{15} 159^{2}	...	...	δ	s	...	...	s	to, chl, DMF s	**B26**², 17
b2150	**1,2,3-Benzoxadiazole, 5,7-dinitro-**	...	210.11	ye pl (al)	158	...	...	...	...	s^h	...	...	...	...	**B16**², 287
Ω b2151	**Benzoxazole**	...	119.12	pr (dil al) λ^{al} 231 (3.90), 270 (3.53), 276 (3.51)	31	182.5^{760} 45^{4}	...	1.5594^{20}	i	s	...	...	...	sulf s	**B27**², 17
b2152	**—,2-chloro-**	C_7H_4ClNO. *See* b2151	153.57		7	201–2	1.3453_4^{18}	1.5678^{20}	i	...	...	...	...	...	**B27**², 17
b2153	**—,2-hydroxy-**	2(3)-Benzoxazolone. $C_7H_5NO_2$. *See* b2151	135.13	nd (bz or w + 1)	141–2 (anh) 97–8 (+w)	230^{30}	...	...	δ	s	s	...	...	...	**B27**², 223
b2154	**—,4-hydroxy-2-phenyl-**	$C_{13}H_9NO_2$. *See* b2151	211.22	nd (bz)	138–9	...	...	...	δ	v	v	...	s^h	aa v lig δ	**B27**², 88
b2155	**—,5-hydroxy-2-phenyl-**	$C_{13}H_9NO_2$. *See* b2151	211.22	nd (lig or dil al)	175	...	...	...	...	...	...	...	v	chl v aa v	**B27**², 88
b2156	**—,6-hydroxy-2-phenyl-**	$C_{13}H_9NO_2$. *See* b2151	211.22	nd	216–7	...	...	...	δ	δ	δ	v	s^h	aa v δ	**B27**, 117
b2157	**—,7-hydroxy-2-phenyl**	$C_{13}H_9NO_2$. *See* b2151	211.22	nd (bz or dil al)	191–2	...	...	...	...	v	v	...	v^h	chl v lig v	**B27**², 91
b2158	**—,2(2-hydroxyphenyl)-**	$C_{13}H_9NO_2$. *See* b2151	211.22	pink nd (al or aa) λ^{cy} 231 sh (4.06), 322 (4.24)	123–4	338	...	...	δ	v	s	s	s	os v	**B27**², 91
b2159	**—,2-mercapto-**	C_7H_5NOS. *See* b2151	151.19	nd (w)	196	...	...	...	δ s^h	δ	v	...	...	aa v NH_3 s	**B27**², 224
Ω b2160	**—,2-methyl-**	C_8H_7NO. *See* b2151	133.16	...	8.5–10	$200\text{–}1^{760}$ $59\text{–}60^{12}$	1.1211_4^{20}	1.5497^{20}	i	v	∞	...	...	...	**B27**², 20
b2161	**2,3-Benzoxazin-1-one**	...	147.14	cr (bz)	120d	...	...	...	...	...	...	...	...	...	**B27**¹, 278
—	**2(3)-Benzoxazolone**	*see* **Benzoxazole, 2-hydroxy-**													
—	**Benzoyl chloride**	*see* **Benzoic acid**, chloride													
—	**Benzyl alcohol**	*see* **Toluene, α-hydroxy-**													
—	**Benzylamine**	*see* **Toluene, α-amino-**													
—	**Benzyl bromide**	*see* **Toluene, α-bromo-**													
—	**Benzyl chloride**	*see* **Toluene, α-chloro-**													
—	**Benzyl cyanide**	*see* **Acetic acid, phenyl-**, nitrile													
—	**Benzyl iodide**	*see* **Toluene, α-iodo-**													
b2162	**Berbamine**	$C_{37}H_{40}N_2O_6$	608.74	lf (+2w, al) cr (peth) $[\alpha]_D^{25}$ +109 (chl) λ^{al} 380 (1.9)	197–210 (anh) 156 (hyd)	...	...	...	δ	s	s	...	...	peth s chl s	**B27**², 891
b2163	**Berberine**	$C_{20}H_{19}NO_5$	353.38	red ye nd (w + 6), cr (chl + 1) λ^{w} 262.5 (4.7), 347.5 (4.3), 425 (3.6)	145 (anh) 110 (+6w)	...	...	anh→ hyd→	i s	s δ	s δ	...	δ ...	chl δ	**B27**², 567
b2164	**—,compound with chloroform**	$C_{21}H_{18}Cl_3NO_4$	454.74	ye tcl ta (chl − al)	179 (145)	...	...	...	δ	δ	...	...	...	chl v	**B27**, 492
b2165	**—,hydrochloride**	$C_{20}H_{22}ClNO_6$	407.86	ye cr (w + 2), nd (w + 4), λ^{al} 270 (4.03), 350 (4.58)	...	...	...	...	δ v^h	i	i	...	...	chl i	**B27**¹, 514
b2166	**—,nitrate**	$C_{20}H_{18}N_2O_7$	398.38	red ye nd (al)	155d	...	...	...	v^h	δ	...	...	...	chl i	**B27**, 500
b2167	**—,sulfate** (trihydrate)	$C_{40}H_{42}N_2O_{15}S$	822.88	red-ye nd		...	...	...	s	s	...	...	...	chl i	**B27**, 500
b2168	**—,tetrahydro- (*dl*)**	$C_{20}H_{21}NO_4$	339.40	mcl nd (al)	173–4	...	...	...	δ	s	...	...	...	CS_2 v chl v	**B27**², 557
—	**Berberonic acid**	*see* **2,4,5-Pyridinetricarboxylic acid**													
b2169	**Berbine (*dl*)**	...	235.33	nd (eth or MeOH)	89	...	...	...	i	v	s	s	...	os v lig, peth δ	**B20**², 311
Ω b2170	**Betaine**	Lycine. Oxyneurine $\overset{+}{(CH_3)_3N}CH_2COO^-$	117.15	(w + 1), pr or lf (al)	293d (anh) (310d)	...	...	...	v	s	δ	...	...	MeOH v	**B4**², 785

For explanations, symbols and abbreviations see beginning of table. For structural formulas see end of table.

No.	Name	Synonyms and Formula	Mol. wt.	Color, crystalline form, specific rotation and λ_{max} (log ε)	m.p. °C	b.p. °C	Density	n_D	Solubility w	al	eth	ace	bz	other solvents	Ref.
	Betol														
—	**Betol**	*see* **Benzoic acid, 2-hydroxy-2-naphthyl ester**													
b2172	**Betonicine**	$C_7H_{13}NO_3$	159.19	pr (dil al + 1w) $[\alpha]_D^{21}$ −37 (w,c = 4.8)	252d (anh)				δ	s	...	...	...		**B22**[1], 547
b2173	**Betulin**	Lupenediol. $C_{30}H_{50}O_2$	442.73	nd (al + 1) $[\alpha]_D^{15}$ +20(Py, c = 2)	251–2	$170–80^{0.08}$ δd			i	δ s^h	s^h	...	δ s^h	chl δ, s^h aa s^h AcOEt s lig δ	**E14**, 568
b2174	**Betulinic acid**	Betulic acid. $C_{30}H_{48}O_3$	456.71	pl or nd (al + 1) $[\alpha]_{546}^{22}$ +7.9 (Py)	316–8				δ	δ	δ	δ	δ	chl δ Py s	**B6**[2], 939
b2175	**Biacene**	Biacenaphthene. Biacenaphthylidene.	304.39	red-ye pl or nd (bz) λ^{bz} 254 (3.54), 298 (3.54), 340 (3.99), 403 (4.77)	277				...	...	...	...	s	CS_2 s^h	**E13**, 187
—	**Biacetyl**	*see* **2,3-Butanedione***													
—	**Biacetylene**	*see* **Butadiyne***													
—	**Bianiline**	*see* **Biphenyl, diamino-**													
—	**Bianisole**	*see* **Biphenyl, dimethoxy-**													
b2176	**10,10′-Bianthronyl**	Bianthrone.	386.45	pl (ace)	256–8d				i	δ	i	...	...	chl v alk i	**E13**, 766
b2177	**Biarsine, tetraethyl-**	Ethyl cacodyl. $(C_2H_5)_2AsAs(C_2H_5)_2$	266.09			185–7	1.1388_4^{24}	1.4709	i	s	s	...	...		**B4**, 616
b2178	**—,tetrakis(trifluoromethyl)-**	Perfluorocacodyl. $(CF_3)_2AsAs(CF_3)_2$	425.87			$106–7^{760}$		1.372^{19}	i	...	...	...	...		**J1952**, 2552
—	**—,tetramethyl-**	*see* **Cacodyl**													
—	**2,2′-Bicamphane-2,2′-diol**	*see* **Camphor pinacol (*l*)**													
b2179	**3,3′-Bicoumarin**		290.28	nd (aa)	> 330				...	i	i	...	i	chl δ	**B19**, 183
—	**Bicresol**	*see* **Biphenyl, dihydroxy-dimethyl-**													
b2180	**Bicyclo[2,2,1]-hepta-2,5-diene**		92.15	λ^{al} 205 (3.31), 214 (3.14), 220 (2.74), 230 (2.30)	−19.1	89.5^{760}	0.9064_4^{20}	1.4702^{20}	i	s	s	s	s	to ∞ lig ∞	**C54**, 12073
b2181	**Bicyclo[2,2,1]-heptane**	Norbornylane. Norbornane. Norcamphane.	96.17		87.5–7.8	sub			i	s	s	s	s	PrOH s	**B5**[2], 45
b2182	**Bicyclo[4,1,0]-heptane, 7-aza-**		97.16		20–2	$48–51^{22}$			i	s	s	...	s		**B7**[1], 7
—	**Bicyclo[2,2,1]-heptane, 1,7,7-trimethyl-**	*see* **Camphane**													
Ω b2183	**Bicyclo[2,2,1]-heptane-2,3-dicarboxylic acid**	2,3-Norcamphane-dicarboxylic acid	184.19		192–5				δ s^h	δ	δ	...	...		**E12A**, 982
b2184	**—,anhydride**	$C_9H_{10}O_3$. *See* b2183	166.18		165–7				...	...	...	...	s^h		**E12A**, 982
b2185	**Bicyclo[2,2,1]-heptane-2-carboxaldehyde**		122.17	cr (w)		$70–2^{22}$	1.0227_4^{19}	1.4760^{25}	i	...	s				**E12A**, 705
b2186	**Bicyclohexyl** (*cis, cis*)	Dodecahydrobiphenyl.	166.31		4	238	0.8914_4^{20}	1.4766^{20}	i	∞	∞	...	...		**B5**[3], 273
b2187	—(*trans, trans*)	$C_{12}H_{22}$. *See* b2186	166.31		4.2	217–9 $95–6^9$	0.8592_4^{20}	1.4663^{20}	i	∞	∞	...	...		**B5**, 108
b2188	**1,1′-Bicyclohexyldicarboxylic acid,** dinitrile		216.33	lf (ace)	224.5–5.5				...	...	s	...	...		**B9**[2], 571
b2189	**Bicyclo[3,3,1]-nonane**		124.23	cr (MeOH)	145–6	169–70 sub			i	s	...	...	...	MeOH s^h aa s^h	**E12A**, 1057
b2190	**Bicyclo[2,2,2]-octane**		110.20		169–71				i	...	...	...	...		**E12A**, 1068
b2191	**Bicyclo[3,2,1]-octane**		110.20		139.5–41				i	...	...	...	...		**E12A**, 1048
b2192	**Bicyclo[3,3,0] octane** (*cis*)		110.20		< −80	137^{765}	0.8638_4^{25}	1.4595^{25}	i	s	...	...	...		**E12A**, 80
b2193	—(*trans*)	C_8H_{14}. *See* b2192	110.20		−30	132^{755}	0.8626_4^{18}	1.4625^{18}	i	s	...	...	...		**E12A**, 80
b2194	**Bicyclo[3,3,0]-octane-2,6-dione**		138.17		45	$86–8^{0.2}$	1.1290_4^{60}	1.4877^{54}	...	...	...	...	...		**E12A**, 84
b2195	**Bicyclo[2,2,2]-octane, 2-methyl-**	C_9H_{16}. *See* b2190	124.23		33–4	158–8.5^{749}	$0.8664_4^{40.5}$	$1.4608^{40.5}$	i	...	...	...	...		**E12A**, 1068
b2196	**Bicyclo[3,3,0]-octan-2-one** (*cis*)		124.19			72^{12} $50^{2.3}$	1.0097_4^{20}	1.4790^{20}	δ	s	...	s	...		**E12A**, 82
Ω b2197	**9,9′-Bifluorenyl-**		330.43	nd (bz-al) λ^{al} 266 (4.53), 291.5 (3.99), 302.5 (4.09)	247				i	δ	δ	...	δ	Py s^h aa s^h	**E13**, 123

For explanations, symbols and abbreviations see beginning of table. For structural formulas see end of table.

No.	Name	Synonyms and Formula	Mol. wt.	Color, crystalline form, specific rotation and λ_{max} (log ε)	m.p. °C	b.p. °C	Density	n_D	Solubility w	al	eth	ace	bz	other solvents	Ref.
	9,9′-Bifluorenyl														
b2198	—,9,9′-diphenyl-	$C_{38}H_{26}$. *See* b2197	482.63	pl (bz)	256 (under CO_2)	...	1.266_4^0	...	i	...	...	...	δ	os δ	**B5²**, 638
b2199	**Bifluorenylidene**	...	328.42	red nd (bz) λ^{diox} 244 (4.89), 272 (4.60), 308 (3.38), 340 (4.50), 458 (4.37)	194–5	...	...	...	i	...	s	...	s	chl s	**E13**, 123
—	**Bigitaligenin**	*see* **Gitoxigenin**													
—	**Bigitaline**	*see* **Gitoxine**													
b2200	**Biguanide**	$H_2NC(:NH)NHC(:NH)NH_2$	101.11	pr or nd (al) λ^{w} 231 (3.98)	136	d at 142	...	...	v	s	...	...	i	chl i	**B3²**, 76
b2201	—,1(2-tolyl)-	$C_9H_{13}N_5$. *See* b2200	191.24	nd or pl (w + ½) $\lambda^{aq\ HCl}$ 236 (4.19)	144	...	...	...	δ v^h	v	i	v	i	chl i	**B12**, 803
b2202	**2,2′-Biindane, 1,1′,3,3′-tetraoxo-**	Bisdiketohydrindene	290.28	vt nd ($PhNO_2$) red nd (bz)	297 (270)	...	...	...	...	...	...	...	s^h	$PhNO_2$ δ alk s	**B7²**, 863
b2203	**Bikhaconitine**	$C_{36}H_{51}NO_{11}$	673.81	$[\alpha]_D$ +12(al)	118–23 (anh) 113–16 (+w)	...	...	...	i	s	s	...	...	chl s peth i	**C19**, 2104
b2204	**Bilifucsin**	$C_{16}H_{20}N_2O_4$	304.35	dk br pw	183	...	...	...	i	s	i	...	...	chl i aa s alk s	**C30**, 1936
Ω b2205	**Bilirubin**	Haematoidine. $C_{33}H_{36}N_4O_6$	584.68	red mcl pr or pl (chl) $\lambda^{w,pH=8}$ 437 (4.67)	...	...	...	...	i	δ	δ	...	s	CS_2 s chl s	**C38**, 1230
Ω b2206	**Biliverdin**	Dehydrobilirubin. Oöcyan. $C_{33}H_{34}N_4O_6$. *See* b2205	582.66	dk gr pl or pr (MeOH) λ^{MeOH} 378 (4.61), 640–50 (4.11)	>300 (206–9)	...	...	...	i	s	δ	...	s	MeOH δ chl δ CS_2 δ alk s	**J1961**, 2264
—	**Binaphthol**	*see* **Binaphthyl, dihydroxy-**													
b2207	**1,1-Binaphthyl**	α,α′-Dinaphthyl	254.34	(i) pl (aa) (ii) rh (peth) λ^{al} 220 (5.0), 283 (4.1), 295 (4.1)	(i) 144.5–5.0 (ii) 160.5	>360 $240–4^{12}$	...	...	i	δ s^h	s	s	s	CS_2 s	**B5²**, 642
Ω b2208	**2,2-Binaphthyl**	β,β-Dinaphthyl.	254.34	bl fluor pl (al) λ^{al} 215 (4.6), 255 (5.0), 307 (4.3)	187–8	452^{753} sub	...	...	i	δ	s	...	s^h	aa δ CS_2 s	**B5²**, 643
b2209	**1,1-Binaphthyl, 4,4′-diamino-3,3′-dimethyl-**	$C_{22}H_{20}N_2$. *See* b2207	312.42	...	213	...	...	...	...	s	...	...	v	...	**B13²**, 140
Ω b2210	—,2,2′-dihydroxy-	1,1′-Bi-2-naphthol. β-Dinaphthol. $C_{20}H_{14}O_2$. *See* b2207	286.33	nd (al), cr (to)	220	...	...	...	i	s	s	...	...	chl δ alk s	**B6²**, 1026
b2211	—,4,4′-dihydroxy-	4,4′-Bi-1-naphthol. α-Dinaphthol. $C_{20}H_{14}O_2$. *See* b2207	286.33	pl	300	sub	...	...	i	s	s	...	δ	chl δ alk s	**B6**, 1053
Ω b2212	**Biotin**	Vitamin H. Coenzyme R. $C_{10}H_{16}N_2O_3S$.	244.32	nd (w) $[\alpha]_D^{22}$ +92 (0.1N NaOH, c = 1)	232.3 d	...	...	...	s	s	δ	...	...	chl δ	**Am 67**, 2096
b2213	—,methyl ester	$C_{11}H_{18}N_2O_3S$. *See* b2213	258.34	pl (MeOH-eth) $[\alpha]_D^{22}$ +57 (chl)	166.7	sub	...	...	δ	s	i	s	δ	chl s peth i	**C43**, 1810
—	**Bioxirane**	*see* **Butane, 1,2,3,4-diepoxy-**													
—	**Biphenol**	*see* **Biphenyl, dihydroxy-**													
Ω b2214	**Biphenyl**	Diphenyl. Phenylbenzene.	154.21	lf (dil al) λ^{al} 247 (4.24)	71	255.9^{760} 145^{22}	0.8660_4^{20}	1.588^{77} 1.475^{20}	i	s	s	...	v	CCl_4, CS_2, MeOH v	**B5²**, 479
b2215	—,2-acetamido-	$C_{14}H_{13}NO$. *See* b2214	211.27	pr or nd (dil al or peth) λ^{al} 231 (4.1)	121	355^{760}	...	...	i	v	v	...	...	...	**B12²**, 747
b2216	—,3-acetamido-	$C_{14}H_{13}NO$. *See* b2214	211.27	nd (al) λ^{al} 242 (4.4)	149	...	...	...	i	s^h	...	...	v^h	...	**B12²**, 751
b2217	—,4-acetamido-	$C_{14}H_{13}NO$. *See* b2214	211.27	cr (dil MeOH) λ^{al} 237 (4.29)	172	...	...	...	i	v	...	v	...	MeOH v	**B12²**, 755
b2218	—,4-acetamido-3-nitro-	$C_{14}H_{12}N_2O_3$. *See* b2214	256.26	ye nd (al)	132	...	...	...	i	s^h	s	...	...	aa s^h	**B12²**, 760
Ω b2219	—,4-acetyl-	4-Phenylacetophenone. $C_{14}H_{12}O$. *See* b2214	196.25	pr (ace), cr (al) λ^{al} 284 (4.30)	121	325–7	...	...	i	v	...	v	...	...	**B7²**, 337
b2220	—,2-amino-	$C_{12}H_{11}N$. *See* b2214	169.23	lf (dil al) λ^{al} 300 (3.5)	51–3	299^{760} 170^{15}	...	...	i	s	s	...	s	peth δ	**B12²**, 747
b2221	—,3-amino-	$C_{12}H_{11}N$. *See* b2214	169.23	nd λ^{al} 234 (4.37), 302 (3.39)	30	254^{135} 195^{15}	...	...	δ	s	s	s	s	...	**B12²**, 751
b2222	—,4-amino-	Xenylamine. $C_{12}H_{11}N$. *See* b2214	169.23	lf (dil al) λ^{al} 278 (4.24)	53–4	302 191^{15}	...	...	δ	s	s	...	...	chl s	**B12²**, 753

For explanations, symbols and abbreviations see beginning of table. For structural formulas see end of table.

No.	Name	Synonyms and Formula	Mol. wt.	Color, crystalline form, specific rotation and λ_{max} (log ε)	m.p. °C	b.p. °C	Density	n_D	Solubility w	al	eth	ace	bz	other solvents	Ref.
	Biphenyl														
b2223	**—,3-amino-4-hydroxy-**	$C_{12}H_{11}NO$. *See* b2214	185.23	nd (chl)	208	...	...	...	...	v	v	...	v	...	**B13**², 419
b2224	**—,4-amino-2′-hydroxy-**	$C_{12}H_{11}NO$. *See* b2214	185.23	nd (to)	181–2	...	...	...	δ	s	...	...	...	to δ alk s aa s	**B13**², 419
Ω b2225	**—,4-amino-4′-hydroxy-**	$C_{12}H_{11}NO$. *See* b2214	185.23	pl (dil al) λ^{sulf} 285 (4.26)	273	...	...	...	i	δ	i	i	δ	aa δ	**B13**², 420
b2226	**—,5-amino-2-hydroxy-**	$C_{12}H_{11}NO$. *See* b2214	185.23	nd (al or bz)	201	...	...	...	i	s^h	δ	...	s	chl δ lig i	**B13**¹, 280
b2227	**—,2-amino-4′-nitro-**	$C_{12}H_{10}N_2O_2$. *See* b2214	214.23	og-red nd (al)	159	...	...	...	...	s^h	...	...	...	...	**B12**², 750
b2228	**—,2-amino-5-nitro-**	$C_{12}H_{10}N_2O_2$. *See* b2214	214.23	ye nd (al)	125	...	...	...	...	s^h	...	...	...	...	**B12**², 750
b2229	**—,3-amino-4-nitro-**	$C_{12}H_{10}N_2O_2$. *See* b2214	214.23	og nd (dil al)	116	...	...	...	...	s	...	...	...	...	**B12**², 752
b2230	**—,3-amino-4′-nitro-**	$C_{12}H_{10}N_2O_2$. *See* b2214	214.23	og nd (al)	137	...	...	...	...	s^h	...	...	...	aa s	**B12**², 753
b2231	**—,4-amino-2′-nitro-**	$C_{12}H_{10}N_2O_2$. *See* b2214	214.23	red mcl pr (al)	99	...	...	...	δ^h i	v	...	...	...	...	**B12**¹, 547
b2232	**—,4-amino-3-nitro-**	$C_{12}H_{10}N_2O_2$. *See* b2214	214.23	red nd (al) λ^{al} 268 (4.4), 425 (3.6)	170–1	...	...	...	...	s^h	s	...	...	chl s aa s	**B12**¹, 760
b2233	**—,4-amino-4′-nitro-**	$C_{12}H_{10}N_2O_2$. *See* b2214	214.23	red nd (al) λ^{al} 247.5 (4.10), 379 (4.19)	203–4	...	...	...	i^h	s^h	...	...	...	aa s^h	**B12**², 761
b2234	**—,2-benzyl-**	*o*-Biphenylylphenylmethane. $C_{19}H_{16}$. *See* b2214	244.34	mcl nd (al)	54–6	283–7[110]	...	...	i	s	s	...	v	...	**B5**, 708
Ω b2235	**—,4-benzyl-**	$C_{19}H_{16}$. *See* b2214	244.34	lf	85	285–6[110]	1.171^0_4	...	i	s	v	...	v	...	**B5**², 618
b2236	**—,4,4′-bis-(diethylamino)-**	$C_{20}H_{28}N_2$. *See* b2214	296.46	nd (al)	85	...	...	...	...	s v^h	s	...	...	...	**B13**², 98
b2237	**—,2,4′-bis(di-methylamino)-**	$C_{16}H_{20}N_2$. *See* b2214	240.35	pl (al)	51–2	206–7[11]	...	...	...	s v^h	s	...	...	...	**B13**², 88
b2238	**—,4,4′-bis(di-methylamino)-**	$C_{16}H_{20}N_2$. *See* b2214	240.35	nd (al or bz-lig)	198	>360	...	...	...	δ s^h	δ	...	v^h	chl v AcOEt s^h lig δ^h	**B13**², 97
b2239	**—,4,4′-bis-(ethylamino)-**	$C_{16}H_{20}N_2$. *See* b2214	240.35	nd or pl (al)	120.5 (116)	...	...	...	...	v^h	v^h	...	v	lig δ	**B13**, 222
b2240	**—,4,4′-bis-(methylamino)-**	$C_{14}H_{16}N_2$. *See* b2214	212.30	lf (al or w or lig) λ^{cy} 288 (4.8)	91–1.5	...	...	...	δ^h	s^h	...	...	...	lig s	**B13**², 97
b2241	**—,4,4′-bis-(phenylamino)-**	$C_{24}H_{20}N_2$. *See* b2214	336.44	lf (to) λ^{al} 334.5 (4.66)	244–5	...	...	...	i	δ	δ	...	δ	to v^h aa v^h	**B13**², 98
b2242	**—,2-bromo-**	$C_{12}H_9Br$. *See* b2214	233.12	...	fp 1.5–2.0	296–8 160[11]	1.2175^{26}	1.6248^{25}	i	s	s	...	...	...	**B5**², 485
Ω b2243	**—,3-bromo-**	$C_{12}H_9Br$. *See* b2214	233.12	...	...	299–301 169–73[17]	...	1.6411^{20}	i	...	...	...	...	...	**B5**², 485
Ω b2244	**—,4-bromo-**	$C_{12}H_9Br$. *See* b2214	233.12	pl (al) λ^{al} 256 (4.41)	91.2	310	0.9327^{25}_4	...	i	s	s	...	s	CS_2, aa s	**B5**², 485
b2245	**—,4(bromo-acetyl)-**	*p*-Phenylphenacyl bromide. $C_{14}H_{11}BrO$. *See* b2214	275.15	nd (95 % al)	127	...	...	...	i	δ s^h	...	...	...	CCl_4 s^h peth s^h	**Am 52,** 3715
b2246	**—,3-bromo-4-hydroxy-**	$C_{12}H_9BrO$. *See* b2214	249.11	nd (chl-peth)	96	...	...	...	δ	v	...	...	...	chl v^h aa v	**B6**², 625
b2247	**—,4-bromo-4′-hydroxy-**	$C_{12}H_9BrO$. *See* b2214	249.11	pl (al)	164–6	...	...	...	δ	v	v	v	v	chl v	**B6**², 625
Ω b2248	**—,2-chloro-**	$C_{12}H_9Cl$. *See* b2214	188.66	mcl (dil al)	34	274 154[12]	$1.1499^{32.5}$	...	i	s	v	...	...	CCl_4 s lig v	**B5**², 483
b2249	**—,3-chloro-**	$C_{12}H_9Cl$. *See* b2214	188.66	...	16	284–5 150–60[6]	1.1579^{25}_4	1.6181^{25}	i	s	s	v	...	...	**B5**², 483
Ω b2250	**—,4-chloro-**	$C_{12}H_9Cl$. *See* b2214	188.66	lf (lig or al) λ^{al} 203 (4.61), 253 (4.32)	77.7	291 180–95[20–30]	...	...	i	s	s	...	...	lig s	**B5**², 483
b2251	**—,4-(chloro)-acetyl)-**	*p*-Phenylphenacyl chloride. $C_{14}H_{11}ClO$. *See* b2214	230.70	pl (al)	125	...	...	...	...	s^h	...	...	...	...	**B7**, 443
b2252	**—,3-chloro-2-hydroxy**	$C_{12}H_9ClO$. *See* b2214	204.66	...	6	317–8d	1.24^{25}_4	1.6237^{30}	i	s	s	s	s	os s	**C29,** 3688
b2253	**—,4-chloro-4′-hydroxy-**	$C_{12}H_9ClO$. *See* b2214	204.66	cr (dil al)	146–7	...	...	...	δ	s	s	s	s	peth δ	**B6**², 625
b2254	**—,2,2′-diaceta-mido-**	$C_{16}H_{16}N_2O_2$. *See* b2214	268.32	pr (al)	164.2–5.3	...	...	...	δ	v	δ	...	v	peth δ aa v	**B13**, 210
b2255	**—,2,4-diaceta-mido-**	$C_{16}H_{16}N_2O_2$. *See* b2214	268.32	nd (al)	202	...	...	...	...	s	...	...	...	...	**B13**, 212
b2256	**—,4,4′-diaceta-mido-**	$C_{16}H_{16}N_2O_2$. *See* b2214	268.32	nd (aa)	328.30	...	...	...	i	δ	δ	...	...	...	**B13**², 102
Ω b2257	**—,2,2′-diamino-**	*o*-Benzidine. *o,o′*-Bianiline. $C_{12}H_{12}N_2$. *See* b2214	184.24	mcl pr or nd (al) λ^{al} 295 (3.76)	81	162[4]	...	...	s v^h	...	...	...	s	...	**B13**², 87
b2258	**—,2,4′-diamino-**	Diphenyline. $C_{12}H_{12}N_2$. *See* b2214	184.24	nd (dil al) λ^{al} 235(4.2), 300(3.9)	54.5	363[760]	...	...	i	s	s	...	...	...	**B13**², 88
b2259	**—,3,3′-diamino-**	*m*-Benzidine. $C_{12}H_{12}N_2$. *See* b2214	184.24	nd (w), pr (bz) λ^{al} 228 (4.56), 302 (3.68), 314 sh (3.55)	93.5–4	205 –15[0.001]	...	...	δ s^h	...	v	...	s^h	...	**B13**², 90

For explanations, symbols and abbreviations see beginning of table. For structural formulas see end of table.

No.	Name	Synonyms and Formula	Mol. wt.	Color, crystalline form, specific rotation and λ_{max} (log ε)	m.p. °C	b.p. °C	Density	n_D	Solubility w	al	eth	ace	bz	other solvents	Ref.
	Biphenyl														
b2260	—,3,4-diamino-	$C_{12}H_{12}N_2$. *See* b2214	184.24	lf (eth or al)	103	...	...	...	...	s v^h	s	...	...	...	**B13²**, 89
Ω b2261	—,4,4′-diamino-	Benzidine. $C_{12}H_{12}N_2$. *See* b2214	184.24	nd (w) λ^{al} 287 (4.4)	(i) 125 (unst) (ii) 122 (unst) (iii) 128 (st)	400^{740}	...	...	δ^h	s	δ	...	...	...	**B13²**, 90
Ω b2262	—,4,4′-diamino-3,3′-dimethoxy-	*o*-Dianisidine. $C_{14}H_{16}N_2O_2$. *See* b2214	244.30	lf or nd (w)	137	...	...	...	i s^h	s	s	s	s	chl s	**B13²**, 502
b2263	—,4,4′-diamino-2,2′-dimethyl-	Bitoluidine. *m*-Tolidine. $C_{14}H_{16}N_2$. *See* b2214	212.30	pr (w) λ^{al} 248 (4.2), 290 sh (3.6)	108–9	...	...	...	s^h	v	v	...	...	...	**B13**, 255
Ω b2264	—,4,4′-diamino-3,3′-dimethyl-	*o*-Tolidine. $C_{14}H_{16}N_2$. *See* b2214	212.30	lf (dil al) λ^{al} 284 (4.4)	131–2	...	...	...	δ	v	v	...	...	...	**B13**, 256
b2265	—,4,4′-diamino-3-ethoxy-	$C_{14}H_{16}N_2O$. *See* b2214	228.30	nd (w)	134	...	...	...	δ	v^h	δ	...	δ	...	**B13²**, 419
Ω b2266	—,4,4′-dibromo-	$C_{12}H_8Br_2$. *See* b2214	312.03	mcl pr (MeOH) λ^{al} 262 (4.37)	164	355–60	...	...	i	δ^h	...	...	s	...	**B5²**, 485
Ω b2267	—,4,4′-dichloro-	$C_{12}H_8Cl_2$. *See* b2214	223.11	pr or nd (al or to-peth) λ^{al} 201 (4.63), 259 (4.40), 283 sh (3.95)	148–9	315–9	...	...	i	δ^h	...	...	s	...	**B5²**, 484
b2268	—,4,4′-dichloro-2,2′-dinitro-	$C_{12}H_6Cl_2N_2O_4$. *See* b2214	313.10	ye cr (al)	140	...	...	...	i	δ s^h	...	...	s	aa s^h lig δ	**B5²**, 584
b2269	—,4,4′-diethoxy-3,3′-diethyl-	$C_{20}H_{26}O_2$. *See* b2214	298.43	lf (al)	120	...	...	...	i	δ	...	...	...	...	**B6**, 1015
b2270	—,2,2′-diethoxy-3,3′-dimethyl-	$C_{18}H_{22}O_2$. *See* b2214	270.38	lf (al)	85	...	...	...	i	δ	...	...	...	...	**B6²**, 974
b2271	—,2,4′-diethoxy-3,3′-dimethyl-	$C_{18}H_{22}O_2$. *See* b2214	270.38	nd (al)	53	...	...	...	i	δ	...	...	...	...	**B6²**, 974
b2272	—,4,4′-diethoxy-3,3′-dimethyl-	$C_{18}H_{22}O_2$. *See* b2214	270.38	pl (al)	156	...	...	...	i	δ	δ	...	...	...	**B6**, 1010
b2273	—,3,3′-diethyl-6,6′-dihydroxy-	$C_{16}H_{18}O_2$. *See* b2214	242.32	nd (dil al)	131	...	...	...	i	δ	...	...	...	...	**B6²**, 981
Ω b2274	—,4,4′-difluoro-	$C_{12}H_8F_2$. *See* b2214	190.21	mcl pr (al) lf (w) λ^{al} 245 (4.18), 271.5 sh (3.60)	94–5	254–5 119^{14}	...	...	i s^h	v	s	s	v^h	chl v aa v^h MeOH v^h	**B5²**, 482
b2275	—,2,2′-dihydroxy-	*o,o′*-Biphenol. $C_{12}H_{10}O_2$. *See* b2214	186.21	lf (w + 1), pr (to) λ^{hp} 270 (3.7), 280 (3.65)	110–2 (anh) 71–3 (hyd)	325–6^{760}	...	...	s^h	s	s	s	s	peth δ aa s alks	**B6²**, 960
b2276	—,2,4′-dihydroxy-	*o,p′*-Biphenol. $C_{12}H_{10}O_2$. *See* b2214	186.21	mcl pr or nd (dil al)	162–3	342 206–10^{11}	...	...	δ^h	s	s	...	...	...	**B6²**, 961
b2277	—,2,5-dihydroxy-	Phenylhydroquinone. $C_{12}H_{10}O_2$. *See* b2214	186.21	nd (dil al)	97–8	...	...	...	i	s	...	...	...	...	**B6**, 989
b2278	—,3,3′-dihydroxy-	*m,m′*-Biphenol. $C_{12}H_{10}O_2$. *See* b2214	186.21	nd (w) λ^{al} 255 (4.08)	123–4	247^{18}	...	...	s^h δ	s	s	...	s	chl s	**B6²**, 961
b2279	—,3,4-dihydroxy-	4-Phenylpyrocatechol. $C_{12}H_{10}O_2$. *See* b2214	186.21	...	145	>360	...	...	δ^h	v	v	s	s	CS_2 δ chl v	**B6²**, 961
Ω b2280	—,4,4′-dihydroxy-	*p,p′*-Biphenol. $C_{12}H_{10}O_2$. *See* b2214	186.21	nd or pl (al) λ^{al} 210 sh (4.3), 268 (4.3)	274–5	sub	...	...	δ	s	s	...	δ	...	**B6²**, 962
b2281	—,4,4′-dihydroxy-3,3′-diethyl-5,5′-dimethyl-	$C_{18}H_{22}O_2$. *See* b2214	270.38	nd (aa)	148	...	...	...	...	v	...	v	v	peth δ aa v	**B6²**, 984
b2282	—,2,2′-dihydroxy-3,3′-dimethyl-	$C_{14}H_{14}O_2$. *See* b2214	214.27	nd (peth)	113	sub	...	...	...	v	v	...	v	peth δ	**B6²**, 974
b2283	—,2,2′-dihydroxy-5,5′-dimethyl-	$C_{14}H_{14}O_2$. *See* b2214	214.27	nd (bz or w) $\lambda^{?}$ 293 (3.91)	153.5	sub	...	...	δ^h	s	v	s	s	os s	**B6²**, 974
b2284	—,2,2′-dihydroxy-6,6′-dimethyl-	$C_{14}H_{14}O_2$. *See* b2214	214.27	pl (dil al)	164	...	...	...	i	v	...	...	...	...	**C35**, 2138
b2285	—,2,5′-dihydroxy-2′,5-dimethyl-	$C_{14}H_{14}O_2$. *See* b2214	214.27	...	158	...	...	...	i	v	v	...	δ δ^h	...	**B6²**, 973
b2286	—,4,4′-dihydroxy-3,3′-dimethyl-	$C_{14}H_{14}O_2$. *See* b2214	214.27	lf (w), nd λ^{al} 265 (4.3)	161	...	...	...	δ^h	v	v	...	...	...	**B6²**, 974
b2287	—,—,diacetate	$C_{18}H_{18}O_4$. *See* b2214	298.34	wh nd (al or aa)	135.5	...	...	...	i	δ	...	...	...	...	**B6¹**, 492
b2288	—,5,5′-dihydroxy-2,2-dimethyl-	$C_{14}H_{14}O_2$. *See* b2214	214.27	pr (al)	229	...	...	...	i δ^h	v	v	...	δ^h	peth i	**B6²**, 973
b2289	—,4,4′-dihydroxy-3,3′,5,5′-tetra-methoxy-	Hydrocerulignone. $C_{16}H_{18}O_6$. *See* b2214	306.32	mcl pr (al)	190	...	...	...	δ	s	δ	...	...	CS_2 i	**B6¹**, 593
b2290	—,2,2′-dihydroxy-3,3′,5,5′-tetra-methyl-	$C_{16}H_{18}O_2$. *See* b2214	242.32	nd or pl (eth or lig)	137.5–8.5	140–$60^{0.05}$	...	...	i	v	v	...	...	peth δ^h lig v^h	**B6²**, 981
b2291	—,4,4′-dihydroxy-3,3′,5,5′-tetra-methyl-	$C_{16}H_{18}O_2$. *See* b2214	242.32	pa ye nd or pr (aa)	222–3	sub	...	...	...	s^h	...	...	δ^h	aa s^h	**B6**, 1015

For explanations, symbols and abbreviations see beginning of table. For structural formulas see end of table.

No.	Name	Synonyms and Formula	Mol. wt.	Color, crystalline form, specific rotation and λ_{max} (log ε)	m.p. °C	b.p. °C	Density	n_D	Solubility						Ref.
									w	al	eth	ace	bz	other solvents	
	Biphenyl														
b2292	—,**4,4′-dihydroxy-3,3′,5,5′-tetra-nitro-**	$C_{12}H_6N_4O_{10}$. *See* b2214	366.20	ye nd λ^{MeOH} 355 (3.8)	223	....	....	....	...	...	...	...	...	....	**B6**^{2}, 963
b2293	—,**2,2′-dimethoxy-**	*o,o′*-Bianisole. $C_{14}H_{14}O_2$. *See* b2214	214.27	rh bipyr pr (al) λ 278 (3.8)	155	$307–8^{766}$	1.268	....	i	v^h	δ^h	...	v^h	CCl_4, chl, CS_2 v^h lig δ^h	**B6**^{2}, 960
Ω b2294	—,**3,3′-dimethoxy-**	*m,m′*-Bianisole. $C_{14}H_{14}O_2$. *See* b2214	214.27	nd (dil al) λ^{cy} 258 (4.08), 290 (3.81)	36	328 $211–20^{15}$	....	....	i	v	v	s	v	CS_2 v aa v chl v	**B6**^{2}, 961
b2295	—,**4,4′-dimethoxy-**	*p,p′*-Bianisole. $C_{14}H_{14}O_2$. *See* b2214	214.27	lf (bz) λ^{hx} 263 (4.34)	173	sub	....	....	i	v^h	δ	...	v	peth i chl v	**B6**^{2}, 962
b2296	—,**2,2′-dimethoxy-5,5′-dimethyl-**	$C_{16}H_{18}O_2$. *See* b2214........	242.32	nd (dil al)	71	188^{12}	....	....	δ^h	s	s	s	s	os v	**B6**^{2}, 975
b2297	—,**2,5′-dimethoxy-2′,5-dimethyl-**	$C_{16}H_{18}O_2$. *See* b2214........	242.32	pr (al)	86	168^4	....	....	i	δ v^h	...	...	s	peth s	**B6**^{2}, 973
b2298	—,**4,4′-dimethoxy-3,3′-dimethyl-**	$C_{16}H_{18}O_2$. *See* b2214........	242.32	pr (al)	145.5	....	....	....	i	δ s^h	δ	...	...	....	**B6**, 1009
Ω b2299	—,**2,2′-dimethyl-**...	*o,o′*-Bitolyl. $C_{14}H_{14}$. *See* b2214	182.27	cr (al) λ^{peth} 228 sh (3.78), 256 sh (2.89), 266.5 sh (2.83)	19.5–20.2	256^{760}	0.9906^{20}	1.5752^{20}	i	v	v	s	v	....	**B5**^{2}, 512
b2300	—,**2,3′-dimethyl-**...	*o,m′*-Bitolyl. $C_{14}H_{14}$. *See* b2214	182.27	λ^{peth} 238 (3.98), 271 sh (3.04)	....	270^{760}	0.9924^{20}	1.5810^{20}	i	v	v	s	v	....	**B5**^{2}, 512
b2301	—,**2,4′-dimethyl-**...	*o,p′*-Bitolyl. $C_{14}H_{14}$. *See* b2214	182.27	λ^{peth} 239 (4.07), 275 (3.01)	....	$273–6^{760}$ $137^{12.5}$	0.9924^{20}	1.5826^{20}	i	v	v	s	v	....	**B5**^{2}, 512
Ω b2302	—,**3,3′-dimethyl-**...	*m,m′*-Bitolyl. $C_{14}H_{14}$. *See* b2214	182.27	λ^{peth} 208 (4.69), 250.5 (4.21)	9–9.5	280^{760} 150^{18}	0.9993_4^{20}	1.5946^{20}	i	v	v	s	v	....	**B5**, 513
Ω b2303	—,**4,4′-dimethyl-**...	*p,p′*-Bitolyl. $C_{14}H_{14}$. *See* b2214	182.27	mcl pr (eth) λ^{peth} 203 (4.64), 254.5 (4.32), 274 sh (4.00)	125 (121.5)	295^{760}	0.917_4^{121}	....	i	δ	s	s	s	CS_2 s	**B5**^{2}, 514
b2304	—,**3,3′-dimethyl-4,4′-dipropoxy-**	$C_{20}H_{26}O_2$. *See* b2214........	298.43	lf	115	....	....	....	i	δ v^h	...	...	...	....	**B6**, 1010
Ω b2305	—,**2,2′-dinitro-**	$C_{12}H_8N_2O_4$. *See* b2214	244.22	ye mcl pr or nd (al) λ^{iso} 253 (4.1)	127–8	....	1.45 (so)	....	i	v^h	s	...	s^h	aa s^h lig δ	**B5**^{2}, 490
b2306	—,**2,4′-dinitro-**	$C_{12}H_8N_2O_4$. *See* b2214	244.22	mcl pr (al) λ^{iso} 265 (4.2)	93–4	....	1.474	....	i	v^h	s	...	s^h	aa s^h	**B5**^{2}, 491
b2307	—,**3,3′-dinitro-**	$C_{12}H_8N_2O_4$. *See* b2214	244.22	ye-og nd (al or aa) λ^{al} 244 (4.48), 300 sh (3.46)	200	....	....	....	i	δ	δ	...	s^h	aa s^h	**B5**^{2}, 491
Ω b2308	—,**4,4′-dinitro-**	$C_{12}H_8N_2O_4$. *See* b2214	244.22	nd (al) λ^{al} 306 (4.40)	240–3	....	....	....	i	δ s^h	...	...	s^h	aa s^h	**B5**^{2}, 491
—	—,**diphenyl-**.......	*see* **Quaterphenyl**													
—	—,**dodecahydro-**	*see* **Bicyclohexyl**													
b2310	—,**2-ethoxy-**	$C_{14}H_{14}O$. *See* b2214	198.27	pr (peth)	34	276 132^6	....	....	...	s	s	s	s	chl s	**B6**, 672
b2311	—,**3-ethoxy-**	$C_{14}H_{14}O$. *See* b2214........	198.27	(peth)	35	305 158^8	....	....	...	s	s	s	s	os s	**B6**, 674
Ω b2312	—,**2-hydroxy-**.....	*o*-Phenylphenol. $C_{12}H_{10}O$. *See* b2214	170.21	nd (peth) λ^{MeOH} 247 (4.1), 286 (3.7)	58–60 (67–5)	286^{760} (275) 145^{14}	1.213_4^{25}	....	i	s	v	s	s	lig s Py v, ∞^{h}	**B6**^{2}, 623
Ω b2313	—,**3-hydroxy-**.....	*m*-Phenylphenol. $C_{12}H_{10}O$. *See* b2214	170.21	nd (w or peth) λ^{al} 250 (4.2) 285 (3.5)	78	> 300	....	....	δ	v	v	...	v	peth s Py v	**B6**^{2}, 624
Ω b2314	—,**4-hydroxy-**.....	*p*-Phenylphenol. $C_{12}H_{10}O$. *See* b2214	170.21	nd or pl (dil al) λ^{al} 260 (4.24)	165–7	305–8 sub	....	....	δ	v	v	...	...	peth δ chl v Py v	**B6**^{2}, 624
b2315	—,**2-hydroxy-2′-methoxy-5,5′-dimethyl-**	$C_{15}H_{16}O_2$. *See* b2214........	228.29	....	....	205^{12}	....	....	...	v	v	...	v	chl v	**B6**^{2}, 974
b2316	—,**2-iodo-**	$C_{12}H_9I$. *See* b2214	280.11	λ^{hx} 226 (4.3)	....	$189–92^{36}$ 158^6	1.6038_{25}^{25}	1.6620^{20}	i	s^h	s	...	s	aa s	**B5**^{2}, 486
b2317	—,**4-iodo-**	$C_{12}H_9I$. *See* b2214	280.11	nd (al or aa) λ^{al} 258.6 (4.41)	113–4	320 δd 183^{11}	....	....	i	s^h	s	...	s	aa s	**B5**^{2}, 486
Ω b2318	—,**2-methoxy-**.....	$C_{12}H_{12}O$. *See* b2214........	184.24	pr (peth) λ^{al} 264 (4.11), 285 (3.74)	29	274 150^{13}	1.0233_4^{99}	1.5641^{99}	i	s^h	...	...	...	peth s	**B6**^{2}, 623
b2319	—,**4-methoxy-**.....	$C_{13}H_{12}O$. *See* b2214........	184.24	pl (al) λ^{al} 261 (4.32)	90	157^{10}	1.0278_4^{100}	1.5744^{100}	i	s^h	s	...	...	....	**B6**^{2}, 625
Ω b2320	—,**2-methyl-**	$C_{13}H_{12}$. *See* b2214	168.24	λ^{al} 237 (4.00)	−0.2	255.3^{760} $130–6^{27}$	1.010_4^{22}	1.5914^{20}	i	s	s	...	...	....	**B5**^{2}, 504
Ω b2321	—,**3-methyl-**	$C_{13}H_{12}$. *See* b2214	168.24	λ^{iso} 248 (4.2)	4.53	272.7^{760} $148–50^{20}$	1.0182_4^{17}	1.5972^{20}	i	s	s	...	...	....	**B5**^{2}, 504
Ω b2322	—,**4-methyl-**	$C_{13}H_{12}$. *See* b2214	168.24	pl (lig or MeOH) λ^{iso} 250 (4.22)	49–50	267–8 $134–6^{15}$	1.015^{27}	....	i	s	s	...	...	os v	**B5**^{2}, 504
Ω b2323	—,**2-nitro-**	$C_{12}H_9NO_2$. *See* b2214......	199.21	pl (al or MeOH) λ^{iso} 231 (4.2)	37.2	320 201^{30}	1.44	....	i	s^h	s	...	...	....	**B5**^{2}, 487

For explanations, symbols and abbreviations see beginning of table. For structural formulas see end of table.

No.	Name	Synonyms and Formula	Mol. wt.	Color, crystalline form, specific rotation and λ_{max} (log ε)	m.p. °C	b.p. °C	Density	n_D	Solubility w	al	eth	ace	bz	other solvents	Ref.
	Biphenyl														
Ω b2324	—,**3-nitro-**	$C_{12}H_9NO_2$. *See* b2214	199.21	ye pl or nd (dil al) λ^{iso} 245 (4.4), 303 (3.3)	62	$225–30^{35}$			i	s	s	...	...	aa s lig s	**B5**[2], 487
Ω b2325	—,**4-nitro-**	$C_{12}H_9NO_2$. *See* b2214	199.21	ye nd (al) λ^{iso} 295 (4.2)	114–4.5	340^{760} 224^{30}			i	δ v^h	s	...	s	chl s aa s	**B5**[2], 487
b2326	—,**3,3′,5,5′-tetrahydroxy-**	Diresorcinol. $C_{12}H_{10}O_4$. *See* b2214	218.21	pl or nd (w + 2)	310 (anh)				s^h	s	s	i	i	aa i	**B6**[2], 1129
b2327	—,**2,2′,4,4′-tetranitro-**	$C_{12}H_6N_4O_8$. *See* b2214	334.20	ye pr (bz)	165–6				i	δ	δ	...	v	aa v	**B5**[2], 494
b2328	—,**2,4,4′-triamino-**	$C_{12}H_{13}N_3$. *See* b2214	199.26	nd	134				...	...	...	...	...		**B13**[2], 149
b2329	**2-Biphenylcarboxylic acid**		198.22	lf (dil al) λ^{al} 245 (4.07)	113.5–4.5	$343–4^{760}$ 199^{10}			i	v	...	...	v	aa v	**B9**[2], 463
b2330	—,nitrile		179.22	nd	41 (32)	$170–2^{15}$ 151^5			...	s	v	...	...		**B9**[2], 463
Ω b2331	**3-Biphenylcarboxylic acid**		198.22	lf (al) λ^{al} 231 (4.32), 250 sh (4.10)	165–6 (161)				δ	s	s	...	s	aa s lig s	**B9**[2], 464
Ω b2332	**4-Biphenylcarboxylic acid**		198.22	nd (bz or al)	228	sub			δ^h i	s	s	...	s		**B9**[2], 464
b2333	—,nitrile		179.22	cr (al or peth) λ^{al} 264.5 (4.42)	88				i	v	v	...	...		**B9**[2], 464
b2334	**3-Biphenylcarboxylic acid, 2-hydroxy-**	$C_{13}H_{10}O_3$. *See* b2331	214.22	cr (bz)	186–7				...	s^h	...	...	...		**B10**, 341
Ω b2335	**2,2′-Biphenyldicarboxylic acid**	Diphenic acid.	242.23	mcl pr or lf (w), cr (aa) $\lambda^{w, pH=1.5}$ 284 (3.5), $\lambda^{w, pH=6}$ 275 (3.4)	233.5 (229)	sub			i δ^h	s	s	...	...	os s	**B9**[2], 655
b2336	—,anhydride	$C_{14}H_8O_3$. *See* b2335	224.23	nd (aa or bz)	217	sub			i	...	δ	...	...		**B17**[2], 497
b2337	—,dichloride	$C_{14}H_8Cl_2O_2$. *See* b2335	279.14		94 (97)				...	v	s	...	v	aa v	**B9**[2], 657
b2338	—,diethyl ester	$C_{18}H_{18}O_4$. *See* b2335	298.34		42				i	...	s	...	...		**B9**[2], 656
Ω b2339	—,dimethyl ester	$C_{16}H_{14}O_4$. *See* b2335	270.29	mcl pr (MeOH)	74	$204–6^{14}$			i	v	s	...	s		**B9**[2], 656
b2340	—,imide	$C_{14}H_9NO_2$. *See* b2335	223.23	nd (al)	219–20				δ^h	δ s^h	δ	...	...	chl v	**B21**[2], 392
b2341	**2,3′-Biphenyldicarboxylic acid**	Isodiphenic acid	242.22	nd (w or dil aa)	216				δ^h	v	...	...	...		**B9**[2], 663
b2342	**2,4′-Biphenyldicarboxylic acid**		242.23	lf (al)	272–3				δ	v^h	...	...	s	aa v	**B9**[2], 663
b2343	**3,3′-Biphenyldicarboxylic acid**		242.23	lf (al)	356–7				δ^h	δ s^h	...	...	i	aa δ chl s lig i	**B9**[2], 664
b2344	—,dimethyl ester	$C_{16}H_{14}O_4$. *See* b2343	270.29	lf (MeOH)	104				i	v	v	...	v	lig δ	**B9**, 927
b2345	**3,4′-Biphenyldicarboxylic acid**		242.23	nd ($PhNO_2$)	334–5				i	δ	...	δ	...	$PhNO_2$ s	**B9**, 927
b2346	—,dimethyl ester	$C_{16}H_{14}O_4$. *See* b2345	270.29	nd (MeOH or lig)	98.5–9.5				i	...	...	...	...	MeOH s lig s	**B9**, 927
b2347	**3,5-Biphenyldicarboxylic acid**		242.23	lf (aa)	>310				i	v	v	v	v	aa s^h	**B9**, 926
b2348	—,dimethyl ester	$C_{16}H_{14}O_4$. *See* b2347	270.29	lf (MeOH)	214				i	δ	...	...	δ		**B9**[2], 665
b2349	**2,2′-Biphenyldicarboxylic acid, 3,3′-dimethyl-**	$C_{16}H_{14}O_4$. *See* b2325	270.29		230				...	s	s	...	s		**B9**[1], 407
b2350	—,**4-nitro-**	$C_{14}H_9NO_6$. *See* b2335	287.23	lf, pr or wh nd (w)	217 (250)				i s^h	v	v	...	δ	lig δ	**B9**[2], 659
b2351	—,**5-nitro-**	$C_{14}H_9NO_6$. *See* b2335	287.23	lf (w or dil al)	267				δ v^h	v	v	...	s	MeOH v	**B9**[2], 659
b2352	—,**6-nitro-(*dl*)**	$C_{14}H_9NO_6$. *See* b2335	287.23	lf (w)	248–50 d				i v^h	s	s	s	δ	chl δ lig δ aa v	**B9**[2], 659
Ω b2353	—,**2,2′-Biphenyldisulfonic acid,** dichloride		351.24	pr (chl), cr (aa)	142–4				d^h	d^h	v	...	v	chl v	**B11**[2], 123
b2354	**3,3-Biphenyldisulfonic acid,** diamide		312.37	nd (ace)	285				...	s	...	...	...	MeOH s	**B11**, 219
b2355	—,dichloride		351.24	nd (chl)	128				d^h	d^h	s	...	s	AcOEt s	**B11**, 219
b2356	**4,4′-Biphenyldisulfonic acid**		314.34	pr	72.5	>200			s	...	...	...	...		**B11**, 219
b2357	—,diamide	$C_{12}H_{12}N_2O_4S_2$. *See* b2356	312.37	nd (w)	300				v^h	δ	s	...	δ	CS_2 s	**B11**, 220
b2358	—,dichloride	$C_{12}H_8Cl_2O_4S_2$. *See* b2356	351.24	pr (aa)	205–7				d^h	d^h	s	...	s	CS_2 δ aa s	**B11**[2], 124
b2359	**2,2′-Biphenyldisulfonic acid, 4,4′-diamino-**	$C_{12}H_{12}N_2O_6S_2$. *See* b2353	344.37	lf	175d				δ	i	i	...	...		**B14**[1], 743
b2360	—,**4,4′-diamino-5,5′-dimethyl-**	$C_{14}H_{16}N_2O_6S_2$. *See* b2359	372.42	nd (w + 1½)					s^h	i	i	...	...	aa i	**B14**, 796
Ω b2361	**2,2′-Bipyridyl**		156.20	pr (peth) λ^{MeOH} 235 (4.01), 280 (3.85)	71–3	$272–5^{760}$			δ	v	v	...	v	chl v lig v	**B23**[2], 211

For explanations, symbols and abbreviations see beginning of table. For structural formulas see end of table.

No.	Name	Synonyms and Formula	Mol. wt.	Color, crystalline form, specific rotation and λ_{max} (log ε)	m.p. °C	b.p. °C	Density	n_D	Solubility w	al	eth	ace	bz	other solvents	Ref.
	2,3′-Bipyridyl														
b2362	**2,3′-Bipyridyl**	Isonicoteine	156.20	λ^{al} 238 (4.15), 275 (4.07)	...	295.5–6.5 $122–5^{1}$	1.140_4^{20}	1.6223^{20}	i	v	v	...	v	peth δ chl v	**B23**², 212
Ω b2363	**2,4′-Bipyridyl**	...	156.20	...	61.5	280–2 $148–50^{1}$	...	...	δ	v	v	...	...	chl v	**B23**, 200
b2364	**3,3′-Bipyridyl**	...	156.20	λ^{al} 225 (3.64), 274 (2.89)	68	$291–2^{760}$ $190–2^{25}$	1.1635_{20}^{20}	...	v	v	δ	...	...	...	**B23**², 212
b2365	**3,4′-Bipyridyl**	...	156.20	lf (peth)	62	297	...	...	v	s	...	...	...	peth s	**B23**², 212
b2366	**4,4′-Bipyridyl**	...	156.20	nd (w + 2) λ^{w} 238 (4.2), 264 sh (3.9)	114 (171–2) (anh) 73(+2w)	305^{760} sub	...	...	δ s^h	v	s	...	v	chl v lig v	**B23**², 212
b2367	**2,2′-Biquinolyl**	...	256.31	pl or lf (al) λ^{al} 258 (4.87), 302 (4.30), 327 (4.37)	196	...	...	...	i	v^h	s	s	s	os s	**B23**², 267
b2368	**2,3′-Biquinolyl**	...	256.31	lf (al), ye pl or nd (bz)	176–7	>400	...	...	i	s^h	s	...	s^h	chl s^h	**B23**², 267
b2369	**2,6′-Biquinolyl**	...	256.31	pl (al)	144	...	...	...	i	s	δ	s	s	...	**B23**, 294
b2370	**2,7′-Biquinolyl** (higher melting)	...	256.31	mcl pl (al)	160	...	...	...	i	s^h	δ	...	δ	...	**B23**, 294
b2371	—,(lower melting)	$C_{18}H_{12}N_2$. *See* b2370	256.31	tcl	115	...	...	...	...	v	s	...	v	peth δ	**B23**, 294
b2372	**3,4′-Biquinolyl**	...	256.31	pw (peth)	83–4	...	...	...	...	s	...	...	s	peth δ	**B23**², 267
b2373	**3,7′-Biquinolyl**	...	256.31	lf or nd (al or bz)	190	...	...	...	i	s^h	δ	...	s	chl s	**B23**², 268
b2374	**4,4′-Biquinolyl**	...	256.31	pr (peth)	171	...	...	...	i	s	...	s	s	peth δ	**B23**², 268
b2375	**4,6′-Biquinolyl**	...	256.31	cr (bz)	122	...	...	...	i	v	i	...	v	AcOEt v chl v	**B23**, 294
b2376	**6,6′-Biquinolyl**	...	256.31	lf (al)	181	...	...	...	i	s^h	s	...	s	...	**B23**, 295
b2377	**6,8′-Biquinolyl**	...	256.31	lf (al)	148	...	...	...	i	s^h	δ	...	v^h	...	**B23**, 296
b2378	**8,8′-Biquinolyl**	...	256.31	lf or pl (al or aa) λ^{al} 230 (4.92), 293 (4.00), 305 (4.00), 318 sh (3.95)	205–7	...	...	...	i	s^h	δ	s	s	chl v lig δ CCl_4 s	**B23**², 268
b2379	**Biquinone, 3,3′-dihydroxy-5,5′-dimethyl-**	Phenicin	274.23	yesh-br (al) λ^{chl} 268 (4.52), 406 (3.36)	230–1	...	...	...	δ	v^h	...	...	...	chl v aa v	**C54**, 397
—	**Bismark brown**	*see* **Azobenzene, 2,4,3′-triamino-**													
b2380	**Bistibine. tetramethyl-**	$(CH_3)_2SbSb(CH_3)_2$	303.64	br-red nd	17.5	190^{760}	...	...	i	...	s	...	s	...	**J1935**, 366
—	**Bistyryl**	*see* **1,3-Butadiene, 1,4-diphenyl-***													
Ω b2381	**2,2′-Bithiophene**	2,2′-Bithienyl	166.27	lf (al) λ^{al} 246 (3.78), 301 (4.11)	33	260^{760} 103^{3}	...	...	i	v	s	...	...	aa s	**B19**², 26
b2382	—,hexabromo-	$C_8Br_6S_2$. *See* b2381	639.67	nd (bz)	257–8	...	...	...	i	...	...	...	s^h	...	**B19**, 33
—	**Bitolyl**	*see* **Biphenyl, dimethyl-**													
Ω b2383	**Biuret**	Allophanamide. Carbamoylurea. $H_2NCONHCONH_2$	103.08	pl (al), nd (w + 1) $\lambda^{w, pH=13}$ 216 (3.75)	190d (193d)	...	...	...	δ v^h	v	i	...	...	...	**B3**², 60
b2384	—,acetyl-	$CH_3CONHCONHCONH_2$	145.12	nd (w or al)	193–4	...	...	...	s	s	δ	i	δ	...	**B3**¹, 33
b2385	—,1,5-diamino-	$H_2NNHCONHCONHNH_2$	133.11	pr (dil al), nd (aa)	199–200d	...	...	...	v	δ	δ	...	...	aa v^h	**B3**, 101
b2386	**Bixin**	$C_{25}H_{30}O_4$. CH_3O_2CCH:-[CHC(CH_3):CHCH:]$_4$-$CHCO_2H$	394.52	vt pr (ace) $\lambda^{95\%al}$ 448 (3.92), 480 (3.86)	198	...	...	...	i	s^h	δ	s^h	δ	aa δ chl s	**C54**, 2406
b2387	**Boric acid, tributyl ester**	Tributoxyborine. $B(OC_4H_9^n)_3$	230.16	oil	...	230–1 $114–5^{15}$	0.8567_4^{20}	1.4106^{18}	d	s	v	...	s	MeOH v	**B1**², 398
b2388	—,triethyl ester	Triethoxyborine. $B(OC_2H_5)_3$	146.00	...	...	120^{760}	0.8546_4^{28}	1.3749^{20}	d	∞	∞	...	...	...	**B1**², 333
b2389	—,triisopropyl ester	Triisopropoxyborine. $B(OC_3H_7^i)_3$	188.08	...	...	104^{760} $52–3^{32}$	0.8251_4^{20}	1.3772^{20}	d	v	v	...	v	*i*-PrOH v	**B1**², 382
b2390	—,trimethyl ester	Trimethoxyborine. $B(OCH_3)_3$	103.92	...	−29.3	67–9	0.915^{20}	1.3568^{20}	d	v	∞	...	s	MeOH ∞	**B1**³, 1210
b2391	—,tripropyl ester	Tripropylborine. $B(OC_3H_7^n)_3$	188.08	...	...	$179–80^{760}$ 64^{9}	0.8576_4^{20}	1.3948^{20}	d	v	∞	...	...	PrOH s	**B1**², 369
b2392	—,tris(3-methylbutyl) ester	Triisoamyl borate. $B[OCH_2CH_2CH(CH_3)_2]_3$	272.24	...	...	254–5 $132–3^{12}$	0.8518_4^{20}	1.4156^{20}	d	∞	∞	...	...	...	**B1**³, 1645
b2393	—,trithio-, trimethyl ester	Tris(methylthio)borine. $B(SCH_3)_3$	152.11	...	5	218.2^{760}	1.126^{20}	1.5788^{20}	d	d	s	s	...	os ∞	**C55**, 9346
b2394	**Borine, bis(dimethylamino)-fluoro-**	$[(CH_3)_2N]_2BF$	117.96	...	−44.3	106^{760}	...	...	d	d	s	s	...	os ∞	**Am 76**, 3905
b2395	—,bis(methylthio)methyl-	$(CH_3S)_2BCH_3$	120.05	...	−59	100^{147}	...	...	d	d	v	v	...	os ∞	**Am 78**, 1523
b2396	—,difluoro(dimethylamino)-	$(CH_3)_2NBF_2$	92.89	rh	165–8d	sub 132	...	...	d	...	...	...	...	...	**Am 76**, 3904

For explanations, symbols and abbreviations see beginning of table. For structural formulas see end of table.

No.	Name	Synonyms and Formula	Mol. wt.	Color, crystalline form, specific rotation and λ_{max} (log ε)	m.p. °C	b.p. °C	Density	n_D	Solubility w	al	eth	ace	bz	other solvents	Ref.
	Boric acid														
b2397	—,**difluoro(phenyl)-**	$C_6H_5BF_2$	125.92	...	−36.2	$97–8^{747}$	1.087^{25}	1.4441^{25}	d		s	...	s	...	**B16**[2], 638
b2398	—,**difluoro(4-tolyl)-**	...	139.94	...	...	127.8^{747}	1.055^{25}	1.4535^{25}	d		s	...	s	...	**C18,** 992
b2399	—,**dimethylarsino-**, (tetramer)	$[(CH_3)_2AsBH_2]_4$	471.28	...	150	185^{11}	...	...	i	s	v	v	...	os s	**Am 76,** 389
b2400	—,—,(trimer)	$[(CH_3)_2AsBH_2]_3$	353.46	...	50	146^{26}	...	...	i	s	v	v	...	os s	**Am 76,** 386
b2401	—,**dimethyl(dimethylamino)-**	$(CH_3)_2NB(CH_3)_2$	84.99	...	−92	65^{760}	...	...	...	...	v	v	...	os ∞	**Am 78,** 1521
b2402	—,**dimethyl)(dimethylphosphino)-**, (trimer)	$[(CH_3)_2PB(CH_3)_2]_3$	305.78	...	333	200^{10}	...	...	i	s					**C54,** 9767
b2403	—,**dimethyl-(methoxy)-**	$CH_3OB(CH_3)_2$	71.92	...	...	21^{760}	...	...	d	d	v	v	...	os ∞	**Am 75,** 3872
b2404	—,**dimethyl-(methylthio)-**	$CH_3SB(CH_3)_2$	87.98	...	−84	71^{760}	...	...	d	d	v	v	...	os ∞	**Am 76,** 3307
b2405	—,**(dimethylphosphino)-**, (tetramer)	$[(CH_3)_2PBH_2]_4$	295.48	nd	161	190^{29}	...	...	i	s	v	v	...	os s	**C54,** 4166
b2406	—,—,(trimer)	$[(CH_3)_2PBH_2]_3$	221.61	rh	86	90^{4}	...	...	i	s	v	v	...	os s	**C53,** 16062
b2407	—,**methylthio-**, (polymer)	$(CH_3SBH_2)_n$	...	...	...	...	...	...	d	d	v	v	...	os ∞	**C55,** 27015
b2408	—,**triethyl-**	$B(C_2H_5)_3$	98.00	...	−92.9	$95–6^{760}$	0.6961^{23}	...	δ	s	s	...	...	...	**B4**[2], 1022
b2409	—,**triisobutyl-**	$B(C_4H_9^i)_3$	182.16	...	...	188^{766} 86^{20}	0.7380^{25}_{4}	1.4188^{23}	δ	s	s	...	s	...	**B4**[2], 1022
b2410	—,**trimethyl-**	$B(CH_3)_3$	55.92	λ^{hx} 260 (3.78)	−161.5	20	...	...	δ	v	v	...	...	...	**B4**[2], 1022
b2411	—,**triphenyl-**	$B(C_6H_5)_3$	242.13	wh cr λ^{hp} 240 (4.04), 270 (3.35)	142	$245–50^{15}$	...	...	i	d	δ	...	s	lig s	**B16**[2], 636
b2412	—,**tripropyl-**	$B(C_3H_7^n)_3$	140.08	...	−56	159^{760} $43–4^{17}$	0.7204^{25}_{4}	$1.4135^{22.5}$	...	...	...	...	...	...	**B4**[2], 1022
b2413	—,**tris(3-methylbutyl)-**	$B[CH_2CH_2CH(CH_3)_2]_3$	224.24	...	...	119^{14}	0.7600^{23}_{4}	1.4321	v	s	s	v	...	os v	**B4**[2], 1023
—	—,**tris(methylthio)-**	*see* **Boric acid, trithio-,** trimethyl ester													
b2414	**Borneol** (*d*)	$C_{10}H_{18}O$	154.26	lf or hex pl (peth) $[\alpha]^{20}_{D}$ +37.7 (al)	208	212^{760} sub	1.011^{20}_{4}	...	δ	s	s	...	v	lig s	**B6**[2], 82
Ω b2415	—(*dl*)	$C_{10}H_{18}O$. *See* b2414	154.26	lf (lig)	210.5	sub	1.011^{20}_{4}	...	i	v	v	...	v	...	**B6**[2], 82
b2416	—(*l*)	$C_{10}H_{18}O$. *See* b2414	154.26	hex pl $[\alpha]^{20}_{D}$ −37.74 (al)	208.6	210^{779}	1.1011^{20}_{4}	...	i	s	v	...	v	...	**B6**[2], 82
b2417	—,acetate (*d*)	$C_{12}H_{20}O_2$. *See* b2414	196.29	rh, $[\alpha]^{20}_{D}$ +44.4 (al)	29	$223–4^{760}$ 107^{15}	0.9920^{20}_{4}	...	...	...	...	...	...	...	**B6**[2], 83
Ω b2418	—,—(*dl*)	$C_{12}H_{20}$. *See* b2414	196.29	...	< -17	$223–4^{760}$	0.9838^{20}_{4}	1.4630^{20}	...	...	...	...	...	...	**B6**[2], 86
b2419	—,—(*l*)	$C_{12}H_{20}O_2$. *See* b2414	196.29	$[\alpha]^{20}_{D}$ −44.45 (undil)	29	$223–4^{760}$ 107^{15}	0.9920^{20}_{4}	1.4634^{20}	δ	s	s	...	...	...	**B6**[2], 85
Ω b2420	—,formate(*d*)	$C_{11}H_{18}O_2$. *See* b2414	182.27	$[\alpha]_D$ +48.75 (undil)	...	90^{10}	1.009^{22}	1.4708^{15}	...	...	...	...	...	...	**B6**[2], 85
b2421	**Bornyl amine**(*d*)	...	153.27	$[\alpha]^{20}_{D}$ +47.2 (al)	163	200 sub	...	...	i	v	v	v	s	os v	**B12**[2], 39
b2422	**Bornyl chloride**(*d*)	...	172.70	nd $[\alpha]_D$ +34.5 (eth), $[\alpha]^{14}_{D}$ +31 (to)	132	207–8 sub	...	...	i	s	s	...	s	peth s	**B5**[2], 62
b2423	**Bornylene**(*d*)	2-Bornene	136.24	cr (al) $[\alpha]_D$ +30.5 (to)	109–10	146^{750} sub	...	...	...	...	...	...	...	...	**B5**[1], 80
b2424	—(*l*)	$C_{10}H_{16}$. *See* b2423	136.24	cr (al) $[\alpha]^{19}_{D}$ −23.9 (bz)	113	146^{746}	...	...	i	s	s	...	s	...	**B5**[2], 105
—	**Bourbonal**	*see* **Benzaldehyde, 3-ethoxy-4-hydroxy-**													
—	**Brassidic acid**	*see* **13-Docosenoic acid** (*trans*)*													
b2425	**Brazilein**	...	284.27	red-br nd or lf (w + 1)	250	...	...	...	s[h]	s	δ	...	s	chl s, aa s	**B18**[2], 194
—	**British antilewisite**	*see* **1-Propanol,2,3-dimercapto-***													
—	**Bromal**	*see* **Acetaldehyde, tribromo-**													
Ω b2426	**Bromocresol green**	3,3′,5,5′-Tetrabromo-*m*-cresolsulfonphthalein	698.04	wh or red (+7w), ye (aa) $\lambda^{0.01N\ NaOH}$ 613 (4.66)	218–9	...	...	...	δ	v	v	...	s	AcOEt v, aa s	**B19**[2], 108
—	**Bromoform**	*see* **Methane, tribromo-***													
Ω b2427	**Bromophenol blue**	3,3′,5,5′-Tetrabromophenolsulfonphthalein	670.02	hex pr (aa–ace) $\lambda^{w, pH=5.2}$ 440 (3.48), 600 (4.82) $\lambda^{0.01NaOH}$ 595 (4.82)	279d	...	...	...	δ	s	...	...	s	aa s	**B19**[2], 105
	romopicrin	*see* **Methane, nitrotribromo-***													
—	**Bromoprene**	*see* **1,3-butadiene, 2-bromo-***													

For explanations, symbols and abbreviations see beginning of table. For structural formulas see end of table.

No.	Name	Synonyms and Formula	Mol. wt.	Color, crystalline form, specific rotation and λ_{max} (log ε)	m.p. °C	b.p. °C	Density	n_D	Solubility w	al	eth	ace	bz	other solvents	Ref.
	Bromural														
—	**Bromural**	*see* **Urea, 1(2-bromo-3-methylbutanoyl)-**													
Ω b2428	**Brucine**	$C_{23}H_{26}N_2O_4$	394.44	mcl pr (w + 4), $[\alpha]^{20}_{5461}$ −149.5 $[\alpha]^{20}_{5983}$ −120.5 (chl, c = 1) λ^{al} 264 (4.01), 301 (3.94)	178 (anh) 105 (hyd)				δ	v	δ	...	δ	chl v	**B27**[2], 797
b2429	—,hydrochloride	$C_{23}H_{27}ClN_2O_4$. *See* b2428	430.94	pr					s	s	...	...	...		**B27**[2], 797
b2430	—,nitrate, dihydrate	$C_{23}H_{31}N_3O_9$. *See* b2428	493.52	pr	230d	sub			s	s	...	...	...		**B27**[2], 797
b2431	—,sulfate, heptahydrate	$C_{46}H_{68}N_4O_{19}S$. *See* b2428	1013.1	nd, $[\alpha]_D$ −24.4 (w)					s v^h	δ	...	...	i	chl δ MeOH v	**B27**[2], 797
b2432	**Bufotalin**	$C_{26}H_{36}O_6$.	444.58	cr(+ 1 al, al) $[\alpha]^{20}_D$ +5.4 (chl, c = 0.5) λ^{al} 304 (3.78)	223d				i	s	...	...	...	chl s	**C53**, 2281
b2433	**Bulbocapnine** (*d*)	$C_{19}H_{19}NO_4$.	325.37	pr (al) $[\alpha]_D$ +237.1 (chl, c = 4)	199–200				i	s	...	...	...	os s chl v	**B27**[2], 554
b2434	—(*l*)	$C_{19}H_{19}NO_4$. *See* b2433	325.37		209–10				...	...	...	...	...		**B27**[1], 467
b2435	**1,2-Butadiene***	Methylallene. $CH_2:C:CHCH_3$	54.09	λ^{gas} 178 (4.3), 186 sh (3.6)	−136.19	10.85^{760}	0.676^{0}_{4} 0.652^{20}_{4}	$1.4205^{1.3}$	i	∞	∞	...	v		**B1**[2], 224
Ω b2436	**1,3-Butadiene***	Bivinyl. $CH_2:CHCH:CH_2$	54.09	λ^{al} 217 (4.32)	−108.91	−4.41	0.6211^{20}_{4}	1.4292^{-25}	i	s	s	v	s	os s	**B1**[3], 929
b2437	**1,2-Butadiene, 4-bromo-***	$CH_2:C:CHCH_2Br$	133.00			$109–11^{760}$	1.4255^{20}_{4}	1.5248^{20}	...	...	...	s	...		**B1**[3], 929
b2438	**1,3-Butadiene, 2-bromo-***	Bromoprene. $CH_2:CBrCH:CH_2$	133.00	grsh-ye oil		$42–3^{165}$	1.397^{20}_{4}	1.4988^{20}	i	s	s	...			**B1**[3], 955
b2439	**1,2-Butadiene, 4-chloro-***	$CH_2:C:CHCH_2Cl$	88.54			88	0.9891^{20}_{4}	1.4775^{20}	δ	...	s	v	s	os v	**B1**[3], 928
b2440	**1,3-Butadiene, 1-chloro-***	$CH_2:CHCH:CHCl$	88.54			68	0.9606^{20}_{4}	1.4712^{20}	...	s	...	...	...	chl v	**B1**[3], 949
b2441	—,**2-chloro-***	Chloroprene. $CH_2:CClCH:CH_2$	88.54	λ^{hx} 223 (4.15)		59.4^{760} 6.4^{100}	0.9583^{20}_{4}	1.4583^{20}	δ	...	∞	∞	∞	os ∞	**B1**[3], 949
b2442	—,**1-chloro-2-methyl-***	$CH_2:CHC(CH_3):CHCl$	102.57			107 50.4^{100}	0.9710^{20}_{4}	1.4792^{20}	...	v^h	...	s	...		**B1**[3], 974
b2443	—,**1-chloro-3-methyl-***	$CH_2:C(CH_3)CH:CHCl$	102.57			99–100	0.9543^{20}_{4}	1.4719^{20}_{α}	i	s	s	s	...	chl s	**B1**[3], 975
Ω b2444	—,**2-chloro-3-methyl-***	$CH_2:C(CH_3)CCl:CH_2$	102.57			93^{760}	0.9593^{20}_{4}	1.4686^{20}	i	s	v	s	...	chl v	**B1**[3], 975
b2445	—,**1,2-dichloro-***	$CH_2:CHCCl:CHCl$	122.98			$60–5^{105}$ 35^{40}	1.1991^{20}_{4}	1.4960^{20}	...	...	...	...	...	CCl_4 v	**B1**[3], 954
b2446	—,**2,3-dichloro-***	$CH_2:CClCCl:CH_2$	122.98			98^{760} $41–3^{85}$	1.1829^{20}_{4}	1.4890^{20}	...	...	...	...	...	chl v	**B1**[3], 954
Ω b2447	—,**2,3-dimethyl-***	Biisopropenyl. $CH_2:C(CH_3)C(CH_3):CH_2$	82.15	λ^{hx} 226 (4.33)	−76	$68–78^{760}$	0.7267^{20}_{4}	1.4394^{20}_{4}	...	...	...	...	...		**B1**[3], 991
b2448	—,**1,4-diphenyl-*** (*cis, cis*)	*cis, cis*-Bistyryl. $C_6H_5CH:CHCH:CHC_6H_5$	206.29	lf or nd (MeOH or al) λ^{bz} 239 (4.23), 313 (4.48)	70.5		$0.9697^{100.7}$	$1.6183^{100.6}$	i	s^h δ	v	...	v	peth s chl v aa s^h	**B5**[2], 550
Ω b2449	—,—(*trans, trans*)*.	*trans, trans*-Bistyryl. $C_6H_5CH:CHCH:CHC_6H_5$	206.29	lf (al or aa) λ^{bz} 334 (4.68), 352 (4.50)	152.5	350^{720}			i	s	δ	...	s	peth s chl s	**B5**[2], 589
—	—,**1,4-epoxy-***	*see* **Furan**													
b2450	—,**1-ethoxy-***	$CH_2:CHCH:CHOC_2H_5$	98.15			109–12	0.8154^{20}_{4}	1.4529^{20}	...	s	v	v	s	chl s	**B1**[3], 1975
b2451	—,**2-ethoxy-***	$CH_2:CHC(OC_2H_5):CH_2$	98.15			$94.5–5.5^{760}$	0.8177^{20}_{4}	1.4400^{20}	...	v	v	v	s	os v	**B1**[3], 1977
b2452	—,**2-fluoro-***	Fluoroprene. $CH_2:CHCF:CH_2$	72.08			12^{760}	0.843^{4}_{4}	1.400^{4}	...	...	...	...	...		**B1**[3], 948
Ω b2453	—,**hexachloro-***	$CCl_2:CClCCl:CCl_2$	260.76	λ^{hp} 253 (3.7)	−21	215^{760} 101^{20}	1.6820^{20}_{4}	1.5542^{20}	i	s	s	...	...		**B1**[3], 955
b2454	—,**hexafluoro-***	$CF_2:CFCF:CF_2$	162.04		−132.4 to 132.1	6^{760}	1.553^{-20}_{4}	1.378^{-20}	...	...	...	...	...		**B1**[3], 948
b2455	**1,2-Butadiene, 4-iodo-***	$CH_2:C:CHCH_2I$	179.99			130^{760}	1.7129^{20}_{4}	1.5709^{20}	...	...	...	...	...		**B1**[3], 929
b2456	**1,3-Butadiene, 2-iodo-***	Iodoprene. $CH_2:CHCI:CH_2$	179.99			$111–3^{760}$	1.7278^{20}_{4}	1.5616	...	...	...	...	...		**B1**[3], 956
b2457	**1,2-Butadiene, 4-methoxy-***	$CH_2:C:CHCH_2OCH_3$	84.13			87–9	0.8286^{20}_{4}	1.435^{20}	...	s	...	...	...		**B1**[3], 1974
Ω b2458	**1,3-Butadiene, 1-methoxy-***	$CH_2:CHCH:CHOCH_3$	84.13			$91–2^{760}$	0.8296^{20}_{4}	1.4594^{20}	s	s	...	...	...		**B1**[3], 1975
b2459	—,**2-methoxy-***	$CH_2:CHC(OCH_3):CH_2$	84.13			75^{760}	0.8272^{20}_{4}	1.4442^{20}	...	v	v	v	v	os v	**B1**[3], 1976
b2460	**1,2-Butadiene, 3-methyl-***	*asym*-Dimethylallene. $CH_2:C:(CH_3)_2$	68.13		−120	40^{760}	0.6804^{20}_{4}	1.4166^{20}	...	∞	∞	∞	∞	peth, CCl_4 ∞	**B1**[3], 976
Ω b2461	**1,3-Butadiene, 2-methyl-***	Isoprene. $CH_2:CHC(CH_3):CH_2$	68.13	λ^{iso} 217.5 (4.28), 223.5 (4.33), 230 (4.16)	−146	34^{760}	0.6810^{20}_{4}	1.4219^{20}	i	∞	∞	∞	∞		**B1**[3], 966
b2462	—,**pentafluoro-2-trifluoromethyl-***	Perfluoroisoprene. $F_2C:C(CF_3)CF:CF_2$	212.05			39^{760}	1.527^{0}_{4}	1.3000^{0}	...	...	s	s	s		**C49**, 2479

For explanations, symbols and abbreviations see beginning of table. For structural formulas see end of table.

No.	Name	Synonyms and Formula	Mol. wt.	Color, crystalline form, specific rotation and λ_{max} (log ε)	m.p. °C	b.p. °C	Density	n_D	Solubility w	al	eth	ace	bz	other solvents	Ref.
	1,3-Butadiene														
b2463	**—,1-phenyl-, (*trans*)***	$CH_2:CHCH:CHC_6H_5$	130.19	λ^{iso} 233 (3.81), 280 (4.35), 306 (3.65)	4.5	76^{11}	0.9286_4^{20}	1.6089^{25}	i	s	s	s	s	os s	**$B5^2$**, 414
b2464	**1,2,3,4-tetra-chloro-, (liquid)***	ClCH:CClCCl:CHCl	191.87	...	−4	188 67^{10}	1.516_{15}^{15}	1.5455^{20}	i	s	s	s	s	os s	**$B1^3$**, 954
b2465	**—,—,(solid)***	ClCH:CClCCl:CHCl	191.87	λ^{hp} 248 (4.3)	52	...	1.4961^{20}	1.5438^{20}	i	s	s	s	s	chl v os s	**$B1^3$**, 954
b2466	**—,1,2,3-trichloro-***	$CH_2:CClCCl:CHCl$	157.44	...	...	$33–4^{7}$	1.4060_4^{20}	1.5262^{20}	...	...	s	...	...	chl s	**$B1^3$**, 954
b2467	**2,3-Butadien-1-ol***	$CH_2:C:CHCH_2OH$	70.09	...	...	$126–8^{756}$ $68–9^{45}$	0.9164_4^{20}	1.4759^{20}	s	v	s	s	s	os v	**$B1^3$**, 1974
b2468	**Butadiyne***	Biacetylene. $CH\vdots CC\vdots CH$	50.06	λ^{iso} 224 (2.35), 247 (2.26), 251.5 (1.74)	−36.4	10.3^{760}	0.7364_4^{0}	1.4189^{5}	v	s	v	v	...	chl s	**$B1^3$**, 1056
b2469	**—,1,4-bis(1-hydroxycyclo-hexyl)-***	...	246.35	cr (bz) λ^{al} 232.5 (2.51), 243 (2.54), 257 (2.34)	174		...	...	i δ^h	...	...	...	δ^h	MeOH s chl δ^h	**C31**, 4283
b2470	**—,1,4-dichloro-***	$ClC\vdots CC\vdots CCl$	118.95	nd	1–3	...	...	...	...	...	...	...	...	chl v	**$B1^2$**, 246
Ω b2471	**Butanal***	Butyraldehyde. $CH_3CH_2CH_2CHO$	72.12	λ^{w} 225 (1.07), 282.5 (1.13)	−99	75.7^{760}	0.8170_4^{20}	1.3843^{20}	s	∞	∞	v	v	...	**$B1^2$**, 721
Ω b2472	**—,oxime**	$CH_3CH_2CH_2CH:NOH$	87.12	...	−29.5	152^{715}	0.923_4^{20}	...	v	∞	∞	v	v	...	**$B1^2$**, 724
Ω b2473	**—,phenylhydrazone**	$CH_3CH_2CH_2CH:NNHC_6H_5$	162.24	...	93–5	$190–5^{80}$ 152^{14}	...	...	...	...	...	...	...	...	**$B15^2$**, 55
b2474	**—,2-benzylidene-**	α-Ethylcinnamaldehyde. 2-Ethyl-3-phenylpropenal. $C_6H_5CH:C(C_2H_5)CHO$	160.22	ye λ^{al} 253.5 (4.09)	...	$157–8^{5}$	1.0201_4^{22}	1.578^{20}	...	...	...	...	...	...	**C26**, 4684
b2475	**—,3-chloro-,** diethyl acetal	$CH_3CHClCH_2CH(OC_2H_5)_2$	180.68	...	...	$70–1^{12}$	0.9709_4^{20}	1.4210^{20}	i	∞	v	v	v	os ∞	**$B1^2$**, 724
b2476	**—,4-chloro-***	$ClCH_2CH_2CH_2CHO$	106.55	...	...	$50–1^{13}$	$1.107_{15}^{8.5}$	$1.4466^{8.5}$	...	v	v	v	...	...	**$B1^3$**, 2767
b2477	**—,2,3-dichloro-***	$CH_3CHClCHClCHO$	141.00	...	...	$58–60^{20}$	1.2666_4^{21}	1.4618^{21}	i	s	s	v	...	chl s	**$B1^2$**, 724
Ω b2478	**—,2-ethyl-***	$(C_2H_5)_2CHCHO$	100.16	...	...	$117–9^{160}$	0.8110_4^{20}	1.4025^{20}	δ	∞	∞	...	...	...	**$B1^2$**, 749
Ω b2479	**—,3-hydroxy-***	Aldol. $CH_3CHOHCH_2CHO$	88.12	...	...	83^{20} d85	1.103_4^{20}	1.4238^{20}	∞	∞	s	v	...	...	**$B1^2$**, 868
Ω b2480	**—,2-methyl-(*dl*)***	$CH_3CH_2CH(CH_3)CHO$	86.14	λ^{al} 295 (1.32)	...	$92–3^{760}$ 54^{200}	0.8029_4^{20}	1.3869^{20}	i	s	s	v	...		**B1**, 682
Ω b2481	**—,3-methyl-***	Isovaleraldehyde. $(CH_3)_2CHCH_2CHO$	86.14	λ^{vap} 190.1	−51	92.5^{760}	0.7977_4^{20}	1.3902^{20}	δ	s	s	...	...	...	**$B1^2$**, 742
b2482	**—,—,oxime**	$(CH_3)_2CHCH_2CH:NOH$	101.15	...	48.5	161.3^{759}	0.8934_4^{20}	1.4367^{20}	i	s	v	v	...	...	**$B1^2$**, 744
b2483	**—,2,2,3-trichloro-***	*n*-Butyrchloral. $CH_3CHClCCl_2CHO$	175.44	...	...	$163–5^{760}$ 49^{8}	1.3956_4^{20}	1.4755^{20}	s	s	s	...	...	...	**$B1^2$**, 725
Ω b2484	**—,—,hydrate**	$CH_3CHClCCl_2CH(OH)_2$	192.46	rh pl or lf (w)	78	d	1.694_4^{20}	...	s^h	v	s	...	...	...	**$B1^2$**, 725
Ω b2485	**Butane***	$CH_3CH_2CH_2CH_3$	58.12	...	−138.35	−0.5	0.6012_4^{0} 0.5788_4^{20}	1.3543^{-13} 1.3326^{20}	s	v	v	...	...	chl v	**$B1^2$**, 261
Ω b2486	**—,1-amino-***	*n*-Butylamine. $CH_3CH_2CH_2CH_2NH_2$	73.14	...	−49.1	77.8^{760}	0.7414_4^{20}	1.4031^{20}	∞	s	s	...	...	...	**$B4^2$**, 631
b2487	**—,2-amino-(*d*)***	*sec*-Butylamine. $CH_3CH_2CH(CH_3)NH_2$	73.14	$[\alpha]_D^{20}$ +7.4 (w)	−104.5	63^{760}	0.724_4^{20}	1.344^{20}	s	∞	∞	v	...	chl s	**$B4^1$**, 372
Ω b2488	**—,—(*dl*)***	$CH_3CH_2CH(CH_3)NH_2$	73.14	...	< −72	63.5^{764}	0.7246_4^{20}	1.3932^{20}	s	∞	∞	v	...	chl s	**$B4^2$**, 636
b2489	**—,—(*l*)***	$CH_3CH_2CH(CH_3)NH_2$	73.14	$[\alpha]_D^{20}$ −7.4 (w, c = 4.7)	...	63	0.7205_4^{20}	...	v	v	v	v	...	chl s	**$B4^2$**, 372
b2490	**—,2-amino-2,3-dimethyl-***	$(CH_3)_2CHC(CH_3)_2NH_2$	101.19	...	...	$104–5^{760}$	0.7683_0^{0}	1.4096^{17}	...	...	...	...	...	...	**B4**, 193
b2491	**—,3-amino-2,2-dimethyl-***	$(CH_3)_3CCH(CH_3)NH_2$	101.19	...	−20	102^{760}	...	1.4105^{25}	s^h	...	...	...	...	...	**B4**, 193
b2492	**—,—,hydro-chloride***	$(CH_3)_3CCH(CH_3)NH_2.HCl$	137.65	nd (CH_2Cl_2-CS_2)	300–1 (cor)	sub 245	...	...	v	...	...	...	...	...	**B4**, 193
Ω b2493	**—,1-amino-3-methyl-***	$(CH_3)_2CHCH_2CH_2NH_2$	87.17	...	...	$95–7^{761}$	0.7505_4^{20}	1.4083^{20}	∞	∞	∞	s	...	chl s	**$B4^2$**, 644
b2494	**—,2-amino-2-methyl-***	$CH_3CH_2C(CH_3)_2NH_2$	87.17	...	−105	77^{760}	0.731_4^{25}	1.3954^{25}	∞	∞	∞	s	...	...	**$B4^2$**, 644
b2495	**—,2-amino-3-methyl-***	$(CH_3)_2CHCH(CH_3)NH_2$	87.17	...	...	$84–7^{760}$	0.7574^{19}	1.4096^{18}	v	s	...	...	...	...	**$B4^2$**, 644
b2496	**1,4-bis(dicar-bethoxyamino)-**	$C_2H_5O_2CNH(CH_2)_4NHCO_2C_2H_5$	232.28	nd (lig)	85–6	...	...	...	i	s	s		s	chl s lig s^h	**$B4^1$**, 421
b2496[1]	**1,4-bis(di-methylamino)-**	Tetramethylputrescine. $(CH_3)_2NCH_2CH_2CH_2CH_2N(CH_3)_2$	232.28	...	...	168^{740} $78–80^{28}$	0.7942^{15}	1.4621^{25}	s	s	s	...	...	...	**$B4^2$**, 702
b2497	**—,2,2-bis(ethyl-sulfonyl)-***	Trional. $CH_3CH_2C(SO_2C_2H_5)_2CH_3$	144.27 242.36	pl (w)	76	d	1.199_4^{85}	...	δ	v	s	...	s	peth s lig s CCl_4 s	**$B1^2$**, 731
b2498	**—,1,1-bis(4-hydroxyphenyl)-***	...	242.32	nd (to)	137	270^{12}	...	...	s	s	...	...	s^h	to s^h	**$B6^2$**, 980
b2499	**—,2,2-bis(4-hydroxyphenyl)-***	...	242.32	nd or pr (w)	133–4 (128)	$250–3^{12}$	...	...	i	v	v	v	δ	MeOH v	**$B6^2$**, 980
Ω b2500	**—,1-bromo-***	*n*-Butyl bromide. $CH_3CH_2CH_2CH_2Br$	137.03	...	−112.4	101.6^{760} 18.8^{30}	1.2758_8^{20}	1.4401^{20}	i	∞	∞	∞	...	chl s	**$B1^2$**, 290
Ω b2501	**—,2-bromo-(*dl*)***	*sec*-Butyl bromide. $CH_3CH_2CH(CH_3)Br$	137.03	...	−111.9	91.2^{760}	1.2585_4^{20}	1.4366^{20}	i	...	∞	∞	...	chl s	**$B1^3$**, 293
b2502	**—,—(*l*)***	$CH_3CH_2CH(CH_3)Br$	137.03	$[\alpha]_D^{22}$ −23.13 (undil)	...	90–1	1.2536_4^{25}	$1.4359^{19.5}$	i	s	∞	∞	...	chl s	**$B1^3$**, 293
Ω b2503	**—,1-bromo-4-chloro-***	$ClCH_2CH_2CH_2CH_2Br$	171.48	...	...	$174–5^{756}$ $63–4^{10}$	1.488_4^{20}	1.4885^{20}	i	s	s	...	...	chl s	**$B1^3$**, 294

For explanations, symbols and abbreviations see beginning of table. For structural formulas see end of table.

No.	Name	Synonyms and Formula	Mol. wt.	Color, crystalline form, specific rotation and λ_{max} (log ε)	m.p. °C	b.p. °C	Density	n_D	Solubility w	al	eth	ace	bz	other solvents	Ref.
	Butane														
b2504	**—,2-bromo-1-chloro-***	$CH_3CH_2CHBrCH_2Cl$	171.48	...	...	$146–7^{758}$	1.4688^{20}_{4}	1.4880^{20}	i	s	s	...	s	chl s	**B1**^{2}, 83
b2505	**—,1-bromo-3,3-dimethyl-***	$(CH_3)_3CCH_2CH_2Br$	165.08	...	...	138 54^{40}	1.1556^{20}_{4}	1.4440^{20}	i	s	s	...	...	chl v	**B1**^{3}, 409
b2506	**—,2-bromo-2,3-dimethyl-***	$(CH_3)_2CHCBr(CH_3)_2$	165.08	...	24–5	$132–3^{742}$ 87^{180}	1.1772^{20}	1.4517	d^h	...	s	...	...	chl v	**B1**^{3}, 414
b2507	**—,1-bromo-4-fluoro-***	$FCH_2CH_2CH_2CH_2Br$	155.01	...	...	$134–5^{740}$	...	1.4370^{25}	i	v	v	...	...	...	**C51,** 7300
b2508	**—,1-bromo-2-methyl-(*d*)***	*act*-Amyl bromide. $CH_3CH_2CH(CH_3)CH_2Br$	151.05	$[\alpha]^{20}_{D}$ +3.68	...	121.6^{760}	1.2234^{20}_{4}	1.4451^{20}	i	s	s	...	...	chl v	**B1**^{3}, 362
b2509	**—,—(*dl*)***	$CH_3CH_2CH(CH_3)CH_2Br$	151.05	...	...	120.5^{760} 12.3^{10}	1.2205^{20}_{4}	1.4452^{20}	i	s	s	...	...	chl v	**B1**^{3}, 363
Ω b2510	**—,1-bromo-3-methyl-***	Isoamyl bromide. $(CH_3)_2CHCH_2CH_2Br$	151.05	...	−112	120.4^{760} 12.3^{10}	1.2071^{20}_{4}	1.4420^{20}	i	s	s	...	...	chl v	**B1**^{3}, 363
Ω b2511	**—,1-chloro-***	*n*-Butyl chloride. $CH_3CH_2CH_2CH_2Cl$	92.57	...	−123.1	78.44^{760}	0.8862^{20}_{4}	1.4021^{20}	i	∞	∞	...	...	...	**B1**^{3}, 275
Ω b2512	**—,2-chloro-(*dl*)***	*sec*-Butyl chloride. $CH_3CH_2CHClCH_3$	92.57	...	−131.3	68.25^{760}	0.8732^{20}_{4}	1.3971^{20}	i	∞	∞	...	v	chl v	**B1**^{3}, 278
b2513	**—,—(*l*)***	$CH_3CH_2CHClCH_3$	92.57	$[\alpha]^{20}_{D}$ −8.48	−140.5	68^{760}	0.8950^{0}_{4}	...	i	∞	∞	...	v	chl v	**B1**^{2}, 81
b2514	**—,1-chloro-2,2-dimethyl-***	*neo*-Hexyl chloride. $CH_3CH_2C(CH_3)_2CH_2Cl$	120.62	...	...	116^{735}	...	1.4200^{20}	i	v	v	...	...	chl v	**B1**^{3}, 408
b2515	**—,1-chloro-3,3-dimethyl-***	$(CH_3)_3CCH_2CH_2Cl$	120.62	...	...	115^{760} 41^{50}	0.8670^{20}_{4}	1.4161^{20}	i	v	v	...	...	chl v	**B1**^{3}, 408
b2516	**—,2-chloro-2,3-dimethyl-***	$(CH_3)_2CHCCl(CH_3)_2$	120.62	...	−10.4	112^{760}	0.8780^{20}_{4}	1.4191^{20}	d	s	...	s	...	...	**B1**^{3}, 414
b2517	**—,3-chloro-2,2-dimethyl-***	Pinacolyl chloride. $CH_3CHClC(CH_3)_3$	120.62	...	−0.9	111^{760} 7^{10}	0.8767^{20}_{4}	1.4182^{20}	...	...	v	...	...	...	**B1**^{3}, 408
b2519	**—,1-chloro-4-fluoro-***	$FCH_2CH_2CH_2CH_2Cl$	110.57	...	...	114.7^{760}	1.0627^{25}_{4}	1.4020^{25}	i	v	v	...	...	...	**B1**^{3}, 279
b2520	**—,1-chloro-2-methyl-(*d*)***	*act*-Amyl chloride. $CH_3CH_2CH(CH_3)CH_2Cl$	106.60	$[\alpha]^{20}_{5892}$ +1.64	...	100.5^{760} 43^{100}	0.8857^{20}_{4}	1.4126^{20}	i	s	s	...	...	...	**B1**^{3}, 356
b2521	**—,—(*dl*)***	$CH_3CH_2CH(CH_3)CH_2Cl$	106.60	...	...	99.9^{760} 52.2^{50}	0.8818^{15}_{15}	1.4102^{25}	...	s	s	...	...	...	**B1**^{3}, 357
Ω b2522	**—,1-chloro-3-methyl-***	Isoamyl chloride. $(CH_3)_2CHCH_2CH_2Cl$	106.60	...	−104.4	98.8^{760}	0.8704^{20}_{4}	1.4084^{20}	δ	∞	∞	...	...	chl v	**B1**^{3}, 359
Ω b2523	**—,2-chloro-2-methyl-***	*tert*-Amyl chloride. $CH_3CH_2CCl(CH_3)_2$	106.60	...	−73.5	85.6^{760}	0.8653^{20}_{4}	1.4055^{20}	δ	s	s	...	...	...	**B1**^{3}, 357
b2524	**—,2-chloro-3-methyl-(*dl*)***	*sec*-Isoamyl chloride. $(CH_3)_2CHCHClCH_3$	106.60	...	...	92.8^{760} 25.7^{60}	0.8620^{20}_{4}	1.4020^{20}	...	s	v	...	...	...	**B1**^{1}, 46
b2525	**—,2(chloromethyl)-1,3-dichloro-(*dl*)***	$CH_3CHClCH(CH_2Cl)_2$	175.49	...	...	$79–81^{15}$	1.2793^{15}_{4}	...	i	s	s	...	...	chl s	**C31,** 1003
b2526	**—,2(chloromethyl)-1,2,3-trichloro-***	$CH_3CHClCCl(CH_2Cl)_2$	209.94	...	...	$102–3^{13}$	1.3977^{18}_{4}	1.5012^{18}_{α}	i	...	v	...	...	chl s	**B1**^{3}, 362
b2527	**—,1-chloro-2,2,3,3-tetramethyl-***	$(CH_3)_3CC(CH_3)_2CH_2Cl$	148.68	...	52–3	$80–1^{40}$	...	...	i	...	v	...	...	...	**B1**^{3}, 502
b2528	**—,2-chloro-2,3,3-trimethyl-***	$(CH_3)_2CClC(CH_3)_3$	134.65	...	136 (132)	sub	...	...	i	...	v	...	...	MeOH δ^h	**B1**^{1}, 59
b2529	**—,1,4-diamino-***	Putrescine. $H_2NCH_2CH_2CH_2CH_2NH_2$	88.15	lf	27–8	$158–9^{760}$	0.877^{25}_{4}	1.4569^{20}	s	...	...	...	...	...	**B4,** 264
b2530	**—,—,dihydrochloride***	$H_2NCH_2CH_2CH_2CH_2NH_2 . 2HCl$	161.08	nd or lf (al or w)	315d (290)	sub	...	...	v	v^h	i	...	i	MeOH i	**B4**^{2}, 701
Ω b2533	**—,1,2-dibromo-***	$CH_3CH_2CHBrCH_2Br$	215.94	λ212 (3.1)	−65.4	166.3^{760} 53.8^{10}	1.7915^{20}_{4}	1.4025^{20} (1.5150^{20})	i	...	s	...	...	chl s	**B1**^{3}, 295
Ω b2534	**—,1,3-dibromo-***	$CH_3CHBrCH_2CH_2Br$	215.94	...	...	174^{760} 72^{20}	1.800^{20}	1.507^{20}	i	...	s	...	...	chl s	**B1**^{3}, 295
Ω b2535	**—,1,4-dibromo-***	Tetramethylene dibromide. $BrCH_2CH_2CH_2CH_2Br$	215.94	λ212 (2.8)	−16.53	197^{760} 79^{10}	1.7890^{20}_{4}	1.5190^{20}	i	...	...	...	...	chl s	**B1**^{3}, 295
b2536	**—,2,3-dibromo-***	$CH_3CHBrCHBrCH_3$	215.94	λ <225	< −80	161^{760}	1.7893^{22}_{4}	1.5133^{22}	i	...	s	...	...	chl s	**B1**^{3}, 296
b2537	**—,1,1-dichloro-***	Butylidene dichloride. $CH_3CH_2CH_2CHCl_2$	127.03	...	...	113.8^{760}	1.0863^{20}_{4}	1.4355^{20}	i	...	...	...	...	chl s	**B1**^{3}, 280
Ω b2538	**—,1,2-dichloro-***	$CH_3CH_2CHClCH_2Cl$	127.03	...	...	124^{760}	1.1116^{25}_{4}	1.4450^{20}	i	...	s	...	...	chl s	**B1**^{3}, 280
Ω b2539	**—,1,3-dichloro-***	$CH_3CHClCH_2CH_2Cl$	127.03	...	...	134^{760}	1.1158^{20}_{4}	1.4445^{20}	i	...	s	...	...	chl s	**B1**^{3}, 281
Ω b2540	**—,1,4-dichloro-***	Tetramethylene dichloride. $ClCH_2CH_2CH_2CH_2Cl$	127.03	...	−37.3	153.9^{760} 39.7^{10}	1.1408^{20}_{4}	1.4542^{20}	i	...	...	...	...	chl s	**B1**^{2}, 81
b2541	**—,2,2-dichloro-***	$CH_3CH_2CCl_2CH_3$	127.03	...	−74	104	...	1.4295	i	...	...	...	...	chl s	**B1**^{3}, 282
Ω b2542	**—,2,3-dichloro-***	$CH_3CHClCHClCH_3$	127.03	...	−80.0	116^{760} 49.5^{80}	1.1134^{20}_{4}	1.4420^{20}	i	...	...	...	...	chl s	**B1**^{2}, 282
b2543	**—,2,2-dichloro-3,3-dimethyl-***	$(CH_3)_3CCCl_2CH_3$	155.07	...	151–2	...	...	...	i	v^h	s^h	...	...	...	**B1**^{3}, 409
b2544	**—,2,3-dichloro-2,3-dimethyl-***	$(CH_3)_2CClCCl(CH_3)_2$	155.07	pr (dil al)	164	...	...	...	i	v	v	...	...	...	**B1,** 152
b2545	**—,1,1-dichloro-3-methyl-(*dl*)***	$(CH_3)_2CHCH_2CHCl_2$	141.04	...	...	130 $48–9^{40}$	1.0473^{20}	1.4344^{20}	i	s	s	...	...	...	**B1**^{1}, 47
b2546	**—,1,2-dichloro-2-methyl-(*dl*)***	$CH_3CH_2CCl(CH_3)CH_2Cl$	141.04	...	...	133–5 71.5^{100}	1.0785^{20}_{4}	$1.4432^{23.5}$	d	s	s	...	...	chl s	**B1**^{3}, 360
b2547	**—,1,3-dichloro-3-methyl-(*d*)***	$(CH_3)_2CClCH_2CH_2Cl$	141.04	...	...	145–6 39^{10}	1.0654^{20}_{4}	1.4455^{20}	i	...	s	...	...	chl s	**B1**^{3}, 361

For explanations, symbols and abbreviations see beginning of table. For structural formulas see end of table.

No.	Name	Synonyms and Formula	Mol. wt.	Color, crystalline form, specific rotation and λ_{max} (log ε)	m.p. °C	b.p. °C	Density	n_D	Solubility: w	al	eth	ace	bz	other solvents	Ref.
	Butane														
b2548	—,1,4-dichloro-2-methyl-*	$ClCH_2CH_2CH(CH_3)CH_2Cl$	141.04	....	....	$168–9^{760}$ 50^{12}	1.1003^{25}_{4}	1.4562^{21}	i	...	...	...	...	chl s	**B1**^{2}, 101
b2549	—,2,3-dichloro-2-methyl-*	$CH_3CHClCCl(CH_3)_2$	141.04	....	....	138^{760} 37.5^{20}	1.0696^{15}_{4}	1.4450^{18}	i	s	s	...	...	....	**B1**^{3}, 360
Ω b2550	—,1:2,3:4-diepoxy-(*dl*)*	Bioxirane. Butadiene dioxide.	86.09	....	4	144^{760}	1.113^{20}	1.435^{20}	s	s	...	...	...	....	**B19**^{2}, 14
b2551	—,—(*meso*)*	....	86.09	....	−16	138^{767}	1.1157^{20}_{4}	1.4330^{20}_{4}	∞	v	...	...	...	....	**B19**, 15
b2552	—,1(diethylaminomethoxy)-*	Butyl (diethylaminomethyl) ether. $CH_3CH_2CH_2CH_2OCH_2N(C_2H_5)_2$	159.27	....	....	$172–4^{754}$	....	....	...	v	...	...	...	os s	**B4**^{2}, 598
b2553	—,1,4-difluorooctachloro-*	$FCCl_2CCl_2CCl_2CCl_2F$	369.67	....	4–5	152.5^{20}	1.9272^{20}_{4}	1.5256^{20}	...	...	...	...	...	....	**Am 62,** 342
b2554	—,2,2-di(2-furyl)-	....	190.25	....	....	$64–6^{1}$	1.0330^{20}_{4}	1.4970^{20}	i	s	s	...	...	MeOH s	**C52,** 13702
Ω b2555	—,1,4-diiodo-*	Tetramethylene diiodide. $ICH_2CH_2CH_2CH_2I$	309.92	....	5.8	$125–6^{15}$d	2.349^{26}_{4}	1.619^{25}	i	...	...	...	...	os s	**B1**^{3}, 302
Ω b2556	—,2,2-dimethyl-*	*neo*-Hexane. $CH_3CH_2C(CH_3)_3$	86.18	....	−99.87	49.74^{760}	0.6485^{20}_{4}	1.3688^{20}	i	s	s	v	v	peth, CCl_4 v	**B1**^{3}, 405
Ω b2557	—,2,3-dimethyl-*	$(CH_3)_2CHCH(CH_3)_2$	86.18	....	−128.53	58^{760}	0.6616^{20}_{4}	1.3750^{20}	i	s	s	v	v	peth, CCl_4 v	**B1**^{3}, 410
b2558	—,1,4-dinitro-*	$O_2NCH_2CH_2CH_2CH_2NO_2$	148.13	pr (al or MeOH) λ^{al} 277 (1.60)	33–4	$176–8^{13}$	....	....	i	δ	s	...	s	MeOH s^h	**B1**^{3}, 305
b2559	—,2,3-dinitro-(*dl*)*	$CH_3CHNO_2CHNO_2CH_3$	148.13	pr (eth)	48–9	....	....	....	...	δ	s	...	...	MeOH δ^h lig δ	**B1**^{3}, 305
b2560	—,—(*meso*)*	$CH_3CHNO_2CHNO_2CH_3$	148.13	pr (eth)	....	$76–7^{1}$	....	....	...	δ	s	...	...	lig δ	**B1**^{3}, 305
Ω b2561	—,1,2-epoxy-*	1,2-Butylene oxide	72.12	....	....	63.3	0.837^{17}_{4}	1.3851^{20}	d^h	v	∞	v	...	os v	**B17**^{2}, 17
—	—,1,4-epoxy-*	*see* **Furan, tetrahydro-**													
b2562	—,2,3-epoxy-(*cis*)*	*β*-Butylene oxide	72.12	....	−80	59.7^{742}	0.8226^{25}_{4}	1.3802^{20}	d	...	v	v	v	os v	**B17**^{2}, 17
b2563	—,—(*trans*)*	....	72.12	....	−85	$56–7^{760}$	0.8010^{25}_{4}	1.3736^{20}	d	...	v	v	v	os v	**B17**^{2}, 17
b2564	—,1-fluoro-*	Butyl fluoride. $CH_3CH_2CH_2CH_2F$	76.12	....	−134	32.5^{760}	0.7789^{20}_{4}	1.3396^{20}	...	v	...	...	...	....	**B1**^{3}, 273
b2565	—,1,1,2,2,3,4,4-heptachloro-*	$Cl_2CHCHClCCl_2CHCl_2$	299.24	....	....	$137.5^{13.5}$ 97.5^{2}	1.742^{20}_{20}	1.5407^{20}	i	...	...	s	...	CCl_4 v	**B1**^{3}, 289
b2566	—,1,1,2,3,4,4-hexachloro-* (liquid)	$Cl_2CHCHClCHClCHCl_2$	264.79	....	....	111^{10}	1.6460^{20}_{4}	1.5258^{20}	i	...	...	s	...	CCl_4 s	**B1**^{3}, 288
b2567	—,—(solid)	$Cl_2CHCHClCHClCHCl_2$	264.79	nd (al), pr (bz, aa)	109–10	....	....	....	i	...	...	s	...	CCl_4 s	**B1**^{3}, 288
Ω b2568	—,1-iodo-*	Butyl iodide. $CH_3CH_2CH_2CH_2I$	184.02	λ^{MeOH} 254 (2.70)	−103	130.53^{760} 19.2^{10}	1.6154^{20}_{4}	1.5001^{20}	i	∞	∞	...	...	chl v	**B1**^{3}, 294
Ω b2569	—,2-iodo-(*dl*)*	*sec*-Butyl iodide. $CH_3CH_2CHICH_3$	184.02	λ^{MeOH} 258.5 (277)	−104.2	120^{760} 33^{45}	1.5920^{20}_{4}	1.4991^{20}	i	∞	∞	...	...	chl v	**B1**^{3}, 301
b2570	—,—(*l*)*	$CH_3CH_2CHICH_3$	184.02	$[\alpha]_D$ −12.15 (al, c = 20)	....	117.5–8.5	1.585^{20}_{4}	1.4945^{19}	i	∞	∞	...	...	chl v	**B1**^{3}, 301
b2571	—,1-iodo-2-methyl-(*d*)*	*act*-Amyl iodide. $CH_3CH_2CH(CH_3)CH_2I$	198.05	$[\alpha]^{15}_{D}$ +5.78 (undil)	....	148^{760} 47.1^{20}	1.5253^{20}_{4}	1.4977^{20}	i	s	s	...	...	....	**B1**^{3}, 366
b2572	—,—(*dl*)*	$CH_3CH_2CH(CH_3)CH_2I$	198.05	....	....	144–7	....	1.497^{20}	i	s	s	...	...	....	**B1**^{3}, 366
Ω b2573	—,1-iodo-3-methyl-*	Isoamyl iodide. $(CH_3)_2CHCH_2CH_2I$	198.05	....	....	147^{760}	1.5118^{20}_{4}	1.4939^{20}	δ	∞	∞	...	...	....	**B1**^{3}, 367
b2574	—,2-iodo-2-methyl-*	*tert*-Amyl iodide. $CH_3CH_2CI(CH_3)_2$	198.05	....	....	124.5^{760}	1.4937^{20}_{4}	1.4981^{20}	i	∞	∞	...	...	....	**B1**^{3}, 366
b2575	—,2-iodo-3-methyl-*	$(CH_3)_2CHCHICH_3$	198.05	....	....	$138–9^{760}$	....	....	i	v	s	s	...	os s	**B1**^{1}, 49
Ω b2577	—,1-isocyano-*	Butylcarbylamine. Butyl isocyanide. $CH_3CH_2CH_2CH_2NC$	83.13	....	....	118^{760}	....	1.4061^{20}	i	∞	∞	...	...	....	**B4**^{3}, 294
b2578	—,1-isocyano-3-methyl-*	Isoamylcarbylamine. $(CH_3)_2CHCH_2CH_2NC$	97.16	....	....	140^{760}	0.806^{20}_{4}	1.406^{20}	i	∞	∞	...	...	....	**B4**, 184
Ω b2579	—,2-methyl-*	Isopentane. $(CH_3)_2CHCH_2CH_3$	72.15	....	−159.9	27.85^{760}	0.6201^{20}_{4}	1.3537^{20}	i	∞	∞	...	...	....	**B1**^{3}, 352
b2580	—,2-methyl-1,2,3-trichloro-*	$CH_3CHClCCl(CH_3)CH_2Cl$	175.49	....	....	$183–5^{762}$ 65.5^{11}	1.2527^{20}_{4}	....	i	...	...	...	...	chl s	**B1**^{3}, 361
b2581	—,2-methyl-2,3,3-trichloro-*	$CH_3CCl_2CCl(CH_3)_2$	175.49	....	....	182–3	1.215^{15}_{4}	1.472^{21}	i	...	...	...	...	chl s aa v^h	**B1**^{3}, 361
Ω b2582	—,1-nitro-*	$CH_3CH_2CH_2CH_2NO_2$	103.12	λ^{al} 272 (1.40)	....	153^{760}	0.9710^{20}_{4}	1.4103^{20}	δ	∞	∞	...	...	alk s	**B1**^{3}, 303
b2583	—,2-nitro-(*dl*)*	$CH_3CH_2CHNO_2CH_3$	103.12	λ^{al} 279 (1.39)	−132	140^{760}	0.9854^{17}_{4}	1.4044^{20}	...	...	...	...	...	....	**B1**^{3}, 303
b2585	—,1,1,2,2,3,3,4,4-octachloro-*	$Cl_2CHCCl_2CCl_2CHCl_2$	333.69	cr (al)	81	....	....	....	i	v^h	s	s	s	os s	**B1**^{3}, 289
b2586	—,1,1,2,3,4-pentachloro- (liquid)*	$ClCH_2CHClCHClCHCl_2$	230.35	....	....	95.5^{11}	1.561^{18}	1.5140^{18}	i	s	...	...	...	CCl_4 v	**B1**^{3}, 288
b2587	—,—(solid)*	$ClCH_2CHClCHClCHCl_2$	230.35	lf (al)	49	230 102^{11}	1.539^{53}	1.5065^{53}	i	v^h	...	...	...	CCl_4 v	**B1**^{3}, 287
b2588	—,1,2,2,3,4-pentachloro-*	$ClCH_2CHClCCl_2CH_2Cl$	230.35	....	....	85^{10}	1.5543^{20}_{4}	1.5157^{20}	i	...	...	s	...	chl s	**B1**^{3}, 288
b2589	—,1,1,4,4-tetrabromo-*	$Br_2CHCH_2CH_2CHBr_2$	373.73	....	....	$138–45^{10}$	2.529^{20}_{4}	1.6077^{20}	i	δ	∞	...	∞	chl ∞	**B1**, 121

For explanations, symbols and abbreviations see beginning of table. For structural formulas see end of table.

No.	Name	Synonyms and Formula	Mol. wt.	Color, crystalline form, specific rotation and λ_{max} (log ε)	m.p. °C	b.p. °C	Density	n_D	Solubility w	al	eth	ace	bz	other solvents	Ref.
	Butane														
b2590	**—,1,2,2,3-tetra-bromo-***	$CH_3CHBrCBr_2CH_2Br$	373.73	...	−2	$128–30^{14}$ 97.5^{7}	2.5100^{20}_{4}	1.6070^{20}	i	...	...	s	...	...	**B1**^{3}, 298
b2591	**—,1,2,2,4-tetra-bromo-***	$BrCH_2CH_2CBr_2CH_2Br$	373.73	nd (lig)	72–3	...	...	...	i	...	...	...	...	lig s^h	**B1**^{3}, 298
Ω b2592	**—,1,2,3,4-tetra-bromo-(*dl*)***	$BrCH_2CHBrCHBrCH_2Br$	373.73	lf (peth)	40–1	...	...	...	i	v	v	s	...	lig s	**B1**^{3}, 298
b2593	**—,—(*meso*)***	$BrCH_2CHBrCHBrCH_2Br$	373.73	nd (al or lig)	118–9	$180–1^{60}$	...	...	i	s^h	...	s	...	chl s lig δ	**B1**^{3}, 298
b2594	**—,2,2,3,3-tetra-bromo-***	$CH_3CBr_2CBr_2CH_3$	373.73	lf (lig), pr (eth-lig)	243	...	...	...	i	δ	s	s	s	chl v lig s	**B1**^{3}, 298
b2595	**—,1,1,1,2-tetra-chloro-***	$CH_3CH_2CHClCCl_3$	195.91	...	...	$134–5^{742}$	1.3932^{20}_{20}	1.4920^{25}	i	...	...	s	...	chl s	**B1**^{3}, 280
b2596	**—,1,2,2,3-tetra-chloro-***	$CH_3CHClCCl_2CH_2Cl$	195.91	...	−48 to −46	182 85^{10}	1.4276^{18}_{4}	1.491^{20}	i	...	...	s	...	chl s	**B1**^{3}, 286
b2597	**—,1,2,3,3-tetra-chloro-***	$CH_3CCl_2CHClCH_2Cl$	195.91	...	...	90^{32} $55–7^{10}$	1.4204^{20}_{4}	1.4958^{20}	i	...	s	s	...	chl s	**B1**^{3}, 286
b2601	**—,2,2,3,3-tetra-methyl-***	$CH_3C(CH_3)_2C(CH_3)_2CH_3$	114.23	lf (eth)	100.7	106.5^{760} 13.1^{10}	0.8242^{20} 0.6485^{110}_{4}	1.4695^{20}	i	...	s	...	...	...	**B1**^{3}, 501
b2602	**—,1,1,2-tri-bromo-***	$CH_3CH_2CHBrCHBr_2$	294.83	...	...	216.2^{760} 98^{14}	2.1836^{20}_{4}	1.5626^{17}	i	s	s	...	...	chl s	**B1**^{2}, 84
b2603	**—,1,2,2-tri-bromo-***	$CH_3CH_2CBr_2CH_2Br$	294.83	...	...	213.8^{760} 90.1^{14}	2.1692^{20}_{4}	1.5624^{16}	i	s	s	...	...	chl s	**B1**^{2}, 85
b2604	**—,1,2,3-tri-bromo-***	$CH_3CHBrCHBrCH_2Br$	294.83	...	−19	220^{760} 97^{10}	2.1908^{20}_{4}	1.5680^{20}	i	s	s	...	...	chl s	**B1**^{2}, 85
b2605	**—,1,2,4-tri-bromo-***	$BrCH_2CH_2CHBrCH_2Br$	294.83	...	−18	215^{760} 93^{10}	2.170^{20}_{4}	1.5608^{20}	i	s	s	...	...	chl s	**B1**^{2}, 85
b2606	**—,1,3,3-tri-bromo-***	$CH_3CBr_2CH_2CH_2Br$	294.83	...	...	$200–5^{760}$ 70^{8}	1.4464^{20}_{4}	1.5564^{20}	i	s	s	...	...	chl s	**B1**^{3}, 298
Ω b2607	**—,2,2,3-tri-bromo-***	$CH_3CHBrCBr_2CH_3$	294.83	...	1.85	206^{760} 86^{10}	2.1724^{20}_{4}	1.5602^{20}	i	s	s	...	...	chl s	**B1**^{2}, 85
b2608	**—,1,1,3-tri-chloro-***	$CH_3CHClCH_2CHCl_2$	161.46	...	...	152^{753}	1.317^{15}	1.4600^{15}	i	s	s	...	...	chl v	**B1**^{3}, 285
b2609	**—,1,2,3-tri-chloro-***	$CH_3CHClCHClCH_2Cl$	161.46	...	...	$165–8^{725}$ 63^{28}	1.3164^{20}_{4}	1.4790^{20}	i	s	s	...	...	chl v	**B1**^{3}, 285
b2610	**—,2,2,3-tri-chloro-***	$CH_3CHClCCl_2CH_3$	161.46	...	...	$143–5^{760}$	1.2699^{20}_{4}	1.4645^{20}	i	...	...	...	...	chl v	**B1**^{3}, 285
Ω b2611	**—,2,2,3-tri-methyl-***	Triptan. $(CH_3)_2CHC(CH_3)_3$	100.21	...	−24.19	80.88^{760}	0.6901^{20}_{4}	1.3894^{20}	i	s	s	v	v	peth, CCl_4 v	**B1**^{3}, 454
Ω b2612	**1-Butanearsonic acid***	*n*-Butylarsonic acid. $CH_3CH_2CH_2CH_2AsO_3H_2$	182.05	...	160	...	...	...	v	s	i	...	...	...	**B4**^{2}, 997
b2613	**1-Butaneboronic acid, 3-methyl-***	Isoamylboric acid. $(CH_3)_2CHCH_2CH_2B(OH)_2$	115.97	pl (w)	169	...	...	...	v^h	s	s	s	...	os s	**B4**^{2}, 1023
—	**Butanedial***	*see* **Succinaldehyde**													
—	**1,1-Butanedi-carboxylic acid***	*see* **Malonic acid, propyl-**													
—	**Butanedioic acid***	*see* **Succinic acid**													
—	**—,2,3-dihydroxy-***	*see* **Tartaric acid**													
b2614	**1,2-Butanediol (*d*)***	$CH_3CH_2CHOHCH_2OH$	90.12	$[\alpha]^{20}_{D}$ +14.5 (al, c = 6)	...	$192–4$ $68^{0.4}$	$1.0059^{17.5}_{0}$	1.4375^{20}	v	∞	...	s	...	...	**B1**^{2}, 544
Ω b2615	**—(*dl*)***	α-Butyleneglycol. $CH_3CH_2CHOHCH_2OH$	90.12	...	...	190.5^{760} 96.5^{10}	1.0024^{20}_{4}	1.4378^{20}	s	s	...	s	...	...	**B1**^{2}, 545
b2616	**—(*l*)***	$CH_3CH_2CHOHCH_2OH$	90.12	$[\alpha]^{22}_{D}$ −7.4 (al, c = 4)	...	$94–6^{12}$	...	...	s	s	...	s	...	...	**B1**^{2}, 544
b2617	**1,3-Butanediol (*d*)***	β-Butyleneglycol. $CH_3CHOHCH_2CH_2OH$	90.12	$[\alpha]^{22}_{D}$ +18.5 (al, c = 4)	...	204 $60–5^{0.8}$	1.0053^{20}_{4}	1.4418^{20}	s	s	i	...	...	...	**B1**^{2}, 545
Ω b2618	**—(*dl*)***	$CH_3CHOHCH_2CH_2OH$	90.12	...	...	207.5^{760} $103–4^{8}$	1.0053^{20}_{4}	1.4410^{20}	...	...	...	...	...	...	**B1**^{2}, 545
b2619	**—(*l*)***	$CH_3CHOHCH_2CH_2OH$	90.12	$[\alpha]^{25}_{D}$ −18.8 (al, c = 4)	...	$107–10^{23}$	1.005^{20}_{4}	...	...	...	...	...	...	...	**B1**^{3}, 2167
Ω b2620	**—,sulfite**	...	135.16	...	5	185^{760} $76–7^{17}$	1.2352^{20}_{4}	1.4461^{20}	d^h	v	v	v	v	aa v	**B1**^{3}, 2171
Ω b2621	**1,4-Butanediol***	Tetramethyleneglycol. $HOCH_2CH_2CH_2CH_2OH$	90.12	...	20.1	235 (230) 120^{10}	1.0171^{20}_{4}	1.4460^{20}	∞	s	δ	...	...	...	**B1**^{2}, 545
b2622	**2,3-Butanediol (*d*)***	$CH_3CHOHCHOHCH_3$	90.12	$[\alpha]^{25}_{D}$ +12.5 (undil)	34 (25) (anh), 16.8 (+5w)	$180–2^{760}$	0.9872^{25}	1.4306^{25}	∞	∞	s	s	...	...	**B1**^{2}, 546
Ω b2623	**—(*dl*)***	$CH_3CHOHCHOHCH_3$	90.12	cr (Pr^i_2O)	7.6	182.5^{760} 86^{16}	1.0033^{20}_{4}	1.4310^{25}	∞	∞	s	s	...	...	**B1**^{2}, 546
b2624	**—(*l*)***	$CH_3CHOHCHOHCH_3$	90.12	$[\alpha]^{25}_{D}$ −13.0 (undil)	19.7	$178–81^{760}$ 77.5^{10}	0.9869^{25}_{4}	1.4340^{18}	∞	∞	s	s	...	...	**B1**^{2}, 547
b2625	**—(*meso*)***	$CH_3CHOHCHOHCH_3$	90.12	cr (Pr^i_2O)	34.4 (23.4)	181.7 83.5^{10}	1.0003^{20}_{4}	1.4367^{25}	v	s	s	...	...	...	**B1**^{2}, 546
Ω b2626	**—,2,3-dimethyl-***	Pinacol. Tetramethylethylene glycol. $(CH_3)_2COHCOH(CH_3)_2$	118.17	nd (al or eth)	43 (38)	174.4^{760}	...	...	δ v^h	v	v	...	...	CS_2 δ	**B1**^{2}, 553
Ω b2627	**—,—,hexa-hydrate***	Pinacol hydrate. $(CH_3)_2COHCOH(CH_3)_2.6H_2O$	226.27	pl (w + 6)	47	...	0.967^{15}	...	δ v^h	v	v	...	...	CS_2 δ	**B1**^{2}, 553
b2628	**—,2,3-diphenyl-***	Acetophenonepinacol. $CH_3COH(C_6H_5)COH(C_6H_5)CH_3$	242.32	pr (dil al)	121–2	...	...	...	i	v	v	...	...	peth δ	**B6**^{2}, 979

For explanations, symbols and abbreviations see beginning of table. For structural formulas see end of table.

No.	Name	Synonyms and Formula	Mol. wt.	Color, crystalline form, specific rotation and λ_{max} (log ε)	m.p. °C	b.p. °C	Density	n_D	Solubility w	al	eth	ace	bz	other solvents	Ref.
	1,3-Butanediol														
b2629	**1,2-Butanediol, 3-methyl-***	α-Isoamylene glycol. $(CH_3)_2CHCHOHCH_2OH$	104.15			206 $81–3^{5}$	0.9987^{0}_{4}		...	s	s	...	...		**B1**[2], 550
b2630	**1,3-Butanediol, 3-methyl-***	γ-Isoamylene glycol. $(CH_3)_2C(OH)CH_2CH_2OH$	104.15			202–3 108^{16}	0.9448^{20}_{4}	1.4452^{20}	s	s	...	...	...		**B1**[1], 251
b2631	**2,3-Butanediol, 2-methyl-***	β-Isoamylene glycol. $CH_3CH(OH)COH(CH_3)_2$	104.15			175 68^{5}	0.9920^{25}_{4}	1.4375^{20}	s	s	s	...	...		**B1**[2], 550
Ω b2632	**2,3-Butanedione***	Biacetyl. Dimethylglyoxal. $CH_3COCOCH_3$	86.09	λ^{hp} 510 (−3.22)	−2.4	88^{760}	$0.9808^{18.5}_{4}$	1.3951^{20}	v	∞	∞	v	s		**B1**[2], 824
Ω b2633	—,dioxime	Dimethyl glyoxime. $CH_3C(:NOH)C(:NOH)CH_3$	116.13	nd (to or dil al) λ^{al} 227 (4.23)	245–6 (cor) (235)	sub 234–5			i	v	v	...	δ	to δ	**B1**[2], 826
Ω b2634	—,monooxime	Biacetyl monooxime. $CH_3COC(:NOH)CH_3$	101.11	pr (chl), lf (w) λ^{al} 229 (4.11)	77–8	$185–6^{760}$			δ	v	v	...	...	chl v alk s	**B1**[2], 826
Ω b2635	**1,3-Butanedione, 1-phenyl-***	Benzoylacetone. $CH_3COCH_2COC_6H_5$	162.19	pr λ^{al} 310.5 (4.14)	56	$261–2^{760}$	1.0599^{74}_{4}	1.5678^{78}	i	...	s	...	...		**B7**[2], 616
Ω b2636	**1,4-Butane-dithiol-***	$HSCH_2CH_2CH_2CH_2SH$	122.25		−53.9	$195–6^{760}$ $110–2^{50}$	1.0621^{0}_{4}	1.5290^{20}	i	v	...	...	...		**B1**[3], 2177
b2637	**1-Butanephos-phonic acid***	Butylphosphonic acid. $CH_3CH_2CH_2CH_2PO_3H_2$	138.10	pl (bz)	106	d			v	v	v	...	δ v^{h}		**Am 75,** 3379
b2638	**2-Butanephos-phonic acid***	*sec*-Butylphosphonic acid. $CH_3CH_2CH(CH_3)PO_3H_2$	138.10	pl (bz-peth)	56	d			v	v	v	...	s	peth δ	**Am 75,** 3379
b2639	**1-Butanesulfonic acid,** amide	*n*-Butylsulfonamide. $CH_3CH_2CH_2CH_2SO_2NH_2$	137.12	lf (eth-lig)	48				v	v	v	...	v	lig i	**B4,** 8
b2640	—,chloride	$CH_3CH_2CH_2CH_2SO_2Cl$	156.63			75^{10}		1.4559^{20}	d	d^{h}	...	...	...		**B4,** 8
b2641	**1,2,3,4-Butane-tetracarboxylic acid (*dl*)***	$HO_2CCH_2[CH(CO_2H)]_2CH_2CO_2H$	234.16	lf (w) cr (ace)	236–7 (slow htng)				v	v	δ	...	i	chl δ lig i	**B2**[1], 333
Ω b2642	— (*meso*)*	$HO_2CCH_2[CH(CO_2H)]_2CH_2CO_2H$	234.16	nd or pr (w)	192–3d				v	s	...	...	i	lig i	**B2**[2], 702
Ω b2643	**1-Butanethiol***	*n*-Butyl mercaptan. $CH_3CH_2CH_2CH_2SH$	90.19	λ^{iso} 225–30 (2.20)	−115.67	98.46^{760}	0.8337^{20}_{4}	1.4440^{20}	δ	v	v	...	...		**B1**[2], 398
b2644	**2-Butanethiol (*d*)***	$CH_3CH_2CHSHCH_3$	90.19	$[\alpha]^{20}_{D}$ +15.7 λ^{iso} 225–30 (2.13)		85–95 37.4^{134}	0.8299^{20}_{4}	1.43385^{25}	...	s	s	...	s	peth s	**B1**[3], 1549
Ω b2645	—(*dl*)*	*sec*-Butyl mercaptan. $CH_3CH_2CHSHCH_3$	90.19			85^{760}	0.8295^{20}_{4}	1.4366^{20}	...	s	s	...	s	peth s	**B1**[3], 1548
b2646	—(*l*)*	$CH_3CH_2CHSHCH_3$	90.19	$[\alpha]^{17}_{D}$ −17.35		83–4	0.8300^{17}_{4}		...	s	s	...	s	peth s	**B1**[3], 1549
b2647	**1-Butanethiol, 2-methyl-(*d*)***	*act*-Amyl mercaptan. $CH_3CH_2CH(CH_3)CH_2SH$	104.22	$[\alpha]^{23}_{D}$ +3.21		118.2^{760}	0.8420^{20}_{4}	1.4440^{20}	...	...	...	...	...		**B1**[2], 421
Ω b2648	—,3-methyl-*	Isoamyl mercaptan. $(CH_3)_2CHCH_2CH_2SH$	104.22	λ^{iso} 225 sh (2.3)		118^{760}	0.8350^{20}_{4}	1.4418^{20}	i	∞	∞	...	...		**B1**[2], 434
—	**1,2,3-Butanetri-carboxylic acid, 2,3-dimethyl-***	*see* **Camphoronic acid**													
Ω b2649	**1,2,3-Butanetriol***	1-Methylglycerol. $CH_3CH(OH)CH(OH)CH_2OH$	106.12			170^{20} $140–2^{1}$		1.4462^{20}	v	v	...	...	...		**B1**[3], 2343
Ω b2650	**1,2,4-Butanetriol***	$HOCH_2CH(OH)CH_2CH_2OH$	106.12			$172–4^{12}$ $132–3^{1}$	1.018^{20}	1.4688^{20}	v	v	...	...	...		**B1**[3], 2344
Ω b2651	**Butanoic acid***	Butyric acid. $CH_3CH_2CH_2CO_2H$	88.12	λ^{w} 208 (1.8), 270 sh (−0.8)	−4.26 f.p. −19	163.53^{760}	0.9577^{20}_{4}	1.3980^{20}	∞	∞	∞	...	...		**B2**[3], 576
Ω b2652	—,allyl ester	$CH_3CH_2CH_2CO_2CH_2CH:CH_2$	128.17			$142–3^{772}$ 44.5^{15}	0.9017^{20}_{4}	1.4158^{20}	i	∞	∞	...	...		**B2,** 272
Ω b2653	—,amide	Butyramide. $CH_3CH_2CH_2CONH_2$	87.12	lf (bz) λ^{bz} 282 (2.68)	114.8	216	0.8850^{120}	1.4087^{130}	...	s	δ	...	i		**B2**[2], 251
b2654	—,—,*N,N*-diethyl-	$CH_3CH_2CH_2CON(C_2H_5)_2$	143.23			206 97^{16}		1.4403^{25}	∞	s	...	...	...		**B4**[2], 604
b2655	—,—,*N,N*-dimethyl-	$CH_3CH_2CH_2CON(CH_3)_2$	115.18		−40	185–8 $124–5^{100}$	0.9064^{25}_{4}	1.4391^{25}	s	s	∞	∞	∞	peth ∞ CS_2 ∞	**B4**[2], 564
Ω b2656	—,—,*N*-phenyl-	Butyranilide. $CH_3CH_2CH_2CONHC_6H_5$	163.22	mcl pr (al, bz, eth) λ^{al} 242 (4.17)	97	189^{15}	1.134		i	v	v	...	...		**B12**[2], 146
Ω b2657	—,anhydride*	Butyric anhydride. $(CH_3CH_2CH_2CO)_2O$	158.20	λ^{bz} 282.5 (1.90)	−75	199.4–201.4^{760} $85–6^{8}$	0.9668^{20}_{4}	1.4070^{20} (1.4124^{20})	d	d	s	...	...		**B2**[2], 251
Ω b2658	—,benzyl ester	$CH_3CH_2CH_2CO_2CH_2C_6H_5$	178.23			238–40 105^{7}	1.0111^{20}_{4}	1.4920^{20}	i	v	v	...	...		**B6**[2], 417
b2659	—,bromide	Butyryl bromide. $CH_3CH_2CH_2COBr$	151.01			128^{760}	1.4162^{17}_{4}	1.1596^{17}	...	...	...	...	...		**B2**[2], 251
Ω b2660	—,(4-bromo-phenacyl) ester		285.15	rod (dil al)	64				...	...	...	...	...		**B8**[2], 90
Ω b2661	—,butyl ester*	$CH_3CH_2CH_2CO_2(CH_2)_3CH_3$	144.22		−91.5	166.6^{760} 55^{13}	0.8709^{20}_{4}	1.4075^{20}	i	∞	∞	...	...		**B2**[2], 246
b2662	—,*sec*-butyl ester (*d*)	$CH_3CH_2CH_2CO_2CH(CH_3)CH_2CH_3$	144.22	$[\alpha]^{20}_{D}$ +22	−91.5	151.5^{747} 54^{18}	0.8737^{13}_{4}	1.4011^{20}	...	s	...	...	s	CS_2 s Py s	**B2**[2], 247
Ω b2663	—,—(*dl*)	$CH_3CH_2CH_2CO_2CH(CH_3)CH_2CH_3$	144.22			152.5^{760} 52^{16}	0.8609^{20}_{4}	1.4019^{20}_{4}	...	...	...	...	...		**C33,** 5806
b2664	—,*tert*-butyl ester	$CH_3CH_2CH_2CO_2C(CH_3)_3$	144.22			145–7		$1.4007^{17.5}$	...	s	s	s	...		**B2**[3], 601
Ω b2665	—,chloride	Butyryl chloride. $CH_3CH_2CH_2COCl$	106.55		−89	102^{760}	1.0277^{20}_{4}	1.4121^{20}	d	d	∞	...	...		**B2**[2], 251
Ω b2666	—,cyclohexyl ester*		170.25			212^{750}	0.9572^{0}_{4}		i	s	...	...	...	os s	**B6**[1], 6

For explanations, symbols and abbreviations see beginning of table. For structural formulas see end of table.

No.	Name	Synonyms and Formula	Mol. wt.	Color, crystalline form, specific rotation and λ_{max} (log ε)	m.p. °C	b.p. °C	Density	n_D	Solubility						Ref.
									w	al	eth	ace	bz	other solvents	
	Butanoic acid														
b2667	—,2,2-dimethyl-propyl ester*	*neo*-Pentyl butyrate. $CH_3CH_2CH_2CO_2CH_2C(CH_3)_3$	158.24	...	...	165–6^{760}	0.8719^{0}	...	...	...	...	...	...	...	**B2**, 272
Ω b2668	—,ethyl ester*	$CH_3CH_2CH_2CO_2C_2H_5$	116.16	...	−100.8	121–6^{760} 48.8^{50}	0.8785$^{20}_{4}$	1.4000^{20}	δ	s	s	...	...	...	**B2^2**, 244
b2669	—,furfuryl ester	...	168.20	...	...	212–3^{764}	1.0530$^{20}_{4}$	...	δ	s	∞	...	...	...	**B17^2**, 115
b2670	—,heptyl ester*	$CH_3CH_2CH_2CO_2(CH_2)_6CH_3$	186.30	...	−57.5	225.87^{760} 105^{10}	0.8637^{20}	1.4231^{20}	i	s	...	...	...	os s	**B2**, 272
Ω b2671	—,hexyl ester*	$CH_3CH_2CH_2CO_2(CH_2)_5CH_3$	172.27	...	−78	208	0.8652^{20}	1.4160$^{15}_{\alpha}$	i	s	...	...	...	os s	**B2**, 272
Ω b2672	—,isobutyl ester*	$CH_3CH_2CH_2CO_2CH_2CH(CH_3)_2$	144.22	...	...	157^{760}	0.8364$^{18}_{4}$	1.4032^{20}	δ	∞	∞	...	...	...	**B2^2**, 247
Ω b2673	—,isopropyl ester*	$CH_3CH_2CH_2CO_2CH(CH_3)_2$	130.19	...	...	130–1^{760}	0.8588$^{20}_{4}$	1.3936^{20}	i	s	...	...	...	...	**B2**, 271
Ω b2674	—,methyl ester*	$CH_3CH_2CH_2CO_2CH_3$	102.13	...	−84.8	102.3^{760}	0.8984$^{20}_{4}$	1.3878^{20}	δ	∞	∞	...	...	...	**B2^2**, 243
b2675	—,2-methylbutyl ester (*d*)*	$CH_3CH_2CH_2CO_2CH_2CH(CH_3)CH_2CH_3$	158.24	$[\alpha]^{20}_{D}$ +3.5	...	179^{765}	0.8620$^{20}_{4}$	1.4135^{20}	...	...	...	...	...	...	**B2^2**, 247
b2676	—,—(*dl*)	$CH_3CH_2CH_2CO_2CH_2CH(CH_3)CH_2CH_3$	158.24		...	166–6.7^{760}	0.862$^{20}_{4}$	1.4100^{25}	...	...	...	...	...	...	**C39**, 4083
b2677	—,2-methyl-2-butyl ester	*tert*-Amyl butyrate. $CH_3CH_2CH_2CO_2C(CH_3)_2CH_2CH_3$	158.24	...	...	164	0.8646$^{15}_{0}$	...	...	...	...	...	...	...	**B2**, 271
Ω b2678	—,3-methylbutyl ester*	Isoamyl butyrate. $CH_3CH_2CH_2CO_2CH_2CH_2CH(CH_3)_2$	158.24	...	...	178.5^{760} 65–8^{12}	0.8651$^{20}_{4}$	1.4110^{20}	i	v	v	...	...	...	**B2^2**, 247
b2679	—,1-methylpentyl ester (*d*)	*sec*-Hexyl butyrate $CH_3(CH_2)_2COCH(CH_3)CH_2$	172.27	$[\alpha]_D$ +10.16	...	85^{20}	0.8744$^{21}_{4}$	...	i	s	s	...	...	...	**B2^1**, 120
Ω b2680	—,nitrile	Butyronitrile. Propyl cyanide. $CH_3CH_2CH_2CN$	69.11	...	−112	118^{760}	0.7936$^{20}_{4}$	1.3842^{20}	δ	∞	∞	...	s	...	**B2^2**, 252
b2681	—,octyl ester*	$CH_3CH_2CH_2CO_2(CH_2)_7CH_3$	200.33	...	−55.6	244.1^{760}	0.8629^{20}	1.4267$^{15}_{He}$	i	s	...	...	...	os s	**B2^2**, 248
Ω b2682	—,pentyl ester*	Amyl butyrate. $CH_3CH_2CH_2CO_2(CH_2)_4CH_3$	158.24	...	−73.2	186.4 78–80^{18}	0.8713$^{15}_{4}$	1.4123^{20}	i	v	v	...	...	...	**B2^2**, 247
b2683	—,phenyl ester*	$CH_3CH_2CH_2CO_2C_6H_5$	164.21	λ^{al} 259 (2.3), 266 (2.3)	...	227–8^{760} (220–2) 85^{8}	1.0382$^{15}_{4}$	1.0267$^{15}_{15}$	i	s	s	...	...	...	**B6^2**, 155
Ω b2684	—,4-phenyl-phenacyl ester	...	282.34	cr (al)	97	...	...	...	i	...	...	...	...	...	**Am 73**, 5301
b2685	—,piperazinium salt	$2(CH_3CH_2CH_2CO_2H).C_4H_{10}N_2$	262.35	...	121–2	...	...	...	s	s	i	...	...	...	**Am 56**, 1759
Ω b2686	—,propyl ester*	$CH_3CH_2CH_2CO_2CH_3CH_2CH_3$	130.19	...	−97.2	143^{760} 39.2^{14}	0.8730$^{20}_{4}$	1.4001^{20}	δ	∞	∞	...	...	...	**B2^2**, 246
b2687	**—,4(5-ace-naphthyl)-4-oxo-**	...	254.29	nd (al or aa)	208d	...	...	...	...	s^h	...	...	s^h	xyl s^h	**B10^2**, 532
b2688	**—,2-acetyl-3-oxo-**, ethyl ester	Ethyl diacetoacetate. Ethyl α-acetylacetoacetate. $(CH_3CO)_2CHCO_2C_2H_5$	172.19	...	...	209–11 104^{16}	1.1045$^{20}_{4}$	1.4690^{20}	δ	v	v	...	v	...	**B3^2**, 467
b2689	**—,2-amino-(*d*)***	$CH_3CH_2CHNH_2CO_2H$	103.12	lf (dil al) $[\alpha]^{16}_{D}$ +8.4 (w, c = 4)	292d (303 sealed tube)	...	...	...	v	δ	i	...	i	...	**B4^2**, 831
Ω b2690	—,—(*dl*)*	$CH_3CH_2CHNH_2CO_2H$	103.12	lf (w)	304d (307d sealed tube)	sub	...	...	v	δ	i	...	i	...	**B4^2**, 831
Ω b2691	—,—(*l*)*	$CH_3CH_2CHNH_2CO_2H$	103.12	lf (w-al), cr (al) $[\alpha]^{20}_{D}$ −14.9 (w, c = 5) $[\alpha]^{20}_{D}$ +20.6 (5N HCl)	292d (>300)	...	...	...	s	δ	δ	...	i	MeOH s	**B4^2**, 831
b2692	**—,3-amino-(*d*)***	$CH_3CHNH_2CH_2CO_2H$	103.12	pr (MeOH) $[\alpha]^{20}_{D}$ +35.3 w, p = 9.6)	d at 220	...	...	...	...	...	...	...	...	...	**B4^1**, 504
Ω b2693	—,—(*dl*)*	$CH_3CHNH_2CH_2CO_2H$	103.12	nd (al)	193–4	...	...	...	v	i	i	...	i	...	**B4^2**, 833
b2694	—,—(*l*)*	$CH_3CHNH_2CH_2CO_2H$	103.12	pr (MeOH) $[\alpha]^{20}_{D}$ −35.2 (w, p = 10)	d at 220	...	...	...	v	i	i	...	i	...	**B4^1**, 504
Ω b2695	**—,4-amino-***	Piperidinic acid. $H_2NCH_2CH_2CH_2CO_2H$	103.12	pr or nd (dil al), lf (MeOH-eth)	203d	...	...	...	v	δ	i	δ	i	MeOH δ^h	**B4^2**, 837
—	**—,2-amino-3-hydroxy-***	*see* **Threonine**													
b2696	**—,3-amino-2-hydroxy-***	3-Methylisoserine. $CH_3CHNH_2CH(OH)CO_2H$	119.12	pr (dil al)	200d	...	...	...	s	δ	i	...	...	MeOH δ	**B4**, 515
b2697	**—,4-amino-2-hydroxy-***	$H_2NCH_2CH_2CH(OH)CO_2H$	119.12	pr (w or dil al)	214	...	...	...	v	δ^h	i	...	...	...	**B4^1**, 548
b2698	**—,4-amino-3-hydroxy-(*dl*)***	$H_2NCH_2CH(OH)CH_2CO_2H$	119.12	pr (w), cr (dil al)	218	...	...	...	δ v^h	δ	δ	δ	δ	MeOH δ oos δ	**B4^2**, 937
—	**—,2-amino-2-methyl-***	*see* **Isovaline**													
—	**—,2-amino-3-methyl-***	*see* **Valine**													
b2699	**—,3-amino-3-methyl-***	$(CH_3)_2CNH_2CH_2CO_2H$	117.15	pr (w + 1), cr (dil al), nd (eth-al)	217	...	...	...	v	δ	i	...	...	...	**B4^3**, 1364
—	**—,2-amino-4(methylthio)-***	*see* **Methionine**													

For explanations, symbols and abbreviations see beginning of table. For structural formulas see end of table.

No.	Name	Synonyms and Formula	Mol. wt.	Color, crystalline form, specific rotation and λ_{max} (log ε)	m.p. °C	b.p. °C	Density	n_D	Solubility w	al	eth	ace	bz	other solvents	Ref.
	Butanoic acid														
b2700	—,**2-benzoyl-**, ethyl ester	Ethyl benzoylethyl acetate. $CH_3CH_2CH(COC_6H_5)CO_2C_2H_5$	220.27	...	...	152^{7} $122^{1.5}$	1.0706_4^{15}	1.509^{15}	i	...	s	...	...	...	**B10²**, 488
—	—,**2-benzylidene-**	*see* **Cinnamic acid, α-ethyl-**													
b2701	—,**2-benzylidene-3-oxo-**, ethyl ester	Ethyl α-acetylcinnamate. Ethyl α-benzalacetoacetate. $C_6H_5CH{:}C(COCH_3)CO_2C_2H_5$	218.26	rh pl or pyr (dil al)	60–1	$295–7^{760}$	...	...	i	δ	δ	...	δ	chl v CS_2 δ	**B10²**, 503
b2702	—,**2-bromo-**(*d*)*	$CH_3CH_2CHBrCO_2H$	167.01	$[\alpha]_D^{20}$ +35.2 (eth, c = 20), +12.9 (w, c = 4)	...	$105–7^{15}$ $66–9^{0.04}$	1.568_4^{20}	1.4483^{15}	...	s	...	...	...	...	**B2²**, 255
Ω b2703	—,—(*dl*)*	$CH_3CH_2CHBrCO_2H$	167.01	...	−4	217d 108^{13}	1.5669_{20}^{20}	...	s	s	s	...	...	...	**B2²**, 255
b2704	—,—,amide	$CH_3CH_2CHBrCONH_2$	166.03	lf (bz), nd (ace)	112–3	...	...	...	v	v	v	v^h	v^h	...	**B2¹**, 125
b2705	—,—,bromide	$CH_3CH_2CHBrCOBr$	229.91	...	...	172–4 $57–60^{10}$	...	...	d	d	s	...	...	...	**B2²**, 255
Ω b2706	—,—,chloride	$CH_3CH_2CHBrCOCl$	185.45	...	...	150–2 43^{12}	1.5320^{20}	...	d^h	d^h	s	...	...	...	**B2¹**, 125
Ω b2707	—,—,ethyl ester*	$CH_3CH_2CHBrCO_2C_2H_5$	195.06	...	...	177.5^{765} $43–4^{5.5}$	1.3297_{20}^{20}	1.4475^{20}	i	∞	∞	...	...	...	**B2²**, 255
b2708	—,—,methyl ester*	$CH_3CH_2CHBrCO_2CH_3$	181.04	...	...	170–2 $75–8^{18}$	1.4528^{20}	1.4029^{25}	...	s	...	...	...	...	**B2¹**, 125
b2708¹	—,**4-bromo-***	$BrCH_2CH_2CH_2CO_2H$	167.01	...	33	$124–7^{7}$	...	...	...	...	...	...	...	...	**B2²**, 256
b2709	—,—,methyl ester*	$BrCH_2CH_2CH_2CO_2CH_3$	181.04	...	...	186–7 86^{15}	1.371^{25}	1.4567^{25}	...	s	...	...	...	...	**B2²**, 283
Ω b2710	—,—,nitrile	$BrCH_2CH_2CH_2CN$	148.01	...	...	$205–7^{760}$ 91^{12}	1.4967_4^{20}	1.4818^{20}	...	s	s	...	...	...	**B2²**, 256
Ω b2711	—,**2-bromo-2-ethyl-**, amide	Neuronal. $(C_2H_5)_2CBrCONH_2$	194.08	...	67	...	...	...	δ v^h	v	v	...	v	peth δ	**B2²**, 293
b2712	—,**2-bromo-2-ethyl-3-methyl-**, amide	2-Bromo-2-isopropylbutyramide. Neodorme. $(CH_3)_2CHCBr(C_2H_5)CONH_2$	208.11	nd (sub)	50–1	sub	...	...	δ d^h	v	v	s	v	peth s os s	**B2²**, 299
b2713	—,**2-bromo-3-methyl-**(*d*)*	$(CH_3)_2CHCHBrCO_2H$	181.04	pr (peth) $[\alpha]_D^{20}$ +22.8 (bz, p = 4), +4.5 (w, c = 1.6), +7.5 (eth, c = 7)	43.5–5.5	230^{760} $95–100^{2}$	...	...	δ	v	s	s	s	peth v^h	**B2²**, 279
Ω b2714	—,—(*dl*)*	$(CH_3)_2CHCHBrCO_2H$	181.04	pr (eth or chl)	44	230^{760}δd $136–40^{25}$	1.459^{20}	...	δ	v	s		s	...	**B2²**, 279
b2715	—,—(*l*)*	$(CH_3)_2CHCHBrCO_2H$	181.04	cr (peth) $[\alpha]_D^{20}$ −21.6 (bz, p = 4)	43.5–5.5	150^{40} $119–20^{14}$	...	...	δ	v	s	s	s	...	**B2²**, 279
b2716	—,—,amide (*dl*)	$(CH_3)_2CHCHBrCONH_2$	180.05	lf (bz)	133	...	...	...	s^h	...	...	...	...	lig s^h	**B2**, 318
Ω b2717	—,—,ethyl ester (*dl*)*	$(CH_3)_2CHCHBrCO_2CO_2H_5$	209.09	...	...	186 $73–4^{12}$	1.2760_4^{20}	1.4496^{20}	...	v	v	...	...	...	**B2²**, 279
b2718	—,—,methyl ester (*dl*)*	$(CH_3)_2CHCHBrCO_2CH_3$	195.06	...	...	176.4–7.8^760^ $64–5^{11}$	1.353_{13}^{13}	1.4530^{20}	...	s	s	...	...	...	**B2**, 317
b2719	—,**3-bromo-3-methyl-***	$(CH_3)_2CBrCH_2CO_2H$	181.04	nd (lig)	73.5	...	...	...	δ	v	v	...	v	lig i	**B2²**, 278
b2720	—,**2-bromo-3-oxo-**, amide, *N*-phenyl-	α-Bromoacetoacetanilide. $CH_3COCHBrCONHC_6H_5$	256.11	lf (al)	138d	...	...	...	δ	s^h	δ	...	...	chl δ	**B12¹**, 276
b2721	—,—,ethyl ester*	Ethyl α-bromoacetoacetate. $CH_3COCHBrCO_2C_2H_5$	209.05	...	...	210–5d $104–10^{15}$	1.4294_4^{14}	1.463^{14}	...	s	s	...	...	...	**B3²**, 427
b2722	—,**4-bromo-3-oxo**, ethyl ester*	Ethyl γ-bromoacetoacetate. $BrCH_2COCH_2CO_2C_2H_5$	209.05	...	...	$114–7^{14}$ $56^{0.005}$	1.4840_4^{20}	1.5281^{20}	δ	v	v	...	...	...	**B3²**, 427
Ω b2723	—,**2-chloro-***	$CH_3CH_2CHClCO_2H$	122.55	...	...	189^{627} 101^{15}	1.1796_4^{20}	1.4411^{20}	δ v^h	v	v	...	...	...	**B2²**, 253
b2724	—,—,chloride	$CH_3CH_2CHClCOCl$	141.00	...	...	$130–1^{760}$ $51–2^{41}$	1.2360^{17}	1.4475^{20}	d	d	s	...	...	...	**B2¹**, 123
b2725	—,—,ethyl ester*	$CH_3CH_2CHClCO_2C_2H_5$	150.61	...	...	$163–4^{760}$ 63^{70}	1.0560^{20}	1.4248^{20}	...	s	s	...	...	...	**B2**, 277
b2726	—,—,methyl ester*	$CH_3CH_2CHClCO_2CH_3$	136.58	...	...	$145–6^{756}$	1.0979^{14}	1.4247^{20}	...	s	s	...	...	MeOH s	**B2**, 277
Ω b2727	—,**3-chloro-**(*dl*)*	$CH_3CHClCH_2CO_2H$	122.55	cr (eth)	16	116^{22}	1.1898_4^{20}	1.4421^{20}	...	s	v^h	...	...	...	**B2²**, 253
b2728	—,—,chloride	$CH_3CHClCH_2COCl$	141.00	...	...	$40–1^{12}$	1.2163_4^{20}	1.4509^{20}	d	d	...	...	...	CS_2 s	**B2²**, 253
b2729	—,—,ethyl ester*	$CH_3CHClCH_2CO_2C_2H_5$	150.61	...	...	109 65^{15}	1.0517_4^{20}	1.4246^{20}	i	s	...	...	...	CCl_4 s	**B2²**, 253
b2730	—,—,methyl ester*	$CH_3CHClCH_2COCH_3$	136.58	...	...	$155–6^{760}$	1.0996_4^{20}	1.4258^{20}	i	...	v	...	...	...	**B2**, 277
b2731	—,**4-chloro-***	$ClCH_2CH_2CH_2CO_2H$	122.55	...	16	196^{22} $68^{0.2}$	1.2236_4^{20}	1.4642^{20}	δ	...	v	...	...	...	**B2²**, 253
Ω b2732	—,—,chloride	$ClCH_2CH_2CH_2COCl$	141.00	...	...	$173–4^{760}$ $60–1^{12}$	1.2581_4^{20}	1.4616^{20}	d	d	s	...	...	...	**B2²**, 254
b2733	—,—,ethyl ester*	$ClCH_2CH_2CH_2CO_2C_2H_5$	150.61	...	...	186^{760} 77^{10}	1.0756_4^{20}	1.4311^{20}	...	s	s	s	...	...	**B2²**, 254
Ω b2734	—,—,methyl ester*	$ClCH_2CH_2CH_2CO_2CH_3$	136.58	...	...	$175–6^{764}$ 55^{4}	1.1201_4^{20}	1.4321^{20}	i	v^h	v	s	...	...	**B2**, 278
Ω b2735	—,—,nitrile	$ClCH_2CH_2CH_2CN$	103.55	...	...	$189–91^{760}$ (196–7) 75^{11}	1.0934^{15}	1.4413^{20}	i	s	s	...	...	...	**B2²**, 254

For explanations, symbols and abbreviations see beginning of table. For structural formulas see end of table.

No.	Name	Synonyms and Formula	Mol. wt.	Color, crystalline form, specific rotation and λ_{max} (log ε)	m.p. °C	b.p. °C	Density	n_D	Solubility w	al	eth	ace	bz	other solvents	Ref.
	Butanoic acid														
b2736	**—,2-chloro-2-methyl-***	$CH_3CH_2CCl(CH_3)CO_2H$	136.58	...	...	$200\text{–}5^{754}$ δd	1.1204_4^{20}	1.4445^{20}	i	s	s	...	...	...	**B2,** 306
b2737	—,—,chloride	$CH_3CH_2CCl(CH_3)COCl$	155.04	...	...	144^{750}	1.187^{14}	...	d	...	s	...	...	...	**B2,** 307
b2738	—,—,ethyl ester*	$CH_3CH_2CCl(CH_3)CO_2C_2H_5$	164.63	...	...	175^{747}	1.069^{14}	1.4368^{11}	i	s	s	...	...	...	**B2,** 306
b2739	**—,2-chloro-3-methyl-***	2-Chloroisovaleric acid. $(CH_3)_2CHCHClCO_2H$	136.58	...	20–2	$210\text{–}2^{756}$ 126^{32}	1.135^{13}	1.4450^{11}	i	s	s	...	...	...	**B2,** 316
b2740	—,—,chloride	$(CH_3)_2CHCHClCOCl$	155.04	...	...	$148\text{–}9^{760}$	1.135^{13}	...	d	d	s	...	...	...	**B2,** 316
b2741	—,—,ethyl ester*	$(CH_3)_2CHCHClCO_2C_2H_5$	164.63	...	...	$178\text{–}9^{756}$	1.021^{13}	...	i	s	s	...	...	...	**B2,** 316
b2742	**—,2-chloro-2-, methyl-3-oxo-,** ethyl ester*	Ethyl α-chloro-α-acetopropionate. $CH_3COCCl(CH_3)CO_2C_2H_5$	178.62	...	...	$116\text{–}7^{75}$	1.157_4^{18}	1.4382^{18}	...	s	...	...	...	...	**B3**[1], 237
b2743	**—,2-chloro-3-oxo-, amide,** *N*-phenyl-	α-Chloroacetoacetanilide. $CH_3COCHClCONHC_6H_5$	211.65	nd (al)	137.5	...	...	...	i	v	...	...	s	chl v MeOH v	**B12**[2], 267
b2744	—,—,ethyl ester*	Ethyl α-chloroacetoacetate. $CH_3COCHClCO_2C_2H_5$	164.59	...	...	197^{748} δd $86\text{–}9^{12}$	1.19_{17}^{14}	1.4414^{20}	δ	v	v	...	...	...	**B3**[2], 426
b2745	**—,4-chloro-3-oxo-, chloride**	$ClCH_2COCH_2COCl$	154.99	...	...	$117\text{–}9^{17}$	1.4397_4^{20}	1.4860^{20}	d	d	...	...	s	CCl_4 s	**Am 62,** 1147
b2746	—,—,ethyl ester*	Ethyl γ-chloroacetoacetate. $ClCH_2COCH_2CO_2C_2H_5$	164.59	...	−8	220d 115^{14}	1.2157_4^{20}	1.4546^{17}	δ	s	s	s	s	chl s oos s	**B3**[2], 426
b2747	**—,4(4-chlorophenyl)-3-hydroxy-3-oxo-2-phenyl-***	...	304.73	nd (eth-peth)	139–40	...	...	...	i	s^h	s	s	s	peth δ	**B10**[2], 710
b2748	**—,2,3-dibromo-** (high m.p.)*	Dibromocrotonic acid. $CH_3CHBrCHBrCO_2H$	245.91	nd (eth), mcl (CS_2)	87	$100\text{–}10^{20}$	...	...	δ d^h	v	v	...	v	...	**B2**[3], 634
b2749	—,— (low m.p.)*	Dibromoisocrotonic acid. $CH_3CHBrCHBrCO_2H$	245.91	nd (lig)	59–60	...	...	...	δ	v	v	...	...	...	**B2**[1], 125
Ω b2750	—,—,ethyl ester*	$CH_3CHBrCHBrCO_2C_2H_5$	273.96	nd	58–9	113^{30} $103\text{–}4^{17}$	...	...	δ	s	s	...	...	...	**B2,** 285
Ω b2751	**—,2,4-dibromo-,** ethyl ester*	$BrCH_2CH_2CHBrCO_2C_2H_5$	273.96	...	...	$149\text{–}50^{52}$	1.6990_0^{20}	1.4960^{20}	i	s	s	...	...	...	**B2,** 285
b2752	**—,2,2-dibromo-3-oxo-,** ethyl ester*	Ethyl α,α-dibromoacetoacetate. $CH_3COCBr_2CO_2C_2H_5$	287.96	...	...	$120\text{–}4^{13}$	...	...	δ	s	s	...	...	...	**B3**[2], 427
b2753	**—,2,3-dichloro-** (high m.p.)*	Dichloroisocrotonic acid. $CH_3CHClCHClCO_2H$	157.00	pr (dil al)	78	131.5^{20}	...	...	δ	v	v	...	...	peth δ	**B2,** 279
b2754	—,— (low m.p.)*	Dichlorocrotonic acid. $CH_3CHClCHClCO_2H$	157.00	pr (dil al)	63	$124\text{–}5^{20}$	...	...	δ	s	s	...	s	CS_2 v lig δ chl s	**B2,** 279
b2755	**—,2,2-dichloro-3-oxo-,** ethyl ester	Ethyl α,α-dichloroacetoacetate. $CH_3COCCl_2CO_2C_2H_5$	199.04	...	...	$205\text{–}7^{756}$ 91^{11}	1.293_{17}^{16}	1.4492^{17}	...	s	...	...	...	...	**B3**[2], 427
b2756	**—,2,2-diethyl-3-oxo-,** ethyl ester*	$CH_3COC(C_2H_5)_2CO_2C_2H_5$	186.25	...	...	$215\text{–}6^{744}$ 64^3	0.9717_4^{18}	1.4326^{17}	i	∞	∞	...	...	...	**B3**[3], 1958
Ω b2757	**—,2,4-dihydroxy-3,3-dimethyl-,** amide, *N*(3-hydroxypropyl)-	Panthothenyl alcohol. $HOCH_2C(CH_3)_2CHOHCONH(CH_2)_2CH_2OH$	205.26	$[\alpha]_D^{20}$ +29.7 (w, 3%)	...	$118\text{–}20^{0.2}$	...	1.4935^{20}	v	v	δ	...	...	...	**C41,** 6199
Ω b2758	**—,2,2-dimethyl-***	$CH_3CH_2C(CH_3)_2CO_2H$	116.16	...	−14	186^{760} 80^{11}	0.9276_4^{20}	1.4145^{20}	δ	s	s	...	...	...	**B2**[2], 293
b2759	—,—,chloride	$CH_3CH_2C(CH_3)_2COCl$	134.61	...	...	132^{760} 27^{11}	0.9801_4^{20}	1.4245^{20}	d	d	s	...	...	...	**B2,** 336
b2760	**—,2,3-dimethyl-***	$(CH_3)_2CHCH(CH_3)CO_2H$	116.16	...	−1.5	191.7^{760} 90^{16}	0.9275_4^{20}	1.4146^{20}	s	s	s	...	...	...	**B2**[2], 294
b2761	—,—,amide	$(CH_3)_2CHCH(CH_3)CONH_2$	115.18	pl (ace-peth)	130.9	...	...	...	δ	s	s	...	...	...	**B2,** 338
b2762	—,—,chloride	$(CH_3)_2CHCH(CH_3)COCl$	134.61	...	...	$135\text{–}6^{751}$ $38\text{–}9^{18}$	0.9795_4^{20}	...	d^h	d^h	s	...	...	...	**B2**[3], 761
Ω b2763	**—,3,3-dimethyl-***	*tert*-Butylacetic acid. $(CH_3)_3CCH_2CO_2H$	116.16	...	6–7	190^{760} 96^{26}	0.9124_4^{20}	1.4096^{20}	...	s	s	...	...	...	**B2,** 337
Ω b2764	—,—,amide	$(CH_3)_3CCH_2CONH_2$	115.18	lf (w or ace-peth)	134	...	...	...	...	s	...	...	...	...	**B2**[3], 759
b2765	—,—,chloride	$(CH_3)_3CCH_2COCl$	134.61	...	...	$128\text{–}30^{745}$ 68^{100}	0.9696_4^{20}	1.4210^{20}	d	d	v	...	...	...	**B2**[3], 759
b2766	**—,2,2-dimethyl-3-oxo-,** ethyl ester	$CH_3COC(CH_3)_2CO_2C_2H_5$	158.20	...	...	184^{760} $40\text{–}1^3$	0.9773_{20}^{20}	1.4180^{20}	δ	s	s	...	...	...	**B3**[2], 439
b2767	**—,2,3-dioxo-,** 2-phenylhydrazone	$CH_3COC(:NNHC_6H_5)CO_2H$	206.20	ye nd or lf (al-eth)	161	...	...	...	i	s^h	s^h	...	...	...	**B15**[2], 133
b2768	**—,2,4-diphenyl-3-oxo-,** nitrile	Benzyl α-cyanobenzyl ketone. $C_6H_5CH_2COCH(C_6H_5)CN$	235.29	nd (bz-peth)	86	...	...	...	i	s	s	...	s	aa s	**B10,** 762
b2769	**—,2,4-diphenyl-4-oxo-,** nitrile	$C_6H_5COCH_2CH(C_6H_5)CN$	235.29	pl (al) λ^{al} 244 (4.16), 268 sh (3.11), 281 (3.04)	127.5	...	...	...	...	v	v	...	...	chl v	**B10**[2], 528
b2770	**—,3,4-epoxy-,** nitrile	Epicyanohydrin.	83.09	pr	162	...	...	...	δ s^h	s	...	...	...	...	**B18,** 261
b2771	**—,2,3-epoxy-3-phenyl-,** ethyl ester*	...	206.24	...	...	272–5 $147\text{–}9^{12}$	1.0442^{20}	1.5182^{20}	...	...	...	...	...	...	**B18**[1], 442
Ω b2772	**—,2-ethyl-***	Diethylacetic acid. $(C_2H_5)_2CHCO_2H$	116.16	...	−31.8	194^{760} 90^{13}	0.9239_4^{20}	1.4132^{20}	δ	∞	∞	...	...	...	**B2**[2], 291

For explanations, symbols and abbreviations see beginning of table. For structural formulas see end of table.

No.	Name	Synonyms and Formula	Mol. wt.	Color, crystalline form, specific rotation and λ_{max} (log ε)	m.p. °C	b.p. °C	Density	n_D	Solubility w	al	eth	ace	bz	other solvents	Ref.
	Butanoic acid														
b2773	—,—,amide,*N,N*-diethyl-	$(C_2H_5)_2CHCON(C_2H_5)_2$	171.29	...	...	220–1 108^{12}	...	...	δ	s	v	...	v	...	**B4**, 111
b2774	—,—,chloride	$(C_2H_5)_2CHCOCl$	134.61	...	...	140^{760} 40^{20}	0.9825_4^{20}	1.4234^{20}	d	d^h	v	...	...	...	**B2²**, 292
b2775	—,—,nitrile	3-Cyanopentane. $(C_2H_5)_2CHCN$	97.16	...	...	145.1– 5.6^{760} 37^{11}	...	1.3891^{24}	...	∞	∞	...	...	...	**B2²**, 292
b2776	—,**2-ethyl-2-methyl-***	$(C_2H_5)_2C(CH_3)CO_2H$	130.19	...	< −20	208.4– 8.6^{760} 104^{13}	...	1.4250^{20}	i	s	...	...	...	...	**B2²**, 299
Ω b2777	—,**2-ethyl-3-oxo-**, ethyl ester*	Ethyl ethylacetoacetate. $CH_3COCH(C_2H_5)CO_2C_2H_5$	158.20	...	...	190^{760} (195^{743}) 58^{3}	0.9847_4^{16}	1.4214^{25}	...	v	s	...	...	...	**B3²**, 438
b2778	—,—,methyl ester*	$CH_3COCH(C_2H_5)CO_2CH_3$	144.17	...	...	182^{760} $79–80^{14}$	0.995^{14}	...	...	s	s	s	...	...	**B3²**, 437
b2779	—,**4-fluoro-***	$FCH_2CH_2CH_2CO_2H$	106.10	...	...	$76–8^{5}$	...	1.3993^{25}	s	v	v	...	...	...	**B2³**, 621
Ω b2780	—,**2-hydroxy-(*dl*)***	$CH_3CH_2CHOHCO_2H$	104.12	nd (CCl_4)	44–4.5	260d 140^{14}	1.125^{20}	...	s	s	s	...	...	...	**B3²**, 216
b2781	—,—,ethyl ester (*d*)*	$CH_3CH_2CHOHCO_2C_2H_5$	132.16	$[\alpha]_D^{22}$ +8.4	...	165–70	0.978^{15}	1.4101	...	s	...	...	...	...	**B3²**, 215
b2782	—,—,—(*dl*)*	$CH_3CH_2CHOHCO_2C_2H_5$	132.16	...	...	167^{760} 74.5^{25}	1.0069_4^{20}	1.4179^{20}	...	s	...	...	...	...	**B3²**, 217
Ω b2783	—,**3-hydroxy-(*dl*)***	$CH_3CHOHCH_2CO_2H$	104.12	...	48–50	130^{12-14} $94–6^{0.1}$	...	1.4424^{20}	v	v	v	...	i	...	**B3²**, 220
b2784	—,—(*l*)*	$CH_3CHOHCH_2CO_2H$	104.12	$[\alpha]_D^{25}$ −24.5 (w, c = 5)	49–50	...	...	...	v	v	v	...	i	...	**B3²**, 219
Ω b2785	—,—,ethyl ester (*dl*)*	$CH_3CHOHCH_2CO_2C_2H_5$	132.16	...	...	$184–5^{755}$ $76–7^{15}$	1.017_4^{20}	1.4182^{20}	s	s	...	...	...	...	**B3²**, 221
b2786	—,—,methyl ester (*l*)*	$CH_3CHOHCH_2CO_2CH_3$	118.13	$[\alpha]_D^{20}$ −21.09	...	$76–7^{20}$	1.058_{20}^{20}	...	v	v	v	...	v	peth δ	**B3²**, 219
b2787	—,**4-hydroxy-***	$HOCH_2CH_2CH_2CO_2H$	104.12	...	< −17	d at 178–80	...	...	...	...	...	...	...	...	**B3²**, 222
Ω b2788	—,—,lactone*	γ-Butyrolactone	86.09	λ^{MeOH} 209 (1.63)	−42	206^{760} 89^{12}	1.1286_0^{16}	1.4341^{20}	∞	v	v	v	v	...	**B17²**, 286
b2789	—,**2-hydroxy-2-isopropyl-3-methyl-**, nitrile	Diisopropylketone cyanohydrin. $[(CH_3)_2CH]_2C(OH)CN$	141.22	rh (eth or peth)	59	111^{18}	...	...	i	s	s	s	s	os s	**B3²**, 239
b2790	—,**2-hydroxy-3-methyl-(*d*)***	$(CH_3)_2CHCHOHCO_2H$	118.13	cr (eth-peth) $[\alpha]_D^{20}$ −1.81 (w, c = 12)	69.5	$124–5^{13}$	...	...	s	s	s	v	...	...	**B3**, 328
Ω b2791	—,—(*dl*)*	$(CH_3)_2CHCHOHCO_2H$	118.13	rh bipyr	86	...	...	...	s	s	s	s	...	...	**B3**, 328
b2792	—,**3-hydroxy-3-methyl-***	$(CH_3)_2COHCH_2CO_2H$	118.13	...	< −32	162^{12}	0.9384_4^{20}	1.5081^{20}	s	s	s	...	...	...	**B3**, 327
b2793	—,**4-hydroxy-2-methylene-**, lactone*	...	98.10	λ^{al} < 220	...	$85–6^{10}$ $37^{0.35}$	1.1206^{20}	1.4650^{20}	s	v	s	s	s	os s	**Am 68**, 2332
—	—,**4-hydroxy-3-oxo-**, lactone*	*see* **Furan, 2,4-dioxo-, (tetrahydro)-**													
Ω b2794	—,**4(3-indolyl)-**	...	203.24	pl (bz-peth)	124	...	...	...	...	...	...	...	v^h	peth i	**B22²**, 54
b2795	—,**2-isopropyl-3-oxo-**, ethyl ester*	$CH_3COCH(C_3H_7^i)CO_2C_2H_5$	172.23	...	...	201^{758} $97–8^{20}$	0.9648_4^{18}	$1.4256_4^{18.5}$	i	∞	∞	...	...	...	**B3²**, 441
b2795[1]	—,**4-oxo-4(4-methoxyphenyl)-**	...	208.22	nd	150–1	...	...	...	s^h	s	...	...	s	chl s	**B10²**, 681
b2796	—,**2-methyl-(*d*)***	$CH_3CH_2CH(CH_3)CO_2H$	102.13	$[\alpha]_D^{15}$ +19.2 (w)	...	176^{760} 77^{12}	0.9419_4^{20}	1.4058^{20}	δ	∞	∞	...	...	...	**B2²**, 270
Ω b2797	—,—(*dl*)*	$CH_3CH_2CH(CH_3)CO_2H$	102.13	λ 205 (4.1), 255 (2.2)	< −80	177 $71.5^{10.5}$	0.9410_4^{20}	1.4051^{20}	δ	∞	∞	...	...	...	**B2²**, 270
b2798	—,—(*l*)*	$CH_3CH_2CH(CH_3)CO_2H$	102.13	$[\alpha]_D^{14}$ −10.1 (al) $[\alpha]_D^{20}$ −24 (w, c = 0.9)	...	176–7 $71–2^{12}$	0.9340_4^{20}	1.4042^{25}	δ	∞	∞	...	...	...	**B2³**, 685
b2799	—,—,chloride (*dl*)	$CH_3CH_2CH(CH_3)COCl$	120.58	...	...	116^{760}	0.9917_4^{20}	1.4170^{20}	d^h	d^h	...	...	...	...	**B2**, 315
b2800	—,—,ethyl ester (*d*)*	$CH_3CH_2CH(CH_3)CO_2C_2H_5$	130.19	$[\alpha]_{5892}^{26}$ +5.16	...	$131–3^{730}$ 35^{16}	0.8689_4^{25}	1.3964^{20}	i	s	...	...	s	...	**B2**, 306
b2801	—,—,nitrile (*dl*)	$CH_3CH_2CH(CH_3)CN$	83.13	...	...	125^{760}	0.7913_4^{15}	1.3933^{20}	...	s	s	...	...	...	**B2**, 306
Ω b2802	—,**3-methyl-***	Isovaleric acid. $(CH_3)_2CHCH_2CO_2H$	102.13	...	−29.3 (−37)	176.7^{760}	0.9286_4^{20}	14033^{20}	s	∞	∞	...	...	chl ∞	**B2²**, 271
b2803	—,—,amide	$(CH_3)_2CHCH_2CONH_2$	101.15	mcl lf (al)	137	$224–8^{760}$	...	...	s	s	s	...	...	peth v	**B2²**, 277
b2804	—,—,anhydride*	$[(CH_3)_2CHCH_2CO]_2O$	186.25	...	...	215^{762} $102–3^{15}$	0.9327_4^{20}	1.4043^{20}	d^h	d^h	s	...	...	...	**B2²**, 277
b2805	—,—,bromide	Isovaleryl bromide. $(CH_3)_2CHCH_2COBr$	165.03	mcl pl (al)	...	143^{760}	...	...	d	d^h	s	...	...	...	**B2²**, 277
Ω b2806	—,—,chloride	Isovaleryl chloride. $(CH_3)_2CHCH_2COCl$	120.58	...	...	114.5– 5.5^{771}	0.9844_4^{20}	1.4149^{20}	d	d	s	...	...	...	**B2²**, 277
Ω b2806[1]	—,—,ethyl ester	$(CH_3)_2CHCH_2CO_2C_2H_3$	130.19	...	−99.3	134.7^{760}	0.8656_4^{20}	1.3962^{20}	...	v	v	...	...	...	**B2²**, 275
Ω b2807	—,—,isobutyl ester	$(CH_3)_2CHCH_2CO_2C_4H_9^i$	158.24	...	...	171.4^{760} $60–2^{12}$	0.8736_4^{20}	1.4057^{20}	i	∞	∞	v	...	...	**B2²**, 276
b2808	—,—,isopropyl ester*	$(CH_3)_2CHCH_2CO_2C_3H_7^i$	144.22	...	...	142^{756} 68.5– 70^{55}	0.8538^{17}	$1.3960^{20.5}$	i	s	s	v	...	...	**B2²**, 275
Ω b2809	—,—,methyl ester*	$(CH_3)_2CHCH_2CO_2CH_3$	116.16	...	...	116.7^{760}	0.8808_4^{20}	1.3927^{20}	i	v	v	v	...	...	**B2²**, 274

For explanations, symbols and abbreviations see beginning of table. For structural formulas see end of table.

No.	Name	Synonyms and Formula	Mol. wt.	Color, crystalline form, specific rotation and λ_{max} (log ε)	m.p. °C	b.p. °C	Density	n_D	Solubility w	al	eth	ace	bz	other solvents	Ref.
	Butanoic acid														
Ω b2810	—,—,nitrile	Isobutyl cyanide. Isovaleronitrile. $(CH_3)_2CHCH_2CN$	83.13	...	fp → −100.85	130.5 53^{50}	0.7914^{20}_{4}	1.3927^{20}	δ	∞	∞	v	...	...	**B2²**, 278
b2811	—,—,propyl ester*	$(CH_3)_2CHCH_2CO_2C_3H_7^n$	144.22	...	...	155.7^{760} 49.5^{13}	0.8617^{20}_{4}	1.4031^{20}	i	∞	∞	...	...	...	**B2²**, 275
Ω b2812	—,**2-methyl-3-oxo-**, ethyl ester*	Ethyl methylacetoacetate. $CH_3COCH(CH_3)CO_2C_2H_5$	144.17	...	...	187^{760} 44^{2}	0.9941^{20}_{4}	1.4185^{20}	δ	s	s	v	...	...	**B3²**, 432
Ω b2813	—,—,methyl ester*	$CH_3COCH(CH_3)CO_2CH_3$	130.15	...	...	177.4^{760} 80^{20}	1.0247^{25}_{25}	1.416^{24}	...	s	s	...	...	...	**B3³**, 1225
b2814	—,**3-methyl-2-phenyl-***	$(CH_3)_2CHCH(C_6H_5)CO_2H$	178.23	pr (lig)	63	$159–60^{14}$	...	...	...	s	v	...	...	lig δ	**B9²**, 364
b2815	—,—,amide	$(CH_3)_2CHCH(C_6H_5)CONH_2$	177.25	nd (dil al)	111–2	$180–2^{14}$	...	...	i	s	s	...	...	lig δ	**B9²**, 364
b2816	—,—,nitrile	$(CH_3)_2CHCH(C_6H_5)CN$	159.23	...	...	$245–9^{765}$ $123–4^{15}$	$0.967^{15.5}$	1.5038^{25}	i	s	...	...	s	...	**B9²**, 364
Ω b2817	—,**2-oxo-***	α-Ketobutyric acid. $CH_3CH_2COCO_2H$	102.09	pl	31–2	$80–2^{16}$	1.200^{17}_{4}	1.3972^{20}	v	v	δ	...	...	...	**B3²**, 411
b2818	—,—,oxime	$CH_3CH_2C(:NOH)CO_2H$	117.11	nd (w or ba) tcl (diox)	164	...	...	...	δ	v	δ	...	...	...	**B3²**, 412
b2819	—,**3-oxo-***	Acetoacetic acid. $CH_3COCH_2CO_2H$	102.09	syr	...	< 100d	...	...	∞	∞	s	...	...	...	**B3²**, 412
b2820	—,—,allyl ester	Allyl acetoacetate. $CH_3COCH_2CO_2CH_2CH{:}CH_2$	142.16	...	−85	$194–5^{737}$ 66.5^{14}	1.0385^{20}_{20}	1.4398^{20}	s	∞	...	...	∞	lig s	**B3³**, 1203
Ω b2821	—,—,amide, *N*(2-chlorophenyl)-	...	211.65	nd	105	...	...	...	...	s	i	...	...	lig i	**B12²**, 319
Ω b2822	—,—,—,*N*-phenyl-	Acetoacetanilide. $CH_3COCH_2CONHC_6H_5$	177.21	pl or nd (bz or lig) λ^{al} 244 (4.12)	86	...	...	...	δ	s	s	...	s^h	chl s lig, ac, alk s	**B12²**, 266
Ω b2823	—,—,—,*N*(2-tolyl)-	...	191.23	pr (AcOEt) λ^{al} 236 (3.48)	107–8	...	...	...	δ	s	...	...	s	lig δ	**B12²**, 450
b2824	—,—,—,*N*(3-tolyl)-	...	191.23	pl (bz-peth)	57–8	...	...	...	δ	s	...	...	s	lig δ	**B12¹**, 404
Ω b2825	—,—,—,*N*(4-tolyl)-	...	191.23	pr (AcOEt) λ^{al} 248 (4.16)	95	...	...	...	δ	s	...	...	s	lig δ	**B12²**, 521
b2826	—,—,butyl ester*	$CH_3COCH_2CO_2C_4H_9^n$	158.20	...	−35.6	127^{50} 85^{8}	0.9671^{25}_{4}	1.4137^{20}	δ	∞	...	...	∞	lig ∞	**B3³**, 1201
b2827	—,—,chloride	Acetoacetyl chloride. CH_3COCH_2COCl	120.54	...	−50 (−20d)	...	...	...	d	d	v	...	...	...	**B3³**, 1204
Ω b2828	—,—,ethyl ester*	Ethyl acetoacetate. $CH_3COCH_2CO_2C_2H_5$	130.15	λ^{al} 248 (3.2)	< −80	180.4^{760} 74^{14}	1.0282^{20}_{4}	1.4194^{20}	v	∞	∞	...	s	chl s	**B3²**, 415
b2829	—,—,—(enol form)	$CH_3C(OH){:}CHCO_2C_2H_5$	130.15	λ^{vap} 238.5 (4.21)	...	...	1.0119^{10}	1.4432^{20}	...	...	...	...	...	...	**B3²**, 415
b2830	—,—,—(keto form)	$CH_3COCH_2CO_2C_2H_5$	130.15	λ^{vap} 263 (2.47)	−39	...	1.0368^{10}_{4}	1.4171^{20}	...	...	...	...	...	...	**B3²**, 415
b2831	—,—,isopropyl ester*	$CH_3COCH_2CO_2CH(CH_3)_2$	144.17	λ^{al} 244 (2.96)	−27.3	185–7 $75–6^{15}$	0.9835^{25}_{4}	1.4173^{20}	s	∞	∞	...	...	lig ∞	**B3**, 659
Ω b2832	—,—,methyl ester*	Methyl acetoacetate. $CH_3COCH_2CO_2CH_3$	116.12	λ^{al} 240 (3.18)	27–8	171.7^{760} 60^{8}	1.0762^{20}_{4}	1.4184^{20}	v	∞	∞	...	...	...	**B3²**, 414
Ω b2833	—,—,nitrile	Acetoacetonitrile. Cyanoacetone. CH_3COCH_2CN	83.09	...	...	120–5	...	...	...	v	...	v	...	...	**B3²**, 424
b2834	—,**3-oxo-2-phenyl-**, ethyl ester*	$CH_3COCH(C_6H_5)CO_2C_2H_5$	206.24	...	...	156^{22} $110^{0.8}$	1.0855^{20}_{4}	1.5176^{20}	...	s	s	...	...	...	**B10²**, 485
b2835	—,—,nitrile	$CH_3COCH(C_6H_5)CN$	159.19	pr (bz), cr (dil al or ace-peth)	90–1	...	...	...	$δ^h$	v	v	...	v	chl v peth δ	**B10²**, 485
Ω b2836	—,**4-oxo-4-phenyl-***	β-Benzoylpropionic acid. $C_6H_5COCH_2CH_2CO_2H$	178.19	lf (dil al)	116	...	...	...	s^h	s	s	...	s	chl s CS_2 s	**B10**, 696
b2838	—,**3-oxo-2-propyl-**, ethyl ester*	Ethyl α-acetovalerate. $CH_3COCH(C_3H_7^n)CO_2C_2H_5$	172.23	...	...	224^{760} 85^{5}. s	0.9661^{20}_{4}	1.4255^{20}	...	...	...	...	...	...	**B3²**, 441
Ω b2839	—,**2-phenoxy-***	$CH_3CH_2CH(OC_6H_5)CO_2H$	180.21	nd (w), pl (lig) λ^{cy} 270 (3.18)	99 (82–3)	258	...	...	δ s^h	s	s	s	s	lig s^h	**B6**, 163
b2840	—,—,amide	$CH_3CH_2CH(OC_6H_5)CONH_2$	179.22	nd (w or al)	123	...	...	...	d^h	s^h	v	v	s^h	chl v	**B6**, 164
b2841	—,—,ethyl ester*	$CH_3CH_2CH(OC_6H_5)CO_2C_2H_5$	208.24	...	...	$250–1^{748}$ $87–90^{5}$	1.0388^{21}	...	i	∞	∞	s	...	chl s	**B6**, 164
b2842	—,—,nitrile	$CH_3CH_2CH(OC_6H_5)CN$	161.21	...	...	228–30^{748}d	...	...	...	s	s	...	...	...	**B6**, 164
b2843	—,**4-phenoxy-***	$C_6H_5OCH_2CH_2CH_2CO_2H$	180.21	pl (lig), cr (w)	64–5	$192–7^{15}$	...	...	i	v	v	s	s	CS_2 chl s lig s^h	**B6²**, 158
b2844	—,—,amide	$C_6H_5OCH_2CH_2CH_2CONH_2$	179.22	lf (dil al), nd (bz)	80	...	...	...	...	v^h	...	...	...	...	**B6**, 164
b2845	—,—,ethyl ester*	$C_6H_5OCH_2CH_2CH_2CO_2C_2H_5$	208.24	...	...	$170–3^{25}$	1.045^{35}_{25}	1.491^{33}	...	s	s	...	...	...	**B6²**, 159
b2846	—,—,nitrile	$C_6H_5OCH_2CH_2CH_2CN$	161.21	nd	45–6	$287–9^{765}$ 170.5^{22}	...	...	...	s	...	...	...	...	**B6²**, 159
Ω b2847	—,**2-phenyl-***	$CH_3CH_2CH(C_6H_5)CO_2H$	164.21	pl (eth) λ^{hp} 252 (2.15), 258 (2.30), 264 (2.18)	47.5	$270–2^{760}$ $145–50^{14}$	...	...	...	s	...	...	s	...	**B9²**, 356
Ω b2848	—,—,amide	$CH_3CH_2CH(C_6H_5)CONH_2$	163.22	cr	86	185^{16}	...	...	...	s	...	...	δ	...	**B9²**, 356
b2849	—,—,methyl ester*	$CH_3CH_2CH(C_6H_5)CO_2CH_3$	178.23	nd (dil al)	77–8	228^{760}	...	...	i	s	s	...	...	...	**B9²**, 356

For explanations, symbols and abbreviations see beginning of table. For structural formulas see end of table.

No.	Name	Synonyms and Formula	Mol. wt.	Color, crystalline form, specific rotation and λ_{max} (log ε)	m.p. °C	b.p. °C	Density	n_D	Solubility w	al	eth	ace	bz	other solvents	Ref.
	Butanoic acid														
b2850	—,—,nitrile	$CH_3CH_2CH(C_6H_5)CN$	145.21			238–40^{765} $141–3^{8}$	0.977^{14}		i	s	s		s		**B9²**, 356
b2851	—,**3-phenyl-**, amide	$CH_3CH(C_6H_5)CH_2CONH_2$	163.22	nd (dil al)	106–7					s			δ		**B9¹**, 212
Ω b2852	—,**4-phenyl-***	$C_6H_5CH_2CH_2CH_2CO_2H$	164.21	lf (w) λ 242 (1.79), 258 (2.62), 267 (2.12)	52	290^{760} 171^{15}			s^h	s	s				**B9²**, 354
b2853	—,—,amide	$C_6H_5CH_2CH_2CH_2CONH_2$	163.22	pl (w) λ^{al} 260 (2.2)	84.5				s^h	v	v				**B9¹**, 211
Ω b2854	—,—,nitrile	$C_6H_5CH_2CH_2CH_2CN$	145.21			$142–5^{16}$									**B9²**, 354
b2855	—,**4(3-pyrenyl)-**	3-Pyrenebutyric acid	288.35	pl (aa)	190								s	aa s^h	**E14**, 392
b2856	—,**2,2,3,3-tetra-methyl-**, amide	$(CH_3)_3CC(CH_3)_2CONH_2$	143.23	nd (peth-al), cr (dil NH_3)	201–2				δ	s^h				lig i	**B2²**, 305
b2857	—,**3-thioxo-**, ethyl ester	$CH_3CSCH_2CO_2C_2H_5$	146.21	dk red		75^{15}	1.0554_4^{31}	1.4712^{26}		s	s				**B3²**, 427
b2858	—,**2,2,3-trichloro-***	$CH_3CHClCCl_2CO_2H$	191.44	lf or nd (peth)	60	236–8			v		v			peth v^h	**B2²**, 255
b2859	—,—,ethyl ester*	$CH_3CHClCCl_2CO_2C_2H_5$	219.50			212 101.5^{17}	1.3138_{20}^{20}			v	v				**B2**, 281
b2860	—,**2,2,4-tri-chloro-***	$ClCH_2CH_2CCl_2CO_2H$	191.44	cr (peth)	73–5				s	v	v				**B2**, 281
b2861	—,**2,3,3-tri-chloro-***	$CH_3CCl_2CHClCO_2H$	191.44	pl (lig)	52				δ	v	v	v	v	chl v CS_2 v	**B2**, 281
b2862	—,**4,4,4-tri-chloro-***	$Cl_3CCH_2CH_2CO_2H$	191.44	nd (w)	55				s^h	v	v			chl s	**B2³**, 629
—	—,**2,3,4-tri-hydroxy-**	*see* **Threonic acid**													
b2863	—,**2,2,3-tri-methyl-**, chloride	$(CH_3)_2CHC(CH_3)_2COCl$	148.63			148–50			d		s				**B2³**, 300
Ω b2864	**1-Butanol***	$CH_3CH_2CH_2CH_2OH$	74.12		−89.53	117.25^{760}	0.8098_4^{20}	1.39931^{20}	s	∞	∞	v	s		**B1²**, 387
b2865	**2-Butanol** (*d*)	*sec*-Butyl alcohol. $CH_3CH_2CHOHCH_3$	74.12	$[\alpha]_D^{20}$ +13.9 (undil)		99.5^{760}	0.8080_4^{20}	1.3954^{20}	v	∞	∞	s	s		**B1²**, 404
Ω b2866	—(*dl*)*	$CH_3CH_2CHOHCH_3$	74.12	λ^{hp} 170–8 (2.5), 208–14 sh (1.5)	−114.7	99.5^{760} 45.5^{60}	0.8063_4^{20}	1.3978^{20}	v	∞	∞	v	s		**B1²**, 400
b2867	—(*l*)*	$CH_3CH_2CHOHCH_3$	74.12	$[\alpha]_D^{20}$ −13.9		99.5^{760}	0.8070_4^{20}	1.3975^{20}	v	∞	∞	v	s		**B1²**, 405
—	*tert*-**Butanol**	*see* **2-Propanol, 2-methyl-***													
b2868	**1-Butanol, 2-amino-(*d*)***	$CH_3CH_2CHNH_2CH_2OH$	89.14	$[\alpha]_D^{20}$ +9.8 (w)		80^{11}	0.947^{20}	1.4518^{20}	∞	∞	∞				**B4³**, 771
Ω b2869	—,—(*dl*)*	$CH_3CH_2CHNH_2CH_2OH$	89.14		−2	178 (174)	0.9162^{20}	1.4489^{25}	∞	∞	∞				**B4¹**, 438
b2870	—,**3-amino-***	$CH_3CHNH_2CH_2CH_2OH$	89.14			$82–5^{19}$		1.4534^{25}	s	s					**B4¹**, 438
Ω b2871	—,**4-amino-***	$H_2NCH_2CH_2CH_2CH_2OH$	89.14			206^{776} 100^{15}	0.967^{12}	1.4625^{20}	s	s	i				**B4¹**, 439
Ω b2872	**2-Butanol, 3-amino-(*dl*)***	$CH_3CHNH_2CHOHCH_3$	89.14		18–20 (44)	159–60^{745} 70^{20}	0.9299_4^{25}	1.4502^{20}	s	s		s		lig i	**B4³**, 778
b2873	**1-Butanol, 2-amino-1-phenyl-***	$CH_3CH_2CHNH_2CH(C_6H_5)OH$	165.24	pl (bz-eth)	79–80					v			s	chl s peth δ	**B13²**, 390
b2874	**2-Butanol, 3-bromo-(*dl*)***	$CH_3CHBrCHOHCH_3$	153.03			154^{760} $46–50^{8}$	1.4550_4^{20}	1.4780^{20}		s	s				**B1³**, 1542
b2875	**1-Butanol, 2-chloro-***	$CH_3CH_2CHClCH_2OH$	108.57			83^{72} $74–6^{25}$	1.062_4^{25}	1.4438^{20}		v	v				**B1³**, 1516
Ω b2876	—,**4-chloro-***	Tetramethylene chlorohydrin. $ClCH_2CH_2CH_2CH_2OH$	108.57			$84–5^{16}$	1.0883_4^{20}	1.4518^{20}		v	v				**B1²**, 398
Ω b2877	**2-Butanol, 1-chloro-***	$CH_3CH_2CHOHCH_2Cl$	108.57			141^{760} 52^{15}	1.068_4^{25}	1.4400^{20}		s	s				**B1²**, 402
b2878	—,**3-chloro-***	$CH_3CHClCHOHCH_3$	108.57			138–40 $52–4^{30}$	1.0669_4^{20}	1.4432^{20}		v	v				**B1²**, 403
b2879	—,—(*erythro, dl*)*	$CH_3CHClCHOHCH_3$	108.57			135.4^{748} 56.1^{30}	1.0610_4^{25}	1.4397^{25}		v	v			chl v	**B1³**, 1538
b2880	—,**4-chloro-***	$ClCH_2CH_2CHOHCH_3$	108.57			67^{20}		1.4408^{26}		v	v				**B1¹**, 188
b2881	—,**1-chloro-2-methyl-***	$CH_3CH_2COH(CH_3)CH_2Cl$	122.60			150–2	1.0161^{20}	1.4469^{20}		v					**B1²**, 424
b2882	—,**3-chloro-2-methyl-***	$CH_3CHClCOH(CH_3)_2$	122.60			141–2 $55–6^{30}$	1.0295_4^{20}	1.4436^{20}	s	v	v				**B1²**, 424
b2883	—,**1,3-dichloro-***	$CH_3CHClCHOHCH_2Cl$	143.02			$63–4^{10}$	1.2860_4^{15}	1.4766^{20}		v	v				**B1**, 373
Ω b2884	**1-Butanol, 2,2-dimethyl-***	*tert*-Amyl carbinol. $CH_3CH_2C(CH_3)_2CH_2OH$	102.18		< −15	136.7^{760}	0.8283_4^{20}	1.4208^{20}	δ	s	s				**B1³**, 1675
b2885	—,**2,3-dimethyl-(*d*)***	Isopropyl dimethyl carbinol. $(CH_3)_2CHCH(CH_3)CH_2OH$	102.18	$[\alpha]_D^{25}$ +1.9		142^{760}	0.823_4^{25}			s		s			**B1³**, 1677
Ω b2886	—,—(*dl*)*	$(CH_3)_2CHCH(CH_3)CH_2OH$	102.18			$144–5^{761}$	$0.8297_4^{20.5}$	$1.4195^{20.5}$		s	s	s			**B1³**, 1677
Ω b2887	—,**3,3-dimethyl-***	$(CH_3)_3CCH_2CH_2OH$	102.18		−60	143	0.844_4^{15}	1.4323^{15}	δ	s	s	s			**B1³**, 1677
Ω b2888	**2-Butanol, 2,3-dimethyl-***	$(CH_3)_2CHCOH(CH_3)_2$	102.18		−14	118.4^{760}	0.8236_4^{20}	1.4176^{20}	s	∞	∞				**B1²**, 441
Ω b2889	—,**3,3-dimethyl-(*dl*)***	Pinacolyl alcohol. $(CH_3)_3CCHOHCH_3$	102.18		5.6	120.4^{760}	0.8122^{25}	1.4148^{20}	δ	s	∞				**B1²**, 441
Ω b2890	**1-Butanol, 2-ethyl-***	$(C_2H_5)_2CHCH_2OH$	102.18		< −15	146.27^{760}	0.8326_4^{20}	1.4220^{20}	δ	s	s				**B1**, 412

For explanations, symbols and abbreviations see beginning of table. For structural formulas see end of table.

No.	Name	Synonyms and Formula	Mol. wt.	Color, crystalline form, specific rotation and λ_{max} (log ε)	m.p. °C	b.p. °C	Density	n_D	Solubility						Ref.
									w	al	eth	ace	bz	other solvents	
	1-Butanol														
b2891	**—,4-fluoro-***	Tetramethylene fluorohydrin. $FCH_2CH_2CH_2CH_2OH$	92.12	...	...	58^{15}	...	1.3942^{25}	s	v	v	v	...	...	**B1**^{3}, 1516
b2892	**—,2-methyl-(*d*)***	*act*-Amyl alcohol. $CH_3CH_2CH(CH_3)CH_2OH$	88.15	$[\alpha]^{20}_{D}$ −5.90 (undil)	...	128^{760} 65.7^{50}	0.8193^{20}_{4}	1.4102^{20}	δ	∞	∞	v	...	...	**B1**^{2}, 421
Ω b2893	**—,—(*dl*)***	$CH_3CH_2CH(CH_3)CH_2OH$	88.15	...	...	127.5– 8^{760} 70^{60}	0.8152^{25}_{4}	1.4092^{20}	δ	∞	∞	v	...	...	**B1**^{2}, 422
b2894	**—,—(*l*)***	$CH_3CH_2CH(CH_3)CH_2OH$	88.15	$[\alpha]^{18}_{D}$ +3.75	...	129^{760}	0.816^{18}_{4}	1.4098^{20}	δ	∞	∞	v	...	...	**B1**^{3}, 1619
Ω b2895	**—,3-methyl-***	Isopentyl alcohol. $(CH_3)_2CHCH_2CH_2OH$	88.15	...	...	128.5^{750}	0.8092^{20}_{4}	1.4053^{20}_{4}	δ	∞	∞	v	...	...	**B1**^{3}, 1633
Ω b2896	**2-Butanol, 2-methyl-***	*tert*-Amyl alcohol. $CH_3CH_2COH(CH_3)_2$	88.15	$\lambda^{con\ sulf}$ 300 (2.85)	−8.4	102^{760} 50^{60}	0.8059^{25}_{4}	1.4052^{20}	s	∞	∞	v	s	chl s	**B1**^{2}, 422
b2897	**—,3-methyl-(*d*)***	$(CH_3)_2CHCHOHCH_3$	88.15	$[\alpha]^{20}_{D}$ +5.34 (al)	...	112^{734}	0.8225^{16}_{4}	1.4089^{20}	δ	s	s	v	s	chl s	**B1**^{1}, 196
Ω b2898	**—,—(*dl*)***	$(CH_3)_2CHCHOHCH_3$	88.15	...	...	112.9^{760} 61^{60}	0.8180^{20}_{4}	1.4089^{20}	δ	∞	∞	v	s	CCl_4 s	**B1**^{2}, 425
b2899	**1-Butanol, 2-methyl-1-phenyl-***	*sec*-Butyl phenyl carbinol. $CH_3CH_2CH(CH_3)CH(C_6H_5)OH$	164.25	...	...	120^{13}	...	...	i	∞	∞	v	∞	os v	**B6**^{2}, 505
b2900	**—,2-methyl-4-phenyl-(*dl*)***	$C_6H_5CH_2CH_2CH(CH_3)CH_2OH$	164.25	...	...	135^{11}	0.9719^{20}_{4}	1.5173^{16}	i	v	∞	v	v	os v	**B6**^{1}, 269
b2901	**—,3-methyl-1-phenyl-***	$(CH_3)_2CHCH_2CHOHC_6H_5$	164.25	...	...	$235–6^{746}$ 122^{9}	0.9537^{19}_{4}	1.5080^{18}	i	v	v	v	v	os v	**B6**, 548
b2902	**—,3-methyl-2-phenyl-***	$(CH_3)_2CHCH(C_6H_5)CH_2OH$	164.25	...	...	130^{15}	0.9694^{25}_{4}	1.5137^{25}	i	v	v	v	v	os v	**B6**^{2}, 506
b2903	**2-Butanol, 2-methyl-1-phenyl-(*dl*)***	$CH_3CH_2COH(CH_3)CH_2C_6H_5$	164.25	...	...	$215–25^{747}$ $103–5^{11}$	0.9754^{20}_{0}	1.5182^{20}	i	s	s	...	...	lig s	**B6**^{2}, 505
b2904	**—,2-methyl-3-phenyl-***	$CH_3CH(C_6H_5)COH(CH_3)_2$	164.25	...	...	196–8 118^{24}	0.9794^{20}_{4}	1.5193^{20}	i	s	s	s	s	os s	**B6**^{1}, 270
Ω b2905	**—,2-methyl-4-phenyl-***	$C_6H_5CH_2CH_2COH(CH_3)_2$	164.25	nd	24.5	121^{13}	0.9626^{21}_{4}	1.5077^{21}	i	v	v	v	v	os v	**B6**^{2}, 506
b2906	**—,3-methyl-3-phenyl-***	$(CH_3)_2CHCOH(C_6H_5)CH_3$	164.25	...	...	$196–8^{760}$	0.9653^{13}_{4}	1.5161^{13}	i	s	s	s	s	os s	**B6**^{1}, 269
Ω b2907	**1-Butanol-, 2-nitro-***	$CH_3CH_2CHNO_2CH_2OH$	119.12	λ^{al} 279 (1.47)	−47	105^{10}	1.1332^{25}_{4}	1.4390^{20}	s	∞	∞	s	...	aa s	**B1**, 370
b2908	**—,1-phenyl-(*dl*)***	α-Propylbenzyl alcohol. $CH_3CH_2CH_2CHOHC_6H_5$	150.22	...	16	232^{760} $113–5^{12}$	0.9740^{20}_{4}	1.5139^{20}	i	s	s	...	...	...	**B6**^{2}, 486
b2909	**—,2,2,3-trichloro-***	$CH_3CHClCCl_2CH_2OH$	177.46	pr (dil al)	62	199– 200^{760} $97–8^{18}$	...	...	δ^{h}	s	s	s	...	...	**B1**^{2}, 398
b2910	**2-Butanol, 1,1,1-trichloro-***	$CH_3CH_2CHOHCCl_3$	177.46	...	...	$169–71^{738}$ $82–4^{22}$	1.3670^{25}_{25}	1.4800^{20}	δ^{h}	v	v	v	v	chl v CS_2 v	**B1**^{2}, 403
b2911	**—,2,3,3-trimethyl-***	$(CH_3)_3CCOH(CH_3)_2$	116.21	cr (dil al + ½w)	83–4	$131–2^{760}$ $40–1^{15}$	0.8380^{25}_{4}	1.4233^{22}	δ s^{h}	∞	∞	v	...	...	**B1**^{2}, 447
Ω b2912	**2-Butanone***	Ethyl methyl ketone. $CH_3CH_2COCH_3$	72.12	λ^{al} 273.5 (1.24)	−86.35	79.6^{760} 30^{119}	0.8054^{20}_{4}	1.3788^{20}	v	∞	∞	∞	∞	...	**B1**^{2}, 726
Ω b2913	**—,oxime**	$CH_3CH_2C(:NOH)CH_3$	87.12	...	−29.5	$152–3^{760}$ $59–60^{15}$	0.9232^{20}_{4}	1.4410^{20}	s	∞	∞	...	...	...	**B1**^{2}, 730
b2914	**—,1-chloro-***	$CH_3CH_2COCH_2Cl$	106.55	...	...	$137–8^{760}$ $34–5^{10}$	1.0850^{20}_{4}	1.4372^{20}	i	...	...	...	...	MeOH v^{h}	**B1**^{2}, 731
b2915	**—,3-chloro-***	$CH_3CHClCOCH_3$	106.55	...	...	115^{760} (117–9) 40^{30}	1.0554^{0}	1.4219^{20}	i	v	v	...	...	...	**B1**^{2}, 731
b2916	**—,4-chloro-***	$ClCH_2CH_2COCH_3$	106.55	...	...	$120–1^{760}$d 48^{15}	1.0680^{23}	1.4284^{23}	i	v	v	...	...	MeOH v	**B1**^{2}, 731
b2917	**—,3-chloro-3-methyl-***	$(CH_3)_2CClCOCH_3$	120.58	...	...	117.2^{758}	1.0083^{20}_{4}	1.4204^{20}	i	s	s	...	...	...	**B1**^{3}, 2818
Ω b2918	**—,3,4-dibromo-4-phenyl-***	Benzalacetone dibromide. $C_6H_5CHBrCHBrCOCH_3$	306.01	nd (al)	124–5	...	...	...	i	δ s^{h}	...	...	...	chl v	**B7**^{2}, 244
b2919	**—,1,3-dichloro-***	$CH_3CHClCOCH_2Cl$	141.00	...	...	$166–7^{760}$ 55.5^{10}	1.3116^{20}_{4}	1.4686^{20}	i	s	s	s	s	os s	**B1**^{1}, 348
Ω b2920	**—,4(diethyl-amino)-***	$(C_2H_5)_2NCH_2CH_2COCH_3$	143.23	...	...	84^{30} 53^{4}	0.8630^{20}_{4}	1.4333^{24}	δ	s	s	...	...	...	**B4**^{1}, 452
Ω b2921	**—,3,3-dimethyl-***	Pinacolone. $(CH_3)_3CCOCH_3$	100.16	λ^{al} 282 (2.62)	−49.8	106^{760}	0.8012^{25}_{4}	1.3952^{20}	δ	s	s	s	...	...	**B1**^{3}, 354
Ω b2922	**—,3,3-diphenyl-***	$CH_3C(C_6H_5)_2COCH_3$	224.31	pr (al) λ^{al} 260 (2.7), 300 (2.6)	41	310–1 176^{16}	1.069^{20}_{4}	1.5748^{20}	i	s v^{h}	v	...	...	chl v aa v	**B7**^{2}, 393
b2923	**—,1-hydroxy-***	$CH_3CH_2COCH_2OH$	88.12	λ^{al} 276 (1.28)	...	160^{760} (153–4) 48^{9}	1.0272^{20}_{4}	1.4189^{20}	v	v	v	...	...	...	**B1**^{2}, 870
Ω b2924	**—,3-hydroxy-(*dl*)***	Acetoin. $CH_3CHOHCOCH_3$	88.12	λ^{w} 272.5 (−0.5)	−72	143^{760} 37^{11}	1.0062^{20}_{20}	1.4171^{20}	∞	δ	δ	s	...	lig i	**B1**^{2}, 870
b2925	**—,1(1-hydroxy-2-naphthyl)-***	...	214.27	ye nd (eth)	85–6	$145–52^{1}$	...	...	i	s	s	...	...	...	**B8**, 152
Ω b2926	**—,3-methyl-***	Isopropyl methyl ketone. $(CH_3)_2CHCOCH_3$	86.14	λ^{al} 280 (1.32)	−92	$94–4.5^{760}$	0.8051^{20}_{4}	1.38804^{20}	δ	∞	∞	v	...	...	**B1**^{2}, 741
b2927	**—,—,oxime**	$(CH_3)_2CHC(:NOH)CH_3$	101.15	...	...	157–8	...	...	s	∞	∞	...	...	...	**B1**, 683
b2928	**1-Butanone, 3-methyl-1-phenyl-***	Isovalerophenone. $(CH_3)_2CHCH_2COC_6H_5$	162.23	λ^{al} 243 (4.09)	...	236.54^{764} $137–8^{38}$	$0.9701^{16.4}_{4}$	$1.5139^{15.3}$	i	∞	∞	v	...	...	**B7**^{2}, 252

For explanations, symbols and abbreviations see beginning of table. For structural formulas see end of table.

No.	Name	Synonyms and Formula	Mol. wt.	Color, crystalline form, specific rotation and λ_{max} (log ε)	m.p. °C	b.p. °C	Density	n_D	Solubility w	al	eth	ace	bz	other solvents	Ref.
	1-Butanone														
Ω b2929	**—,1-phenyl-***	Butyrophenone. $CH_3CH_2CH_2COC_6H_5$	148.21	λ^{al} 240 (4.06), 276 (3.00)	11.5–3.0	$228–9^{760}$	0.988^{20}_{4}	1.5203^{20}	i	∞	∞	v	...	peth δ	**B7²**, 241
Ω b2930	**2-Butanone, 1-phenyl-***	Ethyl benzyl ketone. $CH_3CH_2COCH_2C_6H_5$	148.21	λ^{al} 284 (2.11)	...	230^{755}	1.002^{0}_{4}	...	i	s	∞	v	...	...	**B7²**, 243
Ω b2931	**—,4-phenyl-***	Benzylacetone. $C_6H_5CH_2CH_2COCH_3$	148.21	λ^{cy} 281 (1.51)	...	233–4 115^{13}	0.9849^{22}_{4}	1.511^{22}	i	s	s	v	...	...	**B7²**, 243
Ω b2932	**1-Butanone, 1(4-tolyl)-**	*p*-Methylbutyrophenone.	162.24	...	12	251.5^{758}	0.9745^{20}_{4}	1.5232^{20}	i	v	v	...	...	...	**B7²**, 254
Ω b2933	**2-Butenal** (*trans*)*	Crotonaldehyde. $CH_3CH{:}CHCHO$	70.09	λ^{al} 218 (4.25)	−74	104–5 $(102–3^{765})$	0.8495^{25}_{4}	1.4366^{20}	s	v	v	v	∞	...	**B1²**, 787
Ω b2934	**—,diethyl acetal**	$CH_3CH{:}CHCH(OC_2H_5)_2$	144.22	...	...	147–8 49^{17}	0.8473^{18}_{4}	1.4097^{20}	i	∞	∞	∞	∞	peth ∞	**B1²**, 789
b2935	**—,2-bromo-,** diethyl acetal	$CH_3CH{:}CBrCH(OC_2H_5)_2$	223.12	...	...	86^{15}	1.2255^{21}	1.4565^{21}	...	...	...	...	...	...	**B1¹**, 380
b2936	**—,2-chloro-***	$CH_3CH{:}CClCHO$	104.54	λ < 210	...	147–8 $53–4^{20}$	1.1404^{23}_{4}	1.4780^{23}	δ	s	s	...	...	CCl_4 s chl s	**B1²**, 789
b2937	**—,3-ethoxy-,** diethyl acetal	1,1,3-Triethoxy-2-butene*. $CH_3C(OC_2H_5){:}CHCH(OC_2H_5)_2$	188.27	...	...	190–5 $79–82^{10}$	0.908^{21}_{0}	1.430^{21}	...	s	...	...	...	...	**B1³**, 3267
b2938	**—,2-methyl-***	Tiglaldehyde. $CH_3CH{:}C(CH_3)CHO$	84.13	...	...	116.5–7.5^{738} $63–5^{119}$	0.8710^{20}_{4}	1.4475^{20}	s	∞	∞	...	...	...	**B1²**, 792
b2939	**—,3-methyl-***	β,β-Dimethylacrolein. Senecio-aldehyde. $(CH_3)_2C{:}CHCHO$	84.13	...	...	133^{730}	0.8722^{20}_{4}	1.4526^{20}	s	s	s	...	...	...	**B1³**, 2990
Ω b2940	**1-Butene-***	α-Butylene. $CH_3CH_2CH{:}CH_2$	56.12	λ^{gas} 162 sh (4.0), 175 (4.2), 187 (4.1)	−185.35	-6.3^{760}	0.5951^{20}_{4}	1.3962^{20} (1.3465^{20})	i	v	v	...	s	...	**B1³**, 715
Ω b2941	**2-Butene** (*cis*)*	*cis*-β-Butylene. $CH_3CH{:}CHCH_3$	56.12	λ^{gas} 160 sh (4.1), 175 (4.3), 196 (3.00), 200 (2.7)	−138.91	3.7^{760}	0.6213^{20}_{4}	1.3931^{-25}	i	v	v	...	s	...	**B1³**, 728
Ω b2942	**—(*trans*)***	*trans*-β-Butylene. $CH_3CH{:}CHCH_3$	56.12	λ^{gas} 163 (3.9), 177 (4.1), 187 (3.8), 202 (2.7)	−105.55	0.88^{760}	0.6042^{20}_{4}	1.3848^{-25}	...	...	...	...	s	...	**B1³**, 730
b2943	**1-Butene, 4-bromo-***	$BrCH_2CH_2CH{:}CH_2$	135.01	...	...	98.5^{758}	1.3230^{20}_{4}	1.4622^{20}	i	s	s	...	s	...	**B1³**, 727
b2944	**—,2-bromo-3-methyl-***	$(CH_3)_2CHCBr{:}CH_2$	149.04	...	...	105^{757}	1.2328^{20}_{4}	1.4504^{20}	i	...	s	...	s	chl s	**B1³**, 800
b2945	**2-Butene, 1-bromo-3-methyl-***	$(CH_3)_2C{:}CHCH_2Br$	149.04	...	...	129–33d $50–1^{40}$	1.2819^{20}_{0}	1.4930^{15}	i d^h	s	s	s	s	CS_2 s chl s	**B1³**, 796
b2946	**—,2-bromo-3-methyl-***	$(CH_3)_2C{:}CBrCH_3$	149.04	...	...	$119–20^{766}$	1.2773^{20}_{0}	1.4738^{20}	i	...	s	s	...	chl s	**B1³**, 796
b2947	**1-Butene, 2-bromo-4-phenyl-***	$C_6H_5CH_2CH_2CBr{:}CH_2$	211.11	...	...	$117–8^{21}$ $90–1^{5}$	1.2901^{20}_{4}	1.5450^{20}	i	...	s	...	...	...	**B5²**, 380
b2948	**2-Butene, 1 bromo-4-phenyl-***	$C_6H_5CH_2CH{:}CHCH_2Br$	211.11	...	...	126–$30^{10.5}$	1.2660^{20}_{4}	1.5678^{20}	i	...	s	...	...	...	**B5²**, 379
b2949	**—,2-bromo-3-phenyl-***	$CH_3C(C_6H_5){:}CBrCH_3$	211.11		...	$120–30^{11}$	1.3348^{20}_{4}	1.5811^{20}	i	...	s	...	s	chl s	**B5**, 488
b2950	**1-Butene, 1-chloro-(*cis*)***	$CH_3CH_2CH{:}CHCl$	90.55	...	...	63.5^{760}	0.9153^{15}_{4}	1.4194^{15}	i	∞	v	s	...	chl s	**B1³**, 723
b2951	**—,—(*trans*)***	$CH_3CH_2CH{:}CHCl$	90.55	...	...	68^{760}	0.9205^{15}_{4}	1.4225^{15}	i	∞	v	s	...	chl s	**B1³**, 723
b2952	**—,2-chloro-***	$CH_3CH_2CCl{:}CH_2$	90.55	...	...	58.5^{760}	0.9107^{15}_{4}	1.4115^{21}	i	∞	v	s	...	chl s	**B1**, 204
b2953	**—,3-chloro-***	$CH_3CHClCH{:}CH_2$	90.55	...	...	$64–5^{766}$	0.8978^{20}_{4}	1.4149^{20}	...	...	v	v	...	chl s	**B1³**, 723
b2954	**—,4-chloro-***	$ClCH_2CH_2CH{:}CH_2$	90.55	...	...	75^{773}	0.9211^{20}_{4}	1.4233^{20}	i	...	v	s	...	chl v	**B1³**, 724
b2955	**2-Butene, 1-chloro-(*cis*)**	$CH_3CH{:}CHCH_2Cl$	90.55	...	...	84.1^{758}	0.9426^{20}_{4}	1.4390^{20}	i	s	...	s	...	chl s	**B1²**, 176
b2956	**—,—(*trans*)***	$CH_3CH{:}CHCH_2Cl$	90.55	...	...	84.8^{752}	0.9295^{20}_{4}	1.4350^{20}	i	...	...	s	...	chl s	**B1²**, 176
b2957	**—,2-chloro-(*cis*)***	$CH_3CH{:}CClCH_3$	90.55	...	−117.3	70.6^{760}	0.9239^{20}_{4}	1.4240^{20}	i	∞	...	s	...	chl s	**B1³**, 740
b2958	**—,—(*trans*)***	$CH_3CH{:}CClCH_3$	90.55	...	−105.8	62.8	0.9138^{20}_{4}	1.4190^{20}	i	∞	...	s	...	chl s	**B1³**, 741
b2959	**1-Butene, 3-chloro-2-chloro-methyl-***	$CH_3CHClC(CH_2Cl){:}CH_2$	139.04	...	...	155^{760} $31–3^{7}$	1.1233^{20}_{4}	1.4724^{20}	i	...	...	s	...	chl s	**B1³**, 787
b2960	**2-Butene, 1-chloro-2,3-dimethyl-***	$(CH_3)_2C{:}C(CH_3)CH_2Cl$	118.61	...	...	$111–2^{756}$	0.9355^{20}_{4}	1.4605^{20}	...	...	v	v	...	chl v	**B1³**, 819
b2961	**1-Butene, 1-chloro-2-methyl-***	$CH_3CH_2C(CH_3){:}CHCl$	104.58	...	...	$96–7^{760}$ $(88–90.5^{718})$	0.9170^{20}_{4}	1.4141^{20}	i	...	v	v	...	...	**B1²**, 187
b2962	**—,1-chloro-3-methyl-***	$(CH_3)_2CHCH{:}CHCl$	104.58	...	...	$86–8^{756}$	...	1.4229^{20}	i	...	s	s	...	chl s	**B1³**, 799
b2963	**—,3-chloro-2-methyl-***	$CH_3CHClC(CH_3){:}CH_2$	104.58	...	...	94^{760}	0.9088^{20}_{4}	1.4304^{20}	i	...	v	s	...	chl s	**B1³**, 787
b2964	**2-Butene, 1-chloro-2-methyl-***	$CH_3CH{:}C(CH_3)CH_2Cl$	104.58	...	...	110^{760} 26.4^{25}	0.9327^{20}_{4}	1.4481^{20}	d^h	v	v	v	...	chl s	**B1²**, 189
b2965	**—,1-chloro-3-methyl-***	$(CH_3)_2C{:}CHCH_2Cl$	104.58	...	...	109^{760} 54.3^{95}	0.9273^{20}_{4}	1.4485^{20}	i	s	s	s	...	chl v	**B1²**, 191

For explanations, symbols and abbreviations see beginning of table. For structural formulas see end of table.

No.	Name	Synonyms and Formula	Mol. wt.	Color, crystalline form, specific rotation and λ_{max} (log ε)	m.p. °C	b.p. °C	Density	n_D	Solubility w	al	eth	ace	bz	other solvents	Ref.
	2-Butene														
b2966	—,**2-chloro-3-methyl-***	Trimethylvinyl chloride. $(CH_3)_2C:CClCH_3$	104.58	...	...	94^{760} (97–8)	0.9324_4^{20}	1.4320^{20}	i	s	v	s	...	chl s	**B1**[2], 189
b2967	**1-Butene, 2(chloromethyl)-1,3-dichloro-***	$CH_3CHClC(CH_2Cl):CHCl$	174.47	...	...	$68–70^8$	1.2775_4^{17}	...	i	...	...	...	...	CCl_4 s	**B1**[3], 788
b2968	**2-Butene, 1,4-dibromo-(*trans*)***	$BrCH_2CH:CHCH_2Br$	213.91	pl (peth)	53–4	205^{760} 85^{10}	...	...	δ	v	...	s	...	MeOH v peth v	**B1**[2], 206
b2969	**1-Butene, 1,3-dichloro-***	$CH_3CHClCH:CHCl$	125.00	...	...	125^{760} $58–60^{25}$	1.1341_4^{24}	1.4647^{20}	i	s	s	s	...	chl s	**B1**[2], 724
b2970	—,**2,3-dichloro-***	$CH_3CHClCCl:CH_2$	125.00	...	...	112	1.1340_4^{20}	1.4580^{20}	i	...	s	s	...	chl s	**B1**[3], 724
b2971	**2-Butene, 1,1-dichloro-***	$CH_3CH:CHCHCl_2$	125.00	...	...	125–7	1.1310^{20}	1.466^{18}	i	...	s	v	...	chl s	**B1**[3], 741
b2972	—,**1,2-dichloro-** (high b.p.)*	$CH_3CH:CClCH_2Cl$	125.00	...	...	130.2–0.8^{760}	1.1601_4^{20}	1.4734^{20}	i	...	s	v	...	chl v	**B1**[3], 741
b2973	—,—(low b.p.)*	$CH_3CH:CClCH_2Cl$	125.00	...	...	$116–8^{765}$	1.1544_4^{20}	1.4642^{20}	...	...	...	v	...	CCl_4 v	**B1**[3], 741
b2974	—,**1,3-dichloro-** (*cis*)*	$CH_3CCl:CHCH_2Cl$	125.00	...	...	129.9^{745} 34^{20}	1.1605_4^{20}	1.4735^{20}	i	s	s	s	s	os s, chl s	**B1**[3], 742
b2975	—,—(*trans*)*	$CH_3CCl:CHCH_2Cl$	125.00	...	...	130^{745} 53^{50}	1.1585_4^{20}	1.4719^{20}	i	s	s	s	s	os s, chl s	**B1**[3], 742
b2976	—,**1,4-dichloro-** (*cis*)*	$ClCH_2CH:CHCH_2Cl$	125.00	...	−48	152.5^{758} 22.5^3	1.188_4^{25}	1.4887^{25}	i	s	s	s	s	os s, chl s	**B1**[3], 743
b2977	—,—(*trans*)*	$ClCH_2CH:CHCH_2Cl$	125.00	...	1–3	155.5^{758} 55.5^{20}	1.183_4^{25}	1.4871^{25}	i	s	s	s	s	os s, chl s	**B1**[3], 743
b2978	—,**2,3-dichloro-** (*cis*)*	$CH_3CCl:CClCH_3$	125.00	...	...	$125–6^{758}$	1.1618_4^{20}	1.4590^{20}	i	s	s	s	s	os s, chl s	**B1**[3], 743
b2979	—,—(*trans*)*	$CH_3CCl:CClCH_3$	125.00	...	...	101.3^{758}	1.1416_4^{20}	1.4582^{20}	i	s	s	s	s	os s, chl s	**B1**[3], 744
b2980	**1-Butene, 3,3-dichloro-2-methyl-***	$CH_3CCl_2C(CH_3):CH_2$	139.04	...	...	$151–3^{760}$	1.1276_4^{19}	1.4737_c^{19}	i	...	s	s	...	chl s	**B1**[3], 787
b2981	**2-Butene, 1,3-dichloro-2-methyl-***	$CH_3CCl:C(CH_3)CH_2Cl$	139.04	...	...	$151–3^{760}$	1.1293_4^{20}	...	i	...	...	v	...	chl s	**B1**[3], 795
b2982	—,**1,4-dichloro-2-methyl-***	$ClCH_2CH:C(CH_3)CH_2Cl$	139.04	...	...	93^{50} 56^{10}	1.1526_4^{20}	1.4932^{20}	i	...	...	v	...	chl v	**B1**[3], 795
—	—,**1,1-diethoxy-**	*see* **2-Butenal,** diethyl acetal*													
Ω b2983	**1-Butene, 2,3-dimethyl-***	$(CH_3)_2CHC(CH_3):CH_2$	84.16	λ^{gas} 169 (3.9), 187 (4.0), 198 (3.6)	−157.3	55.67^{760}	0.6803_4^{20}	1.3995^{20}	i	s	s	s	...	CS_2 s	**B1**[3], 816
Ω b2984	—,**3,3-dimethyl-***	$(CH_3)_3CCH:CH_2$	84.16	λ^{gas} 174 (4.1), 188 (3.6)	−115.2	41.2^{760}	0.6529_4^{20}	1.3763^{20}	i	s	s	...	...	chl s	**B1**[3], 814
b2985	**2-Butene, 2,3-dimethyl-***	$(CH_3)_2C:C(CH_3)_2$	84.16	λ^{al} 198 (4.12)	−74.28	73.2^{760}	0.7080_4^{20}	1.4122^{20}	i	s	s	s	...	chl s	**B1**[3], 817
Ω b2986	**1-Butene, 3,4-epoxy-***	Butadiene mono-oxide. Vinyloxirane.	70.09	λ^{al} <210	...	70^{760}	0.9006^0	1.4168^{20}	...	s	s	...	s	os s	**B17**[1], 13
Ω b2987	—,**2-ethyl-***	3-Methylenepentane. $(C_2H_5)_2C:CH_2$	84.16	λ^{gas} 172 (3.7), 187 (3.9), 195 sh (3.8)	−131.53	64.7^{760}	0.6894_4^{20}	1.3969^{20}	i	...	s	s	s	chl s	**B1**[3], 814
b2988	—,**2-ethyl-3-methyl-***	$(CH_3)_2CHC(C_2H_5):CH_2$	98.19	λ^{iso} <215	...	89	0.7150_4^{20}	1.410^{20}	i	...	s	s	s	chl s	**B1**[3], 833
b2989	**2-Butene, 1,1,2,3,4,4-hexachloro-** (liquid)*	$Cl_2CHCCl:CClCHCl_2$	262.78	...	−19	$97–8^{10}$	1.651_{15}^{15}	1.5331	i	...	...	...	...	chl s	**B1**[3], 744
b2990	—,—(solid)*	$Cl_2CHCCl:CClCHCl_2$	262.78	lf (al)	80 (78)	...	...	...	...	v^h	v	...	v	chl v CCl_4 v	**B1**[3], 744
Ω b2991	**1-Butene, 2-methyl-***	$CH_3CH_2C(CH_3):CH_2$	70.14	λ^{gas} 200 (3.44)	−137.56	31.16^{760}	0.6504_4^{20}	1.3378^{20}	i	s	s	...	s	...	**B1**[3], 784
Ω b2992	—,**3-methyl-***	$(CH_3)_2CHCH:CH_2$	70.14	λ^{gas} 195 (2.50)	−168.5	20^{760}	0.6272_4^{20}	1.3643^{20}	i	∞	∞	...	s	...	**B1**[3], 797
Ω b2993	**2-Butene, 2-methyl-***	$(CH_3)_2C:CHCH_3$	70.14	λ^{gas} 205 (2.93)	−133.77	38.57^{760}	0.6623_4^{20}	1.3874^{20}	i	s	s	...	s	lig v	**B1**[2], 187
b2994	**1-Butene, octafluoro-***	Perfluoro-α-butylene. $F_3CCF_2CF:CF_2$	200.03	...	...	4.8^{764}	1.5443^0 1.615_4^{-20}	...	...	...	...	...	...	...	**B1**[3], 722
Ω b2995	**2-Butene, octafluoro-***	Perfluoro-β-butylene. $F_3CCF:CFCF_3$	200.03	...	−129 (−139)	$0–3^{760}$	1.5297^0	...	...	...	...	...	...	...	**B1**[3], 739
b2996	—,**1,1,1,4,4-pentachloro-***	$Cl_2CHCH:CHCCl_3$	228.33	...	...	$78–80^{11}$	1.612_{21}^{21}	1.5538^{21}	...	s^h	...	...	...	chl s	**B1**[3], 744
b2997	**1-Butene, 1,3,4,4-tetrachloro-***	$ClCH:CHCHClCHCl_2$	193.89	...	...	88^{20}	1.0711_4^{20}	1.4773^{20}	i	...	...	...	...	chl s	**B1**[3], 726
b2998	—,**2,3,3,4-tetrachloro-***	$ClCH_2CCl_2CCl:CH_2$	193.89	...	...	$41–2^7$	1.4602_4^{20}	1.5133^{20}	i	...	...	s	...	chl s	**B1**[3], 726
b2999	—,**2,3,4-trichloro-***	$ClCH_2CHClCCl:CH_2$	159.44	...	...	60^{20}	1.3430_4^{20}	1.4944^{20}	i	...	...	s	...	chl s	**B1**[3], 725
b3000	**2-Butene, 1,2,4-trichloro-***	$ClCH_2CH:CClCH_2Cl$	159.44	...	...	$67–9^{10}$	1.3843_4^{20}	1.5175^{20}	i	s	s	...	s	chl s	**B1**[3], 744
—	—,**1,1,3-triethoxy-***	*see* **2-Butenal, 3-ethoxy-,** diethyl acetal													
b3001	**1-Butene, 2,3,3-trimethyl-***	Triptene. $(CH_3)_3CC(CH_3):CH_2$	98.19	...	−109.9	77.87^{760}	0.7050_4^{20}	1.4029^{20}	i	...	s	...	s	MeOH s	**B1**[3], 834
—	**3-Butene, 1,1-dicarboxylic acid***	*see* **Malonic acid, allyl-**													

For explanations, symbols and abbreviations see beginning of table. For structural formulas see end of table.

No.	Name	Synonyms and Formula	Mol. wt.	Color, crystalline form, specific rotation and λ_{max} (log ε)	m.p. °C	b.p. °C	Density	n_D	Solubility w	al	eth	ace	bz	other solvents	Ref.
	Butenedioic acid														
—	**Butenedioic acid** (*cis*)*	*see* **Maleic acid**													
—	—(*trans*)*	*see* **Fumaric acid**													
b3002	**2-Butene-1,4-diol** (*cis*)*	$HOCH_2CH{:}CHCH_2OH$	88.12	...	4	235^{760} 132^{16}	1.0698_4^{20}	1.4782^{20}	s	v	...	...	...	...	**B1**[3], 2255
b3003	—(*trans*)*	$HOCH_2CH{:}CHCH_2OH$	88.12	...	25	131^{13}	1.0700_4^{20}	1.4755^{20}	v	v	...	...	...	...	**B1**[3], 2256
b3004	**3-Butene-1,2-diol***	Vinylethylene glycol. $CH_2{:}CHCHOHCH_2OH$	88.12	...		196.5 98^{16}	1.0470_4^{20}	1.4628^{21}	s	s	...	...	...	...	**B1**[3], 2252
b3005	**2-Butene-1,4-dione, 1,4-diphenyl-**(*cis*)*	Dibenzoyl ethylene. $C_6H_5COCH{:}CHCOC_6H_5$	236.27	nd (al) λ^{al} 260 (4.26)	134 (130)	...	...	...	i	s^h	s	s	s	chl s oos s	**B7**[2], 741
b3006	—,—(*trans*)*	$C_6H_5COCH{:}CHCOC_6H_5$	236.27	ye nd (al or bz) λ^{al} 269 (4.27)	111	...	...	...	...	δ	...	...	s	chl v aa s lig i	**B7**[2], 741
b3007	**2-Butenoic acid** (*cis*)*	Isocrotonic acid. $CH_3CH{:}CHCO_2H$	86.09	nd or pr (peth) λ^{al} 204 (4.06)	15.5	169.3^{760} 74^{15}	1.0267_4^{20}	1.4483^{14}	v	s	...	...	...	...	**B2**[2], 394
Ω b3009	—(*trans*)*	Crotonic acid. $CH_3CH{:}CHCO_2H$	86.09	mcl nd or pr (w or lig) λ^{al} 205 (4.20)	71.5–1.7	185^{760} (189)	1.018_4^{15} 0.9604_4^{77}	1.4249^{77}	v	v	s	s	...	lig s^h	**B2**[2], 390
Ω b3010	—,amide (*trans*)	Crotonamide. $CH_3CH{:}CHCONH_2$	85.11	nd (ace) λ^{al} <216	161.5 (cor)	sub at 140^{13}	...	1.4420^{165}	δ	s	δ	...	s	...	**B2**[2], 392
Ω b3011	—,anhydride (*trans*)*	Crotonic anhydride. $(CH_3CH{:}CHCO)_2O$	154.17	...	...	246–8 129^{19}	1.0397^{20}	1.4745^{20}	d	d	s	...	...	...	**B2**[3], 1265
Ω b3012	—,chloride (*trans*)	Crotonyl chloride. $CH_3CH{:}CHCOCl$	104.54	λ^{al} <210	...	$124–5^{760}$ 35^{18}	1.0905^{20}	1.460^{18}	d	d	...	s	...	...	**B2**[2], 392
b3012[1]	—,ethyl ester (*cis*)*	Ethyl isocrotonate. $CH_3CH{:}CHCO_2C_2H_5$	114.15	...	...	136^{760}	0.9182_4^{20}	1.4242^{20}	i	s	s	s	...	os s	**B2**[2], 394
Ω b3013	—,—(*trans*)*	Ethyl crotonate. $CH_3CH{:}CHCO_2C_2H_5$	114.15	λ^{al} 210 (4.10)	...	136.5^{760} $58–9^{48}$	0.9175_4^{20}	1.4243^{20}	i	s	s	...	...	...	**B2**[2], 392
Ω b3014	—,methyl ester (*trans*)*	Methyl crotonate. $CH_3CH{:}CHCO_2CH_3$	100.13	λ^{al} 212 (4.16), 250 sh (2.3)	−42	121 (118.8–9.3)	0.9444_4^{20}	1.4242^{20}	i	v	v	...	...	...	**B2**[2], 392
Ω b3015	—,nitrile (*trans*)	Crotononitrile. $CH_3CH{:}CHCN$	67.09	...	−51.5	$120–1^{762}$	0.8239_4^{20}	1.4225^{20}	...	...	s	s	...	...	**B2**[2], 393
Ω b3016	**3-Butenoic acid***	Vinylacetic acid. $CH_2{:}CHCH_2CO_2H$	86.09	...	−35	169^{764} (163) $69–70^{12}$	1.0091_4^{20}	1.4239^{20}	s	∞	∞	...	...	...	**B2**[2], 389
b3017	—,ethyl ester*	$CH_2{:}CHCH_2CO_2C_2H_5$	114.14	λ^{hx} 225 sh (2.0)	...	119	0.9122^{20}	1.4105^{20}	...	s	...	...	...	...	**B2**, 407
b3018	—,nitrile	Allyl cyanide. $CH_2{:}CHCH_2CN$	67.09	...	−84	119	0.8329_4^{20}	1.4060^{20}	δ	∞	∞	...	...	...	**B2**[2], 389
Ω b3019	**2-Butenoic acid, 3-amino-**, ethyl ester (*trans*)*	$CH_3C(NH_2){:}CHCO_2C_2H_5$	129.16	mcl pr λ^{al} 275 (4.26)	34 (st) 20–1 (unst)	$210–5d^{760}$ 105^{15}	1.0219_4^{19}	1.4988^{22}	i	s	s	...	s	chl s CS_2 s lig s	**B3**[2], 423
b3020	—,**3-bromo-** (*trans*)*	$CH_3CBr{:}CHCO_2H$	165.00	nd (lig), lf (w)	97	...	...	...	δ	s	s	...	s	CS_2 s aa s	**B2**, 419
b3021	—,**4-bromo-**, ethyl ester (*trans*)*	$BrCH_2CH{:}CHCO_2C_2H_5$	193.05	...	...	$97–8^{15}$ $66–70^{0.3}$	1.402_4^{16}	1.4925^{20}	...	v	...	...	...	...	**C50**, 10804
b3022	—,**2-chloro-** (*cis*)*	$CH_3CH{:}CClCO_2H$	120.54	nd (w) λ^{al} 228 (3.85)	67	...	...	...	s	v	...	...	...	CS_2 v lig δ v^h	**B2**[2], 396
b3023	—,—(*trans*)*	$CH_3CH{:}CClCO_2H$	120.54	nd (w or peth) λ^{al} 222 (4.02)	100.5	212 $111–2^{14}$	...	...	δ	s	s	...	...	...	**B2**[2], 395
b3024	—,—,ethyl ester (*cis*)*	$CH_3CH{:}CClCO_2C_2H_5$	148.59	...	...	75^{30} 58^{12}	1.1021_4^{18}	...	...	s	s	...	...	...	**B2**[2], 396
b3025	—,—,—(*trans*)*	$CH_3CH{:}CClCO_2C_2H_5$	148.59	...	...	176–8 61^{10}	1.1133_4^{20}	1.4538^{20}	...	s	s	...	...	...	**B2**[2], 395
b3026	—,—,methyl ester (*trans*)*	$CH_3CH{:}CClCO_2CH_3$	134.57	...	...	161.5^{762} 59.5^{16}	1.160_4^{20}	1.4569^{23}	...	...	s	...	...	...	**B2**[2], 395
b3027	—,**3-chloro-** (*cis*)*	$CH_3CCl{:}CHCO_2H$	120.54	cr (w) λ^{hx} 226 (4.64)	61	195 sub	1.1995_4^{66}	1.4704^{66}	δ	v	...	...	...	peth v^h CS_2 v	**B2**[2], 396
b3028	—,—(*trans*)*	$CH_3CCl{:}CHCO_2H$	120.54	λ^{hx} 221 (4.30)	94–5	206–11 δd	...	...	δ	v	...	...	...	CS_2 v	**B2**[2], 396
b3029	—,—,ethyl ester (*cis*)*	$CH_3CCl{:}CHCO_2C_2H_5$	148.59			161.4 50^{10}	1.0860_4^{20}	1.4542_4^{19}	δ	s	s	...	...	...	**B2**[1], 190
b3030	—,—,—(*trans*)*	$CH_3CCl{:}CHCO_2C_2H_5$	148.59			184 66^{10}	1.1062_4^{20}	1.4592^{20}	δ	s	s	...	...	...	**B2**[2], 396
b3031	—,—,methyl ester (*cis*)*	$CH_3CCl{:}CHCO_2CH_3$	134.57	...	...	142.4 $42–3^{13}$	1.138_4^{20}	1.4573^{19}	δ	...	v	...	...	MeOH v	**B2**[2], 396
b3032	—,—,—(*trans*)*	$CH_3CCl{:}CHCO_2CH_3$	134.57	...	...	$64–7^{14}$	1.157_4^{20}	1.4630^{20}	δ	v	v	...	...	...	**B2**[2], 396
b3033	—,**4-chloro-** (*trans*)*	$ClCH_2CH{:}CHCO_2H$	120.54	cr (eth-peth)	83	$117–8^{13}$	...	...	$δ^h$	...	v	...	...	peth δ aa v	**B2**, 418
b3034	—,**2,3-dibromo-4-oxo-***	Mucobromic acid. $O{:}CHCBr{:}CBrCO_2H$	257.88	pl (eth-lig), rh cr (w)	125	...	...	...	s v^h	v	v	...	δ	chl δ CS_2 δ	**B3**[2], 460
Ω b3035	—,**2,3-dichloro-4-oxo-***	Mucochloric acid. $O{:}CHCCl{:}CClCO_2H$	168.97	mcl pr (eth-lig), pl (w) λ^{al} 227 (3.9)	127	...	...	...	δ s^h	s	s	...	s	...	**B3**, 727
b3036	—,**2-ethyl-** (*trans*)*	$CH_3CH{:}C(C_2H_5)CO_2H$	114.15	mcl pr (peth)	45–6	209 109^{13}	0.9578_4^{50}	1.4475^{50}	δ	s	s	...	...	...	**B2**[2], 208
b3037	**3-Butenoic acid, 2-hydroxy-**, ethyl ester*	$CH_2{:}CHCHOHCO_2C_2H_5$	130.14	...	...	173^{756}d 68^{15}	1.0470_4^{15}	1.436^{13}	v	∞	∞	...	...	...	**B3**, 271
b3038	—,**2-hydroxy-4-phenyl-***	Benzallactic acid. $C_6H_5CH{:}CHCHOHCO_2H$	178.19	nd (w)	137	...	...	...	δ v^h	...	i	...	i	CS_2 δ lig δ	**C51**, 10449

For explanations, symbols and abbreviations see beginning of table. For structural formulas see end of table.

PHYSICAL CONSTANTS OF ORGANIC COMPOUNDS (Continued)

No.	Name	Synonyms and Formula	Mol. wt.	Color, crystalline form, specific rotation and λ_{max} (log ε)	m.p. °C	b.p. °C	Density	n_D	Solubility w	al	eth	ace	bz	other solvents	Ref.
	2-Butenoic acid														
b3039	**2-Butenoic acid, 2-methyl- (*cis*)***	Angelic acid. $CH_3CH{:}C(CH_3)CO_2H$	100.13	mcl pr or nd λ^{al} 215.5 (3.97)	45–6	185 (cor) $88–9^{10}$	0.983_4^{49} 0.9539_4^{76}	1.4434^{47}	δ s^h	s	v	...	...	...	**B2**[2], 401
Ω b3040	**—,—(*trans*)***	Tiglic acid. $CH_3CH{:}C(CH_3)CO_2H$	100.13	ta (w) λ^{al} 212.5 (4.09)	64.5–5.0	198.5^{760} (cor)	0.9641_4^{76}	1.4330_2^{76}	δ s^h	s	s	...	...	...	**B2**[2], 401
b3041	**—,—,chloride (*trans*)**	Tiglyl chloride. $CH_3CH{:}C(CH_3)COCl$	118.56	...	...	$146–8^{760}$ 45^{12}	...	...	d	d	s	...	...	...	**B2**, 431
b3042	**—,—,ethyl ester***	$CH_3CH{:}C(CH_3)CO_2C_2H_5$	128.17	...	...	156^{760} 55.5^{11}	0.9200_4^{20}	1.4340^{20}	...	s	...	...	s	...	**B2**[2], 401
b3042[1]	**—,3(4-nitro-phenyl)-***	...	207.19	pa ye nd (aa)	168–9	...	...	...	...	...	...	...	...	aa s^h	**B9**, 615
Ω b3043	**—,4-oxo-4-phenyl-***	β-Benzoylacrylic acid. $C_6H_5COCH{:}CHCO_2H$	176.18	nd or pr (to) λ^{al} 238 (4.05), 260 sh (4.00), 360 (1.88)	99	...	...	...	δ s^h	s	s	...	...	to s^h lig δ	**B10**[2], 499
b3043[1]	**—,—,hydrate***	$C_6H_5COCH{:}CHCO_2H.H_2O$	194.19	lf (w + 1)	65	...	...	...	s^h	s	s	...	...	...	**B10**, 726
b3044	**3-Butenoic acid, 4-phenyl-***	Styrylacetic acid. $C_6H_5CH{:}CHCH_2CO_2H$	162.19	nd (w), pr (CS_2) λ^{al} 250 (4.24), 283 (3.18), 292 (2.99)	87	302	...	...	i δ^h	v	v	...	...	CS_2 v^h	**B9**[2], 407
Ω b3045	**2-Buten-1-ol (*trans*)***	Crotyl alcohol. $CH_3CH{:}CHCH_2OH$	72.12	...	< −30	121.2^{760}	0.8521_4^{20}	1.4288^{20}	v	∞	∞	...	...	...	**B1**[2], 480
—	**—,acetate**	*see* **Acetic acid,** 2-buten-1-yl ester													
b3046	**3-Buten-1-ol***	Allylcarbinol. $CH_2{:}CHCH_2CH_2OH$	72.12	...	...	113.5^{760}	0.8424_4^{20}	1.4224^{20}	s	∞	∞	s	...	...	**B1**[2], 480
b3047	**3-Buten-2-ol(*d*)***	$CH_3CHOHCH{:}CH_2$	72.12	$[\alpha]_D^{20}$ +33.9 (undil)	...	$96.2–.5^{745}$	0.8362_4^{15}	1.4120^{20}	...	...	...	...	...	...	**B1**[2], 480
b3048	**—(*dl*)***	$CH_3CHOHCH{:}CH_2$	72.12	...	< −100	97.3^{760}	0.8318_4^{20}	1.4137^{20}	δ	...	...	...	...	...	**B1**[2], 479
b3049	**2-Buten-1-ol, 2-chloro-***	$CH_3CH{:}CClCH_2OH$	106.55	...	...	159^{760}	1.1180_4^{20}	1.4682^{20}	s	v	...	...	...	...	**B1**[2], 481
b3050	**—,4-chloro-***	$ClCH_2CH{:}CHCH_2OH$	106.55	...	...	$64–5^{2}$	...	1.4845^{20}	d^h	v	v	...	...	...	**B1**[3], 1901
b3051	**3-Buten-1-ol 2-chloro-***	$CH_2{:}CHCHClCH_2OH$	106.55	...	...	$66.5–7^{30}$	1.1044_4^{20}	1.4665^{20}	...	v	v	...	...	...	**B1**[3], 1897
b3052	**3-Buten-2-ol, 1-chloro-***	$CH_2{:}CHCHOHCH_2Cl$	106.55	...	...	144–7 63.3^{30}	1.111_4^{20}	1.4643^{20}	...	...	...	...	...	chl v	**B1**[3], 1893
b3053	**—,3-chloro-***	$CH_3CHOHCCl{:}CH_2$	106.55	...	...	$53–7^{19}$	1.1138_4^{23}	...	...	v	...	...	...	...	**B1**[3], 1893
b3054	**3-Butene-2-one***	Methyl vinyl ketone. $CH_3COCH{:}CH_2$	70.09	λ^{al} 210 (3.81)	...	81.4^{760} $33–4^{130}$	0.8636_4^{20}	1.4081^{20}	s	s	v	v	s	MeOH s aa s	**B1**[2], 786
b3055	**—,4-bromo-4-phenyl-***	(α-Bromobenzal)acetone. $CH_3COCH{:}CBrC_6H_5$	225.09	...	...	$150–1^{10}$ $128–31^{1}$	...	...	...	...	...	...	s	...	**B7**, 367
Ω b3056	**2-Buten-1-one, 1,3-diphenyl-***	Dypnone. β-Methylchalcone. $CH_3C(C_6H_5){:}CHCOC_6H_5$	222.29	λ^{cy} 288.5 (4.24)	...	340–5 δd 225^{22}	1.108_0^{20}	1.6312^{20}	...	...	s	...	...	...	**B7**[2], 433
Ω b3057	**3-Buten-2-one, 4(2-furyl)-**	Furfuralacetone.	138.16	nd λ^{al} 236 (3.28), 316 (4.36)	39–40	229d $112–5^{10}$	1.0496_4^{57}	1.5788^{45}	i	v	v	...	...	chl v peth s con sulf s (ye)	**B17**[2], 326
Ω b3058	**—,4(4-hydroxy-3-methoxy-phenyl)-**	Vanillalacetone. $C_{11}H_{12}O_3$. *See* b3065	192.22	nd (dil al)	129	...	...	...	δ	v	v	...	v	...	**B8**[2], 326
b3059	**—4(2-hydroxy-phenyl)-**	Salicylideneacetone. $C_{10}H_{10}O_2$. *See* b3065	162.19	nd (al or lig), pr (bz) λ^{al} 285 (4.1), 335 (4.0)	140	...	...	...	δ	s^h	v	...	s^h	...	**B8**[2], 153
b3060	**—,4(3-hydroxy-phenyl)-**	$C_{10}H_{10}O_2$. *See* b3065	162.19	ye pr (bz)	97–8	...	...	...	...	...	...	...	s^h	...	**B8**[2], 155
b3061	**—,4(4-hydroxy-phenyl)-**	$C_{10}H_{10}O_2$. *See* b3065	162.19	nd (w) λ^{al} 234 (4.01), 323 (4.37)	114–5 (109)	...	...	...	δ s^h	v	...	...	...	aa v	**B8**[2], 155
Ω b3062	**—,4(4-methoxy-phenyl)-**	Anisylideneacetone. $C_{11}H_{12}O_2$. *See* b3065	176.22	lf (al, MeOH, eth or aa) λ^{al} 233.5 (3.92) 322 (4.27)	73	...	...	...	i	v	v	...	s	aa s sulf s (ye)	**B8**[2], 155
Ω b3063	**—,3-methyl-**	Isopropenyl methyl ketone. $CH_3COC(CH_3){:}CH_2$	84.13	λ^{al} 219.5 (3.80)	−54	98^{760}	0.8527_4^{20}	1.4220^{20}	δ	∞	...	...	...	...	**B1**, 733
Ω b3064	**—,4(3,4-methyl-enedioxy-phenyl)-(*trans*)**	Piperonalacetone. $C_{11}H_{10}O_3$. *See* b3065	190.20	...	111	...	...	...	i δ^h	δ v^h	s	...	s	CCl_4 v peth i	**B19**[2], 157
Ω b3065	**—,4-phenyl-(*trans*)***	Benzalacetone. Methyl styryl ketone.	146.19	pl λ^{al} 221 (4.00), 286.5 (4.32)	42	262^{760} 140^{16}	1.0097^{45}	1.5836^{46}	i	v	s	s	s	chl s peth δ	**B7**[2], 287
b3066	**1-Buten-3-yne***	Vinylacetylene. $HC{:}CCH{:}CH_2$	52.08	λ^{iso} 219 (3.88), 227.5(3.89)	...	5.1^{760}	0.7095_0^0	1.4161^{1}	i	...	...	...	s	...	**B1**[3], 1032
b3067	**—,4-chloro-***	$ClC{:}CCH{:}CH_2$	86.52	...	...	$55–7^{760}$	1.0022_4^{20}	1.4656^{20}	...	...	...	...	...	chl s	**B1**[3], 1038
b3068	**—,1-methoxy-***	$HC{:}CCH{:}CHOCH_3$	82.10	...	...	$122–5^{760}$ d $30–2^{13}$	0.906_4^{20}	1.4818^{20}	i	...	...	...	...	os v	**C53**, 21292
Ω b3069	**—,2-methyl-***	Isopropylideneacetylene. Valylene. $CH{:}CC(CH_3){:}CH_2$	66.10	...	...	34^{760}	0.6801_4^{11}	1.4105^{20}	...	...	...	...	...	...	**B1**[3], 1039

For explanations, symbols and abbreviations see beginning of table. For structural formulas see end of table.

No.	Name	Synonyms and Formula	Mol. wt.	Color, crystalline form, specific rotation and λ_{max} (log ε)	m.p. °C	b.p. °C	Density	n_D	Solubility w	al	eth	ace	bz	other solvents	Ref.
	Butesin														
—	**Butesin**	*see* **Benzoic acid, 4-amino-**, butyl ester													
—	***tert*-Butyl bromide**	*see* **Propane, 2-bromo-2-methyl-***													
—	***tert*-Butyl chloride**	*see* **Propane, 2-chloro-2-methyl-***													
—	**Butylene**	*see* **Butene***													
—	**Butylene oxide**	*see* **Butane, epoxy-***													
b3070	**1-Butyne***	$CH_3CH_2C{:}CH$	54.09	λ^{gas} 172 (3.65)	−125.72	8.1^{760}	0.6784^{0}_{4} 0.650^{30}_{4}	1.3962^{20}	i	s	s	...	...	...	**B1**^{2}, 223
b3071	**2-Butyne***	Dimethylacetylene. $CH_3C{:}CCH_3$	54.09	...	−32.26	27^{760}	0.6910^{20}_{4}	1.3921^{20}	i	s	s	...	...	...	**B1**^{3}, 925
b3072	—,1-chloro-*	$CH_3C{:}CCH_2Cl$	88.54	...	...	$104\text{–}6^{760}$	1.0152^{20}	1.4581^{20}	...	s	v	v			**B1**^{3}, 927
b3073	**1-Butyne, 3-chloro-3-methyl-***	$(CH_3)_2CClC{:}CH$	102.57	...	...	77–9	0.9061^{20}_{4}	...	...	...	...	...	...	...	**B1**^{3}, 965
Ω b3074	**2-Butyne, 1,4-dibromo-***	$BrCH_2C{:}CCH_2Br$	211.90	λ^{al} <217	...	92^{15}	2.014^{18}	1.588^{18}	...	...	s^h	s	...	chl v	**B1**^{3}, 927
Ω b3075	—,1,4-dichloro-*	$ClCH_2C{:}CCH_2Cl$	122.98	...	...	$165\text{–}6^{760}$ 73^{24}	1.258^{20}_{4}	1.5058^{20}	...	...	s	s	...	chl v	**B1**^{3}, 927
b3076	—,1,4-diiodo-*	$ICH_2C{:}CCH_2I$	305.89	nd (al) λ 240 (4.00)	53	$70\text{–}2^{0.1}$	...	...	... δ	s^h	s	s	...	chl v d in air	**B1**^{3}, 927
b3077	**1-Butyne, 3,3-dimethyl-***	*tert*-Butylacetylene. $(CH_3)_3CC{:}CH$	82.15	...	−81.2	$39\text{–}40^{760}$	0.6695^{20}_{4}	1.3738^{20}	...	...	...	...	...	...	**B1**^{3}, 991
Ω b3078	**2-Butyne, hexa-fluoro-***	Perfluoro-2-butyne*. $CF_3C{:}CCF_3$	162.04	...	−117.4	-24.6^{760}	...	...	...	s	s	s	...	CCl_4 s aa s	**B1**^{3}, 926
b3079	**1-Butyne, 3-methyl-***	Isopropylacetylene. $(CH_3)_2CHC{:}CH$	68.13	...	−89.7	29.35^{760}	0.6660^{20}_{4}	1.3723^{20}	i	∞	∞	...	...	...	**B1**^{3}, 965
b3080	—,4-phenyl-*	$C_6H_5CH_2CH_2C{:}CH$	130.19	...	...	190^{760} 83^{15}	0.9258^{20}_{4}	1.5208^{20}	i	...	...	...	...		**B5**^{2}, 413
Ω b3081	**2-Butynedioic acid***	Acetylenedicarboxylic acid. $HO_2CC{:}CCO_2H$	114.06	pl (eth), cr (+2w, aq eth)	179	...	...	...	v	v	v	...	...	...	**B2**^{2}, 670
Ω b3082	—,diethyl ester*	$C_2H_5O_2CC{:}CCO_2C_2H_5$	170.17	...	1–2	184^{200} 108^{13}	1.0675^{20}_{4}	1.4425^{20}	...	s	s	...	...	CCl_4 s	**B2**^{2}, 671
Ω b3083	—,dimethyl ester*	$CH_3O_2CC{:}CCO_2CH_3$	142.11	...	...	$195\text{–}8^{760}$ δd 98^{20}	1.1564^{20}_{4}	1.4434^{20}	...	s	s	...	...	CCl_4 s	**B2**^{2}, 671
Ω b3084	**2-Butyne-1,4-diol***	$HOCH_2C{:}CCH_2OH$	86.09	pl (bz, AcOEt), lf (aa)	58	238^{760} 145^{15}	...	1.4804^{20}	v	v	δ	v	i	chl δ peth i MeOH v	**B1**^{1}, 261
b3085	**3-Butyne-1,2-diol***	$HC{:}CCHOHCH_2OH$	86.09	...	40	$64\text{–}6^{0.2}$	...	...	s	s	δ	...	...		**B1**^{2}, 569
b3086	**2-Butynoic acid***	Tetrolic acid. $CH_3C{:}CCO_2H$	84.08	pl (eth, peth or CS_2) λ^{al} 205.5 (3.80)	78	203^{760} 99–100^{18}	0.9641^{20}_{4}	...	s	s	s	...	...	chl, CS_2 s	**B2**^{2}, 451
b3087	—,ethyl ester*	$CH_3C{:}CCO_2C_2H_5$	112.14	...	...	163^{760} 105^{190}	0.9641^{20}_{4}	1.4372^{20}	...	...	...	...	...	...	**B2**^{1}, 208
b3088	**2-Butyn-1-ol***	$CH_3C{:}CCH_2OH$	70.09	...	−2.2	143^{760} $52\text{–}3^{14}$	0.9370^{20}_{4}	1.4530^{20}	...	s	∞	...	...	...	**B1**^{3}, 1973
Ω b3089	**3-Butyn-1-ol***	$HOCH_2CH_2C{:}CH$	70.09	...	−63.6	129^{760}	0.9257^{20}_{4}	1.4409^{20}	s	v	...	...	...	MeOH v	**B1**^{3}, 1972
b3090	**3-Butyn-2-ol**	$CH{:}CCHOHCH_3$	70.09	...	...	107^{760}	0.8858^{20}	1.4265^{20}	s	v	s	...	...	...	**B1**^{3}, 1971
Ω b3091	—,2-methyl-*	$(CH_3)_2COHC{:}CH$	84.13	...	+3 (−3.5)	104^{760} 56^{97}	0.8618^{20}_{4}	1.4207^{20}	s	s	...	...	...	...	**B1**^{2}, 505
—	**Butyraldehyde**	*see* **Butanal***													
—	**Butyramide**	*see* **Butanoic acid,** amide													
—	**Butyric acid**	*see* **Butanoic acid***													
—	**Butyroin**	*see* **4-Octanone, 5-hydroxy-***													
—	**Butyrolactone**	*see* **Butanoic acid, 4-hydroxy-**, lactone*													
—	**Butyrophenone**	*see* **1-Butanone, 1-phenyl-***													

For explanations, symbols and abbreviations see beginning of table. For structural formulas see end of table.

No.	Name	Synonyms and Formula	Mol. wt.	Color, crystalline form, specific rotation and λ_{max} (log ε)	m.p. °C	b.p. °C	Density	n_D	Solubility: w	al	eth	ace	bz	other solvents	Ref.
	Cacodyl														
c1	**Cacodyl**	Diarsenic tetramethyl. Dimethylarsenic. Tetramethylbiarsine. $(CH_3)_2AsAs(CH_3)_2$	209.98	pl	−6	165^{760}	1.447^{15}		δ	s	s				**B4**[2], 1002
c2	**Cacodyl chloride**	Dimethylarsenic monochloride. Dimethylchlorarsine. $(CH_3)_2AsCl$	140.45		< −45	109	1.5046_4^{12}	1.5203^{12}	i	s	i				**B4**[2], 987
c3	**Cacodyl oxide**	Alkarsine. Bis-dimethylarsenous oxide. Dicacodyl oxide. $(CH_3)_2AsOAs(CH_3)_2$	225.98		−25	150^{760}	1.4816^{15}	1.5255^{9}	δ	s	s				**B4**[2], 989
c4	**—,perfluoro-**	Bis(trifluoromethyl) arsenous oxide. $(CF_3)_2AsOAs(CF_3)_2$	441.87			100^{760}		1.3540^{20}							**J 1953,** 1562
c5	**Cacodyl sulfide**	Bis-dimethylarsine sulfide. Dicacodyl sulfide. $(CH_3)_2AsSAs(CH_3)_2$	242.05		< −40	211^{760}			δ	s	s				**B4**, 608
c6	**Cacodyl trichloride**	Dimethylarsenic trichloride. Dimethylorthoarsenic acid trichloride. $(CH_3)_2AsCl_3$	211.35	cr (eth)	d at 50				d	d	s			CS_2 s	**B4**, 612
—	**Cacodylic acid**	*see* **Dimethylarsinic acid**													
—	**Cadalene**	*see* **Naphthalene, 1,6-dimethyl-4-isopropyl-**													
—	**Cadaverine**	*see* **Pentane, 1,5-diamino-***													
c7	**β-Cadinene(*l*)**	3,9-Cadinadiene	204.36	$[\alpha]_D^{22}$ −15.9 (chl, c = 1) ($[\alpha]_D$ −251) λ^{hx} 250 (2.65), 323 (0.75)		274^{760} 149^{20}	0.9239_4^{20}	1.5059^{20}	i	δ	s			lig s chl s[h]	**B5**[2], 347
—	**Caffeic acid**	*see* **Cinnamic acid, 3,4-dihydroxy-**													
Ω c8	**Caffeine**	Theine. 1,3,7-Trimethylxanthine.	194.20	wh nd (w + 1), hex pr (sub) λ^{al} 227 (4.3), 235 (4.3), 274 (4.3)	238 (anh)	sub 178 sub 89^{15}	1.23^{19}		δ s[h]	δ	i	δ	δ	Py s chl s[h] aa δ CS_2, CCl_4 i	**B26**[2], 266
c9	—,benzoate.	$C_8H_{10}N_4O_2.C_6H_5CO_2H$. *See* c8	316.32	wh so pw					s	s					**B26**[2], 268
c10	—,citrate	$C_8H_{10}N_4O_2.C_6H_8O_7$. *See* c8.	386.32	mcl cr					d	d					**B26**[2], 269
c11	—,hydrobromide	$C_8H_{10}N_4O_2.HBr.2H_2O$. *See* c8.	311.14	ye	d 80–100				s	d	i			chl i	**B26**[1], 137
c12	—,hydrochloride	$C_8H_{10}N_4O_2.HCl.2H_2O$. *See* c8.	266.69	mcl pr	d 80–100				d	d					**B26**[2], 268
c13	—,hydroiodide diiodide	Caffeine iodide. Diiodocaffeine hydroiodide. Caffeine triiodide. $C_8H_{10}N_4O_2.I_2.HI.1\frac{1}{2}H_2O$. *See* c8	603.01	gr pr	171d				i	v	i			chl δ	**B26**, 466
c14	—,2-hydroxybenzoate	Caffeine salicylate. $C_8H_{10}N_4O_2.C_7H_6O_3$. *See* c8	332.32	wh nd (w)	137				δ	s					**B26**[2], 269
c16	—,3-methylbutanoate	Caffeine isovalerate. $C_8H_{10}N_4O_2.C_5H_{10}O_2$. *See* c8	296.33	unst nd					s						**B26**, 467
c17	—,sulfate.	$C_8H_{10}N_4O_2.H_2SO_4$. *See* c8.	292.27	wh nd					d	d					**B26**, 466
c18	**—,8-ethoxy-**	1,3,7-Trimethyl-2,6-dioxo-8-ethoxypurine. $C_{10}H_{14}N_4O_3$. *See* c8	238.25	wh or yesh nd (w)	143				δ	v[h]	δ				**B26**[2], 322
c19	**—,8-methoxy-**	$C_9H_{12}N_4O_3$. *See* c8	224.22	wh nd (al or w)	176		1.399_4^{25}		s[h]	v			s		**B26**[2], 322
Ω c20	**Calciferol**	Irradiated ergosterol. Vitamin D2. $C_{28}H_{44}O$.	396.67	pr (ace) $[\alpha]_D^{20}$ +102.5 (al), +81 (ace) λ^{al} 265 (4.27)	115–8	sub at 0.0006 mm			i	s	s	s			**E12A,** 170
c21	**Camphane**	Bornane. Bornylane. Dihydrocamphene. 1,7,7-Trimethylbicyclo[2,2,1]heptane.	138.25	hex pl (al), pr (MeOH)	158–9	sub 161			i	s[h]	s			AcOEt s MeOH s[h]	**E12A,** 570
—	**—,2,3-dioxo-**	*see* **Camphorquinone**													
c22	**3-Camphanecarboxylic acid**		182.27	pl (dil aa) $[\alpha]_D^{20}$ +56 (al, p = 8)	90–1	153^{13}				s				aa v	**B9**[2], 52
—	**2,3-Camphanedione**	*see* **Camphorquinone**													
c23	**Camphene(*d*)**		136.24	nd, $[\alpha]_D^{17}$ +103.9 (eth, c = 4), +104.7 (al) λ^{al} 206 (3.74)	52	160–2^{760} 52^{17}	0.8450_4^{50}	1.4570^{25}	i	δ	s				**B5**[3], 380
Ω c24	—(*dl*)	$C_{10}H_{16}$. *See* c23	136.24	nd (sub)	51–2	158.5–9.5^{760}	0.879_4^{20}	1.4551^{54}	i	v	v				**B5**[3], 380
Ω c25	—(*l*)	$C_{10}H_{16}$. *See* c23	136.24	$[\alpha]_D^{19}$ −106.1 (eth, c = 4)	52	158^{760}	0.8446_4^{50}	1.4564^{54}			s				**B5**[3], 380

For explanations, symbols and abbreviations see beginning of table. For structural formulas see end of table.

No.	Name	Synonyms and Formula	Mol. wt.	Color, crystalline form, specific rotation and λ_{max} (log ε)	m.p. °C	b.p. °C	Density	n_D	Solubility w	al	eth	ace	bz	other solvents	Ref.
	Camphenilone														
c26	**Camphenilone (*d*)**		138.21	$[\alpha]_D^{20}$ +70.4 (al)	41	193^{751} 76^{12}			i	...	s	...	...		**B7**[1], 59
—	**Camphenol**	*See* **1-Epiborneol**													
—	**Camphol**	*See* **Isoborneol**													
c27	**Campholic acid**		170.25	pr, $[\alpha]_D^{20}$ +59.3 (bz) λ <220	106	255^{768} 146^{12}			δ s^h	s	s	...	...		**B9**[2], 17
c28	—(*dl*)	$C_{10}H_{18}O_2$. *See* c27	170.25	tcl pr	109				i	v	s	...			**B9**[2], 19
c29	—(*l*)	$C_{10}H_{18}O_2$. *See* c27	170.25	pr (dil al) $[\alpha]_D^{15}$ −49.1 (al)	106–7	250^{760}			i	s	s	...			**B9**, 36
c30	**Campholytic acid** (α,*l*)		154.21	$[\alpha]_D^{13}$ −60.4		240 −3 140^{15}	1.0145^{18}	1.4712^{17}	...	...	...	...	...	lig v	**B9**[1], 33
c31	**Camphor (*d*)**	*d*-2-Camphanone. Formosa camphor. Laurel camphor. $C_{10}H_{16}O$.	152.24	pl, $[\alpha]_D^{20}$ +44.26 (al) λ^{al} 290 (1.48)	179.8	sub 204^{760}	0.990_4^{25}	1.5462	i	v	v	s	s	chl, CCl_4, MeOH, aa v	**B7**[2], 93
Ω c32	—(*dl*)	$C_{10}H_{16}O$. *See* c31	152.24	wh λ^{al} 290 (1.48)	178.8	sub			i	v	v	s	s	chl, aa v	**B7**[2], 104
c33	—(*l*)	$C_{10}H_{16}O$. *See* c31	152.24	$[\alpha]_D^{16}$ −44.2 (al, c = 16.5) λ^{al} 290 (1.49)	178.6	204 sub	0.9853^{18}		i	v	v	s	s	chl, aa v	**B7**[2], 103
c34	—,oxime (*d*)		167.25	pr (lig-eth) $[\alpha]_D^{22}$ +42.5 (al)	115				i	v	s	...			**B7**[1], 80
c35	—,—(*dl*)	$C_{10}H_{17}NO$. *See* c34	167.25	cr (peth)	118				i	v	s	...			**B7**[2], 104
c36	—,—(*l*)	$C_{10}H_{17}NO$. *See* c34	167.25	mcl nd or pr (dil al) $[\alpha]_D^{20}$ −42.4 (al)	118	249–54d	1.01_4^{116}		i	v	s	...			**B7**[1], 84
c37	—,**3-amino-(*d*)**	3-Aminocamphor. 3-Camphorylamine. $C_{10}H_{17}NO$. *See* c31	167.25	wx	110 −15d	244			i	s	s	...			**B14**[2], 6
c38	—,**3-bromo-(*d*,α)**	$C_{10}H_{15}BrO$. *See* c31	231.14	pr (al) $[\alpha]_D^{20}$ +129.3 (MeOH, c = 4.6), +122.7 (bz) λ^{bz} 307.5 (1.98)	76	274 δd sub	1.449_4^{20}		i	s v^h	v	...	s	chl, CCl_4 s	**B7**[2], 100
Ω c39	—,—(*dl*,α)	$C_{10}H_{15}BrO$. *See* c31	231.14		51				i	v	v	...	s	chl s	**B7**[2], 105
c40	—,—(*l*,α)	$C_{10}H_{15}BrO$. *See* c31	231.14	mcl nd (al) $[\alpha]_D^{18}$ −138.8 (ace, c = 6)	76				i	v	v	...	s	chl s	**B7**[2], 104
c41	—,—(*d*,α′)	$C_{10}H_{15}BrO$. *See* c31	231.14	nd (dil al) $[\alpha]_D^{20}$ +29.4 λ^{bz} 312 (1.95)	78	265d	1.484^{14}			v	...	...	...	chl, CS_2, aa v	**B7**[2], 101
Ω c42	—,**8-bromo-(*d*)**	π-Bromocamphor. $C_{10}H_{15}BrO$. *See* c31	231.14	tetr pr (lig) $[\alpha]_D^{19}$ +122.2 (chl) λ^{al} 289 (1.77)	93	sub			...	s	v^h	s	s	os s peth s^h	**B7**, 123
c43	—,—(*dl*)	$C_{10}H_{15}BrO$. *See* c31	231.14	pr (eth-peth), pym (eth)	92.7	sub			...	s	v^h	s	s	os s peth s^h	**B7**, 136
Ω c44	—,**10-bromo-(*d*)**	$C_{10}H_{15}BrO$. *See* c31	231.14	pr (peth) $[\alpha]_D^{20}$ +19.2 (abs al), +15.7 (bz) λ^{cy} 293 (1.39)	78	265d			i	v^h	v	s	v	chl, CS_2, aa v	**B7**[2], 101
c45	—,—(*dl*)	$C_{10}H_{15}BrO$. *See* c31	231.14	cr	77				...	...	...	...	...		**B7**[2], 105
c46	—,**3-chloro-(*d*,α)**	$C_{10}H_{15}ClO$. *See* c31	186.68	lf, $[\alpha]_D^{20}$ +71.1 (bz, c = 9.2), +97 (al) λ^{cy} 305 (1.72)	94	244–7d			s^h	s^h	s	s	s	CS_2, chl s	**B7**[2], 100
c47	—,—(*d*,α′)	$C_{10}H_{15}ClO$. *See* c31	186.68	$[\alpha]_D^{20}$ +35 (al, c = 5) λ^{cy} 306 (1.75)	118	231d			δ	δ	s	...	...	CS_2, chl s	**B7**[2], 100
Ω c48	—,**8-chloro-(*d*)**	π-Chlorocamphor. $C_{10}H_{15}ClO$. *See* c31	186.68	pr, $[\alpha]_D$ +99.9 (chl)	139				...	...	...	...	...		**B7**, 119
c49	—,—(*dl*)	$C_{10}H_{15}ClO$. *See* c31	186.68		138	sub			...	...	...	...	...	os v	**B7**, 136
c50	—,**10-chloro-(*d*)**	$C_{10}H_{15}ClO$. *See* c31	186.68	pr (al) $[\alpha]_D^{14}$ +40.7 (al)	132–5				...	s	v	...	s	chl, peth, aa s	**B7**[2], 100
c51	—,**3,3-dibromo-(*d*)**	$C_{10}H_{14}Br_2O$. *See* c31	310.04	wh-ye rh pr (al, peth), $[\alpha]_D^{20}$ +40 (chl), +39.2 (al) λ^{al} 323 (1.88)	64	δ sub δ d	$1.854_4^{21.6}$		i	v	v	...	v	chl, peth v AcOEt s	**B7**[2], 101

No.	Name	Synonyms and Formula	Mol. wt.	Color, crystalline form, specific rotation and λ_{max} (log ε)	m.p. °C	b.p. °C	Density	n_D	Solubility						Ref.
									w	al	eth	ace	bz	other solvents	
	Camphor														
c52	—,**3-nitro-**(*l*)	$C_{10}H_{15}NO_3$. *See* c31	197.24	mcl pr (bz), $[\alpha]_D^{13}$ $-26 \rightarrow -9$ (al) (mut) $[\alpha]_D^{15}$ $-124 \rightarrow -104$ (bz) (mut)	104	...	...	...	i	s	s	...	v	chl v lig s peth δ	**B7**², 103
—	**Camphoramic acid**	*See* **Camphoric acid,** monoamide													
c53	**3-Camphorcarboxylic acid** (*d*)	...	196.25	pr (eth, 50 % al) $[\alpha]_D^{20}$ +34.9 (bz)	128d	...	...	...	s	v	s	...	v	lig δ	**B10**², 442
c54	—(*dl*)	$C_{11}H_{16}O_3$. *See* c53	196.25	cr (bz)	136–7	...	...	...	s	v	sʰ	...	vʰ	AcOEt s	**B10**¹, 308
c55	—(*l*)	$C_{11}H_{16}O_3$. *See* c53	196.25	pr (eth), cr (bz) $[\alpha]_D^{20}$ −64 (al), −57.4 (AcOEt)	127–8d	...	...	...	s	v	sʰ	...	vʰ	AcOEt s	**B10**¹, 308
c56	**Camphoric acid** (*d*)	1,2,2-Trimethylcyclopentane-1,3-dicarboxylic acid.	200.24	pr lf (wʰ) $[\alpha]_D^{20}$ +47.7 (al), +50.8 (ace)	188.2	...	1.186_4^{20}	...	δ sʰ	v	v	s	i	MeOH v to, lig, chl i	**B9**², 534
Ω c57	—(*dl*)	$C_{10}H_{16}O_4$. *See* c56	200.24	pr (al, aa), mcl nd	208 (203)	...	1.228_4^{20}	...	vʰ	s	v	...	...	chl δ	**B9**¹, 332
c58	—(*l*)	$C_{10}H_{16}O_4$. *See* c56	200.24	cr (w), $[\alpha]_D^{16}$ −48.1 (abs al, c = 8)	187.5	...	1.190	...	δ	s	s	s	...	MeOH s aa δ	**B9**², 539
Ω c59	—, anhydride (*d*)	...	182.22	rh (al), pr (bz)	223.5	>270d	1.194^{20}	...	δ	δ	δ	...	s	chl δ	**B17**¹, 238
Ω c60	—,—(*dl*)	$C_{10}H_{14}O_3$. *See* c59	182.22	rh (al)	221	270	1.194_4^{20}	...	δ	δ	δ	...	s	AcOEt, CS_2 s chl δ	**B17**, 459
c61	—,—(*l*)	$C_{10}H_{14}O_3$. *See* c59	182.22	$[\alpha]_D$ −77(bz)	221	>270	1.194_4^{20}	...	...	δ	δ	...	s	...	**B17**, 459
c62	—,diethyl ester (*d*).	$C_{14}H_{24}O_4$. *See* c56	256.35	$[\alpha]_D^{15}$ +7.5 (al), +9 (bz)	...	286^{752} 164^{20}	1.0298_4^{20}	1.4535^{26}	i	s	s	...	s	CCl_4 s	**B9**², 536
c62¹	—,dimethyl ester (*d*)	$C_{12}H_{20}O_4$. *See* c56	228.29	$[\alpha]_D$ +49.07 (al)	< −16	264^{738} 155^{15}	1.0747_4^{20}	1.4627^{19}	i	v	s	...	...	...	**B9**², 535
c63	—, 1-monoamide (*d*)	β-Camphoramic acid.	199.25	pl, nd, $[\alpha]_D^{20}$ +74 (al)	183	...	...	...	δ sʰ	s	δ	s	δ	lig δ	**B9**², 536
c64	—,3-monoamide (*d*)	α-Camphoramic acid.	199.25	nd or lf (w) $[\alpha]_D^{20}$ +25 (al)	176–7	...	...	...	sʰ	sʰ	δ	sʰ	δʰ	MeOH sʰ chl δ	**B9**², 536
c65	—,—(*dl*)	$C_{10}H_{17}NO_3$. *See* c64	199.25	nd (w)	198	...	...	...	sʰ	sʰ	δ	sʰ	δʰ	chl δ	**B9**, 761
c67	**Camphoronic acid** (*d*)	$(CH_3)_2C(CO_2H)C(CH_3)(CO_2H)CH_2CO_2H$	218.21	nd (w) $[\alpha]^{19}$ +27.05 (w)	159d	...	...	...	v	v	v	s	i	chl v peth i	**B2**, 837
c68	—(*dl*)	$(CH_3)_2C(CO_2H)C(CH_3)(CO_2H)CH_2CO_2H$	218.21	nd or pr (w)	172d	...	...	...	δ	v	v	s	i	peth i	**B2**, 839
c69	—(*l*)	$(CH_3)_2C(CO_2H)C(CH_3)(CO_2H)CH_2CO_2H$	218.21	nd (w) $[\alpha]_D^{19}$ −26.9	164–5d (rapid htng)	...	...	...	v	v	s	s	i	chl v peth i	**B2**², 690
c70	**Camphor pinacol** (*l*)	2,2′-Bicamphane-2,2′-diol.	306.49	rh, $[\alpha]_D$ −27.2 (bz)	158	...	...	...	i	s	s	...	...	...	**B6**³, 4767
Ω c71	**Camphorquinone** (*l*)	3-Ketocamphor. 2,3-Camphanedione.	166.22	ye nd (dil al, w), pr (eth) $[\alpha]_D^{20}$ −113.2 (bz), −105.4 (chl) λ^{al} 269 (1.4), 279 (1.4), 468 (1.5)	199	sub	...	...	sʰ	v	v	...	s	chl s	**B7**, 581
c72	**3-Camphorsulfonic acid,** methyl ester (*d*)	...	246.33	cr (MeOH), nd (peth), $[\alpha]_D^{20}$ +98.6 (chl, c = 5) λ^{al} 288 (1.43)	77	201^{20}	...	...	i	v	...	...	...	MeOH v os s	**B2**, 179
c73	**10-Camphorsulfonic acid** (*d*)	...	232.30	pr (aa), $[\alpha]_D^{20}$ +32.8 (AcOEt, c = 3), +24 (w) λ^{w} 285 (1.54)	195d	...	...	...	v	...	i	...	...	aa δ	**B2**², 180
c74	—(*dl*)	$C_{10}H_{16}O_4S$. *See* c73	232.30	cr (aa)	202d	...	...	...	v	...	i	...	...	aa δ	**B2**², 182
c75	—(*l*)	$C_{10}H_{16}O_4S$. *See* c73	232.30	cr (aa), nd (AcOEt), $[\alpha]_D^{20}$ −20.75 (w)	194–5d	...	...	...	v	...	i	...	...	AcOEt, aa δ	**B2**², 182
c76	**α-Camphylamine**	4-(2-Aminoethyl)-1,5,5-trimethylcyclopentene.	153.27	$[\alpha]_D$ +6	...	194–6 95^{12}	0.8688^{20}	1.4728^{18}	...	...	...	...	...	...	**B12**², 35
c77	**β-Camphylamine**	2-(2-Aminoethyl)-1,5,5-trimethylcyclopentine.	153.27	...	...	206	0.8697_{20}^{20}	...	...	...	...	...	...	...	**B12**, 40

For explanations, symbols and abbreviations see beginning of table. For structural formulas see end of table.

No.	Name	Synonyms and Formula	Mol. wt.	Color, crystalline form, specific rotation and λ_{max} (log ε)	m.p. °C	b.p. °C	Density	n_D	Solubility w	al	eth	ace	bz	other solvents	Ref.
	Canadine														
c78	**Canadine** (*d*)	*d*-Tetrahydrobeberine. $C_{20}H_{21}NO_4$.	339.40	ye nd (dil al) $[\alpha]_D^{29}$ +299 (chl, c = 1) λ^{MeOH} 288 (3.83)	132 (140)	....	....	....	i	v	s	...	s	chl s lig δ^h	**B27**[2], 557
Ω c79	—(*dl*)...........	$C_{20}H_{21}NO_4$. *See* c78	339.40	mcl nd (al) λ^{MeOH} 288 (3.83)	174	....	....	....	δ	δ	...	...	...	chl, CS_2 v	**B27**[2], 557
c80	—(*l*)...........	$C_{20}H_{21}NO_4$. *See* c78	339.40	ye nd (al), $[\alpha]_D^{20}$ −299.2 (chl) λ^{MeOH} 288 (3.83)	134	....	....	....	i	v	s	...	s	chl s lig δ^h	**B27**[2], 557
c81	**Canaline**	$H_2NOCH_2CH_2CH(NH_2)CO_2H$	134.14	nd (al) $[\alpha]_D^{21}$ −8.31 (w)	214d	....	....	....	v	s^h	i	...	i	....	**B4**[3], 1636
c82	**Canavanine** (L)....	$NH{:}C(NH_2)NHOCH_2CH_2CH(NH_2)CO_2H$	176.18	cr (al), $[\alpha]_D^{20}$ +7.9 (w, c = 2)	184	....	....	....	s	i	i	...	i	....	**B4**[3], 1636
—	**Canescine**	*see* **Deserpidine**													
c83	**Cannabidiol**.......	....	314.47	rods (peth) λ^{bz} 217 (3.1), 275 (2.0), 280 (2.0)	67	187–90[2]	....	....	i	s	s	..	s	chl s	**Am 63,** 2209
c84	**Cannabinol**	$C_{21}H_{26}O_2$	310.44	pl, lf (peth) $[\alpha]_D^{20}$ −148 (al) λ^{al} 220 (4.55), 285 (4.26)	77	$185^{0.05}$	....	....	i	s	s	s	s	alk, peth s	**B17**[2], 151
Ω c85	**Cantharidin**.......	$C_{10}H_{12}O_4$	196.21	rh pl or sc	218	sub 84	....	....	i	δ	δ	δ	δ	aa, con sulf s chl δ	**B19**[2], 179
—	**Capraldehyde**.....	*see* **Decanal***													
—	**Capric acid**	*see* **Decanoic acid***													
—	**Caproaldehyde**	*see* **Hexanal***													
—	**Caproic acid**	*see* **Hexanoic acid***													
—	**ε-Caprolactam**	*see* **Hexanoic acid, 6-amino-, lactam***													
—	**Caprophenone**.....	*see* **1-Hexanone, 1-phenyl-***													
—	**Caprylaldehyde**....	*see* **Octanal***													
—	**Caprylic acid**......	*see* **Octanoic acid***													
c86	**Capsaicin**.........	$C_{18}H_{27}NO_3$..............	305.42	mcl pl or sc (petn) λ^{al} 227 (3.91), 281 (3.43)	65	$210\text{–}20^{0.01}$	....	....	i	v	s	...	s	peth s con HCl δ	**B13**[2], 482
—	**Captan**...........	*see* **4-Cyclohexene-1,2-dicarboxylic acid,** imide, *N* (trichloromethylthio)-													
c87	**Carbamic acid,** benzyl ester	$H_2NCO_2CH_2C_6H_5$	151.17	pl (to), lf (w)	91 (86–7)	220d	....	....	δ^h	v	δ	...	...	to s	**B6,** 437
Ω c88	—,4-benzylphenyl ester	....	227.27	....	145	....	....	....	δ^h	v^h	...	...	v	NaOH s	**B6**[2], 630
Ω c89	—,butyl ester......	$H_2NCO_2C_4H_9^n$..........	117.15	pr	54	204 δd	....	....	...	v	...	...	...	....	**B3**[2], 25
c90	—,chloride	Carbamoyl chloride. H_2NCOCl	79.49	....	....	62^{760}d	....	....	d	d	...	...	...	....	**B3**[1], 15
Ω c91	—,ethyl ester......	Urethane. $H_2NCO_2C_2H_5$ ...	89.10	pr (bz, to)	48.5–50	185^{760}	0.9862_4^{21}	1.4144^{52}	v	v	v	...	v	chl, py, to v lig δ	**B3**[2], 19
c92	—,isobutyl ester ..	$H_2NCO_2C_4H_9^i$............	117.15	lf	67	207	....	1.4098^{76}	i	s	s	...	...	....	**B3**[2], 26
Ω c93	—,isopropyl ester..	$H_2NCO_2C_3H_7^i$............	103.12	nd	92–4	181^{711}	0.9951^{66}	....	...	...	...	...	...	....	**B3**[2], 25
Ω c94	—,methyl ester	Urethylan. $H_2NCO_2CH_3$	75.07	nd	54	177 82^{14}	1.1361_4^{56}	1.4125^{56}	v	v	s	...	...	....	**B3**[2], 18
c95	—,2-methyl-2-butyl ester	*tert*-Amyl carbamate. Aponal. $H_2NCO_2C(CH_3)_2C_2H_5$	131.18	nd (dil al)	85–7	....	....	....	δ	δ	...	s	s	os v lig δ	**B3**[1], 14
c96	—,3-methylbutyl ester	$H_2NCO_2CH_2CH_2CH(CH_3)_2$	131.18	nd (w^h)	64 (59)	220^{760} 114.5^{16}	0.9438_4^{71}	1.4175^{71}	s^h	s	s	...	...	....	**B3**[2], 26
—	—,nitrile	*see* **Cyanamide**													
c97	—,propyl ester	$H_2NCO_2CH_2CH_2CH_3$......	103.12	pr	60	196 $92\text{–}2.5^{12}$	....	....	s	s	s	s	...	....	**B3**[2], 25
Ω c98	—,*N*-**benzyl-**, ethyl ester	$C_6H_5CH_2NHCO_2C_2H_5$....	179.22	lf (lig)	49 (44)	230δd	....	....	δ	v	v	...	v	chl v lig s^h	**B12**[2], 563
c99	—,*N*-**benzyl-*N*-nitro-**, ethyl ester	$C_6H_5CH_2N(NO_2)CO_2C_2H_5$.	224.22	ye	....	d	1.213_{20}^{20}	1.5203^{20}	...	∞	∞	...	...	....	**C55,** 24616
c100	—,*N*-**butyl-**, butyl ester	$C_4H_9^nNHCO_2C_4H_9^n$	173.26	....	....	88^3	0.9238_{20}^{20}	1.4359^{20}	...	∞	∞	...	...	....	**Am 73,** 5043
c101	—,*N*-**butyl-*N*-nitro-**, butyl ester	$C_4H_9^nN(NO_2)CO_2C_4H_9^n$....	218.26	λ^{al} 238 (3.83)	....	98^3	1.048_{20}^{20}	1.448^{20}	...	∞	∞	...	...	....	**Am 73,** 5449
c102	—,*N*-*tert*-**butyl-*N*-nitro-**, ethyl ester	$C_4H_9^tN(NO_2)CO_2C_2H_5$....	190.20	λ^{al} 239 (3.43)	....	56^2d	1.051_{20}^{20}	1.4331^{20}	...	∞	∞	...	...	....	**Am 83,** 1191
c103	—,*N*-**cyclohexyl-(*N*-ethyl)-dithio-**, hexyl-(ethyl)-, ammonium salt	....	330.60	ye	93	....	....	....	s^h	v	s	...	...	....	**B12**[2], 12

For explanations, symbols and abbreviations see beginning of table. For structural formulas see end of table.

No.	Name	Synonyms and Formula	Mol. wt.	Color, crystalline form, specific rotation and λ_{max} (log ε)	m.p. °C	b.p. °C	Density	n_D	Solubility: w	al	eth	ace	bz	other solvents	Ref.
	Carbamic acid														
c104	—,*N*-**dibenzyldithio-**, dibenzylammonium salt	$(C_6H_5CH_2)_2NCS_2\ NH_2(CH_2C_6H_5)_2$	470.71	ye (eth)	82				s	v					**B12**, 1058
c105	—,*N*,*N*-**diethyl-**	$(C_2H_5)_2NCO_2H$	117.15	nd (eth)	−15d	171	0.9276^{20}_{4}	1.4206^{20}	v	v	δ				**B4**^{2}, 611
Ω c106	—,*N*,*N*-**diethyl-**, chloride	N,N-Diethylchloroformamide. $(C_2H_5)_2NCOCl$	135.60			186 (190)			d^h	d^h					**B4**^{2}, 611
c107	—,*N*,*N*-**diethyldithio-**, benzylidene ester	$[(C_2H_5)_2NCS_2]_2CHC_6H_5$	386.67	ye (al)	110				i	s^h					**C29**, 8408
c108	—,—,diethylammonium salt	$(C_2H_5)_2NCS_2\ \overset{+}{N}H_2(C_2H_5)_2$	222.42	ye pl (eth-al)	88 (82)				v	v	δ				**B4**, 121
Ω c109	—,*N*,*N*-**diethylthio-**, chloride	$(C_2H_5)_2NCSCl$	151.66	pr	48–51	108^{10}									**B4**, 121
c110	—,*N*,*N*-**dimethyldithio-**, dimethylammonium salt	$(CH_3)_2NCS_2^-\overset{+}{N}H_2(CH_3)_2\cdot\frac{1}{2}C_6H_6$	205.37	ye pl (bz + ½) nd (al–eth, unsolv)	125 (+ ½bz), 134–6 (unsolv)				v	v	δ				**B4**^{2}, 577
Ω c111	—,—,2,4-dinitrophenyl ester		287.32	ye	152–3		1.54^{20}_{4}		i	s^h		s	s		**C51**, 3481
c112	—,*N*,*N*-**dimethylthio-**, chloride	$(CH_3)_2NCSCl$	123.61	pr	42.5–3.5	98^{10}					v			peth, chl s	**B4**, 576
c113	—,*N*,*N*-**diphenyl-**, chloride	$(C_6H_5)_2NCOCl$	231.68	lf (al)	85										**B12**^{2}, 241
c114	—,—,ethyl ester	Diphenylurethan. $(C_6H_5)_2NCO_2C_2H_5$	241.29	pr (lig) λ^{al} 238 (7.13)	72	360^{760}			s	δ	s		s	peth s	**B12**^{2}, 240
c115	—,**dithio-**	Aminodithioformic acid. Aminomethanethionothiolic acid*. H_2NCS_2H	93.17	nd $\lambda^{al\text{-}NH_4\ salt}$ 291 (4.30), 345 (2.60)					d	v	v				**B3**^{2}, 155
c116	—,*N*-**ethyl-**, butyl ester	$C_2H_5NHCO_2C_4H_9^n$	145.20			66^3	0.9413^{20}_{20}	1.4301^{20}	s	∞	∞				**C55**, 2334
Ω c117	—,—,ethyl ester	Ethylurethan. $C_2H_5NHCO_2C_2H_5$	117.15			176^{760}, 75^{14}	0.9813^{20}_{4}	1.4215^{20}	s	v	v				**B4**^{2}, 607
c118	—,*N*-**ethyl**-*N*-**nitro-**, butyl ester	$C_2H_5N(NO_2)CO_2C_4H_9^n$	190.20			79^3	1.091^{20}_{20}	1.4455^{20}	s	∞	∞				**Am73**, 5043
c119	—,—,ethyl ester	$C_2H_5N(NO_2)CO_2C_2H_5$	162.15	λ^{al} 237 (3.75)		107^{31}	1.163^{20}_{20}	1.4432^{20}		∞	∞				**Am73**, 5449
c119[1]	—,—,methyl ester	$C_2H_5N(NO_2)CO_2CH_3$	148.13	λ^{al} 235 (3.75)		72^{11}	1.233^{20}_{20}	1.4483^{20}		∞	∞				**Am73**, 5449
Ω c120	—,*N*-**ethylidenedi-**, diethyl ester	Ethylidenediurethan. $CH_3CH(NHCO_2C_2H_5)_2$	204.23	nd (eth)	126	170–8^{20}			v	v	δ	v	δ	chl, MeOH s to, lig δ	**B3**^{1}, 11
c121	—,*N*-**isobutyl-**, ethyl ester	Isobutylurethan. $C_4H_9^iNHCO_2C_2H_5$	145.20		> −65	110^{30}, 96^{17}	0.9432^{20}_{4}	1.4288^{20}	δ	v	v				**B4**^{2}, 640
c122	—,*N*-**isopropyl-**, ethyl ester	Isopropylurethan. $C_3H_7^iNHCO_2C_2H_5$	131.18			79^{15}	0.9548^{20}_{20}	1.4229^{20}		∞	∞				**B4**^{1}, 369
c123	—,*N*-**isopropyl**-*N*-**nitro-**, ethyl ester	$C_3H_7^iN(NO_2)CO_2C_2H_5$	176.17	λ^{al} 240 (3.7)		72^7	1.112^{20}_{20}	1.4381^{20}		∞	∞				**Am 73**, 5449
c124	—,*N*-**methyl-**, ethyl ester	Methylurethan. $CH_3NHCO_2C_2H_5$	103.12			170^{760}, 80^{15}	1.0115^{20}_{4}	1.4183^{20}	v	s					**B4**^{2}, 567
c125	—,*N*-**methyl**-*N*-**phenyl-**, chloride	$CH_3N(C_6H_5)COCl$	169.62	pl (al)	88–9	280			i	v	v		s^h		**B12**^{2}, 235
c126	—,*N*-**nitro-**, ethyl ester	$O_2NNHCO_2C_2H_5$	134.09	pl (eth, lig) λ^{al} 215 (3.9), 260 sh (2.6), 290 sh (2.2)	64	140d	1.0074^{20}_{4}		v	v	v			lig δ	**B3**^{2}, 99
c127	—,*N*-**nitro**-*N*-**propyl-**, ethyl ester	$C_3H_7^nN(NO_2)CO_2C_2H_5$	176.17	λ^{al} 239 (3.80)		66^3	1.123^{20}_{20}	1.4431^{20}		∞	∞				**Am 73**, 5449
c128	—,—,methyl ester	$C_3H_7^nN(NO_2)CO_2CH_3$	162.15				1.1585^{15}_{15}		v	∞	∞				**B4**, 146
c129	—,*N*-**phenyl-**, ethyl ester	Ethyl carbanilate. $C_6H_5NHCO_2C_2H_5$	165.19	wh nd (w), pl (dil al) λ^{al} 235.5 (4.23), 274 (2.93)	53	237^{760} δd, 152^{14}	1.1064^{30}_{4}	1.5376^{30}	i s^h	v	v		s		**B12**^{2}, 184
c130	—,—,isobutyl ester	$C_6H_5NHCO_2C_4H_9^i$	193.25	nd (dil al)	86	216^{760}			δ	v	v		s		**B12**^{1}, 219
c131	—,—,isopropyl ester	$C_6H_5NHCO_2C_3H_7^i$	179.22	wh nd (al) λ^{al} 235.5 (4.23), 274 (2.92)	90 (76)		1.09^{20}	1.4989^{91}	i	s			s		**B12**, 321
c132	—,—,propyl ester	$C_6H_5NHCO_2C_3H_7^n$	179.22	wh nd (dil al)	57–9				δ	s	v		s		**B12**, 321
Ω c133	—,*N*-**propyl-**, ethyl ester	$C_3H_7^nNHCO_2C_2H_5$	131.18			192^{758}, 92^{22}	0.9921^{15}		s						**B4**^{2}, 626
c134	—,**thiolo-**, ethyl ester	$H_2NCOSC_2H_5$	105.16	pl (w)	109	sub d			i s^h	s	s				**B3**^{1}, 64
c135	—,**thiono-**, ethyl ester	Thiourethan. Xanthogenamide. $H_2NCSOC_2H_5$	105.16	mcl lf or pyr	41	d	1.069^{20}_{4}	1.520^{20}	δ	∞	∞			chl v	**B3**^{2}, 07
—	**Carbamyl chloride**	*see* **Carbamic acid**, chloride													

For explanations, symbols and abbreviations see beginning of table. For structural formulas see end of table.

No.	Name	Synonyms and Formula	Mol. wt.	Color, crystalline form, specific rotation and λ_{max} (log ε)	m.p. °C	b.p. °C	Density	n_D	Solubility w	al	eth	ace	bz	other solvents	Ref.
	Carbamyl guanidine														
—	**Carbamyl guanidine**	*see* **Guanidine, 1-ureido-**													
—	**Carbanilic acid**	*see* **Carbamic acid, *N*-phenyl-**													
—	**Carbanilide**	*see* **Urea, 1,3-diphenyl-**													
Ω c136	**Carbazide**	$H_2NNHCONHNH_2$	90.09	nd (dil al)	154	....	1.616^{20}	....	v	v	...	...	...	....	**B3**[2], 96
c137	**—,1,5-diphenyl-**	$CO(NHNHC_6H_5)_2$	242.28	cr (al + 1), cr (aa)	170	d	....	....	δ^h	s^h	δ	s^h	s	aa s	**B15**[2], 107
c138	**—,1-phenyl-**	$C_6H_5NHNHCONHNH_2$	166.18	nd (al)	151	....	....	....	s	s^h	...	...	...	....	**B15**[2], 107
c139	**—,1,1,5,5-tetraphenyl-**	$[(C_6H_5)_2NNH]_2CO$	394.48	(al), nd (aa)	242	....	....	....	...	v^h	...	...	...	aa v^h	**B15**[2], 115
c140	**—,3-thio-**	$H_2NNHCSNHNH_2$	106.15	nd, pl (w)	170d	....	....	....	v^h	δ^h	i	...	...	alk v	**B3**[2], 137
—	**Carbazine**	*see* **Acridine, 9,10-dihydro-**													
Ω c141	**Carbazole**	Dibenzopyrrole. Diphenylenimine.	167.21	pl or lf	247–8	355^{760} 200^{147}	....	....	i	s^h δ	δ	s v^h	s^h δ	Py δ, s^h chl, to, aa, CS_2 δ	**B20**[2], 279
Ω c142	**—,9-acetyl-**	$C_{14}H_{11}NO$. *See* c141	209.25	(eth), nd (w) λ^{al} 230 (4.61), 262.5 (4.23), 286 (4.05), 300 (3.79), 312 (3.82)	69	190^6	1.161^{100}_{24}	1.640^{100}	δ^h	...	...	...	...	....	**B20**[2], 283
c143	**—,9-benzoyl-**	$C_{19}H_{13}NO$. *See* c141	271.32	nd or pr (al)	98.5	....	....	....	...	v^h	...	...	v	....	**B20**[1], 165
c144	**—,9-benzyl-**	$C_{19}H_{15}N$. *See* c141	257.34	nd (al)	118–120 (114)	$267–8^{24}$	....	....	i	δ	...	...	v	....	**B20**[2], 283
c145	**—,9-butyl-**	$C_{16}H_{17}N$. *See* c141	223.32	nd (al)	58	$218–9^{19}$	....	....	i	δ	v	...	...	....	**B20**[1], 164
c146	**—,9-ethenyl-**	$C_{14}H_{11}N$. *See* c141	193.25	cr (al)	66	....	....	....	i	δ	v	...	...	....	**B20**[2], 282
Ω c147	**—,9-ethyl-**	$C_{14}H_{13}N$. *See* c141	195.27	nd (al) λ^{cy} 236 (4.7), 294 (4.35), 329 (3.6), 343 (3.75)	68	190^{10}	1.059^{80}_{4}	1.6394^{80}	i	v^h	v	...	...	....	**B20**[2], 282
c148	**—,9-methyl-**	$C_{13}H_{11}N$. *See* c141	181.24	nd, lf (al) λ^{al} 235 (4.61), 293 (4.23), 329.5 (3.56), 343.5 (4.59)	88	195^{12}	....	....	i	δ	v	...	...	....	**B20**[2], 281
c149	**—,1-nitro-**	$C_{12}H_8N_2O_2$. *See* c141	212.22	ye nd (aa) λ^{al} 223.5 (4.54), 260 (3.97), 300.5 (4.16), 403 (3.88)	187	....	....	....	i	...	...	...	...	aa v CS_2 s peth δ	**B20**[2], 288
c150	**—,3-nitro-**	$C_{12}H_8N_2O_2$. *See* c141	212.22	ye λ^{al} 230.5 (4.45), 280 (4.40), 306 (4.16), 363 (4.01)	214 (205–7)	....	....	....	i	...	δ	...	δ	peth, chl, aa δ	**B20**[2], 288
c151	**—,3-nitro-9-nitroso-**	$C_{12}H_7N_3O_3$. *See* c141	241.21	ye nd (al) λ^{al} 255.5 (4.51), 307 (3.99), 332 (4.02)	169 (164)	....	....	....	...	s^h	...	...	...	chl s lig δ	**B20**[2], 289
c152	**—,1-oxo-1,2,3,4-tetrahydro-**	$C_{12}H_{11}NO$. *See* c141	185.23	nd (dil al) λ^{al} 237 (4.21), 308.5 (4.36)	170	....	....	....	...	s	...	...	s	aa s	**B21**[2], 275
Ω c153	**—,9-phenyl-**	$C_{18}H_{13}N$. *See* c141	243.31	nd or pl (al)	95	....	....	....	...	v^h	v	...	v	aa v	**B20**[2], 282
c154	**—,9-propionyl-**	$C_{15}H_{13}NO$. *See* c141	223.28	....	90	....	....	....	...	v	v	...	...	peth δ	**B20**[1], 165
c155	**—,9-propyl-**	$C_{15}H_{15}N$. *See* c141	209.29	nd (al)	50	....	....	....	i	δ	v	...	...	....	**B20**[1], 164
Ω c156	**—,1,2,3,4-tetrahydro-**	$C_{12}H_{13}N$. *See* c141	171.25	lf (dil al) λ^{al} 227 (4.72), 282 (3.97), 290 (3.92)	120	325–30 190^{10}	....	....	i	s	v	...	v	MeOH v	**B20**[2], 257
c157	**Carbazone, 1,5-diphenyl-**	$C_6H_5N{:}NCONHNHC_6H_5$	240.27	og nd (bz), pr (al) λ^{w} *ca.* 470	157d	....	....	....	i	v	...	...	v	chl v	**B16**[2], 9
—	**Carbinol**	*see* **Methanol***													
—	**Carbitol**	*see* **Diethyleneglycol**, monoethyl ether													
—	**α-Carbocinchomeronic acid**	*see* **2,3,4-Pyridinetricarboxylic acid**													
c158	**Carbodiimide, diphenyl-**	$C_6H_5N{:}C{:}NC_6H_5$	194.24	....	168–170	331^{760} 218^{31}	....	....	δ	δ	δ	...	s	....	**B12**[2], 246
—	**Carbohydrazide**	*see* **Carbazide**													
Ω c159	**Carbon dioxide**	CO_2	44.00	....	−56.6 (at 5.2 atm)	-78.5^{760} sub	0.00198 (gas at 0° 1 atm 1.0310^{-20} (liq)	....	...	...	...	...	...	....	**B3**, 4

For explanations, symbols and abbreviations see beginning of table. For structural formulas see end of table.

No.	Name	Synonyms and Formula	Mol. wt.	Color, crystalline form, specific rotation and λ_{max} (log ε)	m.p. °C	b.p. °C	Density	n_D	Solubility w	al	eth	ace	bz	other solvents	Ref.
	Carbon diselenide														
c160	**Carbon diselenide**	CSe_2	169.93	ye	−45.5	126^{760} 46^{50}	2.6824_4^{20}	1.8454^{20}	...	...	...	...	...	...	**B3**[3], 366
Ω c161	**Carbon disulfide**	CS_2	76.14	λ^{cy} 314 (2.89), 319 (2.83)	−111.53	46.25^{760} 0^{128}	1.2632_4^{20}	1.6319^{20}	s	∞	∞	...	...	chl s	**B3**[2], 139
c162	**Carbonic acid,** bis-(2-chloroethyl) ester*	$OC(OCH_2CH_2Cl)_2$	187.03	...	8	241	1.3506_4^{20}	1.461^{20}	i	...	...	...	...	...	**B3**[2], 5
c163	—,bis(3-chloropropyl) ester*	$OC(OCH_2CH_2CH_2Cl)_2$	215.08	...	...	$265–70^{740}$	...	...	...	...	...	...	...	...	**B3**[2], 5
c164	—,bis(2-ethoxyethyl) ester*	$OC(OCH_2CH_2OC_2H_5)_2$	206.24	...	...	$245–6^{758}$ $112–3^{5}$	1.0439_4^{20}	1.4227^{20}	...	v	v	v	...	...	**B3**[3], 17
Ω c165	—,bis(2-methoxyethyl) ester*	$OC(OCH_2CH_2OCH_3)_2$	178.19	...	...	$230–2^{760}$ $99–100^{5}$	1.0988_4^{20}	1.4204^{20}	...	s	s	s	...	...	**B3**[3], 17
c166	—,bis(2-methoxyphenyl) ester*	Duotal	274.28	cr (al)	89	...	...	...	i	δ	s	...	...	chl v	**B6**[2], 784
c166[1]	—,bis(3-methylbutyl) ester*	$OC[OCH_2CH_2CH(CH_3)_2]_2$	202.30	...	...	232.5^{751} 122^{16}	0.9067_4^{20}	1.4174^{20}	i	...	...	...	...	...	**B3**[2], 7
c167	—,bis(trichloromethyl) ester*	$OC(OCCl_3)_2$	296.75	cr (eth, peth)	79	203^{760} δd	...	...	...	...	δ^h	...	...	peth δ^h	**B3**[1], 8
Ω c168	—,dibutyl ester*	$OC(OC_4H_9^n)_2$	174.24	...	...	207^{760} $96–7^{16}$	0.9251_4^{20}	1.4117^{20}	i	s	s	...	...	...	**B3**[2], 6
c169	—,di-*tert*-butyl ester*	$OC(OC_4H_9^t)_2$	174.24	cr (al)	40 sub	158^{767}	...	...	i	s^h	...	...	...	...	**B3**[3], 11
—	—,dichloride*	*see* **Phosgene**													
Ω c170	—,diethyl ester*	$OC(OC_2H_5)_2$	118.13	...	−43	126	0.9752_4^{20}	1.3845^{20}	i	s	s	...	...	...	**B3**[3], 5
—	—,dihydrazide	*see* **Carbazide**													
c171	—,diisobutyl ester	$OC(OC_4H_9^i)_2$	174.24	...	...	190^{760} 85^{16}	0.9138_4^{20}	1.4072^{20}	i	∞	∞	...	...	...	**B3**[2], 6
c172	—,diisopropyl ester*	$OC(OC_3H_7^i)_2$	146.19	...	...	147^{760} 43^{12}	0.9162_4^{20}	1.3932^{20}	...	s	...	...	...	...	**B3**[2], 6
Ω c173	—,dimethyl ester*	$OC(OCH_3)_2$	90.08	λ^{undil} < 180	2–4	90–1	1.0694_4^{20}	1.3687^{20}	i	s	s	...	...	...	**B3**[2], 3
Ω c175	—,diphenyl ester*	$OC(OC_6H_5)_2$	214.22	nd (al, bz)	83 (88)	306^{760} 168^{15}	1.1215_4^{87}	...	i	s^h	s	...	...	CCl_4 s aa s^h	**B6**[2], 156
c176	—,dipropyl ester*	$CO(OC_3H_7^n)_2$	146.19	...	...	168^{760} 59.5^{11}	0.9435_4^{20}	1.4008^{20}	δ	∞	∞	...	...	...	**B3**[2], 5
Ω c177	—,di-2-tolyl ester	...	242.28	nd (al)	60 (57)	$144–5^{0.5}$	...	...	...	s^h	...	...	...	aa v	**B6**[2], 330
Ω c178	—,di-3-tolyl ester	...	242.28	cr (al)	50.5–1.0	...	...	...	i	s^h	...	...	s	chl s	**B6**[2], 353
Ω c179	—,di-4-tolyl ester	...	242.28	nd (al)	115	...	...	...	δ	δ	...	...	δ s^h	chl s	**B6**[2], 380
—	—,ethylene ester	*see* **1,3-dioxolan-2-one**													
c180	—,monoethyl mono 2-butoxyethyl ester	$C_2H_5OCOOCH_2CH_2OC_4H_9^n$	190.24	...	...	224–4.4^{759}	0.9756_4^{25}	1.4143^{25}	...	s	s	s	...	...	**B3**[3], 14
c181	—,monoethyl monomethyl ester	$CH_3OCOOC_2H_5$	104.12	...	−14	107.2–$.8^{765}$	1.012_4^{20}	1.3778^{20}	i	∞	∞	...	...	...	**B3**[2], 4
c182	—,**dithiolo-**, diethyl ester	$OC(SC_2H_5)_2$	150.26	ye	...	197 $85–7^{19}$	1.085^{19}	1.5237^{18}	i	s	s	...	...	...	**B3**[1], 84
c183	,trithio-	$SC(SH)_2$	110.22	red	−30	57d	1.47_4^{17}	...	d	s	...	...	...	chl, to s	**B3**[2], 161
Ω c184	**Carbon monoxide**	CO	28.01	...	−205.06	-191.47^{760}	0.4220^{-141} 0.7909^{-19}	0.00125 at 0°/1 atm	δ	...	...	...	s	aa s	**B1**, 720
—	**Carbon suboxide**	*see* **Propadiene, 1,3-dioxo-***													
—	**Carbon tetrabromide**	*see* **Methane, tetrabromo-***													
—	**Carbon tetrachloride**	*see* **Methane, tetrachloro-***													
—	**Carbon tetrafluoride**	*see* **Methane, tetrafluoro-***													
—	**Carbon tetraiodide**	*see* **Methane, tetraiodo-***													
c185	**Carbonyl fluoride**	COF_2	66.01	...	−114	-83^{760} -128^{39}	1.139^{-114}	...	d	d	...	...	...	...	**B3**[2], 9
c186	**Carbonyl sulfide**	Carbon oxysulfide. COS	60.08	...	−138	-50^{760}	0.0021 at 1 atm 1.24^{-87} 1.028_4^{17}	...	δ	s	...	...	...	al KOH v	**B3**[2], 104
—	**Carbostyril**	*see* **Quinoline, 2-hydroxy-**													
c187	**Carbothialdine**	...	162.28	cr (al) λ^{cy} 243 (3.70), 288 (4.11)	120d	...	...	...	i	v^h	i	...	...	ac s	**B27**[2], 687
—	**Cardiazol**	*see* **Metrazol**													
Ω c188	**Δ³-Carene(*dl*)**	...	136.24	...	...	167^{732} $44–5^{8}$	0.8602_4^{20}	1.4759^{20}	i	...	s	s	s	aa s	**B5**[3], 362
c189	—(*l*)	Isodiprene. $C_{10}H_{16}$. *See* c188	136.24	$[\alpha]_D^{30}$ −5.72 λ^{iso} 220 (2.3) 250 (0.92)	...	$168–9^{705}$ $123–4^{200}$	0.8586_{30}^{30}	1.4684^{30}	i	...	s	s	s	aa, peth s	**E12A**, 34
c190	**Δ⁴-Carene(*l*)**	...	136.24	$[\alpha]_D^{30}$ +62.2	...	167^{707} 64^{29}	0.8441_4^{30}	1.4740^{30}	i	...	s	s	s	aa, peth s	**E12A**, 34
—	**Caricaxanthin**	*see* **Cryptoxanthin**													

For explanations, symbols and abbreviations see beginning of table. For structural formulas see end of table.

No.	Name	Synonyms and Formula	Mol. wt.	Color, crystalline form, specific rotation and λ_{max} (log ε)	m.p. °C	b.p. °C	Density	n_D	Solubility						Ref.
									w	al	eth	ace	bz	other solvents	
	Carminic acid														
Ω c191	**Carminic acid**	$C_{22}H_{20}O_{13}$	492.40	red mcl pr (aq MeOH) $[\alpha]^{15}_{6450}$ +51.6 (w, p = 1)	d at 136	...	...	...	s^h	s	δ	...	i	chl i	**B10**[2], 776
c192	**Carnaubyl alcohol**	$CH_3(CH_2)_{29}OH$	438.81	lf (al)	69	...	...	...	δ	s	...	s	...	...	**B1**[2], 472
—	**Carnegine**	*see* **Isoquinoline, 6,7-dimethoxy-1,2-dimethyl-1,2,3,4-tetrahydro-**													
c193	**Carnosine**(*D*, −)	Ignotine. β-Alanyl-*D*(−)-histidine	226.24	$[\alpha]^{18}_D$ −20.4 (w, c = 1.5)	260	...	...	...	s	i	...	...	...	...	**B25**[2], 408
Ω c194	—,(*L*, +)	$C_9H_{14}N_4O_3$. *See* c193	226.24	nd (w-al) $[\alpha]^{20}_D$ +24.1 (w, c = 1.5)	246–50d (308)	...	...	...	v	i	...	...	...	...	**B25**[2], 408
c195	**α-Carotene**	α-Carotin. $C_{40}H_{56}$	536.90	red pl or pr (peth, bz–MeOH) $[\alpha]_{Cd}$ +385 (bz, c = 0.08) λ^{bz} 277 (3.26), 340 (3.84), 436 (4.92), 458 (5.09), 488 (5.03)	187.5	...	1.00^{20}_{20}	...	i	δ	s	...	s	chl, CS_2 v peth δ	**B30**, 91
Ω c196	**β-Carotene**	β-Carotin. $C_{40}H_{56}$. Provitamin A.	536.90	red br hex pr (bz–MeOH), red rh (peth) λ^{bz} 278 (4.30), 364 (4.62), 463 (5.10), 494 (4.77)	184	...	1.00^{20}_{20}	...	i	δ	s	s	s	peth s chl δ	**B30**, 87
c197	**γ-Carotene**	γ-Carotin. $C_{40}H_{56}$	536.90	red pr (bz-MeOH), vt pr (bz-eth) λ^{peth} 410 sh, 435, 460, 490.5	178	...	...	...	i	i	δ	...	s	chl s peth δ	**B30**, 92
—	**β-Carotene, dehydro-**	*see* **Isocarotene**													
c199	**Carpaine**(*d*)	$C_{28}H_{50}N_2O_4$	478.72	mcl pr (al or ace) $[\alpha]^{21}_D$ +24.7 (al, c = 1.07)	121	sub $120^{0.05}$	...	...	δ	s	s	s	s	chl s peth i	**B27**[3], 209
c200	—,hydrochloride(*d*)	$C_{28}H_{50}N_2O_4$. HCl. *See* c199	515.18	wh mcl nd or pl	225d	...	...	...	v	s	s	...	...	...	**B27**[2], 210
c201	**Carpiline**	Carpidine. Pilosine. $C_{16}H_{18}N_2O_3$.	286.33	pl (al), pr (dil al or w) $[\alpha]^{20}_D$ +35.9 (al), +40.2 (chl)	187	...	...	...	δ, s^h	v^h	δ	δ	δ	dil ac v chl, AcOEt δ	**B27**[1], 612
—	**Carvacrol**	*see* **Benzene, 2-hydroxy-4-isopropyl-1-methyl-**													
—	***o*-Carvacrotonic acid**	*see* **Benzoic acid, 2-hydroxy-6-isopropyl-3-methyl-**													
—	**Carvacrylamine**	*see* **Benzene, 2-amino-4-isopropyl-1-methyl-**													
c202	**Carvenone**(*dl*)	...	152.24	λ^{al} 312 (1.70)	...	235.5–6.0^{762} 104^{10}	0.9263^{20}_4	1.4826^{20}	i	...	...	s	...	...	**B7**[2], 79
c203	—(*l*)	$C_{10}H_{16}O$. *See* c202	152.24	$[\alpha]_D$ −2.08	...	232–4^{760}	0.9290^{20}_0	1.4805	i	...	...	s	...	...	**B7**[1], 66
c204	**Carveol, dihydro-**(*d*)	...	154.26	$[\alpha]^{16}_D$ +34.2	...	225^{760} 107^{15}	0.9274^{20}_4	1.4780^{20}	i	...	s	...	...	...	**B6**[2], 70
c205	—,—(*l*)	$C_{10}H_{16}O$. *See* c204	154.26	$[\alpha]^{20}_D$ −33.3	...	107^{14} 90^4	0.9368^{15}	1.4836^{20}	i	...	s	...	...	...	**B6**[1], 43
c206	**β-Carveol, dihydro-**	...	154.26	$[\alpha]_D$ +7.64	...	120^{20}	0.9266^{20}_4	1.4809^{20}	...	...	...	...	...	...	**B6**, 64
Ω c207	**Carvomenthane**(*d*)	...	138.25	$[\alpha]_{5780}$ +118	...	175 – 7^{760} 77^{24}	0.8246^{18}_4	1.4563^{18}	i	s	...	...	s	peth s	**B5**[2], 52
c208	**Carvomenthol**(*d*)	Hexahydrocarvacrol	156.27	$[\alpha]^{21}_D$ +31.40	...	222 102^{14}	0.8995^{20}_4	1.4617^{20}	...	v	s	...	...	...	**B6**[2], 39
c209	—(*l*, *neo*)	$C_{10}H_{20}O$. *See* c208	156.27	$[\alpha]^{21}_D$ −41.7	...	217–8^{760} 102^{18}	0.9012^{20}_4	1.4632^{20}	...	v	s	...	...	...	**B6**[1], 19
c210	**Carvomenthone**(*d*)	...	154.26	$[\alpha]^{21}_D$ +17.15 λ^{al} 283 (1.43)	...	218–20^{745} 95–9^{15}	0.9075^{15}	1.4544^{20}	...	v	...	s	...	chl s	**B7**[2], 37
Ω c211	**Carvone**(*d*)	Carvol	150.22	$[\alpha]^{20}_D$ +69.1 λ^{al} 235 (3.93), 318 (1.62)	...	231^{760} 104^{11}	0.9608^{20}_4	1.4999^{18}	δ^h	v	s	...	...	chl s	**B7**[2], 128
c212	—(*dl*)	$C_{10}H_{14}O$. *See* c211	150.22	λ^{al} 235 (3.93), 318 (1.62)	...	231^{760} 85^5	0.9645^{15}_{15}	1.5003^{20}	δ^h	v	s	...	...	chl s	**B7**[2], 130

For explanations, symbols and abbreviations see beginning of table. For structural formulas see end of table.

No.	Name	Synonyms and Formula	Mol. wt.	Color, crystalline form, specific rotation and λ_{max} (log ε)	m.p. °C	b.p. °C	Density	n_D	Solubility w	al	eth	ace	bz	other solvents	Ref.
	Carvone														
Ω c213	—(*l*)	$C_{10}H_{14}O$. *See* c211	150.22	$[\alpha]^{20}_D$ −62.46 λ^{al} 235 (3.93), 318 (1.62)	...	231^{760} 98^{9}	0.9593^{20}_4	1.4988^{20}	δ^h	v	s	...	...	chl s	**B7**², 129
c214	—,oxime(α,*d*)	*D*-Carvoxime(α).	165.24	lf (al) $[\alpha]^{17}_D$ +39.71 (al, p = 8.45) λ^{al} 234 (4.21)	72	...	...	...	...	s	s	s	...	...	**B7**², 129
c215	—,—(α,*l*)	*L*-Carvoxime(α). $C_{10}H_{15}NO$. *See* c214	165.24	mcl (dil al) $[\alpha]^{18}_D$ −39.43 (al, p = 4.33)	73.5	...	1.0140^{73}	...	s	s	s	s	...	...	**B7**², 129
c216	—,—(β,*d*)	$C_{10}H_{15}NO$. *See* c214	165.24	nd (dil al) $[\alpha]_D$ +68.3 (bz)	57–8	...	...	...	s	s	s	s	...	oos v	**B7**¹, 102
c217	—,—(β,*l*)	$C_{10}H_{15}NO$. *See* c214	165.24	nd (al)	56–7	...	...	...	s	s	s	s	...	...	**B7**¹, 102
c218	—,—(*dl*)	*dl*-Carvoxime. $C_{10}H_{15}NO$. *See* c214	165.24	...	93–4	...	...	...	...	s	s	s	...	...	**B7**¹, 103
c219	—,**dihydro-(*d*)**	...	152.24	$[\alpha]_D$ +17.5	...	221–2	0.928^{19}	1.4724	...	...	s	s	...	...	**B7**¹, 69
c220	—,—(*l*)	$C_{10}H_{16}O$. *See* c219	152.24	$[\alpha]^{20}_D$ −19 $[\alpha]^{31}_D$ −46.63	...	221–2 104^{18}	0.9253^{20}	1.4717^{20}	i	...	s	s	...	...	**B7**², 81
—	—,**tetrahydro-**	*see* **Carvomenthone**													
c221	α-**Caryophyllene**	$C_{15}H_{24}$	204.36	$[\alpha]^{20}_4$ +1.0 ± 0.3 (chl, c = 9.26) λ 280 (2.1)	...	123^{10}	0.8905^{20}_4	1.5038^{20}	...	...	...	...	...	...	**Am72,** 5350
c222	β-**Caryophyllene**	$C_{15}H_{24}$	204.36	$[\alpha]^{20}_D$ −9.08	...	$122^{13.5}$	0.9075^{20}_4	1.4988^{20}	i	...	...	...	s	...	**B5**³, 1083
c223	γ-**Caryophyllene**	Isocaryophyllene. $C_{15}H_{24}$	204.36	$[\alpha]^{19}_D$ −26.2	...	$130-1^{24}$	0.8953^{20}_4	1.4967^{19}	...	...	...	...	s	...	**B5**³, 1085
c224	**Caryophyllenic acid** (*l, cis*)	...	186.21	pr (w) $[\alpha]_{546}$ −7.4	77–8	...	...	...	s	s	s	s	s	...	**B9**², 531
c225	—(*d,trans*)	$C_9H_{14}O_4$. *See* c224	186.21	nd (cy) $[\alpha]^{20}_D$ +35.3 (bz)	81–2	...	...	...	v	s	s	s	s	lig δ	**B9**², 531
c226	**Caryophyllin**	Oleanolic acid. $C_{30}H_{48}O_3$	456.72	nd or pr (al) $[\alpha]^{20}_D$ +83.3 (chl, c = 0.9) λ^{al} 207 (3.67)	310d	280–308 sub (vac)	...	...	i	δ	δ	δ	δ	Py v aa v^h MeOH s CCl_4 δ	**B10**², 198
Ω c227	**Catechin**(*cis'd*)	3,5,7,3′,4′-Flavanpentol, (*one form*). $C_{15}H_{14}O_6$.	290.28	nd (w + 4) $[\alpha]^{18}_D$ +18.4 (w, p = 0.9)	96 (hyd) 177 (anh)	240–5	1.344^4_4	...	v^h	v	δ	v	i	aa v lig δ chl i	**B17**², 254
c228	—(*cis,dl*)	$C_{15}H_{14}O_6$. *See* c227	290.28	nd (w + 3)	212–4d	...	...	...	δ	v	δ	v	i	chl i	**B17**², 256
c229	—(*cis,l*)	$C_{15}H_{14}O_6$. *See* c227	290.28	nd (+4, w) $[\alpha]_{5780}$ −16.8 (w-ace, p = 3)	96 (hyd) 177 (anh)	...	...	...	δ	v	δ	v	i	lig δ aa v chl i	**B17**², 255
—	**Catechol**	*see* **Benzene, 1,2-dihydroxy-***													
—	**Caulophylline**	*see* **Cytisine,** ***N*-methyl-**													
c230	**Cedrene**	$C_{15}H_{24}$	204.36	$[\alpha]^{20}_D$ −91.3	...	$262-3^{760}$ $124-6^{12}$	0.9342^{20}_4	1.5034^{20}	i	...	...	...	s	lig s	**B5**², 350
Ω c231	**Cellobiose**(β)	Glucose-β-glucoside	342.30	cr (dil al) $[\alpha]^{20}_D$ +14.2 → +34.6 (w, c = 8, 15 hr) (mut)	d 225	...	...	...	s v^h	i	i	i	i	...	**B31,** 380
c232	—,octa-acetate(α)	Octaacetylcellobiose. $C_{28}H_{38}O_{19}$. *See* c231	678.61	nd (al) $[\alpha]^{20}_D$ +43.6 (chl, c = 6)	229.5	...	...	...	i	i	i	...	...	chl, aa v	**B31,** 382
c233	—,—(β)	$C_{28}H_{38}O_{19}$. *See* c231	678.61	nd (al) $[\alpha]^{20}_D$ −14.7 (chl, c = 5)	202 (192)	...	...	...	...	s^h	...	...	δ	chl v	**B31,** 383
—	**Cellosolve,** acetate	*see* **Acetic acid,** 2-ethoxyethyl ester													
c234	**Cellulose**	Polycellobiose. $(C_6H_{10}O_5)_x$	$(162.14)_x$	wh amor	260–70d	...	1.27–1.61	...	i	i	i	i	i	$Cu(OH)_2.(NH_3)_4$ s oos i	
c235	—,hexanitrate	Chief constituent of guncotton. $(C_{12}H_{14}N_6O_{22})_x$	$(594.27)_x$	wh amor	160–70 ign	...	1.66	...	i	i	i	...	i	$PhNO_2$ s	
c236	—,pentanitrate	$(C_{12}H_{15}N_5O_{20})_x$	$(549.28)_x$	wh amor	...	...	1.66	...	i	i	i	...	i	MeOH, eth-al s	...
c237	—,tetranitrate	Constituent of collodion. $(C_{12}H_{16}N_4O_{18})_x$	$(504.26)_x$	wh amor	...	...	1.66	...	i	i	i	...	i	MeOH, eth-al s	...
c238	—,triacetate	$(C_{12}H_{16}O_8)_x$	$(288.26)_x$	non-inflam yesh fl $[\alpha]_D$ −22.5 (chl)	...	...	...	...	i	i	i	i	...	$ClCH_2CH_2Cl$, chl, $PhNO_2$, aa s	...
c239	—,triethyl ether	Ethyl cellulose. Triethyl cellulose. $(C_{12}H_{22}O_5)_x$.	$(246.31)_x$	wh nd (bz) $[\alpha]^{20}_D$ +26.1 (bz)	240–55	...	...	...	i	δ	s^h	...	...	...	...
c240	—,trinitrate	Constituent of collodion. $(C_{12}H_{17}N_3O_{16})_x$.	$(459.28)_x$	wh	...	...	1.66	...	i	i	...	s	...	MeOH, AcOEt s aa s^h	...
c241	**Cepharanthine**	$C_{37}H_{38}N_2O_6$	606.73	ye amor pw nd (+1½ bz, ace-bz) $[\alpha]^{20}_D$ +277 (chl)	145–55 (anh) 103 (+1¼ bz)	...	...	...	...	s	s	s	s	lig i	**C49,** 1745

For explanations, symbols and abbreviations see beginning of table. For structural formulas see end of table.

No.	Name	Synonyms and Formula	Mol. wt.	Color, crystalline form, specific rotation and λ_{max} (log ε)	m.p. °C	b.p. °C	Density	n_D	Solubility w	al	eth	ace	bz	other solvents	Ref.
	Cerane														
c242	**Cerane**	Isohexacosane. $CH_3(CH_2)_{24}CH_3$	366.72	pl (eth), sc(w)	61	$207^{0.7}$			i δ^h	s	s				**B1**², 143
—	**Cerotic acid**	*see* **Hexacosanoic acid***													
—	**Ceryl alcohol**	*see* **1-Hexacosanol***													
c243	**Cerulignone**		304.30	bl gr					i	i	i	i	i	$PhNO_2$, sulf, PhOH s oos i	**B8**², 573
—	**Cetene**	*see* **1-Hexadecene***													
—	**Cetyl amine**	*see* **Hexadecane, 1-amino-***													
—	**Cetyl bromide**	*see* **Hexadecane, 1-bromo-***													
—	**Cevadine**	*see* **Veratrine**													
c244	**Cevagenine**	$C_{27}H_{43}NO_8$	509.65	nd (MeOH-eth) $[\alpha]^{20}_D$ −52 (chl) $[\alpha]^{20}_D$ −47.5 (al) λ^{al} 280 (1.7)	246–8 (242)										**H35**, 1270
Ω c245	**Chalcone** (*trans*)	Benzalacetophenone. Benzylidene acetophenone. 1,3-Diphenyl-2-propen-l-one.	208.26	pa ye lf, pr, nd (peth) λ^{al} 228 (2.99) 358 (3.39)	(i) 59 (st) (ii) 57 (iii) 49 (unst) (iv) 48 (v) 30 (vi) 18	$345–8^{760}$ δd 208^{25}	1.0712^{62}_4		i	δ	s		s	chl, CS_2, sulf s lig i	**B7**², 423
c246	—,4,4′-dimethyl-	$C_{17}H_{16}O$. *See* c245	236.32	cr (MeOH)	127–9				i	s				MeOH s^h	**B7**², 441
c247	—,3,3′-dinitro-	$C_{15}H_{16}N_2O_5$. *See* c245	298.26	pa ye nd (aa) λ^{al} 247 (4.22), 314 (4.45)	210–1				i	i	i		s	Ac_2O s^h	**B7**², 429
c248	—,2-methoxy-	*o*-Anisylideneacetophenone. $C_{16}H_{14}O_2$. *See* c245	238.29	yesh nd (peth or eth-lig) λ^{al} 252 (4.08), 344 (3.83)	64–5					v	v		v	chl v, con sulf s lig δ	**B8**², 218
c249	—,3-methoxy-	$C_{16}H_{14}O_2$. *See* c245	238.29	yesh pl or pr (MeOH)	65	247^{12}			i	s	s	s	s	oos s	**B8**¹, 579
Ω c250	—,4-methoxy-	$C_{16}H_{14}O_2$. *See* c245	238.29	ye nd (al) λ^{al} 240 (4.09), 344 (4.07)	79 (74)	$187.5–8^{19}$			i	v^h	s			chl, aa, con sulf s MeOH v^h	**B8**², 218
Ω c251	—,3,4-methyl-enedioxy-	Piperonylideneacetophenone. $C_{16}H_{12}O_3$. *See* c245	252.27	ye nd (al)	128 (122)				i	s^h				con sulf s (og-ye) aa s	**B19**², 167
c252	—,2-nitro-	$C_{15}H_{11}NO_3$. *See* c245	253.26	pa br nd (al) λ^{al} 210 (4.17), 276 (4.29)	125					s^h	s			aa s	**B7**², 428
c253	—,2′-nitro-	$C_{15}H_{11}NO_3$. *See* c245	253.26	nd (al) λ^{al} 287 (4.32)	128–9					s^h	s			con sulf → pa ye	**B7**², 429
c254	—,3-nitro-	$C_{15}H_{11}NO_3$. *See* c245	253.26	ye nd (al or bz) λ^{al} 215 (4.11), 290 (4.43)	145–6					s	i		s	chl, aa s peth i	**B7**², 429
c255	—,4-nitro-	$C_{15}H_{11}NO_3$. *See* c245	253.26	pa ye nd (al), pl (bz) λ^{al} 210 (4.18), 315 (4.48)	164					s^h	i			chl s lig i	**B7**², 429
c256	—,α-nitro-	$C_6H_5CH{:}C(NO_2)COC_6H_5$	253.26	ye pl (eth or bz-lig), cr (aa) λ^{al} 257 (4.21), 313 (4.14)	90				i		s	s	s	oos v	**B7**, 483
—	**Chamazulene**	*see* **Azulene, 1,5-dimethyl-8-isopropyl-**													
c257	**Chaulmoogric acid**(*d*)	*d*-13(2-Cyclopentenyl)tri-decanoic acid. Hydnocarpyl-acetic acid. $C_{18}H_{32}O_2$.	280.46	pl or lf (al, aa) $[\alpha]_D$ +62 (chl)	68.5	$247–8^{20}$			i	δ	s			chl s	**B9**², 58
c258	—(*dl*)	$C_{18}H_{32}O_2$. *See* c257	280.46	cr (peth)	68.5	$247–8^{20}$									**B9**², 61
—	**Chavibetol**	*see* **Benzene, 5-allyl-1-hydroxy-2-methoxy-**													
—	**Chavicol**	*see* **Benzene, 1-allyl-4-hydroxy-**													
c259	**Chelerythrine**	$C_{19}H_{21}NO_5$	365.39	cr (chl-MeOH), cr (al + l)	207				δ	δ	δ^h	s^h		chl s AcOEt δ	**B27**², 563
Ω c260	**Chelidonic acid**	Jervasic acid. 4-Pyrone-2,6-dicarboxylic acid.	184.11	rose mcl nd (al-w + 1w) λ 270 (4.0)	262				s^h	δ					**B18**², 367
c261	—,diethyl ester	$C_{11}H_{12}O_6$. *See* c260	240.22	pr, nd λ^{MeOH} 270 (4.0)	69 (63)				δ^h	s^h	s				**B18**², 367
c262	**Chelidonine**(*d*)	$C_{20}H_{19}NO_5$	353.38	mcl pl (al + 1w) $[\alpha]_D$ +151 (al, 1 %) λ^{MeOH} 239 (3.92), 289 (3.87)	136–40d				i^h	s v^h	s v^h			chl s	**B27**², 615
c263	—,hydrochloride(*d*)	$C_{20}H_{19}NO_5$. HCl. *See* c262	389.84	wh (w)					i	δ^h	δ^h				**B27**, 557
—	**Chenodeoxy-cholic acid**	*see* **Cholanic acid, 3α,7α-dihydroxy-**													

For explanations, symbols and abbreviations see beginning of table. For structural formulas see end of table.

No.	Name	Synonyms and Formula	Mol. wt.	Color, crystalline form, specific rotation and λ_{max} (log ε)	m.p. °C	b.p. °C	Density	n_D	Solubility w	al	eth	ace	bz	other solvents	Ref.
	Chimyl alcohol														
—	**Chimyl alcohol**	*see* **Glycerol,** 1-hexadecyl ether													
—	**Chloral**	*see* **Acetaldehyde, trichloro-**													
—	**Chloralacetone-chloroform**	*see* **Acetaldehyde, trichloro-,** β,β,β-trichloro-*tert*-butyl hemiacetal													
—	**Chloral formamide**	*see* **Formic acid,** amide, *N*(1-hydroxy-2,2,2-trichloroethyl)-													
—	**Chloral hydrate**	*see* **Acetaldehyde, trichloro-,** hydrate													
Ω c264	**Chloralide**	2,5-Bis(trichloromethyl)-1,3-dioxolan-4-one.	322.79	pr (al or eth)	116	272–3 $147–8^{12}$	......	......	i	v^h δ	v	...	...	aa v	**B19**[1], 656
—	**Chloramphenicol**	*see* **Chlormycetin**													
—	**Chloranil**	*see* **1,4-Benzoquinone, tetrachloro-***													
—	**Chloranilic acid**	*see* **1,4-Benzoquinone, 2,5-dichloro-3,6-dihydroxy-***													
—	**Chloretone**	*see* **2-Propanol, 2-methyl-1,1,1-trichloro-***													
—	**Chloroform**	*see* **Methane, trichloro-**													
—	**Chloroform-*d***	*see* **Methane, deuterotrichloro-***													
c265	**Chloromycetin**	Chloramphenicol. Syntomycetin. $C_{11}H_{12}Cl_2N_2O_5$.	323.14	pa ye pl or nd (w), $[\alpha]_D^{25}$ +19 (al, c = 5), −25.5 (AcOEt) λ 278 (3.99)	150.5–1.5	sub vac	......	......	δ	v	δ	v	i	chl s	**Am 71,** 2458
c266	**Chlorophyll a**	$C_{55}H_{72}MgN_4O_5$	893.53	bl bk hex pl λ^{hx} 440 (4.0), 660 (3.9)	150–3	......	......	......	i	v^h	v^h	...	...	lig s	**M,** 245
c267	**Chlorophyll b**	$C_{55}H_{70}MgN_4O_6$	907.51	bl bk gr pw λ^{bx} 460 (4.2), 650 (3.6)	120–30	......	......	......	i^h	v^h	v^h	...	...	Py v^h MeOH, lig s	**M,** 245
—	**Chloropicrin**	*see* **Methane, nitrotrichloro-***													
—	**Chloroprene**	*see* **1,3-Butadiene, 2-chloro-***													
—	**Chlorosulfonic acid,** ethyl ester	*see* **Sulfuric acid,** monochloride monoethyl ester													
c268	**Chlorpromazine**	Largactil. Thorazine. $C_{17}H_{19}ClN_2S$.	318.87	λ 250 (4.5), 320 (3.5)	......	$200–5^{0.8}$	......	......	i	v	v	...	v	chl v dil HCl s	**C50,** 1931
c269	—,hydrochloride	$C_{17}H_{19}ClN_2S$. HCl. *See* c268	355.33	λ 250 (4.5), 300 (3.5)	194–7d (180)	......	......	......	s	v	i	...	i	MeOH, PrOH, chl v	**C50,** 1931
—	**Chlortetracyline**	*see* **Aureomycin**													
c270	**Cholanic acid**	Ursocholanic acid. $C_{24}H_{40}O_2$.	360.59	nd (al), cr (aa) $[\alpha]_D^{20}$ +21.7 (chl) $\lambda^{97\%\ sulf}$ 313 (3.88)	163–4	......	......	......	...	s	...	...	...	chl, aa s	**E14,** 170
c271	—,ethyl ester	$C_{26}H_{44}O_2$. *See* c270	388.64	lf, nd (dil al) $[\alpha]_D^{19}$ +21 (chl)	93–4	273^{12}	......	......	...	...	...	...	...	chl s	**Am 77,** 3308
c272	—,methyl ester	$C_{25}H_{42}O_2$. *See* c270	374.61	nd $[\alpha]_D$ +23 ±2 (diox)	87–8	......	......	......	...	...	...	...	...	diox s	**Am 77,** 3308
Ω c273	—,3α,6α-dihydroxy-	Hyodeoxycholic acid. $C_{24}H_{40}O_4$. *See* c270	392.56	cr (AcOEt) $[\alpha]_D^{20}$ +37.2 (MeOH) $\lambda^{65\%\ sulf}$ 276 (3.51), 380 (4.18)	198–9	......	......	......	δ	s	δ	δ	δ	aa s AcOEt δ	**C50,** 5067
c274	—,3α,7α-dihydroxy-	Chenodeoxycholic acid. $C_{24}H_{40}O_4$. *See* c270	392.59	nd (AcOEt-hp) $[\alpha]_D^{20}$ +11.1 (al, c = 2.1) $\lambda^{65\%\ sulf}$ 272 (3.58), 380 (4.35)	143 (119)	......	......	......	i	v	s	v	i	aa s peth i	**C48,** 3377
—	—,3,12-dihydroxy-	*see* **Desoxycholic acid**													
—	—,3,7,12-trihydroxy-	*see* **Cholic acid**													
—	—,3,7,12-trioxo-	*see* **Dehydrocholic acid**													
c275	**Cholanthrene**	Benz[*j*]aceanthrene	254.34	pa ye lf (bz-al) λ^{al} 261.5 (4.60), 274 (4.60), 284.5 (4.80), 295 (4.90)	174.5–5.0d	sub $210^{0.2}$	......	......	i	s	...	...	s	aa, lig, to s	**B5**[3], 2469
c276	—,6,7-dihydro-20-methyl-	$C_{21}H_{18}$. *See* c275	270.37	lf (MeOH)	155	......	......	......	...	...	...	...	...	MeOH s^h	**B5**[3], 2429

For explanations, symbols and abbreviations see beginning of table. For structural formulas see end of table.

No.	Name	Synonyms and Formula	Mol. wt.	Color, crystalline form, specific rotation and λ_{max} (log ε)	m.p. °C	b.p. °C	Density	n_D	Solubility w	al	eth	ace	bz	other solvents	Ref.
	Cholanthrene														
c277	—,11,14-dihydro-20-methyl-	$C_{21}H_{18}$. *See* c275	270.37	nd (PrOH)	138–9									PrOH s^h	$B5^3$, 2429
c278	—,20-methyl-	$C_{21}H_{16}$. *See* c275	268.38	yesh nd (bz) λ^{al} 327 (3.63), 343 (3.79), 359 (3.88)	180				i						$B5^3$, 2484
c279	$\Delta^{2,4}$-Cholestadiene	$C_{27}H_{44}$	368.65	cr (eth-ace) $[\alpha]_D^{23}$ +168.5 (eth, c = 1.5) λ^{al} 276 (4.04)	68.5				i	s	s			chl s	$B5^2$, 1428
c280	$\Delta^{3,5}$-Cholestadiene	Cholesterilene. $C_{27}H_{44}$	368.65	wh nd (al) $[\alpha]_D^{20}$ −129.6 (chl, c = 3) λ^{cy} 193 (2.50), 228 (4.23), 235 (4.27), 244 (4.07)	80	260^{13}	0.9254^{100}_4		i^h	s^h	∞		∞	chl ∞ lig v	$B5^3$, 1429
c281	$\Delta^{5,7}$-Cholesta-dien-3β-ol	7-Dehydrocholesterol. Provitamin D_3.$C_{27}H_{44}O$.	384.65	pl (+1w, eth-MeOH) $[\alpha]_D^{25}$ −115 (chl, c = 2.5) λ^{eth} 262 (3.99)	150–1 (143)				i	δ	s	s		oos s	$B6^3$, 2819
c282	$\Delta^{4,6}$-Cholestadien-3-one	$C_{27}H_{42}O$	382.64	ye pr $[\alpha]_D^{20}$ +31 (chl) λ^{al} 284 (4.36)	78				i						J1956, 627
c283	Cholestane	$C_{27}H_{48}$	372.69	sc or pl (eth-al, ace, AcOEt) $[\alpha]_D^{20}$ +30.2 (chl, c = 2) λ 175	80–80.5	250^1	0.9090^{88}_4	1.4887^{88}	i	δ	v		v	chl v MeOH δ	$B5^3$, 1133
—	—,3β-hydroxy-	*see* **Cholestanol**													
c284	3β-Cholestane-carboxylic acid	$C_{28}H_{48}O_2$. *See* c283	416.70	nd, $[\alpha]_D^{25}$ +28.8 (chl, c = 1.7)	210–1										Am 75, 6234
c285	3,6-Cholestane-dione	$C_{27}H_{44}O_2$. *See* c283	400.65	nd, $[\alpha]_D^{20}$ +8.9 (c = 0.98) λ^{MeOH} 295 (1.75)	171–2										H37, 258
c286	3α-Cholestanol	Epicholestanol. Epidihydro-cholesterol. $C_{27}H_{48}O$. *See* c283	388.69	nd (al) $[\alpha]_D^{20}$ +34 (chl, c = 1.1)	188										$B3^2$, 2135
c287	3β-Cholestanol	Dihydrocholesterol. 3β-Hydroxycholestane. $C_{27}H_{48}O$. *See* c283	388.69	lf (al + 1 w) pr, pl (MeOH) $[\alpha]_D^{22}$ +24.2 (chl, c = 1.3)	141–3 (146)				i	v^h	v			chl v con sulf s MeOH δ	$B6^3$, 2131
c288	3-Cholestanone	Zymostanone. $C_{27}H_{46}O$. *See* c283	386.67	nd or lf (al) $[\alpha]_D$ +42 (chl, c = 2.12) λ^{al} 280 (1.86)	128–30				i	s^h					Am 77, 190
c289	6-Cholestanone, 3β-hydroxy-	6-Ketocholestanol. $C_{27}H_{46}O_2$. *See* c283	402.67	nd (al) $[\alpha]_D$ −3.0 (chl)	150–1				i	s^h					Am 76, 532
c290	7-Cholestanone, 3β-hydroxy-	7-Oxocholestanol. 7-Keto-cholestanol. $C_{27}H_{46}O_2$. *See* c283	402.67	pl, $[\alpha]_D^{22}$ −34 (chl)	165–8				i	s^h					Am 76, 532
c291	2-Cholestene	Neocholestene. $C_{27}H_{46}$	370.67	nd (eth-ace or al) $[\alpha]_D$ +66 (c = 1.65) λ^{al} 205 (3.75)	75–6										$B5^3$, 1520
c292	5-Cholestene	$C_{27}H_{46}$	370.67	wh pr or nd (al) $[\alpha]_D^{18}$ −56.3 (chl) λ^{cy} 193 (3.91) λ^{al} 206 (3.32)	93–4 (90)										$B5^3$, 1323
c293	—,3β-bromo-	Cholesteryl bromide. $C_{27}H_{45}Br$. *See* c292	449.57	mcl lf (al) $[\alpha]_D^{20}$ −19 (bz, c = 0.4) λ^{al} 209 (3.72)	100–2 (99)				i	s^h			s	chl s	$B5^3$, 1327
c294	—,3β-chloro-	Cholesteryl chloride. $C_{27}H_{45}Cl$. *See* c292	405.12	nd (al or ace) $[\alpha]_D^{20}$ −26.4 (bz), −33.3 (chl) λ^{al} 207 (3.52)	96				i	s^h		s^h	s	CS_2 v aa v^h chl s	$B5^3$, 1324
c295	—,3β-iodo-	Cholesteryl iodide. $C_{27}H_{45}I$. *See* c292	496.57	nd (ace or AcOEt) $[\alpha]_D^{20}$ −11.9 (chl, c = 1)	106–7				i	δ		δ	s	chl lig s EtOAc s^h	$B5^3$, 1328
c296	5-Cholestene-3β, 7β-diol	7β-Hydroxycholesterol. $C_{27}H_{46}O_2$. *See* c292	402.67	wh nd (eth) $[\alpha]_D^{20}$ +7.2 (chl, c = 2)	177–8.5	sub $145^{0.005}$									$B6^3$, 1328

For explanations, symbols and abbreviations see beginning of table. For structural formulas see end of table.

1-Cholesten-3-one

No.	Name	Synonyms and Formula	Mol. wt.	Color, crystalline form, specific rotation and λ_{max} (log ε)	m.p. °C	b.p. °C	Density	n_D	Solubility w	al	eth	ace	bz	other solvents	Ref.
c297	**1-Cholesten-3-one**	$C_{27}H_{44}O$	384.65	nd (dil ace or al) $[\alpha]_D^{25}$ +88.2 (chl) λ^{al} 231 (3.99)	99–101 (95)	...	...	...	...	...	...	...	s	...	**Am 75,** 3513
Ω c298	**4-Cholesten-3-one**	...	384.65	nd or pl (al) $[\alpha]_D^{25}$ +92 (chl, c = 2.01) λ^{hx} 241 (4.26), 314 (1.65), 323.5 (1.69), 349 (1.53), 367 (1.15)	81–2	...	...	...	i	δ	s	...	s	peth, CS_2 s AcOEt δ	**C50,** 1568
c299	**5-Cholesten-3-one**	$C_{27}H_{44}O$. *See* c292	384.65	lf (al) $[\alpha]_D^{20}$ −4.3 (chl) λ^{al} 285 (2.00)	127	...	...	...	i	s^h	s	...	...	...	**C50,** 1568
—	**Cholesterilene**	*see* **3,5-Cholestadiene**													
Ω c300	**Cholesterol**	Δ^5-3β-Cholestenol. Cholesterin. $C_{27}H_{46}O$.	386.67	rh or tcl lf (al + 1w), nd (eth) $[\alpha]_D^{20}$ −31.5 (eth, c = 2), −39.5 (chl, c = 2) λ^{al} 206 (3.53)	148.5 (anh) (150.1)	360δd $233^{0.5}$	1.067_4^{20}	...	i	δ^h	v	δ^h	s	diox, chl v CS_2, aa s	**B6³**, 2607
Ω c301	—,acetate	Cholesteryl acetate. $C_{29}H_{48}O_2$. *See* c300	428.71	wh nd (ace or al), $[\alpha]_D^{20}$ −47.7 (chl, c = 2) λ^{al} 197 (3.89)	115–6	...	...	...	δ	...	s	s	s	chl s	**B6³**, 2630
c302	—,benzoate	Cholesteryl benzoate. $C_{34}H_{50}O_2$. *See* c300	490.78	wh nd, [α] −13.7 (chl, c = 0.9) λ^{al} 230 (4.16)	152–3	...	...	...	...	i	s	...	...	chl s	**E14,** 53
c303	—,hexadecanoate	Cholesteryl palmitate. $C_{43}H_{76}O_2$. *See* c300	625.09	wh nd (eth or al), $[\alpha]^{20}$ −25.4 (chl, c = 2)	80	...	...	...	i^h	s^h	s^h	s^h	s	chl s	**B6³**, 2640
—	—,7β-hydroxy-	*see* **5-Cholestene-3β, 7β-diol**													
—	—,7-dehydro-	*see* **5,7-Cholestadien-3β-ol**													
—	**Cholesteryl bromide**	*see* **5-Cholestene, 3β-bromo-**													
—	**Cholesteryl chloride**	*see* **5-Cholestene, 3β-chloro-**													
c304	**Cholestrophane**	Dimethylparabanic acid. Oxalyldimethylurea.	142.12	lf or pl (w or al)	155.5	275–7 148–50^{13}	...	...	s v^h	δ	s	...	...	...	**B24²**, 265
Ω c305	**Cholic acid**	Cholanic acid. 3α, 7α, 12α-Trihydroxy-5β-cholanoic acid. $C_{24}H_{40}O_5$	408.59	rh (eth), tetr rh (w or dil al + 1w), cr (al + 1) $[\alpha]_D^{20}$ +37 (al, c = 0.6) $\lambda^{65\%\ sulf}$ 318 (4.16), 377.5 (2.71)	198 (anh)	...	...	...	δ	s	v	s	...	chl v alk, aa s	**E14,** 191
Ω c306	**Choline**	Trimethyl (2-hydroxyethyl) ammonium hydroxide. $(CH_3)_3\overset{+}{N}CH_2CH_2OH.OH^-$	121.18	syr	...	...	...	...	v	v	i	i	i	chl, CS_2, CCl_4 i	**B4²**, 720
Ω c307	—,*O*-**acetyl-**, bromide	Acetylcholine bromide. Progmoline. $(CH_3)_3\overset{+}{N}CH_2CH_2O_2CCH_3.Br^-$	226.12	hex pr (dil al)	143	d	...	...	v d^h	s	i	...	...	MeOH s	**B4²**, 724
Ω c308	—,—,chloride	Acetylcholine chloride. $(CH_3)_3\overset{+}{N}CH_2CH_2O_2CCH_3.Cl^-$	181.66	yesh nd	151	...	...	...	s d	s^h	i	...	...	...	**B4³**, 656
Ω c309	—,*O*-**benzoyl-**, chloride	$(CH_3)_3\overset{+}{N}CH_2CH_2O_2CC_6H_5.Cl^-$	243.74	pr (ace-al)	200	...	...	...	...	s^h	...	s^h	...	...	**B9¹**, 90
Ω c310	—,**carbamate,** chloride	$(CH_3)_3\overset{+}{N}CH_2CH_2O_2CNH_2.Cl^-$	182.65	hyg pr or pw	210–2 (205)	...	...	...	v	δ	i	...	...	MeOH v chl i	**C38,** 1800
—	**Chondrodendrin**	*see* **Bebeerine**(*d*)													
c311	**Chroman**	Dihydrobenzopyran	134.18	λ^{hx} 276 (3.33), 284 (3.43)	...	$214\text{–}5^{742}$ $98\text{–}9^{18}$	1.0610^{20}	1.5444^{20}	s^h	...	...	...	...	oos ∞	**B17,** 52
Ω c312	**Chromone**	α-Benzopyrone. 4-Oxo-1,4-chromene.	146.15	nd (peth or w) λ^{al} 274.5 (4.20), 312.5 (3.91)	59	sub	...	...	δ^h	s	s	...	s	chl s	**B17¹**, 170
—	—,2-phenyl-	*see* **Flavone**													
—	**Chromotropic acid**	*see* **2,7-Naphthalene disulfonic acid, 4,5-dihydroxy-***													

For explanations, symbols and abbreviations see beginning of table. For structural formulas see end of table.

No.	Name	Synonyms and Formula	Mol. wt.	Color, crystalline form, specific rotation and λ_{max} (log ε)	m.p. °C	b.p. °C	Density	n_D	Solubility w	al	eth	ace	bz	other solvents	Ref.
	Chrysammic acid														
—	**Chrysammic acid** . .	*see* **9,10-Anthraquinone, 1,8-dihydroxy-2,4,5,7-tetranitro-***													
—	**Chrysaniline**	*see* **Acridine, 2-amino-5(4-aminophenyl)-**													
—	**Chrysanthemum-monocarboxylic acid**	*see* **Cyclopropanecarboxylic acid, 2,2-dimethyl-3(2-methylpropenyl)-***													
—	**Chrysanthemumic acid**	*see* **Cyclopropanecarboxylic acid, 2,2-dimethyl-3(2-methylpropenyl)***													
—	**Chrysazin**	*see* **9,10-Anthraquinone, 1,8-dihydroxy-***													
—	**Chrysazol**	*see* **Anthracene, 1,8-dihydroxy-***													
Ω c313	**Chrysene**	1,2-Benzophenanthrene. Benzo[*a*]phenanthrene.	228.30	red bl flr rh pl (bz, aa) λ^{al} 222 (4.55), 258 (4.96), 268 (5.20), 295 (4.08), 320 (4.12), 344 (2.81), 353 (2.57), 361 (2.80)	255–6	448	1.274^{20}	. . . .	i	δ	δ	δ	δ s^h	to s^h CS_2, aa δ	**B5**[3], 2380
c314	—,5,6-dimethyl- . . .	$C_{20}H_{16}$. *See* c313	256.35	pl or nd (bz-al) λ^{al} 228 (4.3), 272 (5.0), 330 (4.1), 345 (2.8)	128–9	$200^{0.5}$ (sub 140 vac)	. . . .	. . . .	i	s	. . .	. . .	. . .	CS_2, aa s	**B5**[3], 2418
c315	—,1-methyl-	$C_{19}H_{14}$. *See* c313	242.32	lf (hx, bz, to) λ^{al} 259 (4.54), 281 (4.81), 292 (4.89), 302 (3.97), 337 (3.81), 352 (3.87), 390 (2.87)	256.5–7.0	sub 130–40 (vac)	. . . .	. . . .	i	s	. . .	. . .	. . .	. . . .	**B5**[3], 2395
c316	—,2-methyl-	$C_{19}H_{14}$. *See* c313	242.32	lf (bz-al)	229.5–30 (225)	. . . .	. . . .	. . . .	i	s	. . .	. . .	. . .	aa s	**B5**[3], 2395
c317	—,3-methyl- . . .	$C_{19}H_{14}$. *See* c313	242.32	lf (bz-peth) λ^{cy} 268 (4.9), 275 (5.1), 297 (4.1), 324 (4.1), 355 (3.0)	172.5–3.5	. . . .	. . . .	. . . .	. . .	s	. . .	. . .	. . .	. . . .	**B5**[3], 2395
—	**Chrysin**	*see* **Flavone, 5,7-dihydroxy-**													
—	**Chrysoidine**	*see* **Azobenzene, 2,4-diamino-**													
—	**Chrysophanic acid**	*see* **9,10-Anthraquinone, 1,8-dihydroxy-3-methyl-***													
c318	**5,6-Chryso-quinone**	. . . .	258.28	red nd (bz or to), lf or pl (aa) λ^{diox} 250 (4.5), 385 (3.75)	239.5	sub	. . . .	. . . .	i	s^h	δ	. . .	s^h	sulf s aa δ to δ, s^h	**B7**[2], 760
c319	**6,12-Chryso-quinone**	. . . .	258.28	red ye nd (aa)	288–90d	. . . .	. . . .	. . . .	i	s^h	. . .	. . .	. . .	aa s^h	**B7**[1], 441
—	**Cinchomeronic acid**	*see* **3,4-Pyridinedicarboxylic acid**													
c320	**Cinchonamine**	$C_{19}H_{24}N_2O$	296.42	rh nd (al), orh pr (MeOH) $[\alpha]^{20}_{D}$ +123 (al, c = 0.66) λ^{al} 230 sh (4.2), 282 (4.0), 292 (4.0)	186 (194)	. . . .	. . . .	. . . .	i	v^h	v^h	. . .	s	chl s	**B23**[2], 358
c321	**Cinchonicine**	Cinchotoxine. $C_{19}H_{22}N_2O$ . .	294.40	nd or pr (eth) $[\alpha]^{15}_{D}$ +48 (al, c = 1)	58–60	. . . .	. . . .	. . . .	i	v	v	v	v	chl v	**B24**[2], 100
c322	**Cinchonidine**	Cinchovatine. α-Quinidine. $C_{19}H_{22}N_2O$.	294.40	orh pl or pr (al) $[\alpha]^{20}_{D}$ − 109.20 (al, p = 1) $\lambda^{w, pH=1}$ 305 sh (3.81), 315 (3.87)	210.5	sub	. . . .	. . . .	i	s	δ	. . .	i	Py, chl, MeOH s peth i	**B23**[2], 373
c323	—,bisulfate	$C_{19}H_{22}N_2O.H_2SO_4.5H_2O$. *See* c322	482.56	mcl pr $[\alpha]_D$ − 110 (al)	. . . .	. . . .	. . . .	. . . .	v	v	. . .	. . .	i	. . . .	**B23**, 441

For explanations, symbols and abbreviations see beginning of table. For structural formulas see end of table.

No.	Name	Synonyms and Formula	Mol. wt.	Color, crystalline form, specific rotation and λ_{max} (log ε)	m.p. °C	b.p. °C	Density	n_D	Solubility w	al	eth	ace	bz	other solvents	Ref.
	Cinchonidine														
c324	—,hydrochloride	$C_{19}H_{22}N_2O.HCl.H_2O$. *See* c322	348.88	wh pr (w) $[\alpha]^{20}_D -117.6$ (w, c = 1.2)	242d (anh)				s	v	δ	...	i	chl v	**B23**[2], 373
c325	—,sulfate	$(C_{19}H_{22}N_2O)_2.H_2SO_4.6H_2O$. *See* c322	794.97	mcl eff nd (al + 2w) $[\alpha]^{18.5}_D -97.9$ (w, c = 1.2)	205 (anh)				δ	δ	i	...	i	chl δ	**B23**[2], 373
c326	**β-Cinchonidine**	$C_{19}H_{22}N_2O$	294.40	pr or lf (al or dil al) $[\alpha]^{25}_D -126.6$ (al, c = 0.5)	241				i	v	δ	...	...	chl v	**B23**[1], 131
Ω c327	**Cinchonine**	$C_{19}H_{22}N_2O$. *See* c322	294.40	nd or mcl cr (al) $[\alpha]^{17}_D +229$ (al) $\lambda^{w, pH=6}$ 285 (3.71), 300 (3.57), 313 (3.40)	255 (265)	sub >220			i	δ	δ	i	i	chl s MeOH, Py δ peth i	**B23**[2], 369
c328	—,bisulfate	$C_{19}H_{22}N_2O.H_2SO_4.4H_2O$. *See* c322	464.53	rh oct (w) $[\alpha]^{18}_D +195$ (w, c = 2)					v	v	i	...	...		**B23**[2], 371
c329	—,dihydrochloride	$C_{19}H_{22}N_2O.2HCl$. *See* c322.	367.32	pl, $[\alpha]^{21}_D +205.5$ (w, c = 3.6)					v	s	i	...	i		**B23**[1], 133
c330	—,hydrochloride	$C_{19}H_{22}N_2O.HCl.2H_2O$. *See* c322	366.89	mcl $[\alpha]^{25}_D +133.6$ (chl, c = 1.4)	*ca.* 215 (anh)		1.234		s v^h	v	δ	...	i	chl s	**B23**[2], 370
c331	—,nitrate	$C_{19}H_{22}N_2O.HNO_3.\frac{1}{2}H_2O$. *See* c322	366.42	mcl (w)					s	v	...	...	i		**B23**, 430
c332	—,sulfate	$(C_{19}H_{22}N_2O)_2.H_2SO_4.2H_2O$. *See* c322	722.91	rh pr (w) $[\alpha]^{15}_D +169$ (w, c = 1.4)	206–7 (198) (anh)				s^h	s v^h	i	...	i	MeOH v chl δ	**B23**[2], 371
—	**Cinchoninic acid**	*see* **4-Quinolinecarboxylic acid**													
—	**Cinchopene**	*see* **4-Quinolinecarboxylic acid, 2-phenyl-**													
c333	**Cinchotine**	Hydrocinchonine. Pseudo-cinchonine. $C_{19}H_{24}N_2O$.	296.42	pr $[\alpha]^{21}_D +203.4$ $\lambda^{w, pH=1}$ 298 (3.67), 308 (3.85), 314 (3.87)	268–9				s^h	δ	i	...	...		**B23**[2], 356
Ω c334	**1,4-Cineole**	*p*-Cineole. 1,4-Epoxy-*p*-menthane.	154.26		1	173–4	0.8997^{20}	1.4562^{20}	δ	∞	∞	...	s	lig s	**B17**[2], 32
c335	**1,8-Cineole**	Cajeputol. Eucalyptol.	154.26		1.5	176.4 61^{14}	0.9267^{20}	1.4586^{20}	i	s	s	...	...	chl s	**B17**[2], 32
c336	**Cineolic acid**(*d*)	Tetrahydro-2,6,6-trimethylpyran-2,5-dicarboxylic acid.	216.24	cr (w + 1) $[\alpha]^{20}_D +18.6$ (w, p = 8.21)	79 (+w) 138–9 (anh)				δ s^h	v	δ	...	δ	chl, AcOEt v lig δ	**B18**, 322
c337	—(*dl*)	$C_{10}H_{16}O_5$. *See* c336	216.24		(i) 197.5 (ii) 208				δ s^h	v^h	v	...	δ	chl δ	**B18**[2], 285
c338	—(*l*)	$C_{10}H_{16}O_5.H_2O$. *See* c336	234.25	rh (w + 1) $[\alpha]^{20}_D -19.1$ (w)	79 (+w) 138.9 (anh)				δ s^h	δ^h	δ	...	δ	chl δ	**B18**, 322
Ω c339	**Cinnamaldehyde** (*trans*)	Cinnamic aldehyde. *β*-Phenylacrolein. 3-Phenylpropenal (*trans*)*	132.17	yesh λ^{al} 291 (4.40)	−7.5	253^{760} δd 127^{16}	1.0497^{20}_4	1.6195^{20}	δ	s	s	...	...	chl s lig i	**B7**[2], 273
Ω c340	—,oxime (*syn*)	$C_6H_5CH:CHCH:NOH$	147.18	nd λ^{al} 290 (4.3)	138.5				δ	δ	δ	...	s		**B7**[2], 278
c341	**—,4-hydroxy-3-methoxy-**	Coniferaldehyde. Ferulaldehyde. $C_{10}H_{10}O_3$. *See* c339	178.19	cr (bz)	84	$157^{2.5}$			δ	s	s	...	s		**B8**, 288
c342	**—,4-methyl-**	$C_{10}H_{10}O$. *See* c339	146.19	ye lf (dil al)	41.5	154^{25}			...	s^h	...	...	...		**B7**, 369
c343	**—,α-methyl-**	$C_6H_5CH:C(CH_3)CHO$	146.19	ye λ^{al} 280 (4.3)		150^{100} 131–2^{16}	1.0407^{17}_4	1.6057^{17} (1.3917^{20})	...	...	...	...	...		**B7**[2], 291
c344	**—,2-nitro-**	$C_9H_7NO_3$. *See* c339	177.16	nd (eth or al) λ^{al} 250 (4.3), 280 sh (4.0)	127.0–7.5				v^h	s^h	s	...	...	chl v	**B7**[2], 281
c345	**—,3-nitro-**	$C_9H_7NO_3$. *See* c339	177.16	ye nd (w), pr (al), cr (aa)	116				δ^h	δ s^h	δ	...	v	aa v	**B7**[2], 282
c346	**—,4-nitro-**	$C_9H_7NO_3$. *See* c339	177.16	nd (w or al) λ^{al} 295 (4.3)	141–2				s^h	v	s	s	s	oos v	**B7**[2], 282
—	**—,α-pentyl-**	*see* **Heptanal, 2-benzylidene-***													
—	**Cinnamalmalonic acid**	*see* **Malonic acid, cinnamylidene** (*trans*)													
—	**Cinnamein**	*see* **Cinnamic acid,** benzyl ester													
c347	**Cinnamic acid** (*cis*) (1st form)	Isocinnamic acid. *cis-β*-Phenylacrylic acid.	148.17	mcl pr (w) λ^{al} 263–5 (4.0)	42				δ s^h	s	...	...	...	aa, lig v CCl_4 s	**B9**[2], 393
c348	—(*cis*) (2nd form)	$C_9H_8O_2$. *See* c347	148.17	lo mcl pr (lig) λ^{al} 267 (3.98)	58	265			δ s^h	s	v	v	...	chl, lig s	**B9**[2], 393

For explanations, symbols and abbreviations see beginning of table. For structural formulas see end of table.

No.	Name	Synonyms and Formula	Mol. wt.	Color, crystalline form, specific rotation and λ_{max} (log ε)	m.p. °C	b.p. °C	Density	n_D	Solubility w	al	eth	ace	bz	other solvents	Ref.
	Cinnamic acid														
c349	—(*cis*) (3rd form) . .	Allocinnamic acid. Isocinnamic acid. $C_9H_8O_2$. *See* c347	148.17	mcl pr λ^{al} 266 (3.99)	68	. . . .	. . . .	. . . .	δ s^h	v^h	v^h	. . .	. . .	lig s	**B9**^{2}, 393
Ω c350	—(*trans*)	*trans*-β-Phenylacrylic acid. *trans*-3-Phenylpropenoic acid.*	148.17	mcl pr (dil al) λ^{al} 210 (4.24), 215 (4.28), 221 (4.18), 268 (4.31)	135–6 (cor) (133)	300^{760} (cor)	1.2475_4^4	. . . .	i δ^h	v	s	s	s	chl, to, MeOH s lig, peth i	**B9**^{2}, 377
Ω c351	—,allyl ester	Allyl cinnamate $C_6H_5CH:CHCO_2CH_2CH:CH_2$	188.23	. . . .	. . . .	286d 163^{17}	1.048^{23}	1.530^{20}	i	v	∞	. . .	. . .	. . . .	**B9**^{2}, 387
Ω c352	—,amide(*trans*)	Cinnamamide. $C_6H_5CH:CHCONH_2$	147.18	nd (bz) λ 273 (4.4)	148.0–8.5 (cor)	. . . .	. . . .	. . . .	δ^h	v	v	. . .	. . .	CS_2 v	**B9**^{2}, 391
c353	—,anhydride(*trans*)	$(C_6H_5CH:CHCO)_2O$	278.31	nd (bz or al), pr (al) λ^{peth} 290 (4.69)	138 (136)	. . . .	. . . .	. . . .	i	δ s^h	. . .	. . .	v^h	. . . .	**B9**^{2}, 390
Ω c354	—,benzyl ester (*trans*)	Benzyl cinnamate. Cinnamein. $C_6H_5CH:CHCO_2CH_2C_6H_5$	238.29	pr	39 (34)	350d 244^5	1.109^{15}	. . . .	i	s	s	. . .	. . .	. . . .	**B9**^{2}, 388
Ω c355	—,chloride(*trans*). .	Cinnamoyl chloride. $C_6H_5CH:CHCOCl$	166.61	. . . .	37–8	257.5^{760} 131^{11}	1.1617_4^{45}	$1.614^{42.5}$	i	s^h	. . .	. . .	. . .	CCl_4, lig s	**B9**^{2}, 390
Ω c356	—,ethyl ester(*trans*)	Ethyl cinnamate. $C_6H_5CH:CHCO_2C_2H_5$	176.22	λ^{al} 216.5 (4.14), 222 (4.09), 277 (4.30)	12 (6.5)	271.5^{760} 144^{15}	1.0491_4^{20}	1.5598^{20}	i	s	v	v	s	oos v	**B9**^{2}, 385
Ω c357	—,isopropyl ester (*trans*)	$C_6H_5CH:CHCO_2CH(CH_3)_2$	190.25	. . . .	. . . .	268–70 $153–5^{20}$	1.0320^{20}	1.5455^{20}	i	s	s	s	. . .	. . . .	**B9**^{2}, 387
c358	—,2-methoxyphenyl ester(*trans*)	*o*-Anisyl cinnamate. Guaiacol cinnamate.	254.29	wh nd (al)	130	. . . .	. . . .	. . . .	i	. . .	. . .	s	s	chl s	**B9**^{2}, 389
Ω c359	—,methyl ester (*trans*)	Methyl cinnamate. $C_6H_5CH:CHCO_2CH_3$	162.19	cr (peth or dil al) λ^{al} 217 (4.22), 223 (4.14), 278 (4.37)	36.5	261.9^{760} 127^{10}	1.0911_4^{20}	1.5766^{21}	i	v	v	s	s	oos v	**B9**^{2}, 384
c360	—,nitrile(*cis*).	Allocinnamonitrile. *cis*-Cinnamonitrile. *cis*-Styryl cyanide. $C_6H_5CH:CHCN$	129.16	λ 273 (4.22)	−4.4	249 139^{30}	. . . .	1.5843^{20}	i	s	. . .	. . .	v		**Am 58,** 2428
c361	—,—(*trans*).	$C_6H_5CH:CHCN$	129.16	. . . .	22	263.8 $134–6^{12}$	1.0304_4^{20}	1.6013^{20}	i	s	. . .	s	. . .	. . . .	**B9**^{2}, 392
c362	—,2-octyl ester (*trans, d*)	$C_6H_5CH:CHCO_2CH(CH_3)(CH_2)_5CH_3$	260.38	$[\alpha]_D^{17}$ +40.19	. . . .	218^{28}	0.9645_4^{20}	1.5145^{20}	i	. . .	. . .	. . .	s	chl s	**B9**^{1}, 229
c363	—,—(*trans, dl*)	$C_6H_5CH:CHCO_2CH(CH_3)(CH_2)_5CH_3$	260.38	. . . .	. . . .	240^{60} 213^{28}	0.9715_4^{17}	. . . .	i	. . .	. . .	. . .	s	chl s	**B9**^{1}, 230
c364	—,—(*trans, l*)	$C_6H_5CH:CHCO_2CH(CH_3)(CH_2)_5CH_8$	260.38	$[\alpha]_D^{17}$ − 39.78	. . . .	211^{28}	0.9692_4^{17}	. . . .	i	. . .	. . .	. . .	s	chl s	**B9**^{1}, 229
c365	—,phenyl ester (*trans*)	$C_6H_5CH:CHCO_2C_6H_5$	224.26	. . . .	72.5	$205–7^{15}$	. . . .	. . . .	. . .	. . .	s	. . .	s	. . . .	**B9**^{2}, 387
Ω c366	—,*p*-phenylphenacyl ester (*trans*)	. . . .	342.40	. . . .	182.5	. . . .	. . . .	. . . .	. . .	. . .	. . .	. . .	. . .	. . . .	**Am 52,** 3719
c367	—,propyl ester (*trans*)	$C_6H_5CH:CHCO_2CH_2CH_2CH_3$	190.25	. . . .	. . . .	285^{760}	1.0435_0^0	. . . .	i	. . .	. . .	. . .	. . .	. . . .	**B9**^{2}, 387
c368	—,4-ureidophenyl ester (*trans*)	. . . .	282.30	wh	204	. . . .	. . . .	. . . .	i	s	. . .	s	. . .	. . . .	**C52,** 10721
Ω c369	—,**α-acetamido-** (*trans*)	$C_6H_5CH:C(NHCOCH_3)CO_2H$	205.22	cr (+2w) λ^{al} 273 (4.36)	185–6 (+2w) 193–4 (anh)	. . . .	. . . .	. . . .	s	. . .	. . .	. . .	. . .	. . . .	**B10**^{2}, 471
c370	—,**2-amino-**(*trans*) .	$C_9H_9NO_2$. *See* c350.	163.18	ye nd (w)	158–9d	. . . .	. . . .	. . . .	s^h	s	s	. . .	. . .	. . . .	**B14**^{2}, 316
c371	—,**3-amino-**(*trans*) .	$C_9H_9NO_2$. *See* c350.	163.18	pa ye nd (w or al)	185 (181)	. . . .	. . . .	. . . .	s^h	v	v	. . .	. . .	ac, alk s	**B14**^{2}, 316
Ω c372	—,**4-amino-**(*trans*) .	$C_9H_9NO_2$. *See* c350.	163.18	ye nd (w or al)	175–6d	. . . .	. . . .	. . . .	s^h	s	s	. . .	. . .	. . . .	**B14**^{2}, 317
c373	—,**2-benzamido-** (*trans*)	$C_{16}H_{13}NO_3$. *See* c350	267.29	nd (al)	191–3	. . . .	. . . .	. . . .	. . .	δ^h	. . .	. . .	. . .	aa s^h oos δ	**B14**^{1}, 617
c374	—,**3-benzamido-** (*trans*)	$C_{16}H_{13}NO_3$. *See* c350	267.29	nd (AcOEt)	229	. . . .	. . . .	. . . .	. . .	s^h	. . .	s	δ	aa s	**B14**^{1}, 618
c375	—,**4-benzamido-** (*trans*)	$C_{16}H_{13}NO_3$. *See* c350	267.29	lf (aa), nd (ace)	274d	. . . .	. . . .	. . . .	. . .	δ	. . .	v^h	s	aa v^h	**B14**^{1}, 619
c376	—,**α-bromo-**(*cis*) . . .	α-Bromoallocinnamic acid. $C_6H_5CH:CBrCO_2H$	227.06	lf (w), pr (chl) λ^{al} 206 (4.03), 257 (4.16)	120–1	. . . .	. . . .	. . . .	s^h	s	. . .	. . .	s	CS_2 s	**B9**^{2}, 399
c377	—,—(*trans*).	$C_6H_5CH:CBrCO_2H$	227.06	nd (w) λ^{al} 206 (4.1), 217 (4.05), 273 (4.19)	131–2	. . . .	. . . .	. . . .	v^h	∞	s	. . .	s	. . . .	**B9**^{2}, 398
c378	—,**β-bromo**(*cis*). . . .	β-Bromoallocinnamic acid. $C_6H_5CBr:CHCO_2H$	227.06	nd (bz), pl (al) λ^{al} 205 (4.3), 230 sh (3.93), 267 (3.74)	159–60	. . . .	. . . .	. . . .	δ^h	δ	s	. . .	δ s^h	chl s	**B9**^{2}, 398
c379	—,—(*trans*).	$C_6H_5CBr:CHCO_2H$	227.06	pa ye nd or pl (w), pr (chl) λ^{al} 205 (4.2), 263 (4.07)	135	. . . .	. . . .	. . . .	δ^h	s	. . .	. . .	s^h	CS_2 δ	**B9**^{2}, 397

For explanations, symbols and abbreviations see beginning of table. For structural formulas see end of table.

No.	Name	Synonyms and Formula	Mol. wt.	Color, crystalline form, specific rotation and λ_{max} (log ε)	m.p. °C	b.p. °C	Density	n_D	Solubility w	al	eth	ace	bz	other solvents	Ref.
	Cinnamic acid														
c380	**—,2-carboxy-** (*trans*)	β(2-Carboxyphenyl)acrylic acid. *o*,β-Styrenedicarboxylic acid. $C_{10}H_8O_4$. *See* c350	192.18	pr or nd (w) λ^{al} 220 (4.18), 272 (4.18)	208–9	...	...	...	δ^h	v	δ	...	i	CS_2 δ chl i	**B9**[2], 641
c381	**—,3-carboxy-** (*trans*)	$C_{10}H_8O_4$. *See* c350	192.18	nd (ace-lig)	275 (264)	...	...	...	i	δ	δ	...	i	aa v chl i	**B9**[2], 642
c382	**—,4-carboxy-** (*trans*)	$C_{10}H_8O_4$. *See* c350	192.18	pw	358d	sub >350	...	...	...	i	i	...	i	aa s^h oos i	**B9**[2], 642
c383	**—,α,β-dibromo-** (*cis*)	$C_6H_5CBr:CBrCO_2H$	305.97	ye pr or pl (chl or lig) λ^{al} 205 (4.3), 236 (4.06), 288 (3.47)	100	$124^{0.5}$	...	...	i	s^h	s	...	...	chl, aa s lig s^h peth δ	**B9**[2], 401
c384	**—,2,4-dihydroxy-** (*trans*)	Umbellic acid. $C_9H_8O_4$. *See* c350	180.17	ye nd or pl λ^{MeOH} 217 (4.1), 286 (4.0), 324 (4.1)	d260 (darkens at 240)	...	...	...	s^h	s	i	...	i	lig i	**B10**[2], 293
c385	**—,2,5-dihydroxy-** (*trans*)	$C_9H_8O_4$. *See* c350	180.17	ye cr (w), cr (dil al + 1w) λ^{MeOH} 220 (4.1), 274 (4.2), 350 (3.8)	207d	...	...	...	...	s	...	...	...	...	**B10**, 435
Ω c386	**—,3,4-dihydroxy-** (*trans*)	Caffeic acid. $C_9H_8O_4$. *See* c350	180.17	ye pr or pl (w) λ^{al} 244 (4.05), 299 (4.18), 326 (4.27)	225d (200)	...	...	...	s^h	s	δ	...	...	...	**B10**[2], 294
c387	**—,3,5-dihydroxy-** (*trans*)	$C_9H_8O_4$. *See* c350	180.17	nd (w + ½)	245–6	...	...	...	v^h	v	v	...	i	...	**B10**[2], 297
c388	**—,2,3-dimethoxy-** (*trans*)	$C_{11}H_{12}O_4$. *See* c350	208.22	nd	180–1	...	...	...	δ	δ	...	...	...	oos δ	**B10**[2], 292
c389	**—,2,4-dimethoxy-** (*cis*)	$C_{11}H_{12}O_4$. *See* c347	208.22	nd (al) λ^{MeOH} 217 (3.9), 271 (3.9), 310 (3.9)	138	...	...	...	...	v	v	...	v	...	**B10**[2], 293
c390	**—,—(*trans*)**	$C_{11}H_{12}O_4$. *See* c350	208.22	nd (w or dil al) λ^{MeOH} 217 (4.1), 244 (4.1), 286 (4.2), 322 (4.2)	187.5–9.0	...	...	...	δ	v	v	...	v	chl v lig i	**B10**[2], 293
c391	**—,2,5-dimethoxy-** (*trans*)	$C_{11}H_{12}O_4$. *See* c350	208.22	pa ye or ye-gr nd (w) λ^{MeOH} 223 (4.1, 274 (4.2, 340 (3.8)	148–9	...	...	...	δ s^h	s	s	...	...	oos, alk v	**B10**[2], 294
Ω c392	**—,3,4-dimethoxy-** (*trans*)	$C_{11}H_{12}O_4$. *See* c350	208.22	nd (w, dil al), pa ye pw (dil aa) λ^{al} 220, 258, 290, 318	183	...	...	...	s^h	v	v	...	...	aa s^h	**B10**[2], 294
c393	**—,3,5-dimethoxy-** (*trans*)	$C_{11}H_{12}O_4$. *See* c350	208.22	nd (w)	175–6	...	...	...	v^h	v	...	...	v	...	**B10**[2], 297
Ω c394	**—,3,5-dimethoxy-4-hydroxy-(*trans*)**	5-Methoxyferulic acid. Sinapic acid. $C_{11}H_{12}O_5$. *See* c350	224.22	pa ye nd (al) λ^{al} 225 (4.0), 240 (4.0), 320 (4.2)	192	...	...	...	δ	s^h	δ	...	...	...	**B10**, 508
c395	**—,α-ethyl-(*cis*)**	α-Benzalbutyric acid. $C_6H_5CH:C(C_2H_5)CO_2H$	176.22	nd (w)	82	...	...	...	v^h	v	v	...	v	CS_2 v	**B9**, 623
c396	**—,—(*trans*)**	$C_6H_5CH:C(C_2H_5)CO_2H$	176.22	nd (w)	106 (114)	...	...	...	i	v	s	...	...	lig δ, s^h	**B9**[1], 258
Ω c397	**—,2-hydroxy-** (*trans*)	*trans-o*-Coumaric acid. $C_9H_8O_3$. *See* c350	164.17	nd (w) λ^w 274 (4.26), 325 (4.00)	217d	...	...	...	δ	s	δ	...	...	chl, CS_2 i	**B10**[2], 174
Ω c398	**—,3-hydroxy-** (*trans*)	*trans-m*-Coumaric acid. $C_9H_8O_3$. *See* c350	164.17	pr (w) λ^{eth} 277 (4.29), 319 (3.70)	193	...	...	...	s^h	s	s	...	s	chl, CS_2 i	**B10**[2], 178
Ω c399	**—,4-hydroxy-** (*trans*)	*trans-p*-Coumaric acid. $C_9H_8O_3$. *See* c350	164.17	nd (w^c + 1), cr (w^h) λ^{al} 227 (4.09), 312 (4.36)	215d	...	...	...	δ	s^h	s	...	δ	lig i	**B10**[2], 178
Ω c400	**—,4-hydroxy-3-methoxy-(*trans*)**	Ferulic acid. $C_{10}H_{10}O_4$. *See* c350	194.19	pr or nd (w) λ^{al} 234 (4.10), 296 (4.09), 323 (4.22)	171	...	...	...	s^h	s	δ	...	δ	AcOEt v chl s lig δ	**B10**[2], 294
c401	**—,4-isopropyl-** (*trans*)	Cumylideneacetic acid. $C_{12}H_{14}O_2$. *See* c350	190.25	pr (bz)	165	...	...	...	δ^h	s	...	...	s^h	aa s^h	**B9**, 629
Ω c402	**—,4-methoxy-** (*trans*)	$C_{10}H_{10}O_3$. *See* c350	178.19	wh nd (al)	172–5	...	...	...	δ	δ	...	...	δ	CCl_4, aa s	**B10**[2], 179

For explanations, symbols and abbreviations see beginning of table. For structural formulas see end of table.

No.	Name	Synonyms and Formula	Mol. wt.	Color, crystalline form, specific rotation and λ_{max} (log ε)	m.p. °C	b.p. °C	Density	n_D	Solubility						Ref.
									w	al	eth	ace	bz	other solvents	
	Cinnamic acid														
c403	—,**α-methyl-(*cis*)**	α-Benzalpropionic acid. β-Phenylmethacrylic acid. $C_6H_5CH:C(CH_3)CO_2H$	162.19	nd (bz or CS_2) λ^{al} 205 (4.3), 257 (4.20)	74	288 190^{21}	...	...	δ	v	v	...	v	CS_2 v peth s	**B9², 409**
c404	—,—(*trans*)	$C_6H_5CH:C(CH_3)CO_2H$	162.19	pr (aa, eth or dil al) λ^{al} 206 (4.2), 261 (4.22)	81–2	...	...	...	δ	v	s	...	v	peth s	**B9, 615**
c405	—,**2-methyl-4-nitro-(*trans*)**	$C_{10}H_9NO_4$. *See* c350	207.19	nd (al)	256	...	...	...	δ^h	δ^h	i	...	...	...	**B9¹, 256**
c406	—,**4-methyl-3-nitro-(*trans*)**	$C_{10}H_9NO_4$. *See* c350	207.19	ye pl or nd (al)	173.5	...	...	...	v^h	s	v	...	...	lig i	**B9¹, 257**
c407	—,**α-methyl-2-nitro-**	$C_{10}H_9NO_4$. *See* c350	207.19	mcl pr (al)	164–5	...	...	...	...	v	v	...	δ	lig δ	**B9¹, 256**
c408	—,**α-methyl-3-nitro-**	$C_{10}H_9NO_4$. *See* c350	207.19	nd or pw λ^{al} 258 (4.35)	203.5 (198)	...	...	...	...	s^h	v	...	v	aa v lig δ	**B9¹, 256**
c409	—,**α-methyl-4-nitro-**	$C_{10}H_9NO_4$. *See* c350	207.19	ye rh (aa), tcl pym (aa, al-eth) λ^{al} 282 (3.98)	208	...	...	...	...	s^h	δ	...	δ	aa s^h oos δ	**B9¹, 256**
c411	—,**2-nitro-(*cis*)**	$C_9H_7NO_4$. *See* c347	193.16	yesh (bz or chl)	146–7	...	...	...	...	v	...	...	v^h	chl δ	**B9¹, 246**
c412	—,—(*trans*)	$C_9H_7NO_4$. *See* c350	193.16	nd (al) λ^{al} 248 (4.1), 290 sh (4.6)	242–3	sub	...	...	i	s^h	...	...	i	...	**B9², 402**
c413	—,—,ethyl ester (*trans*)	$C_{11}H_{11}NO_4$. *See* c350	221.22	ye rh bipym (al)	44	...	...	...	...	v^h	v^h	...	v^h	CS_2 v^h	**B9², 402**
c414	—,—,methyl ester (*trans*)	$C_{10}H_9NO_4$. *See* c350	207.19	wh nd (w)	73	187–9¹⁵	...	...	δ^h	v^h	...	...	...	...	**B9², 402**
c415	—,**3-nitro-(*cis*)**	$C_9H_7NO_4$. *See* c347	193.16	nd	158	...	...	...	...	...	...	...	...	...	**B9¹, 247**
c416	—,—(*trans*)	$C_9H_7NO_4$. *See* c350	193.16	nd (al) λ^{al} 263 (4.4), 305 sh (3.1)	204.5–5.5	...	...	...	δ	s	...	...	s^h	...	**B9², 402**
c417	—,—,ethyl ester (*trans*)	$C_{11}H_{11}NO_4$. *See* c350	221.22	nd (al), pr (aa)	78–9	...	...	...	δ	δ	δ	...	...	...	**B9², 403**
c418	—,—,methyl ester (*trans*)	$C_{10}H_9NO_4$. *See* c350	207.19	pa ye pr (MeOH)	123–4	d	...	...	i	δ	s	...	v	chl v MeOH δ	**B9², 403**
Ω c419	—,**4-nitro-(*trans*)**	$C_9H_7NO_4$. *See* c350	193.16	yesh-wh pr (al) λ^{al} 295 (4.3)	286 (289)	...	...	...	δ^h	δ	δ	...	...	CS_2, lig i	**B9², 404**
c420	—,—,ethyl ester (*trans*)	$C_{11}H_{11}NO_4$. *See* c350	221.22	pl (aa) λ^{al} 300 (4.34)	141–2	...	...	...	i	i	δ	...	...	...	**B9², 404**
c421	—,—,methyl ester (*trans*)	$C_{10}H_9NO_4$. *See* c350	207.19	wh nd (al) λ^{al} 299.5 (4.32)	162	281–6	...	...	...	s^h	δ	...	...	...	**B9², 404**
c422	—,—,piperazinium salt(*trans*)	$2C_9H_7NO_4.C_4H_{10}N_2$. *See* c350	472.46	...	...	248d	...	...	δ^h	s	i	...	...	...	**Am 70, 2758**
—	—,**α-phenyl-**	*see* **Propenoic acid, 2,3-diphenyl-***													
Ω c424	—,**3,4,5-tri-methoxy-(*trans*)**	$C_{12}H_{14}O_5$. *See* c350	238.24	nd (w) $\lambda^{w,pH=10}$ 298 (4.28)	126–7	...	...	...	s^h	...	...	...	...	chl s	**Am 77, 2241**
—	**Cinnamic alcohol**	*see* **2-Propen-1-ol, 3-phenyl-***													
—	**Cinnamylidene chloride**	*see* **Propene, 3,3-dichloro-1-phenyl-***													
—	**Citraconic acid**	*see* **Maleic acid, methyl-**													
Ω c425	**Citral a**	Geranial	152.24	λ^{al} 237 (4.16)	...	229⁷⁶⁰ 118–9²⁰	0.8888²⁰	1.4898²⁰	i	∞	∞	...	...	...	**B1³, 3056**
c426	**Citral b**	Neral	152.24	λ^{al} 237 (4.16)	...	120²⁰ 103¹²	0.8869²⁰	1.4869²⁰	i	∞	∞	...	...	...	**B1³, 3056**
—	**Citramalic acid**	*see* **Succinic acid, 2-hydroxy-2-methyl-**													
c427	**β-Citraurin**	$C_{30}H_{40}O_2$	432.65	pl (bz-peth), cr (al) λ 467, 497	147	...	...	...	i	v	v	v	v	CS_2 v lig δ	**H21, 448**
Ω c428	**Citric acid**	2-Hydroxy-1,2,3-propanetri-carboxylic acid.* $HOC(CH_2CO_2H)_2CO_2H$	192.14	rh (w + 1)	153 (anh) (156–7)	d	1.542^{18}_4 (hyd) 1.665^{18}_4 (anh)	...	v	v	s	...	i	AcOEt s chl, to i	**B3², 359**
c429	—,piperazinium salt	$HOC(CH_2CO_2H)_2CO_2H.C_4H_{10}N_2$	278.27	...	141–2	...	...	...	s	s	i	...	...	...	**Am 70, 2758**
c430	—,triamide	Citramide. Citric triamide. $HOC(CH_2CONH_2)_2CONH_2$	189.17	(w)	210–5d	...	...	...	s^h	i	i	...	...	...	**B3¹, 197**
c431	—,tribenzyl ester	Benzyl citrate. $HOC(CH_2CO_2CH_2C_6H_5)_2CO_2CH_2C_6H_5$	462.51	nd (al)	51	...	...	...	...	s^h	...	...	...	oos s	**B6², 421**
Ω c432	—,triethyl ester	Ethyl citrate. $HOC(CH_2CO_2C_2H_5)_2CO_2C_2H_5$	276.29	...	...	294⁷⁶⁰ 185¹⁷	1.1369^{20}_4	1.4455²⁰	i	s	s	...	...	...	**B3², 370**
c433	—,trimethyl ester	Methyl citrate. $HOC(CH_2CO_2CH_3)_2CO_2CH_3$	234.21	tcl	78.5–79	287d 176¹⁶	...	...	δ	v	v	...	...	...	**B3², 370**
c434	—,triphenyl ester	Phenyl citrate. $HOC(CH_2CO_2C_6H_5)_2CO_2C_6H_5$	420.42	nd (al)	124.5	...	...	...	i	s^h	s	...	...	...	**B6, 170**
c435	—,tripropyl ester	Propyl citrate. $HOC(CH_2CO_2C_3H_7)_2CO_2C_3H_7$	318.37	...	...	198¹⁸	...	...	...	s	s	...	...	...	**B3, 568**

For explanations, symbols and abbreviations see beginning of table. For structural formulas see end of table.

No.	Name	Synonyms and Formula	Mol. wt.	Color, crystalline form, specific rotation and λ_{max} (log ε)	m.p. °C	b.p. °C	Density	n_D	Solubility: w	al	eth	ace	bz	other solvents	Ref.
	Citric acid														
c436	—,**anhydro-methylene-**	1,3-Dioxolan-4-one-5,5-diacetic acid.	204.15	(w)	298	....	....	....	v^h	δ	δ	v	...	chl v alk s	**B19**[2], 324
c437	**Citrinin**	$C_{13}H_{14}O_5$	250.25	ye nd (MeOH) $[\alpha]_D^{11}$ −37 λ^{al} 250 (3.97), 331 (4.02), 400 sh (4.2)	178–9d	....	....	....	i	δ	δ	s	s	alk, chl, diox s	**J1963,** 3777
c438	**Citronellal**(*d*)	*d*-Rhodinal. $(CH_3)_2C{:}CH(CH_2)_2CH(CH_3)CH_2CHO$	154.26	$[\alpha]_D^{18}$ +13.09 λ^{al} 235 (1.93), 290 sh (1.08)	....	207.8^{760} 92^{14}	0.8573_4^{20}	1.4456^{20}	δ	s	s	...	...	....	**B1**[2], 803
Ω c439	—(*dl*)...........	$C_{10}H_{18}O$ *See* c438........	154.26	λ^{al} 235 (1.93), 290 sh (1.08)	....	207–8^{760} 79–81^{10}	0.8535_4^{17}	1.4473^{20}	...	...	...	...	...	....	**B1**[2], 805
c440	—(*l*)............	$C_{10}H_{18}O$. *See* c438........	154.26	$[\alpha]_D^{20}$ −2.5 λ^{al} 235 (1.93), 290 sh (1.08)	....	205–6 87^{10}	0.8567_4^{17}	1.4479^{20}	δ	∞	∞	...	...	....	**B1**[2], 805
c441	**Citronellol**(*d*)	*d*-Rhodinol. $(CH_3)_2C{:}CH(CH_2)_2CH(CH_3)CH_2CH_2OH$	156.27	$[\alpha]_D^{17}$ +6.8 λ^{al} 212 (2.98)	....	244.4^{760} (222) 118^{17}	0.8590^{20}	1.4565^{20}	δ^h	∞	∞	...	...	....	**B1**[2], 495
Ω c442	—(*dl*)...........	Dihydrogeraniol. $C_{10}H_{20}O$. *See* c441	156.27	λ^{al} 212 (2.98)	....	99^{10}	0.8560_4^{20}	1.4543^{20}	δ^h	∞	∞	...	...	....	**B1**[2], 495
c443	—(*l*)............	*l*-Rhodinol. $C_{10}H_{20}O$. *See* c441	156.27	$[\alpha]_D^{18}$ −5.3 λ^{al} 212 (2.98)	....	108–9^{10}	0.859_4^{18}	1.4576^{18}	...	...	...	...	...	....	**B1**[2], 496
Ω c444	**Citrulline**(*L*)	α-Amino-δ-ureidovaleric acid. *N*-δ-Carbamyl-ornithine. δ-Ureidonor-valine. $H_2NCONH(CH_2)_3CH(NH_2)CO_2H$	175.19	pr (aq MeOH) $[\alpha]_D^{20}$ +3.7 (w, c = 2)	234–7 (222)	....	....	....	s	i	...	...	...	MeOH i	**B4**[3], 1347
—	**Civetone**	*see* **9-Cycloheptadecen-1-one***													
—	**Cleve's acid**	*see* **1-Naphthalenesulfonic acid, 3-amino-***													
—	**Clupanodonic acid**	*see* **Docosapentaenoic acid***													
—	**Cocaethyline**......	*see* **Ecgonine**, benzoate, ethyl ester													
c445	**Cocaine**(*d*)	β-Cocaine. Benzoyl methyl ecgonine. $C_{17}H_{21}NO_4$.	303.36	mcl pr (eth) $[\alpha]_D^{20}$ +15.8 (chl, p = 10) λ^{al} 229 (4.2), 274 (3.0), 281 (2.9)	98	....	....	....	i	v	v	s	s	....	**B22**[2], 150
c446	—(*dl*)...........	$C_{17}H_{21}NO_4$. *See* c445	303.36	rh bipym pr (peth) λ^{al} 229 (4.2), 274 (3.0), 281 (2.9)	79–80	....	....	....	i	v	v	s	v	chl, CCl_4, lig v	**B22**[2], 156
c447	—(*l*)............	$C_{17}H_{21}NO_4$. *See* c445	303.36	mcl pr (al) $[\alpha]_D^{20}$ −16.3 (chl, c = 4), $[\alpha]_D^{18}$ −35 50 % al, c = 1) λ^{al} 229 (4.2), 274 (3.0), 281 (2.9)	98	187–8$^{0.1}$ (sub 75–90 (vac))	....	1.5022^{98}	δ^h	v	v	s	v	chl, Py v CS_2 s	**B22**[2], 151
c448	—,chromate(*l*)....	$C_{17}H_{21}NO_4.H_2CrO_4.H_2O$. *See* c445	439.40	og ye lf	127	....	....	....	δ	...	...	...	...	....	**B22**, 200
c449	—,hydro-chloride(*dl*)	$C_{17}H_{21}NO_4$.HCl. *See* c445..	339.82	pl (al) λ^w 233 (4.09), 274 (3.03)	187 (cor)	....	....	....	v	v	i	s	i	chl s	**B22**[2], 156
c450	—,—(*l*)	$C_{17}H_{21}NO_4$.HCl. *See* c445..	339.82	mcl pr (al), cr (w + 2) $[\alpha]_D^{20}$ −71.95 (w) λ^w 233 (4.09), 274 (3.03)	197	....	....	....	v	s	i	s	i	chl s	**B22**[2], 153
c451	"**Cocaine cinna-mate**"(*d*)	*O*-Cinnamoyl-*d*-ecgonine methyl ester.	329.40	pr, $[\alpha]_D$ +2 (al)	68	....	....	....	i	s	s	s	s	chl, lig s	**B22**, 207
c452	—(*l*)............	$C_{19}H_{23}NO_4$. *See* c451	329.40	mcl pr, nd (bz) $[\alpha]_D^{15}$ −4.7 (chl, c = 10)	121	....	....	....	i	s	s	s	s	chl, lig s	**B22**[2], 154
c453	**Coclaurine**(*l*).....	$C_{17}H_{19}NO_3$............	285.35	pl (al) $[\alpha]_D^{29}$ −17.01	220–1	....	....	....	δ	δ v^h	δ	δ v^h	i	chl δ peth i	**C47,** 6430
c454	**Codamine**	$C_{20}H_{25}NO_4$............	343.43	pr (bz or eth)	127	....	....	....	s^h	v	s	...	...	chl v	**B21**[2], 184

For explanations, symbols and abbreviations see beginning of table. For structural formulas see end of table.

No.	Name	Synonyms and Formula	Mol. wt.	Color, crystalline form, specific rotation and λ_{max} (log ε)	m.p. °C	b.p. °C	Density	n_D	Solubility w	al	eth	ace	bz	other solvents	Ref.
	Codeine														
c455	**Codeine**	Morphine 3-methyl ether. $C_{18}H_{21}NO_3$.	299.37	rh oct (+1w, w or dil al), cr (eth or CS_2) $[\alpha]_D^{15}$ −137.8 (al, p = 2) λ^{al} 212 (4.4), 236 sh (3.7), 286 (3.2)	157–8.5 (156)	250^{12} (sub 140–$5^{1.5}$)	1.32	...	s^h	v	s	...	s	chl, to, MeOH s peth i	**B27**^{2}, 137
c456	—,hydrate	$C_{18}H_{21}NO_3.H_2O$. *See* c455	317.39	rh oct (w, aq al) $[\alpha]_D^{25}$ −136 (al, c = 2.8)	...	...	1.31	...	δ	v	s	v	s	chl v	**B27**^{2}, 137
c457	—,hydrochloride	$C_{18}H_{21}NO_3.HCl.2H_2O$. *See* c455	371.87	nd, $[\alpha]_D^{15}$ −108.2 (w) λ^w 238 sh (3.70), 285.5 (3.24)	287d	...	...	...	v	s	...	...	...	...	**B27**^{2}, 143
c458	—,phosphate	$C_{18}H_{21}NO_3.H_3PO_4.1\frac{1}{2}H_2O$. *See* c455	424.39	lf or pr (dil al) λ^w 212 (4.41), 284 (3.22)	220–35d	...	...	...	v	s^h	s	...	...	chl s	**B27**^{2}, 144
c459	—,sulfate	$(C_{18}H_{21}NO_3)_2.H_2SO_4.5H_2O$. *See* c455	786.90	pr, $[\alpha]_D^{15}$ −100.9 (w, p = 3)	278 (anh)	...	...	...	s	δ	i	...	...	chl i	**B27**^{2}, 144
c460	**β-Codeine**	Neopine. $C_{18}H_{21}NO_3$	299.37	nd (peth) $[\alpha]_D^{23}$ −28 (chl, c = 7.5) λaq. HBr 210 (4.6), 232 sh (3.8), 284 (3.2)	127.5	...	...	...	s	s	s	...	s	MeOH, chl v lig δ	**B27**^{2}, 176
c461	**Codeine, dihydro- (*d*)**	Cohydrin. Dihydroneopine. Paracodin. $C_{18}H_{23}NO_3$.	301.39	cr (+1w, dil MeOH) $[\alpha]_D$ +146.4 (al, c = 2) λ^{al} 285.5 (3.23)	112–3	248^{15}	...	...	...	...	...	...	...	...	**B27**^{2}, 103
c462	**Colchiceine**	*N*-Acetyltrimethyl colchicinic acid. $C_{21}H_{23}NO_6$.	385.42	pa ye nd (diox) $[\alpha]_D^{14}$ −252.5 (chl, c = 1.2) λ^{al} 244 (4.47), 350 (4.22)	178–9	...	1.24	...	δ	v	i	...	i	chl v	**B14**^{2}, 190
Ω c463	**Colchicine(*l*)**	$C_{22}H_{25}NO_6$	399.45	ye pl (w + 1½), pa ye nd (AcOEt), ye cr (bz + 1) $[\alpha]_D^{17}$ −121 (chl, c = 0.9), −429 (w, c = 1.72) λ^{al} 244 (4.47), 351 (4.21)	155–7 (anh) 140 (+1 bz)	...	...	...	s	s	δ	...	δ	chl, MeOH s CCl_4 δ peth i	**B14**^{2}, 191
c464	**Colchicinic acid, trimethyl-**	$C_{19}H_{21}NO_5$	343.39	pa ye nd (al) $[\alpha]_D^{25}$ −184.5 (chl, c = 1)	155–7	...	...	...	s	s	...	...	...	...	**Am 75,** 5292
—	**α-Collidine**	*see* **Pyridine, 4-ethyl-2-methyl-**													
—	**β-Collidine**	*see* **Pyridine, 3-ethyl-4-methyl-**													
—	**γ-Collidine**	*see* **Pyridine, 2,4,6-trimethyl-**													
—	**Comenic acid**	*see* **4-Pyrone-2-carboxylic acid, 5-hydroxy-**													
c465	**Conessine**	Neriine. Wrightine. $C_{24}H_{40}N_2$.	356.60	lf or pl (ace) $[\alpha]_D^{20}$ +25.3 (al, c = 0.7)	125–6	165–$7^{0.1}$	...	...	δ	...	...	...	...	aa s	**C28,** 5072
Ω c466	**Congo red**	$C_{32}H_{22}N_6Na_2O_6S_2$	696.68	pw λ^w 350 (4.5), 500 (4.7)	...	...	...	...	δ	s	i	...	...	...	**B16**^{2}, 223
c467	**Conhydrine(*d*)**	α-Hydroxyconiine. 2(1-Hydroxypropyl) piperidine. $C_8H_{17}NO$.	143.23	lf (eth) $[\alpha]_D$ +10 (w)	121	226^{760}	...	...	s	v	v	...	...	chl s Py δ	**B21**^{2}, 11
c467[1]	—(*dl*) (lower m.p.)	$C_8H_{17}NO$. *See* c467	143.23	nd (peth)	69–70	sub	...	...	v	v	v	...	v	...	**B21**^{2}, 12
c467[2]	—(*dl*) (higher m.p.)	$C_8H_{17}NO$. *See* c467	143.23	nd (eth)	98–9	sub	...	...	v	v	v	...	v	...	**B21**^{2}, 11
c468	**α-Coniceine(*d*)**	2-Methylconidine	125.22	$[\alpha]_D^{15}$ +18.4 (al, c = 2)	−16	158^{760}	$0.891_4^{15.5}$	...	δ	s	...	...	...	...	**B20,** 152
c469	—(*dl*)	$C_8H_{15}N$. *See* c468	125.22	...	...	156–9	$0.890_4^{15.5}$	...	δ	...	...	...	...	...	**B20,** 153
c470	**β-Coniceine(*d*)**	2-Propenylpiperidine	125.22	nd, $[\alpha]_D^{45}$ +49.9	38.5–9.0	168–9	...	...	δ	s	s	...	...	...	**B20,** 146
c471	—(*dl*)	$C_8H_{15}N$. *See* c470	125.22	nd	8	168.5–70^{753}	0.8716_4^{15}	...	δ	s	s	...	...	...	**B20,** 146
c472	—(*l*)	$C_8H_{15}N$. *See* c470	125.22	nd, $[\alpha]_D^{45}$ −50.5	41	168.5–9^{760}	0.8520_4^{50}	...	δ	s	s	...	...	...	**B20,** 146
c473	**γ-Coniceine**	...	125.22	...	...	173–4^{760} 64–5^{14}	0.8720_4^{20}	1.4607^{18}	δ	s	...	...	...	...	**B20,** 144
—	**δ-Coniceine**	*see* **Piperolidine**													

For explanations, symbols and abbreviations see beginning of table. For structural formulas see end of table.

No.	Name	Synonyms and Formula	Mol. wt.	Color, crystalline form, specific rotation and λ_{max} (log ε)	m.p. °C	b.p. °C	Density	n_D	Solubility w	al	eth	ace	bz	other solvents	Ref.
	ε-Coniceine														
c474	**ε-Coniceine(*d*)**	*d*-2-Methylconidine. Stereoisomer of c468	125.22	$[\alpha]_D^{15}$ +67.4	...	151.5–4.0	0.8856_4^{15}	...	...	...	...	...	...	...	**B20**, 151
c475	—(*dl*)	$C_8H_{15}N$. *See* c474	125.22	...	...	150–1	0.8836_4^{15}	...	...	...	...	...	...	...	**B20**, 152
c476	—(*l*)	$C_8H_{15}N$. *See* c474	125.22	$[\alpha]_D^{15}$ −87.34	...	151–3.5	0.8642_4^{15}	...	...	...	...	...	...	...	**B20**, 151
Ω c477	**α-Conidendrin**	Tsugalactone. Tsugaresinol. $C_{20}H_{20}O_6$.	356.38	cr (al) $[\alpha]_D^{20}$ −54.5 (ace, c = 2.1) λ^{al} 230 sh (4.13), 284 (3.83)	255–6	...	...	...	i	s	s	...	s	peth i	**Am 77**, 432
—	**Conidine, 2-methyl-**	*see* **α-Coniceine**													
c478	—,**3-methyl-(*d*)**	...	125.22	$[\alpha]_D^{8}$ +16	...	158^{760}	0.8856_4^{15}	...	δ	s	s	...	...	...	**B20**, 153
c479	—,—(*dl*)	$C_8H_{15}N$. *See* c478	125.22	...	...	158^{760}	0.8946_4^{15}	...	δ	s	s	...	...	...	**B20**, 153
c480	—,—(*l*)	$C_8H_{15}N$. *See* c478	125.22	$[\alpha]_D^{17}$ −17.1	...	158^{760}	0.8856_4^{15}	...	δ	s	s	...	...	...	**B20**, 153
—	**Coniferaldehyde**	*see* **Cinnamaldehyde, 4-hydroxy-3-methoxy-**													
c481	**Coniferin**	$C_{16}H_{22}O_8$	342.35	nd (w + 2) $[\alpha]_D^{20}$ −68 (w, c = 0.5) λ^{w} 260 (3.22)	186 (anh)	...	...	...	s^n	δ	i	...	...	Py, con sulf s	**B31**, 221
Ω c482	**Coniferyl alcohol**	...	180.21	pr (eth-lig) λ^{chl} 270 (4.2), 295 sh (3.7)	74	$163–5^3$	...	...	δ^h i	s	v	...	...	alk s	**B6**^{2}, 1093
Ω c483	**Coniine(*d*)**	Conicine. 2-Propylpiperidine.	127.23	$[\alpha]_D^{20}$ +15.6	−2	$166–7^{760}$ 64^{18}	0.8440^{20}	$1.4512^{21.9}$	δ	∞	v	...	s	chl δ	**B20**^{2}, 61
c484	—(*dl*)	$C_8H_{17}N$. *See* c483	127.23	...	...	$166–7^{745}$ $59–63^{17}$	0.8447^{20}	1.4513^{23}	δ	∞	v	...	s	oos s	**B20**^{2}, 62
c485	—(*l*)	$C_8H_{17}N$. *See* c483	127.23	$[\alpha]_D^{15}$ −15.6	...	166^{760} 64^{18}	0.845_4^{15}	1.4512^{22}	δ	∞	v	...	s	oos s	**B20**^{2}, 62
c486	—,hydrobromide (*d*)	$C_8H_{17}N.HBr$. *See* c483	208.15	pr	211 (207)	...	...	...	v	v	s	...	...	oos v chl s	**B20**, 112
c487	—,hydrochloride (*d*)	$C_8H_{17}N.HCl$. *See* c483	163.69	orh (w) $[\alpha]_D^{20}$ +10.1, (liq NH_3, c = 8)	221	...	...	...	v	v	δ	δ	...	chl s	**B20**^{2}, 61
Ω c488	—,—(*dl*)	$C_8H_{17}N.HCl$. *See* c483	163.69	nd (al-eth)	216–7	...	...	...	v	s	δ	...	...	...	**B20**^{1}, 31
c489	—,—(*l*)	$C_8H_{17}N.HCl$. *See* c483	163.69	nd	220–1	...	...	...	v	v	i	...	...	...	**B20**, 118
c490	—,picrate(*d*)	$C_8H_{17}N.C_6H_3N_3O_7$. *See* c483	356.34	ye pr (w)	75	...	...	...	s^h	s	s	...	...	...	**B20**, 112
c491	—,—(*l*)	$C_8H_{17}N.C_6H_3N_3O_7$. *See* c483	356.34	ye pr (w)	74	...	...	...	s^h	s	s	...	...	...	**B20**^{2}, 62
c492	—,***N*-methyl-(*d*)**	$C_9H_{19}N$. *See* c483	141.26	$[\alpha]_D^{24}$ +82.4	...	$173–4^{757}$	0.8326_4^{23}	1.4538^{13}	δ	s	...	s	...	oos s	**B20**^{2}, 61
c493	**Conquinamine**	$C_{19}H_{24}N_2O_2$	312.42	ye tetr $[\alpha]_D^{15}$ +200 (116) (al, c = 0.5) λ^{al} 245 (4.0), 299 (3.4)	123	...	...	...	δ	s	s	...	...	chl s	**B27**^{2}, 667
—	**Conyrine**	*see* **Pyridine, 2-propyl-**													
c496	**Copaene**	$C_{15}H_{24}$	204.36	$[\alpha]_D$ −25.8	...	246–51 119–20^{10}	0.8996_4^{20}	1.4894^{20}	i	...	s	s	...	aa, lig s	**B5**^{2}, 349
c497	**Coproergostane**	Pseudoergostane. $C_{28}H_{50}$	386.71	nd (ace) $[\alpha]_D^{19}$ +25.3 (chl, c = 2)	64	...	...	...	i	...	s	s^h	...	chl s	**B5**^{3}, 1143
c498	**Coprostane**	Pseudocholestane. $C_{27}H_{48}$	372.69	orh nd (al, ace or eth-al) $[\alpha]_D^{20}$ +25.1 (chl, c = 2)	72	...	$0.9119_4^{87.7}$	1.4884_4^{88}	i	δ	v	...	...	chl, oos v	**B5**^{3}, 1132
c499	**3β-Coprostanol**	Coprosterol. Strecorin. $C_{27}H_{48}O$. *See* c498	388.69	nd (MeOH) $[\alpha]_D^{18}$ +28 (chl, c = 1.8)	102 (105)	...	...	...	i	v	v	...	v	chl v MeOH δ	**B6**^{3}, 2128
c500	**Coprostenol**	Allocholesterol. 4-Cholesten-3β-ol. $C_{27}H_{46}O$.	386.67	nd (eth-MeOH) $[\alpha]_D$ +43.7 (bz, c = 1) λ^{al} 207 (3.81)	132	...	...	...	i	s	v	v	v	chl, diox, Py v	**B6**^{3}, 2604
Ω c501	**Coprostenone**	Δ^4-Cholesten-3-one. $C_{27}H_{44}O$.	384.65	pl (MeOH) $[\alpha]_D$ +88.6 (chl) λ^{al} 242.5 (4.18)	81–2	...	...	...	...	δ	v	...	v	lig v AcOEt δ	**E14**, 120
—	**Coprosterol**	*see* **Coprostanol**													
Ω c502	**Coronene**	Hexabenzobenzene	300.36	ye nd (bz) λ^{al} 290 (5.04)	438–40 (cor)	525^{760}	1.377	...	i	...	...	...	δ^h	con sulf i	**J1952**, 2991
c503	**Corticosterone**	11,21-Dihydroprogesterone. $C_{21}H_{30}O_4$.	346.47	nd (+al, al), pl (ace) $[\alpha]_D^{15}$ +223 (al, c = 1.1) λ^{w} 248 (4.20)	180–2 (cor) (177–9)	sub $190^{0.01}$	...	...	i	s	s	...	...	...	**Am 78**, 1414

For explanations, symbols and abbreviations see beginning of table. For structural formulas see end of table.

No.	Name	Synonyms and Formula	Mol. wt.	Color, crystalline form, specific rotation and λ_{max} (log ε)	m.p. °C	b.p. °C	Density	n_D	Solubility w	al	eth	ace	bz	other solvents	Ref.
	Corticosterone														
c504	—,11-dehydro-	$C_{21}H_{28}O_4$	344.46	pr (ace-w, al, ace or ace-eth), $[\alpha]_D^{25}$ +258 (al), +239 (diox)	183–3.5	distb vac	...	...	i	s	...	s	s	...	**E14s**, 2972
Ω c505	—,17-hydroxy-	Cortisol. Hydrocortisone. $C_{21}H_{30}O_5$. *See* c503	362.47	pl (al or *i*-PrOH) $[\alpha]_D^{22}$ +167 (al) λ^{al} 242 (4.19)	220	...	...	...	δ^h	s^h	...	...	...	diox, aa s	**E14s**, 2868
—	**Cortisol**	*see* **Corticosterone, 17-hydroxy-**													
c506	**Cortisone**	17-hydroxy-11-dehydro-corticosterone. $C_{21}H_{28}O_5$. *See* c504	360.46	rhd nd (al), cr (ace) $[\alpha]_D^{25}$ +209 (95% al, c = 1.2) λ^{al} 238 (4.18)	230–1 (224d)	...	...	...	i	v	δ	v	δ	MeOH v sulf s chl δ	**Am 73**, 4055
c507	—,21-acetate	$C_{23}H_{30}O_6$. *See* c504	402.49	nd (ace), rods (chl) $[\alpha]_D^{25}$ +164 (ace, c = 0.5) λ^{al} 237 (4.18)	239–4.1	...	...	...	i	δ	δ	s	...	chl v peth δ	**J1954**, 125
c508	**Corybulbine(*d*)**	Corydalis-G. $C_{21}H_{25}NO_4$	355.44	nd (al) $[\alpha]_D^{20}$ +303 (chl, c = 1.4) λ^{MeOH} 286 (3.76)	237–8 (244 sealed tube)	...	...	...	i	δ	δ	s	s^h	chl, HCl s AcOEt δ	**B21**[2], 200
c509	—(*dl*)	$C_{21}H_{25}NO_4$. *See* c508	355.44	cr (chl-al)	220–2	...	...	...	i	s	δ	s	s^h	chl s	**B21**, 217
c510	**Corycavamine**	$C_{21}H_{21}NO_5$	367.41	pr (eth or al) $[\alpha]_D^{20}$ +166.6 (chl, c = 2.2)	149	...	...	...	...	s	...	...	...	chl s	**B27**[2], 621
c511	**Corycavine**	$C_{21}H_{21}NO_5$	367.41	orh pl (al) λ^{MeOH} 288 (3.99)	221–2	...	...	...	i	δ s^h	...	...	...	chl s alk i	**B27**[2], 621
c512	**Corydaldine**	$C_{11}H_{13}NO_3$	207.23	mcl pr (w or al)	175	...	...	...	s	s	s	...	s	chl s peth i	**B21**[2], 442
c513	**Corydaline(*d*)**	Corydalis-A. $C_{22}H_{27}NO_4$	369.47	pr (al) $[\alpha]_D^{20}$ +311 (al, c = 0.8) λ^{MeOH} 282 (3.76)	136	...	...	...	i	s^h	s	...	s	chl, CS_2 s	**B21**[2], 200
c514	—(*dl*)	$C_{22}H_{27}NO_4$. *See* c513	369.47	cr (al) λ^{MeOH} 282 (3.76)	135–6	...	...	...	δ	s^h	s	...	...	...	**B21**[2], 200
c515	—(*meso, d*)	$C_{22}H_{27}NO_4$. *See* c513	369.47	pr (eth) $[\alpha]_D$ +180 (chl, c = 3) λ^{MeOH} 282 (3.76)	155–6 (152)	...	...	...	δ	s^h	s	...	...	...	**B21**[1], 257
c516	—(*meso, dl*)	$C_{22}H_{27}NO_4$. *See* c513	369.47	cr (al) λ^{MeOH} 282 (3.76)	163–4 (158)	...	...	...	δ	s^h	s	...	...	...	**B21**[2], 200
c517	—(*meso, l*)	$C_{22}H_{27}NO_4$. *See* c513	369.47	pr (eth) $[\alpha]_D^{20}$ −181 (chl, c = 3) λ^{MeOH} 282 (3.76)	155–6 (152)	...	...	...	δ	s^h	s	...	...	...	**B21**[1], 257
c518	**Corynantheine(*β*)**	$C_{22}H_{28}N_2O_3$	384.48	$[\alpha]_D^{18}$ +28.8 (MeOH, c = 1.0) λ^{al} 225 (4.7), 272 (3.9), 282 (3.9), 291 (3.8)	165–6	...	...	...	...	s	...	...	...	...	**H35**, 851
—	**Corynine**	*see* **Yohimbine**													
c519	**Cotarnine**	$C_{12}H_{15}NO_4$	237.26	nd (bz), cr (eth) λ^{al} 245 (3.9), 279 (3.6), 326 (3.8)	132–3d	...	...	...	δ	s	s	...	s	chl, Na_2CO_3, NH_4OH s NaOH i	**B27**[2], 543
c520	—,chloride	Stypticin. $C_{12}H_{14}ClNO_3.2H_2O$. *See* c519	291.74	ye pw or nd (al-AcOEt) λ^{al} 255 (0.65) 338 (0.89)	197	...	...	...	s	s	...	...	...	...	**B27**[1], 456
c521	—,*O*(hydrogen phthalate)	$C_{12}H_{14}NO_3.O_2CC_6H_4CO_2H$. *See* c519	385.38	ye	115d	...	...	...	δ	...	...	...	...	...	**B27**, 476
c522	—,*O*-phthalate	Styptol. $(C_{12}H_{14}NO_3)_2.C_6H_4(CO_2)_2$. *See* c519	604.62	og cr or pw	103–5	...	...	...	s	...	...	...	...	...	**B27**, 476
—	**Cotoin**	*see* **Benzophenone, 2,6-dihydroxy-4-methoxy-**													
Ω c523	**Coumalic acid**	α-Pyrone-5-carboxylic acid	140.10	pr (MeOH) λ^{MeOH} 240 (3.9), 290 (3.6)	205–10δd 207–9 (+1w)	218^{120} partially sub	...	...	δ d^h	s	δ	δ	i	aa s AcOEt δ chl, lig i	**B18**[2], 326

For explanations, symbols and abbreviations see beginning of table. For structural formulas see end of table.

No.	Name	Synonyms and Formula	Mol. wt.	Color, crystalline form, specific rotation and λ_{max} (log ε)	m.p. °C	b.p. °C	Density	n_D	Solubility w	al	eth	ace	bz	other solvents	Ref.
	Coumaran														
c524	**Coumaran**	2,3-Dihydrobenzofuran. Dihydrocoumarone.	120.16	λ^{hx} 283 (3.49), 289 (3.49)	−21.5	188–9 76^{14}	1.0576_4^{24}	1.5426^{20}	i	s	s	...	...	chl, CS_2 s	**B17**[1], 22
c525	**3-Coumaranone**	...	134.14	red nd (al)	102–3	$152–4^{16}$	...	...	δ	s^h	...	...	s	alk s lig δ	**B17**[2], 126
—	**Coumaric acid**	see **Cinnamic acid, hydroxy-**													
—	***o*-Coumaric acid, lactone**	see **Coumarin**													
—	**Coumarilic acid**	see **Coumarone-2-carboxylic acid**													
Ω c526	**Coumarin**	1,2-Benzopyrone. *o*-Coumaric acid lactone. Coumarinic lactone.	146.15	rh pym (eth) λ^{al} 274.5 (4.20), 312.5 (3.91)	71	301.72^{760}	0.935_4^{20}	...	s^h	s	v	...	...	Py, chl v alk s	**B17**[2], 357
c527	**—,6-chloro-**	$C_9H_5ClO_2$. *See* c526	180.59	nd (al) λ^{al} 222 (4.38), 272 (4.02), 321 (3.65)	165	...	...	...	δ	v^h	v^h	...	s	CS_2 v	**B17**[2], 359
Ω c528	**—,7-diethyl-amino-4-methyl-**	$C_{14}H_{17}NO_2$. *See* c526	231.30	cr (al, bz-lig) λ^{al} 243 (4.19), 278 (2.28), 318 (3.60), 375 (4.41)	89 (al) 135 (bz-lig)	...	...	...	δ^h	s	s	s	...	oos s	**B18**, 612
Ω c529	**—,3,4-dihydro-**	Hydrocoumarin. Melilotic lactone. Melilotol. $C_9H_8O_2$. *See* c526	148.17	lf λ^{al} 266 (2.85), 273 (2.86), 312 (1.02)	25	272 145^{18}	1.169^{18}	1.5563^{20}	i	δ	δ	...	...	chl s	**B17**[2], 334
c530	**—,6,7-dihydroxy-**	Aesculetin. Esculetin. $C_9H_6O_4$. *See* c526	178.15	nd (w + 1), pr (aa), lf (sub) λ^{al} 261.5 (3.7), 303 (3.8), 354 (4.1)	276	sub	...	...	δ	s^h	δ	s	...	alk, chl, AcOEt s	**B18**[2], 68
c531	**—,7,8-dihydroxy-**	Daphnetin. $C_9H_6O_4$. *See* c526	178.15	yesh (dil al) λ^{al} 258 (3.8), 335 (4.1)	261–3 (258)	sub	...	...	s^h	s^h	δ	...	δ	aa s^h chl, CS_2 δ	**B18**[2], 69
Ω c532	**—,5,7-dihydroxy-4-methyl-**	$C_{10}H_8O_4$. *See* c526	192.18	nd (al), lf (aa) λ^{al} 250 (3.73), 320 (4.08)	282–4	...	...	...	δ	v^h	δ	...	δ	alk v chl δ	**B18**[1], 351
Ω c533	**—,6,7-dihydroxy-4-methyl-**	4-Methylaesculetin. $C_{10}H_8O_4$. *See* c526	192.18	ye nd (dil al) λ^{al} 290 (3.73), 348 (4.09)	274–6	...	...	...	s^h	s^h	...	...	...	aa s^h	**B18**[1], 351
Ω c534	**—,7,8-dihydroxy-4-methyl-**	4-Methyldaphnetin. $C_{10}H_8O_4$. *See* c526	192.18	nd (w), pr (al)	242–5 (235)	...	...	...	i	s	...	...	...	...	**B18**[1], 352
c534[1]	**—,7,8-dihydroxy-6-methoxy-**	Fraxetin. $C_{10}H_8O_5$. *See* c526	208.18	pl (dil al) λ^{al} 273 (3.5), 348 (4.0)	230–2	...	...	...	δ^h	s	δ	...	...	...	**B18**[2], 152
Ω c535	**—,5,7-dimethoxy-**	Citropten. Limettin. $C_{11}H_{10}O_4$. *See* c526	206.20	pr or nd (al) λ^{al} 250 (3.8), 260 (3.8), 330 (4.2)	148–50	200sd	...	...	δ^h	v^h	i	v	...	chl v aa s lig i	**B18**[1], 348
Ω c536	**—,7-ethoxy-4-methyl-**	Maraniol. $C_{12}H_{12}O_3$. *See* c526	204.23	wh	114	...	...	...	i	s	...	...	...	...	**C28**, 578
Ω c537	**—,4-hydroxy-**	Benzotetronic acid. $C_9H_6O_3$. *See* c526	162.15	nd (w)	213–4 (232)	...	...	...	s^h	s	s	...	...	...	**B17**[2], 469
c538	**—,6-hydroxy-**	$C_9H_6O_3$. *See* c526	162.15	nd (dil HCl)	250	...	1.25	...	δ	s	...	...	...	AcOEt s	**B18**[1], 306
Ω c539	**—,7-hydroxy-**	Umbelliferone. $C_9H_6O_3$. *See* c526	162.15	nd (w) λ^{MeOH} 217 (4.1), 240 sh (3.5), 325 (4.2)	230–1	sub	...	...	δ^h	∞	δ	...	...	chl, aa ∞	**B18**[2], 16
c540	**—,8-hydroxy-**	$C_9H_6O_3$. *See* c526	162.15	nd (dil al) λ^{al} 255 (3.9), 295 (4.0)	160	...	...	...	s^h	s	i	...	i	aa s chl i	**C34**, 3252
c541	**—,7-hydroxy-6-methoxy-**	Chrysatropic acid. Gelseminic acid. Scopoletin. $C_{10}H_8O_4$. *See* c526	192.18	nd or pr (al) λ^{al} 255 (3.8), 299 (3.8), 350 (4.1)	204	...	...	...	δ	δ s^h	δ	...	i	chl s aa s CS_2 i	**B18**[2], 68
Ω c542	**—,7-hydroxy-4-methyl-**	4-Methylumbelliferone. $C_{10}H_8O_3$. *See* c526	176.18	nd (al) λ^{al} 325 (4.20)	185–7	...	...	...	δ^h	s	δ	...	...	aa, alk s chl δ	**B18**[2], 19
c543	**—,4-hydroxy-3(1-phenyl-3-oxo-butyl)-**	Coumadin. Warfarin. $C_{19}H_{16}O_4$.	308.34	cr (al) λ^{al} 271 (3.04), 287 (3.05), 306 (3.04)	161	...	...	...	i	s	...	s	...	diox s	**Am 66**, 900
c543[1]	**—,7-methoxy-**	Herniarin. $C_{10}H_8O_3$. *See* c526	176.18	lf (w or MeOH) λ^{al} 290 (3.85), 323 (4.17)	117–8	...	...	...	δ	s	s	...	...	con sulf, alk s	**B18**[2], 17
Ω c544	**—,3-methyl-**	$C_{10}H_8O_2$. *See* c526	160.18	rh bipym (al) λ^{al} 275 (4.02), 308 (3.80)	91	292.5	...	...	δ^h	s	...	...	...	aq KOH i	**B17**[2], 362
Ω c545	**—,4-methyl-**	$C_{10}H_8O_2$. *See* c526	160.18	nd (w), pr (bz) λ^{al} 270 (4.00), 310 (3.78)	83–4	...	...	...		s	...	...	s	alk s^h	**B17**[2], 362

For explanations, symbols and abbreviations see beginning of table. For structural formulas see end of table.

No.	Name	Synonyms and Formula	Mol. wt.	Color, crystalline form, specific rotation and λ_{max} (log ε)	m.p. °C	b.p. °C	Density	n_D	Solubility w	al	eth	ace	bz	other solvents	Ref.
	Coumarin														
c546	**—,6,7,8-tri-methoxy-**	Fraxetin dimethyl ether. $C_{12}H_{12}O_5$. *See* c526	236.23	rh bipym pl (dil al) λ^{al} 228 (4.2), 296 (4.0), 338 (3.8)	103–4	90–100$^{0.2}$				s	s				**B18**², 152
c547	**3-Coumarin-carboxylic acid**		190.16	nd (w or bz)	190d				s^h	s	i		i	alk s lig i	**B18**², 336
Ω c548	**2-Coumarone-carboxylic acid**	Coumarilic acid	162.15	nd (w) λ^{al} 266 (4.24), 292.5 (3.76)	192–3	310–5δd			s^h	s				CS_2, chl δ	**B18**², 276
c549	—,ethyl ester	$C_{11}H_{10}O_3$. *See* c548	190.20		30–1	274^{720} 161^{15}	1.1656$^{28.5}_4$	1.564$^{27.6}$							**B18**¹, 442
Ω c550	**—,3-methyl-**	$C_{10}H_8O_3$. *See* c548	176.18	nd (dil al) λ^{al} 270 (4.25), 294 (3.86)	188–9					s^h					**B18**, 309
—	**Coumaryl alcohol**	*see* **2-Propen-1-ol, 3-hydroxyphenyl-**													
c551	**Coumestrol**	$C_{15}H_8O_5$	268.24	gy micr rods λ^{MeOH} 247, 342	385d				i^h	δ	i	δ^h			**Am 80,** 3969
c552	—,diacetate	$C_{19}H_{12}O_7$. *See* c551	352.30	pl λ^{MeOH} 237 (4.4), 293 (4.1), 326 (4.4), 342 (4.3)	234				i^h	i	i	δ^h			**Am 80,** 4381
Ω c553	**Creatine**	(α-Methylguanido)acetic acid. $H_2NC(:NH)N(CH_3)CH_2CO_2H$	131.14	mcl pr (w + 1) $\lambda^{w, pH=8}$ 215 (3.1)	303		1.33		s	δ	i				**B4**², 796
Ω c554	**Creatinine**	1-Methylglycocyamidine	113.12	rh pr (w + 2), lf (w) $\lambda^{w, pH=7}$ 235 (3.90)	*ca.* 300d (260)				s^h	δ	i	i		chl i	**B24**², 128
—	**Cresol**	*see* **Toluene, hydroxy-**													
—	**Cresolol**	*see* **Toluene, 4-hydroxy-3-methoxy-**													
—	**Cresorcinol**	*see* **Toluene, 2,4-dihydroxy-**													
—	**Cresotic acid**	*see* **Benzoic acid, 2-hydroxy-3-methyl-**													
—	**Croceic acid**	*see* **1-Naphthalenesulfonic acid, 7-hydroxy-**													
Ω c555	**Crocetin**(*trans*)	Gardenin. $C_{20}H_{24}O_4$	328.41	brick red rh (Ac_2O) λ^{al} 256 (2.92), 415 (3.5), 432 (3.7), 460 (3.65)	285–7 (cor)				δ	δ	i		i	oos i NaOH v Py s	**B30**, 106
—	**Crocic acid**	*see* **Croconic acid**													
c556	**Croconic acid**	Crocic acid	142.07	pa ye nd (+3w, al-diox) λ^{MeOH} 298 (4.2), 360 (3.9)	eff 100 sub d > 150, d *ca.* 120 (+3w)				v	s					**B8**², 532
—	**Crotonaldehyde**	*see* **2-Butenal***													
—	**Crotonic acid**	*see* **2-Butenoic acid**(*trans*)*													
—	**Crotyl alcohol**	*see* **2-Buten-1-ol***													
—	**Cryogenine**	*see* **Benzoic acid, 3(1-semi-carbazido)-, amide**													
c557	**Cryptopine**	Cryptocavine. $C_{21}H_{23}NO_5$	369.42	pr or pl (bz), nd (chl-MeOH) $\lambda^{0.02N\ HCl}$ 234 (4.0), 285 (3.8)	223 (cor)		1.315$^{20}_4$		i	δ	δ		δ	chl, aa s	**B27**², 578
c558	**Cryptoxanthin**	Caricaxanthin. β-Caroten-3-ol. Cryptoxanthol. $C_{40}H_{56}O$.	552.90	garnet red pr (bz-MeOH) α: λ^{hx} 268 (4.41), 333 (3.86), 421 (4.99), 446 (5.16), 475 (5.12), β: λ^{hx} 274 (4.33), 346 (3.84), 452 (5.14), 479 (5.07)	169				i	δ			v	chl, Py v lig δ	**B30**, 93
c559	**Cubebin**	$C_{20}H_{20}O_6$	356.38	nd (al or bz) $[\alpha]^{25}_D$ −45.6 (chl, c = 5)	131–2				i	s	s			chl s con sulf s (red) peth δ	**B19**², 470

For explanations, symbols and abbreviations see beginning of table. For structural formulas see end of table.

No.	Name	Synonyms and Formula	Mol. wt.	Color, crystalline form, specific rotation and λ_{max} (log ε)	m.p. °C	b.p. °C	Density	n_D	Solubility w	al	eth	ace	bz	other solvents	Ref.
	Cumaldehyde														
—	**Cumaldehyde**	*see* **Benzaldehyde, 4-isopropyl-**													
—	**Cumene**	*see* **Benzene, isopropyl-***													
—	**Cumenol**	*see* **Benzene, hydroxy-(isopropyl)-***													
—	**Cumic acid**	*see* **Benzoic acid, 4-isopropyl-**													
—	**Cumic alcohol**	*see* **Benzene, 1(hydroxy-methyl)-4-isopropyl-***													
—	***m*-Cumidic acid**	*see* **1,3-Benzenedicarboxylic acid, 4,6-dimethyl-***													
—	**Cumidine**	*see* **Benzene, 1-amino-4-isopropyl-***													
—	**Cuminic acid**	*see* **Benzoic acid, isopropyl-**													
—	**Cuminylamine**	*see* **Benzene, 1(amino-methyl)-4-isopropyl-***													
c563	**Cupreine**	Hydroxycinchonine. $C_{19}H_{22}N_2O_2$.	310.40	pr (eth) $[\alpha]_D^{17}$ −175.5 (al)	198 (anh)	...	...	...	i	s	δ	...	δ	NaOH s chl, peth δ	**B23**, 510
c564	c-**Curarine-III, hydroxide**	c-Fluorocuranine hydroxide. $C_{20}H_{24}N_2O_2$	328.46	cr (MeOH-eth)	212	...	...	...	s	s	i	...	...	...	**H36**, 102
c565	*ar*-**Curcumene**	$C_{15}H_{22}$	202.34	$[\alpha]_D^{18}$ +35.8 λ^{bz} 212 (3.8), 261 (2.6), 267 (2.7), 274 (2.7)	...	140^{19} $117–8^{7}$	0.8821_{20}^{20}	1.4989^{20}	i	...	...	...	s	...	**B5**[2], 402
Ω c566	**Curcumin**	$C_{21}H_{20}O_6$	368.39	or ye pr, rh pr (MeOH) λ^{diox} 265 (4.18), 420 (4.77)	183	...	...	...	i	s	δ	δ	δ	aa, alk s CS_2 δ lig i	**B8**, 554
—	**Curine**	*see* **Bebeerine(*l*)**													
c567	**Cuscohygrine**	α,α′-Bis(*N*-methyl-α-pyrrolidyl)acetone. Bellaradine. Cuskhygrine. $C_{13}H_{24}N_2O$.	224.35	...	...	185^{32} 152^{14}	0.9782_4^{16}	...	∞	s	s	...	s	...	**B24**, 78
c568	—,**hydrate**	$C_{13}H_{24}N_2O.3\frac{1}{2}H_2O$. *See* c567	287.40	nd (w)	40–1	...	...	...	...	...	s	...	s	...	**B24**, 78
c569	**Cusparine**	$C_{19}H_{17}NO_3$	307.35	(i) wh nd (peth) (ii) ye nd (iii) pr	(i) 92 (ii) 92 (iii) 110–22	...	...	...	i	s	s	v	v	chl v lig δ	**B27**, 483
c570	**Cyamelide**	*sym*-Trioxanetriimine	129.08	amor pw	...	d	1.127_4^{15}	...	i	i	i	i	i	con sulf s d^h NH_4OH δ oos i	**B3**, 35
—	**Cyanacetamide**	*see* **Malonic acid,** mono-amide mononitrile													
c571	**Cyanamide**	Carbamic acid nitrile. Carbamonitrile. Carbodiimide. H_2NCN	42.04	nd	42 (46)	140^{19}	1.0729_4^{18} 1.282_4^{20} (so)	1.4418^{48}	v	v	s	s	s	chl s CS_2 δ	**B3**, 74
c572	—,**benzyl-**	$C_6H_5CH_2NHCN$	132.18	pl (eth)	43	...	...	...	i	s	s	...	...	...	**B12**, 1051
Ω c573	—,**diallyl-**	$(CH_2{:}CHCH_2)_2NCN$	122.17	...	...	$140–5^{90}$ 95^{9}	...	...	i	s	s	s	s	oos s	**B4**[2], 666
c574′	—,**dibutyl-**	$(C_4H_9)_2NCN$	154.26	...	...	$187–91^{190}$ $146–51^{35}$	...	...	i	s	s	s	s	oos s	**B4**[2], 635
Ω c575	—,**diethyl-**	*N*-Cyanodiethylamine. $(C_2H_5)_2NCN$	98.15	...	...	188^{760} 68^{10}	0.854_4^{20}	1.4126^{25}	i	s	s	...	...	HCl d	**B4**, 121
Ω c576	—,**dimethyl-**	*N*-Cyanodimethylamine. $(CH_3)_2NCN$	70.10	...	...	163.5^{760} 56^{15}	...	1.4089^{19}	v	s	s	s	...	...	**B4**[2], 574
c577	—,**diphenyl-**	*N*-cyanodiphenylamine. $(C_6H_5)_2NCN$	194.24	pr (al)	73–4	$235–40^{60}$	...	...	i	s^h	...	...	...	lig s MeOH s^h	**B4**[3], 145
c578	—,**methyl)-(1-naphthyl)-**	*N*-methyl-*N*-cyano-α-naphthylamine. $C_{10}H_7N(CH_3)CN$	182.23	yesh	...	$185–7^{3}$	...	...	i	s	s	...	...	...	**B12**[2], 697
c579	—,**phenyl-**	Carbanilonitrile. Cyananilide. *N*-Cyanoaniline. $C_6H_5NHCN.\frac{1}{2}H_2O$	127.15	cr (w, eth), lf (aa, dil KOH) λ^{al} 252 (4.5), 276 (3.5), 303 (2.9), 330 sh (2.8)	47 (hyd)	...	...	...	δ	v	v	...	...	KOH s	**B12**, 368
c580	**Cyanic acid**	Carbonimid. Isocyanic acid. HOCN	43.03	gas	−81– −79	23.5^{760}	1.140_4^{20} (liq)	...	s	...	s	...	s	chl, aa s	**B3**, 31
c581	—,ethyl ester	C_2H_5OCN	71.08	...	...	162d 30^{12}	0.89_4^{20}	1.3788^{25}	i	∞	∞	...	...	...	**Tet 1964**, 2829

For explanations, symbols and abbreviations see beginning of table. For structural formulas see end of table.

No.	Name	Synonyms and Formula	Mol. wt.	Color, crystalline form, specific rotation and λ_{max} (log ε)	m.p. °C	b.p. °C	Density	n_D	Solubility w	al	eth	ace	bz	other solvents	Ref.
	Cyanoacetanilide														
—	**Cyanoacetanilide**	*see* **Malonic acid,** monoamide mononitrile, *N*-phenyl-													
—	**Cyanoacetic acid**	*see* **Malonic acid,** mononitrile													
c581[1]	**Cyanogen**	Oxalic acid dinitrile. NCCN	52.04	gas λ^{gas} 219	−27.9	-21.17^{760}	0.9537_4^{-21}	....	s	s	s	...	...	....	**B2**^{2}, 511
c582	**Cyanogen bromide**	Bromine cyanide. BrCN	105.93	nd	52	61.4^{760}	2.015_4^{20}	....	s	s	s	...	...	....	**B3,** 39
c583	**Cyanogen chloride**	Chlorine cyanide. ClCN	61.47	gas	−6	12.66^{760}	1.186_4^{20}	....	s	s	v	...	...	....	**B3,** 38
c584	**Cyanogen iodide**	Iodine cyanide. ICN	152.92	nd (al, or eth)	146–7	sub >45	2.84^{18}	....	δ	s	s	...	...	....	**B3,** 41
c585	**Cyanogen sulfide**	Cyanogen thiocyanate. $S(CN)_2$	84.10	rh pl	65	sub 30–40	....	....	s	s	s	...	...	....	**B3,** 180
Ω c586	**Cyanuric acid,** dihydrate	2,4,6-Triazinetriol*. Tricyanic acid. Trihydroxycyanidine.	165.11	mcl (w + 2) $\lambda^{w, pH=8}$ 214 (4.01)	>360d	d	1.768^0 2.500_4^{20} (anh)	....	s	δ s^h	i	i	i	con sulf, Py s chl, aa i	**B26**^{2}, 131
—	—,triamide	*see* **Melamine**													
c587	—,tribenzyl ester	Benzyl cyanurate. $C_{24}H_{21}N_3O_3$. *See* c586	399.46	nd (al)	159	>320	....	....	i	s	δ	...	...	....	**B26**^{1}, 76
Ω c588	—,trichloride	Cyanuric chloride. Trichloro-*s*-triazine*. Trichlorocyanidine. Tricyanogen chloride.	184.41	cr (eth or bz) λ^{al} 241 (3.41)	154	190^{720}	....	....	i	s	...	...	...	....	**B26**^{2}, 16
Ω c589	**—,thio-**	Thiocyanuric acid. Trithiocyanuric acid.	177.27	ye pr	200d	....	....	....	δ	i	δ	...	...	....	**B26,** 259
—	**Cyanurotriamide**	*see* **Melamine**													
—	**Cyclamen aldehyde**	*see* **Propanal, 3(4-isopropylphenyl)-2-methyl-***													
—	**Cyclobarbital**	*see* **Barbituric acid, 5(1-cyclohexenyl)-5-ethyl-**													
Ω c590	**Cyclobutane***	Tetramethylene	56.12	....	−50	12^{760}	0.720_4^5	1.4260^{20}	i	v	∞	v	s	peth s	**B5,** 17
c591	—,ethyl-*	C_6H_{12}. *See* c590	84.16	....	−142.9	70.7^{760}	0.7284_4^{20}	1.4020^{20}	i	∞	∞	s	s	peth s	**B5**^{1}, 11
c592	—,methyl-*	C_5H_{10}. *See* c590	70.14	....	....	36.3^{760}	0.6884_4^{20}	1.3866^{20}	i	∞	∞	s	s	peth s	**B5**^{1}, 5
Ω c593	—,octafluoro-*	Perfluorocyclobutane. C_4F_8. *See* c590	200.03	....	−38.7	-4^{764}	1.724_4^0	....	...	...	s	...	...	....	**Am 70,** 4090
Ω c594	**Cyclobutanecarboxylic acid***	Tetramethylenecarboxylic acid.	100.13	....	−2	190^{754} $74–5^2$	1.0599_4^{20}	1.4400^{20}	δ	∞	∞	...	...	....	**B9**^{2}, 5
Ω c595	**1,1-Cyclobutanedicarboxylic acid***	....	144.14	pr (eth or w)	156.6(cor) (160)	....	....	....	v	s	s	...	s	lig δ	**B9**^{2}, 514
Ω c596	—,diethyl ester*	$C_{10}H_{16}O_4$. *See* c595	200.24	....	....	229^{755} (224) 104^{12}	1.0456_4^{20}	1.4344^{20}	...	v	...	...	...	....	**B9**^{2}, 514
Ω c597	**1,2-Cyclobutanedicarboxylic acid** (*cis, dl*)*	Ethylenesuccinic acid	144.14	pl (w), pr (bz)	138	....	....	....	v	v	v	...	δ	lig δ	**B9**^{2}, 515
c598	—(*trans, d*)*	$C_6H_8O_4$. *See* c597	144.14	$[\alpha]_D^{30}$ +123.3 (w, c = 1.2)	105	....	....	....	s	s	...	...	...	....	**B9**^{2}, 515
Ω c599	—(*trans, dl*)*	$C_6H_8O_4$. *See* c597	144.14	rh nd (bz)	131	....	....	....	s	s	...	...	...	con HCl δ	**B9**^{2}, 515
Ω c600	—(*trans, l*)*	$C_6H_8O_4$. *See* c597	144.14	nd (HCl) $[\alpha]_D^{30}$ −124.3 (w, c = 0.8)	105	....	....	....	s	s	...	...	...	....	**B9**^{2}, 515
c601	**1,3-Cyclobutanedicarboxylic acid** (*cis*)*	....	144.14	pr (w)	143.0–3.5	252	....	....	v	v	δ	...	...	....	**B9,** 726
c602	—(*trans*)*	$C_6H_8O_4$. *See* c601	144.14	pr (w), nd (sub)	171	sub	....	....	v^h	s	δ	...	...	....	**B9,** 726
c603	**Cyclobutene***	Cyclobutylene	54.09	λ^{gas} 175 (4.0)	....	2	0.733_4^0	....	...	...	...	v	s	peth s	**B5**^{1}, 29
c604	**—,hexafluoro-***	Perfluorocyclobutene. C_4F_6. *See* c603	162.04	....	−60	3^{760}	1.602_4^{-20}	1.298^{-20}	...	...	...	...	...	....	**B5**^{3}, 170
c605	**Cyclocamphane**	Epicyclene. Isocyclene. β-Pericyclocamphane.	136.24	....	117–8	$150–1^{743}$	0.7948^{121}	....	...	s	...	...	...	aa s	**B5**^{3}, 393
c606	**1,6-Cyclodecanediol** (*trans*)*	....	172.27	cr (AcOEt, chl)	151–3	....	....	....	...	...	s	...	...	AcOEt, chl s^h	**B6**^{3}, 4105
c607	**1,6-Cyclodecanedione***	....	168.24	cr (eth) λ^{al} 287 (1.54)	100	....	....	....	i	...	s^h	s	...	aa s	**B7**^{2}, 540
Ω c608	**Cyclodecanol***	....	156.27	....	40–1	125^{12}	0.9606_4^{20}	1.4926^{20}	...	s	...	...	...	....	**B6**^{3}, 120
c609	**Cyclodecanone***	....	154.26	amor pw λ^{al} 283 (1.18)	28	106.7^{13}	0.9654_4^{20}	1.4806^{20}	...	...	s	...	s	chl s	**B7**^{2}, 36
c610	**9-Cycloheptadecen-1-one***	Civetone	250.43	....	32.5	342^{742} 159^2	0.9170_4^{33}	1.4830^{33}	δ	s	...	...	s	....	**B7**^{2}, 121
Ω c611	**Cycloheptane***	Heptamethylene. Suberane.	98.19	....	−12	118.48^{760}	0.8098_4^{20}	1.4436^{20}	i	v	v	...	s	lig, chl s	**B5**^{2}, 15
Ω c612	—,1-aza-*	Hexamethylenimine. Homopiperidine.	99.18	....	....	138.0–8.2^{749}	0.8643_4^{22}	1.4631^{20}	s	v	v	...	...	....	**B20,** 94
Ω c613	—,bromo-*	Suberyl bromide. $C_7H_{13}Br$. *See* c611	177.09	....	....	101.5^{40} 75^{12}	1.2887_4^{22}	1.4996^{20}	i	...	v	...	...	chl v	**B5**^{2}, 15
c614	—,methyl-*	C_8H_{16}. *See* c611	112.22	....	....	134^{760}	0.8001_4^{20}	1.4401^{20}	i	v	v	...	s	peth s	**B5**^{2}, 20

For explanations, symbols and abbreviations see beginning of table. For structural formulas see end of table.

No.	Name	Synonyms and Formula	Mol. wt.	Color, crystalline form, specific rotation and λ_{max} (log ε)	m.p. °C	b.p. °C	Density	n_D	Solubility w	al	eth	ace	bz	other solvents	Ref.
	1,2-Cycloheptanedione														
c615	**1,2-Cyclo-heptanedione***	$C_7H_{10}O_2$. *See* c611	126.16	ye λ^{hx} 280 (1.3), 375 (0.8), 405 (0.7), 420 (0.6)	−40	$107–9^{17}$	1.0607^{22}_{22}	1.4689^{22}	...	s	...	...	...	...	**C50,** 6327
Ω c616	**Cycloheptanol***	Suberol. Suberyl alcohol. $C_7H_{14}O$. *See* c611	114.19	...	2	185^{760} 95^{24}	0.9554^{20}	1.4705^{20}	δ	v	v	...	...	...	**B6²,** 16
Ω c617	**Cycloheptanone***	Suberone. $C_7H_{12}O$. *See* c611	112.17	λ^{al} 285 (1.30)	...	$178.5–9.5^{760}$ 71^{19}	0.9508^{20}_4	1.4608^{20}	i	v	s	...	...	...	**B7²,** 14
c618	**Cycloheptasil-oxane, tetradeca-methyl-***	...	519.09	...	−26	154^{20}	0.9703^{20}_4	1.4040^{20}	...	...	...	...	...	...	**Am 68,** 667
c619	**1,3,5-Cyclo-heptatriene***	Tropilidene	92.15	cubic (at −80) λ^{al} 261 (3.54)	−79.49	117^{760} 60.5^{122}	0.8875^{19}_4	1.5343^{20}	i	s	s	...	v	chl v	**B5,** 280
c620	**2,4,6-Cyclohepta-trien-1-one***	Tropone	106.13	λ^{MeOH} 225 (4.30), 232 (4.29), 303 (3.87)	−7−−5	113^{15} $84–5^6$	1.095^{22}_4	1.6172^{22}	∞	...	δ	...	δ	hx s	**Am 73,** 876
c621	**—,2-amino-***	C_7H_7NO. *See* c620	121.14	ye pl (bz) λ^{al} 242 (4.4), 337 (4.1), 397 (4.0)	106–7	...	...	...	s^h	s	...	...	s	chl s	**C46,** 7559
c622	**—,3-bromo-2-hydroxy-***	$C_7H_5BrO_2$. *See* c620	201.03	ye pl or nd λ^{chl} 258 (4.58), 329 (3.87), 364 (3.84), 381.5 (3.86)	107–8	...	...	...	...	s	s	...	...	...	**C49,** 2405
c623	**—,2-hydroxy-***	Tropolone. $C_7H_6O_2$. *See* c620	122.13	nd λ^{al} 228 (4.36), 320 (3.83), 351 (3.76)	51–2	sub 40^4	...	...	s	...	s	s	...	oos s	**Am 74,** 4456
Ω c624	**—,2-hydroxy-4-isopropyl-***	$C_{10}H_{12}O_2$. *See* c620	164.21	pa ye (peth) λ^{al} 344 (4.09), 393 (3.09)	50–1	...	...	...	δ	...	...	...	δ	lig δ	**C45,** 2884
c625	**—,2-hydroxy-4-methyl-***	$C_8H_8O_2$. *See* c620	136.16	nd (peth) λ^{cy} 235 (4.4), 320 (3.7), 352 (3.7) 372 (3.5)	75–6	...	...	...	...	...	s	...	...	chl s	**C46,** 5035
c626	**—,2-methoxy-***	$C_8H_8O_2$. *See* c620	136.16	pa ye nd (+½w) λ^{al} 235 (4.46), 319 (3.88), 350 (3.81), 363 sh (3.64)	41 (+½w)	128^5	...	...	...	s	...	...	s	...	**C46,** 4521
Ω c627	**Cycloheptene***	Suberene. Suberylene	96.17	...	−56	115^{760}	0.8228^{20}_4	1.4552^{20}	i	s	s	...	s	peth, chl s	**B5²,** 42
c628	**1,3-Cyclo-hexadiene***	1,2-Dihydrobenzene	80.14	λ^{al} 259 (4.00)	−89	80.5^{760}	0.8405^{20}_4	1.47548^{20}	i	s	v	...	s	peth, chl s	**B5²,** 79
c629	**1,4-Cyclo-hexadiene***	1,4-Dihydrobenzene	80.14	λ^{hx} 224 (1.5), 270 (−0.5)	−49.2	85.6^{760}	0.8471^{20}_4	1.4725^{20}	i	∞	∞	...	s	peth, chl s	**B5²,** 80
c630	**1,3-Cyclo-hexadiene, 5-methyl-(*dl*)***	1,2-Dihydrotoluene. C_7H_{10}. *See* c628	94.16	λ^{al} 259 (3.69)	...	101.5^{762}	0.8354^{20}_4	1.4763^{20}	i	v	s	...	s	lig s	**B5¹,** 62
c631	**—,octafluoro-***	C_6F_8. *See* c628	224.05	...	...	63–4	1.601^{20}	1.3149^{20}	...	...	...	...	...	...	**J1954,** 3780
c632	**1,4-Cyclo-hexadiene, octafluoro-***	C_6F_8. *See* c629	224.05	...	...	57–8	...	1.318^{18}	...	...	...	...	...	...	**J1954,** 3780
c633	**1,4-Cyclohexa-diene-1,2-dicar-boxylic acid***	3,6-Dihydrophthalic acid	168.16	mcl pr (w)	153	...	...	...	δ s^h	s	...	...	...	...	**B9,** 781
c634	**2,4-Cyclohexa-diene-1,2-dicar-boxylic acid***	2,3-Dihydrophthalic acid	168.16	pr (w or al)	179–80	...	...	...	δ^h	s	δ	...	...	...	**B9,** 781
c635	**2,6-Cyclohexa-diene-1,2-dicar-boxylic acid***	4,5-Dihydrophthalic acid	168.16	tcl (w)	215	...	...	...	δ s^h	s	...	s	...	...	**B9,** 782
c636	**2,5-Cyclo-hexadien-1-one, hexachloro-***	Hexachlorophenol	300.78	λ^{cy} 262 (4.20), 268 (4.20)	107	...	...	...	i	...	...	...	s	CCl_4 s	**B7¹,** 96
Ω c637	**Cyclohexane***	Hexahydrobenzene. Hexamethylene.	84.16	...	6.55	80.74^{760}	0.77855^{20}_4	1.42662^{20}	i	∞	∞	∞	∞	lig, CCl_4 ∞	**B5,** 21
c638	**—,acetyl-**	Cyclohexyl methyl ketone. Hexahydroacetophenone. $C_8H_{14}O$. *See* c637	126.20	...	...	$180–1^{760}$ 69^{12}	$0.9176^{20.7}_4$	1.45652^{16}	i	...	s	...	...	...	**B7²,** 23
Ω c639	**—,allyl-**	3-Cyclohexylpropene.* C_9H_{16}. *See* c637	124.23	...	...	151.5^{757}	0.8135^{20}	1.4500^{20}	...	v	s	s	s	chl, lig s	**B5²,** 49
Ω c640	**—,amino-***	Cyclohexylamine. Hexa-hydroaniline. $C_6H_{13}N$. *See* c637	99.18	...	−17.7	134.5^{760} 30.5^{15}	0.8191^{20}_4	1.43716^{20}	s	v	∞	∞	∞	oos ∞	**B12²,** 4

For explanations, symbols and abbreviations see beginning of table. For structural formulas see end of table.

No.	Name	Synonyms and Formula	Mol. wt.	Color, crystalline form, specific rotation and λ_{max} (log ε)	m.p. °C	b.p. °C	Density	n_D	Solubility w	al	eth	ace	bz	other solvents	Ref.
	Cyclohexane														
c641	—,—,hydrochloride*	$C_6H_{13}N$. HCl. *See* c637	135.64	nd (w or al-eth)	206–7	...	...	...	v	v	s^h	...	i	...	**B12**[2], 4
Ω c642	—,**bromo-***	Cyclohexyl bromide. $C_6H_{11}Br$. *See* c637	163.06	...	−56.5	166.2^{760} 45.5^{10}	1.3359^{20}_{4}	1.4957^{20}	i	∞	∞	∞	∞	lig, CCl_4 ∞	**B5**, 24
c643	—,**1-bromo-1-methyl-***	$C_7H_{13}Br$. *See* c637	177.09	...	...	156–60 $65–9^{16}$	1.2510^{20}	1.4866^{20}	i	s	...	...	...	chl s	**B5**[1], 12
Ω c644	—,**1-bromo-3-methyl-(*dl*)***	$C_7H_{13}Br$. *See* c637	177.09	...	...	181.0–1.2 δd 60^{11}	1.275^{25}_{4}	1.4979^{20}	i	...	v	...	s	...	**B5**[2], 18
Ω c645	—,**1-bromo-4-methyl-***	$C_7H_{13}Br$. *See* c637	177.09	...	...	130^{200} 55^{15}	...	...	i	...	v	...	s	...	**B5**[2], 18
c646	—,**(bromomethyl)-***	$C_7H_{13}Br$. *See* c637	177.09	...	...	$76–7^{26}$	1.2763^{25}_{4}	1.4907^{20}	i	...	s	...	s	chl s	**B5**[2], 18
Ω c647	—,**butyl-***	1-Cyclohexylbutane.* $C_{10}H_{20}$. *See* c637	140.27	...	−74.7	180.95^{760} 59^{10}	0.7992^{20}_{4}	1.4408^{20}	i	...	...	...	...	...	**B5**[2], 25
Ω c648	—,*sec*-**butyl-**	2-Cyclohexylbutane.* $C_{10}H_{20}$. *See* c637	140.27	...	...	179.34^{760}	0.81314^{20}_{4}	1.44673^{20}	i	...	...	s	...	...	**B5**[2], 26
Ω c649	—,*tert*-**butyl-**	$C_{10}H_{20}$. *See* c637	140.27	...	−41.17	171.5^{760}	0.81267^{20}_{4}	1.44694^{20}	i	...	...	...	...	...	**B5**[2], 26
Ω c650	—,**(butylamino)-***	*N*-Butylcyclohexylamine. $C_{10}H_{21}N$. *See* c637	155.29	...	...	207	...	...	δ	v	v	...	...	...	**B12**[2], 7
Ω c651	—,**chloro-***	Cyclohexyl chloride. $C_6H_{11}Cl$. *See* c637	118.61	...	−43.9	143^{760}	1.000^{20}_{4}	1.4626^{20}	i	∞	∞	∞	∞	chl v	**B5**, 21
Ω c652	—,**cyclopentyl-***	...	152.28	...	...	215.1^{760}	0.8758^{20}_{4}	1.4725^{20}	...	...	...	...	...	...	**B5**[3], 268
Ω c653	—,**1,2-dibromo-(*cis*)***	$C_6H_{10}Br_2$. *See* c637	241.97	...	9.7	115^{14}	1.803^{25}_{25}	1.5514^{25}	...	...	s	s	s	chl, CCl_4, lig s	**B5**[2], 12
Ω c654	—,—(*trans, dl*)*	$C_6H_{10}Br_2$. *See* c637	241.97	...	−4(−2)	$145–6^{100}$ 105^{20}	1.7759^{20}_{4}	1.5445^{19}	i	s	s	s	s	oos s	**B5**[2], 12
c655	—,**1,3-dibromo-(*cis*)***	$C_6H_{10}Br_2$. *See* c637	241.97	rods (al)	112	...	...	...	δ	s^h	...	...	v	lig δ	**B5**[3], 13
c656	—,—(*trans*)*	$C_6H_{10}Br_2$. *See* c637	241.97	...	1	116^{16}	...	1.5480^{20}	i	s^h	...	...	v	...	**B5**[2], 13
Ω c657	—,**1,4-dibromo-(*cis*)***	$C_6H_{10}Br_2$. *See* c637	241.97	...	...	$137–8^{25}$	1.7834^{20}_{4}	1.5531^{20}	i	...	s	...	...	...	**B5**[2], 13
Ω c658	—,—(*trans*)*	$C_6H_{10}Br_2$. *See* c637	241.97	cr (eth)	113	...	...	...	i	...	s^h	...	...	...	**B5**[2], 13
c659	—,**1,2-dichloro-(*cis*)***	$C_6H_{10}Cl_2$. *See* c637	153.05	...	−1.5	$206–9^{763}$ 91^{20}	1.2021^{20}_{4}	1.4967^{20}	...	...	...	...	s	CCl_4 s	**B5**[1], 8
Ω c660	—,—(*trans, dl*)*	$C_6H_{10}Cl_2$. *See* c637	153.05	...	−6.3	189^{760} 78^{20}	1.1839^{20}_{4}	1.4902^{20}	...	...	...	...	...	...	**B5**[1], 8
c661	—,**(diethylamino)-***	$C_{10}H_{21}N$. *See* c637	155.29	...	...	$192–3^{740}$ $85–6^{20}$	0.872^{0}_{0}	...	...	s	...	...	...	...	**B12**[2], 6
c662	—,**(difluoroamino)-decafluoro-***	Perfluorocyclohexylamine. $C_6F_{13}N$. *See* c637	333.05	...	...	$75–6^{760}$	1.787^{25}_{4}	1.286^{25}	...	...	...	...	...	...	**J1950**, 1966
Ω c663	—,**1,1-dimethyl-***	C_8H_{16}. *See* c637	112.22	...	−33.50	119.54^{760} 10^{10}	0.7809^{20}_{4}	1.4290^{20}	i	s	s	s	s	CCl_4, lig ∞	**B5**[2], 20
Ω c664	—,**1,2-dimethyl-(*cis*)***	C_8H_{16}. *See* c637	112.22	...	−50.1	129.73^{760} 18.3^{10}	0.7963^{20}_{4}	1.4360^{20}	i	s	∞	∞	s	lig v CCl_4 s	**B5**[2], 21
Ω c665	—,—(*trans*)*	C_8H_{16}. *See* c637	112.22	...	−89.2	123.42^{760} 12.9^{10}	0.7760^{20}_{4}	1.4270^{20}	i	s	s	∞	∞	lig v CCl_4 s	**B5**[2], 21
Ω c666	—,**1,3-dimethyl-(*cis*)***	C_8H_{16}. *See* c637	112.22	...	−75.57	120.1^{760} 11.1^{10}	0.7660^{20}_{4}	1.4229^{20}	i	∞	∞	∞	∞	CCl_4, lig ∞	**B5**, 36
c667	—,—(*trans, d*)*	C_8H_{16}. *See* c637	112.22	$[\alpha]_{549}$ +1.33	−90.10	124.45^{760} 15.0^{10}	0.7847^{20}_{4}	1.4309^{20}	i	∞	∞	∞	∞	CCl_4, lig ∞	**C40**, 6427
Ω c668	—,**1,4-dimethyl-(*cis*)***	C_8H_{16}. *See* c637	112.22	...	−87.44	124.32^{760} 14.4^{10}	0.7829^{20}_{4}	1.4230^{20}	i	∞	∞	∞	∞	CCl_4, lig ∞	**B5**, 38
Ω c669	—,—(*trans, dl*)*	C_8H_{16}. *See* c637	112.22	...	−37.0	119.35^{760} 10.0^{10}	0.7626^{20}_{0}	1.4209^{20}	i	∞	∞	∞	∞	CCl_4, lig ∞	**C30**, 2180
c670	—,**1,2-dimethyl-ene-***	...	108.19	λ^{al} 220 (3.88)	...	124^{740} $60–1^{90}$	0.8229^{25}_{4}	1.4718^{25}	i	s	s	v	s	chl s	**Am 75**, 4780
Ω c671	—,**1,2-epoxy-***	Cyclohexene oxide. 7-Oxabicyclo[4,1,0] heptane. $C_6H_{10}O$. *See* c637	98.15	...	< −10	131.5^{760} $54–5^{10}$	0.9663^{20}	1.4519^{20}	i	v	v	v	v	chl s	**B17**[2], 29
c672	—,**1,2-epoxy-4(epoxyethyl)-***	4-Vinylcyclohexene dioxide. $C_8H_{12}O_2$. *See* c637	140.18	...	< −55	227^{760} 92^{5}	1.0986^{20}_{20}	1.4787^{20}	v	...	...	...	...	...	**C46**, 2574
c673	—,**1,2-epoxy-4-ethenyl-**	3-Vinyl-7-oxabicyclo[4,1,0]-heptane. $C_8H_{12}O$. *See* c637	124.19	...	< −100	169^{760} 20^{2}	0.9598^{20}_{20}	1.4700^{20}	δ	...	...	...	...	...	**Am 81**, 3350
Ω c674	—,**ethyl-***	C_8H_{16}. *See* c637	112.22	...	−111.32	131.78^{760} 20.5^{10}	0.7880^{20}_{4}	1.4330^{20}	i	s	s	s	s	CCl_4 ∞ lig v	**B5**[2], 20
Ω c675	—,**(ethylamino)-***	$C_8H_{17}N$. *See* c637	127.23	...	...	164^{760} $62–5^{15}$	0.868^{0}_{0}	...	δ	∞	∞	...	...	...	**B12**[2], 6
c676	—,**fluoro-***	$C_6H_{11}F$. *See* c637	102.15	...	13	100.2 48^{100}	0.9279^{20}_{4}	1.4146^{20}	i	...	...	...	...	Py s	**Am 77**, 3099
c677	—,**hendecafluoro-(trifluoromethyl)-***	Perfluoromethylcyclohexane. C_7F_{14}. *See* c637	350.06	...	−44.72	76.14^{760}	1.7878^{25}_{4}	1.285^{17}	...	...	...	s	s^h	CCl_4, to, AcOEt s	**J1955**, 1749
c678	—,**1,1,2,3,4,5,6-heptachloro-**, (ε, *dl*)	$C_6H_5Cl_7$. *See* c637	325.28	rods	55.0–5.5	...	...	...	i	δ	δ	...	δ	...	**C46**, 441
c679	—,**1,2,3,4,5,6-hexabromo-** (*β* or *cis*)*	Benzene-*β*-hexabromide. $C_6H_6Br_6$. *See* c637	557.57	pr	253d	...	...	...	i	i	i	i	i	...	**B5**, 25

For explanations, symbols and abbreviations see beginning of table. For structural formulas see end of table.

No.	Name	Synonyms and Formula	Mol. wt.	Color, crystalline form, specific rotation and λ_{max} (log ε)	m.p. °C	b.p. °C	Density	n_D	Solubility w	al	eth	ace	bz	other solvents	Ref.
	Cyclohexane														
c680	—,—(α or *trans*)	$C_6H_6Br_6$. *See* c637	557.57	mcl pr (xyl)	212				i	δ	δ	...	δ		**B5**, 25
Ω c681	—,**1,2,3,4,5,6-hexachloro-**(α, *dl*)*	Benzene-*trans*-hexachloride. $C_6H_6Cl_6$. *See* c637	290.83	mcl pr (al or aa)	159.5–60	288	1.87^{20}	....	i	v^h	...	...	s	Ph NH_2 v chl s	**B5**², 11
Ω c682	—,—(β)*	Benzene-*cis*-hexachloride. $C_6H_6Cl_6$. *See* c637	290.83	cr (bz, al or xyl)	314–5 sub >314	$60^{0.58}$	1.89^{19}	....	i	δ	...	...	δ	chl, aa δ	**B5**², 11
Ω c683	—,—(γ)*	Benzene γ-hexachloride. Gammexane. Lindane. $C_6H_6Cl_6$. *See* c637	290.83	nd (al)	112.5–3	323.4^{760} 176.2^{10}	....	....	i	...	...	s	v	....	**B5**, 8
Ω c684	—,—(δ)*	$C_6H_6Cl_6$. *See* c637	290.83	pl	141.5–2.0	$60^{0.34}$									**B5**, 8
Ω c685	—,**iodo-***	Cyclohexyl iodide. $C_6H_{11}I$. *See* c637	210.06	....	....	180 δd 81.5^{20}	1.6244_4^{20}	1.5477^{20}	i	s	s	s	s	oos v lig,	**B5**², 13
Ω c686	—,**isobutyl-**	$C_{10}H_{20}$. *See* c637	140.27	....	−94.85	171.3^{760}	0.79521_4^{20}	1.43861^{20}	i	s	v	s	v	chl s	**B5**², 26
Ω c687	—,**isopropyl-***	Hexahydrocumene. Normenthane. C_9H_{18}. *See* c637	126.24	....	−89.8	154.5^{760} 38.3^{10}	0.8023_4^{20}	1.4410^{20}	i	v	v	∞	∞	lig, CCl_4 ∞	**B5**, 41
—	—,**isopropyl-(methyl)-***	*see* **Menthane**													
Ω c688	—,**methyl-***	Hexahydrotoluene. C_7H_{14}. *See* c637	98.19	....	−126.59	100.9^{760} 16.3^{10}	0.7694_4^{20}	1.4231^{20}	i	s	∞		∞	lig, CCl_4 ∞	**B5**, 29
Ω c689	—,**(methyl-amino)-**	$C_7H_{15}N$. *See* c637	113.20	....	....	$145–7^{748}$ (154–8) $76–7^{18}$	0.8660^{23}	1.4530^{23}	δ	v	∞	...	...	....	**B12**², 5
Ω c690	—,**(3-methyl-butyl)-***	1-Cyclohexyl-3-methylbutane.* Isoamylcyclohexane. $C_{11}H_{22}$. *See* c637	154.30	....	....	196.5^{760}	0.8023_4^{20}	1.4420^{20}	i	...	...	...	s	lig s	**B5**², 32
Ω c691	—,**methylene-***	....	96.17	....	−106.7	$102–3^{764}$	0.8074_4^{20}	1.4523^{20}	i	s	...	...	s	lig s	**B5**², 44
Ω c692	—,**nitro-***	$C_6H_{11}NO_2$. *See* c637	129.16	λ280 (1.4)	fr. −34	205.5^{768} 95^{22}	1.0610_4^{20}	1.4612^{19}	i	s	...	...	...	lig s	**B5**¹, 10
Ω c693	—,**pentyl-***	$C_{11}H_{22}$. *See* c637	154.30	....	−57.5	202.8^{760} 75.3^{10}	0.8037_4^{20}	1.4437^{20}	i	v	s	s	s	....	**B5**², 32
—	—,**phenyl-***	*see* **Benzene, cyclohexyl-***													
Ω c694	—,**propyl-***	C_9H_{18}. *See* c637	126.24	....	−94.9	156.7^{760} 40.1^{10}	0.7936_4^{20}	1.43705^{20}	i	∞	s	∞	s	CCl_4 ∞ peth v aa s	**B5**², 23
—	—,**(2-propyn-1-yl)***	*see* **Propyne, 3-cyclohexyl-***													
c695	—,**1,3,5-tri-methyl-(*cis*)***	Hexahydromesitylene. C_9H_{18}. *See* c637	126.24	....	−49.7	138.5^{760}	0.7708_4^{20}	1.4269^{20}	i	...	s	...	s	lig s	**B5**, 45
c696	—,—(*trans*)*	C_9H_{18}. *See* c637	126.24	....	−107.4	140.5^{760}	0.7794_4^{20}	1.4307^{20}	i	...	s	...	s	lig s	**B5**³, 122
c697	**Cyclohexanecar-boxaldehyde**	Formylcyclohexane. Hexahydrobenzaldehyde.	112.17	λ^{cy} < 186	....	159.3^{760} 36^{10}	0.9035_4^{20}	1.4496^{20}	s	...	s	...	...	....	**B7**², 20
Ω c698	**Cyclohexanecar-boxylic acid***	Hexahydrobenzoic acid.	128.17	mcl pr	31–2	232–3 $120–1^{13}$	1.0334_4^{22}	1.4599^{22} 1.4520^{38}	δ	v	...	...	v	chl v	**B9**, 7
Ω c699	—,chloride	$C_7H_{11}ClO$. *See* c698	146.62	....	....	180^{760} (184) $75–7^{15}$	1.0962_4^{15}	1.4711^{20}	d	d	...	...	...	....	**B9**, 9
Ω c700	—,ethyl ester*	$C_9H_{16}O_2$. *See* c698	156.23	....	....	196^{760} 83^{12}	0.9362_4^{20}	1.45012^{15}	i	v	v	s	...	AcOEt, chl v	**B9**, 8
c701	—,methyl ester*	$C_8H_{14}O_2$. *See* c698	142.20	....	....	183^{760} 73^{15}	0.9954_4^{15}	1.4433^{20}	i	s	s	s	...	chl s	**B9**¹, 5
c702	—,propyl ester*	$C_{10}H_{18}O_2$. *See* c698	170.25	....	....	215.5	0.9530_4^{15}	1.4486^{15}	i	s	s	s	...	chl s	**B9**, 8
c703	—,**1,1-azobis-**, dinitrile.	1,1-Bicyanoazocyclohexane.	244.34	cr (lig) λ^{MeOH} 350 (1.23)	*ca.* 100	....	....	....	i	...	...	...	...	lig s^h	**B16**², 97
Ω c704	—,**epoxy-**,nitrile	C_7H_9NO. *See* c698	123.16	....	−33	244.5^{760} 110^{10}	1.0929_{20}^{20}	1.4763^{20}	s	...	s	...	...	....	**C46**, 3011
c705	—,**1-hydroxy-**, nitrile	Cyclohexanone cyanohydrin. $C_7H_{11}NO$. *See* c698	125.17	....	34–6	$109–13^9$	....	1.4643^{20}	s	i	s	i	i	oos i	**Am 77**, 4571
Ω c706	—,**2-hydroxy-***	Hexahydrosalicylic acid. $C_7H_{12}O_3$. *See* c698	144.17	nd (AcOEt)	111	....	....	....	v	v	v	...	δ	....	**B10**, 5
Ω c706¹	—,**1,3,4,5-tetra-hydroxy-(*d*)**	*d*-Quinic acid. $C_7H_{12}O_6$. *See* c698	192.17	mcl pr (w) $[\alpha]_D^{20}$ +44 (w, c = 10)	164	d	1.637	....	v^h	δ	i	...	...	....	**B10**, 538
c706²	—(*dl*)	$C_7H_{12}O_6$. *See* c698	192.17	pr (w)	142	....	....	....	v^h	δ	i				**B10**, 538
c706³	—(*l*)	$C_7H_{12}O_6$. *See* c698	192.17	pr (w), $[\alpha]_D^{18}$ −44.1 (w, c = 12)	172	d	1.64	....	v^h	δ	δ	...	...	ac, alk s	**B10**², 377
c707	**1,2-Cyclohexane-dicarboxylic acid (*cis*)***	Hexahydrophthalic acid.	172.19	tcl nd (al)	192	d		....	δ	s	s		s	lig δ chl i	**B9**, 730
c708	—(*trans, d*)*	$C_8H_{12}O_4$. *See* c707	172.19	pw (w) $[\alpha]_D$ +18.2	179–83	....	....	....	v^h	...	δ	...	δ	....	**B9**, 732
c709	—(*trans, dl*)*	$C_8H_{12}O_4$. *See* c707	172.19	lf or pr (w)	222	....	....	....	v^h	...	δ	...	δ	chl, peth i	**B7**, 731
c710	—(*trans, l*)*	$C_8H_{12}O_4$. *See* c707	172.19	pw (w) $[\alpha]_D$ −18.5	179–83	....	....	....	...	...	...	...	...	....	**B9**, 732
c711	—,diethyl ester (*cis*)*	$C_{12}H_{20}O_4$. *See* c707	228.29	....	....	133^{10}	1.0540_4^{20}	1.4551^{14}	...	...	s	...	...	....	**B9**², 521
c712	—,—,(*trans, dl*)*	$C_{12}H_{20}O_4$. *See* c707	228.29	....	....	135^{11}	1.040_4^{20}	$1.4522^{13.8}$	...	...	s	...	...	....	**B9**², 522
c713	**1,3-Cyclohexane-dicarboxylic acid (*cis*)***	*cis*-Hexahydroisophthalic acid.	172.19	nd (con HCl), cr (w)	167–7.8 (188)				v	v	s	...	v	lig δ	**B9**², 522

For explanations, symbols and abbreviations see beginning of table. For structural formulas see end of table.

No.	Name	Synonyms and Formula	Mol. wt.	Color, crystalline form, specific rotation and λ_{max} (log ε)	m.p. °C	b.p. °C	Density	n_D	Solubility w	al	eth	ace	bz	other solvents	Ref.
	1,3-Cyclohexanedicarboxylic acid														
c714	—(*trans, d*)*	$C_8H_{12}O_4$. *See* c713	172.19	cr (w), $[\alpha]^{22}_{D}$ + 23.8 (w, c = 4)	134	...	...	...	s	s	s	...	...	peth δ	**B9**^{2}, 523
c715	—(*trans, dl*)*	$C_8H_{12}O_4$. *See* c713	172.19	nd (w)	150.5	...	...	...	v^h	s	s	...	...	peth δ	**B9**^{2}, 523
c716	—(*trans, l*)*	$C_8H_{12}O_4$. *See* c713	172.19	(w), $[\alpha]^{22}_{D}$ − 23.2 (w, c = 2)	134	...	...	...	v^h	s	s	...	...	peth δ	**B9**^{2}, 523
c717	—,diethyl ester (*cis*)*	$C_{12}H_{20}O_4$. *See* c713	228.29	...	...	288, 142^{11}	1.0450$^{20}_{4}$	1.4521^{20}	...	...	...	...	...	...	**B9**^{2}, 523
c718	—,—(*trans, dl*)*	$C_{12}H_{20}O_4$. *See* c713	228.29	...	...	286^{756}, 142^{12}	1.0485$^{21}_{4}$	1.4530^{20}	...	...	...	...	...	...	**B9**^{2}, 523
c719	**1,4-Cyclohexane-dicarboxylic acid** (*cis*)*	*cis*-Hexahydroterephthalic acid.	178.19	lf (w)	170–1	...	...	...	s^h	v	v	...	...	chl v	**B9**^{2}, 523
Ω c720	—(*trans*)*	$C_8H_{12}O_4$. *See* c719	172.19	pr (w), pl (ace)	312–3	300 sub	...	...	δ^h	v	δ	s	...	chl i	**B9**^{2}, 524
c721	—,diethyl ester (*cis*)*	$C_{12}H_{20}O_4$. *See* c719	228.29	...	...	151^{13}	1.0516$^{21}_{4}$	1.4522^{21}	...	...	s	...	...	...	**B9**^{2}, 524
c722	—,—(*trans*)*	$C_{12}H_{20}O_4$. *See* c719	228.29	nd	43–4	...	1.0110$^{20}_{4}$	1.4337^{64}	...	...	s	...	...	...	**B9**^{2}, 529
Ω c723	**1,2-Cyclohexane-diol** (*cis*)*	...	116.16	cr (eth), pl (bz)	99–101	120^{15}	1.0297$^{101}_{4}$	...	...	s	...	s	s	...	**B6**^{2}, 743
Ω c724	—(*trans, dl*)*	$C_6H_{12}O_2$. *See* c723	116.16	cr (AcOEt or ace)	105	117^{13}	1.147$^{24}_{4}$	...	s	s	...	s	s	...	**B6**^{2}, 744
c725	**1,4-Cyclohexane-diol** (*cis*)*	*cis*-Quinitol.	116.16	pr (ace)	113–4	...	...	...	s	s	i	δ, s^h	...	MeOH s, chl δ	**B6**^{2}, 747
Ω c726	—(*trans*)*	$C_6H_{12}O_2$. *See* c725	116.16	mcl pl (ace)	143	...	1.18$^{20}_{4}$	...	s	s	i	δ, s^h	...	MeOH s	**B6**^{2}, 748
Ω c727	**1,2-Cyclohexane-dione***	Dihydropyrocatechol.	112.14	cr (peth) λ^{al} 266 (3.42)	38–40	193–5^{760}, 96–7^{25}	...	1.4995^{20}	s	s	s	...	s	...	**B7**^{2}, 526
Ω c728	—,dioxime*	Nioxime. $C_6H_{10}N_2O_2$. *See* c727	142.16	nd (w or ace) $\lambda^{w, pH=6.5}$ 232.5 (4.12)	191–3	...	...	...	s^h	...	...	s^h	...	...	**B7**^{2}, 526
Ω c729	**1,3-Cyclohexane-dione***	Dihydroresorcinol.	112.14	pr (bz or AcOEt) λ^{al} 253 (4.35), 282 (4.47)	105–6	...	1.0861^{91}	1.4576^{102}	s	s	δ	s	δ	chl s	**B7**^{2}, 526
c730	—,dioxime*	$C_6H_{10}N_2O_2$. *See* c729	142.16	cr (w)	156–7	...	...	...	v^h	v	...	...	...	aa v^h	**B7**, 555
Ω c731	**1,4-Cyclohexane-dione***	Tetrahydroquinone.	112.14	mcl pl (w), nd (peth)	78	sub 100	...	...	s	s	s	s	s	...	**B7**^{2}, 526
c732	—,dioxime	$C_6H_{10}N_2O_2$. *See* c731	142.16	cr (w)	188	...	...	...	s	...	...	...	...	...	**B7**^{2}, 526
c733	**1,3-Cyclohexane-dione, 2-bromo-***	$C_6H_7BrO_2$. *See* c729	191.03	micr nd	169–70	...	...	...	i	s^h	...	...	...	...	**B7**, 556
c734	**1,2-Cyclohexane-dione, 3,5-di-methyl-***	$C_8H_{12}O_2$. *See* c727	140.19	cr (dil MeOH)	71–2	...	...	...	s	...	...	...	...	alk v, MeOH s^h	**B7**^{1}, 314
Ω c735	**1,3-Cyclohexane-dione, 5,5-di-methyl-***	Dimedone. Methone. $C_8H_{12}O_2$. *See* c729	140.19	yesh nd (w, aq ace), mcl pr (al-eth) λ^{al} 257 (4.22)	150	...	...	...	δ^h	...	δ	s	...	chl, aa v, CCl_4 s, peth δ	**B7**^{2}, 531
Ω c736	**1,2,3,4,5,6-Cyclo-hexanehexacar-boxylic acid***	Hexahydromellitic acid.	348.22	syr	d	...	...	...	v	v	δ	...	...	...	**B9**, 1007
c736[1]	**Cyclohexane-hexone, octa hydrate***	Cyclohexone hydrate.	312.19	mic nd (dil HNO_3)	100–1	...	...	...	s^h	δ	δ	...	...	alk s	**B7**^{2}, 882
c737	**1,2,3,4,5-Cyclo-hexanepentol** (*d*)*	Protoquercitol. *d*-Quercite. *d*-Quercitol.	164.16	pr (w or dil (al), $[\alpha]^{15}_{D}$ + 25 (w, c = 10)	235–7 (240)	...	1.5845^{13}	...	v	i, δ^h	i	...	i	MeOH δ	**B6**^{2}, 1151
c738	—(*l*)*	Viboquercitol. $C_6H_{12}O_5$. *See* c737	164.16	pr (w), nd (al), nd (w + 1) $[\alpha]^{20}_{D}$ − 50 (w, c = 4)	180–1 (anh) (cor)	...	...	...	v	δ	i	...	i	MeOH δ	**B6**, 1188
Ω c739	**Cyclohexane-thiol***	Cyclohexyl mercaptan.	116.23	λ^{cy} 230 (2.14)	...	158^{760}, 41^{12}	0.9782$^{20}_{4}$	1.4921^{20}	i	s	s	s	s	chl s	**B6**^{2}, 14
c740	**1,3,5-Cyclohex-anetrione,** triox-ime*	Phloroglucinol trioxime.	171.16	pw (aa)	155 (exp) (darkens at 140)	...	...	...	δ	δ	...	s	...	chl, alk, ac s	**B15**, 34
c741	—**,2,2-dimethyl-*** (trioxoform)	Filicin. Filicinic acid. Filigic acid.	154.17	cubic or rhd (al)	213–5δd	...	...	...	δ^h	v^h	δ	...	δ	Na_2CO_3 v, aa δ, peth i	**B7**^{1}, 470
Ω c742	**Cyclohexanol***	Hexahydrophenol. Hexalin.	100.16	hyg nd	25.15	161.1^{760}	0.9624$^{20}_{4}$	1.4641^{20}	s	s	s	s	∞	CS_2 ∞	**B6**^{2}, 5
c743	—**,1-acetyl-**	$C_8H_{14}O_2$. *See* c742	142.20	...	...	125–6^{50}, 91^{11}	1.0248$^{25}_{4}$	1.4670^{25}	...	s	s	...	...	...	**B8**^{2}, 7
c744	—**,2-allyl-**(*trans*)	$C_9H_{16}O$. *See* c742	140.23	...	...	94–6^{15}	0.943$^{20}_{4}$	1.4778^{20}	i	...	...	...	...	MeOH, aa s	**Am74**, 399
c745	—**,2-amino-** (*trans, dl*)*	$C_6H_{13}NO$. *See* c742	115.18	hyg	68	105^{10}	...	...	...	...	...	...	s	chl v ac s	**B13**^{2}, 157
Ω c746	—**,2-butyl** (*trans*)*	$C_{10}H_{20}O$. *See* c742	156.27	...	...	111–2^{16}, 75–6^{3}	0.9020$^{20}_{4}$	1.4641^{20}	i	...	s	s	s	oos v	**B6**^{3}, 121
c747	—**,2-chloro-** (*cis, dl*)*	*cis*-Cyclohexene chlorohydrin. $C_6H_{11}ClO$. *See* c742	134.61	hyg (peth)	36–7	93–4^{26}	1.1261^{25}	1.4894^{25}	...	s	...	...	s	chl s	**B6**^{3}, 39
c748	—,—(*cis, l*)*	$C_6H_{11}ClO$. *See* c742	134.61	hyg $[\alpha]_{546}$ 19.5	...	87^{15}	1.137^{15}	1.4894^{25}	s	s	...	...	s	chl s	**B6**, 7

For explanations, symbols and abbreviations see beginning of table. For structural formulas see end of table.

No.	Name	Synonyms and Formula	Mol. wt.	Color, crystalline form, specific rotation and λ_{max} (log ε)	m.p. °C	b.p. °C	Density	n_D	Solubility						Ref.
									w	al	eth	ace	bz	other solvents	
	Cyclohexanol														
Ω c749	—,—(*trans*)*	$C_6H_{11}ClO$. *See* c742	134.61	pr (bz-lig)	29	93^{26}	1.146_4^{16}	1.4899^{20}	...	v	s	...	s	chl s	$B6^2$, 12
Ω c750	—,4-chloro-(*trans*)*	$C_6H_{11}ClO$. *See* c742	134.61	pl (cy)	82–3	106^{14}	1.1435_4^{17}	1.4930^{17}	...	s	s	...	s	chl s	$B6^1$, 12
c751	—,3-(dimethyl-amino)-*	$C_8H_{17}NO$. *See* c742	143.23	...	73	231^{740} $126–7^{22}$	0.9766_{25}^{25}	1.4852^{20}	...	s	...	...	...	...	$B13^2$, 159
Ω c752	—,1-ethyl-*	$C_8H_{16}O$. *See* c742	128.22	pr λ^{sulf} 333 (3.40)	34.5–5.0	166 67^{16}	0.9227^{25}	1.4633^{20}	δ	...	...	...	s	peth s	$B6^2$, 26
c753	—,2-ethyl- (*cis, dl*)*	$C_8H_{16}O$. *See* c742	128.22	...	...	$180–2^{760}$ 74^{12}	0.9274_4^{21}	1.4655^{21}	i	...	s	s	s	oos v peth s	$B6^2$, 26
c754	—,—(*trans, dl*)*	$C_8H_{16}O$. *See* c742	128.22	...	...	79^{12}	0.9193_4^{21}	1.4640^{21}	i	...	s	s	s	oos v peth s	$B6^2$, 26
Ω c755	—,1-ethynyl-*	$C_8H_{12}O$. *See* c742	124.19	cr (peth)	31–2	174^{760} 73^{12}	0.9873_4^{20}	1.4822^{20}	i	s	...	...	s	peth s	$B6^2$, 100
c757	—,2(1-hydroxy-ethyl)-*	$C_8H_{16}O_2$. *See* c742	144.22	...	...	140^{12}	0.976_7^{20}	1.4900^{20}	i	...	...	...	...	oos v	C50, 3299
—	—,2-isopropyl-5-methyl-*	*see* **Neoisomenthol**													
Ω c758	—,1-methyl-*	$C_7H_{14}O$. *See* c742	114.19	...	25	155^{760} 70^{25}	9.9194_4^{20}	1.4595^{20}	i	s	...	...	s	chl s	$B6^2$, 16
Ω c759	—,2-methyl-(*cis, dl*)*	$C_7H_{14}O$. *See* c742	114.19	...	7 (−4)	165 60^{12}	0.9360_4^{20}	1.4640^{20}	δ	∞	s	...	...	...	$B6^2$, 20
c760	—,—(*trans, d*)*	$C_7H_{14}O$. *See* c742	114.19	$[\alpha]_D^{20.1}$ +17.19 (undil)	...	166 78^{20}	0.9454^{20}	1.4610^{20}	δ	∞	s	...	...	...	$B6^2$, 18
Ω c761	—,—(*trans, dl*)*	$C_7H_{14}O$. *See* c742	114.19	...	−4.3 to −3.7	167.2–7.6 78^{20}	0.9247_4^{20}	1.4616^{20}	δ	∞	s	...	...	...	$B6^2$, 18
c762	—,—(*trans, l*)*	$C_7H_{14}O$. *See* c742	114.19	$[\alpha]_D^{19}$ −35.5 (undil)	...	166^{760} 78^{20}	0.9454^{20}	1.4610^{20}	δ	∞	s	...	...	...	$B6^2$, 18
Ω c763	—,3-methyl-(*cis, l*)*	$C_7H_{14}O$. *See* c742	114.19	$[\alpha]_D^{22}$ −4.75 (undil)	−4.7	174–5 94^{12}	0.9155_4^{20}	1.4574^{20}	δ	∞	∞	...	...	...	$B6^2$, 20
Ω c764	—,—(*trans, l*)*	$C_7H_{14}O$. *See* c742	114.19	$[\alpha]_D^{20}$ −7.3 (undil)	−1	$174–5^{762}$ 84^{13}	0.9214_4^{20}	1.4590^{20}	δ	∞	v	...	...	...	$B6^2$, 20
Ω c765	—,4-methyl-(*cis*)*	$C_7H_{14}O$. *See* c742	114.19	...	−9.2	$173–4^{760}$ $78–9^{20}$	0.9170_4^{20}	1.4614^{20}	δ	∞	s	...	...	...	$B6^2$, 22
Ω c766	—,—(*trans*)*	$C_7H_{14}O$. *See* c742	114.19	...	...	$173–4^{760}$ 54^3	0.9118_4^{21}	1.4561^{20}	δ	∞	s	...	...	...	$B6^2$, 22
Ω c767	—,2-phenyl-(*cis, dl*)*	$C_{12}H_{16}O$. *See* c742	176.24	...	41–2 (56)	$140–1^{16}$	1.035^{16}	1.5415^{16}	...	...	...	...	...	...	$B6^2$, 548
Ω c768	—,—(*trans, dl*)*	$C_{12}H_{16}O$. *See* c742	176.24	cr (peth)	56–7	$152–5^{16}$	...	...	...	s	...	...	...	chl s	$B6^2$, 548
Ω c769	—,2,2,6,6-tetra kis(hydroxy-methyl)-*	$C_{10}H_{20}O_5$. *See* c742	220.27	pl (al)	131	...	...	...	v	v	i	i	i	MeOH v Py s	$B6^2$, 1151
c770	—,1,2,2-tri-methyl-(*dl*)*	$C_9H_{18}O$. *See* c742	142.24	cr (+½w)	41 (hyd)	81.4–1.8^{20}	0.9230_4^{20}	1.4682^{20}	i	s	s	...	s	oos s	$B6^2$, 16
c771	—,1,2,6-tri-methyl-*	$C_9H_{18}O$. *See* c742	142.24	...	...	78^{23}	0.9126_4^{15}	1.4598^{15}	i	s	s	s	s	oos s	$B6^1$, 17
c772	—,1,3,3-tri-methyl-*	$C_9H_{18}O$. *See* c742	142.24	pr (dil al)	74	...	...	...	i	v	s	s	s	oos v	$B6^1$, 16
c773	—,1,3,5-tri-methyl-*	$C_9H_{18}O$. *See* c742	142.24	...	...	181 $82–3^{19}$	0.8876_4^{17}	$1.454^{16.3}$	i	s	...	...	...	chl s	$B6^1$, 17
c774	—,1,4,4-tri-methyl-*	$C_9H_{18}O$. *See* c742	142.24	hyg nd (dil al)	58	$79–80^{15}$	...	...	i	s	s	...	...	chl s	$B6^1$, 16
c775	—,2,2,3-tri-methyl-*	$C_9H_{18}O$. *See* c742	142.24	...	...	$85–7^{15}$	...	...	i	s	s	...	...	chl s	$B6^1$, 16
c776	—,2,2,5-tri-methyl-*	Pulenol. $C_9H_{18}O$. *See* c742	142.24	...	...	$187–9^{760}$ $90–2^{23}$	0.8955_4^{20}	1.4569^{20}	i	s	...	...	...	oos s	B6, 22
c777	—,2,2,6-tri-methyl-(liquid)*	$C_9H_{18}O$. *See* c742	142.24	...	...	$186–7^{753}$	0.9128_4^{20}	1.4600^{20}	i	s	s	...	...	chl s	$B6^1$, 16
c778	—,—(solid)*	$C_9H_{18}O$. *See* c742	142.24	cr (peth or al)	51	87^{28}	...	...	i	s	s	...	...	chl s	$B6^1$, 16
c779	—,2,3,3-tri-methyl-*	$C_9H_{18}O$. *See* c742	142.24	nd	28	197 97^{19}	...	...	i	v	...	v	...	oos v	$B6^1$, 16
c780	—,2,3,6-tri-methyl-*	$C_9H_{18}O$. *See* c742	142.24	...	...	$193–5^{747}$	0.9117_4^{17}	...	i	s	...	...	...	chl s	B6, 22
c781	—,2,4,5-tri-methyl-(*cis*)*	$C_9H_{18}O$. *See* c742	142.24	hyg	...	$191–3^{760}$ 84^{17}	0.9120_4^{20}	1.463^{20}	i	s	...	...	...	chl s	$B6^2$, 36
c782	—,—(*trans*)*	$C_9H_{18}O$. *See* c742	142.24	hyg	...	196^{760} 112^{35}	0.906_4^{20}	1.461^{20}	i	s	s	...	...	chl s	$B6^2$, 36
Ω c783	—,3,3,5-tri-methyl-(*cis*)*	*cis*-Dihydroisophorole. $C_9H_{18}O$. *See* c742	142.24	...	37.3	$201–3^{750}$ 92^{12}	0.9006_4^{16}	1.4550^{16}	i	s	...	...	...	chl s	$B6^1$, 16
Ω c784	—,—(*trans*)*	$C_9H_{18}O$. *See* c742	142.24	cr (eth)	55.8	189.2^{760}	0.8643_{20}^{60}	...	i	s	...	...	...	chl s	$B6^1$, 16
Ω c785	**Cyclohexanone***	Ketohexamethylene. Pimelic ketone.	98.15	λ^{al} 284(1.26)	−16.4 (−45)	155.65^{760} 47^{15}	0.9478_4^{20}	1.4507^{20}	s	s	s	s	s	chl s	$B7^2$, 5
Ω c786	—oxime*	$C_6H_{11}NO$. *See* c785	113.16	hex pr (lig)	90	206–10	...	...	s	s	s	...	...	MeOH s	$B7^2$, 10
c787	—,2-acetyl-	$C_8H_{12}O_2$. *See* c785	140.19	λ^{al} 290 (3.95)	...	$111–2^{18}$	1.0782^0	1.5138^{20}	...	...	...	...	...	...	$B7^2$, 530
Ω c788	—,2-butyl-*	$C_{10}H_{18}O$. *See* c785	154.26	...	...	70^2	0.905_4^{20}	1.4545^{20}	i	...	...	...	...	oos v	Am 78, 5339
Ω c789	—,2-butylidene-	$C_{10}H_{16}O$. *See* c785	152.24	...	...	$98–100^{10}$ $(95–100^3)$	0.935_4^{20}	1.4800^{20}	i	s	v	s	v	oos v	C49, 1598
Ω c790	—,2-chloro-*	C_6H_9ClO. *See* c785	132.59	λ^{al} 294(1.38)	23	82^{13}	1.161_{15}^{20}	1.4825^{20}	...	...	s	...	s	diox s	$B7^2$, 11
c791	—,3-chloro-*	C_6H_9ClO. *See* c785	132.59	...	...	$91–2^{14}$	...	...	...	...	s	...	...	...	B7, 10
c792	—,4-chloro-*	C_6H_9ClO. *See* c785	132.59	...	...	95^{17}	...	1.4867^{20}	...	...	s	...	...	...	$B7^2$, 11
c793	—,2,6-dibenzyl-idene-	$C_{20}H_{18}O$. *See* c785	274.37	ye nd (al) λ^{al} 330 (4.40)	117–8	$185–95^{20}$	...	...	...	δ	...	...	s	aa s	$B7^2$, 465

For explanations, symbols and abbreviations see beginning of table. For structural formulas see end of table.

No.	Name	Synonyms and Formula	Mol. wt.	Color, crystalline form, specific rotation and λ_{max} (log ε)	m.p. °C	b.p. °C	Density	n_D	Solubility w	al	eth	ace	bz	other solvents	Ref.
	Cyclohexanone														
c794	—,2,6-dibromo- (*cis*)*	$C_6H_8Br_2O$. *See* c785	255.96	cr (eth or aa)	106–7						s^h	s^h			**B7**^{2}, 12
c795	—,2,4-dimethyl-, oxime (*cis*)*	$C_8H_{15}NO$. *See* c785	141.21	nd (al)	98–9				i	s^h					**B7**^{2}, 26
c797	—,2,5-dimethyl- (*d*)*	$C_8H_{14}O$. *See* c785	126.20	$[\alpha]^{20}$ +11.6 (undil)		$172–4^{750}$ 51^{10}	0.8985_4^{20}	1.4445^{20}	i	s	s				**B71**, 19
c798	—,—(*trans, dl*)*	$C_8H_{14}O$. *See* c785	126.20			171–3 (176) $76–7^{27}$	0.9025^{20}	1.4446^{20}	i	s	s				**B7**^{2}, 27
c799	—,2(dimethylaminomethyl)-*	$C_9H_{17}NO$. *See* c785	155.24			$92^{10.5}$	0.9504_4^{20}	1.4672^{20}	δ	s	s				**B14**^{2}, 3
Ω c800	—,2-ethylidene-*	$C_8H_{12}O$. *See* c785	124.19	λ^{al} 250 (3.60)		92^{20}	0.962_4^{20}	1.4882^{20}	i					oos v	**B7**^{2}, 58
c801	—,2-hydroxy-*	Adipoin. $C_6H_{10}O_2$. *See* c785	114.15	nd (MeOH or al) λ^{al} 279 (1.28)	113(92)			1.4785^{21}	v^h	v^h	i		i	peth i	**B8**^{2}, 3
c802	—,2-isopropyl-*	$C_9H_{16}O$. *See* c785	140.23	λ 230 (2.09), 293 (1.29)		$72–3^9$	0.922_4^{16}	1.4564^{15}	i	s	s	s	s	oos v	**B7**^{2}, 30
c803	—,2-methyl- (*d*)*	$C_7H_{12}O$. *See* c785	112.17	$[\alpha]_D^{25}$ +14.21 (chl)		$167–8^{735}$	0.9262_4^{18}	1.4440^{25}							**B7**^{2}, 15
Ω c804	—,—(*dl*)*	$C_7H_{12}O$. *See* c785	112.17	λ 284 (1.2)	−13.9	165^{757} 90^{20}	0.9250_4^{20}	1.4483^{20}	i	s	s				**B7**^{2}, 16
c805	—,—(*l*)*	$C_7H_{12}O$. *See* c785	112.17	$[\alpha]_D^{25}$ −15.22 (undil)		$59–60^{20}$	0.9230_4^{25}	1.4440^{25}	i	s	s				**B7**^{2}, 15
c806	—,3-methyl-(*d*)*	$C_7H_{12}O$. *See* c785	112.17	$[\alpha]_D^{20}$ +12.7 (undil)		169^{760}	0.9155_4^{20}	1.4493^{20}	i	s	s				**B7**^{2}, 17
Ω c807	—,—(*dl*)*	$C_7H_{12}O$. *See* c785	112.17		−73.5	$168–9^{738}$ 65^{15}	0.9136_4^{20}	1.44566^{20}	i	s	s				**B7**^{2}, 18
Ω c808	—,4-methyl-*	$C_7H_{12}O$. *See* c785	112.17	λ 281 (1.2)	−40.6	170^{761}	0.9138_4^{20}	1.4451^{20}	i	s	s				**B7**^{2}, 19
Ω c809	—,2-propyl-*	$C_9H_{16}O$. *See* c785	140.23			195^{760} 70^4	0.927_4^{20}	1.4538^{20}	i	s	v	s	v	oos v	**B7**^{2}, 30
c810	**Cyclohexasiloxane, dodecamethyl-**		444.93		−3	245 128^{20}	0.9672	1.4015^{20}	i						**Am 68**, 667
Ω c811	**Cyclohexene***	3,4,5,6-Tetrahydrobenzene.	82.15	λ^{al} 207 (2.65)	−103.50	82.98^{760}	0.8102_4^{20}	1.4465^{20}	i	∞	∞	∞	∞	CCl_4, lig ∞	**B5**^{2}, 37
c812	—,1-acetyl-	Tetrahydroacetophenone. $C_8H_{12}O$. *See* c811	124.19			$201–2^{760}$ $63–4^6$	0.9655_4^{20}	1.4881^{20}		s	s				**B7**^{2}, 58
c813	—,3-bromo-*	C_6H_9Br. *See* c811	161.05			$80–2^{40}$ 57.5–58^{12}	1.3890_4^{20}	1.5230^{20}	i		s		s	chl s	**B5**^{2}, 40
Ω c814	—,decafluoro-*	Perfluorocyclohexene. C_6F_{10}. *See* c811	262.05			$52–3^{750}$		1.293^{20}							**J 1952**, 1251
c815	—,1-ethenyl-*	C_8H_{12}. *See* c811	108.19	λ^{al} 230 (4.32)		145^{760} $50–2^{22}$	0.8623_4^{15}	1.4915^{20}	i		s		s	MeOH v	**B5**^{1}, 62
c816	—,4-ethenyl-*	C_8H_{12}. *See* c811	108.19			128.9^{760} $66–7^{100}$	0.8299_4^{20}	1.4639^{20}	i		s		s	peth s	**B5**^{2}, 81
—	—,1-isopropyl-4-methyl-	*see* **Menthene**													
Ω c817	—,1-methyl-*	C_7H_{12}. *See* c811	96.17	λ^{al} 208 (3.15)	−121	110^{760} 24.6^{30}	0.8102_4^{20}	1.4503^{20}	i		s		s	CCl_4 s	**B5**^{2}, 42
c818	—,3-methyl- (*d*)*	C_7H_{12}. *See* c811	96.17	$[\alpha]_D^{20}$ +110		104^{760}	0.8010_4^{20}	1.4444^{20}	i		s		s	peth, chl s	**B5**^{2}, 43
c819	—,—(*dl*)*	C_7H_{12}. *See* c811	96.17		−115.5	104^{760} 0.46^{10}	0.7990_4^{20}	1.4414^{20}	i		s		s	peth, chl s	**B5**^{2}, 43
Ω c820	—,4-methyl-*	C_7H_{12}. *See* c811	96.17		−115.5	102.74^{760} 19^{30}	0.7991_4^{20}	1.4414^{20}	i	s	s				**B5**^{2}, 43
c821	—,1,3,4,5,6-pentachloro-(γ)*	$C_6H_5Cl_5$. *See* c811	254.37			$115–6^4$		1.5630^{20}							**Am 76**, 1244
c822	—,—(δ)	$C_6H_5Cl_5$. *See* c811	254.37		68.2–8.6		1.80			s					**Am 76**, 1244
c823	—,1-phenyl-*	1,2,3,4-Tetrahydrobiphenyl. $C_{12}H_{14}$. *See* c811	158.25	λ^{iso} 250 (4.08)	−11	$251–3^{760}$ $125–6^{14}$	0.9939_4^{20}	1.5718^{20}						MeOH v	**B5**^{2}, 419
c824	—,1-Cyclohexene-1-carboxaldehyde	Δ^1-Tetrahydrobenzaldehyde.	110.16	λ^{al} 231 (4.12)		72^{15}	0.9694_4^{20}	1.5005^{20}		s	s				**J 1955**, 320
Ω c825	**3-Cyclohexene-1-carboxaldehyde**	Δ^3-Tetrahydrobenzaldehyde.	110.16		fr −96.1	164^{760} 52^{13}	0.9709_4^{20}	1.4725^{19}				s		MeOH s	**B7**^{2}, 57
c826	**1-Cyclohexene-1-carboxylic acid***	Δ^1-Tetrahydrobenzoic acid.	126.16	pl λ^{al} 218 (4.01)	38	$240–2^{760}$ 138^{14}	1.109_4^{20}	1.4902_{He}^{20}	δ	s		s			**B9**^{2}, 30
Ω c827	**3-Cyclohexene-1-carboxylic acid***	Δ^3-Tetrahydrobenzoic acid.	126.16		17	237^{748} 132.5–3.0^{20}	1.0815_4^{20}	1.4812^{20}	v	s		s			**B9**^{2}, 30
Ω c828	**2-Cyclohexene-1-carboxylic acid, 2,6-dimethyl-4-oxo-, ethyl ester***	3,5-Dimethyl-4-carboxy-2-cyclohexen-1-one.	196.25			$157–8^{18}$	1.0493_4^{20}	1.4773^{20}				s			**B10**^{2}, 436
c829	**1-Cyclohexene-1,2-dicarboxylic acid***	Δ^1-Tetrahydrophthalic acid.	170.17	mcl pl (w)	126				v						**B9**^{2}, 556
Ω c830	—,anhydride*		152.16	pl (eth) λ^{CH_3CN} 255 (3.10)	74				d^h	s	v	s		chl s	**B17**^{2}, 457

For explanations, symbols and abbreviations see beginning of table. For structural formulas see end of table.

2-Cyclohexene-1,2-dicarboxylic acid

No.	Name	Synonyms and Formula	Mol. wt.	Color, crystalline form, specific rotation and λ_{max} (log ε)	m.p. °C	b.p. °C	Density	n_D	w	al	eth	ace	bz	other solvents	Ref.
c831	**2-Cyclohexene-1,2-dicarboxylic acid, anhydride***	Δ^2-Tetrahydrophthalic anhydride.	152.16	pr (eth)	78–9	...	...	...	d^h	s	s	...	...	chl s	**B17^2, 457**
Ω c832	**4-Cyclohexene-1,2-dicarboxylic acid, anhydride (*cis*)***	*cis*-Δ^4-Tetrahydrophthalic anhydride.	152.16	pl (eth or lig)	103–4	...	...	...	d^h	s	...	s	...	chl, os s lig δ	**B17^2, 457**
c833	**—,—(*trans, d*)***	$C_8H_8O_3$, *See* c832	152.16	lf, $[\alpha]_D^{25}$ +6.6 (al)	128	...	...	...	d^h	v	...	...	v	...	**B17, 462**
c834	**—,—(*trans, dl*)***	$C_8H_8O_3$. *See* c832	152.16	cr (bz-lig)	141 (130)	...	...	...	d^h	s	...	...	s	chl s	**B17, 462**
Ω c835	**—,imide, *N*(trichloromethylthio)-**	Captan.	300.60	cr (bz, CCl_4)	172	...	1.74	...	...	...	...	δ	δ	...	**C46,** 11232
—	**Cyclohexene oxide**	*see* **Cyclohexane, 1,2-epoxy-***													
c836	**2-Cyclohexen-1-ol***	...	98.15	...	...	164–6^{760} 63–5^{12}	0.9923_4^{15}	1.4790^{25}	...	s	...	s	...	...	**B6^2, 60**
c837	**—,5-methyl-(*cis, d*)***	$C_7H_{12}O$. *See* c836	112.17	$[\alpha]_D^{30}$ +6.95	...	83^{25}	0.9391_4^{25}	1.4727^{25}	...	...	s	...	...	lig s	**Am 77,** 4042
c838	**—,—(*cis, l*)***	$C_7H_{12}O$. *See* c836	112.17	$[\alpha]_D^{25}$ −7	...	82^{25}	0.9391_4^{25}	1.4727^{25}	...	...	s	...	...	lig s	**Am 77,** 4042
c839	**—,—(*trans, d*)***	$C_7H_{12}O$. *See* c836	112.17	$[\alpha]_D^{27}$ +127 (ace, c = 19.4)	...	68–9^{24}	0.9430_4^{20}	1.4737^{25}	...	...	s	...	...	lig s	**Am 77,** 4042
c840	**—,—(*trans, l*)***	$C_7H_{12}O$. *See* c836	112.17	$[\alpha]_D^{27}$ −163.9	...	82–3^{24}	0.9430_4^{20}	1.4737^{25}	...	...	s	...	...	lig s	**Am 77,** 4042
Ω c841	**2-Cyclohexen-1-one***	...	96.14	λ^{al} 225 (3.68)	...	169–71^{760} 61–2^{10}	0.9620^{25}	1.4883^{20}	...	v	...	s	...	...	**B7^2, 55**
c842	**—,2,5-dimethyl-***	$C_8H_{12}O$. *See* c841	124.19	...	...	189–90	0.938^{22}	1.4753^{22}	...	s	s	...	...	...	**B7^1, 51**
Ω c843	**—,3,5-dimethyl-***	$C_8H_{12}O$. *See* c841	124.19	λ^{al} 233 (4.16) 310 (1.70)	...	208–9 94^{17}	0.9400_4^{20}	1.4812^{20}	...	s	s	...	...	...	**B7^2, 59**
c844	**—,3,6-dimethyl-***	$C_8H_{12}O$. *See* c841	124.19	λ 236 (4.10)	...	75^{19}	1.008_{18}^{18}	1.4805^{18}	i	s	s	...	...	...	**B7^1, 51**
c845	**3-Cyclohexen-1-one,4,6-dimethyl-***	...	124.19	...	...	194	0.9539^0	...	i	...	s	...	...	...	**B7, 61**
—	**2-Cyclohexen-1-one,3-hydroxy-***	*see* **1,3-Cyclohexanedione***													
c846	**—,5-isopropyl-3-methyl-***	*m*-6-Menten-5-one. Hexeton. $C_{10}H_{16}O$. *See* c841	152.24	pa ye λ^{al} 234.5 (4.10)	...	244 124^{15}	0.9340^{21}	1.4865^{21}	δ	v	...	v	...	oos ∞	**B7^2, 74**
c847	**—,2-methyl-***	$C_7H_{10}O$. *See* c841	110.16	λ^{al} 235 (3.98)	...	178–9^{760} 56^9	0.9667_4^{20}	1.4833^{20}	...	...	...	...	s	...	**B7^2, 57**
Ω c848	**—,3-methyl-***	$C_7H_{10}O$. *See* c841	110.16	λ^{al} 235 (4.10)	−21	200–2^{760} 78–9^{12}	0.9693_4^{20}	1.49475^{20}	∞	...	...	...	s	...	**B7^2, 56**
—	**—,3,5,5-trimethyl-***	*see* **Isophorone**													
—	**Cyclohexone, hydrate**	*see* **Cyclohexanehexone, octahydrate.***													
—	**Cyclone**	*see* **Cyclopentadienone, tetraphenyl-**													
—	**Cyclonite**	*see* **1,3,5-Triazine, hexahydro-1,3,5-trinitro-**													
c850	**Cyclononanone***	...	140.23	λ^{al} 280 (1.20)	34	148.5^{24}	0.9560_4^{20}	1.4729^{20}	...	s	...	...	...	...	**B7^2, 29**
c851	**Cyclononasiloxane, octadecamethyl-**	...	667.40	...	...	188^{20}	...	1.4070^{20}	i	...	...	...	s	lig s	**Am 68,** 358
c852	**Cyclononene (*cis*)***	...	124.23	...	...	167–9^{760} 73–4^{30}	0.8671_4^{20}	1.4805^{20}	i	...	...	...	s	...	**Am 74,** 3643
c853	**—(*trans*)***	C_9H_{16}. *See* c852	124.23	...	...	94–6^{30}	0.8615_4^{20}	1.4799^{20}	i	...	...	...	s	...	**Am 74,** 3643
Ω c854	**1,5-Cyclooctadiene (*cis, cis*)***	...	108.19	...	−70→ −69	150.8^{757} 51–2^{25}	0.8818_4^{25}	1.4905^{25}	i	...	...	...	s	CCl_4 s	**B5, 116**
Ω c855	**Cyclooctane***	Octamethylene.	112.22	...	14.3	148.5–9.5^{749} 63^{45}	0.8349_4^{20}	1.4586^{20}	i	...	...	...	s	lig s	**B5^2, 20**
Ω c856	**Cyclooctanol***	...	128.22	...	25.1	99^{16} 74^3	0.9740_4^{20}	1.4871^{20}	...	s	...	...	...	...	**B6^2, 25**
Ω c857	**Cyclooctanone**	Azelaone.	126.20	λ^{al} 282 (1.34)	28–30	194–8^{761} (202^{713}) 74^{12}	0.9581_4^{20}	1.4694^{20}	i	s	...	s	s	...	**B7^2, 22**
c858	**—,semicarbazone**	$C_9H_{17}N_3O$. *See* c857	183.26	lf (dil MeOH)	170–1	...	...	...	...	...	...	...	...	MeOH s^h	**B7^2, 22**
c859	**Cyclooctasiloxane, hexadecamethyl-**	...	593.25	...	31.5	290^{760} 175^{20}	1.177	1.4060^{20}	i	...	...	...	s	lig s	**Am 67,** 667
Ω c860	**Cyclooctatetraene***	...	104.16	ye or wh λ^{al} 280 (2.64)	−4.68	140.56^{760} 29.1^{10}	0.9206_4^{20}	1.5381^{20}	...	s	s	s	s	...	**B5^1, 228**
Ω c861	**Cyclooctene(*cis*)**	...	110.20	...	−12	138^{760} 42^{18}	0.8472_4^{20}	1.4698^{20}	...	s	s	...	...	CCl_4 s	**B5^2, 45**
Ω c862	**—(*trans*)**	...	110.20	...	−59	143^{760} 75^{78}	0.8483_4^{20}	1.4741^{25}	...	s	...	...	...	chl s	**B5^2, 45**
Ω c863	**Cyclopentadecanone***	Exaltone.	224.39	λ^{al} 282 (1.32)	63	120$^{0.3}$	0.8895	1.4637^{66}	δ	s	...	s	...	...	**B7^2, 50**
c864	**—,3-methyl-***	Muscone. Muskone. $C_{16}H_{30}O$. *See* c863	238.42	ye, $[\alpha]_D^{17}$ −13 (undil)	...	327–30^{752} 130$^{0.5}$	0.9221_4^{17}	1.4802^{17}	δ	∞	s	s	...	...	**B7^2, 51**

For explanations, symbols and abbreviations see beginning of table. For structural formulas see end of table.

No.	Name	Synonyms and Formula	Mol. wt.	Color, crystalline form, specific rotation and λ_{max} (log ε)	m.p. °C	b.p. °C	Density	n_D	Solubility w	al	eth	ace	bz	other solvents	Ref.
	1,3-Cyclopentadiene														
c865	**1,3-Cyclopenta-diene***	Cyclopentadiene*.	66.10	λ^{al} 238 (3.62)	−97.2	40.0^{760}	0.8021^{20}_{4}	1.4440^{20}	i	∞	∞	s	∞	...	**B5²**, 77
c866	—,hexachloro-	Perchlorocyclopentadiene. C_5Cl_6	272.77	ye gr liq λ^{hp} 323 (3.2)	−9	239^{753} 48–$9^{0.3}$	1.7019^{25}_{4}	1.5658^{20}	...	...	...	...	...	...	**Am 69**, 1918
—	**Cyclopentadiene-benzoquinone**	*see* **Naphthalene, 5,8-dioxo-1,4,5,8,9,10-hexahydro-1,4-methylene-***													
Ω c867	**Cyclopenta-dienone, tetra-phenyl-***	Cyclone. Tetracyclone.	384.48	bk-vt lf, cr (aa or xyl) λ^{iso} 260 (4.45), 342 (3.82), 512 (3.11)	220–1	...	...	...	...	s	...	...	s	iso s xyl, aa s^h	**B7²**, 521
Ω c868	**Cyclopentane***	Pentamethylene.	70.14	...	−93.879	49.262^{760}	0.7457^{20}_{4}	1.4065^{20}	i	∞	∞	∞	∞	CCl_4, peth ∞	**B5²**, 4
c869	—,acetyl-	Cyclopentyl methyl ketone. $C_7H_{12}O$. *See* c868	112.17	λ^{al} 279 (1.37)	...	158–9.5 (163)	0.918^{20}_{20}	1.4409^{25}	...	...	s	...	...	...	**B7²**, 20
Ω c870	—,bromo-*	Cyclopentyl bromide. C_5H_9Br. *See* c868	149.04	...	...	136.7 – 7.7^{760} 56^{48}	1.3873^{20}_{4}	1.4886^{20}	...	...	...	...	...	...	**B5²**, 4
c871	—,butyl-*	C_9H_{18}. *See* c868	126.23	...	−107.98	156.7^{760} 41.6^{10}	0.7846^{20}_{4}	1.4316^{20}	i	∞	∞	∞	∞	peth, CCl_4 ∞	**B5²**, 25
Ω c872	—,chloro-*	Cyclopentyl chloride. C_5H_9Cl. *See* c868	104.58	...	...	113.5–4.5^{752}	1.0051^{20}_{4}	1.4510^{20}	i	...	s	s	s	oos s	**B5²**, 4
c873	—,1,2-diethyl- (*trans*)*	C_9H_{18}. *See* c868	124.24	...	−95.6	153.58^{760}	0.7832^{20}_{4}	1.4295^{20}	i	...	s	...	s	peth s	**B5²**, 126
Ω c874	—,ethyl-*	C_7H_{14}. *See* c868	98.19	...	−138.44	103.5^{760} 19.36^{30}	0.7665^{20}_{4}	1.4198^{20}	i	∞	∞	∞	s	CCl_4 ∞ to, peth s	**B5³**, 78
c875	—,isopropyl-*	C_8H_{16}. *See* c868	112.22	...	−111.4	126.4^{760} 16.3^{10}	0.7765^{20}_{4}	1.4258^{20}	i	∞	s	∞	s	CCl_4 ∞ peth v	**B5²**, 22
Ω c876	—,methyl-*	Methylpentamethylene. C_6H_{12}. *See* c868.	84.16	...	−142.4	71.8^{760}	0.7486^{20}_{4}	1.4097^{20}	i	∞	∞	∞	∞	lig, CCl_4 ∞	**B5**, 27
c877	—,1-methyl-2-propyl- (*trans*)*	C_9H_{18}. *See* c868	126.24	...	−104.9	152.58^{760}	0.7921^{20}_{4}	1.4321^{20}	...	...	...	...	...	...	**C45**, 7026
c878	—,propyl-*	1-Cyclopentylpropane. C_8H_{16}. *See* c868	112.22	...	−117.3	130.95^{760} 21.2^{10}	0.7763^{20}_{4}	1.4266^{20}	i	∞	∞	∞	s	peth, CCl_4 v	**B5²**, 22
c879	**Cyclopentanecar-boxaldehyde**	...	98.15	...	...	133–4^{760}	0.9371^{20}_{4}	1.1432^{20}	s	s	s	...	...	...	**B7²**, 14
Ω c880	**Cyclopentanecar-boxylic acid***	...	114.15	...	−7 (−4)	212.5–13.5^{752} 104^{11}	1.0527^{20}_{4}	1.4532^{20}	δ	...	...	...	...	MeOH s	**B9²**, 6
c881	—,3-formyl-2,2,3-trimethyl-, methyl ester (*d*)	$C_{11}H_{18}O_3$. *See* c880	198.27	$[\alpha]_D$ +51.4 (al)	...	130–2^8	1.048^{20}	1.4160^{25}	δ	s	...	...	...	MeOH s	**Am 64**, 1416
Ω c882	—,2-oxo-, ethyl ester	$C_8H_{12}O_3$. *See* c880	156.19	λ^{al} 258 (2.5)	...	218^{704} 110^{16}	1.0781^{20}_{4}	1.4519^{20}	...	...	s	...	s	...	**B10²**, 419
—	—,1,2,2,3-tetra-methyl-	*See* **Campholic acid**													
c883	**1,2-Cyclopentane-dicarboxylic acid (*cis*)***	...	158.16	nd (w)	140	...	...	...	s	...	...	...	...	...	**B9**, 728
c884	—(*trans, d*)*	$C_7H_{10}O_4$. *See* c883.	158.16	cr (w), $[\alpha]_D$ + 87.6 (w, c = 0.9)	181	...	...	...	s^h	s	δ	...	δ	chl δ	**B9¹**, 316
c885	—(*trans, dl*)*	$C_7H_{10}O_4$. *See* c883.	158.16	cr (w)	162–3	...	...	...	s^h	v^h	δ	...	δ	AcOEt s chl, peth δ	**B9²**, 518
c886	—(*trans, l*)*	$C_7H_{10}O_4$. *See* c883.	158.16	cr (w), $[\alpha]_D$ −85.9 (w, c = 1.2)	180–1	...	...	...	s^h	s	δ	...	δ	chl δ	**B9¹**, 316
Ω c887	**1,3-Cyclopentane-dicarboxylic acid (*cis*)***	Norcamphoric acid.	158.16	pr (w)	121	>300d	...	...	s^h	s	s	s	s^h	chl s peth i	**B9²**, 518
c888	—(*trans, d*)*	$C_7H_{10}O_4$. *See* c887.	158.16	cr (CCl_4) $[\alpha]_D$ +5.9 (w, c = 5)	93.5	...	...	...	s	...	...	...	...	CCl_4 s^h	**B9²**, 519
c889	—(*trans, dl*)*	$C_7H_{10}O_4$. *See* c887.	158.16	pr (CCl_4)	88	...	...	...	s	...	...	...	...	CCl_4 s^h	**B9²**, 519
c890	— (*trans, l*)*	$C_7H_{10}O_4$. *See* c887.	158.16	cr (CCl_4) $[\alpha]_D$ −5.3 (w, c = 5)	93	...	...	...	s	...	...	...	...	CCl_4 s^h	**B9²**, 519
c891	—,4,4-dimethyl- (*cis*)*	Apofenchocamphoric acid. $C_9H_{14}C_4$. *See* c887	186.21	mcl	144–5	...	...	...	s^h	v	v	...	δ	...	**B9²**, 529
—	—,1,2,2-trimethyl-*	*See* **Camphoric acid**													
Ω c892	**Cyclopentanol***	...	86.14	...	−19	140.85^{760} 53^{10}	0.9478^{20}_{4}	1.4530^{20}	δ	s	s	s	...	...	**B6³**, 4
c893	—,2-acetyl-1,3,3,4,4-penta-methyl- (α)	Desoxymesityl oxide. $C_{12}H_{22}O_2$. *See* c892	198.31	cr (peth-eth)	45	...	...	...	i	...	...	s	...	...	**B8²**, 11
Ω c894	**Cyclopentanone***	Adipic ketone. Ketopentamethylene.	84.12	λ^{al} 290 (1.28)	−51.3	130.65^{760}	0.94869^{20}_{4}	1.4366^{20}	i	s	∞	s	...	MeOH, hx s	**B7²**, 3
c895	—,2-methyl-*	$C_6H_{10}O$. *See* c894.	98.15	...	−75	139.5^{760} 44^{18}	0.9139^{20}	1.4364^{20}	s	v	v	v	...	...	**B7²**, 13

For explanations, symbols and abbreviations see beginning of table. For structural formulas see end of table.

No.	Name	Synonyms and Formula	Mol. wt.	Color, crystalline form, specific rotation and λ_{max} (log ε)	m.p. °C	b.p. °C	Density	n_D	Solubility w	al	eth	ace	bz	other solvents	Ref.
	Cyclopentanone														
c896	—,3-methyl- (*d*)*	$C_6H_{10}O$. *See* c894	98.15	$[\alpha]_D^{25}$ +143.7 (undil) λ^{al} 288 (1.34)	−58.4	143.5^{742} 42.5–44^{13}	0.9140_4^{19}	1.4340^{19}	v	v	v	v	...	aa v	**B7**², 13
Ω c897	—,—(*dl*)*	$C_6H_{10}O$. *See* c894	98.15	λ^{al} 288 (1.34)	...	144–4.5^{760} 38^{11}	0.913^{22}	1.4329^{20}	s	v	v	v	...	aa v	**B7**², 14
c898	**Cyclopentasiloxane, decamethyl-**	...	370.78	...	−38	210^{760} 101^{20}	0.9593^{20}	1.3982^{20}	i	...	...	...	...	...	**Am 68**, 358
Ω c899	**Cyclopentene***	...	68.13	λ^{gas} 182.8 (4.13)	−135.076	44.242^{760}	0.7720_4^{20}	1.4225^{20}	i	s	s	...	s	peth s	**B5**², 35
c900	—,3-chloro-*	C_5H_7Cl. *See* c899	102.57	...	...	$25–31^{30}$	1.0577^{15}	1.4708^{26}	i	s	s	...	...	chl s	**B5**², 36
Ω c901	—,octachloro-*	Perchlorocyclopentene. C_5Cl_8. *See* c899	943.68	nd (al) λ^{hp} <230 (4.0)	41	283^{760} 140^{10}	1.8200_4^{50}	1.5660^{50}	i	v^h	...	...	...	...	**B5**, 62
—	**3-Cyclopentene-carboxylic acid, 2,2,3-trimethyl-***	*see* **α-Campholytic acid**													
c902	**3-Cyclopenten-1-one, 3,4-bis(4-methoxy-phenyl)-***	...	294.35	ye br	129	...	...	...	i	s	s	s	...	oos s	**B8**, 355
c903	**1,2-Cyclopenteno-phenanthrene**	...	218.30	nd (al), cr (peth) λ^{al} 215 (4.3), 257 (4.75), 300 (4.27), 345 (3.0), 352 (3.0)	135–6	...	...	...	i	s	...	...	...	peth s^h	**E14**, 15
c904	**2,3-Cyclopenteno-phenanthrene**	...	218.30	pl or pr (al), nd (MeOH)	84–5	...	...	...	i	s	...	...	...	MeOH s^h	**E14s**, 19
c905	**9,10-Cyclo-pentenophen-anthrene**	...	218.30	pl (xyl), nd (*i*-PrOH)	155–6	...	...	...	i	s	...	...	s	...	**E14s**, 25
c906	**1,2-Cyclopenteno-phenanthrene, 3-methyl-**	$C_{18}H_{16}$. *See* c903	232.33	cr (aa)	126–7	...	...	...	i	...	...	...	...	aa s^h	**E14**, 16
Ω c907	**Cyclopropane***	Trimethylene.	42.08	...	−127.6	-32.7^{760}	0.720_4^{-79} 0.6769_4^{-30}	$1.3799^{-42.5}$	s	v	v	...	s	peth s	**B5**², 3
Ω c908	—,acetyl-	Cyclopropyl methyl ketone. C_5H_8O. *See* c907	84.13	λ^{al} 206 (3.12), 270 (1.33)	fp −68.4	114^{772}	0.89839_4^{20}	1.42514^{20}	s	s	s	...	...	...	**B7**², 5
Ω c909	—,1,1-dimethyl-*	C_5H_{10}. *See* c907	70.14	...	−108.96	20.63^{760}	0.6589_4^{20}	1.3668^{20}	i	s	v	...	...	sulf v	**B5**, 20
c910	—,methoxy-*	Cyprome ether. C_4H_8O. *See* c907	72.12	...	−119	44.7^{760}	0.8100_4^{20}	1.3802^{20}	s	s	v	...	s	oos s	**Am 69**, 2451
c911	—,methyl-*	C_4H_8. *See* c907	56.12	...	−177.2	4–5	0.6912_4^{-20}	...	δ	v	v	...	...	sulf s	**B5**², 3
Ω c912	**Cyclopropanecar-boxylic acid***	Ethyleneacetic acid.	86.09	...	18–9	182–4	1.0885_4^{20}	1.4390^{20}	s^h	s	s	...	...	...	**B9**², 3
Ω c913	—,nitrile	Cyclopropanecarbonitrile*. Cyclopropyl cyanide.	67.09	λ^{al} 208 (0.82)	...	135^{760} $69–70^{88}$	0.89461_4^{20}	1.42293^{20}	...	...	s	...	...	hx s	**B9**², 4
c914	—,2,2-dimethyl-3(2-methyl-propenyl)- (*cis, d*)*	*d, cis*-Chrysanthemumic acid. $C_{10}H_{16}O_2$. *See* c912.	168.24	pr, $[\alpha]_D^{22}$ +83.3 (chl, c = 1.6)	40–2	$95^{0.1}$	...	...	...	s	s	...	...	chl s	**C47**, 3247
c915	—,—(*cis, dl*)*	$C_{10}H_{16}O_2$. *See* c912	168.24	pr (AcOEt), cr (peth)	115–6	...	...	...	...	s	s	...	...	oos s AcOEt s^h	**B9**², 47
c916	—,—(*cis, l*)*	$C_{10}H_{16}O_2$. *See* c912	168.24	pr, $[\alpha]_D^{19}$ −83.3 (chl, c = 1.6)	41–3	$95^{0.1}$	...	...	...	s	s	...	...	chl s	**C47**, 3247
c917	—,—(*trans, d*)*	*d, trans*-Chrysanthemumic acid. $C_{10}H_{16}O_2$. *See* c912	168.24	pr, $[\alpha]_D^{20}$ +25.8 (chl, c = 2.5)	18–21	245 δd 135^{12}	...	...	δ	v	s	...	...	oos v chl s	**B9**², 45
c918	—,—(*trans, dl*)*	$C_{10}H_{16}O_2$. *See* c912	168.24	pr (AcOEt)	54	$145–6^{13}$	...	...	...	s	s	...	...	chl, AcOEt s	**B9**², 46
c919	—,—(*trans, l*)*	$C_{10}H_{16}O_2$. *See* c912	168.24	$[\alpha]_D^{20}$ −25.8 (chl, 2.9 %)	17–21	99–$100^{0.2-0.3}$	...	...	...	s	...	...	...	AcOEt v chl s	**C47**, 3247
c920	**1,1-Cyclopropane-dicarboxylic acid***	Ethylenemalonic acid. Vinaconic acid.	130.10	pr or nd (chl), pr (w + 1)	140–1	...	...	...	s	...	s	...	...	chl s	**B9**, 721
Ω c921	—,diethyl ester*	$C_9H_{14}O_4$. *See* c920	186.21	...	...	$214–6^{748}$ $99–100^{11}$	1.0566_{25}^{25}	1.4345^{18}	...	v	v	...	...	...	**B9**², 512
Ω c922	**1,2-Cyclopropane-dicarboxylic acid (*cis*)***	...	130.10	pr (eth or w)	139	...	...	...	s	v	v	...	δ	con HCl v chl, peth δ	**B9**², 513
c923	—(*trans, d*)*	$C_5H_6O_4$. *See* c922	130.10	$[\alpha]_D^{27}$ +84.87 (w)	175	...	...	...	v	v	v	...	δ	chl δ	**B9**, 724
Ω c924	—(*trans, dl*)*	$C_5H_6O_4$. *See* c922	130.10	nd (eth), pl (ace-bz)	175	210^{30}	...	...	v	v	v	...	δ	chl δ	**B9**², 514
c925	—(*trans, l*)*	$C_5H_6O_4$. *See* c922	130.10	$[\alpha]_D^{27}$ −84.40 (w)	175	...	...	...	s	v	v	...	δ	chl δ	**B9**, 724
c926	—,diethyl ester (*cis*)*	$C_9H_{14}O_4$. *See* c922	186.21	...	...	106.5–7.5	1.062_4^{12}	1.4450^{20}	...	s	s	...	...	...	**B9**², 513
c927	—,dimethyl ester (*cis*)*	$C_7H_{10}O_4$. *See* c922	158.16	...	...	$219–20^{760}$ 110^3	1.1584_4^{16}	1.4472^{14}	...	s	s	...	...	...	**B9**¹, 315
c928	—,1-bromo-*	$C_5H_5BrO_4$. *See* c922	209.00	pr (eth-chl), cr (ace-bz)	175	...	...	...	...	...	v	v	...	chl δ	**B9**², 514
c929	**1,2,3-Cyclopro-panetricarboxylic acid***	...	174.11	nd (w), cr (HCl)	220	...	...	...	s	s	...	...	...	chl δ	**B9**², 702

For explanations, symbols and abbreviations see beginning of table. For structural formulas see end of table.

No.	Name	Synonyms and Formula	Mol. wt.	Color, crystalline form, specific rotation and λ_{max} (log ε)	m.p. °C	b.p. °C	Density	n_D	Solubility w	al	eth	ace	bz	other solvents	Ref.
	Cyclotetrasiloxane														
Ω c930	**Cyclotetrasiloxane, octamethyl-**	...	296.62	...	17.5	175.8^{760}, 74^{20}	0.9561^{20}	1.3968^{20}	i	...	...	...	...	...	**Am 68,** 358
c931	**—,octaphenyl-**	...	793.20	nd (bz-al or aa)	200–1	$330–4^{1}$	...	...	i	δ	...	...	s	aa s^h	**Am 67,** 2173
c932	**—,1,3,5,7-tetramethyl, 1,3,5,7-tetraphenyl-**	...	544.91	cr (aa)	99	$237^{1.5}$	1.1183^{20}_{4}	1.5461^{20}	i	...	...	∞	...	hp ∞	**Am 70,** 1115
c933	**Cyclotrisiloxane, hexaphenyl-**		594.90	pl (bz-al or aa)	190	$290–300^{1}$	1.23^{25}_{4}	...	i	δ	...	...	s	aa s	**Am 69,** 488
c934	**—,1,3,5-triethyl-1,3,5-triphenyl-**	...	450.77	...	177.5	$166^{0.025}$	1.0952^{25}_{4}	1.5402^{25}	i	...	...	s	...	...	**Am 70,** 3758
c935	**—,1,3,5-trimethyl-1,3,5-triphenyl-** (*cis*)	...	408.68	pl (al)	99.5	$165–85^{1.5}$	...	...	i	s^h	...	...	...	...	**Am 70,** 1115
c936	**—,—(*trans*)**	$C_{21}H_{24}O_3Si_3$. *See* c935	408.68	nd (al)	39.5	$190^{1.5}$	1.1062^{20}_{4}	1.5397^{20}	i	s^h	...	...	...	...	**Am 70,** 1115
c937	**Cymarose**	4,5-Dihydroxy-3-methoxyhexanal*. 3-Methyldigitoxose. $CH_3(CHOH)_2CH(OCH_3)CH_2CHO$	162.19	pr (eth-peth), nd (ace) $[\alpha]^{21}_{D}$ +53.4 (w, c = 2.2) (mut)	100–2 (92–4)		...	...	v	v	δ	v	i	chl, peth δ	**B31,** 19
—	**Cymene**	*see* **Benzene, isopropyl(methyl)-***													
c938	**Cysteic acid** (*d*)	*L*-2-Amino-3-sulfopropanoic acid*. β-Sulfoalanine. $HO_2CCH(NH_2)CH_2SO_3H$	169.16	oct cr or nd (dil al), pr or nd (w + 1) $[\alpha]^{20}_{D}$ +8.66 (w)	260d (anh)		...	...	s^h	i	...	...	i	...	**B4**[3], 1713
c939	**—(*dl*)***	$HO_2CCH(NH_2)CH_2SO_3H$	169.16	pr (w)	272–4d		...	...	s^h	i	...	...	i	...	**B4**[2], 951
Ω c940	**Cysteine(*L*)**	*L*-β-Mercaptoalanine. $HSCH_2CH(NH_2)CO_2H$	121.16	cr (w) $[\alpha]^{30}_{D}$ +9.8 (w, c = 1.3) $\lambda^{w, pH=9}$ 231 (3.49)	*ca.* 240d		...	...	v	v	i	i	i	aa v	**B4**[2], 920
Ω c941	**Cystine(*D*)**	*D*-Dicystine. $[HO_2CCH(NH_2)CH_2S\text{-}]_2$	240.30	cr (dil NH_3) $[\alpha]^{20}_{D}$ +224 (1N HCl, c = 1.00)	247–9		...	...	δ s^h	i	...	...	...	ac, alk s	**B4**[2], 919
Ω c942	**—(*DL*)**	$[HO_2CCH(NH_2)CH_2S\text{-}]_2$	240.30	...	260		...	...	δ s^h	i	...	...	...	...	**B4**[2], 936
Ω c943	**—(*L*)**	$[HO_2CCH(NH_2)CH_2S\text{-}]_2$	240.30	hex pl or pr (w) $[\alpha]^{20}_{D}$ −223.4 (1N HCl, c = 1.00) $\lambda^{0.01N\,HCl}$ 250 (2.5)	260–1d		1.677	...	δ s^h	i	i	...	i	ac, alk s	**B4**[2], 925
Ω c944	**—(*meso*)**	$[HO_2CCH(NH_2)CH_2S\text{-}]_2$	240.30	$\lambda^{w, pH=7}$ 250 sh (2.57)	200–218d		...	...	i	...	...	...	...	...	**B4**[2], 936
Ω c945	**Cytidine**	Cytosine-3β-*D*-ribofuranoside. $C_9H_{13}N_3O_5$.	243.22	nd (dil al) $[\alpha]^{20.5}_{D}$ +35.3 (w, c = 1) λ^{w} 271 (3.95)	230–1d		...	...	v	δ	...	...	...	...	**B31,** 24
c946	**Cytidylic acid b**	3′-Cytosylic acid. $C_9H_{14}N_3O_8P$.	323.20	orh nd $[\alpha]^{20}_{D}$ +49.4 (w, c = 1) λ^{w} 271 (3.96)	233–4d		...	...	s^h	s^h	...	...	...	...	**B31,** 25
Ω c947	**Cytisine** (*l*)	Baptitoxine. Sophorine. Ulexine. $C_{11}H_{14}N_2O$.	190.25	orh pr (aa or ace), $[\alpha]^{17}_{D}$ −120 (w) λ^{MeOH} 235 (3.9), 306 (4.0)	154.5 sub	218^{2}	...	...	s	s	i	s	v	chl s, peth i	**B24**[2], 70
c948	**—,*N*-methyl-**	Caulophylline. $C_{12}H_{16}N_2O$. *See* c947	204.28	cr (w + 2), nd (al, bz or lig), pr (al) $[\alpha]^{18.5}_{D}$ −230 (w) λ^{MeOH} 234 (3.81), 309 (3.90)	137		...	...	s	v	...	v	v	chl v	**B24**[2], 70
—	**Cytosylic acid**	*see* **Cytidylic acid**													
Ω c949	**Cytosine**	4-Amino-1,2-dihydro-1,3-diazin-2-one*. 4-Amino-2-pyrimidone. 6-Aminouracil.	111.10	mcl or tcl pl (w + 1) $\lambda^{w, pH=7}$ 267 (3.79)	320–25d		...	...	s^h	δ	i	...	...	ac s	**B24,** 314
Ω c950	**—,5-methyl-**	$C_5H_7N_3O$. *See* c949	125.13	pr (w + ½) $\lambda^{w, pH=7}$ 278 (3.8) $\lambda^{w, pH=14}$ 289.5 (3.91)	270d		...	...	s	δ	i	...	...	ac s	**B24,** 355
—	**—,3β-D-ribofuranoside-**	*see* **Cytidine**													

For explanations, symbols and abbreviations see beginning of table. For structural formulas see end of table.

No.	Name	Synonyms and Formula	Mol. wt.	Color, crystalline form, specific rotation and λ_{max} (log ε)	m.p. °C	b.p. °C	Density	n_D	Solubility w	al	eth	ace	bz	other solvents	Ref.
	Daidzein														
—	**Daidzein**	*see* **Isoflavone, 4′,7-dihydroxy-**													
—	**Damascenine**	*see* **Benzoic acid, 3-methoxy-2-(methylamino)-, methyl ester**													
—	**Danilone**	*see* **1,3-Indanedione, 2-phenyl-**													
—	**Daphnetin**	*see* **Coumarin, 7,8-dihydroxy-**													
—	**Datiscetin**	*see* **Flavone, 2′,3,5,7-tetrahydroxy-**													
—	**D.D.D.**	*see* **Ethane, 1,1-bis(4-chlorophenyl)-2,2-dichloro-**													
—	**D.D.T.**	*see* **Ethane, 1,1-bis(4-chlorophenyl)-2,2,2-trichloro-**													
d1	**1,3-Decadiene***	$CH_3(CH_2)_5CH:CHCH:CH_2$	138.25			$168–70^{760}$	0.750^{20}		δ				s		**B1**, 260
d2	**3,7-Decadiene-5-yne,4,7-dipropyl-***	$C_2H_5CH:C(C_3H_7^n)C⋮CCH:C(C_3H_7^n)C_2H_5$	218.39			$125–7^{18}$	0.8131^{19}_{4}	1.4890^{20}	i				s		**B1**[1], 129
d4	**4,6-Decadiyne***	Dipentyne. $CH_3CH_2CH_2C⋮CC⋮CCH_2CH_2CH_3$	134.22			88^{12}	0.8695^{19}_{4}		i						**B1**[2], 248
Ω d5	**4,6-Decadiyne-3,8-diol,3,8-dimethyl-***	$C_2H_5COH(CH_3)C⋮CC⋮CCOH(CH_3)C_2H_5$	194.28		89–91				δ, s^h				δ, s^h	CCl_4 δ, MeOH s	**C53**, 18868
d6	**Decalin**(*cis*)	Bicyclo[4,4,0]decane. Decahydronaphthalene. Naphthane	138.25	λ < 210	−43.01	195.65^{760}, 69.4^{10}	0.8965^{20}_{4}	1.4810^{20}	i	∞	v	v	∞	chl v	**B5**[2], 5657
d7	—(*trans*)	$C_{10}H_{18}$. *See* d6	138.25	λ < 210	−30.4	187.25^{760}, 63^{10}	0.8699^{20}_{4}	1.4695^{20}	i	v	v	v	∞	chl v, MeOH δ	**B5**[2], 243
d8	—,**1-amino-**(*cis*)	*cis*-Decalylamine. $C_{10}H_{19}N$. *See* d6	153.27		(i) 8, (ii) −2	100^{12}								os, ac s	**B12**[2], 35
d9	—,—(*trans*)	$C_{10}H_{19}N$. *See* d6	153.27		(i) −18, (ii) −1	106^{16}								os, ac s	**B12**[2], 35
d10	—,**1-chloro-**	$C_{10}H_{17}Cl$. *See* d6	172.70			d^{760}, $114–6^{20}$									**Am 76**, 4420
d11	—,**2-methylene-**(*trans*)	$C_{11}H_{18}$. *See* d6	150.27			$200–1^{756}$, $82–2.5^{10}$	0.8897^{20}_{4}	1.4841^{22}							**E12B**, 106
d12	**1,3-Decalindione** (*cis*)	1,3-Diketodecalin	166.22	nd (bz, ace, dil al) λ^{diox} 245 (4.11), 301 (1.98)	124–5					s		s	s^h	alk s	**E12B**, 2804
d13	—(*trans*)	$C_{10}H_{14}O_2$. *See* d12	166.22	nd (bz, dil al) λ^{diox} 244 (4.02), 295 (2.03)	152–3					s				alk s	**E12B**, 2805
d14	**2,3-Decalindione** (*cis*)	2,3-Diketodecalin	166.22	rh (al, ace, lig)	88–9				δ	s				alk s	**E12B**, 2812
d15	—(*trans*)	$C_{10}H_{14}O_2$. *See* d14	166.22	nd (w), lf (dil al) λ^{al} 269 (3.87)	100–1				δ	s				os s, lig s^h	**E12B**, 2812
d17	**2-Decalincarboxylic acid** (*cis*)	Decahydro-2-naphthoic acid.	182.27	cr (hx)	81	150^{15}				s	s		s	chl s	**E12B**, 4086
Ω d18	**Decanal***	Capraldehyde. *n*-Decylaldehyde. $CH_3(CH_2)_8CHO$	156.27	$\lambda^{C_{10}H_{22}}$ 222 (1.9), 293 (1.4)	*ca.* −5	208–9, 81^7	0.830^{15}_{4}	1.4287^{20}	i	s	s	s			**B1**[2], 764
d19	—,**oxime**	Capraldoxime. $CH_3(CH_2)_8CH{:}NOH$	171.29	lf (dil MeOH)	69				i	s	s				**B1**, 711
Ω d20	**Decane***	$CH_3(CH_2)_8CH_3$	142.29		−29.7	174.1^{760}, 57.6^{10}	0.7300^{20}_{4}	1.41023^{20}	i	∞	s				**B1**[3], 519
Ω d21	—,**1-amino-***	*n*-Decylamine. $CH_3(CH_2)_9NH_2$	157.30		17, fr. 16.1	220.5^{760}, 95.8^{10}	0.7936^{20}_{4}	1.4369^{20}	δ	∞	∞	∞	∞	chl ∞	**B4**, 199
Ω d22	—,**1-bromo-***	*n*-Decyl bromide. $CH_3(CH_2)_9Br$	221.19		−29.2	240.6^{760}, 110^{10}	1.0702^{20}_{4}	1.4557^{20}	i		v			chl v	**B1**[3], 523
Ω d23	—,**2-bromo-**(*dl*)*	$CH_3(CH_2)_7CHBrCH_3$	221.19			111^{11}	1.0512^{20}	1.4526^{25}	i		v			chl v	**B1**[3], 523
d24	—,**1-bromo-10-fluoro-***	$F(CH_2)_{10}Br$	239.18			$131–2^{11}$	1.152^{20}_{4}	1.4512^{25}	i	v	v				**C51**, 7300
Ω d25	—,**1-chloro-***	*n*-Decyl chloride. $CH_3(CH_2)_9Cl$	176.73		−31.3	223.4^{760}, 97^{10}	0.8705^{20}_{4}	1.4379^{20}	i		v			chl v	**B1**[3], 522
d26	—,**1-chloro-10-fluoro-***	$F(CH_2)_{10}Cl$	194.72			115^9	0.957^{20}_{4}	1.4333^{25}	i	v	v				**C51**, 7300
d27	—,**1,10-diamino-***	Decamethylene diamine. $H_2N(CH_2)_{10}NH_2$	172.32		61.5	140^{12}									**B4**[2], 712
Ω d28	—,**1,10-dibromo-***	Decamethylene dibromide. $Br(CH_2)_{10}Br$	300.09	pl (al)	28	160^{15} δd, $127–30^4$	1.335^{30}	1.4905^{20}	i	δ, v^h	s				**B1**[2], 130
d29	—,**1-fluoro-***	*n*-Decyl fluoride. $CH_3(CH_2)_9F$	160.28		−35	186.2^{760}, 69^{10}	0.8194^{20}_{4}	1.4085	i		s				**B1**[2], 129
Ω d30	—,**1-iodo-***	*n*-Decyl iodide. $CH_3(CH_2)_9I$	268.18		−16.3	132^{15}	1.2546^{20}_{4}	1.4858^{20}	i	s	s				**B1**[3], 523
Ω d31	**Decanedioic acid***	Sebacylic acid. Sebacic acid. $HO_2C(CH_2)_8CO_2H$	202.25	lf	134.5	295^{100}, 232^{10}	1.2705^{20}_{4}	1.422^{133}	δ	s	s		i		**B2**[2], 608

For explanations, symbols and abbreviations see beginning of table. For structural formulas see end of table.

No.	Name	Synonyms and Formula	Mol. wt.	Color, crystalline form, specific rotation and λ_{max} (log ε)	m.p. °C	b.p. °C	Density	n_D	Solubility w	Solubility al	Solubility eth	Solubility ace	Solubility bz	Solubility other solvents	Ref.
	Decanedioic acid														
Ω d32	—,diamide	Decanediamide. Sebacamide*. $H_2NCO(CH_2)_8CONH_2$	200.28	pr or pl (aa)	210	...	...	...	i δ^h	i v^h	...	...	...	aa v	**B2,** 720
Ω d33	—,dibutyl ester	Di-*n*-butyl sebacate. $C_4H_9^nO_2C(CH_2)_8CO_2C_4H_9^n$	314.47	...	−10 (1)	344–5 227^{17}	0.9405^{15}	1.4433^{15}	i	...	s	...	...	...	**B2,** 719
Ω d34	—,dichloride	Sebacyl chloride. $ClCO(CH_2)_8COCl$	239.14	...	−2.5	220^{75} 165^{11}	1.1212^{20}_{4}	1.4684^{18}	d	d	...	...	...	...	**B2^2,** 610
Ω d35	—,diethyl ester*	Diethyl sebacate. $C_2H_5O_2C(CH_2)_8CO_2C_2H_5$	258.36	...	5 (1.25)	306^{773} 188^{19}	0.9646^{20}_{4}	1.4366^{20}	δ	s	...	s	i	...	**B2^1,** 293
d36	—,di(2-ethylbutyl) ester*	$(C_2H_5)_2CHCH_2O_2C(CH_2)_8CO_2CH_2CH(C_2H_5)_2$	370.58	...	−22	$344–6^{760}$	0.920^{25}_{4}	...	i	s	...	s	s	...	**B2,** 719
d37	—,di(2-ethylhexyl) ester*	$C_4H_9^nCH(C_2H_5)CH_2O_2C(CH_2)_8CO_2CH_2CH(C_2H_5)(C_4H_9^n)$	426.69	...	−48	256^5	0.912^{25}_{4}	1.451^{25}	i	s	...	s	s	...	**B2^3,** 1810
Ω d38	—,dimethyl ester*	Dimethyl sebacate. $CH_3O_2C(CH_2)_8CO_2CH_3$	230.31	lo pr	38	175^{20} 144^5	0.9882^{28}	1.4355^{28}	i	s	s	s	...	...	**B2,** 719
Ω d39	—,dinitrile	1,8-Dicyanooctane. Sebaconitrile. $NC(CH_2)_8CN$	164.25	...	...	204^{16}	0.9313^{20}_{4}	1.4474^{20}	i	...	...	...	...	...	**B2^2,** 610
d39[1]	—,monomethyl ester, mononitrile	9-Carbomethoxynonanonitrile. $NC(CH_2)_8CO_2CH_3$	197.28	...	3–4	178^{16}	0.934^{20}	1.4398^{25}	i	...	s	...	...	...	**B2^3,** 1816
Ω d40	**1,10-Decanediol***	Decamethyleneglycol. $HO(CH_2)_{10}OH$	174.29	nd (w)	72–5	$175–6^{14}$	...	...	δ	v	δ v^h	...	...	lig i	**B1^2,** 560
Ω d41	**1-Decanethiol***	*n*-Decyl mercaptan. $CH_3(CH_2)_9SH$	174.35	λ^{MeOH} 231 (3.64)	−26	240.6^{760} $125–7^{19}$	0.8443^{20}_{4}	1.4569^{20}	i	s	s	...	...	...	**B1^3,** 761
Ω d42	**Decanoic acid***	Capric acid. *n*-Decylic acid. $CH_3(CH_2)_8CO_2H$	172.27	nd	fr. 31.5	270^{760} $148–50^{11}$	0.8858^{40}_{4}	1.4288^{40}	δ	∞^h	s	v	v	chl, peth v	**B2^2,** 309
d43	—,amide	Capramide. Decanamide*. $CH_3(CH_2)_8CONH_2$	171.29	lf (eth)	108 fr. 98	...	0.999^{20}_{4}	1.4261^{110}	i	s	s	s	δ	CCl_4 δ	**B2,** 356
d44	—,anhydride*	Capric anhydride. $[CH_3(CH_2)_8CO]_2O$	326.53	lf	24.7	...	0.8865^{25}_{4}	1.400^{25}	i	s	s	...	...	...	**B2^2,** 311
Ω d45	—,chloride	Capryl chloride. $CH_3(CH_2)_8COCl$	190.72	...	−34.5	232^{760} 114^{15}	0.973^{8}_{4}	...	d	d	s	...	...	...	**B2,** 356
d46	—,decyl ester*	*n*-Decyl caprate. $CH_3(CH_2)_8CO_2(CH_2)_9CH_3$	312.54	...	9.7	219^{15}	0.8586^{20}	1.4423^{20}	i	...	s	...	...	...	**B2,** 356
Ω d47	—,ethyl ester*	Ethyl caprate. $CH_3(CH_2)_8CO_2C_2H_5$	200.33	...	−20	241.5^{760} $122–4^{13}$	0.8650^{20}_{4}	1.4256^{20}	i	∞	∞	...	...	chl ∞	**B2^3,** 842
d48	—,isopropyl ester*	Isopropyl caprate. $CH_3(CH_2)_8CO_2CH(CH_3)_2$	214.35	...	...	121^{10}	0.8543^{20}	1.4221^{25}	...	...	...	...	...	...	**B2^3,** 842
Ω d49	—,methyl ester*	Methyl caprate. $CH_3(CH_2)_8CO_2CH_3$	186.30	...	−18 (−10)	224^{760} 114^{15}	0.8730^{20}_{4}	1.4259^{20}	i	v	v	...	...	chl ∞	**B2^3,** 841
d50	—,nitrile	Caprinitrile. Nonyl cyanide. $CH_3(CH_2)_8CN$	153.27	...	fr. −17.9	243^{760} 106^{10}	0.8199^{20}_{4}	1.4296^{20}	i	∞	∞	∞	...	chl ∞	**B2,** 356
d51	—,piperazinium salt	$C_4H_{10}N_2.2CH_3(CH_2)_8CO_2H$	430.68	...	93	...	...	...	δ^h	s^h	i	...	...	...	**Am 70,** 2758
d52	—,propyl ester*	Propyl caprate. $CH_3(CH_2)_8CO_2C_3H_7^n$	214.35	...	...	128.5^{10}	0.8623^{20}	1.4280^{20}	...	...	...	...	...	...	**B2^3,** 842
d53	—,**2-bromo-***	$CH_3(CH_2)_7CHBrCO_2H$	251.17	...	4	$140–1^2$	1.1912^{24}	1.4595^{24}	...	...	v	...	...	...	**B2,** 356
d54	—,**10-fluoro-***	$F(CH_2)_9CO_2H$	190.26	...	49	$135–8^{10}$	...	...	δ	v	v	...	...	lig s	**B2^3,** 846
d55	—,**2-octyl-***	9-Heptadecanecarboxylic acid. $[CH_3(CH_2)_7]_2CHCO_2H$	284.49	nd or lf (al)	38.5	$212–8^{13}$	...	...	i	s^h	s	...	...	...	**B2^2,** 368
d56	—,**4-oxo-***	γ-Ketocapric acid. $CH_3(CH_2)_5COCH_2CH_2CO_2H$	186.25	(dil al)	70–1	...	...	...	...	s^h	...	...	...	...	**B3^2,** 449
Ω d57	**1-Decanol***	Decyl alcohol. $CH_3(CH_2)_9OH$	158.29	...	fr. 7	229^{760} $107–8^7$	0.8297^{20}_{4}	1.43719^{20}	i	∞	∞	∞	∞	chl ∞	**B1^2,** 459
Ω d58	**2-Decanol**(*dl*)*	$CH_3(CH_2)_7CHOHCH_3$	158.29	...	−2.4	211 $110–11^{10}$	0.8250^{20}_{4}	1.4326^{25}	...	s	∞	∞	s	...	**B1^3,** 1762
Ω d59	**4-Decanol***	$CH_3CH_2CH_2CHOH(CH_2)_5CH_3$	158.29	...	−11.8 to −8.8	210–1 96^{13}	0.8262^{20}_{0}	1.4320^{20}	i	s	...	...	...	...	**B1,** 426
d60	**1-Decanol, 10-chloro-***	$Cl(CH_2)_{10}OH$	192.73	...	12–3	$185–9^{15}$	0.9630^{25}	1.4578^{20}	i	v	v	...	...	...	**B1^3,** 1761
d61	—,**10-fluoro-***	$F(CH_2)_{10}OH$	176.28	...	ca. 22	$136–7^{15}$	0.919^{20}_{4}	1.4322^{25}	i	v	v	...	...	...	**C51,** 7300
Ω d62	**2-Decanone***	Methyl *n*-octyl ketone. $CH_3CO(CH_2)_7CH_3$	156.27	nd	14 fr. 3.1	$210–1^{767}$ $95–7^{12}$	0.8248^{20}	1.4255^{20}	i	s	s	...	...	...	**B1^2,** 764
Ω d63	**3-Decanone***	Ethyl *n*-heptyl ketone. $CH_3CH_2CO(CH_3)_6CH_3$	156.27	...	1–3.8	203^{754}	0.8251^{20}_{4}	1.4252^{20}	...	s	s	...	...	...	**B1^1,** 367
Ω d64	**4-Decanone***	*n*-Propyl *n*-hexyl ketone. $CH_3CH_2CH_2CO(CH_2)_5CH_3$	156.27	nd	−9	206–7 $87–9^{11}$	0.824^{20}_{0}	1.4240^{21}	i	∞	∞	...	...	...	**B1,** 711
d65	**Decasiloxane, dicosamethyl-**	$CH_3[Si(CH_3)_2O]_9Si(CH_3)_3$	755.63	...	...	183^4	0.925^{20}_{4}	1.3988^{20}	i	δ	...	...	s	lig s	**C47,** 4679
Ω d66	**1-Decene***	*n*-Decylene. $CH_3(CH_2)_7CH:CH_2$	140.27	...	fr. −66.3	170.56^{760} 54.3^{10}	0.7408^{20}_{4}	1.4215^{20}	i	∞	∞	...	...	...	**B1^3,** 858
d67	**5-Decene**(*cis*)*	*cis*-1,2-Dibutylethylene. $CH_3(CH_2)_3CH:CH(CH_2)_3CH_3$	140.27	...	−112	170^{739} 73^{30}	0.74451^{20}_{4}	1.4258^{20}	i	∞	∞	...	...	...	**B1^3,** 859
Ω d68	—(*trans*)*	$CH_3(CH_2)_3CH:CH(CH_2)_3CH_3$	140.27	...	−73	170.2^{739}	0.74012^{20}_{4}	1.4243^{20}_{4}	i	∞	∞	...	...	...	**B1^3,** 859
d69	**1-Decene, 2-bromo-***	$CH_3(CH_2)_7CBr:CH_2$	219.17	...	...	$115–6^{22}$	1.0844^{20}_{4}	1.4629^{20}	...	...	...	...	...	...	**B1^3,** 859
d70	**2-Decene, 1-bromo-***	$CH_3(CH_2)_6CH:CHCH_2Br$	219.17	...	...	121^{17}	1.074^{18}_{4}	1.4716^{18}	...	...	...	...	...	lig s	**B1^3,** 859

For explanations, symbols and abbreviations see beginning of table. For structural formulas see end of table.

No.	Name	Synonyms and Formula	Mol. wt.	Color, crystalline form, specific rotation and λ_{max} (log ε)	m.p. °C	b.p. °C	Density	n_D	Solubility w	al	eth	ace	bz	other solvents	Ref.
	1-Decen-3-yne*														
d71	**1-Decen-3-yne***	$CH_3(CH_2)_5C⋮CCH{:}CH_2$	136.24			76^{20} 45^{4}	0.7873^{20}	1.4620^{20}	...				...		**B1**[3], 1049
d72	**1-Decen-4-yne***	$CH_3(CH_2)_4C⋮CCH_2CH{:}CH_2$	136.24			$73–4^{22}$	0.7880^{20}	1.445^{20}	...				...		**B1**[3], 1049
d73	**2-Decen-4-yne***	$CH_3(CH_2)_4C⋮CCH{:}CHCH_3$	136.24			55^{5}	0.7850^{25}_{4}	1.4609^{25}	...				...		**B1**[3], 1049
Ωd74	**1-Decyne***	*n*-Octylacetylene. $CH_3(CH_2)_7C⋮CH$	138.25		−36	174^{760} 57^{10}	0.7655^{20}_{4}	1.4265^{20}	i	s	s	...	...	os s	**B1**[3], 1016
d75	**3-Decyne***	$CH_3CH_2C⋮C(CH_2)_5CH_3$	138.25			$175–6^{760}$	0.765^{21}_{4}	1.433^{21}_{4}	...				...		**B1**[3], 1017
d76	**4-Decyne***	$CH_3(CH_2)_4C⋮C(CH_2)_2CH_3$	138.25			74.5^{19}	0.772^{17}_{4}	1.436^{17}	...				...		**B1**[3], 1017
d77	**5-Decyne***	Dibutylacetylene. $CH_3(CH_2)_3C⋮C(CH_2)_3CH_3$	138.25		−73	177^{751} 78.8^{25}	0.7690^{20}_{4}	1.4331^{20}	i	s	s	...	...		**B1**[3], 1017
d78	**4-Decyne, 3,3-dimethyl-***	$CH_3CH_2C(CH_3)_2C⋮C(CH_2)_4CH_3$	166.31			86^{20}	0.7731^{20}_{4}	1.4399^{20}	...	...	...		...		**B1**[3], 1026
—	**Dehydroacetic acid**	*see* **4-Hexenoic acid, 2-acetyl-5-hydroxy-3-oxo-,** lactone													
Ωd79	**Dehydrocholic acid**	3,7,12-Trioxocholanic acid.	402.54	(ace), $[\alpha]^{20}_{D}$ +26 (al, c = 1.4) λ^{al} 290 (2.95), $\lambda^{0.1N\,NaOH}$ 285 (3.04)	237				i	δ	i	s	δ	chl s AcOEt s	**E14**, 211
d80	**Dehydro-ergosterol**	$\Delta^{5:6,7:8,9:11,22:23}$-Ergostatetraen-3-ol. $C_{28}H_{42}O$.	394.65	lf (al + 1w), pl (al), nd (eth) $[\alpha]^{15}_{D}$ +149.2 (chl, c = 1.9) λ^{al} 310 (3.95), 324 (4.00), 340 (3.82)	146	$230^{0.5}$				s^{h}	v	s	v	chl v AcOEt s MeOH δ	**J1957**, 93
—	**Dehydromucic acid**	*see* **2,5-Furandicarboxylic acid**													
Ωd81	**Delphinidine chloride**	3,3′,4′,5,5′,7′-Hexahydroxyflavinium chloride. $C_{15}H_{11}ClO_7$.	338.70	br pr, nd or pl (HCl) λ^{al} 275 (4.20), 522 (4.54)	>350 (anh)				v	v			...	MeOH v AcOEt s	**B18**[2], 247
d82	**Delphinine**	$C_{33}H_{45}NO_9$	599.73	orh (al) $[\alpha]^{25}_{D}$ +25 (al) λ^{al} 230 (4.15), 273 (2.96)	198–200d				i	s	s	s	...	chl s	**J1952**, 1750
d82[1]	**Demissine**	$C_{50}H_{33}NO_{20}$	1050.22	nd (al) [α]-20(Py)	276–9 (305–8d)				...	s^{h}	...		...	MeOH s^{h}, dil ac s	**C45**, 3855
d83	**Derritol**	$C_{21}H_{22}O_6$	370.41	ye nd (MeOH) $[\alpha]^{20}_{D}$ −66.2 (chl, c = 3)	164 (161)	220– $5^{0.06}$			i	...	...		...	alk v	**B18**[2], 223
Ωd84	**Deserpidine**	Canescine. 11-Desmethoxyreserpine. $C_{32}H_{38}N_2O_8$.	578.67	nd or pr $[\alpha]^{24.5}_{D}$ −137 (chl) λ^{al} 217 (4.81), 271 (4.30)	229–32				i	s^{h}	...		...	chl s	**Am 77**, 4335
—	**Desmethyl-morphine**	*see* **Normorphine**													
—	**Desoxalic acid**	*see* **1,1,2-Ethanetricarboxylic acid, 1,2-dihydroxy-***													
—	**Desoxybenzoin**	*see* **Ketone, benzyl phenyl**													
Ωd85	**Desoxycholic acid**	3,12-Dihydroxycholanic acid. $C_{24}H_{40}O_4$.	392.59	(al) $[\alpha]^{20}_{D}$ +57 (al) $\lambda^{65\%H_2SO_4}$ 310 (3.79), 380 (3.29)	176				i	v	δ	δ	i	chl δ aa δ	**C48**, 1147
d86	**Desoxycorti-costerone**	Δ^4-Pregnene-3,20-dion-21-ol. $C_{21}H_{30}O_3$.	330.47	pl (eth) $[\alpha]^{20}_{D}$ +178 (al, c = 1) λ^{al} 240 (4.2)	141–2				δ	v	s	v	...		**E14**, 153
d87	**Desthiobiotin**	$C_{10}H_{18}N_2O_3$	214.27	lo nd (w) $[\alpha]^{21}_{D}$ +10.7 (w, c = 2)	156–8				s	...	...	...	...		**J1948**, 1552
d88	—, methyl ester	$C_{11}H_{20}N_2O_3$. *See* d87	228.29	cr (MeOH) $[\alpha]^{28}_{D}$ +2.6 (chl, c = 2)	69–70	$194–7^{0.03}$			...	s^{h}	...		...	MeOH s chl v	**C44**, 4934
—	**Desyl chloride**	*see* **Ketone, benzyl phenyl, α-chloro-**													
—	***O*-Deutero-methanol**	*see* **Methanol-*d***													
—	**Deuteroxy(tri-deutero)methane**	*see* **Methan-d_3-ol-*d***													
—	**Dexedrine**	*see* **Amphetamine**													
Ωd89	**Dextrin** (starch)	Amylin. $(C_6H_{10}O_5)_n$	(162.14)$_n$	amor $[\alpha]_D$ > +200	chars		1.0384^{20}_{4}		s	i	i	...	...		**J1925**, 636
—	**Dextronic acid**	*see* **Gluconic acid(*D*)**													
—	**Dextrose**	*see* **Glucose(*D*)**													
d90	**Dextropimaric acid, methyl ester**	$C_{21}H_{32}O_2$	316.49	$[\alpha]_D$ +60.5 (MeOH)	69	$149^{0.03}$	1.030^{19}_{4}	1.5208^{19}	...	v	v	...	...		**B9**[2], 435

For explanations, symbols and abbreviations see beginning of table. For structural formulas see end of table.

No.	Name	Synonyms and Formula	Mol. wt.	Color, crystalline form, specific rotation and λ_{max} (log ε)	m.p. °C	b.p. °C	Density	n_D	Solubility w	al	eth	ace	bz	other solvents	Ref.
	Diacetamide														
—	**Diacetamide**	*see* **Acetic acid,** amide, *N*-acetyl-													
—	**Diacetanilide**	*see* **Acetic acid,** amide, *N*-acetyl-*N*-phenyl-													
—	**Diacetoacetic acid,** ethyl ester	*see* **Butanoic acid, 2-acetyl-3-oxo-,** ethyl ester													
—	**Diacetonamine**	*see* **2-Pentanone, 4-amino-4-methyl-***													
—	**Dialuric acid**	*see* **Barbituric acid, 5-hydroxy-**													
d91	**Diazoamino-benzene**	1,3-Diphenyltriazene	197.24	ye lf or pr (al) λ^{al} 275 (3.9), 355 (4.3)	98	...	...	...	i	v^h	v	...	v	Py v	**B16²**, 351
d92	—(isomer)	$C_{12}H_{11}N_3$. *See* d91	197.24	ye pr (bz)	80–1	...	...	...	...	...	...	...	s	lig s	**B16²**, 351
d93	—,2,2′-dimethyl-	$C_{14}H_{15}N_3$. *See* d91	225.30	og (al)	51	...	...	...	...	s	s	...	...	lig s	**B16,** 703
d94	—,2,4′-dimethyl-	$C_{14}H_{15}N_3$. *See* d91	225.30	ye nd (lig)	120	...	...	...	...	...	...	...	...	lig v	**B16,** 708
d95	—,2′,3-dimethyl-	$C_{14}H_{15}N_3$. *See* d91	225.30	ye cr (lig)	74	...	...	...	...	...	...	s	s	lig v	**B16,** 705
d96	—,3,3′-dimethyl-	$C_{14}H_{15}N_3$. *See* d91	225.30	ye nd (peth)	52	...	...	...	...	...	s	s	s	os v	**B16¹**, 407
d97	—,3,4′-dimethyl-	$C_{14}H_{15}N_3$. *See* d91	225.30	ye nd (lig)	97	...	...	...	...	...	...	...	...	os s lig s^h	**B16,** 708
d98	—,4,4′-dimethyl-	$C_{14}H_{15}N_3$. *See* d91	225.30	red-ye nd (lig), pr (al) λ^{cy} 238 (4.23), 295 (4.00), 355 (4.34)	118	...	...	...	...	s^h	...	...	...	sulf s lig s^h	**B16²**, 355
d99	—,4,4′-dinitro-	$C_{12}H_9N_5O_4$. *See* d91	287.24	ye nd (al), lf (bz) λ^{al} 273 (3.75), 400 (4.6)	240d	...	...	...	i	δ^h	s	δ	δ	chl δ	**B16²**, 354
d100	—,3-methyl-	$C_{13}H_{13}N_3$. *See* d91	211.27	ye nd (lig)	86	...	...	...	...	...	...	...	...	...	**B16,** 714
d101	—,4-methyl-	$C_{13}H_{13}N_3$. *See* d91	211.27	ye pl (lig)	86–7	...	...	...	i	...	...	...	...	...	**B16²**, 354
d102	**1,1′-Diazoamino-naphthalene**	...	297.36	ye lf (al)	exp >100	...	...	...	...	...	...	...	...	...	**B16,** 716
d103	**2,2′-Diazoamino-naphthalene**	...	297.36	red nd (xyl)	156	...	...	...	...	...	...	...	...	con sulf →vt	**Ber 19,** 1282
—	**Diazoamino-toluene**	*see* **Diazoaminobenzene, dimethyl-**													
—	**Diazomethane**	*see* **Methane, diazo-**													
d104	**1,2:3,4-Dibenz-anthracene**	Dibenz[a,c]anthracene. Naphtho-2′,3′:9,10-phenanthrene.	278.36	nd (aa or al) λ^{al} 248 (4.62), 265 (4.73), 275 (5.01), 286 (5.13), 321 (3.89), 349 (3.45), 374 (2.63)	205 (200–2)	...	...	...	...	...	...	...	s	AcOH s^h peth δ	**B5²**, 668
d105	**1,2:5,6-Dibenz-anthracene**	Dibenz[a,h]anthracene	278.36	pl (dil ace), silv lf (aa) λ^{bz} 290 (5.0), 300 (5.2), 373 (3.0), 385 (2.6), 395 (3.1)	269–70 (262)	...	...	...	i	δ	...	s	s	CS_2 s aa s	**E14s,** 604
d106	**1,2:6,7-Dibenz-anthracene**	1,2-Benzonaphthacene. Isopentaphene.	278.36	ye lf or nd (xyl) λ^{bz} 362 (3.16), 380 (3.50), 401 (3.80), 425 (3.97), 425.5 (4.00)	263–4	sub 275^{2-4}	...	...	...	...	...	...	...	con sulf s	**B5²**, 667
d107	**1,2:7,8-Dibenz-anthracene**	Dibenz[a,j]anthracene. Dinaphthanthracene.	278.36	og lf or nd (bz) λ^{bz} 304 (5.2), 338 (4.3), 351 (3.7), 385 (2.5) 395 (2.4)	197–8	...	...	...	i	δ	δ	...	δ	aa i peth s	**B5²**, 668
d108	**1,2:5,6-Dibenz-anthracene, 4′,4″-dihydroxy-**	$C_{22}H_{14}O_2$. *See* d105	310.36	og (bz) λ^{al} 238 (4.80), 298 (4.88), 324 (4.36), 338 (4.28), 354 (4.35), 365 (3.69), 384 (3.99), 403 (4.15)	415–8	sub $300^{2-3.10^{-4}}$	...	...	i	δ^h	i^h	...	...	...	**E14s,** 620
d109	**Dibenzanthrone**	Violanthrone.	456.51	vt-bl or bk nd ($PhNO_2$ or quinoline) $\lambda^{con\ sulf}$ 380 (4.2), 570 (4.4), 755 (4.3), 850 (4.3)	490–5d	...	...	...	...	i	...	...	i	xyl, Py sulf s (vt) aa i	**E14s,** 907

For explanations, symbols and abbreviations see beginning of table. For structural formulas see end of table.

1,2:5,6-Dibenzofluorene

No.	Name	Synonyms and Formula	Mol. wt.	Color, crystalline form, specific rotation and λ_{max} (log ε)	m.p. °C	b.p. °C	Density	n_D	Solubility w	al	eth	ace	bz	other solvents	Ref.
d110	**1,2:5,6-Dibenzo-fluorene**	13H-Dibenzo[a,g]fluorene	266.35	pl (bz-al or AcOEt) λ^{diox} 255 (4.9), 278 (4.2), 294 (4.2), 334 (4.4), 351 (4.4)	174–5	195–$200^{0.5}$			i	δ	δ		s	to, $PhNO_2$ s	**E14s,** 520
d111	**1,2:7,8-Dibenzo-fluorene**	13H-Dibenzo[b,h]fluorene	266.35	lf (bz), pl (aa) λ^{al} 258 (4.4), 264 (4.8), 278 (5.0), 318 (3.8), 330 (4.0), 342 (4.2), 358 (3.9)	234				i	δ	δ		s		**E14s,** 518
d112	**1,2:6,7-Dibenzo-9-fluorenone**		280.33	og-ye pl or pr (xyl or aa)	214	sub 190–$200^{0.04}$								to s^h sulf s (red)	**E14s,** 514
—	**Dibenzolfulvene**	*see* **Fluorene, 9-methylene-**													
Ωd113	**Dibenzofuran**	Diphenylene oxide	168.21	lf or nd (al) λ^{al} 217 (4.5), 245 (4.0), 250 (4.3), 280 (4.2), 285 (4.2), 295 (4.0), 300 sh (4.2)	86–7	287^{760}	1.0886^{99}_{4}	1.6079^{99}_{4}	δ	s	v	s	s^h	aa v	**B17**[2], 67
d114	**—,1-amino-**	$C_{12}H_9NO$. *See* d113	183.21	br nd (dil MeOH)	85										**Am 75,** 4845
d115	**—,2-amino-**	$C_{12}H_9NO$. *See* d113	183.21	pl (dil al) λ^{al} 218 (4.46), 238 (4.13), 261 (4.11), 314 (4.22)	128				i	v^h	v				**B18**[2], 423
d116	**—,3-amino-**	$C_{12}H_9NO$. *See* d113	183.21	(dil al) λ^{al} 217 (4.44), 237 (4.13), 261 (4.11), 302 sh (4.15), 313 (4.22)	94 (99)										**B18**[1], 557
d117	**—,4-amino-**	$C_{12}H_9NO$. *See* d113	183.21	(al)	85										**Am 61,** 1365
d118	**—,2-bromo-**	$C_{12}H_7BrO$. *See* d113	247.10	nd (al), lf (aa) λ^{al} 222 (4.58), 251 (4.19), 288 (4.25), 310 (3.67)	110	220^{40}				s				aa δ	**B17**[2], 68
d119	**—,3-bromo-**	$C_{12}H_7BrO$. *See* d113	247.10	lf (al) λ^{al} 219 (4.55), 254 (4.31), 290 (4.31), 300 (4.22)	120	220^{40}			i	s^h	s				**J1931,** 529
d120	**—,4-bromo-**	$C_{12}H_7BrO$. *See* d113	247.10		67										**Am 76,** 5783
d122	**—,2,8-dibromo-**	$C_{12}H_6Br_2O$. *See* d113	326.00	lf (al)	199–200					s	v		v	aa v	**B17**[2], 69
d123	**—,3,6-dinitro-**	$C_{12}H_6N_2O_5$. *See* d113	258.19		245										**C49,** 720
d124	**—,3,8-dinitro-**	$C_{12}H_6N_2O_5$. *See* d113	258.19		255–6					δ		s	i	xyl δ lig i	**B17**[2], 69
d125	**—,1-nitro-**	$C_{12}H_7NO_3$. *See* d113	213.20	lt ye nd (al) λ^{al} 210 (4.65), 244 (4.10), 336 (3.92)	120–1										**Am 66,** 1884
Ωd126	**—,3-nitro-**	$C_{12}H_7NO_3$. *See* d113	213.20	ye nd (aa) λ^{al} 322 (4.23)	181–2	180–5^3			i	δ^h	δ			aa s^h	**B17**[2], 69
d127	**—,4-nitro-**	$C_{12}H_7NO_3$. *See* d113	213.20	ye nd	138–9 (126)	190–205^{15}									**Am 75,** 4844
d128	**1-Dibenzofuran-carboxylic acid**		212.22	nd (50% al)	232–3										**C51,** 12065
Ωd129	**2-Dibenzofuran-carboxylic acid**		212.22	nd (dil al) nd (aa)	246–7 (252–5)				δ	s^h	s				**B18,** 313
d130	**4-Dibenzofuran-carboxylic acid**		212.22	nd (al)	209–10										**Am 56,** 1416
d131	**1,2:6,7-Dibenzo-phenanthrene**	Naphtho-2′,1′:1,2-anthracene. Benzo[b]chrysene. 3,4-Benzotetraphene.	278.36	pa gr-ye lf (xyl) λ^{diox} 247 (4.56), 280 (4.90), 288 (5.13), 305 (4.68), 365 (3.93), 384 (3.78), 392 (3.60)	294				i				s	diox s	**B5**[2], 668

For explanations, symbols and abbreviations see beginning of table. For structural formulas see end of table.

No.	Name	Synonyms and Formula	Mol. wt.	Color, crystalline form, specific rotation and λ_{max} (log ε)	m.p. °C	b.p. °C	Density	n_D	Solubility w	al	eth	ace	bz	other solvents	Ref.
	1,2:7,8-Dibenzophenanthrene														
—	**1,2:7,8-Dibenzo-phenanthrene**	*see* Picene													
d132	**2,3:6,7-Dibenzo-phenanthrene**	Dibenzo[b,h]phenanthrene. Pentaphene.	278.36	ye-gr nd or lf (xyl) λ^{al} 245 (4.92), 257.5 (5.06), 314.5 (5.00), 345 (4.47), 356 (4.44), 399 (2.98), 412 (2.56), 423.5 (2.98)	257	...	...	...	i	δ	δ	...	s	xyl δ	**B5**^{3}, 2552
—	**Dibenzo-1,4-pyran**	*see* Xanthene													
d133	**1,2:4,5-Dibenzo-pyrene**	Dibenzo[a,e]pyrene	302.38	pa ye nd (xyl) λ^{bz} 294.5 (4.72), 306.5 (4.83), 343 (4.06), 360 (4.24), 378 (4.30), 396 (3.24), 416 (2.84)	233–4	...	...	...	...	δ	...	δ	δ	to s^h con sulf s (ye-red) aa δ	**E14s**, 805
—	**Dibenzopyrrole**	*see* Carbazole													
Ωd134	**Dibenzothiophene**	Diphenylene sulfide	184.27	nd (dil al or lig) λ^{al} 234 (4.9), 256 (4.3), 262 (4.1), 286 (4.2), 303 (3.5), 325 (3.6)	99–100	332–3 152–4^3	...	...	s v^h	v	...	...	v	MeOH s^h	**B17**^{2}, 70
d135	**—,2-amino-**	$C_{12}H_9NS$. *See* d134	199.28	lf (dil al) λ^{al} 237 (4.62), 262 sh (4.20), 289 (3.85), 296 (3.90), 346 (3.43)	122–3	...	...	...	...	...	...	...	...	...	**B18**^{2}, 423
Ωd136	**—,3-amino-**	$C_{12}H_9NS$. *See* d134	199.28	(dil al) λ^{al} 242 (4.70), 282 (4.17), 303 (4.11)	129–31	...	...	...	...	...	...	...	...	...	**B18**^{2}, 423
Ωd137	**—,2-bromo-**	$C_{12}H_7BrS$. *See* d134	263.16	nd (al)	125–6 (98–9)	...	...	...	...	...	...	...	...	...	**B17**^{2}, 70
d138	**—,—,monoxide**	3-Bromodiphenylene sulfonide.	279.16	...	171–2	...	...	...	...	...	...	...	...	...	**B17**^{2}, 71
d139	**—,3-bromo-, dioxide**	...	295.16	...	224–5	...	...	...	...	...	...	...	...	...	**Am 75,** 3843
d140	**—,4-bromo-**	$C_{12}H_7BrS$. *See* d134	263.16	(al)	84	...	...	...	...	...	...	...	...	...	**Am 76,** 5786
d141	**—,2,8-diamino-**	$C_{12}H_{10}N_2S$. *See* d134	214.29	nd (al)	194–6	...	...	...	...	v^h	...	...	...	...	**B18**^{2}, 427
d142	**—,3,7-diamino-**	$C_{12}H_{10}N_2S$. *See* d134	214.29	pa ye cr	169–70	...	...	...	...	...	...	...	...	...	**Am 74,** 1166
d143	**—,—,dioxide**	Benzidine sulfone	246.29	ye nd (al)	327–8	...	...	...	i^h	i^h	i	...	i^h	...	**B18**, 591
d147	**—,2,8-dibromo-**	$C_{12}H_6Br_2S$. *See* d134	342.06	(aa) λ^{al} 242 (4.68), 290 (4.33), 329 (4.01)	229	...	...	...	...	...	...	...	...	...	**Am 69,** 1920
d148	**—,—,dioxide**	...	374.06	(aa)	361–2	...	...	...	...	...	...	...	...	...	**Am 76,** 5786
d150	**—,3,7-dinitro-, dioxide**	...	306.26	(ace)	273–5	...	...	...	...	...	...	...	...	...	**Am 74,** 1165
Ωd151	**—,2-nitro-**	$C_{12}H_7NO_2S$. *See* d134	229.26	pa ye nd λ^{al} 227 (4.4), 241 (4.4), 253 (4.4), 273 (4.3), 305 (3.9), 338 (3.9)	186	...	...	...	...	...	...	...	...	...	**B17**^{2}, 71
d152	**—,—,dioxide**	...	261.26	(ace) λ^{al} 252 (4.50), 275 sh (3.97), 332 (3.20)	257–8	...	...	...	...	...	...	...	...	...	**Am 74,** 1165
Ωd153	**—,3-nitro-**	$C_{12}H_7NO_2S$. *See* d134	229.26	pa ye (dil al) λ^{al} 222 (4.5), 240 sh (4.2), 320 (4.2)	153–4	...	...	...	...	...	...	...	...	...	**Am 74,** 1165
d154	**—,—,monoxide**	...	245.26	(al)	210	...	...	...	...	...	...	...	...	...	**Am 70,** 1749
d155	**2-Dibenzothio-phenecarboxylic acid**	...	228.28	(al)	255	...	...	...	...	...	...	...	...	...	**B18**^{2}, 279

For explanations, symbols and abbreviations see beginning of table. For structural formulas see end of table.

No.	Name	Synonyms and Formula	Mol. wt.	Color, crystalline form, specific rotation and λ_{max} (log ε)	m.p. °C	b.p. °C	Density	n_D	Solubility w	al	eth	ace	bz	other solvents	Ref.
	4-Dibenzothiophenecarboxylic acid														
d156	**4-Dibenzothiophenecarboxylic acid**	...	228.28	(dil MeOH) λ^{al} 282 (3.85), 340 (3.51)	261–2	...	...	...	...	...	...	...	...	...	**Am 74,** 266
d157	—,dioxide	...	260.28	...	337–8	...	...	...	...	...	...	...	...	...	**Am 75,** 278
—	**Dibenzoxazine**	*see* **Phenoxazine**													
—	**Dibenzyl**	*see* **Ethane, 1,2-diphenyl-**													
d158	**Diborane, methylthio-**	$CH_3SB_2H_5$	73.76	...	−101.5	53^{760} (extrap.) (-35^{9})	...	...	d	d	...	s	s	os ∞	**Am 76,** 3307
—	**Dichloramine B**	*see* **Benzenesulfonic acid,** amide, *N,N*-dichloro-													
—	**Dichloramine T**	*see* **4-Toluenesulfonic acid,** amide, *N,N*-dichloro-													
—	**Dichlorophene**	*see* **Methane, bis(5-chloro-2-hydroxyphenyl)-**													
d159	**Dicoumarin**(*cis*)	...	292.30	lf (aa)	262	...	...	...	i	...	...	...	...	...	**B19,** 181
d160	—(*trans*)	...	292.30	nd or pl (aa)	>275	...	...	...	...	...	...	...	...	aa δ	**B19,** 181
d161	**Dicoumarol**	4,4′-Dihydroxy-3,3′-methylene biscoumarin. Melitoxin. $C_{19}H_{12}O_6$	336.31	nd λ 280 (4.34)	288–92	...	...	...	i	i	i	i	δ	chl δ	**B19,** 197
d162	**Dictamnine**	2,3-(4-Methoxy-2,3-quinolino)-furan.	199.21	pr (al) λ^{al} 236 (4.78), 309 (3.95), 329 (3.88)	133–4	...	...	...	δ	v^h	s	...	...	chl, AcOEt con sulf s	**B27^{2},** 79
d163	**Dicyclohexadiene**	...	160.26	...	...	$229–30^{766}$ 104^{16}	0.9950^{20}_{4}	$1.5267^{20.5}$	i	δ	v	v	v	aa v	**E13,** 1038
Ω d164	**α-Dicyclopentadiene** (*endo* form)	...	132.21	λ^{iso} 217 (1.4)	32	170 δd $64–5^{14}$	0.9302^{35}_{4}	1.5050^{35}	...	s	v	...	...	CCl_4 s aa s peth s	**E13,** 1018
d165	—,**3,4,5,6,7,8,8a-heptachloro-**	Heptachlor	373.32	wh, λ^{al} 236 (4.78), 309 (3.96), 328 (3.88)	95–6 (93)	...	$1.57–9^{9}$	...	i	s	s	...	s	lig s	**B5^{3},** 1286
Ω d166	—,**tetrahydro-** (*endo* form)	Tricyclodecane	136.24	(al or aa)	77	193^{769} $86–7^{12}$	0.9128^{79}	1.4726^{79}	i	s	...	...	...	aa s	**B5^{3},** 390
Ω d167	**Dieldrin**	Octalox	380.93	...	175–6	...	1.75	...	i	δ	...	s	s	...	**B5^{3},** 390
—	**Diethanolamine**	*see* **Amine, diethyl, 2,2′-dihydroxy-**													
Ω d168	**Diethylene glycol**	2,2′-Dihydroxydiethyl ether. 2,2′-Oxydiethanol. $HOCH_2CH_2OCH_2CH_2OH$	106.12	...	−10.5	245^{760} 133^{14}	1.1197^{15}_{4}	1.4472^{20}	s	s	s	...	...	...	**B1^{2},** 520
d169	—,diacetate	$CH_3CO_2CH_2CH_2OCH_2CH_2O_2CCH_3$	190.20	...	...	$245–51$ $110–35^{16}$	1.1078^{15}_{15}	1.4348^{20}	...	v	...	...	...	...	**B2^{2},** 155
d170	—,diethyl ether	Diethyl carbitol. $C_2H_5OCH_2CH_2OCH_2CH_2OC_2H_5$	162.23	...	...	189^{760}	0.9063^{20}_{4}	1.4115^{20}	v	v	s	...	...	os v	**B1^{3},** 2098
d171	—,dioctadecanoate	Diethylene glycol distearate. Glycosterin. $[CH_3(CH_2)_{16}CO_2CH_2]_2O$	611.02	wx	54–5	...	0.9333^{20}_{4}	...	...	i	i	...	...	...	**C50,** 9692
d172	—,dioleate	$[CH_3(CH_2)_7CH{:}CH(CH_2)_7CO_2CH_2CH_2]_2O$	635.04	pa ye liq	...	...	0.9310^{20}_{4}	...	...	∞	∞	...	...	...	**C47,** 4618
d173	—,monobutyl ether	Butyl carbitol. $CH_3(CH_2)_3OCH_2CH_2OCH_2CH_2OH$	162.23	...	−68.1	231^{760} 118^{12}	0.9553^{20}_{4}	1.4321^{20}	∞	v	v	v	s	...	**B1^{2},** 521
d174	—,—,acetate	$CH_3(CH_2)_3OCH_2CH_2OCH_2CH_2O_2CCH_3$	204.27	...	−32	245^{760}	0.985^{20}_{4}	1.4262^{20}	s	∞	∞	∞	...	oos ∞	**B2^{3},** 308
d175	—,monododecanoate	Diethylene glycol monolaurate. Glaurin. $CH_3(CH_2)_{10}CO_2CH_2CH_2OCH_2CH_2OH$	288.43	lt ye	17–8	$>270^{760}$	0.96^{25}_{25}	...	...	∞	∞	∞	s	to s	**C37,** 3202^{1}
Ω d176	—,monoethyl ether	Carbitol. Cellosolve. $CH_3CH_2OCH_2CH_2OCH_2CH_2OH$	134.18	hyg liq	...	195^{760}	0.9881^{20}_{4}	1.4300^{20}	∞	∞	v	∞	∞	oos ∞	**B1^{2},** 520
d177	—,—,acetate	Carbitol acetate. $CH_3CH_2OCH_2CH_2OCH_2CH_2O_2CCH_3$	176.22	hyg liq	−25	218^{760}	1.0096^{20}_{4}	1.4230^{20}	∞	∞	∞	v	...	oos s	**B2^{2},** 155
d178	—,mono(2-hydroxypropyl) ether	$CH_3CHOHCH_2OCH_2CH_2OCH_2CH_2OH$	164.20	...	...	$277–9^{760}$	1.0789^{20}_{4}	1.4498^{20}	∞	v	...	...	v	MeOH, CCl_4 v	**C52,** 17693
d179	—,monomethyl ether	Methyl carbitol. $CH_3OCH_2CH_2OCH_2CH_2OH$	120.15	...	...	193^{760}	1.0270^{20}_{4}	1.4264^{20}	∞	v	v	∞	...	...	**C32,** 2217^{5}
Ω d180	**Diethylenetriamine**	2,2′-Diaminodiethylamine. $(NH_2CH_2CH_2)_2NH$	103.17	ye hyg liq	−39	207^{760}	0.9586^{20}_{20}	1.4810^{25}	∞	∞	i	...	...	lig s	**B4^{2},** 695
d181	***m*-Digallic acid**	Gallic acid 3-monogallate.	332.23	nd (dil al + 1w)	268–70d (280d)	...	...	...	i s^h	s	δ	s	...	MeOH s aa δ	**B10^{2},** 344
—	**Digine**	*see* **Gitogenin**													
d182	**Digitalose**	3-Methyl *D*-fucose	178.19	nd (AcOEt). $[\alpha]^{22}_{D}$ + 109 → + 126 (mut)	106 (fresh) 119 (after 4 months)	...	...	...	s	...	...	...	...	AcOEt s^h	**H32,** 163
d183	**Digitogenin**	5α,22α-Spirostan-2,3,15-triol.	448.65	nd (al) $[\alpha]^{19}_{D}$ − 18 (chl. c = 1.4)	280–3	...	...	...	i	s^h δ	...	...	...	chl s	**E14,** 286

For explanations, symbols and abbreviations see beginning of table. For structural formulas see end of table.

No.	Name	Synonyms and Formula	Mol. wt.	Color, crystalline form, specific rotation and λ_{max} (log ε)	m.p. °C	b.p. °C	Density	n_D	Solubility w	al	eth	ace	bz	other solvents	Ref.
	Digitoxigenin														
d184	**Digitoxigenin**	3β,14-Dihydroxy-5β-card-20(22) enolide. $C_{23}H_{34}O_4$	374.53	(dil MeOH) $[\alpha]_D^{20}$ +19.1 (MeOH, c = 1.36) $\lambda^{50\% al}$ 220 (4.21)	253 (256)					s				MeOH v	**E14,** 225
Ωd185	**Digitoxin**	$C_{41}H_{64}O_{13}$	764.96	wh (chl – eth), pl (dil al) $[\alpha]_D^{20}$ +4.8 (diox, c = 1.2) λ^{al} 220 (4.3)	255–6				δ	v	s			chl, MeOH, Py s	**E14,** 230
—	**Digitoxit**	*see* **1,3,4,5-Hexanetetrol**													
Ωd186	**Digitoxose**	2-Desoxy-*D*-altromethylose.	148.16	cr (MeOH – eth or AcOEt) $[\alpha]_D^{18}$ +27.9 to +43.3 (pyr, c = 1) (mut)	112				v			v		Py s AcOEt s^h	**B31,** 19
—	—**,3-methyl-**	*see* **Cymarose**													
—	**Diglycolic acid**	*see* **Acetic acid, oxydi-**													
d187	**Digoxigenin**	3α,12,14β-Trihydroxy-5β-card-20(22)enolide. $C_{23}H_{34}O_5$	390.53	pr (AcOEt). $[\alpha]_D^{20}$ +23.1 (MeOH, c = 1.7) $\lambda^{50\% al}$ 220 (4.13)	222					v				chl δ MeOH v	**E14,** 239
d190	**Dihydrosamidin**		388.42	$[\alpha]_D$ +19 (al)	117–9				i	s	s				**Am 79,** 3534
d191	**Diisoeugenol**		328.41	nd (bz, lig or al)	180–1					s^h	v		δ	chl v	**B6^2,** 917
d192	**Dilactic acid**	1,1′-Dicarboxydiethyl ether. Dilactylic acid. $HO_2CCH(CH_3)OCH(CH_3)CO_2H$	162.14	rh	112–3 (105–7)				s	δ	s		δ		**B3^4,** 205
—	α,γ**-Dilaurin**	*see* **Glycerol,** 1,3-didodecanoate													
—	**Dilirutic acid**	*see* **Barbituric acid, 5-nitro-**													
—	**Dimedone**	*see* **1,3-Cyclohexanedione, 5,5-dimethyl-**													
d193	**Dimethisoquin,** hydrochloride	Quotane	308.86		146				s	s					**Am 71,** 937
Ωd194	**Dimethylarsinic acid**	Alkargen. Cacodylic acid. $(CH_3)_2AsO_2H$	138.00	tcl	200				v	v	i				**B4^2,** 993
d195	**Dimethyl phosphinic acid,** ethyl ester	$(CH_3)_2PO_2C_2H_5$	122.11			89^{15}	1.0278_4^{25}	1.4261^{25}	δ	s	s				**Am 73,** 5466
—	**Dimite**	*see* **Ethanol, 1,1-bis(4-chlorophenyl)-**													
—	**Dinicotinic acid**	*see* **3,5-Pyridinedicarboxylic acid**													
d197	**1,3-Dioxane**	*m*-Dioxane. Trimethylene glycol methylene ether.	88.12	λ^{vap} 180 (3.8)	–42	105^{755}	1.0342_4^{20}	1.4165_α^{20}	∞	∞	∞	∞	∞		**B19^2,** 4
Ωd198	**1,4-Dioxane**	Diethylene dioxide. *p*-Dioxane. Glycol ethylene ether.	88.12	λ^{vap} 180 (3.8)	11.8	101^{750}	1.0337_4^{20}	1.4224^{20}	∞	∞	∞	∞	∞	os ∞ aa ∞	**B19^2,** 4
Ωd199	—**,2,3-dichloro-**	$C_4H_6Cl_2O_2$. *See* d198	157.00		30	$80–2^{10}$	1.468_4^{20}	1.4928^{20}	i d^h		v	v	v	chl, CCl_4, diox v lig v	**Am 68,** 2046
Ωd200	**1,3-Dioxane, 2,4-dimethyl-**	$C_6H_{12}O_2$. *See* b197	116.16			115–8 (120)	0.9392_4^{20}	1.4136^{20}	δ					os v	**B19^2,** 10
d201	—**,5-ethyl-4-propyl-**	$C_9H_{18}O_2$. *See* d197	158.24			196^{760}	0.9305_4^{20}	1.4370^{20}	i					os s	**C52,** 2795
d202	**1,4-Dioxane, heptachloro-**	$C_4HCl_7O_2$. *See* d198	329.22		54–6	$123–8^8$			i		v	v	v	chl, CCl_4 v lig v	**C54,** 12468
d203	**1,3-Dioxane, 5-hydroxy-2-methyl-**	$C_5H_{10}O_3$. *See* d197	118.13			176^{760}	1.0705_4^{17}	1.4375^{17}	∞						**B19^2,** 70
Ωd204	—**,4-methyl-**	$C_5H_{10}O_2$. *See* d197	102.13			114^{760}	0.9758^{20}	1.4159^{20}	δ					os v	**C52,** 8145
Ωd205	—**,4-methyl-4-phenyl-**	$C_{11}H_{14}O_2$. *See* d197	178.23		35–40	256^{760} 102^4	1.0864^{20}	1.5240^{20}	i					os s	**C52,** 8145
d206	—**,2-phenyl-**	$C_{10}H_{12}O_2$. *See* d197	164.21	nd (peth)	41	252–4 $98–9^4$			d^h	v	v				**B19^2,** 22
d207	—**,4-phenyl-**	$C_{10}H_{12}O_2$. *See* d197	164.21			245^{760} $128–30^{13}$	1.1038_4^{20}	1.5306^{18}	i					os s	**B19^1,** 616
	2,5-*p*-Dioxane-dione	*see* **Glycolide**													
Ωd208	—**,3-Dioxolane**	Glycol methylene ether.	74.08		–95	78^{765}	1.0600_4^{20}	1.3974_α^{20}	∞	s	s	s			**B19^2,** 3
d209	—**,4-(hydroxy-methyl)-2-methyl-**	Glycerol ethylidene ether. $C_5H_{10}O_3$. *See* d208	118.13			187^{760} $68–70^1$	1.1243_4^{17}	1.4413^{17}	δ	s					**B19^2,** 71
d210	—**,2-methyl-**	$C_4H_8O_2$. *See* d208	88.12			$81–2^{760}$	0.9811_4^{20}	1.4035^{17}	v	∞	∞				**B19^2,** 8

For explanations, symbols and abbreviations see beginning of table. For structural formulas see end of table.

No.	Name	Synonyms and Formula	Mol. wt.	Color, crystalline form, specific rotation and λ_{max} (log ε)	m.p. °C	b.p. °C	Density	n_D	Solubility w	al	eth	ace	bz	other solvents	Ref.
	1,3-Dioxolan-4-carboxaldehyde														
d211	**1,3-Dioxolan-4-carboxaldehyde, 2,2-dimethyl-**	....	130.15	....	....	74^{50}	....	1.4189^{25}	s^h	...	...	...	...	....	**B19²**, 136
Ω d212	**1,3-Dioxolan-2-one**	1,2-Ethanediol carbonate*. Ethylene carbonate. Glycol carbonate.	88.06	mcl pl (al)	39–40	248^{760}	1.3218_4^{39}	1.4158^{50}	∞	∞	∞	...	∞	chl ∞ AcOEt ∞ aa ∞	**B19**, 100
Ω d213	—,**4-methyl-**	Propylene carbonate. Isopropylene carbonate. $C_4H_6O_3$. *See* d212	102.09	....	−48.8	242^{760}	1.2069_{20}^{20}	1.4189^{20}	v	**v**	**v**	**v**	s	....	**C49,** 12303
—	α,γ-**Dipalmitin**	*see* **Glycerol**, 1,3-dihexadecanoate													
—	**Dipentene**	*see d,l*-**Limonene**													
d214	**Diphenadione**	Diandin. Dipaxin. 2-Diphenylacetyl-1,3-indanedione.	340.38	pa ye mcl (al)	146–7	....	....	1.670_β	i	δ^h	...	s	δ	CH_3CN, aa, cy s	**C49,** 3264
—	**Diphenic acid**	*see* **2,2′-Biphenyldicarboxylic acid**													
—	**Diphenylene-imine**	*see* **Carbazole**													
—	**Diphenylene oxide**	*see* **Dibenzofuran**													
—	**Diphenylene sulfide**	*see* **Dibenzothiophen**													
—	**Diphenylmethane carboxylic acid**	*see* **Benzoic acid, benzyl-**													
—	**Diphosgene**	*see* **Formic acid, chloro-,** trichloromethyl ester													
d215	**Diphosphine, tetrakis (trifluoromethyl)-**	$(F_3C)_2PP(CF_3)_2$	337.97	....	....	84^{760}	>1.0	....	i δd^h	...	...	...	...	....	**J1953,** 1565
—	**Dipicolinic acid**	*see* **2,6-Pyridinedicarboxylic acid**													
d216	**Diploicin**	$C_{16}H_{10}Cl_4O_5$.	424.07	(bz)	232 (225d)	....	....	....	i	i	i	...	i	lig δ aa i	**B19²**, 238
—	**Dipropionamide** ...	*see* **Propanoic acid,** amide, *N*-propionyl-													
d217	**Dipropylene glycol**	2,2′-Dihydroxydipropyl ether. $(CH_3CHOHCH_2)_2O$	134.18	....	....	229–32	1.0224_{20}^{20}	....	∞	s	...	...	...	....	**B1³**, 2149
—	**Disalicylic acid**	*see* **Benzoic acid, 2-hydroxy-, (2-carboxyphenyl)** ester													
d218	**Diselenide, diphenyl**	$C_6H_5SeSeC_6H_5$	312.13	ye nd λ^{al} 240 (4.24), 265 (3.74), 329 (3.03)	63–4	....	1.557_4^{80} 1.743_4^{20}	....	...	s^h	s	...	...	xyl s MeOH s^h	**B6²**, 319
d219	**Disiloxane, 1,3-diphenyl-1,1,3,3-tetramethyl-**	$C_6H_5Si(CH_3)_2OSi(CH_3)_2C_6H_5$	286.53	λ^{al} 264 (3.01)	....	110^2	0.9763_4^{20}	1.5176^{20}	...	...	...	...	...	....	**Am 79,** 1437
d220	—,**hexaethyl-**	$[(C_2H_5)_3Si]_2O$	246.55	....	....	233^{756} 129^{30}	0.8590_0^0	1.4340^{20}	...	...	...	...	...	....	**B4,** 627
Ω d221	—,**hexakis(2-ethylbutoxy)-**	$([(C_2H_5)_2CHCH_2O]_3Si)_2O$ ..	679.20	....	< −54	220^1	0.9219_{20}^{20}	1.4330^{20}	i	δ	s	...	s	....	**C48,** 3761
d222	—,**hexakis(2-ethylhexoxy)-**	$([CH_3(CH_2)_3CH(C_2H_5)CH_2O]_3Si)_2O$	847.52			$253^{0.9}$	0.9044_{20}^{20}	1.4402^{20}	i	δ	s	...	s	....	**C48,** 3761
—	α,γ-**Distearin**	*see* **Glycerol,** 1,3-dioctadecanoate													
d223	**Distibine, tetrakis-(trifluoromethyl)-**	Perfluoroantimony cacodyl. $(F_3C)_2SbSb(CF_3)_2$	519.53	pa ye or col, slow d	....	*ca.* 136^{760}	....	....	i	...	...	...	...	alk d	**J1957,** 3708
	Disulfide														
Ω d225	—,**bis(dibutylthiocarbamyl)**	$[(C_4H_9^n)_2NCS]_2S_2$	408.76	red liq	....	....	1.03^{20}	....	i	δ	s	...	...	....	**B4³**, 302
Ω d226	—,**bis(diethylthiocarbamyl)**	Antabuse. Antietil. $[(C_2H_5)_2NCS]_2S_2$	296.54	(al)	71–2	1.17^{17}	....	....	i	s^h	δ	...	...	chl v	**B4²**, 613
Ω d227	—,**bis(dimethylthiocarbamyl)**	Arasan. $[(CH_3)_2NCS]_2S_2$	240.43	wh or ye mcl (chl-al) λ^{chl} 243 (4.1), 282 (4.06)	155–6 (151)	129^{20}	....	....	i	δ s^h	δ	...	...	chl s	**B4²**, 557
d228	—,**bis(ethylmethylthiocarbamyl)**	$[CH_3N(C_2H_5)CS]_2S_2$	268.49	ye	72	....	....	....	i	δ	δ	...	...	chl s	**C43,** 6856
Ω d229	—,**bis(1-piperidylthiocarbamyl)**	Dicyclopentamethylenethiuram disulfide.	320.56	....	137–8 (129)	....	....	....	i	s^h	δ	...	...	chl s	**Am 65,** 1267
d230	—,**diacetyl**	$CH_3COSSCOCH_3$	150.22	....	20	$105–6^{18}$	....	....	i d^h	v	v	...	...	CS_2 v	**B2²**, 210
Ω d231	—,**2,2′-dibenzothiazyl**	2,2′-Dithiobisbenzothiazole.	332.50	lt ye sc (bz) λ^{al} 271 (4.32)	180	d	1.50_4^{20}	....	i	i	...	...	v^h	chl δ	**B27¹**, 249

For explanations, symbols and abbreviations see beginning of table. For structural formulas see end of table.

No.	Name	Synonyms and Formula	Mol. wt.	Color, crystalline form, specific rotation and λ_{max} (log ε)	m.p. °C	b.p. °C	Density	n_D	Solubility w	al	eth	ace	bz	other solvents	Ref.
	Disulfide														
d232	**—,dibenzoyl**	$C_6H_5COSSCOC_6H_5$	274.36	pr (al), sc (chl-peth)	136	d	...	...	i	δ^h	δ^h	...	...	CS_2 s^h	**B9**^{2}, 289
d233	**—,dibenzyl**	$C_6H_5CH_2SSCH_2C_6H_5$	246.40	lf (MeOH or al), nd (aa) λ^{al} 265 (3.1), 285 sh (2.4)	(i) 71–2 (ii) 69	...	...	...	δ	s^h	s	...	s	MeOH s	**B6**^{2}, 437
d234	**—,—,4,4′-dimethoxy-**	...	306.45	lf or nd (al)	101	...	...	...	i	i	...	...	v	chl v	**B6**, 901
d235	**—,dibutyl**	$CH_3(CH_2)_3SS(CH_2)_3CH_3$	178.36	λ^{al} 204 (3.32), 251.5 (2.60)	...	226^{760} 85^3	0.9383^{20}_4	1.4926^{20}	i	∞	∞	...	...	...	**B1**^{3}, 1524
d236	**—,—,3,3′-dimethyl-**	Isoamyl sulfide. $(CH_3)_2CHCH_2CH_2SSCH_2CH_2CH(CH_3)_2$	206.42	λ^{al} 251 (2.59)		250	0.9192^{20}_4	1.4864^{20}	d	...	...	...	...	...	**B1**^{3}, 1647
d237	**—,diethoxy**	Ethoxyl disulfide. $C_2H_5OSSOC_2H_5$	154.25	...	...	$67–8^{16}$	1.0913^{20}_4	1.4766^{20}	...	...	...	...	...	...	**B1**^{3}, 1314
d238	**—,diethyl**	$C_2H_5SSC_2H_5$	122.25	λ^{al} 202 (3.32), 251.5 (2.62)	−101.52	154^{760} 40^{10}	0.9931^{20}_4	1.5073^{20}	δ	∞	∞	...	...	...	**B1**^{2}, 345
Ω d239	**—,—,2,2′-diamino-**	Dithiobisethylamine. Cystamine. $H_2NCH_2CH_2SSCH_2CH_2NH_2$	152.28	...	...	d	...	...	s	...	δ	...	...	...	**B4**^{2}, 731
d240	**—,di(2-furfuryl)**	...	226.32	...	10	$112–3^{0.5}$	...	...	...	v	...	...	...	os v lig δ	**B17**^{2}, 116
d241	**—,diisopropyl**	$(CH_3)_2CHSSCH(CH_3)_2$	150.31	λ^{iso} 245 (2.36)	...	177.2^{760} 56.8^{10}	0.9435^{20}_4	1.4916^{20}	...	...	...	...	...	...	**B1**^{3}, 1480
d242	**—,dimethyl**	CH_3SSCH_3	94.20	λ^{iso} 256 (2.5)	−84.72	109.7^{760} 6.4^{10}	1.0625^{20}_4	1.5259^{20}	i	∞	∞	...	...	...	**B1**^{2}, 278
d243	**—,—,hexafluoro-**	F_3CSSCF_3	202.14	λ^{gas} 235 (2.46)	...	34.6^{760}	> 1.26	...	i	s	...	...	...	CS_2 i peth s	**J1952**, 2198
d244	**—,—,1,1,1′,1′-tetraphenyl-**	$(C_6H_5)_2CHSSCH(C_6H_5)_2$	398.60	nd (al) λ^{al} 265 sh (3.5)	152–3	...	...	...	...	s^h	s	...	s	CS_2 v	**B6**^{2}, 636
d245	**—,1,1′-dinaphthyl**	$(C_{10}H_7)^{\alpha}SS(C_{10}H_7)^{\alpha}$	318.46	pl (al), nd (lig) λ^{chl} 287 (4.2)	91	...	1.144^{20}	...	i	δ	v	...	...	...	**B6**^{2}, 588
d246	**—,2,2′-dinaphthyl**	$(C_{10}H_7)^{\beta}SS(C_{10}H_7)^{\beta}$	318.46	nd λ^{chl} 255 (4.6), 288 (4.2), 322 sh (3.7), 335 sh (3.5)	139–40	...	0.8409^{20}_4	1.4555^{20}	i	v	v	...	...	lig i	**B6**^{1}, 317
d247	**—,dipentyl**	*n*-Amyl disulfide. $CH_3(CH_2)_4SS(CH_2)_4CH_3$	206.42	...	...	119^7	0.9221^{20}_4	1.4889^{20}	i	...	s	...	...	...	**B1**^{3}, 1608
d248	**—,diphenyl**	...	218.34	nd (al) or rh λ^{al} 240 (4.23), 270 sh (3.50), 300 sh (3.2)	61–2	310^{760} 192^{15}	1.353^{20}_4	...	i	s	s	...	s	CS_2 s	**B6**, 323
d249	**—,—,2,2′-diamino-**	$C_{12}H_{12}N_2S_2$. *See* d248	248.37	pa ye pl or nd (50 % al) λ^{al} 221 (4.41), 342 (3.78)	93	...	...	...	i	v^h	s	...	...	...	**B13**^{2}, 201
d250	**—,—,3,3′-diamino-**	$C_{12}H_{12}N_2S_2$. *See* d248	248.37	nd (dil al)	62	...	...	...	i	v	v	...	v	lig i	**B13**^{2}, 219
d251	**—,—,4,4′-diamino-**	$C_{12}H_{12}N_2S_2$. *See* d248	248.37	ye or col pr or nd (w, eth or dil al) λ^{al} 256 (4.23), 294 sh (4.11)	85 (77)	...	...	...	s^h	v	v	...	δ^h	chl v lig δ^h	**B13**^{2}, 300
Ω d252	**—,—,4,4′-dichloro-2,2′-dinitro-**	$C_{12}H_6Cl_2N_2O_4S_2$. *See* d248	377.23	ye (ace-al)	212–3	...	...	...	...	δ	...	δ	s	CCl_4 δ peth i lig δ	**B6**^{2}, 313
d253	**—,—,5,5′-dichloro-2,2′-dinitro-**	$C_{12}H_6Cl_2N_2O_4S_2$. *See* d248	377.23	ye nd (90 % aa)	171	...	...	...	i	δ	...	...	v	chl, CS_2 v lig i	**B6**^{2}, 314
d254	**—,—,4,4′-diethoxy-**	$C_{16}H_{18}O_2S_2$. *See* d248	306.45	nd (al)	48–9	...	...	...	i	s^h	...	...	...	...	**B6**^{1}, 421
d255	**—,—,2,2′-dimethoxy-**	$C_{14}H_{14}O_2S_2$. *See* d248	278.40	nd (al)	88–90	...	...	...	i	s^h	...	...	...	...	**B6**, 795
d257	**—,—,2,2′-dimethoxy-5,5′-dimethyl-**	$C_{16}H_{18}O_2S_2$. *See* d248	306.45	pa ye pr (al), pl (al or eth)	74	...	1.160^{77}	...	i	s^h	...	...	...	...	**B6**^{2}, 874
Ω d258	**—,—,2,2′-dinitro-**	$C_{12}H_8N_2O_4S_2$. *See* d248	308.35	ye nd (bz or aa), pl (aa) λ^{al} 240 (4.37)	198–9	...	...	...	i	δ	i	δ	δ	peth i aa δ	**B6**^{2}, 307
Ω d259	**—,—,3,3′-dinitro-**	Nitrophenide. $C_{12}H_8N_2O_4S_2$. *See* d248	308.35	ye nd (al) or rh λ^{al} 242 (4.47), 320 sh (3.41)	84	...	...	...	...	δ	s	...	...	...	**B6**, 339
Ω d260	**—,—,4,4′-dinitro-**	$C_{12}H_8N_2O_4S_2$. *See* d248	308.35	nd (aa), pl (al) λ^{al} 228 sh (4.10), 315 (4.30)	182	...	...	...	...	δ	...	...	...	aa δ	**B6**^{2}, 312
Ω d261	**—,—,2,2′4,4′-tetranitro-**	$C_{12}H_6N_4O_8S_2$. *See* d248	398.35	ye nd (al) λ^{al} 248 (4.19)	303d	...	...	...	i	...	...	...	...	oos i $PhNO_2$, Py s	**B6**^{2}, 316
d262	**—,dipropyl**	$CH_3CH_2CH_2SSCH_2CH_2CH_3$	150.31	λ^{al} 250 (2.7)	...	193.5^{750}	0.9599^{20}_4	1.4981^{20}	...	...	...	...	...	...	**B1**^{3}, 1435

For explanations, symbols and abbreviations see beginning of table. For structural formulas see end of table.

No.	Name	Synonyms and Formula	Mol. wt.	Color, crystalline form, specific rotation and λ_{max} (log ε)	m.p. °C	b.p. °C	Density	n_D	Solubility w	al	eth	ace	bz	other solvents	Ref.
	Disulfide														
d263	—,**di-2-tolyl**	....	246.40	lf (al) λ^{al} 240 (4.21), 280 (3.45), 325 sh (2.95)	38–9	....	....	....	i	s	s	s	...	os s	**B6**^{2}, 342
d264	—,**di-4-tolyl**	....	246.40	nd or lf (al) λ^{al} 242 (4.29), 270 (3.65), 315 sh (3.2)	47–8	210–15^{20}	1.114^{51}	....	i	s	v	s	...	....	**B6**^{2}, 400
—	**Ditaine**	*see* **Echitamine**													
—	**Ditan**	*see* **Methane, diphenyl-**													
Ω d265	**1,4-Dithiane**	Diethylene disulfide	120.25	mcl pr λ^{al} 225 sh (2.54), 292 (2.5)	111–2	199–200	....	....	δ^h	s^h	s	...	...	CS_2 s aa s^h	**B19**^{2}, 6
d266	**1,3,5-Dithiazine, 4,5-dihydro-5-methyl-**	Methylthioformaldine	135.25	nd (eth)	65	ca. 185d	....	....	...	s	s	...	...	aa s	**B27**, 460
d267	**1,4-Dithiine 2,5-diphenyl-**	....	268.40	ye pr (al) λ 260 (4.35), 308 (3.95)	118–9	....	....	....	i	s	...	...	v	con sulf s alk s	**B19**^{2}, 46
—	**Dithioacetic acid**	*see* **Acetic acid, thionothiolo-**													
d268	**1,3-Dithiolane**	Trimethylene 1,3-disulfide.	106.21	λ^{cy} 207 (3.13), 247 (2.56)	−51	175^{760}	1.259^{17}	1.5975^{15}	...	s	s	...	...	xyl s	**B19**^{2}, 3
Ω d269	**Dithizone**	Diphenylthiocarbazone. Formazyl mercaptan. $C_6H_5N{:}NCSNHNHC_6H_5$	256.33	bl-bk (chl-al) λ^{ace} 445 (4.29), 610 (4.51)	165–9d	....	....	....	i	δ	δ	...	...	chl, alk s	**B16**, 26
d270	**Diurea**	Dicarbamide. *p*-Urazine.	116.08	pr (w)	270	....	....	....	δ^h	δ	...	...	...	aa δ	**B26**^{2}, 258
—	**Divarinol**	*see* **Benzene, 1,3-dihydroxy-5-propyl-**													
Ω d271	**Djenkolic acid**	β,β′-Methylenedithiodialanine. $HO_2CCH(NH_2)CH_2SCH_2SCH_2CH(NH_2)CO_2H$	254.33	nd (w, HCl) $[\alpha]^{25}_D$ −47.5 (1 % HCl, c = 2)	300–50d	....	....	....	s^h	...	...	...	...	dil ac s dil alk s	**B4**^{3}, 1591
Ω d272	**Docosane***	$CH_3(CH_2)_{20}CH_3$	310.61	pl (to), cr (eth)	44.4	368.6^{760} 213^{10}	0.7944^{20}_4	1.4455^{20}	i	s^h	v	...	...	chl s	**B1**^{3}, 574
Ω d273	**Docosanoic acid***	Behenic acid. Docosoic acid. $CH_3(CH_2)_{20}CO_2H$	340.60	nd	80	306^{60}	0.8223^{90}	1.4270^{100}	δ	δ	δ	...	...	....	**B2**^{3}, 373
d274	—,ethyl ester	Ethyl behenate. $CH_3(CH_2)_{20}CO_2C_2H_5$	368.65	nd (al), cr (ace)	50	240–2^{10}	0.8820^{51}	....	i	s	s	...	...	....	**B2**^{2}, 374
d275	—,methyl ester	Methyl behenate. $CH_3(CH_2)_{20}CO_2CH_3$	354.62	nd (ace)	54	224–5^{15}	....	1.4339^{60}	i	s	s	...	...	....	**B2**^{2}, 373
Ω d276	**1-Docosanol***	Docosyl alcohol. $CH_3(CH_2)_{21}OH$	326.61	(ace, chl)	71 (87)	$180^{0.22}$	....	....	δ	v	δ v^h	...	...	chl s MeOH v peth v^h	**B1**^{3}, 1846
d277	**4,7,11-Docosatrien-18-ynoic acid**	Clupanodonic acid. $CH_3(CH_2)_2C{:}C(CH_2)_5CH{:}CH\,CH_2$ / $HO_2CCH_2CH{:}CHCH_2CH{:}CHCH_2$ (CH₂ groups joined)	330.52	pa ye λ^{MeOH} 240 (4.28), 275 (4.2), 300 (4.2), 328 (4.2), 346 (3.7)	< −78	236^5	0.9290^{20}	1.4868^{20}	i	...	s	...	...	....	**C49**, 1550
Ω d278	**13-Docosenoic acid***(*cis*)	Erucic acid. $CH_3(CH_2)_7CH{:}CH(CH_2)_{11}CO_2H$	338.58	nd (al)	33–4	265^{15}	0.860^{55}_4	1.4758^{20}	i	s	v	...	...	MeOH v	**B2**^{2}, 445
Ω d279	—(*trans*)*	Brassidic acid. Brassic acid. Isoerucic acid. $CH_3(CH_2)_7CH{:}CH(CH_2)_{11}CO_2H$	338.58	pl (al)	61.5	282^{30}	0.8585^{57}_4	1.4472^{64}	i	δ	δ	...	...	....	**B2**^{2}, 447
d280	—,anhydride (*trans*)*	Brassidic anhydride. $[CH_3(CH_2)_7CH{:}CH(CH_2)_{11}CO]_2O$	659.15	nd (al), pl (eth, ace, peth)	64		0.835^{70}_4	1.4366^{100}	i	δ	s	s	...	peth s	**B2**^{2}, 448
d281	—,13,14-diiodo-, ethyl ester*(*trans*)	Iodobrassid. $CH_3(CH_2)_7CI{:}CI(CH_2)_{11}CO_2C_2H_5$	618.43	nd or sc	37 (40)	d	....	....	i	s v^h	v	...	v	chl v	**B2**^{2}, 448
d282	**13-Docosynoic acid***	Behenolic acid. $CH_3(CH_2)_7C{:}C(CH_2)_{11}CO_2H$	336.57	mcl pr or nd (al)	59.5	....	....	....	i	v	v	...	...	chl s	**B2**^{2}, 462
Ω d282[1]	**Dodecanal***	Lauraldehyde. $CH_3(CH_2)_{10}CHO$	184.33	lf	44.5 (12)	185^{100} $100^{3.5}$	0.8352^{15}_4	1.435^{22}	i	s	s	...	...	....	**B1**^{3}, 2911
d283	—,dimethyl acetal	1,1-Dimethoxydodecane*. $CH_3(CH_2)_{10}CH(OCH_3)_2$	230.40	....	....	132–4^5	....	1.4310^{25}	...	v	v	...	...	MeOH v	**Am 80**, 6613
Ω d284	**Dodecane***	$CH_3(CH_2)_{10}CH_3$	170.34	....	−9.6	216.3^{760} 91.5^{10}	0.7487^{20}_4	1.4216^{20}	i	v	v	v	...	chl, CCl_4, v	**B1**^{3}, 539
Ω d285	—,**1-amino-***	Dodecylamine. Laurylamine. $CH_3(CH_2)_{11}NH_2$	185.36	....	28.3	259^{760} 126.5^{10}	0.8015^{20}_4	1.4421^{20}	δ	∞	∞	...	∞	chl, CCl_4 ∞	**B4**, 200
d286	—,—,acetate	$CH_3(CH_2)_{11}NH_2.CH_3CO_2H$	245.41	....	fr. 69.5	....	....	....	v	v	...	...	δ	....	**C48**, 2525
d287	—,—,hydrochloride	$CH_3(CH_2)_{11}NH_2.HCl$	221.82		98	....	....	....	v	v	i	...	δ s^h	....	**Am 74**, 4287
Ω d288	—,**1-bromo-***	Dodecyl bromide. Lauryl-bromide. $CH_3(CH_2)_{11}Br$	249.24	....	−9.5	276^{760} 139^{10}	1.0399^{20}_4	1.4583^{20}	i	s	s	∞	...	....	**B1**^{3}, 542
d289	—,**1-bromo-12-fluoro-***	$F(CH_2)_{12}Br$	267.23	....	....	85–$6^{0.15}$	....	1.4524^{25}	i	v	v	s	...	....	**C51**, 7300
Ω d290	—,**1-chloro-***	Dodecyl chloride. Lauryl chloride. $CH_3(CH_2)_{11}Cl$	204.79	....	fr. −9.3	260^{760} 126.4^{10}	0.8682^{20}_4	1.4433^{20}	i	v	...	∞	s	CCl_4 ∞ lig ∞	**B1**^{3}, 541

For explanations, symbols and abbreviations see beginning of table. For structural formulas see end of table.

No.	Name	Synonyms and Formula	Mol. wt.	Color, crystalline form, specific rotation and λ_{max} (log ε)	m.p. °C	b.p. °C	Density	n_D	Solubility w	al	eth	ace	bz	other solvents	Ref.
	Dodecane*														
d291	—,**1,12-dibromo-***	$BrCH_2(CH_2)_{10}CH_2Br$	328.14	nd (aa, al)	41	215^{15}	...	...	i	v	s	...	...	chl v aa s	**B1**^{2}, 543
d292	—,**1-iodo-***	$CH_3(CH_2)_{11}I$	296.24	...	0.3	298.2^{760} 153^{10}	1.1999^{20}_{4}	1.4840^{20}	i	s ∞^h	∞	∞	...	chl, CCl_4 ∞ MeOH s	**B1**^{1}, 67
Ω d293	**Dodecanedioic acid,** dimethyl ester*	$CH_3O_2C(CH_2)_{10}CO_2CH_3$	258.36	pr	31.3	$167–9^9$ 150^2	...	...	i	...	...	...	...	AcOEt s^h	**B2**^{3}, 1844
Ω d294	**1-Dodecanethiol***	Dodecyl mercaptan. Lauryl mercaptan. $CH_3(CH_2)_{11}SH$	202.41	...	...	$142–5^{15}$	0.8450^{20}_{20}	1.4589^{20}	i	s	s	...	...	...	**B1**^{3}, 1789
Ω d295	**Dodecanoic acid***	Lauric acid. Undecane-1-carboxylic acid. $CH_3(CH_2)_{10}CO_2H$	200.33	nd (al)	44	131^1	0.8679^{50}_{4}	1.4304^{50}	i	v	v	s	∞^h v	peth s MeOH v	**B2**^{3}, 868
d296	—,amide*	Lauramide. $CH_3(CH_2)_{10}CONH_2$	199.34	nd	110	199^{12}	...	1.4287^{110}	i	s	δ	s	δ	CCl_4 s^h	**B2**^{3}, 894
d297	—,—,*N*-phenyl-	Lauranilide. $CH_3(CH_2)_{10}CONHC_6H_5$	275.44	nd (dil al) λ^{cy} 241 (4.19)	78	...			i	s	s	s	s	CCl_4, chl s	**B12**^{2}, 148
d298	—,anhydride*	Lauric anhydride. $[CH_3(CH_2)_{10}CO]_2O$	382.64	lf (al or eth)	41.8	166	0.8533^{70}_{4}	1.4292^{70}	d	s^h	...	...	...	...	**B2**^{2}, 321
d299	—,benzyl ester*	Benzyl laurate. $CH_3(CH_2)_{10}CO_2CH_2C_6H_5$	290.45	...	8.5	$209–11^{12}$	0.9457^{25}_{25}	1.4812^{24}	i	s	v	...	v	chl v peth s	**B6**^{2}, 417
Ω d300	—,chloride	Lauryl chloride. $CH_3(CH_2)_{10}COCl$	218.77	...	−17	145^{18}	...	1.4458^{20}	d	d	s	...	...	...	**B2**^{2}, 321
Ω d301	—,ethyl ester*	Ethyl laurate. $CH_3(CH_2)_{10}CO_2C_2H_5$	228.38	...	fr. −1.8	273^{764} 154^{15}	0.8618^{20}_{4}	1.4311^{20}	i	v	∞	...	...	...	**B2**^{3}, 884
d302	—,isopropyl ester*	Isopropyl laurate. $CH_3(CH_2)_{10}CO_2C_3H_7^i$	242.41	...	...	196^{60} 117.4^2	0.8536^{20}	1.4280^{25}	...	s	v	...	...	...	**B2**^{3}, 886
Ω d303	—,methyl ester*	Methyl laurate. $CH_3(CH_2)_{10}CO_2CH_3$	214.35	...	fr. 5.2	262^{766} 141^{15}	0.8702^{20}_{4}	1.4319^{20}	i	∞	∞	∞	∞	MeOH, chl, CCl_4, AcOEt s	**B2**^{3}, 883
Ω d304	—,nitrile	Lauronitrile. Undecyl cyanide. $CH_3(CH_2)_{10}CN$	181.33	...	fr. 4	277^{760} 131^{10}	0.8240^{20}_{4}	1.4361^{20}	i	∞	∞	∞	∞	chl ∞	**B2**^{3}, 895
d305	—,phenyl ester*	Phenyl laurate. $CH_3(CH_2)_{10}CO_2C_6H_5$	276.42	lf (al)	24.5	210^{15}	...	...	i	s	s	s	...	...	**B6,** 154
d306	—,4-phenyl-phenacyl ester.	...	394.56	...	86	...	...	...	i	...	...	...	...	...	**C32,** 4943
d307	—,piperazinium salt	$C_4H_{10}N_2 \cdot 2CH_3(CH_2)_{10}CO_2H$	486.79	...	92–2.5	...	...	...	s	s	i	...	...	...	**Am 70,** 2758
d308	—,propyl ester*	Propyl laurate. $CH_3(CH_2)_{10}CO_2CH_2CH_2CH_3$	242.31	...	...	205^{60} 124^2	0.8600^{20}	1.4335^{20}	...	...	...	...	...	...	**B2**^{2}, 885
d309	—,**2-bromo-***	$CH_3(CH_2)_9CHBrCO_2H$	279.23	pl	32	$157–9^2$	1.1474^{24}	1.4585^{24}	i	v	s	...	v	chl, lig s	**B2,** 363
d310	—,**12-fluoro-***	$F(CH_2)_{11}CO_2H$	218.32	...	60–1	...	...	...	i	v	v	...	...	lig s^h	**C51,** 7300
Ω d311	**1-Dodecanol***	Lauryl alcohol. $CH_3(CH_2)_{11}OH$	186.32	lf (dil al)	26 (22)	$255–9^{760}$ 150^{20}	0.8309^{24}_{4}	...	i	s	s	...	...	...	**B1**^{2}, 463
d312	**6-Dodecanol***	$CH_3(CH_2)_4CHOH(CH_2)_5CH_3$	186.34	(peth)	30	119^9	...	...	i	s	s	...	...	...	**B1,** 428
Ω d313	**2-Dodecanone***	*n*-Decyl methyl ketone. $CH_3CO(CH_2)_9CH_3$	184.33	...	21	$246–7^{760}$ 144^{11}	0.8198^{30}_{4}	1.4330^{20}	i	s	s	s	...	os s	**B1**^{2}, 769
Ω d314	**1-Dodecanone, 1-phenyl-***	Laurophenone. Lauroylbenzene. *n*-Undecyl phenyl ketone. $CH_3(CH_2)_{10}COC_6H_5$	260.43	og cr	46–7	$222–3^{21}$ 187^5	0.8969^{52}_{4}	1.4850^{52}	i	...	...	s	...	...	**B7**^{1}, 186
—	**1,6,10-Dodecatrien-3-ol, 3,7,11-trimethyl-***	*see* **Nerolidol**													
Ω d315	**1-Dodecene***	α-Dodecylene. $CH_3(CH_2)_9CH{:}CH_2$	168.33	...	−35.23	213.4^{760} 88.7^{10}	0.7584^{20}_{4}	1.4300^{20}	i	s	s	s	...	peth s	**B1**^{3}, 869
d316	**2-Dodecenedioic acid**(*cis*)*	Traumatic acid. $HO_2CCH{:}CH(CH_2)_8CO_2H$	228.29	(al, ace)	67–8	...	...	...	δ	s	s	...	s	chl s	**B2**^{3}, 1978
d317	—(*trans*)*	$HO_2CCH{:}CH(CH_2)_8CO_2H$	228.29	(al, ace)	165–6	...	...	...	δ	s	s	...	...	chl s	**B2**^{3}, 1979
d318	**1-Dodecen-3-yne***	$CH_2{:}CHC{:}C(CH_2)_7CH_3$	164.29	...	...	78^4	0.7858^{25}_{4}	1.4510^{25}	...	...	...	...	...	...	**B1**^{3}, 1055
d319	**1-Dodecyne***	$CH{:}C(CH_2)_9CH_3$	166.31	...	−19	215^{760} 89^{10}	0.7788^{20}_{4}	1.4340^{20}	...	...	...	...	...	...	**B1**^{3}, 1024
d320	**2-Dodecyne***	$CH_3C{:}C(CH_2)_8CH_3$	166.31	...	−9	105^{15}	0.7917^{15}_{4}	1.4828^{20}	i	...	s	s	...	...	**B1**^{3}, 1025
d321	**3-Dodecyne***	$CH_3CH_2C{:}C(CH_2)_7CH_3$	166.31	...	...	95^{12}	0.7871^{20}_{4}	1.4442^{20}	i	...	s	s	...	...	**B1**^{3}, 1025
d322	**6-Dodecyne***	Di *n*-amylacetylene. $CH_3(CH_2)_4C{:}C(CH_2)_4CH_3$	166.31	...	...	209^{745} 100^{14}	0.7871^{20}_{4}	1.4442^{20}	i	s	s	s	...	...	**B1**^{3}, 1025
Ω d323	**Dotriacontane***	Dicetyl. $CH_3(CH_2)_{30}CH_3$	450.89	pl (bz, chl, aa, eth)	69.7	467^{760} 292.7^{10}	0.8124^{20}_{4} suc	1.4550^{20}	i	δ	s^h	...	v^h δ	CCl_4, aa s^h chl δ, CS_2 s	**B1**^{3}, 587
d324	**1-Dotriacontanol***	$CH_3(CH_2)_{30}CH_2OH$	466.89	pl (bz)	89.4	sub $200–50^1$	...	...	i	...	...	...	...	...	**B1**^{3}, 1851
—	**Dulcitol**	*see* **Galactitol**													
—	**Durene**	*see* **Benzene, 1,2,4,5-tetramethyl-**													
—	**Durenol**	*see* **Benzene, 1-hydroxy-2,3,5,6-tetramethyl-**													
—	**Durohydroquinone**	*see* **Benzene, 1,4-dihydroxy-2,3,5,6-tetramethyl-**													

For explanations, symbols and abbreviations see beginning of table. For structural formulas see end of table.

No.	Name	Synonyms and Formula	Mol. wt.	Color, crystalline form, specific rotation and λ_{max} (log ε)	m.p. °C	b.p. °C	Density	n_D	Solubility w	al	eth	ace	bz	other solvents	Ref.
	Duroquinone														
—	**Duroquinone**	*see* **1,4-Benzoquinone, tetramethyl-**													
—	**Durylic acid**	*see* **Benzoic acid, 2,4,5-trimethyl-**													
—	**Dypnone**	*see* **2-Buten-1-one, 1,3-diphenyl-**													

For explanations, symbols and abbreviations see beginning of table. For structural formulas see end of table.

No.	Name	Synonyms and Formula	Mol. wt.	Color, crystalline form, specific rotation and λ_{max} (log ε)	m.p. °C	b.p. °C	Density	n_D	Solubility w	al	eth	ace	bz	other solvents	Ref.
	Ecgonidine														
e1	**Ecgonidine**(*l*)	Anhydroecgonine(*l*)	167.21	cr (MeOH-eth) $[\alpha]_D^{14}$ −84.6 (w, p = 1.7)	235d	...	...	...	...	s	δ	...	...	...	**B22²**, 26
e2	**Ecgonine**(*dl*)	...	185.23	pl (w + 3) λ^{al} 218 (3.8), 224 sh (3.7) 275 (3.3)	203 (212) (anh) 93–118 (+3w)	...	...	...	v^h	s	i	δ	δ	chl δ	**B22²**, 156
e3	—(*l*)	Tropinecarboxylic acid. $C_9H_{15}NO_3$. *See* e2	185.23	mcl pr (al + 1w) $[\alpha]_D^{15}$ −45.5 (w, c = 5)	205 (anh) 198 (hyd)	...	...	...	v	s	i	δ	δ	MeOH s chl δ	**B22²**, 150
e4	—,hydrate(*l*)	$C_9H_{15}NO_3.H_2O$. *See* e2	203.23	mcl pr (al), eff 120–30	198	...	...	...	v	δ	δ	δ	δ	MeOH s chl δ	**B22**, 196
e5	—,benzoate(*l*)	O-Benzoyl-*l*-ecgonine.	289.34	nd (w) $[\alpha]_D^{14}$ −63.5 (w, p = 1.7)	195	...	...	...	δ s^h	s	i	...	v^h	dil ac, dil alk s	**B22²**, 150
e6	—,—,ethyl ester(*l*)	Cocaethyline. Homococaine. $C_{18}H_{23}NO_4$. *See* e5	317.39	pr (eth)	109	...	...	...	δ	s	s	...	...	...	**B22**, 202
e7	—,—,tetra-hydrate(*l*)	$C_{16}H_{19}NO_4.4H_2O$. *See* e5	361.40	pr (w) $[\alpha]_D^{15}$ −44.6 (abs al, c = 3)	92	...	...	...	δ s^h	s	i	...	v^h	...	**B22**, 197
e8	—,hydrochloride(*l*)	$C_9H_{15}NO_3.HCl$. *See* e2	221.69	rh (al), nd (*i*-AmOH) $[\alpha]_D^{15}$ −59 (w, c = 10)	246	...	...	...	s	s	...	...	...	MeOH v *i*-AmOH δ	**B22²**, 150
e8¹	**Echinochrome A**	$C_{12}H_{10}O_7$	266.21	dk red nd (to) λ^w 400 (3.7), 465 (3.7), 480 (3.8), 550 (3.5)	220d	Sub $120^{10^{-4}}$	...	...	i	s	s	s	...	chl, peth δ	**C55**, 6710
e9	—,3,6,7-trimethyl ether	$C_{15}H_{16}O_7$. *See* e8¹	308.29	dk red lf (aq diox), nd λ^{chl} 323, 476 502, 537	133 vac (130)	...	...	...	i	s	...	...	...	MeOH, NaOH s (bl)	**C55**, 6710
e10	**Echinopsine**	N-Methyl-γ-quinolone.	159.19	α nd(bz), (w + 1) β cr (al) λ^{al} 215 (4.2), 290 (3.6), 320 (4.2), 330 (4.2)	α152 β135	...	...	...	s v^h	v	δ	...	v^h	chl, MeOH v	**B21²**, 259
e11	**Echitamidine**	$C_{20}H_{26}N_2O_3$	342.44	pl (eth) $[\alpha]_D^{16}$ −515 (al)	244d	...	...	...	s	s	...	...	...	con HNO_3 (bl→ye)	**J 1932**, 2628
e12	**Echitamine**	Ditaine. $C_{22}H_{28}N_2O_4.4H_2O$	456.54	pr (al + 4w) eff 105 (−3w) $[\alpha]_D^{20}$ −29 (al) λ^{al}236 (4.1), 296 (4.0), 331 (4.3)	206 (+1w)	...	...	...	s	s	s	...	...	chl, con sulf s (red) peth i	**J 1925**, 1640
e13	—,hydrochloride	$C_{22}H_{28}N_2O_4.HCl$	420.94	nd (w), $[\alpha]_D^{15}$ −58 (w, c = 1) λ^{al} 235 (3.93), 295 (3.55)	295–300d	...	...	...	s	v	δ	...	δ	...	**J 1925**, 1640
e14	**Echitin**	$C_{32}H_{52}O_2$	468.77	lf $[\alpha]_D$ +73 (eth)	170	...	...	...	...	s^h	s	s	s	peth s	**M** 397
e15	**Egonol**	$C_{19}H_{18}O_5$	326.35	pl (BuOH)	118	$228–30^{0.15}$	...	...	...	...	...	...	...	chl s	**C33**, 5394
Ω e16	**Eicosane***	Didecyl. $CH_3(CH_2)_{18}CH_3$	282.56	lf (al)	36.8	343^{760} 195.7^{10}	0.7886_4^{20}	1.4425^{20}	i	...	s	v	s	peth s	**B1³**, 570
Ω e17	**Eicosanedioic acid***	Octadecane-1,18-dicarboxylic acid. $HO_2C(CH_2)_{18}CO_2H$	342.53	cr (bz or al)	125–6.5 (124)	$233–4^2$	...	...	...	...	s	...	...	...	**B2³**, 1880
Ω e18	**Eicosanoic acid***	Arachidic acid. Eicosoic acid. $CH_3(CH_2)_{18}CO_2H$	312.54	pl (al)	77	328 δd $203–5^1$	0.8240_4^{100}	1.425^{100}	i	δ s^h	v	...	s	chl s peth δ	**B2²**, 369
Ω e19	—,ethyl ester*	$CH_3(CH_2)_{18}CO_2C_2H_5$	340.60	...	50(42)	$295–7^{100}$ $186–7^2$	...	...	i	s	s	...	s	chl s	**B2²**, 370
e20	—,methyl ester*	$CH_3(CH_2)_{18}CO_2CH_3$	326.57	lf (MeOH)	54.5 (47)	$215–6^{10}$ 188^2	...	1.4317^{60}	i	s	s	...	s	chl s	**B2²**, 370
Ω e21	**1-Eicosanol***	Arachidic alcohol. *pri-n*-Eicosyl alcohol. $CH_3(CH_2)_{18}CH_2OH$	298.56	wx (al), cr (chl)	72.5–3.0 (67)	369^{760} $220–5^3$	0.8405_4^{20} suc	1.4550^{20} suc	i	δ	...	v	s^h	peth s^h	**B1³**, 1843
e22	**2-Eicosanol***	2-Eicosyl alcohol. $CH_3(CH_2)_{17}CH(OH)CH_3$	298.56	cr (MeOH)	63–4 (60)	357^{760}	0.8378_4^{20} suc	1.4524^{20} suc 1.4312^{80}	i	...	...	v	s^h	...	**B1³**, 1843
e23	**2-Eicosanone***	Methyl *n*-octadecyl ketone. $CH_3(CH_2)_{17}COCH_3$	296.54	lf (MeOH)	58	...	...	...	...	s	s	s	...	...	**B1³**, 2932
e24	**3-Eicosanone***	Ethyl *n*-heptadecyl ketone. $CH_3(CH_2)_{16}COC_2H_5$	296.54	lf (al)	60–1	...	...	...	...	δ	s	s	s	chl, aa s	**B1²**, 774
e25	—,oxime*	$CH_3(CH_2)_{16}C(:NOH)C_2H_5$	311.56	nd (al)	α 55.5–6.5 β 64–5	...	...	...	...	δ	s	s	...	peth δ	**B1**, 719
e26	**7-Eicosanone***	Hexyl *n*-tridecyl ketone. $CH_3(CH_2)_{12}CO(CH_2)_5CH_3$	296.54	cr	52.7–3.4	$210–1^{11}$	...	1.4258^{80}	...	...	s	s	...	...	**C49**, 1546

For explanations, symbols and abbreviations see beginning of table. For structural formulas see end of table.

No.	Name	Synonyms and Formula	Mol. wt.	Color, crystalline form, specific rotation and λ_{max} (log ε)	m.p. °C	b.p. °C	Density	n_D	Solubility w	al	eth	ace	bz	other solvents	Ref.
	5,8,11,14-Eicosatetraenoic acid														
e27	**5,8,11,14-Eicosatetraenoic acid***	Arachidonic acid. $CH_3(CH_2)_3(CH_2CH{:}CH)_4(CH_2)_3CO_2H$	304.48	λ 233 (4.24), 257 (4.09), 268 (4.21), 315 (3.82)	−49.5	d	....	1.4824^{20} 1.5563^{23}	i	s	s	s	...	os s	**B2**[2], 469
Ω e28	**1-Eicosene***	$CH_3(CH_2)_{17}CH{:}CH_2$	280.54	....	28.5	341^{760} $151^{1.5}$	0.7882_4^{30}	1.4440^{30}	i	...	...	...	s	peth s	**B1**[3], 881
e29	**1-Eicosyne***	$CH_3(CH_2)_{17}C{:}CH$	278.53	....	36	340^{760} 191.8^{10}	0.8073_4^{20}	1.4501^{20}	i	...	...	...	s	peth s	**B1,** 262
—	**Elaidic acid**	*see* **9-Octadecenoic acid (*trans*)***													
—	**Elaidyl alcohol**	*see* **9-Octadecen-1-ol (*trans*)***													
e30	**α-Elaterin**	$C_{32}H_{44}O_8$	556.67	cr (chl-MeOH or al) λ 234 (4.0), 267 sh (3.87) $[\alpha]_D^{20}$ −64.3 (chl, c = 1.6)	234	....	....	....	i	δ	s	...	s	chl s	**J95,** 1989
e31	**β-Elaterin**	$C_{20}H_{28}O_5$	348.45	nd (al) $[\alpha]_D^{25}$ +13.9	195.5	....	....	....	i	i δ^h	δ	...	δ	chl s	**J95,** 1989
—	**Eldoral**	*see* **Barbituric acid, 5-ethyl-5-piperidyl-**													
e32	**Elemane**	Dihydroelemene. $C_{15}H_{30}$ ...	210.41	....	....	115–9[10]	0.8509_4^{20}	1.4640^{20}	i	...	s	...	s	peth s	**B5**[2], 117
e33	**Elemenonic acid** ...	Dihydro-β-elemonic acid. $C_{30}H_{48}O_3$.	456.72	nd (al or AcOEt), $[\alpha]_D$ +43.7 (chl, c = 0.9)	249–50 (246)	....	....	....	...	...	...	...	...	....	**J1963,** 2762
e34	**α-Elemol**	....	222.38	cr (al) $[\alpha]_D$ −5.8 (chl, c = 3.4)	52.5–3.5	142–3[12]	0.9345_4^{18}	1.4980^{18}	...	...	...	...	...	....	**B6**[3], 410
—	**Eleostearic acid**	*see* **9,11,13-Octadecatrienoic acid***													
e35	**Ellagene**	Indeno-2′,3′:2,3-fluorene.	254.34	pl (bz) λ^{chl} 309 (4.65), 324 (4.68)	216	....	....	....	...	...	...	...	...	....	**E14s,** 498
e36	**Ellagic acid, dihydrate**	$C_{14}H_6O_8.2H_2O$	338.23	pa ye nd (Py) $\lambda^{aq\text{-}diox}$ 248 (4.64), 305 (3.92), 360 (4.0)	450–80 d without m	....	....	....	δ	δ	i	δ	δ	alk s (ye) os δ	**B19**[2], 284
e37	**Elliptic acid**	$C_{20}H_{18}O_8$	386.36	nd (aq al)	190	....	....	....	...	...	...	...	...	....	**J1942,** 587
—	**Embelin**	*see* **1,4-Benzoquinone, 2,5-dihydroxy-3-dodecyl-**													
e38	**Emeraldine**	$C_{48}H_{40}N_8$	726.90	indigo-bl pw	....	....	....	....	...	i	...	...	i	80 % aa s chl, gl aa i	**B12**[1], 147
e39	**Emetine(*l*)**	Cephaline-O-methyl ether $C_{29}H_{40}N_2O_4$.	480.66	amor pw (ye under light) λ^{al} 235 sh (4.2), 285 (3.8), 360 sh (2.2) $[\alpha]_D^{20}$ −50 (chl, c = 2)	74 (cor)	....	....	....	i	s	s	s	δ	chl δ	**B23**[2], 449
Ω e40	—,hydrochloride(*l*)	$C_{29}H_{40}N_2O_4.2HCl.7H_2O$ ... *See* e39	679.69	nd (w) $[\alpha]_D$ + 11 (w, c = l) +21 (w, c = 8) $\lambda^{dil\ HCl}$ 230 (1.6), 280 (0.65)	269–70d	....	....	....	v	s	i	...	...	....	**B23**[2], 451
e41	**Emicymarin**	$C_{30}H_{46}O_9$	550.70	nd or pr (+MeOH) $[\alpha]_D^{20}$ +12.8 (al, c = 2.5)	*ca.* 207	....	....	....	δ	...	...	...	...	....	**J1963,** 1461
—	**Emodin**	*see* **9,10-Anthracenequinone, 6-methyl-1,3,8-trihydroxy-**													
—	**Enanthic acid**	*see* **Heptanoic acid***													
e42	**Enneaphyllin**	$C_{90}H_{154}$	1236.24	rods (bz)	98	....	....	....	...	s	δ	i	s	AcOEt, to s lig δ Py i	**C32,** 2686
e43	**Eosin**	2,4,5,7-Tetrabromofluorescein.	647.93	ye-red λ^w 520 (5.1)	295–6	....	....	....	i	s	i	...	δ	alk s chl, aa δ	**B19**[2], 254
e44	—,sodium salt decahydrate	Eosin dye. Sodium eosin. $C_{20}H_6Br_4Na_2O_5.10H_2O$. *See* e43	871.90	red nd (dil al) eff 150 λ 519 (4.91)	....	....	....	....	s	s	i	...	...	....	**B19,** 230
e45	**Ephedrine(*d*)**	$C_6H_5CH(OH)CH(CH_3)NHCH_3$	165.24	pl (w) $[\alpha]_D$ +13.4 (4 % w)	40	225^{760}	....	....	s	s	s	...	s	chl s	**Am 69,** 128

For explanations, symbols and abbreviations see beginning of table. For structural formulas see end of table.

No.	Name	Synonyms and Formula	Mol. wt.	Color, crystalline form, specific rotation and λ_{max} (log ε)	m.p. °C	b.p. °C	Density	n_D	Solubility w	al	eth	ace	bz	other solvents	Ref.
	Ephedrine														
e46	—(*dl*)	Racephedrine. $C_{10}H_{15}NO$. *See* e45	165.24	nd (eth or peth) λ^{al} 240 (1.78), 254 (2.26)	76–7	$135–7^{12}$	...	...	s	s	s	...	s	chl s	**B13**[2], 376
e47	—(*l*)	Natural ephedrine. $C_{10}H_{15}NO.H_2O$. *See* e45	183.25	pl (w + 1) $[\alpha]^{20}_D$ −6.3 (al) λ^{al} 252 (2.2), 258 (2.3), 266 (2.2)	40	225^{760}	...	...	s	s	s	...	s	chl s	**C34**, 2852
Ω e48	—,hydrochloride(*d*)	$C_{10}H_{15}NO.HCl$. *See* e45	201.70	pl (abs al) $[\alpha]^{20}_D$ +35.8 (w, c = 11.5) $\lambda^{dil\,HCl}$ 250 (2.2), 257 (2.3), 263 (2.2)	218	...	...	...	v	s	i	...	...	...	**B13**[2], 375
e49	—,—(*dl*)	Ephetonin. $C_{10}H_{15}NO.HCl$. *See* e45	201.70	pl (al) $\lambda^{dil\,HCl}$ 250 (2.2), 257 (2.3), 263 (2.2)	189–90	...	...	...	v	s	i	...	...	...	**B13**[2], 376
e50	—,—(*l*)	$C_{10}H_{15}NO.HCl$. *See* e45	201.70	orh nd, $[\alpha]^{20}_D$ −36.6 (w, c = 5)	218–20	...	...	...	v	s	i	...	...	...	**C51**, 4304
Ω e51	—,sulfate(*l*)	$(C_{10}H_{15}NO)_2.H_2SO_4$. *See* e45	428.54	hex pl or orh nd (w), [α] −31.5 (w, c = 5)	245–8d	...	...	...	v	δ	...	...	...	...	**C21**, 2169
e52	—,*N*-**methyl** (*l*)	$C_6H_5CH(OH)CH(CH_3)N(CH_3)_2$	179.26	nd or pl (al or eth, $[\alpha]_D$ −29.5 (MeOH, c = 4.5)	87–8 (cor)	v volat	...	...	i	s	s^h	...	...	MeOH s	**B13**[2], 378
e53	—,*N*-**(4-nitrobenzoyl)-(*dl*)**	...	314.35	pa ye pl (al)	162	...	...	...	...	...	...	...	...	...	**Am 69**, 128
e54	**Epi-β-amyrin**	...	426.74	cr (MeOH) $[\alpha]_D$ +73.3 (chl)	225 (cor)	...	...	...	...	...	...	...	...	...	**E14s**, 948
e55	**Epi-α-amyrin, acetate**	...	468.77	nd (chl-MeOH) $[\alpha]_D$ +39 (chl)	135 (cor)	...	...	...	...	...	...	...	...	...	**E14s**, 1070
e56	**Epiandrosterone**	Androstan-3β-ol-17-one. Isoandrosterone. $C_{19}H_{30}O_2$.	290.45	cr (bz-peth, ace) $[\alpha]^{20}_D$ +108 (MeOH) λ^{sulf} 405 (2.90)	177–9 (176)	...	...	...	...	...	...	...	...	...	**E14**, 144
e57	**Epiborneol**(*l*)	3-Camphanol	154.26	nd (C_5H_{12}) $[\alpha]^{17}_D$ +11.1 (al)	181–2	213^{742}	...	...	...	...	...	...	...	peth s	**B6**[2], 91
e58	—,acetate	...	196.29	$[\alpha]^{16}_D$ +15.63	< −15	101^{11}	0.9872^{14}_4	1.4651^{14}	...	...	...	...	...	...	**B6**[2], 92
e59	**Epibreinonol**	Breinonol A. Breienonol A.	440.72	pl (MeOH), nd (chl), $[\alpha]^{17}_D$ +37 (chl)	204	...	...	...	...	...	...	...	...	...	**E14s**, 1083
—	**Epibromohydrin**	*see* **Propane, 3-bromo-1,2-epoxy-***													
e60	**Epicamphor**(*d*)	3-Bornanone. $C_{10}H_{16}$. *See* e60	152.24	$[\alpha]^{17}_D$ +45.4 (bz)	182	...	...	...	δ	s	s	...	...	peth v	**B7**[1], 86
e61	—(*dl*)		152.24	cr (peth)	175–7	...	...	...	δ	v	v	...	...	peth v	**B7**[2], 104
e62	—(*l*)	β-Camphor. $C_{10}H_{16}O$. *See* e60.	152.24	$[\alpha]^{19}_D$ −58.21 (bz, c = 13)	184 (cor) (186–7)	213	...	...	δ	v	v	...	...	peth v	**B7**[2], 105
e63	—,oxime(*d*)	...	167.25	nd (MeOH) $[\alpha]_D$ −98.9	103	...	...	...	...	...	s	s	...	...	**B7**[1], 86
e64	—,—(*dl*)	$C_{10}H_{17}NO$. *See* e63	167.25	nd (dil al)	98–100	...	...	...	...	...	s	s	...	...	**B7**[1], 87
e65	—,—(*l*)	$C_{10}H_{17}NO$. *See* e63	167.25	nd (dil MeOH) $[\alpha]_D$ +100.5 (bz, c = 6.3)	103–4	...	...	...	i	s^h	s	s	...	os s	**B7**[1], 86
e66	—,semicarbazone (*d*)	...	209.29	nd (al)	237–8	...	...	...	i	s^h	...	...	...	...	**B7**[1], 86
e67	—,—(*l*)	$C_{11}H_{19}N_3O$. *See* e66	209.29	nd (al), $[\alpha]^{20}_D$ +145 (MeOH, c = 0.73)	237–8d	...	...	...	i	s^h	...	...	...	...	**B7**[1], 86
e68	—,**bromo-**(*l*)	...	231.14	nd or pr (peth) $[\alpha]_D$ −86.6 (AcOEt, c = 3.6)	133–4	...	...	...	...	δ	...	...	v	chl v lig s	**B7**[1], 86
e70	**α-Epicamphylamine**	...	153.27	$[\alpha]_D$ +17.6 (bz, c = 6.5)	...	$127–8^{100}$	...	...	v	...	...	...	...	...	**B12**[2], 35
e71	**Epicatechin**(*dl*)	Epicatechol. $C_{15}H_{14}O_6$	290.28	nd (w + 1), pr (w + 4) λ^{al} 279 (3.56)	224–6d anh (cor)	...	...	...	δ	s	δ	s	...	...	**B17**[2], 258
e72	—(*l*)	$C_{15}H_{14}O_6$. *See* e71	290.28	cr (w + 4) $[\alpha]^{25}_D$ −69 (al) $[\alpha]_D$ −60 (1 : 1 ace − w, c = 2) λ^{al} 279 (3.56)	245d (cor) (242)	...	...	...	δ	s	δ	s	...	...	**B17**[2], 257

For explanations, symbols and abbreviations see beginning of table. For structural formulas see end of table.

No.	Name	Synonyms and Formula	Mol. wt.	Color, crystalline form, specific rotation and λ_{max} (log ε)	m.p. °C	b.p. °C	Density	n_D	Solubility w	al	eth	ace	bz	other solvents	Ref.
	β-Epichlorohydrin														
—	**β-Epichlorohydrin**	*see* **Propane, 2-chloro-1,3-epoxy***													
—	**Epicholestanol**	*see* **3α-Cholestanol**													
e74	**Epicholestan-4-ol**	4α-Hydroxycholestane. $C_{27}H_{48}O$.	388.69	lf (al, MeOH – ace) $[\alpha]_D^{21}$ +29.0 (chl)	187–8	...	...	...	i	...	s	...	...	chl s	**E14,** 65
e75	**Epicholestanol**	ε-Cholestanol. 3α-Hydroxycholestane. $C_{27}H_{48}O$.	388.69	nd (al, MeOH) $[\alpha]_D^{20}$ +34 (chl, c = 1.7)	185–6	...	...	...	i	δ s^h	s	...	...	chl s MeOH δ	**E14,** 59
e76	**Epicholesterol**	5-Cholesten-3α-ol. $C_{27}H_{46}O$.	386.67	cr (al, chl-MeOH) $[\alpha]_D^{30}$ −37.5 (al) λ^{hx} 192 (3.95)	141.5	...	...	...	...	δ s^h	...	...	...	...	**E14,** 56
e77	**Epicoprostanol**	3-α-Coprostanol. Epicoprosterol. $C_{27}H_{48}O$.	388.69	cr (al, ace) $[\alpha]_D^{20}$ +31.6 (chl)	117–8	...	...	...	i	s	s	...	v	chl s	**E14,** 60
e78	**Epicoprostenol**	4-Cholesten-3α-ol. Epiallocholesterol. $C_{27}H_{46}O$	386.67	nd (ace) $[\alpha]_D^{24}$ +120.8 (bz)	84 (82)	...	...	...	...	s	v	δ s^h	v	chl v	**E14,** 37
—	**Epicyclene**	*see* **Cyclocamphane**													
e79	**Epidicentrin(*dl*)**	*dl*-Domesticine methyl ether. $C_{20}H_{21}NO_4$	339.40	pr (MeOH)	142	...	...	...	...	v^h	s	...	...	chl v^h	**B27²,** 553
e80	—(*l*)	$C_{20}H_{21}NO_4$. *See* e79	339.40	pr (MeOH) $[\alpha]_D^{18}$ −101.3 (chl, c = 0.5)	138–9	...	...	...	...	v^h	s	...	...	chl v^h	**B27²,** 553
e81	**Epiergosterol(*D*)**	$\Delta^{7,9(11),22}$-Ergostatrien-3α-ol. $C_{28}H_{44}O$.	396.67	nd (eth-MeOH) $[\alpha]_D^{19}$ +36.2 (chl)	203–4	...	...	...	...	...	...	...	...	...	**E14s,** 1752
e82	**Epifucitol(*d*)**	$C_6H_{14}O_5$	166.18	cr (eth, w) $[\alpha]_D^{21}$ +2.2 (w, c = 1)	105–7	...	...	...	s	...	s	...	...	...	**Am 74,** 4373
e83	—(*l*)	$C_6H_{14}O_5$. *See* e83	166.18	cr (w), $[\alpha]_D^{20}$ −2.3 (w, c = 1)	105–7	...	...	...	s	...	s	...	...	...	**C24,** 2431
e84	**Epifucose(*l*)**	6-Deoxy-*D*-glucose. α-*D*-Glucomethylose. *l*-Talomethylose.	164.16	cr (AcOEt) $[\alpha]_D$ −36.9 (w, c = 6)	135–45	...	...	...	s	...	i	i	...	abs al s	**B31,** 65
—	**Epiiodohydrin**	*see* **Propane, 1,2-epoxy-3-iodo-***													
e85	**Epiisofenchol**	4,6,6-Trimethyl-2-norbornanol.	154.26	nd (sub) $[\alpha]_D^{20}$ −7.35	71–2 (67)	...	...	...	...	...	...	...	...	...	**E12A,** 645
e86	**Epiisofenchone(*d*)**	4,6,6-Trimethyl-2-norbornanone	152.24	$[\alpha]_D$ +13.6	...	195	0.934_4^{20}	1.459^{20}	...	...	s	s	...	...	**E12A,** 732
e87	—(*dl*)	$C_{10}H_{16}O$. *See* e86	152.24	...	...	195–8 $90–3^{21}$	...	1.4625^{25}	...	...	s	s	...	...	**E12A,** 732
e88	**Epilupinine**	*d*-Isolupinine	169.27	nd (peth) $[\alpha]_D^{17}$ +32 (al, c = 1.5)	76–8	...	...	...	s	...	s	s	s	os s peth δ	**B21²,** 30
—	**Epimethylin**	*see* **Ether, methyl propyl, 2′,3′-epoxy-**													
—	**Epinephrine**	*see* **Adrenaline**													
e89	**Epinine**	...	167.21	nd (al)	188–9 (cor)	...	...	...	δ s^h	δ δ^h	...	...	δ	os δ	**B13²,** 487
e90	**Epiquinidine**	$C_{20}H_{24}N_2O_2$	324.43	cr (AcOEt), lf (eth) $[\alpha]_D^{20}$ +103.7 (al, c = 0.86) $\lambda^{w, pH=1}$ 315 (3.62), 350 (3.62)	113	...	...	...	...	v	s	...	...	...	**C31,** 1816
e91	**Epirhodanhydrin**	2,3-Epoxypropylrhodanine. CH_2CHCH_2SCN (epoxide O bridging first two C)	115.16	dk red liq, garlic smell	...	d	...	...	i	s	...	...	...	chl s	**B17,** 106
e92	**Episarsapogenin**	$C_{27}H_{44}O_3$	416.65	nd (ace) $[\alpha]_D$ −71 λ^{al} 215 (3.78)	204–6 (198)	...	...	...	...	...	...	...	...	...	**E14,** 284
e93	**Epitruxillic acid**	2,4-*cis*-Diphenylcyclobutane 1,3-*trans*-dicarboxylic acid*.	296.33	cr (dil al or bz-aa)	285–7	...	...	...	...	...	δ	...	δ	...	**B9²,** 691
e94	**Equilenin(*d*)**	$C_{18}H_{18}O_2$	266.34	nd (dil al) $[\alpha]_D^{16}$ +87 (diox) λ^{MeOH} 235, 281, 340	258–9 (251)	$170–80^{0.01}$ sub	...	...	...	δ s^h	...	δ	...	chl δ	**E14,** 134
e95	—(*dl*)	$C_{18}H_{18}O_2$. *See* e94	266.34	cr (bz) λ^{MeOH} 235, 281, 340	276–8	...	...	...	...	...	...	...	...	...	**Am 67,** 2274
e96	—(*l*)	$C_{18}H_{18}O_2$. *See* e94	266.34	$[\alpha]_D^{30}$ −85 (diox) λ^{MeOH} 235, 281, 340	258–9 (cor)	$170–80^1$ sub	...	...	...	s^h	...	...	...	...	**Am 67,** 2274

For explanations, symbols and abbreviations see beginning of table. For structural formulas see end of table.

No.	Name	Synonyms and Formula	Mol. wt.	Color, crystalline form, specific rotation and λ_{max} (log ε)	m.p. °C	b.p. °C	Density	n_D	Solubility w	al	eth	ace	bz	other solvents	Ref.
	Equilin														
Ω e97	**Equilin**	$C_{18}H_{20}O_2$	268.36	orh sph pl (AcOEt) $[\alpha]_D$ +308 (diox, c = 2) $\lambda^{97\%sulf}$ 232 (4.00), 353 (4.38), 459 (3.62)	238–40	170–200 sub vac	...	...	δ	s	...	s	...	diox, AcOEt s	**E14,** 135
e98	—,α-dihydro-	$\Delta^{1,3,5(10),7}$-Estratetraene-3, 17β-diol. $C_{18}H_{22}O_2$.	270.38	cr (ace) $[\alpha]_D$ +220 (diox, c = 1) λ^{MeOH} 225, 260, 305	174.6	...	...	...	δ	s	...	s	...	sulf, diox s	**E14s,** 1962
e99	**Equisetrin**	Kaempferol-7-diglucoside. $C_{27}H_{30}O_{16}$.	610.53	ye nd (+2w)	195–6	...	...	...	...	s	...	...	...	sulf s	**C34,** 7910
e100	**Equol**	4′,7-Isoflavandiol. $C_{15}H_{14}O_3$.	242.28	cr (aq al) $[\alpha]_{5461}$ −21.5	189–90.5	...	...	...	...	...	...	...	...	...	**C52,** 20511
e101	**Eremophilol**	$C_{15}H_{24}O$.	220.36	visc oil $[\alpha]_{5461}$ −55.6 (MeOH)	...	164.5^{13}	...	1.5202^{20}	...	...	...	...	...	...	**E12B,** 1432
e102	**Eremophilone**	$C_{15}H_{22}O$	218.34	nd (MeOH) $[\alpha]_{5461}$ −207 (MeOH) λ^{al} 241 (3.99), 318 (1.92)	42–3	171^{15}	0.9994^{25}_{25}	1.5182^{25}	...	...	...	...	...	...	**E12B,** 2597
e103	—,8,9-epoxy-	$C_{15}H_{22}O_2$	234.34	nd (peth) λ^{al} 295 (1.7)	63–4	...	...	...	i	s	s	s	v	oos v	**E12B,** 2600
e104	**Ergine**	Ergonovine. Lysergamide. $C_{16}H_{17}N_3O$.	283.33	cr (MeOH +1), pr (aq ace +2w) $[\alpha]^{20}_{5461}$ +598 (chl), $[\alpha]_D$ +479 (Py) λ^{MeOH} 241 (4.30), 311 (3.95)	135d resolidify 140, remelt 190	...	...	...	...	δ	...	δ	...	os δ	**H32,** 506
—	**Ergobasine**	*see* **Ergometrine**													
—	**Ergobasinine**	*see* **Ergometrinine**													
e107	**Ergocornine**	$C_{31}H_{39}N_5O_5$	561.70	cr (MeOH) $[\alpha]^{20}_D$ −188 (chl), c = 1), −105 (Py)	182–4d	...	...	...	i	s	...	s	s^h	chl, AcOEt s	**H34,** 1544
e108	**Ergocorninine**	$C_{31}H_{39}N_5O_5$. *See* e107	561.70	lo pr (al) $[\alpha]^{20}_D$ +409 (chl, c = 1)	228d	...	...	...	i	s	..	v	s^h	chl v AcOEt s	**$B27^2$,** 860
Ω e109	**Ergocristine**	$C_{35}H_{39}N_5O_5$	609.63	rh (bz +2) $[\alpha]^{20}_D$ −183 (chl, c = 1) $[\alpha]_D$ −93 (Py) λ^{MeOH} 240 (4.30), 311 (3.95)	175d (170–90)d	...	...	...	i	s	...	s	...	chl s	**H34,** 1544
Ω e110	**Ergocristinine**	$C_{35}H_{39}N_5O_5$. *See* e109	609.63	pr (al, AcOEt) $[\alpha]^{20}_D$ +366 (chl, c = 0.68), +460 (Py) λ^{MeOH} 241 (4.33), 312 (3.96)	237–8d (226d)	...	...	...	i	δ^h	...	δ	...	chl δ	**$B27^2$,** 860
e111	**Ergocryptine**	$C_{32}H_{41}N_5O_5$	575.72	pr (al, MeOH) $[\alpha]^{20}_D$ −187 (chl, c = 1), −112 (Py) λ^{MeOH} 240 (4.31), 312 (3.97)	212–4d	...	...	...	i	s	...	...	...	chl s	**H34,** 1544
e112	**Ergocryptinine**	$C_{32}H_{41}N_5O_5$. *See* e111	575.72	lo pr (al) $[\alpha]^{20}_D$ +408 (chl, c = 1), +479 (Py) λ^{MeOH} 240 (4.31), 313 (3.96)	245d	...	...	...	δ	s^h	...	v	...	chl v	**$B27^2$,** 860
e113	**Ergometrine(*l*)**	Ergobasine. *l*-Lysergic-*l*(β-hydroxyisopropylamide). $C_{19}H_{23}N_3O_2$.	325.42	nd (bz), $[\alpha]^{20}_D$ −89 (w)	159–62d (cor)	...	...	...	δ	s	...	s	...	chl δ	**Am 69,** 1701
Ω e114	**Ergometrinine(*d*)**	Ergobasinine. *d*-Isolysergic-*d*-(β-hydroxyisopropylamide). $C_{19}H_{23}N_3O_2$.	325.42	pr (ace) $[\alpha]^{20}_D$ +416 (chl, c = 0.26), +413 (MeOH)	195–7d	...	...	...	i	δ	...	δ	...	Py, chl s	**Am 60,** 1701

For explanations, symbols and abbreviations see beginning of table. For structural formulas see end of table.

No.	Name	Synonyms and Formula	Mol. wt.	Color, crystalline form, specific rotation and λ_{max} (log ε)	m.p. °C	b.p. °C	Density	n_D	Solubility w	al	eth	ace	bz	other solvents	Ref.
	Ergometrinine														
e115	—(*l*)	$C_{19}H_{23}N_3O_2$. *See* e114	325.42	pr (ace) $[\alpha]_D^{20}$ −415 (chl, c = 0.26)	196d (cor)				i	δ		δ		Py, chl s	**Am 60,** 1701
e116	**Ergopinacol II**	Bisergostadienol. $C_{56}H_{36}O_2$	791.31	nd (bz-al) $[\alpha]_D$ −155 (Py, c = 0.8)	205										**E14, 74**
e117	**Ergosine**	$C_{30}H_{37}N_5O_5$	547.66	pr (MeOH, AcOEt) $[\alpha]_D^{20}$ +16 (ace), −161 (chl, c = 1) λ^{MeOH} 240 (4.33), 312 (3.99)	228d							s		chl s MeOH δ	**H34,** 1544
e118	**Ergosinine**	$C_{30}H_{37}N_5O_5$. *See* e117	547.66	pr (al, aq ace, bz), nd (MeOH + 1) $[\alpha]_D^{20}$ + 420 (chl, c = 1) λ^{MeOH} 241 (4.32), 312 (3.96)	220d (228d)					δ		s		chl s	**J1937,** 396
e119	**$\Delta^{5:6,7:8}$-Ergostadien-3β-ol**	22,23-Dihydroergosterol. Provitamin D_4. $C_{28}H_{46}O$.	398.68	nd (MeOH-AcOEt) $[\alpha]_D^{19}$ −109 (chl)	152–3										**E14s,** 1754
e120	**$\Delta^{14,22}$-Ergostadien-3β-ol**	$C_{28}H_{46}O$	398.68	cr (MeOH) $[\alpha]_D^{20}$ −9 (chl)	116										**E14s,** 1762
e121	**α-Ergostadienone**	$C_{28}H_{44}O$	396.67	lf (al) $[\alpha]_D^{19}$ +2 (chl)	182–3										**E14, 125**
e122	**Ergostane**	24-Methyl-5α-cholestane. $C_{28}H_{50}$.	386.71	lf or pl (eth − MeOH or ace) $[\alpha]_D^{23}$ +21 (chl, c = 2)	85				i	δ	v	v		chl v	**E14, 23**
e123	**Ergostanol**	Brassicastanol. Ergostan-3β-ol. $C_{28}H_{50}O$.	402.71	nd (MeOH-eth) $[\alpha]_D$ +15.94 (chl, c = 1.8) $\lambda^{97\%\,sulf}$ 321 (3.82), 410 sh (3.07), 480 sh (2.75)	144–5				i		s			chl s	**E14s,** 1767
e124	**$\Delta^{3,5,7,22}$-Ergostatetraene**	$C_{28}H_{42}$	378.65	pl (Ac_2O or al) $[\alpha]_D^{20}$ −40.5 (chl) λ 301.5 (4.20), 316 (4.28), 331.5 (4.15)	104										**E14s,** 1434
e125	**$\Delta^{4,6,22}$-Ergostatrienone**	Isoergosterone. $C_{28}H_{42}O$	394.65	nd (ace-eth) λ^{al} 284 (4.42)	110										**E14, 124**
e126	**α-Ergostenol**	$\Delta^{8:14}$-Ergosten-3β-ol. α-Tetrahydroergosterol. $C_{28}H_{46}O$.	400.70	lf or nd (MeOH) nd (gl aa) $[\alpha]_D^{16}$ +11 (MeOH, c = 0.9) λ^{hx} 208 (4.05)	131 (134–5)					δ	s		s	chl s	**E14s,** 1765
e127	**β-Ergostenol**	$\Delta^{14:15}$-Ergosten-3β-ol. β-Tetrahydroergosterol. $C_{28}H_{46}O$.	400.70	pl or ta (al) $[\alpha]_D^{20}$ +21.2 (chl, c = 0.9) λ^{hx} 196 (3.93)	141–2										**E14s,** 1767
e128	**γ-Ergostenol**	$\Delta^{7:8}$-Ergosten-3β-ol. β-Tetrahydroergosterol. $C_{28}H_{46}O$.	400.70	nd (MeOH), cr (Pr^iOH), lf (w) $[\alpha]_D^{19}$ 0 λ^{al} 195.5 (3.92)	148(152)						s				**E14s,** 1763
e129	**δ-$\Delta^{8:9}$-Ergostenol**	$C_{28}H_{46}O$	400.70	$[\alpha]_D$ +39	155									chl s	**J1949,** 214
Ω e130	**Ergosterol**	$\Delta^{5,7,22}$-Ergostatrien-3β-ol. Ergosterin. $C_{28}H_{44}O$.	396.67	pl (+w, al), nd (eth), $[\alpha]_D^{20}$ −135 (chl, c = 1.2) λ^{al} 271 (3.80), 281 (3.82), 293 (3.57)	168 (+1.5 w)	$250^{0.01}$			i	δ s^h	δ s^h		s	chl s aa, peth δ	**E14s,** 1735
e131	**Ergosterol D**	$\Delta^{7,9,11,22}$-Ergostatrien-3β-ol. $C_{28}H_{44}O$.	396.67	nd (al) $[\alpha]_D^{17}$ +24.6	167					δ s^h			s	chl s	**E14s,** 1752

For explanations, symbols and abbreviations see beginning of table. For structural formulas see end of table.

No.	Name	Synonyms and Formula	Mol. wt.	Color, crystalline form, specific rotation and λ_{max} (log ε)	m.p. °C	b.p. °C	Density	n_D	Solubility w	al	eth	ace	bz	other solvents	Ref.
	Ergosterol														
e133	**Ergosterol,5,6-dihydro-**	α-Dihydroergosterol. $C_{28}H_{46}O$.	398.68	lf (ace), pl (chl-MeOH, AcOEt-MeOH) $[\alpha]_D^{20}$ −19 (chl) $\lambda^{97\%sulf}$ 238 (3.57), 319 (3.95), 419 (3.51), 469 sh (3.28)	176–7	....	....	....	...	...	...	...	...	....	**E14s,** 1758
e134	**Ergosterone**	$\Delta^{4,7,22}$-Ergostatrienone. $C_{28}H_{42}O$.	394.65	nd (ace-MeOH) $[\alpha]_D^{20}$ −0.52 (chl) λ^{al} 242 (4.15)	132–2.5	....	....	....	...	...	...	...	...	....	**E14,** 124
e135	**Ergotamine**	$C_{33}H_{35}N_5O_5$	581.68	nd (al), pr (bz) pl (aq ace) $[\alpha]_D^{20}$ −160 (chl, c = 1) λ^{al} 318 (3.86)	213–4d	....	....	....	i	δ	s	δ	s	$PhNO_2$ v chl, Py s peth i	**B27[2]**, 860
Ω e136	**Ergotaminine**	$C_{33}H_{35}N_5O_5$	581.68	rh pl (MeOH), pl (al) $[\alpha]_D^{20}$ +369 (chl, c = 0.5) λ^{MeOH} 241 (4.28), 309 (393)	252d	....	....	....	i	δ	...	δ	δ	Py v chl s peth i	**B27[2]**, 860
e137	**Ergothioneine**	Thiasine. Thioneine. $C_9H_{15}N_3O_2S$.	229.30	pl (w + 2), nd or lf (dil al) $[\alpha]_D$ +115 (w, c = 1) $\lambda^{w, pH=2-10}$ 258 (2.16)	290d	....	....	....	v^h	δ^h	i	δ	i	MeOH δ^h chl i	**B25[2]**, 413
—	**Eriodictol**	*see* **Flavanone,3′,4′,5,7-tetrahydroxy-**													
—	**Erucic acid**	*see* **13-Docosenoic acid(*cis*)**													
e138	**Erysocine**	$C_{18}H_{21}NO_3$	299.37	nd (eth) $[\alpha]_D$ +238.1	162	....	....	....	i	s	s	...	...	chl s	**Am 62,** 1677
e139	**Erysodine**	$C_{18}H_{21}NO_3$	299.37	nd (al) $[\alpha]_D^{27}$ +248 (al) λ^{ac} 235 (4.4), 285 (3.5)	204–5	....	....	....	i	s	s	...	...	chl s	**Am 73,** 589
e140	**Erysonine**	$C_{17}H_{19}NO_3$	285.35	cr (al), $[\alpha]_D^{25}$ +285–8 (aq HCl), +272 (morpholine)	236–7d	....	....	....	...	δ	...	δ	δ	os δ	**Am 36,** 1544
e141	**Erysopine**	$C_{17}H_{19}NO_3$	285.35	cr (al) $[\alpha]_D^{25}$ +265.2 (al-glycerol)	241–2	....	....	....	δ	δ	...	...	...	chl δ	**Am 73,** 589
e142	**Erysothiopine**	$C_{19}H_{21}NO_7S$	407.45	cr (al-w + w) $[\alpha]_D^{25}$ +194 (al)	168–9	....	....	....	...	...	...	...	...	....	**Am 66,** 1083
e143	**Erysovine**	$C_{18}H_{21}NO_3$	299.37	pr (eth) $[\alpha]_D$ +252 (al)	178–9	....	....	....	...	s	s	...	...	chl s	**Am 73,** 589
e144	**Erythraline**	$C_{18}H_{19}NO_3$	297.34	cr (al) $[\alpha]_D^{27}$ +211.8 (al) λ^{ac} 238 (4.4), 290 (3.6); λ^{alk} 230 (4.2), 290 (3.6)	106–7	....	....	....	...	s	...	...	...	chl s	**Am 73,** 589
e145	**Erythramine**	Dihydroerythraline. $C_{18}H_{21}NO_3$.	299.37	cr (eth-peth) $[\alpha]_D^{30}$ +228 (al, c = 0.19) λ^{ac} 240 (3.6), 290 (3.6)	103–4	$125^{3.9\cdot10^{-4}}$	....	....	...	s	s	s	s	lig i	**Am 73,** 589
e146	**Erythratine**	$C_{18}H_{19}NO_4$	315.37	cr (eth-peth + ½w) $[\alpha]_D^{28}$ +145.5 (al) λ^{al} 293 (0.68)	170	....	....	....	...	...	...	...	...	....	**Am 73,** 589
e147	**Erythritol(*d*)**	*d*-Erythryte. *L*-Threite.	122.12	pr (w), nd (al) $[\alpha]_D$ +11.1 (al, c = 5), −4.4 (w, c = 5–10)	88–9	....	....	....	s	...	i	...	i	....	**B1[3]**, 2360
Ω e148	—(*l*)	*l*-Erythryte. *D*-Threite.	122.12	pr (w), nd (al) $[\alpha]_D$ −11.1 (al, c = 5), −4.33 (w, c = 6)	90–1	....	....	....	s	...	i	...	i	....	**B1[3]**, 2360

For explanations, symbols and abbreviations see beginning of table. For structural formulas see end of table.

No.	Name	Synonyms and Formula	Mol. wt.	Color, crystalline form, specific rotation and λ_{max} (log ε)	m.p. °C	b.p. °C	Density	n_D	Solubility						Ref.
									w	al	eth	ace	bz	other solvents	
	Erythritol														
e148[1]	—(*meso*)		122.12	bipym tetr pr	121.5	329–31 294–6[200]	1.451_4^{20}		v	δ s^h	i	...	...	Py v	**B1**[3], 2356
e148[2]	—,tetranitrate (*meso*)		302.11	pl or lf (al)	61				i	s	s	...	...	glycerol s	**B1**[3], 2358
e149	*β*-**Erythroidine**	$C_{16}H_{19}NO_3$	273.34	cr (abs al) $[\alpha]_D^{25}$ +88.8 (w) $\lambda^{al\text{-}HCl}$ 238 (4.40)	99.5–100				s	v	s	...	v	chl s	**Am 80,** 3905
e150	**Erythronic acid,** γ-lactone(*D*)		118.09	pr $[\alpha]_D^{20}$ −73.2 (w, c = 4)	104–5				...	s^h	...	...	...	...	**B18**[2], 54
e151	—,—(*L*)	$C_4H_6O_4$. *See* e150	118.09	nd (AcOEt) $[\alpha]_D^{20}$ +73 (w, c = 4)	105 (cor)				...	s^h	...	...	...	...	**B18**[2], 54
Ω e152	**Erythrose**(*D*)		120.12	syr $[\alpha]_D^{20}$ +1 → −14.8 (w, c = 11, 3 days) (mut)					s	v	...	...	...	...	**B31,** 12
e153	—(*L*)	$C_4H_8O_4$. *See* e152	120.12	syr $[\alpha]_D^{24}$ +11.5 → +30.5 (w, c = 3, 3 days) (mut)					s	v	...	...	...	...	**B31,** 12
Ω e154	**Erythrosin**	2,4,5,7-Tetraiodofluorescein.	835.91	og-ye (eth) λ^w 530 (5.12)					δ	s	s	...	i	aq ace s lig i	**B19**[2], 258
Ω e155	—(dye)	Disodium salt of erythrosin. Iodeosin B. $C_{20}H_6I_4Na_2O_5$.	879.87	red-br pw λ^{alk} 522.5 (4.86)					v	s	...	...	i	...	**M,** 415
e156	**Erythrulose**(*L*)	$HOCH_2CH(OH)COCH_2OH$.	120.12	syr $[\alpha]_D^{30}$ +11.31 (w, at equilibrium) (mut)		d			s	s	i	...	...	...	**B31,** 14
e157	**Escholerine**	$C_{41}H_{61}NO_{13}$	775.95	pl (ace-w) $[\alpha]_D^{25}$ −30 (Py, c = 1)	235d				i	...	...	s	...	...	**Am 76,** 1152
—	**Esculetin**	*see* **Coumarin, 6,7-dihydroxy-**													
Ω e158	**Esculin**	Aesculin. $C_{15}H_{16}O_9$.	340.29	pr (w + 2) $[\alpha]_D^{18}$ −14.6 (MeOH) $[\alpha]_D^{22}$ −38 (Py, c = 1.5) λ^{al} 227 (4.2), 251 sh (3.7), 298 (3.8), 338 (4.1)	205d	230d			δ s^h	δ s^h	δ	...	...	alk, aa, Py, AcOEt s chl s^h	**B31,** 246
Ω e159	α-**Estradiol**	$C_{18}H_{24}O_2$	272.39	nd (+½w, 80 % al) $[\alpha]_D^{20}$ +56 (diox, c = 0.9)	220–3				i	s	...	s	δ^h	peth s	**E14,** 98
e159[1]	*β*-**Estradiol**	$C_{18}H_{24}O_2$. *See* e159	272.39	pr (80 % al) $[\alpha]_D^{25}$ +76–83 (diox) λ^{al} 280 (3.33)	178–9				i	s	...	s	...	diox s	**E14,** 99
—	**Estragole**	*see* **Benzene,1-allyl-4-methoxy-**													
Ω e160	**Estriol**	$C_{18}H_{24}O_3$	288.39	lf (al), mcl (dil al) $[\alpha]_D$ +61 (al), +30 (Py) λ^{al} 281 (3.34)	288d (cor)		1.27		...	s	δ	...	δ	Py v os δ	**E14s,** 2152
Ω e161	**Estrone**	$C_{18}H_{22}O_2$	270.38	α mcl (al) β, γ orh (al) $[\alpha]_D^{25}$ +158 to +168 (diox) λ^{al} 280 (3.37)	260–2		β 1.236 γ 1.228		i	δ^h	δ	s^h	δ	diox, Py s chl δ^h	**E14,** 135
—	**Ethanal***	*see* **Acetaldehyde**													
Ω e162	**Ethane***	CH_3CH_3	30.07	gas, hex cr	−183.3	−88.63	0.572_4^{-108} 0.509_7^{-60}	1.03769 at 0°/546 mm.	i	δ	...	δ	s	...	**B1**[3], 120
e163	—,amino-*	Ethylamine. $C_2H_5NH_2$	45.09	λ^{gas} 177 (3.20), 213 (2.90)	−81	16.6^{760}	0.6829_4^{20}	1.3663^{20}	∞	∞	∞	...	...	...	**B4**[2], 586

For explanations, symbols and abbreviations see beginning of table. For structural formulas see end of table.

No.	Name	Synonyms and Formula	Mol. wt.	Color, crystalline form, specific rotation and λ_{max} (log ε)	m.p. °C	b.p. °C	Density	n_D	Solubility w	al	eth	ace	bz	other solvents	Ref.
	Ethane														
e164	—,—,hydro-bromide*	$C_2H_5NH_2.HBr$	126.01	mcl nd or pl (al)	159.5	...	...	...	s	v	...	...	i	chl δ	**B4**^{2}, 588
Ω e165	—,—,hydro-chloride*	$C_2H_5NH_2.HCl$	81.56	mcl pl (al)	109–10	d315	...	...	v	v	i	i	i	chl i	**B4**^{2}, 588
e166	—,—,hydroiodide*	$C_2H_5NH_2.HI$	173.01	mcl nd (w)	188.5	...	2.100	...	v	δ	i	...	i	...	**B4**^{2}, 588
e167	—,1-amino-2-bromo-, hydro chloride*	$BrCH_2CH_2NH_2.HCl$	160.45	lf (al-AcOEt)	174–5d (cor)	...	...	...	v	v	i	...	s	...	**B4**^{2}, 618
Ω e168	—,1-amino-2-chloro-, hydro-chloride*	$ClCH_2CH_2NH_2.HCl$	115.99	hyg cr (AmOH or al-eth)	144(148)	...	...	...	v	v	i	s	...	os v	**B4**, 133
—	—,2-amino-1,1-diethoxy-*	*see* **Acetaldehyde, 2-amino-,** diethyl acetal													
Ω e169	—,1-amino-2-diethylamino-*	*N'N*-diethylethylenediamine	116.21	...	...	144^{760} $38–40^{15}$	0.8280_4^{20}	1.4340^{20}	∞	s	s	...	...	to s	**B4**^{2}, 690
Ω e170	—,1-amino-2-ethoxy-*	2-Aminodiethyl ether $C_2H_5OCH_2CH_2NH_2$	89.14	...	...	108^{758} (cor)	0.8512_4^{20}	1.4101^{20}	∞	∞	∞	s	s	...	**B4**^{2}, 718
Ω e171	—,1-amino 2-methoxy-*	$CH_3OCH_2CH_2NH_2$	75.11	...	...	95^{756}	...	...	v	v	...	...	...	...	**B4**^{2}, 718
e176	—,aminoxy-	Ethoxylamine. α-Ethyl-hydroxylamine. $C_2H_5ONH_2$	61.05	...	...	68	0.8827_8^8	...	∞	∞	∞	...	...	os ∞	**B1**^{2}, 333
Ω e177	—,1,2-bis(4-aminophenyl)-*	α,α'-Bi-*p*-toluidine. 4,4'-Diaminobibenzyl.	212.30	pl (w)	135–6 sub	sub	...	...	i s^h	v	...	...	...	...	**B13**^{1}, 75
e178	—,1,2-bis-(benzoyl-amino)-*	*N,N'*-Dibenzoyl-ethylenediamine. $C_6H_5CONHCH_2CH_2NHCOC_6H_5$	268.32	pr or nd (al)	247	...	...	...	i	i	...	...	...	aa v^h	**B9**^{2}, 187
e179	—,1,2-bis(2-bromophenyl)-*	2,2'-Dibromobibenzyl	340.07	pl (al)	84.5	138–$48^{0.013}$	...	...	...	s^h	...	...	...	...	**B5**^{2}, 507
e180	—,1,2-bis(4-bromophenyl)-*	4,4'-Dibromobibenzyl.	340.07	pr (al)	115	*ca.* 198^{10}	...	...	...	i δ^h	...	...	δ	...	**B5**^{2}, 507
Ω e181	—,1,2-bis(2-chloroethoxy)-*	Triglycol dichloride ether. $ClCH_2CH_2OCH_2CH_2OCH_2CH_2Cl$	187.07	...	...	230^{760} 118^{10}	1.197_{20}^{20}	1.4592^{25}	...	...	...	...	...	...	**C52**, 18219
e182	—,1,1-bis(4-chlorophenyl)-2,2-dichloro-*	DDD. Rothane	320.05	λ^{al} 233 (4.3)	109–10	...	...	...	...	...	...	...	...	...	**Am 79**, 5979
Ω e183	—,2,2-bis(4-chlorophenyl)-1,1,1-trichloro-	DDT. Dichlorodiphenyl-trichloroethane.	354.49	nd (al) λ^{al} 240 (4.3), 270 (2.8)	108.5–9	260	...	...	i	δ s^h	v	v	v	Py v chl, peth s	**Am 75**, 4853
Ω e184	—,1,2-bis(ethyl-thio)-*	$C_2H_5SCH_2CH_2SC_2H_5$	150.31	...	...	217 95.5^{13}	0.9815_4^{20}	1.5118^{20}	...	s	s	...	...	...	**B1**^{3}, 2131
Ω e185	—,2,2-bis(4-methoxyphenyl)-1,1,1-trichloro-*	Methoxychlor	345.66	cr (dil al) λ^{bz} 230 (4.3), 238 (4.2), 270 (3.9), 275 (3.8)	94	...	...	...	i	s	v	...	v	os v	**H27**, 892
e186	—,1,2-bis(2-methoxy-phenoxy)-*	Guaiacol ethylene ether	274.32	nd (al)	139–40*	...	...	...	i	v	v	...	...	...	**B6**, 772
Ω e187	—,1,2-bis(methyl-amino)-*	*N,N'*-Dimethylethylene-diamine. $CH_3NHCH_2CH_2NHCH_3$	88.15	...	...	120	0.828_4^{15}	...	...	s	s	...	...	dil HCl s	**B4**^{2}, 689
e189	—,1,2-bis(methyl-thio)-*	$CH_3SCH_2CH_2SCH_3$	122.25	...	...	182.5^{750} $78–80^{11}$	1.0371_4^{20}	1.5292^{20}	s	s	s	s	...	chl s	**B1**^{3}, 2130
Ω e190	—,1,2-bis(2-nitro-phenyl)-*	2,2'-Dinitrobibenzyl	272.26	pr (aa)	127 (122)	...	...	...	...	δ s^h	v	...	v	aa s^h	**B5**, 603
e191	—,1,2-bis(4-nitro-phenyl)-*	4,4'-Dinitrobibenzyl	272.26	yesh nd (al or bz)	180.5 (cor)	...	...	...	...	i δ^h	δ	...	δ s^h	chl δ aa δ^h	**B5**^{3}, 1821
e192	—,1,2-bis(phenyl-amino)-*	*N,N'*-Diphenyl-ethylenediamine. $C_6H_5NHCH_2CH_2NHC_6H_5$	212.30	lf (dil al) λ 250 (4.44), 294 (3.70)	74 68–9 (+1 w)	$178–82^2$	...	...	i	s	s	...	...	...	**B12**^{2}, 287
e193	—,1,2-bis(phenyl-sulfonyl)-*	$C_6H_5SO_2CH_2CH_2SO_2C_6H_5$	310.39	nd or lf (al)	180	...	...	...	δ^h	s^h	...	...	s	aa s^h	**B6**^{3}, 997
e194	—,1,2-bis(phenyl-thio)-*	$C_6H_5SCH_2CH_2SC_6H_5$	246.40	ta (al) λ^{al} 256 (4.00)	70	...	...	...	i	...	...	s	...	...	**B6**^{3}, 994
Ω e195	—,bromo-*	Ethyl bromide. C_2H_5Br	108.97	...	–118.6	38.40^{760}	1.4604_4^{20}	1.4239^{20}	δ	∞	∞	...	...	chl ∞	**B1**^{3}, 171
Ω e196	—,1-bromo-2-chloro-*	Ethylene chlorobromide. $BrCH_2CH_2Cl$	143.42	...	–16.7	107^{760}	1.7392_4^{20}	1.4908^{20}	δ	s	s	...	...	...	**B1**^{3}, 179
e197	—,1-bromo-2-fluoro-*	$BrCH_2CH_2F$	126.96	...	...	71.5–1.8^{760}	1.7044_4^{25}	1.4236^{20}	i	v	v	...	...	...	**B1**^{3}, 178
Ω e199	—,chloro-*	Ethyl chloride. C_2H_5Cl	64.52	...	–136.4	12.27^{760}	0.8978_4^{20}	1.3676^{20}	δ	v	∞	...	...	...	**B1**^{3}, 133
e200	—,1-chloro-2-diethylamino-, hydrochloride*	β-Chlorotriethylammonium chloride. $ClCH_2CH_2N(C_2H_5)_2.HCl$	172.10	nd (al-eth)	210–1 (cor)	...	...	...	s	s	i	...	...	...	**B4**^{2}, 618
e201	—,1-chloro-1,2-dinitro-1,2,2-trifluoro-*	$F_2C(NO_2)CClFNO_2$	208.48	...	...	98.5–100^{760}	...	...	...	...	...	...	...	...	**J1953**, 2081
e202	—,1-chloro-2-fluoro-*	$ClCH_2CH_2F$	82.51	...	...	57^{750}	1.1747_4^{20}	1.3775^{20}	i	v	v	...	...	...	**B1**^{3}, 138
Ω e203	—,1-chloro-1-nitro-*	$CH_3CHClNO_2$	109.51	$\lambda^{50\%\,al-alk}$ 237 (4.00)	...	$124–5^{760}$ (127.5)	1.2860_{20}^{20}	1.4224^{20}	i	s	...	...	...	alk s	**B1**, 101

For explanations, symbols and abbreviations see beginning of table. For structural formulas see end of table.

No.	Name	Synonyms and Formula	Mol. wt.	Color, crystalline form, specific rotation and λ_{max} (log ε)	m.p. °C	b.p. °C	Density	n_D	Solubility w	al	eth	ace	bz	other solvents	Ref.
	Ethane														
Ω e204	—,**chloropenta-fluoro-***	$ClCF_2CF_3$	154.47	gas	−106	-38^{760}	...	...	i	s	s	...	...	...	**B1**³, 139
e205	—,**1(2-chloro-ethoxy)-2-phenoxy-***	β-chloro-β′-phenoxy diethyl ether. $C_6H_5OCH_2CH_2OCH_2CH_2Cl$	200.67	...	...	149^{10} 127^4	1.149^{15}_{15}	...	...	...	...	...	...	...	**B6**², 150
e206	—,**2(2-chloro-phenyl)-2(4-chlorophenyl)-1,1,1-trichloro-***	o,p′-DDT.	354.49	cr (MeOH) $\lambda^{80\%\ al}$ 235 (4.2), 275 (2.8), 280 (2.7)	74–4.5 (cor)	...	...	...	...	...	...	...	...	...	**Am 67**, 1498
Ω e207	—,**1,2-diamino-***	Ethylenediamine. 1,2-Ethanediamine. $H_2NCH_2CH_2NH_2$	60.11	...	8.5	116.5^{760}	0.8995^{20}_{20}	1.4568^{20}	v	∞	i	...	i	...	**B4**², 676
e208	—,—,**hydrate***	$H_2NCH_2CH_2NH_2.H_2O$	78.12	...	10	118^{760}	$0.964^{20.5}_4$	$1.4500^{20.5}$	∞	...	...	...	...	...	**B4**², 677
Ω e209	—,—,**hydro-chloride***	$H_2NCH_2CH_2NH_2.2HCl$	133.02	mcl pr (w)	300–330 sub	sub	...	1.633	s	i	...	...	...	...	**B4**², 677
e210	—,**2-diazo-1,1,1-trifluoro-***	Trifluorodiazoethane. N_2CHCF_3	110.04	ye	13^{752}	...	...	...	s	...	s	...	...	...	**Am 65**, 1460
e211	—,**1,1-dibromo-***	Ethylidene bromide. CH_3CHBr_2	187.87	...	−63	108^{760} 9.0^{10}	2.0555^{20}_4	1.5128^{20}	i	s	v	s	s	...	**B1**³, 181
Ω e212	—,**1,2-dibromo-***	Ethylene bromide. $BrCH_2CH_2Br$	187.87	...	9.79	131.36^{760} 29.1^{10}	2.1792^{20}_4	1.5387^{20}	δ	v	∞	s	s	...	**B1**³, 182
e213	—,**1,2-dibromo-1,1-dichloro-***	$BrCH_2CBrCl_2$	256.76	...	−66.85	178.3^{760} 58.8^{10}	2.2622^{20}_4	1.5567^{20}	i	s	s	s	s	os s	**B1**², 64
e214	—,**1,2-dibromo-1,2-dichloro-***	$BrCHClCHBrCl$	256.76	...	−26	195^{760} 84^{45}	2.135^{20}_4	1.5662^{20}	i	s	s	s	s	os s	**B1**³, 190
Ω e215	—,**1,1-dichloro-***	Ethylidene chloride. CH_3CHCl_2	98.96	...	−96.98	57.28^{760}	1.1757^{20}_4	1.4164^{20}	δ	v	v	s	s	...	**B1**³, 139
Ω e216	—,**1,2-dichloro-***	Ethylene chloride. $ClCH_2CH_2Cl$	98.96	...	−35.36	83.47^{760}	1.2351^{20}	1.4448^{20}	δ	v	∞	s	s	oos, chl s	**B1**³, 141
e217	—,**1,1-dichloro-2,2-difluoro-1,2-dinitro-**	$O_2NCF_2CCl_2NO_2$	224.94	...	...	142^{760}d $81–2^{108}$	...	...	i	...	...	...	...	...	**J1953**, 2081
e218	—,**1,1-dichloro-1,2,2,2-tetra-fluoro-***	Cl_2CFCF_3	170.92	...	−94	3.6^{760} (cor)	1.455^{25}_4	1.3092^0	i	s	s	...	s	chl s	**B1**³, 152
—	—,**1,2-diethoxy-***	*See* **1,2-Ethanediol,** diethyl ether													
—	—,**2,2-diethoxy-1,1,1-trichloro-***	*see* **Acetaldehyde, trichloro-,** diethyl acetal													
Ω e221	—,**1,1-difluoro-***	Ethylidene fluoride. CH_3CHF_2	66.05	gas	−117	−24.7	0.95^{20} (at sat pres-sure)	1.3011^{-72} 1.26^{20} (liq)	...	...	...	...	...	...	**B1**³, 130
e222	—,**1,2-difluoro-***	Ethylene fluoride. FCH_2CH_2F	66.05	...	...	30.7^{760}	...	...	...	...	s	...	s	chl s	**B1**, 82
e223	—,**1,1-difluoro-1,2,2,2-tetra-chloro-***	$F_2CClCCl_3$	203.83	...	40.6	91.5^{760} (cor)	...	...	i	s	s	...	...	chl s	**B1**³, 164
Ω e224	—,**1,2-difluoro-1,1,2,2-tetra-chloro-***	$Cl_2CFCFCl_2$	203.83	...	25	93^{760} -37.5^1	1.6447^{25}_4	1.4130^{25}	i	s	s	...	...	chl s	**B1**³, 165
e225	—,**(difluoro-amino)penta-fluoro-***	Perfluoroethylamine. $F_3CCF_2NF_2$	171.02	...	...	-35^{760}	...	...	i	...	...	...	...	...	**Am 81**, 3599
e227	—,**1,1-diiodo-***	Ethylidene iodide. CH_3CHI_2	281.86	...	...	$179–80^{760}$ $60–1^{12}$	2.84^0	1.673^{20}	i	v	v	s	...	chl s	**B1**³, 198
Ω e228	—,**1,2-diiodo-***	Ethylene iodide. ICH_2CH_2I	281.86	ye mcl pr or rh (eth), d in lt	83	200^{760} 74^{10}	3.325^{20}_4	1.871^{20}	δ	s	s	s	...	chl s	**B1**², 69
—	—,**1,2-di-methoxy-***	*see* **1,2-Ethanediol,** dimethyl ether													
e230	—,**1,2-di-*N*-morpholyl-**	...	200.28	wh-yesh (eth or lig)	75 (71)	$160–3^{25}$ $(153–4^9)$	...	...	v	v	...	s	s	MeOH s	**B27**, 7
e231	—,**1,1-dinitro-***	$CH_3CH(NO_2)_2$	120.07	ye mcl (bz or MeOH) λ^{hx} 280 (1.76)	...	185–6 (cor) 72^{12}	1.3503^{24}_{24}	...	δ	s	s	...	...	...	**B1**², 70
e232	—,**1,2-dinitro-1,2-diphenyl-(*dl*)***	α,α′-Dinitrobibenzyl. $C_6H_5CH(NO_2)CH(NO_2)C_6H_5$	272.26	(al), pr (aa)	154–5	...	...	...	...	v	v	v	v	os, aa, chl v lig s	**B5**², 508
e233	—,—(*meso*)	$C_6H_5CH(NO_2)CH(NO_2)C_6H_5$	272.26	nd (aa)	235–6 (226d)	...	...	...	...	δ	δ	v	...	aa δ, s^h	**B5**², 508
e234	—,**1,2-dinitro-1,1,2,2-tetra-fluoro-***	$O_2NCF_2CF_2NO_2$	192.03	...	−41.5	$58–9^{760}$	1.6024^{25}_4	1.3265^{25}	i	...	...	s	...	...	**B1**³, 203
Ω e235	—,**1,2-diphen-oxy-***	Ethylene glycol diphenyl ether. $C_6H_5OCH_2CH_2OC_6H_5$	214.27	lf (al)	98	$180–5^{12}$	...	...	i	δ s^h	s	...	...	chl s	**B6**², 150
Ω e236	—,**1,1-diphenyl-***	α-Methylditan. $CH_3CH(C_6H_5)_2$	182.27	$\lambda^{97\%\ suf}$ 430 (4.35)	−21.5	286^{760} (270) 148^{15}	0.9997^{20}_4	1.5756^{20}	i	∞	∞	...	s	...	**B5**², 653

For explanations, symbols and abbreviations see beginning of table. For structural formulas see end of table.

No.	Name	Synonyms and Formula	Mol. wt.	Color, crystalline form, specific rotation and λ_{max} (log ε)	m.p. °C	b.p. °C	Density	n_D	Solubility w	al	eth	ace	bz	other solvents	Ref.
	Ethane														
Ω e237	**—,1,2-diphenyl-***	Bibenzyl. Dibenzyl. $C_6H_5CH_2CH_2C_6H_5$	182.27	nd (al) λ^{diox} 265 (2.9), 290 sh(2.6)	52.2	285^{760} $95\text{–}6^{1}$	0.9583_4^{60}	1.5476^{60}	i	s v^h	s	...	...	CS_2 liq SO_2 s liq NH_3 i	**B5**[2], 506
Ω e238	**—,2,2-diphenyl-1,1,1-trichloro-***	$(C_6H_5)_2CHCCl_3$	285.60	lf (al or MeOH) λ^{cy} 260 (2.6), 268 (2.5)	65	...	...	...	...	s v^h	...	...	...	...	**B5**[2], 510
e239	**—,1,1-di-4-tolyl-**	...	210.32	...	< −20	298–9 $153\text{–}6^{11}$	0.974_4^{20}	...	i	...	...	...	s	aa s	**B5**[2], 502
Ω e240	**—,1,2-di-4-tolyl-**	...	210.32	lf (MeOH, or dil al), pl (lig) λ^{cy} 221 (4.2), 254 (2.6), 267 (2.9), 275 (2.9), 343 (0.8)	82–3 (85)	$296\text{–}8^{730}$ 178^{18}	...	...	i	δ	...	...	s	peth s	**B5**[2], 521
e240[1]	**—,2,2-di-4-tolyl-1,1,1-trichloro-**	...	313.66	mcl pr (al or eth-al) $\lambda^{80\%al}$ 235 (4.1), 277 (2.7)	92	...	...	...	i	s v^h	v	s	...	...	**B5**[2], 522
Ω e241	**—,1,2-epoxy-***	Ethylene oxide. Oxirane.	44.05	λ^{gas} 169 (3.58), 171 (3.57)	−111	13.5^{746}	0.8824_{10}^{10}	1.3597^{7}	s	s	s	s	s	...	**B17**[2], 9
e242	**—,1,2-epoxy-1-phenyl-***	Styrene oxide.	120.16	λ^{al} 254 (2.24), 260 (2.28), 265 sh (2.1)	−35.6	194.1^{760} $20^{0.3}$	1.0469_4^{25}	1.5342^{20}	δ	s	∞	...	∞	...	**B17**[2], 49
e243	**—,1-ethoxy-2-methoxy-***	$C_2H_5OCH_2CH_2OCH_3$	104.15	...	...	102^{760}	0.8529_4^{20}	1.3868^{20}	∞	∞	∞	s	s	chl s	**B1**[3], 2078
e244	**—,1-ethoxy-2-methylamino-***	Ethyl β-methylaminoethyl ether. $CH_3NHCH_2CH_2OC_2H_5$	103.17	...	...	$114\text{–}5^{744}$ (cor)	0.8363_4^{20}	1.4147^{20}	∞	∞	∞	s	s	chl s	**B4**, 276
e245	**—,fluoro-***	Ethyl fluoride. C_2H_5F	48.06	gas	−143.2	-37.7^{760}	0.00220 at *ca.* 0° 0.7182_4^{20} (liq)	1.2656^{20}	s	v	v	...	...	...	**B1**[3], 130
e246	**—,fluoropentachloro-***	$FCCl_2CCl_3$	220.29	...	101.3 (97)	$134\text{–}6^{760}$	...	...	i	s	s	...	...	...	**B1**[3], 168
e247	**—,hexabromo-***	Perbromoethane. Br_3CCBr_3	503.48	rh pr (bz)	d 200–10	d	2.823_4^{20}	1.863	i	δ^h	δ	...	...	CS_2 s	**B1**[3], 193
Ω e248	**—,hexachloro-***	Perchloroethane. Cl_3CCCl_3	236.74	rh (al-eth)	186.8–7.4 (sealed tube)	186^{777}	2.091_4^{20}	...	i	v	v	...	s	liq HF δ	**B1**[3], 168
Ω e249	**—,hexafluoro-***	Perfluoroethane. F_3CCF_3	138.01	gas	−94	−79	1.590^{-78}	...	i	δ	δ	...	...	...	**B1**[3], 132
e250	**—,hexaphenyl-***	$(C_6H_5)_3CC(C_6H_5)_3$	486.67	cr (ace)	145–7d	d	...	...	...	s^h	s	s	δ	chl, CS_2, MeOH v CCl_4, lig δ lig i	**B5**[2], 626
e251	**—,*N*-hydroxylamino-***	β-Ethylhydroxylamine. C_2H_5NHOH	61.09	nd (lig)	59–60d	...	0.9079_4^{20}	1.4152^{20} 1.4152^{64}	v	v	δ	...	δ	lig δ	**B4**[2], 953
e252	**—,2-imino-1,1,1-trichloro-***	Chloralimide. 2,2,2-Trichloroethylideneimide. $Cl_3CCH{:}NH$	146.40	...	150–5	...	...	...	δ	v	v	...	...	...	**M**, 233
Ω e253	**—,iodo-***	Ethyl iodide. CH_3CH_2I	155.97	...	−108 (−111)	72.3^{760}	1.9358_4^{20}	1.5133^{20}	δ^d	∞	s	...	...	os ∞	**B1**[3], 193
Ω e254	**—,isocyano-**	Ethylcarbylamine. Ethylisocyanide. C_2H_5NC	55.08	...	< −66	79^{775}	0.7402_4^{20}	1.3622^{20}	v	∞	∞	s	...	...	**B4**[2], 600
e254[1]	**—,1(4-methoxyphenyl)-1-phenyl-***	1-*p*-Anisyl-1-phenylethane.	212.30	...	...	$180\text{–}2^{19}$ $130^{0.5}$	1.0473_4^{20}	1.5725^{20}	i	...	s	s	s	os s	**B6**[2], 639
Ω e255	**—,nitro-***	$CH_3CH_2NO_2$	75.07	λ^{al} 260 (1.63)	−50	115^{760}	1.0448_4^{25}	1.3917^{20}	δ	∞	∞	s	...	dil alk s	**B1**[3], 199
e256	**—,nitropentafluoro-***	$F_3CCF_2NO_2$	165.02	λ^{vap} 281 (1.45)	...	0^{760}	...	...	...	...	s	...	...	...	**J1956**, 3416
e257	**—,1-nitro-2,2,2-trifluoro-***	$F_3CCH_2NO_2$	129.04	...	...	96^{760}	1.3914_4^{20}	1.3394^{20}	...	...	s	...	...	...	**Am 72**, 3579
e258	**—,1(2-nitrophenyl)-2(4-nitrophenyl)-***	2,4′-Dinitrobibenzyl.	272.26	nd (dil al)	74–5	...	...	...	i	v	...	...	...	...	**B5**, 603
e259	**—,nitrosopentafluoro-***	F_3CCF_2NO	149.02	bl λ^{vap} 692 (1.37), 708 (1.37)	...	-42^{760}	...	...	...	...	...	...	...	...	**J1956**, 3416
Ω e260	**—,pentabromo-***	Br_3CCHBr_2	424.58	mcl pr (dil al)	56–7	210^{300}d	3.312_4^{20}	...	i	s	v	...	...	...	**B1**[3], 193
Ω e261	**—,pentachloro-***	Pentalin. Cl_3CCHCl_2	202.30	...	−29	162^{760}	1.6796_4^{20}	1.5025^{20}	i	∞	∞	...	...	...	**B1**[3], 165
e262	**—,pentaiodo-***	I_3CCHI_2	659.55	mcl pr (aa)	182–4	...	...	...	i	s	s	...	s	aa s	**B1**[1], 31
e264	**—,1,1,1,2-tetrabromo-**	Br_3CCH_2Br	345.67	...	0.0	112^{18}d	2.8748_4^{20}	1.6277^{20}	...	s	s	s	s	chl s	**B1**[3], 191
Ω e265	**—,1,1,2,2-tetrabromo-***	Acetylene tetrabromide. $Br_2CHCHBr_2$	345.67	yesh	0	243.5^{760} 114.8^{10}	2.9656_4^{20}	1.6353^{20}	i	∞	∞	s	s	chl, $PhNH_2$, aa ∞	**B1**[3], 192
e266	**—,1,1,1,2-tetrachloro-***	Cl_3CCH_2Cl	167.85	yesh red	−70.2	130.5^{760} 22.1^{10}	1.54064_4^{20}	1.4821^{20}	δ	∞	∞	s	s	chl s	**B1**[3], 158
Ω e267	**—,1,1,2,2-tetrachloro-***	Acetylene tetrachloride. $Cl_2CHCHCl_2$	167.85	...	−36 (fp −43.8)	146.2^{760} 33.9^{10}	1.5953_4^{20}	1.4940^{20}	δ	∞	∞	s	s	chl s	**B1**[3], 159
e268	**—,1,1,1,2-tetrafluoro-***	F_3CCH_2F	102.03	...	...	-26.5^{736}	...	...	i	...	s	...	...	...	**Am 82**, 543
e269	**—,1,1,1,2-tetraphenyl-***	$(C_6H_5)_3CCH_2C_6H_5$	334.47	mcl (eth-peth)	143–4	$277\text{–}80^{21}$	...	...	i	δ	...	δ	s	...	**B5**[2], 674

For explanations, symbols and abbreviations see beginning of table. For structural formulas see end of table.

	No.	Name	Synonyms and Formula	Mol. wt.	Color, crystalline form, specific rotation and λ_{max} (log ε)	m.p. °C	b.p. °C	Density	n_D	Solubility w	al	eth	ace	bz	other solvents	Ref.
		Ethane														
	e270	—,**1,1,2,2-tetra-phenyl-***	$(C_6H_5)_2CHCH(C_6H_5)_2$	334.47	cr (bz + 1), rh nd (chl) λ^{al} 263 (3.05), 270.5 (2.91)	214–5	358–62 260^{16}				δ^h			s^h	aa s	**B5**[2], 673
	e271	—,**triazo-***	Azidoethane. Ethyl azide. $CH_3CH_2N_3$	71.08			49^{760}								peth s	**B1**[2], 71
Ω	e272	—,**1,1,2-tribromo-***	Br_2CHCH_2Br	266.77		−29.3	188.93^{760} 73.1^{10}	2.6211^{20}_{4}	1.5933^{20}	i	s	s		s	chl s	**B1**[3], 191
Ω	e273	—,**1,1,1-trichloro-***	Methylchloroform. CH_3CCl_3	133.41		−30.41	74.1^{760}	1.3390^{20}_{4}	1.4379^{20}	i	∞	∞			chl s	**B1**[2], 55
Ω	e274	—,**1,1,2-trichloro-***	Cl_2CHCH_2Cl	133.41		−36.5	113.77^{760} 9.48^{10}	1.4397^{20}_{4}	1.4714^{20}	δ	s	∞			chl s	**B1**[3], 154
	e275	—,**1,1,1-trichloro-2,2,2-trifluoro-***	Cl_3CCF_3	187.38		14.2	45.8^{760} (cor)	1.5790^{20}_{4}	1.3610^{25}	i	s	s			chl s	**B1**[3], 157
Ω	e276	—,**1,1,2-trichloro-1,2,2-trifluoro-***	$Cl_2CFCClF_2$	187.38		−36.4	47.7^{760} (cor)	1.5635^{25}_{4}	1.3557^{25}	i	s	∞		∞		**B1**[3], 157
	e277	—,**1,1,1-trifluoro-***	CH_3CF_3	84.04	gas	−111.3	-47.3^{760}	0.00378				s			chl s	**B1**[3], 131
	e278	—,**1,1,1-triiodo-***	Methyliodoform. CH_3CI_3	407.76	ye oct (al)	95 (93)					δ	v		v	CS_2 v lig δ	**B1**[3], 199
	e279	—,**1,1,1-triphenyl-***	α-Methyltritan. $(C_6H_5)_3CCH_3$	258.37	nd (al, eth) λ^{al} 262 (2.9)	95	$205–10^{18}$			i	δ s^h	s		s	aa δ	**B5**[1], 350
	e280	—,**1,1,2-triphenyl-***	$(C_6H_5)_2CHCH_2C_6H_5$	258.37	mcl lf (dil al), nd (al) λ^{al} 208 (4.4), 261 (2.8), 270 (2.7)	57	$348–9^{751}$ (cor) (396–400)			i	v	v		v	MeOH δ	**B5**[2], 620
	e281	**Ethanearsonic acid***	Ethylarsonic acid. $C_2H_5AsO_3H_2$	154.00	nd (al), rh nd (w)	99.5	$209–11^{12}$			v	v				alk v	**B4**[3], 1823
	e282	**Ethaneboronic acid***	Ethylboric acid. $C_2H_5B(OH)_2$	73.89	volat pl (eth)	40 sub	d			v	v	v				**B4**, 642
	—	**Ethanedial***	*see* **Glyoxal**													
	—	**Ethanedionic acid***	*see* **Oxalic acid**													
	—	**1,1-Ethanediol,** diacetate	*see* **Acetaldehyde,** diacetate													
Ω	e283	**1,2-Ethanediol***	Dihydroxyethane*. Ethylene glycol. Glycol. $HOCH_2CH_2OH$	62.07		−11.5	198^{760} (cor) 93^{13}	1.1088^{20}_{4}	1.4318^{20}	∞	∞	s	∞	δ	aa ∞ chl s	**B1**[3], 2053
	e284	—,bis(chloro-acetate)	$ClCH_2CO_2(CH_2)_2O_2CCH_2Cl$	215.03	pr (eth-peth)	45–6	$142–4^{2}$			i	d^h	v				**B2**[3], 448
	—	—,carbonate	*see* **1,3-Dioxolan-2-one**													
Ω	e285	—,diacetate	Ethylene acetate. $CH_3CO_2CH_2CH_2O_2CCH_3$	146.14		−31	190^{760}	1.1063^{20}_{20}	1.4159^{20}	v	∞	∞	∞	∞	CS_2, CCl_4, aa ∞	**B2**[3], 309
Ω	e286	—,dibenzoate	Ethylene benzoate. $C_6H_5CO_2CH_2CH_2O_2CC_6H_5$	270.29	rh pr (eth)	73–4	>360d			i		s				**B9**[2], 109
	e287	—,dibutanoate*	Ethylene butyrate. $C_3H_7^nCO_2CH_2CH_2O_2CC_3H_7^n$	202.25			240 $118–21^{11}$	1.0005^{20}_{4}	1.42619^{20}	i	v	v				**B2**[2], 272
	e288	—,didodecanoate*	Ethylene laurate. $CH_3(CH_2)_{10}CO_2CH_2CH_2O_2C(CH_2)_{10}CH_3$	426.69	pl (al)	56.6	188^{20}			i	v	v				**B2**[2], 320
Ω	e289	—,diethyl ether	1,2-Diethoxyethane*. $C_2H_5OCH_2CH_2OC_2H_5$	118.18			123.5^{760}	0.8484^{20}	1.3860^{20}		s	s	s	s	os v	**B1**[2], 519
Ω	e290	—,diformate	$HCO_2CH_2CH_2O_2CH$	118.09			174	1.193^{0}_{4}	1.3580	δ	s	s				**B2**, 23
	e291	—,dihexadecylate*	Ethylene palmitate. $CH_3(CH_2)_{14}CO_2CH_2CH_2O_2C(CH_2)_{14}CH_3$	538.91	lf or nd (al-chl)	72 (69)	226 (vac)	0.8594^{78}		i	i	s^h	v		oos v	**B2**[2], 338
Ω	e292	—,dimethyl ether	1,2-Dimethoxyethane*. $CH_3OCH_2CH_2OCH_3$	90.12		−58	$83–4^{760}$	0.86285^{20}_{4}	1.3796^{20}	s	s	s	s	s	chl s	**B1**[3], 2073
	e293	—,dinitrate*	Ethylene nitrate. $O_2NOCH_2CH_2ONO_2$	152.06	ye	−22.3	197–200^{760}	1.4918^{20}_{4}		i	s	s			alk d	**B1**[3], 2112
	e294	—,dinitrite*	Ethylene nitrite. $ONOCH_2CH_2ONO$	120.07		< −15	98	1.2156^{0}_{4}		i	s	s			glycerol s	**B1**, 469
	e295	—,dioctadecylate*	Ethylene stearate. $CH_3(CH_2)_{16}CO_2CH_2CH_2O_2C(CH_2)_{16}CH_3$	595.02	lf	79 (77)	241^{20}	0.8581^{78}		i	i	v	v			**B2**[2], 354
	e296	—,dipropanoate*	Ethylene propionate $C_2H_5CO_2CH_2CH_2O_2CC_2H_5$	174.20			211	1.020^{15}		δ	∞	∞	v			**B2**, 242
	e297	—,ditetradecylate*	Ethylene myristate. $CH_3(CH_2)_{12}CO_2CH_2CH_2O_2C(CH_2)_{12}CH_3$	482.80	cr (eth or ace)	65				i	i	s	v	v	CCl_4 v	**B2**[2], 327
	e298	—,dithiocyanate	$NCSCH_2CH_2SCN$	144.22	rh pl or nd (w), ta (al or eth)	90	d			δ s^h	s	s	v	δ		**B3**[2], 123
	e301	—,**1,2-bis(4-methoxyphenyl)-*** (lower m.p. form)	*p,p'*-Dimethoxyhydro-benzoin. Isohydroanisoin.	274.32	cr (eth), nd (bz)	125–6					v	v				**B6**[2], 1130
	e302	—,—,(higher m.p. form)	Hydroanisoin. $C_{16}H_{18}O_4$. *See* e301.	274.32	rh ta or lf (al)	173–4				i δ^h	δ s^h	i δ^h				**B6**[2], 1129
	e304	—**1,2-dicyclo-hexyl-(*dl*)***	Cyclohexanonepinacol. Dodecahydrobenzoin.	226.36	nd	129–30								v	peth s	**Am 71,** 829

For explanations, symbols and abbreviations see beginning of table. For structural formulas see end of table.

No.	Name	Synonyms and Formula	Mol. wt.	Color, crystalline form, specific rotation and λ_{max} (log ε)	m.p. °C	b.p. °C	Density	n_D	Solubility w	al	eth	ace	bz	other solvents	Ref.
	1,2-Ethanediol														
e305	—,1,2-diphenyl-(*d*)*	*d*-Hydrobenzoin. *d*-Isohydrobenzoin.	214.27	nd (w), lf or pr (abs al) $[\alpha]_D$ +92 (abs al, c = 1.2), +128 (bz, c = 0.3) λ^{al} 253 (2.51), 258 (2.60), 263 sh (2.5)	148.5–9.5d (cor)	....	....	....	δ^h	s	...	...	δ^h	MeOH, AcOEt s	**B6**^{2}, 970
Ω e306	—,—(*dl*)*	$C_{14}H_{14}O_2$. *See* e305	214.27	nd (w or al), ta (eth) λ^{al} 253 (2.51), 258 (2.60), 263 sh (2.5)	122–3 (anh)	>300^{760} 133$^{0.023}$	....	....	i δ^h	v	v	...	...	chl v lig i	**B6**^{2}, 969
e307	—,—(*l*)*	$C_{14}H_{14}O_2$. *See* e305	214.27	lf (eth, abs al or bz), pr (bz or abs al) $[\alpha]_D$ −92 (abs al, c = 1.2) λ^{al} 253 (2.51), 258 (2.60), 263 sh (2.5)	148.5–9.5d (cor)	....	....	....	...	v	s^h	v	s^h	MeOH, AcOEt v chl s^h	**B6**^{2}, 970
e308	—,—(*meso*)*	Tolylene glycol. $C_{14}H_{14}O_2$. *See* e305	214.27	nd or lf (w or bz-peth), mcl lf (al, w) λ^{al} 253 (2.51), 258 (2.60), 263 sh (2.5)	139–40	>300^{760} 139$^{0.023}$	....	....	i δ^h	v	...	...	...	chl v lig i	**B6**^{2}, 967
Ω e309	—,phenyl-(*dl*)*	Phenyl ethylene glycol. $C_6H_5CH(OH)CH_2OH$	138.17	nd (lig) λ^{al} 252 (2.21), 228 (2.26)	69–70	272–4^{755}	....	....	v	v	v	...	v	lig δ, v^h	**B6**^{2}, 887
Ω e309[1]	—,tetraphenyl-*	Benzopinacol. $(C_6H_5)_2C(OH)COH(C_6H_5)_2$	366.47	pr (bz + 1), cr (ace + 2)	182–3 (slow htng) 193–5 (rapid htng)	....	....	....	i	δ^h	s	s^h	s^h	chl, CS_2 s lig, liq NH_3 i	**B6**^{2}, 1034
—	**1,1-Ethanediol, 2,2,2-tribromo-***	*see* **Acetaldehyde, tribromo-, hydrate**													
Ω e310	**1,2-Ethanedione, 1,2-di(2-furyl)-**	Bipyromucyl. Furil.	190.16	ye nd (al), cr (bz) λ^w 227.5 (3.97), 302.5 (4.24)	165–6 (cor)	....	....	....	δ	s	s	...	s^h	chl s	**B19**^{2}, 183
—	—,1,2-diphenyl-*	*see* **Benzil**													
e311	**1,2-Ethanedisulfonic acid***	Ethylenedisulfonic acid. $HO_3SCH_2CH_2SO_3H$	190.19	hyg nd (gl aa or aa-Ac_2O)	111–2 (+2w) 174 (anh)	....	....	....	v	v	i	...	...	diox s	**B4**^{3}, 36
Ω e312	**1,2-Ethanedithiol***	Dithioglycol. Ethylene mercaptan. $HSCH_2CH_2SH$	94.20	....	−41.2	146^{760} 46–7^6	1.1243^{20}_4	1.5590^{20}	i	s	s	s	s	alk v	**B1**^{2}, 529
e313	**Ethanedithiolic acid, diethyl ester***	Diethyl dithioloxalate. $C_2H_5SCOCOSC_2H_5$	178.28	ye nd (eth)	27	235 80–2^{32}	1.0565^{21}_4	....	...	...	s^h	...	...	....	**B2**^{2}, 515
e314	**Ethanephosphonic acid***	Ethylphosphonic acid. $C_2H_5PO_3H_2$	110.05	pl or nd (w)	61–2 (44)	....	....	....	v	v	v	...	i	lig i	**B4**^{3}, 1779
e315	—,diethyl ester*	$C_2H_5PO(OC_2H_5)_2$	166.16	....	....	198^{760} 83^{13}	1.0259^{20}_4	1.4163^{20}	δ	s	s	...	...	....	**B4**^{2}, 975
e316	—,dimethyl ester*	$C_2H_5PO(OCH_3)_2$	138.10	....	....	82^{18}	1.1029^{30}_4	1.4128^{30}	s	s	...	...	s	....	**J1954**, 3222
e317	**Ethanesulfinic acid***	Ethylsulfinic acid. $C_2H_5SO_2H$	94.13	syr	....	....	....	....	...	...	...	...	...	alk s	**B4**^{2}, 524
e318	**Ethanesulfonic acid***	Ethylsulfonic acid. $C_2H_5SO_3H$	110.13	hyg	−17	123^1	1.3341^{25}	1.4335^{20}	s	s	...	...	i	alk s	**B4**^{2}, 525
e319	—,chloride*	$C_2H_5SO_2Cl$	128.58	pa ye	....	171^{760} 65^{13}	1.357$^{22.5}$	1.4531^{20}	d	d	v	...	...	....	**B4**^{2}, 526
—	—,2-amino-*	*see* **Taurine**													
e320	—,2-bromo-, chloride	$BrCH_2CH_2SO_2Cl$	207.48	pa ye	....	102^{13}	1.921^{20}	1.5242^{20}	d	d	...	...	...	....	**B4**^{2}, 526
e321	—,1-chloro-, chloride*	$CH_3CHClSO_2Cl$	163.02	....	....	80–1^{22} 50–3^3	....	1.4782^{20}	δ	...	...	...	...	....	**B1**^{3}, 2656
e322	—,2-chloro-* chloride	$ClCH_2CH_2SO_2Cl$	163.02	....	....	200–3^{760} 93–7^{17}	1.555^{20}_4	1.4920^{20}	d	d	...	...	...	....	**B4**^{2}, 526
Ω e323	—2-hydroxy-, dihydrate*	Isethionic acid. $HOCH_2CH_2SO_3H.2H_2O$	152.17	hyg cr (aa-Ac_2O)	111–2	....	....	....	v	s	i	i	i	CCl_4 i	**B4**^{2}, 529
e324	**1,1,2,2-Ethanetetracarboxylic acid**, 1,2-diethyl ester*	$C_2H_5O_2CCH(CO_2H)CH(CO_2H)CO_2C_2H_5$	262.22	hyg lf (+½w)	132–3d	....	....	....	s	v	v	...	...	chl, CS_2 δ	**B2**, 858
e325	—,tetraamide	Dimalonamide. Ethane-1,1,2,2-tetracarbonamide*. $(H_2NCO)_2CHCH(CONH_2)_2$	202.17	....	230d	....	....	....	δ^h	δ^h	i	i	i	os i	**B2**, 859

For explanations, symbols and abbreviations see beginning of table. For structural formulas see end of table.

No.	Name	Synonyms and Formula	Mol. wt.	Color, crystalline form, specific rotation and λ_{max} (log ε)	m.p. °C	b.p. °C	Density	n_D	Solubility						Ref.
									w	al	eth	ace	bz	other solvents	
	1,1,2,2-Ethanetetracarboxylic acid														
Ω e326	—,tetraethyl ester*	Ethyl dimalonate. $(C_2H_5O_2C)_2CHCH(CO_2C_2H_5)_2$	318.33	tetr pr (al-peth)	77	305d	1.064^{80}	1.4105^{80}	...	s	...	...	...	...	**B2**[2], 699
e327	—,tetramethyl ester*	Methyl dimalonate. $(CH_3O_2C)_2CHCH(CO_2CH_3)_2$	262.22	cr (eth, al, bz)	138	...	...	...	...	v^h	δ	...	s^h	lig i	**B2**[2], 699
Ω e328	**Ethanethiol***	Ethyl hydrosulfide. Ethyl mercaptan. Ethyl thioalcohol. C_2H_5SH	62.13	λ^{al} 195 (3.15), 225 sh (2.2)	−144.4	35^{760}	0.8391^{20}_{4}	1.43105^{20}	δ	s	s	s	...	dil alk s	**B1**[2], 341
e329	—,sodium salt	Sodium thioethylate. C_2H_5SNa	84.12	wh cr λ^{al} 195 (3.15), 225 sh (2.2)	d	...	...	...	v	s	...	...	i	...	**B1**, 341
Ω e330	—,**2-amino-***	Cysteamine. $H_2NCH_2CH_2SH$	77.15	cr (sub)	99–100	d^{760} sub (vac)	...	...	v	s	...	...	...	dil HCl v	**B4**[1], 431
e331	—,**2-chloro-***	$ClCH_2CH_2SH$	96.58	...	...	113^{760}	1.1826^{20}_{4}	1.4929^{20}	s	v	v	...	...	diox v	**B1**[3], 1381
Ω e332	—,**1-phenyl-(*l*)***	$CH_3CH(C_6H_5)SH$	138.23	$[\alpha]^{20}_{D}$ −89 (al, c = 6)	...	199–200 83^{10}	1.022^{20}_{4}	1.5593^{20}	...	s	s	...	s	...	**B6**[2], 445
e333	**1,1,1-Ethanetricarboxylic acid***	$CH_3C(CO_2H)_3$	162.10	pr	159d	...	...	...	s	s	s	...	...	...	**B2**[3], 2026
e334	**1,1,2-Ethanetricarboxylic acid***	Carboxysuccinic acid. $HO_2CCH_2CH(CO_2H)_2$	162.10	pr (w)	178d	...	...	...	v	v	v	...	δ^h	...	**B2**[2], 681
e335	—,**1,2-dihydroxy-***	Desoxalic acid. $HO_2CCH(OH)COH(CO_2H)_2$	194.10	hyg (+1w)	45d	...	...	...	v d^h	...	δ	...	...	...	**B3**, 586
—	**Ethanoic acid***	*see* **Acetic acid**													
Ω e336	**Ethanol***	Alcohol. Ethyl alcohol. Methyl carbinol. C_2H_5OH	46.07	λ^{gas} 181 (2.51)	−117.3 (−112.3)	78.5 4^{16}	0.7893^{20}_{4}	1.3611^{20}	∞	∞	∞	∞	s	chl, aa ∞	**B1**[3], 1223
e337	—(*o-d*)	o-Deuteroethanol. Deuteroxyethane. C_2H_5OD	47.08	...	...	78.8^{760}	0.801^{25}_{4}	1.3610^{20}	∞	∞	∞	∞	s	aa ∞ CCl_4 s	**B1**[3], 1287
Ω e338	—,**2-allyloxy-**	Allyl cellosolve. Glycol monoallyl ether. $CH_2{:}CHCH_2OCH_2CH_2OH$	102.13	...	...	159^{755} 64^{15}	0.9580^{20}_{4}	1.4358^{20}	∞	v	...	...	s	MeOH, CCl_4 s	**B1**, 468
Ω e339	—,**1-amino-***	Acetaldehyde-ammonia. $CH_3CH(OH)NH_2$	61.09	rh (eth-al)	97	110 δd	...	...	s	...	δ	...	...	...	**C52**, 10081
Ω e340	—,**2-amino-***	Colamine. Ethanolamine. $H_2NCH_2CH_2OH$	61.09	...	10.3	170^{760} 58^{5}	1.0180^{20}_{4}	1.4541^{20}	∞	∞	δ	...	δ	glycerol ∞ chl s lig δ	**B4**[2], 717
—	—,**2-amino-1(3,4-dihydroxyphenyl)-(*l*)***	*see* **Noradrenaline (*l*)**													
Ω e341	—,**2-amino-1-phenyl-***	$H_2NCH_2CH(OH)C_6H_5$	137.18	nd (al-eth-peth)	56–7 (40)	160^{17}	...	...	v	s	...	...	...	...	**B13**[2], 361
e342	—,**1-amino-2,2,2-trichloro-***	Chloral-ammonia. $Cl_3CCH(OH)NH_2$	164.42	nd (al)	72–4	100d	...	...	δ	s	v	...	v	...	**B1**[2], 681
Ω e343	—,**2(2-amino-ethylamino)-***	$H_2NCH_2CH_2NHCH_2CH_2OH$	104.15	hyg liq	...	238–40^{752} (cor) 91.2^{5}	0.9556^{25}_{4}	1.4860^{20}	v	v	δ	...	...	...	**B4**, 286
e344	—,**2(2-amino-phenyl)-***	...	137.18	ye in air	...	152–3^{6}	...	1.5849^{19}	s	...	...	...	...	...	**B13**[2], 362
e345	—,**2(4-amino-phenyl)-***	...	137.18	nd (al)	108	...	...	...	...	δ s^h	...	...	...	...	**B13**[2], 362
e346	—,**2-benzyloxy-**	Glycol monobenzyl ether. $C_6H_5CH_2OCH_2CH_2OH$	152.20	...	< −75	256^{760} 138^{15}	1.0640^{20}_{4}	1.5233^{20}	s	s	s	...	...	...	**B6**[2], 413
e346[1]	—,**1,1-bis(4-chlorophenyl)-***	4,4′-Dichloro-α-methylbenzhydrol. Dimite.	267.16	cr (al-w) λ^{sulf} 240 (3.8), 325 (4.2), 370 sh (3.5), 450 (4.9), 500 sh (3.01)	69–70	...	...	...	i	i	s	...	s	...	**Am 73**, 1856
Ω e347	—,**2-bromo-***	Ethylene bromohydrin. $BrCH_2CH_2OH$	124.97	...	...	149–50^{750} 51^{4}	1.7629^{20}_{4}	1.4915^{20}	∞	∞	∞	...	...	oos v lig δ	**B1**[2], 337
Ω e348	—,**2-butoxy-***	Glycol monobutyl ether. $C_4H_9^nOCH_2CH_2OH$	118.18	...	...	171^{760} 50^{4}	0.9015^{20}_{4}	1.4198^{20}	∞	∞	∞	...	...	...	**B1**[2], 519
Ω e349	—,**2-butylamino-***	$C_4H_9^nNHCH_2CH_2OH$	117.19	...	...	199–200^{760} 91–2^{11}	0.8907^{20}_{4}	1.4437^{20}	v	v	v	...	...	...	**B4**, 283
Ω e350	—,**2-chloro-***	Ethylene chlorohydrin. $ClCH_2CH_2OH$	80.52	...	−67.5	128^{760} 44^{20}	1.20027^{20}_{4}	1.44189^{20}	∞	∞	δ	...	...	os ∞	**B1**[2], 333
e351	—,**2-chloro-1-phenyl-***	Styrene chlorohydrin. $C_6H_5CH(OH)CH_2Cl$	156.61	...	...	128^{17} $91.5^{1.5}$	1.1926^{20}_{4}	1.5523^{20}	...	s	v	...	...	...	**B6**[2], 446
Ω e352	—,**2(2-chloro-ethoxy)-***	β-Chloroethyl cellosolve. $ClCH_2CH_2OCH_2CH_2OH$	124.57	...	...	180–85 91–2^{13}	...	1.4505^{19}	v	∞	∞	...	...	...	**B1**[2], 519
Ω e353	—,**1-cyclohexyl-***	Methylcyclohexylcarbinol.	128.22	λ^{al} 255 (3.8)	...	189 81–2^{15}	0.9250^{20}_{4}	1.4677^{20}	...	v	v	...	...	...	**B6**[2], 27
Ω e354	—,**2-cyclohexyl-***	...	128.22	...	...	207–9^{757} 97–9^{15}	0.9229^{20}_{4}	1.4641^{20}	...	s	s	...	s	...	**B6**[2], 27
Ω e355	—,**2-cyclopentyl-***	...	114.19	...	...	183–4^{770} 96–7^{24}	0.9180^{20}_{4}	1.4577^{20}	i	...	s	...	...	...	**B6**[2], 25
Ω e356	—**2,2-dichloro-***	Cl_2CHCH_2OH	114.96	...	...	146^{760} 37–8^{6}	1.4040^{25}_{4}	1.4626^{25}	δ	s	s	...	...	...	**B1**, 338
—	—,**2,2-diethoxy-***	*see* **Acetaldehyde, 2-hydroxy-,** diethyl acetal													

For explanations, symbols and abbreviations see beginning of table. For structural formulas see end of table.

No.	Name	Synonyms and Formula	Mol. wt.	Color, crystalline form, specific rotation and λ_{max} (log ε)	m.p. °C	b.p. °C	Density	n_D	Solubility w	al	eth	ace	bz	other solvents	Ref.
	Ethanol														
Ω e358	—,2-diethyl-amino-*	β-Hydroxytriethylamine. $(C_2H_5)_2NCH_2CH_2OH$	117.19	hyg	...	163^{760} $56–7^{15}$	0.8921_4^{20}	1.4412^{20}	∞	s	s	s	s	peth s	**B4**[2], 727
Ω e359	—,2-dimethyl-amino-*	$(CH_3)_2NCH_2CH_2OH$	89.14	...	...	134^{760} (cor)	0.8866_4^{20}	1.4300^{20}	∞	∞	∞	...	...	...	**B4**[2], 719
e360	—,1,2-diphenyl-(*d*)*	Benzyl phenyl carbinol. $C_6H_5CH_2CH(OH)C_6H_5$	198.27	nd (eth-peth or dil al) $[\alpha]_D^{25}$ +53 (al)	67–8	$167–70^{10}$	1.0358_4^{70}	...	i	v	v	...	...	...	**B6**[2], 637
e361	—,—(*dl*)*	$C_6H_5CH_2CH(OH)C_6H_5$	198.27	nd (bz-peth)	69	177^{15} $125–30^2$	...	...	i	v	v	...	...	...	**B6**[2], 638
e362	—,—(*l*)*	$C_6H_5CH_2CH(OH)C_6H_5$	198.27	nd (eth-peth) $[\alpha]_D^{20}$ −9.4 (w, c = 10)	67	...	1.0358_4^{70}	...	i	v	v	...	...	...	**B6**[2], 637
Ω e363	—,2-ethoxy-*	Ethyl cellosolve. Glycol monoethyl ether. $C_2H_5OCH_2CH_2OH$	90.12	...	...	135^{760} 35^{10}	0.9297_4^{20}	1.4080^{20}	∞	∞	∞	v	...	...	**B1**[2], 518
Ω e364	—,2-ethylamino-*	$C_2H_5NHCH_2CH_2OH$	89.14	...	...	$169–70^{760}$ $78–80^{27}$	0.914_4^{20}	1.444^{20}	v	v	v	...	...	...	**B4**[2], 727
Ω e365	—,2-(ethylthio)-*	$C_2H_5SCH_2CH_2OH$	106.19	...	*ca.* −100	184 $35^{0.12}$	1.0166_4^{20}	1.4867^{20}	δ	s	...	v	...	...	**B1**[3], 2120
Ω e366	—,2-fluoro-*	Ethylene fluorohydrin. FCH_2CH_2OH	64.06	...	−26.45	103.5	1.1040_4^{20}	1.3647^{18}	∞	∞	∞	v	...	...	**B1**[2], 333
Ω e367	—,2-hexyloxy-*	Hexyl cellosolve. $CH_3(CH_2)_5OCH_2CH_2OH$	146.23	...	−45.1	208^{760} 96^{13}	0.8894_{20}^{20}	1.4291^{20}	δ	v	v	...	...	...	**B1**[3], 2086
e368	—,2-iodo-*	Ethylene iodohydrin. ICH_2CH_2OH	171.97	...	...	176–7δd $85–8^{25}$	2.1968_4^{20} 2.2289_4^0	1.5713^{20}	s	v	v	...	...	...	**B1**[3], 1363
e368[1]	—,isobutoxy-	Isobutyl cellosolve. $C_4H_9^iOCH_2CH_2OH$	118.18	...	...	159^{745}	0.8900_4^{20}	1.4143^{20}	v	v	v	...	...	...	**B1**[3], 2084
Ω e369	—,2-isobutyl-amino-*	$C_4H_9^iNHCH_2CH_2OH$	117.19	...	...	199–200 (cor) 90^{16}	0.8818_4^{20}	1.4402^{20}	v	v	v	...	...	...	**B4**, 283
e370	—,2-isopropoxy-*	Isopropyl cellosolve. $C_3H_7^iOCH_2CH_2OH$	104.15	...	...	144^{743}	0.9030_4^{20}	1.4095^{20}	∞	∞	∞	s	...	...	**B1**[3], 2080
e371	—,2-isopropyl-amino-*	$C_3H_7^iNHCH_2CH_2OH$	103.17	...	...	$172–4^{760}$ $76–7^{15}$	0.8970_4^{20}	1.4395^{20}	∞	∞	∞	...	...	...	**B4**, 282
Ω e372	—,2-mercapto-*	Monothioethylene glycol. Thioglycol. $HSCH_2CH_2OH$	78.13	...	...	157–8^{742} δd 55^{13}	1.1143_4^{20}	1.4996^{20}	s	s	s	...	s	...	**B1**[2], 523
Ω e373	—,2-methoxy-*	Methyl cellosolve. $CH_3OCH_2CH_2OH$	76.11	...	−85.1	125^{768}	0.9647_4^{20}	1.4024^{20}	∞	v	∞	s	∞	...	**B1**[3], 2069
e374	—,1(2-methoxy-phenyl)-*	*o*-Anisylmethylcarbinol.	152.20	...	...	128^{17}	1.0862_4^{15}	1.5312^{25}	i	s	s	...	...	...	**B6**[2], 886
e375	—,1(3-methoxy-phenyl)-*	*m*-Anisylmethylcarbinol.	152.20	...	...	133^{15}	1.0781_4^{19}	1.5325^{20}	...	v	v	...	...	...	**B6**[2], 886
e376	—,1(4-methoxy-phenyl)-*	*p*-Anisylmethylcarbinol.	152.20	...	...	*ca.* 310^{760} δd $140–1^{17}$	1.0794_4^{20}	1.5310^{25}	...	s	s	...	...	con sulf s	**B6**[2], 886
Ω e377	—,2-methyl-amino-*	$CH_3NHCH_2CH_2OH$	75.11	...	...	158^{760} (cor) 52^6	0.937^{20}	1.4385^{20}	∞	∞	∞	...	...	...	**B4**[2], 718
—	—,2-methylamino-1-phenyl-(*l*)*	*see* **Halostachine**													
e378	—,2(methyl-phenyl)amino-*	$C_6H_5N(CH_3)CH_2CH_2OH$	151.21	yesh $\lambda^{95\%\ al}$ 256	...	150^{14}	0.9995_0^{15}	...	s^h	v	v	v	v	...	**B12**[2], 107
e380	—,2(methylthio)-*	$CH_3SCH_2CH_2OH$	92.17	...	...	$68–70^{20}$	1.6640_{20}^{20}	1.4867^{30}	s	v	v	...	...	...	**B1**[2], 524
—	—,morpholinyl-	*see* **Morpholine, 4(2-hydroxy-ethyl)-**													
e381	—,1(1-naphthyl)-(*dl*)*	$C_{10}H_7^{\alpha}CH(OH)CH_3$	172.23	nd(peth)	66	178^{15} $119–25^1$	1.1190_4^{14}	1.6188^{25}	i	s	...	s	s	chl s	**B6**[2], 619
e382	—,—(*l*)*	$C_{10}H_7^{\alpha}CH(OH)CH_3$	172.23	nd, $[\alpha]_D^{20}$ − 78.9 (al, c = 5)	47	166^{11}	1.1190_4^{14}	1.6180^{25}	i	s	...	s	s	chl s	**B6**[2], 619
e383	—,2(2-naphthyl-amino)-*	$C_{10}H_7^{\beta}NHCH_2CH_2OH$	187.24	lf (eth or al)	52	$197–8^3$	...	...	...	...	v	...	...	...	**B12**[2], 717
e384	—,2-nitro-*	$O_2NCH_2CH_2OH$	91.07	λ^{al} 270 (1.45)	−80	194^{765} 102^{10}	1.270_4^{15}	1.4438^{19}	∞	∞	∞	...	i	...	**B1**[3], 1364
Ω e385	—,2-phenoxy-*	Phenoxytol. $C_6H_5OCH_2CH_2OH$	138.17	...	...	237 $134–5^{18}$	1.1020_4^{22}	1.5340^{20}	i	s	s	...	...	alk s	**B6**[3], 567
e386	—,1-phenyl-(*d*)*	$C_6H_5CH(OH)CH_3$	122.17	$[\alpha]_D^{19}$ +42.9 (undil) λ^{sulf} 435 (4.0)	...	203^{760} 100^{18}	1.0129_4^{20}	1.5272^{20}	...	v	...	...	s	chl s	**B6**[2], 444
Ω e387	—,—(*dl*)*	$C_6H_5CH(OH)CH_3$	122.17	glassy λ^{sulf} 435 (4.0)	20	203.4^{760} 87.25^{10}	1.0135_4^{20}	1.5275_4^{20}	i	∞	∞	...	...	...	**B6**[2], 444
e388	—,—(*l*)*	$C_6H_5CH(OH)CH_3$	122.17	$[\alpha]_D^{23}$ − 45.5 (MeOH, c = 5) λ^{sulf} 435 (4.0)	...	$202–4^{760}$ 93^{14}	1.0129_4^{20}	1.5272^{20}	δ	v	v	...	...	...	**B6**[2], 444
Ω e389	—,2-phenyl-*	Benzylcarbinol. Phenethyl-alcohol. $C_6H_5CH_2CH_2OH$	122.17	glass λ^{hx} 271 (3.29)	fr −27	218.2^{760} 97.4^{10}	1.0202_4^{20}	1.5325^{20}	δ	∞	∞	...	...	...	**B6**[2], 448
Ω e390	—,2-phenyl-amino-*	2-Anilinoethanol. $C_6H_5NHCH_2CH_2OH$	137.18	$\lambda^{95\%\ al}$ 248	...	286^{760} 167^{17}	1.0945_4^{20}	1.5760^{20}	δ	v	v	...	...	chl v	**B12**[2], 106
—	—,piperidyl-	*see* **Piperidine, hydroxy-ethyl-**													
Ω e391	—,2-propoxy-*	Propyl cellosolve. $C_3H_7^nOCH_2CH_2OH$	104.15	...	...	150^{743}	0.9112_4^{20}	1.4133^{20}	s	v	v	...	...	...	**B1**[3], 2079

For explanations, symbols and abbreviations see beginning of table. For structural formulas see end of table.

No.	Name	Synonyms and Formula	Mol. wt.	Color, crystalline form, specific rotation and λ_{max} (log ε)	m.p. °C	b.p. °C	Density	n_D	Solubility w	al	eth	ace	bz	other solvents	Ref.
	Ethanol														
e393	—,1(3-tolyl)-		136.20			112^{12}	0.9974^{15}_{4}	1.5240^{20}	i	v	v				**B6**^{2}, 478
e394	—,1(4-tolyl)-		136.20			219^{756} 120^{19}	0.9944^{20}_{4}	1.5246^{20}	i	v	v				**B6**^{2}, 479
—	—,tolylamino-	*see* **Amine, ethyl tolyl, hydroxy-**													
e395	—,2(4-tolylthio)-		168.26			282–3d 174^{30}				s	s		s	aa s chl δ lig i	**B6**^{2}, 396
e396	—,2-triazo-*	$N_3CH_2CH_2OH$	87.08			75^{40} 60^{8}	1.149^{24}_{24}	1.4578^{25}	∞						**B1**^{1}, 171
Ω e397	—,2,2,2-tribromo-*	Avertin. Bromethol. Ethobrom. Renarcol. Br_3CCH_2OH	282.77	nd or pr (peth)	81–1.5	$92–3^{10}$			δ	v	v		s	peth s^h	**B1**^{2}, 338
e398	—,2,2,2-trichloro-*	Cl_3CCH_2OH	149.41	hyg rh ta or pl	19	151^{737} 52^{11}		1.4861^{20}	δ	∞	∞			alk s	**B1**^{3}, 1358
Ω e399	—,1,1,2-triphenyl-*	Benzyldiphenylcarbinol. $(C_6H_5)_2C(OH)CH_2C_6H_5$	274.37	nd (bz-lig), pr (peth)	89–90	222^{11}			i	v	δ			peth, lig δ	**B6**^{2}, 696
e400	—,2,2,2-triphenyl-*	Tritylcarbinol. $(C_6H_5)_3CCH_2OH$	274.37	cr (al, eth, lig)	110.5d				i	s	s		s	lig s	**B6**^{2}, 696
Ω e401	Ethene*	Ethylene. $CH_2:CH_2$	28.05	gas, mcl pr λ^{gas} 161.5 (3.94), 166 (3.8), 174 (3.7)	−169.15 fr −181	-103.71^{760}	0.00126 (at 0° 760 mm) 0.384^{-10}_{4}	1.363^{100}	i	δ	s	δ	δ		**B1**^{1}, 75
e402	—,amino-*	Ethenylamine. Vinylamine. $CH_2:CHNH_2$	43.07			$55–6^{756}$	0.8321^{24}		∞	s	∞				**B4**, 203
e403	—,bromo-*	Vinyl bromide. $CH_2:CHBr$	106.96		−139.54	15.80^{760}	1.4933^{20}_{4} (liq)	1.4410^{20}	i	s	s	s	s	chl s	**B1**^{2}, 162
Ω e404	—,1-bromo-1,2,2-triphenyl-*	$(C_6H_5)_2C:CBrC_6H_5$	335.25	nd (aa) λ^{al} 230 (4.3), 290 (3.9)	116–7				i					aa s^h	**B5**^{1}, 355
Ω e405	—,chloro-*	Vinyl chloride. $CH_2:CHCl$	62.50	gas	−153.8	-13.37^{760}	0.9106^{20}_{4}	1.3700^{20}	δ	s	v				**B1**^{2}, 157
e406	—,1-chloro-2-dichloroarsino-(*trans*)*	Lewisite. $ClCH:CHAsCl_2$	207.32		0.1	196^{760}d 93^{26}	1.888^{20}_{4}		i	s	s			os s alk d	**B4**^{3}, 1810
e407	—,1-chloro-1,2,2-triphenyl-*	$(C_6H_5)_2C:CClC_6H_5$	290.80	λ^{al} cr (al) 228 (4.3), 288 (4.0)	117				i	s	s	s	s	chl v peth s	**B5**^{1}, 355
e408	—,1,2-dibromo-(*cis*)*	*Sym*-Dibromoethylene. BrCH:CHBr.	185.86		−53	112.5^{760}	2.2464^{20}	1.5428^{20}	i	v	v	s	s	chl s	**B1**^{3}, 672
Ω e409	—,—(*trans*)*	BrCH:CHBr	185.86		−6.5	108^{760}	2.2308^{20}	1.5505^{18}	i	v	v	s	s	chl s	**B1**^{3}, 672
e410	—,1,1-dibromo-2-ethoxy-*	$C_2H_5OCH:CBr_2$	229.91			$170–2^{747}$ (cor) $73–5^{15}$	1.7697^{18}_{4}		i		v				**B1**^{2}, 473
Ω e411	—,1,1-dichloro-*	Vinylidene chloride. $CH_2:CCl_2$	96.94	λ^{vap} <200	−122.1	37^{760} (32)	1.218^{20}	1.4249^{20}	i	s	v	s	s	chl v	**B1**^{3}, 647
Ω e412	—,1,2-dichloro-(*cis*)*	*Sym*-Dichloroethylene. ClCH:CHCl	96.94	λ^{vap} <200	−80.5	60.3^{760}	1.2837^{20}_{4}	1.4490^{20}	δ	∞	∞	∞	v	chl v	**B1**^{3}, 651
Ω e413	—,—(*trans*)*	ClCH:CHCl	96.94	λ^{vap} <200	−50	47.5	1.2565^{20}_{4}	1.4454^{20}	δ	∞	∞	∞	v	chl v	**B1**^{3}, 651
Ω e414	—,1,1-difluoro-*	Vinylidene fluoride. $CH_2:CF_2$	64.04	gas		< −84			i	s	v				**B1**^{3}, 638
e415	—,1,1-dinitro-2,2-diphenyl-*	$(C_6H_5)_2C:C(NO_2)_2$	270.25	ye nd (aa or al)	149				s^h	s	s				**B5**^{2}, 545
Ω e416	—,1,1-diphenyl-*	$(C_6H_5)_2C:CH_2$	180.25	λ^{al} 236 sh (4.2), 255 (4.0)	8.2	277^{760} $94–5^{11}$	1.028^{16}_{4}	1.6100^{20}	i					chl s	**B5**^{2}, 543
—	—,1,2-diphenyl-*	*see* **Stilbene**													
Ω e417	—,fluoro-*	Vinyl fluoride. $CH_2:CHF$	46.05	gas	−160.5	−72.2			i	s	s				**B1**^{1}, 77
e418	—,iodo-*	Vinyl iodide. $CH_2:CHI$	153.95			56	2.037^{20}	1.5385^{20}	i	∞	∞				**B1**, 199
Ω e419	—,tetrabromo-*	Perbromoethylene $Br_2C:CBr_2$	343.66	pl (dil al), nd (al) λ^{vap}<200	56.5	226–7 100^{15} sub			i	s	s	s		chl v	**B1**^{3}, 673
Ω e420	—,tetrachloro-*	Perchloroethylene. $Cl_2C:CCl_2$	165.83		−19 fr −22	121^{760} 14^{10}	1.6227^{20}_{4}	1.5053^{20}	i	∞	∞		∞		**B1**^{3}, 664
e421	—,tetrafluoro-*	Perfluoroethylene. $F_2C:CF_2$	100.02	gas	−142.5	-76.3^{760}	$1.519^{76.3}$		i						**B1**^{3}, 638
e422	—,tetraiodo-*	Periodoethylene. $I_2C:CI_2$	531.64	lemon ye lf, pr (eth)	192	sub	2.983^{20}		i	δ	δ		s	CS_2 v chl s	**B1**^{3}, 676
Ω e423	—,tetraphenyl-*	$(C_6H_5)_2C:C(C_6H_5)_2$	332.45	mcl or rh (bz-eth or chl-al) λ^{diox} 310 (4.4)	225 (cor)	$415–25^{760}$	1.155^{0}_{4}		i	δ	δ		v^h		**B5**^{2}, 679
e424	—,tribromo-*	$BrCH:CBr_2$	264.76			$163–4^{760}$ 75^{15}	$2.708^{20.5}_{4}$	$1,6045^{16}$	δ	v	s	s		chl s	**B1**^{2}, 164
Ω e425	—,trichloro-*	$ClCH:CCl_2$	131.39	λ^{vap} <200	−73 fr −86.4	87^{760}	1.4642^{20}_{4}	1.4773^{20}	δ	∞	∞	s		chl s	**B1**^{3}, 656
Ω e426	—,triphenyl-*	$C_6H_5CH:C(C_6H_5)_2$	256.35	lf (al or MeOH) λ^{al} 231 (4.1), 299 (4.2)	72–3	$220–1^{14}$	1.0373^{78}_{4}	1.6292^{78}	i	s	v			MeOH s	**B5**^{2}, 630
Ω e427	**Ethenetetracarboxylic acid**, tetraethyl ester*	$(C_2H_5O_2C)_2C:C(CO_2C_2H_5)_2$	316.31	tcl pr (eth)	58	325–8 δd 210^{22}			i	v	v				**B2**^{2}, 709
e428	**Ether, allyl butyl, 3'-methyl-**	Allyl isoamyl ether. $(CH_3)_2CHCH_2CH_2OCH_2CH:CH_2$	128.22			120–2			i	∞	∞				**B1**^{2}, 477

For explanations, symbols and abbreviations see beginning of table. For structural formulas see end of table.

No.	Name	Synonyms and Formula	Mol. wt.	Color, crystalline form, specific rotation and λ_{max} (log ε)	m.p. °C	b.p. °C	Density	n_D	Solubility						Ref.
									w	al	eth	ace	bz	other solvents	
	Ether														
Ω e429	**—,allyl ethyl**	3-Ethoxypropene*. $C_2H_5OCH_2CH:CH_2$	86.14			66^{761}	0.7651^{20}_{4}	1.3881^{20}	i	∞	∞	s			**B1**^{3}, 1881
e430	**—,allyl isopropyl**	$C_3H_7^iOCH_2CH:CH_2$	100.16			83–4	0.7764^{20}	1.3946^{20}	i	v	∞	s			**B1**^{3}, 1882
e431	**—,allyl methyl**	$CH_3OCH_2CH:CH_2$	72.12			46^{760}	0.77^{11}_{11}	1.3778^{20}	i	∞	∞	s			**B1**^{3}, 1881
e432	**—,allyl 2-naphthyl**	$C_{10}H_7^{\beta}OCH_2CH:CH_2$	184.24		16	d 210		1.600^{25}	i						**B6**^{1}, 313
Ω e433	**—,allyl phenyl**	$C_6H_5OCH_2CH:CH_2$	134.18			191.7^{760} 74^{5}	0.9811^{20}_{4}	1.5223^{20}	i	s	∞				**B6**^{2}, 146
e434	**—,—,4′-chloro-**		168.63			d^{760} $106–7^{12}$	1.131^{15}	1.5348^{25}	i	s	s		s		**B6**^{1}, 101
e434[1]	**—,—,2′,4′,6′-tri-bromo-**		370.88	nd (al)	33–4				i		i				**B6**^{2}, 194
e435	**—,allyl propyl**	$CH_3CH_2CH_2OCH_2CH:CH_2$	100.16			90–2	0.7764^{20}	1.3919^{20}	i	s	∞	s			**B1**^{3}, 1882
Ω e436	**—,allyl 2-tolyl**		148.21			205–8 85^{12}	0.9698^{15}_{4}	1.5188^{15}	i						**B6**^{2}, 329
e437	**—,allyl 3-tolyl**		148.21			211–4 93.5^{13}	0.9564^{20}	1.5179^{20}	i			s			**B6**^{1}, 186
Ω e438	**—,allyl 4-tolyl**		148.21			214.5^{760} $97–8^{16}$	0.9728^{15}_{15}	1.5157^{24}	i			s			**B6**^{2}, 377
e439	**—,benzyl butyl**	$C_6H_5CH_2O(CH_2)_3CH_3$	164.25			223^{758} 92^{10}	0.9227^{20}_{4}	1.4833^{20}	i	∞	∞	s			**B6**^{2}, 410
e440	**—,—,2′-methyl-(*d*)**	$C_6H_5CH_2OCH_2CH(CH_3)CH_2CH_3$	178.28	$[\alpha]^{22}_{D}$ +1.82		231^{722}	0.911^{22}_{4}	1.4854^{22}	i	v	v				**B6**, 431
Ω e441	**—,—,3′-methyl-**	Benzyl isoamyl ether. $C_6H_5CH_2OCH_2CH_2CH(CH_3)_2$	178.28			$236–7^{748}$ $117–9^{19}$	0.9098^{20}_{4}	1.4792^{20}	i	v	v		δ		**B6**^{2}, 410
Ω e442	**—,benzyl ethyl**	α-Ethoxytoluene. $C_6H_5CH_2OC_2H_5$	136.20			185^{760} 70^{15}	0.9490^{20}_{4}	1.4955^{20}	i	∞	∞				**B6**^{2}, 409
e443	**—,benzyl isobutyl**	$C_6H_5CH_2OCH_2CH(CH_3)_2$	164.25			$211–2^{743}$	0.9233^{20}_{4}	1.4826^{20}	i		v			chl v	**B6**^{2}, 410
e444	**—,benzyl methyl**	α-Methoxytoluene. $C_6H_5CH_2OCH_3$	122.17		−52.6	170^{760} $59–60^{12}$	0.9634^{20}_{4}	1.5008^{20}	i	v	v		s	lig i	**B6**^{2}, 409
e445	**—,—,1′-chloro-**	$C_6H_5CH_2OCH_2Cl$	156.61			103^{13}	1.1350^{20}_{4}	1.5192^{20}							**B6**^{2}, 414
e446	**—,benzyl 1-naphthyl**	$C_6H_5CH_2OC_{10}H_7^{\alpha}$	234.30	cr (al)	61 (77)	200^{12}			i	δ v^{h}					**B6**^{2}, 579
e447	**—,benzyl 2-naphthyl**	$C_6H_5CH_2OC_{10}H_7^{\beta}$	234.30	lf or nd (al)	101–2				i	s	s		s	chl s	**B6**^{2}, 599
e448	**—,butyl ethenyl**	Butoxyethene.* $CH_3(CH_2)_3OCH:CH_2$	100.16		−92	93.8^{760}	0.7888^{20}_{4}	1.4026^{20}	i	v	∞	v	s	oos ∞ glycol, glycerol δ	**B1**^{3}, 1860
e449	**—,—,3-methyl-**	Isoamyl vinyl ether. $(CH_3)_2CHCH_2CH_2OCH:CH_2$	114.19			$112–3^{760}$	0.7826^{20}_{4}	1.4072^{20}	i	s	v				**B1**^{3}, 1863
Ω e450	**—,butyl ethyl**	1-Ethoxybutane.* $CH_3(CH_2)_3OC_2H_5$	102.18		−124	96^{760}	0.7490^{20}_{4}	1.3818^{20}	i	∞	∞	v			**B1**^{3}, 1502
e451	**—,*sec*-butyl ethyl**	$CH_3CH(C_2H_5)OC_2H_5$	102.18			81^{760}	0.7503^{20}_{4}	1.3802^{20}	i	v	v				**B1**^{3}, 1533
e452	**—,*tert*-butyl ethyl**	$(CH_3)_3COC_2H_5$	102.18		−94	73.1^{760}	0.7519^{25}	1.3794^{20}	i	v	v				**B1**^{3}, 1577
e453	**—,butyl ethyl, 2′-chloro-**	$CH_3(CH_2)_3OCH_2CH_2Cl$	136.62			154.5^{750}d (cor) $49–50^{11}$	0.9335^{20}_{4}	1.4155^{20}			s				**B1**^{3}, 1502
e454	**—,—,3-methyl-**	Ethyl isoamyl ether. $(CH_3)_2CHCH_2CH_2OC_2H_5$	116.21			$112–3^{760}$	0.7695^{21}_{15}		i	∞	∞				**B1**^{2}, 432
e454[1]	**—,*sec*-butyl ethyl, 2-methyl-**	Ethyl *tert*-amyl ether. $CH_3CH_2C(CH_3)_2OC_2H_5$	116.21			101^{760}	0.7657^{20}_{4}	1.3912^{20}	δ	v	∞				**B1**, 389
e455	**—,butyl ethynyl**	Butoxyacetylene. $CH_3(CH_2)_3OC⋮CH$	98.15			$102–4^{760}$ exp *ca.* 100 $41–2^{78}$	0.8200^{20}_{4}	1.4020^{20}	d^{h}	v	s				**B1**^{3}, 1969
e456	**—,butyl furfuryl**		154.21			189.90^{765}	0.9516^{20}_{4}	1.4522^{20}	i	s	v				**B17**^{2}, 114
e457	**—,butyl isobutyl**	$CH_3(CH_2)_3OCH_2CH(CH_3)_2$	130.23			$148–52^{730}$	0.7980^{22}_{4}	1.4077^{21}	i	s	∞	s			**B1**, 376
e458	**—,butyl isopropyl**	$CH_3(CH_2)_3OCH(CH_3)_2$	116.21			108^{738}	0.7594^{15}_{4}	1.3870^{15}	i	s	s	s		os, con sulf s	**B1**^{3}, 1503
Ω e458[1]	**—,butyl methyl**	$CH_3(CH_2)_3OCH_3$	88.15		−115.5	71^{760}	0.7443^{20}_{4}	1.3736^{20}	i	∞	∞	s			**B1**^{2}, 395
e459	**—,*sec*-butyl methyl**	$CH_3CH_2CH(CH_3)OCH_3$	88.15			60 (cor)	0.7415^{20}_{4}	1.3680^{25}	δ	v	v	s			**B1**^{3}, 1532
e460	**—,*tert*-butyl methyl**	$(CH_3)_3COCH_3$	88.15		−109	55.2^{760} (cor)	0.7405^{20}_{4}	1.3690^{20}	s	v	v	s			**B1**^{3}, 1576
e461	**—,butyl methyl, 3-methyl-**	Isoamyl methyl ether. $(CH_3)_2CHCH_2CH_2OCH_3$	102.18			91^{765}	0.7517^{20}	1.3830^{20}		s	s				**B1**^{2}, 432
e462	**—,*sec*-butyl methyl, 2-methyl-**	Methyl *tert*-amyl ether $CH_3CH_2C(CH_3)_2OCH_3$	102.18			86.3^{760}	0.7703^{20}_{4}	$1,3885^{20}$	δ	v	∞			MeOH v	**B1**, 389
e462[1]	**—,butyl propyl**	$CH_3(CH_2)_3OCH_2CH_2CH_3$	116.21			117.1	0.7773^{0}_{0}		i	v	v				**B1**^{3}, 1503
e463	**—,—,3-methyl-**	Isoamyl propyl ether. $(CH_3)_2CHCH_2CH_2OCH_2CH_2CH_3$	130.23			130			i	s	∞				**B1**^{2}, 432
e464	**—,butyl 2-tolyl**		164.25			223^{760}	0.9437^{0}_{0}		i						**B6**, 353
e465	**—,2-butynyl methyl**	$CH_3C⋮CCH_2OCH_3$	84.13			$99–100^{760}$ 33^{27}	0.8496^{20}_{4}	1.4262^{20}	i	s	s		s		**B1**^{3}, 1973
e466	**—,cyclohexyl 2-furyl**		166.22			118.9^{28}	1.0200^{28}_{4}	1.4861^{28}	i	s	s	s			**C51**, 3550
e467	**—,cyclohexyl methyl**	Hexahydroanisole	114.19		−74.37	133^{760}	0.8756^{20}_{4}	1.4355^{20}	i	s	s			MeOH s	**B6**^{2}, 9
e468	**—,—,2-bromo- (*dl, trans*)**		193.09			$78–9^{12}$	1.3314^{20}_{4}	1.4871^{20}	i	s	s			MeOH s	**B6**^{2}, 13
Ω e469	**—,diallyl**	Allyl ether $CH_2:CHCH_2OCH_2CH:CH_2$	98.15			94	0.8260^{20}_{4}	1.4163^{20}	i	∞	∞	v			**B1**^{2}, 477
Ω e470	**—,dibenzyl**	Benzyl ether. $C_6H_5CH_2OCH_2C_6H_5$	198.27	λ^{cy} 252 (2.5), 259 (2.6), 265 (2.5)	3.60	298^{760} 160^{11}	1.0428^{20}_{4}	1.5168^{20}	i	∞	∞				**B6**^{2}, 412

For explanations, symbols and abbreviations see beginning of table. For structural formulas see end of table.

No.	Name	Synonyms and Formula	Mol. wt.	Color, crystalline form, specific rotation and λ_{max} (log ε)	m.p. °C	b.p. °C	Density	n_D	Solubility w	al	eth	ace	bz	other solvents	Ref.
	Ether														
Ω e471	—,**dibutyl**	Butyl ether. $CH_3(CH_2)_3O(CH_2)_3CH_3$	130.23	...	−95.3	142^{760} 47^{24}	0.7689^{20}_{4}	1.3992^{20}	i	∞	∞	v	...	...	**B1**^{2}, 396
e472	—,**di-*sec*-butyl** (*dl*)	*sec*-Butyl ether. $CH_3CH(C_2H_5)OCH(C_2H_5)CH_3$	130.23	...	...	$120\text{–}1^{760}$	0.756^{25}	1.393^{25}	i	∞	∞	v	...	...	**B1**^{2}, 402
e473	—,**dibutyl, 3,3′-dimethyl-**	Isoamyl ether. $(CH_3)_2CHCH_2CH_2OCH_2CH_2CH(CH_3)_2$	158.29	...	...	$172\text{–}3^{760}$ 60^{10}	0.7777^{20}_{4}	1.4085^{20}	i	∞	∞	...	...	chl ∞	**B1**^{3}, 1638
e474	—,—,**3-methyl-**	Butyl isoamyl ether. $CH_3(CH_2)_3OCH_2CH_2CH(CH_3)_2$	144.26	...	...	157^{756}	...	...	i	s	∞	...	...	...	**B1**, 401
Ω e475	—,**diethenyl**	Ethenyloxyethene*. Vinyl ether. $CH_2{:}CHOCH{:}CH_2$	70.09	λ^{gas} 164 (3.82), 302 (4.19)	−101	28^{760} (39)	0.773^{20}_{4}	1.3989^{20}	i	∞	∞	∞	...	chl ∞	**B1**^{2}, 473
e476	—,—,**hexachloro-**	$Cl_2C{:}CClOCCl{:}CCl_2$	276.76	...	...	210	1.654^{21}	...	...	...	...	...	...	...	**B1**, 725
Ω e477	—,**diethyl**	Ether. Ethyl ether. Ethoxyethane*. $CH_3CH_2OCH_2CH_3$	74.12	λ^{gas} 171 (3.60), 188 (3.30)	fr −116.2	34.51^{760}	0.71378^{20}_{4}	1.3526^{20}	δ	∞	∞	v	∞	chl, oils, lig ∞	**B1**^{2}, 311
e478	—,—,borofluoride	$(C_2H_5)_2O.BF_3$	141.93	...	−60.4	$125\text{–}6^{760}$d 60^{20}	1.125^{25}_{4}	1.348^{20}	d	d	s	...	...	...	**B1**^{3}, 1308
Ω e479	—,—,**2-bromo-**	$BrCH_2CH_2OC_2H_5$	153.03	ye nd (al)	...	$127\text{–}8^{755}$ 40^{24}	1.3572^{20}_{4}	1.4447^{20}	δ	∞	∞	...	...	...	**B1**^{3}, 1361
Ω e480	—,—,**1-chloro-**	$CH_3CHClOC_2H_5$	108.57	λ^{hx} < 180	...	92–5	0.950^{20}_{4}	1.4053^{20}	d	d^h	...	...	...	...	**B1**^{2}, 674
e481	—,—,**2-chloro-**	$ClCH_2CH_2OC_2H_5$	108.57	λ^{hx} < 180	...	107–8	0.9894^{20}_{4}	1.4113^{20}	δ	...	∞	...	...	chl s	**B1**^{3}, 1349
e482	—,—,**decachloro-**	Perchloroether. $Cl_3CCCl_2OCCl_2CCl_3$	418.57	oct or tetr pym	69	d	$1.900^{14.5}$	...	...	...	...	...	...	...	**B2**, 210
e483	—,—,**2,2′-di-benzoyloxy-**	Oxybis(β-ethylbenzoate). $C_6H_5CO_2CH_2CH_2OCH_2CH_2O_2CC_6H_5$	314.34	...	33.5	$279\text{–}81^{24}$ 250^3	1.1701^{15}_{15}	...	s	s	...	...	...	...	**B9**^{2}, 108
e484	—,—,**1,2-dibromo-**	$BrCH_2CHBrOC_2H_5$	231.94	...	...	80^{20} $55\text{–}8^9$	1.7320^{20}_{4}	1.5044^{20}	d	s	...	...	...	chl s	**B1**, 625
—	—,—,**1,1′-di-carboxy-**	*see* **Dilactic acid**													
e485	—,—,**1,1′-dichloro-**	$CH_3CHClOCHClCH_3$	143.02	...	...	116–7	1.1060^{25}_{4}	1.4186^{25}	d	v	∞	...	...	chl s	**B1**, 607
e486	—,—,**1,2-dichloro-**	$ClCH_2CHClOC_2H_5$	143.02	λ^{hx} 230 (1.6), 298 (−0.8)	...	145^{760} $66\text{–}8^{45}$	1.1370^{20}_{4}	1.4435^{20}	d	v	v	...	...	chl s	**B1**^{2}, 676
e487	—,—,**1,2′-dichloro-**	$ClCH_2CH_2OCHClCH_3$	143.02	...	...	d^{760} $55\text{–}7^{17}$	1.1867^{20}_{4}	1.4473^{20}	d	v	v	...	...	chl s	**B1**^{2}, 674
e488	—,—,**2,2′-dichloro-**	$ClCH_2CH_2OCH_2CH_2Cl$	143.02	...	−24.5 (−46.8)	178^{760} 75^{20}	1.2199^{20}_{4}	1.4575^{20}	i δ^h	s	s	s	∞	MeOH ∞ oos s	**B1**^{3}, 1349
—	—,—,**2,2′-di-hydroxy-**	*see* **Diethylene glycol**													
e489	—,—,**2,2′-di-phenoxy-**	$C_6H_5OCH_2CH_2OCH_2CH_2OC_6H_5$	258.32	nd (dil al)	66–7	...	...	...	...	s	...	...	...	...	**B6**^{2}, 150
e490	—,—,**1,1′-di-phenyl-**(*dl*)	$C_6H_5CH(CH_3)OCH(CH_3)C_6H_5$	226.32	...	...	280.2 $167\text{–}8^{23}$	1.0058^{15}_{4}	1.5454^{21}	i	...	s	...	...	chl s	**B6**^{2}, 445
e491	—,—,**2,2′-di-phenyl-**	$C_6H_5CH_2CH_2OCH_2CH_2C_6H_5$	226.32	vt-bl flr	...	$317\text{–}20^{760}$ 194.5^{20}	1.0141^{18}_{4}	1.5488^{18}	i	...	s	...	...	chl s	**B6**^{2}, 450
—	—,—,**2-methoxy-**	*see* **Ethane, 1-ethoxy-2-methoxy-***													
e492	—,—,**1,1′,2,2,2,2′,2′,2′-octachloro-**	$Cl_3CCHClOCHClCCl_3$	349.68	cr (al or MeOH)	40–2	$130\text{–}1^{11}$	...	...	i	δ	...	...	v	peth, MeOH v	**B1**^{2}, 681
e493	—,**difurfuryl**	Furfuryl ether.	178.19	...	...	101^2 $88\text{–}9^1$	1.1405^{20}_{4}	1.5088^{20}	i	...	...	...	...	...	**B17**^{2}, 116
e494	—,**diheptyl**	Heptyl ether. $CH_3(CH_2)_6O(CH_2)_6CH_3$	214.40	...	...	258.5^{769}	0.8008^{20}_{4}	1.4275^{20}	i	s	s	...	...	...	**B1**^{3}, 1683
Ω e495	—,**dihexadecyl**	Dicetyl ether. Hexadecyl ether. $CH_3(CH_2)_{15}O(CH_2)_{15}CH_3$	466.89	lf (al)	55	270 d	0.978^{19}	...	i	s	s	...	...	...	**B1**^{2}, 467
Ω e496	—,**dihexyl**	Hexyl ether. $CH_3(CH_2)_5O(CH_2)_5CH_3$	186.34	...	...	223^{768}	0.7936^{20}_{4}	1.4204^{20}	i	...	s	...	...	...	**B1**^{3}, 1656
e496[1]	—,**diisobutyl, α,β-dichloro-**	$(CH_3)_2CClCHClOC_4H_9^i$	199.12	...	...	192.5 83^{15}	1.031^{5}_{4}	...	d	...	s	s	...	os v	**B1**, 675
Ω e497	—,**diisopropyl**	Isopropyl ether. $(CH_3)_2CHOCH(CH_3)_2$	102.18	...	−85.89	68^{760}	0.7241^{20}_{4}	1.3679^{20}	δ	∞	∞	s	...	dil sulf i	**B1**^{3}, 1459
e498	—,—,**β,β′-dichloro-**	$ClCH_2CH(CH_3)OCH(CH_3)CH_2Cl$	171.07	...	...	187^{760}	1.103^{20}_{4}	1.4505^{20}	i	∞	∞	∞	v	oos ∞	**B1**^{3}, 1470
Ω e499	—,**dimethyl**	Methoxymethane*. Methyl ether. CH_3OCH_3	46.07	gas λ^{gas} 162.5 (3.60), 184 (3.40)	−138.5	-23^{760}	...	...	s	s	s	s	δ	chl s	**B1**^{3}, 1188
e500	—,—,borofluoride	$(CH_3)_2O.BF_3$	113.88	...	−14 (−12)	127^{760} d	1.2410^{20}_{4}	1.302^{20}	d	d	...	...	...	...	**B1**^{3}, 1192
Ω e501	—,—,**chloro-**	$ClCH_2OCH_3$	80.52	...	−103.5	59.15^{760}	1.0605^{20}_{4}	1.3974^{20}	d	s	s	s	...	chl s	**B1**^{2}, 645
e502	—,—,**1,1′-dichloro-**	$ClCH_2OCH_2Cl$	114.96	...	−41.5	104^{760}	1.328^{15}_{4}	1.435^{21}	d	∞	∞	...	...	...	**B1**^{2}, 646
e503	—,—,**1,1,1′,1′-tetraphenyl-**	Benzhydryl ether. $(C_6H_5)_2CHOCH(C_6H_5)_2$	350.47	mcl (bz)	110	315^{745} d 267^{15}	...	...	i	δ^h	δ	...	v	...	**B6**^{2}, 634
e504	—,—,**1,1,1-tri-2-tolyl-**	...	316.45	cr (peth)	108.6	...	...	...	i	s	s	...	...	MeOH s peth s^h	**Am 73**, 3644
e505	—,**1,1′-dinaphthyl**	α-Naphthyl ether. $C_{10}H_7^{\alpha}OC_{10}H_7^{\alpha}$	270.34	lf (al or al-eth)	110	$280\text{–}5^{23}$	...	...	i	δ s^h	s	...	s	aa δ^h	**B6**^{3}, 2926
e506	—,**1,2′-dinaphthyl**	α,β′-Naphthyl ether. $C_{10}H_7^{\alpha}OC_{10}H_7^{\beta}$	270.34	nd (al or al-eth)	81	264^{15}	...	...	i	δ	v	s	v	chl v sulf s aa δ^h	**B6**^{2}, 600
e507	—,**2,2′-dinaphthyl**	β-Naphthyl ether. $C_{10}H_7^{\beta}OC_{10}H_7^{\beta}$	270.34	nd or lf (al)	105	250^{19} δd	...	...	i	δ s^h	v	...	v	aa δ, s^h	**B6**^{3}, 2976

For explanations, symbols and abbreviations see beginning of table. For structural formulas see end of table.

No.	Name	Synonyms and Formula	Mol. wt.	Color, crystalline form, specific rotation and λ_{max} (log ε)	m.p. °C	b.p. °C	Density	n_D	Solubility w	al	eth	ace	bz	other solvents	Ref.
	Ether														
Ω e508	—,**dioctyl**	Octyl ether $CH_3(CH_2)_7O(CH_2)_7CH_3$	242.45	....	....	$286–7^{760}$	0.8063^{20}_{4}	1.4327^{20}	δ	s	s	...	...	....	**B1**, 419
Ω e509	—,**dipentyl**	Amyl ether. $CH_3(CH_2)_4O(CH_2)_4CH_3$	158.29	....	−69	190^{760} 70^{12}	0.7833^{20}_{4}	1.4119^{20}	i	∞	∞	...	...	....	**B1**², 417
Ω e510	—,**diphenyl**	Phenyl ether	170.21	mcl, rh λ^{al} 265 (3.20), 272 (3.28), 279 (3.23)	26.84	257.93^{760} 121^{10}	1.0748^{20}	1.5787^{25}	i	s	s	...	s	aa s	**B6**, 146
—	—,—,**amino-**	*see* **Aniline, phenoxy-**													
Ω e511	—,—,**4-bromo-**	$C_{12}H_9BrO$. *See* e510	249.11	λ^{al} 272 (3.18), 279 (3.18)	18.72	310.14^{760} 163^{10}	1.4208^{20}_{4}	1.6084^{20}	i	...	s	...	...	....	**B6**², 185
e512	—,—,**4,4′-dibromo-**	$C_{12}H_8Br_2O$. *See* e510	328.03	lf (al) λ^{al} 273 (3.26), 280 (3.26), 290 (3.18)	60.5	338–40 210^{11}	1.8 (so)	....	i	s	v	...	s	aa δ	**B6**², 185
e513	—,—,**4,4′-dichloro-**	$C_{12}H_8Cl_2O$. *See* e510	239.11	nd (al)	30	312–4 168.72^{7}	1.1231^{20}	1.611^{20}	i	...	...	...	...	....	**B6**², 176
e514	—,—,**2,2′-dihydroxy-**	*o*-Diphenol ether. 2,2′-Oxydiphenol. $C_{12}H_{10}O_3$. *See* e510	202.21	nd (w)	121	....	....	....	δ^h	s	...	...	s	lig δ	**B6**, 773
e515	—,—,**4,4′-dihydroxy-**	*p*-Diphenol ether. 4,4′-Oxydiphenol. $C_{12}H_{10}O_3$. *See* e510	202.21	lf (w or to)	165–7	....	....	....	δ s^h	v	v	v	v^h i	to i	**B6**, 845
e516	—,—,**2,2′-dimethoxy-**	*o*-Anisyl ether. $C_{14}H_{14}O_3$. *See* e510	230.27	pl (lig)	79–80	330–1	....	....	i	v	v	...	...	lig i, s^h	**B6**, 773
e517	—,—,**2,3′-dimethoxy-**	$C_{14}H_{14}O_3$. *See* e510	230.27	pr (bz-peth)	54	$326–9^{760}$ 152^{2}	....	....	...	s	s	...	s	lig δ	**B6**², 816
e518	—,—,**3,3′-dimethoxy-**	*m*-Anisyl ether. $C_{14}H_{14}O_3$. *See* e510	230.27	pa br liq λ^{diox} 275 (3.6)	....	332–4 258^{16}	....	....	i	δ	v	...	v	lig δ	**B6**², 816
e519	—,—,**2,2′-dinitro-**	$C_{12}H_8N_2O_5$. *See* e510	260.22	pl or nd (al)	119	....	....	....	...	δ v^h	...	...	...	....	**B6**, 219
e520	—,—,**2,4-dinitro-**	$C_{12}H_8N_2O_5$. *See* e510	260.22	pl (al), nd (al-ace) λ^{al} 293 (4.15)	71	$230–50^{27}$	....	....	δ	v	s	...	...	....	**B6**², 242
e521	—,—,**2,4′-dinitro-**	$C_{12}H_8N_2O_5$. *See* e510	260.22	nd (al)	103.5	....	....	....	...	δ s^h	...	...	...	....	**B6**², 222
e522	—,—,**2,6-dinitro-**	$C_{12}H_8N_2O_5$. *See* e510	260.22	lf (al)	99–100	....	....	....	...	δ s^h	...	...	...	....	**B6**², 245
e523	—,—,**3,4-dinitro-**	$C_{12}H_8N_2O_5$. *See* e510	260.22	pa ye lf	89	....	....	....	...	...	s	...	...	....	**B6**¹, 127
e524	—,—,**4,4′-dinitro-**	$C_{12}H_8N_2O_5$. *See* e510	260.22	ye nd (al), pr (bz) λ^{al} 218 (4.20), 301 (4.33)	144 (cor)	....	....	....	i	δ s^h	δ	...	s	aa, bz s	**B6**², 222
e525	—,—,**2-methoxy-**	$C_{13}H_{12}O_2$. *See* e510	200.24	cr (MeOH), nd (lig)	79	288^{745} $91–2^{7}$	....	....	i	s	s	...	s	MeOH, lig s^h	**B6**², 781
Ω e526	—,—,**2-nitro-**	$C_{12}H_9NO_3$. *See* e510	215.21	ye liq λ^{al} 255 (3.79), 315 (3.45)	< −20	235^{60} $183–5^{8}$	1.2539^{22}	1.575^{20}	i	s	s	...	s	chl, aa s	**B6**², 222
Ω e527	—,—,**4-nitro-**	$C_{12}H_9NO_3$. *See* e510	215.21	pl (peth or MeOH) λ^{al} 303 (4.06)	61 (56)	$188–90^{8}$	....	....	i	δ	s	...	s	....	**B6**², 210
e528	—,**dipropyl**	Propyl ether. $CH_3CH_2CH_2OCH_2CH_2CH_3$	102.18	....	fr −122	91^{760}	0.7360^{20}_{4}	1.3809^{20}	δ	∞	∞	...	...	....	**B1**², 367
e529	—,—,**1,2-dichloro-**	$CH_3CHClCHClOC_3H_7^n$	171.07	....	....	176^{760}	1.129^{15}_{4}	1.447^{16}	i	...	s	...	...	....	**B1**¹, 334
e530	—,—,**1,3-dichloro-**	$ClCH_2CH_2CHClOC_3H_7^n$	171.07	....	....	65^{12}	1.112^{20}	1.4476^{20}	i	s	s	...	...	....	**B1**², 690
e531	—,—,**2,2′-dichloro-**	$CH_3CHClCH_2OCH_2CHClCH_3$	171.07	....	....	188^{760}	1.109^{20}_{4}	1.4467^{20}	...	s	s	...	...	....	**B1**², 370
e532	—,—,**3,3′-dichloro-**	$Cl(CH_2)_3O(CH_2)_3Cl$	171.07	....	....	215^{745} 90.5^{11}	1.140^{20}_{20}	1.4158^{20}	...	s	s	...	...	....	**B1**², 370
Ω e534	—,**ethenyl ethyl**	Ethyl vinyl ether. $C_2H_5OCH:CH_2$	72.12	....	−115.8	$35–6^{760}$	0.7589^{20}_{4}	1.3767^{20}	δ	s	∞	...	...	....	**B1**³, 1857
Ω e535	—,—,**2′-chloro-**	$ClCH_2CH_2OCH:CH_2$	106.55	....	....	108^{760}	1.0475^{20}_{4}	1.4378^{20}	...	v	v	...	...	....	**B1**³, 1859
e536	—,—,**1,2-dichloro-**	$C_2H_5OCCl:CHCl$	141.00	....	....	128.2	1.1972^{25}_{4}	1.4558^{17}	d^h s	...	...	...	...	....	**B1**², 780
Ω e537	—,**ethenyl isobutyl**	Isobutyl vinyl ether. $(CH_3)_2CHCH_2OCH:CH_2$	100.16	....	−112	83^{760}	0.7645^{20}_{4}	1.3966^{20}	δ	v	∞	v	v	oos ∞ glycol, glycerol δ	**B1**³, 1862
Ω e538	—,**ethenyl isopropyl**	Isopropyl vinyl ether. $(CH_3)_2CHOCH:CH_2$	86.14	....	....	$55–6^{760}$	0.7534^{20}_{4}	1.3840^{20}	δ	v	v	v	v	....	**B1**³, 1859
Ω e539	—,**ethenyl methyl**	Methoxyethene*. Methyl vinyl ether. $CH_3OCH:CH_2$	58.08	....	−122	12^{760}	0.7725^{0}_{4}	1.3730^{0}	δ	v	v	v	v	os v glycol, glycerol i	**B1**³, 1857
e540	—,**ethenyl phenyl**	Phenyl vinyl ether. $C_6H_5OCH:CH_2$	120.16	λ^{cy} 225 (4.3), 271 (3.2), 274 (3.0)	....	155–6	0.9770^{20}_{4}	1.5224^{20}	i	...	v	...	...	....	**B6**², 146
e541	—,**ethyl ethynyl**	Ethoxyacetylene. $C_2H_5OC:CH$	70.09	....	....	50^{760} exp 100	0.8000^{20}_{4}	1.3796^{20}	...	v	v	...	v	....	**B1**³, 1968
e542	—,**ethyl furfuryl**	....	126.16	....	....	$149–50^{770}$	0.9844^{20}_{4}	1.4523^{20}	i	s	s	...	...	....	**B17**², 114
e543	—,**ethyl heptyl**	$CH_3(CH_2)_6OC_2H_5$	144.26	....	....	166.6	0.790^{16}_{4}	1.4111^{20}	i	s	s	...	...	....	**B1**³, 1682
e544	—,**ethyl hexyl**	$CH_3(CH_2)_5OC_2H_5$	130.23	....	....	$142–3^{773}$ 42^{14}	0.7722^{20}_{4}	1.4008^{20}	i	v	v	...	...	....	**B1**³, 1656

For explanations, symbols and abbreviations see beginning of table. For structural formulas see end of table.

No.	Name	Synonyms and Formula	Mol. wt.	Color, crystalline form, specific rotation and λ_{max} (log ε)	m.p. °C	b.p. °C	Density	n_D	Solubility w	al	eth	ace	bz	other solvents	Ref.
	Ether														
e545	**—,ethyl isobutyl**	$(CH_3)_2CHCH_2OC_2H_5$	102.18			81^{760}	0.751^{20}_{4}	1.3739^{25}	i	∞	∞	s	...	chl s	**B1**^{3}, 1559
e546	**—,ethyl isopropyl**	$(CH_3)_2CHOC_2H_5$	88.15			63–4	0.720^{25}_{4}	1.3698^{25}	s	∞	∞	s	...	chl s	**B1**^{3}, 1458
e547	**—,ethyl methyl**	$CH_3OC_2H_5$	60.10			10.8^{760}	0.7252^{0}_{0}	1.3420^{4}	s	∞	∞	s	...	chl s	**B1**^{3}, 1288
e548	**—,—,1′-bromo-**	$BrCH_2OC_2H_5$	139.00			109^{746}	1.4402^{20}_{4}	1.4515^{20}	...	...	v	...	...	...	**B1**^{2}, 647
e549	**—,—,1-chloro-**	$CH_3OCHClCH_3$	94.54			$72–3^{751}$	0.9902^{20}_{4}	1.4004^{20}	d	...	v	...	...	...	**B1**^{3}, 2654
e550	**—,—,1′-chloro-**	$ClCH_2OC_2H_5$	94.54			83^{763}	1.0372^{0}_{4}	1.4040^{20}	...	s	v	...	...	...	**B1**^{2}, 645
e551	**—,—,2-chloro-**	$CH_3OCH_2CH_2Cl$	94.54			92–3	1.0345^{20}_{4}	1.4111^{20}	s	...	v	...	...	...	**B1**^{2}, 335
e552	**—,—,1′-diethyl-amino-**	$(CH_3CH_2)_2NCH_2OC_2H_5$	131.22			136^{760} $76^{11.5}$			i	v	s	s	...	os v	**B4**^{2}, 598
	—,—,2-methyl-amino-	*see* **Ethane, 1-ethoxy-2-methylamino-**													
e553	**—,ethyl octyl**	$CH_3(CH_2)_7OC_2H_5$	158.29		12.5	186.3^{760} 74^{9}	0.7847^{20}_{4}	1.4127^{20}	i	s	...	...	...	AcOEt s	**B1**^{3}, 1708
e554	**—,ethyl pentyl**	Amyl ethyl ether. $CH_3(CH_2)_4OC_2H_5$	116.21			119–20	0.7622^{20}_{4}	1.3927^{20}	δ	∞	∞	...	...	...	**B1**^{3}, 1602
e555	**—,—,1-chloro-**	$CH_3(CH_2)_4OCHClCH_3$	150.65			$63–6^{8}$	0.9200^{20}_{4}	1.4218^{20}	d	...	s	...	...	...	**Am 53,** 4077
e556	**—,ethyl phenyl, 2-bromo-**	$C_6H_5OCH_2CH_2Br$	201.07		39	240–50 d $125–30^{20}$			i	v	v	...	...	...	**B6**^{2}, 145
e557	**—,ethyl propyl**	$CH_3CH_2CH_2OC_2H_5$	88.15		< −79	63.6^{760}	0.7386^{20}_{4}	1.3695^{20}	δ	∞	∞	...	...	aa ∞	**B1**^{3}, 1414
e558	**—,—,1-chloro-**	$C_3H_7^nOCHClCH_3$	122.60			$112–5^{731}$ δd	0.9322^{20}_{4}	1.4013^{20}	d	...	s	...	...	...	**B1**, 607
e560	**—,—,2′,3′-epoxy-**		102.13			128^{760}	0.9700^{20}	1.4320^{20}	s	s	s	...	...	...	**B17**, 105
e561	**—,ethyl 1-propynyl**	Ethoxymethylacetylene. $CH_3C\vdots COC_2H_5$	84.13			84^{760}	0.8276^{20}_{4}	1.4130^{20}	i d^h	...	v	v	v	os v	**B1**^{3}, 1969
e562	**—,ethyl 3-propynyl**	Ethyl propargyl ether. $C_2H_5OCH_2C\vdots CH$	84.13			82^{760}	0.8326^{20}_{4}	1.4039^{20}	i	s	s	...	...	...	**B1**^{2}, 504
e563	**—,ethynyl methyl**	Methoxyacetylene. $CH_3OC\vdots CH$	56.07			50^{760}	0.8001^{20}_{4}	1.3812^{20}	...	s	v	...	...	...	**B1**^{3}, 1968
e564	**—,ethynyl phenyl**	Phenoxyacetylene. $C_6H_5OC\vdots CH$	118.14		−36	$61–2^{25}$ $43–4^{10}$	1.0614^{20}_{4}	1.5125^{20}	i	s	s	...	...	...	**B6**, 145
e565	**—,ethynyl propyl**	Propoxyacetylene. $CH_3CH_2CH_2OC\vdots CH$	84.13			75	0.8080^{20}_{4}	1.3935^{20}	d^h	v	v	...	...	...	**B1**^{3}, 1969
e566	**—,furfuryl methyl**		112.14			$131–3^{760}$	1.0163^{20}_{4}	1.4570^{20}	i	s	v	...	...	...	**B17**^{2}, 114
e567	**—,2-furyl octyl**		196.29			$129–30^{18}$	0.9214^{28}_{4}	1.4520^{28}	...	...	s	...	...	...	**C51,** 3550
e568	**—,2-furyl phenyl**		160.18	λ^{al} 224 (4.06)		$105–6^{18}$	1.1010^{23}_{4}	1.5418^{23}	...	...	s	...	...	...	**C51,** 3550
e569	**—,heptyl methyl**	$CH_3(CH_2)_6OCH_3$	130.23			151^{760}	0.7869^{15}_{15}	1.4073^{20}	i	∞	∞	s	...	...	**B1**^{3}, 1682
e570	**—,heptyl phenyl**	$CH_3(CH_2)_6OC_6H_5$	192.31			267 $128–30^{12}$	0.9178^{15}_{15}	1.4912^{20}	i	∞	∞	s	...	...	**B6**, 144
—	**—,—,4′-hydroxy-**	*see* **Phenol, 4-heptoxy-**													
e572	**—,hexadecyl phenyl**	Cetyl phenyl ether. $CH_3(CH_2)_{15}OC_6H_5$	318.55	lf (al)	41.8	200^{1}	0.8434^{82}	1.4556^{82}	...	...	...	...	...	...	**B6**, 144
e573	**—,hexyl phenyl**	$CH_3(CH_2)_5OC_6H_5$	178.28		−19	240^{760} 130^{22}	0.9174^{20}_{4}	1.4921^{20}	i	...	s	...	...	...	**B6**^{2}, 146
e575	**—,isobutyl methyl**	$(CH_3)_2CHCH_2OCH_3$	88.15			58^{760}(cor)	0.7311^{20}_{4}		i	s	s	...	...	...	**B1**^{3}, 1559
e576	**—,isobutyl propyl**	$(CH_3)_2CHCH_2OCH_2CH_2CH_3$	116.21			$105–6^{720}$	0.7549^{20}	1.3852^{25}	δ	s	∞	...	...	...	**B1**^{2}, 410
e577	**—,isopropyl methyl**	$(CH_3)_2CHOCH_3$	74.12			32.5^{777}	0.7237^{15}_{4}	1.3576^{20}	δ	∞	∞	...	...	...	**B1**^{2}, 1458
e578	**—,isopropyl propyl**	$(CH_3)_2CHOCH_2CH_2CH_3$	102.18			83	0.7370^{20}_{4}	1.376^{21}	δ	v	s	s	...	...	**B1**^{2}, 381
e579	**—,methyl 2-octyn-1-yl**	$CH_3OCH_2C\vdots C(CH_2)_4CH_3$	140.23			77^{19}	0.8370^{20}_{4}	1.4380^{20}	i	...	s	...	...	...	**B1**^{3}, 1996
e580	**—,methyl pentyl**	$CH_3(CH_2)_4OCH_3$	102.18			$99–100^{760}$	0.767^{19}	1.3855^{19}	...	v	s	...	...	...	**B1**^{2}, 417
e581	**—,methyl propyl**	$CH_3CH_2CH_2OCH_3$	74.12			38.9^{760}	0.738^{20}_{4}	1.3579^{20}	s	∞	∞	s	...	...	**B1**^{3}, 1413
e582	**—,—,1-chloro-**	$CH_3CH_2CH_2OCH_2Cl$	108.57			109^{760}	0.9884^{20}_{4}	1.4125^{20}	d	v	v	...	...	...	**B1**^{1}, 305
e583	**—,—,2′,3′-dibromo-**	$CH_3CHBrCHBrOCH_3$	231.94			185^{760} 84^{15}	1.8320^{12}_{4}	1.5123^{20}	...	...	s	...	...	...	**B1**^{3}, 1428
e585	**—,—,2′,3′-epoxy-**	Epimethylin.	88.12			115–8	0.9890^{20}	1.4320^{20}	s	s	s	s	...	...	**B17**, 104
e586	**—,1-naphthyl pentyl**	$C_{10}H_7^{\alpha}O(CH_2)_4CH_3$	214.31	nd (al)	30	322^{760}			i	s	s	...	s	chl s	**B6**^{3}, 2925
e587	**—,2-naphthyl pentyl**	$C_{10}H_7^{\beta}O(CH_2)_4CH_3$	214.31	lf (al)	24.5	335^{760} (327.5 cor) 159^{10}		1.5587^{30}	i	s	s	...	s	chl s	**B6**^{3}, 2973
e588	**—,octyl phenyl**	$C_6H_5O(CH_2)_7CH_3$	206.33		8	285^{760} $164–7^{20}$	0.9139^{15}_{15}	1.4875^{20}	i	s	s	...	...	...	**B6**, 144
e590	**Ethionic acid, anhydride**	Carbyl sulfate. 1,3,2,4-Dioxadithian 2,2,4,4-tetroxide	188.18	dlq	*ca.* 80				d	...	...	...	...	...	**B19**, 433
—	**Ethyl acetate**	*see* **Acetic acid,** ethyl ester													
—	**Ethyl acetoacetate**	*see* **Butanoic acid, 3-oxo-,** ethyl ester*													
—	**Ethyl alcohol**	*see* **Ethanol***													
—	**Ethylamine***	*see* **Ethane, amino-***													
—	**Ethyl cellosolve**	*see* **Ethanol, 2-ethoxy-***													
—	**Ethylene**	*see* **Ethene***													
—	**Ethylene bromide**	*see* **Ethane, 1,2-dibromo-***													
—	**Ethylene bromo-hydrin**	*see* **Ethanol, 2-bromo-***													
—	**Ethylene chloride**	*see* **Ethane, 1,2-dichloro-***													

For explanations, symbols and abbreviations see beginning of table. For structural formulas see end of table.

No.	Name	Synonyms and Formula	Mol. wt.	Color, crystalline form, specific rotation and λ_{max} (log ε)	m.p. °C	b.p. °C	Density	n_D	Solubility w	al	eth	ace	bz	other solvents	Ref.
	Ethylene chlorohydrin														
—	**Ethylene chlorohydrin**	*see* **Ethanol, 2-chloro-***													
—	**Ethylenediamine**	*see* **Ethane, 1,2-diamino-***													
—	**Ethylene dicyanide**	*see* **Succinic acid,** dinitrile													
—	**Ethylene fluoride**	*see* **Ethane, 1,2-difluoro-***													
—	**Ethylene fluorohydrin**	*see* **Ethanol, 2-fluoro-***													
—	**Ethylene glycol**	*see* **1,2-Ethanediol***													
—	—,monoallyl ether	*see* **Ethanol, 2-allyloxy-**													
—	—,monoethyl ether	*see* **Ethanol, 2-ethoxy-***													
—	**Ethylene iodide**	*see* **Ethane, 1,2-diiodo-***													
—	**Ethylene iodohydrin**	*see* **Ethanol, 2-iodo-***													
—	**Ethylene oxide**	*see* **Ethane, 1,2-epoxy-***													
—	**Ethylene urea**	*see* **2-Imidazolidone**													
—	**Ethylenimine**	*see* **Aziridine**													
—	**Ethyl ether**	*see* **Ethyl, diethyl**													
—	**Ethylidene acetate**	*see* **Acetaldehyde,** diacetate													
—	**Ethyl vanillin**	*see* **Benzaldehyde, 3-ethoxy-4-hydroxy-**													
Ω e591	**Ethyne***	Acetylene. Ethine. HC⋮CH	26.04		−80.8	-84.0^{760} sub	0.6208^{-82}_{4}	1.00051^{0}	δ	δ	...	s	s	chl s CS_2 δ	**B1**[3], 887
e592	—,**bis(1-hydroxy-cyclohexyl)-***		222.33	nd (CCl_4)	112–3	182^{13}			i	v	v	v	s[h]	MeOH v CCl_4 s[h] lig δ	**B6**[2], 909
e593	—,**bromo-***	Bromoacetylene. Ethynyl bromide. HC⋮CBr	104.94			4.7	0.0047 (760 mm.)		δ	...	s	...	...	dil HNO_3 s dil HCl δ	**B1**[3], 919
e594	—,**chloro-***	Chloroacetylene. Ethynyl chloride. HC⋮CCl	60.48		−126	−30	0.002 (760 mm.)		d	δ	...	...	...		**B1**[3], 917
e599	—,**dibromo-***	BrC⋮CBr	183.84	nd	−25–−23	76 (exp)			i	s	s	s	s	oos s	**B1**[3], 919
e600	—,**dichloro-***	ClC⋮CCl	94.93		−66–−64.2	exp			...	s	s	s	...		**B1**[3], 918
e601	—,**diiodo-***	IC⋮CI	277.83	rh nd (lig)	81–2	$32–3^{748}$ exp			i	s	s	s	s	os s lig δ	**B1**[3], 919
Ω e602	—,**diphenyl-***	Tolane. $C_6H_5C⋮CC_6H_5$	178.24	mcl pr or pl (al) λ^{hx} 270 (4.4), 280 (4.5), 290 (4.4), 298 (4.4)	63.5	300^{760} 170^{19}	0.9657^{100}_{4}		i	δ v[h]	v	...	...		**B5**[2], 568
e603	—,**1(2-naphthyl)-2-phenyl-***	$C_{10}H_7^{\beta}C⋮CC_6H_5$	228.30	cr (al)	117				...	s	...	...	...		**B5**[2], 628
e604	α-**Eucaine**	$C_{19}H_{27}NO_4$	333.43	pr (eth or al)	104–5				s	s	s	...	s	peth, lig, chl s	**B22,** 194
e605	—,hydrochloride	$C_{19}H_{27}NO_4$. HCl. *See* e604	369.89	pl (w + 1), pr (MeOH + 2)	*ca.* 200 d				v	v	δ	...	...		**B22,** 194
e606	β-**Eucaine** (*d*)	$C_{15}H_{21}NO_2$.	247.34	pr (peth)	57–8				i	s	s	...	s	chl, peth s	**B21**[2], 14
e607	—(*dl*)	Betacaine. $C_{15}H_{21}NO_2$. *See* e606.	247.34	pl (peth)	70–1 (78)				i	s	s	...	s	chl, peth s	**B21**[2], 13
e608	—(*l*)	$C_{15}H_{21}NO_2$. *See* e606	247.34	pr (peth)	57–8				i	s	s	...	s	chl, peth s	**B21**[2], 14
e609	—,hydrochloride (*dl*)	$C_{15}H_{21}NO_2$. HCl. *See* e606	283.80	pl (w)	277–9				s	s	s	...	...	chl s	**B21**[2], 13
e610	—,lactate (*dl*)	$C_{15}H_{21}NO_2$. $CH_3CH(OH)CO_2H$. *See* e606	337.42	wh pw	*ca.* 152				v	s	δ	...	...	chl s	**B21**[2], 13
—	**Eucalyptol**	*see* **1,8-Cineole**													
Ω e611	**Eugenol**	5-Allylguaiacol	164.21	cr (hx)	−7.5	253.2^{760} 130.5^{18}	1.0652^{20}_{4}	1.5405^{20}	i	∞	∞	...	...	chl, aa, oil s	**B6**[2], 921
Ω e612	—,acetate	$C_{12}H_{14}O_3$. *See* e611	206.25	pl (al) λ^{iso} 229.5 (3.81), 278 (3.45), 281.5 (3.48), 287.5 (3.34)	30–1	$127.1–7.8^{4}$	1.0806^{20}_{4}	1.5205^{20}	i	s	...	...	...		**B6**[3], 5029
—	**Eugenic acid**	*see* **Benzoic acid, 5-allyl-2-hydroxy-3-methoxy-**													
e613	**Eupittone**	Eupittonic acid. $C_{25}H_{26}O_9$	470.48	nd (al-eth) $\lambda^{10\% HCl}$ 560 (4.43)	200				...	δ[h]	...	...	...	aa, alk s (bl)	**B8,** 574
e614	**Euxanthic acid**	$C_{19}H_{16}O$.	404.43	ye nd (w + 1) [α] −108 (+1w)	130 d (+w) 162 δd (anh)				δ s[h]	v[h]	...	...	...	alk s	**B31,** 277
—	**Euxanthone**	*see* **Xanthone, 1,7-dihydroxy-**													
—	**Euxanthonic acid**	*see* **Benzophenone, 2,2′,5,6′-tetrahydroxy-**													

For explanations, symbols and abbreviations see beginning of table. For structural formulas see end of table.

No.	Name	Synonyms and Formula	Mol. wt.	Color, crystalline form, specific rotation and λ_{max} (log ε)	m.p. °C	b.p. °C	Density	n_D	Solubility w	al	eth	ace	bz	other solvents	Ref.
	Evernic acid														
e615	**Evernic acid**	Orselinic acid 4-everitate. Lecanoric acid monomethyl ether. $C_{17}H_{16}O_7$.	332.31	nd (w or ace), pr (al)	170	....	....	....	i δ^h	δ s^h	i δ^h	...	...	....	**B10**^{2}, 274
—	**Everninic acid**	*see* **Benzoic acid, 2-hydroxy-4-methoxy-6-methyl-**													
e616	**Evodiamine** (*d*)	Rhetsine. $C_{19}H_{17}N_3O$.	303.37	yesh lf (al) $[\alpha]_D^{15}$ +352 (ace, c = 0.5)	278	....	....	....	i	δ	δ	s	i	chl, aa δ peth i	**B26**^{2}, 103
e617	—,hydrate (*d*)	$C_{19}H_{17}N_3O \cdot H_2O$	321.38	pl (al)	146–7	....	....	....							**B24**^{2}, 72
—	**Exaltone**	*see* **Cyclopentadecanone***													

For explanations, symbols and abbreviations see beginning of table. For structural formulas see end of table.

No.	Name	Synonyms and Formula	Mol. wt.	Color, crystalline form, specific rotation and λ_{max} (log ε)	m.p. °C	b.p. °C	Density	n_D	Solubility w	al	eth	ace	bz	other solvents	Ref.
	Fagaramide														
f1	**Fagaramide**		247.30	nd (bz, dil al or peth) pl (AcOEt)	119.5				δ	s^h			s	peth i AcOEt s^h	**B19**[2], 299
f2	**β-Fagarine**	Skimmianine. $C_{14}H_{13}NO_4$.	259.27	pym, oct (al) λ^{al} 244 (4.43), 320 (3.90), 331 (3.90)	177				i	s	δ			chl s CS_2 δ peth i	**B27**, 134
f3	**γ-Fagarine**	Haplophine. $C_{13}H_{11}NO_3$.	229.24	pr (al) λ^{al} 238 (4.76), 370 (3.89), 332 (3.88)	142				δ	s^h	s		s	chl s peth δ	**C51**, 4402
f4	**α-Farnesene**	3,7,11-Trimethyl-1,3,6,10-dodecatetraene. $(CH_3)_2C{:}CH(CH_2)_2C(CH_3){:}CHCH_2CH{:}C(CH_3)CH{:}CH_2$	204.36			129–32[12]	0.8410_4^{20}	1.4836^{20}	i		s	s		peth ∞ lig ∞	**B1**[3], 1067
f5	**β-Farnesene**	$CH_2{:}CHC({:}CH_2)(CH_2)_2CH{:}C(CH_3)(CH_2)_2CH{:}C(CH_3)_2$	204.36			121–2[9]	0.8363_4^{20}	1.4899^{20}	i		s	s		chl v aa s os s	**B1**[3], 1067
Ω f6	**Farnesol** (*trans, trans*)	$(CH_3)_2C{:}CH(CH_2)_2C(CH_3){:}CH(CH_2)_2C(CH_3){:}CHCH_2OH$	222.38	λ^{al} 215 (3.30), 235 (2.65), 325 sh (0.45)		160[10] 120[0.3]	0.8846_4^{20}	1.4877^{20}	i	v	s	s		os s	**B1**[3], 2040
f7	**α-Fenchene(*d*)**		136.24	$[\alpha]_D^{14}$ +29 (l = 10 cm)		155–6	0.8660_4^{20}	1.4713^{20}	i	∞	s	s			**B5**[1], 86
f8	—(*dl*)	Isopinene. $C_{10}H_{16}$. *See* f7	136.24			154–6	0.8660_4^{20}	1.4705^{20}	i	∞	s	s			**B5**[1], 86
f9	—(*l*)	$C_{10}H_{16}$. *See* f7	136.24	$[\alpha]_D^{20}$ −43.8		158.5–8.8	0.8670_4^{20}	1.4713^{20}	i	∞	s	s			**B5**[2], 109
Ω f10	**Fenchone(*d*)**	*d*-2-Fenchanone. *d*-1,3,3-Trimethylnorcamphor.	152.24	$[\alpha]_D^{20}$ +66.9 (al)	6	193.5[760] 80[20]	0.9465_4^{20}	1.4623^{20}	i	v	v	s			**E12A**, 724
f11	—(*dl*)	$C_{10}H_{16}O$. *See* f10	152.24	λ^{MeOH} 288 (1.34)	−18 to −16	193–4 72–3[12]	0.9501_{15}^{15}	1.4702^{20}	i	v	s	s			**B7**[2], 93
f12	—(*l*)	$C_{10}H_{16}O$. *See* f10	152.24	$[\alpha]_D^{23}$ −66.94 (al) λ^{al} 285 (1.32)	5 (8.5)	192–4	0.948^{20}	1.4636^{20}	i	v	v	s			**B7**[2], 92
Ω f13	**Fenchyl alcohol** (*dl*)		154.26		α 38–9 β 6	α 202–3 β 201			i	v	v			peth	**B6**, 71
—	**Ferulaldehyde**	*see* **Cinnamaldehyde, 4-hydroxy-3-methoxy-**													
—	**Ferulic acid**	*see* **Cinnamic acid, 4-hydroxy-3-methoxy-**													
—	**Filicinic acid**	*see* **1,3,5-Cyclohexanetrione, 2,2-dimethyl-**													
f15	**Filixic acid BBB**	Filicic acid BBB. Filicin.	652.70	cr (AcOEt-ace)	172–4				i	i d^h	δ		s	CS_2, chl, xyl s	**J1952**, 3102
—	**Fisetin**	*see* **Flavone,3,3′,4′,7-tetrahydroxy-**													
f16	**Flavaniline**	2(*p*-Aminophenyl) lepidine.	234.30	pr (bz)	97	132–41[15]			δ	s			s		**B22**, 469
Ω f17	**Flavanone**	2,3-Dihydro-2-phenyl-1,4-benzopyrone. 4-Oxo-2-phenylchroman.	224.26	nd (lig) λ^{al} 252 (3.94), 320 (3.53)	76				i			s	s		**B17**[2], 387
Ω f18	—,4′-methoxy-3′,5,7-trihydroxy-	Hesperetin. $C_{16}H_{14}O_6$. *See* f17	302.29	pl (dil al + ½w), pl (AcOEt)	227–8 (233d)	sub 205[0.004]			i	v	s		δ	chl δ dil alk s	**B8**[2], 580 **B18**[2], 204
Ω f19	—,3′,4′,5,7-tetrahydroxy-	Eriodictyol. $C_{15}H_{12}O_6$. *See* f17	288.26	pa br nd (dil al + 1.5 w or dil aa + 2.5w), pl (al) λ^{MeOH} 250 (3.28), 292 (4.24)	267d 257d (+ 1.5w)				δ^h	s^h	δ			dil alk s os δ aa s	**B8**, 543 **B18**[2], 204
—	**Flavianic acid**	*see* **2-Naphthalenesulfonic acid, 5,7-dinitro-8-hydroxy-**													
—	**Flavinium, 3,4′,5,7-tetrahydroxy-, chloride**	*see* **Pelargonidine**, chloride													
—	**Flavol**	*see* **Anthracene, 2,6-dihydroxy-**													
Ω f20	**Flavone**	2-Phenyl-γ-benzopyrone. 2-Phenylchromone.	222.25	nd (lig), cr (30 % al), λ^{al} 252 (4.33), 295 (4.41)	100 (97)				i	s	s	s	s	os s chl s lig s	**B17**[2], 395
f21	—,6-bromo-	$C_{15}H_9BrO_2$. *See* f20	301.15	nd (al)	191–2				i	s^h					**B17**, 373
f22	—,5,7-dihydroxy-	Chrysin. $C_{15}H_{10}O_4$. *See* f20	254.25	pa ye pl or pr (MeOH), nd (sub) λ^{al} 270 (4.46), 318 (4.08)	275	sub			i	s^h δ	δ	s	δ	CS_2, chl δ aa s^h lig δ	**B18**[2], 97
f23	—,5,7-dihydroxy-4′-methoxy-	Acacetin, $C_{16}H_{12}O_5$. *See* f20	284.27	pa ye nd (al)	261				i	δ s^h	i	v	δ	AcOEt δ alk s lig δ	**B18**[3], 173
f24	—,5,7-dihydroxy-6-methoxy-	Oroxylin-A. $C_{16}H_{12}O_5$. *See* f20	284.27	ye nd (al)	231–2				i	s	s	s	s^h	alk s aa s	**J1936**, 591

For explanations, symbols and abbreviations see beginning of table. For structural formulas see end of table.

No.	Name	Synonyms and Formula	Mol. wt.	Color, crystalline form, specific rotation and λ_{max} (log ε)	m.p. °C	b.p. °C	Density	n_D	Solubility: w	al	eth	ace	bz	other solvents	Ref.
	Flavone														
Ω f25	—,3-hydroxy-	Flavonol. $C_{15}H_{10}O_3$. *See* f20	238.25	pa ye nd (al, MeOH) λ^{al} 240 (4.30), 307 (4.25)	169–70	...	...	...	...	s^h	...	...	...	...	**B17**[2], 498
—	—,7-methoxy-3,3′,4′,5-tetrahydroxy-	*see* **Rhamnetin**													
f26	—,2′,3,3′,5,7-pentahydroxy-	$C_{15}H_{10}O_7$. *See* f20	302.24	ye nd (aa + 1.5w) λ^{MeOH} 262 (4.5), 365 (4.3)	300	...	...	...	δ^h	s^h	...	...	...	...	**B18**[2], 235
Ω f27	—,2′,3,4′,5,7-pentahydroxy-	Morin. $C_{15}H_{10}O_7$. *See* f20	302.24	pa ye nd (+1w, dil aa) λ^{al} 265 (4.24), 273 (4.24)	303–4 (286)	...	...	...	δ^h	v	δ	...	s	CS_2 i aa δ alk s	**B18,** 239
f28	—,2′,3,5,5′,7-pentahydroxy-	$C_{15}H_{10}O_7$. *See* f20	302.24	red-ye cr (dil al + 1w)	306–8	...	...	...	...	...	...	...	...	...	**C48,** 2055
Ω f29	—,3,3′,4′,5,7-pentahydroxy-	Quercetin. Meletin. Sophoretin. $C_{15}H_{10}O_7$. *See* f20	302.24	ye nd (dil al + 2w) λ^{al} 256 (4.32), 301 (3.89), 373 (4.32)	316–7 (anh)	sub	...	...	δ^h	s^h	δ	s	...	MeOH δ Py s aa s	**B18**[2], 236
f30	—,3,3′,4′,7,8-pentahydroxy-	$C_{15}H_{10}O_7$. *See* f20	302.24	ye nd (dil al + 1w)	308d	...	...	...	δ^h	s^h	...	s	...	...	**B18,** 250
f31	—,3,3′,5,5′,7-pentahydroxy-	$C_{15}H_{10}O_7$. *See* f20	302.24	ye nd	>300	...	...	...	...	...	...	...	...	...	**B18**[2], 239
f32	—,3′,4′,5,5′,7-pentahydroxy-	Tricetin. $C_{15}H_{10}O_7$. *See* f20	302.24	ye nd (dil al + w)	>330 d	...	...	...	...	δ	i	...	i	...	**B18**[1], 423
f33	—,2′,3,5,7-tetrahydroxy-	Datiscetin. $C_{15}H_{10}O_6$. *See* f20	286.24	pa ye nd (al, aq aa)	277–8	...	...	...	δ^h	v	v	s	...	os, alk, con sulf s	**B18**[2], 214
Ω f34	—,3,3′,4′,7-tetrahydroxy-	Fisetin. $C_{15}H_{10}O_6$. *See* f20	286.24	lt ye nd (dil al + 1w)	330 (>360)	sub^{vac}	...	...	i	s	δ	s	δ	peth δ	**B18**[2], 216
f35	—,3′,4′,5,7-tetrahydroxy-	Luteolin. $C_{15}H_{10}O_6$. *See* f20	286.24	ye nd (dil al + 1w) λ^{al} 256 (4.20), 355 (4.25)	329–30d (anh)	sub^{vac}	...	...	δ^h	s	s	...	...	alk s con sulf s	**B18**[2], 212
Ω f36	—,3,4′,5,7-tetrahydroxy-	Kaempferol. $C_{15}H_{10}O_6$. *See* f20	286.24	ye nd (al + 1w) (aa) λ^{al} 235 sh (4.15), 273 (4.20), 320 sh (3.72)	276–8	...	...	...	δ	v^h	v	v	i	chl δ aa s^h alk s	**B18**[2], 214
f37	—,4′,5,7-trihydroxy-	Apigenin. $C_{15}H_{10}O_5$. *See* f20	270.25	ye nd (Py-w or w, +½w), lf (al) $\lambda^{w(hyd)}$ 269 (4.30), 335 (4.32)	347–8 (352)	sub^{vac}	...	...	i	s^h	...	...	...	Py s dil alk v con sulf s	**B18**[2], 172
f38	—,5,6,7-trihydroxy-	Baicalein. $C_{15}H_{10}O_5$. *See* f20	270.25	ye pr (al, MeOH or aa) λ^{al} 276 (4.42), 324 (4.18)	264–5d	...	...	...	δ	s v^h	s	s	δ	AcOEt s alk s(og-red) aa s^h chl δ	**B18**[2], 172
—	**Flavopurpurin**	*see* **9,10-Anthraquinone, 1,2,6-trihydroxy-**													
f39	**Floridoside**	Glycerol-2-α-*D*-monogalactoside.	254.24	pr (al) $[\alpha]_D$ +15.1 (w)	86–7	...	...	...	v	δ	...	...	...	...	**C31,** 3878
Ω f40	**Fluoran**	9-Hydroxy-9-xanthene-*o*-benzoic acid lactone.	300.32	nd (al + 2 al)	182–3	...	...	...	i	s	...	...	...	con sulf s	**B19**[2], 173
f41	—,1,6-dihydroxy-	$C_{20}H_{12}O_5$. *See* f40	332.32	ye nd	>280	...	...	...	...	...	...	...	...	...	**B19**[2], 247
f42	—,2,6-dihydroxy-	$C_{20}H_{12}O_5$. *See* f40	332.32	ye nd (al)	177	...	...	...	...	...	...	...	...	...	**B19**[2], 247
f43	—,3,5-dihydroxy-	$C_{20}H_{12}O_5$. *See* f40	332.32	ye-gr nd (aa)	179	...	...	...	...	...	...	...	...	...	**B19**[2], 248
Ω f44	**Fluoranthene**	1,2-Benzacenaphthene. Idryl.	202.26	pa ye nd or pl (al) λ^{cy} 237 (4.75), 254 (4.22), 263 (4.18), 277 (4.41), 288 (4.73)	111	*ca.* 375 217^{30}	1.252^0_4	...	i	s	s	...	s	chl, CS_2 s aa s	**E14s,** 54
Ω f45	**Fluorene**	2,3-Benzindene. Diphenylenemethane.	166.23	lf (al) λ^{chl} 266 (4.3), 290 (3.8), 301 (4.0)	116–7	$293–5^{760}$	1.203^0_4	...	i	δ s^h	s	s	s	CS_2 CCl_4 s, to s, v^h aa, Py s, v^h MeOH δ	**E13,** 25
Ω f46	—,2-acetamido-	*N*(2-Fluorenyl) acetamide. $C_{15}H_{13}NO$. *See* f45	223.28	nd (50% al or 50% aa) λ^{al} 287 (4.4), 300 sh (4.2)	194	...	...	...	i	s	s	...	...	aa s	**B12,** 1331
f47	—,9-acetamido-	$C_{15}H_{13}NO$. *See* f45	223.28	nd (aa)	262	...	...	...	...	...	...	...	...	con sulf s	**B12**[2], 781
f48	—,2-amino-	$C_{13}H_{11}N$. *See* f45	181.24	lo pl or nd (dil al) λ^{al} 287.5 (4.32), 315 sh (3.94)	131–2	...	...	...	i	s	s	...	...	...	**B12**[2], 779

For explanations, symbols and abbreviations see beginning of table. For structural formulas see end of table.

No.	Name	Synonyms and Formula	Mol. wt.	Color, crystalline form, specific rotation and λ_{max} (log ε)	m.p. °C	b.p. °C	Density	n_D	Solubility w	al	eth	ace	bz	other solvents	Ref.
	Fluorene														
f49	**—,9-amino-**	$C_{13}H_{11}N$. *See* f45	181.24	nd (lig) λ 258 (4.37), 293 (3.82), 304 (3.79)	α 64–5 γ 46–7 (unst)	...	...	...	δ	v	s	s	s	MeOH, chl v con sulf s (gr)	**B12^{2}**, 780
f50	**—,1-amino-9-hydroxy-**	$C_{13}H_{11}NO$. *See* f45	197.24	dk red nd (w)	142	...	...	...	δ	v	v	...	s	con sulf s aa v	**B13^{2}**, 435
f51	**—,2-amino-9-hydroxy-**	$C_{13}H_{11}NO$. *See* f45	197.24	irid nd (al)	200.5–1.0	...	...	...	...	s^h	δ	...	...	chl s^h con sulf s	**B13^{2}**, 435
f52	**—,4-amino-9-hydroxy-**	$C_{13}H_{11}NO$. *See* f45	197.24	ye (60 % al)	183–4	...	...	...	...	s	...	...	...	...	**B13^{2}**, 436
f53	**—,9-benzhydryl-idene-**	β,β-Diphenyl-α,α-biphenylene ethylene. ω,ω-Diphenyl dibenzofulvene. $C_{26}H_{18}$. *See* f45	330.43	ye (bz) λ^{al} 235 (4.6), 260 (4.4), 338 (4.0)	229.5	...	...	...	i	δ	δ	...	...	chl s	**E13**, 37
Ω f54	**—,9-benzylidene-**	ω-Phenyldibenzofulvene. $C_{20}H_{14}$. *See* f45	254.34	lf (al) λ^{al} 227 (4.6), 256 (4.4), 325 (4.1)	76	...	...	...	i	s^h	...	...	s	...	**E13**, 35
Ω f55	**—,2-bromo-**	$C_{13}H_9Br$. *See* f45	245.13	nd or pl (al) λ^{al} 270 (4.39), 295 (3.90), 306 (3.98)	113–4	*ca.* 185^{135}	...	...	i	s^h	...	...	...	chl v aa s	**B5^{2}**, 534
Ω f56	**—,9-bromo-**	$C_{13}H_9Br$. *See* f45	245.13	(lig or al) λ^{hx} 241 (4.7), 279 (4.4)	104–5	...	...	...	i	s^h	...	s	...	con sulf s^h	**B5^{2}**, 534
f57	**—,9(3-bromo-benzylidene)-**	$C_{20}H_{13}Br$. *See* f45	333.24	ye nd (aa)	92–3	...	...	...	i	s	...	...	...	MeOH s con sulf s^h	**B5^{1}**, 358
f58	**—,9(4-bromo-benzylidene)-**	$C_{20}H_{13}Br$. *See* f45	333.24	ye nd (aa, AcOEt-chl) λ^{diox} 258 (4.54), 330 (4.24)	147–8	...	...	...	i	...	...	...	...	con sulf s^h aa s^h	**B5^{2}**, 640
f59	**—,9(2-chloro-benzylidene)-**	$C_{20}H_{13}Cl$. *See* f45	288.78	ye nd (aa, MeOH)	69–70 (176)	$180^{0.7}$	...	...	i	v	...	...	...	con sulf s^h aa s^h	**B5^{1}**, 358
f60	**—,9(3-chloro-benzylidene)-**	$C_{20}H_{13}Cl$. *See* f45	288.78	pa ye pr or pym (MeOH)	90.5	...	...	...	i	v	...	...	...	con sulf s^h lig v	**B5^{1}**, 358
f61	**—,9(4-chloro-benzylidene)-**	$C_{20}H_{13}Cl$. *See* f45	288.78	ye nd (aa, al)	151	...	...	...	i	s	...	...	...	con sulf s^h aa s^h	**B5^{1}**, 358
f62	**—,9-cinnamyl-idene-(*trans*)**	$C_{22}H_{16}$. *See* f45	280.37	pa ye nd (aa) λ^{hx} 240 (4.6), 373 (4.67), 390 sh (4.5)	155	...	...	...	...	s^h	...	...	...	chl s^h sulf s (og)	**E13**, 36
f63	**—,2,7-diamino-**	$C_{13}H_{12}N_2$. *See* f45	196.26	nd (w), pr (bz), pl (eth) λ^{al} 293 (4.46), 330 sh (3.88)	165–7	...	...	...	i s^h	s	...	...	...	chl s	**B13^{2}**, 123
f64	**—,2,7-dichloro-**	$C_{13}H_8Cl_2$. *See* f45	235.13	pl or nd (bz)	128	sub	...	...	i	...	...	...	s^h	chl, CCl_4 s	**B5^{2}**, 533
f65	**—,9,9-dichloro-**	$C_{13}H_8Cl_2$. *See* f45	235.13	rh pr (bz, eth) nd (peth) λ^{hx} 234 (4.6), 244 (4.62)	103	...	...	...	d^h	v	s	v	v	os v con sulf s	**B5^{2}**, 534
f66	**—,1,8-dimethyl-9-(2-tolyl)-**	$C_{22}H_{20}$. *See* f45	299.40	...	168–9	...	...	...	...	s	s	s	...	...	**C50**, 11293
Ω f67	**—,2-hydroxy-**	$C_{13}H_{10}O$. *See* f45	182.23	lf (w), nd (chl or 60 % al) λ^{cy} 271 (4.33), 304 (3.74), 315 (3.76)	171–4	...	...	...	i	v	v	s	...	alk v aa v lig i	**B6^{2}**, 655
Ω f68	**—,9-hydroxy-**	Diphenylenecarbinol. 9-Fluorenol. $C_{13}H_{10}O$. *See* f45	182.23	hex nd (w or peth) λ^{sulf} 480 (3.30), 660 (3.49)	154	...	...	...	δ	δ	s	s	v	peth δ	**E13**, 63
f69	**—,9-hydroxy-9-phenyl-**	$C_{19}H_{14}O$. *See* f45	258.32	ye or col pr (lig) λ^{diox} 242 (3.22), 272 (3.06), 280 (3.08), 311 (2.76)	108–9	...	...	...	...	...	...	...	s	con sulf s (ye) aa s	**E13**, 64
f70	**—,9-methyl-**	$C_{14}H_{12}$. *See* f45	180.25	pr λ 263 (4.40), 290 (3.61), 300 (3.89)	46–7	154–6^{15}	1.0263^{66}_{4}	1.610^{66}	i	s	s	s	s	chl s	**E13**, 29
f71	**—,9(2-methyl-benzylidene)-**	$C_{21}H_{16}$. *See* f45	268.38	nd or pr (aa) λ^{al} 258 (4.62), 298 (4.21), 312 (4.24)	109.5	...	...	...	i	s^h	...	...	s^h	con sulf s^h	**B5^{1}**, 359
f72	**—,9(4-methyl-benzylidene)-**	$C_{21}H_{16}$. *See* f45	268.38	nd (aa), pr (al) λ^{al} 258 (4.55), 330 (4.30)	97.5	...	...	...	i	s^h	...	...	s^h	con sulf s^h	**B5^{1}**, 359
f73	**—,9-methylene-**	Biphenyleneethylene. Dibenzofulvene. $C_{14}H_{10}$. *See* f45	178.24	...	53	...	...	...	i	s	s	s	s	os v	**E13**, 34

For explanations, symbols and abbreviations see beginning of table. For structural formulas see end of table.

No.	Name	Synonyms and Formula	Mol. wt.	Color, crystalline form, specific rotation and λ_{max} (log ε)	m.p. °C	b.p. °C	Density	n_D	Solubility w	al	eth	ace	bz	other solvents	Ref.
	Fluorene														
Ω f74	—,2-nitro-	$C_{13}H_9NO_2$. *See* f45	211.22	nd (50 % aa or ace) λ^{al} 234 (3.98), 332 (4.25)	158	...	...	...	i	...	...	s^h	s^h	...	**B5**[2], 535
f75	—,3-nitro-	$C_{13}H_9NO_2$. *See* f45	211.22	pa ye nd (al, chl-peth)	106	...	...	...	...	s^h	...	s^h	...	...	**E13**, 48
f76	—,9-nitro-	$C_{13}H_9NO_2$. *See* f45	211.22	gr-ye nd (al), lf (bz) λ 248 (4.6), 305 (3.4), 340 (4.0) (*aci* form)	181–2d	...	...	...	i	s^h	s	s	s	chl s aa s, lig i	**E13**, 48
—	—,9-oxo-	*see* **9-Fluorenone**													
f77	—,9-phenyl-	$C_{19}H_{14}$. *See* f45	242.32	nd or lf (al or bz) λ^{al} 265 (4.2), 292 (3.7), 305 (3.9)	148	...	...	...	i	s^h	δ	...	s	aa, chl s peth s	**E13**, 29
—	**Fluorenecarboxylic acid, 9-oxo**	*see* **9-Fluorenonecarboxylic acid**													
f78	**2-Fluorenesulfonic acid**	...	246.29	nd (aa)	155 (+1w)	...	...	...	v	s	...	s	...	chl s alk s	**E13**, 114
—	**Fluorenol**	*see* **Fluorene, hydroxy-**													
Ω f80	**9-Fluorenone**	Diphenylene ketone. 9-Oxofluorene.	180.22	ye rh bipym (al, bz-peth) λ^{diox} 258 (4.90), 294 (3.62), 328 (2.89), 378 (2.43)	84 (86)	341.5^{760}	1.1300^{99}_4	1.6369^{99}	i	s	v	s	s v^h	to v	**E13**, 77
Ω f81	—,oxime	9-Isonitrosofluorene. 9-Oximinofluorene.	195.22	nd (chl-peth or bz)	195–6	...	...	...	i	s	...	...	...	chl s dil alk v	**B7**[2], 407
f82	—,1-amino-	$C_{13}H_9NO$. *See* f80	195.22	ye nd (dil al)	118.5–20	...	...	...	s^h	s	s	v	...	oos v	**B14**, 113
Ω f83	—,2-amino-	$C_{13}H_9NO$. *See* f80	195.22	red-vt pr (al)	163	...	...	...	i	s^h	s^h	...	s^h	peth δ, aa s	**B14**[2], 68
f84	—,3-amino-	$C_{13}H_9NO$. *See* f80	195.22	ye nd (w or dil al)	158–9	...	...	...	s^h	s	...	...	...	...	**B14**[2], 69
f85	—,4-amino-	$C_{13}H_9NO$. *See* f80	195.22	red nd (al)	145	...	...	...	...	v	v	s	...	chl v, aa v	**B14**[2], 69
f86	—,2-bromo-	$C_{13}H_7BrO$. *See* f80	259.11	ye nd (al or aa) λ^{diox} 262 (4.92), 304 (3.63), 380 (2.76)	149	...	...	...	...	s^h	...	s	s	chl, aa s	**E13**, 83
f87	—,2-chloro-	$C_{13}H_7ClO$. *See* f80	214.65	og-ye nd (dil al)	125–6	sub	...	...	...	s	...	...	...	sulf s (vt-red)	**E13**, 83
f88	—,1,8-dimethyl-	$C_{15}H_{12}O$. *See* f80	208.24	ye	197–8	...	...	...	...	i	...	...	...	chl s, aa s	**C62**, 7611
f89	—,2-nitro-	$C_{13}H_7NO_3$. *See* f80	225.21	ye nd or lf (aa) λ^{al} 284 (4.46), 336 (3.79)	222–3	sub	...	...	...	δ	...	s	...	sulf s aa s^h (red-ye)	**E13**, 85
f90	—,2,3,7-trinitro-	$C_{13}H_5N_3O_7$. *See* f80	315.20	pa ye nd (aa)	180–1	...	...	...	...	δ	δ	...	v	chl v	**B7**[1], 254
Ω f91	—,2,4,7-trinitro-	$C_{13}H_5N_3O_7$. *See* f80	315.20	pa ye nd (aa or bz) λ^{aa} 350 sh (3.9), 380 sh (3.5)	176	...	...	...	δ	...	...	v	v	chl v	**B7**[2], 410
f92	**9-Fluorenone-1-carboxylic acid**	...	224.23	og-red nd (dil al)	192–4	...	...	...	i	v	v	...	...	sulf s (red)	**B10**[2], 534
f93	—,amide	...	223.23	ye nd (al)	229–30	...	...	...	δ	δ s^h	...	s	...	...	**B10**, 774
f94	—,chloride	$C_{14}H_7ClO_2$. *See* f92	242.66	pa ye nd (bz)	140	...	...	...	d^h	s	s	...	v^h	...	**B10**, 774
f95	—,ethyl ester	$C_{10}H_{12}O_3$. *See* f92	252.27	ye nd (dil al)	84–5	...	...	...	i	s	s	...	...	sulf s	**B10**, 774
Ω f96	**9-Fluorenone-2-carboxylic acid**	...	224.23	ye nd (al or aa)	338	sub 340	...	...	i	s^h δ	...	...	...	aa s^h	**E13**, 108
f97	—,methyl ester	$C_{15}H_{10}O_3$. *See* f96	238.25	ye nd (MeOH)	181	...	...	...	i	s	s	s	...	sulf s	**B10**, 774
f98	**9-Fluorenone-3-carboxylic acid**	...	224.23	ye (aa, MeOH)	299 (286)	...	...	...	...	s^h	...	...	...	aa s^h	**E13**, 108
Ω f99	**9-Fluorenone-4-carboxylic acid**	...	224.23	ye nd (al)	227	...	...	...	i	s^h	s	...	...	sulf s (red) aa s^h	**B10**[2], 535
Ω f100	**Fluorescein**	3′,4′-Dehydroxyfluoran. Resorcinolphthalein.	332.32	red rh pr (stable form) ye gran (MeOH) (labile form) λ^{diox} 225 (4.8), 227.5 (3.8), 490 (2.1)	314–6 d (sealed tube)	...	...	...	$δ^h$	δ	δ	v	δ	peth i Py s MeOH s	**B19**[2], 249
—	—,sodium salt	*see* **Uranine**													
f102	**Fluorescin**	2(3,6-Dihydroxyxanthyl)-benzoic acid.	334.33	col or ye nd (aa or eth) pl (bz) λ^{w} 490 (4.5)	125–7	...	...	...	i	s	s	s	s^h	aa s^h	**B18**[2], 307

For explanations, symbols and abbreviations see beginning of table. For structural formulas see end of table.

No.	Name	Synonyms and Formula	Mol. wt.	Color, crystalline form, specific rotation and λ_{max} (log ε)	m.p. °C	b.p. °C	Density	n_D	Solubility						Ref.
									w	al	eth	ace	bz	other solvents	
	Fluoroform														
—	**Fluoroform**	see **Methane, trifluoro-**													
f103	**Fluorophosphoric acid,** diisopropyl ester	$(C_3H_7^iO)_2POF$	184.15	λ^{al} 243 (0.28), 255 (0.33), 261 (0.30)	...	62^9	1.055	1.3830^{25}	δ d^h	...	s	...	...	oils v lig δ	**C41,** 1233
—	**Fluoroprene**	see **1,3-Butadiene, 2-fluoro-**													
Ω f104	**Folic acid**	P.G.A. Pteroylglutamic acid. Vitamin Bc. $C_{19}H_{19}N_7O_6$	441.41	ye-og nd (w) $[\alpha]_D^{25}$ +23 (0.1 N NaOH, c = 0.5) $\lambda^{w,pH=13}$ 259 (4.51), 368 (3.87)	darkens 250 d	...	...	...	δ s^h	s	i	i	i	Py, aa s chl i alk δ	**Am 69,** 1476
f105	**Folinic acid** (*d*)	5-Formyl-5,6,7,8-tetrahydro-pteroyl-L-glutamic acid. Leucovorin.	473.45	cr (w + 3) $[\alpha]_D^{25}$ + 16.76 (5 % Na_2CO_3 sol, c = 3.5) $\lambda^{w,pH=13}$ 287	248–50 d	...	...	...	δ	...	...	...	...	...	**Am 73,** 1979
Ω f106	**Formaldehyde**	Methanal*. HCHO	30.03	gas λ^{vap} 155.5 (4.37), 175 (4.26)	−92	-21^{760} -79.6^{20}	0.815_4^{-20}	...	s	s	∞	∞	∞	chl s	**B1**[2], 619 **B6**[2], 1244
f107	—, β,β′-dichloro-isopropyl ethyl acetal	$(ClCH_2)_2CHOCH_2OC_2H_5$	187.07	...	...	$96–8^{16}$	1.182_{17}^{17}	1.4491^{17}	...	...	s	...	...	...	**B1**[2], 640
Ω f108	—, diethyl acetal	Diethoxymethane*. Diethyl formal. Ethylal. $CH_2(OC_2H_5)_2$	104.15	...	−66.5	89	0.8319_4^{20}	1.3748^{18}	s	∞	∞	v	v	...	**B1**[2], 639
Ω f109	—, dimethyl acetal	Methyal. $CH_2(OCH_3)_2$	76.11	λ 288 (−2.4)	−104.8	45.5^{760}	0.8593_4^{20}	1.3513^{20}	v	∞	∞	∞	∞	oos ∞	**B1**[2], 638
Ω f110	—, 2,4-dinitro-phenylhydrazone	...	210.15	ye cr (al), pr (lig) λ^{chl} 346 (4.31)	167	...	...	...	i	s^h	δ	...	...	...	**B15,** 490
f111	—, dipropyl acetal	Dipropoxymethane*. $CH_2(OCH_2CH_2CH_3)_2$	132.21	...	−97.3	140.5^{760}	0.8345^{20}	1.3939^{19}	v	v	v	v	v	...	**B1**[2], 639
f112	—, oxime	Formaldoxime. CH_2:NOH	45.04	...	2.5	109^{15} (84)	1.133	...	s	v	v	...	...	...	**B1,** 590
f113	—, **fluoro-**	Formyl fluoride. FCHO	48.02	gas	...	-24^{760} (−29)	...	...	...	d	...	...	...	...	**Am 77** 5278
—	—, **thio,** trimer	see **1,3,5-Trithiane**													
f114	**Formaldomedone**	...	292.38	nd (al or bz) λ^{al} 257 (4.42)	189–90	...	...	...	i	δ s^h	...	...	δ s^h	peth δ	**B7**[2], 852
—	**Formamide**	see **Formic acid,** amide													
—	**Formamidine**	see **Formic acid,** amidine													
—	**Formanilide**	see **Formic acid,** amide, *N*-phenyl-													
Ω f115	**Formic acid**	Methanoic acid*. HCO_2H	46.03	λ^{undil} 205 (1.65)	8.4	100.7^{760} 50^{120}	1.220_4^{20}	1.3714^{20}	∞	∞	∞	v	s	to s	**B2**[3], 3
Ω f116	—, allyl ester	$HCO_2CH_2CH{:}CH_2$	86.09	...	...	83.6	0.9460_4^{20}	...	δ	s	∞	...	...	...	**B2**[3], 46
Ω f117	—, amide	Formamide. $HCONH_2$	45.04	λ^{undil} 205 sh (2.20)	2.55	111^{20}	1.1334_4^{20}	1.4472^{20}	∞	∞	δ	s	i	chl i peth i	**B2,** 26
Ω f119	—, —, *N,N*-diethyl-	$HCON(C_2H_5)_2$	101.15	...	...	$177–8^{760}$ 68^{15}	0.9080^{19}	1.4321^{25}	∞	v	v	∞	∞	...	**B4,** 109
Ω f120	—, —, *N,N*-dimethyl-	$HCON(CH_3)_2$	73.09	λ^{gas} 162 (3.84), 197.4 (3.94)	−60.48	$149–56^{760}$ 39.9^{10}	0.9487_4^{20}	1.4305^{20}	∞	∞	∞	∞	∞	chl ∞ lig δ	**B4,** 58
Ω f121	—, —, *N,N*-diphenyl-	*N*-Formyldiphenyl amine. $HCON(C_6H_5)_2$	197.24	rh (dil al) λ^{bz} 324 (4.22)	73–4	337.5^{762} 190^{13}	...	...	i s^h	s	s	...	s	...	**B12**[1], 190
f122	—, —, *N*-ethyl-	$HCONHC_2H_5$	73.09	...	...	197–9 109.6^{30}	0.9552_4^{20}	1.4320^{20}	∞	∞	∞	...	...	...	**B4,** 109
f123	—, —, *N*(1-hydroxy-2,2,2-trichloro-ethyl)-	Chloral formamide. $HCONHCHOHCCl_3$	192.43	cr	...	118	...	...	s d^h	v d^h	v	v	...	AcoEt s	**B2**[2], 37
Ω f124	—, —, *N*-methyl-	$HCONHCH_3$	59.07	...	...	180–5 $102–3^{20}$	1.011^{19}	1.4319^{20}	s	s	i	∞	...	...	**B4**[2], 563
Ω f125	—, —, *N*-phenyl-	Formanilide. $HCONHC_6H_5$	121.14	mcl pr (lig-xyl) λ^{al} 242 (4.14)	50	271 166^{14}	1.1437_4^{17} 1.1322_{50}^{50}	...	s	v	s	...	s	...	**B12**[1], 190
Ω f126	—, —, *N*-2-tolyl-	Form-*o*-toluidide. *N*-Formyl-*o*-toluidine	135.17	lf (al)	62	288^{760}	1.086_4^{55}	...	s	v	...	...	...	...	**B12**[2], 439
Ω f127	—, —, *N*-3-tolyl-	Form-*m*-toluidide.	135.17	...	< −18	278^{724} δd $176–8^{17}$	...	...	s	...	...	...	...	NaOH s	**B12**[2], 468
Ω f128	—, —, *N*-4-tolyl-	Form-*p*-toluidide.	135.17	nd	53	...	...	...	s	v	...	s	...	...	**B12**[1], 419
f129	—, amidine	Formamidine. $HN{:}CHNH_2$	44.06	pr	81	d	...	...	v	s	...	...	...	...	**B2,** 90
Ω f130	—, —, *N,N′*-diphenyl-	$C_6H_5N{:}CHNHC_6H_5$	196.26	nd (al) λ^{al} 282 (4.34)	142 (136–7)	> 250	...	...	δ	s	v	s	s	chl s, lig δ CS_2, peth δ	**B12,** 236
f130[1]	—, amidoxime	Isoretin. $HON{:}CHNH_2$	60.06	rh nd (al)	114–5	d	...	...	v	δ	δ	...	i	...	**B2**[2], 89
Ω f131	—, benzyl ester	Benzyl formate. $HCO_2CH_2C_6H_5$	136.16	...	...	$202–3^{747}$ $84–5^{10}$	1.081_4^{20}	1.5154^{20}	i	s	∞	s	...	...	**B6**[2], 415
Ω f132	—, butyl ester	Butyl formate. $HCO_2CH_2CH_2CH_2CH_3$	102.13	...	−91.9	106.8^{760}	0.8885_4^{20}	1.3912^{20}	δ	∞	∞	s	...	...	**B2**[3], 39
f133	—, *sec*-butyl ester. (*dl*)	$HCO_2CH(CH_3)C_2H_5$	102.13	...	...	97	0.8846_4^{20}	1.3865^{20}	δ	∞	∞	s	...	...	**B2**[3], 41
Ω f134	—, cyclohexyl ester	...	128.17	...	...	162.5^{750}	1.0057_4^{0}	1.4430^{20}	i	s	v	...	...	HCO_2H s aa s	**B6**[2], 10

For explanations, symbols and abbreviations see beginning of table. For structural formulas see end of table.

No.	Name	Synonyms and Formula	Mol. wt.	Color, crystalline form, specific rotation and λ_{max} (log ε)	m.p. °C	b.p. °C	Density	n_D	Solubility w	al	eth	ace	bz	other solvents	Ref.
	Formic acid														
Ω f135	—,ethyl ester	$HCO_2C_2H_5$	74.08		−80.5	54.5^{760}	0.9168^{20}	1.3598^{20}	s	∞	∞	v			**B2**^{3}, 31
Ω f136	—,heptyl ester	$HCO_2(CH_2)_6CH_3$	144.22			178.12^{760} 83^{30}	0.8784^{20}	1.4140^{20}	i	∞	∞				**B2**^{3}, 44
Ω f137	—,hexyl ester	$HCO_2(CH_2)_5CH_3$	130.19		−62.65	155.5^{760}	0.8813^{20}_4	1.4071^{20}	i	∞	∞				**B2**^{3}, 44
f138	—,hydrazide	Formyl hydrazide. $HCONHNH_2$	60.06	ye lf or nd (al)	54				i	v	v^h		v	chl v	**B2**, 93
Ω f139	—,isobutyl ester	$HCO_2CH_2CH(CH_3)_2$	102.13		−95.8	98.4^{760} 33^{54}	0.8854^{20}_4	1.3857^{20}	δ	∞	∞	v			**B2**^{2}, 30
Ω f140	—,isopropyl ester	Isopropyl formate. $HCO_2CH(CH_3)_2$	88.12			68.2^{760}	0.8728^{20}_4	1.3678^{20}	δ	∞	∞	v			**B2**^{2}, 29
Ω f141	—,methyl ester	Methyl formate. HCO_2CH_3	60.05		−99	31.5^{760}	0.9742^{20}_4	1.3433^{20}	v	∞	s			MeOH s	**B2**^{2}, 25
Ω f142	—,3-methylbutyl ester	Isoamyl formate. $HCO_2(CH_2)_2CH(CH_3)_2$	116.16		−93.5	124.2^{760}	0.8857^{20}_4	1.3976^{20}	δ	s	∞				**B2**^{3}, 43
f143	—,4-nitrobenzyl ester		181.15	(dil al)	31					s^h					**B6**^{1}, 223
Ω f144	—,octyl ester	$HCO_2(CH_2)_7CH_3$	158.24		−39.1	198.8	0.8744^{20}	1.4208^{15}_{He}	i	s	∞				**B2**, 22
Ω f145	—,pentyl ester	Amyl formate. $HCO_2(CH_2)_4CH_3$	116.16		−73.5	132.1 (130)	0.8853^{20}_4	1.3992^{20}	δ	∞	∞				**B2**^{3}, 42
Ω f146	—,propyl ester	Propyl formate. $HCO_2CH_2CH_2CH_3$	88.12		−92.9	81.3^{760}	0.9058^{20}_4	1.3779^{20}	δ	∞	∞				**B2**^{2}, 28
—	—,**amino-**	*see* **Carbamic acid**													
—	—,**azodi-**	*see* **Azodicarboxylic acid**													
—	—,**chloro-**, amide	*see* **Carbamic acid**, chloride													
Ω f150	—,—,benzyl ester	Carbobenzoxy chloride. $ClCO_2CH_2C_6H_5$	170.60			103^{20}	1.20	1.5160^{20}	d	d^h	s	s	s		**B6**, 437
Ω f151	—,—,butyl ester	$ClCO_2(CH_2)_3CH_3$	136.58			138^{756} 35.5^{13}	1.0513^{20}_4	1.4121^{20}	d	d	∞	s			**B3**^{2}, 111
Ω f152	—,—,2-chloroethyl ester	$ClCO_2CH_2CH_2Cl$	142.97			155.7–6.0	1.3847^{20}_4	1.4465^{20}	i	s	s	s	s	oos s	**B2**^{2}, 193
f153	—,—,chloromethyl ester	$ClCO_2CH_2Cl$	128.94			107^{760}	1.465^{15}	1.4286^{22}	d^h		s	v			**B3**^{2}, 11
f154	—,—,3-chloropropyl ester	$ClCO_2(CH_2)_3Cl$	157.00			177	1.2949^{25}_{20}	1.4456^{20}	i						**B3**^{2}, 10
f155	—,—,cyclohexyl ester		162.62			87.5^{27}					s				**B6**^{2}, 11
f156	—,—,dichloromethyl ester	$ClCO_2CHCl_2$	163.39			$110–1^{760}$			d^h			v			**B3**^{2}, 11
f157	—,—,2-ethoxyethyl ester	Cellosolve chloroformate. $ClCO_2CH_2CH_2OC_2H_5$	152.58			67.2^{14}	1.1341^{25}	1.4169^{25}	i	s					**B3**^{2}, 29
Ω f158	—,—,ethyl ester	Ethyl chloroformate. $ClCO_2C_2H_5$	108.53		−80.6	95^{760} (93.1)	1.1352^{20}_4	1.3974^{20}	d	d	s		s	chl s	**B3**^{2}, 10
Ω f159	—,—,isobutyl ester	$ClCO_2CH_2CH(CH_3)_2$	136.58			128.8	1.0426^{18}_4	1.4071^{18}_{He}	d	s^h	∞		s	chl s	**B3**^{2}, 11
f160	—,—,isopropenyl ester	$ClCO_2C(CH_3):CH_2$	120.53			100^{760}	1.103^{20}_{20}				s				**C28**, 6702
f161	—,—,isopropyl ester	$ClCO_2CH(CH_3)_2$	122.55			104.6–0.9^{761} 66.3^{200}		1.4013^{20}	i		s				**B3**^{2}, 10
f162	—,—,2-methoxyethyl ester	$ClCO_2CH_2CH_2OCH_3$	138.55			58.7^{13}	1.1905^{25}	1.4163^{25}	i		v				**B3**^{3}, 29
Ω f163	—,—,methyl ester	Methyl chloroformate. $ClCO_2CH_3$	94.50			70.4–0.9^{752}	1.2231^{20}_4	1.3868^{20}	d	∞	∞		s	chl s	**B3**^{2}, 9
f164	—,—,3-methylbutyl ester	Isoamyl chloroformate. $ClCO_2CH_2CH_2CH(CH_3)_2$	150.61			154.3^{760} 60^{15}	1.0288^{17}_4	1.4176^{20}	d	∞	∞				**B3**^{2}, 11
Ω f165	—,—,pentyl ester	Amyl chloroformate. $ClCO_2(CH_2)_4CH_3$	150.61			$60–2^{15}$		1.4181^{18}			s				**B3**^{3}, 27
f166	—,—,propyl ester	$ClCO_2CH_2CH_2CH_3$	122.55			115.2^{760} (105^{759})	1.0901^{20}_4	1.4035^{20}	d	∞	∞				**B3**^{2}, 10
f167	—,—,trichloromethyl ester	Diphosgene. Perchloromethyl formate. Superpalite. $ClCO_2CCl_3$	197.83		−57	128^{760} 49^{50}	1.6525^{14}	1.4566^{22}	i	s	v				**B3**^{2}, 16
—	—,**cyano-**, ethyl ester	*see* **Oxalic acid,** monoethyl ester mononitrile													
f168	—,(ethoxy)dithio-	Ethylxanthic acid. $C_2H_5OCS_2H$	122.21		ca. −53	25d			δ					chl, CS_2 s $PhNO_2$ s	**B3**^{2}, 151
—	—,**fluoro-**, fluoride	*see* **Carbonyl fluoride**													
f169	**Formimidic acid, *N*-phenyl-,** ethyl ester	*N*-Phenylformininoethyl ether. $C_6H_5N:CHOC_2H_5$	149.19	λ^{hx} 250 (3.7)		$213–5^{761}$	1.0051^{20}_4	1.5279^{20}			s		s		**C21**, 387
f170	**Formonitrolic acid**	Methyl nitrolic acid. $HON:CHNO_2$	90.04	nd (eth or eth-peth)	68d				s	s	s			alk s (red)	**B2**, 92
f171	**Frangulin A**	$C_{21}H_{20}O_9$	416.39	ye or red (al or AcOEt) λ 225 (4.52), 264 (4.28), 282 (4.15), 300 (3.97), 430 (4.05)	228				i s^h	s^h			s^h	chl s^h aa s	**B31**, 74
	Fraxetin	*see* **Coumarin 7,8-dihydroxy-6-methoxy-**													
f173	**Fraxin**	8-Glucosidofraxetin.	370.32	ye nd (al) ye nd (w + 3)	205 (anh)				δ v^h	δ s^h	i			alk s	**B31**, 249

For explanations, symbols and abbreviations see beginning of table. For structural formulas see end of table.

No.	Name	Synonyms and Formula	Mol. wt.	Color, crystalline form, specific rotation and λ_{max} (log ε)	m.p. °C	b.p. °C	Density	n_D	Solubility: w	al	eth	ace	bz	other solvents	Ref.
	Freon														
—	**Freon**	*see* **Methane, dichloro-* difluoro-***													
—	**Freon 114**	*see* **Ethane, 1,1-dichloro-1,2,2,2-tetrafluoro-***													
f174	**1-Fructosamine** (*D*)	Isoglucosamine.	179.17	syr	...	...	...	...	i	i	...	...	...	dil ac s	**B31,** 342, 345
Ω f175	*β*-**Fructose** (*D*)	Fruit sugar. Levulose.	180.16	pr or nd (w) orh pr (al) $[\alpha]_D^{20}$ −133 → −92 (w, c = 2) λ^w 279 (−0.42)	103–5d	...	1.60_4^{20}	...	v	s	...	v^h δ	...	MeOH s Py s	**B31,** 321
Ω f176	*α*-**Fucose** (*L*)	6-Deoxygalactose.	164.16	nd (al) $[\alpha]_D^{20}$ (mut) −124.1 → 75.6 (w, c = 9)	145	...	...	...	v	δ s^h	i	...	...	...	**B31,** 76, 78
—	—,3-methyl-(*D*)	*see* Digitalose													
f177	**Fucoxanthin**	$C_{40}H_{60}O_6$.	632.89	red-br pl (eth-peth) hex pl (+2w, dil al, dil ace), nd (MeOH + 3) $[\alpha]_D^{18}$ +72.5 (chl) λ^{chl} 475, 492	168	...	...	...	...	s	s	...	...	MeOH δ peth δ con sulf → bl	**B30,** 105
—	**Fulminuric acid**	*see* **Malonic acid, nitro-,** monoamide, mononitrile													
Ω f178	**Fumaric acid**	*trans*-Butenedioic acid*. $HO_2CCH{:}CHCO_2H$	116.07	nd, mcl pr or lf (w) λ^w 208 (4.20)	300–2 (286–7) (sealed tube)	$165^{1.7}$ sub	1.635_4^{20}	...	δ s^h	s	δ	δ	...	chl, CCl_4 δ con sulf s	**B2**[2], 631
Ω f179	—,dichloride	Fumaryl chloride. $ClCOCH{:}CHCOCl$	152.97	pa ye liq	...	$158–60^{760}$ 63^{13}	1.408^{20}	1.5004^{18}	d	d	...	...	...	...	**B2**[3], 1907
f180	—,diisobutyl ester	$(CH_3)_2CHCH_2O_2CCH{:}CHCO_2CH_2CH(CH_3)_2$	228.29	...	...	170^{160} 122^5	0.9760_4^{20}	1.4432^{20}	i	s	s	s	...	...	**B2,** 742
f181	—,diisopropyl ester	$(CH_3)_2CHO_2CCH{:}CHCO_2CH(CH_3)_2$	200.24	...	...	$225–6^{760}$	...	...	...	s	s	s	...	...	**B2,** 742
Ω f182	—,dinitrile	*trans*-1,2-Dicyanoethylene. Fumaronitrile. $NCCH{:}CHCN$	78.07	nd (bz-peth) λ^{al} 232 sh (4.06), 292 (0.60)	96.8	186^{760}	0.9416^{111}	1.4349^{111}	s	s	s	s	s	peth δ	**B2**[1], 302
f183	—,diphenyl ester	$C_6H_5O_2CCH{:}CHCO_2C_6H_5$.	268.27	nd (al)	161–2	219^{14}	...	...	i	δ s^h	...	...	...	...	**B6,** 156
Ω f184	—,dipropyl ester	$CH_3CH_2CH_2O_2CCH{:}CHCO_2CH_2CH_2CH_3$	200.24	...	...	110^5	1.0129_4^{20}	1.4435^{20}	...	s	s	...	...	...	**B2**[2], 639
f185	—,piperazinium salt	$C_4H_{10}N_2 . HO_2CCH{:}CHCO_2H$	202.21	...	240d	...	...	...	s^h	s	i	...	...	...	**Am 70,** 2758
f185[1]	—,**bromo-**	$HO_2CCBr{:}CHCO_2H$.	194.98	pr (AcOEt)	185–6	d 200	...	...	s	s	...	...	...	...	**B2**[2], 640
f185[2]	—,**chloro-**	$HO_2CCCl{:}CHCO_2H$.	150.52	pl (aa)	192–3	sub	...	...	v	v	v	...	δ	...	**B2**[2], 640
f186	—,—,dichloride	$ClCOCCl{:}CHCOCl$.	187.41	pa gr	...	184–7 δd $73–5^{20}$	1.564_4^{20}	1.5206^{20}	d	d	v	...	...	aa s	**B2**[2], 640
f187	—,—,diethyl ester	$C_2H_5O_2CCH{:}CClCO_2C_2H_5$.	206.63	...	...	250 δd 127^{10}	1.1880_4^{20}	1.4571^{20}	i	v	v	...	...	...	**B2**[2], 640
f188	—,—,dimethyl ester	$CH_3O_2CCH{:}CClCO_2CH_3$	178.57	...	...	224^{760} 108^{15}	1.2899_4^{25}	1.4720_{He}^{18}	...	s	s	...	...	...	**B2**[2], 640
f189	—,**dimethyl-**	$HO_2C(CH_3)C{:}C(CH_3)CO_2H$	144.14	nd (w)	241	...	...	...	δ s^h	δ	...	...	i	chl, lig i	**B2**[2], 660
Ω f190	—,**methyl-**	Mesaconic acid. $HO_2CC(CH_3){:}CHCO_2H$	130.10	rh nd or mcl pr (eth, AcOEt, w) rods (sub)	204.5 (cor)	sub	1.466_4^{20}	...	δ v^h	v	s	...	δ	chl, CS_2 δ lig δ	**B2**[2], 651
f191	—,—,diethyl ester	$C_2H_5O_2CC(CH_3){:}CHCO_2C_2H_5$	186.21	...	...	229^{760} $93–5^{10}$	1.0453_{20}^{20}	1.4488^{20}	δ	s	s	s	s	...	**B2**[2], 652
f192	—,—,dimethyl ester	$CH_3O_2CC(CH_3){:}CHCO_2CH_3$	158.16	...	...	203.5 100^{16}	1.0914_4^{20}	1.4512^{20}	δ	s	s	...	...	...	**B2**[2], 652
—	**Fumigatin**	*see* **1,4-Benzoquinone, 3-hydroxy-2-methoxy-5-methyl-***													
Ω f193	**Furan**	1,4-Epoxy-1,3-butadiene*. Furfuran.	68.08	λ^{al} 208 sh (3.9)	−85.65	31.36^{760}	0.9514_4^{20}	1.4214_0^2	i	v	v	s	s	...	**B17**[2], 34
Ω f194	—,**2-acetyl-**	2-Furyl methyl ketone. $C_6H_6O_2$. *See* f193	110.11	cr (lig) λ^{al} 225.5 (3.37), 270 (4.11)	33	173 67^{10}	1.098^{20}	1.5017^{20}	i	s	s	...	...	...	**B17**[2], 314
Ω f194[1]	—,**2-(amino-methyl)-**	Furfurylamine. C_5H_7NO. *See* f193	97.12	...	...	$145–6^{761}$ 80^{84}	1.0995_4^{20}	1.4908^{20}	∞	s	s	...	...	...	**B18**[2], 416
f195	—,**2(amino-methyl)tetra-hydro-**	Tetrahydrofurfuryl amine. $C_5H_{11}NO$. *See* f220	101.15	...	...	$151–2^{735}$	0.9770_{20}^{20}	1.4551^{20}	∞	∞	s	...	...	...	**Am 67,** 693
f196	—,**2-bromo-**	*α*-Furyl bromide. C_4H_3BrO. *See* f193	146.98	λ^{al} 215.5 (3.99)	...	102^{744}	1.6500^{20}	1.4980^{20}	δ	s	s	s	s	...	**Am 52,** 2083

For explanations, symbols and abbreviations see beginning of table. For structural formulas see end of table.

No.	Name	Synonyms and Formula	Mol. wt.	Color, crystalline form, specific rotation and λ_{max} (log ε)	m.p. °C	b.p. °C	Density	n_D	Solubility w	al	eth	ace	bz	other solvents	Ref.
	Furan														
f197	**—,3-bromo-**	β-Furyl bromide. C_4H_3BrO. *See* f193	146.98	...	...	103^{760}	1.6606^{20}_{4}	1.4958^{20}	i	s	s	s	s	...	**B17**, 27
f198	**—,2(bromo-methyl)-**	Furfuryl bromide. C_5H_5BrO. *See* f193	161.00	pa ye	...	$33–4^{2}$	1.560^{20}_{20}	1.5380^{20}	...	...	s	...	...	...	**B17**², 39
Ω f199	**—,2(bromo-methyl)tetra-hydro-**	1-Bromo-2,5-epoxypentane*. C_5H_9BrO. *See* f220	165.03	...	...	$168–70^{744}$ $69–70^{22}$	1.3653^{20}_{4}	1.4850^{20}	...	s	s	...	...	...	**B17**², 21
f200	**—,2-*tert*-butyl-**	$C_8H_{12}O$. *See* f193	124.19	...	...	119–20^{760}	0.869^{20}_{4}	1.4373^{20}	i	s	s	s	...	...	**C49**, 2419
f200¹	**—,2-chloro-**	α-Furyl chloride. C_4H_3ClO. *See* f193	102.52	...	...	77.5^{744}	1.1923^{20}_{4}	1.4569^{20}	i	s	s	s	...	...	**C42**, 2284
f201	**—,3-chloro-**	β-Furyl chloride. C_4H_3ClO. *See* f193	102.52	...	...	79^{742}	1.2094^{20}_{4}	1.4601^{20}	δ	...	s	s	...	...	**C49**, 2419
f201¹	**—,2-(chloro-methyl)-**	Furfuryl chloride. C_5H_5ClO. *See* f193	116.55	...	...	49^{26} 37^{15}	1.1783^{20}_{4}	1.4941^{20}	i	s	s	s	s	os s	**B17**², 39
f202	**—,2,5-dibromo-**	$C_4H_2Br_2O$. *See* f193	225.88	pl	9–10	$164–5^{764}$ 62^{15}	2.27^{20}_{20}	1.5455^{20}	...	...	...	...	...	...	**B17**, 28
f203	**—,2,5-di-*tert*-butyl-**	$C_{12}H_{20}O$. *See* f193	180.29	...	...	210^{760} $61–2^{17}$	0.837^{20}_{4}	1.4369^{20}	i	s	s	...	...	...	**C51**, 12875
f204	**—,2,5-dichloro-**	$C_4H_2Cl_2O$. *See* f193	136.97	...	...	115	1.371^{25}	...	...	...	...	...	...	...	**C42**, 2284
f205	**—,2,2-diethyl-(tetrahydro)-**	$C_8H_{16}O$. *See* f220	128.22	...	...	146	0.8703^{20}_{4}	1.4317^{20}	i	s	s	v	v	os v	**B17**², 25
Ω f206	**—,2,5-dimethyl-**	C_6H_8O. *See* f193	96.14	λ^{al} 220 (3.90)	−62.84	$93–4^{760}$	0.8883^{20}_{4}	1.4363^{20}	i	s	s	s	s	chl s aa s	**B17**¹, 20
f207	**—,2,4-dioxo-(tetrahydro)-**	Tetronic acid. β-Ketobutyro-lactone. $C_4H_4O_3$. *See* f202	100.08	pl (al-lig) λ^{w} 233 (3.41), 310 (4.15)	141	...	...	...	v^h	v^h	δ	...	δ	chl δ lig δ	**B17**, 403
f207¹	**—,2,5-dinitro-**	$C_4H_2N_2O_5$. *See* f193	158.07	nd (w), pr (al)	102–2.5	...	...	...	δ	i	s	...	...	...	**B17**, 29
Ω f207²	**—,2,5-diphenyl-**	$C_{16}H_{12}O$. *See* f193	220.27	nd or lf (dil al) λ^{iso} 324 (4.46)	91	343–5	...	...	i	v	v	s	s	oos s	**C40**, 4705
f208	**—,2-ethoxy-**	Ethyl 2-furyl ether. $C_6H_8O_2$. *See* f193	112.14	...	...	125–6	0.9849^{23}_{4}	1.4500^{23}	...	...	...	...	...	...	**C51**, 3549
f209	**—,2-ethyl-**	C_6H_8O. *See* f193	96.14	...	...	$92–3^{768}$	0.912^{13}_{15}	1.4466^{13}	...	s	s	s	s	...	**C49**, 10918
f210	**—,2-ethyl-(tetrahydro)-**	$C_6H_{12}O$. *See* f220	100.16	...	...	109^{760}	0.8570^{19}_{4}	1.4147^{19}	...	s	s	s	s	os s	**B17**², 23
—	**—,2-(hydroxy-methyl)-**	*see* **Methanol, 2-furyl-**													
f210¹	**—,2-iodo-**	C_4H_3IO. *See* f193	193.97	...	...	$43–5^{15}$	2.024^{20}_{4}	1.5661^{20}	...	...	s	...	...	...	**C51**, 14672
f210²	**—,3-iodo-**	C_4H_3IO. *See* f193	193.97	...	...	132.2^{732} $37–8^{22}$	2.045^{20}_{4}	1.5610^{20}	i	...	s	...	...	...	**C56**, 11536
f211	**—,2-isopropoxy-**	2-Furyl isopropyl ether. $C_7H_{10}O_2$. *See* f193	126.16	...	...	135–6	0.9689^{20}_{4}	1.4419^{20}	...	...	...	...	...	...	**C51**, 3550
f212	**—,2-methoxy-**	2-Furyl methyl ether. $C_5H_6O_2$. *See* f193	98.10	λ^{al} 221 (3.82)	...	110–1	1.0646^{25}_{4}	1.4468^{25}	...	...	...	...	...	...	**C49**, 6217
Ω f213	**—,2-methyl-**	Sylvan. C_5H_6O. *See* f193	82.10	λ^{MeOH} <220	...	(i) 63^{737} (ii) 79^{42} (unst)	0.9132^{20}_{4}	1.4342^{20}	δ	s	s	...	...	...	**B17**², 39
f214	**—,3-methyl-**	C_5H_6O. *See* f193	82.10	...	...	65.5^{749}	0.923^{18}_{4}	1.4330^{19}	i	s	s	...	...	...	**B17**², 40
Ω f215	**—,2-methyl-(tetrahydro)-**	$C_5H_{10}O$. *See* f220	86.14	...	...	80^{761}	0.8552^{20}_{4}	1.40595^{21}	s	v	v	v	v	chl, os v	**B17**², 21
f216	**—,3-methyl-(tetrahydro)-**	$C_5H_{10}O$. *See* f220	86.14	...	...	86–7	0.8642^{20}_{4}	1.4122^{20}	s	v	v	v	s	os v	**B17**², 21
f217	**—,2-nitro-**	$C_4H_3NO_3$. *See* f193	113.07	yesh mcl cr (peth) λ^{w} 225 (3.53), 315 (3.91)	29	$133–5^{123}$	...	...	s	s	s	...	...	...	**B17**, 28
f218	**—,2-phenyl-**	$C_{10}H_8O$. *See* f193	144.19	...	...	$107–8^{18}$	1.083^{20}_{4}	1.5920^{20}	...	...	s	s	...	...	**B17**², 64
Ω f219	**—,2-propyl-**	$C_7H_{10}O$. *See* f193	110.16	...	...	$114–5^{750}$	0.8876^{20}_{4}	1.4549^{20}	...	s	s	s	...	...	**C33**, 1316³
f219¹	**—,2-propyl(tetra-hydro)-**	$C_7H_{14}O$. *See* f220	114.19	...	...	139.4^{758}	0.8547^{20}_{4}	1.4242^{20}	i	...	...	...	...	...	**B17**¹, 11
Ω f220	**—,tetrahydro-**	1,4-Epoxybutane*. Tetra-methylene oxide.	72.12	...	fr −108.56	67 (64.5)	0.8892^{20}_{4}	1.4050^{20}	s	v	v	v	v	os v	**B17**², 15
f220¹	**—,tetraiodo-**	C_4I_4O. *See* f193	571.66	nd	165	...	...	...	i	s	...	...	...	lig δ	**B17**², 34
f221	**—,2,3,5-trichloro-**	C_4HCl_3O. *See* f193	171.41	...	...	147	1.50^{25}	...	...	...	...	...	...	...	**C42**, 2284
—	**2-Furanacrylic acid**	*see* **Propenoic acid, 3(2-furyl)-**													
—	**2-Furancarboxal-dehyde**	*see* Furfural													
Ω f222	**2-Furancar-boxylic acid**	α-Furoic acid. Pyromucic acid.	112.09	mcl nd or lf (w) λ^{al} 214 sh (3.58), 242.5 (4.03)	133–4	230–2 $141–4^{20}$ (sub 130–40^{50})	...	...	s v^h	s	v	...	...	...	**B18**², 265
Ω f223	**—,allyl ester**	$C_8H_8O_3$. *See* f222	152.16	λ^{w} 225 (3.53), 315 (3.91)	...	$206–9^{760}$	1.118^{25}_{25}	1.4945^{20}	...	...	s	s	...	...	**C36**, 5812
f224	**—,benzyl ester**	$C_{12}H_{10}O_3$. *See* f222	202.21	ye	...	$179–81^{18}$	1.1632^{20}_{4}	1.5550^{20}	i	...	s	s	...	...	**B18**², 267

For explanations, symbols and abbreviations see beginning of table. For structural formulas see end of table.

No.	Name	Synonyms and Formula	Mol. wt.	Color, crystalline form, specific rotation and λ_{max} (log ε)	m.p. °C	b.p. °C	Density	n_D	Solubility: w	al	eth	ace	bz	other solvents	Ref.
	2-Furancarboxylic acid														
Ω f225	—,butyl ester	$C_9H_{12}O_3$. *See* f222	168.20	...	...	233 83–4[1]	1.0555^{20}_{4}	1.4740	i	s	s	...	s	peth s	**B18**[2], 266
f226	—,*sec*-butyl ester	$C_9H_{12}O_3$. *See* f222	168.20	...	...	67–9[1]	1.0465^{20}_{4}	...	i	∞	∞	...	...	...	**B18**[2], 266
Ω f227	—,chloride	$C_5H_3ClO_2$. *See* f222	130.53	...	−2 (+1)	173 66[10]	...	...	i d[h]	d	s	...	...	chl s	**B18**[2], 267
Ω f228	—,ethyl ester	$C_7H_8O_3$. *See* f222	140.15	lf or pr λ^{al} 220 sh (3.46), 251 (4.13)	34–5	196.8[760] 128[95]	1.1174^{21}_{4}	1.4797[21]	i	∞	∞	∞	s	peth ∞	**B18**[2], 266
f229	—,furfuryl ester	...	192.18	dimorphic	27.5	122[2]	1.2384^{25}_{25}	1.5280[20]	i	s	s	v	∞	chl ∞	**B18**[2], 267
f230	—,heptyl ester	$C_{12}H_{18}O_3$. *See* f222	210.28	...	...	116–7[1]	1.0005^{20}_{4}	...	i	s	s	...	...	...	**B18**[2], 267
Ω f231	—,hexyl ester	$C_{11}H_{16}O_3$. *See* f222	196.25	...	...	105–7[1]	1.0170^{20}_{4}	...	i	s	s	...	∞	MeOH s	**B18**[2], 267
f232	—,isobutyl ester	$C_9H_{12}O_3$. *See* f222	168.20	...	...	221–3[760] 97[13.5]	1.0388^{28}_{4}	1.4676[28]	i	s	s	s	v	...	**B18**[2], 266
f233	—,isopropyl ester	$C_8H_{10}O_3$. *See* f222	154.17	...	...	198–9[760]	1.0655^{24}_{4}	1.4682[24]	i	s	s	s	s	...	**B18**, 275
Ω f234	—,methyl ester	$C_6H_6O_3$. *See* f222	126.11	λ^{al} 252 (4.13)	...	181.3[760]	1.1786^{21}_{4}	1.4860[20]	i	s	s	...	s	...	**B18**[2], 266
f235	—,3-methylbutyl ester	Isoamyl α-furoate. $C_{10}H_{14}O_3$. *See* f222	182.22	...	...	282[760] (232)	1.030^{20}_{4}	1.4274[20]	i	∞	...	...	s	peth ∞ aa s	**B18**[2], 266
f236	—,nitrile	α-Furyl cyanide.	93.09	...	...	146[738]	1.0822^{20}_{4}	1.4798[20]	...	s	s	...	...	...	**B18**[2], 268
Ω f237	—,octyl ester	$C_{13}H_{20}O_3$. *See* f222	224.30	...	...	126–7[1]	0.9885^{20}_{4}	...	i	s	...	...	...	...	**B18**[2], 267
Ω f238	—,pentyl ester	Amyl furoate. $C_{10}H_{14}O_3$. *See* f222	182.22	...	...	95–7[1]	1.0335^{20}_{4}	...	i	∞	...	...	...	...	**B18**[2], 266
f239	—,piperazinium salt	$2(C_5H_4O_3).C_4H_{10}N_2$. *See* f222	310.31	...	234–6	...	...	...	s	s	δ	...	...	...	**Am 70**, 2758
Ω f240	—,propyl ester	$C_8H_{10}O_3$. *See* f222	154.17	...	...	210.9	1.0745^{26}_{4}	1.4737[26]	i	s	s	s	s	peth s	**B18**[2], 266
f241	**3-Furancarboxylic acid**	β-Furoic acid	112.09	nd (w) λ^{al} 232 (3.36)	122–3	105–10[12] sub	...	...	δ s[h]	s	v	...	...	AcOEt s	**B18**[1], 439
f242	—,methyl ester	$C_6H_6O_3$. *See* f241	126.11	λ^{al} 238 (3.40)	...	160[760] 79[42]	1.1744^{15}_{15}	1.4676[20]	i	s	...	s	...	...	**B18**[1], 439
f244	—,**4-bromo-**	$C_5H_3BrO_3$. *See* f222	190.99	nd (w)	129	...	...	...	s	s	s	...	s	chl s lig δ CS_2 δ	**B18**[2], 268
Ω f245	—,**5-bromo-**	$C_5H_3BrO_3$. *See* f222	190.99	lf (w) λ^{al} 251 (4.08)	190–1	...	...	...	δ	s	v	...	δ	chl δ, lig i peth i	**B18**, 284
f246	—,—,ethyl ester	$C_7H_7BrO_3$. *See* f222	219.04	pr	17	235[767] 134–6[34]	1.528[20]	...	i	s	s	...	...	...	**B18**, 284
f247	—,**3-chloro-**	$C_5H_3ClO_3$. *See* f222	146.53	pl or pr (w)	149.5	...	...	...	δ	s	s	...	s[h]	chl s[h]	**B18**, 282
f248	—,**4-chloro-**	$C_5H_3ClO_3$. *See* f222	146.53	pl or pr (w)	149 (146)	...	...	...	δ	s	s	...	s[h]	chl s[h]	**C31**, 2207
f249	—,**5-chloro-**	$C_5H_3ClO_3$. *See* f222	146.53	lf (w)	179–80	...	...	...	δ s[h]	s	s	...	s[h]	...	**B18**, 282
f250	—,—,methyl ester	$C_6H_5ClO_3$. *See* f222	160.56	λ^{al} 260.5 (4.23)	40–1	...	...	...	i	...	s	...	...	...	**C52**, 1990
f251	—,**5(cyclohexyloxy)**	$C_{11}H_{14}O_4$. *See* f222	210.23	...	138–9	...	...	...	δ d[h]	s	...	...	...	...	**C51**, 3549
f252	**3-Furancarboxylic acid, 2,5-dimethyl-**	Pyrotritaric acid. $C_7H_8O_3$. *See* f241	140.15	nd (w)	135	sub	...	...	s[h]	s	v	...	...	...	**B18**, 297
f253	**2-Furancarboxylic acid, 5-ethoxy-**	$C_7H_8O_4$. *See* f222	156.15	...	140–1	...	...	...	δ d[h]	s	...	...	...	...	**C51**, 3549
f254	—,**5-iodo-**	$C_5H_3IO_3$. *See* f222	237.98	...	197d	...	...	...	...	...	...	...	...	...	**C53**, 12267
f255	—,**5-methoxy-**	$C_6H_6O_4$. *See* f222	142.11	...	136–8d	...	...	...	δ d[h]	s	...	...	...	...	**C51**, 3549
f256	—,**3-methyl-**	$C_6H_6O_3$. *See* f222	126.11	nd (w)	134	sub	...	...	s[h]	...	...	...	...	...	**B18**[1], 439
f257	—,—,ethyl ester	$C_8H_{10}O_3$. *See* f222	154.17	pl	47–8	205[760]	...	...	...	...	...	...	...	...	**B18**[2], 271
f258	—,—,methyl ester	$C_7H_8O_3$. *See* f222	140.15	pl (al)	36–8	72–6[8]	...	...	...	...	...	...	...	...	**B18**, 293
f259	—,**4-methyl-**	$C_6H_6O_3$. *See* f222	126.11	nd (bz-peth)	131–2	...	...	...	...	...	...	...	...	...	**C26**, 1603
Ω f260	—,**5-methyl-**	$C_6H_6O_3$. *See* f222	126.11	pl or nd (w) λ^{w} 257 (4.10)	109–10	105[1]	...	...	δ v[h]	s	s	...	δ	chl s	**B18**[2], 272
Ω f261	—,—,methyl ester	$C_7H_8O_3$. *See* f222	140.14	λ^{w} 272 (4.06)	...	205[760] 98[15]	...	...	...	...	...	...	...	...	**B18**[2], 272
f262	**3-Furancarboxylic acid, 2-methyl-**	$C_6H_6O_3$. *See* f241	126.11	cr (w)	102–3	...	...	...	...	s	s	...	...	peth s aa s	**B18**[2], 271
f263	—,—,ethyl ester	$C_8H_{10}O_3$. *See* f241	154.17	...	...	85–7[20]	1.0102^{25}_{4}	1.4620[25]	i	...	s	...	...	...	**B18**[1], 439
f264	—,**4-methyl-**	$C_6H_6O_3$. *See* f241	126.11	nd (bz-peth)	138–9	...	...	...	s	...	...	...	...	...	**C26**, 1603
f265	—,**5-methyl-**	$C_6H_6O_3$. *See* f241	126.11	(w)	119	sub	...	...	s	...	s	...	...	...	**C33**, 4983
Ω f266	**2-Furancarboxylic acid, 5-nitro-**	$C_5H_3NO_5$. *See* f222	157.08	pa ye pl (w) λ^{al} 212 (3.98), 314 (4.06)	184	sub	...	...	s[h]	s	s	...	δ	chl i	**B18**, 287
f267	—,**5-octyloxy-**	$C_{13}H_{20}O_4$. *See* f222	240.30	...	125–6	...	...	...	δ d[h]	s	...	...	...	...	**C51**, 3549
—	**3-Furancarboxylic acid, 5-oxo(tetrahydro)-**	*see* **Paraconic acid**													

For explanations, symbols and abbreviations see beginning of table. For structural formulas see end of table.

No.	Name	Synonyms and Formula	Mol. wt.	Color, crystalline form, specific rotation and λ_{max} (log ε)	m.p. °C	b.p. °C	Density	n_D	Solubility w	al	eth	ace	bz	other solvents	Ref.
	3-Furancarboxylic acid														
f268	**2-Furancarboxylic acid, 5-phenoxy-**	$C_{11}H_8O_4$. *See* f222	204.19		122–3				δ d^h	s	...	...	...		**C51,** 3549
f269	—**,tetrahydro-**		116.13		21	145^{25}	1.1933^{20}_{20}	1.4612^{20}	s	...	...	...	...		**B18**², 262
f270	**2,3-Furandicarboxylic acid**		156.10	pr (aa or sub) λ^{al} 263 (4.14)	226	sub			v	v	δ	...	...	AcOEt v aa δ v^h	**B18**², 287
f271	—,dimethyl ester	$C_8H_8O_5$. *See* f270	184.16	(MeOH) λ^{al} 254 (3.96)	39				i	s	v	...	...		**B18**², 288
f272	**2,4-Furandicarboxylic acid**		156.10	lf (w + 1) λ^{w} 240 (4.02)	266 (anh)	sub			δ v^h	v	...	v	...	chl δ CS$_2$ δ aa δ	**B18**, 327
f273	—,dimethyl ester	$C_8H_8O_5$. *See* f272	184.16	pr (MeOH) λ^{al} 242.5 (4.09)	109–10				...	...	...	...	...		**B18**, 327
f273¹	**2,5-Furandicarboxylic acid**	Dehydromucic acid.	156.10	nd (w), lf (al) λ^{al} 263 (4.20)	> 320	sub			δ	δ	...	...	...		**B18**², 288
f274	—,dichloride	Dehydromucyl chloride. $C_6H_2Cl_2O_3$. *See* f273¹	192.99	nd (sub)	80	245^{760}			v	v	v	...	...	chl v	**B18**, 330
f275	—,dimethyl ester	$C_8H_8O_5$. *See* f273¹	184.16	nd (w), cr (MeOH) λ^{al} 263 (4.19)	112	$154–6^{15}$			i	s	s	...	...	chl s	**B18**, 329
f275¹	**3,4-Furandicarboxylic acid**		156.10	λ^{al} 240 (3.43)	217–8 (212)				...	...	...	...	...		**C54,** 9876
Ω f276	**Furazan, 3,4-dimethyl-**	3,4-Dimethyl-1,2,5-oxadiazole	98.11		−7	156^{744}	1.0528^{14}_{4}	1.4237^{20}	δ	v	v	...	...		**B27**², 628
Ω f277	**Furfural**	α-Furaldehyde. Fural. 2-Furancarboxaldehyde.	96.09	λ^{sulf} 280 (3.07) (initial)	−38.7	161.7^{760} 90^{65}	1.1594^{20}_{4}	1.5261^{20}	s v^h	v	∞	v	s	chl s	**B17**², 305
f278	—,diacetate	Furfurylidiene diacetate.	198.17	nd or pl (eth-peth)	52–3	220^{760} $143–4^{20}$			δ	s	v	...	v	peth δ	**B17**², 309
f279	—,diethyl acetal	$C_9H_{14}O_3$. *See* f277	170.21			$191–2^{760}$ $62–4^{3}$	0.9994^{20}_{20}	1.4451^{20}	i	s	...	...	...		**B17**², 309
f280	—,oxime (*anti*)	α-Furfuraldoxime. $C_5H_5NO_2$. *See* f277	111.10	nd (lig) λ^{al} 270 (4.2)	75–6				s	v	v	...	v	lig δ s^h	**B17**², 311
f281	—,—(*syn*)	β-Furfuraldoxime. $C_5H_5NO_2$. *See* f277	111.10	nd (lig) λ^{al} 265 (4.23)	91–2	201–8 d 98^{9}			δ	v	v	...	v	CS$_2$ v aa v lig s^h	**B17**², 311
Ω f282	—,phenylhydrazone	$C_{11}H_{10}N_2O$. *See* f277	186.22	ye lf (al) λ 302 (4.0), 345 (4.4)	97–8				i	v	v	...	...	lig δ	**B17**², 312
f283	—**,5-bromo-**	$C_5H_3BrO_2$. *See* f277	174.99	cr (50 % al)	82	112^{16}			δ	s	s	...	...		**C35,** 2508
f284	—**,5-chloro-**	$C_5H_3ClO_2$. *See* f277	130.53		31.5–33	70^{1}			...	...	...	...	...		**C40,** 1825
Ω f285	—**,5-hydroxymethyl-**	$C_6H_6O_3$. *See* f277	126.11	nd (eth-peth) λ^{w} 284 (4.26)	35–5.5	$114–6^{0.5}$ $72^{0.002}$	1.2062^{25}_{4}	1.5627^{18}	s	s	δ	...	s	chl, AcOEt s CCl$_4$ δ	**B18**², 6
f286	—**,5-methyl-**	$C_6H_6O_2$. *See* f277	110.11	λ 225 (3.45), 283.5 (4.23)		187^{760} $79–81^{12}$	1.1072^{18}_{4}	1.5264^{20}	s	v	∞	...	...		**B17**², 315
f287	—**,5-nitro-**	$C_5H_3NO_4$. *See* f277	141.08	pa ye (peth) λ^{w} 297 (4.06)	35–6	$128–32^{10}$			δ	...	...	...	...	peth s	**Am 52,** 2550
f288	—,—,semicarbazone	Furacin. Nitrofurazone. $C_6H_6N_4O_4$. *See* f277	198.14	pa ye nd (w) darkens in light	237 d				i	δ	i	...	...	alk s	**C46,** 4029
f289	—**,tetrahydro-**		100.13			$142–3^{779}$ $45–7^{29}$	1.0727^{20}_{4}	1.4366^{20}	s	...	s	...	...		**C38,** 2336
Ω f290	**Furfurin**	Furin. 2,4,5-Tris(2-furyl) 2-imidazoline.	268.28	lt br nd or rh pr (w or eth)	116–7				δ^h	v	v	...	...		**B27**², 918
—	**Furfuryl alcohol**	*see* **Methanol, 2-furyl-**													
—	**Furfurylamine**	*see* **Furan, 2(aminomethyl)-**													
—	**2-Furfurylfuran**	*see* **Methane, di-2-furyl-**													
—	**Furfuryl mercaptan**	*see* **Methanethiol, 2-furyl-**													
—	**Furil**	*see* **1,2-Ethanedione, 1,2-di(2-furyl)-**													
—	**Furoic acid**	*see* **Furancarboxylic acid**													
Ω f291	**Furoin**	Furoylfurylcarbinol.	192.18	nd (al)	138–9 (cor)				δ^h	δ s^h	s	...	...	MeOH s	**B19**², 224

For explanations, symbols and abbreviations see beginning of table. For structural formulas see end of table.

Gaidic acid

No.	Name	Synonyms and Formula	Mol. wt.	Color, crystalline form, specific rotation and λ_{max} (log ε)	m.p. °C	b.p. °C	Density	n_D	Solubility w	al	eth	ace	bz	other solvents	Ref.
—	**Gaidic acid**	*see* **2-Hexadecenoic acid**													
Ω g1	**Galactitol(*D*)**	*D*-Dulcitol	182.18	mcl pr (w)	189	275–8[1] sub	1.466	...	s v^h	δ	i	...	i	Py δ	**B1**[2], 612
Ω g2	**Galactonic acid, γ-lactone (*D*)**	...	178.14	nd (w + 1), nd (al or AcOEt) $[\alpha]^{20}_D$ −65.5 (w)	112 (anh) 66 (hyd)	...	...	...	s	...	...	...	...	AcOEt v	**B18**[1], 408
Ω g3	**Galactose(*D*)**	...	180.16	pl or pr (al), pr or nd (w + 1) [α] +83.3 (w) $\lambda^{w, pH=11}$ 290 (2.36)	170 (anh) 118–20 (hyd)	...	...	...	v^h	δ	i	...	i	Py s	**B1,** 909 **B31,** 295
g4	**—,3,6-anhydro-(*D*)**	...	162.14	lf (PrOH) $[\alpha]_D$ +24(w)	123–5	...	...	...	s	...	...	...	...	...	**C41,** 3062
g5	**—,2,3,4,6-tetra-O-methyl-(*D*, α)**	...	236.27	$[\alpha]^{20}_D$ +150 → +114 (w) (mut)	71–3	172^{12}	...	...	s	s	s	...	...	...	**B31,** 304
Ω g6	**Galacturonic acid(*D*)**	...	194.14	nd (w + 1) α: $[\alpha]^{20}_D$ +98 → +53.4 (w, c = 10) (mut) β: $[\alpha]^{20}_D$ +27 (w) $\lambda^{79\%\,sulf}$ 277	α: 156–9 d β: 160 d	...	...	...	s	s^h	i	...	...	...	**B3**[1], 306
g7	**Galegine**	4-Guanidino-2-methyl-2-butene. γ,γ-Dimethylallylguanidine. $(CH_3)_2C{:}CHCH_2N{:}C(NH_2)_2$	127.19	hyg	60–5	d	...	...	v	v	δ	...	...	chl, peth i	**B4**[2], 672
g8	**Galipine**	Galipoline methyl ether. $C_{20}H_{21}NO_3$.	323.40	pr (al, eth), nd (peth)	115.5	...	...	...	δ	v	v	v	v	chl v peth δ	**B21**[2], 171
—	**Gallamide**	*see* **Benzoic acid, 3,4,5-trihydroxy-, amide**													
—	**Gallanilide**	*see* **Benzoic acid, 3,4,5-trihydroxy-, amide, *N*-phenyl-**													
g9	**Gallein**	4,5-Dihydroxyfluorescein. Pyrogallophthalein. $C_{20}H_{12}O_7$.	364.32	br-red pw (+1.5w), red (anh)	>300 (darkens 180)	...	...	...	i	v	δ	s	i	alk s aa δ chl i	**B19**[2], 282
—	**Gallic acid**	*see* **Benzoic acid, 3,4,5-trihydroxy-***													
g10	**Gallin**	4,5-Dihydroxyfluorescin. $C_{20}H_{14}O_7$.	365.33	nd (eth), turns red in air	...	...	...	...	δ	v	δ	v	...	aa v	**B18,** 368
—	**Gammexane**	*see* **Cyclohexane, 1,2,3,4,5,6-hexachloro-***													
—	**Gelseminic acid**	*see* **Coumarin, 7-hydroxy-6-methoxy-**													
Ω g12	**Gelsemine(*d*)**	Gelseminin. $C_{20}H_{22}N_2O_2$.	322.41	cr (ace), $[\alpha]^{20}_D$ +15.9 (chl) λ^{al} 252 (3.87), 280 (3.15)	178	...	...	...	i	s	s	s^h	s	chl s	**B27**[2], 720
Ω g13	—,hydrochloride(*d*)	$C_{20}H_{22}N_2O_2$. HCl. *See* g12	358.87	pr (w) $[\alpha]^{20}_D$ +2.6 (w) λ^{al} 255 (3.8), 288 (3.2)	326	...	...	...	s	δ	...	...	...	...	**B27,** 720
—	**Genistein**	*see* **Isoflavone, 4′,5,7-trihydroxy-**													
—	**Gentianin**	*see* **Xanthone, 1,7-dihydroxy-3-methoxy-**													
—	**Gentisic acid**	*see* **Benzoic acid, 2,5-dihydroxy-**													
—	**Gentisic aldehyde**	*see* **Benzaldehyde, 2,5-dihydroxy-**													
—	**Gentisin**	*see* **Xanthone, 1,7-dihydroxy-3-methoxy-**													
—	**Gentisyl alcohol**	*see* **Toluene, α,2,5-trihydroxy-**													
—	**Geranial**	*see* **Citral a**													
Ω g14	**Geraniol**	3,7-Dimethyl-2,6-octadien-1-ol*.	154.26	λ^{al} 215 (3.25), 238 (2.30), 325 sh (−1.99)	< −15	230^{760} 121^{18}	0.8894^{20}	1.4766^{20}	i	s	s	s	...	...	**B1**[2], 508
Ω g15	—,formate	$C_{11}H_{18}O_2$. *See* g14	182.27	...	...	229d $113\text{–}4^{15}$	0.9086^{25}_4	1.4659^{20}	i	v	s	s	...	...	**B2**[2], 33
g16	**—,tetrahydro-(*dl*)**	3,7-Dimethyl-1-octanol*. $(CH_3)_2CH(CH_2)_3CH(CH_3)CH_2CH_2OH$	158.29	λ^{al} 237 (1.1)	...	212–3 106^{12}	0.8308^{20}_4	1.4367^{20}_4	i	s	s	s	s	...	**B1**[2], 460
g17	**Geranyl bromide**	8-Bromo-2,6-dimethyl-2,6-octadiene*. $BrCH_2CH{:}C(CH_3)CH_2CH_2CH{:}C(CH_3)_2$	217.16	...	...	$101\text{–}2^{12}$ $47\text{–}8^{0.005}$	1.0940^{22}_4	1.5027^{20}	d^h	s	s	...	...	...	**B1**[2], 240
g18	**Germanidine**	$C_{37}H_{57}NO_{10}$	675.87	nd, $[\alpha]^{24}_D$ −30 ±2 (al)	221–2 (vac)	...	...	...	i	...	...	...	...	...	**Am 75,** 4925

For explanations, symbols and abbreviations see beginning of table. For structural formulas see end of table.

No.	Name	Synonyms and Formula	Mol. wt.	Color, crystalline form, specific rotation and λ_{max} (log ε)	m.p. °C	b.p. °C	Density	n_D	Solubility w	al	eth	ace	bz	other solvents	Ref.
	Germanitrine														
g19	**Germanitrine**	Germine-3-angelate-7-acetate-15-[(−)2-methylbutyrate]. $C_{39}H_{59}NO_{11}$. *See* g22	717.91	nd (aq ace) $[\alpha]_D^{24}$ −61 (Py, c = 1)	228–9 (vac)	...	...	...	i	...	...	...	...	...	**Am 75,** 4925
g20	**Germerine**	Germine-3[(+)-2-methylbutyrate]-15-[(+)2-hydroxy-2-methylbutyrate]. $C_{37}H_{59}NO_{11}$. *See* g22	693.88	cr (ace), lf (bz, +1w) $[\alpha]_D^{25}$ +15.7 (chl, c = 1.02)	193–5d (cor) (202–5d)	...	...	...	i	s	δ	s	s	MeOH, chl v alk, ac s	**Am 72,** 4625
g21	**Germidine**	Germine-3-acetate-15-[(−)-2-methylbutyrate]. $C_{34}H_{53}NO_{10}$. *See* g22	635.81	pl (aq MeOH) $[\alpha]_D^{25}$ +13 (chl), −11(Py)	(i) 242–4 (ii) 202–3	...	...	...	...	s^h	...	...	s	MeOH s^h ac s	**Am 72,** 4625
g22	**Germine**	$C_{27}H_{43}NO_8$	509.65	pr or cr (MeOH +2) $[\alpha]_D^{25}$ +5 (95 % al) λ^{al} 276 (0.6)	220	...	...	...	...	...	...	...	s	MeOH s^h alk, ac s	**Am 71,** 3260
g23	**Germinitrine**	$C_{39}H_{57}NO_{11}$	715.89	pr (dil ace) $[\alpha]_D^{24}$ −36 (Py, c = 1.12)	175 (vac)	...	...	...	i	...	...	...	...	...	**Am 75,** 4925
g24	**Germitrine**	Germine 3[(−)-2-methylbutyrate]-7-acetate-15-[(+)-2-hydroxy-2-methylbutyrate]. $C_{42}H_{67}NO_{13}$. *See* g22	735.92	cr (dil al) $[\alpha]_D^{25}$ +11 (chl)	197–9	...	...	...	i	d^h	...	s^h	s	chl, ac s	**Am 72,** 4625
g25	**Gitogenin**	Digine. $C_{27}H_{44}O_4$	432.65	lf (bz), nd (eth) $[\alpha]_D^{20}$ −75 (chl, c = 1) λ^{sulf} 399 (3.09)	271–2	...	...	...	i	s^h	δ	...	...	chl s	**Am 77,** 5661
g26	**Gitoxigenin**	Bigitaligenin. Hydroxydigitoxin. $C_{23}H_{34}O_5$.	390.53	pr (AcOEt), pr (+w, dil al) $[\alpha]_{546}^{20}$ +38.5 (MeOH, c = 0.7) λ^{al} 220 (4.12), 278 sh (2.2)	222 (cor) (234 anh)	...	...	...	i	...	δ	...	...	chl s	**E14,** 242
g27	**Gitoxin**	Anhydrogitalin. Bigitalin. $C_{41}H_{64}O_{14}$.	780.96	pr (chl-MeOH) $[\alpha]_D^{24}$ +5 (Py, c = 1)	285d	...	...	...	δ	δ	...	i	...	chl δ	**E14,** 243
—	**Glucitol**	*see* **Sorbitol**													
g28	**Glucoascorbic acid**(*D*)	...	206.15	rods (+1w, ace-MeOH-peth) $[\alpha]_D^{20}$ −22 (MeOH, c = 1), −80 (NaOH, c = 0.75)	191 (anh) 140 (hyd)	...	...	...	s	s	...	...	...	MeOH s	**J1937,** 549
g29	**Gluco-α-heptose**(*D*)	...	210.19	rh pl (w) $[\alpha]_D^{20}$ −19.7 (w) (mut)	193 (210–5)	...	...	...	v	δ	...	...	...	...	**B31,** 359
g30	**Gluco-methylose**(*D*)	Epifucose. *D*-Isorhamnose. *D*-Isorhodeose. Quinovose.	164.16	cr (AcOEt) $[\alpha]_D^{20}$ +73 → +29.7 (w) (mut)	139–40	...	...	...	v	v	i	i	...	AcOEt δ^h	**B31,** 63
g31	**Gluconal**(*d*)	...	146.14	hyg nd $[\alpha]_D^{22}$ −7 (w)	60	...	...	...	δ	δ	i	...	...	MeOH δ	**B31,** 116 **B17²,** 214
Ω g32	**Gluconic acid**(*D*)	Dextronic acid.	196.16	nd (al-eth) $[\alpha]_D^{25}$ −3.49 → +12.95 (w) (mut)	131	...	...	...	s d^h	δ	i	...	i	...	**B3²,** 352
g33	—,γ-lactone	...	178.14	nd (al) $[\alpha]_D$ +67.5 → +6.2 (w) (mut)	134–6	...	...	...	d	v^h	...	...	...	...	**B18,** 203
Ω g34	—,δ-lactone	...	178.14	nd (al) $[\alpha]_D^{25}$ +63.5 → +6.2 (w) (mut)	153	...	1.610⁻⁵	...	d^h v	δ	i	i	...	...	**B18²,** 190
g35	—,nitrile(*d*)	*d*-Gluconназвания	177.16	(i) cr (al, aa) $[\alpha]_D^{24}$ +10.0 (w, c = 1.8) (ii) pl (al) $[\alpha]_D^{21}$ + 8.8 (w) λ^{w} 278 (0.85)	(i) 146–8 (ii) 120.5	...	...	...	v	δ	...	δ	...	MeOH, Py s	**B3²,** 351
g36	—,phenyl-hydrazide(*D*)	...	286.29	pr (w) [α] +12 (+18) (w)	204–5	...	...	...	s	δ^h	i	...	...	...	**B15²,** 122
g37	—,**5-oxo-**(*D*)	...	194.14	cr or syr $[\alpha]_D$ −14.5 (w)	125–6	...	...	...	s	s	δ	...	...	...	**B3,** 883

For explanations, symbols and abbreviations see beginning of table. For structural formulas see end of table.

No.	Name	Synonyms and Formula	Mol. wt.	Color, crystalline form, specific rotation and λ_{max} (log ε)	m.p. °C	b.p. °C	Density	n_D	Solubility w	al	eth	ace	bz	other solvents	Ref.
	Glucosamine														
—	**Glucosamine**	*see* **Glucose, amino-**													
Ω g38	**Glucose(*D*)** (equilibrium mixture)	Dextrose	180.16	$[\alpha]^{20}_D$ +52.7 (w) λ^w 282 (1.2)	146 (150)	...	...	...	s	s^h	...	δ	...	Py, MeOH s^h	**B1**, 879
g39	—(*D*, α)	...	180.16	rods, cubes, orh, nd (al) $[\alpha]^{20}_D$ +112.2 → +52.7 (w, c = 4) (mut)	146d	...	1.5620^{18}_4	...	v	δ s^h	...	i	...	Py s^h MeOH δ AcOEt i	**B31**, 83
g40	—,mono-hydrate(*D*, α)	$C_6H_{12}O_6.H_2O$. *See* g39	198.18	lf or pl or orh (w) $[\alpha]^{20}_D$ +102 → +47.9 (mut)	86	...	1.54^{25}_4	...	v	δ	i	...	...	...	**B31**, 86
g41	—(*D*, β)	...	180.16	nd (al), (w + 1) $[\alpha]^{20}_D$ +17.5 → +52.7 (w, c = 4) (mut)	150	...	1.5620^{18}_4	...	v	δ s^h	i	...	...	Py s^h MeOH δ	**B31**, 87
g43	—,penta-acetate(*D*, α)	$C_{16}H_{22}O_{11}$. *See* g39	390.35	pl or nd (al) $[\alpha]^{20}_D$ +100.9 (al, c = 0.5) λ^{chl} 292 (1.51)	112–3	sub	...	...	δ	δ	s	...	...	chl, aa s CS_2 δ lig δ	**B31**, 119
Ω g44	—,—(*D*, β)	$C_{16}H_{22}O_{11}$. *See* g41	390.35	nd (al) $[\alpha]_D$ +3.9 (chl, c = 6)	134	sub vac	...	...	i	δ	δ	...	s	chl ∞ peth δ	**B31**, 120
g45	—,pentamethyl-ether(*D*)	$C_{11}H_{22}O_6$. *See* g38	250.30	$[\alpha]^{20}_D$ +147.4 (w, p = 10)	...	$180^{0.4}$	1.0944^{20}_4	1.4466^{20}	s	s	s	s	...	...	**B31**, 180
g46	—,phenyl-hydrazone(*D*, α)	...	270.29	pl (al) $[\alpha]^{25}_D$ −8.7 → −52.5 (w, c = 2) (mut) $\lambda^{50\%\,al}$ 237 (3.95), 278 (3.4)	160	...	...	...	s	δ^h	δ^h	...	δ	...	**B31**, 173
g47	—,phenyl-hydrazone(*D*, β)	$C_{12}H_{18}N_2O_5$. *See* g46	270.29	pr, nd (al) $[\alpha]^{19}_D$ −4.5 → −53.7 (w-Py) (mut) $\lambda^{50\%\,al}$ 235 (3.7), 277 (4.14)	140–1	...	...	...	s	δ	i	...	δ	...	**B31**, 173
g48	—,phenyl-osazone(*D*)	*D*-Fructosazone. *D*-Glucosazone. *D*-Mannosazone.	358.40	ye nd (dil al) $[\alpha]_D$ −41 (MeOH), $[\alpha]^{20}_D$ −0.62 → −0.35 (al-Py, c = 2) λ 256 (4.30), 310 (4.02), 397 (4.31)	210	d213	...	...	i	s^h	...	...	...	Py s^h lig, SO_2 i	**B31**, 350
g49	—,—(*DL*)	$C_{18}H_{22}N_4O_4$. *See* g48	358.40	ye nd (al)	...	...	...	...	i	δ	i	...	i	...	**B31**, 355
g50	—,2-amino-(*D*, β)	Chitosamine. *D*-Glucosamine.	179.17	nd (al, MeOH) $[\alpha]_D$ +28 → +47.5 (w, c = 0.4) (mut) λ 530	110d	...	...	...	v	δ	i	...	i	MeOH δ, s^h chl i	**B31**, 167
Ω g51	—,—,hydro-chloride(*D*, β)	$C_6H_{13}NO_5$.HCl. *See* g50	215.64	mcl (w or dil al) $[\alpha]^{20}_D$ +25 → +72.6 (w) (mut)	...	...	...	...	v	s^h	...	...	...	...	**B31**, 169
—	—,6-benzoyl-(*D*)	*see* **Vacciniin**													
g52	—,2(methyl-amino)-	*N*-Methyl-2-glucosamine	193.20	gummy $[\alpha]^{25}_4$ −65 (MeOH, c = 1) (mut)	130±2d	...	...	...	s	...	...	...	...	...	**B31**, 186
g53	—,thio-	Glucothiose	196.22	hyg pw (w + 1) $[\alpha]^{30}_D$ +48.7 (w, c = 1.4) (mut)	70	...	...	...	s	...	...	...	...	...	**B31**, 474
g54	**Glucoside, α-methyl-(*D*)**	Methyl-α-*D*-glucopyranoside	194.18	rh nd (al) $[\alpha]^{20}_D$ +158.9 (w) λ^w 280 (−1.4)	168 (166)	$200^{0.2}$	1.46^{30}_4	...	v	δ	...	...	...	MeOH s	**B31**, 179
g55	—,β-methyl-(*D*)	Methyl-β-*D*-glucopyranoside.	194.19	tetr pr (al) $[\alpha]^{20}_D$ −34.2 (w, p = 10)	115–6 (105)	...	...	...	v	δ	i	...	...	MeOH s	**B31**, 182
—	**Glucothiose**	*see* **Glucose, thio-**													
—	**Glucurone**	*see* **Glucuronic acid, γ-lactone.**													

For explanations, symbols and abbreviations see beginning of table. For structural formulas see end of table.

No.	Name	Synonyms and Formula	Mol. wt.	Color, crystalline form, specific rotation and λ_{max} (log ε)	m.p. °C	b.p. °C	Density	n_D	Solubility w	al	eth	ace	bz	other solvents	Ref.
	Glucuronic acid (*D*)														
Ω g56	**Glucuronic acid(*D*)**	....	194.14	nd (al, AcOEt) $[\alpha]^{20}_D$ +11.7→ +36.3 (w, c = 6) (mut)	165	....	....	....	s	s	...	...	...	....	**B31**, 261
g57	—,γ-lactone	Glucurone...............	176.13	mcl pl (w), cr (al) $[\alpha]^{25}_D$ +19.8 (w, c = 5.2)	177–8	....	1.76^{30}_4	....	s	δ	...	...	i	MeOH, aa δ	**B31**, 264
Ω g58	**Glutamic acid(*D*, −)**	α-Aminoglutaric acid. 2-Aminopentanedioic acid.* $HO_2CCH_2CH_2CH(NH_2)CO_2H$	147.13	lf (w), $[\alpha]^{25}_D$ −31.7 (1.7N HCl)	213d	....	1.538^{20}_4	....	δ s^h	i	i	i	i	MeOH, aa i	**B4**[2], 910
Ω g59	—(*DL*)...........	$HO_2CCH_2CH_2CH(NH_2)CO_2H$	147.13	rh (al, w)	199d (225–7)	....	1.4601^{20}_4	....	δ s^h	i	δ	...	...	CS_2, lig i	**B4**[2], 911
Ω g60	—(*L*, +).........	$HO_2CCH_2CH_2CH(NH_2)CO_2H$	147.13	orh (dil al) $[\alpha]^{22}_D$ +31.4 (6N HCl, c = 1) $\lambda^{w,pH=0}$ 206 (2.0)	224–5d (247–9d)	sub 175^{10}	1.538^{20}_4 (vac)	....	δ s^h	i	δ	i	i	MeOH, aa i	**B4**[2], 902
g61	—,hydrochloride(*L*)	Acidulin. $HO_2CCH_2CH_2CH(NH_2)CO_2H.HCl$	183.59	rh pl (w) $[\alpha]^{19}_D$ +31.1 (dil HCl)	214d	....	....	....	s	s	...	...	...	....	**B4**[3], 1537
Ω g62	—,*N*-**acetyl-(*L*)**....	$HO_2CCH_2CH_2CH(NHCOCH_3)CO_2H$	189.17	pr (w), $[\alpha]_D$ −15.3 (w, c = 2) $\lambda^{w,pH=1}$ 215 sh (2.7)	199	....	....	....	s^h	s^h	...	...	...	....	**B4**[2], 911
g63	—,**3-hydroxy-(*D*)** ..	$HO_2CCH_2CH(OH)CH(NH_2)CO_2H$	163.13	hyg pr (w)	135 (sinters 100)	d	....	....	v	i	i	...	i	aa v	**B4**[2], 946
Ω g64	—,—(*DL*)........	$HO_2CCH_2CH(OH)CH(NH_2)CO_2H$	163.13	rh pr or nd (w)	198d	d	....	....	v^h	i	i	...	i	os i	**B4**[1], 550
—	—,**pteroyl-**.........	*see* **Folic acid**													
Ω g65	**Glutamine(*L*, +)** ..	α-Aminoglutaramic acid. $HO_2CCH(NH_2)CH_2CH_2CONH_2$	146.15	nd (w or dil al) $[\alpha]^{20}_D$ +32.5 (5% HCl) $[\alpha]^{25}_D$ +6.5 (w, c = 2) λ^w 220 sh (2.0)	185–6d (205)	....	....	....	s	i	i	...	i	MeOH i	**B4**[3], 1540
—	**α-Glutamylcysteinylglycine**	*see* **Glutathione**													
—	**Glutaraldehyde**....	*see* **Pentanedial**													
—	**Glutaric acid**.......	*see* **Pentanedioic acid***													
—	**Glutaronitrile**.....	*see* **Pentanedioic acid, dinitrile**													
—	**Glutaryl chloride**...	*see* **Pentanedioic acid, dichloride**													
Ω g66	**Glutathione**.......	α-Glutamylcysteinylglycine. $HO_2CCH(NH_2)CH_2CH_2CONHCH(CH_2SH)CONHCH_2CO_2H$	307.33	rh (w), cr (50% al) $[\alpha]^{27}_D$ −21.3 (w, c = 2) $\lambda^{w,pH=9}$ 236 (3.5)	195	....	....	....	v	i	i	...	...	DMF s	**B4**[2], 931
g67	**Glyceraldehyde(*D*)**.	*D*-2,3-Dihydroxypropanal*. $HOCH_2CH(OH)CHO$	90.08	syr $[\alpha]_D$ +14(w) λ^w 294 (0.1)	....	....	....	....	s	...	...	...	i	....	**B1**[1], 427
Ω g68	—(*DL*)...........	$HOCH_2CH(OH)CHO$	90.08	nd or pr (40% MeOH, al-eth) λ^w 294 (0.1)	145 (142)	140–$50^{0.8}$	1.455^{18}_{18}	....	s	δ	δ	...	i	peth, lig i	**B1**[2], 888
g68[1]	—(*L*)............	$HOCH_2CH(OH)CHO$......	90.08	$[\alpha]$ −13.8 (w, c = 11) λ^w 294 (0.1)	....	....	....	....	s	δ	δ	...	i	....	**B1**[1], 427
g69	—,diethyl acetal (*D*)	3,3-Diethoxy-1,2-propanediol*. $HOCH_2CH(OH)CH(OC_2H_5)_2$	164.20	$[\alpha]^{15}_D$ +21.2 (w, c = 18)	....	127–9^{17}	....	....	∞	∞	∞	v	...	....	**B1**[2], 889
—	**Glyceric acid**......	*see* **Propanoic acid, 2,3-dihydroxy-***													
—	**Glycerin**..........	*see* **Glycerol**													
Ω g70	**Glycerol**..........	Glycerin. 1,2,3-Propanetriol*. $HOCH_2CH(OH)CH_2OH$	92.11	syr, rh pl λ^{undil} 270 (−1.2)	20	290d 182^{20}	1.2613^{20}_4	1.4746^{20}	∞	∞	δ	...	i	chl, CCl_4, CS_2, peth i	**B1**[3], 2297
g71	—,borate..........	$(C_3H_5BO_3)_n$	$(99.88)_n$	glass	*ca.* 150	....	....	....	d^h	...	...	...	...	....	**B1**, 519
g72	—,1(2-chlorophenyl) ether	....	202.64	nd (bz)	71–2	250^{19}	....	....	s^h	...	v	...	s^h	....	**B6**[1], 99
g73	—,1(4-chlorophenyl) ether	....	202.64	nd (eth-peth), lf (bz) λ^{cy} 281 (3.26), 289 (3.17)	76	214–5^{19}	....	....	i	v	v	...	s^h	con sulf s	**B6**[1], 101

For explanations, symbols and abbreviations see beginning of table. For structural formulas see end of table.

No.	Name	Synonyms and Formula	Mol. wt.	Color, crystalline form, specific rotation and λ_{max} (log ε)	m.p. °C	b.p. °C	Density	n_D	Solubility w	al	eth	ace	bz	other solvents	Ref.
	Glycerol														
g74	—,1,3-diacetate	Diacetin. $HOCH(CH_2O_2CCH_3)_2$	176.17	....	40	280 $155–6^{15}$	1.1779^{15}_{4}	1.4395^{20}	v	v	δ	...	...	CS_2 i	**B2**^{2}, 160
g75	—,1,2-dibutanoate (*d*)	α,β-Dibutyrin. $C_3H_7^nCO_2CH_2CH(CH_2OH)O_2CC_3H_7^n$	232.28	$[\alpha]_D$ +1.7 (Py, c = 7)	....	273.5 167^{20}	....	1.4422^{20}	i	s	...	...	...	....	**B2**, 273
g76	—,1,3-didodecanoate	α,γ-Dilaurin. Glycerol-1,3-dilaurate. $HOCH[CH_2O_2C(CH_2)_{10}CH_3]_2$	456.72	pl (al), nd (eth-al) β:56.5 (st)	α:49.5 (unst) β:56.5 (st)	....	....	....	...	s	s	...	s	chl, lig s	**B2**^{2}, 320
g77	—,1,3-dihexadecanoate	α,γ-Dipalmitin. Glycerol-1,3-dipalmitate. $HOCH[CH_2O_2C(CH_2)_{14}CH_3]_2$	568.93	cr (al or chl)	72–4 (α:50 β:63.5)	....	....	....	δ s^h	s^h	s	...	...	chl s^h	**B2**^{3}, 968
g78	—,1,2-dimethyl ether(*dl*)	2,3-Dimethoxy-1-propanol*. $CH_3OCH_2CH(OCH_3)CH_2OH$	120.15	....	....	180^{760} 100^{40}	1.016^{25}_{4}	1.4200^{20}	∞	s	s	...	...	....	**B1**^{3}, 2317
g79	—,1,3-dimethyl ether	1,3-Dimethoxy-2-propanol*. $CH_3OCH_2CH(OH)CH_2OCH_3$	120.15	....	....	169^{760} 88^{40}	1.0085^{20}_{4}	1.4192^{20}	∞	v	s	...	...	....	**B1**^{3}, 2318
g80	—,1,3-dinitrate	$O_2NOCH_2CH(OH)CH_2ONO_2$	191.10	pr (w), cr (eth +1w)	26 (hyd)	148^{15} $116^{0.6}$	1.523^{20}_{4}	1.4715^{20}_{4}	s	v	s	...	...	....	**B1**^{2}, 591
Ω g81	—,1,3-dioctadecanoate	α,γ-Distearin. Glycerol-1,3-distearate. $HOCH[CH_2O_2C(CH_2)_{16}CH_3]_2$	625.04	nd or pl (eth, chl, lig)	79.1	....	....	....	...	δ s^h	δ s^h	s^h	...	os s^h	**B2**^{2}, 356
g82	—,1,3-diphenyl ether	$C_6H_5OCH_2CH(OH)CH_2OC_6H_5$	244.29	lf (al)	81–2	$224.5^{17.5}$ 175^{2}	1.179^{24}_{4}	....	i	v^h	v	...	v	chl v	**B6**^{2}, 152
g83	—,—,2-acetate	$CH_3CO_2CH(CH_2OC_6H_5)_2$	286.33	(dil al)	70–1	190^{160}	....	....	i	v	v	...	v	chl v	**B6**, 149
g84	—,1,3-dipropanoate	α,γ-Dipropioin. $HOCH(CH_2O_2CCH_2CH_3)_2$	204.23	....	....	$170–3^{10}$	....	....	...	s	...	...	...	....	**B2**^{1}, 107
g85	—,1-hexadecyl ether	Chimyl alcohol. 3-Hexadecyloxy-1,3-propanediol. $HOCH_2CH(OH)CH_2O(CH_2)_{15}CH_3$	316.53	lf (hx) $[\alpha]^{20}_D$ +3 (chl)	64	$120^{0.005}$	....	....	...	...	...	s	...	chl, peth s	**H27**, 674
g86	—,1(2-methoxyphenyl) ether	Guaiacol α-glyceryl ether. Guaiamar.	198.22	rh pr (eth, eth-peth)	78.5–9	215^{19} $126–8^{0.2}$	....	....	s v^h	v	...	...	s	chl s peth i	**B6**^{1}, 382
g87	—,1-methyl ether	3-Methoxy-1,2-propanediol*. $HOCH_2CH(OH)CH_2OCH_3$	106.12	hyg liq	....	220^{760} (198–200) $110–2^{13}$	1.1830^{20}_{4}	1.442^{25}	v	v	s	v	...	os s	**B1**^{3}, 2317
g88	—,2-methyl ether	2-Methoxy-1,3-propanediol*. $HOCH_2CH(OCH_3)CH_2OH$	106.12	hyg liq	....	232^{760} $119–20^{9}$	1.124^{25}_{4}	1.4505^{17}	v	v	s	v	...	os v	**B1**^{3}, 2317
Ω g89	—,1-monoacetate(*dl*)	α-Monoacetin. $HOCH_2CH(OH)CH_2O_2CCH_3$	134.13	....	....	158^{165} $129–31^{3}$	1.2060^{20}_{4}	1.4157^{20}	s	s	δ	...	i	....	**B2**^{2}, 159
g90	—,1-monobutanoate(*dl*)	α-Monobutyrin. $HOCH_2CH(OH)CH_2O_2CCH_2CH_2CH_3$	162.19	....	....	269–71 163^{16}	1.129^{18}	1.4531^{20}	v	s	...	...	...	....	**B2**^{2}, 249
g91	—,1-monododecanoate(*dl*)	α-Monolaurin. $HOCH_2CH(OH)CH_2O_2C(CH_2)_{10}CH_3$	274.41	lf (CCl_4 or peth)	α:44 (unst) β':59.5 (unst) β:63 (st)	186^{1}	0.9248^{97}	1.4350^{86}	...	δ s^h	v	v	s	chl v	**B2**^{2}, 320
g92	—,—(*l*)	$HOCH_2CH(OH)CH_2O_2C(CH_2)_{10}CH_3$	274.41	cr (eth or peth) $[\alpha]_D$ −4.9 (Py)	54–5	....	....	....	...	δ s^h	v	v	s	chl v	**B2**^{3}, 890
g93	—,monohexadecanoate(*dl*)	α-Monopalmitin. $HOCH_2CH(OH)CH_2O_2C(CH_2)_{14}CH_3$	330.51	pl or lf (eth, lig)	α:66.5 (unst) β':74.6 (unst) β:77 (st)	....	....	....	i	v	δ	...	...	....	**B2**^{2}, 338
g94	—,—(*l*)	$HOCH_2CH(OH)CH_2O_2C(CH_2)_{14}CH_3$	330.51	cr (eth or peth) $[\alpha]_D$ −4.37 (Py)	71–2	....	....	....	...	...	...	...	...	....	**B2**^{3}, 966
g95	—,1-mono(2-hydroxybenzoate)	Glycosal. α-Glyceryl salicylate.	212.21	nd (eth)	76	....	....	....	δ v^h	v	δ	...	v^h	chl, peth δ	**B10**^{2}, 53
g96	—,1-mono(12-hydroxy-9-octadecenoate)	Glycerol-1-monoricinoleate. α-Monoricinolein. $HOCH_2CH(OH)CH_2O_2C(CH_2)_7CH{:}CHCH_2CH(OH)(CH_2)_5CH_3$	372.55	ye	....	....	1.028^{20}_{4}	....	...	s	s	s	s	chl, AcOEt s lig, MeOH, CS_2 δ	**B3**^{1}, 138
g97	—,1-mononitrate	$HOCH_2CH(OH)CH_2ONO_2$	137.09	pr (w, al or eth)	61	155–60 102^{1}	1.4164^{29}_{4}	1.4698^{20}	v	v	δ	...	...	....	**B1**^{2}, 591
g98	—,2-mononitrate	$HOCH_2CH(ONO_2)CH_2OH$	137.09	lf (w)	54	155–60	1.40^{20}_{4}	....	s	s	s	...	...	....	**B1**^{2}, 591
g99	—,1-mono(9,12-octadecadienoate)	Glycerol-α-monolinoleate. α-Monolinolein. $HOCH_2CH(OH)CH_2O_2C(CH_2)_7CH{:}CHCH_2CH{:}CH(CH_2)_4CH_3$	354.54	cr (bz)	14–5	....	....	1.4758^{20}	...	δ	v	...	v	chl v lig, MeOH, CS_2 δ	**B2**^{1}, 213
Ω g100	—,1-monooctadecanoate(*dl*)	α-Monostearin. $HOCH_2CH(OH)CH_2O_2C(CH_2)_{16}CH_3$	358.57	pl (MeOH)	α:74 (unst) β':79 (unst) β:81 (st)	....	0.9841^{20}_{4}	1.4400^{86}	i	δ s^h	δ v^h	...	...	lig s peth δ	**B2**^{2}, 354
g101	—,—(*l*)	$HOCH_2CH(OH)CH_2O_2C(CH_2)_{16}CH_3$	358.57	cr (eth or peth) $[\alpha]_D$ −3.58 (Py)	76–7	....	....	....	...	...	...	...	...	....	**B2**^{3}, 1024

For explanations, symbols and abbreviations see beginning of table. For structural formulas see end of table.

No.	Name	Synonyms and Formula	Mol. wt.	Color, crystalline form, specific rotation and λ_{max} (log ε)	m.p. °C	b.p. °C	Density	n_D	Solubility w	al	eth	ace	bz	other solvents	Ref.
	Glycerol														
g102	—,1-mono(9-octadecenoate)	Glycerol 1-monooleate. α-Monoolein. $HOCH_2CH(OH)CH_2O_2C(CH_2)_7CH:CH(CH_2)_7CH_3$	356.55	pl (al)	α:25 (unst) β′:32 (unst) β:35.5 (st)	238–40^{3}	0.9420^{20}_{4}	1.4626^{40}	i	s	s	...	...	chl s	**B2**^{2}, 439
—	—,1-monooleate ..	*see* **Glycerol**, 1-mono(9-octadecenoate)													
—	—,1-monoricinoleate	*see* **Glycerol**, 1-mono(12-hydroxy-9-octadecenoate)													
Ω g103	—,1-octadecyl ether (*d*)	Batyl alcohol. 3-Octadecyloxy-1,2-propanediol*. $HOCH_2CH(OH)CH_2O(CH_2)_{17}CH_3$	344.59	pl (bz, ace) $[\alpha]^{20}_{D}$ +1.14 (chl, c = 6.6)	70–1	215–20^{2}	....	....	i	s^h	s	s^h	s^h	....	**B1**^{2}, 590
g104	—,1-phenyl ether ..	Antodyne. $HOCH_2CH(OH)CH_2OC_6H_5$	168.20	nd (eth, peth or bz-lig)	67–8 (62–4)	200^{22} 145–8$^{0.6}$	1.225^{20}_{4}	....	v	v	s^h	...	v	peth δ con sulf s	**B6**^{2}, 152
g105	—,1(2-tolyl)ether ..	....	182.22	nd (bz-peth)	70–1d (cor)	....	....	....	δ	s	δ	...	...	....	**B6**, 354
g106	—,triacetate	Triacetin. $CH_3CO_2CH(CH_2O_2CCH_3)_2$	218.21	cr (al)	4.1	258–60^{760} 130.5^{7}	1.1596^{20}_{4}	1.4301^{20}	δ	∞	∞	v	∞	chl ∞ lig, CS_2 δ	**B2**^{2}, 160
g107	—,tribenzoate.....	Tribenzoin. $C_6H_5CO_2CH(CH_2O_2CC_6H_5)_2$	404.43	nd (MeOH)	76	....	1.228^{12}_{4}	....	i	s^h	v	v	v	chl v	**B9**^{2}, 122
Ω g108	—,tributanoate....	Tributyrin. $CH_3CH_2CH_2CO_2CH(CH_2O_2CCH_2CH_2CH_3)_2$	302.37	....	−75	305–10 (287–8) 190^{15}	1.0350^{20}_{4}	1.4359^{20}	i	s	v	s	s	....	**B2**^{2}, 249
g109	—,tridodecanoate .	Glycerol trilaurate. Trilaurin. $CH_3(CH_2)_{10}CO_2CH[CH_2O_2C(CH_2)_{10}CH_3]_2$	639.03	nd (al)	α:35.6 (unst) β′:32.9 (unst) β:46.4 (st)	....	0.8986^{55}	1.4404^{60}	i	s	s	v	v	chl, peth s	**B2**^{2}, 320
—	—,trielaidate......	*see* **Glycerol**, tri(*trans*-9-octadecenoate)													
Ω g110	—,trihexadecanoate	Glycerol tripalmitate. Tripalmitin. $CH_3(CH_2)_{14}CO_2CH[CH_2O_2C(CH_2)_{14}CH_3]_2$	807.35	nd (eth)	α:44.7 (unst) β′:56.6 (unst) β:66.4 (st)	310–20	0.8752^{70}_{4}	1.4381^{80}	i	δ	v	...	s v^h	chl s	**B2**^{2}, 340
g111	—,trihexanoate ...	Glycerol tricaproate. Tricaproin. $CH_3(CH_2)_4CO_2CH[CH_2O_2C(CH_2)_4CH_3]_2$	386.54	....	−60	>200	0.9867^{20}_{4}	1.4427^{20}	i	∞	∞	v	∞	peth, chl ∞	**B2**^{2}, 285
g112	—,tri(3-methylbutanoate)	Glycerol isovalerate. Triisovalerin. $(CH_3)_2CHCH_2CO_2CH[CH_2O_2CCH_2CH(CH_3)_2]_2$	344.45	....	....	330–5^{763} 194^{15}	0.9984^{20}_{4}	1.4354^{20}	...	s	s	...	...	....	**B2**^{2}, 277
g113	—,trimethyl ether..	1,2,3-Trimethoxypropane*. $CH_3OCH_2CH(OCH_3)CH_2OCH_3$	134.18	....	....	148	0.9460^{15}_{4}	1.4055^{15}	∞	...	s	v	s	sulf v	**B1**^{3}, 2318
g114	—,trinitrate.......	Nitroglycerin. Trinitrin. $O_2NOCH_2CH(ONO_2)CH_2ONO_2$	227.09	pa ye tcl or rh	13 (st) 2 (unst)	256 exp 125^{2}	1.5931^{20}_{4}	1.4786^{12}	δ	s	∞	v	s	MeOH s CS_2, lig, peth δ chl v	**B1**^{3}, 2328
g115	—,trioctadecanoate.	Glycerol tristearate. Tristearin. $CH_3(CH_2)_{16}CO_2CH[CH_2O_2C(CH_2)_{16}CH_3]_2$	891.51	cr (eth, peth)	α:55 β′:64.5 β:73	....	0.8559^{90}_{4}	1.4399^{80}	i	i	...	s	δ s^h	CCl_4 δ, s^h chl, CS_2 s AcOEt i lig, peth i, s^h	**B2**^{3}, 1035
g116	—,tri(*cis*-9-octadecenoate)	Glycerol trioleate. Triolein. $CH_3(CH_2)_7CH:CH(CH_2)_7CO_2CH[CH_2O_2C(CH_2)_7CH:CH(CH_2)_7CH_3]_2$	885.47	polymorphic	α: −32 (unst) β′: −12 (unst) β: −5.5 (st)	235–40^{18}	0.8988^{40}	1.4621^{40}	i	δ	v	...	...	chl, peth s	**B2**^{2}, 440
Ω g117	—,tri(*trans*-9-octadecenoate)	Glycerol trielaidate. Trielaidin. $CH_3(CH_2)_7CH:CH(CH_2)_7CO_2CH[CH_2O_2C(CH_2)_7CH:CH(CH_2)_7CH_3]_2$	885.47	....	α:16.6 β:42.8	....	....	....	i	δ	v	...	s	chl s	**B2**, 470
g118	—,trioctanoate....	Glycerol tricaprylate. Tricaprylin. $CH_3(CH_2)_6CO_2CH[CH_2O_2C(CH_2)_6CH_3]_2$	470.70	....	9.8–10.1 (st) −21 (metast)	233.1	0.9540^{20}_{4}	1.4482^{20}	i	∞	v	...	v	peth, lig, chl v	**B2**^{2}, 303
g119	—,tripropanoate ..	Tripropionin. Tripropin. $CH_3CH_2CO_2CH(CH_2O_2CCH_2CH_3)_2$	260.29	....	....	175–6^{20} 130–2^{3}	1.100^{20}_{18}	1.4318^{19}	i	s	v	...	...	chl s	**B2**^{2}, 222
Ω g120	—,tritetradecanoate	Glycerol trimyristate. trimyristin. $CH_3(CH_2)_{12}CO_2CH[CH_2O_2C(CH_2)_{12}CH_3]_2$	723.19	polymorphic (al-eth)	α:32 (unst) β:44 (unst) β′:56.5 (st)	311	0.8848^{60}_{4}	1.4428^{60}	i	δ	s	s	s	chl s lig, CS_2 δ	**B2**^{2}, 328
—	—,**1,2-dithio-**......	*see* **1-Propanol, 2,3-dimercapto-**													

For explanations, symbols and abbreviations see beginning of table. For structural formulas see end of table.

No.	Name	Synonyms and Formula	Mol. wt.	Color, crystalline form, specific rotation and λ_{max} (log ε)	m.p. °C	b.p. °C	Density	n_D	Solubility: w	al	eth	ace	bz	other solvents	Ref.
	Glycerophosphoric acid														
g121	**Glycerophosphoric acid(*L*)**	$HOCH_2CH(OH)CH_2OPO(OH)_2$	172.08	syr $[\alpha]_D$ −1.45 (Ba salt in 2N HCl, c = 10.3)	...	...	...	...	s	s	i	...	i	MeOH s	**B1**[3], 2335
—	**Glycidic acid**	*see* **Propanoic acid, 2,3-epoxy-***													
g122	**Glycidol(*d*)**	2,3-Epoxy-1-propanol*. Glycide. 3-Hydroxypropylene oxide.	74.08	$[\alpha]_D$ +15 (undil)	...	56.5^{11}	1.117_4^{20}	1.4293^{16}	s	s	s	s	s	chl s	**B17**[1], 50
g123	—(*l*)	$C_3H_6O_2$. *See* g122	74.08	$[\alpha]_D^{18}$ −8.6 (undil)	...	$56–6.5^{11}$	1.1050^{18}	1.4293^{16}	s	s	s	s	s	chl s	**B17**[1], 50
Ω g124	**Glycine**	Aminoacetic acid. Glycocoll. $H_2NCH_2CO_2H$	75.07	mcl or trg pr (dil al), $\lambda^{al, pH=2}$ 285 (3.76)	262d (br at 228)	...	0.828^{17} 1.1607	...	v	i	i	δ	...	Py δ	**B4**[1], 462
g125	—,amide	Glycinamide. $H_2NCH_2CONH_2$	74.08	hyg nd (chl) $\lambda^{w, pH=6}$ 210 sh (2.1)	67–8	...	...	...	v	v	δ	s	δ	MeOH v AcOEt s chl s[h]	**B4**[2], 783
g125[1]	—,—,*N'*(4-ethoxyphenyl)-	Glycine-*p*-phenetidide. Phenocoll.	194.24	nd (+1w)	100.5 (anh) 95 (hyd)	...	...	...	δ	s	s	...	...	...	**B13**[2], 268
g126	—,—,hydrochloride	$H_2NCH_2CONH_2.HCl$	110.54	nd (al)	203–4 (189)	...	...	...	v	δ	...	...	...	...	**B4**[1], 468
—	—,anhydride	*see* **2,5-Piperazinedione**													
g127	—,ethyl ester	Ethyl glycinate. $H_2NCH_2CO_2C_2H_5$	103.12	$\lambda^{w, pH=9}$ 218 sh (1.9)	...	$148–9^{750}$ 57.6^{18}	1.0275_4^{10}	1.4242^{20}	∞	∞	∞	∞	∞	lig v	**B4**[2], 780
Ω g128	—,—,hydrochloride	$H_2NCH_2CO_2C_2H_5.HCl$	139.58	nd (al)	144	sub	...	...	v	v	...	...	...	...	**B4**[2], 781
g129	—,hydrochloride	$H_2NCH_2CO_2H.HCl$	111.53	hyg rh nd (w)	200–1 (187)	...	...	...	v	δ	...	...	...	...	**B4**, 340
g130	—,methyl ester	Methyl glycinate. $H_2NCH_2CO_2CH_3$	89.10	...	...	130d 45^{20}	...	...	...	s	...	...	...	...	**B4**[2], 780
g131	—,nitrile	Aminoacetonitrile. H_2NCH_2CN	56.07	...	...	58^{15} δd	...	...	...	s	...	...	...	ac v	**B4**[3], 1120
g132	—,*N*(*p*-acetamidophenyl)-	...	208.22	...	241–2	...	...	...	s	...	...	...	...	...	**Am 39,** 1457
Ω g133	—,*N*-acetyl-	Acetamidoacetic acid. Aceturic acid. $CH_3CONHCH_2CO_2H$	117.11	lo nd (w, MeOH) $\lambda^{w, pH=1}$ 215 sh (2.4)	206	...	...	...	δ	s	i	δ	i	chl, aa δ	**B4**[2], 789
g134	—,*N*-acetyl-*N*-phenyl-	*N*-Phenylaceturic acid. $CH_3CON(C_6H_5)CH_2CO_2H$	193.21	lf (w)	194	...	...	...	δ	s	δ	...	δ	AcOEt s chl, lig δ	**B12**, 476
g136	—,*N*(4-aminophenyl)-, monohydrate	$C_8H_{10}N_2O_2.H_2O$. *See* g167	184.20	pl (dil aa)	222–3d	...	...	...	s[h]	...	...	...	...	...	**B13**[1], 34
g137	—,*N*(2-arsonophenyl)-, amide	...	274.11	sc (w)	198–9d	...	...	...	v[h]	δ[h]	...	...	...	MeOH δ[h]	**B16**[2], 487
g138	—,*N*(3-arsonophenyl)-, amide	...	274.11	nd (w)	175–7d	...	...	...	v[h]	v	...	i	...	aa v	**B16**[2], 489
—	—,*N*-benzoyl-	*see* **Hippuric acid**													
g142	—,*N*-benzyl-	...	165.19	nd (w)	198–9	...	...	...	δ s[h]	...	...	...	i	alk s	**B12**[2], 567
Ω g143	—,—,ethyl ester	$C_{11}H_{15}NO_2$. *See* g142	193.25	...	...	$175–9^{50}$ 165^{16}	...	1.5041^{20}	...	v	v	...	v	...	**B12**[2], 567
Ω g144	—,*N,N*-bis(2-hydroxyethyl)-	$(HOCH_2CH_2)_2NCH_2CO_2H$	163.18	nd (al)	193–5d	...	...	...	v	i s[h]	...	...	...	...	**B4**[2], 787
g145	—,*N*-bromoacetyl-*N*-phenyl-	$BrCH_2CON(C_6H_5)CH_2CO_2H$	272.11	pl (w)	153d	...	...	...	δ	s	...	...	s	...	**B12**, 477
Ω g148	—,*N*(2-carboxyphenyl)-	*N*-Carboxymethylanthranilic acid. Anthranilidoacetic acid. $C_9H_9NO_4$. *See* g167	195.18	nd (MeOH)	218–20	...	...	...	δ	s	s	...	i	chl i aa s	**B14**[2], 225
g149	—,*N*-chloroacetyl-*N*-phenyl-	$ClCH_2CON(C_6H_5)CH_2CO_2H$	227.65	pl or pr (bz)	132–3	...	...	...	δ	s	...	...	s	...	**B12**, 476
g150	—,—,methyl ester	$ClCH_2CON(C_6H_5)CH_2CO_2CH_3$	241.68	pr (lig)	59–60	...	...	...	δ	v	v	...	v	chl v	**B12**, 477
—	—,cholyl-	*see* **Glycocholic acid**													
g151	—,*N,N*-dicarbethoxy-, ethyl ester	$(C_2H_5O_2C)_2NCH_2CO_2C_2H_5$	247.25	pr (peth)	36.5	$152–3^{10}$	...	...	δ	v	v	...	v	AcOEt v peth δ	**B4**, 365
g153	—,—,nitrile	Diethylaminoacetonitrile. $(C_2H_5)_2NCH_2CN$	112.18	...	...	170 70^{24}	...	1.4260^{20}	s	s	...	...	...	...	**B4**[2], 787
g154	—,*N,N*-dimethyl-	Dimethylaminoacetic acid. $(CH_3)_2NCH_2CO_2H$	103.12	hyg nd (PrOH)	185–6 (157–60)	...	...	...	v	s	s	s[h]	...	MeOH v	**B4**[2], 785
g155	—,—,nitrile	Dimethylaminoacetonitrile. $(CH_3)_2NCH_2CN$	84.13	...	...	137–8 42^{21}	0.8650_0^{20}	1.4095^{20}	s	s	...	...	...	HCl s	**B4**, 346
g156	—,*N*-ethyl-	Ethylaminoacetic acid. $C_2H_5NHCH_2CO_2H$	103.12	pl (al)	181–2d	...	...	...	s	s	...	...	...	...	**B4**[2], 787
g157	—,—,nitrile	Ethylaminoacetonitrile. $C_2H_5NHCH_2CN$	84.13	...	...	$166–7^{760}$ $81–3^{29}$	...	...	...	s	s	...	...	dil HCl v	**B4**[2], 787
g158	—,*N*-formyl-*N*-phenyl-	$C_6H_5N(CHO)CH_2CO_2H$	179.18	nd (w)	125	...	...	...	s	s	s	...	...	...	**B12**, 476

For explanations, symbols and abbreviations see beginning of table. For structural formulas see end of table.

No.	Name	Synonyms and Formula	Mol. wt.	Color, crystalline form, specific rotation and λ_{max} (log ε)	m.p. °C	b.p. °C	Density	n_D	Solubility w	al	eth	ace	bz	other solvents	Ref.
	Glycine														
—	—,*N*(*N*-γ-glutamylcysteinyl)-,	*see* **Glutathione**													
—	—,*N*(*N*-glycylglycine)-	*see* **Triglycine**													
—	—,*N*-guanyl-	*see* **Glycocyamine**													
Ω g159	—,*N*(4-hydroxyphenyl)-	$C_8H_9NO_3$. *See* g167	167.17	pl (w)	245–7d (200d)	...	...	...	δ s^h	δ	i	δ	δ	AcOEt, chl, alk s	**B13**², 259
g160	—,*N*(4-hydroxyphenylacetyl)-	...	209.20	pl (w)	153–5	...	...	...	s^h	s	δ	...	i	AcOEt s	**B10**¹, 83
Ω g161	—,*N*-leucyl-(*DL*)	*DL*-Leucylglycine. $(CH_3)_2CHCH_2CH(NH_2)CONHCH_2CO_2H$	188.23	cr (w) λ^w 300 (0.8) (<220)	243d	...	...	...	s	δ	δ	i	i	chl i	**B4**², 872
Ω g162	—,—(*L*)	$(CH_3)_2CHCH_2CH(NH_2)CONHCH_2CO_2H$	188.23	lf or nd (w-al) $[\alpha]^{20}_D$ +85.8 (w, c = 2.4)	248d	...	...	...	s	δ	δ	i	i	chl i	**B4**, 444
g163	—,*N*,*N'*-methylenedi-	$CH_2(NHCH_2CO_2H)_2$	162.15	pl	199d	...	...	...	...	s	...	...	...	HCl s	**B4**¹, 473
g164	—,*N*(2-naphthyl)-	$C_{10}H_7^{\beta}NHCH_2CO_2H$	201.23	(w)	134–5	...	...	...	v^h	v	v	...	...	aa v	**B12**, 1298
Ω g165	—,*N*(2-nitrophenyl)-	...	196.17	dk red pr (al)	192–3d	...	...	...	δ	v^h	δ	...	...	...	**B12**, 695
g167	—,*N*-phenyl-	Anilinoacetic acid.	151.17	...	127–8	...	...	...	s	s	δ	...	...	...	**B12**, 468
g168	—,—,amide	$C_6H_5NHCH_2CONH_2$	150.18	nd (w), lf (al)	141 (136)	...	...	...	v^h	v	δ	v	...	...	**B12**², 249
g169	—,—,ethyl ester	$C_6H_5NHCH_2CO_2C_2H_5$	179.22	lf (dil al)	58	273–4 163^18	...	...	δ^h	s	s	...	...	...	**C48**, 719
g170	—,—,methyl ester	$C_6H_5NHCH_2CO_2CH_3$	165.19	nd (al)	48	...	...	...	i	s	s	...	...	...	**C49**, 1337
Ω g171	—,—,nitrile	Anilinoacetonitrile. $C_6H_5NHCH_2CN$	132.18	pl (lig-eth)	48	...	...	...	δ	s	δ	...	s	lig δ	**B12**, 472
Ω g173	—,*N*-phthaloyl-	Phthalimide-*N*-acetic acid.	205.17	nd or pr (w or al)	193 (193–6)	...	...	...	δ v^h	s^h	s^h	...	...	peth, lig, chl i	**B21**, 481
g174	—,—,chloride	$C_{10}H_6ClNO_3$. *See* g173	223.62	nd (lig)	84–5	distb vac	...	...	d	v	...	...	v	...	**B21**², 358
g175	—,—,ethyl ester	$C_{12}H_{11}NO_4$. *See* g173	233.23	nd (w, al or eth)	112–3 (104–5)	300	...	...	δ	v^h	v	...	v	chl v lig δ	**B21**², 357
g176	—,*N*(*N*-phthaloylglycyl)-	Phthaloyldiglycine.	262.22	nd (w)	232	...	...	...	δ	s	δ	...	i	chl i aa v	**B21**², 358
g177	—,—,ethyl ester	$C_{14}H_{14}N_2O_5$. *See* g176	290.28	nd (al)	193.5–4.5	...	...	...	i	v^h	s	...	v	chl v	**B21**², 358
g178	—,*N*-succinyl-, ethyl ester	Ethyl succinimide-*N*-acetate.	185.18	nd (eth)	67	198^32	...	...	v	v	δ	...	...	...	**B21**², 305
g179	**Glycocholic acid**	Cholylglycine. $C_{26}H_{43}NO_6$.	465.64	nd (w), $[\alpha]^{23}_D$ +32.3 (al, c = 1) $\lambda^{65\%\ sulf}$ 318 (4.18)	165–8 (anh) 154–5d (+1.5w) 132–4 (+6w)	...	...	...	δ	v	δ	...	...	...	**E14**, 194
—	**Glycocoll**	*see* **Glycine**													
Ω g180	**Glycocyamine**	Guanidoacetic acid. *N*-Guanylglycine. $NH{:}C(NH_2)NHCH_2CO_2H$	117.11	pl or nd (w)	>300 (250–66)	...	...	...	δ s^h	δ	δ	...	...	...	**B4**², 793
Ω g181	**Glycogen**	Animal starch. $(C_6H_{10}O_5)_n$	$(162.14)_n$	wh pw, $[\alpha]^{25}_D$ +196.5 (w) λ^w 420	...	...	...	...	v	i s^h	i	...	...	...	**C25**, 4940
—	**Glycol**	*see* **1,2-Ethanediol**													
—	**Glycolaldehyde**	*see* **Acetaldehyde, hydroxy-**													
—	**Glycolic acid**	*see* **Acetic acid, hydroxy-**													
g182	**Glycolide**	2,5-*p*-Dioxanedione.	116.07	lf (al-chl or al)	86–7 (83)	...	...	...	s^h	δ s^h	δ	v	...	chl s^h	**B19**², 175
Ω g183	**Glycoluril**	Acetylenediurene. Glyoxaldiurene.	142.12	nd or pr (w)	300d	...	...	...	δ^h	i	s	...	...	HCl, NH_4OH s aa i	**B26**², 260
—	**Glycolylurea**	*see* **Hydantoin**													
—	**Glycylglycine**	*see* **Diglycine**													
g184	**18α-Glycyrrhetinic acid**	Glycyrrhetic acid. $C_{30}H_{46}O_4$.	470.70	α: pl (dil al) β: nd (al-peth) pl (chl-MeOH) α: $[\alpha]^{20}_D$ +140 (al) β: $[\alpha]^{20}_D$ +145.5 (diox) λ^{al} 242 (4.09)	α: 283 β: 296 (331–5)	...	...	...	...	s	...	...	...	chl, diox, aa s peth i	**E14**, 561
g185	—,glycoside	Glycyrrhizic acid. Glycyrrhizin. $C_{42}H_{62}O_{16}$. *See* g184	822.96	pl or pr (aa)	220d	...	...	...	v^h	i	δ	...	...	...	**H4**, 100

For explanations, symbols and abbreviations see beginning of table. For structural formulas see end of table.

No.	Name	Synonyms and Formula	Mol. wt.	Color, crystalline form, specific rotation and λ_{max} (log ε)	m.p. °C	b.p. °C	Density	n_D	Solubility w	al	eth	ace	bz	other solvents	Ref.
	Glyoxal														
g186	**Glyoxal**	Ethanedial*. Biformyl. OCHCHO	58.04	ye pr	15	50.4	1.14^{20} (1.26^{20})	1.3826^{20}	v	s	s	...	...		**B1**[3], 3076
Ω g187	—,dioxime	Diisonitrosoethane. Glyoxime. HON:CHCH:NOH	88.07	rh pl (w) λ^{al} 233 (4.23)	178d (181)	sub			v^h	v^h	v^h	...	...		**B1**[2], 818
g188	—,**phenyl-**	Benzoylformaldehyde. C_6H_5COCHO	134.14	nd (+w)	91 (hyd)	142^{125} (anh) $95–7^{25}$			s d^h	s	s	s	s	chl s	**B7**[1], 360
—	**Glyoxaline**	*see* **Imidazole**													
—	**Glyoxime**	*see* **Glyoxal,** dioxime													
—	**Glyoxylic acid**	*see* **Acetic acid, oxo-**													
—	**Gnoscopine**	*see* **Narcotine(*dl*)**													
Ω g189	**Gramine**	3(Dimethylaminomethyl)-indole. Donaxine.	174.25	lf (eth), nd (ace) λ^{al} 280 (3.80), 289 (3.68)	138–9				i	s	s	...	...	chl s peth i	**C30,** 4502
g190	**Griseofulvin**	Fulvicin. $C_{17}H_{17}ClO_6$.	352.78	oct or rh (bz) $[\alpha]_D^{17}$ +376 (chl sat sol) $[\alpha]_D^{21}$ +337 (ace) λ^{al} 291 (4.37), 325 (3.77)	220				i	δ	δ	δ	δ s^h	AcOEt, chl, aa δ	**J1952,** 3949
—	**Guaethol**	*see* **Phenol, 2-ethoxy-***													
—	**Guaiacol**	*see* **Phenol, 2-methoxy-***													
—	**Guaiacol, 4-propenyl-**	*see* **Benzene, 1-hydroxy-3-methoxy-4-propenyl-**													
—	**Guaiazulene**	*see* **Azulene, 1,4-dimethyl-7-isopropyl-**													
—	**Guaiene**	*see* **Naphthalene, 2,3-dimethyl-***													
g191	**Guaiol**	Champacol. 3,8-Dimethyl-5(α-hydroxypropyl)-Δ^9-octahydroazulene. $C_{15}H_{26}O$.	222.38	trg pr (al) $[\alpha]_D^{20}$ −30 (al, c = 4)	91	288^{760} δd 165^{17}	0.9074^{100}_{20}	1.4716^{100}	i	s	s	...	...		**E12A,** 431
g192	**Guanidine**	Aminomethanamidine. Carbamamidine. $HN:C(NH_2)_2$	59.07	cr λ^w 209	*ca.* 50				v	v	...	...	...		**B3**[2], 69
Ω g193	—,acetate	$HN:C(NH_2)_2.CH_3CO_2H$	119.13	nd (al-eth)	229–30				v	v	i	...	...		**B3**[2], 69
Ω g194	—,carbonate	$2[HN:C(NH_2)_2].H_2CO_3$	180.17	oct tetr pr (w)	198		1.24^4		v	i	...	...	...		**B3**[2], 72
Ω g195	—,hydrochloride	$HN:C(NH_2)_2.HCl$	95.53	rh bipym (al) λ^w 262 (1.25)	178–85		1.354^{20}_4		v	v	...	...	...		**B3**[2], 71
Ω g196	—,nitrate	$HN:C(NH_2)_2.HNO_3$	122.08	lf (w)	217	d			v	s v^h	...	δ	...		**B3**[2], 72
g197	—,picrate	$HN:C(NH_2)_2.C_6H_3N_3O_7$	288.19	og-ye pl or nd (w)	333d				δ^h	δ	δ	...	...		**B6**[2], 265
Ω g198	—,thiocyanate	$HN:C(NH_2)_2.HCNS$	118.16	lf	118				v	...	...	...	...		**B3**[2], 121
g199	—,**1-amino-***	Guanylhydrazine. $HN:C(NH_2)NHNH_2$	74.09	cr	d				s	s	i	...	...		**B3**[2], 95
Ω g200	—,**1-cyano-***	Dicyanodiamide. Param. $NH:C(NH_2)NHCN$	84.08	rh lf or pl (al)	211–2	d	1.404^{14}		s	s	i	s	i	AcOEt δ chl, CS_2, CCl_4 i	**B3**[2], 75
Ω g201	—,**1,3-diphenyl-***	Melaniline. $HN:C(NHC_6H_5)_2$	211.27	mcl nd (al or to) λ^{al} 251 (4.4), 286 sh (1.1)	150	d 170	1.13^{20}_4		δ	s	v	...	...	CCl_4, chl s to s^h	**B12**[2], 216
Ω g202	—,**1,3-di(2-tolyl)-**		239.32	cr (dil al) λ^{chl} 252 (4.08)	179		1.10^{20}_4		δ^h	δ s^h	v	...	...	chl s	**B12,** 803
g203	—,**1-methyl-, sulfate***	$[HN:C(NH_2)NHCH_3]_2.H_2SO_4$	244.27	(w)	238–40				v	δ	i	...	i	os i	**B4**[2], 570
Ω g204	—,**1-nitro-***	$HN:C(NH_2)NHNO_2$	104.07	nd or pr (w) λ^w 264 (4.16)	239d (slow htng), 246–7 (rapid htng)				δ s^h	δ	i	...	...	alk v	**B3**[2], 100
g205	—,**1-phenyl-3(2-tolyl)-**		225.30	nd (aa)	123–5 (130)				s	v	...	...	v	chl v	**B12**[2], 445
g206	—,**1,1,3,3-tetraphenyl-***	$HN:C[N(C_6H_5)_2]_2$	363.47	rh (lig)	130–1				i	v	v	...	v		**B12,** 430
g207	—,**1,1,3-triphenyl-***	$HN:C(NHC_6H_5)N(C_6H_5)_2$	287.37	ta (dil al)	134				δ	v	v	...	δ		**B12**[2], 242
Ω g208	—,**1,2,3-triphenyl-***	$C_6H_5N:C(NHC_6H_5)_2$	287.37	nd or pr (al)	146–7	d	1.163^{20}_4		δ^h	s	...	...	...		**B12**[2], 246
g209	—,—,hydrochloride monohydrate	$C_6H_5N:C(NHC_6H_5)_2.HCl.H_2O$	341.85	pr (w+1)	241–2d				δ	...	i	...	i	HCl δ	**B12**[2], 246
g210	—,**1-ureido-***	1-Carbamoylguanidine. Guanylurea. Dicyandiamidine. $HN:C(NH_2)NHCONH_2$	102.10	pr (al) $\lambda^{0.1N\,NaOH}$ 218 (4.29)	105	160d			s^h	δ	i	...	i	Py s chl, CS_2 i	**B3,** 89

For explanations, symbols and abbreviations see beginning of table. For structural formulas see end of table.

No.	Name	Synonyms and Formula	Mol. wt.	Color, crystalline form, specific rotation and λ_{max} (log ε)	m.p. °C	b.p. °C	Density	n_D	Solubility w	al	eth	ace	bz	other solvents	Ref.
	Guanine														
Ω g211	**Guanine**	2-Aminohypoxanthine. 6-hydroxy-2-aminopurine.	151.13	nd or pl (NH_4OH) $\lambda^{w, pH=11}$ 245 (3.81), 273 (3.85)	360d	sub	...	...	i	δ	δ	...	...	NH_4OH, alk, ac s, aa i	**B26**[2], 262
Ω g212	**Guanosine**	9-D-Ribosidoguanine. Vernine. $C_{10}H_{13}N_5O_5$.	283.25	nd (w), $[\alpha]_D^{20}$ −60.5 (0.1N NaOH, p = 3) λ^w 254 (4.13)	239d (250d)	...	...	...	δ s^h	i	i	...	...	aa v^h	**B31,** 28
	Guanosine phosphoric acid	*see* **Guanylic acid**													
g213	**Guanylic acid**	Guanosine 3′-phosphate. Guanosine phosphoric acid. GMP. $C_{10}H_{14}N_5O_8P$.	363.23	nd or pr (w + 2) $[\alpha]_D^{20}$ −7.5 (w, p = 1) $[\alpha]_D^{25}$ −65 (5 % NaOH, c = 2) λ^w 260 (4.07)	208d (anh) (218d)	...	...	...	s v^h	...	...	...	...	...	**B31,** 29
—	**Gulitol**	*see* **Sorbitol**													
g214	**Gulonic acid,** γ-lactone(*d*)		178.14	pr ta (w) $[\alpha]_D^{20}$ +55.1(w)	180–1	...	...	...	s	δ	i	...	i	...	**B18**[2], 193
g215	—,—(*l*)	$C_6H_{10}O_6$. *See* g214	178.14	rh pr (w) $[\alpha]_4^{20}$ −57.1(w)	185	...	...	...	v^h	δ^h	i	...	i	...	**B18**[2], 193
g216	—,phenyl-hydrazide(*D*)	...	286.29	(w), $[\alpha]_D^{20}$ +13.45	147–9	d 195	...	...	v^h	v^h	...	...	...	...	**B15,** 332
g217	—,**1-oxo-**(*L*)	...	194.14	cr (w), $[\alpha]_D^{18}$ −48 (w, c = 1)	171 δd	...	...	...	δ	...	...	δ	...	alk s aa δ	**B3**[2], 522
g218	**Gulose**(*D*)	...	180.16	syr $[\alpha]_D^{20}$ +61.6	...	d	...	...	s	δ	...	...	...	MeOH δ	**B31,** 282
g219	—(*L*)	$C_6H_{12}O_6$. *See* g218	180.16	syr $[\alpha]_D^{20}$ −20.4	...	d	...	...	s	δ	...	...	...	MeOH δ	**B31,** 282
g220	—,osazone(*D*)	D-Idosazone.	358.40	$[\alpha]$ +0.5 (al-Py) (mut)	168 (180d)	...	...	...	...	...	...	...	...	...	**B31,** 355
g221	**Guvacine**	1,2,5,6-Tetrahydronicotinic acid.	127.14	pr (w), rods (+1w, dil al)	295d (285d)	...	...	...	s	i	i	...	i	chl i	**B22**[1], 489

For explanations, symbols and abbreviations see beginning of table. For structural formulas see end of table.

No.	Name	Synonyms and Formula	Mol. wt.	Color, crystalline form, specific rotation and λ_{max} (log ε)	m.p. °C	b.p. °C	Density	n_D	Solubility w	al	eth	ace	bz	other solvents	Ref.
	Halazone														
—	**Halazone**	*see* **Benzoic acid, 4(dichlorosulfamyl)-***													
h1	**Halostachine(*l*)**	$C_6H_5CH(OH)CH_2NHCH_3$	151.21	[α] −47	43–5	....	....	....	s	s	s	i	...	....	**C42,** 2245
Ω h2	**Harmaline**	3,4-Dihydroharmine. $C_{13}H_{14}N_2O$.	214.27	ta (MeOH), rh, pr (al) $\lambda^{0.1\ NHCl}$ 264 (4.07), 380 (4.49)	250d	....	....	....	δ	δ s^h	δ	...	...	chl, Py s	**B23**², 345
Ω h3	**Harmine**	Banisterine. Telepathine. Yageine. $C_{13}H_{12}N_2O$.	212.55	rh (al), pr (MeOH) λ^{MeOH} 240 (4.63), 301 (4.13), 322 (4.11)	272–4 (264–5d)	sub	....	....	δ	δ	δ	...	...	Py s chl δ	**B23**², 348
h4	**Hecogenin**	$C_{27}H_{42}O_4$	430.64	pl (eth), $[\alpha]^{22}_D$ +7 λ^{sulf} 351 (3.86), 396 (4.20)	265–8 (253)	....	....	....		s	s	s	...	lig, MeOH s	**J1955,** 1966
h5	—,acetate	$C_{29}H_{44}O_5$. *See* h4.	472.67	cr (MeOH)	243 (252)	....	....	....	i	s^h	...	...	...	....	**Am 77,** 5196
h6	**Hederagenin**	$C_{30}H_{48}O_4$	472.72	pr (al), $[\alpha]^{21}_D$ +70.1 (chl-MeOH), $[\alpha]^{23}_D$ +80 (Py, c = 0.7)	332–4	....	....	....	i	s	δ	δ	δ	Py, aa, alk s	**B10**², 305
Ω h7	**Helenine**	Alantolactone	232.33	nd	76	275^{760} 197^{10}	....	....	δ^h	v	v	...	s	chl, aa, alk, lig s	**Am 79,** 1009
h8	**Helicin**	Salicylaldehyde-*D*-glucoside.	284.27	nd (w), $[\alpha]^{20}_D$ −60.4 (w)	175	....	....	....	v^h	s	δ s^h	...	...	....	**B31,** 223
h9	**Helvolic acid**	Fumigacin. $C_{33}H_{44}O_8$	554.69	nd (dil aa) $[\alpha]^{18}_D$ −125 (chl, c = 1) λ^{al} 231 (4.29), 322 (1.99)	212d	....	....	....	δ	δ s^h	s	s	s	aa, diox s MeOH s^h peth δ	**Am 78,** 5275
Ω h10	**Hematein**	Haematein. $C_{16}H_{12}O_6$	300.27	red-br cr λ^{diox} 277.5 (3.86), 295 (3.51), 308.5 (3.46), 430 (4.6)	250d	....	....	....	i	δ	i	...	i	NH_3 v aa δ chl i	**B18**², 221
h11	**Hematin**	Hydroxyhemin. Ferriporphyrin hydroxide. $C_{34}H_{32}N_4O_4.FeOH$	633.51	br pw (Py) λ^{w} 550, 575	>200	....	....	....	i	s^h	i	...	...	alk s Py δ^h aa δ	**M,** 508
h12	**Hematommic acid**		244.17	nd (aa)	172–3	....	....	....	...	...	...	...	...	aa s^h	**H16,** 282
Ω h13	**Hematoporphyrin**	Photodyn. $C_{34}H_{38}N_4O_6$	598.71	red λ^{al} 494 (2.7), 529 (2.5)	....	....	....	....	i	s	δ	...	...	chl δ	**M,** 509
Ω h14	**Hematoxylin**	Hydroxybrasilin. Haematoxylin. $C_{16}H_{14}O_6.3H_2O$	356.33	yesh cr, [α] +11 (w, c = 3.7) $\lambda^{0.1NKOH}$ 499, 534.4, 565, 615.5	140 (120)	....	....	....	δ s^h	s^h	δ	...	...	alk s	**B17**², 273
—	**Hemellitic acid**	*see* **Benzoic acid, 2,3-dimethyl-**													
—	**Hemimellitene**	*see* **Benzene 1,2,3-trimethyl-***													
—	**Hemimellitic acid**	*see* **1,2,3-benzenetricarboxylic acid***													
h15	**1,10-Hendecadiyne***	$HC{:}C(CH_2)_7C{:}CH$	148.25	....	−17.0	83^{12}	0.8182^{21}_4	1.453^{21}				s	s	....	**B1**², 248
Ω h16	**Hendecanal***	*n*-Undecylaldehyde. $CH_3(CH_2)_9CHO$	170.30	....	−4	117^{18}	0.8251^{23}_4	1.4520^{20}	i	s	s	...	...	....	**B1,** 712
h17	—,oxime*	$CH_3(CH_2)_9CH{:}NOH$	185.31	wh nd (dil MeOH)	72	....	....	....	s	s	s	...	...	....	**B1,** 713
h18	—,2-methyl-*	$CH_3(CH_2)_8CH(CH_3)CHO$	184.33	....	....	114^{10}	0.830^{15}_4	1.4321^{20}	δ	s	s	...	...	....	**C47,** 831
Ω h19	**Hendecane***	Undecane*. $CH_3(CH_2)_9CH_3$	156.32	....	−25.59	195.9^{760} 75^{10}	0.74017^{20}_4	1.4172^{20}	i	∞	∞	...	...	....	**B1**³, 534
h20	—,1-amino-*	$CH_3(CH_2)_{10}NH_2$	171.33	cr (eth or al)	17 (15.5)	242^{760} 112.3^{10}	0.7979^{20}_4	1.4398^{20}	s^h	s	i	...	...	....	**B4**², 658
h21	—,1-bromo-11-fluoro-*	$F(CH_2)_{11}Br$	253.21	....	....	$95^{0.6}$	....	1.4518^{25}	i	v	v	...	...	....	**C51,** 7300
Ω h22	**Hendecanoic acid**	Undecanoic acid.* Undecylic acid. $CH_3(CH_2)_9CO_2H$	186.30	cr (ace)	28.6 α 13.4 β 16.3	280^{760} 164^{15}	0.8907^{30}_4	1.4294^{45}	i	v	s	v	∞	chl v	**B2**², 314
Ω h23	—,ethyl ester*	$CH_3(CH_2)_9CO_2C_2H_5$	214.35	....	(i) −19.5 (unst) (ii) −15 (st) (iii) −27.8	131^{14}	0.8633^{20}_4	1.4285^{20}	i	s	s	s	s	os s	**B2**¹, 154

For explanations, symbols and abbreviations see beginning of table. For structural formulas see end of table.

No.	Name	Synonyms and Formula	Mol. wt.	Color, crystalline form, specific rotation and λ_{max} (log ε)	m.p. °C	b.p. °C	Density	n_D	Solubility w	al	eth	ace	bz	other solvents	Ref.
	Hendecanoic acid														
h24	—,nitrile	Decyl cyanide. $CH_3(CH_2)_9CN$	167.30	...	...	253^{760} 143^{22}	0.8254_4^{30}	1.4293^{30}	i	s	s	...	...	...	**B2**, 358
h25	**—,2-bromo-***	$CH_3(CH_2)_8CHBrCO_2H$	265.20	...	10	178–183^{14}	...	...	i	...	...	...	...	alk s	**B2**^{1}, 155
h26	**—,10-bromo-***	$CH_3CHBr(CH_2)_8CO_2H$	265.20	cr (peth-eth, bz or ace)	35.7	...	...	...	i	v	...	...	...	os v	**B2**^{2}, 315
Ω h27	**—,11-bromo-***	$Br(CH_2)_{10}CO_2H$	265.20	nd (lig)	57 (52)	188^{18}	...	...	i	v	s	s	s	os v	**B2**^{2}, 315
h28	**—,10,11-di-bromo-***	$BrCH_2CHBr(CH_2)_8CO_2H$	344.10	cr	38	...	...	...	i	s	...	...	...	...	**B2**^{2}, 315
h29	**—,11-fluoro-***	$F(CH_2)_{10}CO_2H$	204.29	...	36	113–$5^{0.25}$	...	...	...	v	v	...	...	lig v	**C52**, 1918
Ω h30	**—,4-hydroxy-, lactone***	γ-Undecalactone.	184.28	...	...	286 162^{13}	0.9494_4^{20}	1.4512^{20}	i	s	...	...	...	...	**B17**^{2}, 294
Ω h31	**1-Hendecanol***	1-Undecyl alcohol. $CH_3(CH_2)_9CH_2OH$	172.31	...	19	243^{760} 131^{15}	0.8298_4^{20}	1.43918^{20}	i	s	v	...	...	...	**J1948**, 1814
h32	**2-Hendecanol(*d*)***	2-Undecanol. $CH_3(CH_2)_8CH(OH)CH_3$	172.31	$[\alpha]_D^{20}$ +10.29 (bz)	12	128^{20}	0.8270_4^{20}	1.4369^{20}	i	s	s	...	...	...	**B1**^{3}, 1774
h33	**—(*dl*)***	$CH_3(CH_2)_8CH(OH)CH_3$	172.31	...	0	228^{760} 119^{12}	0.8268^{19}	1.4369^{20}	i	s	s	...	...	...	**B1**^{3}, 1774
h34	**—(*l*)***	$CH_3(CH_2)_8CH(OH)CH_3$	172.31	$[\alpha]_D$ −0.02	...	231–3	0.8302^{20}	1.4381^{20}	i	s	s	...	...	...	**B1**^{3}, 1774
h35	**3-Hendecanol(*l*)***	3-Undecanol. $CH_3(CH_2)_7CH(OH)CH_2CH_3$	172.31	$[\alpha]_D^{20}$ −6.22 (al), −7.08 (bz)	17	229 117^{16}	0.8295_4^{20}	1.4367^{20}	i	s	s	...	s	...	**B1**^{2}, 462
h36	**5-Hendecanol***	5-Undecanol. $CH_3(CH_2)_5CH(OH)(CH_2)_3CH_3$	172.31	...	−3.5	229^{760} 107^{12}	0.8292_4^{20}	1.4354^{24}	i	s	s	...	...	...	**B1**^{2}, 462
h37	**6-Hendecanol***	6-Undecanol. $CH_3(CH_2)_4CH(OH)(CH_2)_4CH_3$	172.31	cr (ace)	25	228^{760} (235^{754}) 117–8^{16}	0.8334_4^{20}	1.4374^{20}	i	s	...	s	...	...	**B1**^{2}, 462
h38	**1-Hendecanol, 2-methyl-***	$CH_3(CH_2)_8CH(CH_3)CH_2OH$	186.34	...	...	129–31^{12}	0.8300_4^{15}	1.4382^{20}	δ	s	s	...	...	...	**C54**, 2880
Ω h39	**2-Hendecanone***	2-Undecanone. $CH_3(CH_2)_8COCH_3$	170.30	...	15	231.5–2.5^{760} 106^{12}	0.8250_4^{20}	1.42907^{20}	i	s	s	s	s	chl s	**B1**^{3}, 2109
Ω h40	**3-Hendecanone***	3-Undecanone. $CH_3(CH_2)_7COCH_2CH_3$	170.30	...	12	227^{760} 104–6^{11}	0.8272_4^{20}	1.4296^{20}	i	s	s	...	...	...	**B1**^{2}, 767
Ω h41	**4-Hendecanone***	4-Undecanone. $CH_3(CH_2)_6COCH_2CH_2CH_3$	170.30	...	4–5	106^{13}	0.8274_4^{25}	1.4248^{24}	i	s	s	...	...	...	**B1**^{2}, 767
Ω h42	**5-Hendecanone***	5-Undecanone. $CH_3(CH_2)_5CO(CH_2)_3CH_3$	170.30	...	2	227^{760} 105–6^{12}	0.8278_4^{19}	1.4275^{18}	i	s	s	...	...	...	**B1**^{2}, 767
Ω h43	**6-Hendecanone***	*n*-Caprone. 6-Undecanone. $CH_3(CH_2)_4CO(CH_2)_4CH_3$	170.30	lf λ^{MeOH} 278 (1.69)	14–5	228^{760} 145.7^{29}	0.8308_4^{20}	1.4270^{20}	i	v	v	...	...	...	**B1**^{2}, 768
h44	**4-Hendecanone, 7-ethyl-2-methyl-***	$CH_3(CH_2)_3CH(C_2H_5)CH_2CH_2COCH_2CH(CH_3)_2$	212.38	...	...	252–3^{760} 101–3^4	0.8362_4^{20}	1.4370^{20}	i	v	v	v	v	os v	**C31**, 6773
h45	**Hendecasiloxane, tetracosamethyl-**	$(CH_3)_3SiO[Si(CH_3)_2O]_9Si(CH_3)_3$	829.78	...	...	322.8^{760} $202^{4.7}$	0.930^{20}	1.3994^{20}	...	δ	...	...	s	...	**Am 68**, 691
h46	**1-Hendecene***	1-Undecene. α-Undecylene. $CH_3(CH_2)_8CH{:}CH_2$	154.30	...	−49.19	192.7^{760} 72^{10}	0.7503_4^{20}	1.42609^{20}	i	...	s	...	...	chl, lig s	**B1**^{3}, 866
h47	**2-Hendecene(*trans*)***	2-Undecene. $CH_3(CH_2)_7CH{:}CHCH_3$	154.30	...	−48.3	192–3^{760} 75^{10}	0.7528_4^{20}	1.4292^{20}	...	...	...	...	...	...	**B1**, 225
h48	**5-Hendecene(*trans*)***	5-Undecene. $CH_3(CH_2)_4CH{:}CH(CH_2)_3CH_3$	154.30	...	−61.1	192^{760} 73^{10}	0.7497_4^{20}	1.4285^{20}	i	...	s	...	...	chl, lig s	**B1**^{3}, 867
h49	**9-Hendecenoic acid(*trans*)***	$CH_3CH{:}CH(CH_2)_7CO_2H$	184.28	...	19	273 121–$3^{0.7}$	0.9119_0^{25}	1.4519^{20}	...	...	s	...	...	...	**B2**^{2}, 420
Ω h50	**10-Hendecenoic acid***	$CH_2{:}CH(CH_2)_8CO_2H$	184.28	cr	24.5	275^{760} 165^{15}	0.9072_4^{24}	1.4486^{24}	i	s	s	...	...	...	**B2**^{2}, 419
Ω h51	**—,ethyl ester***	$CH_2{:}CH(CH_2)_8CO_2C_2H_5$	212.34	...	−38	263.5–5.5 131.5^{16}	0.88271^{15}	1.4449^{23}	i	s	s	...	...	aa s	**B2**^{2}, 420
Ω h52	**—,methyl ester***	$CH_2{:}CH(CH_2)_8CO_2CH_3$	198.31	...	−27.5	248^{760} 124^{10}	0.889^{15}	1.43928^{20}	i	s	s	...	...	aa s	**B2**^{2}, 420
Ω h53	**10-Hendecen-1-ol***	Undecenyl alcohol. $CH_2{:}CH(CH_2)_8CH_2OH$	170.30	...	−2	250 132–3^{15}	0.8495_4^{15}	1.4500^{20}	i	s	s	...	...	...	**C53**, 21717
h54	**1-Hendecen-3-yne***	$CH_3(CH_2)_6C{:}CCH{:}CH_2$	150.27	...	...	74^9	0.7962_4^{20}	1.4606^{20}	...	...	...	...	...	...	**C35**, 3499
h55	**1-Hendecyne***	1-Undecyne. $CH_3(CH_2)_8C{:}CH$	152.28	...	−25 (−17.7)	195^{760} 73.2^{10}	0.7728_4^{20}	1.4306^{20}	i	s	s	s	s	oos s	**B1**, 240
h56	**5-Hendecyne***	5-Undecyne. $CH_3(CH_2)_4C{:}C(CH_2)_3CH_3$	152.28	...	−74.1	198.1^{760} 78.3^{10}	0.7753_4^{20}	1.4369^{20}	...	...	...	...	...	...	**Am 61**, 2897
h57	**9-Hendecynoic acid***	$CH_3C{:}C(CH_2)_7CO_2H$	182.27	pl (dil al)	61	170^{15} $132^{0.1}$	...	...	v	v	v	...	...	...	**B2**^{2}, 457
h58	**10-Hendecynoic acid***	$HC{:}C(CH_2)_8CO_2H$	182.27	lf	43–4	175^{15} 130^3	0.9060^{25}	...	...	...	...	...	...	...	**B2**^{2}, 456
h59	**Heneicosane***	$CH_3(CH_2)_{19}CH_3$	296.59	cr (w)	40.5	356.5^{760} 203^{10}	0.7917_4^{20}	1.4441^{20}	i	δ	...	...	...	peth s	**B1**^{3}, 573
h60	**Heneicosanoic acid***	$CH_3(CH_2)_{19}CO_2H$	326.57	nd (ace)	82 (75)	...	...	...	δ	...	s	s^h	s	chl s	**B2**^{2}, 372
h61	**Hentriacontane***	$CH_3(CH_2)_{29}CH_3$	436.86	lf (AcOEt)	67.9	458^{760} 302^{15}	0.8111_4^{20} 0.781_4^{68}	1.4543^{20} 1.4278^{90}	...	δ	δ	...	δ	aa s^h peth s chl δ	**B1**^{3}, 585
Ω h61[1]	**Hentriacontanoic acid***	Melissic acid. $CH_3(CH_2)_{29}CO_2H$	466.84	sc or nd (al or ace)	93–3.2	...	...	...	i	δ s^h	δ^h	δ	s	MeOH δ^h	**B2**^{2}, 382
h62	**16-Hentriacontanone**	Palmitone. $CH_3(CH_2)_{14}CO(CH_2)_{14}CH_3$	450.84	lf (al) λ^{hx} 275 (1.50)	83	...	0.7947_4^{91}	1.4297^{94}	i	δ	s	δ	δ	chl δ	**B1**^{2}, 776

For explanations, symbols and abbreviations see beginning of table. For structural formulas see end of table.

No.	Name	Synonyms and Formula	Mol. wt.	Color, crystalline form, specific rotation and λ_{max} (log ε)	m.p. °C	b.p. °C	Density	n_D	Solubility w	al	eth	ace	bz	other solvents	Ref.
	Heptachlor														
—	**Heptachlor**	*see* **Dicyclopentadiene, 3,4,5,6,7,8,9-heptachloro-**													
h63	**Heptacosane***	$CH_3(CH_2)_{25}CH_3$	380.75	cr (al, bz), lf (AcOEt)	59.5	422.1^{760} 270^{15}	0.8050_4^{20} 0.7796_4^{60}	1.4511^{20} 1.4345^{65}	i	i	δ	...	...	...	**B1**^{3}, 581
h64	**7,10-Heptadecadiyne***	$CH_3(CH_2)_5C{:}CCH_2C{:}C(CH_2)_5CH_3$	232.41	...	...	150^6	0.84_4^{19}	1.4700^{19}	...	...	...	...	...	...	**C28,** 4035
h65	**Heptadecanal***	Margaraldehyde. $CH_3(CH_2)_{15}CHO$	254.46	nd (peth), cr (al + 1)	36 52 (+al)	204^{26}	...	...	i	δ	s	δ	s	aa v AcOEt δ	**B1,** 717
Ω h66	**Heptadecane***	$CH_3(CH_2)_{15}CH_3$	240.48	hex lf	22	301.8^{760} 161.75^{10}	0.7780_4^{20}	1.4369^{20}	i	δ	s	...	...	aa δ	**B1**^{3}, 563
Ω h67	—,1-amino-*	Heptadecylamine. $CH_3(CH_2)_{16}NH_2$	255.49	...	49	336^{760} 188.9^{10}	0.8510_4^{20}	1.4510^{20}	i		s	...	...	...	**B4**^{2}, 660
Ω h68	—,1-bromo-*	*n*-Heptadecyl bromide. $CH_3(CH_2)_{16}Br$	319.38	...	32	349^{760} 199^{10}	0.9916_4^{20}	1.4625^{20}	i	...	...	...	...	chl v	**B1**^{3}, 564
h69	—,1,17-dibromo-*	$Br(CH_2)_{17}Br$	398.28	lf (al)	38–8.4	$208–10^3$	...	...	i	δ	...	...	...	chl v	**B1**^{3}, 564
Ω h70	**Heptadecanoic acid***	Margaric acid. $CH_3(CH_2)_{15}CO_2H$	270.46	pl (peth)	62–3	227^{100}	0.8532^{60}	1.4342^{60}	...	δ	s	s	s	chl s	**Am 64,** 2739
h71	—,ethyl ester*	$CH_3(CH_2)_{15}CO_2C_2H_5$	298.52	pl (dil al)	28	185^5	...	...	i	s	v	s	s	os s	**B2**^{2}, 344
Ω h72	—,methyl ester*	$CH_3(CH_2)_{15}CO_2CH_3$	284.49	pl (al)	30	$184–7^9$	...	...	i	s	v	s	v	os s	**B2**^{1}, 169
h73	—,nitrile	Cetyl cyanide. Margaronitrile. $CH_3(CH_2)_{15}CN$	251.46	cr (al)	34	349^{760} 183^{10}	0.8315_4^{20}	1.4467^{20}	i	δ v^h	v	...	...	...	**B2**^{2}, 345
Ω h74	**1-Heptadecanol***	$CH_3(CH_2)_{16}OH$	256.48	lf (al), cr (ace)	54	308^{760}	0.8475_4^{20}	...	i	s	s	...	...	...	**B1**^{3}, 1830
h75	**2-Heptadecanol***	$CH_3(CH_2)_{14}CH(OH)CH_3$	256.48	pl (dil al)	54 (46–7)	$140^{0.5}$	...	1.4407^{37}	i	s	s	...	...	...	**C30,** 8255
h76	**9-Heptadecanol***	$CH_3(CH_2)_7CH(OH)(CH_2)_7CH_3$	256.48	pl (dil al)	61	174^9	...	1.4262^{80}	i	s	s	s	s	os s	**B1,** 430
Ω h77	**2-Heptadecanone***	Methyl heptadecyl ketone. $CH_3(CH_2)_{14}COCH_3$	254.46	pl (dil al)	48	320 246^{110}	0.8140_{48}^{48}	...	i	δ	v	s	v	peth s chl v	**B1**^{2}, 772
Ω h78	**9-Heptadecanone***	Nonylone. Pelargone. $CH_3(CH_2)_7CO(CH_2)_7CH_3$	254.46	pl (MeOH)	53	250–3 142^1	...	...	...	δ	...	...	...	MeOH s^h	**B1**^{2}, 772
h79	**1-Heptadecanone, 1-phenyl-***	Stearophenone. $CH_3(CH_2)_{15}COC_6H_5$	330.56	...	56–6.5	...	...	...	i	...	s	...	...	...	**B7**^{2}, 272
h80	**1-Heptadecene***	$CH_3(CH_2)_{14}CH{:}CH_2$	238.46	...	11.2	300^{760} 160^{10}	0.7852_4^{20}	1.4432^{20}	i		v	...	s	lig ∞	**B1**^{3}, 878
h81	**1,4-Heptadiene***	$CH_3CH_2CH{:}CHCH_2CH{:}CH_2$	96.17	...	...	93^{772} (101)	0.7270_4^{20}	1.4370^{20}	i	...	s	...	s	peth s	**B1**^{3}, 999
h82	**1,5-Heptadiene***	$CH_3CH{:}CHCH_2CH_2CH{:}CH_2$	96.17	...	...	94^{760}	0.7186_4^{20}	1.4200^{20}	i	s	s	s	s	os s	**B1**^{3}, 999
h83	**2,4-Heptadiene***	$CH_3CH_2CH{:}CHCH{:}CHCH_3$	96.17	λ^{al} 227 (4.41)	...	108^{760}	0.7384_4^{20}	1.4578^{20}	i	s	s	s	s	peth s	**B1**^{3}, 1000
h84	**1,5-Heptadien-4-ol***	$CH_3CH{:}CHCH(OH)CH_2CH{:}CH_2$	112.17	...	...	$155–6^{742}$ 68^{24}	0.8598_4^{20}	1.4510^{25}	δ	...	...	s	...	...	**B2**^{3}, 1992
h85	**3,5-Heptadien-2-one***	Crotonylideneacetone. $CH_3CH{:}CHCH{:}CHCOCH_3$	110.16	λ^{al} 271 (4.34)	...	88^{28} 63^{10}	0.8946_4^{19}	1.51767^{19}	...	...	s	...	...	...	**B1**^{2}, 809
Ω h86	**2,5-Heptadien-4-one-2,6-dimethyl-***	Diisopropylidenacetone. Phorone. $CH_3C(CH_3){:}CHCOCH{:}C(CH_3)CH_3$	138.21	ye gr pr	28	197.8^{760}	0.8850_4^{20}	1.49982^{20}	δ	s	s	s	...	...	**B1,** 810
h87	**1,6-Heptadien-3-yne***	$CH_2{:}CHCH_2C{:}CCH{:}CH_2$	92.14	...	...	110^{750}	0.787_4^{25}	1.4694^{25}	...	...	...	...	s	peth s	**Am 58,** 612
h88	**1,5-Heptadiyne***	$CH_3C{:}CCH_2CH_2C{:}CH$	92.14	...	...	26^{30}	0.8100_4^{21}	1.4521^{21}	i	...	...	...	s	peth s	**B1**^{2}, 247
h89	**1,6-Heptadiyne***	$HC{:}CCH_2CH_2CH_2C{:}CH$	92.14	...	−85	112^{760} 36^{20}	0.8164_4^{17}	1.451^{17}	i	...	...	...	s	aa s	**B1**^{3}, 1061
Ω h90	**Heptanal***	Enanthaldehyde. Heptaldehyde. $CH_3(CH_2)_5CHO$	114.19	...	−43.3	152.8^{760} 59.6^{30}	0.8495_4^{20}	1.4113^{20}	δ	∞	∞	...	...	...	**B1**^{2}, 750
h91	—,oxime*	$CH_3(CH_2)_5CH{:}NOH$	129.20	pl (al)	57–8	195^{760} 100.5^{14}	0.8583^{55}	1.4210^{20}	δ	s	s	...	...	...	**B1**^{1}, 752
Ω h92	—,2-benzylidene-	α-Pentylcinnamaldehyde. $CH_3(CH_2)_4C({:}CHC_6H_5)CHO$	202.30	ye oil	80	$174–5^{20}$ 140^5	0.97108^{20}	1.5381^{20}	i			s	...	...	**Am 53,** 3122
Ω h93	**Heptane***	$CH_3(CH_2)_5CH_3$	100.21	...	−90.61	98.42^{760} -2.1^{10}	0.68376_4^{20}	1.38777^{20}	i	v	∞	∞	∞	chl, peth ∞	**B1**^{3}, 415
Ω h94	—,1-amino-*	$CH_3(CH_2)_6NH_2$	115.22	...	−18	156.9^{760} 45.6^{10}	0.7754_4^{20}	1.4251^{20}	δ	∞	∞	...	...	...	**B4**^{2}, 652
h95	—,2-amino-*	$CH_3(CH_2)_4CH(NH_2)CH_3$	115.22			142^{760}	0.7665^{19}	1.41997^{19}	δ	s	s	...		peth s	**B4**^{3}, 374
Ω h96	—,1-bromo-*	$CH_3(CH_2)_6Br$	179.11	...	−56.1	178.9^{760} 59.7^{10}	1.1400_4^{20}	1.4502^{20}	i	v	v	...	...	chl s	**B1**^{3}, 431
h97	—,1-bromo-7-fluoro-*	$F(CH_2)_7Br$	197.10	...	...	85^{11}	...	1.4463^{20}	i	v	v	...	...	...	**JOC, 1956,** 739
Ω h98	—,1-chloro-*	$CH_3(CH_2)_6Cl$	134.65	...	−69.5	159^{760} 45^{10}	0.8758_4^{20}	1.4256^{20}	i	∞	∞	...	...	chl s	**B1**^{2}, 117
h99	—,2-chloro-*	$CH_3(CH_2)_4CHClCH_3$	134.65	...	...	46^{19}	0.8672^{20}	1.4221^{20}	i	...	v	...	s	chl, aa s	**B1**^{3}, 429
h100	—,3-chloro-*	$CH_3(CH_2)_3CHClCH_2CH_3$	134.65	...	...	144^{751} 48.3^{20}	0.8960_4^{20}	1.4228^{20}	i	...	v	...	s	chl v	**B1**^{2}, 430
h101	—,4-chloro-*	$(CH_3CH_2CH_2)_2CHCl$	134.65	...	...	144^{758} 48.9^{21}	0.8710_4^{20}	1.4237^{20}	i	...	v	...	s	chl s	**B1**^{2}, 117
h102	—,3-chloro-2,3-dimethyl-*	$CH_3(CH_2)_3CCl(CH_3)CH(CH_3)_2$	162.71	...	...	54^8 42^2	0.8395^{20}	1.4391^{20}	...	...	s	...	...	chl s	**Am 55,** 812
h103	—,5-chloro-2,5-dimethyl-*	$CH_3CH_2CCl(CH_3)CH_2CH_2CH(CH_3)_2$	162.71	...	...	63^{15}	0.8692_4^{18}	1.43457^{15}	i	...	...	...	s	chl s	**B1**^{1}, 64
h104	—,3-chloro-3-ethyl-*	$CH_3(CH_2)_3CCl(C_2H_5)CH_2CH_3$	162.71	...	...	46^3	0.8856_4^{20}	1.4400^{20}	...	...	s	...	...	chl s	**B1**^{3}, 510
h105	—,4-chloro-4-ethyl-*	$(CH_3CH_2CH_2)_2CClCH_2CH_3$	162.71	...	...	67^{12}	0.8821_4^{20}	1.4438^{20}	i	...	v	...	s	chl s	**B1**^{3}, 511

For explanations, symbols and abbreviations see beginning of table. For structural formulas see end of table.

No.	Name	Synonyms and Formula	Mol. wt.	Color, crystalline form, specific rotation and λ_{max} (log ε)	m.p. °C	b.p. °C	Density	n_D	Solubility w	al	eth	ace	bz	other solvents	Ref.
	Heptane														
h106	**—,1-chloro-7-fluoro-***	$F(CH_2)_7Cl$	152.64	...	...	70^{10}	0.993^{20}	1.4222^{25}	i	v	v	...	s	chl s	C51, 7300
h107	**—,2-chloro-2-methyl-***	$CH_3(CH_2)_4CCl(CH_3)_2$	148.68	...	...	50^{15}	0.8568^{25}_{4}	1.4240^{25}	i	v	v	...	s	chl s	$B1^{3}$, 472
h108	**—,2-chloro-6-methyl-***	$(CH_3)_2CH(CH_2)_3CHClCH_3$	148.68	...	...	74^{35}	...	1.4260^{15}	i	v	v	...	s	chl s	$B1^{3}$, 472
h109	**—,3-chloro-3-methyl-***	$CH_3(CH_2)_3CCl(CH_3)CH_2CH_3$	148.68	...	...	64^{27}	0.8764^{20}_{4}	1.4317^{20}	i	v	v	...	s	chl s	$B1^{3}$, 474
h110	**—,4-chloro-4-methyl-***	$(CH_3CH_2CH_2)_2CClCH_3$	148.68	...	...	50^{12}	0.8690^{20}_{4}	1.43098^{15}	i	v	s	...	s	chl s	$B1^{3}$, 477
Ω h110[1]	**—,3(chloro-methyl)-***	$CH_3(CH_2)_3CH(C_2H_5)CH_2Cl$	148.68	...	...	174^{760}	0.8769^{20}_{4}	1.4319^{20}	i	s	s	s	s	os s	C31, 7408
h111	**—,1,1-dichloro-***	$CH_3(CH_2)_5CHCl_2$	169.10	...	...	187^{760} 82^{20}	1.0008^{20}_{4}	1.4440^{20}	i	...	s	...	s	chl v	$B1^{3}$, 430
h112	**—,1,2-dichloro-***	$CH_3(CH_2)_4CHClCH_2Cl$	169.10	...	...	$68–72^{7}$	1.064^{20}_{4}	1.4490^{20}	i	...	s	...	s	chl v	$B1^{3}$, 430
h113	**—,2,2-dichloro-***	$CH_3(CH_2)_4CCl_2CH_3$	169.10	...	...	77^{25}	1.012^{20}_{4}	1.4440^{20}	i	...	s	...	s	chl s	$B1^{3}$, 430
h114	**—,4,4-dichloro-***	$(CH_3CH_2CH_2)_2CCl_2$	169.10	...	...	86^{27}	1.008^{17}	1.448^{17}	i	...	s	...	...	chl s	B1, 154
h115	**—,2,6-dichloro-2,6-dimethyl-***	$(CH_3)_2CCl(CH_2)_3CCl(CH_3)_2$	197.15	...	43	93^{16}	...	...	i	...	s	...	s	chl s	$B1^{3}$, 513
h116	**—,2,3-dimethyl-***	$CH_3(CH_2)_3CH(CH_3)CH(CH_3)_2$	128.26	...	−116	140.5^{760} 29.4^{10}	0.7260^{20}_{4}	1.4088^{20}	i	∞	∞	∞	∞	chl, peth ∞ os v	$B1^{3}$, 511
h117	**—,2,4-dimethyl-***	$CH_3CH_2CH_2CH(CH_3)CH_2CH(CH_3)_2$	128.26	...	...	133.5^{760} 23.8^{10}	0.7143^{20}_{4}	1.4031^{20}	i	∞	∞	∞	∞	os, peth, chl ∞	$B1^{3}$, 512
h118	**—,2,5-dimethyl (*d*)-***	$CH_3CH_2CH(CH_3)CH_2CH_2CH(CH_3)_2$	128.26	$[\alpha]^{27}_{D}$ +4.2	...	136^{760}	0.7198^{20}_{4}	1.4033^{20}	i	∞	∞	∞	∞	os, chl ∞ peth v	$B1^{2}$, 128
Ω h119	**—,2,6-dimethyl-***	Diisobutylmethane. Isobutylisoamyl. $(CH_3)_2CH(CH_2)_3CH(CH_3)_2$	128.26	...	−102.9	135.2^{760} 25.5^{10}	0.7089^{20}_{4}	1.4011^{20}	...	...	...	...	...	...	$B1^{3}$, 512
h120	**—,3,3-dimethyl-***	$CH_3(CH_2)_3C(CH_3)_2CH_2CH_3$	128.26	...	...	137.3^{760} 26^{10}	0.7254^{20}_{4}	1.4087^{20}	i	∞	s	v	v	os v	$B1^{3}$, 513
h122	**—,4-ethyl-***	Ethyldipropylmethane. $CH_3(CH_2)_2CH(C_2H_5)(CH_2)_2CH_3$	128.26	...	...	141.2^{760} 31^{10}	0.7270^{20}_{4}	1.4096^{20}	i	∞	s	∞	∞	chl, lig ∞	C37, 2631
h123	**—,1-fluoro-***	$CH_3(CH_2)_5CH_2F$	118.20	...	−73	117.9^{760}	0.8062^{20}_{4}	1.3854^{20}	i	...	s	s	s	peth v	B1, 56
h124	**—,hexadeca-fluoro-***	Perfluoroheptane. $CF_3(CF_2)_5CF_3$	388.05	...	−78	82.43^{760}	1.7333^{20}	1.2618^{20}	...	...	...	...	...	...	$B1^{3}$, 429
Ω h125	**—,1-iodo-***	$CH_3(CH_2)_5CH_2I$	226.10	...	−48.2	204^{760} 76.16^{10}	1.3791^{20}_{4}	1.4904^{20}	i	s	...	s	...	chl s	$B1^{3}$, 433
Ω h126	**—,2-methyl-***	$CH_3(CH_2)_4CH(CH_3)_2$	114.23	...	−109	117.65^{760} 12.3^{10}	0.6980^{20}_{4}	1.39494^{20}	i	∞	s	∞	∞	chl, peth ∞	$B1^{3}$, 471
h127	**—,3-methyl-(*d*)***	$CH_3(CH_2)_3CH(CH_3)CH_2CH_3$	114.23	$[\alpha]^{26}_{D}$ +9.34	...	115–8	0.7075^{16}_{4}	1.4002^{18}	i	s	s	∞	∞	chl ∞	$B1^{3}$, 475
Ω h128	**—,—(*dl*)***	$CH_3(CH_2)_3CH(CH_3)CH_2CH_3$	114.23	...	−120.5	119 13.3^{10}	0.70583^{20}_{4}	1.3985^{20}	i	∞	s	∞	∞	chl, peth ∞	$B1^{3}$, 476
h129	**—,—(*l*)***	$CH_3(CH_2)_3CH(CH_3)CH_2CH_3$	114.23	$[\alpha]^{25}_{D}$ −3.87	...	$117–8^{745}$	...	1.3990^{20}	i	∞	s	∞	∞	chl ∞	C40, 2048
Ω h130	**—,4-methyl-***	$(CH_3CH_2CH_2)_2CHCH_3$	114.23	...	−121	117.7^{760} 12.4^{10}	0.70463^{20}_{4}	1.39792^{20}	i	∞	s	∞	∞	chl, peth ∞	$B1^{3}$, 476
h131	**—,2(methyl amino)-**	$CH_3(CH_2)_4CH(NHCH_3)CH_3$	129.25	...	...	155^{760}	...	...	δ	...	...	...	...	...	C36, 222
Ω h132	**Heptanedioic acid***	Pimelic acid. $HO_2C(CH_2)_5CO_2H$	160.17	pr (w)	106	272^{100} sub 212^{10}	1.329^{15}	...	s	s	s	...	i s^{h}	...	$B2^{2}$, 586
h133	—,dichloride	$ClCO(CH_2)_5COCl$	197.06	...	...	137^{15}	...	...	d	d	...	...	...	...	B2, 671
Ω h134	—,diethyl ester*	$C_2H_5O_2C(CH_2)_5CO_2C_2H_5$	216.28	...	−24	$252–5^{748}$ $139–41^{15}$	0.99448^{20}	1.43052^{20}	i	s	s	...	...	AcOEt s	$B2^{1}$, 282
Ω h135	—,dimethyl ester*	$CH_3O_2C(CH_2)_5CO_2CH_3$	188.23	...	−21	120^{10} 80^{1}	1.0625^{20}_{4}	1.4309^{20}	δ	s	s	...	...	...	$B2^{2}$, 587
Ω h136	—,dinitrile	1,5-Dicyanopentane. $NC(CH_2)_5CN$	122.17	...	−31.4	175^{14}	0.949^{18}	1.4472^{20}	i	∞	∞	...	...	chl ∞	$B2^{2}$, 587
h137	—,monoethyl ester	$HO_2C(CH_2)_5CO_2C_2H_5$	188.23	cr (eth)	10	182^{18} 160^{4}	...	1.4415^{20}	δ	δ	δ	...	δ	chl δ	$B2^{1}$, 282
h138	**—,4-oxo-***	$OC(CH_2CH_2CO_2H)_2$	174.16	rh pl (w)	143 (141.5)	...	...	...	s^{h}	s	δ	δ	i	...	$B3^{2}$, 487
h139	**1,7-Heptanediol***	Heptamethyleneglycol. $HOCH_2(CH_2)_5CH_2OH$	132.21	...	22	262 151^{14}	0.9569^{25}_{4}	1.4520^{25}	s	s	δ	...	...	...	$B1^{3}$, 2214
Ω h140	**2,4-Heptanediol, 3-methyl-***	$CH_3CH_2CH_2CH(OH)CH(CH_3)CH(OH)CH_3$	146.23	...	...	115^{3}	0.928^{20}_{4}	1.4459^{20}	i	s	...	...	...	...	B1, 491
h141	**2,4-Heptane-dione***	$CH_3CH_2CH_2COCH_2COCH_3$	128.17	...	...	174^{760} 70^{20}	0.9411^{15}_{4}	...	...	...	...	...	...	...	B1, 792
Ω h142	**1-Heptanethiol***	$CH_3(CH_2)_6SH$	132.27	λ^{cy} 225 sh (2.2)	−43	177^{760}	0.8427^{20}_{4}	1.4521^{20}	i	∞	∞	...	...	...	$B1^{3}$, 1684
h143	**2,4,6-Heptane-trione***	Diacetylacetone. $CH_3COCH_2COCH_2COCH_3$	142.16	lf	49	121^{10}	1.0681^{40}_{40}	1.4930^{20}	s d^{h}	s	s	...	...	alk s	C41, 1917
Ω h144	**Heptanoic acid***	Enanthic acid. *n*-Heptylic acid. $CH_3(CH_2)_5CO_2H$	130.19	...	−7.5	223^{760} 116^{11}	0.9200^{20}_{4}	1.4170^{20}	δ	s	s	s	...	...	$B2^{2}$, 294
h145	—,amide	$CH_3(CH_2)_5CONH_2$	129.20	nd (al), lf (w)	96	250–8	0.8521^{110}_{4}	1.4217^{110}	s	s	s	...	...	...	$B2^{2}$, 296
h146	—,anhydride*	$[CH_3(CH_2)_5CO]_2O$	242.36	λ^{MeOH} 274 (3.93), 314 (3.68)	−12.4	268–71 $164^{12.5}$	0.9321^{20}_{4}	1.4335^{15}	i	s	s	...	...	...	B2, 340

For explanations, symbols and abbreviations see beginning of table. For structural formulas see end of table.

No.	Name	Synonyms and Formula	Mol. wt.	Color, crystalline form, specific rotation and λ_{max} (log ε)	m.p. °C	b.p. °C	Density	n_D	Solubility w	al	eth	ace	bz	other solvents	Ref.
	Heptanoic acid														
h147	—,butyl ester*	$CH_3(CH_2)_5CO_2C_4H_9^n$	186.30		−67.5	226.18^{760}	0.86382^{20}	1.4204^{25}	i	s	s	s	s	os s	**B2**, 340
h148	—,chloride*	$CH_3(CH_2)_5COCl$	148.63		−83.8	125.2^{760}	0.9590^{20}	1.4345^{15}	d	d	s	...	...	lig v	**J87**, 93
Ω h149	—,ethyl ester*	$CH_3(CH_2)_5CO_2C_2H_5$	158.24		−66.1	187^{760} 78^{14}	0.8817^{20}_{4}	1.4100^{20}	δ	s	s	...	...	...	**B2**², 295
h150	—,heptyl ester*	$CH_3(CH_2)_5CO_2(CH_2)_6CH_3$	228.38		−33	$276–8^{760}$	0.86492^{19}_{4}	1.4320^{20}	i	s	s	...	...	...	**B2**², 145
h151	—,hexyl ester*	$CH_3(CH_2)_5CO_2(CH_2)_5CH_3$	214.35		−48	261^{760}	0.86114^{20}	1.429^{15}	i	s	s	s	s	os s	**B2**², 295
Ω h152	—,methyl ester*	$CH_3(CH_2)_5CO_2CH_3$	144.22		−56	172^{760}	0.8815^{20}_{4}	1.4152^{20}	δ	s	s	s	...	...	**B2**², 296
h153	—,2-methylpropyl ester*	$CH_3(CH_2)_5CO_2CH_2CH(CH_3)_2$	186.30			209^{759}	0.8593^{20}		i	s	s	s	s	os s	**B2**¹, 145
h154	—,nitrile*	$CH_3(CH_2)_5CN$	111.19			183^{765} (187) $70–2^{10}$	0.8107^{20}_{0}	1.41037^{30}	i	...	s	s	s	aa s	**B2**², 296
h155	—,octyl ester*	$CH_3(CH_2)_5CO_2(CH_2)_7CH_3$	242.41		−22.5	290^{760}	0.85961^{20}	1.43488^{15}	i	s	s	s	s	os s	**B2**, 340
h156	—,pentyl ester*	$CH_3(CH_2)_5CO_2(CH_2)_4CH_3$	200.33		−50	245.44^{760}	0.86232^{20}	1.42627^{15}	i	s	s	s	s	os s	**B2**³, 766
h157	—,*p*-phenylphenacyl ester		324.43		62										**Am 52**, 3715
h158	—,propyl ester*	$CH_3(CH_2)_5CO_2CH_2CH_2CH_3$	172.27	cr (peth)	−63.5	207.9^{760}	0.86412^{15}_{4}	1.41835^{15}	i	s	s	s	s	os s	**B2**, 340
h159	—,**7-amino-***	$H_2N(CH_2)_6CO_2H$	145.20	cr (w, MeOH-peth)	195 (188)				s^h	s^h	i	s^h	...		**B4**, 459
h160	—,**2-bromo-***	$CH_3(CH_2)_4CHBrCO_2H$	209.09			250d 147^{12}	1.319^{18}	1.471^{18}	...	...	s	s	...	...	**B2**², 296
h161	—,**7-bromo-***	$Br(CH_2)_6CO_2H$	209.09	wh cr (dil al)	31	280			i	s	s	s	s	os s lig i	**B2**, 341
h162	—,**7-fluoro-***	$F(CH_2)_6CO_2H$	148.18			133^{10}	1.039^{20}	1.4207^{25}	δ	δ	δ	...	...	...	**C50**, 14528
h163	—,**3-hydroxy-4-methyl-***	$CH_3CH_2CH(CH_3)CH(OH)CH_2CH_2CO_2H$	160.22	ye pw (w)	95–100				δ	s	i	v	i	aa s peth i	**B3**¹, 127
h164	—,**7-iodo-***	$I(CH_2)_6CO_2H$	256.09	lf (dil al)	49–51				δ	s	s	s	s	os s	**B2**, 341
h165	—,**6-oxo-***	$CH_3CO(CH_2)_4CO_2H$	144.17		40–2	$250–3^{280}$ 135^{1}		1.4306^{25}	s	s	s	s	...	...	**B3**¹, 242
Ω h166	**1-Heptanol***	$CH_3(CH_2)_6OH$	116.21		−34.1	176^{760}	0.8219^{20}_{4}	1.4249^{20}	δ	∞	∞	...	...	...	**B1**³, 1679
h167	**2-Heptanol(*d*)***	$CH_3(CH_2)_4CH(OH)CH_3$	116.21	$[\alpha]^{20}_{D}$ +11.4 (al), +13.7 (bz)		160–2 73^{20}	0.8190^{20}_{4}	1.4209^{20}	δ	s	s	...	...	...	**B1**³, 1687
Ω h168	—(*dl*)*	$CH_3(CH_2)_4CH(OH)CH_3$	116.21			160^{760} 66^{20}	0.8167^{20}_{4}	1.4210^{20}	δ	s	s	...	...	...	**B1**³, 1686
h169	—(*l*)*	$CH_3(CH_2)_4CH(OH)CH_3$	116.21	$[\alpha]^{12}_{D}$ −10.5		74^{23}	0.8184^{20}_{4}	1.4208^{20}	δ	s	s	...	...	...	**B1**³, 1687
Ω h170	**3-Heptanol(*d*)***	$CH_3(CH_2)_3CH(OH)CH_2CH_3$	116.21	$[\alpha]^{20}_{D}$ +6.7 (w) $[\alpha]^{25}_{D}$ −3.78 (undil)	−70	157^{750} 66^{18}	0.8227^{20}_{4}	1.4201^{20}	δ	s	s	...	...	...	**B1**², 444
Ω h171	**4-Heptanol***	$(CH_3CH_2CH_2)_2CHOH$	116.21		−41.2	161^{760} (154) 63.8^{16}	0.8183^{20}_{4}	1.4205^{20}	i	s	s	...	...	...	**B1**³, 1685
h172	**2-Heptanol, 6-amino-2-methyl-**, hydrochloride*	$CH_3CH(NH_2)(CH_2)_3COH(CH_3)_2.HCl$	181.71	cr	154–5				v	s	i	...	i	...	**C54**, 8627
h173	**1-Heptanol, 7-chloro-***	$Cl(CH_2)_7OH$	150.65	cr (peth or bz)	11	150^{20} 120^{13}	0.9998^{15}_{4}	1.4537^{25}	δ	v	...	...	...	peth s	**B1**³, 1684
Ω h174	**4-Heptanol, 2,6-dimethyl-***	$[(CH_3)_2CHCH_2]_2CHOH$	144.26			$176–7^{760}$	0.809^{21}_{4}	1.4242^{20}	i	s	s	...	...	...	**B1**³, 1753
h175	**4-Heptanol, 4-ethyl-***	$(CH_3CH_2CH_2)_2C(C_2H_5)OH$	144.26			182^{760}	0.8350^{20}	1.4332^{20}	i	s	s	...	...	...	**B1**², 457
h176	**1-Heptanol, 7-fluoro-***	Heptamethylene fluorohydrin. $F(CH_2)_7OH$	134.20			$98–9^{12}$	0.956^{20}_{4}	1.4197^{25}	δ	v	v	...	...	...	**C51**, 7299
h177	—,**4-methyl-***	$CH_3CH_2CH_2CH(CH_3)(CH_2)_3OH$	130.23			183^{760} 71.6^{20}	0.8065^{25}_{4}	1.4258^{20}	i	s	s	...	...	os s	**B1**³, 1734
h178	—,**6-methyl-***	$(CH_3)_2CH(CH_2)_5OH$	130.23		−106	188^{764} 95.8^{20}	0.8176^{25}_{4}	1.4251^{25}	i	s	s	...	...	...	**Am 63**, 3100
Ω h179	**2-Heptanol, 2-methyl-***	$CH_3(CH_2)_4COH(CH_3)_2$	130.23			156^{760} $66–8^{15}$	0.8142^{20}_{4}	1.4250^{20}	i	s	s	...	...	...	**B1**³, 1725
Ω h180	**3-Heptanol, 3-methyl-***	$CH_3(CH_2)_3COH(CH_3)CH_2CH_3$	130.23		−83	163^{760} (159) $64–5^{16}$	0.8282^{20}_{4}	1.4279^{20}	i	s	s	...	...	...	**B1**³, 1729
Ω h181	**4-Heptanol, 4-methyl-***	$(CH_3CH_2CH_2)_2C(CH_3)OH$	130.23		−82	161^{760} $61–3^{12}$	0.8248^{20}_{4}	1.4258^{20}_{4}	i	s	s	...	...	...	**B1**³, 1735
h182	**1-Heptanol, 1-phenyl-***	$CH_3(CH_2)_5CH(OH)C_6H_5$	192.31			275 $153–5^{18}$	0.946	1.5024^{20}	i	...	...	...	...	...	**B6**², 513
h183	**4-Heptanol, 4-propyl-***	$(CH_3CH_2CH_2)_3COH$	158.29			190–2 $89–90^{15}$	0.8338^{21}_{0}	1.43551^{21}	i	s	s	...	s	...	**B1**³, 1769
Ω h184	**2-Heptanone***	$CH_3(CH_2)_4COCH_3$	114.19	λ^{MeOH} 274 (1.34)	−35.5	151.45^{760} 111^{21}	0.8111^{20}_{4}	1.4088^{20}	λ	s	s	...	...	...	**B1**², 753
Ω h185	**3-Heptanone***	$CH_3(CH_2)_3COCH_2CH_3$	114.19	λ^{al} 277.5 (1.34)	−39	147^{765} (149)	0.8183^{20}_{4}	1.4057^{20}	i	∞	∞	...	...	...	**B1**², 754
Ω h186	**4-Heptanone***	$(CH_3CH_2CH_2)_2CO$	114.19	λ^{al} 279.5 (1.38)	−33	144^{760}	0.8174^{20}_{4}	1.4069^{20}	i	∞	∞	...	...	...	**B1**², 754
h187	**2-Heptanone, 1-chloro-***	$CH_3(CH_2)_4COCH_2Cl$	148.63	λ^{al} 284 (1.43)		83^{16}	0.802^{20}	1.4371^{20}	δ	s	s	...	...	chl s	**Am 67**, 1944
Ω h187[1]	**4-Heptanone, 2,6-dimethyl-***	Isovalerone. $[(CH_3)_2CHCH_2]_2CO$	142.24	λ^{vap} 179, 192, 195		168^{760} $60–1^{18}$	0.8053^{20}_{4}	1.412^{21}	i	∞	∞	...	...	...	**B1**², 763
h188	**3-Heptanone, 6-dimethylamino-4,4-diphenyl-(*l*)***	*l*-Amidon. *l*-Methadone. $(CH_3)_2NCH(CH_3)CH_2C(C_6H_5)_2COC_2H_5$	309.46	$[\alpha]^{20}_{D}$ −32 (al) λ^{al} 295 (2.66)	99–100				...	s	...	...	...	dil HCl s	**Am 71**, 1648

For explanations, symbols and abbreviations see beginning of table. For structural formulas see end of table.

No.	Name	Synonyms and Formula	Mol. wt.	Color, crystalline form, specific rotation and λ_{max} (log ε)	m.p. °C	b.p. °C	Density	n_D	Solubility w	al	eth	ace	bz	other solvents	Ref.
	3-Heptanone														
h189	—,—hydrochloride (*dl*)*	Physopeptone. $(CH_3)_2NCH(CH_3)CH_2C(C_6H_5)_2COC_2H_5 . HCl$	345.92	pl (al-eth) λ^{al} 292	236–6.5	....	....	....	v	s	i	...	...	chl s	**Am 69,** 2941
h190	—,—,—(*l*)*	$(CH_3)_2NCH(CH_3)CH_2C(C_6H_5)_2COC_2H_5 . HCl$	345.92	$[\alpha]^{20}_{D}$ −169 (al, c = 2.1)	245–6	....	....	....	s	s	i	...	...	chl s	**Am 71,** 1648
Ω h191	**2-Heptanone, 4-isopropyl-***	$CH_3(CH_2)_2CH[CH(CH_3)_2]CH_2COCH_3$	156.27	....	....	$82–4^{14}$	....	....	δ	s	v	s	s	os v	**B1**^{2}, 2902
Ω h192	**—,3-methyl-***	$CH_3(CH_2)_3CH(CH_3)COCH_3$	128.22	....	....	167^{760}	0.8218^{20}_{4}	1.4172^{20}	δ	s	s	s	s	os s	**B1**^{2}, 759
Ω h193	**3-Heptanone, 6-methyl-***	$(CH_3)_2CHCH_2CH_2COCH_2CH_3$	128.22	....	....	163–3.5^{734}	0.8304^{20}	1.4209^{20}	i	s	s	...	s	os s	**B1**^{2}, 759
h194	**4-Heptanone, 2-methyl-***	$CH_3CH_2CH_2COCH_2CH(CH_3)_2$	128.22	λ^{al} 282 (1.40)	....	155^{750}	0.813^{22}_{0}	....	i	s	s	...	...	....	**B1**^{1}, 363
h194[1]	**1-Heptanone, 1-phenyl-***	Enanthophenone. $CH_3(CH_2)_5COC_6H_5$	190.29	lf	16.4	283.3 155^{15}	0.9516^{20}_{4}	1.5060^{20}_{4}	i	s	s	s	...	....	**B7**^{2}, 264
h195	**Heptasiloxane, hexadecamethyl-**	$CH_3[Si(CH_3)_2O-]_6Si(CH_3)_3$	533.06	....	−78	270^{760} 165^{20}	0.9012^{20}_{4}	1.3965^{20}	i	δ	...	...	s	peth s lig s	**Am 68,** 2284
h196	**1,4,6-Heptatrien-3-one, 1(4-methoxyphenyl)-7-phenyl-***	....	290.37	ye lf (al)	139	....	....	....	i	s^{h}	s	...	s	chl s lig s	**B8,** 208
Ω h197	**1-Heptene***	α-Heptylene. $CH_3(CH_2)_4CH:CH_2$	98.19	λ^{hx} 237 sh (0.3), 263 (−1.4)	−119	93.64^{760}	0.6970^{20}_{4}	1.3998^{20}	i	s	s	...	...	....	**B1**^{3}, 820
h198	**2-Heptene(*cis*)***	β-Heptylene. $CH_3(CH_2)_3CH:CHCH_3$	98.19	....	....	98.5^{760}	0.708^{20}_{4}	1.406^{20}	i	s	s	s	s	chl s peth s	**B1**^{3}, 824
h199	**—(*trans*)***	$CH_3(CH_2)_3CH:CHCH_3$	98.19	....	−109.48	98^{760}	0.7012^{20}_{4}	1.4045^{20}	i	s	s	s	s	chl s peth s	**B1**^{3}, 825
h200	**3-Heptene(*cis*)***	$CH_3CH_2CH_2CH:CHCH_2CH_3$	98.19	....	....	95.75^{760}	0.7030^{20}_{4}	1.4059^{20}	i	s	s	s	s	chl s peth s	**B1**^{3}, 826
h201	**—(*trans*)***	$CH_3CH_2CH_2CH:CHCH_2CH_3$	98.19	....	−136.63	95.67^{760}	0.6981^{20}_{4}	1.4043^{20}	i	s	s	s	s	chl s peth s	**B1**^{3}, 827
h202	**1-Heptene, 1-chloro-***	$CH_3(CH_2)_4CH:CHCl$	132.64	....	....	155 $78–82^{75}$	0.8948^{20}	1.4380^{20}	i	...	s	s	s	chl v to v	**B1**^{2}, 196
h203	**—,2-chloro-***	$CH_3(CH_2)_4CCl:CH_2$	132.64	....	....	138^{748}	0.8895^{20}_{4}	1.4349^{20}	i	s	s	...	...	chl s	**B1**^{3}, 823
h204	**2-Heptene, 4-chloro-***	$CH_3CH_2CH_2CHClCH:CHCH_3$	132.64	....	....	140–145 49^{21}	0.879^{18}_{4}	1.4430^{21}	i	s	s	...	...	MeOH v chl v	**B1**^{3}, 825
h205	**3-Heptene, 4-chloro-***	$CH_3CH_2CH_2CCl:CHCH_2CH_3$		....		139	0.883^{14}	1.437^{14}	i	s	s	...	...	chl s	**B1**^{3}, 827
h206	**2-Heptene, 6-chloro-2-methyl-***	$CH_3CHCl(CH_2)_2CH:C(CH_3)_2$	147.67	....	....	$60–61^{15}$	0.8931^{18}_{4}	1.4458^{18}	i	s	s	s	s	os s	**B1**^{2}, 200
h207	**2-Heptene, 2-methyl-6-methylamino-***	$CH_3CH(NHCH_3)CH_2CH_2CH:C(CH_3)_2$	141.26		....	176–8 $58–9^{17}$	....	....	i	v	v	...	...	....	**B4**^{3}, 467
h208	**1-Heptene-2-carboxaldehyde, 1-phenyl-***	α-*m*-Amylcinnamaldehyde. Jasminaldehyde. $CH_3(CH_2)_4C(CHO):CHC_6H_5$	203.30	....	....	$174–5^{20}$ 140^{5}	0.9718^{20}	1.5381^{20}	i	s	s	...	...	....	**B7**^{2}, 310
Ω h209	**2-Hepten-4-ol(*dl*)***	$CH_3CH_2CH_2CH(OH)CH:CHCH_3$	114.19	....	....	$152–4^{760}$ 64^{14}	0.8445^{20}_{4}	1.4373^{20}	...	s	s	...	...	....	**B1,** 447
Ω h210	**1-Hepten-4-ol, 4-methyl-***	$CH_3CH_2CH_2COH(CH_3)CH_2CH:CH_2$	128.22	....	....	159–60	0.8345^{20}_{0}	1.4479^{18}	...	s	s	...	...	....	**B1**^{3}, 1945
h211	**3-Hepten-2-one (*trans*)***	$CH_3CH_2CH_2CH:CHCOCH_3$	112.17	λ^{al} 222 (4.09)	....	62^{15} 52^{11}	0.8496^{20}_{4}	1.4436^{20}	i	v	v	...	...	....	**B1**^{3}, 3003
Ω h212	**5-Hepten-2-one, 6-methyl-***	$(CH_3)_2C:CHCH_2CH_2COCH_3$	126.20	....	−67	173^{760} 58.6^{10}	0.8546^{16}_{4}	1.4445^{20}	i	∞	∞	...	...	....	**B1**^{2}, 797
h213	**1-Hepten-3-yne***	$CH_3CH_2CH_2C{\vdots}CCH:CH_2$	94.16	....	....	110^{760} 44^{75}	0.7603^{20}_{4}	1.4520^{25}	i	s	...	s	s	peth s	**Am 61,** 572
h214	**1-Hepten-4-yne, 6,6-dimethyl-***	$CH_2:CHCH_2C{\vdots}CC(CH_3)_3$	122.21	....	....	125^{760} 68^{100}	0.758^{20}	1.4312^{20}	i	s	...	s	s	peth s	**C55,** 23329
h215	**6-Hepten-4-yne-3-ol,3-ethyl-***	$CH_2:CHC{\vdots}CCOH(C_2H_5)CH_2CH_3$	138.21	....	....	62^{4}	0.8875^{20}_{4}	1.4800^{20}	i	v	v	...	...	....	**B1**^{3}, 2034
—	**Heptyl alcohol**	*see* **Heptanol***													
—	**Heptyl amine***	*see* **Heptane, amino-***													
—	**Heptylene**	*see* **Heptene***													
—	**Heptyl thiocyanate**	*see* **Thiocyanic acid,** heptyl ester													
h216	**1-Heptyne***	$CH_3(CH_2)_4C{\vdots}CH$	96.17	....	−81	99.74^{760} 6.0^{10}	0.7328^{20}_{4}	1.4087^{20}	δ	∞	∞	...	s	chl s peth s	**B1**^{3}, 995
h217	**2-Heptyne***	$CH_3(CH_2)_3C{\vdots}CCH_3$	96.17	....	....	112^{760}	0.7480^{20}_{4}	1.4230^{20}	i	∞	∞	...	s	chl s peth s	**B1**^{3}, 998
h218	**3-Heptyne***	$CH_3CH_2CH_2C{\vdots}CCH_2CH_3$	96.17	....	....	$105–6^{760}$	0.7527^{20}_{4}	1.4220^{20}	i	∞	∞	...	s	chl s peth s	**B1**^{3}, 999
h219	**1-Heptyne, 1-bromo-***	$CH_3(CH_2)_4C{\vdots}CBr$	175.08	....	....	164^{758} 69^{25}	1.2120^{22}_{4}	1.4678^{22}	i	s	s	s	...	chl s	**B1**^{3}, 998
h220	**2-Heptyne, 1-bromo-***	$CH_3(CH_2)_3C{\vdots}CCH_2Br$	175.08	....	....	104^{56} 84^{20}	....	1.4878^{25}	i	s	s	s	...	....	**B1**^{3}, 998
h221	**1-Heptyne, 1-chloro-***	$CH_3(CH_2)_4C{\vdots}CCl$	130.62	....	....	141^{760}	0.9250^{24}_{4}	1.4411^{24}	i	v	v	...	...	MeOH v	**B1**^{3}, 997

For explanations, symbols and abbreviations see beginning of table. For structural formulas see end of table.

No.	Name	Synonyms and Formula	Mol. wt.	Color, crystalline form, specific rotation and λ_{max} (log ε)	m.p. °C	b.p. °C	Density	n_D	Solubility w	al	eth	ace	bz	other solvents	Ref.
	2-Heptyne														
h222	**2-Heptyne, 1-chloro-***	$CH_3(CH_2)_3C⋮CCH_2Cl$	130.62	...	...	167^{760} 73^{24}	...	1.4570^{25}	i	s	s	...	...	...	**B1**^{3}, 998
h223	**—,7-chloro-***	$ClCH_2(CH_2)_3C⋮CCH_3$	130.62	...	...	166^{760}	...	1.4507^{20}	i	s	s	...	...	...	**B1**^{3}, 998
h224	**3-Heptyne, 1-chloro-***	$CH_3CH_2CH_2C⋮CCH_2CH_2Cl$	130.62	...	...	162^{760} $90–93^{70}$	...	1.4520^{25}	i	s	s	...	...	...	**B1**^{3}, 999
h225	**—,7-chloro-***	$Cl(CH_2)_3C⋮CCH_2CH_3$	130.62	...	...	164^{760} $74–5^{31}$	...	1.4517^{20}	i	s	s	...	...	...	**B1**^{3}, 999
h226	**—,2,6-dimethyl-***	$(CH_3)_2CHCH_2C⋮CCH(CH_3)_2$	124.23	...	...	$130–6^{760}$	0.785_4^{20}	...	i	...	s	s	...	...	**B1**^{3}, 1005
h227	**—,5,5-dimethyl-***	$CH_3CH_2C(CH_3)_2C⋮CCH_2CH_3$	124.23	...	...	69^{100}	0.7610_4^{20}	1.4360^{20}	i	...	s	s	...	...	**Am 62**, 1800
h228	**—,5-ethyl-5-methyl-***	$(CH_3CH_2)_2C(CH_3)C⋮CCH_2CH_3$	138.25	...	...	88^{100}	0.7714_4^{20}	1.4386^{20}	i	...	s	s	...	...	**Am 62**, 1800
h229	**2-Heptyn-1-ol***	Butyl propargyl alcohol. $CH_3(CH_2)_3C⋮CCH_2OH$	112.17	...	...	94^{22}	...	1.4523^{25}	...	...	...	...	...	...	**Am 71**, 1294
—	**Herniarin**	*see* **Coumarin, 7-methoxy-**													
—	**Heroin**	*see* **Morphine, diacetyl-**													
—	**Hesperetin**	*see* **Flavanone, 4′-methoxy-,3′,5,7-trihydroxy-**													
—	**Hesperetol**	*see* **Styrene, 4-hydroxy-3-methoxy-**													
h232	**Hesperidin**	Hesperetin-1-rhamnosido-*D*-glucose. $C_{28}H_{34}O_{15}$.	610.58	wh nd (dil MeOH or aa) $[\alpha]_D^{30}$ −77.5 (Py, c = 4.28)	261–3 (254d)	...	...	...	δ	s	i	...	i	Py v aa s chl, CS_2 i	**Am 68**, 2108
—	**Heteroxanthine**	*see* **Xanthine, 7-methyl-**													
h233	**Hexacene**	Anthraceno-2′:3′,2:3-anthracene.	328.42	dk bl-gr cr (sub)	*ca.* 380	sub	...	...	i	i	...	...	...	...	**E14s**, 789
—	**Hexachlorophene**	*see* **Methane, bis(2-hydroxy-3,5,6-trichlorophenyl)-***													
h234	**Hexacosane***	Cerane. $CH_3(CH_2)_{24}CH_3$	366.72	mcl, tcl or rh (bz), cr (eth)	56.4	412.2^{760} 248.22^{10}	0.8032_4^{20} 0.7783^{60}	1.4501^{20} 1.4357^{60}	i	s^h δ	...	...	s	lig v chl s	**B1**^{2}, 143
Ω h235	**Hexacosanoic acid***	Cerotic acid. Phthioic acid. $CH_3(CH_2)_{24}CO_2H$	396.71	nd (al)	88–9	...	0.8198_4^{100}	1.4301^{100}	i	v^h	v	...	...	...	**B2**^{2}, 380
Ω h236	**1-Hexacosanol***	Cerotin. Ceryl alcohol. $CH_3(CH_2)_{25}OH$	382.72	rh pl (dil al)	80	305^{20}d	...	...	i	s	s	...	...	...	**B1**^{2}, 470
h237	**1,15-Hexadecadiyne***	$HC⋮C(CH_2)_{12}C⋮CH$	218.39	fl (al)	44–5	$152–5^{12}$	...	...	...	...	...	...	...	...	**B1**^{2}, 249
h238	**6,9-Hexadecadiyne***	$CH_3(CH_2)_5C⋮CCH_2C⋮C(CH_2)_4CH_3$	218.39	...	...	169^{15}	0.845_4^{18}	1.4694^{18}	...	...	...	...	...	...	**C28**, 4035
h239	**6,10-Hexadecadiyne***	$CH_3(CH_2)_4C⋮CCH_2CH_2C⋮C(CH_2)_4CH_3$	218.39	...	...	157^{10} 130.6^6	0.7907_4^{20}	1.4523^{20}	...	...	...	...	...	...	**C45**, 3638
h240	**Hexadecanal***	Palmitaldehyde. $CH_3(CH_2)_{14}CHO$	240.43	pl (eth), nd (peth) λ^{cy} 293 (1.45)	34	$200–2^{29}$	...	...	i	s	s	s	s	os s	**B1**^{2}, 771
h241	**—,dimethyl acetal**	$CH_3(CH_2)_{14}CH(OCH_3)_2$	286.50	...	10	144^2	0.8542^{20}	1.4382^{25}	v	s	v	s	...	MeOH v	**C49**, 161
h242	**—,oxime**	$CH_3(CH_2)_{14}CH{:}NOH$	255.45	nd (dil al)	88	...	...	...	i	s	...	...	δ	chl s peth δ	**B1**^{2}, 771
Ω h243	**Hexadecane***	Cetane. $CH_3(CH_2)_{14}CH_3$	226.45	lf (ace)	18.17	287^{760} 149^{10}	0.77331_4^{20}	1.4345	i	$δ^h$	∞	...	...	...	**B1**^{3}, 555
Ω h244	**—,1-amino-***	Cetylamine. $CH_3(CH_2)_{15}NH_2$	241.47	lf	46.77	322.5^{760} 177.9^{10} 144^2	0.8129_4^{20}	1.4496^{20}	i	v	v	s	v	chl s	**B4**^{2}, 660
Ω h245	**—,1-bromo-***	Cetyl bromide. $CH_3(CH_2)_{15}Br$	305.35	...	17–9	336^{760} 188^{10}	0.9991_4^{20}	1.4618^{20}	i	...	s	...	...	...	**B1**^{2}, 138
Ω h246	**—,1-chloro-***	Cetyl chloride. $CH_3(CH_2)_{15}Cl$	260.90	...	17.9	322^{760} 177^{10}	0.8652_4^{20}	1.4505^{20}	i	...	...	...	...	...	**B1**^{3}, 558
h247	**—,1,16-dibromo-***	$Br(CH_2)_{16}Br$	384.25	lf (al)	56	204^4	...	...	i	δ v^h	...	...	...	chl v	**B1**^{2}, 138
h248	**—,1-fluoro-***	Cetyl fluoride. $CH_3(CH_2)_{15}F$	244.44	...	18	289^{760} 152.63^{10}	0.8321_4^{20}	1.4317^{20}	...	...	s	...	...	lig s	**B1**^{2}, 138
h249	**—,1-iodo-***	Cetyl iodide. $CH_3(CH_2)_{15}I$	352.35	lf (al)	24.7	357^{760} 202^{10}	1.1257_4^{20}	1.4818^{20}	i	δ	s	s	∞	chl v	**B1**^{3}, 560
h250	**—,1-phenyl-***	Cetylbenzene. $CH_3(CH_2)_{15}C_6H_5$	302.55	...	27	237^{16}	0.8560_4^{20}	1.4814^{20}	i	δ	v	...	v	CS_2, lig v	**B5**, 472
h251	**Hexadecanedioic acid***	Thapsic acid. $HO_2C(CH_2)_{14}CO_2H$	286.42	pl (al, AcOEt)	126	...	...	...	i	s^h	δ	s^h	i s^h	AcOEt s^h CS_2 i	**B2**, 733
Ω h252	**1-Hexadecanethiol***	Cetyl mercaptan. $CH_3(CH_2)_{15}SH$	258.52	(lig) λ^{hx} 225 sh (2.2)	18–20	$123–8^{0.5}$	...	...	i	δ	s	...	...	...	**B2**^{2}, 330
Ω h253	**Hexadecanoic acid***	Palmitic acid. $CH_3(CH_2)_{14}CO_2H$	256.43	nd (al) λ^{al} 210 (1.70)	63	390^{760} 267^{100}	0.8527_4^{62}	1.4335^{60}	i	s	∞	s	s	chl v	**Am 64**, 2739
Ω h254	**—,amide**	Palmitamide. $CH_3(CH_2)_{14}CONH_2$	255.45	lf	107	236^{12}	...	...	i	δ	δ	δ	δ	...	**B2**^{2}, 341
h255	**—,—,*N*-phenyl-***	Palmitanilide. $CH_3(CH_2)_{14}CONHC_6H_5$	331.55	nd (al)	90.5	$282–4^{17}$ 132^{10}	...	...	i	v	δ	v	v	chl v aa v^h	**B12**^{2}, 148
h256	**—,anhydride***	$[CH_3(CH_2)_{14}CO]_2O$	494.85	lf (peth)	64	...	0.8383_4^{83}	1.4364^{68}	i	δ	v	...	...	peth $δ^h$	**B2**^{2}, 341
h257	**—,benzyl ester***	$CH_3(CH_2)_{14}CO_2CH_2C_6H_5$	346.56	cr (al)	36	...	0.9136_{25}^{35}	1.4689^{50}	i	s	s	...	s	chl v	**B6**^{2}, 417
h258	**—,butyl ester***	$CH_3(CH_2)_{14}CO_2C_4H_9^n$	312.54	cr (dil al)	16.9	...	...	1.4312^{50}	i	s	s	...	...	...	**B2**^{2}, 336
Ω h259	**—,chloride**	$CH_3(CH_2)_{14}COCl$	274.88	...	12	199^{20}	...	1.4514^{20}	d	d	v	...	...	...	**B2**^{2}, 341

For explanations, symbols and abbreviations see beginning of table. For structural formulas see end of table.

No.	Name	Synonyms and Formula	Mol. wt.	Color, crystalline form, specific rotation and λ_{max} (log ε)	m.p. °C	b.p. °C	Density	n_D	Solubility w	al	eth	ace	bz	other solvents	Ref.
	Hexadecanoic acid														
h260	—,ethyl ester*	$CH_3(CH_2)_{14}CO_2C_2H_5$	284.49	nd	α: 24 β: 19.3	191^{10}	0.8577^{25}_{4}	1.4347^{34}	i	s	s	s	s	chl s	**B2**[2], 336
h261	—,hentriacontyl ester*	Myricyl palmitate. $CH_3(CH_2)_{14}CO_2(CH_2)_{30}CH_3$	691.28	cr (eth)	72	...	...	...	i	i	s	...	...	...	**B2**, 373
h262	—,hexadecyl ester*	Cetyl palmitate. $CH_3(CH_2)_{14}CO_2(CH_2)_{15}CH_3$	480.87	pl (eth)	53–4	360	0.8324^{50}_{4}	1.4425^{50}	i	s^h	s	...	s	chl, CS_2 s	**B2**[2], 337
h262[1]	—,2-hydroxyethyl ester*	$CH_3(CH_2)_{14}CO_2CH_2CH_2OH$	300.49	...	51	$173–4^{3}$	0.8786^{60}_{4}	...	...	v	s^h	...	...	...	**C42**, 2927
h263	—,isopropyl ester*	$CH_3(CH_2)_{14}CO_2C_3H_7^i$	298.52	...	13–4	160^{2}	0.8404^{38}	1.4364^{25}	i	v	s	v	v	chl v	**B2**[2], 336
Ω h264	—,methyl ester*	$CH_3(CH_2)_{14}CO_2CH_3$	270.46	...	30	$415–8^{747}$ 148^{2}	...	...	i	v	s	v	v	chl v	**B2**, 372
Ω h265	—,nitrile	$CH_3(CH_2)_{14}CN$	237.43	hex	31	333^{760} 251^{100} 142^{1}	0.8303^{20}_{4}	1.4450^{20}	i	v	v	v	v	chl v	**B2**[2], 341
h266	—,propyl ester*	$CH_3(CH_2)_{14}CO_2C_3H_7^n$	298.52	nd	20.4	190^{12}	0.8455^{38}	1.4392^{25}	...	...	...	...	...	...	**B2**[2], 336
Ω h267	—,16-hydroxy-*	$HO(CH_2)_{15}CO_2H$	272.43	cr (bz-eth)	95	...	...	...	i	s	δ s^h	s	$δ^h$	...	**B3**, 362
Ω h268	—,9,10,16-trihydroxy*	Aleuritic acid. $HO(CH_2)_6CH(OH)CH(OH)(CH_2)_7CO_2H$	304.43	lf (dil al), nd(w)	102	...	...	...	δ	...	...	...	...	...	**B3**[2], 272
Ω h269	**1-Hexadecanol***	Cetyl alcohol. $CH_3(CH_2)_{15}OH$	242.45	fl (AcOEt)	50	344^{760} 190^{15}	0.8176^{50}_{4}	1.4283^{79}	i	δ	v	s	v	chl v	**B1**[2], 466
Ω h270	**1-Hexadecene***	Cetene. Cetylene. $CH_3(CH_2)_{13}CH{:}CH_2$	224.44	lf	4.1	284.4^{760} 155^{15}	0.7811^{20}_{4}	1.4412^{20}	i	s	s	...	...	peth s	**B1**[2], 206
h271	**2-Hexadecenoic acid** (form I)*	Gaidic acid. $CH_3(CH_2)_{12}CH{:}CHCO_2H$	254.42	lf (al)	39	...	...	...	...	v	...	...	...	chl v peth s	**B2**, 461
h272	**2-Hexadecenoic acid** (form II)*	Δ-α,β-Hypogeic acid. $CH_3(CH_2)_{12}CH{:}CHCO_2H$	254.42	fl (al)	49	...	...	...	...	s	v	...	...	chl, peth v	**B2**, 460
h273	**7-Hexadecenoic acid***	Hypogeic acid (artificial). $CH_3(CH_2)_7CH{:}CH(CH_2)_5CO_2H$	254.42	cr	33	230^{10}	...	...	i	v	s	...	...	...	**B2**, 460
h274	**1-Hexadecyne***	$CH_3(CH_2)_{13}C{:}CH$	222.42	...	15	284^{760} 147.8^{10}	0.7965^{20}_{4}	1.4440^{20}	i	...	...	...	s	...	**B1**[3], 1028
h275	**2-Hexadecyne***	$CH_3(CH_2)_{12}C{:}CCH_3$	222.42	fl	20	160^{15} (184^{15})	0.8039^{20}_{4}	...	...	...	...	...	...	...	**B1**, 262
h276	**7-Hexadecynoic acid***	$CH_3(CH_2)_7C{:}C(CH_2)_5CO_2H$	252.40	nd (w), cr (al)	47	214^{15}	...	...	i	v	v	...	...	...	**B2**, 494
h277	**2,4-Hexadien-1-al***	Sorbaldehyde. $CH_3CH{:}CHCH{:}CHCHO$	96.14	λ^{al} 271 (4.99)	...	$173–4^{754}$ 76^{30}	0.898^{20}	1.5384^{20}	...	...	...	...	...	...	**B1**[2], 809
h278	**1,2-Hexadiene***	Propylallene. $CH_3CH_2CH_2CH{:}C{:}CH_2$	82.15	...	...	76^{760}	0.7149^{20}	1.4282^{20}	i	...	s	...	...	chl s	**B1**[3], 981
h279	**1,3-Hexadiene***	$CH_3CH_2CH{:}CHCH{:}CH_2$	82.15	...	...	73^{760}	0.7050^{20}_{4}	1.4380^{20}	i	...	s	...	...	...	**B1**[3], 981
h280	**1,4-Hexadiene***	1-Allylpropene. $CH_3CH{:}CHCH_2CH{:}CH_2$	82.15	...	...	65^{760}	0.7000^{20}_{4}	1.4150^{20}	i	...	v	...	...	...	**B1**[3], 982
Ω h281	**1,5-Hexadiene***	Biallyl. $H_2C{:}CHCH_2CH_2CH{:}CH_2$	82.15	λ^{vap} 177 (4.4)	−141	59.5^{760}	0.6880^{20}_{4}	1.4042^{20}	i	s	s	...	s	chl s	**B1**[3], 982
h282	**2,4-Hexadiene***	Bipropenyl. $CH_3CH{:}CHCH{:}CHCH_3$	82.15	...	−79	80^{760}	0.7196^{20}_{4}	1.4500^{20}	i	s	s	...	...	chl s	**B1**[3], 985
h283	**1,3-Hexadiene,3-chloro-***	$CH_3CH_2CH{:}CClCH{:}CH_2$	116.59	...	...	68^{117}	0.9390^{20}_{4}	1.4770^{20}	i	...	s	...	...	chl s	**B1**[3], 981
h284	**1,4-Hexadiene,5-chloro-2-isopropyl-***	$CH_3CCl{:}CHCH_2C[CH(CH_3)_2]{:}CH_2$	158.67	...	...	95^{18}	0.9310^{25}_{4}	1.4730^{25}	i	...	...	s	...	chl s	**C33**, 3362
h285	**2,4-Hexadiene, 6-chloro-2-methyl-***	$ClCH_2CH{:}CHCH{:}C(CH_3)_2$	130.62	...	...	57^{11}	0.9416^{20}_{4}	1.5120^{20}	i	...	...	s	...	chl s	**B1**[3], 1000
h286	**1,5-Hexadiene, decachloro-***	Perchlorodiallyl. $Cl_2C{:}CClCCl_2CCl_2CCl{:}CCl_2$	426.60	cr (ace)	49	$121^{0.03}$	1.905^{51}_{4}	1.6012^{51}	i	...	s	s	...	chl s	**B1**[3], 984
h287	**2,4-Hexadiene, 1,3-dichloro-***	$CH_3CH{:}CHCCl{:}CHCH_2Cl$	151.05	...	...	$80–2^{17}$	1.1456^{20}_{4}	1.5271^{20}	i	...	...	s	...	chl s	**B1**[3], 987
Ω h288	**1,5-Hexadiene, 2,5-dimethyl-***	Diisobutenyl. $CH_2{:}C(CH_3)CH_2CH_2C(CH_3){:}CH_2$	110.20	...	−75.6	134^{760}	0.7512^{20}	1.43995^{21}	i	...	...	s	...	chl s	**B1**[2], 239
Ω h289	**2,4-Hexadiene, 2,5-dimethyl-***	Diisocrotyl. $(CH_3)_2C{:}CHCH{:}C(CH_3)_2$	110.20	...	14.5–.8	134^{760} 75^{100}	0.7625^{20}_{4}	1.4785^{20}	i	s	s	...	s	chl s	**B1**[3], 1011
h290	—,1,3,4,6-tetrachloro-*	$ClCH_2CH{:}CClCCl{:}CHCH_2Cl$	219.93	...	...	$84–9^{2}$	1.4013^{20}_{4}	1.5465^{20}	i	...	...	...	...	chl, MeOH s	**B1**[3], 987
h291	**1,4-Hexadiene, 3,3,6-trichloro-***	$ClCH_2CH{:}CHCCl_2CH{:}CH_2$	185.48	...	...	$100–3^{4}$	1.3036^{20}_{4}	1.5585^{20}	i	...	s	...	...	chl s	**B1**[3], 982
h292	**2,4-Hexadiene-1,6-dioic acid** (*cis, cis*)*	*cis*-Muconic acid. $HO_2CCH{:}CHCH{:}CHCO_2H$	142.11	...	194–5	...	...	...	...	...	...	...	...	aa s	**B2**, 803
Ω h293	—(*trans, trans*)*	*trans*-Muconic acid. $HO_2CCH{:}CHCH{:}CHCO_2H$	142.11	nd (al)	305d	320	...	...	δ	δ	...	...	...	AcOEt, aa s	**B2**, 803
h294	—,dimethyl ester (*trans, trans*)*	$CH_3O_2CCH{:}CHCH{:}CHCO_2CH_3$	170.17	nd (dil al)	157	...	...	...	δ	s	s	...	...	aa s	**B2**[2], 672
h295	**1,5-Hexadiene-3,4-diol(*d*)***	Divinyl glycol. $H_2C{:}CHCH(OH)CH(OH)CH{:}CH_2$	114.15	$[\alpha]^{17}_{D}$ +94.8 (al)	−60	198^{760} 97^{13}	1.006^{20}_{4}	1.4700^{20}	s	s	s	...	...	chl s	**B1**[3], 2272
h296	—(*dl*)*	$H_2C{:}CHCH(OH)CH(OH)CH{:}CH_2$	114.15	hyg	21.7	$90.5–.7^{8}$	1.017^{19}_{4}	1.4790^{19}	s	s	s	...	...	chl s	**B1**[3], 2272
h297	—(*meso*)*	$CH_2{:}CHCH(OH)CH(OH)CH{:}CH_2$	114.15	hyg	18	100^{14}	1.023^{19}_{4}	1.4810^{19}	v	v	v	...	...	chl v	**B1**[3], 2272

For explanations, symbols and abbreviations see beginning of table. For structural formulas see end of table.

No.	Name	Synonyms and Formula	Mol. wt.	Color, crystalline form, specific rotation and λ_{max} (log ε)	m.p. °C	b.p. °C	Density	n_D	Solubility w	al	eth	ace	bz	other solvents	Ref.
	2,4-Hexadienoic acid														
Ω h298	**2,4-Hexadienoic acid** (*trans, trans*)	Sorbic acid. $CH_3CH:CHCH:CHCO_2H$	112.14	nd (dil al)	134.5	228d 153^{50}	1.204_4^{19}	...	s^h δ	s	v	...	...	...	**B2**, 489
h299	—,amide	$CH_3CH:CHCH:CHCONH_2$	111.15	nd (w)	171.5–2.5	...	...	...	s	s	...	...	...	...	**B2**[2], 453
h300	—,chloride	$CH_3CH:CHCH:CHCOCl$	130.58	...	...	78^{15}	1.0666_4^{19}	1.5545^{20}	d	...	...	s	...	...	**B2**[2], 453
Ω h301	—,ethyl ester*	$CH_3CH:CHCH:CHCO_2C_2H_5$	140.19	...	...	195.5 85^{20}	0.9506_4^{20}	1.4951^{20}	i	s	s	...	...	chl s	**B2**[2], 452
h302	—,methyl ester*	$CH_3CH:CHCH:CHCO_2CH_3$	126.16	lf	15	180^{759} 70^{20}	0.9777_4^{20}	1.5025^{22}	i	s	s	...	...	...	**B2**[2], 452
h303	**2,4-Hexadien-1-ol***	Sorbyl alcohol. $CH_3CH:CHCH:CHCH_2OH$	98.15	nd	30.5–1.5	76^{12}	0.8967_4^{23}	1.4981^{20}	i	s	s	...	...	...	**B1**[3], 1988
h304	**3,5-Hexadien-2-ol***	$H_2C:CHCH:CHCH(OH)CH_3$	98.15	λ^{al} 225 (4.33)	...	$77–8^{26}$ $44–5^{8}$	0.8678^{20}	1.4816^{20}	...	v	...	...	...	...	**B1**[3], 1987
h306	**3,5-Hexadien-2-one, 6-phenyl-***	Cinnamylideneacetone. $C_6H_5CH:CHCH:CHCOCH_3$	172.23	wh lf (eth) λ^{al} 234 (3.65), 319 (4.56)	68	$170–2^{15}$	...	...	i	s	s	s	s	chl, MeOH s	**B7**[2], 321
h308	**1,5-Hexadien-3-yne***	Divinyl acetylene. $H_2C:CHC:CCH:CH_2$	78.12	...	−88	85^{760}	0.7851_4^{20}	1.5035^{20}	i	...	...	...	s	...	**B1**[3], 1058
h309	**3,5-Hexadien-1-yne***	$HC:CCH:CHCH:CH_2$	78.12	...	...	83.4^{760} 32^{100}	0.7806_4^{20}	1.5095^{20}	i	...	...	...	s	sulf s	**B1**[3], 1058
h310	**1,5-Hexadien-3-yne, 2,5-dimethyl-***	Diisopropenyl acetylene. $CH_2:C(CH_3)C:CC(CH_3):CH_2$	106.17	ye	...	123^{760}	0.7863_4^{25}	1.4845^{20}	i	...	...	...	s	chl s	**B1**[3], 1062
h311	**1,4-Hexadiyne-**	$HC:CCH_2C:CCH_3$	78.12	...	< −80	$78–83^{760}$	0.825_4^{0}	...	i	...	...	...	s	chl s	**B1**[3], 1057
h312	**1,5-Hexadiyne***	Dipropargyl. $HC:CCH_2CH_2C:CH$	78.12	...	−6	86^{760} 20^{46}	0.8049_4^{20}	1.4380^{23}	i	s	s	s	s	os s	**B1**[3], 1057
h313	**2,4-Hexadiyne***	Dimethyl diacetylene. $CH_3C:CC:CCH_3$	78.12	pr (sub) λ^{al} 218.5 (2.48), 236 (2.52), 250 (2.20)	68.5	129–30	...	...	...	v	v	...	...	...	**B1**[2], 247
h314	**1,5-Hexadiyne,1,6-diamino-**	$H_2NC:CCH_2CH_2C:CNH_2$	108.13	...	104–5	...	...	...	s^h	...	...	...	s	...	**C53**, 10036
—	**Hexahydrobenzyl alcohol**	*see* **Methanol, cyclohexyl-***													
—	**Hexahydro-*p*-cymene**	*see* ***p*-Menthane**													
—	**Hexahydrothymol**	*see* **Menthol**													
—	**Hexamethylene-diamine**	*see* **Hexane, 1,6-diamino-***													
—	**Hexamethylene glycol**	*see* **1,6-Hexanediol***													
—	**Hexamethylene-imine**	*see* **Cycloheptane, 1-aza-***													
Ω h315	**Hexamethylene-tetramine**	Aminoform. Formin. Hexamine. Urotropine.	140.19	rh (al)	285–95 sub	sub	1.331^{-5}	...	v s^h	s	δ	s	δ	chl s	**B1**[2], 648
Ω h316	**Hexanal***	Caproaldehyde. *n*-Caproic aldehyde. $CH_3(CH_2)_4CHO$	100.16	λ^{vap} 184.5	−56	128^{760} (131) 28^{12}	0.81392_4^{20}	1.4039^{20}	δ	v	v	s	s	...	**B1**[2], 745
h317	—,oxime*	$CH_3(CH_2)_4CH:NOH$	115.18	cr (MeOH)	51	...	...	...	...	s	s	...	...	...	**B1**, 689
Ω h318	—,2-ethyl-*	$CH_3(CH_2)_3CH(C_2H_5)CHO$	128.22	...	< −100	163^{760} 65^{15}	0.8540^{20}	1.4142^{20}	i	s	s	...	...	...	**B1**, 707
Ω h319	**Hexane***	Dipropyl. $CH_3(CH_2)_4CH_3$	86.18	...	−95 (−93.5)	68.95^{760}	0.6603_4^{20}	1.37506^{20}	i	v	s	...	...	chl s	**B1**[3], 374
—	*neo*-Hexane	*see* **Butane, 2,2-dimethyl-***													
Ω h320	**Hexane, 1-amino-***	*n*-Hexylamine. $CH_3(CH_2)_5NH_2$	101.19	...	−19	130^{762}	0.7660^{20}	1.4180^{20}	δ	∞	∞	...	...	...	**B4**[2], 649
h321	—,2-amino-(*d*)*	*d*-α-Methylamylamine. $CH_3(CH_2)_3CH(NH_2)CH_3$	101.19	$[\alpha]_D^{27}$ +4.30	...	114–5 64^{90}	0.755_4^{27}	...	...	s	s	...	...	...	**C50**, 4782
Ω h322	—,—(*dl*)*	$CH_3(CH_2)_3CH(NH_2)CH_3$	101.19	...	−19	117–8	0.7534_0^{20}	1.4080^{25}	...	s	s	...	...	...	**B4**, 190
h323	—,2-amino-4-methyl-*	$CH_3CH_2CH(CH_3)CH_2CH(NH_2)CH_3$	115.22	...	...	$130–5^{760}$	0.7655^{20}	1.4150^{25}	δ	v	v	...	...	chl, dil ac v	**Am 66**, 1516
h324	—,3,4-bis(4-hydroxy-3-methylphenyl)-	Dimethylhexestrol. Promethestrol.	298.43	cr (dil aa)	145	...	...	...	...	...	...	...	...	aa δ	**Am 70**, 508
h325	—,2,4-bis(4-hydroxyphenyl)-3-ethyl-	Benzestrol. Octofollin.	298.43	cr (al)	162–6	...	...	...	i	v	v	v	δ	MeOH v aa s chl, peth δ	**Am 68**, 729
Ω h326	—,1-bromo-*	*n*-Hexyl bromide. $CH_3(CH_2)_5Br$	165.08	...	−84.7	155.3^{760} 41^{10}	1.1744_4^{20}	1.44781^{20}	i	∞	∞	s	...	chl v	**B1**[3], 391
Ω h327	—,2-bromo-*	*sec*-Hexyl bromide. $CH_3(CH_2)_3CHBrCH_3$	165.08	...	...	144^{749} 78^{90}	1.1658_4^{20}	1.4832^{25}	i	v	s	s	...	chl v	**B1**[3], 392
Ω h328	—,3-bromo-*	$CH_3CH_2CH_2CHBrCH_2CH_3$	165.08	...	...	$141–3^{760}$	1.1799_4^{20}	1.4472^{20}	i	s	s	s	...	chl v	**B1**, 144
h329	—,1-bromo-6-fluoro-*	$F(CH_2)_6Br$	183.07	...	...	$67–8^{11}$	1.293_4^{20}	1.4435^{25}	i	v	v	s	...	chl s	**C51**, 7300
Ω h330	—,1-chloro-*	*n*-Hexyl chloride. $CH_3(CH_2)_5Cl$	120.62	...	−94.0	134.5^{760}	0.8785_4^{20}	1.41991^{20}	i	s	s	s	s	chl v	**B1**[3], 388
h331	—,2-chloro-*	*sec*-Hexyl chloride. $CH_3(CH_2)_3CHClCH_3$	120.62	...	...	$122–3^{760}$ 61^{100}	0.8694_4^{21}	1.4142^{22}	i	s	s	s	s	chl v	**B1**[3], 389
h332	—,3-chloro-*	$CH_3CH_2CH_2CHClCH_2CH_3$	120.62	...	...	123^{760} 60^{95}	0.8700_{20}^{20}	1.4163^{20}	i	s	v	s	s	chl v	**Am 61**, 940

For explanations, symbols and abbreviations see beginning of table. For structural formulas see end of table.

No.	Name	Synonyms and Formula	Mol. wt.	Color, crystalline form, specific rotation and λ_{max} (log ε)	m.p. °C	b.p. °C	Density	n_D	Solubility w	al	eth	ace	bz	other solvents	Ref.
	Hexane														
h333	**—,2-chloro-2,5-dimethyl-***	$(CH_3)_2CHCH_2CH_2CCl(CH_3)_2$	148.68	...	...	86^{100} 44^{14}	0.8476^{18}_{4}	1.4232^{20}	i	s	s	s	s	...	**B1**^{3}, 485
h334	**—,3-chloro-2,3-dimethyl-***	$CH_3CH_2CH_2CCl(CH_3)CH(CH_3)_2$	148.68	...	...	$41–3^{12}$	0.8869^{20}_{4}	1.4333^{25}	i	s	...	...	...	...	**B1**^{3}, 481
h336	**—,1-chloro-3-ethyl-(*d*)***	$CH_3CH_2CH_2CH(C_2H_5)CH_2CH_2Cl$	148.68	$[\alpha]^{27}_{D}$ +1.15	...	85^{40}	0.879^{21}_{4}	1.4335^{25}	i	...	s	...	...	...	**B1**^{3}, 478
h337	**—,3-chloro-3-ethyl-***	$CH_3CH_2CH_2CCl(C_2H_5)_2$	148.68	...	...	155d $62–3^{24}$	0.9018^{0}	1.4358^{20}	i	...	s	...	...	chl s	**B1**^{3}, 479
h338	**—,1-chloro-6-fluoro-***	$F(CH_2)_6Cl$	138.61	...	...	167^{740} 62^{15}	1.015^{20}_{4}	1.4168^{25}	i	v	v	...	...	chl s	**C51**, 7300
h339	**—,1-chloro-3-methyl-***	$CH_3CH_2CH_2CH(CH_3)CH_2CH_2Cl$	134.65	...	...	$150–2^{758}$	0.8766^{20}_{4}	1.4274^{20}	i	v	v	...	...	chl s	**B1**^{2}, 119
h340	**—,2-chloro-2-methyl-***	$CH_3(CH_2)_3CCl(CH_3)_2$	134.65	...	...	135^{760} δd 59.5^{52}	0.8635^{20}_{4}	1.4200^{20}	i	v	v	...	...	chl s	**B1**, 156
h341	**—,2-chloro-5-methyl-***	$(CH_3)_2CHCH_2CH_2CHClCH_3$	134.65	...	...	138^{735}d	0.863^{20}_{4}	...	i	v	v	...	...	chl s	**B1**, 156
h342	**—,3-chloro-3-methyl-***	$CH_3CH_2CH_2CCl(CH_3)CH_2CH_3$	134.65	...	...	135^{760}	0.8787^{20}_{4}	1.4250^{20}	i	v	v	...	...	chl s	**B1**^{2}, 119
h343	**—,3-chloro-2,2,3-trimethyl-***	$CH_3(CH_2)_2CCl(CH_3)C(CH_3)_3$	162.71	...	...	$64–5^{13}$	0.9010^{20}_{4}	1.4465^{20}	i	v	s	...	...	chl s	**B1**^{3}, 515
b344	**—,1,6-diamino-***	Hexamethylenediamine. $H_2N(CH_2)_6NH_2$	116.21	rh bipym pl	41–2	204–5 100^{20}	...	...	v	s	...	...	s	...	**B4**^{2}, 710
h345	**—,—,dihydrochloride***	$H_2N(CH_2)_6NH_2.2HCl$	189.13	nd (al-eth)	248–50	...	...	...	v	δ	i	...	i	chl i	**B4**^{2}, 710
h346	**—,2,5-diamino-***	$CH_3CH(NH_2)CH_2CH_2CH(NH_2)CH_3$	116.21	...	...	175	...	...	v	v	v	...	...	...	**B4**, 269
h347	**—,1,2-dibromo-***	$CH_3(CH_2)_3CHBrCH_2Br$	243.98	...	...	$103–5^{36}$ 87^{16}	1.5774^{20}_{4}	1.5024^{20}	...	s	∞	...	s	chl s	**B1**^{2}, 109
h348	**—,1,2-dichloro-***	$CH_3(CH_2)_3CHClCH_2Cl$	155.07	...	...	$172–4^{760}$ $73–4^{0.3}$	1.085^{15}	...	...	...	s	...	...	chl s	**B1**, 144
Ω h349	**—,1,6-dichloro-***	$Cl(CH_2)_6Cl$	155.07	...	...	203–5 94^{22}	1.0677^{20}_{4}	1.4572^{20}	i	...	s	...	...	chl s	**B1**^{3}, 389
h350	**—,2,2-dichloro-***	$CH_3(CH_2)_3CCl_2CH_3$	155.07	...	...	68^{49}	1.0150^{25}_{4}	1.4353^{25}	i	...	s	...	...	chl s	**B1**^{3}, 390
h351	**—,2,3-dichloro-***	$CH_3CH_2CH_2CHClCHClCH_3$	155.07	...	...	162–5	1.0527^{11}	...	...	...	s	...	...	chl s	**B1**^{2}, 109
h352	**—,2,5-dichloro-(*dl*)***	$CH_3CHClCH_2CH_2CHClCH_3$	155.07	...	fp −38.36	177^{751} (cor) 106^{91}	1.0474^{20}	1.4491^{20}	...	...	...	...	...	chl s	**B1**, 144
h353	**—,2,5-dichloro-(*meso*)***	$CH_3CHClCH_2CH_2CHClCH_3$	155.07	...	19.88	178^{752} (cor) 109^{99}	1.0474^{20}_{4}	1.4484^{20}	...	...	...	...	...	chl s	**B1**, 144
h354	**—,3,4-dichloro-(*dl*)***	$CH_3CH_2CHClCHClCH_2CH_3$	155.07	...	...	167.7 62^{20}	1.0617^{20}	1.4541^{20}	i	...	...	s	...	chl s	**B1**^{3}, 390
h355	**—,2,5-dichloro-2,5-dimethyl-***	$(CH_3)_2CClCH_2CH_2CCl(CH_3)_2$	183.12	lf, nd	67–8	...	0.9543^{70}	...	...	s	s	...	s	chl s	**B1**^{2}, 127
h356	**—,3,4-dichloro-3,4-dimethyl-***	$CH_3CH_2CCl(CH_3)CCl(CH_3)CH_2CH_3$	183.12	...	...	165^{760} $114–5^{18}$	...	...	...	...	...	...	...	chl s	**B1**^{2}, 62
Ω h357	**—,1,6-diiodo-***	$I(CH_2)_6I$	337.97	nd	10	$141–2^{10}$ 113^{3}	2.03^{22}_{4}	1.585^{20}	i	v	v	...	...	...	**B1**^{2}, 110
h358	**—,2,2-dimethyl-***	$CH_3(CH_2)_3C(CH_3)_3$	114.23	...	−121.18	106.84^{760} 3.1^{10}	0.69528^{20}_{4}	1.39349^{20}	i	v	∞	∞	∞	chl, lig ∞	**B1**^{3}, 479
h359	**—,2,3-dimethyl-(*dl*)***	$CH_3CH_2CH_2CH(CH_3)CH(CH_3)_2$	114.23	...	...	115.6^{760} 9.9^{10}	0.71214^{20}_{4}	1.40113^{20}	i	∞	s	∞	∞	chl, lig ∞	**B1**^{3}, 480
h360	**—,—(*l*)***	$CH_3CH_2CH_2CH(CH_3)CH(CH_3)_2$	114.23	$[\alpha]^{25}_{D}$ −0.92	...	113^{760}	...	...	i	∞	s	∞	∞	chl, lig ∞	**B1**^{3}, 482
h361	**—,2,4-dimethyl-(*d*)***	$CH_3CH_2CH(CH_3)CH_2CH(CH_3)_2$	114.23	$[\alpha]^{30}_{D}$ +2.99	...	111^{760}	0.696^{30}_{4}	1.3810^{20}	i	∞	v	∞	∞	chl, lig ∞	**B1**^{3}, 483
h362	**—,—(*dl*)***	$CH_3CH_2CH(CH_3)CH_2CH(CH_3)_2$	114.23	...	...	109.4^{760} 5.2^{10}	0.70036^{20}_{4}	1.39534^{20}	i	∞	s	∞	∞	chl, lig ∞	**B1**^{3}, 482
h363	**—,—(*l*)***	$CH_3CH_2CH(CH_3)CH_2CH(CH_3)_2$	114.23	$[\alpha]^{21}_{D}$ −10.85	...	110^{760}	0.703^{21}_{4}	...	i	∞	s	∞	∞	chl, lig ∞	**B1**^{3}, 483
Ω h364	**—,2,5-dimethyl-***	$(CH_3)_2CHCH_2CH_2CH(CH_3)_2$	114.23	...	−91.2	109^{760} 5.3^{10}	0.69354^{20}_{4}	1.39246^{20}	i	∞	s	∞	∞	chl, lig ∞	**B1**^{3}, 483
h365	**—,3,3-dimethyl-***	$CH_3CH_2CH_2C(CH_3)_2CH_2CH_3$	114.23	...	−126.1	112^{760} 6.1^{10}	0.71000^{20}_{4}	1.40009^{20}	i	∞	v	v	v	os s	**B1**^{3}, 486
Ω h366	**—,3,4-dimethyl-***	$CH_3CH_2CH(CH_3)CH(CH_3)CH_2CH_3$	114.23	...	...	117.72^{760} 11.3^{10}	0.7200^{20}_{4}	1.4046^{20}	i	∞	s	∞	∞	chl, lig ∞	**B1**^{3}, 488
h367	**—,1,6-dinitro-***	$O_2N(CH_2)_6NO_2$	176.17	cr (MeOH) $\lambda^{95\%al}$ 276 (1.65)	37.5	$100–3^{0.3}$	...	...	i	...	...	...	...	aa s	**B1**^{3}, 396
h368	**—,3-ethyl-***	$CH_3CH_2CH_2CH(CH_2CH_3)_2$	114.23	...	...	118.53^{760} 12.8^{10}	0.7136^{20}_{4}	1.4018^{20}	i	∞	∞	∞	∞	chl, lig ∞	**B1**^{3}, 478
h369	**—,1-fluoro-***	$CH_3(CH_2)_5F$	104.17	...	−103	91.5^{760}	0.7995^{20}	1.3738^{20}	...	...	s	...	s	...	**B1**^{3}, 387
Ω h370	**—,1-iodo-***	$CH_3(CH_2)_5I$	212.08	...	−75	181.33^{760} 58.23^{10}	1.4397^{20}_{4}	1.49290^{20}	i	...	...	...	...	...	**B1**^{3}, 394
h371	**—,2-methyl-***	Isoheptane. $CH_3(CH_2)_3CH(CH_3)_2$	100.21	...	−118.27	90^{760} -9.09^{10}	0.67869^{20}_{4}	1.38485^{20}	i	s	∞	∞	∞	chl, lig ∞	**B1**^{3}, 434
h372	**—,3-methyl-(*d*)***	$CH_3CH_2CH_2CH(CH_3)CH_2CH_3$	100.21	$[\alpha]^{20}_{D}$ +9.5	−119	92^{760}	0.6860^{20}_{4}	1.3887^{20}	i	s	∞	∞	∞	chl, lig ∞	**B1**^{3}, 439

For explanations, symbols and abbreviations see beginning of table. For structural formulas see end of table.

No.	Name	Synonyms and Formula	Mol. wt.	Color, crystalline form, specific rotation and λ_{max} (log ε)	m.p. °C	b.p. °C	Density	n_D	Solubility						Ref.
									w	al	eth	ace	bz	other solvents	
	Hexane														
Ω h373	—,—(*dl*)*	$CH_3CH_2CH_2CH(CH_3)CH_2CH_3$	100.21	...	*ca.* −173	92^{760}	0.6872^{20}	1.3885^{20}	i	∞	∞	∞	∞	chl, lig ∞	**B1**^{3}, 437
h374	—,—(*l*)*	$CH_3CH_2CH_2CH(CH_3)CH_2CH_3$	100.21	$[\alpha]^{21}_{D}$ −7.75	...	92^{760}	0.687^{21}_{4}	1.3854^{25}	i	s	∞	∞	∞	chl, lig ∞	**B1**^{3}, 440
h375	—,**1,1,1,2,2-pentachloro-***	$CH_3(CH_2)_3CCl_2CCl_3$	258.40	...	...	$129–31^{10}$	1.370^{25}	1.4872^{25}	i	...	s	...	...	chl s	**B1**^{3}, 390
h376	—,**1-phenyl-***	Hexylbenzene. $CH_3(CH_2)_5C_6H_5$	162.28	...	−62	227^{760}	0.8613^{20}	1.4900^{20}	i	...	∞	...	s	peth s	**B5**^{2}, 337
h377	—,**1,1,2,2-tetra-chloro-***	$CH_3(CH_2)_3CCl_2CHCl_2$	223.96	...	...	$99–101^{14}$	1.3096^{25}_{4}	1.488^{25}	i	...	s		s	chl s	**B1**^{3}, 390
Ω h378	—,**2,2,5-tri-methyl-***	$(CH_3)_2CHCH_2CH_2C(CH_3)_3$	128.26	...	−105.78	124^{760} 16.17^{10}	0.7072^{20}_{4}	1.3997^{20}	i	v	v	v	v	lig ∞ os v	**B1**^{3}, 516
h379	**1,6-Hexanedial***	Adipaldehyde. $OCH(CH_2)_4CHO$	114.15	...	−8	$92–4^{9}$ 70^{3}	1.003^{19}_{4}	1.4350^{20}	δ	v	v	...	s	aa s	**B1**^{2}, 839
Ω h380	**Hexanedioic acid***	Adipic acid. $HO_2C(CH_2)_4CO_2H$	146.14	mcl pr (w, ace-lig)	153	265^{100} 205^{10} sub	1.360^{25}_{4}		δ s^{h}	v	s	...	...	aa, lig i	**B2**^{2}, 572
Ω h381	—,diamide	Adipamide. $H_2NCO(CH_2)_4CONH_2$	144.18	pl	220	...	...	...	δ	v	δ	...	...	...	**B2**^{2}, 576
h382	—,diazide	$N_3OC(CH_2)_4CON_3$	196.18	...	−1	d	...	...	...	v	v	...	...	...	**B2**^{1}, 278
Ω h383	—,dibutyl ester*	$C_4H^{n}_{9}O_2C(CH_2)_4CO_2C_4H^{n}_{9}$	258.36	...	−32.4	165^{10} 145^{4}	0.9615^{20}_{4}	1.4369^{20}	i	∞	∞	...	...	...	**B2**^{2}, 575
Ω h384	—,dichloride	$ClCO(CH_2)_4COCl$	183.05	...	...	126^{12}	...	...	d	d	...			...	**B2**^{2}, 575
Ω h385	—,diethyl ester*	$C_2H_5O_2C(CH_2)_4CO_2C_2H_5$	202.25	λ^{al} 210 (2.0)	−19.8	245^{760} 127^{13}	1.0076^{20}_{4}	1.4272^{20}	i	s	s	...	...	...	**B2**^{2}, 574
h386	—,di(2-ethylbutyl) ester*	$(C_2H_5)_2CHCH_2O_2C(CH_2)_4CO_2CH_2CH(C_2H_5)_2$	314.47	...	−15	200^{10}	0.934^{25}_{4}	1.4434^{20}	i	s		s	...	aa s	**C41,** 4298
h387	—,di(2-ethylhexyl) ester	$C_4H^{n}_{9}CH(C_2H_5)CH_2O_2C(CH_2)_4CO_2CH_2CH(C_2H_5)C_4H^{n}_{9}$	370.58	...	−67.8	214^{5}	0.922^{25}_{4}	1.4474^{20}	i	s		s	...	aa s	**B2**^{2}, 1715
h388	—,dihydrazide	$H_2NNHOC(CH_2)_4CONHNH_2$	174.20	lf (w)	182–2.5	...	...	...	s	s^{h}	i	...	s	aa i	**B2**^{3}, 1724
Ω h389	—,diisopropyl ester	$C_3H^{i}_{7}O_2C(CH_2)_4CO_2C_3H^{i}_{7}$	230.31	...	−1.1	$120^{6.5}$	0.9659^{20}	1.4247^{20}	i	s	s	s	...	aa s	**B2**^{3}, 1713
Ω h390	—,dimethyl ester*	$CH_3O_2C(CH_2)_4CO_2CH_3$	174.20	cr	10.3	115^{13} 105^{7}	1.0600^{20}_{4}	1.4283^{20}	i	s	s	...	...	aa s	**B2,** 365
Ω h391	—,dinitrile	$NC(CH_2)_4CN$	108.15	nd (eth)	1	295^{760} 180^{20}	0.9676^{20}	1.4380^{20}	δ	s	δ	...	...	chl s	**B2**^{2}, 576
h392	—,dipropyl ester*	$C_3H^{n}_{7}O_2C(CH_2)_4CO_2C_3H^{n}_{7}$	230.31	...	−15.7	151^{11}	0.9790^{20}_{4}	1.4314^{20}	i	s	s	...	...	chl s	**B2**^{2}, 574
h393	—,monoamide*	Adipamic acid. $HO_2C(CH_2)_4CONH_2$	145.16	nd (w)	161–2	...	...	...	...	...	...	...	...	...	**J1934,** 1101
h394	—,monoethyl ester*	$HO_2C(CH_2)_4CO_2C_2H_5$	174.20	hyg (eth-peth)	29	285^{760} 170^{17}	0.9796^{20}_{4}	1.4311^{20}	...	s	s^{h}	...	...	peth s^{h}	**B2**^{2}, 574
Ω h395	—,monomethyl ester*	$HO_2C(CH_2)_4CO_2CH_3$	160.17	lf (Me_3N-MeOH)	9	158^{10}	1.0623^{20}	1.4283^{20}	...	s	...	...	...	...	**B2,** 652
h396	—,piperazinium salt	$HO_2C(CH_2)_4CO_2H.\ C_4H_{10}N_2$	232.28	...	244d	...	...	...	s	...	i	...	...	...	**Am 56,** 1759
Ω h397	—,**2-amino-(*dl*)***	$HO_2C(CH_2)_3CH(NH_2)CO_2H$	161.16	pl (w)	206 (anh)	...	...	...	δ s^{h}	δ	δ	...	...	...	**Am 75,** 1994
h398	—,**2-methyl-***	$HO_2C(CH_2)_3CH(CH_3)CO_2H$	160.17	cr (peth-bz)	93	209^{13}	...	...	s	s	s	...	δ	chl s peth δ	**B2**^{3}, 1745
h399	—,**3-methyl-(*d*)***	$HO_2CCH_2CH_2CH(CH_3)CH_2CO_2H$	160.17	cr (chl-bz), $[\alpha]^{25}_{D}$ +9.81 (w)	94–4.5	230^{30}	...	...	v	v	v	v	s	chl, AcOEt v peth δ	**B2**^{2}, 588
Ω h400	—,—(*dl*)*	$HO_2CCH_2CH_2CH(CH_3)CH_2CO_2H$	160.17	nd (bz), cr (ace-bz)	97	$190–200^{12}$	...	...	v	v	v	v	δ	chl, AcOEt v lig δ	**B2**^{2}, 589
h401	—,**2-oxo-***	$HO_2C(CH_2)_3COCO_2H$	160.14	cr (al or eth)	127	...	...	...	v	v	δ	v	i	peth, chl i	**B3**^{2}, 485
h402	—,**3-oxo-***	$HO_2CCH_2CH_2COCH_2CO_2H$	160.14	pl (ace-chl)	125	...	...	...	δ	δ	δ	s	δ	chl δ	**B3**^{2}, 486
—	—,**2,3,4,5-tetra-hydroxy-***	*see* **Mucic acid**													
Ω h406	**1,6-Hexanediol***	Hexamethylene glycol. $HO(CH_2)_6OH$	118.18	nd (w)	43 (59)	250^{760} 132^{9}	...	...	s	s	δ	s	i	...	**B1**^{3}, 2201
h407	**2,3-Hexanediol***	$CH_3CH_2CH_2CH(OH)CH(OH)CH_3$	118.18	cr	60	$204–6^{760}$ $102^{0.8}$	0.9900^{15}	1.4510^{15}	s	s	s	...	...	...	**B1,** 484
Ω h408	**2,5-Hexanediol***	$CH_3CH(OH)CH_2CH_2CH(OH)CH_3$	118.18	cr (eth)	43	$216–8^{750}$ $85–7^{1}$	0.9610^{20}_{4}	1.4475^{20}	s	s	s	...	...	...	**B1**^{2}, 551
h409	**3,4-Hexanediol, 3,4-diethyl-***	$(C_2H_5)_2C(OH)C(OH)(C_2H_5)_2$	174.29	cr (eth)	28	230^{760} 112^{10}	0.9630^{13}_{25}	1.467^{13}	i	v	v	...	...	...	**B1**^{3}, 2235
Ω h410	**2,5-Hexanediol, 2,5-dimethyl-***	$(CH_3)_2C(OH)CH_2CH_2C(OH)(CH_3)_2$	146.23	pr (AcOEt), fl (peth)	92	214 118^{15}	0.898^{20}	...	s	v	...	...	v^{h}	chl v	**B1**^{2}, 557
Ω h411	**1,3-Hexanediol, 2-ethyl-***	$CH_3CH_2CH_2CH(OH)CH(C_2H_5)CH_2OH$	146.23	...	−40	244^{760}	0.9325^{22}_{4}	1.4497^{20}	δ	s	s	...	...	...	**C37,** 5020
h412	**2,3-Hexanedione, 3-oxime***	$CH_3(CH_2)_2C(:NOH)COCH_3$	129.16	cr (al)	60	...	...	...	...	s^{h}	...	...	...	...	**B1**^{1}, 405
Ω h413	**2,5-Hexanedione***	Acetonylacetone. $CH_3COCH_2CH_2COCH_3$	114.15	λ 271 (2.09)	−5.5	194^{754} 89^{25}	0.9737^{20}_{4}	1.4421^{20}	∞	∞	∞	v	s	peth δ^{h}	**B1**^{2}, 841
h414	—,dioxime*	$CH_3C(:NOH)CH_2CH_2C(:NOH)CH_3$	144.18	pl (bz)	137	...	...	...	v^{h}	v^{h}	v^{h}	...	...	...	**B1,** 790
Ω h415	**1,6-Hexanedione, 1,6-diphenyl-***	1,4-Dibenzoylbutane. $C_6H_5CO(CH_2)_4COC_6H_5$	266.34	nd (al), pr (chl-al), lf (dil al)	107	...	...	...	...	s^{h}	...	...	...	MeOH, chl s^{h}	**B7**^{2}, 705
h416	**2,5-Hexanedione, 3-hydroxy-***	$CH_3COCH_2CH(OH)COCH_3$	130.15	...	...	$62–7^{0.5}$	...	1.4497^{25}	δ	...	s	...	s	...	**Am 71,** 3171

For explanations, symbols and abbreviations see beginning of table. For structural formulas see end of table.

No.	Name	Synonyms and Formula	Mol. wt.	Color, crystalline form, specific rotation and λ_{max} (log ε)	m.p. °C	b.p. °C	Density	n_D	Solubility						Ref.
									w	al	eth	ace	bz	other solvents	
	3,4-Hexanedione														
h417	**3,4-Hexanedione, 2,2,5,5-tetra-methyl-***	Di-*tert*-butylglyoxal. Dipivaloyl. Pivalil. $(CH_3)_3CCOCOC(CH_3)_3$	170.25	$\lambda^{95\%\,al}$ 294 (1.67), 365 (1.31)	−2	168^{745}	0.8776^{20}	1.4157^{20}	i	...	∞	...	...	...	**B1**^{3}, 3147
Ω h418	**1,2,3,4,5,6-Hexanehexol***	Dulcite. Dulcitol. Melampyrin. $HOCH_2(CHOH)_4CH_2OH$	182.18	mcl pr	189	$275–80^1$	1.466^{15}	...	s v^h	δ	i	...	...	...	**B1**^{2}, 612
h419	**1,3,4,5-Hexane-tetrol***	Digitoxit. $CH_3(CHOH)_3CH_2CH_2OH$	150.18	pr, $[\alpha]^{15}_D$ −86.2	88	...	...	...	i	v	s^h	...	...	...	**B1**^{2}, 603
Ω h420	**1-Hexanethiol***	*n*-Hexyl mercaptan. $CH_3(CH_2)_5SH$	118.24	λ^{hx} 224 sh (2.1)	−81	151^{760}	0.8424^{20}_4	1.4496^{20}	i	v	v	...	...	...	**B1**^{3}, 1659
h421	**2-Hexanethiol***	*sec*-Hexyl mercaptan. 2-Mercaptohexane*. $CH_3(CH_2)_3CH(SH)CH_3$	118.24	λ^{iso} 223 sh (2.2)	−147	142^{760} $60–6^{50}$	0.8345^{20}_4	1.4451^{20}	i	s	s	...	s	...	**B1**^{3}, 1662
h422	**3-Hexanethiol***	3-Mercaptohexane*. $CH_3CH_2CH_2CH(SH)CH_2CH_3$	118.24	...	...	57^{25}	0.9206^{20}	1.4496^{20}	...	s	...	...	...	...	**B1**^{3}, 1665
h423	**1,2,3-Hexanetriol (*threo*)***	$CH_3CH_2CH_2CH(OH)CH(OH)CH_2OH$	134.18	...	64–4.5	$130^{0.05}$	1.089^{26}	1.472^{26}	s	s	i	...	...	...	**B1**^{3}, 2349
h424	**1,2,4-Hexanetriol***	$CH_3CH_2CH(OH)CH_2CH(OH)CH_2OH$	134.18	...	...	$190–2^{30}$	...	...	s	s	i	...	...	...	**B1**, 521
h425	**1,2,5-Hexanetriol***	$CH_3CH(OH)CH_2CH_2CH(OH)CH_2OH$	134.18	...	...	181^{10} 150^3	1.1012^{20}_4	...	∞	∞	i	...	...	...	**B1**^{3}, 2349
h426	**2,3,4-Hexanetriol***	$CH_3CH_2(CHOH)_3CH_3$	134.18	...	...	256–7 $155–6.5^{20}$	...	...	∞	∞	i	...	...	...	**B1**^{3}, 2349
Ω h427	**Hexanoic acid***	*n*-Caproic acid. $CH_3(CH_2)_4CO_2H$	116.16	...	−2– −1.5	205^{760}	0.9274^{20}_4	1.4163^{20}	i	s	s	...	...	...	**B2**^{2}, 281
Ω h428	—,amide	Caproamide. $CH_3(CH_2)_4CONH_2$	115.18	cr (ace)	101	255^{760}	0.999^{20}_4	1.4200^{110}	δ s^h	s	s	...	s	chl s	**B2**^{2}, 286
h429	—,—,*N*-phenyl-	Hexanilide. Capranilide. $CH_3(CH_2)_4CONHC_6H_5$	191.28	nd (peth), pr (al) λ^{al} 243 (4.18)	95	...	1.112	...	i	v	v	...	...	peth s^h	**B12**^{2}, 147
h430	—,anhydride*	$[CH_3(CH_2)_4CO]_2O$	214.31	...	−41	254–7d (245) 143^{15}	0.9240^{15}_4	1.4297^{20}	...	s	s	...	...	...	**B2**^{2}, 285
Ω h431	—,butyl ester*	$CH_3(CH_2)_4CO_2(CH_2)_3CH_3$	172.27	...	−64.3	208^{760}	0.8653^{20}_4	1.4153^{25}	i	s	∞	...	...	...	**B2**, 323
Ω h432	—,chloride	$CH_3(CH_2)_4COCl$	134.61	...	−87	153^{760}	0.9754^{20}_4	1.4264^{20}	d	d	s	s	...	...	**B2**^{2}, 286
Ω h433	—,ethyl ester*	$CH_3(CH_2)_4CO_2C_2H_5$	144.22	...	−67	168^{760}	0.8710^{20}_4	1.4073^{20}	i	s	s	...	...	...	**B2**^{2}, 285
h434	—,heptyl ester*	$CH_3(CH_2)_4CO_2(CH_2)_6CH_3$	214.35	...	−34.4	261^{760}	0.8611^{20}	1.4293^{15}	i	s	s	s	s	os s	**B2**, 323
h435	—,hexyl ester*	$CH_3(CH_2)_4CO_2(CH_2)_5CH_3$	200.33	...	−55	246^{761}	0.865^{18}	1.4264^{15}	i	s	s	s	s	os s	**B2**, 323
Ω h436	—,methyl ester*	$CH_3(CH_2)_4CO_2CH_3$	130.19	...	−71	151^{760} 52^{15}	0.8846^{20}_4	1.4049^{20}	i	v	v	s	s	...	**B2**^{2}, 284
Ω h437	—,3-methylbutyl ester*	Isoamyl caproate. $CH_3(CH_2)_4CO_2CH_2CH_2CH(CH_3)_2$	186.30	...	...	$224–7^{760}$	0.861^{20}_4	...	i	s	s	...	...	...	**B2**^{3}, 727
Ω h438	—,nitrile	Amyl cyanide. $CH_3(CH_2)_4CN$	97.16	...	−80.3	163.6^{760} 47.3^{10}	0.8051^{20}_4	1.4068^{20}	i	s	s	...	...	...	**B2**^{2}, 286
h439	—,octyl ester*	$CH_3(CH_2)_4CO_2(CH_2)_7CH_3$	228.38	...	−28	275^{760}	0.8603^{20}	1.4326^{15}	i	s	s	s	s	os s	**B2**, 323
Ω h440	—,pentyl ester*	$CH_3(CH_2)_4CO_2(CH_2)_4CH_3$	186.30	...	−47	226^{760} 116.6^{20}	0.8612^{25}_4	1.4202^{25}	...	s	s	s	...	...	**B2**^{2}, 285
h441	—,4-phenyl-phenacyl ester*	...	310.40	cr (dil al)	69–70	...	...	...	...	s^h	...	...	...	...	**Am 52,** 3715
h442	—,piperazinium salt	$2[CH_3(CH_2)_4CO_2H].\ C_4H_{10}N_2$	318.46	...	111	...	...	...	s	s	i	s^h	...	...	**Am 56,** 1759
h443	—,propyl ester*	$CH_3(CH_2)_4CO_2CH_2CH_2CH_3$	158.24	...	−68.7	187^{760}	0.8672^{20}_4	1.4170^{20}	i	s	s	...	...	...	**B2**, 323
Ω h444	—,2-acetyl-, ethyl. ester	Ethyl α-butylacetoacetate. $CH_3(CH_2)_3CH(COCH_3)CO_2C_2H_5$	186.25	...	...	219–24 104^{12}	0.9523^{20}_4	1.4301^{20}	i	...	s	s	...	AcOEt s	**B1**^{3}, 706
Ω h445	—,6-amino-*	$H_2N(CH_2)_5CO_2H$	131.18	lf (eth)	202–3	...	...	...	v	i	...	...	...	MeOH δ	**B4**^{2}, 856
h446	—,—,lactam*	ε-Caprolactam	113.16	lf (lig)	69–71	139^{12}	...	...	s	v	...	...	s	chl v	**B21**^{2}, 216
h447	—,6-amino-3-methyl-, lactam*	...	127.19	cr (bz-peth) $[\alpha]^{20}_D$ −36.1	105–6	...	...	...	v	...	...	...	s	...	**B21**, 243
h448	—,6-amino-5-methyl-, lactam*	...	127.19	cr (peth, bz-lig) $[\alpha]^{20}_D$ −22.2	68–9	...	...	...	v	...	s	...	...	...	**B21**, 242
h449	—,6-benzoyl-amino-	$C_6H_5CONH(CH_2)_5CO_2H$	235.29	nd (al-eth)	79–80	...	...	...	δ^h	...	...	...	δ	AcOEt s lig δ	**B9**^{2}, 181
h450	—,6-benzoyl-amino-2-bromo-(*dl*)	$C_6H_5CONH(CH_2)_4CHBrCO_2H$	314.19	cr (dil al)	166	...	...	...	δ	v^h	...	...	...	...	**B9**^{2}, 182
h451	—,—(*l*)	$C_6H_5CONH(CH_2)_4CHBrCO_2H$	314.19	nd (dil al), $[\alpha]^{18}_D$ −29.2 (al)	129	...	...	...	δ	s^h	...	...	...	...	**B9**^{2}, 181
Ω h452	—,2-bromo-(*dl*)*	$CH_3(CH_2)_3CHBrCO_2H$	195.06	...	4	240^{760} $140–2^{23}$	...	...	...	s	s	...	...	...	**B2**^{2}, 287
h453	—,—(*l*)*	$CH_3(CH_2)_3CHBrCO_2H$	195.06	$[\alpha]^{30}_D$ −27 (eth, c = 5)	...	129^{14}	...	...	...	s	s	...	...	...	**B2**^{2}, 287
Ω h454	—,—,ethyl ester(*dl*)*	$CH_3(CH_2)_3CHBrCO_2C_2H_5$	223.12	...	...	205–10 $95–6^9$	...	...	...	s	...	...	...	...	**B2**^{1}, 141
h455	—,3-bromo-*	$CH_3CH_2CH_2CHBrCH_2CO_2H$	195.06	nd (dil al)	35	...	...	...	...	s	...	...	v	chl, lig v	**B2**, 235
Ω h456	—,6-bromo-*	$Br(CH_2)_5CO_2H$	195.06	cr (peth)	35	$165–70^{20}$	...	...	...	...	...	...	...	peth s^h	**B2**^{2}, 287
h457	—,6-cyclohexyl-*	...	198.31	...	33–3.5	180^{11}	0.9626^{20}_4	1.4750^{20}	...	...	s	...	...	...	**B9**^{2}, 21
—	—,2,6-diamino-*	*see* **Lysine**													

For explanations, symbols and abbreviations see beginning of table. For structural formulas see end of table.

No.	Name	Synonyms and Formula	Mol. wt.	Color, crystalline form, specific rotation and λ_{max} (log ε)	m.p. °C	b.p. °C	Density	n_D	Solubility w	al	eth	ace	bz	other solvents	Ref.
	Hexanoic acid														
h458	—,**2,4-dioxo-***	Propionylpyruvic acid. $CH_3CH_2COCH_2COCO_2H$	144.14	cr (al, w + 1)	83	...	...	...	s	s	s	...	δ	peth i	**B3**[2], 437
h459	—,—,ethyl ester*	$CH_3CH_2COCH_2COCO_2C_2H_5$	172.19	...	...	163–5 108–11^{11}	...	...	...	v	v	v	...	...	**B3**[2], 437
Ω h460	—,**2-ethyl-***	$CH_3(CH_2)_3CH(C_2H_5)CO_2H$	144.22	...	...	228^{755} 120^{13}	0.9031_4^{25}	1.4241^{20}	s^h	δ	s	...	...	...	**B2**[2], 304
h461	—,—,amide	$CH_3(CH_2)_3CH(C_2H_5)CONH_2$	143.23	nd (w)	102–3	...	...	...	s^h	...	...			...	**B2**, 350
h462	—,**6-fluoro-***	$F(CH_2)_5CO_2H$	134.15	...	...	138^{26} 67–$8^{0.6}$	...	1.4166^{25}	δ	v	v	...	...	...	**C52**, 1922
h463	—,**2-hydroxy-(*d*)***	$CH_3(CH_2)_3CH(OH)CO_2H$	132.16	$[\alpha]_D^{20}$ +0.7 (w, c = 14)	60	...	...	...	v	v	s	...	...	...	**B3**[2], 231
Ω h464	—,—(*dl*)*	$CH_3(CH_2)_3CH(OH)CO_2H$	132.16	pr (eth-al, peth)	61	270	...	...	v	v	s	...	...	chl v	**B3**[2], 232
h465	—,—(*l*)*	$CH_3(CH_2)_3CH(OH)CO_2H$	132.16	pr (eth) $[\alpha]_D^{20}$ −3.8 (w, c = 4.5)	60–1	...	...	...	v	v	v	...	...	chl v	**B3**[2], 232
h466	—,**4-hydroxy-, lactone***	γ-Caprolactone	114.15	...	−18	215–6 103^{14}	...	1.4495^{20}	s	s	...	...	...	alk s	**B17**[2], 290
h467	—,**2-methyl-(*d*)***	$CH_3(CH_2)_3CH(CH_3)CO_2H$	130.19	$[\alpha]_D^{22}$ +19.6 (eth)	...	105^5	0.909^{25}	1.4189^{25}	∞	∞	∞	∞	∞	chl ∞	**B2**[2], 296
h468	—,—(*dl*)*	$CH_3(CH_2)_3CH(CH_3)CO_2H$	130.19	...	...	215–6^{760} 100^{12}	0.9612_4^{20}	1.4193^{20}	∞	∞	∞	∞	∞	chl ∞	**B2**[2], 297
h469	—,—(*l*)*	$CH_3(CH_2)_3CH(CH_3)CO_2H$	130.19	$[\alpha]_D^{25}$ −4.3 (w, c = 26) $[\alpha]_D^{25}$ −15.25 (eth)	...	121^{20} 105^5	0.909_4^{25}	1.4189^{25}	∞	∞	∞	∞	∞	chl ∞	**B2**[2], 296
h470	—,**3-methyl-, chloride**	$CH_3CH_2CH_2CH(CH_3)CH_2COCl$	148.63	...	...	163^{751} 82^{50}	0.967_4^{20}	1.4293_4^{25}	d	d	...	s	...	to s^h	**B2**[2], 295
h471	—,**4-methyl-(*d*)***	*act*-Amylacetic acid. $CH_3CH_2CH(CH_3)CH_2CH_2CO_2H$	130.19	$[\alpha]_D^{20}$ +7.6 (MeOH)	...	221^{760} 115^{16}	0.9228_4^{20}	1.4198^{20}	δ	s	...	s	s	MeOH s	**B2**[1], 146
h472	—,—(*dl*)*	$CH_3CH_2CH(CH_3)CH_2CH_2CO_2H$	130.19	...	−80	217–8^{760} 85^2	0.9215_4^{20}	1.4211^{20}	δ	s	s	s	s	os s	**B2**[2], 298
h473	—,—,chloride(*dl*)	$CH_3CH_2CH(CH_3)CH_2CH_2COCl$	148.63	...	...	167^{767}	0.9677_4^{20}	...	...	...	s	...	...	...	**C19**, 464
h474	—,**5-methyl-***	$(CH_3)_2CH(CH_2)_3CO_2H$	130.19	...	< −25	216 (207^{752}) 109^{14}	0.9138_4^{21}	1.4220^{20}	δ	s	s	s	s	peth s	**B2**[2], 297
h475	—,—,chloride	$(CH_3)_2CH(CH_2)_3COCl$	148.63	...	...	168^{739} 76–82^{34}	...	...	i	...	s	...	...	...	**B2**, 342
h476	—,**4-oxo-**	Homolevulinic acid. $CH_3CH_2COCH_2CH_2CO_2H$	130.15	hyg ta or lf (eth-peth)	41–2	183^{20} $89^{0.4}$	...	...	s	s	...	...	...	...	**B3**[2], 435
—	—,**2,3,4,5,6-pentahydroxy-***	*see* **Idonic acid**													
Ω h478	**1-Hexanol***	*n*-Hexyl alcohol. $CH_3(CH_2)_5OH$	102.18	...	−46.7	158^{760}	0.8136_4^{20}	1.4178^{20}	δ	s	∞	s	∞	chl s	**B1**[3], 1650
h479	**2-Hexanol(*d*)***	$CH_3(CH_2)_3CH(OH)CH_3$	102.18	$[\alpha]_D^{25}$ + 14.1 (eth, c = 11) $[\alpha]_D^{20}$ + 12.70 (al)	...	138^{760}	0.81036_4^{25}	1.4126^{25}	δ	s	s	...	...	...	**B1**[2], 437
Ω h480	—(*dl*)*	$CH_3(CH_2)_3CH(OH)CH_3$	102.18	...	...	140^{760}	0.8159_4^{20}	1.4144^{20}	δ	s	s	...	...	...	**B1**[3], 1660
h481	—(*l*)*	$CH_3(CH_2)_3CH(OH)CH_3$	102.18	$[\alpha]_{5780}^{18}$ −12.04	...	136–8^{754}	0.8178_4^{18}	...	δ	s	s	...	...	...	**B1**[3], 1663
h482	**3-Hexanol(*d*)***	$CH_3CH_2CH_2CH(OH)CH_2CH_3$	102.18	$[\alpha]_D^{20}$ +6.81 (chl)	...	131–3	0.8213_4^{20}	1.4150^{20}	δ	s	∞	s	...	...	**B1**[3], 1663
Ω h483	—(*dl*)*	$CH_3CH_2CH_2CH(OH)CH_2CH_3$	102.18	...	...	135^{760}	0.81825_4^{20}	1.4167^{20}	δ	s	∞	s	...	...	**B1**[3], 1663
h484	—(*l*)*	$CH_3CH_2CH_2CH(OH)CH_2CH_3$	102.18	$[\alpha]_D^{18}$ −7.17	...	135^{760}	0.8213_4^{20}	1.4140^{22}	δ	s	∞	s	...	...	**B1**[3], 1665
Ω h485	**1-Hexanol, 6-chloro-***	Hexamethylene chlorohydrin. $Cl(CH_2)_6OH$	136.62	...	...	107^{12} 94^2	1.0241_4^{20}	1.4550^{20}	δ	v	v	...	...	...	**B1**[3], 1658
h486	**2-Hexanol, 1-chloro-***	$CH_3(CH_2)_3CH(OH)CH_2Cl$	136.62	...	...	73–5^{12}	1.0139_4^{20}	1.4478^{20}	δ	v	v	s	...	...	**B1**[2], 1661
h487	**3-Hexanol, 1-chloro-***	$CH_3CH_2CH_2CH(OH)CH_2CH_2Cl$	136.62	...	...	120^{35} 78^6	1.003_4^{25}	1.446^{25}	δ s^h	s	s	s	s	os v	**B1**[3], 1664
h488	—,**2-chloro-***	$CH_3CH_2CH_2CH(OH)CHClCH_3$	136.62	...	...	171^{760}	1.0143^{11}	...	i	s	s	s	s	os s	**B1**[2], 438
h489	—,**5-chloro-***	$CH_3CHClCH_2CH(OH)CH_2CH_3$	136.62	...	...	78–9^{13}	1.0012_4^{19}	1.4433^{19}	δ	s	v	s	s	os ∞	**B1**[3], 1664
Ω h491	—,**2-ethyl-***	$CH_3(CH_2)_3CH(C_2H_5)CH_2OH$	130.23	...	< −76	185 84–6^{15}	0.8328_4^{20}	1.4328^{20}	i	s	s	s	s	...	**Am 58**, 586
h492	**3-Hexanol, 3-ethyl*-**	$CH_3CH_2CH_2COH(C_2H_5)_2$	130.23	λ^{sulf} 224 (1.20), 291 (3.08)	...	160^{760}	0.8373_4^{20}	1.4300^{20}	i	s	s	s	s	os s	**B1**[3], 1736
h493	—,**3-ethyl-5-methyl-***	$(CH_3)_2CHCH_2COH(C_2H_5)_2$	144.26	...	...	272	0.8396_4^{22}	1.43457^{13}	i	s	s	...	...	...	**B1**[3], 1755
h494	**1-Hexanol, 6-fluoro-***	Hexamethylene fluorohydrin. $F(CH_2)_6OH$	120.17	...	...	85–6^{14}	0.975_4^{20}	1.4141^{25}	δ	v	v	...	...	...	**C51**, 7300
Ω h495	—,**2-isopropyl-5-methyl-***	$(CH_3)_2CHCH_2CH_2CH(C_3H_7^i)CH_2OH$	158.29	...	...	211^{760}	0.8345_4^{20}	1.4369^{20}	i	s	v	...	...	...	**B1**[1], 215
h496	—,**2-methyl-(*d*)***	$CH_3(CH_2)_3CH(CH_3)CH_2OH$	116.21	$[\alpha]_D^{25}$ +2.45 (al)	...	164–5 71–2^{15}	0.8313_4^{13}	1.4245^{17}	i	∞	s	...	...	...	**B1**[2], 444

For explanations, symbols and abbreviations see beginning of table. For structural formulas see end of table.

No.	Name	Synonyms and Formula	Mol. wt.	Color, crystalline form, specific rotation and λ_{max} (log ε)	m.p. °C	b.p. °C	Density	n_D	Solubility w	al	eth	ace	bz	other solvents	Ref.
	1-Hexanol														
h497	—,—(*dl*)*	$CH_3(CH_2)_3CH(CH_3)CH_2OH$	116.21	...	...	164^{750}	0.8270^{20}_{4}	1.4226^{20}	i	∞	s	...	...	...	**B1**³, 1689
h498	—,4-methyl-(*d*)*	$CH_3CH_2CH(CH_3)(CH_2)_3OH$	116.21	$[\alpha]^{28}_{D}$ +2.2	...	77^{20}	0.809^{23}_{4}	1.4233^{25}	i	s	s	s	s	os s	**B1**³, 1695
h499	—,—(*dl*)*	$CH_3CH_2CH(CH_3)(CH_2)_3OH$	116.21	...	...	173^{761} 83^{24}	0.8239^{20}_{4}	1.4219^{20}	i	s	s	s	s	os s	**B1**³, 1695
Ω h500	—,5-methyl-*	$(CH_3)_2CH(CH_2)_4OH$	116.21	...	...	170^{755}	0.8192^{25}_{4}	1.4251^{23}	δ	s	s	s	...	...	**B1**³, 1692
Ω h501	2-Hexanol, 2-methyl-*	$CH_3(CH_2)_3COH(CH_3)_2$	116.21	...	...	143^{760} $53\text{–}5^{15}$	0.8119^{20}_{4}	1.4175^{20}	δ	∞	∞	...	...	...	**B1**³, 1690
Ω h502	—,5-methyl-*	$(CH_3)_2CHCH_2CH_2CH(OH)CH_3$	116.21	...	...	150^{744} 78^{28}	0.814^{20}_{4}	1.4180^{20}	i	s	s	...	...	...	**B1**², 1692
Ω h503	3-Hexanol, 3-methyl-*	$CH_3CH_2CH_2C(OH)(CH_3)CH_2CH_3$	116.21	...	...	143^{760} 56^{18}	0.8234^{20}_{0}	1.4231^{20}	δ	s	s	...	...	...	**B1**³, 1693
h504	—,5-methyl-(*d*)*	$(CH_3)_2CHCH_2CH(OH)CH_2CH_3$	116.21	$[\alpha]^{35}_{D}$ +21.23	...	81^{60}	...	1.4171^{25}	...	...	...	...	...	...	**B1**³, 1692
Ω h505	—,—(*dl*)*	$(CH_3)_2CHCH_2CH(OH)CH_2CH_3$	116.21	...	...	$147\text{–}8^{756}$	0.827^{0}	1.4128^{20}	i	s	s	...	...	...	**B1**³, 1692
h506	—,—(*l*)*	$(CH_3)_2CHCH_2CH(OH)CH_2CH_3$	116.21	$[\alpha]^{23}_{D}$ −3.88	...	$93\text{–}6^{105}$ 63^{19}	...	1.4171^{25}	...	...	...	...	...	...	**B1**³, 1691
h507	1-Hexanol, 1-phenyl-*	$CH_3(CH_2)_4CHOH(C_6H_5)$	178.28	λ^{al} 242 (1.78)	...	170^{50} $110^{0.3}$	0.9477^{25}_{4}	1.5105^{20}	...	s	s	...	...	...	**C46,** 8005
h508	3-Hexanol, 2,2,5,5-tetra-methyl-*	$(CH_3)_3CCH_2CHOHC(CH_3)_3$	158.29	cr (peth)	52.5–3.0	166–70	...	...	i	s	s	s	...	peth s	**B1**³, 1772
h509	2-Hexanol, 2,3,4-trimethyl-*	$CH_3CH_2CH(CH_3)CH(CH_3)COH(CH_3)_2$	144.26	...	...	57^{5}	0.853^{15}_{4}	1.4415^{15}	...	s	s	s	...	peth s	**B1**³, 1756
h510	—,2,3,5-tri-methyl-*	$(CH_3)_2CHCH_2CH(CH_3)COH(CH_3)_2$	144.26	...	...	171^{755} 72^{21}	0.8271^{20}_{20}	1.4321^{20}	...	s	s	s	...	...	**B1**, 212
Ω h511	3-Hexanol, 2,2,3-trimethyl-*	$CH_3CH_2CH_2COH(CH_3)C(CH_3)_3$	144.26	...	...	170^{760}	0.8474^{20}_{4}	1.4402^{20}	...	s	s	s	...	...	**B1**³, 1755
h512	—,2,3,5-tri-methyl-*	$(CH_3)_2CHCH_2COH(CH_3)CH(CH_3)_2$	144.26	...	...	72^{21}	0.8271^{20}_{20}	1.4321^{20}	...	s	s	s	...	...	**B1**³, 1756
Ω h513	—,2,4,4-tri-methyl-*	$CH_3CH_2C(CH_3)_2CH(OH)CH(CH_3)_2$	144.26	...	...	170^{760}	0.8489^{20}	1.4395^{20}	...	s	s	s	...	...	**B1**², 458
h514	—,2,5,5-tri-methyl-*	$(CH_3)_3CCH_2CH(OH)CH(CH_3)_2$	144.26	...	...	77^{32}	0.8250^{20}	1.4286^{20}	...	s	s	s	...	...	**B1**³, 1756
Ω h515	—,3,4,4-tri-methyl-*	$CH_3CH_2C(CH_3)_2COH(CH_3)CH_2CH_3$	144.26	...	...	$165\text{–}6^{760}$	0.8323^{21}_{0}	1.43407^{21}	...	s	s	s	...	...	**B1**, 425
h516	—,3,5,5-tri-methyl-*	$(CH_3)_3CCH_2COH(CH_3)CH_2CH_3$	144.26	...	...	62^{14}	0.8350^{20}	1.4352^{20}	...	s	s	s	...	...	**B1**³, 1755
Ω h517	2-Hexanone*	$CH_3(CH_2)_3COCH_3$	100.16	λ^{al} 275.5 (1.31)	−57	128^{760}	0.81127^{20}_{4}	1.4007^{20}	δ	∞	∞	s	...	...	**B1**², 745
Ω h518	3-Hexanone*	$CH_3CH_2CH_2COCH_2CH_3$	100.16	λ^{al} 277.5 (1.34)	...	125^{760}	0.8118^{20}_{4}	1.4004^{20}	δ	∞	∞	s	...	...	**B1**², 746
Ω h519	1-Hexanone, 1(2,4-dihydroxy-phenyl)-*	...	208.24	pl (to-peth)	56	343d	...	...	i	s	s	s	...	chl s	**B8**², 314
h520	2-Hexanone, 3,3-dimethyl-*	$CH_3CH_2CH_2C(CH_3)_2COCH_3$	128.22	...	...	149^{745}	0.838^{0}_{4}	1.4098^{20}	...	s	s	s	...	...	**B1**³, 2881
Ω h521	—,3,4-dimethyl-*	$CH_3CH_2CH(CH_3)CH(CH_3)COCH_3$	128.22	...	...	158^{760} (155)	0.8295^{22}_{4}	1.4193^{20}	...	s	s	s	...	...	**B1**¹, 760
h522	3-Hexanone, 2,2-dimethyl-*	$CH_3CH_2CH_2COC(CH_3)_3$	128.22	λ^{al} 285.5 (1.47)	...	$145\text{–}8^{745}$	0.8105^{25}_{4}	1.4119^{20}	...	s	s	s	...	...	**B1**², 760
h523	—,—,oxime*	$CH_3CH_2CH_2C(:NOH)C(CH_3)_3$	143.23	nd (al)	78–8.5	...	...	...	...	δ	...	...	...	...	**B1**¹, 364
h524	—,2,5-dimethyl-*	$(CH_3)_2CHCH_2COCH(CH_3)_2$	128.22	...	...	$147\text{–}8^{760}$	0.8270^{0}_{0}	1.4049^{20}	...	s	s	s	...	...	**B1**², 760
h525	—,4,4-dimethyl-*	$CH_3CH_2C(CH_3)_2COCH_2CH_3$	128.22	...	...	151	0.8285^{20}	1.4203^{25}	s	s	...	s	...	chl s	**B1**², 760
h526	—,6-dimethyl-amino-4,4-diphenyl-5-methyl-(*l*)*	*l*-Isomethadone. *l*-Isoamidon. Liden. $(CH_3)_2NCH_2CH(CH_3)C(C_6H_5)_2COCH_2CH_3$	309.46	$[\alpha]^{25}_{D}$ −20 (95 % al, c = 1.5) λ^{al} 225 (4.0), 252 sh (2.8), 265 (2.3), 299 (2.8)	...	$162\text{–}5^{0.6}$	...	...	...	s	...	...	...	...	**C51,** 7420
h527	—,4-hydroxy-*	Propioin. $CH_3CH_2CH(OH)COCH_2CH_3$	116.16	...	...	$132\text{–}5^{227}$ 73^{20}	0.956^{21}_{4}	1.4340^{21}	...	s	...	s	...	...	**B1**, 835
h528	—,4-hydroxy-2,2,5,5-tetra-methyl-*	Pivaloin. $(CH_3)_3CCH(OH)COC(CH_3)_3$	172.27	cr (sub)	81 sub	80^{10}	...	...	s^{h}	...	s	...	...	peth i	**B1**², 884
h529	2-Hexanone, 3-methyl-*	$CH_3CH_2CH_2CH(CH_3)COCH_3$	114.19	...	...	$142\text{–}5^{760}$	0.828^{25}	1.4035^{20}	...	s	s	s	s	os v	**B1**¹, 360
h530	—,4-methyl-*	$CH_3CH_2CH(CH_3)CH_2COCH_3$	114.19	...	...	142^{760} $(139^{762}$ $35\text{–}7^{11}$	...	1.4081^{24}	δ	v	v	v	v	os v	**B1**², 756
Ω h531	—,5-methyl-*	$(CH_3)_2CHCH_2CH_2COCH_3$	114.19	λ^{vap} 176, 188, 192, 196 sh	...	144^{760}	0.888^{20}_{4}	1.4062^{20}	δ	∞	∞	v	v	...	**B1**², 756
h532	—,—,oxime*	$(CH_3)_2CHCH_2CH_2C(:NOH)CH_3$	129.20	...	...	$195\text{–}6^{761}$	0.8881^{20}_{4}	1.4448^{20}	...	...	...	...	...	...	**B1**, 701

For explanations, symbols and abbreviations see beginning of table. For structural formulas see end of table.

No.	Name	Synonyms and Formula	Mol. wt.	Color, crystalline form, specific rotation and λ_{max} (log ε)	m.p. °C	b.p. °C	Density	n_D	Solubility w	al	eth	ace	bz	other solvents	Ref.
	3-Hexanone														
h533	**3-Hexanone, 4-methyl-***	$CH_3CH_2CH(CH_3)COCH_2CH_3$	114.19	...	...	134–5	0.8162^{20}	1.4069^{20}	δ	v	v	v	v	os v	**B1**, 701
Ω h534	**—,5-methyl-***	$(CH_3)_2CHCH_2COCH_2CH_3$	114.19	λ^{al} 280 (1.39)	...	134^{735}	0.8090^{20}	1.4047^{20}	i	∞	∞	...	...	...	**B1^{1}**, 360
Ω h535	**1-Hexanone, 1-phenyl-***	Caprophenone. $CH_3(CH_2)_4COC_6H_5$	176.24	fl	27	265^{760} $122–4^{15}$	0.9576^{20}_{4}	1.5027^{25}	δ	s	s	s	...	...	**B7^{2}**, 257
h536	**Hexasiloxane tetradeca-methyl-**	$(CH_3)_3SiO[-Si(CH_3)_2O-]_4Si(CH_3)_3$	443.97	...	−59	245.5^{760} 142^{20}	0.8910^{20}_{4}	1.3948^{20}	i	δ	...	...	s	...	**C41**, 1998
h538	**1,2,3,5-Hexa-tetraene, 4-chloro-***	$CH_2:CHCCl:C:C:CH_2$	112.56	...	...	127^{760}d 55^{54}	0.9997^{20}_{4}	1.5280^{20}	i	...	s	s	...	chl s	**B1^{3}**, 1061
h539	**1,2,4,5-Hexa-tetraene, 3,4-dichloro-***	$CH_2:C:CClCCl:C:CH_2$	147.01	...	...	$38–40^{8}$	1.1819^{20}_{4}	1.5456^{20}	i	...	s	s	...	chl s	**B1^{3}**, 1061
h540	**1,3,5-Hexatriene (*cis*)***	Divinylethylene. $CH_2:CHCH:CHCH:CH_2$	80.14	λ^{hx} 247 (4.21), 256 (4.35), 266 (4.27)	−12	78^{760}	0.7175^{20}_{4}	1.4577^{20}	i	s	...	s	...	peth, chl s	**B1^{2}**, 243
h541	**—(*trans*)***	$CH_2:CHCH:CHCH:CH_2$	80.14	λ^{hx} 247 (4.53), 258 (4.64), 268 (4.56)	−12	78.5^{760}	0.7369^{15}_{4}	1.5135^{20}	i	s	...	s	...	peth, chl s	**B1^{3}**, 1041
h542	**1,3,4-Hexatriene, 3,6-dichloro-***	$ClCH_2CH:C:CClCH:CH_2$	149.02	...	...	$45–6^{3}$	1.1807^{20}_{4}	1.5195^{20}	...	s	...	...	...	AcOEt s	**C27**, 2932
h543	**1,3,5-Hexatriene, 2,5-dimethyl-***	$CH_2:C(CH_3)CH:CHC(CH_3):CH_2$	108.19	λ^{hx} 252 (4.49), 261 (4.62), 272 (4.51)	−9	145^{747}	0.7822^{20}_{4}	1.5122^{20}	i	...	...	s	...	MeOH, lig s	**B1^{3}**, 1047
Ω h544	**—,1,6-diphenyl-***	$C_6H_5CH:CHCH:CHCH:CHC_6H_5$	232.33	lf (ace) λ^{bz} 343 (4.69), 360 (4.83), 380 (4.7)	200–3	...	...	...	i	i	i	s^{h} δ	δ	chl δ aa i	**B5^{2}**, 605
h545	**1,2,4-Hexatriene, 3,4,6-trichloro-***	$ClCH_2CH:CClCCl:C:CH_2$	183.47	...	...	50^{1}	1.3132^{20}_{4}	1.5517^{20}	i	...	...	s	s	chl s	**C27**, 2932
Ω h546	**2-Hexen-1-al (*trans*)***	$CH_3CH_2CH_2CH:CHCHO$	98.15	...	...	$146–7^{760}$ 43^{12}	0.8491^{20}_{4}	1.4480^{20}	...	...	...	...	...	...	**B1**, 382
h546[3]	**1-Hexene***	$CH_3CH_2CH_2CH_2CH:CH_2$	84.16	...	−139.82	63.35^{760}	0.6731^{20}_{4}	1.3837^{20}	i	s	s	...	s	peth, chl s	**B1^{3}**, 800
h547	**2-Hexene(*cis*)***	$CH_3CH_2CH_2CH:CHCH_3$	84.16	λ^{gas} 178 (4.2), 186 (3.9), 194 sh (3.4)	−141.35	68.84^{760}	0.6869^{20}_{4}	1.3977^{20}	i	s	s	...	s	lig, chl s	**Am 63**, 2683
h548	**—(*trans*)***	$CH_3CH_2CH_2CH:CHCH_3$	84.16	λ^{gas} 170 (3.8), 179 (4.1), 201 (3.0)	−133	68^{750}	0.6784^{20}_{4}	1.3935^{20}	i	s	s	...	s	lig, chl s	**Am 63**, 2683
h549	**3-Hexene(*cis*)***	$CH_3CH_2CH:CHCH_2CH_3$	84.16	λ^{gas} 179 (4.17)	−137.82	66.44^{760}	0.6796^{20}_{4}	1.3947^{20}	i	s	s	...	s	lig, chl s	**Am 63**, 2683
Ω h550	**—(*trans*)***	$CH_3CH_2CH:CHCH_2CH_3$	84.16	...	−113.43	67.08^{760}	0.6772^{20}_{4}	1.3943^{20}	i	s	s	...	s	lig, chl s	**Am 63**, 2683
h551	**1-Hexene, 5-amino-4-methyl-***	$CH_3CH(NH_2)CH(CH_3)CH_2CH:CH_2$	113.20	...	...	$133–6^{760}$	0.793^{15}_{0}	...	s	...	...	...	...	...	**B4**, 226
h552	**—,1-chloro-***	$CH_3(CH_2)_3CH:CHCl$	118.61	...	...	121	0.8872^{22}	1.4300^{22}	i	...	v	s	s	chl v	**B1^{3}**, 803
h553	**—,2-chloro-***	$CH_3(CH_2)_3CCl:CH_2$	118.61	...	...	113^{740} 63^{110}	0.8886^{25}_{4}	1.4278^{25}	i	...	...	s	s	chl v	**B1^{3}**, 803
h554	**—,5-chloro-***	$CH_3CHClCH_2CH_2CH:CH_2$	118.61	...	...	120.7^{760} $28–30^{13}$	0.8891^{25}_{4}	1.4305^{20}	i	...	v	s	s	chl v	**B1^{3}**, 803
h555	**2-Hexene, 4-chloro-***	$CH_3CH_2CHClCH:CHCH_3$	118.61	...	...	123^{760} 30^{10}	0.8934^{20}_{4}	1.4400^{20}	i	...	v	v	s	chl v	**B1^{2}**, 192
h556	**3-Hexene, 1-chloro-***	$CH_3CH_2CH:CHCH_2CH_2Cl$	118.61	...	...	$61–1.5^{60}$	0.900^{24}_{4}	1.435^{24}	i	...	v	v	s	chl v	**B1^{3}**, 807
h557	**—,3-chloro-(*cis*)**	$CH_3CH_2CH:CClCH_2CH_3$	118.61	...	...	119.6^{760}	0.9009^{20}_{4}	1.4360^{20}	i	...	s	s	s	chl s	**B1^{3}**, 807
h558	**—,2-chloro-2,5-dimethyl-***	$(CH_3)_2CHCH:CHCCl(CH_3)_2$	146.66	...	...	$45–60^{15}$	...	1.450^{20}	i	...	s	s	s	chl s	**Am 63**, 3474
h559	**—,1-chloro-4-ethyl-***	$CH_3CH_2C(C_2H_5):CHCH_2CH_2Cl$	146.66	...	...	173	0.9102^{20}_{4}	1.4524^{20}	δ	...	...	...	s	chl s	**B1^{2}**, 201
h560	**1-Hexene, 1,2-dichloro-(*cis*)***	$CH_3(CH_2)_3CCl:CHCl$	153.05	...	...	88^{30}	1.0812^{25}_{4}	1.4631^{25}	i	...	...	...	s	CCl_4, chl s	**B1^{3}**, 803
h561	**—,—(*trans*)***	$CH_3(CH_2)_3CCl:CHCl$	153.05	...	...	$63–5^{22}$	1.1167^{25}_{4}	1.4576^{25}	i	...	...	...	s	chl s	**B1^{3}**, 803
h562	**—,dodecafluoro-***	Perfluoro-1-hexene. $F_3C(CF_2)_3CF:CF_2$	300.05	...	...	57	...	...	i	...	...	...	...	chl s	**J1952**, 4259
h563	**3-Hexene, dodecafluoro-***	Perfluoro-3-hexene. $F_3CCF_2CF:CFCF_2CF_3$	300.05	...	...	49	...	...	i	...	...	...	...	chl s	**J1953**, 1592
Ω h564	**1-Hexene, 2-ethyl-***	$CH_3(CH_2)_3C(C_2H_5):CH_2$	112.22	...	...	120^{760}	0.7270^{20}_{4}	1.4157^{20}	i	...	v	...	v	peth v	**Am 77**, 5443
h565	**3-Hexene, 1,2,3,4,5,6-hexa-chloro-***	$ClCH_2CHClCCl:CClCHClCH_2Cl$	290.83	cr (peth)	58–9	$110–12^{2}$	...	...	...	...	...	...	...	MeOH, chl v^{h}	**B1^{3}**, 808
h566	**—,3(4-hydroxy-phenyl)-4(4-methoxy-phenyl)-***	Mestilbol	282.39	(i) nd (bz-lig), (ii) lf 70% al)	(i) 117–8 (120–1) (ii) 114	$185–95^{0.3}$	...	...	i	s	v	v	...	...	**Am 64**, 1625
h567	**1-Hexene, 1,1,2-trichloro-***	$CH_3(CH_2)_3CCl:CCl_2$	187.50	...	...	$90–93^{10}$	1.225^{25}	1.4760^{25}	i	...	s	...	...	...	**B1^{3}**, 803

For explanations, symbols and abbreviations see beginning of table. For structural formulas see end of table.

No.	Name	Synonyms and Formula	Mol. wt.	Color, crystalline form, specific rotation and λ_{max} (log ε)	m.p. °C	b.p. °C	Density	n_D	Solubility w	al	eth	ace	bz	other solvents	Ref.
	3-Hexenoic acid														
h568	**3-Hexenoic acid***	Hydrosorbic acid. $CH_3CH_2CH{:}CHCH_2CO_2H$	114.15	...	12	208 81–2[2]	0.9640^{23}_{4}	1.4935^{20}	...	...	...	...	...	...	**B2**, 435
Ω h569	**4-Hexenoic acid, 2-acetyl-5-hydroxy-3-oxo-, lactone**	Dehydracetic acid.	168.16	nd (w), rh nd or pr (al)	109	270^{760} 132–3[5]	...	...	v^h	δ s^h	v	...	...	...	**B17**[2], 524
h570	**2-Hexenoic acid, 3-phenyl-**	$CH_3(CH_2)_2C(C_6H_5){:}CHCO_2H$	190.25	cr (peth)	94	183–4[14]	...	...	...	...	...	...	s	peth δ	**B9**[2], 477
h571	**1-Hexen-3-ol***	$CH_3CH_2CH_2CH(OH)CH{:}CH_2$	100.16	...	...	134^{760} 50^{20}	0.834^{22}_{4}	1.4297^{18}	...	s	s	s	...	...	**B1**[3], 1924
h572	**3-Hexen-1-ol(*cis*)***	$CH_3CH_2CH{:}CHCH_2CH_2OH$	100.16	...	...	156–7 58^{12}	0.8478^{22}_{4}	1.4380^{20}	s	v	s	...	...	...	**B1**[3], 1925
h573	**1-Hexen-3-yne***	$CH_3CH_2C{⋮}CCH{:}CH_2$	80.14	lf	...	85^{758}	0.7492^{20}_{4}	1.4522^{20}	i	...	s	...	s	peth, chl s	**B1**[3], 1041
h574	**1-Hexen-5-yne***	Diallylene. $HC{⋮}CCH_2CH_2CH{:}CH_2$	80.14	...	...	70^{760}	0.7650^{20}_{4}	1.4318^{20}	i	...	s	...	s	peth, chl s	**C54**, 3272
h575	**1-Hexen-3-yne, 5-chloro-5-methyl-***	$(CH_3)_2CClC{⋮}CCH{:}CH_2$	128.60	...	...	48^{28}	0.9375^{15}	1.4778^{20}	i	s	s	v^h	s	peth s	**B1**[3], 1045
h576	**3-Hexen-1-yne, 3-propyl-***	$CH_3CH_2CH{:}C(C_3H_7^n)C{⋮}CH$	122.21	...	...	136^{760}	0.7799^{25}_{4}	1.4432^{25}	...	...	...	...	s	peth, chl s	**Am 64**, 365
h577	**4-Hexen-1-yn-3-ol(*d*)***	$CH_3CH{:}CHCH(OH)C{⋮}CH$	96.14	$[\alpha]^{20}_{D}$ +16.06	...	157–9	0.9090^{20}_{4}	1.4645^{17}	...	s	...	s	s	...	**J1945**, 273
h578	—(*dl*)*	$CH_3CH{:}CHCH(OH)C{⋮}CH$	96.14	...	...	154–6 60^{18}	0.9148^{23}_{4}	1.4651^{23}	...	s	...	s	s	...	**Am 66**, 1289
—	**Hexestrol, dimethyl-**	*see* **Hexane, 3,4-bis(4-hydroxy-3-methyl-phenyl)-***													
—	**Hexeton**	*see* **2-Cyclohexen-1-one, 5-isopropyl-3-methyl-***													
Ω h579	**1-Hexyne***	*n*-Butylacetylene. $CH_3(CH_2)_3C{⋮}CH$	82.15	...	−131.9	71.3^{760}	0.7155^{20}_{4}	1.3989^{20}	i	s	s	...	s	peth, chl s	**B1**[3], 977
h580	**2-Hexyne***	$CH_3CH_2CH_2C{⋮}CCH_3$	82.15	...	−89.58	84^{760}	0.73146^{20}_{4}	1.41382^{20}	i	∞	∞	...	s	peth, chl s	**B1**[3], 980
h581	**3-Hexyne***	Diethylacetylene. $CH_3CH_2C{⋮}CCH_2CH_3$	82.15	...	−103	81.5^{744}	0.7231^{20}_{4}	1.4115^{20}	i	s	s	...	s	peth, chl s	**B1**[3], 980
h582	—,2,5-dichloro-2,5-dimethyl-*	$(CH_3)_2CClCH{:}CHCCl(CH_3)_2$	181.11	...	29	175–8[745]	...	...	...	...	...	...	...	chl s	**B1**[3], 847
h583	—,1:2,5:6-diepoxy-	...	110.11	...	−16	98^{20}	1.1189^{23}	1.4871^{23}	...	...	...	...	...	chl s	**B19**[2], 19
h584	**1-Hexyne, 5-methyl-***	Isoamylacetylene. $(CH_3)_2CHCH_2CH_2C{⋮}CH$	96.17	...	−125	92^{760}	0.7274^{20}_{4}	1.4059^{20}	i	s	s	...	s	peth, chl s	**B1**[3], 1000
h585	**2-Hexyne, 5-methyl-***	$(CH_3)_2CHCH_2C{⋮}CCH_3$	96.17	...	−92.91	102.46^{760}	0.73776^{20}_{4}	1.41762^{20}	i	...	s	s	s	peth, chl s	**B1**[3], 1000
h586	**3-Hexyne, 2-methyl-***	$CH_3CH_2C{⋮}CCH(CH_3)_2$	96.17	...	−116.7	95.2^{760}	0.7263^{20}_{4}	1.4114^{20}	i	...	s	...	s	peth, chl s	**B1**[3], 1000
Ω h587	**3-Hexyne-2,5-diol, 2,5-dimethyl-***	Acetylenepinacol. $(CH_3)_2C(OH)C{⋮}CC(OH)(CH_3)_2$	142.20	nd (w)	95	205^{759}	0.949^{20}_{20}	...	s	v	v	v	v	chl s	**B1**[2], 572
h588	**1-Hexyn-3-ol, 3-methyl-***	$CH_3CH_2CH_2COH(CH_3)C{⋮}CH$	112.17	...	...	137^{760}	0.8620^{20}_{4}	1.4338^{20}	s	s	s	...	...	...	**B1**[3], 1994
h589	**3-Hexyn-2-ol, 2-methyl-***	$CH_3CH_2C{⋮}CCOH(CH_3)_2$	112.17	...	...	145–7 46–7[7]	0.962^{0}	1.4392^{25}	s	s	s	...	...	...	**B1**[3], 1993
Ω h589[1]	**Hippuric acid**	Benzamidoacetic acid. *N*-benzoylglycine.	179.18	pr (w or al) λ 226 (4.03), 269 (2.8)	190–3	...	1.371^{20}_{4}	...	s	s	δ	...	δ	AcOEt s chl δ peth, CS_2 i	**B9**[2], 174
h589[2]	—,4-phenyl-phenacyl ester	...	373.41	...	163	...	...	...	...	...	...	...	...	...	**Am 52**, 3715
h589[3]	—,piperazinium salt	$2(C_6H_5CONHCH_2CO_2H).C_4H_{10}N_2$	444.49	wh cr	182–4d	...	...	...	s^h	s^h	i	...	...	...	**Am 70**, 2758
Ω h589[4]	—,*p*-**amino-**	$C_9H_{10}N_2O_3$. *See* h589[1]	194.19	pr or nd (w) $\lambda^{1N\ NaOH}$ 273 (4.21)	198–9	...	...	...	i	s	i	...	...	...	**B14**[2], 258
h589[5]	—,*m*-**bromo-**	$C_9H_8BrNO_3$. *See* h589[1]	258.09	nd (w)	146–7	...	...	...	s^h	s	...	...	δ	MeOH s	**B9**[2], 233
h589[6]	—,*o*-**bromo-**	$C_9H_8BrNO_3$. *See* h589[1]	258.09	nd (w)	192–3	...	...	...	s^h	...	i	...	...	AcOEt v	**B9**[2], 231
Ω h589[7]	—,*p*-**bromo-**	$C_9H_8BrNO_3$. *See* h589[1]	258.09	nd (w)	162	...	...	...	s^h	v	...	...	...	...	**B9**, 353
Ω h590	**Histamine**	4-Imidoazolethylamine. β-Aminoethylglyoxaline.	111.15	wh nd (chl)	86 (83–4)	209^{18} $167^{0.8}$	...	...	s	s	δ	...	...	chl s^h	**B25**[2], 302
Ω h591	—,**dihydrochloride**	$C_5H_9N_3.2HCl$. *See* h590	184.07	pl (eth-ace), pr (w)	249–52	...	...	...	v	δ	i	...	i	MeOH v	**B25**[2], 303
h592	**Histidine(*d*)**	...	155.16	ta (w), $[\alpha]^{20}_{D}$ +40.2 (w)	287d	...	...	...	s	i	i	i	i	chl i	**B25**[2], 404
Ω h593	—(*dl*)	$C_6H_9N_3O_2$. *See* h592	155.16	ta, tetr pr (w)	285d	...	...	...	s	δ	i	i	i	os i	**B25**[2], 409
Ω h594	—(*l*)	$C_6H_9N_3O_2$. *See* h592	155.16	nd or pl (dil al) $[\alpha]^{20}_{D}$ −39.7 (w, c = 1.13) $\lambda^{w, pH=0}$ 211 (3.8)	287d	...	...	...	s	δ	i	i	i	chl i	**B25**[2], 409

For explanations, symbols and abbreviations see beginning of table. For structural formulas see end of table.

No.	Name	Synonyms and Formula	Mol. wt.	Color, crystalline form, specific rotation and λ_{max} (log ε)	m.p. °C	b.p. °C	Density	n_D	Solubility w	al	eth	ace	bz	other solvents	Ref.
	Histidine														
h595	—,bis(3,4-dichloro-benzene-sulfonate) (*l*)	$C_6H_9N_3O_2 . 2C_6H_4Cl_2O_3S$. *See* h592	609.29	rh nd (w)	280d				s^h	i					**C51,** 5184
h596	—,diflavianate(*l*)	$C_6H_9N_3O_2 . 2C_{10}H_6N_2O_8S$. *See* h592	783.62	nd (w)	251–4d				δ	i	i				**B25**², 427
h597	—,dihydro-chloride(*dl*)	$C_6H_9N_3O_2 . 2HCl$. *See* h592	228.08		237d										**B25**¹, 718
Ω h598	—,—(*l*)	$C_6H_9N_3O_2 . 2HCl$. *See* h592	228.08	rh pl	252d				s	s	i				**B25**, 513
h599	—,monohydro-chloride mono-hydrate(*l*)	$C_6H_9N_3O_2 . HCl. H_2O$. *See* h592	209.63	pl (w) $[\alpha]_D^{26}$ +8.0 (3N HCl, c = 2)	259d (254d)				s	i	i				**B25**², 407
—	—,β-alanyl-	*see* **Carnosine**													
h600	**Holocaine**	*N,N*-Bis(*p*-ethoxyphenyl)-acetamidine. $C_{18}H_{22}N_2O_2$.	298.39	nd (al)	117–8				δ	v	v	s	v	lig δ	**B13**², 247
h601	—,hydrochloride	Phenacaine. $C_{18}H_{22}N_2O_2 . HCl$. *See* h600	334.85	cr (w + 1)	190–2 (anh)				v^h	v	i			chl s	**B13**², 248
Ω h602	**Homatropine**	Homotropine. Mandelyltropine. $C_{16}H_{21}NO_3$.	275.35	pr (al or eth)	99–100				δ	s	s	s	δ	chl s	**B21**², 18
Ω h603	—,hydrobromide	$C_{16}H_{21}NO_3 . HBr$. *See* h602	356.27	rh pym or pl (w)	217–8d				s	s	i			chl δ	**B21**, 23
h604	—,hydrochloride	$C_{16}H_{21}NO_3 . HCl$. *See* h602	311.81	wh pr (w) λ^{al} 266	220–27d				s	s				chl δ	**B21**², 18
—	**Homoanisic acid**	*see* **Acetic acid, (4-methoxyphenyl)-**													
—	**Homocaine**	*see* **Ecgonine,** benzoate ethyl ester													
—	**Homogentisic acid**	*see* **Acetic acid, (2,5-dihydroxyphenyl)-**													
—	**Homoisophthalic acid**	*see* **Acetic acid, (3-carboxyphenyl)-**													
—	**Homoisovanillic acid**	*see* **Acetic acid, (3-hydroxy-4-methoxyphenyl)-**													
—	**Homophthalic acid**	*see* **Acetic acid, (2-carboxyphenyl)-**													
—	**Homopyro-catechol**	*see* **Toluene, 3,4-dihydroxy-**													
—	**Homosaligenine**	*see* **Benzene, 1-hydroxy-2-hydroxymethyl-4-methyl-**													
—	**Homotere-phthalic acid**	*see* **Acetic acid, (4-carboxyphenyl)-**													
—	**Homoveratric acid**	*see* **Acetic acid, (3,4-dimethoxyphenyl)-**													
h605	**Hordenine**	Anhaline. 1(Dimethylamino)-2(4-hydroxyphenyl)ethane*. $C_{10}H_{15}NO$.	165.24	rh pr (al or bz-peth), nd (w)	117–8	$173–4^{11}$ sub 140–50			s^h	v	v		δ	chl, lig s	**B13**², 356
Ω h606	—,sulfate	$2(C_{10}H_{15}NO). H_2SO_4$. *See* h605	428.54	fl	210–11 (205)				s	δ	i				**B13**¹, 236
Ω h607	—,sulfate, dihydrate	$2(C_{10}H_{15}NO). H_2SO_4 . 2H_2O$. *See* h605	464.58	pr or pl	197				s	δ	i				**B13**¹, 236
h608	**Humulon**	α-Lupulic acid. $C_{21}H_{30}O_5$.	362.47	yesh cr (eth) $[\alpha]_D^{20}$ −232 (bz), −212 (al) $\lambda^{aq\ ac}$ 260 (3.70)	66.5				δ^h	s	s	s	s	os, alk s	**B8**², 537
Ω h609	**Hydantoic acid**	*N*-Carbamoylglycine. Ureidoacetic acid. $H_2NCONHCH_2CO_2H$	118.09	mcl pr	180d (163)				v^h s	v^h δ	δ				**B4**², 792
h610	—,ethyl ester	$H_2NCONHCH_2CO_2C_2H_5$	146.15	nd (w)	135				v^h	s^h	i				**B4**², 794
h611	—,phenylthio-	*N*-Phenylpseudothio-hydantoic acid. $C_6H_5N{:}C(NH_2)SCH_2CO_2H$	210.26	wh (al)	159–61	175^{760}			s^h	δ	δ		δ		**B12**¹, 248
Ω h612	**Hydantoin**	Glycolylurea	100.18	nd (MeOH), lf (w) $\lambda^{0.02N\ NaOH}$ 223 (3.9)	220				v^h δ	s^h	δ			alk s peth i	**B24**², 127
Ω h613	—,1-acetyl-2-thio-	$C_5H_6N_2O_2S$. *See* h612	158.18	pl (al) $\lambda^{w.pH=3}$ 235 (4.2), 278 (4.2)	175–6					s^h					**B24**¹, 293
Ω h614	—,1-benzoyl-2-thio-	$C_{10}H_8N_2O_2S$. *See* h612	220.26	pr (al) λ^{al} 240 (4.3), 287 (3.7)	165d					s^h					**B24**¹, 294
h615	—,5-benzylidene-2-thio-	$C_{10}H_8N_2OS$. *See* h612	204.26	ye nd (al) λ^{al} 240 (4.08), 295 sh (3.86), 365 (4.53)	258d					s^h				aa δ	**B24**¹, 355
Ω h616	—,5,5-dimethyl-	$C_5H_8N_2O_2$. *See* h612	128.14	pr (w-al) $\lambda^{al\ KOH}$ 219 (3.82)	178	sub			v	v	v	v	v	chl v	**B24**², 157

For explanations, symbols and abb. eviations see beginning of table. For structural formulas see end of table.

No.	Name	Synonyms and Formula	Mol. wt.	Color, crystalline form, specific rotation and λ_{max} (log ε)	m.p. °C	b.p. °C	Density	n_D	Solubility w	al	eth	ace	bz	other solvents	Ref.
	Hydantoin														
Ω h617	—,5,5-diphenyl-	$C_{15}H_{12}N_2O_2$. *See* h612	252.28	nd (al) λ^{w} 252 sh (2.99), 258 (2.85), 269 sh (2.38)	286 (cor)	...	...	...	i	s	δ	s	δ	aa s chl δ	**B24**[2], 227
h618	—,3,5-diphenyl-2-thio-	$C_{15}H_{12}N_2OS$. *See* h612	268.34	λ^{al} 266 (4.23), 271 (4.25)	233	...	...	...	δ	v	v	...	v	...	**B24**, 385
h619	—,5-ethyl-5-phenyl-(*d*)	*d*-Nirvanol. $C_{11}H_{12}N_2O_2$. *See* h612	204.23	pl (10% al) $[\alpha]_D$ +123 (al), +169 (aq alk)	237	...	...	...	...	s	...	...	...	...	**Am 54**, 4697
Ω h620	—,—(*dl*)	*dl*-Nirvanol. $C_{11}H_{12}N_2O_2$. *See* h612	204.23	pr (dil al)	199–200	...	...	...	δ^h	s	δ	s^h	i	aa s	**B24**[2], 206
h621	—,5(2-hydroxy-benzylidene)-2-thio-	$C_{10}H_8N_2O_2S$. *See* h612	220.26	nd (aa)	248	...	...	...	...	...	...	...	...	aa s^h	**B25**[1], 502
Ω h622	—,1-methyl-	$C_4H_6N_2O_2$. *See* h612	114.11	cr (w), pl (al)	157–9	...	...	...	s	s^h	δ	...	...	chl s	**B24**[2], 128
Ω h623	—,5-methyl-(*dl*)	α-Lactylurea. $C_4H_6N_2O_2$. *See* h612	114.11	pr (+w, w)	145–6 (anh)	...	...	...	s	s	δ	...	...	...	**B24**[2], 155
h624	—,—(*l*)	$C_4H_6N_2O_2$. *See* h612	114.11	(w) $[\alpha]_D^{20}$ −50.6	175	...	...	...	v	s	δ	...	...	...	**B24**[1], 304
h625	—,—,hydrate(*dl*)	$C_4H_6N_2O_2 . H_2O$. *See* h612	132.12	rh (w)	155–6	...	...	...	v	v	...	v	...	...	**B24**[2], 155
Ω h626	—,2-thio-	Glycolylthiourea. $C_3H_4N_2OS$. *See* h612	116.14	wh nd (w) λ^{al} 222 (3.96), 264 (4.23)	229–31d	...	...	...	v^h	v^h	s	...	...	alk s	**B24**[2], 138
—	—,5-ureido-	*see* **Allantoin**													
—	**Hydracrylic acid**	*see* **Propanoic acid, 3-hydroxy-***													
Ω h627	β(-)-**Hydrastine**	$C_{21}H_{21}NO_6$	383.41	yesh pr (al) $[\alpha]_D^{17}$ −67.8 (chl, c = 2.5) λ^{MeOH} 295 (3.88)	132 (135)	...	...	...	i	δ	s	...	s	chl v AcOEt s CCl_4 δ peth i	**B27**[2], 603
Ω h628	—,hydrochloride	$C_{21}H_{21}NO_6 . HCl$. *See* h627	419.87	micr pw $[\alpha]_D$ +158.0 (w, c = 2)	116	...	...	...	v^h	...	δ	...	...	chl δ	**B27**[2], 604
h629	**Hydrastinine**	$C_{11}H_{13}NO_3$	207.23	nd (lig), cr (eth) $\lambda^{aq\ HCl}$ 250 (4.3), 307 (3.8)	116–7	...	...	...	s^h	v	v	...	...	chl v HCl s	**B27**[2], 530
h630	—,bisulfate	$C_{11}H_{13}NO_3 . H_2SO_4$. *See* h629	287.29	gr-flr ye cr (al)	216d	...	...	...	s	s	...	...	...	...	**B27**, 465
h631	—,hydrochloride	$C_{11}H_{13}NO_3 . HCl$. *See* h629	225.68	pa ye nd	212d	...	...	...	v	v	i	...	...	chl δ	**B27**[2], 530
—	**Hydratrop-aldehyde**	*see* **Propanal, 2-phenyl-***													
Ω h633	**Hydrazine, 1-acetyl-2-phenyl-**	Acetic acid β-phenyl hydrazide. Hydracetin. $C_6H_5NHNHCOCH_3$	150.18	hex pr (eth) λ 246 (3.9), 290 (3.3)	130–2	...	...	...	v^h δ	v	δ	...	s	chl s	**B15**[2], 92
h634	—,allyl-	$CH_2{:}CHCH_2NHNH_2$	72.12	...	...	122–4[757]	...	...	s	...	s	...	...	chl s	**B4**[1], 562
—	—,benzoyl-	*see* **Benzoic acid,** hydrazide													
h635	—,benzyl-	$C_6H_5CH_2NHNH_2$	122.17	fl or pr (al)	26	103[41]	...	...	∞	∞	∞	...	...	...	**B15**[2], 244
h636	—,1-benzyl-2(4-tolyl)-	...	212.30	...	...	212[17]	...	...	...	...	...	...	...	...	**B15**, 533
h637	—,1,2-bis(3-aminophenyl)-*	*m,m'*-Hydrazodianiline. $C_{12}H_{14}N_4$. *See* h669	214.27	pym (al)	151	...	...	...	i	δ	i	...	...	...	**B15**[2], 309
h638	—,1,2-bis(4-aminophenyl)-*	*p,p'*-Hydrazodianiline. Diphenine. $C_{12}H_{14}N_4$. *See* h669	214.27	ye cr	145	...	...	...	s^h	v	v	...	...	...	**B15**, 653
—	—,1,2-bis(car-boxyphenyl)-*	*see* **Benzoic acid, hydrazodi-**													
h640	—,1,2-bis(1-cyanocyclo-hexyl)-	...	246.36	cr (al)	145d	...	...	...	...	s^h	...	...	...	...	**B15**[2], 294
h641	—,2-(bromo-phenyl)-*	$C_6H_7BrN_2$. *See* h694	187.05	nd	48	...	...	...	...	...	...	...	...	...	**B15**[1], 117
h643	—,(4-bromo-phenyl)-*	$C_6H_7BrN_2$. *See* h694	187.05	nd (w), lf (lig), cr (al)	108	...	...	...	s^h	s	s	...	...	lig v^h peth δ	**B15**[2], 160
h644	—,1-butyl-1-phenyl-*	$H_2NN(C_6H_5)C_4H_9^n$	164.25	...	...	250[763]	...	...	...	...	...	...	...	...	**B15**[1], 28
h645	—,1,2-diallyl-	$CH_2{:}CHCH_2NHNHCH_2CH{:}CH_2$	112.18	...	...	145[752]	...	...	...	...	...	...	...	...	**B4**[2], 963
Ω h646	—,1,2-dibenzoyl-	$C_6H_5CONHNHCOC_6H_5$	240.27	nd (al) λ^{MeOH} 232 (4.27)	241	...	...	...	δ^h	i	i	...	...	MeOH s^h chl i	**B9**[2], 216
Ω h647	—,1,2-dibenzoyl-1,2-dimethyl-	$C_6H_5CON(CH_3)N(CH_3)COC_6H_5$	268.32	pr (al)	85–6	...	...	...	i	s^h	i	...	...	...	**B9**[2], 217
h648	—,1,1-dibenzyl-	$H_2NN(CH_2C_6H_5)_2$	212.30	cr (peth)	65	...	...	...	i	v	v	...	...	...	**B15**[2], 245
h649	—,1,2-dibenzyl-	$C_6H_5CH_2NHNHCH_2C_6H_5$	212.30	lf (dil al)	47	...	...	...	i	...	δ	...	...	...	**B15**[2], 246
h650	—,(2,4-dichloro-phenyl)-*	$C_6H_6Cl_2N_2$. *See* h694	177.03	nd (eth or peth)	94	...	...	...	s^h	v	v	...	...	aa v	**B15**, 431
h651	—,1,1-diethyl-*	$H_2NN(C_2H_5)_2$	88.15	...	...	98–9.6[756]	0.8804_4^{20}	1.4214^{20}	v	v	v	...	v	chl v	**B4**, 550
h652	—,1,2-diethyl-*	$C_2H_5NHNHC_2H_5$	88.15	...	...	85–6 (106–7)	0.797^{26}	1.4204^{20}	...	s	s	...	s	...	**B4**, 550
Ω h653	—,1,2-diformyl-	Hydrazodiformic acid. OCHNHNHCHO	88.07	pr (al)	160	...	...	...	v	δ	i	...	...	...	**B2**[1], 38

For explanations, symbols and abbreviations see beginning of table. For structural formulas see end of table.

No.	Name	Synonyms and Formula	Mol. wt.	Color, crystalline form, specific rotation and λ_{max} (log ε)	m.p. °C	b.p. °C	Density	n_D	Solubility w	al	eth	ace	bz	other solvents	Ref.
	Hydrazine														
h654	—,**1,2-diisobutyl-**	$(C_4H_9^i)NHNH(C_4H_9^i)$	144.26	...	...	170^{735}, 63^{10}	0.8002_4^{20}	1.4276	δ	v	v	v	v	os s	**B4^2**, 962
h655	—,**1,2-di-isopropyl-***	$(C_3H_7^i)NHNH(C_3H_7^i)$	116.21	...	...	125^{760}, 63^{84}	0.7894_4^{20}	1.4173^{20}	δ	v	v	v	v	os ∞	**B4^2**, 960
Ω h656	—,**1,1-dimethyl-***	$H_2NN(CH_3)_2$	60.11	...	...	63^{752}	0.7914^{22}	1.4075^{22}	v	v	v	...	...	MeOH v	**B4^2**, 958
h657	—,**1,2-dimethyl-***	$CH_3NHNHCH_3$	60.11	λ^{al} 235 (4.0), 280 (3.1)	...	81^{753}	0.8274_4^{20}	1.4209^{20}	∞	∞	∞	...	...	...	**B4^2**, 958
Ω h658	—,—,**dihydro-chloride***	$CH_3NHNHCH_3 . 2HCl$	133.02	pr (w)	170d	...	...	...	v	v	...	...	...	...	**B4^1**, 560
h659	—,**(2,3-dimethyl-phenyl)-***	$C_8H_{12}N_2$. *See* h694	136.20	nd (al)	111–2	...	...	...	...	v^h	s	...	...	...	**B15^1**, 171
h660	—,**(2,4-dimethyl-phenyl)-***	$C_8H_{12}N_2$. *See* h694	136.20	nd (eth)	85	...	...	...	δ	v	s	...	...	...	**B15^1**, 173
h661	—,**(2,5-dimethyl-phenyl)-***	$C_8H_{12}N_2$. *See* h694	136.20	nd (dil aa)	78	...	...	...	i	v	s	s	s	os s	**B15^1**, 175
h662	—,**(2,6-dimethyl-phenyl)-***	$C_8H_{12}N_2$. *See* h694	136.20	nd (lig)	46	...	...	...	...	...	...	...	...	lig s	**B15^1**, 172
h663	—,**(3,4-dimethyl-phenyl)-***	$C_8H_{12}N_2$. *See* h694	136.20	yesh nd (eth)	57	...	...	...	...	...	...	...	...	...	**B15^1**, 172
h664	—,**1,2-di(1-naphthyl)-***	1,1′-Hydrazonaphthalene. $C_{10}H_7^{\alpha}NHNHC_{10}H_7^{\alpha}$	284.37	lf (bz), pl (peth)	153	...	...	...	i	...	s	...	s	peth δ	**B15^2**, 256
h665	—,**1,2-di(2-naphthyl)-***	2,2′-Hydrazonaphthalene. $C_{10}H_7^{\beta}NHNHC_{10}H_7^{\beta}$	284.37	red pl (bz)	140–1 (133)	...	...	...	...	i	δ	...	...	os s	**B15**, 569
Ω h666	—,**(2,4-dinitro-phenyl)-***	$C_6H_6N_4O_4$. *See* h694	198.14	blsh-red (al) λ^{diox} 261 (3.86), 349 (4.07)	194 (198d)	...	...	...	i	s^h	δ	...	δ	AcOEt s^h chl δ	**B15**, 489
h667	—,**(2,6-dinitro-phenyl)-***	$C_6H_6N_4O_4$. *See* h694	198.14	red nd (dil al) λ^{chl} 409 (3.81)	145	...	...	...	i	s^h	δ	...	δ	chl δ	**B15^1**, 146
Ω h668	—,**1,1-diphenyl-***	$H_2NN(C_6H_5)_2$	184.24	ta (lig) λ^{al} 287 (4.08)	49–52	220^{40-50}	1.190_4^{16}	...	δ	v	v	...	v	chl v con sulf s	**B15^2**, 52
Ω h669	—,**1,2-diphenyl-***	Hydrazobenzene	184.24	ta (al-eth) λ^{al} 245 (4.3), 290 (3.6), 340 sh (2.8)	131	...	1.158_4^{16}	...	...	v	...	...	δ	aa i	**B15^2**, 52
h670	—,**1,1-di(4-tolyl)-**	...	212.30	lf (al)	93	...	...	...	...	s^h	...	...	...	sulf s	**B15^1**, 154
Ω h671	—,**1,2-di(2-tolyl)-**	*o*-Hydrazotoluene	212.30	lf (al) λ^{al} 245 (4.3), 285 (3.6)	165	...	...	...	i	s	s	...	s	...	**B15^2**, 223
h672	—,**1,2-di(3-tolyl)-**	*m*-Hydrazotoluene	212.30	cr (peth) λ^{al} 250 (4.2), 300 (3.6)	38	224	...	...	i	s	s	...	s	...	**B15**, 506
h673	—,**1,2-di(4-tolyl)-**	*p*-Hydrazotoluene	212.30	lf (lig), cr (bz-al), pl (al-eth)	135	...	0.957_4^{20}	...	i	s	s	...	s	...	**B15^2**, 234
h674	—,**ethyl-***	$H_2NNHC_2H_5$	60.11	...	...	101^{760}	...	...	v	v	v	s	v	chl v	**B4^2**, 959
h675	—,**1-ethyl-1-phenyl-**	$H_2NN(C_2H_5)C_6H_5$	136.20	...	...	237^{760}, $115–9^{19}$	1.0181_4^{21}	1.5711^{21}	δ	v	v	s	v	chl v	**B15^2**, 50
h676	—,**1-ethyl-2-phenyl-**	$C_6H_5NHNHC_2H_5$	136.20	...	...	240^{750}, 110^{14}	1.0150_4^{20}	1.5676^{20}	δ	v	v	...	v	chl v	**B15^2**, 50
h680	—,**1-isobutyl-1-phenyl-**	$H_2NN(C_6H_5)C_4H_9^i$	164.25	λ^{al} 235 (4.0), 280 (3.1)	...	240–5	0.9633_4^{15}	...	...	...	...	...	...	...	**B15**, 121
h681	—,**isopropyl-***	$H_2NNHCH(CH_3)_2$	74.13	...	...	107^{750}	...	...	∞	∞	δ	...	∞	AcOEt ∞	**B4^2**, 960
h682	—,**1-isopropyl-2-methyl-***	$(CH_3)_2CHNHNHCH_3$	88.15	cr (eth)	...	100^{760}	...	...	...	...	...	...	...	...	**B4^2**, 960
Ω h683	—,**methyl-***	H_2NNHCH_3	46.07	...	< −80	87^{745}	...	...	s	∞	s	...	...	lig i	**B4^2**, 957
h684	—,**1(3-methyl-butyl)-1-phenyl-***	$H_2NN(C_6H_5)CH_2CH_2CH(CH_3)_2$	178.28	...	...	236	0.9588^{15}	...	...	...	...	...	...	...	**B15**, 121
Ω h685	—,**1-methyl-1-phenyl-***	$H_2NN(C_6H_5)CH_3$	122.17	...	...	227^{745}, 131^{35}	1.0404_4^{20}	1.5691^{20}	δ	∞	∞	...	∞	chl ∞	**B15^2**, 49
h686	—,**1-methyl-2-phenyl-***	$C_6H_5NHNHCH_3$	122.17	...	...	230^{728}, 112^{14}	1.0320_4^{20}	1.5733^{20}	δ	v	v	...	v	chl s	**B15^2**, 50
h687	—,**1-methyl-2(3-tolyl)-**	...	136.20	ye	59–61	...	1.0265_4^{100}	...	i	v	δ	...	v	...	**B15^2**, 229
h687^1	—,**1-methyl-2(4-tolyl)-**	...	136.20	ye nd (eth), pl (lig)	91	...	...	...	i	s	s	...	s	...	**B15^1**, 154
h688	—,**(1-naphthyl)-***	$H_2NNHC_{10}H_7^{\alpha}$	158.21	lf (al), pl (w) sc (eth)	117	203^{20}	...	...	δ^h	v^h	s	...	v	chl v	**B15^2**, 256
h689	—,**(2-naphthyl)-***	$H_2NNHC_{10}H_7^{\beta}$	158.21	lf (w)	124–5	...	...	...	s^h	v	δ	...	v	chl δ	**E12B**, 885
h690	—,**(2-nitro-phenyl)-***	$C_6H_7N_3O_2$. *See* h694	153.14	red nd (bz) λ^{al} 280 (3.8), 430 (3.8)	90–2	...	...	...	...	v^h	δ	...	δ	lig δ	**B15^2**, 177
h691	—,**(3-nitro-phenyl)-***	$C_6H_7N_3O_2$. *See* h694	153.14	red nd or pr (ace), ye nd (al) λ^{al} 375 (3.0)	93	...	...	...	δ^h	δ	...	...	δ	chl, aa s	**B15^2**, 182

For explanations, symbols and abbreviations see beginning of table. For structural formulas see end of table.

No.	Name	Synonyms and Formula	Mol. wt.	Color, crystalline form, specific rotation and λ_{max} (log ε)	m.p. °C	b.p. °C	Density	n_D	Solubility: w	al	eth	ace	bz	other solvents	Ref.
	Hydrazine														
Ω h692	**—,(4-nitro-phenyl)-***	$C_6H_7N_3O_2$. *See* h694	153.14	og-red lf or nd (al) λ^{al} 230 (3.85), 384 (4.24)	158d				δ	s^h	s		s^h	chl, AcOEt s	**B15**^{2}, 183
h693	**—,1-pentyl-2-phenyl-(*d*)***	$C_6H_5NHNH(CH_2)_4CH_3$	178.28	$[\alpha]_D$ +4.45		$173–5^{50}$ $142–4^{12}$	0.986^{20}	1.5523^{20}			s				**B15**, 121
Ω h694	**—,phenyl-***		108.15	mcl pr or pl λ^{al} 241 (3.97), 283 (3.2)	19.8	243^{760} 115^{10}	1.0986_4^{20}	1.6084^{20}	s^h	∞	∞	v	∞	chl ∞ lig δ	**B15**^{2}, 44
h695	**—,—,hemihydrate***	$C_6H_5NHNH_2.\frac{1}{2}H_2O$	117.15	fl (w)	24	120^{12}	1.0970_{25}^{25}	1.6081^{20}	δ						**B15**, 68
h696	**—,—,hemihydro-chloride***	$2(C_6H_5NHNH_2).HCl$	252.75	nd	225				s	s	δ				**B15**, 108
Ω h697	**—,—,hydro-chloride***	$C_6H_5NHNH_2.HCl$	144.61	lf (al)	243–6d	sub			v	s	i				**B15**^{2}, 48
h698	**—,1-phenyl-2(2-tolyl)-**	2-Methylhydrazobenzene	198.27	pl (al)	101–2				i	δ	v		v	peth δ	**B15**, 497
h699	**—,1-phenyl-2(3-tolyl)-**	3-Methylhydrazobenzene	198.27	ye cr (peth)	61		1.0265_4^{100}		i	v	δ		s	lig s	**B15**^{2}, 229
h700	**—,1-phenyl-2(4-tolyl)-**	4-Methylhydrazobenzene	198.27	pl (lig), cr (al)	91				i	v			v		**B15**^{1}, 154
h701	**—,propyl-***	$H_2NNHCH_2CH_2CH_3$	74.13		119										**B4**^{2}, 960
h702	**—,tetraphenyl-**	$(C_6H_5)_2NN(C_6H_5)_2$	336.44	pr (chl-al) λ^{al} 258.5 (4.16), 292.5 (4.31)	149d				i	δ	s	v	v	chl v	**B15**^{1}, 29
h703	**—,(2-tolyl)-**		122.17	nd (dil al)	59					v	v			chl v lig δ	**B15**^{2}, 222
Ω h704	**—,(3-tolyl)-**		122.17			244δd	1.057_4^{20}		δ	v	v		v		**B15**^{2}, 229
Ω h705	**—,(4-tolyl)-**		122.17	lf (w or eth)	66	244δd			δ	v	v		v		**B15**^{2}, 233
h706	**—,1(3-tolyl)-2(4-tolyl)-**		212.30	ta (lig)	74										**B15**, 511
h707	**—,(2,4,6-tri-bromophenyl)-***	$C_6H_5Br_3N_2$. *See* h694	344.85	nd (peth), cr (lig)	146					$δ^h$			v	chl v lig s aa s^h	**B15**^{1}, 126
h708	**—,(2,4,6-tri-chlorophenyl)-***	$C_6H_5Cl_3N_2$. *See* h694	211.48	cr (bz)	143				s^h				s^h		**B15**^{2}, 156
Ω h709	**—,(2,4,6-trinitro-phenyl)-***	$C_6H_5N_5O_6$. *See* h694	243.14	red pl (al) λ^{chl} 337 (4.38), 385 (3.72)	186				δ	s	δ		δ	aa s chl δ	**B15**^{2}, 221
h710	**—,triphenyl-***	$C_6H_5NHN(C_6H_5)_2$	260.34	nd (bz-peth), cr (al)	142d		0.869_4^{70}		i	s	δ		v		**B15**^{2}, 54
h711	**Hydrazinecar-boxylic acid, ethyl ester***	*N*-Aminourethane. $H_2NNHCO_2C_2H_5$	104.12	cr	46	198d 93^9				s	s				**B3**^{2}, 79
h712	**1,1-Hydrazine-dicarboxylic acid,** diethyl ester*	$H_2NN(CO_2C_2H_5)_2$	176.17	pr (w)	29	138^{12}				s					**B3**^{2}, 79
Ω h713	**1,2-Hydrazine-dicarboxylic acid,** diamide	Dicarbonamide. $H_2NCONHNHCONH_2$	118.10	pl (w)	257–9 (248)				s^h	i	i			30 % KOH s	**B3**^{2}, 95
h714	—,diethyl ester*	$C_2H_5O_2CNHNHCO_2C_2H_5$	176.17	nd (chl), pr (w)	135	250d	1.324^8		s^h	v	v			chl s^h	**B3**^{2}, 79
—	**Hydrazobenzene**	*see* **Hydrazine, 1,2-diphenyl-***													
—	**Hydrazo compounds**	*see* **Hydrazine derivatives**													
h715	**Hydrindane** (*trans, l*)	Hexahydroindane. Octahydroindene.	124.23	$[\alpha]_D^{21}$ −5.98		161.08^{760} 71.7^{40}	0.86268_4^{20}	1.46363^{20}	i		s		s	peth s	**B5**^{2}, 50
h716	**2-Hydrindanone** (*cis*)	$C_9H_{14}O$. *See* h715	138.21		10	225^{754} 109^{23}		1.4830^{20}		s			s	lig s	**E12A**, 221
h717	—(*trans*)	$C_9H_{14}O$. *See* h715	138.21		−12	218^{754}	0.9807_4^{17}	1.4769^{17}		s			s	lig s	**E12A**, 222
—	**Hydrindene**	*see* **Indan**													
—	**Hydrindene, 1,2,3-triketo-,** monohydrate	*see* **Ninhydrin**													
Ω h718	**Hydrobenzamide**	Tribenzaldiamine. $C_6H_5CH(N{:}CHC_6H_5)_2$	298.39	nd (bz), cr (al, w)	110	130^{760}			i	s	v				**B7**^{2}, 166
—	**Hydrobenzoin**	*see* **1,2-Ethanediol, 1,2-diphenyl-***													
h719	**Hydroberberine**(*d*)	*d*-Canadine(*D*). Tetrahydroberberine. $C_{20}H_{21}NO_4$	339.40	nd (dil al) $[\alpha]_D^{20}$ +297.4 (chl, c = 1)	132				i	v	s		s	chl s	**B27**^{2}, 556
h720	—(*l*)	$C_{20}H_{21}NO_4$. *See* h719	339.40	nd (al), $[\alpha]_D^{20}$ −298.2 (chl, c = 1)	134				i	v	s		s	chl s lig $δ^h$	**B27**^{2}, 557
—	**Hydrocerulignone**	*see* **Biphenyl, 4,4′-di-hydroxy-3,3′,5,5′-tetramethoxy-**													
h721	**Hydrocin-chonidine**	Cinchamidine. Dihydrocin-chonidine. $C_{19}H_{24}N_2O$.	296.42	lf (al), $[\alpha]_D^{20}$ −98.4 (al)	229				i	s	δ	δ	δ		**B23**^{2}, 357
—	**Hydrocinna-maldehyde**	*see* **Propanal, 3-phenyl-***													

For explanations, symbols and abbreviations see beginning of table. For structural formulas see end of table.

No.	Name	Synonyms and Formula	Mol. wt.	Color, crystalline form, specific rotation and λ_{max} (log ε)	m.p. °C	b.p. °C	Density	n_D	Solubility w	al	eth	ace	bz	other solvents	Ref.
	Hydrocinnamic acid														
—	**Hydrocinnamic acid**	*see* **Propanoic acid, 3-phenyl-***													
—	**Hydrocortisone**	*see* **Corticosterone, 17-hydroxy-**													
h722	**Hydrocotarnine,** hemihydrate	$C_{12}H_{15}NO_3.\frac{1}{2}H_2O$	230.27	pr (eth) λ 283 (3.23)	56				i	v	v	v	...	chl, aa v	**B27**[2], 541
h723	**Hydrocupreine**	$C_{19}H_{24}N_2O_2$	312.42	pl (dil al) $[\alpha]^{23}_D$ −159.2 (abs al)	230				δ	...	s	...	...	chl s	**B23**[1], 151
h724	**Hydrocyanic acid***	Hydrogen cyanide. Formonitrile. HCN	27.04		−13.24	25.7^{760}	0.6876^{20}_4 0.925^{-40} (sol)	1.2614^{20}	∞	∞	∞	...	...		**B2**[2], 37
Ω h725	**Hydrofuramide**	Furfuralhydramide. Furfuramide. Trifurfuraldiamine.	268.28	nd (al)	117				i	v	v	...	...		**B17**[2], 311
—	**Hydrogen cyanide**	*see* **Hydrocyanic acid***													
h726	**Hydrohydras-tinine**	$C_{11}H_{13}NO_2$	191.23	nd (lig), cr (peth)	66	303^{752}			i	v	v	v	s	AcOEt, CS_2, aa v	**B27**[2], 528
h727	**Hydrolapachol**	$C_{15}H_{16}O_3$	244.29	ye nd (al), cr (peth) λ^{al} 252 (4.32), 282 (4.15)	94				...	s	...	...	...	peth s[h]	**E12B**, 3082
—	**Hydroperoxide, acetyl**	*see* **Acetic acid, per-**													
h728	**—,*tert*-butyl-**	$(CH_3)_3COOH$	90.12		6	d at 89 $35–7^{17}$	0.8960^{20}	1.4015^{20}	s	s	s	...	...	chl s	**C55**, 22109
h729	**—,cyclohexyl-**		116.16		−20	$42^{0.1}$	1.019^{20}_4	1.4645^{25}	...	s	s	...	...	aa s	**Am 72**, 3333
h730	**—,ethyl-**	C_2H_5OOH	62.07		−100	$93–7^{760}$ exp >100	0.9332^{20}_4	1.3800^{20}	∞	∞	∞	...	s		**B1**[3], 1312
h731	**—,methyl-**	CH_3OOH	48.04			$38–40^{65}$	1.9967^{15}_4	1.3641^{15}	∞	∞	∞	...	s	chl, aa δ	**B1**[2], 270
—	**Hydrophlorone**	*see* **Benzene, 1,4-dihydroxy-2,5-dimethyl-***													
Ω h732	**Hydroquinidine**	$C_{20}H_{26}N_2O_2$	326.44	nd (al) $[\alpha]^{18}_D$ +229.26	168–9				...	s	s	s	...	chl s	**Am 41**, 826
h733	**Hydroquinine(*d*)**	Quinotine. $C_{20}H_{26}N_2O_2$. *See* h732	326.44	$[\alpha]^{16}_D$ +143.5	171				i	s	s	s	...	chl s	**Am 41**, 819
h734	**—(*dl*)**	$C_{20}H_{26}N_2O_2$. *See* h732	326.44	nd (al, chl)	175–7				i	s	s	s	...	chl s	**Am 41**, 819
h735	**—(*l*)**	$C_{20}H_{26}N_2O_2$. *See* h732	326.44	nd (eth, chl) $[\alpha]^{20}_D$ −142.2	172.3 (anh)				i	s	s	s	...	chl s	**B23**[2], 400
—	**Hydroquinol**	*see* **Benzene, 1,2,4-trihydroxy-***													
—	**Hydroquinone**	*see* **Benzene, 1,4-dihydroxy-***													
Ω h736	**Hydro-quinoneph-thalein**	2,7-Dihydroxyfluoran	332.32	nd (eth)	228–9				δ[h]	v	v	v	...	aa v	**B19**[2], 247
—	**Hydrosorbic acid**	*see* **3-Hexenoic acid***													
h737	**Hydroxy-amphetamine**	Paredrine. $C_9H_{13}NO$	151.21		125–6				δ	s	...	...	...	chl s	**C55**, 14473
h738	—,hydrobromide	$C_9H_{13}NO.HBr$. *See* h737	232.13		189				v	s	...	...	...		**C46**, 6603
h739	**Hydroxy-citronellal**	$(CH_3)_2COH(CH_2)_3CH(CH_3)CH_2CHO$	172.27			103^3 $91^{0.3}$	0.9220^{20}	1.4494^{20}	δ	s	...	s			**C52**, 3855
—	**Hydroxyditan**	*see* **Methane, (hydroxy-phenyl)phenyl-***													
—	**Hydroxyhemin**	*see* **Hematin**													
h742	**Hyenic Acid**	$C_{25}H_{50}O_2$	382.68	cr (eth), nd (bz)	77–8				i	δ	v	...	...		**B2**[2], 380
h743	**Hygrine(*l*)**	2-Acetonyl-1-methyl pyrrolidine	141.22	$[\alpha]_D$ −1.3		$193–5^{760}$ $92–4^{20}$	0.935^{17}_4		δ	s	...	...	...	chl, dil ac s	**B21**[2], 218
—	**Hyodeoxycholic acid**	*see* **Cholanic Acid, 3α,6α-dihydroxy-**													
h744	**Hyoscine(*dl*)**	Atroscine. Scopolamine. $C_{17}H_{21}NO_4$	303.36	syr					δ	v	v	v	v	chl s	**B27**[1], 248
h745	—,dihydrate(*dl*)	$C_{17}H_{21}NO_4.2H_2O$. *See* h744	339.39	nd (w + 2)	36–7				...	...	δ				**B27**, 101
h746	—,hydrobromide(*d*)	$C_{17}H_{21}NO_4.HBr$. *See* h744	384.28	$[\alpha]_D$ +26.3 (w)	195										**B27**[1], 247
h747	—,—(*dl*)	$C_{17}H_{21}NO_4.HBr$. *See* h744	384.28	eff (ace)	185 (182)				s	s	i	δ	...	chl δ	**B27**[1], 248
h748	—,—(*l*)	$C_{17}H_{21}NO_4.HBr$. *See* h744	384.28	lf (al), $[\alpha]_D$ −26	209 (194)				v	δ	i				**B27**[2], 64
h749	—,hydrobromide trihydrate(*d*)	$C_{17}H_{21}NO_4.HBr.3H_2O$. *See* h744	438.33	ta (w), $[\alpha]_D$ +26.3 (w, c = 3)	55										**B27**[1], 247
h750	—,—(*dl*)	$C_{17}H_{21}NO_4.HBr.3H_2O$. *See* h744	438.33		55–8										**B27**[1], 248
h751	—,—(*l*)	$C_{17}H_{21}NO_4.HBr.3H_2O$. *See* h744	438.33	ta (w + 3), $[\alpha]_D$ −22.8 (w, c = 2)					v	δ	...				**B27**[2], 64
h752	—,hydrochloride(*l*)	$C_{17}H_{21}NO_4.HCl$. *See* h744	339.82	cr (al)	200				v	v	...				**B27**[2], 64
h753	—,hydrochloride dihydrate(*l*)	$C_{17}H_{21}NO_4.HCl.2H_2O$. *See* h744	375.85	pr (w)	80				v	v	...				**B27**, 101

For explanations, symbols and abbreviations see beginning of table. For structural formulas see end of table.

No.	Name	Synonyms and Formula	Mol. wt.	Color, crystalline form, specific rotation and λ_{max} (log ε)	m.p. °C	b.p. °C	Density	n_D	Solubility w	al	eth	ace	bz	other solvents	Ref.
	Hyoscine (*dl*)														
h754	—,monohydrate(*dl*)	$C_{17}H_{21}NO_4.H_2O$. *See* h744	321.38	cr (w + 1)	56–7	...	...	...	δ	v	v	...	...	chl v	**B27**, 101
h755	—,—(*l*)	$C_{17}H_{21}NO_4.H_2O$. *See* h744	321.38	cr (w + 1), $[\alpha]_D^{20}$ −28 (w, c = 2.7), −18 (abs al, c = 5.2)	59	...	...	...	v^h	v	v	v	v	chl v	**B27**[2], 63
h756	—,nitrate(*l*)	$C_{17}H_{21}NO_4.HNO_3$. *See* h744	366.38	nd (al)	213	...	...	...	...	...	...	...	...	...	**B27**[2], 64
h757	—,sulfate dihydrate	$(C_{17}H_{21}NO_4)_2.H_2SO_4.2H_2O$. *See* h744	740.83	nd (w, ace)	...	...	...	...	...	...	...	...	...	...	**B27**, 101
h758	**Hyoscyamine**(*d*)	$C_{17}H_{23}NO_3$	289.38	nd (dil al) $[\alpha]_D$ +31.3 (c = 4)	106	...	...	...	δ	v	v	...	v	chl v	**B21**[2], 18
h759	—(*l*)	Daturine. *l*-Tropyltropeine. $C_{17}H_{23}NO_3$. *See* h758	289.38	tetr nd (dil al) $[\alpha]_D^{20}$ −1 (al, c = 1) λ^{MeOH} 252 (2.24), 258 (2.30), 264 (2.20)	108.5 (109–13)	...	...	...	δ	v	δ	...	δ	chl v dil ac s	**B21**[2], 18
h760	—,hydrobromide(*l*)	$C_{17}H_{23}NO_3.HBr$. *See* h758	370.30	pr	152	...	...	...	v	v	i	...	...	chl v	**B21**[1], 198
Ω h761	—,hydrochloride(*l*)	$C_{17}H_{23}NO_3.HCl$. *See* h758	325.84	$[\alpha]_D$ −23.2 (w, c = 0.5)	149–51	...	...	...	s	s	...	...	...	...	**B21**, 26
h762	—,sulfate(*l*)	$(C_{17}H_{23}NO_3)_2.H_2SO_4.2H_2O$. *See* h758	712.86	dlq nd (al or w) $[\alpha]_D$ −28.3 (w)	206 (anh)	...	...	...	v	v^h	i	v^h	...	chl δ	**B21**[2], 19
h763	**Hypaphorine**(*d*)	*N,N*-Dimethyl-*L*-tryptophane betaine.	246.31	cr (dil al) $[\alpha]_D^{25}$ +113.4 (w, c = 1.6)	255d (anh)	...	...	...	v	v	i	...	i	os i	**B22**[2], 469
h764	**Hypnal**	Antipyrine chloral hydrate.	353.64	wh	68	...	...	...	v	v	...	...	...	...	**C23**, 141
h765	**Hypochlorous acid,** *tert*-butyl ester	*tert*-Butyl hypochlorite. $(CH_3)_3COCl$	108.57	ye liq λ^{CCl_4} 280 (2.1), 340 sh (1.5)	...	$77–8^{760}$	0.9583_4^{18}	1.403^{20}	i	d	v	s	v	aa d	**B1**[3], 1581
h766	—,ethyl ester*	CH_3CH_2OCl	80.52	ye liq λ^{CCl_4} 260 (1.5), 310 (1.2)	...	36^{752}	1.013_4^{-6}	...	i	s	∞	...	∞	chl ∞	**B1**[2], 325
h767	—,methyl ester*	CH_3OCl	66.49	...	...	12^{726}d	...	...	i	s	v	...	s	chl s	**Am 48,** 2166
h768	**Hypofluorous acid,** trifluoromethyl ester*	F_3COF	104.00	...	< −215	-95^{760}	...	...	...	...	...	...	...	...	**Am 70,** 3986
—	$\Delta^{\alpha,\beta}$-**Hypogeic acid**	*see* **2-Hexadecenoic acid***													
h769	**Hypophosphoric acid,** tetraethyl ester*	Tetraethyl hypophosphate. $(C_2H_5O)_2POOP(OC_2H_5)_2$	274.19	...	...	$116–7^2$ $89^{0.005}$	1.1283^{18}	1.4284^{20}	d	d	...	...	s	...	**B1**[3], 1330
Ω h770	**Hypoxanthine**	6-Hydroxypurine. 6(1)-Purinone. Sarcine. $C_5H_4N_4O$.	136.11	oct nd (w) $\lambda^{0.01\ NaOH}$ 250 (4.04)	d at 150	...	...	...	δ s^h	...	...	...	...	alk, dil ac s	**B26**[2], 252
—	—,riboside	*see* **Inosine**													
—	—,2-amino-	*see* **Guanine**													
—	**Hystarazin**	*see* **9,10-Anthraquinone, 2,3-dihydroxy-***													

For explanations, symbols and abbreviations see beginning of table. For structural formulas see end of table.

No.	Name	Synonyms and Formula	Mol. wt.	Color, crystalline form, specific rotation and λ_{max} (log ε)	m.p. °C	b.p. °C	Density	n_D	Solubility: w	al	eth	ace	bz	other solvents	Ref.
	Iditol														
i1	**Iditol**	L(+)-Idite. 1,2,3,4,5,6-Hexanehexol.	182.18	mcl pr (al) $[\alpha]_D^{20}$ +3.5 (w)	73.5	...	...	...	s	...	i	...	i	...	**B1**^{3}, 2405
i1[1]	**Idonic acid** (L)	2,3,4,5,6-Pentahydroxy-hexanoic acid.	196.16	nd (w, al) $[\alpha]_D^{20}$ +5.2 → −13.7 (mut)	205d	...	...	...	s	...	i	...	i	...	**B3**^{2}, **354**
i2	—,γ-lactone(d)	...	178.14	pl, $[\alpha]_D^{20}$ −52.6 (w)	174	...	...	...	s	...	...	...	...	...	**Am 76**, 3543
i3	—,phenylhydrazide (L)	...	286.29	$[\alpha]_D^{20}$ −15.3 (w, c = 1)	115–7 (117–20)	...	...	...	v	v^h	...	...	...	AcOEt δ^h	**B15**^{1}, 82
i4	**Idose**(D)	...	180.16	syr, $[\alpha]_D^{13}$ +16 (w)	...	...	...	...	s	...	...	...	i	...	**B31**, 294; **B1**, 909
i5	—(L)	$C_6H_{12}O_6$. *See* i4	180.16	syr, $[\alpha]_D^{20}$ +52.7 (w, c = 6.2)	...	...	...	...	s	...	...	...	i	...	**B1**, 909
—	**Ignotine**	*see* **Carnosine**													
i6	**Imesatin**	Isatin-3-imide.	146.15	dk ye pr (dil al)	175–6	...	...	...	i s^h	v	δ	...	i	lig i	**B21**^{2}, 332
Ω i7	**Imidazole**	1,3-Diazole. Glyoxaline. Iminazole.	68.08	mcl pr (bz) λ^{al} 207 (3.70)	90–1	257^{760} 138.2^{12}	1.0303_4^{101}	1.4801^{101}	v	v	s	s	δ	Py s chl s peth δ	**B23**^{2}, 34
—	—,4-ethylamino-	*see* **Histamine**													
i8	—,2-mercapto-1-methyl-	Methimazol. Topazole. $C_4H_6N_2S$. *See* i7	114.17	lf (al) λ^{chl} 267 (4.23)	142	280 δd	...	...	v	s	δ	...	δ	chl s lig δ	**B24**, 17
Ω i9	—,1-methyl-	Oxalmethyline. $C_4H_6N_2$. *See* i7	82.11	λ^{al} 212 (3.63)	−6	195–6 $94–5^{15}$	$1.0325_4^{20.5}$	1.4970^{20}	∞	v	s	s	...	...	**B23**^{2}, 35
—	—,2,4,5-triphenyl-	*see* **Lophine**													
i10	**4,5-Imidazoledicarboxylic acid**	1,3-Diazole-4,5-dicarboxylic acid.	156.10	pr	...	288d	1.749	...	δ^h	i	i	...	i	Py δ con ac s os i	**B25**^{2}, 159
Ω i11	**2-Imidazolidinethione**	N,N′-Ethylenethiourea.	102.16	nd (al), pr (AmOH) λ^{al} 235 (4.18)	200–3	...	...	...	v	s	i	...	i	chl, lig i	**B24**^{2}, 4
i12	**2-Imidazolidone**	Ethylencurea.	86.09	nd (chl)	131–3	...	...	...	v	v^h	δ	...	...	chl δ	**B24**^{2}, 3
Ω i13	**Δ^2-Imidazoline, 2-methyl-**	Lysidine. 2-Methyl-2-glyoxalidine.	84.13	hyg	107 (103)	$195–8^{760}$ (222) ca. 120^{65}	...	...	v	v	i	...	...	chl s	**B23**^{2}, 26
i14	**Imperatorin**	Ammidin. $C_{16}H_{14}O_4$	270.29	cr (al)	102	...	...	...	i	s	s	s	s	os s alk s	**B19**^{2}, 277
i15	**Indaconitine**	Acetylbenzoylpseudoaconine. $C_{34}H_{47}NO_{10}$	629.76	cr	202–3d	...	...	...	i	s	s	...	...	peth i chl s	**J 1905**, 1620
Ω i16	**Indan**	2,3-Dihydroindene. Hydrindene.	118.18	λ^{cy} 262 (2.8), 268 (3.1), 273 (3.0)	−51.4	178 73^{13}	0.9639_4^{20}	1.5378^{20}	i	∞	∞	...	...	...	**B5**^{2}, 376
i17	—,1-amino-	dl-1-Hydrindamine. $C_9H_{11}N$. *See* i16	133.20	oil	...	220.5^{747} $96–7^8$	1.038_4^{15}	1.5613^{20}	δ	...	s	s	s	...	**E12A**, 149
i18	—,5-amino-	5-Hydrindamine. $C_9H_{11}N$. *See* i16	133.20	nd (peth)	37–8	$247–9^{745}$ 131^{15}	...	...	δ	...	s	s	s	ac s os v	**E12A**, 154
i19	—,2,3-dibromo-	Indene dibromide. $C_9H_8Br_2$. *See* i16	275.99	pr (peth)	31.5–2.5	144^{10} $100–5^1$	1.747_4^{25}	1.6290^{25}	d^h	d^h	s	...	...	sulf s (red)	**E12A**, 142
i20	—,2,3-dichloro-	Indene dichloride. $C_9H_8Cl_2$. *See* i16	187.07	...	...	$87–90^2$	1.254_4^{25}	1.5715^{23}	d^h	d^h	...	...	...	...	**E12A**, 142
—	—,hexahydro-	*see* **Hydrindane**													
i21	—,1-hydroxy-	1-Indanol. $C_9H_{10}O$. *See* i16	134.18	pl (peth)	54	255 128^{12}	...	...	δ^h	v	v	...	v	chl v peth δ	**B6**^{2}, 530
Ω i22	—,4-hydroxy-	4-Indanol. $C_9H_{10}O$. *See* i16	134.18	(i) tcl pr (peth) (ii) nd (peth) λ^{MeOH} 269 (2.88), 277 (2.88)	(i) 50 (ii) 40	120^{12}	...	...	...	...	...	...	...	...	**B6**^{2}, 530
Ω i23	—,5-hydroxy-	5-Indanol. $C_9H_{10}O$. *See* i16	134.18	nd (peth) λ^{MeOH} 282 (3.45)	56	255 110^8	...	...	δ	v	v	...	...	sulf s (ye) peth δ	**B6**^{3}, 2428
Ω i24	—,1-methyl-	$C_{10}H_{12}$. *See* i16	132.21	λ^{iso} 260 (3.5), 273 (3.7)	...	188–90 60^{10}	0.9402_4^{20}	1.5260^{20}	i	...	...	...	...	...	**B5**^{3}, 1230
Ω i25	—,2-methyl-	$C_{10}H_{12}$. *See* i16	132.21	λ^{iso} 260 (2.8), 273 (3.0), 292 (1.5)	...	187^{760} 70^{10}	0.9034_4^{20}	1.5070^{20}	i	...	...	...	...	...	**B5**^{3}, 1231
i26	—,4-nitro-	$C_9H_9NO_2$. *See* i16	163.18	wh cr (al)	44–4.5	139^{10}	...	...	i	...	...	...	...	os v	**B5**^{3}, 1204
i27	—,1-phenyl-1,3,3-trimethyl-	$C_{18}H_{20}$. *See* i16	236.36	tcl pr (al, MeOH)	52–3	307–10 $161–5^{12}$	1.0009_4^{20}	1.5681^{20} 1.5433^{70}	i	s^h	...	...	s	MeOH s	**B5**^{3}, 2038
i28	**1,2-Indandione**	α,β-Dioxohydrindene. Oxindone.	146.15	gold ye pl or lf (bz, eth)	114–6	...	...	...	s^h	v	...	...	...	chl v MeOH v	**B7**^{2}, 631
i29	—,β-oxime	...	161.16	nd (al), (bz)	215–20d	...	...	...	δ	s^h	...	...	s^h	dil alk s (ye)	**B7**^{2}, 631
Ω i30	**1,3-Indandione**	1,3-Dioxohydrindene. Oxindone.	146.15	nd (eth, lig) $\lambda^{1\,NHCl}$ 227 (4.59), 255 (3.97), 300 (3.05)	131–2d	...	1.37^{21}	...	δ	v^h	s^h	...	s^h	alk s (ye) lig δ	**B7**^{2}, 632
i31	—,dioxime	$C_9H_8N_2O_2$. *See* i30	176.19	nd (w)	ca. 225d	...	...	...	δ^h	δ	i	...	...	...	**B7**, 695

For explanations, symbols and abbreviations see beginning of table. For structural formulas see end of table.

No.	Name	Synonyms and Formula	Mol. wt.	Color, crystalline form, specific rotation and λ_{max} (log ε)	m.p. °C	b.p. °C	Density	n_D	Solubility w	al	eth	ace	bz	other solvents	Ref.
	1,3-Indandione														
Ω i32	**—,2,2-dimethylpropoxy-**	Pivalyl indandione. Pivalyl valone. $C_{14}H_{14}O_3$. *See* i30	230.27	(dil al)	108.5–10.5	....	....	....	i	s	s	s	...	aq alk s (bt ye)	**C53**, 21830
i33	**—,2(3-methylbutoxy)-**	Valone. $C_{14}H_{14}O_3$. *See* i30 ..	230.27	ye (dil al) λ^{al} 237 (4.24), 274 (4.36), 283 (4.57), 323 (3.91)	67–8	....	....	....	i	s	s	s	...	....	**C55**, 12371
Ω i34	**—,2-phenyl-**	Danilone. $C_{15}H_{10}O_2$. *See* i30	222.25	lf (al or bz) λ^{al} 269 (4.35), 320 (3.53)	149–51	....	....	....	i δ^h	s	s	s	s	chl, MeOH s	**B7**[2], 732
Ω i35	**1-Indanone**	α-Hydrindone. 1-Keto-hydrindene.	132.17	ta, nd (w+3) λ^{al} 244 (4.16), 286 (3.48), 291.5 (3.49)	42	241–2^{739} 129^{13}	1.1028$^{40}_{40}$	1.561^{25}	δ	v	v	v	...	chl v lig s	**B7**[2], 283
i36	**2-Indanone**	β-Hydrindone. 2-Keto-hydrindene.	132.17	nd (al or eth) λ^{hx} 296–316 (1.0)	59 (61)	218d	1.0712$^{69}_{4}$	1.538^{67}	i	v	v	v	...	chl v	**B7**[2], 286
Ω i37	**1-Indanone, 2-nitro-**	$C_9H_7NO_3$. *See* i35	177.16	ye nd (bz-lig)	117d	....	....	....	s	s	s	s	s	os s lig δ^h	**B7**[1], 192
i38	**—,6-nitro-**	$C_9H_7NO_3$. *See* i35	177.16	ye lf or nd (peth, al or AcOEt)	74	....	....	....	δ	v	v	s	v	AcOEt s chl v peth δ	**B7**[2], 285
i39	**2-Indanone, 5-nitro-**	$C_9H_7NO_3$. *See* i36	177.16	br nd (al)	141–1.5	....	....	....	δ	v	v	...	...	aa v	**B7**, 364
i40	**Indanthrene**	Dihydroanthraquinonazine. Indanthrone.	442.43	bl nd λ^{al} 216 (4.98), 255 (4.92), 292.5 (5.17), 371 (4.21)	470–500d	....	....	....	i	i	i	i	i	dil alk s $PhNO_2$ s (gr-bl)	**B24**[2], 317
Ω i41	**Indazole**	1,2-Benzodiazole. 1,2-Benzo-pyrazole	118.14	nd (al or w) λ^{MeOH} 250 (3.65), 285 (3.66)	147–9	267–70^{743}	....	....	s^h	s	s	...	....	....	**B23**[2], 117
i42	**—,3-chloro-**	$C_7H_5ClN_2$. *See* i41	152.58	nd (w or lig) λ^{al} 254 (3.49), 290 (3.72), 300 (3.61)	148–8.5	sub	....	....	s^h	v	v	...	v	lig s^h	**B23**[2], 139
i43	**—,4-chloro-**	$C_7H_5ClN_2$. *See* i41	152.58	nd (to) λ^{al} 256 (3.62), 288 (3.67), 299 (3.61)	156	....	....	....	s	s	s	s	s^h	os s, to s^h	**B23**[2], 139
i44	**—,4-nitro-**	$C_7H_5N_3O_2$. *See* i41	163.14	nd (w) λ^{al} 228 (4.09), 399 (3.83)	205–7	....	....	....	δ	s	s	v	s	aa v lig i	**B23**[2], 144
Ω i45	**—,5-nitro-**	$C_7H_5N_3O_2$. *See* i41	163.14	yesh nd or col nd (al) λ^{al} 232 (4.12), 254 (4.21), 304 (3.85)	208	....	....	....	...	s	s	v	s	aa v lig i	**B23**[2], 145
i46	**—,6-nitro-**	$C_7H_5N_3O_2$. *See* i41	163.14	nd (w, al, aa or xyl) λ^{al} 288 (3.97), 337 (3.44)	181d	....	....	....	s^h	s	s	v	s	NaOH, xyl s^h lig i	**B23**[2], 146
Ω i47	**—,7-nitro-**	$C_7H_5N_3O_2$. *See* i41	163.14	cr (al) λ^{al} 310 (3.48), 356 (3.91)	188–90 sub	....	....	....	...	...	s	v	...	2N NaOH v	**B23**[2], 150
i48	**3-Indazolinone**	Benzopyrazolone. Indazolone.	134.14	nd or lf (MeOH or w), pl or nd (al) λ^{MeOH} 215 (4.32), 306 (3.62)	250–2	....	....	....	δ^h	s^h δ	δ	...	...	MeOH s^h	**B24**[2], 59
Ω i49	**Indene**	Indonaphthene.	116.16	(aa) λ^{al} 249 (3.97), 279 (2.72), 290 (2.26)	−1.8 to −1.5	182.6^{760}	0.9960$^{20}_{4}$	1.5768^{20}	i	∞	∞	s	s	Py s CS_2 s	**B5**[2], 410
i50	**—,1-benzhydrylidene-**	ω,ω-Diphenylbenzofulvene.	280.37	og-ye (al)	114.5	....	....	....	...	δ	s	s	s	os s aa δ	**B5**[3], 2532
—	**—,2,3-dihydro-**	***see* Indan**													
i51	**—,1,2-diphenyl-** ..	$C_{21}H_{16}$. *See* i49	268.38	lo nd (aa)	177–8	....	....	....	i	s^h δ	s	...	δ	sulf s(gr) peth δ aa s^h	**B5**[3], 2473
i52	**—,1,3-diphenyl-** ..	$C_{21}H_{16}$. *See* i49	268.38	(i) nd (aa) (ii) pym (aa)	(i) 68–9 (ii) 85	230^{15}	....	....	i	s^h	s	s	...	aa s^h	**B5**[3], 2474
i53	**—,2,3-diphenyl-** ..	$C_{21}H_{16}$. *See* i49	268.38	pr (aa) λ^{diox} 235 (4.3), 310 (4.2)	108–9	235–40^{12}	....	....	i	s^h	s	s	s	sulf s (gr) aa s^h	**B5**[3], 2473
i54	**—,2-methyl-**	$C_{10}H_{10}$. *See* i49	130.19	oil	....	187^{760} 62–5^{20}	0.9034$^{20}_{4}$	1.5070^{20}	i	...	s	s	s	....	**B5**[3], 1371
i55	**—,3-methyl-**	$C_{10}H_{10}$. *See* i49	130.19	λ^{al} 252 (3.97), 284 (2.5), 291 (2.4)	....	198.5d 70^{10}	0.9640$^{20}_{4}$	1.5591^{27}	i	...	s	s	s	....	**B5**[3], 1372

For explanations, symbols and abbreviations see beginning of table. For structural formulas see end of table.

PHYSICAL CONSTANTS OF ORGANIC COMPOUNDS (Continued)

No.	Name	Synonyms and Formula	Mol. wt.	Color, crystalline form, specific rotation and λ_{max} (log ε)	m.p. °C	b.p. °C	Density	n_D	Solubility w	al	eth	ace	bz	other solvents	Ref.
	1-Indenecarboxylic acid														
Ω i56	**1-Indenecarboxylic acid**	$C_{10}H_8O_2$. *See* i49	160.18	pa ye nd (bz)	161	$193–5^{12}$	....	....	δ	v	δ	...	δ	to, chl δ	**E12A**, 356
i57	**2-Indenecarboxylic acid**	$C_{10}H_8O_2$. *See* i49	160.18	nd or lf (bz)	234 sub	....	....	....	δ	v	v	...	δ	chl δ	**E12A**, 357
i58	**1-Indenone, 2,3-diphenyl-**	....	282.35	og-red (lig or al) λ^{al} 260 (4.50), 440 (3.24)	153–5 (151)	....	....	....	i	s	...	s	s	lig s^h	**B7²**, 501
i59	**Indican**	Indoxyl-β-glucoside. $C_{14}H_{17}NO_6.3H_2O$	349.34	orh nd (w + 3), $[\alpha]^{19}_{546}$ −65.6 (w, c = 1)	57–8 (hyd) 178–80d (anh)	....	....	....	s	s	δ	s	δ	MeOH s AcOEt, CS_2, chl δ	**B31**, 258
—	**Indigo carmine**	*see* **5,5′-Indigotindisulfonic acid**													
—	**Indigo red**	*see* **Indirubin**													
i60	**Indigo white**	2,2′-Diindoxyl. Leucoindigo.	264.29	ye cr (dil al)	....	....	....	....	δ	s	s	...	...	alk s	**B23²**, 429
i61	**4,4′-Indigotindicarboxylic acid**	....	350.29	bl nd	....	....	....	....	i	i	i	...	...	sulf s chl i	**B25**, 273
Ω i62	**5,5′-Indigotindisulfonic acid, sodium salt**	Indigo carmine.	466.4	dk bl amor or br-red cr λ^{w} 260 (3.8), 310 (4.5), 600 (3.8)	....	....	....	....	...	...	...	...	...	....	**B25²**, 298
i63	**Indigotinsulfonic acid**	....	342.33	amor	200d	....	....	....	s	s			...	dil ac i	**B25²**, 296
i64	**Indirubin**	Indigo red.	262.27	red or br rh nd (sub) λ^{chl} 285 (3.3), 603 (3.0)	sub	....	....	....	i	δ	s	...	...	con sulf s aa δ	**B24²**, 246
—	**Indogenic acid**	*see* **2-Indolecarboxylic acid, 3-hydroxy-**													
Ω i65	**Indole**	1-Benzo[b]pyrrole.	117.15	lf (w, peth), cr (eth) λ^{al} 218 (4.44), 271 (3.79), 287 (3.66)	52.5	254 $123–4^5$	1.22	....	s^h	v	v	...	s	lig s to v	**B20²**, 196
i66	**—,1-acetyl-**	$C_{10}H_9NO$. *See* i65	159.19	....	....	$152–3^{14}$ $100^{0.001}$	....	....	...	...	s	s	...	....	**B20²**, 200
i67	**—,3(2-aminoethyl)-**	Tryptamine. $C_{10}H_{12}N_2$. *See* i65	160.22	nd (al-bz or lig) λ^{w} 279 (3.72)	120 (146)	$137^{0.15}$	....	....	i	s	i	s	i	chl i	**B22²**, 346
—	**—,2,3-dihydro-**	*see* **Indoline**													
i68	**—,1,3-dimethyl-**	*N*-Methylskatole. $C_{10}H_{11}N$. *See* i65	145.21	nd λ^{al} 282.5 (3.74), 292 (3.74)	141–3	257–60 119^7	....	....	...	...	s	...	...	....	**B20²**, 204
—	**—,3(dimethylaminomethyl)-**	*see* **Gramine**													
i69	**—,3-hydroxy-**	3-Indolol. Indoxyl. C_8H_7NO. *See* i65	133.15	bt ye pr	85	....	....	....	s	s	s	v	s	chl s lig δ	**B21²**, 42
—	**—,—,β-glucoside**	*see* **Indican**													
i70	**—,2-hydroxy-3-nitroso-**	β-Isatoxime. $C_8H_6N_2O_2$. *See* i65	162.15	gold-ye nd	214	....	....	....	δ	s	...	...	...	KOH s	**B21**, 443
i71	**—,2-methyl-**	C_9H_9N. *See* i65	131.18	pl (dil al), nd or lf (w) λ^{al} 220 (4.51), 270 (3.89), 288 (3.70)	61	272	1.07^{20}_4	....	δ^h	v	v	s	s	sulf s	**B20²**, 201
Ω i72	**—,3-methyl-**	Skatole. C_9H_9N. *See* i65	131.18	lf (lig) wh-br sc λ^{al} 223 (4.51), 281 (3.80), 290 (3.94)	97–8	$265–6^{755}$	....	....	s	s	s	s	s	chl s	**B20²**, 203
Ω i73	**2-Indolecarboxylic acid**	....	161.16	ye pl (bz or eth-peth) λ^{al} 219 (4.40), 291 (4.21)	205–8 (203)	....	....	....	δ	s	s	...	...	aa δ	**B22²**, 45
i74	**—,3-hydroxy-**	Indogenic acid. Indoxylic acid. $C_9H_7NO_3$. *See* i73	177.16	cr (sub)	....	122–3 sub d	....	....	δ d^h	...	...	...	...	....	**B22²**, 168
Ω i75	**Indoline**	2,3-Dihydroindole.	119.17	λ^{MeOH} 243 (3.8), 291 (3.3)	....	228–30 (220–1) $70–5^2$	1.069^{20}_4	1.5923^{20}	δ	...	s	s	s	....	**B20²**, 170
—	**2,3-Indolinedione**	*see* **Isatin**													
—	**Indolol**	*see* **Indole, hydroxy-**													
—	**Indonaphthene**	*see* **Indene**													
i76	**Indone, 2,3-dibromo-**	....	287.95	og-ye nd (al, aa)	123	....	....	....	i	s	v	...	...	chl v	**B7¹**, 205
i77	**Indophenin**	....	426.20	bl nd or pw (PhOH-al) λ^{MeOH} 262 (4.19), 314 (3.97), 503 (3.93)	d	....	....	....	i	δ	δ	...	δ	chl δ aa δ	**B21²**, 330

For explanations, symbols and abbreviations see beginning of table. For structural formulas see end of table.

No.	Name	Synonyms and Formula	Mol. wt.	Color, crystalline form, specific rotation and λ_{max} (log ε)	m.p. °C	b.p. °C	Density	n_D	Solubility w	al	eth	ace	bz	other solvents	Ref.
	Indophenol														
i78	**Indophenol**	...	199.21	red br pl (acepeth) λ^{MeOH} 262 (4.19), 314 (3.97), 503 (3.93)	160	...	...	...	s	s	s	...	s	chl s aa s	**C47,** 7456
i79	**Indoxazene**	4,5-Benzoisoxazol.	119.12	oil	...	100^{26} $86–7^{11}$	1.1727_4^{21}	1.5570^{20}	i	...	s	...	...	...	**B27²**, 15
—	**Indoxylic acid**	*see* **2-Indolecarboxylic acid, 3-hydroxy-**													
Ω i82	**Inosine**	Hypoxanthosine. Hypoxanthine riboside.	268.23	pl (w+2), nd (80 % al) $[\alpha]_D^{20}$ −73 (dil NaOH) λ^{w} 249 (4.13)	90 (+2w) 218d (anh)	...	...	...	δ s^h	s	...	...	...	dil alk s	**B31**, 25
i83	**Inositol**(*D*)	*d*-Inosite. 1,2,3,4,5,6-Cyclohexanehexol*.	180.16	pr (w+2), (al) $[\alpha]_D$ +65.0 (w, 12%)	249–50 (cor)	...	...	...	v	δ	i	...	i	aa v	**B6²**, 1157
Ω i84	—(*DL*)	*i*-Inosite. Mesoinosite. Phaseomannitol. $C_6H_{12}O_6$. *See* i83	180.16	mcl pr (w), cr (gl aa)	253 (225)	319 vac	1.752^{15}	...	v	δ	i	...	i	aa v	**B6²**, 1158
i85	—(*L*)	*l*-Inosite. $C_6H_{12}O_6$. *See* i83	180.16	nd (w+2) $[\alpha]_D^{28}$ −64.1 (w, 4%)	247 (238)	250 vac	1.598^{20}	...	v	δ	i	...	i	aa v	**B6²**, 1157
i86	**Inulin**	Plant starch. $(C_6H_{10}O_5)_n$	7000	wh amor or pw, $[\alpha]_D^{21}$ −38.3	178d	...	1.35_4^{20}	...	δ s^h	i	δ	δ	i	HCO_2H s	**J 1932,** 2384
—	**Iodisan**	*see* **2-Propanol, 1,3-bis(dimethylamino)-, dimethyl iodide**													
—	**Iodival**	*see* **Urea, 1(2-iodo-3-methylbutanoyl)-***													
—	**Iodoalphionic acid**	*see* **Propanoic acid, 3(3,5-diodophenyl-4-hydroxy)-2-phenyl-**													
—	**Iodoform**	*see* **Methane, triiodo-***													
i87	**Iodogorgoic acid**(*d*)	3,5-Diiodotyrosine(*d*). Dityrin.	432.99	yesh nd (w or 70% al), $[\alpha]_D^{20}$ +2.9 (HCl, c=5)	213 (204 cor)	...	...	...	δ	i	i	...	i	...	**B14²**, 378
i88	—(*dl*)	$C_9H_9I_2NO_3$. *See* i87	432.99	nd (50 % aa), pl (w), cr (70 % al) $\lambda^{al-NaOH}$ 315 (3.82)	200d (195d)	...	...	...	δ	i	i	...	i	...	**B14²**, 384
i89	—(*l*)	$C_9H_9I_2NO_3$. *See* i87	432.99	nd (w or 70 % al), $[\alpha]^{20}$ −2.98 (4 % HCl, c=5) $\lambda^{w, pH=12.2}$ 311 (3.76)	213d (204 cor)	...	...	...	δ	i	i	...	i	...	**B14²**, 366
i90	**Iodonium fluoride didiphenyl-**	$(C_6H_5)_2IF$	300.12	dlq rh (ace)	85d	...	...	...	v	v	i	s^h δ	...	...	**J1946,** 1129
Ω i91	**Iodonium iodide, diphenyl-**	$(C_6H_5)_2II$	408.02	ye nd (al)	182	...	...	...	...	s^h	...	...	...	...	**C36,** 5790
—	**Iodophen**	*see* **Phenolphthalein, 3,3″,5,5″-tetraiodo-**													
—	**Iodoprene**	*see* **1,3-Butadiene, 2-iodo-***													
—	**Ionol**	*see* **Benzene, 1,3-di-*tert*-butyl-2-hydroxy-5-methyl-**													
i92	**α-Ionol**	...	194.32	oil	...	127^{15}	0.9474_4^{20}	1.4735^{20}	...	s	s	s	...	...	**C42,** 5428
i93	**β-Ionol**	...	194.32	λ^{al} 234 (3.72)	...	131^{15} $89^{0.7}$	0.9243_4^{20}	1.4969^{20}	...	s	s	s	...	...	**C32,** 2090
i94	**α-Ionone**(*d, trans*)	4(2,6,6-Trimethyl-2-cyclohexenyl)-3-buten-2-one*	192.31	$[\alpha]_D^{23}$ +347	...	...	...	1.5016^{20}	δ	∞	∞	v	...	...	**Am 46,** 119
Ω i95	—(*dl*)	$C_{13}H_{20}O$. *See* i94	192.31	λ^{al} 232 (4.2)	...	146.5–7.5^28^ $73–7^{0.15}$	0.9298^{21}	1.5041^{20}	δ	∞	∞	v	...	chl i	**B7²**, 140
i96	—(*l*)	$C_{13}H_{20}O$. *See* i94	192.31	$[\alpha]_D^{27}$ −406	...	...	...	1.5000^{25}	δ	∞	∞	v	...	...	**Am 65,** 2061
i97	—,semicarbazone (*dl*)	$C_{14}H_{23}N_3O$. *See* i94	249.36	(60 % al) λ^{al} 263.5 (4.48)	(i) 107–8 (ii) 143	...	...	...	...	...	...	...	...	...	**B7**, 169
Ω i98	**β-Ionone**	$C_{13}H_{20}O$	192.31	λ^{al} 228 (3.81), 295 (4.03)	...	140^{18} $72–4^{0.1}$	0.9462_4^{20}	1.5198^{20}	δ	∞	∞	...	...	...	**B7²**, 140
i99	—,semicarbazone	$C_{14}H_{23}N_3O$. *See* i98	249.36	nd (al) λ^{al} 282 (4.47), 291 (4.49)	149	...	...	...	i	s	s	...	s	chl s lig i	**B7**, 168
—	**Ipral**	*see* **Barbituric acid, 5-ethyl-5-isopropyl-**													

For explanations, symbols and abbreviations see beginning of table. For structural formulas see end of table.

No.	Name	Synonyms and Formula	Mol. wt.	Color, crystalline form, specific rotation and λ_{max} (log ε)	m.p. °C	b.p. °C	Density	n_D	Solubility w	al	eth	ace	bz	other solvents	Ref.
	β-Irone														
i100	**β-Irone**	4(2,5,6,6-Tetramethyl-1-cyclohexenyl)-3-buten-2-one. $C_{13}H_{20}O$	192.31	$[\alpha]_D$ +41.6 λ^{al} 226(3.98), 295(4.05)	....	$85–90^{0.1}$	0.9434_4^{21}	1.5017^{20}	δ	v	v	...	v	chl, lig v	**B7**[1], 110
i101	**Isatic acid**	2-Aminobenzoylformic acid	165.15	pw	d	....	....	....	s d^h	...	...	...	...	....	**B14**[2], 410
Ω i102	**Isatin**	2,3-Indolinedione. Isatic acid lactam.	147.14	yesh-red pr (sub) λ^{chl} 245(4.2), 310(3.4), 420(2.7)	203.5	sub	....	....	s^h δ	v^h	δ	s	s	alk s, d^h	**B21**[2], 327
i103	—,chloride	2-chloro-3-indolone	165.58	br nd	180d	....	....	....	i	v	v	...	δ v^h	aa v lig δ	**B21**[2], 259
—	—,3-imide	see **Imesatin**													
i104	—,2-oxime	$C_8H_6N_2O_2$. See i102	162.15	ye-og nd (al or w)	198–200d	....	....	....	δ v^h	s^h	v	v	δ	alk s lig i chl δ aa v	**B21**[1], 353
Ω i105	—,3-oxime	$C_8H_6N_2O_2$. See i102	162.15	gold ye nd	225d	....	....	....	v^h	v	δ	...	...	alk, ac s lig i	**B21**[2], 334
Ω i106	**—,1-acetyl-**	$C_{10}H_7NO_3$. See i102	189.17	ye pr or nd λ^{al} 236(4.36), 272(3.91), 342(3.56)	144–5	....	....	....	δ	s	...	s	...	....	**B21**[2], 338
i107	**—,1-methyl-**	$C_9H_7NO_2$. See i102	161.16	red-ye orh nd (w) λ^{al} 242(4.35), 304(3.32), 426(2.70)	134	....	....	....	s^h	s	s	s	s	MeOH s	**B21**[2], 336
Ω i108	**—,5-methyl-**	$C_9H_7NO_2$. See i102	161.16	red pl (w), nd (w or al) λ^{diox} 294(3.43), 418(2.95)	187	....	....	....	δ	s	δ	...	...	HCl, alk s	**B21**[2], 375
Ω i109	**—,5-nitro-**	$C_8H_4N_2O_4$. See i102	192.13	ye nd (al)	254–5d	....	....	....	δ	v	...	...	...	KOH s	**B21**[2], 346
Ω i110	**Isatoic acid, anhydride**	*N*-Carboxyanthranilic anhydride.	163.13	pr (al or gl aa), cr (al or diox) λ^{diox} 239(3.95), 315(3.58)	243d	....	....	....	δ	δ^h	i	δ	i	chl i	**B27**[2], 299
—	**α-Isatropic acid**	see **1,4-Tetralindicarboxylic acid, 1-phenyl-**													
—	**Isoanthraflavin**	see **9,10-Anthraquinone, 2,7-dihydroxy-***													
i111	**Isoapiol**	....	222.24	mcl pr, lf or nd (al)	56 (fp 46)	303–4 189^{33}	....	....	i	v^h	v	v	v	con sulf s AcOEt s aa v^h	**B19**[2], 97
—	**Isobarbituric acid**	see **Uracil, 5-hydroxy-**													
i112	**Isobergaptene**	$C_{12}H_8O_4$	216.21	cr (al) λ^{diox} 255(4.3), 270(4.2), 310(4.0), 355 sh (2.9)	222–3	....	....	....	...	s	...	...	...	diox, MeOH s	**C50**, 12037
i113	**Isoborneol(*d*)**	2-Hydroxybornane	154.26	cr (peth), $[\alpha]_D^{20}$ −34.6 (al, c = 5)	212 (sealed tube) (214)	....	....	....	δ	s	s	...	s	lig, to s chl s	**B6**[2], 89
Ω i114	—(*dl*)	α,β-Camphol. $C_{10}H_{18}O$. See i113	154.26	ta (peth)	212 (sealed tube)	sub	....	....	i	v	v	...	δ	chl v	**B6**, 87
i115	—(*l*)	β-Camphol. $C_{10}H_{18}O$. See i113	154.26	(peth), $[\alpha]_D$ +33.9 (al)	214 (218)	....	....	....	i	v	v	...	δ	chl v	**B6**, 87
i116	—,acetate(*d*)	$C_{12}H_{20}O_2$. See i113	196.29	$[\alpha]_4^{20}$ −50.2 (al)	....	112^{17}	0.9905_4^{20}	1.4633^{20}	i	s	...	s	...	....	**B6**[2], 90
Ω i117	—,—(*dl*)	$C_{12}H_{20}O_2$. See i113	196.29	....	....	$115–7^{22}$ $102–3^{15}$	0.9841_4^{20}	1.4640^{20}	i	s	...	s	...	....	**B6**[2], 91
i118	—,—(*l*)	$C_{12}H_{20}O_2$. See i113	196.29	....	< −50	225 $123–7^{35}$	1.002_0^0	....	i	s	...	s	...	....	**B6**, 89
i119	—,formate(*d*)	$C_{11}H_{18}O_2$. See i113	182.27	$[\alpha]_D^{20}$ +29.5 (al, c = 5)	....	94^{15}	1.0136_4^{20}	1.4678_{6708}^{22}	i	s	...	s	...	....	**E12A**, 681
—	**Isobornylamine**	see **Neobornylamine**													
—	**Isobutane**	see **Propane, 2-methyl**													
—	**Isobutylphosphonic acid**	see **1-Propanephosphonic acid, 2-methyl-***													
—	**Isobutyranilide**	see **Propanoic acid, 2-methyl-, amide, *N*-phenyl-**													
—	**Isobutyric acid**	see **Propanoic acid, 2-methyl-***													
—	**Isobutyrophenone**	see **1-Propanone 2-methyl-1-phenyl-***													
i120	**Isocalycanthine**	$C_{22}H_{28}N_4.H_2O$	183.26	rh	235–6	....	....	....	i	s	s	s	...	chl s	**Am 31**, 1305

For explanations, symbols and abbreviations see beginning of table. For structural formulas see end of table.

No.	Name	Synonyms and Formula	Mol. wt.	Color, crystalline form, specific rotation and λ_{max} (log ε)	m.p. °C	b.p. °C	Density	n_D	Solubility w	al	eth	ace	bz	other solvents	Ref.
	Isocamphane														
i121	**Isocamphane**(*d*)	Dihydrocamphene. 2,2,3-Trimethylnorcamphane.	138.25	cr (MeOH), $[\alpha]_D^{20}$ +8.68 (bz, p = 20)	62–3	166–6.5^{750} (cor)			i	v	...	v	...		**B5**^{2}, 67
i122	—(*dl*)	$C_{10}H_{18}$. *See* i121	138.25	cr (MeOH)	65–7(cor)	165.5–5.7^{730}	0.8276$^{67}_{4}$	1.4419^{67}	i	v	...	v	v	MeOH s AcOEt v	**B5**^{1}, 52
i123	—(*l*)	$C_{10}H_{18}$. *See* i121	138.25	cr (al) $[\alpha]_D^{20}$ − 8.50 (al)	64	164–5^{757} 62–3^{17}			i	s	...	v	δ	lig δ	**B5**^{2}, 67
i124	**Isocamphoric acid** (*d*)	*α-trans*-1,2,2-Trimethyl-1,3-cyclopentanedicarboxylic acid*	200.24	lf (w), $[\alpha]_D^{20}$ +48.6	171–2				δ	v	...	...	...	aa v	**B9**, 762
Ω i125	—(*dl*)	$C_{10}H_{16}O_4$. *See* i124	200.24	pr (al or gl aa), cr (w)	197 (191)		1.249		δ^h	δ	v	...	i^h	peth i	**B9**^{1}, 334
i126	—(*l*)⁻	$C_{10}H_{16}O_4$. *See* i124	200.24	tetr $[\alpha]_D^{17}$ −48.4 (MeOH, p = 9.9)	173				δ	s	...	...	...	aa v	**B9**^{1}, 333
—	**Isocaproamide**	*see* **Pentanoic acid, 4-methyl-,** amide													
—	**Isocaproic acid**	*see* **Pentanoic acid, 4-methyl-***													
—	—,*α*-**amino-**	*see* **Leucine**													
—	**Isocarbostyril**	*see* **Isoquinoline, 1-hydroxy-**													
Ω i127	**Isocarotene**	Dehydro-*β*-carotene. $C_{40}H_{54}$	534.88	vt pr (bz-MeOH), vt nd + lf (bz) λ^{hx} 220 (2.7), 320 (2.5), 445 (3.2), 470 (3.4), 495 (3.2)	192–3 cor										**B30**, 88
i128	**Isocarvomenthol**(*d*)	$C_{10}H_{20}O$.	156.27	$[\alpha]_D$ +20.2		110^{20}	0.904$^{20}_{4}$	1.4669^{18}	...	s	s	s	...		**B6**^{3}, 131
i129	—(*l*)	$C_{10}H_{20}O$. *See* i128	156.27	$[\alpha]_D^{16}$ −17.7		106^{17}	0.9109$^{20}_{4}$	1.4662^{20}	...	s	s	s	...		**B6**^{2}, 39
—	**Isocholesterol**	*see* **Lanosterol**													
—	**Isocinchomeronic acid**	*see* **2,5-Pyridinedicarboxylic acid**													
—	**Isocinnamic acid**	*see* **Cinnamic acid**(*cis*)													
—	**Isocitric acid**	*see* **1,2,3-Propanetricarboxylic acid, 1-hydroxy-***													
i130	**Isocodeine**	$C_{18}H_{21}NO_3$	299.37	pl (bz), pr (AcOEt or aa) $[\alpha]_D^{15}$ −152 (chl, c = 2)	171–2	d	1.87^4	1.675	...	...	...	...	...		**B27**^{2}, 175
i131	**Isocorybulbine**	$C_{21}H_{25}NO_4$	355.44	lf (al) $[\alpha]_D^{15}$ +301 (chl, c = 1) λ^{MeOH} 283 (3.80)	187.5–8.5 (180)		1.045$^{20}_{4}$		i	s	...	...	...	chl s ac s	**B21**^{2}, 200
i133	**Isocorydine**	Corytuberine methyl ether. Luteanine. $C_{20}H_{23}NO_4$.	341.41	pl, $[\alpha]_D^{20}$ +195.3 (chl) λ^{MeOH} 268 (4.3), 300 (3.8)	185				...	...	δ	...	...	chl s	**B21**^{2}, 192
—	**Isocotin**	*see* **Benzophenone, 2,4-dihydroxy-6-methoxy-**													
i134	**Isocoumarin**	2,1-Benzopyrone. *o*-(*β*-Hydroxyvinyl) benzoic acid lactone.	146.15	pl (bz) λ 239 (4.22), 261 (3.87), 318 (3.58)	47	285–6^{719}			i	v	v	...	v	CS_2 v	**B17**, 333
—	**Isocrotonic acid**	*see* **2-Butenoic acid**(*cis*)*													
i135	**Isocyanic acid,** 4-bromophenyl ester	*p*-Bromophenyl isocyanate. C_7H_4BrNO. *See* i149	198.03	nd	42	226 158^{14}			d^h	d^h	v	...	...		**B12**^{1}, 321
i136	—,*tert*-butyl ester	*tert*-Butyl isocyanate. $(CH_3)_3CNCO$	99.13		85.5(cor)		0.8670^9	1.4061^{20}	...	...	...	...	...	HCl d	**B4**, 175
Ω i137	—,2-chlorophenyl ester	*o*-Chlorophenyl isocyanate. C_7H_4ClNO. *See* i149	153.57	λ^{hx} 231 (3.99), 274 (2.63), 283 (2.56)		114–5^{43}			...	...	...	...	...		**B12**, 601
Ω i138	—,3-chlorophenyl ester	*m*-Chlorophenyl isocyanate. C_7H_4ClNO. *See* i149	153.57	λ^{hx} 229 (3.97), 276 (2.88), 284 (2.71)		113–4^{43}			...	...	...	...	...		**B12**, 606
Ω i139	—,4-chlorophenyl ester	*p*-Chlorophenyl isocyanate. C_7H_4ClNO. *See* i149	153.57	λ^{hx} 235 (4.22), 273 (2.82), 287 (2.64)	30–1	115–7^{45}			...	...	...	...	...		**B12**, 616
Ω i140	—,ethyl ester	Ethyl isocyanate. C_2H_5NCO.	71.08			60	0.9031$^{20}_{4}$	1.3808^{20}	i	∞	∞	...	...		**B4**^{2}, 613
Ω i141	—,hendecyl ester	Undecyl isocyanate. $CH_3(CH_2)_{10}NCO$	197.32			103^3			d^h	...	...	...	...	lig v	**B4**^{2}, 658
i142	—,isobutyl ester	Isobutyl isocyanate. $(CH_3)_2CHCH_2NCO$	99.13			106			...	...	...	...	...		**B4**^{2}, 641

For explanations, symbols and abbreviations see beginning of table. For structural formulas see end of table.

No.	Name	Synonyms and Formula	Mol. wt.	Color, crystalline form, specific rotation and λ_{max} (log ε)	m.p. °C	b.p. °C	Density	n_D	Solubility w	al	eth	ace	bz	other solvents	Ref.
	Isocyanic acid														
i143	—,methyl ester	Methylcarbylamine. Methyl isocyanate. CH_3NCO	57.05		−45	59.6 (43–5)	0.9230_4^{27}	1.3419^{18}	s						**B4²**, 578
Ω i144	—,1-naphthyl ester	α-Naphthylcarbylamine. α-Naphthyl isocyanate. $C_{10}H_7^{\alpha}NCO$	169.19	λ^{cy} 227 (4.86), 291 (3.94), 322 (2.75)		269–70	1.1774_4^{20}				s		s		**E12B**, 478
i145	—,2-naphthyl ester	β-Naphthylcarbylamine. β-Naphthyl isocyanate. $C_{10}H_7^{\beta}NCO$	169.19	lf	55–6						v		v		**B12**, 1297
i146	—2-nitrophenyl ester	*o*-Nitrophenyl isocyanate. $C_7H_4N_2O_3$. *See* i149	164.12	wh nd (peth) λ^{hx} 222 (4.23), 259 (3.64), 314–8 (3.36)	41	135^7			d	d	s		s	chl s	**B12²**, 373
i147	—,3-nitrophenyl ester	*m*-Nitrophenyl isocyanate. $C_7H_4N_2O_3$. *See* i149	164.12	wh lf (lig) λ^{hx} 222 (4.37), 253 (3.84), 300 (3.17), 335 (2.22)	51	$130–1^{11}$			d	d	s		s	chl, to s lig s^h,	**B12²**, 382
Ω i148	—,4-nitrophenyl ester	*p*-Nitrophenyl isocyanate. $C_7H_4N_2O_3$. *See* i149	164.12	pa ye nd λ^{hx} 215 (4.01), 276 (4.14), 324 sh (2.74)	57	$137–8^{11}$			d	d	v		v	chl, to s lig s^h	**B12²**, 394
Ω i149	—,phenyl ester	Phenyl isocyanate. Carbanil. Phenylcarbylamine.	119.12	λ^{hx} 226 (4.04), 263 (2.66), 277 (2.67)		$162–3^{751}$ (cor) 55^{13}	1.0956_4^{20}	1.5368^{20}	d	d	v				**B12²**, 244
i150	—,2-tolyl ester	*o*-Tolyl isocyanate	133.15	λ^{hx} 228 (4.01), 273 (2.80), 280 (2.75)		$184–6^{760}$		1.5282^{20}	i d^h	d^h	s				**B12**, 812
i151	—,3-tolyl ester	*m*-Tolyl isocyanate	133.15	λ^{hx} 229 (3.98), 273 (2.74), 281 (2.69)		195–8			d	d	s		s		**B12**, 864
i152	—,4-tolyl ester	*p*-Tolyl isocyanate	133.15	λ^{hx} 230 (4.12), 274 (2.89), 283 (2.88)		187^{751}			d	d	s		s		**B12¹**, 427
i153	**Isocyanuric acid, trimethyl ester**	Tricarbonimide trimethyl ester.	171.16	mcl pr (w or al)	176–7	274			i δ^h	s					**B26²**, 134
—	**Isocyclene**	*see* **Cyclocamphane**													
i154	**Isoderritol**	$C_{21}H_{22}O_6$	370.41	ye lf (MeOH)	150									alk s	**B18²**, 223
—	**Isodurene**	*see* **Benzene, 1,2,3,5-tetramethyl-***													
—	**Isodurenol**	*see* **Benzene, 1-hydroxy-2,3,4,6-tetramethyl-***													
—	**Isoduridine**	*see* **Benzene, 1-amino-2,3,4,6-tetramethyl-***													
—	**α-Isodurylic acid**	*see* **Benzoic acid, 3,4,5-trimethyl-**													
—	**β-Isodurylic acid**	*see* **Benzoic acid, 2,4,6-trimethyl-**													
—	**γ-Isodurylic acid**	*see* **Benzoic acid, 2,3,5-trimethyl-**													
—	**Isoergosterone**	*see* **$\Delta^{4,6,22}$-Ergostatrienone**													
i155	**8-Isoestradiol**	$\Delta^{1,3,5}$-8-α-Epiestratrien-3,17-β-diol. 8-Epiestradiol. $C_{18}H_{24}O_2$.	272.39	cr (dil MeOH-chl), $[\alpha]_D^{20}$ +18 (diox) λ^{al} 280 (3.31)	181					s				diox s	**E14s**, 1982
i156	**8-Isoestrone**	8-Epiestrone. $C_{18}H_{22}O_2$.	270.38	cr (MeOH or MeOH-eth), $[\alpha]_D^{20}$ +94 (diox)	247				i		s			diox s	**E14s**, 2559
i157	**Iso-β-eucaine(*dl*)**	4-Benzoyloxy-2,2,6-trimethylpiperidine. $C_{15}H_{21}NO_2$.	247.34		< −5	188^{19}	1.0467_{22}^{22}		i	s					**B21²**, 15
i158	—,hydrochloride(*d*)	$C_{15}H_{21}NO_2.HCl$. *See* i157	283.80	nd (w), $[\alpha]_{5461}$ +17 (w, c = 1)	271–3				s						**B21²**, 16
i159	—,—(*dl*)	$C_{15}H_{21}NO_2.HCl$. *See* i157	283.80	unst nd or st ta (w), pl (aq al or al-eth)	269–71				s						**B21²**, 15
i160	—,—(*l*)	$C_{15}H_{21}NO_2.HCl$. *See* i157	283.80	nd, $[\alpha]_{5461}$ −16.3 (w, c = 1)	271–3				s						**B21²**, 15
—	**Isoeugenol**	*see* **Benzene, 1-hydroxy-2-methoxy-4-propenyl-***													
—	**Isoeuxanthone**	*see* **Xanthone, 1,6-dihydroxy-**													
—	**Isoeuxanthonic acid**	*see* **Benzophenone, 2,2′,4,6′-tetrahydroxy-**													
i161	**Isoflavone, 4′,7-dihydroxy-**	Daidzein. $C_{15}H_{10}O_4$.	254.25	pa ye pr (50 % al) λ^{al} 249.5 (4.43), 303 (4.05)	323d	sub				s	s				**B18²**, 100

For explanations, symbols and abbreviations see beginning of table. For structural formulas see end of table.

PHYSICAL CONSTANTS OF ORGANIC COMPOUNDS (Continued)

No.	Name	Synonyms and Formula	Mol. wt.	Color, crystalline form, specific rotation and λ_{max} (log ε)	m.p. °C	b.p. °C	Density	n_D	Solubility w	al	eth	ace	bz	other solvents	Ref.
	Isoflavone														
i162	—,4′,5,7-tri-hydroxy-	Genistein. Prunetol. $C_{15}H_{10}O_5$.	270.25	nd (eth), pr (dil al) λ^{al} 263 (4.50), 325 sh (3.71)	301–2d	...	...	...	δ^h	δ	...	...	...	os s dil alk s aa δ	**B18**², 176
i163	**Isofurfurine**	$C_{15}H_{12}N_2O_3$.	268.28	nd (w)	143	...	...	...	...	...	...	...	...	...	**B27**, 764
i164	**Isogeraniolene**	2,6-Dimethyl-1,3-heptadiene. $CH_2{:}C(CH_3)CH{:}CHCH_2CH(CH_3)_2$	124.23	λ^{al} 230 (4.36)	...	$143–4^{755}$ (140–2) 31^7	0.7561_4^{20}	1.4520^{20}	i	...	s	...	s	...	**B1**³, 1015
—	**Isohemipinic acid**	*see* **1,3-Benzenedicarboxylic acid, 4,5-dimethoxy-**													
—	**Isohydroanisoin**	*see* **1,2-Ethanediol, 1,2-bis(4-methoxyphenyl)-***													
—	**Isoindoledione**	*See* **Phthalic acid,** imide													
i165	**Isolysergic acid** (*d*)	$C_{16}H_{16}N_2O_2$.	268.32	cr (w + 2), $[\alpha]_D^{20}$ +281 (Py, c = 1) λ^{al} 312 (3.97)	218d	...	...	...	δ	δ	...	...	...	Py s	**J 1936,** 1440
—	**α-Isomalic acid**	*see* **Malonic acid, hydroxy-(methyl)-**													
i166	**Isomannide**	1:4, 3:6-Dianhydro-D-mannitol.	146.14	mcl cr, $[\alpha]_D^{26.2}$ +62.2 (chl) $[\alpha]_D$ +91 (w)	87–9.5	274d	...	...	v	δ	i	...	i	chl δ	**B1**³, 2402
i167	**Isomenthol**(*d*)	2-Isopropyl-5-methylcyclo-hexanol. *p*-Menthanol-3.	156.27	nd (dil al) $[\alpha]_D^{20}$ +26.5 (al, c = 4)	82.5 (85)	218.6^{760} 96.5^{10}	...	...	δ	s	s	...	...	aa s	**B6**, 41
Ω i168	—(*dl*)	$C_{10}H_{20}O$. *See* i167	156.27	nd	53–4	218.5^{760} $97.4^{10.5}$	0.9040^{30}	1.4510^{60}	δ	s	s	...	...	aa s	**B6**², 51
i169	—(*l*)	$C_{10}H_{20}O$. *See* i167	156.27	$[\alpha]_D^{15}$ −24.1	82.5	...	...	...	δ	s	s	...	...	aa s	**B6**², 51
—	**Isomethadone**	*see* **3-Hexanone, 6-dimethylamino-4,4-diphenyl-5-methyl-***													
i170	**α-Isomorphine**	$C_{17}H_{19}NO_3$.	285.35	nd (MeOH-AcOEt) $[\alpha]_D^{15}$ −167 (MeOH, c = 3)	248	...	...	...	s^h	s	...	...	...	MeOH v	**B27**², 174
—	**Isonicoteine**	*see* **2,3′-Bipyridyl**													
—	**Isonicotinalde-hyde**	*see* **4-Pyridinecarboxalde-hyde**													
i171	**Isonicotine**	4(4-Pyridyl)piperidine.	162.24	hyg wh nd	80	292	...	...	δ	v	s	...	s	lig s	**B23**, 119
—	**Isonicotinic acid**	*see* **4-Pyridinecarboxylic acid**													
—	**Isonipecotic acid**	*see* **4-Piperidinecarboxylic acid**													
i172	**Isopapaverine,** ***N*-benzyl-**	$C_{27}H_{27}NO_4$.	429.52	ye lf (al)	139–40	...	...	...	...	s^h	s	...	...	...	**B21**, 229
i173	—,*N*-ethyl-	$C_{22}H_{25}NO_4$. *See* i172	367.45	pr (al)	*ca.* 101	...	...	...	...	...	s	...	s	...	**B21**, 229
i174	—,*N*-methyl-	$C_{21}H_{23}NO_4$. *See* i172	353.42	ye hyg mcl pr (al)	129–31	...	...	...	s	δ	...	...	...	os s dil al v	**B21**, 229
i175	**Isopelletierine**(*dl*)	2-Acetonylpiperidine. Isopimicine.	141.22	oil liq	...	$91–2^{14}$ $62^{0.8}$	0.9624_4^{20}	1.4683^{20}	...	s	...	...	...	chl s dil ac s	**B21**², 219
i176	—,*N*-methyl-	$C_9H_{17}NO$. *See* i175	155.24	...	...	$96–8^{13}$	0.9478_4^{20}	1.4674^{20}	s	...	...	...	...	dil ac s lig s	**B21**², 219
—	**Isopentane**	*see* **Butane, 2-methyl-***													
—	**Isopeonol**	*see* **Acetophenone, 4-hydroxy-2-methoxy-**													
i177	**Isophenolph-thalein**	3-[*o*-Hydroxyphenyl]-3-(*p*-hydroxyphenyl)-phthalide.	320.35	(dil aa)	189–90	...	...	...	...	v^h	...	v^h	...	...	**B10**², 322
i178	**3-Isopheno-thiazin-3-one**	Azthione. Thiazone.	213.26	red (dil al) λ^{al} 280 (4.2), 370 (4.0), 510 (3.9)	164	...	...	...	s^h	v	...	...	v^h	chl s	**B27**¹, 251
i179	—,7-hydroxy-	Thionol. $C_{12}H_7NO_2S$. *See* i178	229.26	red br pw or nd (aa) $\lambda^{al-0.04NHCl}$ 248 (4.4), 280 (4.34), 440 (4.08), 520 (4.20)	> 360	...	...	...	i	s	...	...	...	chl i aa s^h Ac_2O s	**B27**², 109
Ω i180	**Isophorone**	Isoacetophorone. 3,5,5-Tri-methyl-2-cyclohexen-1-one	138.21	λ^{al} 236 (4.09)	...	214^{754} 99^{18}	0.9229^{20}	1.4759^{20}	i	s	s	s	...	...	**B7**², 64
—	**Isophthaldehyde**	*see* **1,3-Benzenedicarbox-aldehyde**													
—	**Isophthaldehydic acid**	*see* **Benzoic acid, 3-formyl-**													
—	**Isophthalic acid**	*see* **1,3-Benzenedicar-boxylic acid***													
i181	**Isopilocarpine**	*N*-Methylisopilocarpidine. $C_{11}H_{16}N_2O_2$.	208.26	pr $[\alpha]_D^{18}$ +50 (w, c = 2)	...	261^{10}	...	...	v	v	s	...	s	chl v peth i	**B27**², 697

For explanations, symbols and abbreviations see beginning of table. For structural formulas see end of table.

No.	Name	Synonyms and Formula	Mol. wt.	Color, crystalline form, specific rotation and λ_{max} (log ε)	m.p. °C	b.p. °C	Density	n_D	Solubility: w	al	eth	ace	bz	other solvents	Ref.
	Isopimpinellin														
i182	**Isopimpinellin**	$C_{13}H_{10}O_5$	246.22	ye nd (MeOH) λ^{diox} 245 (4.2), 270 (4.2), 365 sh (3.1)	151									MeOH s	J 1945, 540
—	**Isopinene**	*see* α-**Fenchene**													
i183	**Isopomiferin**	$C_{25}H_{24}O_6$.	270.25	nd (dil al) λ^{al} 268	265d					s					Am 61, 2832
—	**Isoprene**	*see* **1,3-Butadiene, 2-methyl-***													
—	**Isopropyl alcohol**	*see* **2-Propanol***													
—	**Isopropyl bromide**	*see* **Propane, 2-bromo-***													
—	**Isopropyl chloride**	*see* **Propane, 2-chloro-***													
i184	**Isopulegol(*d*)**	$\Delta^{8(9)}$-*p*-Menthenol-3.	154.26	$[\alpha]_{5461}$ +29.3		212, 93–4[14]	0.9110^{20}_4	1.4723^{20}	δ	s	s				B6[2], 70
i185	—(*l*)	$C_{10}H_{18}O$. *See* i184	154.26	$[\alpha]^{20}_{5461}$ −25.90		212, 94[14]	0.9110^{20}_4	1.4723^{20}	δ	s	s				B6[3], 257
i186	**α-Isoquinine**		324.43	(bz-peth) $[\alpha]^{18}_D$ −245 (al, c = 1), −248 (al, c = 0.5)	196.5				i	v	v				B23[2], 414,423
i187	**β-Isoquinine**	$C_{20}H_{24}N_2O_2$. *See* i186	324.43	pr (dil al), or amor $[\alpha]^{17}_D$ −187 (97 % al, c = 1)	190–1				i	v	δ		v	chl v lig s	B23[2], 413
Ω i188	**Isoquinoline**	2-Benzazine*. Benzo(c)pyridine. Leucoline. C_9H_7N	129.16	hyg pl λ^{al} 220 (4.80), 268 (3.56), 280 (3.37), 319 (3.45)	26.5	243.25[743], 142[40]	1.0986^{20}	1.6148^{20}	i	v	∞	v	∞	chl v dil ac s	B20[2], 236
i189	—,hydrochloride	Isoquinolinium chloride. $C_9H_7N.HCl$. *See* i188	165.63	pr or pl (al)	209–9.5				s	s^h					B20, 381
i190	—,hydrogen sulfate	$C_9H_7N.H_2SO_4$. *See* i188	227.15	pr or pl (al)	209–9.5				v	s^h	i				B20[2], 237
i191	—,**1-amino-**	$C_9H_8N_2$. *See* i188	144.19	pl (w) λ^{al} 300 (3.8), 330 (3.7)	123				δ	v	δ				B22[1], 640
Ω i192	—,**5-amino-**	$C_9H_8N_2$. *See* i188	144.19	pa ye nd (peth) $\lambda^{w, pH=9}$ 238 (4.22), 332 (3.70)	128	sub								lig s^h	B22[2], 359
Ω i193	—,**4-bromo-**	C_9H_6BrN. *See* i188	208.09	(peth)	40–3	280–5					v^h				B20[2], 238
i194	—,**6,7-dimethoxy-1,2-dimethyl-1,2,3,4-tetrahydro-**	Carnegine. $C_{13}H_{19}NO_2$. *See* i188	221.30	pa br syr $[\alpha]^{25}_D$ +20 (w)		170[1]			s	s	s			chl, con sulf, HCl s	B21[2], 110
i195	—,**1-hydroxy-**	Isocarbostyril. 1-Isoquinolinol. C_9H_7NO. *See* i188	145.16	mcl (bz), nd (bz, al, w) λ^{al} 270 (3.90), 328 (3.67)	209–10	240 sub			δ	v^h	δ s^h		δ s^h	lig i	B21[2], 58
i196	—,**1-methyl-**	$C_{10}H_9N$. *See* i188	143.19	λ^{al} 217 (4.80), 270 (3.66), 307 (3.45), 320 (3.57)	10.1–10.4	248 (253), 124–5[10]	1.0777^{20}_4	1.6095^{20}	δ		s	s	s		B20[2], 247
Ω i197	—,**3-methyl-**	$C_{10}H_9N$. *See* i188	143.19	cr (eth)	68	246[761]			δ		s	s			B20[2], 247
i198	—,**4-methyl-**	$C_{10}H_9N$. *See* i188	143.19	λ^{hx} 271 (3.64), 321.5 (3.62)	< −75	256			δ		s	s	s		B20[1], 152
i199	—,**6-methyl-**	$C_{10}H_9N$. *See* i188	143.19	cr	85–6 (83)	265.5			δ	s	s	s	s	os s	B20[2], 247
i200	—,**7-methyl-**	$C_{10}H_9N$. *See* i188	143.19		67–8	245			δ		s	s			B20[2], 248
i201	—,**8-methyl-**	$C_{10}H_9N$. *See* i188	143.19			258			δ		s	s	s		B20, 404
Ω i202	—,**5-nitro-**	$C_9H_6N_2O_2$. *See* i188	174.16	nd (w + 1) λ 235 (4.12), 333 (3.70)	110	sub			δ v^h	v	v			CS_2, chl, aa v	B20[2], 238
Ω i203	—,**1,2,3,4-tetrahydro-**	2-Azatetralin. $C_9H_{11}N$. *See* i188	133.20	λ^{al} 266 (2.61), 273 (2.61)	< −15	232–3	1.0642^{24}_4	1.5668^{20}	i	s			s^h	dil ac s xyl s^h	B20[2], 176
i204	**1-Isoquinolinecarboxylic acid, nitrile**	1-Cyanoisoquinoline.	154.17	nd (peth), nd (MeOH)	78 (93)				δ	v	v		v	lig δ peth s^h MeOH s	B22[1], 511
i205	**5-Isoquinolinecarboxylic acid, nitrile**		154.17	nd (w or dil al)	135	100–20 sub			δ	v	v			dil ac v	B22[2], 58
i207	**Isoraunescine**	$C_{31}H_{36}N_2O_8$.	564.64	wh nd, $[\alpha]_D$ −70 (chl) λ^{al} 218 (4.76), 270 (4.24)	241–2.5				i	δ				chl, aa s	C50, 1267
i208	**Isoreserpiline**	$C_{23}H_{28}N_2O_5$.	412.49	wh pr, $[\alpha]^{20}_D$ −84 (Py) λ^{al} 228 (4.58), 304 (4.02)	210–2				i						C53, 14134
—	**Isorhamnose**	*see* **Glucomethylose**													

For explanations, symbols and abbreviations see beginning of table. For structural formulas see end of table.

No.	Name	Synonyms and Formula	Mol. wt.	Color, crystalline form, specific rotation and λ_{max} (log ε)	m.p. °C	b.p. °C	Density	n_D	Solubility w	al	eth	ace	bz	other solvents	Ref.
	Isorubijervine														
i209	**Isorubijervine**	Δ^5-3β-18-Dihydroxysolanidene. $C_{27}H_{43}NO_2$	413.65	pr (al), $[\alpha]_D$ +9.2 (al)	241–4					i s^h			s	ac s alk i chl s	**Am 75,** 2133
i210	**Isorubijervosine**	3-β-D-Glucosylisorubijervine. 3-β-D-Glucosyl-Δ^5-solanidene-18-ol. $C_{33}H_{53}NO_7$. *See* i209	575.80	wh nd, $[\alpha]_D^{24}$ −20 ± 2 (Py, c = 1.45)	279–80				i						**Am 75,** 2134
i211	**Isosaccharic acid**	3,4-Dihydroxytetrahydro 2,5-furandicarboxylic acid.	192.14	rh, $[\alpha]_D^{20}$ +46.1 (w, p = 4.2)	185	d			v	v	δ				**B18**², 309
Ω i212	**Isosafrole**(*trans*)		162.19	λ^{al} 259.5 (4.08), 267 (4.06), 305 (3.73)	fp 6.8	253^{760} 111–2^6	1.1224_4^{20}	1.5782^{20}	i	∞	∞	v	∞	chl s	**B19**², 27
i213	**Isoserine**(*l*)	$H_2NCH_2CH(OH)CO_2H$	105.10	cr (dil al, w) $[\alpha]_D^{20}$ −32.58 (w, p = 1.3)	199–201d				s		i		i		**B4**, 503
—	**Isosuccinic acid**	*see* **Malonic acid, 2-methyl-**													
i214	**Isothebaine**(*d*)	$C_{19}H_{21}NO_3$.	311.39	rh cr (al), $[\alpha]_D^{18}$ 285.1 (al, c = 2) λ 398.5	203–4				i	∞	δ			MeOH s chl ∞	**B21**², 169
i215	—,sulfate(*d*)	$(C_{19}H_{21}NO_3)_2.H_2SO_4$	720.85	nd	120–1d				v						**B21**¹, 250
Ω i216	**Isothiocyanic acid, allyl ester**	Allyl mustard oil. $CH_2{:}CHCH_2NCS$	99.16	λ^{cy} 250 (3.00), 275 sh (1.65)	−80 (−102.5)	152^{760} 44^{12}	1.0126_4^{20}	1.5306^{20}	δ	∞	∞		v		**B4**², 667
i217	—,benzyl ester	Benzyl mustard oil. $C_6H_5CH_2NCS$	149.22	ye oil λ^{al} 247.5 (3.12)		243 124–5^{12}	1.1246_4^{15}	1.6049^{15}	i	∞	s				**B12**², 567
i218	—,4-biphenylyl ester		211.29	nd (eth)	60						v				**B12,** 1319
i219	—,4-bromophenyl ester	C_7H_4BrNS. *See* i236	214.09	nd	60–1				i	s^h					**B12**², 354
i220	—,butyl ester	Butyl mustard oil. *n*-Butyl thiocarbimide. $CH_3(CH_2)_2CH_2NCS$	115.20	λ^{cy} 248 (3.24) 270 sh (2.10)		168 64–6^{12}	0.9546_4^{20}	1.501^{20}	i	v	v				**B4**², 635
i221	—,*sec*-butyl ester(*d*)	*sec*-Butyl mustard oil. $CH_3CH_2CH(CH_3)NCS$	115.20	$[\alpha]_D^{20}$ +61.88		159	0.943_4^{20}								**B4**, 161
i222	—,—(*dl*)	$CH_3CH_2CH(CH_3)NCS$	115.20			159.5	0.944^{12}		i	s	s				**B4**, 162
i223	—,—(*l*)	$CH_3CH_2CH(CH_3)NCS$	115.20	$[\alpha]_D^{20}$ −61.80		159	0.942_4^{20}								**B4**, 161
i224	—,*tert*-butyl ester	*tert*-Butyl mustard oil. $(CH_3)_3CNCS$	115.20		10.5–1.5	140^{770} 64^{12}	0.9187_4^{10}		i	i	s				**B4**, 175
Ω i225	—,4-chloro-phenyl ester	C_7H_4ClNS. *See* i236	169.63	nd (al) λ^{cy} 228 (4.48), 276 (4.18), 288 (4.17), 304 (3.25)	45	249–50			i	s^h					**B12**², 330
Ω i226	—,cyclohexyl ester		141.24	λ^{cy} 249 (3.08), 275 sh (1.65)		219^{746}		1.5375^{20}	i	s	s				**B12**², 12
i227	—,4(dimethyl-amino)phenyl ester	$C_9H_{10}N_2S$. *See* i236	178.26	ye pr λ^{cy} 256 (3.75), 297 (4.46), 310 (4.41)	67										**B13**³, 53
Ω i228	—,ethyl ester	Ethyl mustard oil. C_2H_5NCS.	87.14	λ^{diox} 245 (2.86)	−5.9	131–2 (cor)	0.9990_4^{20}	1.5130^{20}	i	∞	∞				**B4**², 614
i229	—,isobutyl ester	Isobutyl mustard oil. $(CH_3)_2CHCH_2NCS$	115.20			160 (cor)	0.9638_4^{14}	1.5005^{14}	i	s	s				**B4**², 641
i230	—,isopropyl ester	Isopropyl mustard oil. $(CH_3)_2CHNCS$	101.17			138 29–30^{10}			i	s	s				**B4**, 155
Ω i231	—,methyl ester	Methyl mustard oil. CH_3NCS	73.12		36	119^{758}	1.0691_4^{37}	1.5258	δ	∞	v				**B4**², 579
i232	—,3-methylbutyl ester	Isoamyl mustard oil. $(CH_3)_2CHCH_2CH_2NCS$	129.23			182–4	0.9419^{17}		δ	v	v				**B4**², 649
i233	—,1-naphthyl ester	α-Naphthyl mustard oil. $C_{10}H_7^\alpha NCS$	185.25	wh nd (al)	58				i	v^h	v	v	v	chl v CCl_4 s lig δ	**B12**², 698
Ω i234	—,2-naphthyl ester	β-Naphthyl mustard oil. $C_{10}H_7^\beta NCS$	185.25	yesh nd (al)	62–3				i	v^h	v		v	chl s	**B12**², 724
i235	—,pentyl ester	*n*-Amyl mustard oil. $CH_3(CH_2)_4NCS$	129.23	bt ye		193.4 (cor)			δ	v	v				**B4**², 642
Ω i236	—,phenyl ester	Phenyl mustard oil.	135.19	λ^{hx} 221 (4.52), 270 (4.03), 280 (4.04)	−21 (cor)	221 (cor) 95^{12}	1.1303_4^{20}	1.6492^{23}	i	s	s				**B12**², 247
i237	—,propyl ester	*n*-Propyl mustard oil. $CH_3CH_2CH_2NCS$	101.17			153 (cor)	0.9781_4^{16}	1.5085^{16}	δ	∞	∞				**B4**², 627
i238	—,4-propylphenyl ester	$C_{10}H_{11}NS$. *See* i236	177.27			263				v	v				**B12,** 1144
—	**Isothiourea**	*see* **Urea, 2-thio-(*S*-substituted)**													
—	**Isovaleraldehyde**	*see* **Butanal, 3-methyl-***													
—	**Isovaleric acid**	*see* **Butanoic acid, 3-methyl-***													
—	**Isovaleronitrile**	*see* **Butanoic acid, 3-methyl-, nitrile**													

For explanations, symbols and abbreviations see beginning of table. For structural formulas see end of table.

No.	Name	Synonyms and Formula	Mol. wt.	Color, crystalline form, specific rotation and λ_{max} (log ε)	m.p. °C	b.p. °C	Density	n_D	Solubility w	al	eth	ace	bz	other solvents	Ref.
	Isovalerophenone														
—	**Isovalerophenone** . .	*see* **1-Butanone,3-methyl-1-phenyl-***													
—	**Isovaleryl chloride**	*see* **Butanoic acid, 3-methyl-,chloride**													
i241	**Isovaline**(*d*)	2-Amino-2-methylbutanoic acid*. $CH_3CH_2C(CH_3)(NH_2)CO_2H$	117.15	nd (aq al + 1w) $[\alpha]_D$ +13 (w, c = 2)	ca 300	sub	. . . .	. . . .	s	. . .	δ	. . .	i	. . . .	**B4**[2], 851
Ω i242	—(*dl*)	$CH_3CH_2C(CH_3)(NH_2)CO_2H$	117.15	rh nd (al-eth, + 1w), mcl pr	315 (307–8 sealed tube)	sub 300	. . . .	. . . .	s	s	δ	. . .	i	. . . .	**B4**[2], 851
i243	—(*l*)	$CH_3CH_2C(CH_3)(NH_2)CO_2H$	117.15	lo nd (w-ace) $[\alpha]^{20}_D$ −9.1 (w, c = 2)	*ca.* 300	sub	. . . .	. . . .	v	v	δ	. . .	i	. . . .	**B4**[2], 851
—	**Isovanillic acid**	*see* **Benzoic acid, 3-hydroxy-4-methoxy-**													
—	**Isovanillin**	*see* **Benzaldehyde, 3-hydroxy-4-methoxy-***													
i244	**Isoxanthen-3-one, 9-phenyl-2,6,7-trihydroxy-**	9-Phenyl-2,6,7-trihydroxyfluorone.	320.31	og red (al-HCl)	> 300	. . . .	. . . .	. . . .	. . .	δ	δ	. . .	δ	os δ alk s	**B18**[1], 404
i245	**Isoxazole**		69.06	λ^{al} 211 (3.60)	. . . .	95–5.5	1.078^{20}_4	1.4298^{17}	s	. . .	. . .	. . .	. . .	. . . .	**B27**[2], 9
—	**Isoxylic acid**	*see* **Benzoic acid, 2,5-dimethyl-**													
—	**Itaconic acid**	*see* **Succinic acid, methylene-**													

For explanations, symbols and abbreviations see beginning of table. For structural formulas see end of table.

No.	Name	Synonyms and Formula	Mol. wt.	Color, crystalline form, specific rotation and λ_{max} (log ε)	m.p. °C	b.p. °C	Density	n_D	Solubility w	al	eth	ace	bz	other solvents	Ref.
	Jacareubin														
j1	**Jacareubin**	$C_{18}H_{14}O_6$	326.32	ye pr (MeOH) λ^{al} 240 (4.03), 279 (4.60), 334 (4.26)	256–7d	...	...	...	...	s	δ	s^h	δ	AcOEt s MeOH s chl δ	**J1953,** 3932
j2	**Jaconecic acid**	$C_{10}H_{16}O_6$	232.24	nd (eth), $[\alpha]^{25}_D$ +28.1 (95% al) λ^w 211 (2.43)	183–4	...	...	...	...	s	s^h	...	...	...	**Am 78,** 3518
j3	**Japaconitine-A**	Acetyl benzoyl aconitine. $C_{34}H_{47}NO_{11}$	645.76	rh $[\alpha]^{13}_D$ +20.7 (chl)	202–3d	...	...	...	i	s^h	s^h	v	...	chl s peth i	**Ber 57,** 1462
j4	—-A1	$C_{34}H_{47}NO_{11}$	645.76	rh (MeOH) $[\alpha]^{12}_D$ +26.4 (chl)	208–9	...	...	...	i	s	s	...	...	MeOH s chl s peth i	**Ber 57,** 1462
j5	—-B	$C_{34}H_{47}NO_{11}$	645.76	rh (MeOH) $[\alpha]^{11}_D$ +26.9	208–9d	...	...	...	i	s	s	...	...	MeOH, chl s peth i	**Ber 57,** 1462
—	**Jasminaldehyde**	*see* **1-Heptene-2-carboxaldehyde, 1-phenyl-**													
j6	**Jasmone**	3-Methyl-2-(2-pentenyl)-2-cyclopenten-1-one	164.25	ye oil λ^{al} 235 (4.14)	...	$257–8^{755}$ $134–5^{12}$	0.9437^{22}_4	1.4979^{22}	δ	s	s	...	...	CCl_4 s lig s	**B7²**, 135
j7	**Javanicin**	$C_{15}H_{14}O_6$	290.28	red (al) λ^{al} 303 (3.97), 505 (3.90)	208d	...	...	...	...	...	...	...	...	alk s (red)	**J1947,** 1021
—	**Jervasic acid**	*see* **Chelidonic acid**													
Ω j8	**Jervine**	$C_{27}H_{39}NO_3.2H_2O$	461.65	nd (w + 2, MeOH-w) $[\alpha]^{23}_D$ −158.5 (al, c = 0.99) λ^{al} 247 (4.1), 370 (1.8)	243–4d (cor) 247–8	...	...	...	i	s	δ	s	...	chl s	**Am 73,** 2970
—	**Juglone**	*see* **1,4-Naphthoquinone, 5-hydroxy-***													
j9	**Julolidine**	$C_{12}H_{15}N$	173.25	λ^{al} 225 (4.4), 300 (3.4), 405 (3.8)	40	280d $155–6^{17}$	1.003^{20}	1.568^{25}	...	...	...	...	...	...	**B20²**, 214
j10	—,**1,6-dioxo-**	1,6-Diketojulolidine. $C_{12}H_{11}NO_2$	201.23	ye (al) λ^{al} 250 (3.5), 305 (3.5)	145–6	190– $210^{0.3}$	...	...	...	s^h	...	...	...	MeOH s	**J1951,** 1898
j11	**Junipal**	5(α-Propynyl)-2-formyl thiophene. C_8H_6OS	150.20	nd (pet eth or dil al) λ^{al} 286 (3.91), 320 (4.27)	80	...	...	...	δ s^h	...	...	...	...	os s	**C49**, 13354
j12	**Juniperol**	Kuromatsuol. Macrocarpol. $C_{15}H_{26}O$.	220.38	tcl (EtOH) $[\alpha]^{20}_D$ +25.4 (al) +18.4 (chl)	112	286–8d	1.0460^{20}_{20}	1.519	i	...	...	...	...	os s	**B6²**, 516
j13	**Junipic acid**	5(α-Propynyl)-2-thiophenecarboxylic acid. $C_8H_6O_2S$.	166.20	ye nd (pet eth), cr (aq EtOH) λ^{al} 293 (4.02)	180	...	...	...	i	s	...	...	...	$NaHCO_3$ s lig s^h	**C54,** 9880
j14	—,methyl ester	$C_9H_8O_2S$. *See* j13	180.22	(aq MeOH)	62	sub 50–5	...	...	...	...	...	s	...	...	**C49,** 13354

For explanations, symbols and abbreviations see beginning of table. For structural formulas see end of table.

No.	Name	Synonyms and Formula	Mol. wt.	Color, crystalline form, specific rotation and λ_{max} (log ε)	m.p. °C	b.p. °C	Density	n_D	Solubility						Ref.
									w	al	eth	ace	bz	other solvents	
	Kaempferol														
—	**Kaempferol**	see **Flavone, 3,4′,5,7-tetrahydroxy-**													
—	**Kairoline**	see **Quinoline, 1-methyl-1,2,3,4-tetrahydro-**													
—	**Kavahin**	see **Methysticin**													
k1	**Ketene**	Carbomethene. Ethonone*. $CH_2{:}CO$	42.04	λ^{gas} 330 (1.0)	−151	−56 (−41[760])	...	...	d	d	δ	δ	...	NH_3 d	**B1**[2], 779
k2	—,diethyl acetal	1,1-Diethoxyethene*. $CH_2{:}C(OC_2H_5)_2$	116.16	λ^{vap} 250 (2.4), 380 (1.0)	...	68[100] (76–7)	0.7932_4^{20}	1.3643^{21}	d	d	...	...	...	...	**B1**[3], 2948
k3	**—,dimethyl-***	$(CH_3)_2C{:}CO$	70.09	ye	−97.5	34 −49[12]	...	...	d	d	...	...	...	...	**B1**[2], 789
k3[1]	**—,diphenyl-***	$(C_6H_5)_2C{:}CO$	193.24	ye-red	...	265–70d 146[12]	1.1107_4^{14}	1.615^{14}	...	...	s	...	s	...	**B7**[2], 412
k4	**—,methyl-***	$CH_3CH{:}CO$	56.07	...	−80	...	...	...	...	...	s	...	...	...	**B1**[2], 782
—	**Ketine**	see **Pyrazine, 2,5-dimethyl-**													
k5	**Ketone, benzyl 1-naphthyl**	α-Phenyl-1-acetonaphthone. $C_6H_5CH_2COC_{10}H_7^{\alpha}$	246.31	ta, lf (al or MeOH)	66–7	194–6[0.05]	...	...	i	δ v[h]	s	...	...	chl s	**B7**[2], 461
k6	**—,benzyl 2-naphthyl**	β-Phenyl-2-acetonaphthone. $C_6H_5CH_2COC_{10}H_7^{\beta}$	246.31	nd (al)	99.5	...	...	...	...	s[h]	s	...	s	chl s	**B7**[2], 461
Ω k7	**—,benzyl phenyl**	Desoxybenzoin. α-Phenyl acetophenone. $C_6H_5CH_2COC_6H_5$	196.25	pl (al) λ^{al} 240 (4.2)	60	320 177[20]	1.201_4^0	...	δ[h]	s	s	...	...	chl, CCl_4 s	**B7**[2], 368
Ω k8	**—,—,α-chloro-**	Desyl chloride. $C_6H_5COCHClC_6H_5$	230.70	nd (al) λ^{hx} 281 (2.94), 290 (2.75), 332 (2.2)	68.5	d	...	...	...	s[h]	...	...	...	alk i	**B7**[2], 369
k9	**—,—,α-diazo-**	Azibenzil. Diazodesoxybenzoin. $N_2{:}C(C_6H_5)COC_6H_5$	222.25	red pl (eth) λ^{eth} 267 (4.3), 317 sh (3.7), 417 (2.0)	79	...	...	...	...	δ	δ s[h]	...	δ	...	**B7**[2], 683
Ω k10	**—,cyclobutyl phenyl**	Benzoylcyclobutane.	160.22	λ^{iso} 239 (4.11), 278 (2.94), 287 (2.84)	...	260 121.5–2[10]	1.0457_{25}^{25}	1.5472^{20}	...	...	...	...	...	...	**B7**, 374
k11	**—,1,1′-dinaphthyl**	...	282.35	wh nd, yesh pr (aa)	104	...	...	...	...	v[h] s	s	...	v	aa s[h] lig δ[h] chl, sulf s	**B7**[2], 503
k12	**—,1,2′-dinaphthyl**	...	282.35	nd (al, bz-lig)	136–7	235[0.6]	...	...	...	δ	...	...	v	...	**B7**[2], 504
k13	**—,2,2′-dinaphthyl**	$C_{10}H_7^{\beta}COC_{10}H_7^{\beta}$	282.35	(i) nd (eth). (ii) lf (chl-eth)	(i) 125.5 (ii) 164.5	...	...	...	...	δ	δ	...	...	chl s	**B7**[2], 504
k14	**—,1,1′-dinaphthyl,2-methyl**	$C_{22}H_{16}O$. *See* k11	296.37	gr-yesh nd (al)	171	230–5[5]	...	...	i	s[h]	...	...	...	con sulf s	**B7**[2], 506
k15	**—,1,2′-dinaphthyl,2-methyl-**	$C_{22}H_{16}O$. *See* k12	296.37	gr-yesh nd (aa or al)	142–3	...	...	...	i	...	...	...	...	con sulf s aa s[h]	**B7**[2], 506
—	**—,diphenyl**	see **Benzophenone**													
k16	**—,di(2-thienyl)**	Thienone.	194.28	nd (al)	90–0.5	326	...	...	i	s[h]	s	s	...	oos s	**B19**, 135
—	**—,ethyl methyl**	see **2-Butanone***													
—	**—,ethyl styryl**	see **1-Penten-3-one, 1-phenyl-***													
k18	**—,2-furyl phenyl**	2-Benzoylfuran.	172.20	br λ^{al} 289 (4.2)	< −15	285 164[19]	1.1732^{20}	1.6055^{20}	i	s	s	...	s	...	**B17**[2], 372
—	**—,methyl phenyl**	see **Acetophenone**													
k20	**—,1-naphthyl 2-tolyl**	...	246.31	(al)	64	365	...	...	i	s[h]	v	...	...	lig δ	**B7**[2], 462
k21	**—,1-naphthyl 3-tolyl**	...	246.31	nd (al)	74–5	...	...	...	i	s	...	...	...	os s	**B7**[1], 284
k22	**—,phenyl 4′-piperidyl, 1′-methyl-**	4-Benzoyl-*N*-methyl piperidine. $C_{13}H_{17}NO$.	203.29	nd λ^{alHBr} 245 (4.15), 281 (3.05)	35–7	160–3[13] 130–7[2]	...	1.5430^{23}	s[h]	s	v	s	s	...	**C49**, 12472
—	**—,phenyl 4-pyridyl**	see **Pyridine, 4-benzoyl-**													
Ω k23	**—,phenyl 2-thienyl**	2-Benzoylthiophene. $C_{11}H_8OS$.	188.26	nd (dil al) λ^{al} 263 (4.11), 293 (4.13)	56–7	300	1.1890_4^{54}	1.6181_{α}^{54}	i	v[h]	v	s	...	...	**B17**[2], 372
Ω k24	**Khellin**	$C_{14}H_{12}O_5$	260.25	(MeOH or eth) λ^{al} 250 (4.6), 290 (3.6), 340 (3.6)	154–5 (vac)	180–200[0.05]	...	...	s[h] i	s[h]	δ	s	...	MeOH s, v[h]	**B19**[2], 236
—	**Kojic acid**	see **4-Pyrone, 5-hydroxy-2-hydroxymethyl-**													
—	**Kynurenic acid**	see **2-Quinolinecarboxylic acid, 4-hydroxy-**													
k25	**Kynurenine(*l*)**	3-Anthranyloylalanine. $C_{10}H_{12}N_2O_3$.	208.22	lf (+⅓w, w) $[\alpha]_D^{20}$ −29 (w, c = 4) λ^{w} 260 (2.3), 360 (2.0)	191d	...	...	...	δ	...	...	...	...	...	**B14**[2], 415
—	**Kynurine**	see **Quinoline, 4-hydroxy-**													

For explanations, symbols and abbreviations see beginning of table. For structural formulas see end of table.

No.	Name	Synonyms and Formula	Mol. wt.	Color, crystalline form, specific rotation and λ_{max} (log ε)	m.p. °C	b.p. °C	Density	n_D	Solubility w	al	eth	ace	bz	other solvents	Ref.
	Lactamide														
—	**Lactamide**	*see* **Propanoic acid, 2-hydroxy-, amide**													
—	**Lactaric acid**	*see* **Octadecanoic acid, 6-oxo-***													
—	**Lactic acid**	*see* **Propanoic acid, 2-hydroxy-***													
—	**β-Lactic acid**	*see* **Propanoic acid, 3-hydroxy-***													
Ω l1	**Lactide**(*d*)	Dilactide. 2,5-Dimethyl-3,6-dioxo-1,4-dioxane.	144.14	hyg rh (eth) $[\alpha]^{18}_D$ −298 (bz, c = 1.17)	95	150^{25}			...	δ	δ	...	δ	chl δ	**B19**, 154
l2	—(*dl*)	$C_6H_8O_4$. *See* l1	144.14	pa ye tcl or nd (al)	124.5	255^{757} $138–42^{8}$	0.862^{10}_4		δ	δ	δ	s	s	peth δ	**B19**², 176
l3	—(*l*)	$C_6H_8O_4$. *See* l1	144.14	rh (eth) $[\alpha]^{16}_D$ +281.6 (bz, c = 0.82)	95	150^{25}			...	...	δ	...	...	...	**B19**, 154
Ω l4	**Lactobionic acid**		358.30	syr					v	δ	i	...	...	aa δ MeOH δ	**B31**, 415
—	**Lactobiose**	*see* **Lactose**													
—	**Lactoflavin**	*see* **Riboflavin**													
l5	**Lactose**(α)	Lactobiose. Milk sugar	342.30	pw $[\alpha]^{20}_D$ +92.6 (w, c = 4.5) → + 52.3 (final) (mut)	222.8				v	δ	i	...	...	chl i	**B31**, 408
l6	—(β)		342.30	$[\alpha]^{20}_D$ +34.2 (5 min)→ 52.3 (final) (w) (mut)	253		1.59^{20}		v	δ	i	...	...	chl i	**B31**, 408
l7	—, monohydrate(α)	$C_{12}H_{22}O_{11}.H_2O$. *See* l5	360.31	mcl (w) $[\alpha]^{20}_D$ +83.5 (w, 10 min)	201–2 (−w, 130)	d	1.525^{20}		v	i	i	...	...	MeOH i chl i	**B31**, 408
l8	**Lanosterol**	$C_{33}H_{54}O$	426.70	nd (eth), cr (MeOH-ace) $[\alpha]^{20}_D$ +62 (chl, c = 1)	140–1				...	s^h	v	δ	...	aa s^h chl s	**B6**³, 2880
l9	—, benzoate	$C_{37}H_{53}O_2$	530.81	pw or nd (eth) $[\alpha]^{17}_D$ +72.2	191.5				...	s	s	...	...	...	**B4**³, 2881
l10	**Lanthionine**(*D*)	*d*-β,β-Thiodialanine. $[HO_2CCH(NH_2)CH_2]_2S$	208.24	hex pl $[\alpha]^{21}_D$ −8.0 (2.4N NaOH, c = 5)	293–5d darkens 245				δ	i	i	i	i	dil alk s dil ac s	**B4**³, 1618
l11	—(*DL*)	$[HO_2CCH(NH_2)CH_2]_2S$	208.24	hex pl	282–95d chars 240				δ	i	i	i	i	dil alk s dil ac s chl i	**B4**³, 1620
l12	—(*L*)	$[HO_2CCH(NH_2)CH_2]_2S$	208.24	hex pl $[\alpha]^{22}_D$ +8.6 (2.4N NaOH, c = 5)	293–5d darkens 245				δ	i	i	i	i	dil alk s dil ac s	**B4**³, 1593
l13	—(*meso*)	$[HO_2CCH(NH_2)CH_2]_2S$	208.24	hex pl (dil NH_4OH)	304d softens 270				δ	i	i	i	i	dil alk s dil ac s chl i	**B4**³, 1620
l14	**Lanthopine**	$C_{23}H_{25}NO_4$	379.46	pw	200				...	δ	δ	...	...	chl s	**C20**, 2715
l15	**Lapachol**	Taiguic acid. Tecomine	242.28	ye pr (eth or bz), pl (aa or al) λ^{al} 252 (4.38), 278 (4.28), 331 (3.43), 382 sh (3.17)	139.5–140.5 (142–3)				i	v^h s	s	...	v^h s	alk s aa v chl s	**E12 B** 3083
l16	—, δ-hydroxy-	$C_{15}H_{14}O_4$. *See* l15	258.28	ye nd (bz or w)	127				s^h	v	v	...	v^h	...	**E12B**, 3170
l17	**Lappaconitine**	$C_{32}H_{44}N_2O_9$	600.72	hex pl (al) $[\alpha]^{18}_D$ +27 (chl) λ^{MeOH} 224 (4.41), 251 (4.17), 307 (3.74)	223 (215–6)				i	δ	δ	...	s	chl s	**C18**, 2899
l18	**Laudanidine**(*d*)	$C_{20}H_{25}NO_4$	343.43	(MeOH) $[\alpha]^{17}_D$ +93.5 (chl, c = 1)	184–5				...	δ	δ	...	...	...	**B21**², 184
l19	—(*l*)	Tritopine. $C_{20}H_{25}NO_4$. *See* l18	343.43	hex pr (al) $[\alpha]^{17}_D$ −94.8 (chl, c = 2)	184–5				s	δ	δ	...	s	peth δ con sulf s → ye-red	**B21**², 183
l20	**Laudanine**	*dl*-Laudanidine. $C_{20}H_{25}NO_4$. *See* l18	343.43	ye wh pr (dil al or al-chl) λ^{al} 225 sh (4.2), 283 (3.8)	167		1.26^{20}_4		δ	v^h δ	δ	...	s	chl s	**B21**², 184

For explanations, symbols and abbreviations see beginning of table. For structural formulas see end of table.

No.	Name	Synonyms and Formula	Mol. wt.	Color, crystalline form, specific rotation and λ_{max} (log ε)	m.p. °C	b.p. °C	Density	n_D	Solubility: w	al	eth	ace	bz	other solvents	Ref.
	Laudanosine (*d*)														
l21	**Laudanosine(*d*)**	*N*-Methyltetrahydro-papaverine. $C_{21}H_{27}NO_4$	357.46	nd (peth), pr (al) $[\alpha]_D^{16}$ +106 (al, c = 1.6) λ^{MeOH} 231 (4.20), 282 (3.78)	89	...	...	...	i	s	s	s	s^h	chl s	**B21**[2], 184
l22	—(*dl*)	$C_{21}H_{27}NO_4$. *See* l21	357.46	nd (al)	115–6	...	...	...	i	v^h	s	s	s	chl v peth δ^h	**B21**[2], 185
l23	—(*l*)	$C_{21}H_{27}NO_4$. *See* l21	357.46	cr (al) $[\alpha]_D^{15}$ −105.4 (al, c = 3)	89	...	...	...	...	...	...	...	...	...	**B21**[2], 185
—	**Lauramide**	*see* **Dodecanoic acid, amide***													
l24	**Laureline**	$C_{19}H_{19}NO_3$	309.35	ta (al) cubes (peth) $[\alpha]_D^{18}$ −99 (abs al, c = 0.7) λ 350 (1.8), 394 (1.65)	114 (97)	...	...	...	i	s	s	...	...	dil ac s con sulf s→red	**B27**[1], 461
—	**Lauric acid**	*see* **Dodecanoic acid***													
—	**Laurone**	*see* **12-Tricosanone***													
—	**Laurophenone**	*see* **1-Dodecanone, 1-phenyl-**													
—	**Lawsone**	*see* **1,4-Naphthoquinone, 2-hydroxy-***													
—	**Lawsone anilide**	*see* **1,4-Naphthoquinone, 2(phenylamino)-***													
l25	**Lecithin**	α(Dimyristoyl)lecithin	677.92	$[\alpha]_D^{24}$ +7.0	236–7	...	...	...	i	...	s	i	δ	chl s peth s	**Am 74,** 158
l26	**Ledol**	...	222.38	nd (al) $[\alpha]_D^{20}$ +28 (chl, c = 10) $[\alpha]_D^{20}$ +7.98 (al)	105–6.5 δsub	292^{760} (282–3) with sub	0.9094_{20}^{100}	1.4667^{110}	δ^h i	v	s	s	...	oos s	**E13,** 5
—	**Lepidine**	*see* **Quinoline, 4-methyl-**													
—	***p*-Leucaniline, *N,N,N′,N′*-tetra-methyl-**	*see* **Methane, bis(4-di-methylaminophenyl) (4-amino phenyl)-***													
—	**Leucaurine**	*see* **Methane, tris-(4-hydroxyphenyl)-***													
—	**Leucenol**	*see* **Mimosine**													
l27	**Leucic acid(*D*)**	4-Hydroxy-4-methylvaleric acid. Leucinic acid. $(CH_3)_2CHCH_2CHOHCO_2H$	132.16	nd (bz), pr (eth-peth) $[\alpha]_D^{12.5}$ +10.7 (w, c = 5) $[\alpha]_D^{18}$ +26.1 (1N NaOH, c = 2)	80–1	...	...	...	...	...	...	...	...	...	**B3**[2], 233
Ω l28	—(*DL*)	$(CH_3)_2CHCH_2CHOHCO_2H$	132.16	pl (eth-peth)	77	...	...	...	v	v	v	...	...	...	**B3**[2], 234
l29	—(*L*)	$(CH_3)_2CHCH_2CHOHCO_2H$	132.16	rh (eth) $[\alpha]_D^{20}$ −11.3 (w, c = 1)	81–2	<100	...	...	v	v	v	...	...	...	**B3**[2], 233
l30	**Leucinamide(*dl*)**	α-Aminoisocaproamide. $(CH_3)_2CHCH_2CH(NH_2)CONH_2$	130.19	pr (bz)	106–7	...	...	...	s	v	...	v	δ	...	**B4,** 448
Ω l31	**Leucine(*D*)**	*D*-α-Aminoisocaproic acid. $(CH_3)_2CHCH_2CH(NH_2)CO_2H$	131.18	pl (al) $[\alpha]_D^{20}$ +10.34	293 sealed tube	sub	...	...	δ	...	...	...	...	...	**B4,** 446
Ω l32	—(*DL*)	$(CH_3)_2CHCH_2CH(NH_2)CO_2H$	131.18	lf (w) $\lambda^{w,pH=1}$ 207 (1.82)	293–5 sealed tube (332d)	sub	1.293_4^{18}	...	s	δ	i	...	...	...	**B4**[2], 870
Ω l33	—(*L*)	$(CH_3)_2CHCH_2CH(NH_2)CO_2H$	131.18	hex pl (dil al) $[\alpha]_D^{25}$ −10.42 (w, p = 22) $[\alpha]_D^{15}$ +17.3 (20 % HCl)	293–5 sealed tube	sub	1.293_4^{18}	...	δ	i	i	...	...	...	**B4**[2], 859
—	—,amide	*see* **Leucinamide**													
Ω l34	—,*N*-acetyl-(*dl*)	$(CH_3)_2CHCH_2CH(NHCOCH_3)CO_2H$	173.21	nd (dil al) $\lambda^{w,pH=1}$ 215 (2.8)	161	...	...	...	...	v	δ	...	...	...	**B4,** 451
l35	—,*N*-benzoyl-(*dl*)	$(CH_3)_2CHCH_2CH(NHCOC_6H_5)CO_2H$	235.29	nd (dil al)	137–41	...	...	...	s	s	s	...	δ	chl s lig i	**B9,** 253
Ω l36	—,*N*-glycyl-(*dl*)	$(CH_3)_2CHCH_2CH(NHCOCH_2NH_2)CO_2H$	188.23	tetr (dil al)	242d	...	1.181	...	v	i	...	...	...	...	**B4,** 453
Ω l37	—(*l*)	$(CH_3)_2CHCH_2CH(NHCOCH_2NH_2)CO_2H$	188.23	pl (dil al) $[\alpha]_D^{20}$ −35.2 (w)	256d	...	...	...	v	i	...	...	...	...	**Am 74,** 3818
—	**Leuco indigo**	*see* **Indigo white**													

For explanations, symbols and abbreviations see beginning of table. For structural formulas see end of table.

No.	Name	Synonyms and Formula	Mol. wt.	Color, crystalline form, specific rotation and λ_{max} (log ε)	m.p. °C	b.p. °C	Density	n_D	Solubility w	al	eth	ace	bz	other solvents	Ref.
	Leucomethylene blue														
l38	**Leucomethylene blue**	....	285.42	ye nd (eth or al) λ^{w} 256	185 -		....	....	δ	s			...	....	**B27**, 393
—	**Leucylglycine**	*see* **Glycine,** ***N*****-leucyl-,**													
l39	**Levopimaric acid,** methyl ester	Methyl sapietate. $C_{21}H_{32}O_2$.	316.49	(MeOH or eth) $[\alpha]_D$ −190.4 (al) $[\alpha]_D$ −268 (eth)	63–4	$166\text{–}9^{0.5}$	1.0312_4^{22}	1.5232^{22}	...	s	s^h	...	...	MeOH s^h	**B9**², 433
—	**Levulinaldehyde** ...	*see* **Pentanal, 4-oxo-***													
l40	**Levulin**..........	Fructosin. Laevulin. Levulosin. $(C_6H_{19}O_5)_n$	(162.14)	amor [α] −52.1	140–45d	....	....	....	∞	s	i	...	...	....	**B1**, 925
—	**Levulinic acid**	*see* **Pentanoic acid, 4-oxo-***													
—	**Lewisite**..........	*see* **Ethene, 1-chloro-2-dichloroarsino-***													
—	**Limettin**..........	*see* **Coumarin, 5,7-dimethoxy-**													
Ω l41	**Limonene**(*d*)	Citrene. Carvene. 4-Isopropenyl-1-methylcyclohexene.	136.24	$[\alpha]_D^{20}$ +125.6 (undil) λ^{iso} 220 (2.41), 250 (1.36)	−74.35	178^{760} 61^{13}	0.8411_4^{20}	1.4730^{20}	i	∞	∞	...	...	....	**B5**², 88
Ω l42	—(*dl*)...........	Cinene. Dipentene. $C_{10}H_{16}$. *See* l41	136.24	λ^{iso} 220 (2.41), 250 (1.36)	−95.5	178^{760} 64.4^{15}	0.8402_4^{21}	1.4727^{20}	...	...	...	...	...	....	**B5**², 89
l43	— (*l*)	$C_{10}H_{16}$. *See* l41...........	136.24	$[\alpha]_D^{20}$ −122.1 (undil) λ^{iso} 220 (2.4), 250 (1.1)	....	177.6– 7.8^{755} 64.4^{15}	0.8422_4^{20}	1.4746^{20}	...	...	...	...	...	....	**B5**², 89
Ω l44	**Linalool**(*d*)	Coriandrol. 3,7-Dimethyl-1,6-octadien-3-ol. CH_2:CHCOH$(CH_3)CH_2CH_2$CH:C$(CH_3)_2$	154.26	$[\alpha]_D^{20}$ +19.18 λ^{al} 243 sh (1.58), 272 sh (0.44)	....	198– 200^{760} $87\text{–}8^{12}$	0.8700_4^{20}	1.4636^{20}	δ	∞	∞	...	...	....	**B1**³, 2012
Ω l45	—,acetate(*l*)	Bergamol. Linalyl acetate. Licareol acetate. $C_{12}H_{20}O_2$. *See* l44	196.29	$[\alpha]_D^{20}$ −9.45	....	220^{762}	0.8951_4^{20}	1.4544^{21}	i	∞	∞	...	...	....	**B2**², 153
l46	—,formate(*l*)	Linalyl formate. $C_{11}H_{18}O_2$. *See* l44	182.27	....	....	$100\text{–}3^{10}$	0.915_4^{25}	1.456^{20}–	i	s	...	...	...	....	**B2**³, 47
l47	*β*-**Linaloolene**	Dihydromyrcene. CH_2:CH$CH(CH_3)CH_2CH_2$CH:C$(CH_3)_2$	138.25	....	....	165–8 (161–2)	0.7601_4^{20}	1.4362^{20}	...	...	...	...	...	....	**B1**³, 1020
l48	**Linamarin**.........	Acetone cyanohydrin-*β-d*-glycopyranoside. Phaseolunatin.	247.25	nd (w, al or AcOEt) $[\alpha]_D^{18}$ −29.1 (w, p = 5)	145 (142–3)	....	....	....	v	s^h δ	i	v^h	δ	chl δ peth i	**B31**, 197
—	**Lindane**	*see* **Cyclohexane, 1,2,3,4,5,6-hexachloro-**													
—	**Linoleic acid**	*see* **9,12-Octadecadienoic acid***													
—	**Linolenic acid**	*see* **9,12,15-Octadecatrienoic acid***													
—	**Linolenin**.........	*see* **Glycerol,** 1-mono-(9,12-octadecadienoate)													
Ω l49	*α*-**Lipoic acid**(*d*) ...	6,8-Epidithiooctanoic acid*. Thioctic acid.	206.33	pa ye pl $[\alpha]_D^{25}$ +96.7 λ^{MeOH} 333 (2.18)	47.5	....	....	....	δ				v	MeOH v AcOEt v	**Am 76**, 1270
l50	**Lithocholic acid**....	3α-Hydroxycholanic acid. $C_{24}H_{40}O_3$.	376.59	hex lf (al), pr (dil al or aa) $[\alpha]_D^{20}$ +32.14 (al) $\lambda^{con\ sulf}$ 326 (3.53)	186	....	....	....	i	v^h s	δ	...	...	chl s aa s lig i	**C30**, 2575
l52	**Lobelanidine**	$C_{22}H_{29}NO_2$	339.48	sc (al or eth) λ^{MeOH} 252 (2.58), 258 (2.63), 263 (2.55)	150	distb vac	....	....	δ^h i	s	δ	v	v	Py v chl v peth δ	**B21**², 136
l53	**Lobelanine**........	$C_{22}H_{25}NO_2$........	335.45	nd (peth or eth)	99	....	....	....	i	v	δ	v	v	MeOH, aa, chl v	**B21**², 394
Ω l54	**Lobeline**(*dl*)	$C_{22}H_{27}NO_2$........	337.47	ye pl (eth), pr (al)	110	....	....	....	δ	v	v	...	s	chl s	**B21**², 434
l55	—(*l*)............	$C_{22}H_{27}NO_2$. *See* l55........	337.47	nd (al, eth or bz) $[\alpha]_D^{15}$ −42.85 (al, c = 1) λ^{MeOH} 245 (4.12), 280 (3.15)	130–1	....	....	....	δ	s^h	s	v	s	peth δ chl s	**B21**², 434
Ω l56	**Longifolene**(*d*)	$C_{15}H_{24}$........	204.36	$[\alpha]_D^{18}$ +42.73	....	$254\text{–}6^{706}$ $126\text{–}7^{15}$	0.9319_4^{18}	1.5040^{20}	i	...	...	...	s	....	**B5**³, 1097
l57	**Lophine**	2,4,5-Triphenylimidazole...	296.38	nd (al)	275	sub	....	....	i	s	s	...	...	....	**B23**², 280
—	**Loretine**..........	*see* **5-Quinolinesulfonic acid, 8-hydroxy-7-iodo-**													

For explanations, symbols and abbreviations see beginning of table. For structural formulas see end of table.

No.	Name	Synonyms and Formula	Mol. wt.	Color, crystalline form, specific rotation and λ_{max} (log ε)	m.p. °C	b.p. °C	Density	n_D	Solubility w	al	eth	ace	bz	other solvents	Ref.
	Luciculine														
l58	**Luciculine**	$C_{22}H_{35}NO_3$	361.51	cr (+1w, ace) $[\alpha]_D^{11.5}$ −11.4 (al)	148–50	$165^{0.02}$	...	...	i	s	...	...	...	...	**C46,** 1008
l59	—,hydrochloride	$C_{22}H_{35}NO_3.HCl$	397.97	cr (w + 1.5) $[\alpha]_D$ = −9.4 (w)	198–203d (anh)	...	...	...	i^h	s	...	...	...	...	**C46,** 1008
—	**Luminal**	*see* **Barbituric acid, 5-ethyl-5-phenyl-**													
Ω l60	**Luminol**	5-Amino-2,3-dihydro-1,4-phthalazinedione. 3-Aminophthalhydrazide.	177.16	ye nd (al) λ^{ac} 240 (3.9), 280 sh (3.6), 290 (3.7), 350 (3.65), λ^{al} 235 (4.1), 320 (3.8)	329–32	...	...	...	i	δ	δ	...	...	alk v dil ac δ aa s^h	**B25**^{2}, 389
l61	**Lumisterol**	$C_{28}H_{44}O$	396.67	nd (ace-MeOH) $[\alpha]_D^{18}$ +197 (chl) λ^{al} 278.5 (3.95), 294 (3.63)	118	...	...	...	i	s	v	v	...	aa s chl v MeOH s	**E14,** 76
l62	**Lupanine**(*d*)	$C_{15}H_{24}N_2O$.	248.37	hyg nd $[\alpha]_D^{20}$ +82.4 (w, c = 3)	40–44	$190–3^3$	...	1.5444^{24}	s	v	v	...	...	chl v lig i	**B24**^{2}, 53
l63	—(*dl*)	$C_{15}H_{24}N_2O$. *See* l62	248.37	nd (peth), rh pr (ace)	98–9	$233–4^{18}$	...	...	v	v	v	...	...	peth s chl v	**B24**^{2}, 55
Ω l64	**Lupeol**	Lupenol. 2-Hydroxy-$\Delta^{20,29}$-lupene. $C_{30}H_{50}O$.	426.74	nd (al or ace) $[\alpha]_D^{20}$ +27.2 (chl) λ^{al} 206 (3.40)	215–7	...	0.9457_4^{218}	1.4910^{218}	i	v^h	v	v	v	chl v lig v	**E14s,** 1115
—	**Lupetidine**	*see* **Piperidine, dimethyl-**													
Ω l65	**Lupinine**(*l*)	$C_{10}H_{19}NO$.	169.27	rh (peth) $[\alpha]_D^{17}$ −20.35 (al)	70	$269–70^{754}$	...	...	s	s	s	...	s	chl s peth δ	**B21**^{2}, 28
l66	—,hydrochloride	$C_{10}H_{19}NO$. HCl. *See* l65	205.73	pr (aq al) $[\alpha]_D$ −14 (w)	212–3	...	...	...	s	s	...	...	...	...	**B21**^{2}, 29
l67	—,des-N-methyl-	$C_{11}H_{21}NO$	183.30	...	...	$145–6^{15}$	...	...	i	s	s	...	...	...	**B21**^{2}, 27
—	**Lupulic acid**	*see* **Humulon**													
l68	**Lupulone**	β-Bitter acid. β-Lupulinic acid. $C_{26}H_{38}O_4$.	414.59	pr (MeOH) λ^{al} 228 (4.15), 280 (3.79), 338 (4.01)	93	...	...	...	i	s	...	...	...	peth s hx s	**B7**^{2}, 856
—	**Lutein**	*see* **Xanthophyll**													
—	**Luteolin**	*see* **Flavone, 3′,4′,5,7-tetrahydroxy-**													
—	**Lutidine**	*see* **Pyridine, dimethyl-**													
l69	**Lycaconitine**	$C_{27}H_{34}N_2O_6$	566.61	amor $[\alpha]_D$ +31.5	111–4	...	...	...	δ	s	δ	...	s	peth, chl s	**C53,** 9266
l70	**Lycomarasmine**	$HO_2CCH_2CH(CO_2H)NHCH_2CH(CO_2H)NHCH_2CONH_2$	277.24	$[\alpha]_D^{20}$ −48 (w)	227–9d	...	...	...	δ	...	...	...	...	dil ac, dil alk v	**H46,** 60
Ω l71	**Lycopene**	Lycopin. Neolycopene. $C_{40}H_{56}$.	536.90	red pr or nd (peth or CS_2-al) λ^{hx} 412 (4.9), 433 (4.9), 460 sh (4.7), (all *cis*) λ^{hx} 418 sh (4.8), 442 (5.1), 473 (5.3), 504 (5.2), (all *trans*)	175	...	...	...	...	δ^h	s	...	v^h	chl v^h peth δ CS_2 s	**B30,** 81
l72	**Lycorine**(*l*)	Amarylline. Narcissine. $C_{16}H_{17}NO_4$.	287.32	pr (al or Py) $[\alpha]_D^{16}$ −129 (al, c = 0.16) λ^{MeOH} 233 (3.54), 293 (3.68)	280 (262–3)	sub	...	...	i	δ	δ	...	...	chl δ	**B27**^{2}, 547
l73	**Lycoxanthin**	$C_{40}H_{56}O$.	552.93	purple or red br nd (CS_2), (bz-peth), red pl (bz-MeOH) λ^{CS_2} 473, 507, 547	168	...	...	...	i	δ	...	...	s	CS_2 s peth δ	**C30,** 3832
—	**Lysergamide**	*see* **Ergine**													
l74	**Lysergic acid**	$C_{16}H_{16}N_2O_2$.	268.32	lf or hex sc (w) $[\alpha]_D^{20}$ +40 (Py, c = 0.5) $\lambda^{aq\ NaOH}$ 223 (4.3), 238 (4.31), 310 (3.96)	240d	...	...	...	δ	s	δ	...	δ	Py s	**J1955,** 1626

For explanations, symbols and abbreviations see beginning of table. For structural formulas see end of table.

No.	Name	Synonyms and Formula	Mol. wt.	Color, crystalline form, specific rotation and λ_{max} (log ε)	m.p. °C	b.p. °C	Density	n_D	Solubility w	al	eth	ace	bz	other solvents	Ref.
	Lysidine														
—	**Lysidine**	*see* **Δ²-Imidazoline, 2-methyl-**													
Ω 175	**Lysine(*L*)**	*L*-α,ε-Diaminocaproic acid. $H_2N(CH_2)_4CH(NH_2)CO_2H$	146.19	nd (w or dil al) $[\alpha]_D^{20}$ +14.6 (w, c = 6)	224–5 d darkens 210	....	....	....	v	i	i	i	i	....	**B4²**, 857
Ω 176	—,dihydrochloride(*L*)	$H_2N(CH_2)_4CH(NH_2)CO_2H.2HCl$	219.11	(al-eth or aq HCl) $[\alpha]_D^{20}$ +15.29 (w)	201–2 (193)	....	....	....	v	s^h	...	...	i	....	**B4²**, 857
177	—,**N^α-benzoyl-(*dl*)**	$H_2N(CH_2)_4CH(NHCOC_6H_5)CO_2H$	250.30	nd (w)	235 (235–249)	....	....	....	v^h	δ	i	...	i	chl i	**B9²**, 193
178	—,**N^ϵ-benzoyl-(*dl*)**	$C_6H_5CONH(CH_2)_4CH(NH_2)CO_2H$	250.30	cr (w)	268 (254)	....	....	....	s^h	i	i	...	...	chl i	**B9²**, 193
179	—,—(*L*)	$C_6H_5CONH(CH_2)_4CH(NH_2)CO_2H$	250.30	lf (w) $[\alpha]_D^{19}$ +20.1 (aq HCl)	240d	....	....	....	s^h	i	i	...	i	chl i	**B9²**, 193
—	**Lyxitol**	*see* **Arabitol**													

For explanations, symbols and abbreviations see beginning of table. For structural formulas see end of table.

No.	Name	Synonyms and Formula	Mol. wt.	Color, crystalline form, specific rotation and λ_{max} (log ε)	m.p. °C	b.p. °C	Density	n_D	Solubility w	al	eth	ace	bz	other solvents	Ref.
	Malachite green														
—	**Malachite green, leuco-**	*see* **Methane, bis(4-dimethylaminophenyl)-(phenyl)-**													
Ω m1	**Malathion**	*S*(1,2-Dicarboxyethyl)-*O*, *O*-dimethyldithiophosphate.	330.36	ye-br	2.85	$156–7^{0.7}$ δd	1.2076^{20}	1.4960^{20}	δ	s	s	...	s	...	**C48,** 5572
Ω m2	**Maleic acid**	*cis*-Butenedioic acid*. $HO_2CCH{:}CHCO_2H$	116.07	mcl pr (w) λ^{w} 210 (4.14)	139–40 (131)	...	1.590^{20}	...	v	v	s	v	i	con sulf, aa s chl i	**B2**[2], 641
Ω m3	—,anhydride	...	98.06	nd (chl or eth) λ^{iso} 208 (3.90), 290 (1.11)	60 (56)	197–9 82^{14}	1.314^{60}	...	s d^h	...	s	s	...	chl s lig δ	**B17**[2], 445
Ω m3[1]	—,diethyl ester	Ethyl maleate. $C_2H_5O_2CCH{:}CHCO_2C_2H_5$	172.19	λ^{al} 250 sh (2.7)	−8.8	223^{760} $105–6^{14}$	1.0662_4^{20}	1.4416^{20}	i	s	s	...	...	...	**B2**[3], 1923
Ω m3[2]	—,dimethyl ester	Methyl maleate. $CH_3O_2CCH{:}CHCO_2CH_3$	144.14	λ^{al} 205 (3.90)	−19	202^{762} 102^{17}	1.1606_4^{20}	1.4416^{20}	i	...	s	...	...	lig i	**B2**[3], 1921
m4	—,diphenyl ester	$C_6H_5O_2CCH{:}CHCO_2C_6H_5$	268.27	pl (lig)	73	226^{15}	...	...	i	v	v	v	v	chl v	**B6,** 156
—	—,hydrazide	*see* **Pyridazine, 3,6-dioxo-1,2,3,6-tetrahydro-**													
Ω m5	—,imide	Maleimide.	97.07	pl (bz) λ^{al} 216 (4.13), 230 sh (3.70), 280 (2.64)	93–5	sub	...	...	s	s	s	...	...	...	**B21**[2], 311
Ω m6	—,—,*N*-ethyl-	$C_6H_7NO_2$. *See* m5	125.13	cr (bz) λ^{al} 217 (4.10), 224 sh (4.03), 310 (2.71)	45.5	...	...	...	δ	v	v	...	...	...	**B21,** 399
Ω m6[1]	—,—,*N*-phenyl-	Maleanil. $C_{10}H_7NO_2$. *See* m5	173.17	ye nd (bz-lig) λ^{cy} 205 (3.98), 220 (3.97), 275 (3.48)	90–1	162^{12}	...	...	δ	s	s	...	s	chl, CS_2 δ	**B21,** 400
m6[2]	—,monoamide	Maleamic acid $HO_2CCH{:}CHCONH_2$	115.09	lf (w)	178 (d 153)	...	...	...	v	s^h	i	...	i	chl i	**B2**[3], 1927
m6[3]	—,**bromo-**	$HO_2CCH{:}CBrCO_2H$	194.98	nd or pr	140–1 (rapid htng) 136–8 (slow htng)	d	...	...	v	v	v	...	...	...	**B2**[3], 1929
m6[4]	—,**chloro-**	$HO_2CCH{:}CClCO_2H$	150.52	pr (eth-chl)	108 (114)	...	...	...	s^h	v	v	s	δ	aa s chl δ peth i	**B2**[3], 1928
m7	—,—,diethyl ester	$C_2H_5O_2CCH{:}CClCO_2C_2H_5$	206.63	...	...	235δd 125^{19}	1.1741_4^{20}	...	...	v	v	...	...	...	**B2**[2], 646
m8	—,—,dimethyl ester	$CH_3O_2CCH{:}CClCO_2CH_3$	178.57	...	...	106.5^{18}	1.2775_4^{25}	...	...	v	v	...	...	...	**B2**[2], 646
m9	—,**dichloro-**	$HO_2CCCl{:}CClCO_2H$	184.96	nd (lig or eth) λ^{al} 218 sh (3.7), 242 (3.8)	119–20	...	...	...	v	s	s	...	i	chl, CS_2 i	**B2,** 753
Ω m10	—,**dihydroxy-**	$HO_2CC(OH){:}C(OH)CO_2H$	148.07	pl (w + 2) $\lambda^{w, pH=5.8}$ 291 (3.7)	155 (anh)	...	...	...	δ	s	δ	...	...	MeOH δ	**B3**[2], 346
Ω m11	—,**dimethyl-**, anhydride	Pyrocinchonic anhydride. $C_6H_6O_3$. *See* m3	126.11	pl or lf (dil al)	96	223 105^{12}	1.107_4^{100} 0.90912_4^{19}	...	δ	v	v	...	v	chl v	**B17**[2], 450
Ω m12	—,**methyl-**	Citraconic acid. $HO_2CCH{:}C(CH_3)CO_2H$	130.10	tcl pr or pl (eth-bz), nd (eth-lig)	93–3.8	...	1.617	...	v	...	δ	...	i	chl δ CS_2 i	**B2**[2], 652
Ω m13	—,—,anhydride	Citraconic anhydride. $C_5H_4O_3$. *See* m3	112.09	...	7–8	$213–4^{760}$ 99–100^{15}	1.2469_4^{16}	1.4710^{21}	d	v	s	s	...	...	**B17**[2], 448
m14	—,—,diethyl ester	Diethyl citraconate. $C_2H_5O_2CCH{:}C(CH_3)CO_2C_2H_5$	186.21	λ^{al} 230 sh (3.4), 263 sh (2.0)	...	228^{766} 120^{20}	1.0491_4^{20}	1.44676^{20}	δ	s	s	...	...	aa s	**B2**[2], 653
m15	—,—,dimethyl ester	Dimethyl citraconate. $CH_3O_2CCH{:}C(CH_3)CO_2CH_3$	158.16	λ^{al} 233 (3.37)	...	210.5^{758} 92.8^{10}	1.1153_4^{20}	1.44733^{20}	δ	s	s	...	...	aa s	**B2**[2], 653
m16	—,**phenyl-**, anhydride	$C_{10}H_6O_3$. *See* m3	174.16	ye nd (CS_2)	122	...	...	...	...	s	s	...	...	chl s CS_2, lig δ	**B17**[2], 481
—	**Malonanilic acid**	*see* **Malonic acid,** mono-amide, *N*-phenyl-													
Ω m17	**Malonic acid**	Propanedioic acid*. $CH_2(CO_2H)_2$	104.06	tcl (al) λ^{MeOH} (not max) 228 (2.34)	135.6 δ sub	d at 140	1.619^{16}	...	v	s	s	...	i	Py v	**B2**[2], 516
Ω m19	—,diamide	Malonamide. $CH_2(CONH_2)_2$	102.09	mcl st pr (w), unst tetr	171.5–2.5 (st)	...	...	...	s	i	i	...	i	...	**B2**[2], 530
Ω m20	—,—,*N,N'*-diphenyl-	Malonanilide. $CH_2(CONHC_6H_5)_2$	254.29	nd (al)	226	...	...	...	i	v^h	i	...	v	aa s	**B12**[2], 167
Ω m21	—,dibutyl ester	Dibutyl malonate. $CH_2(CO_2CH_2CH_2CH_2CH_3)_2$	216.28	...	−83	$251–2^{760}$ 137^{14}	0.9824_4^{20}	1.4262^{20}	i	s	s	s	s	aa s	**B2**[2], 528
m22	—,dichloride	Malonyl chloride. $CH_2(COCl)_2$	140.95	λ^{diox} 280 (2.76), 293 (2.64), 304 (2.57)	...	58^{26}	1.4509_4^{20}	1.4639_4^{20}	d^h	d^h	s	...	...	AcOEt s	**B2**[2], 529

For explanations, symbols and abbreviations see beginning of table. For structural formulas see end of table.

No.	Name	Synonyms and Formula	Mol. wt.	Color, crystalline form, specific rotation and λ_{max} (log ε)	m.p. °C	b.p. °C	Density	n_D	Solubility: w	al	eth	ace	bz	other solvents	Ref.
	Malonic acid														
Ω m23	—,diethyl ester	Ethyl malonate. Malonic ester. $CH_2(CO_2C_2H_5)_2$	160.17	λ^{MeOH} 235 (1.70)	−48.9	199.3^{760} 96^{22}	1.0551^{20}_{4}	1.4139^{20}	δ	∞	∞	v	v	chl, aa s	**B2**², 524
Ω m24	—,dimethyl ester	Methyl malonate. $CH_2(CO_2CH_3)_2$	132.13	...	−61.9	181.4^{760} 78.4^{15}	1.528^{20}_{4}	1.41348^{20}	δ	∞	∞	v	v	chl s	**B2**², 522
Ω m25	—,dinitrile	Malononitrile. Methylene cyanide. Dicyanomethane. $CH_2(CN)_2$	66.06	λ^{w} 267 (0.5)	32	$218–9^{760}$ 109^{20}	1.1910^{20}_{4}	1.4146^{34}	s	v	v	s	s	chl, aa s	**B2**², 535
m26	—,dipropyl ester	Propyl malonate. $CH_2(CO_2CH_2CH_2CH_3)_2$	188.23	glass	−77.1	229^{760} 113^{13}	1.0097^{20}_{4}	1.42061^{20}	...	v	v	v	v	...	**B2**², 528
m27	—,monoamide, N-phenyl-	Malonanilic acid. $C_6H_5NHCOCH_2CO_2H$	179.18	cr (w, eth, al)	132	...	...	...	s^h	s^h	s^h	...	s^h	...	**B12**¹, 208
Ω m28	—,monoamide mononitrile	Cyanoacetamide. H_2NCOCH_2CN	84.08	pl (w)	121–2	...	...	...	s	δ	...	...	...	...	**B2**², 534
m29	—,—,N-phenyl-	α-Cyanoacetanilide. $C_6H_5NHCOCH_2CN$	160.19	nd (al)	199–200	...	...	...	δ	...	...	...	...	...	**B12**², 167
m30	—,monochloride monoethyl ester	Carbethoxyacetyl chloride. $ClCOCH_2CO_2C_2H_5$	150.56	...	...	170–80d $75–7^{15}$	...	...	d	d	v	...	...	...	**B2**², 529
m31	—,monochloride monomethyl ester	$ClCOCH_2CO_2CH_3$	136.54	...	...	71^{15}	...	...	d	d	v	...	...	...	**B2**¹, 252
Ω m32	—,monoethyl ester mononitrile	Ethyl cyanoacetate. $NCCH_2CO_2C_2H_5$	113.12	...	−22.5	205^{760} 99^{15}	1.0654^{20}_{4}	1.4175^{20}	i	∞	∞	...	...	...	**B2**², 531
m33	—,monoethyl ester piperazinium salt	$C_4H_{10}N_2.2HO_2CCH_2CO_2C_2H_5$	350.37	...	144	...	...	...	s	s^h	i	...	...	...	**Am 56,** 150
Ω m34	—,monomethyl ester mononitrile	Methyl cyanoacetate. $NCCH_2CO_2CH_3$	99.09	...	−22.5	200.4–$.9^{761}$ 115^{36}	1.1128^{20}	1.4176^{20}	i	∞	∞	...	...	...	**B2**², 530
Ω m35	—,mononitrile	Cyanoacetic acid. $NCCH_2CO_2H$	85.06	dlq dimorphic cr λ^{al} 208 (1.6)	70–1	$108^{0.15}$δd	...	...	s	s	s	...	...	chl, aa δ	**B2**², 530
m36	—,piperazinium salt	$C_4H_{10}N_2.CH_2(CO_2H)_2$	190.20	...	180d	...	...	...	s	s^h	i	...	...	...	**Am 56,** 150
Ω m37	—,**acetamido-**, diethyl ester	$CH_3CONHCH(CO_2C_2H_5)_2$	217.22	cr (al or bz-peth)	95–6	185^{20}	...	...	δ^h	s^h	δ	...	...	...	**B4**², 891
m38	—,**acetyl-**, diethyl ester	$CH_3COCH(CO_2C_2H_5)_2$	202.21	...	...	232^{760} 120^{17}	1.0834^{26}_{4}	1.4435^{25}	i	...	...	s	...	Na_2CO_3 s	**B3**², 485
m39	—,**allyl-**	3-Butene-1,1-dicarboxylic acid*. $CH_2:CHCH_2CH(CO_2H)_2$	144.14	tcl (eth)	105	>180d	...	...	s	s	s	...	s^h	...	**B2**², 658
Ω m40	—,—,diethyl ester	$CH_2:CHCH_2CH(CO_2C_2H_5)_2$	200.24	...	...	222.5–3^{766} 93^{6}	1.0098^{20}_{4}	1.4305^{20}	i	v	v	...	...	...	**B2**², 658
m41	—,**amino-**, monohydrate	$H_2NCH(CO_2H)_2.H_2O$	137.09	pr (w + 1)	112d	...	...	...	δ	δ	...	...	...	...	**B4**², 890
m42	—,—,diethyl ester	$H_2NCH(CO_2C_2H_5)_2$	175.19	...	...	$122–3^{16}$ 88^{4}	1.100^{16}_{4}	1.4353^{16}	v	v	v	s	s	os s lig i	**B4**², 890
Ω m43	—,**benzoyl-amino-**, diethyl ester	Ethyl benzamidomalonate. $C_6H_5CONHCH(CO_2C_2H_5)_2$	279.30	nd (peth)	61	...	...	...	i	v	s	...	...	peth δ	**B9**², 185
Ω m44	—,**benzyl-**	$C_6H_5CH_2CH(CO_2H)_2$	194.19	pr (bz, eth, chl-peth)	121–2	...	...	...	v	v	v	...	v^h	...	**B9**², 619
m45	—,—,diamide	$C_6H_5CH_2CH(CONH_2)_2$	192.22	nd (al)	225	...	...	...	...	δ s^h	i	...	...	...	**B9**, 869
Ω m46	—,—,diethyl ester	$C_6H_5CH_2CH(CO_2C_2H_5)_2$	250.30	...	...	300 169^{212}	1.0750^{20}_{4}	1.4872^{20}	i	...	...	...	...	...	**B9**², 620
m47	—,—,dinitrile	Benzylmalononitrile. $C_6H_5CH_2CH(CN)_2$	156.20	pl (al), nd (w or lig)	91	174^{23}	...	...	...	s	s	...	s	AcOEt s lig δ	**B9**, 870
m48	—,**benzyl-(hydroxy)-**	Benzyltartronic acid. $C_6H_5CH_2COH(CO_2H)_2$	210.19	pr	147d	...	...	...	s	s	s	...	...	...	**B10**, 515
m49	—,**benzylidene-**	$C_6H_5CH:C(CO_2H)_2$	192.18	pr (w)	195–6d	...	...	...	δ s^h	s	δ	s	δ	AcOEt s CS_2, chl δ	**B9**², 638
Ω m50	—,—,diethyl ester	$C_6H_5CH:C(CO_2C_2H_5)_2$	248.28	...	32(27)	308–12δd 180^{10}	1.1045^{20}_{4}	1.5389^{20}	i	s	s	s	s	os s	**B9**², 639
m51	—,—,monoethyl ester mononitrile	Ethyl α-cyanocinnamate. Ethyl benzalcyanoacetate. $C_6H_5CH:C(CN)CO_2C_2H_5$	201.23	(i) oil (ii) nd (al)	(ii) 51	(i) $118^{0.4}$ 188^{15}	(i) 1.0762	(i) 1.5033	i	δ s^h	s	...	v	aa v chl s	**B9**², 640
Ω m52	—,—,mononitrile	Benzalcyanoacetic acid. α-Cyanocinnamic acid. $C_6H_5CH:C(CN)CO_2H$	173.17	cr (al)	183	...	...	...	...	s^h	...	...	...	dil NH_3 s	**B9**², 639
m53	—,**bromo-**	$BrCH(CO_2H)_2$	182.96	nd (eth), pl (ace-bz)	113d	...	...	...	s^h	v	v	...	...	...	**B2**², 538
Ω m54	—,—,diethyl ester	$BrCH(CO_2C_2H_5)_2$	239.07	...	−54	233–5d 123^{20}	1.4022^{25}_{4}	1.4521^{20}	i	∞	∞	s	...	...	**B2**², 538
Ω m55	—,**butyl-**	1,1-Pentanedicarboxylic acid.* $CH_3(CH_2)_3CH(CO_2H)_2$	160.17	pr (w)	104–5	...	...	...	v	s	s	...	...	...	**B2**², 588
Ω m56	—,—,diethyl ester	$CH_3(CH_2)_3CH(CO_2C_2H_5)_2$	216.28	...	...	$235–40^{760}$ 122^{12}	...	1.4250^{20}	...	v	v	...	...	...	**B2**², 588
m57	—,—,monoethyl ester mononitrile	Ethyl α-cyanocaproate. $CH_3(CH_2)_3CH(CN)CO_2C_2H_5$	169.23	syr	...	$230–3^{734}$ $128–9^{25}$	0.9537^{25}_{4}	1.4262^{20}	...	s	...	...	s	...	**B2**², 588
Ω m58	—,*sec*-**butyl-**, diethyl ester	$CH_3CH_2CH(CH_3)CH(CO_2C_2H_5)_2$	216.28	...	...	$245–50^{762}$ 105^{9}	0.988^{15}	1.4248^{20}	i	v	v	...	...	...	**B2**², 590

For explanations, symbols and abbreviations see beginning of table. For structural formulas see end of table.

No.	Name	Synonyms and Formula	Mol. wt.	Color, crystalline form, specific rotation and λ_{max} (log ε)	m.p. °C	b.p. °C	Density	n_D	Solubility w	al	eth	ace	bz	other solvents	Ref.
	Malonic acid														
m59	—,**chloro-**	$ClCH(CO_2H)_2$	138.51	pr (w)	133	d	...	...	v	v	v	...	...	...	**B2**[2], 537
Ω m60	—,—,diethyl ester	Ethyl chloromalonate. $ClCH(CO_2C_2H_5)_2$	194.62	...	...	222^{760} 118^{16}	1.2040_4^{20}	1.4327^{20}	i	∞	∞	...	...	chl ∞ CS_2 s	**B2**[2], 537
Ω m61	—,**(3-chloro-propyl)-**, diethyl ester	$ClCH_2CH_2CH_2CH(CO_2C_2H_5)_2$	236.71	...	...	$147–9^{10}$	...	1.4429^{20}	i	s	s	...	...	chl, CS_2 s	**B2**[1], 278
Ω m62	—,**cinnamylidene-**(*trans*)	Cinnamalmalonic acid. $C_6H_5CH{:}CHCH{:}C(CO_2H)_2$	218.21	dk ye nd	212d	...	...	...	i	s	...	...	...	chl, AcOEt s	**B9**[2], 649
m63	—,**1-cyclo-hexenyl-**, mono-nitrile	1-Cyclohexenylcyanoacetic acid	165.19	nd (bz)	109–10	...	...	...	δ v^h	s	...	s	s^h	...	**B9**[2], 560
m64	—,**cyclohexyl-**, diethyl ester	...	242.32	...	...	$163–5^{20}$	1.0281_4^{19}	1.4478^{25}	i	s	s	s	s	os s	**B9**[2], 526
Ω m65	—,**2-cyclopenten-yl-**, diethyl ester	...	226.28	...	...	141^{16}	1.0507_4^{20}	1.4536^{20}	i		s	s	...	...	**B9**[2], 559
m66	—,**cyclopentyli-dene-**, diethyl ester	...	226.28	...	...	140^{10}	1.0616_4^{20}	1.4724^{20}	i	...	s	s	...	...	**B9**[2], 559
Ω m67	—,**dibenzyl-**, diethyl ester	$(C_6H_5CH_2)_2C(CO_2C_2H_5)_2$	340.42	...	14	$234–5^{23}$	1.093_4^{20}	...	i	s	s	...	...	...	**B9**[1], 408
m68	—,**dibromo-**, diethyl ester	$Br_2C(CO_2C_2H_5)_2$	317.97	...	...	250–6δd 154^{28}	...	...	i	s	s	s	...	...	**B2**[2], 539
Ω m69	—,**dibutyl-**, diethyl ester	$(CH_3CH_2CH_2CH_2)_2C(CO_2C_2H_5)_2$	272.39	...	...	$153–4^{14}$	0.9457_4^{20}	1.4341^{20}	i	s	s	...	...	...	**B2**[2], 614
Ω m70	—,**diethyl-**	3,3-Pentanedicarboxylic acid*. $(C_2H_5)_2C(CO_2H)_2$	160.17	pr (w or bz)	127	d	...	...	v	v	v		$δ^h$	chl δ	**B2**[2], 593
Ω m71	—,—,diethyl ester	$(C_2H_5)_2C(CO_2C_2H_5)_2$	216.28	...	...	230 (225) 100^{12}	0.9643^{30}	1.4240^{20}	i	∞	∞	...	...	...	**B2**[2], 593
m72	—,—,monoethyl ester mononitrile	$(C_2H_5)_2C(CN)CO_2C_2H_5$	169.23	...	...	214–5 $100–1^{15}$	...	1.4200^{27}	i	∞	∞	...	...	...	**B2**[2], 594
m73	—,—,piperazinium salt	$(C_2H_5)_2C(CO_2H)_2 . C_4H_{10}N_2$	246.31	...	80–1	...	...	...	s	s	i	s^h	...	...	**Am 56**, 150
—	—,**dihydroxy-**	*see* **Malonic acid, oxo-,** hydrate													
m74	—,**dimethyl-**, diethyl ester	$(CH_3)_2C(CO_2C_2H_5)_2$	188.23	...	−30.4	197^{760} $97–8^{22}$	0.9964_4^{20}	1.4129^{20}	i	∞	∞	...	...	...	**B2**[2], 572
m75	—,**(2,4-dinitro-phenyl)-**, diethyl ester	...	326.27	pr (al)	51	...	...	...	i	v^h	v^h	...	...	alk s	**B9**[1], 378
Ω m76	—,**(ethoxy-methylene)-**, diethyl ester	$C_2H_5OCH{:}C(CO_2C_2H_5)_2$	216.24	...	...	279–81d 165.5^{19} $109^{0.5}$	...	1.4600^{20}	i	s	s	...	...	...	**B3**[2], 300
Ω m77	—,**ethyl-**	1,1-Propanedicarboxylic acid*. $CH_3CH_2CH(CO_2H)_2$	132.13	pr (w + 1)	114 (112) (anh)	160d	...	...	v	s	s	i	s	MeOH, AcOEt, chl s	**B2**[2], 569
Ω m78	—,—,diethyl ester	$CH_3CH_2CH(CO_2C_2H_5)_2$	188.23	...	...	$207–9^{755}$ $98–9^{12}$	1.0047_4^{20}	1.4166^{20}	δ	v	v	s	...	chl s	**B2**[2], 570
m79	—,**ethyl(iso-propyl)-**, diethyl ester	$(CH_3)_2CHC(C_2H_5)(CO_2C_2H_5)_2$	230.31	...	...	$232–4^{742}$ $108–10^{11}$	...	1.4280^{25}	...	s	s	...	...	...	**B2**, 706
m80	—,**ethyl(methyl)-**	2,2-Butanedicarboxylic acid*. $C_2H_5C(CH_3)(CO_2H)_2$	146.14	pr or nd (eth)	122	...	...	...	s	s	s	s	...	...	**B2**[2], 585
Ω m81	—,**ethyl(phenyl)-**, diethyl ester	$C_6H_5C(C_2H_5)(CO_2C_2H_5)_2$	264.33	...	...	170^{19}	1.071_4^{20}	1.4896^{25}	i	s	s	...	...	...	**B9**[2], 630
m82	—,**ethylidene-**, diethyl ester	$CH_3CH{:}C(CO_2C_2H_5)_2$	186.21	λ^{al} 225 (3.75)	...	$115–8^{17}$	1.0194_4^{17}	1.4308^{17}	i	s	s	...	...	...	**B2**[2], 654
m83	—,**(formyl-amino)-**, diethyl ester	Diethyl formamidomalonate. $HCONHCH(CO_2C_2H_5)_2$	203.20	pl (MeOH)	48–9	$173–4^{11}$ $130–2^{2–3}$	...	...	i	s	...	...	...	...	**B4**[2], 891
m84	—,**heptyl-**	1,1-Octanedicarboxylic acid*. $CH_3(CH_2)_6CH(CO_2H)_2$	202.25	pr (bz-peth)	96.5–8.0	...	...	...	i	v	v	v	i	...	**B2**[2], 610
m85	—,**hexadecyl-**	Cetylmalonic acid. 1,1-Hepta-decanedicarboxylic acid*. $CH_3(CH_2)_{15}CH(CO_2H)_2$	328.50	nd (lig), lf (aa)	121–2	...	...	...	i	δ	v	...	v^h	aa v^h lig s^h peth δ	**B2**[2], 627
m86	—,—,diethyl ester	$CH_3(CH_2)_{15}CH(CO_2C_2H_5)_2$	384.61	amor	(i) 25.1 (ii) 12.7	$238–40^{14}$ $204–8^{2}$	...	1.4433^{20}	i	s	...	...	...	...	**B2**[1], 299
m87	—,**hexyl-**, diethyl ester	1,1-Heptanedicarboxylic acid. $CH_3(CH_2)_5CH(CO_2C_2H_5)_2$	244.34	...	...	268–70 143^{15}	0.9577_4^{21}	1.4278^{21}	i	s	s	s	s	os s	**B2**[2], 604
Ω m88	—,**hydroxy-**	Tartronic acid. $HOCH(CO_2H)_2$	120.06	pr (w + 1)	156–8d (rapid htng) 141–2 (slow htng) (anh)	sub	...	...	s	s	δ	...	...	...	**B3**, 415
m89	—,—,diethyl ester	$HOCH(CO_2C_2H_5)_2$	176.17	...	−2.5	222–5 121^{15}	1.152^{15}	...	...	s	s	s	s	os s	**B3**[1], 148
m90	—,—,dimethyl ester	$HOCH(CO_2CH_3)_2$	148.13	cr (eth-peth)	45	122^{19}	...	...	s	s	v	s	s	chl v os s peth δ	**B3**[3], 905

For explanations, symbols and abbreviations see beginning of table. For structural formulas see end of table.

No.	Name	Synonyms and Formula	Mol. wt.	Color, crystalline form, specific rotation and λ_{max} (log ε)	m.p. °C	b.p. °C	Density	n_D	Solubility w	al	eth	ace	bz	other solvents	Ref.
	Malonic acid														
m91	—,**(2-hydroxy-cyclohexyl)-** lactone, monoethyl ester	...	212.25	pa ye	...	199^{30} (195^{10})	1.0735^{17}	...	...	...	...	...	...	...	**B18²**, 322
Ω m92	—,**hydroxy-(methyl)-**	α-Isomalic acid. $CH_3COH(CO_2H)_2$	134.09	...	142d	170d	...	...	v	v	v	...	...	...	**B3**, 440
m93	—,**isobutyl-**	$(CH_3)_2CHCH_2CH(CO_2H)_2$	160.17	cr (bz)	115d (108)	d	...	...	v	v	v	...	...	...	**B2¹**, 284
Ω m94	—,—,diethyl ester	$(CH_3)_2CHCH_2CH(CO_2C_2H_5)_2$	216.28	...	...	255 119–20^{16}	0.9804^{20}_{4}	1.4236^{20}	i	v	v	...	...	chl s	**B2**, 683
Ω m95	—,**isopropyl-**, diethyl ester	$(CH_3)_2CHCH(CO_2C_2H_5)_2$	202.25	...	...	215^{760} 107–9^{18}	0.9970^{20}_{15}	1.4188^{21}	δ	v	v	...	...	chl s	**B2²**, 586
m96	—,**isopropylidene-**	$(CH_3)_2C{:}C(CO_2H)_2$	144.14	cr (ace-chl)	170–1	d	...	...	...	s	...	...	...	alk v chl i	**B2¹**, 312
m97	—,—,diethyl ester	$(CH_3)_2C{:}C(CO_2C_2H_5)_2$	200.24	...	...	175–8 140–1^{20}	1.0282^{18}_{4}	1.4486^{17}	...	s	s	...	...	CCl_4 s	**B2²**, 661
Ω m98	—,**methyl-**	1,1-Ethanedicarboxylic acid*. Isosuccinic acid. $CH_3CH(CO_2H)_2$	118.09	nd (AcOEt-bz), pr (eth-bz)	135d	...	1.455^{20}_{4}	...	v	v	v	...	δ	aa v AcOEt s	**B2²**, 562
Ω m99	—,—,diethyl ester	Ethyl isosuccinate. $CH_3CH(CO_2C_2H_5)_2$	174.20	$\lambda^{MeOH-KOH}$ 268 (2.0)	...	201 94^{16}	1.0225^{20}_{4}	1.4126^{20}	δ	v	v	s	...	chl s	**B2²**, 563
m100	—,—,dimethyl ester	Methyl isosuccinate. $CH_3CH(CO_2CH_3)_2$	146.14	...	...	176.5^{761}	1.0977^{20}_{4}	1.4128^{20}	δ	∞	∞	s	...	chl s	**B2²**, 562
m101	—,**(3-methyl-butyl)-**, diethyl ester	Ethyl isoamylmalonate. $(CH_3)_2CHCH_2CH_2CH(CO_2C_2H_5)_2$	230.31	...	...	240–2^{760} 137.5^{19}	0.9580^{25}_{4}	1.4255^{25}	i	v	v	s	...	...	**B2²**, 599
m102	—,**nitro-**, mono-amide mononitrile	Cyanonitroacetamide. Fulminuric acid. $NCCH(NO_2)CONH_2$	129.08	pr (al)	145 exp (136–49)	...	...	...	s	s	δ	...	i	chl, lig i	**B2¹**, 258
m103	—,**octyl-**	1,1-Nonanedicarboxylic acid*. $CH_3(CH_2)_7CH(CO_2H)_2$	216.28	pr (bz-peth)	116	...	1.173^{17}	...	s^h	s	...	s	s^h	chl s^h	**B2²**, 613
Ω m104	—,**oxo-**, hydrate	Mesoxalic acid. Dihydroxymalonic acid. $(HO)_2C(CO_2H)_2$	136.06	dlq nd (w)	120–1	...	...	...	v	v	v	...	...	...	**B3²**, 472
Ω m105	—,—,diethyl ester	Ethyl mesoxalate. $OC(CO_2C_2H_5)_2$	174.16	pa ye gr oil	*ca* −30	208–10^{220} 105^{19}	1.1419^{16}_{4}	1.4310^{22}	v	s	s	...	...	chl s CS_2 i	**B3¹**, 267
m106	—,—,diethyl ester hydrate	Ethyl dihydroxymalonate. $(HO)_2C(CO_2C_2H_5)_2$	192.17	pl (bz)	57	*ca.* 200	...	...	s	s	v	s	s	chl s lig δ CS_2 i	**B3²**, 474
m107	—,—,diethyl ester, oxime	Ethyl isonitrosomalonate. $HON{:}C(CO_2C_2H_5)_2$	189.17	...	...	172^{12} $151^{4.8}$	1.1821^{18}_{4}	1.4544^{18}	i	s	v	v	s	os ∞, alk s	**B3²**, 475
Ω m108	—,**pentyl-**, diethyl ester	$CH_3(CH_2)_4CH(CO_2C_2H_5)_2$	230.31	...	...	134–6^{14}	0.9652^{20}_{4}	1.4253^{20}	i	v	v	...	...	...	**B2²**, 597
m109	—,**(3-pentyl)-**, diethyl ester	Ethyl *sec*-amylmalonate. $(C_2H_5)_2CHCH(CO_2C_2H_5)_2$	230.31	...	...	130^{16}	...	1.4291^{20}	i	v	v	...	...	...	**B2³**, 1781
Ω m111	—,**phenyl-**, diethyl ester	$C_6H_5CH(CO_2C_2H_5)_2$	236.27	...	16–7	205δd 168^{12}	1.0950^{20}_{4}	1.4977^{20}	i	s	...	s	...	...	**B9²**, 615
m112	—,—,dimethyl ester	$C_6H_5CH(CO_2CH_3)_2$	208.22	cr (lig)	51	145–7^{13}	...	...	...	s	s	...	...	lig δ^h	**B9²**, 615
m113	—,—,dinitrile	$C_6H_5CH(CN)_2$	142.16	cr (dil al)	70–1	152–3^{21}	...	...	δ	v	δ v^h	...	...	lig δ	**B9²**, 616
m114	—,—,monoethyl ester mononitrile	Ethyl phenylcyanoacetate. $C_6H_5CH(CN)CO_2C_2H_5$	189.22	oil	...	275δd 165^{20}	1.091^{20}_{4}	1.5012^{25}	i	∞	∞	v	v	os ∞	**B9²**, 615
m115	—,**(phenyl-amino)-**, diethyl ester	Ethyl anilinomalonate. $C_6H_5NHCH(CO_2C_2H_5)_2$	251.29	cr (al or lig)	45	...	...	...	i	v	v	...	v	CS_2, chl v	**B12¹**, 271
m116	—,**phthalimido-**, diethyl ester	...	305.29	pr (al)	74	...	...	...	δ	v^h	s	v	v	chl v CS_2 s peth δ	**B21¹**, 379
m117	—,**(4-phthali-midobutyl)-**, diethyl ester	...	361.40	nd (peth)	46	...	...	...	i	...	s	v	v	peth s^h	**B21**, 489
m118	—,**propyl-**	1,1-Butanedicarboxylic acid*. $CH_3CH_2CH_2CH(CO_2H)_2$	146.14	pl (bz)	96–7.5	d	...	...	v	s	s	...	δ	chl s	**B2²**, 581
Ω m119	—,—,diethyl ester	$CH_3CH_2CH_2CH(CO_2C_2H_5)_2$	202.25	...	...	221–1.5^{767} 114^{22}	0.9873^{20}	1.4197^{20}	δ	v	v	...	...	...	**B2²**, 582
—	**Maltobiose**	*see* **Maltose**													
—	**Maltol**	*see* **4-Pyrone, 3-hydroxy-2-methyl-**													
m120	**α-Maltose,** (pyranose form) anhydrous	Glucose-4-α-glucoside. Maltobiose.	342.30	nd (abs al) $[\alpha]_D$ +140.7 (w, c = 10)	160–5	...	...	...	s	i	i	...	i	...	**B31**, 386
Ω m121	—,monohydrate	$C_{12}H_{22}O_{11}\cdot H_2O$. *See* m120	360.32	nd (w), $[\alpha]^{20}_D$ +111.7 → +130.4 (c = 4) (mut)	102–3	...	1.540	...	v	δ	i	...	...	...	**B31**, 386

For explanations, symbols and abbreviations see beginning of table. For structural formulas see end of table.

No.	Name	Synonyms and Formula	Mol. wt.	Color, crystalline form, specific rotation and λ_{max} (log ε)	m.p. °C	b.p. °C	Density	n_D	Solubility w	al	eth	ace	bz	other solvents	Ref.
	Malvidine chloride														
Ω m122	**Malvidine chloride**	Syringidine chloride. $C_{17}H_{15}ClO_7$.	366.76	rh ta or pr (al-aq HCl + 1w) λ^{al-HCl} 560 (1.77)	>300	....	....	....	δ	s	...	...	...	MeOH s (red)	**B18**[2], 247
—	**Mandelic acid**	*see* **Acetic acid, hydroxy-(phenyl)-**													
—	***l*-Mandelonitrile-β-gentiobioside**	*see* **Neoamygdalin**													
m123	**Mannitane**	3,6(1,4)-Anhydro-*D*-mannitol.	164.16	(i) amor (ii) pl $[\alpha]^{20}_D$ −26.2 (w)	146–7 (137)	....	....	....	v	(i) s (ii) i	i	...	i	....	**B1**[1], 284
—	**Mannite**	*see* **Mannitol**													
Ω m124	**Mannitol**(*D*)	*D*-Mannite	182.18	rh nd or pr (w) $[\alpha]^{25}_D$ −0.49 (w)	168	$295^{3.5}$ sub 130	1.489^{20}_4	1.3330	v	δ^h	i	...	...	Py δ	**B1**[3], 2392
m125	—(*DL*)	α-Acritol. $C_6H_{14}O_6$. *See* m124	182.18	cr (al)	168	....	....	....	s^h	δ	...	...	...	....	**B1**[3], 2405
m126	—(*L*)	*L*-Mannite. $C_6H_{14}O_6$. *See* m124	182.18	nd (al)	163–4	....	....	....	v	δ	...	...	...	MeOH δ^h	**B1**[3], 2405
m127	—,hexanitrate(*D*). .	Manexin. Nitranitol. Nitromannitol. $C_6H_8N_6O_{18}$. *See* m124	452.17	nd (al), $[\alpha]^{22}_{546}$ +46.8 ($ClCH_2CH_2Cl$, c = 0.33)	112–3	120 exp	1.8^{20}_4 1.604^0	....	i	s^h	s^h	...	s^h	aa s	**B1**[3], 2404
—	**—,1,4:3,6-dianhydro-(*D*)**	*see* **Isomannide**													
m128	**Mannoheptose-(*d*, α)**	....	210.19	nd (al), $[\alpha]^{20}_D$ +85 to +68.6 (w, c = 10) (mut)	134–5	....	....	....	v	δ	...	...	...	....	**B31**, 361
m129	—(*d*, β), monohydrate	....	244.20	cr (w + 1), $[\alpha]^{20}_D$ +45.7	83	....	....	....	s	δ	...	...	...	....	**B31**, 362
m132	**Mannonic acid,** γ-lactone(*D*)	....	178.14	nd (96% al), pr (abs al) $[\alpha]^{20}_D$ +54 (w, c = 2)	151	....	....	....	v	δ^h	...	...	i	MeOH δ	**B18**[1], 407
m133	—,—(*L*)	$C_6H_{10}O_6$. *See* m132	178.14	rh nd (w) $[\alpha]_D$ −51.8	151	....	....	....	v	δ	...	...	...	....	**B18**[2], 199
m134	**Mannosaccharic acid,** γ,γ'-dilactone(*d*)	....	174.11	nd (w + 2), $[\alpha]^{23}_D$ +202 (w) $\lambda^{aq\ NaOH}$ 263 (3.72)	190d	....	....	....	δ v^h	s^h	i	...	...	....	**B19**[2], 266
m135	—,—(*l*)	$C_6H_6O_6$. *See* m134	174.11	nd (w or al) $[\alpha]^{20}_D$ −202.5 (w)	183–5d (anh) 68 (hyd)	....	....	....	s v^h	δ	i	...	...	....	**B19**[2], 266
Ω m136	**Mannose**(*D*)	Carubinose. Seminose	180.16	nd or orh pr (al or aa) $[\alpha]^{20}_D$ −17 to +14.6 (w, p = 3) (mut) λ^w <220	132d	....	1.539^{20}_4	....	v	δ	i	...	i	MeOH δ	**B31**, 284
m137	—(*DL*)	$C_6H_{12}O_6$. *See* m136	180.16	cr (al)	132–3	....	....	....	v	δ	i	...	i	MeOH δ	**B31**, 294
m138	—(*L*)	$C_6H_{12}O_6$. *See* m136	180.16	cr (al) $[\alpha]_D$ +14 → −14 (mut)	132	....	....	....	v	δ	i	...	i	MeOH δ	**B31**, 293
m139	—,phenylhydrazone(*D*)	$C_{12}H_{18}N_2O_5$. *See* m136	270.29	ye pr (w), nd (al), $[\alpha]_D$ +26.3 → +33.8 (Py) (mut)	199–200	....	....	....	i s^h	s^h	...	...	i	....	**B31**, 290
m140	—,—(*L*)	$C_{12}H_{18}N_2O_5$. *See* m136	270.29	ye pr (w)	195	....	....	....	i s^h	s^h	...	...	i	....	**B31**, 294
m141	**Mannuronic acid**(*D*, α)	....	194.14	hyg nd (al-eth) $[\alpha]^{25}_D$ +16 → −6 (w) (mut)	120–30d	....	....	....	s	...	i	...	i	....	**C27**, 3451
m142	—(β)	$C_6H_{10}O_7$. *See* m141	194.14	cr (w + ½, w-ace-eth) $[\alpha]^{25}_D$ −47.9 → −23.9 (w) (mut)	165–7	....	....	....	s	...	i	...	i	....	**C27**, 3457
—	**Maraniol**	*see* **Coumarin, 7-ethoxy-4-methyl-**													
—	**Maretine**	*see* **Semicarbazide, 1(3-tolyl)-**													
—	**Margaraldehyde** ...	*see* **Heptadecanal***													
—	**Margaric acid**	*see* **Heptadecanoic acid***													
m144	**Matrine**	Sophocarpidine. $C_{15}H_{24}N_2O$.	248.37	(4 forms) α: nd or pl, $[\alpha]^{15}_D$ +40.93 (w) β: orh pr $[\alpha]^{10}_D$ −28.73 (w) γ: liq δ: pr or lf (peth)	α 76–7 β 87 δ 84	$\gamma 223^6$	$\gamma 1.088^{20}_4$	$\gamma 1.52865^{85.4}$	s δ^h	v	s	s	v	CS_2, chl v peth δ	**B24**[2], 58

For explanations, symbols and abbreviations see beginning of table. For structural formulas see end of table.

No.	Name	Synonyms and Formula	Mol. wt.	Color, crystalline form, specific rotation and λ_{max} (log ε)	m.p. °C	b.p. °C	Density	n_D	Solubility w	al	eth	ace	bz	other solvents	Ref.
	Meconic acid														
Ω m145	**Meconic acid**	3-Hydroxy-γ-pyrone-2,6-dicarboxylic acid.	200.11	rh pl (w or dil HCl + 3w), pr (con HCl + 1w) λ^{w} 212 (4.2), 235 (4.15), 305 (3.95)	−w at 100	d at 120	....	....	δ v^{h}	s	δ	δ	s	MeOH δ	**B18**^{2}, 372
m146	**Meconidine**	$C_{21}H_{23}NO_4$	353.42	ye amor	58	....	....	....	i	s	s	s	s	chl, peth s	**M**, 641
m147	**Meconin**	6,7-Dimethoxyphthalide. Opianyl.	194.19	wh nd (w) λ^{al} 213 (4.40), 308 (3.58)	102–3	155 sub	....	....	s^{h} δ	s	s	s	s	chl, aa s	**B18**^{2}, 62
m148	**Medicagenic acid**	2β,3β-Dihydroxy-Δ^{12}-oleanene-23,28-dioic acid. $C_{30}H_{46}O_6$.	502.70	pr or nd $[\alpha]^{25}_{D}$ +106	352–3	....	....	....	i	δ^{h}	δ	...	...	....	**Am 76**, 2271
m149	—,diacetate	$C_{34}H_{50}O_8$. *See* m148	586.78	mcl, $[\alpha]^{22}_{D}$ +92 (chl)	210–2 (207)	....	1.190^{25}_{25}	....	i^{h}	δ^{h}	i	...	...	....	**Am 79**, 5292
m150	**Melam**	$C_6H_8N_{11}$	235.21	pw	....	....	....	....	i	δ	...	...	...	ac s	**B3**^{2}, 121
Ω m151	**Melamine**	Cyanurotriamide. Cyanuramide. 2,4,6-Triamino-1,3,5-triazine*	126.12	mcl pr (w) $\lambda^{0.1N HCl}$ 235 (4.01) $\lambda^{w, pH=7}$ 236 (3.4)	354d (cor)	sub	1.573^{14}	1.872^{20}	δ v^{h}	δ^{h}	i	...	...	....	**B26**^{2}, 132
—	**Melaniline**	*see* **Guanidine, 1,3-diphenyl-**													
m152	**Melene**	$C_{30}H_{60}$	420.82	nd (ace), cr (peth)	62–3	380	0.9037^{25}_{25}	1.4228^{90}	i	δ^{h}	δ^{h}	δ^{h}	δ	to δ	**B1**^{3}, 885
m153	**Melezitose**	Melizitose. $C_{18}H_{32}O_{16}$	504.45	cr (w + 2) $[\alpha]^{20}_{D}$ +88 (w, c = 4)	153–4 (anh)	....	1.5565^{0}	....	s	δ	i	...	i	peth i	**B31**, 466
m154	**Melibiose**(α)	Glucose-6-α-galactoside. $C_{12}H_{22}O_{11}$.	342.30	amor (anh) $[\alpha]^{15}$ +145.8 → +141.6 (w, c = 2) (mut)	....	....	....	....	v	δ	i	...	i	peth i	**B31**, 422
m155	—(β), dihydrate	$C_{12}H_{22}O_{11}.2H_2O$. *See* m154	378.33	mcl cr (w + 2) $[\alpha]^{20}_{D}$ +111.7 → +129.5 (w, c = 4) (mut)	84–5	....	....	....	v	δ	i	...	i	MeOH s	**B31**, 421
—	**Melilotic acid**	*see* **Propanoic acid, 3(2-hydroxyphenyl)-***													
—	**Melitotol**	*see* **Coumarin, 3,4-dihydro-**													
—	**Melissic acid**	*see* **Hentriacontanoic acid***													
—	**Melitoxin**	*see* **Dicoumarol**													
—	**Mellitene**	*see* **Benzene, hexamethyl-***													
—	**Mellitic acid**	*see* **Benzenehexacarboxylic acid***													
—	**Mellophanic acid**	*see* **1,2,3,4-Benzenetetracarboxylic acid***													
—	**Menadione**	*see* **1,4-Naphthoquinone, 2-methyl-***													
Ω m157	*p*-**Menthane**(*cis*)	Hexahydro-*p*-cymene. 1-Isopropyl-4-methylcyclohexane.	140.27	λ^{iso} 220 (−0.67), 250 (−1.38)	−89.84	170.9^{725}	0.8039^{20}_{4}	1.4431^{20}	i	v	v	...	s	peth s	**B5**^{2}, 26
Ω m158	—(*trans*)	$C_{10}H_{20}$. *See* m157	140.27	λ^{iso} 220 (−1.15), 250 (−2.40)	....	170.6^{760} (161) 58.5^{15}	0.7928^{20}_{4}	1.4366^{20}	i	v	v	...	s	lig s	**B5**^{2}, 27
—	**Menthanone**	*see* **Menthone**													
m159	Δ^{3}-*p*-**Menthene**(*d*) (one form)	1-Isopropyl-4-methylcyclohexane	138.25	$[\alpha]^{20}_{D}$ +115.6 (undil)	....	168	0.8118^{20}_{4}	1.4524^{20}	i	s	s	...	s	aa s	**B5**^{2}, 52
m160	—(*d*) (one form)	$C_{10}H_{18}$. *See* m159	138.25	$[\alpha]_D$ +29.6 → +54.4 (mut)	....	167–8	0.8078^{20}_{4}	....	i	s	s	...	s	peth s	**B5**^{3}, 237
Ω m161	—(*dl*)	$C_{10}H_{18}$. *See* m159	138.25	....	....	168^{754} 60.5^{12}	0.8069^{20}_{4}	1.4503^{20}	i	s	s	...	s	peth s	**B5**^{2}, 53
m162	—(*l*)	$C_{10}H_{18}$. *See* m159	138.25	$[\alpha]_D$ −13.46 (al)	....	167–8	0.8112^{19}_{19}	1.4511^{20}	i	s	s	...	s	peth s	**B5**, 89
—	$\Delta^{8(9)}$-*p*-**Menthenol**	*see* **Isopulegol**													
—	**Menthenone**	*see* **Piperitone**													
m163	**Menthol**(*d*)	3-*p*-Menthanol. Hexahydrothymol.	156.27	$[\alpha]_D$ +49.2 (al, c = 5)	42–3	$103–4^{9}$	....	....	...	s	s		s	os s	**B6**^{2}, 49
m164	—(*dl*)	$C_{10}H_{20}O$. *See* m163	156.27	nd (peth)	(i) 28 (ii) 38	216^{760} $103–5^{16}$	0.904^{15}_{15}	1.4615^{20}	i	v	v	v	v	os, chl v	**B6**^{2}, 49
Ω m165	—(*l*)	$C_{10}H_{20}O$. *See* m163	156.27	nd (MeOH) $[\alpha]^{20}_{D}$ −48 (al, c = 2.5)	(i) 44 (ii) 35 (iii) 33 (iv) 31	216.4^{760} 111^{20}	0.904^{15}_{15}	1.460^{22}	δ	v	v	v	v	peth, aa s chl v	**B6**^{2}, 39
Ω m166	—,3-methylbutanoate	Menthyl isovalerate. $C_{15}H_{28}O_2$. *See* m163	240.39	$[\alpha]^{20}_{D}$ −64.02 (bz, c = 10)	....	129^{9}	0.9089^{15}	1.4486^{20}	i	s	...	s	...	....	**B6**^{1}, 22
m167	**Menthone**(*d*)	2-Isopropyl-5-methylcyclohexanone. *d*-3-*p*-Menthanone.	154.26	$[\alpha]^{18}_{D}$ +24.85 λ^{al} 236 (1.40), 287 (1.38)	....	204^{750} 85^{14}	0.8963^{20}_{20}	1.4503^{20}	δ	s	s	s	∞	aa s	**B7**^{1}, 35

For explanations, symbols and abbreviations see beginning of table. For structural formulas see end of table.

No.	Name	Synonyms and Formula	Mol. wt.	Color, crystalline form, specific rotation and λ_{max} (log ε)	m.p. °C	b.p. °C	Density	n_D	Solubility w	al	eth	ace	bz	other solvents	Ref.
	Menthone														
Ω m168	—(*dl*)	$C_{10}H_{18}O$. *See* m167	154.26	...	...	210.5^{760}	0.911^{0}	...	δ	s	s	s	∞	aa s	**B7**[2], 41
m169	—(*l*)	$C_{10}H_{18}O$. *See* m167	154.26	$[\alpha]_D^{20}$ −29.6 λ^{hx} 293 (1.3)	−6.6	209.6^{760} 96^{20}	0.8954_4^{20}	1.4505^{20}	δ	∞	∞	s	∞	CS_2 ∞	**B7**[1], 35
—	**Mepheridine**	*see* **4-Piperidinecarboxylic acid, 1-methyl-4-phenyl-,** ethyl ester, hydrochloride													
—	**Mesaconic acid**	*see* **Fumaric acid, methyl-**													
m170	**Mescaline**	Mezcaline. 3,4,5-Trimethoxy-*β*-phenethylamine. $C_{11}H_{17}NO_3$.	211.26	cr λ^{al} 225 sh (4.2), 270 (3.2)	35–6	180^{12} 120–$30^{0.05}$	...	...	s	s	i	...	s	chl s lig δ peth i	**B13**[2], 521
—	**Mesidine**	*see* **Benzene, 2-amino-1,3,5-trimethyl-***													
—	**Mesitoic acid**	*see* **Benzoic acid, 2,4,6-trimethyl-**													
—	**Mesitol**	*see* **Benzene, 2-hydroxy-1,3,5-trimethyl-***													
—	**Mesitylaldehyde**	*see* **Benzaldehyde, 2,4,6-trimethyl-**													
—	**Mesitylene**	*see* **Benzene, 1,3,5-trimethyl-***													
—	**Mesitylenic acid**	*see* **Benzoic acid, 3,5-dimethyl-**													
—	**Mesityl oxide**	*see* **3-Penten-2-one, 4-methyl-***													
—	**Mesorcinol**	*see* **Benzene, 2,4-dihydroxy-1,3,5-trimethyl-***													
—	**Mesoxalic acid**	*see* **Malonic acid, oxo-**													
—	**Mestilbol**	*see* **3-Hexene, 3(4-hydroxyphenyl)-4(4-methoxyphenyl)-***													
—	**Metacetaldehyde,** (*tetramer*)	*see* **Metaldehyde II**													
m171	**Metacrolein**	2,4,6-Triethenyl-1,3,5-trioxane.	168.20	pl (al)	50 (45)	170^{760}	...	...	i δ[h]	s	s	...	...	...	**B1**[1], 378
—	**Metahemipinic acid**	*see* **Phthalic acid, 4,5-dimethoxy-**													
Ω m172	**Metaldehyde**	Metacetaldehyde. $(C_2H_4O)_{4-6}$	$(44.05)_{4-6}$	tetr nd or pr (al)	246.2 (sealed tube)	sub 112–5	...	...	i	δ[h]	δ[h]	i	δ	chl δ, s[h] CS_2, aa i	**B1**[2], 671
m173	**Metaldehyde II**	**Metacetaldehyde** (tetramer).	176.22	...	47	710^{760} δd 65^{15}	...	...	δ	s	s	s	s	os s	**B1**[3], 2640
m174	**Metameconin**	5,6-Dimethoxyphthalide.	194.19	cr (dil al) λ^{al} 220 (4.38), 258 (3.97), 294 (3.86)	155–7	...	...	...	i s[h]	...	δ	...	...	KOH v	**B18**[2], 62
—	**Metanilic acid**	*see* **Benzenesulfonic acid, 3-amino-***													
—	**Methacrylic acid**	*see* **Propenoic acid, 2-methyl-***													
—	**Methadone**	*see* **3-Heptanone, 6-dimethylamino-4,4-diphenyl-***													
—	**Methanal***	*see* **Formaldehyde**													
Ω m175	**Methane***	Marsh gas. CH_4	16.04	gas	−182.48	-164^{760}	0.466^{-164} 0.5547^{0}	...	s	s	s	δ	s	MeOH, to s	**B1**[3], 1
m176	**—,amino-***	Methylamine. CH_3NH_2	31.06	gas λ^{gas} 190.5 (3.51), 215 (2.77)	−93.5	-6.3^{760}	0.699^{-11} 0.6628^{20}	...	v	s	∞	s	s	...	**B4**[2], 546
Ω m177	**—,—,hydrochloride**	$CH_3NH_2.HCl$	67.52	dlq tetr ta (al)	227–8	sub^{760} 225–30^{15}	...	...	s	s	i	i	...	AcOEt, chl i	**B4**[3], 90
Ω m178	**—,amino-(diphenyl)-***	Benzhydrylamine. $(C_6H_5)_2CHNH_2$	183.26	hex pl	34	304^{763} (288) 176^{23}	1.0635_0^{20}	1.5963^{22}	δ	...	...	s	...	...	**B12**[2], 768
—	**—,amino(2-furyl)-**	*see* **Furan, 2(aminomethyl)-**													
m180	**—,aminooxy-,** hydrochloride.	Methoxylamine hydrochloride. *O*-Methylhydroxylamine hydrochloride. $CH_3ONH_2.HCl$	83.52	pr	149	...	...	...	s	s	...	...	...	...	**B1**[2], 275
m181	**—,(2-amino-phenyl)-bis(4-aminophenyl)-***	$C_{19}H_{19}N_3$. *See* m312	289.38	cr (al)	165	...	...	...	s[h]	v	i	...	...	lig i	**B13**, 311
m182	**—,(3-amino-phenyl)-bis(4-aminophenyl)-***	$C_{19}H_{19}N_3$. *See* m312	289.38	nd (eth or eth-lig)	150	...	...	...	i	s	δ	...	...	lig i	**B13**, 312
m183	**—,(3-amino-phenyl)-diphenyl***	3-Aminotritan. $C_{19}H_{17}N$. *See* m312	259.36	nd (eth)	120	...	...	...	i	s	s	...	...	...	**B12**[2], 790

For explanations, symbols and abbreviations see beginning of table. For structural formulas see end of table.

No.	Name	Synonyms and Formula	Mol. wt.	Color, crystalline form, specific rotation and λ_{max} (log ε)	m.p. °C	b.p. °C	Density	n_D	Solubility w	al	eth	ace	bz	other solvents	Ref.
	Methane														
m184	—,(4-amino-phenyl)-diphenyl-*	4-Aminotritan. *p*-Benzhydrylaniline. $C_{19}H_{17}N$. *See* m312	259.36	pr or lf (eth, lig), ta (path), pr (bz + 1)	84.5	*ca.* 248^{12}	....	....	i	...	s	...	s	lig s	**B12**^{2}, 790
m185	—,(3-amino-phenyl)phenyl-*	3-Aminoditan. *m*-Benzyl-aniline. $C_{13}H_{13}N$. *See* m271	183.26	cr (lig)	46	....	....	....	...	...	...	...	...	lig s	**B12**, 1323
m186	—,(4-amino-phenyl)phenyl-*	4-Aminoditan. *p*-Benzyl-aniline. $C_{13}H_{13}N$. *See* m271	183.26	mcl (lig)	34–5	300	1.038^{55}	....	i	v	v	...	...	lig s	**B12**, 1323
—	—,biphenylyl-(phenyl)	*see* **Biphenyl, benzyl-**													
m187	—,bis(4-amino-cyclohexyl)-(*cis, cis*)	4,4′-Methylenebiscyclo-hexylamine.	210.37	....	60.5–1.9	141^2	....	1.5014^{27} (mixture of isomers)	i	...	...	...	...	....	**Am 73**, 741
m188	—,—(*cis, trans*)....	$C_{13}H_{26}N_2$. *See* m187	210.37	cr	36–7	$127–8^{1.2}$	0.9608^{25}_4	1.5046^{25}	i	...	...	...	...	....	**Am 73**, 741
m189	—,—(*trans, trans*)..	$C_{13}H_{26}N_2$. *See* m187	210.37	cr (peth)	64–5.4	$130–1^{0.8}$	....	1.5032^{25}	i	...	...	...	...	....	**Am 73**, 741
m190	—,bis(4-amino-2-nitrophenyl)-	$C_{13}H_{12}N_4O_4$. *See* m271.....	288.27	og pl (al)	205	....	....	....	...	v^h	δ	...	δ	aa v lig δ	**B13**^{2}, 113
m191	—,bis(4-amino-3-nitrophenyl)-	$C_{13}H_{12}N_4O_4$. *See* m271.....	288.27	red nd (PhOH-al)	232–3	....	....	....	...	i	...	...	i	$PhNO_2$ s^h chl i	**B13**^{2}, 113
Ω m192	—,bis(4-amino-phenyl)-*	4,4′-Diaminoditan. $C_{13}H_{14}N_2$. *See* m271	198.27	pl or nd (w), pl (bz)	92–3	$398–9^{768}$ 257^{18}	....	....	δ	v	v	...	v	....	**B13**^{2}, 111
m193	—,bis(4-amino-phenyl)phenyl-*	4,4′-Diaminotritan. $C_{19}H_{18}N_2$. *See* m312	274.37	pr (bz or eth)	139–40	....	....	....	$δ^h$	v	v	...	...	chl, lig s	**B13**^{2}, 134
Ω m194	—,bis(3-carboxy-4-hydroxy-phenyl)-*	5,5′-Methylenedisalicylic acid.	288.26	nd (bz)	243–4	....	....	....	δ	s	s	s	δ	....	**B10**^{2}, 397
m195	—,bis(5-chloro-2-hydroxyphenyl)-*	Dichlorophene. Dihydroxane $C_{13}H_{10}Cl_2O_2$. *See* m271	269.13	cr (bz, to, path)	177–8	....	....	....	i	s	...	s	...	....	**B6**^{3}, 5406
m196	—,bis(4-chloro-phenoxy)-*	Neotran. Oxythane	269.13	....	69.7–70.2	$189–94^6$	....	....	i	δ	v	v	s	....	**Am 73**, 1872
m197	—,bis(4-dimethyl-aminophenyl)-*	$C_{17}H_{22}N_2$. *See* m271	254.38	pl or ta (al, lig)	91–2	390d $182–5^3$	....	....	i	δ s^h	v	...	v	CS_2 v ac s	**B13**^{2}, 111
m198	—,bis(4-dimethyl-aminophenyl)(4-aminophenyl)-*	*N,N,N′,N′*-Tetramethyl-*p*-leucaniline. $C_{23}H_{27}N_3$. *See* m312	345.49	(al)	151–2	....	....	....	...	δ	...	...	...	....	**B13**, 314
m198[1]	—,[2,4-bis(di-methylamino)phenyl]diphenyl-*	$C_{23}H_{26}N_2$. *See* m312	330.48	pl (peth)	122–3	....	....	....	i	δ	s	s	s	os s	**B13**, 273
m199	—,bis(4-dimethyl-aminophenyl)(2-hydroxy-phenyl)-*	$C_{23}H_{26}N_2O$. *See* m312......	346.48	nd (al)	127–8	....	....	....	δ	δ v^h	...	...	v	lig δ	**B13**^{2}, 440
m200	—,bis(4-dimethyl-aminophenyl)-(3-hydroxy-phenyl)-*	$C_{23}H_{26}N_2O$. *See* m312......	346.48	cr (al)	149	....	....	....	δ	δ	...	...	v	....	**B13**^{2}, 440
m201	—,bis(4-dimethyl-aminophenyl)-(4-hydroxy-phenyl)-*	$C_{23}H_{26}N_2O$. *See* m312......	346.48	cr (al)	165	....	....	....	δ	v^h	...	...	v	alk s lig δ	**B13**^{2}, 440
m202	—,bis(4-dimethyl-aminophenyl)-phenyl-	Leucomalachite green. $C_{23}H_{26}N_2$. *See* m312	330.48	nd or lf (al, bz)	102 (94)	....	....	....	i	δ	v	...	v	to v lig δ	**B13**^{2}, 135
Ω m203	—,bis(2,4-dinitro-phenyl)-*	2,2′,4,4′-Tetranitroditan. $C_{13}H_8N_4O_8$. *See* m271	348.24	ye pr (aa)	181	....	....	....	...	i	i	...	δ	alk s aa δ	**B5**^{2}, 503
m204	—,bis(3-formyl-4-hydroxy-phenyl)-	5,5′-Methylene-disalicylaldehyde. $C_{15}H_{12}O_4$. *See* m271	256.26	pa ye (aa or ace)	142–3	....	....	....	...	...	...	...	...	aa s^h	**B8**, 436
m205	—,bis(3-hydroxy-phenyl)-*	*m,m′*-Methylenediphenol. $C_{13}H_{12}O_2$. *See* m271	200.24	nd (dil aa)	102–3	$230–40^3$	....	....	s^h	v	v	...	δ	aa v lig δ	**B6**, 995
m206	—,bis(4-hydroxy-phenyl)-*	*p,p′*-Methylenediphenol. $C_{13}H_{12}O_2$. *See* m271	200.24	lf or nd (w) λ^{al} 225 (4.7), 280 (3.6)	162–3	sub	....	....	...	s	s	...	...	chl, alk s CS_2 i	**B6**^{2}, 964
m207	—,bis(2-hydroxy-3,5,6-trichloro-phenyl)-*	Gamophen. Hexa-chlorophen. Hexosan. $C_{13}H_6Cl_6O_2$. *See* m271	406.91	nd (bz, to)	166–7	....	....	....	i	s	s	s	...	dil alk, chl s	**B6**^{3}, 5407
m209	—,bis(2-nitro-phenyl)-*	2,2′-Dinitroditan. $C_{13}H_{10}N_2O_4$. *See* m271	258.24	cr	83.5	....	....	....	...	s	s	...	...	lig i	**B5**^{3}, 1796
m210	—,bis(3-nitro-phenyl)-*	3,3′-Dinitroditan. $C_{13}H_{10}N_2O_4$. *See* m271	258.24	lf (aa)	175.5	....	....	....	...	s	δ	...	v^h	aa v^h	**B5**^{2}, 503
m211	—,bis(4-nitro-phenyl)-*	4,4′-Dinitroditan. $C_{13}H_{10}N_2O_4$. *See* m271	258.24	nd (bz, peth, aa)	188 (184)	....	....	....	...	i	δ	...	v^h	aa v^h	**B5**, 595
—	—,bis(phenyl-imino)-*	*see* **Carbodiimide, *N,N′*-diphenyl-**													
m212	—,bis(trichloro-silyl)-	$Cl_3SiCH_2SiCl_3$	282.92	....	....	182.7^{745} 64^{10}	1.5567^{20}_4	1.4740^{20}	...	s	v	...	...	....	**B1**^{3}, 2587
Ω m213	—,bromo-*	Methyl bromide. CH_3Br	94.94	λ^{gas} 204 (2.26)	−93.6	3.56^{760}	1.6755^{20}_4	1.4218^{20}	δ	∞	∞	...	...	chl, CS_2 ∞	**B1**^{3}, 79

For explanations, symbols and abbreviations see beginning of table. For structural formulas see end of table.

No.	Name	Synonyms and Formula	Mol. wt.	Color, crystalline form, specific rotation and λ_{max} (log ε)	m.p. °C	b.p. °C	Density	n_D	Solubility						Ref.
									w	al	eth	ace	bz	other solvents	
	Methane														
Ω m214	**—,bromochloro-***	$ClCH_2Br$	129.39	...	−86.5	68.11^{760}	1.9344$^{20}_{4}$	1.4838^{20}	i	s	s	s	s	os s	**B1**^{3}, 83
m215	**—,bromochloro-dinitro-***	$BrCCl(NO_2)_2$	219.38	...	9.2–9.3	75–6^{15}	2.0394$^{20}_{4}$	1.4793	...	v	...	...	...	...	**B1**^{3}, 115
m216	**—,bromochloro-fluoro-***	$BrCHClF$	147.38	...	−115	36.1^{756}	1.9771$^{0}_{4}$	1.4144^{25}	i	...	s	s	...	chl s	**B1**^{3}, 84
Ω m217	**—,bromodi-chloro-***	$BrCHCl_2$	163.83	...	−57.1	90^{760}	1.980$^{20}_{4}$	1.4964^{20}	i	v	v	v	v	chl ∞ os v	**B1**^{3}, 84
m218	**—,bromodi-fluoro-***	$BrCHF_2$	130.93	λ^{gas} <208	−14.5	...	1.55^{16}	...	s	v	...	...	...	...	**B1**^{3}, 83
m219	**—,bromodi-fluoronitroso-***	$BrCF_2(NO)$	159.92	bl gas	...	*ca.* −12^{760}	...	...	...	...	...	...	...	...	**J1953,** 2075
m220	**—,bromodiiodo-***	$BrCHI_2$	346.74	ye (peth)	60	110^{25}	...	...	i	...	...	...	...	peth δ	**B1**, 72
Ω m221	**—,bromo-(diphenyl)-***	Benzhydryl bromide. α-Bromoditan. $(C_6H_5)_2CHBr$	247.14	tcl (peth)	45	193^{26}δd 111$^{0.3}$	...	...	d^h	s	...	...	v		**B5**^{2}, 502
m222	**—,bromofluoro-***	$BrCH_2F$	112.94	...	...	18–20	...	...	s	...	...	...	...	chl v	**B1**^{3}, 83
m223	**—,bromoiodo-***	$BrCH_2I$	220.84	...	...	138–41^{760}	2.926^{17}	1.6410^{20}	...	...	...	...	...	chl v	**B1**^{3}, 99
m224	**—,bromonitro-***	$BrCH_2NO_2$	139.94	...	...	148–9^{742} 70–2^{45}	...	1.4880^{20}	i	v	...	...	...	alk s	**B1**^{3}, 115
Ω m225	**—,bromotri-chloro-***	$BrCCl_3$	198.28	λ^{iso} 240 (2.3)	−5.65	104.7^{760} 0.6^{10}	2.0122$^{20}_{4}$	1.5063^{20}	i	∞	∞	...	...	...	**B1**^{3}, 85
Ω m226	**—,bromotri-fluoro-***	$BrCF_3$	148.92	gas λ^{gas} 208 (1.52)	...	−59^{740}	...	...	...	...	...	...	...	chl v	**B1**^{3}, 83
m227	**—,bromotri-nitro-***	$BrC(NO_2)_3$	229.94	...	17–8	56^{10}	2.0313$^{20}_{4}$	1.4808^{20}	δ	s	...	...	...	chl s	**B1**^{3}, 116
Ω m228	**—,chloro-***	Methyl chloride. CH_3Cl	50.49	gas	−97.73	−24.2^{760}	0.9159$^{20}_{4}$	1.3661^{-10} 1.3389^{20}	s	s	∞	∞	∞	chl, aa ∞	**B1**^{3}, 36
m229	**—,chlorodi-bromo-***	$ClCHBr_2$	208.29	...	...	119–20^{748}	2.451$^{20}_{4}$	1.5482^{20}	i	s	s	s	s	os s	**B1**^{3}, 87
Ω m230	**—,chlorodi-fluoro-***	Freon 22. $ClCHF_2$	86.47	gas	−146–7 (−160)	−40.8^{760}	1.4909^{-69}	...	v	...	s	s	...	chl s	**B1**^{3}, 41
m231	**—,chlorodifluoro-nitro-***	$ClCF_2(NO_2)$	131.47	...	...	25^{760}	...	...	...	...	...	...	...	chl s	**J1953,** 2075
m232	**—,chlorodifluoro-nitroso-***	$ClCF_2(NO)$	115.47	bl gas	...	*ca.* −35^{760}	...	...	...	...	...	...	...	...	**J1953,** 2075
m233	**—,chlorodiiodo-***	$ClCHI_2$	302.28	...	−4	200d^{760} 88^{30}	...	...	...	...	s	s	...	chl v	**B1**, 72
m234	**—,chlorodinitro-***	$ClCH(NO_2)_2$	140.48	...	...	34–6^{13}	1.6123^{20}	1.4575^{20}	...	...	...	...	...	alk s	**B1**^{3}, 115
Ω m235	**—,chloro-(diphenyl)-***	Benzhydryl chloride. α-Chloroditan. $(C_6H_5)_2CHCl$	202.69	nd	20.5 (18)	173^{19} 141^{4}	1.1398$^{20}_{4}$	1.5959^{20}	...	...	...	...	...	...	**B5**^{2}, 500
m237	**—,chlorofluoro-***	$ClCH_2F$	68.48	gas	...	−9.1	...	...	...	...	...	...	...	chl v	**B1**^{3}, 41
m238	**—,chloroiodo-***	$ClCH_2I$	176.39	...	...	109^{760}	2.422$^{20}_{4}$	1.5822^{20}	i	s	s	s	s	chl v liq HF δ	**B1**^{2}, 99
m239	**—,chloronitro-***	$ClCH_2NO_2$	95.49	...	...	122–3	1.466^{15}	...	s	...	...	...	...	alk v	**B1**^{3}, 112
m239[1]	**—,(4-chloro-phenyl)phenyl-**	4-Chloroditan. $C_{13}H_{11}Cl$. *See* m271	202.69	...	7.5	298^{742} 147–8^{8}	1.1247$^{20}_{0}$	...	i	...	...	s	...	...	**B5**^{2}, 500
m240	**—,chlorotri-bromo-***	$ClCBr_3$	287.19	lf (eth)	55	158–9	2.71^{15}	...	i	...	s	...	...	...	**B1**^{3}, 91
Ω m241	**—,chlorotri-fluoro-***	Freon −13. $ClCF_3$	104.46	gas	−181	−81.1^{760}	...	...	...	...	...	...	...	...	**B1**^{3}, 42
m242	**—,chlorotri-nitro-***	$ClC(NO_2)_3$	185.48	...	4.5	133–5δd 56^{40}	1.6769$^{20}_{4}$	1.4500^{20}	δ	s	s	...	...	chl s	**B1**^{3}, 116
Ω m243	**—,chloro(tri-phenyl)-***	Trityl chloride. $(C_6H_5)_3CCl$	278.79	nd or pr (bz-peth) $\lambda^{98\%\,sulf}$ 430 (4.58)	113–4	310^{760} 230–5^{20}	...	...	i d^h	δ d^h	v	s	v	chl, CCl_4, CS_2 v	**B5**^{2}, 615
—	**—,cyano-**	*see* **Acetic acid,** nitrile													
m243[1]	**—,deuterotri-chloro-***	Chloroform-*d*. $CDCl_3$	120.39	...	−64.12	61.3–1.5	1.5004$^{20}_{4}$	1.4450^{20}	i	...	...	...	...	...	**B1**^{3}, 63
m244	**—,diazo-***	Azomethylene. CH_2N_2	42.04	ye gas λ^{gas} 380 (0.5), 408 (0.5), 435 (0.5)	−145	*ca.* 0	...	...	d	s^h	s	...	...	...	**B1**^{2}, 650
m245	**—,diazo-(diphenyl)-***	$(C_6H_5)_2CN_2$	194.24	bl-red nd (peth)	30–2	exp	...	...	...	s	v	v	v	os v	**B7**^{2}, 358
—	**—,dibenzoyl-**	*see* **1,3-Propanedione, 1,3-diphenyl-***													
Ω m246	**—,dibromo-***	Methylene bromide. CH_2Br_2	173.85	...	−52.55	97^{760}	2.4970$^{20}_{4}$	1.5420^{20}	δ	∞	∞	∞	...	...	**B1**^{3}, 85
Ω m247	**—,dibromodi-chloro-***	Br_2CCl_2	242.74	nd	38 (22)	150.2	2.42$^{25}_{0}$	...	i	s	s	s	s	os s	**B1**^{3}, 88
Ω m248	**—,dibromodi-fluoro-***	F_2CBr_2	209.83	λ^{gas} 227 (2.86)	...	24.5^{760}	...	...	s	...	s		s	...	**B1**^{1}, 16
m249	**—,dibromo-dinitro-***	$Br_2(CNO_2)_2$	263.84	nd	5.5	158d 77^{21}	2.4440$^{20}_{4}$	1.5280^{20}	i	∞	...	...	...	...	**B1**^{3}, 115
m250	**—,dibromo-fluoro-***	$FCHBr_2$	191.84	...	...	64.9^{757}	2.421$^{20}_{4}$	1.4685^{20}	i	s	s	s	s	chl s	**B1**^{3}, 87
m251	**—,dibromoiodo-***	$ICHBr_2$	299.74	pl (peth)	22.5	91^{42}	...	...	...	...	...	...	...	peth δ	**B1**^{3}, 99
Ω m252	**—,dichloro-***	Methylene chloride. CH_2Cl_2	84.93	λ^{vap} <200	−95.1	40^{760}	1.3266$^{20}_{4}$	1.4242^{20}	δ	∞	∞	...	...	...	**B1**^{2}, 13
Ω m253	**—,dichlorodi-fluoro-***	Freon 12. Cl_2CF_2	120.91	λ^{vap} <200	−158	−29.8^{760}	1.75^{-115} 1.1834^{57}	...	s	s	s	...	...	aa s	**B1**, 61

For explanations, symbols and abbreviations see beginning of table. For structural formulas see end of table.

No.	Name	Synonyms and Formula	Mol. wt.	Color, crystalline form, specific rotation and λ_{max} (log ε)	m.p. °C	b.p. °C	Density	n_D	Solubility w	al	eth	ace	bz	other solvents	Ref.
	Methane														
m254	**—,dichloro-dinitro-***	$Cl_2C(NO_2)_2$	174.93	...	...	121–2.5 46^{20}	1.6124_4^{20}	1.4575^{20}	i	s	s	...	s	chl, CS_2 s	**B1**^{3}, 115
Ω m255	**—,dichloro-(diphenyl)-***	Benzophenone dichloride. $(C_6H_5)_2CCl_2$	237.13	...	...	305d 190^{21}	1.235^{18}	...	...	...	s	...	s	...	**B5**^{2}, 501
Ω m256	**—,dichloro-fluoro-***	Freon 21. $FCHCl_2$	102.92	...	−135	9^{760}	1.405^9	1.3724^9	i	s	s	...	...	aa, chl s	**B1**^{3}, 47
m257	**—,dichloroiodo-***	$ICHCl_2$	210.83	...	...	132^{760} 40^{30}	2.392_4^{20}	1.5840^{20}	i	s	s	s	s	chl v	**B1**^{3}, 99
m258	**—,dichloronitro-***	Cl_2CHNO_2	129.93	...	...	107^{760}	...	...	...	...	...	...	...	...	**B1**^{3}, 113
m259	**—,difluoro-***	Methylene fluoride. CH_2F_2	52.02	...	...	-51.6^{760}	0.909^{20}	1.190^{20}	i	s	...	...	...	...	**B1**, 59
m260	**—,(difluoro-amino)trifluoro-**	Perfluoromethylamine. F_3CNF_2	121.01	...	...	-75^{760}	...	...	...	...	...	...	...	...	**J1950**, 1966
m261	**—,difluoroiodo-***	$ICHF_2$	177.92	...	−122	21.6^{760}	3.238^{-19}	...	...	...	...	...	...	...	**B1**^{3}, 98
m262	**—,di(2-furyl)-**	2-Furfurylfuran	148.17	...	...	$94^{22.5}$ 75^{13}	1.102_4^{20}	1.5049^{20}	i	s	s	s	...	MeOH s	**Am 55**, 3302
Ω m263	**—,diiodo-***	Methylene iodide. CH_2I_2	267.84	ye nd or lf	6.1	182^{760} 60^{10}	3.3254_4^{20}	1.7425^{20}	δ	s	s	...	s	chl s	**B1**^{2}, 99
m264	**—,diiodofluoro-***	$FCHI_2$	285.83	pa ye	−34.5	$100–1^{760}$ 50^{50}	3.1969^{22}	...	δ	s	s	...	...	...	**B1**^{3}, 102
m265	**—,[3(dimethyl-amino)phenyl] [4(dimethyl-amino)-phenyl]phenyl-***	$C_{23}H_{26}N_2$. *See* m312	330.48	cr (abs al)	83–4	...	...	...	...	δ^h	δ^h	...	...	...	**B13**, 274
m267	**—,di(1-naphthyl)-**	$C_{10}H_7^{\alpha}CH_2C_{10}H_7^{\alpha}$	268.38	pr or nd (al) $\lambda^{liq\,NH_3-KNH_2}$ 645 (4.9)	109	>360 270^{14}	...	...	δ s^h	δ s^h	s	...	s	chl s peth δ	**B5**^{2}, 650
m268	**—,di(2-naphthyl)-**	$C_{10}H_7^{\beta}CH_2C_{10}H_7^{\beta}$	268.38	nd (al or eth)	93	...	...	...	i	δ	...	...	s	...	**B5**^{1}, 360
m269	**—,dinitro-***	$CH_2(NO_2)_2$	106.04	ye nd λ^w 360 (4.22)	< −15	100 exp	...	...	i δ^h	s	s	...	...	...	**B1**^{3}, 115
m270	**—,dinitro-(diphenyl)-***	$(C_6H_5)_2C(NO_2)_2$	258.24	pl (dil al)	79–80	...	...	...	i	v^h	v	...	v	chl v	**B5**, 596
Ω m271	**—,diphenyl-***	Ditan	168.24	pr nd λ^{al} 260 (2.70), 268.5 (2.61)	25.35	264.3^{760} 125.5^{10}	1.0060_4^{20}	1.5753^{20}	i	s	s	...	...	chl s	**B5**^{2}, 498
m272	**—,diphenyl(3-tolyl)-**	3-Methyltritan	258.37	pr (al or MeOH)	62	354^{706}	1.07^{16}	...	i	δ	v	...	v	chl, aa v lig s	**B5**, 710
m273	**—,di(α-pyrryl)-**	2,2′-Methylenedipyrrole. 2,2′-Pyrromethane.	146.19	lf or nd (al)	73 (66)	$163–7^{12}$	...	...	i	s	s	...	s	lig s^h	**B23**, 167
m274	**—,disilano-**	2-Carbatrisilane. Disilylmethane. $CH_2(SiH_3)_2$	76.25	...	...	14.7^{754}	0.6979_4^4	1.4115^4	...	...	...	...	...	...	**C48**, 5080
m275	**—,di(4-tolyl)-phenyl-***	4,4′-Dimethyltritan	272.40	nd (MeOH) λ^{al} 270 (4.0), 358 (4.5)	56	$218–20^{12}$	...	...	i	s	v	...	v	chl, CS_2 v aa, lig s	**B5**^{1}, 352
m276	**—,fluoro-***	Methyl fluoride. CH_3F	34.03	...	−141.8	-78.4^{760}	0.8428^{-60} 0.5786_4^{20}	1.1727^{20}	v	v	v	...	s	chl s	**B1**^{3}, 33
m277	**—,fluoroiodo-***	FCH_2I	159.93	...	...	53.4	2.366_4^{20}	1.4911^{20}	i	...	s	s	s	chl v	**B1**^{3}, 98
m278	**—,fluorotri-bromo-***	$FCBr_3$	270.74	...	...	106^{760}	2.7648_4^{20}	1.5256^{20}	i	...	s	s	s	chl v	**B1**^{3}, 91
m279	**—,hydroxyl-amino-***	*N*-Methylhydroxylamine. CH_3NHOH	47.06	hyg nd	42 (rapid htng) (87–8)	62.5^{15}	1.0003_4^{20}	1.4164^{20}	v	v	δ	...	δ	MeOH v lig δ	**B4**^{2}, 952
m280	**—,(2-hydroxy-phenyl) (4-hydroxy-phenyl)-***	2,4′-Dihydroxyditan. $C_{13}H_{12}O_2$. *See* m271	200.24	nd (dil al, bz or w) λ^{al} 277.5 (3.65)	119–20	...	...	...	δ	v	v	...	...	...	**B6**, 994
Ω m281	**—,(4-hydroxy-phenyl)-phenyl-***	4-Benzylphenol. p-Hydroxy-ditan. $C_{13}H_{12}O$. *See* m271	184.24	nd or pl (al)	84	325–30 198–200^{10}	...	...	δ^h	s v^h	s	...	s	chl, aa, alk s	**B6**^{2}, 629
Ω m282	**—,iodo-***	Methyl iodide. CH_3I	141.94	...	−66.45	42.4^{760}	2.279_4^{20}	1.5380^{20}	δ	∞	∞	s	s	...	**B1**^{2}, 35
m283	**—,iodotrichloro-***	$ICCl_3$	245.27	...	...	142^{760}	2.355_4^{20}	1.5854^{20}	i	...	s	s	s	chl v	**B1**^{3}, 99
m284	**—,(1-naphthyl) phenyl-***	α-Benzylnaphthalene. $C_{10}H_7^{\alpha}CH_2C_6H_5$	218.30	mcl lf or ta (al) λ^{al} 273 (3.8), 283 (3.9), 290 (3.75), 312 (2.6)	59	350^{760} 220^{20}	1.166^{17}	...	i	δ	v	...	s	CS_2 v chl s	**B5**^{2}, 604
m285	**—,(2-naphthyl) phenyl-***	β-Benzylnaphthalene. $C_{10}H_7^{\beta}CH_2C_6H_5$	218.30	mcl pr (al or MeOH)	58	350^{760}	1.176^0	...	i	δ v^h	v	...	v	chl s	**B5**, 690
Ω m286	**—,nitro-***	CH_3NO_2	61.04	mcl λ^{al} 260 (1.59)	fr −17 (−28.5)	100.8^{760}	1.1371_4^{20}	1.3817^{20}	s	s	s	s	...	alk s	**B1**^{2}, 40
m287	**—,(2-nitro-phenyl) (4-nitro-phenyl)-***	2,4′-Dinitroditan. $C_{13}H_{10}N_2O_4$. *See* m271	258.24	ye mcl pr (bz)	118	...	...	...	i	...	δ	...	δ s^h	...	**B5**, 595
m288	**—,(3-nitro-phenyl) (4-nitro-phenyl)-***	3,4′-Dinitroditan. $C_{13}H_{10}N_2O_4$. *See* m271	258.24	nd (al)	103–4	...	...	...	i	δ s^h	...	...	...	...	**B5**, 595
m289	**—,nitrosotri-fluoro-***	F_3CNO	99.01	λ^{chl} 680 (1.3), 700 (1.3)	−197 (−205)	-84^{760} (-93^{760})	...	...	...	...	...	...	...	...	**B1**^{3}, 105

For explanations, symbols and abbreviations see beginning of table. For structural formulas see end of table.

No.	Name	Synonyms and Formula	Mol. wt.	Color, crystalline form, specific rotation and λ_{max} (log ε)	m.p. °C	b.p. °C	Density	n_D	Solubility w	al	eth	ace	bz	other solvents	Ref.
	Methane														
m290	**—,nitrotribromo-***	Bromopicrin. Nitrobromoform. Br_3CNO_2	297.74	pr λ^{peth} (not max) 240 (3.28), 260 (3.00), 300 (1.90)	10.25	exp^{760} 89–90^{20}	2.7930_4^{20}	1.5790^{20}	i	s	s	v	v	aa v	**B1**^{3}, 115
m291	**—,nitrotrichloro-***	Chloropicrin. Nitrochloroform. Cl_3CNO_2	164.38	λ^{al} 276.5 (1.79)	−64.5	111.84^{760} -8.9^{10}	1.6566_4^{20}	1.4622^{20}	s^h	∞	...	∞	∞	MeOH, aa ∞	**B1**^{3}, 113
m291[1]	**—,nitrotrifluoro-***	Fluoropicrin. F_3CNO_2	115.01	λ^{gas} 277.5 (1.05)	....	−31.1 (−20)	....	....	...	...	...	...	...	....	**J1953**, 2075
m292	**—,(pentafluorothio)trifluoro-**	$F_3C(SF_5)$	196.06	....	....	-20^{760}	....	....	i	i	...	...	...	CS_2 s	**J1953**, 2372
Ω m293	**—,phenyl(3-tolyl)-**	*m*-Benzyltoluene. 3-Methylditan.	182.27	....	(i) −27.83 (ii) −34.46	279.24^{760} $120^{0.2}$	0.99135_4^{20}	1.5712^{20}	i	s	s	...	s	aa, chl s	**B5**, 607
Ω m294	**—,phenyl(4-tolyl)-**	*p*-Benzyltoluene. 4-Methylditan.	182.27	λ^{iso} 260.5 (2.77), 266 (2.80), 274.5 (2.75)	−30	286^{760} 114–5^3	0.9976_4^{20}	1.5712^{20}	i	s	s	...	s	chl, aa s	**B5**^{2}, 511
Ω m295	**—,tetrabromo-***	Carbon tetrabromide. CBr_4	331.65	mcl ta (dil al) λ^{Bu^tOH} 228	α 90–1 β 94.3	α 189–90^{760} 102^{50}	α 2.9609_4^{100} 3.273^{18}	α 1.59419_α^{100}	i	s	s	...	...	CS_2 v chl s liq HF δ	**B1**^{3}, 92
Ω m296	**—,tetrachloro-***	Carbon tetrachloride. CCl_4	153.82	λ^{vap} <200	−22.99	76.54^{760}	1.5940_4^{20}	1.4601^{20}	i	s	∞	s	∞	chl ∞	**B1**^{3}, 65
Ω m297	**—,tetrafluoro-***	Carbon tetrafluoride. CF_4	88.01	....	−150	-129^{754}	3.034^0	....	δ	...	...	...	s	chl s	**B1**^{3}, 35
m298	**—,tetraiodo-***	Carbon tetraiodide. CI_4	519.63	red lf (bz or chl)	171d	130–40^{1-2} sub	4.23^{20}	....	i	i	...	...	...	chl, CCl_4, Py s	**B1**^{3}, 104
	—,tetrakis(hydroxymethyl)-*	*see* **Pentaerythritol**													
Ω m299	**—,tetranitro-***	$C(NO_2)_4$	196.03	λ^{cy} <220	14.2	126 21–3^{22}	1.6380_4^{20}	1.4384^{20}	i	s	s	...	...	....	**B1**^{2}, 47
Ω m300	**—,tetraphenyl-***	$C(C_6H_5)_4$	320.44	rh nd (bz, sub, Ac_2O) λ^{cy} 262 (3.11), 272.5 (2.95)	285 (282)	431^{760} (sub)	....	....	i	i	i	...	s^h	to s^h aa i, s^h lig i	**B5**^{2}, 672
m301	**—,tribenzoyl-** (enol form)	$(C_6H_5CO)_3CH$	328.37	....	155 (in Jena, 240–5 (in soft glass)	....	....	....	...	...	...	...	...	chl v	**B7**^{1}, 485
m302	**—,—,(keto form)**	$(C_6H_5CO)_3CH$	328.37	nd (al or ace)	231 245–50 (in soft glass)	sub	....	....	i	i δ^h	...	δ	δ	chl δ	**B7**^{2}, 842
Ω m303	**—,tribromo-***	Bromoform. $CHBr_3$	252.75	hex sc λ^{Bu^tOH} 224	8.3	149.5^{760} 46^{15}	2.8899_4^{20}	1.5976^{20}	δ	∞	∞	...	s	chl, lig s	**B1**^{3}, 88
Ω m306	**—,trichloro-***	Chloroform. $CHCl_3$	119.38	λ^{vap} <200	−63.5	61.7^{760}	1.4832_4^{20}	1.4459^{20}	δ	∞	∞	s	∞	lig ∞	**B1**^{2}, 14
—	**—,—,*d*.**	*see* **Methane, deuterotrichloro-***													
Ω m308	**—,trifluoro-***	Fluoroform. CHF_3	70.01	....	−160	-82.2^{760}	1.52^{-100}	....	s	v	...	s	s	chl δ	**B1**^{3}, 34
Ω m310	**—,triiodo-***	Iodoform. CHI_3	393.73	ye hex pr or nd (ace) λ^{chl} 244 (3.0), 307 (3.2), 347 (3.2)	123	*ca.* 218^{760}	4.008_4^{20}	....	i	s^h	s	s	i	chl, CS_2, aa s	**B1**^{3}, 102
m310[1]	—, trimethyl	*see* Propane, 2-methyl													
m311	**—,trinitro-***	Nitroform. $CH(NO_2)_3$	151.04		19 (15)	exp^{760} 45–7^{22}	1.479_4^{20}	1.44511_{He}^{24}	s	...	...	s	...	alk s	**B1**^{3}, 116
Ω m312	**—,triphenyl-***	Tritan.	244.34	rh (al) λ^{cy} 256 (2.86), 262.5 (2.92), 269.5 (2.81)	(i) 94 (st) (ii) 81 (unst)	358–9^{754} 190–215^{10}	1.014_4^{99}	1.5839^{99}	i	δ v^h	v	...	s v^h	chl, CS_2, Py v	**B5**^{2}, 613
m313	**—,tris(4-aminophenyl)-***	*p*-Leucaniline. $C_{19}H_{19}N_3$. *See* m312	289.38	lf (w, al or bz)	208	....	....	....	i	s	s	...	...	....	**B13**^{2}, 150
m314	**—,tris(4-dimethylaminophenyl)-***	Leucocrystal violet. $C_{25}H_{31}N_3$. *See* m312	373.55	lf (al), nd (bz or lig)	175	....	....	....	i	δ v^h	v	...	v	chl, aa v	**B13**^{2}, 150
	—,tris(ethylthio)-*	*see* **Orthoformic acid, trithio-**, triethyl ester													
m315	**—,tris(4-hydroxyphenyl)-***	Leucoaurin. $C_{19}H_{16}O_3$. *See* m312	292.34	pr (aa or dil al)	240 (235)	....	....	....	δ	v	s	...	...	aa v	**B6**^{2}, 1106
m316	**—,tris(4-nitrophenyl)-***	$C_{19}H_{13}N_3O_6$. *See* m312	379.33	sc (bz) λ^{hp} 382 (3.2)	212.5 (207)	....	....	....	...	...	δ	...	δ	aa δ	**B5**^{2}, 618
m317	**—,tri(2-tolyl)-**		286.42	nd (al)	130.5–1.5	....	....	....	...	s	s	...	...	MeOH s	**B5**^{3}, 2347
m319	**Methanearsonic acid***	Methylarsonic acid. $CH_3AsO(OH)_2$	139.97	lf (al)	160–1	....	....	....	s	s	...	...	...	....	**B4**^{2}, 996
m320	**Methanedisulfonic acid***, dihydrate	Methionic acid. $CH_2(SO_3H)_2.2H_2O$	212.18	hyg nd (w+2)	....	220–70^{15-20}d	....	....	s	s^h	...	...	...	....	**B1**^{2}, 644
m323	**Methanephosphonic acid***	Methylphosphonic acid. $CH_3PO(OH)_2$	96.02	hyg pl	108–9 (105)	d	....	....	v	v	v	...	i	peth i	**Am 74**, 5540

For explanations, symbols and abbreviations see beginning of table. For structural formulas see end of table.

No.	Name	Synonyms and Formula	Mol. wt.	Color, crystalline form, specific rotation and λ_{max} (log ε)	m.p. °C	b.p. °C	Density	n_D	Solubility w	al	eth	ace	bz	other solvents	Ref.
	Methanephosphonic acid														
m324	—,diethyl ester*	Diethyl methylphosphonate. $CH_3PO(OC_2H_5)_2$	152.13	...	...	194^{763} 85^{15}	1.0406_4^{30}	1.4101^{30}	s	s	s	...	i	...	**J1954,** 3222
m325	—,dimethyl ester*	Dimethyl methylphosphonate. $CH_3PO(OCH_3)_2$	124.08	λ^{al} 217 (1.12)	...	181^{754} 79.5^{20}	1.1507_4^{20}	1.4099^{30}	s	s	s	...	...	...	**J1954,** 3222
m326	**—,trifluoro-, diammonium salt***	$F_3CPO(ONH_4)_2$	184.06	...	212–6d	...	...	...	s	...	...	...	...	...	**J1954,** 3598
m328	**Methanesiliconic acid***	Silicoacetic acid. CH_3SiO_2H.	76.13	amor	...	...	...	...	i	...	s	...	...	KOH s	**B4,** 629
Ω m329	**Methanesulfenic acid, trichloro-, chloride**	Perchloromethylmercaptan. Trichloromethanesulfenyl chloride. Cl_3CSCl	185.89	ye oil λ^{chl} 324 (1.08)	...	$147–8^{760}$ 51^{25}	1.6947_4^{20}	1.5484^{20}	...	...	s	...	...	...	**B3**[2], 106
m330	**—,trifluoro-, chloride**	Trifluoromethylsulfenyl chloride. F_3CSCl	136.52	ye λ^{gas} 214 (2.37), 333 (1.40)	...	-0.7^{760}	...	...	i d	...	...	...	...	...	**J1953,** 3219
Ω m331	**Methanesulfinic acid, amino-(imino)-***	$HN{:}C(NH_2)SO_2H$	108.12	nd (al)	144d	...	...	...	v	...	i	...	i	os i	**B3**[1], 36
m332	**Methanesulfonic acid***	Methylsulfonic acid. CH_3SO_3H	96.11	...	20	167^{10} 122^{1}	1.4812_4^{18}	1.4317^{16}	v	s	v	...	...	...	**B4**[2], 524
Ω m333	—,chloride	Methanesulfonyl chloride*. CH_3SO_2Cl	114.55	...	...	161^{730} 55^{11}	1.48053_4^{18}	1.4573^{20}	i	s	s	...	...	...	**B4**[2], 525
m334	—,**phenyl-**, amide, *N*-methyl-	*N*-Methylbenzylsulfonamide. $C_6H_5CH_2SO_2NHCH_3$	185.25	nd or lf (aa-lig)	108–9	...	...	...	s^h	s	s	...	...	alk s	**B11**[2], 73
Ω m335	**—,trichloro-, chloride**	Cl_3CSO_2Cl	217.89	cr (al-w)	140–1	170 (δsub)	...	...	i d^h	s d^h	s	...	...	CS_2 s	**B3**[2], 16
m336	**—,trifluoro-***	F_3CSO_3H	150.08	hyg liq	...	162^{760} $81^{37.5}$	...	...	...	d	∞ d^h	...	...	...	**J1955,** 2901
m337	—,—,amide	$F_3CSO_2NH_2$	149.09	...	119	...	...	...	s	...	...	...	...	chl s	**J1956,** 173
m338	—,—,—,*N,N*-diethyl-	$F_3CSO_2N(C_2H_5)_2$	205.20	...	...	55^{7}	...	...	i	...	s	...	...	...	**J1956,** 173
m339	—,—,anhydride*	$(F_3CSO_2)_2O$	282.14	...	...	84^{760}	...	...	d	d	...	...	...	...	**J1957,** 4069
m340	—,—,chloride	F_3CSO_2Cl	168.52	...	...	31.6^{760}	...	...	i	...	...	...	...	...	**J1955,** 2901
m341	—,—,ethyl ester	$F_3CSO_3C_2H_5$	178.13	...	...	115^{760} 42^{40}	...	...	d	...	s	...	...	...	**J1956,** 173
m342	—,—,fluoride	F_3CSO_2F	152.07	...	...	-21.7^{760}	...	...	d	i^h	...	...	...	...	**J1956,** 173
m343	—,—,potassium salt	CF_3SO_3K	188.19	cr (ace)	230		...	...	...	...	...	...	...	...	**J1957,** 4069
m344	—,—,sodium salt	CF_3SO_3Na	172.06	cr (ace)	248		...	...	...	...	...	s	...	...	**J1957,** 4069
m345	**Methanethiol***	Methylmercaptan. CH_3SH	48.11	λ^{cy} 228.5 (2.15)	−123	6.2	0.8665_4^{20}	...	δ^h	v	v	...	...	...	**B1**[3], 1212
m346	**—,(2-furyl)-**	Furfurylmercaptan	114.17	...	...	155^{760} 47^{12}	1.1319_4^{20}	1.5329^{20}	i	...	...	...	...	...	**B17**[2], 116
m348	**Methanetricarboxylic acid, trimethyl ester***	Tricarbomethoxymethane. $CH(CO_2CH_3)_3$	190.15	pr (MeOH)	46–7	242.7^{760} 128^{15}	...	...	...	v	v	...	v	chl v MeOH s^h	**B2**[2], 680
—	**Methanal***	*see* **Formaldehyde**													
—	**Methanoic acid***	*see* **Formic acid**													
Ω m349	**Methanol***	Carbinol. Methyl alcohol. Wood alcohol. CH_3OH	32.04	λ^{gas} 183.3 (2.18)	−93.9	64.96^{760} 15^{73}	0.7914_4^{20}	1.3288^{20}	∞	∞	∞	∞	v	chl s	**B1**[3], 1147
m349[1]	**Methanol-*d***	*O*-Deuteromethanol. CH_3OD.	33.05	...	−100	65.5^{760}	0.8127_4^{20}	...	∞	∞	∞	∞	v	chl s	**B1**[3], 1186
m349[2]	**Methan-d^3-ol-*d***	Deuteroxy(trideutero)-methane*. Tetradeuteromethanol. D_3COD	36.07	...	...	65.4^{760}	...	...	∞	∞	∞	∞	v	chl s	**B1**[3], 1187
m350	**Methanol, (2-amino-3-methylphenyl)-***	2-Amino-3-methylbenzyl alcohol.	137.18	nd (bz)	71	$135–45^{12}$	...	...	...	...	...	...	δ s^h	...	**B13**[2], 367
m351	**—,(2-amino-4-methylphenyl)-***	2-Amino-4-methylbenzyl alcohol.	137.18	nd (bz)	141	$140–50^{13}$	...	...	...	...	...	...	δ s^h	...	**B13**[2], 369
m352	**—,(2-amino-5-methyl phenyl)-***	2-Amino-5-methylbenzyl alcohol.	137.18	nd (bz)	123	$145–50^{13}$	...	...	...	...	...	...	δ s^h	...	**B13**[2], 367
m353	**—,(4-amino-3-methylphenyl)-bis(4-aminophenyl)-***	Rosaniline. $C_{20}H_{21}N_3O$. *See* m381	319.41	nd (w) $\lambda^{aq\ HCl}$ 540 (3.1)	186d	...	...	...	δ	s	i	...	...	...	**B13,** 763
m355	**—,(4-aminophenyl)phenyl-***	*p*-Aminobenzhydrol. $C_{13}H_{13}NO$. *See* m366	199.26	nd (w or bz)	121	...	...	...	...	v	δ	s	...	MeOH, AcOEt v lig δ	**B13,** 696
m356	**—,bis[4(dimethylamino)phenyl]-***	Michler's hydrol. $C_{17}H_{22}N_2O$. *See* m366	270.38	lt gr lf or pr (bz) λ^{aa} 607.5 (5.17)	98 (102–3)	...	...	...	i	v^h	s	...	s	aa s	**B13**[2], 423
m357	**—,bis[2(dimethylamino)phenyl] phenyl-***	$C_{23}H_{26}N_2O$. *See* m381	346.48	pr (lig)	105	...	...	...	...	...	δ	...	...	ac s lig δ	**B13,** 741

For explanations, symbols and abbreviations see beginning of table. For structural formulas see end of table.

No.	Name	Synonyms and Formula	Mol. wt.	Color, crystalline form, specific rotation and λ_{max} (log ε)	m.p. °C	b.p. °C	Density	n_D	Solubility w	al	eth	ace	bz	other solvents	Ref.
	Methanol														
m358	**—,bis[3(dimethyl-amino) phenyl] phenyl-***	$C_{23}H_{26}N_2O$. *See* m381	346.48	cr (eth)	128–9	...	...	...	...	...	v^h	...	...	aa s	**B13**, 742
m359	**—,bis[4(dimethyl-amino)phenyl] phenyl-***	$C_{23}H_{26}N_2O$. *See* m381	346.48	cr (eth, bz, lig, MeOH or peth) λ^{aa} 320 (4.25), 428 (4.28), 620 (5.00)	121–3 (110)	...	...	...	...	...	v	...	v^h	ac s lig v^h	**B13**[2], 442
m360	**—,bis(4-hydroxy-phenyl)phenyl-***	Benzaurin. $C_{19}H_{16}O_3$. *See* m381	292.34	ye-red pw λ^{chl} 414 (4.3)	110–20	...	...	...	i	δ	δ	...	δ	...	**B8**[2], 245
m362	**—,cyclohexyl-**	Hexahydrobenzyl alcohol. Hydroxymethyl-cyclohexane.	114.19	...	−43	183^{760} 83^{14}	0.9297^{20}_{4}	1.46439^{20}	...	s	s	...	...	...	**B6**[3], 76
m363	**—,[2(dimethyl-amino)phenyl]-[3(dimethyl-amino)phenyl]-phenyl-***	$C_{23}H_{26}N_2O$. *See* m381	346.48	pl (bz)	183–4	...	...	...	...	δ	δ	...	s	HCl s	**B13**, 742
m364	**—,[2(dimethyl-amino)phenyl]-[4(dimethyl-amino)phenyl]-phenyl-***	$C_{23}H_{26}N_2O$. *See* m381	346.48	(al)	169–70	...	...	...	...	δ	δ	...	v	ac s	**B13**, 742
m365	**—,[3(dimethyl-amino)phenyl]-[4(dimethyl-amino)phenyl]-phenyl-***	$C_{23}H_{26}N_2O$. *See* m381	346.48	pr (bz-al)	140	...	...	...	...	δ	δ	...	v	ac s	**B13**, 742
Ω m366	**—,diphenyl-***	Benzhydrol	184.24	nd (lig) λ^{MeOH} 220 sh (4.0), 259 (2.67) λ^{sulf} 442 (4.60)	69	$297–8^{748}$ 180^{20}	...	...	δ^h	v	v	...	...	CCl_4, chl v aa s lig δ	**B6**[2], 631
m367	**—,diphenyl-(1-naphthyl)-***	$C_{10}H_7^{\alpha}COH(C_6H_5)_2$	310.40	cr (lig or bz)	136.5	d	...	...	i	s^h	s	...	s	lig i, s^h	**B6**[2], 721
m368	**—,diphenyl-(2-naphthyl)-***	$C_{10}H_7^{\beta}COH(C_6H_5)_2$	310.40	pr (eth-lig)	118	...	...	...	i	s	s	s	v	oos s lig δ	**B6**[2], 722
m368[1]	**—,di(4-tolyl)-**		212.30	nd (al) λ^{sulf} 472 (4.87)	69	...	...	...	i	s	s	s	...	chl, aa s	**B6**, 688
m369	**—,(2-furyl)-**	Furfuryl alcohol. 2-Hydroxymethylfuran.	98.10	col-ye λ^{w} 217 (3.9)	...	171^{750} $68–9^{20}$	1.1296^{20}_{4}	1.4868^{20}	∞ d	v	v	...	...	...	**B17**[2], 113
m370	**—,(4-methoxy-phenyl)phenyl-***	*p*-Anisylphenylcarbinol. *p*-Methoxybenzhydrol. $C_{14}H_{14}O_2$. *See* m366	214.27	nd (w, lig or dil al)	66–8	...	...	...	s^h	v	...	...	s	chl s lig s^h	**B6**[2], 965
m371	**—,(5-methyl-2-furyl)-**	5-Methylfurfuryl alcohol	112.14	...	...	$194–6^{744}$ δd 81^{13}	1.0769^{20}_{4}	1.4853^{20}	δ^h	v	v	...	...	...	**Am 54**, 2554
m373	**—,(1-naphthyl)-***	1(Hydroxymethyl)-naphthalene*. $C_{10}H_7^{\alpha}CH_2OH$	158.20	nd (w or al), cr (bz-lig)	64 (61)	301^{715} 163^{12}	1.1039^{80}_{4}	...	δ^h	v	v	...	...	...	**B6**[2], 617
m374	**—,(1-naphthyl)-phenyl-(*dl*)***	$C_{10}H_7^{\alpha}CH(OH)C_6H_5$	234.30	cr (al or lig)	86.5	*ca.* 360	...	...	i	v	v	...	v	lig δ	**B6**[2], 681
m375	**—,(2-naphthyl)-phenyl-(*dl*)***	$C_{10}H_7^{\beta}CH(OH)C_6H_5$	234.30	nd (al or lig)	87–8 (83)	...	...	...	i	s	s	...	s	to s lig δ	**B6**[2], 681
Ω m376	**—,(2-tetrahydro-furyl)-(*dl*)**	Tetrahydrofurfuryl alcohol	102.13	hyg	...	$177–8^{750}$ $80–2^{20}$	1.0544^{20}_{4}	1.4517^{20}	...	...	s	s	...	...	**B17**[2], 106
m377	**—,(2-thienyl)-**	...	114.17	...	...	207^{760} 96^{12}	...	1.5280^{20}	i	v	v	...	...	...	**B17**, 113
m378	**—,(2-tolyl)-**	*o*-Methylbenzyl alcohol	122.17	nd λ^{iso} 263 (2.4), 275 (2.3)	37–9	223^{750} $117–9^{20}$	1.023^{40}	...	δ v^h	v	v	...	...	chl v	**B6**[2], 457
Ω m379	**—,(3-tolyl)-**	*m*-Methylbenzyl alcohol	122.17	...	< −20	$215–6^{760}$ $(218–9^{743})$ $110–2^{17}$	0.9157^{17}	...	δ s^h	v	v	...	...	...	**B6**[2], 465
Ω m380	**—,(4-tolyl)-**	*p*-Methylbenzyl alcohol	122.17	nd (hp)	61–2.1	217^{760} $116–8^{20}$	0.978^{22}	...	δ s^h	v	v	...	...	...	**B6**[2], 469
Ω m381	**—,triphenyl-***	Triphenylcarbinol. Tritanol	260.34	pl (al), trig (bz), rh (CCl_4) λ^{diox} 240 (3.16), 253 (3.26), 260 (3.28) λ^{sulf} 405 (4.6), 435 (4.5)	164.2	380^{760}	1.199^{0}_{4}	...	i	v	v	s	s	aa s peth i	**B6**[2], 686
m382	**—,tris(4-amino-phenyl)-***	Pararosaniline. $C_{19}H_{19}N_3O$. *See* m381	305.38	col to red lf $\lambda^{w, pH=1.3}$ 260 sh (3.0), 410 (2.7), 540 (3.0)	189 *ca.* 205)	...	...	...	δ	s	δ	...	...	...	**B13**[2], 447
m383	**—,tris-(4-biphenylyl)-**	Tri-p-xenylcarbinol	488.64	nd (aa or bz)	212	...	...	...	...	δ	...	...	v	aa s^h	**B6**[2], 741

For explanations, symbols and abbreviations see beginning of table. For structural formulas see end of table.

No.	Name	Synonyms and Formula	Mol. wt.	Color, crystalline form, specific rotation and λ_{max} (log ε)	m.p. °C	b.p. °C	Density	n_D	Solubility w	al	eth	ace	bz	other solvents	Ref.
	Methanol														
m384	—,**tris(3-nitrophenyl)-***	$C_{19}H_{13}N_3O_7$. *See* m381	395.33	rhd (MeOH, chl or AcOEt-lig)	167				...	δ^h	δ	...	s	aa s CS_2 δ	**B6**[1], 352
m385	—,**tris(4-nitrophenyl)-***	$C_{19}H_{13}N_3O_7$. *See* m381	395.33	mcl pr (bz or aa) λ^{sulf} 265 (4.4), 425 (4.6), 450 sh (4.5)	(i) 190–1 (ii) 167				i	δ^h	δ	...	s	aa s CS_2 δ	**B6**[1], 352
m386	**Methantheline bromide**	Banthine bromide	420.36	cr (*i*-PrOH) λ^{al} 246, 282	172–7				s	s	i	...	...	chl s	**Am 65,** 1582
m387	**Methapyrilene,** (base)	Histadyl base. Tenalin base. Thenylene base.	261.39	λ^{w} 238 (4.17), 304 (3.60)		$173–5^{3}$		1.5915^{20}	...	...	...	...	...		**Am 71,** 333
—	**Methionic acid**	*see* **Methanedisulfonic acid***													
Ω m388	**Methionine(*DL*)**	*dl*-2-Amino-4(methylthio)-butanoic acid. $CH_3SCH_2CH_2CH(NH_2)CO_2H$	149.21	pl (al) λ^{w} <200	281d (272)		1.340		v	δ	i	...	i	peth i	**B4**[2], 938
Ω m389	—(*L*)	$CH_3SCH_2CH_2CH(NH_2)CO_2H$	149.21	hex pl (dil al) $[\alpha]^{25}_{D}$ −8.2 (w, c = 1) $[\alpha]^{25}_{D}$ +22.5 (1N HCl) $\lambda^{0.01N\,HCl}$ 208 sh (3.2)	283d	sub 186			s	i	i	i	i	aa δ peth i	**B4**[2], 938
—	**Methone**	*see* **1,3-Cyclohexanedione, 5,5-dimethyl-***													
—	**Methoxychlor**	*see* **Ethane, 2,2-bis(4-methoxyphenyl)-1,1,1-trichloro-***													
—	**Methyl alcohol**	*see* **Methanol***													
—	**Methylamine**	*see* **Methane, amino-***													
—	**Methyl cellosolve**	*see* **Ethanol, 2-methoxy-***													
m390	**Methylene blue**	3,9-Bisdimethylamino-phenazothionium chloride.	319.86	dk gr cr or pw (chl-eth) λ^{al} 655 (4.95)					s	s	i	...	...	chl s Py δ	**B27**[2], 448
—	**Methylene bromide**	*see* **Methane, dibromo-***													
—	**Methylene chloride**	*see* **Methane, dichloro-***													
—	**Methylene cyanide**	*see* **Malonic acid,** dinitrile													
—	**5,5′-Methylene disalicylaldehyde**	*see* **Methane, bis(3-formyl-4-hydroxyphenyl)-**													
—	**Methylene fluoride**	*see* **Methane, difluoro-***													
Ω m391	**Methyl green**	Paris green. Heptamethyl pararosaniline chloride.	458.48	gr pw (al)					s	δ	i	...	...		**B13**[2], 451
—	**Methylhydroxylamine**	*see* **Methane, hydroxylamino-***													
—	**Methyl mercaptan**	*see* **Methanethiol***													
Ω m392	**Methyl orange**	Sodium 4′-dimethyl-aminoazobenzene-4-sulfonate. Helianthin-*β*. Orange III.	327.34	og, ye pl or sc (w) λ^{al} 265 (4.2), 305 (3.9), 420 (4.6)	d				δ	δ	i	...	...	Py δ	**B16,** 331
Ω m393	**Methyl red**	4′-Dimethylaminoazo-benzene-2-carboxylic acid.	269.31	vt or red pr (to or bz), nd (aq aa), lf (dil al) $\lambda^{MeOH-0.4N\,HCl}$ 521 (4.76) $\lambda^{MeOH-NaOMe}$ 410 (4.4)	183				δ	s	...	v^h	v^h	chl, aa v lig δ, peth δ	**B16**[2], 164
m394	**Methysticin**	Kavahin. Kavatin. $C_{15}H_{14}O_5$.	274.28	nd (MeOH), pr (ace) $[\alpha]^{20}_{D}$ +94.3 (ace, p = 5)	137				i δ^h	v^h	δ	s	s^h	chl s^h peth δ alk i	**B19**[2], 431
Ω m395	**Metrazol**	Cardiazole. Leptazole.	138.17	cr (bz-lig)	59–60	194^{12}			v	v	s	v	s	os s	**B26**[2], 213
m396	**Metycaine**	Piperocaine hydrochloride.	292.78		172–5				s	s	i	...	...	chl s	**C25,** 1037
—	**Michler's ketone**	*see* **Benzophenone, 4,4′-bis(dimethylamino)-**													
m397	**Mimosine(*l*)**	*l*-Leucenol	198.18	ta (w) $[\alpha]^{22}_{D}$ −21 (w, c = 0.5) λ^{w} 215 (4.6), 280 (4.6)	228–9d				δ	i	i	i	i	dil ac, dil alk s oos, aa, diox i	**Am 71,** 705

For explanations, symbols and abbreviations see beginning of table. For structural formulas see end of table.

No.	Name	Synonyms and Formula	Mol. wt.	Color, crystalline form, specific rotation and λ_{max} (log ε)	m.p. °C	b.p. °C	Density	n_D	Solubility w	al	eth	ace	bz	other solvents	Ref.
	Monolaurin														
—	**Monolaurin**	*see* **Glycerol,** monododecanoate													
—	**Monoolein**	*see* **Glycerol,** mono(9-octadecenoate)													
—	**Monopalmitin**	*see* **Glycerol,** monohexadecanoate													
—	**Monoricinolein**	*see* **Glycerol,** mono(12-hydroxy-9-octadecenoate)													
—	**Monostearin**	*see* **Glycerol,** monooctadecanoate													
—	**Morin**	*see* **Flavone, 2′,3,4′,5,7,8-pentahydroxy-**													
—	**Morindone**	*see* **9,10-Anthraquinone, 6-methyl-1,2,5-tri-hydroxy-***													
m399	**Morphine**	$C_{17}H_{19}NO_3$	285.35	pr (PhOMe) λ^{al} 210 sh (4.4), 236 sh (3.7), 287 (3.2)	(i) 254–6.4d (st) (ii) 197 (unst)	....	....	....	i δ^h	δ	i	i	i	MeOH, Py s chl, peth i	**B27²**, 118
m400	—,acetate tri-hydrate(*l*)	$C_{17}H_{19}NO_3 . CH_3CO_2H . 3H_2O$. *See* m399	398.44	col to ye cr (dil al) $[\alpha]^{15}_D$ −77 (w)	200d	....	....	....	v	s v^h	i	...	...	chl δ	**B27²**, 134
m401	—,hydrate	$C_{17}H_{19}NO_3 . H_2O$. *See* m399	303.36	orh pr (dil al) $[\alpha]^{25}_D$ −132 (MeOH, c = 1)	−w, 130; 254–6.4 (anh)	d	1.32^{20}_4	1.58–1.64	s^h	δ	i	...	...	....	**B27²**, 122
m402	—,hydrochloride trihydrate	$C_{17}H_{19}NO_3 . HCl . 3H_2O$. *See* m399	375.85	nd or fl (dil HCl) $[\alpha]^{25}_D$ −113.5 (w, c = 2.2) λ^w 230 (3.85), 286 (3.2)	200d (250d)	....	....	....	s v^h	s	i	...	...	chl i	**B27²**, 132
m403	—,*N*-oxide	Genomorphine. $C_{17}H_{19}NO_4$. *See* m399	301.35	pr (50% al) λ^w 208 (4.4), 235 sh (3.7), 284 (3.2)	274–5	....	....	....	δ	δ	...	i	i	NH_4OH v chl i	**B27²**, 159
m404	—,sulfate penta-hydrate	$2(C_{17}H_{19}NO_3) . H_2SO_4 . 5H_2O$. *See* m399	758.85	pw or cubes $[\alpha]^{25}_D$ −107.8 (w, c = 4) λ^{al} 285	*ca.* 250d	....	....	....	s v^h	δ	i	...	i	chl i	**B27²**, 133
m405	—,*O,O*-**diacetyl-**	Diamorphine. Heroin. $C_{21}H_{23}NO_5$. *See* m399	369.42	rh $[\alpha]^{15}_D$ −166 (MeOH) λ^{al} 230 sh (3.8), 281 (3.3)	173	$272–4^{12}$	1.56–1.61	....	i	s^h	δ	...	s	chl v MeOH s	**B27²**, 151
m406	—,—,hydrochloride monohydrate	$C_{21}H_{23}NO_5 . HCl . H_2O$. *See* m399	423.90	$[\alpha]^{20}_D$ −153 (w, c = 1.17) λ^w 227 (3.85), 279.5 (3.25)	231–2 (243–4)	....	....	....	v	s	i	...	...	chl v	**B27²**, 153
m407	—,3-ethyl ether, hydrochloride dihydrate	Dionin. $C_{19}H_{23}NO_3 . HCl . 2H_2O$. *See* m399	385.89	cr $\lambda^{0.02N\ HCl}$ 244 (3.5), 286 (3.2)	123–5d (170 anh)	....	....	....	s	s	δ	...	...	chl δ	**B27²**, 148
—	**Morphol**	*see* **Phenanthrene, 3,4-dihydroxy-***													
Ω m408	**Morpholine**	Diethylenimide oxide. Tetrahydro-1,4-isoxazine.	87.12	hyg	−4.75	128.3^{760} 24.86^{10}	1.0005^{20}_4	1.4548^{20}	∞	s	s	s	s	os s	**B27²**, 3
Ω m409	—,**4-acetyl-**	$C_6H_{11}NO_2$. *See* m408	129.16	....	14.5	152^{50} 118^{12}	1.1165^{20}_{20}	1.4827^{20}	∞	s	...	s	...	....	**C50,** 7112
Ω m410	—,**4(2-amino-ethyl)-**	$C_6H_{14}N_2O$. *See* m408	130.19	....	25.6	116^{50}	0.9915^{20}_{20}	1.4715^{20}	∞	∞	...	s	∞	lig ∞	**C42,** 6747
Ω m411	—,**4(3-amino-propyl)-**	$C_7H_{16}N_2O$. *See* m408	144.22	....	−15	219^{733} 134^{50}	0.9872^{20}_{20}	1.4762^{20}	∞	∞	...	s	∞	lig ∞	**Am 66,** 725
m412	—,**4-benzyl-**	$C_{11}H_{15}NO$. *See* m408	177.25	....	....	260–1 $128–9^{13}$	1.0387^{20}_4	1.5302^{20}	δ	...	...	s	s	ac s	**B27¹**, 203
m413	—,**4-butyl-**	$C_8H_{17}NO$. *See* m408	143.23	....	−57.1	$213–4^{760}$ $67–8^{10}$	0.9068^{20}_4	1.4451^{20}	v	s	...	s	s	....	**Am 61,** 171
Ω m414	—,**2,6-dimethyl-**	$C_6H_{13}NO$. *See* m408	115.18	....	fr −85	146.6^{760} 58^{30}	0.9346^{20}_{20}	1.4460^{20}	∞	∞	...	s	∞	lig ∞	**Am 80,** 1257
m415	—,**4(2-ethoxy-ethyl)-**	$C_8H_{17}NO_2$. *See* m408	159.23	....	−100	206^{760} $93–7^{14}$	0.963^{20}	....	∞	...	s	s	s	....	**Am 63,** 298
Ω m416	—,**4-ethyl-**	$C_6H_{13}NO$. *See* m408	115.18	....	....	$138–9^{763}$	0.8996^{20}_4	1.4400^{20}	∞	∞	∞	s	s	....	**B27¹**, 203
Ω m417	—,**4(2-hydroxy-ethyl)-**	4-Morpholineethanol. $C_6H_{13}NO_2$. *See* m408	131.18	....	....	227^{757}	1.0710^{20}_4	1.4763^{20}	s	s	...	...	...	....	**B27,** 7
m418	—,**4(2-hydroxy-propyl)-**	$C_7H_{15}NO_2$. *See* m408	145.20	....	....	$92–4^{13}$	1.0174^{20}_4	1.4638^{20}	s	s	s	s	s	MeOH s	**Am 64,** 970
Ω m419	—,**4-methyl-**	$C_5H_{11}NO$. *See* m408	101.15	....	....	$115–6^{750}$	0.9051^{20}_4	1.4332^{20}	s	s	s	...	...	....	**B27,** 6
Ω m420	—,**4-phenyl-**	$C_{10}H_{13}NO$. *See* m408	163.22	cr (al-eth)	57–8	$259–60^{745}$ $165–70^{45}$	....	....	i	i	v	...	...	....	**B27²**, 3
m421	—,**4(4-tolyl)-**	$C_{11}H_{15}NO$. *See* m408	177.25	cr (dil al)	51	167^{30}	....	....	...	v	s	...	...	....	**B27²**, 4

For explanations, symbols and abbreviations see beginning of table. For structural formulas see end of table.

No.	Name	Synonyms and Formula	Mol. wt.	Color, crystalline form, specific rotation and λ_{max} (log ε)	m.p. °C	b.p. °C	Density	n_D	Solubility w	al	eth	ace	bz	other solvents	Ref.
	Mucic acid														
Ω m422	**Mucic acid**	Galactosaccharic acid. 2,3,4,5-Tetrahydroxy-hexanedioic acid*.	210.14	pr (w)	255 (rapid htng) 225 (208) (slow htng)	....	....	....	δ s^h	i	i	...	...	alk s Py i	**B3**[3], 1122
—	**Mucobromic acid**	*see* **2-Butenoic acid, 2,3-dibromo-4-oxo-***													
—	**Mucochloric acid**	*see* **2-Butenoic acid, 2,3-dichloro-4-oxo-***													
—	**Muconic acid**	*see* **2,4-Hexadienedioic acid***													
Ω m423	**Murexide**	Ammonium purpurate	302.21	red-gr pr (aq NH_4Cl) λ^w 510 (0.7)	....	....	....	....	δ s^h	i	i	...	...	alk s (bl)	**B25**, 499
—	**Muscone**	*see* **Cyclopentadecanone, 3-methyl-***													
—	**Musk baur**	*see* **Benzene, 2-*tert*-butyl-4-methyl-1,3,5-trinitro-**													
—	**Musk ketone**	*see* **Acetophenone, 4-*tert*-butyl-2,6-dimethyl-3,5-dinitro-**													
—	**Musk xylene**	*see* **Benzene, 1-*tert*-butyl-3,5-dimethyl-2,4,6-trinitro-**													
—	**Mustard gas**	*see* **Sulfide, diethyl, 2,2′-dichloro-**													
m424	**Mycophenolic acid**	$C_{17}H_{20}O_6$	320.35	nd (w)	141	....	....	....	i δ^h	v	v	...	δ	chl v to δ, v^h	**B18**[2], 393
m425	**Myrcene**	7-Methyl-3-methylene-1,6-octadiene. $(CH_3)_2C{:}CH(CH_2)_2C({:}CH_2)CH{:}CH_2$	136.24	λ^{al} 225 (4.30)	....	167^{760} 65^{20}	0.8013^{15}_{4}	1.4722^{20}	i	s	s	...	s	chl, aa s	**B1**, 264
—	**Myricyl alcohol**	*see* **1-Triacontanol***													
—	**Myristaldehyde**	*see* **Tetradecanal***													
—	**Myristic acid**	*see* **Tetradecanoic acid***													
m426	**Myristicin**	....	192.22	....	< −20	$276\text{–}7^{760}$ 157^{21}	1.1437^{20}_{20}	1.5403^{20}	i	δ	s	...	s	....	**B19**[2], 84
—	**Myristyl bromide**	*see* **Tetradecane, 1-bromo-***													

For explanations, symbols and abbreviations see beginning of table. For structural formulas see end of table.

No.	Name	Synonyms and Formula	Mol. wt.	Color, crystalline form, specific rotation and λ_{max} (log ε)	m.p. °C	b.p. °C	Density	n_D	Solubility w	al	eth	ace	bz	other solvents	Ref.
	Naphthacene														
Ω n1	**Naphthacene**	2,3-Benzanthracene. Benz[b]anthracene. Tetracene.	228.30	og-ye lf (bz or xyl) λ^{bz} 396 (3.35), 418 (3.64), 444.5 (3.91), 475.5 (3.98)	357 (341)	sub	...	...	δ	...	...	...	δ s^h	con sulf s oos δ	**B5**^{3}, 2372
n2	—,9,10-dihydro-	$C_{18}H_{14}$. *See* n1	230.31	nd (xyl), lf (bz)	212	*ca.* 400	...	...	...	δ^h	δ^h	...	s^h	$PhNO_2$ v sulf s (ye) aa s^h	**E14s**, 76
n3	—,9,10-diphenyl-	$C_{30}H_{20}$. *See* n1	380.50	og (eth + ½) λ^{bz} 407 (3.32), 434 (3.73), 461 (3.96), 493 (3.98)	207–8	...	...	...	...	...	s^h	...	...	os v	**E14s**, 77
n4	—,9,11-diphenyl-	$C_{30}H_{20}$. *See* n1	380.50	ye λ^{bz} 408 (3.3), 435 (3.8), 462 (4.1), 493 (4.1)	301–2	...	...	...	i	...	s	...	s	CS_2 s	**E14**, 313
n5	—,9,10,11-triphenyl-	$C_{36}H_{24}$. *See* n1	456.60	og (eth), (bz + 1) λ^{bz} 417 (3.48), 442 (3.80), 472 (4.08), 504 (4.14)	236–7 177–8 (+1 bz)	...	...	...	i	...	s	...	s	CS_2 s	**E14**, 313
Ω n6	**9,10-Naphthacenequinone**	2,3-Benzanthraquinone.	258.28	ye nd ($PhNO_2$ or aa)	294 (285)	sub	...	...	...	s	...	δ^h	δ^h	sulf s (red-vt) aa δ^h	**E14**, 319
n7	**9,11-Naphthacenequinone**	...	258.28	dk red (aa or xyl)	322 part δ	...	...	...	i	...	...	...	...	os δ xyl, aa δ	**E14s**, 86
Ω n8	**1-Naphthaldehyde**	1-Formylnaphthalene. α-Naphthaldehyde. 1-Naphthalenecarbonal*.	156.20	pa ye	33–4	292^{760} 160^{15}	1.1503^{20}_{4}	1.6507^{20}	i	s	s	s	s	sulf s	**E12B**, 2197
Ω n9	**2-Naphthaldehyde**	2-Formylnaphthalene. β-Naphthaldehyde. 2-Naphthalenecarbonal.*	156.20	lf (w) λ^{chl} 384 (4.51)	61–3	160^{19}	1.0775^{99}_{4}	1.6211^{99}	δ^h	v	v	s	...	...	**E12B**, 2204
n10	**1-Naphthaldehyde, 2-ethoxy-**	$C_{13}H_{12}O_2$. *See* n8	200.24	yesh nd (al or aa)	115	$185–7^{25}$	...	...	...	s^h	...	...	...	aa s^h	**B8**^{1}, 564
n11	—,4-ethoxy-	$C_{13}H_{12}O_2$. *See* n8	200.24	yesh cr (aa or AcOEt)	75 (72)	...	...	...	...	s^h	...	...	...	aa s^h	**B8**^{2}, 174
n12	—,2-hydroxy-	β-Naphthol-1-aldehyde. $C_{14}H_8O_2$. *See* n8	172.20	pr (al), nd (AcOEt) $\lambda^{MeOH-HCl}$ 225 (4.7), 230 (3.8), 270 sh (3.7)	82	192^{27}	...	...	i	s	s	...	...	aq alk s sulf s(ye) peth s	**B7**^{2}, 171
n13	**2-Naphthaldehyde, 1-hydroxy-**	α-Naphthol-2-aldehyde. $C_{14}H_8O_2$. *See* n9	172.20	grsh-ye nd (dil aa, dil al or lig) $\lambda^{MeOH-HCl}$ 260 (4.6), 295 (3.8), 375 (3.8)	60	...	...	...	δ s^h	s	s	...	...	os v alk s aa s^h	**E12B**, 2369
Ω n14	**Naphthalene***	...	128.19	mcl pl (al) λ^{al} 221 (5.04), 275.5 (3.76), 286 (3.59), 311 (2.38)	80.55	218^{760} 87.5^{10}	1.0253^{20}_{4} 0.9625^{100}_{4}	1.4003^{24} 1.5898^{85}	i	s v^h	v^h	v	v	CS_2, CCl_4, chl v aa s, v^h	**B5**^{3}, 1549
Ω n15	—,picrate	$C_{10}H_8.C_6H_3N_3O_7$. *See* n14	357.28	ye pr or pl (aa), cr (eth) $\lambda^{C_2H_4Cl_2}$ 230 (4.30), 240 (4.30), 340 (3.60)	152	...	1.53	...	δ d^h	d	s	...	s	...	**B5**^{3}, 1568
n16	—,1-acetamido-2-nitro-*	$C_{12}H_{10}N_2O_3$. *See* n14	230.23	lt ye nd (aa or al)	200	...	...	...	...	s^h	...	...	...	aa s	**E12B**, 746
n17	—,1-acetamido-4-nitro-*	$C_{12}H_{10}N_2O_3$. *See* n14	230.23	pa ye nd (ace)	192.5–3.5	...	...	...	...	s^h	...	s^h	...	...	**E12B**, 751
n18	—,1-acetamido-5-nitro-*	$C_{12}H_{10}N_2O_3$. *See* n14	230.23	br pr (aa), ye cr (al)	220	...	...	...	δ	s^h	δ	...	δ	...	**E12B**, 755
n19	—,1-acetamido-8-nitro-*	$C_{12}H_{10}N_2O_3$. *See* n14	230.23	nd (w)	191	...	...	...	s^h	...	...	...	...	...	**E12B**, 756
n20	—,2-acetamido-1-nitro-*	$C_{12}H_{10}N_2O_3$. *See* n14	230.23	ye rh bipym nd or pl (al)	123.5	...	...	...	δ	s	δ	...	s	aa s lig δ	**B12**^{2}, 731
n21	—,2-acetamido-5-nitro-*	$C_{12}H_{10}N_2O_3$. *See* n14	230.23	ye-br rh (al), ye nd (bz)	186	...	...	...	...	s	...	...	δ	aa s	**E12B**, 764
n22	—,2-acetamido-6-nitro-*	$C_{12}H_{10}N_2O_3$. *See* n14	230.23	lt ye nd (dil al)	224	...	...	...	i	v	i	...	s	aa v xyl s	**E12B**, 765
n23	—,2-acetamido-8-nitro-*	$C_{12}H_{10}N_2O_3$. *See* n14	230.23	ye nd (al)	195.5	...	...	...	...	δ s^h	...	...	δ	aa s	**E12B**, 766
Ω n24	—,1-acetyl-	1-Acetonaphthone. Methyl 1-naphthyl ketone. $C_{12}H_{10}O$. *See* n14	170.21	λ^{al} 242 (4.4), 290 (3.9)	34	$296–8^{760}$ $170–0.5^{20}$	$1.1171^{21.5}_{4}$	1.6280^{22}	i	s	s	s	...	os s	**B7**^{2}, 337

For explanations, symbols and abbreviations see beginning of table. For structural formulas see end of table.

No.	Name	Synonyms and Formula	Mol. wt.	Color, crystalline form, specific rotation and λ_{max} (log ε)	m.p. °C	b.p. °C	Density	n_D	Solubility w	al	eth	ace	bz	other solvents	Ref.
	Naphthalene														
Ω n25	**—,2-acetyl-**	2-Acetonaphthone. Methyl 2-naphthyl ketone. $C_{12}H_{10}O$. *See* n14	170.21	nd (lig, dil al or xyl) λ^{al} 247 (4.8), 280 (3.9), 290 (3.9)	56	301–3 171–3[17]	...	...	...	δ	...	...	...	CS_2 s lig δ, s^h	**B7²**, 338
n26	**—,2-acetyl-1-amino-, hydrochloride**	$C_{12}H_{11}NO.HCl$. *See* n14	186.21	cr (w)	220d	...	...	...	s	...	...	...	...	chl, CS_2, aa s	**C34**, 5086
n27	**—,2-acetyl-4-bromo-1-hydroxy-**	$C_{12}H_9BrO_2$. *See* n14	265.11	ye nd (al)	126–7	...	...	...	i	s	s	...	s	CS_2, chl, lig s	**B8²**, 178
n28	**—,1-acetyl-2-hydroxy-**	1-Acetyl-2-naphthol. $C_{12}H_{10}O_2$. *See* n14	186.21	pa ye lf (peth), rh (lig), nd or pl (gasoline) λ^{al} 227 (4.71), 299 (3.60), 337 (3.58)	64–5	...	...	...	...	v	v	...	v	os, con sulf v	**B8²**, 175
n29	**—,1-acetyl-4-hydroxy-**	4-Acetyl-1-naphthol. $C_{12}H_{10}O_2$. *See* n14	186.21	yesh pr (aa, to or al)	198	...	...	...	...	s	...	...	s	alk s aa s^h	**B8²**, 176
Ω n30	**—,2-acetyl-1-hydroxy-**	2-Acetyl-1-naphthol. $C_{12}H_{10}O_2$. *See* n14	186.21	(i) pr (bz or lig) (ii) gr-ye nd (al) λ^{al} 256 (4.38), 284 (3.71), 366 (3.66)	(i) 98 (ii) 103	325δd	...	...	i	δ	...	...	s	chl, CS_2, aa s	**B8¹**, 567
n31	**—,2-acetyl-3-hydroxy-**	3-Acetyl-2-naphthol. $C_{12}H_{10}O_2$. *See* n14	186.21	ye lf or nd (al, peth) λ^{al} 250 (4.46), 304 (3.82), 390 (3.18)	112	...	...	...	...	δ s^h	...	v	v	dil NaOH s lig δ	**B8²**, 179
n32	**—,2-acetyl-6-hydroxy-**	6-Acetyl-2-naphthol. $C_{12}H_{10}O_2$. *See* n14	186.21	pr (bz)	172	...	...	...	...	s	...	...	s^h	NaOH s	**B8²**, 179
n33	**—,3-acetyl-1-hydroxy-**	3-Acetyl-1-naphthol. $C_{12}H_{10}O_2$. *See* n14	186.21	nd (bz)	173–4	...	...	...	...	s	...	...	δ s^h	alk, aa, con sulf s	**B8**, 150
n34	**—,2-acetyl-1-hydroxy-4-nitro-**	$C_{12}H_9NO_4$. *See* n14	231.21	ye nd (al)	159	...	...	...	i	δ s^h	s	...	s	...	**B8²**, 179
Ω n35	**—,1-allyl-**	4-α-Naphthylpropene. $C_{13}H_{12}$. *See* n14	168.24	λ^{al} 228 (4.73), 296 (3.97)	...	265–7 129–30[10]	1.0228^{20}_4	1.6140^{20}	...	s	...	...	v	chl s	**E12B**, 118
Ω n36	**—,1-amino-***	1-Naphthylamine*. α-Naphthylamine. $C_{10}H_9N$. *See* n14	143.19	nd (dil al, eth) λ^{al} 242 (4.27), 320 (3.71)	50	300.8[760] 160[12] sub	1.1229^{25}_{25}	1.67034^{51}	δ	v	v	...	...	...	**B12²**, 675
Ω n37	**—,—,hydrochloride**	$C_{10}H_9N.HCl$. *See* n14	179.65	nd $\lambda^{w, 1N HCl}$ 302 (2.7), 311 (2.5), 317 (2.3)	...	sub	...	...	δ	s	s	...	...	...	**B12²**, 678
Ω n38	**—,2-amino-***	2-Naphthylamine* β-Naphthylamine. $C_{10}H_9N$. *See* n14	143.19	lf (w) λ^{al} 236 (4.78), 280 (3.82), 292 (3.73), 340 (3.28)	113	306.1 (294)	1.0614^{98}_4	1.64927^{98}	s v^h	s	s	...	...	...	**B12²**, 710
Ω n39	**—,—,hydrochloride**	$C_{10}H_9N.HCl$. *See* n14	179.65	lf	254	...	...	...	v	v	...	...	i	...	**B12²**, 712
Ω n42	**—,1-amino-4-bromo-***	$C_{10}H_8BrN$. *See* n14	222.10	nd (al, bz or peth)	102	...	...	...	...	s	...	s	s	lig s	**E12B**, 713
n43	**—,1-amino-5-bromo-***	$C_{10}H_8BrN$. *See* n14	222.10	lf or pl (w or lig)	69 (cor)	sub	...	...	δ^h	s	s	s	v	chl v lig s	**E12B**, 717
n44	**—,2-amino-1-bromo-***	$C_{10}H_8BrN$. *See* n14	222.10	rh nd (dil al or lig)	63–4	...	...	...	δ	v	s	...	v	chl v	**B12**, 1310
n45	**—,2-amino-3-bromo-***	$C_{10}H_8BrN$. *See* n14	222.10	pl (al)	169	...	...	...	...	s	...	...	...	aa s	**E12B**, 729
n46	**—,2-amino-6-bromo-***	$C_{10}H_8BrN$. *See* n14	222.10	lf (al, w or peth)	128	...	...	...	δ^h	s^h	...	s	s	peth s^h	**E12B**, 731
n47	**—,3-amino-1-bromo-***	$C_{10}H_8BrN$. *See* n14	222.10	nd (bz-peth or 90 % aa)	72	...	...	...	i	v	v	...	v	peth δ	**B12**, 1311
n48	**—,6-amino-1-bromo-***	$C_{10}H_8BrN$. *See* n14	222.10	cr λ^{cy} 218 (4.82), 247 (4.19), 331 (3.76)	38	207–10[16]	...	...	i	v	...	s	...	os v lig δ	**E12B**, 730
n49	**—,1-amino-4-bromo-2-nitro-***	$C_{10}H_7BrN_2O_2$. *See* n14	267.09	og cr (al or aa)	200	...	...	...	...	v^h	...	...	v	to, chl v aa s^h CS_2 δ	**E12B**, 775
n50	**—,1-amino-2-chloro-***	$C_{10}H_8ClN$. *See* n14	177.65	nd (peth or dil al)	60 (56)	...	...	...	δ^h	s	...	s	...	peth s^h	**E12B**, 709
Ω n51	**—,1-amino-4-chloro-***	$C_{10}H_8ClN$. *See* n14	177.65	nd (al, bz or lig)	99–100	...	...	...	...	s	s	...	s^h	lig s^h	**E12B**, 711
n52	**—,2-amino-1-chloro-***	$C_{10}H_8ClN$. *See* n14	177.65	nd (al or peth)	60	...	...	...	...	s	...	...	...	aa s lig s^h	**E12B**, 720
n53	**—,1-amino-2,4-dibromo-***	$C_{10}H_7Br_2N$. *See* n14	300.99	nd or pl (dil al)	118–9	...	...	...	i	v	v	...	v	chl, lig v	**E12B**, 735
n54	**—,2-amino-1,4-dibromo-***	$C_{10}H_7Br_2N$. *See* n14	300.99	nd (al or bz)	106–7	...	...	...	...	v	v	...	s	os v	**E12B**, 737

For explanations, symbols and abbreviations see beginning of table. For structural formulas see end of table.

No.	Name	Synonyms and Formula	Mol. wt.	Color, crystalline form, specific rotation and λ_{max} (log ε)	m.p. °C	b.p. °C	Density	n_D	Solubility w	al	eth	ace	bz	other solvents	Ref.
	Naphthalene														
n55	—,2-amino-1,6-dibromo-*	$C_{10}H_7Br_2N$. *See* n14	300.99	nd (al or peth)	121	...	...	...	δ	v	...	...	v	aa s	**E12B,** 737
n56	—,1-amino-2,4-dichloro-*	$C_{10}H_7Cl_2N$. *See* n14	212.08	nd (al)	83–4	...	...	...	i	v	...	...	...	to s^h	**E12B,** 734
n57	—,5-amino-1,4-dihydro-*	$C_{10}H_{11}N$. *See* n14	145.21	pl or nd (to), turns pink in air	37.5 (cor)	247^{408}	...	...	...	s	...	...	...	chl, ac s	**E12B,** 667
n58	—,1-amino-2,4-dinitro-*	$C_{10}H_7N_3O_4$. *See* n14	233.19	ye nd (al), pr (aa), cr (ace) λ^{HClO_4} 400 (4.04)	242	...	...	...	i	δ	δ	v	δ	chl, aa δ lig i Py s^h	**E12B,** 784
n59	—,2-amino-1,6-dinitro-*	$C_{10}H_7N_3O_4$. *See* n14	233.19	gold-ye nd (al or aa), ye pw (Py)	248	...	...	...	δ	δ	δ	δ	δ	Py s^h chl δ lig i	**E12B,** 793
n60	—, 1-amino-4-fluoro-*	$C_{10}H_8FN$. *See* n14	161.19	lt ye	48	162^{16}	...	...	...	...	...	...	...	ac s	**E12B,** 710
n61	—,1-amino-2-hydroxy-*	1-Amino-2-naphthol. $C_{10}H_9NO$. *See* n14	159.19	silvery lf (bz, eth)	150d (175d)	...	...	...	δ^h	s	δ s^h	...	...	dil alk, dil ac v	**B13^2,** 412
n62	—,1-amino-6-hydroxy-*	5-Amino-2-naphthol. $C_{10}H_9NO$. *See* n14	159.19	nd or og pr (w)	190.6 (185)	...	...	...	s	s	s	s	...	NH_3, os s	**E12B,** 1688
Ω n63	—,1-amino-7-hydroxy-*	8-Amino-2-naphthol. $C_{10}H_9NO$. *See* n14	159.19	nd (w, al)	205–7	sub	...	...	s^h	v^h	s	...	δ	lig δ	**E12B,** 1696
n64	—,2-amino-3-hydroxy-*	3-Amino-2-naphthol. $C_{10}H_9NO$. *See* n14	159.19	silvery lf (bz), nd (al, w)	235	...	...	...	δ s^h	v	δ s^h	...	δ	dil ac, dil alk v	**E12B,** 1677
n65	—,2-amino-6-hydroxy-*	6-Amino-2-naphthol. $C_{10}H_9NO$. *See* n14	159.19	br (dil al), pr (w)	192–4 212–3d (sealed tube)	...	...	...	s	s	...	...	...	...	**B13^2,** 412
n66	—,2-amino-7-hydroxy-*	7-Amino-2-naphthol. $C_{10}H_9NO$. *See* n14	159.19	nd or lf (al)	201 (208)	...	...	...	δ	v	v	...	...	...	**E12B,** 1691
n67	—,2-amino-3-iodo-*	$C_{10}H_8IN$. *See* n14	269.10	cr (al)	137	...	...	...	...	s	...	...	s	chl, aa s	**E12B,** 729
n68	—,1-amino-2-methyl-*	$C_{11}H_{11}N$. *See* n14	157.22	nd (peth), turns red in air	32	...	...	...	δ	...	s	...	...	os v lig s	**B12^2,** 742
n69	—,1-amino-3-methyl-*	$C_{11}H_{11}N$. *See* n14	157.22	cr (peth) λ^{al} 234 (4.6), 299 (3.78)	51–2	...	...	...	...	s	s	...	...	os, lig s	**B12^2,** 743
n70	—,1-amino-4-methyl-*	$C_{11}N_{11}N$. *See* n14	157.22	nd (peth)	51–2	176^{12}	...	...	δ	...	s	...	...	os v lig s^h	**B12^2,** 740
n71	—,2-amino-1-methyl-*	$C_{11}H_{11}N$. *See* n14	157.22	nd (lig), pr (peth)	51	...	...	...	δ^h	v	v	s	v	chl v lig s	**B12^2,** 740
n72	—,2-amino-6-methyl-*	$C_{11}H_{11}N$. *See* n14	157.22	lf (peth, w) turns red in air	129–30	...	...	...	δ	...	...	...	...	os, min ac v	**B12^2,** 743
n73	—,1-amino-2-nitro-*	$C_{10}H_8N_2O_2$. *See* n14	188.20	ye-red mcl pr (al)	144	...	...	...	...	s^h	...	...	...	...	**B12^2,** 703
n74	—,1-amino-3-nitro-*	$C_{10}H_8N_2O_2$. *See* n14	188.20	og-ye nd (50 % al)	137	...	...	...	...	v	...	...	s	chl s	**E12B,** 748
Ω n75	—,1-amino-4-nitro-*	$C_{10}H_8N_2O_2$. *See* n14	188.20	og-ye nd (al) λ^{al} 269 (3.94), 443 (4.18)	195 (192)	...	...	...	δ^h	v	...	...	...	aa v	**B12^2,** 704
n76	—,1-amino-5-nitro-*	$C_{10}H_8N_2O_2$. *See* n14	188.20	red nd (w)	118–9	...	...	...	s^h	...	s	...	...	aa s	**B12^2,** 705
n77	—,1-amino-6-nitro-*	$C_{10}H_8N_2O_2$. *See* n14	188.20	og-red nd (chl) λ^{al} 290 (4.11) 353 (3.49), 429 (3.29)	172–3	...	...	...	...	s	...	...	...	chl s^h	**B12^2,** 705
n78	—,1-amino-8-nitro-*	$C_{10}H_8N_2O_2$. *See* n14	188.20	red lf (peth)	96–7	...	...	...	...	...	s	...	...	dil sulf s lig s^h	**B12^2,** 705
Ω n79	—,2-amino-1-nitro-*	$C_{10}H_8N_2O_2$. *See* n14	188.20	og-ye nd (al)	126–7	...	...	...	s^h	v	...	v	...	aa v	**B12,** 1313
n80	—,2-amino-6-nitro-*	$C_{10}H_8N_2O_2$. *See* n14	188.20	lt og pl (al), ye pl (aa) λ^{al} 292 (3.81), 429 (4.14)	207.5	...	...	...	δ	...	...	...	s	diox s aa s^h	**E12B,** 765
n81	—,6-amino-1-nitro-*	$C_{10}H_8N_2O_2$. *See* n14	188.20	red nd (al) λ^{al} 277 (4.20), 352 (3.55)	143.5	...	...	...	...	v^h	...	...	v	aa v lig i	**B12,** 1314
n82	—,7-amino-1-nitro-*	$C_{10}H_8N_2O_2$. *See* n14	188.20	red nd (aa) λ^{al} 275 (4.22), 347 (3.42), 427 (3.46)	104.5	...	...	...	...	s	s	...	s	os v aa s^h lig i	**B12,** 1315
n83	—,2-amino-1-nitroso-*	$C_{10}H_8N_2O$. *See* n14	172.20	gr nd (dil al), dk gr nd (bz)	150–2	...	...	...	δ^h	v^h	...	...	...	dil ac v os s^h	**B7,** 717
n84	—,2-amino-1,3,6-tribromo-*	$C_{10}H_6Br_3N$. *See* n14	379.89	pa red cr (chl, al or al-eth)	143	...	...	...	δ^h	δ^h	v	...	...	chl v peth δ	**E12B,** 741
n85	—,1(2-amino-ethyl)-*	$C_{12}H_{13}N$. *See* n14	171.25	...	...	$182–3^{18}$	...	...	...	v	...	...	...	xyl s	**E12B,** 425
n86	—,2(2-amino-ethyl)-*	$C_{12}H_{13}N$. *See* n14	171.25	...	...	174^{25}	...	...	...	s	...	...	...	aa s	**E12B,** 426

For explanations, symbols and abbreviations see beginning of table. For structural formulas see end of table.

No.	Name	Synonyms and Formula	Mol. wt.	Color, crystalline form, specific rotation and λ_{max} (log ε)	m.p. °C	b.p. °C	Density	n_D	Solubility w	al	eth	ace	bz	other solvents	Ref.
	Naphthalene														
n87	**—,1(2-amino-ethylamino)-***	*N*-α-Naphthylethylenediamine $C_{10}H_7NHCH_2CH_2NH_2$	186.26	ye	....	320d 204^{9}	1.114_4^{25}	1.6648^{25}	δ	s	...	v	...	os v lig i	**E12B,** 518
n88	**—,1(amino-methyl)-***	$C_{11}H_{11}N$. *See* n14	157.22	yesh turns red in air	....	294–5 $162–3^{12}$	....	....	...	s	s	...	...	sulf s (bl) CS_2 s	**B12², 741**
n89	**—,2(amino-methyl)-***	$C_{11}H_{11}N$. *See* n14	157.22	pr (eth)	59–60	180^{24} 149^{12}	....	....	δ	v	v	...	...	....	**E12B,** 421
n90	**—,1-benzyl-2-hydroxy-**	1-Benzyl-2-naphthol. $C_{17}H_{14}O$. *See* n14	234.30	nd (bz, dil HCO_2H)	115 (110)	$247–50^{15}$	....	....	i	s	v	v	s	chl s	**B6²**, 680
n91	**—,1-benzyl-4-hydroxy-**	4-Benzyl-1-naphthol. $C_{17}H_{14}O$. *See* n14	234.30	pl or nd (dil aa-lig), pr (bz)	125–6	237^{9-10}	....	....	δ	...	...	...	...	os s lig δ	**B6²**, 680
n92	**—,2-benzyl-1-hydroxy-**	2-Benzyl-1-naphthol. $C_{17}H_{14}O$. *See* n14	234.30	nd (lig), pr (bz) λ^{MeOH} 220 (4.52), 236 (4.60), 270 (3.48), 295 (3.64)	73.5–4.0	$237–40^{12}$	....	....	i	...	...	...	s^h	lig δ	**B6²**, 681
n92[1]	**—,1(benzylidene-amino)-**	Benzaldehyde α-naphthylimide. $C_{17}H_{13}N$. *See* n14	231.30	ye lf (al)	73.5	....	....	....	i	s^h	s	...	s	MeOH s	**E12B,** 507
n92[2]	**—,2(benzylidene-amino)-**	Benzaldehyde β-naphthylimide. $C_{17}H_{13}N$. *See* n14	231.30	yesh nd (al) $\lambda^{decalin}$ 335 (4.1)	103	....	....	....	i	s^h	...	...	s	chl, aa s	**E12B,** 602
Ω n93	**—,1-bromo-***	$C_{10}H_7Br$. *See* n14	207.08	pr (β form)	−6.2(α) 2–2.7(β)	281 139^{16}	1.4826_4^{20}	1.658^{20}	s^h	∞	∞	s	∞	chl s	**B5³**, 1580
Ω n94	**—,2-bromo-***	$C_{10}H_7Br$. *See* n14	207.08	pl or rh lf (al) λ^{Me-cy} 410 (−1.2)	59, fp 55.2	$281–2^{760}$ 147^{18}	1.605^{0}	1.6382^{60}	i	s	s	...	s	chl, CS_2 s	**B5³**, 1583
n95	**—,1-bromo-2(bromo-methyl)-***	$C_{11}H_8Br_2$. *See* n14	300.02	nd (al), cr (peth)	107–8	....	....	....	d^h	s	...	...	s	lig s^h	**B5³**, 1633
Ω n96	**—,1-bromo-2-hydroxy-***	1-Bromo-2-naphthol. $C_{10}H_7BrO$. *See* n14	223.08	rh pr (bz-lig), nd (aa, lig)	84	d130	....	....	i	s	s	...	s	aa v lig s	**E12B,** 1484
n97	**—,1-bromo-4-hydroxy-***	4-Bromo-1-naphthol. $C_{10}H_7BrO$. *See* n14	223.08	nd (dil al), nd (hx)	128	....	....	....	...	s^h	...	...	...	chl, aa s	**B6³**, 2935
n98	**—,1-bromo-5-hydroxy-***	5-Bromo-1-naphthol. $C_{10}H_7BrO$. *See* n14	223.08	nd (w)	137	....	....	....	s^h	...	s	...	...	....	**B6³**, 2936
n99	**—,1-bromo-6-hydroxy-***	5-Bromo-2-naphthol. $C_{10}H_7BrO$. *See* n14	223.08	nd (w)	105	....	....	....	s^h	s	...	...	...	....	**B6²**, 605
n100	**—,1-bromo-8-hydroxy-***	8-Bromo-1-naphthol. $C_{10}H_7BrO$. *See* n14	223.08	pl (peth)	61	....	....	....	...	s	s	s	...	oos s lig s^h	**B6**, 614
n101	**—,2-bromo-3-hydroxy-***	3-Bromo-2-naphthol. $C_{10}H_7BrO$. *See* n14	223.08	nd (lig)	84–5	....	....	....	δ^h	v	...	...	v	oos s lig s^h	**B6²**, 605
Ω n102	**—,2-bromo-6-hydroxy-***	6-Bromo-2-naphthol. $C_{10}H_7BrO$. *See* n14	223.08	nd (bz)	127 (130)	....	....	....	...	s	...	...	s^h	....	**B6²**, 605
n103	**—,2-bromo-7-hydroxy-***	7-Bromo-2-naphthol. $C_{10}H_7BrO$. *See* n14	223.08	cr (peth)	132–3	....	....	....	...	...	s	s	...	lig s^h	**B6²**, 605
n104	**—,6-bromo-1-hydroxy-***	6-Bromo-1-naphthol. $C_{10}H_7BrO$. *See* n14	223.08	nd (w)	129.5–30	....	....	....	v^h	...	...	...	...	....	**B6³**, 2936
n105	**—,7-bromo-1-hydroxy-***	7-Bromo-1-naphthol. $C_{10}H_7BrO$. *See* n14	223.08	cr (w)	105.6–6.5	....	....	....	s^h	...	...	...	...	....	**B6²**, 583
n106	**—,6-bromo-2-hydroxy-1-methyl-***	$C_{11}H_9BrO$. *See* n14	237.10	nd (bz)	129	....	....	....	i	v^h s	v^h s	v	δ^h	chl v aa s, v^h peth δ	**E12B,** 1501
n107	**—,1(bromo-methyl)-***	$C_{11}H_9Br$. *See* n14	221.10	cr (peth or al)	56	183^{18} $145–50^{1}$	....	....	i^h	s	s	v	s	aa s^h	**B5²**, 462
n108	**—,2(bromo-methyl)-***	$C_{11}H_9Br$. *See* n14	221.10	lf (al)	56	213^{100} $165–9^{14}$	....	....	δd	s	s	...	...	chl s aa s^h	**B5²**, 464
Ω n109	**—,1-chloro-***	$C_{10}H_7Cl$. *See* n14	162.62	cr (al, ace, CCl_4) λ^{cy} 225 (4.83), 285.5 (3.83), 295 (3.61)	−2.3	258.8^{753} 106.5^{5}	1.1938_4^{20}	1.6326^{20}	i	s	s	...	s	CS_2 s	**B5³**, 1570
Ω n110	**—,2-chloro-***	$C_{10}H_7Cl$. *See* n14	162.62	pl (dil al), lf λ^{MeOH} 226 (4.91), 270 (3.67), 288 (3.47), 322 (2.25)	61	256 $121–2^{12}$	1.1377_4^{71}	1.6079^{71}	i	s	s	...	s	chl, CS_2 s	**B5³**, 1573
n111	**—,1-chloro-2-hydroxy-***	1-Chloro-2-naphthol. $C_{10}H_7ClO$. *See* n14	178.62	nd (lig), pr (chl), pl (w)	71	....	....	....	δ v^h	v^h	...	...	v^h	chl v lig s^h	**B6³**, 2990
n112	**—,1-chloro-2-nitro-***	$C_{10}H_6ClNO$. *See* n14	207.62	pa ye nd (al, lig) λ^{MeOH} 220 (4.73), 257 (4.13), 284 (3.73)	81	....	....	....	...	s^h	...	...	...	lig s^h	**E12B,** 374
n113	**—,1-chloro-3-nitro-***	$C_{10}H_6ClNO$. *See* n14	207.62	ye nd (HCO_2H, al)	129.5	....	....	....	...	s^h	...	...	...	HCO_2H s^h	**E12B,** 376

For explanations, symbols and abbreviations see beginning of table. For structural formulas see end of table.

No.	Name	Synonyms and Formula	Mol. wt.	Color, crystalline form, specific rotation and λ_{max} (log ε)	m.p. °C	b.p. °C	Density	n_D	Solubility w	al	eth	ace	bz	other solvents	Ref.
	Naphthalene														
n114	**—,1-chloro-4-nitro-***	$C_{10}H_6ClNO_2$. *See* n14	207.62	br-ye nd (peth or al) λ^{MeOH} 230 (4.24), 250 (4.05)	87–7.5 (85)	...	...	...	i	s	s	...	...	peth s^h	**E12B,** 370
n115	**—,1-chloro-5-nitro-***	$C_{10}H_6ClNO_2$. *See* n14	207.62	nd (dil al or aa)	111	>360 181^{2}	...	...	i	s	...	...	...	aa s^h	**E12B,** 372
n116	**—,1-chloro-6-nitro-***	$C_{10}H_6ClNO_2$. *See* n14	207.62	ye nd (dil al)	131 (120)	...	...	...	...	v	δ	v	v	chl, lig v	**E12B,** 377
Ω n117	**—,1-chloro-8-nitro-***	$C_{10}H_6ClNO_2$. *See* n14	207.62	lt ye nd (gl aa bz or lig)	94–5	>360 175^{2}	...	...	...	s	...	...	s^h	aa s lig δ, s^h	**E12B,** 373
n118	**—,2-chloro-1-nitro-***	$C_{10}H_6ClNO_2$. *See* n14	207.62	nd (peth), ye nd (al) λ^{MeOH} 226 (4.93), 265 (3.60)	99–100	>360	...	...	...	v	v	v	v	aa v peth δ, s^h	**E12B,** 368
n119	**—,2-chloro-3-nitro-***	$C_{10}H_6ClNO_2$. *See* n14	207.62	br cr or nd (al)	94.5 (79)	...	...	...	...	...	...	...	...	...	**E12B,** 375
n120	**—,2-chloro-6-nitro-***	$C_{10}H_6ClNO_2$. *See* n14	207.62	ye nd	170	$180–90^{15}$	...	...	...	...	...	...	...	...	**C41,** 5492
n121	**—,2-chloro-7-nitro-***	$C_{10}H_6ClNO_2$. *See* n14	207.62	pa ye nd	136	...	...	...	...	...	...	...	...	...	**C41,** 5492
n122	**—,3-chloro-1-nitro-***	$C_{10}H_6ClNO_2$. *See* n14	207.62	grsh-br nd ($PhNO_2$ or 90 % HCO_2H)	105	...	...	...	δ	...	...	...	...	$PhNO_2$ s^h	**E12B,** 369
n123	**—,6-chloro-1-nitro-***	$C_{10}H_6ClNO_2$. *See* n14	207.62	nd (aq ace)	100.5	...	...	...	δ	...	...	s	...	...	**E12B,** 373
n124	**—,7-chloro-1-nitro-***	$C_{10}H_6ClNO_2$. *See* n14	207.62	ye nd (al)	116	...	...	...	i	s	s	...	...	...	**E12B,** 373
Ω n126	**—,1(chloro-methyl)-***	$C_{11}H_9Cl$. *See* n14	176.65	pr	32	291–2 $135–9^{6}$	...	...	...	...	...	...	...	...	**B5³,** 1622
n127	**—,2(chloro-methyl)-***	$C_{11}H_9Cl$. *See* n14	176.65	lf (al)	48–9	170^{20}	...	...	$δ^h$	s^h	...	...	s	...	**B5³,** 1632
—	**—,decahydro-***	*see* **Decalin**													
Ω n128	**—,1,2-diamino-***	1,2-Naphthylenediamine $C_{10}H_{10}N_2$. *See* n14	158.21	lf (w), rose to br in air	98–5	214^{13} $150–1^{0.5}$	...	...	$δ^h$	v	v	...	...	chl v	**E12B,** 805
n129	**—,1,4-diamino-***	1,4-Naphthylenediamine. $C_{10}H_{10}N_2$. *See* n14	158.21	ye nd (w)	120	...	...	1.6441^{18}	$δ^h$	v	v	...	v	chl v	**B13²,** 82
Ω n130	**—,1,5-diamino-***	1,5-Naphthylenediamine. $C_{10}H_{10}N_2$. *See* n14	158.21	pr (eth, al, w)	190	sub	1.4	...	s^h	s v^h	s v^h	...	...	chl v	**E12B,** 824
n131	**—,1,6-diamino-***	1,6-Naphthylenediamine. $C_{10}H_{10}N_2$. *See* n14	158.21	nd (w, eth)	85–6	...	1.1477^{99}_{4}	1.7083^{99}	δ s^h	s s^h	δ s^h	...	s	...	**E12B,** 828
n132	**—,1,7-diamino-***	1,7-Naphthylenediamine. $C_{10}H_{10}N_2$. *See* n14	158.21	lf (bz), nd (w)	117.5	...	...	...	$δ^h$	s	δ	...	s^h	lig δ	**B13,** 204
Ω n133	**—,1,8-diamino-***	1,8-Naphthylenediamine. $C_{10}H_{10}N_2$. *See* n14	158.21	nd (dil al)	66.5	205^{12} sub	1.1265^{99}_{4}	1.6828^{99}	δ s^h	v	v	...	...	...	**B13²,** 85
Ω n134	**—,2,3-diamino-***	2,3-Naphthylenediamine. $C_{10}H_{10}N_2$. *See* n14	158.21	lf (eth or w)	199	...	1.0968^{26}_{4}	1.6342^{26}	δ	v	s	...	...	...	**B13²,** 86
n135	**—,2,6-diamino-***	2,6-Naphthylenediamine. $C_{10}H_{10}N_2$. *See* n14	158.21	nd or lf (w), lf (al)	222 d	...	...	...	$δ^h$	δ	δ	...	...	...	**B13²,** 86
n136	**—,1,4-diamino-2-methyl-***	$C_{11}H_{12}N_2$. *See* n14	172.23	ye cr (peth)	113–4	...	...	...	δ	...	...	...	...	aa, ac s peth s^h	**E12B,** 846
n137	**—,—,dihydro-chloride**	Vitamin K_6. $C_{11}H_{12}N_2$.2HCl. *See* n14	245.15	cr (dil HCl)	300d	...	...	...	v	...	...	...	...	...	**E12B,** 846
n138	**—,1,5-diamino-2-methyl-***	$C_{11}H_{12}N_2$. *See* n14	172.23	red-ye lf (dil al)	136 (128d)	...	...	...	δ	v	v	...	v	...	**E12B,** 847
n139	**—,1,3-diamino-2-phenyl-***	$C_{16}H_{14}N_2$. *See* n14	234.30	pl (MeOH or bz) turns red in air	116	...	...	...	...	v	δ	...	s^h	lig i	**E12B,** 843
n140	**—,2,6-dibromo-1,5-dihydroxy-, diacetate**	$C_{14}H_{10}Br_2O_4$. *See* n14	402.05	nd	147.5	...	...	...	...	...	...	...	...	lig s^h	**B6²,** 951
Ω n141	**—,1,6-dibromo-2-hydroxy-***	1,6-Dibromo-2-naphthol. $C_{10}H_6Br_2O$. *See* n14	301.98	nd (peth, bz), lf (al), nd (aa + 1)	106 84 (+1 aa)	...	...	...	i	s	s	s	...	...	**E12B,** 1512
n142	**—,1,2-dichloro-***	$C_{10}H_6Cl_2$. *See* n14	197.07	pl (al) λ^{hx} 273, 284, 316, 325	35–7	295–8 $151–3^{19}$	1.3147^{49}_{4}	1.6338^{49}	...	s	s	...	...	...	**E12B,** 302
n143	**—,1,3-dichloro-***	Dichloronaphthalene. $C_{10}H_6Cl_2$. *See* n14	197.07	nd or pr (al) λ^{hx} 272, 277, 312, 319, 326	61.5	291^{775}	...	...	...	s	...	...	...	...	**E12B,** 304
n144	**—,1,4-dichloro-***	β-Dichloronaphthalene. $C_{10}H_6Cl_2$. *See* n14	197.07	nd or pr (al, aa or ace)	68	$286–7^{740}$ 147^{12}	1.2997^{76}_{4}	1.6228^{76}	i	δ	s	v	s	aa s	**E12B,** 306
n145	**—,1,5-dichloro-***	γ-Dichloronaphthalene. $C_{10}H_6Cl_2$. *See* n14	197.07	nd or lf (al, MeOH, aa), pr (sub) λ^{hx} 280, 292, 320, 324	107	sub	...	...	i	δ	s	...	...	...	**E12B,** 309

For explanations, symbols and abbreviations see beginning of table. For structural formulas see end of table.

No.	Name	Synonyms and Formula	Mol. wt.	Color, crystalline form, specific rotation and λ_{max} (log ε)	m.p. °C	b.p. °C	Density	n_D	Solubility						Ref.
									w	al	eth	ace	bz	other solvents	
	Naphthalene														
n146	**—,1,6-dichloro-***	η-Dichloronaphthalene. $C_{10}H_6Cl_2$, *See* n14	197.07	nd or pr (al, peth, sub) λ^{hx} 276, 288, 317, 325	49	sub	...	...	...	...	...	...	...	...	**E12B,** 310
n147	**—,1,7-dichloro-***	θ-Dichloronaphthalene. $C_{10}H_6Cl_2$. *See* n14	197.07	nd or pr (al, aa) λ^{hx} 267, 273, 296, 320, 326	63.5–4.5	285–6	1.2611_4^{100}	1.6092^{100}	...	s	s	...	s	aa s	**E12B,** 312
n148	**—,1,8-dichloro-***	ζ-Dichloronaphthalene. *peri*-Dichloronaphthalene. $C_{10}H_6Cl_2$. *See* n14	197.07	rh pl (hx), nd (al, sub) λ^{hx} 296, 308 319, 324, 329	89–9.5	d sub	1.2924_4^{100}	1.6236^{100}	...	s^h	...	...	...	peth s^h	**E12B,** 313
n149	**—,2,3-dichloro-***	ι-Dichloronaphthalene. $C_{10}H_6Cl_2$. *See* n14	197.07	rh lf (al) λ^{hx} 272, 283, 312, 326	120	...	...	...	i	δ s^h	v	...	...	...	**E12B,** 315
n150	**—,2,6-dichloro-***	ε-Dichloronaphthalene. $C_{10}H_6Cl_2$. *See* n14	197.07	pr (aa), nd or lf (al), pl (eth, bz) λ^{hx} 277, 301, 313, 330	140–1	285	...	...	...	δ	s	...	s	chl, aa s	**E12B,** 317
n151	**—,2,7-dichloro-***	δ-Dichloronaphthalene. $C_{10}H_6Cl_2$. *See* n14	197.07	pl or lf (al) λ^{hx} 260, 270, 281, 311, 327	114	...	...	...	...	v^h	...	...	...	hx, aa s^h	**E12B,** 318
Ω n152	**—,1(diethyl-amino)-***	*N,N*-Diethyl-1-naphthylamine*. $C_{14}H_{17}N$. *See* n14	199.30	...	...	285 $155–65^{30}$	1.015_{20}^{20}	1.5961^{20}	i	s	s	...	...	aa s	**C37,** 1397
n153	**—,2,3-di-hydrazino-***	$C_{10}H_{12}N_4$. *See* n14	188.23	red-br (al or w), col nd (bz), br-ye (AmOH)	167–8d (w, AmOH) 155–6 (bz, al)	...	...	...	...	v	...	...	...	dil ac s sulf s(red) aa δ	**E12B,** 908
Ω n154	**—,1,2-dihydro-***	Δ^1-Dialin. Δ^1-Dihydro-naphthalene. $C_{10}H_{10}$. *See* n14	130.19	lf, pl on cooling	−8– −7	206–7 78^9	0.9974_4^{20}	1.58137^{20}	...	...	...	...	...	...	**E12B,** 53
n155	**—,1,4-dihydro-***	Δ^2-Dialin. $C_{10}H_{10}$. *See* n14	130.19	pl	25 (30)	$211–2^{760}$ 94^{17}	0.9928_4^{33}	1.5577^{20}	...	...	...	...	...	aa s^h	**E12B,** 54
n156	**—,1,2-dihydroxy-***	β-Naphthohydroquinone. 1,2-Naphthalenediol*. $C_{10}H_8O_2$. *See* n14	160.18	lf or nd (CS_2), lf (w + 1), nd (lig)	103–4 (anh) 58–60 (+1w)	...	...	...	δ		s	...	...	sulf v alk s lig s^h	**E12B,** 1962
Ω n157	**—,1,4-dihydroxy-***	1,4-Naphthalenediol*. α-Naphthohydroquinone. $C_{10}H_8O_2$. *See* n14	160.18	mcl nd (bz, w)	192 (176)	...	...	...	s^h	s	s	...	i	aa s CS_2, lig i	**E12B,** 1973
Ω n158	**—,1,5-dihydroxy-***	1,5-Naphthalenediol*. $C_{10}H_8O_2$. *See* n14	160.18	pr (w), nd (sub)	265d	sub	...	...	δ	δ	v	v	i	aa s lig i	**E12B,** 1980
n159	**—,—,diacetate**	$C_{14}H_{12}O_4$. *See* n14	244.25	nd (bz)	161	...	...	...	...	...	...	...	s^h	aa s^h	**E12B,** 1982
Ω n160	**—,1,6-dihydroxy-***	1,6-Naphthalenediol*. $C_{10}H_8O_2$. *See* n14	160.18	pr (bz)	138	sub	...	...	δ	δ	s	s	s	MeOH s chl δ lig i	**E12B,** 1987
Ω n161	**—,1,7-dihydroxy-***	1,7-Naphthalenediol*. $C_{10}H_8O_2$. *See* n14	160.18	nd (bz or sub)	178–81	sub	...	...	δ s^h	v	v	...	s	aa s	**E12B,** 1989
n162	**—,1,8-dihydroxy-***	1,8-Naphthalenediol*. *peri*-Dihydroxy-naphthalene. $C_{10}H_8O_2$. *See* n14	160.18	lf or nd (w)	144 (140)	...	...	...	δ^h	s^h	v	...	v	to v lig δ	**E12B,** 1991
Ω n163	**—,2,3-dihydroxy-***	2,3-Naphthalenediol*. $C_{10}H_8O_2$. *See* n14	160.18	lf (w)	163.5–4	...	...	...	s^h	s	s	...	s	aa, lig s	**E12B,** 1994
Ω n164	**—,2,6-dihydroxy.***	2,6-Naphthalenediol*. $C_{10}H_8O_2$. *See* n14	160.18	rh pl (w)	222	sub	...	...	δ s^h	s	s	s	δ	aa s lig i	**E12B,** 2003
Ω n165	**—,2,7-dihydroxy-***	2,7-Naphthalenediol*. $C_{10}H_8O_2$. *See* n14	160.18	nd (w or dil al), pl (dil al)	190 (nd) 194 (pl)	sub	...	...	s^h	s	s	...	s	chl, to s lig, CS_2 i	**E12B,** 2007
Ω n166	**—,1,4-dihydroxy-2-methyl-, diacetate**	$C_{15}H_{14}O_4$. *See* n14	258.28	pr (al)	113	...	...	...	i	s^h	...	...	δ	...	**B6**^{2}, 958
n167	**—,1,5-di-mercapto-***	1,5-Naphthalenedithiol*. $C_{10}H_8S_2$. *See* n14	192.31	ye lf (al, eth, bz)	119	...	...	...	...	s	s		s	dil HCl v	**B6**^{2}, 952
n168	**—,1,4-dimethyl-***	α-Dimethylnaphthalene. $C_{12}H_{12}$. *See* n14	156.23	...	7.66	268^{760} 129^{10}	1.0166_4^{20}	1.6127^{20}	i	v	∞	∞	∞	CCl_4, lig ∞	**B5**^{3}, 1645
Ω n169	**—,2,3-dimethyl-***	Guaiene. $C_{12}H_{12}$. *See* n14	156.23	lf (al, MeOH)	105	268^{760} 128.7^{10}	1.003_4^{20}	1.5060^{20}	i	δ s^h	s	...	s	aa δ	**B5**^{3}, 1650
Ω n170	**—,1(dimethyl-amino)-***	*N,N*-Dimethyl-1-naphthylamine*. $C_{12}H_{13}N$. *See* n14	171.25	vt flr λ^{al} 236 (4.13), 305 (3.68)	...	274.5^{711} $139–40^{13}$	1.0423_4^{20}	1.624^{15}	i	s	s	...	...	...	**B12**^{1}, 521
Ω n171	**—,2(dimethyl-amino)-***	*N,N*-Dimethyl-2-naphthylamine*. $C_{12}H_{13}N$. *See* n14	171.25	dk red nd λ^{al} 240 (4.65), 282 (3.86), 345 (3.35)	52–3 (46–7)	305 $160–2^{12}$	1.0455_{40}^{40}	1.6443^{53}	i	s	s	...	...	...	**B12,** 1273
n172	**—,1,6-dimethyl-4-isopropyl-***	Cadalene. $C_{15}H_{18}$. *See* n14	198.31	...	...	$291–2^{720}$ 165^{20}	0.9792_4^{19}	1.5851^{19}	...	...	...	...	...	...	**B5**^{3}, 1683
n173	**—,1,3-dinitro-***	γ-Nitronaphthalene. $C_{10}H_6N_2O_4$. *See* n14	218.17	ye nd (bz or Py-w)	147–9	sub	...	...	i	s	...	s	...	...	**B5**^{3}, 1605

For explanations, symbols and abbreviations see beginning of table. For structural formulas see end of table.

No.	Name	Synonyms and Formula	Mol. wt.	Color, crystalline form, specific rotation and λ_{max} (log ε)	m.p. °C	b.p. °C	Density	n_D	Solubility w	al	eth	ace	bz	other solvents	Ref.
	Naphthalene														
Ω n174	**—,1,5-dinitro-***	$C_{10}H_6N_2O_4$. *See* n14	218.17	hex nd (aa or ace) λ^{al} 233 (4.32), 327 (3.81)	219	sub	...	...	i	δ	v	δ	s^h	Py s^h, CS_2 δ	**B5**³, 1606
Ω n175	**—,1,8-dinitro-***	$C_{10}H_6N_2O_4$. *See* n14	218.17	ye rh pl (chl) λ^{al} 231 (4.44), 313 (3.81)	173–3.5	445d	...	...	i	δ	...	s	δ	Py s chl δ	**B5**³, 1607 **B14**², 653
n176	**—,1,6-dinitro-2-hydroxy-***	$C_{10}H_6N_2O_5$. *See* n14	234.17	pa ye nd (chl)	195d	...	...	...	i	s^h	v	...	...	Py, chl v lig i	**E12B,** 1581
n177	**—,2,4-dinitro-1-triazo-***	$C_{10}H_5N_5O_4$. *See* n14	259.18	ye rh nd (al)	105d	...	...	...	...	s^h	s	...	s	chl, to s lig s^h	**B5**², 460
n178	**—,5,8-dioxo-1,4,5,8,9,10-hexahydro-1,4-methylene-***	Cyclopentadienebenzoquinone.	174.20	gr-ye lf (MeOH) λ 285 (1.40)	77–8	...	...	...	...	s	s	s	s	os s	**E13,** 1033
Ω n179	**—,1-ethoxy-***	Ethyl α-naphthyl ether. $C_{12}H_{12}O$. *See* n14	172.23	nd	5.5	280.5 136–8[14]	1.060_4^{20}	1.5953[25]	i	v	v	...	...	...	**B6**³, 2924
Ω n180	**—,2-ethoxy-***	Ethyl β-naphthyl ether Nerolin II. $C_{12}H_{12}O$. *See* n14	172.23	pl (al)	37–8	282 148[10]	1.0640_{20}^{20}	1.5975[36]	i	s	s	...	...	to, lig, CS_2, s	**B6**³, 2972
Ω n181	**—,1-ethyl-***	$C_{12}H_{12}$. *See* n14	156.23	λ^{iso} 224 (4.9), 282 (3.8), 323 (1.4)	−13.88	258.67[760] 120[10]	1.00816_4^{20}	1.6062[20]	i	∞	∞	...	...	...	**B5**³, 1639
Ω n182	**—,2-ethyl-***	$C_{12}H_{12}$. *See* n14	156.23	...	−7.4	257.9[760] 119[10]	0.9922_4^{20}	1.5999[20]	i	∞	∞	...	...	...	**B5**³, 1641
n183	**—,1(ethylamino)-***	Ethyl α-naphthylamine. $C_{12}H_{13}N$. *See* n14	171.25	...	...	303[723] 191[16]	1.060[20]	1.6477[15]	i	s	s	...	...	...	**B12**², 682
n184	**—,2(ethylamino)-***	Ethyl β-naphthylamine. $C_{12}H_{13}N$. *See* n14.	171.25	...	<15	316–7 191[25]	1.0545[21]	1.6544[21]	...	...	...	...	...	...	**B12**², 715
Ω n185	**—,1-fluoro-***	$C_{10}H_7F$. *See* n14.	146.17	...	−9	215[756] 80[11]	1.1322[20]	1.5939[20]	i	s	s	...	s	chl, aa s	**B5**³, 1569
Ω n186	**—,2-fluoro-***	$C_{10}H_7F$. *See* n14.	146.17	nd (al)	61	211.5[737] 90[16] sub	...	...	i	s	s	...	s	chl, aa s	**B5**³, 1569
n187	**—,1(formylamino)-***	$C_{11}H_9NO$. *See* n14	171.20	nd (w)	137.5	...	...	...	s^h	s	s	s	...	os s	**E12B,** 459
n188	**—,2(formylamino)-***	$C_{11}H_9NO$. *See* n14	171.20	lf (bz-peth)	129	...	...	...	...	s	...	...	s	peth δ	**E12B,** 562
n189	**—,1,2,3,4,9,10-hexahydro-***	Naphthalene hexahydride. $C_{10}H_{14}$. *See* n14	134.22	...	...	200 82[2]	0.934[23]	1.5260[16]	...	...	s	...	s	...	**B5**, 433
Ω n190	**—,1-hydroxy-***	1-Naphthol. α-Naphthol. $C_{10}H_8O$. *See* n14	144.19	ye mcl nd (w) λ^{al} 292 (3.67), 308 (3.52), 322 (3.31)	96 (94)	288[760] sub	1.0989_4^{99}	1.6224[99]	i δ^h	v	v	s	s	chl v CCl_4 δ	**E12B,** 1148
Ω n191	**—,2-hydroxy-***	2-Naphthol. β-Naphthol. $C_{10}H_8O$. *See* n14	144.19	mcl lf (w), pl (CS_2) λ^{al} 226 (4.86), 265 (3.59), 274 (3.67), 285 (3.52)	123–4	295[760]	1.28[20]	...	i δ^h	v	v	...	s	chl s SO_2, CCl_4 δ lig δ^h	**E12B,** 1210
n192	**—,—,acetate**	2-Acetoxynaphthalene. $C_{12}H_{10}O_2$. *See* n14	186.21	nd (al)	71–2	132–4[2]	...	...	i	v	v	...	...	chl v	**E12B,** 1256
Ω n193	**—,—,benzoate**	2-Benzoyloxynaphthalene. $C_{17}H_{12}O_2$. *See* n14	248.29	nd or pr (al), cr (lig) λ^{al} 221 (4.88), 274 (3.71), 303 (2.59), 317 (2.50)	108	...	...	...	i	v^h	δ	...	...	...	**E12B,** 1260
Ω n197	**—,2-hydroxy-1-methyl-***	1-Methyl-2-naphthol. $C_{11}H_{10}O$. *See* n14	158.20	nd (w, bz-lig, dil aa)	112	180[12] sub	...	...	δ s^h	v	v	v	v	aa, chl v peth s	**E12B,** 1388
n198	**—,1-hydroxy-8-nitro-***	8-Nitro-1-naphthol. $C_{10}H_7NO_3$. *See* n14	189.17	grsh ye nd (al, chl, bz-hx)	130–3	...	...	...	s	v	v	v	v	alk s	**E12B,** 1547
Ω n199	**—,2-hydroxy-1-nitro-***	1-Nitro-2-naphthol. $C_{10}H_7NO_3$. *See* n14	189.17	ye nd, lf or pr (al) λ^{al} 330 (3.50)	104	115[0.05]	...	...	s^h	s	v	...	...	...	**E12B,** 1547
n199[2]	**—,2-hydroxy-5-nitro**	5-Nitro-2-naphthol. $C_{10}H_7NO_3$. *See* n14	189.17	lt ye nd (w)	147–9	...	...	...	v^h	δ	v	v	...	oos v	**E12B,** 1556
n201	**—,1-hydroxy-4-nitroso-***	1,4-Naphthoquinone 1-oxime*. 4-Nitroso-1-naphthol. $C_{10}H_7NO_2$. *See* n14	173.17	pa ye nd (bz), nd (dil al) λ^{al} 263 (4.08), 372.5 (3.70)	198	...	...	...	i	v	v	v	δ^h	MeOH v chl, CS_2 δ	**E12B,** 2786
Ω n202	**—,2-hydroxy-1-nitroso-***	1,2-Naphthoquinone 1-oxime*. 1-Nitroso-2-naphthol. $C_{10}H_7NO_2$. *See* n14	173.17	ye nd (bz), og pr or pl (al) $\lambda^{w, pH=7}$ 260 (3.78)	112 (109.5)	...	...	...	i	s v^h	v	s	v	aa v lig δ	**E12B,** 2731
n203	**—,2-hydroxy-1,3,6-tribromo-***	1,3,6-Tribromo-2-naphthol. $C_{10}H_5Br_3O$. *See* n14	380.88	nd (aa or al)	133	...	...	...	...	s	...	...	s	MeOH, CCl_4 s	**B6**³, 3000
n204	**—,2-hydroxy-1,4,6-tribromo-***	Providoform. 1,4,6-Tribromo-2-naphthol. $C_{10}H_5Br_3O$. *See* n14	380.88	nd (bz)	157–8	...	...	...	...	s	...	...	s^h	chl, aa s	**E12B,** 1522

For explanations, symbols and abbreviations see beginning of table. For structural formulas see end of table.

No.	Name	Synonyms and Formula	Mol. wt.	Color, crystalline form, specific rotation and λ_{max} (log ε)	m.p. °C	b.p. °C	Density	n_D	Solubility w	al	eth	ace	bz	other solvents	Ref.
	Naphthalene														
n205	**—,3-hydroxy-1,2,7-tribromo-***	3,4,6-Tribromo-β-naphthol. $C_{10}H_5Br_3O$. *See* n14	380.88	nd (bz)	127–8	....	....	....	...	s	...	...	v^h	....	**B6^2**, 607
n206	**—,1-hydroxy-2,3,4-trichloro-***	2,3,4-Trichloro-α-naphthol. $C_{10}H_5Cl_3O$. *See* n14	247.51	nd (aa or lig)	168	....	....	....	i	s^h	s	...	...	aa, lig δ^h	**B6^2**, 582
Ω n207	**—,2-hydroxy-1,3,4-trichloro-***	1,3,4-Trichloro-β-naphthol. $C_{10}H_5Cl_3O$. *See* n14	247.51	nd	162	....	....	....	...	s	...	...	...	aa s	**B6^2**, 604
Ω n208	**—,1-iodo-***	$C_{10}H_7I$. *See* n14	254.07	λ^{hp} 225 (4.6), 280 (3.9), 287 (4.0), 299 (3.6)	4.2	302	1.7399^{20}_{4}	1.70256^{20}	i	∞	∞	...	∞	CS_2 ∞	**B5^3**, 1590
n209	**—,2-iodo-***	$C_{10}H_7I$. *See* n14	254.07	lf (dil al) λ^{hp} 236 (4.4), 275 (3.6), 292 (3.5)	54.5	308 172^{21}	1.6319^{99}_{4}	1.6662^{99}	i	v	v	...	...	aa v	**B5^3**, 1591
n210	**—,1-mercapto-***	1-Naphthalenethiol*. 1-Thionaphthol. $C_{10}H_8S$. *See* n14	160.24	λ^{iso} 297 (4.34)	....	285d 208.5^{200} 161^{20}	1.1607^{20}_{4}	1.6802^{20}	δ	v	v	...	...	dil alk δ	**B6^3**, 2943
Ω n211	**—,2-mercapto-***	2-Naphthalenethiol*. 2-Thionaphthol. $C_{10}H_8S$. *See* n14	160.24	pl (al) λ^{chl} 273 (3.8), 292 (3.7), 336 (2.8)	81	288 162.7^{20}	1.550	....	δ	v	v	...	...	lig v	**B6^3**, 3006
Ω n215	**—,1-methoxy-***	Methyl α-naphthyl ether $C_{11}H_{10}O$. *See* n14	158.20	λ^{cy} 293 (3.5), 306 (3.3), 320 (3.1)	< −10	269 (cor) 135^{10}	1.0964^{14}_{2}	1.6940^{25}	i	s	s	...	s	CS_2 v chl s	**B6^3**, 2922
Ω n216	**—,2-methoxy-***	Methyl β-naphthyl ether. Nerolin. $C_{11}H_{10}O$. *See* n14	158.20	lf (eth), pl (peth) λ^{al} 227 (4.85), 271 (3.65), 314 (3.19), 328 (3.30)	73–4	274 138^{10} sub	....	....	δ	δ	s	...	v	chl v CS_2 s MeOH δ	**B6^3**, 2926
Ω n217	**—,1-methyl-***	$C_{11}H_{10}$. *See* n14	142.20	λ^{diox} 267 (3.6), 279 (3.6), 314 (2.4)	−22 (−30.5)	244.64^{760} 107.4^{10}	1.0202^{20}_{4}	1.6170^{20}	i	v	v	...	s	....	**B5^3**, 1620
Ω n218	**—,2-methyl-***	$C_{11}H_{10}$. *See* n14	142.20	mcl (al) λ^{al} 224 (4.38), 274 (3.72), 305 (2.66), 319 (2.65)	34.58	241.05^{760} 104.7^{10}	1.0058^{20}_{4}	1.6019^{40}	i	v	v	...	s	....	**B5^3**, 1627
n219	**—,1(methyl-amino)-***	Methyl α-naphthylamine. $C_{11}H_{11}N$. *See* n14	157.22	oil	....	293 $165–75^{15}$	....	1.6722^{20}	i	s	s	...	...	CS_2 s	**B12^2**, 681
n220	**—,2(methyl-amino)-***	Methyl β-naphthylamine. $C_{11}H_{11}N$. *See* n14	157.22	dk in air	....	317 (308–10) $165–70^{12}$	....	....	i	...	s	...	...	....	**B12**, 1273
n221	**—,1(3-methyl-butoxy)-***	Isoamyl α-naphthyl ether. $C_{15}H_{18}O$. *See* n14	214.31	....	....	$317–9^{742}$ $148–53^{3}$	1.0069^{14}_{4}	1.5705^{14}	...	...	...	...	...	....	**B6**, 607
n222	**—,2(3-methyl-butoxy)-***	Isoamyl β-naphthyl ether. $C_{15}H_{18}O$. *See* n14	214.31	lf	26.5	323–6d $155–60^{6}$	1.0155^{12}_{4}	1.5768^{12}	i	s	s	...	...	....	**B6**, 642
n223	**—,1-methyl-2-nitro-***	$C_{11}H_9NO_2$. *See* n14	187.20	lt ye nd (al)	58–9	....	....	....	i	s	...	...	...	....	**E12B**, 363
n224	**—,1-methyl-3-nitro-***	$C_{11}H_9NO_2$. *See* n14	187.20	ye nd (al)	81–2	....	....	....	i	v	...	...	...	....	**E12B**, 364
n225	**—,1-methyl-4-nitro-***	$C_{11}H_9NO_2$. *See* n14	187.20	pa ye nd (al)	71–2	$182–3^{18}$	....	....	...	s	s	s	...	os s	**E12B**, 362
n226	**—,1-methyl-5-nitro-***	$C_{11}H_9NO_2$. *See* n14	187.20	brsh nd (al)	82–3	....	....	....	i	s	...	s	...	os s	**E12B**, 363
n227	**—,1-methyl-6-nitro-***	$C_{11}H_9NO_2$. *See* n14	187.20	ye nd (dil al)	76–7	....	....	....	i	s	...	...	...	....	**E12B**, 364
n228	**—,1-methyl-7-nitro-***	$C_{11}H_9NO_2$. *See* n14	187.20	ye nd (al)	98–9	....	....	....	i	s^h	...	...	...	....	**E12B**, 364
n229	**—,1-methyl-8-nitro-***	$C_{11}H_9NO_2$. *See* n14	187.20	br lf (al)	65	....	....	....	i	s^h	...	...	...	....	**E12B**, 363
Ω n230	**—,2-methyl-1-nitro-***	$C_{11}H_9NO_2$. *See* n14	187.20	yesh pr or nd (al)	81–2	188^{20}	....	....	i	s	...	v	...	os v	**E12B**, 362
n231	**—,2-methyl-3-nitro-***	$C_{11}H_9NO_2$. *See* n14	187.20	yesh pl (al)	117–8	....	....	....	i	s^h	...	...	...	....	**E12B**, 364
n232	**—,2-methyl-6-nitro-***	$C_{11}H_9NO_2$. *See* n14	187.20	ye nd (al)	119	....	....	....	i	s^h	...	...	...	....	**E12B**, 364
n233	**—,2-methyl-7-nitro-***	$C_{11}H_9NO_2$. *See* n14	187.20	yesh pl (al)	105 (102)	....	....	....	i	s^h	...	...	...	....	**E12B**, 364
n234	**—,3-methyl-1-nitro-***	$C_{11}H_9NO_2$. *See* n14	187.20	pa ye nd (al)	49–50	....	....	....	i	v^h	...	...	...	....	**E12B**, 362
n235	**—,6-methyl-1-nitro-***	$C_{11}H_9NO_2$. *See* n14	187.20	ye nd (al)	61–2	....	....	....	i	s^h	...	...	...	....	**E12B**, 363
n236	**—,7-methyl-1-nitro-***	$C_{11}H_9NO_2$. *See* n14	187.20	ye nd (al)	36–8	....	....	....	i	s^h	...	...	...	....	**E12B**, 363
n237	**—,1-nitramino-***	1-Diazonaphthalenic acid. $C_{10}H_7^{\alpha}NHNO_2$	188.20	lt ye nd (w)	123–4	....	....	....	s^h	...	...	s	...	alk, os v lig i	**E12B**, 862
n238	**—,2-nitramino-***	2-Diazonaphthalenic acid. $C_{10}H_7^{\beta}NHNO_2$	188.20	lf or nd	131.5–6	....	....	....	...	...	...	...	...	....	**E12B**, 862

For explanations, symbols and abbreviations see beginning of table. For structural formulas see end of table.

No.	Name	Synonyms and Formula	Mol. wt.	Color, crystalline form, specific rotation and λ_{max} (log ε)	m.p. °C	b.p. °C	Density	n_D	Solubility w	al	eth	ace	bz	other solvents	Ref.
	Naphthalene														
Ω n239	**—,1-nitro-***	$C_{10}H_7NO_2$. *See* n14	173.17	ye nd (al) λ^{al} 243 (4.02), 343 (3.60)	61.5	304 sub 30–40$^{0.01}$	1.332_4^{20}	...	i	v	v	...	v	chl, CS_2, Py v	**B5^3**, 1593
n240	**—,2-nitro-***	$C_{10}H_7NO_2$. *See* n14	173.17	ye rh nd or pl (al) λ^{al} 258 (4.40), 306 (3.93), 352 (3.44)	79	312.5^{734} 165^{15}	...	...	i	v	v	...	...	...	**B5^3**, 1596
n241	**—,1-nitro-5-triazo-***	$C_{10}H_6N_4O_2$. *See* n14	214.19	gold-ye nd (al)	121	...	...	...	δ s^h	s^h	...	s	...	...	**B5^2**, 459
n242	**—,2-nitro-1-triazo-***	$C_{10}H_6N_4O_2$. *See* n14	214.19	ye nd (dil ace)	103–4d	...	...	...	...	v	...	v	v	aa v lig δ^h	**B5**, 565
n243	**—,1(nitroso-hydroxyl-amino)-***	$C_{10}H_7^{\alpha}N(NO)OH$. *See* n14	188.20	nd (peth)	54–5	...	...	...	...	...	...	...	...	chl, NH_3 s peth s^h	**E12B**, 863
n244	**—,2(nitroso-hydroxyl-amino)-***	$C_{10}H_7^{\beta}N(NO)OH$. *See* n14	188.20	nd (AcOEt-peth)	88–92	...	...	...	...	...	s	...	...	NH_3, aa s	**E12B**, 863
n245	**—,octachloro-***	Perchloronaphthalene. $C_{10}Cl_8$. *See* n14	403.74	nd (bz-CCl_4) λ^{al} 275 (4.67), 332 (3.87), 345 sh (3.79)	197.5–8 (cor)	440–2$^{7.4}$ δd 246$^{0.5}$	...	...	...	δ	...	...	v	chl, lig v	**B5^3**, 1579
Ω n246	**—,1-phenyl-***	$C_{16}H_{12}$. *See* n14	204.28	cr λ^{al} 225 (4.70), 288 (3.95)	*ca.* 45	334^{760} 190^{12}	1.096_4^{20}	1.6664^{20}	i	v	v	...	v	aa v	**B5^3**, 2230
Ω n247	**—,2-phenyl-***	$C_{16}H_{12}$. *See* n14	204.28	lf (al) λ^{al} 250 (4.70), 285 (4.00)	103–4	345–6 (357–8) 185–90^5	...	...	...	s	v	...	s	aa, chl s	**B5^3**, 2231
n248	**—,1-propoxy-***	α-Naphthyl propyl ether. $C_{13}H_{14}O$. *See* n14	186.26	...	...	293.5^{760} 167^{18}	1.0447_4^{18}	1.5928^{18}	...	...	...	...	...	...	**B6^3**, 2924
n249	**—,2-propoxy-***	β-Naphthyl propyl ether. $C_{13}H_{14}O$. *See* n14	186.26	nd (al)	41	305^{760} 144^{10}	...	...	...	s^h	...	...	...	...	**B6^3**, 2973
—	**—,1,2,3,4-tetra-hydro-***	*see* **Tetralin**													
n251	**—,1,3,5,8-tetra-nitro-***	γ-Tetranitronaphthalene. $C_{10}H_4N_4O_8$. *See* n14	308.17	lt ye tetr (ace)	194–5	...	...	...	...	δ	...	v	...	HNO_3 s chl, aa δ	**E12B**, 410
n252	**—,1,3,6,8-tetra-nitro-***	β-Tetranitronaphthalene. $C_{10}H_4N_4O_8$. *See* n14	308.17	ye nd (al, bz)	207	exp	...	...	i	δ s^h	...	...	s^h	aa, to s^h	**B5^3**, 1614
n253	**—,1-triazo-***	α-Naphthyl azide. $C_{10}H_7N_3$. *See* n14	169.19	pa ye pr λ^{al} 300 (3.85)	12	d	1.1713^{25}	1.6550^{25}	...	v	∞	∞	...	...	**E12B**, 1041
n254	**—,2-triazo-***	β-Naphthyl azide. $C_{10}H_7N_3$. *See* n14	169.19	pr (al), nd (peth) ye in air	33	...	...	...	δ^h	s	s	s	s	MeOH, os s peth s^h	**E12B**, 1041
n255	**—, 1,2,5-tri-methyl-***	$C_{13}H_{14}$. *See* n14	170.26	nd (al) λ^{hx} 231 (4.98), 289 (3.80), 325 (2.94)	33.5	140^{12}	1.0103_4^{22}	1.6093^{22}	i	δ	v	...	v	MeOH,	**B5^3**, 1661
n256	**—, 1,2,6-tri-methyl-***	$C_{13}H_{14}$. *See* n14	170.26	lf λ^{peth} 229 (4.92), 282 (3.69), 327 (3.03), 355 (0.94)	14	146^{10}	...	1.6010^{20}	i	...	s	...	s	...	**B5^3**, 1662
n257	**—, 1,2,7-tri-methyl-***	Sapotalin. $C_{13}H_{14}$. *See* n14	170.26	λ^{eth} 230.5 (5.0), 280-5 (3.74), 326 (2.36)	...	147–8^{16}	1.0087_4^{20}	1.6097^{20}	i	...	s	...	s	...	**B5^3**, 1662
Ω n258	**—, 2,3,6-tri-methyl-***	$C_{13}H_{14}$. *See* n14	170.26	λ 229 (5.04), 275 (3.76), 309 (2.78), 323 (2.83)	100–2 (92–3)	263–4^{760} 146–8^{14}	...	...	i	...	s	...	s	...	**B5**, 572
n259	**—, 1,2,5-trinitro-***	$C_{10}H_5N_3O_6$. *See* n14	263.17	lt ye nd (al)	112–3	...	...	...	i	s	...	...	...	CCl_4 s	**B5^2**, 457
n260	**—, 1,3,5-trinitro-***	$C_{10}H_5N_3O_6$. *See* n14	263.17	ye rh (chl)	122	364 exp	...	...	i	v	δ	v	...	aa v chl s	**B5^3**, 1612
n261	**—, 1,3,8-trinitro-***	$C_{10}H_5N_3O_6$. *See* n14	263.17	yesh mcl pr (al, ace or aa), wh nd (80 % HNO_3)	218	...	...	...	i^h	i δ^h	δ	s	δ	Py s to, chl δ	**E12B**, 405
n262	**—,1,4,5-trinitro-***	$C_{10}H_5N_3O_6$. *See* n14	263.17	ye lf, nd or rh pl (al, aa, bz or HNO_3)	154	...	...	...	i	δ	δ	v^h	δ	AcOEt s Py v^h chl δ lig i	**E12B**, 406
—	**Naphthalene car-boxylic acid***	*see* **Naphthoic acid**													
n263	**1,2-Naphthalene-dicarboxylic acid***	...	216.21	nd (al), cr (w)	175d	...	...	...	s^h	s	s	...	δ	aa s lig, chl, CS_2 δ	**E12B**, 4681
n264	**1,4-Naphthalene-dicarboxylic acid***	...	216.21	rods (aa or $PhNO_2$)	309 (320)	...	...	...	i δ^h	v	...	...	...	os δ aa, $PhNO_2$ s^h	**E12B**, 4690
n265	**1,5-Naphthalene-dicarboxylic acid***	...	216.21	nd ($PhNO_2$)	320–2 d (cor)	...	...	...	i	δ^h	δ^h	...	i^h	chl δ^h lig i^h os i	**E12B**, 4693

1,6-Naphthalenedicarboxylic acid

No.	Name	Synonyms and Formula	Mol. wt.	Color, crystalline form, specific rotation and λ_{max} (log ε)	m.p. °C	b.p. °C	Density	n_D	w	al	eth	ace	bz	other solvents	Ref.
n266	**1,6-Naphthalene-dicarboxylic acid***	...	216.21	nd (aa)	310 (sinters *ca.* 290)	...	...	...	i	s^h	i	...	...	aa s^h, os δ	**E12B,** 4695
n267	**1,7-Naphthalene-dicarboxylic acid***	...	216.21	ye pw (dil al or aa)	308d (294–6d)	...	...	...	...	s	s	s	...	aa, os s	**E12B,** 4696
Ω n268	**1,8-Naphthalene-dicarboxylic acid***	Naphthalic acid	216.21	nd (al)	260 (270d)	...	...	...	i	δ s^h	δ	...	...	...	**E12B,** 4697
Ω n269	—,anhydride	Naphthalic anhydride. $C_{12}H_6O_3$. *See* n268	198.18	nd (al), pr (aa), lf (sub)	274	sub	...	...	i	δ^h	i	...	i	aa s^h	**E12B,** 4705
n270	—,dichloride	Naphthaloyl chloride. $C_{12}H_6Cl_2O_2$. *See* n268	253.09	pr (CS_2)	84–6	$195–200^{0.2}$	...	...	...	...	...	...	s	chl s, CS_2 s^h, peth δ	**E12B,** 4713
n271	—,diethyl ester	Diethyl naphthalate. $C_{16}H_{16}O_4$. *See* n268	272.30	yesh mcl pr or lf (dil al)	59–60	$238–9^{19}$ $202–4^{11}$	1.1399_4^{70}	1.5586^{70}	i	s	s	...	...	AcOEt s, con sulf s (bl flr)	**B9**[2], 652
n272	—,dimethyl ester	Dimethyl naphthalate. $C_{14}H_{12}O_4$. *See* n268	244.25	nd (al), pr (MeOH)	104	...	...	...	i	s	...	...	...	MeOH v, aa s	**E12B,** 4703
n273	—,imide	Naphthalimide. $C_{12}H_7NO_2$. *See* n268	197.20	nd (chl-al)	300	...	...	...	i	δ	i	...	i	...	**E12B,** 4724
Ω n274	**2,3-Naphthalene-dicarboxylic acid***	...	216.21	pr (aa or w), pr (sub)	246 (sub) 239–41 (aa)	...	...	...	i δ^h	δ s^h	δ	...	i	aa δ^h, CS_2, chl, lig i	**E12B,** 4723
n275	**2,6-Naphthalene-dicarboxylic acid***	...	216.21	nd (al or sub)	>300d	...	...	...	δ^h	s^h	...	...	i	to δ^h, aa i	**E12B,** 4728
n276	**2,7-Naphthalene-dicarboxylic acid***	...	216.21	nd (w, al or dil HCl)	>300d	...	...	...	i^h	s^h	...	...	i	aa, to i	**E12B,** 4730
n277	**1,2-Naphthalene-dicarboxylic acid, 3,4-dihydro-, anhydride**	$C_{12}H_8O_3$. *See* n263	200.21	pa ye nd (lig or al)	126–7	$227–30^{23}$	...	...	...	...	...	...	v	MeOH, aa s, lig δ, s^h	**B17**[2], 487
n278	**1,8-Naphthalene-dicarboxylic acid, 3,6-dinitro-***	3,6-Dinitronaphthalic acid. $C_{12}H_6N_2O_3$. *See* n268	306.19	silvery lf (w)	212	...	...	...	δ^h	v^h	i	...	i	AcOEt, $PhNO_2$, aa v^h	**E12B,** 4775
n279	**—,3-nitro-, anhydride**	3-Nitronaphthalic anhydride. $C_{12}H_5NO_5$. *See* n268	243.18	yesh nd (aa or $PhNO_2$)	252–3	...	...	...	i^h	i^h	i	...	i^h	aa s^h	**E12B,** 4771
n280	**—,4-nitro-***	4-Nitronaphthalic acid. $C_{12}H_7NO_6$. *See* n268	261.19	ye nd	d 140–50	...	...	...	...	δ	δ	...	...	aa s^h, lig δ	**E12B,** 4772
n281	**1,5-Naphthalene-disulfonic acid***	γ-Naphthalenedisulfonic acid.	288.31	pl (+4w, dil aa or dil HCl) λ^{al} 277 (4.82), 298 (3.98), 318 (2.93)	240–5d (anh)	...	1.493	...	v	s	i	...	...	...	**B11**[2], 119
n282	**1,6-Naphthalene-disulfonic acid***	δ-Naphthalenedisulfonic acid	288.31	og pr (+4w, aa or w) $\lambda^{5\%HCl-al}$ 237 (4.89), 273 (3.62), 342 (3.23)	125d (anh)	...	...	...	v	s	i	...	...	...	**B11**[2], 121
n283	**2,7-Naphthalene-disulfonic acid***	...	288.31	hyg nd (con HCl) λ^{al} 233 (4.94), 266 (3.59), 308 (2.41), 323 (2.25)	199	...	...	...	s	...	...	...	...	con HCl δ	**B11**[2], 122
n284	**1,3-Naphthalene-disulfonic acid, 7-amino-***	β,γ-Acid. Amino-G-acid.	303.32	mcl pr or nd (w + 4)	273–5	...	...	...	s	s	...	...	...	...	**B14**, 784
n285	**2,7-Naphthalene-disulfonic acid, 4-amino-5-hydroxy-***	H acid. $C_{10}H_9NO_7S_2$. *See* n283	319.31	...	...	...	...	...	δ	δ	δ	...	...	...	**B14**[2], 502
Ω n286	**—,4,5-dihydroxy-***	Chromotropic acid. $C_{10}H_8O_8S_2.2H_2O$. *See* n283	356.33	nd or lf (w + 2) λ^{sulf} 235 (4.7), 316 (3.7), 335 (3.8), 351 (4.0)	...	...	...	...	s	i	i	...	...	alk s (bl-vt flr)	**B11**[2], 72
n287	**1,3-Naphthalene-disulfonic acid, 7-hydroxy-***	G Acid. β-Naphthol-γ-disulfonic acid.	304.31	$\lambda^{5\%HCl-al}$ 237 (4.75), 287 (3.75), 338 (3.48)	...	...	...	...	s	...	...	...	...	...	**B11**[2], 165
n288	**2,7-Naphthalene-disulfonic acid, 3-hydroxy-***	R-Acid. $C_{10}H_8O_7S_2$. *See* n283	304.31	dlq nd	d	...	...	...	s	s	i	...	...	...	**B11**[2], 164

For explanations, symbols and abbreviations see beginning of table. For structural formulas see end of table.

No.	Name	Synonyms and Formula	Mol. wt.	Color, crystalline form, specific rotation and λ_{max} (log ε)	m.p. °C	b.p. °C	Density	n_D	Solubility w	al	eth	ace	bz	other solvents	Ref.
	1-Naphthalenephosphonic acid														
n289	**1-Naphthalene-phosphonic acid***		208.16	cr (w)	189				δ	v	...	...	...		**B16**², 392
n290	—,dichloride	$C_{10}H_7Cl_2OP$. *See* n289	245.05		*ca.* 60				d	d^h	...	...	...		**B16**², 392
n291	**1-Naphthalene-sulfinic acid***		192.25	nd (w)	104 (96)				s	s	δ	...	...	dil HCl δ	**B11**¹, 5
n292	**2-Naphthalene-sulfinic acid***		192.25	nd (w)	98(105)				s	s	s	...	...		**B11**², 10
n293	**1-Naphthalene-sulfonic acid***		208.25	pr (+2w, dil HCl) λ^{w} 228 (4.69), 275 (3.98), 320 (3.32)	90 (hyd) 139–40 (anh)				s	s	δ	...	...		**B11**², 91
n294	—,chloride	$C_{10}H_7ClO_2S$. *See* n293	226.68	lf (eth)	68	195^{13} $147.5^{0.9}$			i	s	v	...	v	peth i	**B11**², 93
Ωn295	**2-Naphthalene-sulfonic acid***		208.25	dlq pl (+1w), cr (+3w, dil HCl) $\lambda^{5\%HCl-al}$ 227 (4.98), 275 (3.69), 320 (2.62)	124–5 (+1w) 83 (+3w)	d	1.441^{25}_{4}		v	s	s	...	δ^h		**B11**², 96
Ωn296	—,chloride	$C_{10}H_7ClO_2S$. *See* n295	226.68	pw or lf (bz-peth)	79	201^{13} $148^{0.6}$			i	s	v	...	s	chl, CS_2 s peth δ	**B11**², 99
Ωn297	**1-Naphthalene-sulfonic acid, 4-amino-***	Naphthionic acid. $C_{10}H_9NO_3S.\frac{1}{2}H_2O$. *See* n293	232.26	wh nd (w + ½), red-br cr $\lambda^{1N HCl}$ 283 (3.9)	d		1.6703^{25}_{4}		i δ^h	δ	...	...	...	MeOH, Py s	**E12B**, 5033
n298	—,7-amino-*	δ-Acid. Amino-F-acid. Bayer's acid. Cassela's acid. $C_{10}H_9NO_3S.H_2O$. *See* n293	241.27	nd (w + 1), pl (aq ace)					i δ^h	δ	δ	...	...	aa s	**E12B**, 5106
n299	**2-Naphthalene-sulfonic acid, 4-amino-***	Cleve's acid. $C_{10}H_9NO_3S.H_2O$. *See* n295	241.27	nd (w + 1)					...	...	...	...	...		**E12B**, 5031
n300	**1-Naphthalene-sulfonic acid, 4-amino-5-hydroxy-***	S-acid. $C_{10}H_9NO_4S$. *See* n293	239.25	nd					δ	δ	i	...	...		**B14**, 835
n301	**2-Naphthalene-sulfonic acid, 5,7-dinitro-8-hydroxy-***	Flavianic acid. $C_{10}H_6N_2O_8S$. *See* n295	314.23	pa ye nd (con HCl, +3w), cr (w) λ^{w} 358 (4.30), 380 (3.99)	100(+3w) 151 (anh)				v	v	...	...	...	BuOH s con HCl δ	**B11**², 156
n302	**1-Naphthalene-sulfonic acid, 4-hydroxy-***	Nevile-Winther acid. NW-Acid. $C_{10}H_8O_4S$. *See* n293	224.25	ta or pl (w) $\lambda^{5\%HCl-al}$ 234 (4.50), 300 (3.85), 320 sh (3.64)	170 d (rapid htng)				v	...	i	...	...		**B11**, 271
n303	—, 5-hydroxy-*	α-Naphtholsulfonic acid L. $C_{10}H_8O_4S$. *See* n293	224.25	dlq $\lambda^{5\%HCl-al}$ 241 (4.46), 311.5 (3.66), 323 sh (3.62)	120(112)				s	...	...	...	...	aa s	**B11**², 155
n304	—,7-hydroxy-*	Croceic acid. $C_{10}H_8O_4S$. *See* n293	224.25	λ^{w} 278(3.6), 330(3.1)					s	...	...	...	...		**B11**², 163
n305	—,8-hydroxy-*	α-Naphtholsulfonic acid S. $C_{10}H_8O_4S$. *See* n293	224.25	cr (w + 1)	106–7 (+1w)				s	...	...	...	...		**B11**, 275
n306	—,—,lactone.	1-Naphthol-8-sulfonic acid, sultone. Naphthosultone. $C_{10}H_6O_3S$. *See* n293	206.22	pr (bz)	154	>360			δ	δ	...	...	s v^h	chl v CS_2 i	**B19**, 43
n307	**2-Naphthalene-sulfonic acid, 1-hydroxy-***	α-Naphtholsulfonic acid. $C_{10}H_8O_4S$. *See* n295	224.25	pl (w) $\lambda^{5\%HCl-al}$ 237 (4.58), 294 (3.64), 328 (3.57)	>250				s^h δ	s	i	...	...	dil HCl δ	**E12B**, 5262
n308	—,6-hydroxy-*	Schaeffer acid. β-Naphtholsulfonic acid S. $C_{10}H_8O_4S$. *See* n295	224.25	lf, cr (w + 1) $\lambda^{5\%HCl-al}$ 233 (4.85), 280 (3.71), 332 (3.13)	167 (anh) 129 (+1w) 118 (+2w)				v	v	i	...	...	aa s	**B11**, 282
n309	—,7-hydroxy-*	F acid. β-Naphtholsulfonic acid. $C_{10}H_8O_4S$. *See* n295	224.25	nd (HCl), cr (w + 1,2 or 4)	115–6 (anh) 108–9 (+1w) 95 (+2w), 67 (+4w)	150d			v	v	i	...	i		**B11**, 285
n309¹	—,6-hydroxy-5-nitroso-*	$C_{10}H_7NO_5S$. *See* n295	253.23	og	d				v	...	...	...	...		**B11**², 190
—	—,5,6,7,8-tetrahydro-	*see* **5-Tetralinsulfonic acid**													

For explanations, symbols and abbreviations see beginning of table. For structural formulas see end of table.

No.	Name	Synonyms and Formula	Mol. wt.	Color, crystalline form, specific rotation and λ_{max} (log ε)	m.p. °C	b.p. °C	Density	n_D	Solubility w	al	eth	ace	bz	other solvents	Ref.
	1,4,5,8-Naphthalenetetracarboxylic acid														
Ω n310	**1,4,5,8-Naphthalenetetracarboxylic acid***	....	304.23	lf or nd (w, dil HCl or dil HNO_3)	320	....	....	....	δ s^h	δ^h	...	v	δ^h	aa s^h CS_2, chl δ^h	**E12B,** 4833
Ω n311	—,1,8:4,5-dianhydride*	$C_{14}H_4O_6$. *See* n310	268.19	nd (aa), pr ($PhNO_2$)	> 300	sub 320[3]	....	....	i	...	...	...	...	Na_2CO_3 s aa s^h	**E12B,** 4835
n312	**1,2,5-Naphthalenetricarboxylic acid***	....	260.22	nd (MeOH or sub)	270–2	....	....	....	s	...	δ	...	...	MeOH s	**E12B,** 4829
n313	**1,4,5-Naphthalenetricarboxylic acid***	....	260.22	cr (eth or con HCl)	266–8	....	....	....	δ	v	s^h	...	...		**E12B,** 4830
—	**Naphthalic acid**	*see* **1,8-Naphthalenedicarboxylic acid***													
—	**Naphthazarin**	*see* **1,4-Naphthoquinone, 5,8-dihydroxy-***													
—	**Naphthionic acid** ...	*see* **1-Naphthalenesulfonic acid, 4-amino-***													
n314	**Naphthocaine,** hydrochloride	....	322.84	pa ye	212–6	....	....	....	δ	...	...	...	...	....	**C35,** 4003
—	**Naphthohydroquinone**	*see* **Naphthalene, dihydroxy-***													
Ω n315	**1-Naphthoic acid** ...	α-Naphthalenecarboxylic acid.* α-Naphthoic acid.	172.20	nd (aa-w, w or al) λ^{al} 293 (3.9)	161	> 300 231[50]	1.398	....	δ	s v^h	s	...	...	chl s lig δ	**E12B,** 3966
Ω n316	—,amide	1-Naphthalenecarbonamide* $C_{11}H_9NO$. *See* n315	171.20	nd or pl (al or gl aa)	204–5	sub	....	....	δ^h	δ s^h	...	...	...	con HCl, aa s^h	**E12B,** 3992
n317	—,amidine	1-Naphthamidine*. $C_{11}H_{10}N_2$. *See* n315	170.22	lf (dil al)	154	....	....	....	δ^h	s	δ	s	δ	chl s lig δ	**E12B,** 4002
n318	—,anhydride	1-Naphthoic acid anhydride. $C_{22}H_{14}O_3$. *See* n315	326.36	pr (bz)	145–6	....	....	....	i	δ	s	...	s	....	**E12B,** 3988
n319	—,chloride	1-Naphthoyl chloride. $C_{11}H_7ClO$. *See* n315	190.63	....	20	297.5 172[15]	....	....	d^h	d^h	...	...	...	....	**B9**[2], 450
n320	—,ethyl ester	$C_{13}H_{12}O_2$. *See* n315	200.24	....	....	310[760] 183–6[20]	1.1274_{15}^{15}	1.5966^{15}	i	s	...	...	...	....	**B9**[2], 449
n321	—,methyl ester	$C_{12}H_{10}O_2$. *See* n315	186.21	....	59.5	167–9[20] 100–2[0.04]	1.1290_4^{20}	1.6086^{20}	...	s	...	...	s	....	**E12B,** 3981
Ω n322	—,nitrile	1-Cyanonaphthalene. 1-Naphthonitrile. $C_{11}H_7N$. *See* n315	153.19	nd (lig)	37.5	299 148[12]	1.1113_{25}^{25}	1.6298^{18}	i	v	v	...	...	lig s^h	**B9**[2], 450
Ω n324	**2-Naphthoic acid** ...	2-Naphthalenecarboxylic acid*. β-Naphthoic acid.	172.20	nd (lig, chl, sub), pl (ace)	185.5	> 300	1.077_4^{100}	....	δ^h	s	s	...	...	chl s lig δ^h	**E12B,** 4021
n325	—,amide	2-Naphthalenecarbonamide.* $C_{11}H_9NO$. *See* n324	171.20	lf (al)	195 (cor)	....	....	....	δ	s	s	s	...	chl, CS_2, lig s	**E12B,** 4039
n326	—,amidine	2-Naphthamidine.* $C_{11}H_{10}N_2$. *See* n324	170.22	cr (bz)	133–6	....	....	....	δ	v	δ	...	v^h	aa v^h	**E12B,** 4047
n327	—,anhydride......	2-Naphthoic acid anhydride. $C_{22}H_{14}O_3$. *See* n324	326.36	nd (eth)	135	....	....	....	i	...	δ s^h	...	v^h	aa v	**E12B,** 4037
n328	—,chloride	2-Naphthoyl chloride. $C_{11}H_7ClO$. *See* n324	190.63	cr (peth)	51	304–6 142–3[5]	....	....	i d^h	...	s	...	s	chl, aa s	**B9**, 657
n329	—,ethyl ester.....	$C_{13}H_{12}O_2$. *See* n324	200.24	....	32	308–9 (315–6) 224[74]	1.1143_4^{23}	1.5951^{23}	i	s	s	...	...	chl s aa s^h	**B9**[2], 453
n330	—,methyl ester	$C_{12}H_{10}O_2$. *See* n324	186.21	lf (MeOH)	77	290 141–3[4]	....	....	...	v	v	...	v	MeOH, chl v	**E12B,** 4031
Ω n331	—,nitrile	2-Cyanonaphthalene. 2-Naphthonitrile. $C_{11}H_7N$. *See* n324	153.19	lf (lig)	66	306.5[760] 156–8[12]	1.0939_{60}^{60}	....	δ	s	s	...	...	lig s^h	**B9**, 659
n333	**2-Naphthoic acid, 4-acetyl-3-hydroxy-**	$C_{13}H_{10}O_4$. *See* n324	230.22	ye pr (aa)	194	....	....	....	δ^h	v	...	...	δ	chl δ aa δ, s^h	**E12B,** 4589
n334	**1-Naphthoic acid, 2-amino-**	$C_{11}H_9NO_2$. *See* n315	187.20	nd (dil al)	126d	....	....	....	δ	v	v	...	s^h	aa v^h	**E12B,** 4184
n335	—,3-amino-.......	$C_{11}H_9NO_2$. *See* n315	187.20	ye or pksh nd (eth)	181–2	....	....	....	...	s	...	...	...	MeOH s	**E12B,** 4185
n336	—,4-amino-.......	$C_{11}H_9NO_2$. *See* n315	187.20	brsh nd (w or al)	177	....	....	....	δ s^h	s	s	s	δ	aa s lig δ	**E12B,** 4188
n337	—,5-amino-.......	$C_{11}H_9NO_2$. *See* n315	187.20	og nd (w or dil al), nd (sub)	211–2 (217–20) 196 (sub)	sub	....	....	δ^h	s	δ	...	s	aa s	**E12B,** 4198
n338	—,6-amino-.......	$C_{11}H_9NO_2$. *See* n315	187.20	pa ye nd (al or w)	205–6	....	....	....	δ	v	s	s	δ	AcOEt, aa s	**E12B,** 4200
n339	—,7-amino-.......	$C_{11}H_9NO_2$. *See* n315	187.20	pa br pr (al)	223–4	....	....	....	...	v	s	s	δ	aa v AcOEt s	**B14**[2], 322
n340	**2-Naphthoic acid, 1-amino-**	$C_{11}H_9NO_2$. *See* n324	187.20	nd (dil al or aa)	205 (rapid htng)	....	....	....	δ	s	s	...	s	....	**B14**[2], 323
Ω n341	—,3-amino-.......	$C_{11}H_9NO_2$. *See* n324	187.20	ye lf (dil al)	216–7	....	....	....	...	s	s	...	...	....	**B14**[2], 323
n342	—,4-amino-.......	$C_{11}H_9NO_2$. *See* n324	187.20	nd (dil al)	215–6 (204–6)	....	....	....	...	s^h	s	s	s	AcOEt s	**B14**[2], 323

For explanations, symbols and abbreviations see beginning of table. For structural formulas see end of table.

No.	Name	Synonyms and Formula	Mol. wt.	Color, crystalline form, specific rotation and λ_{max} (log ε)	m.p. °C	b.p. °C	Density	n_D	Solubility w	al	eth	ace	bz	other solvents	Ref.
	2-Naphthoic acid														
n343	**—,5-amino-**	$C_{11}H_9NO_2$. *See* n324	187.20	ye lf (al)	291–2 (232)	...	...	...	δ	s^h	s	s	s	aa i	**B14**^{2}, 324
n344	**—,6-amino-**	$C_{11}H_9NO_2$. *See* n324	187.20	pa ye nd (w or dil al)	225	...	...	...	δ s^h	v	v	s	v	dil ac, aa v	**B14**^{2}, 324
n345	**—,7-amino-**	$C_{11}H_9NO_2$. *See* n324	187.20	pa ye nd or lf (al)	245	...	...	...	...	v	s	s	δ	NH_3, aa v AcOEt s	**B14**^{2}, 324
n346	**—,8-amino-**	$C_{11}H_9NO_2$. *See* n324	187.20	grsh-ye nd (aa)	220	...	...	...	...	s	s	s	δ	AcOEt s	**B14**^{2}, 324
n347	**1-Naphthoic acid, 4-bromo-**	$C_{11}H_7BrO_2$. *See* n315	251.09	nd (aa, dil al or xyl) λ^{diox} 305 (3.97)	220	...	...	...	δ^h	v	s	s	s	xyl, aa s^h	**E12B,** 4121
n348	**—,5-bromo-**	$C_{11}H_7BrO_2$. *See* n315	251.09	nd (aa or al) λ^{diox} 305 (3.9)	261	sub	...	...	δ^h	δ	δ	...	v	aa δ	**E12B,** 4123
n349	**—,8-bromo-**	$C_{11}H_7BrO_2$. *See* n315	251.09	pr (w or bz) λ^{diox} 282.5 sh (3.79), 292 (3.89), 302.5 sh (3.81)	178	...	...	...	s^h	v	δ	...	s^h	aa v chl, lig δ	**E12B,** 4130
n350	**2-Naphthoic acid, 1-bromo-**	$C_{11}H_7BrO_2$. *See* n324	251.09	nd (aa or bz)	191	...	...	...	δ	s	...	...	s^h	aa s	**E12B,** 4133
n351	**—,5-bromo-**	$C_{11}H_7BrO_2$. *See* n324	251.09	nd (al, sub)	270 (cor)	sub	...	...	δ^h	s	s	...	s	aa s	**E12B,** 4138
n352	**1-Naphthoic acid, 2-chloro-**	$C_{11}H_7ClO_2$. *See* n315	206.63	cr (w, bz) λ^{diox} 282 (3.75), 292.5 (3.78)	153	...	...	...	s^h	s	s	...	s^h	...	**E12B,** 4120
n353	**—,4-chloro-**	$C_{11}H_7ClO_2$. *See* n315	206.63	nd (al) λ^{diox} 298 (3.73)	210 (223–4)	...	...	...	δ	v	δ	...	δ	aa v chl δ	**E12B,** 4121
n354	**—,5-chloro-**	$C_{11}H_7ClO_2$. *See* n315	206.63	nd (dil al) λ^{diox} 303 (3.85)	245	sub	...	...	δ	v	...	...	δ	aa δ	**E12B,** 4122
n355	**—,8-chloro-**	$C_{11}H_7ClO_2$. *See* n315	206.63	pl (al, w or bz) λ^{diox} 280 sh (3.78), 290 (3.85), 312.5 sh (3.74)	171–2	sub	...	...	s^h	s^h	...	v	s^h	aa s xyl s^h	**E12B,** 4128
n356	**2-Naphthoic acid, 1-chloro-**	$C_{11}H_7ClO_2$. *See* n324	206.63	nd (bz)	196	...	...	...	δ	s	...	s	...	...	**E12B,** 4132
n357	**—,3-chloro-**	$C_{11}H_7ClO_2$. *See* n324	206.63	cr (dil MeOH)	216.5	...	...	...	...	s	s	v	s	MeOH v chl, aa, AcOEt s	**E12B,** 4134
n358	**—,5-chloro-**	$C_{11}H_7ClO_2$. *See* n324	206.63	nd (al or aa)	270 (cor)	...	...	...	...	s^h	...	...	s	aa s	**E12B,** 4137
n359	**—,4-chloro-1-hydroxy-**	$C_{11}H_7ClO_3$. *See* n324	222.63	nd (al or aa)	234 (229)	...	...	...	i^h	s	δ	s	δ	aa s^h	**B10**^{1}, 146
n360	**—,4-chloro-3-hydroxy-**	$C_{11}H_7ClO_3$. *See* n324	222.63	ye nd	231d	...	...	...	...	...	...	...	...	...	**B10,** 336
n361	**1-Naphthoic acid, 1,2-dihydro-**	$C_{11}H_{10}O_2$. *See* n315	174.20	cr (50 % al)	138	...	...	...	s	...	...	...	...	sulf s (red)	**B9**^{2}, 443
n362	**—,1,4-dihydro-**	$C_{11}H_{10}O_2$. *See* n315	174.20	nd or pl (lig)	91	...	...	...	δ	v	s	v	δ	AcOEt, CS_2 v lig δ	**B9**^{2}, 442
n363	**—,3,4-dihydro-**	$C_{11}H_{10}O_2$. *See* n315	174.20	nd (w, AcOEt), cr (peth) λ^{al} 223 (4.10) 274 (3.78)	125	305–6^{748}	...	...	δ	s	...	...	...	MeOH, chl, CS_2 s	**B9**^{2}, 442
Ω n364	**2-Naphthoic acid, 1,2-dihydro-(*dl*)**	$C_{11}H_{10}O_2$. *See* n324	174.20	nd or pr (dil al, w or peth) λ^{al} 211.5 (4.37), 260.5 (3.99)	105–6	...	...	...	i v^h	s	...	...	...	chl v aa s	**B9**^{2}, 443
Ω n365	**—,1,4-dihydro-**	$C_{11}H_{10}O_2$. *See* n324	174.20	pl (bz, w or dil al) λ^{al} 273(2,61)	162–3	...	...	...	i s^h	s	s	...	δ	lig δ CS_2 i	**B9**^{2}, 443
Ω n366	**—,3,4-dihydro-**	$C_{11}H_{10}O_2$. *See* n324	174.20	nd (dil aa, dil al, or dil MeOH) λ^{al} 225 (4.30), 291 (4.18)	120	...	...	...	δ s^h	...	...	...	s	chl v aa s	**B9**^{2}, 443
n367	**—,1,3-dihydroxy-,** ethyl ester	$C_{13}H_{12}O_4$. *See* n324	232.24	nd (dil al or dil aa)	83–4	...	...	...	i	v	v	...	...	alk s lig v^h	**B10**^{2}, 310
n368	**1-Naphthoic acid, 4,5-dinitro-**	$C_{11}H_6N_2O_6$. *See* n315	262.18	yesh nd or pl (al)	266.7 sub	...	...	...	δ^h	v^h	δ	...	δ	aa v	**B9,** 654
n369	**—,5-hydroxy-**	$C_{11}H_8O_3$. *See* n315	188.20	pr or nd (w), cr (bz)	236	...	...	...	δ^h	v	s	...	...	chl, aa s	**B10**^{2}, 208
n370	**—,6-hydroxy-**	$C_{11}H_8O_3$. *See* n315	188.20	nd or pr (w)	212.5–13	...	...	...	s^h	s	s	s	δ	aa s^h chl δ	**B10**^{2}, 208
n371	**—,7-hydroxy-**	$C_{11}H_8O_3$. *See* n315	188.20	nd (w or gl aa)	256	...	...	...	δ s^h	v	δ	...	i	aa s	**B10**^{2}, 208
Ω n372	**2-Naphthoic acid, 1-hydroxy-**	$C_{11}H_8O_3$. *See* n324	188.20	cr (dil al, w, aa), nd (al, eth, bz), pw (chl)	195(188)	...	...	...	δ s^h	v	v	...	s	...	**B10**^{2}, 210
n373	**—,—,phenyl ester**	$C_{17}H_{12}O_3$. *See* n324	264.28	...	96	...	...	...	...	s^h	...	...	s	...	**B10,** 332

For explanations, symbols and abbreviations see beginning of table. For structural formulas see end of table.

No.	Name	Synonyms and Formula	Mol. wt.	Color, crystalline form, specific rotation and λ_{max} (log ε)	m.p. °C	b.p. °C	Density	n_D	Solubility w	al	eth	ace	bz	other solvents	Ref.
	2-Naphthoic acid														
Ω n374	**—,3-hydroxy-**	$C_{11}H_8O_3$. *See* n324	188.20	ye lf (dil al), ye lf (w), nd (dil al) λ^{al} 232 (4.80), 266 (3.78), 328 (3.28)	222–3	...	...	...	δ	v	v	...	s	to, chl s	**B10**[2], 211
n375	**—,—,ethyl ester**	$C_{13}H_{12}O_3$. *See* n324	216.24	nd or mcl pr (aa)	85	291	...	...	i^h	s^h	...	s	...	chl s aa s^h	**B10**[2], 213
n376	**—,—,hexadecyl ester**	$C_{27}H_{40}O_3$. *See* n324	412.62	grsh-wh pr	72–3	...	...	...	i	δ	...	...	s	lig s	**C42**, 2108
n377	**—,—,methyl ester**	$C_{12}H_{10}O_3$. *See* n324	202.21	pa ye rh nd (dil MeOH) λ^{al} 240 (4.6), 290 (3.8), 365 (3.3)	75–6	205–7	...	...	i	s	...	...	...	os s	**B10**[2], 213
n378	**—,5-hydroxy-**	$C_{11}H_8O_3$. *See* n324	188.20	wh nd (w, dil al)	213	...	...	...	i s^h	s	s	s	i	aa s chl i	**B10**[2], 216
n379	**—,7-hydroxy-**	$C_{11}H_8O_3$. *See* n324	188.20	wh lf or pl (dil al), pa ye nd (al)	269–70	...	...	...	i s^h	s	s	s	i	aa s chl i	**B10**[2], 217
n380	**1-Naphthoic acid, 8-iodo-**	$C_{11}H_7IO_2$. *See* n315	298.08	br pr (w) λ^{diox} 299 (3.91), 312.5 sh (3.84)	164.5	...	...	...	s^h	s	s	...	s	aa s	**E12B**, 4131
n381	**—,3-nitro-**	$C_{11}H_7NO_4$. *See* n315	217.18	cr (al) λ^{diox} 308 (3.86), 351 (3.40)	271.5	...	...	...	$δ^h$	s^h	...	...	...	...	**B9**[2], 451
n382	**—,4-nitro-**	$C_{11}H_7NO_4$. *See* n315	217.18	yesh nd (al or aa) λ^{diox} 335 (3.65)	225–6 (221)	...	...	...	$δ^h$	v	...	...	δ	chl, aa s lig δ	**B9**[2], 451
n383	**—,5-nitro-**	$C_{11}H_7NO_4$. *See* n315	217.18	yesh nd (al) λ^{diox} 310 (3.81), 328 (2.70)	241–2	sub	...	...	i s^h	s^h	δ	...	δ	aa s chl, CS_2 δ	**B9**[2], 452
n384	**—,8-nitro-**	$C_{11}H_7NO_4$. *See* n315	217.18	nd (w), pr (al) λ^{diox} 281 (3.74), 308 (2.85)	217	...	...	...	i	s	δ	...	δ	aa s^h	**B9**[2], 452
n385	**2-Naphthoic acid, 5-nitro-**	$C_{11}H_7NO_4$. *See* n324	217.18	yesh nd (al)	295 (287)	...	...	...	i	$δ^h$	δ	s	δ	chl, aa, peth δ	**B9**[2], 455
n386	**—,8-nitro-**	$C_{11}H_7NO_4$. *See* n324	217.18	yesh nd (al)	295	sub	...	...	δ	δ	...	...	...	...	**B9**[2], 455
—	**α-Naphthol**	*see* **Naphthalene, 1-hydroxy-***													
—	**β-Naphthol**	*see* **Naphthalene, 2-hydroxy-***													
—	**β-Naphthol-1-aldehyde**	*see* **1-Naphthaldehyde, 2-hydroxy-**													
—	**α-Naphtholsulfonic acid**	*see* **2-Naphthalenesulfonic acid, 1-hydroxy-***													
—	**α-Naphtholsulfonic acid L**	*see* **1-Naphthalenesulfonic acid, 5-hydroxy-***													
—	**α-Naphtholsulfonic acid S**	*see* **1-Naphthalenesulfonic acid, 8-hydroxy-***													
—	**β-Naphtholsulfonic acid**	*see* **2-Naphthalenesulfonic acid, 7-hydroxy-***													
—	**β-Naphtholsulfonic acid S**	*see* **2-Naphthalenesulfonic acid, 6-hydroxy-***													
—	**1-Naphthol-8-sulfonic acid, sultone**	*see* **1-Naphthalenesulfonic acid, 8-hydroxy-, lactone***													
—	**Naphthonitrile**	*see* **Naphthoic acid,** nitrile													
n387	**Naphtho-(2′,3′:5,6)quinoline**	β-Anthraquinoline.	229.29	lf or ta (al)	170	446	...	...	i	v^h	v	...	v	...	**B20**, 506
Ω n388	**1,2-Naphthoquinone-***	1,2-Dihydro-1,2-diketonaphthalene*. β-Naphthoquinone.	158.16	ye-red nd (eth), og lf (bz) λ^{al} 240 (4.4), 340 (3.4), 415 (3.3)	146 (125)	...	1.450	...	s	s	s	...	...	sulf s (gr) lig δ	**B7**[2], 645
n389	**—,dioxime***	$C_{10}H_8N_2O_2$. *See* n388	188.20	ye nd (bz-lig or dil al) λ^{al} 260 (4.30), 270 (4.27), 300 (3.50), 330 (3.83), 350 (3.95)	169 (152)	...	...	...	...	s^h	...	...	s	diox s sulf s(red) alk s (ye)	**B7**[2], 649
—	**—,1-oxime***	*see* **Naphthalene, 2-hydroxy-1-nitroso-**													
Ω n390	**1,4-Naphthoquinone***	1,4-Dihydro-1,4-diketonaphthalene*. α-Naphthoquinone.	158.16	bt ye nd (al or peth), ye (sub) λ^{al} 255 (4.2), 330 (3.4)	128.5 (125)	sub	...	...	δ	v^h	s	...	s	CS_2, chl, aa s lig δ	**B7**[2], 651

For explanations, symbols and abbreviations see beginning of table. For structural formulas see end of table.

No.	Name	Synonyms and Formula	Mol. wt.	Color. crystalline form. specific rotation and λ_{max} (log ε)	m.p. °C	b.p. °C	Density	n_D	Solubility w	al	eth	ace	bz	other solvents	Ref.
	1,4-Naphthoquinone														
n391	—,dioxime*	$C_{10}H_8N_2O_2$. *See* n390	188.20	nd (dil al)	207d	...	...	...	s	s	...	...	...	alk s	**B7**^{2}, 653
—	—,1-oxime*	*see* **Naphthalene, 1-hydroxy-4-nitroso-***													
n392	**2,6-Naphthoquinone***	2,6-Dihydro-2,6-diketonaphthalene.*	158.16	ye-red pr (bz or pz-peth)	135d	...	...	...	...	v	δ	d	δ	MeOH v Py, aa d	**B7**^{2}, 656
n393	**1,4-Naphthoquinone, 4-anil-, 2-anilino-**	Naphthoquinone dianilide	324.39	ye-red nd (bz-or al)	182.5–3	...	...	...	i	δ	...	s	s	os s dil ac i	**B14**^{2}, 91
n394	**1,2-Naphthoquinone, 3-bromo-***	$C_{10}H_5BrO_2$. *See* n388	237.06	red nd or pl (aa or al)	178	sub	...	...	...	s	...	...	s^h	aa s^h	**B7**, 721
n395	—,4-bromo-*	$C_{10}H_5BrO_2$. *See* n388	237.06	red nd (bz-lig)	154	...	...	...	...	s	...	...	v	sulf s (red) aa s lig δ	**E12B**, 2890
n396	—,6-bromo-*	$C_{10}H_5BrO_2$. *See* n388	237.06	og-red or ye pr or pl (bz or AcOEt), nd (w)	168d (150d)	...	...	...	...	s	s^h	s	...	xyl, aa, lig s^h	**B7**, 722
n397	**1,4-Naphthoquinone, 2-bromo-***	$C_{10}H_5BrO_2$. *See* n390	237.06	ye pl or nd (al or dil aa)	132	...	...	...	i	s^h	δ	s	v	chl, aa s CS_2 δ	**B7**^{2}, 655
n398	—,5-bromo-2,3-dichloro-*	$C_{10}H_3BrCl_2O_2$. *See* n390	305.95	ye pr (al)	180	...	...	...	...	s^h	...	v	v	...	**B7**^{2}, 655
n399	—,2-bromo-3-hydroxy-*	$C_{10}H_5BrO_3$. *See* n390	253.06	ye mcl pr (al), nd (al, w or aa)	202	sub	...	...	i δ^h	δ s^h	δ	s	δ	...	**B8**^{1}, 636
n400	—,2-bromo-3-methyl-*	$C_{11}H_7BrO_2$. *See* n390	251.09	ye-br nd (al)	151	sub 100	...	...	i	s	v	s	s	chl v aa s alk i	**B7**^{2}, 657
n401	**1,2-Naphthoquinone, 3-chloro-***	$C_{10}H_5ClO_2$. *See* n388	192.60	red nd (al, aa, bz or chl)	172d	...	...	...	i	s^h	...	...	s	aa, chl s^h Na_2CO_3 i	**B7**, 720
Ω n402	**1,4-Naphthoquinone, 2-chloro-***	$C_{10}H_5ClO_2$. *See* n390	192.60	ye nd (w, al or aa)	117–8	...	...	...	s^h	s	δ	s	s	aa s^h Na_2CO_3 i	**B7**^{2}, 653
n403	—,5-chloro-*	$C_{10}H_5ClO_2$. *See* n390	192.60	ye nd (lig)	163	sub	...	...	i	s	...	...	...	aa, lig s	**B7**^{2}, 653
n404	—,6-chloro-*	$C_{10}H_5ClO_2$. *See* n390	192.60	red-br (al or eth), ye cr (dil MeOH)	109–10	...	...	...	v	...	δ	...	v	sulf s (red) chl δ	**E12B**, 2896
n405	—,2-chloro-3-hydroxy-*	$C_{10}H_5ClO_3$. *See* n390	208.60	ye nd (al or aa)	215	sub	...	...	δ s^h	s	s	...	s	...	**B8**^{2}, 347
n406	**1,2-Naphthoquinone, 3,4-dibromo-***	$C_{10}H_4Br_2O_2$. *See* n388	315.96	red lf or pl (aa or bz)	172–4	...	...	...	...	δ	δ	...	s^h	alk s aa s^h	**B7**, 722
n407	—,3,6-dibromo-*	$C_{10}H_4Br_2O_2$. *See* n388	315.96	red pr (AcOEt)	176	...	...	...	...	s	...	...	...	aa, AcOEt s^h	**B7**^{2}, 650
n408	—,4,6-dibromo-*	$C_{10}H_4Br_2O_2$. *See* n388	315.96	og-red pr (bz, peth), nd (AcOEt)	153	...	...	...	...	δ	...	...	v	aa v lig δ, s^h	**B7**^{2}, 650
n409	**1,4-Naphthoquinone, 2,3-dibromo-***	$C_{10}H_4Br_2O_2$. *See* n390	315.96	ye nd (aa)	218	...	...	...	...	δ	δ	...	...	aa s^h lig δ	**B7**, 731
n410	—,5,8-dibromo-*	$C_{10}H_4Br_2O_2$. *See* n390	315.96	ye nd (al)	171–3	...	...	...	...	δ	...	...	...	...	**B7**, 732
n411	**1,2-Naphthoquinone, 3,4-dichloro-***	$C_{10}H_4Cl_2O_2$. *See* n388	227.05	red lf or pl (aa or bz), nd (bz or chl)	184	sub	...	...	...	δ	...	...	s^h	chl s aa s^h	**B7**^{2}, 650
Ω n412	**1,4-Naphthoquinone, 2,3-dichloro-***	$C_{10}H_4Cl_2O_2$. *See* n390	227.05	ye nd (al)	195	...	...	...	i	δ s^h	δ	...	δ s^h	chl s^h	**B7**^{2}, 653
n413	—,2,6-dichloro-*	$C_{10}H_4Cl_2O_2$. *See* n390	227.05	dk ye nd (al)	148–9	...	...	...	i	s^h	...	...	...	...	**B7**, 730
n414	—,5,6-dichloro-*	$C_{10}H_4Cl_2O_2$. *See* n390	227.05	ye nd (al)	181	sub	...	...	i	δ s^h	s	...	...	...	**B7**, 730
n415	—,5,8-dichloro-*	$C_{10}H_4Cl_2O_2$. *See* n390	227.05	ye nd (al)	173–4	sub	...	...	i	s^h	s	...	...	...	**B7**, 730
n416	**2,6-Naphthoquinone, 1,5-dichloro-***	$C_{10}H_4Cl_2O_2$. *See* n392	227.05	og pr (chl), gold-ye nd (al)	206.5d	...	...	...	i	s^h	δ	v	v	chl, aa v AcOEt s lig i	**B7**, 733
n417	**1,4-Naphthoquinone, 2,3-dihydroxy-***	Isonaphthazarin. $C_{10}H_6O_4$. *See* n390	190.16	red og nd or lf (sub) λ^{al} 268 (4.32), 330 (3.32), 440 (3.26)	282	sub	...	...	δ^h	δ	δ	s	δ	alk s chl δ	**B8**^{2}, 461
n418	—,5,8-dihydroxy-*	Naphthazarin. $C_{10}H_6O_4$. *See* n390	190.16	dk red mcl pr (bz), red-br nd (al) λ^{diox} 270 (3.9), 520 (3.8)	276–80	sub	...	...	δ^h	δ^h	δ^h	...	...	aa s^h	**B8**^{2}, 463

For explanations, symbols and abbreviations see beginning of table. For structural formulas see end of table.

No.	Name	Synonyms and Formula	Mol. wt.	Color, crystalline form, specific rotation and λ_{max} (log ε)	m.p. °C	b.p. °C	Density	n_D	Solubility						Ref.
									w	al	eth	ace	bz	other solvents	
	1,4-Naphthoquinone														
n419	**—,3,5-di-hydroxy-2-methyl-***	Droserone. $C_{11}H_8O_4$. *See* n390	204.19	og-ye nd (al or aa) λ^{al} 232 (4.12), 285 (4.08), 409 (3.66)	181	sub 100^3	...	...	s^h	s	s	...	...	peth s aa s^h	**B8**², 465
n420	**—,5,8-dihydroxy-2-methyl-***	$C_{11}H_8O_4$. *See* n390	204.19	gr pl	173	...	...	...	i	s^h	...	...	...	...	**E12B**, 3195
Ω n421	**—,2,3-dimethyl-***	$C_{12}H_{10}O_2$. *See* n390	186.21	ye pr (al) λ^{al} 249 (4.26), 267 (4.24), 330 (3.38)	127	...	...	...	i	s^h	...	...	s	aa s	**B7**², 658
n422	**—,2,5-dimethyl-***	$C_{12}H_{10}O_2$. *See* n390	186.21	ye nd (peth, eth or MeOH)	95	...	...	...	...	...	...	...	...	diox s lig δ	**B7**¹, 386
n423	**—,2,6-dimethyl-***	$C_{12}H_{10}O_2$. *See* n390	186.21	ye pr or nd (AcOEt)	136–7	...	...	...	...	s	s	...	s	...	**B7**¹, 386
n424	**—,2,8-dimethyl-***	$C_{12}H_{10}O_2$. *See* n390	186.21	pr (peth), nd (MeOH)	135–6	...	...	...	δ^h	s	...	...	...	...	**E12B**, 2850
n425	**—,2-ethyl-***	$C_{12}H_{10}O_2$. *See* n390	186.21	pr (al), nd (aa, MeOH, peth)	88–9	...	...	...	...	s^h	...	...	...	MeOH, aa s^h	**B7**², 658
n426	**—,2-ethyl-3-hydroxy-***	$C_{12}H_{10}O_3$. *See* n390	202.21	ye nd (dil MeOH)	141	...	...	...	...	...	s	s^h	...	...	**E12B**, 3075
n427	**—,2-ethyl-3,5,6,7,8-penta-hydroxy-***	Echinochrome A. $C_{12}H_{10}O_7$. *See* n390	266.21	red nd (diox-w, to) $\lambda^{w,pH=7}$ 400 (3.7), 465 (3.7), 480 (3.8), 550 (3.5)	220 δd	sub 120^{10}	...	...	δ	s	v	s	v	con sulf s	**E12B**, 3230
n428	**1,2-Naphtho-quinone, 6-hy-droxy-***	$C_{10}H_6O_3$. *See* n388	174.16	red lf (ace), og-ye nd λ^{chl} 280 (4.18), 370 (3.79)	165d	...	...	...	δ	s	s	s^h	δ	aa s	**B8**¹, 634
n429	**—,7-hydroxy-***	$C_{10}H_6O_3$. *See* n388	174.16	br nd λ^{chl} 265 (4.47), 335 (3.11)	194	...	...	...	...	v	i	...	i	chl i	**B8**¹, 634
Ω n430	**1,4-Naphtho-quinone, 2-hy-droxy-***	Lawsone. $C_{10}H_6O_3$. *See* n390	174.16	redsh-br (aa) λ^{MeOH} 275 (4.5), 335 (3.8)	192d (194–6)	...	...	...	...	v	i	...	i	aa s^h chl i	**B8**², 344
n431	**—,—,acetate**	$C_{12}H_8O_4$. *See* n390	216.21	ye lf (al)	131	...	...	...	δ	s	s	...	...	CS_2 s^h	**B8**², 346
Ω n432	**—,5-hydroxy-***	Juglon. Nucin. $C_{10}H_6O_3$. *See* n390	174.16	redsh-ye nd or pr (chl or bz) λ^{al} 245 (4.1), 330 sh (3.0), 420 (3.5)	154 (161–3)	sub	...	...	i	s	s	...	s^h	chl, aa v^h lig δ	**B8**², 347
n433	**—,6-hydroxy-***	$C_{10}H_6O_3$. *See* n390	174.16	gold-ye or red ye nd (w, bz-al, bz-ace) λ^{chl} 270 (4.37), 390 (3.60)	170d	...	...	...	...	v	v	v	δ	MeOH v sulf s (red) lig δ	**B8**², 348
n434	**—,2-hydroxy-3-methyl-***	Phthiocol. $C_{11}H_8O_3$. *See* n390	188.20	ye pr (eth-peth, MeOH)	173–4	sub	...	...	δ	...	s	s	...	os s lig i	**E12B**, 3069
n435	**—,5-hydroxy-2-methyl-***	Plumbagin. $C_{11}H_8O_3$. *See* n390	188.20	gold pr or og-ye nd (dil al)	78–9	sub	...	...	i	s v^h	v	v	v	chl, CS_2, lig v	**B8**², 350
Ω n436	**—,2-hydroxy-3-phenyl-***	$C_{16}H_{10}O_3$. *See* n390	250.26	gold-ye or og pr or nd (al, bz, MeOH) $\lambda^{0.1N\ NaOH}$ 480 (3.41)	147	...	...	...	...	v^h	v	...	v	chl v alk, lig s	**B8**², 409
n437	**1,2-Naphtho-quinone, 3-methyl-***	$C_{11}H_8O_2$. *See* n388	172.20	red or og nd (abs al), lf (bz, peth) λ^{al} 253 (4.4), 340 (3.3)	116 (122–2.5)	...	...	...	...	s	...	...	s^h	alk, sulf s (gr)	**E12B**, 2814
n438	**—,4-methyl-***	$C_{11}H_8O_2$. *See* n388	172.20	og nd (MeOH), nd (aa) λ^{al} 252 (4.4), 345 (3.4)	248–50d	...	...	...	...	s	...	...	...	alk s (bl) MeOH s^h	**B8**¹, 565
Ω n439	**1,4-Naphtho-quinone, 2-methyl-***	Menadione, Vitamin K_3. $C_{11}H_8O_2$. *See* n390	172.20	ye nd (al, peth) λ^{al} 246 (4.28), 262 sh (4.23), 333 (3.42)	107	...	...	...	i	δ	s	...	s	sulf s aa, lig δ	**B7**², 656
n440	**1,2-Naphtho-quinone, 3-nitro-***	$C_{10}H_5NO_4$. *See* n388	203.16	red pl (aa)	158	...	...	...	s^h	s^h	δ	...	s	aa v^h CS_2, lig i	**B7**², 651
n441	**1,4-Naphtho-quinone, 2-phenyl-***	$C_{16}H_{10}O_2$. *See* n390	234.26	gold-ye nd (al)	111	...	...	...	...	v	v	...	v	chl v sulf s (red) lig δ	**B7**, 822
n442	**—,2(phenyl-amino)-***	Lawsone anilide. $C_{16}H_{11}NO_2$. *See* n390	249.27	red nd (dil al)	193	sub	...	...	...	s^h	s	...	s	sulf s (red) lig δ	**B14**², 91
n443	**—,5,6,7,8-tetra-hydro-***	$C_{10}H_{10}O_2$. *See* n390	162.19	gold-ye nd (peth)	55–6	...	...	...	...	s	s	...	...	aa s	**B7**², 625

For explanations, symbols and abbreviations see beginning of table. For structural formulas see end of table.

No.	Name	Synonyms and Formula	Mol. wt.	Color, crystalline form, specific rotation and λ_{max} (log ε)	m.p. °C	b.p. °C	Density	n_D	Solubility w	al	eth	ace	bz	other solvents	Ref.
	1,4-Naphthoquinone														
n444	—,2,5,8-tri-hydroxy-*	Naphthopurpurin. $C_{10}H_6O_5$. *See* n390	206.16	red nd (bz or MeOH) λ^{al} 480 (3.8)	195	....	....	....	δ v^h	v	...	...	s^h	aa, NH_4OH s sulf s (red)	**B8**², 537
—	**Naphthosultone**....	*see* **1-Naphthalenesulfonic acid, 8-hydroxy-, lactone***													
—	**α-Naphthylamine**..	*see* **Naphthalene, 1-amino-***													
—	**β-Naphthylamine**..	*see* **Naphthalene, 2-amino-***													
—	**α-Naphthylazide**...	*see* **Naphthalene, 1-triazo-***													
—	**Naphthylene-diamine**	*see* **Naphthalene, diamino-***													
—	**Naphthyl red**......	*see* **Azo, benzene 1-naphthalene, 4′-amino-**													
n445	**Narceine**.........	Narcein. Narcéne. Pseudonarcéne. $C_{23}H_{27}NO_8 . 3H_2O$	499.52	nd or pr (w + 3) $\lambda^{w, pH=2}$ 278 (3.0)	145.2 (155) (anh) 176–7 (+3w)	....	....	....	δ s^h	s^h	i	...	i	chl, peth i	**B19**², 386
n447	—,bisulfate decahydrate	$C_{23}H_{27}NO_8 . H_2SO_4 . 10H_2O$. *See* n445	723.71	nd (sulf), pw	d	....	....	....	s	s^h	s	...	...	....	**B19**², 386
n448	—,hydrochloride trihydrate	$C_{23}H_{27}NO_8 . HCl . 3H_2O$. *See* n445	535.98	pr (HCl) λ^w 208 (4.76), 277 (4.19)	192 (anh)	....	....	....	δ^h	s^h	δ	...	...	....	**B19**¹, 372
n449	**α-Narcotine(*dl*)**....	$C_{22}H_{23}NO_7$..............	413.43	nd (al, chl, MeOH) λ^{chl} 290 (3.6), 309 (3.7)	232–3	....	....	....	...	δ	...	...	δ	chl v^h peth i	**B27**², 607
n450	—(*l*)...........	$C_{22}H_{23}NO_7$. *See* n449	413.43	pr or nd (al), [α] −200 (chl) λ^{MeOH} 291 (3.60), 309 (3.69)	176	....	....	....	i	s	δ	v	s	chl s CCl_4 δ	**B27**², 605
n451	—,hydrochloride..	$C_{22}H_{23}NO_7 . HCl$	449.89	(w + 3), [α] +100 λ^w 211 (4.76), 291 (3.40), 313 (3.55)	193 (anh)	....	....	....	v	δ					**B27**², 605
n452	**β-Narcotine(*dl*)**....	β-Gnoscopine. $C_{22}H_{23}NO_7$. *See* n449	413.43	nd, pr (MeOH, al)	180	....	....	....	...	s^h	...	...	...	MeOH s^h	**B27**¹, 559
n452¹	—,hydrochloride..	$C_{22}H_{23}NO_7 . HCl$. *See* n449..	449.89	pr	86–8 (224–6 on standing)	....	....	....	...	...	...	...	...	....	**B27**¹, 559
n453	**Naringin**........	Naringenin-7-rhamno-glucoside. Naringoside. $C_{27}H_{32}O_{14} . 2H_2O$.	616.58	nd (w + 8) $[\alpha]_D^{19}$ −82.11 (al) λ^{al} 283 (4.37)	82 (+8w), 171 (+2w)	....	....	....	δ s^h	δ v^h	i	...	i	aa s^h chl i	**C54**, 25366
n454	**Neoabietic acid**, methyl ester	Methyl neoabietate. $C_{21}H_{32}O_2$.	316.49	cr (MeOH)	61.5–2	....	....	....	...	...	...	...	...	MeOH v	**B9**², 433
n455	**Neoamygdalin**.....	*l*-Mandelonitrile-β-gentobioside. $C_{20}H_{27}NO_{11}$.	457.44	cr (al), $[\alpha]_D^{25}$ −61.4 (w, c = 8.5)	212	....	....	....	v	v	...	...	...	....	**B31**, 404
n456	**Neobornyl amine**	Isobornylamine...........	153.27	pw $[\alpha]_D$ −47.7 (4 % al) −27 (bz)	184	....	....	....	i	...	s	s	...	os s	**B12**², 40
n457	**Neocarvo-menthol(*dl*)**	....	156.27	....	....	....	....	1.4637^{20}	...	s	...	...	...	....	**B6**³, 130
n458	—(*l*)...........	$C_{10}H_{20}O$. *See* n457......	156.27	$[\alpha]_D^{21}$ −41.7 λ^{eth} 270 (2.5)	....	102^{18} 90^9	0.9012$^{20}_4$	1.4632^{20}	...	s	...	...	...	....	**B6**³, 130
—	**Neocholestene**....	*see* **2-Cholestene**													
n458¹	**Neoergosterol**....	$C_{26}H_{38}O$.............	380.62	pr or nd (al), $[\alpha]_D^{17}$ −12 (chl, c = 2)	154.5 (157)	....	....	....	i	...	s	s	...	os s	**B6**³, 3474
n459	**Neogermitrine**....	$C_{38}H_{55}NO_{11}$...........	677.84	wh nd $[\alpha]_D^{25}$ −79.2 (Py)	237–9	....	....	....	i	d^h	...	...	...	chl s	**C54**, 6780
—	**Neohexane**.......	*see* **Butane, 2,2-dimethyl-*** ...													
n460	**Neoisocarvo-menthol(*l*)**	....	156.27	$[\alpha]_D^{17}$ −34.7	< −25	87–8^4	0.9102$^{20}_4$	1.4676^{20}	i	s	...	s	...	....	**B6**³, 130
n461	**Neoisomenthol(*d*)**..	*p*-Menthol-3..........	156.27	$[\alpha]_D^{15}$ +2.2 (al, c = 2)	−8	214.6^{760} 91.5^{11}	0.9131$^{18}_4$	1.4670^{20}	i	s	...	s	...	....	**B6**³, 132
n462	—(*dl*)...........	$C_{10}H_{20}O$. *See* n461........	156.27	....	14	214.5^{760} 81^6	0.8854$^{55}_4$	1.4649^{20}	i	s	...	s	...	....	**B6**³, 132
n463	**Neomenthol(*d*)**....	*d*-β-Pulegomenthol.	156.27	$[\alpha]_D^{20}$ +19.6 (al)	−15	211.7^{760} 95^{12}	0.897$^{22}_4$	1.4600^{20}	i	s	...	s	...	oos s	**B6**³, 139
n464	—(*dl*)...........	$C_{10}H_{20}O$. *See* n463........	156.27	pl or pr (peth)	52	211.7^{760} 103–5^{16}	0.903$^{15}_{15}$	1.4600^{20}	i	s	...	s	...	oos s	**B6**³, 140

For explanations, symbols and abbreviations see beginning of table. For structural formulas see end of table.

No.	Name	Synonyms and Formula	Mol. wt.	Color, crystalline form, specific rotation and λ_{max} (log ε)	m.p. °C	b.p. °C	Density	n_D	Solubility w	al	eth	ace	bz	other solvents	Ref.
	Neomenthol														
n465	—(*l*)	$C_{10}H_{20}O$. *See* n463	156.27	$[\alpha]_D^{18}$ −19.62 (al)	...	211.7^{760} 97.6^{10}	...	1.4603^{20}	i	s	...	s	...	oos s	**B6**[3], 140
—	**Neonicotine**	*see* **Anabasine**													
—	**Neopentane**	*see* **Propane, 2,2-dimethyl-***													
—	**Neopentyl alcohol**	*see* **1-Propanol, 2,2-dimethyl-***													
—	**Neopentyl chloride**	*see* **Propane, 1-chloro-2,2-dimethyl-***													
—	**Neopine**	*see* **β-Codeine**													
—	**Neral**	*see* **Citral b**													
—	**Neriine**	*see* **Conessine**													
Ω n466	**Nerol**	3,7-Dimethyl-2,6-octadien-1-ol*. $C_{10}H_{18}O$. $(CH_3)_2C{:}CH(CH_2)_2C(CH_3){:}CHCH_2OH$	154.26	λ^{hp} 189–94 (4.26)	< −15	$224–5^{745}$ 125^{25}	0.8756_4^{20}	1.4746^{20}	i	s	...	...	...	...	**B1**[3], 2008
n467	**Nerolidol**(*d*)	Peruviol. α-3,7,11-Trimethyl-1,6,10-dodecatrien-3-ol*. $C_{15}H_{26}O$. $(CH_3)_2C{:}CH(CH_2)_2C(CH_3){:}CH(CH_2)_2C(CH_3)(OH)CH{:}CH_2$	222.38	$[\alpha]_D^{20}$ +15.5 (undil)	...	276^{760} $128–9^{6}$	0.8778_4^{20}	1.48982^{20}	...	v	s	s	...	os, aa s	**B1**[3], 2041
Ω n468	—(*dl*)	$C_{16}H_{25}O$. *See* n467	222.38	λ^{al} 273 sh (1.74), 280 sh (1.70)	...	$145–6^{12}$ $75–6^{0.1}$	0.8756_4^{19}	1.4801^{16}	...	s	s	s	...	os, aa s	**B1**[3], 2042
n469	—(*l*)	$C_{15}H_{26}O$. *See* n467	222.38	$[\alpha]_D$ −6.5 (undil)	...	$124–6^{3}$	0.8881_{15}^{15}	1.4799^{20}	...	s	s	s	...	os, aa s	**B1**[3], 2042
n470	**Neurine**	Trimethylethenylammonium hydroxide*. $CH_2{:}CHN^+(CH_3)_3OH^-$	103.17	syr	...	...	...	...	s	s	s	...	...	...	**B4**[3], 442
—	**Neuronal**	*see* **Butanoic acid, 2-bromo-2-ethyl-,** amide													
—	**Nevile-winther acid**	*see* **1-Naphthalenesulfonic acid, 4-hydroxy-**													
—	**Niacinamide**	*see* **3-Pyridinecarboxylic acid,** amide													
—	**Nicotinamide**	*see* **3-Pyridinecarboxylic acid,** amide													
n471	**Nicotine**(*d*)	α-*N*-Methyl-*d*-β-pyridylpyrrolidine. $C_{10}H_{14}N_2$	162.24	hyg, $[\alpha]_D^{20}$ +163.2 λ^{al} 262 (3.47)	...	245.5–6.5^{729}	1.0094_4^{20}	1.5280^{20}	∞	v	v	...	...	chl v lig s	**B23**, 117
n472	—(*dl*)	Tetrahydronicotyrine. $C_{10}H_{14}N_2$. *See* n471	162.24	λ^{al} 262 (3.47)	...	242–3	1.0082_4^{20}	1.5289^{20}	∞	v	v	...	...	chl v lig s	**B23**, 117
Ω n473	—(*l*)	$C_{10}H_{14}N_2$. *See* n471	162.24	hyg, br in air, $[\alpha]_D^{20}$ −169 λ^{al} 262 (3.47)	−79	246.7^{745} $124–5^{18}$	1.0097_4^{20}	1.5282^{20}	∞	v	v	...	...	chl v lig s	**B23**, 110
n474	—,hydrochloride(*d*)	$C_{10}H_{14}N_2 \cdot HCl$. *See* n471	198.70	dlq, $[\alpha]_D^{20}$ +104 (w, p = 10)	...	...	1.0337	...	...	...	...	...	...	...	**B23**, 114
—	**Nicotinic acid**	*see* **3-Pyridinecarboxylic acid**													
—	**Nicotinonitrile**	*see* **3-Pyridinecarboxylic acid,** nitrile													
n475	**2,2′-Nicotyrine** (solid)	*N*-Methyl-2(2-pyridyl)-pyrrole. α-Nicotyrine. $C_{10}H_{10}N_2$	158.21	cr, br in air λ^{al} 288 (3.99)	43.4–4.5	...	...	...	i	v	s	...	v	dil HCl s	**B23**[2], 192
n476	—(liquid)	$C_{10}H_{10}N_2$. *See* n475	158.21	λ^{al} 288 (3.99)	−28	273^{764} $149–50^{22}$	...	...	δ	...	...	...	s	os s	**B23**[2], 191
n477	**3,2′-Nicotyrine**	*N*-Methyl-2(3-pyridyl)-pyrrole. β-Nicotyrine.	158.21	br in air λ^{al} 288 (3.99)	...	$280–1^{744}$ 150^{15}	1.2411_4^{20}	1.6057^{20}	δ s^h	s	s	s	...	os s	**B23**[2], 192
Ω n478	**Ninhydrin**	1,2,3-Triketohydrindene monohydrate.	178.15	pr (w) λ^{al} 228 (4.64) 357 (2.00)	241–3d (250d)	...	...	...	v^h	s	δ	...	...	alk s	**B7**[2], 831
—	**Nioxime**	*see* **1,2-Cyclohexanedione, dioxime*.**													
—	**Nirvanol**	*see* **Hydantoin, 5-ethyl-5-phenyl-**													
Ω n479	**Nitranilic acid**	2,5-Dihydroxy-3,6-dinitro-*p*-benzoquinone.	230.09	gold-ye pl (+w, dil HNO_3)	86–7d (hyd) (100d)	exp	...	...	v	v	i	...	...	...	**B8**[3], 433
Ω n480	**Nitric acid,** butyl ester*	*n*-Butyl nitrate. $CH_3(CH_2)_3ONO_2$	119.12	...	...	135.5^{763} (cor) $70–1^{86}$	1.0228^{30}	1.4013^{23}	i	s	s	...	...	...	**B1**[2], 397
n481	—,*sec*-butyl ester (*dl*)	*sec*-Butyl nitrate. $CH_3CH_2CH(CH_3)ONO_2$	119.12	λ^{al} 270 sh (1.23)	...	124 59^{80}	1.0264_4^{20}	1.4015^{20}	...	∞	∞	...	...	...	**B1**[3], 1533
n482	—,decyl ester*	*n*-Decyl nitrate. $CH_3(CH_2)_9ONO_2$	203.28	...	...	$127–8^{11}$ $88–9^{1}$	0.951_4^{0}	...	...	s	s	...	...	...	**B1**, 425
n483	—,ethyl ester*	Ethyl nitrate. $CH_3CH_2ONO_2$.	91.07	inflam λ^{al} 266 sh (1.17)	−94.6	87.2^{762}	1.1084_4^{20}	1.3852^{20}	s	∞	∞	...	...	...	**B1**[2], 328

For explanations, symbols and abbreviations see beginning of table. For structural formulas see end of table.

No.	Name	Synonyms and Formula	Mol. wt.	Color, crystalline form, specific rotation and λ_{max} (log ε)	m.p. °C	b.p. °C	Density	n_D	Solubility w	al	eth	ace	bz	other solvents	Ref.
	Nitric acid														
n484	—,isopropyl ester*	Isopropyl nitrate. $(CH_3)_2CHONO_2$	105.10	...	...	100–1.4^{760}	1.036$^{19}_{19}$	1.3912^{16}	...	s	s	...	...	...	**B1^3**, 1465
n485	—,methyl ester*	Methyl nitrate. CH_3ONO_2	77.04	explosive vapor	−82.3 exp	64.6^{760} exp	1.2075$^{20}_4$	1.3748^{20}	δ	s	s	...	...	...	**B1^1**, 284
Ω n486	—,3-methylbutyl ester*	Isoamyl nitrate. $(CH_3)_2CHCH_2CH_2ONO_2$	133.15	...	...	147–8	0.9961$^{22}_4$	1.4122^{22}	δ	∞	∞	...	...	...	**B1^3**, 1642
n487	—,octyl ester*	*n*-Octyl nitrate. $CH_3(CH_2)_7ONO_2$	175.23	λ^{al} 270 sh (1.18)	...	110–2^{20}	0.8419$^{17}_{17}$	...	δ	s	s	...	...	...	**B1**, 419
n488	—,propyl ester*	*n*-Propyl nitrate. $CH_3CH_2CH_2ONO_2$	105.10	...	...	110^{762}	1.0538$^{20}_4$	1.3973^{20}	δ	s	s	...	...	...	**B1^2**, 369
—	**Nitroform**	*see* **Methane, trinitro-***													
—	**Nitrofurazone**	*see* **Furfural, 5-nitro-,** semicarbazone													
—	**Nitrogen mustard**	*see* **Amine, triethyl, 2,2′2″-trichloro-**													
—	**Nitroglycerin**	*see* **Glycerol,** trinitrate													
—	***N*-Nitroguanidine**	*see* **Guanidine,1-nitro-***													
Ω n489	**Nitron**	4,5-Dihydro-1,4-diphenyl-3,5-phenylimino-1,2,4-triazole.	312.38	ye lf (al), nd (+chl)	189d	...	...	...	i	s	δ	s	s	chl, CCl_4, dil ac s	**B26^2**, 76, 199
n490	**Nitrous acid,** butyl ester*	*n*-Butyl nitrite. $CH_3(CH_2)_3ONO$	103.12	λ^{cy} 223 (3.15), 333 (1.72), 356 (1.93), 384 (1.51)	...	77.8^{760} 27^{88}	0.8823$^{20}_4$	1.3762^{20}	...	∞	∞	...	...	...	**B1^2**, 397
n491	—,*sec*-butyl ester (*dl*)	*sec*-Butyl nitrite. $C_2H_5CH(CH_3)ONO$	103.12	...	...	68–9 28^{180}	0.8726$^{20}_4$	1.3710^{20}	δ	v	v	...	...	CCl_4, CS_2, chl s	**B1^2**, 402
Ω n492	—,*tert*-butyl ester	*tert*-Butyl nitrite. $(CH_3)_3CONO$	103.12	pa ye λ^{cy} 221 (3.23), 353 (1.68), 382 (1.91), 397 (1.72)	...	62.8–3.2 34^{250}	0.8670$^{20}_4$	1.3687^{20}	δ	s	s	...	...	chl, CS_2 s	**B1^2**, 415
n493	—,decyl ester*	*n*-Decyl nitrite. $CH_3(CH_2)_9ONO$	187.29	yesh λ^{hp} 357 (1.92), 384 (1.53)	...	105–8^{12} 95^4	...	1.4247^{20}	...	s	s	...	...	...	**B1**, 425
n494	—,ethyl ester*	Ethyl nitrite. CH_3CH_2ONO	75.07	yesh or col	...	16.5–17^{725}	0.90$^{15}_{15}$	1.3418^{10}	d	∞	∞	...	...	...	**B1^2**, 328
n495	—, heptyl ester*	*n*-Heptyl nitrite. $CH_3(CH_2)_6ONO$	145.20	λ^{eth} 315 (1.38), 356 (1.93), 382.5 (1.52)	...	155–8 44^{18}	0.8939$^0_4$	1.4032^{20}	i	...	s	...	...	...	**B1^2**, 443
n496	—, hexyl ester*	*n*-Hexyl nitrite. $CH_3(CH_2)_5ONO$	131.18	ye λ^{hp} 357 (1.91), 387 (1.52)	...	129–30^{774} (cor) 52^{44}	0.8778$^{20}_4$	1.3987^{20}	i	s	s	...	...	...	**B1^3**, 1657
n497	—, isopropyl ester*	Isopropyl nitrite. $(CH_3)_2CHONO$	89.10	λ^{eth} 327 (1.20), 359 (1.70), 387 (1.45)	...	45^{762}	0.8684$^{15}_4$	...	i	s	s	...	...	...	**B1^3**, 1464
n498	—, methyl ester*	Methyl nitrite. CH_3ONO	61.04	gas	−16	−12	0.991^{15} (liq)	...	...	s	...	...	...	...	**B1**, 284
Ω n499	—, 3-methylbutyl ester*	Isoamyl nitrite. $(CH_3)_2CHCH_2CH_2ONO$	117.15	yesh, inflam λ^{cy} 224 (3.29), 333 (1.71), 369 (1.83), 384 (1.52)	...	99.2^{760} 30^{60}	0.8828$^{20}_4$	1.3918^{20}	δ	∞	∞	...	...	...	**B1^3**, 1641
n500	—, octyl ester*	*n*-Octyl nitrite. $CH_3(CH_2)_7ONO$	159.23	ye λ^{hp} 357 (1.92), 384 (1.52)	...	174–5^{760} 60^{10}	0.862^{17}	1.4127^{20}	δ	v	v	...	...	...	**Am 78**, 1501
n501	—, pentyl ester*	*n*-Amyl nitrite. $CH_3(CH_2)_4ONO$	117.15	ye λ^{cy} 224 (3.25), 334 (1.66), 356 (1.87), 383 (1.45)	...	104.5^{763} (cor) 29^{40}	0.8817$^{20}_4$	1.3851^{20}	δ	∞	∞	...	...	...	**B1^3**, 1604
n502	—, propyl ester*	*n*-Propyl nitrite. $CH_3CH_2CH_2ONO$	89.10	...	...	79^{760} (57)	0.935^{21}	1.3604^{20}	δ	s	s	...	...	...	**B1^2**, 369
Ω n502^1	**Nonacosane***	$CH_3(CH_2)_{27}CH_3$	408.80	rh cr (peth)	63.7	440.8^{760} 271.4^{10}	0.7630$^{100}_4$ 0.8083$^{20}_4$	1.4529^{20}	i	v^h	v	v	s	chl δ	**B1^2**, 144
n503	—, **2-methyl-***	$(CH_3)_2CH(CH_2)_{26}CH_3$	422.83	pl (lig)	73–4	222$^{0.3}$	...	...	i	...	s	s	s	chl δ	**B1^1**, 72
n504	**1-Nonacosanol***	$CH_3(CH_2)_{27}CH_2OH$	424.80	cr (al)	84.6–5.0	sub 200–50^1	...	...	...	s^h	...	...	...	...	**B1^2**, 471
Ω n505	**Nonadecane***	$CH_3(CH_2)_{17}CH_3$	268.53	wax	32.1	329.7^{760} 193^{15}	0.7774$^{32}_4$ 0.7855$^{20}_4$	1.4409^{20}	i	δ	s	s	...	...	**B1^3**, 568
Ω n506	**1,2,3-Nonadecanetricarboxylic acid, 2-hydroxy-***	Agaric acid. Agaricin. Laricic acid. $CH_3(CH_2)_{15}CH(CO_2H)C(OH)(CO_2H)CH_2(CO_2H)$	416.56	lf (+1.5w, dil al) $[\alpha]^{19}_D$ −8.8 (NaOH)	142d	...	...	...	s^h	δ	δ	...	i	chl i	**B3^2**, 372
n507	—, triethyl ester	$C_{28}H_{52}O_7$. *See* n506.	500.73	nd	36–7	...	...	...	...	s^h	...	...	s	...	**B3^2**, 373
n508	—, trimethyl ester	$C_{25}H_{46}O_7$. *See* n506.	458.64	nd (al)	63–4	...	...	...	...	s^h	...	...	s	...	**B3^2**, 373
Ω n509	**Nonadecanoic acid***	*n*-Nonadecylic acid. $CH_3(CH_2)_{17}CO_2H$	298.52	lf (al)	69.4	297–8^{100} 227–30^{10}	...	...	i	v^h	v	...	v	chl, lig v	**B2^2**, 368
n510	**1-Nonadecanol***	*n*-Nonadecyl alcohol. $CH_3(CH_2)_{17}CH_2OH$	284.53	cr (ace)	62–3	166–7$^{0.32}$	...	1.4328^{75}	...	...	s	s^h	...	...	**B1^3**, 1841

For explanations, symbols and abbreviations see beginning of table. For structural formulas see end of table.

No.	Name	Synonyms and Formula	Mol. wt.	Color. crystalline form. specific rotation and λ_{max} (log ε)	m.p. °C	b.p. °C	Density	n_D	Solubility: w	al	eth	ace	bz	other solvents	Ref.
	2-Nonadecanone														
Ω n511	**2-Nonadecanone***	Methyl *n*-heptadecyl ketone. $CH_3(CH_2)_{16}COCH_3$	282.52	pl (al)	57	266.5^{110} 165^{2}	0.8108^{56}	....	i	δ s^h	v	s v^h	...	chl, CCl_4 v to v^h bz s, v^h	**B1**[2], 773
n512	—, oxime*	$CH_3(CH_2)_{16}C(:NOH)CH_3$	297.53	cr (al)	76.5–7.5	....	....	....	...	s^h	...	...	...	....	**B1**, 718
n513	**4-Nonadecanone***	*n*-Propyl *n*-pentadecyl ketone. $CH_3(CH_2)_{14}CO(CH_2)_2CH_3$	282.52	lf (al)	50.5	211^{11}δd	....	....	i	δ	v	v	...	....	**B1**[2], 773
Ω n514	**10-Nonadecanone***	Caprinone. Dinonyl ketone. $[CH_3(CH_2)_7CH_2]_2CO$	282.52	lf (al)	65.5 (58)	>350 $155.6^{1.1}$	....	....	i	δ s^h	s	s	v	chl, lig s	**B1**, 718
n515	**1-Nonadecyne***	$CH_3(CH_2)_{16}C⋮CH$	264.50	....	37–8 (33)	327^{760} 181.6^{10}	0.8054_4^{20}	1.4488^{20}	...	...	s	s	s	....	**B1**[3], 1030
n516	**1,8-Nonadiyne***	$HC⋮C(CH_2)_5C⋮CH$	120.20	....	−27.3	162^{760} 55^{13}	0.8158_4^{20}	1.4490^{20}	...	...	s	s	...	....	**B1**[3], 1063
Ω n517	**Nonanal***	*n*-Nonylaldehyde. Pelargonaldehyde. $CH_3(CH_2)_7CHO$	142.24	....	....	190–2 93.5^{23}	0.8264_4^{22}	1.4273^{20}	...	...	s	...	...	....	**B1**[2], 761
n518	—, oxime*	$CH_3(CH_2)_7CH:NOH$	157.26	lf (dil al)	64	....	....	....	i	s	s	s	...	os s	**B1**[2], 761
Ω n519	**Nonane***	$CH_3(CH_2)_7CH_3$	128.26	....	−51	150.798^{760} 39^{10}	0.7176_4^{20}	1.4054^{20}	i	v	v	∞	∞	chl, hp ∞	**B1**[3], 502
Ω n520	—, **1-amino-***	*n*-Nonylamine. $CH_3(CH_2)_7CH_2NH_2$	143.28	....	−1	202.2^{760} 80.8^{10}	0.7886_4^{20}	1.4336^{20}	δ	s	s	...	...	....	**B4**[3], 393
n521	—, **1-chloro-***	*n*-Nonyl chloride. $CH_3(CH_2)_7CH_2Cl$	162.71	....	−39.4	203.4^{760} 80.5^{10}	0.8720_4^{20}	1.4345^{20}	i	...	s	...	...	chl s	**B1**[2], 128
n522	—, **2-chloro-***	2-Nonyl chloride. $CH_3(CH_2)_6CHClCH_3$	162.71	....	....	190^{764}	0.8790^{20}	1.4420^{20}	i	...	...	...	...	chl s	**B1**, 166
n523	—, **5-chloro-***	5-Nonyl chloride. $(CH_3CH_2CH_2CH_2)_2CHCl$	162.71	....	....	$85–7^{14}$	0.8639_4^{15}	1.4314^{15}	i	...	v	...	...	....	**B1**[2], 128
n524	—, **1-chloro-9-fluoro-***	$FCH_2(CH_2)_7CH_2Cl$	180.70	....	....	$102–2.5^{11}$	0.966_4^{20}	1.4301^{25}	i	v	v	...	...	....	**C51**, 7300
Ω n525	**Nonanedioic acid***	Azelaic acid. 1,7-Heptanedicarboxylic acid*. $HO_2C(CH_2)_7CO_2H$	188.23	lf or nd	106.5	>360d 287^{100} 225.5^{10}	1.225_4^{25}	1.4303^{111}	δ s^h	s	δ	...	δ v^h	....	**B2**[2], 602
Ω n526	—, dichloride	Azelayl dichloride. $ClCO(CH_2)_7COCl$	225.12	....	....	166^{18} $140^{0.4}$	....	1.4680^{20}	d	d^h	s	...	v^h	....	**B2**, 709
Ω n527	—, diethyl ester*	Ethyl azelate. $C_2H_5O_2C(CH_2)_7CO_2C_2H_5$	244.34	λ^{al} 255 (0.7), 262 (0.8), 273 (0.7)	−18.5	291–2 $174–5^{20}$	0.97294_4^{20}	1.43509^{20}	i	s	s	...	...	....	**B2**[2], 603
Ω n528	—, di(2-ethylbutyl) ester*	Di-2-ethylbutyl azelate. $CH_2[(CH_2)_3CO_2CH_2CH(C_2H_5)_2]_2$	356.55	....	−45	230^{5}	0.928_4^{25}	1.443^{25}	i	s	...	s	s	....	**B2**[3], 1787
n529	—, di(2-ethylhexyl) ester*	Di-2-ethylhexyl azelate. $CH_2[(CH_2)_3CO_2CH_2CH(C_2H_5)C_4H_9]_2$	412.66	....	−78	237^{5}	0.915_4^{25}	1.446^{25}	i	s	...	s	s	....	**B2**[3], 1787
Ω n530	—, dimethyl ester*	Dimethyl azelate. $CH_3O_2C(CH_2)_7CO_2CH_3$	216.28	....	....	156^{20} 128^{5}	1.0082_4^{20}	1.4367^{20}	i	s	...	s	s	os s	**B2**[1], 290
Ω n531	—, dinitrile	Azelanitrile. $NC(CH_2)_7CN$	150.23	....	....	$198–9^{23}$ 160^{3}	0.9290^{19}	1.4518^{19}	i	v	v	...	v	....	**B2**[1], 290
n532	—, diphenyl ester*	Diphenyl azelate. $C_6H_5O_2C(CH_2)_7CO_2C_6H_5$	340.42	nd (al)	59–60 (49)	....	....	....	i	δ s^h	v	...	v	....	**B6**[3], 606
Ω n533	**1,9-Nonanediol***	Nonamethylene glycol. $HOCH_2(CH_2)_7CH_2OH$	160.26	cr (bz)	45.8	$173–5^{20}$	....	....	δ	v	v	...	s^h	lig i	**B1**[3], 2226
Ω n534	**Nonanoic acid***	*n*-Nonylic acid. Pelargonic acid. $CH_3(CH_2)_7CO_2H$	158.24	....	fp 12.24 15	255^{760} 150^{20}	0.9057_4^{20}	1.4343^{19}	i	s	s	...	...	chl s	**B2**[2], 360
n535	—, amide	Nonanamide*. Pelargonamide. $CH_3(CH_2)_7CONH_2$	157.26	....	99.2–9.7	sub	0.8394^{110}	1.4248^{110}	i	δ	δ	...	...	....	**B2**[3], 822
Ω n536	—, chloride	Nonanoyl chloride*. Pelargonyl chloride. $CH_3(CH_2)_7COCl$	176.69	....	−60.5	215.35^{760} 98^{15}	0.9463_4^{15}	....	d	d	s	s	...	....	**B2**[2], 308
Ω n537	—, ethyl ester*	$CH_3(CH_2)_7CO_2C_2H_5$	186.30	....	−36.7	227^{760} $96–8^{10}$	0.8657_4^{20}	1.4220^{20}	i	s	s	s	...	....	**B2**[2], 307
Ω n538	—, methyl ester*	$CH_3(CH_2)_7CO_2CH_3$	172.27	....	....	$213–4^{757}$ $104–6^{23}$	0.8799^{15}	1.4214^{20}	i	s	s	...	...	....	**B2**[2], 307
Ω n539	—, nitrile	Nonanonitrile*. Octyl cyanide. Pelargonitrile. $CH_3(CH_2)_7CN$	139.24	....	−34.2	224.4^{760} 91.9^{10}	0.8178_4^{20}	1.4255^{20}	i	s	s	...	...	....	**B2**[3], 828
n540	—, piperazinium salt	$2[CH_3(CH_2)_7CO_2H].C_4H_{10}N_2$	402.62	wh	95.1–6.2	....	....	....	s^h	s	i	...	...	....	**Am70**, 2758
n541	—, **9-amino***	$H_2NCH_2(CH_2)_7CO_2H$	173.26	....	185.6–6.6	....	....	....	s	s	...	...	...	....	**B4**[3], 1479
n542	—, **9-fluoro-***	ω-Fluoropelargonic acid. $FCH_2(CH_2)_7CO_2H$	176.23	....	*ca.* 18	$89–90^{0.2}$	....	1.4289^{25}	δ	v	v	...	...	....	**C52**, 1918
Ω n543	**1-Nonanol***	Nonyl alcohol. $CH_3(CH_2)_7CH_2OH$	144.26	....	−5.5	213.5^{760} 118^{15}	0.8273_4^{20}	1.4333^{20}	i	s	s	...	...	....	**B1**[3], 1743
n549	**5-Nonanol, 5-butyl-***	Tri-*n*-butyl carbinol. $[CH_3(CH_2)_3]_3COH$	200.37	....	20	230–5d 118–20^{17}	0.8408_4^{20}	1.4445^{20}	i	s	...	...	...	....	**B1**[3], 1802
n550	**1-Nonanol, 9-chloro-***	Nonamethylene chlorohydrin. $ClCH_2(CH_2)_7CH_2OH$	178.71	....	28	$146–8^{14}$	....	1.4575^{20}	i	v	v	...	...	....	**C51**, 7300
n551	—, **9-fluoro-***	Nonamethylene fluorohydrin. $FCH_2(CH_2)_7CH_2OH$	162.25	....	....	$125–6^{15}$	0.928_4^{20}	1.4279^{25}	i	v	v	...	...	....	**C51**, 7300
Ω n552	**2-Nonanone***	Heptyl methyl ketone. $CH_3(CH_2)_6COCH_3$	142.24	λ<200	−7.46	195.3^{760} 73.8^{10}	0.8208_4^{20}	1.42096^{20}	i	s	s	v	s	chl, MeOH v	**B1**[3], 2887

For explanations, symbols and abbreviations see beginning of table. For structural formulas see end of table.

No.	Name	Synonyms and Formula	Mol. wt.	Color, crystalline form, specific rotation and λ_{max} (log ε)	m.p. °C	b.p. °C	Density	n_D	Solubility w	al	eth	ace	bz	other solvents	Ref.
	4-Nonanone														
Ω n553	**4-Nonanone***	*n*-Propyl *n*-amyl ketone. $CH_3(CH_2)_4CO(CH_2)_2CH_3$	142.24	λ^{MeOH} 279 (1.42)	...	187–8 $75\text{–}6^{20}$	0.8190_4^{25}	1.4189^{20}	i	s	s	v	...	chl s	**B1**[3], 2888
n554	**5-Nonanone***	Dibutyl ketone. $CH_3(CH_2)_3CO(CH_2)_3CH_3$	142.24	λ^{MeOH} 278 (1.46)	−4.8	188.4 88^{22}	0.8217_4^{20}	1.4195^{20}	i	s	v	...	...	chl v	**B1**[2], 762
n555	**Nonasiloxane, eicosamethyl-**	$CH_3[Si(CH_3)_2O\text{-}]_8Si(CH_3)_3$	681.47	...	...	307.5^{760} 198.8^{16}	0.9173^{20}	1.3980^{20}	i	δ	...	...	s	...	**C43**, 4402
Ω n556	**1,3,6,8-Nonatetraen-5-one, 1,9-diphenyl-***	Dicinnamylidene acetone, $(C_6H_5CH{:}CHCH{:}CH)_2CO$	286.38	ye nd (abs al) λ^{al} 266 (4.74)	144	...	...	...	i	δ s^h	δ	...	...	con sulf s	**B7**, 524
n557	**1-Nonen-3-yne***	$CH_3(CH_2)_4C{:}CCH{:}CH_2$	122.21	...	...	$27.2\text{–}8.2^{4}$	0.7602_4^{25}	1.4487^{25}	i	...	s	s	...	...	**B1**[3], 1048
n558	**1-Nonen-4-yne***	$CH_3(CH_2)_3C{:}CCH_2CH{:}CH_2$	122.21	...	...	58.0^{22}	0.777_4^{25}	1.4413^{25}	i	...	s	s	...	...	**B1**[3], 1048
n559	**2-Nonen-4-yne***	$CH_3(CH_2)_3C{:}CCH{:}CHCH_3$	122.21	...	...	$70\text{–}0.5^{29}$	0.7832_4^{25}	1.4590^{25}	i	...	s	s	...	...	**B1**[3], 1048
n560	**1-Nonyne***	$CH_3(CH_2)_6C{:}CH$	124.23	...	−50	150.8^{760} 33.3^{10}	0.7568_4^{20}	1.4217^{20}	i	...	s	...	s	os s	**B1**[3], 1012
n561	**2-Nonyne***	$CH_3(CH_2)_5C{:}CCH_3$	124.23	...	...	$158\text{–}9^{760}$	0.7690_4^{20}	1.4331^{20}	i	...	s	...	...	lig s	**B1**[3], 1013
n562	**3-Nonyne***	$CH_3(CH_2)_4C{:}CCH_2CH_3$	124.23	...	...	$153\text{–}5^{745}$ 92^{97}	0.7616_4^{20}	1.4299^{20}	i	...	s	...	...	lig s	**B1**[3], 1013
n563	**4-Nonyne***	$CH_3(CH_2)_3C{:}CCH_2CH_2CH_3$	124.23	...	...	$150\text{–}4^{750}$	0.757_4^{25}	1.4296^{25}	i	...	s	s	...	...	**B1**[3], 1013
n564	**1-Nonyne, 1-chloro-***	$CH_3(CH_2)_6C{:}CCl$	158.67	...	...	$75\text{–}7^{15}$	0.906^{20}	1.450^{20}	i	...	v	...	...	...	**B1**[3], 1012
n565	**4-Nonyne,3,3-dimethyl-***	$CH_3(CH_2)_3C{:}CC(CH_3)_2CH_2CH_3$	152.28	...	...	82^{40}	0.7667_4^{20}	1.4317^{20}	...	...	s	...	s	...	**B1**[3], 1024
n566	**—,8-methyl-***	2-Methyl-5-nonyne. $(CH_3)_2CHCH_2CH_2C{:}CCH_2CH_2CH_3$	138.25	...	...	104.5^{97}	0.7681_4^{20}	1.4311^{20}	...	...	s	...	s	...	**B1**[3], 1018
—	**Nopinene**	*see* **β-Pinene**													
n567	**Nopinone(*d*)**	...	138.21	$[\alpha]_D^{20}$ +34 (chl), +11.5 (eth)	0	209^{760} $87\text{–}8^{14}$	0.9807_4^{20}	1.4787^{20}	s	s	s	...	...	...	**B7**[2], 69
n568	**Noradrenaline(*l*)**	*l*-Arterenol. *l*-Norepinephrine.	169.18	$[\alpha]_D^{25}$ −37.3 (dil HCl) $\lambda^{w, pH=7}$ 280	216.5–8d	...	...	...	δ	δ	δ	...	...	alk, dil HCl v	**B13**[2], 523
—	**Norbornane**	*see* **Bicyclo(2.2.1)heptane**													
—	**Norbornylane**	*see* **Bicyclo(2.2.1)heptane**													
—	**Norcamphane**	*see* **Bicyclo(2.2.1)heptane**													
—	**Norcamphane dicarboxylic acid**	*see* **Bicyclo(2.2.1)heptane-2,3-dicarboxylic acid**													
—	**Norcamphoric acid**	*see* **1,3-Cyclopentane-dicarboxylic acid***													
n569	**Nordihydro-guaiaretic acid**	NDGA	302.38	nd (w, al, aa)	185–6	...	...	...	δ^h	s	s	s	i	alk s chl δ^h to i	**B6**[1], 577
Ω n570	**Norephedrine,** hydrochloride(*dl*)	2-Amino-1-phenyl-1-propanol hydrochloride. $C_6H_5CH(OH)CH(CH_3)NH_2.HCl$	187.67	pl (abs al), cr (dil HCl, al or al-AcOEt)	194	...	...	...	v	s	i	...	i	chl i	**B13**[2], 371
n571	**—,*N,N*-diethyl-hydro-chloride(*dl*)**	$C_6H_5CH(OH)CH(CH_3)N(C_2H_5)_2.HCl$	243.78	cr (al-ace, AcOEt-al)	205–6	...	...	...	...	s	...	...	...	...	**B13**[2], 380
n572	**—,*N*-ethyl-**	$C_6H_5CH(OH)CH(CH_3)NHCH_2CH_3$	179.26	cr (lig)	51.5	143^{18}	...	...	...	...	...	...	s	...	**B13**[2], 380
n573	**Normorphine**	Desmethylmorphine. $C_{16}H_{17}NO_3$.	271.32	(w + 1½), (MeOH + 1)	273 (w + 1½) 263–4 (anh) (277)	...	...	...	i^h	i^h	...	...	i	chl i	**B27**[2], 117
n574	**Nornicotine(*l*)**	*l*-3(2-Pyrrolidyl)pyridine. $C_9H_{12}N_2$.	148.21	hyg $[\alpha]_D^{22}$ −88.8 (undil) λ^{al} 262 (3.47)	...	270 130.5–1.3^{11}	$1.0737^{19.5}$	$1.5378^{18.5}$	∞	v	v	v	...	chl, peth v	**B23**[2], 107
—	**Norvaline**	*see* **Pentanoic acid, 2-amino-***													
—	**Novocaine**	*see* **Benzoic acid, 4-amino-** 2-diethylaminoethyl ester, hydrochloride													
—	**Novonal**	*see* **4-Pentenoic acid, 2,2-diethyl-**, amide													

For explanations, symbols and abbreviations see beginning of table. For structural formulas see end of table.

No.	Name	Synonyms and Formula	Mol. wt.	Color, crystalline form, specific rotation and λ_{max} (log ε)	m.p. °C	b.p. °C	Density	n_D	Solubility w	al	eth	ace	bz	other solvents	Ref.
	Octacosane														
o1	**Octacosane***	$CH_3(CH_2)_{26}CH_3$	394.78	mcl or rh (bz-al)	64.5 fr 61.4	431.6^{760} 264^{10}	0.8067_4^{20} 0.7750^{70}	1.4520^{20} 1.4330^{70}	i	...	...	∞	s	chl s	**B1**^{3}, 582
o2	**Octacosanoic acid***	$CH_3(CH_2)_{26}CO_2H$	424.76	(ace or aa)	90.4	...	0.8191^{100}	1.4313^{100}	...	...	...	...	...	...	**B2**^{3}, 1095
o3	**1-Octacosanol***	$CH_3(CH_2)_{27}OH$	410.78	(ace or peth)	83.3	sub $200–50^1$	...	...	...	...	...	...	...	...	**B1**^{3}, 1849
Ω o4	**9,12-Octadecadienoic acid** (*cis,cis*)*	Linoleic acid. $CH_3(CH_2)_4CH{:}CHCH_2CH{:}CH(CH_2)_7CO_2H$	280.46	pa ye or col λ^{al} 232 (3.8), 275 (2.9)	−5	$229–30^{16}$	0.9022_4^{20}	1.4699^{20}	i	∞	∞	∞	∞	chl, CCl_4, MeOH ∞	**B2**^{2}, 459
o5	—,ethyl ester*	Ethyl linoleate. $CH_3(CH_2)_4CH{:}CHCH_2CH{:}CH(CH_2)_7CO_2C_2H_5$	308.49	ye or col λ^{hx} 198 (4.6), λ^{al} 233 sh (2.1), 270 (1.79)	...	$270–5^{180}$ 212^{12}	0.8865_4^{20}	...	i	s	s	...	...	...	**B2**^{2}, 461
Ω o6	—,methyl ester*	Methyl linoleate. $CH_3(CH_2)_4CH{:}CHCH_2CH{:}CH(CH_2)_7CO_2CH_3$	294.48	ye or col λ^{al} 232 (3.78)	−35	215^{20} $168–70^1$	0.8886_4^{18}	1.4638^{20}	i	s	s	...	...	...	**B2**^{2}, 461
o7	**10,12-Octadecadienoic acid** (*trans,trans*)*	10,12-Linoleic acid. $CH_3(CH_2)_4CH{:}CHCH{:}CH(CH_2)_8CO_2H$	280.45	(bz or al) λ^{al} 233 (3.51)	56–7.5	...	0.8686_4^{70}	1.4689^{60}	...	...	...	...	...	...	**B2**^{3}, 1476
o8	**7,11-Octadecadiyne***	$CH_3(CH_2)_5C{:}CCH_2CH_2C{:}C(CH_2)_5CH_3$	246.44	...	...	$167–8^7$	0.841_4^{19}	1.4698^{19}	i	...	...	...	...	...	**B2**^{3}, 1068
o9	**Octadecanal***	Stearaldehyde. $CH_3(CH_2)_{16}CHO$	268.49	nd (peth)	55 (38–9)	261 $212–3^{12}$	...	...	i	...	...	...	...	...	**B1**^{2}, 772
o10	—,dimethyl acetal	1,1-Dimethoxyoctadecane*. $CH_3(CH_2)_{16}CH(OCH_3)_2$	314.56	...	...	$168–70^3$	...	1.4410^{25}	...	v	v	...	...	MeOH v	**Am 80**, 6613
Ω o11	**Octadecane***	$CH_3(CH_2)_{16}CH_3$	254.51	nd (al or eth-MeOH)	28.18	316.1^{760} 173.5^{10}	0.7768_4^{28}	1.4390^{20}	i	δ	s	s	...	lig s	**B1**^{3}, 565
Ω o12	—,**1-amino-***	$CH_3(CH_2)_{17}NH_2$	269.52	(w)	fr 52.86	348.8^{760} 199.5^{10}	0.8618_4^{20}	1.4522^{20}	i	s	s	δ	s	chl v	**B4**^{3}, 431
o13	—,—,acetate	$CH_3(CH_2)_{17}NH_2 . CH_3CO_2H$	329.57	nd (al), cr (bz)	fr 84.5	...	...	...	i s^h	v			s	...	**B4**^{3}, 433
o14	—,—,hydrochloride	$CH_3(CH_2)_{17}NH_2 . HCl$	305.98	orh pl (al)	162–3	...	...	...	i v^h	δ v^h	i	...	i		**B4**^{3}, 432
Ω o15	—,**1-bromo-***	$CH_3(CH_2)_{17}Br$	333.41	cr (al)	28.2	210^{10}	0.9848_4^{20}	1.4631^{20} 1.4594^{30}	i	s	s		...	peth, AcOEt s	**B1**^{3}, 567
Ω o16	—,**1-chloro-***	$CH_3(CH_2)_{17}Cl$	288.95	...	28.6	348^{760} 199^{10}	0.8641^{20}	1.4531^{20}	i	δ	...	...	...	...	**B1**^{3}, 566
o17	—,**1,18-dibromo-***	$Br(CH_2)_{18}Br$	412.31	nd or lf (al)	64	$205–7^{15}$	...	...	i	δ v^h	...	...	...	chl v	**B1**^{2}, 139
o18	—,**1-iodo-***	$CH_3(CH_2)_{17}I$	380.40	lf (lig), nd (ace or al-ace)	34	383^{760} 223^{10}	1.0994_4^{20}	1.4810^{20}	i	δ	δ			...	**B1**^{3}, 567
o19	**Octadecanedioic acid,** diethyl ester*	Diethyl eicosanedioate. $C_2H_5O_2C(CH_2)_{16}CO_2C_2H_5$	370.58	...	54.5–5.0	240^{12}	...	...	i	s	s		...	...	**B2**^{2}, 626
o20	**1,18-Octadecanediol***	$HO(CH_2)_{18}OH$	286.50	lf (al, AcOEt or bz), nd (bz, diox)	97–9	$210–1^2$	...	...	i	...	...	...	...	...	**B1**^{3}, 2247
o21	**1-Octadecanethiol***	*n*-Octadecyl mercaptan. $CH_3(CH_2)_{17}SH$	286.57	...	(i) 24–6 (ii) 28	188^{1-2}	0.8475^{20}	1.4645^{20}	i	δ	s	...	...	...	**B1**^{3}, 1838
Ω o22	**Octadecanoic acid***	Stearic acid. $CH_3(CH_2)_{16}CO_2H$	284.50	mcl lf (al) λ^{al} 210 (1.69)	71.5–2.0	360d 232^{15}	0.9408_4^{20}	1.4299^{80}	i	s^h δ	v	s	δ	chl, CCl_4, CS_2, to s	**B2**^{2}, 346
Ω o23	—,amide	Stearamide. $CH_3(CH_2)_{16}CONH_2$	283.50	lf (al)	109	$250–1^{12}$	...	...	i	s^h δ	s	δ	δ	chl s	**B2**, 384
o24	—,—,*N*-phenyl-	Stearanilide. $CH_3(CH_2)_{16}CONHC_6H_5$	359.60	nd (al)	94	153.5^{10}	...	...	i	v	v	s	v	chl s peth δ	**B12**^{2}, 148
o24[1]	—,anhydride*	Stearic anhydride. $[CH_3(CH_2)_{16}CO]_2O$	550.96	λ^{al} 243 (4.17)	72	...	0.8368_4^{82}	1.4362^{80}	i	i	δ		δ	...	**B2**^{2}, 360
o25	—,benzyl ester	Benzyl stearate. $CH_3(CH_2)_{16}CO_2CH_2C_6H_5$	374.61	pa ye	28 (45–6)	...	0.9075_{25}^{50}	1.4663^{50}	i	δ	δ		...	chl δ	**B6**^{2}, 418
Ω o26	—,butyl ester*	*n*-Butyl stearate. $CH_3(CH_2)_{16}CO_2(CH_2)_3CH_3$	340.60	...	27.5	223	0.855_4^{20}	1.4328^{50}	i	s	...	v	...	...	**B2**^{2}, 352
Ω o27	—,chloride	Stearoyl chloride. $CH_3(CH_2)_{16}COCl$	302.93	...	23	215^{15} $202–3^6$	...	1.4523^{24}	...	s^h	...	...	...	...	**B2**, 384
o28	—,cyclohexyl ester*	...	366.64	...	44 (28–9)	...	0.890_{25}^{25}	...	i	i	s	...	...	...	**B6**^{2}, 11
Ω o29	—,ethyl ester*	Ethyl stearate. $CH_3(CH_2)_{16}CO_2C_2H_5$	312.54	...	(i) 33.4 (ii) 30.9	199^{10}	1.057_4^{20} (0.8973^{25})	1.4349^{40}	i	s	s	v	...	...	**B2**^{2}, 379
o30	—,hexadecyl ester*	Cetyl stearate. $CH_3(CH_2)_{16}CO_2(CH_2)_{15}CH_3$	508.92	lf or pl (eth, aa)	57	...	...	1.4410^{70}	i	i	s	s	i s^h	chl, CS_2 s aa i	**B2**^{2}, 353
o30[1]	—,2-hydroxyethyl ester*	Glycol monostearate. $CH_3(CH_2)_{16}CO_2CH_2CH_2OH$	328.54	(peth)	60–1	$189–91^3$	0.8780_4^{60}	1.4310^{60}	...	δ	s^h	...	...	...	**B2**^{3}, 1020
o31	—,isobutyl ester	Isobutyl stearate. $CH_3(CH_2)_{16}CO_2CH_2CH(CH_3)_2$	340.60	wax	(i) 22.5 (ii) 28.9	223^{15} 199^5	0.8498_4^{20}	...	i	...	s	...	...	...	**B2**^{2}, 353

For explanations, symbols and abbreviations see beginning of table. For structural formulas see end of table.

No.	Name	Synonyms and Formula	Mol. wt.	Color, crystalline form, specific rotation and λ_{max} (log ε)	m.p. °C	b.p. °C	Density	n_D	Solubility w	al	eth	ace	bz	other solvents	Ref.
	Octadecanoic acid														
o32	—,isopropyl ester*	Isopropyl stearate. $CH_3(CH_2)_{16}CO_2CH(CH_3)_2$	326.57	...	28	207^{6}	0.8403^{38}	...	...	s	s	v	...	chl s	**B2**^{2}, 352
Ω o33	—,methyl ester*	Methyl stearate. $CH_3(CH_2)_{16}CO_2CH_3$	298.52	...	39.1	$442–3^{747}$ 215^{15}	0.8498^{40}_{4}	1.4367^{40}	i	s	s	v	...	chl v	**B2**^{3}, 1011
Ω o34	—,3-methylbutyl ester*	Isoamyl stearate. $CH_3(CH_2)_{16}CO_2CH_2CH_2CH(CH_3)_2$	354.62	...	25.5	192^{2}	0.855^{20}_{4}	1.433^{50}	i	δ	s	s	...	...	**B2**^{2}, 353
Ω o35	—,nitrile	Stearonitrile. $CH_3(CH_2)_{16}CN$	265.49	...	41	362^{760} 193^{10}	0.8325^{20}_{4}	1.4389^{45} 1.4481^{20}	i	s	v	v	...	chl v	**B2**, 384
o36	—,pentyl ester*	*n*-Amyl stearate. $CH_3(CH_2)_{16}CO_2(CH_2)_4CH_3$	354.62	pl	30	...	...	1.4342^{50}	i	s	v	...	...	...	**B2**^{2}, 353
o37	—,phenyl ester*	Phenyl stearate. $CH_3(CH_2)_{16}CO_2C_6H_5$	360.59	...	51.5–3.0	267^{15}	...	...	i	s	s	...	...	...	**B6**^{2}, 155
o38	—,*p*-phenyl-phenacyl ester	...	478.72	...	97	...	...	...	...	...	s	...	...	...	**C32**, 4944
o39	—,propyl ester*	Propyl stearate. $CH_3(CH_2)_{16}CO_2CH_2CH_2CH_3$	326.57	cr (eth or PrOH), pr (peth)	fr 28.9	186.8^{2}	0.8452^{38}	1.4400^{30}	i	s	v	v	...	chl, CCl_4 v	**B2**^{2}, 352
o40	—,tetrahydro-furfuryl ester	...	368.61	...	22	...	0.917^{25}_{25}	...	i	s	s	...	...	...	**Am 50**, 134
o41	—,**9,10-dibromo-*** (*trans, dl*)	Elaidic acid dibromide. $CH_3(CH_2)_7CHBrCHBr(CH_2)_7CO_2H$	442.29	col or ye nd (eth)	29–30 (27)	...	1.2458^{30}_{4}	1.4893^{42}	i	i	s	...	...	...	**B2**^{2}, 362
o42	—,**2,3-dihydroxy-***	$CH_3(CH_2)_{14}CH(OH)CH(OH)CO_2H$	316.49	nd (aa)	α: 107 β: 126	...	...	...	δ^{h}	δ	δ	δ	...	chl δ^{h} peth δ	**B3**, 406
o43	—,**9,10-di-hydroxy-***	$CH_3(CH_2)_7CH(OH)CH(OH)(CH_2)_7CO_2H$	316.49	lf λ^{al} 220 (2.1)	95	...	...	...	i	δ	δ	...	...	...	**B3**^{3}, 876
o44	—,**9,10-dioxo-***	Stearoxylic acid. $CH_3(CH_2)_7COCO(CH_2)_7CO_2H$	312.46	ye pl (al) λ^{al} 270 (1.7)	86	...	...	...	i	v^{h}	v^{h}	v	...	lig v	**B3**^{3}, 1350
Ω o45	—,**9,10,12,13,15,16-hexabromo-***	α-Linolenic acid hexa-bromide. $CH_3[CH_2CHBrCHBr]_3(CH_2)_7CO_2H$	757.89	cr (diox)	182	...	...	...	i	δ	i	...	δ	chl i aa δ xyl s^{h}	**B2**^{2}, 364
o46	—,**2-hydroxy-***	$CH_3(CH_2)_{15}CH(OH)CO_2H$	300.49	flat nd (chl, AcOEt)	93	...	...	...	...	s	s	s	s^{h}	chl, AcOEt s	**B3**, 364
o47	—,**3-hydroxy-***	$CH_3(CH_2)_{14}CH(OH)CH_2CO_2H$	300.49	pl (chl)	89–90	...	...	...	...	s^{h}	s	...	...	chl s^{h}	**B3**^{2}, 249
o48	—,**10-hydroxy-***	$CH_3(CH_2)_7CH(OH)(CH_2)_8CO_2H$	300.49	pl	84	...	...	...	i	s	δ	...	...	...	**B3**^{2}, 249
o49	—,**11-hydroxy-***	$CH_3(CH_2)_6CH(OH)(CH_2)_9CO_2H$	300.49	ta or pl (al)	81–2	...	...	...	i	δ	δ	...	...	lig δ	**B3**^{3}, 670
Ω o50	—,**12-hydroxy-***	$CH_3(CH_2)_5CH(OH)(CH_2)_{10}CO_2H$	300.49	(al)	82	...	...	...	i	s	s	...	...	chl s	**B3**^{2}, 250
o51	—,**10-methyl-*** (D, −)	Tuberculostearic acid. $CH_3(CH_2)_7CH(CH_3)(CH_2)_8CO_2H$	298.52	$[\alpha]^{22}_{D}$ −0.11 (ace), $[\alpha]^{19}_{D}$ −0.41 (Py, c = 17)	12–3	$175–8^{0.7}$	0.8771^{25}_{4}	1.4512^{25}	...	s^{h}	s^{h}	s^{h}	...	...	**B2**^{2}, 369
o52	—,**3-oxo-***	Palmitoylacetic acid. $CH_3(CH_2)_{14}COCH_2CO_2H$	298.47	...	102–3	...	...	...	...	s^{h}	...	...	...	...	**B3**^{2}, 457
o53	—,—,ethyl ester*	Ethyl palmitoylacetate. $CH_3(CH_2)_{14}COCH_2CO_2C_2H_5$	326.53	(al or peth)	37–8	...	...	...	i	s^{h}	...	...	...	peth s^{h}	**B3**^{2}, 457
o54	—,**6-oxo-***	Lactaric acid. $CH_3(CH_2)_{11}CO(CH_2)_4CO_2H$	298.47	pl (al or peth-al)	87	...	...	...	i	...	s	...	δ	chl s	**B3**^{2}, 457
o55	—,—,ethyl ester*	$CH_3(CH_2)_{11}CO(CH_2)_4CO_2C_2H_5$	326.53	cr	41	...	...	...	i	v	v	...	...	...	**B3**^{1}, 253
Ω o56	—,**10-oxo-***	$CH_3(CH_2)_7CO(CH_2)_8CO_2H$	298.47	pl (al)	76 (82)	...	...	...	...	δ	...	...	...	...	**B3**^{2}, 457
o57	—,—,ethyl ester*	$CH_3(CH_2)_7CO(CH_2)_8CO_2C_2H_5$	326.53	pl (al)	41	...	...	...	i	s^{h}	...	...	...	...	**B3**, 725
o58	—,**12-oxo-***	$CH_3(CH_2)_5CO(CH_2)_{10}CO_2H$	298.47	lf (aa), cr (lig)	82	...	...	...	...	s^{h}	...	...	...	aa s^{h}	**B3**^{2}, 458
o59	—,—,ethyl ester*	$CH_3(CH_2)_5CO(CH_2)_{10}CO_2C_2H_5$	326.53	lf (al)	38	$199–200^{5}$	...	...	i	s^{h}	...	...	...	...	**B3**^{2}, 458
Ω o60	—,**9,10,12,13-tetrabromo-***	Linoleic acid tetrabromide. $CH_3(CH_2)_4[CHBrCHBrCH_2]_2(CH_2)_6CO_2H$	600.09	pl or lf (aa)	114–5	...	...	...	i	v	v	...	v	chl, aa v	**B2**^{2}, 362
o61	—,—,ethyl ester*	$CH_3(CH_2)_4[CHBrCHBrCH_2]_2(CH_2)_6CO_2C_2H_5$	628.15	nd	63	...	...	...	i	δ	...	...	...	peth δ	**B2**^{2}, 363
o62	—,—,methyl ester*	$CH_3(CH_2)_4[CHBrCHBrCH_2]_2(CH_2)_6CO_2CH_3$	614.12	lf	63	215^{15}	...	1.4346^{45}	i	s	s	...	...	chl s	**B2**^{2}, 351
Ω o63	**1-Octadecanol***	Stearyl alcohol. $CH_3(CH_2)_{17}OH$	270.50	lf (al)	59.4–9.8 fr 57.95	210.5^{15}	0.8124^{59}_{4}	...	i	s	s	δ	δ	chl s	**B1**^{3}, 1833
o64	**9,11,13-Octa-decatrienoic acid** (*cis*)*	α-Eleostearic acid. $CH_3(CH_2)_3[CH{:}CH]_3(CH_2)_7CO_2H$	278.44	nd (al) λ^{al} 261 (4.56), 271 (4.67), 281 (4.58)	49	235^{12} δd 170^{1}	0.9028^{50}_{4}	1.5112^{50}	i	s	s	...	...	...	**B2**^{2}, 465
o65	—(*trans*)*	β-Eleostearic acid. $CH_3(CH_2)_3[CH{:}CH]_3(CH_2)_7CO_2H$	278.44	lf (al, MeOH or CCl_4) λ^{al} 259 (4.67), 268 (4.78), 279 (4.69)	71–2	188^{1}	0.8839^{80}_{4}	1.5000^{80}	i	s^{h}	i	...	...	MeOH s	**B2**^{2}, 467
Ω o66	**9,12,15-Octade-catrienoic acid** (*cis, cis, cis*)*	α-Linolenic acid. $CH_3[CH_2CH{:}CH]_3(CH_2)_7CO_2H$	278.44	λ 195 (4.4), 233 (2.1)	−11.3	$230–2^{17}$ $125^{0.05}$	0.9164^{20}_{4}	1.4800^{20}	i	s	s	...	δ	...	**B2**^{2}, 463
Ω o67	—,ethyl ester (*cis, cis, cis*)*	Ethyl linolenate. $CH_3[CH_2CH{:}CH]_3(CH_2)_7CO_2C_2H_5$	306.49	λ^{al} 260 (1.7)	...	218^{15} $174^{2.5}$	0.8919^{20}_{4}	1.4694^{20}	i	s	s	...	...	...	**B2**^{2}, 465

For explanations, symbols and abbreviations see beginning of table. For structural formulas see end of table.

No.	Name	Synonyms and Formula	Mol. wt.	Color, crystalline form, specific rotation and λ_{max} (log ε)	m.p. °C	b.p. °C	Density	n_D	Solubility w	al	eth	ace	bz	other solvents	Ref.
	9-Octadecenal*														
Ω o68	**9-Octadecenal***	Olealdehyde. $CH_3(CH_2)_7CH:CH(CH_2)_7CHO$	266.47	ye nd		$168–9^{3–4}$	0.8509^{20}_4	1.4558^{20}							**B1**, 749
Ω o69	**1-Octadecene***	$CH_3(CH_2)_{15}CH:CH_2$	252.49		17.5	179^{15} 145^8	0.7891^{20}_4	1.4448^{20}	i			s^h			**B1**^{3}, 878
o70	**9-Octadecene***	$CH_3(CH_2)_7CH:CH(CH_2)_7CH_3$	252.49		−30.5	162^9	0.7916^{20}_4	1.4470^{20}	i						**B1**^{3}, 879
Ω o71	**9-Octadecenoic acid(*cis*)***	Oleic acid. $CH_3(CH_2)_7CH:CH(CH_2)_7CO_2H$	282.47	nd λ^{hx} 185 (3.8)	16.3 (13.4)	286^{100} $228–9^{15}$	0.8935^{20}_4	1.4582^{20}	i	∞	∞	∞	∞	chl, CCl_4, MeOH ∞	**B2**^{3}, 1387
Ω o72	—(*trans*)*	Elaidic acid. $CH_3(CH_2)_7CH:CH(CH_2)_7CO_2H$	282.47	pl (al) λ^{hx} 189 (3.9)	45	288^{100} 234^{15}	0.8734^{45}	1.4499^{45} 1.4308^{100}	i	s	s		s	chl s	**B2**^{2}, 441
o73	—,amide(*cis*)	Oleamide. $CH_3(CH_2)_7CH:CH(CH_2)_7CONH_2$	281.49		76				i	s^h	s				**B2**^{3}, 1425
o74	—,—,(*trans*)	Elaidamide. $CH_3(CH_2)_7CH:CH(CH_2)_7CONH_2$	281.49		93–4				i	s^h					**B2**^{3}, 1433
o75	—,—,*N*-phenyl-(*cis*)	Oleanilide. $CH_3(CH_2)_7CH:CH(CH_2)_7CONHC_6H_5$	357.59	nd	41	143.5^{10}			i	s^h	v	s^h	v^h	MeOH, aa s	**B12**^{1}, 198
o76	—,benzyl ester(*cis*)	Benzyl oleate. $CH_3(CH_2)_7CH:CH(CH_2)_7CO_2CH_2C_6H_5$	372.60			237^1	0.9330^{25}_{25}	1.4875^{25}	i	s	v				**B6**^{2}, 418
Ω o77	—,butyl ester(*cis*)*	Butyl oleate. $CH_3(CH_2)_7CH:CH(CH_2)_7CO_2(CH_2)_3CH_3$	338.58	ye	−26.4	$227–8^{15}$	0.8704^{15}	1.4480^{25}	i	s	s				**B2**^{2}, 439
Ω o78	—,ethyl ester(*cis*)*	Ethyl oleate. $CH_3(CH_2)_7CH:CH(CH_2)_7CO_2C_2H_5$	310.53	λ^{al} 237 (1.5)		207^{13}	0.8720^{20}	1.4515^{20}	i	∞	∞				**B2**^{3}, 1410
Ω o79	—,—(*trans*)*	Ethyl elaidate. $CH_3(CH_2)_7CH:CH(CH_2)_7CO_2C_2H_5$	310.53		5.8	$217–9^{15}$	0.8664^{25}	1.4480^{25}	i	s	s				**B2**^{2}, 443
Ω o80	—,methyl ester (*cis*)*	Methyl oleate. $CH_3(CH_2)_7CH:CH(CH_2)_7CO_2CH_3$	296.50	λ^{al} 230 (3.5)	−19.9	218.5^{20}	0.8739^{20}	1.4522^{20}	i	∞	∞				**B2**^{3}, 1407
Ω o81	—,—(*trans*)*	Methyl elaidate. $CH_3(CH_2)_7CH:CH(CH_2)_7CO_2CH_3$	296.50			$213–5^{15}$	0.8730^{20}	1.4513^{20}	i	s	s				**B2**^{2}, 443
o82	—,3-methylbutyl ester(*cis*)*	Isoamyl oleate. $CH_3(CH_2)_7CH:CH(CH_2)_7CO_2CH_2CH_2CH(CH_3)_2$	352.61			$223–4^{10}$	0.897^{15}		i	s	v				**B2**^{2}, 439
o83	—,nitrile(*cis*)	Oleonitrile. $CH_3(CH_2)_7CH:CH(CH_2)_7CN$	263.47		−1	330–5d 204^{12}	0.848^{17}_{17}	1.4566^{20}	i	s					**B2**^{2}, 443
o84	**11-Octadecenoic acid(*trans*)***	Vaccenic acid. $CH_3(CH_2)_5CH:CH(CH_2)_9CO_2H$	282.47	pl (ace) λ <210	44			1.4439^{60}				s^h			**B2**^{3}, 1384
o85	**6-Octadecenoic acid, 6,7-diiodo-***	6,7-Diiodotariric acid. Iodostarin. $CH_3(CH_2)_{10}CI:CI(CH_2)_4CO_2H$	534.27	nd (al)	48.5	d			i	δ v^h	v		v	chl, CS_2 v alk δ	**B2**^{2}, 428
Ω o86	**9-Octadecenoic acid, 12-hydroxy-(*cis*)***	Ricinoleic acid. $CH_3(CH_2)_5CH(OH)CH_2CH:CH(CH_2)_7CO_2H$	298.47	$[\alpha]^{22}_D$ +5.05 (+7.86) λ <210	α: 7.7 β: 16 γ: 5.5	$226–8^{10}$	0.9450^{21}_4	1.4716^{21}	i	s	s				**B3**, 385
o87	—,—,butyl ester (*cis*)*	Butyl ricinoleate. $CH_3(CH_2)_5CH(OH)CH_2CH:CH(CH_2)_7CO_2(CH_2)_3CH_3$	354.58	$[\alpha]^{22}_D$ +3.73		278^{13}	0.9058^{22}	1.4566^{22}	i		s				**B3**, 388
o88	—,—,ethyl ester (*cis*)*	Ethyl ricinoleate. $CH_3(CH_2)_5CH(OH)CH_2CH:CH(CH_2)_7CO_2C_2H_5$	326.53	$[\alpha]^{22}_D$ +5.28		258^{13}	0.9145^{22}	1.4618^{22}	i	s^h					**B3**^{2}, 259
o89	—,—(*trans*)*	Ethyl ricinelaidate. $CH_3(CH_2)_5CH(OH)CH_2CH:CH(CH_2)_7CO_2C_2H_5$	326.53		16										**B3**, 389
o90	—,—,isobutyl ester(*cis*)*	Isobutyl ricinoleate. $CH_3(CH_2)_5CH(OH)CH_2CH:CH(CH_2)_7CO_2C_4H_9^i$	354.58	[α] +4.01		282^9	0.9078^{22}	1.4538^{22}	i	s	s				**B3**, 388
Ω o91	**9-Octadecen-1-ol(*cis*)***	Olelyl alcohol. $CH_3(CH_2)_7CH:CH(CH_2)_8OH$	268.49		6–7	$205–10^{15}$	0.8489^{20}_4	1.4606^{20}	i	s	s				**B1**^{3}, 1962
o92	—(*trans*)*	Elaidyl alcohol. $CH_3(CH_2)_7CH:CH(CH_2)_8OH$	268.49	(al or ace)	36–7 (34)	*ca.* 333 198^{10}	0.8338^{40}_4	1.4552^{40}	i	s^h		s^h		dil NaOH s	**B1**^{3}, 1965
o93	**1-Octadecyne***	$HC:C(CH_2)_{15}CH_3$	250.47	(al)	22.5 fr 27	313^{760} 180^{15}	0.8025^{20}_4	1.4774^{20}							**B1**^{3}, 1029
o94	**2-Octadecyne***	$CH_3(CH_2)_{14}C:CCH_3$	250.47		30	184^{15}	0.8016^{30}_4								**B1**, 262
o95	**9-Octadecyne***	$CH_3(CH_2)_7C:C(CH_2)_7CH_3$	250.47		3	$163–4^7$	0.8012^{20}_4	1.4488^{25}							**B1**^{3}, 1029
o96	**9-Octadecynoic acid***	Stearolic acid. $CH_3(CH_2)_7C:C(CH_2)_7CO_2H$	280.46	pr (al, peth) nd (dil al) λ 220 (1.8), 265 (1.0)	48	260 $189–90^{1.8}$		1.4510^{54}	i	s^h	v				**B2**^{3}, 1474
o97	—,ethyl ester*	Ethyl stearolate. $CH_3(CH_2)_7C:C(CH_2)_7CO_2C_2H_5$	308.51			$180^{2.5}$		1.4555^{20}	i	s	s				**B2**^{2}, 458
o98	—,methyl ester*	Methyl stearolate. $CH_3(CH_2)_7C:C(CH_2)_7CO_2CH_3$	294.48			175^3		1.4562^{20}							**B2**^{3}, 1475
o99	**1,3-Octadiene, 3-chloro-***	$CH_3(CH_2)_3CH:CClCH:CH_2$	144.65			$64–5^{18}$	0.9366^{20}_4	1.4794^{20}	i	s	s				**B1**^{3}, 1007
Ω o100	**2,6-Octadiene, 2,6-dimethyl-***	Dihydroocimene. $CH_3CH:C(CH_3)CH_2CH_2CH:C(CH_3)_2$	138.25	λ^{hx} 223 (2.0)		168 75^{30}	0.775^{21}_4	1.4498^{20}	i	s	s	s		aa s	**B1**^{3}, 1018
o101	**2,7-Octadiene, 2,6-dimethyl-***	Linaloolen. $(CH_3)_2C:CHCH_2CH_2CH(CH_3)CH:CH_2$	138.25			168 58^{12}	0.7882^{20}	1.4561^{20}		s					**B1**, 261
o102	**2,4-Octadiene, 7-methyl-***	$(CH_3)_2CHCH_2CH:CHCH:CHCH_3$	124.23			149	0.7521^{18}_4	1.4543^{18}	i						**B1**^{2}, 239
—	**1,6-Octadien-2-ol, 3,7-dimethyl-**	*see* **Linalool**													
—	**2,6-Octadien-1-ol, 3,7-dimethyl-**	*see* **Geraniol, Nerol**													
o103	**1,5-Octadien-3-yne, 5-propyl-***	$CH_3CH_2CH:C(C_3H_7^n)C:CCH:CH_2$	148.25			$57–8^6$	0.8047^{20}_4	1.4949^{20}	i	s	s				**B1**^{3}, 1065

For explanations, symbols and abbreviations see beginning of table. For structural formulas see end of table.

No.	Name	Synonyms and Formula	Mol. wt.	Color, crystalline form, specific rotation and λ_{max} (log ε)	m.p. °C	b.p. °C	Density	n_D	Solubility w	al	eth	ace	bz	other solvents	Ref.
	2,6-Octadien-4-yne														
o104	**2,6-Octadien-4-yne, 3,6-diethyl-***	$CH_3CH:C(C_2H_5)C{:}C(C_2H_5)C:CHCH_3$	162.28			$169–71^{760}$ 99^{12}	0.8196^{20}_{4}	1.4965^{20}	...	...	s	s	...		**B1**^{3}, 1066
o105	—,3,6-dimethyl-*	$CH_3CH:C(CH_3)C{:}C(CH_3)C:CHCH_3$	134.22		−45	170^{760}	0.8071^{22}_{4}	1.4998^{20}	...	...	s	s	...		**B1**^{3}, 1065
o106	**1,7-Octadiyne***	$HC{:}C(CH_2)_4C{:}CH$	106.17			$135–6^{760}$ $93–5^{16}$	0.8169^{21}_{4}	1.4521^{18}	...	...	s	...	...		**B1**^{3}, 1061
o107	**2,6-Octadiyne***	$CH_3C{:}CCH_2CH_2C{:}CCH_3$	106.17		27	62^{19}	0.828^{80}_{4}	1.4658^{30}	...	...	s	...	...		**B1**^{2}, 248
o108	**3,5-Octadiyne***	$CH_3CH_2C{:}CC{:}CCH_2CH_3$	106.17	λ^{al} 227.5 (2.56), 238.5 (2.53), 253 (2.36)		$163–4^{760}$ 78^{34}	0.826^{0}_{4}	1.4968^{0}	...	...	s	...	...		**B1**^{3}, 1064
o109	—,2,7-dimethyl-*	$(CH_3)_2CHC{:}CC{:}CCH(CH_3)_2$	134.22			74^{12}	0.8090^{20}_{4}		...	...	s	...	...		**B1**, 128
—	**Octalene**	*see* **Aldrin**													
Ω o110	**Octanal***	Capryl aldehyde. $CH_3(CH_2)_6CHO$	128.22	λ^{hx} 295 (1.11)		171^{760} 72^{20}	0.8211^{20}_{4}	1.4217^{20}	δ	s	∞	∞	v	os s	**B1**^{2}, 757
o111	—,oxime	Caprylaldoxime. $CH_3(CH_2)_6CH:NOH$	143.23	nd (peth, MeOH, dil al)	60	112^{9}			δ	s	...	s	...	MeOH s	**B1**^{2}, 758
Ω o112	**Octane***	$CH_3(CH_2)_6CH_3$	114.23		−56.79	125.66^{760} 19.2^{10}	0.7025^{20}_{4}	1.3974^{20}	i	∞	s	∞	∞	chl, peth ∞	**B1**^{3}, 457
Ω o113	—,1-amino-*	Octylamine. $CH_3(CH_2)_7NH_2$	129.25		0.0	179.6^{760} 63.2^{10}	0.7826^{20}_{4}	1.4924^{20}	δ	v	v	...	...		**B4**^{3}, 379
o114	—,2-amino-(*d*)*	$CH_3(CH_2)_5CH(NH_2)CH_3$	129.25	$[\alpha]^{17}_{D}$ +8.62 (undil)		70^{25}	0.771^{25}_{4}	1.4220^{25}	...	v	v	...	...		**B4**^{3}, 384
o115	—,—(*dl*)	$CH_3(CH_2)_5CH(NH_2)CH_3$	129.25			$163–5^{760}$ $58–9^{13}$	0.7745^{20}_{0}	1.4232^{25}	i	v	v	...	...		**B4**, 196
—	—,1(4-amino-phenyl)-	*see* **Benzene, 1-amino-4-octyl-**													
Ω o116	—,1-bromo-*	$CH_3(CH_2)_7Br$	193.13		−55	200.8^{760} 77.3^{10}	1.1122^{20}_{4}	1.4524^{20}	i	∞	∞	...	...		**B1**^{3}, 466
o117	—,2-bromo-(*d*)*	$CH_3(CH_2)_5CHBrCH_3$	193.13	$[\alpha]^{25}_{D}$ +34.2		71^{14}	1.0982^{25}_{4}	1.4500^{20}	i	∞	∞	...	...		**B1**^{2}, 125
Ω o118	—,—(*dl*)	$CH_3(CH_2)_5CHBrCH_3$	193.13			$188–9^{760}$ 72^{14}	1.0878^{25}_{4}	1.4442^{25}	i	∞	∞	...	...		**B1**^{2}, 125
o119	—,—(*l*)	$CH_3(CH_2)_5CHBrCH_3$	193.13	$[\alpha]^{25}_{D}$ −37.46		$72–3^{18}$ 46^{1}	1.0920^{20}_{4}	1.4475^{25}	i	∞	∞	...	...		**B1**^{2}, 125
o120	—,1-bromo-8-fluoro*	$F(CH_2)_8Br$	211.13			$118–20^{22.5}$		1.4500^{20}	i	v	v	...	...		**C51**, 7300
Ω o121	—,1-chloro-*	$CH_3(CH_2)_7Cl$	148.68		−57.8	182^{760} 78^{15}	0.8738^{20}_{4}	1.4305^{20}	i	v	v	...	...		**B1**, 159
o122	—,2-chloro-(*d*)*	$CH_3(CH_2)_5CHClCH_3$	148.68	$[\alpha]^{20}_{D}$ +33.7		$171–3$ 75^{28}	0.8658^{17}_{4}	1.4273^{21}	i	v	s	...	...		**B1**, 160
o123	—,4-chloro-(*d*)*	$CH_3(CH_2)_3CHClCH_2CH_2CH_3$	148.68	$[\alpha]^{25}_{D}$ +0.28 (undil. l = 10 dm)		92^{50}			i	v	s	...	...	chl s	**B1**^{3}, 466
o124	—,1-chloro-8-fluoro-*	$F(CH_2)_8Cl$	166.67			87^{10}	0.978^{20}_{4}	1.4266^{25}	i	v	v	s	...		**C51**, 7300
o125	—,3-chloro-3-methyl-*	$CH_3(CH_2)_4CCl(CH_3)CH_2CH_3$	162.71			$73–4^{15}$	0.8680^{25}_{4}	1.4351^{20}	...	...	s	s	s	chl s	**B1**^{3}, 508
o126	—,4-chloro-4-methyl-*	$CH_3(CH_2)_3CCl(CH_3)CH_2CH_2CH_3$	162.71			$71^{14.5}$	0.8723^{20}_{4}	1.4360^{20}	...	...	s	...	...	chl s	**B1**^{3}, 509
o127	—,1,2-dibromo-*	$CH_3(CH_2)_5CHBrCH_2Br$	272.04			$240–2$ 118.5^{15}	1.4580^{20}_{4}	1.4970^{20}	...	...	...	...	...		**B1**^{2}, 125
o128	—,2,7-dimethyl-*	Di-isoamyl. $(CH_3)_2CH(CH_2)_4CH(CH_3)_2$	142.29		−54.6	159.6^{760}	0.7240^{20}_{4}	1.4092^{20}	...	...	s	...	...	aa s	**B1**^{2}, 131
o129	—,1-fluoro-*	$CH_3(CH_2)_7F$	132.22			$142–3^{760}$	0.8103^{20}_{4}	1.3935^{20}	...	...	...	...	...		**B1**^{2}, 124
Ω o130	—,1-iodo-*	$CH_3(CH_2)_7I$	240.13		−45.7	225.5^{760} 86.5^{5}	1.3297^{20}_{4}	1.4889^{20}	...	s	s	...	...		**B1**^{2}, 185
o131	—,2-iodo-(*D*, −)*	*sec*-Octyl iodide. $CH_3(CH_2)_5CHICH_3$	240.13	$[\alpha]^{28}_{D}$ −45.47		92^{12}	1.3219^{20}_{4}	1.4863^{25}	i	s	s	...	...	lig s	**B1**^{3}, 470
o132	—,—(*DL*)	$CH_3(CH_2)_5CHICH_3$	240.13			210 $95–6^{16}$	1.3251^{20}_{4}	1.4896^{20}	i	s	s	...	...	lig s	**B1**^{3}, 470
o133	—,—(*L*, +)*	$CH_3(CH_2)_5CHICH_3$	240.13	$[\alpha]^{26}_{D}$ +46.33		101^{22}	1.3314^{17}_{4}	1.4877^{22}	i	s	s	...	...	lig s	**B1**^{3}, 470
o134	—,2-methyl-*	$CH_3(CH_2)_5CH(CH_3)_2$	128.26		−80.1	142.8^{760}	0.7107^{20}_{4}	1.4029^{20}	i	s	s	...	...	peth, lig v	**B1**^{3}, 507
o135	—,3-methyl-(*d*)*	$CH_3(CH_2)_4CH(CH_3)CH_2CH_3$	128.26	$[\alpha]^{17}_{D}$ +9.38	−107.6	$143–4^{760}$	0.7206^{17}	1.4068^{20}	...	...	...	s	s		**B1**, 166
o136	—,—(*l*)	$CH_3(CH_2)_4CH(CH_3)CH_2CH_3$	128.26	$[\alpha]^{27}_{D}$ −8.5		143^{760}	0.714^{27}_{4}	1.4052^{25}	...	...	...	s	s		**B1**^{3}, 509
Ω o137	—,4-methyl-(*dl*)*	$CH_3(CH_2)_3CH(CH_3)CH_2CH_2CH_3$	128.26		−113.2	142.4^{760} 32^{10}	0.7199^{20}_{4}	1.4061^{20}	...	s	v	v	v	os v	**B1**^{3}, 509
o138	—,—(*l*)	$CH_3(CH_2)_3CH(CH_3)CH_2CH_2CH_3$	128.26	$[\alpha]^{19}_{D}$ −1.06		141^{760}	0.717^{19}_{4}		i	s	v	v	v	os v	**B1**^{3}, 510
o139	—,1(4-nitro-phenyl)-*	1-Nitro-4-octylbenzene*.	235.33		−5	200^{10}			i	...	∞	...	...		**B5**, 454
Ω o140	—,1-phenyl-*	Octylbenzene*. $C_6H_5(CH_2)_7CH_3$	190.33		−7	$264–5^{760}$ $131–4^{12}$	0.8582^{20}_{4}	1.4851^{20}	i	...	∞	...	∞		**B5**^{2}, 343
o141	**Octanedial***	Suberaldehyde. $OCH(CH_2)_6CHO$	142.20			230–40d $96–8^{3}$		1.4439^{20}	v	v	...	...	...		**B1**^{2}, 845
o142	—,dioxime	Suberaldoxime. $HON:CH(CH_2)_6CH:NOH$	172.23	pr (al)	155–6				δ^h	s^h	...	...	...		**B1**^{2}, 845
Ω o143	**Octanedioic acid***	Suberic acid. $HO_2C(CH_2)_6CO_2H$	174.20	lo nd or pl (w)	144	300 sub 219.5^{10}			δ s^h	s	δ	...	...	MeOH s chl i	**B2**^{2}, 595

For explanations, symbols and abbreviations see beginning of table. For structural formulas see end of table.

No.	Name	Synonyms and Formula	Mol. wt.	Color, crystalline form, specific rotation and λ_{max} (log ε)	m.p. °C	b.p. °C	Density	n_D	Solubility w	al	eth	ace	bz	other solvents	Ref.
	Octanedioic acid														
Ω o144	—,diethyl ester	Diethyl suberate. $C_2H_5O_2C(CH_2)_6CO_2C_2H_5$	230.31		5.9	$282–6^{760}$ $140–1^{8}$	0.9811_4^{20}	1.4328^{20}	i	s	s				$B2^2$, 596
Ω o145	—,dimethyl ester	Dimethyl suberate. $CH_3O_2C(CH_2)_6CO_2CH_3$	202.25		−3.1	268 120^{6}	1.0217_4^{20}	1.4341^{20}	i	s	s	s		os s	B2, 693
Ω o146	**1,8-Octanediol***	$HO(CH_2)_8OH$	146.23	nd (bz-lig), pr	63	172^{20}			δ	v	δ		s	lig δ	$B1^2$, 556
Ω o147	**4,5-Octanediol** (*dl*)*	1,2-Dipropylethylene glycol. $CH_3CH_2CH_2CH(OH)CH(OH)CH_2CH_2CH_3$	146.23		28	110^{8}		1.4419^{25}							$B1^3$, 2219
o148	—(*meso*)*	$CH_3CH_2CH_2CH(OH)CH(OH)CH_2CH_2CH_3$	146.23	lf (bz, al, lig)	123–4				i	s	δ	δ	δ	lig δ	$B1^3$, 2219
o149	**2,3-Octanedione***	Acetylcaproyl. $CH_3COCO(CH_2)_4CH_3$	142.20			$172–3^{733}$				s				lig δ	B1, 795
o150	—,dioxime*	Methylpentylglyoxime. $CH_3C(:NOH)C(:NOH)(CH_2)_4CH_3$	172.23	nd (dil al) λ^{al} 229 (4.15)	173				i	s		s			B1, 795
o151	—,3-oxime*	$CH_3COC(:NOH)(CH_2)_4CH_3$	157.22	(lig)	59	133^{11}					v			lig s^h	B1, 795
o152	**2,7-Octanedione***	1,4-Diacetylbutane. $CH_3CO(CH_2)_4COCH_3$	142.20	pl (bz)	44	114^{10}			δ	s					$B1^1$, 408
o153	—,dioxime*	$CH_3C(:NOH)(CH_2)_4C(:NOH)CH_3$	172.23	(al)	158				i	s^h					B1, 795
o154	**3,6-Octanedione***	*sym*-Dipropionylethane. $CH_3CH_2COCH_2CH_2COCH_2CH_3$	142.20	pl (al)	35–6	98^{14} sub			$δ^h$	s					$B1^2$, 845
o155	**4,5-Octanedione***	Dibutyryl. Dipropylglyoxal. $CH_3CH_2CH_2COCOCH_2CH_2CH_3$	142.20	ye oil λ^{al} 268 (1.67), 425 (1.16)		168^{760} 60^{12}	0.934_4^0			v	v	v			$B1^2$, 845
o156	—,dioxime*	Dipropylglyoxime. $CH_3CH_2CH_2C(:NOH)C(:NOH)CH_2CH_2CH_3$	172.23		186–7	sub			i	s		s		lig i	$B1^2$, 846
o157	**3,6-Octanedione, 2,2,7,7-tetra-methyl-**	Dipivaloylethane. $(CH_3)_3CCOCH_2CH_2COC(CH_3)_3$ Dipivaloylethane.	198.31	nd	2.5	$115–7^{17}$ $55–60^{0.5}$	0.900_4^{27}	1.4400^{20}	i	s	s				$B1^3$, 3151
Ω o158	**1-Octanethiol***	Octyl mercaptan. $CH_3(CH_2)_7SH$	146.30		−49.2	199.1^{760} 86^{15}	0.8433_4^{20}	1.4540^{20}		s					$B1^3$, 1710
o159	**2-Octanethiol** (*dl*)*	$CH_3(CH_2)_5CH(SH)CH_3$	146.30		−79	186.4^{760} 88.9^{30}	0.8366_4^{20}	1.4504^{20}	s	s			s		$B1^3$, 1717
o160	—(*l*)*	$CH_3(CH_2)_5CH(SH)CH_3$	146.30	$[\alpha]_{546}^{25}$ −36.4		$78–80^{22}$	0.830_4^{25}			s	s		s		$B1^3$, 1722
Ω o161	**Octanoic acid***	Caprylic acid*. $CH_3(CH_2)_6CO_2H$	144.22		16.5	239.3 140^{23}	0.9088^{20}	1.4285^{20}	$δ^h$	∞				chl, CH_3CN ∞	$B2^2$, 301
Ω o162	—,amide	Caprylamide. $CH_3(CH_2)_6CONH_2$	143.23	lf lf, pl	fr 105.9 110	239^{760}	0.8450^{110}		$δ^h$	v	s	s	δ	chl δ	$B2^2$, 303
o163	—,anhydride*	Caprylic anhydride. $[CH_3(CH_2)_6CO]_2O$	270.42		−1	280–5 186^{15}	0.9065_4^{18}	1.4358^{18}	d	s	∞	s		oos s	$B2^2$, 303
o164	—,butyl ester*	Butyl caprylate. $CH_3(CH_2)_6CO_2(CH_2)_3CH_3$	200.33		−42.9	240.5 $121–2^{20}$	0.8628^{20}	1.4232^{25}	i	s	s	s		os s	B2, 348
Ω o165	—,chloride	Caprylyl chloride. $CH_3(CH_2)_6COCl$	162.66		−63	195.6^{760} 89^{20}	0.9535_4^{15}	1.4335^{20}	d	d	s				$B2^2$, 303
Ω o166	—,ethyl ester*	Ethyl caprylate. $CH_3(CH_2)_6CO_2C_2H_5$	172.27		−43.1	208.5^{760} 104^{80}	0.8693_4^{20}	1.4178^{20}	i	v	v				$B2^2$, 302
o167	—,heptyl ester*	$CH_3(CH_2)_6CO_2(CH_2)_6CH_3$	242.41		−10.6	290.6 160^{14}	0.8596^{20}	1.4340^{20}	i	s	s	s		os s	B2, 348
o168	—,hexyl ester*	$CH_3(CH_2)_6CO_2(CH_2)_5CH_3$	228.38		−30.6	277.4	0.8603^{20}	1.4323^{15}	i	s	s	s		os s	$B2^3$, 794
o169	—,isopropyl ester*	Isopropyl caprylate. $CH_3(CH_2)_6CO_2CH(CH_3)_2$	186.30			93.8^{10}	0.8555^{20}	1.4147^{25}							$B2^3$, 794
Ω o170	—,methyl ester*	Methyl caprylate. $CH_3(CH_2)_6CO_2CH_3$	158.24		−40	192.9^{760} 83^{15}	0.8775_4^{20}	1.4170^{20}	i	v	v				$B2^2$, 302
Ω o171	—,2-methylallyl ester*	Isoamyl caprylate. $CH_3(CH_2)_6CO_2CH_2C(CH_3):CH_2$	198.31			147.8^{50}	0.8703	1.4308	δ	s	s				$B2^3$, 795
o172	—,nitrile	Caprylonitrile. $CH_3(CH_2)_6CN$	125.22		−45.6	205.2^{760} 77.8^{10}	0.8136_4^{20}	1.4203^{20}	i	δ	s				$B2^2$, 303
o173	—,octyl ester*	$CH_3(CH_2)_6CO_2(CH_2)_7CH_3$	256.43		−18.1	306.8^{760} 192.5^{30}	0.8554_4^{20}	1.4352^{20}	i	s	s	s		os s	B2, 348
Ω o174	—,pentyl ester*	Amyl caprylate. $CH_3(CH_2)_6CO_2(CH_2)_4CH_3$	214.35		−34.8	260.2 $124–6^{20}$	0.8613^{20}	1.4262^{25}	i	s	s	s		os s	$B2^3$, 794
o175	—,piperazinium salt	$C_4H_{10}N_2 \cdot 2[CH_3(CH_2)_6CO_2H]$	374.59		98				s	s	i				Am 70, 2758
o176	—,propyl ester*	Propyl caprylate. $CH_3(CH_2)_6CO_2CH_2CH_2CH_3$	186.30	(peth)	−46.2	226.4 112^{20}	0.8659^{20}	1.4191^{25}	i	s	s	s		os s	B2, 348
o177	—,2-amino-(*d*)*	$CH_3(CH_2)_5CH(NH_2)CO_2H$	159.23	$[\alpha]_D^{26}$ +23.5 (6N HCl, c = 1) $[\alpha]_D^{20}$ +12.3 (1N NaOH, c = 0.5)					δ	δ	δ		δ	aa s	$B4^2$, 886
Ω o178	—,—(*dl*)	$CH_3(CH_2)_5CH(NH_2)CO_2H$	159.23	lf (w)	270 (263–5)	sub d			δ	δ	δ		δ	aa s	$B4^2$, 886
o179	—,—(*l*)	$CH_3(CH_2)_5CH(NH_2)CO_2H$	159.23	$[\alpha]_D$ −13 (1N NaOH, c = 2) $[\alpha]_D$ −23 (5N HCl)	276				δ	δ	δ		δ	aa s	$B4^2$, 886
o180	—,8-amino-*	$H_2N(CH_2)_7CO_2H$	159.23		172				δ v^h	v					$B4^1$, 527
o181	—,2-bromo-*	$CH_3(CH_2)_5CHBrCO_2H$	223.12			140^{5}	1.2785^{24}	1.4613^{24}							$B2^2$, 303

For explanations, symbols and abbreviations see beginning of table. For structural formulas see end of table.

No.	Name	Synonyms and Formula	Mol. wt.	Color, crystalline form, specific rotation and λ_{max} (log ε)	m.p. °C	b.p. °C	Density	n_D	Solubility w	al	eth	ace	bz	other solvents	Ref.
	Octanoic acid														
o182	—,8-fluoro-*	$F(CH_2)_7CO_2H$	162.21	....	35	$132–3^{4}$	....	....	δ	v	v	...	...	lig s^h	**C51,** 7300
Ω o183	—,2-hydroxy-*	$CH_3(CH_2)_5CH(OH)CO_2H$	160.22	pl	70	$160–5^{10}$	....	....	δ	v	v	...	...	....	**B3²,** 237
Ω o184	—,4-hydroxy-, lactone*	γ-Caprylolactone.	142.20	....	....	$132–3^{20}$	0.9796^{19}_{4}	1.4451^{19}	...	s	...	...	...	....	**B17¹,** 133
o185	—,2-methyl-3-oxo-, ethyl ester*	$CH_3(CH_2)_4COCH(CH_3)CO_2C_2H_5$	200.28	....	....	$128–9^{12}$	0.963^{0}_{4}	....	...	s	...	...	...	....	**B3,** 713
Ω o186	**1-Octanol***	Octyl alcohol. $CH_3(CH_2)_7OH$	130.23	λ^{vap} 197	−16.7	194.45^{760} 98^{19}	0.8270^{20}_{4}	1.4295^{20}	i	∞	∞	...	...	....	**B1³,** 1703
o187	**2-Octanol(*d*)***	$CH_3(CH_2)_5CH(OH)CH_3$	130.23	$[\alpha]^{17}_{D}$ +9.9	....	86^{20}	0.8216^{20}_{4}	1.4264^{20}	δ	s	s	s	...	....	**B1¹,** 208
Ω o188	—(*dl*)*	$CH_3(CH_2)_5CH(OH)CH_3$	130.23	λ^{vap} 196 λ^{hx} 175 (2.5)	−31.6	180^{760} 87^{20}	0.8193^{20}_{4}	1.4203^{20}	δ	s	s	s	...	....	**B1²,** 449
o189	—(*l*)*	$CH_3(CH_2)_5CH(OH)CH_3$	130.23	$[\alpha]^{17}_{D}$ −9.9	....	86^{20}	0.8201^{20}_{4}	1.4264^{20}	δ	s	s	s	...	....	**B1²,** 451
o190	**1-Octanol, 8-chloro-***	Octamethylene chlorohydrin. $Cl(CH_2)_8OH$	164.68	....	....	139^{19}	....	1.4563^{25}	δ	v	v	...	...	....	**B1³,** 1710
o191	—,3,7-dimethyl-(*d*)*	Tetrahydrogeraniol. $(CH_3)_2CH(CH_2)_3CH(CH_3)CH_2CH_2OH$	158.29	λ^{al} 237 (1.09) $[\alpha]^{20}_{D}$ +4.09 (+5.57)	....	$212–3^{760}$ $105–6^{10}$	0.8285^{20}_{4}	1.4355^{20}	...	...	v	...	...	....	**B1²,** 460
o192	—,—(*l*)*	$(CH_3)_2CH(CH_2)_3CH(CH_3)CH_2CH_2OH$	158.29	$[\alpha]^{11}_{546}$ −3.67 λ^{al} 237 (1.09)	....	$212–3^{760}$ 109^{15}	0.830^{18}	1.4370^{15}	...	...	v	...	...	....	**B1²,** 461
Ω o193	**3-Octanol, 3,7-dimethyl-(*dl*)***	Tetrahydrolinalool. $(CH_3)_2CH(CH_2)_3C(OH)(CH_3)CH_2CH_3$	158.29	....	31–2	196–7 $87–8^{10}$	0.8280^{20}	1.4335^{20}	i	s	...	...	...	....	**B1³,** 1767
o194	—,3-ethyl-*	$CH_3(CH_2)_4C(OH)(C_2H_5)_2$	158.29	....	....	199 84^{12}	0.8361^{25}_{4}	1.4390^{20}	i	s	...	...	...	....	**B1,** 426
o195	**1-Octanol, 8-fluoro-***	Octamethylene fluorohydrin. $F(CH_2)_8OH$	148.22	....	....	$106–7^{10}$	0.945^{20}_{4}	1.4248^{25}	δ	v	v	...	...	....	**C51,** 7300
Ω o196	**2-Octanol, 2-methyl-***	$CH_3(CH_2)_5C(OH)(CH_3)_2$	144.26	....	....	178 $81–3^{16}$	0.8210^{20}_{4}	1.4280^{20}	i	s	s	...	...	....	**B1²,** 457
Ω o197	**2-Octanone***	Hexyl methyl ketone. $CH_3(CH_2)_5COCH_3$	128.22	....	−16	173^{760} $59–60^{11}$	0.8202^{20}_{4}	1.4151^{20}	δ	∞	∞	...	...	....	**B1²,** 758
Ω o198	**3-Octanone***	Amyl ethyl ketone. $CH_3(CH_2)_4COC_2H_5$	128.22	λ^{MeOH} 278 (1.38)	....	167^{749}	0.8221^{20}_{4}	1.4153^{20}	i	∞	∞	...	...	....	**B1¹,** 362
Ω o199	**4-Octanone***	Butyl propyl ketone. $CH_3(CH_2)_3COCH_2CH_2CH_3$	128.22	λ^{al} 280 (1.36)	....	163^{760} 70^{26}	0.8146^{25}_{4}	1.4173^{14}	i	∞	∞	...	...	....	**B1²,** 759
o200	—,5-hydroxy-*	Butyroin. $CH_3CH_2CH_2CH(OH)COCH_2CH_2CH_3$	144.22	....	−10	$180–90^{760}$ 95^{20}	0.9231^{20}	1.4290^{20}	...	s	s	s	...	....	**B1²,** 880
o201	—,7-methyl-*	Isoamyl propyl ketone. $(CH_3)_2CHCH_2CH_2COCH_2CH_2CH_3$	142.24	....	....	$177–9^{760}$	0.8239^{20}_{4}	1.4210^{20}	i	s	s	...	...	....	**B1¹,** 366
o202	**Octasiloxane, octadecamethyl-**	$CH_3[-Si(CH_3)_2O-]_7Si(CH_3)_3$	607.32	....	....	$153^{5.1}$	0.913	1.3970^{20}	i	δ	...	...	s	peth, lig s	**Am 68,** 362
o203	**1,3,5,7-Octatetraene***	$CH_2:CHCH:CHCH:CHCH:CH_2$	106.17	(bz) λ^{cy} 278, 290, 302	*ca.* 50	sub	....	....	...	...	...	...	...	peth s^h aa s	**B1³,** 1062
o204	**2,4,6-Octatriene (*trans, trans, trans*)***	$CH_3CH:CHCH:CHCH:CHCH_3$	108.19	lf λ^{iso} 253 (4.7), 263 (4.22), 274 (4.7)	52	$147–8^{764}$ 43^{10}	0.7961^{23}_{4}	1.5131^{27}	...	s	...	...	...	chl, lig s	**B1³,** 1047
o205	**1,3,7-Octatriene, 3,7-dimethyl-***	Ocimene. $CH_2:C(CH_3)CH_2CH_2CH:C(CH_3)CH:CH_2$	136.24	λ^{iso} 235 (4.22)	....	176–8d $73–4^{21}$	0.8000^{20}	1.4862^{20}	i	s	s	...	...	chl, aa s	**B1³,** 1052
o206	**2,4,6-Octatriene, 2,6-dimethyl-** (*4-trans, 6-trans*)*	Alloocimene A. $CH_3CH:C(CH_3)CH:CHCH:C(CH_3)_2$	136.24	λ^{iso} 270 (4.50), 278 (4.60), 289 (4.48)	−35.4	188^{750} 91^{20}	0.8118^{20}_{4}	1.5446^{20}	...	...	...	...	...	....	**B1³,** 1050
o207	—,—(*4-trans, 6-cis*)*	Alloocimene B. $CH_3CH:C(CH_3)CH:CHCH:C(CH_3)_2$	136.24	λ^{iso} 265 (4.52), 273 (4.63), 285 (4.53)	−20.6	89^{20}	0.8060^{20}	1.5446^{20}	...	...	...	...	...	....	**B1³,** 1050
Ω o208	**1-Octene***	$CH_3(CH_2)_5CH:CH_2$	112.22	λ^{hp} 177 (4.1), λ^{al} 210 (2.60)	−101.73	121.3^{760} 15.4^{10}	0.7149^{20}_{4}	1.4087^{20}	i	∞	s	s	s	chl, os v	**B1³,** 835
o209	**2-Octene(*cis*)***	$CH_3(CH_2)_4CH:CHCH_3$	112.22	λ^{hx} 179 (4.15)	−100.2	125.6^{760} 16.5^{10}	0.7243^{20}_{4}	1.4150^{20}	i	s	s	s	s	chl s	**B1³,** 839
o210	—(*trans*)*	$CH_3(CH_2)_4CH:CHCH_3$	112.22	λ^{hx} 183 (4.11)	−87.7	125.0^{760} 16.0^{10}	0.7199^{20}_{4}	1.4132^{20}	i	s	s	s	s	chl v	**B1³,** 839
o211	**3-Octene(*cis*)***	$CH_3(CH_2)_3CH:CHCH_2CH_3$	112.22	....	−126	122.9^{760} 14.3^{10}	0.7189^{20}_{4}	1.4135^{20}	i	s	s	s	s	lig s	**B1³,** 840
Ω o212	—(*trans*)	$CH_3(CH_2)_3CH:CHCH_2CH_3$	112.22	λ^{al} 185 (3.91)	−110	123.3^{760} 14.6^{10}	0.7152^{20}_{4}	1.4126^{20}	i	s	s	s	s	lig s	**B1³,** 841
o213	**4-Octene(*cis*)***	$CH_3CH_2CH_2CH:CHCH_2CH_2CH_3$	112.22	....	−118.7	122.5^{760} 14^{10}	0.7212^{20}_{4}	1.4148^{20}	i	s	s	s	s	lig s	**B1³,** 841
Ω o214	—(*trans*)*	$CH_3CH_2CH_2CH:CHCH_2CH_2CH_3$	112.22	....	−93.8	122.3^{760} 13.7^{10}	0.7141^{20}_{4}	1.4118^{20}	i	s	s	s	s	lig s	**B1³,** 841
o215	**1-Octene, 2-chloro-***	$CH_3(CH_2)_5CCl:CH_2$	146.66	....	....	168–70	0.9274^{0}_{0}	....	...	...	s	s	s	....	**B1³,** 840
o216	**2-Octene, 2-chloro-***	$CH_3(CH_2)_4CH:CClCH_3$	146.66	....	....	167–8	0.8923^{16}_{16}	1.4424^{16}	...	s	s	s	s	....	**B1³,** 840

For explanations, symbols and abbreviations see beginning of table. For structural formulas see end of table.

No.	Name	Synonyms and Formula	Mol. wt.	Color, crystalline form, specific rotation and λ_{max} (log ε)	m.p. °C	b.p. °C	Density	n_D	Solubility w	al	eth	ace	bz	other solvents	Ref.
	2-Octene														
o217	—,4-chloro-*	$CH_3(CH_2)_3CHClCH:CHCH_3$	146.66			153 $65–6^{15}$	0.8924_4^{20}	1.4452^{20}	i		s	s	s	chl s	$B1^3$, 840
o218	**4-Octene, 4-chloro-(*cis*)***	$CH_3CH_2CH_2CH:CClCH_2CH_2CH_3$	146.66			165.33^{760}	0.8912_4^{20}	1.4447^{20}			s			chl s	**Am 73**, 3329
o219	**1-Octen-3-yne***	$CH_3(CH_2)_3C\vdots CCH:CH_2$	108.19			62^{60}	0.7830_4^{20}	1.4592^{20}			s				$B1^3$, 1046
o220	**1-Octen-3-yn-5-ol, 5-methyl-***	$CH_2:CHC\vdots CC(CH_3)(OH)CH_2CH_2CH_3$	138.21			80^{13}	0.8851_4^{15}	1.4735^{20}		s					$B1^3$, 2034
o221	**1-Octyne***	*n*-Hexylacetylene. $CH_3(CH_2)_5C\vdots CH$	110.20	λ^{hp} 185 (3.3), 222.5 (2.1)	−79.3	125.2^{760} 19.7^{10}	0.7461^{20}	1.4159^{20}	i	s	s				**B1**, 258
o222	**2-Octyne***	Methylpentylacetylene. $CH_3(CH_2)_4C\vdots CCH_3$	110.20	λ^{hp} 177.5 (4.0), 196 sh (3.3), 222.5 (2.2)	−61.6	138^{760}	0.7596_4^{20}	1.4278^{20}	i	s	s				**B1**, 258
o223	**3-Octyne***	*n*-Butylethylacetylene $CH_3(CH_2)_3C\vdots CC_2H_5$	110.20		−103.9	133^{760} 85^{169}	0.7529_4^{20}	1.4250^{20}	i	s	s				$B1^3$, 1006
o224	**4-Octyne***	Di-*n*-propylacetylene. $CH_3CH_2CH_2C\vdots CCH_2CH_2CH_3$	110.20		−102.55	131.5^{760}	0.7509_4^{20}	1.4248^{20}	i	s	s				$B1^3$, 1006
o225	**1-Octyne, 1-chloro-***	$CH_3(CH_2)_5C\vdots CCl$	144.65			$61–2^{17}$	0.912^{20}	1.445^{20}		v^h	v				$B1^3$, 1005
o226	**3-Octyne, 2-chloro-2-methyl-***	$CH_3(CH_2)_3C\vdots CCCl(CH_3)_2$	158.67			68^{15}	0.8929_4^{20}	1.4480^{20}	i		v				$B1^3$, 1014
o227	**—,2,2-dimethyl-***	$CH_3(CH_2)_3C\vdots CC(CH_3)_3$	138.25			79^{60}	0.7491_4^{20}	1.4270^{20}	i		s				$B1^3$, 1018
o228	**—,7-methyl-***	$(CH_3)_2CH(CH_2)_2C\vdots CCH_2CH_3$	124.23			87^{99}	0.7599_4^{20}	1.4280^{20}	i		v				$B1^3$, 1014
o229	**2-Octyn-1-ol***	Pentylpropargyl alcohol. $CH_3(CH_2)_4C\vdots CCH_2OH$	126.20		−18	$98–9^{15}$	0.8805_4^{20}	1.4556^{20}			∞				$B1^2$, 506
—	**Octhracene**	see **Anthracene, 1,2,3,4,5,6,7,8-octahydro-**													
—	**Olealdehyde**	see **9-Octadecenal***													
—	**Oleanilide**	see **9-Octadecenoic acid,** amide, *N*-phenyl-(*cis*)													
—	**Oleanolic acid**	see **Caryphyllin**													
—	**Oleic acid**	see **9-Octadecenoic acid(*cis*)***													
—	**Oleyl alcohol**	see **9-Octadecen-1-ol(*cis*)***													
—	**Olivetol**	see **Benzene, 1,3-dihydroxy-5-pentyl-***													
—	**Olivetolcarboxylic acid**	see **Benzoic acid, 2,4-dihydroxy-6-pentyl-**													
o230	**Orcein**	Orcin. $C_{28}H_{24}N_2O_7$	500.51	br-red pw					i	s	i	s	i	chl, CS_2 i	$B6^2$, 876
—	**Orcinol**	see **Toluene, 3,5-dihydroxy-**													
o231	**Ornithine(*L*)**	*L*-2.5-Diaminopentanoic acid*. $H_2N(CH_2)_3CH(NH_2)CO_2H$	132.16	cr (al-eth) $[\alpha]_D^{25}$ +11.5 (w, c = 6.5)	140				s	s	δ				$B4^2$, 844
o232	—,monohydro-chloride(*L*)	$H_2N(CH_2)_3CH(NH_2)CO_2H.HCl$	168.63	nd, $[\alpha]_D^{23}$ + 11.0 (w, c = 5.5)	215 (230–2)				s						$B4^3$, 1347
o233	—,sulfate(*L*)	$H_2N(CH_2)_3CH(NH_2)CO_2H.H_2SO_4$	230.24	$[\alpha]_D^{25}$ +8.4 (w)	234d				v	δ	i				$B4^3$, 1347
—	**Orotic acid**	see **6-Uracilcarboxylic acid**													
—	**Oroxylin**	see **Flavone, 5,7-dihydroxy-6-methoxy-**													
—	**Orsellinic acid**	see **Benzoic acid, 2,4-dihydroxy-6-methyl-**													
—	**Orthanilic acid**	see **Benzenesulfonic acid, 2-amino-***													
Ω o234	**Orthoacetic acid,** triethyl ester	1,1,1-Triethoxyethane*. $CH_3C(OC_2H_5)_3$	162.23			$144–6^{740}$ 66.5^{41}	0.8847_4^{25}	1.3980^{20}	i	∞	∞			chl, CCl_4 ∞	$B2^2$, 137
o235	—.trimethyl ester	1,1,1-Trimethoxyethane*. $CH_3C(OCH_3)_3$	120.15			107–9	0.9438_4^{25}	1.3859^{25}		v	v				$B2^2$, 128
Ω o236	**Orthocarbonic acid,** tetraethyl ester	Tetraethoxymethane*. $C(OC_2H_5)_4$	192.26			160–1 62^{28}	0.9186_4^{20}	1.3928^{20}		∞	∞				$B3^1$, 4
Ω o237	—.tetrapropyl ester	Tetrapropoxymethane*. $C(OCH_2CH_2CH_3)_4$	248.37			224.2	0.897_4^{20}	1.4100^{20}		v	v				**B3**, 6
—	**Orthoform**	see **Benzoic acid, 4-amino-3-hydroxy-, methyl ester**													
Ω o238	**Orthoformic acid,** triethyl ester	Triethoxymethane*. $HC(OC_2H_5)_3$	148.20			143^{765} 60^{30}	0.8909_4^{20}	1.3922^{20}	d	s	s				$B2^3$, 35
o239	—,triisobutyl ester	Triisobutoxymethane. $HC[OCH_2CH(CH_3)_2]_3$	232.37			224–6	0.8582_4^{20}	1.4120^{20}		s	s				$B2^2$, 31
o240	—,triisopropyl ester	Triisopropoxymethane*. $HC[OCH(CH_3)_2]_3$	190.29			166–8	0.8621_4^{20}	1.4000^{20}		s	s				$B2^3$, 39
Ω o241	—,trimethyl ester	Trimethoxymethane*. $HC(OCH_3)_3$	106.12			$103–5^{760}$	0.9676_4^{20}	1.3793^{20}		s	s				**B2**, 19
o242	—,tri(3-methyl-butyl) ester	Triisoamyl orthoformate. $HC[OCH_2CH_2CH(CH_3)_2]_3$	274.45			267–9d 166^{25}	0.8628_4^{20}	1.4233^{20}		s	s				$B2^2$, 31
o243	—,triphenyl ester	Triphenoxymethane.* $HC(OC_6H_5)_3$	292.34		76–7	$269–70^{50}$ d				s^h	s		s^h	chl s	$B6^3$, 595
o244	—,tripropyl ester	Tripropoxymethane.* $HC(OCH_2CH_2CH_3)_3$	190.29			$190–1^{745}$ 93^{30}	0.8805_4^{20}	1.4072^{20}		s	s				$B2^3$, 38

For explanations, symbols and abbreviations see beginning of table. For structural formulas see end of table.

No.	Name	Synonyms and Formula	Mol. wt.	Color, crystalline form, specific rotation and λ_{max} (log ε)	m.p. °C	b.p. °C	Density	n_D	Solubility w	al	eth	ace	bz	other solvents	Ref.
	Orthoformic acid														
o245	—,**trithio-**, triethyl ester	Ethyl orthothioformate. Tris(ethylthio)methane*. $HC(SC_2H_5)_3$	196.40	...	...	235^{760}d; $127–8^{12}$	1.053^{20}_4	1.5410^{15}	...	s	s	...	...	...	**B2**^{1}, 39
Ω o246	**Orthopropionic acid,** triethyl ester	1,1,1-Triethoxypropane*. $CH_3CH_2C(OC_2H_5)_3$	176.26	...	...	161^{766}; 44^9	...	1.4000^{25}	...	v	v	...	...	...	**B2**^{2}, 220
o247	**Orthosilicic acid,** tetraethyl ester	$(OC_2H_5)_4Si$	208.33	...	−82.5	168.8^{760}; 77^{32}	0.9320^{20}_4	1.3928^{20}	d	v	∞	...	...	...	**B1**, 334
o248	—,tetrakis(2-ethylbutyl) ester	$[(C_2H_5)_2CHCH_2O]_4Si$	432.77	liq	...	358^{760}	0.8920^{20}_4	1.4307^{20}	i	δ	s	...	s	...	**C54,** 23706
o249	—,tetrakis(2-ethylhexyl) ester	$[CH_3(CH_2)_3CH(C_2H_5)CH_2O]_4Si$	544.99	...	−90	419^{760}; 227^5	0.8803^{20}_4	1.4388^{20}	i	δ	s	...	s	...	**C54,** 23706
o250	—,tetramethyl ester	Methyl silicate. $(CH_3O)_4Si$	152.22	nd	−2	121^{760}; $25–7^{12}$	1.0232^{20}	1.3683^{20}	...	s	...	...	...	...	**B1**^{2}, 274
o253	**Orthovanadic acid,** triisopropyl ester	$[(CH_3)_2CHO]_3VO$	244.21	pa ye to col	−180	124^{21}	1.088^{15}	...	d	s	s	...	s	to s	**B51,** 751
o254	**1,3,4-Oxadiazole, 2,5-dimethyl-**	...	98.11			178–9	...	...	∞	∞	∞	...	...	...	**B27,** 565
Ω o255	**Oxalic acid**	Ethanedioic acid*. HO_2CCO_2H	90.04	mcl ta or pr (+2w, w) orh (anh) λ^w 250 sh (1.8)	α: 189.5; β: 182 (anh); 101.5 (hyd)	157 sub	α: 1.900^{17}_4; β: 1.895; 1.650^{20}_4 (hyd)	...	s	v	δ	...	i	chl, peth i	**B2**^{3}, 1534
Ω o256	—,diallyl ester	$CH_2{:}CHCH_2O_2CCO_2CH_2CH{:}CH_2$	170.17	...	...	217; 86^3	1.1582^{20}	1.4481^{20}	i	s	...	s	s	chl δ	**B2**^{3}, 1581
Ω o257	—,diamide	Oxamide. $H_2NCOCONH_2$	88.07	nd (w)	419d	...	...	...	δ	δ	i	...	...	...	**B2**^{3}, 1586
o258	—,—,*N,N′*-diethyl-	$C_2H_5NHCOCONHC_2H_5$	144.18	nd (al)	175 (180)	...	1.169^4	...	δ	s	i	...	...	...	**B4**^{2}, 605
o259	—,—,*N,N′*-diisopropyl-	$(CH_3)_2CHNHCOCONHCH(CH_3)_2$	172.23	nd (al)	212	...	...	...	...	v^h	...	...	...	...	**B4,** 154
o260	—,—,*N,N′*-dimethyl-	$CH_3NHCOCONHCH_3$	116.13	pl or nd (al)	217	sub	1.3^4_4	...	v^h	δ	i	...	s^h	chl s^h	**B4**^{2}, 564
Ω o261	—,—,*N,N′*-diphenyl-	Oxanilide. $C_6H_5NHCOCONHC_6H_5$	240.27	lf (bz or $PhNO_2$)	254	>360	...	...	i	δ^h	i	...	s	...	**B12**^{2}, 165
Ω o262	—,dibutyl ester	$CH_3(CH_2)_3O_2CCO_2(CH_2)_3CH_3$	202.25	...	−30.5	242^{773}; 96^2	0.9873^{20}_4	1.4234^{20}	i	s	s	...	...	...	**B2**^{2}, 507
Ω o263	—,dichloride	Oxalyl chloride. ClCOCOCl	126.93	nd (eth or peth) λ^{hp} 306 (1.36), 316 (1.37), 320 (1.40), 330 (1.35), 340 (1.24), 351 (1.20), 369 (0.74)	−16 (−12)	$63–4^{763}$	1.4785^{20}_4	1.4316^{20}	d	d	s	...	...	...	**B2**^{2}, 508
o264	—,di(2-chloroethyl) ester	$ClCH_2CH_2O_2CCO_2CH_2CH_2Cl$	215.04	lf (dil al)	45	132^3	...	...	...	s	...	...	s	...	**B2**^{3}, 1579
o265	—,dicyclohexyl ester	...	254.33	(MeOH)	42 (47)	$190–1^{73}$	...	...	i	v	v	...	...	MeOH s	**B6**^{1}, 6
Ω o266	—,diethyl ester	Oxalic ester. $C_2H_5O_2CCO_2C_2H_5$	146.14	λ^{al} 225 (2.6)	−38.5	185.7^{760}; 97^{20}	1.0785^{20}_4	1.4101^{20}	δ^h	∞	∞	∞	...	...	**B2**^{2}, 504
o267	—,dihydrazide	$H_2NNHCOCONHNH_2$	118.10	nd (w)	243d	...	$1.458^{22.5}$	...	s^h	δ	δ	...	δ	chl δ	**B2**^{2}, 514
o268	—,diisobutyl ester	$(CH_3)_2CHCH_2O_2CCO_2CH_2CH(CH_3)_2$	202.25		...	229^{758}; 143^{20}	0.9737^{20}_4	1.4180^{20}	i	s	s	s	...	...	**B2**^{2}, 507
o269	—,diisopropyl ester	$(CH_3)_2CHO_2CCO_2CH(CH_3)_2$	174.20		...	191^{765}	1.0010^{20}_4	1.4100^{20}	...	s	s	...	...	...	**B2**^{1}, 234
Ω o270	—,dimethyl ester	$CH_3O_2CCO_2CH_3$	118.09	mcl ta	54	164.5^{760}	1.148^{15}; 1.1716^{60}	1.379^{82}	δ; s^h	s	s	s	...	...	**B2**^{2}, 503
o271	—,di(3-methylbutyl) ester	Isoamyl oxalate. $(CH_3)_2CHCH_2CH_2O_2CCO_2CH_2CH_2CH(CH_3)_2$	230.30			267–8; 144^{14}	0.968^{11}_{11}	...	i	v	v	...	...	...	**B2**^{3}, 1580
—	—,dinitrile	*see* **Cyanogen**													
o273	—,dipropyl ester	$CH_3CH_2CH_2O_2CCO_2CH_2CH_2CH_3$	174.20	...	−44.3	211^{749}; $78–80^3$	1.0188^{20}_4	1.4168^{20}	δ	∞	s	...	...	...	**B2**^{3}, 1579
o274	—,di(2-tolyl) ester	...	270.29	nd (al)	91	distb	...	...	i	v	v	v	v	chl, CS_2, os v	**B6**^{2}, 330
o275	—,di(3-tolyl) ester	...	270.29	nd (al)	106	distb	...	...	i	v	v	v	s	os v, chl v	**B6**^{2}, 353
o276	—,di(4-tolyl) ester	...	270.29	lf or pl (al-eth)	148	...	...	...	...	s	v	v	s	os s, chl v	**B6**^{2}, 379
o277	—,imide	Oximide.	71.04	pr (al)	...	...	...	...	d^h	δ	...	...	...	...	**B21,** 368
Ω o278	—,monoamide	Oxamic acid. HO_2CCONH_2	89.05	cr (w)	210d	...	...	...	δ	i	i	...	...	...	**B2**^{2}, 509
o279	—,—,*N-sec*-butyl-	$CH_3CH_2CH(CH_3)NHCOCO_2H$	145.16	cr (eth)	88–9	...	...	...	...	...	s^h	...	...	...	**B4,** 162
o280	—,—,*N*-phenyl-	Oxanilic acid. $C_6H_5NHCOCO_2H$	165.15	nd (bz)	150	...	...	...	s^h	v	s	...	δ	chl s lig δ	**B12**^{2}, 164
Ω o281	—,monoamide monoethyl ester	Ethyl oxamate. $C_2H_5O_2CCONH_2$	117.11	...	114–5	...	...	...	s	...	s	...	i	...	**B2**^{2}, 509
o282	—,—,*N*-acetyl-	$C_2H_5O_2CCONHCOCH_3$	159.14	pl (eth)	54–5	...	...	...	...	s	s	...	...	...	**B2**^{2}, 509
o283	—,—,*N*-phenyl-	$C_2H_5O_2CCONHC_6H_5$	193.21	pl or pr (al), nd (w)	65–6	260–300	...	...	δ	s	s	s	s	oos v	**B12**^{2}, 164
Ω o284	—,monoamide monohydrazide	Semioxamazide. $H_2NNHCOCONH_2$	103.08	lf	221d (232)	...	...	...	δ; s^h	i	i	...	...	alk, ac v	**B2**^{3}, 1594
o285	—,monoamide monoureide	Oxalam. Oxaluramide. $H_2NCOCONHCONH_2$	131.09	...	>310	...	...	...	i	...	...	...	...	sulf s	**B3**^{2}, 54

For explanations, symbols and abbreviations see beginning of table. For structural formulas see end of table.

No.	Name	Synonyms and Formula	Mol. wt.	Color, crystalline form, specific rotation and λ_{max} (log ε)	m.p. °C	b.p. °C	Density	n_D	Solubility w	al	eth	ace	bz	other solvents	Ref.
	Oxalic acid														
o286	—,monochloride monoethyl ester	$ClCOCO_2C_2H_5$	136.54	hyg	...	135^{760} 30^{10}	1.2226^{20}_4	...	d	d	s	...	s	...	**B2**^{2}, 508
o287	—,monoethyl-monomethyl ester	$CH_3O_2CCO_2C_2H_5$	132.13	...	...	173.7	1.5505^0_0	...	i	v	v	...	...	...	**B2**^{1}, 232
Ω o288	—,monoethyl ester mononitrile	Ethyl cyanoformate. $C_2H_5O_2CCN$	99.09	...	...	116.5–$.8^{765}$	1.0034^{20}_4	1.3821^{20}	i	s	s	...	...	...	**B2**^{2}, 510
o289	—,monoureide	Oxaluric acid. $H_2NCONHCOCO_2H$	132.08	cr	208–10d	...	...	...	s	i	δ	...	δ	oos δ	**B3**^{2}, 54
o290	—,piperazinum salt	$C_2H_2O_4 . C_4H_{10}N_2$	176.17	(w)	>300	...	...	...	s^h	v	i	...	...	...	**Am 56,** 1759
o291	—,**dithiono-, diamide**	Dithiooxamide. $H_2NCSCSNH_2$	120.20	og-red	sub	...	...	...	δ	δ	...	...	...	con sulf s	**B2**^{2}, 515
—	**Oxaluramide**	*see* **Oxalic acid,** monoamide monoureide													
—	**Oxaluric acid**	*see* **Oxalic acid,** monureide													
—	**Oxamic acid**	*see* **Oxalic acid,** monoamide													
—	**Oxanilic acid**	*see* **Oxalic acid,** monoamide, *N*-phenyl-													
—	**Oxanthranol**	*see* **Anthracene,9,10-dihydroxy-**													
o292	**Oxazole**	...	69.06	λ^{cy} 303 (4.48)	...	69–70	...	$1.4285^{17.5}$	...	...	...	...	...	...	**B27**^{2}, 9
o292[1]	—,**2,4-dimethyl-**	C_5H_7NO. *See* o292	97.12	...	...	108	0.9352^{15}_4	$1.4166^{15.3}$	v	v	v	...	...	...	**B27**^{2}, 10
o293	—,**2,5-dimethyl-**	C_5H_7NO. *See* o292	97.12	...	...	$117–8^{760}$	0.9958^{21}_4	1.4385^{21}	∞	...	...	...	...	...	**B27**^{2}, 10
o294	—,**2,4-diphenyl-**	$C_{15}H_{11}NO$. *See* o292	221.26	lf (al) λ^{al} 240 sh (4.2), 280 (4.2)	103	338–40	...	...	...	v^h	v	...	v	...	**B27**, 78
o295	—,**2,5-diphenyl-**	$C_{15}H_{11}NO$. *See* o292	221.26	nd (lig) λ^{cy} 223 (4.31), 303 (4.48)	74	360	1.0940^{100}_4	1.6231^{100}	i	v	v	...	...	...	**B27**^{2}, 43
Ω o296	—,**4,5-diphenyl-**	$C_{15}H_{11}NO$. *See* o292	221.26	pl or pr (lig) λ^{MeOH} 220 sh (4.27), 275 (4.09)	44	$192–5^{15}$	...	1.6283^{100}	δ	v	v	...	...	con ac s	**B27**, 79
Ω o297	—,**2,4,5-triphenyl-**	Azobenzil. Benzilam. $C_{21}H_{15}NO$. *See* o292	297.36	pr λ^{cy} 235 (4.40), 310 (4.35)	116	...	...	...	...	δ	...	...	s	...	**B27**^{2}, 56
o298	**2,4-Oxazolidinedione, 5,5-dipropyl-**	...	185.23	...	42–3	$148–50^3$	...	...	i	...	...	...	...	...	**Am 67,** 522
o299	**Oxindole**	...	133.15	nd (w) λ^w 248.5 (3.89)	127	227^{23}	...	...	s^h	s	s	...	...	...	**B21**^{2}, 249
o300	—,**1-ethyl-**	$C_{10}H_{11}NO$. *See* o299	161.21	nd (ace or w)	97–8	...	...	...	δ	...	s	s^h	...	...	**B21**^{1}, 291
o301	—,**3-ethyl-1-methyl-**	$C_{11}H_{13}NO$. *See* o299	175.23	...	...	$280–5^{745}$ $103–7^{0.5}$	...	1.557^{25}	...	s	s	...	...	...	**B21**^{2}, 258
o302	—,**3-hydroxy-**	Dioxindole. $C_8H_7NO_2$. *See* o299	149.15	cr (w, al) $\lambda^{w, 0.1 N HCl}$ 252 (3.72), 292 (3.16)	180 (19s d)	...	...	...	...	...	...	...	...	...	**B21**^{2}, 415
o303	**Oxonium, dimethyl-, bromide**	$[(CH_3)_2OH]^+Br^-$	126.99	...	−13	...	...	...	...	...	s	...	...	liq HBr v	**B1**^{3}, 1192
o304	—,—,chloride	$[(CH_3)_2OH]^+Cl^-$	82.53	gas	−97	−2	...	...	...	...	s	...	...	liq HCl v	**B1**^{3}, 1192
o305	—,**trimethyl-, fluoborate**	$[(CH_3)_3O]^+BF_4^-$	147.91	hyg nd	148d	...	...	...	d	...	...	v^h	...	chl, CH_3NO_2, $PhNO_2$ v	**B1**^{3}, 1193
o306	**Oxyacanthine**	Vinetine. $C_{37}H_{40}N_2O_6$	608.71	nd (al or eth) $[\alpha]^{20}_D$ +131.5 (chl, c = 1) λ^{MeOH} 281 (4.04)	216–7	...	...	...	i	s	s	...	s	chl, min ac s lig l	**B27**^{2}, 892
o307	—,hydrochloride	$C_{37}H_{40}N_2O_6 . HCl$. *See* o306.	645.17	nd, $[\alpha]^{15}_D$ +163.8 (w, c = 3)	270–1	...	...	...	s	...	...	...	...	...	**B27**^{2}, 893
o308	—,nitrate dihydrate	$C_{37}H_{40}N_2O_6 . HNO_3 . 2H_2O$. *See* o306	705.74	nd	195–200	...	...	...	δ	...	...	...	...	...	**B27**^{2}, 893
o309	**Oxynarcotine**	α-Narcotine-*N*-oxide. $C_{22}H_{23}NO_8$.	429.43	hyg nd $[\alpha]_D$ +135 (chl)	...	...	...	...	v	v	i	δ	...	chl v	**B27**^{2}, 607
o311	**Oxysparteine**	Isolupanine. $C_{15}H_{24}N_2O$.	248.36	ye to col hyg nd (peth) $[\alpha]^{18}_D$ −10.0 (al, c = 18) λ^w 230 sh (3.0)	111	209^{12}	...	...	v	v	v	...	...	chl s	**B24**^{2}, 56
o312	—,monohydrochloride tetrahydrate	$C_{15}H_{24}N_2O . HCl . 4H_2O$. *See* o311	356.89	wh cr (w)	48–50	...	...	...	s	s	...	...	...	...	**B24**^{2}, 57
—	**Oxytetracycline**	*see* **Terramycin**													

For explanations, symbols and abbreviations see beginning of table. For structural formulas see end of table.

No.	Name	Synonyms and Formula	Mol. wt.	Color, crystalline form, specific rotation and λ_{max} (log ε)	m.p. °C	b.p. °C	Density	n_D	Solubility w	al	eth	ace	bz	other solvents	Ref.
	Palmitaldehyde														
—	**Palmitaldehyde**	*see* **Hexadecanal***													
—	**Palmitanilide**	*see* **Hexadecanoic acid, amide, *N*-phenyl-***													
—	**Palmitic acid**	*see* **Hexadecanoic acid***													
p1	**Paludrine**	Chloroguanide. Proguanil. BPC. $C_{11}H_{16}ClN_5$.	253.74	pl (dil al) λ^{w} 240 (4.16)	130–1	...	...	...	i	s	...	...	...	to s^h	**C40,** 2931
p2	—, hydrochloride	$C_{11}H_{16}ClN_5.HCl$. *See* p1	290.20	nd (w)	245 (248–52)	...	...	...	δ	s	i	...	...	chl i	**J1946,** 729
p3	**Panthesin**	$C_{18}H_{32}N_2O_5S$.	388.53	pa ye pw (al) $\lambda^{w,pH=9.5}$ 289 (4.7)	157–9	...	...	...	v	s	...	...	...	...	**C27,** 808
p4	**Pantothenic acid** (*d*)	Chick antidermatitis factor. (*D*, +)-*N*(α,γ-Dihydroxy-β,β-dimethylbutyryl)-β-alanine. $C_9H_{17}NO_5.HOCH_2C(CH_3)_2CH(OH)CONH(CH_2)_2CO_2H$	219.24	ye visc oil $[\alpha]_D^{25}$ +37.5 (w)	...	...	...	...	v	...	s	...	v	AmOH, aa v diox s	**C40,** 7163
p5	—, calcium salt (*d*)	Calcium pantothenate. $[C_9H_{16}NO_5]_2Ca$. *See* p4	476.55	wh (MeOH), $[\alpha]_D^{26}$ + 28.2 (w)	195–6	...	...	...	s	i	...	i	i	MeOH s	**C40,** 7163
Ω p6	—, — (*l*)	$[C_9H_{16}NO_5]_2Ca$. *See* p4.	476.55	cr (MeOH), $[\alpha]_D^{26}$ − 27.8 (w)	187.5–9	...	...	...	s	i	...	i	i	...	**C36,** 406
Ω p7	**Pantothenyl alcohol**	Penthenol. $C_9H_{19}NO_4$. $HOCH_2C(CH_3)_2CH(OH)CONH(CH_2)_3OH$	205.26	slightly hyg oil $[\alpha]_D^{20}$ +29.5 (w, c = 5)	...	d 118–$20^{0.02}$	1.2_{20}^{20}	1.497^{20}	v	v	δ	...	...	MeOH v alk d	**C40,** 374
p8	**Papaveraldine**	Xanthaline. $C_{20}H_{19}NO_5$	353.38	nd (al), cr (bz, peth) λ^{al} 265 (4.78), 305 (4.15), 385 (4.1)	210–1	...	...	...	i	δ	δ	...	s	min ac v aa v^h chl s peth δ	**B21**[2], 479
p9	**Papaverine**	Papaveroline tetramethyl ether. $C_{20}H_{21}NO_4$.	339.40	wh pr (al-eth), nd (chl-peth) λ^{al} 240 (4.9), 280 (3.9), 315 (3.7), 327 (3.7)	147–8	d sub 135–40^{11}	1.337_4^{20}	1.625	δ s^h	v^h	δ	s	s^h	chl v Py s CCl_4, lig δ	**B21**[2], 202
p10	—, hydrochloride	$C_{20}H_{21}NO_4.HCl$. *See* p9	375.86	wh mcl pr (w) λ^{w} 250 (4.69), 281 (3.80), 311 (3.82)	224–5	...	...	...	s	s	δ	...	...	chl s	**B21**[2], 203
—	—, **tetrahydro-*N*-methyl-**	*see* **Laudanosine**													
Ω p11	**Parabanic acid**	Imidazoletrione. Oxalylurea.	114.06	mcl nd (w) $\lambda^{w,pH=7.45}$ 272 (2.9)	243–5 d	part sub 100	...	...	s	v	...	...	...	...	**B24**[2], 263
p12	**Parabutyr-aldehyde**	...	216.32	...	...	98–100^{35} (105–8^{12})	0.918	...	...	...	...	...	...	...	**B19**[2], 402
p13	**Paraconic acid**	Hydroxymethylsuccinic acid γ-lactone. Itamalic acid γ-lactone. Tetrahydro-5-oxo-3-furancarboxylic acid.	130.10	dlq	57–8	...	...	...	s	...	...	...	...	...	**B18**[2], 311
p14	**Paracyanogen**	$(CN)_x$	...	br pw	...	sub	...	...	i	i	...	...	...	KOH s	**B2**[3], 1590
p14[1]	**Paraisobutyr-aldehyde**	2,4,6-Triisopropyl-1,3,5-trioxane.	216.32	nd (al)	59–60	195 sub δd	...	...	i	s	v	...	...	...	**B19,** 390
Ω p15	**Paraldehyde**	Paraacetaldehyde. 2,4,6-Trimethyl-1,3,5-trioxane.	132.16	...	12.6	128.0	0.9943_4^{20}	1.4049^{20}	δ s^h	∞	∞	...	...	chl ∞	**B19**[2], 394
p16	—, **trichloro-**	Chloroacetaldehyde trimer.	235.50	nd (eth)	87–7.5 (cor)	142–4^{10}	...	...	i	δ v^h	v	...	...	...	**B19**[1], 807
p17	**Paraldol**	...	176.22	wh tcl pr	89–91	90^{15}	1.116^{20}	1.4610^{20}	v	v	s	...	...	...	**B1**[2], 869 **B19**[2], 93
—	**Pararosaniline**	*see* **Methanol, tris(4-aminophenyl)-***													
p18	**Parasorbic acid**	2-Hexen-5,1-olide. 5-Hydroxy-2-hexenoic acid lactone*	112.14	oily liq $[\alpha]_D^{19}$ +210 (al, c = 2)	...	110^{15}	1.079_4^{18}	1.4730^{19}	s	v	v	...	...	...	**B17,** 225
Ω p19	**Parathion**	Diethyl *p*-nitrophenyl monothiophosphate.	291.27	ye liq	6.1	375^{760} 157–$62^{0.6}$ $115^{0.04}$	1.2704_{20}^{20}	1.5370^{25}	i	∞	s	s	...	chl, AcOEt v lig i	**Am 70,** 3943
—	**Paredrine**	*see* **Benzene, 1(2-amino-propyl)-4-hydroxy-***													
—	**Paris green**	*see* **Methyl green**													
—	**α-Parvoline**	*see* **Pyridine,3,5-dimethyl-2-ethyl-**													
—	**β-Parvoline**	*see* **Pyridine, tetramethyl-**													
p20	**Patulin**	Clavacin.	154.12	pr or pl (eth, chl) λ^{al} 277 (4.3)	111	...	...	...	s	s	s	s	s	AcOEt v os s peth i	**C38,** 1321
p21	**Paucine,** hydrate	$C_{27}H_{39}N_5O_5.6\frac{1}{2}H_2O$	630.75	ye lf	126d	...	...	...	s^h	s^h	i	...	...	alk s^h chl i	**M775**
—	**Peganine**	*see* **Vasicine** (*DL*)													
—	**Pelargonaldehyde**	*see* **Nonanal***													
—	**Pelargonic acid**	*see* **Nonanoic acid***													

For explanations, symbols and abbreviations see beginning of table. For structural formulas see end of table.

PHYSICAL CONSTANTS OF ORGANIC COMPOUNDS (Continued)

No.	Name	Synonyms and Formula	Mol. wt.	Color, crystalline form, specific rotation and λ_{max} (log ε)	m.p. °C	b.p. °C	Density	n_D	Solubility w	al	eth	ace	bz	other solvents	Ref.
	Pelargonidin														
p22	**Pelargonidin, chloride**	$C_{15}H_{11}ClO_5$	306.70	red br hyg (anh), pr or pl (dil HCl), nd (al-HCl, +w), lf (MeOH-eth) λ^{al} 275 (4.2), 330 sh (3.7)	>350 (anh)	...	...	...	s v^h	v	...	...	...	conc sulf s MeOH, chl δ	**B18**², 200
p25	**Pellotine**	*N*-Methylanhalonidine. $C_{13}H_{19}NO_3$.	237.30	pl (al, peth)	111.5	...	...	...	δ	s	s	s	...	chl v peth s	**B21**¹, 249
p26	**Penicillic acid**	CH_2:C(CH_3)COC(OCH_3):CHCO_2H	170.17	rh or hex pl (+1w), nd (peth) λ^{al} 225.5 (4.02)	64–5 (+1w) 87 (anh)	...	...	...	s v^h	v	v	s	v	chl v peth δ^h	**B3**², 519
Ω p27	**Pentacene***	Benzo[b]naphthacene. 2,3,6,7-Dibenzanthracene.	278.36	deep vt-bl nd or lf ($PhNO_2$), cr (bz) λ^{MeOH} 310 (5.4), 575 (4.1)	270–1 (>300 d)	290–300 sub (vac)	...	...	i	...	...	...	δ^h	$PhNO_2$ s^h os δ	**E14s**, 582
p28	—,6,13-diphenyl-*	$C_{34}H_{22}$. *See* p27	430.56	vt-bl nd ($PhNO_2$ + 1)	...	318–20	...	...	...	s^h	...	...	s	aa s CCl_4 s^h	**E14s**, 583
p29	**6,9-Pentadecadiyne***	$CH_3(CH_2)_4C$:CCH_2C:$C(CH_2)_4CH_3$	204.34	...	...	135–6⁴	0.840^{21}_{4}	1.4693^{21}	...	...	...	...	...	...	**B1**³, 1067
p30	**Pentadecanal***	*n*-Pentadecylaldehyde. $CH_3(CH_2)_{13}CHO$	226.41	nd	24–5	185^{25}	...	...	i	s	v	v	...	os v	**B1**³, 2920
p31	—,oxime*	$CH_3(CH_2)_{13}CH$:NOH	241.42	nd (dil al)	86	...	...	...	i	δ	s	...	δ	peth δ	**B1**, 716
p32	**Pentadecane***	$CH_3(CH_2)_{13}CH_3$	212.42	...	10	270.63^{760} 136^{10}	0.7685^{20}_{4}	1.4315^{20}	i	v	v	...	...	...	**B1**³, 553
p33	—,1-amino-*	*n*-Pentadecylamine. $CH_3(CH_2)_{14}NH_2$	227.44	fl	37.3 (40)	307.6^{760} 165.8^{10}	0.8104^{20}_{4} (suc)	1.4480^{20} (suc)	i	s	s	...	...	...	**B4**³, 422
Ω p34	—,1-bromo-*	*n*-Pentadecyl bromide. $CH_3(CH_2)_{14}Br$	291.33	...	19.0	322^{760} 177^{10}	1.0675^{20}_{4}	1.4611^{20}	i	...	...	s	...	chl v	**B1**³, 553
p35	—,1,15-dibromo-*	Pentadecamethylene bromide. $Br(CH_2)_{15}Br$	370.23	lf (al)	27.2–7.5	$215–25^{15}$ 192^{2}	...	...	i	δ v^h	...	...	...	chl v	**B1**³, 554
p36	—,1(2,3-dihydroxyphenyl)-*	Tetrahydrourushiol.	320.52	nd (to, xyl, eth or peth)	59	...	...	...	...	v	v	...	v	aa, chl v lig δ, v^h	**B6**², 911
Ω p37	**Pentadecanoic acid***	Pentadecylic acid. $CH_3(CH_2)_{13}CO_2H$	242.41	pl (aq al, ace), cr (peth)	53–4	257^{100} 158^{1}	0.8423^{80}	1.4254^{80}	i	v	s	v	v	CS_2, peth, chl s	**B2**², 329
Ω p38	—,methyl ester	$CH_3(CH_2)_{13}CO_2CH_3$	256.43	nd (dil al)	18.5	153.5	0.8618^{25}_{4}	1.4390^{25}	...	s	s	...	...	os s	**B2**², 330
Ω p39	**8-Pentadecanone**	Caprylone. Diheptyl ketone. $CH_3(CH_2)_6CO(CH_2)_6CH_3$	226.41	cr (MeOH, al)	43	291^{760} 178^{20}	...	...	...	s	s	...	s	chl, CCl_4 s	**B1**³, 2922
p40	**1-Pentadecene***	$CH_3(CH_2)_{12}CH$:CH_2	210.41	...	2–8 (−3.73)	268.17^{760} 133.7^{10}	0.7764^{20}_{4}	1.4389^{20}_{4}	i	...	...	s	...	...	**B1**³, 874
p41	**1-Pentadecyne***	$CH_3(CH_2)_{12}C$:CH	208.39	...	10	268^{760} 129.8^{10}	0.7928^{20}_{4}	1.4419^{20}	i	...	...	s	...	...	**B1**³, 1028
p43	**2,4-Pentadienal, 5-phenyl-***	Cinnamylideneacetaldehyde. C_6H_5CH:CHCH:CHCHO	158.20	nd λ^{bz} 320 (4.58)	42–3	$155–65^{3}$ $92–5^{0.05}$	...	...	i	∞	v	...	∞	...	**B7**², 320
p44	**1,2-Pentadiene***	Ethylallene. CH_3CH_2CH:C:CH_2	68.13	λ^{gas} 177 (4.4), 181 (4.4)	−137.26	44.86^{760}	0.6926^{20}_{4}	1.4209^{20}	...	∞	∞	∞	∞	hp, CCl_4 ∞	**B1**³, 958
Ω p45	**1,3-Pentadiene***	Piperylene. CH_3CH:CHCH:CH_2	68.13	λ^{al} 223.5 (4.36)	−87.47	42.03^{760}	0.6760^{20}_{4}	1.4301^{20}	...	∞	∞	∞	∞	hp, CCl_4 ∞	**B1**³, 958
Ω p46	**1,4-Pentadiene***	Allylethylene. Divinylmethane. CH_2:$CHCH_2CH$:CH_2	68.13	λ^{gas} 178 (4.23)	−148.28	25.97^{760}	0.6608^{20}_{4}	1.3888^{20}	i	v	v	v	v	...	**B1**³, 963
p47	**2,3-Pentadiene***	*sym*-Dimethylallene. CH_3CH:C:$CHCH_3$	68.13	λ^{gas} 171 (4.3), 180 (4.2), 194 (3.6), 202 sh (3.5)	−125.65	48.25^{760}	0.6950^{20}_{4}	1.4284^{20}	i	∞	∞	∞	∞	hp, CCl_4 ∞	**B1**³, 964
p48	**1,3-Pentadiene, 3-chloro-***	3-Methylchloroprene. CH_3CH:C(Cl)CH:CH_2	102.57	...	...	99.5–101.5^{759}	0.9576^{20}_{4}	1.4785^{20}	i	v	v	v	...	chl s	**B1**³, 962
p49	**1,2-Pentadiene, 1-chloro-3-ethyl-***	$(C_2H_5)_2C$:C:CHCl	130.62	...	...	$85–8^{100}$	0.9297^{19}_{4}	...	i	...	s	...	...	...	**B1**³, 1002
p50	—,1-chloro-3-methyl-*	$CH_3CH_2C(CH_3)$:C:CHCl	116.59	...	...	$68–70^{100}$	0.9562^{20}_{4}	...	i	...	s	...	...	...	**B1**³, 990
p51	**1,3-Pentadiene, 1-chloro-3-methyl-***	CH_3CH:C(CH_3)CH:CHCl	116.59	...	...	$62–3^{100}$	0.9574^{20}_{4}		i	...	s	...	...	chl s	**B1**³, 990
p52	—,2-chloro-3-methyl-*	CH_3CH:C(CH_3)CCl:CH_2	116.59	...	...	$57–60^{95}$	0.9437^{20}_{4}	1.4671^{20}	i	...	s	...	...	chl s	**B1**³, 991
p53	**2,4-Pentadienoic acid***	β-Vinylacrylic acid. CH_2:CHCH:$CHCO_2H$	98.10	hyg pr (eth) λ^{al} 242 (4.39)	80	d110–5	...	...	v^h	v	v	...	v	peth s	**B2**, 451
p54	—,4-hydroxy-, lactone*	Protoanemonin.	96.09	pa ye oil λ^{w} 260 (4.15)	...	73^{11} 45^{11}	...	...	δ	...	...	...	...	chl s	**C54**, 7673
p55	—,5(3,4-methylenedioxyphenyl)-	Piperic acid.	218.21	ye in light, nd (al), ye nd (sub) λ^{al} 340 (4.5)	215	sub	...	...	i	s^h	δ	...	δ	...	**B19**², 300
Ω p56	—,5-phenyl-*	Cinnamylideneacetic acid. C_6H_5CH:CHCH:$CHCO_2H$	174.20	pl (al), pr (bz) λ^{al} 307 (4.56)	166–7	...	...	...	...	s^h	s	...	s v^h	peth δ	**B9**², 441

For explanations, symbols and abbreviations see beginning of table. For structural formulas see end of table.

No.	Name	Synonyms and Formula	Mol. wt.	Color, crystalline form, specific rotation and λ_{max} (log ε)	m.p. °C	b.p. °C	Density	n_D	Solubility w	al	eth	ace	bz	other solvents	Ref.
	2,4-Pentadienoic acid														
p57	—,—,ethyl ester	$C_6H_5CH{:}CHCH{:}CHCO_2C_2H_5$	202.26	ye oil	25–6.7	149–50^{4}	1.0469$_4^{20}$	1.5768^{80}	...	s	s	...	...	...	**B9^2**, 441
p58	—,—,methyl ester	$C_6H_5CH{:}CHCH{:}CHCO_2CH_3$	188.23	lf or pl λ^{eth} 360 (4.57)	71	185^{20}	...	...	i	v	...	...	...	MeOH v	**B9^2**, 441
p59	**1,4-Pentadien-3-one, 1,5-bis-(2-ethoxyphenyl)-***	$C_{21}H_{22}O_3$. *See* p72	322.41	ye lf (dil al)	89	...	...	...	i	v^h	s^h	...	...	...	**B8**, 352
p61	—,1,5-bis(2-hydroxyphenyl)-*	Disalicylidene acetone. $C_{17}H_{14}O_3$. *See* p72	266.30	ye nd (dil al)	168d	...	...	...	δ	v	s	v	s	Py v ac, chl, aa s	**B8^2**, 404
p62	—,1,5-bis(4-hydroxyphenyl)-*	Bis(4-hydroxystyryl) ketone. $C_{17}H_{14}O_3$. *See* p72	266.30	(i) ye-og nd or lf (dil al) (ii) gr lf	(i) 237–8 (st)	...	...	...	...	v	δ	v	δ	aa, MeOH, alk, acs, chl δ	**B8^2**, 405
p63	—,1,5-bis(2-methoxyphenyl)-*	Bis (2-methoxystyryl)ketone. $C_{19}H_{18}O_3$. *See* p72	294.35	ye nd or lf (al)	127	...	...	...	i	δ s^h	...	...	...	...	**B8^2**, 405
p64	—,1,5-bis(3-methoxyphenyl)-*	Bis(3-methoxystyryl) ketone. $C_{19}H_{18}O_3$. *See* p72	294.35	nd (chl-MeOH)	55–6	...	...	...	i	...	...	s	...	os v chl s lig δ	**B8^1**, 666
Ω p65	—,1,5-bis(4-methoxyphenyl)-*	Dianisalacetone. $C_{19}H_{18}O_3$. *See* p72	294.35	ye lf (aa, AcOEt)	129–30	...	...	...	...	δ	δ	...	v	chl v aa s	**B8^2**, 406
Ω p66	—,1,5-bis(3,4-methylenedioxyphenyl)-*	Dipiperonylideneacetone. $C_{19}H_{14}O_5$. *See* p72	322.32	ye nd (bz or AcOEt)	185	...	...	...	i	δ	...	v	s^h	chl v con sulf s lig i	**B19^2**, 463
p67	—,1,5-bis(2-nitrophenyl)-*	Bis(2-nitrostyryl) ketone. $C_{17}H_{12}N_2O_5$. *See* p72	324.30	ye nd (aa)	170.5–1	...	...	...	...	δ	...	...	...	chl v con sulf s	**B7^2**, 455
p68	—,1,5-bis(3-nitrophenyl)-*	Bis(3-nitrostyryl) ketone. $C_{17}H_{12}N_2O_5$. *See* p72	324.30	ye, br (Ac_2O)	238	...	...	...	...	...	...	s	...	os s con sulf, Ac_2O δ	**B7^2**, 455
p69	—,1,5-bis(4-nitrophenyl)-*	Bis(4-nitrostyryl) ketone. $C_{17}H_{12}N_2O_5$. *See* p72	324.30	ye (Ac_2O)	254 (248)	...	...	...	i	...	...	s	...	os v Ac_2O δ	**B7^2**, 455
p70	—,1(2-chlorophenyl)-5(3-chlorophenyl)-	$C_{17}H_{12}Cl_2O$. *See* p72	303.19	ye nd (dil al)	67–8	...	...	...	s	...	δ	...	...	CS_2, lig δ	**B7^2**, 454
p71	—,1(2-chlorophenyl)-5(4-chlorophenyl)-	$C_{17}H_{12}Cl_2O$. *See* p72	303.19	ye nd (al)	109	...	...	...	δ	...	...	...	...	...	**B7^2**, 454
p71[1]	—,1,5-di(2-furyl)-	Difurfurylidene acetone.	214.22	dlq pr (peth), ye pr (lig)	60–1	d	...	...	i	v	v	...	...	chl v con HCl, sulf s lig δ^h	**B19^2**, 162
p72	—,1,5-diphenyl-*	Dibenzalacetone. Cinnamone. Distyryl ketone.	234.30	pl or lf (ace or AcOEt) λ^{al} 331 (4.55)	113d	d	...	...	i	δ	δ	s	...	chl s	**B7^2**, 452
p72[1]	**2,4-Pentadiene-1-one, 1,5-diphenyl-**	Cinnamalacetophenone. $C_6H_5CH{:}CHCH{:}CHCOC_6H_5$	234.30	(i) ye nd (al) (ii) ye cr	(i) 102–3 (ii) 235	...	...	...	i	δ s^h	δ	...	...	con sulf s (red)	**B7^2**, 451
p73	**1,3-Pentadiyne***	$CH_3C{:}CC{:}CH$	64.09	λ^{al} 227 (2.43), 236 (2.59), 249 (2.32)	...	55–6	0.7375$_4^{21}$	1.4431^{21}	i	...	s	...	s	chl s	**B1^2**, 247
Ω p74	**Pentaerythritol**	Tetrakis(hydroxymethyl)-methane*. Tetramethylol-methane. $C(CH_2OH)_4$	136.15	cr (dil HCl)	269 (260)	sub	...	1.548	s	...	i	...	i	...	**B1^2**, 601
p75	—,tetraacetate	$C(CH_2O_2CCH_3)_4$	304.30	tetr nd (w or bz)	83–4 (84–6)	...	1.273$_4^{18}$	...	s	v	v	...	...	...	**B2^2**, 162
Ω p76	—,tetranitrate	Penthrit. PETN. $C(CH_2ONO_2)_4$	316.15	tetr (ace), pr (ace-al)	140–1	...	1.773$_4^{20}$	...	δ	δ	δ	v	s	to s Py s^h MeOH δ	**B1^2**, 602 **B9^2**, 887
—	**Pentalin**	*see* **Ethane, pentachloro-***													
Ω p77	**Pentanal***	Valeral. Valeraldehyde. $CH_3(CH_2)_3CHO$	86.14	λ^{gas} 178, 182, 184	−91.5	103^{760}	0.8095$_4^{20}$	1.3944^{20}	δ	s	s	...	...	...	**B1^3**, 2797
p78	—,oxime*	Valeraldoxime. $CH_3(CH_2)_3CH{:}NOH$	101.15	cr	52	...	...	...	...	...	s	...	...	... –	**B1**, 676
p78[1]	—,4-hydroxy-(*dl*)*	$CH_3CH(OH)CH_2CH_2CHO$	102.13	...	...	63–5^{10}	1.019$_4^{20}$	1.4359^{17}	s	...	...	s	...	os s	**B1^2**, 871
p78[2]	—,—(*l*)*	$CH_3CH(OH)CH_2CH_2CHO$	102.13	$[\alpha]_D^{23}$ − 7.8	...	43–6^{1}	...	...	s	...	...	s	...	os s	**B1^2**, 872
p79	—,3-hydroxy-2-methyl-*	Propionaldol. $C_2H_5CH(OH)CH(CH_3)CHO$	116.16	...	...	94–6^{23}	0.986$_4^{25}$	1.4502^{20}	s	v	v	v	...	...	**B1^1**, 423
Ω p80	—,2-methyl-*	$CH_3CH_2CH_2CH(CH_3)CHO$	100.16	...	...	116^{737}	...	...	s	...	s	s	...	...	**B1^1**, 355
p81	—,4-methyl-2-oxo-*	Formyl isobutyl ketone. $(CH_3)_2CHCH_2COCHO$	114.15	ye-gr	...	45–6^{12}	...	...	...	...	...	...	...	os s	**B1^1**, 406
p82	—,2-oxo-*	Butyrylformaldehyde. $CH_3(CH_2)_2COCHO$	100.13	...	...	112^{760} 36^{16}	...	1.4043^{25}	∞	∞	∞	...	...	...	**B1^2**, 830
p83	—,4-oxo-*	γ-Ketovaleraldehyde. Levulinaldehyde. Levulinic aldehyde. $CH_3COCH_2CH_2CHO$	100.13	...	< −21	186–8^{760} δd 70^{12}	1.0184$_4^{21}$	1.4257^{22}	∞	∞	∞	s	s	...	**B1^2**, 830
Ω p84	**Pentane***	$CH_3(CH_2)_3CH_3$	72.15	...	−129.72	36.07	0.6262$_4^{20}$	1.3575^{20}	δ	∞	∞	∞	∞	chl, hp ∞	**B1^3**, 238
Ω p85	—,1-amino-*	*n*-Amylamine. Pentylamine*. $CH_3(CH_2)_4NH_2$	87.17	...	−55	104.4^{760} 5.9^{10}	0.7547$_4^{20}$	1.4118^{20}	∞	∞	∞	v	v	...	**B4^2**, 641

For explanations, symbols and abbreviations see beginning of table. For structural formulas see end of table.

No.	Name	Synonyms and Formula	Mol. wt.	Color, crystalline form, specific rotation and λ_{max} (log ε)	m.p. °C	b.p. °C	Density	n_D	Solubility w	al	eth	ace	bz	other solvents	Ref.
	Pentane														
p86	**—,2-amino-***	*sec n*-Amylamine. $CH_3(CH_2)_2CH(NH_2)CH_3$	87.17	...	...	91.5^{755}	0.7384^{20}	1.4027^{20}	∞	∞	∞	v	v	...	**B4**^{2}, 643
p87	**—,3-amino-***	$(C_2H_5)_2CHNH_2$	87.17	...	...	91^{760}	0.7487_{4}^{20}	1.4063^{20}	...	s	...	...	...	...	**B4**^{3}, 339
p88	**—,3(amino-methyl)-***	2-Ethyl-*n*-butylamine. $(C_2H_5)_2CHCH_2NH_2$	101.19	...	...	125.3	...	...	...	...	...	...	...	...	**B4**, 192
—	**—,2,2-bis-(hydroxymethyl)-***	*see* **1,3-Propanediol, 2-methyl-2-propyl-***													
p89	**—,3,3-bis(ethyl-sulfonyl)-***	Tetronal. $(C_2H_5)_2C(SO_2C_2H_5)_2$	256.39	lf (dil al)	85	...	...	...	δ	s	s	...	...	...	**B4**, 681
Ω p90	**—,1-bromo-***	Amyl bromide. $CH_3(CH_2)_4Br$	151.05	...	−87.9	129.6^{760} 21^{10}	1.2182_{4}^{20}	1.4447^{20}	i	s	∞	...	s	chl s	**B1**^{3}, 344
Ω p91	**—,2-bromo-(*dl*)***	*sec*-Amyl bromide. $CH_3(CH_2)_2CHBrCH_3$	151.05	...	−95.5	117.4^{760} 58.4^{100}	1.20754_{4}^{20}	1.4413^{20}	i	s	s	...	s	chl s	**B1**^{3}, 345
Ω p92	**—,3-bromo-***	$(C_2H_5)_2CHBr$	151.05	...	−126.2	118.6^{760} 10.8^{10}	1.2124_{4}^{20}	1.4441_{4}^{20}	i	s	s	...	s	chl s	**B1**^{3}, 346
p93	**—,1-bromo-5-fluoro-***	$F(CH_2)_5Br$	169.04	...	...	162	1.3604^{25}	1.4406^{25}	i	v	v	...	...	...	**C51,** 7300
p94	**—,1-bromo-2-methyl-***	$CH_3(CH_2)_2CH(CH_3)CH_2Br$	165.08	...	...	$142–5^{748}$ $51–3^{25}$	1.1624_{4}^{20}	1.4495^{20}	i	...	v	...	...	chl v	**B1**^{3}, 399
Ω p95	**—,1-bromo-3-methyl-***	$CH_3CH_2CH(CH_3)CH_2CH_2Br$	165.08	...	...	148.6–9.4^{766}	1.1829_{4}^{20}	1.4496^{20}	i	...	s	...	...	chl v	**B1**^{3}, 403
p96	**—,1-bromo-4-methyl-***	Isohexyl bromide. $(CH_3)_2CH(CH_2)_3Br$	165.08	...	...	$147–8^{759}$	1.1683^{20}	1.4490	i	...	v	...	...	chl v	**B1**^{3}, 399
p97	**—,2-bromo-2-methyl-***	$CH_3CH_2CH_2CBr(CH_3)_2$	165.08	...	...	$142–3^{760}$ 70^{100}	...	1.442^{23}	d^{h}	...	s	...	...	chl v	**B1**^{3}, 399
p98	**—,3-bromo-3-methyl-***	$(C_2H_5)_2CBrCH_3$	165.08	...	...	129–31 $82–3^{145}$	1.1835^{20}	1.4525^{20}	d^{h}	...	s	...	...	chl v	**B1**^{1}, 54
Ω p99	**—,1-chloro-***	*n*-Amyl chloride. Pentyl chloride. $CH_3(CH_2)_4Cl$	106.60	...	−99	107.8^{760} 4.9^{10}	0.8818_{4}^{20}	1.4127^{20}	i	∞	∞	...	s	chl v	**B1**^{3}, 339
p100	**—,2-chloro-(*d*)***	*sec*-Amyl chloride. $CH_3CH_2CH_2CHClCH_3$	106.60	$[\alpha]_D$ +34.07	−137	96.86^{760}	0.8698_{4}^{20}	1.4069^{20}	i	s	s	...	s	chl v	**B1**^{3}, 340
Ω p101	**—,3-chloro-***	$(C_2H_5)_2CHCl$	106.60	...	−105	97.8^{760}	0.8731_{4}^{20}	1.4082^{20}	i	s	s	...	s	chl v	**B1**^{3}, 340
p102	**—,2-chloro-2,3-dimethyl-***	$CH_3CH_2CH(CH_3)CCl(CH_3)_2$	134.65	...	...	$38–9^{20}$	...	1.4264^{20}	i		s		s	...	**B1**^{3}, 447
p103	**—,2-chloro-2,4-dimethyl-***	$(CH_3)_2CHCH_2CCl(CH_3)_2$	134.65	...	...	$127–8^{733}$d $33–4^{20}$	0.861_{4}^{20}	1.4180^{20}	...	...	v	...	...	...	**B1**^{3}, 451
p104	**—,3-chloro-2,3-dimethyl-***	$CH_3CH_2C(CH_3)ClCH(CH_3)_2$	134.65	...	...	$135–8^{757}$ δd $41–2^{20}$	0.884_{22}^{22}	1.4318^{20}	i	...	s	...	...	chl s	**B1**^{3}, 448
p105	**—,4-chloro-2,2-dimethyl-***	$CH_3CHClCH_2C(CH_3)_3$	134.65	...	...	93^{250}	0.855_{4}^{20}	1.4180^{20}	i	...	...	...	...	chl s	**B1**^{3}, 444
p106	**—,3-chloro-2,2-dimethyl-3-ethyl-***	$(C_2H_5)_2CClC(CH_3)_3$	162.71	...	...	d$53–4^{6}$	...	1.4528^{25}	d	...	s	...	...	chl s	**B1**^{3}, 517
p107	**—,2-chloro-3-ethyl-***	$(C_2H_5)_2CHCHClCH_3$	134.65	...	...	83.5^{100}	0.8951_{25}^{25}	1.4318^{20}	i	...	s	...	...	chl s	**B1**^{2}, 120
p108	**—,3-chloro-3-ethyl-***	$(C_2H_5)_3CCl$	134.65	...	...	$143–4^{760}$ $43–4^{20}$	0.8856_{4}^{20}	1.4400^{20}	...	...	v	...	...	...	**B1**^{3}, 510
p109	**—,3-chloro-3-ethyl-2-methyl-***	$(C_2H_5)_2CClCH(CH_3)_2$	148.68	...	...	150–5d $98–100^{98}$	...	1.4405^{25}	d	s	...	...	...	chl s	**B1**^{3}, 490
p110	**—,1-chloro-5-fluoro-***	$F(CH_2)_5Cl$	124.59	...	...	143.2	1.0325^{25}	1.4120^{23}	i	v	v	...	...	...	**C51,** 7300
p111	**—,2-chloro-2-methyl-***	$CH_3(CH_2)_2CCl(CH_3)_2$	120.62	...	...	$110–1^{734}$d $36–7^{15}$	0.863_{4}^{20}	1.4126^{20}	d^{h}	...	v	...	...	...	**B1**^{3}, 399
p111[1]	**—,2-chloro-4-methyl-***	$(CH_3)_2CHCH_2CHClCH_3$	120.62	...	...	$111–2^{733}$	0.8610_{4}^{20}	1.4113^{20}	d	...	v	...	...	...	**B1**^{3}, 399
p112	**—,3-chloro-2-methyl-***	$CH_3CH_2CHClCH(CH_3)_2$	120.62	...	...	$115–6.5^{752}$ d	...	...	i	...	s	...	...	chl s	**B1**^{2}, 111
p113	**—,3-chloro-3-methyl-***	$(C_2H_5)_2CClCH_3$	120.62	...	...	116^{760} 35^{25}	0.8900_{4}^{20}	1.4210^{20}	d^{h}	...	v	...	s	chl v	**B1**^{3}, 403
p114	**—,3(chloro-methyl)-***	$(C_2H_5)_2CHCH_2Cl$	120.62	...	...	125–7	0.8914_{4}^{20}	1.4222^{20}	i	...	v	...	s	chl v	**B1**^{3}, 403
p115	**—,2-chloro-2,4,4-trimethyl-***	$(CH_3)_2CClCH_2C(CH_3)_3$	148.68	...	−26	145–50d 44^{16}	0.8746^{20}	1.4308^{20}	d	s	...	...	...	alk d^{h}	**B1**^{3}, 498
p116	**—,1,5-diamino-***	Cadaverine. Pentamethylene diamine. $H_2N(CH_2)_5NH_2$	102.18	...	9	178–80	0.867^{25}	1.4561^{25}	s	s	δ	...	...	...	**B4**^{3}, 588
Ω p117	**—,1,5-dibromo-***	Pentamethylene bromide. $Br(CH_2)_5Br$	229.95	...	−39.5	222.3^{760} 98.6^{10}	1.7018_{4}^{20}	1.5126^{20}	i	...	...	...	s	chl s	**B1**^{3}, 374
p118	**—,1,2-dichloro-***	$CH_3CH_2CH_2CHClCH_2Cl$	141.04	...	...	148.4–.8 $58–9^{28}$	1.0872_{4}^{20}	1.4485^{20}	i	s	...	...	...	chl v alk d	**B1**^{3}, 341
p119	**—,1,3-dichloro-***	$CH_3CH_2CHClCH_2CH_2Cl$	141.04	...	...	80.4^{60}	1.0834_{4}^{20}	1.4485^{20}	i	...	...	v	...	chl v	**B1**^{3}, 342
p120	**—,1,4-dichloro-***	$CH_3CHCl(CH_2)_2CH_2Cl$	141.04	...	...	$161–3^{760}$ $58–60^{15}$	1.0840_{4}^{20}	1.4503^{20}	d^{h}	...	...	s	...	chl s	**B1**^{3}, 342
Ω p121	**—,1,5-dichloro-***	$Cl(CH_2)_5Cl$	141.04	...	−72.8	180^{760} 59^{10}	1.1006_{4}^{20}	1.4564^{20}	i	s	s	...	s	chl s	**B1**^{3}, 342
p122	**—,2,2-dichloro-***	$CH_3CH_2CH_2CCl_2CH_3$	141.04	...	...	128–9 (cor) $36–7^{20}$	1.040^{20}	1.434^{20}	i	...	s	...	s	chl s	**B1**^{3}, 342
p123	**—,2,4-dichloro-***	$CH_3CHClCH_2CHClCH_3$	141.04	...	...	$147–50^{760}$ 62^{12}	1.0634_{4}^{15}	1.447^{18}	i	...	v	...	s	chl v alk d	**B1**^{3}, 343

For explanations, symbols and abbreviations see beginning of table. For structural formulas see end of table.

No.	Name	Synonyms and Formula	Mol. wt.	Color, crystalline form, specific rotation and λ_{max} (log ε)	m.p. °C	b.p. °C	Density	n_D	Solubility w	al	eth	ace	bz	other solvents	Ref.
	Pentane														
p124	**—,3,3-dichloro-***	$CH_3CH_2CCl_2CH_2CH_3$	141.04			$131–2^{750}$ 32^{14}	1.053^{20}	1.442^{20}	i				s	chl s	**B1**^{3}, 343
p125	**—,1,2-dichloro-4,4-dimethyl-***	$(CH_3)_3CCH_2CHClCH_2Cl$	169.10			$173–5^{745}$ $58–9^{12}$	1.0259_4^{20}	1.4489^{20}	i				s	chl s	**B1**^{3}, 445
p126	**—,1,5-dichloro-3,3-dimethyl-***	$(ClCH_2CH_2)_2C(CH_3)_2$	169.10			135^{80} $58–9^{8}$	1.0563_4^{20}	1.4652^{20}	i	s^h				chl s	**B1**^{3}, 454
p127	**—,2,4-dichloro-2,4-dimethyl-***	$(CH_3)_2CClCH_2CCl(CH_3)_2$	169.10		23–4	$51.5–.7^{8}$	1.0292_4^{20}	1.4537^{20}	i	δ				chl s	**B1**^{3}, 451
p128	**—,3,3-dichloro-2,4-dimethyl-***	$(CH_3)_2CHCCl_2CH(CH_3)_2$	169.10			118–20δd	0.9513^{9}		i		s			chl s	**B1**, 158
p129	**—,3,3-diethyl-***	Tetraethylmethane. $C(C_2H_5)_4$	128.26		−33.110	146.17^{760} 30.71^{10}	0.75359_4^{20}	1.4206^{20}	i		s		s	os s	**B1**^{3}, 518
Ω p130	**—,1,5-diiodo-***	$I(CH_2)_5I$	323.94		9	149^{20} $101–2^{3}$	2.1903^{15}	1.6046^{15}	i		s			chl s	**B1**^{3}, 350
Ω p131	**—,2,2-dimethyl-***	$CH_3CH_2CH_2C(CH_3)_3$	100.21		−123.82	79.197^{760}	0.6739_4^{20}	1.3822^{20}	i	s	s	∞	∞	chl, hp ∞	**B1**^{3}, 442
Ω p132	**—,2,3-dimethyl-***	$CH_3CH_2CH(CH_3)CH(CH_3)_2$	100.21			89.8	0.6951_4^{20}	1.3919^{20}	i	s	s	∞	∞	chl, hp ∞	**B1-**, 445
Ω p133	**—,2,4-dimethyl-***	$(CH_3)_2CHCH_2CH(CH_3)_2$	100.21		−119.24	80.5^{760}	0.6727_4^{20}	1.3815^{20}	i	s	s	∞	∞	chl, hp ∞	**B1**^{3}, 449
Ω p134	**—,3,3-dimethyl-***	$CH_3CH_2C(CH_3)_2CH_2CH_3$	100.21		−134.46	86.064	0.6936_4^{20}	1.3909^{20}	i	s	s	∞	∞	chl, hp ∞	**B1**^{3}, 452
p135	**—,1,5-dinitro-***	$O_2NCH_2(CH_2)_3CH_2NO_2$	162.15	λ^{al} 275 (1.69)		$134^{1.2}$		1.461^{20}	i				s		**B1**^{3}, 350
p136	**—,1,1-diphenyl-***	$CH_3(CH_2)_3CH(C_6H_5)_2$	224.35	λ^{iso} 262 (2.7), 269 (2.6)	−12.06	307.89	0.96594^{20}	1.5511^{20}	i						**B5**^{2}, 523
p137	**—,1,5-diphenyl-***	$C_6H_5(CH_2)_5C_6H_5$	224.35	λ^{iso} 254 (2.6), 261.5 (2.7), 268 (2.6)		330.6 (cor) $187–9^{10}$	0.9814_0^{19}	1.559^{19}	i						**B5**^{3}, 1895
p138	**—,1,5-dithio-cyanato-**	$NCS(CH_2)_5SCN$	186.30	yesh		221–2									**B3**^{1}, 72
p139	**—,1,2-epoxy-2,4,4-trimethyl-***		128.22		−64	140.9 $20^{5.4}$	0.8287_{20}^{20}	1.4097^{20}	δ		s		s		**B1**^{3}, 850
p140	**—,3-ethyl-***	Triethylmethane. $(C_2H_5)_3CH$	100.21		−118.604	93.5^{760}	0.6982_4^{20}	1.3934^{20}	i	s	s	∞	∞	chl, hp ∞	**B1**^{3}, 441
p141	**—,3-ethyl-2-methyl-***	$(CH_3)_2CHCH(C_2H_5)_2$	114.23		−114.960	115.65^{760} 9.5^{10}	0.7193_4^{20}	1.4040^{20}	i	∞	s	∞	∞	chl, hp ∞	**B1**^{3}, 489
p142	**—,3-ethyl-3-methyl-***	$(C_2H_5)_3CCH_3$	114.23		−90.87	118.26^{760} 9.9^{10}	0.7274_4^{20}	1.4078_4^{20}	i	∞	s	∞	∞	chl, hp ∞	**B1**^{3}, 490
p143	**—,1-fluoro-***	*n*-Amyl fluoride. $CH_3(CH_2)_4F$	90.14		−120	62.8^{760}	0.7907_4^{20}	1.3591^{20}	i	v	v				**B1**^{3}, 338
Ω p144	**—,1-iodo-***	*n*-Amyl iodide $CH_3(CH_2)_4I$	198.05		−85.6	157^{760} 39.31^{10}	1.5161_4^{20}	1.4959^{20}	δ	s	s				**B1**^{3}, 348
p145	**—,2-iodo-(*dl*)***	$CH_3(CH_2)_2CHICH_3$	198.05			$144–5^{760}$	1.5096_4^{20}	1.4961^{20}	i		s	s	s	os s	**B1**^{3}, 349
p146	**—,3-iodo-***	$(C_2H_5)_2CHI$	198.05			145–6 68^{50}	1.5176_4^{20}	1.4974^{20}	i		s	s	s	os s	**B1**^{3}, 349
p147	**—,1-isocyano-**	*n*-Amylcarbylamine. Amyl isocyanide. $CH_3(CH_2)_4NC$	97.16		−51.1	155.5 (130^{760}) 50^{45}	0.806_4^{20}		i	s					**C42**, 869
Ω p148	**—,2-methyl-***	$CH_3CH_2CH_2CH(CH_3)_2$	86.18		−153.67	60.271	0.6532_4^{20}	1.3715^{20}	i	s	s	∞	∞	hp, chl ∞	**B1**^{3}, 396
Ω p149	**—,3-methyl-***	$(C_2H_5)_2CHCH_3$	86.18			63.282	0.6645_4^{20}	1.3765^{20}	i	s	∞	∞	∞	hp, chl ∞	**B1**^{3}, 401
p150	**—,3-nitro-***	$(C_2H_5)_2CHNO_2$	117.15			153–5	0.957_4^{0}		i	s	s	s			**B1**^{3}, 350
p151	**—,2,2,4,4-tetra-methyl-***	$(CH_3)_3CCH_2C(CH_3)_3$	128.26		−66.54	122.68^{760} 12.5^{10}	0.7195_4^{20}	1.4069^{20}	i	v			v	os v	**B1**^{3}, 519
p152	**—,2,2,3-tri-methyl-***	$CH_3CH_2CH(CH_3)C(CH_3)_3$	114.23		−112.27	109.841^{760} 3.9^{10}	0.7161_4^{20}	1.4030_4^{20}	i	∞	∞	∞	s	chl, hp ∞	**B1**^{3}, 491
Ω p153	**—,2,2,4-tri-methyl-***	Isooctane. $(CH_3)_2CHCH_2C(CH_3)_3$	114.23		−107.38	99.238^{760} -4.3^{10}	0.6919_4^{20}	1.3915^{20}	i	∞	s	∞	∞	chl, hp ∞	**B1**^{3}, 492
p154	**—,2,3,3-tri-methyl-***	$CH_3CH_2C(CH_3)_2CH(CH_3)_2$	114.23		−100.7	114.76^{760} 6.9^{10}	0.7262_4^{20}	1.4075^{20}	i	v	∞	∞	∞	chl, hp ∞ os v	**B1**^{3}, 499
Ω p155	**—,2,3,4-tri-methyl-***	$(CH_3)_2CHCH(CH_3)CH(CH_3)_2$	114.23		−109.21	113.467^{760} 7.1^{10}	0.7191_4^{20}	1.4042^{20}	i	v	∞	∞	∞	chl, hp ∞ to, os v	**B1**^{3}, 500
p156	**1,5-Pentanedial***	Glutaraldehyde. Glutaric aldehyde. $OCH(CH_2)_3CHO$	100.13			187–9d $71–2^{10}$			∞	∞			s		**B1**^{2}, 831
Ω p157	—,dioxime	Glutaraldoxime. $HON{:}CH(CH_2)_3CH{:}NOH$	130.15	nd (w or Py)	178	sub			s^h					Py s^h	**B1**^{2}, 831
Ω p158	**Pentanedioic acid***	Glutaric acid. $HO_2C(CH_2)_3CO_2H$	132.13	nd (bz)	99	302–4δd 200^{20}	1.424_4^{25}	1.4188^{106}	v	v	v		i	chl, con sulf s lig δ	**B2**^{3}, 1685
Ω p159	—,dichloride	Glutaryl chloride. $ClCO(CH_2)_3COCl$	169.01			216–8 (cor) 100^{15}	1.324_4^{20}	1.4728^{20}	d	d	s				**B2**^{3}, 1691
Ω p160	—,diethyl ester*	Diethyl glutarate. $C_2H_5O_2C(CH_2)_3CO_2C_2H_5$	188.23	syr	−24.1 (cor)	236.5–7 $103–4^{7}$	1.0220_4^{20}	1.4241^{20}	δ	δ	s				**B2**^{3}, 1689
Ω p161	—,dimethyl ester*	Dimethyl glutarate. $CH_3O_2C(CH_2)_3CO_2CH_3$	160.17		−42.5	214^{751} 109^{21}	1.0876_4^{20}	1.4242^{20}		v	v				**B2**^{3}, 1688
Ω p162	—,dinitrile	Glutaronitrile. Pentanedi-nitrile*. $NC(CH_2)_3CN$	94.12		−29	286 160.4^{22}	0.99112_4^{15}	1.4295^{20}	s	s	i			chl s CS_2 i	**B2**^{3}, 1692
p163	—,diphenyl ester*	Diphenyl glutarate. $C_6H_5O_2C(CH_2)_3CO_2C_6H_5$	284.32	nd (lig)	54	236.5^{15}			i	s		s		os, lig v	**B6**, 156
p164	—,piperazinium salt	$2[HO_2C(CH_2)_3CO_2H].C_4H_{10}N_2$	350.37		152				s	s^h	i				**Am 56**, 1759
Ω p165	**—,2-acetyl-,** diethyl ester	$CH_3COCH(CO_2C_2H_5)CH_2CH_2CO_2C_2H_5$	230.26			271–2δd $110–2^{0.1}$	1.0712_4^{20}	1.4420^{19}	δ	s	s				**B3**^{2}, 488
—	**—,2-amino-***	*see* **Glutamic acid**													
Ω p166	**—,2,2-dimethyl-***	$HO_2CCH_2CH_2C(CH_3)_2CO_2H$	160.17	nd (con HCl, bz-lig)	85				v	v				chl, aa s lig δ	**B2**^{3}, 1750

For explanations, symbols and abbreviations see beginning of table. For structural formulas see end of table.

No.	Name	Synonyms and Formula	Mol. wt.	Color, crystalline form, specific rotation and λ_{max} (log ε)	m.p. °C	b.p. °C	Density	n_D	Solubility w	al	eth	ace	bz	other solvents	Ref.
	Pentanedioic acid														
Ω p167	—,3,3-dimethyl-*	$HO_2CCH_2C(CH_3)_2CH_2CO_2H$	160.17	mcl pl, nd (bz)	103–4	$126–7^{4.5}$	1.4278_4^{20}	...	v	v	v	...	δ s^h	lig i	**B2**^{3}, 1758
p168	—,3[2(3,5-dimethyl-2-oxocyclohexyl)2-hydroxyethyl]-, imide	Actidione. Cycloheximide.	281.36	wh pl (w) $[\alpha]_D^{29}$ −3.38 (al) λ^{al} 235 sh (2.3), 255 sh (2.6), 290 (1.5)	119.5–21	...	...	...	δ	s	s	s	...	...	**Am 69,** 474
p169	—,2-ethyl-3-methyl-*	$HO_2CCH_2CH(CH_3)CH(C_2H_5)CO_2H$	175.20	(i) pr (w) (ii) pr (chl-lig)	(i) 100–1 (ii) 88	...	...	...	v	...	v	...	...	...	**B2**^{3}, 1779
p170	—,2-ethyl-4-methyl (*dl*)*	Paramethylethylglutaric acid. $HO_2CCH(CH_3)CH_2CH(C_2H_5)CO_2H$	175.20	nd (w)	107	...	...	...	s^h	...	v	...	i	os v lig s CS_2 i	**B2**^{3}, 1775
p171	—,—(*meso*)*	Mesomethylethylglutaric acid. $HO_2CCH(CH_3)CH_2CH(C_2H_5)CO_2H$	175.20	nd (w)	83.5–4.5	...	...	...	δ s^h	...	v	...	...	os v lig δ xy i	**B2**^{3}, 1775
Ω p172	—,3-ethyl-3-methyl-*	$HO_2CCH_2C(CH_3)(C_2H_5)CH_2CO_2H$	175.20	nd (w or bz), pl (bz-peth)	87	260^{740}	...	...	s^h	s	s	...	s^h	...	**B2**^{3}, 1781
p173	—,2-hydroxy-(*d*)*	$HO_2CCH_2CH_2CH(OH)CO_2H$	148.13	cr (eth), $[\alpha]_D^{19}$ +1.76 (w, c = 3)	72	...	...	...	...	...	...	...	...	...	**B3**^{3}, 929
p174	—,—(*dl*)*	$HO_2CCH_2CH_2CH(OH)CO_2H$	148.13	pr (AcOEt)	72	...	...	...	s	s	...	...	...	...	**B3**^{2}, 293
p175	—,—(*l*)*	$HO_2CCH_2CH_2CH(OH)CO_2H$	148.13	$[\alpha]_D$ −1.98 (w)	72–3	...	...	...	s	s	...	...	...	...	**B3**^{2}, 93
Ω p177	—,2-oxo-*	2-Oxoglutaric acid. α-Ketoglutaric acid. $HO_2CCH_2CH_2COCO_2H$	146.10	cr (ace-bz) $\lambda^{w, pH=8}$ 315 (1.49)	115–6	...	...	...	v	v	v	s	...	...	**B3**^{2}, 481
Ω p178	—,3-oxo-*	Acetonedicarboxylic acid. β-Ketoglutaric acid. $HO_2CCH_2COCH_2CO_2H$	146.10	nd (al, AcOEt), rh (w)	135d (138)	d	...	...	s d^h	s	δ	...	i	AcOEt s^h chl, lig i	**B3**^{2}, 482
Ω p179	—,diethyl ester*	$C_2H_5O_2CCH_2COCH_2CO_2C_2H_5$	202.21	...	...	250 140^{13}	1.113_4^{20}	...	δ	∞	...	...	...	...	**B3**^{2}, 484
p181	—,2-phenyl-*	$C_6H_5CH(CO_2H)CH_2CH_2CO_2H$	208.22	cr (bz or eth-peth)	82–3	...	...	...	...	...	s^h	...	s^h	...	**B9**, 877
p182	—,—,anhydride*	...	190.20	nd (eth)	95	$218–30^{13}$	...	...	...	s^h	...	...	...	...	**B17**, 494
p183	—,3-phenyl-, anhydride*	...	190.20	cr (bz)	105	$217–9^{15}$	...	...	d^h	...	s	...	s	chl s peth i	**B17**^{2}, 476
p184	—,2,3,4-trihydroxy-(*d*)*	*D*-Arabotrihydroxyglutaric acid. $HO_2C(CHOH)_3CO_2H$	180.13	cr (ace), pl (w) $[\alpha]_D^{20}$ +22.2	128	...	...	...	v	s	i	s^h	...	chl i	**B3**^{1}, 192
p185	—,—(*dl*)*	$HO_2C(CHOH)_3CO_2H$	180.13	cr (ace)	154.5d	...	...	...	v	v	i	s	...	chl i	**B3**, 553
p186	**1,2-Pentanediol (*d*)***	1,2-Amylene glycol. $CH_3(CH_2)_2CH(OH)CH_2OH$	104.15	$[\alpha]_D^{20}$ +0.95	...	210.5–211.5^{751} $99–102^{13}$	0.9802_{20}^{20}	1.4412^{19}	...	...	...	...	...	...	**B1**^{3}, 2190
p187	**1,4-Pentanediol (*dl*)***	γ-Pentylene glycol. $CH_3CH(OH)CH_2CH_2CH_2OH$	104.15	...	...	220^{713} $124–6^{10}$	0.9883_4^{20}	1.4452^{25}	∞	∞	...	...	...	chl ∞ lig i	**B1**^{3}, 2191
Ω p188	**1,5-Pentanediol***	Pentamethylene glycol. $HO(CH_2)_5OH$	104.15	...	−18	260 (240) $137–8^{12}$	0.9939_{20}^{20}	1.4494_4^{20}	s	s	δ	...	δ	...	**B1**^{3}, 2192
Ω p189	**2,3-Pentanediol***	β-*n*-Amylene glycol. $CH_3CH_2CH(OH)CH(OH)CH_3$	104.15	...	...	187.5 97^{17}	0.9800_0^{19}	1.4402^{25}	s	s	δ	...	...	...	**B1**^{3}, 2194
p190	**1,3-Pentanediol, 2,2-dimethyl-***	$CH_3CH_2CH(OH)C(CH_3)_2CH_2OH$	132.21	(eth)	60–3	212–4 119^{21}	...	...	...	s	...	...	...	...	**B1**, 190
p191	**1,5-Pentanediol, 2,2-dimethyl-***	$HO(CH_2)_3C(CH_3)_2CH_2OH$	132.21	...	...	130^{12}	...	...	...	s	s	...	...	...	**B1**^{1}, 254
Ω p192	**2,4-Pentanediol, 2-methyl-***	$CH_3CH(OH)CH_2C(OH)(CH_3)_2$	188.18	...	...	197^{760}	0.9254_4^{17}	1.4250^{20}	s	s	s	...	...	...	**B1**^{1}, 252
Ω p193	—,3-methyl-*	$CH_3CH(OH)CH(CH_3)CH(OH)CH_3$	118.18	...	...	$211–2^{760}$ 91^3	0.9640^{20}	1.4433^{20}	s	s	...	...	...	...	**B1**^{3}, 2209
Ω p194	**1,2-Pentanediol, 2,4,4-trimethyl-***	$(CH_3)_3CCH_2C(OH)(CH_3)CH_2OH$	146.23	pr or pl (bz)	62–3	...	...	...	s	v	v	...	s^h	...	**B1**^{3}, 2226
Ω p195	**1,3-Pentanediol, 2,2,4-trimethyl-***	$(CH_3)_2CHCH(OH)C(CH_3)_2CH_2OH$	146.23	pl (bz)	51.8–2.2	234^{737} $81–2^1$	0.937_{15}^{15}	1.4513^{15}	δ	v	v	...	s^h	...	**B1**^{3}, 2225
p196	**1,4-Pentanediol, 2,2,4-trimethyl-***	$(CH_3)_2C(OH)CH_2C(CH_3)_2CH_2OH$	146.23	cr (eth)	86	209–11 $114–5^{13}$	...	...	...	v	s	...	...	...	**B1**, 493
Ω p197	**2,3-Pentanediol, 2,4,4-trimethyl-***	$CH_3C(CH_3)_2CH(OH)C(CH_3)(OH)CH_3$	146.23	mcl pr (lig)	65–6	...	...	...	...	v	s	...	...	lig s^h	**B1**^{3}, 2225
p198	**2,3-Pentanedione***	Acetylpropionyl. Ethyl methylglyoxal. $CH_3CH_2COCOCH_3$	100.13	dk ye liq	...	108	0.9565_4^{19}	1.4014^{19}	s	∞	∞	∞	...	...	**B1**^{3}, 3112
p199	—,dioxime*	2,3-Diisonitrosopentane. $CH_3CH_2C(:NOH)C(:NOH)CH_3$	130.15	ye nd (al), pl (to, al)	172–3	sub	...	...	i	s	δ	...	...	to s^h	**B1**^{2}, 831
p200	—,2-oxime*	Isonitrosodiethyl ketone. $CH_3CH_2COC(:NOH)CH_3$	115.13	lf (dil al)	69–72	...	...	...	...	v	v	...	...	...	**B1**, 776
p201	—,3-oxime*	Isonitrosopropyl methyl ketone. $CH_3CH_2C(:NOH)COCH_3$	115.13	pl (lig)	58–9	183–7 part d	...	...	δ	v	v	...	...	chl v alk s (ye)	**B1**^{2}, 831
Ω p202	**2,4-Pentanedione***	Acetylacetone. Diacetylmethane. $CH_3COCH_2COCH_3$	100.13	λ^{iso} 271 (4.00), λ^{chl} 274 (4.03)	−23	139^{746}	0.9721_4^{25}	1.4494^{20}	v	∞	∞	∞	...	chl ∞	**B1**, 777
p203	—,dioxime*	2,4-Diisonitrosopentane. $CH_3C(:NOH)CH_2C(:NOH)CH_3$	130.15	pr (eth)	149–50	...	...	...	δ	s	δ	...	...	...	**B1**^{2}, 838
p204	—,monoimide	$CH_3COCH_2C(:NH)CH_3$	99.13	...	43	209	...	...	s d^h	...	v	...	...	...	**B1**^{2}, 838

For explanations, symbols and abbreviations see beginning of table. For structural formulas see end of table.

No.	Name	Synonyms and Formula	Mol. wt.	Color, crystalline form, specific rotation and λ_{max} (log ε)	m.p. °C	b.p. °C	Density	n_D	Solubility: w	al	eth	ace	bz	other solvents	Ref.
	2,4-Pentanedione														
p205	—,3,3-dimethyl-*	$CH_3COC(CH_3)_2COCH_3$	128.17	...	19	173 58^{10}	0.9575^{20}_{4}	1.4306^{20}	...	...	s	...	...	...	B1³, 3137
p206	—,3-ethyl-*	$CH_3COCH(C_2H_5)COCH_3$	128.17	...	...	177–80 $69–70^{13}$	0.9531^{19}_{4}	1.4408^{19}_{4}	...	s	s	...	...	chl s	B1², 844
p207	**1,5-Pentanedione, 1,2,3,4,5-pentaphenyl-***	Benzamaron. Benzylidene-bis-desoxybenzoin. $C_6H_5CH[CH(C_6H_5)COC_6H_5]_2$	480.61	lf (w)	218–9	...	...	...	δ	δ	...	...	δ s^h	...	B7², 803
p208	**1,4-Pentanedione, 1-phenyl-***	γ-Oxovalerophenone. Phenacylacetone. $C_6H_5COCH_2CH_2COCH_3$	176.22	ye oil	...	162^{12}	...	1.5250^{30}	s^h	...	...	s	...	alk i	B7¹, 368
Ω p209	**1-Pentanethiol***	n-Amyl mercaptan. n-Thioamyl alcohol. $CH_3(CH_2)_3CH_2SH$	104.22	λ^{iso} 230 (2.2)	−75.7	126.638^{460} 19.505^{10}	0.84209^{20}_{4}	1.4469^{20}	i	∞	∞	...	...	...	B1³, 1607
p210	**2-Pentanethiol***	(α-Methylbutyl) mercaptan. $CH_3CH_2CH_2CH(SH)CH_3$	104.22	λ^{iso} 227 (2.0)	−169	112.9^{760} 63.9^{150}	0.8327^{20}_{4}	1.4412^{20}	...	s	...	...	...	lig s	B1³, 1615
p211	**3-Pentanethiol***	(α-Ethylpropyl) mercaptan. $(C_2H_5)_2CHSH$	104.22	...	−110.8	105	0.8410^{20}_{4}	1.4447^{20}	...	s	...	...	...	...	B1³, 1619
p212	**1,3,5-Pentanetricarboxylic acid, 3-acetyl-**, 3-ethyl ester, 1,5-dinitrile*	$CH_3COC(CH_2CH_2CN)_2CO_2C_2H_5$	236.27	cr	83	$190–200^{2}$	...	...	i	s^h	...	...	...	...	B3², 512
p213	**1,2,3-Pentanetriol***	$CH_3CH_2CH(OH)CH(OH)CH_2OH$	120.15	syr	...	192^{63}	1.0851^{34}_{0}	...	∞	∞	s	...	...	...	B1³, 2346
Ω p214	**Pentanoic acid***	Propylacetic acid. Valerianic acid. Valeric acid. $CH_3(CH_2)_3CO_2H$	102.13	...	−33.83	186.05^{760} 82.7^{10} $86–8^{15}$	0.9391^{20}_{4}	1.4085^{20}_{4}	s	s	s	...	...	...	B2³, 664
Ω p215	—,amide	Pentanamide.* n-Valeramide. $CH_3(CH_2)_3CONH_2$	101.15	mica-like mcl pl (peth, al)	106 (114–6)	...	1.023 0.8735^{110}	1.4183^{110}	v	v	v	...	...	...	B2³, 674
p216	—,—,N,N-dimethyl-	$CH_3(CH_2)_3CON(CH_3)_2$	129.20	...	−51	141^{100}	0.8962^{25}_{4}	1.4419^{25}	∞	∞	∞	...	...	...	Am 59, 4012
p217	—,—,N-phenyl-	Valeranilide. $CH_3(CH_2)_3CONHC_6H_5$	177.25	mcl pr (al), cr (peth)	63	...	...	...	...	...	s	...	...	...	C51, 251
p218	—,anhydride*	Valeric anhydride. $[CH_3(CH_2)_3CO]_2O$	186.25	...	−56.1	218^{754} 111^{15}	0.924^{20}_{4}	1.4171^{26}	d^h	s	v	...	...	...	B2³, 674
Ω p219	—,butyl ester*	n-Butyl valerate. $CH_3(CH_2)_3CO_2(CH_2)_3CH_3$	158.24	...	−92.8	185.8^{760} $84–5^{8}$	0.8710^{15}_{4}	1.4128^{20}	δ	s	s	...	...	...	B2³, 671
p220	—,sec-butyl ester (d)*	d-sec-Butyl valerate. $CH_3(CH_2)_3CO_2CH(CH_3)C_2H_5$	158.24	$[\alpha]^{20}_{D}$ +20.72	...	174.5 67^{18}	0.8605^{20}_{4}	1.4070^{20}	i	s	s	...	s	Py s	B2¹, 131
p221	—,chloride	Pentanoyl chloride*. Valeryl chloride. $CH_3(CH_2)_3COCl$	120.58	...	−110.0	$107–10^{756}$	1.0155^{15}	1.4200^{20}	...	...	...	...	...	...	B2³, 674
Ω p222	—,ethyl ester*	Ethyl valerate. $CH_3(CH_2)_3CO_2C_2H_5$	130.19	...	−91.2	144.6^{736} 50.5^{29}	0.8770^{20}_{4}	1.4120^{20}	i	∞	∞	...	...	...	B2³, 669
p223	—,furfuryl ester*	Furfuryl valerate.	182.22	...	...	$228–9^{764}$ $82–3^{1}$	1.0284^{20}_{4}	...	i	s	s	...	...	...	B17², 115
p224	—,heptyl ester*	n-Heptyl valerate. $CH_3(CH_2)_3CO_2(CH_2)_6CH_3$	200.33	...	−46.4	245.2	0.8623^{20}	1.42536^{15}_{He}	i	s	s	s	...	os s	B2, 301
p225	—,hexyl ester*	n-Hexyl valerate. $CH_3(CH_2)_3CO_2(CH_2)_5CH_3$	186.30	...	−63.1	226.3	0.8635^{20}	1.42286^{15}	i	s	s	s	...	os s	B2, 301
p226	—,isobutyl ester	Isobutyl valerate. $CH_3(CH_2)_3CO_2CH_2CH(CH_3)_2$	158.24	...	...	179^{760}	0.8625^{25}_{4}	1.4046^{20}	i	∞	s	s	...	...	B2³, 671
p227	—,isopropyl ester*	Isopropyl valerate. $CH_3(CH_2)_3CO_2CH(CH_3)_2$	144.22	...	...	153.5 (164.9)	0.8579^{20}_{4}	1.4061^{20}	i	s	s	s	...	...	B2², 266
Ω p228	—,methyl ester*	Methyl valerate. $CH_3(CH_2)_3CO_2CH_3$	116.16	...	...	126.5^{750}	0.8947^{20}_{4}	1.4003^{20}	δ	∞	∞	s	...	...	B2³, 669
Ω p229	—,nitrile	Pentanonitrile*. Valeronitrile. $CH_3(CH_2)_3CN$	83.13	λ^{cy} 341 (2.1)	−96	141.3^{760} 30.9^{10}	0.8008^{20}_{4}	1.3971^{20}	...	...	s	s	s	...	B2², 267
p230	—,octyl ester*	n-Octyl valerate. $CH_3(CH_2)_3CO_2(CH_2)_7CH_3$	214.35	...	−42.3	261.6	0.8615^{20}	1.4273^{15}	i	s	s	s	...	os s	B2, 301
Ω p231	—,pentyl ester*	n-Amyl valerate. $CH_3(CH_2)_3CO_2(CH_2)_4CH_3$	172.27	...	−78.8	203.7^{760} 103^{23}	0.8638^{20}_{4}	1.4164^{20}	δ	∞	∞	...	...	...	B2³, 671
p232	—,p-phenylphenacyl ester	...	296.37	cr (dil al)	67.5–8.0	...	...	...	δ	s	...	...	...	...	Am 52, 3715
p233	—,piperazinium salt	$2C_4H_9CO_2H.C_4H_{10}N_2$	290.41	wh (diox)	112–5.3	...	...	...	s	s	i	...	...	diox s^h	Am 56, 1759
p234	—,propyl ester*	n-Propyl valerate. $CH_3(CH_2)_3CO_2CH_2CH_2CH_3$	144.22	...	...	167.5	0.8699^{20}_{4}	1.4065^{20}	i	s	s	...	...	chl s	B2³, 670
p235	—,2-acetyl-, ethyl ester	Ethyl propylacetoacetate. $CH_3COCH(C_3H_7)CO_2C_2H_5$.	172.23	...	...	$210–2^{749}$ $90.2–.5^{10}$	0.9682^{15}_{4}	1.4271^{20}	i	s	s	...	...	...	B3², 441
p236	—,2-amino- (D, +)*	$CH_3CH_2CH_2CH(NH_2)CO_2H$	117.15	lf (w) $[\alpha]^{23}_{D}$ +32 (6N HCl, c = 2)	ca. 307	sub	...	...	s v^h	i	i	...	...	chl, AcOEt, lig i	B4³, 1532
Ω p237	—,—(DL)*	Norvaline. $CH_3CH_2CH_2CH(NH_2)CO_2H$.	117.15	lf (al, w)	303 (sealed tube)	sub	...	...	s v^h	i	i	...	...	chl, AcOEt, lig i	B4², 843
p238	—,—(L, −)*	$CH_3CH_2CH_2CH(NH_2)CO_2H$	117.15	cr (dil al, w) $[\alpha]^{20}_{D}$ −23 (20% HCl, c = 10)	ca. 305	sub	...	...	v^h	i	i	...	...	chl, AcOEt, lig i	B4³, 1549

For explanations, symbols and abbreviations see beginning of table. For structural formulas see end of table.

No.	Name	Synonyms and Formula	Mol. wt.	Color, crystalline form, specific rotation and λ_{max} (log ε)	m.p. °C	b.p. °C	Density	n_D	Solubility w	al	eth	ace	bz	other solvents	Ref.
	Pentanoic acid														
p239	—,**4-amino-(*D*)***	$CH_3CH(NH_2)CH_2CH_2CO_2H$	117.15	cr (dil al), $[\alpha]_D^{20}$ +12.0 (w, p = 10)	214 (cor)	...	...	...	v	i	i	...	i	lig i	**B4^2**, 843
p240	—,—(*DL*)*	$CH_3CH(NH_2)CH_2CH_2CO_2H$	117.15	cr (w)	199 (214)	d	...	...	v	i	i	...	i	lig i	**B4^2**, 843
Ω p241	—,**5-amino-***	$H_2N(CH_2)_4CO_2H$	117.15	lf (dil al)	157–8d	d	...	...	s	δ	i	...	i	lig i	**B4^2**, 844
—	—,—,lactam*	*see* **2-Piperidone**													
—	—,**2-amino-5-guanidino-***	*see* **Arginine**													
—	—,**2-amino-4-methyl-***	*see* **Leucine**													
—	—,**2-amino-3-sulfo-***	*see* **Cysteic acid**													
Ω p242	—,**2-bromo-(*dl*)***	$CH_3CH_2CH_2CHBrCO_2H$	181.04	...	...	118^{12}	1.381^{20}	...	δ	v	s	...	...	...	**B2^3**, 680
Ω p243	—,—,ethyl ester*	$CH_3CH_2CH_2CHBrCO_2C_2H_5$	209.09	...	...	190–2 92–4^{18}	1.226$_4^{18}$	1.4496^{20}	i	s	s	...	...	...	**B2^3**, 681
p244	—,**2-bromo-4-methyl-(*d*)***	$(CH_3)_2CHCH_2CHBrCO_2H$	195.06	oil, $[\alpha]_D^{20}$ +29.8 (eth, c = 6) 26.8 (20% dil al, c = 8)	...	131–1.5^{16}	...	...	...	v	v	...	...	...	**B2^2**, 290
p245	—,—(*l*)*	$(CH_3)_2CHCH_2CHBrCO_2H$	195.06	oil, $[\alpha]_D^{20}$ −12.1	...	94$^{0.2-0.4}$ (cor)	...	...	...	v	v	...	...	...	**B2^2**, 291
p246	—,—,amide(*d*)*	$(CH_3)_2CHCH_2CHBrCONH_2$	194.08	(w), $[\alpha]_D^{20}$ −48.3 (al, c = 5.9)	118 (cor)	...	...	...	v^h	s	s	...	...	chl, AcOEt s	**B2^2**, 291
p247	—,—,ethyl ester*	$(CH_3)_2CHCH_2CHBrCO_2C_2H_5$	223.12	pa ye	...	202–4 86–7^{11}	...	...	...	...	...	...	...	...	**B2^1**, 142
p248	—,**2-chloro-(*dl*)***	$CH_3CH_2CH_2CHClCO_2H$	136.58	...	−15	222^{763} 133–5^{30}	1.141^{13}	1.4481^{11}	i	s	s	...	...	...	**B2**, 302
p249	—,—,chloride(*dl*)	$CH_3CH_2CH_2CHClCOCl$	155.03	...	...	155–7^{763} 61–2^{28}	1.1765^{20}	1.4465^{20}	d	d	s	...	...	...	**B2^3**, 677
p250	—,—,ethyl ester	$CH_3CH_2CH_2CHClCO_2C_2H_5$	164.63	...	...	185^{752}	1.040^{12}	1.4307^{11}	i	s	s	...	...	...	**B2**, 302
p251	—,**3-chloro-***	$CH_3CH_2CHClCH_2CO_2H$	136.58	cr (bz)	33	112^{10}	1.1484$_4^{20}$	1.4462^{20}	i	s	s	...	...	...	**B2^3**, 678
p252	—,—,ethyl ester*	$CH_3CH_2CHClCH_2CO_2C_2H_5$	164.63	...	...	189 66–7^{10}	1.0330$_4^{20}$	1.4278^{20}	i	s	s	...	...	...	**B2^3**, 678
p253	—,**4-chloro-***	$CH_3CHClCH_2CH_2CO_2H$	136.58	...	...	116^{10}	1.1514^{20}	1.4458^{20}	i	s	s	...	...	...	**B2^3**, 678
p254	—,—,chloride	$CH_3CHClCH_2CH_2COCl$	155.03	...	...	61^8	...	...	d	d	v	...	...	...	**B2^3**, 679
p255	—,—,ethyl ester*	$CH_3CHClCH_2CH_2CO_2C_2H_5$	164.63	...	...	196^{760} 70.5^9	1.0393$_4^{20}$	1.4310^{20}	i	s	s	...	...	...	**B2^3**, 678
p256	—,**5-chloro-***	$Cl(CH_2)_4CO_2H$	136.58	...	18	230^{760} 141–9^{12} δd	1.3416$_4^{25}$	1.4555^{20}	s v^h	s	v	...	...	...	**B2^3**, 679
p257	—,—,chloride	$Cl(CH_2)_4COCl$	155.03	...	...	83^{12}	1.210^{18}	1.4639^{20}	d	d	s	...	...	...	**B2^3**, 679
Ω p258	—,—,ethyl ester	$Cl(CH_2)_4CO_2C_2H_5$	164.63	...	...	205–6^{760} 93^{16}	1.0561^{20}	1.4355^{20}	i	s	s	...	...	...	**B2^3**, 679
—	—,**2,5-diamino-***	*see* **Ornithine**													
p259	—,**4,4-dimethyl-**, chloride	Neopentylacetyl chloride. $(CH_3)_3CCH_2CH_2COCl$	148.63	...	...	150–2^{760} 58^2	...	1.4294^{20}	d	d	s	...	...	...	**C47**, 6438
p260	—,**2,4-dioxo-***	Acetoneoxalic acid. Acetylpyruvic acid. $CH_3COCH_2COCO_2H$	130.10	nd (bz-chl), pr (bz) $\lambda^{w, pH = 1.5}$ 285 (3.5) λ^{eth} 290 (3.95)	55–63 (+1w) 101(anh)	130^{37} sub par d	...	...	s	s	s	s	s	AcOEt, chl s lig i alk d	**B3^2**, 465
Ω p261	—,—,ethyl ester*	Ethyl acetoneoxalate. $CH_3COCH_2COCO_2C_2H_5$	158.16	...	18	213–5 111–2^{16}	1.1251^{20}	1.4757^{17}	...	s	s	...	...	...	**B3^2**, 465
p262	—,**2-ethyl-(*dl*)***	3-Hexanecarboxylic acid* $CH_3CH_2CH_2CH(C_2H_5)CO_2H$	130.19	...	...	209.2 105–7^{18}	0.9361$_{33}^{33}$	...	i	s	s	...	...	...	**B2^3**, 778
p263	—,—,chloride	$CH_3CH_2CH_2CH(C_2H_5)COCl$	148.63	...	...	158–60 50^{11}	...	...	d	...	s	...	...	...	**B2**, 344
p264	—,**3-ethyl-***	$(C_2H_5)_2CHCH_2CO_2H$	130.19	...	...	212 104–5^{13}	...	1.4250^{20}	...	...	...	...	...	...	**B2^3**, 779
p265	—,**5-fluoro-***	$F(CH_2)_4CO_2H$	120.13	...	...	83^2	...	1.4080^{25}	δ	v	v	...	...	...	**C50**, 4063
Ω p266	—,**2-hydroxy-***	Valerolactic acid. $CH_3CH_2CH_2CH(OH)CO_2H$	118.13	hyg pl	34	sub	...	...	s	s	s	...	...	...	**B3^2**, 225
p267	—,**4-hydroxy-**, lactone(*d*)*	γ-Valerolactone. Tetrahydro-5-methyl-2-furanone.	100.12	$[\alpha]_D^{20}$ +13.5 (undil)	...	86–90^{14}	...	...	∞	s	...	s	...	...	**B17^2**, 288
Ω p268	—,—,—(*dl*)*	$C_5H_8O_2$. *See* p267	100.12	...	−31	206^{760} 83–4^{13}	1.0465^{25}	1.4328^{20}	∞	s	...	s	...	...	**B17^2**, 288
p269	—,—,—(*l*)*	$C_5H_8O_2$. *See* p267	100.12	$[\alpha]_D^{20}$ +4.6 (eth, c = 10)	...	78–80^8	...	1.4322^{20}	∞	s	...	s		...	**B17^2**, 288
Ω p270	—,**5-hydroxy-**, lactone	δ-Valerolactone	100.12	...	−12.5	218–20 113–4^{14}	1.0794^{20}	1.4503^{20}	s	∞	∞	...	...	...	**B17^2**, 287
—	—,**2-hydroxy-4-methyl-***	*see* **Leucic acid**													
p271	—,**2-hydroxy-2-propyl-**, nitrile	4-Heptanone cyanohydrin. $(CH_3CH_2CH_2)_2C(OH)CN$	141.22	...	...	119–20^{21}	0.9077^{18}	1.4337^{18}	...	s	...	...	...	...	**B3^2**, 238
p282	—,**2-isobutyl-4-methyl-**, amide	Diisobutylacetamide. $[(CH_3)_2CHCH_2]_2CHCONH_2$	171.29	lf or nd (eth)	74–5	...	...	...	δ	v	...	...	...	os v lig δ	**B2^1**, 153
p283	—,**2-methyl-(*d*)***	Methylpropylacetic acid. $CH_3CH_2CH_2CH(CH_3)CO_2H$	116.16	$[\alpha]_D^{25}$ +5.58 (eth) $[\alpha]_D^{16}$ +18.5 (undil)	...	96^{15}	0.927$_4^{16}$	1.4112^{25}	s	s	s	...	...	...	**B2^3**, 739

For explanations, symbols and abbreviations see beginning of table. For structural formulas see end of table.

PHYSICAL CONSTANTS OF ORGANIC COMPOUNDS (Continued)

No.	Name	Synonyms and Formula	Mol. wt.	Color, crystalline form, specific rotation and λ_{max} (log ε)	m.p. °C	b.p. °C	Density	n_D	Solubility w	al	eth	ace	bz	other solvents	Ref.
	Pentanoic acid														
Ω p284	—,—(*dl*)*	$CH_3CH_2CH_2CH(CH_3)CO_2H$	116.16	...	...	$195–6^{760}$ $102–5^{12}$	0.9230_4^{20}	1.413^{20}	s	s	s	...	...	...	**B2**^{3}, 740
p285	—,—(*l*)*	$CH_3CH_2CH_2CH(CH_3)CO_2H$	116.16	$[\alpha]_D^{25}$ −7.08 (eth)	...	$190–3$ 96^{15}	0.9781_4^{20}	1.4117^{25}	s	s	s	...	...	...	**B2**^{2}, 288
p286	—,—,chloride(*dl*)	$CH_3CH_2CH_2CH(CH_3)COCl$	134.61	...	...	140.0–0.8^{745}	0.9781_4^{20}	1.4330^{27}	d^h	d^h	s	...	...	...	**B2**^{3}, 741
p287	—,3-methyl-(*d*)*	*see*-Butylacetic acid. $CH_3CH_2CH(CH_3)CH_2CO_2H$	116.16	$[\alpha]_D^{20}$ +8.5 (undil)	...	199 $92–3^{10}$	0.9276_4^{20}	1.4158^{20}	...	s	s	...	...	...	**B2**^{3}, 749
Ω p288	—,—(*dl*)*	$CH_3CH_2CH(CH_3)CH_2CO_2H$	116.16	...	−41.6	197.2–$.8^{760}$	0.9262^{20}	1.4159^{20}	...	s	s	...	...	...	**B2**^{3}, 750
p289	—,—(*l*)*	$CH_3CH_2CH(CH_3)CH_2CO_2H$	116.16	$[\alpha]_D^{20}$−8.9 (al)	...	196–7 105^{30}	0.923_4^{25}	1.4152^{20}	...	s	s	...	...	...	**B2**^{3}, 749
p290	—,—,chloride(*dl*)	*sec*-Butylacetyl chloride. $CH_3CH_2CH(CH_3)CH_2COCl$	134.61	...	...	142.5–3^{749}	0.9781_4^{20}	...	d	d	v	...	...	CS_2 v	**B2**^{3}, 750
Ω p291	—,4-methyl-*	Isobutylacetic acid. Isocaproic acid. $(CH_3)_2CHCH_2CH_2CO_2H$	116.16	...	−33.0	$200–1^{760}$ $86–8^{11}$	0.9225_4^{20}	1.4144^{20}	δ	s	s	...	...	...	**B2**^{3}, 743
p292	—,—,amide	Isocaproamide. $(CH_3)_2CHCH_2CH_2CONH_2$	115.18	nd (al)	120–1	...	...	...	s	s^h	...	...	...	...	**B2**^{2}, 290
p293	—,—,—,*N*-phenyl-	Isocaproanilide. $(CH_3)_2CHCH_2CH_2CONHC_6H_5$	191.28	nd (bz, dil al)	112	...	...	...	i	s^h	s	...	s^h	...	**B12**^{2}, 147
p294	—,—,chloride	Isocaproyl chloride. $(CH_3)_2CHCH_2CH_2COCl$	134.61	...	...	143.8–4.5^{745}	0.9725^{20}	...	d^h	d^h	...	...	...	...	**B2**^{3}, 745
Ω p295	—,—,nitrile	Isoamyl cyanide. Isocapronitrile. $(CH_3)_2CHCH_2CH_2CN$	97.16	...	−51	$156–7^{761}$ 50^{13}	0.8030_4^{20}	1.4059^{20}	i	s	∞	...	...	...	**B2**^{3}, 745
p296	—,4-methyl-2-phenyl-*	α-Phenylisocaproic acid. $(CH_3)_2CHCH_2CH(C_6H_5)CO_2H$	192.24	pr (peth)	78–9	$178–80^{30}$	...	...	...	s	...	...	...	alk s lig δ	**B9**^{2}, 369
p297	—,—,nitrile	α-Phenylisocapronitrile. $(CH_3)_2CHCH_2CH(C_6H_5)CN$	173.26	...	...	$263–6^{765}$ $136–8^{15}$	0.942^{16}	...	i	s	s	...	s	aa s	**B9**^{1}, 220
p298	—,3-oxo-, ethyl ester*	Ethyl propionylacetate. $CH_3CH_2COCH_2CO_2C_2H_5$	144.17	...	...	191 90^{13}	...	1.4230^{20}	...	s	s	...	s	...	**B3**^{2}, 430
Ω p299	—,4-oxo-*	β-Acetopropionic acid. Levulinic acid. $CH_3COCH_2CH_2CO_2H$	116.13	lf or pl λ^w 270 (1.40)	37.2	245–6 δd $139–40^8$	1.1335_4^{20}	1.4396^{20}	v	v	v	...	...	...	**B3**^{3}, 1214
p299[1]	—,—,benzyl ester	Benzyl levulinate. $CH_3COCH_2CH_2CO_2CH_2C_6H_5$	206.24	...	...	$132–4^2$	1.0935_4^{20}	1.5090^{20}	...	...	...	...	...	to s	**Am 55,** 4727
p300	—,—,butyl ester*	Butyl levulinate. $CH_3COCH_2CH_2CO_2(CH_2)_3CH_3$	172.23	...	...	237.8^{760}	0.9735_4^{20}	1.4290^{20}	...	s	s	s	s	CS_2 s	**B3**^{3}, 1220
p301	—,—,*sec*-butyl ester	*sec*-Butyl levulinate. $CH_3COCH_2CH_2CO_2CH(CH_3)CH_2CH_3$	172.23	...	...	225.8^{760}	0.9669_4^{20}	1.4249^{20}	...	s	s	s	...	...	**B3**^{3}, 1221
p302	—,—,isobutyl ester.	Isobutyl levulinate. $CH_3COCH_2CH_2CO_2CH_2CH(CH_3)_2$	172.23	...	...	230.9^{760} (222–4)	0.97047_4^{20}	1.4268^{20}	...	s	s	s	s	CS_2 s	**B3**^{3}, 1221
p303	—,—,isopropyl ester*	Isopropyl levulinate. $CH_3COCH_2CH_2CO_2CH(CH_3)_2$	158.20	...	...	209.3^{760}	0.98422_4^{20}	1.4420^{20}	δ	s	s	s	s	...	**B3**^{3}, 1220
Ω p304	—,—,methyl ester*.	Methyl levulinate. $CH_3COCH_2CH_2CO_2CH_3$	130.15	...	...	196^{760} $85–6^{14}$	1.05113_4^{20}	1.4233^{20}	δ	s	∞	s	s	...	**B3**^{3}, 1217
Ω p305	—,—,propyl ester*.	Propyl levulinate. $CH_2COCH_2CH_2CO_2CH_2CH_2CH_3$	158.20	...	...	221.2^{760}	0.9896_4^{20}	1.4258^{20}	δ	s	s	s	s	CS_2 s	**B3**^{3}, 1219
p306	—,5-oxo-5-phenyl-*	γ-Benzoylbutyric acid. $C_6H_5CO(CH_2)_3CO_2H$	192.22	pl (w) λ^{al} 281 (1.41)	128–9	...	...	...	s^h	...	...	...	...	...	**B10**^{2}, 488
p307	—,2-phenyl-(*d*)*	α-Phenylvaleric acid. $CH_3CH_2CH_2CH(C_6H_5)CO_2H$	178.23	$[\alpha]_D$ +58.81 (chl), +33.1 (eth)	...	165^{14}	1.047_4^{20}	...	...	s	s	...	...	chl s lig s^h	**B9**, 557
p308	—,—(*dl*)*	$CH_3CH_2CH_2CH(C_6H_5)CO_2H$	178.23	nd (lig)	58	280	...	...	...	s	s	...	...	chl s lig s^h	**B9**, 557
p309	—,—,nitrile	$CH_3CH_2CH_2CH(C_6H_5)CN$	159.23	...	...	$254–5^{750}$ $125–8^{13}$	0.9425	1.5000^{20}	i	∞	...	...	∞	...	**B9**^{1}, 216
Ω p310	—,4-phenyl-*	γ-Phenylvaleric acid. $C_6H_5CH(CH_3)CH_2CH_2CO_2H$	178.23	...	*ca.* 13	210^{85} 165^{12}	1.0554_4^{15}	1.5167^{20}	δ	s	s	...	...	...	**B9**^{2}, 362
Ω p311	—,5-phenyl-*	δ-Phenylvaleric acid. $C_6H_5(CH_2)_4CO_2H$	178.23	pl (w), pr (peth)	57–8 (61)	$190–3^{30}$	...	...	δ^h	v	...	...	...	os s	**B9**^{1}, 216
p312	—,2,2,4-trimethyl-, amide	$(CH_3)_2CHCH_2C(CH_3)_2CONH_2$	143.23	lf (lig)	71	...	...	...	...	s^h	...	...	...	lig δ	**B2**^{2}, 305
Ω p313	**1-Pentanol***.	Butyl carbinol. *n*-Amyl alcohol. $CH_3(CH_2)_4OH$	88.15	λ^{undil} 178 (2.4), 215 (1.55)	−79	137.3^{748} 50^{13}	0.8144_4^{20}	1.4101^{20}	i δ^h	∞	∞	s	...	...	**B1**^{3}, 1598
Ω p314	**2-Pentanol***.	Methyl *n*-propyl carbinol. $CH_3CH_2CH_2CH(OH)CH_3$	88.15	...	...	118.9 62^{60}	0.8103_4^{20}	1.4053^{20}	v	s	s	...	...	...	**B1**^{3}, 1609
Ω p315	**3-Pentanol***.	Diethyl carbinol. $(CH_3CH_2)_2CHOH$	88.15	...	...	116.1^{760} 30^{12}	0.8212_4^{20}	1.4104^{20}	δ	s	s	s	...	...	**B1**, 385
p317	—,4-amino-2-methyl-*	$(CH_3)_2C(OH)CH_2CH(NH_2)CH_3$	117.19	...	35–6	174^{745}	...	...	s^h	s	s	s	s	lig i	**B4**, 296
p318	**1-Pentanol, 5-chloro-***	Pentamethylene chlorohydrin $Cl(CH_2)_5OH$	122.60	...	...	112^{12}	...	1.4518^{20}	δ	v	v	...	...	...	**B1**^{3}, 1606
p319	**2-Pentanol, 1-chloro-***	$CH_3CH_2CH_2CH(OH)CH_2Cl$	122.60	...	...	$157–60^{735}$ $59–62^{14}$	1.037_{20}^{20}	1.4404^{25}	...	s	v	...	...	...	**B1**^{2}, 419
p320	**3-Pentanol, 1-chloro-***	$CH_3CH_2CH(OH)CH_2CH_2Cl$	122.60	...	...	173 77^{20}	1.0327_4^{25}	1.448^{25}	δ s^h	v	v	...	...	...	**B1**^{3}, 1618
p321	**1-Pentanol,2,4-dimethyl-(*dl*)***	$(CH_3)_2CHCH_2CH(CH_3)CH_2OH$	116.21	...	...	$160–2^{760}$ $65–7^{18}$	0.793_4^{20}	1.427^{20}	i	s	s	...	...	os s	**B1**^{3}, 1700

For explanations, symbols and abbreviations see beginning of table. For structural formulas see end of table.

No.	Name	Synonyms and Formula	Mol. wt.	Color, crystalline form, specific rotation and λ_{max} (log ε)	m.p. °C	b.p. °C	Density	n_D	Solubility w	al	eth	ace	bz	other solvents	Ref.
	1-Pentanol														
p322	—,—(*l*)*	$(CH_3)_2CHCH_2CH(CH_3)CH_2OH$	116.21	$[\alpha]_D^{22}$ −1.1 (undil)	...	157^{760}	0.816_4^{25}	...	i	s	s	...	...	os s	**B1**^{3}, 1700
Ω p323	**2-Pentanol, 2,4-dimethyl-***	Dimethyl isobutyl carbinol. $(CH_3)_2CHCH_2C(OH)(CH_3)_2$	116.21	...	< −20	133.1^{760} 53.5–4.5^{25}	0.8103_4^{20}	1.4172^{20}	i	s	s	...	...	os s	**B1**^{3}, 1700
Ω p324	**3-Pentanol, 2,3-dimethyl-***	Ethyl isopropyl methyl carbinol. $(CH_3)_2CHC(OH)(CH_3)CH_2CH_3$	116.21	...	< −30	139.7^{760} $44–5^{14}$	0.833_4^{20}	1.4287^{20}	i	s	s	...	...	...	**B1**^{3}, 1698
Ω p325	**—,2,4-dimethyl-***	Diisopropyl carbinol. $(CH_3)_2CHCH(OH)CH(CH_3)_2$	116.21	...	< 70	138.7^{760} 87.5^{125}	0.8288_4^{20}	1.4250^{20}	δ	s	s	...	...	...	**B1**^{3}, 1701
p326	**—,2,4-dimethyl-3-phenyl-***	Diisopropylphenyl carbinol. $[(CH_3)_2CH]_2C(OH)C_6H_5$	192.31	ye	...	229^{760} 157^{60}	0.9755_4^{20}	1.5239^{20}	δ	i	s	...	...	...	**B6**^{1}, 273
Ω p327	**—,3-ethyl-***	Triethyl carbinol. $(C_2H_5)_3COH$	116.21	...	...	143.1–3.2 73^{52}	0.8407_4^{22}	1.4294^{20}	i	s	s	...	...	...	**B1**^{3}, 1696
Ω p328	**—,3-ethyl-2-methyl-***	Diethylisopropyl carbinol. $(CH_3)_2CHCOH(C_2H_5)_2$	130.23	...	...	159.5–61^{750} $55–7^{48}$	0.8295_{20}^{20}	1.4372^{20}	i	s	s	...	...	...	**B1**^{3}, 1740
p329	**1-Pentanol, 5-fluoro-***	Pentamethylene fluorohydrin $F(CH_2)_5OH$	106.14	...	...	$70–1^{11}$	...	1.4057^{25}	s	v	v	...	...	...	**C51**, 7300
Ω p330	**—,2-methyl-***	*sec*-Amyl carbinol. $CH_3CH_2CH_2CH(CH_3)CH_2OH$	102.18	...	...	148^{760}	0.8263_4^{20}	1.4182^{20}	...	s	s	s	...	...	**B1**^{3}, 1665
Ω p331	**—,3-methyl-(*dl*)***	$CH_3CH_2CH(CH_3)CH_2CH_2OH$	102.18	...	...	152.44^{760} $51–2.5^8$	0.8242_4^{20}	1.4112^{23}	i	s	s	...	...	...	**B1**^{3}, 1671
Ω p332	**—,4-methyl-***	Isoamyl carbinol. Isohexyl alcohol. $(CH_3)_2CH(CH_2)_3OH$	102.18	...	...	151.6^{760}	0.8131_4^{20}	1.4134^{25}	i	s	s	...	...	...	**B1**^{3}, 1671
Ω p334	**2-Pentanol, 2-methyl-***	Dimethyl *n*-propyl carbinol. $CH_3CH_2CH_2COH(CH_3)_2$	102.18	col	−103	120.5–1.5 $49.5^{27.5}$	0.8350_4^{16}	1.4100^{20}	δ	s	s	...	...	...	**B1**^{3}, 1666
Ω p335	**—,3-methyl-***	*sec*-Butyl methyl carbinol. $CH_3CH_2CH(CH_3)CH(OH)CH_3$	102.18	...	...	134.32^{760} 75.6^{50}	0.8307_4^{20}	1.4182^{20}	δ	s	s	...	...	...	**B1**^{3}, 1672
Ω p336	**—,4-methyl-***	Methyl isobutyl carbinol. $(CH_3)_2CHCH_2CH(OH)CH_3$	102.18	...	...	133^{760} $50–5^{25}$	0.8075_4^{20}	1.4100^{20}	δ	s	s	...	...	...	**B1**^{3}, 1669
Ω p337	**—,—,acetate**	4-Methyl-2-pentyl acetate. $(CH_3)_2CHCH_2CH(CH_3)O_2CCH_3$	144.22	...	...	147–8 76.5^{47}	0.8805_0^0	1.4066^{20}	...	...	s	s	...	...	**C50**, 4817
Ω p338	**3-Pentanol, 2-methyl-***	Ethyl isopropyl carbinol. $(CH_3)_2CHCH(OH)CH_2CH_3$	102.18	...	...	126.68^{760}	0.8243_4^{20}	1.4175^{20}	δ	∞	∞	...	...	...	**B16**, 1037
Ω p339	**—,3-methyl-***	Diethyl methyl carbinol. $(CH_3CH_2)_2C(OH)CH_3$	102.18	...	−23.6	122.4^{760}	0.8286_4^{20}	1.4186^{20}	δ	∞	∞	...	...	...	**B1**, 411
p340	**1-Pentanol, 1-phenyl-(*d*)***	Butyl phenyl carbinol. $C_6H_5CH(OH)(CH_2)_3CH_3$	164.25	$[\alpha]_D^{20}$ +40.8	...	$140–2^{25}$	0.9672_{20}^{20}	1.4086^{25}	i	s	s	s	...	...	**B6**^{2}, 504
p341	**—,5-phenyl-***	$C_6H_5(CH_2)_5OH$	164.25		...	155^{20} $128–30^{2.5}$	0.9725^{20}	1.5156^{20}	i	v	v	...	...	...	**B6**^{2}, 505
p342	**2-Pentanol, 2-phenyl-***	Methyl phenyl propyl carbinol. $C_6H_5C(OH)(CH_3)(CH_2)_2CH_3$	164.25	...	...	216 112^{14}	0.9723_4^{22}	...	i	v	...	...	...	...	**B6**^{2}, 505
p343	**3-Pentanol, 1-phenyl-(*d*)***	$C_6H_5CH_2CH_2CH(OH)CH_2CH_3$	164.25	$[\alpha]_D^{20}$ +26.8 (al), +30.2 (CS_2)	38	143^{19}	0.9687_4^{20}	...	i	v	...	...	...	CS_2 v	**B6**^{2}, 504
p344	**—,3-phenyl-***	Diethyl phenyl carbinol. $(C_2H_5)_2C(OH)C_6H_5$	164.25	...	< −17	$223–4^{762}$ 110^{12}	0.9831_4^{20}	1.51655^{20}	i	v	...	...	...	...	**B6**^{2}, 506
Ω p345	**2-Pentanol,2,4,4-trimethyl-***	Isodibutol. $(CH_3)_3CCH_2C(OH)(CH_3)_2$	130.23	...	−20	147.5 $42–4^7$	0.8225_4^{20}	1.4284^{20}	i	δ	s	...	...	...	**B1**^{3}, 1741
Ω p346	**2-Pentanone***	Methyl propyl ketone. $CH_3CH_2CH_2COCH_3$	86.14	λ^{al} 280.3 (1.26)	−77.8	102	0.8089_4^{20}	1.3895^{20}	δ	∞	∞	...	...	...	**B1**^{3}, 2800
p347	**—,oxime***	Methyl propyl ketoxime. $CH_3CH_2CH_2C(:NOH)CH_3$	101.15	...	...	167^{748}	0.9095_4^{20}	1.4450^{20}	s	∞	∞	...	...	...	**B1**^{3}, 2803
Ω p348	**3-Pentanone***	Diethyl ketone. Propione. $CH_3CH_2COCH_2CH_3$	86.14	λ^{al} 275 (1.30)	−39.8	101.7^{760}	0.8138_4^{20}	1.3924^{20}	v	∞	∞	...	...	...	**B1**^{3}, 2806
p349	**2-Pentanone, 4-amino-4-methyl-***	Diacetonamine. $(CH_3)_2C(NH_2)CH_2COCH_3$	115.18	...	...	$25^{0.14}$	< 1	...	s	∞	∞	...	...	...	**B4**^{2}, 767
Ω p350	**1-Pentanone, 1(4-amino-phenyl)-***	3-Aminovalerophenone	177.25	cr (bz-peth)	74–5	$160–3^3$	...	...	i	s	s	...	...	...	**B14**^{2}, 43
p351	**2-Pentanone, 1-chloro-***	$CH_3CH_2CH_2COCH_2Cl$	120.58	...	...	154.5–6 δd $58–9^{17}$	...	...	δ	...	...	...	...	MeOH s	**B1**^{3}, 2804
Ω p352	**—,5-chloro-***	$Cl(CH_2)_3COCH_3$	120.58	...	...	76^{34} (75^{23})	1.0523_4^{20}	1.4375^{20}	...	...	s	s	...	...	**B1**^{3}, 2804
p353	**3-Pentanone, 1-chloro-***	$CH_3CH_2COCH_2CH_2Cl$	120.58	...	...	68^{20} $33^{2.5}$	...	1.4361^{20}	...	s	v	...	...	MeOH s alk d^h	**B1**^{3}, 2812
p354	**—,2-chloro-***	$CH_3CH_2COCHClCH_3$	120.58	...	...	135 (145)	...	...	i	s	s	...	...	...	**B1**^{3}, 2812
Ω p355	**—,2,4-dimethyl-***	Diisopropyl ketone. Isobutyrone. Tetramethyl acetone. $(CH_3)_2CHCOCH(CH_3)_2$	114.19	λ^{al} 286 (1.45)	−69.03	124–5	0.8108_4^{20}	1.39995^{20}	i	∞	∞	...	s	...	**B1**^{3}, 2868
Ω p356	**2-Pentanone, 3-ethyl-4-methyl-***	$(CH_3)_2CHCH(C_2H_5)COCH_3$	128.22	...	...	$154–5^{760}$	0.812_4^{20}	1.4105^{20}	s	∞	∞	...	∞	aa, chl ∞ os v	**B1**^{3}, 2883
p357	**—,1-hydroxy-***	$CH_3CH_2CH_2COCH_2OH$	102.13	...	...	152 $62–4^{18}$	0.9860_4^{20}	1.4234^{20}	s	s	s	...	...	...	**B1**^{3}, 3219
p358	**—,3-hydroxy-***	$CH_3CH_2CH(OH)COCH_3$	102.13	...	...	$147–8^{761}$ 59^{27}	0.9500_4^{20}	1.4350^{20}	∞	∞	s	s	s	...	**B1**^{3}, 3220

For explanations, symbols and abbreviations see beginning of table. For structural formulas see end of table.

No.	Name	Synonyms and Formula	Mol. wt.	Color, crystalline form, specific rotation and λ_{max} (log ε)	m.p. °C	b.p. °C	Density	n_D	Solubility w	al	eth	ace	bz	other solvents	Ref.
	2-Pentanone														
p359	**—,4-hydroxy-***	$CH_3CH(OH)CH_2COCH_3$	102.13	λ^{al} 277.5 (1.38)	....	177^{760} $62–4^{12}$	1.0071^{20}_{4}	1.4265^{20}	s	s	s	...	...	....	**B1**³, 3220
Ω p360	**—,5-hydroxy-***	$HOCH_2CH_2CH_2COCH_3$	102.13	....	....	208^{730} $116–8^{33}$	1.0071^{20}_{4}	1.4390^{20}	∞	s	s	...	...	....	**B1**³, 3221
p361	**3-Pentanone, 2-hydroxy-***	$CH_3CH_2COCH(OH)CH_3$	102.13	....	....	152.5^{761} 63^{20}	0.9742^{20}_{4}	1.4128^{20}	s	s	s	...	...	....	**B1**², 873
Ω p362	**2-Pentanone, 4-hydroxy-4-methyl-***	Diacetone alcohol. $(CH_3)_2C(OH)CH_2COCH_3$	116.16	λ^{al} 282 (1.40)	−44	164^{760} (168) $67–9^{19}$	0.9387^{20}_{4}	1.4213^{20}	∞	∞	∞	...	...	....	**B1**³, 3234
p363	**1-Pentanone, 5-hydroxy-1-phenyl-***	*δ*-Hydroxy-*n*-valerophenone. $C_6H_5CO(CH_2)_4OH$	178.23	pl (w)	40–1	....	....	....	s^h	v	v	...	v	MeOH v peth *δ*	**B8**, 123
Ω p364	**2-Pentanone, 3-methyl-(*dl*)***	*sec*-Butyl methyl ketone. $CH_3CH_2CH(CH_3)COCH_3$	100.16	λ^{al} 282 (1.40)	....	118^{758}	0.8130^{20}_{4}	1.4002^{20}	*δ*	∞	∞	...	...	chl s	**B1**³, 2837
Ω p365	**—,4-methyl-***	Isobutyl methyl ketone. Isopropyl acetone. $(CH_3)_2CHCH_2COCH_3$	100.16	....	−84.7	116.85^{760} $35–40^{16}$	0.7978^{20}	1.3962^{20}	*δ*	∞	∞	∞	∞	chl s	**B1**³, 2882
Ω p366	**3-Pentanone, 2-methyl-***	Ethyl isopropyl ketone. $CH_3CH_2COCH(CH_3)_2$	100.16	λ^{al} 281 (1.41)	....	114.5– 5^{745}	0.830^{0}_{0}	1.3975^{20}	*δ*	v	∞	∞	v	chl s	**B1**³, 2831
p367	**1-Pentanone, 4-methyl-1-phenyl-***	Isocaprophenone. Isopentyl phenyl ketone $(CH_3)_2CHCH_2CH_2COC_6H_5$	176.24	....	−2	255–6 $145–7^{30}$	0.9623^{15}_{4}	1.533^{20}	i	v	v	v	v	chl s	**B7**², 258
Ω p368	**—,1-phenyl-***	*n*-Butyl phenyl ketone. Valerophenone. $CH_3(CH_2)_3COC_6H_5$	162.23	λ^{MeOH} 240 (4.11)	....	248.5 (cor) $131–3^{13}$	0.988^{20}_{20}	1.5158^{20}	i	v	v	...	...	....	**B7**², 251
Ω p369	**3-Pentanone, 2,2,4,4-tetra-methyl-***	Hexamethylacetone. Pivalone. $(CH_3)_3CCOC(CH_3)_3$	142.24	....	....	152 (154) 70^{43}	0.8240^{18}	1.4194^{20}	i	s	s	s	...	chl, aa s	**B7**², 251
—	**Pentaphene**	*see* **2:3, 6:7-Dibenzophenanthrene**													
p370	**Pentaquine**	$C_{18}H_{27}N_3O$	301.44	....	....	165– $70^{0.02}$	....	1.5785^{25}	...	...	...	...	...	dil HCl s	**Am 68**, 1529
p371	**Pentasiloxane, dodecamethyl-**	$CH_3[-Si(CH_3)_2O-]_4Si(CH_3)_3$	384.85	....	−80	229^{710} $103–7^{12}$	0.8755^{20}	1.3925^{20}	*δ*	...	...	...	s	lig s	**Am 68**, 2284
p372	**Pentatri-acontane***	$CH_3(CH_2)_{33}CH_3$	492.97	cr (al)	75	490^{760} 311.36^{10}	0.8157^{20}_{4}	1.4568^{20} (suc)	i	...	δ^h	s	...	....	**B1**, 177
Ω p373	**18-Pentatri-acontanone***	Diheptadecyl ketone. Stearone. $CH_3(CH_2)_{16}CO(CH_2)_{16}CH_3$	506.95	lf (lig)	88.4	....	0.793^{95}_{4}	....	i	δ^h	δ^h	δ^h	*δ*	chl, lig *δ*	**B1**, 720
p374	**2-Pentenal, 2-methyl-***	Methylethyl acrolein. $CH_3CH_2CH:C(CH_3)CHO$	98.15	λ^{MeOH} 227.5 (4.14), 311 (1.51)	....	136.4– 7.1^{762} $38–9^{18}$	0.8581^{20}_{4}	1.4488^{20}	i	s	s	...	s	MeOH s	**B1**³, 2995
Ω p375	**1-Pentene***	1-Pentylene. Propylethylene. $CH_3CH_2CH_2CH:CH_2$	70.14	λ^{gas} 177 (4.22), 181.3 (4.14), 187.1 sh (3.65)	−138 (−165.2)	29.968^{760}	0.6405^{20}_{4}	1.3715^{20}	i	∞	∞	...	s	dil sulf v	**B1**³, 770
p376	**2-Pentene(*cis*)***	*cis*-*β*-*n*-Amylene. $CH_3CH_2CH:CHCH_3$	70.14	λ^{gas} 166.7 (4.03), 177 (4.26), 185 (4.1), 204.8 (2.95)	−151.39	36.9^{760}	0.6556^{20}_{4}	1.3830^{20}	i	∞	∞	...	s	dil sulf s	**B1**³, 776
p377	**—(*trans*)***	*trans*-*β*-*n*-Amylene. $CH_3CH_2CH:CHCH_3$	70.14	λ^{gas} 157.7 (4.08), 180.8 (4.15), 200.9 (3.11)	−136	36.353^{760}	0.6482^{20}_{4}	1.3793^{20}	i	∞	∞	...	s	dil sulf v	**B1**³, 777
p378	**1-Pentene, 1-bromo-***	$CH_3CH_2CH_2CH:CHBr$	149.04	....	....	121.5– 22^{760} 43.5^{30}	1.2606^{25}_{4}	1.4572^{20}	i	...	v	...	s	chl s	**B1**³, 775
p379	**—,2-bromo-***	$CH_3CH_2CH_2CBr:CH_2$	149.04	....	....	107–8	1.228^{20}	1.4535^{20}	i	...	s	...	s	chl, lig s	**B1**², 183
p380	**—,3-bromo-***	$CH_3CH_2CHBrCH:CH_2$	149.04	....	....	30.5^{30}	1.2417^{25}_{4}	1.4626^{25}	i	...	...	s	s	chl v	**B1**³, 775
p380[1]	**2-Pentene, 1-bromo-***	$CH_3CH_2CH:CHCH_2Br$	149.04	....	....	123–4 (122) 35^{25}	1.2545^{20}	1.4731^{20}	i	...	...	s	s	chl v	**B1**³, 783
p381	**—,2-bromo-***	$CH_3CH_2CH:CBrCH_3$	149.04	....	....	110.5^{750}	1.277^{20}_{20}	1.4580^{20}	i	...	...	s	s	chl v	**B1**³, 784
p382	**—,3-bromo-***	$CH_3CH_2CBr:CHCH_3$	149.04	....	....	115.2^{750}	1.273^{20}_{20}	1.4628^{20}	i	...	...	s	s	chl v	**B1**³, 784
p383	**—,4-bromo-***	$CH_3CHBrCH:CHCH_3$	149.04	....	....	116.7– 9.2d 22^{8}	1.2312^{21}	1.4752^{21}	d	...	...	s	s	chl v	**B1**³, 784
p384	**—,5-bromo-***	$BrCH_2CH_2CH:CHCH_3$	149.04	....	....	121.7^{621}	1.2715^{20}_{4}	1.4695^{20}	i	...	...	s	s	chl v	**B1**³, 784
p385	**1-Pentene, 2-chloro-***	$CH_3CH_2CH_2CCl:CH_2$	104.58	....	....	95–7	0.872^{5}	....	...	s	v	...	...	....	**B1**³, 774
p386	**—,3-chloro-***	$CH_3CH_2CHClCH:CH_2$	104.58	....	....	$93–4^{764}$	0.8978^{20}_{4}	1.4254^{20}	...	s	v	s	...	....	**B1**³, 774
p387	**—,4-chloro-***	$CH_3CHClCH_2CH:CH_2$	104.58	λ^{undil} 233 (1.25)	....	97–100	0.934^{15}	1.417^{15}	i	...	v	...	...	chl v	**B1**³, 774
p388	**—,5-chloro-***	$ClCH_2CH_2CH_2CH:CH_2$	104.58	....	....	103.5– 4.5^{773} 36.5^{61}	0.9125^{20}_{4}	1.4297^{20}	*δ*	...	s	v	...	....	**B1**³, 775
p389	**2-Pentene, 1-chloro-***	$CH_3CH_2CH:CHCH_2Cl$	104.58	....	....	109.5 62^{148}	0.908^{22}_{4}	1.4352^{22}	i	v^h	v	s	...	chl s	**B1**², 184
p390	**—,2-chloro-***	$CH_3CH_2CH:CClCH_3$	104.58	....	....	95–7 45^{130}	0.9067^{20}_{4}	1.4261^{20}	i	...	v	v	...	xyl v^h	**B1**², 185
p391	**—,3-chloro-***	$CH_3CH_2CCl:CHCH_3$	104.58	....	....	$90–2^{781}$	1.423^{20}	0.9125^{20}_{20}	i	...	v	v	...	xyl v^h	**B1**², 185

For explanations, symbols and abbreviations see beginning of table. For structural formulas see end of table.

No.	Name	Synonyms and Formula	Mol. wt.	Color, crystalline form, specific rotation and λ_{max} (log ε)	m.p. °C	b.p. °C	Density	n_D	Solubility: w	al	eth	ace	bz	other solvents	Ref.
	2-Pentene														
p392	**—,4-chloro-***	Piperylene hydrochloride. $CH_3CHClCH:CHCH_3$	104.58	...	...	103 $18–20^{12}$	0.9004^{20}_{20}	1.4322^{20}	i	...	v	v	...	chl s	**B1**[3], 782
p393	**—,5-chloro-***	$ClCH_2CH_2CH:CHCH_3$	104.58	...	...	107.0– $.6^{755}$	0.9043^{20}_{4}	1.4310^{20}	i	...	v	v	...	chl s	**B1**[3], 782
p394	**—,3-chloro-2,4-dimethyl-***	$(CH_3)_2CHCCl:C(CH_3)_2$	132.64	...	...	118–20 $44–5^{30}$	0.9513^{9}_{9}	...	i	...	v	...	s	chl v	**B1**, 221
p395	**1-Pentene, 2-chloro-3-ethyl-3-methyl-***	$(C_2H_5)_2C(CH_3)CCl:CH_2$	146.66	...	...	147^{743} 53^{20}	0.9147^{20}_{4}	1.4450^{25}	i	...	...	...	s	chl s	**B1**[3], 847
p396	**—,3-chloro-2-methyl-***	$CH_3CH_2CHClC(CH_3):CH_2$	118.61	...	...	121–4	...	1.4422^{20}	...	...	v	...	s	chl v	**B1**[3], 809
p397	**2-Pentene, 5-chloro-2-methyl-***	$ClCH_2CH_2CH:C(CH_3)_2$	118.61	...	...	$132–3^{758}$	0.9135^{20}_{4}	1.4419^{20}	i	...	...	s	...	chl s	**B1**[3], 810
p398	**1-Pentene, decafluoro-***	Perfluoropentene. $F_3CCF_2CF_2CF:CF_2$	250.04	...	...	$29–30^{740}$	...	1.2571^{25}	i	...	...	...	...	chl s	**Am 73**, 4054
p399	**2-Pentene, 2,5-dichloro-***	$ClCH_2CH_2CH:CClCH_3$	139.04	...	...	$40–1^{8}$	1.1182^{15}_{4}	...	i	...	...	...	...	chl s	**B1**[3], 783
p400	**1-Pentene, 2,3-dimethyl-***	$CH_3CH_2CH(CH_3)C(CH_3):CH_2$	98.19	...	−134.8	84.26^{760}	0.7051^{20}_{4}	1.4033^{20}	i	∞	∞	...	...	dil sulf v	**B1**[3], 832
Ω p401	**—,2,4-dimethyl-***	$(CH_3)_2CHCH_2C(CH_3):CH_2$	98.19	...	−123.8	81.64^{760}	0.6943^{20}_{4}	1.3986^{20}	i	∞	∞	...	s	chl s	**B1**[3], 833
p402	**—,3,3-dimethyl-***	$CH_3CH_2C(CH_3)_2CH:CH_2$	98.19	...	−134.3	77.54^{760}	0.6974^{20}_{4}	1.3984^{20}	i	∞	∞	...	s	chl s	**B1**[3], 834
Ω p403	**—,4,4-dimethyl-***	Vinylneopentane. $(CH_3)_3CCH_2CH:CH_2$	98.19	...	−136.6	72.49^{760}	0.6827^{20}_{4}	1.3918^{20}	i	∞	∞	...	s	os, chl s	**B1**[3], 832
p404	**2-Pentene, 2,3-dimethyl-***	$CH_3CH_2C(CH_3):C(CH_3)_2$	98.19	...	−118.3	97.46^{760}	0.7277^{20}	1.4208^{20}	i	s	s	...	s	chl s	**B1**[3], 832
Ω p405	**—,2,4-dimethyl-***	$(CH_3)_2CHCH:C(CH_3)_2$	98.19	...	−127.7	83.44^{760}	0.6954^{20}_{4}	1.4040^{20}	i	s	s	...	s	chl s	**B1**[3], 833
p406	**—,3,4-dimethyl-***	$(CH_3)_2CHC(CH_3):CHCH_3$	98.19	λ^{iso} <215	...	87^{760}	0.7126^{20}_{4}	1.4070^{20}	i	s	s	...	s	chl s	**B1**[3], 833
p407	**—,4,4-dimethyl-(*trans*)***	$(CH_3)_3CCH:CHCH_3$	98.19	...	−115.235	76.75^{760}	0.6889^{20}_{4}	1.3982^{20}	i	s	s	...	s	chl s	**B1**[3], 831
p408	**1-Pentene, 2-ethyl-***	$CH_3CH_2CH_2C(C_2H_5):CH_2$	98.19	...	...	94^{760}	0.7079^{20}_{4}	1.405^{20}	i	s	s	...	s	chl s	**B1**[3], 830
p409	**2-Pentene, 3-ethyl-***	$(C_2H_5)_2C:CHCH_3$	98.19	...	...	96.01^{760}	0.7204^{20}_{4}	1.4148^{20}	i	s	s	...	s	chl s	**B1**[3], 832
Ω p410	**1-Pentene, 2-methyl-***	$CH_3CH_2CH_2C(CH_3):CH_2$	84.16	λ^{gas} 184 (3.8), 189 (3.9), 194 sh (3.8)	−135.72	60.7^{760}	0.6799^{20}_{4}	1.3920^{20}	i	s	...	...	s	chl, peth s	**B1**[3], 808
p411	**—,3-methyl-***	$CH_3CH_2C(CH_3)CH:CH_2$	84.16	λ^{gas} 177 (4.2), 183 sh (4.0), 186 (3.7)	−153.0	51.14^{760}	0.6675^{20}_{4}	1.3841^{20}	i	s	...	...	s	chl, peth s	**B1**[3], 812
p412	**—,4-methyl-***	1-Isohexene. $(CH_3)_2CHCH_2CH:CH_2$	84.16	λ^{gas} 173 (3.9), 178 (4.0)	−153.63	53.88^{760}	0.6642^{20}_{4}	1.3828^{20}	i	s	...	...	s	chl, peth s	**B1**[3], 811
Ω p413	**2-Pentene, 2-methyl-***	$CH_3CH_2CH:C(CH_3)_2$	84.16	λ^{hp} 192 (4.0)	−135.07	67.29^{760}	0.6863^{20}_{4}	1.4004^{20}	i	s	...	...	s	chl, peth s	**B1**[3], 809
p414	**—,3-methyl-(*cis*)***	$CH_3CH_2C(CH_3):CHCH_3$	84.16	λ^{gas} 179 (4.0), 184 (4.0), 187 (3.9), 199 sh (3.6)	−138.445	70.45^{760}	0.6986^{20}_{4}	1.4045^{20}	i	s	...	...	s	chl, peth s	**B1**[3], 813
Ω p415	**—,—(*trans*)***	$CH_3CH_2C(CH_3):CHCH_3$	84.16	λ^{gas} 175 sh (4.0), 180 (4.1), 186 sh (4.0)	−134.84	67.63^{760}	0.6942^{20}_{4}	1.4016^{20}	i	s	...	...	s	chl, peth s	**B1**[3], 813
Ω p416	**—,4-methyl-(*cis*)***	2-Isohexene. $(CH_3)_2CHCH:CHCH_3$	84.16	λ^{gas} 177 (4.2), 197 sh (3.2), 204 (2.9)	−134.43	56.3^{760}	0.6690^{20}_{4}	1.3880^{20}	i	s	...	...	s	chl, peth s	**B1**[3], 810
p417	**—,—(*trans*)***	$(CH_3)_2CHCH:CHCH_3$	84.16	λ^{gas} 180 (4.1), 186 sh (4.0)	−140.81	58.55^{760}	0.6686^{20}_{4}	1.3889^{20}	i	s	...	...	s	chl, peth s	**B1**[3], 810
p418	**1-Pentene, nonafluoro-2-trifluoromethyl-***	Perfluoro-2-methylpentene. $F_3C(CF_2)_2C(CF_3):CF_2$	300.05	...	...	60	...	...	i	...	...	...	s	...	**J1955**, 3005
Ω p419	**—,2,4,4-trimethyl-***	"Diisobutylene." $(CH_3)_3CCH_2C(CH_3):CH_2$	112.22	λ^{al} 206 (2.91)	−93.48	101.44	0.7150^{20}_{4}	1.4086^{20}	i	...	s	...	s	lig, chl s	**B1**[3], 849
Ω p420	**2-Pentene, 2,4,4-trimethyl-***	"Diisobutylene." $(CH_3)_3CCH:C(CH_3)_2$	112.22	λ^{al} 207 (3.40)	−106.33	104.91^{760} 2.33^{10}	0.7218^{20}_{4}	1.4160^{20}	i	...	s	...	s	lig v chl s	**B1**[3], 848
p421	**2-Pentenedioic acid**, diethyl ester(*trans*)*	Ethyl glutaconate. $C_2H_5O_2CCH_2CH:CHCO_2C_2H_5$	186.21	...	...	236–8 125^{12}	1.0496^{20}_{4}	1.4411^{20}	...	s	s	...	...	...	**B2**[3], 1931
p422	**—,3-methyl-**, 5-ethyl 1-methyl ester*	Ethyl methyl 3-methylglutaconate. $C_2H_5O_2CCH_2C(CH_3):CHCO_2CH_3$	186.21	...	...	118^{12}	...	...	...	s	...	...	...	...	**B2**, 778
Ω p423	**4-Pentenoic acid***	Allylacetic acid. $CH_2:CHCH_2CH_2CO_2H$	100.13	...	−22.5	188–9 93^{20}	0.9809^{20}_{4}	1.4281^{20}	δ	v	v	...	...	...	**B2**[3], 1297
p424	—,nitrile	Allylacetonitrile. Allylmethyl cyanide. $CH_2:CHCH_2CH_2CN$	81.12	...	...	140 $60–1^{40}$	0.8239^{24}	1.4213^{14}	i	∞	∞	...	...	...	**B2**, 426
Ω p425	**—,2-acetyl-**, ethyl ester	Ethyl allylacetoacetate. $CH_2:CHCH_2CH(COCH_3)CO_2C_2H_5$	170.21	...	...	211–2δd 102^{12}	0.9898^{20}_{4}	1.4388^{18}	i	∞	∞	...	∞	...	**B3**, 738
p426	**—,2,2-diethyl-**, amide	Novonal. $CH_2:CHCH_2C(C_2H_5)_2CONH_2$	155.24	wh pw, cr (eth-peth)	75–6	155^{10}	...	...	δ	v	v	...	...	...	**B2**[3], 1951
Ω p428	**2-Pentenoic acid, 4-hydroxy-, lactone***	β-Angelica lactone. Δ^1-Angelica lactone.	98.10	λ^{al} <214	< −17	$208–9^{751}$ 98^{15}	1.0810^{20}_{4}	1.4454^{20}_{He}	∞	s	s	...	...	...	**B17**[2], 297

For explanations, symbols and abbreviations see beginning of table. For structural formulas see end of table.

No.	Name	Synonyms and Formula	Mol. wt.	Color, crystalline form, specific rotation and λ_{max} (log ε)	m.p. °C	b.p. °C	Density	n_D	Solubility w	al	eth	ace	bz	other solvents	Ref.
	3-Pentenoic acid														
p429	**3-Pentenoic acid, 2-hydroxy-, nitrile (*cis*)**	Angelactinonitrile. *cis*-Crotonaldehyde cyanohydrin. $CH_3CH:CHCH(OH)CN$	97.12	....	....	139^{70} 112^{12}	0.9675^{15}_{4}	1.4460^{21}	s	s	s	...	s	chl s CS_2, lig i	**B3²**, 256
Ω p430	**—,4-hydroxy-, lactone***	γ-Angelica lactone. Δ^2-Angelica lactone.	98.10	nd λ^{peth} 217 (3.4)	18	167^{760} 53^{12}	1.084^{20}_{4}	1.4476^{20}	s^d	s	s	...	...	CS_2 s	**B17²**, 297
p431	**2-Pentenoic acid, 4-methyl-***	Isobutylideneacetic acid. α-Isohexenoic acid. $(CH_3)_2CHCH:CHCO_2H$	114.15	....	35	217 (211) $115–6^{20}$	0.9529^{21}_{4}	1.4489^{21}	...	v	v	v	...	os v	**B2²**, 406
p432	**1-Penten-3-ol***	Ethyl vinyl carbinol. $CH_2:CHCH(OH)CH_2CH_3$	86.14	....	....	$114–6^{760}$ 37^{20}	0.8395^{22}_{4}	1.4239^{20}	δ	∞	∞	...	...	....	**B1³**, 1909
Ω p433	**4-Penten-1-ol***	Allylethanol. $CH_2:CHCH_2CH_2CH_2OH$	86.14	....	....	140–2 (137)	0.8457^{20}_{4}	1.4309^{20}	δ	...	s	...	...	....	**B1³**, 1910
Ω p434	**4-Penten-2-ol***	Allyl methyl carbinol. $CH_2:CHCH_2CH(OH)CH_3$	86.14	....	....	$115–6^{750}$ (139–42)	0.8367^{20}_{4}	1.4225^{20}	v	∞	∞	...	...	....	**B1³**, 1910
p435	**3-Penten-2-ol, 2-methyl-***	Dimethyl propenyl carbinol. $CH_3CH:CHC(OH)(CH_3)_2$	100.16	....	....	121.6–2^{757} 37^{13}	0.8347^{20}_{4}	1.4302^{20}	s	∞	∞	...	...	....	**B1³**, 1932
Ω p436	**4-Penten-2-ol, 2-methyl-***	Dimethyl allyl carbinol. $CH_2:CHCH_2C(OH)(CH_3)_2$	100.16	....	....	119.5	0.8300^{20}_{4}	1.4263^{20}	δ	s	s	...	...	....	**B1³**, 1932
p437	**3-Penten-2-one (*trans*)***	Ethylideneacetone. Methyl propenyl ketone. $CH_3CH:CHCOCH_3$	84.13	λ 220 (4.12), 310 (1.61)	....	122	0.8624^{20}_{4}	1.4350^{20}	s	...	s	s	...	....	**B1²**, 791
p438	**1-Penten-3-one, 5,5-dimethyl-1-phenyl-**	Benzalpinacolone. 5-Methyl-1-phenyl-1-hexen-3-one*. $C_6H_5CH:CHCOCH_2CH(CH_3)_2$	188.27	cr	43	154^{25}	0.9509^{46}	1.5523^{25}	δ	s	...	...	s	chl s	**B7²**, 308
Ω p439	**—,1(4-methoxy-phenyl)-***	Ethyl 4-methoxystyryl ketone.	190.24	col-lt ye pl (eth-peth)	60	....	....	....	i	s	s	...	...	peth δ	**B8²**, 159
p440	**—,2-methyl-***	Ethyl isopropenyl ketone. $C_2H_5COC(CH_3):CH_2$	98.15	....	69.5	118.5	0.8530^{20}_{4}	1.4289^{20}	δ	s	...	s	...	....	**B1³**, 2994
Ω p441	**3-Penten-2-one,4-methyl-**	Mesityl oxide. $(CH_3)_2C:CHCOCH_3$	98.15	λ 227.5 (3.87), 253 (3.88), 295 (3.59)	−52.85	129.76^{760} 41^{23}	0.8653^{20}_{4}	1.4440^{20}	s	∞	∞	s	...	....	**B1³**, 2995
Ω p441¹	**1-Penten-3-one,1-phenyl-***	$C_6H_5CH:CHCOC_2H_5$	160.22	lf (lig)	38–9	142^{12}	0.8697^{20}_{4}	1.5684^{50}	δ	v	v	...	v	....	**B7²**, 298
p442	**1-Penten-3-yne***	Pyrylene. $CH_2:CHC:CCH_3$	66.10	....	....	59.2–60.1	0.7401^{20}_{4}	1.4496^{20}	i	...	s	...	s	....	**Am 55**, 1622
p443	**1-Penten-4-yne***	Allylacetylene. $CH_2:CHCH_2C:CH$	66.10	....	....	42–3	0.777^{22}_{22}	1.3653^{22}	i	...	s	...	s	....	**C47**, 2682
p444	**3-Penten-1-yne, 3-ethyl-***	$CH_3CH:C(C_2H_5)C:CH$	94.16	....	....	96.5 $41–3^{100}$	0.7886^{25}	1.4338^{25}	...	...	s	...	s	....	**C34**, 7844
p445	**1-Penten-3-yne, 2-methyl-***	$CH_2:C(CH_3)C:CCH_3$	80.14	....	....	$81–2^{100}$	....	1.4002^{20}	...	...	s	...	s	....	**C42**, 1871
p446	**3-Penten-1-yne, 3-methyl-***	$CH_3CH:C(CH_3)C:CH$	80.14	λ^{hx} 219 (3.95)	....	66–7 (67–9)	0.739^{20}	1.4332^{20}	...	...	s	...	s	....	**C51**, 4292
—	**Pentobarbital**	*see* **Barbituric acid, 5-ethyl-5(2-pentyl)-**													
p447	**1-Pentyne***	*n*-Propyl acetylene. $CH_3CH_2CH_2C:CH$	68.13	....	−90.0 (−105.7)	40.18^{760}	0.6901^{20}_{4}	1.3852^{20}	i	v	∞	...	s	chl s	**B1³**, 957
p448	**2-Pentyne***	Ethyl-methyl acetylene. $CH_3CH_2C:CCH_3$	68.13	....	−101 (−109.3)	56.07^{760}	0.7107^{20}_{4}	1.4039^{20}	i	v	∞	...	s	chl s	**B1³**, 958
p449	**1-Pentyne, 3-chloro-3-ethyl-***	$(C_2H_5)_2CClC:CH$	130.62	....	....	$73–6^{100}$	0.9230^{19}_{4}	1.4437^{19}	i	...	s	...	s	chl s	**B1³**, 1002
p450	**—,3-chloro-3-methyl-***	$CH_3CH_2CCl(CH_3)C:CH$ ...	116.59	....	....	102.5–7^{760} 55^{130}	0.9163^{20}_{4}	1.4330^{20}	i	...	s	...	s	chl s	**B1³**, 990
p451	**2-Pentyne, 4-chloro-4-methyl-***	Trimethylpropargyl chloride. $CH_3CCl(CH_3)C:CCH_3$	116.59	....	....	55^{70}	....	1.4143^{20}	...	...	...	v^h	s	chl s	**C25**, 1931
p452	**1-Pentyne, 4,4-dimethyl-***	$(CH_3)_3CCH_2C:CH$	96.17	....	−75.68	76.08^{760}	0.7142^{20}_{4}	1.3983^{20}	i	...	s	...	s	chl s	**C50**, 776
p453	**2-Pentyne, 4,4-dimethyl-***	$(CH_3)_3CC:CCH_3$	96.17	....	−82.37	82.9–3.0^{760}	0.7176^{20}_{4}	1.4071^{20}	i	...	s	...	s	chl s	**C50**, 776
p454	**1-Pentyne, 3-ethyl-***	$(C_2H_5)_2CHC:CH$	96.17	....	....	87–8.5	0.7246^{25}_{4}	1.4043^{25}	i	...	s	...	s	chl s	**B1³**, 1002
p455	**—,3-ethyl-3-methyl-***	$(C_2H_5)_2C(CH_3)C:CH$	110.20	....	....	101.5–2.5^{760}	0.7422^{20}_{4}	1.4110^{20}	i	...	s	...	s	chl s	**C51**, 16284
p456	**—,4-methyl-***	Isobutyl acetylene. $(CH_3)_2CHCH_2C:CH$	82.15	....	−105.1	61.1–2	0.7092^{15}_{4}	1.3936^{15}_{α}	i	...	...	...	s	chl s	**B1³**, 987
p457	**2-Pentyne, 4-methyl-***	Isopropyl methyl acetylene. $(CH_3)_2CHC:CCH_3$	82.15	....	−110.37	72.0–.5	0.716^{19}_{4}	1.4078^{19}	i	...	...	...	s	chl s	**B1³**, 987
p458	**2-Pentynoic acid***	Ethylpropiolic acid. $CH_3CH_2C:CCO_2H$	98.10	cr (peth)	50	122^{10} $81–2^{3}$	0.978^{20}	1.4619^{20}	s	...	...	...	...	peth s^h	**B2**, 481
p459	**4-Pentynoic acid***	Ethynylpropionic acid. $HC:CCH_2CH_2CO_2H$	98.10	....	57.7	102^{17}	....	....	v	v	v	...	...	....	**B2³**, 1450
p460	**1-Pentyn-3-ol, 3,4-dimethyl-***	Ethynyl methyl isopropyl carbinol. $(CH_3)_2CHC(OH)(CH_3)C:CH$	112.17	....	....	133^{735}	0.8691^{20}_{4}	1.4372^{20}	s	s	s	...	...	....	**B1³**, 1996

For explanations, symbols and abbreviations see beginning of table. For structural formulas see end of table.

No.	Name	Synonyms and Formula	Mol. wt.	Color, crystalline form, specific rotation and λ_{max} (log ε)	m.p. °C	b.p. °C	Density	n_D	Solubility w	al	eth	ace	bz	other solvents	Ref.
	1-Pentyn-3-ol														
Ω p461	—,3-methyl-*	$CH_3CH_2C(OH)(CH_3)C{:}CH$	98.15	...	30–1	120–1 61^{70}	0.8688_4^{20}	1.4310^{20}	...	...	...	...	...	...	**B1**^{3}, 1990
—	**Peonol**	*see* **Acetophenone, 2-hydroxy-4-methoxy-**													
—	**Peracetic acid**	*see* **Acetic acid, per-**													
—	**Perbenzoic acid**	*see* **Benzoic acid, per-**													
p462	**Perchloric acid, ethyl ester***	Ethyl perchlorate*. $CH_3CH_2OClO_3$	128.51	oil	...	89	...	...	d	∞	∞	...	...	...	**C30,** 8057
p463	—,methyl ester*	Methyl perchlorate*. CH_3OClO_3	114.49	oil	...	*ca.* 52	...	...	d	∞	∞	...	...	...	**C30,** 8057
p464	—,trichloromethyl ester*	Trichloromethyl perchlorate*. Cl_3COClO_3	217.82	...	−55	exp (40d)	...	...	d	...	∞	...	...	bz, al exp CCl_4 ∞	**C24,** 4506
p465	**Pereirine**	$C_{20}H_{26}N_2O.\frac{1}{2}H_2O$	323.44	pa ye amor pw, [α] +137.5 (al) λ^{al} 245 (3.93), 300 (3.47)	135d	...	...	...	i	v	v	...	...	chl v	**C28,** 5459
Ω p466	**Perimidine**	Peri-naphthimidazole	168.21	gr cr (dil al)	222	...	...	...	i	s	s	s	s	os, ac s	**B23**^{1}, 53
p467	**Peroxide, acetyl benzoyl**	$C_6H_5COOOCOCH_3$	180.17	wh nd (lig)	40–1	85–100 exp 130^{19}	...	...	d	d	s	...	...	chl, oils s	**B9**^{2}, 157
p468	—,diacetyl	Acetyl peroxide. $CH_3COOOCOCH_3$	118.09	nd (eth), lf	30 (26)	63^{21}	...	...	δ	s	s^h	...	...	CCl_4 v NaOH d	**B2**^{3}, 380
Ω p469	—,dibenzoyl	Benzoyl peroxide	242.23	rh (eth), pr	106–8 (104)	exp	...	1.545	δ	s	s	s	s	CS_2 s MeOH δ	**B9**^{2}, 157
p470	—,—,3,3′-dinitro-	$C_{14}H_8N_2O_8$. *See* p469	332.24	nd (al)	139–40d	...	...	...	...	...	v	v	v	AcOEt s^h	**B9**^{2}, 252
p471	—,—,4,4′-dinitro-	$C_{14}H_8N_2O_8$. *See* p469	332.24	ye cr (ace), nd (to)	156d	...	...	...	...	...	...	...	...	AcOEt, to s^h	**B9**^{2}, 270
Ω p472	—,di(*tert*-butyl)	$(CH_3)_3COOC(CH_3)_3$	146.23	...	−40	111^{760} 70^{197}	0.794^{20}	1.3890^{20}	i	...	...	∞	...	os, lig s	**Am 72,** 337
Ω p473	—,didodecanoyl	Alperox c. Lauroyl peroxide. $CH_3(CH_2)_{10}COOOCO(CH_2)_{10}CH_3$	398.63	wh pl	49	...	...	...	i	...	...	...	...	...	**C42,** 1252
p474	—,diethyl	Ethyl peroxide. $C_2H_5OOC_2H_5$	90.12	...	−70	65^{760}	0.8240^{19}	1.3715^{17}	δ	∞	∞	...	...	...	**B1**^{3}, 1313
p475	—,dimethyl, hexafluoro-	Perfluorodimethyl peroxide. F_3COOCF_3	170.01	...	...	−32 (−37)	...	...	d	...	...	...	...	aq KOH i	**C52,** 6154
p476	—,di(1-naphthyl)	$C_{10}H_7^{\alpha}COOOCOC_{10}H_7^{\alpha}$	342.36	yesh pr (chl, ace, eth)	91d exp	...	...	...	...	s	...	s	s	chl s	**E12B,** 3989
p477	—,disuccinyl	$(HO_2CCH_2CH_2CO)_2O_2$	234.16	cr pw	127d	...	...	...	s	s	δ	s	i	chl i	**B2,** 613
Ω p478	**Perseitol**	α-Mannoheptitol. Perseite.	212.20	nd, $[\alpha]^{20}$ +4.53 (w)	188 (cor)	...	...	...	s	δ s^h	δ	...	...	...	**B1**^{3}, 2412
Ω p479	**Perylene**	*peri*-Dinaphthalene	252.32	gold-br ye pl (bz, to, aa) λ^{al} 251 (4.70), 368 (3.68), 387 (4.08), 406 (4.42), 434 (4.56)	277–9 (273–4)	sub 350–400	1.35	...	i	δ	δ	v	s	chl, CS_2 v peth i	**B5**^{2}, 655
p480	**3-Perylene-carboxylic acid**	$C_{21}H_{12}O_2$. *See* p479	296.33	og-br nd ($PhNO_2$)	330	...	...	...	i	...	...	...	...	sulf s (vt, red flr) os s (gr flr)	**E14s,** 743
—	**Petn**	*see* **Pentaerythritol, tetranitrate**													
p481	**Peucedanin**	...	258.28	pr or pl (bz-peth), yesh cr (eth) λ^{MeOH} 255 (4.4), 298 (4.1), 340 (3.7)	109	$276–81^{17}$	...	...	δ	s^h	s	...	δ	chl, CS_2 v aa s peth δ	**B19**^{2}, 228
Ω p482	**α-Phellandrene(*d*)**	5-Isopropyl-2-methyl-1,3-cyclohexadiene*. $\Delta^{1,5\theta}$-*p*-Menthadiene.	136.24	$[\alpha]_D$ +49.1 (undil) $\lambda^{n-C_8H_{18}}$ 263 (3.52)	...	$175–6^{760}$ 61^{11}	0.8463_4^{25}	1.4777^{20}	i	i	s	...	...	...	**B5,** 129
p483	**β-Phellandrene**	3-Isopropyl-6-methylene-cyclohexene*. 2-*p*-Menthadiene.	136.24	$[\alpha]_D^{20}$ +65.2 (undil)	...	$171–2^{760}$ 57^{11}	0.8520_4^{20}	1.4788^{20}	i	i	s	...	...	...	**B5,** 131
—	**Phenacetin**	*see* **Acetic acid,** amide, *N*(4-ethoxyphenyl)-													
—	**Phenaceturic acid**	*see* **Glycine,** *N*-acetyl-*N*-phenyl-													
—	**Phenacyl alcohol**	*see* **Acetophenone,** α-hydroxy-													
—	**Phenacyl bromide**	*see* **Acetophenone,** α-bromo-													
—	**Phenacyl chloride**	*see* **Acetophenone,** α-chloro-													

For explanations, symbols and abbreviations see beginning of table. For structural formulas see end of table.

No.	Name	Synonyms and Formula	Mol. wt.	Color, crystalline form, specific rotation and λ_{max} (log ε)	m.p. °C	b.p. °C	Density	n_D	Solubility w	al	eth	ace	bz	other solvents	Ref.
	Phenanthrene														
Ω p484	**Phenanthrene***	...	178.24	mcl pl (al), lf (sub) λ^{hp} 187.5 (4.48), 211.5 (4.57)	101	340 210–15[12]	0.9800^{4}	1.59427	i	s	s	s	s	CS_2, chl, aa s Py s^h MeOH δ, s^h	**B5**, 667
p485	**—,2-acetyl-***	$C_{16}H_{12}O$. *See* p484	220.27	nd (MeOH) λ^{al} 218 (4.26), 266 (4.77), 304 (3.98), 368 (3.98)	144–5	...	...	...	i	s	...	...	v	MeOH s^h	**E13**, 878
p486	**—,3-acetyl-***	$C_{16}H_{12}O$. *See* p484	220.27	nd (MeOH) λ^{al} 210 (4.11), 252 (4.61), 314 (4.07)	72	...	...	...	i	s	...	...	s	MeOH v, chl, aa s lig δ	**E13**, 879
p487	**—,9-acetyl-***	$C_{16}H_{12}O$. *See* p484	220.27	bl flr nd (MeOH), lf (al)	74.5	...	...	...	i	s	v	...	s	MeOH s^h	**E13**, 879
p488	**—,2-amino-***	2-Phenanthrylamine. $C_{14}H_{11}N$. *See* p484	193.25	lt ye (lig)	85	...	...	...	i	i s^h	i s^h	...	...	lig s^h	**E13**, 819
p489	**—,3-amino-***	3-Phenanthrylamine. $C_{14}H_{11}N$. *See* p484	193.25	α lf (lig), β cr (lig)	α 143 β 87.5 (st)	...	...	...	δ	s	...	...	...	HCl s lig s^h xyl δ	**E13**, 820
p490	**—,9-amino- (a form)***	9-Phenanthrylamine. $C_{14}H_{11}N$. *See* p484	193.25	lt ye cr (al)	137–8	sub	...	...	i	v	v	...	v	chl v	**E13**, 819
p491	**—,—(b form)***	$C_{14}H_{11}N$. *See* p484	193.25	lt ye nd (al)	104	sub	...	...	i	v	v	...	v	chl v	**E13**, 819
Ω p492	**—,9-bromo-***	$C_{14}H_9Br$. *See* p484	257.14	pr (al)	63.8–4.8	>360 190[1.2] sub	1.4093_4^{10}	...	i	s	s	...	...	CS_2, aa s	**B5**[2], 583
p493	**—,9,10-diamino-***	$C_{14}H_{12}N_2$. *See* p484	208.27	pa ye lf	166	...	...	...	...	...	...	...	...	...	**E13**, 822
Ω p494	**—,9,10-dihydro-***	$C_{14}H_{12}$. *See* p484	180.25	nd (MeOH) λ^{al} 267 (4.2), 297 sh (3.7)	34.5–5	168–9[15]	1.0757_4^{40}	1.6415_4^{20}	...	s	v	...	...	MeOH δ, s^h	**E13**, 794
p495	**—,3,4-dihydroxy-***	3,4-Phenanthrenediol*. Morphol. $C_{14}H_{10}O_2$. *See* p484	210.24	col nd (peth), dk in air	143	sub 130 (vac)	...	...	i	s	s	...	...	alk s peth s^h	**E13**, 846
p496	**—,3,4-di-methoxy-***	Dimethylmorphol. $C_{16}H_{14}O_2$. *See* p484	238.29	bt ye lf (MeOH)	45	298–303[112]	...	...	i	v	v	...	...	MeOH s^h	**C34**, 4072
p497	**—,4,5-dimethyl-***	$C_{16}H_{14}$. *See* p484	206.29	pr (MeOH)	76–7.2	...	...	...	δ	...	...	...	...	$PhNO_2$, chl s	**E14**, 309
p498	**—,9,10-dimethyl-***	$C_{16}H_{14}$. *See* p484	206.29	lt red pr (aa), nd (MeOH)	144	sub	...	...	...	δ	...	...	s	chl, aa s	**E13**, 803
p499	**—,9,10-diphenyl-***	$C_{26}H_{18}$. *See* p484	330.44	nd (eth, bz)	240 (235)	sub 270	...	...	i	δ	s	...	s	...	**E13**, 810
p500	**—,1,2,3,4,5,6,7,8,9,10,11,12-dodeca-hydro-***	$C_{14}H_{22}$. *See* p484	190.33	...	...	81–2[1.5]	0.9674_4^{20}	1.5102^{20}	i	...	...	s	s	os s	**E13**, 795
p501	**—,9-ethyl-***	$C_{16}H_{14}$. *See* p484	206.29	λ^{al} 252 (4.81), 274 (4.18), 296 (4.12), 350 (2.70)	62.5–3.0 (66)	199–200[11]	1.0603_4^{78}	1.6582^{78}	i	s	...	...	s	peth δ	**E13**, 799
p502	**—,1,2,3,4,11,12-hexahydro-***	$C_{14}H_{16}$. *See* p484	184.28	λ^{al} 247 (3.75)	(i) −3 (ii) −8	(i) 307 167[13] (ii) 290	1.045^{20}	1.5810^{15}	i	δ	s	...	s	os, aa, CS_2, chl, peth s	**E13**, 794
p503	**—,1-hydroxy-***	1-Phenanthrol*. $C_{14}H_{10}O$. *See* p484	194.24	nd (peth, bz-lig, eth) λ^{al} 245 (4.08), 308 (4.1), 335 (3.5)	157	...	...	...	i	...	s^h	...	s^h	sulf s (red)	**E13**, 829
Ω p504	**—,2-hydroxy-***	2-Phenanthrol*. $C_{14}H_{10}O$. *See* p484	194.24	pl, lf (al, eth, lig) λ^{al} 255 (4.82), 292 (4.15), 355 (3.00)	168	...	...	...	δ	s	s	...	s	lig δ	**B6**, 704
Ω p505	**—,3-hydroxy-***	3-Phenanthrol*. $C_{14}H_{10}O$. *See* p 484	194.24	nd (al, lig) λ^{al} 248 (4.85), 305 (4.28), 355 (3.28)	122–3	...	...	...	i s^h	v	s	...	s^h	lig s^h	**B6**, 705
Ω p506	**—,9-hydroxy-***	9-Phenanthrol*. $C_{14}H_{10}O$. *See* p484	194.24	nd (lig, bz) λ^{al} 248 (4.6), 305 (3.8), 355 (3.2)	158	...	...	...	δ	v	v	...	v	chl v lig s	**E13**, 831
Ω p506[1]	**—,7-isopropyl-1-methyl-***	Retene. $C_{18}H_{18}$. *See* p484	234.34	pl (al) λ^{al} 259, 277, 301, 334, 351	100.5–1	390[760] 290[10]	1.035	...	i	s^h	s^h	...	s	CS_2, lig s aa s^h	**B5**, 683
p507	**—,1-methyl-***	$C_{15}H_{12}$. *See* p484	192.26	lf, pl (dil al)	123	...	...	...	i	s	...	...	...	...	**E13**, 798
Ω p508	**—,3-methyl-***	$C_{15}H_{12}$. *See* p484	192.26	pr or nd (al)	65	140–50[6]	...	...	i^h	s^h	...	s	...	os s	**B5**, 675
p509	**—,9-nitro-***	$C_{14}H_9NO_2$. *See* p484	223.23	ye nd (al)	116–7	...	...	...	...	s	s	...	s	sulf s (red) MeOH s lig δ	**E13**, 817

For explanations, symbols and abbreviations see beginning of table. For structural formulas see end of table.

No.	Name	Synonyms and Formula	Mol. wt.	Color, crystalline form, specific rotation and λ_{max} (log ε)	m.p. °C	b.p. °C	Density	n_D	Solubility: w	al	eth	ace	bz	other solvents	Ref.
	Phenanthrene														
Ω p510	**—,1,2,3,4,5,6,7,8-octahydro-***	Octanthrene. $C_{14}H_{18}$. *See* p484	186.30	.....	16.7	295, 169^{15}	1.026_4^{20}	1.5569^{17}	i	...	...	s	s	CS_2, aa s	**E13**, 795
p511	**—,1,2,3,4,9,10,11,12-octahydro- (*cis*)***	$C_{14}H_{18}$. *See* p484	186.30	.....	.....	129^6, $88–90^{0.1}$	1.0072_4^{25}	1.5549^{21}	i	...	...	...	s	.....	**E13**, 795
p512	**—,—(*trans*)***	$C_{14}H_{18}$. *See* p484	186.30	nd	23–4	$94–5^{15}$	1.0060_4^{20}	1.5528^{21}	i	...	...	...	s	.....	**E13**, 795
Ω p513	**—,tetra-decahydro-***	Perhydrophenanthrene. $C_{14}H_{24}$. *See* p484	192.35	.....	.....	$86–9^2$	0.9447_4^{20}	1.5011^{20}	i	...	s	s	s	os s	**E13**, 796
p514	**—,1,2,3,4-tetra-hydro-***	Tetranthrene. $C_{14}H_{14}$. *See* p484	182.27	lf (MeOH) λ^{al} 229 (4.87), 251 (3.82), 280 (3.77), 301 (3.95), 322 (2.87)	33–4	173^{11}	1.0601_4^{40}	.....	i	s	s	s	s	chl, aa, lig s	**E13**, 794
p515	**—,3,4,5-tri-hydroxy-***	3,4,5-Phenanthrenetriol*. $C_{14}H_{10}O_3$	226.24	lf or pl (w)	148	.....	.....	.....	i	v	s	...	...	chl s	**E13**, 862
p516	**1-Phenanthrene-carboxylic acid***	1-Phenanthroic acid. $C_{15}H_{10}O_2$. *See* p484	222.25	nd (al) λ^{al} 232 (4.3), 251 (4.76), 275 (4.07), 300 (4.00)	232–3	.....	.....	.....	...	s	...	...	s	.....	**E13**, 953
p517	**2-Phenanthrene-carboxylic acid***	2-Phenanthroic acid. $C_{15}H_{10}O_2$. *See* p484	222.25	nd (aa) λ^{al} 251 sh (4.69), 255 (4.72)	258.5–60	.....	.....	.....	...	s	...	...	s	aa s	**E13**, 953
p518	**—,nitrile**	$C_{15}H_9N$. *See* p484	203.25	cr (bz-lig, al)	108–9.5	.....	.....	.....	...	v	s	s	...	os v	**B9**, 706
p519	**3-Phenanthrene-carboxylic acid***	3-Phenanthroic acid. $C_{15}H_{10}O_2$. *See* p484	222.25	nd (aa) λ^{al} 245 (4.07), 257 (4.11)	270	sub	.....	.....	i	s	s	...	...	aa s	**E13**, 954
p520	**—,nitrile**	$C_{15}H_9N$. *See* p484	203.27	nd (abs al)	102	.....	.....	.....	...	s^h	s	s	...	os v	**B9**, 706
p521	**9-Phenanthrene-carboxylic acid***	9-Phenanthroic acid. $C_{15}H_{10}O_2$. *See* p484	222.25	nd (aa), lf (sub) λ^{al} 211 (4.66), 254 (4.88), 299 (4.24), 355 (3.61)	256.5–7	sub	.....	.....	i	s	s	...	s	aa s	**E13**, 955
Ω p522	**—,nitrile**	$C_{15}H_9N$. *See* p484	203.25	nd (al) λ^{al} 229 (4.65), 258 (4.73), 278 (4.08), 311 (4.16), 356 (3.00)	103	.....	.....	.....	...	s^h	s	s	...	os v	**B9**, 707
p523	**2-Phenanthrene-sulfonic acid***	$C_{14}H_{10}O_3S$. *See* p484	258.30	cr (bz), cr (w + 1)	*ca.* 150	.....	.....	.....	v	v	...	...	v^h	to, $PhNO_2$ s	**E13**, 995
p524	**3-Phenanthrene-sulfonic acid***	$C_{14}H_{10}O_3S$. *See* p484	258.30	lf (bz), cr (w + 1 or w + 2)	88–9 (+2w) 120–1 (+1w) 175–6 (anh)	.....	.....	.....	s	...	...	...	s^h	.....	**E13**, 995
p525	**9-Phenanthrene-sulfonic acid***	$C_{14}H_{10}O_3S$. *See* p484	258.30	lf or nd (bz or w + 2)	134 (hyd) 174 (anh)	.....	.....	.....	s	s	...	...	δ s^h	aa s	**E13**, 996
Ω p526	**Phenanthridine**	9-Azaphenanthrene. 3,4-Benzoquinoline.	179.22	nd (dil al)	106–7	349^{769}	.....	.....	δ	v	v	s	v	chl, CS_2 v	**B20**, 466
p527	**—,6-hydroxy-**	9-Phenanthridone. $C_{13}H_9NO$. *See* p526	195.22	nd (al, sub) λ^{al} 230 (4.7), 250 (4.1), 260 (4.3), 320 (4.0), 338 (3.9)	293–4	sub	.....	.....	i	δ s^h	δ	...	...	con sulf s $PhNO_2$ s^h aa δ	**B21**[2], 79
—	**Phenanthrol***	*see* **Phenanthrene, hydroxy-***													
Ω p528	**1,7-Phenan-throline**	1,7-Diazaphenanthrene*. *m*-Phenanthroline.	180.22	pl (anh), nd (w + 2)	78–8.5 (anh) 65.5 (+2w)	> 360	.....	.....	s^h	v	i	...	i δ^h	lig i	**B23**[1], 61
Ω p529	**1,10-Phenan-throline**	4,5-Diazaphenanthrene*. *o*-Phenanthroline.	180.22	wh nd (bz), cr (w + 1)	117 (anh)	> 300	.....	.....	v^h	s	...	s	s	peth i	**B23**, 227
p530	**4,7-Phenan-throline**	1,8-Diazaphenanthrene*. *p*-Phenanthroline.	180.22	nd (w) $\lambda^{dil\ HCl}$ 221 (4.6), 275 (4.5), 304 (3.8), 336 (3.6)	177	sub 100	.....	.....	s^h	v	δ	...	δ	chl v lig s^h CS_2 δ	**B23**[1], 61
Ω p531	**—,hydrate**	$C_{12}H_8N_2 . H_2O$. *See* p530	198.23	wh nd (w)	100–3 (94)	.....	.....	.....	δ s^h	δ	δ	...	...	.....	**B23**[2], 235
p532	**1,7-Phenan-throline, 9-nitro-**	$C_{12}H_7N_3O_2$. *See* p528	225.21	nd (dil al)	168	.....	.....	.....	δ s^h	s^h	...	...	...	os s	**B23**[2], 236
Ω p533	**9,10-Phenan-throquinone**	9,10-Phenanthrenequinone*.	208.23	og nd (to), og-red pl (sub) λ^{al} 252 (4.5), 283 (3.0)	208.5–10	> 360 sub	1.405_4^{20}	.....	i	δ	s	...	δ	aa s^h AcOEt δ	**E13**, 911

For explanations, symbols and abbreviations see beginning of table. For structural formulas see end of table.

No.	Name	Synonyms and Formula	Mol. wt.	Color, crystalline form, specific rotation and λ_{max} (log ε)	m.p. °C	b.p. °C	Density	n_D	Solubility w	al	eth	ace	bz	other solvents	Ref.
	9,10-Phenanthroquinone														
p534	**—,—,2-bromo-***	$C_{14}H_7BrO_2$. *See* p533	287.12	red-ye cr (aa)	233–4	...	...	...	...	δ	δ	...	δ	sulf s (gr) aa s^h	**E13,** 924
p535	**—,—,3-bromo-***	$C_{14}H_7BrO_2$. *See* p533	287.12	ye nd (aa)	268–9	...	...	...	i	δ	δ	...	s	sulf s (dk br) aa δ, s^h $PhNO_2$, chl s	**E13,** 925
p536	**—,2-chloro-***	$C_{14}H_7ClO_2$. *See* p533	242.66	ye-red nd (aa)	252–3	...	...	...	i	s	...	...	...	sulf s (gr-br) aa s^h	**E13,** 924
p537	**—,3-chloro-***	$C_{14}H_7ClO_2$. *See* p533	242.66	og-ye nd (aa, bz-al)	264–5 (246)	...	...	...	i	s	...	...	s	aa s	**E13,** 925
p538	**—,1,2-dihydroxy-***	$C_{14}H_8O_4$. *See* p533	240.23	dk red nd (ace)	d	...	...	...	δ	s	...	s	...	alk, aa, con sulf s	**E13,** 940
p539	**—,2,5-dihydroxy-***	$C_{14}H_8O_4$. *See* p533	240.23	dk red nd (w)	400d	...	...	...	s^h	...	...	...	...	alk, sulf s	**E13,** 941
p540	**—,2,7-dihydroxy-***	$C_{14}H_8O_4$. *See* p533	240.23	dk red or br nd	>400d (293)	...	...	...	...	s	s	s	δ	NaOH, aa s	**E13,** 941
p541	**—,4,5-dihydroxy-***	$C_{14}H_8O_4$. *See* p533	240.23	dk red (al), nd (w)	d >400	...	...	...	s^h	s	...	...	...	...	**B8**[2], 508
p542	**—,2,5-dinitro-***	$C_{14}H_6N_2O_6$. *See* p533	298.21	red ye pr (aa)	228	...	...	...	i	...	...	...	...	aa s	**E13,** 929
p543	**—,2,7-dinitro-***	$C_{14}H_6N_2O_6$. *See* p533	298.21	gold ye nd (aa)	301–3	...	...	...	i	δ	...	...	...	aa δ, s^h	**E13,** 929
p544	**—,2-hydroxy-***	$C_{14}H_8O_3$. *See* p533	224.23	br-red or bk-vt nd (aa)	283 (cor)	sub	...	...	s	...	...	...	...	Ac_2O s al-KOH s (bl)	**E13,** 937
p545	**—,3-hydroxy-***	$C_{14}H_8O_3$. *See* p533	224.23	ye-red or red nd (aa, MeOH, sub)	330d	sub	...	...	s	s	...	...	...	alk s (red) Ac_2O s	**E13,** 937
p546	**—,7-isopropyl-1-methyl-***	Retenoquinone. $C_{18}H_{16}O_2$. *See* p533	264.33	og nd (chl-al)	197–9	sub	...	...	i	s	s	...	i	aa, chl i	**B7,** 819
p547	**—,2-nitro-***	$C_{14}H_7NO_4$. *See* p533	253.22	ye lf or nd (aa)	260	...	...	...	...	i	...	δ	...	aa δ	**E13,** 927
p548	**—,1,2,4-tri-hydroxy-***	$C_{14}H_8O_5 \cdot C_2H_5OH$. *See* p533	302.29	red (al + 1)	d	...	...	...	...	v	...	...	...	alk, con sulf, Py s os δ	**E13,** 943
p549	**—,2,3,4-tri-hydroxy-***	$C_{14}H_8O_5$. *See* p533	256.23	red br pw	185d	...	...	...	s	δ	...	...	...	...	**E13,** 943
—	**Phenanthryl-amine***	*see* **Phenanthrene, amino-***													
Ω p550	**Phenazine**	Azophenylene.	180.22	ye-red nd (aa) λ^{MeOH} 248 (5.10), 350 sh (4.03)	176–7	>360 sub	...	...	δ	s^h	δ	...	s	...	**B23,** 223
p551	**—,9,10-dihydro-**	Hydrazophenylene. $C_{12}H_{10}N_2$. *See* p550	182.23	rh lf	317 (sealed tube)	...	...	...	i	δ	...	...	i	...	**B23,** 209
p552	**—,1,4-dihydroxy-, di(*N*-oxide)**	$C_{12}H_8N_2O_4$. *See* p550	244.22	purp (chl)	236d	...	...	...	i	...	...	...	...	chl, dil alk, con sulf s	**C50,** 358
p553	**—,1-hydroxy-**	Hemipyocyanine. 1-Phenazinol. $C_{12}H_8N_2O$. *See* p550	196.22	ye nd (bz, dil MeOH), lf (dil MeOH) λ^{al} 240 (4.2), 265 (4.7), 370 (3.9)	158	sub	...	...	δ s^h	δ	...	...	s^h	os v dil alk, con ac, Py s	**B23**[2], 360
Ω p554	**—,2-methyl-**	$C_{13}H_{10}N_2$. *See* p550	194.24	lt ye nd or pr λ^{al} 251.5 (5.2), 364.5 (4.2)	117	350d	...	...	δ^h	v	v	...	...	chl v lig δ	**B23**[2], 239
—	**Phenazinol**	*see* **Phenazine, hydroxy-**													
—	**Phenethyl alcohol**	*see* **Ethanol, 2-phenyl-***													
—	**Phenethyl amine**	*see* **Benzene, (1-amino-ethyl)-***													
—	**Phenethyl chloride**	*see* **Benzene, (chloro-ethyl)-***													
—	**Phenetidine**	*see* **Aniline, ethoxy-***													
—	**Phenetole**	*see* **Benzene, ethoxy-***													
—	**Phenetsal**	*see* **Benzoic acid, 2-hydroxy-, 4-acetamidophenyl ester**													
—	**Phenicin**	*see* **Biquinone, 3,3′-dihydroxy-5,5′-dimethyl-**													
—	**Phenobarbital**	*see* **Barbituric acid, 5-ethyl-5-phenyl-**													
—	**Phenocoll**	*see* **Glycine, amide, N′-ethoxyphenyl-**													
Ω p555	**Phenol***	Carbolic acid. Benzenol*. Hydroxybenzene*.	94.11	nd λ^{al} 218.5 (3.78), 266 sh (3.18), 271 (3.28), 276 sh (3.16)	43 (fr 41)	181.75^{760} 70.86^{10}	1.0722^{20}_{20} 1.0576^{20}_{4}	1.5509^{21} 1.5418^{41}	s ∞^h	s	v	∞	∞^h	CCl_4 ∞ chl, CS_2 s	**B6**[2], 116
Ω p556	**—,2-amino-***	*o*-Hydroxyaniline. C_6H_7NO. *See* p555	109.13	wh rh bipym nd (bz) λ^w 229 (3.79), 281 (3.43)	174	sub 153^{11}	1.328	...	s	v	s	...	δ	...	**B13**[2], 164

For explanations, symbols and abbreviations see beginning of table. For structural formulas see end of table.

No.	Name	Synonyms and Formula	Mol. wt.	Color, crystalline form, specific rotation and λ_{max} (log ε)	m.p. °C	b.p. °C	Density	n_D	Solubility w	al	eth	ace	bz	other solvents	Ref.
	Phenol														
Ω p557	—,3-amino-*	*m*-Hydroxyaniline. C_6H_7NO. *See* p555	109.13	pr (to) $\lambda^{0.1N\,HCl}$ 270 (326)	123	164[11]	...	...	s^h	v	v	...	δ	alk s to s^h lig δ	**B13**[2], 209
p558	—,—,hydrobromide*	$C_6H_7NO \cdot HBr$. *See* p555	190.06	pr (w)	224	...	...	...	...	...	...	...	...	...	**B13**, 403
p559	—,—,hydrochloride*	$C_6H_7NO \cdot HCl$. *See* p555	145.60	pr (w)	229	...	...	...	s^h	...	...	...	...	...	**B13**, 403
p560	—,—,hydroiodide	$C_6H_7NO \cdot HI$. *See* p555	237.05	pr (w)	209	...	...	...	s^h	...	...	...	...	...	**B13**, 403
Ω p561	—,4-amino-*	*p*-Hydroxyaniline. C_6H_7NO. *See* p555	109.13	wh pl (w) λ^w 229 (3.78), 294 (3.30)	186–7	sub par d $110^{0.3}$	...	...	δ v^h	v	...	...	i	alk s chl i	**B13**[2], 220
p562	—,2-amino-3-chloro-*	C_6H_6ClNO. *See* p555	143.57	nd (aq $NaHSO_3$)	122	...	...	...	s	...	...	...	...	...	**B13**[2], 182
p563	—,2-amino-5-chloro-*	C_6H_6ClNO. *See* p555	143.57	nd (al), pr (dil al)	154–5	...	...	...	s^h	s	s	...	δ	...	**B13**[2], 184
p564	—,3-amino-2-chloro-*	C_6H_6ClNO. *See* p555	143.57	...	85–7	...	...	...	δ	s	s	...	...	...	**B13**, 420
p565	—,4-amino-2-chloro-*	C_6H_6ClNO. *See* p555	143.57	nd (al, eth, w)	153	...	...	...	δ	s	s	...	...	alk s	**B13**[2], 272
p566	—,4-amino-3-chloro-*	C_6H_6ClNO. *See* p555	143.57	nd (dil $NaHSO_3$)	160	...	...	...	δ	s	s	...	...	...	**B13**[2], 273
p567	—,2-amino-4-chloro-5-nitro-*	$C_6H_5ClN_2O_3$. *See* p555	188.57	ye nd λ^w 285 (3.29)	225d (dk at 200)	...	...	...	δ^h	v	...	...	...	alk s	**B13**[2], 196
p568	—,2-amino-4-chloro-6-nitro-*	$C_6H_5ClN_2O_3$. *See* p555	188.57	...	152	...	...	...	δ^h	v	...	...	...	...	**B13**[2], 196
p569	—,2-amino-6-chloro-4-nitro-*	$C_6H_5ClN_2O_3$. *See* p555	188.57	ye nd (w + l)	160	...	...	...	δ^h	v	v	...	...	...	**B13**[2], 195
p570	—,4-amino-2-chloro-6-nitro-*	$C_6H_5ClN_2O_3$. *See* p555	188.57	...	130	...	...	...	δ^h	v	...	...	...	...	**B13**, 524
p571	—,2-amino-3,5-dibromo-*	$C_6H_5Br_2NO$. *See* p555	266.93	nd (lig)	145 (142–3)	...	...	...	s	...	...	...	...	alk s lig s^h	**B13**[2], 187
p572	—,2-amino-4,6-dibromo-*	$C_6H_5Br_2NO$. *See* p555	266.93	ye nd (dil al)	99	...	...	...	δ	s	s	...	s	chl s	**B13**[2], 188
p573	—,4-amino-2,6-dibromo-*	$C_6H_5Br_2NO$. *See* p555	266.93	nd (al, bz) λ^w 315 (3.51)	192.5–3	...	...	...	i	s	δ	...	s^h	...	**B13**[2], 280
p574	—,2-amino-3,5-dichloro-*	$C_6H_5Cl_2NO$. *See* p555	178.02	nd (bz, w)	132–3	...	...	...	δ	s	...	s	s^h	oos, chl s peth i	**B13**[2], 185
p575	—,3-amino-4,6-dichloro-*	$C_6H_5Cl_2NO$. *See* p555	178.02	ye br pr (w)	135–6	...	...	...	i s^h	s	v	v	s	chl s	**B13**[1], 135
p576	—,4-amino-2,5-dichloro-*	$C_6H_5Cl_2NO$. *See* p555	178.02	cr (bz)	178–9	...	...	...	d^h	s	v	...	s^h	aa v alk s chl i	**B13**[2], 274
Ω p577	—,4-amino-2,6-dichloro-*	$C_6H_5Cl_2NO$. *See* p555	178.02	nd or lf (w, bz)	167	sub	...	...	i δ^h	v	v	...	δ s^h	aa δ	**B13**[2], 274
p578	—,4-amino-3,5-dichloro-*	$C_6H_5Cl_2NO$. *See* p555	178.02	nd (w, bz)	154	...	...	...	s	s	s	...	...	chl, CS_2, CCl_4, aa s	**B13**[2], 276
Ω p579	—,2-amino-4,6-dinitro-*	Picramic acid. $C_6H_5N_3O_5$. *See* p555	199.12	dk red nd (al), pr (chl)	169	...	...	...	s	s	δ	...	s	aa s chl δ	**B13**[2], 196
p580	—,2-amino-3-nitro-*	$C_6H_6N_2O_3$. *See* p555	154.13	red nd (w)	216–7	sub	...	...	v^h	...	...	...	...	...	**B13**[1], 191
Ω p581	—,2-amino-4-nitro-*	$C_6H_6N_2O_3$. *See* p555	154.13	og pr (+w)	80–90 (+w) 145–7 (anh)	...	...	...	δ	v	s	...	s^h	MeOH, aa s	**B13**[2], 192
p582	—,2-amino-6-nitro-*	$C_6H_6N_2O_3$. *See* p555	154.13	red nd (dil al)	111–2	...	...	...	δ	s	v	...	v	chl, aa v	**B13**[1], 122
p583	—,3-amino-4-nitro-*	$C_6H_6N_2O_3$. *See* p555	154.13	og nd (w)	185–6	...	...	...	δ^h	s	v	...	v	chl v	**B13**[1], 136
Ω p584	—,4-amino-2-nitro-*	$C_6H_6N_2O_3$. *See* p555	154.13	dk red pl or nd (w, al)	131	...	...	...	s^h	s	s	...	...	...	**B13**[2], 839
p585	—,4-amino-3-nitro-*	$C_6H_6N_2O_3$. *See* p555	154.13	dk red pr (eth)	154	...	...	...	s	s	s	...	...	chl s	**B13**[1], 186
Ω p586	—,4(2-amino-propyl)-*	$C_9H_{13}NO$. *See* p555	151.21	cr (bz)	125–6	...	...	...	s	s	...	...	s^h	chl, AcOEt s	**B13**[1], 251
—	—,anilino-	*see* **Amine, diphenyl, hydroxy-**													
p587	—,2-benzylidene-amino-	$C_{13}H_{11}NO$. *See* p555	197.24	lf (al) λ^{al} 251 (4.1), 345 (3.8)	89	...	...	...	...	δ	v	...	v	con sulf s	**B13**[1], 112
p588	—,2-benzyloxy-	Pyrocatechol monobenzyl ether. $C_{13}H_{12}O_2$. *See* p555	200.24	...	...	173–4[13]	1.154[22]	1.5906[18]	...	s	s	...	...	alk s	**C53**, 14106
p589	—,3-benzyloxy-	Resorcinol monobenzyl ether. $C_{13}H_{12}O_2$. *See* p555	200.24	cr (CCl_4)	69.2	200[5]	...	...	s	...	...	...	...	5 % KOH s CCl_4 s^h	**Am 53**, 3997
Ω p590	—,4-benzyloxy-	Hydroquinone monobenzyl ether. $C_{13}H_{12}O_2$. *See* p555	200.24	pl (w)	122–2.5	...	...	...	δ v^h	v	v	...	v	...	**B6**, 845
p591	—,4[bis(2-hydroxyethyl) amino]-*	*N-p*-Phenoldiethanolamine. $C_{10}H_{15}NO_3$. *See* p555	197.24	cr (w or dil Na_2CO_3)	140	...	...	...	s	...	...	...	...	ac, alk s	**B13**[2], 233

For explanations, symbols and abbreviations see beginning of table. For structural formulas see end of table.

No.	Name	Synonyms and Formula	Mol. wt.	Color, crystalline form, specific rotation and λ_{max} (log ε)	m.p. °C	b.p. °C	Density	n_D	Solubility w	al	eth	ace	bz	other solvents	Ref.
	Phenol														
Ω p592	—,2-bromo-*	C_6H_5BrO. *See* p555	173.02	λ^{al} 214 sh (3.93), 275.5 (3.44), 281 sh (3.87)	5.6	194–5^{760} 87.3^{13}	1.4924$^{20}_{4}$	1.5892^{20}	δ	s	s	...	...	alk s	**B6**^{3}, 735
Ω p593	—,3-bromo-*	C_6H_5BrO. *See* p555	173.02	lf λ^{cy} 210 (3.88), 272 (3.36), 279 (3.35)	33	236.5^{760} 135–40^{12}	...	...	δ	v	v	...	...	chl, alk s	**B6**^{3}, 738
Ω p594	—,4-bromo-*	C_6H_5BrO. *See* p555	173.02	tetr λ^{cy} 224 (4.0), 279.5 (3.22), 287 (3.18)	66.4	238 118.2^{11}	1.840^{15}	...	s	v	v	...	...	chl s	**B6**^{3}, 739
p595	—,3-bromo-5-chloro-*	C_6H_4BrClO. *See* p555	207.46	nd (peth)	70	256–60^{756}	...	...	...	s	...	...	...	10 % alk, aa s	**B6**^{2}, 187
Ω p596	—,4-bromo-2-chloro-*	C_6H_4BrClO. *See* p555	207.46	nd (lig, bz, aa)	50–1	233–4^{760} 127–30^{12}	1.6170$^{20}_{4}$	1.5859^{20}	...	v	v	v	v	to v lig δ^h	**B6**^{3}, 750
p597	—,3-bromo-2,4-dinitro-*	$C_6H_3BrN_2O_5$. *See* p555	263.02	pa ye nd (w)	175	...	...	...	v	...	s	...	...	...	**B6**^{2}, 249
p598	—,3-bromo-2,6-dinitro-*	$C_6H_3BrN_2O_5$. *See* p555	263.02	nd (peth)	131	...	...	...	δ	s	s	...	...	lig δ, v^h	**B6**^{2}, 250
p599	—,4-bromo-2,6-dinitro-*	$C_6H_3BrN_2O_5$. *See* p555	263.02	pa ye nd (w), nd (al), pr (aa)	78	sub	...	...	δ s^h	s	v	...	v	chl v aa s^h lig δ	**B6**^{3}, 872
p600	—,5-bromo-2,4-dinitro-*	$C_6H_3BrN_2O_5$. *See* p555	263.02	pr (al, eth)	92	...	...	...	s	v	v	...	...	con HNO_3 s	**B6**^{3}, 871
p601	—,6-bromo-2,4-dinitro-*	$C_6H_3BrN_2O_5$. *See* p555	263.02	pa ye cr (al), pr (eth), col nd (w, al)	118–9	sub	...	...	δ^h	s^h	v	...	v	lig s	**B6**^{3}, 871
p602	—,2-bromo-3-nitro-*	$C_6H_4BrNO_3$. *See* p555	218.01	pa ye nd (HCl)	147–8	sub	...	...	i δ^h	v	v	...	...	CS_2, lig δ	**B6**, 244
p603	—,2-bromo-4-nitro-*	$C_6H_4BrNO_3$. *See* p555	218.01	cr (to, w), nd (chl, eth, dil al)	114	...	...	...	δ^h	v	v	...	...	CCl_4, chl s	**B6**^{3}, 845
Ω p604	—,2-bromo-5-nitro-*	$C_6H_4BrNO_3$. *See* p555	218.01	pa ye nd (w), cr (peth)	129–30 (124)	...	...	...	δ	v	v	s	s	os s lig δ	**B6**^{2}, 234
p605	—,2-bromo-6-nitro-*	$C_6H_4BrNO_3$. *See* p555	218.01	pa ye nd (chl, al)	68	...	...	...	s^h	v	...	...	...	aa s	**B6**^{3}, 844
p606	—,3-bromo-2-nitro-*	$C_6H_4BrNO_3$. *See* p555	218.01	ye nd (peth), col nd (+w)	65–7 (anh) 35 (hyd)	...	...	...	s	v	...	...	...	alk s lig δ, v^h	**B6**^{3}, 842
p607	—,3-bromo-4-nitro-*	$C_6H_4BrNO_3$. *See* p555	218.01	pa ye nd (bz, peth)	129–30	...	...	...	...	v	v	...	s v^h	lig s^h	**B6**^{2}, 234
p608	—,3-bromo-5-nitro-*	$C_6H_4BrNO_3$. *See* p555	218.01	cr (w)	145	...	...	...	δ s^h	v	s	...	δ	...	**B6**^{2}, 233
p609	—,4-bromo-2-nitro-*	$C_6H_4BrNO_3$. *See* p555	218.01	ye nd or lf (al), pr (eth)	92	sub	...	...	i	s	v	...	v	chl v lig δ	**B6**^{3}, 842
p610	—,4-bromo-3-nitro-*	$C_6H_4BrNO_3$. *See* p555	218.01	ye nd (w)	147	...	...	...	i δ^h	...	s	...	s	...	**B6**^{3}, 845
p611	—,5-bromo-2-nitro-*	$C_6H_4BrNO_3$. *See* p555	218.01	ye pr or nd (lig)	44	...	...	...	s^h	v	v	...	...	lig s	**B6**^{3}, 844
p612	—,2-butoxy-*	Pyrocatechol monobutyl ether. $C_{10}H_{14}O_2$. *See* p555	166.22	...	...	231–4 159^{69}	1.026^{25}	1.5113^{25}	...	...	...	...	...	...	**B6**, 772
—	—,butyl-*	*see* **Benzene, butyl-(hydroxy)-***													
Ω p613	—,2-chloro-*	C_6H_5ClO. *See* p555	128.56	λ^{al} 216 (3.81), 274.5 (3.37), 280 sh (3.30)	9.0	174.9^{760} 56.4^{10}	1.2634$^{20}_{4}$	1.5524^{20}	δ	s	s	...	v	...	**B6**^{3}, 671
Ω p614	—,3-chloro-*	C_6H_5ClO. *See* p555	128.56	nd λ^{cy} 216 (3.81), 272 (3.31), 279 (3.30)	33 (35–6)	214^{760}	1.268^{25}	1.5565^{40}	δ s^h	s	s	...	v	...	**B6**^{3}, 681
Ω p615	—,4-chloro-*	C_6H_5ClO. *See* p555	128.56	nd λ^{cy} 224 (3.94), 279 (3.27), 287 (3.23)	43.2–3.7	219.75 (217) 125^{18}	1.2651$^{40}_{4}$	1.5579^{40}	i	v	v	...	v	alk s	**B6**^{3}, 684
p616	—,4-chloro-2,3-dinitro-*	$C_6H_3ClN_2O_5$. *See* p555	218.55	pr	127	...	...	...	δ^h	s	v	...	v	chl s	**B6**^{2}, 248
p617	—,5-chloro-2,4-dinitro-*	$C_6H_3ClN_2O_5$. *See* p555	218.55	nd (al, peth)	92	...	1.74^{22}	...	δ	s v^h	s	...	...	chl s peth s^h	**B6**^{2}, 247
Ω p618	—,2-chloro-4-nitro-*	$C_6H_4ClNO_3$. *See* p555	173.55	wh nd (w, 50 % al) $\lambda^{0.1N\ HCl}$ 240 (4.11), 317 (3.93)	111	...	...	...	s^h	s	s	...	δ	chl s	**B6**^{3}, 839
p619	—,2-chloro-5-nitro-*	$C_6H_4ClNO_3$. *See* p555	173.55	ye nd or pr (w), rosettes (bz)	121–2	...	...	...	s^h	...	...	...	s^h	...	**B6**^{3}, 838
Ω p620	—,4-chloro-2-nitro-*	$C_6H_4ClNO_3$. *See* p555	173.55	ye mcl pr (al)	88–9 (87)	...	...	...	i	s v^h	s	...	...	chl s	**B6**^{3}, 834
Ω p621	—,5-chloro-2-nitro-*	$C_6H_4ClNO_3$. *See* p555	173.55	ye pr or nd (w)	41	sub	...	...	δ	s	s	...	...	aa s	**B6**^{3}, 836
p622	—,2,4-diamino-*	$C_6H_8N_2O$. *See* p555	124.14	lf	78–80d	...	...	...	...	s	δ	s	...	ac, alk s chl, lig δ	**B13**^{2}, 308

For explanations, symbols and abbreviations see beginning of table. For structural formulas see end of table.

No.	Name	Synonyms and Formula	Mol. wt.	Color, crystalline form, specific rotation and λ_{max} (log ε)	m.p. °C	b.p. °C	Density	n_D	Solubility w	al	eth	ace	bz	other solvents	Ref.
	Phenol														
Ω p623	—,—,dihydro-chloride*	Amidol. $C_6H_8N_2O.2HCl$. *See* p555	197.07	nd $\lambda^{w, pH=3}$ 235 sh (3.6), 280 (3.3)	230–40d	...	...	...	v	δ	δ	...	...	...	**B13**[2], 308
p624	—,2,5-diamino-*	$C_6H_8N_2O$. *See* p555	124.15	nd	68	...	...	...	v	...	...	...	...	...	**B13**[2], 312
p625	—,3,4-diamino-*	$C_6H_8N_2O$. *See* p555	124.15	nd	170–2	...	...	...	...	...	...	...	...	...	**B13**[1], 210
p626	—,3,5-diamino-*	$C_6H_8N_2O$. *See* p555	124.15	nd or pr (chl)	168–70 (180)	...	...	...	v	...	δ	...	...	chl s^h	**B13**, 567
Ω p627	—,2,4-dibromo-*	$C_6H_4Br_2O$. *See* p555	251.92	nd (peth)	40	238–9 177^{17}	...	...	δ s^h	v	v	...	v	CS_2 v	**B6**[2], 188
p628	—,2,6-dibromo-*	$C_6H_4Br_2O$. *See* p555	251.92	nd (w) $\lambda^{0.1 N\ NaOH}$ 367.5 (3.57)	56–7	162^{21} sub	...	...	s^h	v	v	...	...	...	**B6**[2], 189
Ω p629	—,2,6-dibromo-4-nitro-*	$C_6H_3Br_2NO_3$. *See* p555	296.92	pa ye pr or lf (al)	145–6	d > 144	...	...	i	v^h	v^h	...	δ	CS_2 v^h aa δ lig i	**B6**[2], 234
Ω p630	—,2,3-dichloro-*	$C_6H_4Cl_2O$. *See* p555	163.00	cr (lig, bz) $\lambda^{0.1 N\ HCl}$ 277 (3.52), 280 (3.30), 283 (3.29)	57–9	...	...	...	...	s	s	...	s^h	lig s^h	**B6**[3], 699
Ω p631	—,2,4-dichloro-*	$C_6H_4Cl_2O$. *See* p555	163.00	hex nd (bz) $\lambda^{0.1 N\ NaOH}$ 304 (3.50)	45	210^{760} 145–7^{110}	...	...	δ	s	s	...	s	chl s	**B6**[3], 699
Ω p632	—,2,5-dichloro-*	$C_6H_4Cl_2O$. *See* p555	163.00	pr (bz, peth) $\lambda^{0.1 N\ HCl}$ 280 (3.41)	59	211^{744}	...	...	δ	v	v	...	s	peth s^h	**B6**[3], 712
Ω p633	—,2,6-dichloro-*	$C_6H_4Cl_2O$. *See* p555	163.00	nd (peth) $\lambda^{0.1 N\ NaOH}$ 238 (3.85), 301 (3.71)	68–9	219–20^{740} 80–5^{4}	...	...	...	v	v	...	s	peth s^h	**B6**[3], 713
Ω p634	—,3,4-dichloro-*	$C_6H_4Cl_2O$. *See* p555	163.00	nd (bz-peth) $\lambda^{0.1 N\ NaOH}$ 244 (4.12), 302 (3.51)	68	253.5^{767}	...	...	δ	v	v	...	s	peth s^h	**B6**[3], 715
Ω p635	—,3,5-dichloro-*	$C_6H_4Cl_2O$. *See* p555	163.00	pr (peth) λ^{al} 220 (3.86), 278 (3.23)	68	233^{757} 122–4^{8}	...	...	δ	v	v	...	...	peth s^h	**B6**[3], 715
Ω p637	—,2,6-dichloro-4-nitro-*	$C_6H_3Cl_2NO_3$. *See* p555	208.00	br nd (w) λ^{hp} 285 (3.81)	127d (exp <100)	...	1.822	...	δ s^h	s^h	s	...	δ	chl s	**B6**[3], 841
Ω p638	—,3(diethyl-amino)-*	$C_{10}H_{15}NO$. *See* p555	165.24	rh bipym (CS_2-lig)	78	276–80 170^{15}	...	...	s	s	s	...	...	CS_2 s lig δ	**B13**[2], 212
p639	—,2,4-diiodo-*	$C_6H_4I_2O$. *See* p555	345.91	nd (w)	72–3	sub 100	...	...	δ	v	v	...	δ	chl δ	**B6**[3], 786
p640	—,2,3-dimeth-oxy-*	Pyrogallol 1,2-dimethyl ether. $C_8H_{10}O_3$. *See* p555	154.17	...	...	233–4^{760} 124–5^{17}	...	1.5392^{20}	...	...	...	...	...	...	**B6**, 1081
p641	—,2,6-dimeth-oxy-*	Pyrogallol 1,3-dimethyl ether. $C_8H_{10}O_3$. *See* p555	154.17	mcl pr (w) λ^{MeOH} 267.5 (3.04)	56–7	262–7	...	...	δ	s	v	...	...	aq alk s	**B6**[2], 1065
Ω p642	—,3,5-dimeth-oxy-*	Phloroglucinol dimethyl ether. $C_8H_{10}O_3$. *See* p555	154.17	cr (bz-lig)	36–8	198–200^{35} 172–5^{17}	...	...	...	...	s	...	s	lig δ	**B6**[2], 1078
—	—,dimethyl-*	*see* **Benzene, dimethyl-(hydroxy)-***													
Ω p643	—,3(dimethyl-amino)-*	$C_8H_{11}NO$. *See* p555	137.18	nd (lig) λ^{cy} 252 (4.11), 294 (3.50)	87	265–8^{760} 152–3^{15}	...	1.5895^{26}	i δ^h	s	s	s	s	CS_2 s lig s^h	**B13**[2], 211
p644	—,2,3-dinitro-*	$C_6H_4N_2O_5$. *See* p555	184.11	ye nd (w)	144–5	...	1.681^{20}	...	δ s^h	v^h	v	...	s	...	**B6**[3], 854
Ω p645	—,2,4-dinitro-*	$C_6H_4N_2O_5$. *See* p555	184.11	pa ye pl or lf (w) λ^{AcOEt} 260 (3.91), 280 (3.76)	115–6 (113)	sub	1.683^{24}	...	δ s^h	s	s	s	s	chl, to, MeOH s Py s, v^h	**B6**[3], 854
Ω p646	—,2,5-dinitro-*	$C_6H_4N_2O_5$. *See* p555	184.11	ye mcl pr or nd (dil al, w, lig) λ^{Py} 360 (3.5), 465 (2.5)	108	...	...	...	δ v^h	δ v^h	v	...	s v^h	MeOH δ	**B6**[3], 866
Ω p647	—,2,6-dinitro-*	$C_6H_4N_2O_5$. *See* p555	184.11	pa ye rh nd or lf (dil al)	63–4	...	...	...	i δ^h	v^h	v	v	s v^h	chl, Py s CS_2, CCl_4 δ	**B6**[3], 867
p648	—,3,4-dinitro-*	$C_6H_4N_2O_5$. *See* p555	184.11	tcl nd (w)	134	...	1.672	...	δ s^h	s	s	...	s	...	**B6**[3], 868
p649	—,3,5-dinitro-*	$C_6H_4N_2O_5$. *See* p555	184.11	lf (w)	126.1	...	1.702	...	δ s^h	s	s	...	s	chl s lig δ	**B6**[3], 869
Ω p650	—,2-ethoxy-*	Guaethol. Pyrocatechol monoethyl ether. $C_8H_{10}O_2$. *See* p555	138.17	...	29	217 68^{4}	...	...	δ	∞	∞	...	...	...	**B6**, 771
Ω p651	—,3-ethoxy-*	*m*-Hydroxyphenetole. Resorcinol monoethyl ether. $C_8H_{10}O_2$. *See* p555	138.17	ye	...	246–7 $117^{5.5}$	1.0705^{4}_{4}	...	i	s	s	...	s	...	**B6**, 814
Ω p652	—,4-ethoxy-*	*p*-Hydroxyphenetole. Hydroquinone monoethyl ether. $C_8H_{10}O_2$. *See* p555	138.17	pr or lf (w)	66–7	246–7	...	...	δ s^h	v	v	...	...	...	**B6**, 843

For explanations, symbols and abbreviations see beginning of table. For structural formulas see end of table.

No.	Name	Synonyms and Formula	Mol. wt.	Color, crystalline form, specific rotation and λ_{max} (log ε)	m.p. °C	b.p. °C	Density	n_D	Solubility: w	al	eth	ace	bz	other solvents	Ref.
	Phenol														
—	—,**ethyl-***	*see* **Benzene, ethyl-(hydroxy)-**													
p653	—,**2(ethyl-amino)-***	$C_8H_{11}NO$. *See* p555	137.18	pl (bz)	113–4	...	...	...	i	v	δ	...	s^h	chl δ	**B13,** 364
p654	—,**3(ethyl-amino)-***	$C_8H_{11}NO$. *See* p555	137.18	cr (bz-peth)	62	176^{12} 143^{5}	...	...	s^h	s	s	...	s	chl v peth δ	**B13,** 408
p655	—,**4(ethyl-amino)-***	$C_8H_{11}NO$. *See* p555	137.18	nd (w)	110–2 (104)	...	...	...	s^h	s	s	...	...	...	**B13,** 443
p655[1]	—,**4(heptyloxy)-***	Hydroquinone monoheptyl ether. $C_{13}H_{20}O_2$. *See* p555	208.30	cr (lig)	60	...	...	...	δ	...	...	...	...	lig s^h	**Am 54,** 298
p655[2]	—,**4(hexyloxy)-***	Hydroquinone monohexyl ether. $C_{12}H_{18}O_2$. *See* p555	194.28	cr (lig)	48	...	...	...	δ	...	...	...	...	lig s^h	**Am 54,** 298
p656	—,**2-iodo-***	C_6H_5IO. *See* p555	220.01	nd λ^{cy} 218 (3.97), 229 (3.81), 274 (3.48), 281.5 (3.45)	43	$186\text{–}7^{160}$ $91\text{–}2^{2}$	1.8757^{80}	...	s^h	v	v	...	...	CS_2 v	**B6**[3], 768
p657	—,**3-iodo-***	C_6H_5IO. *See* p555	220.01	nd (lig) λ^{w} 277 (3.26), λ^{alk} 297 (3.45)	118	d	...	...	δ	s	s	...	...	...	**B6**[3], 771
Ω p658	—,**4-iodo-***	C_6H_5IO. *See* p555	220.01	nd (w or sub) λ^{cy} 230.5 (4.18), 273.5 (3.11), 279.5 (3.16), 288 (3.04)	93–4	138–40^{5}d	1.8573^{112}	...	δ	v	v	...	...	...	**B6**[3], 774
p659	—,**2-mercapto-***	Monothiopyrocatechol. C_6H_6OS. *See* p555	126.18	oil	5–6	$216\text{–}7^{751}$ $88\text{–}90^{8}$	1.2373^{0}_{0}	...	δ	...	s	...	...	alk s	**B6,** 793
p660	—,**3-mercapto-***	Monothioresorcinol. C_6H_6OS. *See* p555	126.18	cr	16–7	168^{35}	...	...	δ	s	...	...	...	os v alk, con sulf s	**B6**[2], 827
p661	—,**4-mercapto-***	Monothiohydroquinone. C_6H_6OS. *See* p555	126.18	cr λ^{al} 255 (4.04)	29–30	$166\text{–}8^{45}$ $133\text{–}7^{11}$	1.1285^{25}_{4}	1.5101^{25}	s	s	...	...	...	alk, con sulf s	**B6**[1], 419
Ω p662	—,**2-methoxy-***	Pyrocatechol monomethyl ether. Guaiacol. $C_7H_8O_2$. *See* p555	124.15	hex pr λ^{al} 274 (3.41)	32 (fr 28.65)	205.05^{760} 106.5^{24}	1.1287^{21}_{4}	1.5429^{20}	δ	s	s	...	...	os v chl s	**B6**[2], 776
p663	—,—,acetate	$C_9H_{10}O_3$. *See* p555	166.18	...	...	$123\text{–}4^{13}$ (107^{22})	1.1285^{25}_{4}	1.5101^{25}	i	∞	∞	...	...	...	**B6**[2], 783
Ω p664	—,**3-methoxy-***	*m*-Hydroxyanisole. Resorcinol monomethyl ether. $C_7H_8O_2$. *See* p555	124.15	λ^{cy} 220 (3.73), 271.5 (3.22), 278 (3.19)	< −17.5	244.3^{760} 144^{25}	...	1.5520^{20}	δ	∞	∞	...	...	...	**B6,** 813
Ω p665	—,**4-methoxy-***	Hydroquinone monomethyl ether. 4-Hydroxyanisole. $C_7H_8O_2$. *See* p555	124.15	pl (w) λ^{cy} 224 (3.85), 288 (3.50), 298 (3.37)	57 (66)	243^{760}	...	...	s	v	v	...	s	...	**B6,** 843
p666	—,**2-methoxy-3-nitro-***	6-Nitroguaiacol. $C_7H_7NO_4$. *See* p555	169.14	yesh rh pr (CS_2, peth)	102–3	...	...	...	δ	v	...	...	...	peth s^h CS_2 δ, s^h	**B6**[2], 789
p667	—,**2-methoxy-4-nitro-***	5-Nitroguaiacol. $C_7H_7NO_4$. *See* p555	169.14	ye nd (w)	103–4	...	...	...	s^h	v	v	...	...	...	**B6**[2], 790
p668	—,**2-methoxy-5-nitro-***	4-Nitroguaiacol. $C_7H_7NO_4$. *See* p555	169.14	pa ye nd (w)	105	...	...	...	s^h	s	s	...	...	...	**B6**[2], 790
p669	—,**2-methoxy-6-nitro-***	3-Nitroguaiacol. $C_7H_7NO_4$. *See* p555	169.14	og-ye nd (sub)	62	sub	...	...	s	s	...	...	...	peth δ	**B6**[2], 789
p670	—,**3-methoxy-4-nitro-***	$C_7H_7NO_4$. *See* p555	169.14	ye nd (al) λ^{al} 329.5 (3.98)	144	...	...	...	...	s^h	...	...	s^h	...	**B6**[2], 822
p671	—,**3-methoxy-5-nitro-***	$C_7H_7NO_4$. *See* p555	169.14	ye cr (al)	144	...	...	...	...	s	...	...	...	...	**B6,** 825
p672	—,**4-methoxy-2-nitro-***	$C_7H_7NO_4$. *See* p555	169.14	og ye nd or mcl cr (al, lig)	80 (83)	...	...	...	...	s^h	...	...	...	lig s^h	**B6**[2], 849
p673	—,**4-methoxy-3-nitro-***	$C_7H_7NO_4$. *See* p555	169.14	pa ye nd (w), cr (bz)	98–100	...	...	...	δ v^h	s^h	...	...	δ v^h	...	**B6**[2], 848
p674	—,**5-methoxy-2-nitro-***	$C_7H_7NO_4$. *See* p555	169.14	yesh nd (al)	95	...	...	...	...	s^h	...	...	...	...	**B6**[2], 822
—	—,**methyl-***	*see* **Toluene, hydroxy-**													
p675	—,**2(methyl-amino)-***	C_7H_9NO. *See* p555	123.16	pl (bz-peth)	96–7	...	...	...	i	s	...	...	s	peth δ	**B13,** 362
Ω p676	—,**4(methyl-amino)-, sulfate***	Metol. Photol. Pictol. $2(C_7H_9NO).H_2SO_4$. *See* p555	344.39	wh nd (w)	250–60d	...	...	...	v	s	...	...	...	...	**B13,** 441
p677	—,**4(methyl-amino)-2-nitro-***	$C_7H_8N_2O_3$. *See* p555	168.15	dk red-br nd (al)	113–4	...	...	...	...	s^h	...	...	...	alk s (red)	**B13**[1], 186
p678	—,**4(2-methyl-aminopropyl)-***	$C_{10}H_{15}NO$. *See* p555	165.24	cr (MeOH)	164	...	...	...	δ	s	s	...	...	dil ac v MeOH s^h	**C43,** 1741
Ω p679	—,**2-nitro-***	$C_6H_5NO_3$. *See* p555	139.11	ye nd or pr (eth, al) λ^{w} 230 (3.57), 276 (3.80), 346 (3.48)	45.3–5.7	216^{760} 96.4–$.8^{10}$	1.485^{14} 1.2942^{40}	1.5723^{50}	δ s^h	v^h	v	v	v	chl, Py v alk, to, CS_2 s	**B6**[3], 794

For explanations, symbols and abbreviations see beginning of table. For structural formulas see end of table.

No.	Name	Synonyms and Formula	Mol. wt.	Color, crystalline form, specific rotation and λ_{max} (log ε)	m.p. °C	b.p. °C	Density	n_D	Solubility w	al	eth	ace	bz	other solvents	Ref.
	Phenol														
Ω p680	**—,3-nitro-***	$C_6H_5NO_3$. *See* p555	139.11	ye mcl (aq HCl, eth) λ^{w} 228 (3.88), 272 (3.77), 328 (3.29)	97	194^{70}	1.2797_4^{100}	...	δ v^h	v	v	v	v^h	dil ac, alk s chl s^h	**B6**³, 805
Ω p681	**—,4-nitro-***	$C_6H_5NO_3$. *See* p555	139.11	ye mcl pr (to) λ^{MeOH} 310 (4.00)	114.9–5.6	279d sub	1.479^{20}	...	δ v^h	v	v	v	δ s^h	chl, to, Py s chl, CS_2 δ	**B6**³, 811
Ω p682	**—,4-nitroso-***	1,4-Benzoquinone monoxime. $C_6H_5NO_2$. *See* p555	123.11	pa ye rh nd (ace, bz) λ^{MeOH} 303 (4.09)	d144 (br at 124)	...	...	...	δ s^h	s	s	s	s^h	dil alk s	**B6**², 205 **B7**², 574
p682[1]	**—,4-octyloxy-***	Hydroquinone monooctyl ether. $C_{14}H_{22}O_2$. *See* p555	222.33	cr (lig)	60–1	...	...	...	...	s	...	...	...	lig s^h	**Am 54**, 298
Ω p683	**—,pentabromo-***	C_6HBr_5O. *See* p555	488.62	mcl pr (aa), nd (al, CS_2)	229.5	sub	...	...	i	s^h	δ	...	s^h	CS_2, aa s^h	**B6**³, 766
Ω p684	**—,pentachloro-***	C_6HCl_5O. *See* p555	266.34	mcl pr (al+1w), nd (bz) λ^{al} 300.5 (3.4), 308 (3.4)	174 (+1w), 191 (anh)	309–10^{754} d	1.978_4^{22}	...	δ	v	v	...	s^h	lig δ	**B6**³, 731
p685	**—,2-propoxy-***	Pyrocatechol monopropyl ether. $C_9H_{12}O_2$. *See* p555	152.20	...	...	228–9^{760} 80–3^4	1.0523^{25}	1.5176^{25}	...	s	...	...	...	lig s^h	**Am 54**, 1204
p686	**—,3-propoxy-***	Resorcinol monopropyl ether. $C_9H_{12}O_2$. *See* p555	152.20	cr (w, al)	55	256–7 120^5	...	...	s^h	s^h	...	...	...	lig s^h	**Am 53**, 3397
p687	**—,4-propoxy-***	Hydroquinone monopropyl ether. $C_9H_{12}O_2$. *See* p555	152.20	cr (w, lig, al)	56–7	...	...	...	s^h	s^h	...	...	...	lig s^h	**Am 54**, 298
—	**—,propyl-***	*see* **Benzene, hydroxy-(propyl)-***													
p688	**—,seleno-***	Selenylbenzene. C_6H_5SeH	157.08	λ^{al} 241 (3.94), 270 sh (3.35), 320–350 (3.67)	...	183.6^{760}	1.4865^{15}	...	i	s	v	...	...	CCl_4 v	**B6**³, 766
p689	**—,2,3,4,6-tetra-bromo-***	$C_6H_2Br_4O$ *See* p555	409.72	nd (al, aa)	113–4 (120)	sub	...	...	...	v	...	...	s	alk s aa s^h	**B6**³, 766
p690	**—,2,3,4,5-tetra-chloro-***	$C_6H_2Cl_4O$. *See* p555	231.89	nd (peth, sub)	116–7	sub	...	...	...	v	...	...	...	alk, MeOH v peth s^h	**B6**³, 729
p691	**—,2,3,4,6-tetra-chloro-***	$C_6H_2Cl_4O$. *See* p555	231.89	nd (lig, aa)	70	150^{15}	...	...	i δ^h	s	...	...	s	NaOH v chl, lig s aa s^h	**B6**³, 729
p692	**—,2,3,5,6-tetra-chloro-***	$C_6H_2Cl_4O$. *See* p555	231.89	lf (lig)	115	...	...	...	δ	...	...	...	v	lig s^h	**B6**³, 730
p693	**—,2,3,4,6-tetra-nitro-***	$C_6H_2N_4O_9$. *See* p555	274.10	lt ye nd (chl)	140d	exp	...	...	v	...	...	...	...	chl s^h	**B6**³, 973
p694	**—,2,4,6-tri-amino-***	$C_6H_9N_3O$. *See* p555	139.16	...	...	257	...	...	v	s	s	...	...	...	**B13**², 317
p695	**—,2,3,5-tri-bromo-***	$C_6H_3Br_3O$. *See* p555	330.82	nd or pl (w, lig)	94–5	...	...	...	δ	v	v	v	...	alk v lig v^h	**B6**², 192
p696	**—,2,4,5-tri-bromo-***	$C_6H_3Br_3O$. *See* p555	330.82	nd (lig)	87	...	...	...	...	s	...	...	...	alk, lig s	**B6**², 192
Ω p697	**—,2,4,6-tri-bromo-***	Bromol. $C_6H_3Br_3O$. *See* p555	330.82	nd (al), pr (bz), cr (aa+1)	95–6	282–90^{764} sub	2.55_{20}^{20}	...	δ	v	s	...	s^h	chl s aa s^h	**B6**, 203
p698	**—,3,4,5-tri-bromo-***	$C_6H_3Br_3O$. *See* p555	330.82	tcl (bz-lig) $\lambda^{0.1N\ NaOH}$ 246 (3.85), 309 (3.45)	129	...	...	...	...	s	...	...	s	alk v aa s	**B6**², 195
p699	**—,2,3,4-tri-chloro-***	$C_6H_3Cl_3O$. *See* p555	197.45	nd (sub, bz, lig)	83.5	sub	...	...	...	s	s	...	s	alk, aa s	**B6**³, 716
p700	**—,2,3,5-tri-chloro-***	$C_6H_3Cl_3O$. *See* p555	197.45	nd (al) $\lambda^{0.1N\ HCl}$ 280 (3.28), 290 (3.28)	62	248.5–9.5^{250}	...	...	δ	s	s	...	...	...	**B6**³, 716
p701	**—,2,3,6-tri-chloro-***	$C_6H_3Cl_3O$. *See* p555	197.45	nd (dil al, lig) $\lambda^{0.1N\ HCl}$ 280 (3.28), 289 (3.29)	58	...	...	...	δ^h	v	v	...	v	aa, lig s	**B6**³, 716
Ω p702	**—,2,4,5-tri-chloro-***	$C_6H_3Cl_3O$. *See* p555	197.45	nd (al, peth) $\lambda^{0.1N\ NaOH}$ 244 (3.95), 310 (3.62)	68–70.5	sub	...	...	δ	s	...	...	...	os, lig s	**B6**³, 717
Ω p703	**—,2,4,6-tri-chloro-***	$C_6H_3Cl_3O$. *See* p555	197.45	rh nd (aa) $\lambda^{0.1N\ NaOH}$ 244 (3.95), 311 (3.69)	69.5	246^{760}	1.4901_4^{75}	...	δ	s	s	...	...	aa s^h	**B6**³, 722
p704	**—,3,4,5-tri-chloro-***	$C_6H_3Cl_3O$. *See* p555	197.45	nd (lig) $\lambda^{0.1N\ NaOH}$ 204 (4.03), 307 (3.70)	101	271–7^{746}	...	...	δ	...	s	...	...	lig δ, v^h	**B6**³, 729
p705	**—,2,3,5-triiodo-***	$C_6H_3I_3O$. *See* p555	471.80	nd (peth, bz-lig)	114	...	...	...	...	s	s	s	s	os v	**B6**, 211

For explanations, symbols and abbreviations see beginning of table. For structural formulas see end of table.

No.	Name	Synonyms and Formula	Mol. wt.	Color, crystalline form, specific rotation and λ_{max} (log ε)	m.p. °C	b.p. °C	Density	n_D	Solubility w	al	eth	ace	bz	other solvents	Ref.
	Phenol														
p706	**—,2,4,6-triiodo-***	$C_6H_3I_3O$. *See* p555	471.80	nd (dil al) λ^{al} 295 (3.51)	158–9	sub d	...	...	i	δ	s	s	...	...	**B6**[3], 789
p707	**—,2,3,6-tri-nitro-***	$C_6H_3N_3O_7$. *See* p555	229.11	ye nd (w)	119	...	...	...	δ v^h	s	v	...	s	aa s	**B6**, 265
p708	**—,2,4,5-trinitro-***	$C_6H_3N_3O_7$. *See* p555	229.11	wh nd (w, dil al)	96	...	...	...	δ s^h	v	v	...	v	aa v	**B6**[2], 253
Ω p709	**—,2,4,6-trinitro-***	Picric acid. $C_6H_3N_3O_7$. *See* p555	229.11	ye lf (w), pr (eth), pl (al), col cr (lig) λ^{al} 360 (4.20)	122–3	sub exp >300	...	1.763	δ s^h	s	s	v	s v^h	aa, MeOH, Py s chl s^h	**B6**[3], 873
Ω p710	**Phenolphthalein**	2,2-Bis(*p*-hydroxyphenyl) phthalide.	318.33	wh rh nd λ^{alk-al} 550 (3.90)	262–3	...	1.277^{32}_4	...	i	v	s	v	i	Py v chl, to s CS_2 δ peth i	**B18**, 143
Ω p711	**—,3′,3″,5′,5″-tetrabromo-**	$C_{20}H_{10}Br_4O_4$. *See* p710	633.94	nd (al, eth, aa)	295–7	...	...	...	i	$δ^h$	v	...	...	alk s aa s^h	**B18**[2], 124
p712	**—,3′,3″,5′,5″-tetrachloro-**	$C_{20}H_{10}Cl_4O_4$. *See* p710	456.11	cr (bz, aa)	225	...	...	...	i	s	...	s	v	$PhNO_2$ s aa s^h	**B18**[2], 123
Ω p713	**—,3′,3″,5′,5″-tetraiodo-**	Iodophen. Nosophen. $C_{20}H_{10}I_4O_4$. *See* p710	821.92	amor	277–9d	...	2.0246^{22}_{22}	...	i	i	i	...	...	chl δ	**B18**[2], 119
p714	**Phenolphthalin**	...	320.35	nd (w)	229–32 (225)	...	...	...	δ	s	...	...	...	NaOH s	**B10**[2], 322
Ω p715	**Phenolsulfon-phthalein**	Phenol red.	354.37	dk red nd or pl	>300	...	...	...	δ	δ	i	δ	δ	KOH s chl i	**B19**[2], 102
Ω p716	**Phenothiazine**	Thiodiphenylamine.	199.28	ye pr (al), ye lf or pl (to) λ^{MeOH} 254 (4.5), 285 (3.1), 318 (3.6)	185.5–.9	371^{760} 290^{40} (sub 130^1)	...	...	δ	δ	s	v	v	aa s^h lig δ CCl_4 i	**B27**, 63
p717	**—,2,8-dinitro-, 5-oxide**	$C_{12}H_7N_3O_5S$. *See* p716	305.27	ye-red lf (aa), pr ($PhNH_2$)	d	...	...	...	...	...	...	...	...	$PhNO_2$ s^h os δ	**B27**[1], 229
p718	**—,10(2-diethyl-aminoethyl)-**	$C_{18}H_{22}N_2S$. *See* p716	298.46	ye oil	...	195–208^{4-5} 161–$5^{0.5}$	...	...	i	...	...	...	...	dil HCl s	**Am 66**, 888
p719	**—,10(2-dimethyl-aminopropyl)-**	Phenergan base. Prometha-zine. $C_{17}H_{20}N_2S$. *See* p716	284.43	...	60 (55)	190–$3^{0.5}$ (190–2^3)	...	...	i	...	...	...	...	dil HCl v	**C42**, 575
p720	**—,—,hydrochloride**	$C_{17}H_{20}N_2S$. HCl. *See* p716	320.89	λ 249	230–2	...	...	...	v	s	...	...	...	chl s	**C42**, 575
Ω p721	**—,10-phenyl-**	$C_{18}H_{13}NS$. *See* p716	275.38	ye pr (al)	89–90	...	...	...	i	s^h	...	...	...	con sulf s	**B27**, 227
Ω p722	**Phenoxanthin**	...	200.27	nd, cr (MeOH)	59–60	311^{745} 183–4^{12}	...	...	...	s	s	s	...	os v con sulf s	**B19**[2], 33
p723	**Phenoxazine**	Dibenzoxazine	183.21	lf (dil al, bz)	156	δd	...	...	...	v	v	...	s^h	chl, aa v con sulf, lig s	**B27**[1], 223
—	**Phenoxytol**	*see* **Ethanol, 2-phenoxy-***													
Ω p724	**Phenylalanine** (*D*).	*l*-α-Aminohydrocinnamic acid. *l*-β-Phenyl-α-aminopropionic acid. $C_6H_5CH_2CH(NH_2)CO_2H$	165.19	nd or pr (w), $[\alpha]^{20}_D$ −69.5 (w) λ^w 242 (1.88), 257 (2.25), 267 (1.91)	283 d	sub at 295	...	1.600	s	$δ^h$	i	...	...	alk s dil ac δ os i	**B14**[2], 297
Ω p725	**—(*DL*)**	$C_6H_5CH_2CH(NH_2)CO_2H$	165.19	red-br lf (aq al), pr or nd (w) λ^w 242 (1.88), 257 (2.25), 267 (1.91)	284–8d (271–3)	sub par d	...	...	v	i	i	...	i	as i	**B14**[2], 299
Ω p726	**—(*L*)**	$C_6H_5CH_2CH(NH_2)CO_2H$	165.19	pr (w), $[\alpha]^{18}_D$ +70 (w) λ^w 242 (1.88), 2.57 (2.25), 267 (1.91)	283–4 d	sub at 295	...	...	s	i	i	...	...	MeOH δ os i	**B14**[2], 296
p727	**—,*N*-acetyl-(*d*, +)**	$C_6H_5CH_2CH(NHCOCH_3)CO_2H$	207.23	cr (w), $[\alpha]^{26}_D$ −50.9 λ^w 260 (2.3)	172	...	...	...	s^h	s	...	...	...	...	**B14**[2], 297
p728	**—,—(*dl*)**	$C_6H_5CH_2CH(NHCOCH_3)CO_2H$	207.23	nd or pl (w), hex tab (ace) λ^w 260 (2.3)	152.5–3	...	...	...	s	s	...	...	...	...	**B14**[2], 302
p729	**—,—(*l*, −)**	$C_6H_5CH_2CH(NHCOCH_3)CO_2H$	207.23	$[\alpha]^{26}_D$ +51.4 λ^w 260 (2.3)	172	...	...	...	...	s	...	...	...	...	**B14**[2], 298
—	**Phenylenediamine**	*see* **Benzene, diamino-***													
—	**Phenylglycidol**	*see* **1-Propanol,2,3-epoxy-3-phenyl-***													
—	**Phenylglycine**	*see* **Acetic acid, amino-(phenyl)-***													
—	**Phenylisocyanide**	*see* **Benzene, isocyano-***													
—	**Phloramin**	*see* **Benzene, 1-amino-3,5-dihydroxy-***													
Ω p730	**Phloretin**	Dihydronaringenin. $C_{15}H_{14}O_5$.	274.28	nd (dil al), cr (dil ace)	262–4d (274d)	...	...	...	$δ^h$	∞	i	s^h	∞	MeOH s chl δ	**B8**[2], 542

For explanations, symbols and abbreviations see beginning of table. For structural formulas see end of table.

No.	Name	Synonyms and Formula	Mol. wt.	Color, crystalline form, specific rotation and λ_{max} (log ε)	m.p. °C	b.p. °C	Density	n_D	Solubility w	al	eth	ace	bz	other solvents	Ref.
	Phlorhizin														
p731	**Phlorhizin,** dihydrate	Asebotin. Asebotoside. $C_{21}H_{24}O_{10}.2H_2O$	472.45	nd (w) $[\alpha]_D^{25}$–52 (96% al)	108 (+2w) 170d (anh)		1.4298		δ	s	i	s	i	Py s chl i	**B31**, 231
—	**Phloroglucinol**	*see* **Benzene, 1,3,5-trihydroxy-***													
—	**Phlorol**	*see* **Benzene, 1-ethyl-2-hydroxy-***													
—	**Phorone**	*see* **2,5-Heptadiene-4-one, 2,6-dimethyl-***													
p732	**Phosgene**	Carbonic acid dichloride. Carbonyl chloride. Chloroformyl chloride. Cl_2CO.	98.92	λ^{gas} 295, 297, 298	–118 (–127.76)	7.56^{760}	1.381_4^{20}		d	d	...	...	s	to, CCl_4, chl, aa s	**B3**[3], 31
p733	—,thio-	Thiocarbonyl chloride. $SCCl_2$	114.98	red		73	1.508^{15}	1.5442^{20}	i	s	s	...	...		**B3**[3], 246
p734	**Phosphine, bis-(trifluoro-methyl)-**	$(F_3C)_2PH$	169.99	spontaneously inflam	–137	1									**J1954,** 3896
p735	**—,bis(trifluoro-methyl)chloro-**	$(F_3C)_2PCl$	204.44	spontaneously inflam		21			d	...	...	...	...		**J1953,** 1552
p736	**—,bis(trifluoro-methyl)cyano-**	$(F_3C)_2PCN$	195.00	spontaneously inflam		48		1.3248^{20}							**J1953,** 1552
p737	**—,bis(trifluoro-methyl)iodo-**	$(F_3C)_2PI$	295.89			73		1.403^{15}	d	...	...	...	...	Bu_2O s	**J1953,** 1552
p738	**—,dichloro(2,4-dimethyl-phenyl)-**	$C_8H_9Cl_2P$. *See* p741	207.04			256–8			d^h	...	...	...	...		**B16**, 773
p739	**—,dichloro(2,5-dimethyl-phenyl)-**	$C_8H_9Cl_2P$. *See* p741	207.04		–30	253–4	1.25_{18}^{18}		d^h	...	...	...	...		**B16**, 773
p740	**—,dichloro(4-ethylphenyl)-**	$C_8H_9Cl_2P$. *See* p741	207.04			250–2 $85^{0.4}$	1.237_4^{20}	1.584^{20}	d^h	...	...	...	...		**B16**, 772
p741	**—,dichloro-(phenyl)-**		178.99			224.6^{760} $99–101^{11}$	1.356_4^{20}	1.6030^{20}	d	...	...	...	∞	CS_2 ∞	**B16**[2], 373
p742	**—,dichlorotri-fluoromethyl-**	CF_3PCl_2	170.89			37			d	...	...	...	...		**J1953,** 1552
p743	**—,diethyl-**	$(C_2H_5)_2PH$	90.11			85									**B4**[3], 1761
p744	**—,diiodo(tri-fluoromethyl)-**	F_3CPI_2	353.79	ye fum		d^{760} 73^{37}		1.630^{20}	i d		s	...	...	Bu_2O s	**J1953,** 1552
p745	**—,dimethyl-**	$(CH_3)_2PH$	62.05	λ^{gas} 189 (3.80)		25	<1		i	s	s	...	...		**B4**[3], 1759
p746	**—,diphenyl-(ethyl)-**	$(C_6H_5)_2PC_2H_5$	214.25			293			δ	...	...	...	δ		**B16**[2], 371
p747	**—,ethyl-**	$CH_3CH_2PH_2$	62.05			25	<1		d	...	...	...	...		**B4**[3], 1761
p748	**—,methyl-**	CH_3PH_2	48.02	gas		-14^{759}			i	δ	v	...	...		**B4**[3], 1759
p749	**—,phenyl-**	$C_6H_5PH_2$	110.10	λ^{al} 234 (3.54), 260.5 (2.8), 265.5 (2.84), 272 (2.72)		$160–1^{760}$ 40^{10}	1.001^{15}	1.5796^{20}	...	...	...	...	...		**B16**[2], 369
p750	**—,triethyl-**	$(C_2H_5)_3P$	118.16		–88– –85	129^{762}	0.8006_4^{19}	1.458^{15}	i	∞	∞				**B4**[3], 1761
p751	—,—,oxide	$(C_2H_5)_3PO$	134.16	wh hyg nd	50	243			s	s	s				**B4**[3], 1755
p752	—,—,sulfide	$(C_2H_5)_3PS$	150.23	cr (al)	94				s						**B4**[3], 1755
p753	**—,(trifluoro-methyl)-**	F_3CPH_2	102.00	spontaneously inflam		–26.5									**J1954,** 3896
p754	**—,trimethyl-**	$(CH_3)_3P$	76.08	λ^{gas} 201 (4.27)	–85.3– –84.3	37.8	<1		i	...	s	...	...		**B4**[3], 1759
Ω p755	**—,triphenyl-**	$(C_6H_5)_3P$	262.29	$\lambda^{dil\ al}$ 260 (4.0)	80	>360 188^1	1.0749_4^{80}	1.6358^{80}	i	s	v	...	s	chl s	**B16**[2], 371
Ω p756	—,—,oxide	$(C_6H_5)_3PO$	278.29	pr (hyd) λ^{MeOH} 228 (4.3), 267 (3.3)	156–7	>360	1.2124_4^{23}		δ^h	v	δ	...	v		**B16**[2], 379
p757	**—,tris(chloro-methyl)-**	$P(CH_2Cl)_3$	179.41			100^7	1.414^{20}		i	s	v	v	v		**B4**[3], 1760
p758	**—,tris(trichloro-methyl)-**	$(CCl_3)_3P$	386.08		53				d	...	...	...	...		**C47,** 2685
p759	**—,tris(trifluoro-methyl)-**	$(CF_3)_3P$	237.99	spontaneously inflam	–112	17.3			i	...	...	...	...	w d^{200}	**J1953,** 1565
p760	—,—,oxide	$(CF_3)_3PO$	253.99			23.6			d	...	...	...	...		**J1955,** 574
p761	**Phosphinic acid, bis(trifluoro-methyl)-***	$(F_3C)_2PO_2H$	201.99	visc liq		182^{760} $137–8^{238}$	<1.9		s	...	...	...	...	con sulf i	**J1955,** 563
p762	**—,diethyl-***	$(C_2H_5)_2PO_2H$	122.11		18.5	320 $134^{0.7}$			v	v	v	...	...		**B4**[3], 1776
p763	—,—,anhydride*	$[(C_2H_5)_2PO]_2O$	226.20			188^{14}	1.1053_4^{23}	1.4720^{23}	d	d	s	...	...		**Am 73,** 5466
p764	—,—,chloride	$(C_2H_5)_2POCl$	140.55			104^{15}	1.1394_4^{20}	1.4647^{20}	d	d	...	...	...		**Am 73,** 5466
p765	—,—,ethyl ester*	$(C_2H_5)_2PO(OC_2H_5)$	150.16			95^{14}	0.9908_4^{20}	1.4337^{20}	δ	s	s	...	...		**Am 73,** 5466
p766	**—,dimethyl-***	$(CH_3)_2PO_2H$	94.05	cr (bz)	92	377			v	v	v	...	s^h		**B4**[3], 1776

For explanations, symbols and abbreviations see beginning of table. For structural formulas see end of table.

No.	Name	Synonyms and Formula	Mol. wt.	Color, crystalline form, specific rotation and λ_{max} (log ε)	m.p. °C	b.p. °C	Density	n_D	Solubility w	al	eth	ace	bz	other solvents	Ref.
	Phosphinic acid														
p767	—,—,anhydride*	$[(CH_3)_2PO]_2O$	170.09	nd (bz)	119–21	192^{15}			d	d	...	...	s^h		**Am 73**, 5466
p768	—,—,chloride	$(CH_3)_2POCl$	112.50	hyg nd (bz-peth)	66.8–8.4	204^{760}			d	d	...	...	s^h	peth δ	**B4**[3], 1776
—	—,phenyl-*	*see* **Benzenephosphinic acid**													
—	**phosphonic acid ethyl-***	*see* **Ethanephosphonic acid**													
—	—,methyl-*	*see* **Methanephosphonic acid**													
—	—,phenyl-*	*see* **Benzenephosphonic acid**													
p770	**Phosphoric acid,** diethyl ester*	$(C_2H_5O)_2POOH$	154.10	syr		203.3d	1.186^{25}_{4}	1.4170^{20}	i	...	s	...	...	CCl_4 i	**B1**[3], 1328
p771	—,dimethyl ester*	$(CH_3O)_2POOH$	126.05			172–6d	1.335^{25}	1.408^{25}	s	s	i	s	...	to, CCl_4 i	**B1**[3], 1205
p772	—,monoethyl ester*	$(C_2H_5O)PO(OH)_2$	126.05	hyg cr		d	1.430^{25}_{4}	1.427	s d^h	s	s	s	...	CCl_4, to i	**B1**[3], 1327
p773	—,monophenyl ester*	$(C_6H_5O)PO(OH)_2$	174.09	pl (chl), nd (w)	99.5				v	v	v	...	v	chl s	**B6**[3], 657
p774	—,triamide, hexamethyl-*	$[(CH_3)_2N]_3PO$	179.20			$98–100^{6}$ $66^{0.5}$	1.024^{25}_{25}	1.4579^{20}	...	s	s	...	...		**C53**, 4132
Ω p775	—,tributyl ester*	$(C_4H^n_9O)_3PO$	266.32			289 $160–2^{15}$	0.9727^{25}_{4}	1.4224^{25}	s	∞	s	...	s	CS_2 s	**B1**[3], 1511
Ω p776	—,triethyl ester*	$(C_2H_5O)_3PO$	182.16		−56.4	$215–6^{760}$ 103^{25}	1.0695^{20}_{4}	1.4053^{20}	$s^{\delta d}$	v	s	...	s		**B1**[3], 1328
p777	—,triisobutyl ester	$(C_4H^i_9O)_3PO$	266.32			264^{760} 138^{10}	0.9681^{20}_{4}	1.4193^{20}	v	s	s	...	s		**B1**[3], 1563
p778	—,triisopropyl ester*	$(C_3H^i_7O)_3PO$	224.24			$218–20^{763}$ $95–6^{8}$	0.9867^{20}_{4}	1.4057^{20}	...	s	...	...	...		**B1**[3], 1466
p779	—,trimethyl ester*	$(CH_3O)_3PO$	140.08	λ^{al} 229 (1.74), 263 (0.97)	α: −46 (st) β: −62 (unst)	197.2^{760} 85^{24}	1.2144^{20}_{4}	1.3967^{20}	v	δ	s	...	...		**B1**[3], 1205
p780	—,tripentyl ester*	$[CH_3(CH_2)_4O]_3PO$	308.40			225^{50} 167^{5}	0.9608^{20}_{4}	1.4319^{20}	i	s	s	...	...	to, CS_2 s	**B1**[3], 1605
Ω p781	—,triphenyl ester*		326.29	cr (abs al-lig), pr (al), nd (eth-lig)	50–1	245^{11}	1.2055^{50}_{4}		i	s	v	...	v	chl, CCl_4 v	**B6**[3], 658
Ω p782	—,tripropyl ester*	$(C_3H^n_7O)_3PO$	224.24			252^{760} 107.5^{5}	1.0121^{20}_{4}	1.4165^{20}	δ	s	s	...	...	to, CS_2 s	**B1**[3], 1422
p783	—,tris(2,4-dimethylphenyl) ester*	$C_{24}H_{27}O_4P$. *See* p781	410.46	glassy on fr		232–5	1.142^{38}_{4}	1.5550^{20}	i	...	...	...	s	hx s	**B6**, 488
p784	—,tris(2,5-dimethylphenyl) ester*	$C_{24}H_{27}O_4P$. *See* p781	410.46	cr (dil al)	78.6–81 (77)	$260–5^{8}$	1.197^{25}		i	δ	s	...	s	to, CCl_4 s hx δ	**B6**[3], 1773
p785	—,tris(2,6-dimethylphenyl) ester*	$C_{24}H_{27}O_4P$. *See* p781	410.46		136–8	$262–4^{6}$			i	δ	...	...	s	hx̂ δ	**J1956**, 3043
p786	—,tris(3,4-dimethylphenyl) ester*	$C_{24}H_{27}O_4P$. *See* p781	410.46	wax	71.5–2.5	$260–3^{7}$			i	δ	...	...	s	hx δ	**B6**, 482
p787	—,tris(3,5-dimethylphenyl) ester*	$C_{24}H_{27}O_4P$. *See* p781	410.46	wax	46–6.5	290^{10}			i	δ	...	...	...	aa s hx δ	**C31**, 187
p788	—,tris(2,2,2-trichloroethyl) ester*	$(Cl_3CCH_2O)_3PO$	492.16		73–4				...	...	v	...	...	lig δ	**B1**, 338
Ω p789	—,tri(2-tolyl) ester	Tri-*o*-cresyl phosphate. $C_{21}H_{21}O_4P$. *See* p781	368.37	col or pa ye	11	410 $283–5^{20}$	1.1955^{20}_{4}	1.5575^{20}	i	v	v	...	...	CCl_4, to v aa s	**B6**[3], 1261
Ω p790	—,tri(3-tolyl) ester	Tri-*m*-cresyl phosphate. $C_{21}H_{21}O_4P$. *See* p781	368.37	wax	25–6	260^{15}	1.150^{25}	1.5575^{20}	i	δ	s	...	...	to, CCl_4 v aa s	**B6**[3], 1312
Ω p791	—,tri(4-tolyl) ester	T.P.C. Tri-*p*-cresyl phosphate $C_{21}H_{21}O_4P$. *See* p781	368.37	nd (al), ta (eth)	77–8	$244^{3.5}$	1.247^{25}		...	s	s	...	s	chl, aa s	**B6**[3], 1372
p794	—,**thiono-**, tri(2-tolyl) ester		384.44	nd (al)	45–6	$260–5^{1}$			i	s	...	...	...	aa v	**B6**[3], 1262
p795	—,—,tri(3-tolyl) ester		384.44	nd (al)	40–1	$270–2^{12}$			i	δ s^h	...	...	...	aa v MeOH, lig δ	**B6**[2], 354
p796	—,—,tri(4-tolyl) ester		384.44	nd (al)	93–4				i	δ s^h	...	...	δ	chl, aa v lig δ	**B6**[3], 1372
Ω p797	**Phosphorous acid,** diethyl ester*	$(C_2H_5O)_2POH$	138.10			87^{20} $51–2^{2}$	1.0720^{20}_{4}	1.4101^{20}	d^h	s	s	...	...		**B1**[3], 1324
Ω p798	—,dimethyl ester*	$(CH_3O)_2POH$	110.05			170–1 70^{25}	1.2004^{20}_{0}	1.4036^{20}	...	s	...	...	...	Py s	**B1**[3], 1203
Ω p799	—,tributyl ester*	$[CH_3(CH_2)_3O]_3P$	250.32			122^{12}	0.9259^{20}_{4}	1.4321^{19}	d^h	s	v	...	...		**B1**[3], 1510
Ω p800	—,triethyl ester*	$(C_2H_5O)_3P$	166.16	λ^{al} 260 (0.16)		157.9 49^{12}	0.9629^{20}_{4}	1.4127^{20}	i	v	v	...	...		**B1**[3], 1325
Ω p801	—,triisopropyl ester*	$[(CH_3)_2CHO]_3P$	208.24	λ^{al} 261 (0.60)		$60–1^{10}$ 43.5^{1}	$0.9687^{18.5}_{0}$	1.4085^{25}	...	s	s	...	...		**B1**[3], 1465
Ω p802	—,trimethyl ester*	$(CH_3O)_3P$	124.08	λ^{al} 251 (2.76), 257 (2.80), 263 (2.66)		111–2 22^{23}	1.0520^{20}_{0}	1.4095^{20}	d^h	v	v	...	...		**B1**[3], 1203
Ω p803	—,triphenyl ester*	$(C_6H_5O)_3P$	310.29		*ca.* 25	360^{760} $200–1^{5}$	1.1844^{20}_{0}	1.5900^{20}	i d^h	v	...	...	...	os v	**B6**[3], 656
p804	—,tripropyl ester*	$(CH_3CH_2CH_2O)_3P$	208.24			$206–7^{760}$ 92^{14}	0.9417^{20}	1.4282^{20}	...	s	s	...	...		**B1**[3], 1420
p805	—,tris(2,2,2-trichloroethyl) ester*	$(Cl_3CCH_2O)_3P$	476.16			263 $127–31^{0.1}$		1.5174^{20}	...	...	...	...	...		**B1**, 338

For explanations, symbols and abbreviations see beginning of table. For structural formulas see end of table.

No.	Name	Synonyms and Formula	Mol. wt.	Color. crystalline form. specific rotation and λ_{max} (log ε)	m.p. °C	b.p. °C	Density	n_D	Solubility w	al	eth	ace	bz	other solvents	Ref.
	Phosphorous acid														
p806	—,tri(2-tolyl) ester.	Tri-o-cresyl phosphite.....	352.38	ye	....	238^{11} $197–9^{2}$	1.1423_4^{20}	1.5740^{28}	δd	δd	s	...	...	....	**B6³**, 1260
p807	—,tri(4-tolyl) ester..	Tri-*p*-cresyl phosphite......	352.38	pa ye	....	$250–5^{10}$	1.1313_{25}^{25}	1.5703^{28}	...	...	s	...	...	....	**B6³**, 1371
p808	**Phthalaldehyde**....	1,2-Benzene dicarbonal*. *o*-Phthalic aldehyde.	134.14	ye cr or nd (lig) λ^{w} 210 (3.9), 219 (3.7), 260 (3.0), 300 (2.3)	(i) 53.2 (ii) 56–7	....	....	....	s	s	s	...	...	os s peth δ	**B7²**, 605
—	**Phthalaldehydic acid**	*see* **Benzoic acid, 2-formyl-**													
—	**Phthalamic acid** ...	*see* **Phthalic acid,** monoamide													
—	**Phthalamide**	*see* **Phthalic acid,** diamide													
—	**Phthalanil**	*see* **Phthalic acid,** imide, *N*-phenyl-													
—	**Phthalazine, 6-amino-1,4-dioxo-1,2,3,4-tetrahydro-**	*see* **Phthalic acid, 4-amino-,** hydrazide													
Ω p810	**—,1,2-dihydro-1-oxo**	1(2H)-Phthalazinone. Phthalazone.	146.15	nd (w), pr (sub)	184–5	337^{755} sub	....	....	v^h	s v^h	...	...	s v^h	alk, ac s	**B24²**, 70
Ω p812	**Phthalic acid**	1,2-Benzenedicarboxylic acid*.	166.14	pl (w) λ^{al} 225 (3.88), 275 (3.08)	210–11d 231 (rapid htng) 191 (sealed tube)	d	1.593	....	δ v^h	s	δ	...	...	chl i	**B9²**, 580
Ω p813	—,anhydride	Phthalandione. Phthalic anhydride. $C_8H_4O_3$. *See* p812	148.12	wh nd (al, bz) λ^{CCl_4} 287 (3.2), 295 (3.3)	131.61	295.1 (284^{740}) sub	1.527^{4}	....	δ	s	δ	...	s^h	....	**B17²**, 463 **B21²**, 466
Ω p814	—,diamide	Phthalamide. Phthalic diamide. $C_8H_8N_2O_2$. *See* p812	164.18	cr	222	d	....	....	δ	δ	i	...	...	....	**B9²**, 601
p815	—,—,*N,N,N′,N′*,-tetraethyl-	$C_{16}H_{24}N_2O_2$. *See* p812	276.38	....	36	204^{16}	....	....	s	...	...	...	...	os v	**B9²**, 602
p816	—,dibenzyl ester...	$C_{22}H_{18}O_4$. *See* p812.......	346.39	pr (al) λ^{al} 276 (3.1)	42–3	277^{15}	....	....	i	v	v	...	...	lig i	**B9²**, 595
Ω p817	—,dibutyl ester....	$C_{16}H_{22}O_4$. *See* p812.......	278.35	λ^{al} 226 (3.98), 272 (3.18)	....	340 206^{20}	1.047_{20}^{20}	1.4911^{20}	i	∞	∞	...	∞	....	**B9²**, 586
p818	—,dichloride......	Phthalyl chloride. $C_8H_4Cl_2O_2$. *See* p812	203.03	λ^{eth} 235 (3.8), 272 sh (2.9), 278 (3.0), 286 (3.0)	15–6 (fr 12)	281.1 $131–3^{9}$	1.4089_4^{20}	1.5684^{20}	d	d	...	...	...	....	**B9²**, 599
Ω p819	—,dicyclohexyl ester	$C_{20}H_{26}O_4$. *See* p812.......	330.43	pr (al)	66	....	1.383_4^{20}	1.451^{20}	i	s	s	...	...	....	**B9**, 799
p820	—,di(2-ethoxy-ethyl) ester	$C_{16}H_{22}O_6$. *See* p812.......	310.35	....	34	345 $233–5^{23}$	1.1229^{21}	....	...	...	...	...	...	....	**B9²**, 597
Ω p821	—,diethyl ester....	$C_{12}H_{14}O_4$. *See* p812.......	222.24	λ^{al} 225 (3.9), 275 (3.1)	....	298 (290) 172^{12}	1.1175_4^{20}	1.5000^{20}	i	∞	∞	s	s	....	**B9²**, 584
Ω p822	—,diisobutyl ester .	$C_{16}H_{22}O_4$. *See* p812.......	278.35	....	....	295–8 $182–4^{10}$	1.0490^{15}	....	...	...	...	...	...	....	**B9²**, 587
p823	—,di(2-methoxy-ethyl) ester	$C_{14}H_{18}O_6$. *See* p812.......	282.30	....	....	230^{10}	1.1708^{15}	....	...	...	...	...	...	....	**B9²**, 597
Ω p824	—,dimethyl ester ..	$C_{10}H_{10}O_4$. *See* p812.......	194.19	pa ye λ^{al} 225 (3.92), 274 (3.10)	0–2	283.8^{760}	$1.1905_4^{20.7}$	1.5138^{20}	i	∞	∞	...	s	....	**B9²**, 584
p825	—,di(3-methyl-butyl) ester	$C_{18}H_{26}O_4$. *See* p812.......	306.41	....	....	330–8d 225^{40}	1.0220_{16}^{16}	1.4871^{20}	i	s	...	...	...	os s	**B9²**, 587
Ω p826	—,dinitrile	Phthalonitrile. $C_8H_4N_2$. *See* p812	128.14	nd (w, lig) λ^{MeOH} 237 (4.02), 276 (3.05), 280.5 (3.22), 290 (3.23)	141	....	....	....	δ	v	s	s	v	lig δ, s^h	**B9²**, 602
Ω p827	—,diphenyl ester ..	$C_{20}H_{14}O_4$. *See* p812.......	318.33	pr (al, lig)	73	$250–7^{14}$ sub	....	....	i	δ	δ	...	...	....	**B9²**, 594
Ω p828	—,dipropyl ester...	$C_{14}H_{18}O_4$. *See* p812.......	250.30	....	....	304–5	....	....	i	s	s	...	...	....	**B9²**, 586
p829	—,hydrazide.....	Phthalhydrazide. 1,4-Dioxo-1,2,3,4-tetra-hydrophthalazine.	162.15	mcl nd (w, dil al, aa) λ^{al} 251 (3.6), 262 (3.64), 300 (3.78)	342–4 (cor)	....	....	....	i δ^h	i δ^h	...	...	i	aa, dil alk s aa s^h chl i	**B24²**, 194
Ω p830	—,imide	1,3-Isoindoledione. Phthalimide.	147.14	nd (w), pr (aa), lf (sub) λ^{al} 215 (3.61), 229.5 (4.21), 238 (4.03), 291 (3.15)	238	....	....	....	δ	...	...	...	i	alk s aa s^h lig i	**B21²**, 348
p831	—,—,*N*-acetyl-....	$C_{10}H_7NO_3$. *See* p830.......	189.17	nd (bz), cr (al, aa)	135–6	....	....	....	i	...	s	...	v^h	chl v	**B21²**, 357

For explanations, symbols and abbreviations see beginning of table. For structural formulas see end of table.

No.	Name	Synonyms and Formula	Mol. wt.	Color, crystalline form, specific rotation and λ_{max} (log ε)	m.p. °C	b.p. °C	Density	n_D	Solubility w	al	eth	ace	bz	other solvents	Ref.
	Phthalic acid														
Ω p832	—,—,N-benzyl-	$C_{15}H_{11}NO_2$. *See* p830	237.26	ye nd (al), cr (aa) λ^{al} 222 (4.65), 240 (3.74), 297 (3.28)	116		1.343^{16}		...	s^h	...	...	...	aa s^h	**B21**[2], 351
Ω p833	—,—,N(2-bromoethyl)-	$C_{10}H_8BrNO_2$. *See* p830	254.10	nd (w) λ 291 (3.4)	82.0–3.5				d^h	...	v	...	...		**B21**[2], 349
p834	—,—,N(2-bromoisobutyl)-	$C_{12}H_{12}BrNO_2$. *See* p830	282.14	nd (al), lf (chl)	97				...	s v^h	...	...	v	AcOEt, chl v	**B21**[2], 349
Ω p835	—,—,N(bromomethyl)-	$C_9H_6BrNO_2$. *See* p830	240.06	pr (chl, bz, aa, AcOEt)	151.5				d^h	...	...	s	δ	AcOEt v^h chl δ	**B21**[2], 354
p836	—,—,N(2-bromopropyl)-	$C_{11}H_{10}BrNO_2$. *See* p830	268.12	nd (MeOH, al)	110–11				δ^h	v	v	...	...	lig δ	**B21**[2], 349
Ω p837	—,—,N(3-bromopropyl)-	$C_{11}H_{10}BrNO_2$. *See* p830	268.12	nd (lig)	72–3				i	v^h	v^h	...	...	lig δ peth i	**B21**[2], 349
p838	—,—,N-carbethoxy-	Ethyl N,N-phthalylcarbamate. $C_{11}H_9NO_4$. *See* p830	219.20	(bz-peth)	87–9				...	...	v	...	v	chl v con ac s peth δ	**B21**[2], 35
p839	—,—,N-ethyl-	$C_{10}H_9NO_2$. *See* p830	175.19	nd (al) λ^{al} 219 (4.6)	79	285.6^{758} (cor)			...	s^h	s	...	...		**B21**[2], 349
Ω p840	—,—,N(2-hydroxyethyl)-	$C_{10}H_9NO_3$. *See* p830	191.19	nd (al), lf (w)	129.5 (cor)				δ	...	...	...	...		**B21**[2], 352
Ω p841	—,—,N(hydroxymethyl)-	$C_9H_7NO_3$. *See* p830	177.16	lf, pr (to)	141–2				i s^h	δ s^h	i	...	δ s^h	to s^h CCl_4 i	**B21**[2], 354
p842	—,—,N-isobutyl-	$C_{12}H_{13}NO_2$. *See* p830	203.24		93	293–5			...	...	...	...	...		**B21**, 463
p843	—,—,N-methyl-	$C_9H_7NO_2$. *See* p830	161.16	nd (al), lf (sub)	134	285–7 (cor) sub			i	δ v^h	...	...	...		**B21**[2], 348
p844	—,—,N(1-naphthyl)-	$C_{18}H_{11}NO_2$. *See* p830	273.30	pr, pl (al-aa)	180–1				...	s^h	s^h	...	...		**B21**, 469
p845	—,—,N(2-naphthyl)-	$C_{18}H_{11}NO_2$. *See* p830	273.30	nd (aa)	216				...	s	...	...	...	aa v^h	**B21**[2], 352
Ω p846	—,—,N(4-nitrophenyl)-	$C_{14}H_8N_2O_4$. *See* p830	268.24	cr (aa) λ^{al} 222 (4.36), 317 (4.15)	271–2 (266)				...	i^h	...	...	...	xyl i aa i, s^h	**B21**[2], 350
Ω p847	—,—,N-phenyl-	$C_{14}H_9NO_2$. *See* p830	223.23	wh nd (al) λ^{al} 215 (4.54), 230 (4.22), 295 (3.26)	210	sub			i	δ	...	...	...	chl ∞	**B21**[2], 350
p849	—,monoamide	Phthalamic acid. $C_8H_7NO_3$. *See* p812	165.15	pr	148–9				s^h	s	δ	...	δ	lig i	**B9**[2], 600
Ω p850	—,—,N(1-naphthyl)-	Alanap-1. $C_{18}H_{13}NO_3$. *See* p812	291.31		185				i	δ	...	δ	δ		**B12**[1], 525
p851	—,monoamide mononitrile	o-Cyanobenzamide. Phthalamic nitrile. $C_8H_6N_2O$. *See* p812	146.15	nd (MeOH), cr (aa)	173				i	v	...	v	...	MeOH, aa s^h AcOEt δ	**B9**, 815
p852	—,monobutyl ester	$C_{12}H_{14}O_4$. *See* p812	222.24	pl (al, ace)	73–4				...	s^h	...	s^h	...	os s	**B9**[2], 586
p853	—,mono-sec-butyl ester(d)	$C_{12}H_{14}O_4$. *See* p812	222.24	$[\alpha]^{20}_D$ +38.4 (al)	48				...	s	...	...	...	chl s	**B9**[2], 586
p854	—,—(dl)	$C_{12}H_{14}O_4$. *See* p812	222.24	cr (peth)	63–3.5				...	...	...	...	...	peth s^h	**B9**[2], 587
p855	—,monoethyl ester	$C_{10}H_{10}O_4$. *See* p812	194.19		2	d	1.1877^{22}_{4}	1.509^{22}	δ	s	s	...	...		**B9**[2], 584
p856	—,mononitrile	$C_8H_5NO_2$. *See* p812	147.14	nd (al)	187d (192d)				s^h	v^h	δ	s	δ	peth δ	**B9**[2], 602
p857	—,mono(2-octyl) ester(d)	$C_{16}H_{22}O_4$. *See* p812	278.35	pr (peth), $[\alpha]^{20}_D$ +48.7 (al)	75		1.027^{22}_{4}		...	v	...	...	v	AcOEt, chl v CS_2 s peth s^h	**B9**[2], 587
p858	—,—(dl)	$C_{16}H_{22}O_4$. *See* p812	278.35	pr (peth)	55				...	v	...	v	v	chl v peth s^h	**B9**[1], 353
p859	—,—(l)	$C_{16}H_{22}O_4$. *See* p812	278.35	pr (peth), $[\alpha]_D$ −48.26 (al)	75				...	v	...	...	v	chl v peth s^h	**B9**[2], 587
p860	**—,3-amino-**	$C_8H_7NO_4$. *See* p812	181.15	nd	231–2				δ	δ	δ	...	i	chl i	**B14**[2], 336
p861	**—,4-amino-,** dimethyl ester	$C_{10}H_{11}NO_4$. *See* p812	209.20	pl (al, bz), pr (w)	84 (cor)				δ^h	s	δ	...	s^h	chl, Py s peth δ	**B14**, 554
p862	—,—,hydrazide	$C_8H_7N_3O_2$. *See* p829	177.16	ye nd (w + 1)	> 305				...	δ	i	...	i	chl, lig i	**B25**, 487
p863	**—,3-benzoyl-**	2,3-Benzophenone dicarboxylic acid. $C_{15}H_{10}O_5$. *See* p812	270.25	pl, nd (w + 1)	140–1	d			s^h	s	δ	...	s		**B10**, 880
p864	**—,4-benzoyl-**	3,4-Benzophenone dicarboxylic acid. $C_{15}H_{10}O_5$. *See* p812	270.25	lf (xyl)	177				s	s	...	...	...	xyl s^h	**H29**, 1413
p865	**—,3-bromo-**	$C_8H_5BrO_4$. *See* p812	245.04	nd	188d				s	s	s	...	...	chl i	**B9**[2], 604
p866	**—,4-bromo-**	$C_8H_5BrO_4$. *See* p812	245.04		173–5				δ^h	v	s	...	...		**B9**[2], 605
p867	**—,3-chloro-**	$C_8H_5ClO_4$. *See* p812	200.58	nd (w)	186–7				δ	v	v	...	...		**B9**[2], 603
p868	—,—,anhydride	$C_8H_3ClO_3$. *See* p812	182.56	nd (sub)	124–5	sub			...	...	...	...	...		**B17**[2], 466
p869	**—,4-chloro-**	$C_8H_5ClO_4$. *See* p812	200.58	nd (dil al)	157				s	s	s	...	...		**B9**[2], 603
p870	—,—,anhydride	$C_8H_3ClO_3$. *See* p812	182.56	pr	98.5	$294–5^{720}$			...	s	s	...	...	chl s	**B17**[2], 466
p871	**—,3,4-dichloro-**	$C_8H_4Cl_2O_4$. *See* p812	235.03	pl (w)	195				v	s	v	...	...		**B9**, 817
p871[1]	—,—,anhydride	$C_8H_2Cl_2O_3$. *See* p812	217.01	pl	121	329			...	s	...	...	...	chl, to s	**B17**[2], 467
p872	**—,3,5-dichloro-**	$C_8H_4Cl_2O_4$. *See* p812	235.03	nd, ta (aq HCl)	164d	sub			s	v	v	v	δ	chl δ	**B9**, 817
p873	—,—,anhydride	$C_8H_2Cl_2O_3$. *See* p812	217.01	nd	89				d^h	...	...	...	s	chl s	**B17**, 483

For explanations, symbols and abbreviations see beginning of table. For structural formulas see end of table.

No.	Name	Synonyms and Formula	Mol. wt.	Color, crystalline form, specific rotation and λ_{max} (log ε)	m.p. °C	b.p. °C	Density	n_D	Solubility w	al	eth	ace	bz	other solvents	Ref.
	Phthalic acid														
p874	—,3,6-dichloro-	$C_8H_4Cl_2O_4$. *See* p812	235.03	pl (w)	d at 100 → anh	...	...	...	v^h	v	v	...	...	...	**B9**, 817
p875	—,—,anhydride	$C_8H_2Cl_2O_3$. *See* p812	217.01	nd	194.5	339^{760}	...	...	...	...	...	...	...	...	**B17²**, 467
p876	—,4,5-dichloro-	$C_8H_4Cl_2O_4$. *See* p812	235.03	nd (w)	*ca.* 200d	...	...	...	s v^h	...	v	...	...	...	**B9²**, 604
p877	—,—,anhydride	$C_8H_2Cl_2O_3$. *See* p812	217.01	ta or pr (to, CCl_4)	187.5–8.8 (cor)	313	...	...	...	s	s	...	δ v^h	to v alk s CCl_4 δ	**B17²**, 467
—	—,dihydro-	*see* **Cyclohexadiene-dicarboxylic acid***													
Ω p878	—,3,6-dihydroxy-, dinitrile	$C_8H_4N_2O_2$. *See* p812	160.13	yesh lf (w + 2)	230	...	...	...	δ	v	v	...	δ	chl δ	**B10²**, 383
Ω p879	—,—,imide	$C_8H_5NO_4$. *See* p830	179.13	gr-ye nd (w + 3)	273–4	...	...	...	v^h	...	...	...	...	...	**B21¹**, 478
p880	—,4,5-dimethoxy-	*m*-Hemipic acid. *m*-Hemipinic acid. $C_{10}H_{10}O_6$. *See* p812	226.19	nd (w) or pr (w + 2)	174–5 (206) (179–82, rapid htng)	...	...	...	δ	...	...	...	...	...	**B10²**, 382
p881	—,3-hydroxy-	$C_8H_6O_5$. *See* p812	182.13	nd, pr (eth-peth, w) λ^{al} 235 (3.4), 311 (3.3)	150d→ anh	sub	...	...	s^h	v	v	...	...	peth δ	**B10²**, 352
p882	—,4-hydroxy-	$C_8H_6O_5$. *See* p812	182.13	rosettes (w)	204–5d→ anh	...	...	...	δ s^h	s	s	...	δ	peth δ	**B10¹**, 255
p882¹	—,monoper-	$C_8H_6O_5$. *See* p812	182.13	nd (w)	110–2d	...	...	...	v	v	...	...	s	chl s	**B9²**, 599
Ω p883	—,3-nitro-	$C_8H_5NO_6$. *See* p812	211.13	pa ye pr (w)	218	...	...	...	δ s^h	s^h	δ	...	i	chl, peth i	**B9**, 823
Ω p884	—,—,anhydride	$C_8H_3NO_5$. *See* p812	193.12	nd (aa, ace, al)	164	...	...	...	i	s^h	...	s	δ	aa s^h	**B17²**, 468
p885	—,—,diethyl ester	$C_{12}H_{13}NO_6$. *See* p812	267.24	pr (al), nd (peth)	46	...	...	...	i	v	v	...	...	...	**B9²**, 607
Ω p887	—,4-nitro-	$C_8H_5NO_6$. *See* p812	211.13	pa ye nd (w, eth)	165–6	...	...	...	s	s	...	...	i	AcOEt s^h chl, peth, CS_2 i	**B9²**, 606
Ω p888	—,—,imide	$C_8H_4N_2O_4$. *See* p830	192.13	col nd (w), ye lf (al-ace) λ^{al} 231 (4.25)	202	...	...	...	δ^h	δ s^h	...	s	...	aa s	**B21²**, 373
p889	—,—,dimethyl ester	$C_{10}H_9NO_6$. *See* p812	239.19	cr (dil al)	69–71 (66)	...	...	...	i	s	...	...	...	MeOH s	**B9²**, 607
p890	—,—,piperazinium salt	$C_8H_5NO_6 . C_4H_{10}N_2$. *See* p812	297.27	wh cr	201.5–4.5 d	...	...	...	s^h	s^h	δ	...	...	...	**Am 70**, 2758
p891	—,tetrabromo-	$C_8H_2Br_4O_4$. *See* p812	481.74	nd (w)	266d→ anh	...	...	...	δ s^h	δ	...	...	...	os δ	**B9¹**, 367
Ω p892	—,—,anhydride	$C_8Br_4O_3$. *See* p812	463.72	nd (aa-xyl) λ^{cy} 255 (1.55), 326 (0.15), 345 (0.25)	279.5–80.5	...	...	...	i	i	...	...	δ	$PhNO_2$ s os δ^h	**B17²**, 468
Ω p893	—,tetrachloro-	$C_8H_2Cl_4O_4$. *See* p812	303.92	pl (w)	98d→ anh 250d	...	...	...	δ d^h	...	...	v	δ	chl δ	**B9²**, 604
Ω p894	—,—,anhydride	$C_8Cl_4O_3$. *See* p812	285.90	pr, nd (sub) $\lambda^{Pr_2O-Me-Cy}$ 240 (4.7), 310 sh (3.1), 335 (3.6)	255–6.5	sub	...	...	d^h	...	δ	...	...	...	**B17²**, 467
p895	—,—,monoethyl ester	$C_{10}H_6Cl_4O_4$. *See* p812	331.97	pr (dil al)	94–5	d 150	...	...	δ	v	v	...	...	...	**B9²**, 604
p896	—,tetraiodo-, anhydride	$C_8I_4O_3$. *See* p812	651.71	ye pr, nd (aa, $PhNO_2$), nd (sub) λ^{cy} 270 (0.9), 322 (0.15), 342 (0.1)	327–8 (325)	sub	...	...	i	i	...	...	i	aa δ^h oos i	**B17²**, 468
—	**Phthalic anhydride**	*see* **Phthalic acid**, anhydride													
Ωp897	**Phthalide**	α-Hydroxy-*o*-toluic acid lactone. 1(2)-Isobenzo-furanone.	134.14	(i) nd or pl (w) λ^{al} 272 (3.0), 285 (3.0)	(i) 75 (st) (ii) 65.8 (unst)	290	1.1636^{99}_4	1.536^{99}	s^h	v	v	...	...	...	**B17²**, 332
Ωp898	—,3-benzylidene- (*trans*)	$C_{15}H_{10}O_2$. *See* p897	222.25	mcl pr λ^{al} 239 (4.13), 296 (4.31), 340 (4.32)	108	...	...	...	i	s^h	...	...	...	...	**B17²**, 399
Ωp899	—,3,3-diphenyl-	Phthalophenone. Triphenyl-carbinol-*o*-carboxylic lactone.	286.33	lf (al)	120	235^{15}	...	...	...	...	...	...	...	...	**B17**, 391
Ωp900	—,6-nitro-	$C_8H_5NO_4$. *See* p897	179.13	ye nd (al, aa)	145	...	...	...	i	s	s	...	s	aa s chl s^h	**B17²**, 334
—	**Phthalimide**	*see* **Phthalic acid**, imide													

For explanations, symbols and abbreviations see beginning of table. For structural formulas see end of table.

No.	Name	Synonyms and Formula	Mol. wt.	Color, crystalline form, specific rotation and λ_{max} (log ε)	m.p. °C	b.p. °C	Density	n_D	Solubility w	al	eth	ace	bz	other solvents	Ref.
	Phthalocyanine														
Ωp901	**Phthalocyanine**	Tetrabenzoporphyrazine.	514.55	grsh-bl mcl with purp luster (quinoline) $\lambda^{1\text{-Bromonaphthalene}}$ 549 (3.85), 575 (4.06), 600 (4.66), 632 (4.86)	does not melt	sub 550d (vac)	...	...	i	i	i	...	...	$PhNH_2$ s os i	**J1938,** 1151
—	**Phthalonic acid**	*see* **Acetic acid, (2-carboxyphenyl)-2-oxo-***													
—	**Phthalyl chloride**	*see* **Phthalic acid,** dichloride													
—	**Phthiocol**	*see* **1,4-Naphthoquinone, 2-hydroxy-3-methyl-***													
Ωp902	**Physostigmine**	Eserine. $C_{15}H_{21}N_3O_2$	275.35	orth pr (eth, bz) $[\alpha]_D^{17}$ −82 (chl, c = 1.3), $[\alpha]_D^{25}$ −120 (bz, c = 1) λ^{MeOH} 252 (4.11), 309 (3.50)	(i) 105–6 (st) (ii) 86–7 (unst)	...	...	...	δ	s	s	...	s	chl s	**B23²,** 330
Ωp903	—,2-hydroxybenzoate	Eserin salicylate. $C_{15}H_{21}N_3O_2 . C_7H_6O_3$. *See* p902	413.48	pr (al), lt, air, heat → red pr	185–7	...	...	...	δ s^h	s v^h	δ	...	...	chl s	**B23²,** 332
p904	—,sulfate	$2(C_{15}H_{21}N_3O_2) . H_2SO_4$. *See* p902	648.79	dlq sc (ace-eth) $[\alpha]_D$ −130 (w)	140–2 (anh)	...	...	...	v	v	δ	s	...	...	**B23²,** 332
p905	**Phytadiene**(*d*)	2,6,10,14-Tetramethyl-13,15-hexadecadiene. $C_{20}H_{38}$	278.53	$[\alpha]_D$ +0.89 (undil) λ^{al} 230 (4.20)	...	$186–8^{14}$	0.826^{0}_{4}	...	...	...	...	...	...	peth, MeOH, aa ∞	**B1³,** 1030
Ωp906	**Phytol** (*dl*)	3,7,11,15-Tetramethyl-2-hexadecen-1-ol*. $C_{20}H_{40}O$	296.54	λ^{al} 212 (3.04)	...	$202–4^{10}$ $140–1^{0.03}$	0.8497^{25}_{4}	1.4595^{25}	i	...	...	...	...	os s	**B1³,** 1966
p907	**Picein**	*p*-Hydroxyacetophenone-*D*-glucoside. $C_{14}H_{18}O_7$	298.30	nd (w + 1), nd (MeOH) $[\alpha]_D$ −86.5	195–6	...	...	...	s^h δ	s	s	...	...	aa s AcOEt δ chl i	**B31,** 226
Ωp908	**Picene**	Dibenzo[*a*,*i*]phenanthrene. 3,4-Benzochrysene.	278.36	lf, pl (xyl, Py cumene, sub), cr (w + 1) λ^{bz} 287 (2.99). 304 (4.53), 315 (4.28), 329 (4.35), 358 (3.92), 376 (2.92)	367–9	518–20 sub 300²	...	...	i	δ	...	...	δ^h	cumene s con sulf s (bl flr) chl, aa δ^h	**B5³,** 2555
—	**Picoline**	*see* **Pyridine, methyl-**													
—	**Picolinic acid**	*see* **2-Pyridinecarboxylic acid***													
—	**Picramic acid**	*see* **Phenol, 2-amino-4,6-dinitro-***													
—	**Picric acid**	*see* **Phenol, 2,4,6-trinitro-***													
Ωp909	**Picrolonic acid**	...	264.21	ye nd (al)	116–7 (d125)	...	...	...	δ^h	s^h	s^h	...	...	MeOH s^h	**B24²,** 25
Ωp910	**Picropodophyllin**	$C_{22}H_{22}O_8$	414.42	col nd (al, MeOH, bz) λ^{al} 208.5 (4.76), 289.5 (3.66)	228	...	...	...	i	v^h	s	v	s	chl, MeOH v alk s peth i	**B19,** 424
Ωp911	**Picrotoxin**	Cocculin. $C_{30}H_{34}O_{13}$	602.60	rh lf, $[\alpha]_D^{16}$ −29.3 (absal, c = 4)	203–4	...	...	...	δ v^h	s v^h	δ	...	...	Py v chl δ	**Am 79,** 5550
—	**Picryl chloride**	*see* **Benzene, 2-chloro-1,3,5-trinitro-***													
p912	**Pilocarpidine** (*d*)	$C_{10}H_{14}N_2O_2$	194.24	syr, $[\alpha]_D^{20}$ +81.3 (w, al, chl)	...	...	...	...	s	v	δ	...	...	...	**B27²,** 694
p913	—,nitrate (*d*)	$C_{10}H_{14}N_2O_2 . HNO_3$. *See* p912	257.25	pr (w), $[\alpha]_D^{20}$ +73.2 (w)	137	...	...	...	s	s	...	...	...	...	**B27²,** 694
p914	**Pilocarpine**	$C_{11}H_{16}N_2O_2$	208.26	nd, $[\alpha]_D^{20}$ +100.5 (w) λ^{MeOH} 216 (3.75)	34	260^5	...	...	s	s	δ	...	δ	chl v peth i	**B27²,** 694
Ωp915	—,hydrochloride	$C_{11}H_{16}N_2O_2 . HCl$. *See* p914	244.72	hyg cr	204–5 (198)	...	...	...	v	s	i	...	...	chl δ	**B27²,** 695
p916	—,2-hydroxybenzoate	Pilocarpine salicylate. $C_{11}H_{16}N_2O_2 . C_7H_6O_3$. *See* p914	346.39	nd or lf (al) $[\alpha]_D$ +63	120	...	...	...	v	s	s	...	...	...	**B27,** 635
p917	—,nitrate	$C_{11}H_{16}N_2O_2 . HNO_3$. *See* p914	271.28	wh pw or cr (al) $[\alpha]_D$ +80 (w, c = 4)	178 (cor)	...	...	...	v	s^h δ	i	...	...	chl i	**B27²,** 695
p918	—,sulfate	$(C_{11}H_{16}N_2O_2)_2 . H_2SO_4$. *See* p914	514.60	hyg cr (al-eth) $[\alpha]_D^{20}$ +85 (w, c = 7)	132	...	...	...	s	s	...	...	...	...	**B27,** 635
—	**Pilosine**	*see* **Carpiline**													

For explanations, symbols and abbreviations see beginning of table. For structural formulas see end of table.

No.	Name	Synonyms and Formula	Mol. wt.	Color, crystalline form, specific rotation and λ_{max} (log ε)	m.p. °C	b.p. °C	Density	n_D	Solubility w	al	eth	ace	bz	other solvents	Ref.
	Pimaric acid														
p919	**Pimaric acid** (*d*)	Dextropimaric acid. $C_{20}H_{30}O_2$	302.46	orh (ace), pr (al) $[\alpha]_D^{20}$ +87.3 (chl) λ <215	218–9 (215.4 cor)	282^{18}	...	...	i	s	s	...	...	Py v MeOH δ peth δ^h	**B9^2**, 433
—	**Pimelic acid**	*see* **Heptanedioic acid***													
—	**Pinacol**	*see* **2,3-Butanediol, 2,3-dimethyl-***													
—	**Pinacolone**	*see* **2-Butanone, 3,3-dimethyl-***													
p920	**Pinane** (*d, cis*)	...	138.26	$[\alpha]_D^{20}$ +23.3 (undil) λ <220	−53	169^{760} 60.1^{18}	0.8560_4^{20}	1.4629^{20}	i	...	s	...	s	os s	**B5^3**, 254
Ω p921	—(*dl*)	Pincocampane. $C_{10}H_{18}$. *See* p920	138.26	λ <220	...	164.5–5.0^{760}	0.8551_4^{20}	1.4609^{20}	i	...	s	...	s	os s	**B5^1**, 47
p922	—(*l, cis*)	$C_{10}H_{18}$. *See* p920	138.26	$[\alpha]_D^{20}$ −47 (undil) λ <220	...	167–8^{757}	0.8556^{21}	1.4645^{20}	i	...	s	...	s	os s	**B5^3**, 254
Ω p923	α-**Pinene** (*dl*)	$C_{10}H_{16}$	136.24	oil λ^{al} 210 (3.64)	−55	156.2^{760} 51.4^{20}	0.8582_4^{20}	1.4658^{20}	i	∞	∞	...	...	chl ∞	**B5^3**, 366
Ω p924	β-**Pinene** (*d*)	Nopinene. Pseudopinene. $C_{10}H_{16}$	136.24	$[\alpha]_D$ +28.6 λ^{al} 208 (3.72)	...	164–6^{760} 59.7^{20}	0.8654_4^{20}	1.4789^{20}	i	s	s	...	s	chl v os s	**B5^3**, 378
p925	—(*l*)	$C_{10}H_{16}$. *See* p924	136.24	$[\alpha]_D^{25}$ −21.5 λ^{al} 208 (3.72)	...	164^{760} 59.7^{20}	0.8694_4^{20}	1.4762^{20}	i	s	s	...	s	chl v os s	**B5^3**, 376
p926	**Pinic acid** (*dl*) (*a* form)	3-Carboxy-2,2-dimethylcyclobutylacetic acid.	186.21	wh pr (w)	101–2.5	214–6^{9}	1.0925_4^{109}	1.4458_4^{109}	δ	...	...	...	...	...	**B9^2**, 529
p927	—(*dl*) (*b* form)	$C_9H_{14}O_4$. *See* p926	186.21	...	68	214–6^{9}	...	...	...	...	...	...	...	...	**B9^2**, 529
p928	—(*dl*) (*c* form)	$C_9H_{14}O_4$. *See* p926	186.21	pl	58	214–6^{9}	...	...	...	...	...	...	...	...	**B9^2**, 529
p929	—,—(*l*)	$C_9H_{14}O_4$. *See* p926	186.21	nd (w)	135–6	...	...	...	s^h	...	...	...	...	...	**B9^1**, 320
p930	**Pinol**(*dl*)	*dl*-Sobrerone. 6,8-Epoxy-1-*p*-menthene.	152.24	$[\alpha]_D$ −7.1 (ace)	...	183–4^{760} 76–7^{14}	0.9515_4^{20}	1.4695^{20}	...	s	s	...	...	...	**B17^2**, 44
p931	—,hydrate (*trans,dl*)	$C_{10}H_{16}O.H_2O$. *See* p930	170.25	pl or nd (w)	132 (128)	270–1 157–8^{12}	...	...	s	v	v	...	...	...	**C55**, 4569
—	**Pinosylvin**	*see* **Stilbene, 3,5-dihydroxy-** (*trans*)													
—	**Pipecoline**	*see* **Piperidine, methyl-**													
—	**Pipecolyl carbinol**	*see* **Piperidine, 2(2-hydroxyethyl)-**													
Ω p932	**Piperazine**	Diethylenediamine. Hexahydropyrazine.	86.14	hyg pl or lf (al) λ^{gas} 196 (3.7)	106 (106–9)	146^{760}	...	1.446^{113}	v	s	i	...	...	glycols v	**B23^2**, 3
p933	—,dihydrobromide	$C_4H_{10}N_2.2HBr$. *See* p932	247.97	wh nd	d	...	...	...	v	i	i	...	...	...	**J1957**, 1881
p934	—,dihydrochloride monohydrate.	$C_4H_{10}N_2.2HCl.H_2O$. *See* p932	177.07	nd (dil al)	82.5–3	...	...	...	v	i	i	...	...	...	**B23^2**, 4
Ω p935	—,hexahydrate.	Arthriticine. $C_4H_{10}N_2.6H_2O$. *See* p932	194.23	wh (w)	44–5	125–30	...	...	s	s	i	...	...	...	**C47**, 10014
Ω p936	—,1-benzyl-	$C_{11}H_{16}N_2$. *See* p932	176.26	...	...	145–7^{12} 88–9$^{0.3}$	...	1.5430^{28}	s	s	s	...	...	...	**Am 77**, 753
p937	—,1,4-bis(4-methoxybenzoyl)-	1,4-Dianisoylpiperazine. $C_{20}H_{22}N_2O_4$. *See* p932	354.41	wh	192.5–3.5	...	...	...	i	s^h	i	...	...	...	**Am 56**, 150
p938	—,1,4-bis-(phenylacetyl)-	*N,N'*-Di-α-toluylpiperazine. $C_{20}H_{22}N_2O_2$. *See* p932	322.41	wh	150–1	...	...	...	δ	s^h	δ	...	...	...	**Am 56**, 150
p939	—,1,4-bis-(3-phenylpropanoyl)-	1,4-Bis(hydrocinnamyl)-piperazine. $C_{22}H_{26}N_2O_2$. *See* p932	350.47	wh	122.5–3	...	...	...	i	s^h	δ	...	...	...	**Am 56**, 150
p940	—,1,4-dimethyl-	$C_6H_{14}N_2$. *See* p932	114.19	...	...	131–2^{764}	0.8600_4^{20}	1.4474^{20}	v	v	v	...	...	...	**B23^2**, 5
Ω p941	—,2,5-dimethyl- (*cis*)	$C_6H_{14}N_2$. *See* p932	114.19	rh bipym nd or pr (chl) λ^{iso} 277 (1.3), 308 (0.4), 313 (0.5), 319 (0.4)	114	162 sub	...	1.4720^{20}	v	v	δ	...	δ	chl v	**B23^2**, 21
Ω p942	—,—(*trans*)	$C_6H_{14}N_2$. *See* p932	114.19	mcl pl or pr (bz, chl)	118–9	162 sub	...	...	v	v	δ	...	δ s^h	chl s	**B23^2**, 19
Ω p943	—,2,6-dimethyl- (*cis*)	$C_6H_{14}N_2$. *See* p932	114.19	lf or pl (bz)	111–3	162	...	...	s	s	i	...	δ	peth v chl s	**B23^1**, 8
Ω p944	—,1,4-dinitroso-	$C_4H_8N_4O_2$. *See* p932	144.14	pa ye pl (w)	158	...	...	...	δ	v^h	δ	...	...	...	**B23^1**, 7
p945	—,1-ethyl-	$C_6H_{14}N_2$. *See* p932	114.19	...	...	154^{753}	...	...	s	s	s	...	...	...	**B23^2**, 5
Ω p946	—,1-methyl-	$C_5H_{12}N_2$. *See* p932	100.17	...	...	138^{760}	...	1.4378^{20}	v	v	v	...	...	...	**C51**, 10538
Ω p947	—,2-methyl-	$C_5H_{12}N_2$. *See* p932	100.17	hyg lf (al)	62	155^{763}	...	...	v	s	s	...	s	chl s	**B23^2**, 16
Ω p948	—,1-phenyl-	$C_{10}H_{14}N_2$. *See* p932	162.24	pa ye oil	...	286.5^{760} 156–7^{10}	1.0621_4^{20}	1.5875^{20}	i	∞	∞	...	...	...	**C49**, 11667
Ω p949	**1-Piperazinecarboxylic acid,** ethyl ester	...	158.20	...	...	237 116–7^{12}	...	1.4760^{25}	s	s	s	...	...	...	**B23^2**, 9
Ω p950	**1,4-Piperazinedicarboxylic acid,** diethyl ester	...	230.27	nd (hx)	49	315 131–3^{3}	...	...	...	s	s	...	...	...	**B23**, 12
Ω p951	**2,5-Piperazinedione**	Glycine anhydride.	114.11	ta or pl (w)	318–20d sub 260	...	...	...	δ	δ^h	...	...	...	HCl s	**B24^2**, 141

For explanations, symbols and abbreviations see beginning of table. For structural formulas see end of table.

No.	Name	Synonyms and Formula	Mol. wt.	Color, crystalline form, specific rotation and λ_{max} (log ε)	m.p. °C	b.p. °C	Density	n_D	Solubility w	al	eth	ace	bz	other solvents	Ref.
	Piperic acid														
—	**Piperic acid**	*see* **2,4-Pentadienoic acid, 5(3,4-methylene-dioxyphenyl)-***													
Ωp952	**Piperidine**	Hexahydropyridine. Pentamethyleneimine.	85.15	λ^{gas} 198 (3.5)	−9	106.0, 17.7^{20}	0.8606_4^{20}	1.4530^{20}	∞	∞	s	s	s	chl s	**B20², 6**
—	**—,2-acetonyl-**	*see* **Isopelletierine**													
p953	**—,1-acetyl-**	$C_7H_{13}NO$. *See* p952	127.19			$226–7^{760}$, 109^{18}	1.011^9	1.4790^{25}	∞	s					**B20², 33**
p954	**—,2-allyl-**	$C_8H_{15}N$. *See* p952	125.22			170–1	0.8823_4^{15}								**B20, 147**
Ωp955	**—,1-benzoyl-**	$C_{12}H_{15}NO$. *See* p952	189.26	tcl	49	320–1, 180^{15}			i	s	s				**B20², 34**
—	**—,4-benzoxy-2,2,6-trimethyl-**	*see* **Iso-β-eucaine**													
p956	**—,1-butyl-**	$C_9H_{19}N$. *See* p952	141.26			175–7, $47–8^{20}$	0.8245_4^{20}	1.4467^{20}							**B20², 13**
p957	**—,2-butyl-(*d*)**	$C_9H_{19}N$. *See* p952	141.26	$[\alpha]_D$ +15.7			0.8512								**B20, 127**
p958	**—,—(*dl*)**	$C_9H_{19}N$. *See* p952	141.26			191–3, 75^{14}	0.8529_4^{15}		i					os s	**B20, 128**
p959	**—,—(*l*)**	$C_9H_{19}N$. *See* p952	141.26	$[\alpha]_D^{16}$ − 18.7			0.8533								**B20, 127**
p960	**—,3-butyl-**	$C_9H_{19}N$. *See* p952	141.26			196–7			i	s	s			chl s	**B20¹, 33**
p961	**—,1-*sec*-butyl-(*dl*)**	$C_9H_{19}N$. *See* p952	141.26			$175–6^{768}$	0.8378_4^{20}	1.4506^{20}		s	s				**B20², 13**
p962	**—,—(*l*)**	$C_9H_{19}N$. *See* p952	141.26	$[\alpha]_D^{25}$ − 54.59		175^{760}, $75–6^{25}$	0.835_4^{25}	1.4486^{21}		s	s				**J1935, 1072**
p963	**—,1-*tert*-butyl-**	$C_9H_{19}N$. *See* p952	141.26			166^{763}	0.8465_4^{20}	1.4532^{20}	δ	∞	∞				**B20², 13**
p964	**—,2,4-diethyl-**	$C_9H_{19}N$. *See* p952	141.26			174–9	0.8722^0		δ						**B20, 128**
p965	**—,2,5-diethyl-**	$C_9H_{19}N$. *See* p952	141.26			190, $100–5^{22}$	0.8722^0				v			chl v	**B20, 128**
p966	**—,3,4-diethyl-(*cis*)**	$C_9H_{19}N$. *See* p952	141.26	$[\alpha]_D^{22}$ +26.0 (95 % al, c = 4.35)		70^{12}									**Am 66, 1881**
p967	**—,—(*trans*)**	$C_9H_{19}N$. *See* p952	141.26			193^{720}									**B20, 128**
p969	**—,1,2-di-methyl-(*d*)**	*N*-Methyl-α-pipecoline. $C_7H_{15}N$. *See* p952	113.20	$[\alpha]_D^{15}$ +68.8 (undil)		127	0.825^{16}	1.4395^{20}							**C24, 3240**
p970	**—,—(*dl*)**	$C_7H_{15}N$. *See* p952	113.20			127.9^{760}	0.824_4^{15}	1.4395^{20}	s	s	s			CS_2 s	**B20, 95**
p971	**—,1,3-di-methyl-(*dl*)**	*N*-Methyl-β-pipecoline. $C_7H_{15}N$. *See* p952	113.20			124–6	0.818^{15}								**B20, 100**
p972	**—,2,3-dimethyl-**	α,β-Lupetidine. $C_7H_{15}N$. *See* p952	113.20			$138–40^{720}$			s	v					**B20¹, 29**
p973	**—,2,4-di-methyl-(*d*)**	α,γ-Lupetidine. $C_7H_{15}N$. *See* p952	113.20	$[\alpha]_D$ +23.2		140–2	0.845								**B20, 108**
p974	**—,—(*dl*)**	$C_7H_{15}N$. *See* p952	113.20			140–2	0.8615^0	1.4366^{25}		v	v			ac s	**B20¹, 29**
p975	**—,2,5-dimethyl-**	α,β′-Lupetidine. $C_7H_{15}N$. *See* p952	113.20			138–40 (cor)		1.4452^{25}		v				ac s	**B20, 108**
Ωp976	**—,2,6-dimethyl-**	α,α′-Lupetidine. $C_7H_{15}N$. *See* p952	113.20			127.5–8.3^{768}	0.8158_4^{25}	1.4377^{20}	∞	∞	∞			ac s	**B20², 60**
p977	**—,3,3-dimethyl-**	β,β-Lupetidine. $C_7H_{15}N$. *See* p952	113.20			137, $45–6^{33}$		1.4452^{25}		v					**B20¹, 29**
p978	**—,4,4-dimethyl-**	γ,γ-Lupetidine. $C_7H_{15}N$. *See* p952	113.20			145–6, $30–2^{12}$		1.4489^{25}	∞					os v	**B20¹, 29**
p979	**—,4(dimethyl-amino)-1-phenyl-**	Irenal. $C_{13}H_{20}N_2$. *See* p952	204.32	lf (bz)	47.5–8.5	$123–6^{0.5}$							s^h	dil HCl v	**H26, 1132**
p980	**—,1(2,2-dimethyl-propyl)-**	*N*-Neopentylpiperidine. $C_{10}H_{21}N$. *See* p952	155.29			188	0.8608_4^{20}	1.4593^{20}		s	s				**B20², 13**
p981	**—,1-dodecyl-**	$C_{17}H_{35}N$. *See* p952	253.48	pa ye		161^5, $114–6^{0.4}$	0.8378_4^{20}	1.4588^{20}							**J1952, 1057**
Ωp982	**—,1-ethyl-**	$C_7H_{15}N$. *See* p952	113.20			130.8	0.8237_4^{20}	1.4480^{20}							**B20², 13**
p983	**—,2-ethyl-(*d*)**	$C_7H_{15}N$. *See* p952	113.20	$[\alpha]_D$ +17.1		142–3.5	0.8680^4		δ						**B20, 104**
Ωp984	**—,—(*dl*)**	$C_7H_{15}N$. *See* p952	113.20			$142–3^{719}$, $73–5^{52}$	0.8651_0^0	1.4494^{21}	δ						**B20¹, 28**
p985	**—,—(*l*)**	$C_7H_{15}N$. *See* p952	113.20	$[\alpha]_D$ − 14.9		143^{720}	0.8680^4	1.4544^{20}	δ						**B20¹, 28**
p986	**—,3-ethyl-(*dl*)**	$C_7H_{15}N$. *See* p952	113.20	fum in air		152.6	0.8565_4^{23}	1.4531^{20}	δ			s			**B20¹, 29**
p987	**—,—(*l*)**	$C_7H_{15}N$. *See* p952	113.20	$[\alpha]_D^{15}$ − 4.5		155			δ			s			**B20, 106**
p988	**—,4-ethyl-**	$C_7H_{15}N$. *See* p952	113.20			$156–8^{760}$	0.8759^0	1.4503^{25}	δ						**B20, 108**
—	**—,1,4-ethylene-**	*see* **Quinuclidine**													
p989	**—,1-heptyl-**	$C_{12}H_{25}N$. *See* p952	183.34			239.5, $100–3^9$	0.8316^{20}	1.4531^{20}							**C45, 2940**
p990	**—,1-hexyl-**	$C_{11}H_{23}N$. *See* p952	169.31			219.2, $103–4^{20}$	0.8292_4^{20}	1.4522^{20}							**C55, 2465**
p991	**—,1(2-hydroxy-ethyl)-**	2-*N*-Piperidylethanol. $C_7H_{15}NO$. *See* p952	129.20			$200–2^{742}$, 90^{12}	0.9732_{25}^{25}	1.4749^{20}	∞	v					**B20², 18**
p992	**—,2(1-hydroxy-ethyl)-**	Methyl-2-piperidylcarbinol. $C_7H_{15}NO$. *See* p952	129.20			$106–10^{18}$									**B21², 9**
p993	**—,2(2-hydroxy-ethyl)-**	Pipecolyl carbinol. $C_7H_{15}NO$. *See* p952	129.20		39–40	234.5, $145–6^{36}$	1.01^{17}		v	v	v				**B21², 9**
p994	**—,3(2-hydroxy-ethyl)-**	$C_7H_{15}NO$. *See* p952	129.20			$121–3^6$	1.0106_4^{25}	1.4888^{25}	v	v	v				**B21², 10**
p995	**—,4(2-hydroxy-ethyl)-**	$C_7H_{15}NO$. *See* p952	129.20	syr	132–3.5	227–8, $120–5^{15}$	1.0059_4^{15}	1.4907^{20}	s	v	v				**B21², 10**
—	**—,2(1-hydroxy-propyl)-**	*see* **Conhydrine**													

For explanations, symbols and abbreviations see beginning of table. For structural formulas see end of table.

No.	Name	Synonyms and Formula	Mol. wt.	Color, crystalline form, specific rotation and λ_{max} (log ε)	m.p. °C	b.p. °C	Density	n_D	Solubility w	al	eth	ace	bz	other solvents	Ref.
	Piperidine														
p996	**—,1-isobutyl-**	$C_9H_{19}N$. *See* p952	141.26	...	...	$160–1^{740}$ 87^{43}	0.8161^{25}	1.4382^{25}	v	...	...	...	...	...	B20, 20
p997	**—,2-isobutyl-**	$C_9H_{19}N$. *See* p952	141.26	...	...	181–2	0.8510_4^{22}	1.4553^{22}	...	v	v	...	...	...	B20, 128
p998	**—,1-isopropyl-**	$C_8H_{17}N$. *See* p952	127.23	...	...	$149–50^{757}$	0.8389_4^{20}	1.4491^{20}	s	...	...	...	...	...	$B20^2$, 13
p999	**—,2-isopropyl-** (*dl*)	$C_8H_{17}N$. *See* p952	127.23	...	...	162	0.8668^0	...	δ i^h	...	...	...	...	...	B20, 120
p1000	**—,—(*l*)**	$C_8H_{17}N$. *See* p952	127.23	$[\alpha]_D$ −13.1	...	161.5	0.8503^{19}	...	δ i^h	...	...	...	...	...	B20, 121
p1001	**—,4-isopropyl-**	$C_8H_{17}N$. *See* p952	127.23	...	...	168–71 $66–70^{15}$	...	...	δ i^h	...	...	...	...	...	B20, 121
p1002	**—,1-methyl-**	$C_6H_{13}N$. *See* p952	99.18	λ^{eth} 213 (3.20)	...	107	0.8159_4^{20}	1.4355^{20}	v	∞	∞	...	...	...	$B20^2$, 12
Ωp1003	**—,2-methyl-(*d*)**	*d*-α-Pipecoline. $C_6H_{13}N$. *See* p952	99.18	$[\alpha]_D^{15}$ +18.7 (chl), $(\alpha)_D^{22}$ +5.6 (al)	...	117–7.5^{745}	...	1.4459^{20}	s^h	...	...	...	...	...	$B20^2$, 57
p1004	**—,—(*dl*)**	α-Pipecoline. $C_6H_{13}N$. *See* p952	99.18	...	−4.9	$117–8^{747}$	0.8436_4^{24}	1.4459^{20}	v	s	s	...	...	dil KOH i	$B20^2$, 57
Ωp1005	**—,3-methyl-(*dl*)**	β-Pipecoline. $C_6H_{13}N$. *See* p952	99.18	λ^{iso} 268 (2.63)	0–5	$125–6^{763}$	0.8446_4^{24}	1.4470^{20}	v	...	...	...	...	...	$B20^2$, 58
p1006	**—,—(*l*)**	$C_6H_{13}N$. *See* p952	99.18	$[\alpha]_D^{25}$ −4 λ^{iso} 268 (2.63)	...	124	...	...	v	...	...	...	...	...	$B20^2$, 58
Ωp1007	**—,4-methyl-**	γ-Pipecoline. $C_6H_{13}N$. *See* p952	99.18	...	...	132–4 (127–9)	0.8674^0	1.4458^{20}	v	...	...	...	...	...	B20, 101
p1008	**—,1(3-methyl-butyl)-**	*N*-Isoamylpiperidine. $C_{10}H_{21}N$. *See* p952	155.29	...	...	188–9 $76–9^{20}$	...	...	δ	...	...	s	...	...	$B20^2$, 14
p1009	**—,1(2-methyl-2-pentyl)-**	$C_{11}H_{23}N$. *See* p952	169.31	...	...	205–7	0.8517_4^{20}	1.4592^{20}	...	...	...	s	...	...	$B20^2$, 14
p1010	**—,1(3-methyl-3-pentyl)-**	$C_{11}H_{23}N$. *See* p952	169.31	...	...	214^{752}	0.8614_4^{20}	1.4637^{20}	...	...	...	...	...	...	$B20^2$, 14
—	**—,1-methyl-2-propyl-**	*see* **Coniine, *N*-methyl-**													
Ωp1011	**—,1-nitroso-**	$C_5H_{10}N_2O$. *See* p952	114.15	pa ye λ^{hx} 354 (1.91), 366 (2.04), 377 (2.00)	...	217^{721} 109^{20}	$1.0631_4^{18.5}$	$1.4933^{18.5}$	s	...	...	...	...	HCl s	$B20^1$, 24
p1012	**—,1-nonyl-**	$C_{14}H_{29}N$. *See* p952	211.40	...	...	$135–7^{11}$	0.8313^{25}	1.4538^{25}	...	...	...	...	...	...	Am 56, 2419
p1013	**—,1-octyl-**	$C_{13}H_{27}N$. *See* p952	197.37	...	...	$136–8^{13}$ 89^1	0.8324_4^{20}	1.4544^{20}	...	...	...	...	...	...	C43, 3378
p1014	**—,1-pentyl-**	$C_{10}H_{21}N$. *See* p952	155.29	...	...	198.2^{760} 80^8	0.8282^{20}	1.4498^{20}	...	...	...	...	...	...	$B20^2$, 13
p1015	**—,1-phenyl-**	$C_{11}H_{15}N$. *See* p952	161.25	λ^{MeOH} 253 (4.0)	...	$257–8^{752}$ 126.5^{15}	...	...	...	v	v	...	v	chl v	$B20^2$, 14
p1016	**—,2-propenyl-(*d*)**	β-Coniceine. $C_8H_{15}N$. *See* p952	125.22	nd, $[\alpha]_D^{45}$ +49.9	39	168–9	...	...	δ	s	s	...	...	...	B20, 147
p1017	**—,—(*dl*)**	$C_8H_{15}N$. *See* p952	125.22	nd	8	168.5–70^{753}	0.8716_4^{15}	...	δ	s	s	...	...	...	B20, 147
p1018	**—,—(*l*)**	$C_8H_{15}N$. *See* p952	125.22	nd, $[\alpha]_D^{15}$ −50.47	39.5–40	168–9	0.8672_4^{15}	...	...	...	...	...	...	...	B20, 147
p1019	**—,1-propyl-**	$C_8H_{17}N$. *See* p952	127.23	...	...	151.2	0.8231^{20}	1.4446^{20}	s	v	v	...	...	...	$B20^2$, 13
—	**—,2-propyl-**	*see* **Coniine**													
p1020	**—,3-propyl-(*d*)**	$C_8H_{17}N$. *See* p952	127.23	$[\alpha]_D^{16}$ +5.9	...	174^{752}	0.8517_4^{19}	...	s	s	...	...	...	...	B20, 120
p1021	**—,—(*dl*)**	$C_8H_{17}N$. *See* p952	127.23	...	...	174^{758}	0.8475_4^{26}	...	s	s	...	...	...	...	B20, 119
p1022	**—,—(*l*)**	$C_8H_{17}N$. *See* p952	127.23	oil, $[\alpha]_D^{16}$ −6.6	...	174^{752}	0.8517_4^{19}	...	s	s	...	...	...	...	B20, 120
Ωp1023	**—,4-propyl-**	$C_8H_{17}N$. *See* p952	127.23	...	...	172^{748} (179)	0.864^{23}	1.4465^{23}	...	s	...	...	...	...	B20, 120
Ωp1024	**—,2,2,6,6-tetra-methyl-**	$C_9H_{19}N$. *See* p952	141.26	...	...	155.5–6.5	0.8367_4^{16}	1.4455^{20}	...	...	s	...	...	...	B20, 129
p1025	**—,2,2,4-tri-methyl-**	$C_8H_{17}N$. *See* p952	127.23	...	...	148	0.832^{15}	1.4458^{20}	δ	s	s	...	...	...	B20, 124
p1026	**—,2,3,6-tri-methyl-**	$C_8H_{17}N$. *See* p952	127.23	...	...	36^8	0.8302_4^{20}	1.4434^{20}	...	...	...	...	...	...	C43, 2959
p1027	**—,2,4,6-tri-methyl-**	$C_8H_{17}N$. *See* p952	127.23	...	...	165–6 (151–3)	0.8315_4^{19}	1.4412^{20}	δ	∞	∞	...	...	...	$B20^2$, 64
p1028	**1-Piperidinecar-boxaldehyde**	*N*-Formylpiperidine	113.16	pa ye	...	222 106–10^{17}	1.0205_4^{25}	1.4700^{20}	∞	∞	∞	...	∞	chl, aa, lig ∞	$B20^2$, 33
p1029	**1-Piperidinecar-bodithioic acid,** piperidinium salt	Piperidinium cyclopenta-methylenedithiocarbamate.	246.44	pa ye lf (al), cr (CS_2-eth)	174	...	...	...	v^h	v	...	...	v	...	$B20^2$, 37
p1030	**4-Piperidinecar-boxylic acid**	Hexahydroisonicotinic acid. Isonipecotic acid.	129.16	nd (w)	*ca.* 326	...	...	...	s	i	...	...	...	...	$B22^1$, 486
p1031	**—,ethyl ester**	Ethyl isonipecotate. $C_8H_{15}NO_2$. *See* p1030	157.22	col oil	...	$100–1^{10}$	...	1.4591^{20}	s	s	s	...	s	...	C49, 12473
p1032	**—,methyl ester**	Methyl isonipecotate. $C_7H_{13}NO_2$. *See* p1030	143.19	col oil	...	$107–10^{22}$ $88–9^9$	...	1.4635^{25}	s	s	s	...	s	...	C49, 12473
p1033	**—,1-methyl-,** methyl ester	$C_8H_{15}NO_2$. *See* p1030	157.22	λ^{al} 204 (3.0)	...	$96–100^{20}$ $82–6^{10}$	...	1.4539^{24}	s	s	s	...	s	...	B22, 486
p1034	**—,1-methyl-4-phenyl-,** ethyl ester hydro-chloride	Demerol hydrochloride. Meperidine hydrochloride. Pethidine hydrochloride.	283.80	cr (al)	186–9	...	...	...	s	δ	i	s	i	AcOEt s	C33, 9442

For explanations, symbols and abbreviations see beginning of table. For structural formulas see end of table.

No.	Name	Synonyms and Formula	Mol. wt.	Color, crystalline form, specific rotation and λ_{max} (log ε)	m.p. °C	b.p. °C	Density	n_D	Solubility w	al	eth	ace	bz	other solvents	Ref.
	2,4-Piperidinedione														
p1034^{1}	**2,4-Piperidine-dione, 3,3-diethyl-**	Piperidione	169.23	nd (w)	103–5 (102–7)	...	...	...	v	v	...	...	...	MeOH, chl v	**C44,** 1506
—	**Piperidinic acid**	*see* **Butanoic acid 4-amino-***													
—	**Piperidinium,** cyclopenta-methylene dithio-carbamate	*see* **1-Piperidinecarbodithoic acid,** piperidinium salt													
Ω p1035	**2-Piperidone**	5-Aminopentanoic acid lactam*. δ-Valerolactam.	99.13	hyg λ^{al} <217 (3.0)	39–40	256^{760} 137^{14}	...	...	v	v	v	...	...	dil ac s con alk i	**B21,** 238
p1036	**4-Piperidone,** hydrochloride	...	135.60	cr (+1.5 al, al-eth), cr (w+1)	147–9 139–41d (+al) 140–5d (+w)	...	...	...	v	...	...	...	...	...	**B21**^{2}, 215
p1037	**2-Piperidone, 3-hydroxy-(*dl*)**	$C_5H_9NO_2$. *See* p1035.	115.13	nd or pr (AcOEt)	141–2	...	...	...	v	v	δ	...	...	AcOEt, chl δ	**B21**^{2}, 411
p1038	**—,5-hydroxy-**	$C_5H_9NO_2$. *See* p1035.	115.13	(al, al-AcOEt)	145–6 (cor)	...	...	...	v	v	i	...	i	AcOEt, chl δ	**B21**^{2}, 412
Ω p1039	**4-Piperidone, 2,2,6,6-tetra-methyl-**	Triacetoneamine. $C_9H_{17}NO.H_2O$. *See* p1036	173.26	rh pl (moist eth, +1w), nd (eth)	58 (hyd), 34.9 (anh)	205	...	...	s	s	s	...	...	...	**B21**^{2}, 222
Ω p1040	**Piperine**	1-Piperylpiperidine. $C_{17}H_{19}NO_3$	285.35	pr (AcOEt), pl or mcl pr (al), cr (bz-lig) λ^{al} 345 (4.5)	130–2.5 (129.5)	...	...	...	i	s v^h	δ	...	s	chl v Py, aa, MeOH s peth i	**B20**^{2}, 53
p1041	**—,hydroiodide** diiodide	$2(C_{17}H_{19}NO_3).HI.I_2$. *See* p1040	952.42	bl nd	145	...	...	...	s	v^h	v	...	...	chl, CS_2 v	**B20,** 80
—	**Piperitenone**	*see* **3-Terpinolenone**													
p1042	**Piperitone(*d*)**	Δ^1-*p*-Methanone-3. 1-Methyl-4-isopropylcyclohexenone-3. $C_{10}H_{16}O$	152.24	yesh in air $[\alpha]^{20}_D$ +49.13 (undil) λ^{al} 234 (4.18)	...	222–30 116–8.5^{20}	0.9344^{20}_4	1.4848^{20}	...	...	s	...	...	...	**B7**^{2}, 74
p1043	**—(*dl*)**	$C_{10}H_{16}O$. *See* p1042.	152.24	λ^{al} 234 (4.18)	...	$232–3^{769}$ 113^{18}	0.9331^{20}_4	1.4845^{20}	...	...	...	...	...	...	**B7**^{2}, 75
p1044	**—(*l*)**	$C_{10}H_{16}O$. *See* p1042.	152.24	$[\alpha]^{20}_D$ −51.53 (undil) λ^{al} 234 (4.18)		235 109.5–10.5^{15}	0.9324^{20}_4	1.4848^{20}	...	...	...	...	...	...	**B7**^{2}, 75
Ω p1045	**Piperolidine(*dl*)**	δ-Coniceine. Indolizidine. Octahydropyrrocoline.	125.22	...	...	161.5^{750} $65–7^{18}$	0.9012^{15}_4	1.4748	δ	s	s	...	...	...	**B20,** 150
p1046	**—(*l*)**	$C_8H_{15}N$. *See* p1045.	125.22	$[\alpha]^{27}_D$ −7.89 (c = 1)	...	158^{729} $65–7^{18}$	0.8976^{23}_4	1.4748^{20}	δ	s	s	...	...	...	**B20,** 150
—	**Piperonal**	*see* **Benzaldehyde,3,4-methylenedioxy-**													
—	**4-Piperone**	*see* **Verbenone**													
—	**Piperonyl alcohol**	*see* **Toluene,** α**-hydroxy-3,4-methylenedioxy-**													
—	**Piperonylic acid**	*see* **Benzoic acid, 3,4-methylenedioxy-**													
—	**Pivalaldehyde**	*see* **Propanal,2,2-dimethyl-***													
—	**Pivalic acid**	*see* **Propanoic acid, 2,2-dimethyl-***													
—	**Pivalil**	*see* **3,4-Hexanedione, 2,2,5,5-tetramethyl-***													
—	**Pivalilindandione**	*see* **1,3-Indandione, 2-*tert*-butyl-**													
—	**Pivaloin**	*see* **3-Hexanone, 4-hydroxy-2,2,5,5-tetramethyl-***													
—	**Pivalophenone**	*see* **1-Propanone, 2,2-dimethyl-1-phenyl-***													
	Plasmocid	*see* **Antimalarine**													
—	**Plumbagin**	*see* **1,4-Naphthoquinone, 5-hydroxy-2-methyl-***													
Ω p1047	**Podophyllotoxin**	$C_{22}H_{22}O_8$	414.42	nd (MeOH, dil ace), wh cr (w+1 or 2) $[\alpha]^{20}_D$ −132.7 (chl, c = 2) λ^{al} 291 (3.64)	114–8 (hyd) 183–4 (anh)	...	...	...	δ	v	i	s	s^h	chl v aa s lig i	**B19**^{2}, 443
p1048	**Polyglycolid**	$[\text{-O-CH}_2\text{CO-}]_x$	...	wh ($PhNO_2$)	223	...	...	...	...	...	i	...	i	...	**B19**^{1}, 679
p1049	**Populin**	Salicin benzoate.	390.40	nd (w+2), pr (al) $[\alpha]_D$ −2.0 (Py, c = 5)	180	...	...	...	δ s^h	s^h	v	...	...	aa s	**B31,** 216

For explanations, symbols and abbreviations see beginning of table. For structural formulas see end of table.

No.	Name	Synonyms and Formula	Mol. wt.	Color, crystalline form, specific rotation and λ_{max} (log ε)	m.p. °C	b.p. °C	Density	n_D	Solubility w	al	eth	ace	bz	other solvents	Ref.
	Porphin														
p1050	**Porphin**	Tetramethenetetrapyrrole. $C_{20}H_{14}N_4$	310.36	met red or og lf (chl-MeOH) λ^{bz} 396.5 (5.42), 489.5 (4.20), 519.5 (3.42), 563 (3.70), 569 (3.62), 616.5 (2.93)	darkens 360	300^{12} sub	1.336	....	i	i	i	i	s	diox s chl, aa δ Py	**B26**[2], 228
p1051	**5β-Pregnane**	17β-Ethyletiocholane. $C_{21}H_{36}$	288.52	mcl sc or pl (MeOH) $[\alpha]_D^{19}$ +21.2 (chl, c = 2)	83.5	....	1.032_4^{15}	....	i	...	...	...	...	chl s MeOH s^h	**E14**, 17
p1052	**5β-Pregnane-3α, 20α-diol**	$C_{21}H_{36}O_2$. See p1051	320.52	pl (ace) $[\alpha]_D^{20}$ +27.4 (al, c = 0.7) λ^{sulf} 320 (4.94), 430 (4.73)	243–4	....	1.15	....	...	δ	δ	s^h	...	os δ	**E14**, 95
Ωp1053	**5β-Pregnane-3α-20β-diol**	$C_{21}H_{36}O_2$. See p1051	320.52	cr (al), $[\alpha]_D^{20}$ + 10 $\lambda^{97\,\%sulf}$ 239 (3.56), 323 (3.84), 428 (3.69), 496 sh (3.10)	244–6	....	....	....	...	s^h	...	...	...	....	**E14**, 95
p1054	**5β-Pregnane-3β, 20α-diol**	$C_{21}H_{36}O_2$. See p1051	320.52	cr (al, ace)	182	....	....	....	...	s^h	...	s^h	...	....	**E14**, 95
p1055	**5β-Pregnane-3β, 20β-diol**	$C_{21}H_{36}O_2$. See p1051	320.52	cr (AcOEt-peth or dil al)	174–6	....	....	....	δ	s^h	...	...	...	AcOEt s peth δ	**E14**, 95
p1056	**5β-Pregnane-3,20-dione**	$C_{21}H_{32}O_2$. See p1051	316.49	nd (dil al), cr (dil ace) $\lambda^{97\,\%sulf}$ 237sh (3.70)	123	....	....	....	i	v^h	s	s	...	os v	**E14**, 152
Ωp1057	**5β-Pregnan-3α-ol-20-one**	$C_{21}H_{34}O_2$. See p1051	318.51	nd (bz), cr (dil al), rosetts or nd (hx) $\lambda^{97\,\%sulf}$ 233 sh (3.62), 331 (3.86), 396 sh (2.76)	149.5	....	....	....	i	s^h	...	...	s^h	peth s^h	**E14**, 115
p1058	**5β-Pregnan-3β-ol-20-one**	$C_{21}H_{34}O_2$. See p1051	318.51	cr (dil al)	149	....	....	....	i	s^h	...	...	s^h	....	**E14**, 115
p1059	**5β-Pregnan-20α-ol-3-one**	$C_{21}H_{34}O_2$. See p1051	318.51	pr (ace)	152	....	....	....	...	...	...	s^h	...	....	**E14**, 132
p1060	**5β-Pregnan-20β-ol-3-one**	$C_{21}H_{34}O_2$. See p1051	318.51	cr (dil MeOH)	172	....	....	....	...	...	...	...	...	MeOH s^h	**E14**, 132
Ωp1061	**5-Pregnene-3β-ol-20-one**	$C_{21}H_{32}O_2$. See p1051	316.49	nd (dil al) $[\alpha]_D^{20}$ + 28 (al) λ^{al} 205 (3.52)	193	....	....	....	δ	δ	...	δ	δ	CCl_4 δ lig i	**E14**, 114
p1062	**4-Pregnene-11β, 17α,20β,21-tetrol-3-one**	Reichstein's substance E. $C_{21}H_{32}O_5$. See p1051	364.49	cr (aq ace + w) $[\alpha]_D^{20}$ + 87 (al) $\lambda^{97\,\%sulf}$ 240 (4.31), 283 (4.23), 330 sh (3.73), 405 (3.66), 468 (3.63), 515 sh (3.51)	*ca.* 125d	....	....	....	...	s	...	s^h	...	....	**E14**, 150
p1063	**4-Pregnene-17α, 20β,21-triol-3-one**	$C_{21}H_{32}O_4$. See p1051	348.49	cr (MeOH) $[\alpha]_D$ +63 (diox, c = 1) λ^{al} 240 (4.1)	190	....	....	....	...	...	...	...	...	diox, chl MeOH s	**H22**, 755
—	**Prehnitene**	see **Benzene,1,2,3,4-tetramethyl-***													
—	**Prehnitic acid**	see **1,2,3,4-Benzenetetra-carboxylic acid***													
—	**Prehnitylic acid**	see **Benzoic acid, 2,3,4-trimethyl-**													
p1064	**Primeverose**	6(β-*D*-Xylosido)-*D*-glucose. $C_{11}H_{20}O_{10}$.	312.28	cr (MeOH, 80% al), $[\alpha]_D^{20}$ + 23 → − 3.2 (w, c = 5) (mut)	210 (darkens 190)	....	....	....	s	δ	...	...	...	MeOH s	**B31**, 372
—	**Procaine**	see **Benzoic acid, 4-amino-, (2-diethylaminoelthyl) ester**													
Ωp1065	**α-Progesterone**	17α-Progesterone. $C_{21}H_{30}O_2$.	314.47	orh pr (dil al) $[\alpha]_D$ + 192 λ^{w} 249 (4.23)	128.5–31	....	1.166^{23}	....	δ	...	...	...	...	os δ	**E14**, 151
p1066	**β-Progesterone**	17β-Progesterone. $C_{21}H_{30}O_2$. See p1065	314.47	nd (peth) $[\alpha]_D^{20}$ + 172–82 (diox, c = 2)	121–2	....	1.171^{20}	....	i	s	...	s	...	con sulf, diox s	**E14**, 151

For explanations, symbols and abbreviations see beginning of table. For structural formulas see end of table.

No.	Name	Synonyms and Formula	Mol. wt.	Color, crystalline form, specific rotation and λ_{max} (log ε)	m.p. °C	b.p. °C	Density	n_D	Solubility w	al	eth	ace	bz	other solvents	Ref.
	Progesterone														
p1067	**Progesterone,** dioxime	$C_{21}H_{32}N_2O_2$. *See* 1065	344.50	pl (dil al)	243	...	...	...	i	s^h	...	...	...	...	**E14,** 151
p1068	**Proline**(*D*)	*d*-2-Pyrrolidinecarboxylic acid.	115.13	hyg pr (al-eth) $[\alpha]_D^{20}$ +81.9 (w)	215–20d	...	...	...	v	s^h	i	δ	δ	...	**B22,** 1
p1069	—(*DL*)	$C_5H_9NO_2$. *See* p1068	115.13	hyg nd (al-eth), cr (+w)	205d (anh), 190 (hyd)	...	...	...	v	s v^h	i	δ	δ	chl δ	**B22**², 5
Ωp1070	—(*L*)	$C_5H_9NO_2$. *See* p1068	115.13	nd (al-eth), pr (w), $[\alpha]_D^{20}$ −80.9 (w, c = 1)	220–2d	...	...	...	v	δ	i	δ	δ	BuOH, PrOH i	**B22**², 3
p1071	—,**4-hydroxy-** (*D, cis*)	*Allo*-4-hydroxyproline. $C_5H_9NO_3$. *See* p1068	131.13	nd (w + 1) $[\alpha]_D^{18}$ +58.6 (w, p = 5)	237–41	...	...	...	v	...	i	...	...	...	**B22**¹, 546
p1072	—,—(*DL, cis*)	*Allo*-4-hydroxyproline. $C_5H_9NO_3$. *See* p1068	131.13	cr (w, dil al)	250	...	...	...	s	...	i	...	...	MeOH δ	**B22**², 144
p1073	—,—(*L, cis*)	*Allo*-4-hydroxyproline. $C_5H_9NO_3$. *See* p1068	131.13	nd (w + 1), $[\alpha]_D^{18}$ −58.1 (w, p = 5.2)	238–41	...	...	...	v	δ	...	...	...	MeOH δ	**B22,** 546
Ωp1074	—,—(*D, trans*)	α-4-Hydroxyproline. $C_5H_9NO_3$. *See* p1068	131.13	lf (dil al) $[\alpha]_D^{21}$ +75.2 (w)	274	...	...	...	v	δ	...	...	...	...	**B22**¹, 545
p1075	—,—(*DL, trans*)	α-4-Hydroxyproline. $C_5H_9NO_3$. *See* p1068	131.13	pl (MeOH)	270 (261)	...	...	...	v	δ	...	...	...	MeOH s^h	**B22**², 143
p1076	—,—(*L, trans*)	α-4-Hydroxyproline. $C_5H_9NO_3$. *See* p1068	131.13	lf (dil al), pr (w) $[\alpha]_D$ −76.5 (w, c = 2.5)	274	...	...	...	v	δ	...	...	...	...	**B22**², 143
p1077	—,—,betaine(*d*)	$C_7H_{13}NO_3$. *See* p1068	159.19	pr (w + 1) $[\alpha]_D^{21}$ +36	249d	...	...	...	δ s^h	δ	...	...	...	...	**B22**¹, 546
—	**Prominal**	*see* **Barbituric acid, 5-ethyl-1-methyl-5-phenyl-**													
p1078	**Prontosil**	$C_{12}H_{13}N_5O_2S.HCl$.	327.80	og-red pw $\lambda^{w.pH=6.8}$ 273 (3.86) $\lambda^{w.pH=1.8}$ 262 (4.10), 297 (3.67)	248–51	...	...	...	δ	s		s	...	oils, fats s	**C29,** 4135
p1079	**Propadiene***	Allene. Dimethylenemethane. $CH_2:C:CH_2$	40.07	gas	−136	-34.5^{760}	1.787	1.4168	i	...	...	...	s	peth s	**B1**³, 922
p1080	—,**1,3-dioxo-***	Carbon suboxide. Dioxoallene. O:C:C:C:O	68.03	gas	−107	6.8	1.114_4^0	1.4538^0	d	d	s	...	s	CS_2 s HCO_2H, MeOH d	**B1**³, 3163
p1081	—,**tetrafluoro-***	Perfluoroallene. $F_2C:C:CF_2$	112.03	gas		−38									**Am 81,** 606
Ωp1082	—,**tetraphenyl-***	$(C_6H_5)_2C:C:C(C_6H_5)_2$	344.46	nd, pr (dil ace, al),	166	...	...	...	i	δ	s	s	v	CS_2, chl v	**B5**², 692
—	**Propaesin**	*see* **Benzoic acid, 4-amino-,** propyl ester													
Ωp1083	**Propanal***	Propionaldehyde. Propional. CH_3CH_2CHO	58.08	λ^w 282 (0.9)	−81	48.8	0.8058_4^{20}	1.3636^{20}	s	∞	∞				**B1**³, 2682
p1084	—,diethyl acetal*	1,1-Diethoxypropane*. Propylal. $CH_3CH_2CH(OC_2H_5)_2$.	132.21	...	...	122.8^{744}	0.8232_4^{20}	1.3924^{19}	s	v	v	s	s	...	**B1**³, 2689
p1085	—,oxime*	Propionaldoxime. $CH_3CH_2CH:NOH$	73.10	...	40	131.5 $77^{0.5}$	0.9258_4^{20}	1.4287^{20}	...	...	...	...	...	...	**B1**³, 2689
p1086	—,**2-bromo-***	$CH_3CHBrCHO$	136.98	...	...	109–10 $52–4^{80}$	1.592^{20}	1.4813^{20}	...	...	v	...	...	...	**B1**³, 2693
p1087	—,**2-chloro-***	$CH_3CHClCHO$	92.53	...	...	86	1.182_4^{15}	1.431^{17}	δ	...	∞	...	∞	...	**B1**³, 2691
p1088	—,**3-chloro-***	$ClCH_2CH_2CHO$	92.53	...	...	130–1 40^{19} (40–50^{10})	1.268^{15}	1.475^{15}	i	s	s	...	...	...	**B1**³, 2691
Ωp1089	—,—,diethyl acetal*	$ClCH_2CH_2CH(OC_2H_5)_2$	166.65	...	...	84^{25} 52^9	0.9951_4^{19}	1.4268^{20}	...	s	s	...	...	...	**B1**³, 2692
p1090	—,**2-chloro-2-methyl-***	2-Chloroisobutyraldehyde. $(CH_3)_2CClCHO$	106.55	...	...	90	1.053_4^{15}	1.4160^{16}	...	v	v	...	...	...	**B1,** 675
Ωp1091	—,**2,3-dibromo-***	Acrolein dibromide. $BrCH_2CHBrCHO$	215.88	pa ye fum liq	...	$73–5^{10}$	2.198^{15}	1.5082^{20}	...	...	v	...	...	...	**B1**³, 2694
p1092	—,**2,2-dichloro-***	CH_3CCl_2CHO	126.97	(peth)	38–9	77.8–8.0	...	...	...	...	...	...	...	peth δ	**B1**³, 2692
p1093	—,**2,3-dichloro-***	$ClCH_2CHClCHO$	126.97	hyd (w)	...	73^{50} 48^{14}	1.400^{20}	1.4762^{20}	...	d	...	...	...	CCl_4 s	**B1**³, 2693
—	—,**2,3-dihydroxy-***	*see* **Glyceraldehyde**													
p1094	—,**2,2-dimethyl-***	Pivalaldehyde. $(CH_3)_3CCHO$	86.14	λ^{al} 193 (1.65)	6	77–8	0.7923^{17}	1.3791^{20}	...	s	s	...	...	...	**B1**², 742
p1095	—,—,oxime*	Pivaldoxime. $(CH_3)_3CCH:NOH$	101.15	...	48	65^{20}									**B1**², 744
Ωp1096	—,**3-ethoxy-,** diethyl acetal*	1,1,3-Triethoxypropane*. $C_2H_5OCH_2CH_2CH(OC_2H_5)_2$	176.26	...	...	184–6 δd 78^{14}	0.898_4^{15}	1.4067^{20}	δ	s					**B1**³, 3190
—	—,**2-ethyl-3-phenyl-***	*see* **Butanal, 2-benzylidene-**													
Ωp1097	—,**3-hydroxy-2-oxo-***, enol form	Propanolonal. Reductone. HOCH:C(OH)CHO	88.06	ye nd (w)	200–20d	...	...	...	s	s	...	...	...	...	**B1**³, 3310

For explanations, symbols and abbreviations see beginning of table. For structural formulas see end of table.

No.	Name	Synonyms and Formula	Mol. wt.	Color, crystalline form, specific rotation and λ_{max} (log ε)	m.p. °C	b.p. °C	Density	n_D	Solubility w	al	eth	ace	bz	other solvents	Ref.
	Propanal														
p1098	**—,3(4-isopropyl-phenyl)-2-methyl-**	Cyclamen aldehyde. Cyclomal.	190.29	...	...	133–7^{99} 115^{5}	0.951^{15}	1.5068^{20}	i	s	s	...	s	...	**C55,** 25834
p1099	**—,2-oxo-***	Methylglyoxal. Pyruvaldehyde. CH_3COCHO	72.06	ye hyg liq	...	72	1.0455^{24}	1.4002^{18}	...	s	s	...	s	...	**B1^2,** 819
p1100	**—,—,dioxime***	Methyl glyoxime. $CH_3C(:NOH)CH:NOH$	102.09	nd (to, w, sub), pr (al) λ^{al} 231 (4.32)	157	sub	...	...	δ s^h	s	s	...	s^h	to s^h	**B1^2,** 822
Ωp1101	**—,—,1-oxime***	Isonitrosoacetone. $CH_3COCH:NOH$	87.08	nd (CCl_4), lf (eth-peth)	69	sub	1.0744^{67}	...	s	...	s	...	δ	chl, CCl_4 δ	**B1^3,** 3092
p1102	**—,2-phenoxy-***	$C_6H_5OCH(CH_3)CHO$	150.18	...	...	229–30 99–101^{16}	...	...	i	v	v	...	v	...	**B6,** 151
p1103	**—,—,oxime***	$C_6H_5OCH(CH_3)CH:NOH$	165.19	nd (dil al)	110	...	...	...	δ	v	v	...	...	...	**B6,** 151
Ωp1104	**—,2-phenyl-***	Hydratropic aldehyde. $C_6H_5CH(CH_3)CHO$	134.18	λ 239 (2.87), 259 (2.34), 287 (2.18), 314 (2.00),	...	202–5 92.5^{10}	1.0089$^{20}_4$	1.5176^{20}	i	s	...	...	...	...	**B7^2,** 237
Ωp1105	**—,3-phenyl-***	Hydrocinnamaldehyde. $C_6H_5CH_2CH_2CHO$	134.18	mcl λ^{al} 252 (2.3), 259 (2.4)	47	223^{745} 104–5^{13}	...	...	i	v	∞	...	...	...	**B7^2,** 236
p1106	**—,2,2,3-tri-chloro-***	$ClCH_2CCl_2CHO$	161.42	...	...	63–5^{45}	1.470^{25}	1.473^{25}	...	...	v	...	...	CCl_4 s	**B1^2,** 691
Ωp1107	**Propane***	$CH_3CH_2CH_3$	44.11	...	−189.69	−42.07	0.5005^{20} (lig) 0.5853$^{-45}_4$	1.2898^{20}	s	s	v	δ	v	chl v	**B1^3,** 204
Ωp1108	**—,1-amino-***	Propylamine. $CH_3CH_2CH_2NH_2$	59.11	...	−83.0	47.8^{760}	0.7173$^{20}_4$	1.3870^{20}	s	v	v	v	s	chl s	**B4^3,** 250
Ωp1109	**—,2-amino-***	Isopropylamine. $(CH_3)_2CHNH_2$	59.11	...	−95.2	32.4^{760}	0.6891^{20}	1.3742^{20}	∞	∞	∞	v	s	chl s	**B4^3,** 271
p1110	**—,1-amino-3-cyclohexyloxy-***	...	159.28	...	...	115^{20}	...	...	i	...	...	s	s	...	**C55,** 22136
p1111	**—,1-amino-2,2-dimethyl-***	Neopentylamine. $(CH_3)_3CCH_2NH_2$	87.17	...	...	81–2^{741}	0.7455$^{20}_4$	1.4023^{20}	δ	...	s	...	...	...	**B4^3,** 355
p1112	**—,1-amino-3-dodecyloxy-***	$CH_3(CH_2)_{11}O(CH_2)_3NH_2$	243.44	...	...	140^{3}	0.8439$^{20}_4$	1.4487^{20}	i	...	...	s	s	MeOH, chl s	**C55,** 22136
Ωp1113	**—,1-amino-3-methoxy-***	$CH_3OCH_2CH_2CH_2NH_2$	89.14	...	...	116–9	0.8727$^{20}_4$	1.4191^{20}	s	...	...	s	s	MeOH, chl, CCl_4 s	**C55,** 22136
Ωp1114	**—,2-amino-2-methyl-***	*tert*-Butylamine. $(CH_3)_3CNH_2$	73.14	...	−67.5	44.4^{760}	0.6958$^{20}_4$	1.3784^{20}	∞	∞	∞	...	...	...	**B4^3,** 323
—	**—,2-amino-1-phenyl-***	*see* **Amphetamine**													
Ωp1114[1]	**—,2,2-bis(bromo-methyl)-1,3-dibromo-***	Pentaerythrityl tetrabromide. $C(CH_2Br)_4$	387.76	cr (ace), nd (lig), tcl (bz + 1)	163	305–6	2.596^{15}	...	...	s^h	δ	...	s^h	to s^h os δ	**B1^2,** 105
Ωp1115	**—,2,2-bis(chloro-methyl)-1,3-dichloro-***	Pentaerythrityl tetrachloride. $C(CH_2Cl)_4$	209.94	...	97	110^{12}	...	...	i	...	s	...	...	chl s	**Am 67,** 942
p1116	**—,1,1-bis(ethyl-sulfonyl)-2-methyl-***	$(C_2H_5SO_2)_2CHCH(CH_3)_2$	242.36	nd (al)	94	...	...	...	s^h	...	...	...	...	...	**B1,** 676
p1117	**—,1,1-bis(4-hydroxy-phenyl)-***	4,4′-Propylidenediphenol.	228.29	nd (w)	130	275^{20}	...	...	i	v	v	...	...	alk v aa s	**B6^2,** 977
Ωp1118	**—,2,2-bis(4-hydroxy-phenyl)-***	4,4′-Isopropylidenediphenol.	228.29	pr (dil aa), nd (w) λ^{al} 227 (4.4), 275 (3.8)	152–3	250–2^{13}	...	...	i	v	v	...	v	alk v aa s	**B6^2,** 978
Ωp1118[1]	**—,2,2-bis(iodo-methyl)-1,3-diiodo-***	Pentaerythrityl tetraiodide. $C(CH_2I)_4$	575.75	nd (to)	225	...	...	...	...	δ	δ	...	...	to s^h	**B1^3,** 373
Ωp1119	**—,1-bromo-***	*n*-Propyl bromide. $CH_3CH_2CH_2Br$	123.00	...	−109.85	71.0^{760}	1.3537$^{20}_4$	1.4343^{20}	δ	s	s	s	s	chl s	**B1^3,** 239
Ωp1120	**—,2-bromo-***	Isopropyl bromide. $CH_3CHBrCH_3$	123.00	...	−89.0	59.38^{760}	1.3140$^{20}_4$	1.4251^{20}	δ	∞	∞	s	s	chl s	**B1^3,** 242
p1121	**—,1-bromo-2-chloro-***	$CH_3CHClCH_2Br$	157.44	...	...	118	1.531$^{20}_4$	1.4745^{20}	δ	v	v	s	s	...	**B1^3,** 245
Ωp1122	**—,1-bromo-3-chloro-***	$ClCH_2CH_2CH_2Br$	157.44	...	−58.87	143.36^{760} 32.4^{10}	1.5969$^{20}_4$	1.4864^{20}	i	v	v	...	...	chl v	**B1^3,** 245
Ωp1123	**—,2-bromo-1-chloro-***	$CH_3CHBrCH_2Cl$	157.44	...	...	118^{756}	1.537$^{20}_4$	1.4795^{20}	i	v	v	s	s	chl v	**B1^3,** 245
p1124	**—,2-bromo-2-chloro-***	$(CH_3)_2CBrCl$	157.44	...	...	93.0–5.5^{745}	1.474^{21}	1.4575^{20}	i	s	v	s	s	chl v	**B1^3,** 246
Ωp1125	**—,1-bromo-2,2-dimethyl-***	Neopentyl bromide. $(CH_3)_3CCH_2Br$	151.05	...	...	106^{760} 34.6^{100}	1.1997$^{20}_4$	1.4370^{20}	i	s	s	s	s	chl v	**B1,** 141
Ωp1126	**—,1-bromo-2,3-epoxy-*(*d*)***	Epibromohydrin.	136.98	$[\alpha]^{16}_D$ +45.4 (undil); +6.5 (al, p = 5.5)	...	134–6^{50}	...	...	i	s^h	s	...	s	chl s	**B17^2,** 14
p1127	**—,—(*dl*)**	C_3H_5BrO. *See* p1126	136.98	...	...	138–40 (134–6) 61–2^{50}	1.615^{14}	1.4841^{20}	i	s^h	s	...	s	chl s	**B17^2,** 14

For explanations, symbols and abbreviations see beginning of table. For structural formulas see end of table.

No.	Name	Synonyms and Formula	Mol. wt.	Color, crystalline form, specific rotation and λ_{max} (log ε)	m.p. °C	b.p. °C	Density	n_D	Solubility w	al	eth	ace	bz	other solvents	Ref.
	Propane														
p1128	**—,1-bromo-3-fluoro-***	$FCH_2CH_2CH_2Br$	140.99	...	...	101.4	1.542$^{25}_{4}$	1.4290^{25}	i	v	v		s	chl v	**B1^3, 244**
Ω p1129	**—,2-bromo-2-methyl-***	*tert*-Butyl bromide. $(CH_3)_3CBr$.	137.03	...	−16.2	73.25^{760}	1.2209$^{20}_{4}$	1.4278^{20}	i	...	...	...	...	...	**B1^3, 322**
p1130	**—,2(bromomethyl)-1,2,3-tribromo-***	$(BrCH_2)_3CBr$	373.73	...	25	150–1^{14}	2.5595$^{20}_{4}$	1.6246^{20}	i	...	s	s	...		**B1^2, 91**
p1131	**—,1-bromo-1-nitro-***	$CH_3CH_2CHBrNO_2$	168.00	...	...	160–5 82.5^{50}	...	...	...	s	s		...	KOH s	**B1^3, 260**
p1132	**—,2-bromo-2-nitro-***	$(CH_3)_2CBrNO_2$	168.00	...	...	151.7 −.8^{745} 73–5^{50}	1.6562^{0}	...	...	s	s	...	...	alk i	**B1^3, 260**
Ω p1133	**—,1-chloro***	*n*-Propyl chloride. $CH_3CH_2CH_2Cl$.	78.54	...	−122.8	46.60^{760}	0.8909$^{20}_{4}$	1.3879^{20}	δ	∞	∞	...	s	chl s	**B1^3, 219**
Ω p1134	**—,2-chloro-***	Isopropyl chloride. $CH_3CHClCH_3$.	78.54	...	−117.18	35.74^{760}	0.8617$^{20}_{4}$	1.3777^{20}	δ	∞	∞	...	s	chl s	**B1^3, 221**
p1135	**—,1-chloro-2,2-dimethyl-***	Neopentyl chloride. $(CH_3)_3CCH_2Cl$.	106.60	...	−20	84.3^{760}	0.8660$^{20}_{4}$	1.4044^{20}	δ	s	v	...	s	chl s	**B1^3, 370**
Ω p1136	**—,1-chloro-2,3-epoxy-(*dl*)***	α-Epichlorohydrin.	92.53	...	−48	116.5^{760} 60–1^{100}	1.1801$^{20}_{4}$	1.4361^{20}	δ d^h	∞	∞	...	s	...	**B17^2, 13**
p1137	**—,—(*l*)**	C_3H_5ClO. *See* p1136	92.53	$[\alpha]^{18}_{D}$ −25.6	...	92–3^{360}	1.2007	...	δ d^h	∞	∞	...	s	...	**B17^1, 4**
p1138	**—,2-chloro-1,3-epoxy-***	3-Chloro-1-oxacyclobutane. 3-Chlorooxetane. β-Epichlorohydrin.	92.53	...	132–4	...	...	...	...	v	...	...	...	chl s	**B17, 6**
p1139	**—,3-chloro-1,2-epoxy-2-methyl-***	...	106.55	...	...	122.0 51^{55}	1.1011$^{20}_{4}$	1.4340^{20}	s	...	v	...	...	...	**C55, 18694**
p1140	**—,1-chloro-3-fluoro-***	$FCH_2CH_2CH_2Cl$	96.53	...	...	79.5^{740}		1.3871^{25}	i	v	v	...	s	chl s	**C51, 260**
Ω p1141	**—,2-chloro-2-methyl-***	*tert*-Butyl chloride. $(CH_3)_3CCl$.	92.57	...	−25.4	52	0.8420$^{20}_{4}$	1.3857^{20}	δ	∞	∞	...	s	chl s	**B1^3, 316**
p1142	**—,(2-chloromethyl)-1,1,2,3-tetrachloro-***	$(ClCH_2)_2CClCHCl_2$	230.35	...	...	226.3–.4^{737} 95^{9}	1.5686$^{25}_{4}$	1.5165^{25}		s			s	chl s	**B1^3, 321**
p1143	**—,(2-chloromethyl)-1,2,3-trichloro-***	$(ClCH_2)_3CCl$	195.91	...	...	209.8–10.5^{737} 87^{9}	1.5036$^{25}_{4}$	1.508^{20}	...	...	...	...	...	chl s	**B1^3, 321**
Ω p1144	**—,1-chloro-1-nitro-***	$CH_3CH_2CHClNO_2$	123.54	λ^{al} 280.5 (1.48)	...	141–3 67^{56}	1.209$^{20}_{20}$	1.4251^{20}	δ	s	s	...	...	glycols, oils s	**B1^3, 259**
p1145	**—,1-chloro-2-nitro-***	$CH_3CH(NO_2)CH_2Cl$	123.54	...	...	172–3 94^{46}	1.245^{22}	1.4432^{25}	...	s	s	...	...	chl s	**B1^3, 259**
p1146	**—,1-chloro-3-nitro-***	$O_2NCH_2CH_2CH_2Cl$	123.54	...	...	197 δd 115–6^{40}	1.267^{20}	...	...	s	s	...	...	chl s	**B1, 116**
p1147	**—,2-chloro-1-nitro-***	$CH_3CHClCH_2NO_2$	123.54	...	...	172^{760} 75^{15}	1.2361^{15}	1.4447^{20}	δ	s	s	...	...	...	**B1^3, 259**
Ω p1148	**—,2-chloro-2-nitro-***	$CH_3CCl(NO_2)CH_3$	123.54	...	...	134 δd 57^{50}	1.230^{19}	1.4378^{19}	δ	s	s	...	...	glycerol, oils s KOH i	**B1^3, 259**
Ω p1149	**—,1-chloro-3-phenyl-***	$C_6H_5CH_2CH_2CH_2Cl$	154.64	...	...	219–20 110^{21}	1.056$^{25}_{4}$	1.5160^{25}	...	...	...	...	...	...	**B5^2, 305**
p1150	**—,1,2-diacetamido-**	$CH_3CONHCH(CH_3)CH_2NHCOCH_3$	158.20	nd (bz)	138–9	190^{18}	...	...	v	v	i	...	δ s^h	chl v lig i	**B4, 261**
Ω p1151	**—,1,2-diamino-(*d*)***	1,2-Propanediamine*. $CH_3CH(NH_2)CH_2NH_2$	74.13	$[\alpha]^{25}_{D}$ +29.78	...	120.5	0.8584$^{25}_{4}$	...	v	...	i	...	...	chl v	**B4^2, 697**
Ω p1152	**—,1,3-diamino-***	Trimethylenediamine. $H_2NCH_2CH_2CH_2NH_2$	74.13	...	...	135.5^{738}	0.884$^{25}_{4}$	1.4600^{20}	s	∞	∞	...	...	...	**B4^3, 552**
p1153	**—,1,1-dibromo-***	Propylidene bromide. $CH_3CH_2CHBr_2$	201.90	...	...	133.5^{760} 28.4^{10}	1.982$^{20}_{4}$	1.5100^{20}	...	s	s	...	...	chl s	**B1^3, 246**
Ω p1154	**—,1,2-dibromo-***	Propylene bromide. $CH_3CHBrCH_2Br$	201.90	...	−55.25	140.0^{760} 35.7^{10}	1.9324$^{20}_{4}$	1.5201^{20}	...	s	s	...	...	chl s	**B1^3, 246**
Ω p1155	**—,1,3-dibromo-***	Trimethylene bromide. $BrCH_2CH_2CH_2Br$	201.90	...	fr −34.2	167.3^{760} 56.6^{10}	1.9822$^{20}_{4}$	1.5232^{20}	...	s	s	...	...	chl s	**B1^3, 248**
p1156	**—,2,2-dibromo-***	Isopropylidene bromide. $CH_3CBr_2CH_3$	201.90	...	...	114–5^{740}	1.7825$^{20}_{4}$	...	...	s	s	...	...	chl s	**B1^3, 250**
p1157	**—,1,1-dibromo-2,2-dimethyl-***	Neopentylidene bromide. $(CH_3)_3CCHBr_2$	229.95	...	14	180^{760} 66^{10}	1.6695$^{20}_{4}$	1.5047^{20}	i	...	s	...	s	chl s	**B1^3, 371**
p1158	**—,1,3-dibromo-2,2-dimethyl-***	$(CH_3)_2C(CH_2Br)_2$	229.95	...	...	185–90d 72^{14}	1.6934$^{20}_{4}$	1.5050^{20}	i	s	s	...	s	chl s	**B1^3, 371**
Ω p1159	**—,1,2-dibromo-2-methyl-***	Isobutylene bromide. $(CH_3)_2CBrCH_2Br$	215.94	...	9–12	149–51 61^{40}	1.759$^{20}_{4}$	1.509	...	s	s	...	s	chl s	**B1^3, 324**
p1160	**—,1,1-dichloro-***	Propylidene dichloride. $CH_3CH_2CHCl_2$	112.99	...	...	88.1^{760}	1.1321$^{20}_{4}$	1.4289^{20}	...	s	s	...	s	chl s	**B1^3, 225**
Ω p1161	**—,1,2-dichloro-***	Propylene chloride. $CH_3CHClCH_2Cl$	112.99	...	−100.44	96.37^{760} −3.69^{10}	1.1560$^{20}_{4}$	1.4394^{20}	δ	s	s	...	s	chl s	**B1^3, 225**
Ω p1163	**—,1,3-dichloro-***	Trimethylene chloride. $ClCH_2CH_2CH_2Cl$	112.99	...	−99.5	120.4^{760} 14^{10}	1.1878$^{20}_{4}$	1.4487^{20}	δ	v	v	...	s	chl s	**B1^3, 227**
Ω p1164	**—,2,2-dichloro-***	Acetone dichloride. Isopropylidene chloride. $CH_3CCl_2CH_3$	112.99	...	−33.8	69.3^{760}	1.1120$^{20}_{4}$	1.4148^{20}	i	s	∞	...	s	chl s	**B1^3, 228**

For explanations, symbols and abbreviations see beginning of table. For structural formulas see end of table.

No.	Name	Synonyms and Formula	Mol. wt.	Color, crystalline form, specific rotation and λ_{max} (log ε)	m.p. °C	b.p. °C	Density	n_D	Solubility w	al	eth	ace	bz	other solvents	Ref.
	Propane														
p1165	**—,1,1-dichloro-2-methyl-***	Isobutylidene chloride. $(CH_3)_2CHCHCl_2$	127.03	...	...	105–6 (108–10)	1.0111_{12}^{12}	1.4330^{25}	i	s	s	...	s	chl s	**B1**^{3}, 319
p1166	**—,1,2-dichloro-2-methyl-***	Isobutylene chloride. $(CH_3)_2CClCH_2Cl$	127.03	...	> −130	108; $38–9^{70}$	1.093_4^{20}	1.4370^{20}	i	∞	∞	∞	∞	CCl_4 ∞	**B1**^{3}, 319
p1167	**—,1,3-dichloro-2-methyl-***	$CH_3CH(CH_2Cl)_2$	127.03	...	...	134.6^{760}; 60^{49}	1.1325_4^{25}	1.4488^{25}	i	s	v	...	v	CCl_4 s	**B1**^{3}, 319
—	**—,1,1-diethoxy-***	see **Propanal,** diethyl acetal*													
p1168	**—,2,2-di(ethylsulfonyl)-***	Acetone diethyl sulfone. Sulfonal. $(CH_3)_2C(SO_2C_2H_5)_2$	228.33	mcl (w), pr (al)	125.8	300d	...	...	δ s^h	s	δ	...	s	to, chl, AcOEt s CCl_4 δ CS_2 i	**B1**^{1}, 345
p1169	**—,1,3-difluoro-***	Trimethylene fluoride. $FCH_2CH_2CH_2F$	80.08	...	...	41.6^{760}	1.0057_4^{25}	1.3190^{26}	...	...	...	...	s	...	**B1**^{3}, 218
p1170	**—,2,2-difluoro-***	Isopropylidene fluoride. $CH_3CF_2CH_3$	80.08	gas	−104.8	-0.4^{760}	0.9205_4^{20} (liq at sat pressure)	1.2904^{20} (liq at sat pressure)	...	...	...	...	...	...	**B1**^{3}, 218
p1171	**—,1,2-diiodo-***	Propylene iodide. CH_3CHICH_2I	295.89	...	...	...	...	$2.490^{18.5}$	...	s	s	...	...	MeOH s	**B1**, 115
Ω p1172	**—,1,3-diiodo-***	Trimethylene iodide. $ICH_2CH_2CH_2I$	295.89	...	fr −20	227^{760} δd; 110^{19}	2.5755_4^{20}	1.6423^{20}	i	...	s	...	...	CCl_4, chl s	**B1**^{3}, 255
p1173	**—,2,2-diiodo-***	Isopropylidene iodide. $CH_3CI_2CH_3$	295.89	...	...	173^{760}; 53^{10}	2.5755_4^{20}	1.651^{20}	i	...	s	...	...	CCl_4, chl s	**B1**^{3}, 255
Ω p1174	**—,2,2-dimethyl-***	Neopentane. $C(CH_3)_4$	72.15	gas	−16.55	9.50	0.61350^{20}	1.3476^{6}	i	s	s	...	...	...	**B1**^{3}, 369
p1175	**—,1(dimethylamino)-***	Dimethyl propylamine*. $CH_3CH_2CH_2N(CH_3)_2$	87.17	...	...	65.5^{752}	0.7152_4^{20}	1.3860^{20}	...	s	s	...	s	chl s	**B4**^{2}, 621
p1176	**—,1(dimethylamino)-2-methyl**	Dimethyl isobutylamine*. $(CH_3)_2CHCH_2N(CH_3)_2$.	101.19	...	...	81.0^{753}	0.7097_4^{20}	1.3907^{20}	...	s	s	...	s	chl s	**B4**^{2}, 638
p1177	**—,1,1-dinitro-***	$CH_3CH_2CH(NO_2)_2$	134.09	λ^{hx} 280 (1.76)	−42	184	1.2610^{25}	1.4339^{20}	...	...	...	...	...	alk s	**B1**^{3}, 260
p1178	**—,1,3-dinitro-***	$O_2NCH_2CH_2CH_2NO_2$	134.09	λ^{al} 273 (1.60)	−21.4	103^{1}	1.353_4^{26}	1.4654^{20}	i	...	s	...	...	...	**B1**^{3}, 261
p1179	**—,2,2-dinitro-***	$CH_3C(NO_2)_2CH_3$	134.09	λ^{al} 278 (1.72)	53	185.5; $48–50^{2}$	1.30^{25}	...	δ	...	...	...	...	...	**B1**^{3}, 261
p1179[1]	**—,1,2-diphenoxy-***	$CH_3CH(OC_6H_5)CH_2OC_6H_5$	228.29	rh (MeOH)	32	$175–8^{12}$	$1.0748_4^{33.3}$	$1.5542^{33.3}$	i	s	s	s	s	MeOH, chl s	**B6**^{2}, 151
p1179[2]	**—,1,3-diphenoxy-***	$C_6H_5OCH_2CH_2CH_2OC_6H_5$	228.29	lf (al)	61	$338–40^{760}$; 160^{25}	...	...	i	s	s	...	...	...	**B6**^{2}, 151
Ω p1180	**—,1,2-epoxy-(*dl*)***	Methyloxiran. Propylene oxide.	58.08	...	...	34.3	0.859_4^{0}	1.3670^{20}	∞	∞	∞	...	...	...	**B17**, 6
p1181	**—,1,3-epoxy-***	Oxetane. Trimethylene oxide.	58.08	λ^{gas} 167 (3.04), 170.5 (3.29), 174.2 (3.44), 184 (3.1), 187.5 (3.30)	...	47.8 (cor)	0.8930_4^{25}	1.3961^{20}	∞	∞	s	v	...	os v	**B17**^{2}, 12
p1182	**—,1,2-epoxy-3-iodo-***	Epiiodohydrin.	183.98	...	...	160–2	1.982^{24}	...	i	s	s	...	...	...	**B17**^{2}, 15
p1183	**—,1,2-epoxy-2-methyl-***	Isobutylene oxide. 2,2-Dimethyl oxiran.	72.12	...	...	52	0.8650^{0}	1.3712^{22}	...	s	s	...	...	...	**B17**^{2}, 17
Ω p1184	**—,1,2-epoxy-3,3,3-trichloro-***	...	161.42	...	...	149^{764}; $44–5^{13}$	1.495_4^{20}	1.4737^{25}	...	...	v	...	...	...	**B17**^{2}, 14
p1184[1]	**—,1-ethoxy-3-phenoxy-**	$C_6H_5OCH_2CH_2CH_2OC_2H_5$	180.25	oil	...	328–30	...	...	...	s	s	...	...	...	**B6**, 147
p1185	**—,1-fluoro-***	*n*-Propyl fluoride. $CH_3CH_2CH_2F$	62.09	...	−159	2.50^{760}	0.7956_4^{20}	1.3115^{20}	δ	v	v	...	...	...	**B1**^{3}, 217
Ω p1186	**—,1,1,1,2,2,3,3-heptachloro-***	$Cl_2CHCCl_2CCl_3$	285.21	amor	29.4	247–8; 132^{30}	1.8048_4^{34}	...	...	...	...	...	...	chl s	**B1**^{3}, 238
p1187	**—,1,1,1,2,3,3,3-heptachloro-***	$Cl_3CCHClCCl_3$	285.21	...	11–1.5	249; 93^{3}	1.7921_4^{34}	1.5427^{21}	...	...	...	...	...	chl s	**B1**^{3}, 239
p1188	**—,heptafluoro-1-nitro-***	$F_3CCF_2CF_2NO_2$	215.03	λ^{vap} 282 (1.56)	...	25	...	...	...	...	...	...	...	...	**C51**, 14790
p1189	**—,heptafluoro-1-nitroso-***	$F_3CCF_2CF_2NO$	199.03	deep bl λ^{vap} 686 (1.34)	−151	−12	...	...	...	...	...	...	...	...	**C51**, 14790
p1190	**—,1,1,1,2,2,3-hexafluoro-***	$FCH_2CF_2CF_3$	152.04	...	...	1.2	...	...	...	...	...	...	...	...	**B4**, 537
p1191	**—,1(hydroxylamino)-***	$CH_3CH_2CH_2NHOH$	75.11	nd (eth)	46	...	...	...	v	...	s	...	...	lig δ	**B4**, 537
Ω p1192	**—,1-iodo-***	*n*-Propyl iodide. $CH_3CH_2CH_2I$	169.99	λ^{al} 254.5 (2.69)	−101.3	102.45	1.7489_4^{20}	1.5058^{20}	δ	∞	∞	...	...	...	**B1**^{3}, 252
Ω p1193	**—,2-iodo-***	Isopropyl iodide. CH_3CHICH_3	169.99	...	−90.1	89.45	1.7033_4^{20}	1.5026^{20}	δ	∞	∞	...	∞	chl ∞	**B1**^{3}, 253
p1194	**—,1-iodo-2,2-dimethyl-***	Neopentyl iodide. $(CH_3)_3CCH_2I$	198.05	...	...	127–9d; $42–4^{20}$	1.4940^{20}	1.4890^{20}	i	s	s	...	...	...	**B1**^{3}, 373
p1195	**—,2-iodo-2-methyl-***	*tert*-Butyl iodide. $(CH_3)_3CI$.	184.02	...	−38.20	100 δd; 20.8^{30}	1.5445_4^{20}	1.4918^{20}	d	∞	∞	...	...	...	**B1**^{3}, 326
Ω p1196	**—,1-isocyano-***	Propylcarbylamine. Propyl isocyanide. $CH_3(CH_2)_2NC$	69.11	...	...	99.5	...	...	i	∞	∞	...	...	...	**B4**^{2}, 625
p1197	**—,2-isocyano-***	Isopropylcarbylamine. Isopropyl isocyanide $(CH_3)_2CHNC$	69.11	...	...	87	0.7596^{0}	...	i	∞	∞	...	...	...	**B4**, 154
p1198	**—,2-isocyano-2-methyl-***	*tert*-Butyl carbylamine. *tert*-Butyl isocyanide $(CH_3)_3CNC$	83.13	...	...	167–70; 91^{38}	...	...	...	∞	∞	...	...	HCl d	**B4**^{2}, 641

No.	Name	Synonyms and Formula	Mol. wt.	Color, crystalline form, specific rotation and λ_{max} (log ε)	m.p. °C	b.p. °C	Density	n_D	Solubility: w	al	eth	ace	bz	other solvents	Ref.
	Propane														
p1198^1	–, 2-methyl	Isobutane	58.12	Col. gas	–159.42	–11.7	0.549^{30}		s	v	v			chl v	B1, 124
p1198^2	—,1-methoxy-3 phenoxy-	$CH_3OCH_2CH_2CH_2OC_6H_5$	166.22	oil		230–1									B6, 147
p1199	—,1(2-methoxy-phenyl)-2(methyl-amino)-, hydrochloride.		215.73	cr (al-eth)	128				v	v	δ		δ	chl v	C55, 24677
p1200	—,2-methyl-1-nitro-*	Nitroisobutane. $(CH_3)_2CHCH_2NO_2$	103.12			140.5^{760} $61–2^{45}$	0.9625^{25}_{25}	1.4066^{20}	δ	∞	∞				B1^3, 327
p1201	—,2-methyl-1,1,1,2,3-penta-chloro-*	$ClCH_2CCl(CH_3)CCl_3$	230.35	(al)	73.5	213^{737} $90–3^{10}$	1.5686^{25}_4	1.5165^{25}	i					chl s	B1^3, 321
p1202	—,2-methyl-1,1,1,2-tetra-bromo-*	$(CH_3)_2CBrCBr_3$	373.73	lf	217				i	v	v				B1, 128
p1203	—,2-methyl-1,1,2,3-tetra-bromo-*	$BrCH_2CBr(CH_3)CHBr_2$	373.73			134^{11}	2.4545^{20}	1.5990^{20}	i					chl s	B1^3, 325
p1204	—,2-methyl-1,1,1,2-tetra-chloro-*-	$(CH_3)_2CClCCl_3$	195.91	cr (al)	178.6–9.6	$192^{117.5}$ (cor) sub			i	v	v			chl s	B1^3, 320
p1205	—,2-methyl-1,1,2,3-tetra-chloro-*	$ClCH_2CCl(CH_3)CHCl_2$	195.91		–46	190.6–1.3^{760} 69^{12}	1.4393^{25}_4	1.4963^{20}	i		s		s	chl s	B1^3, 321
p1206	—,2-methyl-1,1,2-tribromo-*	$(CH_3)_2CBrCHBr_2$	294.83			208–15d 96^{34}	2.0169^{20}_4		i		s			chl s	B1^3, 325
p1207	—,2-methyl-1,2,3-tribromo-*	$(BrCH_2)_2CBrCH_3$	294.83			88.5^9	2.1750^{20}_4	1.5652^{20}	i		s			chl s	B1^3, 325
p1208	—,2-methyl-1,1,2-trichloro-*	$(CH_3)_2CClCHCl_2$	161.46		–6	144.5–5.5^{760} $46–7^{18}$	1.2588^{20}_4	1.4666^{20}	i					chl v	B1^3, 320
p1209	—,2-methyl-1,2,3-trichloro-*	$CH_3CCl(CH_2Cl)_2$	161.46			$162–3^{760}$ 81^{50}	1.3012^{25}_4	1.4765^{20}	i					chl s	B1^3, 320
Ω p1210	—,1-nitro-*	$CH_3CH_2CH_2NO_2$	89.09	λ^{al} 270 (1.59)	–108	130.5–1.5^{760}	1.0081^{24}_4	1.4016^{20}	δ	∞	∞			chl s	B1^3, 256
Ω p1211	—,2-nitro-*	$CH_3CH(NO_2)CH_3$	89.09	λ^{al} 260 (1.66)	–93	120^{760}	0.9876^{20}_4	1.3944^{20}	δ					chl s	B1^3, 258
p1212	—,1(nitro-amino)-*	*n*-Propylnitramine. $CH_3CH_2CH_2NHNO_2$	104.12		–21	$128–9^{40}$	1.1046^{15}	1.4610^{20}	δ	∞	∞				B4^1, 569
p1213	—,3-nitro-1,1,1-trifluoro-*	$O_2NCH_2CH_2CF_3$	143.07			135^{760}	1.4203^{20}_{20}	1.3549^{20}	i		s				C51, 1868
Ω p1214	—,octachloro-*	Perchloropropane. $Cl_3CCCl_2CCl_3$	319.66	wh cr (dil al)	160	$268–9^{734}$			i	v	v				B1^3, 239
p1215	—,1,1,1,2,3-penta-chloro-*	$ClCH_2CHClCCl_3$	216.32	nd (al)	179–80	$84–5^{20}$ sub		1.513^{20}	i	s	s	s		os, chl s	B1, 107
Ω p1216	—,1,1,2,3,3-penta-chloro-*	$Cl_2CHCHClCHCl_2$	216.32			198–200^{760} 98–100^{20}	1.6086^{34}_4	1.5131^{17}	i	v	s			chl s	B1^3, 236
Ω p1218	—,1,1,1,2-tetra-chloro-*	$CH_3CHClCCl_3$	181.88		–64	150^{760} (152) 37^{10}	1.473^{20}_4	1.4867^{20}	i	v	s			chl s	B1^3, 233
p1219	—,1,1,2,2-tetra-chloro-*	$CH_3CCl_2CHCl_2$	181.88			153^{760}	1.47^{13}	1.4850^{25}	i	∞	∞			chl s	B1^3, 234
p1220	—,1,1,2,3-tetra-chloro-*	$ClCH_2CHClCHCl_2$	181.88			179–80^{760} (cor) 61^{12}	1.513^{17}	1.5037^{17}	i	s	v			chl s alk d*	B1^3, 234
p1221	–,1,2,2,3-tetra-chloro-*	$ClCH_2CCl_2CH_2Cl$	181.88			165^{760} 51^{12}	1.500^{18}	1.4940^{18}	i	v	v			chl s	B1^3, 234
Ω p1222	—,1,1,2-tri-bromo-*	$CH_3CHBrCHBr_2$	280.80			200–1 83^6	2.3548^{20}_4	1.5790^{20}	i	s	v			chl, aa s	B1^3, 251
p1223	—,1,2,2-tri-bromo-*	$CH_3CBr_2CH_2Br$	280.80			190–1 81^{20}	2.2985^{20}_4	1.5670^{20}	i	s	v			chl, aa s	B1^2, 77
Ω p1224	—,1,2,3-tri-bromo-*	$BrCH_2CHBrCH_2Br$	280.80		16.9	222.16^{760} 98.47^{10}	2.4209^{20}_4	1.5862^{20}	i	v	v				B1^3, 251
p1225	—,1,1,1-tri-chloro-*	$CH_3CH_2CCl_3$	147.43			106.5–8.5	1.287^{23}_4		i	s	v			chl v	B1^3, 230
Ω p1226	—,1,1,2-tri-chloro-*	$CH_3CHClCHCl_2$	147.43			140 (131–3)	1.372^{25}		i	s	v			chl s	B1^3, 230
p1227	—,1,1,3-tri-chloro-*	$ClCH_2CH_2CHCl_2$	147.43		–58.98	145.55^{760} 33.42^{10}	1.3557^{20}_4	1.4718^{20}	i	s	v			chl v aa s	B1^3, 330
Ω p1228	—,1,2,2-tri-chloro-*	$CH_3CCl_2CH_2Cl$	147.43			$123–5^{762}$	1.318^{25}	1.4609^{25}	i	s	s			chl v	B1^3, 231
Ω p1229	—,1,2,3-tri-chloro-*	Trichlorohydrin. $ClCH_2CHClCH_2Cl$	147.43		–14.7	156.85^{760} 41.92^{10}	1.3889^{20}_4	1.4832^{20}	δ	s	s			chl v	B1^3, 231
—	—,1,3,3-tri-ethoxy-*	*see* **Propanal, 3-ethoxy-, diethyl acetal***													
p1230	—,1,1,1-tri-phenyl-*	$(C_6H_5)_3CCH_2CH_3$	272.40	pr	51										B5^3, 2340

For explanations, symbols and abbreviations see beginning of table. For structural formulas see end of table.

No.	Name	Synonyms and Formula	Mol. wt.	Color, crystalline form, specific rotation and λ_{max} (log ε)	m.p. °C	b.p. °C	Density	n_D	Solubility w	al	eth	ace	bz	other solvents	Ref.
	1-Propanearsonic acid														
Ω p1231	**1-Propanearsonic acid***	$CH_3CH_2CH_2AsO_3H_2$	168.03	nd (al), pl (w)	134.6–5.2	...	...	...	v	v	i	...	...	...	**B4^{2}**, 997
p1232	**1-Propaneboronic acid***	*n*-Propylboric acid. $CH_3CH_2CH_2B(OH)_2$	87.92	wh nd	107	d	...	...	v	s	s	...	...	...	**B4^{2}**, 1023
p1233	**—,2-methyl-***	Isobutylboric acid. $(CH_3)_2CHCH_2B(OH)_2$	101.94	lo pl (w), cr ($ClCH_2CH_2Cl$)	112	...	...	...	δ	s	s	...	...	...	**B4^{3}**, 1965
—	**Propanedioic acid***	*see* **Malonic acid**													
Ω p1234	**1,2-Propanediol***	Propylene glycol. $CH_3CH(OH)CH_2OH$	76.11	...	...	189, $96–8^{21}$	1.0361_4^{20}	1.4324^{20}	∞	∞	s	...	s	...	**B1^{3}**, 2142
Ω p1235	**—,carbonate**	4-Methyl-1,3-dioxolan-2-one.	102.09	...	−48.8	240, 110^{10}	1.2041_4^{20}	1.4189^{20}	v	v	v	v	s	...	**C49,** 12303
p1236	**—,diacetate**	$CH_3CO_2CH(CH_3)CH_2O_2CCH_3$	160.17	...	...	$190–1^{762}$	1.059_4^{20}	1.4173^{20}	v	s	s	...	...	...	**B2^{3}**, 312
p1243	**—,sulfite**	...	122.14	...	< −60	175, 85^{25}	1.2960_4^{20}	1.4370^{20}	v d^h	v	v	v	v	AcOEt v	**C51,** 1036
Ω p1244	**1,3-Propanediol***	Trimethylene glycol. $HOCH_2CH_2CH_2OH$.	76.11	...	...	213.5^{760}, 110^{12}	1.0597_4^{20}	1.4398^{20}	∞	∞	v	...	δ s^h	...	**B1^{2}**, 540
p1245	**—,diacetate**	$CH_3CO_2CH_2CH_2CH_2O_2CCH_3$	160.17	...	...	$209–10^{760}$, 84.5^{10}	1.070^{19}	1.4192	v	s	...	...	...	...	**B2^{2}**, 156
Ω p1250	**—,2-amino-2-ethyl-***	$CH_3CH_2C(CH_2OH)_2NH_2$	119.17	ye	37.5–8.5	$143–5^{10}$	1.099_4^{20}	1.490^{20}	∞	...	...	...	...	...	**C34,** 1305
Ω p1251	**—,2-amino-2(hydroxymethyl)-***	$H_2NC(CH_2OH)_3$	121.14	nd of fl (MeOH)	170.5–1.5	$219–20^{18}$ ($230–4^{12}$)	...	...	v	...	...	...	...	MeOH s^h	**C49,** 1357
Ω p1252	**—,2-amino-2-methyl-***	$CH_3C(CH_2OH)_2NH_2$	105.14	...	109–11	151.2^{10}	...	...	v	s	...	...	...	...	**C34,** 1305
Ω p1253	**—,2-butyl-2-ethyl-***	$CH_3(CH_2)_3C(C_2H_5)(CH_2OH)_2$	160.26	wh	43.8	262, 123^{15}	0.929_{20}^{50}	1.4587^{25}	δ	s	...	...	...	...	**Am 70,** 3121
Ω p1254	**1,2-Propanediol, 3-chloro-***	α-Chlorohydrin. $ClCH_2CH(OH)CH_2OH$	110.54	yesh liq	...	213d, 116^{11}	1.326_{15}^{18}	1.4809^{20}	s	s	s	...	...	...	**B1^{3}**, 2150
p1255	**—,—,diacetate**	$ClCH_2CH(O_2CCH_3)CH_2O_2CCH_3$	194.62	...	...	245, 116^{12}	1.199_4^{25}	1.4407^{20}	...	s	s	...	...	...	**B2^{3}**, 313
p1256	**1,3-Propanediol, 2-chloro-***	$HOCH_2CHClCH_2OH$	110.54	...	...	146^{18}	1.3219_4^{20}	1.4831^{20}	v	v	...	v	...	...	**B1^{2}**, 542
p1257	**1,2-Propanediol. 3-chloro-2-methyl-***	$ClCH_2C(OH)(CH_3)CH_2OH$	124.57	...	...	$114–7^{20}$, $80^{1.6}$	1.2362_4^{20}	1.4748^{20}	∞	∞	∞	...	...	...	**B1^{3}**, 2188
—	**—,3,3-diethoxy-***	*see* **Glyceraldehyde,** diethyl acetal													
Ω p1258	**1,3-Propanediol, 2,2-diethyl-***	$HOCH_2C(C_2H_5)_2CH_2OH$	132.21	wh	61.3–.6	240–1, 131^{13}	1.052_{20}^{20}	...	v	v	v	...	...	os s, lig i	**Am 70,** 946
p1259	**1,2-Propanediol, 3(diethylamino)-***	$(C_2H_5)_2NCH_2CH(OH)CH_2OH$	147.22	syr	...	233–5	...	...	s	s	s	...	...	chl s	**B4,** 302
Ω p1260	**1,3-Propanediol, 2,2-dimethyl-***	$HOCH_2C(CH_3)_2CH_2OH$	104.15	nd (bz)	130	206^{747}, $120–30^{15}$	...	...	s	v	v	...	s^h	...	**B1^{3}**, 2199
p1261	**1,2-Propanediol, 3(dimethylamino)-***	$(CH_3)_2NCH_2CH(OH)CH_2OH$	119.17	...	...	220^{749}	...	...	s	s	s	...	...	chl s	**B4,** 302
p1262	**1,3-Propanediol, 2,2-dinitro-***	$HOCH_2C(NO_2)_2CH_2OH$	166.09	wh pl (bz)	142	...	...	...	s^h	s^h	...	...	s^h	diox, $PhNO_2$ s^h	**C45,** 9473
Ω p1263	**—,2-ethyl-2-hydroxymethyl-***	TMP. Trimethylolpropane. $CH_3CH_2C(CH_2OH)_3$	134.18	wh pw or pl	58	160^5	...	...	∞	∞	...	...	i	CCl_4 i	**C54,** 2177
Ω p1264	**—,2-ethyl-2-nitro-***	$CH_3CH_2C(NO_2)(CH_2OH)_2$	149.15	nd (w) λ^{al} 280 sh (1.61)	57–8	d	...	...	v	v	v	...	...	...	**B1,** 483
—	**1,2-Propanediol, 3-hexadecyloxy-***	*see* **Glycerol,** 1-hexadecyl ether													
Ω p1265	**1,3-Propanediol, 2(hydroxymethyl)-2-methyl-***	Pentaglycerol. Trimethylolethane. $CH_3C(CH_2OH)_3$.	120.15	wh pw or nd (al)	204	$135–7^{15}$	...	...	∞	∞	i	...	i	aa v^h	**B1^{3}**, 2348
Ω p1266	**—,2(hydroxymethyl)-2-nitro-***	$O_2NC(CH_2OH)_3$	151.12	nd or pr	165 (144)	d	...	...	v	v	s	...	...	...	**B1^{2}**, 596
p1267	**1,2-Propanediol, 3-mercapto-***	1-Thioglycerol. $HSCH_2CH(OH)CH_2OH$.	108.17	visc	...	$100–1^1$	1.2455^{20}	1.5268^{20}	δ	∞	δ	v	δ	...	**B1^{3}**, 2339
p1268	**—,2-methyl-***	Isobutylene glycol. $(CH_3)_2C(OH)CH_2OH$	90.12	...	...	176^{760}, $79–80^{12}$	1.0024_4^{20}	1.4350^{20}	s	s	v	...	...	...	**B1^{3}**, 2187
p1269	**1,3-Propanediol, 2-methyl-2-nitro-***	$CH_3C(NO_2)(CH_2OH)_2$	135.12	mcl	149–50	d	...	...	v	v	...	...	...	...	**B1^{3}**, 2190
p1270	**—,2-methyl-2-propyl-***	2,2-bis(hydroxymethyl)-pentane. $(HOCH_2)_2C(CH_3)(CH_2)_2CH_3$	132.21	cr (hx)	62–3	234	...	...	s	...	...	...	...	os s, hx s^h	**Am 72,** 3716
—	**1,2-Propanediol, 3-octadecyloxy-***	*see* **Glycerol,** 1-octadecyl ether													
p1271	**1,3-Propanedione, 2,2-dibromo-1,3-diphenyl-***	$C_6H_5COCBr_2COC_6H_5$	382.06	pr (eth)	95	...	...	...	...	δ	s^h	...	...	os δ	**B7,** 772

For explanations, symbols and abbreviations see beginning of table. For structural formulas see end of table.

No.	Name	Synonyms and Formula	Mol. wt.	Color, crystalline form, specific rotation and λ_{max} (log ε)	m.p. °C	b.p. °C	Density	n_D	Solubility: w	al	eth	ace	bz	other solvents	Ref.
	1,2-Propanedione														
p1272	**1,2-Propanedione, 1(3,5-dimethoxy-4-hydroxy-phenyl)-**	Syringoyl methyl ketone	224.22	ye nd	80–1	...	...	...	δ^h	s	...	...	s	peth s^h	**Am 62,** 986
p1273	**1,3-Propanedione, 1,3-diphenyl-**(one enol form)*	Dibenzoylmethane. $C_6H_5COCH{:}C(OH)C_6H_5$	224.26	unst nd λ^{al} 245 (3.65), 342 (4.38)	70–1	...	...	...	...	...	s	...	...	peth, chl, dil NaOH s	**B7², 689**
p1274	—,—(one enol form)*	$C_6H_5COCH{:}C(OH)C_6H_5$	224.26	metast mcl	72–3	...	...	...	...	...	...	...	...	...	**B7², 689**
p1275	—,—(one enol form)*	$C_6H_5COCH{:}C(OH)C_6H_5$	224.26	st rh bipym (eth)	78–9	$219–21^{18}$ $165–70^{3}$	...	...	...	s	v	...	...	chl v NaOH s Na_2CO_3 i	**B7², 689**
p1276	—,—(keto form)	$C_6H_5COCH_2COC_6H_5$	224.26	metast nd or pl (eth)	81	...	...	...	...	s	s	...	...	chl, dil NaOH s	**B7², 690**
p1277	**1,2-Propanedione, 1-phenyl-***	Acetyl benzoyl. $C_6H_5COCOCH_3$	148.17	ye oil	...	222^{760} 101^{12}	1.0065_4^{20}	1.537^{10}	s	s	s	...	...	...	**B7², 608**
p1278	—,—,dioxime*	Methyl phenyl glyoxime. $C_6H_5C({:}NOH)C({:}NOH)CH_3$	178.19	nd (dil al)	238–40	...	...	...	i	s	...	...	...	...	**B7², 609**
p1279	—,—,1-oxime*	$C_6H_5C({:}NOH)COCH_3$	163.18	pa ye nd or lf (aa, al)	166–7	...	...	...	δ	s^h	...	...	δ	aa s^h	**B7², 608**
Ω p1280	—,—,2-oxime*	α-Isonitrosopropiophenone. $C_6H_5COC({:}NOH)CH_3$	163.18	wh nd (w)	115	...	...	...	s^h	...	...	...	...	to s^h	**B7², 608**
p1281	**1,3-Propanedithiol***	$HSCH_2CH_2CH_2SH$	108.24	...	−79	172.9 63^{15}	1.0783_4^{20}	1.5392^{20}	δ	∞	∞	...	∞	chl ∞	**B1³, 2164**
p1282	**1-Propanephosphonic acid***	*n*-Propylphosphonic acid. $CH_3CH_2CH_2PO_3H_2$	124.08	pl (bz)	73	d	...	...	v	v	v	...	δ	peth δ	**Am 75,** 3379
p1283	**2-Propanephosphonic acid***	Isopropylphosphonic acid. $(CH_3)_2CHPO_3H_2$	124.08	pl (bz)	74–5	d	...	...	v	v	v	...	δ s^h	...	**Am 75,** 3379
p1284	**1-Propanephosphonic acid, 2-methyl-***	Isobutylphosphonic acid. $(CH_3)_2CHCH_2PO_3H_2$	138.10	pl (xyl)	119	d	...	...	v	v	s	...	δ	xyl s^h	**Am 75,** 3379
p1285	**2-Propanephosphonic acid, 2-methyl-***	*tert*-Butylphosphonic acid. $(CH_3)_3CPO_3H_2$	138.10	wh nd (xyl, aa-lig)	192	d	...	...	v	v	...	...	...	aa s xyl s^h peth δ	**Am 75,** 3379
p1286	**1-Propanesulfonic acid,** amide	1-Propansulfonamide. $CH_3CH_2CH_2SO_2NH_2$	123.18	pr (eth), cr (bz)	53.5	...	...	...	s	s	δ s^h	...	s^h	...	**B4³, 18**
p1287	—,chloride	1-Propanesulfonyl chloride*. $CH_3CH_2CH_2SO_2Cl$	142.61	...	...	180d 77^{13}	1.2826_4^{15}	1.452^{20}	d^h	d^h	...	...	...	...	**B4³, 18**
p1288	**2-Propanesulfonic acid***	Isopropylsulfonic acid. $(CH_3)_2CHSO_3H$	124.17	...	−37	$159^{1.4}$	1.187^{25}	1.4332^{20}	v	...	...	...	...	...	**B4³, 19**
p1289	—,amide	Isopropylsulfonamide*. $(CH_3)_2CHSO_2NH_2$	123.18	cr (eth-peth)	67.5 (60)	...	...	...	v	s	s	...	i	peth i	**B4³, 19**
p1290	**1-Propanesulfonic acid, 2-methyl-, chloride**	Isobutanesulfonyl chloride*. $(CH_3)_2CHCH_2SO_2Cl$	156.63	...	...	$189–91^{760}$ 87^{15}	...	1.4520^{25}	d	d^h	...	...	...	...	**B4³, 22**
p1291	**1,1,2,3-Propanetetracarboxylic acid,** tetraethyl ester*	$C_2H_5O_2CCH_2CH(CO_2C_2H_5)CH(CO_2C_2H_5)_2$	332.36	...	...	$203–4^{18}$ 187^{5}	1.1184_4^{20}	1.4393^{20}	...	s	...	...	...	...	**B2³, 2077**
p1292	**1,1,3,3-Propanetetracarboxylic acid,** tetraethyl ester*	Ethyl methanedimalonate. $(C_2H_5O_2C)_2CHCH_2CH(CO_2C_2H_5)_2$	332.36	...	−30	300–10d 195^{8}	1.116^{20}	1.4398^{20}	...	s	...	...	...	...	**B2³, 2077**
Ω p1293	**1-Propanethiol***	*n*-Propylmercaptan. $CH_3CH_2CH_2SH$	76.17	$\lambda^{w, pH = 10.2}$ 240 (3.20)	−113.3	$67–8^{760}$	0.8411^{20}	1.4380^{20}	δ	s	s	s	s	...	**B1³, 1430**
Ω p1294	**2-Propanethiol***	Isopropylmercaptan. $(CH_3)_2CHSH$	76.17	...	−130.54	52.56^{760}	0.8143_4^{20}	1.4255^{20}	δ	∞	∞	v	...	...	**B1³, 1477**
Ω p1295	**1-Propanethiol, 2-methyl-***	Isobutylmercaptan. $(CH_3)_2CHCH_2SH$	90.19	λ^{iso} 225–30 (2.20)	< −79	88.72^{760}	0.8339_4^{20}	1.4387^{20}	δ	v	v	v	...	...	**B1³, 1565**
Ω p1296	**2-Propanethiol, 2-methyl-***	*tert*-Butylmercaptan. $(CH_3)_3CSH$	90.19	λ^{cy} 226 sh (2.2)	1.11	64.22^{760}	0.8002_4^{20}	1.4232^{20}	i	...	...	...	...	hp s	**B1³, 1589**
Ω p1297	**1,2,3-Propanetricarboxylic acid***	Tricarballylic acid. $HO_2CCH_2CH(CO_2H)CH_2CO_2H$	176.14	orh (w, eth)	166	...	...	...	v	v	δ	...	...	...	**B2³, 2027**
p1298	—,1,2-dihydroxy-*(*l*)	$HO_2CCH_2C(OH)(CO_2H)CH(OH)CO_2H$	208.14	nd, $[\alpha]_D$ −17.7 (ac)	159–60	...	1.39^{35}	...	v	δ	v	...	...	...	**B3³, 1127**
p1299	—,1-hydroxy-*	Isocitric acid. $HO_2CCH_2CH(CO_2H)CH(OH)CO_2H$	192.14	yesh syr	105	...	...	...	δ	δ	δ	...	...	...	**B3², 359**
—	—,2-hydroxy-*	*see* **Citric acid**													
—	**1,2,3-Propanetriol***	*see* **Glycerol**													
p1301	**Propanetrione, diphenyl-***	Diphenyltriketone. $C_6H_5COCOCOC_6H_5$	238.25	ye nd (lig)	68–70	289^{175} 248^{60}	...	...	i	δ	s	...	...	...	**B7, 871**
Ω p1302	**Propanoic acid***	Propionic acid. $CH_3CH_2CO_2H$	74.08	...	−20.8	140.99^{760} 41.65^{10}	0.9930^{20}	1.3869^{20}	∞	∞	s	...	...	...	**B2³, 502**
Ω p1303	—,allyl ester	$CH_3CH_2CO_2CH_2CH{:}CH_2$	114.15	...	...	$124–4.5^{774}$	0.9140^{20}	1.4105^{20}	...	s	s	s	...	...	**B2³, 532**
p1304	—,amide	Propanamide*. Propionamide $CH_3CH_2CONH_2$	73.10	rh, pl (bz) $\lambda^{hx-C_2H_5CN}$ 175 (3.85)	81.3	213	0.9262^{110}	1.4160^{110}	v	v	v	...	...	chl v	**B2³, 542**

For explanations, symbols and abbreviations see beginning of table. For structural formulas see end of table.

No.	Name	Synonyms and Formula	Mol. wt.	Color, crystalline form, specific rotation and λ_{max} (log ε)	m.p. °C	b.p. °C	Density	n_D	Solubility w	al	eth	ace	bz	other solvents	Ref.
	Propanoic acid														
p1305	—,—,*N*,*N*-diethyl-	$CH_3CH_2CON(C_2H_5)_2$	129.20	...	...	191 $81\text{–}5^{20}$	...	1.4425^{20}	...	v	...	...	...	ac v	**B4**³, 210
p1306	—,—,*N*-phenyl-	Propionanilide. $CH_3CH_2CONHC_6H_5$	149.19	pl (eth, al, bz) λ^{al} 243 (4.16)	105–6	222.2	1.175	...	δ	v	v	...	...	...	**B12**, 250
p1307	—,—,*N*-propionyl-	Dipropionamide. $CH_3CH_2CONHCOCH_2CH_3$	129.16	nd (w, eth) λ^{chl} 281 (2.12)	154 (sub 100)	210–20	...	...	δ s^h	...	δ	...	...	...	**B2**², 224
p1308	—,—,*N*(2-tolyl)-	...	163.22	nd (bz)	89.5	298–9	...	...	...	v	v	...	s^h	chl, aa v	**B12**², 440
p1309	—,—,*N*(3-tolyl)-	...	163.22	nd (eth)	81	...	...	...	...	v	s	...	...	lig s	**B12**, 861
Ω p1310	—,anhydride	Propionic anhydride. $(CH_3CH_2CO)_2O$	130.15	...	−45	168.1–$.4^{712}$ 67.5^{18}	1.0110_4^{20}	1.4038^{20}	d	d	∞	...	...	...	**B2**³, 539
p1311	—,bromide	Propionyl bromide. CH_3CH_2COBr	136.98	...	...	103.0–$.6^{770}$	1.5210_4^{16}	1.4578^{16}	...	...	s	...	...	...	**B2**², 108
Ω p1312	—,butyl ester*	$CH_3CH_2CO_2(CH_2)_3CH_3$	130.19	...	−89.55	145.5^{760}	0.8754_4^{20}	1.4014^{20}	δ	∞	∞	...	...	...	**B2**³, 526
p1313	—,*sec*-butyl ester	$CH_3CH_2CO_2CH(CH_3)CH_2CH_3$	130.19	...	...	132.0–$.5^{760}$	0.8657^{20}	1.3952^{20}	...	s	s	...	...	...	**B2**, 241
Ω p1314	—,chloride	Propionyl chloride. CH_3CH_2COCl	92.53	...	−94	80^{760}	1.0646^{20}	1.4032^{20}	d	d	s	...	...	...	**B2**³, 540
Ω p1315	—,cyclohexyl ester*	...	156.23	...	...	193^{750} 93^{35}	0.9359_4^{20}	1.4403^{20}	i	s	s	s	...	...	**B6**³, 24
Ω p1316	—,ethyl ester*	$CH_3CH_2CO_2C_2H_5$	102.13	...	−73.9	99.10^{760}	0.8917^{20}	1.3839^{20}	δ	∞	∞	s	...	...	**B2**³, 520
p1317	—,fluoride	Propionyl fluoride. CH_3CH_2COF	76.07	...	...	44	0.972_4^{15}	1.329^{13}	d	...	...	...	...	...	**B2**³, 540
Ω p1318	—,furfuryl ester	...	154.17	λ^{w} <220	...	195–6^{762}	1.1085_4^{20}	...	δ	s	∞	s	...	...	**B17**², 115
p1319	—,heptyl ester*	$CH_3CH_2CO_2(CH_2)_6CH_3$	172.27	...	−50.9	210.04^{760} $124\text{–}5^{14}$	0.8679_4^{20}	1.4201^{15}	i	s	s	s	...	os s	**B2**³, 530
p1320	—,hexyl ester*	$CH_3CH_2CO_2(CH_2)_5CH_3$	158.24	...	−57.5	190^{760} $73\text{–}4^{10}$	0.8698^{20}	1.4162^{15}	i	s	s	s	...	AcOEt s	**B2**³, 529
p1321	—,iodide	Propionyl iodide. CH_3CH_2COI	183.98	...	...	127–8	...	...	...	...	...	...	...	...	**B2**, 243
Ω p1322	—,isobutyl ester	$CH_3CH_2CO_2CH_2CH(CH_3)_2$	130.19	...	−71.4	136.8^{760} 66.5^{60}	0.8687_4^{20}	1.3973^{20}	δ	...	...	...	...	...	**B2**³, 527
Ω p1323	—,isopropyl ester*	$CH_3CH_2CO_2CH(CH_3)_2$	116.16	...	...	109–10^{760}	0.8660_4^{20}	1.3872^{20}	δ	∞	∞	...	...	...	**B2**³, 526
Ω p1324	—,methyl ester*	$CH_3CH_2CO_2CH_3$	88.12	...	−87.5	79.85^{760}	0.9150_4^{20}	1.3775^{20}	δ	∞	∞	s	...	os s	**B2**³, 519
Ω p1325	—,3-methylbutyl ester*	Isoamyl propionate. $CH_3CH_2CO_2CH_2CH_2CH(CH_3)_2$	144.22	...	...	160.7^{760}	0.8697_4^{20}	1.4069^{20}	δ	s	s	...	...	...	**B2**³, 529
Ω p1326	—,nitrile	Ethyl cyanide. Propionitrile. CH_3CH_2CN	55.08	...	−92.89	97.35^{760} 1.32^{10}	0.7818_4^{20}	1.3655^{20}	v	s	s	s	s	os s	**B2**³, 547
Ω p1327	—,octyl ester*	$CH_3CH_2CO_2(CH_2)_7CH_3$	186.30	...	−41.6	227.93^{760}	0.8663^{20}	1.4221_{α}^{15}	i	s	s	...	s	os s	**B2**³, 531
Ω p1328	—,pentyl ester*	$CH_3CH_2CO_2(CH_2)_4CH_3$	144.22	...	−73.1	168.65	0.8761_4^{15}	1.4096^{15}	i	∞	∞	...	s	...	**B2**², 221
Ω p1329	—,phenyl ester*	$CH_3CH_2CO_2C_6H_5$	150.18	pr	20	211^{760} 100^{16}	1.0467_{25}^{25}	1.4980^{20}	i	v	v	...	s	...	**B6**³, 599
Ω p1330	—,4-phenylphenacyl ester*	...	268.32	cr	102	...	...	...	i	...	...	...	...	...	**Am 52**, 3715
p1331	—,piperazinium salt	$2(CH_3CH_2CO_2H).C_4H_{10}N_2$	234.30	cr (diox)	124–5	...	...	...	...	...	...	...	...	diox s^h	**Am 56**, 1759
Ω p1332	—,propyl ester*	$CH_3CH_2CO_2CH_2CH_2CH_3$	116.16	...	−75.9	122.3^{760}	0.8809_4^{20}	1.3935^{20}	δ	∞	∞	s	...	chl v	**B2**³, 524
p1333	—,tetrahydrofurfuryl ester*	...	158.20	...	...	$204\text{–}7^{756}$ $85\text{–}7^{3}$	1.044_0^{20}	...	i	∞	∞	...	...	chl ∞	**B17**², 107
—	**—,2-amino-***	*see* **Alanine**													
—	**—,3-amino-***	*see* **β-Alanine**													
p1335	**—,2(2-aminoacetamido)-2-methyl-**	$(CH_3)_2C(CO_2H)NHCOCH_2NH_2$	160.17	nd (al)	260d	...	...	...	...	s^h	...	...	...	...	**B4**², 841
—	**—,2-amino-3-hydroxy-***	*see* **Serine**													
—	**—,2-amino-3(4-hydroxyphenyl)-***	*see* **Tyrosine**													
Ω p1336	**—,2-amino-2-methyl-***	$(CH_3)_2C(NH_2)CO_2H$	103.12	ta or pr (w)	337 (cor)	sub 280	...	...	v	δ	i	...	...	...	**B4**³, 1322
—	**—,2-amino-3-phenyl-***	*see* **Phenylalanine**													
p1337	**—,3-amino-3-phenyl-(*d*)***	β-Aminohydrocinnamic acid. $C_6H_5CH(NH_2)CH_2CO_2H$	165.19	pl (w), $[\alpha]_D$ −9.2 (1N NaOH, p = 9) $[\alpha]_D^{20}$ +7 (w, p = 1)	234–5d	...	...	...	δ s^h	δ s^h	δ	...	...	...	**B14**¹, 602
p1338	**—,—(*dl*)***	$C_6H_5CH(NH_2)CH_2CO_2H$	165.19	cr (w)	231d	...	...	...	δ s^h	δ s^h	δ	...	...	dil ac s	**B14**², 295
p1339	**—,—(*l*)***	$C_6H_5CH(NH_2)CH_2CO_2H$	165.19	cr (w), $[\alpha]_D^{25}$ −7.5 (w, c = 1) $[\alpha]_D$ +8.9 (1N NaOH, p = 10)	234–5d	...	...	...	δ s^h	δ s^h	δ	...	...	...	**B14**¹, 602

For explanations, symbols and abbreviations see beginning of table. For structural formulas see end of table.

No.	Name	Synonyms and Formula	Mol. wt.	Color, crystalline form, specific rotation and λ_{max} (log ε)	m.p. °C	b.p. °C	Density	n_D	Solubility w	al	eth	ace	bz	other solvents	Ref.
	Propanoic acid														
—	—,3(2-amino-phenyl)-, lactam*	see **Quinoline, 2-oxo-1,2,3,4-tetrahydro-**													
—	—,2-benzylidene-	see **Cinnamic acid, α-methyl-**													
p1340	—,2-benzyl-3-phenyl-	Dibenzylacetic acid. $(C_6H_5CH_2)_2CHCO_2H$	240.31	pl (peth, dil aa), nd (w)	89	235^{18}	...	...	δ^h	s	s	...	s	chl, aa s	**B9**^{2}, 475
p1341	—,—,amide	$(C_6H_5CH_2)_2CHCONH_2$	239.32	nd (al, w)	129	259^{18}	...	...	...	s	s	...	...	...	**B9**, 683
p1342	—,—,methyl ester	$(C_6H_5CH_2)_2CHCO_2CH_3$	254.33	nd (al)	42–3	...	...	...	i	δ	...	...	...	os v peth δ	**B9**^{2}, 475
p1343	—,—,nitrile	$(C_6H_5CH_2)_2CHCN$	221.31	lf or pl (al)	89–91	...	...	...	i	s	s	...	...	...	**B9**, 683
Ω p1344	—,2-bromo-(*dl*)*	$CH_3CHBrCO_2H$	152.98	pr	(i) 25.7(st) (ii) −3.9 (unst)	203.5 96^{10}	1.7000$^{20}_{4}$	1.4753^{20}	v	v	v	...	...	...	**B2**^{3}, 565
p1345	—,—,amide, *N*(2-tolyl)-	...	242.12	nd	131	...	...	...	...	s	v	...	s	chl v	**B12**, 794
Ω p1346	—,—,bromide	$CH_3CHBrCOBr$	215.88	...	...	152–4 (cor) 59^{15}	2.0612$^{16}_{4}$	...	d	d	...	...	...	...	**B2**^{3}, 569
p1347	—,—,chloride	$CH_3CHBrCOCl$	171.43	...	...	131–3	1.697^{11}	...	d	d^h	s	...	...	chl s	**B2**^{3}, 569
Ω p1348	—,—,ethyl ester*	$CH_3CHBrCO_2C_2H_5$	181.04	...	...	159–61 δd 71^{26}	1.4135$^{20}_{4}$	1.4490^{20}	i	∞	∞	...	...	chl s	**B2**^{3}, 568
p1349	—,—,methyl ester(*d*)*	$CH_3CHBrCO_2CH_3$	167.01	$[\alpha]^{17}_{D}$ +42.65	...	144^{760} 61–2^{36}	1.482^{17}	...	...	s	s	...	...	chl s	**B2**, 253
p1350	—,—,—(*dl*)*	$CH_3CHBrCO_2CH_3$	167.01	...	...	143–5^{760} 51.5^{19}	1.4966$^{21}_{4}$	1.4451^{22}	...	s	s	...	...	MeOH s	**B2**^{3}, 567
p1351	—,—,—(*l*)*	$CH_3CHBrCO_2CH_3$	167.01	$[\alpha]^{20}_{578}$ −55.5	...	61–3^{32}	1.484^{20}	...	...	s	s	...	...	...	**B2**^{2}, 229
p1352	—,—,nitrile	$CH_3CHBrCN$	133.98	...	...	59^{24}	1.5505$^{20}_{4}$	1.4585^{20}	...	...	s	s	...	...	**B2**^{3}, 567
p1353	—,—,piperazinium salt	$2(CH_3CHBrCO_2H).C_4H_{10}N_2$	392.10	wh cr	195d (cor)	...	...	...	s	s	i	...	...	...	**Am 70**, 2758
Ω p1354	—,3-bromo-*	$BrCH_2CH_2CO_2H$	152.98	pl (CCl_4)	62.5	140–2^{45}	1.48	...	s	s	s	...	s	chl s	**B2**^{3}, 569
p1355	—,—,amide	$BrCH_2CH_2CONH_2$	152.00	cr (w)	111	...	...	...	s^h	s	s	s	...	...	**B2**^{2}, 231
p1356	—,—,butyl ester*	$BrCH_2CH_2CO_2(CH_2)_3CH_3$	209.09	...	...	130^{26}	1.4549^{20}	1.3051^{20}	...	s	s	s	...	os s	**B2**^{2}, 231
p1357	—,—,ethyl ester*	$BrCH_2CH_2CO_2C_2H_5$	181.04	...	...	179 70^{12}	...	1.4516^{20}	...	s	s	s	...	os v	**B2**^{3}, 570
Ω p1358	—,—,methyl ester*	$BrCH_2CH_2CO_2CH_3$	167.01	...	...	105.5^{60} 80^{27}	1.4897^{15}	1.4542^{20}	...	s	s	s	...	os v	**B2**^{3}, 570
p1359	—,—,3-methylbutyl ester*	Isoamyl β-bromopropionate. $BrCH_2CH_2CO_2CH_2CH_2CH(CH_3)_2$	223.12	...	...	110–1^{11}	1.2217$^{15}_{4}$	1.4556^{9}	...	s	s	...	...	...	**B2**^{2}, 231
Ω p1360	—,—,nitrile	$BrCH_2CH_2CN$	133.98	...	...	92^{25} 69^{7}	1.6152$^{20}_{4}$	1.4800^{20}	...	v	v	...	...	...	**B2**^{3}, 571
p1361	—,2-bromo-2-methyl-*	2-Bromoisobutyric acid. $(CH_3)_2CBrCO_2H$	167.01	cr (peth)	48–9	198–200^{760} 115^{24}	1.5225$^{60}_{60}$	...	v	s	s	...	...	os v peth s^h	**B2**^{3}, 659
p1362	—,—,amide	$(CH_3)_2CBrCONH_2$	166.03	pr (chl)	148	145^{17}	...	...	s	s	...	...	...	chl s	**B2**^{2}, 263
p1363	—,—,—,*N*-methyl-*N*-phenyl-	$(CH_3)_2CBrCON(CH_3)C_6H_5$	256.15	cr (lig)	44	...	...	...	...	...	...	...	...	to s lig s^h	**B12**, 254
p1364	—,—,bromide	$(CH_3)_2CBrCOBr$	229.91	...	...	162–4 91–8^{100}	1.4067$^{14}_{4}$	1.4552^{14}	d	d^h	...	s	...	AcOEt, CS_2 s	**B2**^{3}, 661
Ω p1365	—,—,ethyl ester*	$(CH_3)_2CBrCO_2C_2H_5$	195.06	...	...	164^{762} 70^{20}	1.3182$^{20}_{4}$	1.4446^{20}	i	s	∞	...	...	...	**B2**^{3}, 660
p1366	—,—,nitrile	$(CH_3)_2CBrCN$	148.01	...	...	139–40 61.2–.6^{5}	1.4796$^{15}_{4}$	1.4739^{15}	...	s	s	...	s	...	**B2**^{3}, 661
p1367	—,2-bromo-2-phenyl-(*dl*)*	$C_6H_5CBr(CH_3)CO_2H$	229.08	pl (CS_2)	93–4 (90–1)	...	...	...	i	...	...	...	v	CS_2 s lig δ	**B9**, 525
p1368	—,3-bromo-2-phenyl-(*dl*)*	$C_6H_5CH(CH_2Br)CO_2H$	229.08	pr (CS_2)	93–4	...	...	...	i	v	v	...	v	CS_2 s	**B9**, 526
p1369	—,3-bromo-3-phenyl-(*dl*)*	$C_6H_5CHBrCH_2CO_2H$	229.08	mcl pr (al)	137	...	...	...	s	s	...	...	...	...	**B9**^{2}, 343
p1370	—,—,methyl ester*	$C_6H_5CHBrCH_2CO_2CH_3$	243.11	pr	37.5–8.5	...	...	...	s	s	δ	...	...	os s lig δ	**B9**^{1}, 201
p1371	—,3(2-carboxy-phenyl)-*	...	194.19	nd (w)	167	...	...	...	s	...	s	...	δ	...	**B9**^{2}, 622
p1372	—,3(4-carboxy-phenyl)-*	...	194.19	nd (al)	294	sub	...	...	s^h	s^h	...	...	...	...	**B9**^{2}, 622
Ω p1373	—,2-chloro-*	$CH_3CHClCO_2H$	108.53	...	...	186 84^{12}	1.2585$^{20}_{4}$	1.4380^{20}	∞	∞	∞	s	...	...	**B2**^{3}, 554
p1374	—,—,amide, *N*(2-tolyl)-	...	197.67	nd (abs al)	111	...	...	...	...	v^h	...	...	...	...	**B12**, 794
p1375	—,—,butyl ester*	$CH_3CHClCO_2C_4H_9$	164.63	...	...	183.5–5.0 71.6–2.6^{10}	1.0253$^{20}_{4}$	1.4263^{20}	...	...	s	...	...	...	**B2**^{3}, 555
p1376	—,—,chloride(*d*)	$CH_3CHClCOCl$	126.97	$[\alpha]^{18}_{D}$ +0.2 (l = 10)	...	110^{744}	1.2394$^{7.5}$	...	d	d	...	...	...	...	**B2**^{1}, 110
Ω p1377	—,—,ethyl ester*	$CH_3CHClCO_2C_2H_5$	136.58	...	...	147–8 52–4^{18}	1.0793$^{20}_{4}$	1.4178^{20}	i	∞	∞	...	...	...	**B2**^{3}, 555
p1378	—,—,isobutyl ester(*d*)*	$CH_3CHClCO_2CH_2CH(CH_3)_2$	164.63	$[\alpha]_D$ +5.21	...	175–7	1.0312$^{20}_{4}$	1.4247^{20}	...	s	s	s	...	os s	**B2**, 248

For explanations, symbols and abbreviations see beginning of table. For structural formulas see end of table.

No.	Name	Synonyms and Formula	Mol. wt.	Color, crystalline form, specific rotation and λ_{max} (log ε)	m.p. °C	b.p. °C	Density	n_D	Solubility w	al	eth	ace	bz	other solvents	Ref.
	Propanoic acid														
p1379	—,—,isopropyl ester*	$CH_3CHClCO_2CH(CH_3)_2$	150.61	...	...	$151.5–2.5^{760}$ $46.1–.9^{12}$	1.0315_{4}^{20}	1.4149^{20}	i	s	s	...	...	...	**B2^{3}**, 555
p1380	—,—,methyl ester(*d*)*	$CH_3CHClCO_2CH_3$	122.55	$[\alpha]_D$ +19.88 (undil)	...	133–4 50^{35}	1.1815_{4}^{20}	...	...	s	...	...	...	...	**B2**, 248
Ω p1381	—,—,—(*dl*)*	$CH_3CHClCO_2CH_3$	122.55	...	...	132.5^{760}	1.209_{4}^{20}	1.4182^{20}	...	s	...	...	...	0.1N HCl s	**B2^{3}**, 554
p1382	—,—,—(*l*)*	$CH_3CHClCO_2CH_3$	122.55	$[\alpha]_D$ −26.83 (undil)	...	$79–80^{120}$	1.158_{4}^{5}	...	...	s	...	...	...	...	**B2**, 248
p1383	—,—,nitrile	$CH_3CHClCN$	89.53	...	...	123–4 73^{144}	1.0792^{10}	...	...	...	...	...	...	...	**B2^{3}**, 556
Ω p1384	—,**3-chloro-***	$ClCH_2CH_2CO_2H$	108.53	lf (w), hyg cr (lig)	41 (61)	204 δd	...	...	s	s	∞	...	...	...	**B2^{3}**, 556
p1385	—,—,amide, *N*(2-tolyl)-	...	197.67	(w, dil al)	78	...	...	...	s^{h}	v^{h}	...	...	...	...	**B12^{2}**, 441
p1386	—,—,butyl ester*	$ClCH_2CH_2CO_2C_4H_9^n$	164.63	...	...	104^{22} 77^{6}	1.0370_{4}^{20}	1.4321^{20}	s	...	s	...	...	os s	**B2^{3}**, 559
Ω p1387	—,—,chloride	$ClCH_2CH_2COCl$	126.97	yesh	...	$143–5^{763}$ 82^{102}	1.3307^{13}	1.4549^{20}	δ	v	v	...	...	chl v	**B2^{3}**, 560
p1388	—,—,ethyl ester*	$ClCH_2CH_2CO_2C_2H_5$	136.58	...	...	162^{760} 56^{11}	1.1086_{4}^{20}	1.4254^{20}	δ	d^{h} ∞	∞	...	...	os s	**B2^{3}**, 558
p1389	—,—,isobutyl ester	$ClCH_2CH_2CO_2CH_2CH(CH_3)_2$	164.63	...	...	191–3	1.0323_{4}^{20}	1.4295^{20}	i	s	s	...	...	...	**B2^{3}**, 559
Ω p1390	—,—,methyl ester*	$ClCH_2CH_2CO_2CH_3$	122.55	...	...	$155–7^{760}$ (150) $40–2^{10}$	1.1861_{4}^{15}	1.4263^{20}	...	s	...	...	...	...	**B2^{3}**, 558
p1391	—,—,3-methylbutyl ester*	Isoamyl β-chloropropionate. $ClCH_2CH_2CO_2CH_2CH_2CH(CH_3)_2$	178.66	...	...	$207–8^{740}$ 87^{12}	1.0171_{4}^{20}	1.4343^{20}	i	s	s	...	...	...	**B2^{3}**, 559
Ω p1392	—,—,nitrile	$ClCH_2CH_2CN$	89.53	...	...	175–6 $85–7^{20}$	1.1573^{20}	1.4360^{20}	...	...	...	...	...	...	**B2^{2}**, 227
p1393	—,—,propyl ester*	$ClCH_2CH_2CO_2CH_2CH_2CH_3$	150.61	...	...	180 $77–8^{12}$	1.0656_{4}^{20}	1.4290^{20}	i	s	s	...	...	...	**B2^{3}**, 559
p1394	—,**3-chloro-2(chloromethyl)-2-hydroxy-***	$(ClCH_2)_2C(OH)CO_2H$	173.00	ta (chl)	93	...	...	...	...	v	v	...	...	chl s^{h} peth δ	**B3^{2}**, 224
Ω p1395	—,**3-chloro-2,2-dimethyl-***	Chloropivalic acid. $ClCH_2C(CH_3)_2CO_2H$	136.58	...	41–2	$126–9^{30}$ $108–12^{10}$	...	...	...	...	...	...	...	CCl_4 v	**C47**, 11126
p1396	—,—,chloride	$ClCH_2C(CH_3)_2COCl$	155.04	...	...	$85–6^{60}$	...	1.4539^{20}	d	d	...	...	...	CCl_4 s	**C34**, 3677
p1397	—,**3-chloro-2-hydroxy-2-methyl-**, nitrile	Chloroacetone cyanohydrin. $ClCH_2C(CH_3)(OH)CN$	119.55	...	...	110^{27}	1.2027^{15}	1.45362^{11}	v	v	...	v	...	os v peth i	**B3**, 317
p1398	—,**2-chloro-2-methyl-***	2-Chloroisobutyric acid. $(CH_3)_2CClCO_2H$	122.55	...	31	118^{50} $80–2^{12}$	...	1.450^{20}	v	v	...	...	...	...	**B2^{3}**, 657
p1399	—,—,chloride	$(CH_3)_2CClCOCl$	141.00	...	...	126–7 (113–4)	...	1.4369^{20}	d	d	s	...	...	...	**B2^{3}**, 657
p1400	—,—,ethyl ester*	$(CH_3)_2CClCO_2C_2H_5$	150.61	...	...	148.5–9 (cor)	1.062^{0}	1.4109^{16}	...	s	s	...	...	...	**B2^{3}**, 657
p1401	—,—,methyl ester*	$(CH_3)_2CClCO_2CH_3$	136.58	...	...	135 $42–4^{17}$	1.0893_{15}^{15}	1.4122^{21}	...	...	s	...	...	...	**B2^{3}**, 657
p1402	—,**3-chloro-2-methyl-***	$ClCH_2CH(CH_3)CO_2H$	122.55	...	...	$128–33^{50}$	1.0153^{20}	1.4310^{20}	...	s	v	...	...	CCl_4 s	**B2^{3}**, 658
p1403	—,—,chloride	$ClCH_2CH(CH_3)COCl$	141.00	...	...	$151–2^{765}$	...	1.4542^{20}	d	d	...	...	...	CCl_4 s	**B2^{3}**, 658
p1404	—,**3-cyclohexyl-***	...	156.23	...	16	275–8 (268 (cor)) 143.5^{11}	0.9966_{4}^{20}	1.4634^{20}	s	...	s	...	...	...	**B9^{2}**, 13
p1405	—,**2,3-diamino-***	$H_2NCH_2CH(NH_2)CO_2H$	104.12	hyg rosettes	*ca.* 110–20	...	...	...	s	i	i	...	...	...	**B4^{1}**, 500
Ω p1406	—,**2,3-dibromo-***	Acrylic acid dibromide. $BrCH_2CHBrCO_2H$	231.88	(i) st pl (ii) labile pr	(i) 66.5–7 (ii) 51	220–40d 160^{20}	...	...	s	s	s	...	s	CS_2 s	**B2**, 258
Ω p1407	—,—,ethyl ester*	$BrCH_2CHBrCO_2C_2H_5$	259.95	...	...	214–5 112^{23}	1.7966_{4}^{20}	1.5007^{20}	...	s	s	...	...	...	**B2^{3}**, 572
Ω p1408	—,—,methyl ester*	$BrCH_2CHBrCO_2CH_3$	245.91	...	...	206^{760} 115^{25}	1.9333_{4}^{20}	1.5127^{20}	...	s	...	...	...	...	**B2^{3}**, 572
p1410	—,**2,3-dibromo-3-phenyl-(*d*)***	*d*-Cinnamic acid dibromide. $C_6H_5CHBrCHBrCO_2H$	307.99	pr (chl), $[\alpha]_D^{15}$ +45.8 (abs al)	182	...	...	...	...	v	v	...	...	chl s^{h}	**B9^{2}**, 344
Ω p1411	—,—(*dl*)*	$C_6H_5CHBrCHBrCO_2H$	307.99	mcl pr (chl)	240d	sub	...	...	d^{h}	v	v	...	...	chl s^{h} CS_2 δ	**B9^{2}**, 344
p1412	—,—(*meso*)*	$C_6H_5CHBrCHBrCO_2H$	307.99	nd	91–3	...	...	...	d^{h}	...	...	...	v	CS_2 v lig i	**B9^{2}**, 344
p1413	—,—,ethyl ester(*d*)*	$C_6H_5CHBrCHBrCO_2C_2H_5$	336.04	cr (CS_2), $[\alpha]_D$ +59.1	71	...	...	...	...	s	...	...	...	CS_2 s^{h}	**B9**, 518
p1414	—,—,—,(*dl*)*	$C_6H_5CHBrCHBrCO_2C_2H_5$	336.04	mcl pr or pl	75–6	...	...	...	i	s	v	...	...	chl v	**B9^{2}**, 345
Ω p1415	—,**2,2-dichloro-***	$CH_3CCl_2CO_2H$	142.97	...	...	185–90 $90–2^{14}$	$1.389_{4}^{22.8}$	...	v	v	s	...	...	alk v	**B2^{2}**, 228
p1416	—,—,chloride	CH_3CCl_2COCl	161.42	...	...	$117.4–.8^{753}$ $68–73^{89}$	1.4062_{4}^{20}	1.4524^{20}	d	d	...	...	...	...	**B2^{3}**, 561
Ω p1417	—,**2,3-dichloro-***	$ClCH_2CHClCO_2H$	142.97	hyg nd (peth)	50	210 part d 113^{12}	...	1.4650^{20}	s	s	s	...	...	peth δ, s^{h}	**B2^{3}**, 561
p1418	—,—,chloride	$ClCH_2CHClCOCl$	161.42	...	...	$52–4^{16}$	1.4757_{4}^{20}	1.4764^{20}	d	d	s	...	...	...	**B2^{3}**, 562

For explanations, symbols and abbreviations see beginning of table. For structural formulas see end of table.

No.	Name	Synonyms and Formula	Mol. wt.	Color, crystalline form, specific rotation and λ_{max} (log ε)	m.p. °C	b.p. °C	Density	n_D	Solubility w	al	eth	ace	bz	other solvents	Ref.
	Propanoic acid														
p1419	—,—,ethyl ester*	$ClCH_2CHClCO_2C_2H_5$	171.03			183–4 76–7[15]	1.2461_4^{20}	1.4482^{20}		s	s				B2[3], 561
p1420	—,—,methyl ester(*d*)*	$ClCH_2CHClCO_2CH_3$	157.00	$[\alpha]_D^{20}$ +1.70		92^{50}	1.3282_4^{20}			s	s	s			B2[1], 111
p1421	**—,3,3-dichloro-***	$Cl_2CHCH_2CO_2H$	142.97	pr	56				v	v	v		v	chl v	B2, 252
p1422	—,—,chloride	Cl_2CHCH_2COCl	161.42			43–4[10]	1.4557_4^{20}	1.4738^{20}	d	d	s			diox s	B2[3], 562
p1423	**—,3,3-dichloro-2-hydroxy-2-methyl-***	$Cl_2CHC(CH_3)(OH)CO_2H$	173.00	pr (al-eth)	82–3	d			s	s	s				B3[2], 224
Ω p1423[1]	**—,3(3,5-diiodo-4-hydroxyphenyl)-2-phenyl-***	Iodoalphionic acid	494.07	pa ye cr (aa, bz, dil al)	163.5–.8				i	s	s		δ	alk s aa s^h chl δ	C40, 1883
p1423[2]	**—,2,3-dihydroxy-***	Glyceric acid. $HOCH_2CH(OH)CO_2H$	106.08	syr		d			∞	∞	i	∞			B3[3], 845
p1424	—,—,ethyl ester(*D*)*	*D*-Ethyl glycerate. $HOCH_2CH(OH)CO_2C_2H_5$	134.13	$[\alpha]_D^{11}$ −22.73 (1 = 19.84 cm)											B3[1], 141
p1425	—,—,—(*DL*)*	$HOCH_2CH(OH)CO_2C_2H_5$	134.13			230–40 120–1[14]	1.1909_{15}^{15}		s	v	v				B3[3], 854
p1426	—,—,methyl ester (*DL*)*	*DL*-Methyl glycerate. $HOCH_2CH(OH)CO_2CH_3$	120.12			239–44 119–20[14]	1.2814_{15}^{15}	1.4502^{20}	∞	∞	δ				B3[1], 141
p1427	—,—,—(*L*)*	$HOCH_2CH(OH)CO_2CH_3$	120.12	$[\alpha]_D^{15}$ −6.44		119–20[14] 74.5[0.5]	1.2798_{15}^{15}		∞	∞	δ				B3[3], 853
p1428	**—,3(2,5-dimethoxyphenyl)-2-oxo-***		224.22	yesh cr (aa)	166–70d							s	δ	MeOH s aa s^h	B10[2], 723
p1429	**—,3(3,4-dimethoxyphenyl)-2-oxo-***		224.22	lf (aa)	*ca.* 187d				s^h	v	δ	v	δ	chl δ	B10[2], 723
Ω p1430	**—,2,2-dimethyl-***	Pivalic acid. Trimethylacetic acid. $(CH_3)_3CCO_2H$	102.13	nd	35.3–.5	163.7–.8[760] 70[14]	0.905^{50}	$1.3931^{36.5}$	δ	v	v				B2[3], 708
p1431	—,—,amide,*N*,*N*-diethyl-	$(CH_3)_3CCON(C_2H_5)_2$	157.26			203	0.891^{15}		s	s	s				B4, 111
Ω p1432	—,—,chloride	Pivalyl chloride. $(CH_3)_3CCOCl$	120.58			107[760] 48[100]	1.003^{20}	1.4139^{20}	d	d	v				B2[3], 712
p1433	—,—,ethyl ester*	$(CH_3)_3CCO_2C_2H_5$	130.19		−89.55	118–8.2[760]	0.856_4^{20}	1.3906^{20}		s	s				B2[3], 710
p1434	—,—,methyl ester*	$(CH_3)_3CCO_2CH_3$	116.16			100.8–1.4[760]	0.891_4^{0}	1.3880^{20}	δ	∞	∞				B2[3], 709
Ω p1435	—,—,nitrile	*tert*-Butyl cyanide. Pivalonitrile. $(CH_3)_3CCN$	83.13		15–6	105–6	0.7586_4^{25}	1.3774^{20}							B2[3], 713
p1436	**—,2(dimethylamino)-, nitrile**	*N*,*N*-Dimethyl-*dl*-alanine nitrile. $(CH_3)_2NCH(CH_3)CN$	98.15			144				s	s				B4, 392
Ω p1437	**—,2,2-diphenyl-***	$(C_6H_5)_2C(CH_3)CO_2H$	226.28	pl (bz-peth), nd (w), lf (dil al)	175–7	sub >300			δ^h	s v^h	v		s	chl, to v peth δ	B9[2], 474
p1438	**—,2,3-diphenyl-(*d*)***	$C_6H_5CH_2CH(C_6H_5)CO_2H$	226.28	cr (dil al) $[\alpha]_D^{20}$ +94 (bz)	83–9				δ^h	s	s		s		B9[1], 284
p1439	—,—(*dl*)*	$C_6H_5CH_2CH(C_6H_5)CO_2H$	226.28	(i) pr (chl), (ii) pl (chl), (iii) cr (MeOH)	(i) 88–9 (ii) 95–6 (iii) 82	330–40	(i) 1.1481 (ii) 1.1495 (iii) 1.1430		i δ^h	v	v		s	CS_2 s chl s^h	B9[1], 285
p1440	—,—(*l*)*	$C_6H_5CH_2CH(C_6H_5)CO_2H$	226.28	nd (dil al) $[\alpha]_D^{20}$ −85.1 (bz)	83–9				δ^h	s	s		s		B9[1], 285
Ω p1441	**—,3,3-diphenyl-***	Benzhydrylacetic acid. $(C_6H_5)_2CHCH_2CO_2H$	226.28	nd (dil al)	155				δ	v	s				B9[2], 473
p1442	**—,2,3-diphenyl-2-hydroxy-***	α-Benzylmandelic acid. $C_6H_5CH_2C(OH)(C_6H_5)CO_2H$	242.28	nd (bz), cr (dil al)	165–6				δ^h	v	v		s^h	aa v	B10[2], 227
p1443	**—,3,3-diphenyl-2-hydroxy-***	β,β-Diphenyllactic acid. $(C_6H_5)_2CHCH(OH)CO_2H$	242.28	nd (w)	159d				δ s^h	s	s				B10[2], 228
Ω p1444	**—,3,3-diphenyl-3-hydroxy-***	$(C_6H_5)_2C(OH)CH_2CO_2H$	242.28	nd (dil al)	212				δ^h	v	δ	v	δ	aa v	B10[1], 156
p1445	—,—,ethyl ester*	$(C_6H_5)_2C(OH)CH_2CO_2C_2H_5$	270.33	pr (dil al)	87				δ s^h	v				os v con sulf s	B10[2], 228
p1446	**—,2,3-epoxy-***	Acrylic acid oxide. Glycidic acid.	88.06						∞	∞	∞				B18[1], 435
p1447	**—,2-ethoxy-,** nitrile	$CH_3CH(OC_2H_5)CN$	99.13			131^{765}	0.8743_4^{20}	1.3890^{22}	δ	v	v			aa v	B3[3], 498
Ω p1448	**—,3-ethoxy-,** nitrile	$C_2H_5OCH_2CH_2CN$	99.13			171.3–.5[760] 65[15]	0.9285_4^{15}	1.4068^{20}		v	v			ac d	B3[3],538
p1449	**—,3-fluoro-***	$FCH_2CH_2CO_2H$	92.07			83–4[14] 51–2[2]		1.3889^{25}	s	v	v				C42, 866
p1451	**—,3(2-furyl)-**	Furfurylacetic acid.	140.15	cr (chl-lig, w, peth), ye in HCl	58	229 108–10[10]			s		s			chl s lig δ	B18[2], 272
p1452	—,—,ethyl ester	$C_9H_{12}O_3$. *See* p1451	168.20			212 108–10[10]		1.4812^{25}							B18[2], 272

For explanations, symbols and abbreviations see beginning of table. For structural formulas see end of table.

No.	Name	Synonyms and Formula	Mol. wt.	Color, crystalline form, specific rotation and λ_{max} (log ε)	m.p. °C	b.p. °C	Density	n_D	Solubility w	al	eth	ace	bz	other solvents	Ref.
	Propanoic acid														
Ω p1452[1]	**—,3(2-furyl)-3-oxo-**, ethyl ester	$C_9H_{10}O_4$. *See* p1451	182.18	lt ye		$142–3^{10}$	1.165^{17}		i	s	s				**B18**[2], 327
p1453	—,—,methyl ester	$C_8H_8O_4$. *See* p1451	168.16	ye in air		$144–5^{20}$ $96–8^{7}$				s	s				**B18**[2], 327
p1454	**—,2-hydroxy-(*D*)***	*l*-(−)-Lactic acid. Sarcolactic acid. $CH_3CH(OH)CO_2H$	90.08	pl (chl or aa) $[\alpha]_D$ −2.26 (w, c = 1.24)	53	103^{2}			v	v	δ				**B3**[3], 442
Ω p1455	—,—(*DL*)*	$CH_3CH(OH)CO_2H$	90.08	ye	18	122^{15}	1.2060^{25}_{4}	1.4392^{20}_{α}	v	v	δ				**B3**[3], 454
p1456	—,—(*L*)*	$CH_3CH(OH)CO_2H$	90.08	hyg pr (eth-Pr^i_2O), $[\alpha]^{15}_D$ +3.82 (w, c = 10.5)	52.8	119^{12}d			∞	∞	δ				**B3**[3], 448
p1457	—,—,acetate (*dl*)	$CH_3CO_2CH(CH_3)CO_2H$	132.13	dlq	57–60	$167–70^{78}$ 127^{11}	1.1758^{20}_{4}	1.4240^{20}	d^h	s			s	peth δ	**B3**[3], 467
p1458	—,—,acetate chloride(*d*, +)	$CH_3CO_2CH(CH_3)COCl$	150.56	$[\alpha]^{18}_{578}$ +32.4		$51–3^{11}$	1.177^{20}		d^h	d^h					**B3**[2], 189
p1459	—,—,—(*dl*)	$CH_3CO_2CH(CH_3)COCl$	150.56			150δd 56^{11}	1.1920^{17}_{4}	1.4241^{17}	d^h	d^h					**B3**[3], 495
Ω p1460	—,—,allyl ester	Allyl lactate. $CH_3CH(OH)CO_2CH_2CH{:}CH_2$	130.15			$56–60^{8}$	1.0452^{20}_{4}	1.4369^{20}	d^h	d^h				Py s	**B3**[3], 486
p1461	—,—,amide (*d*)	Lactamide. Lactic amide. $CH_3CH(OH)CONH_2$	89.09	cr (AcOEt) $[\alpha]^{18}_{578}$ +22.2	49–51				s	s					**B3**[3], 450
Ω p1462	—,—,—(*dl*)	$CH_3CH(OH)CONH_2$	89.09	pl (AcOEt)	75.5 (78.5–9.5)		1.1381^{80}_{4}		s	s				AcOEt s^h	**B3**[3], 495
p1463	—,—,—,*N*(4-ethoxyphenyl)-	Lactophenin. *N*-Lactyl-*β*-phenetidide.	209.25	nd (w) λ^{MeOH} 249 (4.20)	118				s^h	v	δ		δ	peth δ	**B13**[2], 262
p1464	—,—,anhydride*	Lactic anhydride. $[CH_3CH(OH)CO]_2O$	162.14	pa ye amor or syr		d250			δ	v	v				**B3**, 282
p1465	—,—,butyl ester(*d*)*	*d*-Butyl lactate. $CH_3CH(OH)CO_2C_4H_9^n$	146.19	$[\alpha]^{27.3}_D$ +13.63		77^{10}	$0.9744^{27.6}_{4}$		s	∞	∞				**B3**[2], 188
Ω p1466	—,—,—(*dl*)*	$CH_3CH(OH)CO_2C_4H_9^n$	146.19		−43	83^{13}	0.9803^{22}_{4}	1.4217^{20}	s	∞	∞				**B3**[3], 480
p1467	—,—,ethyl ester(*D*)*	Ethyl lactate. $CH_3CH(OH)CO_2C_2H_5$	118.13	$[\alpha]^{19}_D$ +14.52		58^{20}	$1.0324^{20.4}_{4}$	1.4125^{20}	∞	v	v				**B3**[3], 446
Ω p1468	—,—,—(*DL*)*	$CH_3CH(OH)CO_2C_2H_5$	118.13			154.5 (cor) 58^{19}	1.0302^{20}_{4}	1.4124^{20}	∞	v	v				**B3**[3], 473
p1469	—,—,—(*L*)*	$CH_3CH(OH)CO_2C_2H_5$	118.13	$[\alpha]^{19}_D$ −11.33		$69–70^{36}$	1.0314^{20}_{4}	1.4156^{20}	∞	v	v				**B3**[3], 449
p1470	—,—,isopropyl ester*	Isopropyl lactate. $CH_3CH(OH)CO_2CH(CH_3)_2$	132.16			166–8 $75–80^{32}$	0.9980^{20}_{4}	1.4082^{25}	s	s	s		s		**C24**, 1843
p1471	—,—,methyl ester(*D*)*	*d*-Methyl lactate. $CH_3CH(OH)CO_2CH_3$	104.12	$[\alpha]^{20}_D$ +7.46		40^{13}	1.0857^{26}_{4}		∞	s	s				**B3**[2], 187
Ω p1472	—,—,—(*DL*)*	$CH_3CH(OH)CO_2CH_3$	104.12			144.8 (cor) 58^{19}	1.0928^{20}_{4}	1.4141^{20}	∞	s	s				**B3**[3], 471
p1473	—,—,—(*L*)*	*l*-Methyl lactate. $CH_3CH(OH)CO_2CH_3$	104.12	$[\alpha]^{20}_D$ −8.25		58^{19}	1.0895^{20}_{4}	1.4139^{20}	∞	s	s				**B3**[3], 444
p1474	—,—,3-methyl-butyl ester*	Isoamyl lactate. $CH_3CH(OH)CO_2CH_2CH_2CH(CH_3)_2$	160.22			202.4 82^{7}	0.9617^{25}_{25}	1.4240^{25}	i	s	s				**B3**[3], 483
Ω p1475	—,—,nitrile	Acetaldehyde cyanohydrin. Lactonitrile. $CH_3CH(OH)CN$	71.08	ye liq λ^{undil} 270 (−0.7)	−40	182–4δd 102^{30}	0.9877^{20}_{4}	1.4058^{18}	∞	∞	s			CS_2, peth i	**B3**[2], 209
p1476	—,—,—,acetate	$CH_3CO_2CH(CH_3)CN$	113.12			172–3 $76–7^{25}$	1.0278^{20}_{4}	1.4027^{20}	s	δ	δ			aa δ	**B3**[3], 499
p1477	—,—,4-phenyl-phenacyl ester*		284.32	cr	145				i						**C24**, 3051
p1478	—,—,piperazinium salt	$2[CH_3CH(OH)CO_2H]$. $C_4H_{10}N_2$	266.30		96–6.5				s	s^h	i			cellosolve s^h	**Am 56**, 1759
Ω p1479	**—,3-hydroxy-***	Hydracrylic acid. *β*-Lactic acid. $HOCH_2CH_2CO_2H$	90.08	syr		d		1.4489^{20}	v	s	∞				**B3**[2], 212
Ω p1480	—,—,lactone*	*β*-Propiolactone	72.06		−33.4	162d 51^{10}	1.1460^{20}_{4}	1.4105^{20}	d	d	∞			chl s	**B17**[1], 130
Ω p1481	—,—,nitrile	Ethylene cyanohydrin. Hydracrylonitrile. $HOCH_2CH_2CN$	71.08			230^{760} 110^{15}	1.0588^{0}	1.4240^{20}	∞	∞	δ			CS_2 i	**B3**[2], 213
p1483	**—,3(4-hydroxy-3-methoxy-phenyl)-***	Hydroferulic acid	196.21	pl (w)	89–90				s^h	s	s				**B10**, 424
Ω p1484	**—,2-hydroxy-2-methyl-***	Acetonic acid. 2-Hydroxyisobutyric acid. $(CH_3)_2C(OH)CO_2H$	104.12	hyg pr (eth), nd (bz)	82–3	212 $108–11^{8}$			v	v	v		δ v^h		**B3**[3], 587
Ω p1485	—,—,ethyl ester*	$(CH_3)_2C(OH)CO_2C_2H_5$	132.16			150 (cor) 46^{14}	0.987^{20}	1.4080^{20}	∞	∞					**B3**[3], 591
Ω p1486	—,—,methyl ester*	$(CH_3)_2C(OH)CO_2CH_3$	118.13			137 $62–4^{12}$		1.4056^{20}	v	v					**B3**[3], 590
Ω p1487	—,—,nitrile	Acetone cyanohydrin. $(CH_3)_2C(OH)CN$	85.11	λ^{al} 263 (0.1)	−19	82^{23}	0.932^{20}_{4}	1.3996^{20}	v	v	v	s	s	os v peth i	**B3**[3], 597
p1488	**—,2-hydroxy-3-oxo-3-phenyl-***	Benzoylglycolic acid. $C_6H_5COCH(OH)CO_2H$	180.17	lo pr (lig)	112				δ s^h	s	s			chl s	**B9**[2], 147

For explanations, symbols and abbreviations see beginning of table. For structural formulas see end of table.

No.	Name	Synonyms and Formula	Mol. wt.	Color. crystalline form. specific rotation and λ_{max} (log ε)	m.p. °C	b.p. °C	Density	n_D	Solubility w	al	eth	ace	bz	other solvents	Ref.
	Propanoic acid														
p1490	—,**2-hydroxy-2-phenyl-(*D*)***	Atrolactic acid. $C_6H_5C(OH)(CH_3)CO_2H$	166.18	pr (w) $[\alpha]_D^{16.5}$ +37.7 (al, c = 3.5)	116.5–7	...	...	...	s	s^h	...	v	v^h	...	$B10^2$, 156
p1491	—,—(*DL*)*	$C_6H_5C(OH)(CH_3)CO_2H$	166.18	nd, pl (lig)	93–5	...	...	...	s	s^h	...	v	v^h	lig s^h	$B10^2$, 157
p1492	—,—(*L*)*	$C_6H_5C(OH)(CH_3)CO_2H$	166.18	nd (bz, w) $[\alpha]_D^{13.8}$ −37.7 (al, c = 3.4)	116–7	...	...	...	s	v s^h	...	v	v^h	...	$B10^2$, 157
p1493	—,—,hemi-hydrate (*DL*)*	$C_6H_5C(OH)(CH_3)CO_2H.\frac{1}{2}H_2O$	175.19	nd or pr (w)	67–8	...	...	...	s ∞^h	s^h	...	...	...	...	$B10^2$, 157
p1494	—,**2-hydroxy-3-phenyl-(*d*)***	α-Hydroxyhydrocinnamic acid. $C_6H_5CH_2CH(OH)CO_2H$	166.18	nd (w) $[\alpha]_D^{20}$ +22.2 (w, c = 2.2)	124–6	...	...	...	v^h	s	s^h	s	s^h	MeOH s CS_2, chl, peth δ	$B10^2$, 152
p1495	—,—(*dl*)*	$C_6H_5CH_2CH(OH)CO_2H$	166.18	cr (chl, bz), pr (w)	98	148–50[15]	...	...	δ	s	s	s	s^h	CCl_4, chl s^h	$B10^2$, 154
p1496	—,—(*l*)*	$C_6H_5CH_2CH(OH)CO_2H$	166.18	nd (w) $[\alpha]_D^{20}$ −19.9 (w, c = 3.2)	124–5	...	...	...	s^h	s	s	s	s^h	...	$B10^2$, 153
p1497	—,**3-hydroxy-2-phenyl-(*d*)***	$HOCH_2CH(C_6H_5)CO_2H$	166.18	nd (w, bz), pr (eth or w) $[\alpha]_D^{16}$ +72.2 (al, c = 2.7)	130	...	...	...	s^h	s	s	...	δ	...	$B10^2$, 158
Ω p1498	—,—(*dl*)*	Tropaic acid. Tropic acid. $HOCH_2CH(C_6H_5)CO_2H$	166.18	nd, pl (al, bz, w)	118	d	...	...	s ∞^h	s	s	...	δ	peth, CS_2 i	$B10^2$, 158 $B23^2$, 534
p1499	—,—(*l*)*	$HOCH_2CH(C_6H_5)CO_2H$	166.18	pl (AcOEt), nd (w), $[\alpha]_D^{15}$ −81.2 (w, c = 1.5)	130	...	...	...	...	s	s	...	δ	AcOEt s	$B10^2$, 158
p1500	—,**3-hydroxy-3-phenyl-(*d*)***	$C_6H_5CH(OH)CH_2CO_2H$	166.18	cr (bz), $[\alpha]_D^{18}$ +20.6 (MeOH, c = 5.04)	116	...	...	...	δ	s	s	...	...	MeOH s chl δ	$B10^2$, 148
p1501	—,—(*dl*)*	$C_6H_5CH(OH)CH_2CO_2H$	166.18	pr (w)	96	...	...	...	δ ∞^h	v	...	v	δ	MeOH v chl s peth δ	$B10^2$, 148
p1502	—,—(*l*)*	$C_6H_5CH(OH)CH_2CO_2H$	166.18	nd (bz) $[\alpha]_D^{18}$ −19.8 (al, c = 4.7)	115–6	...	...	...	δ	...	...	...	δ	chl δ	$B10^2$, 147
p1502[1]	—,**3(2-hydroxy-phenyl)-***	*o*-Hydrocoumaric acid. Melilotic acid.	166.18	pr (w)	82–4	...	...	...	s v^h	v	v	...	...	...	$B10^2$, 143
p1503	—,**2-hydroxy-3,3,3-trichloro-***	$Cl_3CCH(OH)CO_2H$	193.42	pr (eth)	125	140–70[45]	...	...	v	v	v	...	...	chl v	$B3^2$, 210
p1504	—,—,monohydrate	$Cl_3CCH(OH)CO_2H.H_2O$	211.43	...	105–10 (115–8)	...	...	...	s	...	...	...	...	...	B3, 287
p1505	—,—,nitrile	Chloral cyanohydrin. $Cl_3CCH(OH)CN$	174.42	pl (w, CS_2)	61	215–20d	...	...	v	v	v	...	...	CS_2 s	$B3^2$, 210
p1506	—,**2,2′-iminodi-(*dl*)***	$HN[CH(CH_3)CO_2H]_2$	161.16	(i) nd (w), pr (eth), (ii) cr	(i) 234–5d (ii) 254–5	...	...	...	v	i	i s^h	i	...	os i	$B4^2$, 824
p1507	—,—,dinitrile	$HN[CH(CH_3)CN]_2$	123.16	nd (eth)	68	...	...	...	δ	s	s	...	...	...	$B4^2$, 825
p1508	—,**3,3-iminodi-**, *N*-methyl-, diethyl ester*	$(C_2H_5O_2CCH_2CH_2)_2NCH_3$	231.30	...	...	136–8[4]	1.0190_{20}^{20}	1.4421^{20}	...	s	s	...	...	...	$B4^2$, 829
Ω p1509	—,**β(3-indolyl)-***	...	189.22	pl (w) λ^w 280 (3.7)	134	...	...	...	δ s^h	v	v	v	v	AcOEt, chl v	$B22^2$, 53
p1510	—,—,methyl ester	$C_{12}H_{13}NO_2$. *See* p1509	203.24	pr (MeOH)	79–80	...	...	...	...	s	...	...	...	MeOH s^h	$B22^2$, 53
p1511	—,**2-iodo-(*dl*)***	CH_3CHICO_2H	199.98	nd (bz, w, al)	45–7	93–6[0.2]	2.073_4^{18}	...	δ	v	v	...	...	...	$B2^3$, 573
Ω p1512	—,**3-iodo-***	$ICH_2CH_2CO_2H$	199.98	lf (w)	85	...	...	...	δ s^h	v	s	s	...	...	$B2^3$, 574
p1513	—,—,methyl ester*	$ICH_2CH_2CO_2CH_3$	214.00	...	...	188[756]	1.8408^7	...	...	s	...	...	...	...	$B2^3$, 574
Ω p1514	—,**3-mercapto-***	$HSCH_2CH_2CO_2H$	106.14	amor	17–9	110.5–1.5[15]	1.218^{21}	1.4911^{20}	s	s	s	...	...	...	$B3^2$, 214
p1515	—,**2-methoxy-**, nitrile	$CH_3CH(OCH_3)CN$	85.11	...	...	118[740]	0.8928_4^{20}	1.3818^{20}	...	s	...	...	...	...	$B3^3$, 497
Ω p1516	—,**3-methoxy-**, nitrile	$CH_3OCH_2CH_2CN$	85.11	...	...	165.5[763] 85.5[49]	0.9379_4^{20}	1.4043^{20}	...	s	s	...	...	...	$B3^3$, 538
p1517	—,**2-methoxy-3-methyl-**, 3-*p*-menthyl ester*	...	256.39	...	...	124–6[10]	0.9466	...	...	s	s	...	...	...	C25, 984
Ω p1518	—,**2-methyl-***	Isobutyric acid. $(CH_3)_2CHCO_2H$	88.11	...	−46.1	153.2[760] 53.74[10]	0.96815_4^{20}	1.3930^{20}	v	∞	∞	...	...	...	$B2^3$, 637
p1519	—,—,allyl ester	Allyl isobutyrate. $(CH_3)_2CHCO_2CH_2CH{:}CH_2$	128.17	...	...	133–5[755]	...	...	δ	∞	∞	s	...	...	$B2^3$, 650
Ω p1521	—,—,amide, *N*-phenyl-	Isobutyranilide. $(CH_3)_2CHCONHC_6H_5$	163.22	mcl pr (al, eth), nd (lig)	106–7	...	...	...	δ^h	v	v	...	...	lig s^h	$B12^2$, 147
Ω p1522	—,—,anhydride*	Isobutyric anhydride. $[(CH_3)_2CHCO]_2O$	158.20	...	−53.5	181.5[734] 89–90[32]	0.9535_4^{20}	1.4061^{19}	d	d	∞	...	...	...	$B2^3$, 653
p1523	—,—,bromide	Isobutyryl bromide. $(CH_3)_2CHCOBr$	151.01	...	...	116–8	1.4067_4^{15}	1.4552^{15}	d	d	...	...	...	...	$B2^2$, 262
p1524	—,—,*tert*-butyl ester	*tert*-Butyl isobutyrate. $(CH_3)_2CHCO_2C(CH_3)_3$	144.22	...	...	126.7[760]	...	1.3921^{20}	i	s	s	s	...	...	$B2^3$, 648

For explanations, symbols and abbreviations see beginning of table. For structural formulas see end of table.

No.	Name	Synonyms and Formula	Mol. wt.	Color, crystalline form, specific rotation and λ_{max} (log ε)	m.p. °C	b.p. °C	Density	n_D	Solubility w	al	eth	ace	bz	other solvents	Ref.
	Propanoic acid														
Ω p1525	—,—,chloride*	Isobutyryl chloride. $(CH_3)_2CHCOCl$	106.55	...	−90.0	92^{760}	1.0174_4^{20}	1.4079^{20}	d	d	s	...	...	...	**B2**^{3}, 653
p1526	—,—,cyclohexyl ester*	Cyclohexyl isobutyrate.	170.25	...	...	204^{750} $87–8^{15}$	0.9489_4^0	...	i	s	s	...	...	os s	**B6**^{2}, 11
Ω p1527	—,—,ethyl ester*	Ethyl isobutyrate. $(CH_3)_2CHCO_2C_2H_5$	116.16	...	−88.2	111.0^{760}	0.8693_4^{20}	1.3869^{20}	δ	∞	∞	s	...	...	**B2**^{2}, 260
p1528	—,—,furfuryl ester	Furfuryl isobutyrate	168.20	...	...	$85–6^{15}$	1.0313_4^{20}	...	δ	s	s	...	...	...	**B17**^{2}, 115
p1529	—,—,α,α′,1,2-hydrazodi-, dinitrile	α,α′-Hydrazodiisobutyronitrile. $(CH_3)_2C(CN)NHNHC(CN)(CH_3)_2$	166.23	pl (eth)	92–3	...	...	...	i	v	s	...	...	...	**B4**, 561
Ω p1530	—,—,isobutyl ester	Isobutyl isobutyrate. $(CH_3)_2CHCO_2CH_2CH(CH_3)_2$	144.22	...	−80.66	148.6^{760} $36–40^{11}$	0.8750_4^0	1.3999_4^{20}	δ	s	∞	s	...	...	**B2**^{3}, 647
p1531	—,—,isopropyl ester*	Isopropyl isobutyrate. $(CH_3)_2CHCO_2CH(CH_3)_2$	130.19	...	...	120.76^{760}	0.8471_4^{21}	...	i	s	s	s	...	...	**B2**^{3}, 646
Ω p1532	—,—,methyl ester*	Methyl isobutyrate. $(CH_3)_2CHCO_2CH_3$	102.13	...	−84.7	92.3^{760}	0.8906_4^{20}	1.3840^{20}	δ	∞	∞	s	...	...	**B2**^{3}, 642
Ω p1533	—,—,3-methylbutyl ester*	Isoamyl isobutyrate. $(CH_3)_2CHCO_2CH_2CH_2CH(CH_3)_2$	158.24	...	...	168.9	0.8627^{20}	...	δ	s	s	s	...	...	**B2**^{3}, 649
Ω p1534	—,—,nitrile	Isobutyronitrile. Isopropyl cyanide. $(CH_3)_2CHCN$	69.11	...	−71.5	103.8^{760} (107–8)	0.7608_4^{30}	1.3720^{20}	δ	v	v	s	...	...	**B2**^{3}, 655
p1535	—,—,2-nitropentyl ester	2-Nitroamyl isobutyrate. $(CH_3)_2CHCO_2CH_2CH(NO_2)(CH_2)_2CH_3$	202.24	...	...	$248–51^{760}$ 122^{10}	1.0329_{20}^{20}	1.4315^{20}	δ	...	∞	s	...	...	**B2**^{3}, 648
p1536	—,—piperazinium salt	$2[(CH_3)_2CHCO_2H]. C_4H_{10}N_2$	262.35	wh cr (diox)	89.5–90	...	...	...	s	s	i^h	...	...	diox s^h	**Am 56**, 1759
p1537	—,—,propyl ester*	*n*-Propyl isobutyrate. $(CH_3)_2CHCO_2CH_2CH_2CH_3$	130.19	...	...	135–5.5^{760}	0.8843_4^0	1.3955^{20}	δ	s	v	s	...	...	**B2**^{3}, 646
p1538	—,**2-methyl-3-phenyl-***	$C_6H_5CH_2CH(CH_3)CO_2H$	164.21	pl (dil al)	36.5	272 $155–6^{11}$	...	...	δ v^h	v	v	...	...	...	**B9**^{2}, 357
Ω p1539	—,**3(1-naphthyl)-***	$C_{10}H_7^{α}CH_2CH_2CO_2H$	200.24	cr (bz), nd (al)	156–6.7 (148)	179^{11}	...	...	s^h	s	...	...	s^h	...	**E12B**, 3283
p1540	—,**3(2-naphthyl)-***	$C_{10}H_7^{β}CH_2CH_2CO_2H$	200.24	lf or nd (w, al)	135	...	...	...	v^h	s	...	...	...	...	**E12B**, 3285
Ω p1541	—,**3(2-nitrophenyl)-2-oxo-***	...	209.16	ye nd or lf (w, al, bz)	130	...	...	...	δ s^h	s	v	...	δ s^h	aa v chl δ lig i	**B10**^{2}, 476
Ω p1542	—,**2-oxo-***	Acetylformic acid. Pyruvic acid. CH_3COCO_2H	88.06	...	13.6	165 part d 54^{10}	1.2272_4^{20}	1.4280^{20}	∞	∞	∞	s	...	...	**B3**^{3}, 1146
Ω p1543	—,—,ethyl ester*	Ethyl pyruvate. $CH_3COCO_2C_2H_5$	116.13	...	−50	155 $69–71^{42}$	$1.0596_4^{15.6}$	1.4052^{20}	δ	∞	∞	s	...	...	**B3**^{3}, 1161
Ω p1544	—,—,methyl ester*	Methyl pyruvate. $CH_3COCO_2CH_3$	102.09	...	...	134–7 53^{15}	1.154_4^0	1.4046^{25}	δ	∞	∞	s	...	...	**B3**^{3}, 1160
Ω p1545	—,—,nitrile	Acetyl cyanide. Pyruvonitrile. CH_3COCN	69.06	rh	...	92.3	0.9745_4^{20}	1.3764^{20}	d	d	s	s	...	CH_3CN s	**B3**^{3}, 1165
Ω p1545[1]	—,**3(2-oxocyclohexyl)-**, nitrile	...	151.21	...	...	$138–42^{10}$	1.0181_4^{20}	1.4755^{20}	...	...	...	...	...	...	**Am 73**, 724
p1546	—,**2-oxo-3-phenyl-***	Phenylpyruvic acid. $C_6H_5CH_2COCO_2H$	164.17	lf (chl, bz) λ^{MeOH} 287.5 (4.28)	157–8	...	...	...	$δ^h$	v	v	...	s^h	chl s^h lig i	**B10**^{2}, 471
p1546[1]	—,**3-oxo-2-phenyl-**, ethyl ester*	Ethyl phenylmalonaldehydate. $C_6H_5CH(CHO)CO_2C_2H_5$	192.22	pl (chl)	70–1	136^{16}	1.12045_{20}^{20}	1.532^{21}	i	∞	v	...	...	chl s^h	**B10**^{2}, 478
p1547	—,**3-oxo-3-phenyl-***	Benzoylacetic acid. $C_6H_5COCH_2CO_2H$	164.17	nd (bz-peth)	103–4d	...	...	...	δ v^h	s	s	...	s	lig δ	**B10**^{2}, 466
Ω p1548	—,—,amide, *N*-phenyl-*	α-Benzoylacetanilide. $C_6H_5COCH_2CONHC_6H_5$	239.28	lf (bz) λ^{al} 235 (4.2), 325 (3.4)	108	...	...	...	δ	v	...	...	δ v^h	alk, chl v	**B12**^{2}, 270
Ω p1549	—,—,ethyl ester*	$C_6H_5COCH_2CO_2C_2H_5$	192.22	...	<0	265–70 part d 165^{14}	1.1220_4^{20}	1.5312^{16}	δ	s	s	...	...	...	**B10**^{2}, 467
p1550	—,—,methyl ester (enol form)*	$C_6H_5C(OH):CHCO_2CH_3$	178.19	...	*ca.* 40	...	...	1.5620^{16}	...	...	...	...	...	...	**B10**^{2}, 467
p1551	—,—,—(keto + enol forms)*	$C_6H_5COCH_2CO_2CH_3$	178.19	pa ye	...	265d 151.5^{13}	1.158_4^{29}	1.537^{25}	i	∞	∞	s	...	dil alk s	**B10**^{2}, 467
Ω p1552	—,—,nitrile	Benzoylacetonitrile. α-Cyanoacetophenone. $C_6H_5COCH_2CN$	145.16	pr or lf (w) λ^{al} 245 (4.11), 280 (3.30)	80–1	160^{10}	...	...	δ	s	s	...	s	chl, aq KCN, alk s	**B10**^{2}, 468
p1553	—,**pentachloro-***	$Cl_3CCCl_2CO_2H$	246.31	cr(CCl_4)	200–15d	...	...	...	v	...	...	...	...	CCl_4 δ, v^h	**B2**^{2}, 228
p1554	—,—,chloride*	Cl_3CCCl_2COCl	264.75	nd	42	...	...	...	d	d	...	...	s	...	**B2**^{1}, 112
p1555	—,**2-phenoxy-(*D*)***	$CH_3CH(OC_6H_5)CO_2H$	166.18	nd (w) $[\alpha]_D^{21}$ +39.3 (al, c = 1.2)	87	$265–6^{758}$	...	...	δ v^h	v	v	...	...	...	**B6**^{3}, 614
p1556	—,—(*DL*)*	$CH_3CH(OC_6H_5)CO_2H$	166.18	nd (w)	115–6	$265–6^{758}$ $105–6^5$	1.1865_4^{20}	1.5184^{20}	δ s^h	s	s	...	...	...	**B6**^{3}, 614
p1557	—,—,amide*	$CH_3CH(OC_6H_5)CONH_2$	165.19	nd or pl (to, w)	132–3	...	...	...	s^h	v	v	...	...	aa s to s^h lig v^h	**B6**, 163
Ω p1558	—,—,chloride	$CH_3CH(OC_6H_5)COCl$	184.62	...	...	$146–7^{55}$ $115–7^{10}$	1.1865_4^{20}	1.5178^{20}	d	d	s	...	...	...	**B6**^{3}, 615
p1559	—,—,ethyl ester*	$CH_3CH(OC_6H_5)CO_2C_2H_5$	194.23	...	...	$243–4^{760}$ $120–5^6$	1.360_4^{17}	...	i	s	s	...	...	...	**B6**^{3}, 615

For explanations, symbols and abbreviations see beginning of table. For structural formulas see end of table.

No.	Name	Synonyms and Formula	Mol. wt.	Color, crystalline form, specific rotation and λ_{max} (log ε)	m.p. °C	b.p. °C	Density	n_D	Solubility w	al	eth	ace	bz	other solvents	Ref.
	Propanoic acid														
Ω p1560	—,3-phenoxy-*	$C_6H_5OCH_2CH_2CO_2H$	166.18	nd (w), lf (lig)	97.5–8	$234–45^{771}$ $188–9^{26}$	...	...	s^h	...	...	...	...	lig s^h	**B6^3**, 615
p1561	—,—,amide	$C_6H_5OCH_2CH_2CONH_2$	165.19	nd (w)	119	...	...	...	s^h	v^h	v^h	...	...	...	**B6^3**, 616
p1562	—,—,ethyl ester*	$C_6H_5OCH_2CH_2CO_2C_2H_5$	194.23	nd (peth)	24	170^{40} 142^{11}	1.0821^{25}_{25}	1.5007^{18}	i	s	s	...	...	...	**B6^3**, 616
p1563	—,2-phenyl-(*d*)*	Hydratropic acid. $C_6H_5CH(CH_3)CO_2H$	150.18	$[\alpha]^{20}_D$ +9.14; +81.1 (al, c = 3)	...	152^{16}	...	...	δ	...	...	...	...	...	**B9^2**, 347
Ω p1564	—,—(*dl*)*	$C_6H_5CH(CH_3)CO_2H$	150.18	...	< −20	260–2 160^{25}	1.1^0_4	1.5237^{20}	δ	...	...	...	...	...	**B9^2**, 348
p1565	—,—(*l*)*	$C_6H_5CH(CH_3)CO_2H$	150.18	$[\alpha]^{20}_D$ −7.0; −58 (al)	...	152^{16}	...	...	δ	...	...	...	...	...	**C53**, 5179
p1566	—,—,amide	Hydratropamide. $C_6H_5CH(CH_3)CONH_2$	149.19	lo nd (w, dil al) $[\alpha]^{28}_D$ +57.9 (chl, c = 1.6)	100.5	...	...	...	i δ^h	s	...	...	...	chl s	**B9**, 525
p1567	—,—,nitrile	Hydratroponitrile. $C_6H_5CH(CH_3)CN$	131.18	...	...	230–2 $116–7^{20}$ $(127–8^{12})$	0.9854^{20}_4	1.5095^{25}	i	s	s	...	...	...	**B9^2**, 348
Ω p1568	—,3-phenyl-*	Hydrocinnamic acid. $C_6H_5CH_2CH_2CO_2H$	150.18	pr (peth)	48.6	279.8^{760} $169–70^{28}$	1.0712^{49}_4	...	s	s	s	...	v	CCl_4, CS_2, chl s	**B9^2**, 337
p1569	—,—,amide	Hydrocinnamamide. $C_6H_5CH_2CH_2CONH_2$	149.19	nd (w)	106–8	...	...	...	s^h	v	v	...	...	...	**B9^2**, 340
p1570	—,—,benzyl ester	Benzyl hydrocinnamate. $C_6H_5CH_2CH_2CO_2CH_2C_6H_5$	240.13	...	...	310–40 $198–9^{20}$	1.090^{15}	...	...	...	s	...	...	...	**B9^2**, 339
p1571	—,—,chloride	Hydrocinnamyl chloride. $C_6H_5CH_2CH_2COCl$	168.63	...	...	225 d 105^{10}	1.135^{21}	...	d	d^h	s	...	...	CS_2 s	**B9^2**, 340
p1572	—,—,ethyl ester*	Ethyl hydrocinnamate. $C_6H_5CH_2CH_2CO_2C_2H_5$	178.23	...	...	247.2^{760} 123^{16}	1.0147^{20}	1.4954^{20}	i	s	s	...	...	...	**B9^2**, 339
p1573	—,—,isopropyl ester*	Isopropyl hydrocinnamate. $C_6H_5CH_2CH_2CO_2CH(CH_3)_2$	192.24	...	...	126^{11}	0.9860^{25}_4	...	i	s	s	...	...	...	**B9^2**, 339
Ω p1574	—,—,methyl ester*	Methyl hydrocinnamate. $C_6H_5CH_2CH_2CO_2CH_3$	164.21	...	...	$238–9^{757}$	1.0455^0	...	i	s	s	...	s	AcOEt s	**B9^2**, 338
Ω p1575	—,—,nitrile	Hydrocinnamonitrile. $C_6H_5CH_2CH_2CN$	131.17	...	...	261 $125–6^{15}$	1.0016^{20}	1.5266^{20}	...	s	s	...	...	...	**B9^2**, 341
p1576	—,—,piperazinium salt	$2[C_6H_5CH_2CH_2CO_2H].C_4H_{10}N_2$	386.50	wh cr	122.5–3	...	...	...	δ	s^h	i	...	...	...	**Am 56**, 150
p1577	—,—,propyl ester*	Propyl hydrocinnamate. $C_6H_5CH_2CH_2CO_2CH_2CH_2CH_3$	192.26	...	...	262.1 135^{16}	1.008^{12}	...	...	...	...	...	...	...	**B9^2**, 339
p1578	—,2-propoxy-, nitrile	$CH_3CH_2CH_2OCH(CH_3)CN$	113.16	...	...	150^{727}	0.866^{20}_4	1.398^{20}	...	...	...	...	...	...	**B3**, 285
Ω p1579	—,3(3-pyrenyl)-	...	274.32	pl(aa)	180	...	...	...	i	...	s	s	...	con sulf, dil alk, aa s	**E14s**, 441
p1580	—,2,2,3,3-tetra-chloro-*	$Cl_2CHCCl_2CO_2H$	211.86	cr (CS_2-chl)	76	...	...	...	s	...	...	...	...	CS_2 v	**B2**, 253
p1581	—,3(2-tetra-hydrofuryl)-	...	144.17	...	...	263 $118–20^2$	1.1155^{20}_{20}	1.4578^{25}	...	s	...	...	...	alk v	**B18^2**, 263
p1582	—,—,ethyl ester*	$C_9H_{16}O_3$. *See* p1581	172.23	...	...	$221–2^{750}$ 73^2	1.024^7_{15}	1.440^{20}	...	s	...	s	...	...	**B18^2**, 263
p1584	—,2,2,3-tri-chloro-*	$ClCH_2CCl_2CO_2H$	177.42	hyg pr (CS_2)	65–6	140^{40}	...	...	s	s	...	...	s	CS_2 s^h	**B2^3**, 563
p1585	—,2,3,3-tri-phenyl-*	$(C_6H_5)_2CHCH(C_6H_5)CO_2H$	302.38	nd (dil al, peth)	222–3	...	...	...	i	v	v	...	...	peth δ	**B9**, 715
Ω p1586	—,3,3,3-tri-phenyl-*	$(C_6H_5)_3CCH_2CO_2H$	302.38	pr (al)	179–80	...	...	...	i	s	v	...	...	...	**B9^2**, 504
Ω p1587	1-Propanol-*	*n*-Propyl alcohol. $CH_3CH_2CH_2OH$	60.11	λ^{gas} 183 (2.38)	−126.5	97.4^{760}	0.8035^{20}_4	1.3850^{20}	∞	∞	∞	s	v	...	**B1^3**, 1397
Ω p1588	2-Propanol*	Isopropanol. Isopropyl alcohol. $CH_3CH(OH)CH_3$	60.11	λ^{gas} 181 (2.79)	−89.5	82.4^{760}	0.7855^{20}_4	1.3776^{20}	∞	∞	∞	s	v	...	**B1^3**, 1439
Ω p1589	1-Propanol, 2-amino-(*dl*)*	$CH_3CH(NH_2)CH_2OH$	75.11	...	...	173–6 80^{18}	...	1.4502^{20}	v	v	v	...	...	...	**B4^3**, 736
Ω p1590	—,3-amino-*	$H_2NCH_2CH_2CH_2OH$	75.11	...	...	$187–8^{756}$ (cor)	0.9824^{26}_4	1.4617^{20}	s	s	s	...	...	...	**B4^3**, 738
Ω p1591	2-Propanol, 1-amino-(*dl*)*	$CH_3CH(OH)CH_2NH_2$	75.11	...	1.74	159.46^{760} 59.5^{10}	0.9611^{20}_4	1.4479^{20}	∞	∞	∞	∞	∞	CCl_4 ∞	**B4^3**, 754
p1592	—,—(*l*)*	$CH_3CH(OH)CH_2NH_2$	75.11	...	...	$156–8^{758}$	0.973^{18}	...	∞	∞	∞	∞	∞	...	**B4^2**, 736
p1593	1-Propanol, 2-amino-1(4-aminophenyl)-, dihydrochloride*	*p*-Aminoneoephedrine dihydrochloride.	239.15	lf (al-eth)	192–3d	...	...	...	v	s	i	...	...	...	**C27**, 2762
Ω p1594	2-Propanol, 1-amino-3-(diethylamino)-*	$H_2NCH_2CH(OH)CH_2N(C_2H_5)_2$	146.24	...	...	223 $116–8^{25}$	0.937^{20}_4	1.465^{20}	...	s	s	...	...	...	**B4^3**, 767
p1595	1-Propanol, 2-amino-2-methyl-*	$(CH_3)_2C(NH_2)CH_2OH$	89.14	...	25–6	165.5 $69–70^{10}$	0.934^{20}_4	1.449^{20}	∞	...	...	...	...	...	**C49**, 1537
—	—,2-amino-1-phenyl-*	*see* **Norephedrin**													

For explanations, symbols and abbreviations see beginning of table. For structural formulas see end of table.

No.	Name	Synonyms and Formula	Mol. wt.	Color, crystalline form, specific rotation and λ_{max} (log ε)	m.p. °C	b.p. °C	Density	n_D	Solubility w	al	eth	ace	bz	other solvents	Ref.
	2-Propanol														
p1596	**2-Propanol, 2-benzyl-**	$C_6H_5CH_2C(OH)(CH_3)_2$	150.22	nd	24	214–6 $103–5^{10}$	0.9790^{20}_{25}	1.5174^{20}	δ	s	...	...	...	...	**B6**[2], 489
p1597	**—,1,3-bis(dimethylamino)-***	$(CH_3)_2NCH_2CH(OH)CH_2N(CH_3)_2$	146.24	...	...	178–85 $79–81^{18}$	0.8788^{20}_{4}	1.4418^{20}	v	...	...	...	...	...	**B4**[3], 766
p1598	—,—,dimethiodide	Endoiosin. Iodisan. $HOCH[CH_2N(CH_3)_3I]_2$	430.14	wh cr	270–5d	...	...	...	v	δ	i	i	...	...	**C19**, 1757
p1599	**1-Propanol, 3-bromo-***	Trimethylene bromohydrin. $BrCH_2CH_2CH_2OH$	139.00	...	...	$98–112^{185}$ 62^5	1.5374^{20}_{4}	1.4834^{25}	s	∞	∞	...	...	...	**B1**[3], 1427
Ω p1600	**2-Propanol, 1-bromo-***	Propylene bromohydrin. $CH_3CH(OH)CH_2Br$	139.00	...	...	145–8 49.6^{12}	1.5585^{30}	1.4801^{20}	s	v	v	...	...	...	**B1**[3], 1474
Ω p1601	**—,1-butoxy-***	1,2-Propyleneglycol, 1-monobutyl ether. $CH_3CH(OH)CH_2O(CH_2)_3CH_3$	132.21	...	...	168–75	1.0035^{20}_{4}	1.4168^{20}	...	s	s	...	s	MeOH, CCl_4 s	**B1**[2], 537
p1602	**1-Propanol, 2-chloro-***	Propylene chlorohydrin. $CH_3CHClCH_2OH$	94.54	...	...	133–4	1.103^{20}	1.4390^{20}	s	s	s	...	...	os s	**B1**[3], 1424
Ω p1603	**—,3-chloro-***	Trimethylene chlorohydrin. $ClCH_2CH_2CH_2OH$	94.54	...	...	165 53^6	1.1309^{20}_{4}	1.4459^{20}	v	s	s	...	...	...	**B1**[3], 1425
Ω p1604	**2-Propanol, 1-chloro-***	Propylene chlorohydrin. $CH_3CH(OH)CH_2Cl$	94.54	...	...	$126–7^{750}$ (cor)	1.115^{20}_{20}	1.4392^{20}	∞	∞	∞	...	...	...	**B1**[3], 1469
Ω p1605	**—,1-chloro-3-isopropoxy-***	$ClCH_2CH(OH)CH_2OCH(CH_3)_2$	152.62	...	...	$87–7.5^{20}$	1.0530^{25}	1.4370^{25}	...	s	s	...	...	...	**B1**[3], 2153
p1606	**1-Propanol, 2-chloro-2-methyl-**	β-Isobutylene chlorohydrin. $(CH_3)_2CClCH_2OH$	108.57	visc	...	132–3δd $59–61^{50}$	1.0472^{20}_{4}	1.4388^{20}	d	...	...	...	...	con HCl s	**B1**[3], 1564
p1607	**—,3-chloro-2-methyl-***	$ClCH_2CH(CH_3)CH_2OH$	108.57	...	...	$76–8^{21}$	1.083^{25}_{4}	1.4460^{25}	...	v	v	...	...	...	**B1**[3], 1564
p1608	**2-Propanol, 1-chloro-2-methyl-***	$(CH_3)_2C(OH)CH_2Cl$	108.57	...	−20	128–9 71^{100}	1.0628^{20}_{4}	1.4380^{24}	v δd^h	v	...	...	...	...	**B1**[3], 1584
p1609	**—,1-chloro-3-propoxy-***	$CH_3CH_2CH_2OCH_2CH(OH)CH_2Cl$	152.63	...	...	$92–5^{15}$	1.0526^{25}_{4}	1.4378^{25}	...	...	s	s	...	...	**B1**[3], 2153
Ω p1610	**—,1,3-diamino-***	$H_2NCH_2CH(OH)CH_2NH_2$	90.13	cr	42–5	235 $93–5^2$	...	...	...	...	i	...	i	...	**B4**[2], 739
p1611	—,—,dihydrochloride	$H_2NCH_2CH(OH)CH_2NH_2 . 2HCl$	163.05	hyg pr or nd (dil al)	184.5	...	...	...	s	i	i	...	i	...	**B4**[2], 739
p1612	**1-Propanol, 2,3-dibromo-(*d*)***	$BrCH_2CHBrCH_2OH$	217.90	$[\alpha]_D$ +7.27	...	219δd	2.11	...	δ	∞	∞	∞	∞	...	**B1**[2], 371
Ω p1613	**—,—(*dl*)***	$BrCH_2CHBrCH_2OH$	217.90	...	...	219δd 118^{17}	2.0739^{20}_{4}	1.5466^{20}	δ	∞	∞	∞	∞	...	**B1**[3], 1428
p1614	**2-Propanol, 1,3-dibromo-***	$BrCH_2CH(OH)CH_2Br$	217.90	yesh liq	...	219δd 105^{16}	2.1202^{25}_{4}	1.5495^{25}	s	s	s	s	...	...	**B1**[3], 1475
p1615	**1-Propanol, 2,3-dichloro-***	$ClCH_2CHClCH_2OH$	128.99	visc	...	183–5 $70–80.5^{17}$	1.3607^{20}_{4}	1.4819^{20}	δ	∞	∞	∞	∞	lig δ	**B1**[3], 1426
p1617	**2-Propanol, 1,1-dichloro-***	$CH_3CH(OH)CHCl_2$	128.99	...	...	$146–8^{765}$	1.3334^{22}_{4}	...	δ	v	v	s	...	...	**B1**[2], 383
Ω p1618	**—,1,3-dichloro-***	$ClCH_2CH(OH)CH_2Cl$	128.99	...	...	176^{760} 69^{12}	1.3506^{17}_{4}	1.4837^{20}	v	v	∞	s	...	...	**B1**[3], 1471
p1620	**—,1,1-dichloro-2-methyl-***	$(CH_3)_2C(OH)CHCl_2$	143.02	...	8	150–5 52^{10}	1.2363^{19}_{4}	1.4598^{19}	...	v	...	...	v^h	...	**B1**[3], 1586
p1621	**—,1,3-dichloro-2-methyl-***	$(ClCH_2)_2C(OH)CH_3$	143.02	...	...	174–5 $55–6^{10}$	1.2745^{20}_{4}	1.4744^{21}	s	v	...	s	...	...	**B1**[3], 1586
p1622	**—,1(diethylamino)-***	$CH_3CH(OH)CH_2N(C_2H_5)_2$	131.22	...	...	$158–9^{756}$ (167–72) 63^{22}	0.8511^{20}_{0}	1.4255^{20}	...	s	...	s	...	...	**B4**[2], 737
Ω p1623	**1-Propanol, 2,3-dimercapto-***	BAL. British antilewisite. 1,2-Dithioglycerol. $HSCH_2CH(SH)CH_2OH$	124.24	visc liq	...	120^{15} 100^8	1.2463^{20}_{4}	1.5733^{20}	d	s	s	...	...	oils s	**B1**[3], 2340
p1624	**—,3(3,5-dimethoxy-4-hydroxyphenyl)-***	Hydrosinapyl alcohol. 2-Syringylethanol.	212.25	wh nd (eth-peth)	75.5–6.5	...	...	...	s	...	...	...	δ	lig i	**Am 70**, 57
Ω p1625	**—,2,2-dimethyl-***	*tert*-Butylcarbinol. Neopentyl alcohol. $(CH_3)_3CCH_2OH$	88.15	...	52–3	$113–4^{760}$	0.812	...	δ	v	v	...	...	...	**B1**[3], 1648
p1626	**—,2,2-dimethyl-1-phenyl-***	$C_6H_5CH(OH)C(CH_3)_3$	164.25	nd	45	$114–6^{16}$	...	...	i	s	s	...	...	os s	**B6**[1], 270
p1627	**—,2,2-dimethyl-3-phenyl-***	$C_6H_5CH_2C(CH_3)_2CH_2OH$	164.25	nd	34–5	$125–6^{14.5}$	...	...	i	s	s	...	...	os s	**B6**[2], 507
Ω p1628	**—,2,3-epoxy-***	Glycide. Glycidol. 3-Hydroxypropylene oxide.	74.08	...	...	166–7d $65–6^{2.5}$	1.1143^{25}	1.4287^{20}	∞	∞	s	s	s	chl s xyl, peth δ	**B17**[2], 104
p1629	**—,2,3-epoxy-3-phenyl-***	Phenyl glycidol	150.18	...	26.5	138^3	1.512^{27}	1.5432^{27}	...	s	s	...	...	CCl_4 s	**C26**, 3493
p1630	**—,2-ethoxy-***	$CH_3CH(OC_2H_5)CH_2OH$	104.15	...	...	$140–1^{760}$	0.9044^{20}_{4}	1.4122^{20}	s	s	s	...	...	...	**B1**[3], 2147
p1630[1]	**2-Propanol, 1-ethoxy-***	$CH_3CH(OH)CH_2OC_2H_5$	104.15	...	...	131^{760} (136)	0.9028^{20}_{4}	1.4075^{20}	s	s	s	...	...	...	**B1**[3], 2147
p1631	**1-Propanol, 3-fluoro-***	Trimethylene fluorohydrin. $FCH_2CH_2CH_2OH$	78.09	...	...	127.8	1.0390^{25}_{4}	1.3771^{25}	s	v	v	...	δ	...	**B1**[3], 1424
Ω p1632	**—,3(4-hydroxy-3-methoxyphenyl)-***	Hydroconiferyl alcohol.	182.22	λ^{al} 281.6 (3.45)	65	197^{15}	...	1.5545^{25}	...	s	s	...	...	...	**C22**, 3884 $_{13}$C

For explanations, symbols and abbreviations see beginning of table. For structural formulas see end of table.

PHYSICAL CONSTANTS OF ORGANIC COMPOUNDS (Continued)

No.	Name	Synonyms and Formula	Mol. wt.	Color. crystalline form. specific rotation and λ_{max} (log ε)	m.p. °C	b.p. °C	Density	n_D	Solubility						Ref.
									w	al	eth	ace	bz	other solvents	
	2-Propanol														
Ω $p1632^{3}$	**2-Propanol, 1-methoxy-***	$CH_3CH(OH)CH_2OCH_3$	90.12	....	....	118–8.5^{740}	0.9620^{20}_{4}	1.4034^{20}	...	...	...	...	...	....	**$B1^{3}$**, 2146
p1633	—,2-methyl-*	*tert*-Butyl alcohol. $(CH_3)_3COH$	74.12	rh pr or pl	25.5	82.2	0.7887^{20}_{4}	1.3878^{20}	∞	∞	∞	...	...	....	**$B1^{3}$**, 1568
$p1633^{1}$	1-Propanol, 2-methyl	iso-butyl alcohol $(CH_3)_2CHCH_2OH$	74.12		-108	108	0.7982^{25}	1.3939^{25}	15^{25}	∞	∞				
Ω p1634	**1-Propanol, 2-methyl-2-nitro-***	$(CH_3)_2C(NO_2)CH_2OH$	119.12	nd or pl (MeOH)	89.5–90	94.5–5.5^{10}	....	....	δ	v	v	...	...	....	**$B1^{3}$**, 1564
Ω p1635	—,2-methyl-1-phenyl-*	α-Isopropylbenzyl alcohol. $C_6H_5CH(OH)CH(CH_3)_2$	150.22	d:$[\alpha]^{20}_{D}$ +47.7 l:$[\alpha]^{20}_{D}$ −25.2	....	222–4 112–3^{15}	0.9869^{14}	1.5193^{14}	i	...	s	...	...	....	**$B6^{3}$**, 1859
p1636	**2-Propanol, 2-methyl-1-phenyl-***	$(CH_3)_2C(OH)CH_2C_6H_5$	150.22	nd	24	214–6 104–5^{17}	0.9840^{20}_{4}	1.5170^{20}	i	s	s	...	s	....	**$B6^{3}$**, 1860
Ω p1637	—,2-methyl-1,1,1-tribromo-*	Brometone. $(CH_3)_2C(OH)CBr_3$	310.83	nd (lig), cr (dil al)	168–70 (167–76)	sub	....	....	δ	s	s	...	...	....	**$B1^{3}$**, 1588
Ω p1638	—,2-methyl-1,1,1-trichloro-*	Chloretone. $(CH_3)_2C(OH)CCl_3$	177.46	hyg nd (w + 1)	98.5–9.5 77 (hyd)	167	....	....	i s^{h}	s	s	s	s	chl, lig s	**$B1^{3}$**, 1586
Ω p1639	—,—,hemihydrate*	$(CH_3)_2C(OH)CCl_3 \cdot \frac{1}{2}H_2O$ ..	186.47	wh	80–1	167	....	....	i s^{h}	v	s	s	s	chl, lig s	**$B1^{2}$**, 415
p1640	**1-Propanol, 2-nitro-***	$CH_3CH(NO_2)CH_2OH$	105.10	λ^{al} 275 (1.49)		120–2^{32} 100^{12}	1.1841^{25}_{4}	1.4379^{20}	s	s	s	...	...	....	**$B1^{3}$**, 1429
p1641	**2-Propanol, 3-nitro-1,1,1-trichloro-***	$O_2NCH_2CH(OH)CCl_3$	208.43	pr or pl (chl)	44.7–5.7	105.5–$6.5^{3.5}$	....	....	i	s	s	...	...	....	**$B1^{3}$**, 1477
p1642	**1-Propanol, 2-phenoxy-**	$CH_3CH(OC_6H_5)CH_2OH$ ...	152.20	....	....	244 124–6^{20}	0.9830^{25}_{25}	1.4760^{25}	...	s	s	...	...	....	**$B6^{3}$**, 577
$p1642^{1}$	—,3-phenoxy-.....	$C_6H_5OCH_2CH_2CH_2OH$....	152.20	oil	....	249–50^{764} 170^{60}	....	1.491^{20}	...	...	...	...	...	....	**$B6^{3}$**, 577
$p1642^{2}$	**2-Propanol, 1-phenoxy-**	$CH_3CH(OH)CH_2OC_6H_5$ ...	152.20	....	....	134.5^{20}	1.0622^{20}_{4}	1.5232^{20}	...	...	...	...	...	....	**$B6^{1}$**, 85
p1643	**1-Propanol, 1-phenyl-(*dl*)***	$CH_3CH_2CH(OH)C_6H_5$....	136.20	....	....	213–5^{740} 98^{10}	0.9938^{23}_{4}	1.5210^{23}	i	s	s	...	...	....	**$B6^{3}$**, 1793
Ω p1644	—,3-phenyl-*	$C_6H_5CH_2CH_2CH_2OH$	136.20	....	< −18	236–7^{750} 132^{21}	1.008^{20}_{4}	1.5278^{20}	s	∞	∞	...	...	....	**$B6^{3}$**, 1800
Ω p1645	**2-Propanol, 2-phenyl-***	$C_6H_5C(OH)(CH_3)_2$	138.20	pr	35–7	202 93^{13}	0.9735^{20}_{4}	1.5325^{20}	i	s	s	...	s	aa s	**$B6^{3}$**, 1813
p1646	—,1,1,1,3-tetra-chloro-*	$ClCH_2CH(OH)CCl_3$	197.88	....	....	95–6^{17}	1.610^{20}_{4}	1.5145^{20}	...	...	...	...	...	....	**$B1^{3}$**, 1474
p1647	—,1,1,3,3-tetra-chloro-*	$Cl_2CHCH(OH)CHCl_2$	197.88	....	....	80–90^{14}	1.612^{20}_{4}	1.5133^{20}	...	...	s	...	...	....	**$B1^{3}$**, 1474
Ω p1648	—,1,1,1-tri-chloro-*	Isopral. $CH_3CH(OH)CCl_3$ ..	163.43	....	50–1	161.8^{773} 53–5^{12}	....	....	δ	v	v	s	s	oos s	**$B1^{3}$**, 1474
Ω p1649	**2-Propanone***	Acetone. Dimethyl ketone. CH_3COCH_3	58.08	....	−95.35	56.2^{760}	0.7899^{20}_{4}	1.3588^{20}	∞	∞	∞	∞	∞	chl ∞	**$B1^{3}$**, 2696
p1650	—,azine*	Acetone azine. $(CH_3)_2C:NN:C(CH_3)_2$	112.18	λ^{w} 217 (3.47)	−12.5	133 54^{37}	0.8389^{20}	1.4535^{20}	∞	∞	∞	s	...	min ac d	**$B1^{3}$**, 2745
Ω p1651	—,diethyl acetal* ..	2,2-Diethoxypropane*. $(CH_3)_2C(OC_2H_5)_2$	132.21	....	....	114^{760} 46^{60}	0.8200^{21}_{4}	1.3891^{20}	...	s	v	s	s	....	**$B1^{3}$**, 2741
Ω p1652	—,2,4-dinitro-phenylhydrazone*.	Acetone DNP............	238.21	ye nd or pl (al) λ^{iso} 344 (4.23)	128	....	....	....	i	s^{h}	s^{h}	...	s	AcOEt, chl s	**$B15^{2}$**, 216
p1653	—,4-nitrophenyl-hydrazone*	....	193.21	og-br or og-ye nd or lf (al)	152 (149)	....	....	....	δ^{h}	s^{h}	s^{h}	...	...	aa δ	**$B15^{2}$**, 184
p1654	—,oxime*	Acetoxime. $(CH_3)_2C:NOH$..	73.10	pr (al) λ^{al} 190 (3.70)	61	134.8^{728} 61^{20}	0.9113^{62}_{4}	1.4156^{20}	s	s	s	...	...	lig s	**$B1^{3}$**, 2743
Ω p1655	—,phenyl-hydrazone*	$C_6H_5NHN:C(CH_3)_2$	148.21	rh	42	163^{50} 140^{16}	....	....	...	s	s	...	...	dil ac s	**$B15^{2}$**, 55
p1656	—,—,hydrate*	$C_6H_5NHN:C(CH_3)_2 \cdot H_2O$..	166.23	....	35	....	....	....	δ	s	s	...	...	....	**$B15^{1}$**, 30
Ω p1657	—,semicarbazone*	Acetone semicarbazone. $(CH_3)_2C:NNHCONH_2$	115.14	nd (w, ace)	190–1d	....	....	....	δ s^{h}	v	s	s^{h}	...	....	**$B3^{2}$**, 81
p1658	**1-Propanone, 1(4-acetamido-phenyl)-***	$C_{11}H_{13}NO_2$. *See* p1696	191.23	pa ye nd (w)	172–3	....	....	....	s	v	v	...	...	....	**$B14^{1}$**, 375
p1659	**2-Propanone, l-amino-, hydrochloride***	Acetonylamine hydrochloride $CH_3COCH_2NH_2 \cdot HCl$	109.56	hyg pl (al-eth)	75	....	....	....	v	s	s	...	...	....	**B4**, 314
p1660	**1-Propanone, 1(2-amino-phenyl)-***	*o*-Aminopropiophenone. $C_9H_{11}NO$. *See* p1696	149.19	pa ye lf (peth), pl (dil al)	46–7	$93^{0.8}$	....	....	s	s	s	s	...	ac, os s chl i	**$B14^{2}$**, 37
p1661	**2-Propanone, 1(4-amino-phenyl)-***	*p*-Aminopropiophenone. $C_9H_{11}NO$. *See* p1696	149.19	pl (al, w), nd (w)	140	....	....	....	s	s	δ	...	...	chl s	**$B14^{1}$**, 375
p1662	—,2-amino-1-phenyl-, hycrochloride	α-Aminopropiophenone hydrochloride. $C_6H_5COCH(CH_3)NH_2 \cdot HCl$	185.66	nd (al-eth)	187	....	....	....	δ	δ	i	...	...	....	**$B14^{2}$**, 37
Ω p1663	**2-Propanone, 1-bromo-***	Bromoacetone. CH_3COCH_2Br	136.98	....	−36.5	136.5^{725} 31.5^{8}	1.634^{23}	1.4697^{15}	δ	s	s	s	...	....	**$B1^{3}$**, 2752
p1664	**1-Propanone, 1(4-bromo-1-hydroxy-2-naphthyl)-***	....	279.14	ye nd (al)	98	....	....	....	i	s	s	...	...	....	**B8**, 152

For explanations, symbols and abbreviations see beginning of table. For structural formulas see end of table.

No.	Name	Synonyms and Formula	Mol. wt.	Color, crystalline form, specific rotation and λ_{max} (log ε)	m.p. °C	b.p. °C	Density	n_D	w	al	eth	ace	bz	other solvents	Ref.
	1-Propanone														
p1665	**—,2-bromo-2-methyl-1(2,4,6-trimethylphenyl)-***	α-Bromoisobutyrylmesitylene.	269.19	gold-ye oil	27	$160–70^{24}$	....	....	...	...	s	...	...	....	**Am 52,** 5036
Ω p1666	**—,1(4-bromophenyl)-***	*p*-Bromopropiophenone. C_9H_9BrO. *See* p1696	213.08	nd	48	169^{15}	....	....	i	s	s	s	...	CS_2 s	**B7**, 302
Ω p1667	**—,2-bromo-1-phenyl-***	α-Bromopropiophenone. $C_6H_5COCHBrCH_3$	213.08	ye	....	245–50 $134–5^{18}$	1.4298^{20}_4	1.5720^{20}	i	s	s	s	s	....	**B7**^{2}, 233
Ω p1668	**2-Propanone, 1-chloro-***	Acetonyl chloride. Chloroacetone. CH_3COCH_2Cl	92.53	λ^{cy} 292 (1.49)	−44.5	119^{763}	1.15^{20}	....	s	s	s	...	...	chl s	**B1**^{3}, 2746
Ω p1669	**1-Propanone, 1(4-chlorophenyl)-***	*p*-Chloropropiophenone. C_9H_9ClO. *See* p1696	168.63	....	36–7	$134–7^{31}$ $114–8^2$	....	....	i	s	...	...	...	CS_2 s	**B7**, 301
Ω p1670	**—,—,oxime***	$C_9H_{10}ClNO$. *See* p1696	183.64	pl (al)	62–3	....	....	....	i	s^h	...	...	...	....	**B7**, 301
p1671	**—,3-chloro-1-phenyl-***	ω-Chloropropiophenone. $C_6H_5COCH_2CH_2Cl$	168.63	lf (eth), cr (peth, al)	49–50	....	....	....	...	s^h	s^h	...	...	peth s^h	**B7**^{2}, 233
p1672	**2-Propanone, 1-chloro-3-phenyl-***	$C_6H_5CH_2COCH_2Cl$	168.63	nd (chl)	72–3	$159–61^{17}$	....	....	...	...	...	...	...	chl s^h	**B7**^{2}, 235
p1673	**—,1,3-diamino-, dihydrochloride***	$(H_2NCH_2)_2CO.2HCl$	161.03	(dil al, dil aa), pr (w + 1 or 1½)	180d	....	....	....	v	i	i	...	i	chl δ aa i	**B4**^{2}, 763
p1674	**1-Propanone, 2,3-dibromo-1,3-diphenyl-* (one form)**	*threo*-Chalcone dibromide. $C_6H_5CHBrCHBrCOC_6H_5$	386.08	nd (al) λ^{iso} 253 (4.17)	122–3	....	....	....	δ	s	...	...	...	....	**B7**^{2}, 381
p1675	**—,—(one form)***	*erythro*-Chalcone dibromide. $C_6H_5CHBrCHBrCOC_6H_5$	386.08	pr or nd (al) λ^{iso} 253 (4.14)	159–60	....	....	....	...	δ v^h	...	...	...	....	**B7**^{2}, 381
p1676	**2-Propanone, 1,1-dichloro-***	*asym*-Dichloroacetone. $CH_3COCHCl_2$	126.97	λ^{diox} 294 (1.8)	....	120 47^{76}	1.305^{18}_{15}	....	δ	s	∞	...	...	....	**B1**^{3}, 2749
Ω p1677	**—,1,3-dichloro-***	*sym*-Dichloroacetone. $ClCH_2COCH_2Cl$	126.97	pl or nd λ^{cy} 300 (1.62)	45	173.4 $86–8^{12}$	1.3826^{46}_4	1.4714^{46}	s	s	s	...	...	....	**B1**^{3}, 2749
p1678	**—,1(diethylamino)-***	$CH_3COCH_2N(C_2H_5)_2$	129.20	....	....	155–6δd 64^{16}	0.8620^{20}_4	1.4249^{20}	∞	∞	∞	...	...	....	**B4**^{3}, 877
Ω p1679	**—,1,3-dihydroxy-***	Dihydroxyacetone. $HOCH_2COCH_2OH$	90.08	λ^w 270 (1.35)	89–91.5 (80)	....	....	....	s^h	s^h	s^h	s	...	lig i	**B1**^{3}, 3292
p1680	**1-Propanone, 1(2,4-dihydroxyphenyl)-***	4-Propionylresorcinol. $C_9H_{10}O_3$. *See* p1696	166.18	ye nd (al)	97 (anh) 56 (hyd)	$176–8^6$	....	....	δ	v	s	...	s	aa s peth, chl, CCl_4 δ	**B8**^{2}, 305
p1681	**—,2,2-dimethyl-1-phenyl-***	*tert*-Butyl phenyl ketone. Pivalophenone. $C_6H_5COC(CH_3)_3$	162.23	λ^{hx} 263 (3.91)	....	$219–21^{760}$ $97–8^{16}$	0.963^{26}	1.5086^{19}	...	...	...	s	...	....	**B7**^{2}, 253
Ω p1682	**2-Propanone, 1,3-diphenyl-***	Dibenzyl ketone. $C_6H_5CH_2COCH_2C_6H_5$	210.28	cr (al, peth, eth) λ^{hx} 294 (2.34)	35	331 $112–25^{0.1}$	1.195^0_4	....	i	s	s	...	...	peth s^h	**B7**^{2}, 382
Ω p1683	**1-Propanone, 1(2-furyl)-**	2-Propionylfuran.	124.15	cr λ^{al} 270 (4.0), 315 (2.0)	28	88^{14}	1.0626^{28}	1.4922^{25}	...	...	s	...	...	....	**B17**^{1}, 157
Ω p1684	**2-Propanone, hexachloro-***	Perchloroacetone. $Cl_3CCOCCl_3$	264.75	....	−2 15 (hyd)	202–4 110^{40}	1.7444^{12}_{12}	1.5112^{20}	δ d^h	...	...	s	...	....	**B1**^{3}, 2751
p1685	**—,1-hydroxy-***	Acetol. CH_3COCH_2OH	74.08	λ^{al} 269 (1.67)	−17	145–6d 54^{18}	1.0824^{20}_{20}	1.4295^{20}	v	v	v	...	...	....	**B1**^{3}, 3191
p1686	**1-Propanone, 1(1-hydroxy-2-naphthyl)-***	....	200.26	grsh ye lf or pl (al)	81	....	....	....	i	s	s	...	...	....	**B8**^{2}, 179
Ω p1687	**—,1(2-hydroxyphenyl)-***	*o*-Propionylphenol. $C_9H_{10}O_2$. *See* p1696	150.18	λ^{al} 250 (3.9), 325 (3.5)	....	150^{80} 115^{15}	....	1.5501^{20}	δ	s	s	...	...	alk s	**B8**^{2}, 103
Ω p1688	**—,1(4-hydroxyphenyl)-***	*p*-Propionylphenol. $C_9H_{10}O_2$. *See* p1696	150.18	wh nd or pr (w) λ^{al} 220 (4.0), 277 (4.1)	149	....	....	....	δ v^h	s	s	...	...	alk s	**B8**^{2}, 104
p1689	**2-Propanone, 1-iodo-***	Iodoacetone. CH_3COCH_2I	183.98	yesh liq λ^{hx} 268 (2.6)	....	62^{12}	2.17^{15}	....	...	s	...	...	...	....	**B1**^{3}, 2753
p1690	**—,—,oxime***	Iodoacetoxime. $CH_3C(:NOH)CH_2I$	198.99	pr (peth)	64.5	....	....	....	...	...	...	...	...	peth s^h	**B1**, 660
p1691	**—,1(3-methoxyphenyl)-***	....	164.21	λ^{al} 250 (3.8), 310 (3.3)	....	258–60 $95–7^{0.7}$	1.0812^0	1.5230^{25}	...	s	s	...	...	....	**B8**, 106
p1692	**—,1(4-methoxyphenyl)-***	....	164.21	λ^{al} 270 (4.2)	< −15	267–9 142^{14}	1.0670^{18}_4	1.5253^{20}	δ	s	s	...	...	....	**B8**^{2}, 104
Ω p1693	**1-Propanone, 2-methyl-1-phenyl-***	Isobutyrophenone. $(CH_3)_2CHCOC_6H_5$	148.21	λ^{al} 242 (4.11), 275 (3.09)	....	221^{760} 86^4	0.9863^{17}_4	1.5172^{20}	i	s	s	...	...	cy s	**B7**^{2}, 245
p1694	**2-Propanone, pentachloro-***	$Cl_2CHCOCCl_3$	230.31	cr (w + 4)	2.1 (anh) 15–7 (hyd)	192^{753} $97.5–8.5^{40}$	1.69^{15}_{15}	....	s	...	...	v	...	....	**B1**^{3}, 2751
p1695	**—,1-phenoxy-***	$CH_3COCH_2OC_6H_5$	150.18	....	....	229–30 $117–9^{20}$	1.0903^{20}_4	1.5228^{20}	...	...	s	s	...	....	**B6**^{3}, 589
Ω p1696	**1-Propanone, 1-phenyl-***	Propiophenone.	134.18	λ^{hx} 238 (4.06), 277 (2.94)	18.61	217.48^{760} 91.6^{10}	1.0096^{20}	1.5269^{20}	i	s	s	...	...	....	**B7**^{2}, 231
p1697	**—,—,oxime***	$C_6H_5C(:NOH)CH_2CH_3$	149.19	pl (peth) λ^{al} 245 (4.00)	53–5	245–6d 165^{38}	....	....	i	s	s	...	...	....	**B7**^{2}, 232

For explanations, symbols and abbreviations see beginning of table. For structural formulas see end of table.

No.	Name	Synonyms and Formula	Mol. wt.	Color, crystalline form, specific rotation and λ_{max} (log ε)	m.p. °C	b.p. °C	Density	n_D	Solubility						Ref.
									w	al	eth	ace	bz	other solvents	
	2-Propanone														
Ωp1698	**2-Propanone, 1-phenyl-***	Acetonylbenzene. Phenylacetone. $CH_3COCH_2C_6H_5$	134.18	λ^{al} 285 (2.18)	−15	216.5, 101^{14}	1.0157^{20}_{4}	1.5168^{20}	i	v	v	...	∞	xyl ∞	**B7**[2], 233
p1699	**—,1,1,1,3-tetra-chloro-***	*asym*-Tetrachloroacetone. $ClCH_2COCCl_3$	195.86	liq (anh), pr (w+4), pl (w+2)	46 (+4w), 65 (+2w)	183^{760}, $71\text{–}2^{13}$	1.624^{15}_{4}	1.497^{18}	δ	...	v	v	...		**B1**[3], 2751
p1700	**—,1,1,3,3-tetra-chloro-***	*sym*-Tetrachloroacetone. $Cl_2CHCOCHCl_2$	195.86			$180\text{–}2^{718}$ (cor)			v	v	v	v	v		**B1**[3], 2751
p1701	**—,—,tetrahydrate***	$Cl_2CHCOCHCl_2 \cdot 4H_2O$	267.92	tcl pl (w)	48–9				s	...	...	...	...		**B1**, 656
Ωp1702	**1-Propanone, 1(3-tolyl)-***		148.21	λ^{al} 245 (4.0), 285 (3.0)		234^{745}, 130–5^{33}	1.0059^{0}_{4}		i	s	s	s	s		**B7**[2], 246
p1703	**—,1(4-tolyl)-***		148.21	λ^{al} 255 (4.0), 315 (1.7)		$238\text{–}9^{760}$, 120^{18}	0.9926^{20}_{4}	1.5278^{20}	i	s	s	s	s	CS_2 s	**B7**[2], 246
p1704	**2-Propanone, 1,1,1-trichloro-***	CH_3COCCl_3	161.42			149^{764} (134) 28^{10}	1.435^{20}_{4}	1.4633^{17}	i	v	v				**B1**[3], 2751
Ωp1705	**1-Propanone, 1,3,3-triphenyl-***	$(C_6H_5)_2CHCH_2COC_6H_5$	286.38	nd (al) λ^{al} 252.5 (4.22), 257 (4.22)	96				...	δ v^h	δ	v	v	chl v lig δ	**B7**[2], 485
p1706	**—,1-ureido-***	Propionylurea. $CH_3CH_2CONHCONH_2$	116.13	cr (w)	210–11				s	s^h	...	...	...		**C37**, 5934
—	**Propargyl alcohol**	see **2-Propyn-1-ol***													
—	**Propargyl aldehyde**	see **Propynal***													
Ωp1707	**Propenal***	Acrolein. Acrylaldehyde. $CH_2{:}CHCHO$	56.07	λ^{al} 207 (4.05)	−86.95	52.5–3.5	0.8410^{20}_{4}	1.4017^{20}	v	s	s	s	...		**B1**[3], 2953
—	**—,diacetate**	see **2-Propene-1,1-diol, diacetate**													
Ωp1708	**—,diethyl acetal**	3,3-Diethoxy-1-propene*. $CH_2{:}CHCH(OC_2H_5)_2$	130.19			123.5^{760}	0.8543^{15}	1.4000^{20}	δ	∞	∞	...	...		**B1**[3], 2960
p1709	**—,2-chloro-***	$CH_2{:}CClCHO$	90.51			40^{30}	1.199^{20}	1.463^{20}	...	...	v	...	...	CCl_4 v	**B1**[3], 2960
Ωp1711	**—,3(2-furyl)-***	Furacrolein	122.13	ye or wh nd (lig) λ^{diox} 312 (4.42)	54	>200d, 135^{14}			i s^h	∞	s	...	...		**B17**[2], 325
Ωp1712	**—,2-methyl-***	Methacrolein. $CH_2{:}C(CH_3)CHO$	70.09	λ^{al} 216 (4.04)		68.4	0.837^{20}_{4}	1.4144^{20}	∞	∞	∞	...	...		**B1**[3], 2981
—	**—,3-phenyl-***	see **Cinnamaldehyde**													
Ωp1713	**Propene***	Propylene. $CH_3CH{:}CH_2$	42.08	gas	−185.25	-47.4^{760}	0.5193^{20}_{4} (liq at sat pressure)	1.3567^{-70}	v	v	...	...	...	aa v	**B1**[3], 677
Ωp1714	**—,3-amino-***	Allylamine. $H_2NCH_2CH{:}CH_2$	57.09			58	0.7621^{20}_{4}	1.4205^{20}	∞	∞	∞	...	...	chl s	**B4**[3], 444
Ωp1715	**—,1-bromo-(*cis*)***	Propenyl bromide. $CH_3CH{:}CHBr$	120.98		−113	57.8^{760}	1.4291^{20}_{4}	1.4560^{20}	i	...	s	s	...	chl s	**B1**[3], 710
Ωp1716	**—,2-bromo-***	Isopropenyl bromide. $CH_3CBr{:}CH_2$	120.98		−124.8	48.4^{748}	1.362^{20}_{4}	1.4440^{20}	i	...	s	s	...	chl s	**B1**[3], 710
Ωp1717	**—,3-bromo-***	Allyl bromide. $BrCH_2CH{:}CH_2$	120.98		−119.4	70^{753}	1.398^{20}_{4}	1.4697^{20}	i	∞	∞	...	...	chl, CS_2, CCl_4 s	**B1**[3], 711
p1718	**—,2-bromo-3-cyclohexyl-***		203.13			90^{15}	1.215^{17}	1.495^{17}	...	...	s	...	...		**B5**[3], 224
p1719	**—,1-chloro-(*cis*)***	*cis*-Propenyl chloride. $CH_3CH{:}CHCl$	76.53		−134.8	32.8^{760}	0.9347^{20}_{4}	1.4055^{20}	i	...	s	s	s	chl s	**B1**[3], 697
p1720	**—,—(*trans*)***	*trans*-Propenyl chloride. $CH_3CH{:}CHCl$	76.53		−99	37.4^{760}	0.9350^{20}_{7}	1.4054^{20}	i	...	s	s	s	chl s	**B1**[3], 697
p1721	**—,2-chloro-***	Isopropenyl chloride. $CH_3CCl{:}CH_2$	76.53		−137.4	22.65^{760}	0.9017^{20}_{4}	1.3973^{20}	i	...	s	s	s	chl s	**B1**[3], 698
Ωp1722	**—,3-chloro-***	Allyl chloride. $ClCH_2CH{:}CH_2$	76.53	λ 230	−134.5	45^{760}	0.9376^{20}_{4}	1.4157^{20}	i	∞	∞	∞	∞	lig ∞	**B1**[3], 699
p1723	**—,3-chloro-2-(chloromethyl)-***	$(ClCH_2)_2C{:}CH_2$	125.00		−15 (−13)	138–8.5, $30\text{–}1^{9}$	1.1782^{20}_{3}	1.4754^{20}	...	v	...	...	...	chl v	**B1**[3], 768
p1724	**—,1-chloro-2-methyl-***	Isocrotyl chloride. $(CH_3)_2C{:}CHCl$	90.55			68^{754}	0.9186^{20}_{4}	1.4221^{20}	...	∞	∞	s	...	chl v	**B1**[3], 765
Ωp1725	**—,3-chloro-2-methyl-***	Isobutenyl chloride. $ClCH_2C(CH_3){:}CH_2$	90.55			71.5–2.5	0.9165^{20}_{4}	1.4291^{20}	...	∞	∞	s	...	chl v	**B1**[3], 765
p1726	**—,1-chloro-1-phenyl-***	$C_6H_5CCl{:}CHCH_3$	152.63	λ^{cy} 253 (3.28)		$90.5^{0.9}$	1.085^{20}_{4}	1.5635^{15}	i	...	...	s	s	to s	**B5**[2], 372
p1727	**—,1-chloro-3-phenyl-***	$C_6H_5CH_2CH{:}CHCl$	152.63	λ^{hx} 253 (3.48)		$212\text{–}4^{760}$ (cor) 97^{18}	1.073^{14}_{4}	1.545^{14}	i	...	s	s	s	to s	**B5**[3], 1191
p1728	**—,2-chloro-1-phenyl-***	$CH_3CCl{:}CHC_6H_5$	152.63			$118\text{–}23^{28}$, $61.5\text{–}2.5^{2}$	1.0738^{19}_{4}	1.5565^{19}	i	...	...	s	s	chl v	**B5**[3], 1186
p1729	**—,3-chloro-1-phenyl-(*trans*)***	*trans*-Cinnamyl chloride. $ClCH_2CH{:}CHC_6H_5$	152.63	nd λ^{cy} 253 (4.31)	8–9	106.7^{13}	1.0926^{20}_{4}	1.5851^{20}	i	∞	∞	v	s		**B5**[3], 1186
Ωp1730	**—,2,3-dibromo-***	α-Epidibromohydrin. $BrCH_2CBr{:}CH_2$	199.88			$141\text{–}1.5^{760}$, $37\text{–}7^{11}$	2.0346^{25}_{4}	1.5416^{25}	i	...	s	s	...	chl s	**B1**[3], 713
Ωp1731	**—,1,1-dichloro-***	$CH_3CH{:}CCl_2$	110.97			$76\text{–}7^{760}$ (78)	1.1864^{25}_{4}	1.4430^{25}	i	...	s	s	...	chl s	**B1**[3], 704
Ωp1732	**—,1,2-dichloro-(*trans*)***	$CH_3CCl{:}CHCl$	110.97			77.0^{757}	1.1818^{20}_{4}	1.4471^{20}	i	v	...	...	...	MeOH, CCl_4 v	**B1**[3], 704

For explanations, symbols and abbreviations see beginning of table. For structural formulas see end of table.

No.	Name	Synonyms and Formula	Mol. wt.	Color, crystalline form, specific rotation and λ_{max} (log ε)	m.p. °C	b.p. °C	Density	n_D	Solubility w	al	eth	ace	bz	other solvents	Ref.
	Propene														
Ωp1733	—,**1,3-dichloro-** (*cis*)*	$ClCH_2CH:CHCl$	110.97	...	...	104.3	1.217_4^{20}	1.4730^{20}	i	...	s	...	s	chl s	**B1**[3], 704
p1734	—,—(*trans*)*	$ClCH_2CH:CHCl$	110.97	...	...	112	1.224_4^{20}	1.4682^{20}	i	...	s	...	s	chl s	**B1**[3], 705
Ωp1735	—,**2,3-dichloro-***	$ClCH_2CCl:CH_2$	110.97	...	...	94^{760}	1.211_4^{20}	1.4603^{20}	i	∞	s	...	s	chl s	**B1**[3], 705
p1736	—,**3,3-dichloro-***	Acrolein dichloride. $Cl_2CHCH:CH_2$	110.97	...	...	84.4 (cor)	1.175_4^{20}	1.4510^{20}	i	v	v	...	s	chl s	**B1**[3], 706
p1737	—,**1,1-dichloro-2-methyl-***	$(CH_3)_2C:CCl_2$	125.00	...	...	108.7–9.1^{760} 42–3^{75}	1.1449_6^{20}	1.4580^{20}	i	...	s	...	s	chl s	**B1**[3], 767
p1738	—,**3,3-dichloro-2-methyl-***	$Cl_2CHC(CH_3):CH_2$	125.00	...	...	108–12^{762} 49–50^{120}	1.1363_4^{24}	1.4523_α^{24}	i	...	s	...	s	chl v	**B1**[3], 767
p1739	—,**1,1-dichloro-3-phenyl-***	Cinnamylidene chloride. $C_6H_5CH:CHCHCl_2$	187.08	cr (eth, chl), pl (peth) λ^{al} 258 (4.28), 295 sh (3.30)	59	142–3^{30} 121–2^{14}	...	...	d	...	s	...	s	chl s peth s^h	**B5**[3], 1188
—	**3,3-diethoxy-***	*see* **Propenal**, diethyl acetal													
p1741	—,**1,1-diphenyl-***	$CH_3CH:C(C_6H_5)_2$	194.28	lf (al) λ^{al} 235 (3.87), 248 (3.94)	52	280–1 149^{11}	1.0250_4^{20}	1.5880^{20}	i	s	...	...	s	...	**B5**[3], 1998
p1742	—,**1,2-epoxy-***	Allylene oxide. Methyloxirene.	56.07	...	...	63	...	...	δ	∞	∞	...	...	...	**B17**, 20
p1743	—,**3-fluoro-***	Allyl fluoride. $FCH_2CH:CH_2$	60.07	gas	...	−3	...	...	δ	v	v	...	...	chl s	**B1**[3], 696
Ωp1744	—,**hexafluoro-***	Perfluoropropene. $CF_3CF:CF_2$	150.02	gas	−156.2	−29.4	1.583_4^{-40}	...	...	...	...	...	...	...	**B1**[3], 697
p1745	—,**3-iodo-***	Allyl iodide. $ICH_2CH:CH_2$	167.98	pa ye	−99.3	102^{760}	1.8494_4^{20}	1.5530^{20}	i	s	s	...	...	chl s	**B1**[3], 714
p1746	—,**3-isocyano-**	Allyl carbylamine. Allyl isocyanide. $CNCH_2CH:CH_2$	67.09	...	...	98	0.794_4^{17}	...	δ	∞	∞	...	...	...	**B4**, 208
Ωp1747	—,**2-methyl-***	Isobutylene. $(CH_3)_2C:CH_2$	56.11	gas λ^{gas} 159 (3.9), 184 (4.1), 188 (4.1), 192 sh (3.9), 200 sh (3.9)	−140.35	−6.9	0.5942_4^{20} (liq)	1.3926^{-25}	i	v	v	...	s	sulf, peth s	**B1**[3], 749
p1748	—,—(trimer)	Triisobutylene. $(C_4H_8)_3$	168.33	...	−76	179–81 56^{10}	0.7590_4^{20}	1.4314^{20}	...	...	...	...	...	...	**B1**[3], 762
p1749	—,—(tetramer)	Tetraisobutylene. $(C_4H_8)_4$	224.44	...	−98	243–6 109.5^{15}	0.7944_4^{20}	1.4482^{20}	...	...	...	...	...	...	**B1**[3], 763
p1750	—,**2-methyl-1,1,3-trichloro-***	$ClCH_2C(CH_3):CCl_2$	159.44	...	...	156 45–6^{12}	1.346^{20}	1.4990^{20}	i	...	...	v	s	chl v	**B1**[3], 768
p1751	—,**2-methyl-3,3,3-trichloro-***	$Cl_3CC(CH_3):CH_2$	159.44	...	...	132–4	1.293^{20}	1.4770^{20}	i	...	...	s	s	chl v	**B1**[3], 768
p1752	—,**2-nitro-1-phenyl-***	$CH_3C(NO_2):CHC_6H_5$	163.18	ye nd (peth) $\lambda^{95\%\ al}$ 223 (3.90), 276 (3.80), 283 (3.89), 291 (3.96), 307 (4.02)	65–6	...	...	...	i	v	δ	...	...	aa s peth s^h alk i	**B5**[3], 1189
p1753	—**1,1,2,3,3-pentachloro-***	$Cl_2CHCCl:CCl_2$	214.31	...	...	185^{760} 116^9	1.6317_4^{34}	1.5313^{20}	i	...	s	...	...	to v^h	**B1**[3], 708
Ωp1754	—,**2-phenyl-***	α-Methylstyrene. $C_6H_5C(CH_3):CH_2$	118.18	λ^{al} 242 (4.08)	24.5	163.4^{760} 60^{17}	0.9082_4^{20}	1.5303^{20}	i	...	s	...	s	chl s	**B5**[3], 1192
—	—,**3-phenyl-***	*see* **Benzene, allyl-***													
p1755	—,**1,1,2-trichloro-***	$CH_3CCl:CCl_2$	145.42	...	...	118 41^{52}	1.382_4^{20}	1.4827^{20}	i	s	s	...	s	chl s	**B1**[3], 707
p1756	—,**1,2,3-trichloro-***	$ClCH_2CCl:CHCl$	145.42	...	...	142 (145) 32–3^{14}	1.414_{20}^{20}	1.5020^{20}	i	v	v	...	s	chl s	**B1**[3], 707
p1757	—,**3,3,3-trichloro-***	$Cl_3CCH:CH_2$	145.42	...	−30	114–5 57^{103}	1.369_{20}^{20}	1.4827^{20}	i	s	s	...	s	chl s	**B1**[3], 707
p1758	**2-Propene-1-arsonic acid***	Allylarsonic acid. $CH_2:CHCH_2AsO_3H_2$	166.01	nd (w), pr (al)	130–1	...	...	...	v^h	v^h	...	...	...	...	**B4**[3], 1826
Ωp1758[1]	**2-Propene-1,1-diol,** diacetate	Allylidene diacetate. $CH_2:CHCH(O_2CCH_3)_2$	158.16	...	−37.6	180^{760} 76^{13}	1.0760_4^{20}	1.4193^{20}	δ	∞	∞	...	∞	lig ∞	**B2**[3], 356
Ωp1759	**2-Propene-1-thiol***	Allyl mercaptan. $CH_2:CHCH_2SH$	74.15	...	...	67–8	0.925_4^{23}	1.4832^{20}	i	∞	∞	...	...	chl s	**B1**[2], 478
p1760	**1,2,3-Propenetricarboxylic acid-** (*cis*)*	*cis*-Aconitic acid. $HO_2CCH_2C(CO_2H):CHCO_2H$	174.11	nd (w) λ^w 389 (2.69), 411 (2.78)	130 (125)	...	...	...	s	...	δ	...	...	...	**B2**[3], 2065
Ωp1761	—(*trans*)*	$HO_2CCH_2C(CO_2H):CHCO_2H$	174.11	lf (w), nd (w, eth) λ^w 411 (2.88), 432 (2.88)	198–9 (195 cor)	...	...	...	v	s	δ	...	...	...	**B2**[3], 2063
p1762	—,triamide (*trans*)*	Aconitamide. $H_2NOCCH_2C(CONH_2):CHCONH_2$	171.16	ye nd (w)	sinters 260	...	...	...	v^h	i	i	...	...	chl i	**B2**, 853
p1763	—,triethyl ester (*trans*)*	$C_2H_5O_2CCH_2C(CO_2C_2H_5):CHCO_2C_2H_5$	258.27		...	275d 159^9	1.1064_4^{20}	1.4556^{20}	...	v	s	...	...	...	**B2**[3], 2065
p1764	—,trimethyl ester (*trans*)*	$CH_3O_2CCH_2C(CO_2CH_3):CHCO_2CH_3$	216.19		...	270–1^{760} 161^{14}	...	...	i	s	s	...	...	...	**B2**[3], 2064
p1765	—,tripropyl ester (*trans*)*	$C_3H_7^nO_2CCH_2C(CO_2C_3H_7^n):CHCO_2C_3H_7^n$	300.36		...	195^{13}	1.050_4^{25}	1.4521^{25}	i	s	s	...	...	...	**B2**[3], 2065
Ωp1766	**Propenoic acid***	Acrylic acid. $CH_2:CHCO_2H$	72.06	λ^{MeOH} 252 (1.96)	13	141.6^{760} 48.5^{15}	1.0511_4^{20}	1.4224^{20}	∞	∞	∞	s	s	...	**B2**[3], 1215

For explanations, symbols and abbreviations see beginning of table. For structural formulas see end of table.

No.	Name	Synonyms and Formula	Mol. wt.	Color, crystalline form, specific rotation and λ_{max} (log ε)	m.p. °C	b.p. °C	Density	n_D	Solubility w	al	eth	ace	bz	other solvents	Ref.
	Propenoic acid														
p1767	—,allyl ester	Allyl acrylate. $CH_2:CHCO_2CH_2CH:CH_2$	112.14	...	...	$122–4^{760}$ 47^{40}	0.9441_4^{20}	1.4320_4^{20}	δ	s	s	...	...	ac s	B2³, 1230
Ω p1768	—,amide	Acrylamide. Propenamide*. $CH_2:CHCONH_2$	71.08	lf (bz)	84–5	...	...	...	v	s	s	...	...	chl v	B2², 388
Ω p1769	—,benzyl ester*	Benzyl acrylate. $CH_2:CHCO_2CH_2C_6H_5$	162.19	...	...	228^{760} 94^{6}	1.0573_4^{20}	1.5143^{20}	i	s	s	s	...	...	B6³, 1481
Ω p1770	—,butyl ester*	*n*-Butyl acrylate. $CH_2:CHCO_2C_4H_9^n$	128.17	...	−64.6	$146–8^{760}$ 39^{10}	0.8898_4^{20}	1.4185^{20}	i	s	s	s	...	...	B2³, 1226
Ω p1771	—,chloride	Acrylyl chloride. $CH_2:CHCOCl$	90.51	...	...	75–6	1.1136_4^{20}	1.4343^{20}	d	d	...	...	...	chl v	B2³, 1233
Ω p1772	—,cyclohexyl ester*	Cyclohexyl acrylate.	154.21	...	...	$182–4^{750}$ 88^{20}	1.0275_4^{20}	1.4673^{20}	i	∞	∞	...	...	chl s	C40, 5697
Ω p1773	—,ethyl ester*	Ethyl acrylate. $CH_2:CHCO_2C_2H_5$	100.13	λ^{al} 208 (3.84)	−71.2	99.8	0.9234_4^{20}	1.4068^{20}	δ	∞	∞	...	...	chl s	B2³, 1223
Ω p1774	—,isobutyl ester*	Isobutyl acrylate. $CH_2:CHCO_2CH_2CH(CH_3)_2$	128.17	...	...	132 62^{50}	0.8896_4^{20}	1.4150^{20}	δ	s	s	...	...	MeOH s	B2³, 1227
Ω p1775	—,methyl ester*	Methyl acrylate. $CH_2:CHCO_2CH_3$	86.09	...	< −75	80.5^{760}	0.9535_4^{20}	1.4040^{20}	δ	s	s	s	s	...	B2³, 1218
Ω p1776	—,nitrile	Acrylonitrile. Vinyl cyanide. $CH_2:CHCN$	53.06	λ^{al} 203 (3.79)	−83.5	$77.5–9^{760}$	0.8060_4^{20}	1.3911^{20}	s v^h	∞	∞	s	s	...	B2³, 1234
p1777	—,2-chloro-*	$CH_2:CClCO_2H$	106.51	nd (peth), pr (sub)	65 sub	176–81d	...	...	s	s	s	...	...	...	B2³, 1244
Ω p1778	—,—,ethyl ester*	$CH_2:CClCO_2C_2H_5$	134.56	...	...	$51–3^{18}$	1.1404_4^{20}	1.4384^{20}	...	v	v	...	...	...	B2³, 1245
p1779	—,—,methyl ester*	$CH_2:CClCO_2CH_3$	120.54	...	...	52^{51}	1.189_4^{20}	1.4420^{20}	...	...	v	...	...	...	B2³, 1244
p1780	—,3-chloro-(*cis*)*	$ClCH:CHCO_2H$	106.51	lf or nd (HCl)	63–4	$107^{17.5}$	...	...	...	s	s	...	...	...	B2³, 1243
p1781	—,—(*trans*)*	$ClCH:CHCO_2H$	106.51	lf	86 (92–4)	94^{18}	...	...	...	s	s	...	...	...	B2³, 1243
Ω p1782	—,2,3-dichloro-*	$ClCH:CClCO_2H$	140.95	mcl pr (chl), cr (peth, CS_2)	87–8	...	...	...	v	s	s	s	δ s^h	chl v lig s^h CS_2 δ CCl_4 i	B2³, 1246
p1783	—,3,3-dichloro-*	$Cl_2C:CHCO_2H$	140.95	nd (peth, sub), pr (chl)	76–7	sub	...	...	δ	...	v	...	...	chl v	B2³, 1246
Ω p1783[1]	—,2,3-diphenyl-(*trans*)*	α-Phenylcinnamic acid. $C_6H_5CH:C(C_6H_5)CO_2H$	224.26	nd (lig, dil al) λ^{al} 228 (4.11) 294 (4.38)	172.5–3	sub	...	...	δ s^h	s	s	...	...	...	B9², 482
p1784	—,3(2-furyl)-(*cis*)*	2-Furanacrylic acid	138.12	wh pr or pl (unst) λ^{al} 230 (3.25), 300 (4.36)	103–4	sub vac at 95	...	...	δ s^h	...	s	...	δ	...	B18, 301
p1785	—,—(*trans*)*	$C_7H_6O_3$. *See* p1784	138.12	nd (w) (st)	141	286 (sub vac at 112)	...	1.5286^{20}	δ s^h	s	s	...	s	aa s lig, CS_2 i	B18², 273
p1786	—,—,benzyl ester	$C_{14}H_{12}O_3$. *See* p1784	228.25	pa ye	42–3	$201–3^{12}$ $155–6^{3}$	...	1.5872^{25}	i	s^h	s	s	s	...	B18², 274
p1787	—,—,butyl ester*	$C_{11}H_{14}O_3$. *See* p1784	194.23	...	...	$147–50^{15}$ $117–8^{3}$	1.045^{20}	1.5129^{20}	i	s	...	s	...	...	C48, 2114
p1788	—,—,ethyl ester*	$C_9H_{10}O_3$. *See* p1784	166.18	pa ye λ^{al} 225 (3.33), 302.5 (4.37)	24.5	232^{760} 120^{17}	1.0891_4^{25}	1.5286^{20}	i	∞	∞	...	...	...	B18¹, 440
p1789	—,—,methyl ester*	$C_8H_8O_3$. *See* p1784	152.16	λ^{al} 302 (4.37)	35–7 (27.5)	$227–8^{774}$ 112^{15}	...	1.4447^{20}	i	v	v	...	v	...	B18¹, 440
p1790	—,—,pentyl ester*	$C_{12}H_{16}O_3$. *See* p1784	208.24	...	...	$116.5–8^{2}$	1.0322_4^{20}	1.5289^{24}	i	s	s	...	...	...	Am 69, 460
p1791	—,—,propyl ester*	$C_{10}H_{12}O_3$. *See* p1784	180.21	...	...	$91–4^{3}$	1.0744_4^{20}	1.5392^{24}	i	s	s	...	s	...	Am 69, 460
Ω p1792	—,2-methyl-*	Methacrylic acid. $CH_2:C(CH_3)CO_2H$	86.09	pr λ^{MeOH} 244 (1.8)	16	$162–3^{757}$ 60^{12}	1.0153_4^{20}	1.4314^{20}	s v^h	∞	∞	...	...	...	B2³, 1278
Ω p1793	—,—,amide	Methacrylamide. $CH_2:C(CH_3)CONH_2$	85.11	cr (bz), pl (CH_2Cl_2)	110–1	...	...	...	...	s	δ s^h	...	...	CH_2Cl_2 s^h	B2³, 1293
Ω p1794	—,—,anhydride*	Methacrylic anhydride. $[CH_2:C(CH_3)CO]_2O$	154.17	...	...	89^{5} (87^{13})	...	1.4540^{20}	d	∞	∞	...	...	...	B2³, 1293
Ω p1795	—,—,butyl ester*	Butyl methacrylate. $CH_2:C(CH_3)CO_2(CH_2)_3CH_3$	142.20	λ^{al} 214 (3.83)	...	160^{760} 52^{11}	0.8936_4^{20}	1.4240^{20}	i	∞	∞	...	...	...	B2³, 1286
Ω p1796	—,—,ethyl ester*	Ethyl methacrylate. $CH_2:C(CH_3)CO_2C_2H_5$	114.15	λ^{al} 208 (3.93)	...	117^{760} 30^{18}	0.9135_4^{20}	1.4147^{20}	δ	∞	∞	...	...	...	B2³, 1284
Ω p1797	—,—,isobutyl ester*	Isobutyl methacrylate. $CH_2:C(CH_3)CO_2CH_2CH(CH_3)_2$	142.20			155^{760} 45^{11}	0.8858_4^{20}	1.4199^{20}	i	∞	∞	...	...	...	B2³, 1287
Ω p1798	—,—,isopropyl ester*	Isopropyl methacrylate. $CH_2:C(CH_3)CO_2CH(CH_3)_2$	128.17			125^{760}	0.8847_4^{20}	1.4122^{20}	i	∞	∞	∞	∞	...	B2³, 1286
Ω p1799	—,—,methyl ester*	Methyl methacrylate. $CH_2:C(CH_3)CO_2CH_3$	100.13	...	−48	$100–1^{760}$ 24^{32}	0.9440_4^{20}	1.4142^{20}	δ	∞	∞	∞	...	glycol δ	B2³, 1280
Ω p1800	—,—,nitrile	Methacrylonitrile. $CH_2:C(CH_3)CN$	67.09	λ^{al} 215 (2.83)	−35.8	90.3^{760}	0.7998_4^{20}	1.4003^{20}	i	∞	∞	∞	...	peth, to ∞	B2³, 1294
Ω p1801	—,—,propyl ester*	Propyl methacrylate. $CH_2:C(CH_3)CO_2CH_2CH_2CH_3$	128.17	...	...	141^{760}	0.9022_4^{20}	1.4190^{20}	i	∞	∞	...	...	...	B2³, 1286
p1802	—,3(1-naphthyl)-(*cis*)*	$C_{10}H_7^{\alpha}CH:CHCO_2H$	198.22	col pl (al) λ^{MeOH} 307 (3.84)	156	...	...	...	...	s	...	...	...	...	E12B, 3287
p1803	—,—(*trans*)*	$C_{10}H_7^{\alpha}CH:CHCO_2H$	198.22	nd (al, w, aa) λ^{MeOH} 317 (4.06)	211–2	sub	...	...	δ^h	s^h	s	...	...	chl s	B9², 465

For explanations, symbols and abbreviations see beginning of table. For structural formulas see end of table.

No.	Name	Synonyms and Formula	Mol. wt.	Color, crystalline form, specific rotation and λ_{max} (log ε)	m.p. °C	b.p. °C	Density	n_D	Solubility w	al	eth	ace	bz	other solvents	Ref.
	Propenoic acid														
p1804	—,3(2-naphthyl)-(*trans*)*	$C_{10}H_7^{\beta}CH{:}CHCO_2H$	198.22	nd (al, w) λ^{MeOH} 269 (4.6), 308 (4.4)	210 (203)				s^h	s					**B9**, 672
p1805	—,3(2-nitro-phenyl)-2-phenyl-(*trans*)*	*o*-Nitro-α-phenylcinnamic acid.	269.26	ye cr (al) λ^{al} 262 (4.06)	196–7				δ^h	δ s^h	s		s	to s^h chl δ, s^h	**B9**[2], 483
p1806	—,3(3-nitro-phenyl)-2-phenyl-(*cis*)*	*m*-Nitro-α-phenylcinnamic acid.	269.26	nd (al, aa)	195–6					δ s^h					**B9**[2], 483
p1807	—,—(*trans*)*	$C_{15}H_{11}NO_4$. *See* p1806	269.26	ye pr (eth), nd (dil al)	182					s^h	s		s	CS_2, peth, chl s	**B9**[2], 483
p1808	—,3(4-nitro phenyl)-2-phenyl-(*cis*)*	*p*-Nitro-α-phenylcinnamic acid.	269.26	ye-gr pr (dil al + 1w), cr (abs al + ½), lf (bz + ½) λ^{al} 270 (4.3)	144				δ	v	s		s^h		**B9**[2], 484
p1809	—,—,(*trans*)*	$C_{15}H_{11}NO_4$. *See* p1808	269.26	ye pr or nd (al)	213–4				δ	s^h	δ		δ	chl δ	**B9**, 696
p1810	—,2-phenyl-*	Atropic acid. $CH_2{:}C(C_6H_5)CO_2H$	148.17	lf (al), nd (w) λ^{al} 246 (3.75)	106–7	267d			δ	s	s		s	chl, CS_2 s	**B9**[2], 407
—	—,3-phenyl-*	*see* **Cinnamic acid**													
p1811	—,trichloro-*	$Cl_2C{:}C(Cl)CO_2H$	175.40	pr (CS_2, eth)	76	221–3 133[30] (cor)			s^h	s	s			CCl_4 s CS_2 s^h	**B2**[3], 1247
p1812	—,—,chloride	$Cl_2C{:}C(Cl)COCl$	193.85			158		$1.5271^{18.5}$	d				s	CS_2 v	**B2**[1], 187
p1813	—,triphenyl-, nitrile	$(C_6H_5)_2C{:}C(C_6H_5)CN$	281.36	nd (al), pr	166–7					s^h	s				**B9**[2], 507
Ωp1814	**2-Propen-1-ol***	Allyl alcohol. $CH_2{:}CHCH_2OH$	58.08	$\lambda^{98\%\ sulf}$ 273 (3.67)	−129 (fr−50)	97	0.8540^{20}_{4}	1.4135^{20}	∞	∞	∞				**B1**[3], 1874
p1815	—,2-bromo-*	$CH_2{:}CBrCH_2OH$	136.98			153–4[755] 62[21]	1.621^{18}	1.500^{18}			s			chl s	**B1**[3], 1888
Ωp1816	—,2-chloro-*	$CH_2{:}CClCH_2OH$	92.53			136–40 47[10]	1.1618^{20}_{4}	1.4588^{20}	d^h	s^h					**B1**[3], 1887
p1817	—,3-chloro-*	$ClCH{:}CHCH_2OH$	92.53			(i) 146[746] (ii) 153[756]	(i) 1.1769^{20}_{4} (ii) 1.1729^{20}_{4}	(i) 1.4638^{20} (ii) 1.4664^{20}	δ						**B1**[3], 1887
p1818	—,3(3,5-dimethoxy-4-hydroxy-phenyl)-*	Sinapyl alcohol. Syringenin.	210.23	nd (eth-peth)	66–7				i		s			peth δ	**B6**[3], 6690
—	—,3(4-hydroxy-3-methoxy-phenyl)-*	*see* **Coniferyl alcohol**													
p1819	—,3(4-hydroxy-phenyl)-*	*p*-Coumaryl alcohol.	150.18	pr (dil al) λ^{al} 260 (4.30)	124				δ	s	s	s	s	to, MeOH s	**C45**, 3359
Ωp1820	—,2-methyl-*	Methallyl alcohol. $CH_2{:}C(CH_3)CH_2OH$	72.12			114.49[760]	0.8515^{20}	1.4255^{20}	v	∞	∞				**B1**[3], 1903
p1821	—,1-phenyl-*	α-Vinylbenzyl alcohol. $CH_2{:}CHCH(C_6H_5)OH$	134.18	λ^{al} 242 sh (2.88), 248 (2.92), 251 (2.92), 257 sh (2.88)		215–6 111[18]	1.0251^{21}_{0}	1.5406^{20}	δ	s	s		s	chl s	**B6**[3], 2417
Ωp1822	—,3-phenyl-(*trans*)*	Cinnamic alcohol. Cinnamyl alcohol. $C_6H_5CH{:}CHCH_2OH$	134.18	wh nd (eth-peth) λ^{eth} 252 (4.08), 275 sh (3.3), 282 (3.2), 292 (3.0)	34	257.5 127–8[10]	1.0440^{20}_{4}	1.5819^{20}	δ	v	v				**B6**[3], 2401
p1823	—,—,acetate (*trans*)	$C_6H_5CH{:}CHCH_2O_2CCH_3$	176.22	λ^{al} 250 (4.24)		145–6[15] 114[1]	1.0567^{20}	1.5425^{20}	i	s	s	s	s	chl s	**B6**[3], 2406
p1824	**2-Propen-1-one, 3(2-furyl)-1-phenyl-***	Furfurylideneacetophenone.	198.22	λ^{al} 260 (3.93), 344 (4.43)		317 181–2[9]	1.1140^{20}			s	s				**B17**[2], 377
p1825	—,1(2-hydroxy-naphthyl)-3-phenyl-*		274.32	og-red lf or pl (al)	129				i	s	s			con sulf s	**B8**[2], 246
—	**Propioin**	*see* **3-Hexanone, 4-hydroxy-***													
—	**Propiolactone**	*see* **Propanoic acid, 3-hydroxy-, β-lactone***													
—	**Propiolaldehyde**	*see* **Propynal***													
—	**Propiolic acid**	*see* **Propynoic acid***													
—	**Propionaldehyde**	*see* **Propanal***													
—	**Propionaldol**	*see* **Pentanal, 3-hydroxy-2-methyl-***													
—	**Propionamide**	*see* **Propanoic acid,** amide*													
—	**Propionic acid**	*see* **Propanoic acid***													
—	**Propiophenone**	*see* **1-Propanone, 1-phenyl-***													
—	**Propylal**	*see* **Propynal,** diethyl acetal													
—	**Propyl alcohol**	*see* **1-Propanol***													
—	**Propyl bromide**	*see* **Propane, 1-bromo-***													
—	**Propyl chloride**	*see* **Propane, 1-chloro-***													

For explanations, symbols and abbreviations see beginning of table. For structural formulas see end of table.

No.	Name	Synonyms and Formula	Mol. wt.	Color, crystalline form, specific rotation and λ_{max} (log ε)	m.p. °C	b.p. °C	Density	n_D	Solubility w	al	eth	ace	bz	other solvents	Ref.
	Propylene														
—	**Propylene**	*see* **Propene***													
—	**Propylene glycol**	*see* **1,2-Propanediol***													
—	**Propylene oxide**	*see* **Propane, 1,2-epoxy-***													
p1826	**Propyl hexedrine**		155.29			205^{760} $92–3^{20}$	0.8501^{20}_{4}	1.4600^{20}	δ	s	...	...	...	dil ac s	**Am 69**, 1117
—	**Propyl iodide**	*see* **Propane, 1-iodo-***													
p1827	**Propyl red**	4′-Dipropylaminoazo-benzene-2-carboxylic acid.	325.42	vt-bl or purp-red cr (al)					...	s^h	...	...	...	KOH s	**B16**^{2}, 165
p1828	**Propynal***	Propargylaldehyde. Propiolaldehyde. HC⋮CCHO	54.05	λ^{iso} 212 (3.7), 335 (1.3), 382 (0.7)		59–61 (55–6)		1.4033^{25}	∞	s	s	s	s	to s	**B1**^{3}, 3040
p1829	—,diethyl acetal	3,3-Diethoxypropyne. $CH⋮CCH(OC_2H_5)_2$	128.17			130.45 37^{11}	0.8942^{22}_{4}	1.4140^{20}	...	s	s	s	...	chl s	**B1**^{2}, 808 **B4**^{2}, 1135
Ω p1830	—,1-phenyl-*	$C_6H_5C⋮CCHO$	130.15	λ 240 (4.5), 250 (4.4), 300 (4.05)		$127–8^{28}$ $65^{0.1}$	1.0639^{16}_{4}	1.6079^{18}	...	...	...	...	...	...	**B7**^{2}, 317
p1831	**Propyne***	Methylacetylene. Propine. $CH_3C⋮CH$	40.07	gas	−101.5 (cor)	-23.2^{760}	0.7062^{-50}_{4} 0.66^{-13}	1.3863^{-40}	δ	v	...	...	s	chl s	**B1**^{3}, 919
Ω p1832	—,3-bromo-*	Propargyl bromide. $BrCH_2C⋮CH$	118.97			88–90 33^{130}	1.579^{19}	1.4922^{20}	...	s	s	...	s	chl s	**B1**^{3}, 922
Ω p1833	—,3-chloro-*	Propargyl chloride. $ClCH_2C⋮CH$	74.51			65 (56–7)	1.0297^{20}_{4}	1.4320^{20}	i	∞	∞	...	s	chl s	**B1**^{3}, 922
p1834	—,3-cyclohexyl-*		122.21	λ^{al} 236.5 (3.90)		$157–8^{760}$ 48^{11}	0.8449^{20}_{4}	1.4605^{20}	...	...	s	...	...	...	**B5**^{3}, 329
p1835	—,1,3-dibromo-*	$BrCH_2C⋮CBr$	197.88			$73–4^{30}$ $53–4^{10}$	2.1894^{20}	1.5690^{20}	i	...	s	...	...	chl s	**B1**, 248
—	—,3,3-diethoxy-*	*see* **Propynal**, diethyl acetal													
p1836	—,1-iodo-*	$CH_3C⋮CI$	165.96			110^{760}	2.08^{22}		δ	v	v	...	...	...	**B1**^{3}, 922
p1837	—,3-iodo-*	Propargyl iodide. $ICH_2C⋮CH$	165.96			115	2.0177^{0}		...	...	s	...	...	...	**B1**^{3}, 922
p1838	—,3-methoxy-*	Methyl propargyl ether. $CH_3OCH_2C⋮CH$	70.09			63	0.83^{12}	1.5035^{20}	δ	∞	∞	...	...	...	**B1**^{3}, 1971
p1839	—,1-phenyl-*	Phenylallylene. $CH_3C⋮CC_6H_5$	116.17	λ^{al} 238 (4.22), 249 (4.22)		183 77^{17}	0.9388^{20}_{4}	1.5650^{20}	i	...	s	...	...	...	**B1**^{3}, 1353
p1840	—,3,3,3-trifluoro-*	$F_3CC⋮CH$	94.04	gas		−48.3			...	...	...	...	...	...	**Am 82**, 543
p1841	**Propynoic acid***	Propargylic acid. Propiolic acid. $HC⋮CCO_2H$	70.05	cr (CS_2)	18 (anh) +0.3 (+1w)	144d $83–4^{50}$	1.1380^{20}_{4}	1.4306^{20}	∞	∞	∞	v	...	chl ∞	**B2**^{3}, 1447
p1842	—,ethyl ester*	$HC⋮CCO_2C_2H_5$	98.10			119^{745}	0.9583^{25}_{25}	1.4105^{20}	i	v	v	...	...	chl v	**B2**^{3}, 1448
p1843	—,chloro-*	$ClC⋮CCO_2H$	104.50	cr (peth)	69–70				...	...	v	...	...	peth δ, v^h	**B2**^{3}, 1448
p1844	—,(2-chloro-phenyl)-*	$C_9H_5ClO_2$. *See* p1849	180.59	cr (50% aa, bz)	132.7–3.8				i	...	...	...	v^h	aa v	**Am 77**, 5549
p1845	—,(3-chloro-phenyl)-*	$C_9H_5ClO_2$. *See* p1849	180.59	cr (aa, bz-peth)	144.3–5.1				i	...	...	...	s^h	aa v	**Am 77**, 5549
p1846	—,(4-chloro-phenyl)-*	$C_9H_5ClO_2$. *See* p1849	180.59	cr (aa), pl (bz)	192–3				i	...	...	...	s^h	aa v^h	**Am 77**, 5549
Ω p1847	—,(2-nitro-phenyl)-*	$C_9H_5NO_4$. *See* p1849	191.15	nd or lf (w)	157d	exp			s^h	s	s	...	...	chl δ CS_2, lig i	**B9**, 637
p1848	—,(4-nitro-phenyl)-*	$C_9H_5NO_4$. *See* p1849	191.15	nd (eth, al)	181				δ	s^h	s	...	δ	aa, chl s CS_2 δ peth i	**B9**, 637
Ω p1849	—,phenyl-*		146.15	nd (w, CS_2) λ^{al} 253 (4.1)	137	sub			δ	v	v	...	...	...	**B9**, 633
p1850	—,—,ethyl ester*	$C_6H_5C⋮CCO_2C_2H_5$	174.20	λ 258 (4.18)		260–70δd 144^{13}	1.0550^{25}_{4}	1.5535^{20}	...	...	s	...	...	...	**B9**, 634
Ω p1851	**2-Propyn-1-ol***	Propargyl alcohol. Propiolic alcohol. $HC⋮CCH_2OH$	56.07		−48	113.6^{760} 30^{21}	0.9485^{20}	1.4322^{20}	s	∞	∞	...	...	...	**B2**^{1}, 234
p1852	—,acetate*	Propargyl acetate. $HC⋮CCH_2O_2CCH_3$	98.10			124–5 (110–2)	1.0052^{20}_{4}	1.4205^{20}	δ	s	s	...	...	...	**B2**^{3}, 294
p1853	**Prostigmine bromide**	Neostigmine bromide	303.21	(al-eth)	167d				v	s	...	...	...	...	**C47**, 4908
—	**Protoanemonin**	*see* **2,4-Pentadienoic acid, 4-hydroxy-, lactone***													
—	**Protocatechu-aldehyde**	*see* **Benzaldehyde, 3,4-dihydroxy-**													
—	**Protocatechuic acid**	*see* **Benzoic acid, 3,4-dihydroxy-**													
p1854	**Protopine**	Fumarine. Macleyine. $C_{20}H_{19}NO_5$	353.38	mcl pr (al-chl) λ^{al} 290 (4.0)	208				i	δ	δ	...	δ	chl s peth, CS_2 AcOEt δ	**B27**^{1}, 568
p1855	**Protoveratridine**	$C_{32}H_{51}NO_9$	593.76	cr (al-chl), $[\alpha]^{20}_{D}$ −14 (Py, c = 1)	272–3				...	s^h	i	...	...	aq ac s	**Am 74**, 2382

For explanations, symbols and abbreviations see beginning of table. For structural formulas see end of table.

No.	Name	Synonyms and Formula	Mol. wt.	Color, crystalline form, specific rotation and λ_{max} (log ε)	m.p. °C	b.p. °C	Density	n_D	Solubility w	al	eth	ace	bz	other solvents	Ref.
	Protoveratrine														
p1856	**Protoveratrine**	$C_{39}H_{61}NO_{13}$	751.93	pl (ace) $[\alpha]_D^{24}$ −8.3 (chl) λ^{al} 260 (1.1), 290 (0.8)	225d	...	...	...	i	s^h	...	...	...	chl s^h	**Am 74,** 5107
p1857	**Protoveratrine A**	$C_{41}H_{63}NO_{14}$	793.96	lf (al) $[\alpha]_D^{20}$ −44.1 (Py, c = 1.12), −12.1 (chl, c = 3.2)	305d	...	...	...	i	...	...	...	...	...	**H36,** 718
p1858	**Protoveratrine B**	Neoprotoveratrine. $C_{41}H_{63}NO_{15}$	809.96	lf (chl-al), $[\alpha]_D^{20}$ −39.8 (Py, c = 1.24), −3.5 (chl, c = 3.1) λ 260 (1.2), 290 (0.9)	285–90d (267–9)	...	...	...	i	s^h	...	...	...	chl, Py s	**H36,** 718
p1859	**Protoverine**	$C_{27}H_{43}NO_9$	525.65	nd (MeOH) $[\alpha]_D^{20}$ −15.7 (Py, c = 1.1) λ^{al} 260 (1.0), 330 (0.5)	220–2	...	...	...	i	s	...	...	s	aq ac s MeOH s^h	**C37,**5726
—	**Provitamin A**	*see* **β-Carotene**													
—	**Provitamin D_3**	*see* **5,7-Cholestadien-3β-ol**													
p1860	**Prulaurasin**	*dl*-Mandelonitrile glucoside. $C_{14}H_{17}NO_6$	295.30	wh nd or pr (al) $[\alpha]_D$ −54	122–2.5	...	...	...	s	s	i	...	...	...	**B31,** 240
p1861	**Prunasin**	*d*-Mandelonitrile-β-*d*-glucoside. $C_{14}H_{17}NO_6$. *See* p1860	295.30	nd (chl or AcOEt–CCl_4) $[\alpha]_D^{21}$ −27.0	149–50	...	...	...	s	s	i	s	...	AcOEt s^h chl δ^h	**B31,** 238
—	**Prunetol**	*see* **Isoflavine, 4′,5,7-trihydroxy-**													
—	**Pseudocholestane**	*see* **Coprostane**													
p1862	**Pseudoaconitine**	$C_{36}H_{51}NO_{12}$	687.79	tcl (MeOH–AmOH) $[\alpha]_D^{15}$ +18.36 (al), +22.75 (chl)	214	...	...	...	i	v	s	...	...	...	**J1928,** 1105
p1863	**Pseudocodeine**	Neoisocodeine. $C_{18}H_{21}NO_3$	299.37	wh nd $[\alpha]_D$ −96.6 (al)	181–2	...	1.290^{80}	1.574	δ	s	δ	...	...	...	**B27²,** 112
p1864	**Pseudoconhydrine**	5-Hydroxyconine. $C_8H_{17}NO$	143.23	hyg nd (eth) $[\alpha]_D^{20}$ +11.0 (al, c = 10)	106 (anh) 60 (+1w)	236	...	...	s	s	s	...	...	...	**B21¹,** 191
p1865	**Pseudoconiceine(*L*)**	γ-Coniceine.	125.22	hyg, $[\alpha]_D^{15}$ +122.6	...	171–2	0.8776_4^{15}	1.4607^{18}	δ	s	s	...	...	chl s	**B20,** 146
—	**Pseudocumene**	*see* **Benzene, 1,2,4-trimethyl-***													
—	**Pseudocumenol**	*see* **Benzene, 1-hydroxy-2,4,5-trimethyl-***													
—	**Pseudocumidine**	*see* **Benzene, 1-amino-2,4,5-trimethyl-***													
p1866	**Pseudoephedrine(*d*)**	Isoephedrine. ψ-Ephedrine. $C_{10}H_{15}NO$	165.24	pr or lf (eth) $[\alpha]_D^{20}$ +51.9 (abs al, c = 0.6)	118.7	...	...	...	δ	s	s	...	v	...	**B13²,** 376
Ω p1867	—(*dl*)	$C_{10}H_{15}NO$. *See* p1866	165.24	nd (eth)	118.2	130^{16}	...	...	δ	v	s	...	v	...	**B13²,** 372
p1868	—(*l*)	$C_{10}H_{15}NO$. *See* p1866	165.24	lf or pr (eth) $[\alpha]_D^{20}$ −51.9 (abs al, c = 0.6)	118.7	...	...	...	δ	s	s	...	v	...	**B13²,** 377
p1869	—,hydrochloride(*d*)	$C_{10}H_{15}NO.HCl$. *See* p1866	201.70	rh pl or nd (al) $[\alpha]_D^{20}$ +62 (w, c = 0.8)	181–2	...	...	...	s	s	s	...	...	25 % HCl v	**B13²,** 373
Ω p1870	—,—(*dl*)	$C_{10}H_{15}NO.HCl$. *See* p1866	201.70	nd (abs al)	164	...	...	...	v	s	s	...	...	...	**B13²,** 377
p1871	—,—(*l*)	$C_{10}H_{15}NO.HCl$. *See* p1866	201.70	nd (abs al or AcOEt) $[\alpha]_D^{20}$ −62.1 (w, c = 1.8)	182–2.5	...	...	...	v	...	...	...	...	chl i	**B13²,** 377
—	**Pseudoergostane**	*see* **Coproergostane**													
p1872	**Pseudohyoscyamine**	Norhyoscyamine. $C_{16}H_{21}NO_3$	275.34	nd, $[\alpha]_D$ −21 to −23 (al)	140.5	...	...	...	δ	s	δ	...	...	chl s	**B21¹,** 197
p1873	**Pseudoionone**	Citrylideneacetone. $C_{13}H_{20}O$. $(CH_3)_2C{:}CHCH_2CH_2C(CH_3){:}CHCH{:}CHCOCH_3$	192.31	pa ye oil λ 290 (4.4)	...	$143–5^{12}$ $114–6^2$	0.8984^{20}	1.5335^{20}	...	s	s	...	...	chl, MeOH s	**B1³,** 3067
p1874	**Pseudojervine**	$C_{33}H_{49}NO_8$	587.73	wh nd or hex pl $[\alpha]_D$ −133 (al-chl, 1:3) λ^{MeOH} 250 (4.18)	304–5d	...	...	...	i	s	i	...	i	chl s peth, to i	**C39,**1413

For explanations, symbols and abbreviations see beginning of table. For structural formulas see end of table.

No.	Name	Synonyms and Formula	Mol. wt.	Color, crystalline form, specific rotation and λ_{max} (log ε)	m.p. °C	b.p. °C	Density	n_D	Solubility w	al	eth	ace	bz	other solvents	Ref.
	Pseudomorphine														
p1875	**Pseudomorphine**	2,2′-Bimorphine. $C_{34}H_{36}N_2O_6$	568.68	cr (aq NH_3 + 3w) $[\alpha]^{24}_D$ +44.8 (1N HCl, c = 0.86) $\lambda^{1N\,sulf}$ < 320	282–3 (327d)	...	...	...	i	i	i	...	...	Py s NH_3 s^h sulf, chl i	**B27²**, 886
p1876	—,hydrochloride	$C_{34}H_{36}N_2O_6 \cdot 2HCl \cdot 2H_2O$. *See* p1876	677.64	pw $[\alpha]_D$ − 114.76	...	...	...	...	δ	...	...	...	...	...	**B3**, 910
p1877	**Pseudopelletierine**	Pseudopunicine. $C_9H_{15}NO$	153.23	orh pr (peth)	54	246	1.001^{100}	1.4760^{100}	v	s	s	...	...	chl s	**B21**, 261
p1878	**Pseudoreserpine**	$C_{32}H_{38}N_2O_9$	594.67	$[\alpha]^{24}_4$ −65 (chl) λ^{MeOH} 217 (4.78), 266 (4.23), 293 (4.00)	257–8	...	...	...	i	...	...	s	...	...	**C52**, 2876
Ωp1879	**Pseudo-thiohydantoin**	2-Imino-4-thiazolidinone	116.14	pr or nd (w) λ 220 (4.17), 250 (3.82)	255–8d	...	...	...	δ v^h	i	i	...	...	...	**B27²**, 284
Ωp1880	**Pseudotropine**	Pseudotropanol. $C_8H_{15}NO$	141.22	rh ta or pr (eth), rh bipym (peth-bz)	109	24^{760}	...	...	v	v	δ s^h	...	s	chl s	**C48**, 2724
p1881	**Psicose**	*d*-Allose. Pseudofructose. $C_6H_{12}O_6$	180.16	$[\alpha]^{25}_D$ +4.7 (w, c = 4.3)	58	...	...	...	s	s	...	i	...	MeOH s	**B31**, 349
p1882	**Pteridine**	Pyrimido[4,5-b]pyrazine	132.13	ye pl (bz, sub) λ^{al} 267 sh (3.6), 300 (3.9), 380 sh (1.9)	139.5–40	sub $125–30^{20}$	...	...	v	s	δ	...	δ	...	**J1951**, 474
p1883	—,**2-amino-4,6-dihydroxy-**	Xantopterin. Uropterin. $C_6H_5N_5O_2$. *See* p1882	179.14	og-ye (w + 1) λ^{aa} 270 (4.07), 380 (3.55)	>410d (darkens 360)	...	...	...	i	...	...	...	...	ac, alk v	**Am 73**, 1497
p1884	—,**2-amino-4-hydroxy-**	2-Amino-4-pteridonol. $C_6H_5N_5O$. *See* p1882	163.14	ye $\lambda^{w,\,pH=1}$ 312 (3.91) $\lambda^{w,\,pH=11}$ 252 (4.34), 358 (3.88)	>360	...	...	...	...	...	...	...	...	...	**C47**, 5945
p1885	**Pukateine(*l*)**	4-Hydroxy-5,6-methylene-dioxyaporphin. $C_{18}H_{17}NO_3$	295.34	cr (al, eth) $[\alpha]^{15}_D$ −200 (al, c = 6)	200	$210–5^2$	...	...	l	δ s^h	δ s^h	...	...	Py v chl s peth δ	**B27¹**, 461
p1886	**Pulegenone**	4-Methyl-1-isopropyl-cyclopentenone-5.	138.21	...	...	$188.5–9^{760}$	0.9144^{20}_0	1.4660^{20}	...	s	s	s	...	chl s	**C34**, 92
Ωp1887	**Pulegone**	4(8)-*p*-Menthen-3-one. $C_{10}H_{16}O$	152.24	$[\alpha]^{20}_D$ +23.4 (undil) λ^{hx} 244.5 (390), 331 (1.70)	...	224^{760} 103^{17}	0.9346^{45}	1.4894^{20}	i	∞	∞	...	...	chl ∞	**B7**, 81
Ωp1888	**Purine**	7-Imidazo[4,5-d]pyrimidine	120.11	nd (to or al) $\lambda^{w,\,pH=1}$ 260 (3.79) $\lambda^{w,\,pH=7}$ 263 (3.90)	216–7	sub	...	...	v	v^h	δ	s	...	chl, AcOEt δ	**B26**, 354
Ωp1889	—,**6-amino-**	Adenine. $C_6H_5N_5$. *See* p1888	135.13	nd or lf (w + 3) $\lambda^{w,\,pH=7}$ 260 (4.13)	360–5d (anh)	sub at 220	...	...	δ v^h	δ	i	...	...	ac, alk v chl i	**B26²**, 252
—	—,**1,3-dimethyl-2,6-dihydroxy-**	*see* **Theophylline**													
p1890	—,**6-mercapto-**	6-Purinethiol. $C_5H_4N_4S$. *See* p1888	152.18	ye pr (w + 1) dk ye hyd λ^{al} 330 (4.28)	313–4d (−w140)	...	...	...	i δ^h	...	...	...	...	alk s	**Am 74**, 411
p1891	—,**2,6,8-trichloro-**	$C_5HCl_3N_4$. *See* p1888	223.45	nd (al)	159–61	...	...	...	δ^h	v^h	v	v	v^h	chl v^h	**B26**, 356
—	—,**2,6,8-trihydroxy-**	*see* **Uric acid**													
—	**6(1)-Purinone**	*see* **Hypoxanthine**													
—	**Purpurin**	*see* **9,10-Anthraquinone, 1,2,4-trihydroxy-***													
—	**Purpuroxanthin**	*see* **9,10-Anthraquinone, 1,3-dihydroxy-***													
—	**Putrescine**	*see* **Butane, 1,4-diamino-***													
p1892	**Pyocyanine**	...	210.24	dk bl nd (w + 1 or chl-peth) λ^{al} 231 (4.4) 315 (4.4), 740 (4.0)	133d	sub	...	...	δ s^h	s^h	i	s	δ	chl v Py, AcOEt, $PhNO_2$ s CCl_4, peth i	**J123**, 3279
p1893	**Pyraconitine**	$C_{32}H_{43}O_9N$	585.70	nd	171	...	...	...	δ	s	...	...	...	...	**C51**, 17963
p1893¹	γ-**Pyran**	C_5H_6O	82.10	col oil λ^{al} 222 (3.85), 238 (3.71)	...	80^{760}	...	1.4559^{20}	...	s	s	...	s	...	**B17**, 36

For explanations, symbols and abbreviations see beginning of table. For structural formulas see end of table.

No.	Name	Synonyms and Formula	Mol. wt.	Color, crystalline form, specific rotation and λ_{max} (log ε)	m.p. °C	b.p. °C	Density	n_D	Solubility w	al	eth	ace	bz	other solvents	Ref.
	γ-Pyran														
p1894	—,2,3-dihydro-	C_5H_8O. *See* p1893[1]	84.13	λ^{hx} 195 (3.58)	...	86–7	0.922^{19}_{15}	1.4402^{19}	s	s	...	...	...	...	**C28,** 4056
—	—,2,3-dihydro-3,4-dihydroxy-2-hydroxy-methyl-	*see* **Gluconal**													
—	—,4-oxo-	*see* **4-Pyrone**													
p1895	—,tetrahydro-	Pentamethylene oxide. $C_5H_{10}O$. *See* p1893[1]	86.14	λ^{vap} 175 (3.3), 187 (2.8)	...	88^{760}	0.8810^{20}_4	1.4200^{20}	...	s	s	...	s	...	**B17**[2], 18
p1896	**Pyranthrone**	8,16-Pyranthenedione	406.45	red-ye or red-br nd ($PhNO_2$)	d	sub vac	...	...	i	...	...	...	...	sulf s (bl) $PhNO_2$ s[h] 50 % Py δ	**E14s,** 887
Ω p1897	**Pyrazine**	1,4-Diazine*. Paradiazine	80.09	pr (w) λ^{al} 261 (3.78), 310 (2.93)	54	115.5 $-.8^{768}$	1.0311^{61}_4	1.4953^{61}	s	s	s	s	...	...	**B23,** 91
Ω p1898	—,2,5-dimethyl-	Ketine. $C_6H_8N_2$. *See* p1897	108.15	λ^{cy} 271 (3.76), 314 (2.96)	15	155^{760} (cor)	0.9887^{20}_4	1.4980^{20}	∞	∞	∞	s	...	...	**B23**[2], 80
Ω p1899	—,2-methyl-	$C_5H_6N_2$. *See* p1897	94.12	λ^{cy} 266 (3.76), 320 (2.92)	...	136–7	1.0290^{20}_4	1.5067^{19}	∞	∞	∞	s	...	...	**B23,** 94
Ω p1900	**Pyrazine-carboxylic acid,** amide	Pyrazinamide	123.12	wh nd (w, al) λ^{al} 268 (3.95), 322 (2.75)	191–3	sub	...	...	s[h]	s[h]	...	...	...	...	**C48,** 2074
Ω p1901	**2,3-Pyrazinedi-carboxylic acid**	...	168.11	pr (w + 2) λ^{al} 269 (3.79), 314 (2.84)	193d (188d)	...	...	...	v	δ	δ	s	δ	MeOH s chl d	**B25**[2], 164
p1902	**2,5-Pyrazinedi-carboxylic acid**	...	168.11	nd (w + 2)	255–6d (sealed tube) (272)	...	...	...	δ	δ	δ	...	δ	chl δ	**B25**[2], 164
p1903	**2,6-Pyrazinedi-carboxylic acid**	...	168.11	mcl nd (w + 2)	217–8d	...	...	...	δ s[h]	s	δ	...	...	...	**B25,** 168
Ω p1904	**2,3-Pyrazinedi-carboxylic acid, 5-methyl-**	$C_7H_6N_2O_4$. *See* p1901	182.14	(w, ace, dil al-eth)	174–5	...	...	...	v[h]	s	...	s[h]	...	...	**B25**[2], 165
Ω p1905	**Pyrazole**	1,2-Diazole	68.08	nd or pr (lig) λ^{diox} 212 (3.45)	69.5–70	186–8^{760}	...	1.4203	s	s	s	...	s	...	**B23,** 39
p1906	—,4-bromo-1,3-dimethyl-	$C_5H_7BrN_2$. *See* p1905	175.03	...	...	76–7^{10}	1.4976^{15}_4	1.5214^{15}	...	s	s	s	...	os v	**B23**[2], 54
p1907	—,4-bromo-1,5-dimethyl-	$C_5H_7BrN_2$. *See* p1905	175.03	cr	38.5–9.5	85^{10}	...	...	δ	...	...	...	...	os, dil ac s peth i	**B23**[2], 54
Ω p1908	—,4-bromo-3,5-dimethyl-	$C_5H_7BrN_2$. *See* p1905	175.03	nd (dil al)	123	...	...	...	s[h]	...	v	v	...	os s	**B23**[2], 68
p1909	—,4-bromo-3-methyl-	$C_4H_5BrN_2$. *See* p1905	161.00	cr (dil al)	76–7	...	1.5638^{100}_4	1.5182^{100}	s[h]	s	v	v	...	...	**B23**[2], 54
p1910	—,3-chloro-1,5-dimethyl-	$C_5H_7ClN_2$. *See* p1905	130.58	pl (w)	47–6	210–2 138^{72}	1.0823^{100}_4	1.4648^{100}	δ	v	v	v	v	...	**B23**[2], 49
p1911	—,4-chloro-3,5-dimethyl-	$C_5H_7ClN_2$. *See* p1905	130.58	pr (w), cr (al)	117.5–8.5	220–22	...	...	δ s[h]	v	v	v	v	os v	**B23**[2], 67
p1912	—,5-chloro-1,3-dimethyl-	$C_5H_7ClN_2$. *See* p1905	130.58	...	...	157–8 (cor)	1.1367^{18}_4	1.4877^{18}	s	...	...	...	...	...	**B23**[2], 49
p1913	—,5-chloro-3-methyl-	$C_4H_5ClN_2$. *See* p1905	116.55	cr (eth, lig)	118–9	258 138^{15}	...	...	v[h]	...	...	...	...	os v	**B23**[2], 49
—	—,dihydro-	*see* **Pyrazoline**													
—	—,dihydro(oxo)-	*see* **Pyrazolone**													
p1914	—,1,3-dimethyl-	$C_5H_8N_2$. *See* p1905	96.14	λ^{al} 221 (3.67)	...	136–8 (143–5) 31^1	0.9561^{17}_4	1.4734^{15}_{α}	v	...	...	...	...	...	**B23**[2], 44
p1915	—,1,5-dimethyl-	$C_5H_8N_2$. *See* p1905	96.14	λ^{al} 217 (3.64)	...	153 (cor)	0.9813^{17}_4	1.4782^{16}_{α}	v	v	v	...	...	...	**B23**[2], 44
Ω p1916	—,3,4-dimethyl-	$C_5H_8N_2$. *See* p1905	96.14	(peth) λ^{al} 222 (3.65)	58	111^{10}	...	...	...	s	s	s	s	os s	**B23**[2], 64
Ω p1917	—,3,5-dimethyl-	$C_5H_8N_2$. *See* p1905	96.14	cr (peth, MeOH, al)	107.5–8.5	218^{758}	0.8839^{26}_4	...	s	v	v	s	v	chl, MeOH v	**B23**[2], 65
p1918	—,1-ethyl-	$C_5H_8N_2$. *See* p1905	96.14	...	...	139^{760}	0.9537^{20}_4	1.4700^{20}	v	v	v	v	v	Py v	**C55,** 22291
p1919	—,1-isopropyl	$C_6H_{10}N_2$. *See* p1905	110.16	...	...	143	...	...	s	v	...	...	...	...	**C53,** 21043
p1920	—,3-methyl-	$C_4H_6N_2$. *See* p1905	82.11	λ^{al} 213 (3.60)	36–7	204^{752} (cor) 108^{25}	1.0203^{16}_4	1.4915^{20}	∞	∞	∞	...	...	...	**B23**[2], 44
p1921	**2-Pyrazoline**	4,5-Dihydropyrazole	70.10	...	...	144^{760}	1.0200^{17}_4	1.4796^{17}	∞	∞	s	...	...	...	**B23,** 28
p1922	—,1-phenyl-	$C_9H_{10}N_2$. *See* p1921	146.19	pl (lig) λ^{al} 242 (4.06), 281 (4.41)	52	273^{754} 151^{17}	1.0689^{58}_4	1.6015^{58}	i	s	s	...	s	...	**B23**[2], 25
p1923	**5-Pyrazolone**	2-Pyrazolin-5-one	84.08	nd (to, w, xyl)	165	sub	...	...	s	s	δ	...	...	...	**B24**[1], 186
—	**3-Pyrazolone, 1,5-dimethyl-2-phenyl-**	*see* **Antipyrine**													

For explanations, symbols and abbreviations see beginning of table. For structural formulas see end of table.

No.	Name	Synonyms and Formula	Mol. wt.	Color, crystalline form, specific rotation and λ_{max} (log ε)	m.p. °C	b.p. °C	Density	n_D	Solubility w	al	eth	ace	bz	other solvents	Ref.
	5-Pyrazolone														
Ωp1925	**5-Pyrazolone, 3-methyl-**	$C_4H_6N_2O$. *See* p1923	98.11	pr (w), nd (al), lf (sub) λ^{al} 240 (3.5)	215 (219)	sub	....	....	v^h	δ^h	...	...	...	....	**B24**, 19
p1926	**—,3-methyl-1(3-nitrophenyl)-**	$C_{10}H_9N_3O_3$. *See* p1923	219.20	ye amor (aa)	185	....	....	....	δ	δ	δ	...	...	alk s aa s^h	**B24**[1], 191
p1927	**—,3-methyl-1-phenyl-**	$C_{10}H_{10}N_2O$. *See* p1923	174.20	mcl pr (w) λ^{al} 245 (1.1), 280 sh (3.6)	127	287^{105} 191^{17}	....	1.637	s^h	s^h	...	...	δ	peth i	**B24**, 20
p1928	**—,3-methyl-1(4-sulfophenyl)-**	$C_{10}H_{10}N_2O_4S$. *See* p1923	254.27	nd (w + 1)	290–320d	....	....	....	s^h	i	i	...	...	alk v aa i	**B24**[2], 20
p1929	**5-Pyrazolone-3-carboxylic acid, 1-phenyl-**	....	204.20	nd (w, al)	261	....	....	....	i δ^h	s	i	...	...	....	**B25**[2], 219
Ωp1930	**Pyrene***	Benzo[*d,e,f*]phenanthrene	202.26	pa ye pl (to, sub) λ^{cy} 273 (4.77), 306 (5.07), 320 (4,51), 335 (4.78)	156 (cor) (150)	393^{760} 260^{60}	1.271_4^{23}	....	i	s	s	...	s	CS_2, to, lig s	**B5**[3], 2279
Ωp1931	**—,1-acetyl-**	$C_{18}H_{12}O$. *See* p1930	244.30	gr-ye lf (al, MeOH)	89–90	....	....	....	...	...	s	...	v	to v sulf s (og-red)	**E14s**, 434
p1932	**—,1-amino-**	$C_{16}H_{11}N$. *See* p1930	217.27	ye nd (hx), lf (dil al) λ^{eth} 238 (4.6), 286 (4.4), 357 (4.2), 400 (4.1)	117–8	....	....	....	...	s	...	s	...	hx, ac s	**E14s**, 422
—	**3-Pyrenebutyric acid**	*see* **Butanoic acid, 4(3-pyrenyl)-***													
p1933	**3-Pyrene-carboxylic acid**	$C_{17}H_{10}O_2$. *See* p1930	246.27	ye nd (PhCl, eth-al, sub)	274	sub	....	....	δ	δ	s	...	...	con sulf, dil alk, chl s	**E14s**, 447
p1934	**4-Pyrene-carboxylic acid**	$C_{17}H_{10}O_2$. *See* p1930	246.27	gr lf or nd ($PhNO_2$, aa-PhCl)	327–8	....	....	....	δ	...	...	...	...	chl, con sulf s	**C52**, 11081
p1935	**Pyrethrin I**	Chrysanthemummono-carboxylic acid, pyrethrolone ester. $C_{21}H_{28}O_3$.	328.46	visc liq $[\alpha]_D^{25}$ −32.3 (eth, c = 5.66)	....	$170^{0.1}$ d	....	1.5192^{18}	i	s	s	...	...	peth, CH_3NO_2, CCl_4 s	**B9**[2], 46
p1936	**Pyrethrin II**	Chrysanthemumdicarboxylic acid, methyl ester, pyrethrolone ester. $C_{22}H_{28}O_5$.	372.47	visc liq $[\alpha]_D^{20}$ −6 (eth, c = 5)	....	$200^{0.1}$ d	....	1.5258^{20}	i	s	s	...	...	peth, CH_3NO_2, CCl_4 s	**B9**[2], 565
Ωp1937	**Pyridazine**	1,2-Diazine*. Orthodiazine.	80.09	λ^{al} 261 (3.78), 310 (2.93)	−8	208^{760} $47–8^1$	1.1035_4^{28}	1.5218^{20}	∞	∞	v	v	v	peth i	**B23**, 89
p1938	**—,3,6-dioxo-1,2,3,6-tetra-hydro-**	Maleic acid hydrazide. $C_4H_4N_2O_2$. *See* p1937	112.09	cr (w) $\lambda^{w,\,pH=8}$ 215 (4.2), 340 (3.4)	> 300d (296 −8d)	....	....	....	δ s^h	δ^h	...	...	...	....	**Am 80**, 3790
Ωp1939	**Pyridine**	Azine*	79.10	λ^{cy} 251 (3.1), 256 (3.1), 279 sh (2.0), 284 sh (1.8), 288 sh (1.4)	−42	115.5	0.9819^{20}	1.5095^{20}	∞	∞	∞	∞	∞	chl ∞	**B20**, 181
Ωp1940	**—,hydrochloride**	Pyridinium hydrochloride. $C_5H_5N.HCl$. *See* p1939	115.56	hyg pl or sc (al) λ^{al} 266 (3.7)	82	218–9	....	....	s	s	i	...	i	chl s	**B20**, 189
p1941	**—,nitrate**	Pyridinium nitrate. $C_5H_5N.HNO_3$. *See* p1939	142.12	nd (al)	sub	....	....	....	v	s	i	...	i	....	**B20**[1], 59
Ωp1942	**—,picrate**	Pyridinium picrate. $C_5H_5N.C_6H_3N_3O_7$. *See* p1939	308.22	ye nd (al)	167–8	....	....	....	s^h	s^h	...	...	...	....	**B20**[2], 122
p1943	**—,sulfate**	Pyridinium sulfate. $C_5H_5N.H_2SO_4$. *See* p1939	177.18	cr (al) $\lambda^{2N\,sulf}$ 250 sh (3.7), 255 (3.8), 262 sh (3.6)	....	....	....	....	∞	∞	i	...	...	....	**B20**, 190
p1944	**—,2-acetyl-**	Methyl 2-pyridyl ketone. C_7H_7NO. *See* p1939	121.14	ye in air λ^{hx} 227.5 (3.84), 267.5 (3.50)	....	192^{760} 78^{12}	....	1.5203^{20}	...	s	s	...	...	min ac s	**B21**[2], 247
p1945	**—,3-acetyl-**	Methyl 3-pyridyl ketone. C_7H_7NO. *See* p1939	121.14	ye in air $\lambda^{ac\text{-}al}$ 235 (4.0), 276 (3.6)	13–4	220^{760} 106^{12}	....	1.5341^{20}	s	s	s	...	...	ac s	**B21**[2], 247
p1946	**—,—,oxime**	$C_7H_8N_2O$. *See* p1939	136.17	cr (al, bz)	113 (130.5)	....	....	....	s^h	...	...	...	...	min ac s	**B21**[2], 247
p1947	**—,2-allyl-**	C_8H_9N. *See* p1939	119.17	....	....	190 $58–8.5^{10}$	0.959^{20}	....	δ	∞	∞	...	...	....	**C52**, 15520
Ωp1948	**—,2-amino-**	α-Pyridylamine. $C_5H_6N_2$. *See* p1939	94.12	lf (lig) λ^{al} 234 (4.1), 297 (3.6)	57.5–87 (59–60)	204 $104–6^{20}$ sub	....	....	...	s	s	s	s	os s	**B22**, 428
Ωp1949	**—,3-amino-**	β-Pyridylamine. $C_5H_6N_2$. *See* p1939	94.12	lf (bz-lig) λ^{al} 241 (4.0), 301 (3.4)	64–5	252 $131–2^{12}$	....	....	s	s	s	...	...	lig δ	**B22**, 431

For explanations, symbols and abbreviations see beginning of table. For structural formulas see end of table.

No.	Name	Synonyms and Formula	Mol. wt.	Color, crystalline form, specific rotation and λ_{max} (log ε)	m.p. °C	b.p. °C	Density	n_D	Solubility w	al	eth	ace	bz	other solvents	Ref.
	Pyridine														
Ω p1950	**—,4-amino-**	γ-Pyridylamine. $C_5H_6N_2$. *See* p1939	94.12	nd (bz) λ^{al} 246 (4.2)	158–9	180^{13}	...	...	s	v	s	...	s	lig δ	**B22**, 433
Ω p1951	**—,2-amino-5-bromo-**	$C_5H_5BrN_2$. *See* p1939	173.02	cr (bz)	137	...	...	...	...	v	i	...	s	lig i	**B22**, 431
Ω p1952	**—,5-amino-2-butoxy-**	$C_9H_{14}N_2O$. *See* p1939	166.23	...	...	$148–50^{12}$ $(146–7^2)$	1.037^{25}	1.5373^{20}	...	s	s	...	...	dil ac s	**C46**, 10285
Ω p1953	**—,2-amino-5-chloro-**	$C_5H_5ClN_2$. *See* p1939	128.56	pl	136–8	$127–8^{11}$	...	...	s	s	...	...	...	peth, lig i	**B22**^{2}, 332
p1954	**—,2-amino-3,5-dibromo-**	$C_5H_4Br_2N_2$. *See* p1939	251.92	nd (dil al, peth)	104–4.5	...	...	...	δ	s	s	...	...	lig δ	**B22**^{2}, 333
p1955	**—,2-amino-3,5-dichloro-**	$C_5H_4Cl_2N_2$. *See* p1939	163.01	nd or pr (dil al)	84–5	...	...	...	δ	s	...	s	...	lig s	**B22**^{2}, 333
p1956	**—,4-amino-2,6-dichloro-**	$C_5H_4Cl_2N_2$. *See* p1939	163.01	nd (dil al)	176	...	...	...	i, s^h	s	s	s	s	...	**B22**^{1}, 632
p1957	**—,4-amino-3,5-dinitro-**	$C_5H_4N_4O_4$. *See* p1939	184.11	ye pl (dil al)	170–1	...	...	...	δ^h	δ	...	...	...	...	**B22**^{2}, 341
p1958	**—,2-amino-5-iodo-**	$C_5H_5IN_2$. *See* p1939	220.01	nd or lf (dil al)	129	...	...	...	δ	s	v	...	...	...	**B22**^{2}, 334
Ω p1959	**—,2-amino-3-methyl-**	α-Amino-β-picoline. $C_6H_8N_2$. *See* p1939	108.15	hyg λ^w 230 (3.8), 294 (3.7)	33.5	221.5^{748} 95^8	...	...	v	s	s	s	s	os s lig δ	**B22**^{2}, 342
Ω p1960	**—,2-amino-4-methyl-**	α-Amino-γ-picoline. $C_6H_8N_2$. *See* p1939	108.15	lf or pl (lig)	100–0.5 (cor)	$115–7^{11}$ (cor)	...	...	v	s	s	s	s	os s lig i	**B22**^{2}, 342
Ω p1961	**—,2-amino-6-methyl-**	α′-Amino-α-picoline. $C_6H_8N_2$. *See* p1939	108.15	hyg (lig)	41	208–9	...	...	v	s	s	s	s	lig s^h	**B22**^{2}, 342
p1962	**—,3-amino-4-methyl-**	β-Amino-γ-picoline. $C_6H_8N_2$. *See* p1939	108.15	pr (bz-peth)	106	254^{735}	...	...	v	v	v	v	v	chl v lig i	**B22**^{2}, 343
p1963	**—,2-amino-3-nitro-**	$C_5H_5N_3O_2$. *See* p1939	139.12	ye nd (dil al)	165–7	...	...	...	δ	s	δ	...	...	...	**B22**^{2}, 335
Ω p1964	**—,2-amino-5-nitro-**	$C_5H_5N_3O_2$. *See* p1939	139.12	ye lf (dil al)	188	...	...	...	δ	s	δ	...	δ	lig δ	**B22**^{1}, 631
p1965	**—,4-amino-3-nitro-**	$C_5H_5N_3O_2$. *See* p1939	139.12	ye nd (w)	200	...	...	...	s^h	s	...	...	...	...	**B22**^{2}, 341
Ω p1966	**—,4-benzoyl-**	Phenyl 4-pyridyl ketone. $C_{12}H_9NO$. *See* p1939	183.21	nd (peth), pl (w)	71.5–3	314^{742} $170–2^{10}$	...	...	δ^h	s	s	...	s	...	**B21**^{2}, 277
Ω p1967	**—,2-benzyl-**	$C_{12}H_{11}N$. *See* p1939	169.23	nd λ^{al} 262.5 (3.67), 270 (3.54)	11–4	276^{742} 149^{16}	1.067^0_0	1.5785^{20}	i	s	s	...	...	...	**B20**, 425
p1968	**—,3-benzyl-**	$C_{12}H_{11}N$. *See* p1939	169.23	nd	34	286^{740}	1.061	...	i	s	s	...	...	...	**B20**, 426
Ω p1969	**—,4-benzyl-**	$C_{12}H_{11}N$. *See* p1939	169.23	λ^{al} 259 (3.5)	...	287^{742} $180–1^{31}$	1.0614^{20}_0	1.5818^{20}	i	s	v	...	...	...	**B20**^{2}, 272
Ω p1970	**—,2-bromo-**	C_5H_4BrN. *See* p1939	158.00	λ^{al} 260 sh (3.4), 265 (3.4), 270 sh (3.3)	...	$193–4^{764}$ $74–5^{13}$	1.657^{15}	1.5734^{20}	δ	s	s	...	...	os s	**B20**^{2}, 153
Ω p1971	**—,3-bromo-**	C_5H_4BrN. *See* p1939	158.00	nd (al) λ^{al} 265 sh (3.3), 268 (3.4), 275 sh (3.2)	142–3	$172–3^{752}$ $68–70^{18}$	1.645^0_4	1.5694^{20}	s	v	v	...	...	...	**B20**, 233
p1972	**—,5-bromo-2-hydroxy-**	C_5H_4BrNO. *See* p1939	174.00	pr (w, al)	177–8	...	...	...	δ, v^h	v^h	δ	...	...	dil alk, con ac v	**B21**^{1}, 202
Ω p1973	**—,2-chloro-**	C_5H_4ClN. *See* p1939	113.55	oil λ^{al} 260 sh (3.4), 265 (3.5), 270 (3.3)	...	170^{760} $54–8^{10}$	1.205^{15}	1.5320^{20}	δ	s	s	...	...	...	**B20**, 230
Ω p1974	**—,3-chloro-**	C_5H_4ClN. *See* p1939	113.55	λ^{al} 265 (3.4), 268 (3.4), 275 (3.3)	...	148^{744} $85–7^{100}$	...	1.5304^{20}	δ	...	...	...	...	...	**B20**, 230
Ω p1975	**—,4-chloro-**	C_5H_4ClN. *See* p1939	113.55	λ^{al} 258 (3.30)	−42.5	$147–8^{760}$ $85–7^{100}$	...	...	s	∞	...	...	...	...	**B20**, 231
Ω p1976	**—,2,3-diamino-** (one form)	$C_5H_7N_3$. *See* p1939	109.13	lf or pl (dil al)	122	$148–50^5$	...	...	δ	s	...	...	δ	peth i	**B22**^{1}, 647
Ω p1977	**—,—(one form)**	$C_5H_7N_3$. *See* p1939	109.13	nd (bz)	116	sub	...	...	s	s	...	...	s^h	...	**B22**^{2}, 394
p1978	**—,2,4-diamino-**	$C_5H_7N_3$. *See* p1939	109.13	hyg lf or nd	107	...	...	...	v	...	...	...	δ	peth i	**B22**^{2}, 394
p1979	**—,2,5-diamino-**	$C_5H_7N_3$. *See* p1939	109.13	nd	109–10	$180–5^{12}$	...	...	s	s	...	...	δ	peth i	**B22**^{2}, 394
p1980	**—,3,4-diamino-**	$C_5H_7N_3$. *See* p1939	109.13	nd or lf	218–9	...	...	...	i^h	...	...	...	...	...	**B22**^{2}, 395
p1981	**—,3,5-diamino-**	$C_5H_7N_3$. *See* p1939	109.13	hyg nd or lf	119–20	...	...	...	v	...	...	...	δ	peth i	**B22**^{1}, 648
Ω p1982	**—,3,5-dibromo-**	$C_5H_3Br_2N$. *See* p1939	236.90	nd (al)	112	222 sub 100	...	...	δ^h	s^h	s	...	...	...	**B20**, 233
p1983	**—,1,2-dihydro-1-methyl-2-oxo-**	C_6H_7NO. *See* p1939	109.13	nd λ^{al} 228.5 (3.66), 303 (3.57)	7	250^{740} 121^{10}	...	...	∞	...	...	...	...	peth, lig δ	**B21**^{2}, 229
p1984	**—,2,4-dihydroxy-**	2,4-Pyridinediol. $C_5H_5NO_2$. *See* p1939	111.10	rh bipym (al or w)	260–5 d	...	...	...	s	s	i	...	...	...	**B21**, 160
p1985	**—,2,6-dihydroxy-**, hydrate	2,6-Pyridinediol hydrate $C_5H_5NO_2.H_2O$. *See* p1939	129.12	ye pr	202–3 (195)	...	...	...	s	s	i	...	...	alk, ac δ	**B21**^{2}, 108
p1986	**—,3,5-diiodo-2-hydroxy-**	$C_5H_3I_2NO$. *See* p1939	346.89	bt br nd (AmOH)	261–2	...	...	...	δ^h	...	δ	...	...	AmOH s^h chl δ	**B21**^{2}, 32
p1987	**—,2,3-dimethyl-**	2,3-Lutidine. C_7H_9N. *See* p1939	107.15	λ^{al} 266 (3.56)	...	$163–4^{760}$ (160.7)	0.9319^{25}_4	1.5057^{20}	s	s	s	...	...	...	**B20**, 243

For explanations, symbols and abbreviations see beginning of table. For structural formulas see end of table.

No.	Name	Synonyms and Formula	Mol. wt.	Color, crystalline form, specific rotation and λ_{max} (log ε)	m.p. °C	b.p. °C	Density	n_D	Solubility						Ref.
									w	al	eth	ace	bz	other solvents	
	Pyridine														
Ω p1988	**—,2,4-dimethyl-**	2,4-Lutidine. C_7H_9N. *See* p1939	107.16	λ^{MeOH} 265 (3.4), 270 (3.3)	...	159–9.5^{760}	0.9309^{20}_{4}	1.5010^{20}	v i^h	v	v	s	...	...	**B20**^{2}, 160
Ω p1989	**—,2,5-dimethyl-**	2,5-Lutidine. C_7H_9N. *See* p1939	107.16	λ^{MeOH} 260 sh (2.5), 270 (3.6), 280 (3.4)	−16	156.8–.9^{760} (159–60)	0.9297^{20}_{4}	1.5006^{20}	δ v^h	v	∞	s	...	...	**B20**, 244
Ω p1990	**—,2,6-dimethyl-**	2,6-Lutidine. C_7H_9N. *See* p1939	107.16	λ^{MeOH} 267 (3.64)	−6.1	145.7^{760}	0.9226^{20}_{4}	1.4953^{20}	∞ s^h	δ	s	s	...	chl s	**B20**^{2}, 160
p1991	**—,3,4-dimethyl-**	3,4-Lutidine. C_7H_9N. *See* p1939	107.16	$\lambda^{0.1NH_2SO_4}$ 258 (3.7)	...	163.5–4.5^{760} (178.8)	0.9281^{25}_{4}	1.5096^{20}	δ	s	s	s	...	chl s	**B20**^{2}, 161
Ω p1992	**—,3,5-dimethyl-**	3,5-Lutidine. C_7H_9N. *See* p1939	107.16	$\lambda^{0.1NH_2SO_4}$ 268 (3.8)	...	171.6^{760}	0.9419^{20}_{4}	1.5061^{20}	s	s	s	s	...	...	**B20**^{2}, 161
Ω p1993	**—,2(dimethyl-amino)-**	$C_7H_{10}N_2$. *See* p1939	122.17	λ^{al} 250 (4.2), 312 (3.8)	...	196, 88^{15}	1.0157^{14}_{14}	1.5663^{20}	...	s	s	...	s	dil ac s	**B22**^{1}, 629
Ω p1994	**—,4(dimethyl-amino)-**	$C_7H_{10}N_2$. *See* p1939	122.17	pl (eth) $\lambda^{72\%\,diox}$ 258 (4.39), 290 (3.30)	114	...	...	...	v	v	s	...	v	chl v	**B22**^{2}, 341
p1995	**—,2(dimethyl-amino)-6-methyl-**	$C_8H_{12}N_2$. *See* p1939	136.20	...	...	198–200^{760}, 88^{15}	...	...	δ	s	s	...	...	...	**B22**^{2}, 342
Ω p1996	**—,3,5-dimethyl-2-ethyl-**	α-Parvoline. $C_9H_{13}N$. *See* p1939	135.21	λ^{MeOH} 255 (3.3), 265 (3.2)	...	198–9^{764}, 85–7^{15}	0.9338^{0}	...	s	v	v	v	...	...	**B20**^{2}, 166
Ω p1997	**—,2-ethenyl-**	α-Vinylpyridine. C_7H_7N. *See* p1939	105.14	λ^{al} 238 (4.1), 282 (3.8)	...	159–60, 50–5^{4}	0.9985^{20}_{0}	1.5495^{20}	δ	v	v	v	...	chl v	**B20**, 256
Ω p1998	**—,4-ethenyl-**	γ-Vinylpyridine. C_7H_7N. *See* p1939	105.14	red to dk-br λ^{al} 242.5 (4.12)	...	65^{15}	0.9800^{20}_{4}	1.5449^{20}	s^h	s^h	δ	...	...	...	**B20**^{2}, 170
Ω p1999	**—,2-ethyl-**	C_7H_9N. *See* p1939	107.16	$\lambda^{0.1NHCl}$ 263 (3.85)	−63.1	148.6^{760}	0.9502^{0}	1.4964^{20}	s	∞	v	v	...	...	**B20**, 241
Ω p2000	**—,3-ethyl-**	C_7H_9N. *See* p1939	107.16	λ^{cy} 257 (3.4), 262 (3.4), 268 sh (3.2)	−76.9	165^{760}	0.9539^{0}	1.5021^{20}	s	s	s	v	...	...	**B20**, 242
p2001	**—,4-ethyl-**	C_7H_9N. *See* p1939	107.16	λ^{MeOH} 255 (3.3), 265 (3.2)	−90.5	167.7^{760}	0.9417^{20}	1.5009^{20}	s	s	s	v	...	dil ac s	**B20**, 243
p2002	**—,2-ethyl-4-methyl-**	2-Ethyl-γ-picoline. $C_8H_{11}N$. *See* p1939	121.18	...	...	173–5^{748}	0.9239^{20}_{0}	...	...	s	s	v	...	...	**B20**^{2}, 162
p2003	**—,2-ethyl-6-methyl-**	6-Ethyl-α-picoline. $C_8H_{11}N$. *See* p1939	121.18	...	...	160–1.5^{760}, 73–6^{12}	0.9207^{25}_{4}	1.4920^{25}	δ	s	s	v	...	...	**B20**^{2}, 162
Ω p2004	**—,3-ethyl-4-methyl-**	β-Collidine. 3-Ethyl-γ-picoline. $C_8H_{11}N$. *See* p1939	121.18	...	...	198^{760}, 76^{12}	0.9286^{17}_{4}	...	δ	s	s	v	...	chl s	**B20**^{2}, 163
p2005	**—,4-ethyl-2-methyl-**	α-Collidine. 4-Ethyl-α-picoline. $C_8H_{11}N$. *See* p1939	121.18	...	...	179^{760}	0.9130^{25}_{4}	...	s δ^h	s	s	v	s	...	**B20**^{2}, 162
Ω p2006	**—,5-ethyl-2-methyl-**	Aldehyde collidine. Aldehydine. 5-Ethyl-α-picoline. $C_8H_{11}N$. *See* p1939	121.18	λ^{MeOH} 270 (3.6), 280 (3.5)	...	178.3^{760}, 65–6^{17}	0.9219^{20}_{20}	1.4971^{20}	δ^h	s	s	v	s	con sulf, dil ac s	**B20**, 248
—	**—,hexahydro-**	*see* **Piperidine**													
Ω p2007	**—,2-hydroxy-**	2-Pyridol. α-Pyridone. C_5H_5NO. *See* p1939	95.10	nd (bz) λ^{al} 227 (4.00), 297 (3.80)	106–7	280–1	...	...	s	s	δ	...	s	chl s lig δ	**B21**^{2}, 30
Ω p2008	**—,3-hydroxy-**	3-Pyridol. C_5H_5NO. *See* p1939	95.10	nd (bz) λ^{al} 215 (3.81), 279 (3.65)	129	...	...	...	s	s	δ	...	...	...	**B21**^{2}, 33
Ω p2009	**—,4-hydroxy-**	4-Pyridol. γ-Pyridone. C_5H_5NO. *See* p1939	95.10	pr or nd (w + 1) λ^{al} 260 (4.2)	148.5 (anh) 65 (hyd)	>350, 257–60^{10}	...	...	s	s	i	...	i	...	**B21**^{2}, 34
Ω p2010	**—,2(2-hydroxy-ethyl)-**	2(2-Pyridyl)ethanol. C_7H_9NO. *See* p1939	123.16	hyg	...	118–21^{15}	1.1111^{4}_{0}	1.5368^{20}	v	v	δ	...	...	chl v	**B21**, 50
p2011	**—,2(β-hydroxy-isopropyl)-**	2(2-Pyridyl)-1-propanol. $C_8H_{11}NO$. *See* p1939	137.18	...	...	128–31^{17}	...	...	v	v	δ	...	...	...	**B21**, 57
p2012	**—,3(α-hydroxy-isopropyl)-**	2(3-Pyridyl)-2-propanol. $C_8H_{11}NO$. *See* p1939	137.18	cr	58	140–1^{12}	...	...	s	s	...	...	...	chl s	**B21**^{2}, 37
Ω p2013	**—,2(hydroxy-methyl)-**	2-Pyridylmethanol. C_6H_7NO. *See* p1939	109.13	λ^{al} 260 (4.58), 266.5 (4.58)	...	112–3^{16}, 102.5^{8}	1.1317^{20}_{4}	1.5444^{20}	∞	v	v	v	v	os v	**B21**^{1}, 203
p2014	**—,2(1-hydroxy-propyl)-**	1(2-Pyridyl)-1-propanol. $C_8H_{11}NO$. *See* p1939	137.18	pa ye	...	213–6, 112–3^{13}	1.0501^{20}_{4}	1.5197^{20}	s	s	...	...	...	...	**B21**^{2}, 37
p2015	**—,2(2-hydroxy-propyl)-**	1(2-Pyridyl)-2-propanol. $C_8H_{11}NO$. *See* p1939	137.18	pr	32	123.5^{20}	...	...	v	v	...	...	...	chl v	**B21**^{2}, 37
Ω p2016	**—,2-isopropyl-**	$C_8H_{11}N$. *See* p1939	121.18	$\lambda^{0.1N\,NaOH}$ 261.5 (3.58), 268 (3,43)	...	159.8	0.9342^{0}	1.4915^{20}	δ	∞	∞	v	...	...	**B20**, 247
Ω p2017	**—,4-isopropyl-**	$C_8H_{11}N$. *See* p1939	121.18	λ^{al} 252 sh (3.3), 256 (3.3), 262 (3.2)	−54.9	178 (181.5)	0.9382^{25}_{4}	1.4962^{20}	δ	∞	∞	v	...	...	**B20**, 248
p2018	**—,4-methoxy-**	C_6H_7NO. *See* p1939	109.13	λ^{MeOH} 220 (3.9)	...	191^{738}, 95^{45}	...	...	∞	...	...	...	...	...	**B21**, 49

For explanations, symbols and abbreviations see beginning of table. For structural formulas see end of table.

No.	Name	Synonyms and Formula	Mol. wt.	Color, crystalline form, specific rotation and λ_{max} (log ε)	m.p. °C	b.p. °C	Density	n_D	Solubility w	al	eth	ace	bz	other solvents	Ref.
	Pyridine														
Ωp2019	**—,2-methyl-**	α-Picoline. C_6H_7N. *See* p1939	93.13	λ^{al} 256 (3.31), 262 (3.34), 269 (3.24)	−66.8	128.8^{760}	0.9443^{20}_{4}	1.4957^{20}	v	∞	∞	v	...	....	**B20**[2], 155
Ωp2020	**—,3-methyl**	β-Picoline. C_6H_7N. *See* p1939	93.13	λ^{MeOH} 260 (3.4), 265 (3.5), 270 (3.3)	−18.3	144.1^{760}	0.9566^{20}_{4}	1.5040^{20}	∞	∞	∞	v	...	....	**B20**[2], 157
Ωp2021	**—,4-methyl-**	γ-Picoline. C_6H_7N. *See* p1939	93.13	λ^{cy} 255 (3.20), 260 sh (3.12)	3.6	144.9^{760}	0.9548^{20}_{4}	1.5037^{20}	∞	∞	∞	s	...	....	**B20**[2], 158
p2022	**—,2(methyl-amino)-**	$C_6H_8N_2$. *See* p1939	108.15	λ^{eth} 300 (3.7)	15	200–1 90^{9}	1.052^{29}_{29}	....	s	v	v	...	s	aa v	**B22**[2], 325
p2023	**—,4(methyl-amino)-**	$C_6H_8N_2$. *See* p1939	108.15	pl (eth)	117–8	....	....	....	s	s	s	s	s	aa v	**B22**[2], 340
—	**—,2(*N*-methyl-2-pyrrolidyl)-**	*see* **2,2′-Nicotyrine**													
—	**—,3(*N*-methyl-2-pyrrolidyl)-**	*see* **Nicotine**													
—	**—,3(*N*-methyl-2-pyrrolidyl)-**	*see* **2′,3-Nicotyrine**													
Ωp2024	**—,2-phenyl-**	α-Pyridylbenzene. $C_{11}H_9N$. *See* p1939	155.20	λ^{al} 277 (4.02)	....	$270–2^{760}$ 146^{15}	1.0833^{25}	1.6210^{20}	δ	∞	∞	...	...	....	**B20**, 424
p2025	**—,3-phenyl-**	β-Pyridylbenzene. $C_{11}H_9N$. *See* p1939	155.20	pa ye oil λ^{al} 246 (4.1), 276 sh (3.8)	....	273– 4^{760} 118^{5}	....	1.6123^{25}	δ	s	s	...	...	....	**B20**, 424
Ωp2026	**—,4-phenyl-**	γ-Pyridylbenzene. $C_{11}H_9N$. *See* p1939	155.20	pl (w) λ^{al} 261 (4.22)	77–8	$280–2^{762}$	....	....	s^{h}	s	s	...	...	....	**B20**, 424
p2027	**—,2-propyl-**	Conyrine. $C_8H_{11}N$. *See* p1939	121.18	λ^{al} 262 (3.6)	2	$166–8^{760}$ 60^{11}	0.9119^{20}_{4}	1.4925^{20}	δ	∞	∞	v	...	....	**B20**[2], 161
p2028	**—,4-propyl-**	$C_8H_{11}N$. *See* p1939	121.18	....	....	184–6 80^{20}	0.9381^{15}	1.4966^{20}	δ	v	v	...	...	....	**B20**[2], 161
—	**—,3(2-pyrrolidyl)-**	*see* **Nornicotine**													
p2029	**—,2,3,4,6-tetra-methyl-**	β-Parvoline. Parvuline. $C_9H_{13}N$. *See* p1939	135.21	....	....	203^{750}	0.9322^{25}_{4}	1.5087^{25}	...	s	s	...	...	....	**C46**, 5587
p2030	**—,2,4,6-trihydroxy-**	2,4,6-Pyridinetriol. $C_5H_5NO_3$. *See* p1939	127.10	ye nd or pw	230d	....	....	....	δ	s	s	s	...	....	**B21**, 197
p2031	**—,2,3,5-trimethyl-**	$C_8H_{11}N$. *See* p1939	121.18	....	....	186.7 (184)	0.9352^{19}_{4}	1.5057^{25}	δ	s	s	s	s	....	**B20**[2], 163
p2032	**—,2,3,6-trimethyl-**	$C_8H_{11}N$. *See* p1939	121.18	λ^{cy} 273 (3.5)	....	$176–8^{759}$	0.9220^{25}_{4}	1.5053^{20}	s^{h}	s	s	s	s	....	**B20**[2], 163
p2033	**—,2,4,5-trimethyl-**	$C_8H_{11}N$. *See* p1939	121.18	....	....	188^{760}	0.9330^{25}_{4}	1.5054^{25}	δ	s	s	s	s	....	**B20**[2], 164
Ωp2034	**—,2,4,6-trimethyl-**	γ-Collidine. *sym*-Collidine. $C_8H_{11}N$. *See* p1939	121.18	λ^{MeOH} 264 (3.58)	−44.5	170.5^{762} (175–8)	0.9166^{22}_{4}	1.4959^{25}	s δ^{h}	s	s	s	...	....	**B20**[2], 164
Ωp2035	**4-Pyridinecarbox-aldehyde**	Isonicotinaldehyde	107.11	λ^{al} 258 (3.35)	....	$77.3–8.1^{12}$	....	1.5423^{20}	s	...	s	...	...	....	**B21**, 287
Ωp2036	**2-Pyridine-carboxylic acid**	Picolinic acid	123.11	nd (w, al, bz) λ^{al} 264 (3.51)	136–7	sub	....	....	δ s^{h}	s	i	...	δ^{h}	aa v chl, CS_2 i	**B22**[2], 30
Ωp2037	—,amide	Picolinamide. $C_6H_6N_2O$. *See* p2036	122.13	mcl pr (w) $\lambda^{w,\ pH=6}$ 265 (3.6)	107–8	....	....	....	δ	s	...	...	s	lig i	**B22**[2], 31
p2038	—,ethyl ester	$C_8H_9NO_2$. *See* p2036	151.17	ye in air $\lambda^{95\%\ al}$ 262 (3.55)	0–2	243 122^{13}	1.1194^{20}	1.5104^{20}	∞	∞	∞	...	...	....	**B22**[2], 31
Ωp2039	—,nitrile	Picolinonitrile. $C_6H_4N_2$. *See* p2036	104.11	nd or pr (eth) λ^{cy} 265 (3.44), 278 sh (2.53)	29	222–7 (212–5)	1.0810^{25}_{4}	1.5242^{25}	s	v	v	...	v	chl s lig δ	**B22**, 36
Ωp2040	**3-Pyridine-carboxylic acid**	Nicotinic acid. Niacin	123.11	nd (w, al) λ^{al} 257 (3.36), 262 (3.42)	236–7	sub	1.473	....	δ s^{h}	δ s^{h}	δ	...	...	....	**B22**[2], 32
Ωp2041	—,amide	Niacin amide. Nicotamide. Nicotine amide. Pellagra preventive vitamin. P. P. factor. $C_6H_6N_2O$. *See* p2040	122.13	wh pw, nd (bz) λ^{al} 255 (3.4), 262.5 (3.4)	129–31	150– $60^{5.10^{-4}}$	1.400	1.466	v	v	...	...	...	glycerol v	**B22**[2], 34
Ωp2042	—,—,*N*,*N*-diethyl-	Coramine. Cardiamine. $C_{10}H_{14}N_2O$. *See* p2040	178.24	yesh	24–6	280 d 175^{25}	1.060^{25}_{4}	1.525^{20}	∞	∞	∞	∞	...	chl ∞	**B22**[2], 34
p2043	—,ethyl betaine	....	151.17	hyg pl	84–6	....	....	....	v	δ	i	...	...	os i	**B22**, 43
Ωp2044	—,ethyl ester	$C_8H_9NO_2$. *See* p2040	151.17	λ^{al} 262 (3.46)	8–9.5	224^{760} $103–5^{5}$	1.1070^{20}	1.5034^{20}	v	v	v	...	v	....	**B22**[2], 33
Ωp2045	—,hydrochloride	$C_6H_5NO_2.HCl$. *See* p2040	159.57	pr or pl, rh bipym (w)	274	....	....	....	s	s	...	...	...	....	**B22**[2], 33
Ωp2046	—,methyl ester	$C_7H_7NO_2$. *See* p2040	137.14	cr λ^{al} 260 (3.4)	42–3	204 118.5^{25}	....	....	s	s	...	...	s	....	**B22**[2], 33
Ωp2047	—,nitrile	Nicotinonitrile. $C_6H_4N_2$. *See* p2040	104.11	nd (peth-eth, lig) λ^{cy} 265 (3.35), 279 sh (2.63)	50–2	240–5	....	....	v	v	v	...	v	lig δ, s^{h}	**B22**[2], 35
Ωp2048	**4-Pyridine-carboxylic acid**	Isonicotinic acid.	123.11	nd (w) λ^{al} 271.5 (3.41)	319 (325–6 sealed tube)	sub at 260^{15}	....	....	δ s^{h}	δ^{h}	δ	...	δ	....	**B22**[2], 37

For explanations, symbols and abbreviations see beginning of table. For structural formulas see end of table.

No.	Name	Synonyms and Formula	Mol. wt.	Color, crystalline form, specific rotation and λ_{max} (log ε)	m.p. °C	b.p. °C	Density	n_D	Solubility w	al	eth	ace	bz	other solvents	Ref.
	4-Pyridinecarboxylic acid														
p2049	—,anhydride	$C_{12}H_8N_2O_3$. *See* p2048	228.22	nd	103–4 (302d sealed tube)	...	...	...	...	...	...	...	...	...	**B22**[2], 37
p2050	—,ethyl betaine	...	151.17	nd	241d	...	...	...	v	v	...	...	...	...	**B22**, 47
Ω p2051	—,ethyl ester	$C_8H_9NO_2$. *See* p2048	151.17	nd	23	220^{760} 110^{15}	1.1052^{20}	1.5177^{20}	δ	s	v	...	s	chl v	**B22**[2], 37
Ω p2052	—,hydrazide	Isoniazid. $C_6H_7N_3O$. *See* p2048	137.14	nd (al) λ^{al} 263 (3.6)	171 (163)	...	...	...	...	...	...	...	...	...	**B22**[2], 37
Ω p2053	—,methyl ester	$C_7H_7NO_2$. *See* p2048	137.14	$\lambda^{w,\,pH=6}$ 278 (3.44)	8.5	209^{760} δd 104^{21}	1.1599^{20}_{4}	1.5135^{20}	δ	s	s	...	s	...	**B22**, 46
Ω p2054	—,nitrile	Isonicotinonitrile. $C_6H_4N_2$. *See* p2048	104.11	nd (lig-eth) λ^{cy} 271 (3.45) 290 sh (2.70)	83	...	...	...	s	s	s	...	s	lig δ	**B22**, 46
p2055	**3-Pyridine-carboxylic acid, 6-amino-**	$C_6H_6N_2O_2$. *See* p2040	138.13	cr (dil aa + 2w)	312	...	...	...	...	δ	...	...	...	os δ	**B22**[2], 464
Ω p2056	**4-Pyridine-carboxylic acid, 2,6-dihydroxy-**	Citrazinic acid. $C_6H_5NO_4$. *See* p2048	155.11	yesh-grsh pw (w)	>330d	...	...	...	s^h	...	...	...	...	alk s HCl δ^h	**B22**, 254
p2057	**3-Pyridine-carboxylic acid, 2-hydroxy-**	$C_6H_5NO_3$. *See* p2040	139.11	nd (w) λ^{al} 234 (3.7), 328 (3.8)	α 259–61d β 301–2d	...	...	...	δ s^h	δ	δ	...	...	chl i	**B22**[2], 165
p2058	—,4-hydroxy-	$C_6H_5NO_3$. *See* p2040	139.11	nd (w + 2), cr (al + 2w)	254–5	...	...	...	s^h	δ	...	...	...	chl δ	**B22**, 214
Ω p2059	—,6-hydroxy-	$C_6H_5NO_3$. *See* p2040	139.11	nd (w) $\lambda^{w,\,pH=6.8}$ 250 (4.1), 295 (3.7)	304 d	sub	...	...	δ^h	i	i	...	i	chl i	**B22**[2], 165
p2060	**4-Pyridine-carboxylic acid, 3-hydroxy-5(hydroxy-methyl)-2-methyl-, lactone**	$C_8H_7NO_3$. *See* p2048	165.15	cr (MeOH)	273–3.5d (260d)	...	...	...	s	...	...	...	...	MeOH δ, s^h	**C43**, 3045
p2061	**3-Pyridine-carboxylic acid, 5-hydroxy-4(hydroxy-methyl)-6-methyl-, lactone**	$C_8H_7NO_3$. *See* p2040	165.15	nd (al)	282–3d	...	...	...	s	δ s^h	...	...	...	...	**C53**, 21931
p2062	—,1-methyl-	Trigonelline	137.14	pr (aq al + 1w) λ^{al} 265 (3.6), 272 sh (3.5)	218d (anh) 130 (hyd)	...	...	...	v	δ v^h	δ	...	δ	chl δ	**B22**[2], 35
p2063	**2-Pyridine-carboxylic acid, 1,2,3,6-tetrahydro-**	Baikiain. $C_6H_9NO_2$. *See* p2036	127.14	pr (MeOH) $[\alpha]^{20}_{D}$ −201.6	274d	...	...	...	v	δ	i	i	i	MeOH s^h AcOEt i	**J1950**, 3590
Ω p2064	**2,3-Pyridinedicar-boxylic acid**	Quinolinic acid	167.12	mcl pr (w) λ^{MeOH} 274 (3.57)	228–9 (191) (rapid htng)	...	...	...	δ	i	i	...	i	...	**B22**, 150
Ω p2065	**2,4-Pyridinedicar-boxylic acid**	Lutidinic acid	167.12	lf (w + 1) λ^{MeOH} 219 (3.63), 258 (3.36)	248–50 (anh)	...	0.942	...	δ s^h	s^h	i	...	i	CS_2 i	**B22**[2], 105
Ω p2066	**2,5-Pyridine-dicarboxylic acid**	Isocinchomeronic acid	167.12	lf (w + 1), (al, dil HCl) λ^{MeOH} 223 (3.97), 272.5 (3.85)	256–8d	sub	...	...	s^h	δ^h	i	...	i	HCl s^h	**B22**[2], 105
Ω p2067	**2,6-Pyridine-dicarboxylic acid**	Dipicolinic acid	167.12	nd (w + 1½) λ^{MeOH} 219 (3.85), 270 (3.79)	252 (anh) (228d sealed tube)	...	...	...	δ s^h	δ^h	...	...	...	aa δ	**B22**[2], 106
Ω p2068	**3,4-Pyridine-dicarboxylic acid**	Cinchomeronic acid	167.12	pr, nd or lf (w) λ^{MeOH} 263.5 (3.57)	262d	sub	...	...	δ^h	δ	i	...	δ	chl i	**B22**[2], 106
Ω p2069	**3,5-Pyridine-dicarboxylic acid**	Dinicotinic acid	167.12	cr (ac-w), pr (gl aa) λ^{MeOH} 268.5 (3.48)	325d	sub	...	...	i	...	δ	...	...	HCl s^h aa δ	**B22**[2], 107
p2070	**3,4-Pyridine-dicarboxylic acid, 4,5-dihydro-2,6-dimethyl-,** diethyl ester	$C_{13}H_{19}NO_4$. *See* p2068	253.30	bl flr, pl (eth)	85	...	...	...	i	v	s^h	s	...	os s lig i	**B22**[2], 98

No.	Name	Synonyms and Formula	Mol. wt.	Color, crystalline form, specific rotation and λ_{max} (log ε)	m.p. °C	b.p. °C	Density	n_D	Solubility w	al	eth	ace	bz	other solvents	Ref.
	3,5-Pyridinedicarboxylic acid														
Ω p2071	**3,5-Pyridine-dicarboxylic acid, 1,4-dihydro-2,6-dimethyl-,** diethyl ester	$C_{13}H_{19}NO_4$. *See* p2069	253.30	ye nd or lf (al) λ^{al} 232 (4.3), 375 (3.8)	184–5	...	...	...	δ^h	s^h	δ	...	...	chl v	**B22²**, 98
Ω p2072	**—,1,4-dihydro-2,4,6-trimethyl-,** diethyl ester	$C_{14}H_{21}NO_4$. *See* p2069	267.33	lt bl flr pl (al) λ^{al} 234 (4.3), 350 (3.9)	131	...	...	...	δ	δ v^h	δ	...	...	chl v CS_2 δ	**B22²**, 100
p2073	**3,4-Pyridine-dicarboxylic acid, 2,6-dimethyl-,** diethyl ester	$C_{13}H_{17}NO_4$. *See* p2068	251.29	...	16	270d 163^{13}	...	...	...	...	...	...	...	...	**B22²**, 109
Ω p2074	**3,5-Pyridine-dicarboxylic acid, 2,6-dimethyl-,** diethyl ester	$C_{13}H_{17}NO_4$. *See* p2069	251.29	nd (al), pr (eth) λ^{MeOH} 233 (4.1), 275 (3.6)	75–6 (73)	301–2 180^{16}	...	...	i	s	s	...	s	chl, lig s	**B22²**, 110
p2075	**3,4-Pyridine-dicarboxylic acid, 5-methoxy-6-methyl-**	$C_9H_9NO_5$. *See* p2068	211.18	cr (w-ace)	213–5	...	...	...	s^h	...	...	s^h	...	...	**C40**, 1636
p2076	**2,4-Pyridine-dicarboxylic acid, 6-methyl-**	Uvitonic acid. $C_8H_7NO_4$. *See* p2065	181.15	cr (w)	282d (>330)	...	...	...	δ	...	...	...	i	HCl s aa s^h chl δ^h CS_2 i	**B22²**, 107
—	**Pyridinediol**	*see* **Pyridine, dihydroxy-**													
p2077	**Pyridinepenta-carboxylic acid,** dihydrate	...	335.18	cr (eth, w)	220d	...	...	...	v	δ	i	...	...	...	**B22**, 190
p2078	**3-Pyridinesulfonic acid**	...	159.16	orh $\lambda^{50\%\,al}$ 208 (3.8), 262 (3.45)	357d	...	1.718^{25}_{25}	...	v	δ	i	...	...	...	**B22**, 387
p2079	**2,3,4,5-Pyridine-tetracarboxylic acid**	...	255.14	cr (w + 2 or 3)	160d (−w115)	...	...	...	s	...	...	...	...	...	**B22**, 188
p2080	**2,3,4,6-Pyridine-tetracarboxylic acid**	...	255.14	nd (w + 3)	236d	...	...	...	v	δ	δ	...	...	aa s	**B22²**, 142
p2081	**2,3,5,6-Pyridine tetracarboxylic acid**	...	255.14	cr (w + 2)	200d	...	...	...	v	...	...	...	...	...	**B22¹**, 544
p2082	**2,3,4-Pyridine-tricarboxylic acid**	α-Carbocinchomeronic acid	211.13	lf (w + 1½)	250 (anh)	...	...	...	s^h	δ s^h	i	...	i	...	**B22²**, 136
p2083	**2,3,5-Pyridine-tricarboxylic acid**	Carbodinicotinic acid	211.13	pl, lf or nd (w + 2, dil al + 2w)	323 (anh)	...	...	...	v	δ	δ	...	...	dil al v aa δ	**B22²**, 136
p2084	**2,4,5-Pyridine-tricarboxylic acid**	Berberonic acid	211.13	tcl pr (dil HCl + 2w) λ^{MeOH} 275 (3.04)	243 (anh) 235 (hyd)	...	...	...	δ s^h	δ^h	i	...	i	dil ac s chl i	**B22²**, 136
p2085	**2,4,6-Pyridine-tricarboxylic acid**	Trimesitic acid	211.13	nd (w + 2)	227d	sub	...	...	δ s^h	δ s^h	δ	...	...	...	**B22²**, 136
p2086	**3,4,5-Pyridine-tricarboxylic acid**	β-Carbocinchomeronic acid	211.13	lf or pl (w + 3), ta (w)	261d (−w115)	...	...	...	δ s^h	s	...	...	...	...	**B22**, 186
p2087	**Pyridinium, 1-ethyl-,** bromide	...	188.08	cr (al)	111–2	...	...	...	s	s	i	...	...	...	**B20**, 214
Ω p2088	**—,1-hexadecyl-,** chloride	...	340.00	wh pw	77–83	...	...	...	v	...	δ	...	δ	chl v	**C51**, 4370
—	**Pyridone**	*see* **Pyridine, hydroxy-**													
p2089	**Pyridoxal,** hydrochloride	...	203.63	rh λ^{w} 318 (3.91), 390 (2.30) $\lambda^{0.1N HCl}$ 288 (3.93)	165d	...	...	...	v	δ	...	...	...	...	**C55**, 17657
p2090	**—,oxime**	...	182.18	cr (al)	225–6d	...	...	...	...	s^h	...	...	...	...	**C44**, 10740
Ω p2091	**Pyridoxamine,** dihydrochloride	...	241.12	pl (al) λ^{w} 253 (3.66), 328 (3.89) $\lambda^{0.1N HCl}$ 293 (3.93)	226–7d	...	...	...	v	δ s^h	...	...	...	...	**C51**, 14833
—	**Pyridoxin**	*see* **Vitamin B_6**													
Ω p2092	**Pyrimidine**	1,3-Diazin*. Miazine	80.09	λ^{al} 243 (3.47), 280 (2.57)	22	123.5–4[760]	...	1.4998^{20}	∞	s	...	...	...	...	**B23**, 89
Ω p2093	**—,2-amino-**	$C_4H_5N_3$. *See* p2092	95.11	nd (AcOEt) λ^{MeOH} 227 (4.22), 297 (3.59)	127–8	sub	...	...	s	...	...	...	...	...	**B24**, 80
p2094	**—,4-amino-**	$C_4H_5N_3$. *See* p2092	95.11	pl (AcOEt) λ^{MeOH} 236 (4.30), 272 (3.71)	151–2	...	...	...	s	s	...	...	...	AcOEt s^h	**B24**, 81

For explanations, symbols and abbreviations see beginning of table. For structural formulas see end of table.

No.	Name	Synonyms and Formula	Mol. wt.	Color, crystalline form, specific rotation and λ_{max} (log ε)	m.p. °C	b.p. °C	Density	n_D	Solubility w	al	eth	ace	bz	other solvents	Ref.
	Pyrimidine														
p2095	**—,2-amino-4,6-dihydroxy-**	$C_4H_5N_3O_2$. *See* p2092	127.10	pl (w + 1) λ^{w} 257 (4.14)	>330	...	...	...	i	i	...	...	...	ac, alk s PhNO₂ δ^h	**B24**, 468
p2096	**—,6-amino-2,4-dihydroxy-**	4-Aminouracil. $C_4H_5N_3O_2$. *See* p2092	127.10	cr (w)	d	...	...	...	s^h	...	...	...	...	min ac v aa i	**B24**, 469
p2097	**—,2-amino-4,5-dimethyl-**	$C_6H_9N_3$. *See* p2092	123.16	nd (w)	214–5	sub	...	...	s^h	s	δ	s	s	chl s lig δ	**B24**, 91
Ωp2098	**—,4-amino-2,6-dimethyl-**	Kyanmethin. $C_6H_9N_3$. *See* p2092	123.16	nd (al), pl (bz) λ^{al} 235 (4.02), 267 (3.62)	183 (191–2)	sub	...	...	δ	δ s^h	...	...	δ s^h	...	**B24**, 89
p2099	**—,6-amino-4,5-dimethyl-**	$C_6H_9N_3$. *See* p2092	123.16	nd (w)	230	...	...	...	s^h	s^h	δ	s	s	chl s lig i	**B24**, 92
Ωp2100	**—,2-amino-4-methyl-**	$C_5H_7N_3$. *See* p2092	109.13	pl (w), nd (sub) $\lambda^{w, pH=6.8}$ 290 (3.6)	159–60	sub	...	...	s^h	s	...	...	...	...	**B24**, 84
p2101	**—,2-amino-5-methyl-**	$C_5H_7N_3$. *See* p2092	109.13	pl (sub), pr (w)	193.5	sub	...	...	s^h	s	i	...	s^h	lig s^h	**B24**, 87
p2102	**—,4-amino-2-methyl-**	$C_5H_7N_3$. *See* p2092	109.13	rh (ace)	205	...	...	...	v	s	...	s	s	...	**B24**, 84
p2103	**—,4-amino-5-methyl-**	$C_5H_7N_3$. *See* p2092	109.13	pl (al, AcOEt)	176	...	...	...	v	s^h	δ	...	s	AcOEt s^h chl, peth, lig δ	**B24**, 87
p2104	**—,4-amino-6-methyl-**	$C_5H_7N_3$. *See* p2092	109.13	pr (w), nd, lf (sub) $\lambda^{w, pH=13}$ 234 (3.95), 264 (3.45)	197–7.5	sub	...	...	s	s	...	...	...	...	**B24**, 85
Ωp2105	**—,2-amino-5-nitro-**	$C_4H_4N_4O_2$. *See* p2092	140.10	nd (al) λ^{w} 317.5 (4.18)	236–7	...	...	...	δ	s	i	s	i	alk s chl i	**B24**[1], 231
p2106	**—,2-chloro-4(dimethylamino)-6-methyl-**	$C_7H_{10}ClN_3$. *See* p2092	171.63	br wx so	87	$140–7^{4}$	...	...	i	s	...	...	...	...	**C36**, 911
Ωp2107	**—,2,4-dichloro-5-methyl-**	$C_5H_4Cl_2N_2$. *See* p2092	163.01	pl (al)	25–7	235^{759}	...	...	δ	v	v	...	v	chl v	**B23**, 93
p2108	**—,2,4-dichloro-6-methyl-**	$C_5H_4Cl_2N_2$. *See* p2092	163.01	nd (lig)	46–7	219	...	...	δ	v	v	...	v	chl v lig s^h	**B23**, 92
p2109	**—,2-hydroxy-, hydrochloride**	2-Pyrimidone hydrochloride. $C_4H_4N_2O.HCl$. *See* p2092	132.55	rods (al)	205	...	...	...	v	s^h	...	...	...	...	**B24**[1], 231
p2110	**—,2-methyl-**	$C_5H_6N_2$. *See* p2092	94.12	λ^{hx} 245 (3.39), 249 (3.43), 255 (3.24), 295 (2.55)	−4	138^{758}	...	...	∞	...	...	...	...	...	**B23**, 92
p2111	**—,4(methylamino)-**	$C_5H_7N_3$. *See* p2092	109.13	lf (al), nd (peth) $\lambda^{w, pH=2.1}$ 254 (4.20)	74–5	$142–4^{16}$	...	...	v	s^h	s^h	...	...	...	**B24**[2], 28
—	**2,4-Pyrimidinedione**	*see* **Uracil**													
—	**2(1)-Pyrimidone, 4-amino-**	*see* **Cytosine**													
p2112	**Pyrocalciferol**	9α-Lumisterol. $C_{28}H_{44}O$	396.67	nd (MeOH), $[\alpha]_D^{20}$ +512 (al) λ^{al} 274 (2.44), 284.5 (2.45), 296 (2.47)	93–5	...	...	...	i	s	...	...	...	chl s MeOH s^h	**E14**, 77
—	**Pyrocatechol**	*see* **Benzene, 1,2-dihydroxy-***													
—	***o*-Pyrocatechuic acid**	*see* **Benzoic acid, 2,3-dihydroxy-**													
—	**Pyrocinchonic acid, anhydride**	*see* **Maleic acid, dimethyl-, anhydride**													
p2113	**Pyrocoll**	...	186.17	ye mcl pl (aa)	268	sub	...	...	i	δ	δ	...	δ	chl s aa δ, s^h	**B24**[1], 360
—	**Pyrogallol**	*see* **Benzene, 1,2,3-trihydroxy-***													
—	**Pyrogallophthalein**	*see* **Gallein**													
—	**Pyromellitic acid**	*see* **1,2,4,5-Benzenetetracarboxylic acid***													
—	**Pyromucic acid**	*see* **2-Furancarboxylic acid**													
p2114	**4-Pyrone**	4H-Pyran-4-one. 4-Oxo-1,4-pyran. γ-Pyrone.	96.09	hyg cr λ^{MeOH} 246 (4.2)	32.5	$215–7^{742}$ 105^{23}	1.190	1.5238	v	s	v	...	s	aa, chl v peth, CS_2 δ	**B17**, 271
p2116	**2-Pyrone, 4,6-dimethyl-**	5-Hydroxy-3-methyl-2,4-hexadienoic acid lactone.* Mesitenlactone.	124.15	lf (eth)	51.5	245 126^{11}	...	...	v	v	s	...	...	CS_2 δ	**B17**, 291
p2117	**4-Pyrone, 2,6-dimethyl-**	$C_7H_8O_2$. *See* p2114	124.15	pl, nd (sub) λ^{MeOH} 208 (3.81), 245 (4.09)	132	$248–9^{713}$ $139–40^{25}$ sub	0.9953^{137}	...	s	s	s	s	...	...	**B17**[2], 315
p2118	**2-Pyrone, 3-hydroxy-**	2,5-Dihydroxy-2,4-hexadienoic acid 5-lactone*. Isopyromucic acid.	112.09	nd (w + 2)	95 (anh) 80–5 (hyd)	112^{20}	...	...	s	s	s	...	s^h	chl s lig s^h CS_2 δ	**B17**[1], 233

For explanations, symbols and abbreviations see beginning of table. For structural formulas see end of table.

No.	Name	Synonyms and Formula	Mol. wt.	Color, crystalline form, specific rotation and λ_{max} (log ε)	m.p. °C	b.p. °C	Density	n_D	Solubility w	al	eth	ace	bz	other solvents	Ref.
	4-Pyrone														
p2119	**4-Pyrone, 5-hydroxy-2(hydroxy-methyl)-**	Kojic acid. $C_6H_6O_4$. *See* p2114	142.11	nd, pr (ace) λ^{al} 268 (3.92)	153–5	sub	...	...	δ s^h	s	s	s^h	δ	AcOEt, chl, Py s aa δ	**B18**[2], 57
p2120	**—,3-hydroxy-2-methyl-**	Larixinic acid. Maltol. $C_6H_6O_3$. *See* p2114	126.11	red mcl pr (chl), bipym pr (50 % al, sub), nd (to)	162–4	sub at 93	...	...	δ v^h	...	δ	...	δ	chl v alk, ac s aa δ peth i	**B17**[2], 450
—	**2-Pyrone-5-carboxylic acid**	*see* **Coumalic acid**													
p2121	**4-Pyrone-2-carboxylic acid, 5,6-dihydro-6,6-dimethyl-, butyl ester**	Butopyronoxyl. Indalone...	226.28	ye or pa red-br λ^{al} 283 (4.00)	...	256–70 113–4^{14}	1.052^{25}	1.4745^{25}	i	∞	∞	...	∞	chl ∞	**C53**, 6145
p2122	**—,5-hydroxy-**	Comenic acid	156.10	ye cr λ^{w} 224 (4.32), 290 (3.76)	>270d	...	...	...	s^h	i s^h	...	...	...	...	**B18**[2], 352
—	**4-Pyrone-2,6-dicarboxylic acid**	*see* **Chelidonic acid**													
p2123	**Pyrophosphoric acid, tetraamide, octamethyl-**	OMPA. Scharadan. $[(CH_3)_2N]_4P_2O_3$	286.25	...	...	118–22$^{0.3}$	1.1343^{25}	1.462^{25}	s	s	...	...	...	chl s	**C52**, 243
p2124	**—,tetraethyl ester**	$(C_2H_5O)_4P_2O_3$	290.19	...	...	155^{5} 104–10$^{0.08}$	1.1847^{20}_{4}	1.4180^{20}	$\infty^{\delta d}$	∞	∞	∞	...	chl, xyl ∞ lig i	**B1**[3], 1330
p2125	**—,dithiono-, tetraethyl ester**	Sulfotep. $(C_2H_5O)_4P_2OS_2$	322.32	col oil	...	136–9^{2} 110–3$^{0.2}$	1.196^{25}_{4}	1.4753^{25}	i	s	...	...	...	...	**C51**, 8641
—	**Pyrotartaric acid**	*see* **Succinic acid, methyl-**													
—	**Pyrrocoline, octahydro-**	*see* **Piperolidine**													
Ωp2126	**Pyrrole**	Azole*	67.09	λ^{cy} 324 (4.47)	...	130–1^{761}	0.9691^{20}_{4}	1.5085^{20}	δ	s	s	s	s	...	**B20**, 159
p2127	**—,1-acetyl-**	C_6H_7NO. *See* p2126	109.13	λ^{al} 238.5 (4.03), 288sh (2.88)	...	181–2	...	...	i	...	...	...	...	HCl d	**B20**, 165
p2128	**—,2-acetyl-**	C_6H_7NO. *See* p2126	109.13	mcl nd (w) λ^{MeOH} 251 (3.61), 290 (2.41)	90	220^{760}	...	...	s	s	s	...	...	...	**B21**[2], 236
p2129	**—,1-benzyl-**	$C_{11}H_{11}N$. *See* p2126	157.22	λ^{al} 208 (4.14)	15	247 138^{27}	...	1.5655^{24}	i	v	v	...	...	...	**B20**[1], 39
p2130	**—,1-butyl-**	$C_8H_{13}N$. *See* p2126	123.20	oil	...	170–1^{760} 53–4^{11}	...	1.4727^{20}	...	...	...	...	...	...	**B20**[2], 83
—	**—,dihydro-**	*see* **Pyrroline**													
p2131	**—,2,4-dimethyl-**	C_6H_9N. *See* p2126	95.15	pa bl flr λ^{al} 218 (3.67)	...	171 62–3^{10}	0.9236^{20}_{4}	1.5048^{20}	δ	v	v	...	v	aa δ	**B20**, 172
p2132	**—,2,5-dimethyl-**	C_6H_9N. *See* p2126	95.15	λ^{w} 212 (3.89)	...	170–2^{765} (165) 50–3^{8}	0.9353^{20}_{4}	1.5036^{20}	i	v	v	...	...	...	**B20**, 173
p2133	**—,2,3-dimethyl-4-ethyl-**	Hemopyrrole. $C_8H_{13}N$. *See* p2126	123.20	...	16–7	198^{725} 113^{16}	0.915^{20}_{4}	...	v^h	...	...	...	...	...	**B20**[2], 91
Ωp2134	**—,2,4-dimethyl-3-ethyl-**	Kryptopyrrole. $C_8H_{13}N$. *See* p2126	123.20	pr λ^{al} 213 (3.85)	0	197^{710} 96^{16}	0.913^{20}_{4}	1.4961^{20}	δ	s	s	...	s	chl s	**B20**[2], 91
p2135	**—,1,3-diphenyl-**	$C_{16}H_{13}N$. *See* p2126	219.29	pl (al)	122–3	...	...	...	...	s^h	v	...	v	chl s	**B20**[1], 148
Ωp2136	**—,2,5-diphenyl-**	$C_{16}H_{13}N$. *See* p2126	219.29	lf (aa, dil al) λ^{al} 230 (4.08), 325 (4.42), 330 (4.43)	143.5	...	...	...	i	v	v	...	v	aa v HCl δ (red) alk i	**B20**[2], 313
p2137	**—,1-ethyl-**	C_6H_9N. *See* p2126	95.15	...	...	129–30^{762}	0.9009^{20}_{4}	1.4841^{20}	i	s	...	...	...	...	**B20**, 163
p2138	**—,2-ethyl-**	C_6H_9N. *See* p2126	95.15	...	...	163–5^{760} 59–60^{15}	0.9042^{20}_{4}	1.4942^{20}	i	s	...	...	...	...	**B20**[2], 85
p2139	**—,3-ethyl-4-methyl-**	Opsopyrrole. $C_7H_{11}N$. *See* p2126	109.17	ye oil	3	70^{11}	0.9059^{20}_{4}	1.4913^{20}	i	s	s	...	...	...	**B20**[2], 89
p2140	**—,3-ethyl-2,4,5-trimethyl-**	Phyllopyrrole. $C_9H_{15}N$. *See* p2126	137.23	pl (sub), lf (eth), wh lf (peth)	67.5–8.5	213^{725} 92–3^{12}	...	...	i	s	s	...	.	lig δ	**B20**[2], 93
p2141	**—,2-isopropyl-**	$C_7H_{11}N$. *See* p2126	109.17	...	...	171–2^{741}	0.908^{25}_{4}	1.491^{25}	i	s	s	...	...	...	**B20**, 176
Ωp2142	**—,1-methyl-**	C_5H_7N. *See* p2126	81.12	...	...	114–5^{747} (cor)	0.9145^{15}_{4}	1.4875^{20}	i	∞	∞	...	...	...	**B20**, 163
p2143	**—,2-methyl-**	C_5H_7N. *See* p2126	81.12	...	...	147–8^{750}	0.9446^{15}_{4}	1.5035^{16}	i	∞	∞	...	...	...	**B20**, 170
p2144	**—,3-methyl-**	C_5H_7N. *See* p2126	81.12	...	...	142–3^{743} 45^{11}	...	1.4970^{20}	...	∞	∞	...	...	...	**B20**[2], 85 **B21**[2], 767
—	**—,oxo(tetra-hydro)-**	*see* **Pyrrolidone**													
Ωp2145	**—,1-phenyl-**	$C_{10}H_9N$. *See* p2126	143.19	pl (sub) red in air	62	234 140^{38} sub	...	...	i	s	s	s	s	peth v chl s	**B20**[2], 83
p2146	**—,2-phenyl-**	$C_{10}H_9N$. *See* p2126	143.19	pl (al, sub) $\lambda^{95\%\,al}$ 290 (4.03)	129	271–2^{726}	...	...	i	v	v	...	v	chl v lig δ	**B20**[2], 238

For explanations, symbols and abbreviations see beginning of table. For structural formulas see end of table.

No.	Name	Synonyms and Formula	Mol. wt.	Color, crystalline form, specific rotation and λ_{max} (log ε)	m.p. °C	b.p. °C	Density	n_D	Solubility w	al	eth	ace	bz	other solvents	Ref.
	Pyrrole														
p2147	—,**1-propyl-**	$C_7H_{11}N$. *See* p2126	109.17			145.5–6.5	0.8833_4^{20}		i	s	s				**B20**^{1}, 44
—	—,**tetrahydro**	*see* **Pyrrolidine**													
p2148	—,**2,3,4,5-tetra-iodo-**	Iodol. C_4HI_4N. *See* p2126	570.68	ye nd (al)	150d (162–4d)				i	s^h	s	s		chl, aa s	**B20**^{2}, 84
p2149	—,**2,3,4,5-tetra-methyl-**	$C_8H_{13}N$. *See* p2126	123.20	lf (dil al, peth) λ^{al} 219 (3.74)	111	130^7				∞	∞	∞	v	peth s^h	**B20**^{2}, 92
p2150	**2-Pyrrolcarboxal-dehyde**	2-Formylpyrrole	95.10	rh pr (peth) λ^{chl} 220 (3.9), 240 (3.9), 255 (3.9), 290 (4.7)	46–7	217–9 114^{15}		1.5939^{16}						os v lig δ, s^h	**B21**^{2}, 236
p2151	**2-Pyrrolecar-boxylic acid**		111.10	lf (w) λ^{al} 262 (4.08)	208.5d				s	s	s				**B22**^{2}, 15
p2152	—,**4-acetyl-3,5-dimethyl,** ethyl ester	$C_{11}H_{15}NO_3$. *See* p2151	209.25	wh nd (al) λ^{al} 235 (4.36), 255sh (4.08), 283 (4.07)	143–4				i	δ					**B27**^{2}, 233
p2152[1]	**3-Pyrrolecar-boxylic acid**		111.10	nd (lig) λ^{al} 245 (3.68)	161–2										**C50**, 7830
p2153	**2-Pyrrolecar-boxylic acid, 3,5-dimethyl-,** ethyl ester	$C_9H_{13}NO_2$. *See* p2151	167.21	cr (al) 240 sh (3.70), 276 (4.29)	125					s		s		os s	**B22**^{2}, 21
p2154	**3-Pyrrolecar-boxylic acid, 2,4-dimethyl-,** ethyl ester	$C_9H_{13}NO_2$. *See* p2152[1]	167.21	cr (eth-lig or peth)	78–9	291 $181–2^{35}$			δ	v	v			peth s^h	**B22**^{2}, 18
p2155	—,**2,5-dimethyl-,** ethyl ester	$C_9H_{13}NO_2$. *See* p2152[1]	167.21	rh (al)	117–8	290^{731} 130^{15}			i	s				os s dil ac, alk i	**B22**^{2}, 22
p2156	—,**4,5-dimethyl-,** ethyl ester	$C_9H_{13}NO_2$. *See* p2152[1]	167.21	cr (dil al)	110–1				δ	v	v			chl v	**B22**^{2}, 20
p2156[1]	**2,4-Pyrroledi-carboxylic acid**		155.11	cr (w)	295d				s^h						**B22**^{1}, 525
p2157	—,**3,5-dimethyl-,** diethyl ester	$C_{12}H_{17}NO_4$. *See* p2156[1]	239.27	lo nd (dil al) λ^{al} 221 (4.44), 273 (4.22)	136–7				i	δ	δ	s	s	chl, aa s lig δ	**B22**^{2}, 94
p2158	—,**3,5-dimethyl-1-ethyl-,** diethyl ester	$C_{14}H_{21}NO_4$. *See* p2156[1]	267.33	cr (dil al) λ^{chl} 275 (4.2)	40.5–1.3				δ	s^h					**C52**, 3866
p2159	—,**5-formyl-3-methyl-,** diethyl ester	$C_{12}H_{15}NO_5$. *See* p2156[1]	253.26	nd λ^{al} 237 (4.33), 270 (3.88), 308 (4.04)	124–5				i	s				to s lig i	**B22**^{2}, 278
Ω p2160	**Pyrrolidine**	1-Azacyclopentane. Tetra-hydropyrrole. Tetra-methylenimine.	71.12	λ^{gas} 171 (3.4), 196 (3.3)		$88.5–9^{760}$	0.8520_4^{22}	1.4431^{20}	∞	s	s			chl s	**B20**^{2}, 3
p2161	—,**1-butyl-**	$C_8H_{17}N$. *See* p2160	127.23			$154–5^{758}$	0.816^{25}	1.4373^{25}	δ	s				os s	**B20**^{2}, 4
p2162	—,**2-butyl-**	$C_8H_{17}N$. *See* p2160	127.23			173.5–4.5^{741} 67^{18}	0.8277_4^{20}	1.4490^{20}	s δ^h	∞				os ∞	**B20**^{2}, 65
p2163	—,**1(chloro-acetyl)-**	$C_6H_{10}ClNO$. *See* p2160	147.61		44–6	$112^{0.5}$									**C51**, 1171
p2164	—,**1,2-dimethyl-**	$C_6H_{13}N$. *See* p2160	99.18			96^{760}	0.7994_4^{20}	1.4252^{20}	∞	v	v				**B20**^{2}, 55
p2165	—,**2,4-dimethyl-**	$C_6H_{13}N$. *See* p2160	99.18			$115–7^{760}$	0.8297_4^{20}	1.4325^{20}	v	v	v				**B20**, 102
p2166	—,**2,5-dimethyl-** (*cis*)	$C_6H_{13}N$. *See* p2160	99.18			106–6.7^{760}	0.8205_4^{20}	1.4299^{20}	∞	∞	∞				**B20**^{2}, 60
Ω p2167	—,**1-methyl-**	$C_5H_{11}N$. *See* p2160	85.15			$81–3^{760}$	0.8188_4^{20}	1.4247^{20}	∞		s				**B20**^{2}, 4
—	—,**oxo-**	*see* **Pyrrolidone**													
p2168	—,**1-phenyl-**	$C_{10}H_{13}N$. *See* p2160	147.22	λ^{iso} 260 (4.27)		119–20.5^{12} $110–6^9$	1.0260^{25}	1.5813^{20}			s				**C53**, 1338
—	**2-Pyrrolidine carboxylic acid**	*see* **Proline**													
Ω p2169	**2-Pyrrolidone**	γ-Butyrolactam. 2-Oxopyr-rolidine.	85.11	cr (peth)	24.6	250.5^{742} 133^{12}	1.120_4^{20}	1.4806^{30}	v	v	v		v	CS_2, chl v	**B21**^{2}, 213
Ω p2170	—,**1,5-dimethyl-**	*N*-Methyl-γ-valerolactam. $C_6H_{11}NO$. *See* p2169	113.16			$215–7^{743}$ 87.5^{10}		1.4650^{20}	∞		∞				**B21**, 239
p2171	—,**3,3-dimethyl-**	$C_6H_{11}NO$. *See* p2169	113.16	lf (bz)	65–7	237			v	v	s	s	v^h		**B21**, 242
Ω p2172	—,**1-methyl-**	C_5H_9NO. *See* p2169	99.13	λ^{MeOH} 205 (3.46)	−23	202^{760} 845^{14}	1.0260_{25}^{25}	1.4684^{20}	v		s	s		os s	**B21**^{2}, 213
p2173	**3-Pyrroline**	Dihydropyrrole	69.11			$90–1^{760}$	0.9097_4^{20}	1.4664^{20}	s	∞	∞	s			**B20**^{2}, 67
—	**Pyruvaldehyde**	*see* **Propanal, 2-oxo-***													
—	**Pyruvic acid**	*see* **Propanoic acid, 2-oxo-***													
—	**Pyruvonitrile**	*see* **Propanoic acid, 2-oxo-,** nitrile													

For explanations, symbols and abbreviations see beginning of table. For structural formulas see end of table.

No.	Name	Synonyms and Formula	Mol. wt.	Color, crystalline form, specific rotation and λ_{max} (log ε)	m.p. °C	b.p. °C	Density	n_D	Solubility w	al	eth	ace	bz	other solvents	Ref.
	Quaterphenyl														
Ω q1	*o,o′*-**Quaterphenyl**	*o,o′*-Diphenylbiphenyl	306.41	pr (al) λ^{hx} 203.5 (4.73), 227 (4.50), 248 sh (4.35)	118–9	420	......	....	i	δ	s	s	s	MeOH δ chl v aa s	**B5**², 669
Ω q2	*p,p′*-**Quaterphenyl**	Tetraphenyl	306.41	lf (bz) λ^{hx} 206.5 (4.82), 294 (4.64)	320	428^{18}	......	....	i	i	i	...	s^h	chl i aa s^h $PhNO_2$ s	**B5**², 669
—	**Quercitin**	*see* **Flavone, 3,3′,4′,5,7-pentahydroxy-**													
—	**Quercitol**	*see* **1,2,3,4,5-Cyclohexanepentol***													
q3	**Quercitrin**	Quercetin 3-rhamnoside. $C_{21}H_{20}O_{11}$	448.39	pa ye nd or pl (+2w, dil al) $[\alpha]_{578}$ −73.5 (al, c = 4) λ^{al} 258 (4.30), 350 (4.18)	182–5 (hyd) 250–2 (anh)		......	....	s^h i	s	i	...	...	aa s^h MeOH s AcOEt i alk s (ye)	**B31**, 75
q4	**Quillaic acid**	Quillaja sapogenin. $C_{30}H_{46}O_5$	486.70	nd (dil al) $[\alpha]_D^{20}$ +56.1 (Py, c = 2.9)	294		......	....	i	s	s	s	...	Py, aa s	**B10**², 737
q5	**Quinacrine,** dihydrochloride (*dl*)	Atebrine. $C_{23}H_{30}ClN_3O.2HCl.2H_2O$	508.90	yesh nd (w), ye cr pw $\lambda^{w, pH=6.5}$ 343 (3.7), 424 (3.9), 444 (3.9)	248–50d		......	....	s v^h	δ	i	i	i	MeOH s	**C28**, 2126
—	**Quinaldine**	*see* **Quinoline, 2-methyl-**													
—	**Quinaldinic acid**	*see* **2-Quinolinecarboxylic acid**													
—	**Quinalizarin**	*see* **9,10-Anthraquinone, 1,2,5,8-tetrahydroxy-**													
q6	**Quinamine**	$C_{19}H_{24}N_2O_2$	312.42	pr (bz), nd (80% al) $[\alpha]_D^{15}$ +93 (chl, c = 2), $[\alpha]_D^{15}$ +105 (al, c = 2) λ^{al} 245 (4.0), 299 (3.4)	185–6 (172)		......	....	i	v^h	s	s	v^h	lig v	**B27**², 667
Ω q7	**Quinazoline**	1,3-Benzodiazine. Benzo(a)-pyrimidine.	130.15	ye pl (peth) λ^{cy} 218 (4.64), 259 (3.43), 267 (3.45), 303 (3.29), 312 (3.32)	48.0–8.5	241.5^{764} 117–20^{15}	......	....	v	s	s	s	s	oos v	**B23**, 175
q8	—,3,4-dihydro-4-oxo-	Quinazolinone. $C_8H_6N_2O$. *See* q7	146.15	nd (dil aa)	216–8	360	......	....	s^h	s^h	...	...	...	....	**B24**, 143
q9	—,3,4-dihydro-2-phenyl-	$C_{14}H_{12}N_2$. *See* q7	208.27	lf (dil al)	142		......	....	...	v	v	...	...	chl v aa v	**B23**, 239
q10	—,3,4-dihydro-3-phenyl-	Orexin. Phenzoline. Cedrarine. $C_{14}H_{12}N_2$. *See* q7	208.27	pl (eth-lig)	96	d	1.290^4	....	i	v	v	...	v	chl, CS_2 v	**B23**², 155
q11	—,3,4-dihydro-4-phenyl-	$C_{14}H_{12}N_2$. *See* q7	208.27	pl (al, AcOEt)	166–7		......	....	i	v	v	...	...	....	**B23**, 239
q12	—,2,4-dihydroxy-6,8-dinitro-	$C_8H_4N_4O_6$. *See* q7	252.15	ye gr pr (aa)	274–5d		......	....	s^h	δ	δ	...	δ	chl δ aa s^h	**B24**¹, 344
Ω q13	—,2,4-dioxo-1,2,3,4-tetrahydro-	2,4-Quinazolinedione. Benzoyleneurea. $C_8H_6N_2O_2$. *See* q7	162.15	nd (w, al), lf (aa) λ^{al} 217 (4.63), 310 (3.56)	356 (cor)		......	....	δ^h	δ	δ	...	...	alk s	**B24**², 197
Ω q14	**Quinhydrone**	Benzoquinhydrone	218.21	red br nd $\lambda^{w, pH=2.5}$ 440 (2.95)	171	sub	1.401_4^{20}	....	s^h δ	v	v	...	...	chl δ lig, peth i	**B7**², 572
—	**Quinic acid**	*see* **Cyclohexanecarboxylic acid, 1,3,4,5-tetrahydroxy-***													
q15	**Quinicine**	Quinotoxin. $C_{20}H_{24}N_2O_2$	324.43	red ye amor $[\alpha]_D^{15}$ +44.1 (chl) λ 256, 344	*ca.* 60		......	....	δ	v	v	...	...	chl v	**B25**², 20
q16	—,oxalate(*d*)	$(C_{20}H_{24}N_2O_2)_2.H_2C_2O_4.9H_2O$. *See* q15	901.03	pr (+9w, chl), nd (al), $[\alpha]_D^{15}$ +19.5 (al-chl) (+9.5(w))	149		......	....	v^h	s	...	...	...	chl v	**B25**, 39
q17	**Quinidine**	$C_{20}H_{24}N_2O_2$.	324.43	cr (+2.5w, dil al) cr (+1 al, al) $[\alpha]_D^{15}$ +230 (chl, c = 1.8) +262 (al, c = 1) λ^{MeOH} 230 (4.53), 280 (3.54), 336 (3.69)	174–5 (anh)		......	....	δ	s	δ	...	s	chl v peth i MeOH δ	**B23**², 414

For explanations, symbols and abbreviations see beginning of table. For structural formulas see end of table.

No.	Name	Synonyms and Formula	Mol. wt.	Color, crystalline form, specific rotation and λ_{max} (log ε)	m.p. °C	b.p. °C	Density	n_D	Solubility w	al	eth	ace	bz	other solvents	Ref.
	Quinidine														
q18	—,hydrate	Conquinine. $C_{20}H_{24}N_2O_2 \cdot 2\frac{1}{2}H_2O$. *See* q17	369.47	cr (dil al)	171.5d (168)				δ	s	s		v	chl s	**B23**[2], 414
q19	—,bisulfate	$C_{20}H_{24}N_2O_2 \cdot H_2SO_4 \cdot 4H_2O$. *See* q17	494.57	pr, nd (w) flr in sol, $[\alpha]_D$ + 184 (chl)					s	v	δ				**M**, 887
q20	—,hydrochloride	$C_{20}H_{24}N_2O_2 \cdot HCl \cdot H_2O$. *See* q17	378.91	pr (w), $[\alpha]_D^{20}$ +200 (w, c = 1)	258–9d (anh)				s^h	s^h	i			chl s	**M**, 887
q21	—,sulfate(*d*)	$(C_{20}H_{24}N_2O_2)_2 \cdot H_2SO_4 \cdot 2H_2O$. *See* q17	782.97	pr, nd (w) sol flr bl, $[\alpha]_D^{25}$ +212 (al)					s	v	δ			chl s	**B23**[2], 415
Ω q22	**Quinine**	$C_{20}H_{24}N_2O_2 \cdot 3H_2O$.	378.47	cr (+3w, eth), nd (+3w, al) rh nd (abs al) $[\alpha]_D^{15}$ −145.2 (al), $[\alpha]_D^{17}$ −117 (chl, c = 1.5) λ^w 347.5 (3.74)	57 (hyd) 177 (anh)	sub		1.625	δ	v	s	δ	δ	Py, MeOH v chl s peth i	**B23**[1], 166
q23	—,bisulfate	$C_{20}H_{24}N_2O_2 \cdot H_2SO_4 \cdot 7H_2O$. *See* q22	548.62	pr (+w, w) pr (+5w, al) $[\alpha]_D^{15}$ +168.4 (w)	160d				i	δ	δ			chl s	**B23**, 522
Ω q24	—,o-ethyl carbonate	$C_{23}H_{28}N_2O_4$. *See* q22	396.49	nd (w), cr (dil al)	95 (91)				δ	v	v			chl v	**B23**[2], 424
q25	—,formate	Quinoform. $C_{20}H_{24}N_2O_2 \cdot HCOOH$. *See* q22	370.45	nd, $[\alpha]_D^{20}$ −144.2 (w, c = 1)	149–50 (anh) 126 (+1w)				s	s	δ			chl s CCl_4 δ	**B23**[1], 169
Ω q26	—,hydrobromide	$C_{20}H_{24}N_2O_2 \cdot HBr \cdot H_2O$. *See* q22	423.36	silky efflor nd λ^w 332 (3.70)	*ca.* 200 softens at 152				s	v	δ			chl v	**B23**[1], 168
q27	—,hydrochloride	$C_{20}H_{24}N_2O_2 \cdot HCl$. *See* q22	360.89	silky efflor nd (w), $[\alpha]_D^{15}$ − 145 (al) $\lambda^{w, 0.02N HCl}$ 249 (4.5), 310 (3.6), 345 (3.7)	158–60 (anh)				s	v	δ			chl, alk s	**M**, 891
Ω q28	—,hydrochloride hydrate	$C_{20}H_{24}N_2O_2 \cdot HCl \cdot 2H_2O$. *See* q22	396.92	silky efflor nd (w), $[\alpha]_D^{20}$ −149.8 (w, c = 1.3) λ^{al} 278 (3.40), 331 (3.51)	156–90 (120)				v	v	δ			chl v	**B23**, 168
q29	—,2-hydroxy-benzoate	Quinine salicylate. $C_{20}H_{24}N_2O_2 \cdot C_7H_6O_3 \cdot H_2O$ *See* q22	480.55	wh pr (al) cr (+2w, w)	195				δ	v	δ			chl, glycerol v	**B23**, 526
q30	—,pentanoate	Quinine valerate. $C_{20}H_{24}N_2O_2 \cdot CH_3(CH_2)_3COOH \cdot H_2O$. *See* q22	444.58	wh	*ca.* 95				δ	v	v				**M**, 892
q31	—,sulfate	$2(C_{20}H_{24}N_2O_2) \cdot H_2SO_4$. *See* q22	746.93	silky nd (w)	235.2				s^h	s	δ		i	chl i	**B23**, 522
Ω q32	—,sulfate dihydrate	$2(C_{20}H_{24}N_2O_2) \cdot H_2SO_4 \cdot 2H_2O$. *See* q22	782.97	silky nd (w) $[\alpha]_D^{15}$ −220 (0.5N HCl) λ^w 274	205				s^h	s	δ			chl δ	**B23**, 522
—	**Quininic acid**	*see* **4-Quinolinecarboxylic acid, 6-methoxy-**													
q33	**Quininone**	$C_{20}H_{22}N_2O_2$.	322.41	nd, lf (eth) $[\alpha]_D^{20}$ +75.5 (al, c = 2)	108 (rapid htng) 101 (slow htng)				i	v	v		v	chl v peth i	**B25**[2], 23
—	**Quinitol**	*see* **1,4-Cyclohexanediol***													
—	**Quinizarin**	*see* **9,10-Anthraquinone, 1,4-dihydroxy-***													
Ω q34	**Quinoline**	1-Benzazine*. Benzo[b]-pyridine.	129.16	λ^w 275 (3.51), 299 (3.46), 312 (3.52)	fr −15.6	238.05^{760} 114^{17}	1.0929_4^{20}	1.6268^{20}	s^h i	∞	∞	∞	∞	CS_2 ∞	**B20**[2], 222
q35	—,hydrochloride	Quinolinium chloride. $C_9H_7N \cdot HCl$. *See* q34	165.63	pr (+½w, w) λ^w 233 (4.5), 312 (3.9)	94 (hyd) 134.5 (anh)				v	s	s^h		s^h	chl s	**B20**[2], 226
q36	—,hydrogen sulfate	Quinolinium bisulfate. $C_9H_7N \cdot H_2SO_4$. *See* q34	227.24	cr (al, aa)	164–4.5				s	s^h				aa s^h	**B20**[2], 226
Ω q37	—,**3-acetamido-**	$C_{11}H_{10}N_2O$. *See* q34	186.22	cr (w)	166–7				δ	s					**B22**[2], 352
q38	—,**4-acetamido-**	$C_{11}H_{10}N_2O$. *See* q34	186.22	nd (w + 1) λ^{al} 227 (4.65), 299 (3.98), 314 (3.85)	176 (anh)	sub			δ	s					**B22**, 445

For explanations, symbols and abbreviations see beginning of table. For structural formulas see end of table.

No.	Name	Synonyms and Formula	Mol. wt.	Color, crystalline form, specific rotation and λ_{max} (log ε)	m.p. °C	b.p. °C	Density	n_D	Solubility w	al	eth	ace	bz	other solvents	Ref.
	Quinoline														
q39	—,5-acetamido-	$C_{11}H_{10}N_2O$. *See* q34	186.22	pl (w), pr (dil al)	178	...	...	...	δ	s	...	...	...	...	B22², 353
q40	—,6-acetamido-	$C_{11}H_{10}N_2O$. *See* q34	186.22	nd (w)	138	...	...	...	v^h	v	δ	...	s^h	...	B22², 355
q41	—,7-acetamido-	$C_{11}H_{10}N_2O$. *See* q34	186.22	cr (w, al)	167.5	...	...	...	s^h	s^h	...	...	...	...	B22², 356
q42	—,8-acetamido-	$C_{11}H_{10}N_2O$. *See* q34	186.22	nd (al)	103	...	...	...	...	s^h	...	...	...	...	B22¹, 640
Ω q43	—,2-amino-	α-Quinolylamine. $C_9H_8N_2$. *See* q34	144.19	lf (w) $\lambda^{95\%\ al}$ 219 (4.60), 314 (3.70)	131.5	sub	...	...	v^h	s	s	s	δ	chl, aa s	B22², 350
q44	—,3-amino- (stable form)	β-Quinolylamine. $C_9H_8N_2$. *See* q34	144.19	rh (w, dil al)	94	...	...	...	δ	v	v	...	...	chl v	B22², 352
q45	—,—(unstable form)	$C_9H_8N_2$. *See* q34	144.19	mcl cr (to) $\lambda^{95\%\ al}$ 242 (4.48), 350 (3.60)	84	...	...	...	δ	v	v	...	...	chl v	B22², 352
q46	—,4-amino-	γ-Quinolylamine. $C_9H_8N_2 . H_2O$. *See* q34	144.19	nd (w + 1), nd (dil al, bz) λ^{al} 233 (4.28), 320 (4.01)	70 (hyd) 154 (anh)	180^{12-3}	...	...	s	v	v^h	...	s^h	chl s CS_2 δ	B22², 353
Ω q47	—,5-amino-	5-Quinolylamine. $C_9H_8N_2$. *See* q34	144.19	ye nd (al), lf (eth) λ^{al} 252 (4.42), 352 (3.46)	110	310 sub 184^{10}	...	...	δ	v	v	...	s	MeOH s lig i	B22², 353
q48	—,6-amino-	6-Quinolylamine. $C_9H_8N_2$. *See* q34	144.19	cr (w + 2), pr (eth) λ^{al} 245 (4.61), 285 (3.64), 355 (3.65)	114 (anh)	187^{11} $146^{0.3}$	...	...	δ	s	δ	...	...	NH_3 s	B22², 355
q49	—,7-amino-	7-Quinolylamine. $C_9H_8N_2$. *See* q34	144.19	ye nd (+1w) λ^{al} 246 (4.63), 285 (3.70), 354 (3.68)	93.5–94 (anh) 74–75.5 (hyd)	...	...	...	...	s	...	...	...	ac s	B22², 355
q50	—,8-amino-	8-Quinolylamine. $C_9H_8N_2$. *See* q34	144.19	pa ye nd (sub), cr (al, lig) λ^{al} 250 (4.44), 340 (3.48)	70 (65)	$157–8^{19}$	...	...	s	s	...	...	...	...	B22², 356
q51	—,2-amino-4-hydroxy-	$C_9H_8N_2O$. *See* q34	160.19	nd (w + 1), rh (al)	303–4	...	...	...	...	δ s^h	...	...	δ	AcOEt, chl δ alk, ac s	B22¹, 653
q52	—,5-amino-6-hydroxy-	$C_9H_8N_2O$. *See* q34	160.19	gr nd (w + 2)	185 (anh)	...	...	...	s^h	v	δ	...	δ	chl δ	B22, 501
q53	—,5-amino-8-hydroxy-	$C_9H_8N_2O$. *See* q34	160.19	nd (bz)	143	...	...	...	d^h	d^h	...	...	δ	...	B22¹, 653
q54	—,7-amino-8-hydroxy-	$C_9H_8N_2O$. *See* q34	160.19	br pr (eth, dil al) λ^{bp} 260 (4.7), 345 (3.5)	124	...	...	...	i	s	s	...	s	chl s	B22², 417
Ω q55	—,8-amino-6-methoxy-	$C_{10}H_{10}N_2O$. *See* q34	174.20	cr	51	$137–8^1$	...	...	...	...	...	...	...	...	B22², 413
q56	—,2-amino-4-methyl-	2-Aminolepidine. $C_{10}H_{10}N_2$. *See* q34	158.21	cr pw (bz)	133	320	...	...	$δ^h$	v	v	...	v^h	chl, aa v	B22², 364
q57	—,3-amino-2-methyl-	3-Aminoquinaldine. $C_{10}H_{10}N_2$. *See* q34	158.21	pa ye nd (eth, peth)	160–0.5	278^{760} δd 198^{16}	...	...	i	v	s	...	v	chl v	B22², 359
Ω q58	—,4-amino-2-methyl-	4-Aminoquinaldine. $C_{10}H_{10}N_2$. *See* q34	158.21	nd (bz-lig), pr (eth-bz)	168	333	...	...	δ	v	v	v	v^h	lig δ	B22², 359
q59	—,5-amino-2-methyl-	5-Aminoquinaldine. $C_{10}H_{10}N_2$. *See* q34	158.21	grsh pl or nd (w + 1)	117–8 (anh)	...	...	...	v^h	v	δ	...	v	lig s	B22², 360
q60	—,5-amino-8-methyl-	$C_{10}H_{10}N_2$. *See* q34	158.21	yesh nd (w or dil al)	143	...	...	...	$δ^h$	v	...	...	...	...	B22, 456
q61	—,6-amino-2-methyl-	6-Aminoquinaldine. $C_{10}H_{10}N_2$. *See* q34	158.21	pa br (w or dil al)	187.5	...	...	...	v^h	...	...	...	...	chl v	B22², 361
q62	—,6-amino-4-methyl-	6-Aminolepidine. $C_{10}H_{10}N_2$. *See* q34	158.21	nd (w)	169–70	...	...	...	s^h	v	s	...	...	chl v ac s	B22, 455
q63	—,7-amino-2-methyl-	7-Aminoquinaldine. $C_{10}H_{10}N_2$. *See* q34	158.21	nd (w + 1)	148 (anh)	...	...	...	v^h	...	δ	...	s	lig s	B22², 363
q64	—,7-amino-8-methyl-	$C_{10}H_{10}N_2$. *See* q34	158.21	pr (dil al)	129	304	...	...	δ	v	v	v	v	lig δ	B22, 456
Ω q65	—,8-amino-2-methyl-	8-Aminoquinaldine. $C_{10}H_{10}N_2$. *See* q34	158.21	pr (lig)	57–8	...	...	...	δ	v	v	v	v	lig s^h	C46, 5045
q65¹	—,8-amino-6-methyl-	$C_{10}H_{10}N_2$. *See* q34	158.21	nd	73 (62–4)	sub	...	...	s	v	v	v	v	oos v	B22², 365
—	—,5-benzamido-8-ethoxy-	*see* Analgen													
q66	—,2-bromo-	C_9H_6BrN. *See* q34	208.06	nd (al) $\lambda^{10\%\ al}$ 277 (3.53), 319 (3.62)	49	...	...	...	δ	v^h	v	...	v	chl v	B20, 362
q67	—,3-bromo-	C_9H_6BrN. *See* q34	208.06	ye oil $\lambda^{10\%\ al}$ 279 (3.53), 323 (3.51)	13–5	274–6 $95^{0.5}$	...	1.6641^{20}	...	...	...	...	...	aa v^h	B20, 363
q68	—,4-bromo-	C_9H_6BrN. *See* q34	208.06	cr	29–30	270d	...	...	δ	...	...	...	...	dil ac v	B20, 364

For explanations, symbols and abbreviations see beginning of table. For structural formulas see end of table.

No.	Name	Synonyms and Formula	Mol. wt.	Color, crystalline form, specific rotation and λ_{max} (log ε)	m.p. °C	b.p. °C	Density	n_D	Solubility w	al	eth	ace	bz	other solvents	Ref.
	Quinoline														
q69	**—,5-bromo-**	C_9H_6BrN. *See* q34	208.06	nd $\lambda^{10\%\,al}$ 293 (3.70), 317 (3.46)	52 (48)	280^{756} $105–7^{1.2}$								ac s	**B20**[2], 235
Ω q70	**—,6-bromo-**	C_9H_6BrN. *See* q34	208.06	$\lambda^{10\%\,al}$ 273 (3.58), 320 (3.53)	24	278 $155–6^{15}$				s	s			ac s	**B20**, 364
q71	**—,7-bromo-**	C_9H_6BrN. *See* q34	208.06	nd $\lambda^{10\%\,al}$ 269 (3.61), 320 (3.58)	34 (52)	290				s	s			ac v	**B20**, 365
q72	**—,8-bromo-**	C_9H_6BrN. *See* q34	208.06	$\lambda^{10\%\,al}$ 291 (3.70), 315 (3.46)	< −10 (80)	302–4 $165–6^{18}$				v				ac s	**B20**[2], 235
q73	**—,3-bromo-2-hydroxy-**	β-Bromocarbostyril. C_9H_6BrNO. *See* q34	224.06	pr (al)	253	sub				s^h			s^h		**B21**, 80
q74	**—,4-bromo-2-hydroxy-**	γ-Bromocarbostyril. C_9H_6BrNO. *See* q34	224.06	nd (al)	266–7	sub				v^h					**B21**, 80
q75	**—,5-bromo-2-hydroxy-**	5-Bromocarbostyril. C_9H_6BrNO. *See* q34	224.06	nd (al)	300					δ s^h					**B21**, 80
q76	**—,5-bromo-6-hydroxy-**	C_9H_6BrNO. *See* q34	224.06	nd (dil al)	186					δ				dil ac δ	**B21**[1], 221
Ω q77	**—,5-bromo-8-hydroxy-**	C_9H_6BrNO. *See* q34	224.06	nd (al), nd or lf (sub)	124	sub			v^h	v^h			v	chl v	**B21**[1], 222
q78	**—,6-bromo-2-hydroxy-**	C_9H_6BrNO. *See* q34	224.06	ye nd (al)	269				v	v^h	s			chl, ac, alk s	**B21**, 80
q79	**—,7-bromo-2-hydroxy-**	7-Bromocarbostyril. C_9H_6BrNO. *See* q34	224.06	nd (aa), pl (al)	228	sub				v^h	s			chl s	**B21**, 80
q80	**—,8-bromo-5-hydroxy-**	C_9H_6BrNO. *See* q34	224.06	nd (al)	190d					v				chl s	**B21**, 85
Ω q81	**—,2-chloro-**	C_9H_6ClN. *See* q34	163.61	nd (aq al) $\lambda^{10\%\,al}$ 282 (3.52), 318 (3.66)	38	$265–6^{753}$ $153–4^{22}$	1.2464_4^{25}	1.6342^{25}	i	v	v		s	lig s	**B20**[2], 233
q82	**—,3-chloro-**	C_9H_6ClN. *See* q34	163.61	hyg cr $\lambda^{10\%\,al}$ 283 (3.48), 323 (3.55)		255^{743} 141^{15}									**B20**[2], 234
q83	**—,4-chloro-**	C_9H_6ClN. *See* q34	163.61	cr $\lambda^{10\%\,al}$ 289 (3.69), 316 (3.45)	34–5	261^{744} 130^{15}	1.251		δ	v	v			dil HCl s	**B20**[2], 234
q84	**—,5-chloro-**	C_9H_6ClN. *See* q34	163.61	cr (al) $\lambda^{10\%\,al}$ 292 (3.66), 317 (3.50)	45	$256–7^{756}$				s^h					**B20**, 360
Ω q85	**—,6-chloro-**	C_9H_6ClN. *See* q34	163.61	pr (eth), nd (al) $\lambda^{10\%\,al}$ 276 (3.56), 319 (3.54)	44–5	$262–4^{760}$		1.6110_α^{56}							**B20**[2], 234
q86	**—,7-chloro-**	C_9H_6ClN. *See* q34	163.61	nd or pr $\lambda^{10\%\,al}$ 279 (3.55), 319 (3.56)	31–2	267–8	1.2158_4^{58}	1.6108_α^{58}	δ	v	v	v	v	chl v	**B20**[2], 234
Ω q87	**—,8-chloro-**	C_9H_6ClN. *See* q34	163.61	$\lambda^{10\%\,al}$ 292 (3.65), 315 (3.47)	fr. −20	$288–9^{760}$	1.2834_4^{14}	$1.6408_\alpha^{14.3}$	s	v	v	v	v	chl v	**B20**[2], 234
Ω q88	**—,7-chloro-4-hydroxy-**	C_9H_6ClNO. *See* q34	179.61	nd (al-w) λ^{chl} 246 (4.28), 290 (3.37), 320 (4.01), 333 (4.08)	276–80										**C44**, 2572
Ω q89	**—,5-chloro-8-hydroxy-7-iodo-**	Vioform. C_9H_5ClINO. *See* q34	305.50	ye br nd (al, aa) λ^{chl} 326 (3.41)	178–9					δ^h				aa s^h	**B21**[2], 58
Ω q90	**—,2-chloro-4-methyl-**	2-Chlorolepidine. $C_{10}H_8ClN$. *See* q34	177.65	nd (dil al)	59	296			i	v	v			chl v	**B20**[2], 245
q91	**—,2-chloro-6-methyl-**	$C_{10}H_8ClN$. *See* q34	177.65	nd (dil al)	112 (116)				δ	v	v		v	chl v	**B20**[1], 151
q92	**—,2-chloro-8-methyl-**	$C_{10}H_8ClN$. *See* q34	177.65	nd (eth)	61	286^{734}			δ	v	v		v	chl v	**B20**[1], 152
q93	**—,3-chloro-2-methyl-**	3-Chloroquinaldine. $C_{10}H_8ClN$. *See* q34	177.65	nd (dil al)	71–2				δ	v	v				**B20**, 392
q94	**—,3-chloro-4-methyl-**	3-Chlorolepidine. $C_{10}H_8ClN$. *See* q34	177.65	nd (dil al)	55					s				HCl v	**B20**[1], 150
q95	**—,3-chloro-6-methyl-**	$C_{10}H_8ClN$. *See* q34	177.65	nd (dil MeOH)	85.5				δ	s				MeOH s	**B20**[2], 246
q96	**—,4-chloro-2-methyl-**	4-Chloroquinaldine. $C_{10}H_8ClN$. *See* q34	177.65	nd (+1w)	42–3 (hyd) 25–6 (anh)	269–70			δ	s	s		s	chl, CS_2 s	**B20**[2], 241
q97	**—,6-chloro-2-methyl-**	5-Chloroquinaldine. $C_{10}H_8ClN$. *See* q34	177.65	lf or nd (dil al)	91					s^h	s				**B20**[2], 242
q98	**—,6-chloro-4-methyl-**	6-Chlorolepidine. $C_{10}H_8ClN$. *See* q34	177.65	nd (al)	71–2 (65–6)				δ	v	v	v	v	oos v	**B20**[2], 245

For explanations, symbols and abbreviations see beginning of table. For structural formulas see end of table.

No.	Name	Synonyms and Formula	Mol. wt.	Color, crystalline form, specific rotation and λ_{max} (log ε)	m.p. °C	b.p. °C	Density	n_D	Solubility w	al	eth	ace	bz	other solvents	Ref.
	Quinoline														
q99	**—,7-chloro-2-methyl-**	7-Chloroquinaldine. $C_{10}H_8ClN$. *See* q34	177.65	nd (eth), cr (lig)	75–6	$87^{0.5}$			...	...	s^h	...	...	lig s^h	**B20**[2], 242
q100	**—,8-chloro-5-methyl-**	$C_{10}H_8ClN$. *See* q34	177.65	nd (w)	49				s^h	v	v	...	v		**B20**[2], 246
q101	**—,8(chloro-methyl)-**	$C_{10}H_8ClN$. *See* q34	177.65	nd or pl (peth)	56				...	...	s	...	...	peth s^h	**B20**, 402
q102	**—,decahydro-(*cis*)**	$C_9H_{17}N$. *See* q34	139.24		−40	205^{735} 90^{20}	0.9426_4^{20}	1.4926^{20}	δ	s	s	...	...		**B20**[2], 73
q103	**—,—(*trans, d*)**	$C_9H_{17}N$. *See* q34	139.24	$[\alpha]_D^{25}$ +4.8 (al, c = 3–4)	75	200–2			δ	s	s	v	s	oos v	**B20**[1], 35
q104	**—,—(*trans, dl*)**	$C_9H_{17}N$. *See* q34	139.24	pr (lig, sub)	48	203^{735} sub	0.9610^{22}	1.4692_α^{56}	s^h	v	v	v	s	alk δ oos v	**B20**[2], 72
q105	**—,—(*trans, l*)**	$C_9H_{17}N$. *See* q34	139.24	$[\alpha]_4^{25}$ −4.5 (al, c = 3–7)	74–5	200–1			δ	...	...	...	...		**B20**[1], 35
Ω q106	**—,5,7-dibromo-8-hydroxy-**	$C_9H_5Br_2NO$. *See* q34	302.96	nd (al) λ^{chl} 328 (3.47)	196	sub			i	s	δ	s	s	aa, chl s	**B21**[2], 58
q107	**—,6,8-dibromo-2-hydroxy-**	6,8-Dibromocarbostyril. $C_9H_5Br_2NO$. *See* q34	302.96	nd (dil al)	230				...	v^h	...	...	...		**B21**, 81
q108	**—,2,3-dichloro-**	$C_9H_5Cl_2N$. *See* q34	198.05	(dil al)	104–5				i	v	v	...	s	lig δ	**B20**, 361
q109	**—,2,4-dichloro-**	$C_9H_5Cl_2N$. *See* q34	198.05	nd (dil al)	67–8	280–2			i	v	v	...	s	chl v	**B20**[2], 234
q110	**—,2,6-dichloro-**	$C_9H_5Cl_2N$. *See* q34	198.05	nd (eth)	156 (161.5)				i	s	s	...	...		**B20**, 361
q111	**—,2,7-dichloro-**	$C_9H_5Cl_2N$. *See* q34	198.05	nd (al)	120	sub 100^2			δ	s	s	...	...		**B20**[2], 234
q112	**—,3,4-dichloro-**	$C_9H_5Cl_2N$. *See* q34	198.05	(skellysolve B)	69–70				...	s	s	...	...		**Am 68**, 2570
q113	**—,4,5-dichloro-**	$C_9H_5Cl_2N$. *See* q34	198.05		118	$134^{5.5}$			...	s	s	...	...		**Am 68**, 113
q114	**—,4,6-dichloro-**	$C_9H_5Cl_2N$. *See* q34	198.05	(peth)	104				...	s	s	...	...		**Am 68**, 1277
Ω q115	**—,4,7-dichloro-**	$C_9H_5Cl_2N$. *See* q34	198.05	cr (MeOH), nd(80 % al) λ^{chl} 279 (3.68), 310 (3.46), 324 (3.52)	93	148^{10}			...	...	...	...	...		**Am 68**, 1206
q116	**—,4,8-dichloro-**	$C_9H_5Cl_2N$. *See* q34	198.05		155–6				δ	s	s	...	...		**Am 68**, 1277
q117	**—,5,6-dichloro-**	$C_9H_5Cl_2N$. *See* q34	198.05	nd (al)	85				δ	v	s	...	...	peth v	**B20**, 361
q118	**—,5,7-dichloro-**	$C_9H_5Cl_2N$. *See* q34	198.05	nd (al)	117				δ	s	s	...	...		**B20**, 362
q119	**—,5,8-dichloro-**	$C_9H_5Cl_2N$. *See* q34	198.05	nd (al), pl (eth)	97–8	sub			...	s	s	...	...		**B20**, 362
Ω q120	**—,6,8-dichloro-**	$C_9H_5Cl_2N$. *See* q34	198.05	nd (al)	104–5				δ	s^h	s	...	...		**B20**, 362
q121	**—,7,8-dichloro-**	$C_9H_5Cl_2N$. *See* q34	198.05	nd	85.5				...	s	s	...	...		**Prak 48**, 279
Ω q122	**—,5,7-dichloro-8-hydroxy-**	$C_9H_5Cl_2NO$. *See* q34	214.05	nd (al)	183				...	δ	...	δ	s	peth, alk s chl, aa δ	**B21**[2], 58
—	**—,8(3-diethyl-aminopropyl-amino-6-methoxy-**	*see* **Antimalarine**													
Ω q123	**—,1,2-dihydro-1-methyl-2-oxo-**	*N*-Methyl-2(1H)-quinolone. *N*-Methylcarbostyril. *N*-Methyl-*o*-aminocinnamic acid lactam.	159.19	nd (lig) λ^{w} 228 (4.53), 245 (4.01), 272 (3.84), 325 (3.81)	74	324^{728}			δ	s	s	s	v	chl s lig δ	**B21**[1], 297
Ω q124	**—,5,7-diiodo-8-hydroxy-**	Diodoquin. Embequin. Floraquin. $C_9H_5I_2NO$. *See* q34	396.96	yesh nd (aa, xyl)	210 (200–15d)				δ	v	δ	...	δ	chl, aa δ alk s	**B21**[2], 58
q125	**—,5,8-diiodo-6-hydroxy-**	$C_9H_5I_2NO$. *See* q34	396.96	yesh	191				i	v	v	...	δ	chl, aa δ alk v	**B21**[2], 54
q126	**—,2,3-dimethyl-**	3-Methylquinaldine. $C_{11}H_{11}N$. *See* q34	157.22	ye rh (eth)	68–9	261^{729}	1.1013		i	v	v	...	...	lig s	**B20**, 406
Ω q127	**—,2,4-dimethyl-**	4-Methylquinaldine. $C_{11}H_{11}N$. *See* q34	157.22			$264–6^{758}$ 143^{15}	1.0611^{15}	1.6075^{20}	δ	v	v	...	...		**B20**, 407
Ω q128	**—,2,6-dimethyl-**	6-Methylquinaldine. $C_{11}H_{11}N$. *See* q34	157.22	rh pr (eth) λ^{MeOH} 240 (4.5), 310 (3.5), 320 (3.5), 330 (3.6)	60	266–7 $152–5^{13}$			δ^h	δ	δ	...	v		**B20**, 408
Ω q129	**—,2,8-dimethyl-**	*o*-Toluquinaldine. $C_{11}H_{11}N$. *See* q34	157.22		27	255.3^{760} $103–4^{5}$	1.0394_4^{20}	1.6022^{20}	δ	v	v	...	...		**B20**[1], 154
q130	**—,3,4-dimethyl-**	3-Methyllepidine. $C_{11}H_{11}N$. *See* q34	157.22	cr (eth)	73–4	290^{737}			i	s	s	...	...		**B20**, 410
q131	**—,5,8-dimethyl-**	$C_{11}H_{11}N$. *See* q34	157.22	nd	4–5	265^{736}	1.070^{21}		δ	s	s	...	...		**B20**, 411
q132	**—,6,8-dimethyl-**	*β*-Cytisolidine. $C_{11}H_{11}N$. *See* q34	157.22			$268–9^{760}$ $133–4^{14}$	1.0665^{4}		δ	s	s	...	...		**B20**, 411

For explanations, symbols and abbreviations see beginning of table. For structural formulas see end of table.

No.	Name	Synonyms and Formula	Mol. wt.	Color, crystalline form, specific rotation and λ_{max} (log ε)	m.p. °C	b.p. °C	Density	n_D	Solubility w	Solubility al	Solubility eth	Solubility ace	Solubility bz	Solubility other solvents	Ref.
	Quinoline														
q133	**—,6(dimethyl-amino)-2-methyl-**	6(Dimethylamino)-quinaldine. $C_{12}H_{14}N_2$. *See* q34	186.26	ye pr (aa, AcOEt)	101	319^{760}	...	...	...	v	v	...	v	peth i	**B22**[2], 361
q134	**—,2,3-dimethyl-4-hydroxy-**	$C_{11}H_{11}NO$. *See* q34	173.22	pr (w + 1) λ^{chl} 248 (4.29), 288 (3.35), 324 (4.03), 337 (4.07)	319–20	sub	...	...	δ^h	δ^h	...	...	...	lig δ^h	**B21**[2], 68
q135	**—,2,4-dimethyl-6-hydroxy-**	$C_{11}H_{11}NO$. *See* q34	173.22	pr or pl (al)	214	360d	...	...	i	s	δ	v	i	ac, alk v	**B21**[2], 68
q136	**—,2,4-dimethyl-7-hydroxy-**	$C_{11}H_{11}NO$. *See* q34	173.22	nd (al)	218	...	...	...	...	s	...	...	...	...	**B21**, 116
q137	**—,2,4-dimethyl-8-hydroxy-**	$C_{11}H_{11}NO$. *See* q34	173.22	pr (eth) λ^{cy} 246 (4.70), 309 (3.52)	65	281 sub	...	...	i	v	v	v	v	peth δ chl v	**B21**, 116
q138	**—,2,6-dimethyl-4-hydroxy-**	$C_{11}H_{11}NO$. *See* q34	173.22	nd (w + 1)	279 (anh)	...	...	...	v^h	...	...	...	...	ac s	**B21**[2], 68
q139	**—,2,8-dimethyl-4-hydroxy-**	$C_{11}H_{11}NO$. *See* q34	173.22	lf or pl (w + 1)	260–1 (anh)	sub	...	...	δ	v	δ	...	δ	chl δ dil ac, alk v	**B21**, 116
q140	**—,4,6-dimethyl-2-hydroxy-**	$C_{11}H_{11}NO$. *See* q34	173.22	pr (al)	249–50	...	...	...	δ	s^h	δ	...	δ	chl, alk, dil ac δ	**B21**[2], 225
q141	**—,4,7-dimethyl-2-hydroxy-**	$C_{11}H_{11}NO$. *See* q34	173.22	cr (aa)	220	...	...	...	δ^h	v^h	...	...	...	alk s aa v^h	**B21**[1], 225
q142	**—,4,8-dimethyl-2-hydroxy-**	$C_{11}H_{11}NO$. *See* q34	173.22	pl (aq aa)	217–8	...	...	...	δ s^h	s	...	...	...	alk s	**B21**[1], 225
q143	**—,6,8-dimethyl-2-hydroxy-**	$C_{11}H_{11}NO$. *See* q34	173.22	nd (al)	201–2	...	...	...	s^h	s	...	...	...	dil ac i	**B21**[1], 225
q144	**—,6,8-dimethyl-5-hydroxy-**	Cytisoline. $C_{11}H_{11}NO$. *See* q34	173.22	pl (chl), cr (al)	197–8	sub	...	...	δ s^h	v	...	...	s	chl, ac s	**B21**, 117
q145	**—,3-ethyl-2-hydroxy-**	3-Ethylcarbostyril. $C_{11}H_{11}NO$. *See* q34	173.22	(dil HCl)	168	...	...	...	...	δ	...	...	s	chl s	**B21**, 115
q146	**—,2-hydrazino-**	$C_9H_9N_3$. *See* q34	159.19	(bz)	142–3	...	...	...	...	v	δ	...	...	lig δ	**B22**[1], 690
q147	**—,5-hydrazino-**	$C_9H_9N_3$. *See* q34	159.19	ye nd (w)	150–1	...	...	...	s^h	v	...	...	δ	peth i	**B22**, 565
Ω q148	**—,2-hydroxy-**	*o*-Aminocinnamic acid lactam. Carbostyril. 2-Quinolinol. α-Quinolone. C_9H_7NO. *See* q34	145.16	pr (al, dil al + 1w), nd (sub) $\lambda^{w, pH = 5.5}$ 224 (4.43), 245 (3.93), 270 (3.82), 324 (3.80)	199–200 (anh)	sub	...	...	δ s^h	v	v	...	...	dil HCl s	**B21**[2], 51
q149	**—,3-hydroxy-**	3-Quinolinol. C_9H_7NO. *See* q34	145.16	cr (bz, to or dil al) λ^{al} 272 (3.49), 329 (3.68)	200–1	...	...	...	s^h i	s	δ	...	v^h	chl δ to v^h	**B21**[2], 52
q150	**—,4-hydroxy-**	Kynurine. 4-Quinolinol. 4(1)-Quinolone. C_9H_7NO. *See* q34	145.16	nd (w + 3) λ^{al} 225 (4.3), 315 (4.1), 325 (4.1)	210 (anh) 100 (hyd)	...	...	...	v^h	v^h	δ	...	δ	peth δ	**B21**[2], 53
q151	**—,5-hydroxy-**	5-Quinolinol. C_9H_7NO. *See* q34	145.16	nd (al), pl λ^{al} 242 (4.75), 323 (3.53)	224d	sub	...	...	s^h	δ	δ	...	s^h	chl s^h lig i MeOH v	**B21**, 84
q152	**—,6-hydroxy-**	6-Quinolinol. C_9H_7NO. *See* q34	145.16	pr (al or eth) λ^{al} 227 (4.51), 274 (3.45), 331 (3.61)	193	> 360	...	...	δ^h i	δ	δ	...	i	chl i ac, alk s	**B21**, 85
q153	**—,7-hydroxy-**	7-Quinolinol. C_9H_7NO. *See* q34	145.16	pr (al), nd (dil al-eth)	238–40	sub	...	...	δ	v	...	...	...	alk v chl s^h	**B21**, 91
Ω q154	**—,8-hydroxy-**	8-Quinolinol. Quinophenol. Oxine. C_9H_7NO. *See* q34	145.16	nd (dil al) λ^{al} 240 (4.60), 308 (3.47)	75–6	266.6^{752} sub	1.034^{209}	...	i	v	i	s	v^h	chl v^h alk, ac s	**B21**, 91
Ω q155	**—,—,sulfate**	Chinosol. $2(C_9H_7NO).H_2SO_4$	388.40	ye pw (w)	177.5	...	...	...	v	s	i	...	...	...	**B21**, 92
q156	**—,—,sulfate monohydrate**	$2(C_9H_7NO).H_2SO_4.H_2O$	406.42	ye pr (w-al)	176–9	...	...	...	v	δ^h	...	...	...	...	**B21**[2], 56
Ω q157	**—,2-hydroxy-3-methyl-**	3-Methylcarbostyril. $C_{10}H_9NO$. *See* q34	159.19	yesh nd (dil al or ace)	234–5	sub	...	...	...	s	...	s^h	...	...	**B21**, 107
q158	**—,2-hydroxy-4-methyl-**	2-Hydroxylepidine. $C_{10}H_9NO$. *See* q34	159.19	nd (w)	245 (223)	$> 360^{760}$ 270^{17}	...	...	s^h δ	v^h	δ	...	δ	chl, lig δ	**B21**[2], 65
Ω q159	**—,2-hydroxy-6-methyl-**	6-Methylcarbostyril. $C_{10}H_9NO$. *See* q34	159.19	nd (al)	237	$240–1^{12}$	...	...	...	v	s	s	s	oos s	**B21**[1], 65
q161	**—,—(high m.p.)**	$C_{10}H_9NO$. *See* q34	159.19	nd (aq ace)	260d	...	...	...	δ	v^h	δ	v^h	δ	lig δ	**B21**[2], 59
q162	**—,3-hydroxy-2-methyl-(low m.p.)**	3-Hydroxyquinaldine. $C_{10}H_9NO$. *See* q34	159.19	nd (al)	203–5	...	...	...	δ	s	s	...	...	chl s	**B21**, 104
Ω q163	**—,4-hydroxy-2-methyl-**	4-Hydroxyquinaldine. $C_{10}H_9NO$. *See* q34	159.19	pr (w + 2) λ^{chl} 247 (4.14), 285 (3.43), 329 (4.06), 317 (4.03)	232	$> 360d^{760}$	...	...	s^h	v	i	...	i	lig i	**B21**, 104

For explanations, symbols and abbreviations see beginning of table. For structural formulas see end of table.

No.	Name	Synonyms and Formula	Mol. wt.	Color, crystalline form, specific rotation and λ_{max} (log ε)	m.p. °C	b.p. °C	Density	n_D	Solubility w	al	eth	ace	bz	other solvents	Ref.
	Quinoline														
q164	**—,5-hydroxy-2-methyl-**	5-Hydroxyquinaldine. $C_{10}H_9NO$. *See* q34	159.19	pl (al)	246–7 (233)	....	....	....	i	δ	v	...	...	alk s Na$_2$CO$_3$ i	**B21**, 106
q165	**—,5-hydroxy-6-methyl-**	$C_{10}H_9NO$. *See* q34	159.19	nd (al or sub)	230	sub	....	....	δ^h	v	v	v	...	oos v	**B21**, 111
q166	**—,5-hydroxy-8-methyl-**	$C_{10}H_9NO$. *See* q34	159.19	nd (dil al or sub)	262–3	sub	....	....	...	...	...	...	...	NaOH s^h chl δ	**B21**, 112
q167	**—,6-hydroxy-2-methyl-**	6-Hydroxyquinaldine. $C_{10}H_9NO$. *See* q34	159.19	(w)	213	$304–5^{760}$ 186^{35}	1.1665^{0}	....	δ	v	v	...	...	ac, alk s	**B21**, 106
q168	**—,6-hydroxy-4-methyl-**	6-Hydroxylepidine. $C_{10}H_9NO$. *See* q34	159.19	nd (w or dil al)	222–4	....	....	....	δ s^h	v^h	...	v	...	chl v^h	**B21**, 109
q169	**—,6-hydroxy-8-methyl-**	$C_{10}H_9NO$. *See* q34	159.19	nd (dil al)	200	....	....	....	...	s^h	...	...	...	....	**B21**, 113
q170	**—,7-hydroxy-6-methyl-**	$C_{10}H_9NO$. *See* q34	159.19	nd (al)	244	240^{22} 210^{11} sub	....	....	i	v^h	v	...	v	alk s	**B21**, 111
Ω q171	**—,8-hydroxy-2-methyl-**	8-Hydroxyquinaldine. $C_{10}H_9NO$. *See* q34	159.19	pr (dil al) λ^{cy} 246 (4.72), 309 (3.46)	74–5	267^{760} 145–60^{22-8} sub 100	....	....	i	s^h	s	...	s	....	**B21**, 106
Ω q172	**—,8-hydroxy-4-methyl-**	8-Hydroxylepidine. $C_{10}H_9NO$. *See* q34	159.19	nd (lig) λ^{cy} 242 (4.68), 319 (3.52)	141	....	....	....	s^h	s^h	...	v	v	ac, alk, chl, aa v	**B21**, 109
q173	**—,8-hydroxy-5-methyl-**	$C_{10}H_9NO$. *See* q34	159.19	nd (dil al)	122–4	sub 100	....	....	i	...	...	...	...	ac, alk s	**B21**, 110
q174	**—,8-hydroxy-6-methyl-**	$C_{10}H_9NO$. *See* q34	159.19	nd (chl or bz)	95–6	sub	....	....	δ	v	...	...	...	alk s^h	**B21**, 111
q175	**—,8-hydroxy-7-methyl-**	$C_{10}H_9NO$. *See* q34	159.19	nd (dil al)	72–4	sub 100	....	....	δ	v	...	...	...	....	**B21**, 111
Ω q176	**—,8-hydroxy-5-nitroso-**	$C_9H_6N_2O_2$. *See* q34	174.16	nd (al)	245d	....	....	....	i	...	δ	...	δ	chl δ	**B21**[1], 405
q177	**—,4-hydroxy-2-phenyl-**	$C_{15}H_{11}NO$. *See* q34	221.26	pl or pr (al), nd (aa) λ^{al} 257 (4.55), 325 sh (4.06), 334 (4.08)	256–7	....	....	....	i	v^h	i	...	...	alk s	**B21**[2], 84
q178	**—,2-iodo-**	C_9H_6IN. *See* q34	255.06	nd (dil al)	52–3	....	....	....	δ	v	v	v	...	oos v	**B20**, 370
q179	**—,4-iodo-**	C_9H_6IN. *See* q34	255.06	nd or pr	97 (100)	sub	....	....	i	v	v	...	...	....	**B20**, 370
q180	**—,5-iodo-**	C_9H_6IN. *See* q34	255.06	nd (al or eth)	101–2	sub	....	....	δ^h	s^h	s^h	...	...	....	**B20**[1], 141
q181	**—,6-iodo-**	C_9H_6IN. *See* q34	255.06	lf (w), nd (sub)	91	sub	....	....	s^h	v	v	...	...	....	**B20**, 370
q182	**—,8-iodo-**	C_9H_6IN. *See* q34	255.06	lo nd (al)	36	....	....	....	...	v	v	v	v	lig s	**B20**[1], 141
q183	**—,4-methoxy-**	$C_{10}H_9NO$. *See* q34	159.19	λ^{al} 275 (3.8)	41	245 167^{20}	....	....	i	s	s	...	s	....	**B21**[2], 53
q184	**—,5-methoxy-**	$C_{10}H_9NO$. *See* q34	159.19	....	....	282^{758}	....	....	...	v	v	...	...	....	**B21**, 84
Ω q185	**—,6-methoxy-**	*p*-Quinanisole. $C_{10}H_9NO$. *See* q34	159.19	hyg lf λ^{w} 223 (4.51), 268 (3.47), 325 (3.59)	26.5	305^{740} 153^{12}	1.1542^{20}_{20} 1.000^{209}	....	...	s	s	...	...	dil HCl s	**B21**[2], 53
q186	**—,8-methoxy-**	*o*-Quinanisole. $C_{10}H_9NO$. *See* q34	159.19	nd (peth) λ^{al} 240 (4.6), 360 (3.52)	49–50	282^{750} 164^{14}	1.034^{29}	....	...	s	s	...	s	peth s	**B21**[1], 222
q187	**—,2-methoxy-6-nitro-**	$C_{10}H_8N_2O_3$. *See* q34	204.20	nd (dil aa, bz or sub)	189–90	....	....	....	i	s	s	...	s	con sulf, chl v lig s	**B21**[1], 219
q188	**—,2-methoxy-8-nitro-**	$C_{10}H_8N_2O_3$. *See* q34	204.20	nd (dil al)	124–5	....	....	....	...	v	...	...	v	....	**B21**, 82
q189	**—,6-methoxy-5-nitro-**	$C_{10}H_8N_2O_3$. *See* q34	204.20	cr (al)	104–5	....	....	....	...	s^h	...	v	...	....	**B21**, 90
Ω q190	**—,6-methoxy-8-nitro-**	$C_{10}H_8N_2O_3$. *See* q34	204.20	yesh nd (al)	159–60	....	....	....	...	...	...	...	...	chl s MeOH δ	**B21**[2], 54
q191	**—,8-methoxy-5-nitro-**	$C_{10}H_8N_2O_3$. *See* q34	204.20	cr (al)	151.5	....	....	....	i	v^h	...	...	...	....	**B21**, 98
q192	**—,6-methoxy-1,2,3,4-tetrahydro-**	Thalline. $C_{10}H_{13}NO$. *See* q34	163.22	pr (peth or al), rh pym (w)	42–3	283^{735} $127–30^{1}$	....	1.5718^{50}	s^h δ	v	v	...	v	peth δ	**B21**, 61
Ω q193	**—,2-methyl-**	Quinaldine. $C_{10}H_9N$. *See* q34	143.19	λ^{MeOH} 230 (4.0), 270 (3.5), 300 (3.6), 310 (3.5), 320 (3.6)	−2 to −1	247.6^{760} 118^{10}	1.0585^{20}_{4}	1.6116^{20}	δ	s	s	s	...	chl s	**B20**[2], 238
q194	**—,3-methyl-**	*β*-Methylquinoline. $C_{10}H_9N$. *See* q34	143.19	pr $\lambda^{0.2 N HCl}$ 316 (3.83)	16–7	259.6^{760} $140–2^{25}$	1.0673^{20}_{4}	1.6171^{20}	δ	s	s	s	...	min ac s alk i	**B20**, 394
Ω q195	**—,4-methyl-**	Lepidine. $C_{10}H_9N$. *See* q34	143.19	red br λ^{al} 280 (3.68)	9–10	264.2^{760} 133^{15}	1.0862^{20}_{4}	1.6206^{20}	δ	∞	∞	s	s	lig s	**B20**[2], 244
q196	**—,5-methyl-**	*ana*-Methylquinoline. $C_{10}H_9N$. *See* q34	143.19	$\lambda^{0.2 N HCl}$ 315 (3.76)	19	262.7^{760}	1.0832^{20}_{4}	1.6219^{20}	δ	∞	∞	s	...	....	**B20**[2], 246
Ω q197	**—,6-methyl-**	*p*-Toluquinoline. $C_{10}H_9N$. *See* q34	143.19	$\lambda^{0.2 N HCl}$ 318 (3.85)	*ca.* −22	258.6^{760} 130^{15}	1.0654^{20}_{4}	1.6157^{20}	δ	s	s	s	...	....	**B20**, 397
q198	**—,7-methyl-**	*m*-Toluquinoline. $C_{10}H_9N$. *See* q34	143.19	ye $\lambda^{0.2 N HCl}$ 316 (3.82)	39	257.6^{760} 144^{18}	1.0609^{20}_{4}	1.6150^{20}	δ	s	s	s	...	....	**B20**[2], 246
q199	**—,8-methyl-**	*o*-Toluquinoline. $C_{10}H_9N$. *See* q34	143.19	$\lambda^{0.2 N HCl}$ 315 (3.76)	....	247.8^{760} 143^{34}	1.0719^{20}_{4}	1.6164^{20}	δ	∞	∞	s	...	....	**B20**[2], 247

For explanations, symbols and abbreviations see beginning of table. For structural formulas see end of table.

No.	Name	Synonyms and Formula	Mol. wt.	Color, crystalline form, specific rotation and λ_{max} (log ε)	m.p. °C	b.p. °C	Density	n_D	Solubility w	al	eth	ace	bz	other solvents	Ref.
	Quinoline														
q200	—,2-methyl-5-nitro-	5-Nitroquinaldine. $C_{10}H_8N_2O_2$. *See* q34	188.20	nd (dil al)	82	...	...	...	δ	v	v	...	...	...	**B20²**, 243
q201	—,2-methyl-6-nitro-	6-Nitroquinaldine. $C_{10}H_8N_2O_2$. *See* q34	188.20	cr (80% MeOH)	165 (173–4)	...	...	...	s^h	s	i	...	...	dil HCl v	**B20²**, 243
q202	—,2-methyl-8-nitro-	8-Nitroquinaldine. $C_{10}H_8N_2O_2$. *See* q34	188.20	pa ye nd (dil al)	137	...	...	...	δ	v	v	...	v	...	**B20²**, 244
q203	—,4-methyl-3-nitro-	3-Nitrolepidine. $C_{10}H_8N_2O_2$. *See* q34	188.20	pr (w)	118	...	...	...	s^h	s	...	...	...	...	**B20²**, 245
q204	—,4-methyl-8-nitro-	8-Nitrolepidine. $C_{10}H_8N_2O_2$. *See* q34	188.20	lf (abs al) λ^{al} 283 (3.70), 314 (3.60)	126–7	...	...	...	δ	s^h δ	s	...	...	...	**B20**, 397
Ω q205	—,6-methyl-5-nitro-	$C_{10}H_8N_2O_2$. *See* q34	188.20	pa ye nd (al)	116–7	...	...	...	i	s	s	s	s	oos v	**B20¹**, 151
q206	—,6-methyl-8-nitro-	$C_{10}H_8N_2O_2$. *See* q34	188.20	pa ye nd (w)	122	...	...	...	s^h	v	s	s	s	oos v	**B20**, 400
q207	—,8-methyl-5-nitro-	$C_{10}H_8N_2O_2$. *See* q34	188.20	pa ye nd (al)	93	...	...	...	...	s	s	s	s	oos v	**B20**, 403
q208	—,8-methyl-6-nitro-	$C_{10}H_8N_2O_2$. *See* q34	188.20	cr (al)	129	...	...	...	δ	v	...	...	...	...	**B20**, 403
q209	—,1-methyl-1,2,3,4-tetrahydro-	Kairoline. $C_{10}H_{13}N$. *See* q34	147.22	...	...	$247–50^{758}$ $123–6^{14}$	1.0222^{20}_4	1.5802^{23}	...	v	δ	...	...	...	**B20²**, 174
q210	—,3-nitro-	$C_9H_6N_2O_2$. *See* q34	174.16	nd (dil al) λ^{cy} 242.5 (4.45), 295 (4.0)	128	...	...	...	δ^h	s	...	v	...	...	**B20²**, 235
q211	—,4-nitro-, oxide	...	190.16	ye nd, pl or nd (ace)	154	...	...	...	...	...	...	...	...	...	**C50**, 13925
Ω q212	—,5-nitro-	$C_9H_6N_2O_2$. *See* q34	174.16	pl (w or al), nd (+w, w) λ 305 (3.80)	73–5 (anh)	sub	...	...	δ^h	s^h	...	...	s	...	**B20²**, 235
Ω q213	—,6-nitro-	$C_9H_6N_2O_2$. *See* q34	174.16	ye pl (HCl-aa), nd (w or dil al) λ 249 (4.40), 257 (4.34), 286 (3.98)	153–4	sub	...	...	s^h	s^h	δ	...	v	dil ac v lig δ	**B20²**, 235
q214	—,7-nitro-	$C_9H_6N_2O_2$. *See* q34	174.16	nd or lf (w or al), pl (sub) λ^{cy} 248 (4.37), 256.5 (4.3), 284 sh (3.97)	132–3	sub	...	...	s^h	s^h	s	...	...	chl v	**B20**, 372
Ω q215	—,8-nitro-	$C_9H_6N_2O_2$. *See* q34	174.16	mcl pr (al) λ 275 (3.74), 301 (3.54), 315 (3.52)	91–2	...	...	...	δ	s	s	...	s	dil ac s	**B20**, 373
q216	—,2-oxo-1,2,3,4-tetrahydro-	3(2-Aminophenyl)propionic acid lactam. Hydrocarbostyril. C_9H_9NO. *See* q34	147.18	pr (al or eth) λ^{al} 251 (4.1)	163–4	201^{45}	...	...	i	v	v	...	...	alk δ^h	**B21²**, 253
q218	—,2-phenyl-	$C_{15}H_{11}N$. *See* q34	205.26	nd (dil al) λ^{al} 252 (4.6), 325 (3.9)	86	363^{760} 310^{187}	...	...	δ	v^h	v	v	v	peth δ	**B20²**, 311
q219	—,3-phenyl-	$C_{15}H_{11}N$. *See* q34	205.26	pl (eth) λ^{al} 253 (4.5)	52	$205–7^{12}$	...	...	δ	s	s	v	s	chl s aa δ	**B20²**, 312
q220	—,5-phenyl-	$C_{15}H_{11}N$. *See* q34	205.26	nd (dil al)	83	...	...	...	δ	s	s	v	...	...	**J1943**, 441
q221	—,6-phenyl-	$C_{15}H_{11}N$. *See* q34	205.26	pl (al, bz or $PhNH_2$), pym (eth) λ^{al} 253 (4.64)	111	260^{77}	1.1945^{20}	...	δ	v	δ	v	v	chl v peth δ	**B20**, 483
q222	—,8-phenyl-	$C_{15}H_{11}N$. *See* q34	205.26	ye gr oil λ^{al} 233 (4.50), 300 (3.80)	...	283^{187}	...	...	δ	v	v	v	v	chl v CS_2 s	**B20**, 484
Ω q223	—,1,2,3,4-tetrahydro-	*Py*-Tetrahydroquinoline. $C_9H_{11}N$. *See* q34	133.20	nd λ^{al} 250 (3.85), 300 (3.30)	20	251^{760}	1.0588^{20}_4	1.6062^{19}	s	∞	∞	...	...	...	**B20²**, 173
q224	—,5,6,7,8-tetrahydro-	*Bz*-Tetrahydroquinoline. $C_9H_{11}N$. *See* q34	133.20	λ^{al} 268 (3.66)	...	222^{760} $92–5^{12}$	1.0304^{20}	1.5435^{20}	δ	s	s	s	s	...	**B20²**, 176
q225	—,2,3,4-trimethyl-	3,4-Dimethylquinaldine. $C_{12}H_{13}N$. *See* q34	171.25	...	92 (65)	285^{760} $156–8^{12}$	...	...	...	...	...	...	...	...	**B20²**, 255
q226	—,2,4,6-trimethyl-	4,6-Dimethylquinaldine. $C_{12}H_{13}N$. *See* q34	171.25	nd (w or dil al + 1w)	65.5 (anh) 40 (44) (hyd)	$281–2^{760}$ $146–8^{13.5}$	...	...	δ	v	v	v	v	chl, peth v	**B20²**, 256
q227	—,2,4,7-trimethyl-	4,7-Dimethylquinaldine. $C_{12}H_{13}N$. *See* q34	171.25	nd (w)	63–4 (anh) 48 (hyd)	280–1	1.0337^{20}_4	1.5973^{24}	δ	v	v	v	v	...	**B20²**, 256
q228	—,2,4,8-trimethyl-	4,8-Dimethylquinaldine. $C_{12}H_{13}N$. *See* q34	171.25	...	50–1	287^{758}	...	1.5855^{50}	...	...	...	...	...	...	**B20²**, 257
q229	—,2,5,6- or -2,6,7-trimethyl-	$C_{12}H_{13}N$. *See* q34	171.25	nd	69–70	...	...	...	δ	v	v	...	v	...	**B20**, 415

For explanations, symbols and abbreviations see beginning of table. For structural formulas see end of table.

No.	Name	Synonyms and Formula	Mol. wt.	Color, crystalline form, specific rotation and λ_{max} (log ε)	m.p. °C	b.p. °C	Density	n_D	Solubility w	al	eth	ace	bz	other solvents	Ref.
	Quinoline														
q230	**—,2,5,7-tri-methyl-**	Tetracoline. $C_{12}H_{13}N$. *See* q34	171.25	pr	...	286.6^{746} $107–8^{1.2}$	...	1.5980^{20}	i	v	v	...	...	...	**Am 81,** 152
q231	**—,2,6,8-tri-methyl-**	$C_{12}H_{13}N$. *See* q34	171.25	pr (peth), lf (dil al) λ^{al} 238 (4.63), 285 (3.59), 311 (3.50), 325 (3.46)	46	260^{719}	...	...	i	v	v	...	v	peth v	**B20,** 415
q232	**—,4,5,8-tri-methyl-**	$C_{12}H_{13}N$. *See* q34	171.25	...	73.4	155^{13}	...	...	i	v	v	...	v	...	**Am 68,** 644
Ω q233	**2-Quinoline-carboxylic acid**	Quinaldinic acid.	173.17	nd (w+2w) (bz)	157 (anh)	...	...	...	v^h s	...	...	...	v^h	...	**$B22^2$,** 55
q234	**—,amide**	Quinaldinamide. $C_{10}H_8N_2O$. *See* q233	172.20	nd (dil al, bz-lig)	133	...	...	...	δ^h	v	δ	...	v	chl, dil HCl v aa s	**B22,** 73
q235	**—,chloride**	Quinaldinyl chloride. $C_{10}H_6ClNO$. *See* q233	191.62	nd (eth or lig)	(i) 97–8 (ii) 175–6	...	...	...	δ	d^h	v	...	v	lig δ	**$B22^1$,** 509
q236	**—,methyl ester**	Methyl quinaldinate. $C_{11}H_9NO_2$. *See* q233	187.20	nd (lig)	86	...	...	...	...	s	...	...	...	lig s^h	**$B22^2$,** 55
q237	**—,nitrile**	2-Cyanoquinoline. $C_{10}H_6N_2$. *See* q233	154.17	nd (lig, chl)	94	160– $70^{20–23}$	...	...	s	v	v	...	v	chl v lig s	**$B22^1$,** 509
q238	**3-Quinoline-carboxylic acid**	...	173.17	pl (al or dil al) λ^{al} 233 (4.56), 275 (3.62), 322 sh (3.02)	275δd	...	...	...	s^h δ	s	...	...	...	dil ac, alk v	**$B22^2$,** 56
q239	**—,nitrile**	3-Cyanoquinoline. $C_{10}H_6N_2$. *See* q238	154.17	(al or sub)	108	...	...	...	δ	s^h	s	v	s	oos s	**$B22^2$,** 57
q240	**4-Quinoline-carboxylic acid**	Cinchoninic acid.	173.17	mcl pr (w), nd (+1w), mcl or tcl (+2w)	257–8 (anh)	...	...	...	δ	δ	i	...	...	...	**$B22^2$,** 57
q241	**—,nitrile**	4-Cyanoquinoline. Cinchoninonitrile. $C_{10}H_6N_2$. *See* q240	154.17	cr (chl, eth or lig) nd (sub)	103–4	240–5	...	...	δ	s	v	v	v	oos v	**$B22^2$,** 76
q242	**5-Quinoline-carboxylic acid**	...	173.17	cr (aa or sub)	342	sub <338	...	...	δ	δ	i	...	i	dil ac, dil alk v CS_2 i	**$B22^1$,** 511
q243	**—,nitrile**	5-Cyanoquinoline. $C_{10}H_6N_2$. *See* q242	154.17	nd (lig), nd (+1.5 w, dil al)	89 (anh) 70 (hyd)	...	...	...	δ	v	...	...	v	CS_2 v lig δ	**B22,** 79
q244	**6-Quinoline-carboxylic acid**	...	173.17	nd, pr or pl (sub)	291–2	sub <290	...	...	δ	s^h	...	...	...	dil ac, dil alk v	**$B22^1$,** 511
q245	**7-Quinoline-carboxylic acid**	...	173.17	nd (w or al)	249–50	sub	...	...	δ^h i	s	i	...	δ	...	**B22,** 81
q246	**8-Quinoline-carboxylic acid**	...	173.17	nd (w)	187	sub	...	...	s^h	s	...	...	...	dil ac, dil alk v	**B22,** 81
q247	**—,nitrile**	8-Cyanoquinoline. $C_{10}H_6N_2$. *See* q246	154.17	nd (dil al)	84	...	...	...	i	v	...	...	...	...	**B22,** 81
q248	**4-Quinoline-carboxylic acid, 2(3-carboxy-4-hydroxyphenyl)-**	Hexophan. $C_{17}H_{11}NO_5$. *See* q240	309.28	yesh pw	283–4 δd	...	...	...	i	δ	...	...	...	chl, alk v lig i	**$B22^2$,** 206
Ω q249	**2-Quinoline-carboxylic acid, 4,8-dihydroxy-**	Xanthurenic acid. Xanthuric acid. $C_{10}H_7NO_4$. *See* q233	205.17	ye micr cr (w) λ^{al} 243 (4.4), 345 (3.8)	289 (297)	...	...	...	i	s	δ	...	δ	dil HCl s^h alk, Na_2CO_3 s	**Am 73,** 3520
q250	**—,4-hydroxy-**	Kynurenic acid. 4-Hydroxyquinaldinic acid. $C_{10}H_7NO_3$. *See* q233	189.17	ye nd (+w, dil aa) λ^{al} 243 (4.4), 345 (4.0)	282–3	...	...	...	δ^h	s^h	i	...	...	oos δ alk v	**$B22^2$,** 174
Ω q251	**4-Quinoline-carboxylic acid, 6-methoxy-**	Quininic acid. $C_{11}H_9NO_3$. *See* q240	203.20	pa ye pr (dil al)	285d	sub	...	...	δ	s^h	δ	...	δ	chl i alk, ac s,	**$B22^2$,** 176
q252	**—,8-methoxy-2-phenyl-**	Isatophan. $C_{17}H_{13}NO_3$. *See* q240	279.30	ye nd (al)	216	...	...	...	i	s	i	...	...	chl, alk, ac s	**$B22^1$,** 559
Ω q253	**—,6-methyl-2-phenyl-, ethyl ester**	6-Methylcinchophene ethyl ester. Neocinchophene. Novatophan. $C_{19}H_{17}NO_2$. *See* q240	291.35	ye cr (al)	75–6	...	...	...	i	s^h	v	s	s	chl v oos s	**$B22^1$,** 520
q254	**3-Quinoline-carboxylic acid, 2-phenyl-**	$C_{16}H_{11}NO_2$. *See* q238	249.27	nd (al)	230d (226)	...	...	...	δ^h	s	...	s	δ	aa, MeOH, $phNO_2$ s	**$B22^2$,** 70
Ω q255	**4-Quinoline-carboxylic acid, 2-phenyl-**	Artamin. Atophan. Cinchophene. $C_{16}H_{11}NO_2$. *See* q240	249.27	nd (MeOH or dil al), ye in air λ^{al} 260 (>4.5), 330 (3.9)	218	...	...	...	i	s^h	s	δ^h	δ^h	peth i alk s	**$B22^2$,** 70
q256	**—,—,allyl ester**	Atoquinol. $C_{19}H_{15}NO_2$. *See* q240	289.34	nd (dil al)	30	265^{15} $215^{0.8}$	...	...	i	s	v	v	v	alk i ac δ	**$B22^2$,** 71
q257	**5-Quinoline-sulfonic acid, 6-hydroxy-**	...	225.23	ye nd (+0.5 w, w or al)	270d	...	...	...	s^h	δ	i	...	i	chl i aa δ	**B22,** 407

For explanations, symbols and abbreviations see beginning of table. For structural formulas see end of table.

No.	Name	Synonyms and Formula	Mol. wt.	Color, crystalline form, specific rotation and λ_{max} (log ε)	m.p. °C	b.p. °C	Density	n_D	Solubility w	al	eth	ace	bz	other solvents	Ref.
	5-Quinoline Sulfonic acid														
Ω q258	**—,8-hydroxy-**	$C_9H_7NO_4S$. *See* q257	225.23	ye (+2w, con HCl) lf, nd (+1w, dil HCl) λ^{w} 362 (3.9), $\lambda^{w,pH=10.5}$ 359 (5.0)	322–3	...	...	...	δ^h	...	...	...	...	...	**B22**[2], 313
q259	**7-Quinoline-sulfonic acid, 8-hydroxy-**	...	225.23	ye nd (al)	314–5	...	...	...	s^h	δ	...	...	i	...	**B22,** 408
Ω q260	**5-Quinoline-sulfonic acid, 8-hydroxy-7-iodo-**	Loretin. $C_9H_6INO_4S$. *See* q257	351.12	ye pr or lf (al) $\lambda^{w,pH=1}$ 327 (3.4), 355 (3.4), $\lambda^{w,pH=9}$ 335 (4.2), 362 (4.8)	260d	...	...	...	δ	δ	i	...	i	chl i con sulf s	**B22**[2], 314
q261	**8-Quinoline-sulfonic acid, 5-hydroxy-6-iodo-**	Lorenite	351.13	ye nd or lf	210–30d	...	...	...	...	...	...	...	...	con sulf s	**B22,** 406
—	**Quinolinic acid**	*see* **2,3-Pyridinedicarboxylic acid**													
Ω q262	**Quinolinium, *N*-ethyl-, iodide**	...	285.13	ye pr (al or MeCN) λ^{MeOH} 316 (3.87)	158	...	...	...	v	s^h	i	...	...	chl s	**B20**[2], 231
q263	**—,*N*-methyl-,** chloride	...	179.65	cr (+w, al) λ^{al} 232 (4.5), 315 (4.0), 375 sh (2.2)	126	...	...	...	s	s^h	...	...	...	chl s	**B20**[1], 139
—	**Quinolinol**	*see* **Quinoline, hydroxy-**													
—	**Quinone**	*see* **Benzoquinone**													
q264	**Quinovic acid**	Quinovaic acid. $C_{30}H_{46}O_5$	486.67	pl or nd $[\alpha]^{16}_D$ +87 (aq KOH)	298d	...	...	...	i	δ^h	...	...	...	alk, Ac_2O s aa δ	**E14,** 580
Ω q265	**Quinoxaline**	1,4-Benzodiazine. Benzopyrazine. Quinazine.	130.15	cr (peth) cr (+w, peth-w) λ^{al} 232 (4.4), 305 sh (3.7), 315 (3.8)	28 (anh) 37 (hyd)	229.5^{760} 108–11^{12}	1.1334^{48}_4	1.6231^{48}	s	∞	∞	∞	∞	aa s	**B23**[2], 177
q266	**—,6-amino-**	$C_8H_7N_3$. *See* q265	145.17	ye nd (eth) λ^{al} 236 (4.35), 321 (3.69)	159	sub	...	...	v	v	s	...	δ	aa δ chl s	**B25,** 326
q267	**—,6-chloro-**	$C_8H_5ClN_2$. *See* q265	164.60	nd (w) λ^{al} 236 (4.35), 321 (3.69)	63.8–4.3	$117–9^{10}$ sub	...	...	δ	...	...	...	...	...	**B23**[2], 177
q268	**—,2,3-dichloro-**	$C_8H_4Cl_2N_2$. *See* q265	199.04	cr (al, bz) λ^{al}245 (4.5), 315 sh (3.7), 325 (3.8), 340 (3.8)	151–3	...	...	...	i	v^h	...	...	v	chl, aa v^h	**B23**[2], 177
Ω q269	**—,2,3-dihydroxy-**	$C_8H_6N_2O_2$. *See* q265	162.15	nd (w) $\lambda^{w,pH=5}$ 262 (3.63), 312 (4.07), 326 (4.00)	410	...	...	...	v	δ	δ	...	s^h	MeOH, aa s^h alk v	**B24**[2], 200
Ω q270	**—,2,3-dimethyl-**	$C_{10}H_{10}N_2$. *See* q265	158.21	nd (w +3), (ace) λ^{al} 237 (4.41), 316 (3.86), 322.5 sh (3.75)	106 (anh) 85 (hyd)	...	...	...	...	s	s	s	s	oos, ac s	**B23**[2], 197
q271	**—,2,6-dimethyl-**	$C_{10}H_{10}N_2$. *See* q265	158.21	...	54	267–9	...	...	v	v	v	s	s	...	**B23,** 192
q273	**—,2-hydroxy-**	2-Quinoxalinol. $C_8H_6N_2O$. *See* q265	146.15	lf (al) $\lambda^{w,pH=4}$ 250 (3.79), 287 (3.70), 343 (3.74)	271	sub $200^{0.5}$	...	...	...	...	...	...	...	...	**C40,** 341
q274	**—,5-hydroxy-**	5-Quinoxalinol. $C_8H_6N_2O$. *See* q265	146.15	(eth) λ^{ac} 263 (4.53), 338 (3.68), 416 (3.05)	101–2	184^7 sub 90^{25}	...	...	...	...	...	...	...	...	**C48,** 8232
q275	**—,6-methoxy-**	$C_9H_8N_2O$. *See* q265	161.19	nd (w)	57.5	128^7 sub	...	...	...	...	s	s	...	...	**B23,** 387
Ω q276	**—,2-methyl-**	$C_9H_8N_2$. *See* q265	144.19	ye λ^{al} 234 (4.48), 250 (4.44), 297 sh (3.61), 309 (3.74), 317 (3.77)	180–1	$245–7^{760}$ 118^{16}	...	...	∞	v	∞	∞	∞	...	**B23**[2], 190
q277	**—,6-methyl-**	$C_9H_8N_2$. *See* q265	144.19	...	218–9	248^{748} 141.5^{29}	1.1164^{20}_4	$1.6211^{18.4}$	∞	∞	∞	∞	∞	...	**B23,** 184

For explanations, symbols and abbreviations see beginning of table. For structural formulas see end of table.

No.	Name	Synonyms and Formula	Mol. wt.	Color, crystalline form, specific rotation and λ_{max} (log ε)	m.p. °C	b.p. °C	Density	n_D	Solubility w	al	eth	ace	bz	other solvents	Ref.
	Quinoxaline														
Ω q278	**—,1,2,3,4-tetra-hydro-**	Ethylene-*o*-phenyl-enediamine. $C_8H_{10}N_2$. *See* q265	134.18	lf (w, eth or peth)	99.0–9.5	288.5–9.5 153–4[14]	....	....	s[h]	v	v	...	v	CCl_4 v chl s peth δ[h]	**B23**[2], 106
q279	**Quinuclidine**	1-Azabicyclo[2,2,2]octane. 1,4-Ethylenepiperidine. Nuclidine.	111.19	(eth)	158 (sealed tube)	....	....	....	v	v	v	v	v	oos v	**B20**[2], 71
Ω q280	**—,3-hydroxy-**	$C_7H_{13}NO$. *See* q279	127.19	cr (bz)	221–3	....	....	....	...	...	...	s	...	....	**C48,** 11499
q281	**2-Quinuclidine-carboxylic acid**	$C_8H_{13}NO_2$. *See* q279	155.20	col hyg cr (al-ace)	280d	....	....	....	...	...	...	...	...	....	**C32**, 166

For explanations, symbols and abbreviations see beginning of table. For structural formulas see end of table.

No.	Name	Synonyms and Formula	Mol. wt.	Color, crystalline form, specific rotation and λ_{max} (log ε)	m.p. °C	b.p. °C	Density	n_D	Solubility w	al	eth	ace	bz	other solvents	Ref.
	Raffinose														
r1	**Raffinose**	Gossypose. Melitose. Melitriose.	594.50	wh pw, pr or nd (+5w, dil al) $[\alpha]_D^{20}$ +105.2 (w, c = 4) (hyd) $[\alpha]_D$ +123 (w) (anh)	80 (hyd) 118–9 (anh)	d 130	1.465⁰	....	v	δ	i	...	...	MeOH v Py s	**B31,** 462
r2	**Raunescine,** hydrate	$C_{31}H_{36}N_2O_8$.	582.61	wh hex pr (90 % MeOH) $[\alpha]_D^{25}$ −74 (chl, c = 1) λ^{al} 218 (4.77), 271 (3.82)	160–70	....	....	....	i	δ	...	...	...	chl s aa s	**Am 79,** 250
—	**Reductone**	*see* **Propanal, 3-hydroxy-2-oxo-***													
r3	**Resazurin**	Diazoresorcinol. 3-Isophenoxazinone 10-oxide. Resazoin. $C_{12}H_7NO_4$.	229.18	dk red to grsh pr or pl (aa or (AcOEt) $\lambda^{w, pH=4}$ 355 (3.6), 425.5 (3.6), 520 (3.9)	d	sub vac	....	....	i	δ	i	...	...	alk s aa δ	**B27,** 128
r4	**Rescinnamine**	Moderil. $C_{35}H_{42}N_2O_9$.	634.74	nd (bz) $[\alpha]_D^{24}$ −98 (chl, c = 0.1) λ^{al} 229 (4.73), 302 (4.39)	238–9	....	....	....	i	δ	...	s	...	chl, AcOEt s	**Am 77,** 2241
r5	**Reserpic acid**	Reserpinolic acid. $C_{22}H_{28}N_2O_5$.	400.48	(MeOH) $[\alpha]_D^{23}$ (of hydro-chloride) −81 $\lambda^{al\ HCl}$ 225 (4.43), 269 (3.71), 295 (3.81)	241–3	....	....	....	...	...	...	...	...	....	**Am 78,** 2023
Ω r6	**Reserpine**	Rivasin. Serparsin. $C_{33}H_{40}N_2O_9$.	608.70	lo pr (dil ace) $[\alpha]_D$ −117.7 (chl, c = 1), −164 (Py, c = 0.96) λ^{chl} 268 (3.15), 295 (4.02)	264–5, 277–7.5 sealed tube	....	....	....	δ^h	s^h	δ	δ	s	chl s AcOEt s	**H37,** 67
—	**—,11-desmethoxy-**	*see* **Deserpidine**													
r7	**Reserpinine**	Raubasinine. $C_{22}H_{26}N_2O_4$.	382.47	pa ye pl (aq ace) $[\alpha]_D^{20}$ −131 (chl, c = 1.18) $[\alpha]_D^{22}$ −129 (Py, c = 0.5) λ^{al} 229 (4.58), 299 (3.74)	243–4d	....	....	....	i	s	...	...	...	chl s	**C49,** 11672
—	**Resodiaceto-phenone**	*see* **Benzene, 1,5-diacetyl-2,4-dihydroxy-**													
—	**Resorcinol**	*see* **Benzene, 1,3-dihydroxy-***													
—	**Resorcylaldehyde**	*see* **Benzaldehyde, 2,4-dihydroxy-**													
—	**Resorcylic acid**	*see* **Benzoic acid, dihydroxy-**													
r8	**Resorufin**	7-Hydroxy-2-phenoxazone.	213.20	br nd $(PhNH_2)$, red pr (HCl) $\lambda^{w, pH=4}$ 413 sh (3.8), 493 (4.0) $\lambda^{w, pH=8}$ 561 (4.4)	....	....	....	....	i	δ	i	...	...	alk v	**B27²,** 108
—	**Retene**	*see* **Phenanthrene, 7-isopropyl-1-methyl-***													
—	**Retenoquinone**	*see* **9,10-Phenanthro-quinone, 7-isopropyl-1-methyl-***													
r10	**Retronecine**	Senecifolinene.	155.20	pr (ace) $[\alpha]_D$ +27.4 (w), $[\alpha]_D^{26}$ +50.2 (al)	121–2 (130–1)	$80^{0.01}$	....	....	v	v	δ	s	...	....	**J1936,** 744
r11	**Rhamnetin**	$C_{16}H_{12}O_7$.	316.27	ye nd (al, PhOH) $\lambda^{alk\ al}$ 238 sh (4.35)	294–6	....	....	....	δ^h	s^h	...	s	...	dil alk v PhOH s^h	**B18²,** 237
r12	**Rhamnitol**(*D*)	Rhamnite.	166.18	pr (ace) $[\alpha]_D^{20}$ −12.4 (w, c = 0.5)	123	....	....	....	s	s	i	...	...	chl δ Py s	**B1,** 532

For explanations, symbols and abbreviations see beginning of table. For structural formulas see end of table.

No.	Name	Synonyms and Formula	Mol. wt.	Color, crystalline form, specific rotation and λ_{max} (log ε)	m.p. °C	b.p. °C	Density	n_D	Solubility w	al	eth	ace	bz	other solvents	Ref.
	α-Rhamnose														
r13	**α-Rhamnose(*d*), hydrate**	...	182.18	(w + 1), $[\alpha]_D^{20}$ + 3.76 → −0.82 (3 hours), (mut) $[\alpha]_D^{16.5}$ −8.25 (w)	90–1	...	...	...	s	...	i	...	...	...	**B1**[1], 439
r14	—(*dl*)	$C_6H_{12}O_5$. *See* r13	164.16	(w)	151.3–3 (anh)	...	...	...	s	s	i	...	...	...	**C25,** 84
r15	—(*l*), hydrate	$C_6H_{12}O_5 \cdot H_2O$. *See* r13	182.18	mcl pl (al, w + 1) $[\alpha]_D^{20}$ −77 → + 8.9 (mut)	92	105^2 sub	1.4708^{20}	...	s	s	...	...	...	MeOH s	**B1**[2], 901 **B31,** 66
r16	**β-Rhamnose(*l*)**	$C_6H_{12}O_5$. *See* r13	164.16	nd (ace) $[\alpha]_D^{20}$ + 38.4 → + 8.9 (w) (mut)	122–6	...	...	...	s	s	...	s^h δ	...	...	**B1**[1], 439
—	**Rhamnoxanthin**	*see* **Frangulin**													
r17	**Rheadin**	Rhoeadine. $C_{21}H_{21}NO_6$	383.41	nd (chl, eth, al) $[\alpha]_D^{17.5}$ + 232 (chl, c = 0.025) λ^{MeOH} 242 (4.0), 291 (4.0)	256–7.5	sub	...	...	δ	δ	δ	...	...	...	**C35,** 6060
r18	**Rhizopterin**	12-Formylpteroic acid. $C_{15}H_{12}N_6O_4$.	340.31	lt ye pl (aq $AcONH_4$), lf (aq aa) $\lambda^{w, pH=7}$ 248 (4.34), 265 (4.30), 348 (3.69)	> 300 285 darkens	...	...	...	i	i	i	...	...	aq alk s aq NH_3 s Py s	**Am 69,** 2751
Ω r19	**Rhodamine B**	Tetraethylrhodamine.	442.57	gr lf (w + 4), col pr (al, xyl) λ^{al} 546 (4.86)	165 (anh)	...	...	...	s	s	s	...	s	os i xyl s^h	**B19**[2], 373
r20	—,hydrochloride	$C_{28}H_{30}N_2O_3 \cdot HCl$. *See* r19	479.01	gr or red vt pw (w), lf (dil HCl)	...	...	...	...	v	v	...	...	s	w (red) dil HCl δ	**B19**[2], 373
Ω r21	**Rhodanine**	4-Oxo-2-thioxothiazolidine. Rhodanic acid. 4-Thioxo-4-thiazolidone.	133.19	lt ye pr (al, w, aa) λ^{al} 250.5 (4.12), 294 (4.24)	170	...	0.868	...	δ v^h	v	v	...	...	MeOH v aa s^h lig i	**B27**[2], 288
Ω r22	—,5[4(dimethyl-amino)-benzylidene]-	$C_{12}H_{12}N_2OS_2$ *See* r21	264.37	red or og nd (al, Py-w)	296 (282)	...	...	...	i	δ^h	v	s	δ	min ac s lig i CCl_4^v chl δ	**B27**[2], 484
r23	—,5-ethyl-	$C_5H_7NOS_2$. *See* r21	161.25	ye amor (dil al) λ^{al} 253 (4.1), 296 (4.3)	105	...	...	...	i	v	v	v	...	peth i aa v	**B27**[1], 313
Ω r24	—,3-phenyl-	$C_9H_7NOS_2$. *See* r21	209.29	ye pl (aa), nd or pr (al) λ^{al} 220 (3.9), 230 (3.9), 250 (4.0), 295 (4.1)	194–5	...	...	...	i	δ s^h	δ	s^h	...	chl s^h lig δ aa s^h	**B27**[2], 288
—	**Rhodinal**	*see* **Citronellal**													
—	**Rhodinol**	*see* **Citronellol**													
r25	**Rhodizonic acid**	5,6-Dihydroxy-5-cyclohexene-1,2,3,4-tetrone*.	170.08	dk og nd (sub)	155–60 (anh)	...	...	...	d	s	...	...	...	...	**B8**[2], 572
—	**Ribitol**	*see* **Adonitol**													
Ω r26	**Riboflavin**	Lactoflavin. Vitamin B_2. Vitamin G. $C_{17}H_{20}N_4O_6$.	376.37	ye or og-ye nd (w, dil aa) $[\alpha]_D^{25}$ −112 → −122 (0.02N NaOH, c = 0.5) (mut) $[\alpha]_D$ −8.80 (w) λ^w 267 (4.51), 373 (4.02), 447 (4.09)	280d	...	...	...	i	δ^h	i	i	...	chl i	**C28,** 2036
Ω r27	**Ribose(*D*)**	...	150.13	pl (abs al) $[\alpha]_D$ −21.5 (w) $[\alpha]_D^{20}$ −38.2 → + 43.1 (Py) (mut)	95 (87)	...	...	...	s	δ	...	...	...	...	**B1**[1], 434
r29	**Ricinidine**	1-Methyl-2-pyridone-3-carbonitrile.	134.14	nd (sub), (chl, al)	140	243^{28}	...	...	...	s^h	...	...	...	chl s^h	**B22**[2], 222

For explanations, symbols and abbreviations see beginning of table. For structural formulas see end of table.

No.	Name	Synonyms and Formula	Mol. wt.	Color, crystalline form, specific rotation and λ_{max} (log ε)	m.p. °C	b.p. °C	Density	n_D	Solubility w	al	eth	ace	bz	other solvents	Ref.
	Ricinine														
r30	**Ricinine**	Ricidine. $C_8H_8N_2O_2$.	164.18	pr or lf (w, al) λ^{MeOH} 256 (3.50), 313 (3.89)	201.5	sub 170–80[20]	...	...	s^h	δ	...	...	δ^h	chl s^h Py v^h peth i	**B22[1]**, 371
—	**Ricinoleic acid**	*see* **9-Octadecenoic acid, 12-hydroxy-(*cis*)***													
—	**Rivanol**	*see* **Acridine, 6,9-diamino-2-ethoxy-**													
—	**Roccelic acid**	*see* **Succinic acid, 2-dodecyl-3-methyl-**													
—	**Rosaniline**	*see* **Methanol,(4-amino-3-methylphenyl)bis(4-aminophenyl)**													
r31	**Rosinduline,** anhydro base	$C_{22}H_{17}NO_3$.	337.39	red-br lf (eth, bz)	198–9	...	...	...	i	v	v	...	v	...	**B25[2]**, 322
Ω r32	**Rotenone**	Tubotoxine. $C_{23}H_{22}O_6$.	394.43	nd or lf (al, aq ace), (CCl_4 + 1) $[\alpha]_D^{29.5}$ − 225.2 (bz) $[\alpha]_D$ − 132 (al, c = 0.125) λ^{al} 237 (4.15), 293.5 (4.25), 330.5 (3.80)	(i) 163 (ii) 176	210–20[0.5]	...	...	i	s	δ	s	s	chl v, aa s os s	**B19[2]**, 438
r33	**Rubicene**	...	326.40	red nd (xyl or Ph NO_2) λ^{al} 230 (5.10), 251 (5.00)	306	...	...	...	...	i	i	...	δ	$PhNO_2$ v^h CS_2 s peth i con sulf i	**B5[2]**, 707
r34	**Rubijervine**	$C_{27}H_{43}NO_2$.	413.65	nd (+1w, dil al) $[\alpha]_D^{25}$ +19.0 (al, c = 1.0)	242	...	...	...	i	s	δ	...	s	chl s aq ac s aq alk i con sulf (red) peth δ	**C32,** 5996
r35	**Rubixanthin**	3-Hydroxy-γ-carotene. $C_{40}H_{56}O$.	552.90	dk red nd (bz-MeOH), og-red (bz-peth) λ^{CS_2} 461, 494, 533	160	...	...	...	...	δ	...	...	s	chl s peth δ	**B30,** 93
Ω r36	**Rubrene**	9,10,11,12-Tetraphenyl-naphthacene.	532.69	og red (bz-lig) λ^{bz} 439 (3.40), 464 (3.76), 495 (4.06), 529 (4.05)	331	...	...	...	i	δ	δ	δ	s	CS_2 δ Py δ aa i	**B5[2]**, 725
—	**Rufigallic acid**	*see* **9,10-Anthraquinone, 1,2,3,5,6,7-hexahydroxy-**													
—	**Rufiopin**	*see* **9,10-Anthraquinone, 1,2,5,6-tetrahydroxy-***													
—	**Rufol**	*see* **Anthracene, 1,5-dihydroxy-***													
r37	**Rutaecarpine**	$C_{18}H_{13}N_3O$.	287.33	yesh nd (al or AcOEt) λ^{MeOH} 232 (4.45), 276 (3.93), 288 (3.97), 328 (4.50), 343 (4.57), 358 (4.47)	259–60	...	...	...	...	δ	...	δ	δ	ac v con sulf → ye	**B26[2]**, 104
r38	**Rutinose**	6(β-1-*L*-Rhamnosido) *D*-glucose.	326.30	hyg pw (al, eth) $[\alpha]_D^{20}$ + 3.2 → − 0.8 (w, c = 4) (mut) $[\alpha]_D^{20}$ − 10 (al)	189–92d	...	...	...	v	s	i	...	...	...	**B31,** 376
—	**Rutonal**	*see* **Barbituric acid, 5-methyl-5-phenyl-**													

For explanations, symbols and abbreviations see beginning of table. For structural formulas see end of table.

No.	Name	Synonyms and Formula	Mol. wt.	Color, crystalline form, specific rotation and λ_{max} (log ε)	m.p. °C	b.p. °C	Density	n_D	Solubility						Ref.
									w	al	eth	ace	bz	other solvents	
	Sabadine														
s1	**Sabadine**.........	$C_{29}H_{48}NO_8$..............	541.73	nd (eth) $[\alpha]_D -11$ (al)	256–60	....	....	....	δ	s	δ	s	...	....	**C54,** 2386
—	**Sabinane**.........	*see* **Thujane**													
s2	**Sabinene**(*d*).......	4(10)-Thujene.............	136.24	$[\alpha]_D^{20} +101.4$	....	$163–5^{758}$ 49^{13}	0.8437_4^{20}	1.4676^{20}	i	∞	s	...	s	....	**B5²,** 96
s3	—(*l*)............	$C_{10}H_{16}$. *See* s2	136.24	$[\alpha]_D^{15} -42.5$	....	162–6	0.8468^{20}	1.4674^{17}	i	∞	s	...	s	....	**B5²,** 96
Ω s4	**Sabinol**(*d*)........	4(10)-Thujen-3-ol. $C_{10}H_{16}O$	152.24	$[\alpha]_D^{18} +3.94$	....	208 90^{11}	0.9488_4^{19}	1.4871^{25}	...	...	s	...	...	....	**E12A,** 13
s5	**Saccharic acid**(*d*)..	....	210.14	nd (95 % al), $[\alpha]_D$ +6.86→ +20.60 (w, c = 2.5) (mut)	125–6	....	....	....	v	v	δ	...	...	chl δ	**B3²,** 377
Ω s6	**Saccharin**........	2-Sulfobenzoic acid imide.	183.19	mcl (ace), pr (al), lf (w) λ^{MeOH} 230 sh (3.8), 260 sh (2.9)	228.8–9.7d	sub vac	0.828	....	δ s^h	s	δ	s	δ	alk v chl δ	**B27,** 168
Ω s7	**Safrole**...........	1-Allyl-3,4-Methylene-dioxybenzene. Shikimole.	162.19	mcl λ^{al} 236 (3.62), 285 (3.58)	11.2	234.5^{760} $104–5^6$	1.1000_4^{20}	1.5383^{20}	i	v	∞	...	...	chl ∞	**B19²,** 29
—	**Salicin**...........	*see* **Toluene, α,2-dihydroxy-,** glucoside													
—	**Salicylaldehyde**....	*see* **Benzaldehyde, 2-hydroxy-**													
—	**Salicylic acid**......	*see* **Benzoic acid, 2-hydroxy-**													
—	**Saligenin**........	*see* **Toluene, α,2-dihydroxy-**													
—	**Salipyrine**........	*see* **Antipyrine,** 2-hydroxybenzoate													
—	**Salol**............	*see* **Benzoic acid, 2-hydroxy-,** phenyl ester													
—	**Salophen**.........	*see* **Benzoic acid, 2-hydroxy-,** 4-acetamido-phenyl ester													
s8	**Salsoline**........	$C_{11}H_{15}NO_2$...............	193.26	pw or cr (al) $[\alpha]_D^{20}$ +34.5 (0.1 N HCl, c = 1) λ^{MeOH} 226 (3.79), 285 (3.55)	221–2	....	....	....	δ	s	i	...	δ	chl, NaOH s peth i	**C53,** 14039
s9	**Salvarsan**.........	606. Arsphenamine.........	475.03	gy pw	185–95d	....	....	....	v^h	δ	i	...	i	dil HCl s MeOH δ con HCl i	**B16²,** 560
s10	**Sambunigrin**......	*d*-Mandelonitrile glucoside. $C_6H_5CH(CN)O.C_6H_{11}O_5$	295.30	nd (bz-peth), $[\alpha]_D^{18} -75.1$	151.5–2.5	....	....	....	s	s	...	...	...	AcOEt s	**B31,** 239
s11	**Samidin**..........	$C_{21}H_{22}O_7$...............	386.41	cubic, $[\alpha]_D^{21}$ +49.1 (chl, c = 1.59) λ^{al} 322 (3.15)	138–9	....	....	....	i	δ	s	...	...	MeOH s	**Ber 92,** 2338
—	**Sandoptal**........	*see* **Barbituric acid, 5-allyl-5-isobutyl-**													
s12	**Sanguinarine**......	ψ-Chelerythrine. $C_{20}H_{15}NO_5$	351.36	cr (eth)....... cr (al + 1)	266 (242–3 slow htng) 195–7 (+al)	....	....	....	i	s	s	s	s	....	**C54,** 13552
s13	**Santalic acid**......	Guerbet's acid............	234.34	red	....	β:202 γ:189	....	β:1.5136^{20} γ:1.5055^{20}	i	∞	...	...	...	....	**B9,** 571
s14	**α-Santalol**........	Arheol. $C_{15}H_{24}O$..........	220.36	$[\alpha]_D^{20} +17.0$	....	$301–2^{760}$ 167^{14}	0.9769_4^{20}	1.5023^{20}	i	s	...	...	...	....	**B6¹,** 275
s15	**β-Santalol**........	$C_{15}H_{24}O$................	220.36	$[\alpha]_D^{20} -90.5$	....	$167–8^{10}$ 133^1	0.9750_4^{20}	1.5115^{20}	...	...	...	...	...	....	**B6²,** 517
s16	**Santene**..........	2,3-Dimethyl-2-norbornene.	122.21	....	....	140–1 35^{15}	0.8698_4^{17}	1.4688^{17}	...	...	s	s	s	....	**E12A,** 540
s17	**Santenic acid** (*cis, d*)	π-Norcamphenic acid......	186.21	pl (w), $[\alpha]_D^{33}$ +38.3 (al)	170–1	....	....	....	s^h	s	...	...	...	alk s	**B9²,** 529
s18	**β-Santenol** (*cis, exo*)	2,3-Dimethyl-2-norbor-nanol.	140.23	nd (al) tab (lig)	101–2	192	....	....	i	...	s	...	...	....	**E12A,** 632
s19	**α-Santenone**(*cis*)...	1,7-Dimethyl-2-norbor-nanone.	138.21	pl, $[\alpha]_D^{22}$ +11.4 (al)	55	191	....	....	...	s	...	s	...	....	**E12A,** 271
s20	**Santonin**........	Santonic acid. $C_{15}H_{18}O_3$.	246.31	orh (w, al, eth) $[\alpha]_D^{18}$ −173 (al, c = 2) λ^{al} 240 (4.0), 268 (3.8)	174–6 (120 sub)	....	1.187_4^{26}	1.590 (1.640)	δ^h	δ s^h	δ s^h	...	s	chl, Py s peth i	**E12B,** 3736

For explanations, symbols and abbreviations see beginning of table. For structural formulas see end of table.

No.	Name	Synonyms and Formula	Mol. wt.	Color, crystalline form, specific rotation and λ_{max} (log ε)	m.p. °C	b.p. °C	Density	n_D	Solubility w	al	eth	ace	bz	other solvents	Ref.
	Sarmentogenin														
s21	**Sarmentogenin**	$C_{23}H_{34}O_5$	390.53	pr (85% al or MeOH-eth), nd (al), $[\alpha]_D^{19}$ +21.1 ± 4 (MeOH, c = 0.5) λ^{al} 217 (4.24), 288 (1.00)	270–5	....	....	....	i	s	i	δ	i	MeOH, Py s chl δ	**E14,** 237
s22	**Sarpagine**	Raupine. $C_{19}H_{22}N_2O_2$	310.40	nd, $[\alpha]_D^{20}$ +54 (Py) λ^{al} 231 (4.2), 280 (3.95)	320	....	....	....	i	s^h	...	...	...	....	**Am 84,** 622
s23	**Sarsasapogenin**	Parigenin. $C_{27}H_{44}O_3$	416.65	lo pr nd (ace), $[\alpha]_D^{25}$ −75 (chl, c = 0.5) λ^{al} <230 λ^{sulf} 398 (3.44)	200–1.5	....	....	....	...	s	...	s	s	chl s	**E14,** 282
s24	**Scarlet red**	....	380.46	dk br pw or nd	186 δd	d 260	....	....	i	δ	...	δ	δ	chl, peth s	**B16,** 172
s25	**Scilliroside**	$C_{32}H_{44}O_{12}$	620.67	lo pr (dil MeOH), $[\alpha]_D^{20}$ −59.4 (MeOH, c = 1) λ^{al} 302 (3.65)	168–70	d	....	....	δ	v	i	δ	...	diox v chl, AcOEt δ	**H42,** 1620
—	**Scopolamine**	*see* **Hyoscine**													
—	**Scopoletin**	*see* **Coumarin, 7-hydroxy-6-methoxy-**													
s26	**Scopoline**	Oscin.	155.20	hyg nd (lig, eth, peth or chl)	108–9	248	1.089_4^{134}	....	s	s	s^h	s	...	chl, peth δ	**B27²,** 61
s27	**Scyllitol**	$C_6H_{12}O_6$	180.16	pr (+3w)	353d	1.659_4^{19}	....	....	δ	i	i	...	i	chl, MeOH i	**B6²,** 1160
—	**Sebacic acid**	*see* **Decanedioic acid***													
s28	**Sedormide**	$H_2NCONHCOCH(CH_2CH{:}CH_2)CH(CH_3)_2$	184.24	nd (al)	194	....	....	....	δ	s^h	δ	...	...	chl s	**B3²,** 53
s29	**Selenanthrene**	Diphenyldiselenide.	310.13	pr (al), nd (ace) $\lambda^{96\%\,al}$ 240 (4.22), 330 (2.99)	181	223^{11}	....	....	...	δ s^h	δ	...	...	CS_2 s	**B19,** 47
s30	**Selenide, diethyl**	$(C_2H_5)_2Se$	137.06	pa ye	....	108	1.2300_4^{20}	1.4768^{20}	i	v	v	...	v	chl v	**B1³,** 357
s31	**—,dimethyl**	$(CH_3)_2Se$	109.03	....	....	$54\text{–}5^{753}$	1.4077_4^{15}	....	i	s	s	δ	...	chl s to, MeOH δ	**B1,** 291
s32	**—,—,hexachloro-**	$(Cl_3C)_2Se$	315.70	....	37	....	....	....	...	...	...	...	...	CCl_4 s	**J1947,** 1080
s33	**—,diphenyl**	$(C_6H_5)_2Se$	233.17	ye nd (bz) λ^{al} 235 (3.78), 255 (3.97), 275 sh (3.67)	2.5	$301\text{–}2^{760}$ $126\text{–}7^{5}$	1.351_4^{20}	1.6500^{20}	i	∞	∞	...	s	xyl s	**B6²,** 318
s34	**—,di(2-tolyl)**	....	261.23	pl or lf (al)	65	186^{16}	....	....	i	s^h	δ	...	...	chl δ	**B6²,** 343
s35	**—,di(4-tolyl)**	....	261.23	rods or nd (al) λ^{al} 257 (4.00), 275 sh (3.68)	69–70	196^{16}	....	....	i	s^h	δ	...	...	chl δ	**B6²,** 402
s36	**Selenonium, diphenyl-,** dichloride	$(C_6H_5)_2SeCl_2$	304.08	pa ye pr (xyl or ace), nd (al-HCl)	183 (142d)	....	....	....	s	s	i	s	...	chl, CCl_4 i	**B6²,** 318
s37	**—,triphenyl-,** chloride	$(C_6H_5)_3SeCl$	345.71	orh (AcOEt)	230d	....	....	....	v	v	δ	δ	...	chl s	**J1946,** 1126
s38	**—,—,fluoride**	$(C_6H_5)_3SeF$	329.26	oct deliq	145d	....	....	....	v	v	...	s	...	chl v	**J1946,** 1126
s39	**—,—,hydrogen** fluoride	$(C_6H_5)_3SeF.HF$	349.27	nd (ace)	99	....	....	....	...	...	...	s	...	....	**J1946,** 1126
—	**Selenophenol**	*see* **Phenol, seleno-**													
Ω s39[1]	**Semicarbazide***	Aminourea. Carbamylhydrazine. $H_2NCONHNH_2$	75.07	pr (al) λ^{al} 231 (1.2)	96	....	....	....	v	s	i	...	i	chl i	**B3²,** 80
Ω s39[2]	**—,hydrochloride**	$H_2NCONHNH_2 \cdot HCl$	111.52	pr (dil al) λ^{w} 278 sh (−0.7) 357 sh (−1.1)	175–7d	....	....	....	v	i δ^h	i	...	...	....	**B3,** 100
s40	**—,1,1-diphenyl-***	$H_2NCONHN(C_6H_5)_2$	227.27	nd (al or bz)	195	....	....	....	...	v^h	...	...	v^h	con sulf s	**B15,** 304
s41	**—,1,4-diphenyl-***	$C_6H_5NHCONHNHC_6H_5$	227.27	nd or lf (al, bz) λ^{al} 275 (3.4)	177	....	....	....	δ^h	v	i	...	...	....	**B15²,** 106
s42	**—,2,4-diphenyl-***	$C_6H_5NHCON(NH_2)C_6H_5$	227.27	lf (al)	165.5	....	....	....	...	s	s		s	chl v	**B15,** 277
s43	**—,1,1-diphenyl-3-thio-***	$H_2NCSNHN(C_6H_5)_2$	243.33	cr (al or bz)	202	....	....	....	i	v^h	...	v	v	chl v lig i	**B15,** 304
s44	**—,2,4-diphenyl-3-thio-***	$C_6H_5NHCSN(NH_2)C_6H_5$	243.33	lf (al)	139	....	....	....	...	δ v^h	...	v	v^h	....	**B15²,** 103
s45	**—,2-phenyl-***	$H_2NCON(NH_2)C_6H_5$	151.17	nd (bz or al) λ^{al} 242 (4.00)	120	....	....	....	v^h	v	δ	...	δ	chl v	**B15²,** 103
Ω s46	**—,4-phenyl-***	$C_6H_5NHCONHNH_2$	151.17	nd (bz), pl (w) λ^{al} 238 (4.3), 280 (2.9)	128	....	....	....	δ^h	v	i	...	...	chl v	**B12²,** 221

For explanations, symbols and abbreviations see beginning of table. For structural formulas see end of table.

No.	Name	Synonyms and Formula	Mol. wt.	Color, crystalline form, specific rotation and λ_{max} (log ε)	m.p. °C	b.p. °C	Density	n_D	Solubility w	al	eth	ace	bz	other solvents	Ref.
	Semicarbazide														
Ω s47	—,1-phenyl-3-thio-*	$H_2NCSNHNHC_6H_5$	167.23	pr (al) λ^{al} 245 (4.2), 275 sh (3.3)	200–1d				δ	s^h	δ	...	δ	chl δ	**B15**[2], 110
s48	—,2-phenyl-3-thio-*	$H_2NCSN(NH_2)C_6H_5$	167.23	(w) λ^{al} 255 (3.91)	153				δ s^h	s	...	...	δ		**B15**[1], 70
s49	—,4-phenyl-3-thio-*	$C_6H_5NHCSNHNH_2$	167.23	pl (al) λ^{al} 256 (4.3)	140d				...	i	...	...	δ	lig i	**B12**[2], 232
s50	—,3-thio-*	Thiosemicarbazide. $H_2NCSNHNH_2$	91.14	lo nd (w) λ^{al} 244 (4.2), 280 sh (2.4)	183				s	s	...	...	...		**B3**[2], 134
Ω s51	—,1(3-tolyl)-	Maretine	165.19	lf (w or dil al) λ^{al} 234.5 (4.00), 282.5 (3.30)	183–4				δ	v^h	i	...	...		**B15,** 508
s52	—,1,1,4-triphenyl-*	$C_6H_5NHCONHN(C_6H_5)_2$	303.37	nd (al)	206–7				...	δ v^h	...	...	...		**B15**[2], 115
s53	—,1,4,4-triphenyl-*	$(C_6H_5)_2NCONHNHC_6H_5$	303.37	pl (al)	151–2				i	v^h	...	...	...		**B15**[1], 71
s54	—,2,4,4-triphenyl-*	$(C_6H_5)_2NCON(NH_2)C_6H_5$	303.37	cr (dil al)	128				i	v	...	...	...	peth i	**B15,** 277
—	**Seminose**	*see* **Mannose**													
s55	**Serine(*D*)**	2-Amino-3-hydroxypropanoic acid*. $HOCH_2CH(NH_2)CO_2H$	105.10	nd or hex pr (w) $[\alpha]^{20}_D$ +6.87 (w, c = 10)	228d	d			v	i	i	...	i	aa i	**B4**[2], 919
Ω s56	—(*DL*)	$HOCH_2CH(NH_2)CO_2H$	105.10	mcl pr or lf (w)	246d (sealed tube)		$1.603^{22.5}$		s	i	i	...	i	aa i	**B4**[2], 934
s57	—(*L*)	$HOCH_2CH(NH_2)CO_2H$	105.10	hex pl or pr (w) $[\alpha]^{20}_D$ −6.83 (w, c = 10)	228d	sub $150^{10^{-4}}$			v	i	i	...	i	aa i	**B4**[2], 919
Ω s58	**Serpentine**	$C_{21}H_{20}N_2O_3$	348.41	ye rods or lf (al) $[\alpha]^{25}_D$ +292 ± 2 (MeOH) λ^{al} 252 (4.3), 310 (4.35), 370 (3.55)	175				i	s	s	s	...	MeOH v aq aa s	**Am 76,** 2843
s59	**Sesamin**	Asaranin. $C_{20}H_{18}O_6$	354.34	nd (al) $[\alpha]^{20}_D$ +68.4 (chl, c = 24) λ^{al} 235 (3.97), 288 (3.92)	123–4				i	s	δ	s	s	chl s peth δ	**B19**[2], 490
s60	**Shikonine**	d-Alkannin	288.30	br-red nd (bz) $[\alpha]^{20}_{644}$ +135 (bz, c = 1.3) λ^{CCl_4} 488 (3.8), 524 (3.9), 543 sh (3.7), 560 (3.7)	147				i	s	s	s	s	oos s	**B8**[2], 543
s61	**Silane, butyl(trichloro)-**	Butylsilicon trichloride. $CH_3(CH_2)_3SiCl_3$	191.56			148.9^{760}	1.1606^{20}_4	1.4363^{20}	d	d^h	s	...	s	to, AcOEt s	**B4**[1], 582
s62	—,chloro(triethoxy)-	Orthosilicic acid monochloride, triethyl ester. $(C_2H_5O)_3SiCl$	198.72		−51	156–6.5[760] 69^{32}	1.032^{20}_{20}	1.3999^{20}	d	s	...	...	...		**B1**[3], 1336
s63	—,dichloro(diethoxy)-	$(C_2H_5O)_2SiCl_2$	189.12		−130	135.9^{760} 51.6^{32}	1.1290^{20}_4		d	s	...	...	...		**B1**[3], 1336
Ω s63[1]	—,dichloro(diethyl)-	$(C_2H_5)_2SiCl_2$	157.12		−96.5	128–30d	1.0504^{20}	1.4309^{20}	d	d	...	...	...		**B4,** 629
Ω s64	—,dichloro(diphenyl)-	$(C_6H_5)_2SiCl_2$	253.21			302–5[757] 163–5[10]	1.2216^{20}_4	1.5819^{20}	d	d	...	...	...		**B16**[2], 608
s65	—,diethoxy(difluoro)-	Diethyl difluorosilicate. $(C_2H_5O)_2SiF_2$	156.21		−122	83^{760}d			d	...	...	...	...		**Am 68,** 76
s67	—,diethyl(difluoro)-	$(C_2H_5)_2SiF_2$	124.21		−78.7	60.9^{755}	0.9348^{20}_4	1.3385^{20}	i d^h	...	...	...	...		**J1944,** 454
s68	—,difluoro(diphenyl)-	$(C_6H_5)_2SiF_2$	220.30			252^{760} 158^{50}	1.145^{17}_4	1.5221^{25}	i d^h	...	...	...	s		**J1944,** 454
s69	—,dimethyl-	Dimethylsilicane. $(CH_3)_2SiH_2$	60.17	gas	−155.2	-19.6^{760}	0.68^{-80}		...	...	...	...	...		**B4**[1], 579
s70	—,ethoxy(trichloro)-	$(C_2H_5O)SiCl_3$	179.51		−135	101.9^{760}	1.2274^{20}	1.4045^{20}	d	s	...	...	...		**B1,** 335
s71	—,ethoxy(triethyl)-	$(C_2H_5O)Si(C_2H_5)_3$	160.34			154–5	0.8160^{20}	1.4140^{20}	i	∞	∞	...	...	sulf s	**B4,** 627
s72	—,ethoxy(trifluoro)-	$(C_2H_5O)SiF_3$	130.15	gas	−122	-7^{760}			...	...	...	...	...		**J1949,** 1696
Ω s73	—,ethyl(trichloro)-	$C_2H_5SiCl_3$	163.51		−105.6	97.9^{760}	1.2381^{20}_4	1.4257^{20}	d	d	...	...	...		**B4**[1], 582
s74	—,ethyl(triethoxy)-	Triethyl ethaneorthosiliconate. $C_2H_5Si(OC_2H_5)_3$	192.33			158.9	0.8594^{20}_4	1.3955^{20}	i	∞	∞	...	...		**B4,** 630
s74[1]	—,ethyl(trifluoro)-	$C_2H_5SiF_3$	114.14		−105	-4.4^{760}	1.227^{-76}		d	...	...	...	...		**J1944,** 454
s74[2]	—,ethyl(trimethoxy)-	Trimethyl ethaneorthosiliconate. $C_2H_5Si(OCH_3)_3$	150.25			124.3^{760}	0.9488^{20}_4	1.3838^{20}	...	s	...	...	...		**B4,** 629

For explanations, symbols and abbreviations see beginning of table. For structural formulas see end of table.

No.	Name	Synonyms and Formula	Mol. wt.	Color, crystalline form, specific rotation and λ_{max} (log ε)	m.p. °C	b.p. °C	Density	n_D	Solubility w	al	eth	ace	bz	other solvents	Ref.
	Silane														
s75	—,**fluoro(tri-ethoxy)-**	$(C_2H_5O)_3SiF$	182.27	...	...	134.6^{760}	...	...	d	...	...	...	...	...	**J1944,** 1696
s76	—,**fluoro(tri-ethyl)-**	$(C_2H_5)_3SiF$	134.27		...	110^{760}	0.8354_4^{25}	1.3900^{25}	i	...	...	...	...	peth, lig ∞	**Am 58,** 897
—	—,**hydroxy-**	*see* **Silicol**													
s77	—,**methyl-**	Methylsilicane. CH_3SiH_3	46.15	...	−156.5	−57	...	...	i	...	...	...	...	...	**B4**[1], 579
Ω s77[1]	—,**methyl(tri-ethoxy)-**	Trimethyl methaneortho-siliconate. $CH_3Si(OC_2H_5)_3$	178.31	...	...	143^{760}	0.8923^{20}	1.3835^{20}	...	s	...	...	...	...	**B4**[3], 1895
s78	—,**phenyl(tri-fluoro)-**	$C_6H_5SiF_3$	162.19	...	...	101.5^{760} 16.5^{24}	1.2169_4^{20}	1.4110^{20}	d	s	...	...	s	...	**J1944,** 454
s79	—,**tetraethyl-**	$(C_2H_5)_4Si$	144.34	...	...	153^{760} 108^{200}	0.7658_4^{20}	1.4268^{20}	i	...	...	...	...	...	**B4**[2], 1007
Ω s80	—,**tetramethyl-**	$(CH_3)_4Si$	88.23	...	α: −102.2 β: −99.1	26.5^{760}	0.648_4^{19}	1.3587^{20}	i	v	v	...	...	sulf i	**B4**[1], 579
s81	—,**triethyl-**	$(C_2H_5)_3SiH$	116.27	...	...	109^{755}	0.7302^{20}	1.4117^{20}	i	...	...	...	...	sulf i	**B4,** 625
—	**Silicane**	*see* **Silane**													
s82	**Silicol, triethyl-**	$(C_2H_5)_3SiOH$	132.28	...	...	154^{760} 46.6^{9}	0.8647_4^{20}	1.4329^{20}	i	∞	∞	...	...	...	**B4**[1], 581
—	**Sinapic acid**	*see* **Cinnamic acid, 3,5-dimethoxy-4-hydroxy-**													
s83	**Sinapine,** hydrogen sulfate	$C_{16}H_{23}NO_5.H_2SO_4$	407.44	lf (al)	186–7d	...	...	...	s	s^h	i	...	...	...	**B10,** 509
s84	—,hydrogen sulfate trihydrate	$C_{16}H_{23}NO_5.H_2SO_4.3H_2O$. *See* s83	461.48	lf (al)	127d	...	...	...	s	s^h	i	...	...	...	**B10,** 509
s85	—,thiocyanate monohydrate	$C_{16}H_{23}NO_5.HSCN.H_2O$. *See* s83	386.45	ye nd	180–1	...	...	...	δ	δ	...	...	...	...	**B10**[2], 354
—	**Sinapyl alcohol**	*see* **2-Propen-1-ol, 3(3,5-dimethoxy-4-hydroxyphenyl)-***													
s86	**Sinomenine**	Cuculine. $C_{19}H_{23}NO_4$	329.40	nd (bz), $[\alpha]_D^{26}$ −71 (al, c = 2.1) $\lambda^{aq.HCl}$ 230 sh (3.8), 266 (3.7)	162 (182 after melting)	...	...	...	δ	s	δ	s	δ	chl, dil alk s	**B21**[2], 470
s87	α_1-**Sitosterol**	$C_{29}H_{48}O$	412.71	nd (al), $[\alpha]_D^{28}$ −1.7 (chl, c = 2)	166	...	...	...	...	s	...	...	...	chl s	**E14,** 89
s88	α_2-**Sitosterol**	$C_{30}H_{50}O$	426.74	cr (al-peth), $[\alpha]_D^{25}$ +3.5 (chl, c = 2)	156	...	...	...	...	s	...	...	...	oos s peth δ	**E14,** 89
s89	α_3-**Sitosterol**	$C_{29}H_{48}O$	412.71	pl (al), $[\alpha]_D^{20}$ +5.2 (chl)	142–3	...	...	...	...	s	...	...	...	chl s	**E14,** 90
s90	β-**Sitosterol**	Cinchol. Vorosterol. $C_{29}H_{50}O$.	414.72	pl (al), nd (MeOH) $[\alpha]_D^{25}$ −37 (chl, c = 2) λ^{al} 267 (3.61)	140 (137)	...	...	...	...	s	s	...	...	aa s	**E14,** 90
—	**Skatole**	*see* **Indole, 3-methyl-**													
s91	**Skimmin**	7-Hydroxycoumarin-7-*D* glucoside. Umbelliferone-*D* glucoside.	324.29	cr (w + 1) $[\alpha]_D^{18}$ −80 (Py, c = 10) λ^{al} 319 (4.1)	219–21	...	...	...	s^h	s	i	...	...	chl i	**B31,** 245
s92	**Smilagenin**	Isosarsapogenin. $C_{27}H_{44}O_3$.	416.65	silky nd (ace), $[\alpha]_D^{25}$ −69 (chl, c = 0.5)	185	...	...	...	...	s	...	s	s	chl s	**E14,** 284
s94	**Solanidine-T**	Solatubine. $C_{27}H_{43}NO$	397.65	lo nd (chl-MeOH) $[\alpha]_D^{21}$ −29 (chl, c = 0.5)	218–9	sub δd	...	...	i	δ	i	...	v	chl v MeOH δ	**J1936,** 1299
s95	**Solanine-S**	Purapurine. Solatunine. $C_{45}H_{73}NO_{16}$. $C_6H_{11}O_4-O-C_6H_{10}O_4-O-C_6H_{10}O_4-O-C_{27}H_{42}N$ rhamnose galactose glucose solanidine-T	884.09	nd (al), fl (diox) $[\alpha]_D^{20}$ −69 (al) $[\alpha]_D^{20}$ −60 (Py, c = 1)	*ca.* 190 (284.5d)	...	...	...	i δ^h	s^h	i	...	i	diox s^h chl i	**J1963,** 745
s96	**Solasodine**	Purapuridine. Solanidine-S. $C_{27}H_{43}NO_2$.	413.65	hex pl (sub) $[\alpha]_D^{20}$ −92.4 (bz), $[\alpha]_D^{20}$ +109.8 (chl)	202	...	...	...	...	s^h	δ	s	s	chl v MeOH, diox, aq ac, Py s	**J1942,** 13
—	**Solatubine**	*see* **Solanidine-T**													
—	**Solatunine**	*see* **Solanine-S**													
—	**Sorbic acid**	*see* **2,4-Hexadienoic acid***													
s98	**Sorbierite**(*D*, −)	Iditol.	182.18.	pr (al), $[\alpha]_D$ −3.53 (w, p = 2)	73–4	...	...	...	s	...	...	...	...	...	**B1,** 544
Ω s99	**Sorbitol**(*D*)	*D*-Glucitol. *D*-Sorbite.	182.18	nd (w + 0.5) $[\alpha]_D^{25}$ −1.98 (w) λ^w < 220	110–2 (anh) 75 (hyd)	$295^{3.5}$	1.489_4^{20}	1.3330^{20}	v	δ^h	i	s	...	Py v^h aa s	**B1**[3], 2385
s100	—(*L*)	Gulitol. $C_6H_{14}O_6$. *See* s99.	182.18	nd (w + 0.5) [α] + 1.7 (w)	77	...	...	...	v	δ^h	i	s	...	aa s	**B1,** 534

For explanations, symbols and abbreviations see beginning of table. For structural formulas see end of table.

No.	Name	Synonyms and Formula	Mol. wt.	Color, crystalline form, specific rotation and λ_{max} (log ε)	m.p. °C	b.p. °C	Density	n_D	Solubility w	al	eth	ace	bz	other solvents	Ref.
	Sorbitol														
Ω s101	—,hexaacetate(*D*)	$C_{18}H_{26}O_{12}$. *See* s99	434.38	pr (w), $[\alpha]_D^{18}$ +6.8 (ace)	99.5 (120)	....	....	....	δ^h	v	δ	...	...	AcOEt s	**B2²**, 163
s102	**Sorbose(*D*)**	Pseudotagatose. *D*-Sorbinose.	180.16	rh (al), $[\alpha]_D^{20}$ +42.9 (w)	165	....	1.612^{17}	....	s	δ	δ	...	...	MeOH δ	**B31**, 345
s103	—*DL*)	β-Acrose. $C_6H_{12}O_6$. *See* s102	180.16	rh (dil al)	162–3 (154)	....	1.638^{17}	....	s	δ^h	δ	...	...	MeOH δ	**B31**, 348
Ω s104	—(*L*)	$C_6H_{12}O_6$. *See* s102	180.16	rh (al), $[\alpha]_D^{20}$ −43.2 (w, c = 5) $\lambda^{79\ \%\ sulf}$ 243 (3.34), 299 (3.70)	165	....	1.612^{17}	....	s	δ^h	δ	...	...	MeOH δ	**B31**, 346
—	**Sorbyl alcohol**	*see* **2,4-Hexadien-1-ol***													
—	**Sparassol**	*see* **Benzoic acid, 2-hydroxy-4-methoxy-6-methyl-**, methyl ester													
s105	**Sparteine(*D*)**	Lupinidine. $C_{15}H_{26}N_2$	234.39	$[\alpha]_D^{20}$ −19.5 (al) λ^{eth} 214 (3.71)	30–1	325^{754} 173^{8}	1.0196_4^{20}	1.5312^{20}	δ	v	v	...	...	chl v	**B2²**, 621
s106	—,sulfate pentahydrate	$C_{15}H_{26}N_2 . H_2SO_4 . 5H_2O$. *See* s105	422.53	pr, $[\alpha]_D^{17}$ −15.3	242	....	....	1.5289	i	v	...	...	...	....	**B23²**, 99
—	**Spinacane**	*see* **Squalene, perhydro-**													
—	**Spinacene**	*see* **Squalene**													
—	**Spinulosin**	*see* **1,4-Benzoquinone, 2,5-dihydroxy-3-methoxy-6-methyl-***													
s107	**Spiro[5,5]hendecane, 2,4,8,10-tetroxa-**	Pentaerythritol dimethylene ether. Spiro-bi(1,3-dioxane).	160.17	....	46.5	147^{53}	....	....	v	v	s	s	s	CCl_4 s	**B19²**, 449
s108	**Squalene**	Spinacene. $C_{30}H_{50}$.	410.74	λ^{cy} < 220	< −20	280^{17} 241^{4}	0.8584_4^{20}	1.4990^{20}	i	δ	s	s	...	CCl_4 s	**B1²**, 250
Ω s109	—,**perhydro-**	2,6,10,15,19,23-Hexamethyltetracosane*. Spinacane. $C_{30}H_{62}$. *See* s108	422.83	oil	−38	*ca.* 350^{760} 263^{10}	0.8125_4^{15}	1.4525^{20}	i	s^h δ	s	δ	∞	peth v MeOH, chl s aa δ	**B1²**, 145
s110	**Stachydrine(*L*)**	Hygric acid methylbetaine. *N*-Methylproline methylbetaine. $C_7H_{13}NO_2$.	143.19	cr (w + 1) $[\alpha]_D^{25}$ −40.25 (w, c = 4)	235d (anh) 116–8 (hyd)	....	....	....	s	s	i	...	...	chl i	**B22²**, 4
s111	—,oxalate	$C_7H_{13}NO_2 . H_2C_2O_4$	233.22	nd	105–7	....	....	....	...	i	...	...	...	....	**M**, 974
s112	**Starch**	Amylum. $(C_6H_{10}O_5)_n$	......	amor pw	d	....	....	....	i	i	...	...	...	....	**C51**, 11746
s113	—,triacetate	$C_{216}H_{288}O_{144}$	7801.3	pw $[\alpha]_D^{25}$ +72.5	180 (> 270 δd)	....	....	....	δ	δ	...	s	s	chl i	**C46**, 8401
—	**Stearic acid**	*see* Octadecanoic acid*													
—	**Stearolic acid**	*see* **9-Octadecynoic acid***													
—	**Stearone**	*see* **18-Pentatriacontanone***													
—	**Stearoxylic acid**	*see* **Octadecanoic acid, 9,10-dioxo-***													
s114	**Stibine, bis(trifluoromethyl)-bromo-***	$(F_3C)_2SbBr$	339.67	....	....	113^{760}	....	....	...	...	...	...	...	....	**J1957**, 3708
s115	—,**bis(trifluoromethyl)-chloro-***	$(F_3C)_2SbCl$	295.22	....	....	*ca.* 88^{760} 17^{20}	....	....	d^h	...	...	...	...	....	**J1957**, 3708
s116	—,**bis(trifluoromethyl)iodo-***	$(F_3C)_2SbI$	386.67	....	−42	*ca.* 129 16^{8}	....	....	...	...	...	...	...	....	**J1957**, 3708
s117	—,**dibromo(trifluoromethyl)-***	F_3CSbBr_2	350.57	....	....	*ca.* 157^{760} $34^{2.5}$	....	....	...	...	...	...	...	....	**J1957**, 3708
s118	—,**dichloro-(phenyl)-***	$C_6H_5SbCl_2$	269.71	nd or ta	69–70	$110–5^{10}$	....	....	d	v	...	s	...	MeOH, aa s	**B16²**, 574
s119	—,**diiodo(trifluoromethyl)-***	F_3CSbI_2	444.57	bt ye	4–8	> 200^{760}d	....	....	i	...	...	...	...	....	**J1957**, 3708
s120	—,**triethyl-***	$(C_2H_5)_3Sb$	208.94	....	−98 (−29)	161.4^{760}	1.3224^{16}	....	i	s	s	...	...	....	**B4**, 618
s121	—,**trimethyl-***	$(CH_3)_3Sb$	166.85	....	−62.0	80.6^{760}	1.523^{15}	1.42^{15}	i	s	s	...	...	CS_2 s	**B4²**, 1004
Ω s122	—,**triphenyl-***	$(C_6H_5)_3Sb$	353.07	pr (peth) λ^{al} 255 (4.11)	53.5	> 360 > 220^{1}	1.4343_4^{25}	1.6948^{42}	i	δ	v	v	v	chl, CS_2, aa v	**B16²**, 573
s123	—,**tris(trifluoromethyl)-**	$(F_3C)_3Sb$	328.77	....	−58	73^{760}	....	....	i	...	...	...	...	....	**J1957**, 3708
s124	—,**tri(2-tolyl)-**	....	395.15	(al)	102	....	....	....	...	s^h	v	...	v	chl, peth v	**B16**, 892
s125	—,**tri(3-tolyl)-**	....	395.15	(peth)	72	....	1.3957_4^{16}	....	...	s	v	...	v	chl, aa v	**B16**, 892
s126	—,**tri(4-tolyl)-**	....	395.15	rh (eth, MeOH)	127–8	....	1.3595^{16}	....	...	δ	s	...	s	chl s peth δ	**B16²**, 574
s127	**Stigmastanol**	Fucostanol. β-Sitostanol. $C_{29}H_{52}O$.	416.74	pl (al), cr (+1 w), $[\alpha]_D^{20}$ +24.8 (chl, c = 1.1)	144–5 (anh) 138–9 (hyd)	....	....	....	...	...	...	...	...	chl s	**E14**, 91

No.	Name	Synonyms and Formula	Mol. wt.	Color, crystalline form, specific rotation and λ_{max} (log ε)	m.p. °C	b.p. °C	Density	n_D	Solubility w	al	eth	ace	bz	other solvents	Ref.
	Stigmasterol														
Ω s128	**Stigmasterol**	$C_{29}H_{48}O$.	412.71	cr (al + 1 w), $[\alpha]_D^{22}$ −51 (chl, c = 2) λ^{al} 205 (3.74)	170	......	....	....	i	s^h	s	s	s	chl, oos s	**E14**, 88
Ω s129	**Stilbene**(*trans*)	*trans*-1,2-Diphenylethene.*	180.25	cr (al) λ^{al} 244 (4.39), 280 (4.02)	124.5–4.8	305^{720} 166–7^{12}	0.9707	1.6264^{17}	i	δ s^h	v	...	v	....	**B5**[2], 537
s130	—,**2,2′-diamino-** (*cis*)	$C_{14}H_{14}N_2$. *See* s129	210.28	red nd (w)	123 (107)	......	....	....	...	s^h	...	s	...	....	**B13**[1], 85
s131	—,—(*trans*)	$C_{14}H_{14}N_2$. *See* s129	210.28	gold-ye pr (al) λ^{al} 228 (4.21), 295 (4.46), 307 (4.45)	176	......	....	....	...	s^h	...	s	...	....	**B13**[1], 85
s132	—,**4,4′-diamino-** (*trans*)	$C_{14}H_{14}N_2$. *See* s129	210.28	ye nd or lf (al) λ^w 328 (4.60)	231	sub	....	....	δ^h	...	...	...	δ	MeOH s CS_2 δ	**B13**, 267
s133	—,**2,2′-dihydroxy-** (α-form)	$C_{14}H_{12}O_2$. *See* s129	212.25	nd (al)	95	......	....	....	...	v	v	...	...	....	**B6**[2], 987
s134	—,—(β-form)	$C_{14}H_{12}O_2$. *See* s129	212.25	flat nd (al)	197	......	....	....	...	δ	s	...	s	MeOH s^h	**B6**, 1022
s135	—,**3,5-dihydroxy-** (*trans*)	Pinosylvin. $C_{14}H_{12}O_2$. *See* s129	212.25	nd (aa)	156	......	....	....	i	...	...	s	s	chl, aa s	**C33**, 8989
s136	—,**4,4′-dihydroxy-** (*trans*)	Stilbestrol. $C_{14}H_{12}O_2$. *See* s129	212.25	nd (aa), ta (al) λ^{al} 228 (4.10), 300 (4.46), 307 (4.46), 325 (4.40)	284	......	....	....	...	δ s^h	δ	v	s	aa δ	**B6**, 1022
Ω s137	—,**4,4′-di-methoxy-**	Bianisal. Photoanethole. $C_{16}H_{16}O_2$. *See* s129	240.31	lf (bz or aa) λ^{al} 228 (4.11), 306 (4.54), 325 (4.53)	214–5	sub	....	....	i	...	...	s	s^h	aa s^h	**B6**[2], 988
s138	—,**α,β-dinitro-** (*cis*)	$C_{14}H_{10}N_2O_4$. *See* s129	270.25	ye pym pr (al) λ^{al} 265 (4.5)	108–9	d	....	....	i	s	s	v	v	chl v MeOH s	**B5**[2], 542
s139	—,—(*trans*)	$C_{14}H_{10}N_2O_4$. *See* s129	270.25	pa ye nd or pr (al) λ^{al} 262 (4.4)	187–8	......	....	....	i	δ s^h	δ	s	s	chl, aa s MeOH δ	**B5**[2], 541
s140	—,**2,2′-dinitro-** (*cis*)	$C_{14}H_{10}N_2O_4$. *See* s129	270.25	ye nd (aa)	126	......	....	....	i	...	...	...	...	aa s^h	**B5**, 637
s141	—,—(*trans*)	$C_{14}H_{10}N_2O_4$. *See* s129	270.25	pa ye nd (chl)	199	420 exp	....	....	...	δ	δ	...	s^h	CS_2s chl s^h	**B5**[1], 306
s142	—,**2,4-dinitro-** (*cis*)	$C_{14}H_{10}N_2O_4$. *See* s129	270.25	ye pl (aa)	127	......	....	....	...	...	...	...	v	chl s aa s^h	**B5**[2], 541
Ω s143	—,—(*trans*)	$C_{14}H_{10}N_2O_4$. *See* s129	270.25	pa ye (aa) λ^{Py} 368 (4.40)	143–5	412 exp	....	....	...	i	...	...	δ	CS_2, chl, xyl s	**B5**[2], 540
s144	—,**2,6-dinitro-** (*trans*)	$C_{14}H_{10}N_2O_4$. *See* s129	270.25	ye nd (bz, aa) λ^{bz} 330 sh (3.7)	114 (86)	......	....	....	i	...	...	...	s^h	chl s	**B5**[1], 306
s145	—,**3,4′-dinitro-** (*cis*)	$C_{14}H_{10}N_2O_4$. *See* s129	270.25	ye nd (aa)	155	......	....	....	...	δ	...	s	s	chl s aa s^h	**B5**[2], 541
s146	—,—(*trans*)	$C_{14}H_{10}N_2O_4$. *See* s129	270.25	ye nd (aa) $PhNO_2$ or Py)	220–2	......	....	....	...	δ^h	...	δ	δ^h	chl c, AcOEt δ	**B5**[2], 541
s147	—,**4,4′-dinitro-** (*cis*)	$C_{14}H_{10}N_2O_4$. *See* s129	270.25	ye nd (aa, chl or $PhNO_2$) λ^{al} 224 (4.15), 330 (4.18)	186–6.5	......	....	....	i	δ	δ	δ	s	chl s lig δ	**B5**[2], 541
s148	—,—(*trans*)	$C_{14}H_{10}N_2O_4$. *See* s129	270.25	pa ye lf (aa), nd (aa or $PhNO_2$) λ^{al} 235 (4.20), 356 (4.45)	303–4 (294–5)	......	....	....	i	s	s	s	s	chl s aa s^h	**B5**[2], 541
—	**Stilbestrol**	*see* **Stilbene, 4,4′-dihydroxy-(*trans*)**													
s151	**Strophanthidin**	Corchorgenin. $C_{23}H_{32}O_6$	404.51	orh ta (MeOH-w), lf (w + 2) $[\alpha]_D^{25}$ +43.1 (MeOH, c = 2.8) λ^{al} 216 (4.23)	235 (anh) 171–5 (hyd)	......	....	....	i	s	i	s	s	chl, aa s peth i	**E14**, 252
Ω s152	**g-Strophanthin**	Ouabain. $C_{29}H_{44}O_{12}$	584.67	hyg pl (+9w) $[\alpha]_D^{25}$ −34	200 (180)	......	....	....	δ s^h	v	...	...	...	....	**E14**, 251
Ω s153	**Strychnine**	$C_{21}H_{22}N_2O_2$	334.42	orh pr (al), $[\alpha]_D^{18}$ −139.3 (chl, c = 1) λ^{al} 255 (4.10), 280 (3.63), 290 (3.53)	286–8 (270 slow htng)	270^5	1.36_4^{20}	....	δ^h	δ	i	δ	δ	chl s peth i to, MeOH δ	**B27**[2], 723
Ω s154	—,**hydrochloride**	$C_{21}H_{22}N_2O_2.HCl.2H_2O$. *See* s153.	406.91	nd (w) $[\alpha]_D$ −28.3 (w, c = 0.7) λ^{al} 254 (4.1), 279 (3.6), 287 (3.5)	......	......	....	....	s	δ	i	...	...	chl δ	**B27**[2], 730

For explanations, symbols and abbreviations see beginning of table. For structural formulas see end of table.

No.	Name	Synonyms and Formula	Mol. wt.	Color, crystalline form, specific rotation and λ_{max} (log ε)	m.p. °C	b.p. °C	Density	n_D	Solubility w	al	eth	ace	bz	other solvents	Ref.
	Strychnine														
Ω s155	—,nitrate	$C_{21}H_{22}N_2O_2.HNO_3$. *See* s153.	397.44	nd (w) $\lambda^{w, pH=2}$ 249 (4.5), 310 (3.6), 345 (3.7)	280–310	...	1.627	...	v^h s	s^h δ	i	...	...	MeOH s chl δ	**B27²**, 732
s156	—,sulfate	$(C_{21}H_{22}N_2O_2)_2.H_2SO_4.5H_2O$ *See* s153.	857.00	wh mcl $[\alpha]_D$ 13.25 λ^w 255 (4.03)	200d (anh)	...	...	...	s v^h	s^h	i	...	...	MeOH s chl δ	**C48,** 4963
—	**Styphnic acid**	*see* **Benzene ,2,4-dihydroxy-1,3,5-trinitro-***													
—	**Styptol**	*see* **Cotarnine**, phthalate													
Ω s157	**Styracin**	Cinnamyl cinnamate. $C_6H_5CH:CHCO_2CH_2CH:CHC_6H_5$	264.33	nd	44	...	1.1565^{4}	...	i	s	v	...	...	...	**B9²**, 388
s158	**Styracitol**	$C_6H_{12}O_5$	164.16	pr (90 % al) $[\alpha]^{20}_D$ − 71.72 (w)	157	...	...	...	v	δ	i	i	i	MeOH δ, s^h	**B17²**, 235 **B18²**, 620
Ω s159	**Styrene**	Ethenylbenzene*. Phenylethylene. Vinylbenzene.	104.16	λ^{al} 245.3 (4.18)	−30.63	145.2^{760} 33.6^{10}	0.9060^{20}_4	1.5468^{20}	i	s	s	s	∞	peth ∞ MeOH, CS_2 s	**B5²**, 362
s160	—,**α-bromo-**	$C_6H_5CBr:CH_2$	183.05	...	−44	$86–7^{14}$	1.4025^{23}	1.5881^{20}	...	...	...	...	...	...	**B5²**, 367
Ω s161	—**β-bromo-(*trans*)**	$C_6H_5CH:CHBr$	183.05	λ^{al} 253 (4.34), 256 (4.36), 260 (4.34)	7	219 δd 108^{20}	1.4269^{16}	1.6093^{20}	i	∞	∞	...	...	...	**B5²**, 368
Ω s162	—,**2-bromo-**	C_8H_7Br. *See* s159	183.05	λ^{chl} 248 (1.84)	fr − 52.8	206.2^{760} 98^{20}	1.4160^{20}_4	1.5927^{20}	...	...	...	...	...	...	**B5³**, 1176
s163	—,**3-bromo-**	C_8H_7Br. *See* s159	183.05	λ^{chl} 248 (1.84)	...	$90–4^{20}$ $48.3^{0.5}$	1.4059^{20}_4	1.5933^{20}	...	...	...	...	...	...	**Am 59,** 1476
Ω s164	—,**4-bromo-**	C_8H_7Br. *See* s159	183.05	lf λ^{chl} 259 (2.19)	4.5 (−7.7)	103^{20} $50^{2.5}$	1.3984^{20}_4	1.5947^{20}	i	...	...	...	...	chl v aa s	**B5²**, 367
s165	—,**α-chloro-**	$C_6H_5CCl:CH_2$	138.60	...	−23	199^{760} 73^{16}	1.1016^{18}_4	1.5612^{20}	i	s	s	...	...	...	**B5²**, 367
s166	—,**β-chloro-(*trans*)**	$C_6H_5CH:CHCl$	138.60	λ^{al} 254 (4.22)	...	199^{760} 90^{18}	1.1095^{18}_4	1.5648^{20}	i	s	s	s	...	...	**B5²**, 367
Ω s167	—,**2-chloro-**	C_8H_7Cl. *See* s159	138.60	λ^{chl} 246.5 (1.92)	−63.15	188.7^{760} 64.6^{10}	1.1000^{20}	1.5649^{20}	...	s	s	s	...	peth ∞ CCl_4, aa s	**Am 66,** 1295
Ω s168	—,**3-chloro-**	C_8H_7Cl. *See* s159	138.60	λ^{chl} 250 (1.98)	...	$62–3^{6}$ 51^{3}	1.1168^{20}_4	1.5625^{20}	i	s	s	...	...	...	**Am 66,** 1296
Ω s169	—,**4-chloro-**	C_8H_7Cl. *See* s159	138.60	λ^{al} 253 (4.29)	−15.90	192.0^{760} 66.3^{10}	1.0868^{20}_4	1.5660^{20}	i	s	s	∞	∞	CCl_4, peth ∞	**B5²**, 367
s170	—,**2-fluoro-**	C_8H_7F. *See* s159	122.14	...	...	46^{32} $32–4^{3}$	1.0282^{20}	1.5200^{20}	i	s	s	...	s	...	**Am 66,** 1295
s171	—,**3-fluoro-**	C_8H_7F. *See* s159	122.14	λ^{chl} 248 (1.99)	...	$30–1^{4}$	1.0177^{20}	1.5170^{20}	i	s	s	...	s	...	**Am 66,** 1295
Ω s172	—,**4-fluoro-**	C_8H_7F. *See* s159	122.14	...	−34.5	67.4^{50} $29–30^{4}$	1.0220^{20}_4	1.5150^{20}	i	s	s	...	s	...	**Am 68,** 1159
s173	—,**2-hydroxy-**	*o*-Vinylphenol. C_8H_8O. *See* s159	120.16	nd	29–9.5	101^{14} 77^{15}	1.0609^{18}_4 1.0468^{36}	1.5851^{20}	s	v	v	...	...	...	**B6**, 560
s174	—,**3-hydroxy-**	*m*-Vinylphenol. C_8H_8O. *See* s159	120.16	...	...	$114–6^{16}$	1.0353^{31}_4	1.5804^{31}	...	...	...	...	...	...	**B6**, 561
s175	—,**4-hydroxy-3-methoxy-**	5-Vinylguaiacol. Hesperetol. $C_9H_{10}O_2$. *See* s159	150.18	cr	57	...	...	...	δ	v	v	...	...	...	**B6**, 954
s176	—,**2-methoxy-**	*o*-Vinylanisole. $C_9H_{10}O$. *See* s159	134.18	nd	29–9.5	195–200 $83–4^{12}$	1.0049^{17}_4 18	1.5388^{20}	i	s	s	s	s	oos s	**B6¹**, 277
s177	—,**3-methoxy-**	*m*-Vinylanisole. $C_9H_{10}O$. *See* s159	134.18	λ^{al} 250 (3.93), 293 (3.33)	...	$114–6^{16–7}$ $90–3^{15}$	0.999^{16}_4	1.5586^{23}	i	s	s	...	s	...	**B6**, 561
Ω s178	—,**4-methoxy-**	*p*-Vinylanisole. $C_9H_{10}O$. *See* s159	134.18	λ^{al-HCl} 260 (4.17), 292 sh (3.3)	...	$204–5^{756}$ 91^{13}	1.0001^{13}_4	1.5642^{13}	i	s	s	...	s	...	**B6²**, 520
Ω s179	—,**β-nitro-(*trans*)**	$C_6H_5CH:CHNO_2$	149.15	ye pr (peth or al) λ^{al} 228 (3.89), 311 (4.22)	60 (58)	250–60^{760}d 150^{14}	...	...	i δ^h	s	v	s	...	chl, CS_2 v peth s	**B5²**, 368
s180	—,**4-nitro-**	$C_8H_7NO_2$. *See* s159	149.15	pr (lig) λ^{al} 301 (4.16)	29	d	...	...	...	v^h	v	...	...	aa, chl, lig s	**B5³**, 1180
—	**Styrene oxide**	*see* **Benzene, (1,2-epoxyethyl)-***													
s181	**β-Styrenesulfonic acid,** chloride (*trans*)	$C_6H_5CH:CHSO_2Cl$	202.66	fl (bz) λ^{iso} 273.7 (4.33)	88	...	...	...	...	d	...	...	s^h	...	**B11²**, 87
s182	**Subboric acid,** tetramethyl ester	$(CH_3O)_2BB(OCH_3)_2$	145.76	...	−24.3	93^{760} (extrap 18^{35})	...	...	d	...	...	...	...	...	**Ber 94,** 509
—	**Suberaldehyde**	*see* **Octanedial***													
—	**Suberic acid**	*see* **Octanedioic acid***													
s183	**Succinaldehyde**	1,4-Butanedial*. $OCHCH_2CH_2CHO$	86.09	...	...	$169–70^{760}$ δd 56.5^{9}	1.064^{20}_4	1.4262^{18}	s	s	s	v	s	os v	**B1²**, 824
—	**Succinamic acid**	*see* **Succinic acid,** monoamide													
—	**Succinamide**	*see* **Succinic acid,** diamide													
—	**Succinanil**	*see* **Succinic acid,** imide, *N*-phenyl-													

For explanations, symbols and abbreviations see beginning of table. For structural formulas see end of table.

No.	Name	Synonyms and Formula	Mol. wt.	Color, crystalline form, specific rotation and λ_{max} (log ε)	m.p. °C	b.p. °C	Density	n_D	Solubility: w	al	eth	ace	bz	other solvents	Ref.
	Succinanilic acid														
—	**Succinanilic acid** ...	*see* **Succinic acid,** monoamide, *N*-phenyl-													
Ω s185	**Succinic acid**	Butanedioic acid*. $HO_2CCH_2CH_2CO_2H$	118.09	tcl or mcl pr λ^{w} 208 (2.0)	188 (185)	235d (→anhydride)	1.572^{25}_{4}	1.450	δ v^h	s	s	s	i	MeOH s to i	**B2**^{2}, 540
Ω s186	—,anhydride	Succinic anhydride.	100.08	nd (al), rh pym (chl or CCl_4) λ^{bz} 278 (3.15)	119.6	261 139^{15}	1.2340^{20}_{4}	...	i	s	δ	...		chl s	**B17**^{2}, 429
s187	—,diamide	Succinamide. $H_2NCOCH_2CH_2CONH_2$	116.12	orh nd (w)	268–70d (260)	125.5 sub	...	...	δ s^h	i	i	...	...	...	**B2**^{2}, 554
Ω s188	—,dibenzyl ester	$C_6H_5CH_2O_2CCH_2CH_2CO_2CH_2C_6H_5$	298.33	orh pl or ta (PrOH)	49–50 (45)	245^{15}	1.256	1.596	i				s	...	**B6**^{2}, 418
Ω s189	—,dibutyl ester	$C_4H_9^nO_2CCH_2CH_2CO_2C_4H_9^n$	230.31	...	−29.25	274.5^{760} 145^{4}	0.9752^{20}_{4}	1.4299^{20}	i	s	s	...	s	...	**B2**^{2}, 551
s190	—,di-*sec*-butyl ester	$C_4H_9^sO_2CCH_2CH_2CO_2C_4H_9^s$	230.31	...	...	256^{760}	0.9735^{20}_{4}	1.4238^{25}	i	s	s	...	s	...	**B2**^{2}, 559
s191	—,di(2-carboxyphenyl) ester	Diaspirin. Succinylsalicylic acid.	358.29	nd (aa or al)	176–8	...	...	...	δ	δ s^h	δ	...	...	aa δ, s^h	**B10**^{2}, 43
Ω s192	—,dichloride	Succinyl chloride. $ClCOCH_2CH_2COCl$	154.98	pl or lf	20	193.3 88.5^{19}	1.3748^{20}_{4}	1.4683^{20}	d	d	s	s	s	...	**B2,** 613
Ω s193	—,diethyl ester	Ethyl succinate. $C_2H_5O_2CCH_2CH_2CO_2C_2H_5$	174.20	...	−20.6	216.5^{760} 105^{15}	1.0402^{20}_{4}	1.4198^{20}	i	∞	∞	s	...	...	**B2**^{2}, 550
Ω s195	—,dimethyl ester	Methyl succinate. $CH_3O_2CCH_2CH_2CO_2CH_3$	146.14	...	19	196.4^{760} 80^{11}	1.1198^{20}_{4}	1.4197^{20}	δ	s	v	s	...	...	**B2**^{2}, 549
Ω s196	—,dinitrile	Succinonitrile. Ethylene dicyanide. $NCCH_2CH_2CN$	80.09	...	57.15–.20	$265–7^{760}$ 124^{5}	0.9867^{60}_{4}	1.4173^{60}	v	s	δ	s	s	chl s	**B2,** 615
s197	—,diphenyl ester	$C_6H_5O_2CCH_2CH_2CO_2C_6H_5$	270.29	lf (al)	121	222.5^{15}	...	...	i	s^h	s	s	s	...	**B6,** 155
s198	—,dipropyl ester	$C_3H_7^nO_2CCH_2CH_2CO_2C_3H_7^n$	202.25	...	−5.9	250.8^{760} 101.5^{3}	1.0020^{20}_{4}	1.4250^{20}	i	...	s	s	s	os s	**B2**^{2}, 551
s199	—,di(4-tolyl)ester	...	298.34	nd or lf (al)	121	...	...	...	i	s	s	s	s	aa, os s lig δ	**B6**^{2}, 379
Ω s200	—,imide	Succinimide.	99.09	pl (+1w, al), rh (ace) λ^{al} 218 sh (2.27), 243 (1.96)	126–7	287–8d	1.418	...	s v^h	δ s^h	δ	...	...	...	**B21,** 369
s201	—,—,*N*-benzyl-	$C_{11}H_{11}NO_2$. *See* s200	189.22	nd (w), pr (al)	103	390–400	...	...	v^h	v	s	...	v^h	chl v CS_2 δ peth i	**B21**^{2}, 304
Ω s202	—,—,*N*-bromo-	NBS. $C_4H_4BrNO_2$. *See* s200	177.99	cr (bz)	...	173.5d	2.098	...	δ	...	δ	v	...	AcOEt v Ac_2O δ	**B21**^{2}, 306
s203	—,—,*N*(2-bromophenyl)-	$C_{10}H_8BrNO_2$. *See* s200	254.10	(dil al)	91	...	...	...	i	s	...	...	...	...	**B21**^{2}, 304
s204	—,—,*N*(3-bromophenyl)-	$C_{10}H_8BrNO_2$. *See* s200	254.10	nd (al)	118	...	...	...	s^h	v^h	...	...	...	...	**B21**^{2}, 304
s205	—,—,*N*(4-bromophenyl)-	$C_{10}H_8BrNO_2$. *See* s200	254.10	pr (al)	172	...	...	...	i^h	...	...	s	s	os v	**B21**^{2}, 304
s206	—,—,*N*-carbethoxy-	$C_7H_9NO_4$. *See* s200	171.15	(eth-lig)	44	...	...	...	v	...	...	s	s	os v	**B21**^{2}, 305
Ω s207	—,—,*N*-chloro-	NCS. $C_4H_4ClNO_2$. *See* s200.	133.53	pl (CCl_4)	150	...	1.65	...	δ	δ	...	s	δ	aa s, lig δ	**B21**^{2}, 306
s208	—,—,*N*(chloromethyl)-	$C_5H_6ClNO_2$. *See* s200	147.56	...	58	$158–60^{12}$	...	...	v	...	...	s	v	oos v	**B21**^{2}, 305
s209	—,—,*N*(4-chlorophenyl)-	$C_{10}H_8ClNO_2$. *See* s200	209.64	nd (dil al)	170	...	...	...	...	v	...	s	v	os v	**B21**^{2}, 303
s210	—,—,*N*(ethoxymethyl)-	$C_7H_{11}NO_3$. *See* s200	157.17	nd	31–2	262 $151\text{-}2^{14}$	...	...	v	v	δ	...	...	...	**B21**^{2}, 304
s211	—,—,*N*(4-ethoxyphenyl)-	$C_{12}H_{13}NO_3$. *See* s200	219.24	pr (al)	155	...	...	...	δ^h	s^h	i	...	...	aa s^h	**B21,** 377
s212	—,—,*N*-ethyl-	$C_6H_9NO_2$. *See* s200	127.14	(eth)	26	236	...	...	v	v	v	...	...	...	**B21,** 373
s213	—,—,*N*(hydroxymethyl)-	$C_5H_7NO_3$. *See* s200	129.12	lf (bz)	66	...	...	...	v	...	...	...	...	os δ	**B21**^{2}, 304
Ω s214	—,—,*N*-iodo-	$C_4H_4INO_2$. *See* s200	224.99	(ace)	135	...	2.245	...	v	s	δ	s	...	...	**B21**^{2}, 306
s215	—,—,*N*-methyl-	$C_5H_7NO_2$. *See* s200	113.12	nd (eth-peth, al, ace) λ^{MeOH} 224 (2.7), 252 (2.0)	71 (66)	234	...	...	s	s	v	...	...	...	**B21**^{2}, 303
s216	—,—,*N*(1-naphthyl)-	$C_{14}H_{11}NO_2$. *See* s200	225.25	nd (dil al)	153	...	...	...	i	v^h	...	...	...	MeOH v^h	**B21,** 376
s217	—,—,*N*(2-naphthyl)-	$C_{14}H_{11}NO_2$. *See* s200	225.25	nd (al or AcOEt)	183	...	...	...	i	s	...	...	s	AcOEt δ	**B21,** 376
s218	—,—,*N*-phenyl-	Succinanil. $C_{10}H_9NO_2$. *See* s200	175.19	mcl pr or nd (w or al) λ^{hx} 262 (2.90), 273 (1.97)	156	*ca.* 400	1.356	...	i δ^h	s^h	s	...	...	...	**B21**^{2}, 303
s219	—,monoamide	Succinamic acid. $HO_2CCH_2CH_2CONH_2$	117.11	nd (w or ace)	156.3–7.8	...	...	...	s	δ	...	s^h	δ	lig δ	**B2**^{2}, 554
Ω s220	—,—,*N*- phenyl-	Succinanilic acid. $HO_2CCH_2CH_2CONHC_6H_5$	193.20	nd (w) λ^{al} 242 (4.11)	148.5	...	...	...	δ s^h	s	v	...	...	...	**B12,** 295
s221	—,monobenzyl ester	$HO_2CCH_2CH_2CO_2CH_2C_6H_5$	208.22	cr (bz)	59 (47)	...	...	...	i	...	...	...	δ	lig δ	**B6**^{2}, 418

For explanations, symbols and abbreviations see beginning of table. For structural formulas see end of table.

No.	Name	Synonyms and Formula	Mol. wt.	Color, crystalline form, specific rotation and λ_{max} (log ε)	m.p. °C	b.p. °C	Density	n_D	Solubility w	al	eth	ace	bz	other solvents	Ref.
	Succinic acid														
s222	—,monochloride monomethyl ester	$CH_3O_2CCH_2CH_2COCl$	150.56			102^{35} $53\text{–}4^{1}$		1.4412^{20}	d^h	d^h					$B2^2$, 553
s223	—,monoethyl ester(*dl*)	$HO_2CCH_2CH_2CO_2C_2H_5$	146.14	pr or nd	8	172^{42} 119^{3}	1.1466_4^{20}	1.4327^{20}	v	v	v			chl δ peth i	$B2^2$, 549
s224	—,monoethyl ester monomethyl ester	$CH_3O_2CCH_2CH_2CO_2C_2H_5$	160.17		< −20	208.2^{760} $90\text{–}5^{3}$	1.076_4^{20}		i	v	v				B2, 609
s226	—,piperazinium salt	$HO_2CCH_2CH_2CO_2H.C_4H_{10}N_2$	204.23		205–6d				s	s^h	i				Am 56, 1759
Ω s227	—,**acetyl-**, diethyl ester	$C_2H_5O_2CCH(COCH_3)CH_2CO_2C_2H_5$	216.24			$254\text{–}6^{760}$ 139^{12}	1.081_4^{20}	1.438^{16}	i	s	s		s	CS_2 s	$B3^2$, 486
—	—,**amino-**	*see* **Aspartic acid**													
s228	—,**benzyl-**(*dl*)	$C_6H_5CH_2CH(CO_2H)CH_2CO_2H$	208.22	nd or lf $\lambda^{a\,c}$ 261 (2.35)	163–4				s^h	s	s		δ		$B9^2$, 628
Ω s229	—,**bromo-**(*dl*)	$HO_2CCH_2CHBrCO_2H$	196.99	(w)	161		2.073		s	s				aa δ	$B2^2$, 560
s230	—,**2,3-diacetyl-**, diethyl ester (diketo form)	Ethyl diacetosuccinate. $C_2H_5O_2CCH(COCH_3)CH(COCH_3)CO_2C_2H_5$	258.27	mcl pr or nd (al) λ^{MeOH} 250 (3.7)	90				δ	s		s	s	chl v lig δ	$B3^2$, 505
s231	—,**2,3-dibromo-**(*d*)	$HO_2CCHBrCHBrCO_2H$	275.89	pl (aa-CCl_4) $[\alpha]_D^{24}$ +147.8 (AcOEt) $[\alpha]_D^{18}$ +64.4 (w, c = 5)	157–8				s	v		v			$B2^2$, 561
Ω s232	—,—(*dl*)	$HO_2CCHBrCHBrCO_2H$	275.89	(w or AcOEt)	171 (166–7)				s	v		v			B2, 561
s233	—,—(*l*)	$HO_2CCHBrCHBrCO_2H$	275.89	nd (bz), $[\alpha]_D^{18}$ −148 (AcOEt, c = 5.79)	157–8d				v	v		v		AcOEt, MeOH v, chl, CCl_4, peth δ	$B2^1$, 268
s234	—,—(*meso*)	$HO_2CCHBrCHBrCO_2H$	275.89		255 (sealed tube)	sub 275			δ v^h	v	v			chl δ	$B2^2$, 561
s235	—,**2,3-dichloro-** (*d*)	$HO_2CCHClCHClCO_2H$	186.98	mcl, $[\alpha]_D^{9}$ +3.6 (w, c = 3)	168		1.820^{15}		s	δ	s	s	δ	chl s	$B2^2$, 556
s236	—,—(*dl*)	$HO_2CCHClCHClCO_2H$	186.98	pr (w or eth-peth)	175d		1.844^{15}		s	δ	s	s	δ	chl s	$B2^2$, 557
Ω s237	—,—(*meso*)	$HO_2CCHClCHClCO_2H$	186.98	hex pr (w)	221d				s	v	v	v	δ	chl v lig δ	$B2^2$, 558
s238	—,—,dichloride(*dl*)	$ClCOCHClCHClCOCl$	223.87		39	78.5^{7}			d	d	s				$B2^2$, 558
s239	—,—,—(*meso*)	$ClCOCHClCHClCOCl$	223.87			$105\text{–}6^{45}$ $79\text{–}80^{15}$			d	d				CCl_4 s	$B2^2$, 558
s240	—,—,diethyl ester(*dl*)	$C_2H_5O_2CCHClCHClCO_2C_2H_5$	243.09			132^{15}	1.1963_4^{77}	1.4512^{20}			v				$B2^2$, 558
s241	—,—,—(*meso*)	$C_2H_5O_2CCHClCHClCO_2C_2H_5$	243.09	nd (dil al)	63	$125.5^{12.5}$	1.1490_4^{69}	1.4266^{65}	i	v	v				$B2^2$, 558
s242	—,**2,3-diethyl-**, diphenyl ester	3,4-Dicarbophenoxyhexane. $C_6H_5O_2CCH(C_2H_5)CH(C_2H_5)CO_2C_6H_5$	326.40	nd (lig)	107–8				i	δ		s	s	to, CS_2 s	B6, 156
—	—,**2,3-dihydroxy-**	*see* **Tartaric acid**													
s243	—,**2,3-dimethoxy-**, dimethyl ester(*d*)	$CH_3O_2CCH(OCH_3)CH(OCH_3)CO_2CH_3$	206.20	pr, $[\alpha]_D^{18}$ +79.9 (MeOH, c = 2)	53–4	$130\text{–}2^{12}$	1.1317_4^{60}	1.4340^{20}						MeOH s	$B3^2$, 328 $B16^2$, 871
—	—,**dioxo-**, dihydrate	*see* **Succinic acid, tetrahydroxy-**													
s244	—,**2-dodecyl-3-methyl-**	2,3-Pentadecanedicarboxylic acid*. Roccellic acid. $CH_3(CH_2)_{11}CH(CO_2H)CH(CH_3)CO_2H$	300.44	lf (aa, bz or al), rods (ace) wh pl (peth) $[\alpha]_D^{20}$ +17.4 (al)	132–3				i	s	v	s	s^h	$NaHCO_3$ s peth i	B2, 734
s245	—,**ethyl-**(*d*)	$HO_2CCH_2CH(C_2H_5)CO_2H$	146.14	pr or nd, $[\alpha]_D^{16}$ +20.6 (aa, c = 3.7) $[\alpha]_D^{17}$ +26 (ace, c = 4)	96	180–3	1.0017_4^{20}		v	v	v			chl δ peth i	$B2^2$, 584
s246	—,—(*l*)	$HO_2CCH_2CH(C_2H_5)CO_2H$	146.14	pr or nd $[\alpha]_D^{24}$ −20.8 (ace, c = 4.6)	96	180–3	1.0018^{20}		v	v	v			chl δ peth i	$B2^2$, 584
s247	—,**formyl-**, diethyl ester (*aldo* form)	$C_2H_5O_2CCH_2CH(CHO)CO_2C_2H_5$	202.21			$130\text{–}4^{15}$ $114\text{–}9^{5}$		1.4486^{25}	i	∞	∞				$B3^2$, 485
—	—,**(hydroxy-methyl)-**,γ-lactone	*see* **Paraconic acid**													
s248	—,**2-hydroxy-3-methyl-**(*d*)-	d-Citramalic acid. $HO_2CC(CH_3)OHCH_2CO_2H$	148.13	$[\alpha]_D^{14}$ +34.67 (w)	108–9				v	s		s	i	peth i	B3, 443
s249	—,—(*dl*)	$HO_2CC(CH_3)OHCH_2CO_2H$	148.13	mcl pr (AcOEt)	123	sub			v	s		s	i	AcOEt s peth i	$B3^2$, 294
s250	—,**isopropylidene-**	Teraconic acid. $(CH_3)_2C{:}C(CO_2H)CH_2CO_2H$	158.16	tcl nd (eth)	160–1 (165–6)				s^h	s	s		δ		B2, 786

For explanations, symbols and abbreviations see beginning of table. For structural formulas see end of table.

No.	Name	Synonyms and Formula	Mol. wt.	Color, crystalline form, specific rotation and λ_{max} (log ε)	m.p. °C	b.p. °C	Density	n_D	Solubility w	al	eth	ace	bz	other solvents	Ref.
	Succinic acid														
s250[1]	**—,mercapto-(*d*)**	*d*-Thiomalic acid. $HO_2CCH_2CH(SH)CO_2H$	150.15	cr (AcOEt-bz) $[\alpha]_D^{17}$ +64.4 (al), +76.1 (ace)	154	....	....	....	s	s	...	s	...	....	**B3**[2]**,** 287
Ω s250[2]	—,—(*dl*)	$HO_2CCH_2CH(SH)CO_2H$	150.15	cr (eth)	151	....	....	....	v	v	s	v	i	....	**B3**[2]**,** 291
s250[3]	—,—(*l*)	$HO_2CCH_2CH(SH)CO_2H$	150.15	$[\alpha]_D^{17}$ −64.8 (al), −75.8 (ace)	152–3	....	....	....	s	s	δ	s	δ	....	**B3**[2]**,** 287
Ω s251	**—,methyl-(*dl*)**	Pyrotartaric acid. $HO_2CCH_2CH(CH_3)CO_2H$	132.13	pr	115	d	....	1.4303	v	v	s	...	...	MeOH v chl δ	**B2,** 637
s252	**—,2-methyl-3-oxo-, diethyl ester**	Diethyl methyloxaloacetate. $C_2H_5O_2CCOCH(CH_3)CO_2C_2H_5$	202.21	....	....	$137–8^{23}$ $75–8^{2}$	1.0970_4^{20}	1.4313^{20}	i	∞	∞	...	...	....	**B3**[2]**,** 484
Ω s253	**—,methylene-**	Itaconic acid. $CH_2{:}C(CO_2H)CH_2CO_2H$	130.10	rh (bz)	175 (162–4)	d	1.632	....	s	s	δ	s	δ	chl s peth, CS_2 δ	**B2**[2]**,** 650
Ω s254	—,—,anhydride	Itaconic anhydride.	112.09	rh bipym pr (eth or chl), sc (aa)	68.5 (70)	$139–40^{30}$ $114–5^{18}$	....	....	d^h	d^h	δ	...	...	chl v	**B17**[2]**,** 449
Ω s255	—,—,dichloride	Itaconyl chloride. $CH_2{:}C(COCl)CH_2COCl$	166.99	....	....	89^{17} 72^{2}	....	1.4919^{20}	d^h	d^h	...	s	...	....	**B2,** 762
s256	—,—,diethyl ester	Diethyl itaconate. $CH_2{:}C(CO_2C_2H_5)CH_2CO_2C_2H_5$	186.21	λ^{al} 265 sh (2.1)	58–9	228 111^{13}	1.0467_4^{20}	1.4377^{20}	...	∞	s	v	s	....	**B2**[2]**,** 651
Ω s257	—,—,dimethyl ester	Dimethyl itaconate. $CH_2{:}C(CO_2CH_3)CH_2CO_2CH_3$	158.16	hyg mcl (MeOH) λ^{al} 205 (3.88), 240 sh (2.2)	38	208^{760} 108^{11}	1.1241_4^{18}	1.4457^{20}	...	s	s	v	...	MeOH s	**B2,** 762
s258	**—,oxo-, diethyl ester**	Diethyl oxaloacetate. $C_2H_5O_2CCH_2COCO_2C_2H_5$	188.18	λ^{w} 265 (2.25)	....	$131–2^{24}$	1.131_4^{20}	1.4561^{17}	i	∞	∞	v	∞	....	**B3**[2]**,** 479
s259	**—,2-oxo-3-phenyl-, 1-ethyl ester 4-nitrile**	Ethyl phenylcyanopyruvate. $C_6H_5CH(CN)COCO_2C_2H_5$	217.23	(eth-lig)	130	206^{20}	....	....	...	v	δ	...	...	chl, alk s	**B10**[2]**,** 607
s260	**—,phenyl-(*d*)**	$HO_2CCH_2CH(C_6H_5)CO_2H$	194.19	pr (w), $[\alpha]_D^{16.5}$ +148.3 (al, c = 1.5) λ^{al} 260 (2.05)	173–4	....	....	....	δ v^h	s	v	v	δ	MeOH s	**B9**[1]**,** 380
Ω s261	—,—(*dl*)	$HO_2CCH_2CH(C_6H_5)CO_2H$	194.19	lf or nd (w) λ^{al} 260 (2.06)	168	d	....	....	δ v^h	v	v	v	i	aa v chl δ CS_2, peth i	**B9**[2]**,** 619
s262	—,—(*l*)	$HO_2CCH_2CH(C_6H_5)CO_2H$	194.19	$[\alpha]_D^{15}$ −173.3 (ace) λ^{al} 260 (2.06)	173–4	....	....	....	...	v	v	v	i	MeOH s	**B9**[1]**,** 381
s263	—,—,anhydride(*d*)	....	176.18	nd (bz-peth), $[\alpha]_D^{15}$ +100.9 (bz) λ^{al} 258 (2.2)	83.5–4.5	....	....	....	d^h	s	...	...	v	chl v peth, CCl_4 δ	**B17**[1]**,** 259
s264	—,—,—(*dl*)	$C_{10}H_8O_3$. *See* s263	176.18	mcl pr or nd (eth)	54	$204–6^{22}$ $191–2^{12}$	....	....	i	v	s	v	v	oos v	**B17**[2]**,** 473
s265	—,—,—(*l*)	$C_{10}H_8O_3$. *See* s263	176.18	$[\alpha]_D^{14}$ − 100.9 (bz)	83.5–4.5	....	....	....	d^h	...	...	...	v	chl v	**B17**[1]**,** 259
s266	**—,(3-phenyl-propenyl)-**	$C_6H_5CH_2CH{:}CHCH(CO_2H)CH_2CO_2H$	234.25	lf (eth), (bz)	112	....	....	....	s	...	v	v	v	....	**B9,** 909
Ω s267	**—,tetrahydroxy-**	$HO_2CC(OH)_2C(OH)_2CO_2H$	182.09	....	114–5	....	....	....	d	...	...	...	...	....	**B3**[2]**,** 500
s268	**—,tetramethyl-**	$HO_2CC(CH_3)_2C(CH_3)_2CO_2H$	174.20	tcl (60% MeOH, lig or AcOEt), mcl and tcl (eth or ace)	200	sub	1.30	....	δ	v	...	...	v	chl s	**B2**[2]**,** 601
s269	—,—,dinitrile	$NCC(CH_3)_2C(CH_3)_2CN$	136.20	mcl pl, lf and pr (dil al)	170.5–1.5	....	1.070	....	...	s^h	...	...	...	....	**B2**[1]**,** 290
—	**Succinimide**	*see* **Succinic acid,** imide													
Ω s273	**Sucrose**	Cane sugar. Saccharose.	342.30	mcl, $[\alpha]_D^{20}$ +66.37 (w)	185–6	....	$1.5805^{17.5}$	1.5376	s v^h	δ	i	...	...	Py s	**B31,** 424
Ω s274	—,octaacetate	$C_{28}H_{38}O_{19}$. *See* s273	678.61	nd (al), $[\alpha]_D^{20}$ +59.6 (chl)	86–87	d 285 260^{1}	1.27^{16}	1.4660	δ^h	s^h	s	s	s	chl, oos s	**B31,** 453
s275	**Sudan III**	Tetrazobenzene-β-naphthol.	352.40	br lf with gr lustre (aa) $\lambda^{cellosolve}$ 345 (4.22), 505 (4.49)	195	....	....	....	i	s	s	s	s	xyl, chl, aa, peth s	**B16**[2]**,** 75
—	**Sudan G**	*see* **Azobenzene, 2,4-dihydroxy-**													
—	**Sudan yellow**	*see* **Azobenzene 1-naphthalene, 2′-hydroxy-**													
Ω s276	**Sulfadiazine**	2-Sulfanilamidopyrimidine. Sulfapyrimidine.	250.28	cr (w), wh pw $\lambda^{0.1N\,HCl}$ 244 (4.15)	255–6d (cor)	....	....	....	δ	δ	...	δ	...	ac s	**C55,** 25956
s277	**Sulfaguanidine**	Sulfoguenil.	214.25	nd (w)	190–3 (anh)	....	....	....	δ s^h	δ	...	δ	...	dil ac s	**C55,** 22204

For explanations, symbols and abbreviations see beginning of table. For structural formulas see end of table.

No.	Name	Synonyms and Formula	Mol. wt.	Color, crystalline form, specific rotation and λ_{max} (log ε)	m.p. °C	b.p. °C	Density	n_D	Solubility w	al	eth	ace	bz	other solvents	Ref.
	Sulfaguanidine														
s278	—monohydrate	$C_7H_{10}N_4O_2S.H_2O$	232.27	nd (w)	143 (sealed tube)				δ s^h	δ	i	δ			**M**, 993
Ω s279	**Sulfamerazine**	Sulfamethyldiazine	264.31	cr $\lambda^{w, pH=2.1}$ 239 (4.16), 315 (3.53)	234–8				δ	δ	i	δ		chl i	**C55**, 5501
Ω s280	**Sulfamethazine**		278.34	pa ye (w + ½), cr (diox-w) λ^{w} 241 (4.20), 260 (4.23)	198–9 (cor) (205–7)				s					ac, alk s	**Am 64**, 567
—	**Sulfanilamide**	*see* **Benzenesulfonic acid, 4-amino-, amide**													
—	**Sulfanilic acid**	*see* **Benzenesulfonic acid, 4-amino-***													
Ω s281	**Sulfapyrazine**	*N'*(2-Pyrazinyl)sulfanilamide.	250.28	nd ($PhNO_2$)	251 (251–4)				i	i	i	δ	i	Py s diox δ chl i	**Am 63**, 3153
s282	**Sulfapyridine**		249.29	ye og (al) λ^{w} 262 (4.2), 310 (4.0)	191–3				s^h i	s^h			i	CCl_4 i	**C34**, 1814
Ω s283	**Sulfaquinoxaline**		300.34	$\lambda^{w, pH=6.6}$ 255 (4.52), 362 (3.93)	247–8				δ	δ		δ		aq alk s	**C49**, 5525
s284	**Sulfathiadiazole**		256.32		218				δ	s				Py s	**C40**, 5411
Ω s285	**Sulfathiazole**		255.32	br pl, rods or pw (45 % al) λ^{al} 260 (4.16), 290 (4.27)	202.5 (form I) 175 (form II)				δ	δ					**C40**, 7518
Ω s286	—,**phthalyl-**		403.44	λ^{w} 270 (4.30)	272–7d				i	δ	i			ac, alk s chl i	**C51**, 9689
s287	—,**4-nitro-**	Nisulfazole.	285.30	pa ye pw	255–62				i	δ	i		i	dil alk v chl i	**C53**, 12281
Ω s288	—,**succinyl-**		355.39	cr λ^{w} 257 (4.26), 275.5 (4.23)	192–5 (184–6)				i	δ	i	δ		alk s chl i	**C51**, 12148
Ω s289	**Sulfide, benzyl phenyl**	Benzylthiobenzene. $C_6H_5CH_2SC_6H_5$	200.31	lf (al) λ^{al} 255 (3.83)	42–3.5	197^{27}			i	s^h	s			con sulf s	**B6**², 428
—	—,**bis(2-amino-2-carboxyethyl)**	*see* **Lanthionine**													
—	—,**bis(dimethyl-arsine)**	*see* **Cacodyl sulfide**													
s290	—,**butyl ethyl**	1(Ethylthio)butane*. $CH_3(CH_2)_3SC_2H_5$	118.24		−95.13	144.2^{760} 33.3^{10}	0.8376^{20}_{4}	1.4491^{20}		v				chl s	**B1**³, 1522
s291	—,**butyl methyl**	1(Methylthio)butane*. $CH_3(CH_2)_3SCH_3$	104.22		−97.8	123.2^{760}	0.8426^{20}_{4}	1.4477^{20}		v				MeOH v^h	**B1**³, 1521
Ω s292	—,**diallyl**	Allyl sulfide. $(CH_2:CHCH_2)_2S$	114.21	λ^{al} 221 (3.33)	−83	139^{758} 35^{6}	0.8877^{27}_{4}	1.4870^{25}	δ	∞	∞			CCl_4, CS_2 s	**B1**², 478
s293	—,**dibenzoyl**	Thiobenzoic anhydride. $C_6H_5COSCOC_6H_5$	242.30	pr (al, eth)	48	d			i		v				**B9**², 289
Ω s294	—,**dibenzyl**	Benzyl sulfide. $(C_6H_5CH_2)_2S$	214.33	pl (eth or chl) λ^{al} 238 sh (3.30), 260 (2.85), 266 sh (2.66)	49–50	d	1.0712^{50}_{50}		i	s	s				**B6**², 429
Ω s295	—,**dibutyl** (α-form)	1(Butylthio)butane*. Butyl sulfide. $CH_3(CH_2)_3S(CH_2)_3CH_3$	146.30		−79.7	185–5.5	0.8386^{20}_{4}	1.4530^{20}	i	v	v			chl, CCl_4 v	**B1**², 399
s296	—,—(β-form)	$CH_3(CH_2)_3S(CH_2)_3CH_3$	146.30	λ^{iso} 251 sh (1.0)		190–230d			i	s	s	s		oos s	**Ber 62**, 2168
s297	—,—,**2,2'-dimethyl-(*d*)**	*d*-Di-*act*-amyl sulfide. $CH_3CH_2CH(CH_3)CH_2SCH_2CH(CH_3)CH_2CH_3$	174.35	$[\alpha]^{20}_{D}$ +24.5	98		0.8362^{20}_{0}								**B1**, 387
Ω s298	—,—,**3,3'-dimethyl-**	Isoamyl sulfide. $(CH_3)_2CHCH_2CH_2SCH_2CH_2CH(CH_3)_2$	174.35	col-pa ye		216^{760} 85.5^{5}	0.8323^{20}_{4}	1.4520^{20}	i	∞	v				**B1**³, 1646
s299	—,**di-*sec*-butyl**	*sec*-Butyl sulfide. $CH_3CH_2CH(CH_3)SCH(CH_3)CH_2CH_3$	146.30			165	0.8348^{20}_{4}	1.4506^{20}	i	v	v				**B1**, 373
Ω s300	—,**di(dimethyl-thiocarbamyl)**	$(CH_3)_2NCSSCSN(CH_3)_2$	208.37	ye mcl λ^{MeOH} 279 (4.21)	109–10		1.37		i	s	δ	s	s	chl s	**B4**, 76
s301	—,**diethenyl**	Vinyl sulfide. $CH_2:CHSCH:CH_2$	86.16	λ^{al} 240 (4.62), 255 (4.58)		84^{759}	0.9174^{15}_{4}		δ	∞	∞	s			**B1**², 474
s302	—,**diethoxy**	Ethoxyl sulfide. $C_2H_5OSOC_2H_5$	122.19	λ < 308		117^{733}	0.9940^{20}_{4}	1.4234^{20}	s	v	v	v	v	os v	**B1**³, 1314
Ω s303	—,**diethyl**	Ethyl sulfide. $C_2H_5SC_2H_5$	90.19	λ^{al} 210 (3.25), 229 sh (2.14)	−103.9	92.1^{760}	0.8362^{20}_{4}	1.4430^{20}	δ	s	s				**B1**², 343
s305	—,—,**2-chloro-**	$ClCH_2CH_2SC_2H_5$	124.63			156^{760} $63–5^{47}$	1.0663^{20}_{4}	1.4878^{20}						chl s	**B1**², 348
s306	—,—,**2,2'-diamino-**	$H_2NCH_2CH_2SCH_2CH_2NH_2$	120.22	ye		$231–3^{755}$ $118–20^{17}$			∞						**B4**, 287

For explanations, symbols and abbreviations see beginning of table. For structural formulas see end of table.

No.	Name	Synonyms and Formula	Mol. wt.	Color, crystalline form, specific rotation and λ_{max} (log ε)	m.p. °C	b.p. °C	Density	n_D	Solubility w	al	eth	ace	bz	other solvents	Ref.
	Sulfide														
—	—,—,2,2′-diamino-2,2′-dicarboxy-	*see* **Lanthionine**													
s307	—,—,2,2′-dichloro-	Mustard gas. Yperite. $ClCH_2CH_2SCH_2CH_2Cl$	159.09	ye pr	13–4	217^{760} 95^{10}	1.2741^{20}_{4}	1.5312^{20}	i	v	s	v	v	lig ∞	**B1**², 349
Ω s308	—,—,2,2′-dihydroxy-	2,2′-Thiodiethanol. Thiodiglycol. $HOCH_2CH_2SCH_2CH_2OH$	122.19	λ^{al} <210	−10 (−16)	$164–6^{20}$ 133^{1}	1.1819^{20}_{4}	1.5203^{20}	∞	∞	s	...	δ	chl, AcOEt ∞	**B1**³, 2122
s309	—,—,2,2′-diphenoxy-	$C_6H_5OCH_2CH_2SCH_2CH_2OC_6H_5$	274.39	nd (al)	42	...	...	...	i	s	s	...	...	...	**B6**², 150
s310	—,diheptyl	Heptyl sulfide. $CH_3(CH_2)_6S(CH_2)_6CH_3$	230.46	...	...	298 164^{20}	0.8416^{20}_{4}	1.4606^{20}	i	...	s	...	...	...	**B1**³, 1685
s311	—,dihexyl	Hexyl sulfide. $CH_3(CH_2)_5S(CH_2)_5CH_3$	202.41	...	...	230 113.5^{4}	0.8411^{20}_{4}	1.4586^{20}	...	...	...	...	...	...	**B1**³, 1659
s312	—,diisopropyl	Isopropyl sulfide. $(CH_3)_2CHSCH(CH_3)_2$	118.24	λ^{iso} 213 (3.4)	−78.08	120.02^{760}	0.8142^{20}_{4}	1.4438^{20}	i	s	s	...	...	...	**B1**³, 1479
Ω s313	—,dimethyl	Methyl sulfide. CH_3SCH_3	62.13	λ^{al} 215 (3.0)	−98.27	37.3^{760}	0.8483^{20}_{4}	1.4355^{20}	i	s	s	...	...	MeOH s	**B1**, 288
s313¹	—,—,hexafluoro-	F_3CSCF_3	170.08	λ^{vap} 210 (0.84)	...	-22.2^{760}	...	...	...	...	...	...	...	...	**J1952**, 2198
s314	—,1,1′-dinaphthyl	α-Naphthyl sulfide. $C_{10}H_7^{\alpha}SC_{10}H_7^{\alpha}$	286.40	nd or pr (al)	110	$289–90^{15}$	...	...	...	δ s^h	...	...	v	CS_2, aa v	**B6**², 588
Ω s315	—,2,2′-dinaphthyl	β-Naphthyl sulfide. $C_{10}H_7^{\beta}SC_{10}H_7^{\beta}$	286.40	pl (al), lf (bz)	151	$295–6^{15}$	...	...	...	δ^h i	...	...	v	CS_2, aa v	**B6**², 611
s316	—,dipentyl	Amyl sulfide. $CH_3(CH_2)_4S(CH_2)_4CH_3$	174.35	...	−51.3	230^{760} 84.5^{4}	0.8409^{20}_{4}	1.4556^{20}	i	...	s	...	...	...	**B1**³, 1608
Ω s317	—,diphenyl	Phenyl sulfide.	186.28	λ^{hx} 240 (3.8), 255 (4.1), 280 (3.7)	−25.9	296 145^{8}	1.1136^{20}_{4}	1.6334^{20}	i	s^h	∞	...	∞	CS_2 ∞	**B6**, 299
s318	—,—,2-amino-	$C_{12}H_{11}NS$. *See* s317	201.29	pl (al)	35–6	257.5^{100} 212^{25}	...	...	...	s v^h	...	...	...	...	**B13**², 200
s319	—,—,4-amino-	$C_{12}H_{11}NS$. *See* s317	201.39	nd (dil al), cr (lig)	95.8	282.3^{100} 242.5^{29}	...	...	δ^h	v	v	...	...	...	**B13**², 298
s320	—,—,2,4′-diamino-	$C_{12}H_{12}N_2S$. *See* s317	216.31	nd (w or al), pr (dil al)	62.5	...	...	...	s^h	v	v	...	v	peth δ	**B13**², 299
Ω s321	—,—,4,4′-diamino-	*p,p′*-Thiodianiline. $C_{12}H_{12}N_2S$. *See* s317	216.31	nd (w) λ^{al} 264 (4.39)	108–9	...	...	...	δ^h	v	v	...	v^h	...	**B13**², 299
s322	—,—,4,4′-dichloro-2,2′-dinitro-	$C_{12}H_6Cl_2N_2O_4S$. *See* s317	345.16	br-ye nd (90% aa)	149–50	...	...	...	...	i	...	...	v	...	**B6**², 312
s323	—,—,4,4′-dihydroxy-	$C_{12}H_{10}O_2S$. *See* s317	218.28	mcl pr or lf (al)	151	...	...	...	i s^h	v	v	...	...	...	**B6**, 860
s324	—,—,2,2′-dimethoxy-	o-Anisyl sulfide. $C_{14}H_{14}O_2S$. *See* s317	246.33	lf (al)	73	$252–3^{10}$	...	...	s^h	v	v	...	v	aa s peth i	**B6**, 794
s325	—,—,2,2′-dinitro-	$C_{12}H_8N_2O_4S$. *See* s317	276.28	gold-ye pl (aa)	122–3	sub δd	...	...	...	δ	...	...	...	aa δ	**B6**², 305
s326	—,—,2,4-dinitro-	$C_{12}H_8N_2O_4S$. *See* s317	276.28	pa ye nd (ace or bz-al) λ^{al} 270 sh (3.89), 331 (4.07)	121	...	...	...	i	δ	...	s	v	aa v	**B6**², 315
Ω s327	—,—,4,4′-dinitro-	$C_{12}H_8N_2O_4S$. *See* s317	276.28	og pl (aa) λ^{al} 235 (4.08), 250 (4.06), 341 (4.18)	160–1	...	...	...	i	δ s^h	...	...	...	con sulf s	**B6**², 311
s328	—,—,2-nitro-	$C_{12}H_9NO_2S$. *See* s317	231.28	ye-og nd (lig or al-eth) λ^{sulf} 414 (4.28), 575 (3.73), 611 (3.75)	82	210^{15}	...	...	...	s	s	...	...	con sulf s peth i	**B6**², 305
Ω s329	—,—,4-nitro-	$C_{12}H_9NO_2S$. *See* s317	231.28	pa ye mcl pr (lig) λ^{sulf} 480 (4.30)	55	288.2^{100} 240^{25}	...	...	...	s	s	...	...	...	**B6**², 311
Ω s330	—,—,2,2′4,4′-tetranitro-	$C_{12}H_6N_4O_8S$. *See* s317	366.27	ye nd or pl (aa)	197	...	...	...	...	i	...	...	i	con HNO_3 s CS_2 i	**B6**², 315
s332	—,dipropyl	Propyl sulfide. $CH_3CH_2CH_2SCH_2CH_2CH_3$	118.24	...	−102.5	142.38^{760} 32.31^{10}	0.8377^{20}_{4}	1.4487^{20}	i	s	s	...	...	...	**B1**², 372
s333	—,—,2,2′-dichloro-	$CH_3CHClCH_2SCH_2CHClCH_3$	187.13	...	−40	122^{23} $95–7^{4.5}$	1.1569^{25}_{4}	1.5020^{20}	i	s	...	...	...	...	**B1**³, 1437
s334	—,—,3,3′-dichloro-	$Cl(CH_2)_3S(CH_2)_3Cl$	187.13	...	...	162^{43} $111–2^{7}$	1.1774^{25}_{4}	1.5075^{20}	...	s	s	...	...	to s	**B1**³, 1438
s335	—,2,2′-ditolyl	*o*-Tolyl sulfide.	214.33	pl (al) λ^{al} 248 (4.11), 274 (3.70)	64	285 174^{15}	...	...	i	s^h δ	v	...	...	chl, CS_2 v	**B6**², 342
s336	—,2,4′-ditolyl	...	214.33	...	...	173^{11}	1.0774^{15}_{4}	...	...	s	s	...	...	...	**B6**, 418
s337	—,3,4′-ditolyl	...	214.33	nd (al)	28	179^{11}	...	...	...	s	s	...	...	...	**B6**, 418
Ω s338	—,4,4′-ditolyl	*p*-Tolyl sulfide.	214.33	nd (al) λ^{al} 252 (4.16), 276 (3.83)	57.3	>300 179^{11} (cor)	...	...	i	s^h	v	s	s^h	aa s^h	**B6**², 395
s340	—,ethyl isobutyl	$C_2H_5SCH_2CH(CH_3)_2$	118.24	...	...	134.22^{760} 24.8^{10}	0.8306^{20}_{4}	1.4450^{20}	...	...	...	...	...	chl s	**B1**³, 1566
s341	—,ethyl methyl	$CH_3SC_2H_5$	76.16	λ^{iso} 212 (3.0), 240 sh (1.6)	−105.91	66.65^{760}	0.8422^{20}_{4}	1.4404^{20}	i	∞	s	...	...	...	**B1**, 343

For explanations, symbols and abbreviations see beginning of table. For structural formulas see end of table.

No.	Name	Synonyms and Formula	Mol. wt.	Color, crystalline form, specific rotation and λ_{max} (log ε)	m.p. °C	b.p. °C	Density	n_D	Solubility w	al	eth	ace	bz	other solvents	Ref.
	Sulfide														
Ω s342	—,—,2-chloro-	$CH_3SCH_2CH_2Cl$	110.61	...	...	140 44^{20}	1.1097^{25}_{4}	1.4902^{20}	...	s	s	s	...	...	**B1**^{2}, 347
—	—,ethyl phenyl	*see* **Benzene, (ethylthio)-***													
s343	—,ethyl propyl	$C_2H_5SCH_2CH_2CH_3$	104.22	λ^{cy} 216 (2.49)	−117.04	118.5^{760} 13.49^{10}	0.8370^{20}_{4}	1.4462^{20}	...	s	...	...	...	...	**B1**^{3}, 1432
s345	—,2-furyl phenyl	...	176.23	...	...	$119–20^{8}$ $97–8^{2.5}$	1.1341^{26}	1.5976^{20}	...	s	s	...	...	...	**Am 81,** 4927
s346	—,isobutyl methyl	$CH_3SCH_2CH(CH_3)_2$	104.22	...	...	112.5^{760}	0.8335^{20}_{4}	1.4433^{20}	...	s	s	s	...	...	**B1**^{3}, 1565
s347	—,isopropyl methyl	$CH_3SCH(CH_3)_2$	90.19	λ^{iso} 215 (2.96)	−101.51	84.75^{760}	0.8291^{20}_{4}	1.4932^{20}	...	s	s	s	...	...	**B1,** 367
—	—,methyl phenyl	*see* **Benzene, (methylthio)-***													
s348	—,methyl propyl	$CH_3SCH_2CH_2CH_3$	90.19	λ^{cy} 210–15 (3.00)	−113	95.54^{760} -4.09^{10}	0.8424^{20}	1.4442^{20}	s	s	s	s	...	...	**B1**^{3}, 1432
Ω s350	—,pentamethylene	Tetrahydrothiopyran. Thiacyclohexane. Thiane.	102.20	λ^{al} 230 sh (2.2)	19	141.75^{760} 93^{82}	0.9861^{20}_{4}	1.5067^{20}	i	s	s	s	s	os s	**B17**^{2}, 18
s351	—,trimethylene	Thiacyclobutane.	74.15	λ^{al} 272 (1.4)	−73.25	94.67^{760} 14.01^{30}	1.0200^{20}_{4}	1.5102^{20}	i	v	...	s	v	os v	**B17**^{2}, 12
s352	**Sulfone, benzyl phenyl**	$C_6H_5CH_2SO_2C_6H_5$	232.30	nd (al) λ^{al} 219 (4.12), 259 (2.85), 265 (3.03), 272 (2.90)	146–6.5	...	1.126^{153}_{4}	...	i s^h	δ	δ	...	δ	...	**B6**^{2}, 428
s353	—,benzyl (4-tolyl)	...	246.33	nd (al)	144–5	...	...	...	...	v^h	...	...	v	aa v	**B6,** 455
Ω s354	—,dibenzyl	Benzyl sulfone. $C_6H_5CH_2SO_2CH_2C_6H_5$	246.33	nd (al-bz) λ^{al} 219 (4.32), 259 (2.68), 265 (2.61), 269 (2.41)	155	290 δd	...	...	s^h i	δ	...	v	s	aa s	**B6**^{2}, 430
Ω s355	—,dibutyl	Butyl sulfone. $CH_3(CH_2)_3SO_2(CH_2)_3CH_3$	178.30	pl (w or al)	46	...	...	...	i	s	s	...	...	...	**B1**^{2}, 400
s356	—,diethyl	Ethyl sulfone. $C_2H_5SO_2C_2H_5$	122.19	rh pl λ < 208	73–4	248	1.357^{20}_{4}	...	s	...	s^h	...	v	peth i	**B1**^{2}, 345
s356[1]	—,—2,2′-diphenoxy-	$C_6H_5OCH_2CH_2SO_2CH_2CH_2OC_6H_5$	306.38	pink lf (al)	180	...	...	...	i	s	...	...	...	...	**B6**^{2}, 150
Ω s357	—,diisopropyl	Isopropyl sulfone. $(CH_3)_2CHSO_2CH(CH_3)_2$	150.24	(eth)	36	...	...	...	v	...	s^h	...	...	...	**B1**^{2}, 387
Ω s358	—,dimethyl	Methyl sulfone. $CH_3SO_2CH_3$	94.13	pr	110	238^{760}	1.1702^{110}_{0}	1.4226	s	s	...	...	s	...	**B1,** 289
Ω s359	—,diphenyl	Phenyl sulfone.	218.28	mcl pr (bz), pl (al), nd (w) λ^{al} 235 (4.20), 261 (3.23), 267 (3.30), 274 (3.11)	128–9	379 232^{18}	1.252^{20}_{4}	...	i δ^h	s^h	s	...	s	...	**B6**^{3}, 992
s360	—,—,2-amino-	$C_{12}H_{11}NO_2S$. *See* s359	233.29	lf (dil al) λ^{cy} 228 (4.33), 274 (2.94), 280 (2.40), 318–20 (3.62)	122–4	...	...	...	i	v	...	...	v	aa v	**B13,** 399
s361	—,—,4-amino-	$C_{12}H_{11}NO_2S$. *See* s359	233.29	nd (al) λ^{cy} 270 (4.27)	176	...	...	...	i	s	...	...	v^h	aa v	**B13**^{2}, 298
Ω s362	—,—,4-chloro-	Sulphenone. $C_{12}H_9ClO_2S$. *See* s359	252.72	cr (al) λ^{al} 239 (4.21), 273 sh (3.19)	98 (95)	...	...	...	i	δ	s	v	v	os s	**B6**^{3}, 1036
s363	—,—,4,4′-diacetamido-	$C_{16}H_{16}N_2O_4S$. *See* s359	332.38	pa ye nd (eth or dil aa), lf (aq al) λ^{al} 257 (4.41), 285 (4.55)	282–5	...	...	...	i	s	...	...	...	...	**B13**^{2}, 303
s364	—,—,2,4-diamino-	$C_{12}H_{12}N_2O_2S$. *See* s359	248.31	nd (al)	188	...	...	...	i	v^h	i	...	δ	...	**B13,** 553
Ω s365	—,—,3,3′-diamino-	$C_{12}H_{12}N_2O_2S$. *See* s359	248.31	pr λ^{al} 231 (4.52), 315 (3.67)	168	...	...	...	v^h	v^h	...	...	...	...	**B13**^{2}, 219
Ω s366	—,—,4,4′-diamino-	$C_{12}H_{12}N_2O_2S$. *See* s359	248.31	lf (dil al, MeOH) λ^{al} 261 (4.25), 295 (4.43)	178	...	...	...	...	s	...	...	...	...	**B13**^{2}, 300
Ω s367	—,—,4,4′-dichloro-	$C_{12}H_8Cl_2O_2S$. *See* s359	287.18	mcl	148–9	sub	...	...	i	δ s^h	...	...	...	...	**B6**^{2}, 297
s368	—,—,4,4′-diethoxy-	$C_{16}H_{18}O_4S$. *See* s359	306.38	pl (al or aa)	163	...	...	...	...	v^h	v	...	...	aa s^h	**B6,** 861
Ω s369	—,—,2,2′-dihydroxy-	$C_{12}H_{10}O_4S$. *See* s359	250.28	nd (bz)	164–5 (179)	...	...	...	s	v	v	...	δ	aa v peth i	**B6**^{1}, 396
s370	—,—,2,5-dihydroxy-	$C_{12}H_{10}O_4S$. *See* s359	250.28	pr (w), nd (dil al)	196	...	...	...	δ s^h	v	δ	...	...	...	**B6**^{2}, 1072
s371	—,—,3,3′-dihydroxy-	$C_{12}H_{10}O_4S$. *See* s359	250.28	...	190–1	...	...	...	δ s^h	s	s	...	...	aa δ	**B6**^{2}, 827
Ω s372	—,—,4,4′-dihydroxy-	$C_{12}H_{10}O_4S$. *See* s359	250.28	nd (w), rh bipym λ^{al} 236 (4.12), 262 (4.30), 295 (3.35)	240–1	...	1.3663^{15}	...	i s^h	s	s	...	δ	alk s	**B6**^{2}, 853

For explanations, symbols and abbreviations see beginning of table. For structural formulas see end of table.

No.	Name	Synonyms and Formula	Mol. wt.	Color, crystalline form, specific rotation and λ_{max} (log ε)	m.p. °C	b.p. °C	Density	n_D	Solubility w	al	eth	ace	bz	other solvents	Ref.
	Sulfone														
Ω s373	**—,—,2,2′-dimethoxy-**	$C_{14}H_{14}O_4S$. *See* s359	278.33	nd (bz) λ^{al} 233 sh (4.01), 289 (3.88)	157–8	...	...	...	...	v^h	...	...	δ^h	aa s con sulf s^h xyl δ^h peth i	**B6,** 794
s374	**—,—,4,4′-dimethoxy-**	Anisyl sulfone. $C_{14}H_{14}O_4S$. *See* s359	278.33	lf or pr (al), nd (al-eth) λ^{al} 237 (4.16), 260 (4.35)	130	sub	...	...	i	v^h	...	...	...	...	**B6,** 861
Ω s375	**—,dipropyl**	Propyl sulfone. $CH_3CH_2CH_2SO_2CH_2CH_2CH_3$	150.24	sc	29.5–30.5	...	1.0278^{50}_4	1.4456^{30}	δ	s	s	...	...	...	**B1²,** 373
s376	**—,2,2′-ditolyl**	*o*-Tolyl sulfone.	246.33	nd (al) λ^{al}233 (4.16), 273 (3.43), 280 (3.40)	134–5 (104.5–5.5)	...	...	...	i	v	v	...	v	chl, CS_2 s	**B6²,** 342
s377	**—,3,4′-ditolyl**	...	246.33	nd (aa) λ^{al} 233 (4.00), 245 (4.30)	116	...	...	...	...	...	...	...	...	...	**B6¹,** 208
s378	**—,4,4′-ditolyl**	*p*-Tolyl sulfone.	246.33	pr (bz), nd (w or al), pl (al)	159	405^{714}	...	...	δ s^h	s^h	δ	...	s	chl, CS_2 s	**B6²,** 395
s381	**—,phenyl (2-tolyl).**	...	232.30	pl (al) λ^{al} 234 (4.15), 266 (3.31), 273 (3.35)	81	...	...	...	i	s	v	...	v	lig δ	**B6,** 371
Ω s382	**—,phenyl (4-tolyl)**	...	232.30	pl (al)	127–8 (124.5)	...	...	...	i	δ	...	...	δ	aa δ	**B6²,** 395
Ω s383	**Sulfoxide, dibenzyl**	$C_6H_5CH_2SOCH_2C_6H_5$	230.33	lf (al or w) λ^{al} 222 (4.29), 260 (2.68), 265 (2.55)	134–5	210d	...	...	i s^h	v	s	...	...	...	**B6²,** 429
Ω s384	**—,dibutyl**	Butyl sulfoxide. $CH_3(CH_2)_3SO(CH_2)_3CH_3$	162.30	nd (dil al)	32.6	d	0.8317^{23}_4	1.4669^{20}	i	s	s	...	...	...	**B1²,** 400
s385	**—,diethyl**	Ethyl sulfoxide. $C_2H_5SOC_2H_5$	106.19	syr	14 (4–6)	104^{25} $88–9^{15}$	...	...	s	s	s	...	...	...	**B1²,** 345
Ω s386	**—,dimethyl**	Methyl sulfoxide. CH_3SOCH_3	78.13	...	18.45	189^{760} $85–7^{20}$	1.1014^{20}_4	1.4770^{20}	s	s	s	s	...	AcOEt s	**B1,** 289
Ω s387	**—,diphenyl**	Phenyl sulfoxide.	202.28	pr (lig) λ^{al} 233 (4.15), 265 (3.32)	70.5	340 δd 210^{15}	...	...	...	v	v	...	v	aa v peth i	**B6²,** 290
s388	**—,—,4-amino-**	$C_{12}H_{11}NOS$. *See* s387	217.29	nd (w) λ^{al} 278 (4.01)	152	...	...	...	s^h	v	v	...	...	...	**B13,** 534
s389	**—,—4-amino-4′-nitro-**	$C_{12}H_{10}N_2O_3S$. *See* s387	262.29	ye (al) λ^{al} 277 (4.33)	132	...	...	...	δ	v	...	...	...	...	**Am 73,** 4356
s390	**—,—,4,4′-diamino-**	*p,p′*-Sulfinyldianiline. $C_{12}H_{12}N_2OS$. *See* s387	232.31	pr (w or al) λ^{al} 264 (4.17), 300 (4.25)	175d	...	...	...	s^h	s^h	...	...	...	...	**B13,** 53
s391	**—,—,4,4′-dihydroxy-**	$C_{12}H_{10}O_3S$. *See* s387	234.28	nd (ace)	195	...	...	...	...	s	...	s	...	...	**B6,** 860
s392	**—,dipropyl**	Propyl sulfoxide. $CH_3CH_2CH_2SOCH_2CH_2CH_3$	134.24	nd	22–3	80^2	0.9654^{20}_4	1.4663^{20}	δ	s	s	...	...	...	**B1²,** 373
Ω s393	**—,4,4′-ditolyl**	*p*-Tolyl sulfoxide.	230.33	cr (lig, peth) λ^{al} 275 (3.35)	94–4.5	...	...	...	...	v	v	...	v	chl, aa v lig δ, s^h	**B6²,** 395
s394	**Sulfuric acid,** diamide, *N,N,N′,N′*-tetra-methyl-	Tetramethylsulfamide. $(CH_3)_2NSO_2N(CH_3)_2$	152.22	pl or nd (dil al)	73	225^{760} (extrap $126^{30.6}$)	...	...	δ	s	...	...	...	...	**Am 76,** 220
s395	—,dibutyl ester*	Butyl sulfate. $[CH_3(CH_2)_3O]_2SO_2$	210.30	...	...	109.5^4	1.0616^{20}	1.4192^{20}	i	...	...	...	...	...	**J1943,** 16
s396	—,didecyl ester*	1-Decanol sulfate. Decyl sulfate. $[CH_3(CH_2)_9O]_2SO_2$	378.62	...	37.6–7.8	...	...	...	...	...	...	...	...	...	**Am 56,** 1204
s397	—,didodecyl ester*	Dodecyl sulfate. $[CH_3(CH_2)_{11}O]_2SO_2$	434.73	...	48.5	...	...	...	...	...	...	...	...	...	**Am 56,** 1204
Ω s398	—,diethyl ester*	Ethyl sulfate. $(C_2H_5O)_2SO_2$	154.19	...	−24.5	208 δd 96^{15}	1.1774^{20}_4	1.4004^{20}	i d^h	∞ d^h	∞	...	...	...	**B1²,** 327
s399	—,diheptyl ester*	Heptyl sulfate. $[CH_3(CH_2)_6O]_2SO_2$	294.46	cr (peth)	13	$146.6^{1.5}$	0.9819^{25}_{25}	1.4362^{25}	...	...	...	...	...	...	**B1³,** 1683
s400	—,dihexadecyl ester*	Cetyl sulfate. $[CH_3(CH_2)_{15}O]_2SO_2$	546.95	...	66.2	...	...	...	s	...	...	...	...	...	**B1³,** 1823
s401	—,dihexyl ester*	Hexyl sulfate. $[CH_3(CH_2)_5O]_2SO_2$	266.40	...	...	125.3^2	1.0036^{21}_0	1.433^{21}	...	...	...	...	...	...	**B1³,** 1656
Ω s402	—,dimethyl ester*	Methyl sulfate. $(CH_3O)_2SO_2$	126.13	λ^{undil} 265	−31.75 fr −27	188.5^{760}d 76^{15}	1.3283^{20}	1.3874^{20}	s	∞	s	...	s	CS_2 i	**B1,** 283
s403	—,di(3-methyl-butyl) ester*	Isoamyl sulfate. $[(CH_3)_2CHCH_2CH_2O]_2SO_2$	238.35	...	−20	$139–41^{12}$	...	...	...	...	...	...	...	...	**B1²,** 434
s404	—,dioctadecyl ester*	Octadecyl sulfate $[CH_3(CH_2)_{17}O]_2SO_2$	603.06	...	70.5	...	...	...	...	...	...	...	...	...	**B1³,** 1837
s405	—,dipentyl ester*	Amyl sulfate. $[CH_3(CH_2)_4O]_2SO_2$	238.35	...	14	$117^{3.5}$	1.029^{20}_0	1.4290^{20}	...	...	...	...	...	...	**B1²,** 418
s406	—,dipropyl ester*	Propyl sulfate. $(CH_3CH_2CH_2O)_2SO_2$	182.24	...	...	121^{20} $93–4^4$	1.1064^{20}_4	1.4135^{20}	i δd	...	...	...	...	peth v	**B1²,** 368

For explanations, symbols and abbreviations see beginning of table. For structural formulas see end of table.

No.	Name	Synonyms and Formula	Mol. wt.	Color, crystalline form, specific rotation and λ_{max} (log ε)	m.p. °C	b.p. °C	Density	n_D	Solubility						Ref.
									w	al	eth	ace	bz	other solvents	
	Sulfuric acid														
s407	—,ethylene ester	Ethylene sulfate. Glycol sulfate.	124.12	nd or pr (bz-lig)	99 (97)	sub	1.735_4^{13}	...	d	s	v	v	s	os v	**B1**^{3}, 2110
Ω s408	—,monoamide, *N*-cyclohexyl-	Cyclohexylsulfamic acid.	179.24	(al)	169–70	...	...	...	δd^h	...	...	...	...	alk v	**C53,** 5157
Ω s409	—,mono(2-aminoethyl) ester*	$H_2NCH_2CH_2OSO_3H$	141.15	mcl pr (w-al)	230d	...	...	...	s	i	...	...	...	...	**B4**^{2}, 718
s410	—,monobutyl ester*	Butyl hydrogen sulfate. $CH_3(CH_2)_3OSO_3H$	154.19	syr	...	d	...	...	v	s	s	...	...	...	**B1**^{2}, 397
s411	—,monochloride monoethyl ester	Ethyl chlorosulfonate*. $C_2H_5OSO_2Cl$	144.58	...	...	151–4, 52^{14}	1.3502_4^{25}	1.416^{20}	d	d	s	...	...	chl, lig s	**B1**^{2}, 327
s412	—,monochloride monomethyl ester	Methyl chlorosulfonate*. CH_3OSO_2Cl	130.55	...	...	133–5^{760}d, 48^{29}	1.4805_4^{25}	1.4138^{18}	i, d^h	d^h	s	v	s	os s	**B1**^{3}, 1199
s413	—,monoethyl ester*	Ethylsulfuric acid. Ethyl hydrogen sulfate. $C_2H_5OSO_3H$	126.13	...	...	280d	1.3657_4^{20}	1.4105^{20}	v	δ	δ	...	...	...	**B1**^{2}, 326
s414	—,monomethyl ester*	Methyl bisulfate. Methyl hydrogen sulfate. CH_3OSO_3H	112.11	...	< −30	130–40d	...	...	v	s	∞	...	...	CCl_4 s	**B1**, 283
s415	**Sulfurous acid,** diamide, *N,N,N′,N′*-tetramethyl-	Tetramethylthionamide. $(CH_3)_2NSON(CH_3)_2$	136.22	...	31	209^{760}, 70^{10}	...	...	...	...	s	...	...	oos s	**Am 76,** 220
Ω s416	—,dibutyl ester*	Butyl sulfite. $[CH_3(CH_2)_3O]_2SO$	194.30	λ^{iso} <230	...	230, 116^{19}	0.9957_4^{20}	1.4310^{20}	...	s	s	...	...	...	**B1**^{3}, 1506
s417	—,diethyl ester*	Ethyl sulfite. $(C_2H_5O)_2SO$	138.19	...	...	157^{768}, 55^{15}	1.0829_4^{20}	1.4144^{20}	d	s	s	...	...	...	**B1**^{2}, 326
s418	—,diisobutyl ester*	Isobutyl sulfite. $[(CH_3)_2CHCH_2O]_2SO$	194.30	...	...	209^{741}, 92–4^{13}	0.9862_4^{20}	1.4268^{20}	...	...	...	...	...	...	**B1**^{3}, 1561
s419	—,dimethyl ester*	Methyl sulfite. $(CH_3O)_2SO$	110.13	λ^{al} 280 (0.90)	...	126^{760}, 52^{45}	1.2129_4^{20}	1.4093^{20}	s^h	s	s	...	...	...	**B1**, 282
Ω s420	—,ethylene ester	Ethylene sulfite. Glycol sulfite.	108.12	...	−11	173^{760}, 70–1^{20}	1.4402_4^{20}	1.4463^{20}	v, d^h	v	v	v	v	AcOEt v	**B1**^{3}, 2110
s421	—,monochloride, monoethyl ester	Ethyl chlorosulfinate. $C_2H_5OS(O)Cl$	128.58	...	...	52.5^{44}, 32^{16}	1.2766_4^{25}	1.4550^{25}	d	d	s	...	...	...	**B1**^{3}, 1316
s422	**Sylvestrene**(*d*)	Carvestrene. 3-Isopropenyl-1-methylcyclohexene *d*-1,8(9)-*m*-Menthadiene.	136.24	$[\alpha]_D^{18}$ +83.18 (undil), +66.3 (chl, c = 4.3)	...	175^{751}	0.8479_4^{18}	1.4760^{18}	i	∞	∞	...	...	...	**B5**^{2}, 84
—	**Syntomycetin**	*see* **Chloromycetin**													
—	**Syringaldehyde**	*see* **Benzaldehyde, 3,5-dimethoxy-4-hydroxy-**													
—	**Syringenin**	*see* **2-Propen-1-ol, 3(3,5-dimethoxy-4-hydroxyphenyl)-**													
—	**Syringic acid**	*see* **Benzoic acid, 3,5-dimethoxy-4-hydroxy-**													
—	**Syringidin**	*see* **Malvidin**													
s423	**Syringin**	Methoxyconiferine	372.38	cr (w), nd (al), nd (w+1), $[\alpha]_D$ −17.1	192	...	...	...	δ, s^h	s	i	...	...	con HNO_3, con sulf s	**B31**, 222

For explanations, symbols and abbreviations see beginning of table. For structural formulas see end of table.

No.	Name	Synonyms and Formula	Mol. wt.	Color, crystalline form, specific rotation and λ_{max} (log ε)	m.p. °C	b.p. °C	Density	n_D	Solubility w	al	eth	ace	bz	other solvents	Ref.
	Tachysterol														
t1	**Tachysterol**	$C_{28}H_{44}O$	396.67	$[\alpha]_D^{18}$ −70 (bz) λ^{eth} 280 (4.44)	....	220 vac	....	....	i	s	s	s	s	os s MeOH i	**E12A,** 175
t2	**Tagatose**(*D*)	$C_6H_{12}O_6$	180.16	cr (dil al), $[\alpha]_D^{20}$ −5 (w, c = 1)	134–5	....	....	....	s	δ	i	...	i	MeOH δ	**B31,** 348
t3	**Talitol**(*D*)	*D*-Altritol. *D*-Talite. $C_6H_{14}O_6$	182.18	pr (al, MeOH) $[\alpha]_D^{18}$ +3.2 (w)	87–8	....	....	....	s	s	i	...	i	....	**B1**[3], 2384
t4	**Talonic acid,** hemihydrate(*D*)	$C_6H_{12}O_7 \cdot \frac{1}{2}H_2O$	205.17	cr (aq al + ½w) $[\alpha]_D^{25}$ +16.7 → −21.6 (w, c = 4) (mut)	138–9	....	....	....	v	...	i	...	i	....	**B3**[3], 1068
Ω t5	—,γ-lactone(*D*)	$C_6H_{10}O_6$	178.14	pr (al), $[\alpha]_D^{25}$ −34.65 → −28.4 (w) (mut)	135–7	....	....	....	...	...	...	...	...	....	**Am 54,** 1593
t6	**Talose** (*D*)	$C_6H_{12}O_6$	180.16	α: cr (al), β: cr (MeOH) $[\alpha]_D^{27}$ +29 → +19.7 (w, c = 1) (mut) β: $[\alpha]_D^{20}$ +11.5 → +21 (w, c = 4) (mut)	α: 130–5 β: 120–1	....	....	....	s	s^h	i	...	i	....	**B31,** 283
Ω t7	**Tannic acid**	Gallotannic acid. Tannin. $C_{76}H_{52}O_{46}$.	1701.24	pa ye-br amor or fl $\lambda^{w, pH=8}$ 280 (2.2), 420 (1.8), 500 (1.8)	210–5 δd	....	....	....	s	v	i	v	i	chl, CS_2 i	**B31,** 133
—	**Tannin**	*see* **Tannic acid**													
t8	**Taraxanthin**	$C_{40}H_{56}O_4$	600.89	ye pr (MeOH) $[\alpha]_{Cd}^{20}$ +200 (AcOEt) λ^{bz} 428.5 (4.96), 455 (5.14), 485 (5.12)	185–6	....	....	....	...	s^h	v	...	s	CS_2, peth s MeOH δ	**B30,** 100
t9	**Tartaric acid**(*d*)	L-2,3-Dihydroxybutanedioic acid*. *d*-2,3-Dihydroxy-succinic acid. $HO_2CCH(OH)CH(OH)CO_2H$	150.09	mcl (anh), rh pr (w + 1), $[\alpha]_D^{20}$ +12.7 (w, c = 17.4)	171–4	....	1.7598^{20}	1.4955	v	v	δ	s	i	....	**B3**[3], 994
Ω t10	—(*dl*)	Racemic acid. $HO_2CCH(OH)CH(OH)CO_2H$	150.09	mcl pr (w or al + 1w)	206 (210d) (anh) 203–4 (+1w)	....	1.788	....	s	s^h	δ	...	i	....	**B3**[3], 1025
t11	—(*meso*)	$HO_2CCH(OH)CH(OH)CO_2H$	150.09	tcl pl (w)	146–8 (140)	....	1.666_4^{20}	1.5–1.6	v	s	δ	...	...	....	**B3**[3], 1029
t12	—,anhydride diacetate(*d*)	α,α′-Diacetoxysuccinic anhydride.	216.16	nd (bz)	135	....	....	....	...	v	v	s	...	aa δ	**B18**[2], 143
t13	—,dibenzyl ester(*d*)	$C_6H_5CH_2O_2CCH(OH)CH(OH)CO_2CH_2C_6H_5$	330.34	$[\alpha]_D^{15}$ +19.26	50	$250–70^4$	1.2036^{72}	....	...	s	...	...	...	Py s	**B6**[3], 1537
t14	—,dibutyl ester(*d*)	$CH_3(CH_2)_3O_2CCH(OH)CH(OH)CO_2(CH_2)_3CH_3$	262.31	pr, $[\alpha]_D^{20}$ +11.3 (al)	22–2.5	320^{760} 178^{12}	1.0909_4^{20}	1.4451^{20}	s	s	...	s	...	os s	**B3**[3], 1021
Ω t15	—,—(*dl*)	$CH_3(CH_2)_3O_2CCH(OH)CH(OH)CO_2(CH_2)_3CH_3$	262.31	....	....	320^{765} 185^{12}	1.0879_4^{18}	1.4474^{15}	s	s	...	s	...	os s	**B3**[3], 1032
t16	—,diethyl ester(*d*)	$C_2H_5O_2CCH(OH)CH(OH)CO_2C_2H_5$	206.20	$[\alpha]_D^{16}$ +7.9 (undil)	18.7	280^{760} 142^8	1.2036_4^{20}	1.4468^{20}	s	s	s	s	...	aa ∞ os s	**B3**[2], 329
Ω t17	—,—(*dl*)	Diethyl racemate. $C_2H_5O_2CCH(OH)CH(OH)CO_2C_2H_5$	206.20	....	18.7	281^{765} 158^{14}	1.2046_4^{20}	1.4438^{20}	δ	∞	∞	s	...	oos s	**B3**[2], 337
t18	—,—(*l*)	$C_2H_5O_2CCH(OH)CH(OH)CO_2C_2H_5$	206.20	$[\alpha]_D^{20}$ −7.55 (undil)	18	280^{760} 162^{19}	1.2054_4^{20}	1.4468^{20}	...	s	s	s	...	os s	**B3**[3], 1020
t19	—,—(*meso*)	$C_2H_5O_2CCH(OH)CH(OH)CO_2C_2H_5$	206.20	....	60 (55)	157.5^{14}	1.1350_4^{99}	1.4315^{65}	...	...	s	s	...	....	**B3**[3], 1031
t20	—,diethyl ester diacetate(*d*)	$C_2H_5O_2CCH(O_2CCH_3)CH(O_2CCH_3)CO_2C_2H_5$	290.27	mcl cr (lig) $[\alpha]_D^{100}$ +6.3 (undil) $[\alpha]_{546}^{20}$ −17.3 (bz, c = 5)	67.3–7.6 (cor)	296^{764} 163^{10}	1.1149_4^{66}	....	δ	s	v	...	...	lig s^h	**B3**[3], 1021
t21	—,diisobutyl ester (*d*)	$(CH_3)_2CHCH_2O_2CCH(OH)CH(OH)CO_2CH_2CH(CH_3)_2$	262.31	$[\alpha]_D$ +11.8 (al) $[\alpha]_{546}^{22}$ +67.1 (Py, p = 7)	72.7–2.9	323–5 183^{11}	1.0265_4^{81}	....	s^h	s	...	...	...	os s	**B3**[3], 1021
t22	—,—(*dl*)	$(CH_3)_2CHCH_2O_2CCH(OH)CH(OH)CO_2CH_2CH(CH_3)_2$	262.31	cr (bz)	63	311^{768} 195^{13}	1.0386_4^{68}	....	s^h	v	...	...	...	os s	**B3**[3], 1028

For explanations, symbols and abbreviations see beginning of table. For structural formulas see end of table.

No.	Name	Synonyms and Formula	Mol. wt.	Color, crystalline form, specific rotation and λ_{max} (log ε)	m.p. °C	b.p. °C	Density	n_D	Solubility w	al	eth	ace	bz	other solvents	Ref.
	Tartaric acid														
t23	—,diisopropyl ester(*d*)	$(CH_3)_2CHO_2CCH(OH)CH(OH)CO_2CH(CH_3)_2$	234.25	$[\alpha]_D^{20}$ +14.9	....	275^{765} 152^{12}	1.1300_4^{20}	....	...	s	s	s	...	os s	**B3²**, 331
t24	—,—(*dl*)	$(CH_3)_2CHO_2CCH(OH)CH(OH)CO_2CH(CH_3)_2$	234.25	....	34	275^{765} 154^{12}	1.1166_4^{20}	....	...	s	s	s	...	os s	**B3²**, 337
t25	—,dimethyl ester(*d*)	$CH_3O_2CCH(OH)CH(OH)CO_2CH_3$	178.14	(i) cr (bz) (ii) cr (bz) (iii) cr (w) $[\alpha]_D^{50}$ +6.72 (MeOH, c = 16), $[\alpha]_D^{20}$ −9.1 (bz, c = 1)	(i) 48 (ii) 50 (iii) 61	280^{760} 166^{12}	1.306_4^{45}	....	s	v	s	s	...	chl v	**B3³**, 1018
t26	—,—(*dl*)	$CH_3O_2CCH(OH)CH(OH)CO_2CH_3$	178.14	orh nd (bz), ta (chl)	90 (st) 80 (unst)	282^{760} 169^{20}	1.2604^{90}	....	s	v	s	s	...	os s	**B3³**, 1027
t27	—,—(*meso*)......	$CH_3O_2CCH(OH)CH(OH)CO_2CH_3$	178.14	nd (chl), cr (MeOH)	114	$98^{0.01}$ sub	....	....	s	v	...	s	...	os v	**B3³**, 1031
t28	—,di(2-methylbutyl) ester(*d*)	$C_2H_5CH(CH_3)CH_2O_2CCH(OH)CH(OH)CO_2CH_2CH(CH_3)C_2H_5$	290.36	$[\alpha]_D^{20}$ +14.10	....	208^{20}	1.0636_4^{20}	....	...	s	...	s	...	....	**B3**, 519
t29	—,dinitrate(*d*)	$HO_2CCH(ONO_2)CH(ONO_2)CO_2H$	240.08	nd (eth-bz) $[\alpha]_D^{20}$ +13.7 (MeOH, p = 9)	d	....	....	....	d	v	v	v	i	peth i	**B3²**, 328
t30	—,dipropyl ester(*d*)	$CH_3CH_2CH_2O_2CCH(OH)CH(OH)CO_2CH_2CH_2CH_3$	234.25	$[\alpha]_D^{20}$ +12.44 (w)	....	303^{760} 181^{23}	1.1390_4^{20}	....	s	v	v	s	...	os s	**B3**, 516
t31	—,—(*dl*)	$CH_3CH_2CH_2O_2CCH(OH)CH(OH)CO_2CH_2CH_2CH_3$	234.25	pr (al-eth)	25	286^{765} 167^{11}	1.1256_4^{20}	....	s	v	v	s	...	os s	**B3²**, 337
t32	—,monoethyl ester (*d*)	$HO_2CCH(OH)CH(OH)CO_2C_2H_5$.	178.14	$[\alpha]_D$ +21.8 (w)	90	....	....	....	s	s	i	...	...	....	**B3**, 512
t33	—,piperazinium salt (*d*)	$HO_2CCH(OH)CH(OH)CO_2H.C_4H_{10}N_2$	236.23	cr	248–54 (cor)	....	....	....	s^h	s	δ	...	...	....	**Am 70**, 2758
t34	—,—(*meso*)......	$HO_2CCH(OH)CH(OH)CO_2H.C_4H_{10}N_2$	236.23	cr	140–1 (cor)	....	....	....	s	s	δ	...	...	....	**Am 70**, 2758
—	**Tartronic acid**	*see* **Malonic acid, hydroxy-**													
Ω t35	**Taurine**	2-Aminoethanesulfonic acid*. $H_2NCH_2CH_2SO_3H$	125.15	mcl pr (w)	328 (317d)	....	....	....	s v^h	i	i	...	...	....	**B4**, 528
t36	—,***N,N*-dimethyl-** ..	2(Dimethylamino)ethanesulfonic acid*. $(CH_3)_2NCH_2CH_2SO_3H$	153.20	pr (MeOH), pl (w + 1)	315–6 (anh) 270–80d (hyd)	....	....	....	v	i	i	...	...	aa v MeOH δ^h	**B4²**, 951
t37	—,***N*-methyl-**	2(Methylamino)ethanesulfonic acid*. $CH_3NHCH_2CH_2SO_3H$	139.18	pr	241–2	....	....	....	v	i	i	...	...	....	**B4**, 529
t38	—,***N*-methyl-*N*-phenyl-**	$C_6H_5N(CH_3)CH_2CH_2SO_3H$.	215.27	pa vt (al)	239–40	....	....	....	...	v^h	...	...	...	....	**B12²**, 285
t39	—,***N*-phenyl-**	$C_6H_5NHCH_2CH_2SO_3H$	201.25	lf (w), pr (al)	277–80d	....	....	....	v	i	i	...	...	....	**B12²**, 284
t40	**Taurocholic acid** ...	Cholaic acid. Cholyltaurine. $C_{26}H_{45}NO_7S$	515.72	pr (al-eth) $[\alpha]_D^{18}$ +38.8 (al, c = 2) $\lambda^{con\ sulf}$ 303 (3.64), 389 (4.45), 480 (3.46)	*ca.* 125d	....	....	....	v	v^h	δ	...	...	AcOEt δ	**E14**, 195
t41	**Telluride, diethyl**...	Ethyl telluride. $C_2H_5TeC_2H_5$	185.73	red-ye	....	$137–8^{760}$	1.599_4^{15}	1.5182^{15}	δ	s	...	...	...	....	**B1²**, 359
t42	—,**dimethyl**	Methyl telluride. CH_3TeCH_3	157.67	pa ye	....	93.5^{749}	>1	....	d	v	i	...	...	....	**B1**, 291
t43	**Tephrosin**	Hydroxydeguelin. $C_{23}H_{22}O_7$.	410.43	pr (chl-MeOH)	198 (218–20)	....	....	....	i	...	s	s	...	chl s MeOH δ	**Am 55**, 759
—	**Teraconic acid**.....	*see* **Succinic acid, isopropylidene-**													
t44	**Terebic acid**(*dl*)....	3,3-Dimethylparaconic acid. Teribinic acid. $C_7H_{10}O_4$	158.16	mcl pr (al)	176	....	0.8155_4^{24}	....	δ s^h	s^h	δ	...	...	....	**B18²**, 314
—	**Terephthalaldehyde**	*see* **1,4-Benzenedicarboxaldehyde**													
—	**Terephthalaldehydic acid**	*see* **Benzoic acid, 4-formyl-**													
—	**Terephthalic acid** ..	*see* **1,4-Benzenedicarboxylic acid***													
t45	**Terpenolic acid**	Terpenylic acid...........	172.19	lf or pr (w + 1)	90 (anh) 57 (+1w)	sub 130–40	....	....	s v^h	...	...	...	...	....	**B18²**, 316
Ω t46	***o*-Terphenyl**......	1,2-Diphenylbenzene.......	230.31	mcl pr (MeOH) λ^{al} 231.5 (4.42), 250.5 (4.06)	58	332^{760} $160–70^2$	....	....	i	...		s	s	MeOH, chl s	**B5³**, 2292
Ω t47	***m*-Terphenyl**	1,3-Diphenylbenzene.......	230.31	ye nd (al) λ^{al} 246.8 (4.59), 290.7 (3.24)	89	365^{760}	....	....	i	s	s	...	s	aa s	**B5³**, 2294
Ω t48	**α-Terpinene**.......	*p*-Mentha-1,3-diene. $C_{10}H_{16}$.	136.24	λ^{al} 265 (3.84)	....	177.2^{760} $68–70^{12}$	0.8502_4^{20}	1.4784^{20}	i	∞	∞	...	...	....	**B5³**, 337

For explanations, symbols and abbreviations see beginning of table. For structural formulas see end of table.

No.	Name	Synonyms and Formula	Mol. wt.	Color, crystalline form, specific rotation and λ_{max} (log ε)	m.p. °C	b.p. °C	Density	n_D	Solubility w	al	eth	ace	bz	other solvents	Ref.
	α-Terpineol														
Ω t49	α-**Terpineol**(*dl*)	*dl-p*-Menth-1-en-8-ol. $C_{10}H_{18}O$.	154.26	cr (peth) λ^{al} 215 (2.25), 260 (0.26)	40–1	220^{760} 85^{3}	0.9337_4^{20}	1.4831^{20}	δ	v	v	s	s	chl s peth s^h	**B6**³, 247
Ω t50	—,acetate	$C_{12}H_{20}O_2$. *See* t49	196.29	*d*:$[\alpha]_D$ +52.5 (undil) *l*:$[\alpha]_D^{20}$ −73 (undil)	....	140^{40} $104\text{–}6^{11}$	0.9659_4^{20}	1.4689^{21}	i	s	s	...	s	....	**B6**², 66
t51	**Terpinol,** hydrate (*cis*)	*cis*-Terpin hydrate	214.31	rh cr	123d	sub 100	....	1.51–1.52	δ	s	δ	...	...	....	**B6**², 754
Ω t52	**Terpinolene**	....	136.24	....	....	185^{760} 76^{10}	0.8623_4^{20}	1.4883^{20}	i	∞	∞	...	s	....	**B5**³, 345
t53	**3-Terpinolenone**	Piperitenone. $C_{10}H_{14}O$	150.22	$[\alpha]_{546}$ −0.1 λ^{al} 242 (4.10), 278 (3.90), 353 sh (2.05)	....	$120\text{–}2^{14}$	0.9774_4^{20}	1.5294^{20}	...	v	v	...	...	....	**J1939,** 1496
t54	**Terramycin**	Oxytetracycline. $C_{22}H_{24}N_2O_9$	498.49	....	184.5–5.5	1.634^{20}	....	....	δ	δ	...	...	...	....	**Am 87,** 134
Ω t55	**Testosterone**	17β-Hydroxy-4-androsten-3-one. $C_{19}H_{28}O_2$.	288.44	nd (dil ace) $[\alpha]_D^{24}$ +109 (al, c = 4) λ^{al} 240.5 (4.23)	155	....	....	....	i	s	s	s	...	os s	**Am 76,** 1962
t56	—,**4,5-dihydro-17-methyl-**	$C_{20}H_{32}O_2$. *See* t55	304.48	cr (AcOEt) λ 283 (4.15)	192–3 (cor)	....	....	....	...	...	...	...	...	AcOEt δ	**E14,** 145
t57	—,**17-ethenyl-**	17-Vinyltestosterone. $C_{21}H_{30}O_2$. *See* t55	314.47	pr nd (peth-eth) $[\alpha]_D$ +87.6	140–1	....	....	....	...	...	s	s	...	MeOH, chl s	**C31,** 2224
t58	—,**17-ethyl-**	$C_{21}H_{32}O_2$. *See* t55	316.49	nd (AcOEt) $[\alpha]_D^{20}$ +71.2	143–4	....	....	....	i	...	...	s^h	...	AcOEt s^h	**E14,** 145
t59	—,**17-ethynyl-**	$C_{21}H_{28}O_2$. *See* t55	312.46	cr (chl-MeOH, AcOEt) $[\alpha]_D^{20}$ +22.5 (diox) λ^{al} 238.5	270–2 (cor)	sub vac	....	....	i	δ	...	...	...	diox, Py s AcOEt s^h	**C34,** 18213
t60	—,**17-methyl-**	$C_{20}H_{30}O_2$. *See* t55	302.46	nd (hx) $[\alpha]_D^{20}$ +82 (al) $\lambda^{97\%\,sulf}$ 298 (4.26), 497 (2.86)	165–6 (cor)	....	....	....	i	s	s	...	...	MeOH, os s	**E14s,** 2645
t61	**Tetrabenzotriaza-porphyrin**	....	513.57	purple nd and pl (quinoline)	....	....	....	....	...	...	...	...	...	Py δ	**J1939,** 1809
—	**Tetracaine,** hydrochloride	*see* **Benzoic acid, 4(butyl-amino)-, (2-dimethyl-aminoethyl) ester, hydrochloride**													
—	**Tetracene**	*see* **Naphthacene**													
Ω t62	**Tetracosane***	$CH_3(CH_2)_{22}CH_3$	338.67	cr (eth)	54 (51)	391.3^{760} 231.3^{10}	0.7665_4^{70} 0.7991_4^{20} (suc)	1.4283^{70} 1.4480^{20} (suc)	i	δ	v	...	...	....	**B1**³, 575
—	**Tetracyclone**	*see* **Cyclopentadienone, tetraphenyl-***													
t63	**6,8-Tetradeca-diyne***	$CH_3(CH_2)_4C{:}CC{:}C(CH_2)_4CH_3$	190.33	....	2	$118\text{–}9^{4}$	0.8699_4^{16}	....	...	...	s	...	...	....	**B1**³, 1067
Ω t64	**Tetradecanal***	Myristaldehyde. Tetradecyl aldehyde. $CH_3(CH_2)_{12}CHO$	212.38	lf	30 (23)	166^{24}	....	....	i	s	s	s	...	oos s	**B1**³, 2919
t65	—,dimethyl acetal	1,1-Dimethoxytetradecane*. $CH_3(CH_2)_{12}CH(OCH_3)_2$	258.45	....	....	$134\text{–}6^{4}$	....	1.4342^{25}	...	v	v	...	...	MeOH v	**C53,** 12168
t66	—,oxime	Myristaldoxime. $CH_3(CH_2)_{12}CH{:}NOH$	227.39	fl or nd (al)	82–3	....	....	....	i	v	...	...	δ	chl δ	**B1,** 716
Ω t67	**Tetradecane***	$CH_3(CH_2)_{12}CH_3$	198.40	....	5.86	253.7^{760} 121.9^{10}	0.7628_4^{20}	1.4290^{20}	i	v	v	...	...	....	**B1**³, 549
Ω t68	—,**1-amino-***	Myristylamine. $CH_3(CH_2)_{13}NH_2$	213.41	....	83.1	291.2^{760} 162^{15}	0.8079_4^{20}	1.4463^{20}	i	v	v	s	v	chl v	**B4**³, 419
Ω t69	—,**1-bromo-***	Myristyl bromide. $CH_3(CH_2)_{13}Br$	277.30	....	5.6	307^{760} 181^{21}	1.0170_4^{20}	1.4603^{20}	i	s	...	∞	∞	chl v	**B1**², 136
Ω t70	—,**1-chloro-***	Myristyl chloride. $CH_3(CH_2)_{13}Cl$	232.84	....	4.9	292^{760} 153^{10}	0.8665_4^{20}	1.4473^{20}	i	s	...	v	v	chl, os s	**B1**³, 550
t71	—,**1,14-dibromo-***	Tetradecamethylene bromide. $Br(CH_2)_{14}Br$	356.20	lf (al-eth), cr (al)	50.4	$190\text{–}2^{8}$	....	....	i	v^h	s	...	...	chl v	**B1**³, 551
t72	—,**1-phenyl-***	Tetradecylbenzene*. $CH_3(CH_2)_{13}C_6H_5$	274.50	....	16.1	358.9^{760} 210^{12}	0.8559_4^{20}	1.4818^{20}	...	...	...	...	...	....	**C50,** 4869
t73	**Tetradecanedioic acid,** dimethyl ester*	$CH_3O_2C(CH_2)_{12}CO_2CH_3$	286.42	nd (MeOH)	43–5	$191\text{–}2^{10}$	....	....	i	...	δ	...	...	MeOH i, s^h	**B2**³, 1858
t74	**1,14-Tetrade-canediol***	Tetradecamethylene glycol. $HO(CH_2)_{14}OH$	230.40	nd (bz)	84.5	200^{9}	....	....	...	s	s	...	s^h	....	**B1**³, 2241
t75	**1-Tetradecane-sulfonic acid***	$CH_3(CH_2)_{13}SO_3H$	278.46	....	65.5 (anh) 55–6 (+1w)	....	0.9996_4^{25}	....	s	...	...	...	...	....	**Am 57,** 1905
Ω t76	**1-Tetradecane-thiol***	*n*-Tetradecyl mercaptan. $CH_3(CH_2)_{13}SH$	230.45	....	....	$176\text{–}80^{22}$	0.8484_{20}^{20}	1.4597^{20}	i	s	s	...	...	....	**Am 55,** 1090

For explanations, symbols and abbreviations see beginning of table. For structural formulas see end of table.

No.	Name	Synonyms and Formula	Mol. wt.	Color, crystalline form, specific rotation and λ_{max} (log ε)	m.p. °C	b.p. °C	Density	n_D	Solubility w	al	eth	ace	bz	other solvents	Ref.
	Tetradecanoic acid														
Ω t77	**Tetradecanoic acid***	Myristic acid. $CH_3(CH_2)_{12}CO_2H$	228.38	lf (eth, 80 % aa) λ^{hp} 210 (1.85)	58	250.5^{100} 149.3^{1}	0.8439_4^{80}	1.4305^{60}	i	s	δ	s	v	chl s MeOH s, v^h	**B2**³, 911
Ω t78	—,amide	Myristamide. $CH_3(CH_2)_{12}CONH_2$	227.39	lf (ace) λ^{al} < 200	105–7	217^{12}	....	....	i	s	δ	δ s^h	δ	CCl_4 δ	**B2**³, 929
t79	—,anhydride*	$[CH_3(CH_2)_{12}CO]_2O$	438.74	lf (peth)	53.4	vac distb	0.8502_4^{70}	1.4335^{70}	i	s	s	...	...	peth s^h	**B2**³, 929
t80	—,benzyl ester	Benzyl myristate. $CH_3(CH_2)_{12}CO_2CH_2C_6H_5$	318.51	....	20.5	229.3^{11}	0.9321_{25}^{25}	....	i	s	v	...	v	chl v	**B6**², 417
Ω t81	—,chloride	Myristoyl chloride. $CH_3(CH_2)_{12}COCl$	246.83	....	−1	174^{16}	....	....	d	d	s	...	...	....	**B2**³, 929
Ω t82	—,ethyl ester*	Ethyl myristate. $CH_3(CH_2)_{12}CO_2C_2H_5$	256.43	....	12.3	295^{760} 162.5^{9}	0.8573_4^{25}	1.4362^{20}	i	s	δ	...	...	lig s	**B2**³, 922
Ω t83	—,isopropyl ester*	Isopropyl myristate. $CH_3(CH_2)_{12}CO_2CH(CH_3)_2$	270.46	....	....	192.6^{20} 140.2^{2}	0.8532^{20}	1.4325^{25}	i	s	s	v	v	chl s	**B2**³, 923
Ω t84	—,methyl ester*	Methyl myristate. $CH_3(CH_2)_{12}CO_2CH_3$	242.41	λ^{hp} 212 (1.85)	19	295^{751} (323) $155\text{–}7^{7}$	....	1.425^{45}	i	∞	∞	∞	∞	chl, CCl_4, MeOH, AcOEt ∞	**B2**³, 921
Ω t85	—,nitrile	Myristonitrile. $CH_3(CH_2)_{12}CN$	209.38	....	19.25	226.5^{100} 119^{1}	0.8281_4^{19}	1.4392^{25}	i	∞	∞	∞	∞	chl ∞	**B2**³, 930
t86	—,propyl ester*	Propyl myristate. $CH_3(CH_2)_{12}CO_2CH_2CH_2CH_3$	270.46	....	....	147^{2}	0.8592^{20}	1.4356^{25}	i	v	v	v	v	....	**B2**³, 923
Ω t87	**1-Tetradecanol***	Myristyl alcohol. $CH_3(CH_2)_{13}OH$	214.40	lf	39–40	263.2 167^{15}	0.8236_4^{38}	....	i	v	v	v	v	chl v	**B1**³, 1803
Ω t88	**2-Tetradecanone***	*n*-Dodecyl methyl ketone. $CH_3(CH_2)_{11}COCH_3$	212.38	cr (dil al)	33–4	$205\text{–}6^{100}$ 134^{13}	....	....	i	s	...	s	...	os s	**B1**³, 2919
Ω t89	**3-Tetradecanone***	Ethyl *n*-undecyl ketone. $CH_3(CH_2)_{10}COC_2H_5$	212.38	cr (MeOH)	34	152^{16}	....	....	i	s	...	s	...	os s	**B1**³, 2919
Ω t90	**1-Tetradecene***	1-Tetradecylene. $CH_3(CH_2)_{11}CH{:}CH_2$	196.38	....	−12	$232\text{–}4^{760}$ 125^{15}	0.7745_4^{15}	1.4351^{20}	i	v	v	...	s	....	**B1**³, 873
t91	**5,9-Tetradeca-dien-7-yne, 6,9-dimethyl-***	$CH_3(CH_2)_3CH{:}C(CH_3)C{:}CC(CH_3){:}CH(CH_2)_3CH_3$	218.39	....	....	$95\text{–}8^{0.5}$	0.8241_4^{20}	1.4866^{20}	i	...	s	...	s	....	**Am 55**, 1655
t92	**2-Tetradecyne***	$CH_3(CH_2)_{10}C{:}CCH_3$	194.36	....	6.5	252.5 134^{15}	0.8000_4^{20}	....	i	v	v	...	...	....	**B1**, 262
t93	**7-Tetradecyne***	$CH_3(CH_2)_5C{:}C(CH_2)_5CH_3$	194.36	....	....	144^{30}	0.7991_4^{25}	1.4330^{25}	i	v	v	...	...	....	**B1**³, 1067
Ω t94	**Tetraethylene glycol**	$HOCH_2(CH_2OCH_2)_3CH_2OH$	194.23	....	−6.2	328^{760} 198^{14}	1.1285_4^{15}	1.4577^{20}	v	s	s	...	...	diox s	**B1**³, 2106
Ω t95	—,dimethyl ether	$CH_3OCH_2(CH_2OCH_2)_3CH_2OCH_3$	222.29	....	....	275.8	1.0132_{20}^{20}	....	∞	s	s	...	...	....	**B1**³, 2107
t96	—,monoocta-decanoate	Tetraethylene glycol monostearate. $CH_3(CH_2)_{16}CO_2CH_2(CH_2OCH_2)_3CH_2OH$	476.72	....	40 (35)	328^{760}	1.1285_4^{15}	1.4593^{20}	...	...	...	...	...	....	**C37**, 3202
Ω t97	**Tetraethylene-pentamine**	$H_2NCH_2(CH_2NHCH_2)_3CH_2NH_2$	189.31	....		340.3^{760} 186–92^{14}	....	1.5042^{20}	...	...	...	...	...	....	**C51**, 5821
Ω t98	**Tetralin**	1,2,3,4-Tetrahydronaphthalene*.	132.21	λ^{al} 259 (2.69), 266 (2.84), 274 (2.90), 286 (2.28)	−35.79 (−31)	207.57^{760} 79.36^{10}	0.9702_4^{20}	1.54135^{20}	i	v	v	...	...	$PhNH_2$ s	**B5**³, 1219
t99	—,6-acetyl-	$C_{12}H_{14}O$. *See* t98	174.25	....	....	289–91d 182^{20}	....	....	...	...	...	...	...	....	**B7**², 305
t100	—,2-amino-(*dl*)	$C_{10}H_{13}N$. *See* t98	147.22	....	38	249^{710} 140^{20}	1.0295_4^{22}	1.5604^{22}	δ s^h	v	s	s	s	....	**B12**, 1200
Ω t101	—,5-amino-	$C_{10}H_{13}N$. *See* t98	147.22	....	....	276.8^{760} 155^{22}	1.0625^{16}	1.6050^{20}	δ	s	s	...	...	ac s	**C50**, 4112
t102	—,1,4-dioxo-	2,3-Dihydro-1,4-naphthoquinone. $C_{10}H_8O_2$. *See* t98	160.18	lf (hx), nd (peth) λ^{al} 228 (4.0), 253 (4.0), 294 (3.2)	98–9	....	....	....	...	s	...	...	...	....	**E12B**, 2806
t103	—,1,2,3,4-tetra-bromo-	Naphthalene tetrabromide. $C_{10}H_8Br_4$. *See* t98	447.81	mcl pr (chl)	111d	....	....	....	i	i δ^h	i	...	s^h	CS_2 s chl s^h	**B5**³, 1229
t104	**1-Tetralincar-boxylic acid (*dl*)**	1,2,3,4-Tetrahydro-1-naphthoic acid.	176.22	tcl pr (AcOEt)	85	....	....	....	δ s^h	v	v	v	s	os v AcOEt s^h	**B9**², 415
Ω t105	**2-Tetralincar-boxylic acid (*dl*)**	1,2,3,4-Tetrahydro-2-naphthoic acid.	176.22	nd (dil al) λ^{al} 266 (2.65), 273.5 (2.70)	97	$168\text{–}70^{15}$	....	....	δ	s	s	...	s	CS_2, chl s	**B9**², 416
t106	**5-Tetralincar-boxylic acid**	5,6,7,8-Tetrahydro-1-naphthoic acid.	176.22	pr (w), wh nd (dil aa) λ^{al} 235 (3.78), 285 (3.11)	150	....	....	....	δ s^h	v	...	...	s	chl s	**E12B**, 4079
t107	**6-Tetralincar-boxylic acid**	5,6,7,8-Tetrahydro-2-naphthoic acid.	176.22	nd (al), cr (aa or bz) λ^{al} 240 (4.1), 272 (3.2), 286 (3.1)	154	216^{14}	....	....	...	v	...	...	s	aa s peth i	**E12B**, 4082
t108	**1,4-Tetralindicar-boxylic acid, 1-phenyl-(*d*, α)**	*d*-α-Isatropic acid.	296.33	pr, $[\alpha]_D^{20}$ +9.44 (al, c = 12.6)	239d	....	....	....	δ^h	δ	i	...	i	....	**B9**¹, 417
t109	—,—(*dl*,α)	$C_{18}H_{16}O_4$. *See* t108	296.33	cr (chl-peth)	238–9	....	....	....	δ^h	δ	i	...	i	CS_2, aa s lig i	**B9**¹, 416

For explanations, symbols and abbreviations see beginning of table. For structural formulas see end of table.

No.	Name	Synonyms and Formula	Mol. wt.	Color, crystalline form, specific rotation and λ_{max} (log ε)	m.p. °C	b.p. °C	Density	n_D	Solubility w	al	eth	ace	bz	other solvents	Ref.
	1,4-Tetralindicarboxylic acid														
t110	—,—(*dl,β*)	$C_{18}H_{16}O_4$. *See* t108	296.33	pl (w)	208–9	...	...	...	i s^h	s	...	...	...	aa s	**B9**[1], 417
t111	—,—(*l,β*)	$C_{18}H_{16}O_4$. *See* t108	296.33	$[\alpha]_D$ −8.8 (al, c = 5)	197	...	...	...	s^h	s	...	...	...	aa s	**B9**[1], 417
—	**Tetralindione**	*see* **Tetralin, dioxo-**													
t112	**5-Tetralinsulfonic acid**	...	212.27	cr (chl + 1w), cr (w or min ac + 2w)	105–10 (+1w)	...	...	...	v^h	...	...	...	...	chl δ^h	**B11**[2], 87
t113	—,amide	$C_{10}H_{13}NO_2S$. *See* t112	211.29	lf (al or 30 % aa)	139–40	...	...	...	...	s^h	...	...	...	dil NaOH v	**B11**[2], 87
t114	—,chloride	$C_{10}H_{11}ClO_2S$. *See* t112	230.72	pl (peth)	70.5	...	...	...	d^h	d^h	s	...	...	peth δ	**B11**[2], 87
t115	**6-Tetralinsulfonic acid**	...	212.26	cr (chl or dil sulf)	75	...	...	...	s	...	s	...	...	chl v	**B11**[2], 88
t116	—,chloride	$C_{10}H_{11}ClO_2S$. *See* t115	230.72	pl (eth)	58	197–200[18]	...	...	d^h	...	s	...	...	...	**B11**[2], 88
Ω t117	**1-Tetralone**	1-Oxo-1,2,3,4-tetrahydronaphthalene*.	146.19	λ^{hx} 247.5 (4.06), 290 (3.28)	8	255–7[760] 129[12]	1.0988^{16}_4	1.5672^{20}	...	...	...	...	...	...	**B7**[2], 292
Ω t118	**2-Tetralone**	2-Oxo-1,2,3,4-tetrahydronaphthalene*.	146.19	λ^{al} 245 (4.0), 290 (3.2)	18	234–40[760] 138[16]	1.1055^{17}_4	1.5598^{20}	i	...	s	...	s	...	**B7**[2], 295
t119	**1-Tetralone, 7-ethyl-**	$C_{12}H_{14}O$. *See* t117	174.25	...	...	152–3[12]	1.0556^{17}_4	1.5599^{17}	i	s	s	...	s	...	**B7**[2], 305
—	**Tetranthrene**	*see* **Phenanthrene, 1,2,3,4-tetrahydro-***													
t120	**Tetraphenylene**	Tetrabenzocyclooctatetraene.	304.40	cr (al or AcOEt) λ^{cy} 300 sh (2.3)	233	200[0.2] sub	...	...	...	s^h	δ	...	...	AcOEt, $PhNO_2$ s^h oos δ	**E14s**, 752
t121	**Tetraphosphoric acid**, hexaethyl ester	$[(C_2H_5O)_2P(O)O]_3PO$	506.26	hyg	*ca.* −40	>150d	1.2917^{27}_4	1.4273^{27}	d	∞	...	∞	∞	oos ∞ peth i	**C41**, 2692
t122	**Tetrasiloxane, decamethyl-**	$(CH_3)_3Si(—OSi(CH_3)_2—)_3CH_3$	310.69	...	−76	194[760] 88[20]	0.8536^{20}_4	1.3895^{20}	i	δ	...	...	s	peth s	**Am 68**, 358
t123	**1-Tetratriacontanol***	*n*-Carnatyl alcohol. $CH_3(CH_2)_{33}OH$	494.94	nd (ace)	91.9	...	...	...	...	...	...	s^h	...	...	**B1**[3], 1852
t124	**1,2,4,5-Tetrazine**	*s*-Tetrazine	82.07	dk red pr λ^w 253 (3.4), 511 (2.56)	99	sub	...	...	s	s	s	...	...	sulf s (red)	**B26**[2], 212
t125	**1,2,3,4-Tetrazole**	...	70.05	pl (al)	156	sub	...	...	s	s	δ	s	δ	aa s	**B26**[2], 196
t126	**1,2,3,4-Tetrazolium, 2,3,5-triphenyl-**, chloride	Tetrazolium salt. T.T.C.	334.81	nd (al or chl) λ^{al} 250, 290, 300	243d	...	...	...	s	s	i	s	...	chl s^h	**C51**, 11334
—	**Tetrolic acid**	*see* **2-Butynoic acid***													
—	**Tetronal**	*see* **Pentane, 3,3-bis(ethylsulfonyl)-***													
—	**Tetronic acid**	*see* **Furan, 2,4-dioxotetrahydro-**													
—	**Tetryl**	*see* **Aniline, *N*-methyl-*N*, 2,4,6-tetranitro-**													
—	**Thalline**	*see* **Quinoline, 6-methoxy-1,2,3,4-tetrahydro-**													
—	**Thapsic acid**	*see* **Hexadecanedioic acid***													
Ω t127	**Thebaine**	Paramorphine. $C_{19}H_{21}NO_3$	311.39	pl (eth), pr (dil al), $[\alpha]^{25}_D$ −218.5 (al, p = 2) λ^{al} 225 sh (4.2), 285 (3.9)	193	sub 91[0.01]	1.305^{20}_4	...	i	v^h	δ	...	s	chl, Py, $PhNH_2$ v	**B27**[2], 177
t128	—,hydrochloride monohydrate	$C_{19}H_{21}NO_3 . HCl. H_2O$	365.86	orh pr (al) $[\alpha]^{20}_D$ −157 (al)	...	...	...	...	s	s	s	...	...	...	**C55**, 8764
Ω t129	**Thebainone A**	$C_{18}H_{21}NO_3$	299.37	nd (dil al + ½w) nd or pr (al, aa, AcOEt) $[\alpha]^{28}_D$ −47 (95 %) al, c = 1.16) ($[\alpha]^{20}_D$ +9.6 (al)) λ 283 (3.25)	151–2	...	...	...	δ	δ s^h	δ	s	s	chl s AcOEt s^h MeOH δ	**B21**[2], 448
—	**Theine**	*see* **Caffeine**													
t130	**Theobromine**	3,7-Dimethylxanthine. $C_7H_8N_4O_2$.	180.18	rh or mcl nd (w) $\lambda^{0.1N\,HCl}$ 274 (5.57)	351 (cor) (357)	sub 290	...	...	δ	δ	i	...	i	CCl_4, chl, lig i	**B26**[2], 264
t131	**Theophylline**	1,3-Dimethyl-2,6-dihydroxypurine. 1,3-Dimethylxanthine. $C_7H_8N_4O_2$.	180.18	nd or pl (w + 1) λ^{al} 270 (3.98)	272–4 (264)	...	...	...	v^h	δ	δ	...	...	chl δ	**B26**[2], 263

For explanations, symbols and abbreviations see beginning of table. For structural formulas see end of table.

No.	Name	Synonyms and Formula	Mol. wt.	Color, crystalline form, specific rotation and λ_{max} (log ε)	m.p. °C	b.p. °C	Density	n_D	Solubility w	al	eth	ace	bz	other solvents	Ref.
	Thevetin														
t132	**Thevetin**	$C_{42}H_{66}O_{18}$	858.99	nd (al), pl or nd (*i*-PrOH + 3w), $[\alpha]_D^{26}$ −62.5 (MeOH, c = 2)	210	....	....	....	δ	s	i	i	i	MeOH, Py s chl i	**E14,** 231
t134	**Thialdine**	Thioacetaldehyde ammonia	163.31	pl (al-eth)	46 (43)	d	1.0632^{50}_{20}	....	δ	s	s	...	...	....	**B27²,** 525
—	**Thiamine hydrochloride**	*see* **Vitamin B₁**													
Ω t135	**Thianthrene**	Diphenylene disulfide	216.34	mcl pr or pl (al) λ^{al} 242 (4.22), 257 (4.63), 275 sh (3.35)	158–9	366^{760} 204^{11}	....	....	i	δ s^h	s	...	s	CS_2 s conc sulf s (vt)	**B19²,** 34
t136	**Thiazole**	....	85.13	λ^{al} 240 (3.60)	....	116.8	1.1998^{17}_4	1.5969^{20}	δ	s	s	s	...	....	**B27²,** 9
Ω t137	**—,2-amino-**	Abadol. $C_3H_4N_2S$. *See* t136	100.14	ye pl (al) λ^{al} 257 (3.79)	93	140^{11}	....	....	δ s^h	δ	δ	...	...	dil HCl v	**B27²,** 205
Ω t138	**—,2-amino-4-methyl-**	$C_4H_6N_2S$. *See* t136	114.17	hyg cr λ^{al} 257 (3.74)	45–6	281–2δd 136^{30-40}	....	....	v	v	v	...	...	....	**B27²,** 206
t139	**—,2-amino-5-methyl-**	$C_4H_6N_2S$. *See* t136	114.17	pl (w) λ^{al} 260 (3.84)	96	....	....	....	δ s^h	v	s	...	...	....	**B27,** 162
t140	**—,2-amino-5-sulfanilyl-**	....	255.32	nd (al)	219–21δd	....	....	....	i	s	s	v	...	diox v AcOEt, dil ac s	**C41,** 447
—	**—,dihydro-**	*see* **Thiazoline**													
t141	**—,2,4-dimethyl-**	C_5H_7NS. *See* t136	113.18	λ 253 (3.65)	....	$144–5^{719}$ $70–3^{50}$	1.0562^{15}_4	1.5091^{20}	δ^h	s	s	...	...	....	**B27²,** 10
t142	**—,4,5-dimethyl-**	C_5H_7NS. *See* t136	113.18	....	83–4	158 $81–3^{59}$	....	....	...	s	s	...	...	....	**Am 74,** 5778
t143	**—,2,4-diphenyl-**	$C_{15}H_{11}NS$. *See* t136	237.33	lf (al)	92–3	>360	1.1554^{98}_4	....	...	v	v	...	...	ac s	**B24²,** 43
t144	**—,5(2-hydroxyethyl)-4-methyl-**	C_6H_9NOS. *See* t136	143.21	col to pa ye λ^w 249 (4.61)	....	135^7 103^1	1.196^{24}_4	....	v	s	s	...	s	chl s	**Am 71,** 2931
t145	**—,2-mercapto-4-methyl-**	$C_4H_5NS_2$. *See* t136	131.22	ye (dil al)	88–9	188^3	....	....	δ	s	...	...	...	....	**B27²,** 208
t146	**—,2-methyl-**	C_4H_5NS. *See* t136	99.16	λ^{MeOH} 234 (3.8), 275 (3.0)	....	$128–9^{760}$ $65–70^{80}$	....	1.510	∞	s	...	s	...	....	**B27,** 16
t147	**—,4-methyl-**	C_4H_5NS. *See* t136	99.16	λ^{al} 250 (3.54)	....	132^{743} 70^{90}	1.112^{25}	....	s	s	s	...	...	....	**B27²,** 9
—	**—,tetrahydro-**	*see* **Thiazolidine**													
t149	**5-Thiazolecarboxylic acid**	....	129.14	lt ye nd (dil HCl)	217–18 (196–7)	....	....	....	s^h	...	s	...	...	dil HCl s	**C48,** 2688
t150	**—,2-amino-,** ethyl ester	$C_6H_8N_2O_2S$. *See* t149	172.22	....	163–4	213–5d	....	....	δ	...	...	...	...	....	**C47,** 7453
t151	**—,4-methyl-**	$C_5H_5NO_2S$. *See* t149	143.17	pr or pl (w), nd (al) λ 255 (3.78)	280d	sub >250	....	....	δ^h	s^h	s^h	...	δ	lig δ	**B27²,** 376
t152	**—,—,ethyl ester**	$C_7H_9NO_2S$. *See* t149	171.22	pr	28	$232–3^{735}$ (215–20) $110–5^{15}$	....	....	...	s	s	s	...	oos v	**B27,** 316
t153	**—,—,—,hydrochloride**	$C_7H_9NO_2S$. HCl. *See* t149	207.68	nd (al)	155d	....	....	....	s	s^h	...	...	...	....	**J1939,** 443
t154	**—,4-methyl-2-sulfanilamido-**	....	313.36	cr	241–2	....	....	....	δ	...	...	...	...	....	**C38,** 2250
t155	**Thiazolidine**	Tetrahydrothiazole	89.16	....	....	$164–5^{760}$	1.131^{25}_4	1.551^{30}	∞	s	v	v	...	....	**Am 59,** 200
Ω t156	**2,4-Thiazolidinedione**	2,4-Dioxothiazolidine	117.13	pl (w), pr (al)	128	179^{19}	....	....	δ s^h	δ v^h	s	...	...	....	**B27²,** 284
t157	**2-Thiazolidinethione**	2-Thiothiazolidone	119.21	nd (w or MeOH) λ 270 (4.11)	106–7	....	....	....	s^h	δ	i	...	s^h	alk v chl δ CS_2 i	**B27²,** 198
—	**4-Thiazolidone, 2-thioxo-**	*see* **Rhodanine**													
t158	**2,3-Thiazoline**	4,5-Dihydrothiazole. Δ^2-Thiazoline.	87.14	....	....	138^{750}	....	....	...	...	s	s	s	....	**B27¹,** 206
Ω t159	**—,2-amino-**	*S,N*-Ethyleneisothiourea. $C_3H_6N_2S$. *See* t158	102.16	nd or fl (bz)	84–5	d	....	....	v	v	...	...	v^h	chl v	**B27²,** 194
Ω t160	**—,2-mercapto-**	2-Thiazolinethiol. $C_3H_5NS_2$. *See* t158	119.21	nd (w or MeOH)	106–7	....	....	....	s^h	s^h	δ	...	s^h	chl s^h CS_2 i	**B27²,** 198
—	**Thiazone**	*see* **3-Isophenothiazin-3-one**													
—	**Thienone**	*see* **Ketone, di(2-thienyl)**													
t161	**Thiirane**	Ethylene sulfide	60.12	λ^{al} 258 (1.5)	....	55–6d	1.0368^0_4	1.4937^{20}	...	δ	δ	s	...	chl s	**B17²,** 12
—	**Thioacetic acid**	*see* **Acetic acid, thio-**													
—	**Thioanisole**	*see* **Benzene, (methylthio)-**													
—	**Thiocarbazide**	*see* **Carbazide, 3-thio-**													
t162	**Thiochrome**	....	262.34	ye pr (chl) $\lambda^{w, pH=7}$ 367 (3.97) $\lambda^{w, pH=3}$ 392 (4.13)	227–8	sub vac	....	....	s	δ	δ	δ	...	MeOH s chl δ	**C29,** 6242
t163	**Thiocyanic acid**	Sulfocyanic acid. HSCN	59.09	....	5	d	....	....	∞	∞	∞	...	...	....	**B3³,** 251
t164	**—,allyl ester**	Allyl rhodanate $CH_2{:}CHCH_2SCN$	99.16	....	....	161	1.056^{15}	....	δ	v	v	...	...	....	**B3,** 177

For explanations, symbols and abbreviations see beginning of table. For structural formulas see end of table.

No.	Name	Synonyms and Formula	Mol. wt.	Color, crystalline form, specific rotation and λ_{max} (log ε)	m.p. °C	b.p. °C	Density	n_D	Solubility: w	al	eth	ace	bz	other solvents	Ref.
	Thiocyanic acid														
t165	—,4-aminophenyl ester	p-Thiocyanatoaniline. $C_7H_6N_2S$. *See* t180	150.20	nd (w), cr (dil al) λ^{al} 262 (4.20), 280 sh (3.75)	57–8	......	......	......	δ	v	s	...	s		**B13**², 299
Ω t166	—,benzyl ester	Benzyl rhodanate. Benzyl thiocyanate. $C_6H_5CH_2SCN$	149.22	pr (al) λ^{cy} 222 (3.91), 258 (2.84)	43	256	......	......	i	s^h	s	...	...	CS_2 s	**B6**³, 1600
Ω t167	—,butyl ester	*n*-Butyl thiocyanate. $CH_3(CH_2)_3SCN$	115.20	......	......	185^{743}	0.9563^{25}	1.4630^{20}	i	s	s	...	...	......	**B3**³, 281
t168	—,*tert*-butyl ester	$(CH_3)_3CSCN$	115.20	......	10.5	140^{770}d, $39–40^{10}$	0.9187^{10}	......	...	...	...	...	...	......	**B3**³, 282
t169	—,4-chlorophenyl ester	C_7H_4ClNS. *See* t180	169.63	nd (al) λ^{al} 238 (4.09)	35–6	......	......	......	i	v	...	...	...	os v	**B6**³, 1038
t170	—,4(dimethyl-amino)phenyl ester	$C_9H_{10}N_2S$. *See* t180	178.26	nd (lig, w, al)	73–4	......	......	......	s^h	δ s^h	s^h	...	...	lig s^h	**B13**², 301
Ω t171	—,ethyl ester	Ethyl rhodanate. Ethyl thiocyanate. C_2H_5SCN	87.14	......	−85.5	145^{758}	1.0071^{22}_{4}	1.4684^{15}	i	∞	∞	...	...	......	**B3**³, 281
t172	—,heptyl ester	*n*-Heptyl rhodanate. $CH_3(CH_2)_6SCN$	157.28	......	......	234–6, 136^{28}	0.92^{20}	......	i	∞	∞	...	...	......	**B3**², 122
t173	—,isobutyl ester	Isobutyl rodanate. $(CH_3)_2CHCH_2SCN$	115.20	......	−59	175.4^{760}, 66^{15}	......	......	i	v	v	...	...	......	**B3**, 177
Ω t174	—,isopropyl ester	Isopropyl rodanate. $(CH_3)_2CHSCN$	101.17	......	......	$152–3^{754}$	0.9784^{20}	......	i	∞	∞	...	...	......	**B3**³, 281
Ω t175	—,methyl ester	Methyl rhodanate. Methyl thiocyanate. CH_3SCN	73.12	......	−51	132.9^{757}	1.0678^{25}_{4}	1.4669^{25}	δ	∞	∞	...	...	......	**B3**³, 280
t176	—,3-methylbutyl ester	Isoamyl rhodanate. Isoamyl thiocyanate. $(CH_3)_2CHCH_2CH_2SCN$	129.23	......	......	197^{760}	......	......	i	s	s	...	...	......	**B3**³, 282
t177	—,1-naphthyl ester	α-Naphthyl thiocyanate. $C_{10}H_7^{\alpha}SCN$	185.25	cr (peth) λ^{al} 225 (4.75), 280 sh (3.83), 287 (3.93), 315 (2.98)	55	......	......	......	...	...	...	...	...	peth s^h	**B6**², 588
t178	—,2-naphthyl ester	β-Naphthyl thiocyanate. $C_{10}H_7^{\beta}SCN$	185.25	......	35	......	......	......	...	...	...	...	...	......	**B6**², 611
Ω t179	—,octyl ester	*n*-Octyl rhodanate. $CH_3(CH_2)_7SCN$	171.31	......	105	$141–2^{19}$, $122–4^{11}$	0.9149^{25}_{4}	1.4649^{20}	i	v	v	...	s^h	......	**B3**³, 282
t180	—,phenyl ester	Thiocyanatobenzene. Phenyl rhodanide. Phenyl thiocyanate.	135.19	λ^{cy} 226 (3.97), 270 (3.13)	......	$232–3^{760}$, $71–3^{1.5}$	1.155^{18}_{18}	......	i	s	s	...	...	......	**B6**³, 1011
t181	—,propyl ester	Propyl rhodanate. Propyl thiocyanate. $CH_3CH_2CH_2SCN$	101.17	......	......	163	......	......	i	s	s	...	...	......	**B3**³, 281
t182	—,4-tolyl ester	p-Tolyl rhodanide	149.22	......	......	240–5, $116–8^{10}$	......	......	i	s	...	...	s	chl s	**B6**², 398
—	**Thiocyanuric acid**	*see* **Cyanuric acid, thio-**													
Ω t183	**Thiodiglycolic acid**	Dimethyl sulfide α,α-dicarboxylic acid. Thiodiacetic acid. $S(CH_2CO_2H)_2$	150.15	cr (w, AcOEt-bz)	129	......	......	......	v	v	...	...	s^h	......	**B3**³, 422
—	**Thioglycolic acid**	*see* **Acetic acid, mercapto-**													
—	**Thiohydantoin**	*see* **Hydantoin, thio-**													
t184	**Thioindigo**	......	296.38	br-red nd (xyl), red mcl nd (bz) λ^{chl} 283 sh (4.1), 300 (4.4), 490 (4.3)	359	sub	......	......	i	i δ^h	...	...	s^h	xyl s, $PhNO_2$ v^h, CS_2, chl δ^h	**B19**², 192
—	**Thioisatin**	*see* **2,3-Benzothiophene-quinone**													
—	**Thiomalic acid**	*see* **Succinic acid, 2-mercapto-**													
t185	**Thiomorpholine**	1,4-Thiazan	103.19	......	......	174^{746}, 110^{100}	1.0882^{20}_{4}	1.5386^{20}	∞	s	s	s	s	oos ∞	**B17**², 4
—	**Thionol**	*see* **3-Isophenothiazin-3-one, 7-hydroxy**													
—	**Thionaphthene**	*see* **2,3-Benzothiophene**													
t186	**Thionin**, hydro-chloride	7-Amino-3-imino-3H-2-phenothiazine hydro-chloride. Lauth's violet.	263.76	dk br or gr pl or nd λ^{al} 602 (4.74)	......	......	......	......	δ s^h	δ	δ	...	s	chl, ac s	**B27**², 447
Ω t187	**Thiophene**	Thiofuran	84.14	λ^{al} 215 (3.8), 231 (3.87)	−38.25	84.16^{760}	1.06494^{20}_{4}	1.5289^{20}	...	∞	∞	∞	∞	oos, CCl_4, hp, Py, diox, to ∞	**B17**², 35
t188	—,2-acetamido-	C_6H_7NOS. *See* t187	141.19	lf (w)	161–2	......	......	......	δ^h	v	δ	v	δ	......	**B17**¹, 136
Ω t189	—,2-acetyl-	Methyl 2-thienyl ketone. C_6H_6OS. *See* t187	126.18	λ^{al} 260 (4.01), 283 (3.87)	10–11	213.5^{760}, $94.5–6.5^{13}$	1.1679^{22}_{4}	1.5667^{20}	δ	∞	∞	...	...	......	**B17**², 314

For explanations, symbols and abbreviations see beginning of table. For structural formulas see end of table.

No.	Name	Synonyms and Formula	Mol. wt.	Color. crystalline form. specific rotation and λ_{max} (log ε)	m.p. °C	b.p. °C	Density	n_D	Solubility w	al	eth	ace	bz	other solvents	Ref.
	Thiophene														
Ω t190	**—,2-acetyl-5-bromo-**	C_6H_5BrOS. *See* t187	205.08	nd (al) λ^{al} 268 (3.92), 293 (4.09)	94–5	$105-7^{4.5}$				δ v^h					**B17**, 288
Ω t191	**—,2-acetyl-5-chloro-**	C_6H_5ClOS. *See* t187	160.62	ta (al or eth) λ^{al} 264.5 (3.94), 292 (4.05)	52	$88-9^{4.5}$			v	v	v				**B17**, 287
Ω t192	**—,2-amino-**	Thiophenine. C_4H_5NS. *See* t187	99.16	pa ye tab (al)		$77-9^{11}$ $61-2^{1}$			v	v	i				**B17**, 248
t192[1]	**—,2-benzoyl-**	Phenyl 2-thienyl ketone. $C_{11}H_8OS$. *See* t187	188.26	nd (dil al or peth) λ^{al} 263 (4.11), 293 (4.13)	56.5–7	300^{760}	1.1890_4^{54}		i	s	s				**B17[2]**, 372
Ω t193	**—,2-bromo-**	C_4H_3BrS. *See* t187	163.04	λ^{al} 236 (3.90)		$149-51$ $42-6^{13}$	1.684_4^{20}	1.5868^{20}	i		v	v			**B17**, 39
t194	**—,2-bromo-5-chloro-**	C_4H_2BrClS. *See* t187	197.49		−20	70^{18}	1.803_{25}^{25}	1.5925^{25}			s	s			**Am 70**, 2379
t195	**—,2-bromo-5-iodo-**	C_4H_2BrIS. *See* t187	288.94			116^{13}									**C32**, 3391
t196	**—,2-bromo-3-methyl-**	C_5H_5BrS. *See* t187	177.07			175^{729} $27^{1.8}$	1.5844_4^{18}	1.5714^{20}			s		s		**B17[2]**, 40
t197	**—,2-bromo-5-methyl**	C_5H_5BrS. *See* t187	177.07	col to pa ye		177^{740} $29^{1.8}$	1.5529_4^{20}	1.5673^{20}			s		s		**B17[2]**, 39
Ω t198	**—,2-chloro-**	C_4H_3ClS. *See* t187	118.59	λ^{iso} 236 (3.94)	−71.9	128.3^{760}	1.2863_4^{20}	1.5487^{20}	i	∞	∞				**B17**, 32
Ω t199	**—,2-chloro-5-butyl-**	$C_8H_{11}ClS$. *See* t187	174.69			$117-8^{88}$	1.0842_4^{17}	1.5162^{20}	i				s		**B17**, 44
t200	**—,2-chloro-5-iodo-**	C_4H_2ClIS. *See* t187	244.48		−25	$95-6^{14}$									**C32**, 3391
t201	**—,2-chloro-5-methyl-**	C_5H_5ClS. *See* t187	132.61			$154-5^{742}$ 55^{19}	1.2147_4^{25}	1.5372^{20}		s	s	s	s	os s	**B17**, 37
Ω t202	**—,2,5-dibromo-**	$C_4H_2Br_2S$. *See* t187	241.94	λ^{iso} 252 (3.96)	−6	210.3^{760} $76-80^{10}$	2.147_{23}^{23}	1.6288^{20}	i	v	v				**B17**, 33
t203	**—,2,5-dibromo-3,4-dinitro-**	$C_4Br_2N_2O_4S$. *See* t187	331.94	cr (al)	139–40					δ					**B17**, 36
Ω t204	**—,2,5-dichloro-**	$C_4H_2Cl_2S$. *See* t187	153.03	λ^{iso} 252 (3.87)	−40.5	162^{760}	1.4422_4^{20}	1.5626^{20}	i	∞	∞				**B17**, 33
t205	**—,2,5-diiodo-**	$C_4H_2I_2S$. *See* t187	335.93	lf (al) λ^{iso} 266 (4.15), 315 (2.43)	40.5–1.5	$139-40^{15}$			i	v					**B17**, 35
t206	**—,2,3-dimethyl-**	2,3-Thioxene. C_6H_8S. *See* t187	112.21	λ^{iso} 233 (3.77)	−49.0	141.6^{760} 29.3^{10}	1.0021_4^{20}	1.5192^{20}	i	v	v		s		**B17**, 40
t207	**—,2,4-dimethyl-**	2,4-Thioxene. C_6H_8S. *See* t187	112.21	λ^{iso} 234.5 (3.80)		140.7^{760} 29.9^{10}	0.9956_{20}^{20}	1.5104^{20}	i	s	s		s		**B17**, 41
t208	**—,2,5-dimethyl-**	α,α-Thioxene. C_6H_8S. *See* t187	112.21	λ^{iso} 238 (4.8)	−62.6	136.7^{760} 26.2^{10}	0.985_4^{20}	1.5129^{20}	i	s	s		s		**B17**, 41
t209	**—,2,5-dinitro-**	$C_4H_2N_2O_4S$. *See* t187	174.14	(i) ye lf (al) (ii) ye nd (al, w)	(i) 52 (ii) 80–2	290^{760}			s^h	s	v				**B17**, 35
Ω t210	**—,2-ethyl-**	C_6H_8S. *See* t187	112.21	λ^{iso} 233 (3.89)		134^{760} 24^{10}	0.9930_4^{20}	1.5122^{20}	i	v	v				**B17**, 39
t211	**—,3-ethyl-**	C_6H_8S. *See* t187	112.21	λ^{iso} 235 (3.6)	−89.1	136^{760} 26^{10}	0.9980_4^{20}	1.5146^{20}	i	s	v				**B17**, 40
t212	**—,2-hydroxy-5-methyl-**	2,5-Thiotenol. C_5H_6OS. *See* t187	114.17			85^{40}			δ	v	v				**B17**, 252
Ω t213	**—,2-iodo-**	C_4H_3IS. *See* t187	210.04	λ^{iso} 243 (3.97), 285 sh (2.4)	−40	$180-2^{760}$ 73^{15}		1.6465^{25}		v	v				**B17**, 34
t214	**—,2-iodo-5-nitro-**	$C_4H_2INO_2S$. *See* t187	255.04	ye pr (al)	74					s					**B17**, 35
Ω t215	**—,2-methyl-**	α-Thiotolene. C_5H_6S. *See* t187	98.17	λ^{w} 234 (3.58)	−63.4	112.56^{760} 9.2^{10}	1.0193_4^{20}	1.5203^{20}	i	∞	∞	∞	∞	CCl_4, hp ∞	**B17**, 37
Ω t216	**—,3-methyl-**	β-Thiotolene. C_5H_6S. *See* t187	98.17	λ^{cy} 234.5 (3.75)	−69	115.4^{760} 10.96^{10}	1.0218_4^{20}	1.5204^{20}	i	∞	∞	∞	∞	chl v	**B17**, 38
t217	**—,2(methyl-amino)-**	*N*-Methyl-2-thiophenine. C_5H_7NS. *See* t187	113.18			$88-92^{15}$					s	s			**B17[1]**, 136
t218	**—,2-methyl-5-phenyl-**	$C_{11}H_{10}S$. *See* t187	174.27	nd λ 225 sh (3.79), 291 (4.21)	51	$270-2^{760}$				v	v			lig v	**B17**, 67
Ω t219	**—,2-nitro-**	$C_4H_3NO_2S$. *See* t187	129.14	lt ye mcl nd (peth) λ^{al} 270 (3.80), 296 (3.78)	46.5	224–5	1.3644_4^{43}		i	v				alk s^h peth δ, s^h	**B17**, 35
t220	**—,4-nitro-2,3,5-tribromo-**	$C_4Br_3NO_2S$. *See* t187	365.84	red-ye nd (al)	106					δ	v				**B17**, 35
t221	**—,3-nitro-2,4,5-trichloro-**	$C_4Cl_3NO_2S$. *See* t187	232.47	red-ye nd (al)	86					s	v		v		**B17**, 35
Ω t222	**—,tetrabromo-**	C_4Br_4S. *See* t187	399.74	nd (al)	117–8	326 $170-3^{13}$			i	s^h	v				**B17**, 34
Ω t223	**—,tetrachloro-**	C_4Cl_4S. *See* t187	221.92	nd (dil al)	30–1	233.4^{760} $75-7^{2}$	1.7036_4^{30}	1.5915^{30}	i	v	∞				**B17**, 33
Ω t224	**—,tetrahydro-**	Tetramethylene sulfide	88.18	λ^{al} 240 sh (1.7)	−96.16	121.12^{760} 14.48^{10}	0.9987_4^{20}	1.5048^{20}	i	∞	∞	∞	∞	os ∞	**B17[2]**, 15
t225	**—,2,3,5-tribromo-**	C_4HBr_3S. *See* t187	320.84	nd (al)	29	260^{760}			i	v^h	v				**B17**, 34
t227	**—,2,3,5-trichloro-**	C_4HCl_3S. *See* t187	187.48		−16.1	198.7^{760}	1.5856_4^{20}	1.5791^{20}	i	∞	∞				**B17**, 33

For explanations, symbols and abbreviations see beginning of table. For structural formulas see end of table.

No.	Name	Synonyms and Formula	Mol. wt.	Color, crystalline form, specific rotation and λ_{max} (log ε)	m.p. °C	b.p. °C	Density	n_D	Solubility w	al	eth	ace	bz	other solvents	Ref.
	Thiophene														
t228	—,**2,3,5-tri-methyl-**	$C_7H_{10}S$. *See* t187	126.22	λ^{hx} 237 (3.83)	...	164.5^{760} 46.8^{10}	0.9753_4^{20}	1.5112^{20}	i	v	v	v	v	os s	**B17**[2], 43
—	**Thiopheneacetic acid**	*see* **Acetic acid, thienyl-**													
Ω t229	**2-Thiophene-carboxaldehyde**	2-Formylthiophene. 2-Thiophenealdehyde.	112.15	pa ye liq λ^{iso} 265 (4.12), 278.5 (3.81)	...	198^{760} $85–6^{16}$	1.215_{21}^{21}	1.5920^{20}	i	v	s	...	...	...	**B17**, 285
Ω t230	—,oxime(*syn*)	2-Thiophenealdoxime	127.17	nd λ 265 (4.1)	133	...	...	...	...	...	v	...	...	...	**B17**, 285
t231	—,phenylhydrazone	...	202.28	ye nd (al)	134.5	...	...	...	i	s	...	...	...	...	**B17**, 286
Ω t232	**2-Thiophene-carboxylic acid**	α-Thiophenic acid	128.15	nd (w) λ^{al} 246 (3.96), 260 (3.84)	129–30	260d	...	...	v^h	v	v	...	...	chl s peth δ	**B18**, 289
Ω t233	—,chloride	C_5H_3ClOS. *See* t232	146.60	...		201 (206–8) 77^{10}	...	...	d	d^h	...	...	...	...	**B18**[2], 269
Ω t234	—,ethyl ester	$C_7H_8O_2S$. *See* t232	156.22	λ^{al} 249 (3.95), 269 (3.89)	...	218 94^{10}	1.1623_4^{16}	1.5248^{20}	...	s	...	s	...	os s	**B18**[2], 269
t235	—,**3-methyl-**	$C_6H_6O_2S$. *See* t232	142.18	nd (dil al, w)	144	...	...	...	δ v^h	s	s	s	...	...	**B18**, 293
t236	**2,3-Thiophenedi-carboxylic acid**	...	172.16	pr or nd (w)	272–4	...	...	...	δ^h	...	v	...	...	...	**B18**, 327
t237	**2,4-Thiophenedi-carboxylic acid**	...	172.16	cr	280d (>300)	sub >200	...	...	δ v^h	...	...	...	...	...	**B18**, 327
Ω t238	**2,5-Thiophenedi-carboxylic acid**	...	172.16	λ^{iso} 271 (4.2)	358.5–9.5 (sealed tube)	sub at 150–300	...	...	δ	s	s	...	...	...	**B18**, 330
t239	—,diethyl ester	$C_{10}H_{12}O_4S$. *See* t238	228.27	nd (al)	51.5	...	...	...	...	s	...	...	...	...	**B18**, 331
t240	**2-Thiophene-sulfonic acid,** amide	...	163.22	nd (w)	147	...	...	...	δ s^h	...	...	...	...	alk s Na_2CO_3 i	**B18**, 567
t241	—,chloride	...	182.65	...	28	$99–101^6$ sub	...	...	d	d^h	s	...	...	...	**B18**, 567
t242	**3-Thiophene-sulfonic acid,** amide	...	163.22	pl (w)	152–3 (146)	...	...	...	δ	...	...	...	...	...	**B18**, 568
t243	—,chloride	...	182.65	cr (eth)	43	$98–9^{0.5}$	...	...	d^h	d^h	v	...	...	lig i	**B18**, 568
—	**Thiophenine**	*see* **Thiophene, amino-**													
—	**Thiophenol**	*see* **Benzene, mercapto-***													
—	**Thiophosphoric acid**	*see* **Phosphoric acid, thiono-***													
t244	**Thiophthene** (solid)	Thieno[3,2-b]thiophene	140.23	bipym orh (lig) λ^{iso} 259 (4.1), 268 (4.0), 278 (4.0)	56	$221–4^{760}$	...	...	i	...	s	...	...	lig δ, s^h	**J1953**, 1837
t245	—(liquid)	Thieno[2,3-b]thiophene	140.23	nd (eth-so CO_2)	6.5	$224–6^{760}$ 106^{16}	...	...	i	...	s	...	...	...	**B19**[1], 612
—	**Thiopyran, tetrahydro-**	*see* **Sulfide, pentamethylene**													
t246	**Thiopyrine**	1,5-Dimethyl-2-phenyl-3-thio-3-pyrazolone.	204.30	cr (w)	166	...	...	...	δ s^h	s	s	...	...	...	**B24**[2], 28
—	**Thiosemi-carbazide**	*see* **Semicarbazide, 3-thio-**													
—	**Thiosinamine**	*see* **Urea, 1-allyl-2-thio-**													
—	**Thiourea**	*see* **Urea, 2-thio-**													
—	**Thiourethan**	*see* **Carbamic acid, thiono-,** ethyl ester													
t248	**Thioxanthene**	Thiaxanthene. Dibenzthiopyran.	198.29	nd (al-chl)	128–9	340^{730} sub	...	...	i	δ s^h	δ s^h	...	...	chl, peth s	**B17**, 74
Ω t249	**Thioxanthone**	9-Oxothioxanthene	212.28	ye nd (chl) λ^{al} 219 (3.87), 257 (4.28), 286 (3.37), 299 (3.20), 379 (3.48)	209 (212)	$371–3^{715}$ sub	...	...	i	δ	...	...	s	CS_2, aa s chl s^h peth i	**B17**[1], 191
—	**Threite**(*D*)	*see* **Erithritol** (*l*)													
t250	**Threonic acid**(*D*)	2,3,4-Trihydroxybutyric acid*.	136.12	nd (al), $[\alpha]_D$ *ca.* −30 (w)	197–8	...	...	...	s	s^h	...	s^h	...	AcOEt s^h	**B3**[2], 272
t251	—(*DL*)	$C_4H_8O_5$. *See* t250	136.12	cr (al, ace)	99	...	...	...	s	s^h	...	s^h	...	AcOEt s^h	**B3**[3], 895
t252	—(*L*)	$C_4H_8O_5$. *See* t250	136.12	nd (al-eth) $[\alpha]_D$ +9.54 (w)	169.5–70.5	...	...	...	s	s^h	...	s	...	AcOEt s^h	**B3**[3], 894
t253	**Threonine**(*D*)	*D*-2-Amino-3-hydroxybutyric acid. $CH_3CH(OH)CH(NH_2)CO_2H$	119.12	cr (80% al) $[\alpha]_D^{26}$ −28.3 (w, c = 1.1)	255–7d	...	...	...	s	i	i	...	...	chl i	**C31**, 6199
Ω t254	—(*DL*)	$CH_3CH(OH)CH(NH_2)CO_2H$	119.12	cr (dil al)	229–30 (+½w) 234–5d	...	...	...	s v^h	i	i	...	...	chl i	**B4**, 514
t255	**Threose**(*D*)	Trihydroxybutyraldehyde	120.12	hyg syr or nd (w), $[\alpha]_D^{22}$+29.1 → +19.6 (w) (mut)	126–32	...	...	...	v	δ	i	...	...	peth i	**B1**, 855 **B31**, 13

For explanations, symbols and abbreviations see beginning of table. For structural formulas see end of table.

No.	Name	Synonyms and Formula	Mol. wt.	Color, crystalline form, specific rotation and λ_{max} (log ε)	m.p. °C	b.p. °C	Density	n_D	Solubility						Ref.
									w	al	eth	ace	bz	other solvents	
	Thujane														
Ω t256	**Thujane(*d*)**	Sabinane	138.26	$[\alpha]_D$ +73.1 (+62)	....	157^{758}	0.8139_4^{20}	1.4376^{20}	...	s	s	s	s	....	**B5**[3], 253
t257	**4(10)-Thujene(*d*)**	Sabinene	136.24	$[\alpha]_D$ +80.17	....	$163–5^{760}$ 66^{30}	0.842^{20}	1.4678^{20}	...	s	s	s	s	chl s	**B5**[3], 365
t258	—(*l*)	$C_{10}H_{16}$. *See* t257	136.24	$[\alpha]_D^{15}$ −42.5	....	$163–5^{760}$	0.8464_4^{20}	1.4515^{20}	...	s	s	s	s	chl s	**B5**[3], 365
Ω t259	**α-Thujone**	....	152.24	$[\alpha]_D^{18}$ −19.94 (undil)	....	$200–1^{760}$ 75^{9}	0.9152^{20}	1.4490^{25}	i	∞	∞	s			**B7**, 92
Ω t260	**Thymidine**	Thymine-2-desoxyriboside. $C_{10}H_{14}N_2O_5$.	242.23	nd (AcOEt) $[\alpha]_D^{25}$ +30.6 (w, c = 1.03) λ^{w} 269 (3.97)	186–7	....	....	....	s	s^h	...	s^h	...	Py, MeOH, aa s AcOEt s^h chl δ^h	**C48**, 226
—	**Thymine**	*see* **Uracil, 5-methyl-**													
—	**Thymohydro-quinone**	*see* **Benzene, 1,4-dihydroxy-2-isopropyl-5-methyl-***													
—	**Thymol**	*see* **Benzene, 2-hydroxy-1-isopropyl-4-methyl-***													
Ω t263	**Thymol blue**	Thymolsulfonephthalein. $C_{27}H_{30}O_5S$	466.60	gr-red (al, eth, aa) λ 450, 600	221–4d	....	....	....	δ	s	...	δ	δ	aniline, aa s chl, CCl_4, to δ	**B19**[2], 112
Ω t264	**Thymolphthalein**	$C_{28}H_{30}O_4$	430.55	pr or nd (al)	253	....	....	....	i	s^h	s	s	...	alk s (bl)	**B18**[2], 130
—	**Thymoquinone,** dioxime	*see* **1,4-Benzoquinone, 2-isopropyl-5-methyl-,** dioxime*													
—	***o*-Thymotic acid**	*see* **Benzoic acid, 2-hydroxy-3-isopropyl-6-methyl-**													
—	**Thymylamine**	*see* **Benzene, 2-amino-1-isopropyl-4-methyl-***													
t266	**Thyroxine(*d*)**	$C_{15}H_{11}I_4NO_4$	776.88	nd, $[\alpha]_{546}^{25}$ +2.97 (al-NaOH) $\lambda^{0.02N\ KOH}$ 325 (3.8)	237d	....	....	....	δ	i	...	...	i	....	**B14**[2], 366
t267	—(*l*)	$C_{15}H_{11}I_4NO_4$. *See* t266	776.88	nd, $[\alpha]_D^{20}$ −4.4 (al-NaOH) $\lambda^{0.02N\ KOH}$ 325 (3.8)	235–6	....	....	....	δ	i	...	...	i	....	**B14**[2], 378
—	**Tiglaldehyde**	*see* **2-Butenal, 2-methyl-***													
—	**Tiglic acid**	*see* **2-Butenoic acid, 2-methyl-(*trans*)***													
t268	**Tigogenin**	5α, 22α-Spirostan-3 β-ol. $C_{27}H_{44}O_3$.	416.65	lf (al + 1w), pr (ace) $[\alpha]_D^{25}$ −49 (Py, c = 1) $[\alpha]_D^{18}$ −67.2 (chl) λ^{sulf} 394 (3.50)	205–6	....	....	....	...	s^h	s	s	...	MeOH, peth s CCl_4 s^h	**E14**, 280
—	**TNT**	*see* **Toluene, 2,4,6-trinitro-**													
Ω t269	**α-Tocopherol**	Vitamin E. 5,7,8-Trimethyltocol. $C_{29}H_{50}O_2$.	430.69	pa ye visc oil $[\alpha]_D^{25}$ +0.65 (al) λ^{MeOH} 292 (3.54)	2.5–3.5	350d $140^{10^{-6}}$	....	....	i	s	s	s	...	os s	**H21**, 520
t270	**β-Tocopherol**	Vitamin E. 5,8-Dimethyltocol. $C_{28}H_{48}O_2$.	416.70	pa ye visc oil $[\alpha]_D^{20}$ +6.37 λ 296 (3.57)	....	$200–10^{0.1}$	....	....	i	∞	∞	∞	...	chl ∞	**H21**, 1234
t271	**γ-Tocopherol**	Vitamin E. 7,8-Dimethyltocol. $C_{28}H_{48}O_2$.	416.70	pa ye visc oil $[\alpha]_D^{20}$ −2.4 (al) λ 298 (3.58)	−3– −2	$200–10^{0.1}$	....	....	i	∞	∞	∞	...	chl ∞	**H22**, 260
t272	**δ-Tocopherol**	8-Methyltocol. $C_{27}H_{46}O_2$	402.67	pa ye visc oil $[\alpha]_{546}^{25}$ +3.4 (al, c = 15.5) λ 298 (3.54)	....	$150^{0.001}$	....	....	i	v	v	v	...	chl v	**J1959**, 3374
t273	**α-Tocopherol-quinone**	α-Tocoquinone	446.72	ye oil λ 259, 269	....	$120^{0.002}$	....	....	...	...	s	...	...	peth s MeOH δ, s^h	**Am 73**, 5148
—	**Tolan**	*see* **Ethyne, diphenyl-***													
t274	**Tolbutamide**	Artosin. 1-Butyl-3(*p*-tolyl-sulfonyl)urea.	270.35	orh cr λ^{al} 228 (4.11), 256 (2.8), 263 (2.8), 274 (2.8)	128.5–9.5	....	1.245^{25}	....	δ^h	s	s	...	...	chl s	**C53**, 13084
—	**Tolidine**	*see* **Biphenyl, diamino(dimethyl)-**													
—	**α-Tolualdehyde**	*see* **Acetaldehyde, phenyl-**													
—	**Tolualdehyde**	*see* **Benzaldehyde, methyl-**													
Ω t275	**Toluene**	Methylbenzene*. Phenylmethane*.	92.15	λ^{al} 207 (3.97), 260 (2.48)	−95	110.6^{760} $14.5^{14.5}$	0.8669_4^{20}	1.4961^{20}	i	∞	∞	s	∞	CS_2, lig s	**B5**[3], 651
t276	**—,2-acetamido-3-bromo-**	6-Bromo-*o*-acetotoluidide. $C_9H_{10}BrNO$. *See* t275	228.10	nd (bz)	166	....	....	....	...	s	...	...	s^h	....	**B12**[2], 455
t277	**—,2-acetamido-4-bromo-**	5-Bromo-*o*-acetotoluidide. $C_9H_{10}BrNO$. *See* t275	228.10	nd (bz)	165.5	....	....	....	...	s	...	...	v^h	....	**B12**[2], 456

For explanations, symbols and abbreviations see beginning of table. For structural formulas see end of table.

No.	Name	Synonyms and Formula	Mol. wt.	Color, crystalline form, specific rotation and λ_{max} (log ε)	m.p. °C	b.p. °C	Density	n_D	Solubility: w	al	eth	ace	bz	other solvents	Ref.
	Toluene														
Ω t278	—,**2-acetamido-5-bromo-**	4-Bromo-*o*-acetotoluidide. $C_9H_{10}BrNO$. *See* t275	228.10	nd (dil al or lig)	159–60	...	...	...	s^h	v	...	...	...	lig s^h	**B12**[2], 456
Ω t279	—,**4-acetamido-2-bromo-**	3-Bromo-*p*-acetotoluidide. $C_9H_{10}BrNO$. *See* t275	228.10	nd (bz or dil al)	113	...	...	...	s	v	...	...	v^h	...	**B12**[1], 436
t280	—,**4-acetamido-3-bromo-**	2-Bromo-*p*-acetotoluidide. $C_9H_{10}BrNO$. *See* t275	228.10	nd (al)	118	...	...	...	...	v^h	...	...	...	xyl v^h	**B12**[2], 532
Ω t281	—,**α-amino-**	Benzylamine. $C_6H_5CH_2NH_2$	107.16	λ 255 (2.1), 262 (2.2), 270 (2.0)	...	185^{770} 90^{12}	0.9813^{20}_{4}	1.5401^{20}	∞	∞	∞	v	s	...	**B12**[2], 540
Ω t282	—,—,hydrochloride	Benzylamine hydrochloride. $C_6H_5CH_2NH_2 . HCl$	143.62	pl (al)	255–8 (248)	...	...	...	s	s^h	...	...	...	chl i	**B12**[2], 540
Ω t283	—,**2-amino-**	2-Methylaniline. *o*-Toluidine. C_7H_9N. *See* t275	107.16	$\lambda^{w, pH=10}$ 232.5 (3.88), 281.5 (3.16)	α: −23.7 (unst) β: −14.7 (st)	200.23^{760} 80.14^{10}	0.9984^{20}_{4}	1.5725^{20}	δ	∞	∞	...	...	CCl_4 ∞	**B12**[2], 429
Ω t285	—,—,hydrochloride	*o*-Toluidine hydrochloride. C_7H_9N. HCl. *See* t275	143.62	mcl pr (w^c), rh pym (w^h)	215	242.2^{760}	...	...	v	s	i	...	i	...	**B12**[2], 432
Ω t286	—,**3-amino-**	3-Methylaniline. *m*-Toluidine. C_7H_9N. *See* t275	107.16	...	−30.4	203.35^{760} 82.33^{10}	0.9889^{20}_{4}	1.5681^{20}	δ	∞	∞	∞	∞	CCl_4, hp ∞	**B12**[2], 463
t287	—,—,hydrochloride	*m*-Toluidine hydrochloride. C_7H_9N. HCl. *See* t275	143.62	lf (w)	228	250^{760}	...	...	v	v	...	...	...	...	**B12**, 856
Ω t288	—,**4-amino-**	4-Methylaniline. *p*-Toluidine. C_7H_9N. *See* t275	107.16	lf (w + 1) λ^{iso} 237 (3.98), 293 (3.28)	43.7 (45)	200.55^{760} 79.63^{10}	0.9619^{20}_{4}	1.5534^{45} 1.5636^{20}	δ	v	s	s	...	Py v	**B12**[2], 482
Ω t289	—,—,hydrochloride	*p*-Toluidine hydrochloride. C_7H_9N. HCl. *See* t275	143.62	mcl nd (aa-eth)	243	257.5^{760}	...	...	v	v	i	...	i	CS_2 i	**B12**[2], 487
—	—,**4-amino-*N*-benzylidene-**	*see* **Benzaldehyde,** imine, *N*(4-tolyl)-													
t291	—,**α-amino-3-bromo-**	*m*-Bromobenzylamine. C_7H_8BrN. *See* t275	186.07	...	...	244–5 84^{15}	...	...	...	...	v	...	...	ac s	**B12**[2], 575
t292	—,**α-amino-4-bromo-**	*p*-Bromobenzylamine. C_7H_8BrN. *See* t275	186.07	...	20	250–1 $126–7^{15}$	...	...	...	...	v	...	...	ac s	**B12**[2], 575
t293	—,**2-amino-3-bromo-**	6-Bromo-*o*-toluidine. C_7H_8BrN. *See* t275	186.07	...	...	130^{16} 107^{3}	...	...	...	...	v	...	...	ac s	**B12**[2], 455
t294	—,**2-amino-4-bromo-**	5-Bromo-*o*-toluidine. C_7H_8BrN. *See* t275	186.07	lf λ^w 293 (3.24)	33	253–7δd 139^{17}	...	...	...	s	v	...	...	ac s	**B12**[2], 456
Ω t295	—,**2-amino-5-bromo-**	4-Bromo-*o*-toluidine. C_7H_8BrN. *See* t275	186.07	cr (al)	59.5	240	...	...	δ	s	v	...	...	aa v	**B12**[2], 456
t296	—,**3-amino-4-bromo-**	6-Bromo-*m*-toluidine. C_7H_8BrN. *See* t275	186.07	pr	46	$129–30^{15}$	1.474^{25}_{25}	1.5990^{25}	...	v	s	...	...	ac s	**B12**[2], 474
t297	—,**3-amino-5-bromo-**	5-Bromo-*m*-toluidine. C_7H_8BrN. *See* t275	186.07	...	37–8	255–60 $150–1^{15}$	1.1422^{19}	...	...	s	...	...	...	ac s	**B12**[2], 474
t298	—,**4-amino-2-bromo-**	3-Bromo-*p*-toluidine. C_7H_8BrN. *See* t275	186.07	...	26	254–7	...	...	...	...	v	...	...	con ac s	**B12**[2], 532
Ω t299	—,**4-amino-3-bromo-**	2-Bromo-*p*-toluidine. C_7H_8BrN. *See* t275	186.07	lf λ^w 298 (3.33)	26	240 $120–2^{30}$	1.510^{20}	1.5999^{20}	i	s	s	...	...	...	**B12**[2], 532
t300	—,**5-amino-2-bromo-**	4-Bromo-*m*-toluidine. C_7H_8BrN. *See* t275	186.07	pl (50 % al), cr (al)	81	240	...	...	i	v	...	...	...	...	**B12**[2], 474
t301	—,**4-amino-5-bromo-2-nitro-**	$C_7H_7BrN_2O_2$. *See* t275	231.06	br to pa ye nd (al or aa)	121	...	...	...	...	...	...	...	...	...	**B12**[1], 441
Ω t302	—,**2-amino-4-chloro-**	5-Chloro-*o*-toluidine. C_7H_8ClN. *See* t275	141.61	...	26 (22)	237^{722} 140^{38}	...	...	...	s^h	...	...	...	...	**B12**[2], 453
Ω t303	—,**2-amino-5-chloro-**	4-Chloro-*o*-toluidine. C_7H_8ClN. *See* t275	141.61	lf (al)	29–30	241^{760}	...	...	...	s^h	...	...	...	...	**B12**[2], 455
Ω t304	—,**2-amino-6-chloro-**	3-Chloro-*o*-toluidine. C_7H_8ClN. *See* t275	141.61	...	0–2	245^{760} $96–9^{10}$	...	1.5880^{20}	s^h	s	i	...	i	...	**B12**[2], 456
t305	—,**3-amino-4-chloro-**	6-Chloro-*m*-toluidine. C_7H_8ClN. *See* t275	141.61	pl	29–30	228–30	...	...	...	s^h	...	...	...	...	**B12**[2], 473
Ω t306	—,**4-amino-2-chloro-**	3-Chloro-*p*-toluidine. C_7H_8ClN. *See* t275	141.61	...	26	$242–4^{760}$ (238) $112–3^{13}$	...	...	...	s	...	...	...	...	**B12**[2], 530
t307	—,**4-amino-3-chloro-**	2-Chloro-*p*-toluidine. C_7H_8ClN. *See* t275	141.61	...	7	219^{732}	1.151^{20}	1.5748^{22}	...	δ	...	...	δ	...	**B12**[2], 531
t308	—,**5-amino-2-chloro-**	4-Chloro-*m*-toluidine. C_7H_8ClN. *See* t275	141.61	nd (peth)	83–4	241	...	...	δ	v	...	s	s	os s lig δ	**B12**[1], 404
t309	—,**4-amino-2,6-dinitro-**	3,5-Dinitro-*p*-toluidine. $C_7H_7N_3O_4$. *See* t275	197.15	ye nd (w or aa)	171	...	...	...	δ^h	v	s	s	s	chl s aa s^h CS_2 δ	**B12**[2], 538
t310	—,**2-amino-3-hydroxy-**	2-Amino-*m*-cresol. C_7H_9NO. *See* t275	123.16	pl (w)	150	sub	...	...	s^h	...	...	...	...	...	**B13**[2], 324
t311	—,**2-amino-4-hydroxy-**	3-Amino-*p*-cresol. C_7H_9NO. *See* t275	123.16	cr (w or eth), lf (sub)	156–7 (144–5)	sub	...	...	s^h	...	s	...	...	...	**B13**[2], 337
Ω t312	—,**2-amino-5-hydroxy-**	4-Amino-*m*-cresol. C_7H_9NO. *See* t275	123.16	pr (dil al), cr (bz)	179	...	...	...	δ	v	v	...	...	...	**B13**[2], 330
t313	—,**2-amino-6-hydroxy-**	3-Amino-*o*-cresol. C_7H_9NO. *See* t275	123.16	nd (w)	129	...	...	...	δ	s	δ	...	...	...	**B13**, 579
t314	—,**3-amino-2-hydroxy-**	6-Amino-*o*-cresol. C_7H_9NO. *See* t275	123.16	pl (w)	89	...	...	...	s^h	s	s	s	s	MeOH, chl, CCl_4 s peth i	**B13**[1], 212

For explanations, symbols and abbreviations see beginning of table. For structural formulas see end of table.

No.	Name	Synonyms and Formula	Mol. wt.	Color, crystalline form, specific rotation and λ_{max} (log ε)	m.p. °C	b.p. °C	Density	n_D	Solubility w	al	eth	ace	bz	other solvents	Ref.
	Toluene														
t315	**—,3-amino-4-hydroxy-**	2-Amino-*p*-cresol. C_7H_9NO. *See* t275	123.16	cr (w), rh (bz), sc (eth), lf or nd (sub)	137	sub	...	...	δ	s	s	...	δ	chl s lig i	**B13**², 338
t316	**—,3-amino-5-hydroxy-**	5-Amino-*m*-cresol. C_7H_9NO. *See* t275	123.16	cr (dil MeOH)	139	245	...	...	δ	s^h	...	...	...	MeOH s^h	**C47**, 9300
t318	**—,4-amino-3-hydroxy-**	6-Amino-*m*-cresol. C_7H_9NO. *See* t275	123.16	nd (bz, dil al)	162d	...	...	...	s^h	s	s	s	...	MeOH s lig i	**B13**², 326
Ω t319	**—,5-amino-2-hydroxy-**	4-Amino-*o*-cresol. C_7H_9NO. *See* t275	123.16	nd or lf (bz)	175	sub	...	...	δ	s	s	...	δ s^h	...	**B13**², 319
t320	**—,α-amino-2-hydroxy-5-nitro-**	2-Hydroxy-5-nitrobenzylamine. $C_7H_8N_2O_3$. *See* t275	168.16	ye nd or lf (w or dil NH_3)	253d	...	...	...	s^h	s^h	...	...	...	...	**B13**, 587
t321	**—,α-amino-4-hydroxy-3-nitro-**	4-Hydroxy-3-nitrobenzylamine. $C_7H_8N_2O_3$. *See* t275	168.16	og red nd (w + 1)	225d (−w 115)	...	...	...	s^h	...	...	...	...	...	**B13**, 610
t322	**—,3-amino-2-hydroxy-5-nitro-**	$C_7H_8N_2O_3$. *See* t275	168.16	red-br nd (bz)	176	...	...	...	...	s^h	...	...	s^h	...	**B13**², 319
t323	**—,3-amino-4-hydroxy-5-nitro-**	$C_7H_8N_2O_3$. *See* t275	168.16	red-br (al)	119	...	...	...	...	s^h	...	...	...	...	**B13**², 345
t324	**—,5-amino-2-hydroxy-3-nitro-**	$C_7H_8N_2O_3$. *See* t275	168.16	br-red nd (al)	118	...	...	...	...	s^h	...	...	...	...	**B13**, 578
t325	**—,5-amino-4-hydroxy-2-nitro-**	$C_7H_8N_2O_3$. *See* t275	168.16	ye-og cr (al)	199–200d	...	...	...	s^h	s^h	...	...	...	dil HCl s	**B13**², 346
t326	**—,6-amino-3-hydroxy-2-nitro-**	$C_7H_8N_2O_3$. *See* t275	168.16	red-br nd (al)	201	...	...	...	...	s^h	...	...	...	dil ac v dil alk v	**B13**, 595
t327	**—,2-amino-4-iodo-**	5-Iodo-*o*-toluidine. C_7H_8IN. *See* t275	233.06	nd (aq al)	48–9	273d	...	...	δ	s	...	...	...	dil ac, aa s	**B12**¹, 391
Ω t328	**—,2-amino-5-iodo-**	4-Iodo-*o*-toluidine. C_7H_8IN. *See* t275	233.06	nd (dil al), pr (lig)	91–2 (87.2)	...	...	...	s^h	v	v	...	v	aa, lig s	**B12**², 457
t329	**—,3-amino-2-iodo-**	2-Iodo-*m*-toluidine. C_7H_8IN. *See* t275	233.06	pr	41–2	...	...	...	i	s	s	s	...	os s	**B12**², 475
t330	**—,3-amino-4-iodo-**	6-Iodo-*m*-toluidine. C_7H_8IN. *See* t275	233.06	nd (dil al), br in air	48 (38)	...	...	...	...	s	s	...	...	chl s	**B12**², 475
t331	**—,3-amino-5-iodo-**	5-Iodo-*m*-toluidine. C_7H_8IN. *See* t275	233.06	nd (peth)	78.5	...	...	...	...	s	s	s	...	oos s peth s^h	**B12**¹, 406
t332	**—,4-amino-2-iodo-**	3-Iodo-*p*-toluidine. C_7H_8IN. *See* t275	233.06	nd (dil al or peth)	39–40	...	...	...	...	s	s	s	...	aa v oos s	**B12**², 533
t333	**—,4-amino-3-iodo-**	2-Iodo-*p*-toluidine. C_7H_8IN. *See* t275	233.06	pr	40	d	...	...	...	v	v	v	v	os, chl, peth v	**B12**², 533
t334	**—,5-amino-2-iodo-**	4-Iodo-*m*-toluidine. C_7H_8IN. *See* t275	233.06	lf or pl (al or peth)	46	...	...	...	δ^h	s	s	...	v	aa, lig v	**B12**², 474
t335	**—,α-amino-3-methoxy-**	*m*-Methoxybenzylamine. $C_8H_{11}NO$. *See* t275	137.18	...	...	141^{30} $81^{0.8}$	...	...	i	s	s	...	...	...	**B13**², 335
t336	**—,α-amino-4-methoxy-**	*p*-Methoxybenzylamine. $C_8H_{11}NO$. *See* t275	137.18	...	...	$236–7^{760}$ (220–3) $133–4^{33}$	1.050^{15}	1.5462^{20}	s	s	s	...	...	...	**B13**², 347
t337	**—,2-amino-3-methoxy-**	6-Methyl-*o*-anisidine. $C_8H_{11}NO$. *See* t275	137.18	nd (w)	31	$119–21^{16}$	...	...	s^h	v	...	...	...	os v	**B13**², 324
t338	**—,2-amino-4-methoxy-**	6-Methyl-*m*-anisidine. $C_8H_{11}NO$. *See* t275	137.18	nd (w)	47	253 140^{20}	...	...	s^h	...	s	...	...	...	**B13**², 337
Ω t339	**—,2-amino-5-methoxy-**	2-Methyl-*p*-anisidine. $C_8H_{11}NO$. *See* t275	137.18	cr (lig)	29–30	248–9 $146–7^{23}$	...	1.5647^{20}	...	s	...	...	...	lig s^h	**B13**², 330
Ω t340	**—,3-amino-4-methoxy-**	Cresidine. 5-Methyl-*o*-anisidine. $C_8H_{11}NO$. *See* t275	137.18	nd or lf (al, lig or peth)	93–4 (52–4)	235^{760}	...	...	δ^h	s	s	...	s	peth s^h	**B13**², 338
t341	**—,4-amino-2-methoxy-**	4-Methyl-*m*-anisidine. $C_8H_{11}NO$. *See* t275	137.18	...	58	250–2	...	...	δ	v	s	...	v	lig v	**B13**¹, 213
t342	**—,4-amino-3-methoxy-**	4-Methyl-*o*-anisidine. $C_8H_{11}NO$. *See* t275	137.18	pa ye	...	237–9 $179–80^{46}$	...	...	...	s	s	s	...	oos s	**B13**², 326
t343	**—,5-amino-2-methoxy-**	3-Methyl-*p*-anisidine. $C_8H_{11}NO$. *See* t275	137.18	cr (dil al)	59–60	...	...	...	δ s^h	s	s	s	s	...	**B13**², 320
Ω t344	**—,2-amino-3-nitro-**	6-Nitro-*o*-toluidine. $C_7H_8N_2O_2$. *See* t275	152.16	og-ye pr (dil al) λ^{al} 208 (3.62), 283.5 (3.65), 330 (4.17)	97	...	1.1900_4^{100}	...	δ	s	s	...	s	chl s	**B12**², 458
Ω t345	**—,2-amino-4-nitro-**	5-Nitro-*o*-toluidine. $C_7H_8N_2O_2$. *See* t275	152.16	ye mcl pr (al) λ^{w} 385 (3.78)	107–8	...	...	...	δ	s	s	s	s	chl s	**B12**², 459
Ω t346	**—,2-amino-5-nitro-**	4-Nitro-*o*-toluidine. $C_7H_8N_2O_2$. *See* t275	152.16	ye mcl pr or nd (w, al or lig) λ^{al} 231 (4.08), 253 (4.08), 288.5 (3.69), 373 (3.23)	134–5 (129)	...	1.1586_4^{140}	...	δ^h	s	...	...	s	aa s	**B12**², 459
Ω t347	**—,2-amino-6-nitro-**	3-Nitro-*o*-toluidine. $C_7H_8N_2O_2$. *See* t275	152.16	ye rh nd (w), ye lf (al) λ^{al} 235 (4.20), 352 (3.14)	92 (97)	305d	...	...	δ^h	s	s	...	s	chl s	**B12**², 460

For explanations, symbols and abbreviations see beginning of table. For structural formulas see end of table.

No.	Name	Synonyms and Formula	Mol. wt.	Color, crystalline form, specific rotation and λ_{max} (log ε)	m.p. °C	b.p. °C	Density	n_D	Solubility: w	al	eth	ace	bz	other solvents	Ref.
	Toluene														
t348	**—,3-amino-2-nitro-**	2-Nitro-*m*-toluidine. $C_7H_8N_2O_2$. *See* t275	152.16	ye-og pr or nd (bz-peth) λ^{al} 282.5 (3.53), 404 (3.50)	108	...	...	...	δ	v	s	...	...	...	**B12²**, 476
t349	**—,3-amino-4-nitro-**	6-Nitro-*m*-toluidine. $C_7H_8N_2O_2$. *See* t275	152.16	ye lf (w), pl (dil al) λ^{al} 233 (4.11), 373.5 (3.87)	112	...	...	...	s^h	s	s	...	s	chl s	**B12²**, 476
t350	**—,3-amino-5-nitro-**	5-Nitro-*m*-toluidine. $C_7H_8N_2O_2$. *See* t275	152.16	ye-red or red-br nd (al) λ^{al} 235 (4.18), 287 (3.59), 375 (3.07)	98	...	...	...	δ	s	v	...	s	...	**B12²**, 476
Ω t352	**—,4-amino-2-nitro-**	3-Nitro-*p*-toluidine. $C_7H_8N_2O_2$. *See* t275	152.16	ye nd (w) λ^{al} 230 (4.30), 280 (3.80), 417.5 (3.76)	78–9	...	...	...	δ s^h	s^h	s	...	s	CS_2 δ	**B12²**, 534
Ω t353	**—,4-amino-3-nitro-**	2-Nitro-*p*-toluidine. $C_7H_8N_2O_2$. *See* t275	152.16	red lf (dil al), mcl pr (al) $\lambda^{dil\ MeOH}$ 227.5 (4.26), 287 (3.75), 428 (3.63)	117	...	1.164_4^{121}	...	δ^h	s	...	...	...	...	**B12²**, 535
Ω t354	**—,5-amino-2-nitro-**	4-Nitro-*m*-toluidine. $C_7H_8N_2O_2$. *See* t275	152.16	lt ye nd (w or dil al) λ^{al} 233 (3.87), 373.5 (4.12)	138 (134)	...	...	...	δ s^h	s	s	...	...	ac v	**B12²**, 476
Ω t355	**—,3-amino-α,α,α-trifluoro-**	*m*-Aminobenzotrifluoride. $C_7H_6F_3N$. *See* t275	161.13	λ^{al} 296 (3.45)	...	187.5^{764} $74–5^{10}$	...	1.4787^{20}	δ	s	s	...	...	...	**B12²**, 473
t356	**—,α-azido-**	Benzyl azide. $C_6H_5CH_2N_3$	133.15	...	...	108^{23} 74^{11}	1.0655^{25}	1.5341^{25}	i	∞	∞	...	...	...	**B5³**, 773
Ω t357	**—,α-bromo-**	Benzyl bromide. $C_6H_5CH_2Br$	171.04	pr λ^{iso} 224.5 (3.87)	−3–−1	201^{760} 114^{15}	1.4380_0^{22}	1.5752^{20}	i	∞	∞	...	...	...	**B5³**, 709
Ω t358	**—,2-bromo-**	*o*-Tolyl bromide. C_7H_7Br. *See* t275	171.04	λ^{chl} 260 sh (2.6), 272 (2.7), 280 (2.6)	−27.73 (−26)	181.7^{760} 59.1^{10}	1.4232_4^{20}	1.5565^{20}	i	v	v	...	v	CCl_4 ∞	**B5³**, 704
Ω t359	**—,3-bromo-**	*m*-Tolyl bromide. C_7H_7Br. *See* t275	171.04	λ^{hx} 267 (2.59), 276 (2.53)	−39.8	183.7	1.4099_4^{20}	1.5510^{20}	i	s	∞	s	...	chl s	**B5³**, 706
Ω t360	**—,4-bromo-**	*p*-Tolyl bromide. C_7H_7Br. *See* t275	171.04	cr (al) λ^{al} 220 (4.02), 262 (2.59), 269 (2.68), 277 (2.59)	28.5 (24.8)	184.35^{760} 61.9^{10}	1.3995_4^{20}	1.5477^{20}	i	s	s	s	s	chl s	**B5³**, 707
t361	**—,α-bromo-2-chloro-**	*o*-Chlorobenzyl bromide. C_7H_6BrCl. *See* t275	205.49	...	...	120^{10}	...	...	d^h	...	...	...	s	chl s	**B5³**, 714
t362	**—,2-bromo-α-chloro-**	*o*-Bromobenzyl chloride. C_7H_6BrCl. *See* t275	205.49	...	...	$124–6^{20}$	...	...	i	v	v	...	...	...	**B5³**, 713
t363	**—,3-bromo-α-chloro-**	*m*-Bromobenzyl chloride. C_7H_6BrCl. *See* t275	205.49	...	22–3	119^{18}	...	...	d	s	...	...	...	...	**B5³**, 714
Ω t364	**—,4-bromo-α-chloro-**	*p*-Bromobenzyl chloride. C_7H_6BrCl. *See* t275	205.49	nd (al or peth)	50	236^{760} $110–1^{9}$	...	...	i	v^h	v	...	...	peth s^h	**B5³**, 714
t365	**—,5-bromo-α,2-dihydroxy-**	5-Bromosaligenin. Bromosalisol. $C_7H_7BrO_2$. *See* t275	203.04	lf (bz)	113 (109)	...	...	...	δ v^h	v	v	...	s	AcOEt v chl s	**B6²**, 879
t366	**—,2-bromo-α-hydroxy-**	*o*-Bromobenzyl alcohol. C_7H_7BrO. *See* t275	187.04	nd (lig)	80	...	...	...	s^h	v	v	...	...	lig s^h	**B6²**, 423
t367	**—,2-bromo-4-hydroxy-**	3-Bromo-*p*-cresol. C_7H_7BrO. *See* t275	187.04	nd (peth)	56	245–7	...	...	δ	v	s	s	v	MeOH v peth s^h	**B6³**, 1377
t368	**—,2-bromo-5-hydroxy-**	4-Bromo-*m*-cresol. C_7H_7BrO. *See* t275	187.04	nd (peth or w)	63.5	$137–43^{16}$	...	...	δ	...	s	...	...	Py v peth, lig δ	**B6³**, 1321
t369	**—,2-bromo-6-hydroxy-**	3-Bromo-*o*-cresol. C_7H_7BrO. *See* t275	187.04	nd (peth)	95	$55–7^{4}$	...	...	δ	s	s	s	s	peth s^h	**B6**, 360
Ω t370	**—,3-bromo-4-hydroxy-**	2-Bromo-*p*-cresol. C_7H_7BrO. *See* t275	187.04	nd (peth)	56–7	213–4 (218–9)	1.5468_{25}^{25}	1.5772^{20}	δ	s	...	...	s	...	**B6³**, 1378
t371	**—,3-bromo-5-hydroxy-**	5-Bromo-*m*-cresol. C_7H_7BrO. *See* t275	187.04	nd (w)	56–7	$161–2^{28}$	...	...	δ	s	s	...	...	...	**B6²**, 357
t372	**—,4-bromo-α-hydroxy-**	*p*-Bromobenzyl alcohol. C_7H_7BrO. *See* t275	187.04	nd (lig)	77	...	...	...	i	v	v	...	v	CS_2 v	**B6³**, 1560
t373	**—,4-bromo-2-hydroxy-**	5-Bromo-*o*-cresol. C_7H_7BrO. *See* t275	187.04	nd (lig or peth)	80	...	...	...	...	s	s	s	...	oos s lig δ, s^h	**B6²**, 333
t374	**—,4-bromo-3-hydroxy-**	6-Bromo-*m*-cresol. C_7H_7BrO. *See* t275	187.04	cr (peth)	38	$206–8^{731}$ $81–2^{4}$	...	...	...	v	s	s	...	os s	**B6³**, 1320
t375	**—,5-bromo-2-hydroxy-**	4-Bromo-*o*-cresol. C_7H_7BrO. *See* t275	187.04	nd (al or peth)	64	235 $137–43^{18}$ sub	...	...	...	v	s	s	...	os s	**B6³**, 1269
Ω t376	**—,α-bromo-2-nitro-**	*o*-Nitrobenzyl bromide. $C_7H_6BrNO_2$. *See* t275	216.04	pl (dil al)	46–7	...	...	...	i	s	s	...	s	lig s	**B5³**, 752
Ω t377	**—,α-bromo-3-nitro-**	*m*-Nitrobenzyl bromide. $C_7H_6BrNO_2$. *See* t275	216.04	nd or pl (al)	58–9	$153.5–4.5^{8}$	...	...	i	s	...	...	...	...	**B5³**, 752

For explanations, symbols and abbreviations see beginning of table. For structural formulas see end of table.

No.	Name	Synonyms and Formula	Mol. wt.	Color, crystalline form, specific rotation and λ_{max} (log ε)	m.p. °C	b.p. °C	Density	n_D	Solubility w	al	eth	ace	bz	other solvents	Ref.
	Toluene														
Ω t378	**—,α-bromo-4-nitro-**	*p*-Nitrobenzyl bromide. $C_7H_6BrNO_2$. *See* t275	216.04	nd (al) λ^{al} 270.5 (4.08)	99–100	....	....	....	δ	v	v	...	...	aa s	**B5**³, 753
t379	**—,2-bromo-3-nitro-**	$C_7H_6BrNO_2$. *See* t275......	216.04	ye pr (al)	41–2	157^{22} $135–6^{8}$	....	....	i	s	...	...	...	....	**B5**³, 752
Ω t380	**—,2-bromo-4-nitro-**	$C_7H_6BrNO_2$. *See* t275......	216.04	nd (al)	78	$150–1^{20}$	....	....	i	δ v^h	s	...	...	CS_2 s	**B5**³, 753
t381	**—,2-bromo-5-nitro-**	$C_7H_6BrNO_2$. *See* t275......	216.04	cr (al)	78	$140–3^{17}$	....	....	...	δ v^h	s	...	...	CS_2 s	**B5**³, 752
t382	**—,2-bromo-6-nitro-**	$C_7H_6BrNO_2$. *See* t275......	216.04	pa ye nd (dil al)	42	143^{22}	....	....	i	s v^h	...	...	...	....	**B5**³, 751
t383	**—,3-bromo-2-nitro-**	$C_7H_6BrNO_2$. *See* t275......	216.04	pa ye nd	28	$129–30^{10}$	....	....	i	s	s	...	...	....	**B5**³, 751
t384	**—,3-bromo-4-nitro-**	$C_7H_6BrNO_2$. *See* t275......	216.04	pa ye pr or nd (MeOH)	37	$154–5^{20}$	....	....	i	s	s	...	...	....	**B5**³, 753
t385	**—,3-bromo-5-nitro-**	$C_7H_6BrNO_2$. *See* t275......	216.04	pa ye nd or pr (MeOH)	84	$269–70^{760}$	....	....	i	s	s	...	...	MeOH δ	**B5**³, 752
t386	**—,4-bromo-2-nitro-**	$C_7H_6BrNO_2$. *See* t275......	216.04	pa ye nd (dil al)	47	$256–7^{760}$ 130^{12}	....	....	i	s	s	...	...	con sulf s	**B5**³, 751
t387	**—,4-bromo-3-nitro-**	$C_7H_6BrNO_2$. *See* t275......	216.04	pa ye nd (MeOH)	35	....	....	1.5682^{50}	i	...	...	...	...	con sulf s MeOH s^h	**B5**³, 752
t388	**—,5-bromo-2-nitro-**	$C_7H_6BrNO_2$. *See* t275......	216.04	cr (al)	56	267 143^{10}	....	....	i	δ s^h	...	...	...	....	**B5**³, 751
t389	**—,3-butoxy-**......	Butyl *m*-tolyl ether. $C_{11}H_{16}O$. *See* t275	164.25	....	....	229.2^{760}	0.9407^{0}_{0}	1.4970^{20}	...	...	s	...	...	....	**B6**³, 1300
t390	**—,4-butoxy-**......	Butyl *p*-tolyl ether. $C_{11}H_{16}O$. *See* t275	164.25	λ^{al} 218 (3.83), 254 (2.31), 260 (2.39), 266 (2.34)	....	229.5^{760} 88^{3}	0.9232^{25}_{25}	1.4970^{20}	...	...	s	...	...	....	**B6**³, 1354
Ω t391	**—,α-chloro-**......	Benzyl chloride. $C_6H_5CH_2Cl$	126.59	λ^{al} 217 (3.85)	−39	179.3^{760} 66^{11}	1.1002^{20}_{20}	1.5391^{20}	i d^h	∞	∞	...	...	chl ∞	**B5**³, 685
Ω t392	**—,2-chloro-**......	*o*-Tolyl chloride. C_7H_7Cl. *See* t275	126.59	λ^{cy} 265 (2.48)	−35.1	159.15^{760} 42.6^{10}	1.0825^{20}_{4}	1.5268^{20}	i	s	∞	∞	s	chl, CCl_4, hp ∞	**B5**³, 680
Ω t393	**—,3-chloro-**......	*m*-Tolyl chloride. C_7H_7Cl. *See* t275	126.59	λ^{iso} 254 (2.13), 261 (2.33), 268 (2.47), 275 (2.45)	−47.8	162^{760}	1.0722^{20}_{4}	1.5214^{19}	i	s	∞	...	s	chl s	**B5**³, 682
Ω t394	**—,4-chloro-**......	*p*-Tolyl chloride. C_7H_7Cl. *See* t275	126.59	λ^{iso} 264 (2.54), 270 (2.68), 277.5 (2.70)	7.5	162^{760} 44^{10}	1.0697^{20}_{4}	1.5150^{20}	i	s	∞	...	...	chl, aa s	**B5**³, 683
t395	**—,α-chloro-α,α-difluoro-**	Benzodifluorochloride. $C_6H_5CClF_2$	162.57	....	−49.8	140.6^{760} $79–81^{100}$	1.2509^{20}	1.4648^{20}	i	s	s	...	...	....	**B5**³, 692
t397	**—,2-chloro-α-hydroxy-**	*o*-Chlorobenzyl alcohol. C_7H_7ClO. *See* t275	142.59	lf or nd (dil al)	74	230 $100–5^{28}$	....	....	δ	v	v	...	...	lig v^h	**B6**³, 1554
t398	**—,2-chloro-3-hydroxy-**	2-Chloro-*m*-cresol. C_7H_7ClO. *See* t275	142.59	pr (peth)	55–6	196^{760} $53–7^{4}$	....	....	δ	...	...	...	...	aa v peth s^h	**B6**³, 1315
t399	**—,2-chloro-4-hydroxy-**	3-Chloro-*p*-cresol. C_7H_7ClO. *See* t275	142.59	nd (al)	55–6	228^{760}	....	....	s^h	s	s	...	s	aa v	**B6**³, 1374
Ω t400	**—,2-chloro-5-hydroxy-**	4-Chloro-*m*-cresol. C_7H_7ClO. *See* t275	142.59	nd (peth)	66–8	235^{760}	....	....	δ	s	s	...	...	peth s^h	**B6**³, 1315
t401	**—,2-chloro-6-hydroxy-**	3-Chloro-*o*-cresol. C_7H_7ClO. *See* t275	142.59	lo nd (w)	86	225	....	....	δ s^h	s	s	...	s	....	**B6**³, 1267
Ω t402	**—,3-chloro-2-hydroxy-**	6-Chloro-*o*-cresol. C_7H_7ClO. *See* t275	142.59	λ^{peth} 275, 282	....	$188–9^{740}$ $80–1^{20}$	....	1.5449^{20}	...	...	s	...	...	....	**B6**³, 1263
t403	**—,3-chloro-4-hydroxy-**	2-Chloro-*p*-cresol. C_7H_7ClO. *See* t275	142.59	....	....	$195–6^{760}$	1.1785^{25}_{4}	1.5200^{27}	δ	s	s	...	s	aa s	**B6**³, 1374
t404	**—,4-chloro-α-hydroxy-**	*p*-Chlorobenzyl chloride. C_7H_7ClO. *See* t275	142.59	nd (w), pr (bz or bz-lig), cr (al) λ^{al} 266 (2.42)	75	235	....	....	δ^h	s	s	...	s^h	lig δ	**B6**³, 1555
t405	**—,4-chloro-2-hydroxy-**	5-Chloro-*o*-cresol. C_7H_7ClO. *See* t275	142.59	nd (peth or gasoline)	73–4	....	....	....	...	v	...	...	v	alk, aa v	**B6**³, 1263
t406	**—,4-chloro-3-hydroxy-**	6-Chloro-*m*-cresol. C_7H_7ClO. *See* t275	142.59	pr (peth)	45–6	196	1.215^{15}	....	s	...	...	...	...	peth s^h	**B6**³, 1315
Ω t407	**—,5-chloro-2-hydroxy-**	4-Chloro-*o*-cresol. C_7H_7ClO. *See* t275	142.59	nd (peth) λ^{peth} 286, 292	51	223	....	....	δ	...	...	...	...	peth s^h	**B6**³, 1264
t408	**—,α-chloro-4-hydroxy-3-nitro-**	$C_7H_6ClNO_3$. *See* t275......	187.58	ye nd (bz, lig or al), lf (peth)	75	....	....	....	...	s^h	...	...	s^h	aa, lig s^h	**B6**, 413
Ω t409	**—,α-chloro-4-methoxy-**	*p*-Methoxybenzyl chloride. C_8H_9ClO. *See* t275	156.61	....	....	$116–20^{15}$ $83–4^{2}$	1.159^{20}	1.553	...	...	...	...	...	....	**B6**³, 1375
Ω t410	**—,α-chloro-2-nitro-**	*o*-Nitrobenzyl chloride. $C_7H_6ClNO_2$. *See* t275	171.58	cr (lig)	50–2	....	....	1.5557^{62}	i	s	s	v	v	AcOEt v aa s	**B5**³, 745
t411	**—,α-chloro-3-nitro-**	*m*-Nitrobenzyl chloride. $C_7H_6ClNO_2$. *See* t275	171.58	pa ye nd (lig)	45–7	175– $83^{30–5}$	....	1.5577^{62}	i	s	s	v	s	AcOEt v aa s	**B5**³, 746
Ω t412	**—,α-chloro-4-nitro-**	*p*-Nitrobenzyl chloride. $C_7H_6ClNO_2$. *See* t275	171.58	pl or nd (al) λ^{al} 265 (4.08)	71	....	....	1.5647^{62}	i	s	s	v	v	AcOEt v MeOH s	**B5**³, 747
Ω t413	**—,2-chloro-4-nitro-**	$C_7H_6ClNO_2$. *See* t275......	171.58	nd (al)	68	260^{760}	....	1.5470^{69}	δ^h	s	s	...	...	aa s	**B5**³, 747
Ω t414	**—,2-chloro-6-nitro-**	$C_7H_6ClNO_2$. *See* t275......	171.58	nd (dil al)	37–40	238^{760}	....	1.5377^{69}	i	s	...	...	...		**B5**³, 745
Ω t415	**—,3-chloro-4-nitro-**	$C_7H_6ClNO_2$. *See* t275......	171.58	pa ye nd	24	146^{19}	....	....	i	...	...	...	...	....	**B5**², 253

For explanations, symbols and abbreviations see beginning of table. For structural formulas see end of table.

No.	Name	Synonyms and Formula	Mol. wt.	Color, crystalline form, specific rotation and λ_{max} (log ε)	m.p. °C	b.p. °C	Density	n_D	Solubility w	al	eth	ace	bz	other solvents	Ref.
	Toluene														
t416	**—,3-chloro-5-nitro-**	$C_7H_6ClNO_2$. *See* t275	171.58	ye nd (al)	61	...	...	1.5404^{69}	i	s^h	...	...	...	...	**B5**^{3}, 746
Ω t417	**—,4-chloro-2-nitro-**	$C_7H_6ClNO_2$. *See* t275	171.58	mcl nd $\lambda^{dil\ MeOH}$ 213 (4.26), 262 (3.68), 315 (3.23)	38	240^{720} 115.5^{11}	1.2559^{80}	...	i	s^h	s	...	...	...	**B5**^{2}, 251
Ω t418	**—,4-chloro-3-nitro-**	$C_7H_6ClNO_2$. *See* t275	171.58	...	7	260^{745} 118^{11}	...	1.5572^{20}	i	...	...	...	...	...	**B5**^{2}, 252
t419	**—,5-chloro-2-nitro-**	$C_7H_6ClNO_2$. *See* t275	171.58	ye	24.9	...	...	1.5496^{65}	i	...	...	...	...	...	**B5**^{1}, 162
t420	**—,2,3-diamino-**	$C_7H_{10}N_2$. *See* t275	122.17	cr	63–4	255	...	...	s	s	s	...	...	oos s	**B13**^{2}, 60
Ω t421	**—,2,4-diamino-**	$C_7H_{10}N_2$. *See* t275	122.17	nd (w), or (al)	99	292 $148–50^{8}$	...	...	v^h	v	v	...	v^h	...	**B13**^{2}, 60
t422	**—,2,5-diamino-**	$C_7H_{10}N_2$. *See* t275	122.17	pl (bz)	64	273–4	...	...	s	s	s	...	δ s^h	aa δ, v^h	**B13**^{2}, 62
Ω t423	**—,2,6-diamino-**	$C_7H_{10}N_2$. *See* t275	122.17	pr (bz or w)	106	...	...	...	s	s	...	...	s^h	...	**B13**^{2}, 64
Ω t424	**—,3,4-diamino-**	$C_7H_{10}N_2$. *See* t275	122.17	lf (lig)	89–90	265 sub	...	...	v	...	...	...	...	lig s^h	**B13**^{2}, 64
t425	**—,3,5-diamino-**	$C_7H_{10}N_2$. *See* t275	122.17	...	<0	283–5	...	...	v	s	s	...	...	...	**B13**, 164
t426	**—,—,hydrochloride**	$C_7H_{10}N_2 . 2HCl$. *See* t275	195.09	nd (dil al)	255–60d	...	...	...	v	...	i	...	i	...	**B13**, 164
t427	**—,α,α-dibromo-**	Benzal bromide. Benzylidene bromide. $C_6H_5CHBr_2$	249.94	λ^{iso} 276 sh (2.85)	...	156^{23}	1.51^{15}	1.6147^{20}	i	∞	∞	...	...	...	**B5**^{3}, 717
t428	**—,α,2-dibromo-**	*o*-Bromobenzyl bromide. $C_7H_6Br_2$. *See* t275	249.94	cr (al, lig)	31	129^{19}	...	...	d^h	∞	∞	...	...	CS_2, aa ∞	**B5**^{3}, 717
t429	**—,α,3-dibromo-**	*m*-Bromobenzyl bromide. $C_7H_6Br_2$. *See* t275	249.94	nd or lf	41	...	...	...	d^h	s	v	...	...	CS_2, aa v	**B5**^{3}, 717
Ω t430	**—,α,4-dibromo-**	*p*-Bromobenzyl bromide. $C_7H_6Br_2$. *See* t275	249.94	nd (MeOH or al)	63	...	...	...	δ d^h	s^h	v	...	s	CS_2, aa v	**B5**^{3}, 717
t431	**—,2,5-dibromo-**	$C_7H_6Br_2$. *See* t275	249.94	...	fr 5.62	236^{760} $135–6^{35}$	1.8127^{19}	1.5982^{18}	i	...	...	...	...	...	**B5**^{3}, 716
t432	**—,3,5-dibromo-**	$C_7H_6Br_2$. *See* t275	249.94	nd	39	246^{760}	...	...	i	...	...	...	...	...	**B5**^{3}, 717
t433	**—,2,4-dibromo-6-hydroxy-**	3,5-Dibromo-*o*-cresol. $C_7H_6Br_2O$. *See* t275	265.94	nd (peth)	98–101	$283–7^{758}$	...	...	...	...	...	...	...	peth s^h	**B6**^{3}, 1271
t434	**—,3,5-dibromo-2-hydroxy-**	4,6-Dibromo-*o*-cresol. $C_7H_6Br_2O$. *See* t275	265.94	nd (peth or 50% al)	58	$263–6^{745}$d	...	...	δ	s	s	...	s	oos, aa s	**B6**^{2}, 334
t435	**—,3,6-dibromo-2-hydroxy-**	3,6-Dibromo-*o*-cresol. $C_7H_6Br_2O$. *See* t275	265.94	cr	38	255–60	...	...	...	...	...	...	...	os v	**B6**^{1}, 176
Ω t436	**—,α,α-dibromo-4-nitro-**	*p*-Nitrobenzylidene bromide. $C_7H_5Br_2NO_2$. *See* t275	294.94	nd (al) λ^{hx} 263.5 (4.23)	84 (82)	...	...	...	i	v	v	...	...	...	**B5**^{3}, 755
Ω t437	**—,α,α-dichloro-**	Benzal chloride. Benzylidene chloride. $C_6H_5CHCl_2$	161.03	λ^{cy} 260 (2.4), 265 (2.5), 272 (2.3)	−16.4	205.2^{760}	1.2557^{14}	1.5502^{20}	i	s	s	...	...	...	**B5**^{3}, 696
Ω t438	**—,α,2-dichloro-**	*o*-Chlorobenzyl chloride. $C_7H_6Cl_2$. *See* t275	161.03	...	−17	217^{760} $94–5^{10}$	1.2699^{0}_{4}	1.5530^{20}	...	...	...	...	...	...	**B5**^{3}, 694
t439	**—,α,3-dichloro-**	*m*-Chlorobenzyl chloride. $C_7H_6Cl_2$. *See* t275	161.03	...	...	$215–6^{753}$ $110–1^{25}$	1.2695^{15}_{4}	1.5554^{20}	i d^h	v	...	...	...	...	**B5**^{3}, 695
Ω t440	**—,α,4-dichloro-**	*p*-Chlorobenzyl chloride. $C_7H_6Cl_2$. *See* t275	161.03	nd (dil al)	31	222 (214) 117^{20}	...	...	i	δ s^h	v	...	v	aa v CS_2 s	**B5**^{3}, 695
t441	**—,2,3-dichloro-**	$C_7H_6Cl_2$. *See* t275	161.03	...	...	$207–8^{760}$ $61–2^{3}$	...	1.5511^{20}	i	...	...	...	v	...	**B5**^{3}, 693
Ω t442	**—,2,4-dichloro-**	$C_7H_6Cl_2$. *See* t275	161.03	...	−13.5	$196–7^{760}$	1.2498^{20}_{20}	1.5511^{20}	i	...	...	...	...	...	**B5**^{3}, 693
t443	**—,2,5-dichloro-**	$C_7H_6Cl_2$. *See* t275	161.03	...	5	200^{770}	1.2535^{20}	1.5449^{20}	i	...	...	...	s	...	**B5**^{3}, 694
Ω t444	**—,2,6-dichloro-**	$C_7H_6Cl_2$. *See* t275	161.03	λ^{MeOH} 267 (2.26)	...	198^{760}	...	1.5507^{20}	i	...	...	...	...	chl s	**B5**^{3}, 694
Ω t445	**—,3,4-dichloro-**	$C_7H_6Cl_2$. *See* t275	161.03	...	−15.25	208.92^{760} 81.8^{10}	1.2564^{20}_{4}	1.5471^{20}	i	∞	∞	∞	∞	CCl_4, lig ∞	**B5**^{3}, 694
t446	**—,3,5-dichloro-**	$C_7H_6Cl_2$. *See* t275	161.03	...	26	$201–2^{760}$	...	...	i	...	...	...	...	...	**B5**^{3}, 694
t447	**—,α,α-dichloro-α-fluoro-**	Benzodichlorofluoride. $C_6H_5CCl_2F$	179.02	...	...	178–80	1.3138^{11}	1.5180^{11}	d	s^h	...	...	...	...	**B5**^{3}, 698
t448	**—,2,4-dichloro-3-hydroxy-**	2,6-Dichloro-*m*-cresol. $C_7H_6Cl_2O$. *See* t275	177.03	...	58–9	$235–6^{745}$ $75–80^{4}$	...	...	...	...	s	...	...	chl v	**B6**^{3}, 1319
Ω t449	**—,2,4-dichloro-5-hydroxy-**	4,6-Dichloro-*m*-cresol. $C_7H_6Cl_2O$. *See* t275	177.03	pr (peth)	72–4	235–6 110^{18}	...	1.572^{20}	...	...	...	...	...	chl v peth s^h	**Am 55,** 4216
t450	**—,2,6-dichloro-3-hydroxy-**	2,4-Dichloro-*m*-cresol. $C_7H_6Cl_2O$. *See* t275	177.03	pr (peth)	27	241–2.5	...	...	...	...	s	...	...	chl v peth s^h	**B6**^{3}, 1319
Ω t451	**—,3,5-dichloro-2-hydroxy-**	4,6-Dichloro-*o*-cresol. $C_7H_6Cl_2O$. *See* t275	177.03	nd (w or peth)	55	266.5 $73–8^{4}$	...	...	δ s^h	v	v	...	...	chl, CS_2 v peth s^h	**B6**^{3}, 1267
t452	**—,3,5-dichloro-4-hydroxy-**	2,6-Dichloro-*p*-cresol. $C_7H_6Cl_2O$. *See* t275	177.03	nd (lig)	39 (42)	$138–9^{28}$	...	...	δ	s	s	...	...	aa s lig s^h	**B6**^{3}, 1377
t453	**—,4,5-dichloro-2-hydroxy-**	4,5-Dichloro-*o*-cresol. $C_7H_6Cl_2O$. *See* t275	177.03	nd (peth)	101	...	...	...	δ	s	...	...	s	aa v peth δ	**B6**^{2}, 333
t454	**—,α,α-dichloro-3-nitro-**	*m*-Nitrobenzylidene chloride. $C_7H_5Cl_2NO_2$. *See* t275	206.03	mcl (al)	65	...	...	...	i	s^h	s	...	...	...	**B5**^{3}, 750
t455	**—,α,α-dichloro-4-nitro-**	*p*-Nitrobenzylidene chloride. $C_7H_5Cl_2NO_2$. *See* t275	206.03	pr (al) λ^{hx} 256 (4.16)	46	...	...	...	i	v	v	...	...	...	**B5**^{3}, 750
t456	**—,2,5-diethoxy-**	$C_{11}H_{16}O_2$. *See* t275	180.25	nd (lig)	24–5	$247–9^{760}$	1.0134^{15}	...	...	∞	∞	...	∞	chl ∞	**B6**^{3}, 4499
t457	**—,2(diethyl-amino)-**	*N,N*-Diethyl-*o*-toluidine. $C_{11}H_{17}N$. *See* t275	163.26	...	...	$208–9^{755}$	...	...	δ	s	s	...	...	...	**B12**^{2}, 436
t458	**—,4(diethyl-amino)-**	*N,N*-Diethyl-*p*-toluidine. $C_{11}H_{17}N$. *See* t275	163.26	...	...	229^{770}	0.9242^{16}	...	δ	∞	∞	...	...	...	**B12**^{2}, 492

For explanations, symbols and abbreviations see beginning of table. For structural formulas see end of table.

No.	Name	Synonyms and Formula	Mol. wt.	Color, crystalline form, specific rotation and λ_{max} (log ε)	m.p. °C	b.p. °C	Density	n_D	Solubility						Ref.
									w	al	eth	ace	bz	other solvents	
	Toluene														
t459	—,α,α-**difluoro-**....	Benzal fluoride. Benzylidene fluoride. $C_6H_5CHF_2$	128.12	λ^{iso} 239 (2.02), 261 (2.47), 284 (0.78)	....	139.9^{760}	1.1357^{20}	1.4577^{20}	i	s	...	...	...	....	**B5**^{2}, 224
Ω t460	—,**α,2-dihydroxy-**..	Saligenin. Salicyl alcohol. $C_7H_8O_2$. *See* t275	124.15	lf (bz), nd or pl (w, eth), pl (sub)	87	sub	1.1613^{25}	....	s v^h	s	s	...	s	chl v	**B6**^{3}, 4537
Ω t461	—,—,glucoside....	Salicin. Saligenin-β,*D*-glucoside.	286.29	rh nd or lf (w) $[\alpha]_D^{20}$ −62.6 (w, c = 3) $\lambda^{w, pH=4}$ 271 (3.0)	204.7–8.7	240d	1.434^{26}	....	s v^h	s	i	...	i	aa, Py s chl i	**B31**, 214
Ω t461[1]	—,**α,3-dihydroxy-**..	$C_7H_8O_2$. *See* t275.........	124.15	nd (bz), cr (CCl_4)	73	300^{760}d	1.161^{25}	....	v^h	v	v	...	...	chl δ	**B6**^{3}, 4545
Ω t462	—,**α,4-dihydroxy-**..	$C_7H_8O_2$. *See* t275.........	124.15	pr or nd (w) λ^w 289 (3.43)	124.5–5.5	252^{760}	....	....	v	v	v	...	δ	chl, peth i	**B6**^{3}, 4546
Ω t463	—,**2,3-dihydroxy-**..	Isohomocatechol. $C_7H_8O_2$. *See* t275	124.15	lf (bz) λ^{cy} 275 (3.26)	68 (47)	241^{760} 127^{12}	....	....	v	v	s	...	v	chl v	**B6**^{3}, 4492
t464	—,**2,4-dihydroxy-**..	Cresorcinol. $C_7H_8O_2$. *See* t275	124.15	cr (bz-peth) λ^{al} 283 (3.48)	105–7	267–70	....	....	s	s	s	...	δ	peth δ	**B6**^{3}, 4495
Ω t465	—,**2,5-dihydroxy-**..	Toluhydroquinone. 2-Methylhydroquinone. $C_7H_8O_2$. *See* t275	124.15	rh pl (bz)	126–7	163^{11} sub	....	....	v	v	v	...	δ	lig δ	**B6**^{3}, 4498
t466	—,—,diacetate....	2,5-Diacetoxytoluene. $C_{11}H_{12}O_4$. *See* t275	208.22	nd (aa), pr (lig) nd or pr (w)	52	....	....	....	...	v	v	...	δ	aa s lig s^h	**B6**^{3}, 4501
Ω t467	—,**3,4-dihydroxy-**..	4-Homopyrocatechol. $C_7H_8O_2$. *See* t275	124.15	lf (bz-lig), pr (bz) λ^{cy} 283 (3.45)	65	251^{766} $143–6^{26}$ sub	1.1287_4^{74}	1.5425^{74}	s	s	s	s	...	os s lig δ	**B6**^{3}, 4514
Ω t468	—,**3,5-dihydroxy-**..	Orcinol. $C_7H_8O_2$. *See* t275..	124.15	pr (w + 1), lf (chl) λ^{al} 278 (2.97)	107–8 (anh) 58 (hyd)	289–90 147^5	1.290^4	....	s	s	s	...	s	lig, peth δ	**B6**^{3}, 4531
Ω t469	—,**α,4-dihydroxy-3-methoxy-**	Vanillyl alcohol. $C_8H_{10}O_3$. *See* t275	154.17	pr (w), nd (bz) λ^{al} 230 (3.8), 280 (3.6)	115	d	....	....	s^h	s^h	s^h	...	s^h	....	**B6**^{3}, 6323
t470	—,**2,4-di-mercapto-**	Dithiocresorcinol. $C_7H_8S_2$. *See* t275	156.28	....	36–7	263^{760}	....	....	...	...	...	...	...	....	**B6**, 873
t471	—,**2,3-di-methoxy-**	Isohomoveratrol. $C_9H_{12}O_2$. *See* t275	152.20	....	....	$202–3^{760}$ (214) $92–3^{18}$	1.0335_4^{20}	1.5121^{25}	...	s	s	...	...	....	**B6**^{3}, 4493
Ω t472	—,**2,4-di-methoxy-**	$C_9H_{12}O_2$. *See* t275.........	152.20	....	....	211 110–20^{30}	....	....	δ	s	s	...	...	....	**B6**^{3}, 4496
Ω t473	—,**3,4-di-methoxy-**	Homoveratrol. $C_9H_{12}O_2$. *See* t275	152.20	pr (eth)	24	$219–21^{760}$ $122–4^{27}$	1.0509_4^{25}	1.5257^{25}	i	...	...	...	...	os v	**B6**^{3}, 4516
t474	—,**3,5-di-methoxy-**	$C_9H_{12}O_2$. *See* t275.........	152.20	....	....	244^{760} 102^8	1.0478^{15}	1.5234^{29}	i	v	v	...	v	CS_2, aa v	**B6**^{3}, 4533
Ω t475	—,**α(dimethyl-amino)-**	*N,N*-Dimethylbenzylamine. $C_6H_5CH_2N(CH_3)_2$	135.21	....	....	$180–2^{760}$ $73–4^{15}$	0.915_4^0	1.5011^{20}	δ^h	∞	∞	...	...	....	**B12**^{2}, 545
Ω t476	—,**2(dimethyl-amino)-**	*N,N*-Dimethyl-*o*-toluidine. $C_9H_{13}N$. *See* t275	135.21	λ^{iso} 248 (3.80)	−60	185.3^{760} $70–2^{15}$	0.9286_4^{20}	1.5153^{20}	δ	∞	∞	...	...	....	**B12**^{2}, 435
Ω t477	—,**3(dimethyl-amino)-**	*N,N*-Dimethyl-*m*-toluidine. $C_9H_{13}N$. *See* t275	135.21	λ^{cy} 253 (4.12), 298 (3.38)	....	212^{760}	0.9410_4^{20}	1.5492^{20}	...	∞	∞	...	...	....	**B12**^{2}, 466
Ω t478	—,**4(dimethyl-amino)-**	*N,N*-Dimethyl-*p*-toluidine. $C_9H_{13}N$. *See* t275	135.21	λ^{hx} 255 (5.0), 305 (3.8)	....	211^{760}	0.9366_4^{20}	1.5366^{20}	i	∞	∞	...	...	....	**B12**^{2}, 491
Ω t479	—,**2,4-dinitro-**.....	$C_7H_6N_2O_4$. *See* t275.......	182.14	ye nd or mcl pr (CS_2) $\lambda^{5\%\,al}$ 252 (4.15)	71	300 δd	1.3208^{71}	1.442 (1.756)	i	s	s	v	s	Py v to, chl, CS_2, AcOEt s	**B5**^{3}, 759
t480	—,**2,5-dinitro-**.....	$C_7H_6N_2O_4$. *See* t275.......	182.14	nd (al) $\lambda^{5\%\,al}$ 266.5 (4.06)	52.5	....	1.282^{111}	....	...	s	...	...	s	CS_2 v	**B5**^{3}, 760
Ω t481	—,**2,6-dinitro-**.....	$C_7H_6N_2O_4$. *See* t275.......	182.14	rh nd (al) $\lambda^{5\%\,al}$ 241 (3.95)	66	....	1.2833^{111}	1.479 (1.734)	...	s	...	...	...	....	**B5**^{3}, 761
Ω t482	—,**3,4-dinitro-**.....	$C_7H_6N_2O_4$. *See* t275.......	182.14	ye nd (CS_2) $\lambda^{5\%\,al}$ 219 (4.00)	58.3	....	1.2594^{111}	....	i	s	...	...	...	CS_2 s	**B5**^{3}, 761
t483	—,**3,5-dinitro-**.....	$C_7H_6N_2O_4$. *See* t275.......	182.14	ye rh nd (aa) $\lambda^{5\%\,al}$ 248 (4.08)	93	sub	1.2772^{111}	....	δ	s	s	...	s	CS_2, chl s	**B5**^{3}, 762
t484	—,**2,4-dinitro-6-hydroxy-**	3,5-Dinitro-*o*-cresol. Sinox. $C_7H_6N_2O_5$. *See* t275	198.14	ye pr (al) $\lambda^{0.1N\,HCl}$ 268 (4.10)	85.8	....	....	....	δ	s	s	s	...	peth δ	**C51**, 10414
Ω t484[1]	—,**3,5-dinitro-2-hydroxy-**	4,6-Dinitro-*o*-cresol. $C_7H_6N_2O_5$. *See* t275	198.14	ye pr or nd (al) $\lambda^{0.1N\,HCl}$ 268 (4.10)	86.5	....	....	....	δ	s	s	s	...	peth δ	**B6**^{3}, 1276
t485	—,**3,5-dinitro-4-hydroxy-**	2,6-Dinitro-*p*-cresol. $C_7H_6N_2O_5$. *See* t275	198.14	ye nd (eth or peth) λ^{al} 241 (4.02), 354 (3.72)	85	....	....	....	i	s	s	...	s	....	**B6**^{3}, 1390
—	—,**α-ethoxy-**......	*see* **Ether, benzyl ethyl**													

For explanations, symbols and abbreviations see beginning of table. For structural formulas see end of table.

No.	Name	Synonyms and Formula	Mol. wt.	Color, crystalline form, specific rotation and λ_{max} (log ε)	m.p. °C	b.p. °C	Density	n_D	Solubility w	al	eth	ace	bz	other solvents	Ref.
	Toluene														
t486	**—,2-ethoxy-**	Ethyl *o*-tolyl ether. *o*-Cresyl ethyl ether. $C_9H_{12}O$. *See* t275	136.20	λ^{hx} 260 sh (2.90), 272 (3.24), 278 (3.22)	...	184^{760} 70^{12}	0.9592_4^{13}	1.508^{13}	i	s	s	...	...	...	**B6**[3], 1246
t487	**—,3-ethoxy-**	Ethyl *m*-tolyl ether. *m*-Cresyl ethyl ether. $C_9H_{12}O$. *See* t275	136.20	...	...	192^{760}	0.949^{20}	1.513^{20}	i	s	s	...	...	...	**B6**[3], 1299
Ω t488	**—,4-ethoxy-**	Ethyl *p*-tolyl ether. *p*-Cresyl ethyl ether. $C_9H_{12}O$. *See* t275	136.20	...	...	$188–9^{760}$	0.9509_4^{18}	1.5058^{18}	i	s	s	...	...	...	**B6**[3], 1353
Ω t489	**—,2(ethylamino)-**	*N*-Ethyl-*o*-toluidine. $C_9H_{13}N$. *See* t275	135.21	λ^{al} 244 (3.86), 291 (3.33)	< −15	218^{760} 95.5^{10}	0.948_4^{25}	1.5456^{20}	...	s	s	...	...	...	**B12**[2], 435
Ω t490	**—,3(ethylamino)-**	*N*-Ethyl-*m*-toluidine. $C_9H_{13}N$. *See* t275	135.21	...	...	221 $111–2^{20}$	...	1.5451^{20}	...	s	s	...	...	...	**B12**[2], 466
Ω t491	**—,4(ethylamino)-**	*N*-Ethyl-*p*-toluidine. $C_9H_{13}N$. *See* t275	135.21	...	...	217	0.9391^{16}	...	...	s	s	...	...	...	**B12**[2], 492
t492	**—,α-fluoro-**	Benzyl fluoride. $C_6H_5CH_2F$	110.13	nd (fr) λ^{iso} 244 (2.27), 250 (2.31), 264 (2.23), 289 (1.27)	−35	139.8^{753} 40^{14}	1.0228_4^{25}	1.4892^{25}	d	...	...	...	...	...	**B5**[3], 678
Ω t493	**—,2-fluoro-**	*o*-Tolyl fluoride. C_7H_7F. *See* t275	110.13	λ^{cy} 206 (3.90), 262 (3.00), 268 (3.00)	−62	114 30^{26}	1.0041_4^{13}	1.4704^{20}	i	v	v	...	...	...	**B5**[3], 676
Ω t494	**—,3-fluoro-**	*m*-Tolyl fluoride. C_7H_7F. *See* t275	110.13	λ^{cy} 207 (3.88), 262 (2.93), 269 (2.98)	−87.7	116^{760}	0.9986^{20}	1.4691^{20}	i	v	v	...	...	...	**B5**[3], 676
Ω t495	**—,4-fluoro-**	*p*-Tolyl fluoride. C_7H_7F. *See* t275	110.13	λ^{cy} 206 (3.86), 264 (3.04), 273 (3.22)	−56.8	116.6^{760}	1.0007_4^{16}	1.4699^{20}	i	v	v	...	...	...	**B5**[3], 677
Ω t496	**—,α,α,2,3,4,5,6-heptachloro-**	Pentachlorobenzylidene chloride. C_7HCl_7. *See* t275	333.26	lf (al) λ^{al} 307 (2.95)	119.5	334 199^{13}	...	...	...	δ v^h	...	...	...	...	**B5**[1], 153
Ω t497	**—,α-hydroxy-**	Benzyl alcohol. $C_6H_5CH_2OH$	108.15	λ^{al} 243 (1.91), 258.5 (2.26), 268 (1.95)	−15.3	205.35^{760} 93^{10}	1.0419_4^{20}	1.5396^{20}	s	s	s	s	s	MeOH, chl s	**B6**[3], 1445
Ω t498	**—,2-hydroxy-**	*o*-Cresol. 2-Methylphenol*. C_7H_8O. *See* t275	108.15	λ^{w} 219 (3.71), 275 (3.22)	30.94	190.95^{760} 74.9^{10}	1.02734_4^{20}	1.5361^{20}	s	v	v	∞	∞	CCl_4 ∞ oos s	**B6**[3], 1233
Ω t499	**—,3-hydroxy-**	*m*-Cresol. 3-Methylphenol*. C_7H_8O. *See* t275	108.15	λ^{hx} 214 (3.79), 271 (3.20), 277 (3.27)	11.5	202.2^{760} 86^{10}	1.0336_4^{20}	1.5438^{20}	δ s^h	∞	∞	∞	∞	CCl_4 ∞ os s	**B6**[3], 1286
Ω t500	**—,4-hydroxy-**	*p*-Cresol. 4-Methylphenol*. C_7H_8O. *See* t275	108.15	pr λ^{cy} 280 (3.23)	34.8	201.9^{760} 85.7^{10}	1.0178_4^{20}	1.5312^{20} (suc)	δ s^h	∞	∞	∞	∞	CCl_4 ∞ os s	**B6**[3], 1341
t502	**—,α-hydroxyl-amino-**	*N*-Benzylhydroxylamine. $C_6H_5CH_2NHOH$	123.16	nd (peth or lig) λ^{al} 209 (3.86), 258 (2.31), 264 (2.21)	57	...	...	...	δ	s	...	...	...	lig s^h	**B15**[2], 17
t503	**—,2-hydroxyl-amino-**	*o*-Tolylhydroxylamine. C_7H_9NO. *See* t275	123.16	nd (eth-bz)	44	...	...	...	i	s	s	...	s	lig δ	**B15**[2], 14
t504	**—,3-hydroxyl-amino-**	*m*-Tolylhydroxylamine. C_7H_9NO. *See* t275	123.16	lf (bz-peth)	68.5	...	...	...	s^h	s	s	...	s^h	chl s lig δ	**B15**[2], 15
Ω t505	**—,4-hydroxyl-amino-**	*p*-Tolylhydroxylamine. C_7H_9NO. *See* t275	123.16	lf (bz)	98 (94)	115–20d	...	...	δ	s	s	...	s^h	chl s lig i	**B15**[2], 16
t506	**—,2-hydroxyl-amino-6-nitro-**	$C_7H_8N_2O_3$. *See* t275	168.18	col or ye (bz)	120–1	...	...	...	δ	v	v	...	δ s^h	peth δ	**B15**[2], 14
t507	**—,4-hydroxyl-amino-2-nitro-**	$C_7H_8N_2O_3$. *See* t275	168.18	ye (bz)	108–9	...	...	...	...	...	...	...	s^h	...	**B15**[2], 16
Ω t508	**—,α-hydroxy-2-methoxy-**	*o*-Methoxybenzyl alcohol. Saligenin 2-methyl ether. $C_8H_{10}O_2$. *See* t275	138.17	...	...	249^{760} 119^{8}	1.0395_{15}^{15}	1.5455^{20}	i	s	∞	...	...	...	**B6**[2], 878
Ω t509	**—,α-hydroxy-4-methoxy-**	Anisyl alcohol. $C_8H_{10}O_2$. *See* t275	138.17	nd	25	259.1^{760} $134–5^{12}$	1.109_4^{26}	1.5420^{25}	s	v	v	...	...	...	**B6**[2], 883
t511	**—,4-hydroxy-3-methoxy-**	Cresolol. $C_8H_{10}O_2$. *See* t275	138.17	pr	5.5	221 113.5^{22}	1.098_4^{20}	1.5353^{25}	δ	∞	∞	...	∞	chl, aa ∞	**B6**[2], 865
t512	**—,4-hydroxy-2-(methylamino)-**	$C_8H_{11}NO$. *See* t275	137.18	cr (bz-lig)	108	...	...	...	...	s	s	...	s	lig δ	**B13**, 599
Ω t513	**—,α-hydroxy-3,4-methylenedioxy-**	Piperonyl alcohol. $C_8H_8O_3$. *See* t275	152.16	nd (peth)	58	157^{16}	...	...	δ	s	s	...	s	chl, MeOH s lig i	**B19**[2], 77
Ω t514	**—,α-hydroxy-2-nitro-**	*o*-Nitrobenzyl alcohol. $C_7H_7NO_3$. *See* t275	153.14	nd (w) $\lambda^{0.1N\,HCl}$ 267.5 (3.75)	74	270^{760} 168^{20}	...	...	δ	s	s	...	...	...	**B6**[3], 1563
t515	**—,α-hydroxy-3-nitro-**	*m*-Nitrobenzyl alcohol. $C_7H_7NO_3$. *See* t275	153.14	rh nd (w)	30.5 (27)	$175–80^{3}$	1.296_{15}^{19}	...	s	s	s	...	...	...	**B6**[3], 1565
Ω t516	**—,α-hydroxy-4-nitro-**	*p*-Nitrobenzyl alcohol. $C_7H_7NO_3$. *See* t275	153.14	nd (w) $\lambda^{0.1N\,HCl}$ 276 (3.96)	96–7	250–60d 185^{12}	...	...	δ s^h	s	s	...	...	...	**B6**[3], 1567
Ω t517	**—,2-hydroxy-3-nitro-**	6-Nitro-*o*-cresol. $C_7H_7NO_3$. *See* t275	153.14	ye pr (dil al or peth)	70	$102–3^{9}$	...	...	i	v	v	...	...	peth s^h	**B6**[3], 1273
t518	**—,2-hydroxy-4-nitro-**	5-Nitro-*o*-cresol. $C_7H_7NO_3$. *See* t275	153.14	ye nd (lig) λ^{MeOH} 320 (3.99)	118	...	...	...	δ	s	s	...	s	CS_2, lig δ	**B6**[3], 1273

For explanations, symbols and abbreviations see beginning of table. For structural formulas see end of table.

PHYSICAL CONSTANTS OF ORGANIC COMPOUNDS (Continued)

No.	Name	Synonyms and Formula	Mol. wt.	Color, crystalline form, specific rotation and λ_{max} (log ε)	m.p. °C	b.p. °C	Density	n_D	Solubility w	al	eth	ace	bz	other solvents	Ref.
	Toluene														
Ω t519	**—,2-hydroxy-5-nitro-**	4-Nitro-*o*-cresol. $C_7H_7NO_3$. *See* t275	153.14	ye or col nd (w or aq al + w), pl (bz)	96 (anh) 30–40 (hyd)	$186–90^{9}$	....	....	δ	v	v	...	v	aa v	**B6**[3], 1274
Ω t521	**—,2-hydroxy-6-nitro-**	3-Nitro-*o*-cresol. $C_7H_7NO_3$. *See* t275	153.14	pa ye nd (w)	147	....	....	....	δ^h	s	s	...	...	....	**B6**[1], 178
t522	**—,3-hydroxy-4-nitro-**	6-Nitro-*m*-cresol. $C_7H_7NO_3$. *See* t275	153.14	ye mcl pl (eth or bz)	56	....	....	....	δ	s	s	...	s	....	**B6**[3], 1326
t523	**—,3-hydroxy-5-nitro-**	5-Nitro-*m*-cresol. $C_7H_7NO_3$. *See* t275	153.14	pa ye cr (bz)	90–1	....	....	....	δ^h	s	s	...	s^h	....	**B6**[2], 361
t524	**—,—,monohydrate**	$C_7H_7NO_3 . H_2O$. *See* t275 ...	171.16	pa ye nd (w)	60–2	....	....	....	δ^h	v	v	...	s	....	**B6**, 385
Ω t525	**—,4-hydroxy-2-nitro-**	3-Nitro-*p*-cresol. $C_7H_7NO_3$. *See* t275	153.14	ye pr (eth)	79	....	....	....	δ	v	v	...	δ	CS_2, lig δ	**B6**[3], 1384
Ω t526	**—,4-hydroxy-3-nitro-**	2-Nitro-*p*-cresol. $C_7H_7NO_3$. *See* t275	153.14	ye nd (al or w)	36.5	125^{22}	1.2399^{39}_{4}	1.574^{40}	δ^h	v	v	v	s	os s	**B6**[3], 1384
Ω t527	**—,5-hydroxy-2-nitro-**	4-Nitro-*m*-cresol. $C_7H_7NO_3$. *See* t275	153.14	nd or pr (w)	129	....	....	....	δ^h	s	s	...	s	chl s	**B6**[3], 1327
Ω t528	**—,2-hydroxy-5-nitroso-**	1,4-Toluquinone 4-oxime. $C_7H_7NO_2$. *See* t275	137.14	nd (w) λ^{MeOH} 305 (4.26)	134–5d	....	....	....	δ v^h	v	v	...	s	chl, alk v CCl_4 s	**B7**[2], 589
Ω t529	**—,5-hydroxy-2-nitroso-**	1,4-Toluquinone 1-oxime. $C_7H_7NO_2$. *See* t275	137.14	nd (w or bz), pr (aa)	165d	....	....	....	δ s^h	s	s	...	s	aa s	**B7**[2], 589
t530	**—,2-hydroxy-3,4,5,6-tetrabromo-**	3,4,5,6-Tetrabromo-*o*-cresol. $C_7H_4Br_4O$. *See* t275	423.75	ye nd (chl or aa)	208	d	....	....	i	s	s	...	s	chl s aa, lig δ	**B6**[3], 1272
t531	**—,3-hydroxy-2,4,5,6-tetrabromo-**	2,4,5,6-Tetrabromo-*m*-cresol. $C_7H_4Br_4O$. *See* t275	423.75	nd (chl, aa)	194	....	....	....	...	...	s	...	...	alk s chl, aa s^h	**B6**[3], 1324
t532	**—,4-hydroxy-2,3,5,6-tetrabromo-**	2,3,5,6-Tetrabromo-*p*-cresol. $C_7H_4Br_4O$. *See* t275	423.75	nd (al or chl)	198–9 (209)	....	....	....	...	s	v	...	...	chl, alk v aa v^h	**B6**[3], 1383
t533	**—,2-hydroxy-3,4,5,6-tetrachloro-**	3,4,5,6-Tetrachloro-*o*-cresol. $C_7H_4Cl_4O$. *See* t275	245.92	nd (lig)	190	....	....	....	...	s	s	...	s	aa s lig s^h	**B6**[2], 333
t534	**—,3-hydroxy-2,4,5,6-tetrachloro-**	2,4,5,6-Tetrachloro-*m*-cresol. $C_7H_4Cl_4O$. *See* t275	245.92	nd (peth)	189–90	....	....	....	...	s	s	s	s	os, KOH s	**B6**[3], 1320
t535	**—,4-hydroxy-2,3,5,6-tetrachloro-**	2,3,5,6-Tetrachloro-*p*-cresol. $C_7H_4Cl_4O$. *See* t275	245.92	nd (dil al, aa, bz-lig) λ^{chl} 295 (3.4)	190	....	....	....	...	s	...	...	s	chl, alk v aa s lig δ	**B6**[3], 1377
t536	**—,2-hydroxy-3,4,5-trinitro-**	4,5,6-Trinitro-*o*-cresol. $C_7H_5N_3O_7$. *See* t275	243.13	og-ye pr (ace)	102	....	....	....	δ^h	s	s	s	...	chl, AcOEt v	**B6**, 369
t537	**—,3-hydroxy-2,4,6-trinitro-**	Methylpicric acid. $C_7H_5N_3O_7$. *See* t275	243.13	pa ye nd (w or al) λ^{iso} 257 (4.00), 335 (3.54)	109–10	150 exp	....	....	δ^h	s	s	s	s	chl s	**B6**[3], 1331
t538	**—,α-iodo-**	Benzyl iodide. $C_6H_5CH_2I$ ...	218.04	col or ye nd (MeOH)	24.5	93^{10}	1.7335^{25}	1.6334^{25}	i	s	s	...	s	MeOH s^h CS_2 δ	**B5**[3], 724
Ω t539	**—,2-iodo-**	*o*-Tolyl iodide. C_7H_7I. *See* t275	218.04	λ^{cy} 232 (4.07)	....	$211–2^{760}$ $73–5^{7}$	1.713^{20}_{4}	1.6079^{20}	i	∞	∞	...	...	....	**B5**[3], 720
Ω t540	**—,3-iodo-**	*m*-Tolyl iodide. C_7H_7I. *See* t275	218.04	λ^{hx} 260 (2.93), 297 (2.78)	−27.2	213^{760} $80–2^{10}$	1.705^{20}	1.6053^{20}	i	∞	∞	...	...	....	**B5**[3], 721
Ω t541	**—,4-iodo-**	*p*-Tolyl iodide. C_7H_7I. *See* t275	218.04	lf (al) λ^{hx} 262 (3.06), 280 (2.85)	36–7	211 sub	1.678^{40}_{4}	....	i	s	s	...	...	CS_2 s	**B5**[3], 722
t542	**—,α-isocyano-**	Benzyl isocyanide. Benzyl carbylamine. $C_6H_5CH_2NC$	117.15	λ^{al} 210 (3.90)	....	198–200 δd $93–4^{55}$	0.972^{15}	....	...	...	...	...	...	....	**B12**, 1041
Ω t543	**—,α-mercapto-**	Benzyl mercaptan. $C_6H_5CH_2SH$	124.22	....	....	194–5	1.058^{20}	1.5751^{20}	i	v	v	...	...	CS_2 s	**B6**[3], 1573
Ω t544	**—,2-mercapto-**	*o*-Toluenethiol. *o*-Thiocresol. C_7H_8S. *See* t275	124.22	pl or lf λ^{iso} 236 (3.88), 271 (2.83), 279 (2.86), 286.5 (2.76)	15	194.2^{760} 67.1^{10}	1.041^{20}_{4}	1.570^{20}_{4}	i	s	v	...	...	....	**B6**[3], 1279
Ω t545	**—,3-mercapto-**	*m*-Toluenethiol. *m*-Thiocresol. C_7H_8S. *See* t275	124.22	λ^{iso} 237 (4.89), 281 (3.80), 289.5 (3.52)	< −20	195.1^{760} 67.8^{10}	1.044^{20}_{4}	1.572^{20}_{4}	i	s	∞	...	...	....	**B6**[3], 1332
Ω t546	**—,4-mercapto-**	*p*-Toluenethiol. *p*-Thiocresol. C_7H_8S. *See* t275	124.22	lf (eth or dil al) λ^{cy} 271 (3.3), 278 (3.3)	44	195^{760} 67.6^{10}	....	....	...	s	v	...	...	....	**B6**[3], 1391
—	**—,α-methoxy-**	*see* **Ether, benzyl methyl**													
Ω t547	**—,2-methoxy-**	*o*-Methylanisole. $C_8H_{10}O$. *See* t275	122.17	λ^{hx} 272 (3.26), 278 (3.25)	....	171.3^{760}	0.9851^{15}_{15}	1.5161^{20}	i	s	s	...	...	....	**B6**[3], 1244
Ω t548	**—,3-methoxy-**	*m*-Methylanisole. $C_8H_{10}O$. *See* t275	122.17	λ^{cy} 273 (3.3), 280 (3.3)	....	177.2^{760}	0.9697^{25}_{25}	1.5164^{13}	i	s	s	s	s	....	**B6**[3], 1297

For explanations, symbols and abbreviations see beginning of table. For structural formulas see end of table.

No.	Name	Synonyms and Formula	Mol. wt.	Color, crystalline form, specific rotation and λ_{max} (log ε)	m.p. °C	b.p. °C	Density	n_D	Solubility w	al	eth	ace	bz	other solvents	Ref.
	Toluene														
Ω t549	—,4-methoxy-	*p*-Methylanisole. $C_8H_{10}O$. *See* t275	122.17	λ^{cy} 270.5 (3.2), 280 (3.4), 286 (3.3)	...	176.5^{760}	0.9689^{25}_{25}	1.5124^{19}	i	s	s	...	...	...	**B6**[3], 1351
Ω t550	—,2(methyl-amino)-	*N*-Methyl-*o*-toluidine. $C_8H_{11}N$. *See* t275	121.18	λ^{al} 244 (4.1), 294 (3.4)	...	$207\text{–}8^{760}$, 99^{17}	0.9769^{20}_{4}	1.5649^{20}	i	∞	∞	s	...	...	**B12**[2], 435
t551	—,3(methyl-amino)-	*N*-Methyl-*m*-toluidine. $C_8H_{11}N$. *See* t275	121.18	...	...	$206\text{–}7^{760}$, $120\text{–}1^{40}$	...	1.5557^{25}	i	∞	∞	s	...	...	**B12**[1], 398
Ω t552	—,4(methyl-amino)-	*N*-Methyl-*p*-toluidine. $C_8H_{11}N$. *See* t275	121.18	λ^{iso} 246 (4.12), 302 (3.35)	...	$209\text{–}11^{761}$, 102^{20}	0.9348^{55}_{4}	1.5568^{20}	i	∞	∞	s	...	...	**B12**[2], 491
t553	—,α-nitro-	Phenylnitromethane*. $C_6H_5CH_2NO_2$	137.14	ye liq	...	225–7, $118\text{–}9^{16}$	1.1598^{20}_{0}	1.5323^{20}	...	...	s	s	...	...	**B5**[3], 741
Ω t554	—,2-nitro-	$C_7H_7NO_2$. *See* t275	137.14	(i) nd (ii) cr λ^{al} 259 (3.72)	(i) −9.55 (ii) −2.9	221.7^{760}, 118^{16}	1.1629^{20}	1.5450^{20}	i	∞	∞	...	...	...	**B5**[3], 730
Ω t555	—,3-nitro-	$C_7H_7NO_2$. *See* t275	137.14	pa ye $\lambda^{5\%\,al}$ 274 (3.86)	16	232.6^{760}	1.1571^{20}_{4}	1.5466^{20}	i	s	∞	...	s	...	**B5**[3], 734
Ω t556	—,4-nitro-	$C_7H_7NO_2$. *See* t275	137.14	orh cr (al, eth) λ^{hx} 265 (4.02)	54.5	238.3^{760}, 105^{9}	1.1038^{75}_{4}	...	i	s	v	v	v	CCl_4, Py, to, chl, CS_2 v	**B5**[3], 736
t557	—,2-nitroso-	C_7H_7NO. *See* t275	121.14	nd or pr λ 286 (3.80)	72.5	...	...	...	δ	v	v	...	...	chl v	**B5**[3], 728
t558	—,3-nitroso-	C_7H_7NO. *See* t275	121.14	nd λ 282 (3.76)	53.5	...	...	...	i	δ	s	...	...	chl s, MeOH s^h, lig δ, s^h	**B5**[3], 728
t559	—,4-nitroso-	C_7H_7NO. *See* t275	121.14	nd (lig) λ^{chl} 745 (1.66)	48.5	...	...	...	i	...	...	...	v	chl v, lig s^h	**B5**[3], 729
t560	—,3-nitro-4-triazo-	$C_7H_6N_4O_2$. *See* t275	178.15	ye nd or pl (lig or dil al)	38	...	...	...	i	s	...	...	...	lig s^h	**B5**[1], 174
Ω t561	—,2-nitro-α,α,α-trifluoro-	*o*-Nitrobenzotrifluoride. $C_7H_4F_3NO_2$. *See* t275	191.11	cr (al)	32.5	216.3^{765}	...	...	i	v	...	...	v	aa v	**B5**[2], 251
Ω t564	—,3-nitro-α,α,α-trifluoro-	*m*-Nitrobenzotrifluoride. $C_7H_4F_3NO_2$. *See* t275	191.11	λ^{al} 247 (3.84)	−2.4	202.75^{760}, 81.6^{10}	1.4357^{15}_{4}	1.4719^{20}	i	s	s	...	...	...	**B5**[2], 251
t565	—,α,α,2,3,4-pentachloro-	$C_7H_3Cl_5$. *See* t275	264.37	cr (lig)	84	275–85	...	...	i	...	...	...	δ v^h	lig δ, s^h	**B5**[1], 153
t566	—,α,α,2,3,6-pentachloro-	$C_7H_3Cl_5$. *See* t275	264.37	nd (MeOH)	83	$145\text{–}50^{12}$	...	...	...	...	...	...	s	MeOH δ	**B5**[3], 703
t567	—,α,α,2,4,5-pentachloro-	$C_7H_3Cl_5$. *See* t275	264.37	...	<0	280–1, $153\text{–}5^{15}$	1.5956^{20}_{4}	1.5992^{20}	i	...	...	...	s	...	**B5**[1], 153
t568	—,α,α,2,4,6-pentachloro-	$C_7H_3Cl_5$. *See* t275	264.37	cr (MeOH)	27	158^{15}	...	...	i	...	...	...	...	MeOH s^h	**B5**[3], 704
t569	—,2,3,4,5,6-pentachloro-	$C_7H_3Cl_5$. *See* t275	264.37	nd (bz or peth) λ^{al} 280 (2.38)	224	301	...	...	...	δ^h	δ^h	...	s^h	to s, peth s^h, CS_2 δ	**B5**[3], 703
t570	—,2-propoxy-	Propyl *o*-tolyl ether. *o*-Cresyl propyl ether. $C_{10}H_{14}O$. *See* t275	150.22	...	...	204.1	0.9517^{0}_{0}	...	...	...	...	...	...	...	**B6**[3], 1246
t571	—,3-propoxy-	Propyl *m*-tolyl ether. *m*-Cresyl propyl ether. $C_{10}H_{14}O$. *See* t275	150.22	...	...	210.6	0.9484^{0}_{0}	...	...	...	...	...	...	...	**B6**[3], 1299
t572	—,4-propoxy-	Propyl *p*-tolyl ether. *p*-Cresyl propyl ether. $C_{10}H_{14}O$. *See* t275	150.22	...	...	210.4	0.9497^{0}_{0}	...	...	...	...	...	...	...	**B6**[3], 1354
t573	—,2,3,5,6-tetra-bromo-	$C_7H_4Br_4$. *See* t275	407.75	nd	116–7	...	...	...	i	δ s^h	...	s	s	os s	**B5**, 310
Ω t574	—,α,α,α,2-tetra-chloro-	*o*-Chlorobenzotrichloride. $C_7H_4Cl_4$. *See* t275	229.92	...	30	264.27^{760}, 129.5^{13}	1.5187^{20}_{4}	1.5836^{20}	i d^h	...	s	s	...	os s	**B5**[3], 703
Ω t575	—,α,α,α,3-tetra-chloro-	*m*-Chlorobenzotrichloride. $C_7H_4Cl_4$. *See* t275	229.92	...	...	255 (247–50)	1.495^{14}	1.4461^{20}	i d^h	...	s	s	...	os s	**B5**, 303
Ω t576	—,α,α,α,4-tetra-chloro-	*p*-Chlorobenzotrichloride. $C_7H_4Cl_4$. *See* t275	229.92	...	...	245, $108\text{–}12^{8}$	...	1.4463^{20}	i	...	s	s	...	os s	**B5**, 303
t577	—,α,α,2,5-tetra-chloro-	$C_7H_4Cl_4$. *See* t275	229.92	cubic cr (chl)	42	...	...	...	i	v	v	...	v	os s	**B5**[3], 702
t578	—,α,α,3,4-tetra-chloro-	$C_7H_4Cl_4$. *See* t275	229.92	...	...	257	1.518^{22}_{22}	...	i	s	s	...	s	aa s	**B5**[3], 702
t579	—,α,α,3,5-tetra-chloro-	$C_7H_4Cl_4$. *See* t275	229.92	cr (MeOH or dil aa)	36.5	...	...	...	...	...	s	s	...	oos s	**B5**[3], 703
t580	—,α,2,4,5-tetra-chloro-	$C_7H_4Cl_4$. *See* t275	229.92	...	...	273	1.547^{20}	...	i	s	s	s	...	os s	**B5**, 302
t581	—,2,3,4,5-tetra-chloro-	$C_7H_4Cl_4$. *See* t275	229.92	nd (MeOH or dil al)	98.1	...	...	...	i	v	s	s	...	os s	**B5**[2], 233
t582	—,2,3,4,6-tetra-chloro-	$C_7H_4Cl_4$. *See* t275	229.92	nd (al or eth)	96 (92)	276.5	...	...	i	s	...	...	v	CS_2 v	**B5**[3], 702
t583	—,2,3,5,6-tetra-chloro-	$C_7H_4Cl_4$. *See* t275	229.92	nd (MeOH)	93–4	sub	...	...	i	s	v	...	...	MeOH s^h	**B5**[3], 702
—	—,triacetoxy-	*see* **Toluene, trihydroxy-,** triacetate													
t586	—,3,4,5-triamino-	$C_7H_{11}N_3$. *See* t275	137.19	col nd (bz), red in air	105	...	...	...	v	v	...	...	s	...	**B13**[2], 147
t587	—,α-triazo-	Benzyl azide. $C_6H_5CH_2N_3$	133.15	...	...	108^{23}, 74^{11}	1.0655^{25}_{4}	1.53414^{25}	i	∞	∞	...	...	...	**B5**[3], 773

For explanations, symbols and abbreviations see beginning of table. For structural formulas see end of table.

No.	Name	Synonyms and Formula	Mol. wt.	Color, crystalline form, specific rotation and λ_{max} (log ε)	m.p. °C	b.p. °C	Density	n_D	Solubility w	al	eth	ace	bz	other solvents	Ref.
	Toluene														
t588	**—,2-triazo-**	*o*-Tolyl azide. $C_7H_7N_3$. *See* t275	133.15	pa ye liq	< −10	90.5^{30}	...	...	...	...	s	...	s^h	...	**B5**^{3}, 773
t589	**—,3-triazo-**	*m*-Tolyl azide. $C_7H_7N_3$. *See* t275	133.15	...	...	92.5^{31}	...	...	...	...	s	...	...	...	**B5**^{2}, 273
t590	**—,4-triazo-**	*p*-Tolyl azide. $C_7H_7N_3$. *See* t275	133.15	...	...	d 180 80^{10}	1.0527^{23}_4	...	i	s	s	...	...	...	**B5**^{3}, 773
t591	**—,α,2,4-tribromo-**	$C_7H_5Br_3$. *See* t275	328.85	...	40–1	...	...	...	i	s	...	...	...	oos v	**B5**^{2}, 240
t592	**—,α,3,5-tribromo-**	$C_7H_5Br_3$. *See* t275	328.85	pl or nd (al)	96	173^{19}	...	...	i	v	...	...	...	os s	**B5**^{3}, 719
t593	**—,2,3,4-tribromo-**	$C_7H_5Br_3$. *See* t275	328.85	rh pl (lig-CS_2, aa)	45–6	...	2.456^{20}	...	i	...	...	...	...	aa s^h	**B5**^{1}, 156
t594	**—,2,3,5-tribromo-**	$C_7H_5Br_3$. *See* t275	328.85	mcl pr (eth-to)	53–4	...	2.467^{17}	...	i	...	v	...	...	...	**B5**^{3}, 719
t595	**—,2,3,6-tribromo-**	$C_7H_5Br_3$. *See* t275	328.85	mcl pr or lf (lig or chl)	60.5	...	2.471^{17}	...	i	...	...	...	...	lig s^h chl δ	**B5**^{1}, 156
t596	**—,2,4,5-tribromo-**	$C_7H_5Br_3$. *See* t275	328.85	mcl pr (eth-al), nd (al)	113.5	...	2.472^{17}	...	i	δ	...	...	...	...	**B5**^{3}, 719
t597	**—,2,4,6-tribromo-**	$C_7H_5Br_3$. *See* t275	328.85	lo nd or mcl pr (eth-AcOEt)	70	290	2.479^{17}	...	i	δ	s	...	...	...	**B5**^{3}, 719
Ω t598	**—,α,α,α-trichloro-**	Benzotrichloride. $C_6H_5CCl_3$.	195.48	λ^{cy} 262 (2.6), 267 (2.7), 274 (2.6)	−4.75	220.6^{760} (214) 150^{100}	1.3723^{20}	1.5580^{20}	i	s	s	...	s	...	**B5**^{3}, 699
t599	**—,α,α,2-trichloro-**	*o*-Chlorobenzylidene chloride. $C_7H_5Cl_3$. *See* t275	195.48	...	...	228.5	1.399^{15}	...	i	...	...	...	...	os s	**B5**, 300
Ω t600	**—,α,2,6-trichloro-**	$C_7H_5Cl_3$. *See* t275	195.48	cr (lig, eth, al-eth)	39–40	$117–9^{14}$	...	...	...	v	v	...	...	lig s^h	**C53**, 21809
Ω t601	**—,α,3,4-trichloro-**	$C_7H_5Cl_3$. *See* t275	195.48	...	37–7.5	241	...	...	i	s	...	...	...	os s	**B5**, 300
t602	**—,α,3,5-trichloro-**	$C_7H_5Cl_3$. *See* t275	195.48	cr (MeOH)	36	$60^{0.35}$	...	...	...	...	...	...	...	MeOH s^h	**C28**, 754
t603	**—,2,3,4-trichloro-**	$C_7H_5Cl_3$. *See* t275	195.48	nd (al or MeOH)	43–4	244^{760}	...	...	i	s	s	s	...	oos s	**B5**^{2}, 232
t604	**—,2,3,5-trichloro-**	$C_7H_5Cl_3$. *See* t275	195.48	nd (al)	45–6	$229–31^{757}$	...	...	i	s	...	s	...	os s	**B5**, 299
Ω t605	**—,2,3,6-trichloro-**	$C_7H_5Cl_3$. *See* t275	195.48	nd (al)	45–6	...	...	...	i	s	...	s	...	os s	**B5**, 299
t606	**—,2,4,5-trichloro-**	$C_7H_5Cl_3$. *See* t275	195.48	nd or lf (al)	82.4	$229–30^{716}$	...	...	i	s	...	s	...	os s	**B5**^{2}, 232
t607	**—,2,4,6-trichloro-**	$C_7H_5Cl_3$. *See* t275	195.48	nd (al)	38	...	...	...	i	s	...	s	...	os s	**B5**^{2}, 232
t608	**—,3,4,5-trichloro-**	$C_7H_5Cl_3$. *See* t275	195.48	...	45–5.5	$246–7^{768}$	...	...	i	s	...	s	...	os s	**B5**, 299
Ω t609	**—,α,α,α-trifluoro-**	Benzotrifluoride. $C_6H_5CF_3$.	146.11	λ^{iso} 250 (2.34), 260 (2.70), 266.5 (2.58)	−29.11	102.06^{760} 10^{10}	1.1884^{20}	1.4146^{20}	d	∞	∞	∞	∞	CCl_4 ∞	**B5**^{2}, 224
t610	**—,α,α,2-trihydroxy-, triacetate**	Salicylaldehyde triacetate. $C_{13}H_{14}O_6$. *See* t275	266.25	nd or pl (al or Ac_2O), pr (al)	107	...	...	...	i	δ v^h	s	...	s	chl, CCl_4 s	**B8**^{2}, 41
t610^1	**—,α,α,3-trihydroxy-, triacetate**	*m*-Acetoxybenzal acetate. $C_{13}H_{14}O_6$. *See* t275	266.25	lf (w-al)	76	...	...	...	δ	s	s	...	...	...	**B8**, 60
t611	**—,α,α,4-trihydroxy-, triacetate**	*p*-Acetoxybenzal acetate. $C_{13}H_{14}O_6$. *See* t275	266.25	pr (eth or lig)	94	...	...	...	i s^h	δ s^h	v	...	...	lig s^h	**B8**^{1}, 530
t611^1	**—,α,2,5-trihydroxy-**	Gentisyl alcohol. $C_7H_8O_3$. *See* t275	140.15	nd (chl)	100	sub vac	...	...	v	v	v	...	δ	chl v lig i	**H30**, 124
Ω t612	**—,2,4,5-trihydroxy-, triacetate**	2,4,5-Triacetoxytoluene. $C_{13}H_{14}O_6$. *See* t275	266.25	cr (al)	114–5	...	...	...	s	s	...	...	s	...	**B6**, 1109
t613	**—,2,4,6-trihydroxy-, triacetate**	2,4,6-Triacetoxytoluene. $C_{13}H_{14}O_6$. *See* t275	266.25	(i) cr (eth-peth) (ii) nd (peth)	(i) 76 (ii) 58	...	...	...	δ^h	v^h	...	...	v	AcOEt v lig s^h	**B6**^{3}, 6320
t614	**—,3,4,5-trihydroxy-**	5-Methylpyrogallol. $C_7H_8O_3$. *See* t275	140.15	pa br nd (bz) λ^w 280 (1.6)	129 (125)	sub	...	...	...	...	...	...	s^h	...	**B6**^{3}, 6320
t615	**—,2,3,4-triiodo-**	$C_7H_5I_3$. *See* t275	469.83	pa br nd (al)	92	...	...	...	i	s	...	...	s	...	**B5**^{1}, 157
t616	**—,2,3,5-triiodo-**	$C_7H_5I_3$. *See* t275	469.83	og pl (al)	72–3	...	...	...	i	δ s^h	...	...	...	...	**B5**^{1}, 157
t617	**—,2,3,6-triiodo-**	$C_7H_5I_3$. *See* t275	469.83	nd (al)	80.5	...	...	...	i	δ	...	...	...	...	**B5**^{1}, 158
t618	**—,2,4,5-triiodo-**	$C_7H_5I_3$. *See* t275	469.83	pa br nd (al)	118–20	...	...	...	i	δ	...	...	...	...	**B5**^{1}, 158
t619	**—,2,4,6-triiodo-**	$C_7H_5I_3$. *See* t275	469.83	nd (al or bz)	118–9	300d	...	...	i	δ	...	...	...	...	**B5**, 317
t620	**—,3,4,5-triiodo-**	$C_7H_5I_3$. *See* t275	469.83	nd (al)	122–3	...	...	...	i	s	...	...	...	...	**B5**, 317
t621	**—,2,3,4-trinitro-**	$C_7H_5N_3O_6$. *See* t275	227.13	tcl lf (al), pr (ace) $\lambda^{5\%\,al}$ 238.5 (4.09)	112	...	1.62	...	i	δ	s	s	s	...	**B5**^{3}, 766
t622	**—,2,4,5-trinitro-**	$C_7H_5N_3O_6$. *See* t275	227.13	yesh pl (ace), pa ye rh bipym (al) $\lambda^{5\%\,al}$ 237 (4.15)	104	exp 290–310	...	...	i	δ	s	s	s	aa s^h	**B5**^{3}, 767
Ω t623	**—,2,4,6-trinitro-**	TNT. $C_7H_5N_3O_6$. *See* t275	227.13	orh (al) λ^{al} 225 (4.36)	82	240 exp	1.654	...	i	δ s^h	s	v	v	to, Py v	**B5**^{3}, 767
t624	**α-Toluenearsonic acid**	Benzylarsonic acid. Phenylmethanearsonic acid*. $C_6H_5CH_2AsO(OH)_2$	216.07	nd (al)	167–8	...	...	...	δ s^h	δ s^h	...	...	...	chl, AcOEt v CS_2 δ CCl_4 δ, s^h	**B16**^{2}, 461
t625	**2-Toluenearsonic acid**	*o*-Tolylarsinic acid	216.07	nd (w)	163–4 (160)	...	...	...	δ^h	v	...	...	...	...	**B16**^{2}, 460
t626	**3-Toluenearsonic acid**	*m*-Tolylarsinic acid	216.07	nd (w)	150	...	...	...	s^h	s	...	...	...	aa s^h	**B16**^{2}, 460

For explanations, symbols and abbreviations see beginning of table. For structural formulas see end of table.

No.	Name	Synonyms and Formula	Mol. wt.	Color, crystalline form, specific rotation and λ_{max} (log ε)	m.p. °C	b.p. °C	Density	n_D	Solubility w	al	eth	ace	bz	other solvents	Ref.
	4-Toluenearsonic acid														
t627	**4-Toluenearsonic acid**	*p*-Tolylarsinic acid	216.07	nd (w)	d (−w 105–10)				δ v^h	s				aa s^h	**B16**[2], 460
t628	**α-Tolueneboronic acid**	Benzylboric acid. $C_6H_5CH_2B(OH)_2$	135.96	cr (w or bz)	140 (195–215) 104 (hyd)				δ	s			s		**B16**[2], 639
t629	**2-Tolueneboronic acid**	*o*-Tolylboronic acid	135.96	nd or pl (w)	165–8	d			δ	s	s		s		**B16**, 921
t630	**3-Tolueneboronic acid**	*m*-Tolylboronic acid	135.96	cr (w)	137–40				δ s^h	v	v				**B16**, 921
t631	**4-Tolueneboronic acid**	*p*-Tolylboronic acid	135.96	nd (w)	245 (226)				δ v^h		s				**B16**[2], 638
t632	**2-Toluenesulfinic acid**	*o*-Tolylsulfinic acid	156.22	nd (w)	80	d			s	v	s	s		os v	**B11**[2], 6
t633	**4-Toluenesulfinic acid**	*p*-Tolylsulfinic acid	156.22	rh pl or lo nd (w) λ^{al} 223 (4.05), 262 (3.14)	86–7				s^h	v	v		δ^h		**B11**[2], 6
t634	—,chloride	*p*-Toluenesulfinyl chloride	174.65	nd	54–8	115–20[4]			d^h	d^h				chl s	**B11**[2], 9
Ω t635	**α-Toluenesulfonic acid**, amide	Benzylsulfonamide. $C_6H_5CH_2SO_2NH_2$	171.22	pr or nd (w), nd (al)	105				v^h	v					**B11**[2], 73
t636	—,—,*N*(2-tolyl)-		261.35	cr (dil al)	83					v				alk s	**B12**[2], 45
t637	—,—,*N*(3-tolyl)-		261.35	cr (al)	75					v				alk s	**B12**[2], 47
t638	—,—,*N*(4-tolyl)-		261.35	pr (al)	113					s^h	s			alk s	**B12**[2], 52
Ω t639	—,chloride	Benzylsulfonyl chloride. $C_6H_5CH_2SO_2Cl$	190.65	pr (eth), nd (bz)	93				d^h	d^h	v		v		**B11**[2], 73
t640	**2-Toluenesulfonic acid**	*o*-Toluenesulfonic acid	172.21	hyg pl (w + 2) λ^w 222 (4.0)	67.5	128.8[25]			v	s	i				**B11**[2], 39
Ω t641	—,amide	*o*-Toluenesulfonamide	171.22	oct (al), pr (w) $\lambda^{0.1N\ HCl}$ 270 (3.1), 276 (3.0)	156.3				δ	s	δ				**B11**[2], 39
t642	—,—,*N*-methyl-	$C_8H_{11}NO_2S$. *See* t641	185.25	pl (bz-lig)	74–5				δ^h	v		v	i	chl v lig δ	**B11**, 87
t643	—,—,*N*-phenyl-	$C_{13}H_{13}NO_2S$. *See* t641	247.32	pr (dil al)	136				i	v					**B12**, 566
t644	—,bromide	*o*-Toluenesulfonyl bromide. $C_7H_7BrO_2S$. *See* t640	235.11	cr	13	138[10]			i d^h	d^h				os ∞	**B11**, 86
Ω t645	—,chloride	*o*-Toluenesulfonyl chloride. $C_7H_7ClO_2S$. *See* t640	190.65		10.2	154[36]	1.3383_4^{20}	1.5565^{20}	i d^h	s^h	s		s		**B11**[2], 39
t646	—,2-tolyl ester	$C_{14}H_{14}O_3S$. *See* t640	262.33	cr (al) λ^{al} 260 (3.01), 271.5 (3.27), 279.5 (3.16)	50–1					s^h			s		**B11**, 85
t647	—,3-tolyl ester	$C_{14}H_{14}O_3S$. *See* t640	262.33	cr (al) λ^{al} 272 (3.27), 280 (3.17)	60					s^h			s		**B11**, 85
t648	—,4-tolyl ester	$C_{14}H_{14}O_3S$. *See* t640	262.33	cr (al) λ^{al} 272 (3.27), 279.5 (3.14)	70–1					s^h			s		**B11**, 85
t649	**3-Toluenesulfonic acid**	*m*-Toluenesulfonic acid	172.21	oil λ^w 222 (3.8), 277 (2.8)					v	s	s				**B11**[2], 23
Ω t650	—,amide	*m*-Toluenesulfonamide	171.22	pl (w), mcl pr (al) $\lambda^{0.1N\ HCl}$ 262 (2.9), 270 (3.0), 279 (3.0)	108				δ s^h	s					**B11**[2], 42
t651	—,—,*N*-phenyl-	*m*-Toluenesulfonanilide. $C_{13}H_{13}NO_2S$. *See* t650	247.32	pr (al)	96				i	s	s				**B12**, 566
Ω t652	**4-Toluenesulfonic acid**	*p*-Toluenesulfonic acid $C_7H_8O_3S \cdot H_2O$	190.19	hyg pl (w + 1), mcl lf or pr λ^w 222 (4.0), 261 (2.5), 272 (2.2)	104–5 38 (anhyd)	140[20]			v	s	s				**B11**[2], 43
Ω t653	,amide	*p*-Toluenesulfonamide	171.22	mcl pl (w + 2) λ^{al} 224 (4.08), 263 (2.7), 275 sh (2.5)	138.5–9 (anh) 105 (hyd)				δ	s	δ				**B11**[2], 55
Ω t654	, ,*N,N*-dichloro-	Dichloramine T. $C_7H_7Cl_2NO_2S$. *See* t653	240.11	pa ye pr (peth-chl)	83				i	s d^h	s		s	aa, chl, CCl_4 s peth δ	**B11**[2], 63
Ω t655	, ,*N*-ethyl-	$C_9H_{13}NO_2S$. *See* t653	199.27	pl (dil al or lig)	64					s					**B11**[2], 56
Ω t656	, ,*N*-methyl-	$C_8H_{11}NO_2S$. *See* t653	185.25	pl (dil al)	78–9		1.340		δ	v	s				**B11**[2], 56
Ω t657	, ,*N*-methyl-*N*-nitroso-	$C_8H_{10}N_2O_3S$. *See* t653	214.25	cr	60				i	v^h	v				**B11**[1], 29
t658	, ,*N*-methyl-*N*-phenyl-	$C_{14}H_{15}NO_2S$. *See* t653	261.35	pl or mcl pr (AcOEt)	95				i	v	v			AcOEt s^h	**B12**[2], 305
t659	, ,*N*-methyl-*N*(2-tolyl)-	$C_{15}H_{17}NO_2S$. *See* t653	275.37	pr (al)	119–20				i	v^h					**B12**[1], 388
t660	, ,*N*-methyl-*N*(4-tolyl)-	$C_{15}H_{17}NO_2S$. *See* t653	275.37	pr	60				i	δ				oos δ	**B12**[2], 529

For explanations, symbols and abbreviations see beginning of table. For structural formulas see end of table.

No.	Name	Synonyms and Formula	Mol. wt.	Color, crystalline form, specific rotation and λ_{max} (log ε)	m.p. °C	b.p. °C	Density	n_D	w	al	eth	ace	bz	other solvents	Ref.
	4-Toluenesulfonic acid														
t661	—,—,*N*-phenyl-	*p*-Toluenesulfonanilide. $C_{13}H_{13}NO_2S$. *See* t653	247.32	dimorphic, α: tcl β: mcl pr (dil al or bz)	103–4	...	...	...	i	v	...	...	s^h	aa s	**B12**^{2}, 298
t662	—,—,*N*(2-tolyl)-	$C_{14}H_{15}NO_2S$. *See* t653	261.35	rh bipym (al), nd (dil aa)	110	...	...	...	...	s	v	v	v	chl v	**B12**^{2}, 452
Ω t663	—,—,*N*(4-tolyl)-	$C_{14}H_{15}NO_2S$. *See* t653	261.35	tcl pr or nd (aa)	118–9	...	...	...	...	v^h	...	...	s	chl s	**B12**^{2}, 528
Ω t664	—,butyl ester	$C_{11}H_{16}O_3S$. *See* t652	228.31	λ^{al} 225.5 (4.09), 263 (2.78), 274 (2.70)	...	164–6^6	1.1319^{20}_4	1.5050^{20}	i	...	s	...	...	...	**B11**^{2}, 46
Ω t665	—,chloride	*p*-Toluenesulfonyl chloride. $C_7H_7ClO_2S$. *See* t652	190.65	tcl (eth or peth)	71	145–6^{15}	...	...	i	s	s	...	v	...	**B11**^{2}, 54
Ω t666	—,2-chloroethyl ester	$C_9H_{11}ClO_3S$. *See* t652	234.70	...	...	210^{21}	...	...	i	...	...	...	...	...	**B11**^{2}, 45
t667	—,2-chloropropyl ester	$C_{10}H_{13}ClO_3S$. *See* t652	248.73	...	...	216–7^{17}	1.2674^{20}_4	1.5225^{21}	...	...	...	...	...	...	**B11**^{2}, 45
Ω t668	—,ethyl ester	$C_9H_{12}O_3S$. *See* t652	200.26	mcl pr (aa, AcOEt) λ^{hp} 257 (2.5), 261 (2.6), 267 (2.6)	34–5	173^{15}	1.166^{48}_4	...	i	s	s	...	...	aa, AcOEt s^h	**B11**^{2}, 45
t669	—,ethylene ester	Ethylene *p*-toluenesulfonate.	370.44	cr (bz)	128	...	...	...	...	...	...	...	...	...	**Am** 77, 4899
Ω t670	—,isopropyl ester	$C_{10}H_{14}O_3S$. *See* t652	214.29	...	21	...	...	1.5065^{20}	...	...	...	...	...	...	**Am** 77, 4899
Ω t671	—,methyl ester	$C_8H_{10}O_3S$. *See* t652	186.23	mcl lf or pr (eth-lig) λ^{hp} 261 (2.7), 273 (2.6)	28–9	292^{760} 140^{20}	...	...	i	v	s	...	v	chl v lig δ	**B11**^{2}, 44
t672	—,propyl ester	$C_{10}H_{14}O_3S$. *See* t652	214.29	λ^{al} 225 (4.09), 263 (2.77), 272 (2.93), 274 (2.69)	< −20	189^9 155^3	1.144^{20}_4	1.4998^{20}	...	...	...	...	...	...	**B11**^{2}, 45
Ω t673	—,2-tolyl ester	$C_{14}H_{14}O_3S$. *See* t652	262.33	nd λ^{al} 228 (4.13), 264 (3.03), 275 (2.81)	54–5	...	...	...	...	s^h	...	...	...	...	**B11**, 100
t674	—,2,4,6-tribromophenyl ester	$C_{13}H_9Br_3O_3S$. *See* t652	485.01	cr (al)	113	...	...	...	...	s^h	...	...	...	...	**B11**^{2}, 47
t675	**3-Toluenesulfonic acid, 6-acetamido-, chloride**	$C_9H_{10}ClNO_3S$. *See* t649	247.70	nd (bz)	159	...	...	...	d	d^h	...	...	δ s^h	...	**B14**^{2}, 448
Ω t676	**2-Toluenesulfonic acid, 4-amino-**	$C_7H_9NO_3S$. *See* t640	187.22	mcl pr (w + 1)	d	...	...	...	δ	i	...	...	...	...	**B14**^{2}, 446
t677	—,—,amide	$C_7H_{10}N_2O_2S$. *See* t641	186.23	nd or pl (w)	164	...	...	...	δ v^h	v^h	...	...	...	...	**B14**, 721
t678	—,5-amino-	$C_7H_9NO_3S$. *See* t640	187.22	pl (w + 1)	>275	...	...	...	δ	...	...	...	...	...	**B14**^{2}, 446
t679	—,6-amino-	$C_7H_9NO_3S$. *See* t640	187.22	nd (+½w)	130d	...	...	...	δ	i	...	...	...	...	**B14**, 723
t680	**3-Toluenesulfonic acid, 4-amino-**	$C_7H_9NO_3S$. *See* t649	187.22	lt ye nd (w + ½)	132d (hyd)	...	...	...	s	...	...	...	...	...	**B14**^{2}, 447
t681	**4-Toluenesulfonic acid, 2-amino-**	$C_7H_9NO_3S$. *See* t652	187.22	pl, nd or pr (w + 1)	...	...	...	...	δ	i	...	...	...	...	**B14**^{2}, 450
t682	—,—,amide	$C_7H_{10}N_2O_2S$. *See* t653	186.23	pr (w)	176	...	...	...	δ s^h	δ	i	...	i	...	**B14**, 729
t683	**2-Toluenesulfonic acid, 5-isopropyl-, amide**	$C_{10}H_{15}NO_2S$. *See* t641	213.30	pl (al), pr or lf (dil al)	75	...	...	...	δ^h	s^h	...	...	...	lig s	**B11**, 140
t684	**3-Toluenesulfonic acid, 4-isopropyl-, amide**	$C_{10}H_{15}NO_2S$. *See* t650	213.30	fl (dil al)	149.9	...	...	...	δ^h	v^h	v	...	...	...	**B11**^{2}, 84
t685	**4-Toluenesulfonic acid, 3-isopropyl-, amide**	$C_{10}H_{15}NO_2S$. *See* t653	213.30	nd (w)	162	...	...	...	δ^h	v	...	...	...	...	**B11**, 139
t686	**2-Toluenesulfonic acid, 4-nitro-, dihydrate**	$C_7H_7NO_5S.2H_2O$. *See* t640	253.23	pl (w)	133.5	...	...	...	v	s	s	...	...	chl s	**B11**^{1}, 23
t687	—,5-nitro-	$C_7H_7NO_5S$. *See* t640	217.20	pr or pl (w + 2)	133.5 (anh) 130 (+2w)	...	...	...	v	v	v	...	...	chl v	**B11**^{2}, 41
t688	—,4-propyl-, amide	$C_{10}H_{15}NO_2S$. *See* t641	213.30	pl (dil al or bz)	101–2	...	...	...	v	s	...	...	δ s^h	...	**B11**, 138
—	**α-Toluic acid**	*see* **Acetic acid, phenyl-**													
—	**Toluic acid**	*see* **Benzoic acid, methyl-**													
—	**Toluidine**	*see* **Toluene, amino-**													
—	**Toluquinone**	*see* **Benzoquinone, methyl-***													
t689	**Tomatidine**	$C_{27}H_{45}NO_2$	415.67	pl, $[\alpha]^{20}_D$ +5 (MeOH) $\lambda^{al\text{-}con\ sulf}$ 405, 470	210–1	...	...	...	...	s^h	s	...	...	...	**Am** 73, 4018

For explanations, symbols and abbreviations see beginning of table. For structural formulas see end of table.

No.	Name	Synonyms and Formula	Mol. wt.	Color, crystalline form, specific rotation and λ_{max} (log ε)	m.p. °C	b.p. °C	Density	n_D	Solubility w	al	eth	ace	bz	other solvents	Ref.
	Tomatine														
t690	**Tomatine**	Lycopersicin. $C_{50}H_{83}NO_{21}$	1034.22	nd (MeOH) $[\alpha]_D^{20}$ − 30 (Py)	270	...	...	...	i	s	i	...	...	MeOH, diox s peth i	**C41,** 3502
t691	**Torularhodin**	Torulene. $C_{37}H_{48}O_2$	524.80	red nd (MeOH-eth), vt-bk (bz-MeOH)	201–3d	...	...	...	...	δ^h	...	s	...	chl, CS_2, Py s peth i	**H42,** 867
t692	**α-Toxicarol(*dl*)**	Hydroxydequelin. $C_{23}H_{22}O_7$	410.43	gr ye pl (al) λ^{al} 273 (4.53), 297 (4.04), 315 sh (4.00)	219–23 (231–2)	...	...	1.580	...	δ	...	...	...	chl s^h	**J1941,** 878
t693	—(*l*)	$C_{23}H_{22}O_7$. *See* t691	410.43	gr-yesh pl or nd (AcOEt-al) $[\alpha]_D^{20}$ − 66 (bz) $[\alpha]_D^{20}$ + 58 (ace) λ^{al} 273 (4.53), 297 (4.04), 315 sh (4.00)	125–7	...	...	...	...	...	...	s	...	AcOEt s	**J1941,** 878
t694	**β-Toxicarol(*dl*)**	$C_{23}H_{22}O_7$	410.43	pa ye pl (al) λ 675 (4.08)	169–70	...	...	...	...	s^h	...	...	...	...	**J1938,** 528
t695	**α-Toxicarol, dihydro-(*l*)**	$C_{23}H_{24}O_7$	412.45	pa ye pl or rods $[\alpha]_D^{20}$ − 30 (bz, c = 5)	179 (206)	...	...	...	...	...	...	...	...	...	**J1940,** 1178
t696	**Trasentin, hydrochloride**	Adiphenine. 2(Diethylamino)ethyl diphenylacetate hydrochloride. $(C_6H_5)_2CHCO_2CH_2CH_2N(C_2H_5)_2 \cdot HCl$	347.89	nd	114–5	...	...	...	s	δ	i	...	i	...	**C54,** 12062
—	**Traumatic acid**	*see* **2-Dodecenedioic acid***													
t697	**α,α-Trehalose**	Mycose. $C_{12}H_{22}O_{11}$	342.30	orh cr $[\alpha]_D^{20}$ + 199 (w, c = 6)	214–6 (anh) 97 (hyd)	...	1.58_{24}^{24}	...	v^h	s^h	i	...	i	...	**B31,** 378
t698	—,dihydrate	$C_{11}H_{22}O_{11} \cdot 2H_2O$. *See* t697	378.33	cr (dil al), $[\alpha]_D^{20}$ + 178.3 (w, c = 7)	103 (97)	...	...	...	v	s^h	i	...	...	...	**B31,** 378
—	**Triacetamide**	*see* **Acetic acid,** amide, *N,N*-diacetyl-													
—	**Triacetin**	*see* **Glycerol,** triacetate													
t699	**Triacontane***	$CH_3(CH_2)_{28}CH_3$	422.83	orh (eth, bz)	65.8	449.7^{760} 304^{15}	0.7750_4^{78} 0.8097_4^{20} (suc)	1.4352^{70} 1.4536^{20} (suc)	i	δ^h	s	...	v^h	...	**B1**^{3}, 584
t700	**1-Triacontanol***	Myricyl alcohol. $CH_3(CH_2)_{29}OH$	438.83	nd (eth), pl (bz)	88	...	0.777^{95}	...	i	s	v	...	v	oos s^h	**B1**^{3}, 1850
t701	**1,3,5-Triazine***	...	81.08	rh λ^{iso} 272 (3.0)	86	114^{760}	1.38	...	...	s	s	...	...	...	**Am 76,** 5646
Ω t701[1]	—,**2,4-diamino-***	Formoguanamine. Guanamine. $C_3H_5N_5$. *See* t701	111.11	nd (w) λ < 205 (< 4.54), 258 (3.56)	329d (315)	...	...	...	s^h	δ	...	...	...	...	**B26**^{1}, 65
Ω t702	—,**2,4-diamino-6-phenyl-***	Benzoquanamine. $C_9H_9N_5$. *See* t701	187.21	nd or pr (al) λ^{al} 249 (4.40)	226–8	...	...	...	...	s	s	...	...	...	**B26**^{1}, 69
t703	—,**hexahydro-1,3,5-trinitro-***	Cyclonite. Hexogen. $C_3H_6N_6O_6$. *See* t701	222.12	orh cr (ace)	205–6	...	1.82^{20}	...	i	i	δ	s	i	aa s MeOH, to, AcOEt δ CCl_4, CS_2 i	**B26**^{2}, 5
Ω t703[1]	—,**hexahydro-1,3,5-triphenyl-***	$C_{21}H_{21}N_3$. *See* t701	315.42	nd (lig), pr (eth or chl-al)	143	185	...	...	i	δ	s	s	s	chl, to s	**B26**^{2}, 3
—	—,**trichloro-***	*see* **Cyanuric acid,** trichloride													
—	—,**trihydroxy-***	*see* **Cyanuric acid**													
t704	**1,3,5-Triazine-2,4,6-tricarboxylic acid**	Cyanuric tricarboxylic acid. Paracyanoformic acid.	213.11	pw	> 250d	...	...	...	δ	i	i	...	i	...	**B26**^{2}, 168
t705	—,triethyl ester	$C_{12}H_{15}N_3O_6$. *See* t704	297.27	nd (al)	168	d	...	...	i	i δ^h	i	...	i	...	**B26**^{2}, 168
t706	—,trinitrile	...	156.11	mcl pr (bz + l)	119 sub	262^{771} 119^1	...	...	d^h	...	...	...	s v^h	CCl_4 s	**B26**^{1}, 91
Ω t707	**1,2,3-Triazole**	Osotriazole	69.07	hyg cr λ^{al} 210 (3.64)	23	203^{739}	1.1861_4^{25}	1.4854^{25}	s	...	s	s	...	oos s lig i	**B26**^{1}, 5
Ω t708	**1,2,4-Triazole**	Pyrrodiazole	69.07	pr (w), nd (al, eth, chl or bz) λ^{THF} 216.5 (3.66)	120–1	260	1.132^{153}	1.4854^{25}	v	v	δ	...	δ	...	**B26**^{2}, 7
Ω t709	—,**3-amino-**	Amizol. ATA. $C_2H_4N_4$. *See* t708	84.08	cr (w, al or AcOEt)	159	...	...	...	v	v	i	i	...	chl s AcOEt δ	**B26**^{2}, 76
Ω t710	—,**4-amino-**	$C_2H_4N_4$. *See* t708	84.08	hyg nd (al or chl)	82–3	...	...	...	s	s	...	...	...	chl, peth δ	**B26**^{2}, 7
—	**Tribenzoin**	*see* **Glycerol,** tribenzoate													
—	**Tributyrin**	*see* **Glycerol,** tributanoate													
—	**Tricaproin**	*see* **Glycerol,** trihexanoate													
—	**Tricaprylin**	*see* **Glycerol,** trioctanoate													

For explanations, symbols and abbreviations see beginning of table. For structural formulas see end of table.

No.	Name	Synonyms and Formula	Mol. wt.	Color, crystalline form, specific rotation and λ_{max} (log ε)	m.p. °C	b.p. °C	Density	n_D	Solubility w	al	eth	ace	bz	other solvents	Ref.
	Tricarballylic acid														
—	**Tricarballylic acid**	*see* **1,2,3-Propanetricarboxylic acid***													
—	**Tricarbonimide,** trimethyl ester	*see* **Isocyanuric acid,** trimethyl ester													
—	**Tricetin**	*see* **Flavone, 3′,4′,5,5′,7-pentahydroxy-**													
Ω t711	**Tricosane***	$CH_3(CH_2)_{21}CH_3$	324.64	lf (eth-al)	47.6	380.2^{760} 243^{15}	0.7785_4^{48} 0.7969_4^{20} (suc)	1.4468^{20} (suc)	i	δ	s	...	...	...	**B1**[2], 141
Ω t712	**12-Tricosanone***	Diundecyl ketone. Laurone. $[CH_3(CH_2)_{10}]_2CO$	338.63	lf (al)	69.3	...	0.8086_4^{69}	1.4283^{80}	i	δ	s	δ	s	chl s	**B1**[2], 774
—	**Tricyclodecane**	*see* **α-Dicyclopentadiene, tetrahydro-**													
t713	**1,12-Tridecadiyne***	$HC{:}C(CH_2)_9C{:}CH$	176.31	...	−3	115.5^{12}	0.8262_4^{21}	1.454^{20}	...	...	...	...	...	...	**B1**[2], 284
t714	**Tridecanal***	$CH_3(CH_2)_{11}CHO$	198.35	...	14	156^{23} 128^{10}	0.8356_4^{18}	1.4384^{18}	i	s	...	...	...	os v	**B1**[3], 2916
t715	—,oxime*	$CH_3(CH_2)_{11}CH{:}NOH$	213.37	nd (dil al)	80.5	...	...	...	i	δ	v	...	δ	chl v peth δ	**B1**[2], 769
Ω t716	**Tridecane***	$CH_3(CH_2)_{11}CH_3$	184.37	...	−5.5	235.4^{760} 107^{10}	0.7564_4^{20}	1.4256^{20}	i	v	v	...	...	...	**B1**[3], 547
Ω t717	—,1-amino-*	$CH_3(CH_2)_{12}NH_2$	199.38	...	27.4	275.8^{760} 140.1^{10}	0.8049_4^{20}	1.4443^{20}	δ	s	s	...	...	...	**B4**[1], 388
Ω t718	—,1-bromo-*	Tridecyl bromide. $CH_3(CH_2)_{12}Br$	263.27	...	6.2	296^{760} 162^{16}	1.0177_4^{20}	1.4593^{20}	i	...	...	...	...	chl v	**B1**[3], 548
t719	—,1,13-dibromo-*	$Br(CH_2)_{13}Br$	342.17	...	8–10	$188\text{–}92^{13}$ $155\text{–}7^{1}$	1.276^{15}	1.4880^{27}	i	...	s	...	...	chl v	**B1**[3], 548
t720	**1,12-Tridecanediol***	$CH_3CH(OH)(CH_2)_{11}OH$	216.37	cr (dil al)	60–1	$188\text{–}90^{8}$	...	...	...	v^h	...	...	...	peth δ	**B1**[2], 563
t721	**1,13-Tridecanediol***	Tridecamethylene glycol. $HO(CH_2)_{13}OH$	216.37	cr (bz)	76.5	$195\text{–}7^{10}$	...	...	...	s	...	...	s^h	aa v^h	**B1**[3], 2240
Ω t722	**Tridecanoic acid***	$CH_3(CH_2)_{11}CO_2H$	214.35	cr (peth, ace)	44.5–5.5	236^{100} 140.5^{1}	...	...	i	v	v	s	...	aa v peth s^h	**B2**[3], 904
t723	—,amide	$CH_3(CH_2)_{11}CONH_2$	213.37	lf (al)	100	...	...	...	...	v	v	...	...	...	**B2**[1], 159
Ω t724	—,methyl ester*	$CH_3(CH_2)_{11}CO_2CH_3$	228.38	...	fr 6.5	$90\text{–}5^{1}$	...	1.4405^{20}	...	∞	...	...	...	...	**B2**[3], 905
Ω t725	—,nitrile	1-Cyanododecane. $CH_3(CH_2)_{11}CN$	195.35	...	9.7	293^{760} 142^{10}	0.8257_4^{20}	1.4378^{20}	v	v	...	...	...	...	**B2**[3], 906
t726	**1-Tridecanol***	Tridecyl alcohol. $CH_3(CH_2)_{12}OH$	200.37	cr (al)	32.5–3.5	152^{14}	0.8223_4^{31}	...	i	s	s	...	...	...	**B1**[3], 1799
Ω t727	**2-Tridecanone***	Hendecyl methyl ketone. $CH_3(CH_2)_{10}COCH_3$	198.35	...	30.5	263^{760} 160^{16}	0.8217_4^{30}	1.4318^{20}	i	v	v	v	v	CCl_4, chl, to, MeOH v	**B1**[3], 2916
Ω t728	**3-Tridecanone***	Ethyl decyl ketone. $CH_3(CH_2)_9COCH_2CH_3$	198.35	pl	31	140^{17}	...	...	i	...	...	s	...	...	**B1**[3], 2916
Ω t729	**7-Tridecanone***	Dihexyl ketone. Enanthone. $CH_3(CH_2)_5CO(CH_2)_5CH_3$	198.35	lf (al)	33	261^{760} 138^{12}	0.825^{30}	...	...	s	v	...	...	chl, lig s	**B1**[3], 2917
t730	**1-Tridecene***	Tridecylene. $CH_3(CH_2)_{10}CH{:}CH_2$	182.35	...	−13 (−23)	232.8^{760} 104^{11}	0.7658_4^{20}	1.4340^{20}	i	v	v	...	s	...	**B1**[3], 872
—	**Tridecylene**	*see* **1-Tridecene**													
t731	**1-Tridecyne**	$CH_3(CH_2)_{10}C{:}CH$	180.34	...	...	94.5^{25}	0.7729_4^{20}	1.4309^{20}	i	...	s	...	s	...	**J1960**, 4719
—	**Triethanolamine**	*see* **Amine, triethyl, 2,2′,2″-trihydroxy-**													
Ω t732	**Triethylene glycol**	$HOCH_2CH_2OCH_2CH_2OCH_2CH_2OH$	150.18	hyg liq	−5 (−7.2)	278.3^{760} 165^{14}	1.1274_4^{15}	1.4531^{20}	∞	∞	δ	...	∞	to ∞ peth i	**B1**[3], 2102
t733	—,3-aminopropyl ether	$HOCH_2CH_2OCH_2CH_2OCH_2CH_2OCH_2CH_2CH_2NH_2$	207.27	glassy	−50	184^{10}	1.0682_{20}^{20}	1.4668^{20}	∞	∞	...	...	...	...	**C55**, 5935
t734	—,diacetate	$CH_3CO_2CH_2CH_2OCH_2CH_2OCH_2CH_2O_2CCH_3$	234.25	...	...	300	...	...	∞	∞	∞	...	...	...	**B2**[3], 309
t735	—,monobutyl ether	$HOCH_2CH_2OCH_2CH_2OCH_2CH_2O(CH_2)_3CH_3$	206.29	...	...	278^{760}	0.9890_4^{20}	1.4389^{20}	∞	v	...	...	...	MeOH, CCl_4 v	**C53**, 1154
Ω t736	**Triethylenetetramine**	$H_2NCH_2(CH_2NHCH_2)_2CH_2NH_2$	146.24	...	12	$266\text{–}7^{760}$ 157^{20}	...	1.4971^{20}	s^h	s	...	...	...	ac s	**B4**[3], 542
Ω t737	**Triglycine**	*N*(*N*-Glycylglycyl)glycine. $H_2NCH_2CONHCH_2CONHCH_2CO_2H$	189.17	ns (dil al) λ 186 (4.1)	246d	...	...	...	s^h	i	i	...	...	...	**B4**[3], 1198
t738	—,*N*-phthalyl-	...	319.28	nd (al)	234–5d	...	...	...	v^h	v^h	i	...	i	chl, peth i	**B21**[2], 358
—	**Triglycolamic acid**	*see* **Acetic acid, nitrilotri-**													
—	**Trigonelline**	*see* **3-Pyridinecarboxylic acid, 1-methyl-**													
—	**Triisovalerin**	*see* **Glycerol,** tri(3-methylbutanoate)													
—	**Trilaurin**	*see* **Glycerol,** tridodecanoate													
—	**Trimellitic acid**	*see* **1,2,4-Benzenetricarboxylic acid***													
—	**Trimesic acid**	*see* **1,3,5-Benzenetricarboxylic acid***													

For explanations, symbols and abbreviations see beginning of table. For structural formulas see end of table.

No.	Name	Synonyms and Formula	Mol. wt.	Color, crystalline form, specific rotation and λ_{max} (log ε)	m.p. °C	b.p. °C	Density	n_D	Solubility w	al	eth	ace	bz	other solvents	Ref.
	Trimesitic acid														
—	**Trimesitic acid**	*see* **2,4,6-Pyridinetricarboxylic acid**													
t739	**Trimethadione**	3,5,5-Trimethyl-2,4-oxazolidinedione. Tridione.	143.14	cr (50 % MeOH) λ^{MeOH} 226 (2.11)	46–6.5	$78–80^{5}$ $71–3^{1}$	...	...	s	v	v	v	v	chl v peth i	**C45,** 9528
—	**Trimethylene glycol**	*see* **1,3-Propanediol***													
—	**Trimethyleneimine**	*see* **Azetidine**													
—	**Trimethylethenylammonium hydroxide**	*see* **Neurine**													
—	**Trimyristin**	*see* **Glycerol,** tritetradecanoate													
—	**Triolein**	*see* **Glycerol,** tri(*cis*-9-octadecenoate)													
—	**Trional**	*see* **Butane, 2,2-bis(ethylsulfonyl)-***													
Ω t741	**1,3,5-Trioxane**	Metaformaldehyde	90.08	rh nd (eth)	64	114.5^{759} sub 46^{1}	1.17^{65}	...	v	s	s	...	s	CCl_4, chl, CS_2 s peth i	**B19**², 392
—	**—,triimine**	*see* **Cyamelide**													
—	**—,2,4,6-triethenyl-**	*see* **Metacrolein**													
—	**—,2,4,6-triisopropyl-**	*see* **Paraisobutyraldehyde**													
—	**—,2,4,6-trimethyl-**	*see* **Paraldehyde**													
—	**—,2,4,6-tripropyl-**	*see* **Parabutyraldehyde**													
—	**Tripalmitin**	*see* **Glycerol,** trihexadecanoate													
—	**Triphenylene**	*see* **9,10-Benzophenanthrene**													
—	**Triphenyl methane**	*see* **Methane, triphenyl-***													
—	**Triphosgene**	*see* **Carbonic acid,** bis(trichloromethyl) ester*													
—	**Tripropionin**	*see* **Glycerol,** tripropanoate													
t743	**Trisiloxane, octamethyl-**	$(CH_3)_3SiOSi(CH_3)_2OSi(CH_3)_3$	236.54	...	−80	153^{774} $50–2^{17}$	0.8200_4^{20}	1.3840^{20}	...	δ	...	...	s	peth s	**C54,** 8602
t744	**—,1,1,3,5,5-pentamethyl-1,3,5-triphenyl-**	$C_6H_5Si(CH_3)_2OSi(CH_3)(C_6H_5)OSi(CH_3)_2C_6H_5$	422.75	...	...	$169^{0.7}$	1.0227_4^{20}	1.5280^{20}	i	...	...	...	s	lig s	**Am 70,** 1116
—	**Tristearin**	*see* **Glycerol,** trioctadecanoate													
t745	**Trisulfide, diallyl**	$CH_2:CHCH_2S_3CH_2CH:CH_2$	178.34	...	...	$112–22^{16}$	1.0845^{15}	...	i	i	∞	...	...	...	**B1,** 441
t746	**—,diethyl**	$CH_3CH_2S_3CH_2CH_3$	154.32	ye λ^{hx} 250 (2.63)	...	$96–7^{26}$	1.114^{20}	1.5689^{13}	...	...	...	...	...	...	**B1**³, 1378
t747	**—,dimethyl, hexabromo-**	$Br_3CS_3CBr_3$	599.67	rh pr (eth)	125d	d	...	...	i	δ	δ	...	s	CS_2 v chl, peth s	**B3**², 107
t748	**—,diphenyl,2,2′-dinitro-**	...	340.41	ye nd (al)	175–6	...	...	...	...	δ	...	...	...	KOH s os δ	**B6**³, 1062
—	**Tritan**	*see* **Methane, triphenyl-***													
Ω t749	**1,3,5-Trithiane**	Trithioformaldehyde	138.27	hex (bz), pr (w), nd (al) λ^{al} 240 (3.16)	220 (cor)	sub	1.6374_4^{24}	...	δ s^h	δ	δ	...	s	...	**B19**², 393
t750	**—,2,4,6-trimethyl-(α-form)**	Trithioacetaldehyde. $C_6H_{12}S_3$. *See* t749	180.36	mcl (al, ace) λ 238 (2.91)	101	245–8	...	...	i	s	s	s	v	chl v CS_2 s	**B19**², 399
t751	**—,—(β-form)**	$C_6H_{12}S_3$. *See* t749	180.36	rh nd (ace)	126–7	245–8	...	...	i	s	s	s	v	chl v	**B19**², 399
Ω t753	**—,2,4,6-trimethyl-2,4,6-triphenyl-**	Trithioacetophenone. $C_{24}H_{24}S_3$. *See* t749	408.65	nd (al)	122	...	...	...	i	δ s^h	v	v	...	chl v	**B19,** 398
t754	**—,2,4,6-triphenyl-(α-form)**	Trithiobenzaldehyde. $C_{21}H_{18}S_3$. *See* t749	366.57	nd (bz-al)	167	d	...	...	i	δ	δ	δ	s	chl, AcOEt, CS_2 s MeOH δ	**B19**¹, 808
t755	**—,—(β-form)**	$C_{21}H_{18}S_3$. *See* t749	366.57	cr (bz or thiophene)	229–30	...	...	...	i	i	δ	δ	s^h	chl, CS_2, AcOEt δ MeOH i	**B19**², 408
—	**Trithioacetaldehyde**	*see* **1,3,5-Trithiane, 2,4,6-trimethyl-**													
—	**Trithiobenzaldehyde**	*see* **1,3,5-Trithiane, 2,4,6-triphenyl-**													
—	**Trithiocyanuric acid**	*see* **Cyanuric acid, thio-**													
t756	**α-Tritisterol**	$C_{30}H_{50}O$	426.74	nd (MeOH-ace), $[\alpha]_D^{20}$ +54.3 (al)	114–5	...	...	...	...	s	...	...	...	peth, chl s	**H20,** 424

For explanations, symbols and abbreviations see beginning of table. For structural formulas see end of table.

No.	Name	Synonyms and Formula	Mol. wt.	Color, crystalline form, specific rotation and λ_{max} (log ε)	m.p. °C	b.p. °C	Density	n_D	Solubility w	al	eth	ace	bz	other solvents	Ref.
	β-Tritisterol														
t757	*β*-**Tritisterol**	$C_{30}H_{50}O$	426.74	nd (MeOH) $[\alpha]_D$ +49.2 (al)	97	...	...	...	...	s	...	...	...	...	**H20**, 424
t758	**Tropacocaine**	Benzoyl-*ψ*-tropeine. $C_{15}H_{19}NO_2$	245.33	pl or tab	49	d	1.0426_4^{100}	1.5080^{100}	δ^h	v	v	...	v	chl, peth, dil ac v	**B21**², 23
t759	—,hydrochloride	$C_{15}H_{19}NO_2 . HCl$. *See* t758	281.79	pl (al)	283d	...	...	...	s	δ	i	...	...	...	**B21**, 39
t760	**Tropane**	2,3-Dihydro-8-methylnor-tropidine.	125.22	...	...	167^{760}	0.9259_{15}^{15}	1.4732^{20}	δ^h	...	...	...	...	...	**B20**², 69
t761	**Tropeine, benzoyl-**	...	245.33	cr (eth)	41–2 37 (+$\frac{5}{3}$w)	175–80	...	...	s	s	s	...	...	...	**B21**², 17
t763	—,dihydrate	$C_{15}H_{19}NO_2 . 2H_2O$. *See* t761	281.36	pl (w)	58	...	...	...	s	s	s	...	...	...	**B21**, 19
—	**Tropic acid**	*see* **Propanoic acid, 3-hydroxy-2-phenyl-**													
—	**Tropilidene**	*see* **1,3,5-Cycloheptatriene***													
Ω t764	**Tropine**	3-Tropanol	141.22	hyg pl (eth)	64	229^{760}	...	...	v	v	s	...	...	chl s	**B21**², 17
t765	**Tropinic acid**(*d*)	...	187.20	cr (w or dil al) $[\alpha]_D$ +14.8 (w)	253d (rapid htng) 247–8 (slow htng)	...	...	...	s	δ s^h	i	...	i	aa i	**B22**, 123
t766	—(*dl*)	$C_8H_{13}NO_4$. *See* t765	187.20	nd (dil al)	251d	...	...	...	s	δ s^h	i	...	i	aa i	**B22**, 124
t767	—(*l*)	$C_8H_{13}NO_4$. *See* t765	187.20	cr (w), $[\alpha]_D^{20}$ −14.8 (w)	243	...	...	...	s	δ s^h	i	...	i	aa i	**B22**, 124
Ω t768	**Tropinone**	3-Tropanone	139.20	lo nd (peth)	42–4	$224–5^{714}$ 113^{25}	1.9872_4^{100}	1.4598^{100}	...	s	s	s	s	oos v peth s^h	**B21**², 225
—	**Tropolone**	*see* **2,4,6-Cycloheptatrien-1-one, 2-hydroxy-**													
—	**Tropone**	*see* **2,4,6-Cycloheptatrien-1-one***													
t769	**Truxane**	Diindene	232.33	pl (peth) λ^{diox} 273 (4.92), 297 (4.82), 335 (2.84)	116	...	...	...	...	v^h	v	...	...	aa v peth s^h	**B5**², 606
t770	—,**tetraphenyl-**	Dimer of triphenylallene	536.73	lf (bz-al)	210	...	...	...	...	δ	...	...	s^h	chl v	**E14s**, 494
t771	*α*-**Truxilline**	Cocamine. $C_{36}H_{46}N_2O_6$	658.80	amor pw $[\alpha]_D$ < 0	*ca.* 80	...	...	...	δ	s	s	...	s	chl s peth δ	**B22**, 202
Ω t772	**Tryptophan**(*D*)	*α*-Amino-*β*-indolylpropionic acid. $C_{11}H_{12}N_2O_2$	204.23	pl (dil al) λ^w 280 (3.8) $[\alpha]_D^{20}$ +33 (w)	281–2	...	...	...	δ s^h	s^h	i	...	...	aa δ	**C43**, 2656
Ω t773	—(*DL*)	$C_{11}H_{12}N_2O_2$. *See* t772	204.23	pl (50 % al) λ^w 280 (3.8)	282 (293)	...	...	...	δ s^h	δ	i	...	...	...	**B22**², 470
Ω t774	—(*L*)	$C_{11}H_{12}N_2O_2$. *See* t772	204.23	lf or pl (dil al), $[\alpha]_D^{20}$ +6.1 (1N NaOH, p = 11), $[\alpha]_D^{20}$ −31.5 (w, c = 0.5) λ^w 280 (3.7)	290–2.5d (278)	...	...	...	δ s^h	s^h	i	...	...	aa δ chl i	**B22**², 466
Ω t775	—,*N*-**acetyl-**(*DL*)	$C_{13}H_{14}N_2O_3$. *See* t772	246.27	pl (dil al) λ^w 279 (3.72)	206–7	...	...	...	...	s	v	...	...	NaOH v	**B22**², 469
Ω t776	—,—(*L*)	$C_{13}H_{14}N_2O_3$. *See* t772	246.27	nd (dil MeOH) $[\alpha]_D^{15}$ +25 (95 % al, c = 1) λ^w 279 (3.72)	189–90	...	...	...	s	s	...	...	...	alk s	**Am 66**, 350
t777	—,*N*-**methyl-**(*DL*)	$C_{12}H_{14}N_2O_2$. *See* t772	218.26	nd (dil al)	297d	...	...	...	δ s^h	s	i	...	...	...	**J1938**, 1910
t778	—,—(*L*)	Abrin. $C_{12}H_{14}N_2O_2$. *See* t772	218.26	pr (w) $[\alpha]_D^{21}$ +44.4 (dil HCl)	295d	...	...	...	δ s^h	s	i	...	...	...	**J1938**, 1910
—	**Tuberculostearic acid**	*see* **Octadecanoic acid, 10-methyl-***													
t779	**Tuduranine**	$C_{18}H_{19}NO_3$	297.36	nd (eth) $[\alpha]_D^{20}$ −127.5 (al)	204	...	...	...	...	s	s	s	...	os v	**C49**, 8317
t780	**Turanose**	3*α*-*D*-Glucosido-*D*-fructose. $C_{12}H_{22}O_{11}$	342.30	pr (w-al or MeOH + ½) $[\alpha]_D^{20}$ +27.2 → +75.8 (w, c = 4) (mut)	168 (157d) 65–70 (+½ MeOH)	...	...	...	v	s	...	...	...	MeOH s^h	**B31**, 454
Ω t781	**Tyramine**	*p*-(*β*-Aminoethyl)pehenol	137.18	pl or nd (bz), cr (al), nd (w) λ 535 (2.78)	164–5	$205–7^{25}$ $165–7^2$	...	...	δ s^h	s v^h	...	...	δ	xyl s to i	**B13**², 354

For explanations, symbols and abbreviations see beginning of table. For structural formulas see end of table.

No.	Name	Synonyms and Formula	Mol. wt.	Color, crystalline form, specific rotation and λ_{max} (log ε)	m.p. °C	b.p. °C	Density	n_D	Solubility w	al	eth	ace	bz	other solvents	Ref.
	Tyrosine														
t782	**Tyrosine(*D*)**	2-Amino-3(4-hydroxyphenyl) propanoic acid*	181.19	cr (w) $[\alpha]^{25}_{D}$ +10.3 (1N HCl, c = 4) $\lambda^{w, pH=6}$ 274.5 (3.15)	310–4d	...	...	...	δ^h	...	...	...	...	dil HCl s	**B14**[2], 365
Ω t783	—(*DL*)	$C_9H_{11}NO_3$. *See* t782	181.19	nd $\lambda^{w, pH=6}$ 274.5 (3.15)	340d (rapid htng) 290–5d (slow htng)	...	...	...	δ	i δ^h	i	...	...	...	**B14**[2], 379
Ω t784	—(*L*)	$C_9H_{11}NO_3$. *See* t782	181.19	nd (w) $[\alpha]^{22}_{D}$ −10.6 (1N HCl, c = 4) $\lambda^{w, pH=6}$ 274.5 (3.15)	342–4d (sealed tube, rapid htng) 290–5 (slow htng)	sub	...	...	δ^h	i	i	...	...	aa δ	**B14**[2], 366
t785	—,amide(*L*)	$C_9H_{12}N_2O_2$. *See* t782	180.21	pl or pr (al) $[\alpha]^{20}_{D}$ +19.5 (w)	153–4	...	...	...	s	v	...	...	...	MeOH v	**B14**, 612
Ω t786	—,ethyl ester(*L*)	$C_{11}H_{15}NO_3$. *See* t782	209.25	pr (AcOEt) $[\alpha]^{20}_{D}$ +20.4 (MeOH)	108–9	...	...	...	δ^h	s	δ	...	v^h	MeOH v AcOEt, alk s	**B14**[2], 372
t787	—,methyl ester(*L*)	$C_{10}H_{13}NO_3$. *See* t782	195.21	pr (AcOEt) $[\alpha]^{20}_{D}$ +25.75 (MeOH)	136–7	...	...	...	δ s^h	s	δ	...	δ	MeOH v AcOEt, alk s	**B14**[2], 372
t788	—,**3-bromo-(*L*)**	$C_9H_{10}BrNO_3$. *See* t782	260.09	cr (w + 1), nd (w + 2)	246–9d	...	...	...	δ v^h	...	...	...	...	...	**B14**[2], 377
Ω t789	—,**3,5-dibromo-(*L*)**	Bromotiren. $C_9H_9Br_2NO_3$. *See* t782	339.00	nd or pl (w + 2) $[\alpha]^{20}_{D}$ −5.5 (1N HCl, c = 5)	245 (anh)	...	...	...	δ^h	δ s^h	i	...	...	ac, alk s	**B14**[2], 377
—	—,**diiodo-**	*see* **Iodogorgoic acid**													
t790	—,***N*-methyl-(*L*)**	Andirine. Surinamine. $C_{10}H_{13}NO_3$. *See* t782	195.21	nd, $[\alpha]^{21}_{D}$ +19.8 (dil HCl)	257 (d280)	...	...	...	i	δ	i	...	...	NH_4OH v dil ac s	**B14**[1], 665

For explanations, symbols and abbreviations see beginning of table. For structural formulas see end of table.

No.	Name	Synonyms and Formula	Mol. wt.	Color, crystalline form, specific rotation and λ_{max} (log ε)	m.p. °C	b.p. °C	Density	n_D	Solubility w	al	eth	ace	bz	other solvents	Ref.
	Umbellatine														
u1	**Umbellatine**	$C_{21}H_{21}NO_8$	415.71	ye nd	206–7	...	...	...	δ s^h	s	...	δ	...	chl δ	**C48,** 10034
—	**Umbellic acid**	*see* **Cinnamic acid, 2,4-dihydroxy-**													
—	**Umbelliferone**	*see* **Coumarin, 7-hydroxy-**													
—	**Undecane***	*see* **Hendecane***													
—	**Undecanoic acid***	*see* **Hendecanoic acid***													
—	**Undecanol***	*see* **Hendecanol***													
—	**Undecanone***	*see* **Hendecanone***													
—	**Undecenoic acid***	*see* **Hendecenoic acid***													
—	**Undecyl alcohol**	*see* **1-Hendecanol***													
—	**Undecylic acid**	*see* **Hendecanoic acid***													
—	**Undecyne***	*see* **Hendecyne***													
Ω u2	**Uracil**	...	112.10	nd (w) $\lambda^{w, pH=2.4}$ 205 (3.9), 260 (3.9) $\lambda^{w, pH=7.2}$ 259.5 (3.91) $\lambda^{w, pH=12}$ 284 (3.79)	338	...	...	...	δ v^h	v	v	...	...	dil NH_3 s	**B24**[2], 168
Ω u3	—,**5-amino-**	$C_4H_5N_3O_2$. *See* u2	127.11	nd (w) λ^w 225 (4.02), 290 (3.93)	d	...	...	...	i δ^h	...	...	...	...	alk, ac s^h	**B24**[2], 266
Ω u4	—,**5,6-dihydro-**	Hydrouracil. β-Lactylurea. $C_4H_6N_2O_2$. *See* u2	114.10	nd (w) λ^{NaOH} 230 (3.91)	275–6	...	...	...	v	s	...	...	...	MeOH, chl s	**B24**[2], 140
u4[1]	—,**5,6-dihydro-5-methyl-**	Dihydrothymine. $C_5H_8N_2O_2$. *See* u2	128.14	(w or al) λ^{NaOH} 230 (3.91)	264–5	...	...	...	v^h	δ	...	...	...	...	**B24**[1], 306
Ω u5	—,**5-hydroxy-**	Isobarbituric acid. $C_4H_4N_2O_3$. *See* u2	128.09	pr (w) λ^w 210 (3.82), 278 (3.72)	d > 300	...	...	...	s	...	...	...	...	alk s	**B24**[1], 408
u6	—,**1-methyl-**	$C_5H_6N_2O_2$. *See* u2	126.12	pr (w), λ^w 267.5 (3.99)	179	...	...	...	v	v	δ	...	i	alk s	**B24**[2], 170
Ω u7	—,**3-methyl-**	$C_5H_6N_2O_2$. *See* u2	126.12	pr (al), nd (w) λ^w 278.5 (3.82)	232	...	...	...	v	v^h	...	...	...	dil alk s	**B24**[2], 170
Ω u7[1]	—,**5-methyl-**	Thymine. $C_5H_6N_2O_2$. *See* u2	126.12	nd (al), pl (w) λ^w 265 (3.90)	326	sub	...	...	δ	δ	δ	...	...	...	**B24,** 353
Ω u8	—,**6-methyl-**	$C_5H_6N_2O_2$. *See* u2	126.12	oct, pr or nd (w or al) λ^w 260 (4.0)	270–80d	...	...	...	s	s^h	δ	...	...	NH_3 v alk s	**B24**[2], 182
Ω u9	—,**6-methyl-2-thio-**	$C_5H_6N_2OS$. *See* u2	142.18	pl (w)	299–303d	...	...	...	i s^h	δ	δ	...	δ	MeOH δ	**B24**[2], 183
u10	—,**6-methyl-4-thio-**	$C_5H_6N_2OS$. *See* u2	142.18	ye pr (w) λ^{MeOH} 277 (4.20)	>250d	...	...	...	δ^h	δ	...	...	...	...	**B24,** 352
Ω u11	—,**5-nitro-**	$C_4H_3N_3O_4$. *See* u2	157.09	gold nd (al) λ^w 235 (3.90), 338 (4.02)	exp >300	...	...	...	δ	s^h	...	...	...	...	**B24**[2], 171
Ω u12	—,**6-propyl-2-thio-**	$C_7H_{10}N_2OS$. *See* u2	170.24	pw (w)	219	...	...	...	δ s^h	δ	i	...	i	chl δ	**C43,** 674
Ω u13	—,**2-thio-**	$C_4H_4N_2OS$. *See* u2	128.15	pr (w or al) λ^{MeOH} 273 (4.14)	>340d	...	...	...	δ	δ	...	...	...	anh HF s	**B24**[2], 171
u14	—,**4-thio-**	$C_4H_4N_2OS$. *See* u2	128.15	yesh nd or pr (w) λ^w 332 (4.25)	328d (289d)	...	...	...	δ s^h	δ	...	...	...	...	**B24,** 323
u15	**6-Uracilcarboxylic acid**	Orotic acid.	156.10	pr (w + 1) λ^{NaOH} 215 (4.40), 284 (3.77)	347d (322–5)	...	...	...	δ	...	...	...	...	os i	**B25**[2], 249
Ω u16	—,**monohydrate**	$C_5H_4N_2O_4 . H_2O$. *See* u15	174.11	pr (w)	125–30d	...	...	...	δ s^h	...	...	i	i	os i	**B25**[2], 249
—	**Uramil**	*see* **Barbituric acid, 5-amino-**													
Ω u17	**Uranin**	Sodium salt of fluorescein.	376.28	ye pw λ^w 478 (3.51), 450 sh (3.59)	...	...	...	...	s	s	...	...	...	glycerol, dil ace s	**B19**[2], 252
—	*p*-**Urazine**	*see* **Diurea**													
Ω u18	**Urea***	Carbamide. Carbonyl diamide. H_2NCONH_2	60.06	tetr pr (al) λ^{NaOH} <220	135	d	1.3230^{20}_4	1.484 (1.602)	v	v	i	...	i	chl i MeOH v abs al, aa, Py s	**B3**[2], 35
Ω u19	—,**nitrate***	$H_2NCONH_2 . HNO_3$	123.07	mcl lf (w)	157d	...	1.690^{20}_4	...	δ v^h	s	i	...	i	chl i HNO_3 δ	**B3**[2], 45
u20	—,**oxalate**	$2(H_2NCONH_2). HO_2CCO_2H$	210.15	mcl (w + 1)	173d	...	1.585	...	s	δ	i	...	...	...	**B3**[2], 48
u21	—,**1-acetyl-**	$CH_3CONHCONH_2$	102.09	nd (al) λ^{NaOH} 222 (3.11)	218	180–90 sub	...	...	δ s^h	s	δ	...	...	...	**B3**[2], 49
u22	—,**1-acetyl-3-methyl-**	$CH_3CONHCONHCH_3$	116.13	tcl (w, al or AcOEt), pr (w)	180–1	d	...	...	s v^h	δ	δ	...	...	...	**B4**[2], 568

For explanations, symbols and abbreviations see beginning of table. For structural formulas see end of table.

No.	Name	Synonyms and Formula	Mol. wt.	Color, crystalline form, specific rotation and λ_{max} (log ε)	m.p. °C	b.p. °C	Density	n_D	Solubility w	al	eth	ace	bz	other solvents	Ref.
	Urea														
Ω u23	—,**1-acetyl-2-thio-**	*N*(Thiocarbamyl)acetamide. $CH_3CONHCSNH_2$	118.16	pr (w), rh (al)	165	...	...	...	δ s^h	s	δ	...	...	dil NaOH	**B3^2**, 131
u24	—,***S*-acetyl-2-thio-**, hydrochloride	$H_2NC(:NH)SCOCH_3.HCl$	154.62	pl	109d	...	...	...	v	δ	...	...	...	...	**B3^2**, 133
Ω u25	—,**1-allyl-**	$H_2NCONHCH_2CH:CH_2$	100.13	nd (al)	85	...	...	...	∞	∞	δ	...	...	chl δ peth i	**B4^2**, 664
u26	—,**1-allyl-3-phenyl-**	$C_6H_5NHCONHCH_2CH:CH_2$	176.22	nd (bz)	115–6	...	...	...	δ	s	...	...	s	...	**B12**, 350
Ω u27	—,**1-allyl-2-thio-**	$H_2NCSNHCH_2CH:CH_2$	116.20	mcl or rh pr (w) λ^{MeOH} 248 (2.82)	(i) 71 (ii) 78.4	...	1.219^{20}_{20}	1.5936^{78}_{α}	s	s	δ	...	i	...	**B4**, 211
u28	—,**1(4-aminobenzenesulfonyl)-***	1-Sulfanylurea. Sulfacarbamide.	215.23	(w)	146–8δd	sub 320	...	...	s^h	s^h	...	...	...	alk s	**Am 64**, 1683
u29	—,**1(4-aminobenzenesulfonyl)-2-thio-**	1-Sulfamylthiourea. Sulfathiourea.	231.30	pw, la pl (w) λ^w 257 (4.3)	182 (200d)	...	...	...	i	δ s^h	...	...	...	...	**Am 68**, 761
Ω u30	—,**1-benzoyl-**	$H_2NCONHCOC_6H_5$	164.18	fl or lf (al)	214.5	...	...	...	s^h	s^h	i	...	...	...	**B9^2**, 172
u31	—,**1-benzoyl-2-thio-**	$H_2NCSNHCOC_6H_5$	180.24	pr (dil al)	171	...	...	...	δ	s	i	...	...	...	**B9^2**, 173
Ω u32	—,**1-benzyl-**	Phthalamic acid. $H_2NCONHCH_2C_6H_5$	150.18	nd (al)	149	d200	...	...	δ s^h	s	i	s	δ	...	**B12^2**, 563
u33	—,**1-benzyl-2-thio-**	$H_2NCSNHCH_2C_6H_5$	166.25	pr (w) λ^w 238 (4.15)	164.5–5.5	...	...	...	i	δ	...	...	...	...	**B12^2**, 564
u33[1]	—,**1,3-bis(2-ethoxyphenyl)-***	$C_{17}H_{20}N_2O_3$. *See* u71	300.36	pr (dil al)	125	...	...	...	...	s^h	...	...	...	...	**B13^2**, 179
u33[2]	—,**1,3-bis(4-ethoxyphenyl)-***	$C_{17}H_{20}N_2O_3$. *See* u71	300.36	nd (aa), pr (abs al)	255–6	...	...	...	i	s^h	...	...	...	aa s^h	**B13^2**, 254
Ω u34	—,**1,3-bis(2,4-dinitrophenyl)-***	$C_{13}H_8N_6O_9$. *See* u71	392.25	lf or pr (al), nd (con HNO_3)	204	...	...	...	i	δ	δ	...	...	...	**B12^2**, 410
Ω u35	—,**1,3-bis-(hydroxymethyl)-***	*N,N'*-Dimethylolurea. $HOCH_2NHCONHCH_2OH$	120.12	pr (abs al), pl (w-al)	126 (137–8)	d260	149^{25}	...	s	s	i	...	...	MeOH s	**B3^2**, 49
u36	—,**1,3-bis(1-hydroxy-2,2,2-trichloroethyl)-***	Dichloralurea. EH2. Crag herbicide. $Cl_3CCH(OH)NHCONHCH(OH)CCl_3$	354.83	...	196	...	...	...	i	s	δ	s	i	chl i	**B3^2**, 49
Ω u37	—,**1(2-bromo-2-ethylbutanoyl)-***	Adalin. Carbromal. Uradal. $H_2NCONHCOCBr(C_2H_5)_2$	237.10	rh (dil al)	118	...	1.544^{25}	...	δ s^h	...	...	s	s	...	**B3^2**, 52
u38	—,**1(2-bromo-3-methylbutanoyl)-***	Adabine. Bromoisovalum. Bromural. Pivadorm. $H_2NCONHCOCHBrCH(CH_3)_2$	223.08	nd or lf (to)	154 (160)	sub	1.56^{15}	...	δ v^h	s v^h	v	s	s	alk v lig i	**B3^2**, 51
u39	—,**1(2-bromophenyl)-***	$C_7H_7BrN_2O$. *See* u110	215.06	nd (al)	202	...	...	...	i	δ	...	...	s	chl s	**B12**, 632
u40	—,**1(3-bromophenyl)-***	$C_7H_7BrN_2O$. *See* u110	215.06	nd (abs al-eth-lig)	164–5	...	...	...	...	s	...	...	...	os i	**B12**, 634
Ω u41	—,**1(4-bromophenyl)-***	$C_7H_7BrN_2O$. *See* u110	215.06	nd (bz), (al) λ^{al} 247 (4.40), 276 (3.14)	225–7	d *ca.* 260 (d 296)	...	...	δ^h	v^h	v	...	v	aa v lig δ	**B12^2**, 351
Ω u42	—,**1-butyl-***	$H_2NCONH(CH_2)_3CH_3$	116.16	ta (w), nd (bz)	96	...	...	...	v	v	...	...	...	...	**B4^2**, 634
u43	—,**1-*sec*-butyl-**	$H_2NCONHCH(CH_3)CH_2CH_3$	116.16	*d* forms nd $[\alpha]^{20}_D$ +24.1 (al, c = 1.5), +27.6 (chl, c = 0.3) *dl* form: pr. *l* form: $[\alpha]^{20}_D$ = −27.3 (w, c = 2)	166(*d*) 169–70 (*dl*) 168–9(*l*)	...	...	...	s	s	...	...	...	...	**B4**, 160
Ω u44	—,**1-*tert*-butyl-**	$H_2NCONHC(CH_3)_3$	116.16	nd (w or dil al)	191d	sub >100	...	...	s	v	...	...	δ	...	**B4^1**, 377
u45	—,**1-butyl-3-phenyl-2-thio-***	$C_6H_5NHCSNH(CH_2)_3CH_3$	208.33	pr	85(63)	...	...	...	...	...	...	...	s	CCl_4 s	**B12^2**, 226
u48	—,**1(2-chlorophenyl)-***	$C_7H_7ClN_2O$. *See* u110	170.60	pr (w) λ^{al} 269 (2.8), 283 (2.8)	152	...	...	...	s	v	δ	v	δ	...	**B12**, 600
u49	—,**3(4-chlorophenyl)-1,1-dimethyl-***	CMU Weed killer (Du Pont).	198.65	gy-wh pl (MeOH)	170.5–1.5	...	...	...	i	δ	...	δ	...	...	**C51**, 16534
u50	—,**1(2-chlorophenyl)-2-thio-***	$C_7H_7ClN_2S$. *See* u112	186.66	nd or pl	146–6.5	...	...	...	...	v	...	...	s	NH_3 δ	**B12^2**, 318
u51	—,**1(4-chlorophenyl)-2-thio-***	$C_7H_7ClN_2S$. *See* u112	186.66	pl or nd (al)	178	...	...	...	...	s^h	...	...	s	...	**B12^2**, 329
u52	—,**1,3-diacetyl-**	$CH_3CONHCONHCOCH_3$	144.14	nd (al, 50 % aa)	154–5	sub d 179–80	...	...	δ	δ	s	s	s	os s	**B3^2**, 50
Ω u55	—,**1,1-diethyl-***	$H_2NCON(C_2H_5)_2$	116.16	pl (abs eth), nd (eth)	75	$94–6^{0.02}$	...	...	v	v	s	...	v	lig v	**B4^2**, 611
Ω u56	—,**1,3-diethyl-***	$CH_3CH_2NHCONHCH_2CH_3$	116.16	ta (lig), dlq nd (al)	112.5	263	1.0415	1.4616^{49}	v	v	v	...	...	...	**B4^2**, 608
Ω u57	—,**1,3-diethyl-1,3-diphenyl-***	$C_6H_5N(C_2H_5)CON(C_2H_5)C_6H_5$	268.36	(al or w) λ^{al} 247 (3.94)	79	...	...	...	i	v	...	...	...	...	**B12^2**, 238

For explanations, symbols and abbreviations see beginning of table. For structural formulas see end of table.

No.	Name	Synonyms and Formula	Mol. wt.	Color, crystalline form, specific rotation and λ_{max} (log ε)	m.p. °C	b.p. °C	Density	n_D	Solubility w	al	eth	ace	bz	other solvents	Ref.
	Urea														
u58	**—,1,3-diethyl-1,3-diphenyl-2-thio-***	$C_6H_5N(C_2H_5)CSN(C_2H_5)C_6H_5$	284.43	rh pl (lig), pr (al)	75.5	....	....	....	i	v^h	...	...	...	lig s^h	**B12**, 424
Ω u59	**—,1,3-diethyl-2-thio-***	$CH_3CH_2NHCSNHCH_2CH_3$	132.23	λ^{MeOH} 234 (3.80), 265 (3.86)	144 (78)	d	....	....	s	s	v	...	...	....	**B4²**, 610
Ω u60	**—,1,1-dimethyl-*** ..	$H_2NCON(CH_3)_2$	88.12	mcl pr (al or chl)	182	....	1.255	....	s	δ	i	...	...	....	**B4²**, 573
Ω u61	**—,1,3-dimethyl-*** ..	$CH_3NHCONHCH_3$	88.12	rh bipym (chl-eth)	108	268–70	1.142	....	v	v	i	...	...	....	**B4²**, 568
Ω u62	**—,1,3-dimethyl-1,3-diphenyl-***	$C_6H_5N(CH_3)CON(CH_3)C_6H_5$	240.31	pl (al) λ^{al} 244 (3.95)	121	350	....	....	v	v	δ	v	δ	chl v, CS_2 δ	**B12²**, 236
Ω u64	**—,1,3-dimethyl-2-thio-***	$CH_3NHCSNHCH_3$	104.19	dlq pl	62 (58)	....	....	....	v	v	δ	v	δ	chl v CS_2, peth i	**B4²**, 573
u65	**—,1,1-di(2-naphthyl)-***	$H_2NCON(C_{10}H_7^{\beta})_2$	312.38	nd (al)	192–3	....	....	....	...	δ	...	...	δ	aa δ	**B12**, 1297
u66	**—,1,3-di(1-naphthyl)-***	$C_{10}H_7^{\alpha}NHCONHC_{10}H_7^{\alpha}$	312.38	nd ($PhNO_2$, Py or aa)	296 (280)	sub	....	....	δ^h	...	...	...	...	Py s $PhNO_2$ s	**B12²**, 692
u67	**—,1,3-di(2-naphthyl)-***	$C_{10}H_7^{\beta}NHCONHC_{10}H_7^{\beta}$	312.38	nd (ace, $PhNO_2$, aa)	310	....	....	....	δ^h	δ^h	δ	...	δ	$PhNO_2$ δ^h *i*-AmOH s aa s^h	**B12²**, 723
u68	**—,1,3-di(1-naphthyl)-2-thio-***	$C_{10}H_7^{\alpha}NHCSNHC_{10}H_7^{\alpha}$.....	328.44	nd ($PhNO_2$), lf (to)	207.5	....	....	....	...	δ^h	δ	...	δ	to, $PhNO_2$ s^h CS_2 δ	**B12²**, 696
u69	**—,1,3-di(2-naphthyl)-2-thio-***	$C_{10}H_7^{\beta}NHCSNHC_{10}H_7^{\beta}$.....	328.44	lf (aa, $PhNO_2$) λ^{al} 256 (4.56)	203 (192–3)	....	....	....	δ^h	δ^h	δ^h	...	δ	os δ	**B12²**, 723
Ω u70	**—,1,1-diphenyl-*** ..	Acardite. $H_2NCON(C_6H_5)_2$.	212.25	ta (al) λ^{al} 242 (4.06)	189	d	1.276	....	δ	s	s	...	...	chl s	**B12²**, 241
Ω u71	**—,1,3-diphenyl-*** ..	Carbanilide.	212.25	rh bipym, pr (al) λ^{al} 256 (4.56)	238	262^{760}d	1.239	....	δ	δ	s	δ	i	chl δ aa s	**B12²**, 207
u72	**—,1,3-diphenyl-1-methyl-***	*N*-Methylcarbanilide. $C_6H_5NHCON(CH_3)C_6H_5$	226.28	nd (al or xyl), cr (lig) λ^{al} 242 (4.2), 262 sh (3.8)	106	203–5^{760}d	....	....	δ^h	δ	v	...	s	chl s bz-aa v lig i	**B12¹**, 251
u73	**—,1,3-diphenyl-*S*-methyl-2-thio-**	*S*-Methyl-*N,N'*-diphenylisothiourea. $C_6H_5N{:}C(SCH_3)NHC_6H_5$	242.35	nd (al)	109–10	....	....	....	...	s^h	...	...	...	....	**B12²**, 248
u74	**—,1,1-diphenyl-2-thio-*** ..	$H_2NCSN(C_6H_5)_2$.........	228.32	pr (al)	210d (218)	....	....	....	i	s	...	...	...	....	**B12²**, 242
Ω u75	**—,1,3-diphenyl-2-thio-***	$C_6H_5NHCSNHC_6H_5$	228.32	lf (al) λ^{al} 273 (4.3)	154 (189)	....	....	....	δ	v	v	...	...	chl, olive oil v	**B12²**, 227
u76	**—,1,1-dipropyl-2-thio-***	$H_2NCSN(CH_2CH_2CH_3)_2$...	160.28	....	67	....	....	....	...	...	...	...	...	....	**B4**, 144
u77	**—,1,3-dipropyl-2-thio-***	$CH_3CH_2CH_2NHCSNHCH_2CH_2CH_3$	160.28	lf (w)	71	....	....	....	δ s^h	...	s	...	...	....	**B4**, 143
Ω u78	**—,1,3-diiso-propyl-2-thio-***	$(CH_3)_2CHNHCSNHCH(CH_3)_2$	160.28	nd (w)	141	....	....	....	δ^h	...	...	...	...	....	**B4**, 155
Ω u79	**—,1,3-di(2-tolyl)-2-thio-**	....	256.37	nd (al, sub)	165–6	216–8^{760} (sub)	....	....	i	s^h	i	...	s	chl, aa s	**B12²**, 446
u80	**—,1,3-di(4-tolyl)-2-thio-**	....	256.37	rh bipym pr	176(cor) (178)	....	....	....	i	δ	s	...	...	....	**B12²**, 514
u81	**—,1(3-ethoxy-phenyl)-***	$C_9H_{12}N_2O_2$. *See* u110......	180.21	nd λ^{al} 240.5 (4.13), 280 (3.41)	112	....	....	....	δ	δ s^h	...	...	δ	....	**B13**, 418
u82	**—,1(4-ethoxy-phenyl)-***	Dulcin. $C_9H_{12}N_2O_2$. *See* u110	180.21	lf (dil al), pl (w) λ^{al} 242 (4.26), 290 (3.30)	173–4	d	....	....	δ	s	...	...	...	AcOEt v	**B13²**, 253
Ω u83	**—,1-ethyl-***.......	$H_2NCONHCH_2CH_3$........	88.12	nd (al-eth), (bz)	92.1–0.4	d	....	....	v	v	s	...	v^h	abs eth, CS_2 i chl v	**B4²**, 607
Ω u84	**—,1-ethyl-1-phenyl-***	$H_2NCON(C_2H_5)C_6H_5$	164.21	ta (peth), cr (ace) λ^{al} 236 (2.54)	62.3–0.5	....	....	....	s	...	s	s	...	oos s lig i	**B12²**, 237
Ω u85	**—,1-ethyl-3-phenyl-***	$C_6H_5NHCONHCH_2CH_3$...	164.21	nd (dil al) λ^{al} 240 (4.30), 275 (3.02)	104 (99)	....	....	....	...	s	...	...	...	....	**B12²**, 205
u86	**—,1-ethyl-3-phenyl-2-thio-***	$C_6H_5NHCSNHCH_2CH_3$ ...	180.27	....	107 (102)	....	....	....	...	s^h	...	...	s^h	....	**B12²**, 226
u87	**—,1-ethyl-2-seleno-**	$H_2NCSeNHCH_2CH_3$	151.08	nd (al-peth or w)	*ca.* 125	....	....	....	δ v^h	s^h	...	...	...	....	**B4²**, 610
u88	**—,1-ethylidene-2-thio-***	$H_2NCSN{:}CHCH_3$.........	102.16	(al)	*ca.* 212d	....	....	....	i	δ s^h	δ	...	...	oos i	**B3¹**, 76
u89	**—,1-hydroxy-***	Carbamide oxide. $H_2NCONHOH$	76.06	nd (al)	141	d	....	....	v	δ v^h	i	...	i	....	**B3²**, 78
u90	**—,1(hydroxy-methyl)-***	Methylolurea. $H_2NCONHCH_2OH$	90.08	pr (al)	111	....	....	....	v	s^h	i	...	...	MeOH, aa s	**B3¹**, 27
u91	**—,1(2-iodo-3-methyl-butanoyl)***	Iodival. α-Iodoisovalerylurea. $H_2NCONHCOCHICH(CH_3)_2$	270.07	lf (al)	180–1	....	....	....	δ s^h	v	i	...	δ	dil NaOH s	**B3¹**, 29

For explanations, symbols and abbreviations see beginning of table. For structural formulas see end of table.

No.	Name	Synonyms and Formula	Mol. wt.	Color, crystalline form, specific rotation and λ_{max} (log ε)	m.p. °C	b.p. °C	Density	n_D	Solubility w	al	eth	ace	bz	other solvents	Ref.
	Urea														
u92	**—,isobutyl-**	$H_2NCONHCH_2CH(CH_3)_2$	116.16	pr (w), nd (ace)	141				...	...	i	δ	δ	...	**B4**[1], 376
u93	**—,1(2-isopropyl-4-pentenoyl)-***	$H_2NCONHCOCH(C_3H_7^i)CH_2CH:CH_2$	184.24	nd (al)	194–4.5				i δ^h	δ v^h	s	s	s	chl, aa v CCl_4, CS_2 δ	**B3**[2], 53
u94	**—,1(4-methoxy-phenyl)-***	$C_8H_{10}N_2O_2$. *See* u110	166.18	pl (w) λ^{al} 243 (4.21), 289 (3.23)	168				δ	...	s	...	...	...	**B13**[2], 252
Ω u95	**—,1-methyl-***	$H_2NCONHCH_3$	74.08	rh pr (w or al)	103	d			v	v	i	...	i	CS_2, lig s	**B4**[2], 567
Ω u95[1]	**—,1(2-methyl-2-butyl)-***	*tert*-Amylurea. $H_2NCONHC(CH_3)_2CH_2CH_3$	130.19	mcl (w)	162 (151)				δ	...	...	...	...	...	**B4**[1], 379
u96	**—,1(3-methyl-butyl)-***	Isoamylurea. $H_2NCONHCH_2CH_2CH(CH_3)_2$	130.19	pl (dil al)	96 (150)				δ	s	δ	...	...	...	**B4**[1], 383
u97	**—,1-methyl-3(1-naphthyl)-2-thio-***	$C_{10}H_7^{\alpha}NHCSNHCH_3$	216.31	pl (al)	198 (191–2)				...	s^h	...	...	...	...	**B12**[2], 696
u98	**—,1-methyl-1-nitroso-***	$H_2NCON(NO)CH_3$	103.08	col or yesh pl (eth)	123–4d				i v^h	v	v	v	s	chl s	**B4**[1], 342
u99	**—,1-methyl-3-phenyl-2-thio-***	$C_6H_5NHCSNHCH_3$	166.25	ta, pl	112–3				...	s	...	...	...	...	**B12**[2], 225
u100	**—,*S*-methyl-2-thio-**	2-Methylthiopseudourea. $H_2NC(:NH)SCH_3$	90.15	lf (ace) $\lambda^{pH=7}$ 220 (4.58), $\lambda^{pH=12}$ 238 (3.74)	79				s	s	...	s	...	...	**B3**[2], 132
u100[1]	—,—,hydroiodide	*S*-Methylisothiouronium iodide. $H_2NC(:NH)SCH_3.HI$	218.06	pr	117				v	v	...	...	...	...	**B3**, 192
u100[2]	—,—,nitrate	*S*-Methylisothiouronium nitrate. $H_2NC(:NH)SCH_3.HNO_3$	153.16	(NHO_3)	109–10				s	v	...	...	...	MeOH v HNO_3 s^h	**B3**[1], 78
u101	—,—,sulfate	*S*-Methylisothiouronium sulfate. $2[H_2NC(:NH)CSH_3].H_2SO_4$	278.37	nd (al or w)	244d				δ s^h	i s^h	...	...	...	...	**B3**[2], 132
Ω u102	**—,1-methyl-2-thio-***	$H_2NCSNHCH_3$	90.15	pr λ^w 235 (4.08)	120.5–1.0				v	v	δ	s	...	...	**B4**[2], 572
Ω u103	**—,1(1-naphthyl)-***	$H_2NCONHC_{10}H_7^{\alpha}$	186.22	nd (al)	219–20				i	v	s	...	...	...	**B12**, 1238
Ω u104	**—,1(2-naphthyl)-***	$H_2NCONHC_{10}H_7^{\beta}$	186.22	nd (al)	219				i	v^h	...	...	...	...	**B12**, 1292
u105	**—,1(1-naphthyl)-3-phenyl-2-thio-***	$C_6H_5NHCSNHC_{10}H_7^{\alpha}$	278.38	lf	162–3				...	i	...	...	...	...	**B12**, 1241
Ω u106	**—,1(1-naphthyl)-2-thio-***	$H_2NCSNHC_{10}H_7^{\alpha}$	202.28	pr (al)	198				i	δ	δ	δ	...	...	**C42**, 4560
Ω u107	**—,1(2-naphthyl)-2-thio-***	$H_2NCSNHC_{10}H_7^{\beta}$	202.28	lf (al)	186				...	s^h	...	...	...	os δ	**B12**[2], 723
u108	**—,1-nitro-***	$H_2NCONHNO_2$	105.05	pl (al-peth), lf or pr (al)	158.4–.8d	exp			δ d^h	v	v	v	δ	peth, chl δ aa v	**B3**[2], 99
—	**—,1-oxalyl-**	*see* **Parabanic acid**													
u109	**—,1(2-phenoxy-ethyl)-***	$H_2NCONHCH_2CH_2OC_6H_5$	180.21	nd (w + 2), cr (50 %al)	120–1				s^h	s	...	...	...	...	**B6**[1], 91
Ω u110	**—,1-phenyl-**	Phenylcarbamide.	136.17	nd or pl (w), tab (al) λ^{al} 238 (4.19)	147	238			δ s^h	s^h	δ	...	...	AcOEt s	**B12**[2], 204
Ω u111	**—,1(phenyl-acetyl)-**	$H_2NCONHCOCH_2C_6H_5$	178.19	(al)	209 (212–6)				...	s^h	...	...	...	...	**Am 70**, 4189
u112	**—,1-phenyl-2-thio-***	Phenylthiocarbamide.	152.23	nd (w), pr (al) λ^{al} 245 sh (4.0), 266 (4.19)	154				δ s^h	s	...	...	...	NaOH s	**B12**[2], 225
u112[1]	**—,*S*-phenyl-2-thio-***	*S*-Phenylisothiourea. $C_6H_5SC(:NH)NH_2$	152.23	nd (bz)	96–7d				...	...	...	...	s	...	**B6**[1], 146
Ω u113	**—,1-propyl-***	$H_2NCONHCH_2CH_2CH_3$	102.14	pr (al)	110				δ	s	...	...	...	...	**B4**[3], 261
u114	**—,2-seleno-**	Selenourea. $H_2NCSeNH_2$	123.02	pr or nd (w)	200d (213d)				s v^h	δ s^h	...	...	...	abs al s^h	**B3**[1], 87
—	**—,1(sulfonamyl-phenyl)-***	*see* **Benzenesulfonic acid, ureido-**, amide													
u115	**—,1,1,3,3-tetra-ethyl-***	$(C_2H_5)_2NCON(C_2H_5)_2$	172.27			209^{760} $94–5^{12}$	0.919_4^{20}	1.4474^{20}	i	...	...	...	...	alk, ac i	**B4**[2], 611
Ω u116	**—,1,1,3,3-tetra-methyl-***	$(CH_3)_2NCON(CH_3)_2$	116.16	λ^{cy} 217.5 (3.29)	−1.2	166–7.2^{760} 63–4^{12}	0.9687^{20}	1.4496^{23}	...	δ	δ	...	...	...	**B4**[2], 574
u117	**—,1,1,3,3-tetra-methyl-2-thio-***	$(CH_3)_2NCSN(CH_3)_2$	132.23	λ^{al} 255.5 (4.23)	78–9	245			s v^h	s	δ	...	...	...	**B4**[2], 576
Ω u118	**—,1,1,3,3-tetra-phenyl-***	$(C_6H_5)_2NCON(C_6H_5)_2$	364.45	rh (bz) λ^{al} 266.5 (4.31)	183		1.222		i	δ^h	...	...	...	...	**B12**[2], 242
u119	**—,1,1,3,3-tetra-phenyl-2-thio-***	$(C_6H_5)_2NCSN(C_6H_5)_2$	380.52	nd (al or MeOH)	194.5–5.5				i	δ	s	...	v	con sulf s	**B12**[2], 242

For explanations, symbols and abbreviations see beginning of table. For structural formulas see end of table.

No.	Name	Synonyms and Formula	Mol. wt.	Color, crystalline form, specific rotation and λ_{max} (log ε)	m.p. °C	b.p. °C	Density	n_D	Solubility w	al	eth	ace	bz	other solvents	Ref.
	Urea														
Ω u120	**—,2-thio-***	Thiourea*. H_2NCSNH_2	76.12	rh (al) $\lambda^{w, pH=7.4}$ 238 (4.1)	182	...	1.405	...	s v^h	s	i	...	...	...	**B3²**, 128
Ω u121	**—,2-thio-1(2-tolyl)-**	...	166.25	nd (dil al), (w)	162	...	...	...	v^h	v	δ	...	...	...	**B12²**, 445
u122	**—,2-thio-1(3-tolyl)-**	...	166.25	pr (al)	110–1	...	...	...	s^h	δ	δ	...	...	...	**B12²**, 470
Ω u123	**—,2-thio-1(4-tolyl)-**	...	166.25	pl (al)	188–9	...	...	...	δ s^h	s^h	...	...	...	...	**B12²**, 514
u123[1]	**—,2-thio-1,1,3-trimethyl-***	$CH_3NHCSN(CH_3)_2$	118.20	pr (bz-lig)	87–8	...	...	...	v	v	...	...	s	chl s	**B4²**, 576
u124	**—,2-thio-1,1,3-triphenyl-***	$C_6H_5NHCSN(C_6H_5)_2$	304.42	nd (al)	152	...	...	...	...	δ	...	...	...	...	**B12**, 432
Ω u125	**—,1(2-tolyl)-**	...	150.18	lf (al or w) λ^{al} 236 (4.02), 270 (2.82)	195–6	...	...	...	i δ^h	v	v	δ	...	...	**B12²**, 444
Ω u126	**—,1(3-tolyl)-**	...	150.18	lf (w) λ^{al} 240 (4.25), 277 (2.89)	142	...	...	...	s	v	δ	...	...	...	**B12²**, 470
Ω u127	**—,1(4-tolyl)-**	...	150.18	nd (w), pl (w-aa) λ^{al} 240 (4.26), 281 (3.04)	183 (187)	...	...	...	δ^h	v	i	s	i	...	**B12²**, 511
u128	**—,1,1,3-trimethyl-***	$CH_3NHCON(CH_3)_2$	102.14	pr (eth)	75.5(cor)	232.5^{764}	...	...	s	s	δ	...	δ	...	**B4¹**, 335
Ω u129	**1-Ureacarboxylic acid**, ethyl ester*	Allophanic acid ethyl ester. $H_2NCONHCO_2C_2H_5$	132.13	nd (w), (bz)	195	d	...	...	i s^h	δ	i	...	δ s^h	...	**B3²**, 56
u130	**—,3-phenyl-**, nitrile	1-Cyano-3-phenylurea. $C_6H_5NHCONHCN$	161.17	nd	125	...	...	...	...	s	...	...	...	...	**B12²**, 210
—	**Urethan**	*see* **Carbamic acid,** ethyl ester													
—	**Urethytan**	*see* **Carbamic acid,** methyl ester													
Ω u131	**Uric acid**	2,6,8-Purinetrione. 2,6,8-Trihydroxypurine. $C_5H_4N_4O_3$.	158.12	rh pr or pl $\lambda^{w, pH=5.5}$ 235 (4.01), 290 (4.04) $\lambda^{w, pH=9.4}$ 292 (4.10)	d	d	1.89	...	i δ^h	i	i	...	...	alk v aq Na_2CO_3, glycerol s ac δ	**B26²**, 293
u132	**—,1-methyl-**	$C_6H_6N_4O_3$. *See* u131	182.14	nd	400	...	...	...	i^h	...	...	...	...	...	**B26²**, 299
u133	**—,3-methyl-**	$C_6H_6N_4O_3$. *See* u131	182.14	pr (w + 1) λ^{w} 241 (2.36), 296 (1.80)	>350	...	1.6104^{25} (anh) 1.6334^{25} (hyd)	...	δ^h	δ	...	...	...	alk s	**B26²**, 299
u134	**—,7-methyl-**	$C_6H_6N_4O_3$. *See* u131	182.14	pl (w) $\lambda^{w, pH=1}$ 234 (2.07), 286 (1.95) $\lambda^{w, pH=7}$ 238 (1.99), 294 (1.92) $\lambda^{w, pH=14}$ 297 (1.87)	370–80d	...	1.706_4^{25}	...	δ^h	...	...	...	...	NaOH s	**B26²**, 299
Ω u135	**Uridine**	3(*d*)-Ribofuranosidouracil. $C_9H_{12}N_2O_6$.	244.21	nd (aq al) $[\alpha]_D^{20}$ +4 (w) λ^{w} 262 (4.00) $\lambda^{w, pH=11}$ 262 (3.87)	165	...	...	...	s	s^h	...	...	...	Py s	**J1947,** 338
u136	**Uridylic acid**	Uridine-3′-phosphate. $C_9H_{13}N_2O_9P$.	324.19	pr (MeOH) $[\alpha]_D^{20}$ +10.5(w) λ^{w} 262 (4.02)	202d	...	...	...	v	...	...	...	...	MeOH s^h	**J1940,** 746
u137	**Urocanic acid**	4-Imidazolylacrylic acid. Urocaninic acid. $C_6H_6N_2O_2$.	138.13	pr (w + 2) λ^{w} 277 (4.27)	*cis*: 175–6 *trans*: 218–24 (hyd)	...	...	...	s^h	i	i	s^h	...	...	**B25²**, 121
u138	**Urochloralic acid**	β,β,β-Trichloroethyl-*D*-glucuronide. $C_8H_{11}Cl_3O_7$.	325.23	nd	142	...	...	...	v	v	δ	...	...	...	**B31**, 266
—	**Uronium, *S*-methyl-2-thio-**, inorganic salt	*see* **Urea,** ***S*-methyl-2-thio-**, inorganic salt													
—	**Uropterin**	*see* **Xanthopterin**													
—	**Urotropine**	*see* **Hexamethylenetetramine**													
—	**Ursocholanic acid**	*see* **Cholanic acid**													
u139	**Ursodesoxycholic acid**	3α,7β-Dihydroxy-5β-cholanic acid. $C_{24}H_{40}O_4$.	392.58	pl (al) $[\alpha]_D^{20}$ +57 (abs al, c = 2)	203	...	...	...	...	v	δ	...	...	...	**E14**, 180

For explanations, symbols and abbreviations see beginning of table. For structural formulas see end of table.

No.	Name	Synonyms and Formula	Mol. wt.	Color, crystalline form, specific rotation and λ_{max} (log ε)	m.p. °C	b.p. °C	Density	n_D	Solubility w	al	eth	ace	bz	other solvents	Ref.
	Ursolic acid														
u140	**Ursolic acid**	Malol. Prunol. Urson. $C_{30}H_{48}O_3$.	456.72	(i) pr (eth) $[\alpha]_D^{21}$ +66 (1N al KOH) (ii) (al) $[\alpha]_D^{21}$ +72.4 (MeOH)	(i) 291 (ii) 284.5 –5	....	....	....	i	δ^h	s	s	...	peth, CS_2 i MeOH, chl s aa s^h	**B10[2]**, 202
u141	**Usnic acid**(*d*).	Usninic acid. $C_{18}H_{16}O_7$.	344.33	ye orh pr (ace) $[\alpha]_D^{20}$ +469 $[\alpha]_D^{16}$ +509 (chl, c = 0.7)	204	....	....	....	i	s	s	δ	...	os δ chl s^h aa δ, s^h	**B18[2]**, 241
Ω u142	—(*dl*).	$C_{18}H_{16}O_7$. *See* u141.	344.33	λ^{al} 234 (4.52), 282 (4.39)	193–4	....	....	1.710_β, 1.611_α	i	s	s	...	...	chl s	**B18[2]**, 241
u143	—(*l*).	$C_{18}H_{16}O_7$. *See* u141.	344.33	$[\alpha]_D^{20}$ –480 $[\alpha]_D^{20}$ –367 (chl)	203	....	....	....	i	s	s	...	...	chl v	**B18[2]**, 241
—	**Uvitic acid**	*see* **1,3-Benzenedicarboxylic acid, 5-methyl-***													
—	**Uvitonic acid**	*see* **2,4-Pyridinedicarboxylic acid, 6-methyl-**													
u144	**Uzarin**	$C_{35}H_{54}O_{14}$.	698.82	pr $[\alpha]_D$ –27 (Py)	268–70	....	....	....	δ	...	i	i	...	Py v chl i	**E14**, 233

For explanations, symbols and abbreviations see beginning of table. For structural formulas see end of table.

No.	Name	Synonyms and Formula	Mol. wt.	Color, crystalline form, specific rotation and λ_{max} (log ε)	m.p. °C	b.p. °C	Density	n_D	Solubility w	al	eth	ace	bz	other solvents	Ref.
	Vaccenic acid														
—	**Vaccenic acid**	*see* **11-Octadecenoic acid(*trans*)***													
v1	**Vacciniin**	6-Benzoyl-*D*-glucose. $C_{13}H_{16}O_7$.	284.27	amor (aq ace +1w) $[\alpha]_D^{21}$ +48 (al)	104–6				s	s	i	s	δ	peth i chl δ AcOEt s	**B31**, 123
—	**Valeraldehyde**	*see* **Pentanal***													
—	**Valeric acid**	*see* **Pentanoic acid***													
—	**Valerolactone**	*see* **Pentanoic acid, hydroxy-**, lactone*													
—	**Valeronitrile**	*see* **Pentanoic acid,** nitrile													
—	**Valerophenone**	*see* **1-Pentanone, 1-phenyl-***													
v2	**Valine(*D*)**	*l*-α-Aminoisovaleric acid. *l*-2-Amino-3-methyl-butanoic acid*. $(CH_3)_2CHCH(NH_2)CO_2H$	117.15	pl (aq al) $[\alpha]_D^{20}$ −29.04 (20% al)	156–7.5 (293d, sealed tube)				s	δ	δ	...	δ		**B4**², 853
Ω v3	—(*DL*)	$(CH_3)_2CHCH(NH_2)CO_2H$.	117.15	tcl pl (al)	298d (sealed tube)	sub	1.316		s	δ	δ	...	δ		**B4**², 854
v4	—(*L*)	*d*-α-Aminoisovaleric acid. $(CH_3)_2CHCH(NH_2)CO_2H$	117.15	lf (w-al) $[\alpha]_D^{23}$ +22.9 (20% al, c = 0.8)	93–6 (315 sealed tube)		1.230		s	δ	δ	...	δ		**B4**², 852
v5	—, β-hydroxy-(*dl*)	*dl*-2-Amino-3-hydroxy-3-methylbutanoic acid. $(CH_3)_2C(OH)CH(NH_2)CO_2H$	133.15	pl (dil al)	240d				s	i	i	...	i	AcOEt i	**B4**², 942
—	**Valone**	*see* **1,3-Indandione, 2(3-methylbutyl)-**													
—	**Valylene**	*see* **1-Buten-3-yne, 2-methyl-***													
—	**Vanillic acid**	*see* **Benzoic acid, 4-hydroxy-3-methoxy-**													
—	**Vanillyl alcohol**	*see* **Toluene, α,4-dihydroxy-3-methoxy-**													
—	**Vanillin**	*see* **Benzaldehyde, 4-hydroxy-3-methoxy-**													
v7	**Vasicine(*dl*)**	Peganine. $C_{11}H_{12}N_2O$.	188.23	nd (al) λ^{al} 285 (3.51)	209–10				δ	s	δ	s	δ	chl s	**B23**², 342
v8	—(*l*)	$C_{11}H_{12}N_2O$. *See* v7	188.23	nd (al) $[\alpha]_D^{14}$ −62 (al, c = 2.4), −254 (chl)	211–2				δ	s	δ	...	δ	chl s	**B23**², 342
—	**Veratraldehyde**	*see* **Benzaldehyde, 3,4-dimethoxy-**													
v9	**Veratramine**	$C_{27}H_{39}NO_2$.	409.62	nd $[\alpha]_D^{19}$ −70 (MeOH) λ^{al} 268 (2.8)	209.5–10.5 (hyd)				...	s^h	...	...	s	dil ac s dil alk i chl s	**C55**, 17675
—	**Veratric acid**	*see* **Benzoic acid, 3,4-dimethoxy-**													
v10	**Veraridine**	3-Veratroylveracerive. $C_{36}H_5NO_{11}$.	673.80	yesh amor $[\alpha]_D^{22}$ +8 (al)	130 (anh) 180 (hyd)				s^h	s^h	δ	...	...		**C49**, 569
v11	**Veratrine**	Cevadine. $C_{32}H_{49}NO_9$.	591.75	rh (+2 al) $[\alpha]_D^{17}$ +12.5 (al) λ^{al} 290.5 (2.1)	205d				i δ^h	s	δ	...	...	chl s Py v	**C53**, 18148
—	**Veratrole**	*see* **Benzene, 1,2-dimethoxy-***													
—	**Veratrophenone**	*see* **Benzophenone, 3,3′,4,4′-tetramethoxy-**													
v13	**Veratrosine**	Veratramine *D*-glucoside. $C_{33}H_{49}NO_7$. *See* v9	571.76	nd (aqu MeOH) $[\alpha]_D^{24}$ −55 (al-chl, c = 0.94)	242–3d				i	...	...	...	...		**C39**, 1413
v14	**Verbenone (*d*)**	*d*-4-Piperone. $C_{10}H_{14}O$.	150.22	$[\alpha]_D^{18}$ +249.6 λ^{MeOH} 253 (3.81)	9.8	227–8 $103–4^{16}$	0.9978^{20}	1.4993^{18}	s	s	...	s	s	oos s	**B7**¹, 104
v15	—(*l*)	$C_{10}H_{14}O$. *See* v14	150.22	$[\alpha]_D$ −144 (l = 10)	6.5	253.5 (227–8) 100^{16}	0.9731^{20}	1.4961^{20}	s	s	...	...	s		**E12A**, 515
—	**Vernine**	*see* **Guanosine**													
—	**Veronal**	*see* **Barbituric acid, 5,5-diethyl-**													
—	**Vetivazulene**	*see* **Azulene, 4,8-dimethyl-2-isopropyl-**													
v16	**α-Vetivone**	α-Vetiverone. $C_{15}H_{22}O$	218.34	(peth) $[\alpha]_D$ +238.2 (al, c = 10) λ^{hx} 328 (2.22)	51.5	144.0–4.5^2	1.0035_4^{20} (suc)	1.5370^{20}	...	...	...	s	...		**E12A**, 436

For explanations, symbols and abbreviations see beginning of table. For structural formulas see end of table.

No.	Name	Synonyms and Formula	Mol. wt.	Color, crystalline form, specific rotation and λ_{max} (log ε)	m.p. °C	b.p. °C	Density	n_D	Solubility						Ref.
									w	al	eth	ace	bz	other solvents	
	β-Vetivone														
v17	**β-Vetivone**	β-Vetiverone. $C_{15}H_{22}O$. *See* v16	218.34	(pentane) $[\alpha]_D^{20}$ −38.9 (al, c = 10.6) λ^{hx} 330 (1.52)	44.5	$141–2^{2}$	1.0001_4^{20}	1.5309^{20}	...	...	...	s	...	...	**E12A,** 436
v18	**Vicianose**	6(β-Arabinoside)*D*-glucose. $C_{11}H_{20}O_{10}$	312.28	nd (dil al) $[\alpha]_D^{20}$ +56.5 → +39.7 (w) (mut)	ca. 210d	...	...	...	s	δ	...	...	...	...	**B31,** 371
v19	**Vicine**	$C_{10}H_{16}N_4O_7$.	304.26	nd (w or dil al, +1w) $[\alpha]_D^{25}$ −12 (w, c = 10) $\lambda^{0.1N\,HCl}$ 274 (4.21)	239–42d (anh)	...	...	...	δ	δ	...	...	...	ac v alk v	**B31,** 163
—	**Vinaconic acid**	*see* **1,1-Cyclopropanedicarboxylic acid***													
—	**Vinetine**	*see* **Oxyacanthine**													
—	**Vinyl acetate**	*see* **Acetic acid,** ethenyl ester													
—	**Vinyl amine**	*see* **Ethene, amino-***													
—	**Vinyl bromide**	*see* **Ethene, bromo-***													
—	**Vinyl chloride**	*see* **Ethene, chloro-***													
—	**Vinyl ether**	*see* **Ether, diethenyl***													
—	**Vinyl fluoride**	*see* **Ethene, fluoro-***													
—	**Vinyl iodide**	*see* **Ethene, iodo-***													
—	**Violanthrone**	*see* **Dibenzanthrone**													
v20	**Violaxanthin**	Zeaxanthin diepoxide. $C_{40}H_{56}O_4$	600.89	red pr (MeOH, al-eth or CS_2) $[\alpha]_{Cd}^{20}$ +35 (chl, c = 0.08) λ^{bz} 428 (4.95), 453.5 (5.13), 483 (5.11)	208	...	...	...	...	s	s	...	...	CS_2 s peth i	**B30,** 99
v21	**Violuric acid**	Alloxan-5-oxime. 5-Isonitrosobarbituric acid. Oximidomesoxalylurea. $C_4H_3N_3O_4$	157.09	pa ye rh λ^{al} 310 (3.3)	203–4d	...	...	...	δ	s	...	...	...	...	**B24**[2]**,** 304
v22	**Visnadin**	$C_{21}H_{24}O_7$.	388.42	nd $[\alpha]_D$ +9 (al) λ^{al} 323 (4.14)	85–6	...	...	...	i	s	s	...	...	...	**Am 79,** 3534
v23	**Visnagin**	5-Methoxy-2-methylfuranochromone. $C_{13}H_{10}O_4$	230.22	nd (w or MeOH) λ^{al} 245 (3.5), 320 (2.6)	144–5	...	...	...	δ	δ	...	...	...	chl v	**C53,** 6220
Ω v24	**Vitamin A_1**	Axerophythol. $C_{20}H_{30}O$	286.46	ye pr (peth) λ^{al} 326 (4.66)	63–4	$137–8^{10^{-6}}$	...	...	i	s	s	s	s	oos v	**B6**[3]**,** 2787
Ω v25	**Vitamin B_1**	Aneurin. Thiamine hydrochloride. Thiamine chloride. $C_{12}H_{18}Cl_2N_4OS$.	337.27	(i) mcl pl (MeOH-al) (ii) pl (MeOH-al, w-al) $\lambda^{w, pH=8}$ 235 (4.1), 268 (3.9)	(i) 233–4 (ii) 250	...	...	...	v	δ	i	δ	i	chl i	**Am 74,** 2409
—	**Vitamin B_2**	*see* **Riboflavin**													
v26	**Vitamin B_6**	Adermin. Pyridoxin. $C_8H_{11}NO_3$.	169.18	nd (ace) λ^{al} 286 (3.76) $\lambda^{w, pH=6.8}$ 254 (3.59), 324 (3.86)	160	sub	...	...	v	s	δ	s	...	chl δ	**C53,** 2225
Ω v27	—, hydrochloride	Adermine hydrochloride. Pyridoxin hydrochloride. $C_8H_{11}NO_3.HCl$. *See* v26	205.64	pl (al, ace) $\lambda^{w, pH=7}$ 253 (3.57), 325 (3.85)	206–8 (205–12d)	sub	...	...	v	δ	i	δ	...	chl i	**Am 61,** 1245
—	**Vitamin C**	*see* **Ascorbic acid(*l*)**													
—	**Vitamin D_2**	*see* **Calciferol**													
—	**Vitamin E**	*see* **Tocopherol**													
—	**Vitamin K**	*see* **1,4-Naphthoquinone, 2-hydroxy-3-methyl-***													
Ω v28	**Vitamin K_1**	Natural vitamin K. Phylloquinone. $C_{31}H_{46}O_2$.	450.71	ye visc oil $[\alpha]_D^{20}$ −0.4 (bz, 57.5 %) λ^{MeOH} 263 (4.20), 270 (4.22), 328 (3.50)	−20	140–$45^{10^{-3}}$ >100–120d	0.967_{25}^{25}	1.5250^{25}	i	s	s	s	s	peth, chl s	**Am 76,** 4529
—	**Vitamin K_3**	*see* **1,4-Naphthoquinone, 2-methyl-***													
—	**Vitamin K_6**	*see* **Naphthalene, 1,4-diamino-2-methyl-, dihydrochloride***.													

For explanations, symbols and abbreviations see beginning of table. For structural formulas see end of table.

No.	Name	Synonyms and Formula	Mol. wt.	Color, crystalline form, specific rotation and λ_{max} (log ε)	m.p. °C	b.p. °C	Density	n_D	Solubility w	al	eth	ace	bz	other solvents	Ref.
	Vomicine														
v29	**Vomicine**	$C_{22}H_{24}N_2O_4$.	380.45	nd (80 % al), pr (ace) $[\alpha]^{22}_D$ +80.4 (al, p = 0.4) λ266 (3.95), 291 (3.62)	282	....	....	....	...	δ s^h	δ	δ	...	chl v AcOEt s	**B27**[2], 795

For explanations, symbols and abbreviations see beginning of table. For structural formulas see end of table.

No.	Name	Synonyms and Formula	Mol. wt.	Color, crystalline form, specific rotation and λ_{max} (log ε)	m.p. °C	b.p. °C	Density	n_D	Solubility w	al	eth	ace	bz	other solvents	Ref.
	Xanthaline														
—	**Xanthaline**	*see* **Papaveraldine**													
Ω x1	**Xanthene**	Dibenzo-1,4-pyran. Diphenylmethane oxide. $C_{13}H_{10}O$.	182.23	wh ye lf (al) λ^{cy} 248 (3.9), 283 (3.4), 287 (3.4), 293 (3.3)	100.5	310–2 (315 sub)	...	...	i	δ	s	...	s	aa, chl, lig s	**B17^2**, 72
Ω x2	—,9-hydroxy-	Dibenz-γ-pyranol. Xanthydrol. $C_{13}H_{10}O_2$. *See* x1	198.22	nd (aq al) λ^{al} 290 (3.59), 336 (2.69)	*ca.* 125	...	...	...	δ	s	s	...	...	chl s	**B17^1**, 72
Ω x3	—,9-phenyl-	$C_{19}H_{14}O$. *See* x1	258.32	(al) λ^{al} 250 (3.91), 284 (3.47)	145.1–5.5	...	...	...	...	s^h	...	...	s	aa s lig s^h	**B17^1**, 38
Ω x4	**9-Xanthene-carboxylic acid**	Xanthanoic acid. $C_{14}H_{10}O_3$.	226.24	nd (dil al or MeOH)	223–4	...	...	...	...	s^h	s	...	...	peth s^h	**B18^2**, 279
Ω x5	***peri*-Xantheno-xanthene**	1,1-Binaphthylene, 2:8′,2:8-dioxide. $C_{20}H_{10}O_2$.	282.30	ye pr (chl)	242	400^{20-25}	...	...	...	...	...	...	s	chl s^h	**B19^2**, 52
—	**Xantic acid**	*see* **Xanthogenic acid**													
Ω x6	**Xanthine**	2,6-Purinedione. $C_5H_4N_4O_2$	152.11	yesh pl (w) $\lambda^{w,pH=2}$ 266 (3.96) $\lambda^{w,pH=10}$ 240 (3.91), 277 (3.95)	d	sub	...	...	δ	δ	i	...	...	ac, alk s chl i	**B26^2**, 260
—	—,3,7-dimethyl-	*see* **Theobromine**													
Ω x7	—,7-methyl-	Heteroxanthine. $C_6H_6N_4O_2$. *See* x6	166.14	nd (w)	380d	...	...	...	δ	i	i	...	...	chl i NH_3 s HCl s^h	**B26^2**, 263
—	—,1,3,7-trimethyl-	*see* **Caffeine**													
x9	**Xanthione**	9-Xanthenethione. $C_{13}H_8OS$.	212.28	red nd (al)	156	...	...	...	i	δ	s	...	s	...	**B17^2**, 382
—	**Xanthogenamide**	*see* **Carbamic acid, thiono-, ethyl ester**													
Ω x10	**Xanthogen, diethyl-**	Auligen. Lenisarin. $C_2H_5OC(:S)SSC(:S)OC_2H_5$	242.40	ye nd or pl (al)	31.5–32	$107–9^{0.05}$	1.2604^{25}_4	...	i	δ v^h	v	s	v	peth v chl, CS_2 s	**B3^2**, 154
x11	**Xanthogenic acid**	Ethoxydithioformic acid. Ethylxanthogenic acid. Xanthic acid. C_2H_5OCSSH	122.21	unst	fr −53	25d	...	...	δ	...	...	...	...	...	**B3^2**, 151
Ω x12	**Xanthone**	Diphenylene ketone oxide. Dibenzopyrone. 9-Xanthenone. $C_{13}H_8O_2$	196.22	nd (al) λ^{al} 260 (4.11) 288 (3.66), 366 (3.85)	174	$349–50^{730}$ $143–6^3$	...	...	i	s	s	...	s	chl s peth δ	**B17^2**, 378
x13	—,1,2-dihydroxy-	$C_{13}H_8O_4$. *See* x12	228.22	pa ye nd (aq al + 3w)	166–7	...	...	...	...	s	...	...	...	Py s	**J1958,** 1790
x14	—,1,3-dihydroxy-	$C_{13}H_8O_4$. *See* x12	228.22	nd (aq al + 1w)	259	sub	...	...	...	s	δ	...	...	alk v	**B18^2**, 85
x15	—,1,6-dihydroxy-	Isoeuxanthone. $C_{13}H_8O_4$. *See* x12	228.22	ye nd (dil al)	245–6	...	...	...	i	s	s	...	...	...	**B18**, 113
Ω x16	—,1,7-dihydroxy-	Euxanthone. $C_{13}H_8O_4$. *See* x12	228.22	yel nd (to), pl (al) λ^{al} 263 (4.0), 305 (4.1), 355 (3.8)	240	sub, δd	...	...	i	s^h	δ	...	...	con alk s	**B18^2**, 87
x17	—,1,8-dihydroxy-	$C_{13}H_8O_4$. *See* x12	228.22	ye lf (bz) λ^{al} 252 (3.90), 334 (3.30), 380 (2.90)	187	...	...	...	i	s^h	δ	...	s^h	alk, chl s	**B18^2**, 87
x18	—,2,3-dihydroxy-	$C_{13}H_8O_4$. *See* x12	228.22	ye nd (al)	294	...	...	...	i	v	δ	v	v	chl δ aa v	**B18**, 116
x19	—,2,7-dihydroxy-	β-Isoeuxanthone. $C_{13}H_8O_4$. *See* x12	228.22	ye nd (al or eth)	>330	sub	...	...	i	s	s	...	...	alk s	**B18^2**, 88
x20	—,3,4-dihydroxy-	$C_{13}H_8O_4$. *See* x12	228.22	pa red-ye nd (dil al + 3w)	240 (anh)	...	...	...	i	v	...	...	...	alk s sulf s (ye)	**B18^2**, 88
x21	—,3,6-dihydroxy-	$C_{13}H_8O_4$. *See* x12	228.22	nd (dil al), pr (sub)	300–50d	sub	...	...	i	v	...	...	i	chl δ aa s	**B18^1**, 357
x22	—,1,7-dihydroxy-3-methoxy-	Gentianin. Gentisin. $C_{14}H_{10}O_5$. *See* x12	258.23	ye rh λ^{al} 260 (4.7), 315 (4.2)	266–7	400 sub	...	...	i	v	...	i	...	Py s	**B28^2**, 158
x23	**Xanthophyll**	Lutein. Luteol. $C_{40}H_{56}O_2$	568.89	ye or vt pr (eth-MeOH) $[\alpha]_{Cd}$ + 160 (chl)	196	...	...	...	i	s	s	...	s	peth s	**B30**, 94
Ω x24	**Xanthopterin**	2-Amino-4,6-pteridinediol. Uropterin. $C_6H_5N_5O_2$.	179.14	hyg ye amor or og (aa) λ^{alk} 255 (4.25), 392 (3.82)	>410d	$98–100^{18}$	1.559	...	i	δ	δ	...	...	NH_3 s alk s (ye)	**B26^2**, 313
—	**Xanthopurpurin**	*see* **9,10-Anthraquinone, 1,3-dihydroxy-**													
x25	**Xanthotoxin**	Ammoidin. 8-Methoxypsoralen. Zanthotoxin. $C_{12}H_8O_4$	216.21	pr (dil al), nd (peth or bz-peth) λ^{al} 249 (4.35), 300 (4.06)	148	...	...	...	δ	v^h	δ	δ	...	peth δ	**B19^1**, 711

For explanations, symbols and abbreviations see beginning of table. For structural formulas see end of table.

No.	Name	Synonyms and Formula	Mol. wt.	Color, crystalline form, specific rotation and λ_{max} (log ε)	m.p. °C	b.p. °C	Density	n_D	Solubility w	al	eth	ace	bz	other solvents	Ref.
	Xanthotoxol														
x26	**Xanthotoxol**	8-Hydroxy-4′:5′, 6:7-furocoumarin. $C_{11}H_6O_4$	202.17	....	251–2	....	....	....	i	...	...	...	...	diox s^h	**Am 72,** 4826
x27	**Xanthoxyletin**	Alloxanthylethin. Xanthoxylin-N. $C_{15}H_{14}O_4$.	258.28	pr (MeOH or peth) λ^{al} 272 (4.33), 348 (4.05)	133	....	....	....	i	s^h	δ	s	v	alk v chl v	**J1936,** 627
—	**Xanthurenic acid**	*see* **2-Quinolinecarboxylic acid, 4,8-dihydroxy-**													
—	**Xanthuric acid**	*see* **2-Quinolinecarboxylic acid, 4,8-dihydroxy-**													
—	**Xanthydrol**	*see* **Xanthene, 9-hydroxy-**													
x28	**Xanthyletin**	2,2-Dimethyl-chromenocoumarin. $C_{14}H_{12}O_3$.	228.25	pr λ^{al} 266 (4.34), 348 (4.15)	131.5	140–5$^{0.1}$	....	....	...	s^h	...	...	...	peth s^h	**J1936,** 1828
—	**Xenylamine**	*see* **Biphenyl, 4-amino-**													
—	**Xylene**	*see* **Benzene, dimethyl-***													
—	**Xylenol**	*see* **Benzene, dimethyl(hydroxy)-***													
—	**Xylic acid**	*see* **Benzoic acid, 2,4-dimethyl-***													
—	**Xylidine**	*see* **Benzene, amino(dimethyl)-***													
Ω x29	**Xylitol**	Xylit. $C_5H_{12}O_5$.	152.15	(i) rh (al) (metast) (ii) mcl (al) (st)	(i) 61–1.5 (ii) 93–4.5	(ii) 215–7^1	....	....	v	s	...	...	...	Py s	**B31^2,** 604
—	**Xyloquinone**	*see* **Benzoquinone, dimethyl-***													
Ω x30	**Xylose(*D*)** (*aldehydo form*)	Wood sugar. $C_5H_{10}O_5$.	150.13	mcl nd $[\alpha]_D^{20}$ +22.5 (chl)	90–1	....	1.525_4^{20}	....	v	s^h	δ	...	...	....	**B31,** 47
x31	—,osazone	$C_{11}H_{20}N_4O_3$.	328.38	pa ye nd $[\alpha]_D$ −40.9 (al) λ 310 (4.02), 397 (4.31)	159 (167d)	....	....	....	δ	s	s	s	...	....	**B31,** 61

For explanations, symbols and abbreviations see beginning of table. For structural formulas see end of table.

No.	Name	Synonyms and Formula	Mol. wt.	Color, crystalline form, specific rotation and λ_{max} (log ε)	m.p. °C	b.p. °C	Density	n_D	Solubility w	al	eth	ace	bz	other solvents	Ref.
	Yamogenin														
y1	**Yamogenin**	$C_{27}H_{42}O_3$.	414.64	pl $[\alpha]_D^{25}$ −123 λ^{sulf} 412 (4.13), 512 (3.40)	201	......	......	......	...	...	s^h	...	...	MeOH s^h	Am 77, 3086
—	**Yellow AB**	*see* **Azo, benzene 1-naphthalene, 2-amino-**													
y2	**Yobyrine**	Yobirine. 1-*o*-Methylbenzyl-9H-pyrid-[3,4-b]indole. $C_{19}H_{16}N_2$.	272.35	nd (dil al) λ^{al} 236 (4.53), 289 (4.23), 338 (3.66), 351 (3.65)	218–9	$150^{0.01}$	......	......	...	s	s^h	...	s^h	chl s	B23[2], 263
Ω y3	**Yohimbine**	Corynine. Quebrachine. Hydroergotocin. Aphrodine. $C_{21}H_{26}N_2O_3$	354.45	nd (dil al) $[\alpha]_D^{20}$ +108 (Py, c = 1) λ^{MeOH} 225 (4.7), 275 (3.9)	241	$159^{0.01}$ sub	......	......	δ	s	s	...	δ s^h	chl s	B25[2], 201
Ω y4	—,hydrochloride ..	$C_{21}H_{26}N_2O_3 . HCl$. *See* y3 ...	390.92	orh nd or pl (w or dil HCl), pr (al) $[\alpha]_D^{22}$ +103.3 (w, c = 2)	302	......	......	......	s^h	δ	i	i	...	....	B25[2], 201
y5	—,nitrate	$C_{21}H_{26}N_2O_3 . HNO_3$. *See* y3	417.47	pr (w)	276	......	......	......	v	...	...	...	...	....	B25[2], 201
—	**Yperite**	*see* **Sulfide, diethyl, 2,2′-dichloro-**													
y6	**Yuccagenin**	$C_{27}H_{42}O_4$.	430.64	nd (al) $[\alpha]_D^{25}$ −113 λ^{sulf} 345 (3.67), 404 (3.91)	252	......	......	......	...	s^h	s^h	...	...	diox s	J1956, 1167

For explanations, symbols and abbreviations see beginning of table. For structural formulas see end of table.

No.	Name	Synonyms and Formula	Mol. wt.	Color, crystalline form, specific rotation and λ_{max} (log ε)	m.p. °C	b.p. °C	Density	n_D	Solubility w	al	eth	ace	bz	other solvents	Ref.
	Zagadinine														
z1	**Zagadinine**	$C_{27}H_{43}NO_7$	493.65	orh (al), nd (bz) [α] −45 (chl)	201–4	....	....	....	δ	s^h	δ	...	s^h	ac, chl s alk, lig i	**Am 35,** 258
z2	**Zeaxanthin**	Zeaxanthol. $C_{40}H_{56}O_2$	568.89	ye pr (MeOH), rh (chl-eth) λ^{al} 275 (4.34), 453 (5.12), 480 (5.07)	215.5 (cor)	$226–9^{0.06}$	....	....	i	δ	s	s	s	chl, Py s peth i^h	**B30, 97**
—	**Zymostanone**	*see* **3-Cholestanone**													
z3	**Zymosterol**	$\Delta^{8(14),24(25')}$-Cholestadienol. $C_{27}H_{44}O$	384.65	pl (MeOH), nd $[\alpha]^{20}_D$ +49 (chl) λ^{al} 209 (3.54)	110	$160^{0.001}$ sub	....	....	...	...	...	s^h	...	MeOH s^h chl s	**E14s,** 1559

For explanations, symbols and abbreviations see beginning of table. For structural formulas see end of table.

STRUCTURAL FORMULAS OF ORGANIC COMPOUNDS

In Alpha Numeric as they occur in Organic Compounds Table

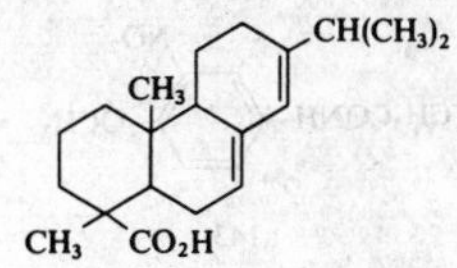

a1

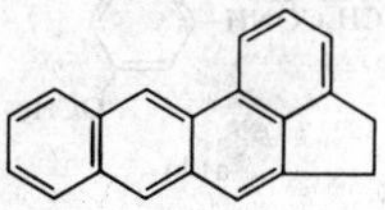

a3

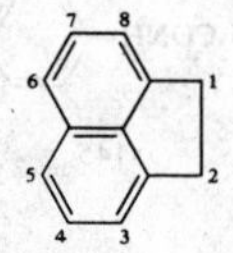

a4

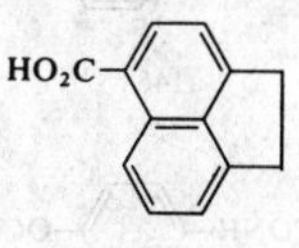

a13

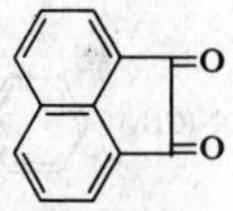

a14

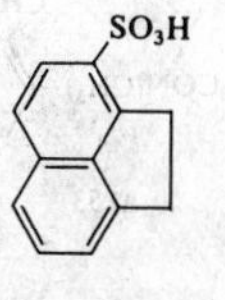

a15

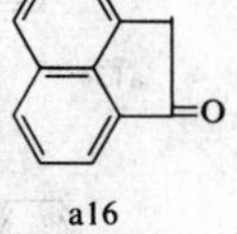

a16

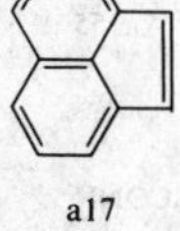

a17

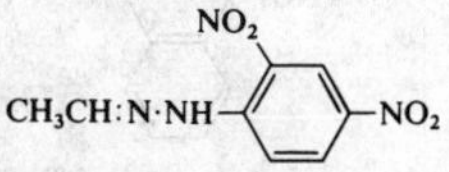

a23

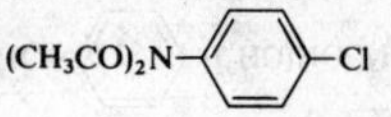

a63

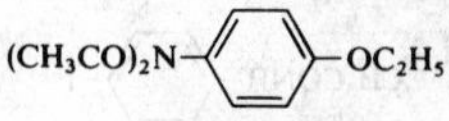

a64

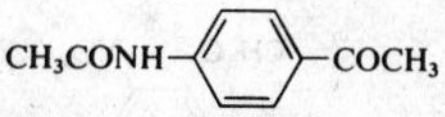

a68

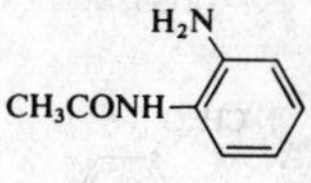

a70

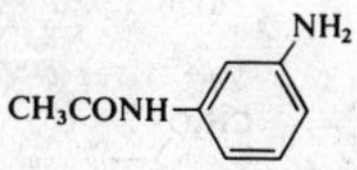

a71

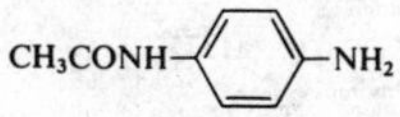

a73

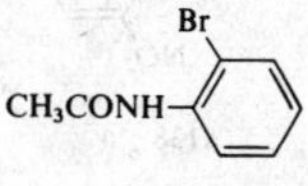

a77

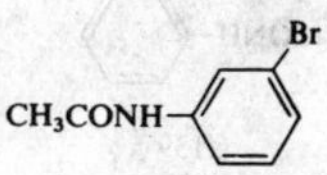

a78

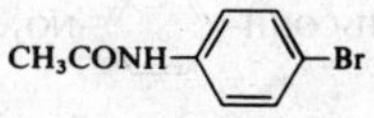

a79

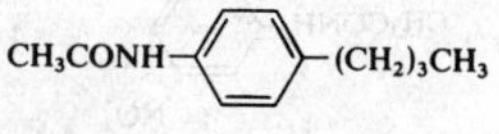

a82

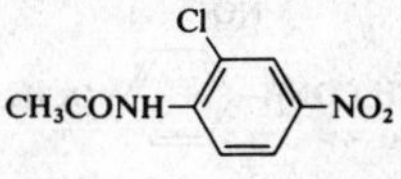

a83

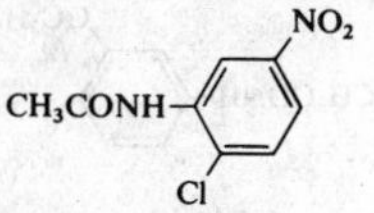

a84

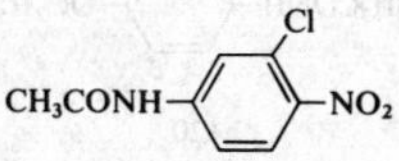

a85

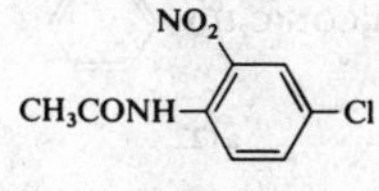

a86

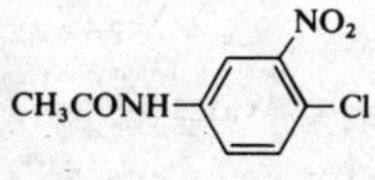

a87

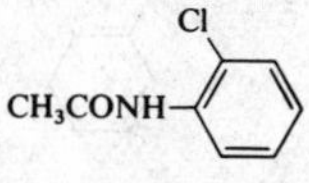

a89

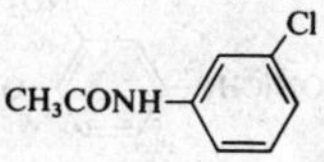

a90

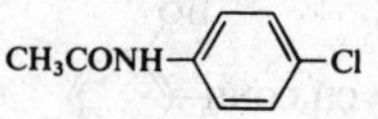

a91

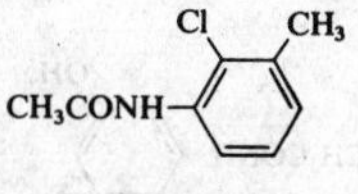

a92

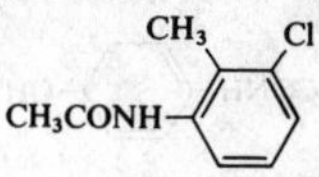

a93

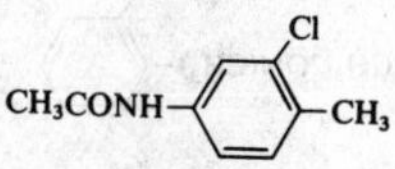

a94

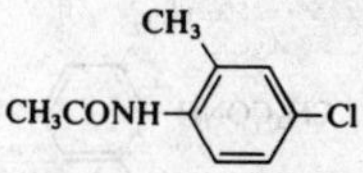

a95

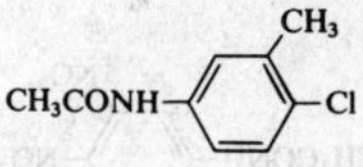

a96

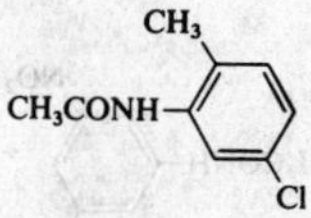

a97

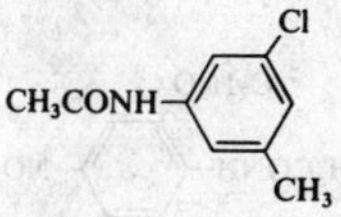

a98

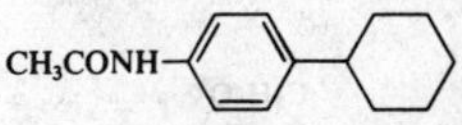

a99

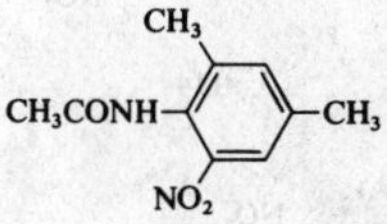

a103

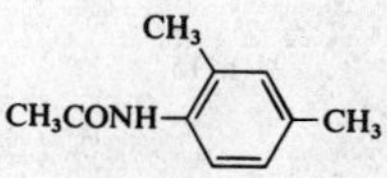

a104

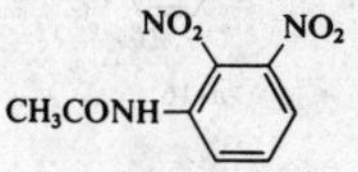

a105

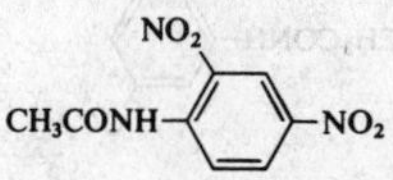

a106

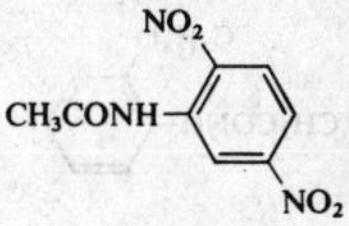

a107

In Alpha Numeric as they occur in Organic Compounds Table

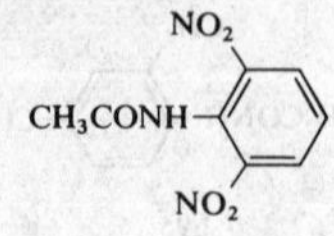

a108

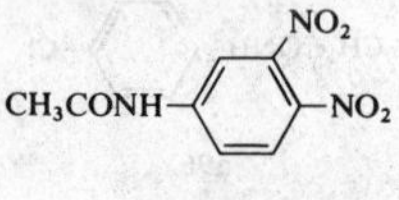

a109

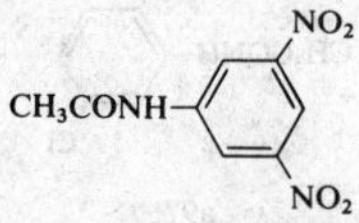

a110

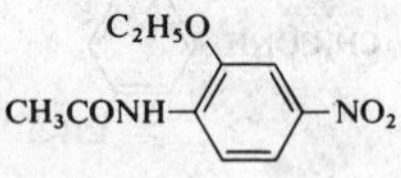

a113

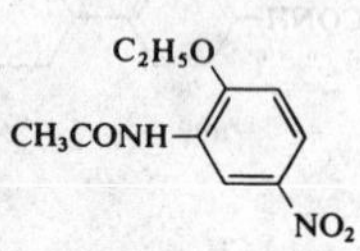

a114

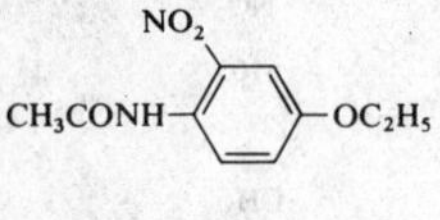

a115

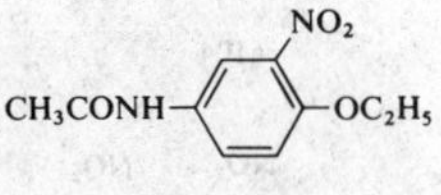

a116

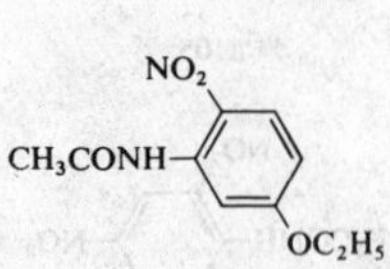

a117

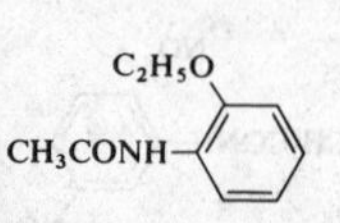

a118

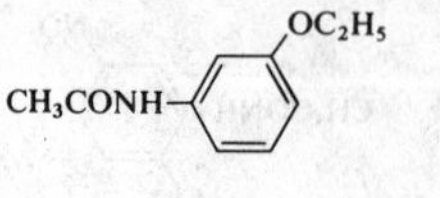

a119

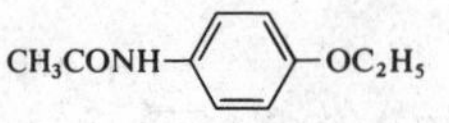

a120

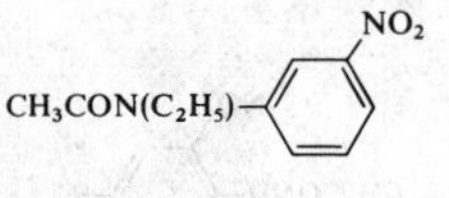

a122

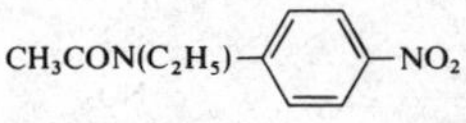

a123

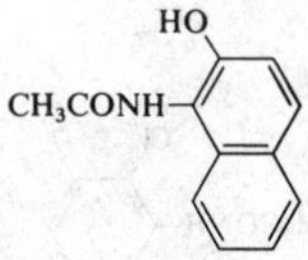

a126

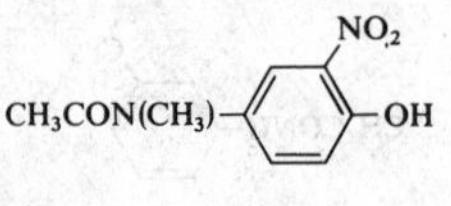

a127

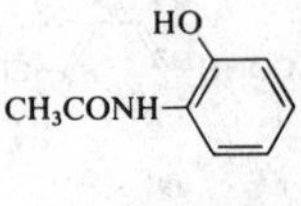

a128

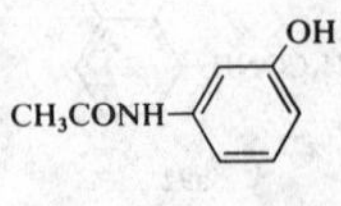

a129

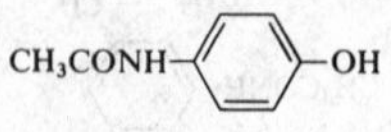

a130

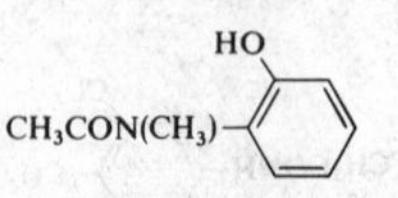

a131

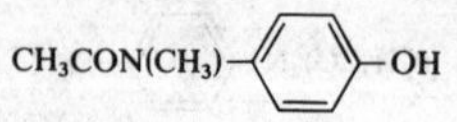

a132

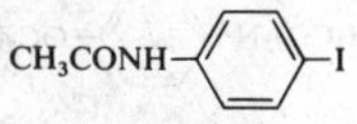

a133

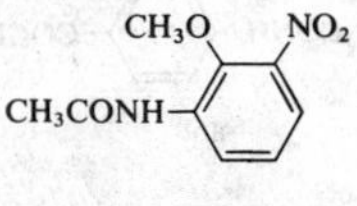

a135

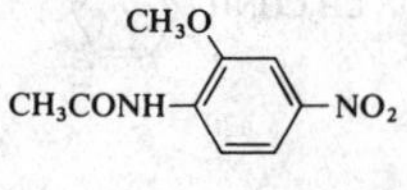

a136

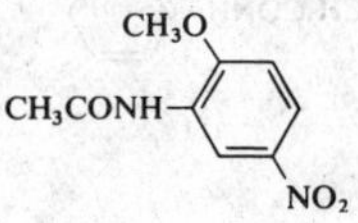

a137

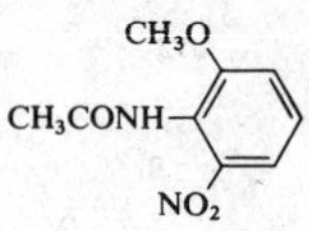

a138

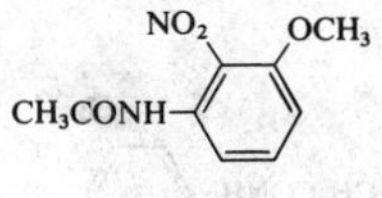

a139

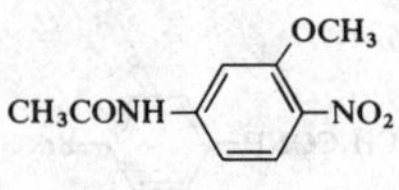

a140

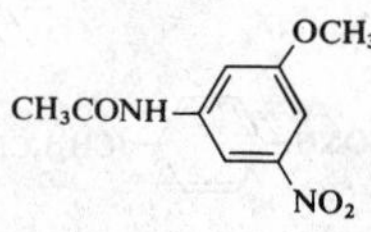

a141

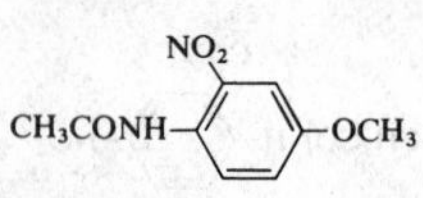

a142

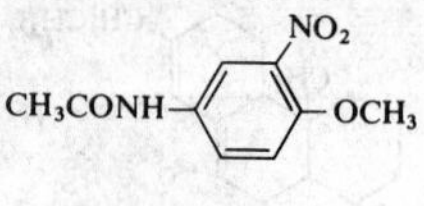

a143

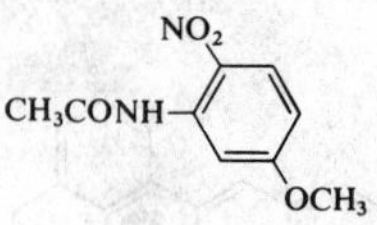

a144

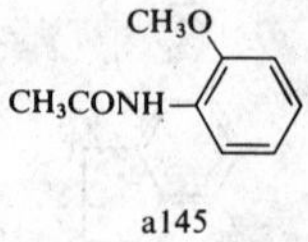

a145

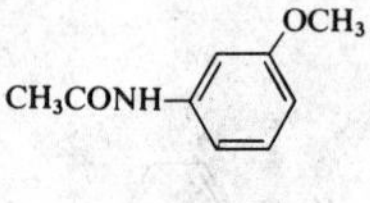

a146

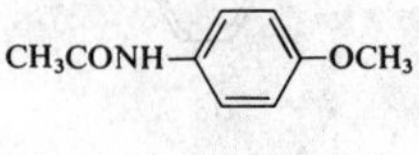

a147

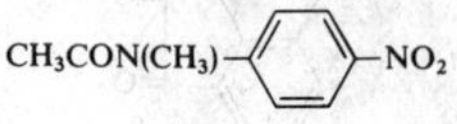

a151

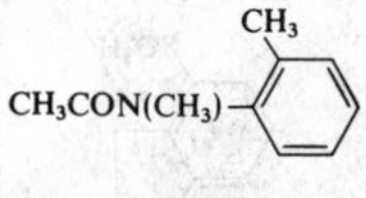

a153

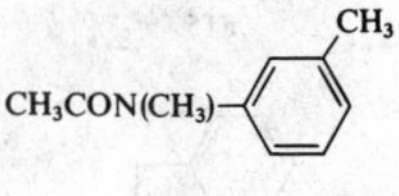

a154

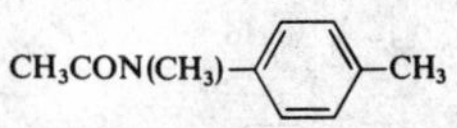

a155

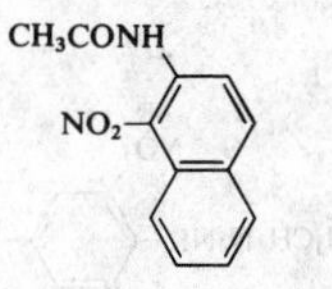

a158

In Alpha Numeric as they occur in Organic Compounds Table

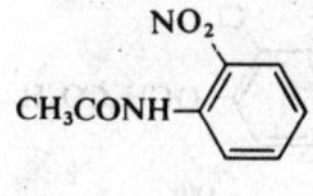

a159

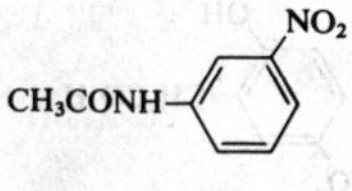

a160

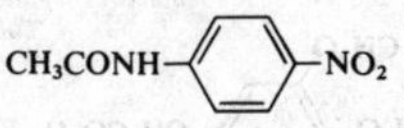

a161

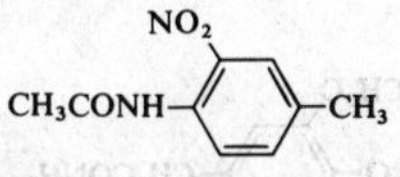

a162

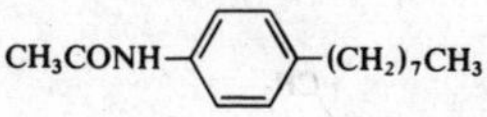

a163

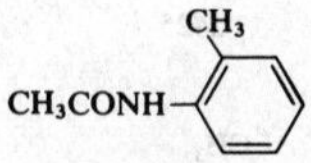

a166

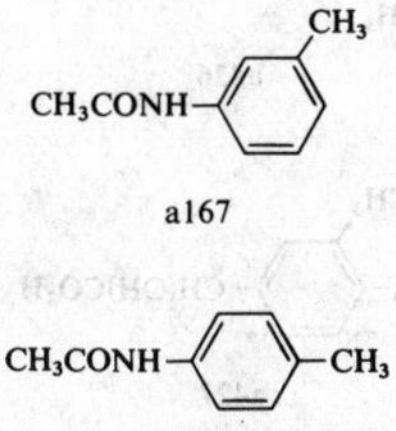

a167

a168

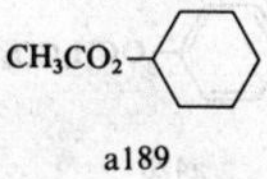

a189

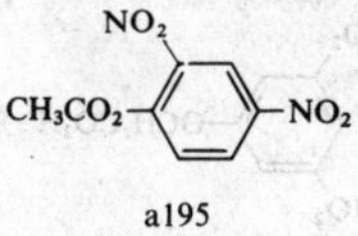

a195

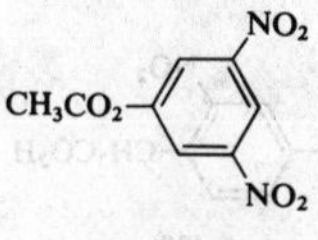

a196

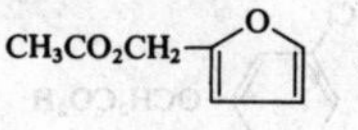

a204

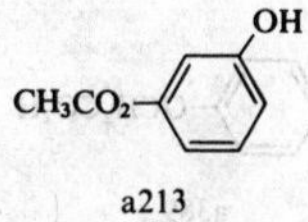

a213

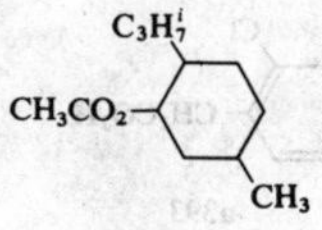

a218

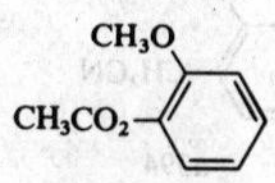

a220

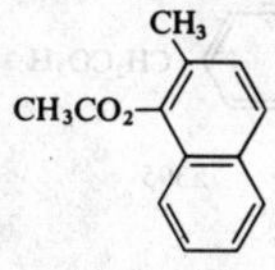

a229

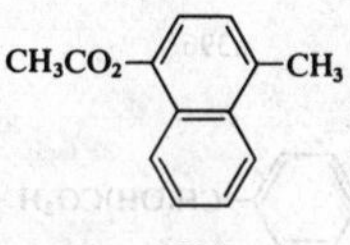

a230

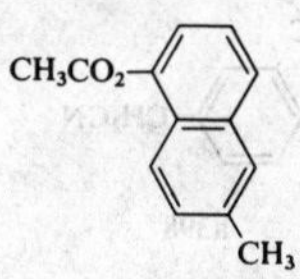

a231

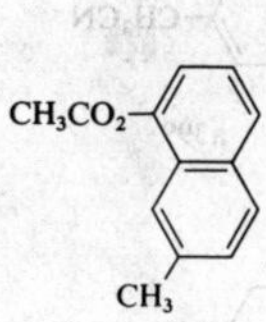

a232

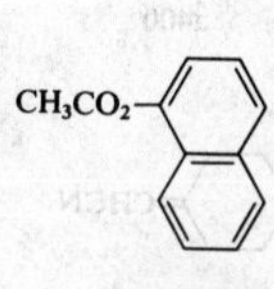

a239

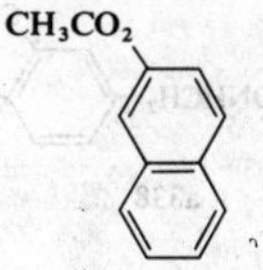

a240

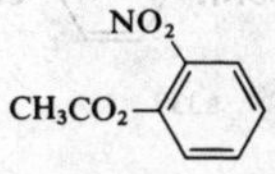

a243

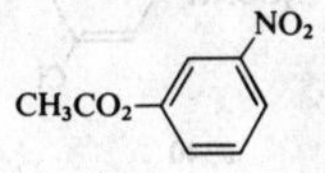

a244

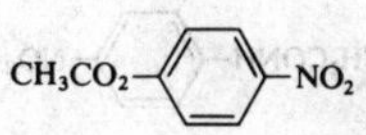

a245

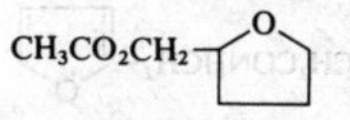

a260

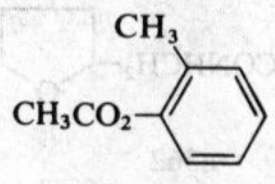

a261

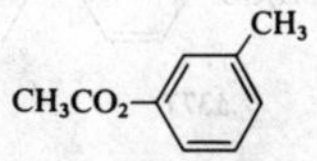

a262

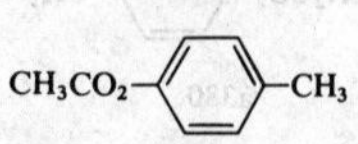

a263

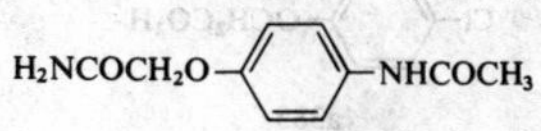

a266

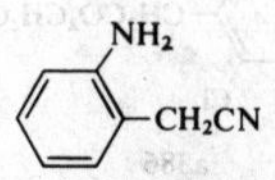

a269

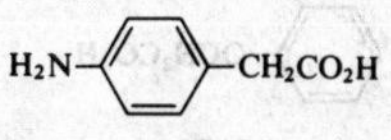

a270

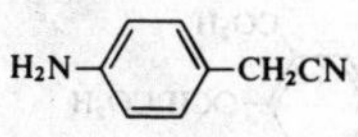

a271

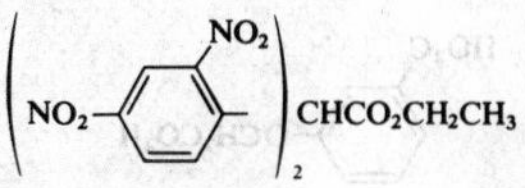

a273

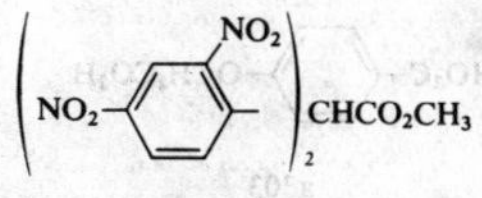

a274

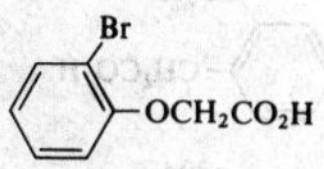

a292

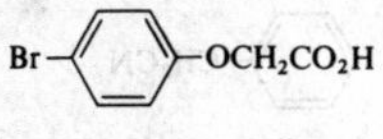

a293

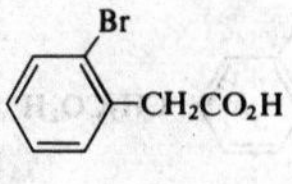

a295

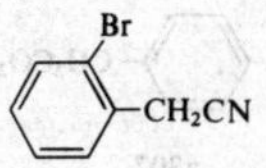

a296

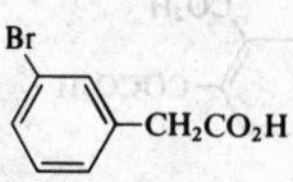

a297

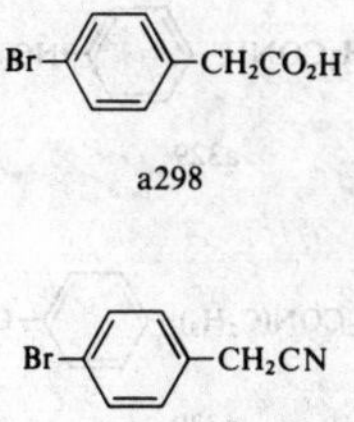

a298

a299

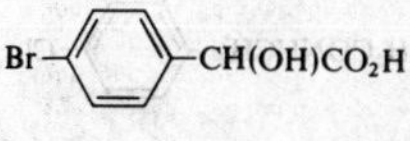

a300

In Alpha Numeric as they occur in Organic Compounds Table

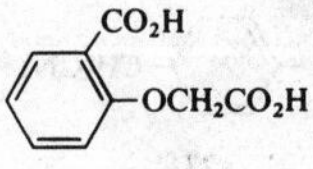

a301

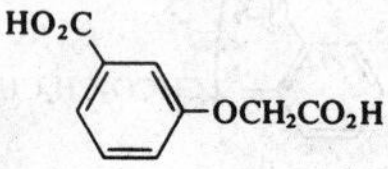

a302

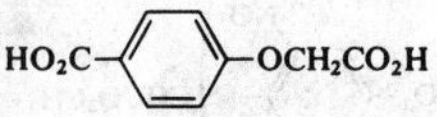

a303

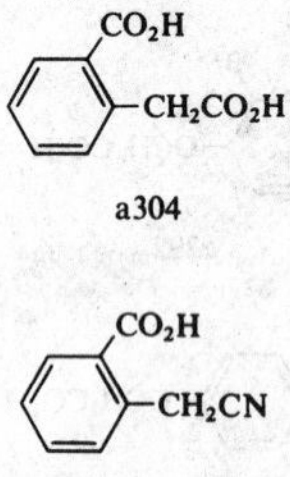

a304

a305

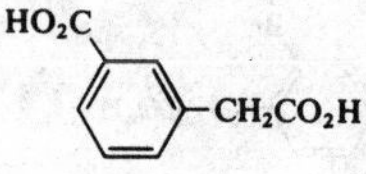

a306

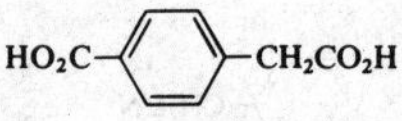

a307

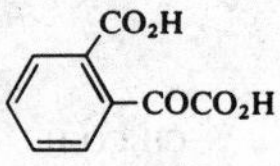

a308

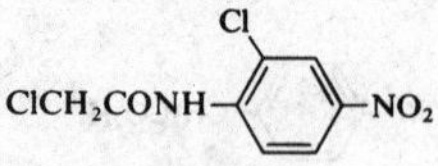

a329

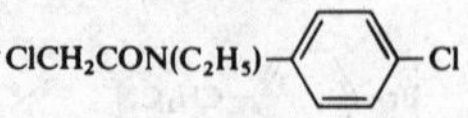

a330

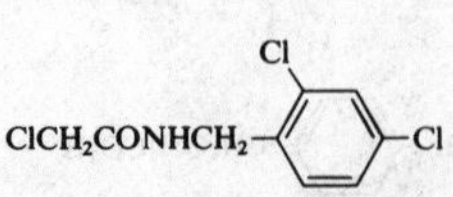

a337

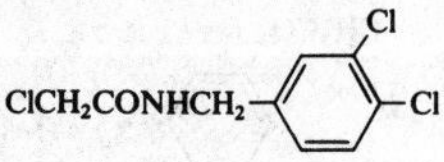

a338

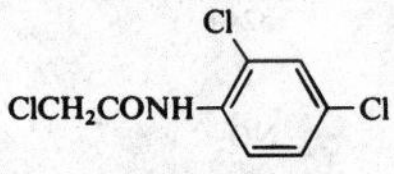

a339

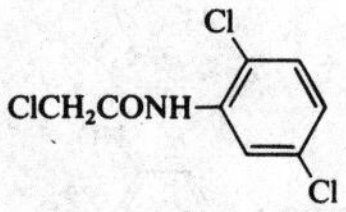

a340

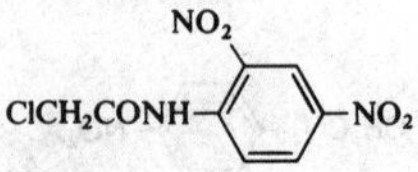

a347

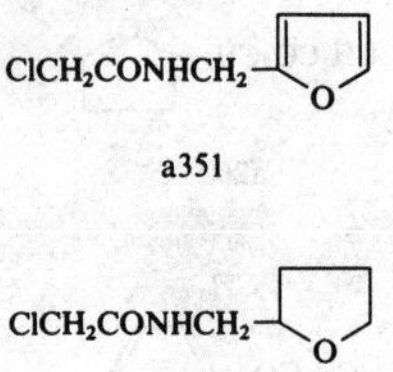

a351

a362

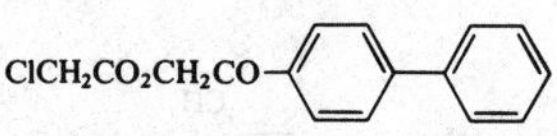

a377

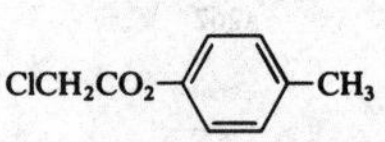

a380

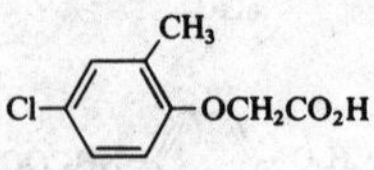

a385

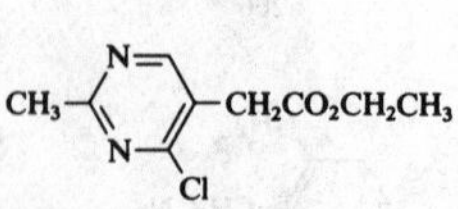

a386

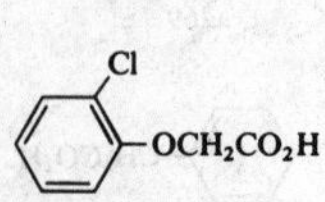

a387

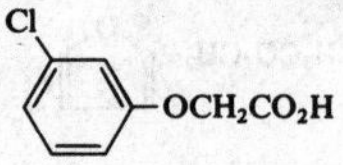

a388

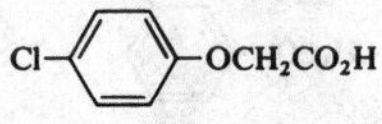

a389

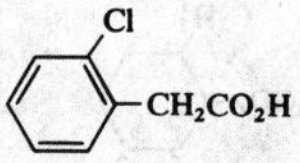

a393

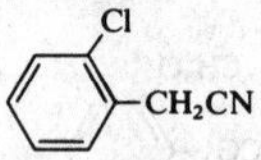

a394

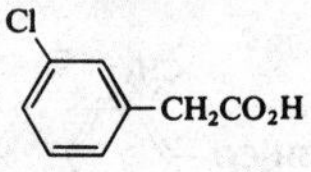

a395

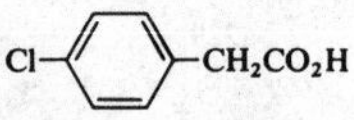

a396

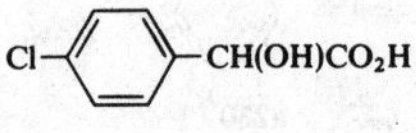

a397

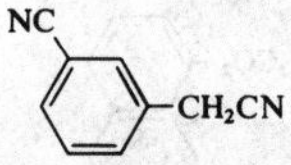

a398

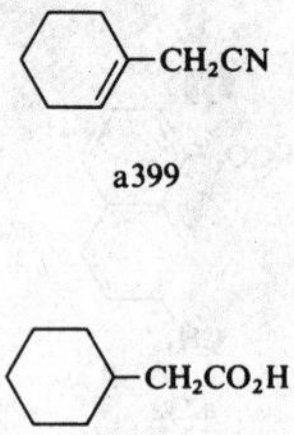

a399

a400

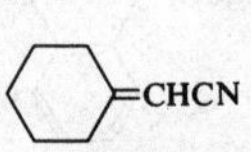

a401

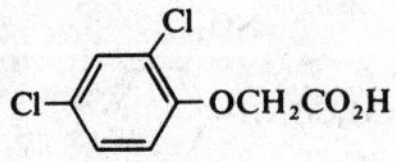

a418

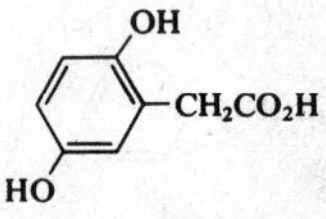

a421

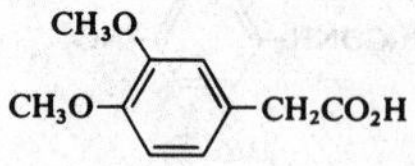

a423

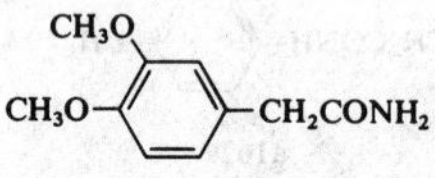

a424

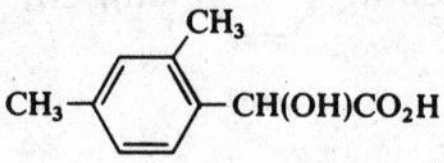

a425

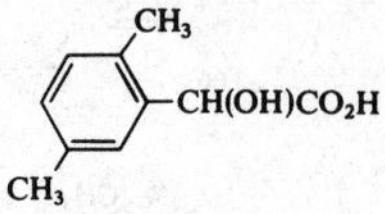

a426

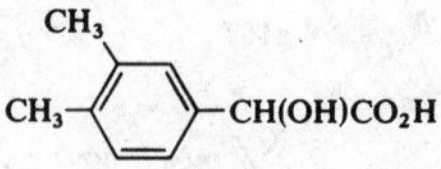

a427

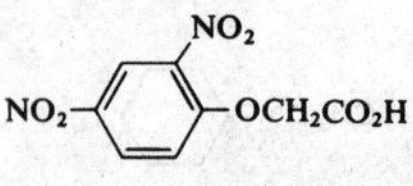

a428

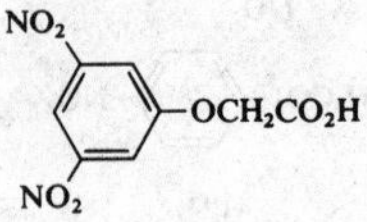

a429

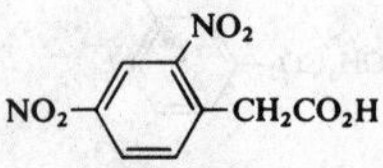

a430

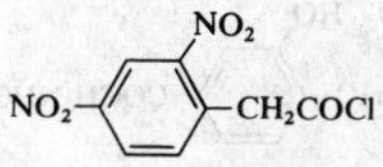
a431

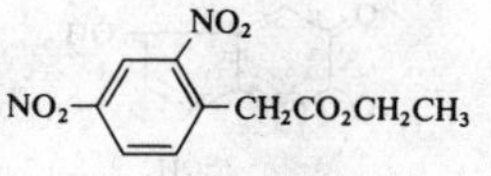
a432

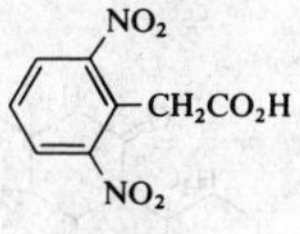
a433

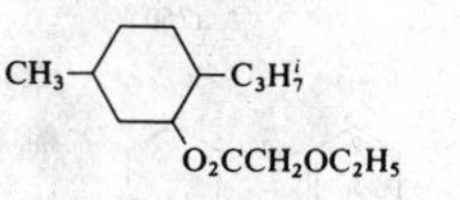
a447

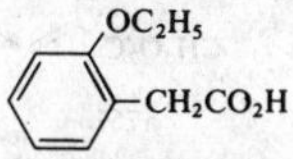
a450

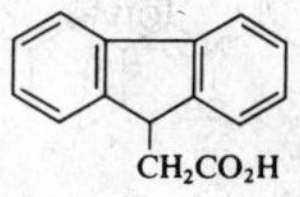
a452

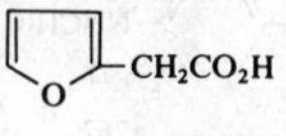
a456

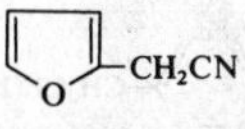
a457

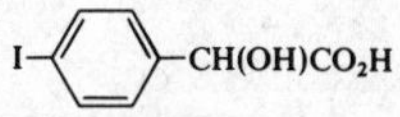
a466

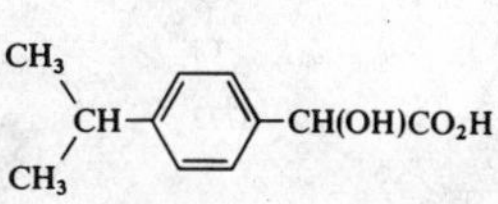
a467

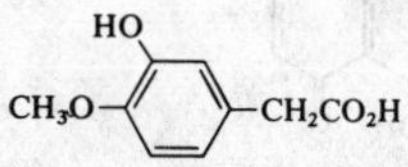
a489

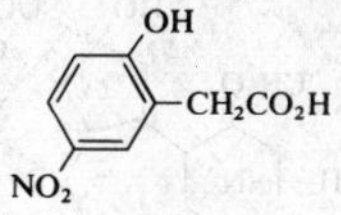
a490

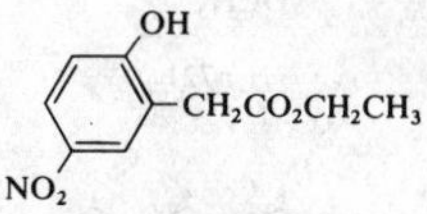
a491

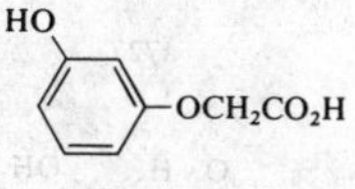
a492

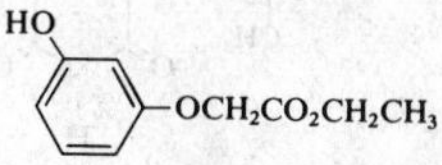
a493

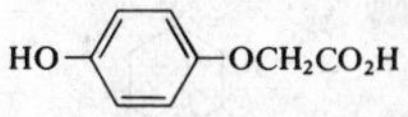
a494

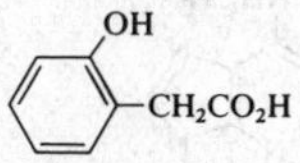
a495

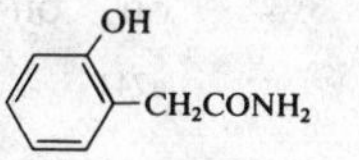
a496

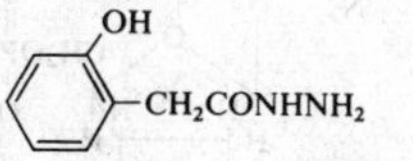
a497

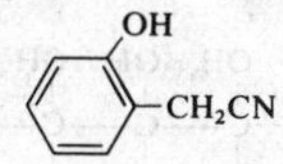
a498

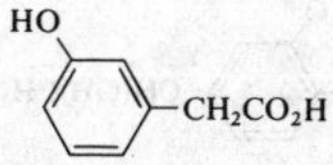
a499

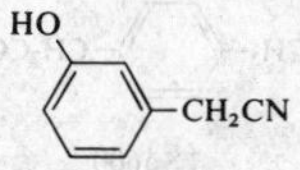
a500

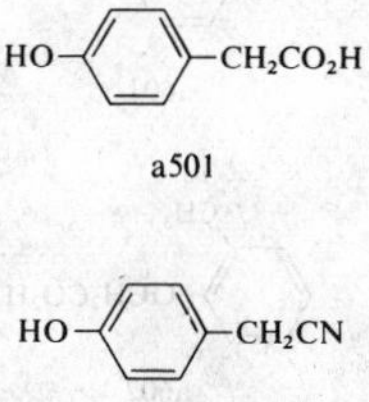
a501

a502

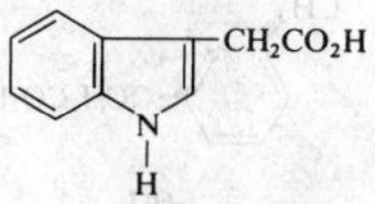
a505

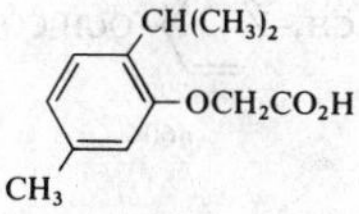
a509

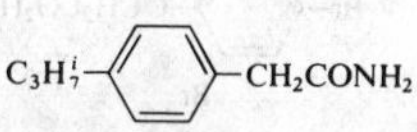
a510

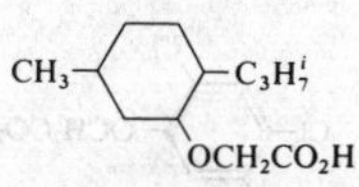
a513

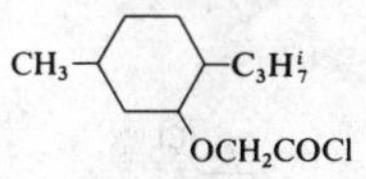
a514

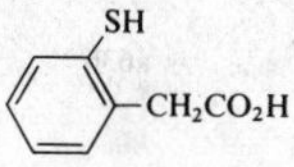
a519

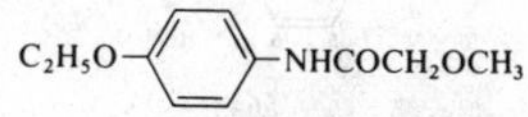
a521

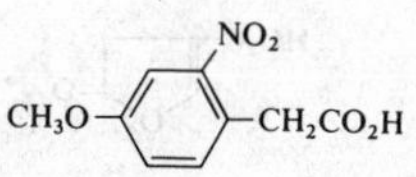
a529

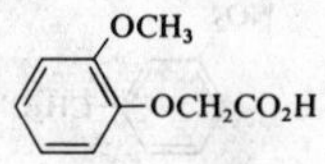
a530

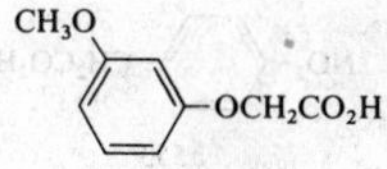
a531

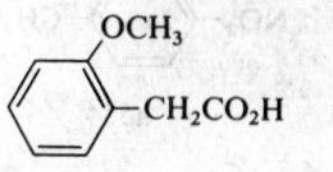
a532

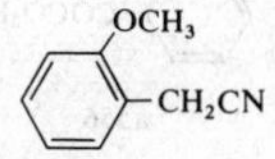
a533

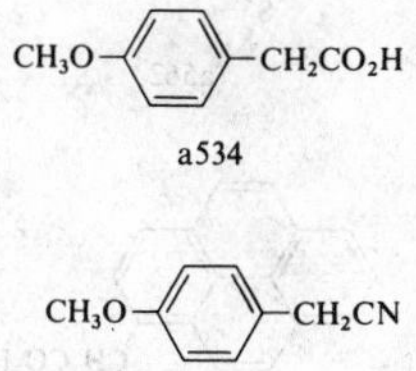
a534

a535

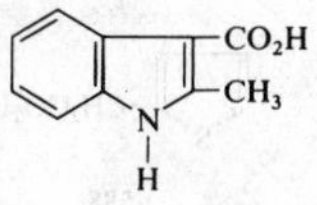
a537

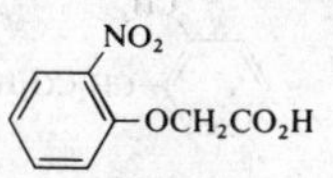
a547

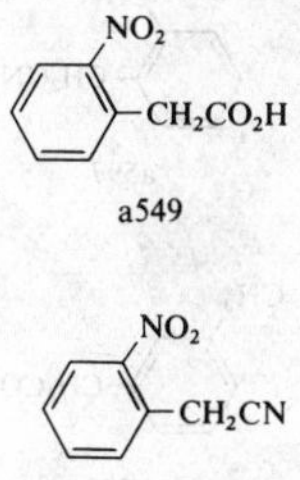
a549

a550

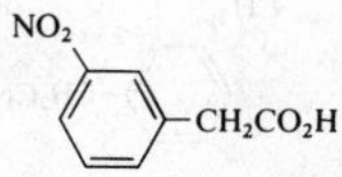
a551

STRUCTURAL FORMULAS OF ORGANIC COMPOUNDS (Continued)

In Alpha Numeric as they occur in Organic Compounds Table

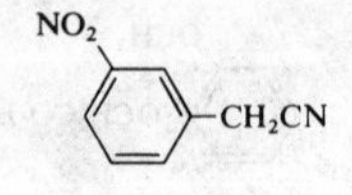

a552

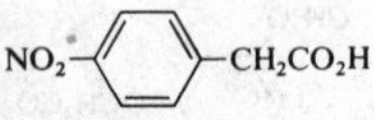

a553

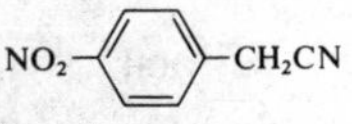

a554

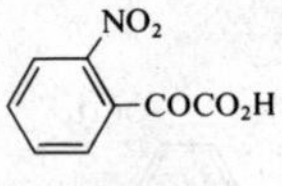

a556

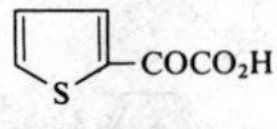

a562

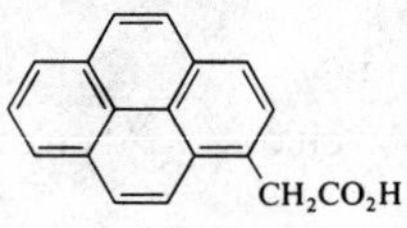

a586

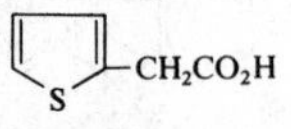

a588

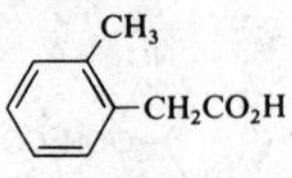

a595

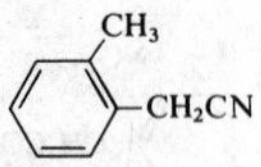

a597

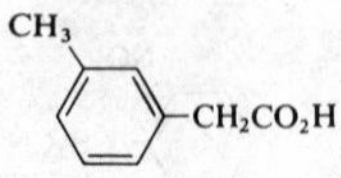

a598

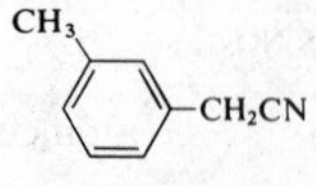

a599

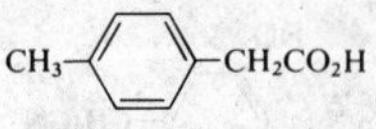

a600

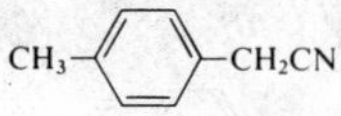

a601

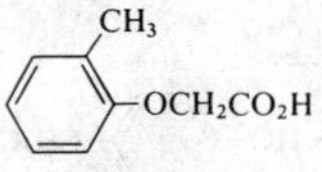

a602

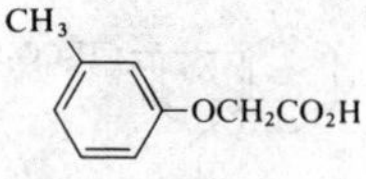

a603

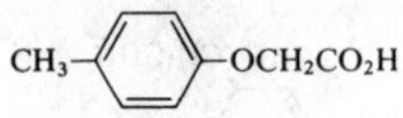

a604

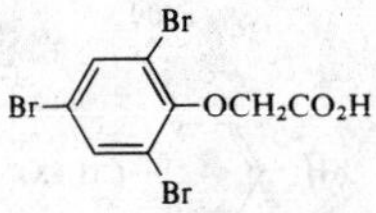

a609

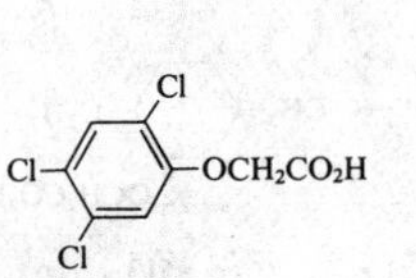

a635

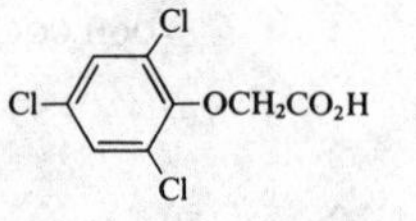

a636

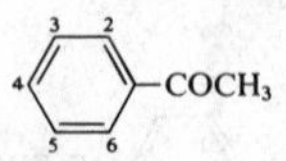

a645

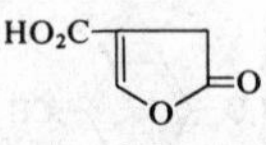

a719

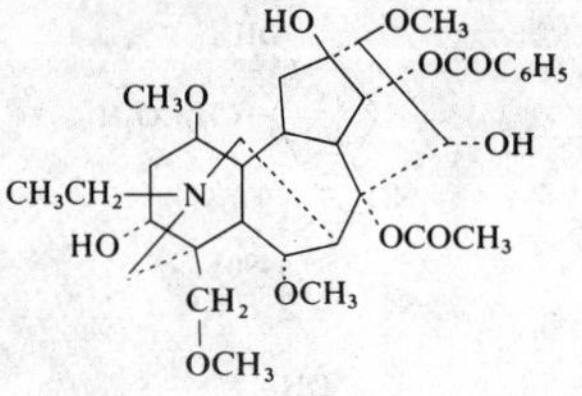

a721

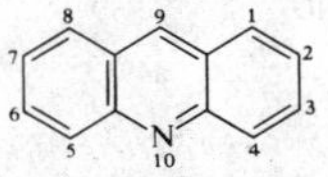

a727

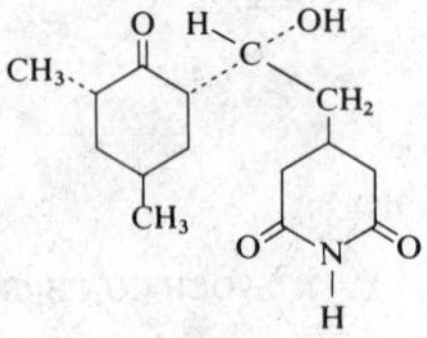

a741

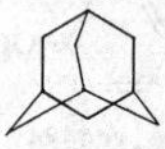

a742

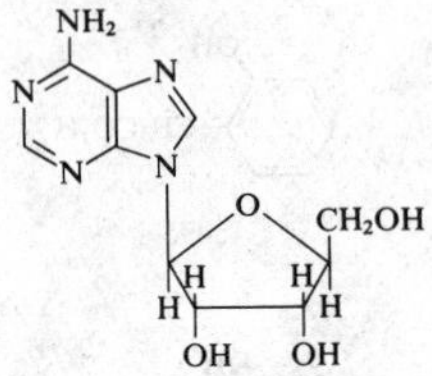

a743

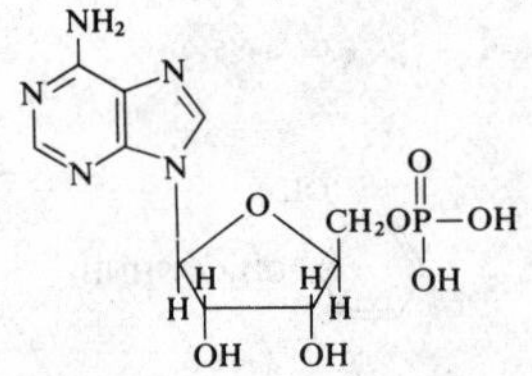

a744

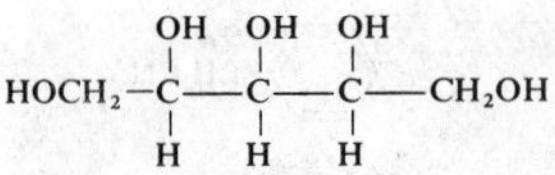

a745

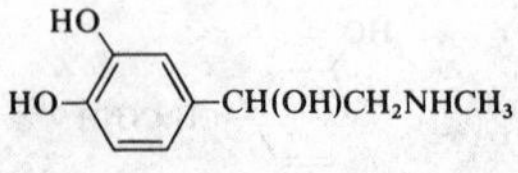

a746

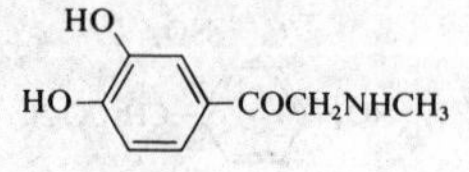

a748

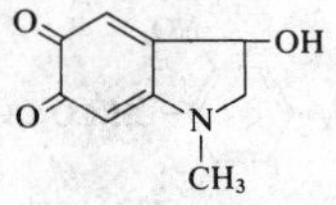

a749

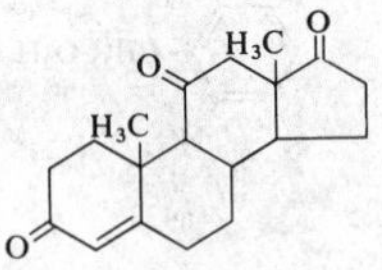

a750

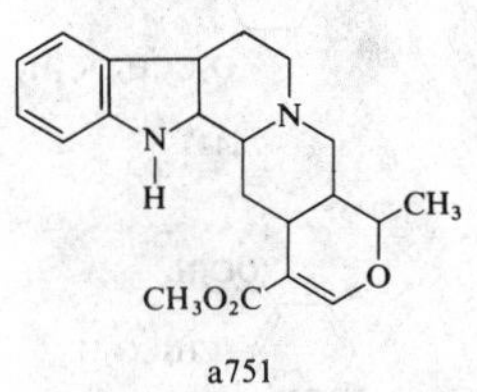

a751

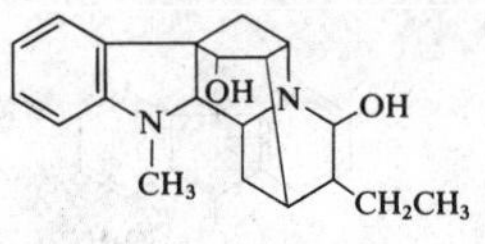

a752

a762

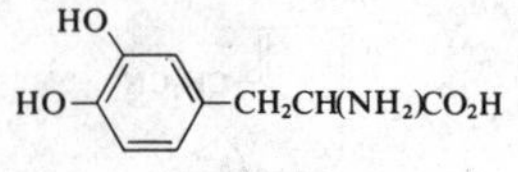

a764

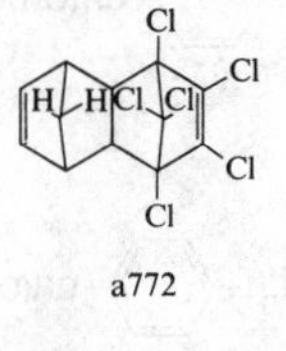

a772

a773

STRUCTURAL FORMULAS OF ORGANIC COMPOUNDS (Continued)

In Alpha Numeric as they occur in Organic Compounds Table

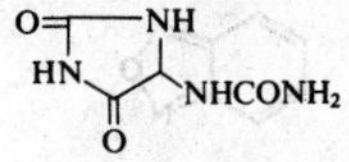

a775

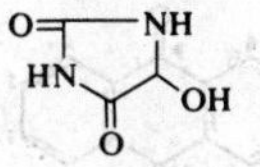

a776

$CH_2{:}CHCH_2S(=O)SCH_2CH{:}CH_2$

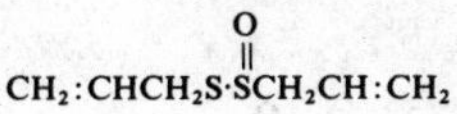

a777

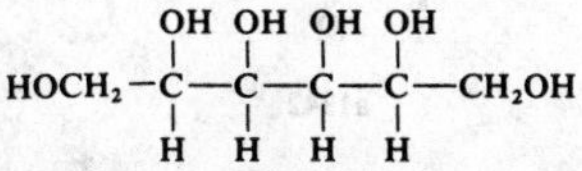

a778

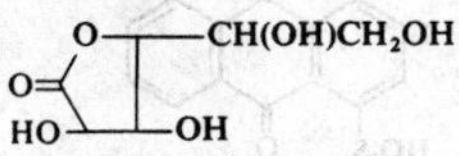

a781

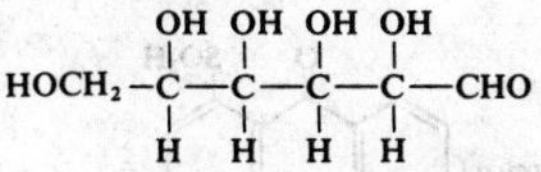

a782

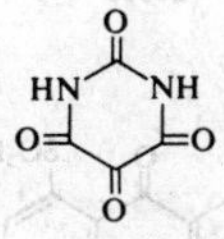

a784

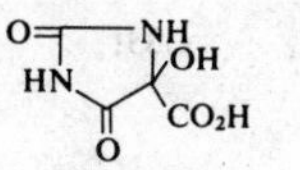

a785

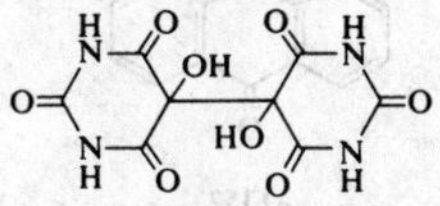

a786

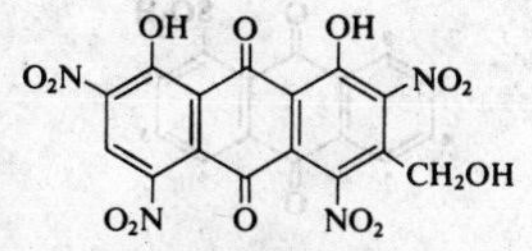

a787

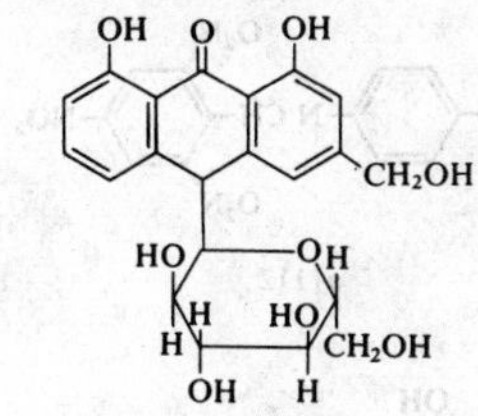

a788

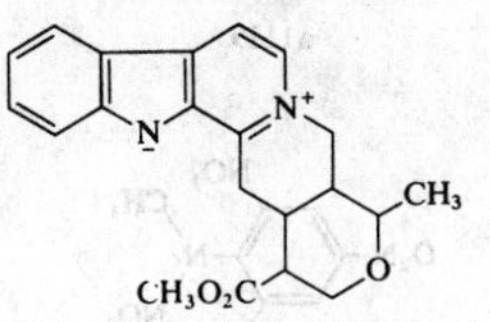

a789

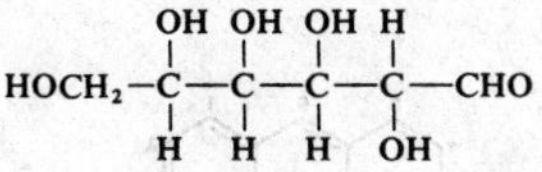

a790

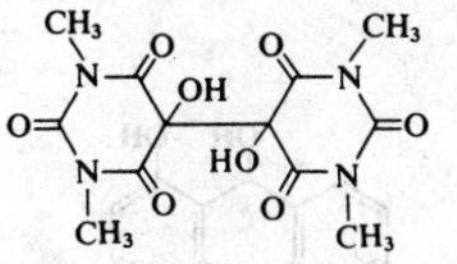

a792

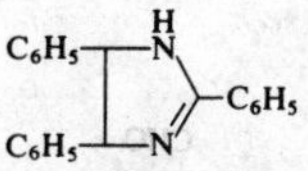

a793

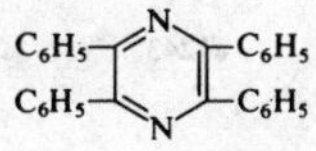

a795

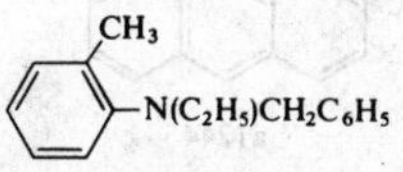

a799

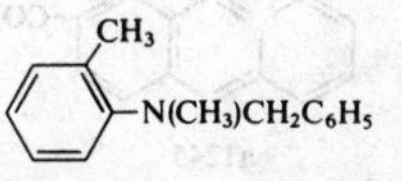

a800

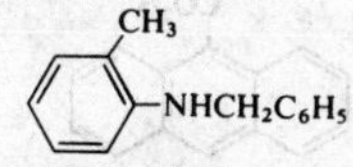

a803

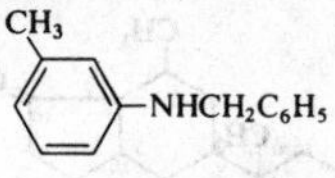

a804

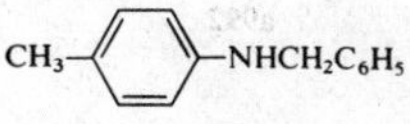

a805

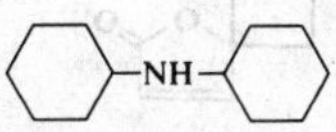

a825

a840

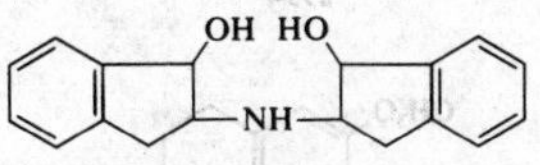

a844

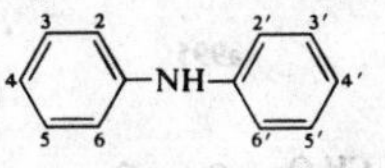

a863

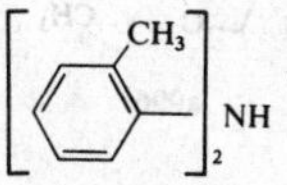

a900

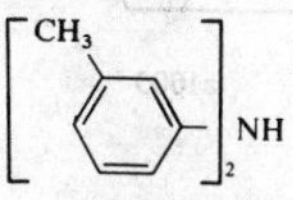

a901

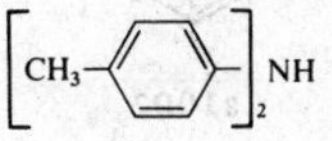

a902

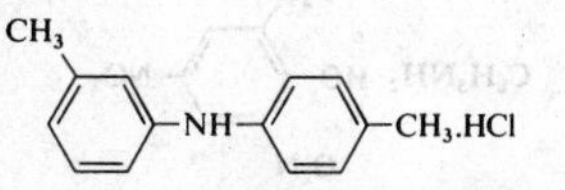

a903

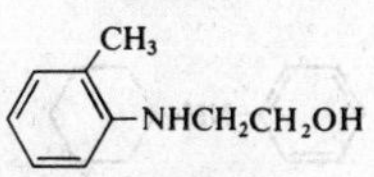

a908

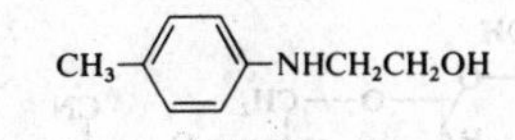

a909

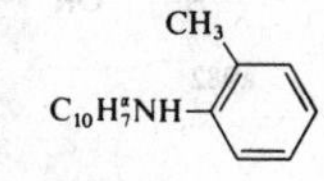

a914

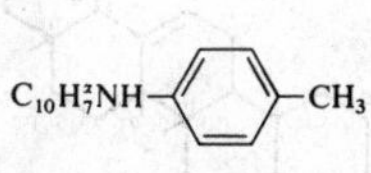

a915

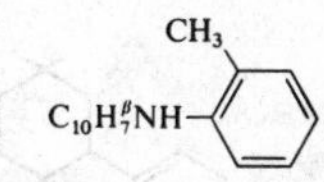

a916

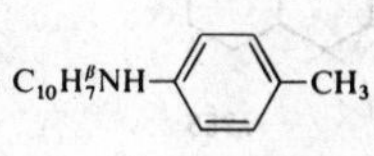

a917

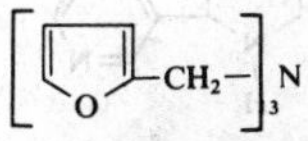

a935

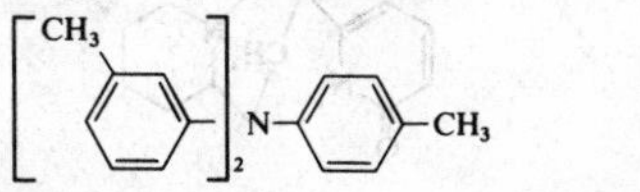

a951

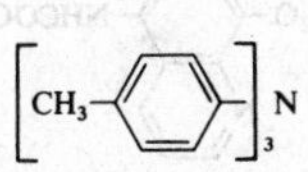

a952

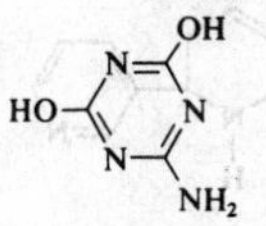

a953

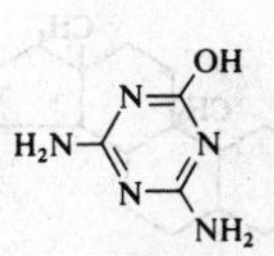

a954

STRUCTURAL FORMULAS OF ORGANIC COMPOUNDS (Continued)

In Alpha Numeric as they occur in Organic Compounds Table

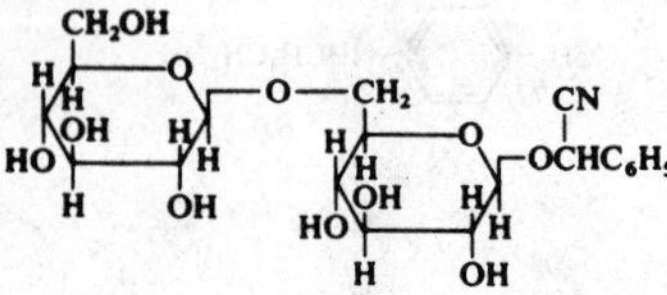

a982

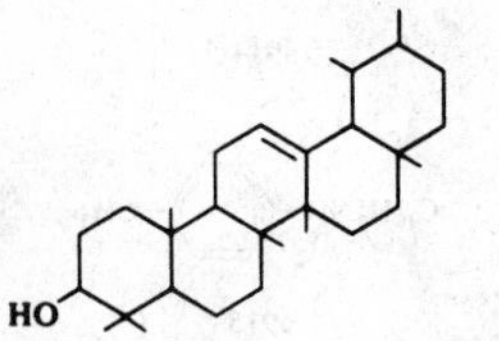

a984

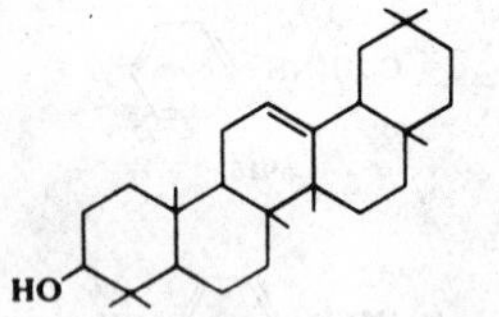

a985

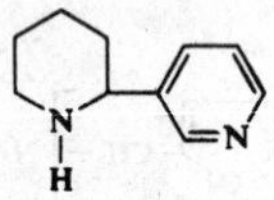

a986

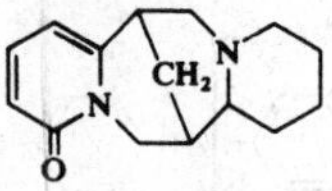

a988

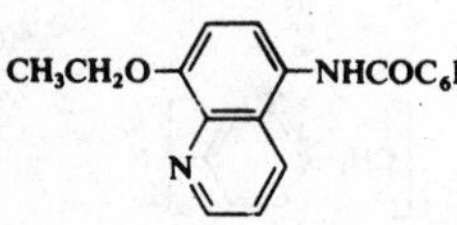

a989

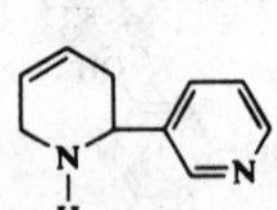

a990

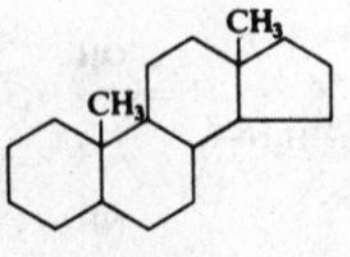

a991

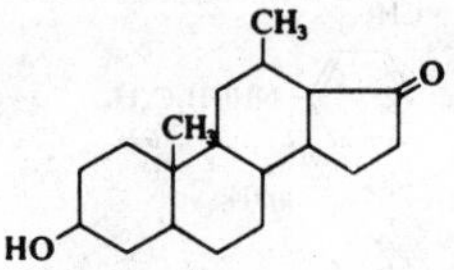

a992

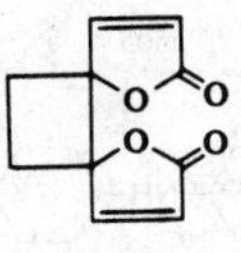

a993

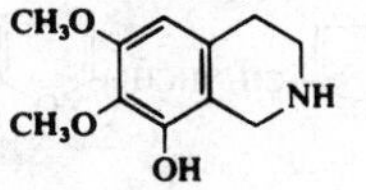

a994

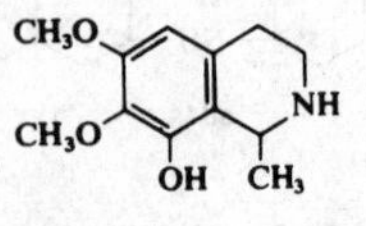

a995

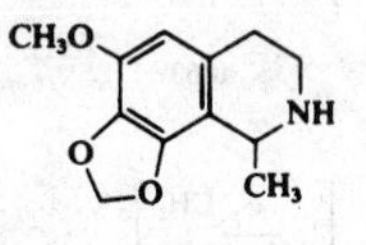

a996

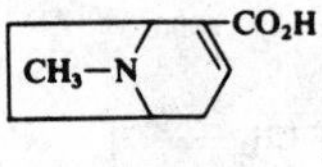

a1000

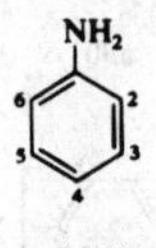

a1002

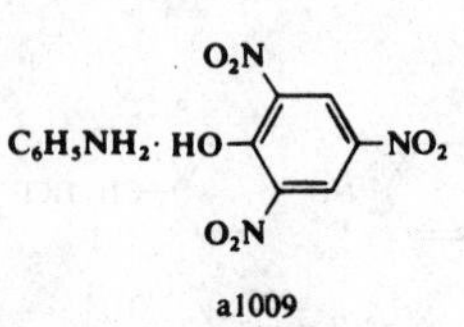

a1009

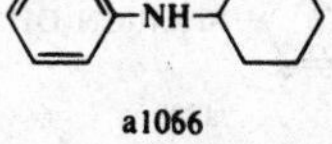

a1066

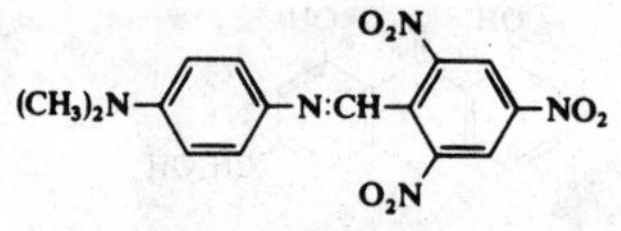

a1112

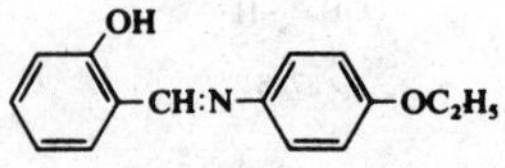

a1119

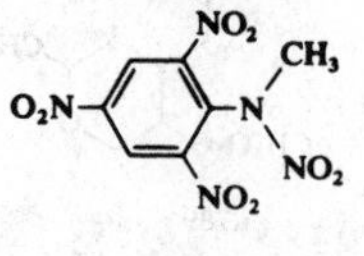

a1165

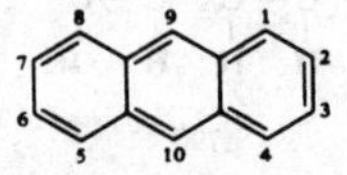

a1193

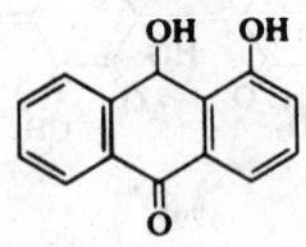

a1241

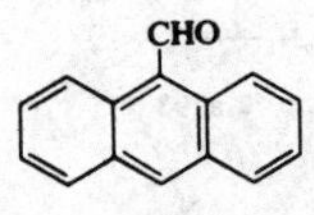

a1243

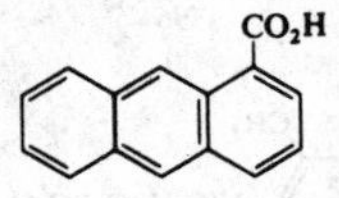

a1244

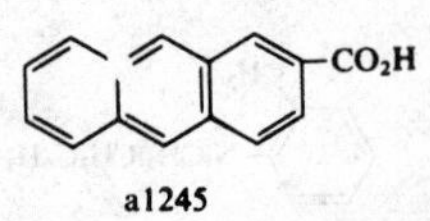

a1245

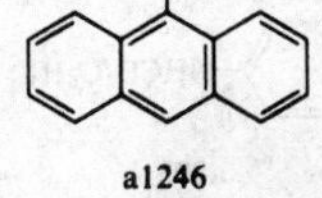

a1246

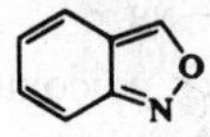

a1247

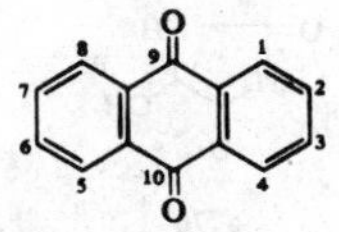

a1248

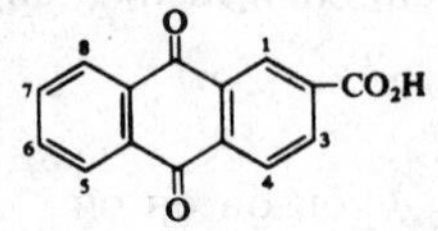

a1342

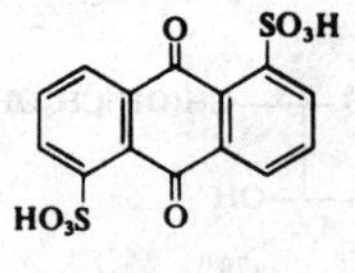

a1349

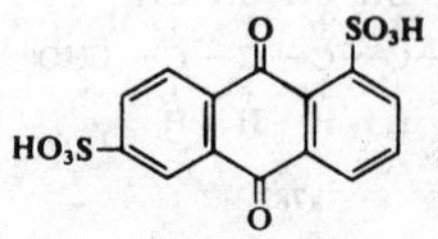

a1350

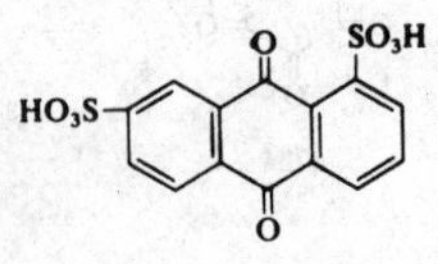

a1351

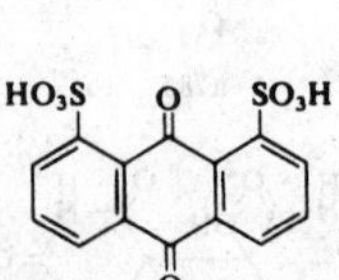

a1352

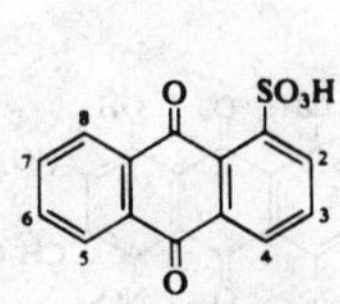

a1353

STRUCTURAL FORMULAS OF ORGANIC COMPOUNDS (Continued)

In Alpha Numeric as they occur in Organic Compounds Table

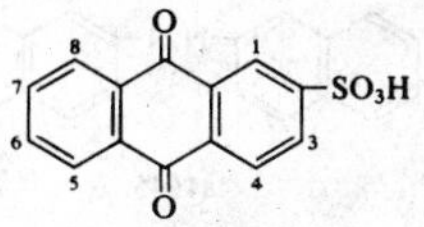

a1354

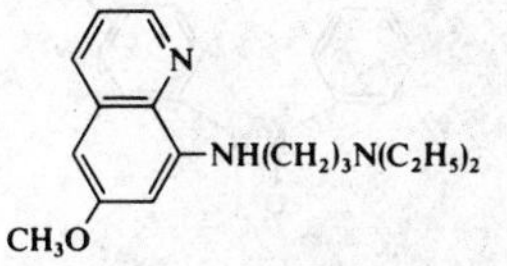

a1358

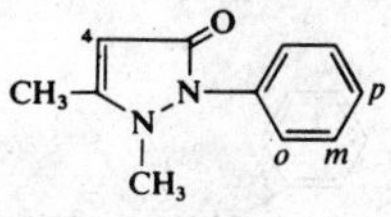

a1359

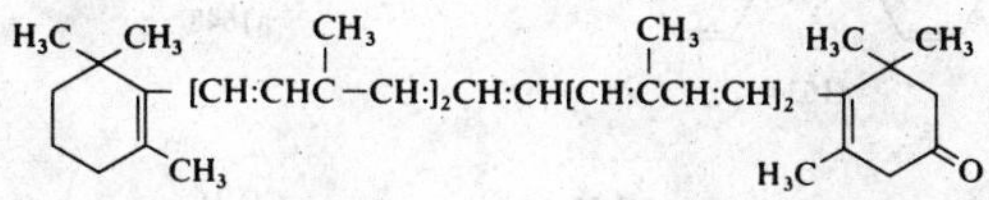

a1366

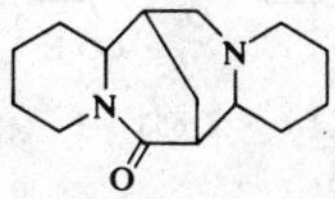

a1367

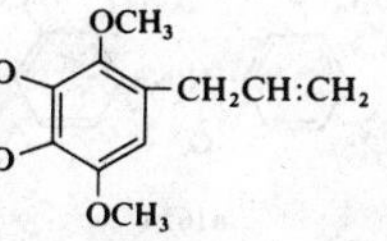

a1368

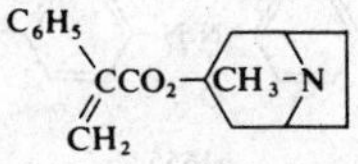

a1369

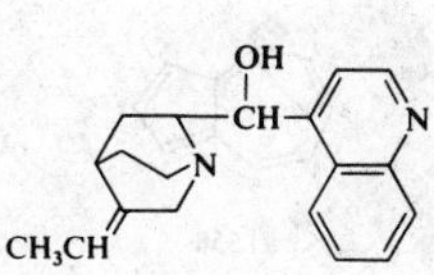

a1372

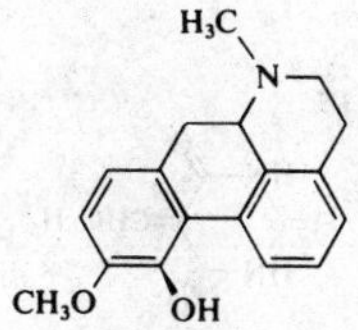

a1374

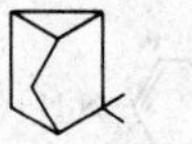

a1377

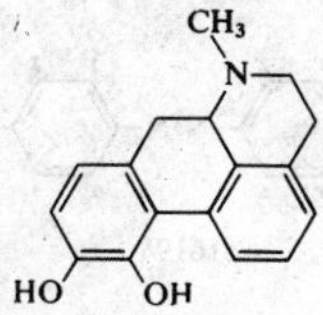

a1378

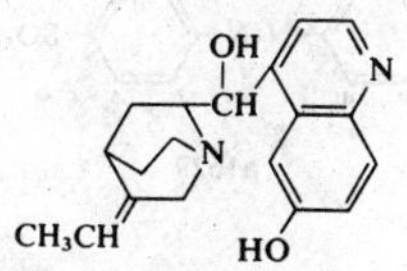

a1380

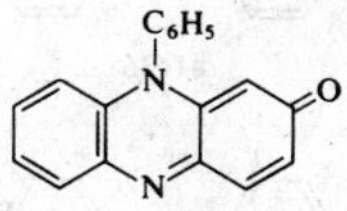

a1382

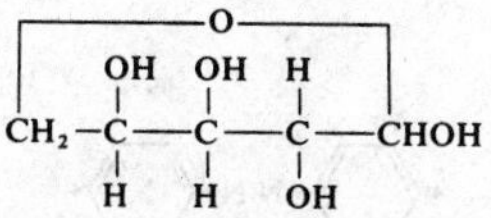

a1383

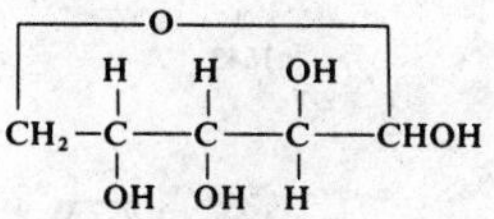

a1386

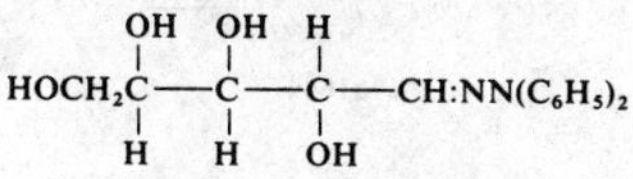

a1388

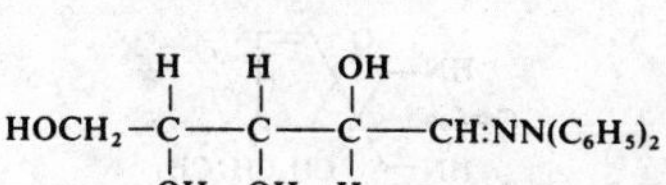

a1390

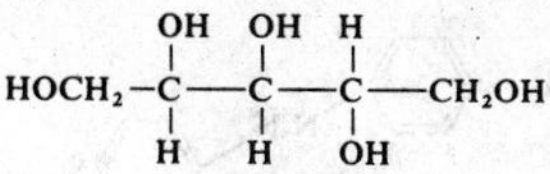

a1391

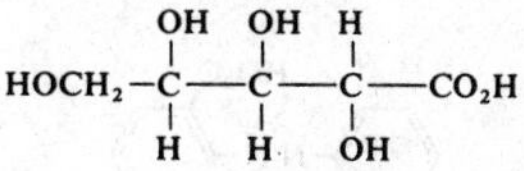

a1394

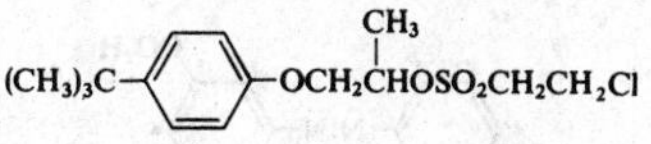

a1397

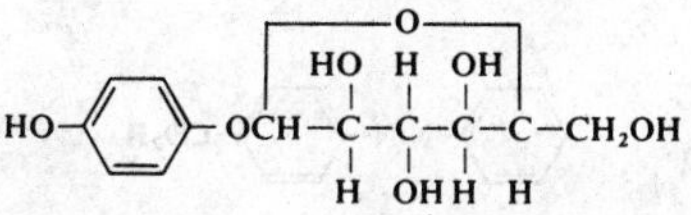

a1398

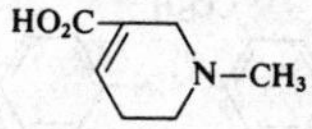

a1400

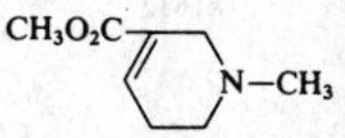

a1401

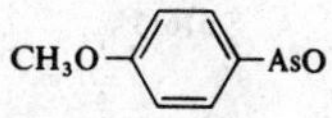

a1444

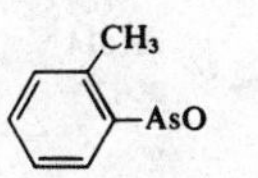

a1447

a1448

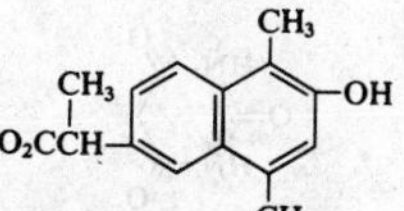

a1453

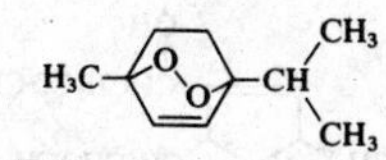

a1454

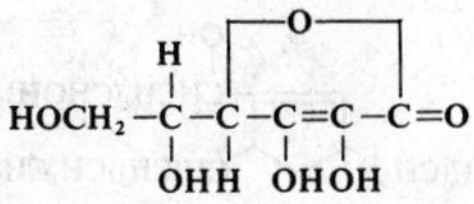

a1455

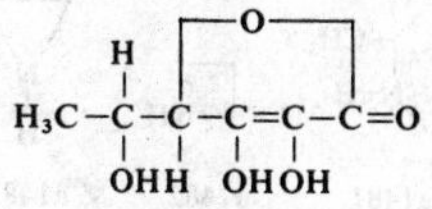

a1458

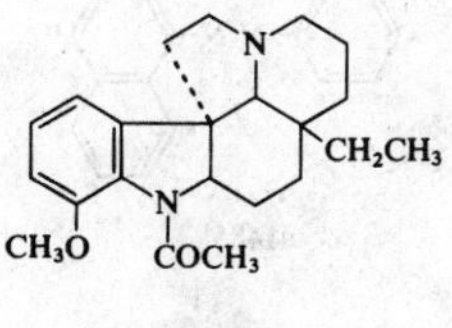

a1466

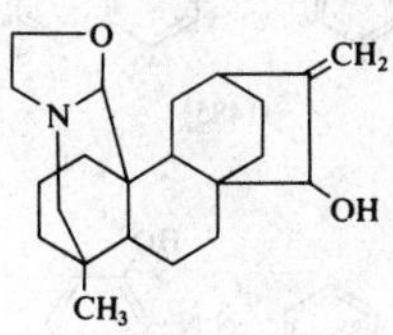

a1467

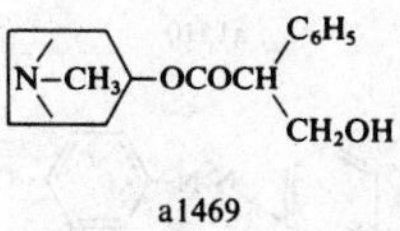

a1469

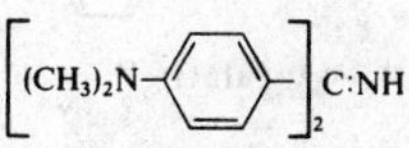

a1473

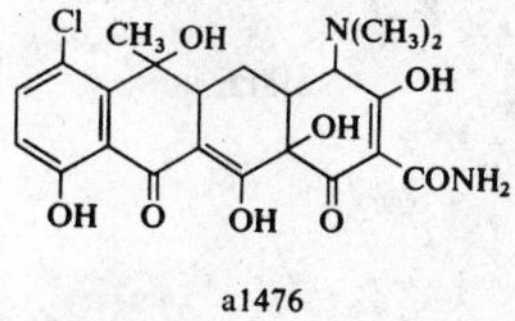

a1476

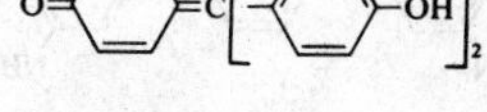

a1478

In Alpha Numeric as they occur in Organic Compounds Table

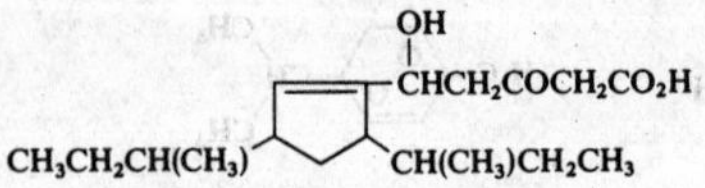
a1479

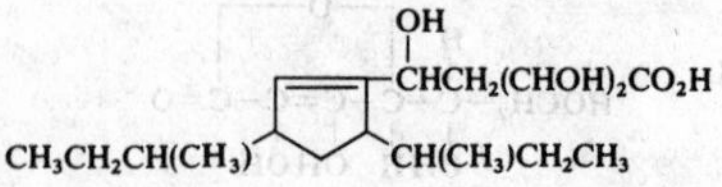
a1480

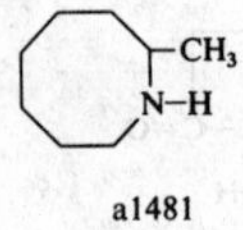
a1481

a1482

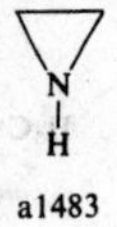
a1483

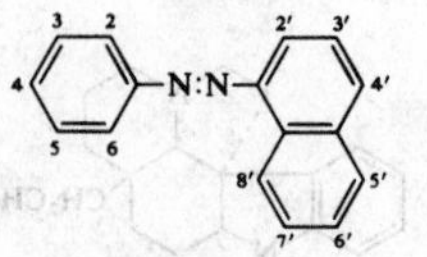
a1485^1

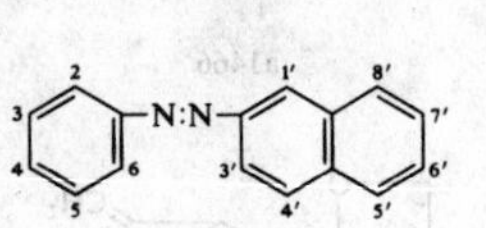
a1485^2

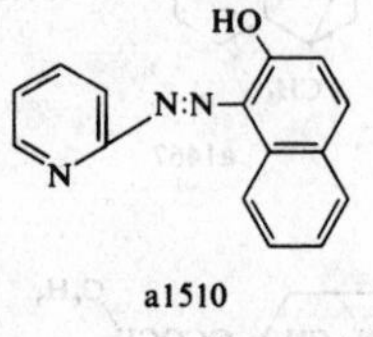
a1510

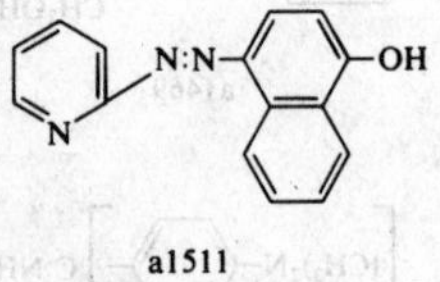
a1511

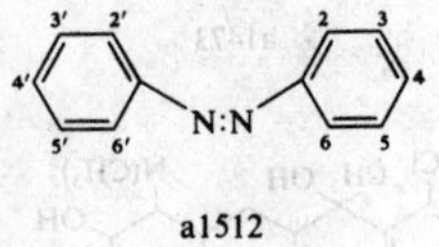
a1512

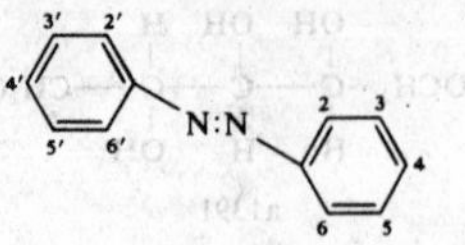
a1513

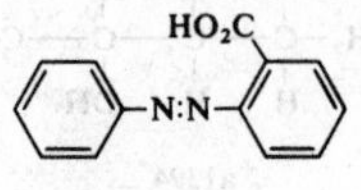
a1604

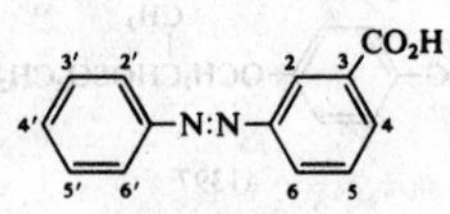
a1605

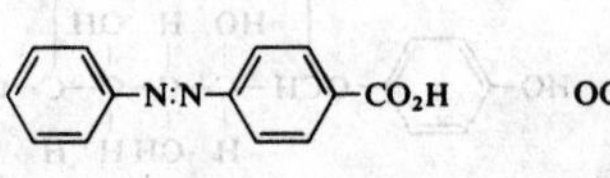
a1606

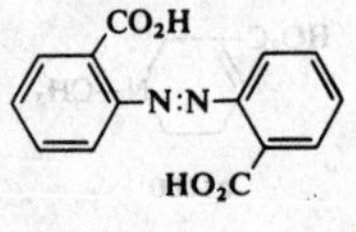
a1612

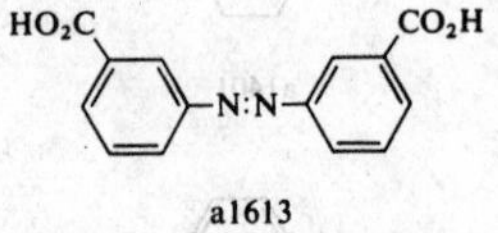
a1613

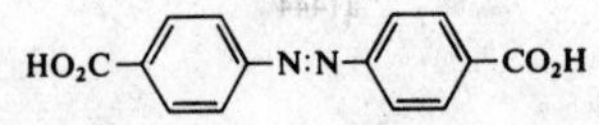
a1614

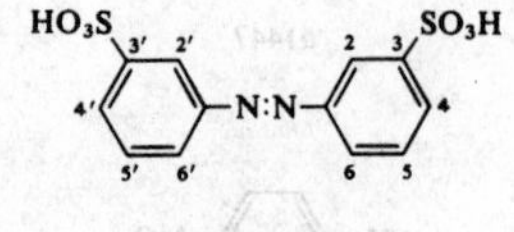
a1614^1

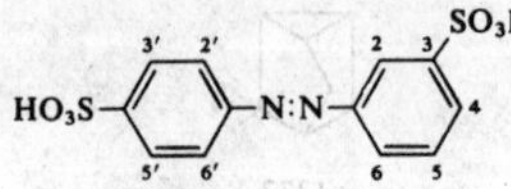
a1617^1

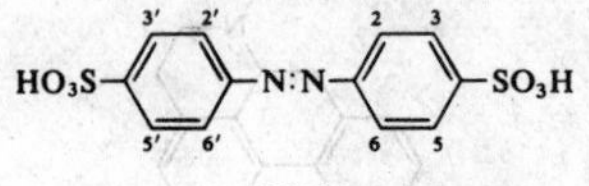
a1619^1

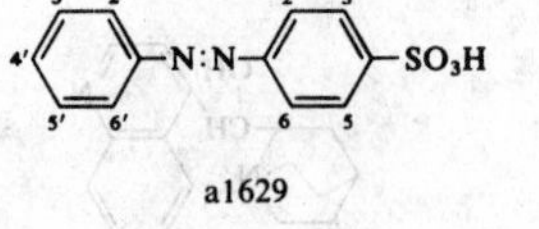
a1629

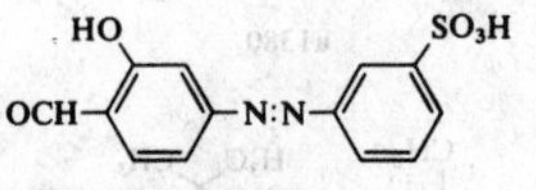
a1634

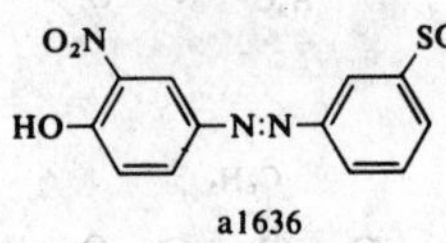
a1636

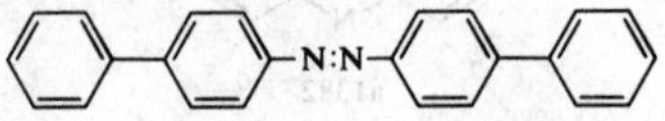
a1638

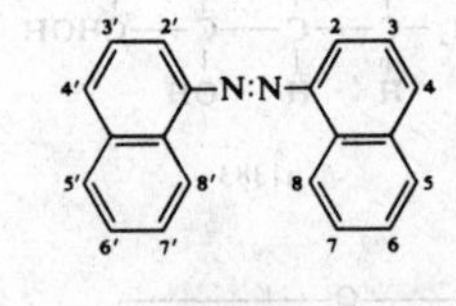
a1643

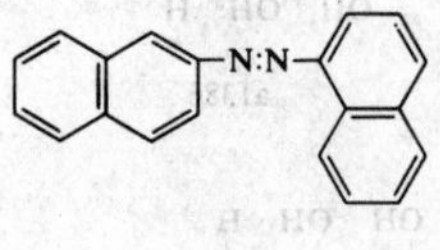
a1644

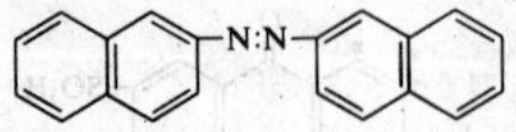
a1645

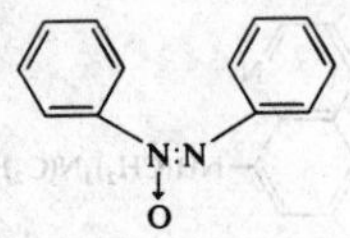
a1647

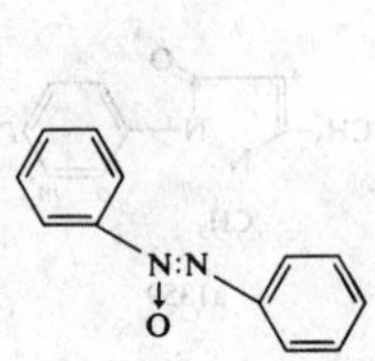
a1648

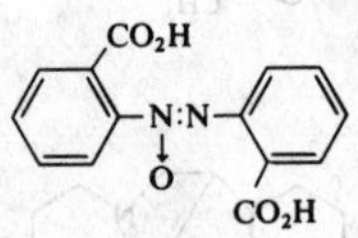
a1653

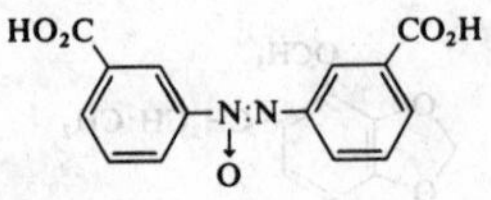
a1654

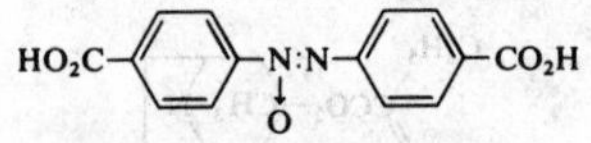
a1655

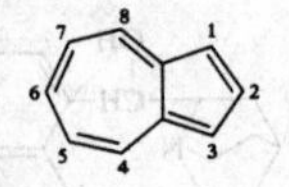
a1658

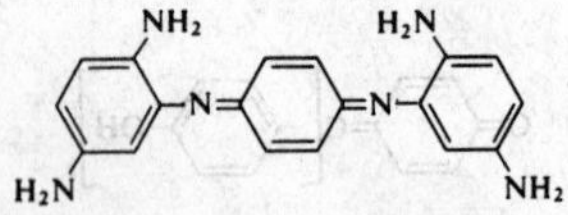
b1

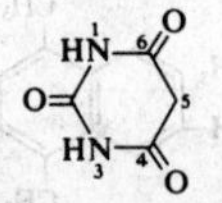
b2

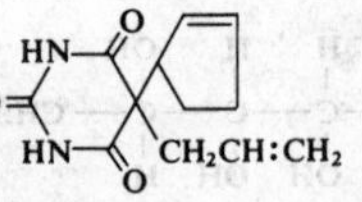
b4

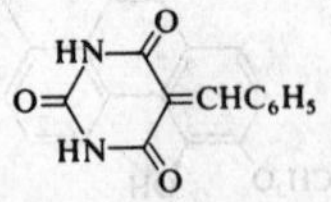
b9

STRUCTURAL FORMULAS OF ORGANIC COMPOUNDS (Continued)

In Alpha Numeric as they occur in Organic Compounds Table

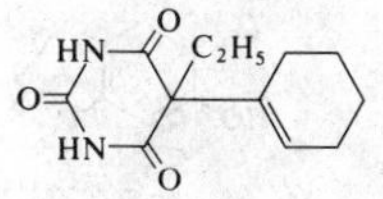

b11

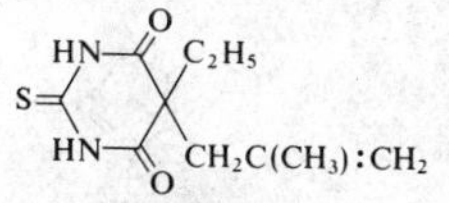

b19

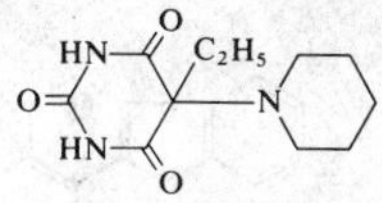

b25

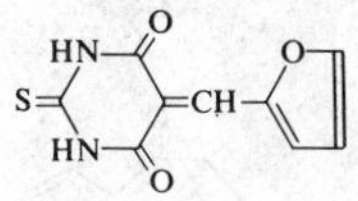

b26

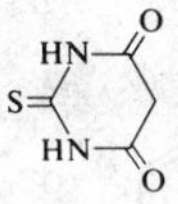

b31

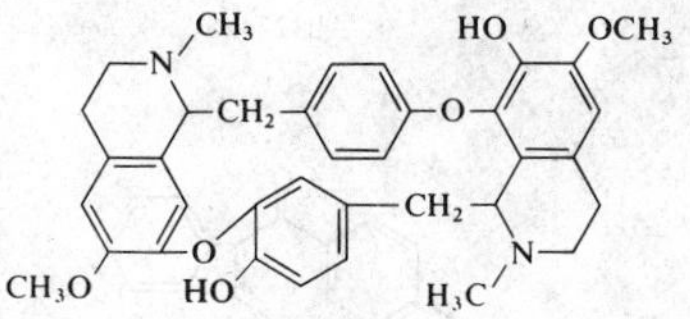

b32

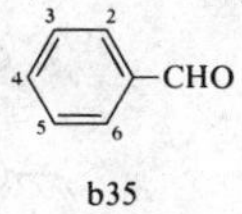

b35

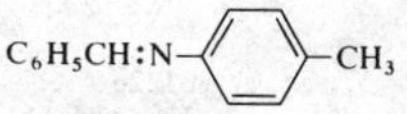

b41

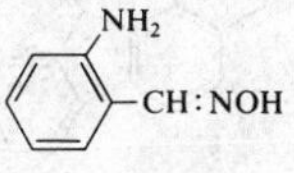

b49

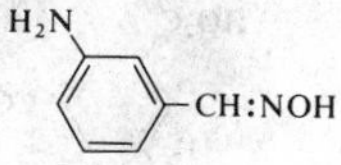

b51

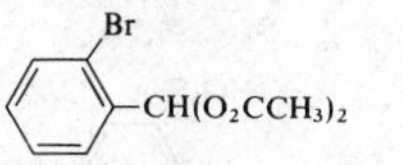

b55

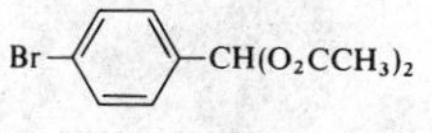

b58

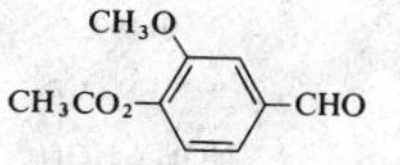

b140

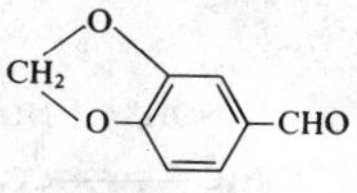

b160

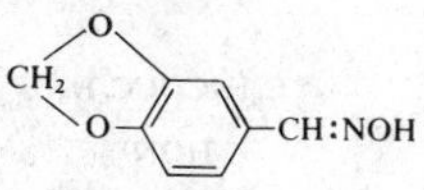

b161

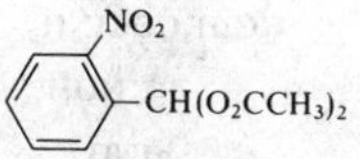

b163

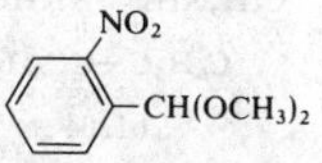

b164

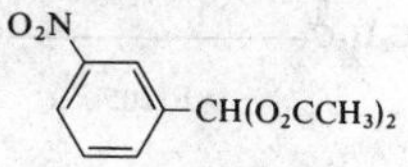

b166

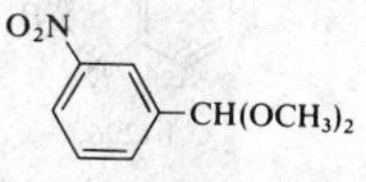

b167

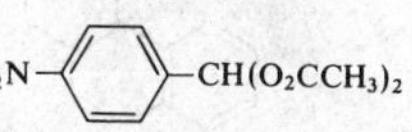

b169

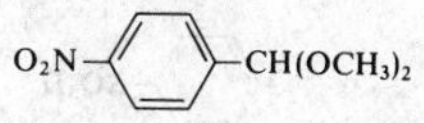

b170

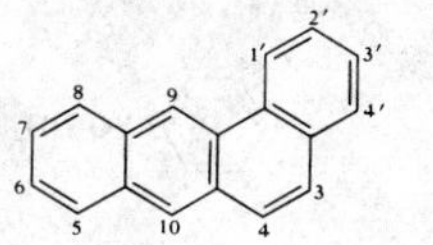

b181

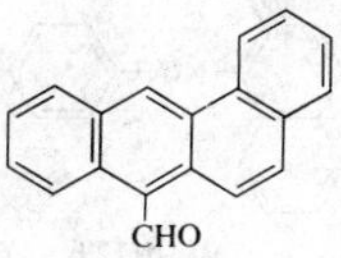

b193

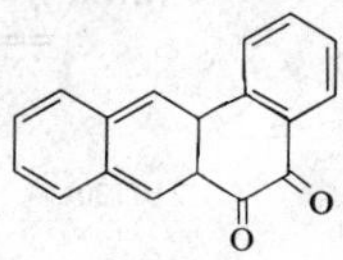

b194

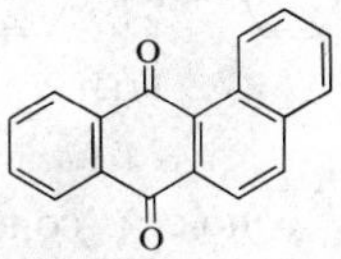

b195

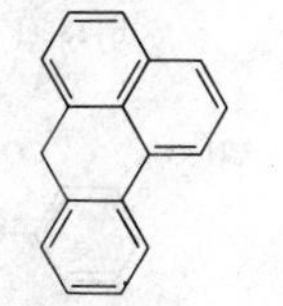

b196

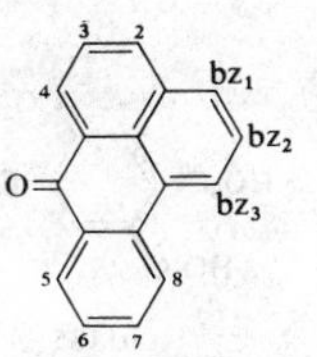

b197

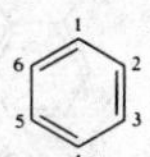

b202

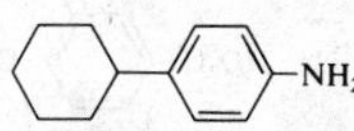

b216

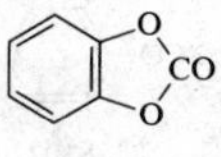

b570

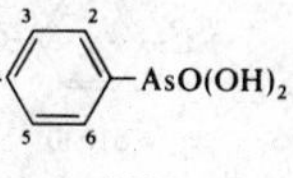

b985

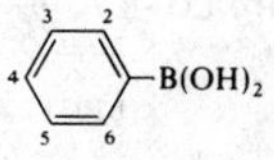

b1006

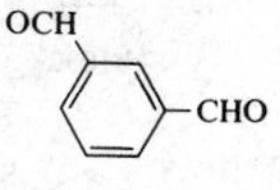

b1014

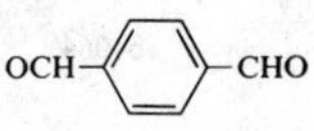

b1015

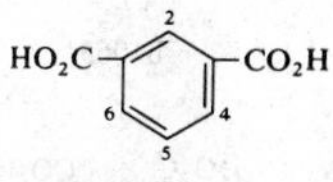

b1016

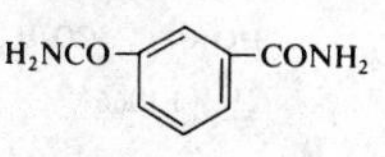

b1016[1]

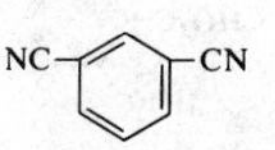

b1021

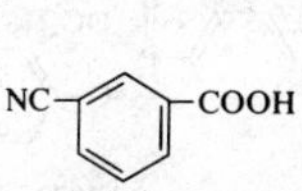

b1022

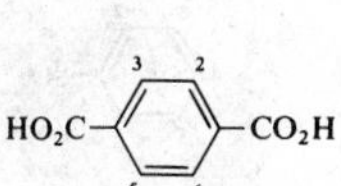

b1024

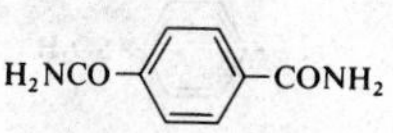

b1025

STRUCTURAL FORMULAS OF ORGANIC COMPOUNDS (Continued)

In Alpha Numeric as they occur in Organic Compounds Table

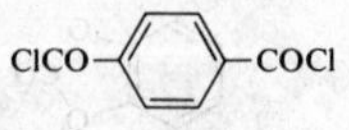

b1027

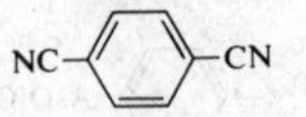

b1030

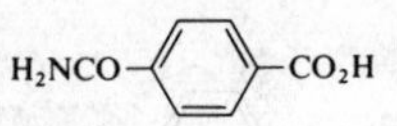

b1031

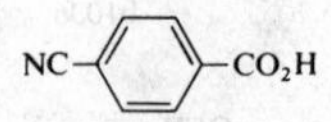

b1032

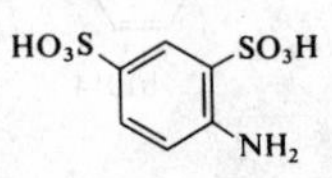

b1064

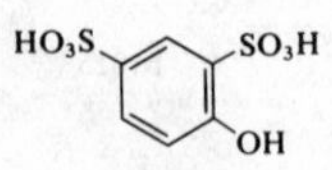

b1065

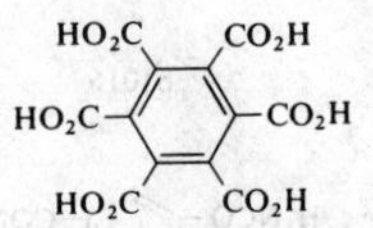

b1066

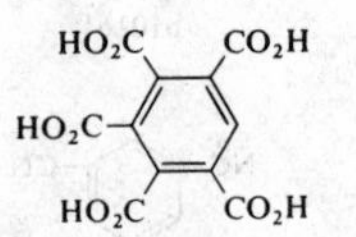

b1067

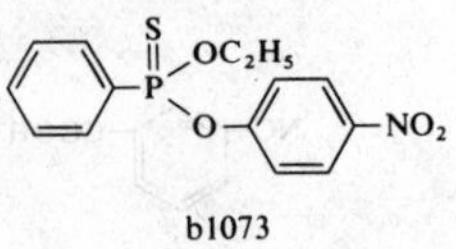

b1073

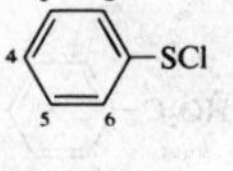

b1078

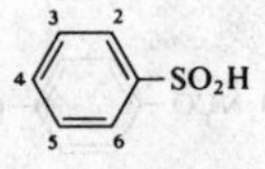

b1082

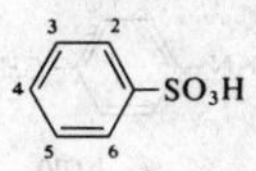

b1091

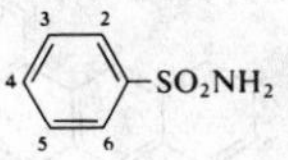

b1093

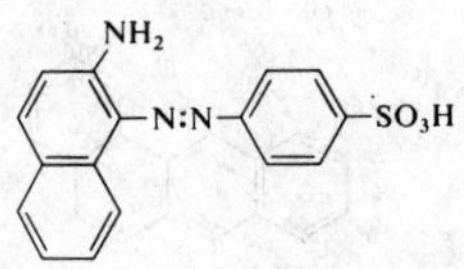

b1129

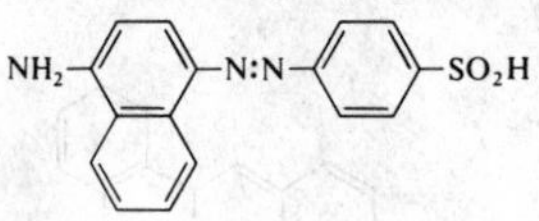

b1130

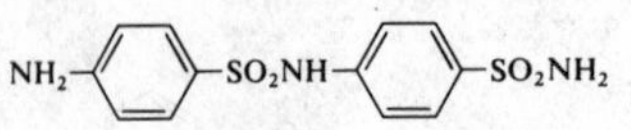

b1131

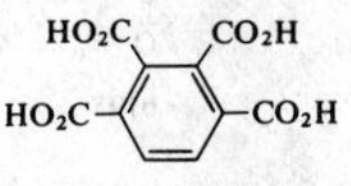

b1182

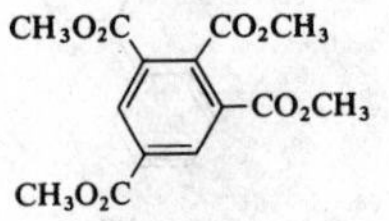

b1184

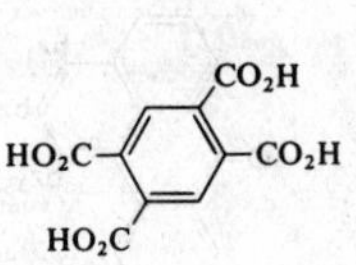

b1185

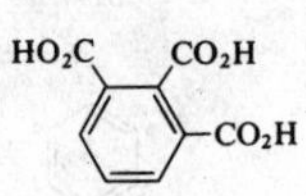

b1188

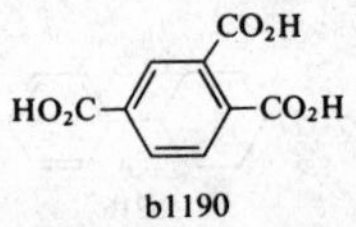

b1190

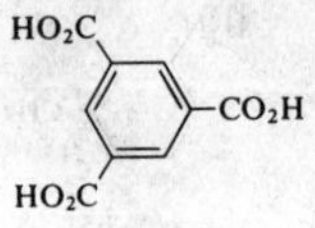

b1191

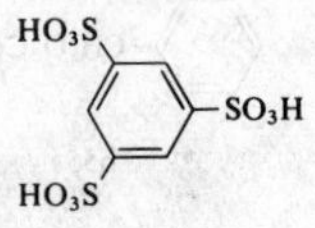

b1197

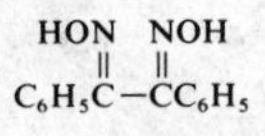

b1199

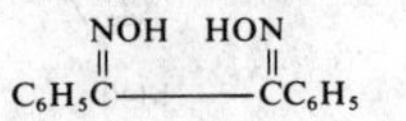

b1200

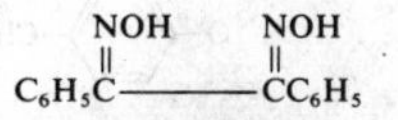

b1201

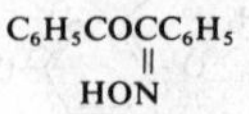

b1202

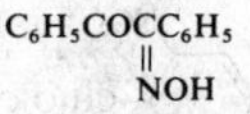

b1203

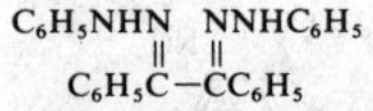

b1204

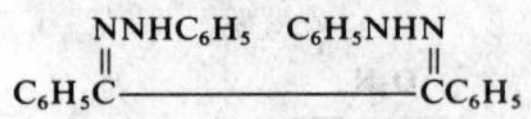

b1205

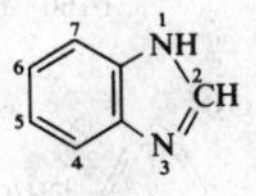

b1206

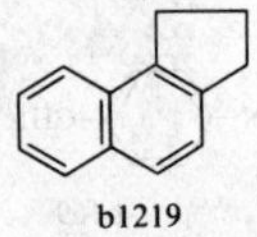

b1219

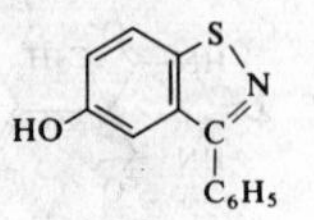

b1220

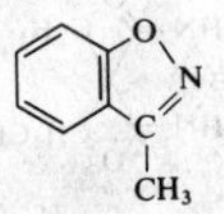

b1221

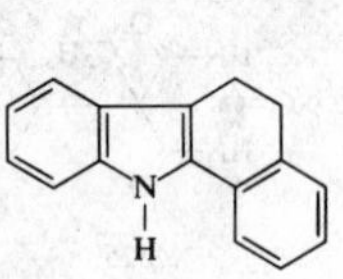

b1222

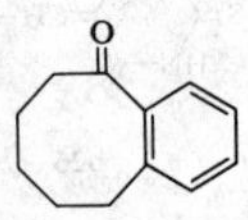

b1223

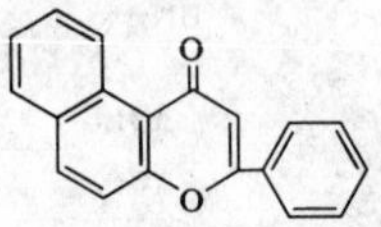

b1224

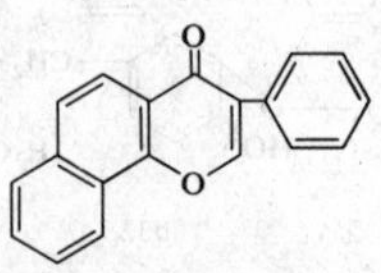

b1225

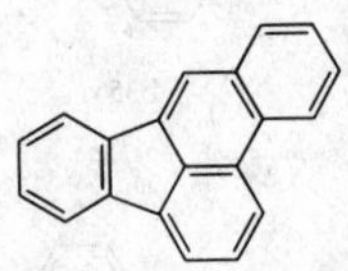

b1226

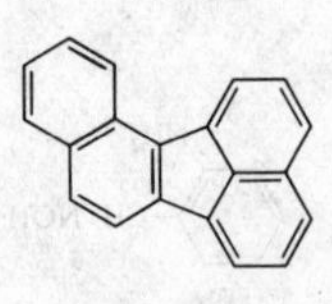

b1227

In Alpha Numeric as they occur in Organic Compounds Table

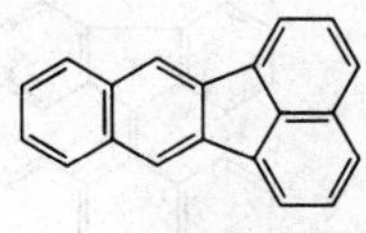
b1228

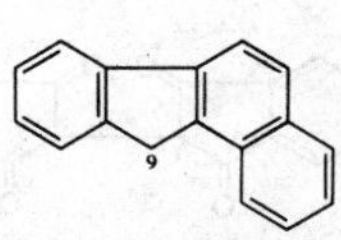

b1229

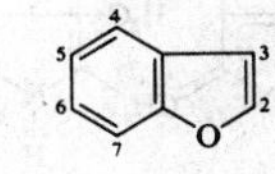

b1231

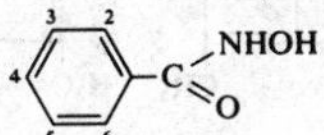

b1237

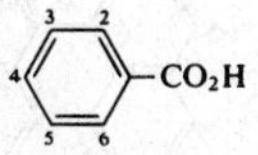

b1239

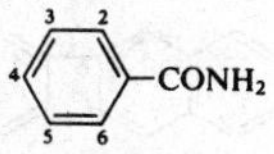

b1244

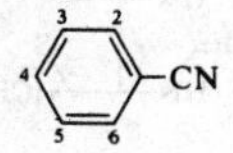

b1291

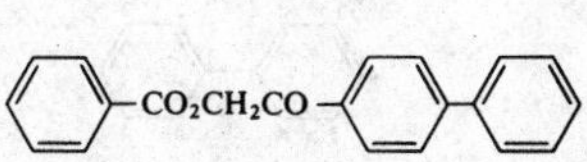

b1298

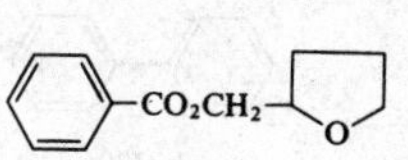

b1300

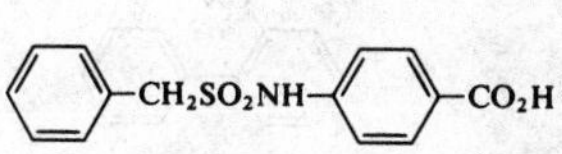

b1409

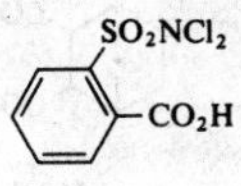

b1547

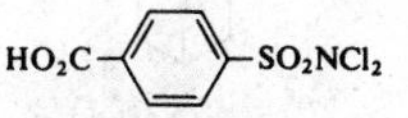

b1548

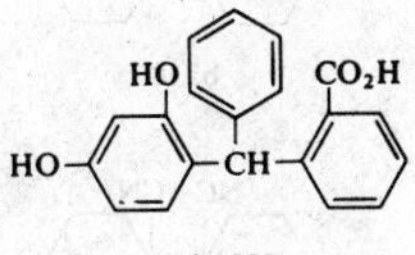

b1558

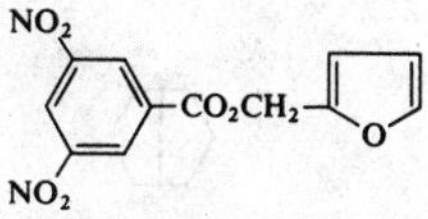

b1598

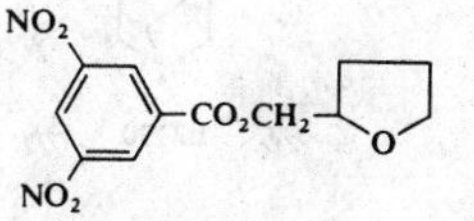

b1605

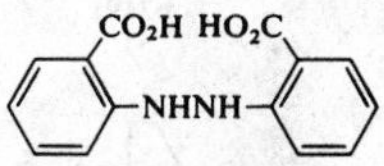

b1646

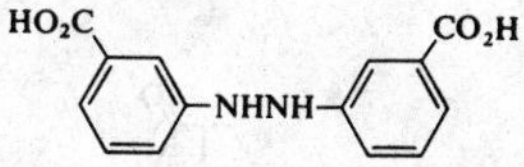

b1647

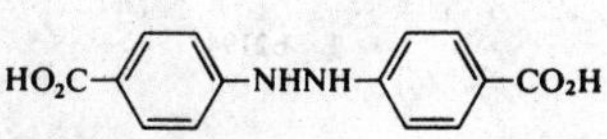

b1648

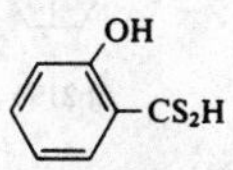

b1699

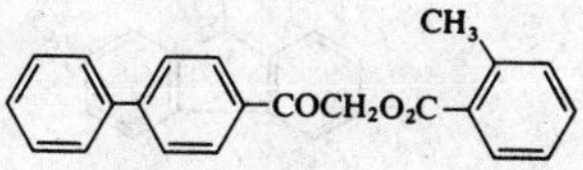

b1791

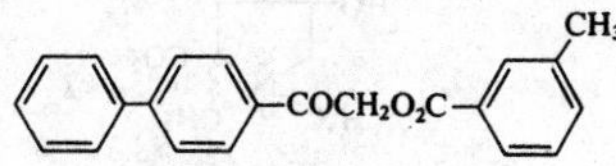

b1799

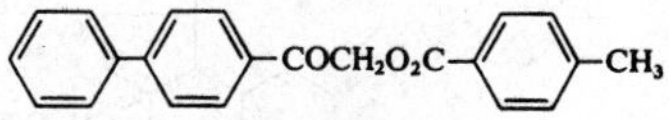

b1807

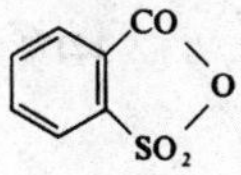

b1880

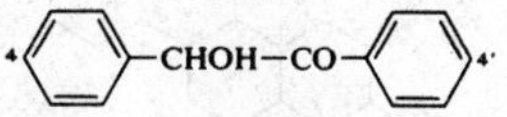

b1944

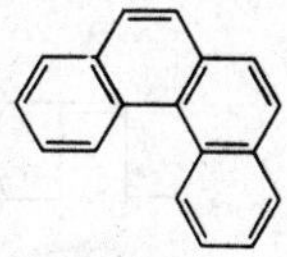
b1956

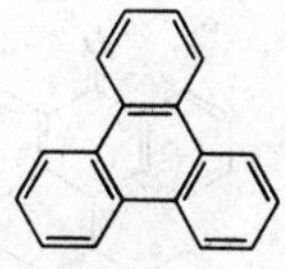
b1957

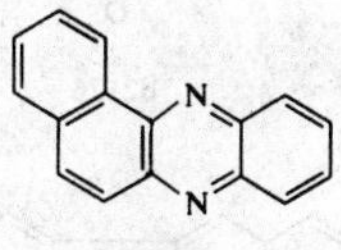

b1958

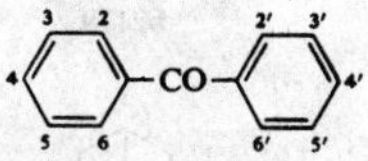

b1959

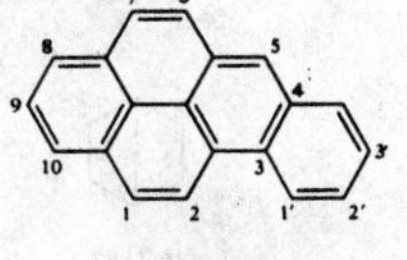
b2062

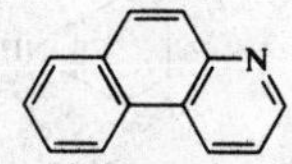

b2066

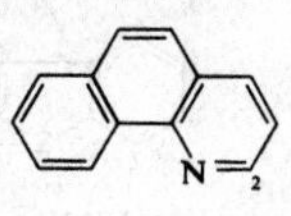

b2067

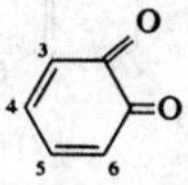

b2068′

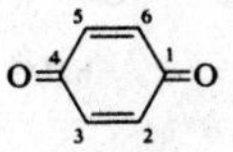

b2069

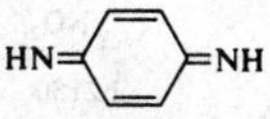

b2070

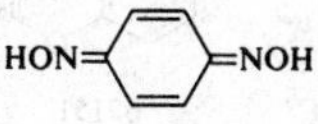

b2072

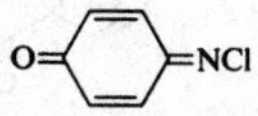

b2073

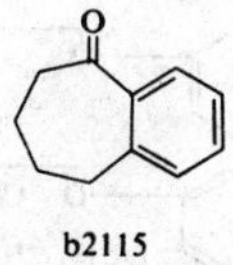

b2115

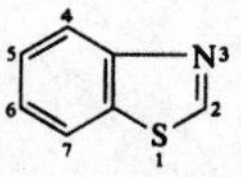

b2116

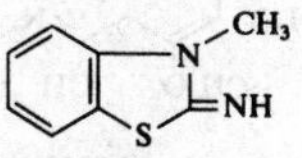

b2143

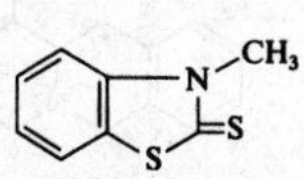

b2144

STRUCTURAL FORMULAS OF ORGANIC COMPOUNDS (Continued)

In Alpha Numeric as they occur in Organic Compounds Table

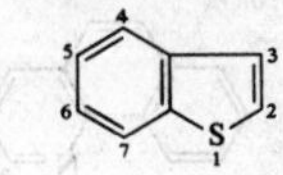
b2145

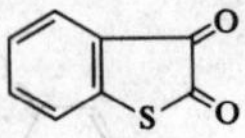
b2148

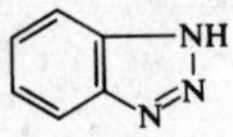
b2149

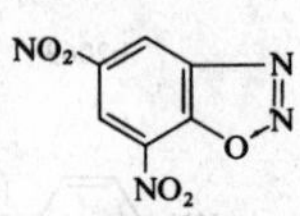
b2150

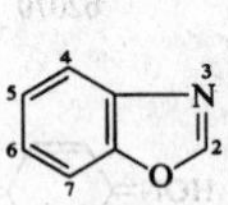
b2151

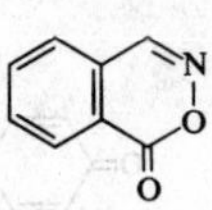
b2161

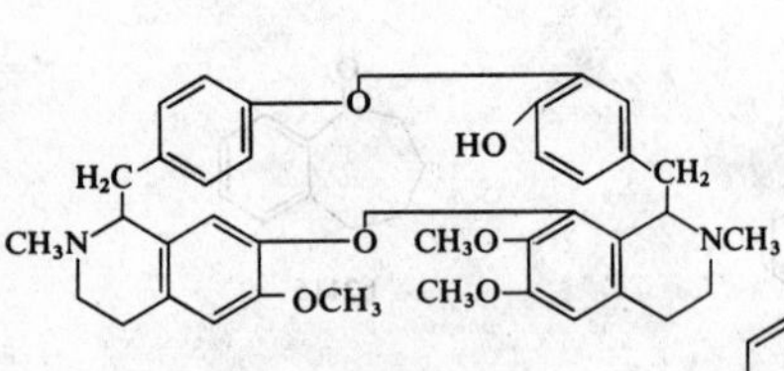
b2162

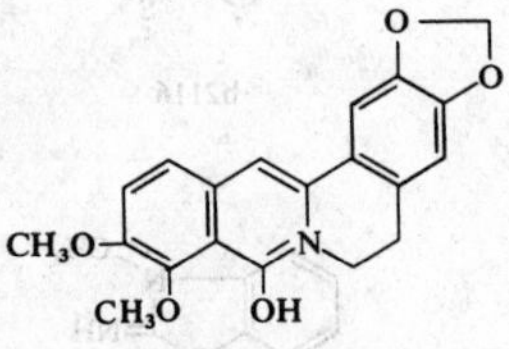
b2163

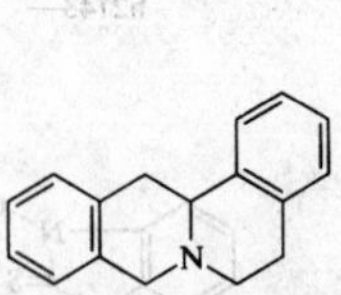
b2169

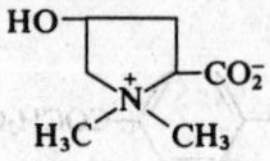
b2172

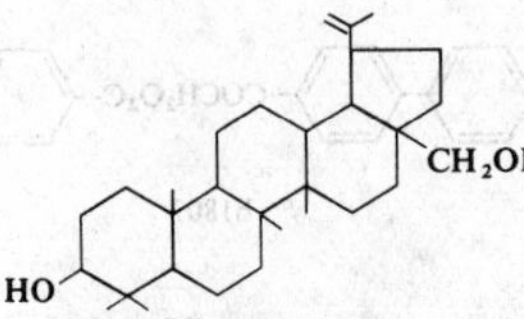
b2173

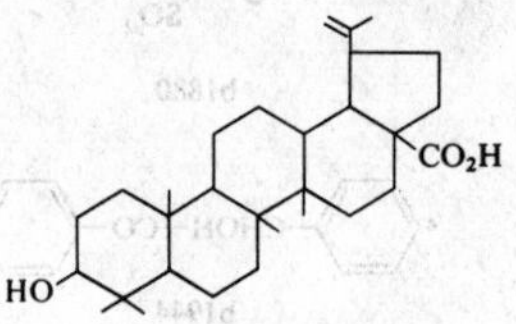
b2174

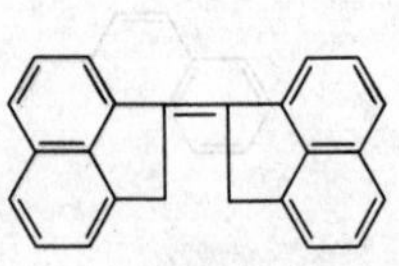
b2175

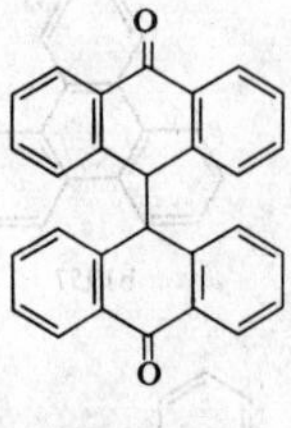
b2176

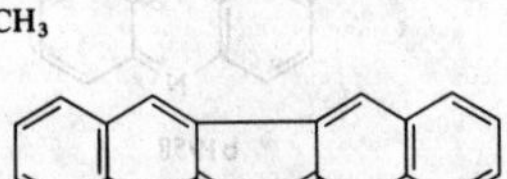
b2179

b2180

b2181

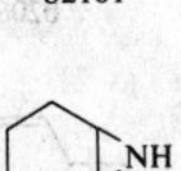
b2182

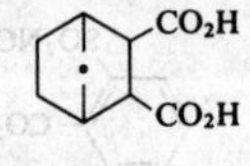
b2183

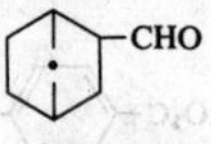
b2185

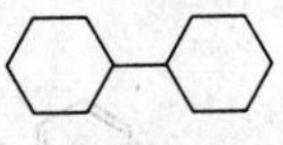
b2186

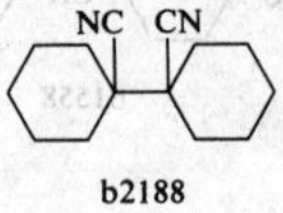
b2188

b2189

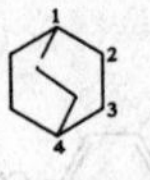
b2190

b2191

b2192

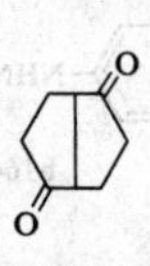
b2194

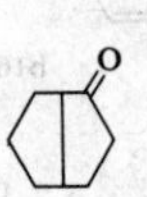
b2196

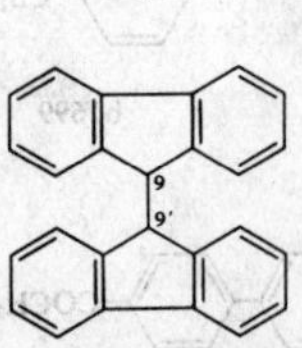
b2197

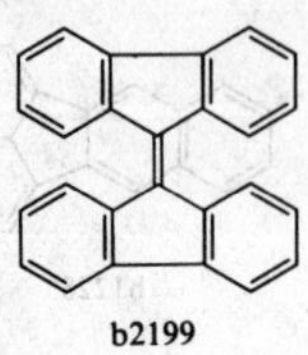
b2199

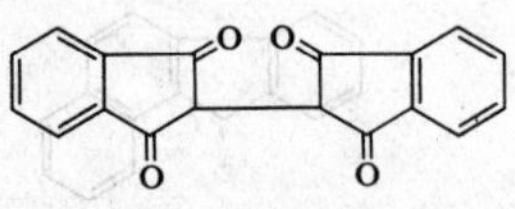
b2202

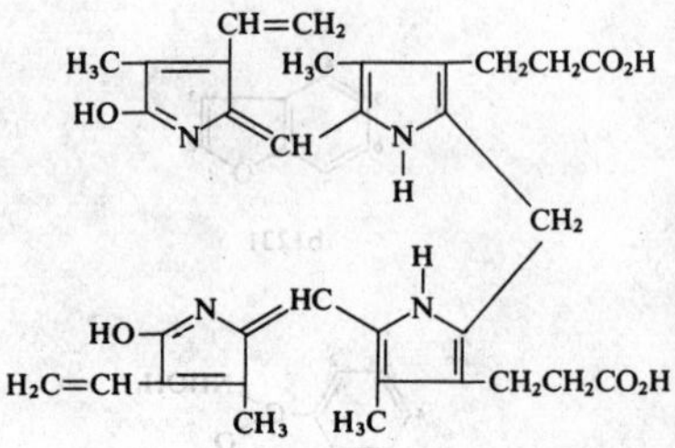
b2205

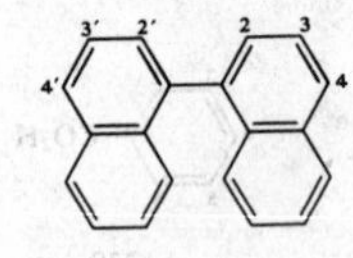
b2207

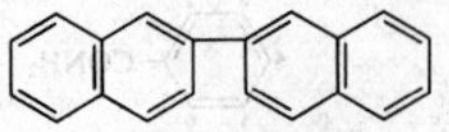
b2208

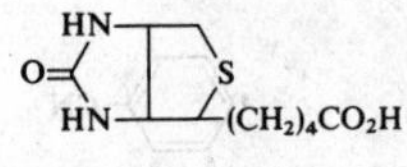
b2212

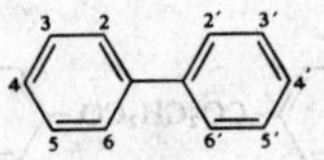
b2214

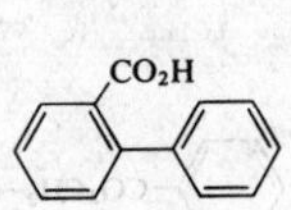
b2329

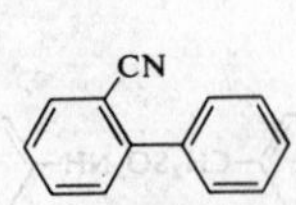
b2330

STRUCTURAL FORMULAS OF ORGANIC COMPOUNDS (Continued)

In Alpha Numeric as they occur in Organic Compounds Table

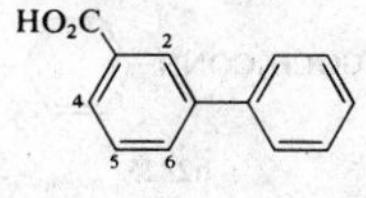

b2331

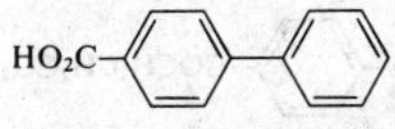

b2332

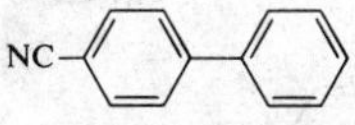

b2333

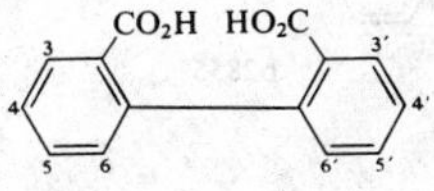

b2335

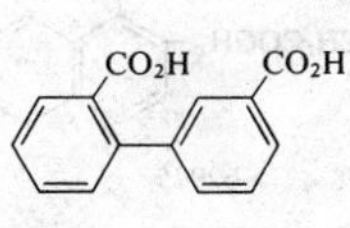

b2341

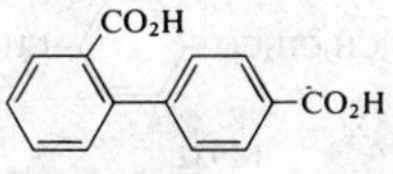

b2342

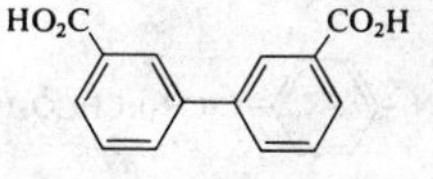

b2343

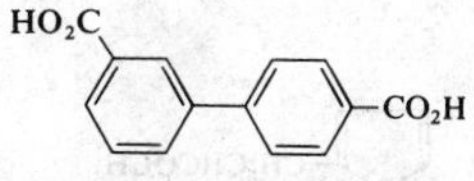

b2345

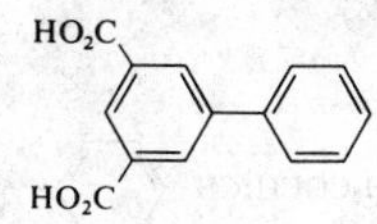

b2347

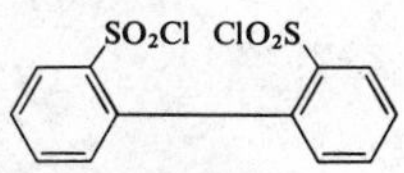

b2353

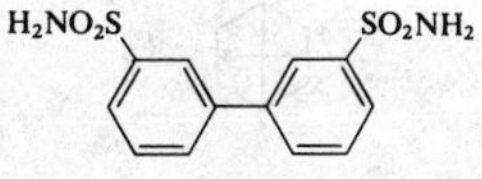

b2354

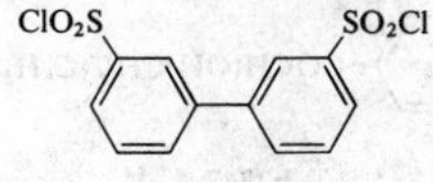

b2355

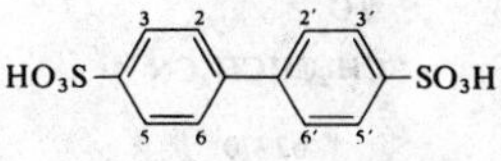

b2356

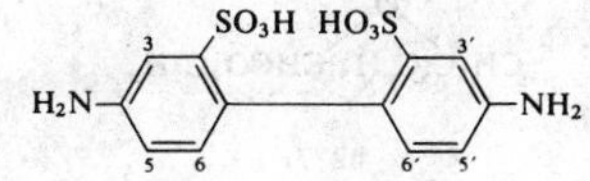

b2359

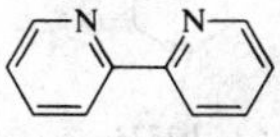

b2361

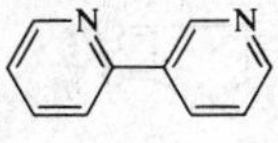

b2362

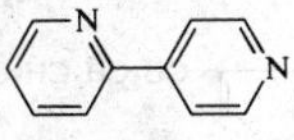

b2363

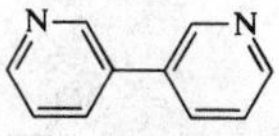

b2364

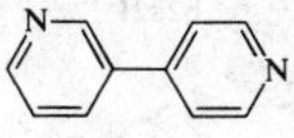

b2365

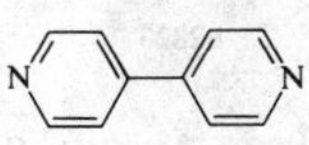

b2366

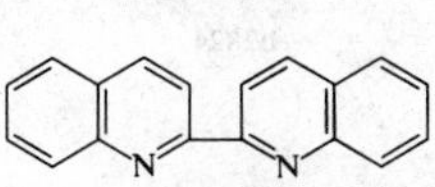

b2367

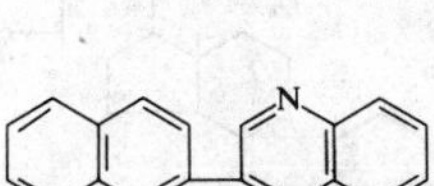

b2368

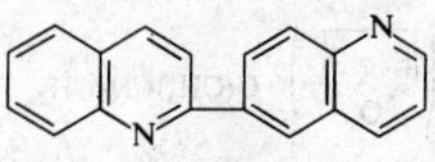

b2369

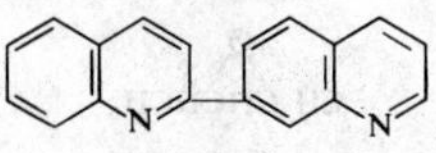

b2370

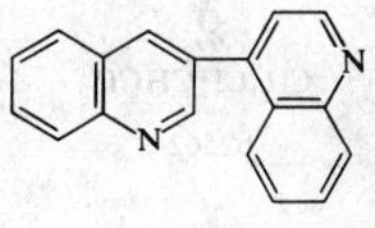

b2372

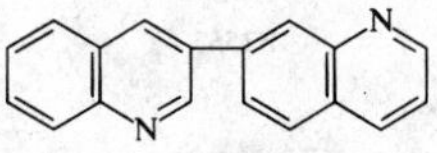

b2373

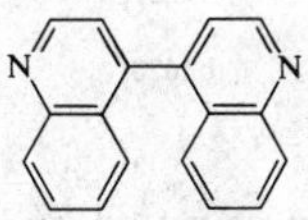

b2374

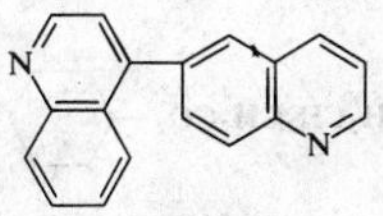

b2375

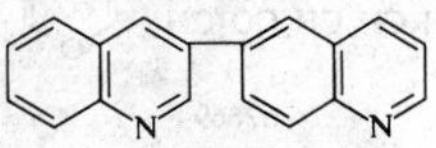

b2376

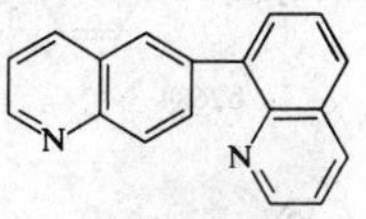

b2377

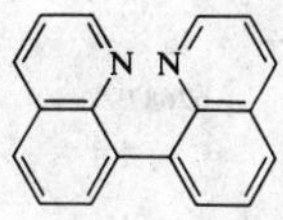

b2378

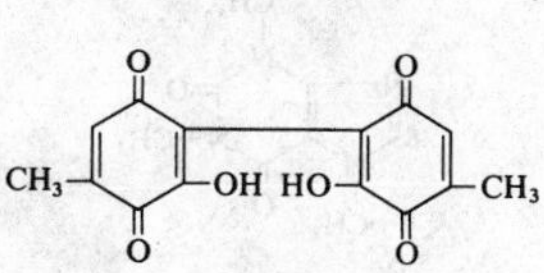

b2379

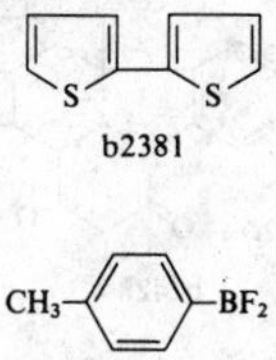

b2381

b2398

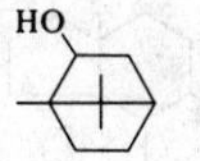

b2414

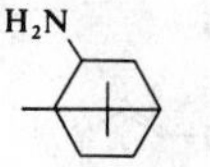

b2421

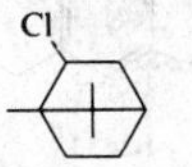

b2422

b2423

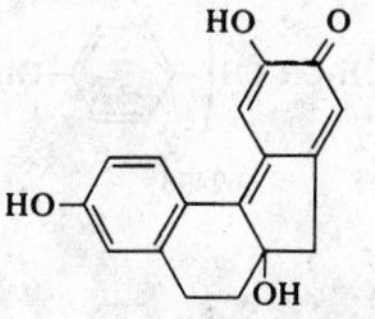

b2425

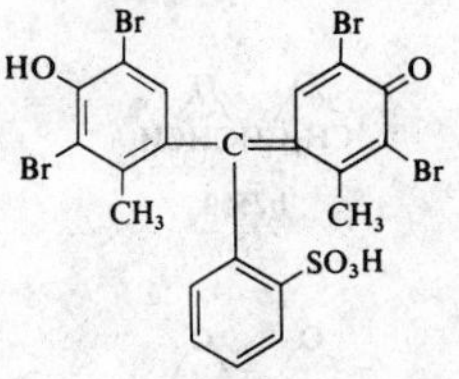

b2426

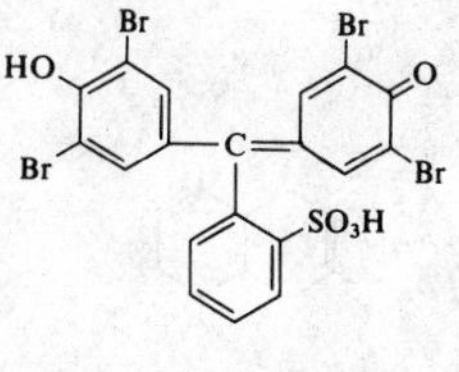

b2427

STRUCTURAL FORMULAS OF ORGANIC COMPOUNDS (Continued)

In Alpha Numeric as they occur in Organic Compounds Table

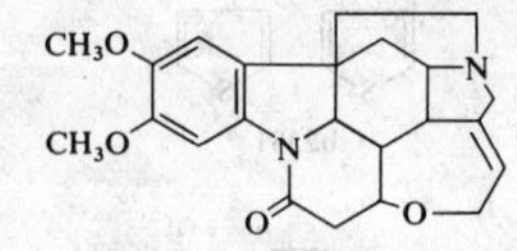

b2428

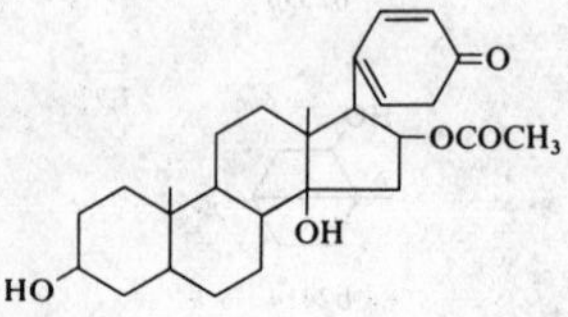

b2432

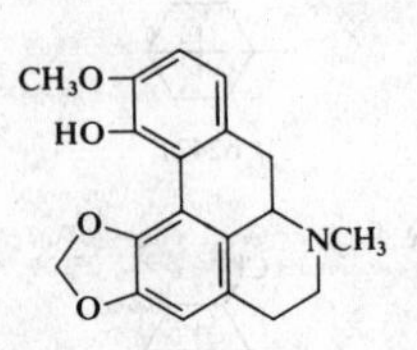

b2433

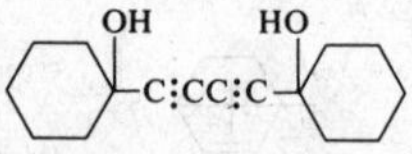

b2469

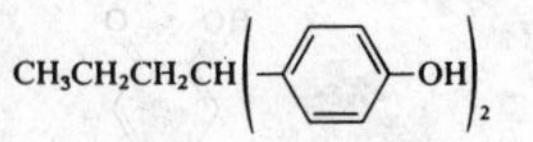

b2498

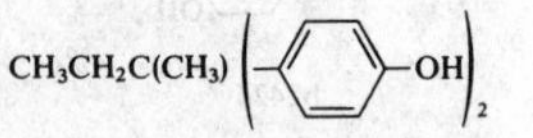

b2499

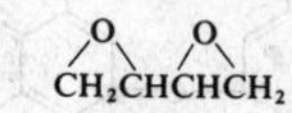

b2550

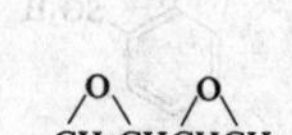

b2551

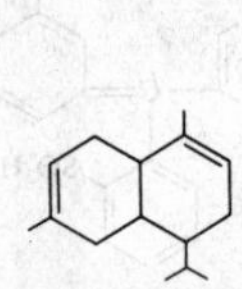

c7

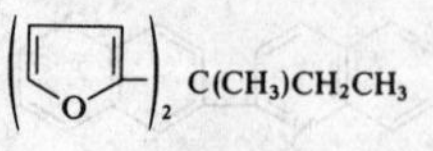

b2554

b2561

b2562

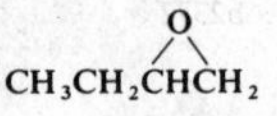

b2563

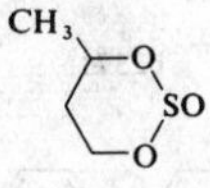

b2620

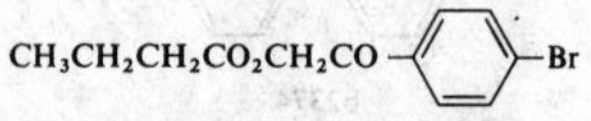

b2660

b2666

b2669

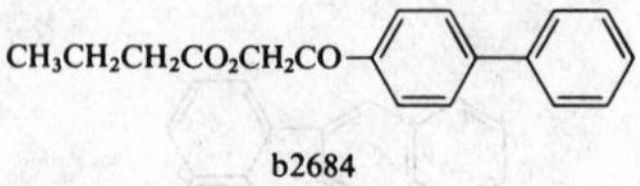

b2684

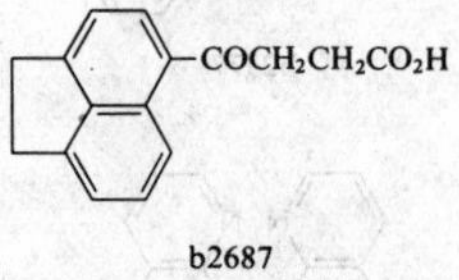

b2687

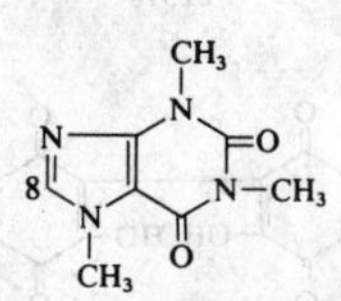

c8

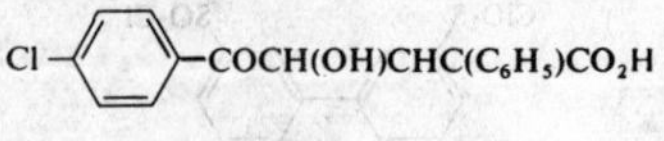

b2747

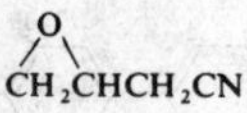

b2770

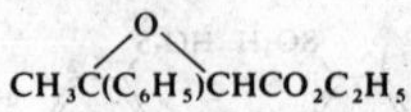

b2771

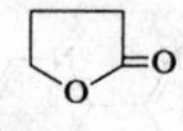

b2778

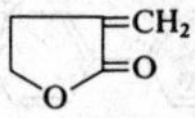

b2793

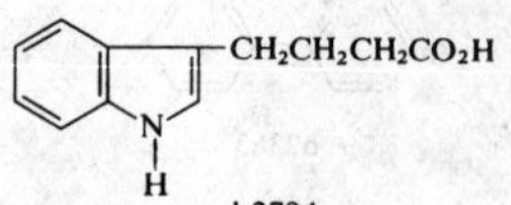

b2794

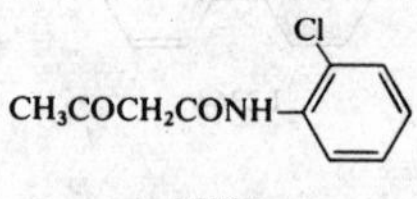

b2821

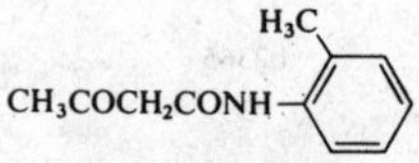

b2823

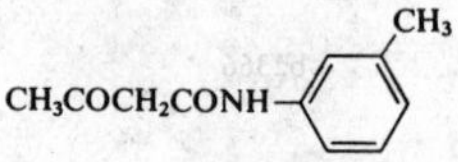

b2824

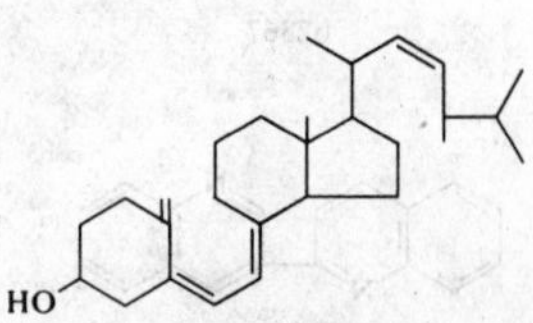

c20

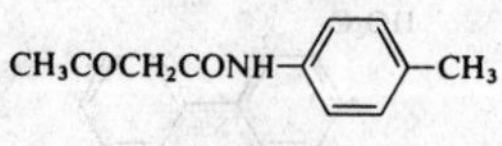

b2825

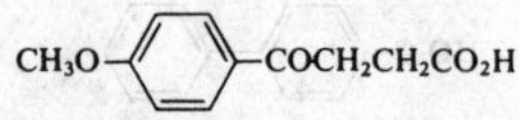

b2833[1]

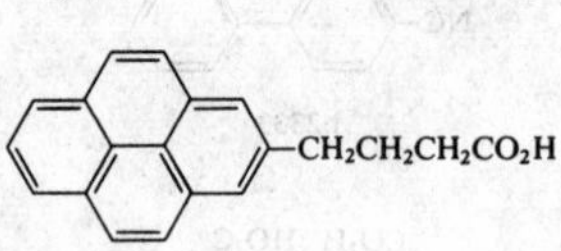

b2855

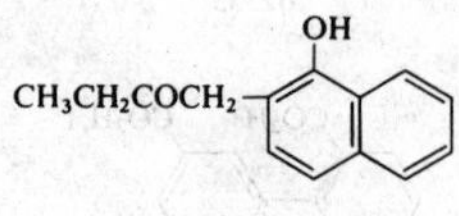

b2925

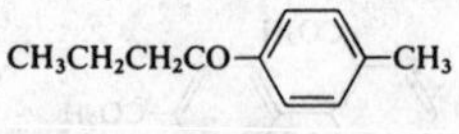

b2932

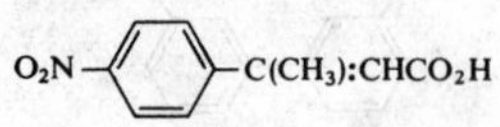

b3042[1]

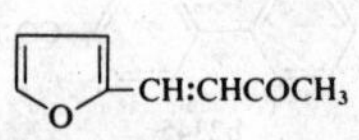

b3057

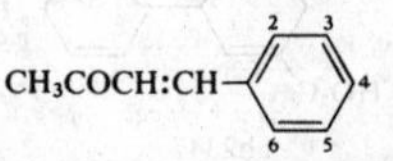

b3065

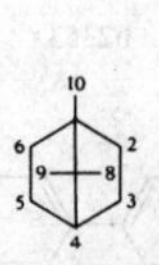

c21

STRUCTURAL FORMULAS OF ORGANIC COMPOUNDS (Continued)

In Alpha Numeric as they occur in Organic Compounds Table

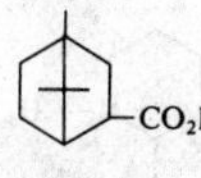
c22

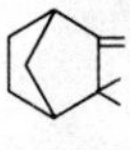
c23

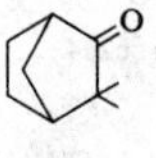
c26

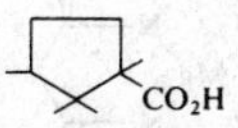
c27

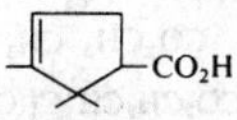
c30

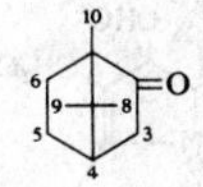
c31

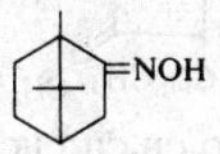
c34

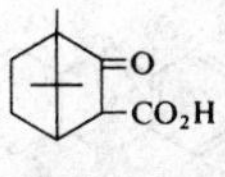
c53

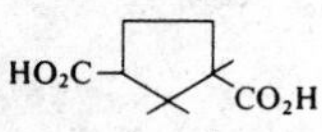
c56

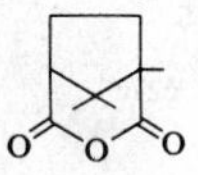
c59

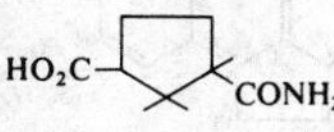
c63

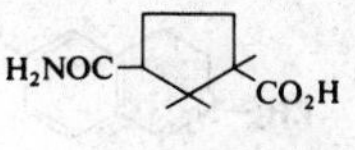
c64

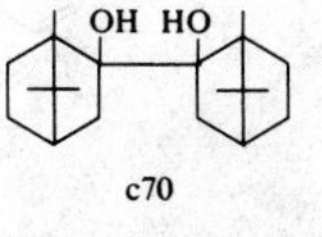
c70

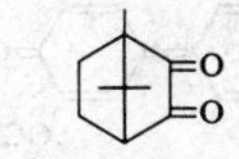
c71

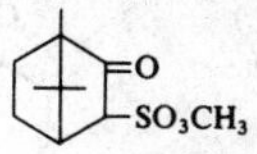
c72

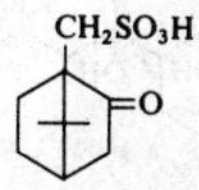
c73

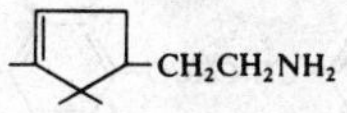
c76

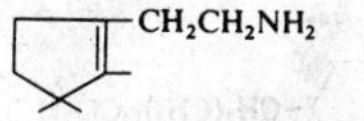
c77

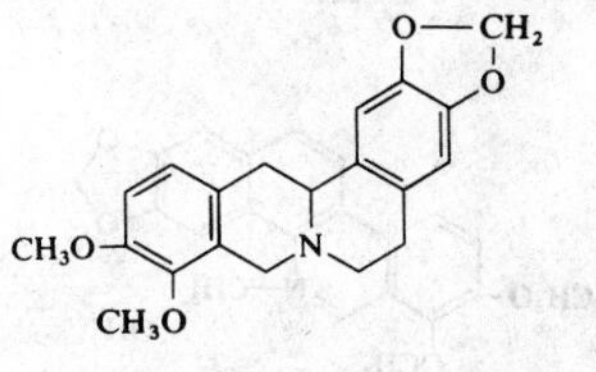
c78

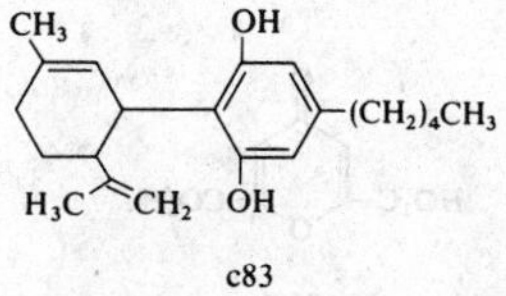
c83

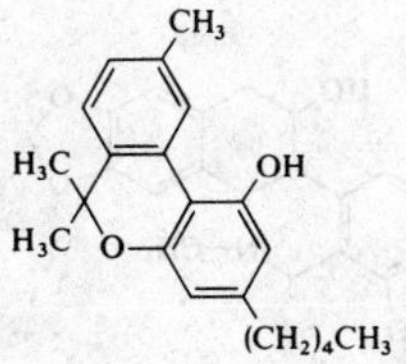
c84

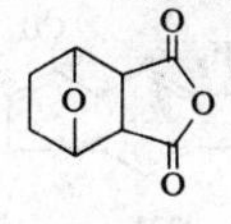
c85

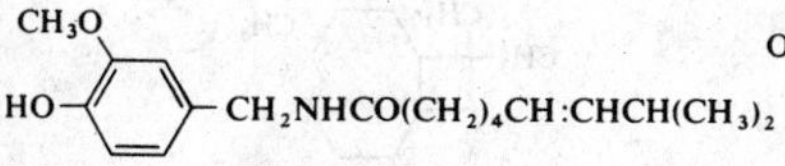
c86

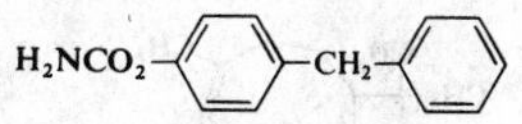
c88

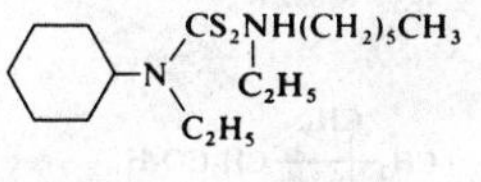
c103

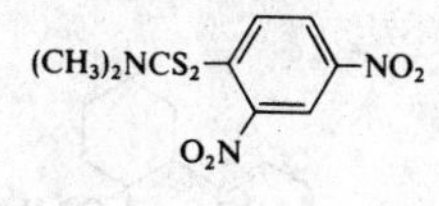
c111

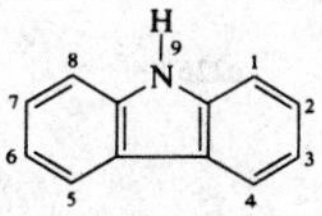
c141

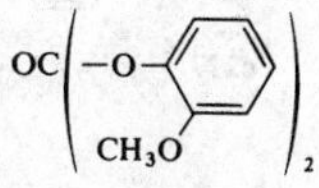
c166

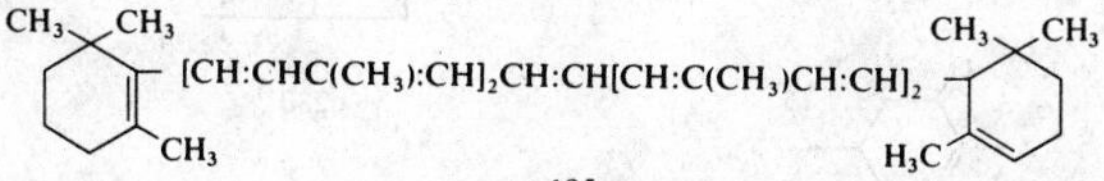
c195

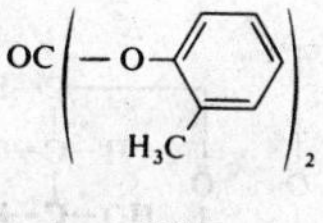
c177

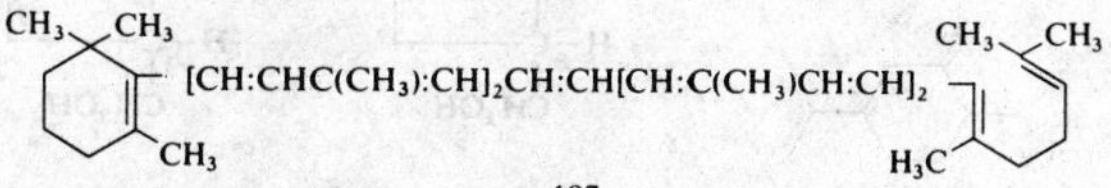
c197

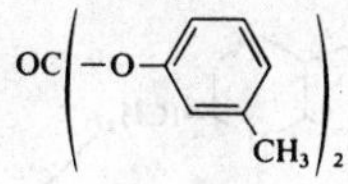
c178

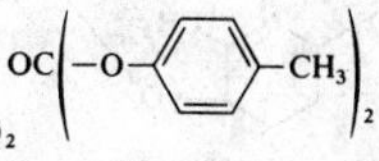
c179

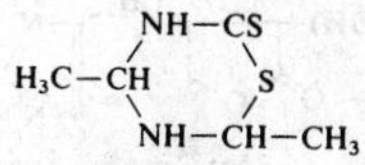
c187

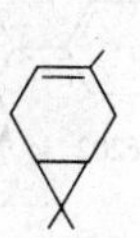
c188

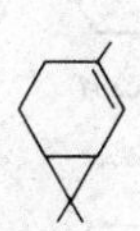
c190

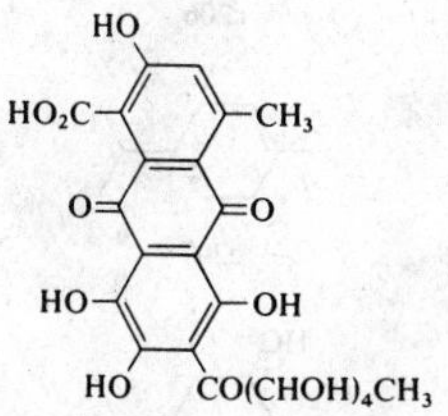
c191

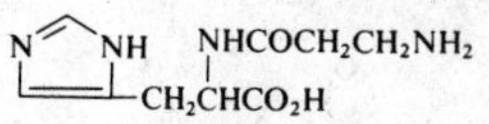
c193

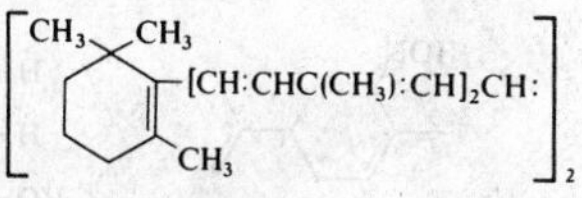
c196

STRUCTURAL FORMULAS OF ORGANIC COMPOUNDS (Continued)

In Alpha Numeric as they occur in Organic Compounds Table

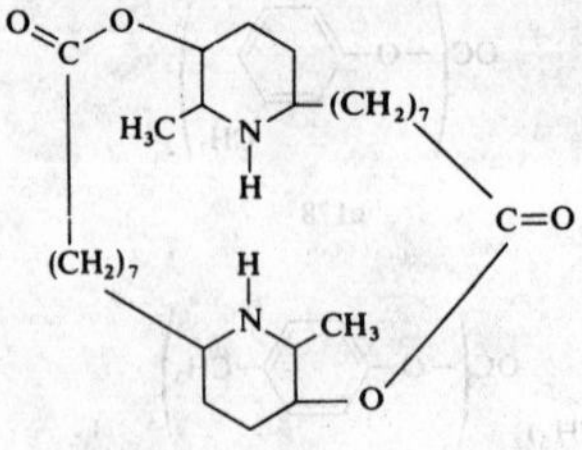
c199

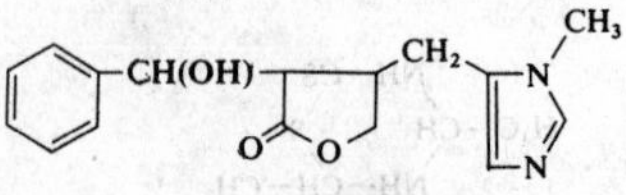
c201

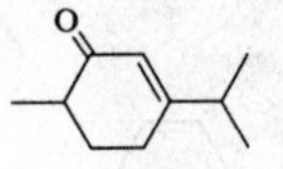
c202

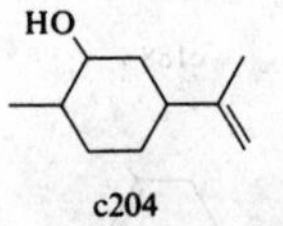
c204

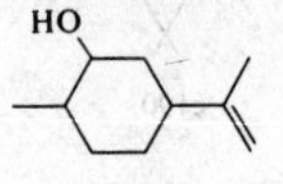
c206

c207

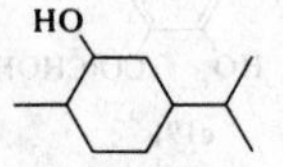
c208

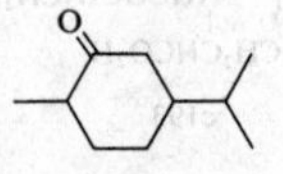
c210

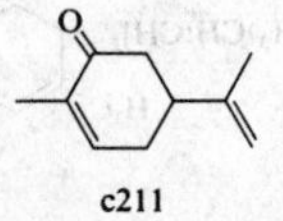
c211

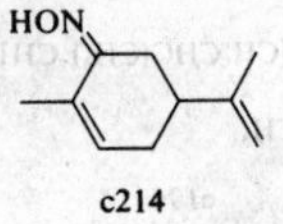
c214

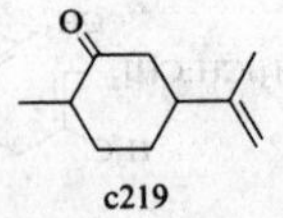
c219

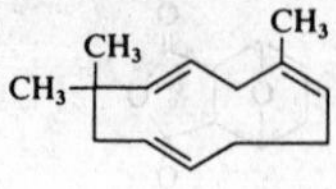
c221

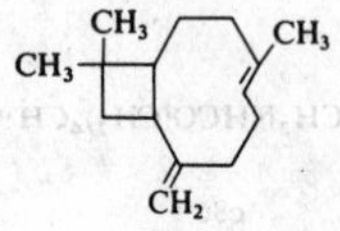
c222

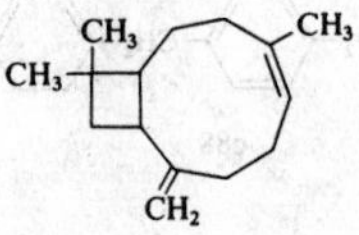
c223

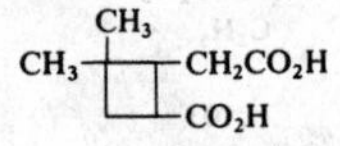
c224

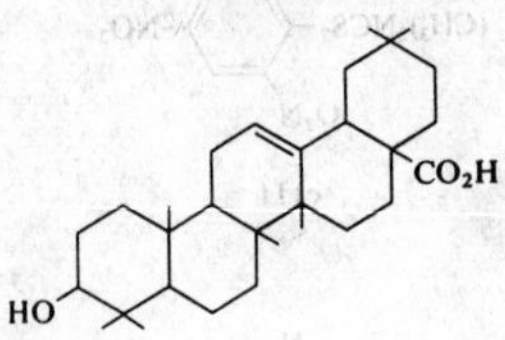
c226

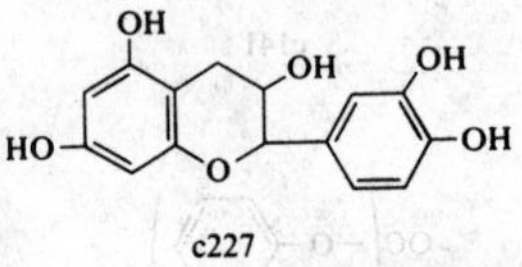
c227

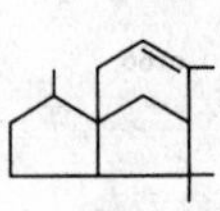
c230

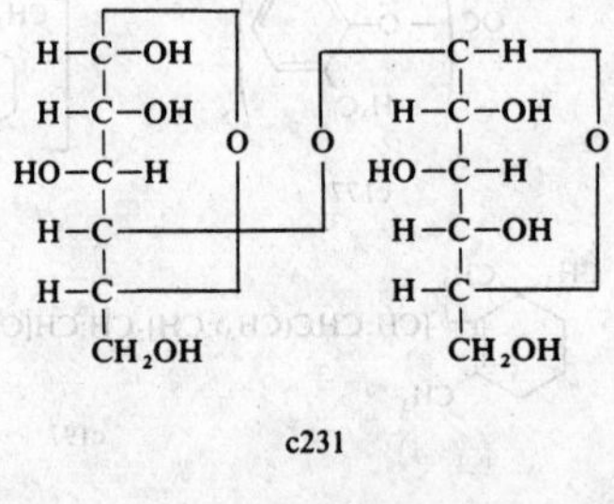
c231

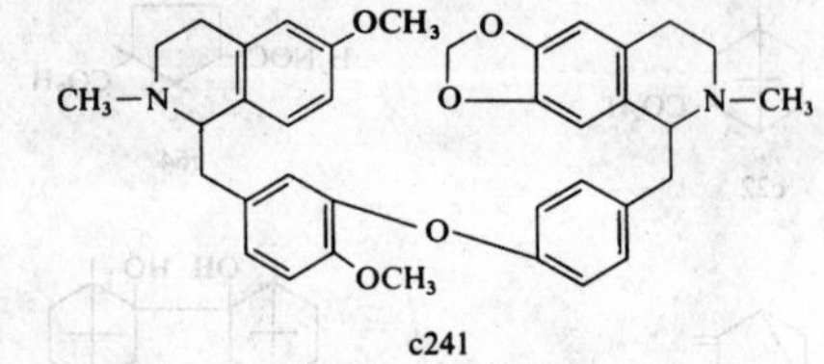
c241

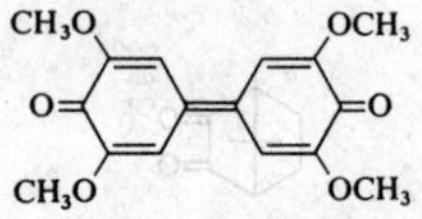
c243

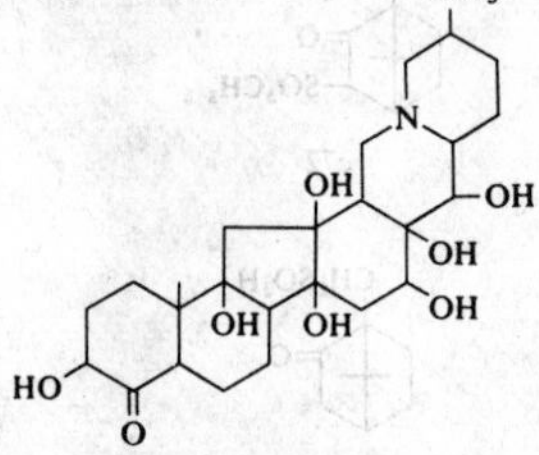
c244

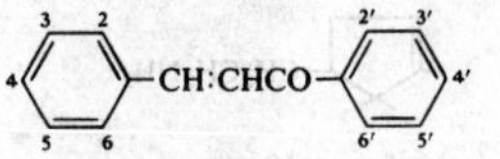
c245

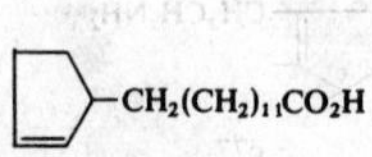
c257

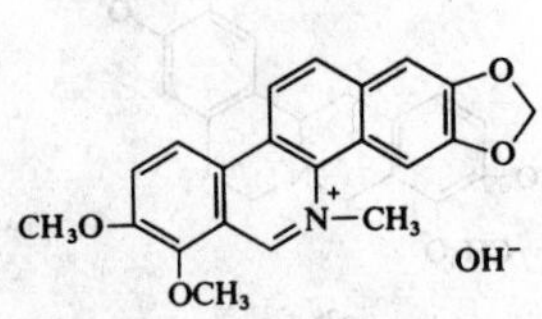
c259

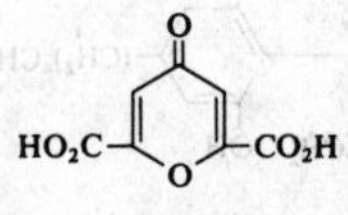
c260

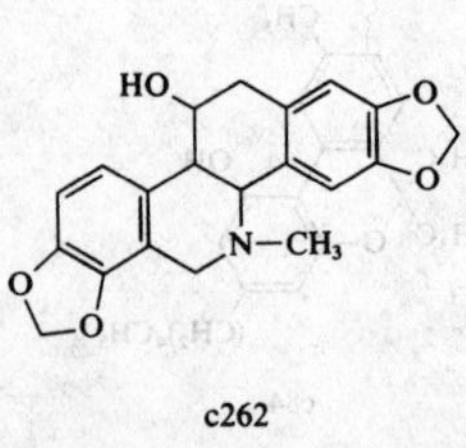
c262

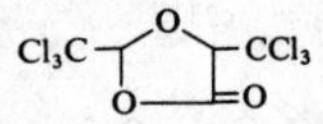
c264

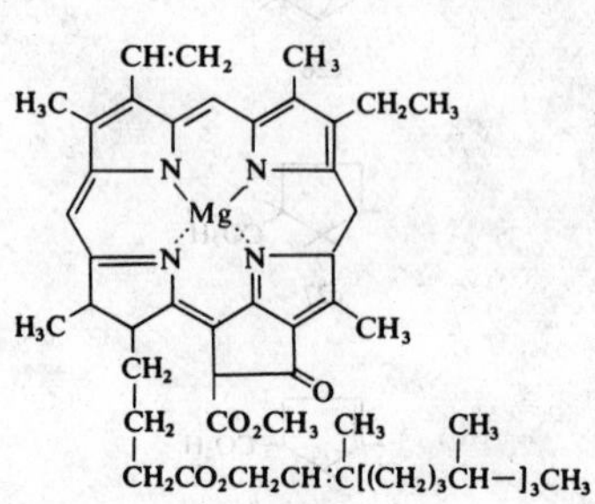
c266

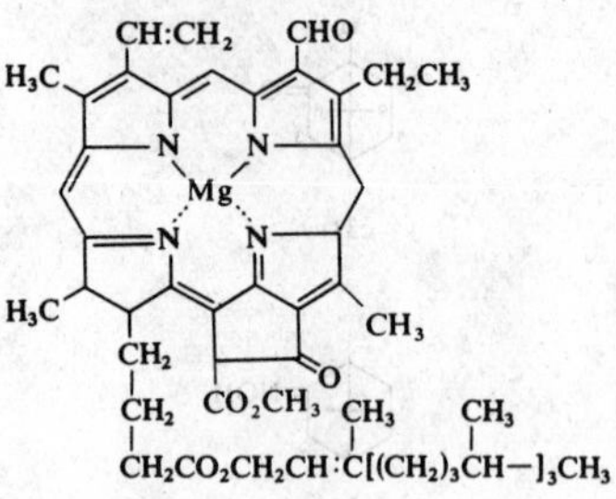
c267

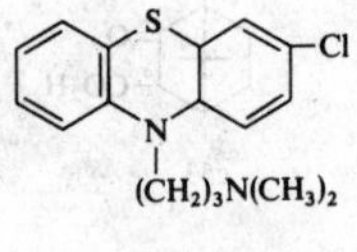
c268

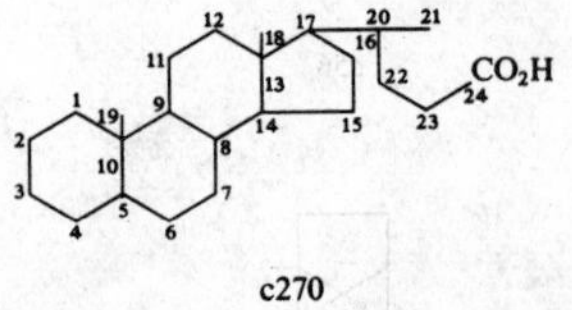
c270

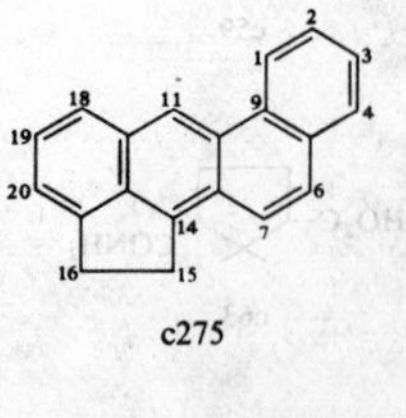
c275

STRUCTURAL FORMULAS OF ORGANIC COMPOUNDS (Continued)

In Alpha Numeric as they occur in Organic Compounds Table

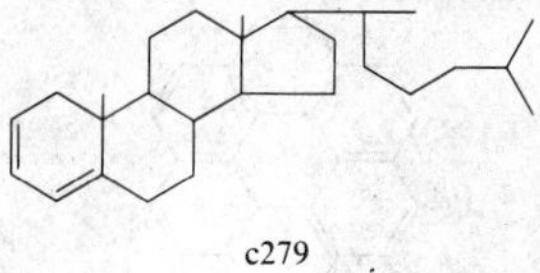

c279

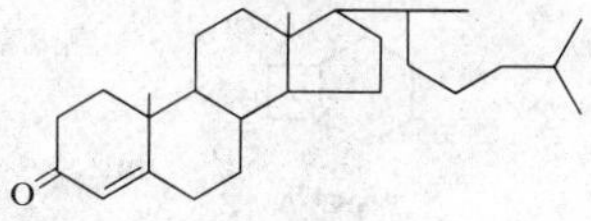

c298

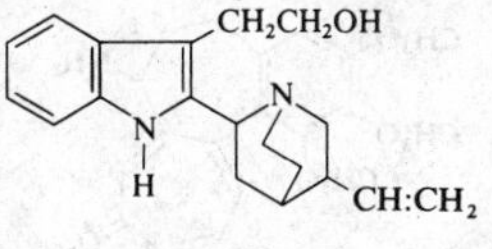

c320

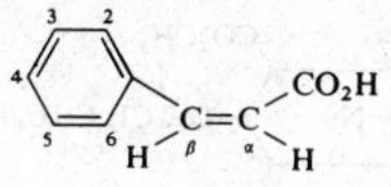

c347

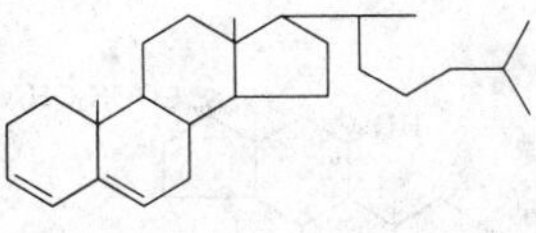

c280

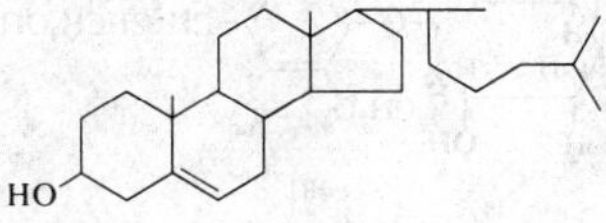

c300

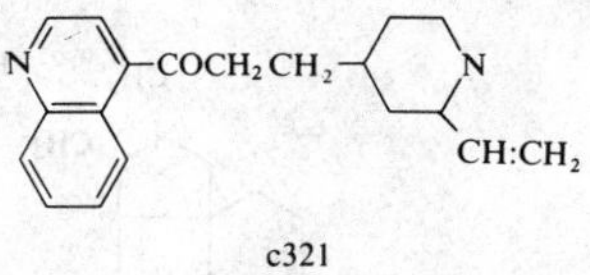

c321

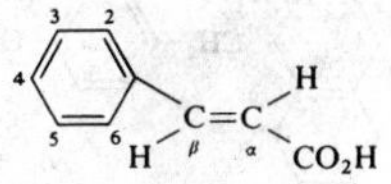

c350

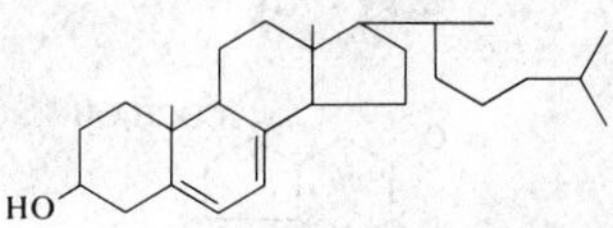

c281

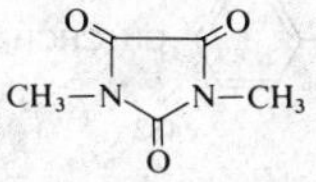

c304

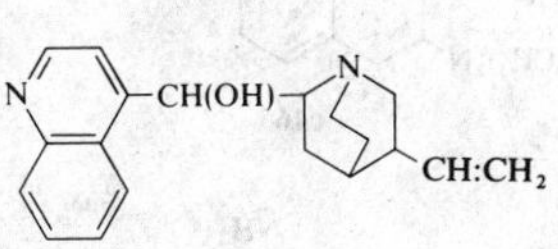

c322

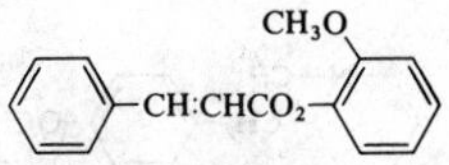

c358

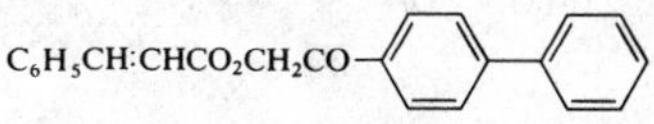

c366

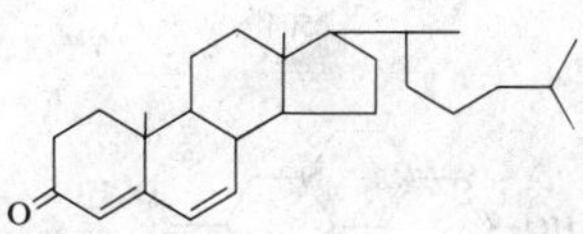

c282

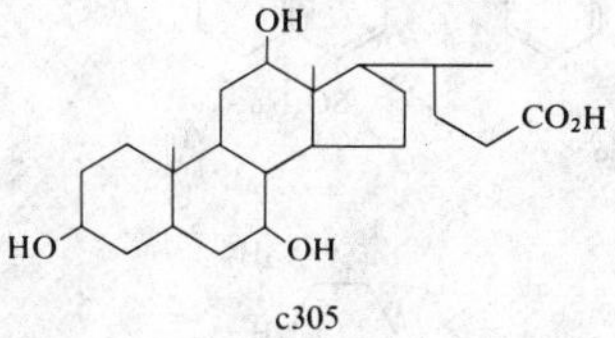

c305

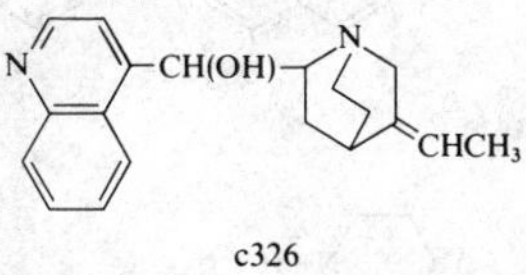

c326

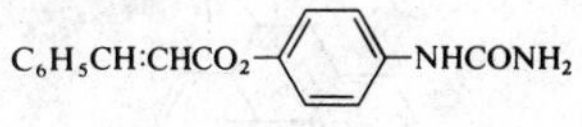

c368

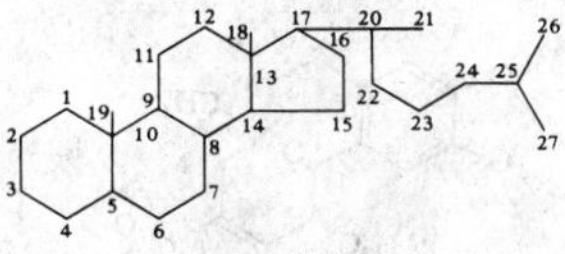

c283

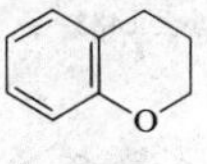

c311

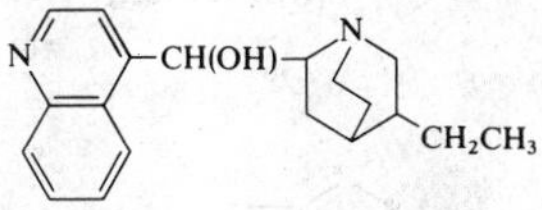

c333

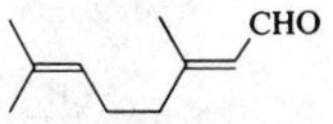

c425

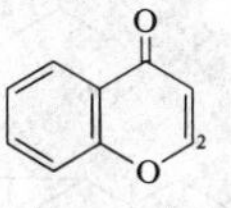

c312

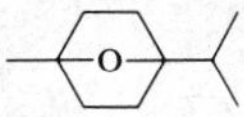

c334

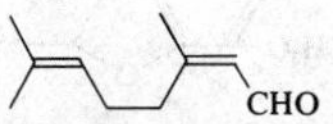

c426

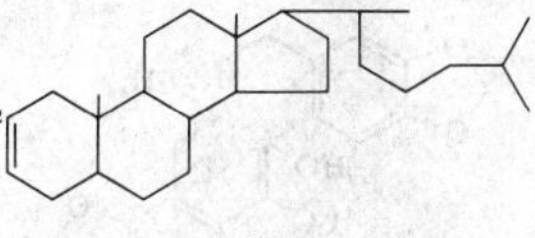

c291

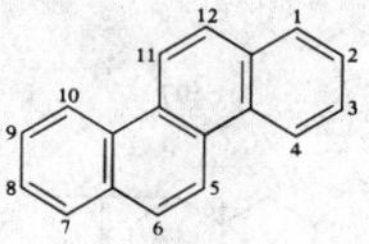

c313

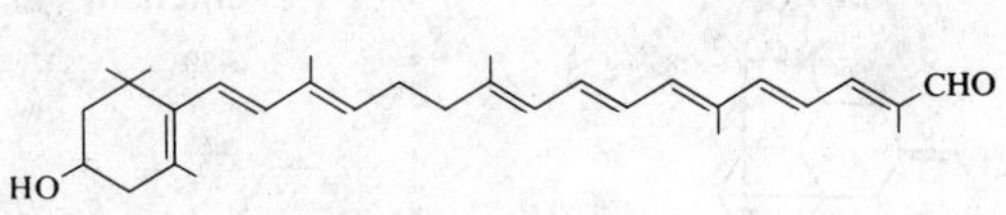

c427

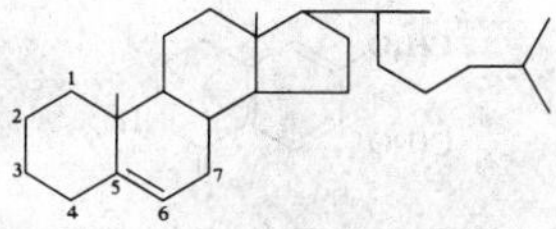

c292

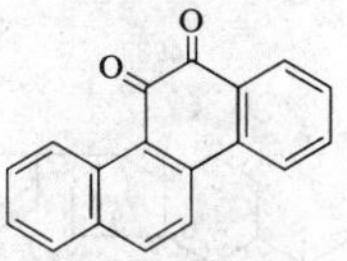

c318

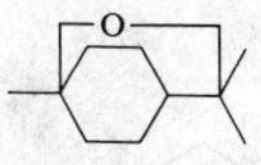

c335

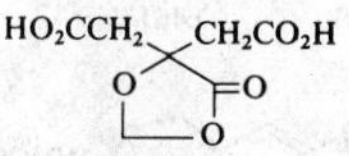

c436

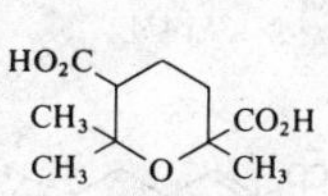

c336

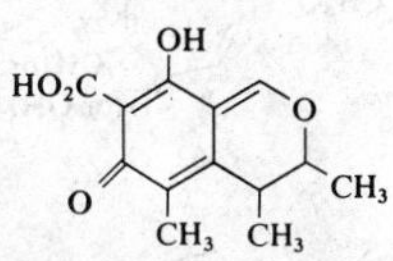

c437

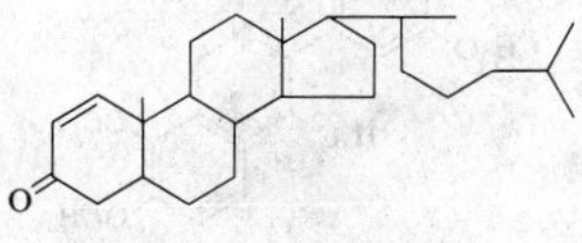

c297

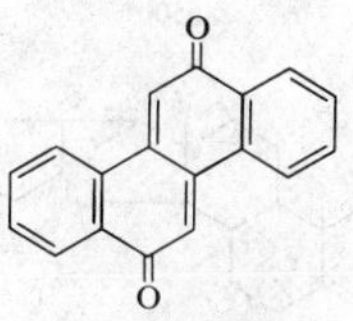

c319

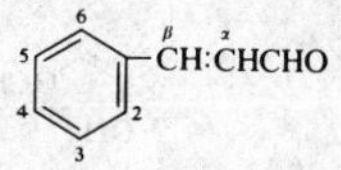

c339

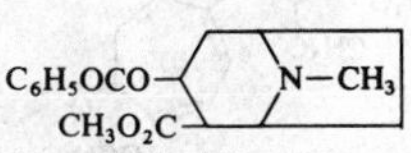

c445

In Alpha Numeric as they occur in Organic Compounds Table

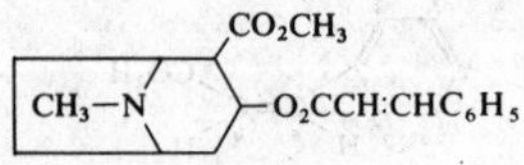
c451

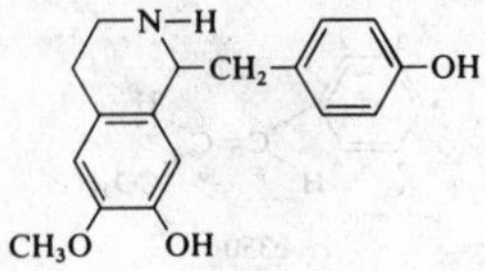
c453

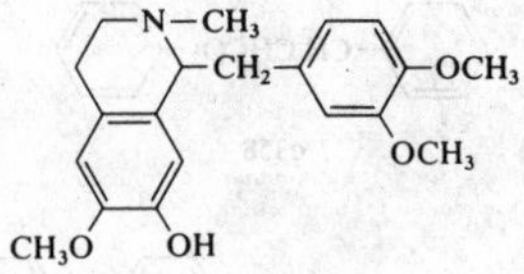
c454

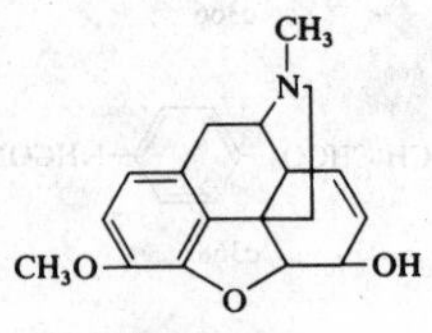
c455

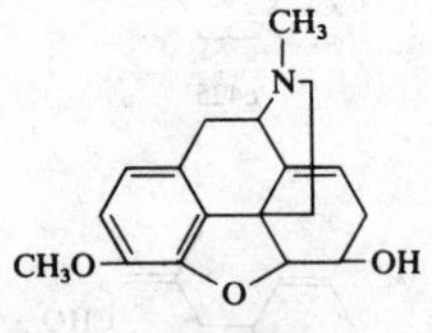
c460

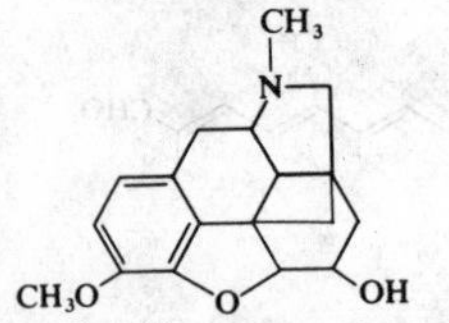
c461

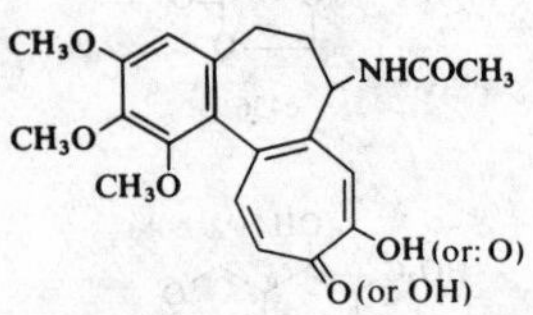
c462

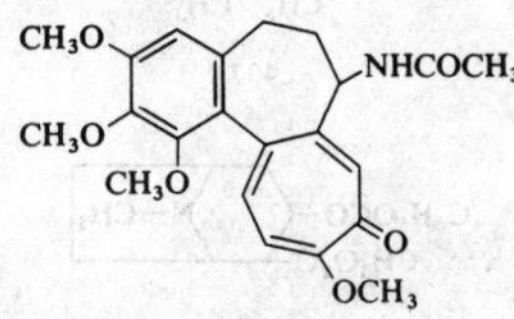
c463

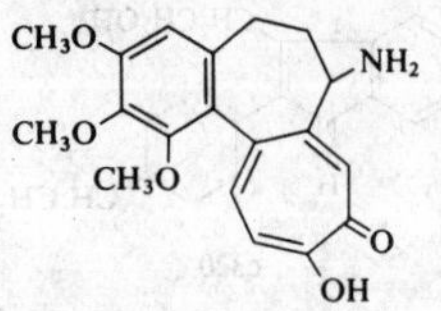
c464

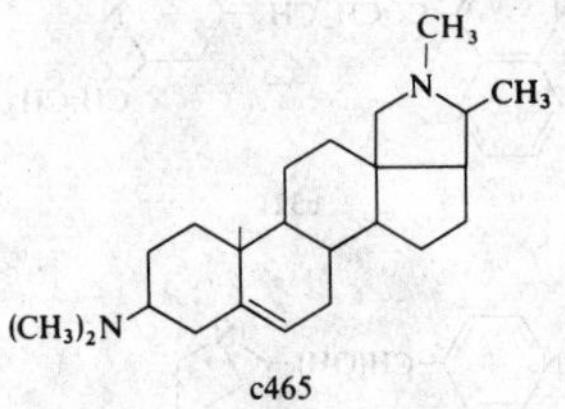
c465

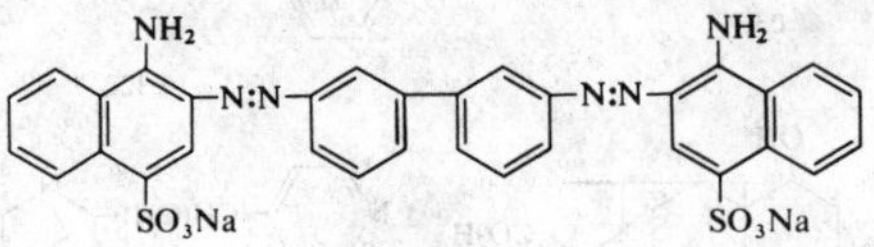
c466

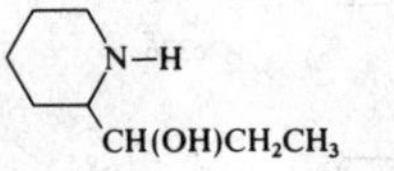
c467

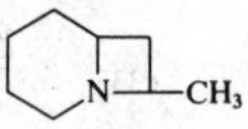
c468

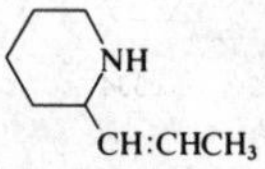
c470

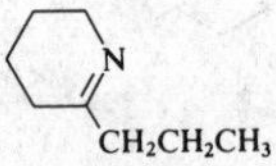
c473

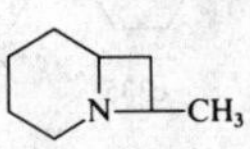
c474

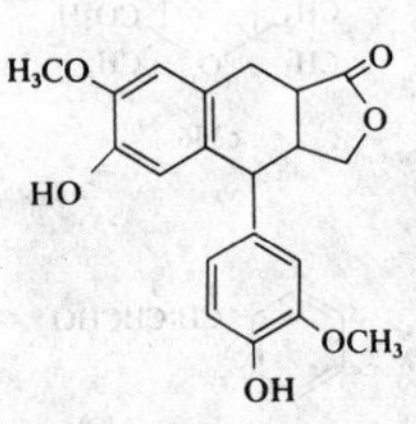
c477

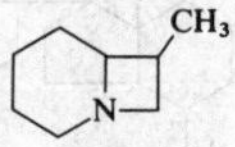
c478

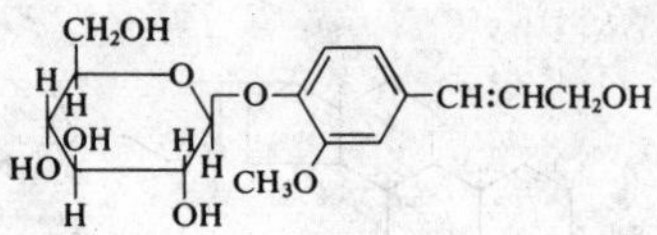
c481

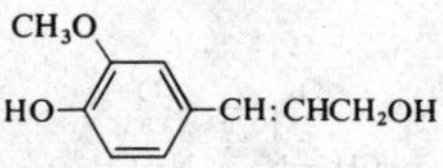
c482

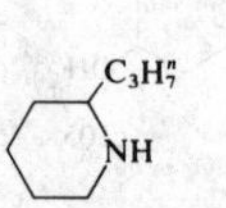
c483

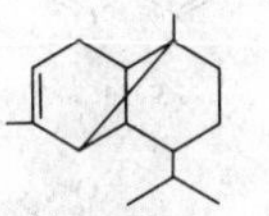
c496

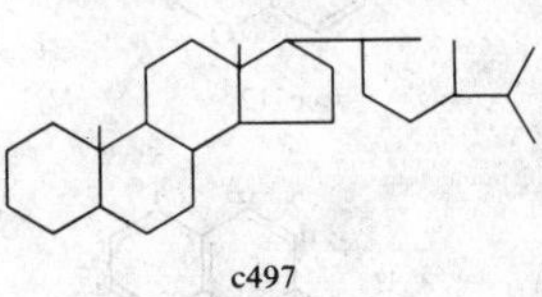
c497

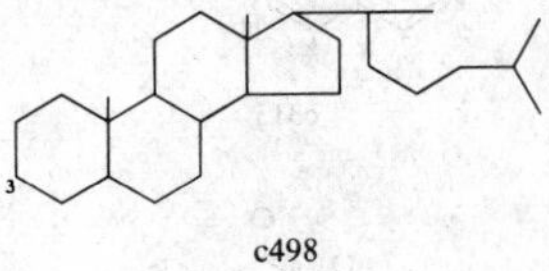
c498

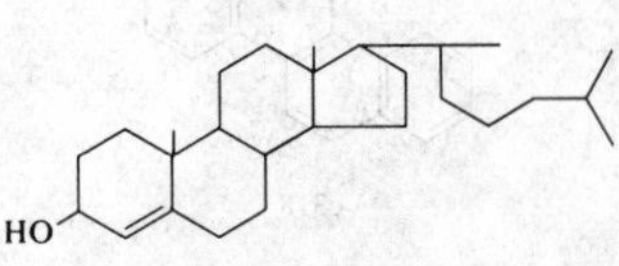
c500

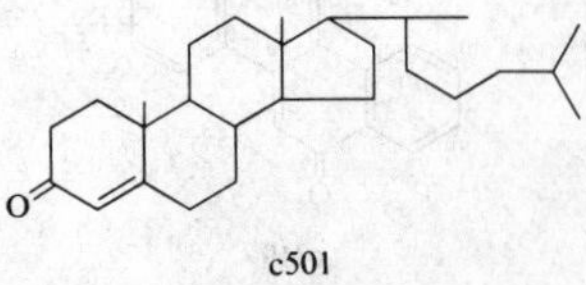
c501

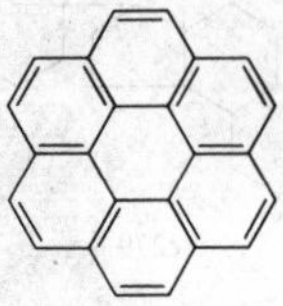
c502

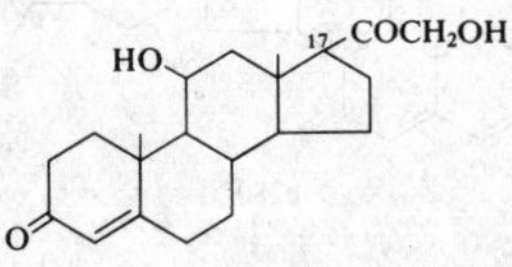
c503

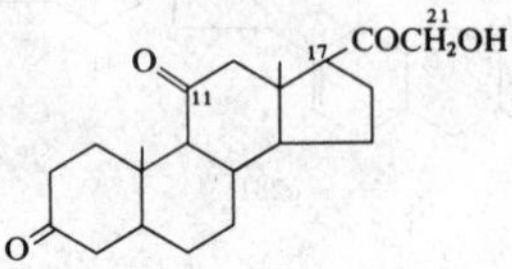
c504

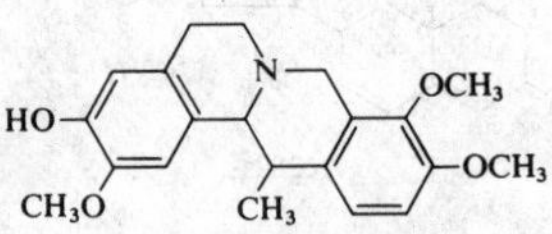
c508

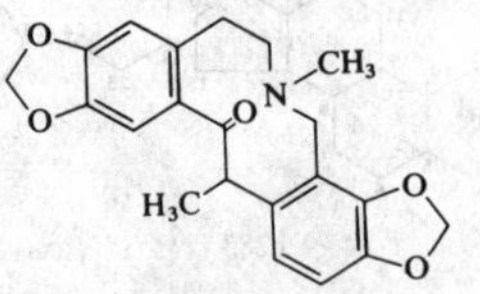
c510

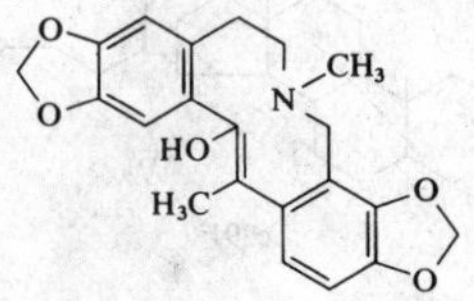
c511

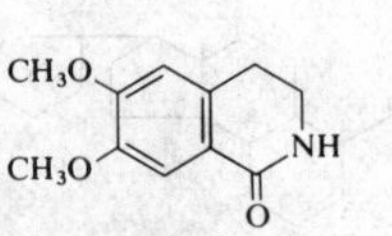
c512

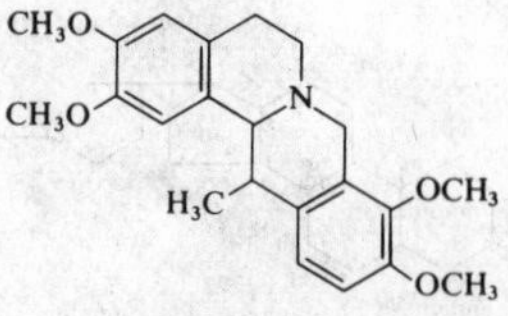
c513

STRUCTURAL FORMULAS OF ORGANIC COMPOUNDS (Continued)

In Alpha Numeric as they occur in Organic Compounds Table

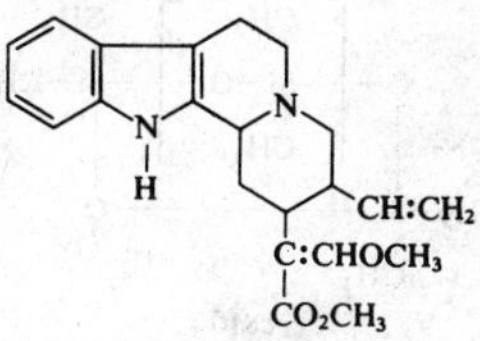

c518

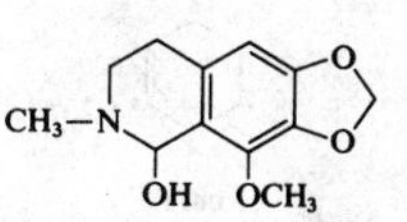

c519

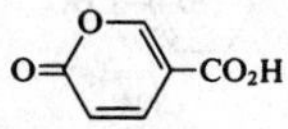

c523

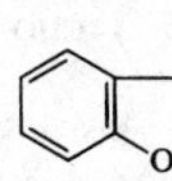

c524

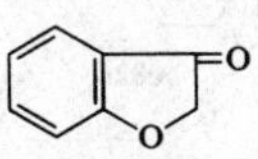

c525

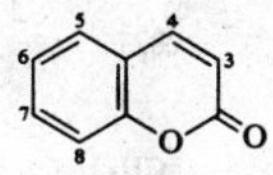

c526

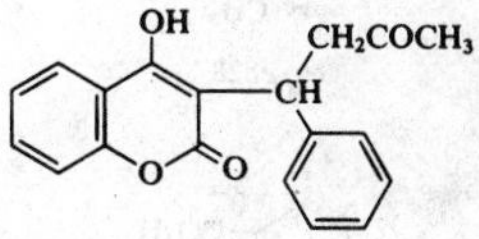

c543

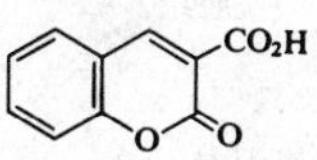

c547

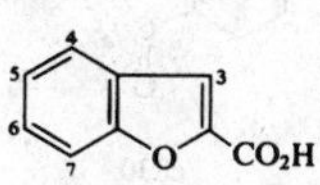

c548

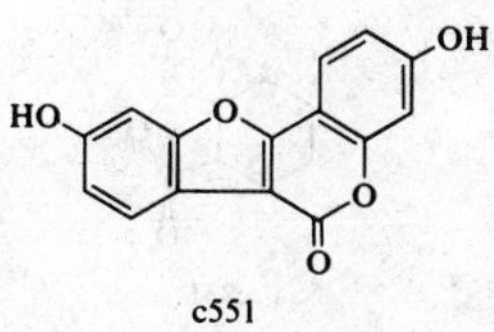

c551

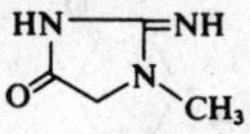

c554

$\left[HO_2C\overset{CH_3}{\overset{|}{C}}:CHCH:CH\overset{CH_3}{\overset{|}{C}}:CHCH=\right]_2$

c555

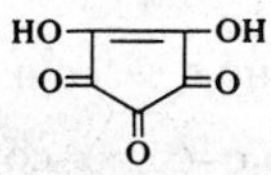

c556

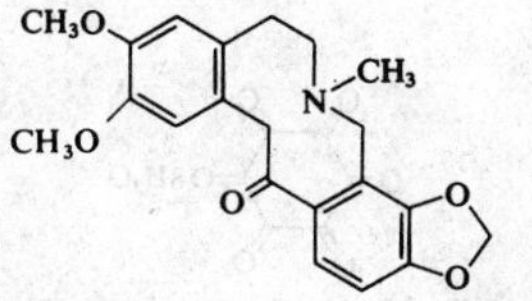

c557

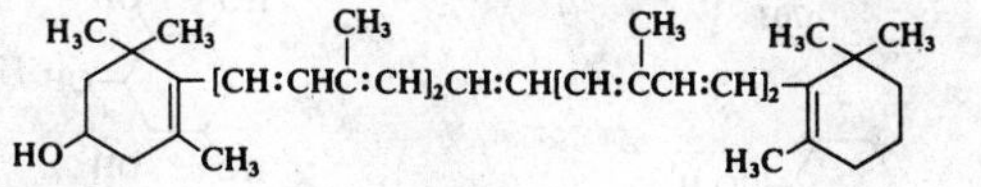

c558

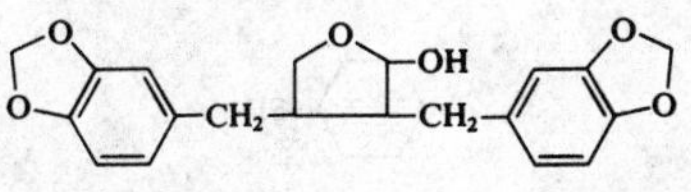

c559

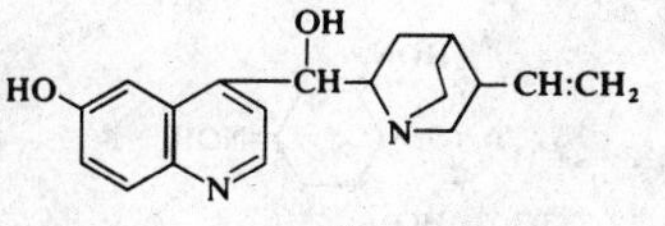

c563

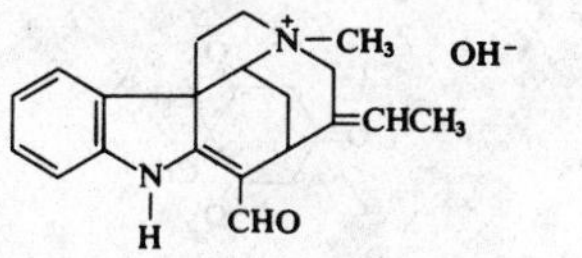

c564

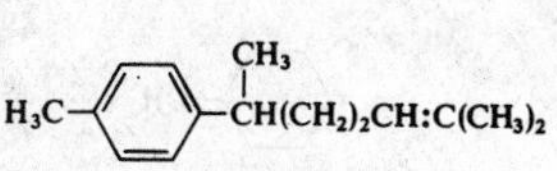

c565

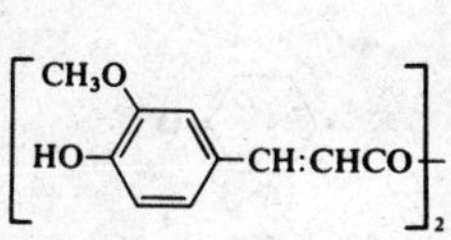

c566

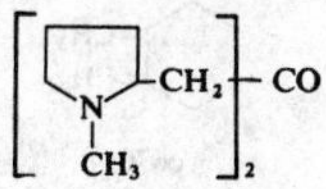

c567

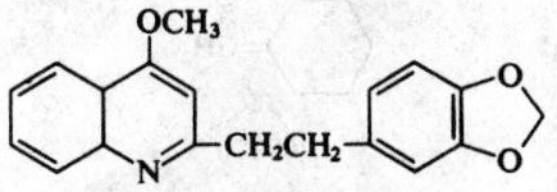

c569

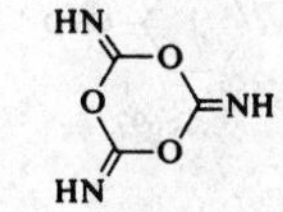

c570

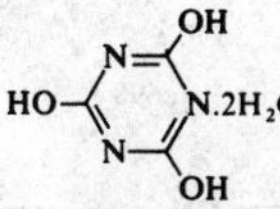

c586

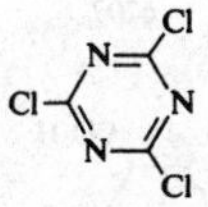

c588

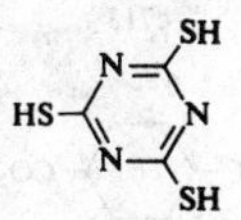

c589

c590

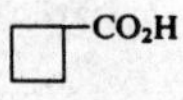

c594

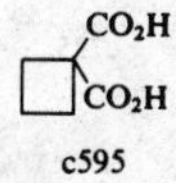

c595

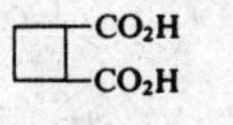

c597

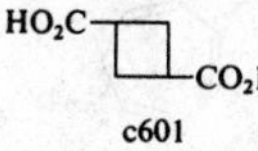

c601

c603

c605

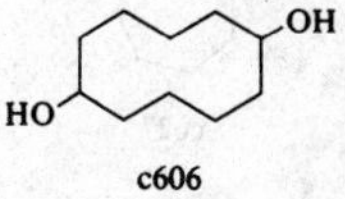

c606

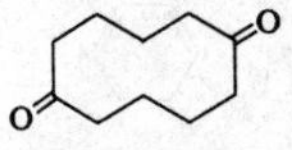

c607

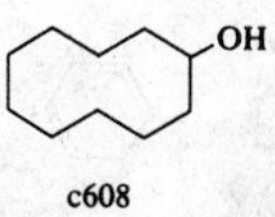

c608

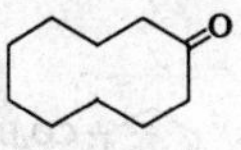

c609

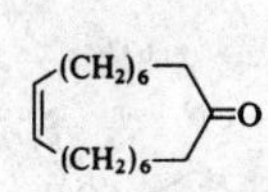

c610

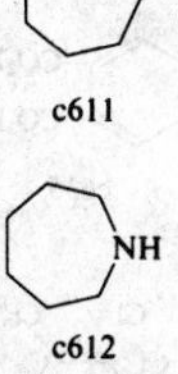

c611

c612

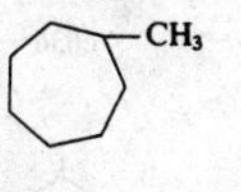

c614

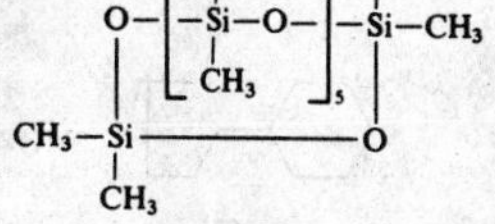

c618

STRUCTURAL FORMULAS OF ORGANIC COMPOUNDS (Continued)

In Alpha Numeric as they occur in Organic Compounds Table

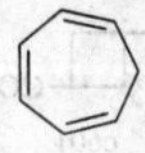
c619

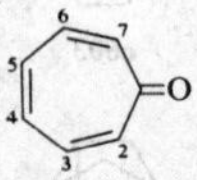
c620

c627

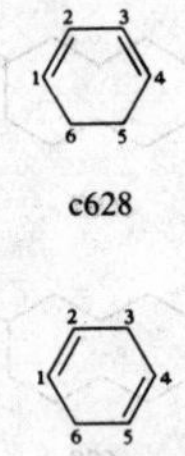
c628

c629

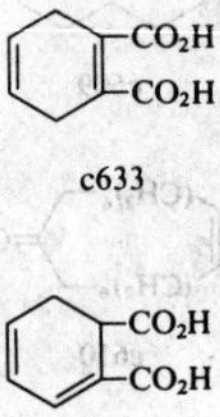
c633

c634

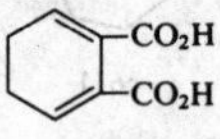
c635

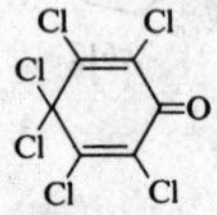
c636

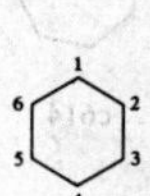
c637

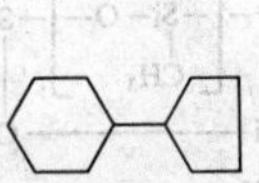
c652

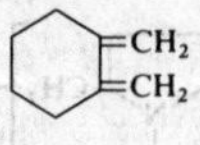
c670

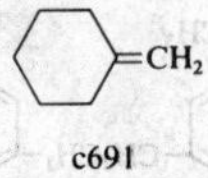
c691

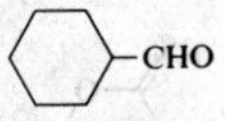
c697

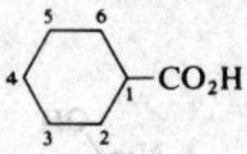
c698

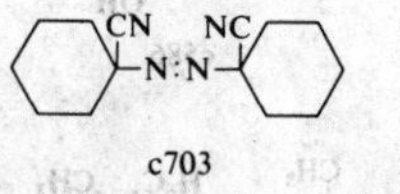
c703

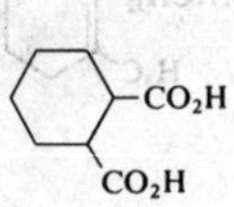
c707

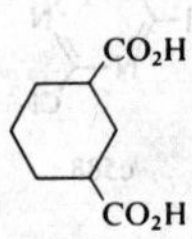
c713

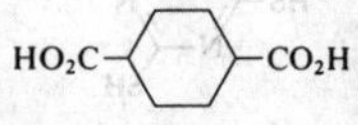
c719

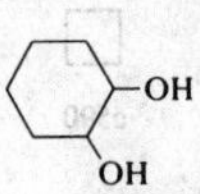
c723

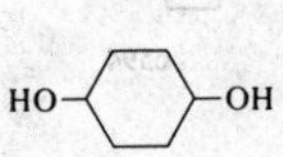
c725

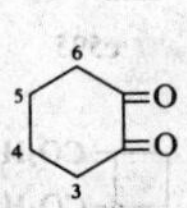
c727

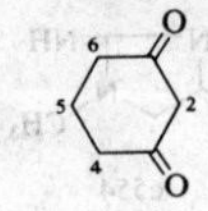
c729

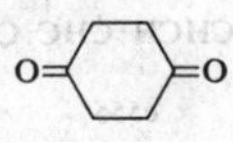
c731

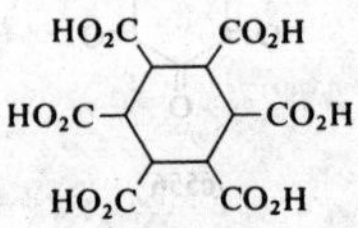
c736

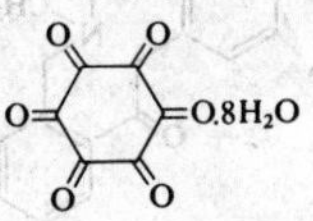
c736[1]

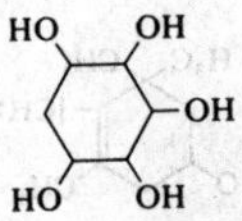
c737

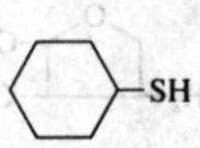
c739

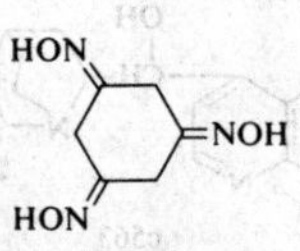
c740

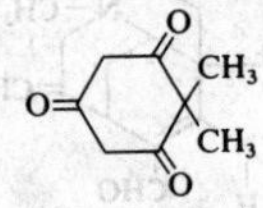
c741

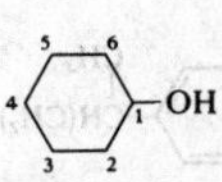
c742

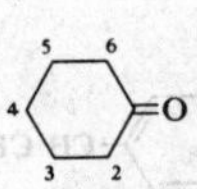
c785

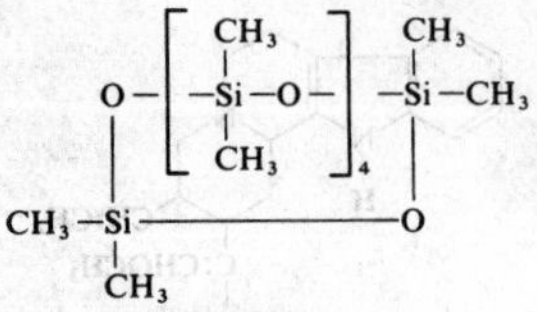
c810

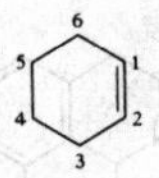
c811

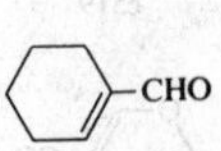
c824

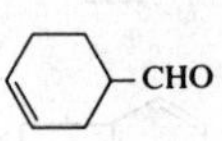
c825

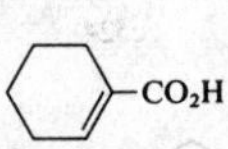
c826

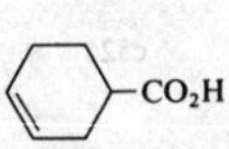
c827

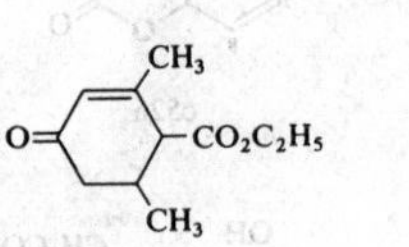
c828

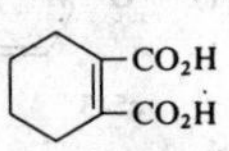
c829

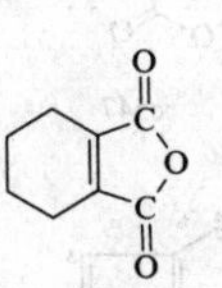
c830

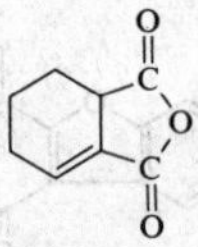
c831

In Alpha Numeric as they occur in Organic Compounds Table

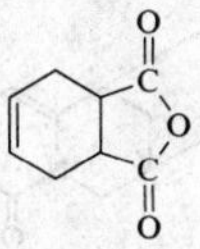

c832

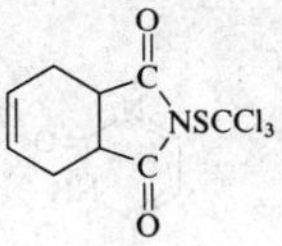

c835

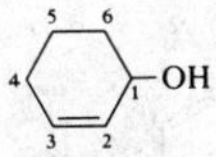

c836

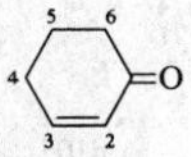

c841

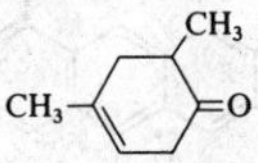

c845

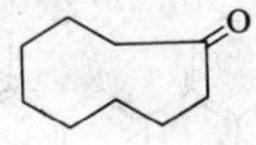

c850

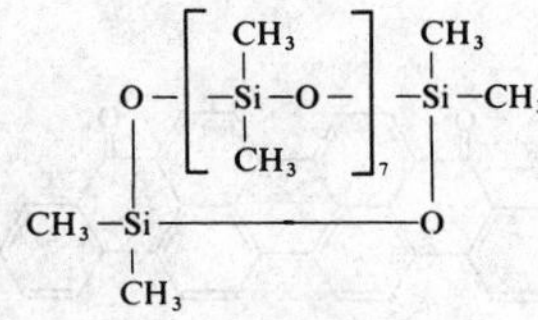

c851

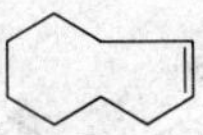
c852

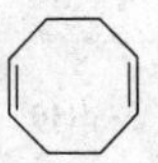
c854

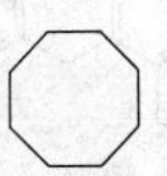
c855

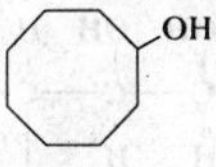

c856

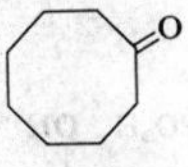

c857

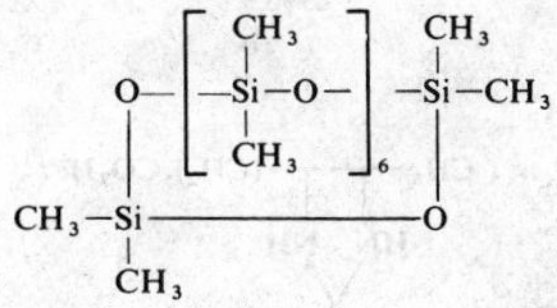

c859

c860

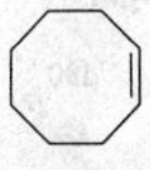
c861

c862

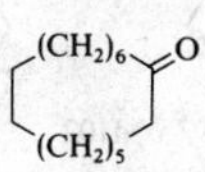

c863

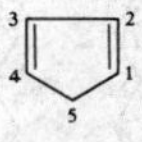
c865

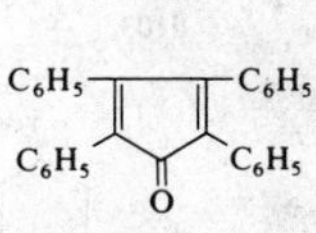

c867

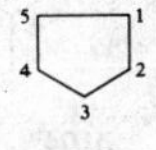
c868

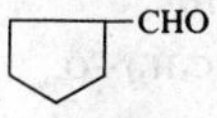

c879

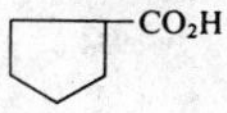

c880

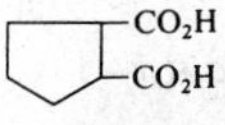

c883

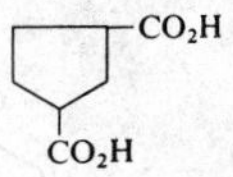

c887

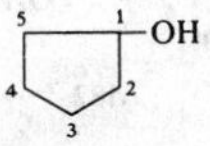

c892

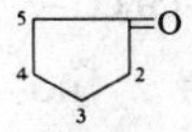

c894

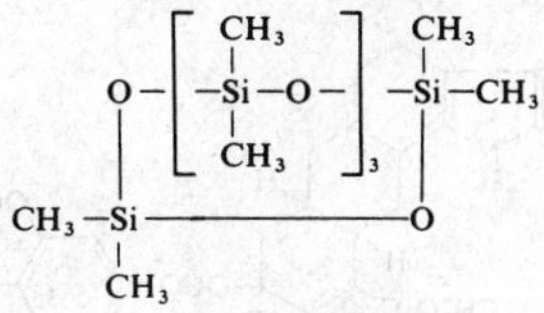

c898

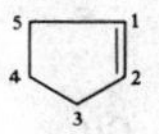
c899

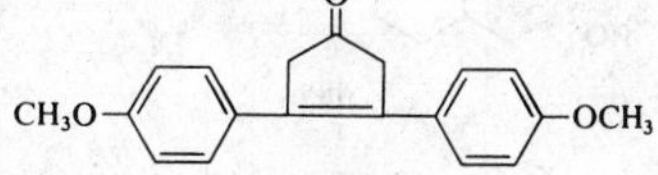

c902

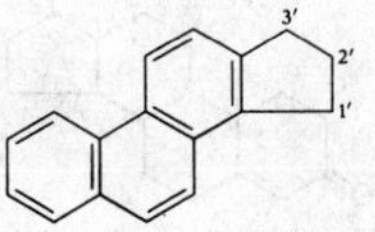
c903

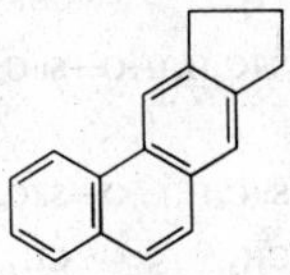
c904

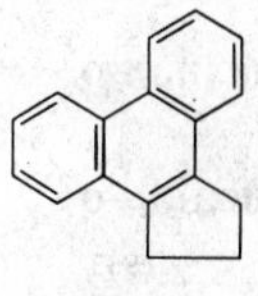
c905

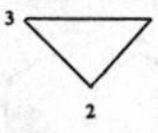
c907

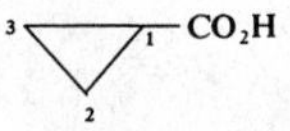

c912

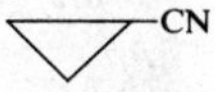

c913

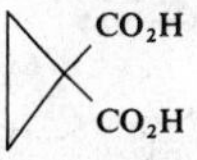

c920

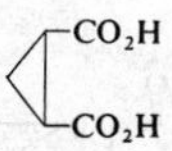

c922

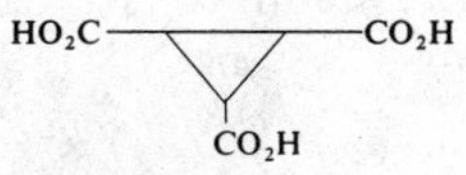

c929

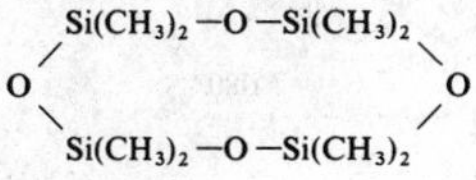

c930

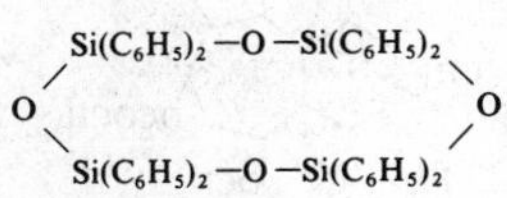

c931

In Alpha Numeric as they occur in Organic Compounds Table

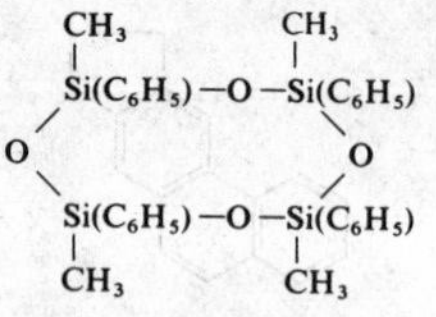

c932

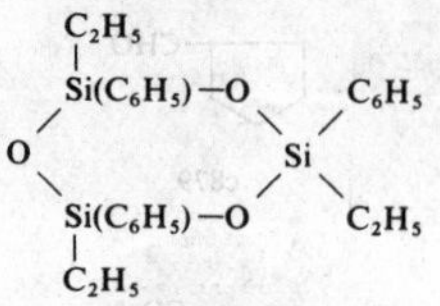

c934

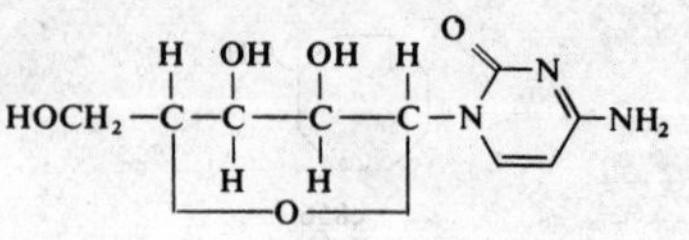

c945

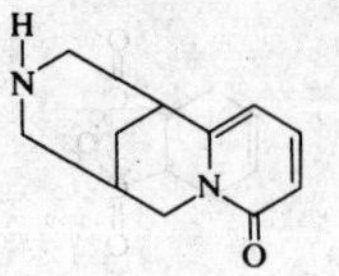

c947

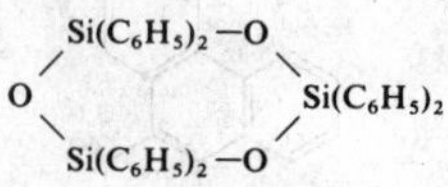

c933

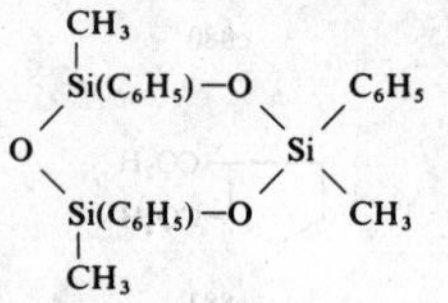

c935

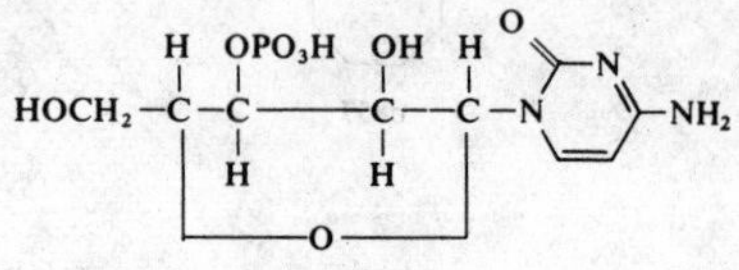

c946

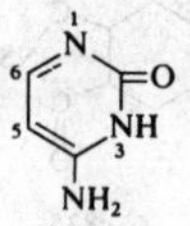

c949

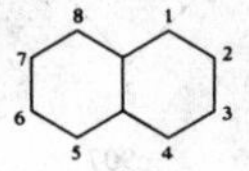

d6

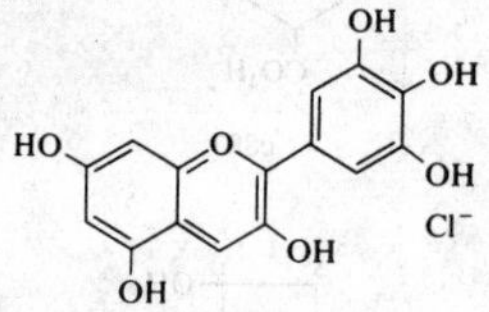

d82

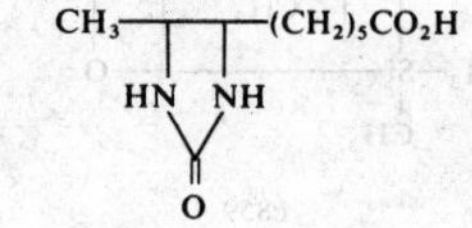

d87

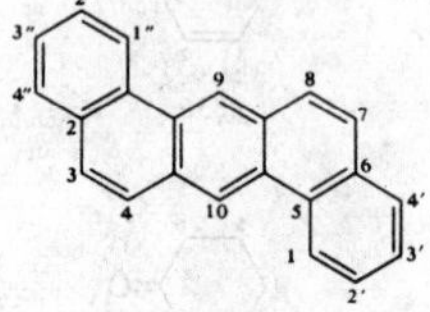

d105

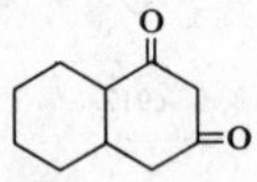

d12

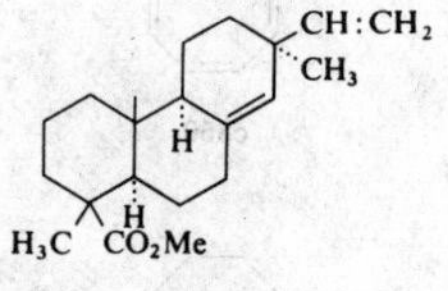

d90

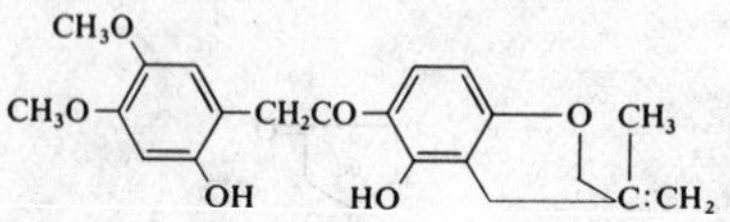

d83

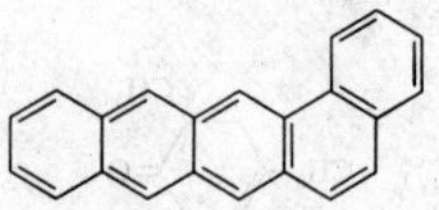

d106

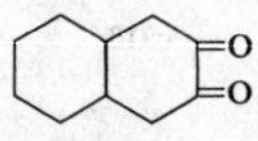

d14

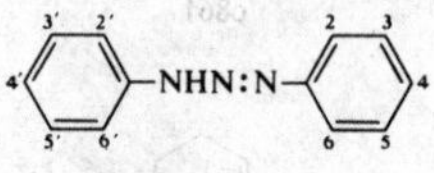

d91

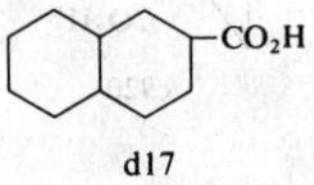

d17

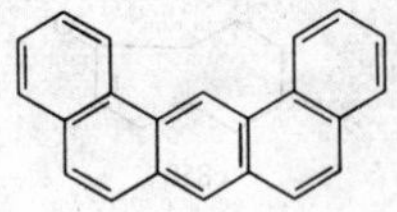

d107

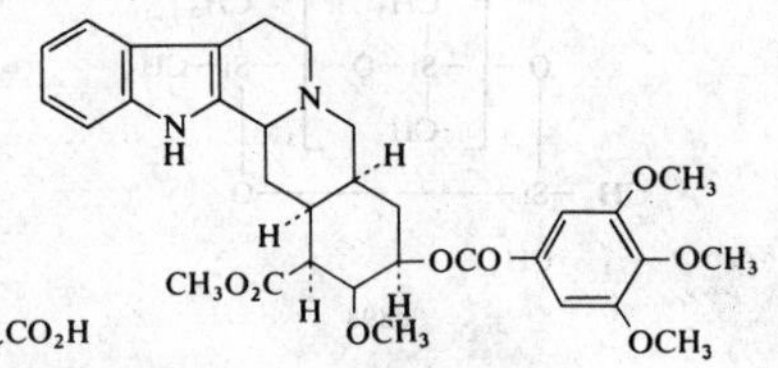

d84

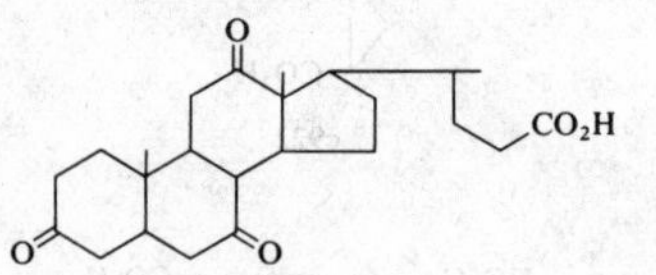

d79

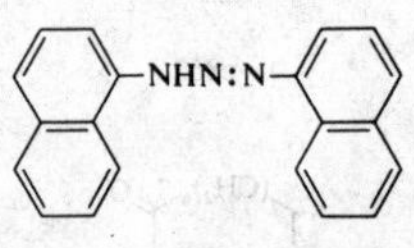

d102

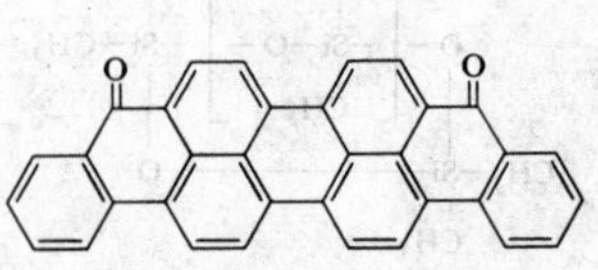

d109

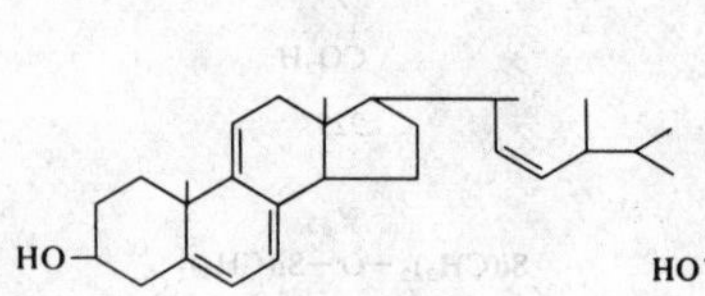

d80

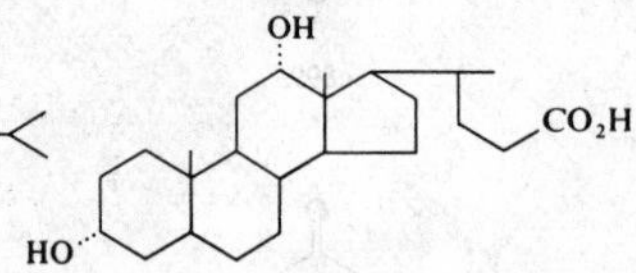

d85

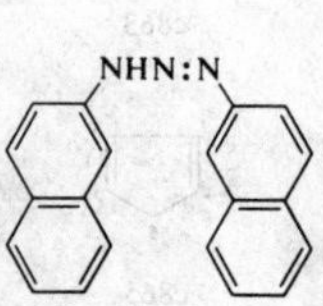

d103

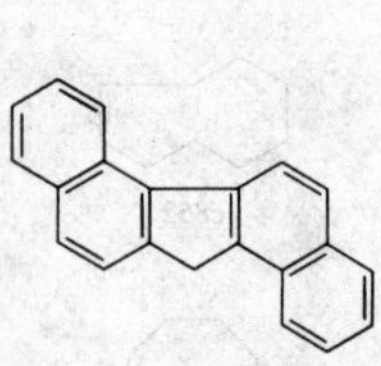

d110

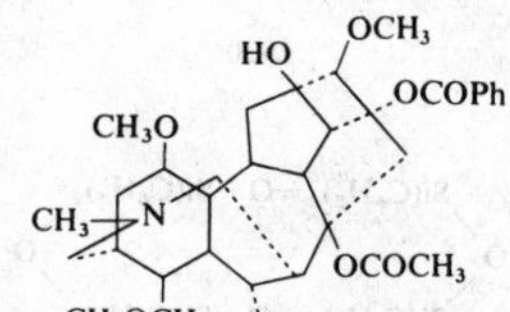

d81

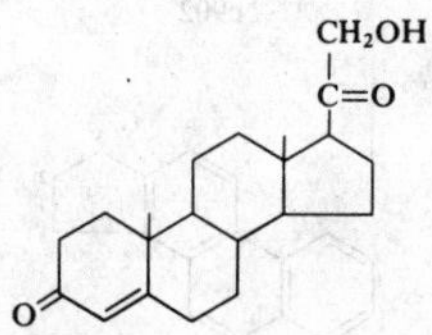

d86

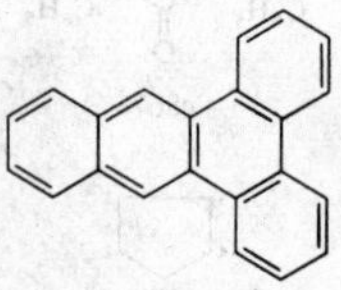

d104

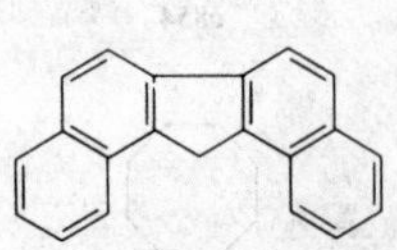

d111

STRUCTURAL FORMULAS OF ORGANIC COMPOUNDS (Continued)

In Alpha Numeric as they occur in Organic Compounds Table

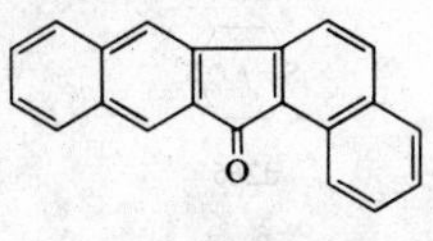

d112

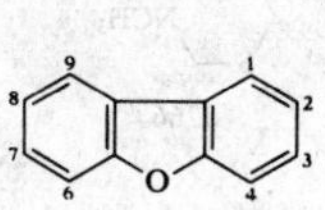

d113

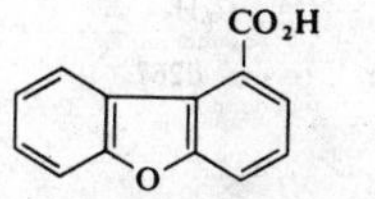

d128

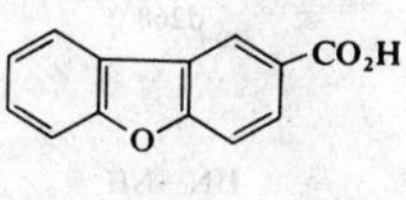

d129

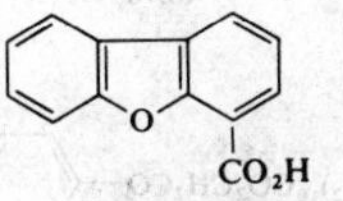

d130

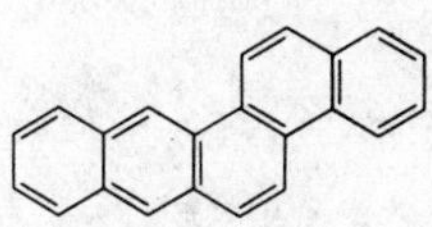

d131

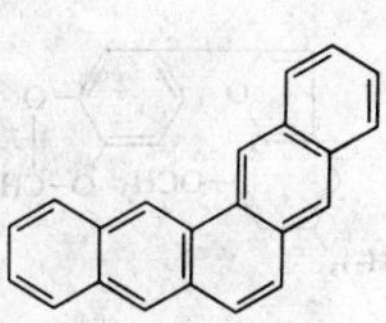

d132

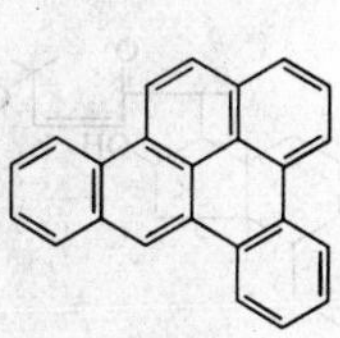

d133

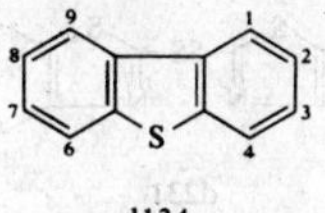

d134

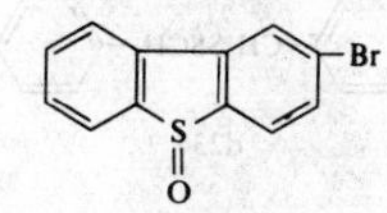

d138

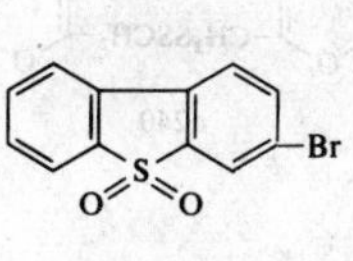

d139

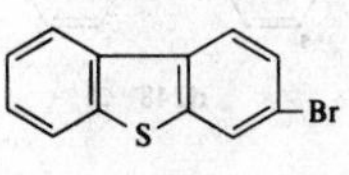

d140

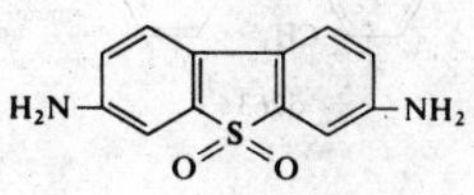

d143

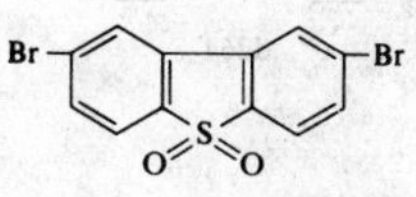

d148

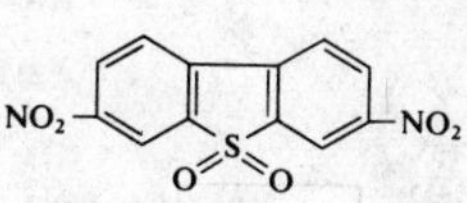

d150

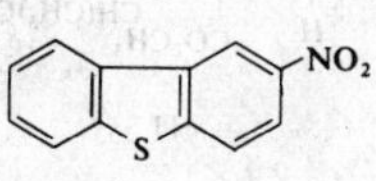

d151

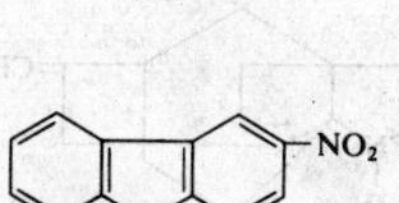

d152

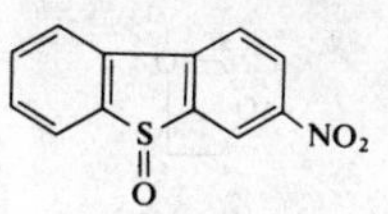

d154

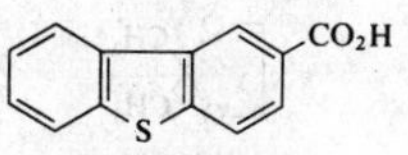

d155

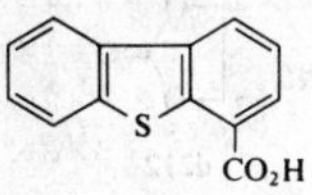

d156

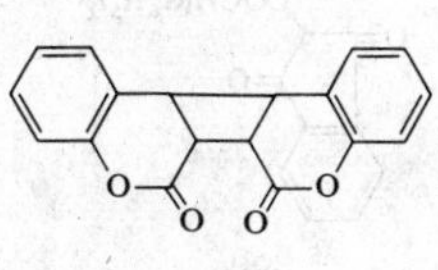

d159

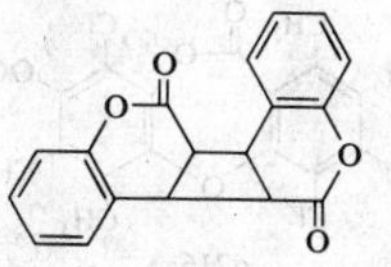

d160

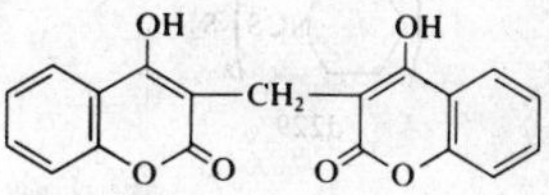

d161

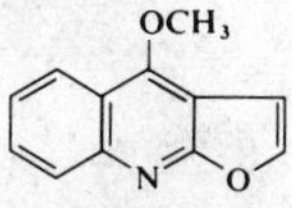

d162

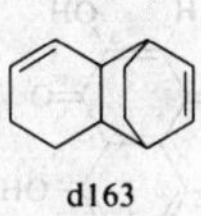

d163

d164

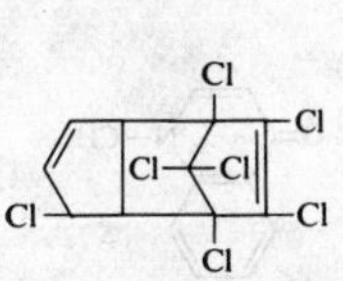

d165

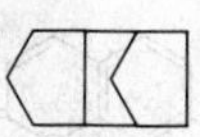

d166

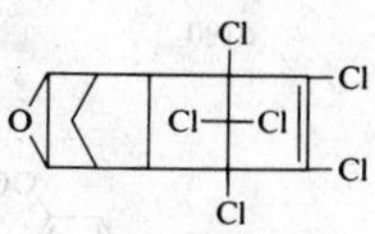

d167

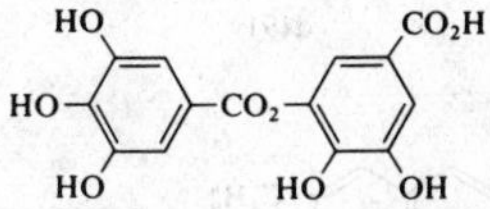

d181

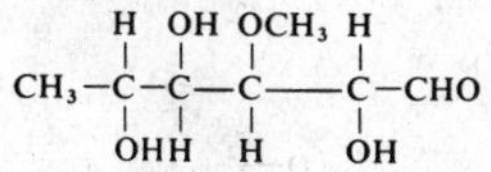

d182

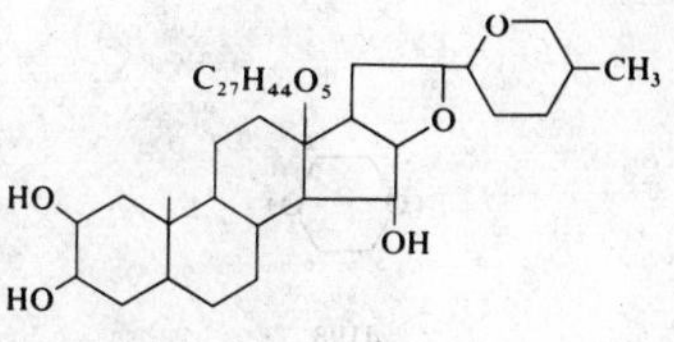

d183

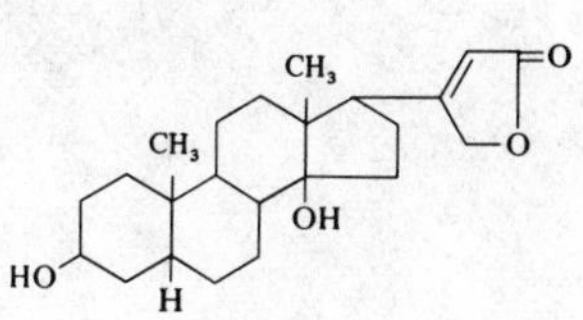

d184

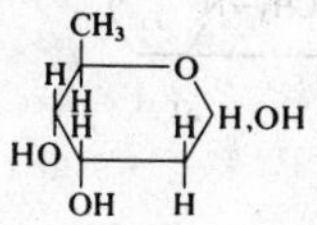

d186

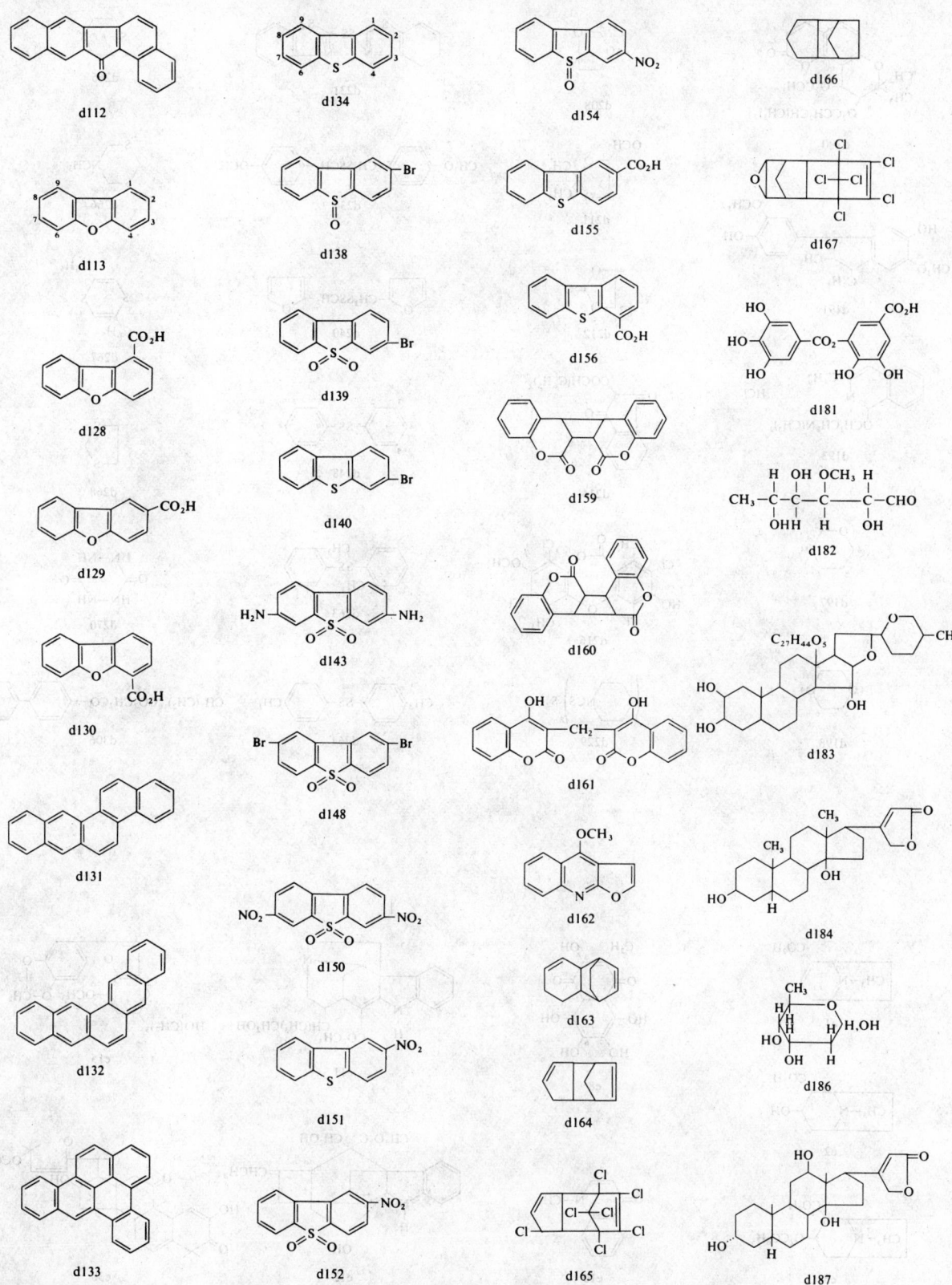

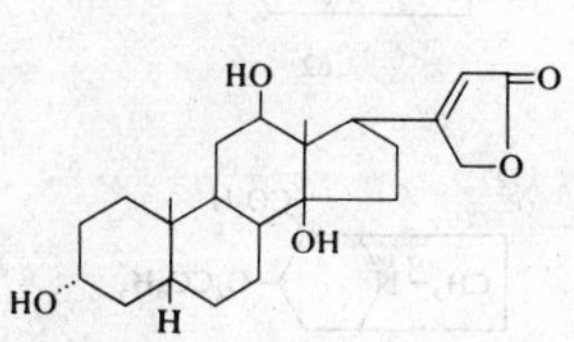

d187

STRUCTURAL FORMULAS OF ORGANIC COMPOUNDS (Continued)

In Alpha Numeric as they occur in Organic Compounds Table

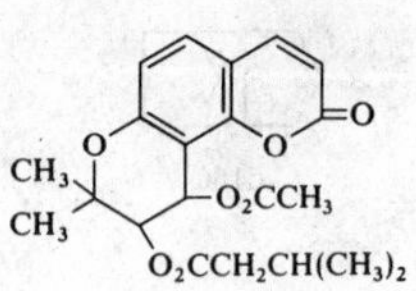
d190

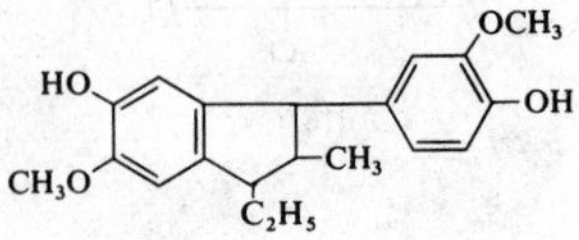
d191

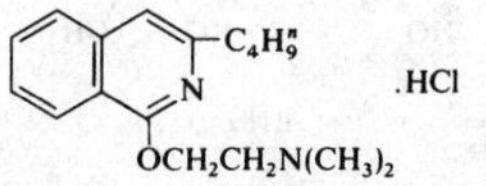
d193

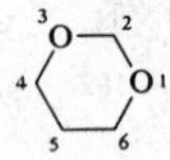
d197

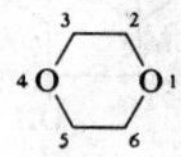
d198

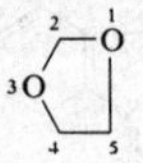
d208

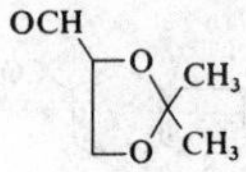
d211

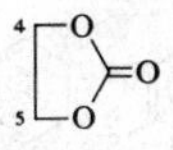
d212

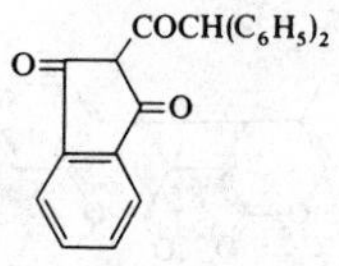
d214

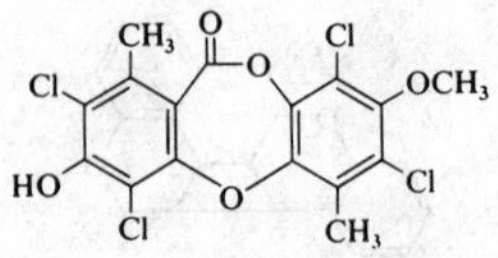
d216

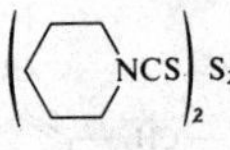
d229

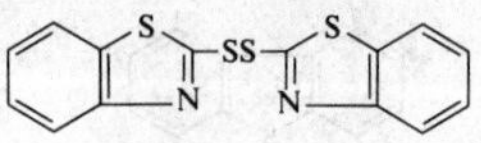
d231

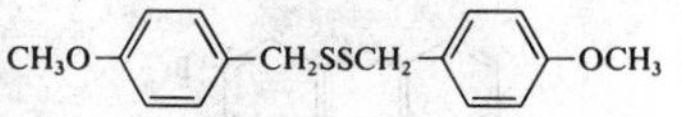
d234

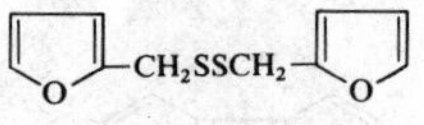
d240

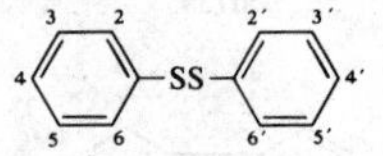
d248

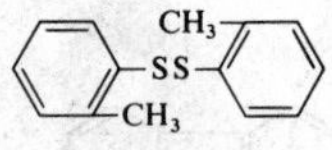
d263

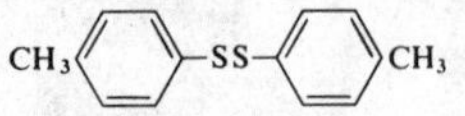
d264

d265

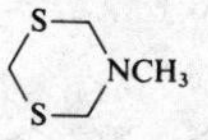
d266

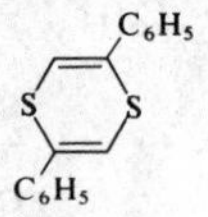
d267

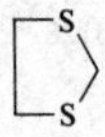
d268

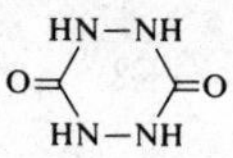
d270

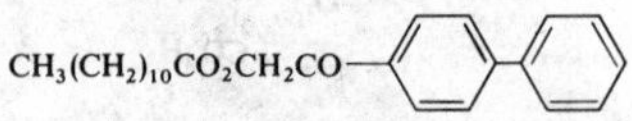
d306

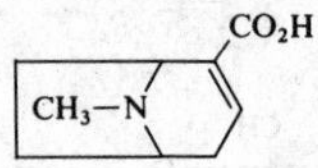
e1

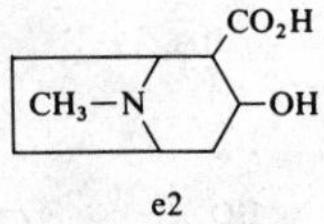
e2

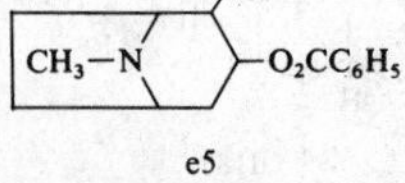
e5

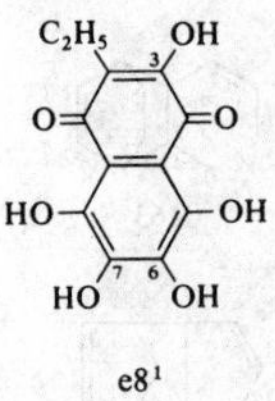
e81

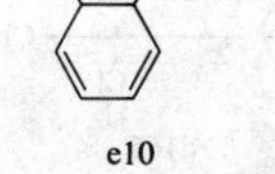
e10

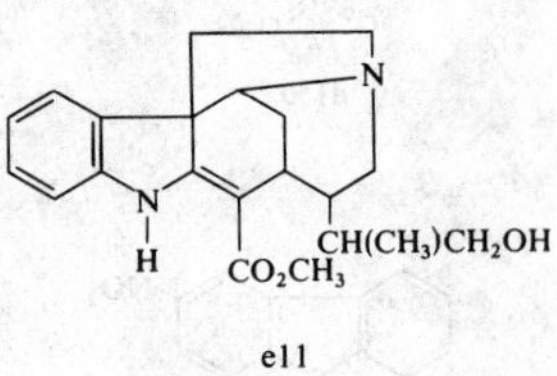
e11

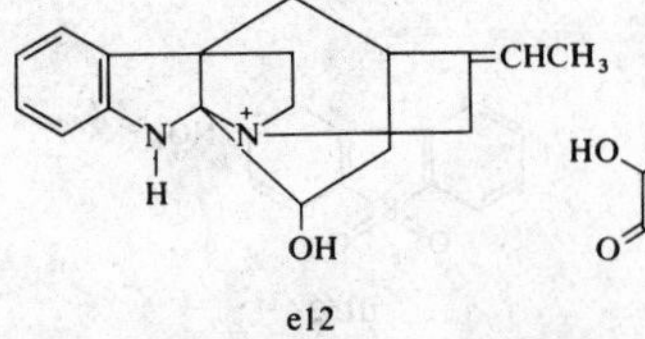
e12

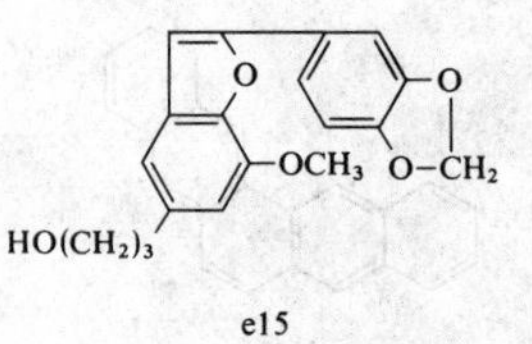
e15

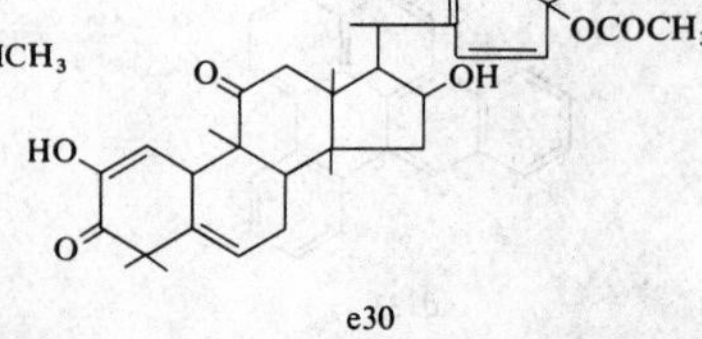
e30

STRUCTURAL FORMULAS OF ORGANIC COMPOUNDS (Continued)

In Alpha Numeric as they occur in Organic Compounds Table

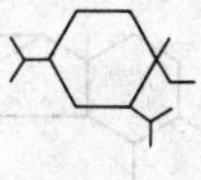
e32

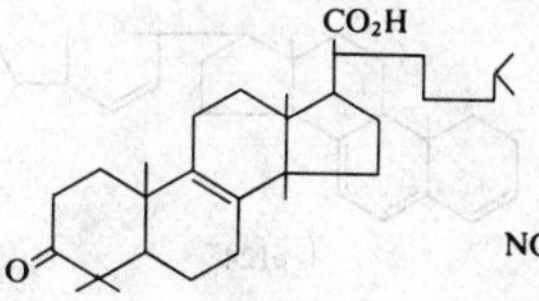
e33

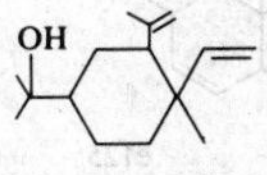
e34

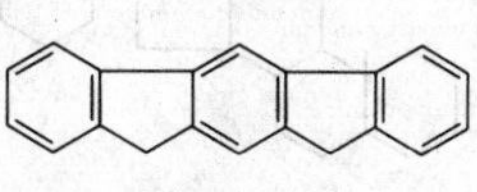
e35

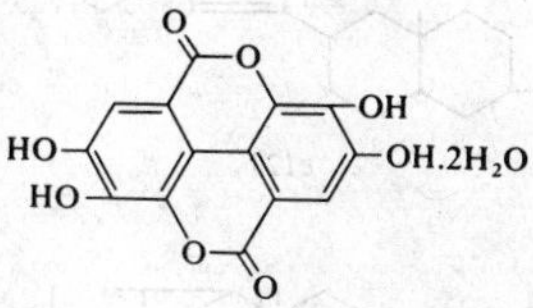
e36

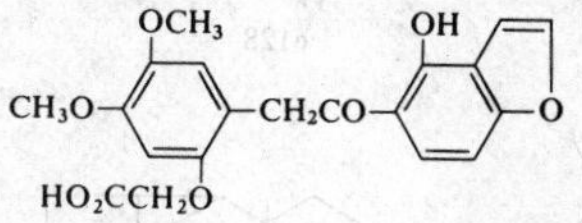
e37

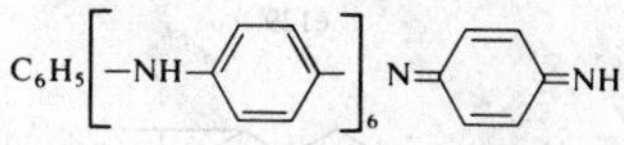
e38

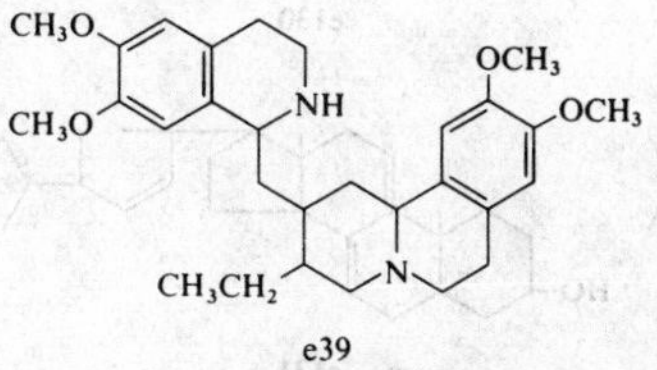
e39

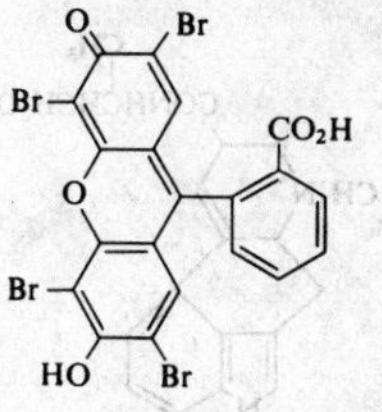
e43

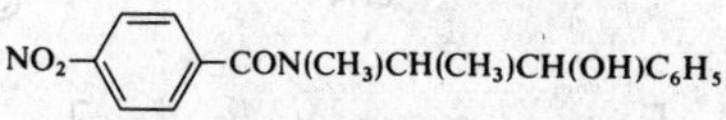
e53

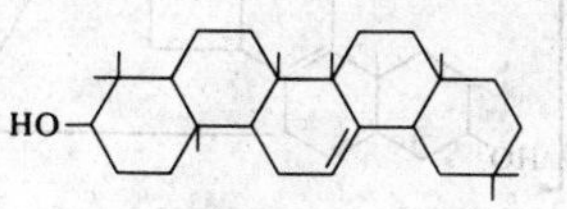
e54

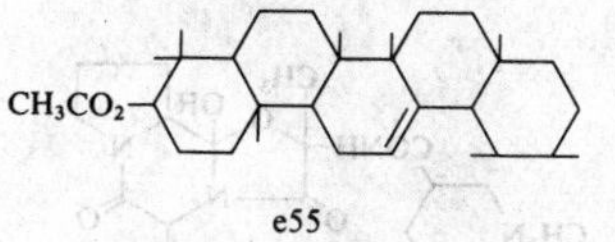
e55

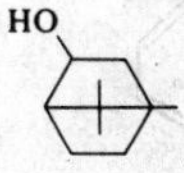
e57

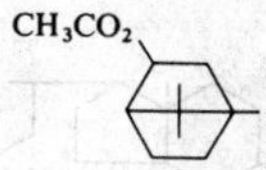
e58

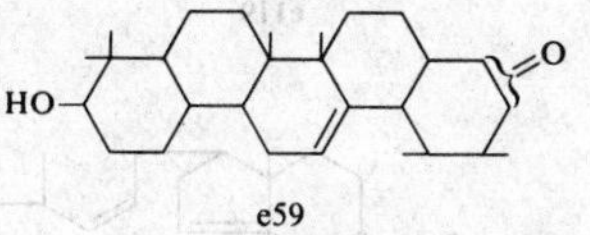
e59

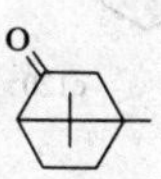
e60

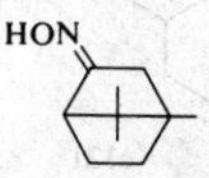
e63

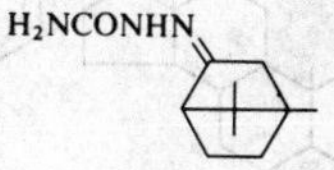
e66

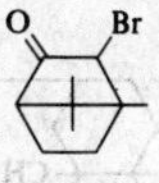
e68

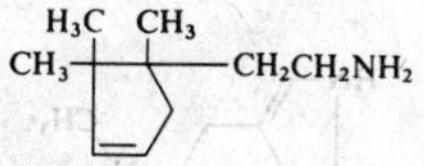
e70

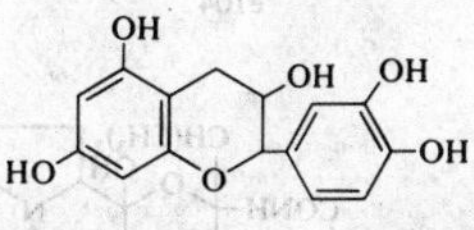
e71

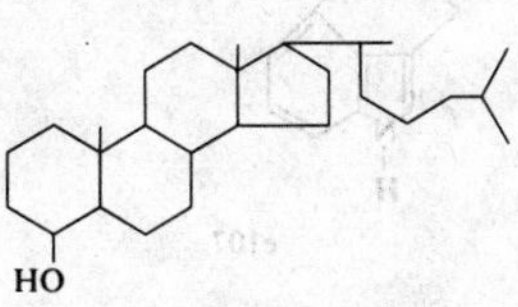
e74

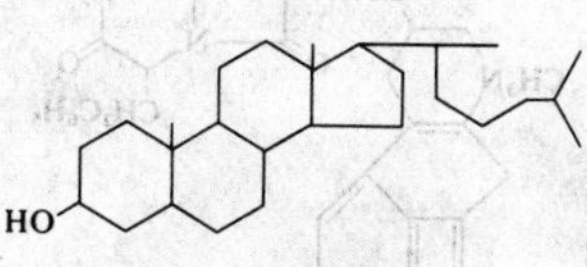
e75

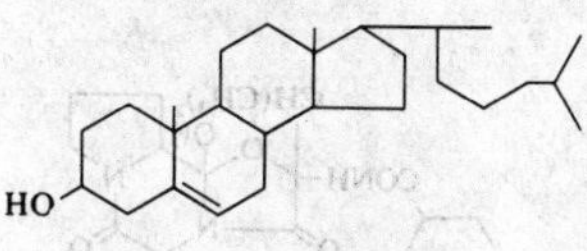
e76

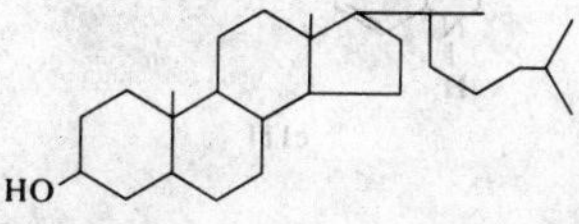
e77

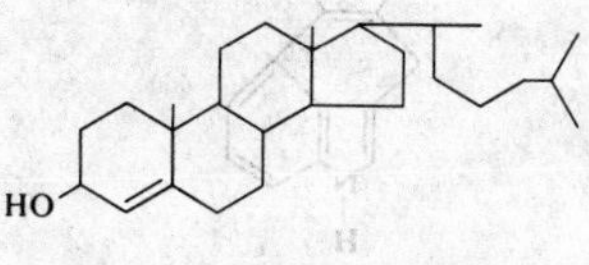
e78

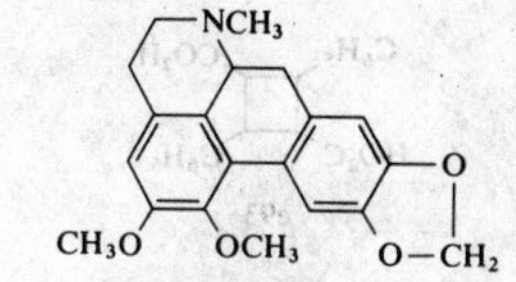
e79

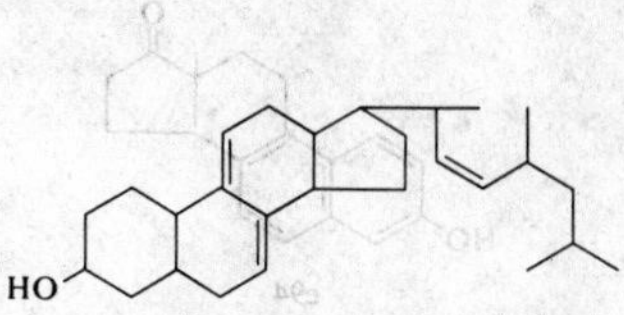
e81

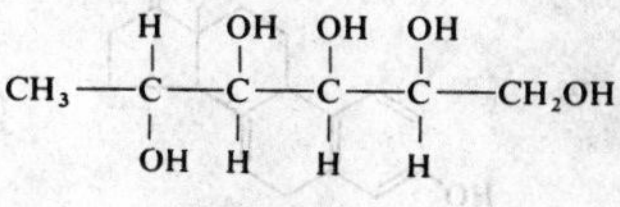
e82

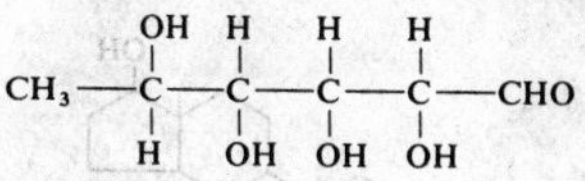
e84

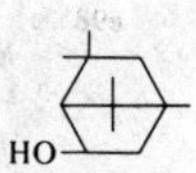
e85

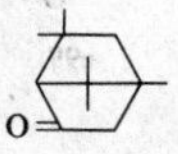
e86

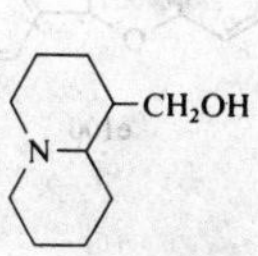
e88

e89

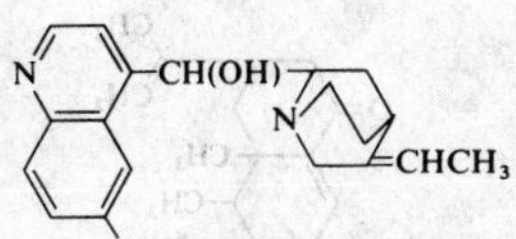
e90

STRUCTURAL FORMULAS OF ORGANIC COMPOUNDS (Continued)

In Alpha Numeric as they occur in Organic Compounds Table

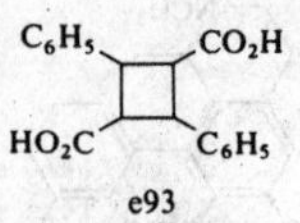

e93

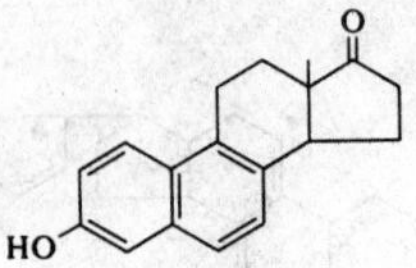

e94

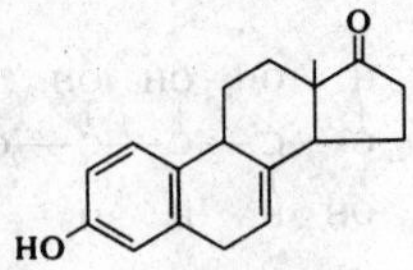

e97

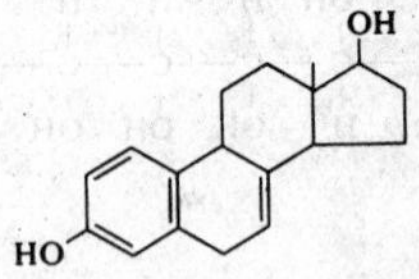

e98

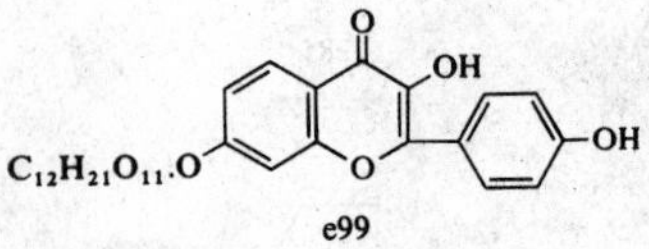

e99

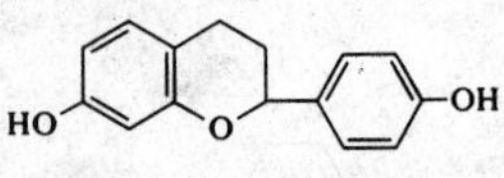

e100

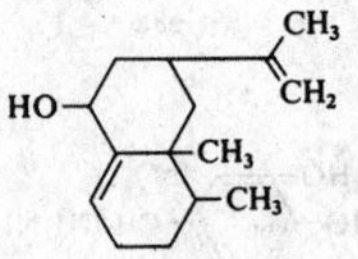

e101

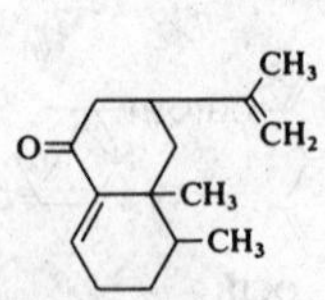

e102

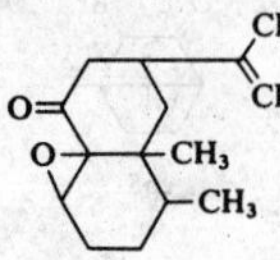

e103

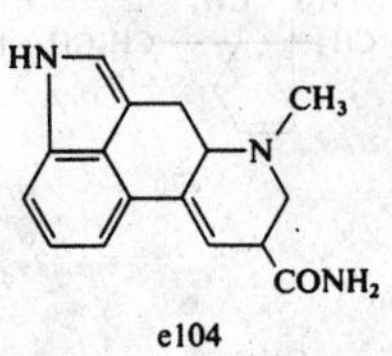

e104

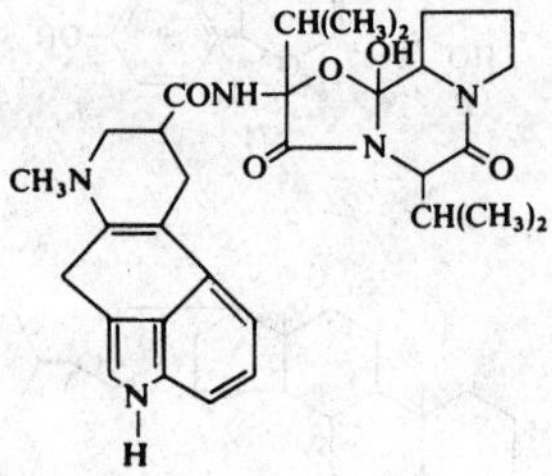

e107

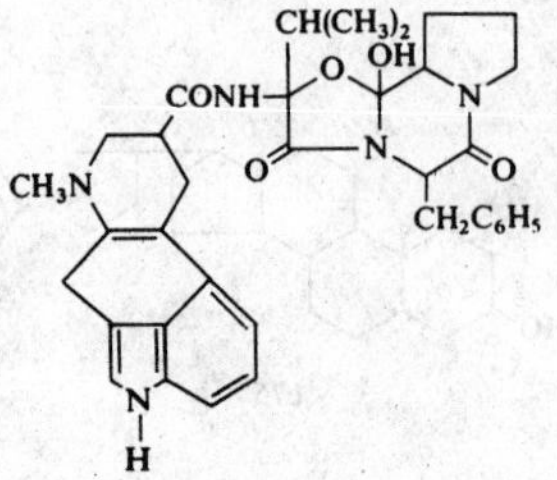

e109

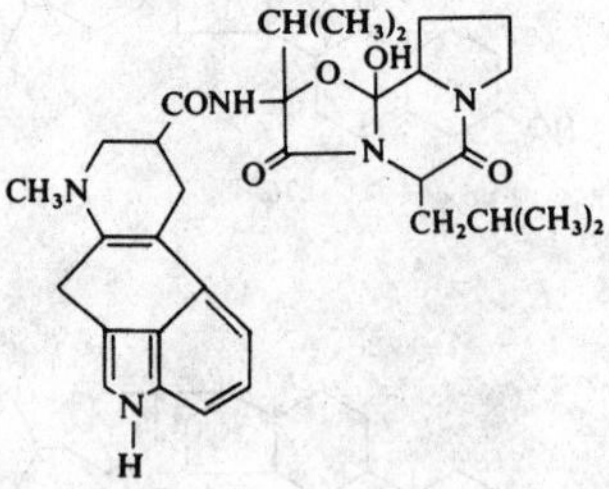

e111

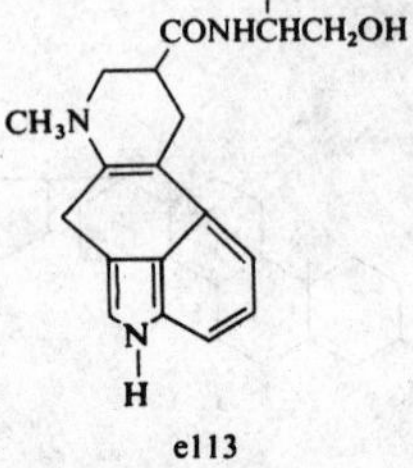

e113

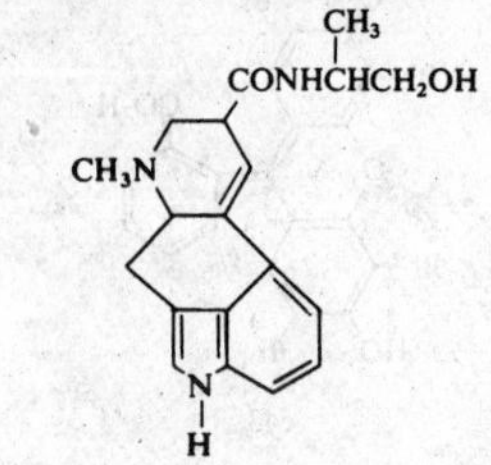

e114

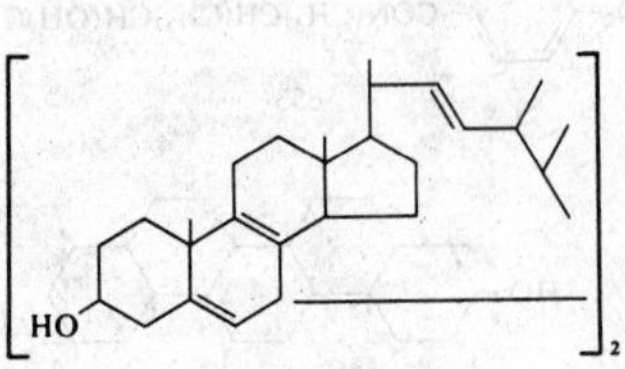

e116

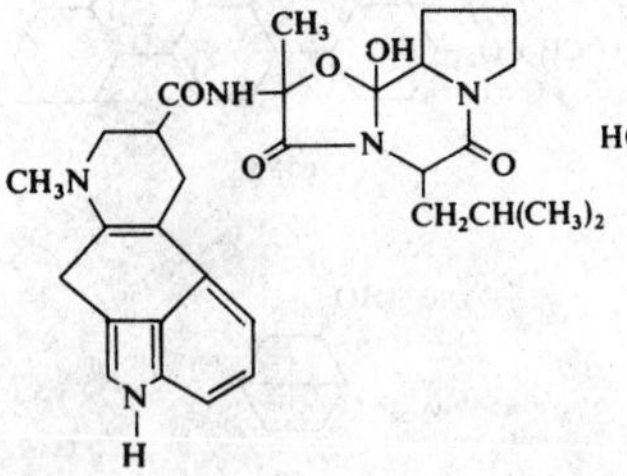

e117

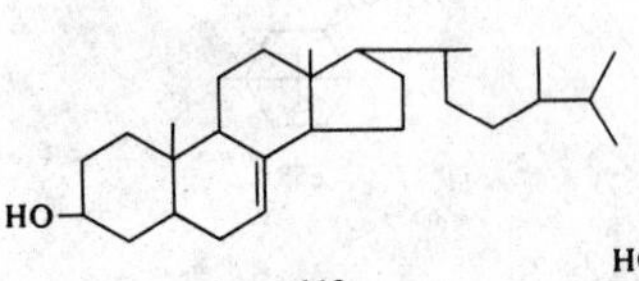

e119

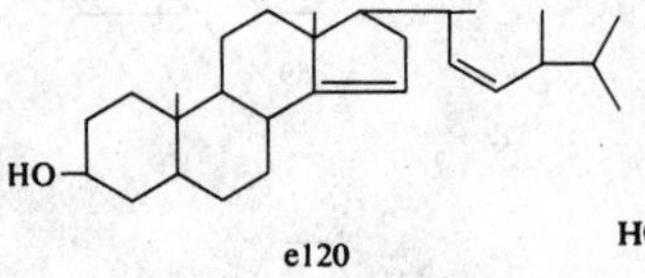

e120

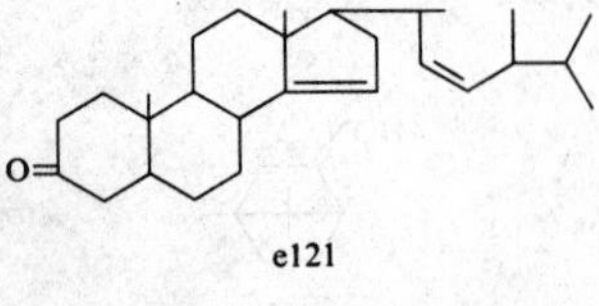

e121

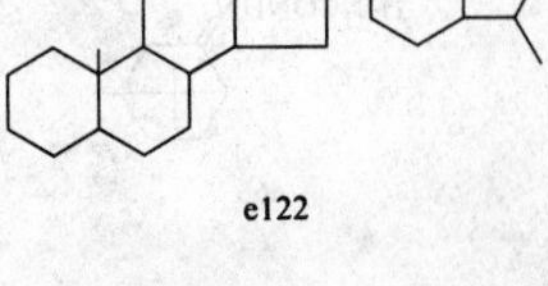

e122

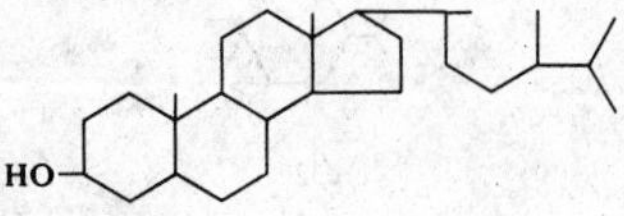

e123

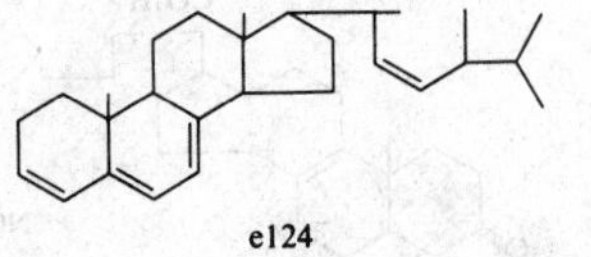

e124

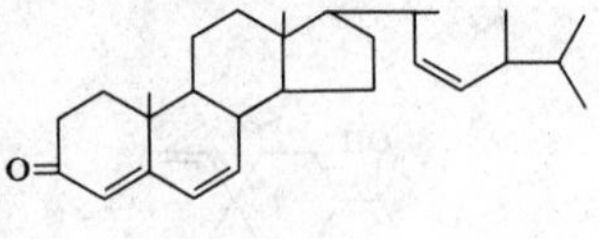

e125

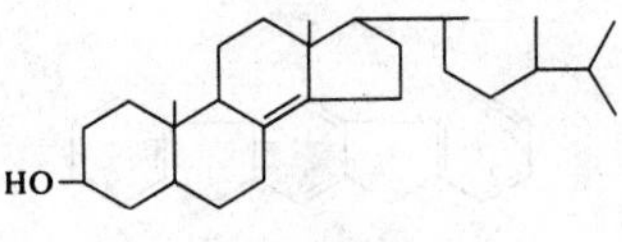

e126

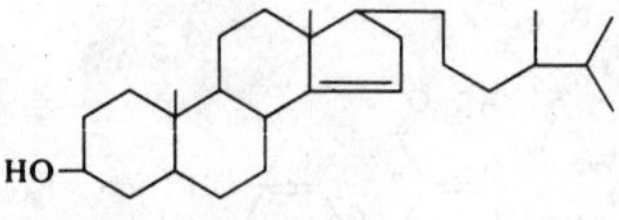

e127

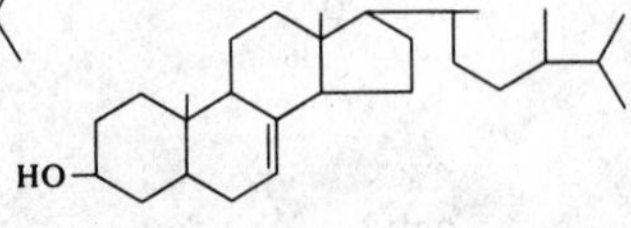

e128

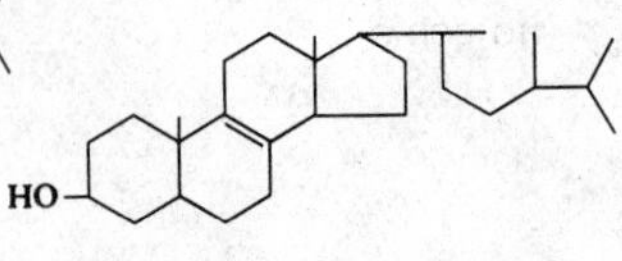

e129

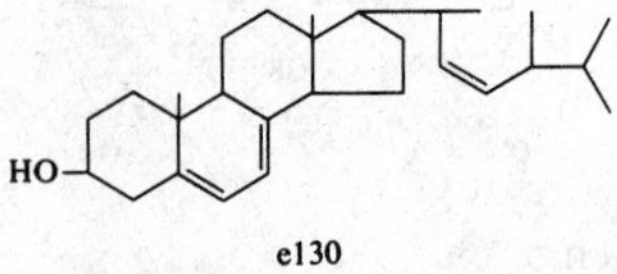

e130

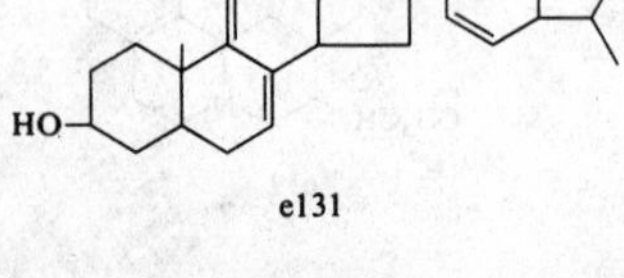

e131

STRUCTURAL FORMULAS OF ORGANIC COMPOUNDS (Continued)

In Alpha Numeric as they occur in Organic Compounds Table

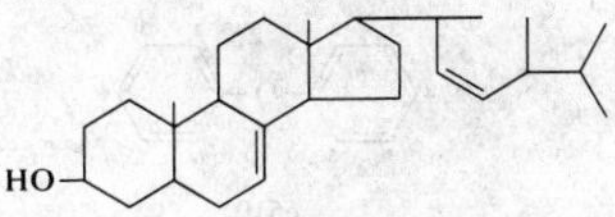

e133

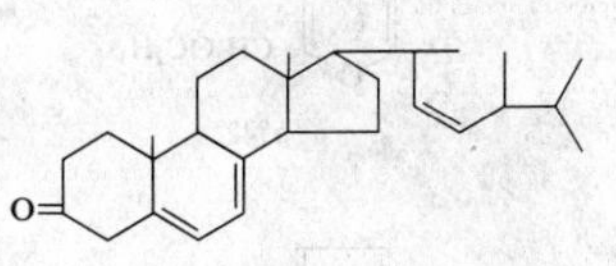

e134

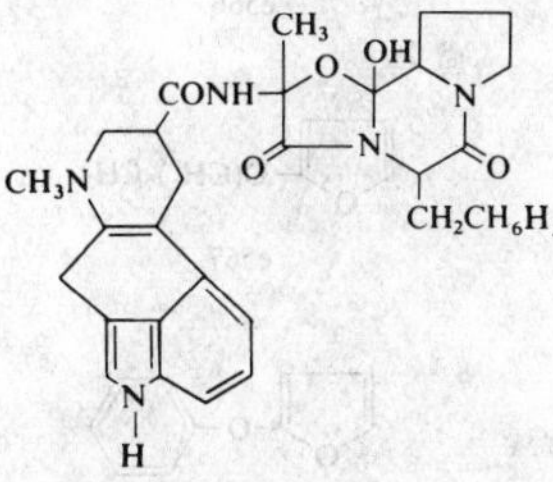

e135

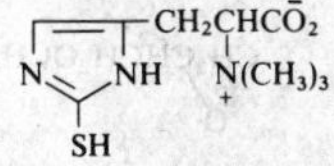

e137

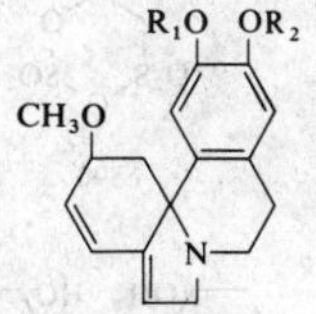

$R_1 = CH_3$ or H
$R_2 = H$ or CH_3

e139

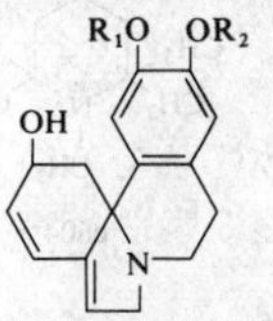

$R_1 = CH_3$ or H
$R_2 = H$ or CH_3

e140

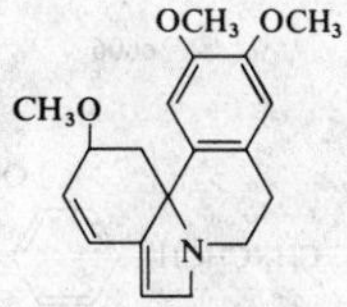

e141

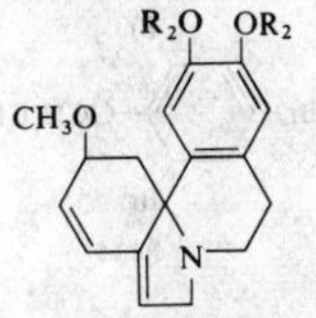

$R_1 = CH_3$ or H
$R_2 = H$ or CH_3

e142

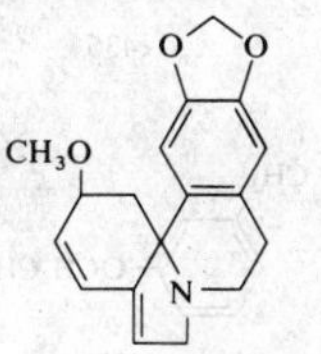

e144

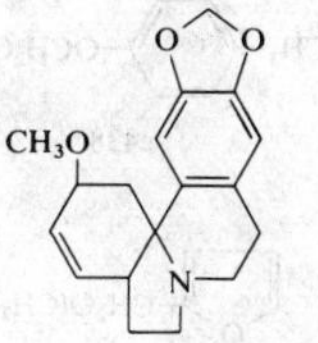

e145

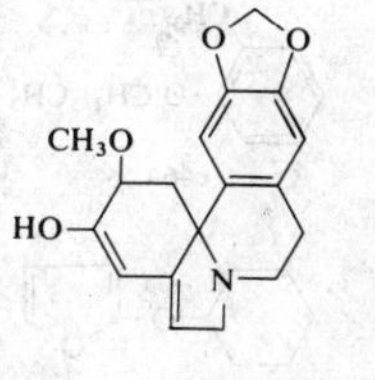

e146

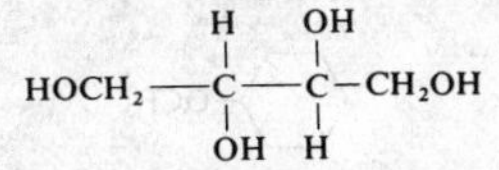

e147

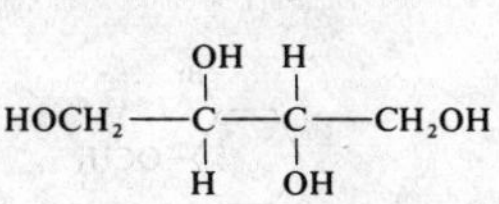

e148

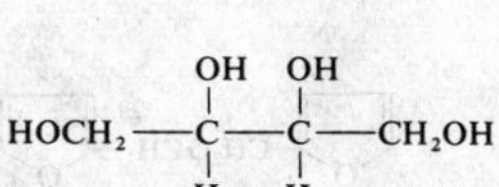

e148[1]

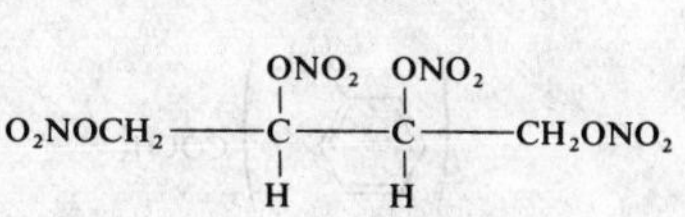

e148[2]

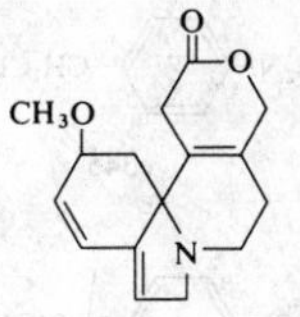

e149

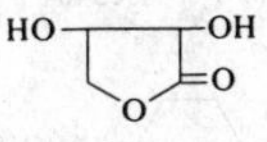

e150

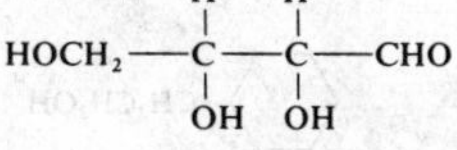

e152

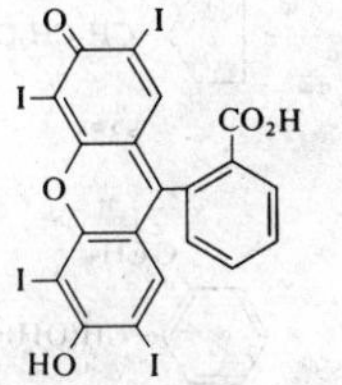

e154

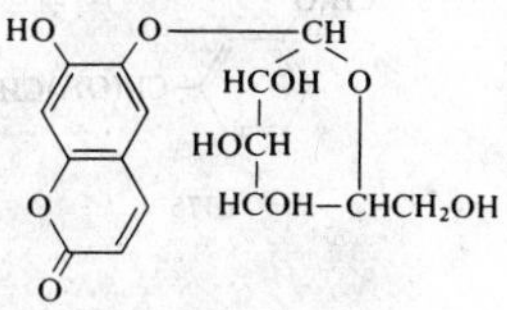

e158

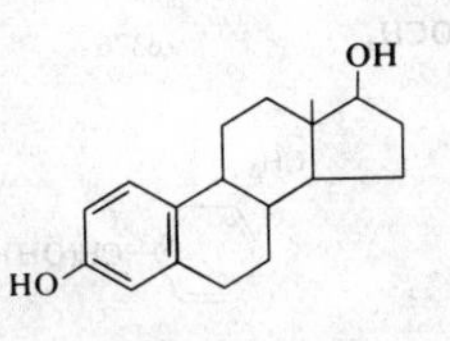

e159

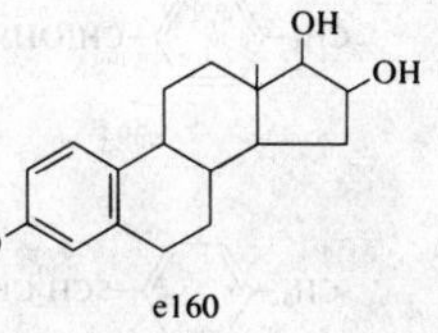

e160

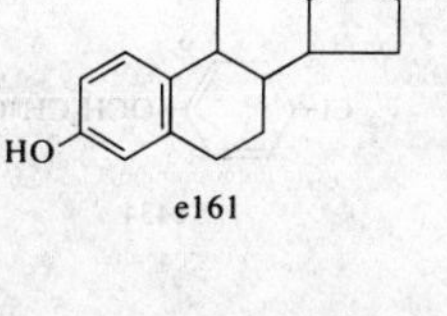

e161

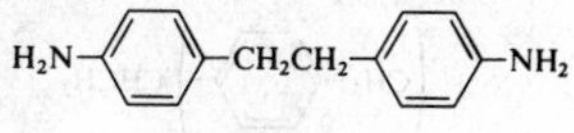

e177

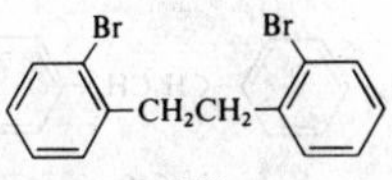

e179

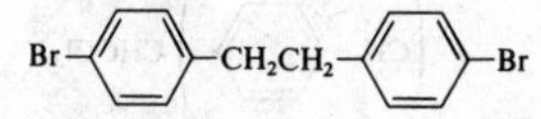

e180

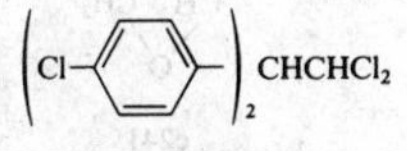

e182

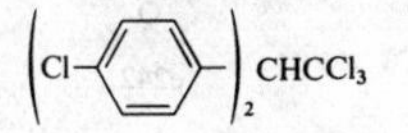

e183

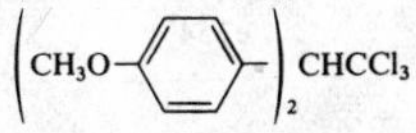

e185

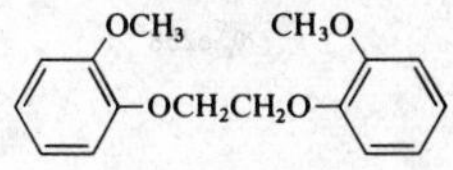

e186

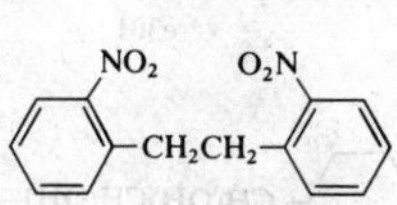

e190

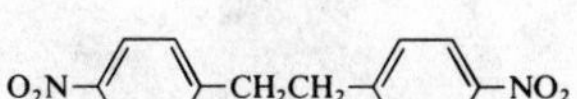

e191

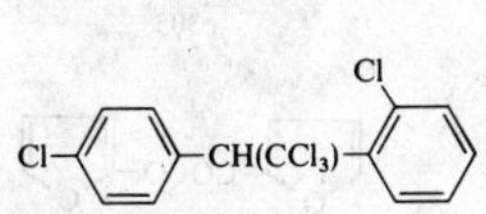

e206

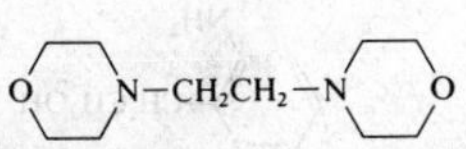

e230

STRUCTURAL FORMULAS OF ORGANIC COMPOUNDS (Continued)

In Alpha Numeric as they occur in Organic Compounds Table

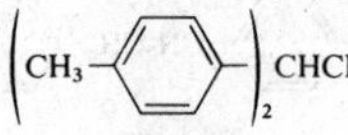

e239

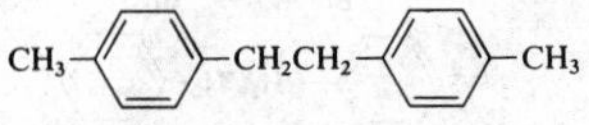

e240

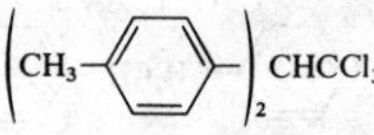

e240[1]

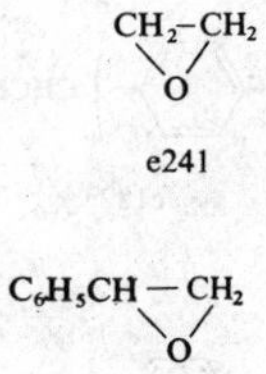

e241

e242

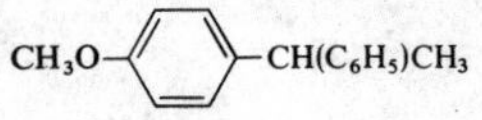

e254[1]

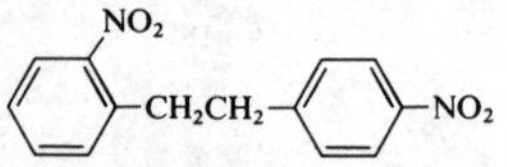

e258

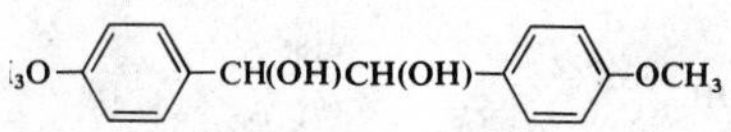

e301

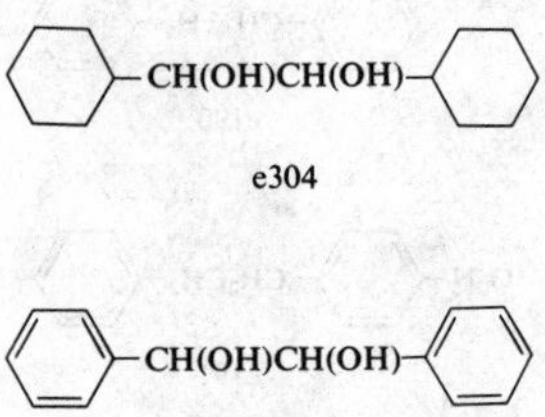

e304

e305

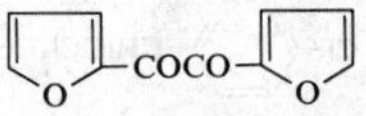

e310

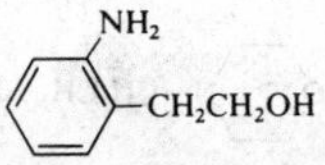

e344

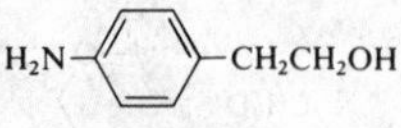

e345

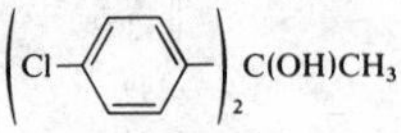

e346[1]

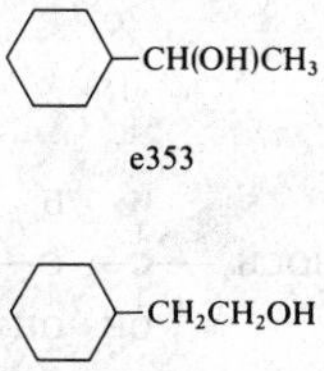

e353

e354

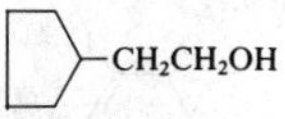

e355

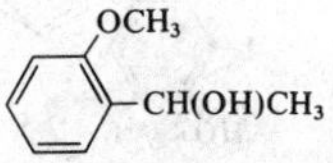

e374

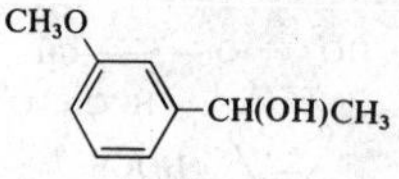

e375

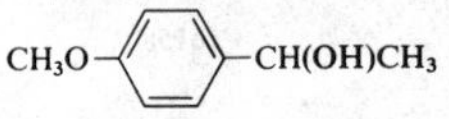

e376

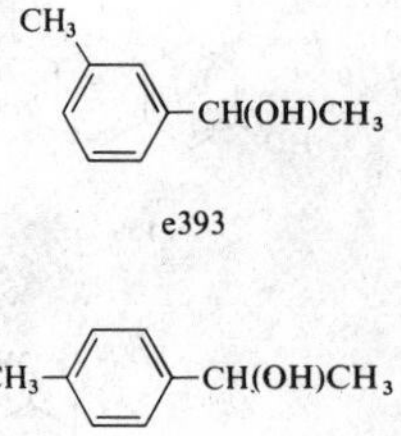

e393

e394

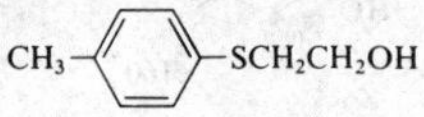

e395

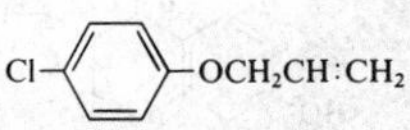

e434

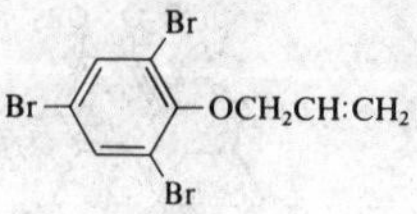

e434[1]

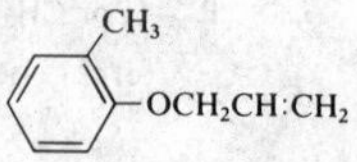

e436

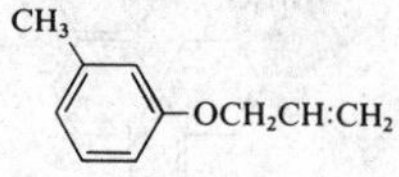

e437

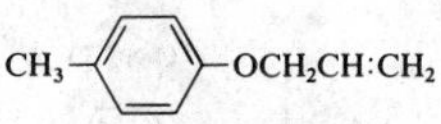

e438

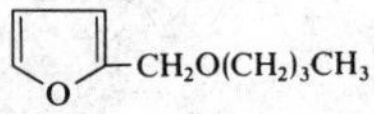

e456

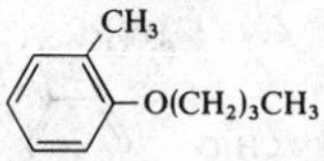

e464

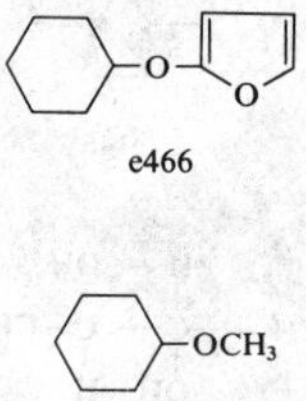

e466

e467

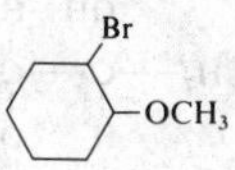

e468

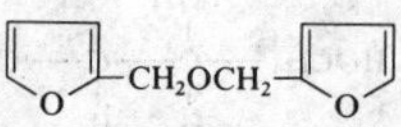

e493

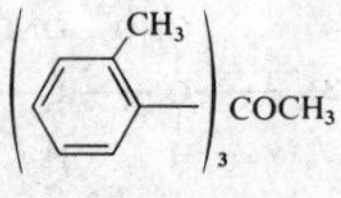

e504

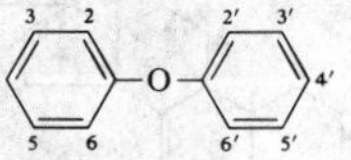

e510

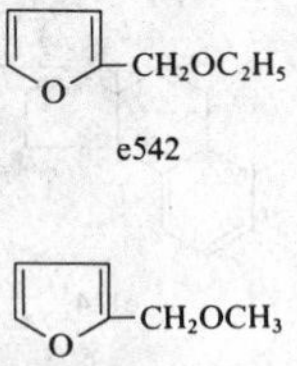

e542

e566

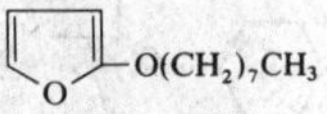

e567

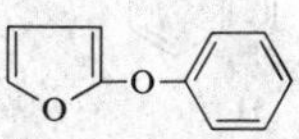

e568

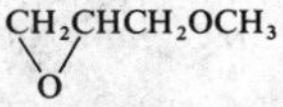

e585

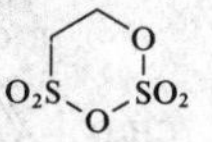

e590

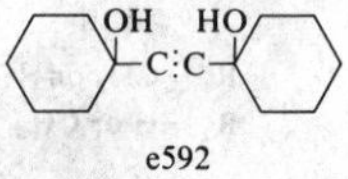

e592

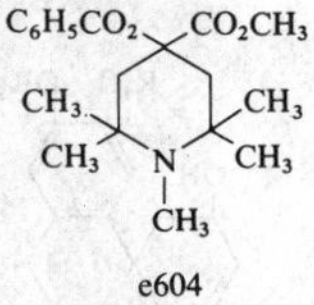

e604

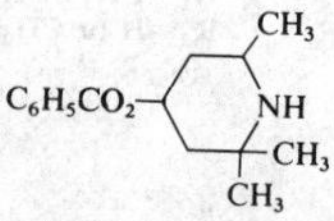

e606

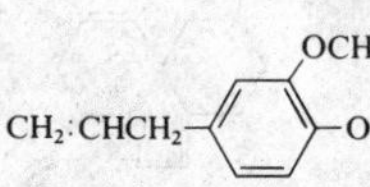

e611

STRUCTURAL FORMULAS OF ORGANIC COMPOUNDS (Continued)

In Alpha Numeric as they occur in Organic Compounds Table

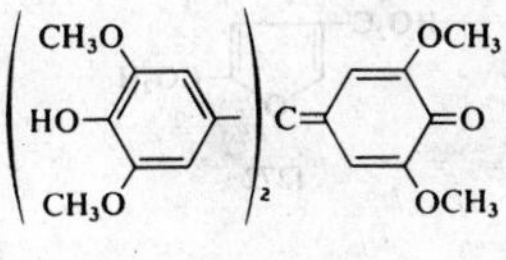

e613

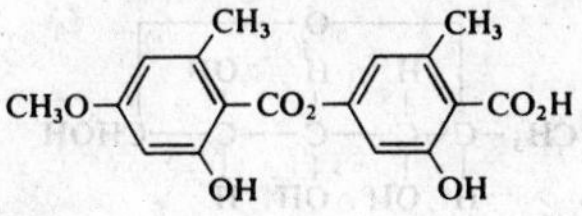

e615

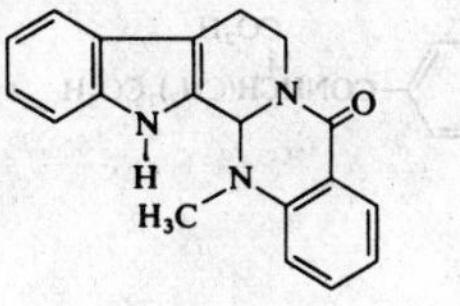

e616

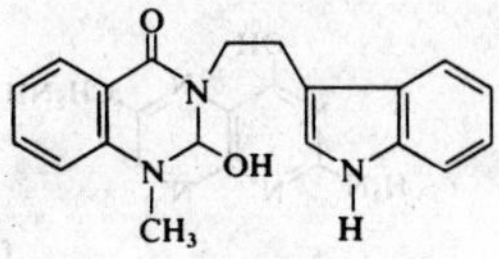

e617

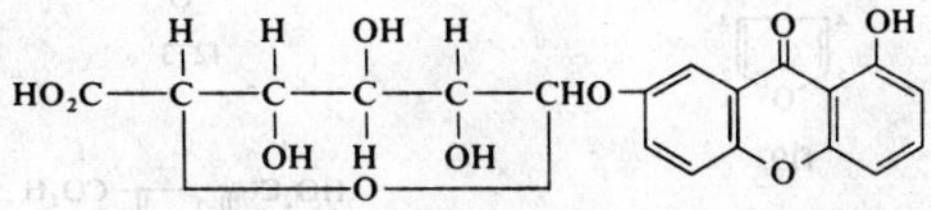

e614

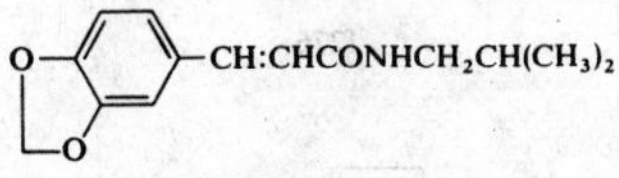

f1

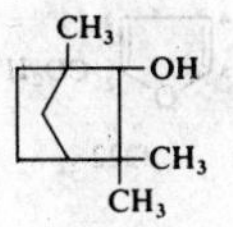

f13

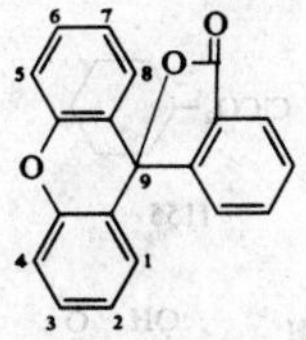

f40

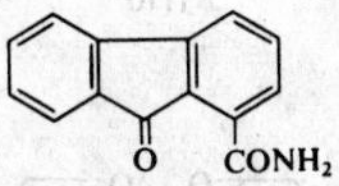

f93

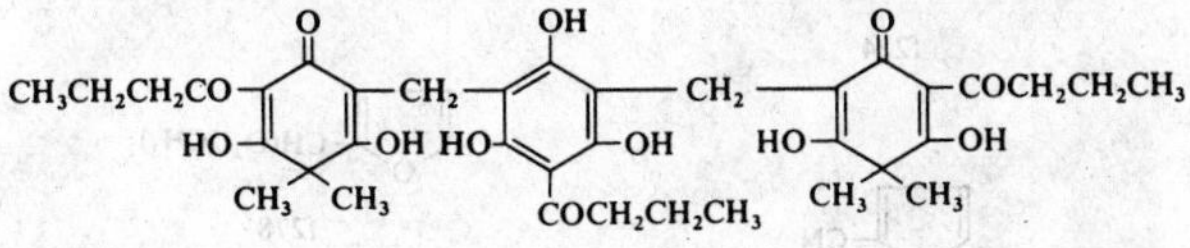

f15

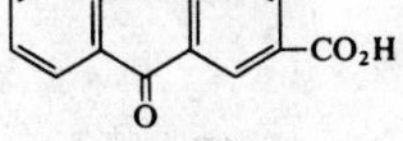

f96

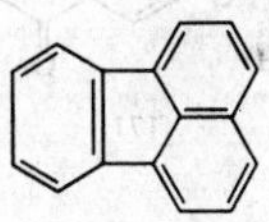

f44

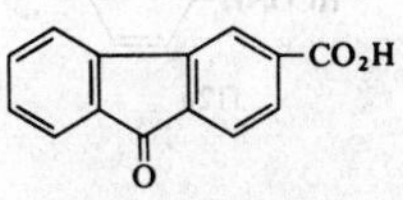

f98

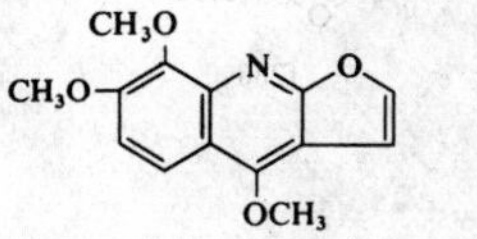

f2

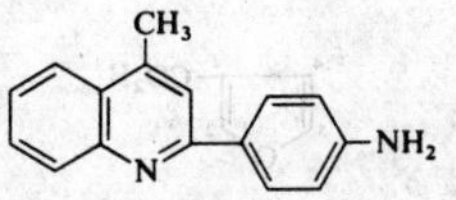

f16

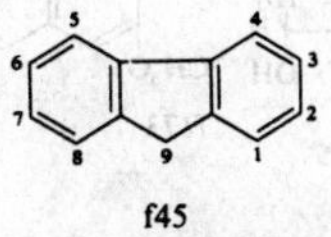

f45

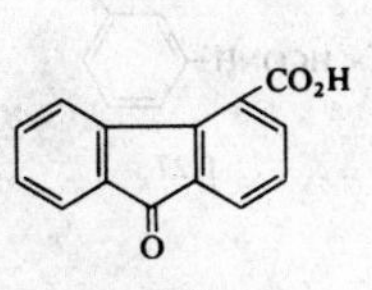

f99

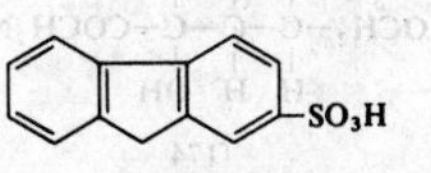

f78

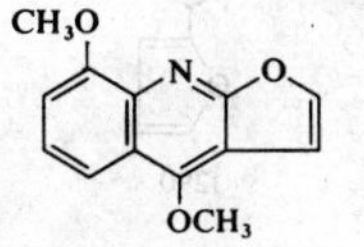

f3

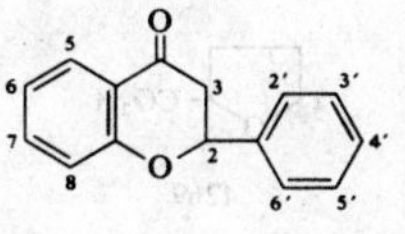

f17

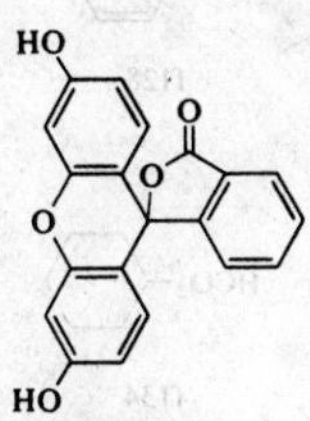

f100

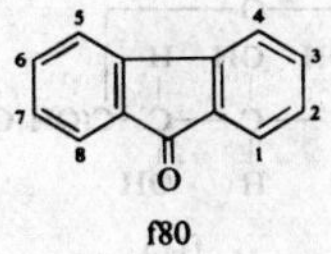

f80

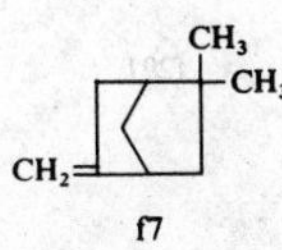

f7

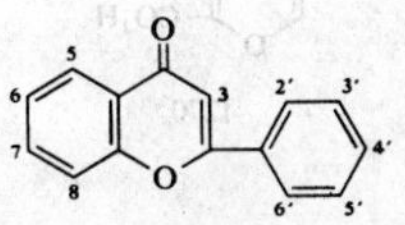

f20

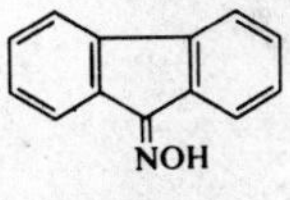

f81

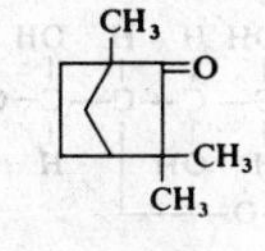

f10

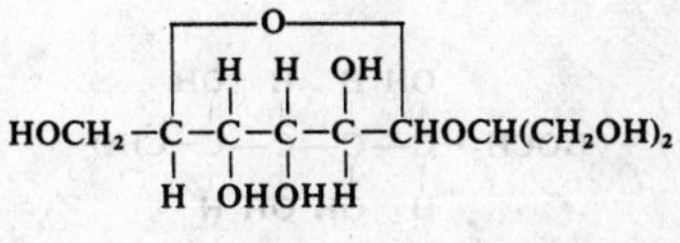

f39

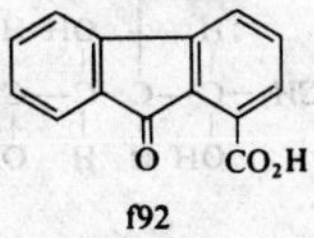

f92

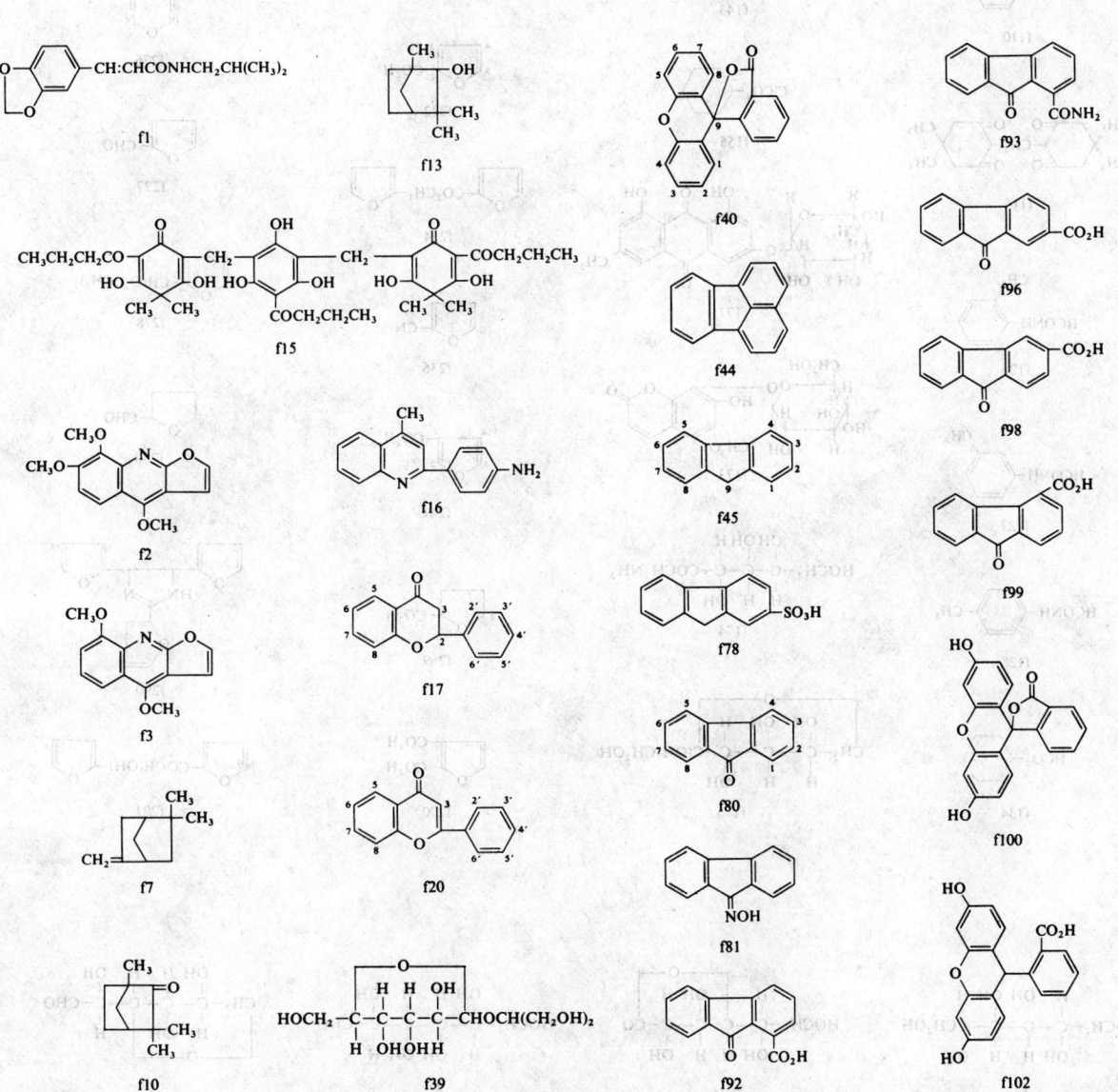

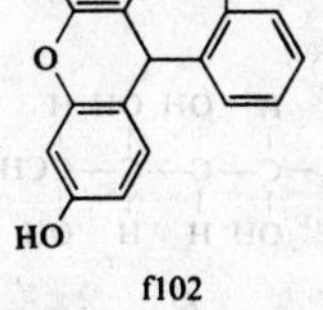

f102

STRUCTURAL FORMULAS OF ORGANIC COMPOUNDS (Continued)

In Alpha Numeric as they occur in Organic Compounds Table

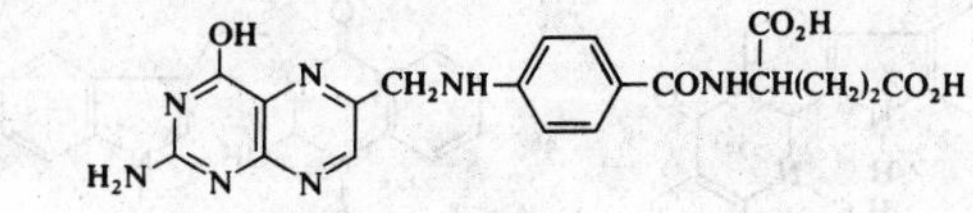

f104

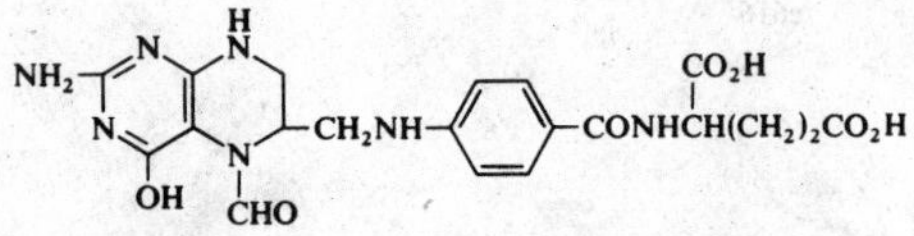

f105

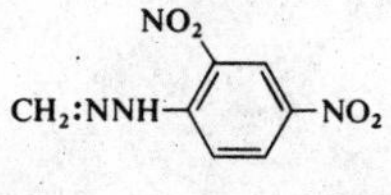

f110

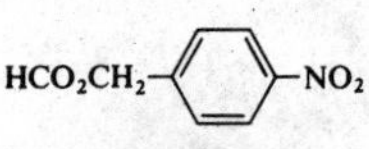

f143

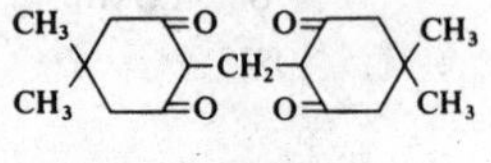

f114

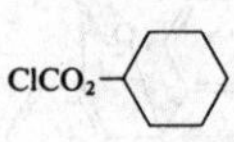

f155

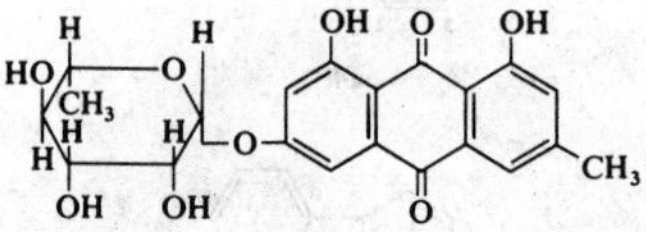

f171

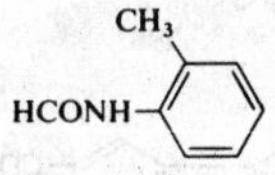

f126

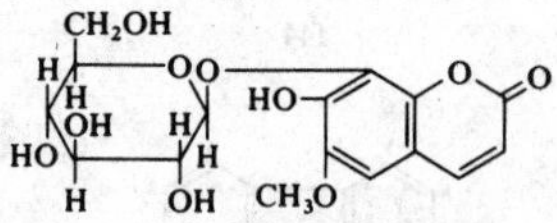

f173

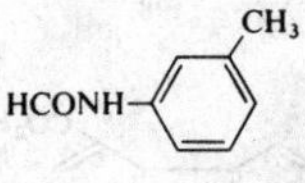

f127

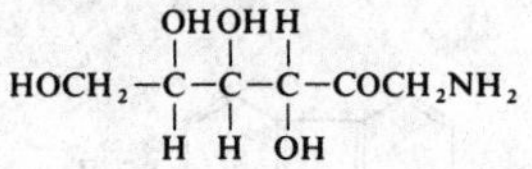

f174

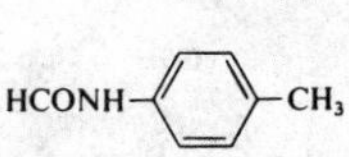

f128

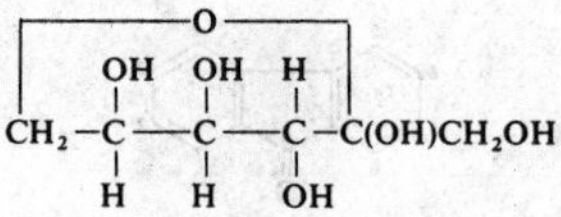

f175

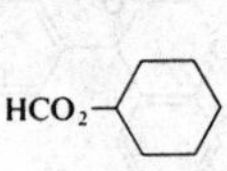

f134

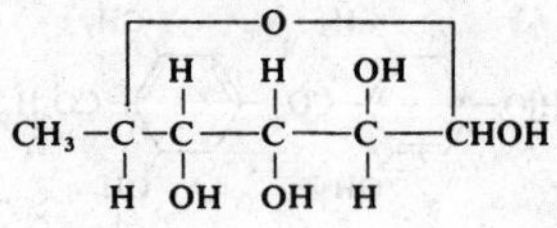

f176

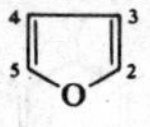

f193

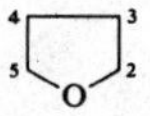

f220

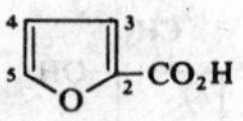

f222

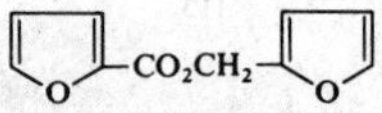

f224

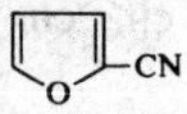

f236

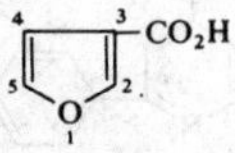

f241

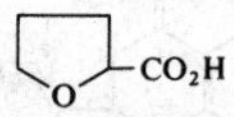

f269

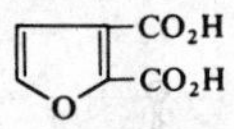

f270

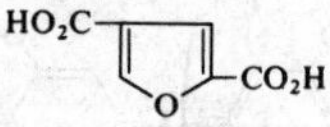

f272

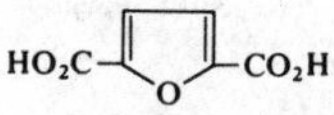

f273[1]

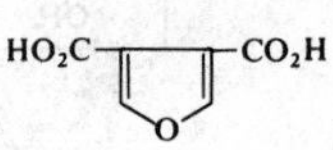

f275[1]

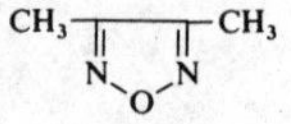

f276

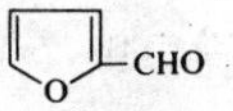

f277

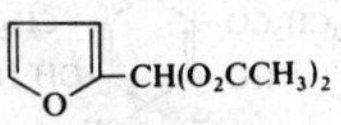

f278

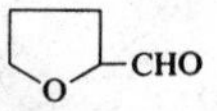

f288

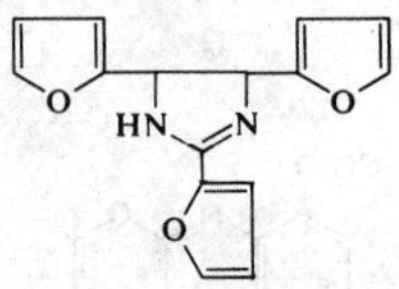

f290

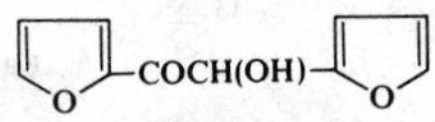

f291

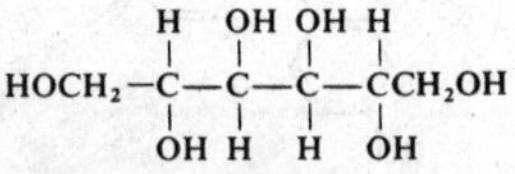

g1

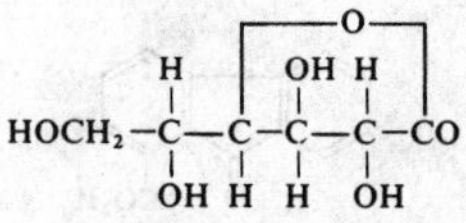

g2

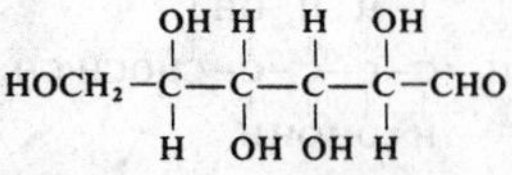

g3

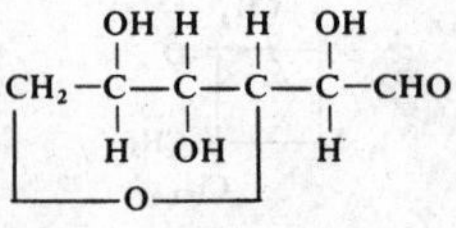

g4

STRUCTURAL FORMULAS OF ORGANIC COMPOUNDS (Continued)

In Alpha Numeric as they occur in Organic Compounds Table

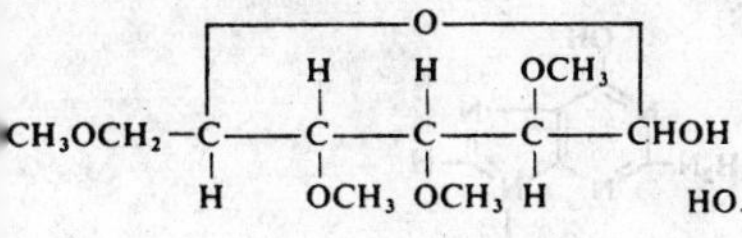

g5

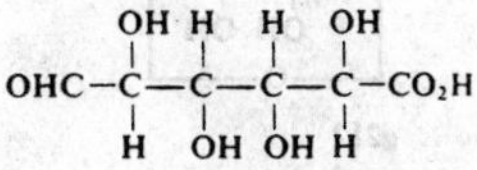

g6

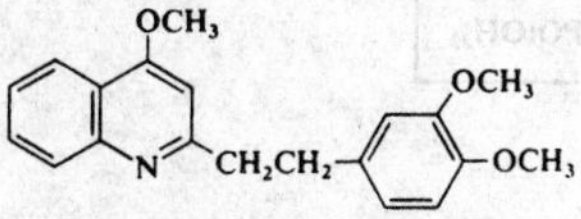

g8

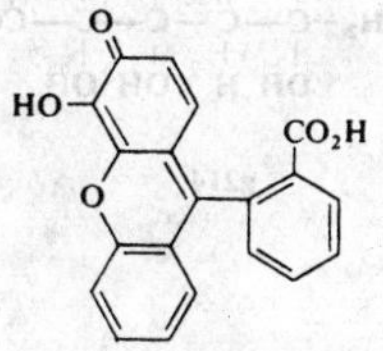

g9

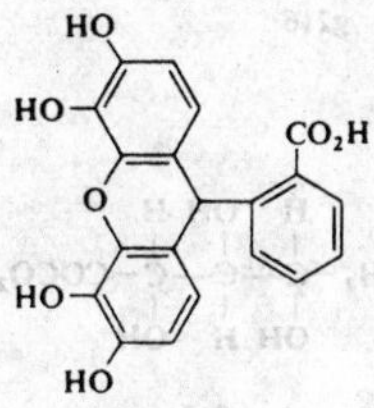

g10

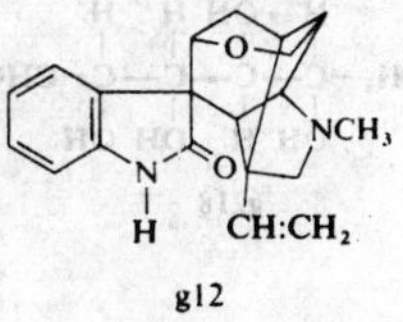

g12

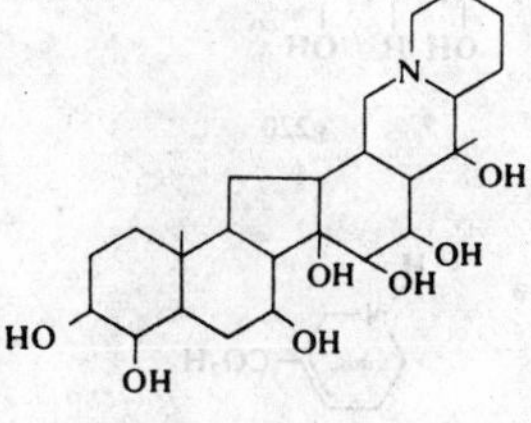

g22

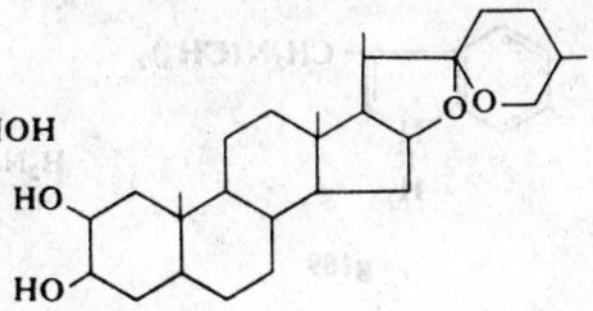

g25

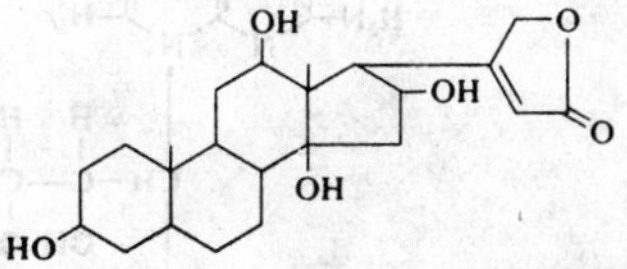

g26

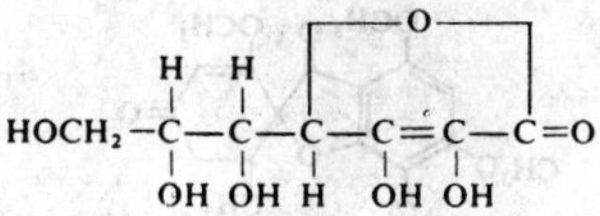

g28

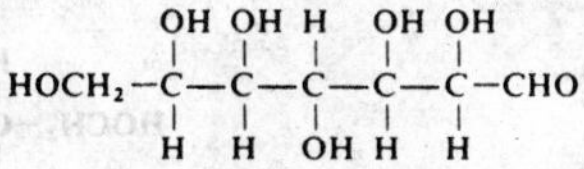

g29

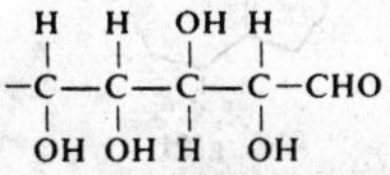

g30

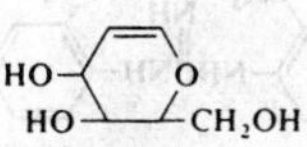

g31

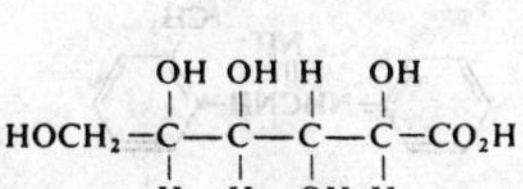

g32

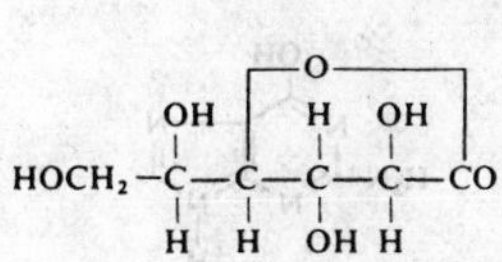

g33

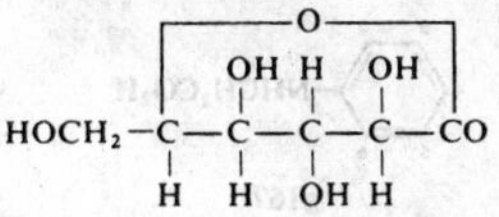

g34

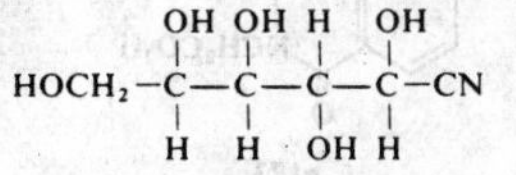

g35

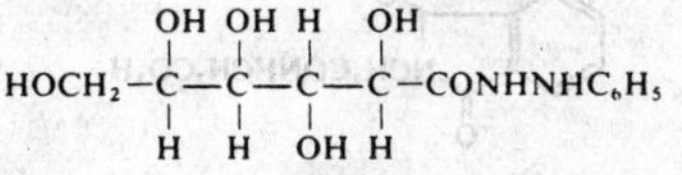

g36

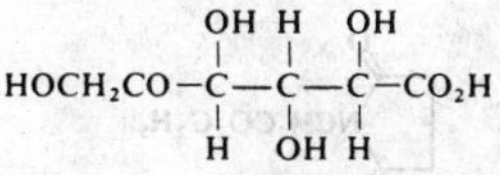

g37

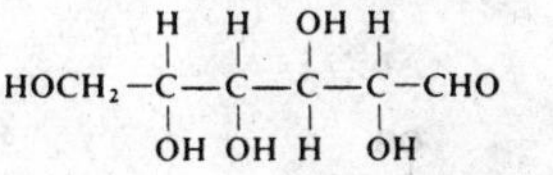

g38

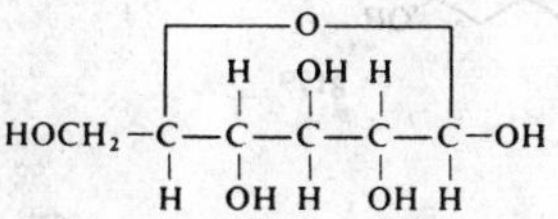

g39

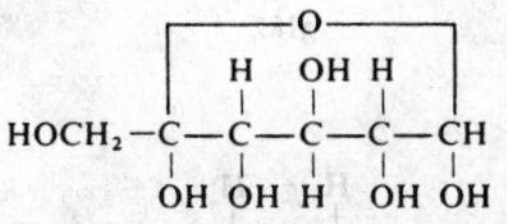

g41

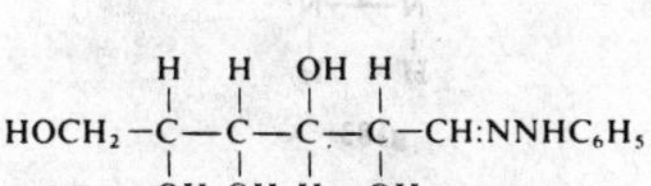

g46

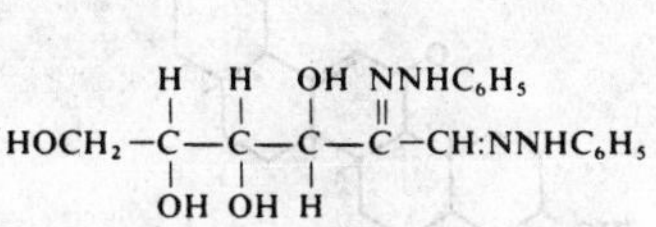

g48

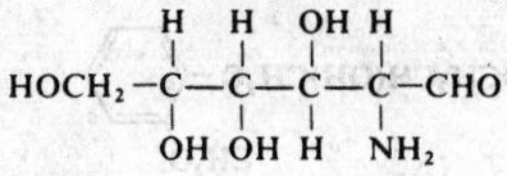

g50

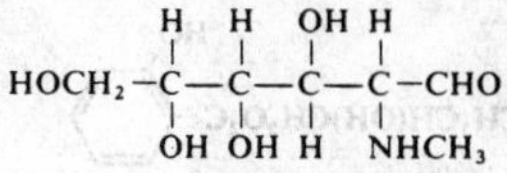

g52

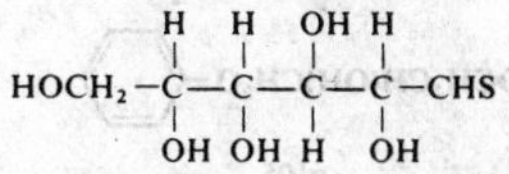

g53

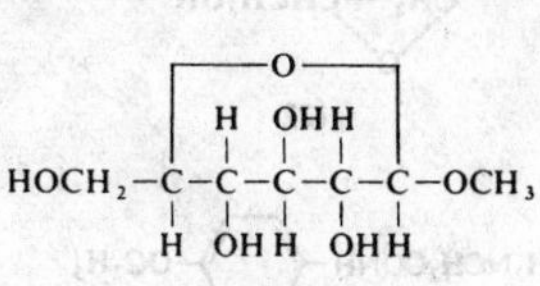

g54

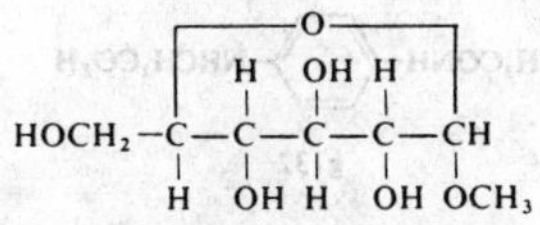

g55

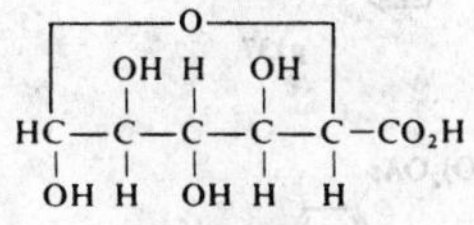

g56

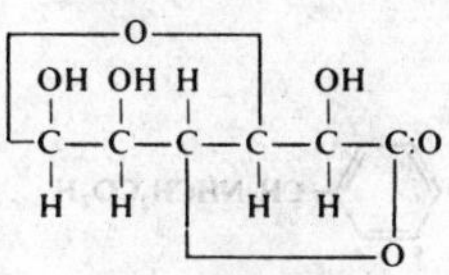

g57

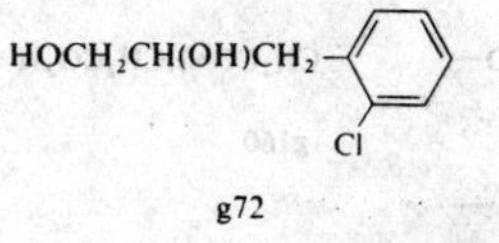

g72

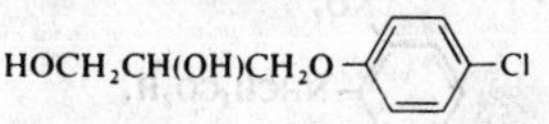

g73

STRUCTURAL FORMULAS OF ORGANIC COMPOUNDS (Continued)

In Alpha Numeric as they occur in Organic Compounds Table

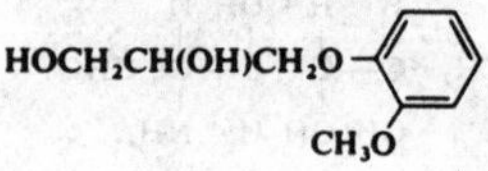

g86

g95

g105

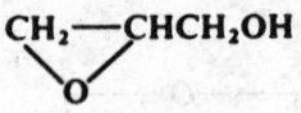

g122

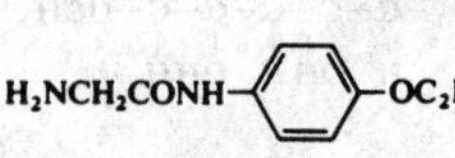

g125[1]

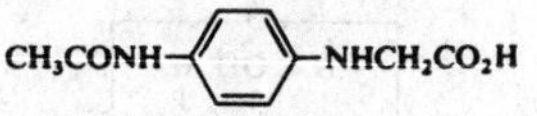

g132

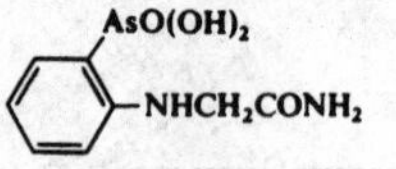

g137

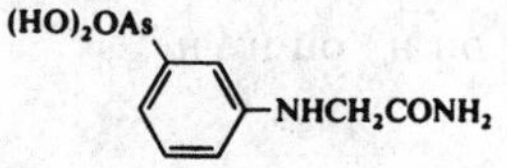

g138

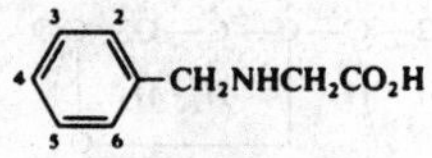

g142

g160

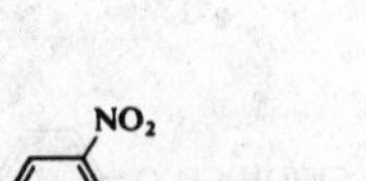

g165

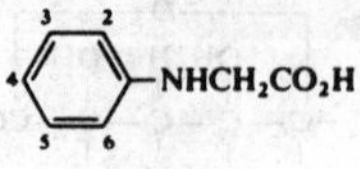

g167

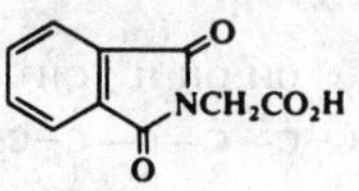

g173

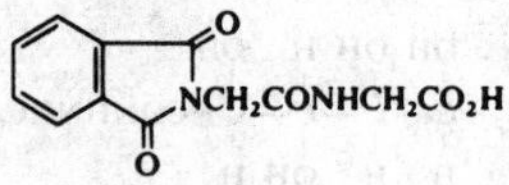

g176

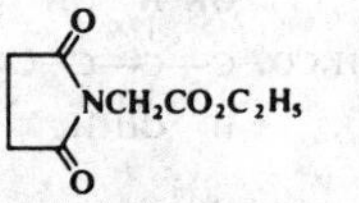

g178

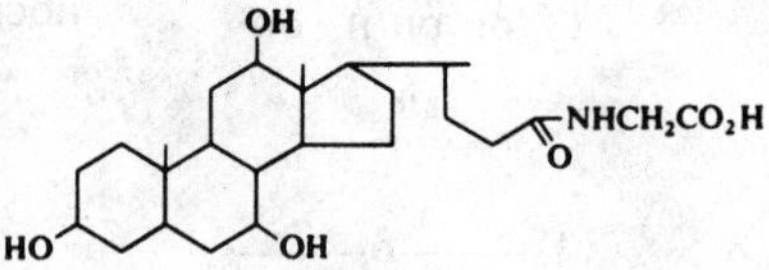

g179

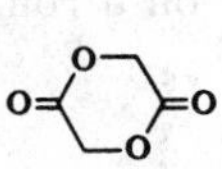

g182

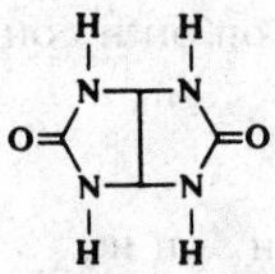

g183

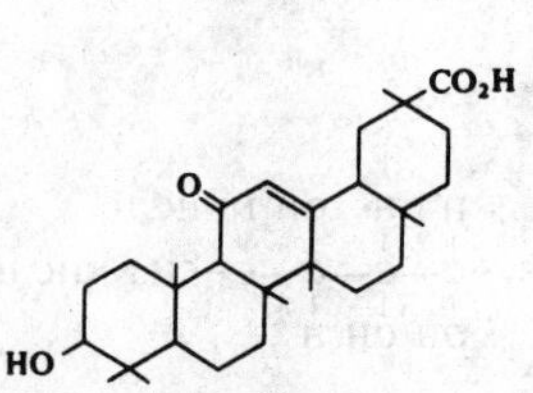

g184

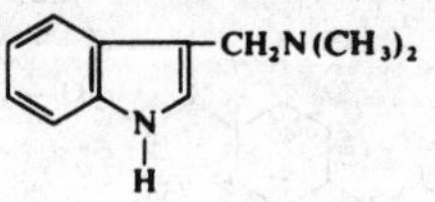

g189

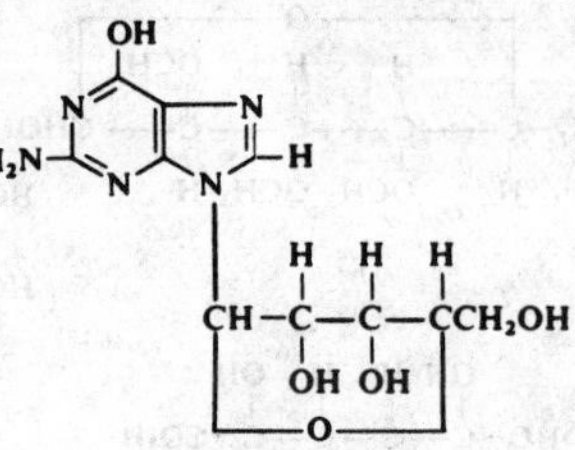

g212

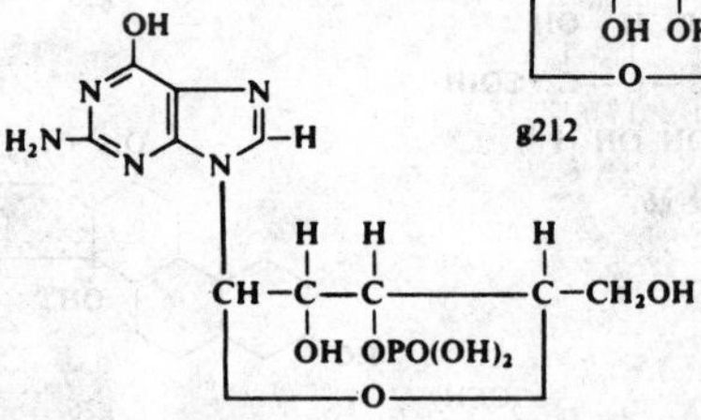

g213

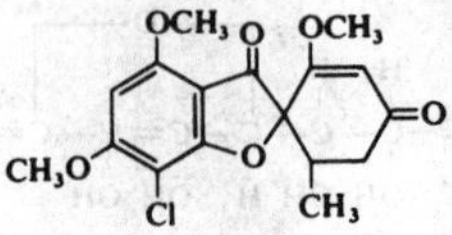

g190

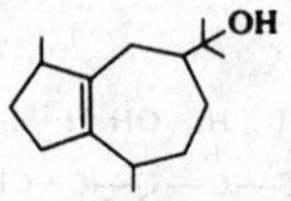

g191

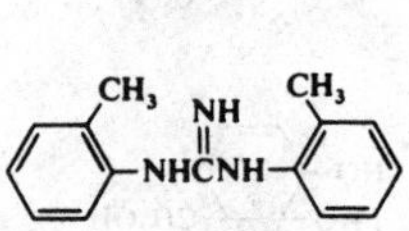

g203

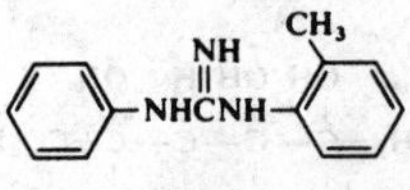

g206

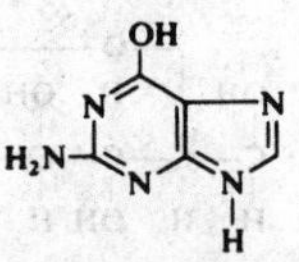

g211

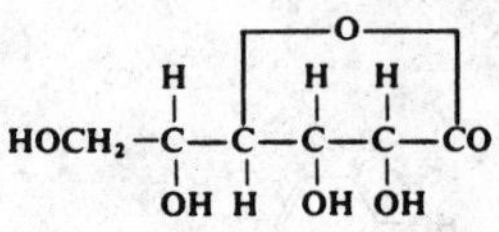

g214

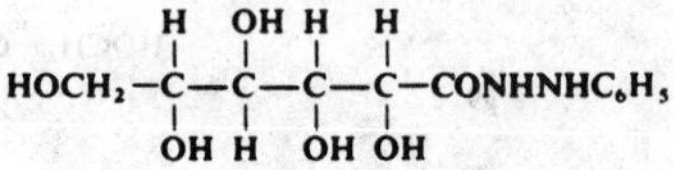

g216

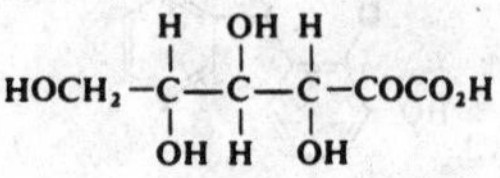

g217

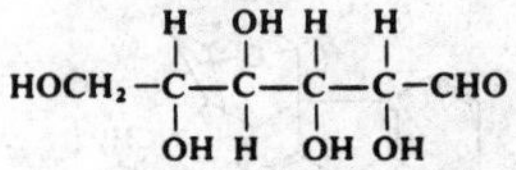

g218

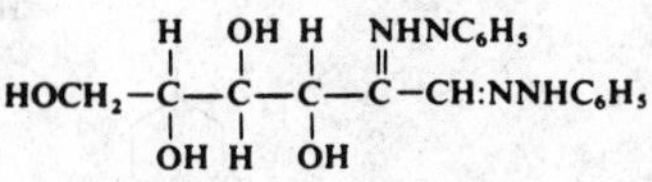

g220

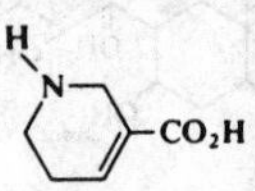

g221

In Alpha Numeric as they occur in Organic Compounds Table

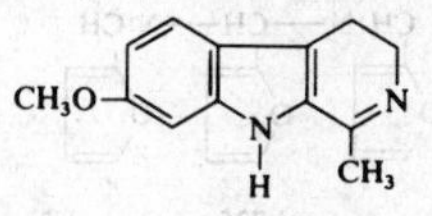

h2

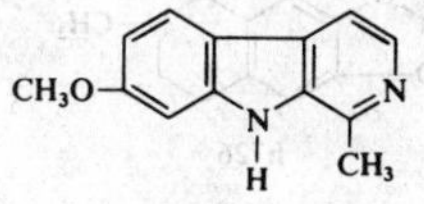

h3

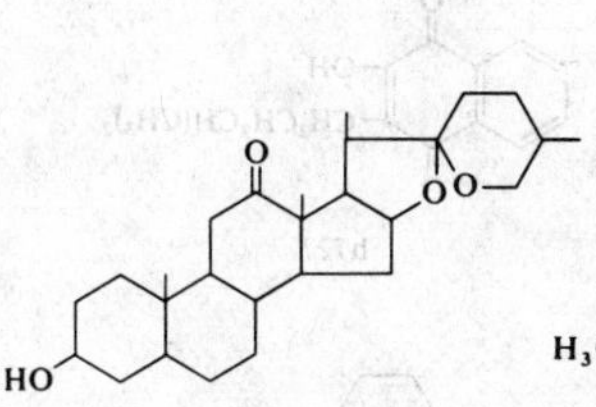

h4

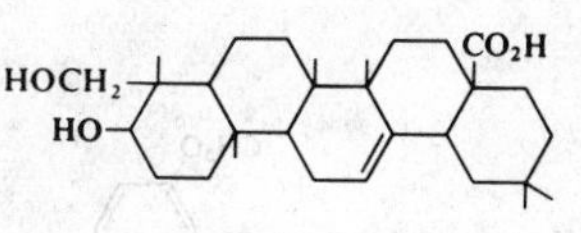

h6

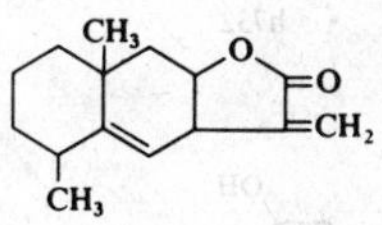

h7

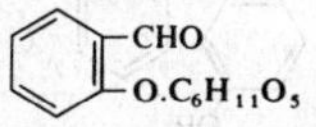

h8

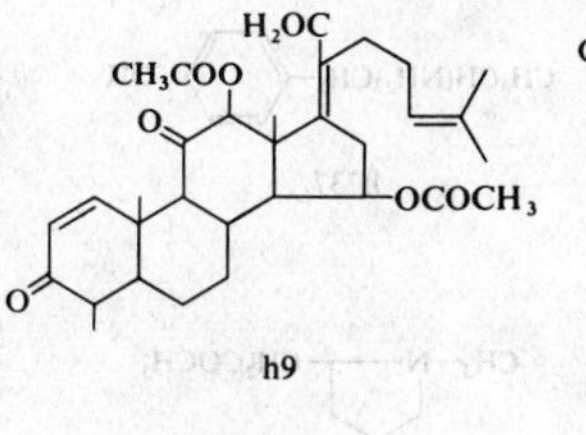

h9

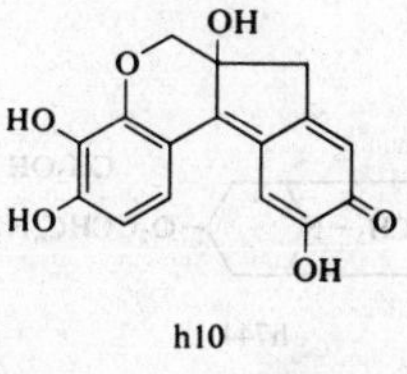

h10

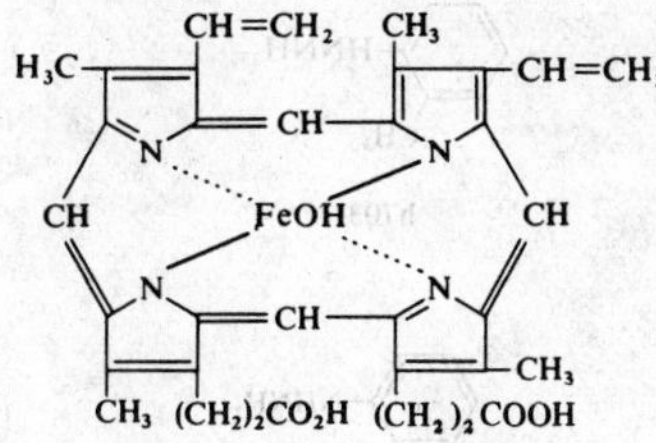

h11

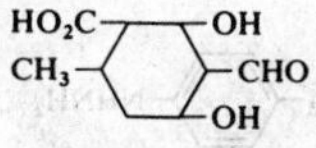

h12

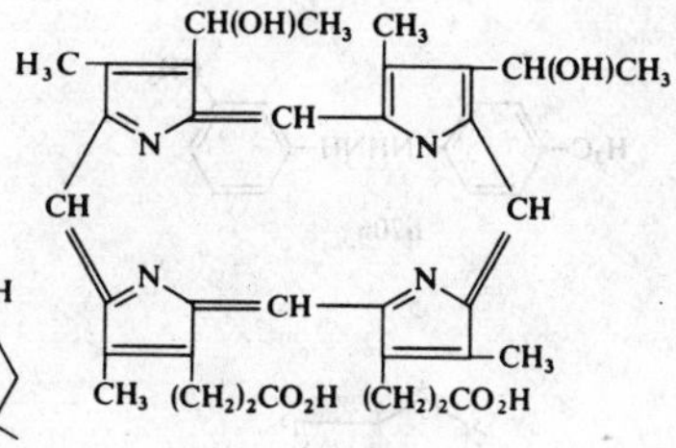

h13

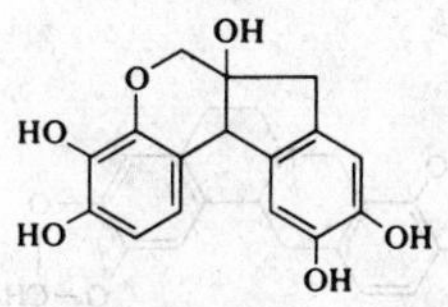

h14

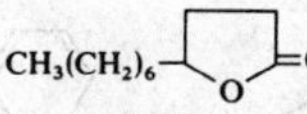

h30

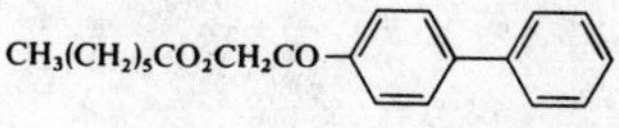

h157

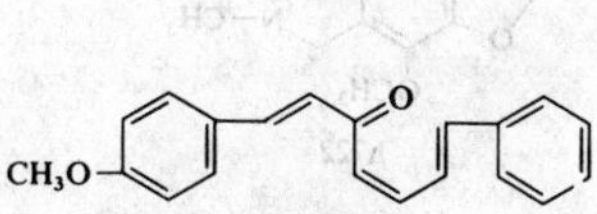

h196

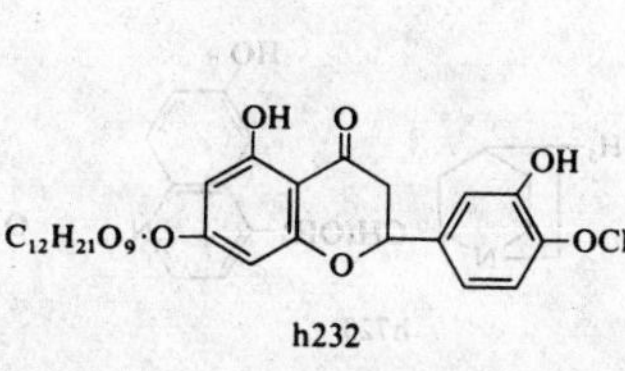

h232

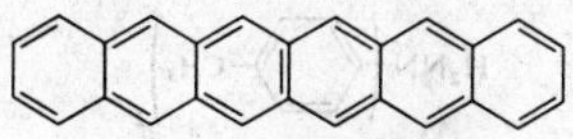

h233

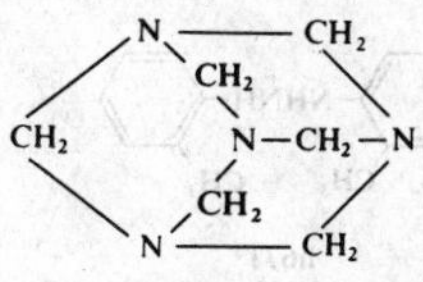

h315

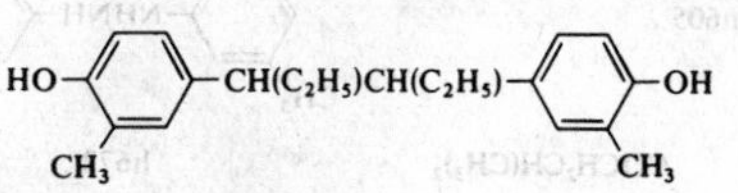

h324

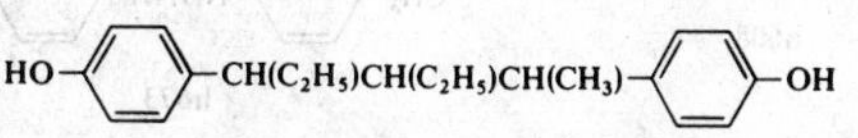

h325

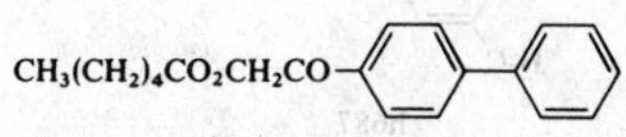

h441

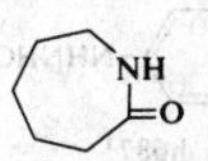

h446

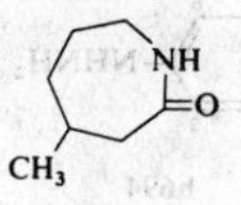

h447

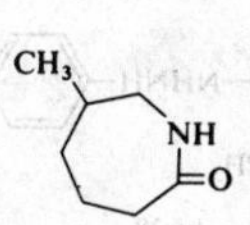

h448

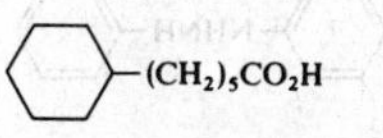

h457

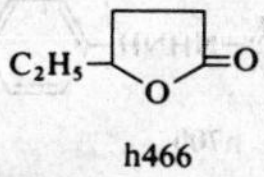

h466

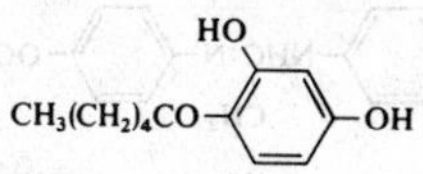

h519

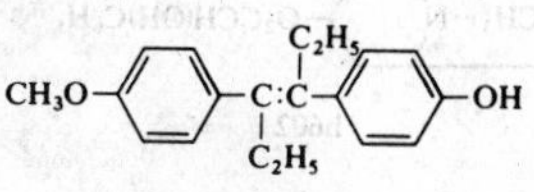

h566

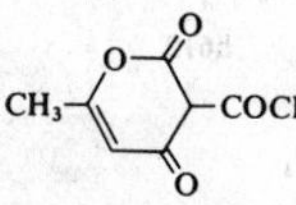

h569

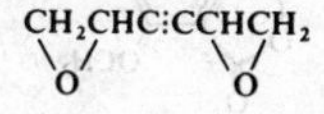

h583

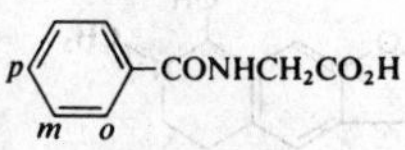

h589[1]

h589[2]

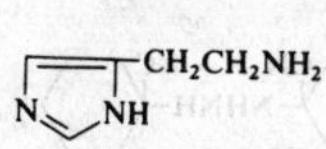

h590

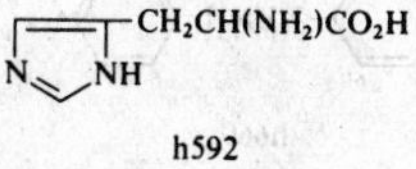

h592

STRUCTURAL FORMULAS OF ORGANIC COMPOUNDS (Continued)

In Alpha Numeric as they occur in Organic Compounds Table

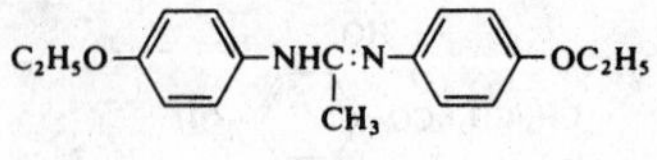

h600

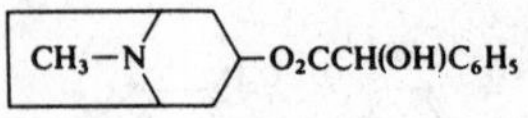

h602

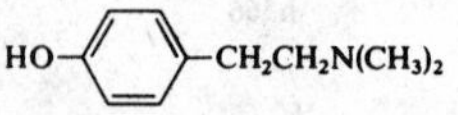

h605

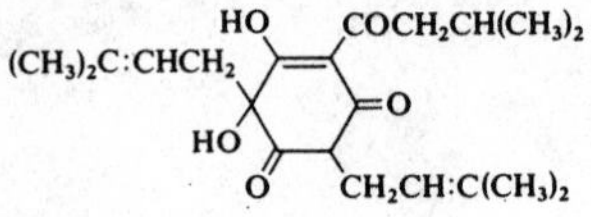

h608

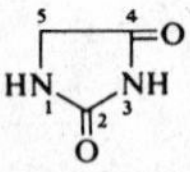

h612

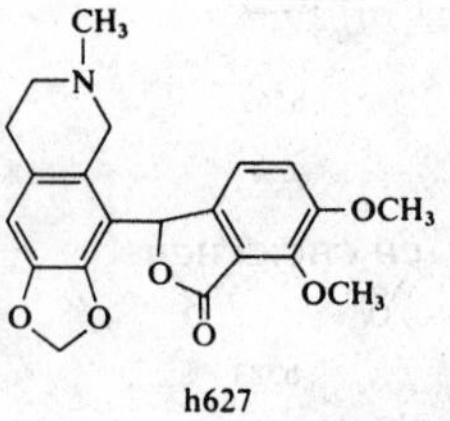

h627

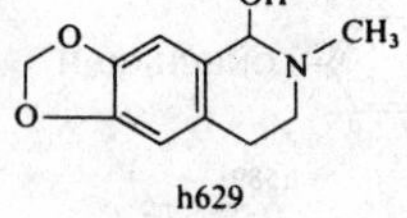

h629

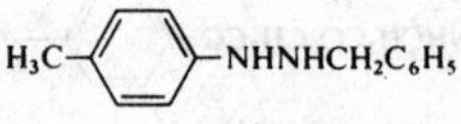

h636

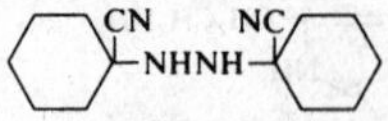

h640

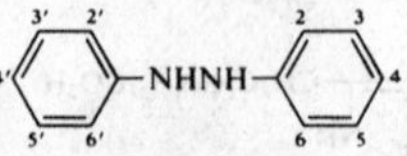

h669

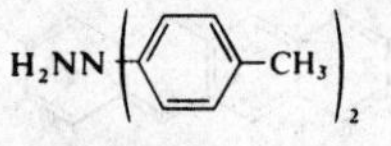

h670

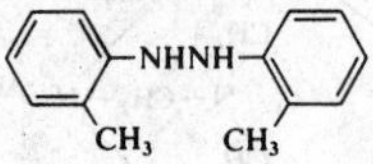

h671

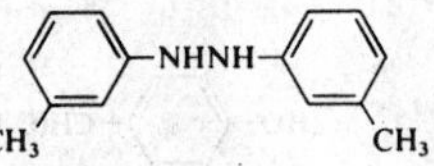

h672

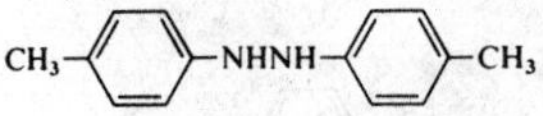

h673

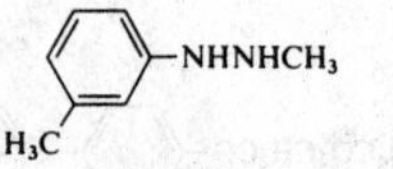

h687

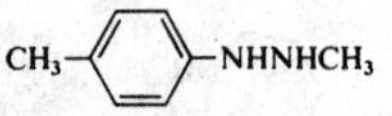

h687'

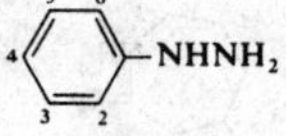

h694

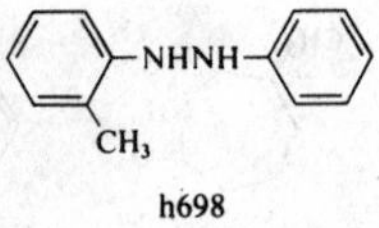

h698

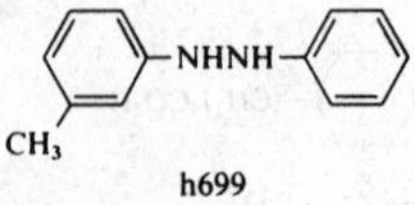

h699

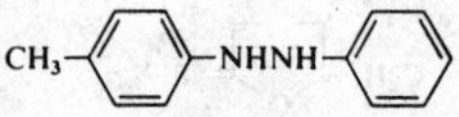

h700

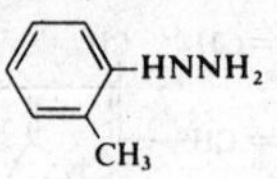

h703

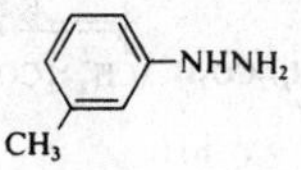

h704

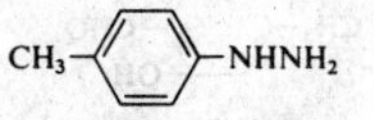

h705

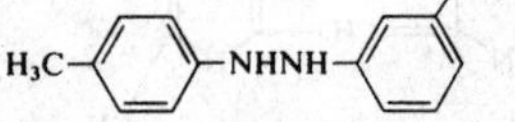

h706

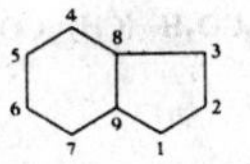

h715

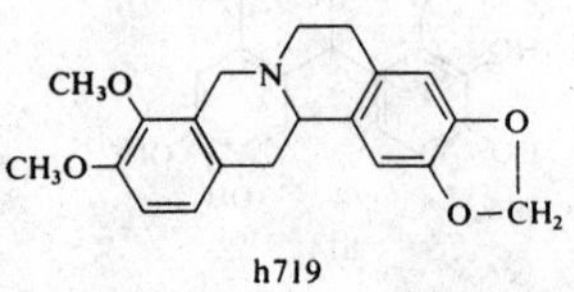

h719

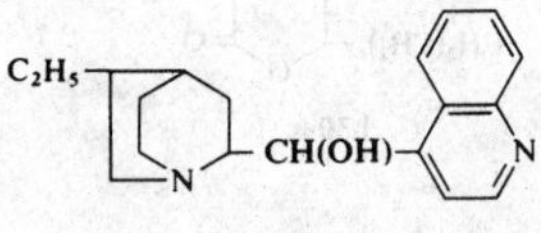

h721

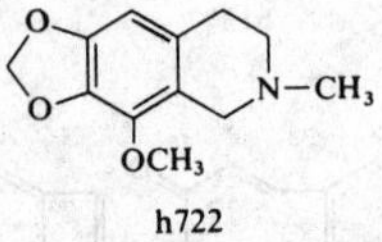

h722

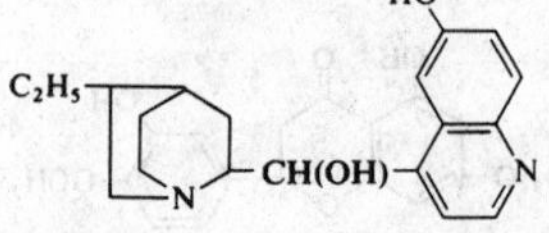

h723

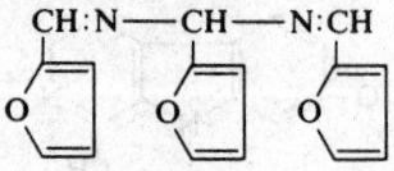

h725

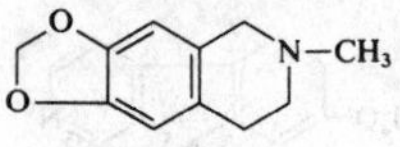

h726

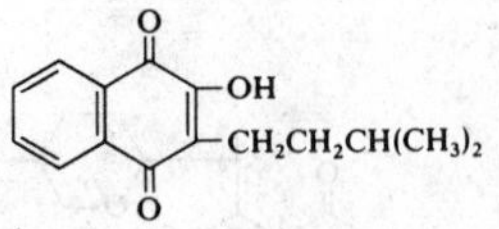

h727

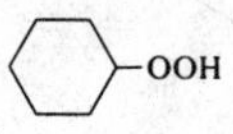

h729

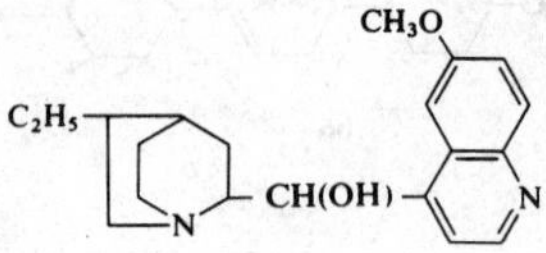

h732

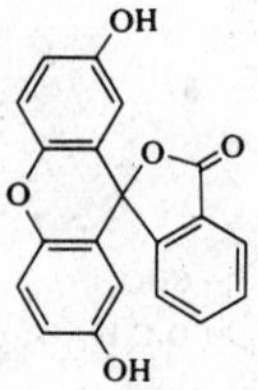

h736

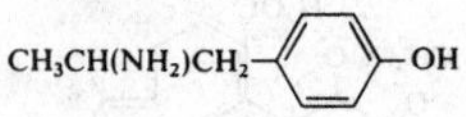

h737

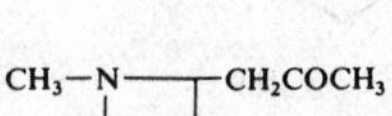

h743

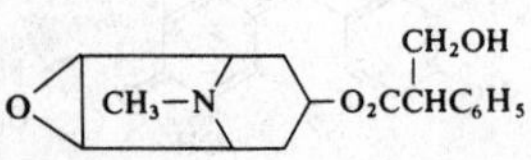

h744

STRUCTURAL FORMULAS OF ORGANIC COMPOUNDS (Continued)

In Alpha Numeric as they occur in Organic Compounds Table

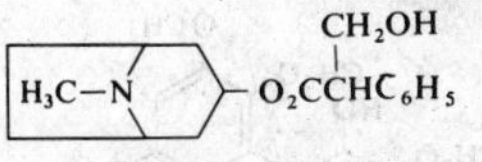
h758

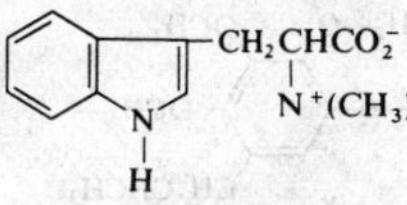
h763

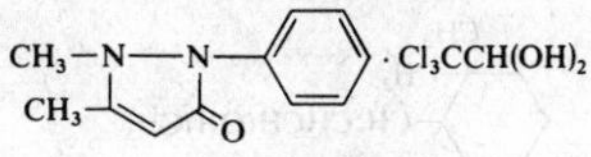
h764

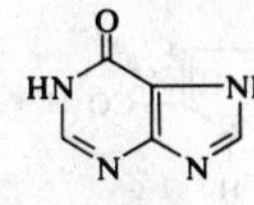
h770

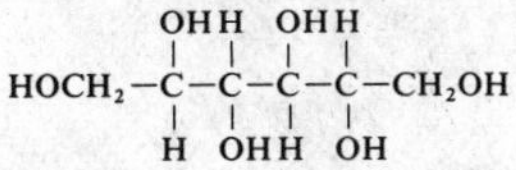
i1

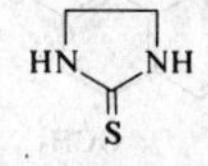
i11

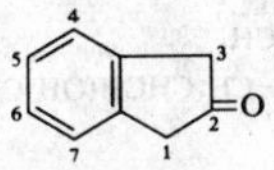
i36

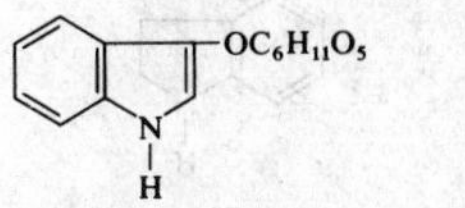
i59

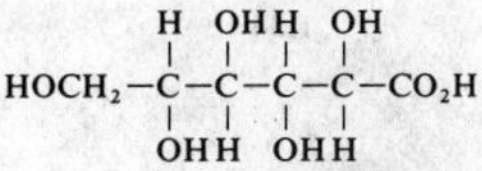
i1'

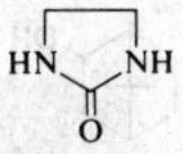
i12

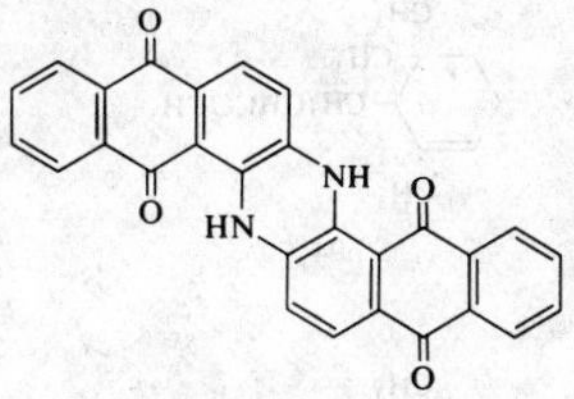
i40

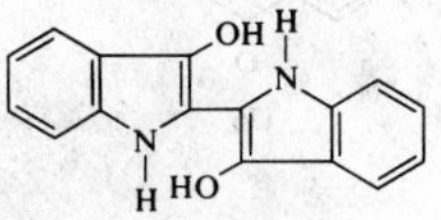
i60

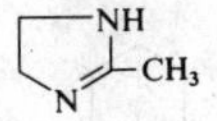
i13

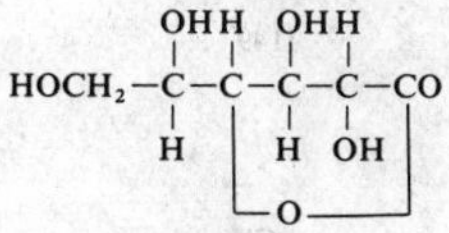
i2

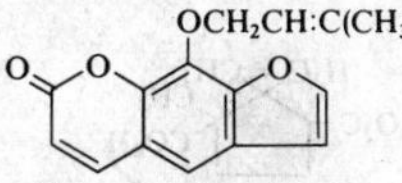
i14

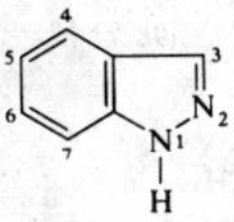
i41

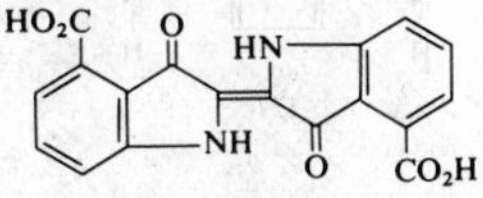
i61

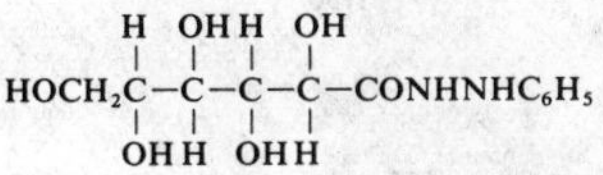
i3

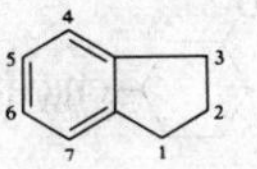
i16

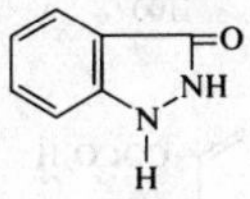
i48

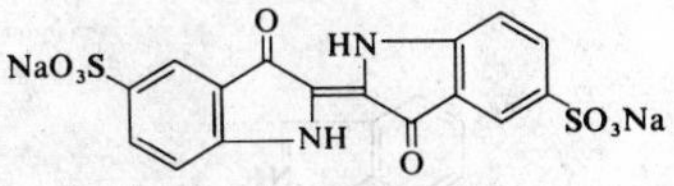
i62

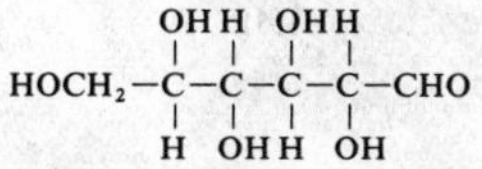
i4

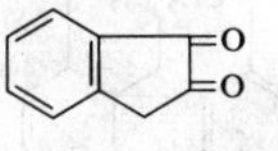
i28

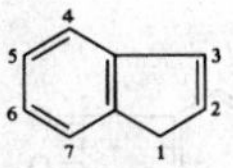
i49

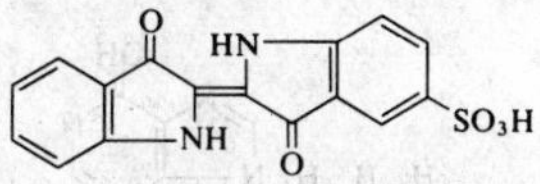
i63

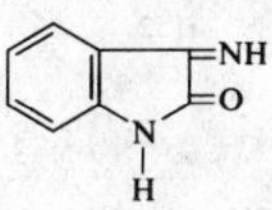
i6

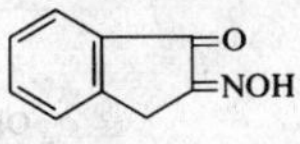
i29

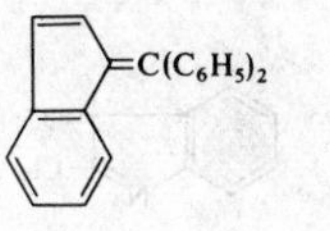
i50

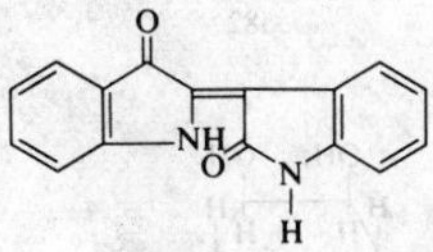
i64

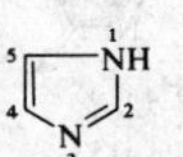
i7

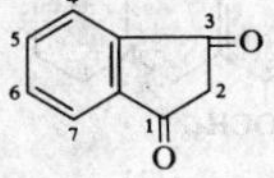
i30

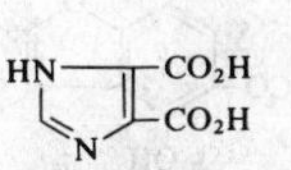
i10

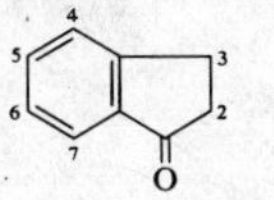
i35

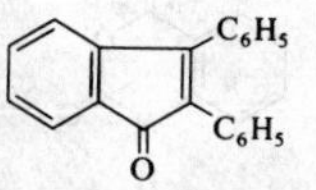
i58

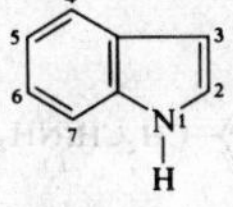
i65

In Alpha Numeric as they occur in Organic Compounds Table

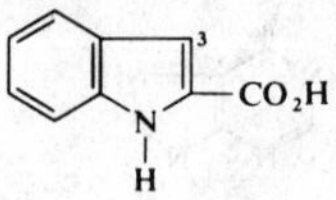
i73

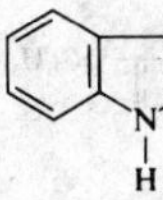
i75

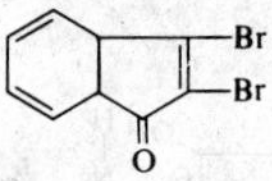
i76

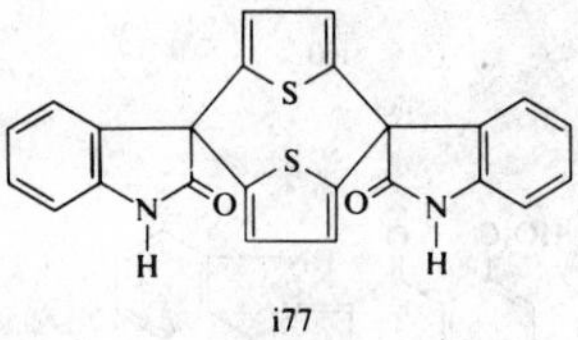
i77

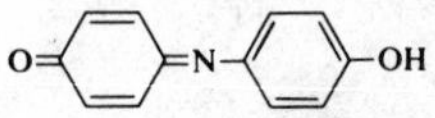
i78

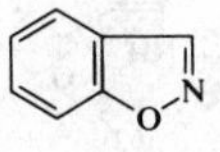
i79

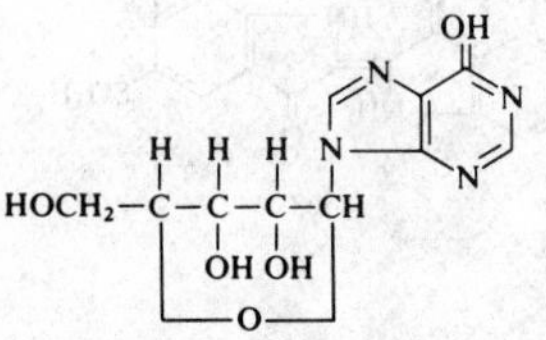
i82

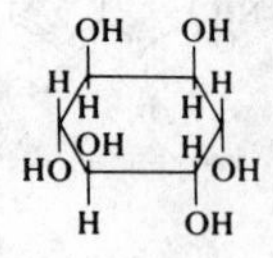
i83

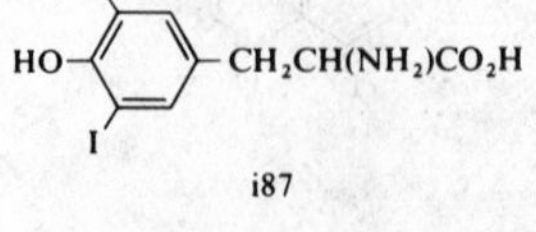
i87

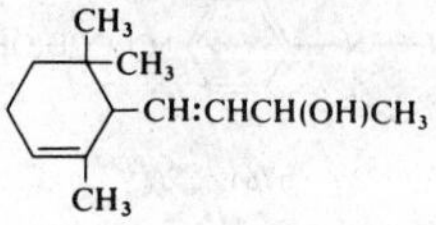
i92

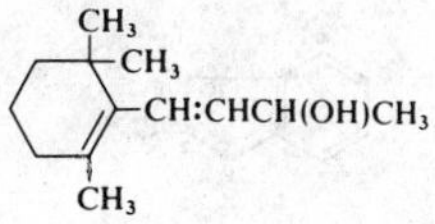
i93

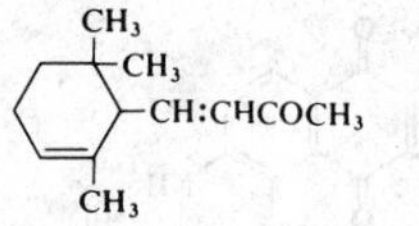
i94

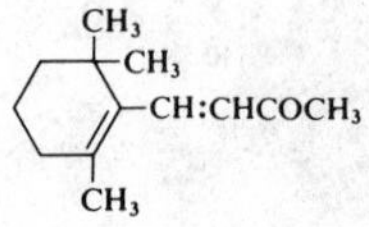
i98

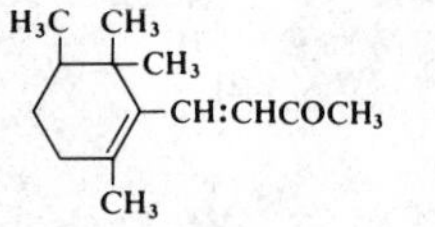
i100

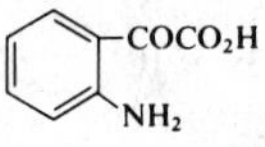
i101

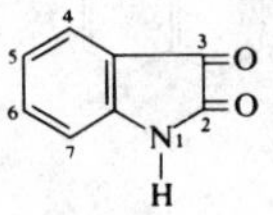
i102

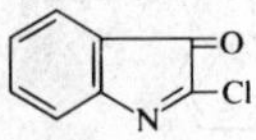
i103

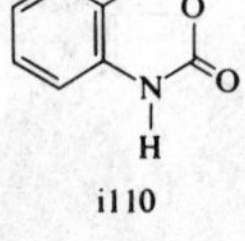
i110

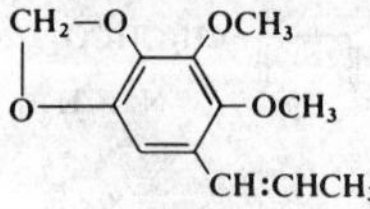
i111

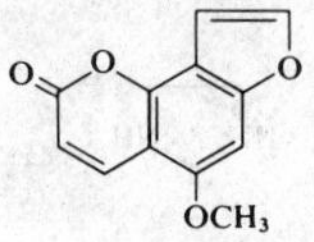
i112

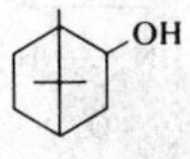
i113

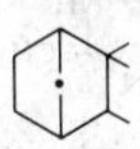
i121

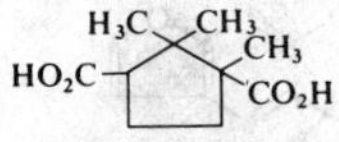
i124

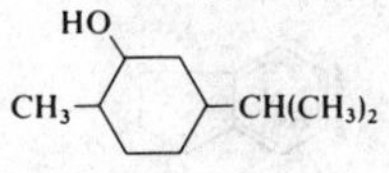
i128

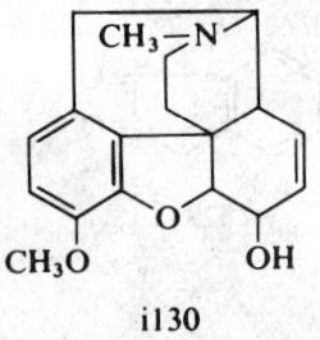
i130

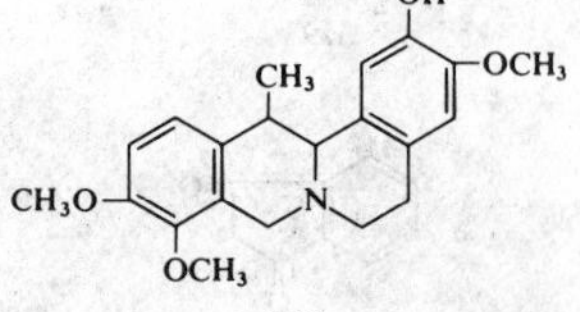
i131

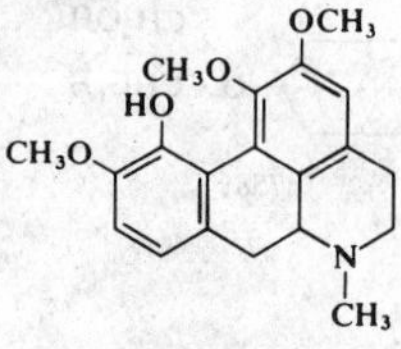
i133

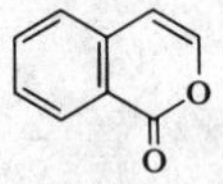
i134

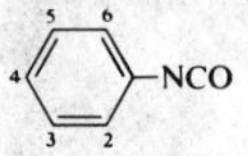
i149

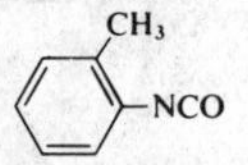
i150

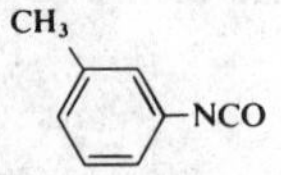
i151

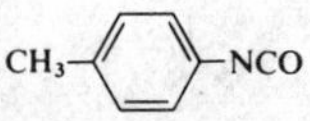
i152

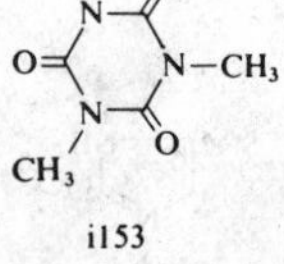
i153

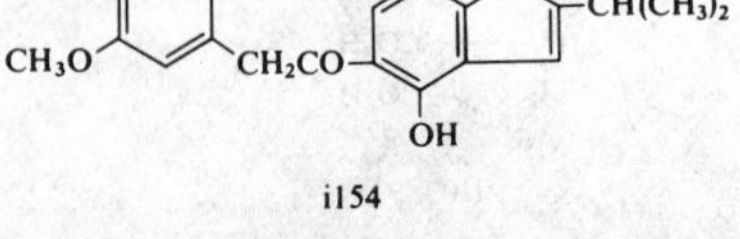
i154

STRUCTURAL FORMULAS OF ORGANIC COMPOUNDS (Continued)

In Alpha Numeric as they occur in Organic Compounds Table

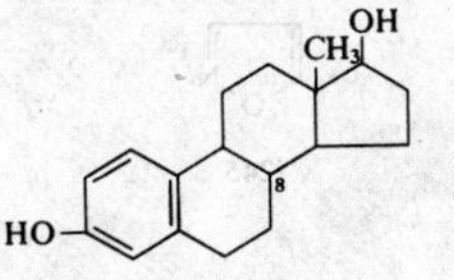

i155

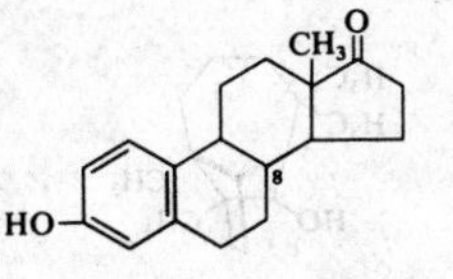

i156

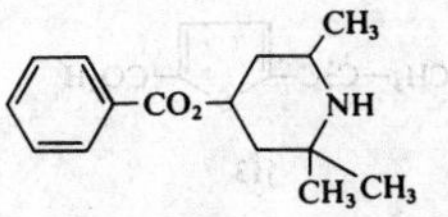

i157

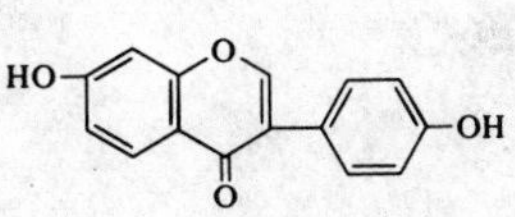

i161

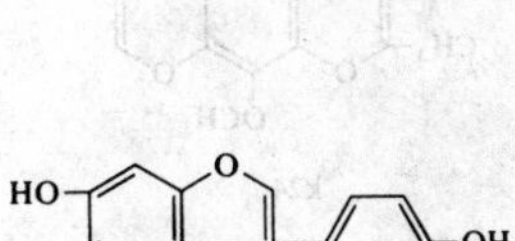

i162

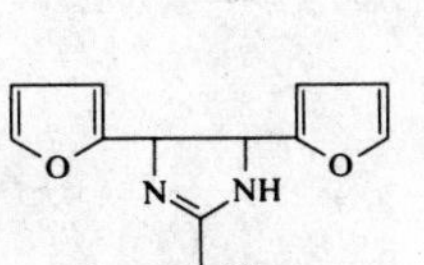

i163

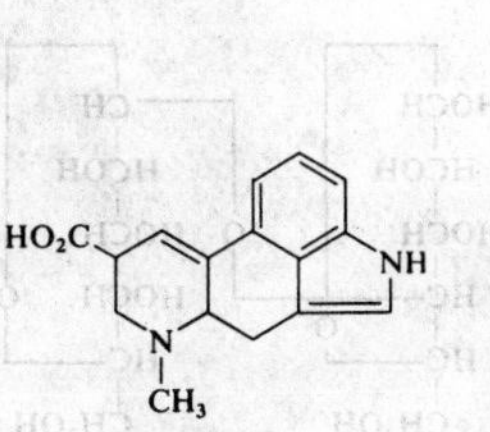

i165

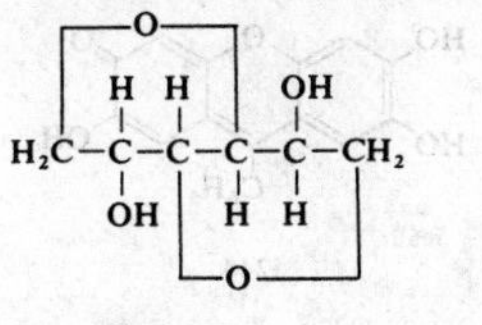

i166

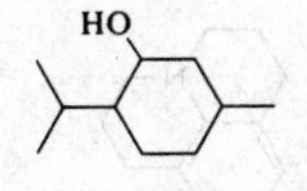

i167

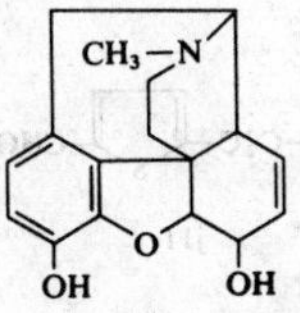

i170

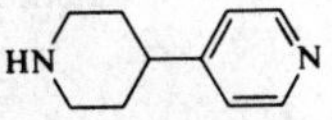

i171

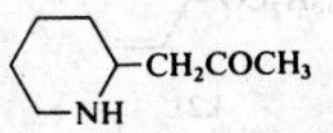

i175

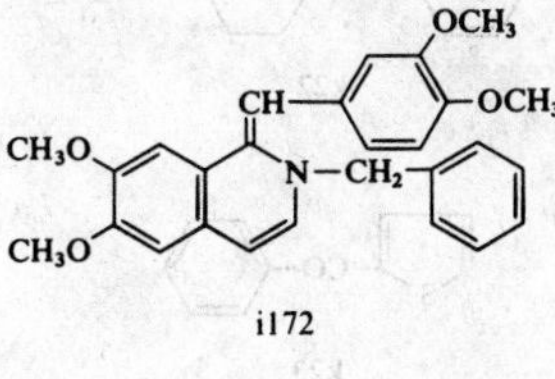

i172

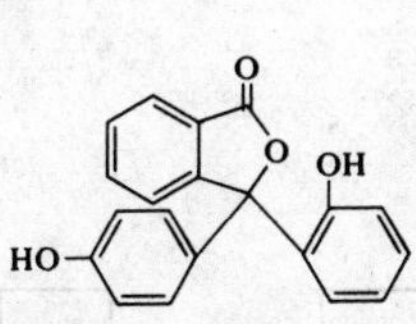

i177

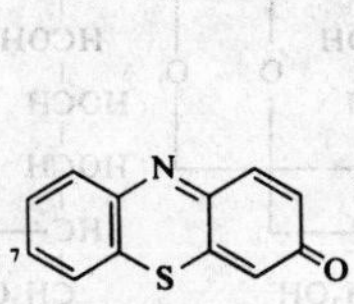

i178

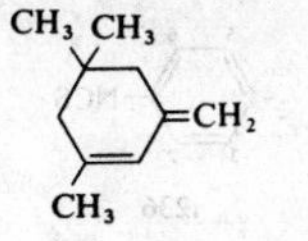

i180

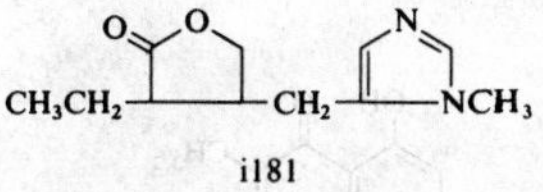

i181

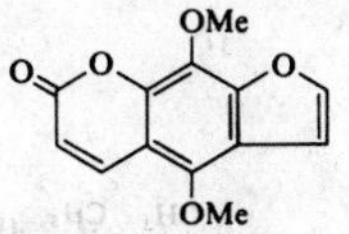

i182

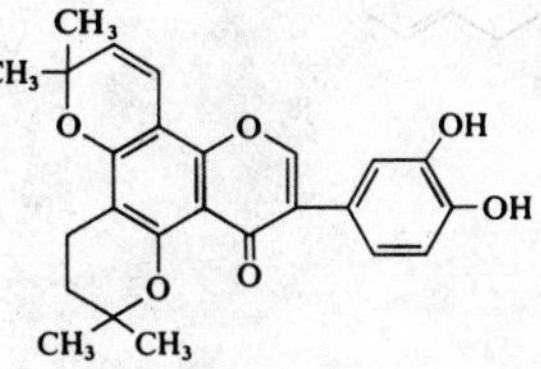

i183

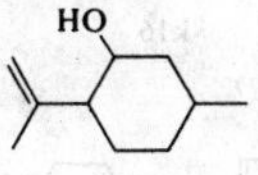

i184

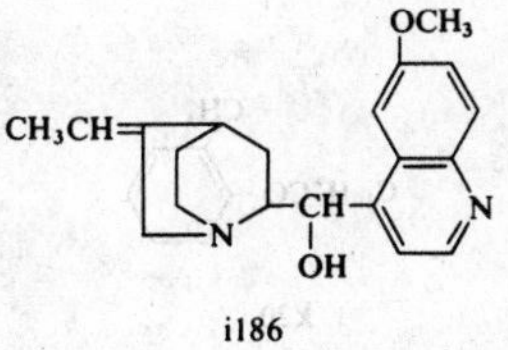

i186

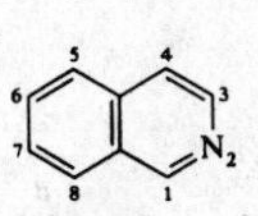

i188

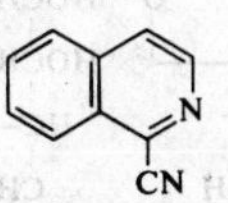

i204

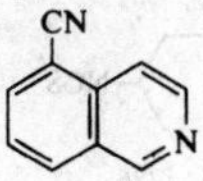

i205

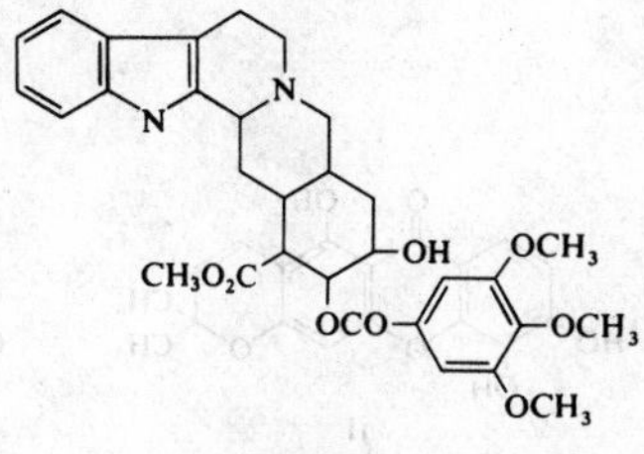

i207

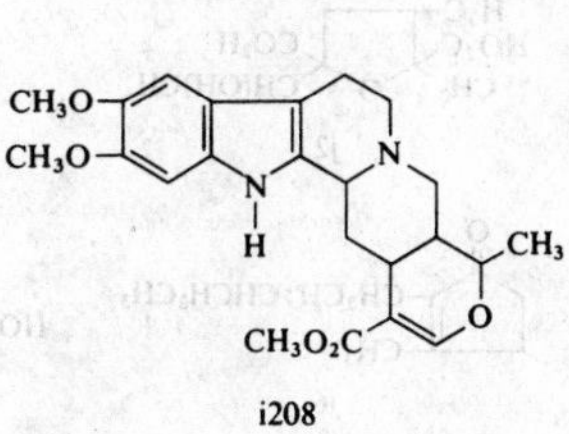

i208

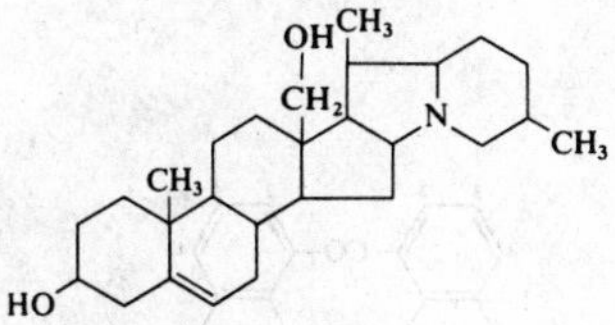

i209

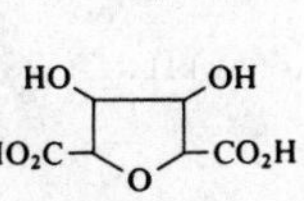

i211

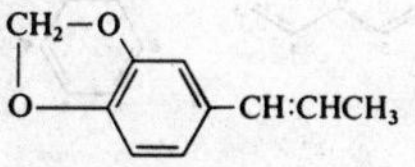

i212

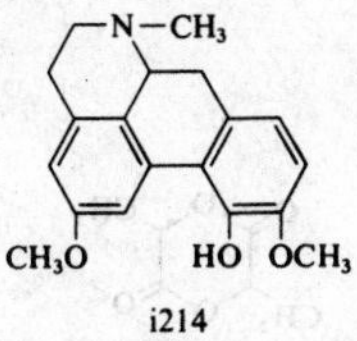

i214

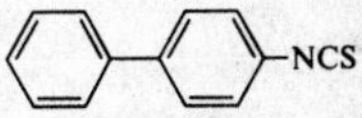

i218

STRUCTURAL FORMULAS OF ORGANIC COMPOUNDS (Continued)

In Alpha Numeric as they occur in Organic Compounds Table

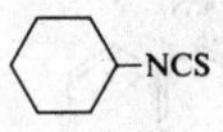

i226

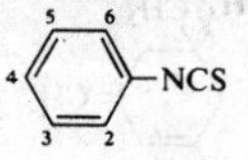

i236

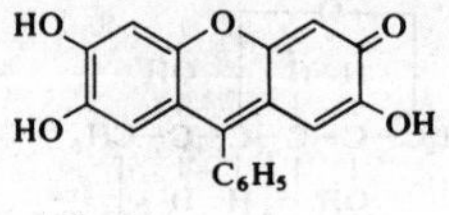

i244

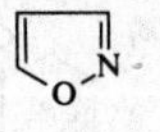

i245

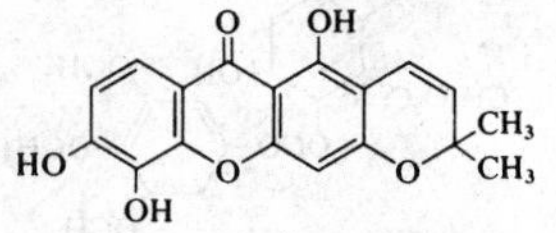

j1

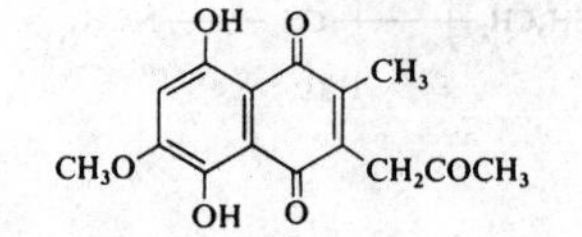

j7

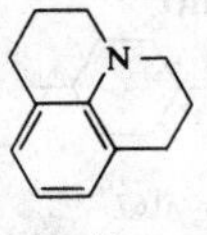

j9

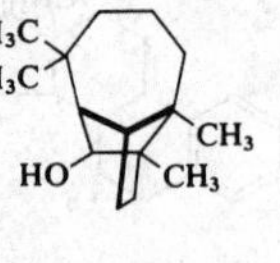

j12

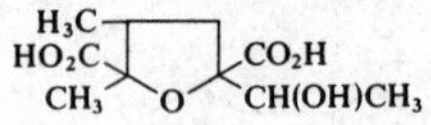

j2

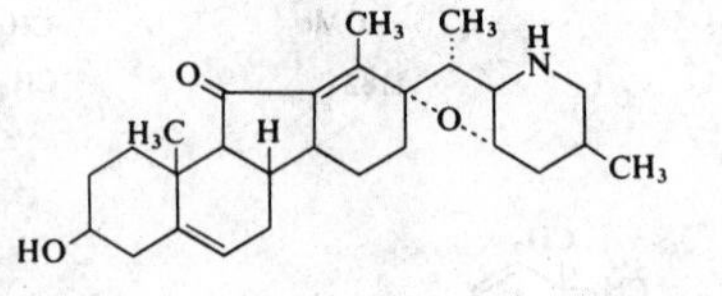

j8

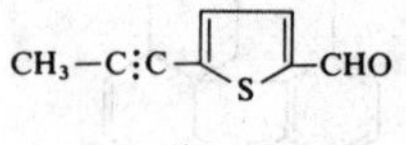

j11

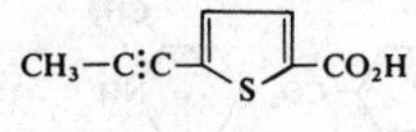

j13

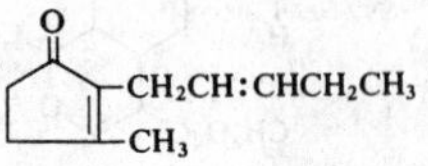

j6

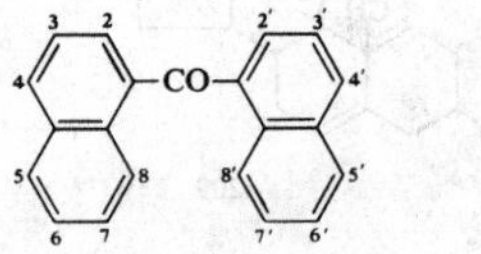

k11

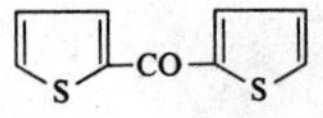

k16

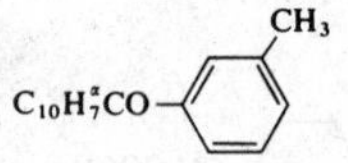

k21

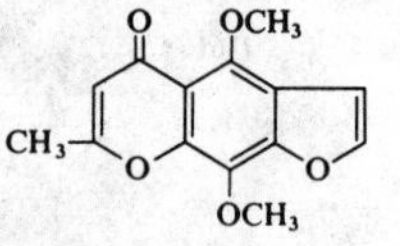

k24

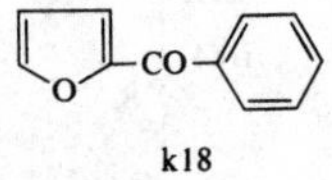

k18

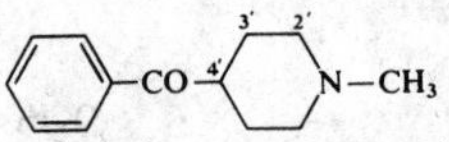

k22

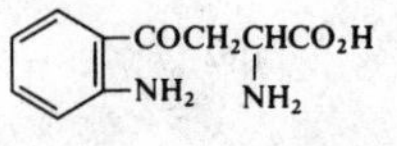

k25

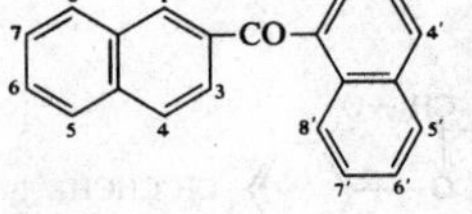

k12

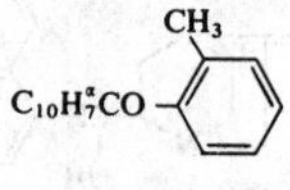

k20

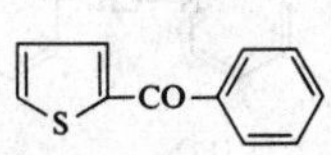

k23

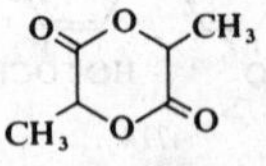

l1

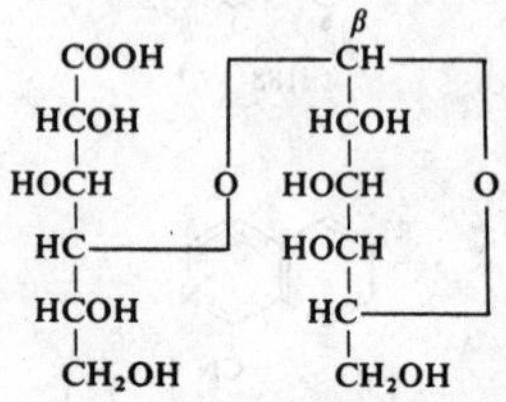

l4

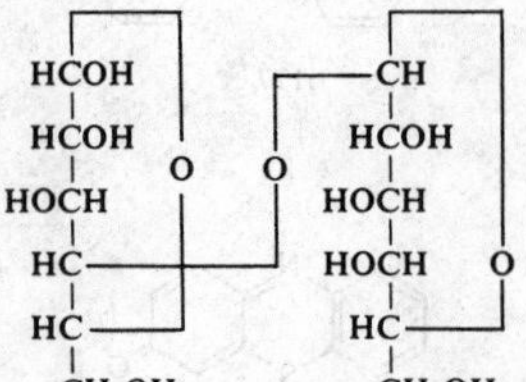

l5

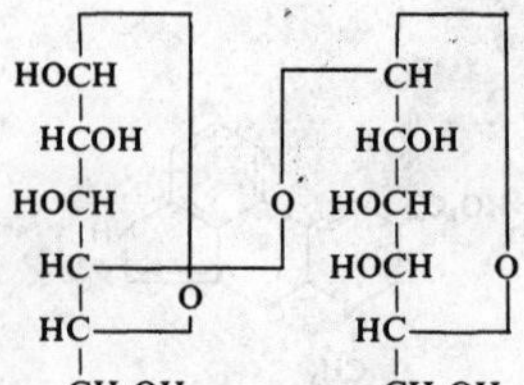

l6

STRUCTURAL FORMULAS OF ORGANIC COMPOUNDS (Continued)

In Alpha Numeric as they occur in Organic Compounds Table

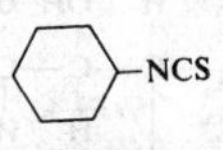

i226

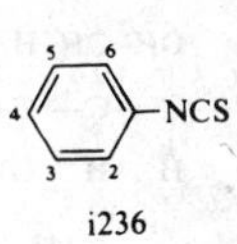

i236

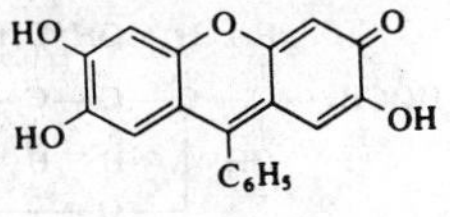

i244

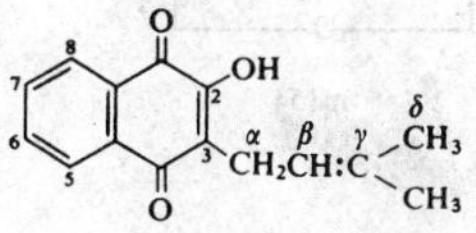

l15

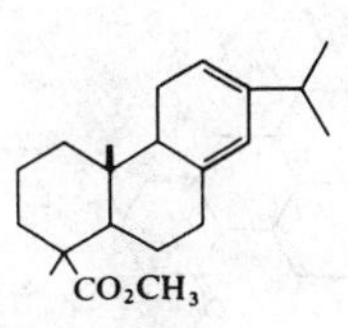

l39

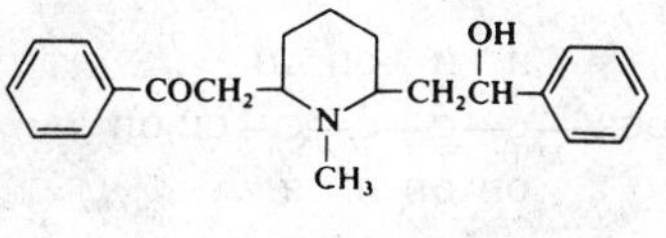

l54

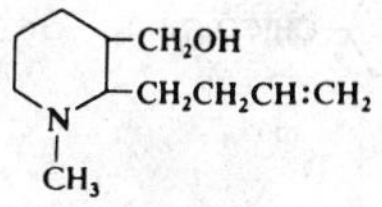

l67

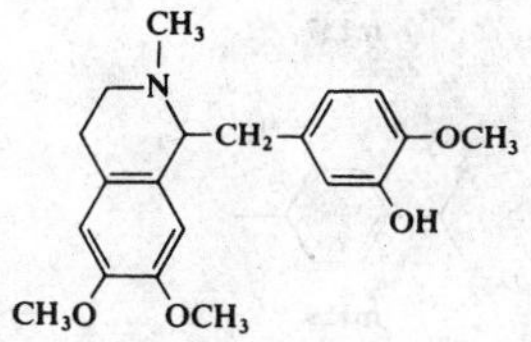

l18

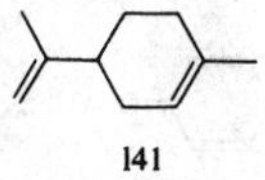

l41

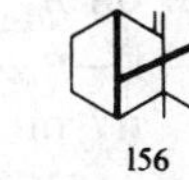

l56

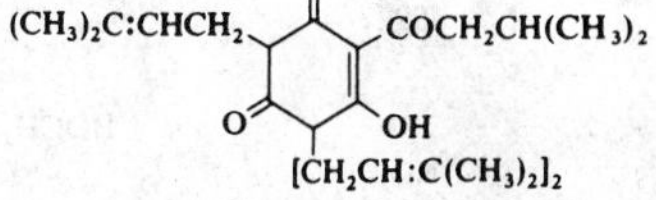

l68

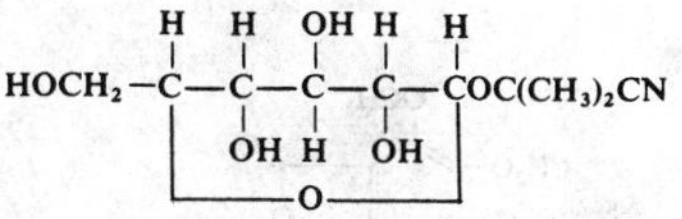

l48

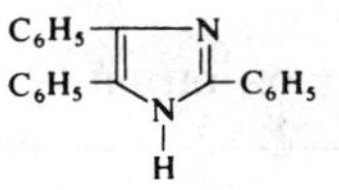

l57

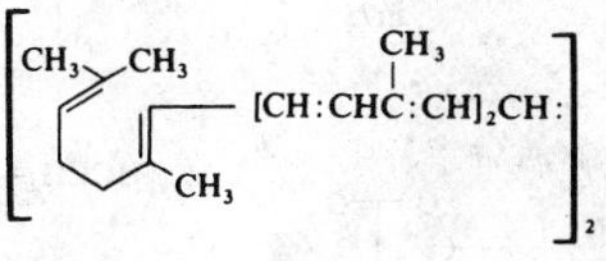

l71

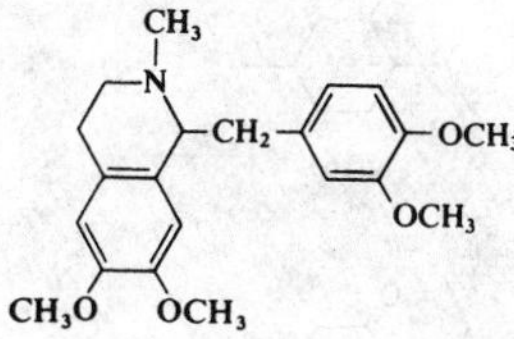

l21

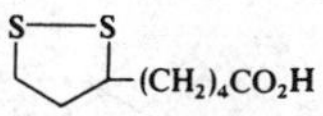

l49

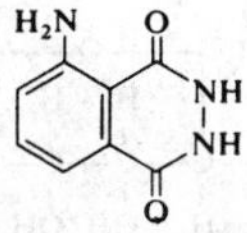

l60

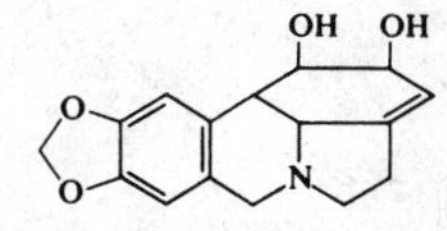

l72

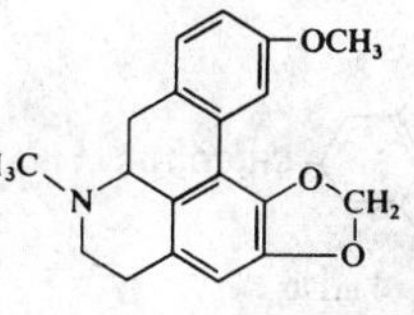

l24

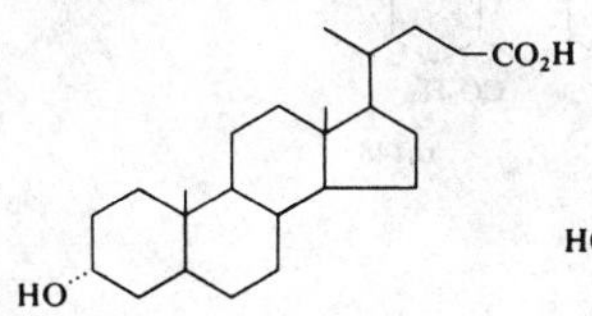

l50

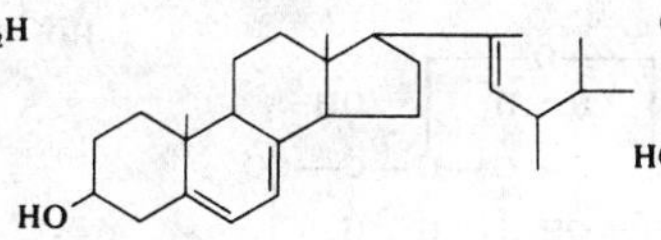

l61

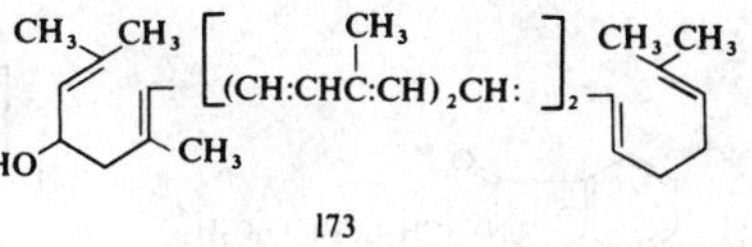

l73

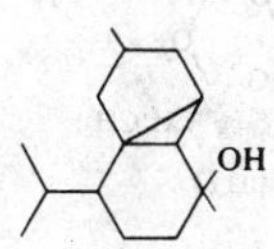

l26

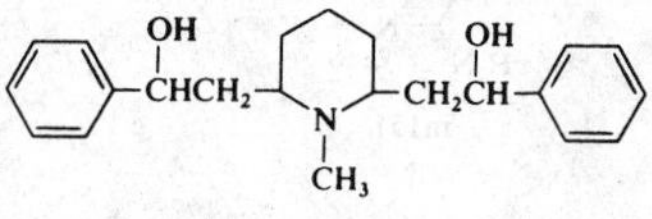

l52

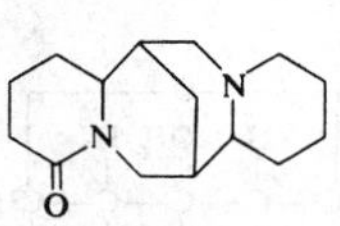

l62

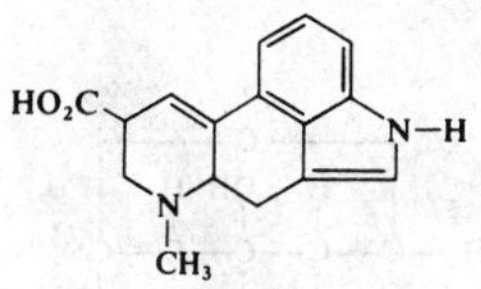

l74

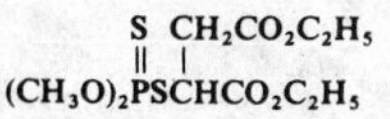

m1

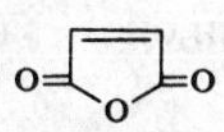

m3

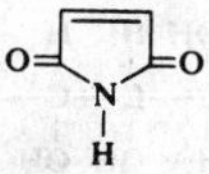

m5

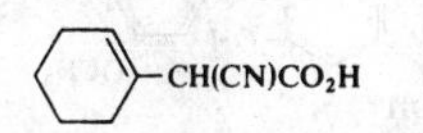

m63

STRUCTURAL FORMULAS OF ORGANIC COMPOUNDS (Continued)

In Alpha Numeric as they occur in Organic Compounds Table

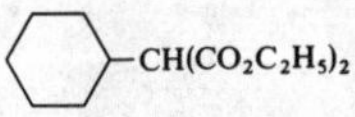

m64

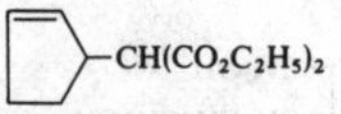

m65

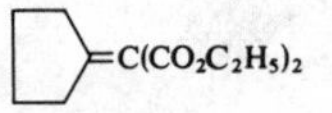

m66

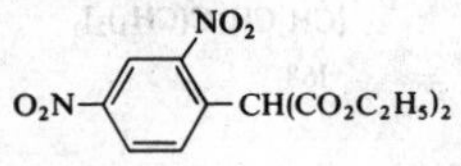

m75

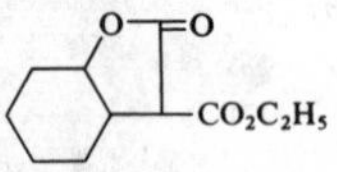

m91

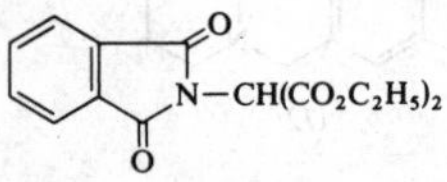

m116

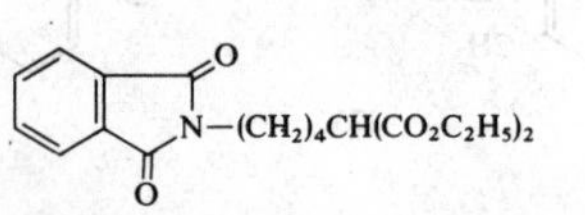

m117

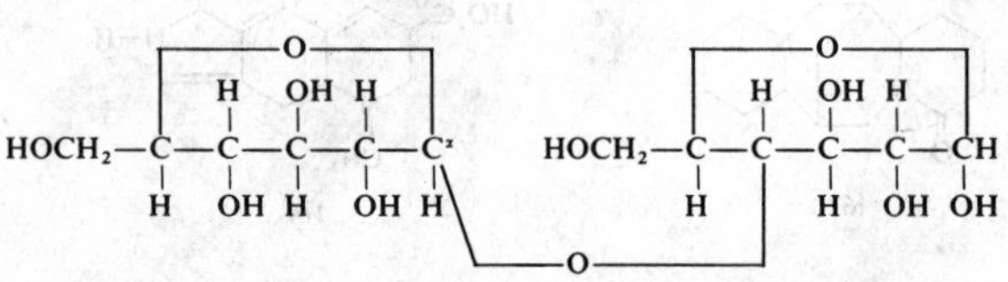

m120

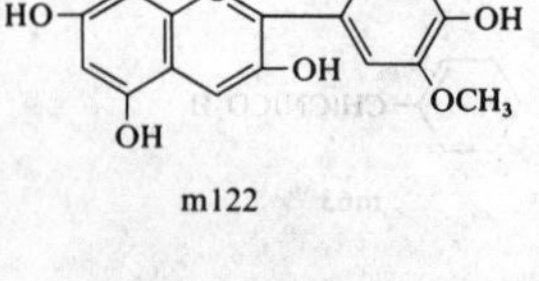

m122

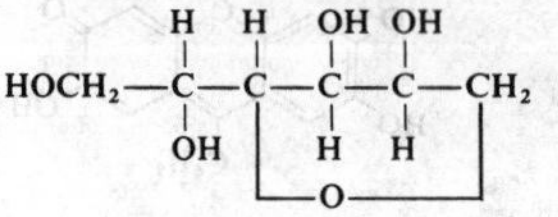

m123

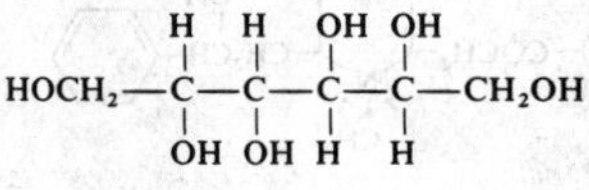

m124

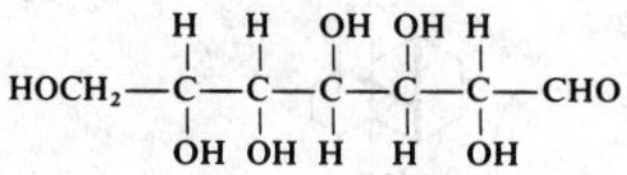

m128

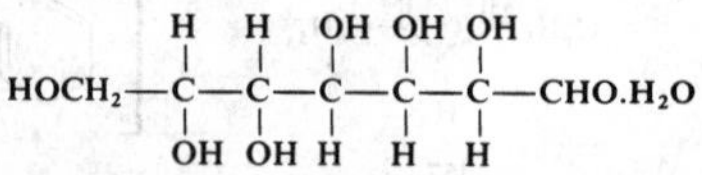

m129

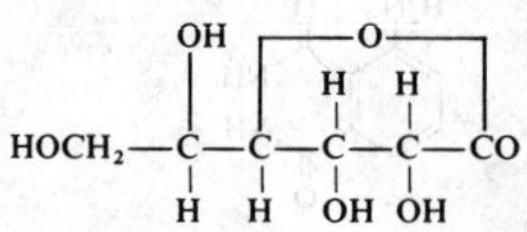

m132

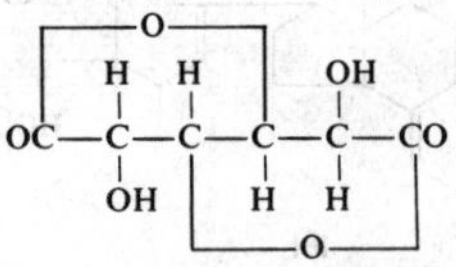

m134

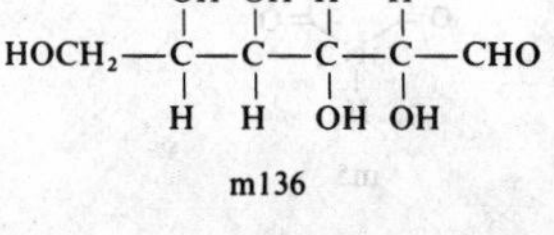

m136

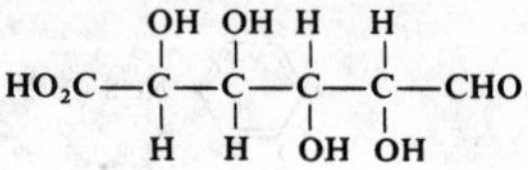

m141

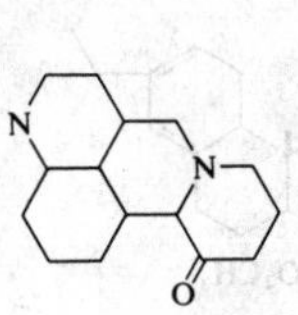

m144

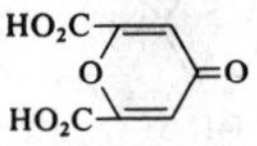

m145

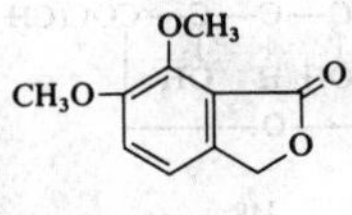

m147

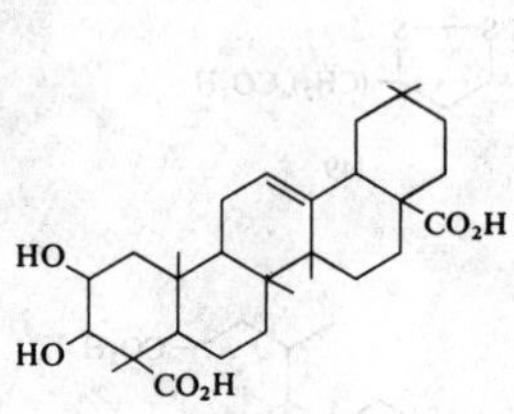

m148

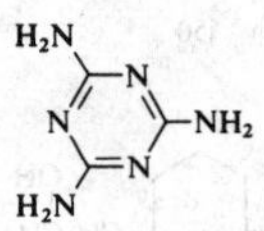

m151

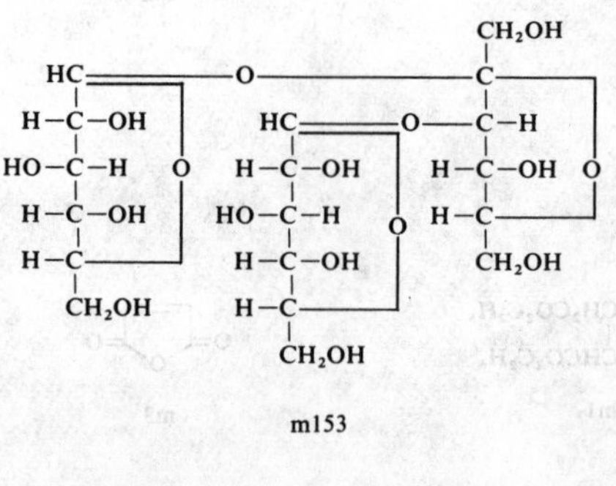

m153

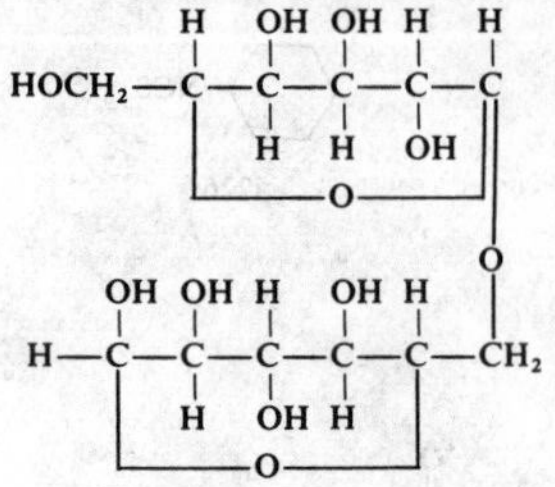

m154

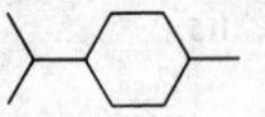

m157

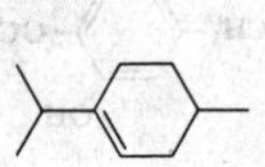

m159

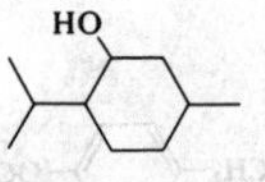

m163

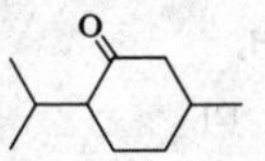

m167

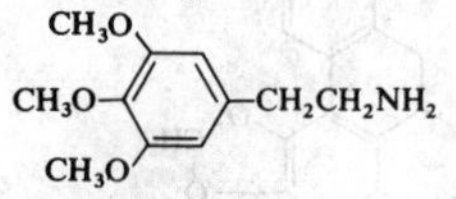

m170

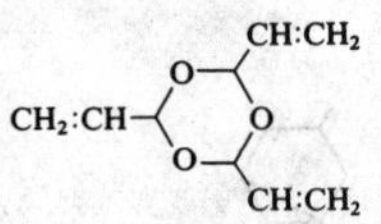

m171

STRUCTURAL FORMULAS OF ORGANIC COMPOUNDS (Continued)

In Alpha Numeric as they occur in Organic Compounds Table

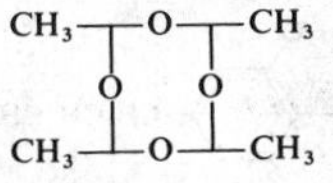

m173

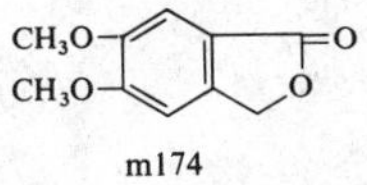

m174

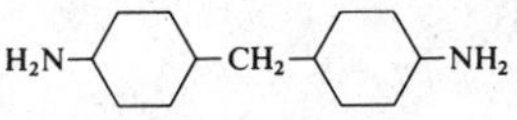

m187

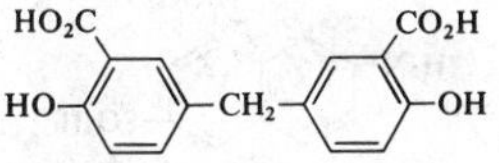

m194

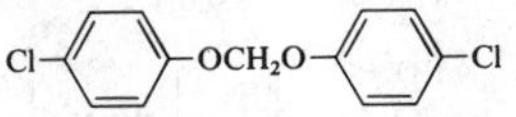

m196

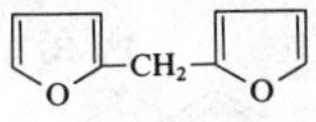

m262

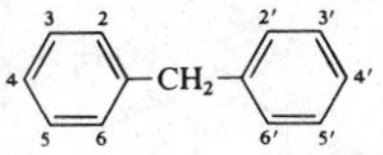

m271

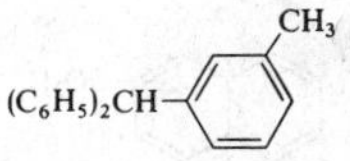

m272

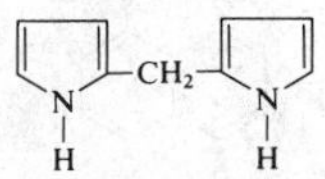

m273

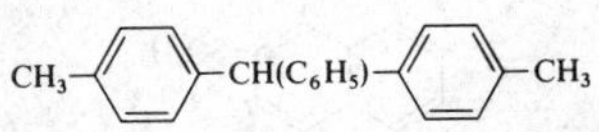

m275

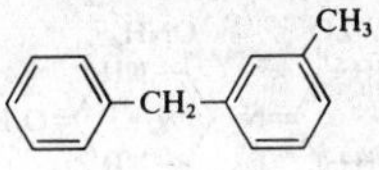

m293

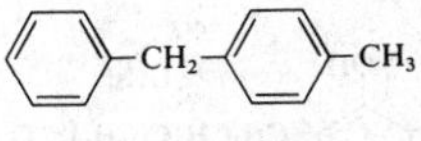

m294

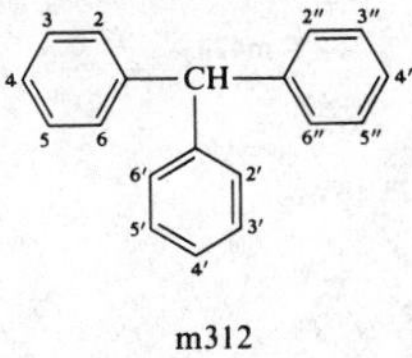

m312

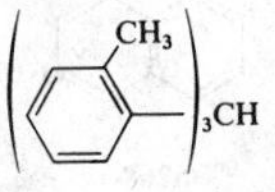

m317

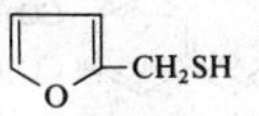

m346

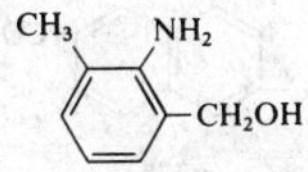

m350

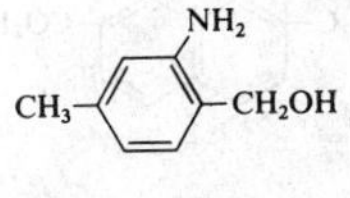

m351

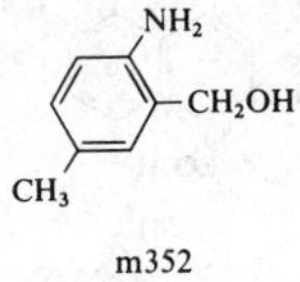

m352

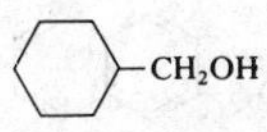

m362

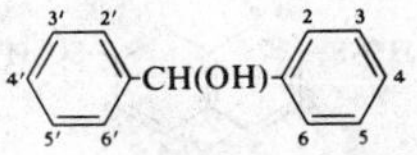

m366

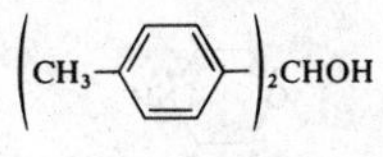

m368[1]

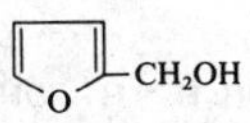

m369

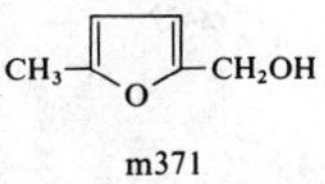

m371

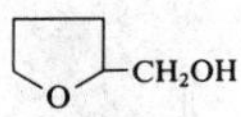

m376

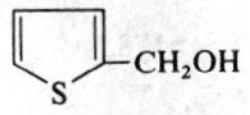

m377

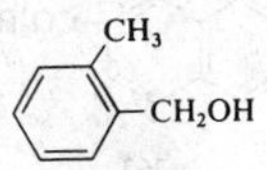

m378

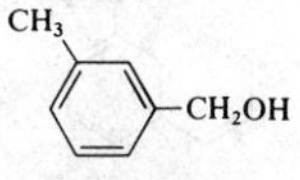

m379

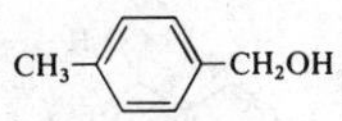

m380

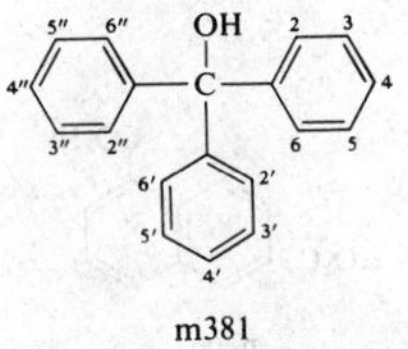

m381

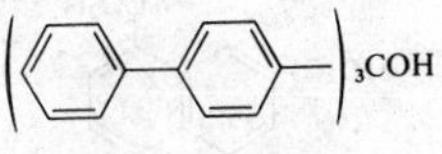

m383

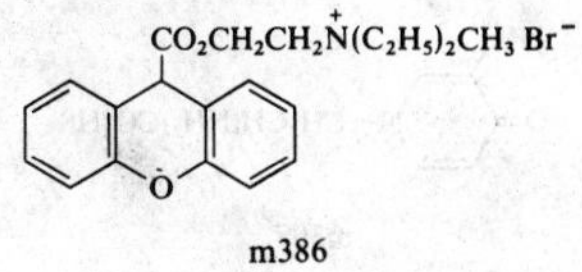

m386

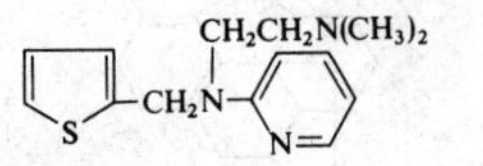

m387

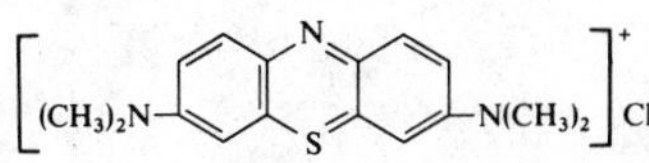

m390

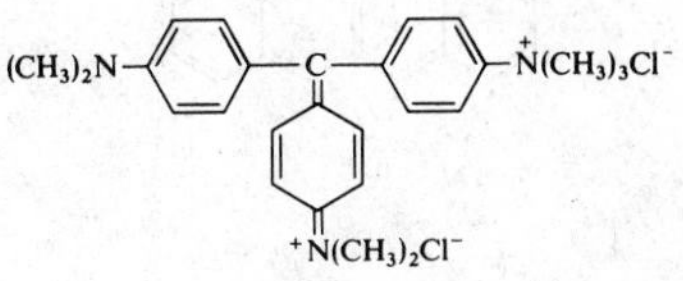

m391

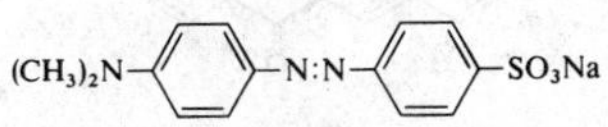

m392

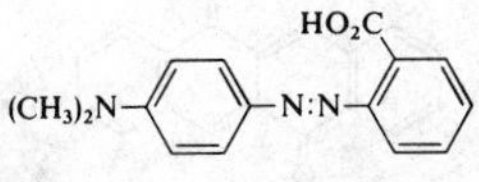

m393

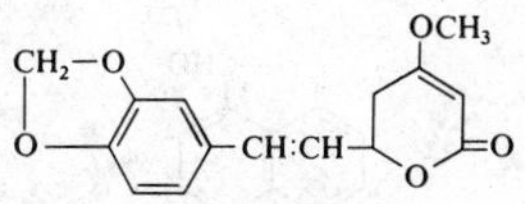

m394

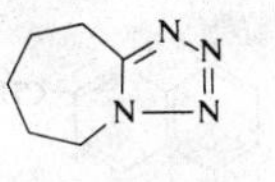

m395

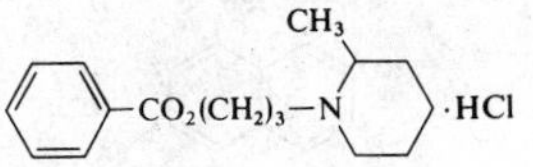

m396

STRUCTURAL FORMULAS OF ORGANIC COMPOUNDS (Continued)

In Alpha Numeric as they occur in Organic Compounds Table

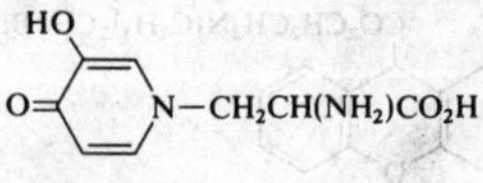

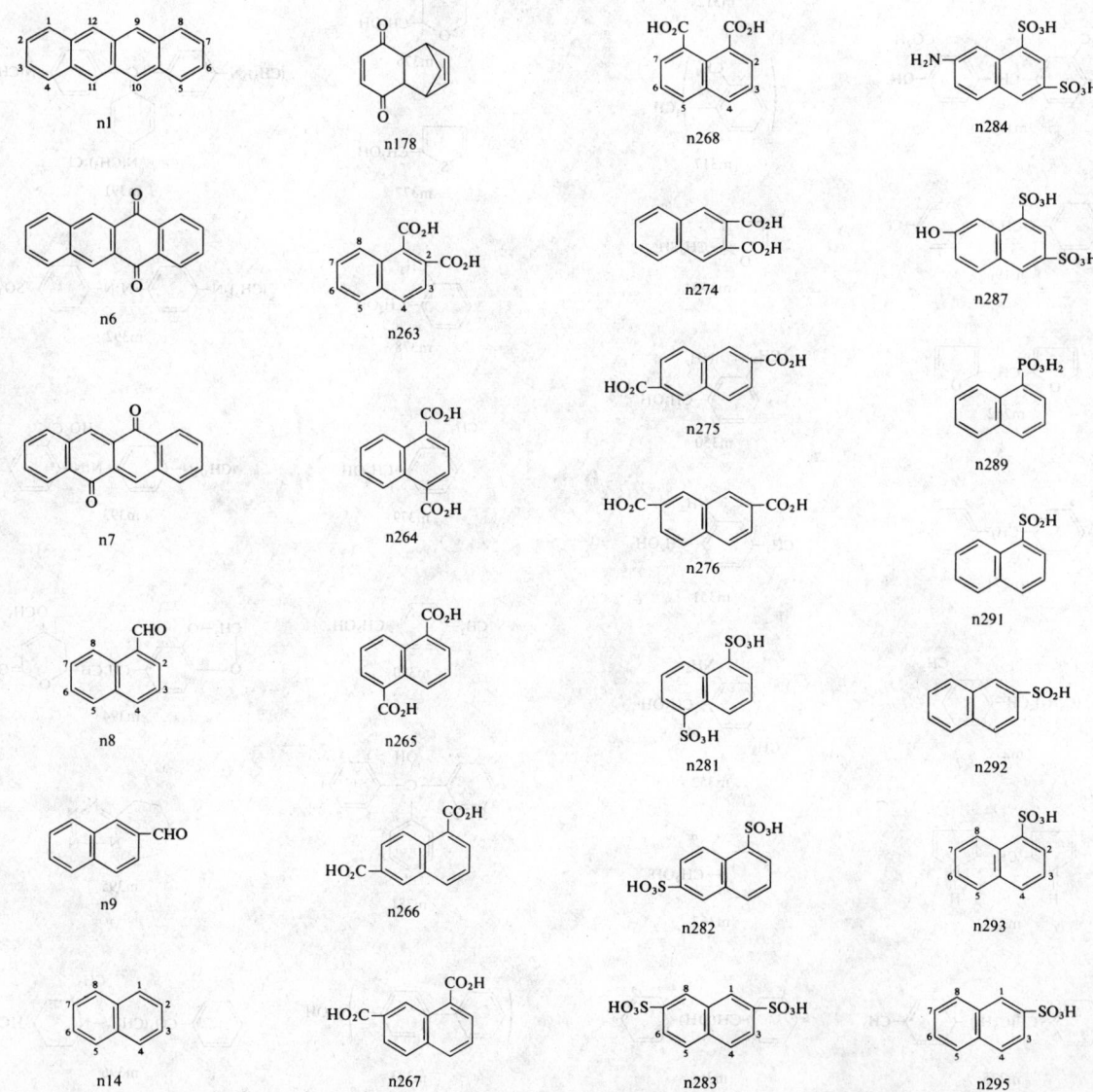

m397

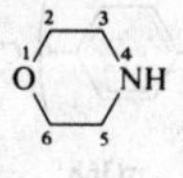

m408

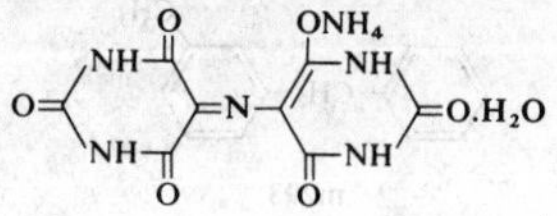

m423

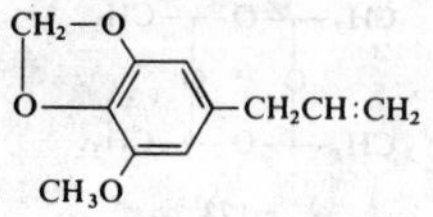

m426

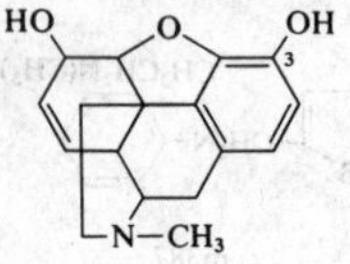

m399

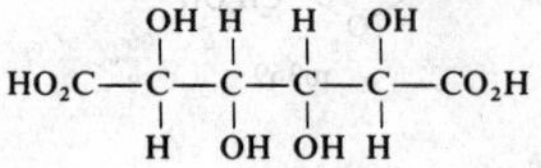

m422

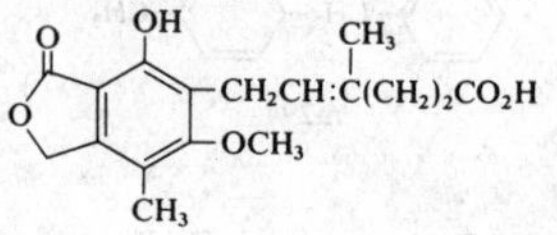

m424

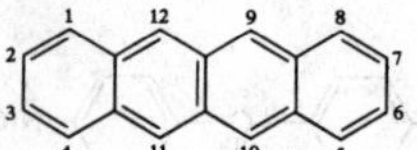

n1

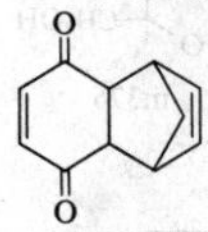

n178

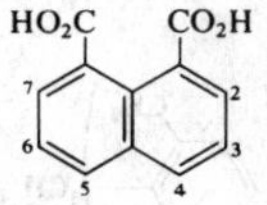

n268

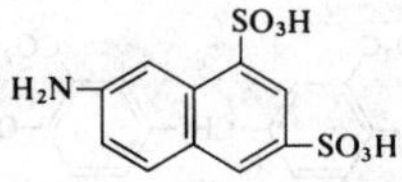

n284

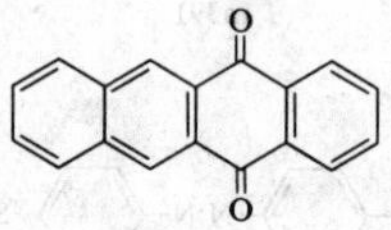

n6

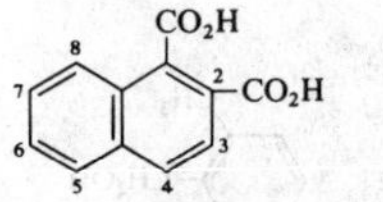

n263

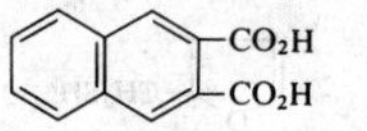

n274

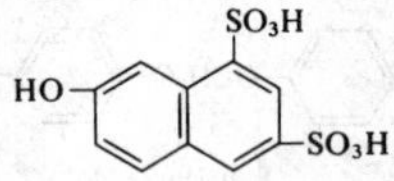

n287

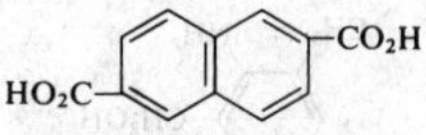

n275

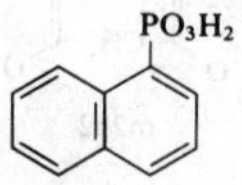

n289

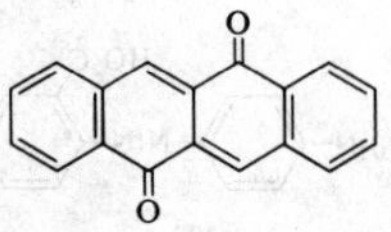

n7

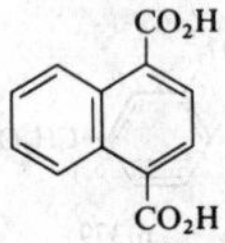

n264

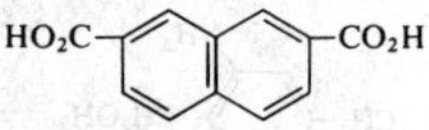

n276

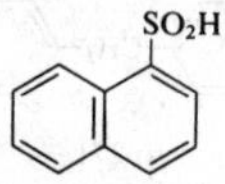

n291

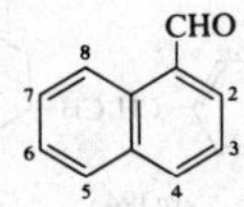

n8

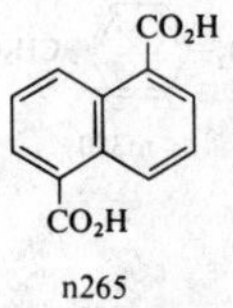

n265

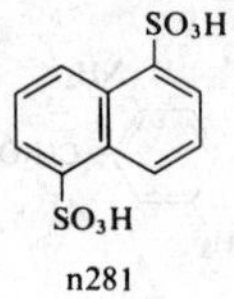

n281

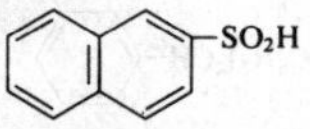

n292

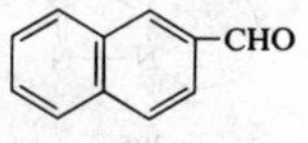

n9

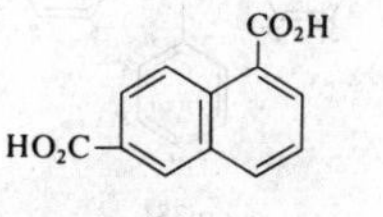

n266

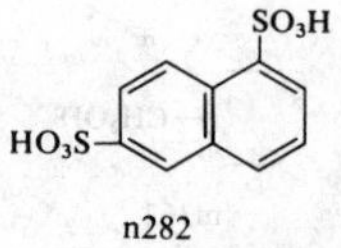

n282

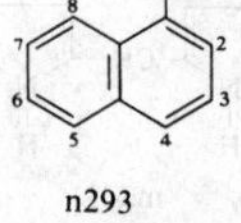

n293

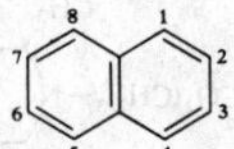

n14

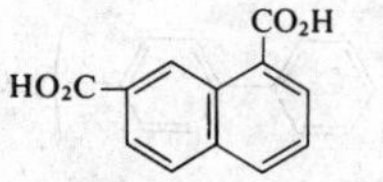

n267

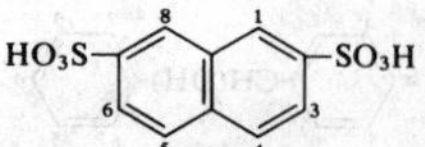

n283

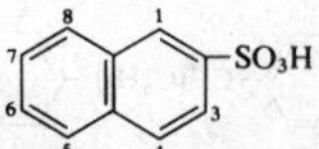

n295

STRUCTURAL FORMULAS OF ORGANIC COMPOUNDS (Continued)

In Alpha Numeric as they occur in Organic Compounds Table

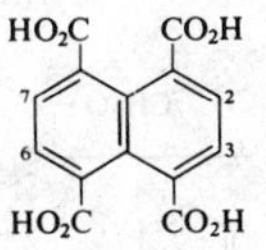

n310

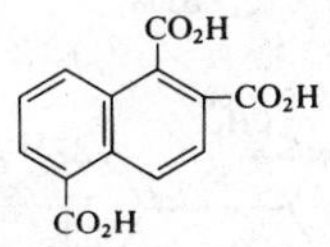

n312

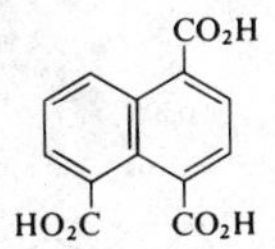

n313

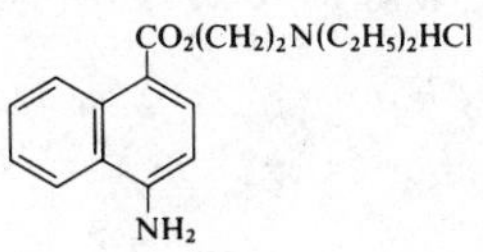

n314

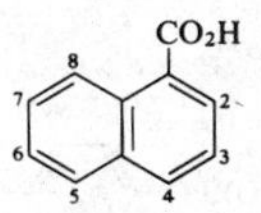

n315

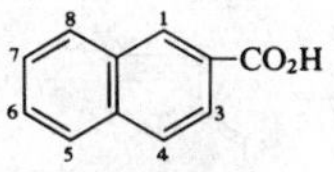

n324

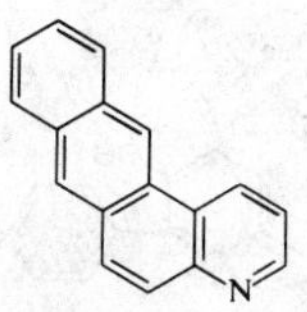

n387

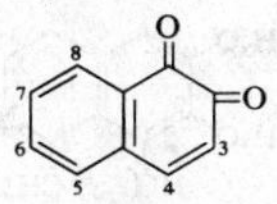

n388

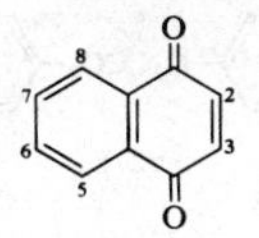

n390

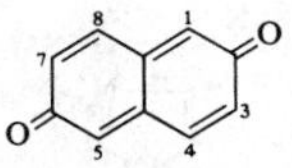

n392

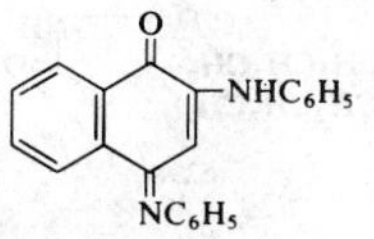

n393

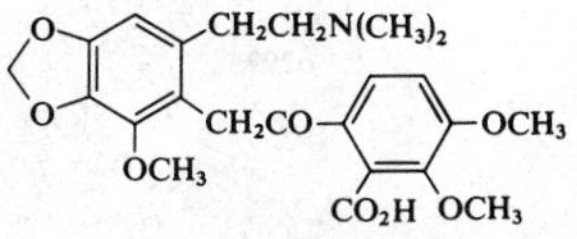

n445

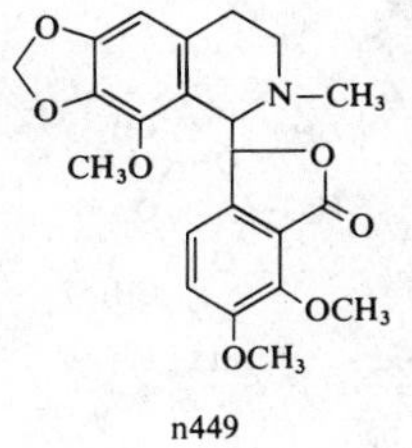

n449

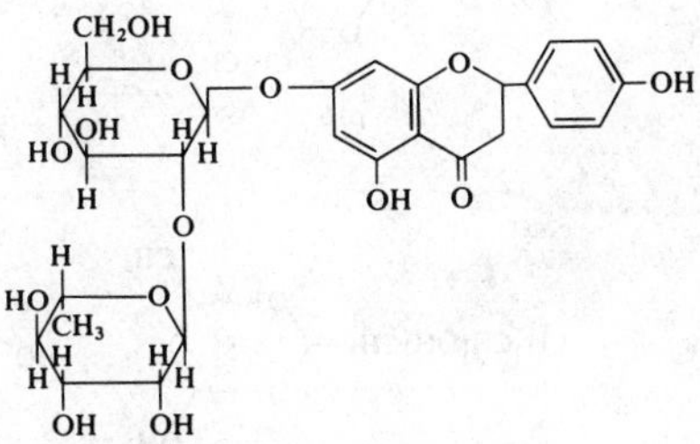

n455

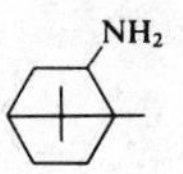

n456

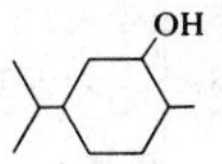

n457

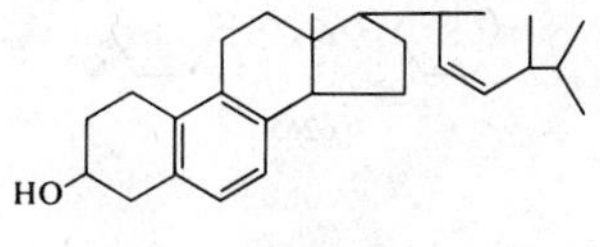

n458[1]

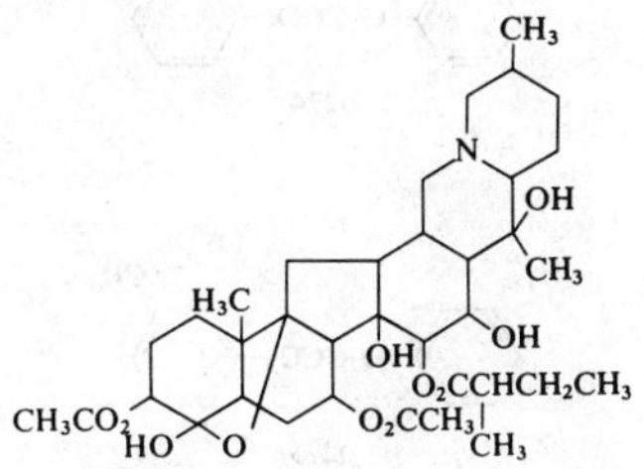

n459

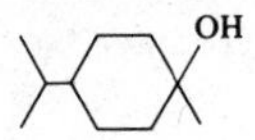

n460

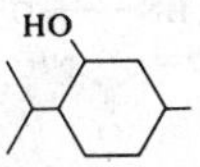

n461

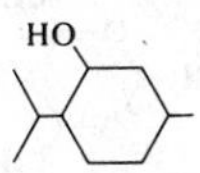

n463

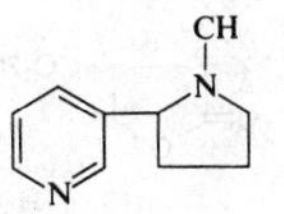

n471

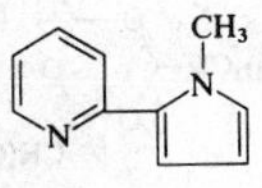

n475

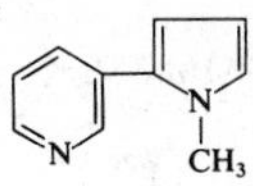

n477

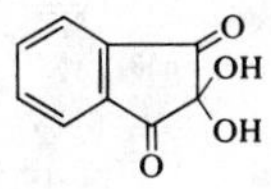

n478

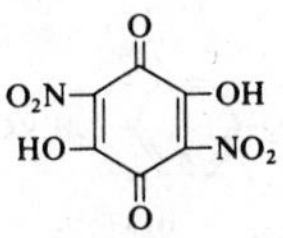

n479

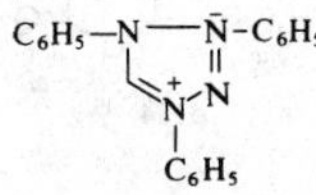

n489

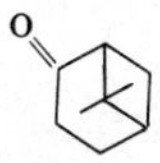

n567

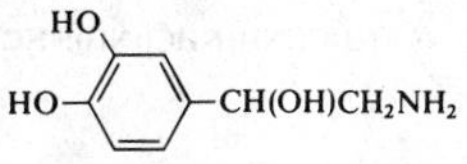

n568

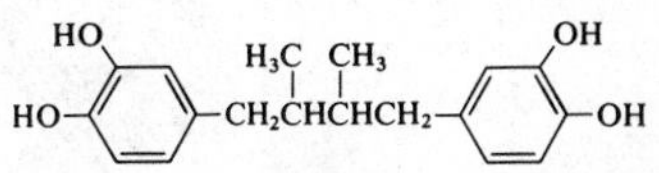

n569

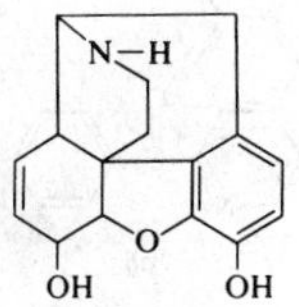

n573

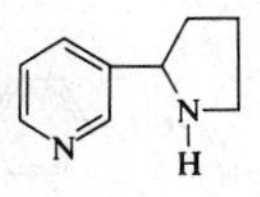

n574

In Alpha Numeric as they occur in Organic Compounds Table

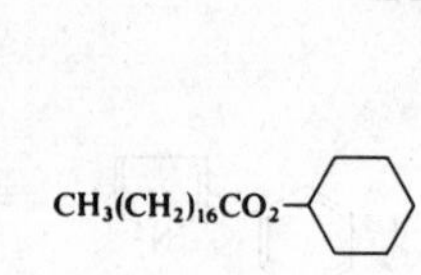

o28

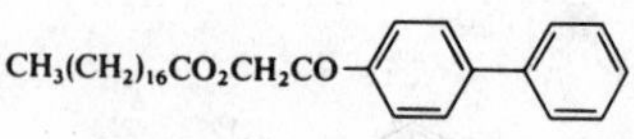

o38

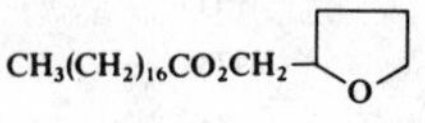

o40

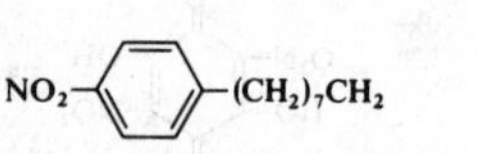

o139

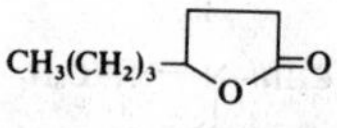

o184

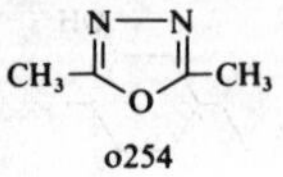

o254

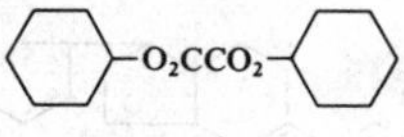

o265

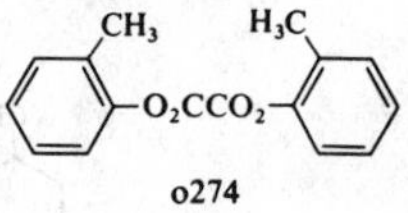

o274

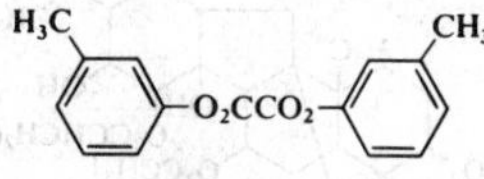

o275

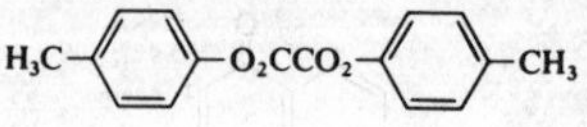

o276

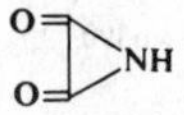

o277

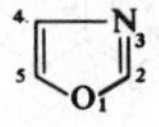

o292

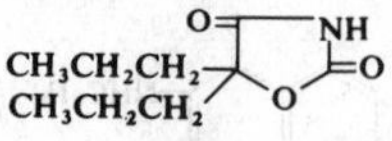

o298

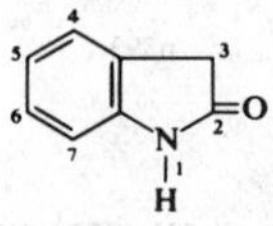

o299

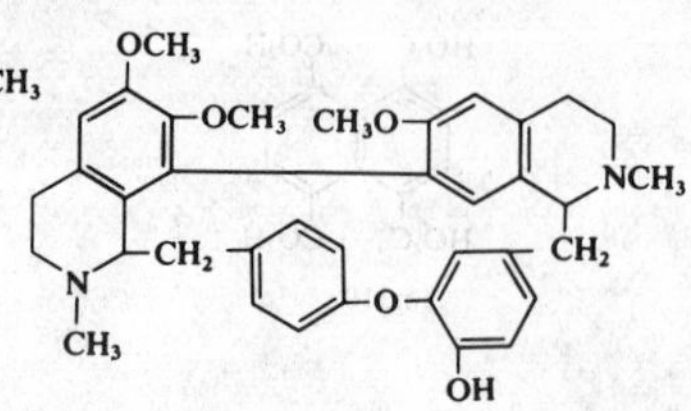

o306

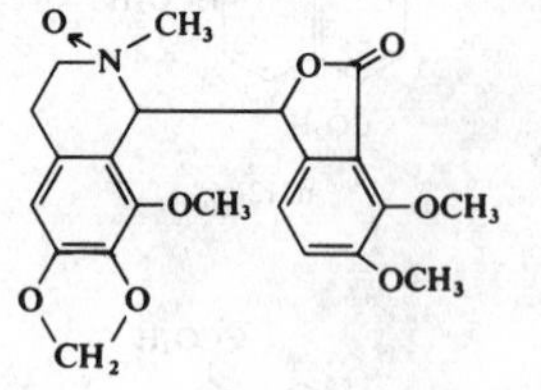

o309

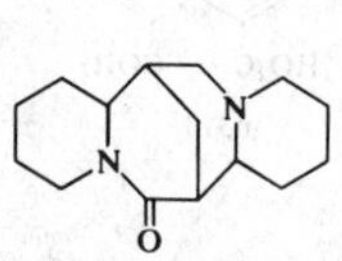

o311

p1

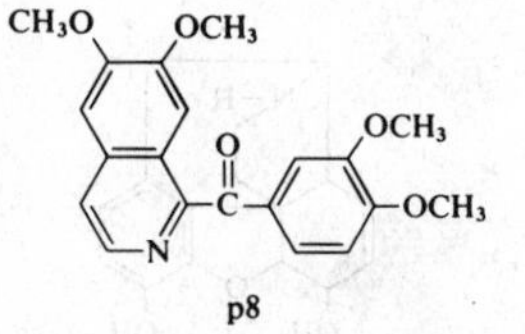

p8

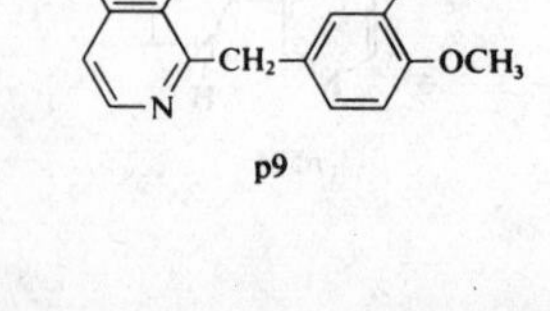

p9

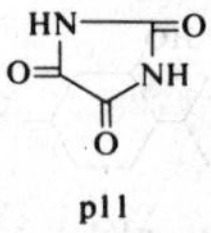

p11

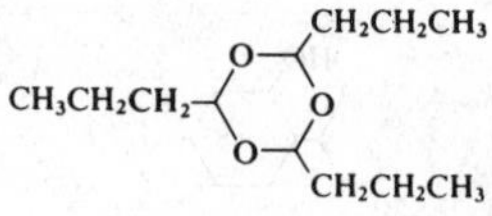

p12

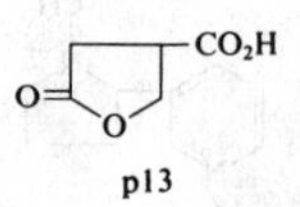

p13

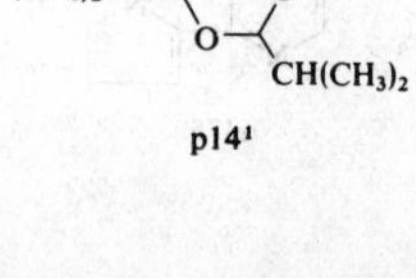

p14[1]

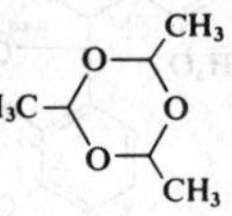

p15

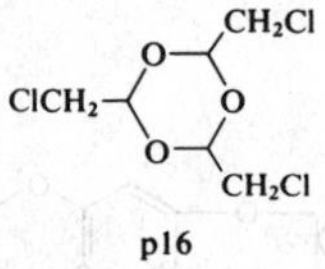

p16

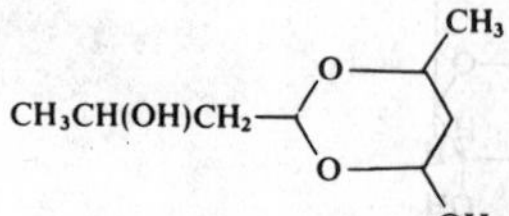

p17

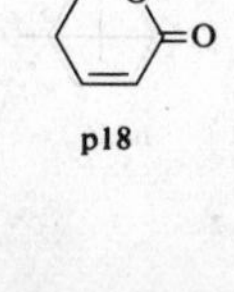

p18

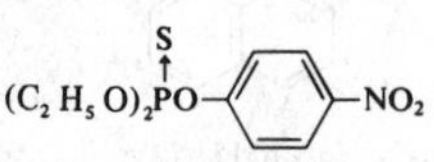

p19

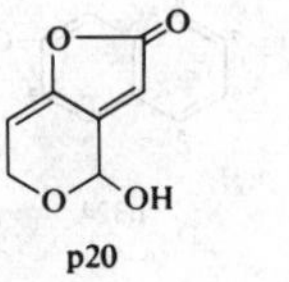

p20

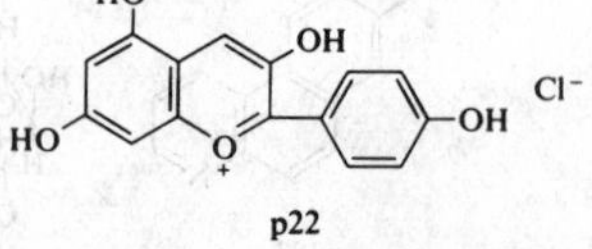

p22

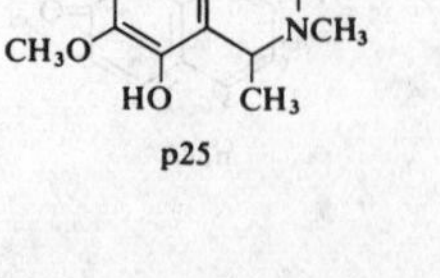

p25

In Alpha Numeric as they occur in Organic Compounds Table

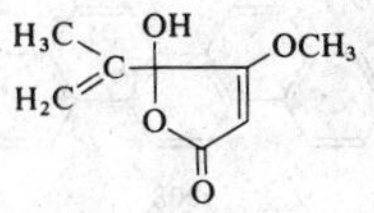

p26

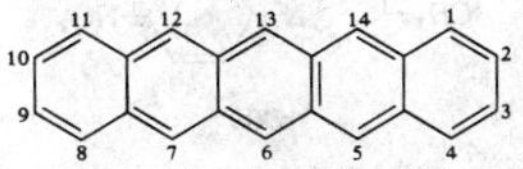

p27

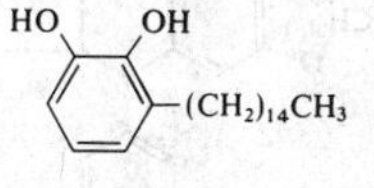

p36

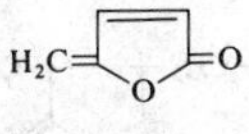

p54

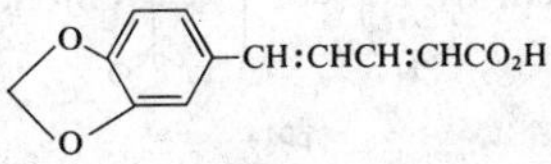

p55

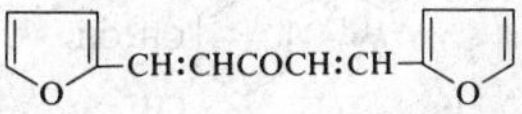

p71[1]

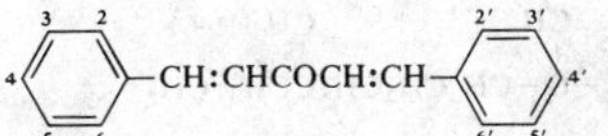

p72

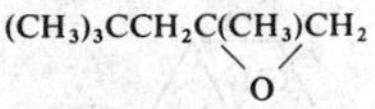

p139

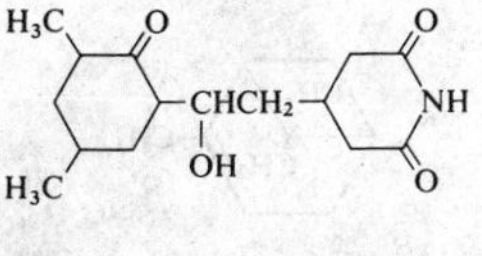

p168

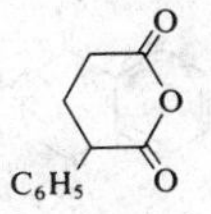

p182

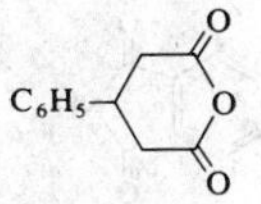

p183

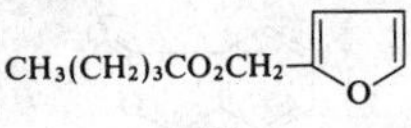

p223

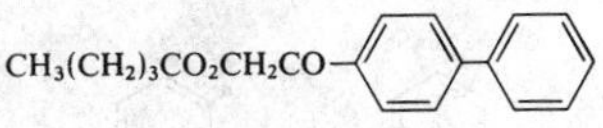

p232

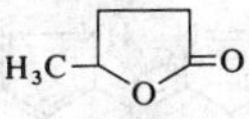

p267

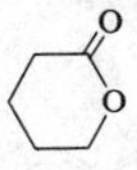

p270

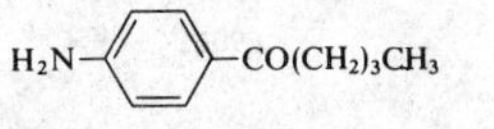

p350

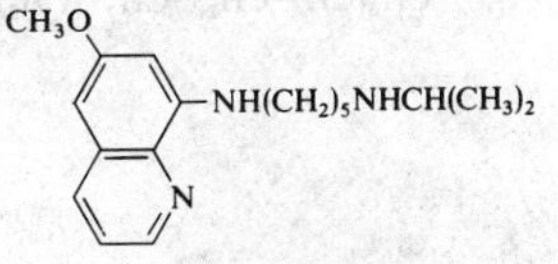

p370

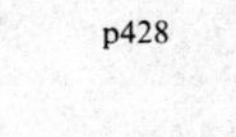

p428

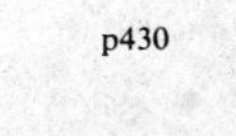

p430

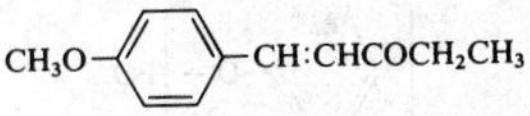

p439

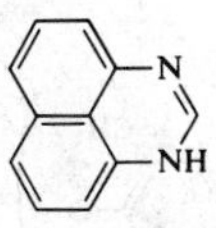

p466

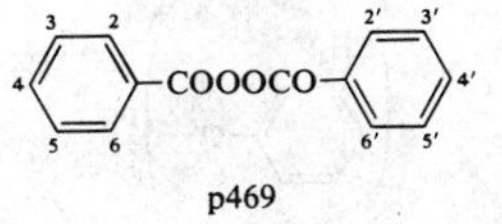

p469

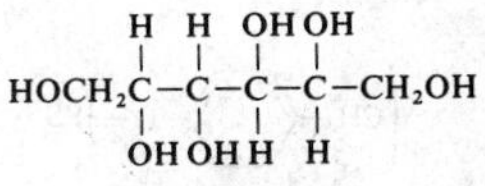

p478

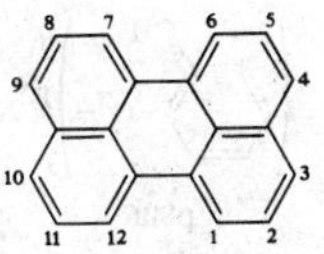

p479

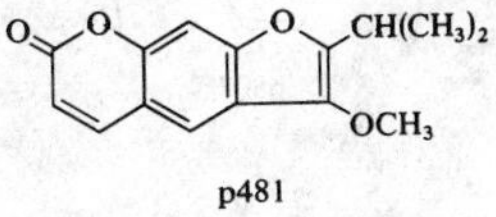

p481

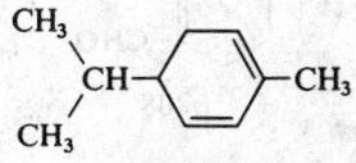

p482

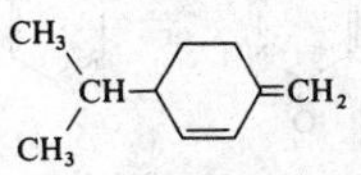

p483

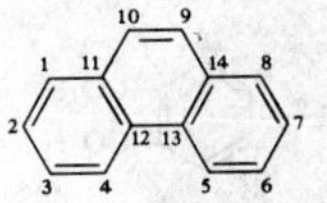

p484

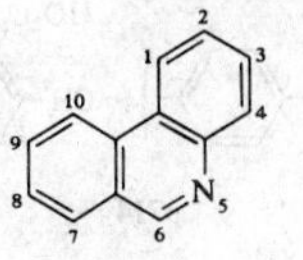

p526

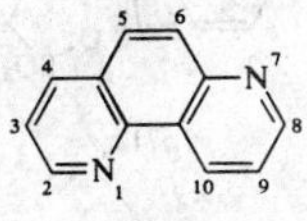

p528

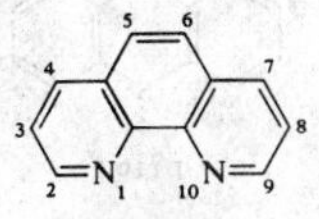

p529

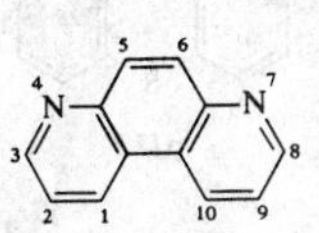

p530

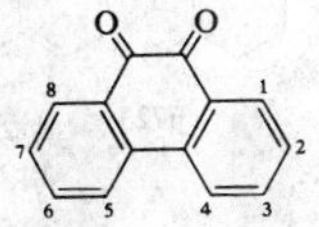

p533

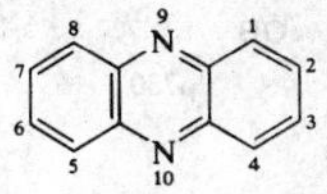

p550

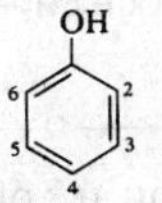

p555

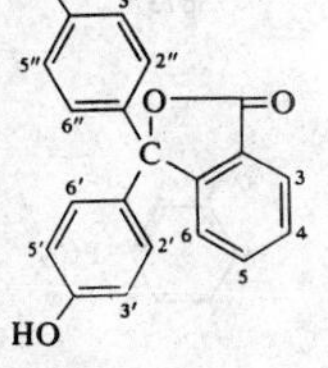

p710

In Alpha Numeric as they occur in Organic Compounds Table

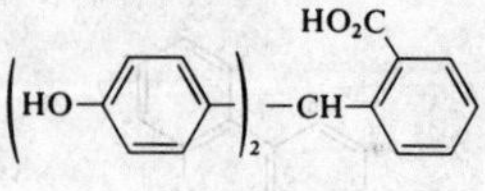

p714

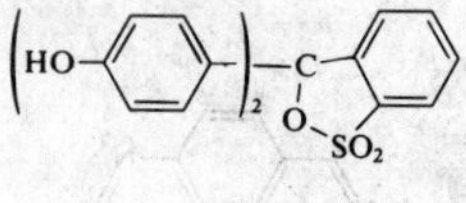

p715

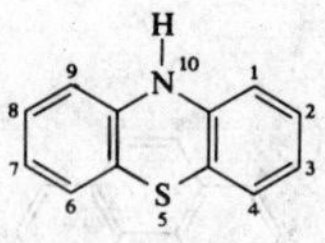

p716

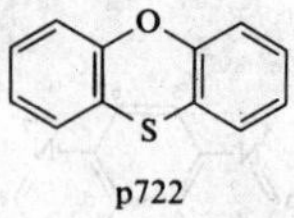

p722

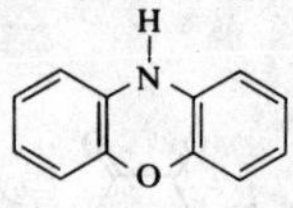

p723

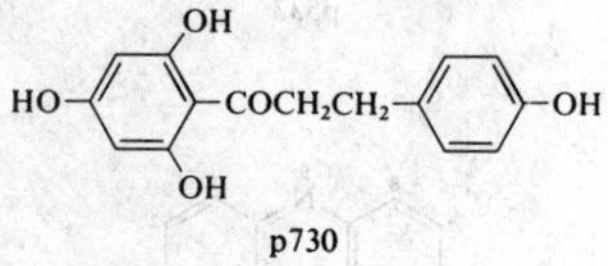

p730

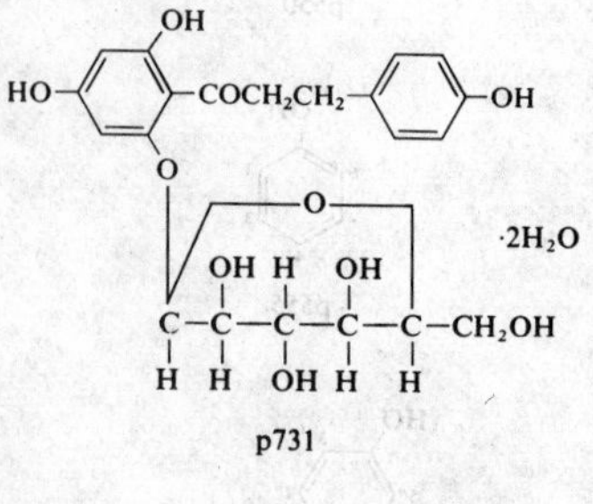

p731

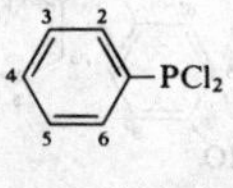

p741

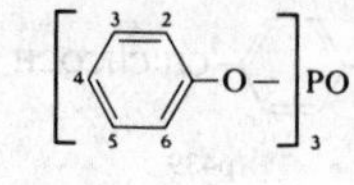

p781

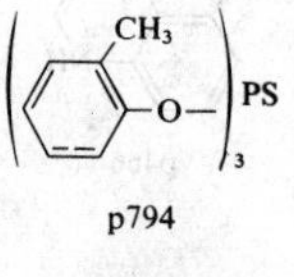

p794

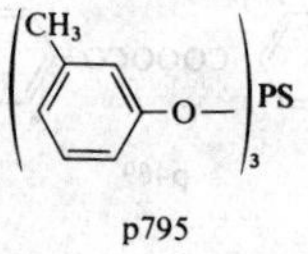

p795

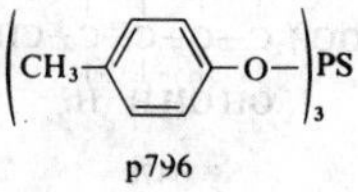

p796

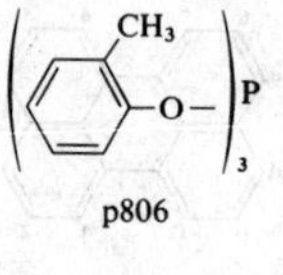

p806

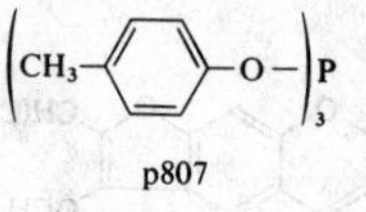

p807

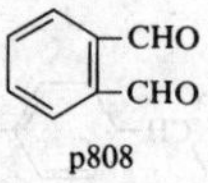

p808

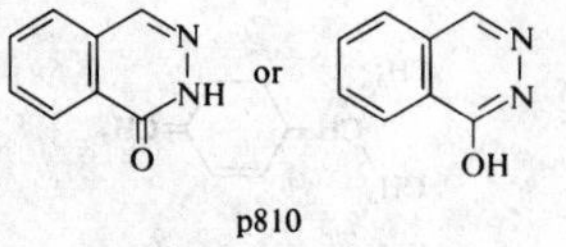

p810

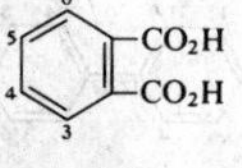

p812

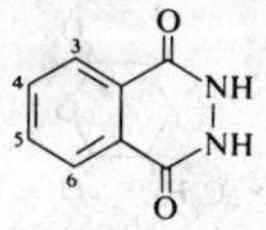

p829

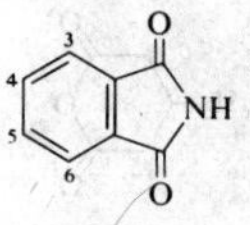

p830

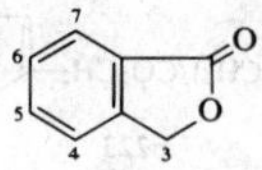

p897

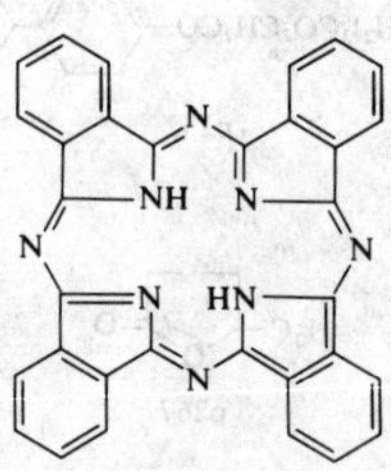

p901

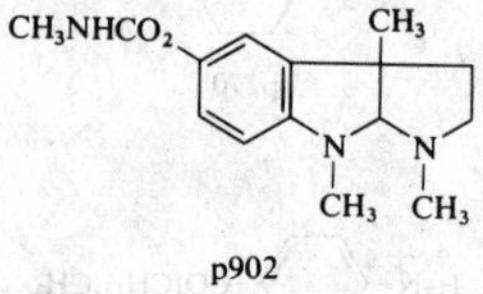

p902

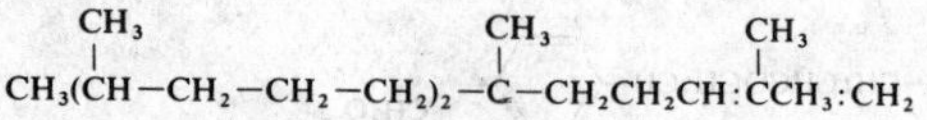

p905

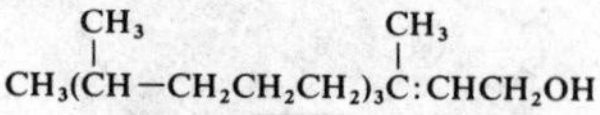

p906

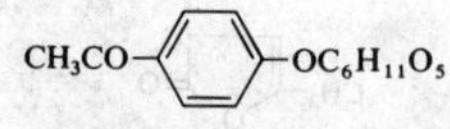

p907

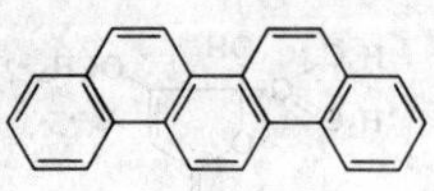

p908

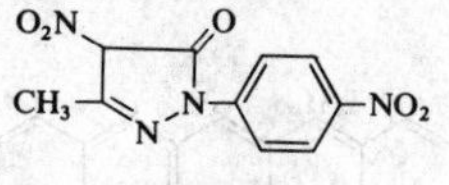

p909

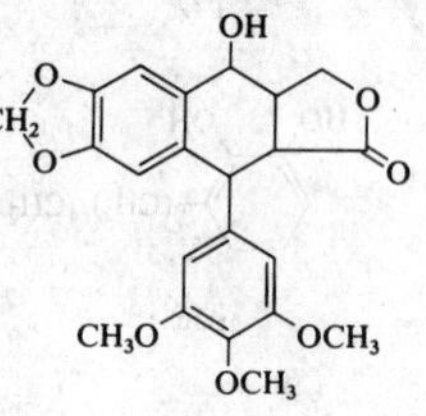

p910

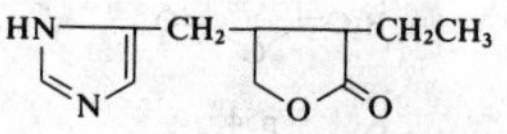

p912

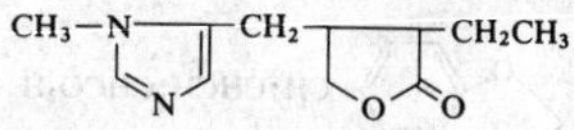

p914

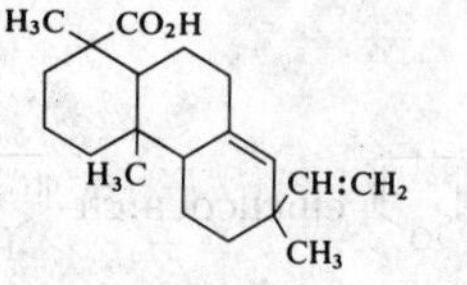

p919

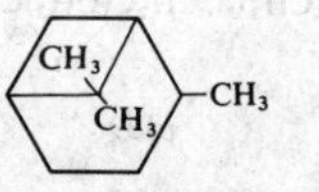

p920

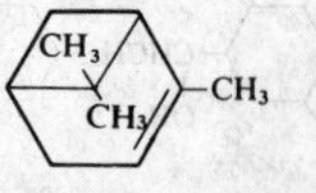

p923

STRUCTURAL FORMULAS OF ORGANIC COMPOUNDS (Continued)

In Alpha Numeric as they occur in Organic Compounds Table

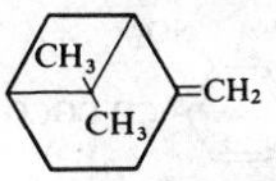
p924

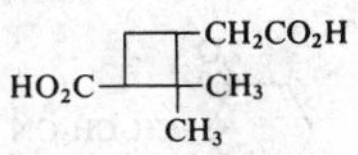
p926

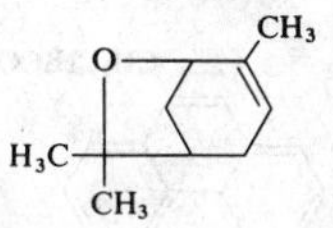
p930

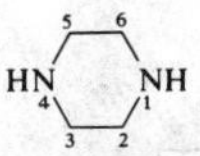
p932

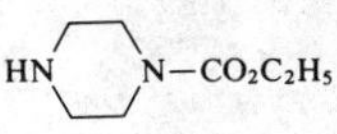
p949

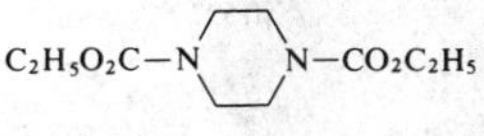
p950

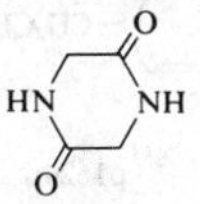
p951

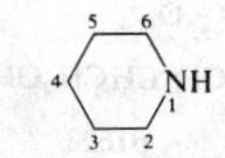
p952

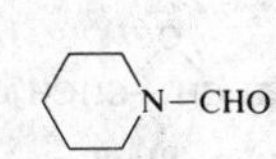
p1028

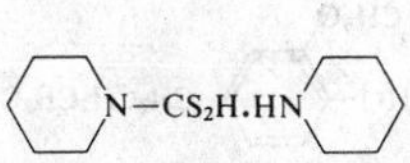
p1029

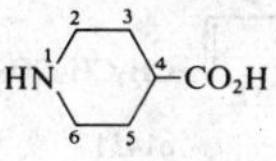
p1030

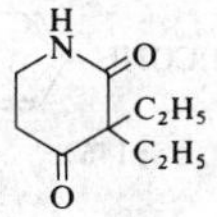
p1034[1]

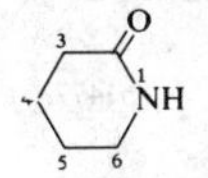
p1035

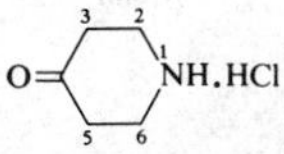
p1036

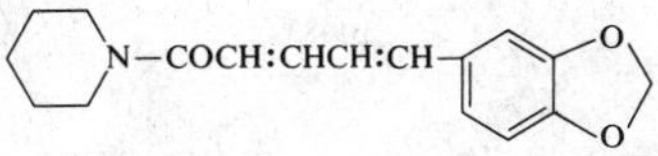
p1040

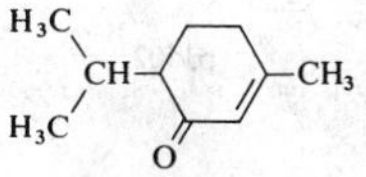
p1042

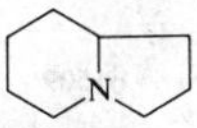
p1045

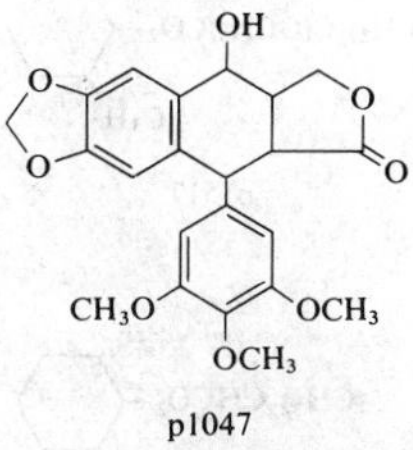
p1047

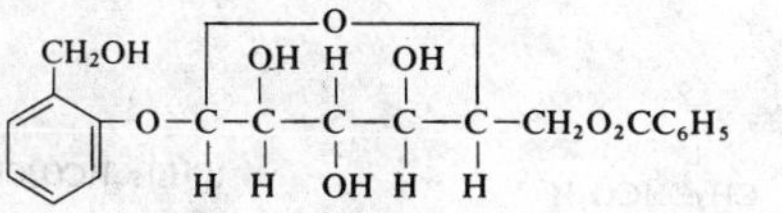
p1049

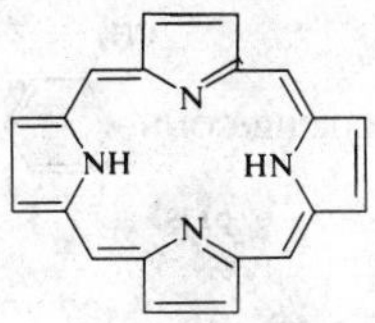
p1050

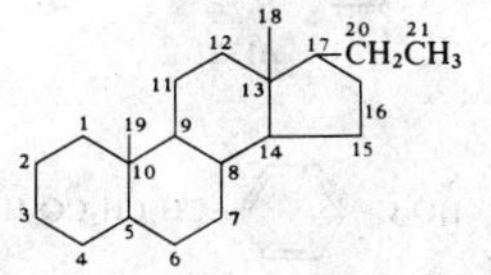
p1051

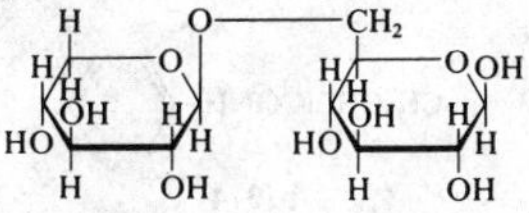
p1064

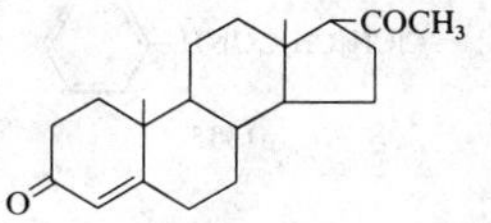
p1065

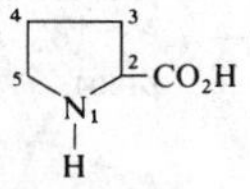
p1068

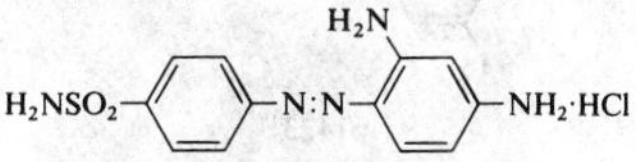
p1078

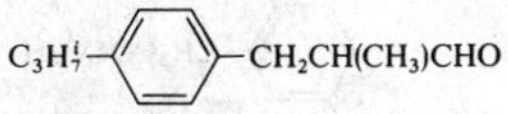
p1098

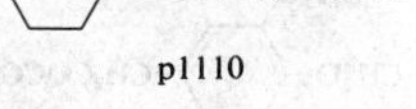
p1110

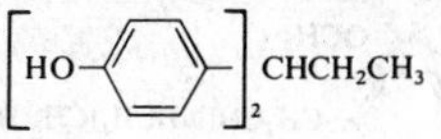
p1117

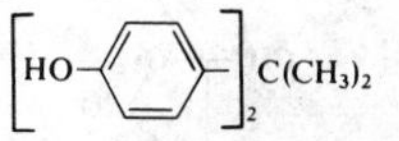
p1118

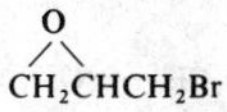
p1126

p1136

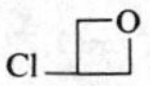
p1138

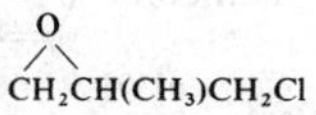
p1139

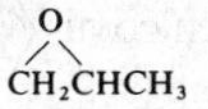
p1180

p1181

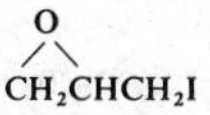
p1182

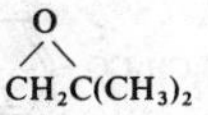
p1183

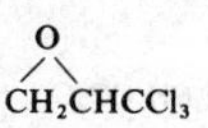
p1184

STRUCTURAL FORMULAS OF ORGANIC COMPOUNDS (Continued)

In Alpha Numeric as they occur in Organic Compounds Table

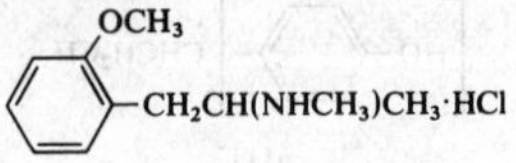
p1199

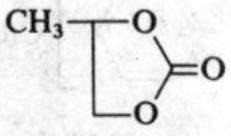
p1235

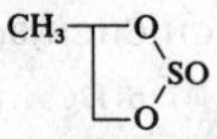
p1243

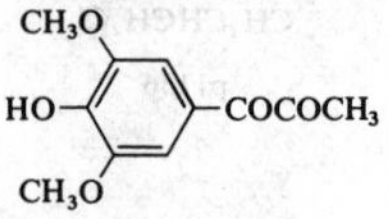
p1272

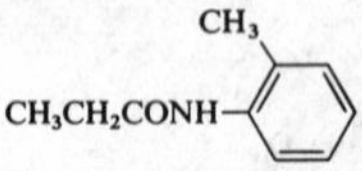
p1308

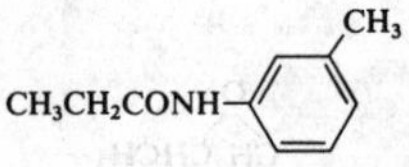
p1309

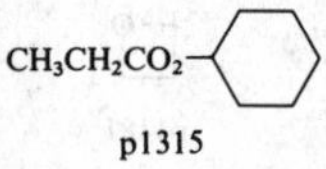
p1315

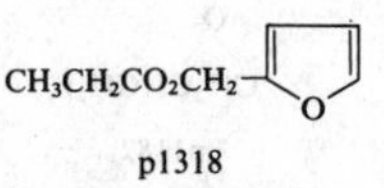
p1318

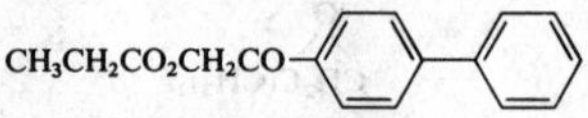
p1330

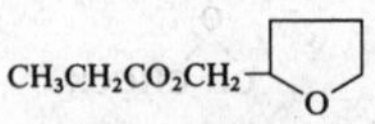
p1333

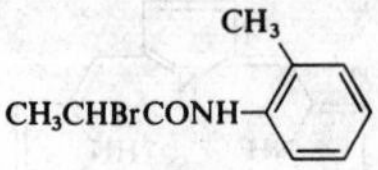
p1345

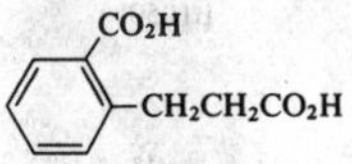
p1371

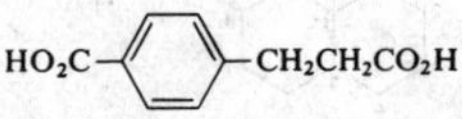
p1372

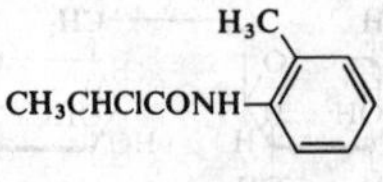
p1374

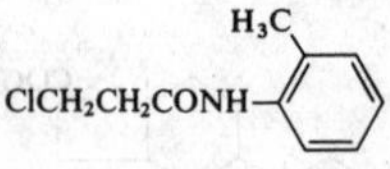
p1385

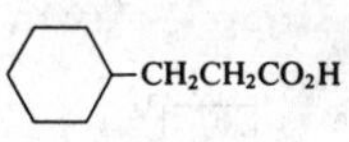
p1404

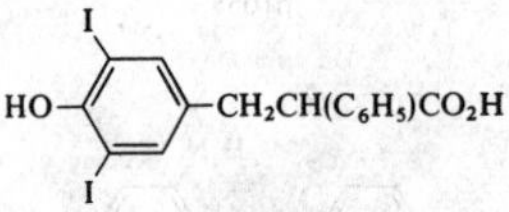
p1423[1]

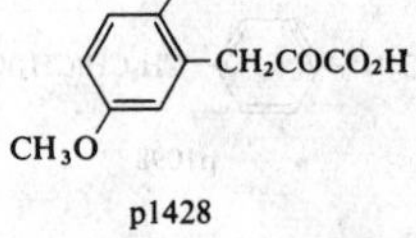
p1428

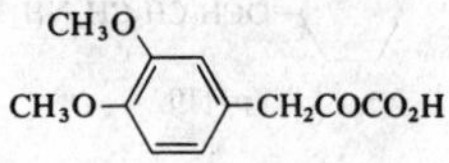
p1429

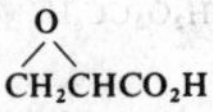
p1446

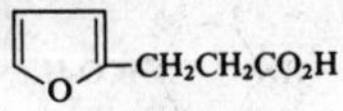
p1451[1]

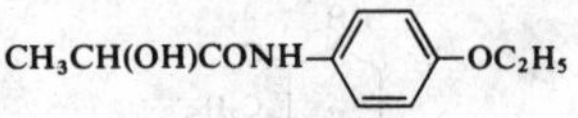
p1463

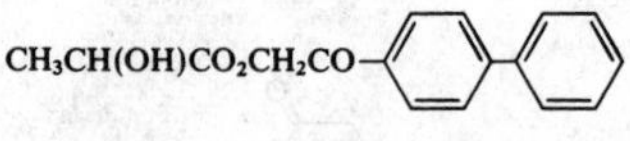
p1477

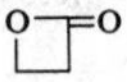
p1480

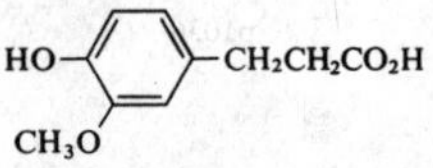
p1483

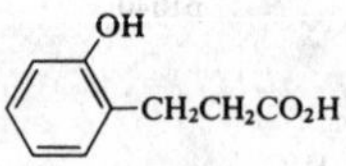
p1502[1]

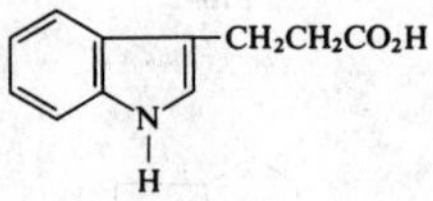
p1509

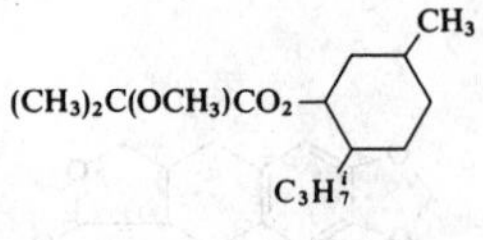
p1517

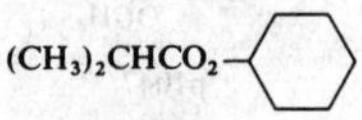
p1526

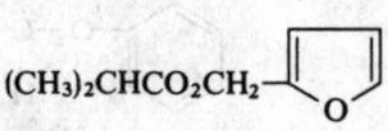
p1528

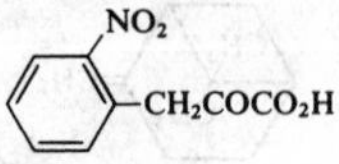
p1541

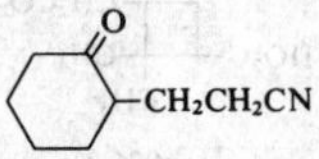
p1545[1]

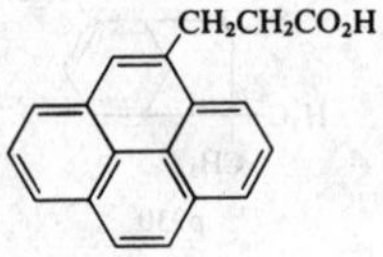
p1579

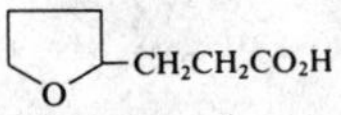
p1581

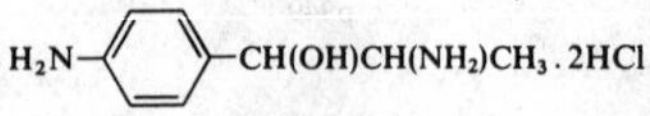
p1593

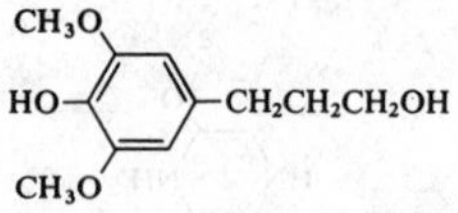
p1624

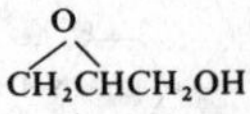
p1628

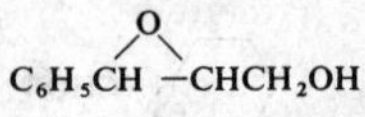
P1629

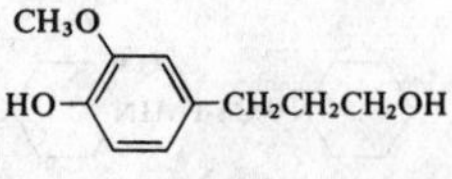
p1632

STRUCTURAL FORMULAS OF ORGANIC COMPOUNDS (Continued)

In Alpha Numeric as they occur in Organic Compounds Table

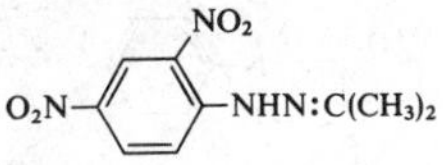

p1652

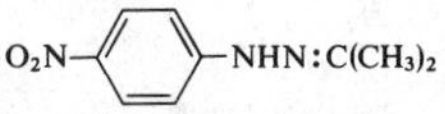

p1653

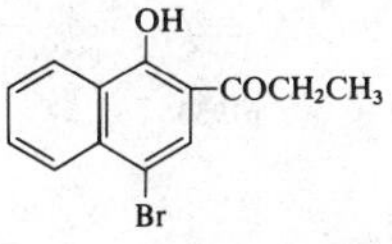

p1664

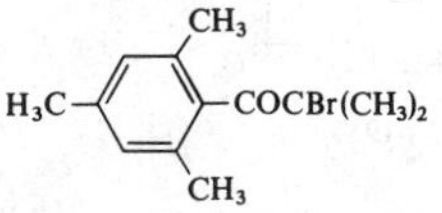

p1665

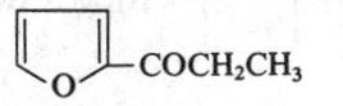

p1683

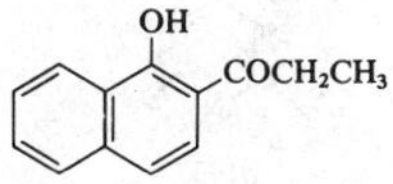

p1686

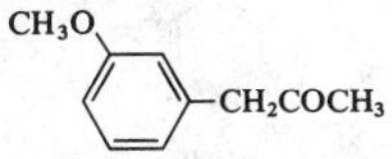

p1691

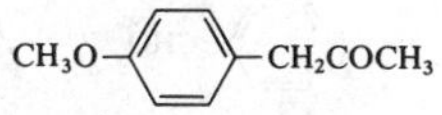

p1692

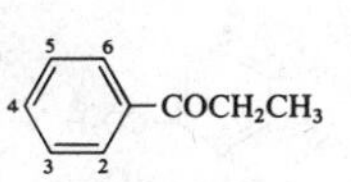

p1696

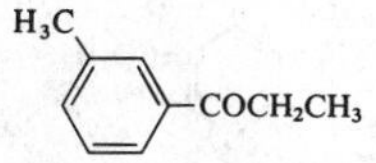

p1702

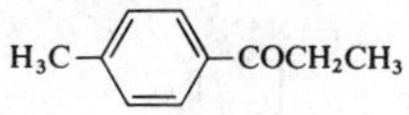

p1703

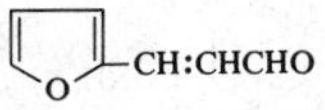

p1711

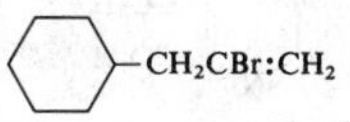

p1718

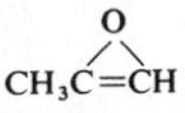

p1742

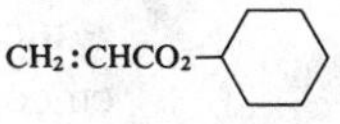

p1772

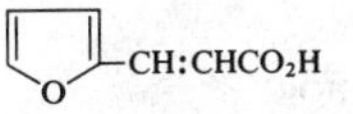

p1784

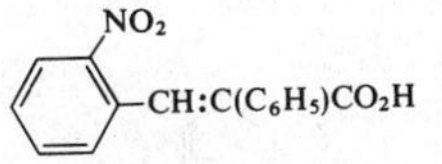

p1805

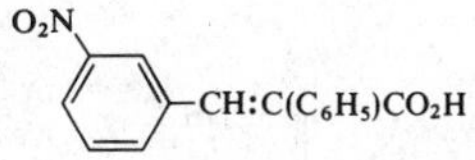

p1806

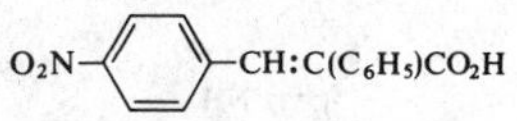

p1808

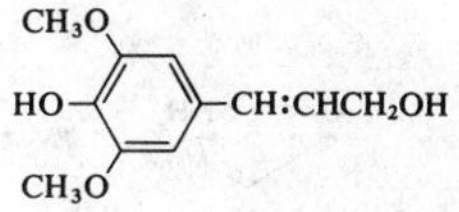

p1818

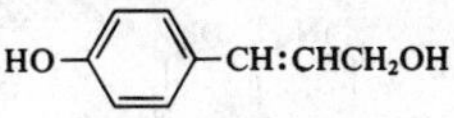

p1819

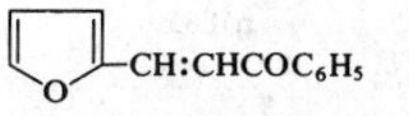

p1824

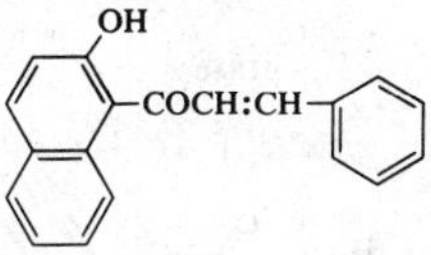

p1825

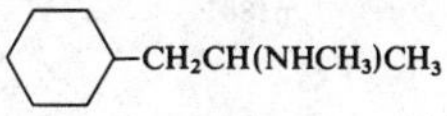

p1826

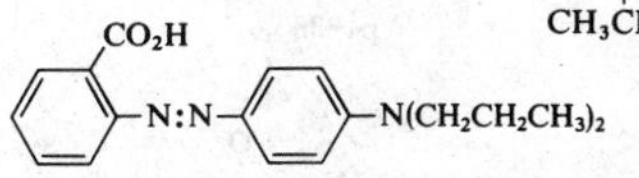

p1827

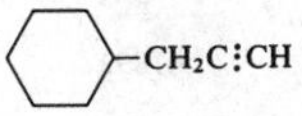

p1834

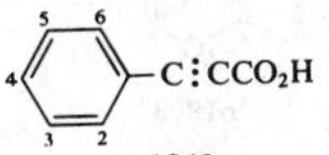

p1849

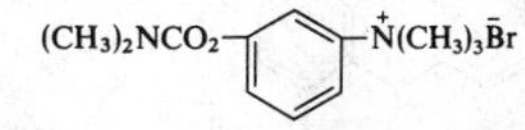

p1853

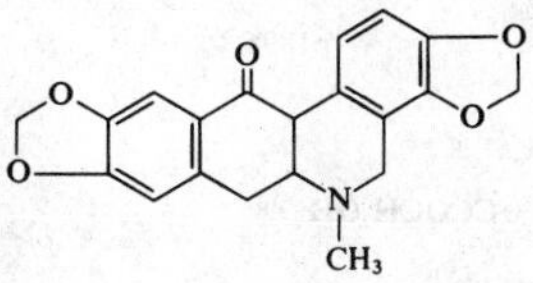

p1854

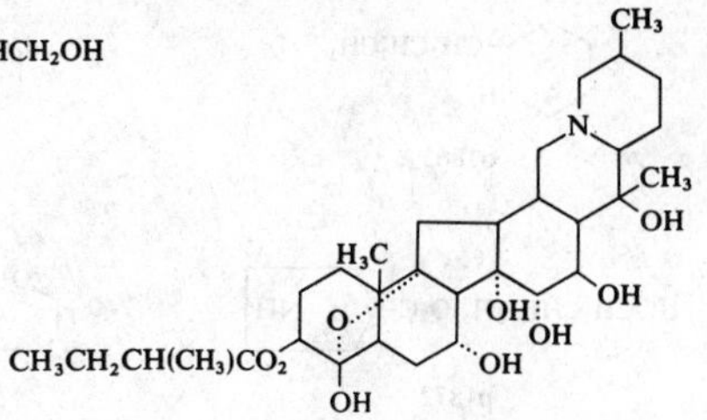

p1855

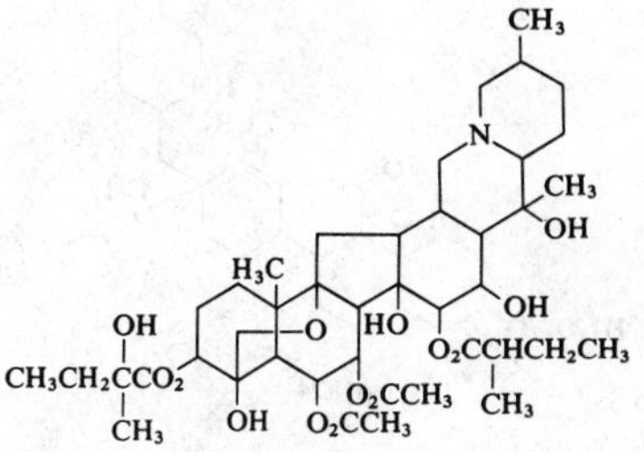

p1857

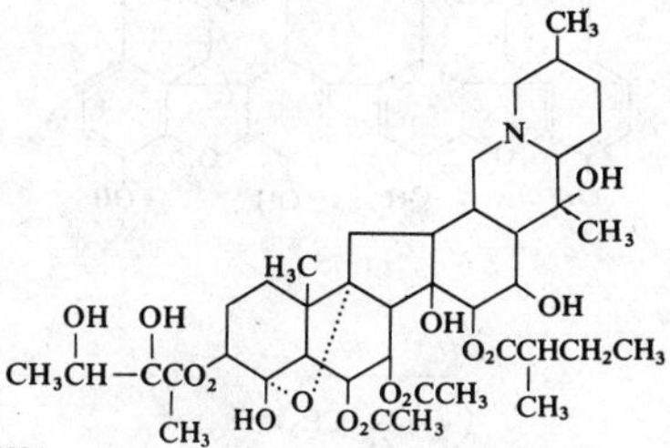

p1858

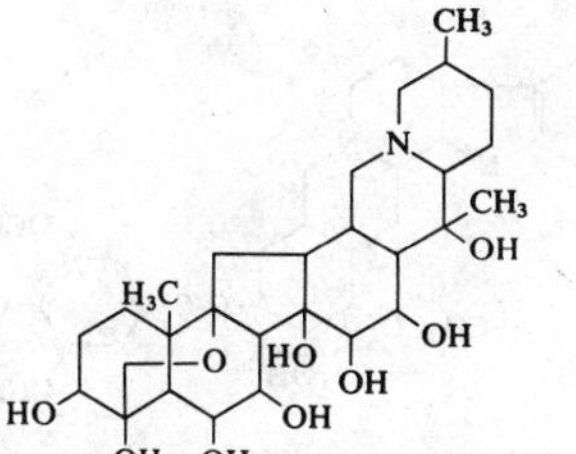

p1859

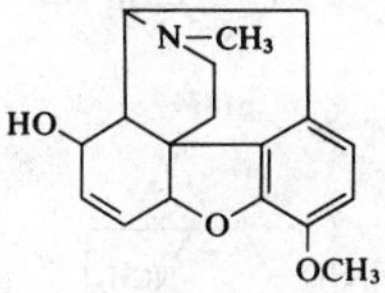

p1863

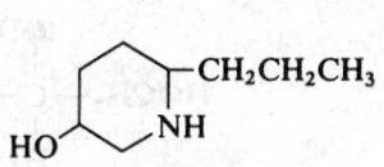

p1864

STRUCTURAL FORMULAS OF ORGANIC COMPOUNDS (Continued)

In Alpha Numeric as they occur in Organic Compounds Table

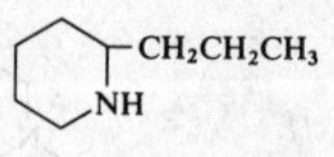

p1865

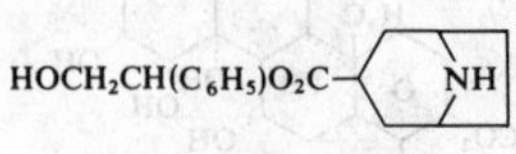

p1872

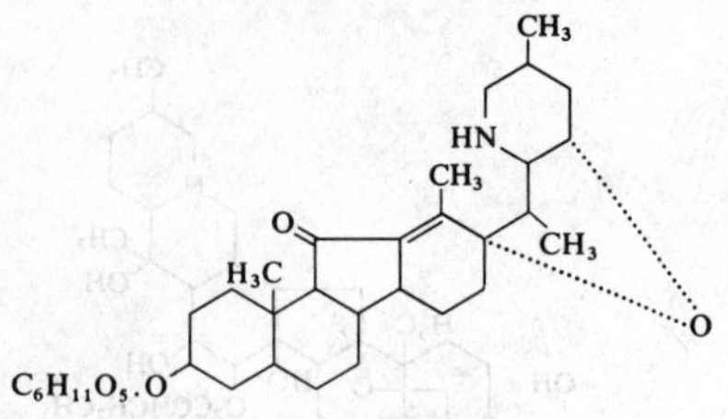

p1874

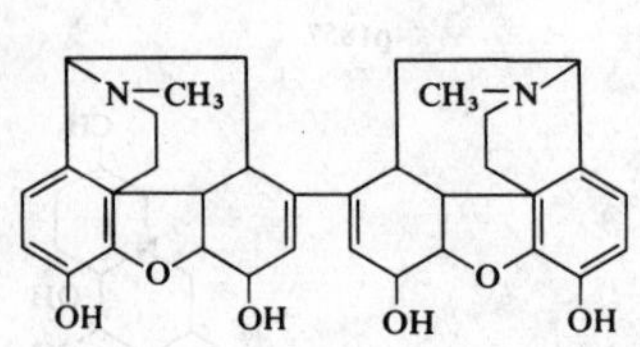

p1875

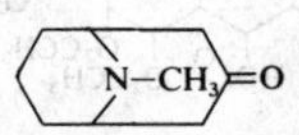

p1877

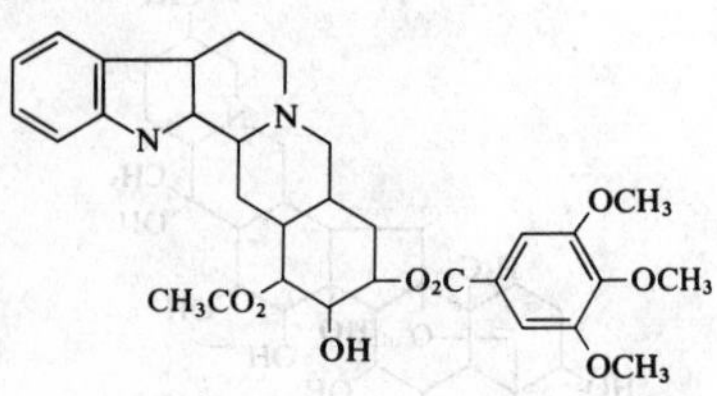

p1878

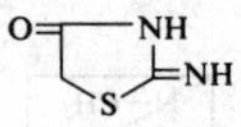

p1879

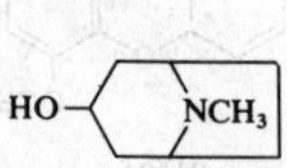

p1880

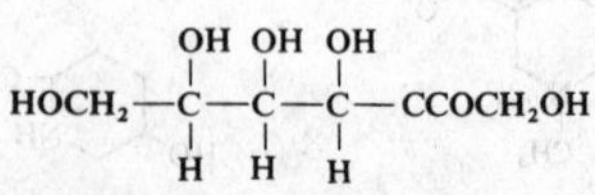

p1881

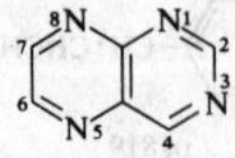

p1882

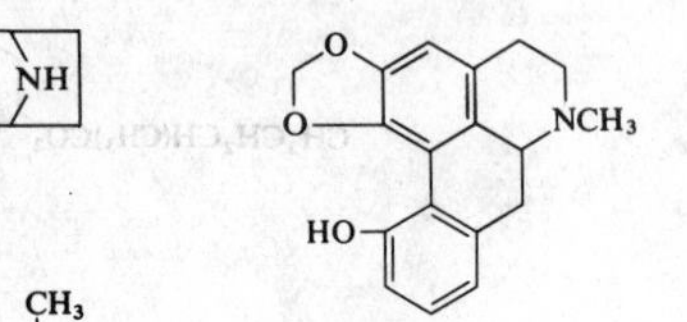

p1885

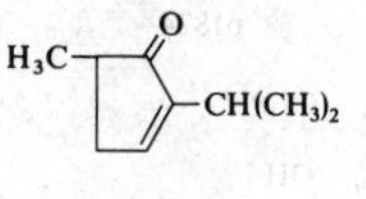

p1886

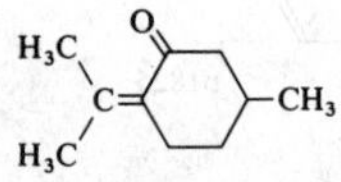

p1887

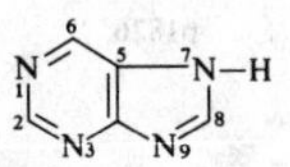

p1888

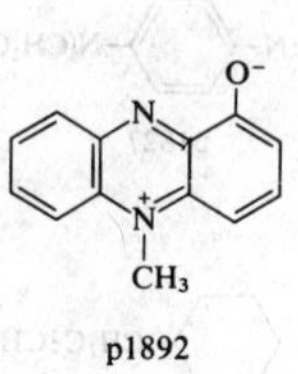

p1892

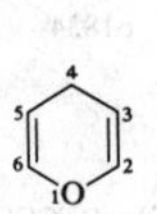

p1893[1]

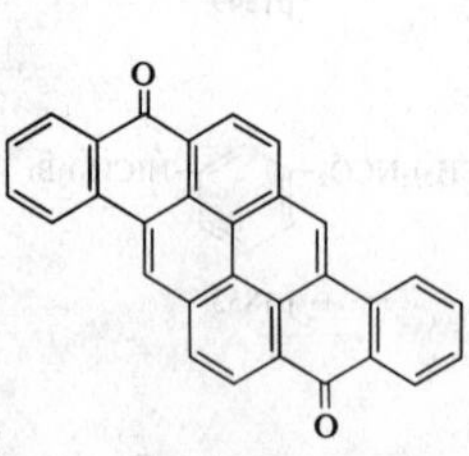

p1896

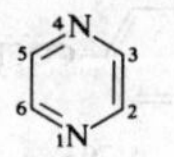

p1897

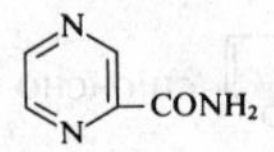

p1900

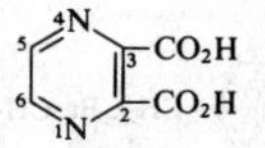

p1901

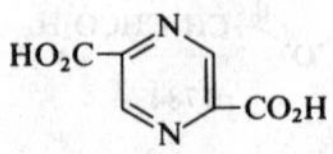

p1902

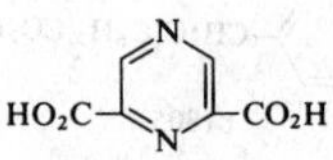

p1903

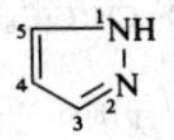

p1905

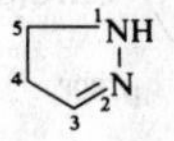

p1921

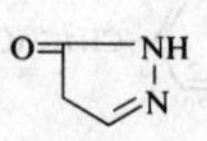

p1923

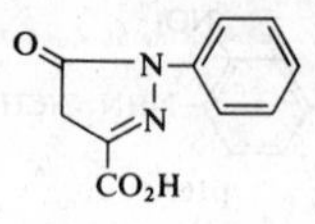

p1929

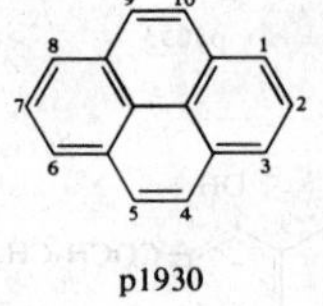

p1930

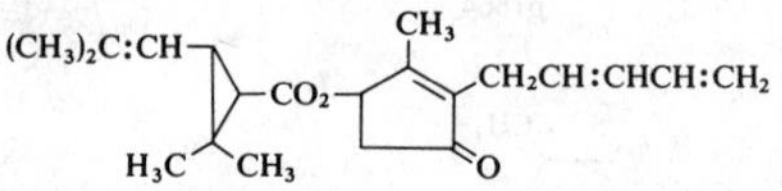

p1935

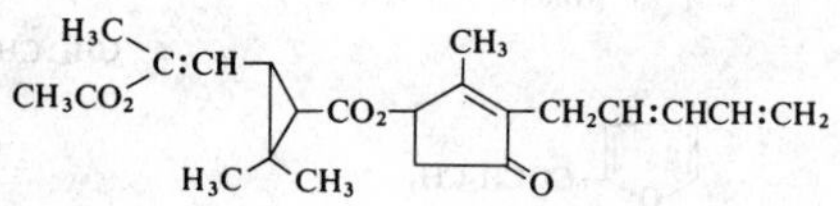

p1936

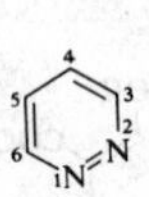

p1937

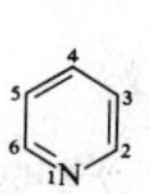

p1939

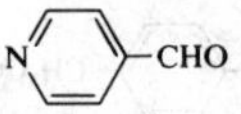

p2035

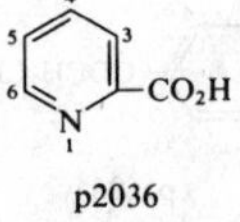

p2036

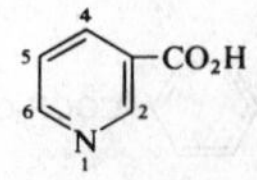

p2040

In Alpha Numeric as they occur in Organic Compounds Table

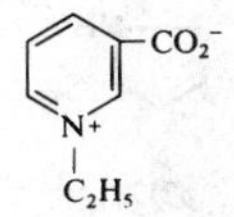

p2043

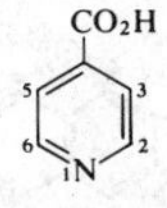

p2048

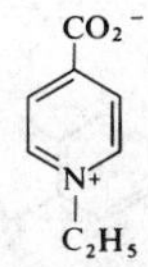

p2050

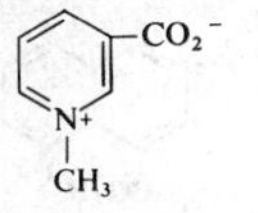

p2062

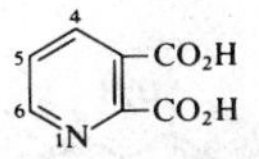

p2064

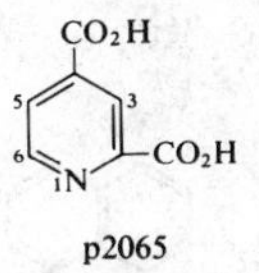

p2065

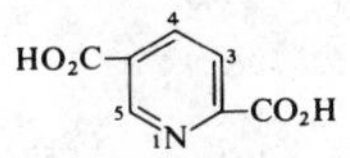

p2066

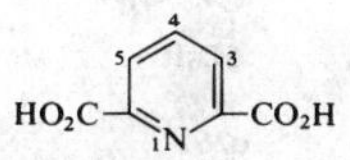

p2067

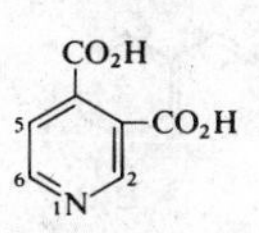

p2068

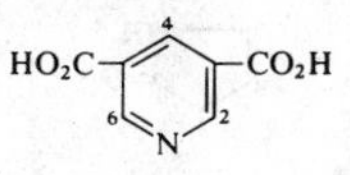

p2069

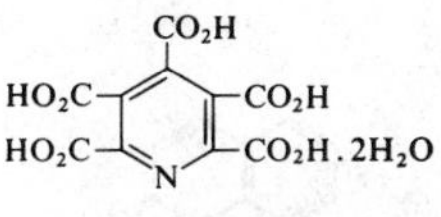

p2077

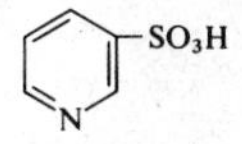

p2078

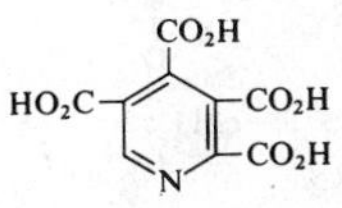

p2079

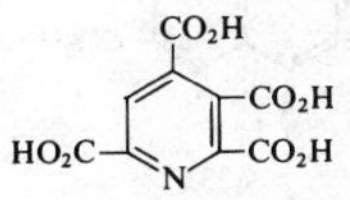

p2080

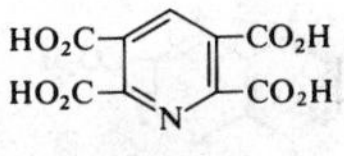

p2081

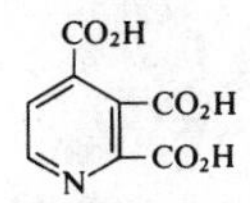

p2082

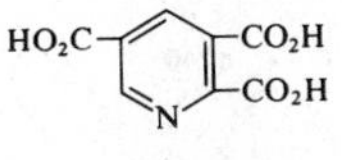

p2083

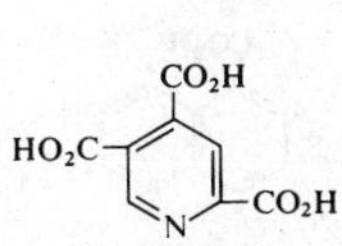

p2084

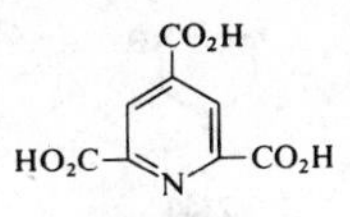

p2085

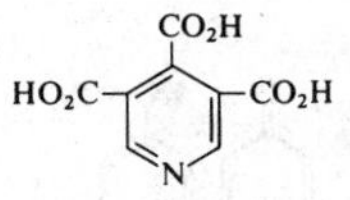

p2086

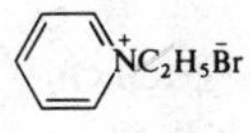

p2087

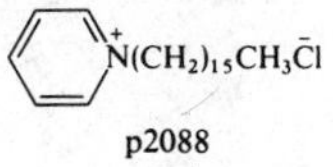

p2088

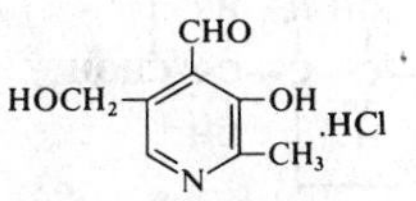

p2089

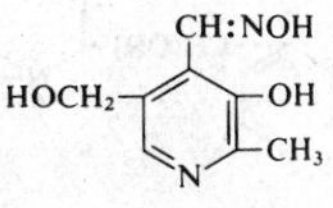

p2090

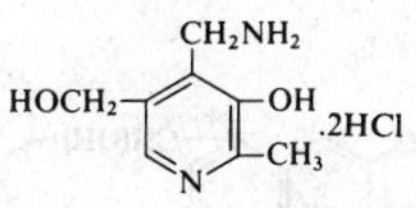

p2091

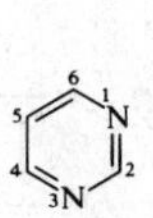

p2092

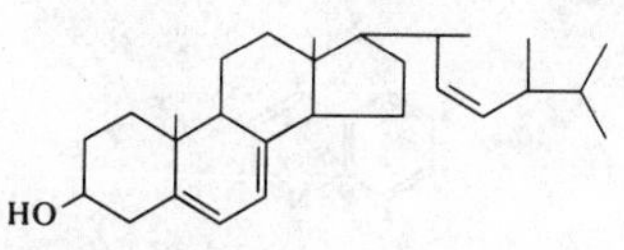

p2112

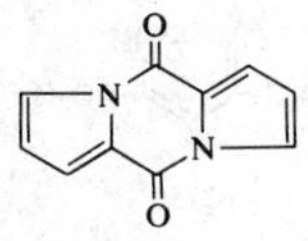

p2113

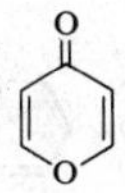

p2114

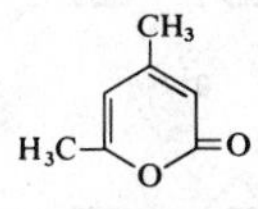

p2116

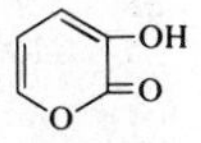

p2118

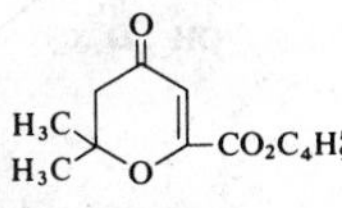

p2121

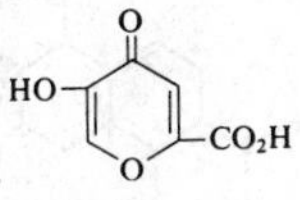

p2122

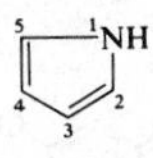

p2126

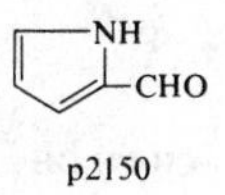

p2150

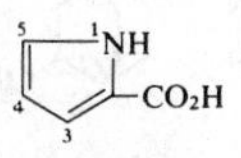

p2151

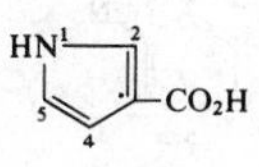

p2152[1]

STRUCTURAL FORMULAS OF ORGANIC COMPOUNDS (Continued)

In Alpha Numeric as they occur in Organic Compounds Table

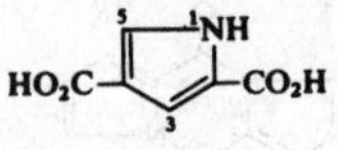

p2156[1]

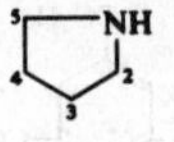

p2160

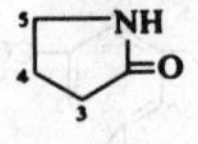

p2169

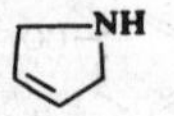

p2173

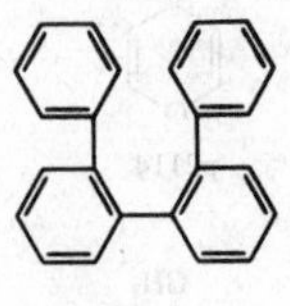

q1

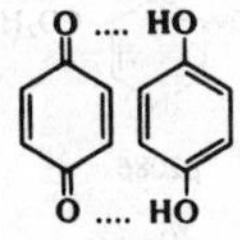

q14

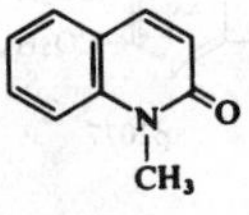

q123

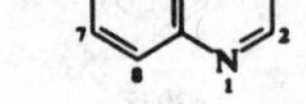

q244

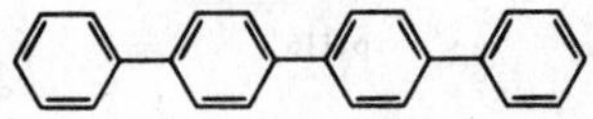

q2

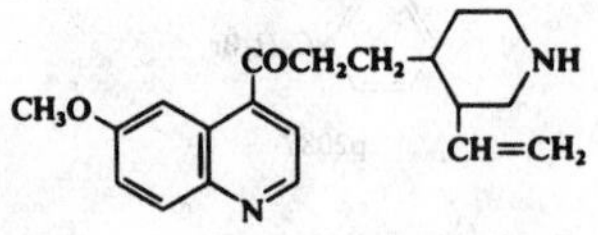

q15

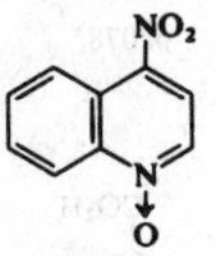

q211

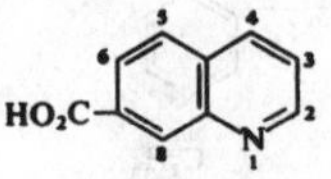

q245

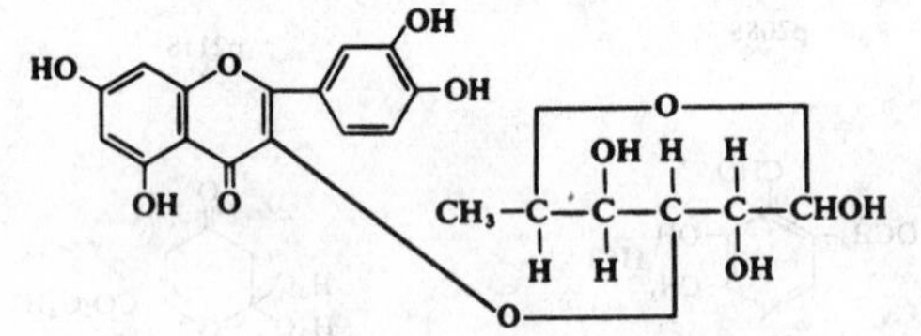

q3

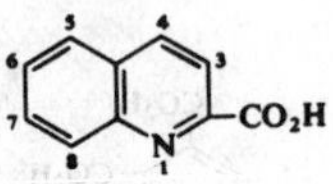

q233

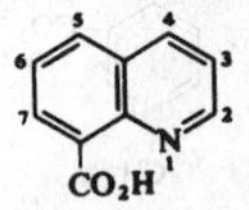

q246

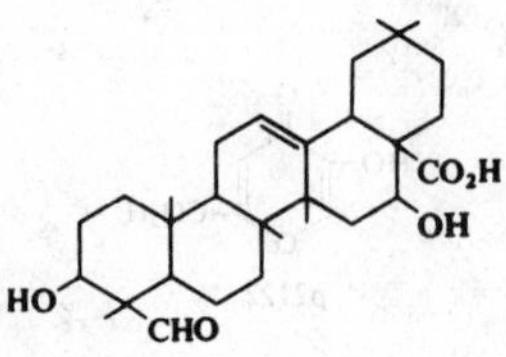

q4

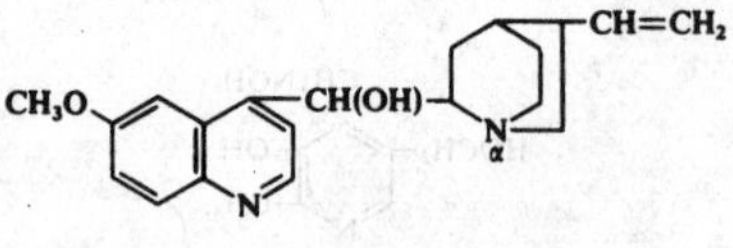

q17

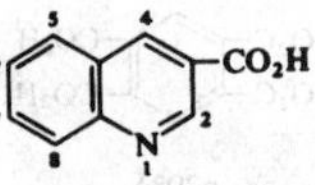

q238

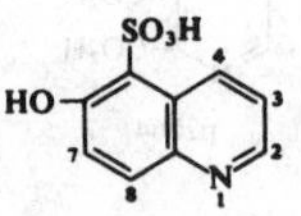

q257

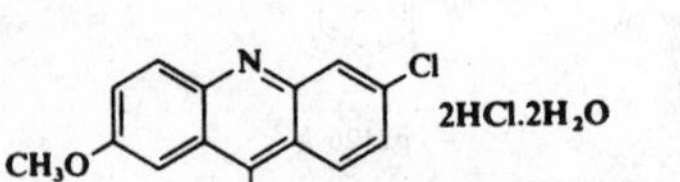

q5

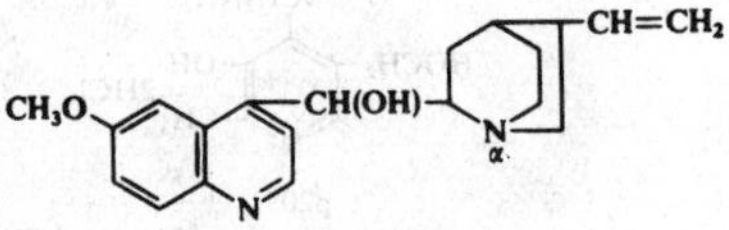

q22

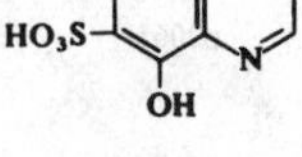

q259

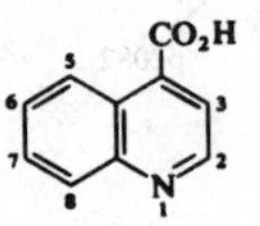

q240

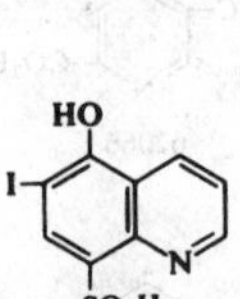

q261

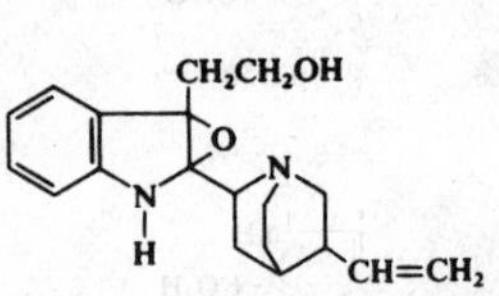

q6

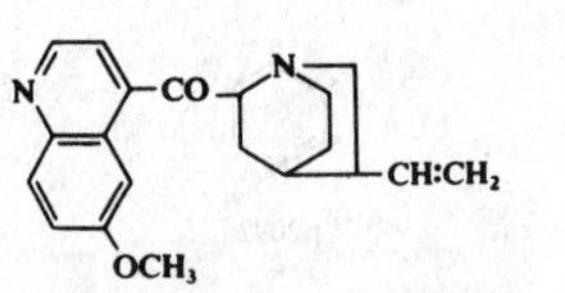

q33

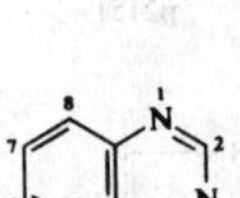

q7

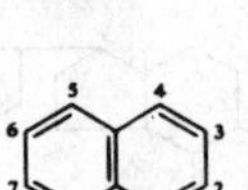

q34

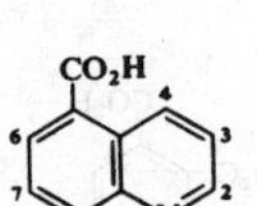

q242

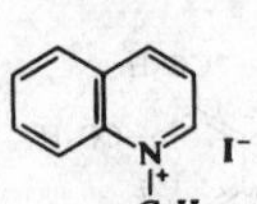

q262

STRUCTURAL FORMULAS OF ORGANIC COMPOUNDS (Continued)

In Alpha Numeric as they occur in Organic Compounds Table

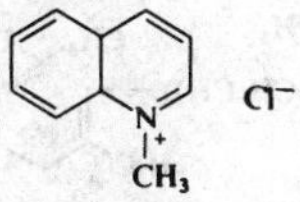

q263

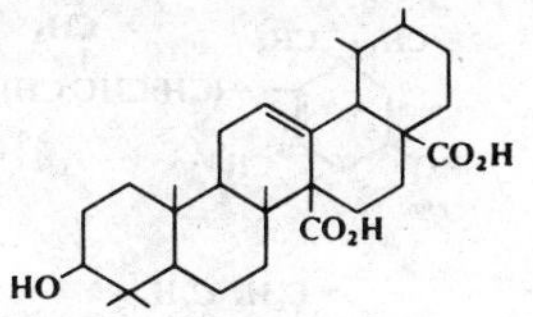

q264

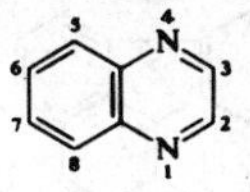

q265

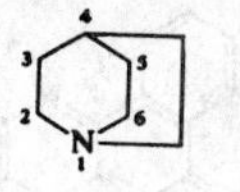

q279

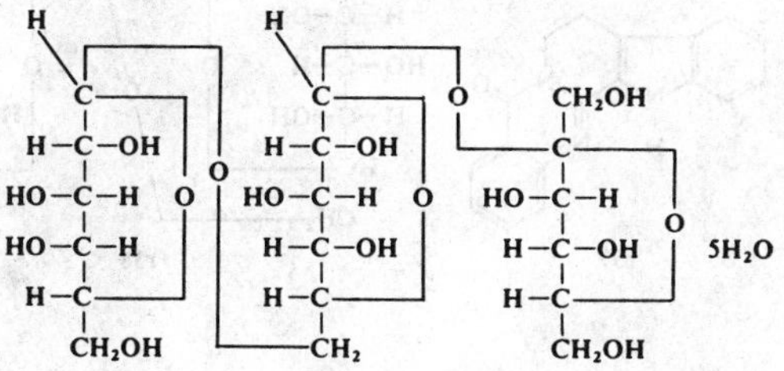

r1

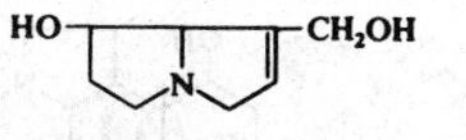

r10

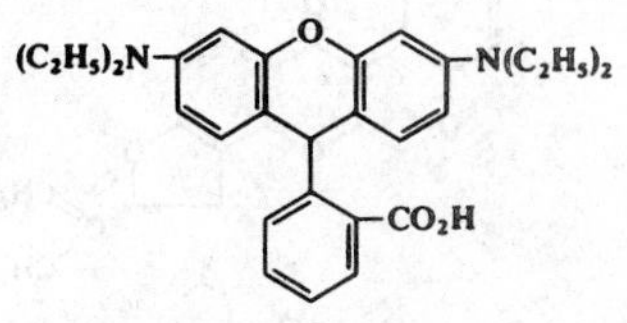

r19

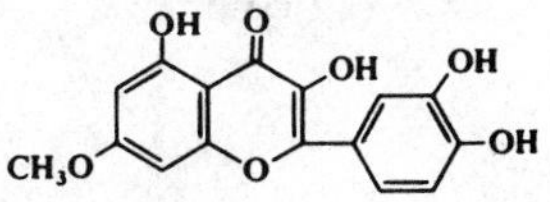

r11

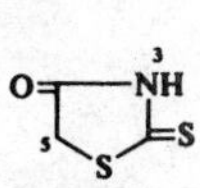

r21

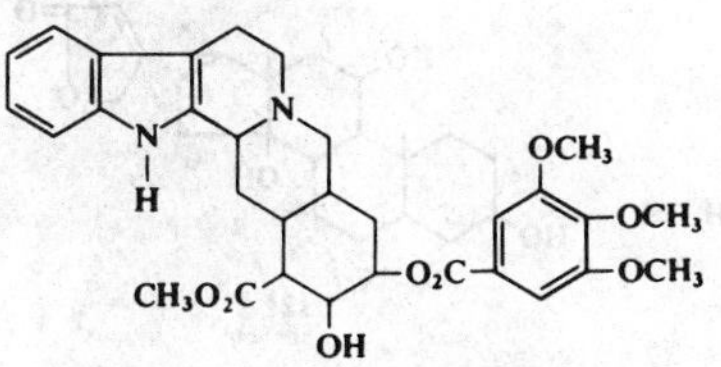

r2

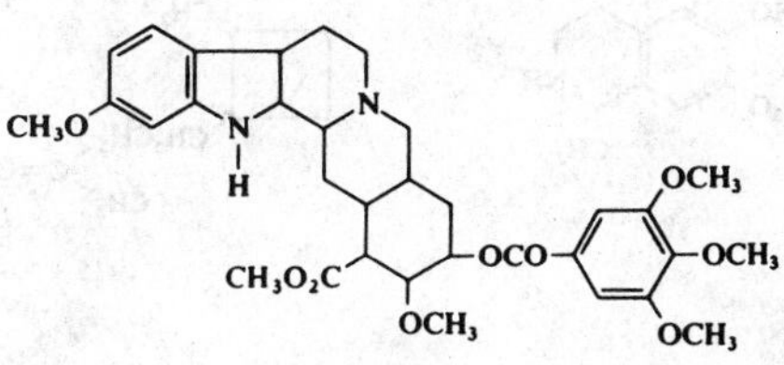

r6

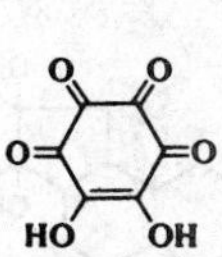

r25

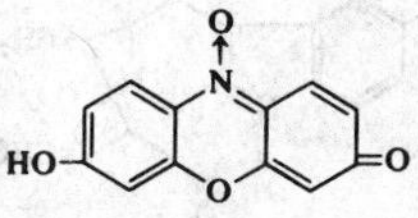

r3

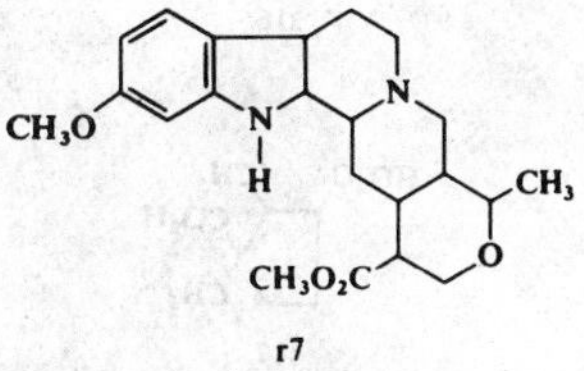

r7

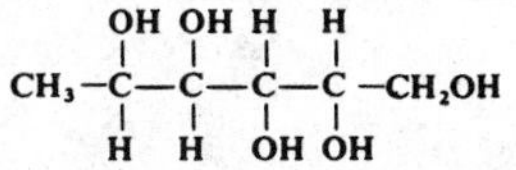

r12

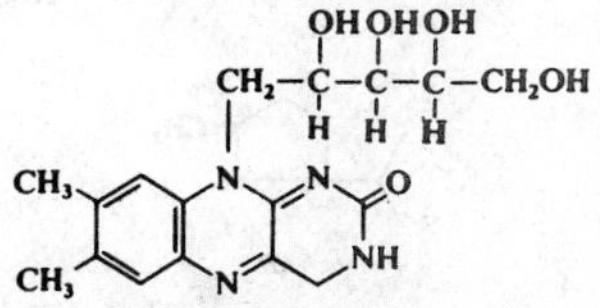

r26

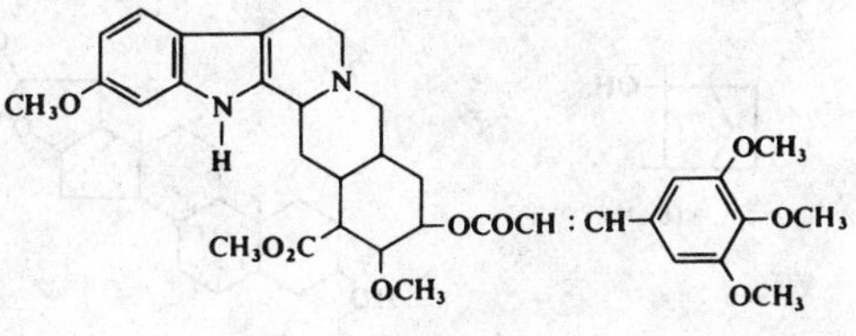

r4

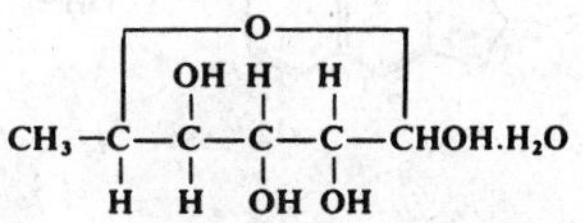

r13

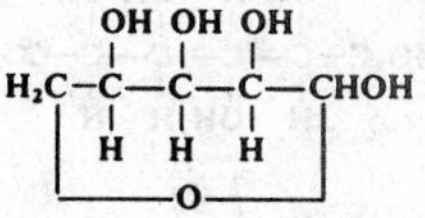

r27

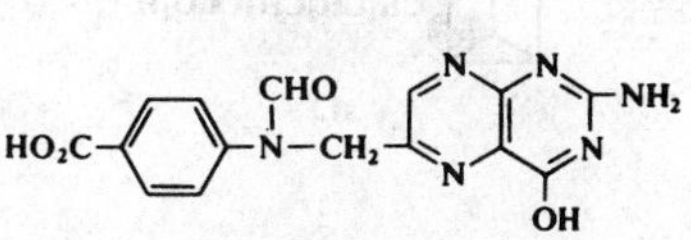

r18

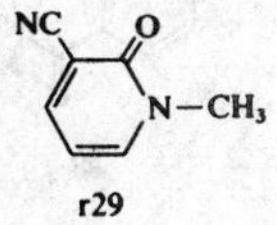

r29

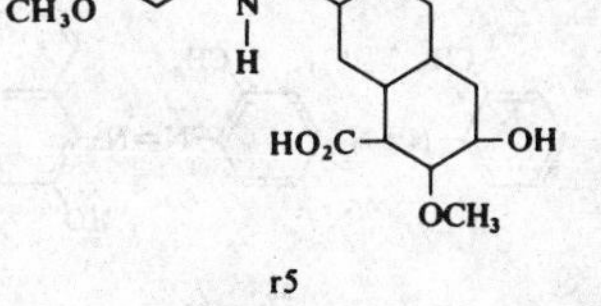

r5

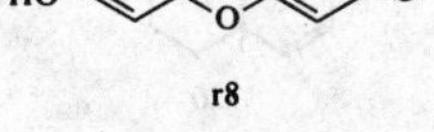

r8

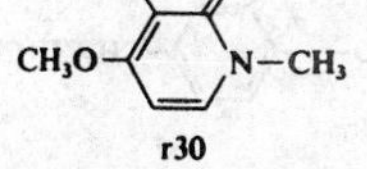

r30

STRUCTURAL FORMULAS OF ORGANIC COMPOUNDS (Continued)

In Alpha Numeric as they occur in Organic Compounds Table

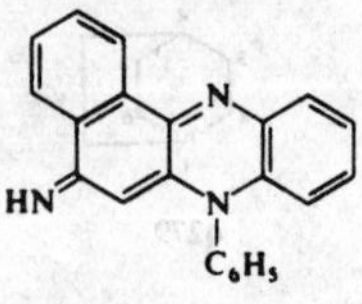

r31

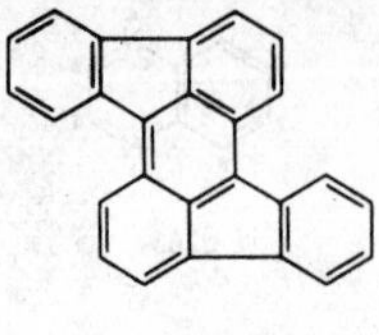

r33

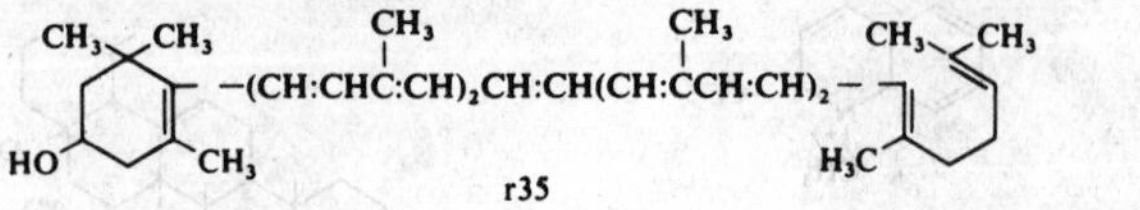

r35

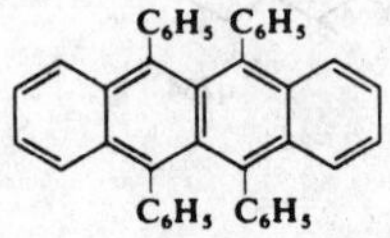

r36

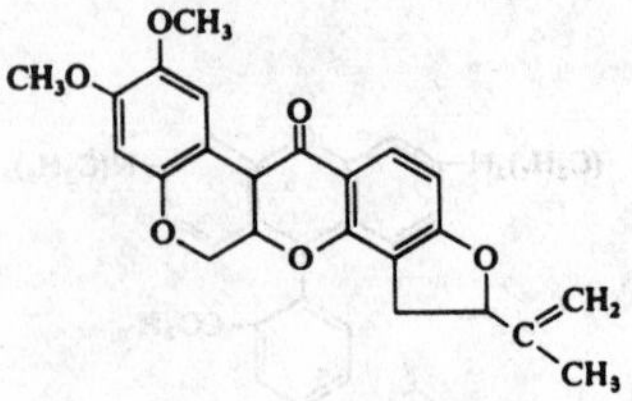

r32

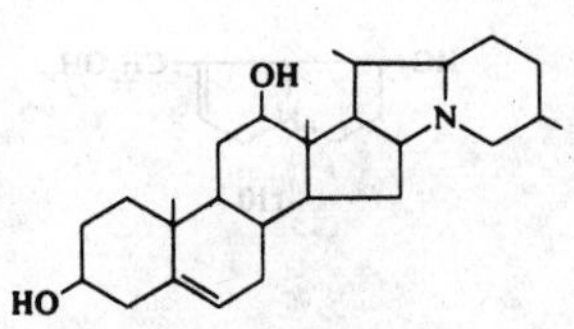

r34

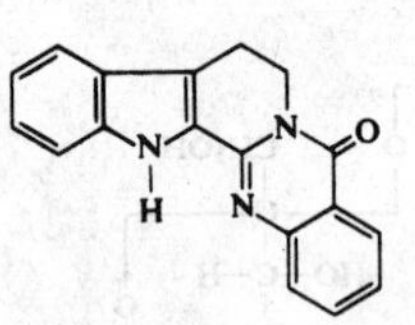

r37

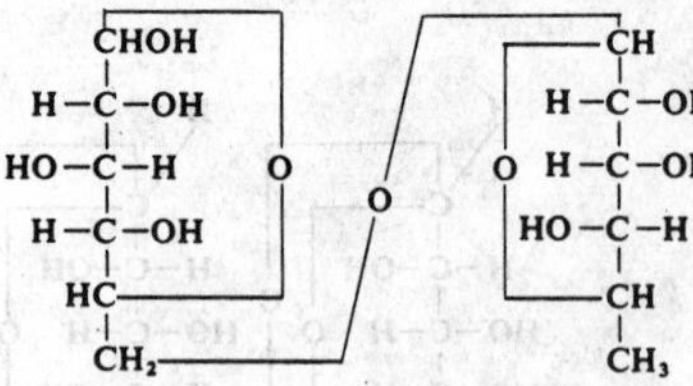

r38

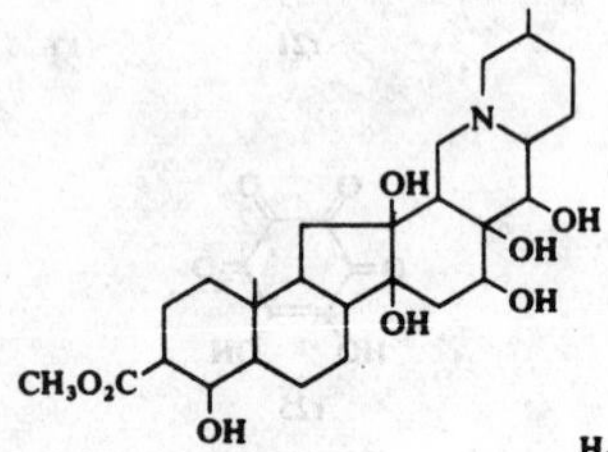

s1

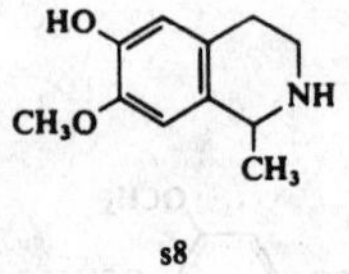

s8

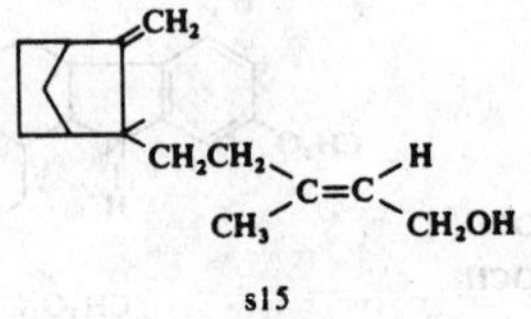

s15

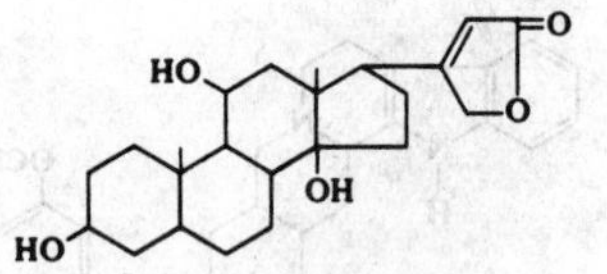

s21

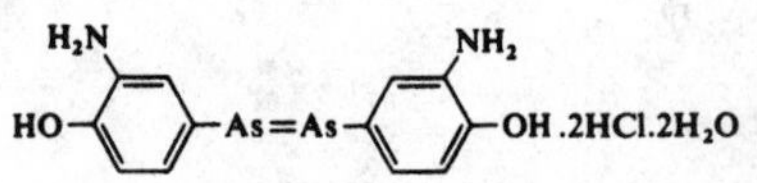

s9

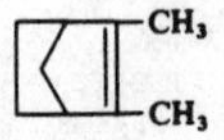

s16

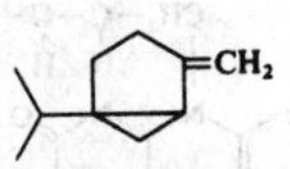

s2

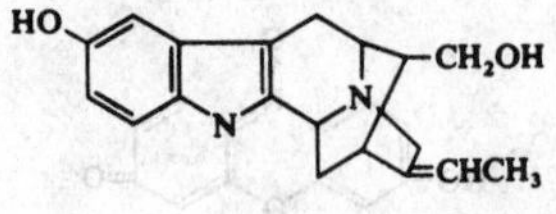

s22

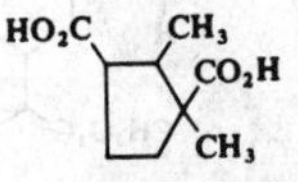

s17

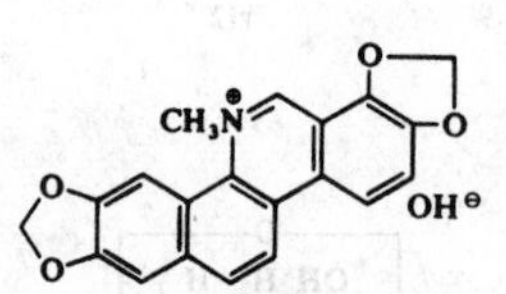

s12

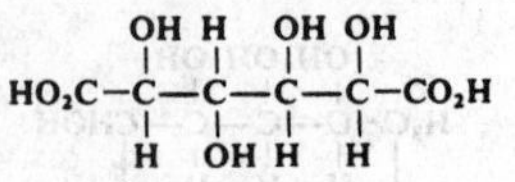

s5

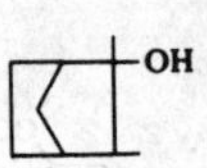

s18

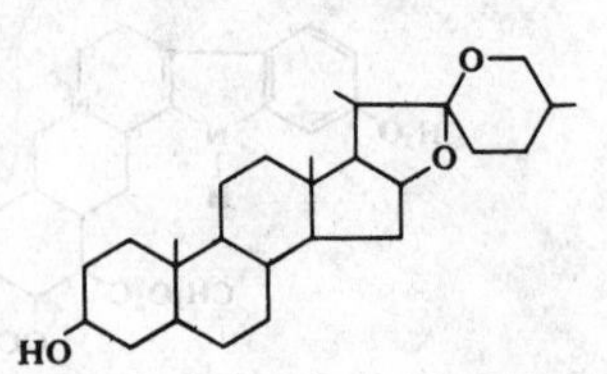

s23

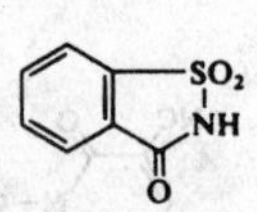

s6

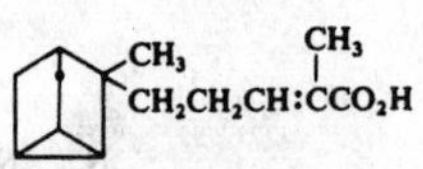

s13

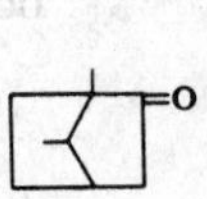

s19

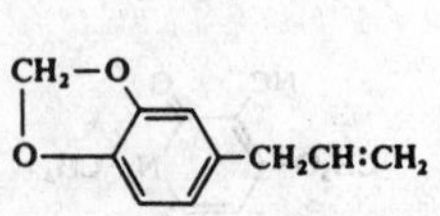

s7

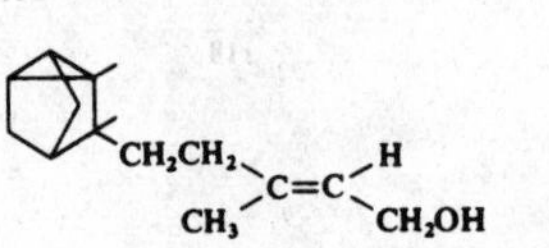

s14

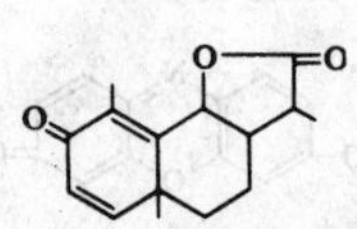

s20

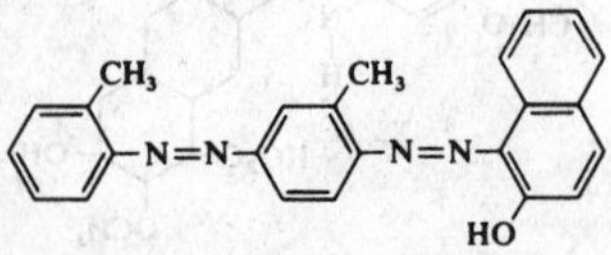

s24

STRUCTURAL FORMULAS OF ORGANIC COMPOUNDS (Continued)

In Alpha Numeric as they occur in Organic Compounds Table

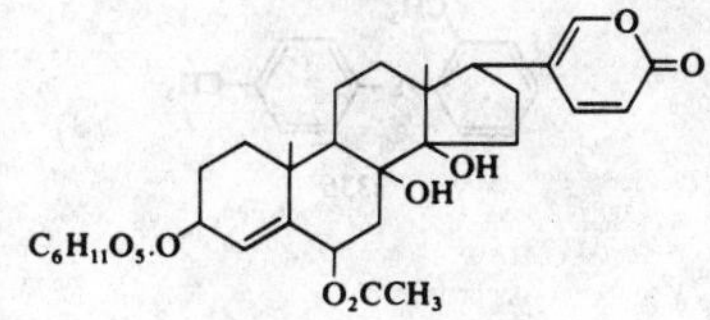

s25

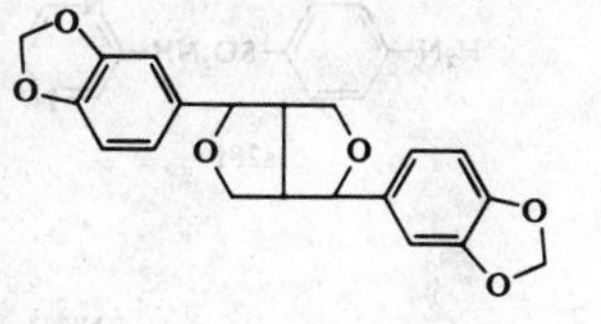

s59

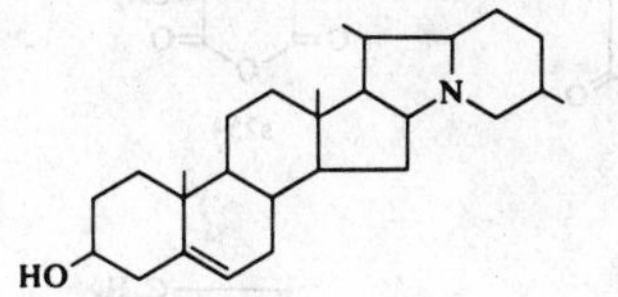

s94

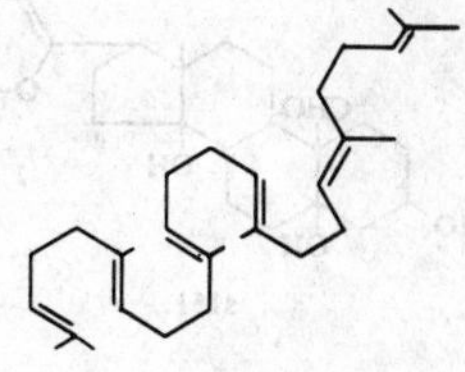

s108

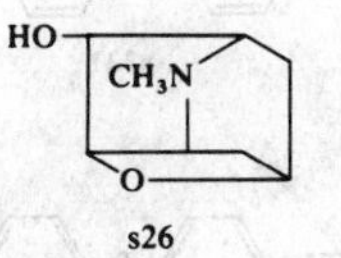

s26

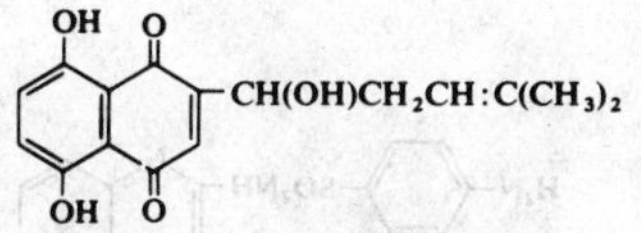

s60

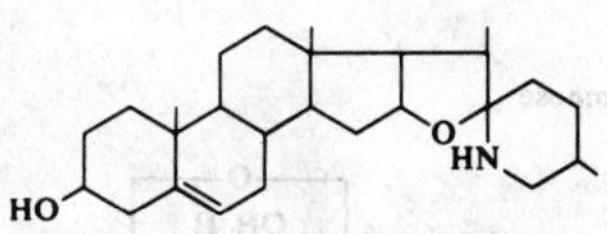

s96

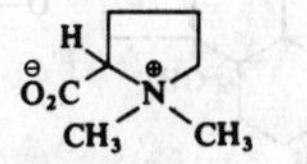

s110

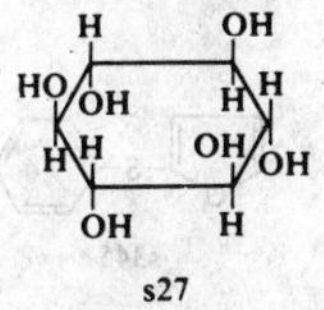

s27

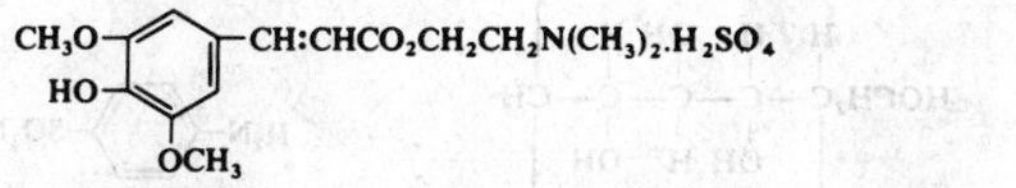

s83

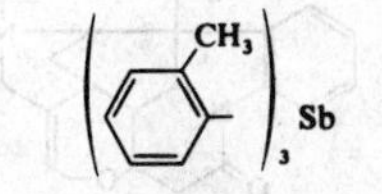

s124

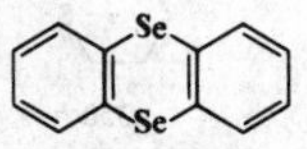

s29

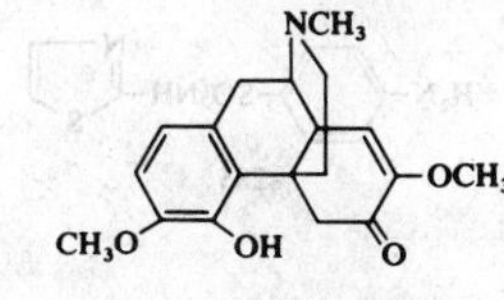

s86

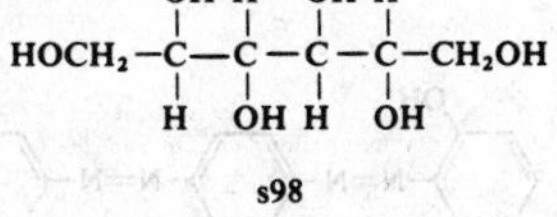

s98

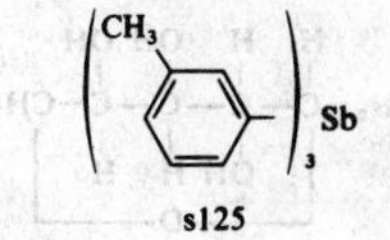

s125

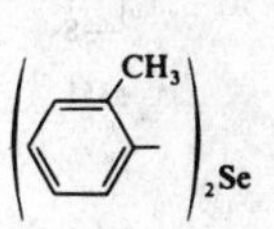

s34

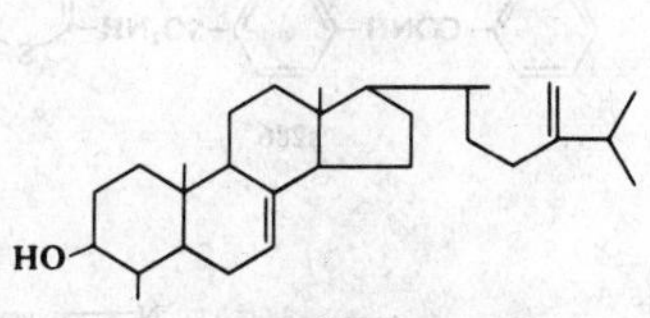

s87

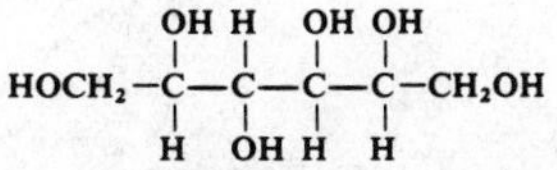

s99

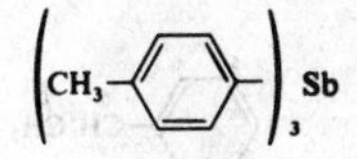

s126

s35

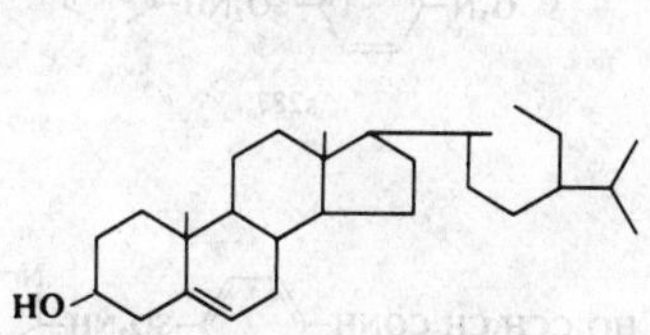

s90

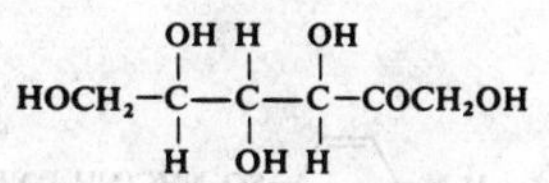

s102

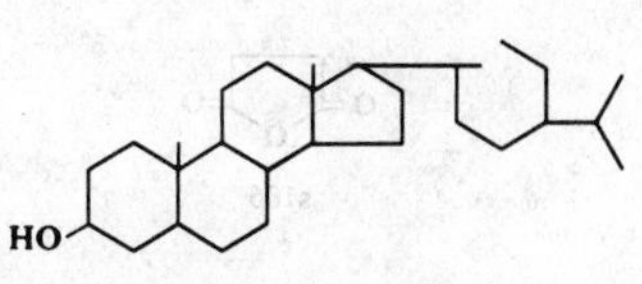

s127

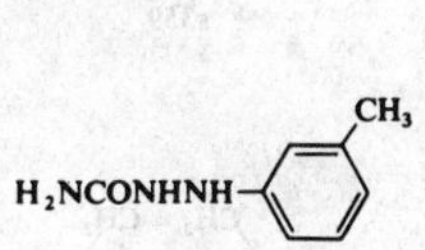

s51

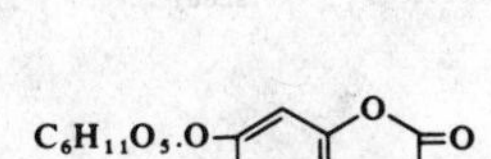

s91

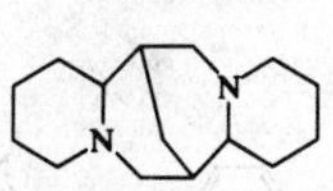

s105

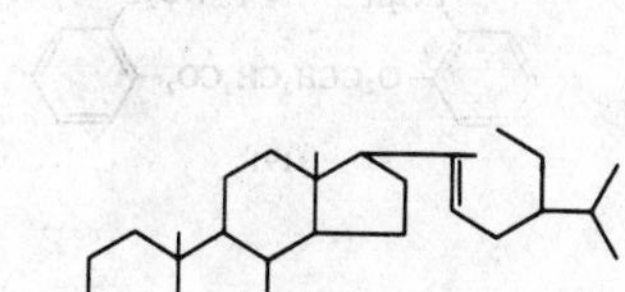

s128

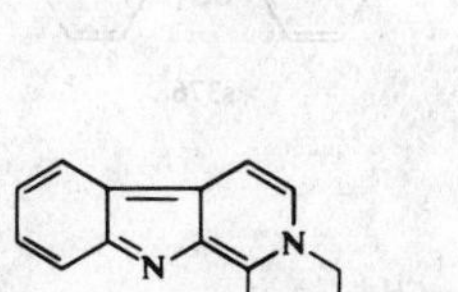

s58

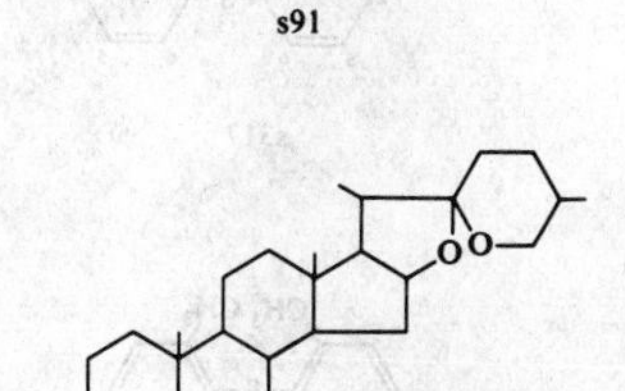

s92

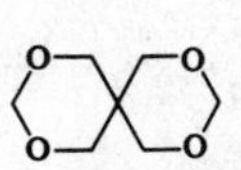

s107

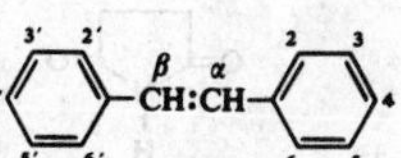

s129

STRUCTURAL FORMULAS OF ORGANIC COMPOUNDS (Continued)

In Alpha Numeric as they occur in Organic Compounds Table

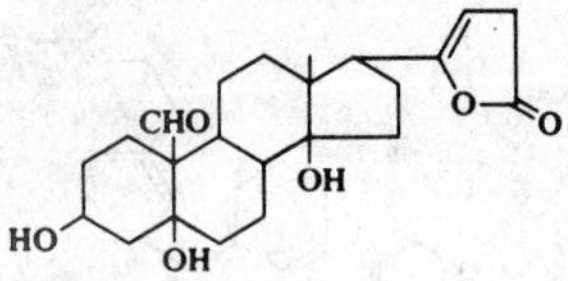

s151

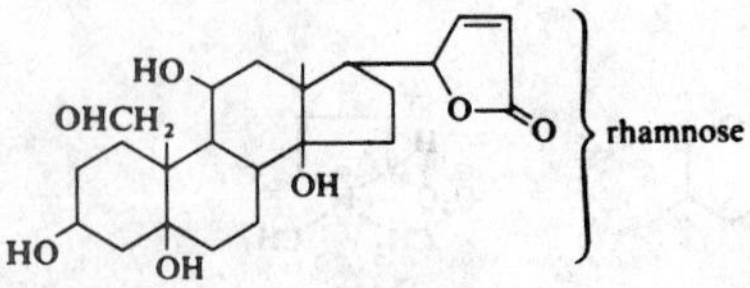

s152

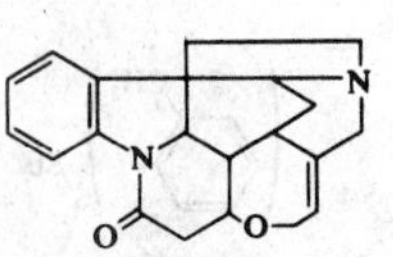

s153

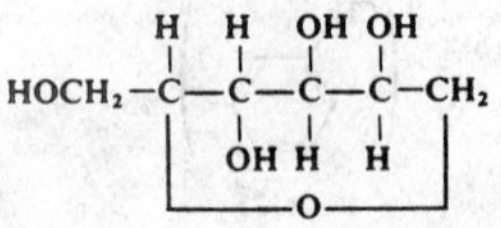

s158

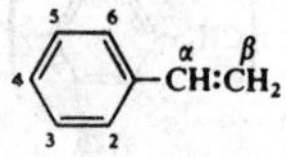

s159

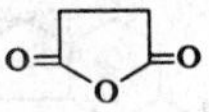

s186

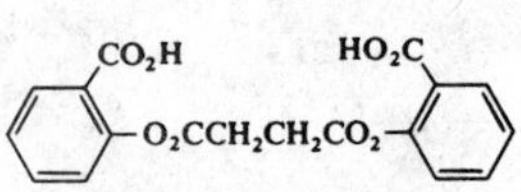

s191

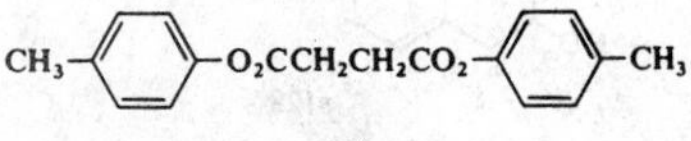

s199

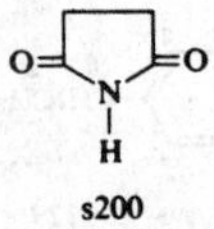

s200

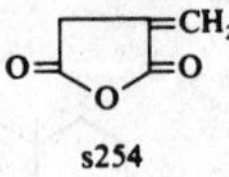

s254

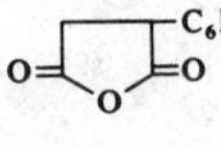

s263

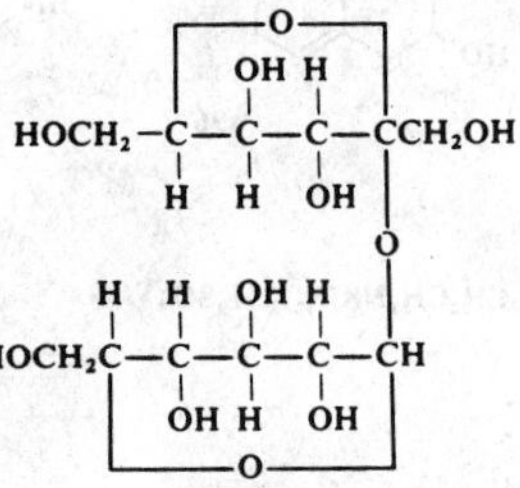

s273

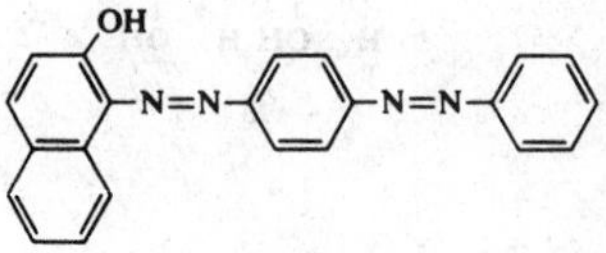

s275

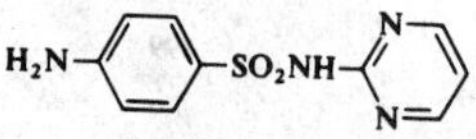

s276

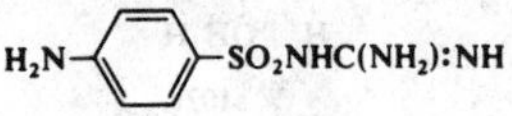

s277

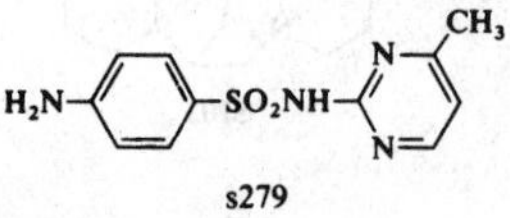

s279

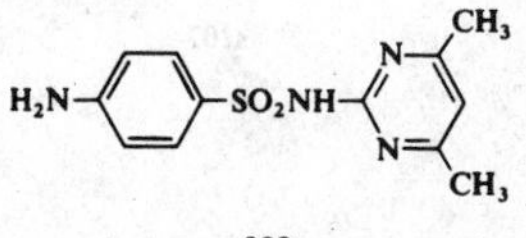

s280

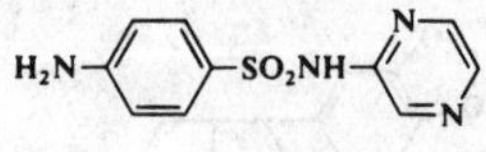

s281

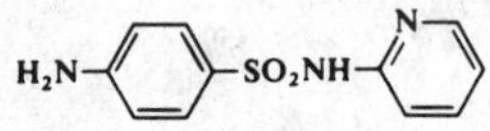

s282

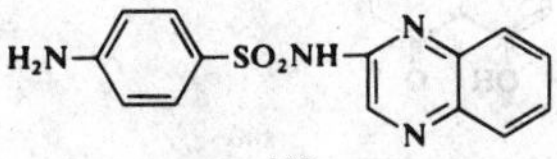

s283

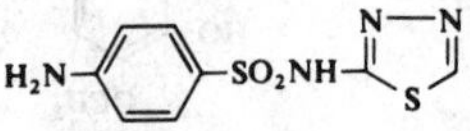

s284

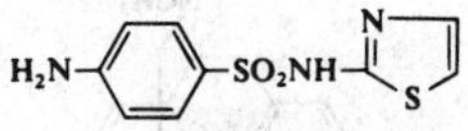

s285

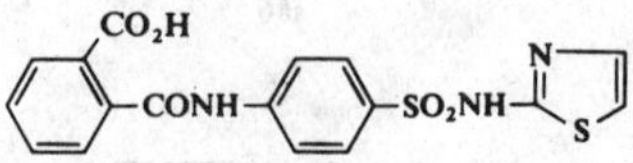

s286

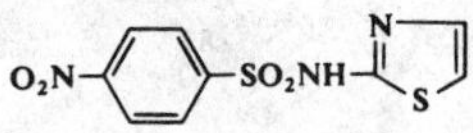

s287

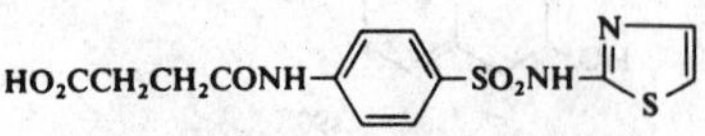

s288

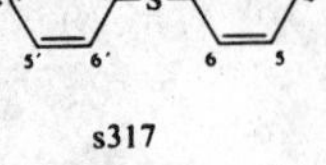

s317

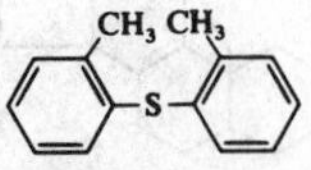

s335

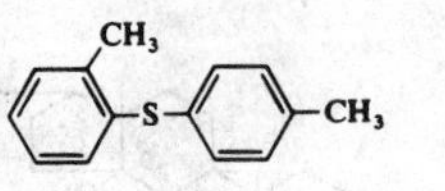

s336

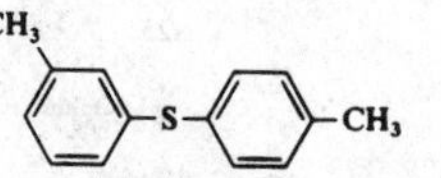

s337

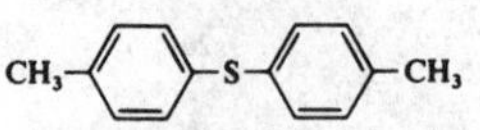

s338

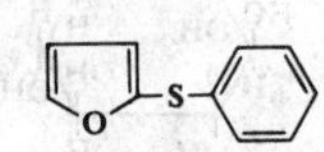

s345

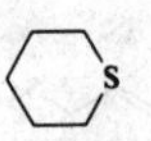

s350

s351

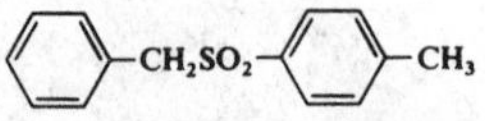

s353

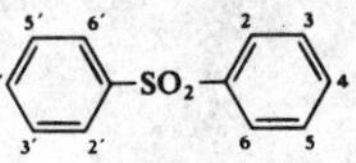

s359

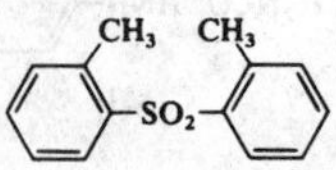

s376

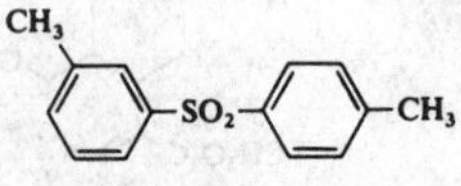

s377

In Alpha Numeric as they occur in Organic Compounds Table

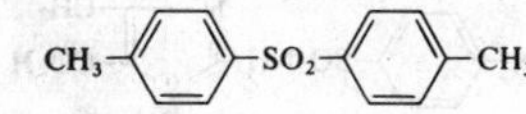
s378

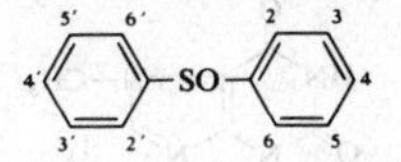
s387

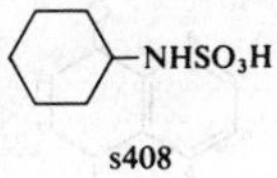
s408

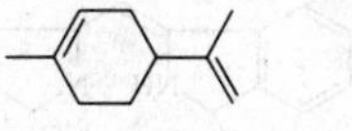
s422

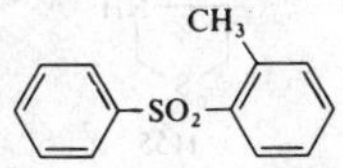
s381

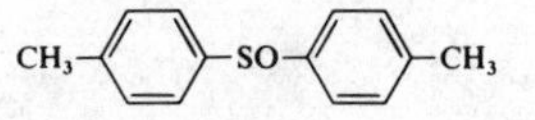
s393

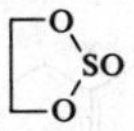
s420

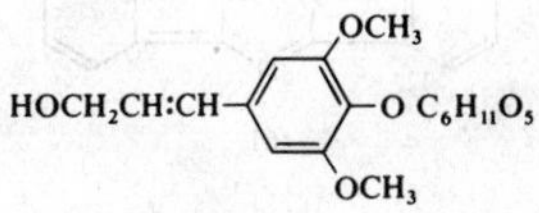
s423

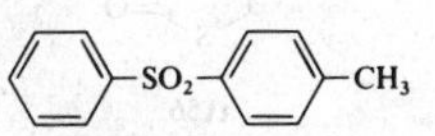
s382

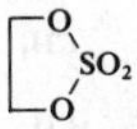
s407

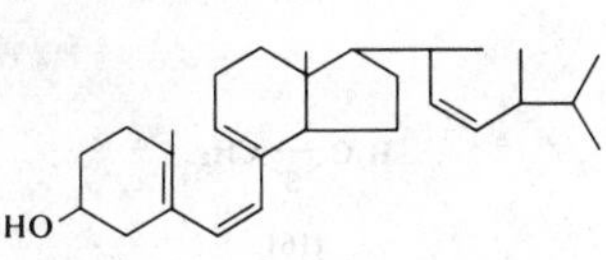
t1

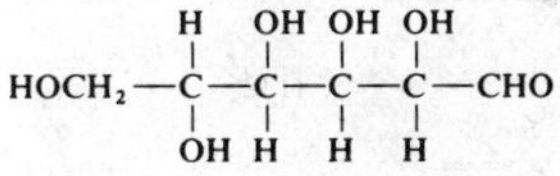
t6

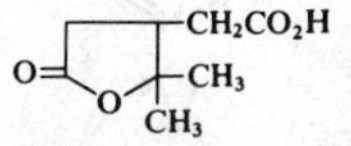
t45

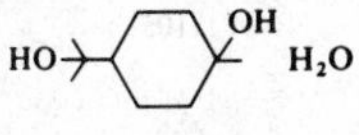
t51

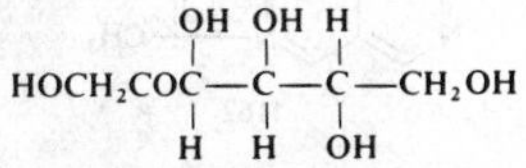
t2

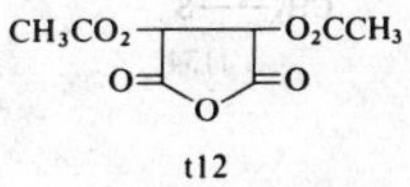
t12

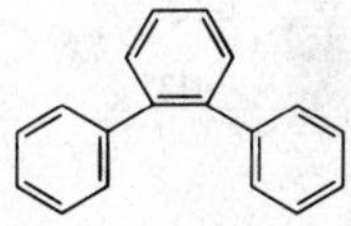
t46

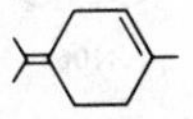
t52

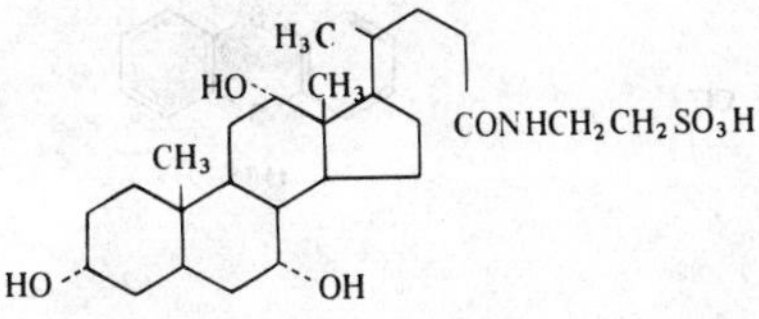
t40

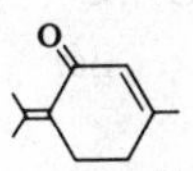
t53

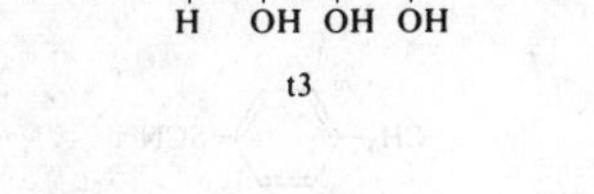
t3

t47

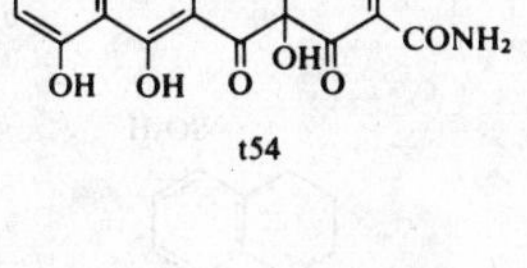
t54

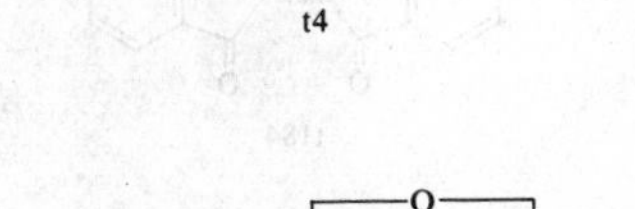
t4

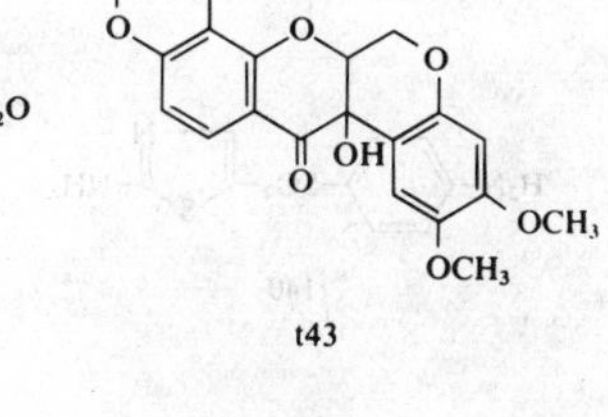
t43

t48

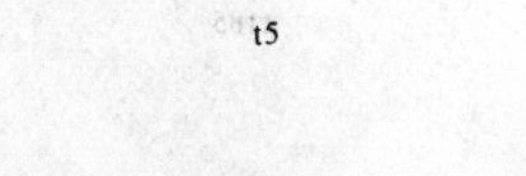
t5

t44

t49

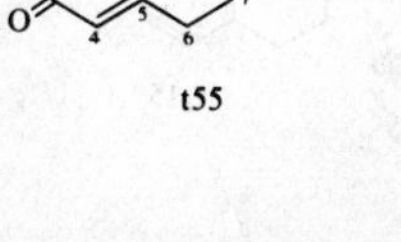
t55

STRUCTURAL FORMULAS OF ORGANIC COMPOUNDS (Continued)

In Alpha Numeric as they occur in Organic Compounds Table

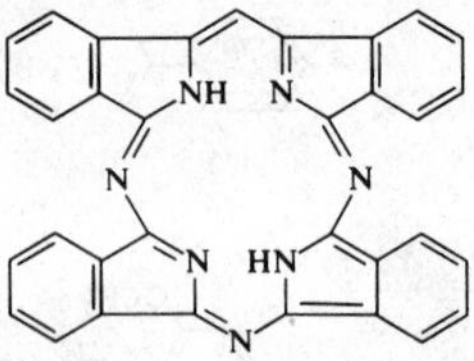
t61

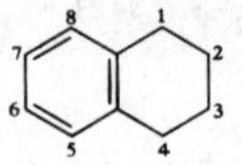
t98

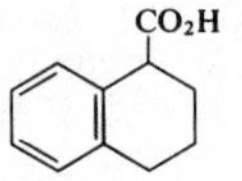
t104

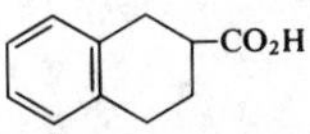
t105

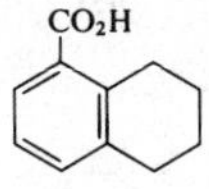
t106

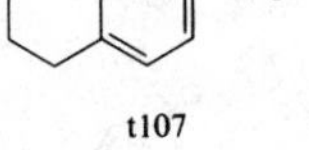
t107

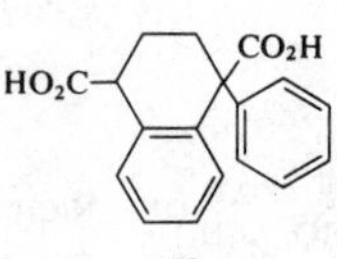
t108

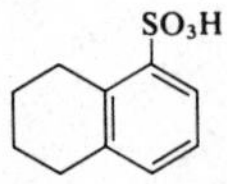
t112

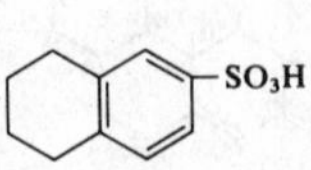
t115

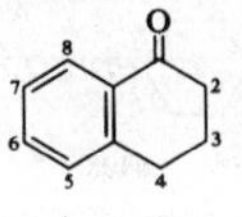
t117

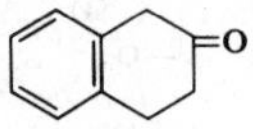
t118

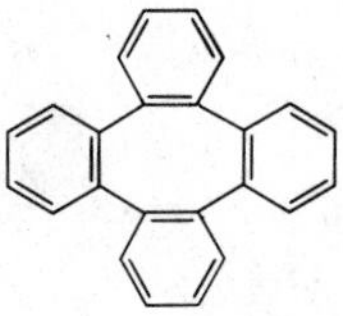
t120

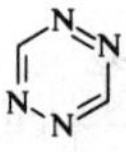
t124

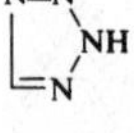
t125

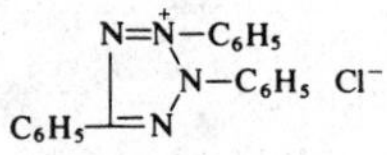
t126

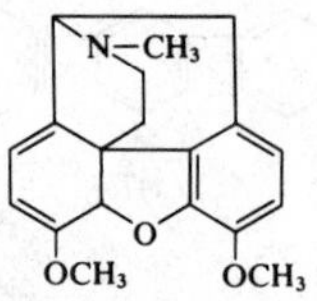
t127

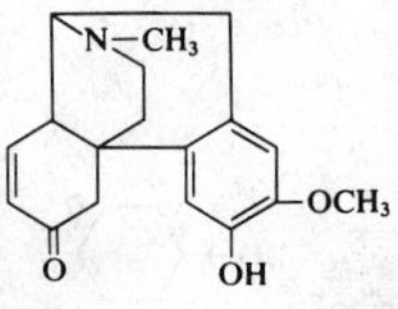
t129

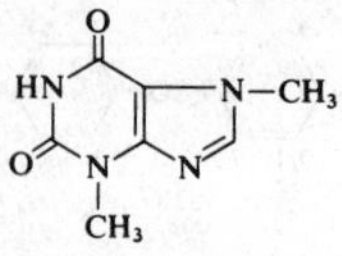
t130

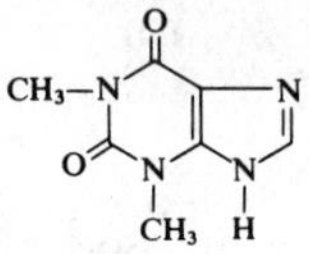
t131

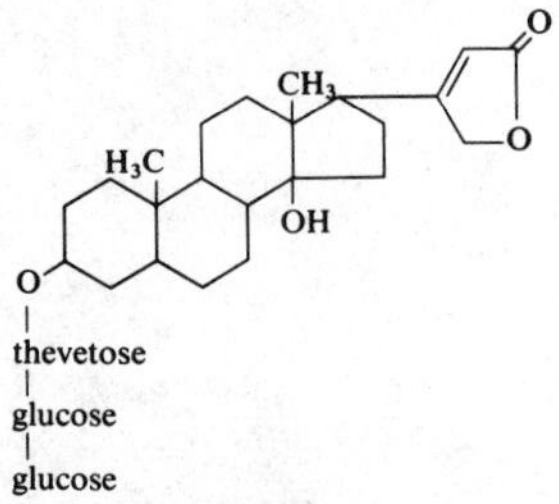
t132

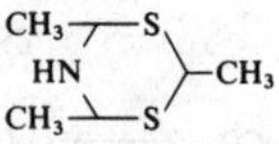
t134

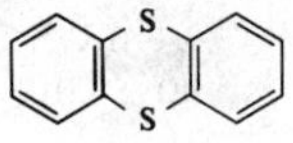
t135

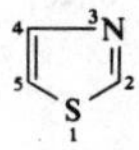
t136

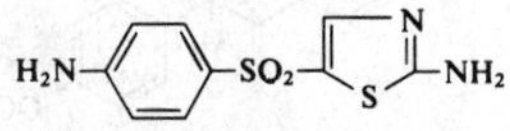
t140

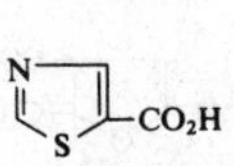
t149

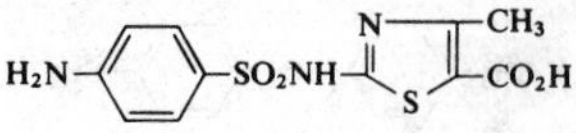
t154

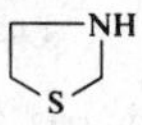
t155

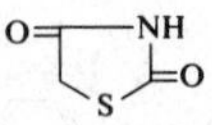
t156

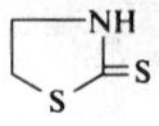
t157

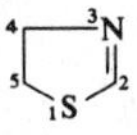
t158

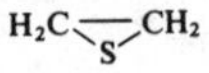
t161

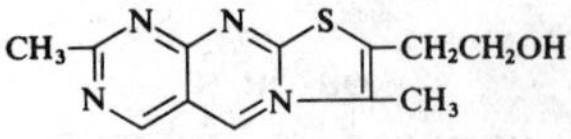
t162

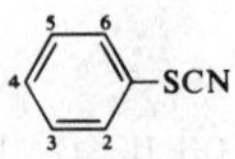
t180

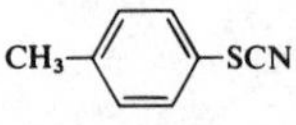
t182

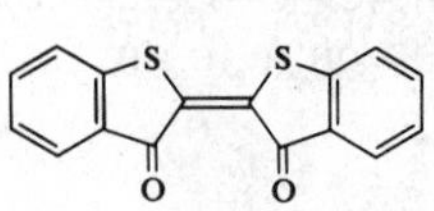
t184

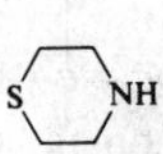
t185

STRUCTURAL FORMULAS OF ORGANIC COMPOUNDS (Continued)

In Alpha Numeric as they occur in Organic Compounds Table

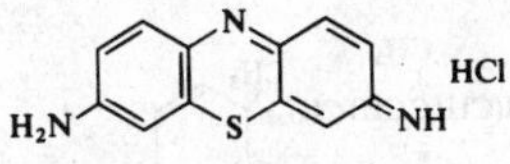

t186

t187

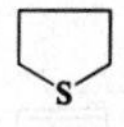

t224

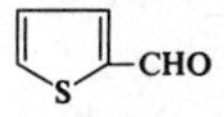

t229

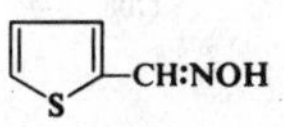

t230

t231

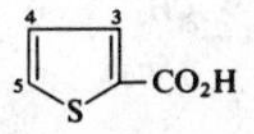

t232

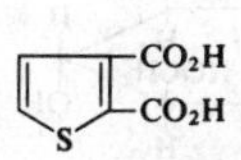

t236

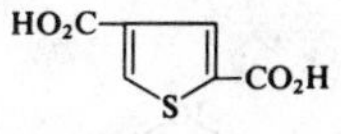

t237

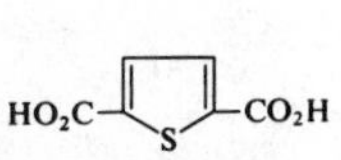

t238

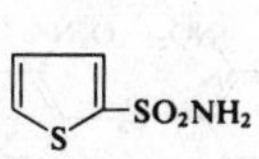

t240

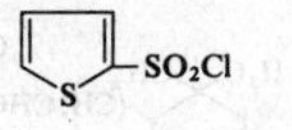

t241

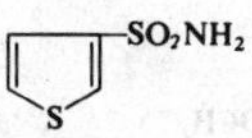

t242

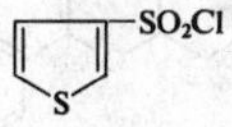

t243

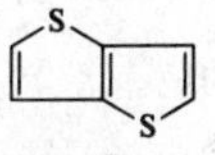

t244

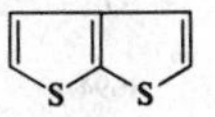

t245

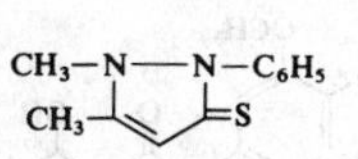

t246

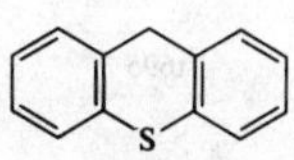

t248

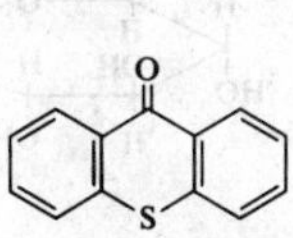

t249

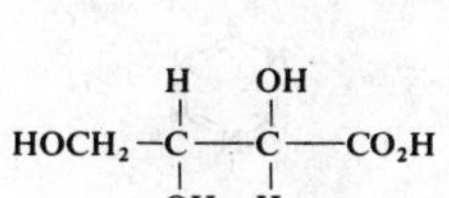

t250

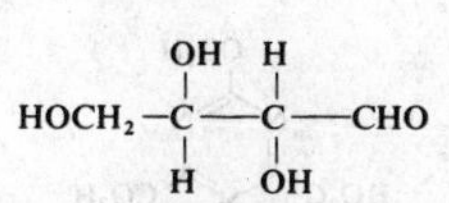

t255

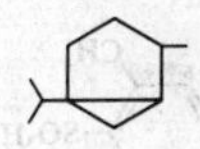

t256

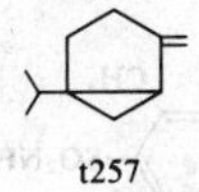

t257

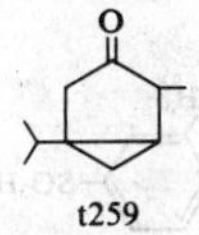

t259

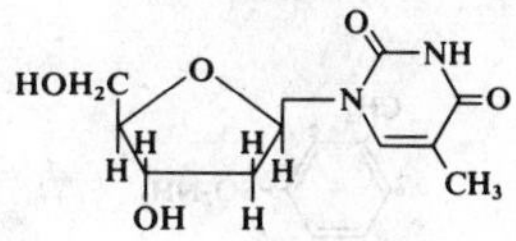

t260

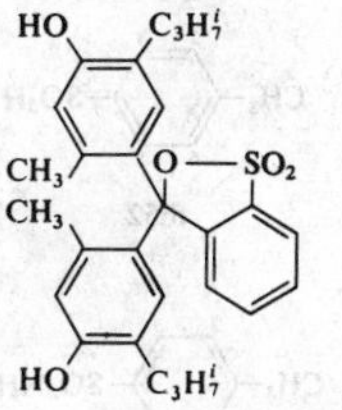

t263

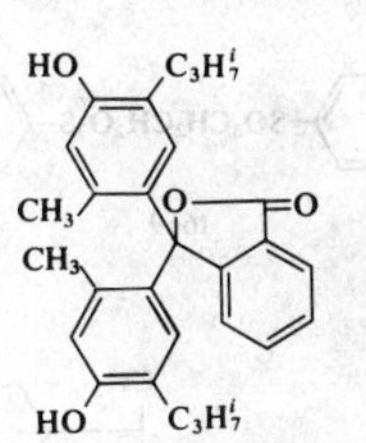

t264

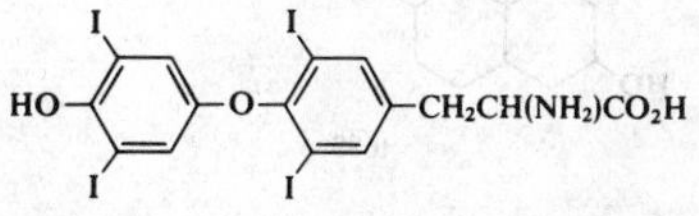

t266

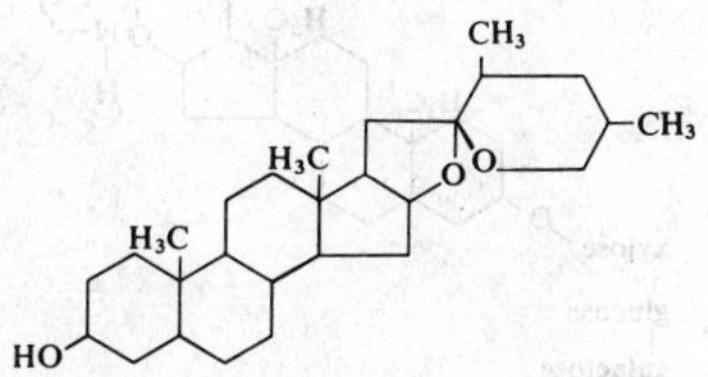

t268

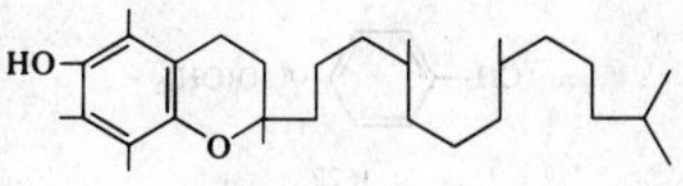

t269

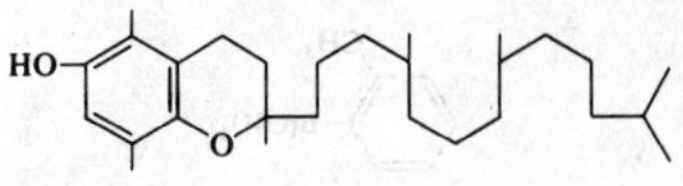

t270

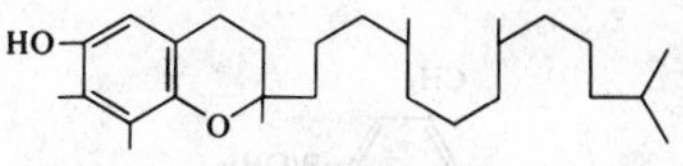

t271

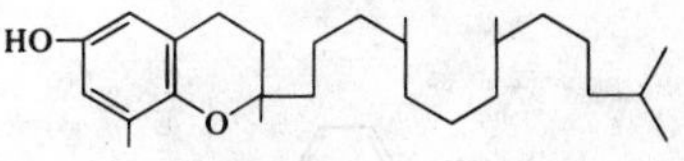

t272

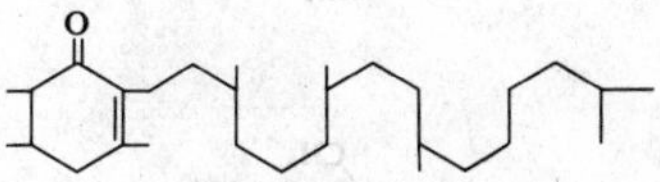

t273

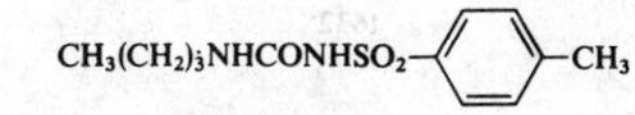

t274

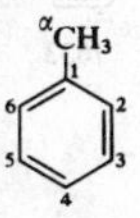

t275

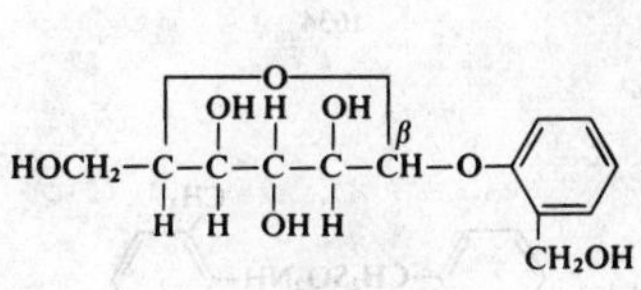

t461

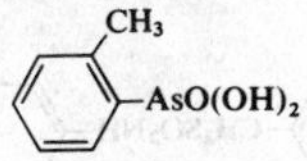

t625

t626

STRUCTURAL FORMULAS OF ORGANIC COMPOUNDS (Continued)

In Alpha Numeric as they occur in Organic Compounds Table

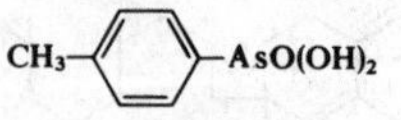
t627

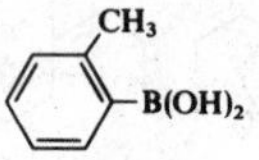
t629

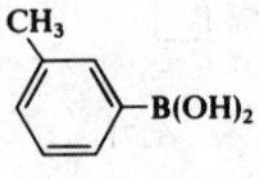
t630

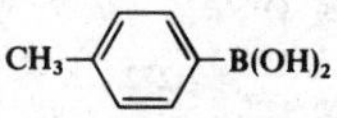
t631

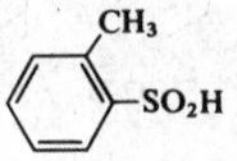
t632

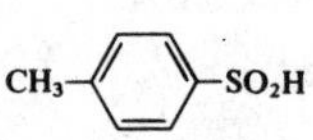
t633

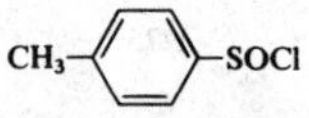
t634

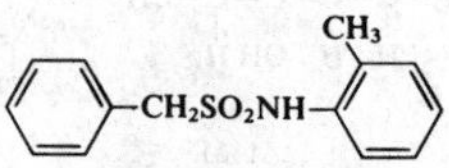
t636

t637

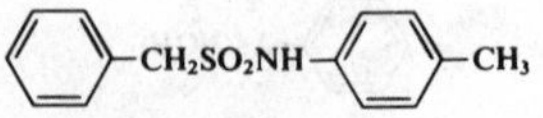
t638

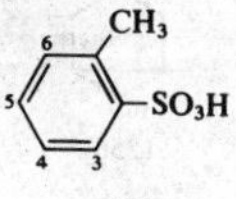
t640

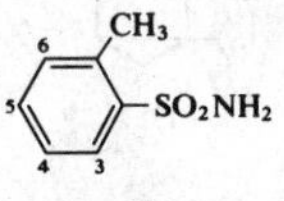
t641

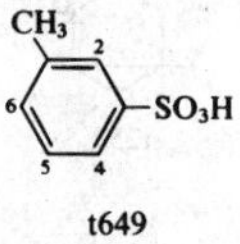
t649

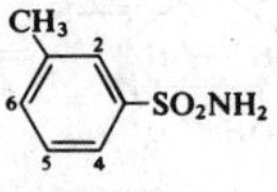
t650

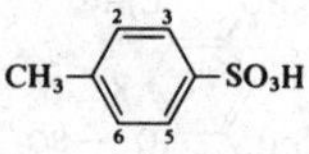
t652

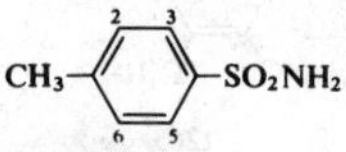
t653

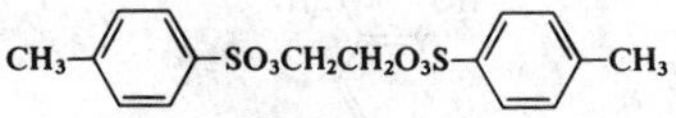
t669

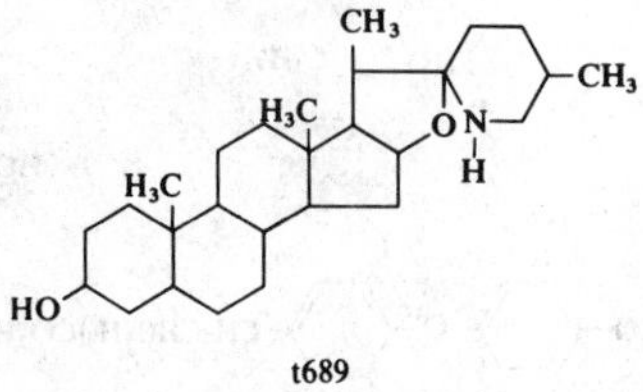
t689

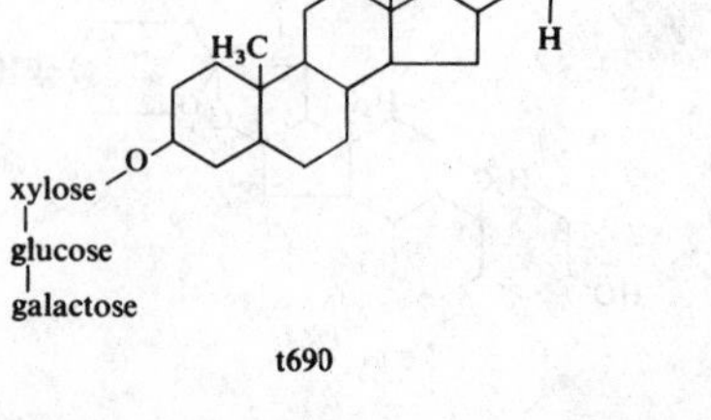
t690

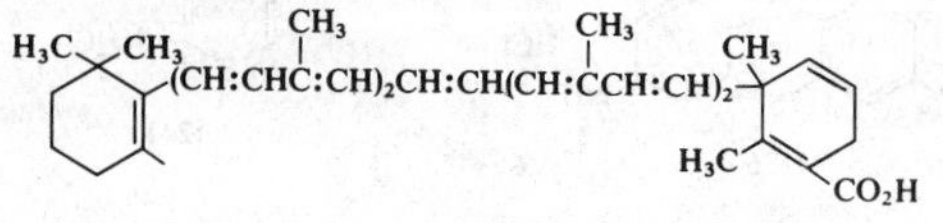
t691

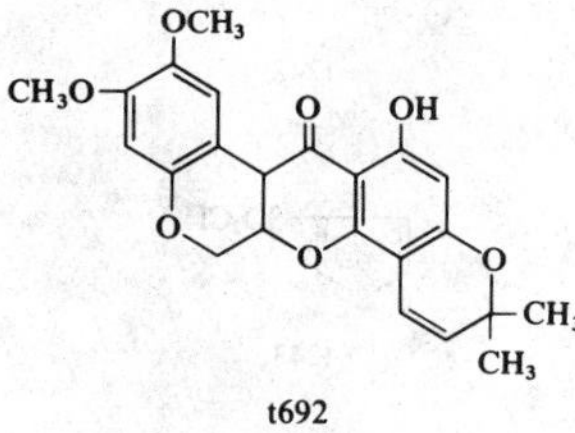
t692

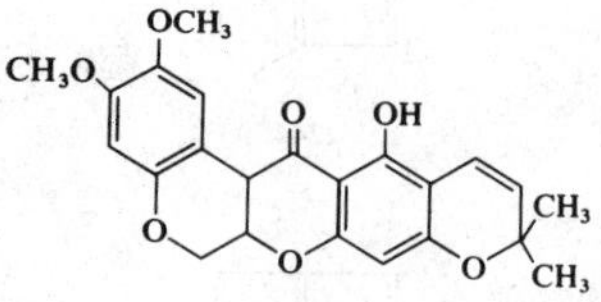
t694

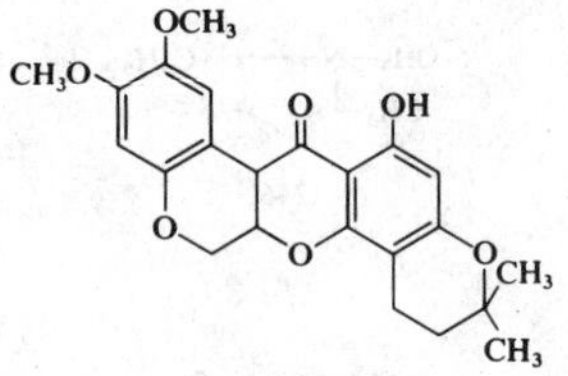
t695

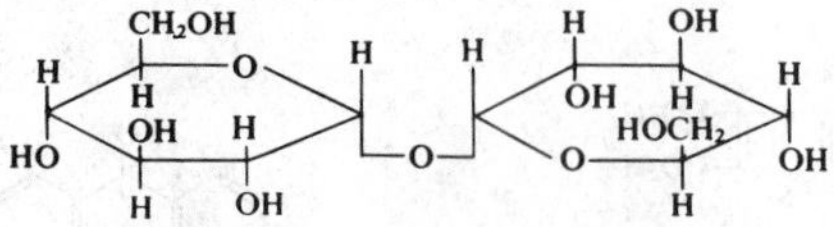
t697

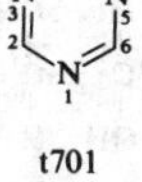
t701

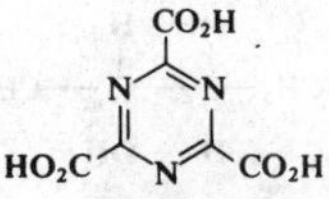
t704

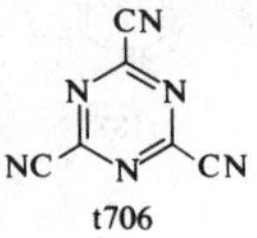
t706

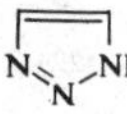
t707

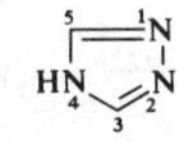
t708

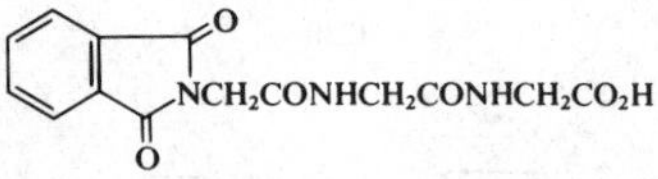
t738

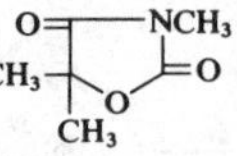
t739

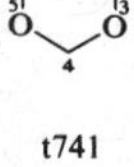
t741

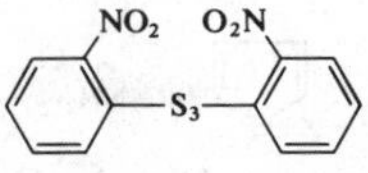
t748

In Alpha Numeric as they occur in Organic Compounds Table

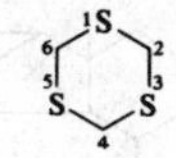
t749

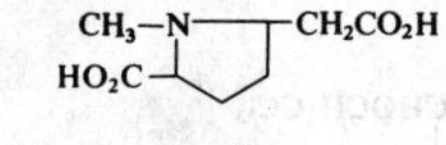
t765

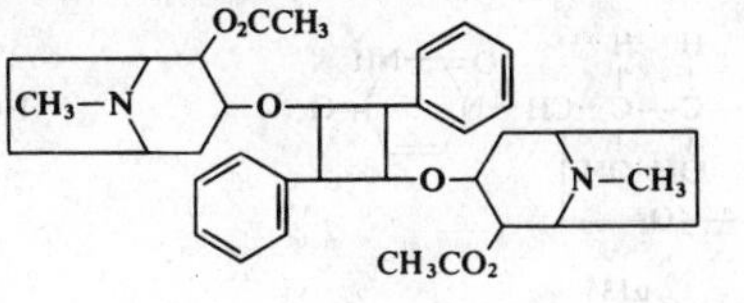
t771

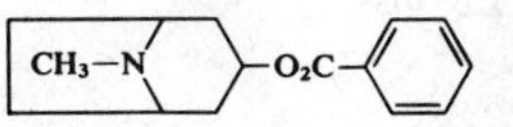
t758

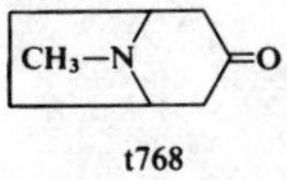
t768

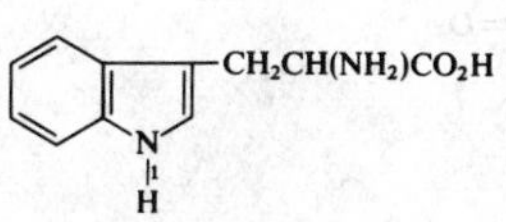
t772

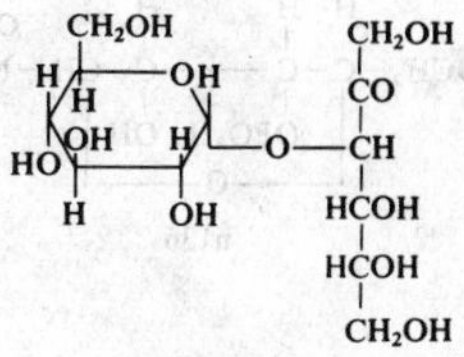
t780

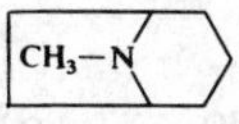
t760

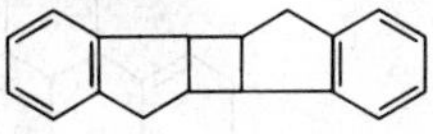
t769

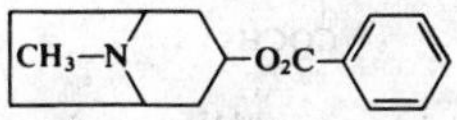
t761

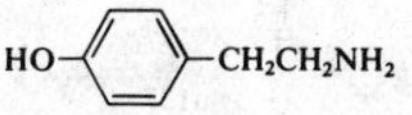
t781

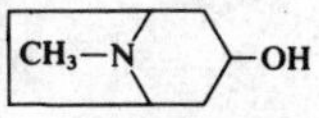
t764

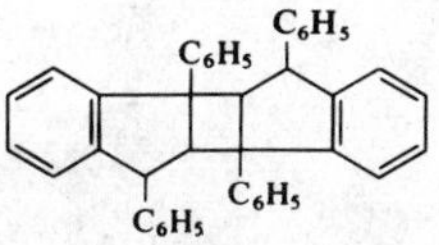
t770

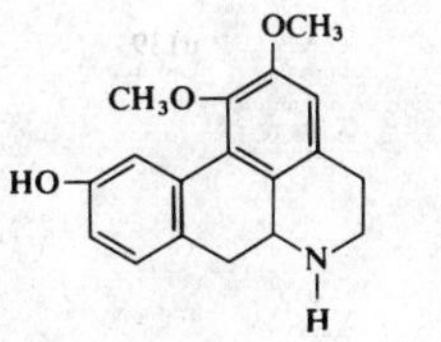
t779

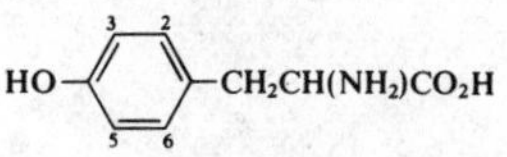
t782

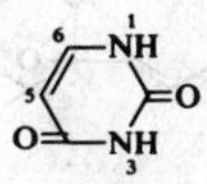
u2

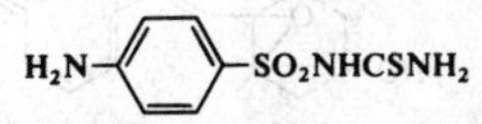
u29

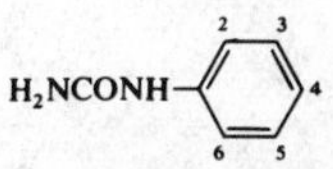
u110

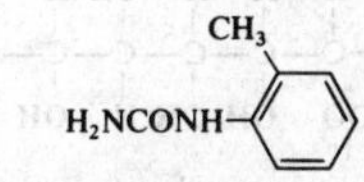
u125

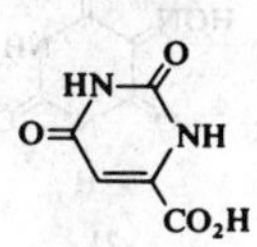
u15

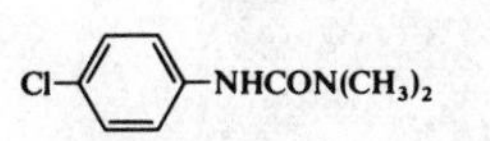
u49

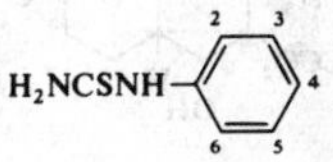
u112

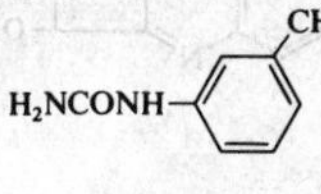
u126

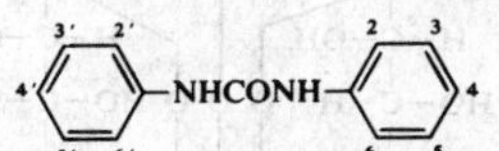
u71

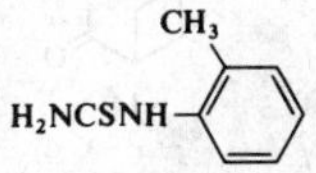
u121

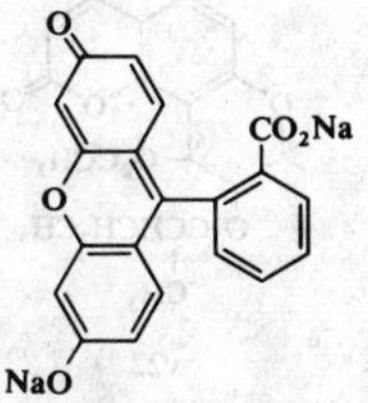
u17

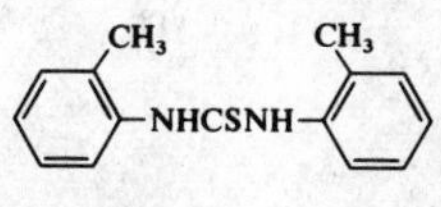
u79

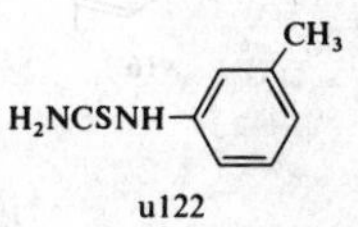
u122

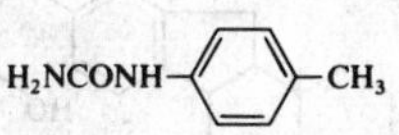
u127

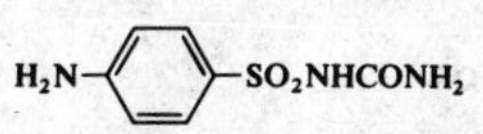
u28

u80

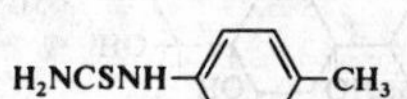
u123

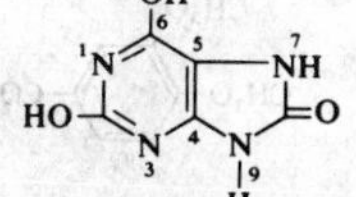
u131

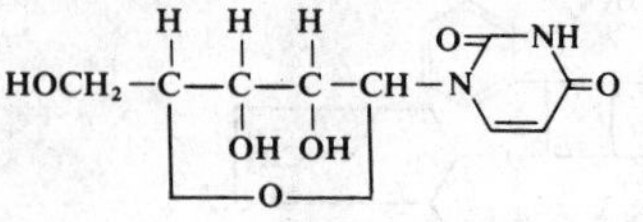

u135

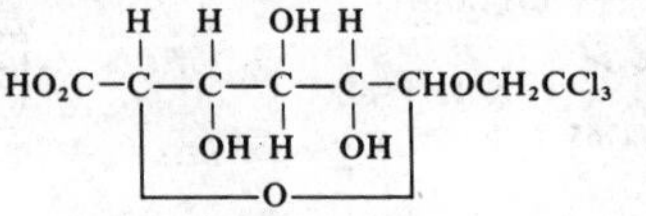

u138

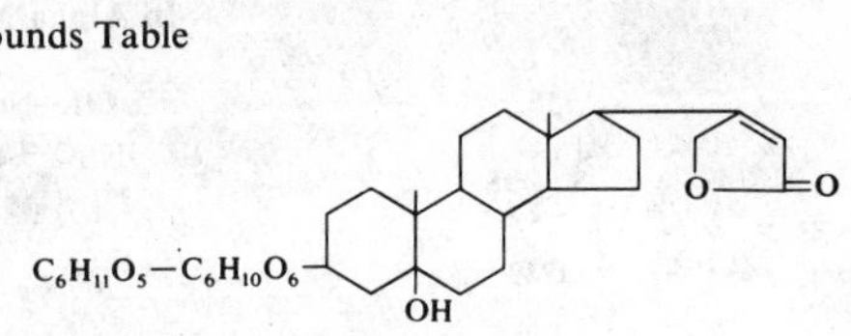

u144

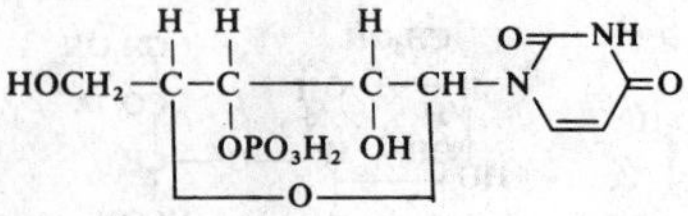

u136

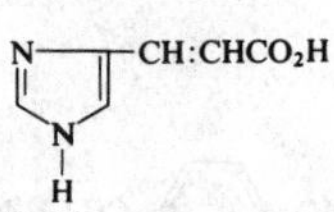

u137

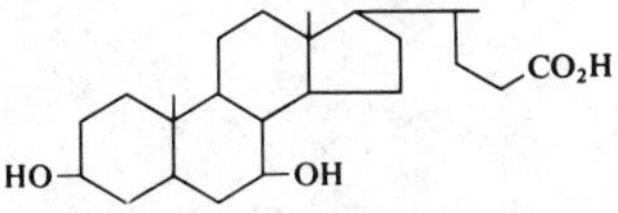

u139

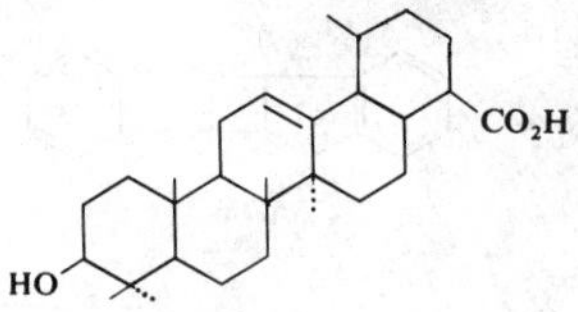

u140

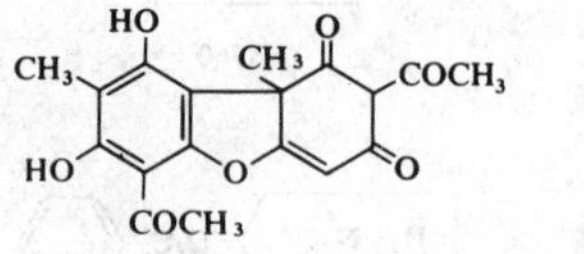

u141

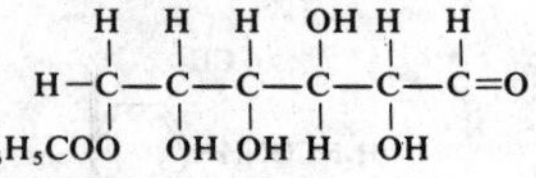

v1

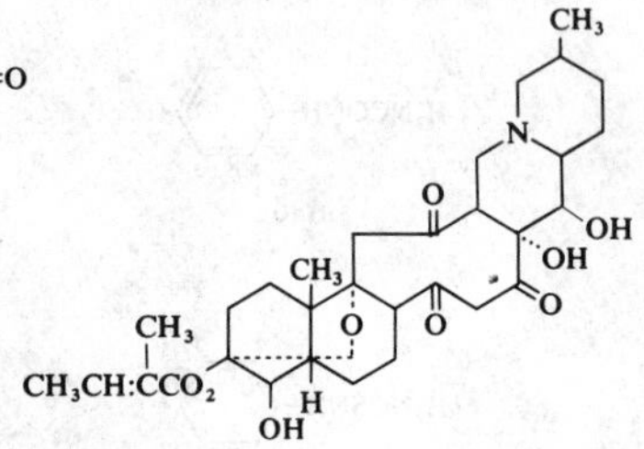

v11

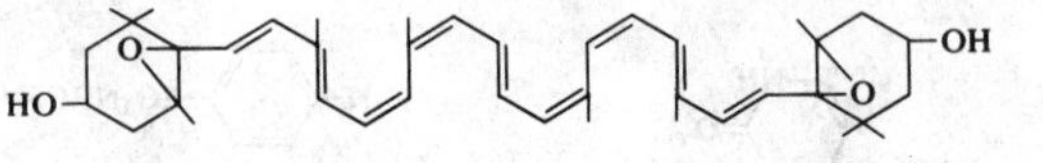

v20

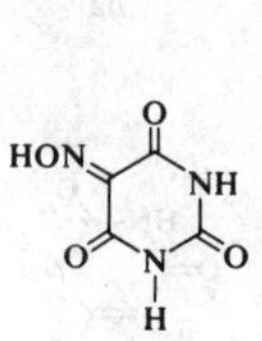

v21

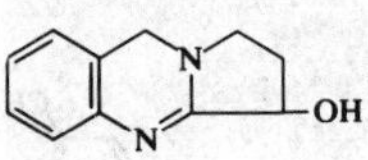

v7

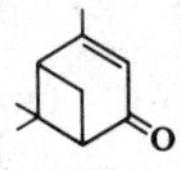

v14

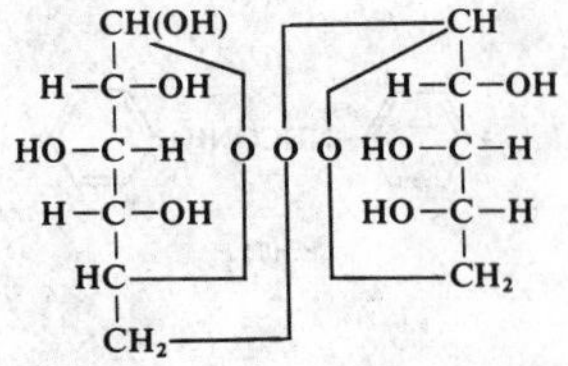

v18

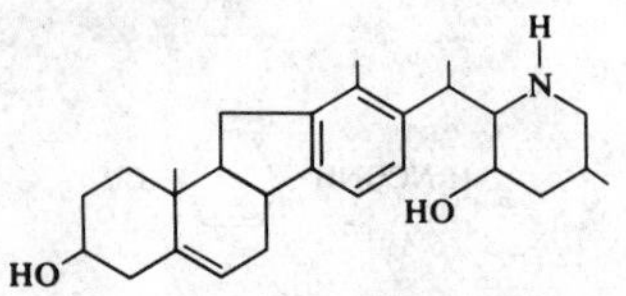

v9

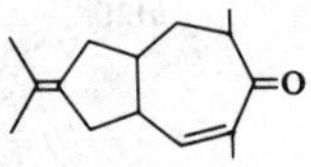

v16

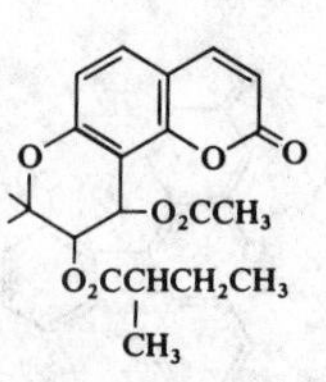

v22

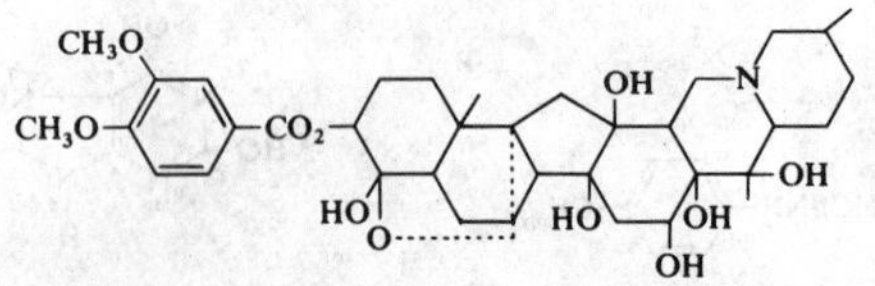

v10

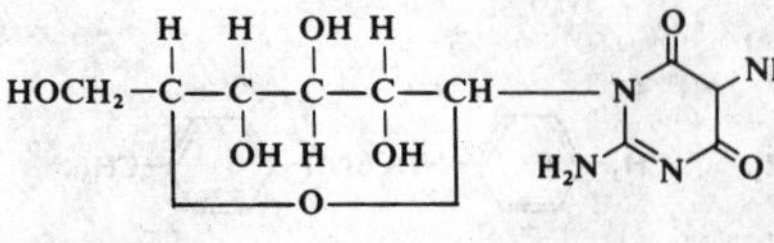

v19

STRUCTURAL FORMULAS OF ORGANIC COMPOUNDS (Continued)

In Alpha Numeric as they occur in Organic Compounds Table

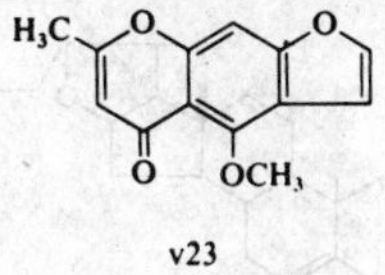

v23

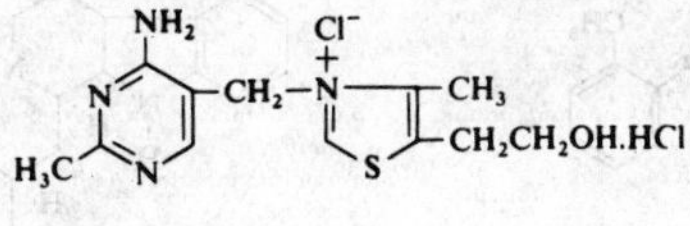

v25

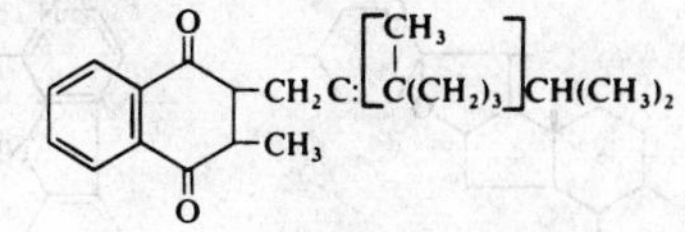

v28

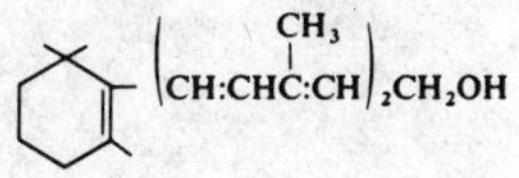

v24

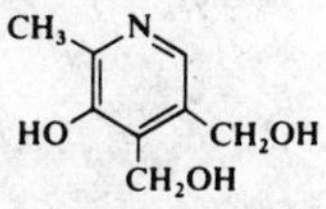

v26

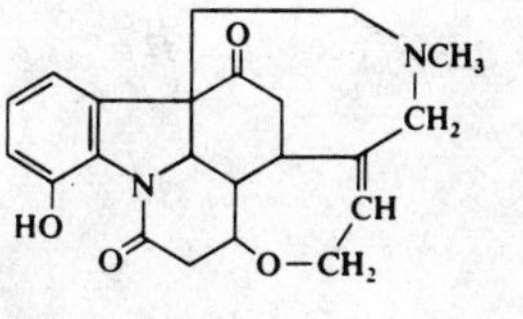

v29

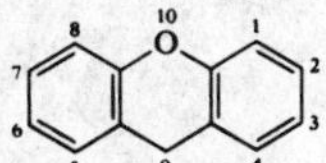

x1

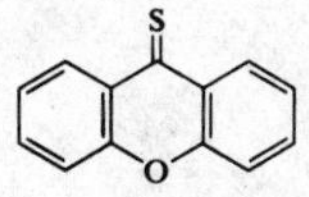

x9

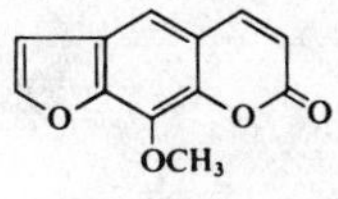

x25

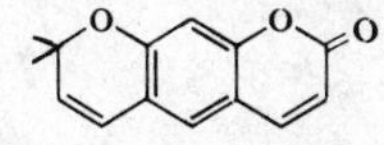

x28

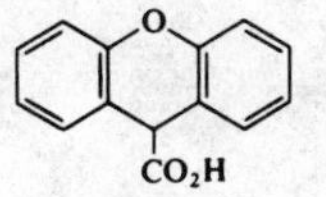

x4

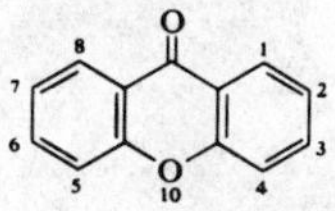

x12

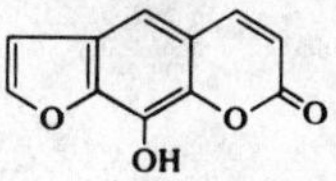

x26

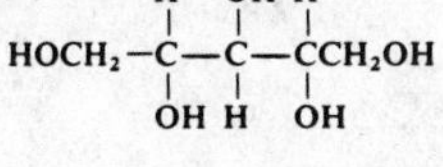

x29

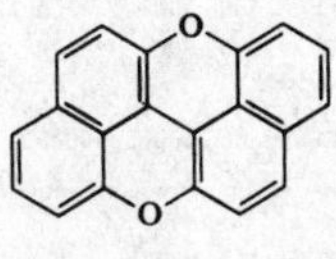

x5

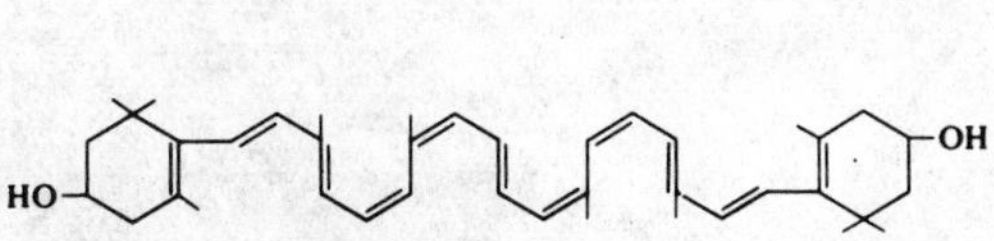

x23

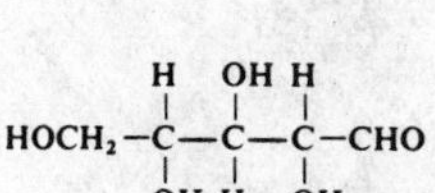

x30

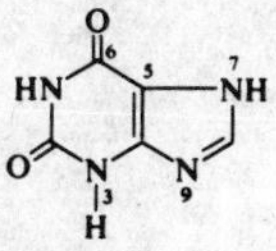

x6

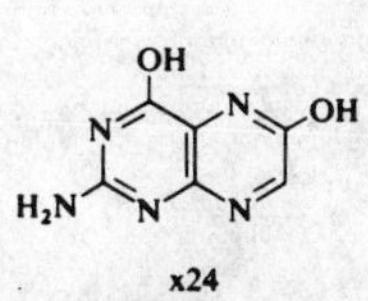

x24

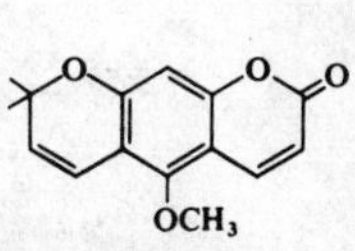

x27

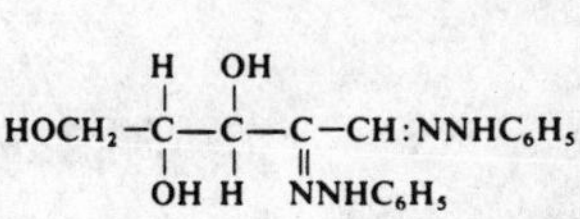

x31

STRUCTURAL FORMULAS OF ORGANIC COMPOUNDS (Continued)

In Alpha Numeric as they occur in Organic Compounds Table

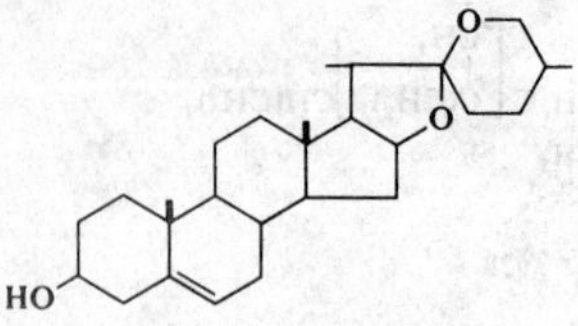

y1

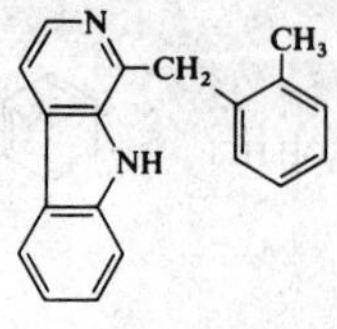

y2

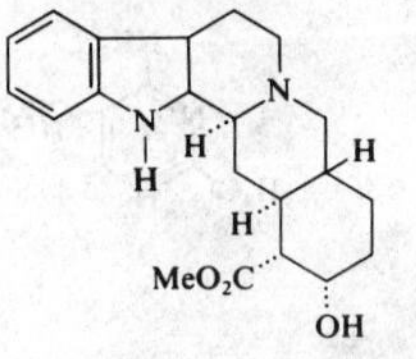

y3

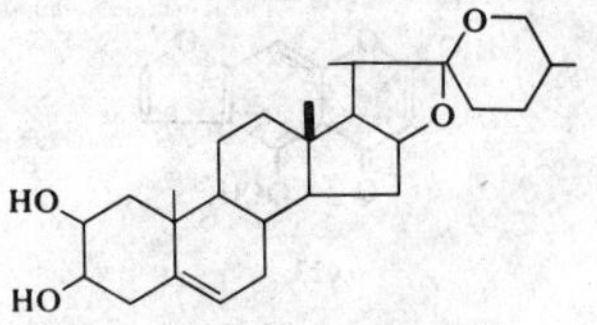

y6

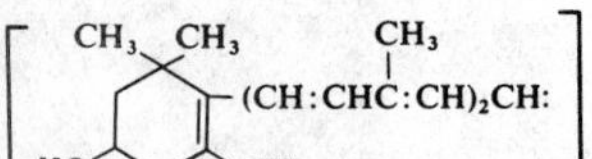

z2

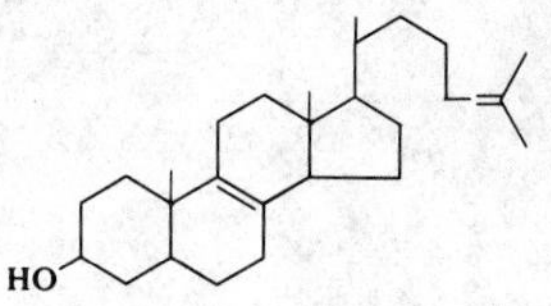

z3

FORMULA INDEX OF ORGANIC COMPOUNDS

Hydrates and other solvates appear under the parent compound

C_2

C_3

c888, c889, c890, f192, m15, p261, p1758[1], s250, s257, t44
$C_7H_{10}O_5$: h138, m105
$C_7H_{10}O_6$: m348
$C_7H_{10}O_7$: g28
$C_7H_{10}S$: t228
$C_7H_{11}Br$: h219, h220
$C_7H_{11}BrO_4$: m54
$C_7H_{11}Cl$: h221, h222, h223, h224, h225, h285, p49, p449
$C_7H_{11}ClO$: c699
$C_7H_{11}ClO_2$: f155
$C_7H_{11}ClO_3$: b2742
$C_7H_{11}ClO_4$: m60, p1255
$C_7H_{11}Cl_3O_2$: a628, a629, a631
$C_7H_{11}N$: p2139, p2141, p2147
$C_7H_{11}NO$: c705
$C_7H_{11}NO_2$: a1400
$C_7H_{11}NO_3$: s210
$C_7H_{11}NO_5$: g62, m107
$C_7H_{11}NO_7S$: t686
$C_7H_{11}NS$: i226
$C_7H_{11}N_3$: t586
C_7H_{12}: b2181, c627, c691, c817, c818, c819, c820, h81, h82, h83, h216, h217, h218, h584, h585, h586, p452, p453, p454
$C_7H_{12}ClNO_2$: a362
$C_7H_{12}Cl_2N_2$: t426
$C_7H_{12}Cl_2O_3$: c163
$C_7H_{12}O$: c617, c697, c803, c804, c805, c806, c807, c808, c837, c838, c839, c840, c869, h84, h211, h229, h588, h589, p460
$C_7H_{12}O_2$: b2652, b3042, c698, f134, h141, p205, p206, p1519, p1770, p1774, p1798, p1801, p1829
$C_7H_{12}O_3$: a260, b2778, b2812, b2831, c706, h165, p298, p1581
$C_7H_{12}O_4$: h132, h395, h398, h399, h400, m23, m55, m70, m93, p161, p166, p167, p1236, p1245, s107, s224
$C_7H_{12}O_5$: g74, m89
$C_7H_{12}O_6$: c706[1], c706[2], m106
$C_7H_{13}Br$: c613, c643, c644, c645, c646
$C_7H_{13}BrN_2O_2$: u37
$C_7H_{13}BrO$: e468
$C_7H_{13}BrO_2$: b2717, h160, h161, p243, p1356
$C_7H_{13}Cl$: h202, h203, h204, h205, p394
$C_7H_{13}ClO$: b2863, h148, h187, h470, h473, h475, p259, p263
$C_7H_{13}ClO_2$: b2738, b2741, p250, p252, p255, p258, p1375, p1378, p1386, p1389
$C_7H_{13}FO_2$: h162
$C_7H_{13}IO_2$: h164
$C_7H_{13}N$: h154, q279
$C_7H_{13}NO$: h447, h448, p953, q280
$C_7H_{13}NO_2$: p1032, s110
$C_7H_{13}NO_3$: b2172, p1077
$C_7H_{13}NO_4$: m42
C_7H_{14}: b2988, b3001, c611, c688, c874, h197, h198. h199, h200, h201, p400, p401, p402, p403, p404, p405, p406, p407, p408, p409
$C_7H_{14}BrF$: h97
$C_7H_{14}BrNO$: b2712
$C_7H_{14}ClF$: h106
$C_7H_{14}ClNO$: a358, a359
$C_7H_{14}Cl_2$: h111, h112, h113, h114, p125, p126, p127, p128
$C_7H_{14}N_2$: a763
$C_7H_{14}N_2O_2$: p949, p1150
$C_7H_{14}N_2O_4$: c102, c118, m36
$C_7H_{14}N_2O_4S_2$: d271
$C_7H_{14}O$: c616, c758, c759, c760, c761, c762, c763, c764, c765, c766, e355, e449, e467, f219[1], h90, h184, h185, h186, h209, h529, h530, h533, h534, m362, p355
$C_7H_{14}O_2$: a194, a222, a223, a226, a228, a253, a253[1], a253[2], a253[3], a253[4], b2673, b2686, b2776, b2800, b2806[1], f137, h144, h436, h467, h468, h469, h471, h472, h474, p222, p262, p264, p1312, p1313, p1322, p1433, p1531, p1537, p1708
$C_7H_{14}O_3$: c172, c173, c176, p1465, p1466
$C_7H_{14}O_4$: c937, g90
$C_7H_{14}O_5$: c165, d182
$C_7H_{14}O_6$: g54, g55
$C_7H_{14}O_7$: g29, m128, m129
$C_7H_{15}Br$: h96
$C_7H_{15}Cl$: b2528, h98, h99, h100, h101, h339, h340, h341, h342, p102, p103, p104, p105, p107, p108
$C_7H_{15}ClO$: e555, h173
$C_7H_{15}ClO_2$: p1089
$C_7H_{15}F$: h123
$C_7H_{15}FO$: h176
$C_7H_{15}I$: h125
$C_7H_{15}N$: c689, h551, p969, p970, p971, p972, p973, p974, p975, p976, p977, p978, p982, p983, p984, p985, p986, p987, p988
$C_7H_{15}NO$: h91, h145, h532, p216, p991, p992, p993, p994, p995, p1305, p1678
$C_7H_{15}NO_2$: c116, c121, h159, m418, n495
$C_7H_{15}NO_5$: g52
C_7H_{16}: b2611, h93, h371, h372, h373, h374, p131, p132, p133, p134, p140
$C_7H_{16}BrNO_2$: c307
$C_7H_{16}ClNO_2$: c308
$C_7H_{16}N_20$: m411
$C_7H_{16}N_2S$: u76, u77, u78
$C_7H_{16}O$: b2911, e454, e454[1], e458, e462, e554, e576, h166, h167, h168, h169, h170, h171, h496, h497, h498, h499, h500, h501, h502, h503, h504, h505, h506, p321, p322, p323, p324, p325, p327
$C_7H_{16}O_2$: f111, h139, p190, p191, p1084, p1258, p1270, p1601, p1651
$C_7H_{16}O_3$: o238
$C_7H_{16}O_4$: d178, g69
$C_7H_{16}O_4S_2$: p1168
$C_7H_{16}O_7$: p478
$C_7H_{16}O_9$: m129
$C_7H_{16}S$: h142
$C_7H_{16}S_3$: o245
$C_7H_{17}N$: a857, h94, h95, h323
$C_7H_{17}NO$: e552, p1622, q280
$C_7H_{17}NO_2$: a898, p1259
$C_7H_{18}N_2O$: p1594, p1597
$C_7H_{18}O_3Si$: s77[1]

C_8

$C_8Br_4O_3$: p892
$C_8Br_6S_2$: b2382
$C_8Cl_4O_3$: p894
$C_8H_2Br_4O_4$: b1062, b1063, p891
$C_8H_2Cl_4O_4$: p893
$C_8H_2N_2O$: h654
$C_8H_3ClO_3$: b1543, p868, p870
$C_8H_3NO_5$: p884
C_8H_4ClNO: i103
$C_8H_4Cl_2N_2$: q268
$C_8H_4Cl_2O_2$: b1018, b1027, p818
$C_8H_4Cl_2O_3$: p871[1], p873, p875, p877
$C_8H_4Cl_2O_4$: b1048, b1049, p871, p872, p874, p876
$C_8H_4N_2$: b1021, b1030, p826
$C_8H_4N_2O_2$: p878
$C_8H_4N_2O_4$: i109, p888
$C_8H_4N_4O_6$: q12
$C_8H_4O_2S$: b2148
$C_8H_4O_3$: p813
$C_8H_5BrO_4$: b1043, b1044, p865, p866
$C_8H_5Br_3O_3$: a609
$C_8H_5ClN_2$: q267
$C_8H_5ClN_2O_5$: a431
$C_8H_5ClO_3$: b1820
$C_8H_5ClO_4$: b1045, b1046, b1047, p867, p869
$C_8H_5Cl_3O_3$: a635, a636
C_8H_5NO: a560, b1635, b1637
$C_8H_5NO_2$: b1022, b1032, b2161, i102, p830, p856
$C_8H_5NO_3$: i110
$C_8H_5NO_4$: p879, p880, p900
$C_8H_5NO_5$: a556
$C_8H_5NO_6$: b1058, b1061, p883, p887, p2082, p2083, p2084, p2085, p2086
C_8H_6: b778
C_8H_6BrClO: a656, a657, a658
C_8H_6BrN: a292, a296, a299
$C_8H_6Br_2O$: a668
$C_8H_6Br_2O_2$: b2085
$C_8H_6Br_4$: b293, b294, b295, b721, b722
C_8H_6ClN: a394, b1512, b1513

$C_8H_6ClNO_2$: a666
$C_8H_6ClNO_4$: b1522, b1523
C_8H_6ClNS: b2126
$C_8H_6ClN_3O_5$: a347
$C_8H_6Cl_2O$: a669, a670, a671, a672
$C_8H_6Cl_2O_3$: a418
$C_8H_6Cl_3NO$: a341, a614
$C_8H_6F_3NO$: a638
$C_8H_6I_2O_3$: b101
$C_8H_6N_2$: q7, q265
$C_8H_6N_2O$: a561, i6, p810, p851, q8, q273, q274
$C_8H_6N_2O_2$: a548, a550, a552, a554, b1823, b1824, b1825, b1826, b1827, b1828, b1829, b1830, b1831, b1832, i69[2], i104, i105, p829, q11, q13, q269
$C_8H_6N_2O_6$: a195, a196, a430, a433, b1601
$C_8H_6N_2O_7$: a428, a429
$C_8H_6N_4O_8$: a786
C_8H_6O: b1231, e564
C_8H_6OS: b2146, b2147, j11
$C_8H_6O_2$: b1014, b1015, c525, g188, p808, p897
$C_8H_6O_2S$: j13
$C_8H_6O_3$: a558, b160, b1633, b1634, b1636
$C_8H_6O_4$: b1016, b1024, b1638, b1639, b1640, b1641, b1819, p812
$C_8H_6O_5$: b1053, b1054, b1055, p881, p882, p882[1],
$C_8H_6O_6$: b1050
C_8H_6S: b2145
$C_8H_6S_2$: b2381
C_8H_7Br: s160, s161 s162, s163, s164
C_8H_7BrO: a653, a654, a655
$C_8H_7BrO_2$: a283, a295, a297, b1414, b1420, b1427
$C_8H_7BrO_3$: a293, a294, a300, b66
$C_8H_7Br_3O$: b694, b695
C_8H_7Cl: s165, s166, s167, s168, s169
$C_8H_7ClN_2O_3$: a83, a84, a85, a86, a87
C_8H_7ClO: a579, a661, a662, a663, a664, b1787, b1795, b1803
$C_8H_7ClO_2$: a376, a390, a391, a392, a393, a395, a396, a570, b1470, b1478, b1485, b1503, b1504, b1505, b1506, b1507, b1508, b1509, b1510, b1511, b1720, b1721, b1765, b1771, b1778, f150
$C_8H_7ClO_2S$: s181
$C_8H_7ClO_3$: a387, a388, a389, a397, b79, b1495, b1501, b1502
$C_8H_7Cl_3O$: b696, b697, b698, b699, b700
C_8H_7F: s170, s171, s172
$C_8H_7IO_2$: b1747, b1749, b1751
$C_8H_7IO_3$: a466, b131, b132
C_8H_7N: a583, b1790, b1798, b1806, c483, i65, t542
C_8H_7NO: a486, a487, a488, a498, a500, a502, a573, b1221, b1768, b1781, b2160, i69, i150, i151, i152, o299
$C_8H_7NO_2$: a47, o302, s179, s180
$C_8H_7NO_3$: a706, a707, a708, b161, b1031, b1632, i101, o280, p849, p2060, p2061
$C_8H_7NO_4$: a243, a244, a245, a549, a551, a553, b156, b1034, b1036, b1038, b1040, b1842, b1851, b1860, f143, p860, p2076
$C_8H_7NO_5$: a490, a547
C_8H_7NS: b2139, i217, t166, t182
$C_8H_7NS_2$: b2134, b2135, b2136, b2137, b2140, b2144
$C_8H_7N_3$: q266
$C_8H_7N_3O$: u130
$C_8H_7N_3O_2$: l60, p862
$C_8H_7N_3O_5$: a105, a106, a107, a108, a109, a110
$C_8H_7N_3O_6$: b723, b724
C_8H_8: c860, s159
C_8H_8BrNO: a76, a77, a78, a79
$C_8H_8Br_2$: b281, b282, b283, b515
C_8H_8ClNO: a88, a89, a90, a91, a650, a651, c125
$C_8H_8ClNO_3S$: b1105, b1107
$C_8H_8Cl_2$: b287, b288, b289, b551, h582
$C_8H_8Cl_2O$: b544, b545, b546, b547, b548, b549, b550
C_8H_8INO: a133
$C_8H_8N_2$: a268, a269, a271, b1211, b1212, b1213, b1214, b1368, b1370, b1372, b1373, b1375, b1376, b1378, b1380, b1382, c572, g171
$C_8H_8N_2OS$: u31
$C_8H_8N_2O_2$: a46, b1016[1], b1025, p814, r30, u30
$C_8H_8N_2O_3$: a159, a160, a161
$C_8H_8N_2O_4$: b663, b664, b665, b666, b667, b668, b669, b670, g165
$C_8H_8N_2O_5$: b728, b729, b730, b731
$C_8H_8N_2O_6$: b650
$C_8H_8N_2S$: b2122, b2123, b2124, b2143
$C_8H_8N_4O_2S_2$: s284
$C_8H_8N_4O_4$: a23, a24
$C_8H_8N_5O_6$: m423
C_8H_8O: a48, a645, b157, b158, b159, b745, c524, e242, e540, s173, s174
$C_8H_8O_2$: a255, a574, a685, a687, a688, a689, b152, b153, b154, b1286, b1785, b1792, b1800, b2098, b2099, b3057, c625, c626, f131
$C_8H_8O_2S$: a519
$C_8H_8O_3$: a45, a213, a470, a471, a472, a473, a495, a499, a501, a567, a673, a674, a675, a676, a677, b100, b133, b134, b135, b136, b137, b138, b139, b573, b1671, b1686, b1693, b1719, b1721, b1722, b1723, b1724, b1725, b1726, b1727, b1728, b1729, b1730, b1763, b1770, b1775, c830, c831, c832, c833, c834, f223, p1547, p1789, t513
$C_8H_8O_4$: a421, a492, a494, a710, a711, b1555, b1713, b1714, b2097, b2101, c633, c634, c635, h569, p1453
$C_8H_8O_5$: b1914, b1918, b1924, b2096, f271, f273, f275
$C_8H_8O_7$: t12
C_8H_9Br: b327, b328, b329, b330, b338[1], b339, b340, b353, b354, b355
C_8H_9BrO: b336, b337, b338, e556
$C_8H_9Br_2NO$: b217
C_8H_9Cl: b420, b421, b422, b423, b443, b444, b445, b446, b447, b468, b469, b470
$C_8H_9ClN_2$: b1383
$C_8H_9ClN_2O_2$: a1051
C_8H_9ClO: b424, b425, b426, b427, b428, b429, b430, b431, b432, b433, b439, b440, b441, b442, e351, e445, t409
$C_8H_9ClO_2S$: b1153, b1155, b1156
C_8H_9ClS: b448
$C_8H_9Cl_2NO$: a1081, a1082
$C_8H_9Cl_2P$: p738, p739, p740
C_8H_9FO: b748, b749, b750
C_8H_9I: b701, b702, b703, b704, b705, b706, b762, b763, b833, b834
C_8H_9IO: b751, b752, b829
$C_8H_9IO_2$: b651
C_8H_9N: a807, b40, i75, p1947
C_8H_9NO: a164, a575, a646, a647, a648, a649, a1135, b1786, b1793, b1801, f126, f127, f128
C_8H_9NOS: a517
$C_8H_9NO_2$: a128, a129, a130, a267, a270, a477, a496, a568, b155, b713, b714, b715, b716, b717, b718, b773, b774, b775, b1319, b1327, b1338, b1367, b1369, b1371, b1374, b1377, b1379, b1381, b1764, b1776, b1809, b1812, b1814, c87, g167, p2038, p2043, p2044, p2050, p2051
$C_8H_9NO_3$: b685, b686, b687, b688, b689, b690, b691, b692, b693, b755, b757, b765, b1360, b1363, g159
$C_8H_9NO_3S$: b1085, b1086
$C_8H_9NO_4$: a1006, b653, b654, b655, b656, b657, b658
C_8H_9NS: a594

C_{11}

$C_{12}H_{10}N_2O$: a887, a892, a1577, a1579, a1582, a1647, a1648, p531
$C_{12}H_{10}N_2O_2$: a884, a885, a886, a1541, a1541[1], a1542, b2227, b2228, b2229, b2230, b2231, b2232, b2233
$C_{12}H_{10}N_2O_2S$: d144
$C_{12}H_{10}N_2O_3$: a158, n16, n17, n18, n19, n20, n21, n22, n23
$C_{12}H_{10}N_2O_3S$: a1629, s389
$C_{12}H_{10}N_2O_4S$: a1635
$C_{12}H_{10}N_2O_5$: g176
$C_{12}H_{10}N_2O_6S_2$: a1614[1], a1617[1], a1619[1]
$C_{12}H_{10}N_2S$: d145, d146
$C_{12}H_{10}N_4O_7$: a1009
$C_{12}H_{10}O$: b2312, b2313, b2314, e510, n24, n25
$C_{12}H_{10}OS$: s387
$C_{12}H_{10}O_2$: a239, a240, a540, b2275, b2276, b2277, b2278, b2279, b2280, n28, n29, n30, n31, n32, n33, n192, n321, n330, n421, n422, n423, n424, n425
$C_{12}H_{10}O_2S$: s323, s359
$C_{12}H_{10}O_3$: a538, a539, e514, e515, f224, n377, n426
$C_{12}H_{10}O_3S$: a15, s391
$C_{12}H_{10}O_4$: b2326, m62, p55, q14
$C_{12}H_{10}O_4S$: s369, s370, s371, s372
$C_{12}H_{10}O_6S_2$: b2356
$C_{12}H_{10}O_7$: e8[1], n427
$C_{12}H_{10}S$: s317
$C_{12}H_{10}S_2$: d248
$C_{12}H_{10}Se$: s33
$C_{12}H_{10}Se_2$: d218
$C_{12}H_{11}As$: a1435
$C_{12}H_{11}AsO_2$: a1450
$C_{12}H_{11}N$: a5, a6, a7, a8, a863, b2220, b2221, b2222, p1967, p1968, p1969
$C_{12}H_{11}NO$: a156, a157, a541, a543, a881, a882, a883, a1175, a1176, a1177, b2223, b2224, b2225, b2226, c152
$C_{12}H_{11}NOS$: a516[1], s388
$C_{12}H_{11}NO_2$: a126, g164, j10, m51
$C_{12}H_{11}NO_2S$: b1096, s360, s361
$C_{12}H_{11}NO_3$: s259
$C_{12}H_{11}NO_3S$: b1180[1]
$C_{12}H_{11}NO_4$: g175
$C_{12}H_{11}NS$: s318, s319
$C_{12}H_{11}N_3$: a1520, a1521, a1522, d91, d92
$C_{12}H_{12}$: n168, n169, n181, n182
$C_{12}H_{12}BrN$: a864
$C_{12}H_{12}BrNO_2$: p834
$C_{12}H_{12}BrN_3$: a1523
$C_{12}H_{12}ClNO$: n26
$C_{12}H_{12}ClN_3$: a1524
$C_{12}H_{12}N_2$: a865, a866, b2257, b2258, b2259, b2260, b2261, h668, h669
$C_{12}H_{12}N_2OS$: s390
$C_{12}H_{12}N_2O_2S$: s364, s365, s366
$C_{12}H_{12}N_2O_3$: a1543, b25, b30
$C_{12}H_{12}N_2O_4S_2$: b2354, b2357
$C_{12}H_{12}N_2O_6S_2$: b2359
$C_{12}H_{12}N_2O_7$: b1605
$C_{12}H_{12}N_2S$: s320, s321, u97
$C_{12}H_{12}N_2S_2$: d249, d250, d251
$C_{12}H_{12}N_4$: a1536, a1537, a1538
$C_{12}H_{12}N_4O_4S_2$: a1615, a1618, a1620
$C_{12}H_{12}N_4O_7$: a792
$C_{12}H_{12}O$: e381, e382, h306, n179, n180
$C_{12}H_{12}O_2$: c351, p58
$C_{12}H_{12}O_3$: b944, c536
$C_{12}H_{12}O_5$: c546
$C_{12}H_{12}O_6$: b962, b964 b967, b1193
$C_{12}H_{12}O_{12}$: c736
$C_{12}H_{13}BrO_3$: b2660
$C_{12}H_{13}N$: c156, n85, n86, n170, n171, n183, n184, q225, q226, q227, q228, q229, q230, q231, q232
$C_{12}H_{13}NO$: e383
$C_{12}H_{13}NO_2$: b2794, p842, p1510
$C_{12}H_{13}NO_3$: s211
$C_{12}H_{13}NO_3S$: a1003
$C_{12}H_{13}NO_6$: b1059, p885
$C_{12}H_{13}N_3$: a868, b2328
$C_{12}H_{13}N_3O_4S_2$: b1131
$C_{12}H_{13}N_5$: a1603
$C_{12}H_{14}$: c823
$C_{12}H_{14}As_2Cl_2N_2O_2$: s9
$C_{12}H_{14}ClNO_3$: c520
$C_{12}H_{14}ClN_5O_2S$: p1078
$C_{12}H_{14}N_2$: n87, q133
$C_{12}H_{14}N_2O_2$: t777, t778
$C_{12}H_{14}N_2O_3$: b4
$C_{12}H_{14}N_2O_6$: b1602
$C_{12}H_{14}N_4$: h637, h638
$C_{12}H_{14}N_4OS$: t162
$C_{12}H_{14}N_4O_2S$: s280
$C_{12}H_{14}N_6O_{22}$: c235
$C_{12}H_{14}O$: b1223, t99, t119
$C_{12}H_{14}O_2$: b2554, c357, c367, c401, h570, p439
$C_{12}H_{14}O_3$: b807, b809, b1300, b2771, b2834, e612, p299[1]
$C_{12}H_{14}O_4$: a1368, b1019, b1028, i111, p821, p852, p853, p854
$C_{12}H_{14}O_5$: c424
$C_{12}H_{15}Br$: b315
$C_{12}H_{15}ClO_2$: b1818
$C_{12}H_{15}N$: j9, p297
$C_{12}H_{15}NO$: p955
$C_{12}H_{15}NO_3$: a64, a996, a997, a998, h722
$C_{12}H_{15}NO_4$: c519
$C_{12}H_{15}NO_5$: p2159
$C_{12}H_{15}N_3O_6$: b383, p890, t705
$C_{12}H_{15}N_5O_{20}$: c236
$C_{12}H_{16}$: b484, d163
$C_{12}H_{16}ClNO_3$: a999
$C_{12}H_{16}N_2O$: c948
$C_{12}H_{16}N_2O_3$: b11, p212
$C_{12}H_{16}N_2O_4$: b1023, b1033
$C_{12}H_{16}N_2O_4S$: a1011
$C_{12}H_{16}N_2O_5$: b385
$(C_{12}H_{16}N_4O_{18})_n$: c237
$C_{12}H_{16}O$: a698, b486, b487, c767, c768, h535, p367
$C_{12}H_{16}O_2$: a581, b485, b798, b801, b1287, b1869, p296, p1573, p1577
$C_{12}H_{16}O_3$: a509, b926, b1673, b1678, b1777, b1816, b1817, b2841, b2845, h519, p1790
$C_{12}H_{16}O_4$: b1557
$C_{12}H_{16}O_7$: a1398
$(C_{12}H_{16}O_8)_n$: c238
$C_{12}H_{17}ClN_4$: h696
$C_{12}H_{17}N$: a1066, b216
$C_{12}H_{17}NO$: a81, a82, h429, p293
$C_{12}H_{17}NO_3$: a995
$C_{12}H_{17}NO_4$: p2158
$C_{12}H_{17}N_3O_{16}$: c240
$C_{12}H_{17}N_7O_9$: a1411
$C_{12}H_{18}$: b641, b642, b643, b764, b791, b886, b959, b960, h376, o104
$C_{12}H_{18}ClNO_2$: c309
$C_{12}H_{18}Cl_2N_4OS$: v25
$C_{12}H_{18}N_2O_3S$: t274
$C_{12}H_{18}N_2O_5$: g46, g47, m139, m140
$C_{12}H_{18}N_2O_6$: g36, g216, i3
$C_{12}H_{18}O$: b378, b379, b380, b381, b382, b387, b388, b744, b860, e440, e441, e573, h507
$C_{12}H_{18}O_2$: b595, b612, b742, b743, d5, p655[2]
$C_{12}H_{18}O_3$: b958, f230
$C_{12}H_{18}O_4$: m65, m66, p2121
$C_{12}H_{18}O_6$: p1763, s230
$C_{12}H_{18}O_8$: a1399, t20
$C_{12}H_{19}BrN_2O_2$: p1853
$C_{12}H_{19}N$: a1115
$C_{12}H_{19}NO$: a1086, a1087
$C_{12}H_{20}$: d318
$C_{12}H_{20}N_2O_3$: b17
$C_{12}H_{20}N_2S_4$: d229
$C_{12}H_{20}O$: f203
$C_{12}H_{20}O_2$: b2417, b2418, b2419, e58, e567, i116, i117, i118, 145, t50
$C_{12}H_{20}O_4$: c62[1], c711, c712, c717, c718, c721, c722, d316, d317, f180
$C_{12}H_{20}O_6$: g119
$C_{12}H_{20}O_7$: c432
$C_{12}H_{21}ClO_2$: a514
$C_{12}H_{21}NO_3$: a950
$C_{12}H_{21}N_7O_{11}$: a1410, a1412
$C_{12}H_{22}$: b2186, b2187, d78, d319, d320, d321, d322
$C_{12}H_{22}O_2$: a218, c893, h52, h457, o157
$C_{12}H_{22}O_3$: a513, h430
$C_{12}H_{22}O_4$: d38, h389, h392, m79, m101, m108, m109, o144, o271, s189, s190
$(C_{12}H_{22}O_5)_n$: c239
$C_{12}H_{22}O_6$: t14, t15, t21, t22
$C_{12}H_{22}O_{10}$: r38
$C_{12}H_{22}O_{11}$: c231, l5, l6, l7, m120, m154, s273, t697, t698, t780
$C_{12}H_{22}O_{12}$: l4
$C_{12}H_{23}BrO_2$: d309
$C_{12}H_{23}ClO$: d300
$C_{12}H_{23}FO_2$: d310
$C_{12}H_{23}N$: a825, d304
$C_{12}H_{23}NO$: i141
$C_{12}H_{23}NO_4$: a894
$C_{12}H_{24}$: d315, p1748
$C_{12}H_{24}BrF$: d289
$C_{12}H_{24}Br_2$: d291
$C_{12}H_{24}ClNO$: a320, a348
$C_{12}H_{24}O$: d282[1], d313, h18

$C_{12}H_{24}O_2$: a190, b2681, d47, d295, h156, h435, o164, p224
$C_{12}H_{24}O_3$: p12, p14[1]
$C_{12}H_{24}O_6$: m172
$C_{12}H_{24}O_{12}$: m121
$C_{12}H_{25}Br$: d288
$C_{12}H_{25}Cl$: d290
$C_{12}H_{25}I$: d292
$C_{12}H_{25}N$: p989
$C_{12}H_{25}NO$: d296
$C_{12}H_{26}$: d284
$C_{12}H_{26}N_2O_4$: b2685, p1536
$C_{12}H_{26}N_2O_6$: a449
$C_{12}H_{26}O$: d311, d312, e496, h38
$C_{12}H_{26}O_4S$: s401
$C_{12}H_{26}O_{13}$: m155
$C_{12}H_{26}S$: d294, s311
$C_{12}H_{27}B$: b2409
$C_{12}H_{27}BO_3$: b2387
$C_{12}H_{27}N$: a843, a920, a938, d285
$C_{12}H_{27}NO_2$: a862
$C_{12}H_{27}O_3P$: p799
$C_{12}H_{27}O_4P$: p775, p777
$C_{12}H_{28}BrN$: a976
$C_{12}H_{28}ClN$: d287
$C_{12}H_{28}IN$: a977
$C_{12}H_{30}OSi_2$: d220
$C_{12}H_{30}O_{13}P_4$: t121
$C_{12}H_{36}B_3P_3$: b2402
$C_{12}H_{36}O_4Si_5$: p371
$C_{12}H_{36}O_6Si_6$: c810

C_{13}

$C_{13}H_5N_3O_7$: f90, f91
$C_{13}H_6Cl_6O_2$: m207
$C_{13}H_7BrO$: f86
$C_{13}H_7ClN_2O_5$: b1982
$C_{13}H_7ClO$: f87
$C_{13}H_7NO_3$: f89
$C_{13}H_7N_3O_4S_2$: b2128
$C_{13}H_8Br_2O$: b1986
$C_{13}H_8ClN$: a734
$C_{13}H_8Cl_2$: f64, f65
$C_{13}H_8Cl_2O$: b1987, b1988
$C_{13}H_8I_2O$: b1999
$C_{13}H_8N_2O_5$: b2009
$C_{13}H_8N_2O_6$: b1603
$C_{13}H_8N_4O_8$: m203
$C_{13}H_8N_6O_9$: u34
$C_{13}H_8O$: f80
$C_{13}H_8OS$: t249, x9
$C_{13}H_8O_2$: x12
$C_{13}H_8O_2S$: d155, d156
$C_{13}H_8O_3$: d128, d129, d130
$C_{13}H_8O_4$: x13, x14, x15, x16, x17, x18, x19, x20, x21
$C_{13}H_8O_4S$: d157
$C_{13}H_8O_6$: n312, n313
$C_{13}H_9Br$: f55, f56
$C_{13}H_9BrO$: b1976, b1977, b1978
$C_{13}H_9Br_3O_3S$: t674
$C_{13}H_9ClO$: b1979, b1980, b1981
$C_{13}H_9N$: a727, b2066, b2067, b2330, b2333, p526
$C_{13}H_9NO$: a738, f81, f82, f83, f84, f85, p527
$C_{13}H_9NOS$: b1220, b2130, b2131
$C_{13}H_9NO_2$: b2154, b2155, b2156, b2157, b2158, f74, f75, f76
$C_{13}H_9NO_3$: b2027, b2028, b2029
$C_{13}H_9NS$: b2141, i218
$C_{13}H_9N_3O_5$: a1608, a1609, a1610
$C_{13}H_{10}$: f45
$C_{13}H_{10}ClNO$: c113
$C_{13}H_{10}Cl_2$: m255
$C_{13}H_{10}Cl_2O_2$: m195, m196
$C_{13}H_{10}N_2$: a728, a729, a730, a732, b1218, c158, c577, m245, p554
$C_{13}H_{10}N_2O$: p1892
$C_{13}H_{10}N_2O_2$: a1604, a1605, a1606
$C_{13}H_{10}N_2O_3$: a1607, b1248, b1249, b1250, b1838, b1846, b1855
$C_{13}H_{10}N_2O_4$: m209, m210, m211, m270, m287, m288
$C_{13}H_{10}N_2O_5S$: a1611, a1634
$C_{13}H_{10}N_2S$: b2142
$C_{13}H_{10}O$: b1959, f67, f68, x1
$C_{13}H_{10}O_2$: a13, b1295, b2010, b2011, b2012, b2329, b2331, b2332, p1802, p1803, p1804, p1824, x2
$C_{13}H_{10}O_3$: b1278, b1279, b1280, b1679, b1871, b1989, b1990, b1991, b1992, b1993, b1994, b1995, b1996, b2334, c175, p71[1]
$C_{13}H_{10}O_3S$: f78
$C_{13}H_{10}O_4$: b2050, b2051, b2052, b2053, b2054, n333, v23
$C_{13}H_{10}O_5$: b2030, b2031, b2032, b2033, b2034, b2035, b2036, b2037, i182
$C_{13}H_{10}S$: b2049, t248
$C_{13}H_{11}Br$: m221
$C_{13}H_{11}BrO_2$: p1664
$C_{13}H_{11}Cl$: m235, m239[1]
$C_{13}H_{11}ClN_2$: a731
$C_{13}H_{11}ClN_2O_2S$: a1637
$C_{13}H_{11}N$: a737, a1015, b1960, c148, f48, f49
$C_{13}H_{11}NO$: b1251, b1961, b1963, b1964, b1965, f50, f51, f52, f121, p587
$C_{13}H_{11}NO_2$: b1321, b1340, b1653, b1683, b1690, b1872
$C_{13}H_{11}NO_3$: f3
$C_{13}H_{11}NO_4$: b1921
$C_{13}H_{11}NS$: b1892
$C_{13}H_{11}N_3$: a735
$C_{13}H_{12}$: b1219, b2320, b2321, b2322, m271, n35
$C_{13}H_{12}N_2$: b44, f63, f130
$C_{13}H_{12}N_2O$: a801, a1587, a1589, a1590, a1591, a1592, a1594, a1595, a1597, a1598, a1599, b1297, b1332, b1983, b1984, b1985, h3, u70, u71
$C_{13}H_{12}N_2O_3$: b7
$C_{13}H_{12}N_2S$: u74, u75
$C_{13}H_{12}N_4O$: c157
$C_{13}H_{12}N_4O_4$: m190, m191
$C_{13}H_{12}N_4S$: d269
$C_{13}H_{12}O$: b279, b280, b2318, b2319, e432, m281, m366
$C_{13}H_{12}O_2$: a229, a230, a231, a232, e535, m205, m206, m280, n10, n11, n320, n329, p588, p589, p590, p1539, p1540
$C_{13}H_{12}O_2S$: s352, s381, s382
$C_{13}H_{12}O_3$: n375
$C_{13}H_{12}O_4$: n367
$C_{13}H_{12}S$: s289
$C_{13}H_{13}N$: a802, a891, m178, m185, m186
$C_{13}H_{13}NO$: a150, m355
$C_{13}H_{13}NO_2S$: t643, t651, t661
$C_{13}H_{13}N_3$: d100, d101, g201
$C_{13}H_{13}N_3O$: s40, s41, s42
$C_{13}H_{13}N_3O_5S_2$: s288
$C_{13}H_{13}N_3S$: s43, s44
$C_{13}H_{14}$: n255, n256, n257, n258
$C_{13}H_{14}N_2$: h698, h699, h700, m192
$C_{13}H_{14}N_2O$: h2
$C_{13}H_{14}N_2O_3$: b21, t775, t776
$C_{13}H_{14}N_2O_8$: m75
$C_{13}H_{14}N_4O$: c137
$C_{13}H_{14}O$: n248, n249
$C_{13}H_{14}O_2$: p57, p1686
$C_{13}H_{14}O_3$: b2701
$C_{13}H_{14}O_4$: s266
$C_{13}H_{14}O_5$: c437
$C_{13}H_{14}O_6$: t610, t611, t612, t613
$C_{13}H_{15}Cl_3N_2O_3$: h764
$C_{13}H_{15}N$: a913
$C_{13}H_{16}BrNO_3$: h450, h451
$C_{13}H_{16}O$: p438
$C_{13}H_{16}O_2$: b1271
$C_{13}H_{16}O_3$: b2700
$C_{13}H_{16}O_4$: m111
$C_{13}H_{16}O_7$: h8, v1
$C_{13}H_{17}BrO$: p1665
$C_{13}H_{17}NO$: k22
$C_{13}H_{17}NO_3$: h449, l35
$C_{13}H_{17}NO_4$: m115, p2073, p2074
$C_{13}H_{17}N_3O$: a1365
$C_{13}H_{18}N_2O_3$: l77, l78, l79
$C_{13}H_{18}N_2O_4$: b386
$C_{13}H_{18}N_4O_2$: a1413
$C_{13}H_{18}O$: h194[1], p1098
$C_{13}H_{18}O_2$: b1275, b1833
$C_{13}H_{18}O_7$: t461
$C_{13}H_{19}NO_2$: i194
$C_{13}H_{19}NO_3$: p25
$C_{13}H_{19}NO_4$: p2070, p2071
$C_{13}H_{20}$: t713
$C_{13}H_{20}N_2$: p979
$C_{13}H_{20}N_2O_2$: b1335
$C_{13}H_{20}N_4O_4$: c16
$C_{13}H_{20}O$: b810, h182, i94, i95, i96, i98, i100, e570, p326, p1873
$C_{13}H_{20}O_2$: p655[1]
$C_{13}H_{20}O_3$: f237
$C_{13}H_{20}O_4$: f267
$C_{13}H_{20}O_8$: p75
$C_{13}H_{21}ClN_2O_2$: b1336
$C_{13}H_{22}ClNO$: n571
$C_{13}H_{22}ClN_3O$: b1331
$C_{13}H_{22}O$: i92, i93
$C_{13}H_{22}O_4$: m64
$C_{13}H_{24}$: t731
$C_{13}H_{24}N_2O$: c567, c568
$C_{13}H_{24}O_2$: h51
$C_{13}H_{24}O_4$: m87, n527
$C_{13}H_{25}N$: t725
$C_{13}H_{26}$: t730
$C_{13}H_{26}Br_2$: t719
$C_{13}H_{26}N_2$: m187, m188, m189
$C_{13}H_{26}O$: t714, t727, t728, t729
$C_{13}H_{26}O_2$: d47[1], d52, d303, h23, h151, h434, o171, o174, p230, t722
$C_{13}H_{27}Br$: t718
$C_{13}H_{27}N$: p1013

b2264, e177, e192, h636, h648, h649, h670, h671, h672, h673, h706
$C_{14}H_{16}N_2O$: b2265
$C_{14}H_{16}N_2O_2$: b2262
$C_{14}H_{16}N_2O_4$: a1007
$C_{14}H_{16}N_2O_6S_2$: b2360
$C_{14}H_{16}N_4O_4S_2$: a1624
$C_{14}H_{16}O_4$: m50
$C_{14}H_{17}N$: n152
$C_{14}H_{17}NO_2$: c528
$C_{14}H_{17}NO_3$: f1
$C_{14}H_{17}NO_5$: m43
$C_{14}H_{17}NO_6$: p1860, p1861, s10
$C_{14}H_{18}$: a1233, p510, p511, p512
$C_{14}H_{18}N_2O_2$: h763
$C_{14}H_{18}N_2O_5$: a660
$C_{14}H_{18}N_2O_6$: f239
$C_{14}H_{18}N_4O_9$: c10
$C_{14}H_{18}O$: h92, h208
$C_{14}H_{18}O_4$: m46, p828
$C_{14}H_{18}O_6$: b286, p823
$C_{14}H_{18}O_7$: p907
$C_{14}H_{19}NO$: a99
$C_{14}H_{19}N_3S$: m387
$C_{14}H_{20}N_2$: b2188
$C_{14}H_{20}N_2O_4$: b384
$C_{14}H_{20}N_2O_6S$: p676
$C_{14}H_{20}N_4$: c703
$C_{14}H_{20}N_4O_7$: c490, c491
$C_{14}H_{20}O_2$: b2087
$C_{14}H_{20}O_8$: e427
$C_{14}H_{21}NO_2$: o139
$C_{14}H_{21}NO_4$: p2072, p2157, p2158
$C_{14}H_{22}$: b525, b934, b935, o140, p499, p500, t63
$C_{14}H_{22}N_4$: h640
$C_{14}H_{22}O$: b529, b530, e588
$C_{13}H_{22}O_2$: b524, b527, e592, p682[1]
$C_{14}H_{22}O_4$: o265
$C_{14}H_{22}O_8$: e326
$C_{14}H_{23}N$: a1075, b269
$C_{14}H_{23}NO_9$: i59
$C_{14}H_{23}N_3O$: i97, i99
$C_{14}H_{24}$: p513
$C_{14}H_{24}N_2$: b296, b297, b298
$C_{14}H_{24}O_4$: c62
$C_{14}H_{25}NO_2$: c199
$C_{14}H_{26}$: t92, t93
$C_{14}H_{26}ClNO_2$: c200
$C_{14}H_{26}N_2O_8$: m33, p164
$C_{14}H_{26}O_2$: e304
$C_{14}H_{26}O_3$: a447, h146
$C_{14}H_{26}O_4$: d35, d293, h383, h389
$C_{14}H_{26}O_6$: t28
$C_{14}H_{27}ClO$: t81
$C_{14}H_{27}N$: t85
$C_{14}H_{28}$: t90
$C_{14}H_{28}Br_2$: t71
$C_{14}H_{28}ClNO$: a344
$C_{14}H_{28}O$: h44, t64, t88, t89
$C_{14}H_{28}O_2$: d301, h150, h439, o168, t77, t724
$C_{14}H_{29}Br$: t69
$C_{14}H_{29}Cl$: t70
$C_{14}H_{29}N$: p1012
$C_{14}H_{29}NO$: t66, t78
$C_{14}H_{30}$: t67
$C_{14}H_{30}N_2O_4$: p233
$C_{14}H_{30}O$: e494, t87
$C_{14}H_{30}O_2$: d283, t74
$C_{14}H_{30}O_3S$: t75
$C_{14}H_{30}O_4S$: s399
$C_{14}H_{30}S$: s310, t76
$C_{14}H_{31}N$: a841, t68
$C_{14}H_{31}NO_2$: d286
$C_{14}H_{42}O_5Si_6$: h536
$C_{14}H_{42}O_7Si_7$: c618

C_{15}

$C_{15}H_6ClNO_5$: a1347
$C_{15}H_6N_4O_{13}$: a787
$C_{15}H_7NO_6$: a1346, a1348
$C_{15}H_8Cl_4O_3$: b1405
$C_{15}H_8O_5$: c551
$C_{15}H_8O_6$: a1342[1], a1343, a1344
$C_{15}H_9BrO_2$: f21
$C_{15}H_9N$: p518, p520, p522
$C_{15}H_9NO_4$: a1317
$C_{15}H_{10}BrNO_2$: a1261, a1262
$C_{15}H_{10}Br_2O_2$: p1271
$C_{15}H_{10}N_2O_5$: c247
$C_{15}H_{10}N_4O_{10}$: a274
$C_{15}H_{10}O$: a1243
$C_{15}H_{10}O_2$: a1244, a1245, a1246, a1316, f20, i34, p516, p517, p519, p521, p898
$C_{15}H_{10}O_3$: f25, f97, p1301
$C_{15}H_{10}O_4$: a1297, a1298, f22, i161
$C_{15}H_{10}O_5$: a1296, a1318, a1319, b1042, b1462, f37, f38, i162, i183, p863, p864
$C_{15}H_{10}O_6$: f33, f34, f35, f36
$C_{15}H_{10}O_7$: f26, f27, f28, f29, f30, f31, f32
$C_{15}H_{10}ClO_5$: p22
$C_{15}H_{11}ClO_7$: d81
$C_{15}H_{11}I_4NO_4$: t266, t267
$C_{15}H_{11}N$: q218, q219, q220, q221, q222
$C_{15}H_{11}NO$: o294, o295, o296, q177
$C_{15}H_{11}NO_2$: a1320, a1321, p832
$C_{15}H_{11}NO_3$: c252, c253, c254, c255, c256
$C_{15}H_{11}NO_4$: p1805, p1806, p1807, p1808, p1809
$C_{15}H_{11}NS$: t143
$C_{15}H_{11}N_3O$: a1510, a1511
$C_{15}H_{12}$: a1229, a1230, a1231, p507, p508
$C_{15}H_{12}Br_2O$: p1674, p1675
$C_{15}H_{12}I_2O_3$: p1423[1]
$C_{15}H_{12}N_2OS$: h618
$C_{15}H_{12}N_2O_2$: h617
$C_{15}H_{12}N_2O_3$: f290, h725, i163
$C_{15}H_{12}N_6O_4$: r18
$C_{15}H_{12}O$: c245, f88
$C_{15}H_{12}O_2$: a452, c365, f17, p1273, p1274, p1275, p1276, p1783[1]
$C_{15}H_{12}O_3$: b1241, b1242, b1264, b1401, b1893, b1894, b1895, b1897, b1898
$C_{15}H_{12}O_4$: b1288, b1672, m204
$C_{15}H_{12}O_6$: f19, m194
$C_{15}H_{13}NO$: c154, f46, f47
$C_{15}H_{13}NO_2$: p1548
$C_{15}H_{13}NO_3$: b1240
$C_{15}H_{13}NO_4$: b1650
$C_{15}H_{13}N_5O_6$: a1112
$C_{15}H_{14}$: p1741
$C_{15}H_{14}N_2O_2$: a1517, a1518, a1519, m20
$C_{15}H_{14}O$: b2006, p1682
$C_{15}H_{14}O_2$: a439, b1296, b1950, p1437, p1438, p1439, p1440, p1441
$C_{15}H_{14}O_3$: a443, b2000, b2001, b2002, b2003, b2004, b2005, c177, c178, c179, e100, 115, p1442, p1443, p1444
$C_{15}H_{14}O_4$: l16, n166, p481, x27
$C_{15}H_{14}O_5$: b1783, c166, m394, p730
$C_{15}H_{14}O_6$: c227, c228, c229, e71, e72, j7
$C_{15}H_{15}N$: c155
$C_{15}H_{15}NO$: b2007, b2008
$C_{15}H_{15}NO_2$: a1119, c114
$C_{15}H_{15}NO_3$: a935
$C_{15}H_{15}NO_6$: m116
$C_{15}H_{15}N_3O$: a736
$C_{15}H_{15}N_3O_2$: m393
$C_{15}H_{16}$: b278
$C_{15}H_{16}N_2O$: a1572, a1573, a1574, a1575, a1576, u62
$C_{15}H_{16}N_2S$: u79, u80
$C_{15}H_{16}N_4O_4$: c9
$C_{15}H_{16}N_4O_5$: c14
$C_{15}H_{16}O$: e254[1], m368[1]
$C_{15}H_{16}O_2$: b2315, p1117, p1118, p1179[1]. p1179[2]
$C_{15}H_{16}O_3$: a1453, g82, h727
$C_{15}H_{16}O_7$: e9
$C_{15}H_{16}O_8$: s91
$C_{15}H_{16}O_9$: e158
$C_{15}H_{17}N$: a798, a800
$C_{15}H_{17}NO_2S$: t659, t660
$C_{15}H_{17}N_3$: g202
$C_{15}H_{18}$: a1660, a1661, a1662, n172
$C_{15}H_{18}O$: e586, e587, n221, n222
$C_{15}H_{18}O_3$: s20
$C_{15}H_{18}O_6$: b1192
$C_{15}H_{19}NO_2$: t758, t761, t763
$C_{15}H_{20}ClNO_2$: t759
$C_{15}H_{20}NO_2$: a988
$C_{15}H_{20}O_2$: h7
$C_{15}H_{20}O_4$: m81
$C_{15}H_{21}NO_2$: e606, e607, e608, i157
$C_{15}H_{21}N_3O_2$: p902
$C_{15}H_{22}$: c565
$C_{15}H_{22}ClNO_2$: e609, i158, i159, i160, p1034
$C_{15}H_{22}O$: e102, v16, v17
$C_{15}H_{22}O_2$: b1866, e103, s13
$C_{15}H_{23}ClO_4S$: a1397
$C_{15}H_{23}NO_4$: a741, p168, t763
$C_{15}H_{24}$: c7, c221, c222, c223, c230, c496, f4, f5, 156, p29
$C_{15}H_{24}N_2O$: a1367, l 62, l 63, m144, o311
$C_{15}H_{24}O$: b531, b532, b533, b534, e101, s14, s15
$C_{15}H_{24}O_6$: p1765
$C_{15}H_{24}O_8$: p1291, p1922
$C_{15}H_{25}ClN_2O$: o312
$C_{15}H_{25}ClN_2O_2$: b1461
$C_{15}H_{26}N_2$: s105
$C_{15}H_{26}O$: e34, f6, g191, 126, n467, n468, n469
$C_{15}H_{26}O_6$: g108
$C_{15}H_{26}O_7$: c435
$C_{15}H_{28}$: p41
$C_{15}H_{28}N_2O_4S$: s106
$C_{15}H_{28}O$: c863
$C_{15}H_{28}O_2$: m166
$C_{15}H_{28}O_3$: p1517
$C_{15}H_{28}O_4$: m69
$C_{15}H_{30}$: e32, p40
$C_{15}H_{30}Br_2$: p35

$C_{17}H_{20}N_2O$: b1974, u57
$C_{17}H_{20}N_2O_3$: u33[1], u33[2]
$C_{17}H_{20}N_2O_4$: a1388, a1390
$C_{17}H_{20}N_2S$: b1975, p719, u58
$C_{17}H_{20}N_4O_3$: x31
$C_{17}H_{20}N_4O_6$: r26
$C_{17}H_{20}O_6$: m424
$C_{17}H_{21}ClN_2S$: p720
$C_{17}H_{21}NO_2$: a1369
$C_{17}H_{21}NO_4$: c445, c446, c447, h744, h745, h754, h755, m400
$C_{17}H_{21}N_3$: a1473
$C_{17}H_{22}BrNO_4$: h746, h747, h748, h749, h750, h751
$C_{17}H_{22}ClNO_2$: a1370
$C_{17}H_{22}ClNO_4$: c449, c450, h752, h753
$C_{17}H_{22}ClN_3$: a1474
$C_{17}H_{22}N_2$: m197
$C_{17}H_{22}N_2O$: m356
$C_{17}H_{22}N_2O_7$: h756
$C_{17}H_{23}ClN_2O_2$: n314
$C_{17}H_{23}CrNO_8$: c448
$C_{17}H_{23}NO_3$: a1469, h758, h759
$C_{17}H_{24}BrNO_3$: h760
$C_{17}H_{24}ClNO_3$: a1470, h761
$C_{17}H_{24}N_2O_5S$: s85
$C_{17}H_{24}O_2$: c362, c363, c364
$C_{17}H_{24}O_3$: b1668
$C_{17}H_{24}O_4$: f114
$C_{17}H_{24}O_9$: s423
$C_{17}H_{25}ClN_2O$: d193
$C_{17}H_{25}N_3O$: a1358
$C_{17}H_{26}N_2S_4$: c107
$C_{17}H_{28}$: h64
$C_{17}H_{28}O$: b301
$C_{17}H_{30}O$: c610
$C_{17}H_{32}O_4$: s244
$C_{17}H_{33}N$: h73
$C_{17}H_{34}$: h80
$C_{17}H_{34}Br_2$: h69
$C_{17}H_{34}N_2S_2$: c103
$C_{17}H_{34}O$: h65, h77, h78
$C_{17}H_{34}O_2$: h70, h264, t83, t86
$C_{17}H_{35}Br$: h68
$C_{17}H_{35}N$: p981
$C_{17}H_{36}$: h66
$C_{17}H_{36}O$: h74, h75, h76
$C_{17}H_{37}N$: h67

C_{18}

$C_{18}H_{10}N_2O_6$: i61
$C_{18}H_{10}O_2$: b194, b195, c318, c319, n6, n7
$C_{18}H_{10}O_4$: b2179, b2202
$C_{18}H_{11}NO_2$: p844, p845
$C_{18}H_{12}$: b181, b1956, b1957, c313, e603, n1
$C_{18}H_{12}N_2$: b2367, b2368, b2369, b2370, b2371, b2372, b2373, b2374, b2375, b2376, b2377, b2378
$C_{18}H_{12}N_2O$: a1382
$C_{18}H_{12}O$: b184, p1931
$C_{18}H_{12}O_2$: a586, b2100
$C_{18}H_{12}O_3$: b1834, b1835
$C_{18}H_{12}O_4$: d159, d160
$C_{18}H_{13}N$: c153
$C_{18}H_{13}NO_3$: p850
$C_{18}H_{13}NS$: p721
$C_{18}H_{13}N_3O$: r37
$C_{18}H_{13}N_3O_2$: b1215
$C_{18}H_{14}$: b182, n2, t46, t47
$C_{18}H_{14}N_4O_2$: b306, b307
$C_{18}H_{14}O$: k5, k6, k20, k21
$C_{18}H_{14}O_3$: b1786[1], c353
$C_{18}H_{14}O_6$: j1
$C_{18}H_{14}O_8$: s191
$C_{18}H_{15}As$: a1441
$C_{18}H_{15}B$: b2411
$C_{18}H_{15}ClSe$: s37
$C_{18}H_{15}FSe$: s38
$C_{18}H_{15}N$: a947
$C_{18}H_{15}N_3$: a1602
$C_{18}H_{15}OP$: p756
$C_{18}H_{15}O_3P$: p803
$C_{18}H_{15}O_4P$: p781
$C_{18}H_{15}P$: p755
$C_{18}H_{15}Sb$: s122
$C_{18}H_{16}$: c906, h544, t769
$C_{18}H_{16}F_2Se$: s39
$C_{18}H_{16}N_2$: h710
$C_{18}H_{16}N_2O$: a1489, a1490, a1491, a1492, a1493
$C_{18}H_{16}N_2O_2$: a989
$C_{18}H_{16}N_2O_6S$: q155, q156
$C_{18}H_{16}O_2$: p546, s157
$C_{18}H_{16}O_4$: e93, t108, t109, t111
$C_{18}H_{16}O_7$: u141, u142, u143
$C_{18}H_{17}Cl_4N_3O_8S_2$: h595
$C_{18}H_{17}NO_3$: p1885
$C_{18}H_{17}N_3$: a1494
$C_{18}H_{18}$: p506[1]
$C_{18}H_{18}N_2O_4$: a1361
$C_{18}H_{18}N_6$: b1
$C_{18}H_{18}O_2$: e94, e95, e96, h415
$C_{18}H_{18}O_3$: b2684
$C_{18}H_{18}O_4$: b2287, b2338, s188, s199
$C_{18}H_{18}O_5$: e483
$C_{18}H_{18}O_6$: t13
$C_{18}H_{19}NO_2$: a844, a845, a846, a1374
$C_{18}H_{19}NO_3$: e144, t779
$C_{18}H_{19}N_3O_2$: a1535
$C_{18}H_{20}$: i27
$C_{18}H_{20}Br_2N_2O_4$: b1416, b1422, b1429
$C_{18}H_{20}Cl_2N_2O_4$: b1472, b1487
$C_{18}H_{20}N_{10}O_{16}$: a1407, a1408
$C_{18}H_{20}O_2$: e97
$C_{18}H_{21}NO_3$: c455, c456, c460, e138, e139, e143, e145, i130, p1863, t129
$C_{18}H_{21}NO_4$: e146
$C_{18}H_{22}ClNO_3$: c457
$C_{18}H_{22}N_2O_2$: h600
$C_{18}H_{22}N_2O_5$: p916
$C_{18}H_{22}N_2S$: p718
$C_{18}H_{22}N_4O_4$: g48, g49, g220
$C_{18}H_{22}O_2$: b2270, b2271, b2272, b2281, e98, e161, i156
$C_{18}H_{22}O_4$: n569
$C_{18}H_{22}O_8$: b1186
$C_{18}H_{23}ClN_2O_2$: h601
$C_{18}H_{23}NO_3$: c461
$C_{18}H_{23}NO_4$: c456, e6
$C_{18}H_{23}N_3$: a1475
$C_{18}H_{24}NO_7P$: c458
$C_{18}H_{24}O_2$: e159, e159[1], i155
$C_{18}H_{24}O_3$: e160
$C_{18}H_{26}O_4$: p825
$C_{18}H_{26}O_{12}$: s101
$C_{18}H_{27}NO_3$: c86
$C_{18}H_{27}NO_5$: e610
$C_{18}H_{27}N_3O$: p370
$C_{18}H_{28}N_2O_4S$: a980, a981
$C_{18}H_{28}O$: d314
$C_{18}H_{28}O_2$: d305
$C_{18}H_{28}O_4$: b2095
$C_{18}H_{29}NO$: d297
$C_{18}H_{30}$: b788, o8
$C_{18}H_{30}Br_6O_2$: o45
$C_{18}H_{30}O$: b824
$C_{18}H_{30}O_2$: o64, o65, o66
$C_{18}H_{30}O_4$: a1480
$C_{18}H_{32}Br_4O_2$: o60
$C_{18}H_{32}CaN_2O_{10}$: p5, p6
$C_{18}H_{32}I_2O_2$: o85
$C_{18}H_{32}N_2O_5S$: p3
$C_{18}H_{32}O_2$: c257, c258, o4, o7, o96
$C_{18}H_{32}O_4$: o44
$C_{18}H_{32}O_5$: a1479
$C_{18}H_{32}O_6$: g112
$C_{18}H_{32}O_{16}$: m153, r1
$C_{18}H_{33}N$: o83
$C_{18}H_{34}$: o93, o94, o95
$C_{18}H_{34}Br_2O_2$: o41
$C_{18}H_{34}O$: o68
$C_{18}H_{34}O_2$: o71, o72, o84
$C_{18}H_{34}O_3$: o52, o54, o56, o58, o86
$C_{18}H_{34}O_4$: d33, h386
$C_{18}H_{35}ClO$: o27
$C_{18}H_{35}N$: o35
$C_{18}H_{35}NO$: o73, o74
$C_{18}H_{36}$: o69, o70
$C_{18}H_{36}Br_2$: o17
$C_{18}H_{36}ClNO$: a318
$C_{18}H_{36}N_2S_4$: d225
$C_{18}H_{36}O$: o9, o91, o92
$C_{18}H_{36}O_2$: a207, d55, h72, h260, o22
$C_{18}H_{36}O_3$: h262[1], o46, o47, o48, o49, o50
$C_{18}H_{36}O_4$: o42, o43
$C_{18}H_{37}Br$: o15
$C_{18}H_{37}Cl$: o16
$C_{18}H_{37}I$: o18
$C_{18}H_{37}NO$: o23
$C_{18}H_{38}$: o11
$C_{18}H_{38}O$: o63
$C_{18}H_{38}O_2$: h241, o20
$C_{18}H_{38}S$: o21
$C_{18}H_{39}N$: a937, o12
$C_{18}H_{40}ClN$: o14
$C_{18}H_{54}O_7Si_8$: o202
$C_{18}H_{54}O_9Si_9$: c851

C_{19}

$C_{19}H_{10}Br_4O_5S$: b2427
$C_{19}H_{12}O$: b193
$C_{19}H_{12}O_2$: b1224, b1225
$C_{19}H_{12}O_5$: i244
$C_{19}H_{12}O_6$: d161
$C_{19}H_{12}O_7$: c552
$C_{19}H_{13}N$: a740
$C_{19}H_{13}NO$: c143
$C_{19}H_{13}N_3O_6$: m316
$C_{19}H_{13}N_3O_7$: m384, m385
$C_{19}H_{14}$: b185, b186, b187, b188, b189, b190, b191, b192, c315, c316, c317, f77
$C_{19}H_{14}N_2O_2$: a1578, a1581, a1584
$C_{19}H_{14}O$: f69, x3
$C_{19}H_{14}O_2$: p1579, p1825
$C_{19}H_{14}O_3$: a1478
$C_{18}H_{14}O_5$: p66
$C_{19}H_{14}O_5S$: p715
$C_{19}H_{15}Cl$: m243
$C_{19}H_{15}ClN_4$: t126
$C_{19}H_{15}N$: c144
$C_{19}H_{15}NO$: b1246
$C_{19}H_{15}NO_2$: q256
$C_{19}H_{15}N_3$: a733
$C_{19}H_{16}$: b2234, b2235, m312
$C_{19}H_{16}N_2$: b1257, b1259, b1962, y3
$C_{19}H_{16}N_2S$: u124
$C_{19}H_{16}O$: m381
$C_{19}H_{16}O_3$: m315, m360, o243

C_{20}

C_{21}

C_{38}

$C_{38}H_{26}$: b2198
$C_{38}H_{30}$: e250
$C_{38}H_{44}N_2O_{10}S$: i215
$C_{38}H_{46}N_2O_8$: t771
$C_{38}H_{50}N_4O_8S$: c332
$C_{38}H_{53}NO_{13}$: a722
$C_{38}H_{58}N_4O_{12}S$: c325
$C_{38}H_{74}O_4$: e295
$C_{38}H_{74}O_5$: d171

C_{39}

$C_{39}H_{57}NO_{11}$: g23
$C_{39}H_{59}NO_{11}$: g19
$C_{39}H_{61}NO_{12}$: g24
$C_{39}H_{61}NO_{13}$: p1856
$C_{39}H_{74}O_6$: g109
$C_{39}H_{76}O_5$: g81

C_{40}

$C_{40}H_{42}N_2O_{15}S$: b2167
$C_{40}H_{50}N_4O_8S$: q21, q31, q32
$C_{40}H_{54}$: i127
$C_{40}H_{54}O$: a1366
$C_{40}H_{56}$: c195, c196, c197, l71
$C_{40}H_{56}O$: c558, r35
$C_{40}H_{56}O_2$: x23, z2
$C_{40}H_{56}O_4$: t8, v20
$C_{40}H_{56}O_6$: f177
$C_{40}H_{60}O$: l73
$C_{40}H_{74}O_5$: d172

C_{41}

$C_{41}H_{61}NO_{13}$: e157
$C_{41}H_{63}NO_{14}$: p1857
$C_{41}H_{63}NO_{15}$: p1858
$C_{41}H_{64}O_{13}$: d185
$C_{41}H_{64}O_{14}$: g27

C_{42}

$C_{42}H_{28}$: r36
$C_{42}H_{32}$: t770
$C_{42}H_{46}N_4O_8S$: s156
$C_{42}H_{50}N_4O_8$: q16
$C_{42}H_{62}O_{16}$: g185
$C_{42}H_{66}O_{18}$: t132

C_{43}

$C_{43}H_{76}O_2$: c303

C_{44}

$C_{44}H_{82}O_3$: d280

C_{45}

$C_{45}H_{86}O_6$: g120
$C_{45}H_{73}NO_{15}$: s93
$C_{45}H_{73}NO_{16}$: s95

C_{46}

$C_{46}H_{68}N_4O_{19}S$: b2431

C_{47}

$C_{47}H_{94}O_2$: h261

C_{48}

$C_{48}H_{38}N_8$: e38
$C_{48}H_{40}O_4Si_4$: c931
$C_{48}H_{102}O_7Si_2$: d222

C_{50} to C_{216}

$C_{50}H_{83}NO_{22}$: d196
$C_{51}H_{98}O_6$: g110
$C_{53}H_{83}NO_{21}$: t690
$C_{54}H_{111}N$: a945
$C_{55}H_{70}MgN_4O_6$: c267
$C_{55}H_{72}MgN_4O_5$: c266
$C_{56}H_{86}O_2$: e116
$C_{57}H_{104}O_6$: g116, g117
$C_{57}H_{110}O_6$: g115
$C_{68}H_{96}N_2O_{26}S$: a726
$C_{76}H_{52}O_{46}$: t7
$C_{90}H_{154}$: e42
$C_{216}H_{288}O_{144}$: s113

MELTING POINT INDEX OF ORGANIC COMPOUNDS

Temperatures in °C; where values are not precisely known, or where there is a range of melting points, the compound is listed according to the lower temperature.

−215: h768
−205: c184, m289
−197: m289
−190: p1107
−185: b2940, p1713
−183: e162
−182: m175
−181: m241
−180: o253
−177: c911
−173: h373
−169: b2992, e401, p210
−165: p375
−162: b2410
−160: b2759, e417, m230, m253, m308
−159 p1185, p1198′
−157: b2983, s77
−156: p1744
−155: s69
−154: e405, p148, p412
−153: p411
−151: k1, p376, p1189
−148: p46
−147: h421
−146: b2461, m230
−145: m244
−144: e328
−143: a1437, e245
−142: c876, e421, m276
−141: b2513, h54, h281, p417
−140: e403, h558, p1747
−139: b2941, b2995
−138: b2485, b2991, c186, c874, e499, h549, p375, p414
−137: h201, p44, p100, p734, p1721
−136: b2435, e199, p377, p403, p410, p1079
−135: c899, m256, p400, p413, p415, p1719, p1722, s70
−134: b2564, b2993, p134, p137, p401, p402
−133: a1642, h548
−132: b2454, b2583, b2987
−131: b2512, p1294
−130: a852, p84, s63
−129: b2557, b2995, p1814
−128: c907, p405, p732
−127: c688, p1587
−126: b3070, e544, h365, o211, p47, p92
−125: h584, p1716
−124: e450, p131
−123: b2511, m345, p1133
−122: e411, e528, e539, m261, s65, s72
−121: a18, c817
−120: b2460, h128, p143
−119: c910, e195, o213, p133, p140, p1717
−118: p404
−117: a939, b2957, b3078, c878, e221, e336, p1134, s343
−116: b2643, c819, c820, e477, e534, p407
−115: a925, b2866, b2984, e458[1], m216, p141
−114: c185
−113: a22, e592, o137, p1293, p1715, s348
−112: a183, b2500, b2501, b2510, b2680, c161, d67, e537, p152, p759
−111: c674, c875, e241, e253, e277, p211
−110: b3001, o212, p221, p457, p1119
−109: b2436, c909, e460, p115, f220
−108: c871, e253, o135, p1210
−107: c691, c696, p153, p1080
−106: b2942, b2958, e204, p420, s73, s341
−105: a813, b2487, b2494, c877, f109, p101, p456, p1170, s74[1]
−104: b2522, b2569, c811, o223, s303
−103: b2568, e501, i216, o224, p334
−102: d238, o208, p1831, s80, s332, s347
−101: b2668, e475, p154, p448, p1192
−100: b2556, b2810, c673, d158, e365, h318, o209, p1161, p1163
−99: a215, b928, b2471, b2806[1], f141, p99, p1720, p1745, s80
−98: a221, m228, p1749, s120, s291, s313
−97: b2686, c150, c865, e215, f111, k3, o304, s63[1]
−96: a174, b771, b847, c825, c873, f139, l42, p91, p229, t224
−95: a208, a259, b758, c686, c694, m252, p1109, p1649, s290, t275
−94: a948, c868, e218, e249, e452, m176, m213, m349, n483, o214, p1314
−93: a197, a201, a216, a850, b2408, f142, f146, h319, p219, p419, p1211, p1326
−92: b2401, b2661, b2662, b2926, e448, f106, f132, p76, p77
−91: c667, p142, p222, p2001
−90: b2864, b3079, c687, m157, o249, p447, p1193, p1312, p1433, p1525, p1588
−89: b2665, c628, c665, p1120, t211
−88: b372, o210, p90, p750, p1324, p1527
−87: a1440, c668, p45, p1707
−86: b2912, e497, f139, m214, p144, t171
−85: b401, b2563, b2674, b2820, e373, p365, p754, p1532
−84: a199, b566, b672, b2404, b3018, d242, p1766
−83: b374, m21, p1108, s292
−82: n485, o247, p453
−81: a208, b770, b3077, e163, e591, f158, p1083, p1530
−80: b2542, b2562, e384, e412, f135, h311, h683, i216, k4, o134, p371, s295, t743
−79: c619, n473, o159, o221, p231, p313, p1281, s67
−78: a178, a223, a1641, b2671, n529, p346, s312
−77: m26, p2000
−76: b2447, c666, h491, p209, p452, p1332, p1748, t122
−75: b373, b374, b924, b2657, c647, c895, g108
−74: a847, b568, b2523, b2541, b2933, b2985, c807, e467, l41, p1316
−73: a172, a217, b2682, d68, d77, e425, f145, p121, p1328, s351
−72: b848, b2924, p1534, t198
−71: a253, p1322, p1773
−70: c854, e266, p474
−69: e509, h98, p355, t216
−68: b850, c908, d173, p1114
−67: b671, e213, e350, h147, p151, p2019
−66: b960, d66, e600, f108, m282
−65: a639, b2533, p1770
−64: a1128, b247, b849, b889, b3089, m243[1], m291, m306, p139, p1218, t215
−63: a893, b257, b447, b642, b674, e211, f137, f206, h158, o165, p225, p1999, s167, t208
−62: a198, b772, m24, o222, s121, t493
−61: a848, n536
−60: a820, b887, b925, b2887, c604, e478, f120, g111, t476
−59: b2395, c862, p1122, p1227, t173
−58: b376, b2670, c693, c896, e292, o121, p1320, s123
−57: a52, a1157, b541, c159, c642, f167, m217, m413, o112, t495
−56: b35, b2412, b2681, c627, p218, p776
−55: o116, o128, p85, p464, p923, p1154, p2017
−54: b673, b3063, m54, p1522
−53: e408, e444, f168, m246, p441, p920, s162, x11
−52: a173, a415, b403, b648, b1299, b3015
−51: b843, b2481, c894, d268, i16, n519, p147, p216, p295, p1319, s62, s316, t175
−50: a205, b2827, b2921, c664, c695, e255, e413, n560, p1543, t395, t733
−49: b2486, c629, d213, e27, m23, o158, t206
−48: a827, b661, b2596, b2976, d37, p1136, p1799, p1851, t393
−47: b2907, c160, e488
−46: a241, b407, o130, o172, o176, p224, p1205, p1518
−45: b778, b782, b945, c677, c785, e367, e471, i143, n528, o105, p1310, p1668, p1974, p2034
−44: a823, b974, b2394, c651, e267, o273, s160
−43: a1424, b246, b449, b567, c170, d6, m362, o164, o166, p161, p1466
−42: a630, b2788, b3014, d197, e234, e502, p230,

a669, b114, b269, b366, b480, b914, b1885, b2182, b2621, b2624, b2739, b3019, d230, g70, h266, h275, h353, h694, m235, m332, n319, n549, q223, s192, t80, t292

21: a933, b54, b353, b441, b552, b785, b864, b1265, b1478, b1494, b1609, d313, f269, t670

22: a487, b380, b391, b647, b677, b821, b1151, c361, d61, h66, h139, h296, m247, m251, o31, o40, o93, p2092, s392, t14, t302, t363

23: b170, b275, b387, b817, b2625, c790, o27, p127, p512, p2051, t64, t68, t707

24: a81, a207, a394, a1076, b473, b1319, b1543, b1768, b1778, b1891, b2125, b2506, b2905, d305, e587, h50, h260, h695, o21, p30, p1562, p1596, p1636, p1754, p1788, p2042, q70, t415, t456, t473, t538

25: a619, b321, b1500, b2622, b3003, c748, c758, c856, d44, e224, g102, h249, h371, m86, m271, m410, n155, o34, o157, p57, p790, p1595, p1633, p2107, p2169, q96, t31, t360, t419, t509

26: a584, a1141, b260, b732, b795, b820, b1657, b1959, d311, g80, h635, n222, p468, p1344, p1629, q25, q185, s212, t298, t299, t302, t306, t446

27: a612, b398, b432, b435, b638, b645, b682, b785, b1720, b1784, b2529, e313, e510, f229, h250, h535, i188, m50, o26, o41, o107, p35, p1665, p1789, q129, t450, t515, t568, t717

28: a148, a486, a705, a706, a830, a832, b50, b442, b553, b644, b744, b812, b2832, c609, c779, c857, d28, d285, e28, h71, h86, h409, m164, n550, o11, o15, o21, o25, o28, o32, o86, o147, p1683, q265, s337, t152, t241, t360, t671

29: a294, a1016, b707, b752, b1806, b2318, b2417, b2419, c749, f217, h22, h394, h582, h712, o16, o31, o39, o41, p650, p661, p662, p1186, p2039, q68, s173, s176, s180, s375, t225, t303, t305, t339

30: a356, a841, b291, b714, b1840, b2221, d199, e312, e513, e586, e612, h72, h264, h303, i139, m245, n155, o36, o94, p461, p468, q256, s105, t64, t223, t515, t574, t727

31: a220, a1660, b829, b2151, b2817, c698, c755, c859, d42, d293, f143, f284, h161, h265, m165, o29, o193, p1389, q86, s210, s415, t337, t428, t440, t498, t728, x10

32: a283, a285, a380, a542, a560, a1016, a1662, b1263, b1420, b2145, b2330, c610, d164, d309, g102, g120, h68, i19, m25, m50, n68, n126, n329, n505, p1179[1], p2015, p2114, t561

33: a48, a400, a472, a482, a671, a1024, a1102, a1137, a1598, b472, b808, b833, b1805, b2195, b2381, b2558, b2708[1], d278, e434[1], e483, f194, g109, h273, h457, m165, n8, n254, n255, n515, o29, p251, p514, p593, p614, p1959, s384, t88, t294, t726, t729

34: a246, a442, a634, a798, b261, b288, b335, b559, b689, b826, b2248, b2310, b2622, b2625, b3019, c354, c705, c752, c850, f228, h73, h240, m178, m186, n24, o18, o92, p266, p820, p914, p1627, p1822, p1968, q71, q83, t24, t89, t668

35: a386, a453, a480, a860, a1050, a1552, a1597, b41, b119, b320, b355, b651, b810, b1631, b1848, b1856, b2311, d205, f285, f287, g102, h455, h456, k22, m165, m170, n142, o154, o182, p317, p431, p494, p606, p614, p1039, p1645, p1656, p1682, p1789, t96, t169, t178, t387, t500

36: a332, a1648, b43, b265, b438, b639, b756, b1073, b1162, b2294, c359, c747, e29, f258, g109, g151, h26, h29, h65, h257, h745, i231, m188, n236, n507, o92, p815, p1538, p1669, p1920, q182, s318, s357, t470, t526, t541, t579, t602

37: a67, a432, a481, a802, a1176, b152, b160, b396, b438, b1665, b1839, b2014, b2019, b2323, c355, c783, d281, e16, h367, i18, m378, n57, n180, n322, n515, o53, p33, p299, q265, s32, t297, t384, t414, t601, t761

38: a285, a475, a701, a1160, b836, b862, b1083, b1149, b1495, b1786[1], b2626, c470, c727, c826, d38, d55, d263, h28, h69, h672, m164, m247, n48, o9, o59, p343, p441[1], p1092, p1250, p1370, q81, s257, s396, t100, t374, t417, t435, t560, t607

39: a134, a232, a548, a684, a808, a1102, b48, b529, b1327, b1421, b1810, b2010, b3057, c354, c936, d212, e556, h271, o33, p993, p1016, p1035, p1907, q198, s238, t87, t332, t432, t452, t600

40: a243, a287, a656, b134, b232, b352, b438, b639, b1479, b2592, b3085, c169, c608, c914, e45, e47, e341, e492, f250, g74, h59, h165, i22, i193, j9, l62, p363, p467, p627, p795, p1018, p1085, t49, t96, t333, t591

41: a642, a1125, b94, b310, b394, b617, b784, b1232, b1870, b2013, b2330, b2922, c26, c135, c342, c472, c626, c767, c770, c901, c916, d206, d291, e223, h344, h476, i146, n249, o35, o55, o57, o75, p621, p1384, p1395, p1961, p2158, q55, q183, t205, t329, t379, t429, t761

42: a909, a1054, a1431, a1471, b286, b1174, b1262, b1425, b1976, b2338, b3065, c112, c347, d298, e19, e102, e572, i135, m163, m279, o265, o298, p41, p817, p1342, p1554, p1610, p1655, p1786, p2046, q96, q192, s289, s309, t382, t452, t557, t768

43: a37, a384, a681, a683, a1067, a1167, a1556, b162, b356, b435, b555, b562, b606, b616, b958, b1018, b1471, b1679, b2626, b2713, b2715, c722, g117, h1, h58, h115, h406, h408, p39, p204, p438, p555, p615, p656, t73, t134, t166, t243, t603

44: a50, a291, a376, a895, a1175, a1422, a1569, b84, b103, b133, b397, b526, b916, b949, b1028, b1485, b1715, b2014, b2714, b2780, c98, c413, d272, d282[1], d295, e314, g91, g120, h237, i26, m165, n475, o28, o84, o152, o296, p935, p1253, p1363, p2163, q85, s157, s206, t288, t503, t546, v17

45: a322, a950, a1049, b876, b1092, b1277, b2194, b2846, b3036, b3039, c893, e281, e335, g110, i225, m6, m90, m115, m171, m221, n247, o25, o72, o264, p496, p631, p679, p794, p1511, p1626, p1641, p1677, q84, s188, t138, t288, t406, t411, t593, t604, t605, t608, t722

46: a271, a363, a556, a1081, a1144, a1546, a1554, b37, b262, b303, b381, b396, b414, b473, b834, b1602, d314, f70, g109, h75, h662, h711, m117, m185, m348, n533, p33, p787, p885, p1191, p1660, p1699, p1910, p2108, p2150, q231, s107, s355, t134, t219, t296, t334, t376, t455, t739

47: a25, a299, b69, b305, b377, b534, b609, b761, b815, b921, b1153, b1623, b1849, b2007, b2627, b2847, d264, e20, e382, f257, h244,

h276, h649, i134, l49, m173, n171, o265, p1105, s221, t338, t386, t463

48: a61, a290, a346, a403, a849, b205, b605, b971, b1177, b1523, b1699, b1959, b2482, b2559, b2639, b2783, c91, c109, d254, g170, g171, h641, m83, n60, n127, o85, o96, o312, p655^{2}, p853, p979, p1059, p1361, p1666, p1701, q7, q69, q104, q227, s293, s397, t25, t327, t330, t559, t711

49: a53, a165, a239, a255, a686, a968, a1171, b61, b431, b452, b617, b678, b799, b1780, b1950, b2322, b2587, b2784, c98, d54, g76, h67, h143, h164, h272, h286, h668, n146, n234, n532, o64, p473, p950, p955, p1461, p1568, p1671, q66, q100, q186, s188, s294, t758

50: a55, a275, a383, a610, a653, a655, a991, a1078, b502, b1501, b2400, b2712, c6, c155, c178, c624, d274, e19, f125, g77, g192, h269, i22, m171, n36, n513, o203, p458, p596, p751, p781, p1417, p1648, p2047, q228, t13, t25, t71, t364, t410, t646

51: a69, a659, a677^{1}, a1015, a1080, a1651, b120, b229, b299, b532, b619, b793, b1134, b1155, b1165, b1169, b1320, b1335, b2002, b2220, b2237, c24, c39, c431, c623, c778, d93, h262^{1}, h317, i147, m51, m75, m112, n69, n70, n71, n328, n572, o37, p1230, p1406, p2116, t62, t218, t239, t407, v16

52: a9, a311, a327, a500, a692, a693, a900, a1056, a1572, b63, b75, b217, b298, b532, b545, b797, b1081, b1401, b1979, b2067, b2140, b2465, b2527, b2852, b2861, c23, c25, d96, e34, e237, e383, f278, h27, i27, n171, n464, o204, p78, p195, p1625, p1922, q69, q71, q178, q219, t191, t208, t242, t340, t466, t480

53: a51, a64, a479, a513, a691, a901, a1069, a1573, b232, b435, b537, b832, b896, b952, b1327, b1328, b1822, b2049, b2222, b2271, b2968, b3076, c129, e26, f73, f128, h78, h262, h508, i65, i168, o12, p37, p808, p1179, p1286, p1454, p1456, s122, s243, t79, t558, t594

54: a863, a1015, a1547, b460, b561, b835, b863, b922, b932, b972, b1186, b1749, b2234, b2258, c89, c94, c918, d171, d202, d275, e20, e517, f138, g92, g98, h74, h75, i21, n209, n243, o19, o270, o282, p163, p1711, p1877, p1897, q271, s264, t62, t556, t634, t673

55: a124, a153, a244, a268, a483, a485, a493, a826, a945, a1027, a1545, b73, b216, b287, b397, b733, b906, b1144, b1302, b1415, b1940, b2018, b2055, b2080, b2098, b2862, c678, c784, e25, e495, g115, h749, h750, i145, m240, n443, o9, p64, p260, p686, p719, p858, q94, s19, s329, t19, t75, t177, t398, t399, t451

56: a37, a55, a165, a310, a437, a661, a1599, b6, b174, b238, b357, b464, b486, b530, b558, b882, b897, b1173, b1289, b2635, b2638, c217, c767, c768, e260, e341, e419, e527, g76, g120, h79, h234, h247, h519, h722, h754, i23, i111, k23, m3, m275, n25, n50, n107, n108, o7, p628, p641, p687, p808, p1421, p1680, q101, t244, t367, t370, t371, t388, t522

57: a26, a56, a667, a1072, a1147, a1156, a1467, b303, b456, b573, b684, b757, b906, b1858, b2078, b2824, c132, c177, c216, e280, e288, e606, e608, g110, h27, h61, h91, h663, i59, i148, m106, n511, o30, p13, p311, p630, p665, p1264, p1457, q22, q65, q275, s175, s196, s338, t45, t165, t192^{1}, t502

58: a351, a484, a562, a610, a801, a839, a855, a1027, b82, b103, b165, b177, b649, b856, b1285, b1333, b1400, b1874, b2004, b2312, b2750, b3084, c145, c321, c348, c774, e23, e427, g169, h565, i233, m146, m285, n156, n223, n514, p201, p308, p459, p701, p928, p1039, p1263, p1451, p1881, p1948, p2012, s179, s208, s256, t46, t77, t116, t341, t377, t434, t448, t513, t613, t763, u64

59: a438, a1222, a1520, b290, b335, b436, b451, b517, b728, b805, b1521, b1989, b2018, b2749, b2789, c96, c245, c312, d282, e251, g91, g150, h63, h406, h687, h703, h755, i36, m284, m395, n89, n94, n271, n321, n532, o63, o151, p14^{1}, p36, p632, p722, p1739, q90, s221, t295, t343

60: a390, a439, a646, a803, a973, a1109, a1122, a1136, a1574, b316, b495, b618, b757, b1697, b2068^{1}, b2701, b2858, c97, c177, d310, e22, e24, e512, g7, g31, h407, h412, h463, h465,. i218, i219, k7, m3, m187, m220, n13, n50, n52, n170, o30^{1}, p71^{1}, p190, p326, p439, p655^{1}, p682^{1}, p719, p1289, p1864, q15, q128, s179, t19, t524, t595, t647, t657, t660, t720

61: a74, a392, a1441, b304^{1}, b392, b460, b554, b856, b877, b980, b981, b1335, b1619, b1781, b1896, b2015, b2363, b3027, c242, d27, d248, d279, e148^{2}, e143, e446, e527, g97, h76, h464, h699, i36, i71, m43, m380, n9, n100, n110, n143, n235, n239, p311, p1179^{2}, p1258, p1384, p1505, p1654, q92, t25, t416, x29

62: a355, a360, a362, a588, a598, a665, a911, a974, b106, b521, b533, b600, b796, b926, b1595, b1948, b2324, b2365, b2909, d250, f126, g104, h70, h157, i121, i234, j14, m152, m272, n454, n510, p.194, p654, p669, p700, p947, p1270, p1354, p1670, p2145, q65^{1}, s320, u64, u84

63: a125, a309, a527, a552, a1077, b124, b493, b510, b522, b954, b1013, b1154, b1370, b2754, b2814, c261, c863, d218, e22, e103, e602, g77, g91, h253, l39, n44, n509, o61, o62, o146, p501, p647, p854, p1697, p1780, q227, s241, t22, t368, t420, t430, u45, v24

64: a361, a1550, a1553, a1570, a1599, b117, b121, b155, b412, b571, b653, b868, b1178, b1716, b1760, b2660, b2843, b3040, c51, c96, c126, c248, d280, e25, g85, g115, h256, h424, i123, k20, m189, m373, n28, n147, n502^{1}, n518, o1, o17, p26, p492, p897, p1690, p1949, q267, s335, t375, t422, t655, t741, t764

65: a11, a57, a167, a289, a342, a478, a1535, a1575, b80, b85, b237, b267, b302, b319, b475, b523, b557, b877, b1091, b3043^{1}, c86, c249, d266, e238, e297, h648, i122, n229, n514, o283, p197, p503, p528, p606, p640, p1584, p1632, p1752, p1777, p2009, p2171, q50, q98, q109, q137, q225, q226, s34, t75, t454, t467

66: a63, a154, a558, a866, a892, a1018, b218, b284, b676, b692, b814, b1211, b1896, b1967, c146, e381, e489, g2, g93, g110, h608, h705, h726, k5, m273, m370, n133, n331, p594, p651, p652, p665, p819, p889, p1818, s213, s400, t400, t481, t699

67: a459, a569, a694, a1138, a1223, a1580, b57, b148, b263, b503, b599, b926, b1020, b1142, b1512, b1569, b1718, b1987, b2092, b2312, b2711, b3022,

c83, c92, d120, d316, e21, e85, e360, e362, g104, g125, g178, h355, i33, i200, i227, p70, p232, p768, p1289, p1406, p1493, q109, t20, t640, u76

68: a456, a533, a1161, a1185, a1513, a1517, a1586, b276, b314, b515, b595, b609, b611, b679, b882, b894, b1176, b1622, b1691, b1847, b1956, b2076, b2364, c147, c257, c258, c279, c349, c451, c745, c822, e314, f170, h306, h313, h448, h764, i52, i197, k8, m135, n144, n294, p605, p624, p633, p634, p635, p702, p927, p1507, p2140, q126, s254, t413, t463, t504

69: a502, a881, a1521, b1011, b2108, b2126, b2790, c142, c192, c261, c467[1], d19, d88, d90, d233, d286, e309, e346[1], e361, e482, f59, h441, h446, m366, m368[1], n43, n509, p200, p440, p589, p703, p889, p1101, p1301, p1843, q112, q229, s35, s118, t712

70: a10, a75, a240, a330, a557, a1223, a1485[1], b45, b105, b274, b334, b1175, b1203, b1321, b1670, b1686, b1877, b2448, d56, d323, e194, e607, g53, g83, g103, g105, h61, l65, m35, m113, m196, o183, p595, p691, p1273, p1546[1], p1905, q46, q50, q243, s387, s404, t114, t517, t597, t648

71: a528, a578, a819, a1168, a1512, b52, b83, b104, b379, b531, b612, b735, b908, b915, b1150, b1293, b1303, b1686, b1794, b1862, b2146, b2214, b2275, b2296, b2361, b3009, c734, d226, d233, d276, e85, e230, e520, g5, g72, g94, m350, n111, n192, n225, o65, p58, p312, p1413, q93, q98, t412, t479, t665, u77

72: a195, a269, a440, a1048, a1079, b81, b102, b112, b166, b429, b476, b556, b564, b616, b657, b825, b827, b909, b1256, b1303, b1586, b1813, b2099, b2356, b2591, c114, c214, c365, d40, d228, e21, e342, e426, g77, h17, h261, n11, n47, n376, o22, o24[1], p173, p174, p175, p486, p487, p639, p786, p837, p1274, p1672, p1966, q175, s125, s274, t449, t557, t616

73: a685, a859, a883, a894, a1185, a1551, a1555, b121, b368, b580, b691, b1072, b1604, b2042, b2056, b2366, b2719, b2860, b3062, c215, c414, c751, e286, f121, g115, i1, m4, m273, n92[1], n216, n503, p788, p827, p852, p1201, p1282, p2074, q65[1], q130, q212, q232, s98, s324, s356, s394, t21, t170, t405, t461[1]

74: a395, a594, a1040, a1094, a1168, a1233, b111, b244, b427, b1596, b1685, b1816, b2339, c250, c403, c772, c830, d95, d257, e39, e192, e258, g100, h706, i38, k21, m116, n92, o295, p282, p350, p1283, p2111, q49, q105, q123, q171, t214, t397, t514, t642

75: a440, a443, a866, a884, a1103, a1535, b332, b348, b505, b675, b680, b716, b898, b1080, b1341, b1376, b1857, b1949, c291, c625, e230, f280, g93, h60, n11, n377, p372, p426, p857, p859, p897, p1414, p1462, p1659, p2074, q99, q103, q154, q253, s99, t115, t404, t408, t637, t683, u55, u58, u128

76: a588, a672, a1064, a1151, a1196, a1212, b242, b413, b658, b1094, b1139, b2497, c38, c40, c131, e46, e88, f17, f54, g73, g95, g101, g107, h7, m144, n227, o56, o73, o243, p1580, p1624, p1783, p1811, p1909, t610[1], t613, t721

77: a395, a431, a574, a1020, a1089, b118, b139, b282, b476, b664, b857, b1138, b1977, b1981, b2634, b2849, c45, c72, c84, c224, d166, e18, e295, e326, e446, g93, h742, l28, n178, n330, n512, p784, p791, p2026, p2088, s100, t372

78: a391, a504, a1187, a1233, b140, b235, b434, b601, b608, b1172, b1598, b1851, b2076, b2147, b2250, b2313, b2484, b2753, b2990, b3086, c41, c44, c282, c417, c433, c731, c831, d297, e607, g86, h523, h661, i204, n435, p296, p528, p599, p622, p638, p1275, p1385, p2154, t331, t352, t380, t381, t656, u27, u59, u117

79: a62, a90, a100, a118, a404, a475, a902, a915, a1068, b323, b857, b943, b1180, b1324, b1513, b1635, b2873, c167, c250, c336, c338, c446, e295, e516, e525, g81, h449, k9, m270, n119, n240, p784, p839, p1510, t525, u57, u100

80: a423, a458, a708, a865, a1181, a1526, b525, b809, b822, b1255, b1261, b1458, b1467, b1820, b1828, b2844, b2990, c11, c12, c280, c283, c303, d92, d273, e590, f274, h236, h753, i171, j11, l27, m73, p53, p581, p672, p755, p1272, p1552, p1639, p2118, r1, t209, t366, t373, t617, t632, t715

81: a6, a146, a229, a245, a707, a882, a1051, a1071, a1231, a1518, a1650, b139, b235, b240, b900, b1375, b1427, b1502, b1977, b2257, b2585, c225, c298, c404, d17, e397, e506, e601, f129, g82, g100, h528, i167, i169, l29, m312, n14, n112, n211, n224, n230, o49, p1276, p1304, p1309, p1686, s381, t300

82: a60, a1577, a1602, a1630, b318, b382, b620, b663, b870, b1068, b1565, b1978, b1980, b2839, c104, c108, c395, c750, e78, e240, f283, h60, n12, n226, o50, o56, o58, p181, p833, p934, p1423, p1439, p1484, p1502[1], p1940, q200, s328, t66, t436, t606, t623, t710

83: a155, a506, a1019, a1070, a1201, a1220, a1577, b433, b483, b541, b547, b619, b632, b804, b866, b1027, b1059, b1522, b1605, b1674, b1687, b1947, b2059, b2084, b2372, b2911, b3033, c175, e228, g182, h62, h458, h590, i199, m129, m209, m265, m375, n56, n295, n367, o3, p75, p212, p672, p699, p1051, p1438, p1440, p2054, q220, s263, s265, t142, t308, t566, t636, t654

84: a323, a398, a550, a587, a996, a1057, b46, b55, b196, b280, b385, b474, b518, b546, b874, b1082, b1827, b1833, b1890, b1893, b2853, c341, c904, d140, d259, e78, f80, f95, g174, m144, m155, m184, m281, n96, n101, n141, n270, o13, o48, p75, p171, p861, p1768, p1955, p2043, q45, q247, t26, t74, t159, t385, t436, t565

85: a997, a1106, a1177, a1181, a1229, a1571, b411, b424, b516, b520, b607, b613, b730, b899, b1017, b1175, b1312, b2073, b2083, b2235, b2236, b2270, b2469, b2925, c95, c113, d114, d117, d251, e122, e179, e240, h647, h660, i52, i69[1], i90, i136, i167, i199, n114, n131, n375, n504, p89, p116, p488, p564, p1512, p2070, q95, q117, q121, q270, t104, t485, u25, u45, v22

86: a230, a534, a544, a998, b175, b572, b731, b741, b919, b1145, b1339, b1913, b2297, b2406, b2768, b2791, b2822, b2824, c87, c130, d100, d101, d113, d306, f39, f80, g40, g182, h590, m374, n479, o44, p31, p196, p902, p1781, q218,

q236, s144, t221, t401, t484, t633, t701
87: a7, a15, a71, a78, a89, a145, a644, a756, a1012, a1070, a1090, a1104, a1647, b323, b326, b453, b514, b581, b613, b686, b739, b1342, b1599, b1868, b1964, b2000, b2118, b2181, b2748, b3044, c272, e52, i166, m144, m279, m375, n114, o54, p16, p26, p172, p489, p620, p643, p696, p838, p1445, p1555, p1782, p2106, r27, t3, t328, t460, t484[1], u123[1]
88: a76, a122, a595, a638, a1092, b51, b437, b872, b948, b1105, b1171, b1241, b1382, b2333, c108, c125, c148, c175, c889, d14, d255, e147, h235, h242, h419, n244, n425, n453, o279, p169, p373, p524, p620, p1439, s181, t145, t700
89: a545, a951, a1538, b156, b415, b506, b656, b669, b867, b907, b1014, b2169, c166, d5, d324, e399, e523, l21, l23, n148, o47, p17, p59, p587, p721, p873, p1308, p1340, p1343, p1483, p1679, p1931, q243, t47, t314, t424
90: a685, a1104, a1521, a1534, a1564, b127, b163, b178, b419, b426, b544, b726, b1012, b1378, b1581, b2082, b2144, b2319, b2835, c22, c131, c154, c256, c292, c786, e148, e298, f60, h255, h690, k16, i7, m6[1], m295, n293, o2, p1536, p2128, r13, s230, t26, t32, t45, t523, x30
91: a88, a96, a276, a562, a600, a694, a919, a1021, a1120, b173, b409, b693, b873, b969, b1168, b2240, b2244, c87, d245, f207[2], f281, g188, g191, h687, h700, m47, m197, h362, o274, p476, p1412, q24, q97, q181, q215, s203, t328
92: a17, a545, a651, a770, a1111, a1119, a1581, b344, b477, b596, b597, b1076, b1337, b1405, b1672, b2008, c43, c93, c801, c937, d307, e7, e240, f57, h92, h410, i82, m192, n258, p600, p609, p617, p766, p1529, p1781, q225, r15, t123, t143, t347, t582, t615, u83
93: a162, a1578, b36, b667, b811, b818, b1088, b1398, b1597, b1832, b2107, b2259, b2306, b2473, c42, c103, c218, c271, c292, c888, c890, d51, d165, d249, e2, e278, h61[1], h398, h670, h691, i204, 168, m5, m12, m268, o46, o74, p658, p796, p842, p1367, p1368, p1394, p1491, p2112, q49, q115, q207, t137, t340, t483, t583, t639, v4, x29
94: a159, a163, a914, a1119, a1227, a1649, b58, b623, b739, b809[1], b1156, b1368, b1486, b1791, b1886, b2066, b2088, b2274, b2337, b3028, c46, d116, e185, h399, h570, h650, h727, m202, m295, m312, n117, n119, n190, o24, p531, p695, p752, p895, p1116, q35, q44, q237, s393, t190, t505, t611
95: a117, a150, a506, a614, a700, a797, a916, a1091, a1445, a1565, a1604, b88, b281, b360, b901, b946, b1079, b1179, b1198, b1311, b1373, b1380, b1475, b1579, b1675, b1783, b1802, b1815, b1973, b2006, b2028, b2074, b2825, c153, c297, d165, e278, e279, g125[1], h163, h267, h429, h587, l1, l3, m37, n422, n540, o43, p182, p674, p697, p1271, p1439, p2118, q24, q30, q174, r27, s133, s362, t369, t658
96: a4, a162, a337, a393, a422, a474, a519, a657, a688, a941, a1121, a1186, b404, b730, b1563, b1680, b1695, b1860, b1966, b2089, b2246, c227, c229, c294, h145, m11, m84, m118, n78, n190, n291, n373, p675, p708, p1478, p1501, p1705, q10, s39[1], s245, s246, s319, t139, t519, t582, t592, t651, u42, u96, u112[1]
97: a42, a119, a673, a1604, b405, b417, b729, b738, b964, b1677, b1793, b2057, b2061, b2153, b2277, b2337, b2656, b2684, b3020, b3060, d97, e339, f16, f72, f282, h400, i72, o20, o38, o300, p680, p834, p1115, p1680, q119, q179, q235, s407, t105, t344, t516, t697, t698,
98: a26, a423, a436, a557, a567, a648, a1519, a1588, b292, b359, b425, b465, b594, b646, b654, b878, b1089, b1248, b1633, b1676, b1802, b1882, b2060, b2346, c143, c445, c447, c467[2], c795, d91, d137, d287, e42, e64, e235, l63, m356, n30, n128, n228, n292, o175, p260, p673, p870, p1495, p1560, p1664, s297, s362, t102, t350, t433, t505, t581
99: a77, a408, a521, a702, a767, b77, b97, b687, b895, b929, b946, b1079, b1088, b1288, b1953, b2231, b2839, b3043, c293, c297, c723, c932, c935, d116, d134, e149, e281, e330, e522, h188, h602, k6, l53, n51, n118, n535, p158, p572, p773, p1638, q278, s39, s101, s407, t124, t251, t378, t421, u85
100: a297, a1617, a1658, b289, b496, b499, b1201, b1519, b1823, b1866, b1916, b1930, b2001, b2089, b2149, b3023, c283, c293, c607, c703, c736[1], c937, d15, f20, g125[1], n123, n258, n301, n479, p169, p531, p1566, p1960, q150, q179, t611[1], t723, x1
101: a339, a568, a670, b331, b395, b666, b1292, b1350, b1637, b1763, b1830, b2043, b2601, d234, e246, e447, h428, h698, i173, p484, p506[1], p704, p926, q33, q133, q180, q274, s18, t453, t688, t750
102: a152, a652, a1393, a1486, a1527, b140, b324, b492, b549, b1136, b1505, b1892, b1913, b2030, f207[1], f262, h268, h461, i14, m121, m147, m202, m205, m356, n42, n233, o52, p72[1], p520, p666, p1034[1], p1330, s124, t536, u86
103: a111, a135, a450, a603, a790, a917, a1061, a1391, b322, b945, b1035, b2003, b2660, c832, e63, e65, e145, e521, f65, f175, i13, m288, n30, n92[2], n156, n242, n247, o294, p167, p522, p667, p1034[1], p1547, p1784, p2049, q42, q241, s201, t661, t698, u95
104: a86, a115, a690, a745, a772, a1034, a1243, a1548, b865, b966, b1829, b2344, c52, e124, e150, e604, f56, g175, h314, k11, n82, n199, n272, n291, p469, p491, p1954, q108, q114, q120, q189, s376, t622, t628, t652, u85, v1
105: a82, a94, a295, a338, a396, a682, a916, a1082, a1126, a1405, b65, b410, b415, b569, b597, b603, b737, b871, b967, b1823, b1825, b1963, b2027, b2081, b2428, b2821, c598, c600, c724, c729, d192, e82, e83, e151, e507, g55, g210, h447, l26, m39, m323, m357, n99, n122, n169, n177, n233, n292, n364, p183, p668, p902, p1299, p1306, p1504, r23, s111, t78, t112, t179, t464, t586, t635, t653
106: a649, a794, a1203, a1392, a1559, b125, b168, b172, b519, b792, b869, b881, b2075, b2637, b2851, c27, c29, c295, c396, c621, c794, d182, e144, f75, h132, h758, l30, n54, n105, n141, n305, n525, o275, p215, p469, p526, p932, p1521, p1569, p1810, p1864, p1962, p2007, q270, t157, t160, t220, t423, u72

107: a791, a872, a1095, a1591, b294, b333, b880, b1290, b1785, b1831, b2041, b2823, c622, c636, h254, h415, i13, i97, n95, n145, n439, o42, p170, p953, p1232, p1978, p2037, s130, s242, t345, t468, t610, u86

108: a8, a75, a352, a730, a912, a1058, a1180, a1194, a1589, b418, b574, b1157, b1407, b2263, d43, e183, e345, e504, f69, h643, h759, i32, i53, m6[4], m93, m323, m334, n193, n477, p646, p731 p898, p1548, p1917, q33, q239, s26, s138, s248, s321, t348, t507, t512, t650, t786, u61

109: a689, a1590, b141, b685, b1179, b1334, b1826, b1968, b2039, b2423, b2567, b3061, c28, c134, e6, e165, e182, f71, f260, f273, h569, h759, m63, m267, n202, n404, o23, p71, p481, p518, p1252, p1880, p1979, q25, s300, t365, t537, u24, u73, u100[2]

110: a166, a388, a422, a668, a1095, a1593, a1600, b72, b325, b383, b622, b1096, b1315, b1770, b1935, b1963, b2163, b2275, c37, c107, d118, d296, e125, e400, e503, g50, h718, i202, l54, m359, m360, n90, o162, p655, p836, p882[1], p1103, p1405, p1793, p2156, q47, s99, s314, s358, t662, u113, u122, z3

111: a516[1], a517, a727, a834, a1037, a1210, b234, b273, b574, b1184, b1360, b1792, b1843, b1969, b2024, b2113, b2815, b3006, b3064, c706, d265, e311, e323, f44, h442, h659, l69, n115, n441, o311, p20, p25, p582, p618, p943, p1355, p1374, p2087, p2149, q221, t103, u90

112: a1052, a1557, a1567, b17, b176, b182, b612, b904, b1257, b1594, b1601, b1626, b1970, b2704, c461, c654, c683, d186, d192, f275, g2, g43, g175, j12, m41, m77, m127, n31, n197, n202, n259, n303, p233, p293, p1233, p1488, p1982, q91, s266, t349, t621, u56, u81, u99

113: a1124, b106[1], b542, b865, b1203, b1276, b1324, b1624, b1694, b1824, b1871, b1955, b2203, b2282, b2317, b2329, b2424, c658, c725, c801, e90, f55, m53, m243, n136, n166, p645, p653, p677, p689, p1946, t279, t365, t596, t638, t674

114: a164, a344, a347, a762, a885, a1169, a1394, a1444, a1560, b233, b383, b490, b858, b938, b1087, b1163, b1214, b1309, b1338, b1428, b1645, b1751, b2048, b2110, b2141, b2325, b2366, b3061, c144, c396, f130[1], h566, i28, i50, l24, m6[4], m77, n151, o60, o281, p215, p603, p705, p941, p1047, p1994, q48, s144, s267, t27, t612, t696, t756

115: a164, a382, a425, a593, a676, a697, a1127, a1562, b304, b430, b587, b598, b631, b2058, b2104, b2304, b2371, b2563, c20, c34, c179, c301, c915, e180, g8, g55, i3, l22, m93, n10, n90, n139, n309, p177, p645, p681, p692, p1280, p1502, p1556, p2086, t321, t469, u26

116: a142, a298, a305, a340, a377, a426, a554, a658, a867, a1060, a1188, a1525, a1636, b136, b293, b482, b538, b968, b1015, b1583, b1692, b2011, b2069, b2101, b2229, b2239, b2836, c264, c345, e120, e404, f45, f290, h628, h629, m103, n124, n437, o297, p509, p690, p832, p909, p1492, p1500, p1977, q91, q205, s110, s377, t573, t769

117: a498, a577, a952, a1189, a1537, a1549[1], a1579, b128, b416, b576, b965, b1761, b2086, c605, c793, d190, e77, e407, e603, h566, h600, h605, h688, h725, i3, i37, n132, n231, n402, p529, p554, p1490, p1932, p2023, p2155, q59, q118, t222, t353, u100[1]

118: a123, a300, a312, a329, a381, a496, a531, a1150, a1164, a1179, a1396, a1469, a1532, b100, b190, b243, b665, b725, b736, b823, b1264, b1406, b1566, b1852, b2203, b2593, c35, c36, c47, c144, d98, d267, e15, f82, g3, g198, l61, m287, m368, n53, n76, p246, p601, p657, p942, p1463, p1498, p1867, p1911, p1913, q1, q113, q203, r1, s204, t280, t324, t518, t618, t619, t663, u37

119: a397, a425, a471, a805, a867, a1197, a1652, b180, b610, b734, b900, c274, e519, f1, f265, h701, m9, m280, m337, n167, n232, p707, p767, p1284, p1561, p1866, p1868, p1981, t323, t496, t659, t706

120: a385, a449, a461, a781, a1154, b570, b685, b1064, b1152, b1363, b1730, b1741, b1742, b1744, b2091, b2161, b2239, b2269, c156, c187, c267, c376, d94, d119, d125, g35, h14, i67, i215, m104, m141, m183, n116, n129, n149, n303, n366, p168, p292, p524, p689, p899, p916, q111, s20, s45, s101, s186, t6, t506, t708, u102, u109

121: a16, a106, a107, a312, a471, a606, a632, a1065, a1576, b76, b950, b981, b2148, b2215, b2219, b2628, b2685, c199, c452, c467, c887, e148[2], e514, m28, m44, m85, m355, m359, n55, n241, p619, p871, p1066, r10, s197, s199, s326, t301, u62

122: a551, a677[2], a734, a1093, a1123, a1158, a1182, a1587, b183, b690, b1239, b1600, b2012, b2025, b2045, b2261, b2375, c251, d135, e190, e306, f241, m16, m80, m198[1], n260, n437, p505, p562, p590, p709, p1674, p1860, p1976, p2135, q173, q206, r16, s325, s360, t620, t753

123: a116, a158, a530, a532, a556, a736, a1184, a1619, b113, b236, b268, b578, b670, b1060, b1562, b1841, b1923, b2158, b2278, b2840, c418, g4, g205, i76, i191, m310, m352, m407, n20, n191, n237, o148, p507, p557, p939, p1056, p1576, p1908, r12, s59, s130, s249, t51, u98

124: a878, a1152, a1529, b53, b62, b100, b223, b500, b910, b1074, b1629, b1908, b1965, b2070, b2794, b2918, c434, d12, e17, h689, l2, n295, p604, p868, p1331, p1494, p1496, p1819, p2159, q54, q77, q188, s129

125: a106, a144, a749, a983, a1096, a1487, a1489, b586, b903, b1253, b1371, b1993, b2044, b2228, b2251, b2261, b2303, b3034, c110, c252, c465, e17, e301, f87, f102, f267, g37, g158, h402, h737, k13, n91, n282, n363, n387, n390, p586, p909, p1062, p1503, p1760, p2153, s5, t462, t614, t693, t747, u16, u33[1], u87, u130, x2

126: a158, a196, a305, a905, a1062, b142, b186, b1095, b1161, b1628, b1990, b2021, b2041, b2071, c120, c424, c829, c906, d127, d137, h251, n27, n79, n277, n334, o42, p21, p649, p1168, q25, q204, q263, s140, s200, t255, t465, t751, u35

127: a561, a947, a1446, a1522, a1528, a1629, a1656, b169, b258, b485, b979, b1026, b1312, b1398, b1459, b1581, b1708, b1812, b1934, b2245, b2305, b2769, b3035, c55, c246, c299, c344, c448, c454, c460, e190, g167, h401, l16, m70, m199, n102, n205, n421,

o299, p63, p477, p616, p637, p1927, p2093, s84, s126, s142, s382

128: a104, a462, a782, a783, a1073, a1530, a1531, a1592, b3, b78, b146, b816, b1372, b1730, b1731, b2143, b2261, b2355, b2499, c53, c251, c253, c288, c314, c833, d115, f64, f243, i192, m358, n46, n97, n138, n390, p184, p306, p931, p1199, p1652, q210, s46, s54, s359, t156, t248, t669

129: a47, a99, a530, a756, a1022, a1155, a1228, b788, b1764, b1880, b1917, b1954, b3058, c902, d136, e304, f244, h451, i174, n72, n106, n113, n188, p65, p604, p607, p698, p840, p1065, p1341, p1825, p1928, p2008, p2041, p2146, q64, q208, t183, t232, t274, t313, t346, t527, t614

130: a147, a489, a695, a943, a972, a1197, a1633, a1637, b23, b42, b91, b137, b502, b511, b602, b650, b1157, b1170, b1278, b1893, b1998, b2023, b2038, b3005, c358, c834, e9, e614, g52, g206, h633, l7, l55, m317, n102, n104, n198, p1, p570, p1040, p1117, p1260, p1497, p1499, p1541, p1758, p1760, p2062, r10, s259, s374, t6, t679, t687, v10

131: a171, a499, a1165, a1485[2], a1539, b64, b264, b501, b640, b655, b824, b1037, b1237, b1693, b2264, b2273, b2761, c559, c769, e126, f48, f259, g32, h669, i12, i30, m2, n116, n431, p584, p598, p953, p1345, q43, x28

132: a70, a720, a1038, a1558, b384, b1244, b1556, b1582, b1610, b2117, b2131, b2133, b2218, b2422, b2528, c50, c78, e134, e324, g149, g179, h579, h627, h719, m27, m136, m137, m138, n103, n238, n323, n397, p574, p813, p918, p931, p995, p1138, p1557, p2117, q214, s244, s389, t680

133: a92, a288, a470, a540, a605, a666, a886, a1063, a1475, a1499, b155, b245, b487, b614, b961, b1183, b1192, b1560, b1884, b1944, b1946, b2050, b2499, b2716, c350, d162, e9, e68, f222, h665, m59, n203, n326, p1844, p1892, q56, q234, t230, t686, t687, x27

134: a157, a473, a477, a739, a1536, b253, b631, b1041, b1983, b2020, b2265, b2328, b2764, b3005, c80, c110, c714, c716, d31, e126, f256, g33, g44, g164, g207, h298, h720, i107, m128, p525, p648, p843, p1509, q35, s376, s383, t2, t231, t346, t354, t528

135: a5, a427, a466, a605, a660, a1241, a1453, a1561, a1601, b22, b49, b60, b386, b1057, b1242, b1279, b1332, b1474, b1651, b2012, b2287, c350, c379, c903, e10, e55, e84, e104, e176, f252, g63, h610, h627, h673, h714, i205, m17, m98, n327, n392, n424, p178, p465, p575, p831, p929, p1231, p1540, s214, t5, t12, u18

136: a93, a604, a793, a1473, a1644, b498, b587, b630, b938, b957, b1653, b1711, b1799, b2200, b2528, c54, c191, c262, c353, c377, d232, f130, f255, g168, k12, m6[3], m102, m367, n121, n138, n423, p785, p1953, p2036, p2157, t643, t787

137: a120, a1510, b912, b1131, b1202, b1352, b1613, b1774, b1945, b1962, b2230, b2262, b2290, b2498, b2743, b2803, b3038, c14, c948, d229, h414, l35, m123, m394, n67, n74, n98, n187, p490, p913, p1369, p1849, p1951, q202, s90, t315, t630, u35

138: a452, a696, a1501, a1584, b5, b191, b308, b2029, b2129, b2154, b2720, c49, c277, c336, c338, c340, c353, c389, c597, d127, e80, e327, f251, f264, f291, g189, n160, n361, p178, p1150, q40, s11, s127, t4, t354

139: a83, a97, a876, a1149, b4, b70, b74, b724, b933, b1077, b1536, b1541, b1614, b1866, b2193, b2747, c48, c922, d246, e186, e308, g30, h196, i172, m2, m193, n293, p470, p1036, s44, t113, t203, t316, t653

140: a95, a418, a1213, a1306, b239, b494, b963, b1457, b1514, b1525, b1844, b2051, b2268, b3059, c78, c883, c920, f94, f253, g47, h14, h665, l8, l15, l40, m335, m365, n150, n162, n280, o231, p76, p591, p693, p863, p904, p1661, p1872, p1882, r29, s49, s90, t11, t34, t57

141: a549, a1035, a1114, b86, b192, b625, b1029, b1169, b1260, b1564, b1817, b1875, b1877, b1882, b2153, c287, c346, c420, c429, c684, c834, d86, e76, e127, f207, g168, h138, i39, i68, i207, m89, m351, m424, n426, p826, p841, p1037, p1785, q172, u78, u89, u92

142: a292, a424, a611, a1399, b87, b668, b1075, b1111, b1363, b1465, b1466, b1518, b1652, b1958, b2124, b2353, b2411, c706[2], c726, e48, e79, f3, f50, f130, g68, h710, i8, k15, l15, m20, m92, n506, p571, p1262, p1971, q9, q146, s36, s89, u126, u138

143: a887, a1152, a1603, b604, b1187, b1360, b1430, b1753, b1786, b1845, b2095, c18, c274, c281, c307, c601, c726, e269, h138, h708, i97, i163, n81, n84, n489, n495, p2136, p2152, q53, q60, s143, s278, t58, t703[1], u28

144: a109, a741, a1446, a1514, a1549, a1594, a1644, b149, b896, b1193, b1535, b1541, b1580, b1654, b1931, b1961, b1991, b2201, b2207, b2369, c891, e123, e168, e524, g128, i106, m33, m331, n73, n162, n556, o143, p485, p498, p644, p670, p671, p682, p1266, p1808, s127, s353, t235, u59, v23

145: a85, a378, a424, a1097, a1199, a1447, b252, b283, b1085, b1143, b1165, b1213, b1252, b1603, b1724, b2026, b2047, b2106, b2122, b2123, b2163, b2164, b2189, b2279, b2298, c88, c241, c254, e250, f85, f176, g68, h324, h623, h638, h640, h667, j10, l48, m102, n318, n388, n445, p571, p581, p608, p629, p900, p1038, p1041, p1477, p1845, s38, x3

146: a24, a308, a767, a1232, a1492, b15, b161, b222, b543, b1259, b1314, b1547, b1753, b1895, b1991, b2005, b2046, b2253, c287, c411, d80, d193, d214, e617, f248, g35, g38, g39, g208, h589[5], h707, i67, m123, s352, t11, t242, u50

147: a428, a495, a585, a674, a677, a1053, b71, b589, b668, b1461, b1517, b1557, b1664, b1709, b1786, b1836, b1988, c427, f58, g216, i41, m48, n140, n173, n200, n436, p9, p602, p610, p1036, s60, t240, t521, u110

148: a129, a168, a387, a434, a563, a650, a788, a966, a1200, a1213, a1631, b193, b939, b1141, b1354, b1356, b1488, b1570, b1731, b2267, b2281, b2377, c300, c352, c391, e128, e168, e305, e307, f77, i42, l58, n413, o276, o305, p515, p849, p1362, p1539, p2009, q63, s220, s367, x25

149: a501, a505, a773, a1097, a1100, a1170, a1235, b509, b582, b636, b1008, b1107,

d138, d253, f67, g217, h299, h733, i124, i130, k14, m298, n355, n410, n453, p67, p1251, p1893, p2052, q14, q18, s151, s232, t9, t309, u31

172: a103, a858, a963, a964, a1402, a1541, b11, b145, b295, b540, b921, b1367, b1740, b1782, b1876, b2217, c68, c317, c402, c706[3], c835, f15, h12, h735, m386, m396, n32, n77, n401, n406, o180, p199, p727, p729, p1060, p1658, q6, s205

173: a1, a710, a774, a1172, b577, b694, b990, b1206, b1245, b1340, b1499, b1509, b1706, b1723, b1984, b2168, b2295, c406, e302, i126, m405, n33, n175, n415, n420, n434, o150, p851, p866, p1783[1], q201, s260, s262, u20, u82

174: a1007, a1195, a1494, b12, b25, b698, b727, b837, b1005, b1323, b1531, b1607, b1728, b1918, b2469, c79, c275, d110, e167, e311, i2, p525, p556, p684, p880, p1029, p1055, p1904, q17, s20, x12

175: a1036, b700, b802, b992, b1369, b1574, b1634, b1758, b2155, b2359, c372, c393, c923, c924, c925, c928, d167, e61, e98, e109, g23, g138, h8, h613, h624, h734, i6, l71, m210, m314, n263, o258, p524, p597, p1437, q235, s39[2], s58, s236, s253, s285, s390, t319, t748

176: a780, a1129, b14, b21, b29, b982, b1039, b1167, b1212, b1348, b1555, b1620, b1834, b2062, b2085, c19, c31, c64, d85, e56, e133, f91, i153, n157, n407, n445, n450, p550, p1956, p2103, q38, q156, r32, s131, s191, s361, t44, t322, t682, u80

177: a170, a636, b699, b1331, b1445, b1590, b1627, b1727, b1837, b1897, b1986, b2054, b2368, c227, c229, c296, e56, f2, f42, g57, i51, m191, n5, n161, n336, p530, p864, p1972, q22, q155, s41

178: a137, a1032, a1129, a1239, a1402, a1450, a1596, b198, b201, b254, b911, b1491, b1508, b1721, b1736, b2428, c197, c437, c462, c934, e143, e334, g12, g187, g195, h616, i59, i86, m6[2], n349, n394, p157, p576, p917, q39, q89, s366, u51, u80

179: a91, a430, a934, b637, b1481, b1638, b1809, b1974, b2161, b3081, c32, c33, c634, c708, f43, f249, g202, p1204, p1215, p1586, s369, t312, t695, u6

180: a46, a541, a724, a957, a1008, a1113, a1219, a1693, b29, b132, b696, b1004, b1246, b1323, b1404, b1451, b1493, b1736, b2132, c31, c269, c278, c388, c738, c886, d191, d231, e191, e193, f90, g214, g220, h609, i103, i131, j13, m36, n398, n452, o258, o302, p626, p844, p1049, p1579, p1673, q276, s85, s113, s152, s356[1], u22, u91, v10

181: a927, a1009, a1028, b10, b1395, b1520, b1567, b2136, b2224, b2376, c371, c884, d126, e57, f76, f97, g156, g187, i46, i155, m203, n335, n414, n419, p1848, p1863, p1869, s29

182: a715, a888, a1253, a1460, a1509, b199, b697, b930, b1002, b1114, b1493, b1607, b1800, c366, d260, e60, e107, e121, e262, e309[1], f40, h388, h589[3], h747, i91, o45, p1054, p1410, p1807, p1871, q3, s86, u29, u60, u120

183: a1480, a1646, b14, b189, b1330, b1589, b2204, c63, c392, f52, j2, m52, m363, m393, n393, p2098, q122, s36, s50, s51, s217, u118, u127

184: a133, a306, a740, a1380, b723, b970, b1387, b1444, b1723, b1872, b2079, b2137, c82, c196, e62, f266, l18, l19, l38, m211, n411, n456, p810, p1611, p2071, s288

185: a304, a992, a1190, b489, b491, b902, b955, b977, b1304, b1443, b1630, b1729, b1775, c369, c371, e75, f185[1], g65, g154, g215, h747, i133, i211, m135, n63, n324, n569, p66, p549, p583, p850, p903, p1926, q6, q52, s9, s92, s185, s273, t8, t54

186: a984, a1029, a1030[1], a1196, a1493, a1516, b984, b991, b1133, b1572, b1644, b1739, b1971, b2134, b2334, c320, d151, e62, h709, l50, m353, n21, n541, o156, p561, p716, p867, p1034, q76, s24, s83, s147, t269, u107

187: a105, a888, a994, a1278, b1437, b1558, b1650, b1748, b1906, b2208, c58, c149, c195, c201, c390, c449, e74, i108, i131, p856, q61, q246, s139, u127, x17

188: a1113, a1258, b1140, b1329, b1544, b2009, c56, c286, c713, c732, e89, e166, h159, i147, m211, n372, p6, p478, p865, p877, p1964, s185, s364, u123

189: a1455, a1457, a1489, b1229, b1710, e49, e100, f114, g1, g126, h418, h738, i177, m382, n289, q187, r38, t534, t776, u70, u75

190: a301, a538, a1005, a1195, a1251, a1381, a1643, b14, b994, b1152, b1308, b1524, b1769, b2289, b2373, b2383, b2855, c933, e37, f245, h589[1], h601, i187, m134, m385, n130, n165, p1063, p1067, p1657, q80, s95, s277, s371, t533, t535

191: a110, a877, a1084, b14, b1449, b2157, c373, c728, f21, g28, i125, k25, l9, n19, n62, n350, p684, p812, p1900, p2064, p2098, q125, s282, u44, u97

192: a3, a1192, a1455, a1457, b96, b621, b1424, b1754, b2183, b2642, c233, c394, c707, e422, f92, f185, g165, h589[2], i127, n65, n75, n157, n430, n448, p856, p1285, p1593, p1846, s288, s423, t56, u65, u69

193: a1292, b983, b1482, b1878, b1902, b2383, b2384, b2693, c369, c398, e309[1], g20, g29, g144, g173, g177, 176, n17, n442, n451, p573, p937, p1061, p1901, p2101, q152, t127, u142

194: a1261, a1472, a1622, b623, b1643, b1889, b1941, b2199, c75, c296, c320, d141, f46, g134, h292, h666, h748, n165, n251, n332, n333, n429, n430, n570, p875, r24, s28, t531, u93, u119

195: a537, a744, a1183, a1378, b51, b89, b285, b1230, b1903, c73, e5, e31, e99, e114, f81, g66, h159, h746, i88, m49, m140, n23, n75, n176, n325, n372, n412, n444, o308, p5, p871, p907, p1353, p1806, q29, s40, s275, s391, u125, u129

196: a537, a1207, a1298, a1407, a1479, b408, b1381, b2031, b2159, b2367, e115, i186, n356, p1805, q106, s370, t149, u36, x23

197: a108, a985, b1188, b1453, b1576, b1702, b1901, b2162, c337, c450, d107, f88, f254, g24, h607, i125, m399, p546, p2104, q144, s134, s330, t111, t250,

198: a780, a1004, b1403, b1616, b1690, b1707, b1742, b1904, b2238, b2386, c65, c273, c305, c332, c408, d82, d258, e4, e92, g64, g137, g142, g194, h589[4], h666, i104, l59, n29, n201, n245, p915, p1761, r31, s280, t43, t532, u97, u106

199: a114, a270, a756[1], b508, b1250, b1307, b1700, b1957, b2385, b2433, c71, d122, g59,

g62, g163, h620, i213, m29, m139, n134, n283, p240, q148, s141, t325

200: a609, a725, a983, a1225, a1238, a1288, a1379, a1398, a1409, a1411, a1544, a1635, b1002, b1507, b1526, b1553, b2032, b2033, b2052, b2307, b2696, c309, c931, c944, d104, d194, e247, e605, e613, f51, g129, g159, h11, h544, h752, i11, i63, i88, l14, m400, m402, n16, n49, p1097, p1553, p1885, p1965, p2081, q26, q124, q149, q169, s23, s47, s152, s156, s268, u29, u114

201: a141, a433, a889, b95, b990, b1681, b1690, b1854, b2226, b2856, l7, l76, n66, p876, q143, q150, r30, t326, t691, t750, y1, z1

202: a113, a543, a903, a1282, a1406, b171, b1436, b1578, b1591, b1924, b2034, b2096, b2225, c74, c233, g20, h445, i15, j3, n399, p888, p890, p1985, s43, s96, s285, u39, u136

203: a1280, b18, b1456, b1808, b2233, b2695, c57, c408, e2, e81, g126, i73, i102, i204, p911, p1804, q162, t10, u69, u139, u143, v21

204: a675, a721, a1240, a1259, a1390, a1413, b579, b1549, b1606, b1975, b2040, c368, c416, e59, e92, e139, f190, g36, i87, n316, n342, p882, p915, p1265, p1411, t779, u34, u141

205: a555, a789, a845, a1488, a1500, b1209, b1217, b1551, b1593, b1646, b2358, b2378, c150, c310, c325, d104, e3, e116, e158, f173, g65, i1[1], i44, i73, m190, n63, n338, n340, n571, p1069, p2102, p2109, q32, s226, s280, t268, t461, t703, v7, v11, v27

206: a302, a714, a989, a1389, a1504, a1523, b1180[1], b1181, b1617, b1647, b1752, b1995, b2206, c332, e12, g133, h397, h762, l51, m641, n416, p880, s52, t10, t695, t775, u1, v27

207: a161, a429, a756[1], a1074, a1388, a1542, b1200, b1497, b1912, b1921, b2064, c259, c385, e41, m149, m316, n3, n80, n252, n391, u68

208: a258, a266, a744, a1236, a1645, b999, b1365, b1384, b1542, b1725, b1743, b1867, b2223, b2414, b2687, c57, c380, c409, g213, i45, j4, j5, j7, m313, n66, o289, p1854, p2151, t110, t530, v20

209: a128, a723, a928, a1230, a1277, b976, b1217, b1390, b1510, b2416, b2434, d130, h148, i189, i190, i195, p533, p560, t249, t532, v9

210: a1307, a1498, b838, b1205, b1310, b1351, b1359, b1366, b1726, b1863, b1869, b1911, b1996, b2036, b2415, c247, c284, c310, c322, c430, d32, d154, e200, g29, g48, h606, i208, m149, n353, o278, p8, p812, p847, p1064, p1804, q124, q261, t7, t10, t132, t689, t770, v18

211: a747, a1264, a1632, b740, b1855, b2094, g200, n337, p1803, v8

212: a1209, a1285, a1466, b30, b591, b1438, b1759, c228, c680, d252, e2, e111, f275[1], h9, h631, i113, i114, l66, m62, m316, m326, m383, n2, n65, n278, n314, n455, o259, p1444, t249, u88, u111

213: a728, a960, a1281, b527, b1346, b1440, b1548, b1714, b1743, b1755, b2209, c741, g58, h756, i87, i89, n370, n378, p1809, q167, u114

214: b27, b123, b1446, b1688, b1859, b1863, b2100, b2348, b2697, c81, c150, d112, e270, g61, i113, i115, i170, p239, p240, p1862, p2097, q135, s137, t697, u30

215: a746, a1202, a1608, b24, b987, b1433, b1437, b1527, b1705, b1912, b1939, c215, c330, c399, i29, l17, l64, n342, n405, o232, p55, p919, p1068, p1925, s251, t285, z2

216: a161, a874, a962, a982, a1193, a1477, a1543, b1104, b1362, b2120, b2156, b2341, e35, h630, n341, n357, o306, p580, p845, p1888, q252

217: a873, a1246, a1404, a1410, a1412, b307, b585, b1022, b1228, b1472, b1915, b2336, b2350, b2699, c397, f275[1], g196, h603, n384, n568, o260, p1202, p1903, q142, t149

218: a942, a1003, a1625, b249, b965, b1006, b1434, b2426, b2698, c85, e48, e50, g148, g213, i82, i115, i165, n261, n409, p207, p883, p919, p1980, p2062, q136, q255, q277, s94, s284, t43, u21, u74, u137, y2

219: a871, b1032, b1106, b1487, b1498, b1550, b2340, n174, p1925, s91, t140, t692, u12, u103, u104

220: a13, a586, a1607, a1623, b28, b1172, b1545, b1645, b2210, b2692, b2694, c453, c458, c867[1], c929, e8[1], e118, e159, g22, g185, g190, h381, h604, h612, n18, n26, n346, n347, n427, p1070, p1859, p2077, q141, s146, t749

221: a1047, b32, b33, c60, c61, g18, n382, o284, q280, s8, s237, t263

222: a711, a750, a761, a871, a1621, a1626, b306, b1030, b2291, c444, d187, f89, g26, g136, i112, n135, n164, n374, p466, p814, p1585, q168

223: a844, a846, a982, a1202, a1296, a1405, a1411, b1188, b1258, b1545, b1757, b2292, b2432, c59, l5, l17, n339, n353, p1048, q158, x4

224: a729, a962, b27, b584, b1207, b1429, b1926, b2188, d139, e71, g60, l75, n22, p10, p558, q151, t569

225: a503, a1217, b143, b184, b1639, c200, c231, c386, e54, e423, g59, h696, i31, i84, i105, m45, n344, n382, p567, p712, p714, p1118[1], p2090, t321, u33[2], u41

226: a1000, a1208, a1257, a1312, b1403, b1704, b2065, e110, e233, f270, m20, p2091, q254, t631, t702

227: a828, b951, b1416, b1550, b2037, b2130, c709, f18, f99, l70, m177, m398, p2085, t162

228: b917, b1000, b1067, b1750, b1898, b2332, e108, e117, e118, f171, g19, h736, m397, p542, p910, p2064, q79, s55, s57, t287

229: a765, a1296[1], b1409, b1733, b1819, b1942, b2288, b2335, c232, c316, c374, d84, d147, f53, f59, g193, h626, h721, n359, p559, p683, p714, s6, t254, t755

230: a864, a969, b592, b787, b1204, b1439, b1659, b1701, b1864, b2127, b2349, b2379, b2430, c945, e325, h723, m343, o232, p623, p720, p878, p2030, p2099, q107, q165, q254, s37, s409

231: a1495, b1344, f24, m302, m406, n360, p812, p860, p1338, s132, t692

232: a1143, a1174, a1400, a1627, b626, b937, b988, b1349, b1554, b2212, d128, d216, g176, m191, n449, o284, p516, q163, u7

233: b628, b2335, c946, d133, h618, p534, q164, t120, v25

234: a731, a1459, b1534, b1640, c444, d111, e30, g26, i57, n359, o233, p1337, p1339, p1506, q157, s279, t738

235: a126, a743, a748, a955, a1045, a1636, b13, b31, b1001, b1158, b1358, b1393, b1559, b1732, b1737, b2103, b2109, b2633, c737, e1, e40, e157, e233,

BOILING POINT INDEX OF ORGANIC COMPOUNDS

Temperatures in °C; where there is a range of values, the compound is listed according to the lower value.

−191: c184
−164: m175
−127: m297
−123: m345
−104: e101
−100: m349[1]
−95: h768
−93: m289
−89: e162
−84: e591, m289
−83: c185
−82: m308
−81: m241
−79: c159, e249
−78: m276
−76: e421
−75: m260
−72: e417
−64: a640
−63: m176
−59: m226
−57: s77
−56: k1
−52: m259
−50: c186
−48: p1713, p1840
−47: e277
−42: e259, p1107
−41: k1, m230
−39: e204
−38: e245, p1081
−37: a853
−35: e225, m232, p1079
−32: a1064, c907
−31: m291[1]
−30: e594, m253
−29: f113, m297, p1744
−27; p753
−26: e268
−25: b3078, e221
−24: f113, m228
−23: e499, p1831
−22: m342, s313[1]
−21: f106
−20: m291[1], m292, s69
−14: p748
−13: e405
−12: a1439, m219, n498, p1189, p1198′
−9: m237
−7: a852, a944, p1747, s72
−6: b2940, m176
−4: b2436, s74[1]
−3: p1743
−2: o304
−1: b2485, m330
0: b2995, e256, p1170
1: a1641, b2942, p1190
2: a1437, p1185
3: a939, c604, e546
4: b2941, c911, e218
5: b2994, b3066
6: b2454, m345
7: a850, e253, p1080
8: b3070, p732
9: m256
10: b2468
11: b2435, e547
12: b2452, e199, e539, h767
13: e241
15: m274
16: e403
17: e163, p759
19: a1416[1]
20: b2410, b2992, c909
21: a18, a203, b2403, p735
22: m261
23: p1721
24: m248, p760
25: a1419, f25, p745, p747, p1188, x11
26: h724, p46, s80
27: b3071
28: b2579, e475
29: b3079, p398
30: e222, p375
31: b2991, f141, f193
32: b2564, e411, e577, e601, m340, p1109
33: a1442, f198, p1719
34: b2461, b3069, k3, p1180
35: d243, e328, e477, e534
36: a904, a1433, a1436, h766, m216, p84, p377, p1134
37: p376, p742, s313
38: e195, p754, p1720
39: b2462, b2993, b3077
40: a639, b2460, c865, e581, m252, p447
41: b2984, p1169
42: m282, p45, p443
43: i143
44: c899, c910, p1114, p1317
45: f109, n497, p44, p1722
46: a1418, c161, e275, e431
47: e276, e413, p1133
48: p47, p736, p1108, p1181, p1716
49: c868, e271, h563, n502, p1083
50: b2556, e541, e563, g186, p448
51: a183, p411
52: a1421, a1440, c814, p1141, p1183, p1294
53: d158, f158, m277, p1707
54: f135, p412, s31
55: b3067, e402, e460, e538, p73, p1828, t161
56: a827, a1483, b2563, b2983, e418, p416, p1649, p1833
57: a221, c183, c632, e202, e215, h562
58: a1434, b2557, b2562, b2952, e234, e575, p1714, p1715
59: a1417, b2441, e501, f162, h281, p417, p1120, p1828
60: e412, e459, i140, i143, p148, p418, p442
61: a910, m306, p410, p456, s67
62: c90
63: a1482, b2487, b2488, b2489, b2561, b2950, b2958, c631, f213, h456[1], h656, n492, o263, p143, p149, p1742, p1838
64: a22, b2953, e557, f220, n485, p1296
65: a796, b2401, b2987, f214, h280, m250, m349, m349[1], m349[2], p474, p1175, p1833
66: a833, e429, h549, p466
67: b2390, f220, h550, p413, p1293, p1759, s341
68: b2440, b2512, b2513, b2951, e497, f140, h548, m214, n491, p415, p1712, p1724
69: b2447, h319, h547, o292, p1164
70: a931, b2986, h574, p414, p1717
71: a40, a197, a1426, b2404, b2957, e458[1], f163, h579, p1119
72: a197, a637, c876, e549, p403, p457, p1099, p1725
73: b2985, e452, h279, p733, p737, p1129, s123
74: e273
75: b2459, b2954, c662, e565
76: a174, a1430, b2471, c677, e599, h278, k2, m296, p452, p1731
77: a199, b2494, b3073, f200[1], h765, p407, p1094, p1732
78: b2486, b2511, b3001, e336, e337, d208, h311, h540, h541, n490, p402, p1092, p1731, p1776
79: e254, f201, p131, p1140
80: a854, b202, b2106, b2912, c628, f215, h282, p133, p1314, p1324, p1775, p1893[1]
81: a174, b2611, b3054, d210, e451, e545, f146, h581, h657, p1111, p1176, p2167, s121
82: a241, e562, h124, p401, p1588, p1633
83: b2646, c811, e216, e292, e430, e537, e550, e578, h309, p453, s65
84: a848, b2495, b2955, d215, e561, f116, h580, m339, p400, p405, p1135, p1736, s301, t187
85: a30, a630, b779, b2644, b2645, b2956, c628, h308, h373, h652, p743, s347
86: b2523, b2962, e459[1], f216, h312, p134, p1087, p1894
87: a589, b2457, e425, h683, n483, p406, p454, p1197
88: b2439, b2632, b2961, c637, h93, p1160, p1832, p1895, s115
89: a925, b2180, b2988, f108, p462, p1193, p1295, p2160
90: a33, a217, b2502, c173, e435, h371, m217, p132, p1090, p1800, p2173
91: b2458, b2501, e223, e461, e528, h369, p86, p87, p391
92: a44, a216, a1420, b2480, b2481, e450, 480, e551, f209, h372, h373, h374, h584, p1525, p1532, p1545, s303
93: a45, b2444, b2524, e224, e448, f206, h81, h730, p140, p386, p1124, s182, t42
94: a1429, b2451, b2926, b2963, b2966, e448, e469, h82, h197, p408, p1735
95: a810, b2408, b2493, e171, f158, h586, i245, p385, p390, s351
96: a56, b2961, b3047, e257, h200, h201, p409, p444, p1161, p2164, s348
97: a37, a182, a813, a1427, a1443, b2966, b3048, f133, p100, p387, p404, p1326, p1587, p1814
98: a52, a812, b2397, b2446, b2643, b2943,

Temperatures in °C; where there is a range of values, the compound is listed according to the lower value.

b3063, e201, e294, f139, h198, h199, h651, m246, p101, p1746, s73

99: a522, b2443, b2522, b2865, b2866, b2867, e465, e580, n499, p153, p1196, p1316

100: a1012, b1102, b2520, b2521, c4, e342, f160, h216, h682, m264, m269, p48, p1195, p1773, p1799

101: b2979, c630, c688, d198, e454[1], f115, h81, h674, m286, n484, p419, p1128, p1434, s78

102: a259, b2491, b2500, b2526, b2665, b2674, b2896, b2933, c691, e243, e455, f196, h585, p345, p348, p450, p455, p1192, p1745, s70, t609

103: a21, a59, c820, c874, e366, f150, f197, o241, p77, p392, p1311

104: b2389, b2490, b2541, b2933, b3072, b3091, c818, c819, e502, n501, p85, p388, p1534

105: a566, a1427, b2944, d197, e576, f161, h218, m225, p211, p420, p1165, p1435

106: b2178, b2394, b2601, b2921, h652, i142, m278, p952, p1125, p2166

107: a367, b2442, b3090, e181, e196, e481, f132, f153, h358, h681, m258, o235, p221, p379, p393, p1002, p1225, p1432, p1534

108: a214, a367, a411, a811, e170, e211, e409, e458, e535, h83, o292[1], p198, p760, p1165, p1166, p1738, s30

109: a36, a893, b2437, b2450, b2729, b2965, c2, e548, e582, f210, h362, h364, m238, p389, p1086, p1323, p1737, s81

110: b2964, c817, d242, e339, f156, f212, h87, h213, h363, n488, p53, p111, p152, p381, p1376, p1836, s76

111: a816, b2456, b2517, b2960, h361, p111[1], p472, p802, p1527, t275

112: a179, a180, a416, b2516, b2897, b2970, c408, e449, e454, e558, h89, h217, h365, m291, p82, p1733, p1734, s346

113: b2898, b3046, c872, e274, e331, h360, h553, p155, p210, p1625, s114, t215

114: b2537, b2806, c908, d204, e244, f219, h321, p432, p1156, p1651, p1757, p1820, p1851, p2142, t493, t741

115: a25, a55, a187, b2515, b2519, b2915, c627, d200, e255, e585, f166, f200, f204, h127, m341, p112, p154, p366, p382, p434, p1837, p1939, p2165, t216, t505

116: a181, a591, a836, b2514, b2542, b2799, b2938, b2973, e207, e485, h359, o288, p80, p113, p141, p315, p365, p383, p1113, p1523, p1897, p1990, t494

117: a215, b2478, b2570, b2809, b2864, b2915, b2917, c619, e462, h129, h322, o293, p1003, p1004, p1796, s302, t136, t495

118: a37, a58, a525, a1423, b2577, b2647, b2648, b2680, b2888, c611, e208, f123, h123, h126, h130, h366, p128, p142, p364, p394, p440, p1121, p1123, p1433, p1515, p1632[1], p1755, s343

119: b2946, b3014, b3017, b3018, c669, e554, h128, h368, i231, m229, p92, p314, p436, p1668, p1842, t224

120: a1416, b2388, b2509, b2510, b2569, b2833, b2889, b2916, b3015, c663, c666, e187, e428, e472, h557, h564, m127, p461, p1151, p1163, p1211, p1676, s312

121: a184, b3014, b3045, e420, h552, h554, o208, o250, p334, p396, p1531

122: b2508, b2668, b3068, h331, h634, o213, o214, p378, p380[1], p384, p435, p437, p1139, p1332, p1767

123: a445, a857, c665, e289, h332, h555, m239, o211, o212, p151, p380[1], p1084, p1228, p1383, p1708, s291

124: a222, b2538, b2574, b3012, c667, c668, c670, e203, f142, h378, k2, p355, p971, p1006, p1303, p1852, p2092, s74[2]

125: a178, b2801, b2969, b2971, b2978, e373, e478, f208, h148, h214, h518, h655, o210, o221, p88, p114, p935, p1005, p1798

126: a178, a375, b2467, c160, c170, c875, m299, o112, o209, p228, p441, p1604, s419

127: a32, b2893, e203, e479, e500, h538, p103, p209, p338, p969, p1007, p1321, p1194, p1524

128: b281, b2398, b2659, b2765, b2892, b2895, c816, e350, e536, e560, f167, h316, h517, m408, p122, p970, p976, p1608, p1631, s63[1], t146, t198

129: a53, b2894, b2945, b3089, f159, h313, n496, p98, p750, p2019, p2137, s116

130: a176, a253[1], a374, a949, b450, b2455, b2545, b2673, b2810, b2972, b2974, b2975, c664, c878, c894, e266, e463, f145, g130, h226, h320, h323, h718, p90, p147, p1088, p1829, p2126, s414, t45

131: a524, b2568, b2800, b2911, c671, e212, e566, h316, h482, i228, o224, p124, p940, p982, p1085, p1210, p1226, p1347, p1447, p1630[1]

132: a177, a254, b407, b2192, b2396, b2506, b2759, c674, f145, f210[2], m257, p397, p1007, p1313, p1381, p1606, p1751, p1774, t147

133: a1424, b2546, b2939, c879, e467, h117, h551, p336, p460, p1153, p1380, p1519, p1602, p1650, s412, t175

134: a253[2], a420, b2265, b2507, b2539, b2595, c614, c640, e246, e359, h288, h289, h330, h533, h534, h571, p335, p1148, p1544, p1704, s340, t210

135: a821, b2762, b2806[1], b2879, c913, e363, f211, h119, h340, h342, h483, h484, n480, o106, o286, p104, p354, p1152, p1167, p1213, p1401, p1537, p1654, s75

136: b758, b3012[1], b3013, d223, e552, h118, h481, h576, p374, p1663, p1816, p1899, p1914, s63, t211

137: b2191, b2884, b2914, c870, g155, h120, h588, p313, p433, p977, p1322, p1486, t41, t208

138: a1438, b449, b662, b2505, b2549, b2550, b2575, b2878, c612, c695, c861, f151, h203, h341, h479, i230, m223, m416, o222, p946, p972, p975, p1127, p1723, p2110, t158

139: a847, b661, c895, f219[1], h205, h530, p202, p325, p1366, p1918, s292

140: a172, a402, b2274, b2578, b2583, c126, c696, c892, f111, h116, h204, h480, i164, i224, p286, p324, p424, p433, p973, p974, p1154, p1200, p1226, p1630, s342, t168, t459, t492

141: a264, b2877, b2882, c860, h122, h221, h328, o138, p139, p229, p1144, p1302, p1730, s350, t207, t395

142: a226, a253[2], b778, b2652, b2808, b2885, b3031, e217, e471, e544, f289, h95, h421, h529, h530, m283, o129, o137 p94, p97, p290, p983, p984, p1756, p1766, p2144, s332, t206

143: a265, a415, a616, b2610, b2686, b2805,

Temperatures in °C; where there is a range of values, the compound is listed according to the lower value.

Temperatures in °C; where there is a range of values, the compound is listed according to the lower value.

f119, f154, h142, h352, m376, o201, p206, p359, t48, t197, t548

178: a508, b848, b2624, b2678, b2741, b2787, b2869, c8, c617, c847, e213, e488, f136, h353, i16, l41, l42, l43, o196, o254, p116, p1597, p2006, p2017, t447

179: a41, a460, a921, b2391, b2675, c648, e227, p227, p1220, p1357, p1748, p2005, t391, u52

180: a29, a441, a1134, b535, b2622, b2828, c638, c685, c699, e352, h302, o113, o188, o200, p121, p1157, p1287, p1393, p1700, p1758[1], s245, s246, t213, t475, t590, u21

181: a406, b566, c93, c644, c647, c773, h370, m24, m325, p555, p997, p1522, p2017, p2127

182: b709, b780, b888, b1316, b2151, b2581, b2596, b2623, b2625, b2778, c912, e189, h175, i232, m263, o121, p761, p1475, p1772, t358

183: a35, a290, a365, a465, a730, a1432, b372, b565, b889, b2580, c701, e355, h154, h177, i148, m211, m362, p201, p930, p1419, p1615, p1699, p1839

184: a228, a1002, b245[3], b447, b567, b672, b2766, b2785, b3030, b3082, c699, e365, f186, i150, p688, p1096, p1177, p1375, p2028, p2031, t359, t360, t486

185: a101, a226, a807, b887, b2177, b2620, b2634, b2655, b2831, b3009, b3039, c91, c92, c616, d266, e442, e583, h491, p250, p1158, p1179, p1415, p1753, s295, t52, t167, t281, t476, t703[1]

186: b767, b1858, b2682, b2709, b2717, b2733, b2758, c106, c777, d22, e248, e553, f182, o159, o266, p83, p214, p219, p1373, p1905

187: a212, a227, a624, a633, a855, a1164, a1133, b245[1], b245[2], b420, b421, b422, b432, b673, b2812, c776, d7, d209, e498, f286, h111, h149, h154, h443, i25, i54, i152, n553, p156, p189, p1590, p2031, t355

188: a280, a309, a310, a311, a823, b346, b674, b828, b2409, b2464, c842, e272, e531, h178, i24, n554, o118, o206, p423, p980, p1008, p1513, p2033, s402, t488

189: b1631, b2735, b3009, c660, c784, d170, e353, e456, m295, p252, p1234, p1290, p1886, s13, s167, s386

190: b671, b927, b1236, b2380, b2615, b2763, b2777, b2937, b3080, c106, c171, e285, e509, h183, n517, n522, o244, p243, p285, p965, p1205, p1223, p1236, p1320, p1947, s296

191: a251, a628, a634, a938, b1291, b2760, c781, f279, p298, p958, p1305, p2018, s19, t498

192: a205, a844, b422, b712, b765, b884, b2614, c133, c662, e433, e496[1], f12, h47, h48, p1389, p1694, p1944, s18, s169, t487

193: a410, b403, b462, b890, c26, c727, c780, d166, d179, d262, f10, f11, f16, h46, h743, i235, o170, p284, p967, p1315, p1970, s192

194: a19, a66, a249, a405, a407, a849, a929, a1105, b371, b423, b710, b711, b745, b2772, b2820, c76, c845, c857, e242, h413, m324, m371, o186, p592, t122, t543, t544

195: a48, a65, a806, b469, b769, b2636, b2777, b3027, b3038, c809, d176, e86, e87, e214, g216, h55, h91, h301, h532, i9, i13, i151, n552, p283, p1318, s176, t403, t545, t546

196: a248, a250, a256, a1157, b313, b2904, b2906, b3004, c97, c690, c700, c782, d6, d201, e406, h19, o165, o193, p255, p289, p304, p960, p1993, s195, t398, t406, t442

197: a610, b122, b248[1], b402, b446, b463, b468, b766, b943, b1233, b1235, b1268, b2535, b2744, c182, e293, f122, f228, m3, m74, p192, p288, p779, p1146, t176

198: a793, b457, b461, b784, b924[1], b942, b3040, e283, e315, g87, h56, h86, h295, h711, l44, p1014, p1216, p1361, p1995, p2001, p2004, p2133, t229, t444, t542

199: a201, a419, a1229, b158, b329, b707, b885, b2657, b2900, d265, e332, e349, e369, f144, m23, o194, p287, s165, s166, t227

200: b43, b157, b470, b789, b1202, b1286, b1763, b2160, b2421, b2606, b2736, c848, d11, e228, e322, f185[1], m34, m106, m233, n189, p291, p991, p1222, p2022, q103, q105, t259, t283, t443, u32

201: b2152, b2795, c783, c812, f281, m99, o116, t233, t288, t357, t446

202: a191, a368, a645, a707, a861, b339, b2630, c693, e388, f131, m3[2], n520, p247, p1104, p1474, p1645, p1684, p2172, s13, t471, t500

203: a363, a520, a978, a979, a1130, a1140, b338[1], b642, b1099, b3086, c167, d63, e386, e387, f192, h349, n521, p770, p1344, p1431, p2029, g104, t286, t499, t564, t707, u72

204: a148, a260, b159, b312, b440, b481, b641, b708, b2617, c31, c33, c89, h125, h344, h407, m167, p231, p768, p1333, p1384, p1526, p1920, p1948, p2046, s178, t570

205: a54, a121, a1128, b328, b329, b400, b458, b459, b460, b568, b776, b777, b924, b941, b1790, b2710, b2755, b2968, c440, c692, e436, f257, f261, h427, h454, h587, m111, n377, o172, p258, p662, p1009, p1039, p1826, q102, t437, t497

206: a444, a560, a837, a899, b647, b785, b2607, b2629, b2654, b2788, b2871, b3028, c77, c93, c659, c786, d64, f223, m415, n154, p268, p804, p1408, s162, t233, t374, t551

207: b401, b759, b2422, b2618, c92, c168, c438, c439, c650, d180, e354, h474, m78, m377, p1391, t441, t550

208: a261, a275, b399, b954, b2671, b2776, c843, e367, h158, h431, h568, o166, p360, p428, p1206, p1937, p1961, s4, s224, s257, s398, t98, t457

209: a112, a1043, a1044, a1401, b246, b763, b2688, b3036, d322, h153, n567, p196, p204, p262, p303, p1245, s415, s418, t445, t552, u115

210: a247, a607, a1083, b350, b370, b439, b643, b682, b1635, b1996, b2416, b2721, b2739, b3019, c898, d59, d62, e432, f203, m15, m168, m169, o132, p235, p631, p1143, p1307, p1319, p1417, p1910, s383, t202, t572

211: b680, b747, b764, b891, c5, d58, e296, e437, e443, f240, h495, n155, n186, p425, p632, p1329, t472, t478, t539, t571

212: a262, a263, b67, b354, b441, b649, b678, b2414, b2666,

Temperatures in °C; where there is a range of values, the compound is listed according to the lower value.

Temperatures in °C; where there is a range of values, the compound is listed according to the lower value.

b2816, c164, c810, c917, d169, d174, d207, f274, h156, h385, h536, i200, m58, m217, p224, p299, p1255, p1667, p1697, p2116, q183, q276, t304, t316, t367, t576, t750, t751, u117

246: a559, a680, a712, a835, a930, b430, b472, b564, b932, b3011, d313, h435, i197, h471, n473, p651, p652, p703, p1877, t432, t608

247: a581, a661, b84, b114, b135, b823, b1015, i18, p1186, p1572, p2129, q209, t456, t575

248: a1117, b382, b390, b391, b2125, b2148, c422, d212, h52, p188, p368, p1535, p2117, q153, q199, q277, s26, s356, t339

249: a648, a669, a895, b116, b154, b380, b752, b812, b1239, b1501, c36, c360, i225, p1187, p1642[1], t100, t508

250: a484, a571, a647, b288, b387, b817, b818, b935, b1031, b1267, b1680, b2841, c29, d236, f187, h53, h78, h145, h160, h406, h644, h714, m21, m68, p179, p740, p1464, p1983, p2169, s179, t287, t292, t341, t516

251: a394, a1017, a1078, a1146, b743, b934, b1097, b1415, b1611, b2932, c823, q223, s198, t467

252: a1076, b343, b344, b875, b1773, d206, h44, h134, p782, p1949, s68, t92, t462

253: a152, a243, a481, b209, c339, e611, h24, i196, i212, p634, p739, t67, t294, t338, v15

254: a1118, a1159, b206, b542, b714, b931, b1270, b1284, b1413, b1765, b2274, b2392, h430, l56, p309, p1962, s227

255: a655, a1425, b275, b515, b552, b555, b811, b916, b972, b1419, b1768, b1811, b2320, c27, d311, h428, i21, i23, i65, l2, n534, p367, q82, q129, t117, t297, t420, t435, t575

256: a709, a825, b358, b492, b744, b1319, b1780, b1781, b2214, b2299, d205, e346, g114, h426, i198, n110, p595, p686, p738, p1035, p2121, s190, t166, t386

257: b265, b342, b388, b554, i7, i67, j6, p694, p1015, p1822, t289, t578

258: a124, a701, b103, b359, b559, b862, b863, b1071, b1512, b1619, b1785, b2839, c355, e414, e494, e510, g106, i201, n182, p1691, p1913, q198

259: a1026, b274, b378, b415, b451, b1622, d285, n109, n181, q197, t509

260: a153, a667, a1056, a1080, b2, b221, b299, b1272, b1495, b1561, b1612, b1658, b1666, b1772, b2381, b2780, d290, e183, k10, o96, o174, o283, p172, p188, p1564, p1850, q194, q231, s24, t225, t232, t413, t418, t708, u35, u41

261: b211, b1419, b1766, b2635, c359, h151, h434, o9, p1575, q83, q126, s186

262: a134, a1070, a1187, b232, b809[1], b1287, b1426, b1778, b3065, c230, d303, h139, p641, p1253, p1577, q85, q196, s210, t706, t729, u71

263: a937, b117, b160, b418, b530, b1320, b1513, b1792, c361, i238, n258, p297, p805, p1581, t87, t434, t470, t727, u56

264: a1027, b379, b610, b756, b1615, c62[1], m271, o140, p777, q127, q195, t574

265: a17, a574, a600, b133, b357, b506, b531, b603, b791, b809[1], b815, b915, c41, c44, c163, c348, h535, i72, i199, k3[1], n35, p643, p1549, p1551, p1555, p1556, q81, q131, s196, t424

266: a165, b1810, q128, t451, t736

267: a73, b494, b558, b755, b816, b1034, b1276, b1581, b2322, e570, i41, o242, o271, p1692, p1810, q154, q171, q271, t388, t464

268: a1087, b231, b560, b1317, b1582, c357, h146, m87, n168, n169, o145, p40, p1214, q132, u61

269: b532, b1494, b1779, g90, i144, l65, q96, t385

270: a255, a670, a686, a1024, a1186, a1658, b601, b605, b659, b1322, b1658, b1693, b1784, b2300, b2847, c60, d42, e495, h464, n574, p931, p1764, p2021, p2073, q68, t218, t514

271: a830, a841, b950, c356, f125, p32, p165, p704, p2146

272: a1079, a1166, b526, b1275, b1663, b1874, b2361, b2771, c264, e309, h493, i71, p1538

273: a664, a809, a1019, b716, b1822, b2301, b2321, d301, g75, g169, h49, n476, p125, p205, p320, p1173, p1581, p1922, t327, t422, t580

274: a493, b164, b863, b1583, b2248, b2318, c7, c38, i153, i166, n170, n216, q67, s189

275: a1075, a1081, a1445, b636, b842, b949, b1618, b1775, b1800, b1840, b1842, b1848, c304, h7, h50, h182, h439, m114, p1404, p1763, t23, t24, t565

276: a986, b220, b574, b586, b1018, b1673, b1749, b2310, d288, h150, n467, p638, p1967, t95, t582, t717

277: b270, b583, b861, b919, b920, b1747, d178, d304, e416, o168, t101

278: b93, b284, b575, f127, q57, q70, t732, t735

279: a4, a1066, t1851, m76, m293, p681

280: a1109, a1137, b298, b836, b1686, b2302, b2363, c125, e490, g47, h22, h151, i8, i193, j9, n179, n477, o163, o301, p308, p1568, p1741, p2007, p2026, p2042, q69, q109, q227, s413, t16, t18, t25, t567

281: a81, a860[1], c421, n93, n94, p818, q137, q226, t17, t138

282: b493, b534, b809, b1020, b1960, c203, e395, n8, n180, o144, p697, q184, q186, t26

283: a155, a213, b301, b486, b757, h194[1], q192, t425, t433

284: b756, b2249, h270, h274, p813, p824

285: a567, a798, a908, b139, b523, b577, b639, b640, b1271, b1716, b1821, c367, e237, e588, g27, h394, i134, k18, n147, n210, p839, p843, q225, s274, s335

286: a536, a677[1], a909, b638, b1211, b2312, c62, c351, c718, e236, e390, e508, h30, j12, m294, n144, p162, p948, p1968, q92, t31

287: a1168, b2846, c433, h248, p1969, q228, q230, s200

288: a1087, a1088, b835, b906, b1328, c403, c681, c717, e525, f126, g191, i10, m288, n190, n211, q87, q278

289: a648, b837, h248, n215, p775, t99, t468

290: b604, b1244, b2852, c859, g70, h155, n330, p27, p897, p2155, q71, q130, s354, t209, t597

291: b726, b1715, b2250, b2364, n126, n172, n375, n527, o167, p39, p2154, t68

292: a1185, b2303, i171, t70, t421, t671

293: a649, a891, a1233, a1513, b487, b522, f45, n248, p746, p842, t725

294: b170, b1219, b1326, c432, n38, n88, p870

295: a697, a831, a890, b1685, b1776, b2303, b2362, b2701, h391, n191, p510, p813, p823, t82, t84

296: a166, a688, b517,

Temperatures in °C; where there is a range of values, the compound is listed according to the lower value.

b732, b926, b1849, b2242, e240, n24, q90, s317, t20, t718
297: a860, b278, b872, b1692, b2365, m366, n319
298: b788, d292, e239, e470, m239[1], p821, p1308, s310
299: b727, b1295, b2220, b2243, n322
300: a737, a803, a817, a1182, b873, b1300, b1315, b1931, c350, e602, g175, h80, k23, m46, m186, o143, p1168, p1292, t461, t479, t619, t734
301: b1540, m373, n25, n36, p2074, s14, s33
302: a863, b1019, b1028, b2222, b3044, h66, n208, p33, p158, s64
303: a145, a167, h726, i111, t30
304: a164, m178, n239, n328, p828, q64, q167
305: a1210, a1169, b870, b2311, b2314, b2366, e326, m255, n171, n249, n363, q185, s129, t347
306: a818, b871, b1959, c175, d35, n38, n331
307: a801, a1175, b1301, b2293, i27, n556, o173
308: h74, m50, n209, n220, n329
309: b961, b2016, p684
310: a1015, b269, b1337, b1566, b1845, b2244, b2922, d248, e376, e511, g110, m243, n320, p1570, q47, x1
311: g120, p722, t22
312: a271, a804, a865, a900, b219, b279, e513
313: o93, p877
314: b1295, b1297, b1302, b2017, p1966
315: a39, a1176, b435, b437, b438, b981, b2267, e165, e502, n329, p950, x1
316: a913, b1303, n184, o11
317: b1837, b2252, e491, n220, n221, p1824
318: p28
319: a10, a805, a901, b725, q133
320: a828, a1211, b280, b331, b1656, b2317, b2323, h77, h293, k7, n87, p955, q56, t14, t15, u28
321: a802
322: b787, e586, h244, h246, p34
323: b1265, h45, m272, n222, t21
324: b2068, q123
325: b2219, b2275, c156, e427, m281, n30, s105
326: e517, k16, t222
327: b2018, c864, n515
328: b2294, e18, p1184[1], t94, t96
329: b929, e148[1], n505, p871[1]
330: a502, a883, a902, a936, b1979, b1989, e516, g112, o83, p137, p825, p1439
331: c158, p1682
332: a1170, b1981, d134, e518, t46
333: a1422, b595, b2006, h265, o92, q58
334: n246, t496
335: a9, b1533, b2144, e587
336: b945, h67, h245
337: f121, p810
338: b2067, b2158, e512, o294, p1179[2]
339: a1571, p875
340: a882, a1193, a1599, b865, b946, b2325, b3056, e29, f96, p484, p817, s387, t97, t248
341: e28, f80
342: b2014, b2276, c610
343: b2329, e16, f207[2], h519
344: b1945, d33, d36, h269
345: a727, b1976, c245, h73, n247, p820
346: b1198
348: e280, o16
349: h68, o12, p526, x12
350: b1401, b1978, b2066, b2449, c354, m284, m285, s109, t269, u62
351: t130
353: b1988
354: a866, b2015
355: b1871, b2215, b2266, c141
356: h59
357: e22, h249, t130
358: e270, m312, o248
359: t72
360: a1441, b1262, b2129, c114, c300, h262, m376, o22, o295, p803, q8, q135, q152, q158, q163
362: o35
363: b2258, q218
364: n260
365: a947, k20, t47
366: t135
369: d272, e21
371: p716, t249
374: b721
375: f44, p19
379: s359
380: a919, b1248, m152, m381, t711
383: o18
385: b1672
390: h253, m197, s201, p506[1]
391: t62
393: p1930
395: a912, b1986
398: b1229, m192
400: a916, b2261, s218
405: s378
410: b724, p789
412: h234, s143
413: b1229
415: b978, e423, h264
417: a1234
418: b418
419: o249
420: q1, s141
422: h63
425: b1957
430: a1286
431: m300, o1
435: b181
441: n502[1]
442: o33
445: n175
446: n387
448: c313
450: t699
452: b2208
458: h61
459: b982
467: d323
471: a858
480: b1228
518: p908
546: s132

Spectra Index of Organic Compounds

This index is a listing of the Sadtler Standard Spectra, infrared prism, and also contains the infrared grating (HR), the UV, and NMR numbers of many compounds listed in the Table of Physical Constants of Organic Compounds of the Handbook of Chemistry and Physics. In the NMR column of this index, the Varian numbers are preceded by the letter V and the JEOL numbers either by the symbol * which represents 100 Mc spectra, or the letter F (F-19 spectra) or J (60 Mc spectra). The various media in which some spectra were run are designated at the end of the reference number by one of the following abbreviations:

B, Between salts (Neat)	K, KBr
C, Cell	M, Mull
F, Film	S, Solution
G, Gas	

HCP No.	Sadtler Reference Number			
	IR		UV	NMR
	Prism	HR		
a1	3963			
a2	5151		1399	
a4	875	548	272	65
	7458	175		
a9	20181			
a14	14633		4140	
a17	13704		3631	
a18	5824			V6
a21	5274			V143
a22	5273			2949
a23	4128	8324	1193	
a25	15391			V373
a31	2395			3001
a32	8000			3827
a33	8040			
a39	18393			
a48	5176			
a52	4626			
a58	76			V8
a59	6476			
a60	69			4280
	13207			
	2217			
a62	26009	1854		427
a69	19170			
a72	5756			
a73	1818	281	510	
a79	4590	416	1271	
a80	1903			
a81	6633		1831	V601
a85	34756	12739		
a87	34758	12741		
a89	7335(K)		2070	
a89	2729(M)		2070	
a90	20856		8039	3015
a105	11270		3078	
a106	20128		6707	
a121	2296			146
a124	24023			1345
a128	24126	794	8443	2921
a129	19561		6428	
a130	1602	261	2913	
a142	13934		3810	
a145	9706		2573	
a147	22087			
a148	15060			
a150	9768			
a152	20972	10488	8130	2403
a156	2164	297	571	
a157	2165	298	572	131
a159	5855	8419	1638	3184
a160	12384			
a161	11426(K)		3138	5129
a162	7489			
a164	160	49	63	5235
a166	2899	350	830	175
a167	2900	351	831	
a168	2901		832	V239
a170	6474			
a172	70			
a173	859	167	260	V530
a174	1867			4980
a178	1041	10971		V140
a180	5250			
a182	18668			V141
a183	6030(gas)			
a185	6640			
a188	14890	10036		4924
a189	22788	10579		
a190	2973			
a197	982	8112		V65
a198	5284			
a199	4	3		V79
a200	58			
a201	108			
a203	11912			
a204	1475	249	420	V167
a205	5249			
a207	5166			
a208	4574	10946		217
a212	15649			
a213	5552		1546	
a215	164	50		
a216	772	10769		V440
a217	1473	10939		99
a218	11299			
a219	18372	10112	5831	4226
a221	2228			
a223	6650			281
a240	9683		2571	
a241	269			
a242	13213			
a245	17371		7321	
a247	3875			
a253	1911	10975		
a255	19118			
a256	5152		1400	
a259	1643	10940		115
a260	5968			
a262	8043		2175	
a263	1649	267	1458	
a275	2325			
a277	9717			
a279	6338			
a281	6334	10988		4205
a309	2094			
a312	2975			
a365	1302			92
a367	2993			
a369	1301	234		91
a374	1318			95
a375	5183			
a387	13137		3424	
a388	13138		3425	
a389	13139	259	457	V500
a391	25003	1003		
a395	26035	1878	9443	411
a402	17518			
a403	23661			
a407	2806			166
a408	12551			4967
a411	6264			
a412	5248			V59
a415	4561			
a418	13141	8174	3427	
a420	14891			
a421	25710	1584	9298	
a423	11174	132	176	
a430	1947	8213	11441	
a434	2112		560	129
a435	19246			
a437	10958			
a440	10650		2918	
a441	1938	289	534	
a444	7392			V417
a446	22898	10593		
a451	9766			
a453	17442			
a456	3990			
a458	5829			
a464	695			
a465	5141			
a471	13457		3518	4410
a475	23517			V554
a476	19191	708		
a481	2346			
a484	25017	1010		
a492	26717	2468	10026	1603
a495	17420		7333	
a499	796	8096	11410	
a503	10591			
a504	30353	6346		4312
a506	4562			214
a507	26608	2373		1754
a508	22423			
a509	20584		7915	

HCP No.	Sadtler Reference Number IR Prism	IR HR	UV	NMR
a515	7248			
a516[1]	21997			
a517	29288	5284	12029	1330
a518	22483			
a520	6533			
a522	25206			
a523	6953	8490		2863
a524	6265			
a525	19936			V28
a534	6263	8454	1727	3192
a535	18504		5928	4141
a538	5870	8422	1646	
a539	5871	504	1647	
a540	5874	585	2965	
a542	17537			
a549	24457		8611	
a553	8441	505	2237	3223
a554	1801		503	V495
a558	22237	10524		2512
a559	23016			
a560	32069	9067		4324
a567	5852	500	1637	264
a570	4723			
a574	1655		474	117
a575	2236	308	599	
a579	4575			J106
a580	882	10770	275	66
a581	13594	8186	3579	
a582	165	4764	64	19
a583	2231	307	598	135
a584	1633	265	467	720
a589	417			V7
a591	7865			
a593	5970	8440	11494	
a597	20772	10425	7993	3052
a600	6517		1810	272
a601	31679	7665		2398
a602	8336			
a603	17479		5550	
a605	22002			
a608	7387			
a610	11346			6
a623	22016			
a625	927			
a635	13148	8175	3432	
a636	13149		3433	
a640	2503			
a643	20686		7953	
a645	3226	8290	953	V192
a646	5780		1618	
a647	18351		5845	4987
a648	1843		518	119
a649	5111	8370	1387	242
a650	5023		1384	
a652	24148	807	8459	2673
a653	8489		2228	240
a654	8626	561		1889
a655	4580	414	1267	219
a657	6642	8483	1835	3206
a661	4576	8338	1264	3174
a664	1769	272	494	V188
a668	6641	531	1834	279
a669	27877	3603		866
a671	28837	4539	11355	
a672	28838	4540	11356	376
a674	8764	8379		1555
a675	19180	10815	6276	4286
a678	3884			
a685	9770			
a687	6011	512	1695	532

HCP No.	Sadtler Reference Number IR Prism	IR HR	UV	NMR
a688	15183		4491	
a689	3415		1066	
a692	27912	3638	10858	
a693	32461	9457		4363
a697	19571		6437	V234
a699	20308		6817	
a700	20309	10362	6818	2314
a701	3883		1177	
a702	20467	10387	6914	2340
a703	23784			
a704	22460	10575		2758
a705	872	174	270	
a706	21154		8267	
a707	5857		1639	531
a708	5112			
a710	20551		7894	
a711	20552		7895	
a712	20573			
a713	20574		7908	2368
a721	7536			
a727	1153		342	81
a732	7591		2125	
a737	10786	10790	13286	
a738	7560		2117	
a742	27864	3590		835
a743	5446		1497	
a744	5473			
a745	10784			
a752	23488			
a754	794	164		
a756	23489			
a756[1]	795			
a759	23283	10910	13329	
a764	10623			
a772	1935			
a773	23495			
a775	6488			
a779	21948			
a788	3555			
a795	18248			
a798	288		114	1178
a802	2323	8251	618	148
a816	11301			
a817	10943	8585	2978	3277
	17113			
a819	17392	680	5526	
a820	3195			
a821	3196			
a825	1033	208		
a827	3194			
	8960			
a828	5451			257
a830	5830			
a831	24492	4752	8634	
a832	277		110	
a836	34412	12397		
a841	24958			
a843	24957			
a847	2246			136
a848	1059			
a850	1140			
a851	5452			
a856	20605			V375
a860	3198			
a861	9728			
a863	68	8009	30	11
a865	20665		6963	3588

HCP No.	Sadtler Reference Number IR Prism	IR HR	UV	NMR
a870	8963			
a871	19698			
a880	10524			
a882	20343	10374	6837	2328
a884	1292	232	388	681
a886	34522	12506		5762
a887	9744			
a892	390		153	
a893	2243			
a907	24683			
a911	179	58	72	21
a912	155	48	61	
a918	15374			
a919	18406		5864	1899
a920	1187			84
a921	30298	6294		
a923	23726			
a925	227	76		29
a926	6605			
a928	23641	10622		3630
a929	18732	8882		2068
a931	1322			
a939	9718			
a940	6377			
a946	6375	521		
a947	24568	4808		
a948	2244			
a949	1313			
a953	8887			
a954	8962			
a955	25746	1615	9315	
a956	23636			
a957	11279			
a960	23618	10615	13339	3622
a966	10266			
a967	21884			
a969	5957			
a970	5958			
a975	5959			
a977	21995			
a982	16450		5152	
a992	718			
a1002	3695	392	1129	191
a1003	8948		2299	
a1004	10645	582		
a1008	6291		1731	
a1011	5447	8397	1498	1217
a1014	13702			
a1015	2731	339	743	1208
a1016	9779			
a1017	16154			
a1018	6630	8481	1829	278
a1022	25596	1473	9229	
a1024	12473			
a1025	12410			
a1027	12442			
a1030	32013	9014		4317
a1034	25572	1449	11540	
a1035	1517			
a1041	6643	532	1836	280
a1042	21769			V568
a1046	1262		376	88
a1047	16545		5193	
a1048	1317	8140	393	V123
a1049	9752			
a1053	21003		8154	
a1057	8926		2291	
a1058	13917	8634	3799	3257
a1059	21922			
a1060	10564	8344	2909	3237

HCP No.	Sadtler Reference Number			
	IR		UV	NMR
	Prism	HR		
a1061	21913			
a1068	23709			2378
a1069	24256	884	8517	2568
a1070	19145		6258	
a1074	21004		8155	
a1075	2304	321	612	147
a1076	19095		6233	5005
a1078	780	8093	253	58
a1079	6459		1794	271
a1080	19146	8973	6259	2164
a1084	6529		1814	
a1085	216	72	83	716
a1090	5890	508	1655	3186
a1098	24026			
a1102	19181		7705	5012
a1104	18927	8944	6158	2138
a1105	6	4	3	1
a1109	25194	1108	8994	
a1110	14894		4313	
a1111	5891		2908	
a1113	1287	231	387	3290
a1114	17483		5551	
a1115	26402	2180	9792	431
a1116	4674	8340	1285	236
a1118	4675	8341	1286	V208
a1128	9167		2378	1890
a1132	19615			5031
a1133	19616	715	9650	5032
a1134	19617		6459	5033
a1135	8418		2234	
a1136	17451			
a1137	17800		5637	
a1138	18976		6181	
a1140	30297	6293	12762	1335
a1141	1568		449	
a1157	6365	8463	1754	3203
a1158	10647			
a1162	18329		5832	V489
a1164	20537			V488
a1165	21134	750	9663	
a1166	19425			
a1168	1276	229	385	
a1169	205	70	80	
a1170	8451		2243	
a1175	4493		1246	
a1176	4494		1247	
a1177	4495		1248	
a1182	9769		2576	
a1186	20524		7882	
a1193	1319	239	394	
a1208	3506	381	1084	V618
a1210	19205	709	6290	V307
a1214	5436			254
a1220	19880		6579	
a1221	19875		6574	
a1224	11383			
a1230	21292		9681	
a1231	20504	10402	6940	V317
a1232	20358		6850	
a1233	15221	628	4511	550
a1246	24157	814	8465	
a1248	1815	279	508	
a1249	13893	8629	3787	
a1250	13894	8630	3788	
a1261	5770		1611	
a1263	2984		859	182
a1264	1855			
a1267	14895		4314	
a1268	2983	8283	858	
a1273	14896			

HCP No.	Sadtler Reference Number			
	IR		UV	NMR
	Prism	HR		
a1279	19109	8962	6238	
a1282	19122		6244	
a1288	8810		2284	
a1292	5420		4318	
a1294	3585			
a1295	3588			
a1298	3579			
a1307	22405	10560		2738
a1314	11414			
a1315	11256		3068	
a1316	6342	8459	1746	3200
a1317	18319	10108	5830	4221
a1333	21103			
a1336	3598			
a1337	3596			
a1342	14897	8695	4315	
a1353	18664			
a1354	18163	8824		
a1370	15657	4702	4443	
a1379	7520			
a1391	25687	1563		
a1393	15309			
a1398	15662		4704	4444
a1402	9603			
a1403	18700			2078
a1404	25689	1565		
a1405	13418			
a1454	2384			V276
a1457	13217	470	3455	3126
a1459	15266			
a1461	17219			
a1463	5425			
	7919			
a1465	6793			
a1469	18670	8857	5999	2070
a1477	4816			
a1478	21993			
a1599	23694		9711	
a1601	4847		1345	
a1643	11578		3203	
a1645	11577		3203	
a1652	6649		1837	
a1655	19108	8961		
a1658	10783			V551
b2	11337	8346	3094	
b8	22214	10511		
b9	6471		1798	
b11	24419	990		
b14	17178		5423	1173
b15	27744	3479	11668	1121
b21	27981	3705	10887	
b25	483		171	
b26	31693	7677		
b31	5971		1679	
b35	3010	364	868	V151
b36	5702	485	1584	702
b43	25698	1574	9290	
b44	20678		7948	3606
b48	15304		4569	
b50	15643			
b54	9988			
b56	19509			
b57	18083	10081	5724	4047
b65	5767	495	1609	
b67	625	8076	208	50
b68	13491			

HCP No.	Sadtler Reference Number			
	IR		UV	NMR
	Prism	HR		
b69	624	8075	207	V146
b81	620	10845	205	3151
b84	622		2066	
b94	7203	542	2044	534
b95	14539			
b97	19175		6271	
b98	19176	10196	6272	4283
b102	17472		7344	
b103	3887	8313	1179	V236
b111	6652		1838	V238
b118	5178	211	318	1320
b121	16462	10048	5155	4218
b122	9989	8567	2646	3232
b123	5705		1586	
b124	5985			V156
b125	2126		562	
b128	4579		1266	
b133	1273	9997	383	528
b136	18980		6183	
b139	8703			197
b151	5973	510	1680	
b152	13224	8614	3456	1895
b154	1946	10848	536	126
b157	32149	9147	13722	3936
b159	21549			
b160	3885		1178	V187
b162	3516		1091	V148
b165	9990(K)		626	
b168	3517	384	1092	
b181	1124		335	
b185	9268		2412	
b186	9269			
b187	9259			
b188	9260		2404	
b190	9261		2405	
b191	9262		2406	
b192	9263		2407	
b197	11260		3071	
b202	6402	136	1765	3429
b203	13701	10873	3630	3248
b206	17520		5557	4257
b209	3880			V260
b215	15603		4680	
b224	17119	663	5403	893
b225	4677	10778	1288	237
b226	34116	12106		5580
b227	7555	10785		
b228	17112	8718	5399	3276
b229	17391	679	5525	5577
b236	18768	8896	6054	2099
b245[2]	5859	8420	1640	V207
b246	1601		3239	111
b248	28862	4564	11849	493
b248[1]	280	8027	113	35
b256	15599		4676	1896
b260	1568		449	
b262	18932		6159	
b276	5340		1462	
b277	5555		1549	
b278	8240			
b281	4560			1882
b282	19167		6269	
b283	4887		1364	
b287	28918	4620		497
b288	32737	9732		4612
b289	13674		3624	3254
b293	20546	10412	7890	2362
b309	3227	366	954	1212
b310	32858	9852		4585
b311	2107			

SPECTRA INDEX OF ORGANIC COMPOUNDS (Continued)

HCP No.	Sadtler Reference Number			
	IR		UV	NMR
	Prism	HR		
b312	20098	10311	6684	2268
b313	16343			
b314	4555	10776	3519	3251
b325	16566		5203	
b329	9736			
b330	20131		6710	
b338	2268	317	607	V198
b339	4617	424	1279	223
b340	4682	434	1289	
b341	5727			V121
b342	5728		1590	
b344	6387	522	1764	
b345	21358			
b348	19440		7785	
b351	2258	8249	605	141
b353	4737	441	1305	239
b354	18666		5997	J114
b355	6345	8460	1748	3201
b356	20969	10485	8128	2426
b357	16554		5198	
b358	7897(K)		505	
b368	18806		6078	
b370	2720	334	735	
b372	6318	8457	1739	3425
b374	247	83	98	31
b376	679	152	228	53
b385	18719	8877	6029	2085
b389	20023		6648	
b390	16264		5046	
b391	20162	10331	6731	2285
b392	13668		3618	J130
b393	11626			
b394	21107		8229	J131
b395	522	133	179	47
b396	257	89	103	525
b397	14511	622	4124	V288
b398	1061	8118	322	527
b403	8311		2215	V586
b407	34	19	16	714
b413	13916	8633	3798	
b415	5553	8404	1547	
b417	26968	2714	10217	3299
b418	1625	264	9787	
b422	13675		3625	
b423	23015			
b435	964	196	5204	
b437	25163	1087	8976	
b439	4671	431	1283	234
b441	4672	432	1284	235
b442	13922		3802	
b446	8637			
b449	21856			
b450	21858			
b451	6627	10855	1827	3205
b454	2892	348	825	690
b456	6871	535	10886	
b458	21703			
b459	21704			
b460	6760			
b461	8802		2281	
b463	2727	338	741	164
b465	4597	419	1274	
b470	895	178	279	68
b472	3231			
b473	3232		958	
b474	4683	435	1290	V122
b483	18318	10107	7579	4220
b484	8033	8520	2712	J188
b485	19704		6492	
b486	17169		5417	

HCP No.	Sadtler Reference Number			
	IR		UV	NMR
	Prism	HR		
b487	6455		1791	
b490	1996		6636	3839
b492	1604	262	460	112
b493	6522		1813	V458
b494	4111	8322	1187	
b503	2971		855	
b507	25940	1788	9405	
b508	5862		1641	
b512	12459	605		
b513	12394	324		
b514	2393	323	625	
b520	24259	7989	8520	2571
b523	22989	10598		2548
b525	25092	1067		
b527	708		233	
b529	7244	8498	2053	3127
b530	24266	887	8523	2574
b535	1003	201	303	746
b536	5934		1671	
b537	146	44	55	715
b542	11926			
b554	19172	10195	6270	4282
b555	3655	387	1114	687
b558	3656	388	1115	688
b559	17387		5521	
b562	2890	347	823	171
b564	711	157	235	309
b565	24283		8528	
b566	18713		6025	4993
b567	24284		8529	
b568	19517			
b569	268		108	V124
b571	2861		802	
b574	467	125	2572	
b575	5551	481	1545	699
b576	1624(C)	263	2103	
b577	153	47	60	
b578	5189	455	2078	252
b594	8318		2216	
b595	8339	10856	2223	3219
b601	20311		6819	
b605	13502			
b606	20276	10355		
b609	19974			
b610	20540	10410	7885	
b613	9807			
b620	24849			
b621	21378			
b626	7847	552	2154	
b637	9534			
b640	4544		1258	207
b642	2958	356		
b643	2957	355		
b647	1658	8189	476	1112
b648	11030		2999	1891
b649	1508			
b654	19576			5028
b657	25106	1074		
b658	19579		7797	
b660	11	9	7	V201
b661	1045	210	317	V202
b662	2276	319	609	V203
b675	2342			
b676	3819	393	1162	193
b678	294	100		36
b679	8080	349	83	
b680	1065	10846	323	82
b682	521		178	
b688	23629			
b692	19411		7773	

HCP No.	Sadtler Reference Number			
	IR		UV	NMR
	Prism	HR		
b699	19653	10253	7802	2228
b706	12172	601	11527	
b707	19395			
b708	19398		9642	
b709	18712	8874	6024	2082
b710	18926	8943	6157	2139
b711	19397		9641	
b712	19396		9640	
b713	20289	10357	6805	2311
b714	20290		6806	
b715	23688	775	9710	320
b717	17346		5498	
b718	2397			
b725	12397			
b726	81	27	33	
b727	3657			
b731	33630	11621	14277	5794
b732	3511	383	1087	712
b739	7033		1983	V149
b740	20321	10365	7864	
b742	2888	346	821	
b744	19706		6494	
b745	9730(C)	10788	2303	V193
b746	192	4826	77	26
b748	26404	2182	9794	434
b750	27954	3679	10871	335
b757	8446		2240	
b758	246	82	97	V505
b759	2340			
b760	2339			
b761	2341			
b768	31977	7957		3943
b770	19677	722	9655	
b773	1579		455	109
b775	19659			5040
b778	4615	8339	1278	222
b779	3509			
b780	21875			
b781	15345	638	4587	3271
b782	4670			233
b783	21876			
b784	19614	714	9649	5030
b785	4618	425	1280	224
b787	4545	410		
b788	26949	2695	11601	
b789	33623	11614	14272	
b791	16015	648	4937	
b794	11624			1893
b796	1073		326	
b797	1916	286	531	V270
b800	2470	325	653	684
b808	1017	8115	311	76
b816	487			
b820	16263		5045	
b823	486		172	
b824	488	130	173	707
b825	3821			
b827	295	8029	115	
b828	2891	8270	824	172
b830	9737			
b831	20115		6696	
b832	19431	710	7778	V153
b835	20984		8140	
b836	2894			
b837	6869		1919	
b838	6704			
b840	6703		1863	
b843	26669	2427	9995	
b844	1075			80
b846	11619			

SPECTRA INDEX OF ORGANIC COMPOUNDS (Continued)

HCP No.	Sadtler Reference Number IR Prism	IR HR	UV	NMR
b847	242	8023	95	V240
b850	1066		324	V268
b854	317		2919	
b855	19428			
b859	2988	362	861	J101
b861	5864		1643	1887
b862	9757(C)			
b863	5865	8421	1644	3185
b864	1467	8156	419	V258
b872	4598	420	1275	696
b877	13471		3520	
b882	34871	12853		
b884	1574			*160
b891	12	10	8	4
b893	24813	4969	8786	
b894	21426			
b895	20450		7878	2341
b896	1879		525	
b897	1884			
b898	1885		526	
b905	28878	4580	11853	
b906	20523	10406	6944	2356
b916	21414			
b922	9755			V287
b924	23608	772	9709	*157
b925	12455			
b926	34263			
b927	34920	12890		
b928	5654			
b931	5935			
b933	2976	360	856	1179
b941	20568	10417	6956	815
b942	23781			
b943	1199	224	355	85
b944	24595	4825	8694	2698
b950	17183		5424	
b952	5936	509	1673	266
b953	4600	422	1277	221
b954	2493			
b960	24553	4796	8670	
b961	8766	8542	2278	3225
b965	3845			
b971	3844			
b972	26606	2371		
b974	1038	8116	316	3417
b975	4539	407	1256	V241
b981	1086		328	
b982	15226	629	4513	
b985	5822		1628	672
b986	5785		1622	
b988	5474		1505	
b992	15166		4477	
b993	19961		7842	
b996	478		169	
b1001	20253		6784	
b1003	20254			
b1005	17321	10057	5482	4441
b1006	11874			
b1007	11867			
b1014	31886	7865	13666	3790
b1015	16651		5240	4508
b1016	9179		2384	
b1018	13750			
b1019	10551			
b1020	10550		2905	
b1021	25026	1017	8891	
b1024	9178		2383	
b1027	7467			
b1028	10642		2917	
b1029	11093(K)	352	3015	3242

HCP No.	Sadtler Reference Number IR Prism	IR HR	UV	NMR
b1032	19477	8999	6376	2177
b1038	18164		5766	
b1039	28929	4631	11386	
b1043	32444	9440	13794	
b1044	2910		838	
b1047	16400		5128	
b1049	16401		7079	
b1054	19977		6623	
b1056	20141		6714	
b1058	20285		6801	
b1060	28932	4634	11875	873
b1061	9716			
b1066	2668			
b1067	2667			
b1068	201	69	2285	
b1079	6165			
b1080	6166			
b1083	13396			
b1086	19598	10240	6455	5029
b1091	13662		3616	
b1092	5733		1593	
b1093	5760	493	1606	1174
b1094	7646		2142	3214
b1095	5670			
b1097	415	118	154	3150
b1098	15653		4700	
b1099	1379			
b1101	15277		4550	
b1103	5735	488	1594	
b1108	18667			
b1110	17193		5427	
b1112	2435		3235	
b1113	301	102	117	
b1139	6623		1826	
b1143	14531	8692	4127	3263
b1144	6457			
b1145	34349	12335	14458	
b1157	22724			
b1162	20100	731	11530	898
b1163	15766		4769	
b1164	9712			
b1174	14532		4128	
b1177	7360		2084	
b1178	5876			
b1179	1830		515	
b1180	1817			
b1182	2664			
b1183	2669			
b1184	2671			
b1185	2666		721	
b1187	2672			
b1188	17449		5545	
b1191	18429	8842	5881	
b1193	33611	11602	14261	5780
b1198	922	186	286	73
b1199	6624			
b1206	1685		491	4270
b1209	24484		8626	
b1212	4578(C)			
b1214	25489		9158	
b1217	20269			
b1218	14331		4063	
b1225	20375		6862	
b1229	25473	1357		
b1231	3739			
b1233	3740			
b1234	3741			
b1235	3743			
b1236	3745			
b1237	8125	10011	2190	

HCP No.	Sadtler Reference Number IR Prism	IR HR	UV	NMR
b1239	779	162	252	57
b1243	13700		3626	
b1244	2732	340	744	689
b1247	3630		1102	
b1251	3624	385		
b1256	18879	8918	6124	2117
b1262	2702	565	2267	155
b1265	1509	253	427	V627
b1267	2711	331	732	158
b1268	3	5402	2	3417
b1273	1506	252	426	1204
b1274	34847	12830		5819
b1276	15671		4706	
b1278	29349	5315	12058	1009
b1279	3888			
b1282	1006	202	304	75
b1283	9724			
b1286	1052	212	319	78
b1287	1866	283	524	
b1290	8453			
b1291	2255	312	602	138
b1295	3856	8311	1171	3173
b1298	2881		814	
b1299	20990			
b1303	921	185	285	1187
b1306	5736	489	1595	1218
b1314	2703	330	729	3159
b1315	10971	8588	2987	
b1316	910	182	283	655
b1317	3868			
b1318	949	193	295	
b1319	897	280	1200	
b1323	6493	526	1802	2822
b1326	28938	4640		801
b1329	1542		441	
b1333	5154		1402	243
b1335	7605		2131	V302
b1336	7542	550	2115	1219
b1337	3059	8286	888	1856
b1338	5433	473	1490	585
b1344	19445	8980	6353	
b1349	9715			
b1353	22220	10514		
b1355	10876			
b1359	30548	6537	13226	
b1360	30543	6531	12849	
b1361	1477	8560	3162	3244
b1364	13598			
b1366	18123	8803	5739	2003
b1367	21254		8331	
b1379	27666	3401	10684	3045
b1385	24015			
b1390	20669		6966	
b1394	18489		5920	2052
b1398	3973			
b1400	5157			
b1401	3974			
b1403	18582	10153	5952	2038
b1406	26061	1899	9458	467
b1409	18599	10159	596	2060
b1410	4563(M)	8337	3297	
b1411	26471	2241	9845	457
b1412	22162			
b1415	19969		6617	3831
b1417	20992	10501	8146	
b1421	20110		6691	
b1423	4564(M)		3296	
b1424	12372			
b1425	21935			
b1428	20111	10315	6692	2272

SPECTRA INDEX OF ORGANIC COMPOUNDS (Continued)

HCP No.	Sadtler Reference Number			
	IR		UV	NMR
	Prism	HR		
b1448	4568(M)		3294	
b1460	5955	8436	1676	3287
b1465	251	85	101	
b1466	8398			
b1468	618			
b1471	9756			
b1473	6617	8479	1824	
b1476	15045		4412	
b1480	147	45	56	
b1481	12373			
b1483	2949		847	3167
b1484	6499			
b1486	2893	349	826	V483
b1488	13912		3795	
b1496	30542	6530	12848	
b1499	3248		968	
b1514	17166		5416	
b1515	4565(M)	599	3298	
b1518	18476	10151	5913	2047
b1519	5331	8389	11477	
b1520	18616		7649	
b1521	21911			
b1527	10865		2962	
b1528	19506	711	7790	587
b1534	20993	10502		
b1538	585	138	200	3340
b1539	2950	8277	848	3164
b1540	2157		2972	
b1541	584			
b1542	583	137	199	
b1543	2951	8278	849	3165
b1546	19149	8975	6262	2166
b1550	17377	8722	7324	
b1551	7173	8495	2028	3210
b1552	10967	8587	2986	540
b1553	11298			
b1554	19530	10214	6407	2197
b1555	31749	7730		
b1559	6540		1817	
b1566	19512	10208	6398	2189
b1567	19540	10216	6412	2194
b1568	19599	10241	6456	2218
b1569	19486		6383	2183
b1577	1961	8216	542	3158
b1584	24490	4750	8632	
b1585	17476		7345	
b1587	13931	8637	3807	
b1589	17106		5398	
b1590	23753	779	9713	
b1592	19635		6462	4502
b1593	15328	8705	4583	3273
b1594	14553		4129	
b1595	9488			
b1596	460			
b1597	9480			
b1598	17372		7322	
b1599	21126		8246	
b1600	20991	10500	8145	
b1601	21117			
b1602	21120		8240	
b1604	21118		8238	
b1605	10730			
b1607	17383		5518	
b1610	19676		6477	
b1616	1623		462	
b1618	26711	2462	10020	476
b1628	11953			
b1629	17454		5546	
b1630	4566		3299	
b1631	25468	1353	9141	906

HCP No.	Sadtler Reference Number			
	IR		UV	NMR
	Prism	HR		
b1633	19471	8995	7787	
b1636	24355	934	8545	
b1637	19652		6467	
b1642	8393			
b1645	15120			
b1649	17414	328	671	
b1651	51		25	
b1652	3253	373	973	
b1656	1523	256	433	719
b1658	2253	311	601	
b1662	3870	396	1174	196
b1664	23213	10757	13315	
b1665	14024			
b1666	1007	203	305	724
b1668	1008	204	306	
b1671	2238			
b1673	1522	255	432	1189
b1679	2931	353	841	178
b1681	3247	368	967	1624
b1682	17370		5510	
b1683	5972	8441	2903	3235
b1685	23763	782	11533	2773
b1686	4081	8320	11462	514
b1687	19484	8713	6381	2182
b1688	5252	9999	3277	3160
b1691	5434	474	972	1214
b1692	3250	370	970	1192
b1693	18045	369	969	1191
b1694	19970		6618	
b1695	18051	371	971	1213
b1714	1975	8217	546	
b1715	1954		538	
b1716	11795(K)			
b1719	17527	8578	2910	3281
b1721	13925		3804	
b1737	24707	4891	8760	
b1739	30541	6973	12847	
b1746	4567	413	1262	3092
b1747	9734			
b1748	12374	602	9629	
b1749	9767			
b1750	20994		8147	2411
b1751	9804			
b1757	18820		6084	
b1761	30118	6118	12615	
b1762	327		125	
b1763	5427		1487	V195
b1764	25854	1706	9342	825
b1765	1805			
b1770	23695			
b1775	11181	8258	730	V196
b1778	2948	10849	846	180
b1779	8018			
b1780	1936	288	533	124
b1781	14300		4049	
b1785	139	41	51	17
b1786	2153	8236	566	3163
b1790	2898	8271	829	174
b1792	140	42	52	18
b1793	2154		567	692
b1797	23802			2788
b1798	2886	8268	819	168
b1800	2152	293	565	1113
b1801	2155	294	568	693
b1803	32045	9044		3882
b1804	11320		3085	
b1805	11319			
b1806	2887	8269	820	169
b1814	24526	4783	8661	2652
b1817	862			

HCP No.	Sadtler Reference Number			
	IR		UV	NMR
	Prism	HR		
b1819	20547		7891	
b1836	11480(K)	8237	3158	3243
b1840	2681			
b1841	21423			
b1842	24737	10692		5742
b1843	17570		5573	
b1844	11481(K)	295	3159	703
b1848	1258			
b1849	30544	6532	12850	1635
b1850	12462(M)			
b1851	8425	559	11507	
b1852	17357		5502	4948
b1853	1826	8203	513	
b1854	6252		1725	
b1857	11648(K)			
b1858	9475			
b1859	8689		2265	
b1860	4899	444	1366	584
b1861	24453	4723	8608	
b1867	582			
b1871	8633			
b1875	30124	6124	12621	
b1879	12935			
b1882	18665		7670	4033
b1887	20116	10318	6697	
b1889	10912		2971	
b1906	10909		2968	
b1907	10910		2969	
b1909	10908		2967	
b1912	4577		1265	
b1916	1122		334	
b1922	17496		7347	
b1923	23625			
b1924	17494			
b1925	9163			
b1926	32151	9149	13723	
b1928	20532		6951	
b1932	8449		2242	
b1937	11925			
b1942	1112	217	331	
b1945	2722		736	
b1947	17366			
b1952	13599	333	734	1207
b1955	6616	8478	1823	V655
b1956	9271		2414	
b1957	7334		2069	
b1959	14800	326	659	153
b1961	5781		1619	
b1963	18349	8832	5843	2020
b1965	18120	8802	5737	2001
b1974	5974	8541	1681	V329
b1978	16552		5197	
b1981	13913	8632	4051	3260
b1987	13914		3797	
b1988	1014	206	309	
b1991	19552	10220	6420	2199
b2001	8460			
b2005	6294			
b2006	5662		1570	
b2010	6779		1909	
b2012	8332		2220	
b2016	4734			
b2017	5656		1566	
b2018	4868			
b2024	12920			
b2029	25210	1117	9004	
b2031	20586		6958	
b2062	1189	8245	594	704
b2066	7521			
b2067	14260	8691	4023	3261

SPECTRA INDEX OF ORGANIC COMPOUNDS (Continued)

HCP No.	Sadtler Reference Number			
	IR		UV	NMR
	Prism	HR		
b2069	751		248	
b2071	10758		2947	
b2072	18459		7603	
b2078	29004	5004	11908	
b2083	5144			
b2087	713	159	237	V314
b2090	29005	5005	11909	
b2091	29006	5006	11910	3392
b2092	23677			
b2093	11917	594		
b2094	10874		2963	
b2096	22029			
b2097	10283			
b2098	34084	12074		5927
b2099	34086	12076		5929
b2100	8194			
b2111	6634	530	1832	
b2112	18321	10109		
b2113	22413			
b2116	2256	313	603	139
b2117	5467		1502	V150
b2119	24145	806	8456	
b2124	7529		2112	
b2125	1796			
b2131	20190		6751	
b2132	11493	8139	390	
b2139	6162		1722	V191
b2140	20127	10322	6706	2275
b2141	8634		2251	
b2149	1631		465	
b2151	9732			V147
b2160	19932		7835	
b2170	16563	479		
b2183	5921			
b2197	11824		3275	
b2205	6555	527		
b2206	25728	1600		
b2208	23539	765		
b2210	20074	730	6677	
b2212	25729	1601		
b2214	783	163	255	V289
b2219	22092			
b2225	32556	9551		4401
b2235	20073	10303	6676	2261
b2243	12171			
b2244	13650		3610	142
b2248	8048		3351	
b2250	15043		4410	
b2257	32555	9550		4400
b2261	6632	8482	1830	
b2262	6316	517	1738	3193
b2264	6379	8468	1760	269
b2266	14554(K)		3282	
b2267	15044		4411	
b2274	13493		3526	541
b2280	23754	10654		3661
b2294	19573		6438	2210
b2299	25471			
b2302	18935	8947	6161	2141
b2303	21289		9678	
b2305	20180	10335	6742	3083
b2308	22809	10580	13303	
b2312	1192(K)		341	
b2313	5832		1629	262
b2314	1107		330	
b2318	6367	8465	1756	3196
b2320	12194			
b2321	21078		8208	
b2322	9393	8550	13281	2860
b2323	1253		373	86
b2324	20340		6834	
b2325	8448	8531		3084
b2331	18852		6110	
b2332	18853			
b2335	7081	538	2008	
b2339	20998		8151	2415
b2353	16067			
b2361	13298	8616	3468	3247
b2363	15551		4661	
b2381	25973	1819		3113
b2383	12988			
b2415	2143			
b2418	2180			
b2420	2177			
b2426	9529			
b2427	9595			
b2428	2252	8248	600	683
b2436	893			
b2444	2427			
b2447	14901			1859
b2449	3512	8298	1088	5140
b2453	3229		956	
b2458	11018		2995	
b2461	688	154		
b2471	333			V78
b2472	13643			V420
b2473	5779			
b2478	942			74
b2479	5719			
b2480	9181			
b2481	648			
b2484	9549			
b2485	2286			
b2486	6648	8485		V89
b2488	13484			V88
b2493	10987			
b2500	4619	10947		V418
b2501	4620	10948		*124
b2503	20087	10310		2266
b2510	1858	10941		121
b2511	4621			
b2512	13902			
b2522	1859	10942		122
b2523	6477			V447
b2533	1844			V74
b2534	3210			
b2535	3211			
b2538	23668			
b2539	4623			
b2540	508			
b2542	12165			
b2550	21359			
b2555	19097	8957		2150
b2556	683	5000		3406
b2557	687	153		
b2561	15718			
b2568	4624	10949		V81
b2569	4625	10950		V82
b2573	4556			
b2577	10661			5086
b2579	691	8087		*141
b2582	23683			
b2592	5421			
b2607	23711			2761
b2611	11835			
b2612	5786			
b2615	5314			
b2618	265	91		V86
b2620	18671			
b2621	4112	10982		5127
b2623	4660			V87
b2626	14254			
b2627	5979			
b2632	2977		857	
b2633	7201		2042	
b2634	5782		1620	V72
b2635	5437	476	1492	
b2636	20090			2267
b2642	5737			
b2643	6613			274
b2645	14801			3437
b2648	6362			
b2649	4662			
b2650	4663			
b2651	125	37		
b2652	10155			
b2653	5761			
b2656	5762	494	1607	261
b2657	5148			
b2658	930	188	288	726
b2660	2867			
b2661	338	112		
b2663	10545			
b2665	6031			
b2666	22381	10552		2726
b2668	341	114		42
b2671	23774	10662		2779
b2672	2713			159
b2673	10546			
b2674	337			
b2678	10675			
b2680	261			33
b2682	886			
b2684	2879		812	
b2686	336			
b2690	2954			
b2691	23502			
b2693	15270			
b2695	15271			
b2703	2259			V66
b2706	2262			
b2707	4698			V137
b2710	2261			V58
b2711	34876			
b2714	2710			
b2717	4704			
b2723	21372			
b2727	18484	8850		2488
b2732	16159			
b2734	18502			
b2735	5186			
b2750	4005			
b2751	19622			
b2757	17635			
b2758	18782	8901		2102
b2763	9187			
b2764	24493	4753		
b2772	67			
b2777	14640			
b2780	17506			
b2783	5280			
b2785	17507			4253
b2788	5330			*129
b2791	18968			
b2794	5430		1488	
b2797	9199			317
b2802	1835			
b2806	4736			
b2806[1]	7394			
b2807	6622			277

SPECTRA INDEX OF ORGANIC COMPOUNDS (Continued)

HCP No.	Sadtler Reference Number: IR Prism	IR HR	UV	NMR
b2809	17502			
b2810	17382			
b2812	19648			5039
b2813	22464			
b2817	17740			
b2821	14638		4144	
b2822	14636	8693	4142	V256
b2823	6494	8475	1803	
b2825	20657		6959	3782
b2828	101	32	40	
b2832	65			
b2833	269			
b2836	12553			
b2839	7796		2150	
b2847	5873	8423	1648	3147
b2848	26059	1897	9457	443
b2852	2774			
b2854	24821		8790	
b2864	75	10934		
b2866	5278	10984		
b2869	3683			
b2871	15640			
b2872	18162			
b2876	4891			241
b2877	2250			
b2884	3431			
b2886	3432			
b2887	3433			
b2888	3435	12976		4944
b2889	3436			
b2890	103	10963		14
b2893	805	9996		
b2895	806			
b2896	919			
b2898	9195			
b2905	18438			
b2907	16558	8716	5200	4018
b2912	297	101		V76
b2913	5784			
b2918	18584		5953	
b2920	24821			
b2921	5975	10008		
b2922	4817			
b2924	7362	545	2086	536
b2926	5287			1885
b2929	5772	8415	1613	3199
b2930	14662	624	4153	547
b2931	13494		3527	
b2932	19862		6568	
b2933	1457		418	V60
b2934	22378			
b2940	7857			
b2941	7859			
b2942	7860			
b2983	1606			3418
b2984	1605			5318
b2986	299			
b2987	14643			5315
b2991	6320			3426
b2992	7856	553		J140
b2993	681	12975		3411
b2995	14551			
b3009	749	8091	247	V61
b3010	10732			
b3011	17150	10049		4494
b3012	14587			
b3013	14585			J143
b3014	14586			J142
b3015	11902			
b3016	15368			

HCP No.	Sadtler Reference Number: IR Prism	IR HR	UV	NMR
b3019	19628			1861
b3035	23595	770		3608
b3040	22001			
b3043	6530			1815
b3045	14583			V414
b3054	13667			
b3056	8687		2264	
b3057	9053(K)			
b3058	9536			
b3062	9760			
b3063	14584			
b3064	13044			5416
b3065	1466			V251
b3069	17302			V99
b3074	23663			2793
b3075	19168			5010
b3078	31117			
b3081	19603			
b3082	22098			
b3083	14652		4151	5095
b3084	5332			
b3089	32794	9789		4548
b3091	8008	10990		V636
c8	1036	209	315	V204
c20	14471			
c24	11919			
c25	1474	248		
c32	244		96	30
c39	4582			
c42	11035			
c44	11037			
c48	11038			
c57	1499			
c59	2701			
c60	15618			
c71	9614			
c79	26825	2573	10103	
c85	13498			
c88	18455		5903	
c89	5747			
c91	11180			245
c93	3402			
c94	12994	607		5240
c98	30325	6320		3482
c106	11592			
c109	3224			
c111	14806			
c117	22419	10565		2748
c120	19184	707		4288
c133	19967			3830
c136	5701			
c141	1311	237	13550	
c142	10962			
c147	14645		4148	
c153	13268			
c156	8396			
c159	1924			
c161	2223	10977	596	
c165	5748			
c168	13481	10026		V243
c170	8024			290
c173	13477	10022		4466
c175	12993			1894
c177	9759	8562	2575	3231
c178	15324	8703	4581	3269
c179	15325	8704	4582	3270
c184	1142			
c188	31914	7892		
c191	23370			
c194	21933			

HCP No.	Sadtler Reference Number: IR Prism	IR HR	UV	NMR
c196	11591			
c207	4608			
c211	6626			V271
c213	21371			
c227	19913		6600	
c231	1020	207		
c245	2723	335	737	160
c250	18778	8897	6063	2100
c251	13124		3419	
c260	20112	10316	6693	
c264	20799			V425
c273	721			
c298	7178		2031	J217
c300	723			J218
c301	5199			J225
c305	722			
c306	5450			
c307	6480			
c308	5956			
c309	25704	1579	9293	
c310	21			
c312	34477	12461		5745
c313	1188	221	351	
c327	9521	8554	175	
c334	8580			
c339	1490		424	101
c340	20882		8601	
c350	29		3452	V230
c351	13697		3628	
c352	19485	8838	6382	
c354	1520		430	104
c355	6033			
c356	884	177	276	718
c357	18097	10092	5728	4246
c359	167	52	66	20
c366	2882		815	
c369	19543		6413	
c372	18350		5844	
c386	17668		5609	
c392	1386			
c394	10002			
c397	17173	8719	5419	
c398	19985		6628	
c399	9915		2612	
c400	9919		2615	
c402	8405	10013		4945
c419	23684	10633	13349	
c424	10535			
c425	5275	8385	1448	
c428	765			
c432	2402			150
c439	5179			
c442	16021			
c444	14889			
c463	17152		5410	V689
c466	31646	7632		
c477	3276			
c482	5948		1674	
c483	17131			
c488	17158			
c501	7178		2031	J217
c502	10787			
c505	8758			
c523	203		79	
c526	1691	270	492	V225
c528	7622			
c529	13593		3578	
c532	8192			
c533	8417			
c534	3157		926	

SPECTRA INDEX OF ORGANIC COMPOUNDS (Continued)

HCP No.	Sadtler Reference Number			
	IR		UV	NMR
	Prism	HR		
c535	11616		3223	
c536	19701		6489	V294
c537	13757			
c539	20137		6712	
c542	465	8041	2232	3220
c544	1260		375	87
c545	8403		2231	
c548	1463			
c550	3747		1147	
c553	21934			
c554	7198			
c555	26680	2436	10002	
c566	13460			
c573	953	195		1034
c575	5794			
c576	5715			
c586	7027			
c588	380			
c589	27902	3628	10845	
c590	2496			
c593	16925			
c594	2385			
c595	19621	10248		
c596	14903			4926
c597	15453			
c599	15454			
c600	15455			
c608	22379			
c611	12046			
c612	23773	10661		2778
c613	20681			
c616	17154			
c617	11956			
c624	21999			1902
c627	23593			
c637	180	59		*150
c639	31987	7967		3949
c640	845			
c642	2263	8250		313
c644	4628			
c645	4629			
c647	11838			
c648	11837			
c649	11836			
c650	6449			
c651	191	8019		25
c652	3386			
c653	19155			
c654	17125			
c657	23664			
c658	23664			
c660	24271			
c663	11843			
c664	12999	8610		3407
c665	12998	8609		
c666	21745			
c668	31992	7972		3954
c669	31993	7973		
c671	3449			
c674	4540			315
c675	6445			
c681	9699			
c682	9698			
c683	9697	574		
c684	9696			
c685	22448			
c686	12047			
c687	11841			
c688	194	68		27
c689	5967			
c690	12054			
c691	31997			
c692	6315			
c693	12051			
c694	11842			
c698	26686	2441		2830
c699	22380			
c700	23760			
c704	13511			V173
c706	13199(CIS)			
c706[1]	16546			
c720	22276	10545		
c723	7837	8512		
c724	7838	8513		
c726	22861			
c727	17130			
c728	3472	378	1072	710
c729	17349		5500	
c731	17177	670	5422	559
c735	14904	8697	4319	V512
c736	256			
c739	19137			5009
c742	41	10960		7
c746	20168	10332		2287
c749	18468			
c750	2540			
c752	27910	3636		486
c755	6464	10989		V211
c758	11911			
c759	13370			
c761	13371			
c763	13372			
c764	13362			
c765	13373			
c766	13363			
c767	468	126	161	
c768	3439			
c769	9510			
c783	13369			
c784	13368			
c785	166	51	65	
c786	276	95		
c788	18889	8925		2120
c789	1887		7682	
c790	19503			5023
c800	19663		7804	
c804	8401			
c807	8402			
c808	42		19	8
c809	20475			
c811	196			3409
c814	30248	6247		
c817	3387			3420
c820	7853			
c825	5721			
c827	25777	164		
c828	20179			
c830	24579	4814	8682	2696
c832	6217			
c835	10592			
c841	19138		7699	
c843	18785		6066	
c848	25016			
c854	18835			V511
c855	17144			
c856	17148			
c857	17139	665	5408	557
c860	13043		3365	
c861	4452			
c862	4451			
c863	1314			
c867	21399			
c868	680			3435
c870	438			
c872	6025			
c874	11829			
c876	689	155		3436
c880	19102	8958		2151
c882	13513			
c887	22825			
c892	376	12974		44
c894	171	55	67	1195
c897	23787			
c899	19781			5162
c901	23852	10920		
c907	30064			
c908	1306			
c909	11831(gas)			
c912	13924			
c913	11885			
c921	2333			
c922	22060			
c924	22059			
c930	1710			
c940	13419			
c941	17161			
c942	17180			
c943	323	111		
c944	17160			
c945	16947		5343	
c947	10785			
c949	17162		5414	V402
c950	9355	8549	2483	
d5	19594		7799	
d18	1524			
d20	4527	8332		*149
d21	7170			
d22	2981			
d23	31985	7965		3948
d25	2982			
d28	2980(C)			
d30	22449	10571		2754
d31	7037	537		
d32	17136			
d33	983			
d34	14296			
d35	5174		250	
d38	999			
d39	18811	8914		V570
d40	2979			
d41	3668			
d42	2705			
d45	6032			
d47	16160			
d49	16347			
d57	159	10965		V282
d58	23657	774		565
d59	32088	9086		3909
d62	22387	10557		2733
d63	22388			2743
d64	23658			
d66	1309			
d68	31991	7971		3953
d74	1671	8193		359
d79	7177		2030	
d81	4509			
d84	10561			
d85	7175			
d89	5416			
d113	5986			V589

SPECTRA INDEX OF ORGANIC COMPOUNDS (Continued)

HCP No.	Sadtler Reference Number			
	IR		UV	NMR
	Prism	HR		
d126	13221			652
d129	12456			
d134	2455		640	
d136	5817		1625	
d137	7261		2059	
d151	13222		642	
d153	6994			
d164	1143			
d166	24551			
d167	15769			
d168	2945	10980		179
d176	2290	10943		
d180	5318			
d185	2588			
d186	17466			
d194	5787			
d198	40	21		1193
d199	9761			
d200	18812			
d204	19960			
d205	8031	8518	11505	V75
d207	2348			
d208	9684			
d212	3296	375		
d213	3297	376		4885
d221	14704			
d225	24574	4812		
d226	1244			
d227	10588			
d229	24276	892		
d231	1344			
d239	17637			4438
d252	9709			
d258	9242			
d259	9241		2395	
d260	9240		2394	
d261	13660		7007	
d265	27950	3675		330
d269	3293		988	
d271	17422			
d272	17453			
d273	15645	644		
d276	15646			
d278	3504	379		711
d279	16560			
d282[1]	1534			106
d284	7103	8492		3433
d285	367			
d288	11630			
d290	4684	10954		
d293	24956			
d294	6363			
d295	8486			
d300	5675			
d301	3851	10775		194
d303	1937			
d304	20585			
d311	1492	251		103
d313	23756	10655		3662
d314	28849	4551	11366	1138
d315	4528			202
d323	10567			
e16	16457	10045		4217
e17	11722			
e18	8465			260
e19	24967			
e21	24968			1317
e28	24951			
e40	936			
e48	7550	8507		

HCP No.	Sadtler Reference Number			
	IR		UV	NMR
	Prism	HR		
e51	7559			
e97	1235			
e105	2351	8252	11446	
e109	4835	8351	11468	
e110	4836	8352		
e114	2229	8247		
e130	7385			
e136	4840	8353		
e148[1]	1030			
e152	6169	8449		
e154	9547			
e155	9522			
e158	19728		7811	
e159	9512			
e160	773			
e161	1288			
e162	7989			
e165	7373			
e168	24369	948		
e169	26610	2375		836
e170	14480	10028		4391
e171	22453			2757
e177	20065	10299	6670	2258
e181	77			
e183	127		47	15
e184	19633			
e185	17345		5497	
e187	18783			
e190	24657	4867	8733	
e195	4631	10951		V10
e196	4685	10983		V5
e199	533			V11
e203	944	190	292	
e204	2440			
e207	5319			
e209	8183			
e212	996			
e215	3205			J118
e216	35	20		
e221	30866	6838		
e224	6524			
e228	22906			
e235	6245			
e236	9982			
e237	3203	365	950	184
e238	1488			
e240	30454	6445		
e241	1109			
e248	4546	411		
e249	30863	6836		
e253	57			V13
e254	23483			
e255	7	5	4	2
e260	5244			
e261	178			
e265	1465	247		
e267	182	60		J119
e272	4633			1883
e273	19461			
e274	9721			V2
e276	23717	10998		
e283	2166	299		
e285	8012			
e286	7092	539	2016	
e289	8803			
e290	2230			
e292	10604			
e306	8333	557	2221	
e309	22919			

HCP No.	Sadtler Reference Number			
	IR		UV	NMR
	Prism	HR		
e309[1]	7506		2102	
e310	3665	389	1120	
e312	15284			
e323	17524			
e326	14647			
e328	6306			3424
e330	17171			
e332	9061		2307	
e336	188	64		V14
e338	19613			
e339	5776			
e340	123			
e341	22917			
e343	97			
e347	4724			
e348	2292	10979		4023
e349	16342			
e350	73			V12
e352	8187			
e353	23654			3635
e354	2978			
e355	32087	9085		3906
e356	17127			
e358	143			
e359	5618			V91
e363	5291	10992		
e364	14666			V92
e365	14802			
e366	270			
e367	383			
e369	18392			
e372	6351			
e373	254	87		32
e377	19906			
e385	248	84	99	V506
e387	2191	8242	582	133
e389	1632	8184	466	113
e390	8636		2252	295
e391	15307			
e397	18320			4224
e399	24183	824	8474	2678
e401	1131			
e404	34471	12456		5738
e405	1119	10973		
e409	7190			
e411	11632			
e412	3645			
e413	3646			
e414	30862			
e416	19518		6401	*111
e417	30864			
e419	5422	469	1485	
e420	237	79		
e423	20510	10404	7881	2355
e425	185	62		
e426	21396			
e427	15279	630	4552	548
e429	22899			
e433	1863		522	
e436	34462	12447		5733
e438	34463	12448		5760
e441	1861		521	
e442	11515			
e450	5297			
e458[1]	22452			
e469	1862			V134
e470	1656	8187	3530	118
e471	106			
e475	8021			
e477	1335(liquid)	241		

SPECTRA INDEX OF ORGANIC COMPOUNDS (Continued)

HCP No.	Sadtler Reference Number: IR Prism	IR HR	UV	NMR
e479	13379			
e480	16348			
e495	8691			
e496	95			
e497	102			
e499	1128			
e501	9682			
e508	5867			
e509	2487			151
e510	4492		1245	200
e511	2269		608	
e526	12234			
e527	4497		1250	
e534	228	10936		*110
e535	221	73		
e537	968			
e538	220			
e539	292			
e591	3961			
e602	4537	406	1255	205
e611	3880			
e612	5257	8384	1447	3183
f6	17489			
f10	6395			522
f13	4994			
f17	34500	12484		5759
f18	20108		6689	
f19	596			
f20	7526			
f25	9412		2519	
f27	21430			
f29	594			
f34	17433		7337	
f36	31626	7613		
f40	10563			
f44	901	180	281	
f45	7540	240	2114	288
f46	2458	8590	4150	
f54	9994			
f55	11938			
f56	7618		2139	
f67	20350		6843	
f68	11386			
f74	8442			
f80	7390		2095	287
f81	25075	1051	8923	
f83	10466		2890	
f91	3518		1093	
f96	22427	10568	8354	
f99	24783	4961		
f100	9792			
f104	17431			
f106	2538			
f108	17522			561
f109	9685	572		
f110	4127	400	1192	
f115	25	15		
f116	9775			
f117	2232			
f119	5051			
f120	156			V39
f121	1989	8218		360
f124	10961			
f125	1031		314	
f126	19981		6626	
f127	19657		6470	
f128	19658		7803	
f130	17511		5556	
f131	1531		436	
f132	2077			
f134	15273			
f135	2515			
f136	2529			
f137	2513			
f139	2716			
f140	5259			V415
f141	693			V9
f142	5258			
f144	1641			114
f145	885			
f146	2514			
f150	16344			
f151	5750			
f152	1552	8166		1552
f158	8312			
f159	5751			
f163	13474			4463
f165	5749			
f175	2755			
f176	1105			
f178	472	127	165	4407
f179	1876			
f182	7236		2051	538
f184	5163			
f190	5520			4982
f193	3664			V50
f194	3135			
f194[1]	8081	8523	2184	V104
f199	10520			
f206	9256			
f207[2]	14146		3936	
f213	210		82	
f215	581	10970		
f219	12862			
f220	74			V77
f222	126	38	46	
f223	13691			
f225	10516			
f227	3667	390	1121	
f228	3129			
f231	10514			
f234	1423			V125
f237	10513			
f238	10515			
f240	10517			
f245	18729		6036	
f260	3130			
f261	3131			
f266	1424	244	410	
f276	18702	694	7673	1257
f277	113	35	42	V95
f282	17226		5442	
f285	22440	757	8361	
f290	3682	391		
f291	17441		5543	V247
g1	1044			
g2	21697			
g3	989	8113		
g6	17499			
g12	12213			
g13	509	8047		
g14	8087			V279
g15	18088	10086		
g32	8335			
g34	5145			
g38	5455	199		
g44	17512			4255
g51	5122			
g56	8951			
g58	8657	482		
g59	17530			
g60	2154	8050		
g62	23515			
g64	4517	8330		
g65	11489	591		
g66	7391			
g68	5180			
g70	169	10001		4979
g81	8654	8012		2400
g89	7594			
g100	6408	8046		
g103	18673			
g108	10269			
g110	8653	719		
g117	6852			
g120	6354			
g124	6475			
g128	6702			
g133	6490	40		
g143	26030	1873	9440	
g144	25751	1618		
g148	8632			
g159	8688			
g161	31643	7629		
g162	31642	7628		
g165	22814			
g166	5892			
g171	7404		2096	
g173	22588			
g180	2151			
g181	8671			
g183	19605	10242		
g187	24880	5992		
g189	3824		1163	
g193	5194			
g194	4588			
g195	8676	544		
g196	452			
g198	8323		2218	
g200	13632			
g201	1337	8143	396	97
g202	2218		592	
g204	21412			
g208	6374	8467	1759	
g211	17446			
g212	17443		7339	
h2	2969	8281	11453	
h3	3358	8292	11456	
h7	11498	8599		
h10	3885		1178	V187
h13	2928	8274	11450	
h14	2968	8280	11452	
h16	2717			
h19	7851			289
h22	5579			
h23	5260			3181
h27	3805			
h30	1525	257		
h31	7039			
h39	5310	461	1454	663
h40	23728	10643		3656
h41	23729	10644		3657
h42	26812	2561		2835
h43	16753			
h50	5580			
h51	6874			
h52	5583			
h53	20576			
h61[1]	19912	10275		
h66	24326	7997		2596

HCP No.	Sadtler Reference Number IR Prism	IR HR	UV	NMR
h67	21345			
h68	21351			
h70	24327	926		2594
h72	3448	8295		1549
h74	7321			
h77	5311			
h78	16752			
h86	24837			
h90	306			
h92	952	194		
h93	271	93		679
h94	12004			
h96	4547			208
h98	4548			209
h110[1]	32154	9152		3938
h119	33859			5425
h125	4549			210
h126	5110	8369		
h128	11833			5320
h130	33851			5056
h132	5961			
h134	22398			
h135	18494	8853		2065
h136	9060			316
h140	19984			3836
h142	8686			297
h144	544			49
h149	6873			283
h152	5255			
h166	4571			215
h168	111			
h170	389	10767		45
h171	9192			
h174	378	10968		
h179	6191			
h180	28947	4649		876
h181	32007	9008		3962
h184	340	10966		41
h185	2034	290	548	1311
h186	9191			
h187[1]	6230	31	39	
h191	20169			
h192	34863	12846		5874
h193	22462	10576		2742
h197	542			
h209	32120	9118		
h210	32128	9126		3929
h212	1645			
h235	1399			
h236	3522	8300		
h243	15329			5313
h244	16685			
h245	11631			
h246	9725			
h252	3558	8301		
h253	11336	8007		
h254	2293			
h259	6035			
h264	5172			248
h265	24948			
h267	12045			
h268	17311	10050		
h269	222			5175
h270	4529	8393		1343
h281	15330			4419
h288	15617			
h289	5315	464	1457	V515
h293	20258			
h298	3674	8305		V462
h301	5271			
h315	2150			
h316	342			
h318	5720			
h319	678	151		
h320	433			46
h322	211			28
h326	4550	412		
h327	16033	10041		4212
h328	16032			
h330	4551	10945		211
h349	24273	889		2577
h357	22400			2736
h364	23741	10649		2765
h366	11832			
h370	4552			212
h373	19929			5312
				*145
h378	684	8085		3413
h380	281			4406
h381	14642			
h383	14660			
h384	6478			
h385	5251			
h389	181			
h390	7571			
h391	1875			
h395	18131			
h397	23499			
h400	5423			
h406	15337			
h408	4573			216
h410	5299	10985		
h411	4226			
h413	121		43	2709
h415	19119	8964	6242	2155
h418	1044			
h420	6364			
h427	308			37
h428	2733			
h431	16162			
h432	18325			
h433	931			
h436	6339			
h437	1642			
h438	2902			176
h440	23609			3614
h444	8013			
h445	15194(K)			4445
h452	16346			
h454	6141			
h456	2985			
h460	11287			
h464	18824			
h478	4064			198
h480	9196			
h483	5293			
h485	18478			
h491	1472	10974		98
h495	20480	10392		
h500	9185			
h501	6192			
h502	9188			
h503	32087	9028		3875
h505	32008	9009		4503
h511	8503			
h513	8507			
h515	8508			
h517	6343			
h518	9197			
h519	24544			
h521	8510			
h531	17523	10876		5997
h534	23482			
h535	9701	575	6986	415
h544	7383	8502	2093	
h546	22438			
h550	15061			5321
h564	15300	633		
h569	7172			V504
h579	23775			
h587	6746	10944		
h589[1]	2802	344	773	
h589[4]	1819	8202	511	
h589[7]	23272		5115	
h590	5860			V443
h591	16250		5040	4939
h593	5456	8396	11479	
h594	13420			
h598	24332	7998	8540	
h602	20211		6764	
h603	7162			
h606	19587		6448	2492
h607	33858		14368	
h609	8334			
h612	13497			
h613	18400	10137	7597	4050
h614	8601			
h616	2095			
h617	5470	478	2168	3215
h620	8086		2186	
h622	26019	1863	9435	831
h623	19939			
h626	18405			
h627	3774	8308	6093	
h628	7163			
h633	6492	8474	1801	3285
h646	19113		6240	
h647	24250	881		
h653	2470	4735		
h656	7647			
h658	23745			
h666	18317	377	989	
h668	17447		7340	
h669	15427	8707	4611	
h671	20362			
h683	7650			
h685	19921		6603	
h692	21427			
h694	3670			
h697	8631			
h704	25023	1015	8889	
h705	20543		7888	
h709	9538			
h713	5700			
h718	20378		6864	
h725	3019			
h732	3837	8310	11458	1551
h736	33119	11112	14024	
h761	7533			
h770	7007			
i1a	8335			
i7	7001			V20
i9	7205			
i11	5619		4571	
i13	19909		6599	

SPECTRA INDEX OF ORGANIC COMPOUNDS (Continued)

HCP No.	Sadtler Reference Number			
	IR		UV	NMR
	Prism	HR		
i16	15219	8701	4509	V527
i22	20144		6717	
i23	17526		5558	
i24	10926			
i25	10927			
i30	3505	380	1083	V224
i32	11597			
i34	23613	10612	13337	3618
i35	4890			V229
i37	3333		1022	
i41	18970		6176	
i45	23545			
i47	23546			V485
i49	1012	10938	308	V227
i56	3400			
i62	6219			
i65	4586	415	1269	473
				J189
i66	21265		8340	
i72	3823			V231
				J190
i73	24575	10688		2692
i75	22446			
i82	9340			
i84	1060	215		
i91	8193			
i95	5906			
i98	5907			
i102	2214	304	590	
i105	11603		3213	
i106	19212	10825	6295	4298
i108	21694			
i109	20294		6809	
i110	10143		2746	
i114	1383			
i117	207	71		524
i125	1499			
i127	14675			
i137	22269			
i138	22270			2527
i139	22271			
i140	23483			
i141	15366			
i144	6251		2052	
	7237			
i148	8436			
i149	1075			80
i168	586	9995		V281
i180	122	36	44	
i188	1144			V520
i192	33672	11663	14296	5447
i193	18728		6035	
i197	15220		4510	
i202	20300		6810	
i203	15886		4845	
i212	9062		2308	V252
i216	1603	8182	459	3155
i225	31857	7836		3771
i226	19174			5011
i228	15275		4548	
i231	19922			
i234	5908	8434		
i236	9063	8545	2309	3228
i242	24151	7983		2674
j8	8162			
k7	7197		2041	
k8	20060	10297	6666	2257
k10	31982	7962		
k23	1627	9998	463	
k24	2986			V628
l4	31634	7620		
l11	21949			
l28	18821			
l31	19911	729		
l32	5459			
l33	790			
l34	10916	586		
l36	564	8065		
l37	563	8064		
				2852
l41	12958	10018		
l42	15333			
l44	1005	8114		
l45	8341			
l49	17458			
l54	31656	7642		
l56	7217			
l60	8410			
l64	10338			
l65	3100			
			4158	
l71	14679			
l75	788			
l76	8699			
m1	11587			
m2	6253			
			163	V48
m3	462	8040		V212
m3[1]	1097		300	723
m3[2]	986		632	
m5	13219		6387	5021
m6	19491			
m6[1]	6225			
m10	17393		7343	
m11	17471			
m12	17143		3514	V427
m13	13445			
m17	2960			
m19	18323	10110		4222
m20	8698		2271	
m21	19907	728		3594
m23	7367			V181
m24	5261			
m25	5188			
m28	2237			
m32	2235	10773		4600
m34	5247	10780		V57
m35	7625			
m37	5169	453	1406	730
m40	3212			
m43	15281		4554	
m44	5739	491	1597	
m46	8380	10012		4983
m50	857			
m52	24407	979	8581	
m54	3217			1880
m55	31692	7676		
m56	7998			
m58	14909			4927
m60	24285	895		
m61	32812	9807		4558
m62	20036	10287	6654	
m65	18735			
m67	3214		951	
m69	27239	2983		519
m70	32100	9098		
m71	3213			1879
m76	6008			V573
m77	15280	631		891
m78	5262			
m81	264	8025	1728	302
m82	19125			
m83	19183			4287
m88	22014			
m92	19753			
m94	3216			
m95	15615	10809		5076
m98	22463			3607
m99	5263			
m104	19915			
m105	8567			
m108	14912			4417
m111	7183	8497	2034	3211
m119	21816			
m121	1019			
m122	4513			
m124	992			
m136	993			
m145	20113	10317	6694	
m151	5460	477	1499	
m157	20221			
m158	1846			
m161	4099			
m165	9599	995		V281
	586			
m166	17529			
m168	5190			
m172	6400	8469		
m175	4063			
m177	6369	520		
m178	9251		2402	
m192	7846			
m194	11341		3097	
m203	21006	747	8157	
m213	1098	10972		
m214	204			
m217	1898			
m221	1897		528	V606
m225	1845			
m226	2478			
m228	842			
m230	3701			
m235	9252	8548	2403	V176
m241	4083			
m243	4557	8336	1259	213
m246	5939	10956		
m247	3204			
m248	6523			
m252	6620			
m253	30865	6837		
m255	24272		8525	2385
m256	1129			2429
m263	4636	426		226
m271	3389			
m281	7608	551	2133	537
m282	59			
m286	61	25	29	
m293	8236			
m294	8237			
m295	13900			
m296	2210	302		
m297	4082			
m299	2492			
m300	11997			

HCP No.	Sadtler Reference Number			
	IR		UV	NMR
	Prism	HR		
m303	10	8		
m306	2224	305		
m308	3722			
m310	4553			
m312	4584		2945	220
m329	6740			
m331	17513			
m333	13515			
m335	14535			
m349	1918	287		V1
m366	3508	382	1085	V607
m376	112	10964		1554
m379	894		278	67
m380	32840	9835		4573
m381	6347	8461	1749	3202
m388	131	39		
m389	5565			
m391	7401	8504		
m392	8110			
m393	7156			
m395	4982			5951
m408	100	10962		V83
m409	879			
m410	11310			
m411	900			69
m414	881			
m416	11309			
m417	3295			185
m419	15541			
m420	9170		2380	
m422	8432			
m423	5938			
n1	1211		360	
n6	11258		3069	
n8	8444		2239	
n9	21406			
n14	865	169	265	62
n15	21007		8158	
n24	8695		2268	298
n25	1245		371	1202
n30	7525		2109	
n35	12169			
n36	9508	571	2545	539
n37	6370	8466	1757	
n38	46		22	
n39	9739		2574	
n42	18124	8804	5740	2004
n51	18361	8837	5850	2024
n63	5441	8394		1494
n75	18934	8946	6160	2140
n79	18674	8859	6000	
n93	2267	316	606	143
n94	14913	625	4321	546
n96	18803		6076	2109
n102	18804	8909	7677	
n109	3659		1117	
n110	23646	10624	13346	3632
n117	18424	8839	5877	2031
n126	23645	10997		1905
n128	19994		7845	
n130	19995		7846	
n133	17146	668	9632	
n134	34395	12381		5860
n141	19457	8989	6364	2174
n152	16249	8714	5039	3275
n154	21667			
n157	21419			
n158	1519		429	
n160	994		301	
n161	19547		6415	
n163	6759	8486	1889	3207
n164	19536		6410	2193
n165	19575	10231	6440	2206
n166	20206		6761	2298
n169	1145	219	340	
n170	1926	8210	11440	517
n171	22408			
n174	9038		2305	
n175	9037		2304	
n179	9810			
n180	918		284	72
n181	8207			
n182	19700		6488	V292
n185	12168			
n186	25445	1330	9120	797
n190	30	8004	2045	5
n191	18	8001	2551	3230
n193	11517		3168	
n197	8424			
n199	1416			
n202	1417			
n207	31481	7472		
n208	9735		11510	
n211	331		126	
n215	6870			
n216	1026		313	77
n217	2175			
n218	1277	230	12432	
n230	11628		3229	
n239	5850	499	1636	263
n246	21084		8214	
n247	21105		8227	
n258	1168			
n268	18841		6099	
n269	10072	8568	2710	
n274	24452	4722	8607	2933
n286	8045			
n295	6333		1742	
n296	21408			4311
n297	5853			
n310	20357		6849	2330
n311	25857	1709	13422	
n315	2162			
n316	17374		5512	
n322	31625	7612		
n324	11483	8238	3161	
n331	19488	10202	6385	2185
n341	18876	8916	6121	
n364	2036			
n365	2037			
n366	2038			
n372	16455	8715	5154	
n374	1829		514	
n388	8458			
n390	5879		1651	V550
n402	31711	7695	13586	1574
n412	11563	8600	3190	
n421	8288		2212	
n430	20150		6722	
n432	20315		6820	
n436	17626		5585	
n439	8077	8522	2183	3217
n466	23805	10669		2790
n468	20208			
n473	1355	8148	401	V269
n478	6361		1753	
n479	11584			
n480	13495			4411
n486	6486	525		
n489	21415			
n492	16165			
n499	5147		1398	
n502[1]	10566			
n505	24949			5311
n506	7534			
n509	24461	4730		2634
n511	5312	462	1455	664
n514	23691	10635		2760
n517	1489			100
n519	4530	8333		*148
n520	12000			
n525	142			
n526	14295			
n527	25695	1571		
n528	7565			
n530	5676			
n531	18731	8881		2089
n533	20608			
n534	60			9
n536	4888			
n537	887	10771		5218
n538	5173			249
n539	24463	10678		2635
n543	1491			102
n552	543			48
n553	25681	1557		
n556	8198		2198	
n570	27968	3692	10881	
o4	914			
o6	913			
o11	4531			5310
o12	20374			
o15	4637			
o16	9056			
o22	50			
o23	17362			
o26	1321			96
o27	14298			
o29	16467			
o33	24966			
o34	12504			
o35	21883			
o45	3451	8296		
o50	507			
o56	22762			
o60	20507			
o63	170	54		
o66	916	184		
o67	31652	7638		
o68	20372			2335
o69	4532			
o71	915			
o72	5329			
o77	7597			
o78	20816			3583
o79	2074	8232		1560
o80	917			71
o81	2113	8233		1632
o86	21992			
o91	20372			2335
o100	25044			
o110	1526			105
o112	5245	457		V216
o113	5861			
o116	4638			
o118	4639			227
o121	13423			4505
o130	23779			
o137	33852			
o140	32043	9042		3881

SPECTRA INDEX OF ORGANIC COMPOUNDS (Continued)

HCP No.	Sadtler Reference Number: IR Prism	IR HR	UV	NMR
o143	18977			4448
o144	18982	10173		4260
o145	24955			
o146	23599			3610
o147	240			
o158	14804			4388
o161	2707			157
o162	16555			
o165	6034			
o166	5159			244
o170	6340	518		
o171	18078	10078		4234
o174	4699			
o178	23503			
o183	4198	8327		512
o184	20213	10344		2500
o186	161			
o188	6188			
o193	3879			
o196	6194			
o197	1634			
o198	10632		2915	
o199	26813	2562		2836
o208	4533	8334		3422
o212	32000	9001		3959
o214	26412	2190		
o234	6269			
o236	5799			
o237	5802			
o238	3273			
o241	5265			
o246	6270			
o255	2962	8279		
o256	1167			
o257	479			
o261	5856			
o262	5266			
o263	6653			
o266	5267			
o270	5171			
o278	5854			
o281	24311	915		2589
o284	24467	4732		
o288	26016	1860		410
o295	14056		3868	3355
o297	14103	8666	3904	3371
p6	524			
p7	17635			
p11	21014			
p15	5181			V474
p19	503			
p27	15223			
p34	23639			
p37	24814			5721
p38	24965			
p39	23604	10608		3612
p45	15377			
p46	23605			
p56	26671	10753	9997	
p65	8046			
p66	8028			5185
p74	2480			
p76	10553			
p77	9490			
p80	14530	10034		4387
p84	690	8086		*140
p85	10988			
p90	1857			120
p91	4641			
p92	21355			

HCP No.	Sadtler Reference Number: IR Prism	IR HR	UV	NMR
p95	2265			
p99	63			
p101	23648			
p117	3654	386		
p121	5851			
p130	9726			
p131	31996	7976		
p132	7862			5317
p133	677	8084		52
p134	11834			
p144	4642			228
p148	686			*142
p149	682			3412
p153	4681	433		3432
p155	18508	12981		
p157	17452			
p158	7386			4409
p159	24323	923		2593
p160	17521			4258
p161	692			
p162	5187			251
p165	19116			5008
p166	17484			4252
p167	10897			
p172	19665	10256		2231
p177	8680		2261	
p178	14649			
p179	14650		4149	
p188	1907			
p189	22964			
	22965			
p192	239			
p193	20266	10353		
p194	17549			4019
p195	6527			V217
p197	17550			4020
p202	5774	497	1615	
p209	6612			
p214	304	10844		5995
p215	22009			
p219	2075			
p222	22010			
p228	24944	10712		
p229	20785	12982		3696
p231	1530			
p237	2326			
p241	15269			
p242	2709			
p243	6145			
p258	34406	12392		5839
p261	23514			
p266	4831	8349		
p268	3407			
p270	15064			
p284	10606	10789		5433
p288	23560			
p291	2706			156
p295	2728			165
p299	17491			
p304	8694			
p305	9058			
p310	21072		8203	
p311	2775			
p313	52	205		
p314	803			60
p315	804			

HCP No.	Sadtler Reference Number: IR Prism	IR HR	UV	NMR
p323	27948	3673		329
p324	8522			4274
p325	8512			
p327	9189			
p328	2400			
p330	10607			
p331	9194			
p332	9193			
p334	6195	8451		
p335	23795			
p336	66	10933		4279
p337	144	43		521
p338	9186			
p339	19759	10274		3976
p345	3445			
p346	2903	8272	833	177
p348	880		274	
p350	20804		8006	
p352	19115			5007
p355	163			
p356	20134			
p360	24137	801		
p362	356	116	139	303
p364	9190			
p365	44	23	21	V139
p366	9183			
p368	11920			
p369	5681			
p373	16751			
p375	685			3414
p401	32004	9005		4315
p403	3390			5319
p405	32005	9006		4316
p410	7988			3393
p413	7987			5316
p415	15315			5314
p416	23796			V471
p419	3391			3421
p420	298			
p423	23090			
p425	6009		1693	
p428	18902			
p430	15287		4558	
p433	27133	2879		968
p434	8511			
p436	32030	9031		3878
p439	3882			
p441	109		41	
p441[1]	15302	634	4568	572
p461	5300			
p466	32038	9037		
p469	332	127		
p472	1333			
p473	334	9994		3845
p478	34418	12403		
p479	5253	458		
p482	12959			2853
p484	810	8035	256	61
p492	7603		2130	
p494	19206			V622
p504	23258			5112
p505	23259			5113
p506	13119		3414	
p506[1]	25474	1358	9145	
p508	20364	10380	6855	2332
p510	15222			
p513	27896	3622		
p522	19725		6504	
p526	15224		4512	
p528	20577		6957	

SPECTRA INDEX OF ORGANIC COMPOUNDS (Continued)

HCP No.	Sadtler Reference Number			
	IR		UV	NMR
	Prism	HR		
p529	21074		8204	
p531	5894			
p533	21389			
p550	14219			
p554	28702	4404	11797	
p555	843	165	258	3152
p556	6764		1894	1176
p557	709	8088	3509	
p561	712	158	236	717
p577	18173	8829	5772	2018
p579	5982	8443	1685	3189
p581	2161	296	570	2824
p584	20670		6967	
p586	30633	6615	13258	
p590	474		167	
p592	4665		1281	
p593	21356			
p594	4892	443	1365	678
p596	18724		6032	
p604	1515			
p613	2800	343	772	1209
p614	4666	10852	1282	232
p615	1418	243	409	
p618	6456	8471	1792	270
p620	1419	8155	1982	3209
p621	15185			
p623	6766	8488	1896	
p627	22392		8249	2735
p629	7191		2037	
p630	17464			
p631	1315	238	9786	
p632	22395	755	9704	
p633	17465	681	9634	
p634	17124		5405	
p635	17468			
p637	19142	8971	6255	2162
p638	13923	8636	3803	
p642	33636	11627	14278	
p643	5444		1496	255
p645	253	86	3234	
p646	17486		5553	
p647	17463		5547	4250
p650	2889		822	170
p651	5980		1683	267
p652	85		35	12
p658	24717	4901	8770	
p662	90	8013	36	3149
p664	5554		1548	259
p665	1482	10768	423	3154
p676	6765	8487	1895	3208
p679	13456	10853	3517	3250
p680	2160		569	
p681	5887	506	1684	
p682	23689	10634	13350	3644
p683	5277	459	1449	
p684	279	96	112	
p697	5323	466	1459	667
p702	14019	616	3859	
p703	29001	5001	11905	
p709	3891	397	1181	1753
p710	8113		2188	
p711	9171	568		
p713	9395			
p715	7157			
p716	321	109	122	
p721	8642			
p722	8623		2248	
p724	5556			
p725	11496			
p726	5557			

HCP No.	Sadtler Reference Number			
	IR		UV	NMR
	Prism	HR		
p730	23620			
p755	12192			
p756	12191		4852	4473
p775	138			
p776	377			V482
p781	9066			
p782	3008			
p789	22			
p790	20	8002	10	301
p791	676	150	227	
p797	110	34		
p798	3003			
p799	104	33		
p800	105			
p801	6739			
p802	13950			
p803	18408		5865	
p810	13930		3806	
p812	6272	8442	1730	
p813	8656		18	
p814	21064		8200	
p817	1902	285	529	721
p819	37	3837	225	1199
p821	375	117	150	705
p822	3228	367	955	709
p824	1265	8135	378	89
p826	9204		2385	
p827	1304	235	391	93
p828	9778			
p830	6227	516	9886	
p832	1812	278	507	
p833	16548		5195	V246
p835	6635		1833	
p837	18797	8906	6071	2106
p840	18823	699	6086	
p841	20138		6713	
p846	27760	3495	10766	
p847	16607	658	5220	
p850	5962	8438	1678	3198
p878	20019			
p879	19545	10217	6414	
p883	21424	10837		5579
p884	8445			
p887	26607	2372		
p888	8700		2272	
p892	13001	8611	3355	
p893	9394			
p894	1271	228	382	
p897	13929	10874	3805	V496
p898	14538			
p899	11032		3001	
p900	31895	7873	13668	3795
p901	8760			
p902	6218	8452	11498	
p903	6236	8453	11499	
p906	18334			V346
p908	15225			
p909	9517			
p910	1342		398	
p911	7548			
p915	7161			
p921	22476			
p923	928			V272
p924	2188			V274
p932	15000			
p935	7194			
p936	34387	12373		5828
p941	368			
p942	9507			
p943	439			

HCP No.	Sadtler Reference Number			
	IR		UV	NMR
	Prism	HR		
p944	7480			
p946	11022			V119
p947	14478			
p948	8770		2279	300
p949	18497			4139
p950	23277			5117
p951	18335			
p952	5991			*135
p955	23634	10620	13344	3628
p976	8303			
p982	15285			
p984	24312			
p1003	8304			V477
p1005	8032	8519		
p1007	23799	10761		V479
p1011	20247			
p1023	34421	12406		5818
p1024	31908	7886		3800
p1035	20522			
p1039	25053			
p1040	8771			V328
p1045	29192	5189		
p1047	1343			
p1053	724			
p1057	763			
p1061	9600			
p1065	2518			
p1070	1905			
p1074	14542			
p1082	8269			
p1083	8348			
	302			
p1089	18482			4496
p1091	26036			
p1096	22883	10590		2541
p1097	4607			
p1101	7014			
p1104	16244	653	5037	
p1105	17410			V529
	17410		5535	*127
p1107	6404			
p1108	2242			
p1109	1081			
p1113	1130			
p1114	4719			V90
p1114[1]	20244			2300
p1115	26024	1868		400
p1118	1070	216	325	
p1118[1]	20246	10349		
p1119	11352			1892
p1120	4644	10952		V392
p1122	11042	10959		V29
p1123	16034			
p1125	1856			
p1126	22414			
p1129	4643	427		
p1133	193	66		
p1134	243			
p1136	129	10935		16
p1141	4646	429		
p1144	943	189	291	V385
p1148	946	192	294	
p1149	7245		2054	
p1151	9059			
p1152	8904			
p1154	3649			187
p1155	3650			188

HCP No.	Sadtler Reference Number			
	IR		UV	NMR
	Prism	HR		
p1159	9777			V412
p1161	3208	10981		V30
p1163	3209			V31
p1164	17517			
p1172	23736	10648		2762
p1174	7986			5309
p1180	387			V32
	2211			V32
	2212			V32
p1184	34404	12390		5837
p1186	5143	448		1215
p1192	4647			230
p1193	4651			V392
p1196	10652			V408
p1210	43	22	20	V42
p1211	1	1	1	
p1214	23596			
p1216	5246			V377
p1218	23630			
p1222	11351			
p1224	4652			231
p1226	23714	10638		3652
p1228	23715			V383
p1229	4653	10777		
p1231	9707			
p1234	267	92		V45
p1235	3297	376		4885
p1244	2804			
p1250	7898			
p1251	7246			
	11184(K)			
p1252	7899			
p1253	18722			
p1254	13919			
p1258	11297			
p1260	5085			
p1263	10749			
	6350			
p1264	5884	8430	1652	3188
p1265	1921			
p1266	1346	242	399	1203
p1280	22451	10572		2756
p1293	5983			
	328			
p1294	329			V396
p1295	330			3410
p1296	14803			3438
p1297	11296			
p1302	307	104		5996
p1303	9801			
p1310	107			
p1312	2170	300		
p1314	5965			
	5966			
p1315	22385	10556		2732
p1316	303	103		
p1318	1432			
p1322	698			55
p1323	5969			
p1324	5270			1884
p1325	2350			149
p1326	190			1877
p1327	23602			3611
p1328	339	113		40
p1329	8759			
p1330	2828		811	
p1332	309	105		38
p1336	9783			
p1344	2271	318		144
p1346	9776			
p1348	6144			

HCP No.	Sadtler Reference Number			
	IR		UV	NMR
	Prism	HR		
p1354	2251			137
p1358	16157			
p1360	2272			145
p1365	16156			
p1373	6618			V25
p1377	23758	10656		2770
	5268			
p1381	4707			
p1384	3647			
p1387	6553			
p1390	1596	8180		358
p1392	363			304
p1395	27940	3665		846
p1406	3648			
	10903(K)			
p1407	4001			
p1408	27238	2982		
p1411	23665	10628	13347	3637
p1415	5673			
p1417	18755			
p1423[1]	19593		6452	
p1430	6355			
p1432	24888	6000		
p1435	22127			
p1437	24668	4873	8738	5333
p1441	20769	10422	7990	3049
p1444	11349		3099	
p1448	15334	10807		4933
p1452[1]	3133			
p1455	5338	467		
p1460	768			
p1462	22444			2755
p1466	5167	10955		246
p1468	14027			
p1472	24952			
p1475	5276			
p1479	832	8098		
p1480	115			V409
p1481	388	10766		306
p1484	14541			
p1485	4703			
p1486	20096			
p1487	385			V70
	385			305
p1498	23727	10642		3655
p1509	5431		1489	
	10719		1489	
p1512	28899	4601		V27
p1514	15547			
p1516	1057			V69
p1518	697			54
p1521	27611	3346	10648	
p1522	11317			V516
p1525	14299			
p1527	18085	10083		4239
p1530	4702			238
p1532	8668			
p1533	17525			
p1534	10652			V408
p1539	11278		3080	
p1541	24458	4727	8612	
p1542	5963			
p1543	16153	651		
p1544	25732	1604		
p1545	27899	3625		
p1545[1]	19188			5016
p1548	24986	10731		
p1549	8017			
p1552	8905			

HCP No.	Sadtler Reference Number			
	IR		UV	NMR
	Prism	HR		
p1558	26058			
p1560	24818	4972	8787	5628
p1564	2495			
p1568	8697		2270	299
p1574	3866			
p1575	24832		8797	5726
p1579	23627			
p1586	26028	1871	9438	428
p1587	64			10
	64	10961		V43
p1588	189	65		V44
p1589	23490			
p1590	5826			
p1591	5977			4273
p1594	7558			
p1600	8765			
p1601	25767			
p1603	9489			
p1604	2249	310		
p1605	25770	1636		
p1610	8050			
p1613	16246	10042		4214
p1618	17455			V386
p1623	17122			
p1625	15720			1897
p1628	15765			
p1632	2089		558	
p1632[3]	6371			
p1633	2	2		V423
	2219	2		V423
p1634	8678	8539	2259	V422
	8678			296
p1635	9398			
p1637	6372			
	6356			
p1638	266	10937		
	4570	10937		
p1639	289	99		
p1644	1651	268	470	116
p1645	15283		4556	
p1648	34872			
p1649	233	77	89	
	6309	77	89	
p1651	13948			
p1652	4135			199
p1655	5944			
p1657	2719			
p1663	16158			
p1666	22248	10531		2515
p1667	27936	3661	10863	496
p1668	8049			
p1669	5669	483	1575	700
p1677	17394			
p1679	18249			
p1682	3881		1176	V638
p1683	25975	1821		1483
p1684	15347			
p1687	833	8099	11411	361
p1688	8329		2219	
p1693	5773	496	1614	V559
p1696	272	94	109	34
p1698	2927	8273	839	1855
p1702	9969	8565	11513	2843
p1705	12170			
p1707	6646			
	6645			
p1708	14668			

SPECTRA INDEX OF ORGANIC COMPOUNDS (Continued)

HCP No.	Sadtler Reference Number IR Prism	IR HR	UV	NMR
p1711	10521			V152
p1712	8350			
	8349			
p1713	6403			
p1714	1100			V38
p1715	4654			
p1716	13406			V23
	2270			V23
p1717	2539			V24
p1722	275			
	7848			
p1725	4689			
p1730	3652			V17
p1731	13903			
p1732	23674			
p1733	23676			
p1735	13904			V18
p1744	7366			
p1747	7858	8514		
p1754	195	8020	78	V232
p1758[1]	5192			V492
p1759	871			
	13687			
p1761	23512			
p1766	11017		2994	5298
p1768	2998	8284	3515	3166
	13447	8284	3515	
p1769	14038		3864	
p1770	4694	10779	1294	5141
p1771	11400			
p1772	14564			
p1773	1310			
p1774	14041			
p1775	1117			V64
p1776	386			V15
p1778	14574			
p1782	26553	2318	9911	593
p1783[1]	12487	606	9630	
p1792	5342	8390	1464	V62
p1793	14604			V71
p1794	1123			
p1795	14049			
p1796	1082			V135
p1797	1120			
p1798	15352			
p1799	2226			V113
p1800	14000			V97
p1801	15351			
p1814	284	97		V34
p1816	26703			479
p1820	12158			
p1822	184	8018	74	23
p1830	8639			
p1832	6603	529		273
p1833	18580			5998
p1847	12394			
p1849	21017		8164	
p1851	5333			V21
p1867	520		177	
p1870	7550	8507		
p1879	20514			
p1880	24059			
p1887	23626			1475
p1888	14700		4171	4021
p1889	7006			
p1897	15548	642	4658	553
p1898	14476		4111	V459
p1899	14477		4112	4390
p1900	3068		895	
p1901	20525	10407	6945	

HCP No.	Sadtler Reference Number IR Prism	IR HR	UV	NMR
p1904	3061		890	
p1905	21056		8194	V379
p1908	24207	847	8493	2683
p1916	34266	12255	14418	
p1917	8625			1858
p1925	5457			
p1930	870	173	269	64
p1931	34705	12688	14498	
p1937	4834	8350	11467	
p1939	15	12	9	V96
p1940	21879			
p1942	21019		8166	
p1948	1788	8199	500	V431
p1949	8394		2230	293
p1950	9495			
p1951	18347	8831	5841	2019
p1952	18145	8817	5756	2011
p1953	5465		1501	
p1959	12481	8608	3354	3246
p1960	13897		3791	
p1961	13895	612	3789	
p1964	16588		5210	
p1966	15314		4575	
p1967	13644	608	3607	V593
p1969	13646			
p1970	4728			
p1971	16341		5112	
p1973	1795		501	
p1974	17133		5406	
p1975	18955			
p1976	18956		6170	
p1977	18956		6170	
p1982	3201			
p1988	1042			
p1989	3298			
p1990	8009			V169
p1992	11029		2998	
p1993	16586		5208	
p1994	32021	9022		3873
p1996	6654		1839	V154
p1997	15806			V155
p1998	1249			
p1999	1250		372	
p2000	1251			
p2004	5614		1559	
p2006	4450	404	1241	571
p2007	18960		6173	
p2008	17490		5554	V428
p2009	18961		6174	
p2010	8188			
p2013	20453			
p2016	1283			
p2017	1247			
p2019	9	7	6	566
p2020	32	18	14	890
p2021	31	17	13	889
p2024	21070			
p2026	11397		3126	
p2034	10656		2920	
p2035	20451	10383	6903	1344
p2036	3515		1090	
p2037	32726	9721		4607
p2039	19132	8969	6251	2160
p2040	434	120	161	
p2041	860	8106	261	V453
p2042	8761		2277	
p2044	5170	454	1407	697
p2045	8452			
p2046	23791			
p2047	17192	8720	5426	3278

HCP No.	Sadtler Reference Number IR Prism	IR HR	UV	NMR
p2048	5644	8408	1564	
p2051	18978	10172	6182	4259
p2052	7547			
p2053	18975		6180	
p2054	17175		5421	3283
p2056	8039			
p2059	20252	10351	6783	569
p2064	5964	8439	3163	
p2065	27970	3694	10882	
p2066	13499		3528	
p2067	17440		5542	
p2068	21061	10836	8197	
p2069	168	53		
p2071	8189			
p2072	19121	8966	6243	2157
p2074	7508			
p2088	2419		627	
p2091	21880			
p2092	2121		561	
p2093	1783		498	V401
p2098	15641		4697	5098
p2100	7517		2107	
p2105	24016			
p2107	27244	2988		
p2126	1010		307	V55
p2134	18498		5924	
p2136	14149			
p2142	5851		1911	
p2145	8640		2254	
p2160	1908			
p2167	16581			5091
p2169	7235			V68
p2170	28861	4563		818
p2172	6528			V116
q1	11995	596	6997	1614
q2	4771	442	9885	
q7	34527	12511		
q13	7527		2110	
q14	5976	511	1682	
q22	7456			
	5993			
q24	9525			
q26	7504			
q28	7455			
q32	7546			
q34	1053	213	320	
q37	9719			
q43	7522			
q47	20672		6968	3589
q55	18144	8816	5755	2010
q58	18682		6006	
q65	33628	11619	14276	5791
q70	7610			
q77	34388	12374		
q81	1794			
q85	9762			
q87	8321			
q88	32735	9730	13970	
q89	4585		1268	
q90	18422		5875	2030
q106	7168			
q115	13310			
q120	2813			
q122	19446	8981	6354	5020
q123	19928		6606	
q124	7179	8496	2032	

SPECTRA INDEX OF ORGANIC COMPOUNDS (Continued)

HCP No.	Sadtler Reference Number IR Prism	IR HR	UV	NMR
q127	8010		2163	V578
q128	10544	578	9625	V579
q129	19690			3971
q148	13899		3792	
q154	187	63	76	24
q155	8628			
q157	26313	2109		
q159	26314	2110		
q163	4214	8328		
q171	2573	329	9883	314
q172	11199		3050	
	2575		3050	
q176	33687	11677	14307	5486
q185	8340			V249
q190	6344		1747	
q193	1055		321	*138
q195	2174		575	*139
	2174			132
q197	8407			
q205	2819			
q212	20301		6811	
q213	5893	8433	1656	265
q215	20302		7862	
q223	22130			
q233	16571		5205	
q249	20021	10281		
q251	6519		1812	
q253	11645			
q255	11644	128	166	
	473	128	166	
q258	8682	8540	2262	
q260	10872			
	13309			
q262	26558	2323	9916	1168
q265	10727			V494
q269	3169			
q270	9467		2542	
q276	31609	7596		2855
q278	34823	12806		
q280	31686	7672		
r1	1032			
r6	7379			
	11280			
r19	11293			
r21	255	88	2964	
	10901	88	2964	
r22	8540			
r24	20687			3668
r26	10526			
r27	8952			
r32	373		149	
r36	25477	1361		
s4	4995			
s6	322	110		
s7	5990			V253
s39[1]	29947	5934		
s39[2]	481			
s46	20541		7886	
s47	23617	10614	13338	3621
s51	25024	1016	8890	
s56	10625			
	454			
s58	10560			
s63[1]	1226			
s64	1702			
s73	24004			
s77[1]	11470			V184
s80	1611			
s99	991	200		
s101	9027			
s104	1069			
s109	17336	10755		5411
s122	9758			
s128	5978			
s129	11985	8603		V306
s137	1827			
s143	20058	10295	6665	2256
s152	7549			
s153	3801			
s154	9471			
s155	13218			
	3803			
s157	2181		577	
s159	241	81	94	
s161	1529	8158	435	V497
s162	25739	1609		
s164	20123		6702	
s167	14032		3862	
s168	14033	617	3863	
s169	14034			V498
s172	22434		8360	
s178	14630		4139	
s179	3301		991	
s185	11335			
	2805			4531
s186	5150	449		
s188	2257	314	604	140
s189	3218			
s192	6312			
s193	3001			V215
	3001			183
s195	14918			4930
s196	500			
s200	482	129	170	
s202	7619			
s207	1431			
s214	15719	646	4732	4381
s220	23902		8382	
s227	1831			V277
s229	18586	10155		
s232	7189	540		
s237	19169	10194		4281
s250[2]	6187			
s251	19937			
s253	5519			4408
s254	18984			4261
s255	28869	4571		
s257	1307			
s261	11334			
s267	19500			
s273	8659	563		
s274	21882			
s276	9514			
s279	49		24	
	9024		24	
s280	423			
s281	312			
s283	3685			
s285	311	106	119	
s286	21089			
	21088		8218	
s288	20468		6915	
	20469			
s289	24186		8477	
s292	9781			V136
	14631			V136
s294	18221		5779	1898
s295	6614			275
s298	8037			
s300	1300			
s303	17151			
s308	11300			
s313	6300			
	6299			
s315	18230		7561	
s317	320	108	121	39
s321	18683	8866	6007	2074
s327	9234		2389	
s329	9231		2386	
s330	34489	12473		
s338	9235		2390	
s342	6655			
s350	23704			V118
s354	15577		4668	
s355	422			
	7459			
s357	20107			
s358	14501			
s359	53	24	26	
s362	17818		5649	
s365	13485		3522	
s366	2991		864	
	13393			
	4873		864	
s367	4875		1356	
s369	17827		5655	
s372	9548			
	1352		400	
s373	17826		5654	
s375	15062			
s382	9705			
s383	15576		4667	
s384	18723			
s386	7031	536		535
s387	16313		5090	
	418		5090	
s393	20593		7917	
s398	8023			
s402	15332			
s408	19478	9000		
s409	461			
s416	9765			
s420	19601			V371
t7	656	147	222	
t10	22011			
t15	5272			
t17	20815			1904
t35	10275			
t46	4543	409	1257	V670
t47	4542			206
	4542	408	2171	V669
	8030	408	2171	206
t48	22478			
t49	1369			
t50	1365			
t52	1364			
t55	727			
	13203			J128
t62	10565			
t64	31621	7608		
t67	4535	8335		204
t68	20503			
t69	4687			

SPECTRA INDEX OF ORGANIC COMPOUNDS (Continued)

HCP No.	Sadtler Reference Number			
	IR		UV	NMR
	Prism	HR		
t70	4688			
t76	24576			2690
t77	21400			167
	2807			167
	8501			167
t78	20502			
t81	209			
t82	5264			
t83	7596			
t84	3852			195
t85	23699			3648
t87	8502			294
t88	23700			
t89	23701			
t90	4536			
t94	32144	9142		3934
t95	5612			
	10601			
t97	5320			
t98	12388			V557
t101	18893		6132	
t105	2035			
t117	8618			
t118	1392			
t127	4104	10007		
t129	4107			
t135	7505			
t137	1787		499	V380
t138	7552			
t156	20513			2359
t159	7512			
t160	3121			
t166	18591	8854	5959	J112
	18591			2055
t167	5753	10781	1601	5217
t171	15274			V384
t174	20086			
t175	20002			
t179	32063	9061		3897
t183	6092	514		1554
t187	172			V52
	172			*134
t189	8476	10847	523	123
	1865	10847	523	123
t190	25976	1822		1484
t191	25979	1825		1485
t192	17538		5560	
t193	1800	276	502	1408
t198	1802		504	
t202	13504			
t204	1809	277	506	1319
t210	19745		6521	
t213	8346			V49
t215	25980	1826		V103
t216	23801			
t219	177	57	71	
t222	20606			
t223	16580			
t224	14805			V80
	14805			*133
t229	174	56	69	V94
t230	25971	1817		1482
t232	176	963	70	523
t233	173			
t234	26712	2463	10021	475
t238	23256			
t249	7058		1994	

HCP No.	Sadtler Reference Number			
	IR		UV	NMR
	Prism	HR		
t254	784			
t256	12963			
t259	22000			
t260	21800			
t263	9543			
t264	1532	8159	437	
t269	10277			V366
t275	419	119	155	V157
t278	5757		1605	
t279	20973	10489		2404
t281	866	170	266	63
t282	28773	4475	11822	
t283	1543	8160	442	107
	7460	8160	422	107
t285	6330		1740	
t286	1544	8161	443	108
t288	3817	8309	1161	3172
t289	6341	8458	1745	3288
t295	21022		8169	
t299	21023			
t302	4721		1302	
t303	17155	669	5411	558
t304	4720		1301	
t306	14921	8699	4325	3266
t312	20668		6965	
t319	24010			
t328	11627		3228	
t339	20460		6909	
t340	4716			
t344	24433	4703	8592	
t345	13670	8626	3620	3252
	4722	8626	3620	
t346	13669	8625	3619	
t347	28904	4606	11379	
t352	212		4691	
	15634		4691	
t353	13671	611	3621	543
t354	21815			
t355	863	168	263	
t357	2504	327	660	154
t358	2724	336	738	161
t359	2725	8259	739	162
t360	2726	337	740	163
t364	4559	10851	1261	
t370	18680	8864	6004	2072
t376	32763	9758		4628
t377	20270	6793		
t378	8443	8530	2238	3222
t380	20880	8060		
t391	8	6	5	3
t392	91	29	37	13
t393	3658	8302	1116	3168
t394	629	145	212	51
t400	4726	440	1303	
	9809	440	1303	
t402	15736			
t407	15735			
t409	19934		6608	
t410	20271	11091	6794	
t412	13918	8635	3800	3258
t413	18905		6140	5000
t414	4690	436	1291	675
t415	21917			
t417	4691		3293	
	12065		3293	
t418	18798		6072	2107
t421	8047	8304	2177	3216
	3671	8304	2177	
t423	17467			
t424	13958	614	3815	
t430	13908	8631	3793	3256

HCP No.	Sadtler Reference Number			
	IR		UV	NMR
	Prism	HR		
t436	4558		1260	
t437	2489		657	685
t438	621			1182
	621	141	206	J116
t440	626			1183
	626	142	209	J117
t442	628	144	211	1185
t444	17120	664	5404	894
t445	627	143	210	574
t449	19448	8983	6356	2169
t451	15737		4745	
t460	4572		1263	
t461	20387			
t461[1]	18817		6083	
t462	18827	10812	6088	4996
t463	19974			
t465	707		232	
t467	19975		6622	
t468	24466	4731	8615	
t469	2092			
t472	11031		3000	
t473	24448	4718	8604	2626
t475	4718	8345	1300	3178
t476	3662		1118	
t477	23751			
t478	3663	8303	1119	3171
t479	175	8017	2550	3229
	258	8017	2550	3229
	9596	8017	2550	3229
t481	17378	676	5514	895
t482	1949	8214		
t484[1]	287	98	2907	
	10554	98	2907	3236
t488	7188		2036	284
t489	8090			
t490	15318		4577	
t491	8313			
t493	4658			
t494	4657			
t495	4659			
t496	775		250	
t497	157	985	62	V161
t498	844	166	259	3153
t499	2338	322	622	V160
t500	33	8005	15	
t505	8122		2189	
t508	25712	1585	9299	
t509	6349	8462	1750	3194
t513	20147	10326	6719	2278
t514	20273			
t516	21403			
t517	19409	8977	7771	
t519	19410		7772	
t521	32463	9459	13799	
t525	20272		7860	
t526	11921		3280	
t527	23685			
t528	20534			
t529	20535			
t539	2896		828	
t540	2895	10774	827	
t541	2897		3100	173
	11353		3100	173
t543	6615	8477	1822	J111
	6615			276
t544	21700			
t545	313			

SPECTRA INDEX OF ORGANIC COMPOUNDS (Continued)

HCP No.	Sadtler Reference Number: IR Prism	IR HR	UV	NMR
t546	45	8006	11403	V168
	12448		3353	
t547	28945	4647	11389	505
t548	2987	361	860	1180
t549	8044	10772	2176	V205
	8044	10772	2176	292
	1366	10772	2176	292
t550	21401			
t552	8409			
t554	4692	437	1292	676
t555	183	61	73	22
t556	4693	438	1293	677
t561	28374	4084	11732	1371
t564	236	8022	92	
t574	1768	271	493	1190
t575	18486	8852	5918	2051
t576	619	140	204	1181
t598	1268	227	380	90
t600	25215			
t601	630			
t605	23716			
t609	232		88	
t612	22032			
t623	21886			V486
t635	869	172	268	
t639	868			
t641	6337	8607	1744	
t645	10657		2921	
t650	28350	4064		
t652	19785		6540	
t653	2289	320	611	666
t654	16248		5038	
t655	23373			
t656	6860		1918	
t657	13453		3516	
t663	17363	675	5505	
t664	5754	492	1602	671
t665	6357			
t666	32049	9048	13694	3886
t668	8956	8544	2302	3227
t670	13518		3533	
t671	1572		451	
t673	9753			
t676	22370	10549		
t701[1]	22278	10547		
t702	2511	615	3855	544
	14013	615	3855	544
t703[1]	4729		1304	
t707	4114			
t708	10626			
t709	8667			
t710	10536			
t711	24950			1419
t712	23718	10639		3653
t716	18328	8830		3440
t717	24552			
t718	18686			
t722	22004			
t724	24445	4715		
t725	24946			
t727	5313	463	1456	
t728	22903			
t729	23719			
t732	5686	10986		5543
t736	5321			
t737	13213			
	5940			
t741	466			
t749	310			
t753	2801			

HCP No.	Sadtler Reference Number: IR Prism	IR HR	UV	NMR
t764	21887			
t768	21888			
t772	21889			
t773	432	8612	3453	
	13210	8612	3453	
t774	5568			
t775	455	592		
	11497	592	11525	
t776	535	134	187	
t781	7501			
t783	10624			
t784	5569			
t786	21891			
t789	17385		5520	
u2	6335		1743	
u3	15638		4695	
u4	24877	5989		
u5	1136	8127	11420	
u7	31616	7603		
u7[1]	8683			
u8	7514	8506	2105	2801
u9	8429		2236	
u11	26885	2631	10150	
u12	25719	1591	9305	
u13	9400		2508	
u16	6467		1796	
u17	1836		517	
u18	447	123		
u19	9795			
u23	28912	4614	11381	
u25	7524			
u27	1823			
u30	18601	10160	5968	
u32	3902			
	11739			
u34	3939			
u35	13656			
u37	14717			
u41	34472	12457		5739
u42	6757			282
u44	9751			
u55	1837			
u56	11732			
	3895			
u57	3219		952	
	11751			
u59	3661			190
	11236			190
u60	3892	10864		
	12493	10864		4957
u61	3660			189
u62	3911			
u64	965			
u70	3898			
u71	3899		1182	
u75	21489			
	8029	554	2170	1220
u78	11234			
u79	202			4399
u83	3894			
	11731			
u84	3912			
	11749			
u85	3913			
	11750			
u95	19919			
u95[1]	13891			
u102	19941		6610	
u103	20330		6827	
u104	20331		6828	
u106	21417			

HCP No.	Sadtler Reference Number: IR Prism	IR HR	UV	NMR
u107	20333	10369	7865	
u110	4591	417	1272	
u111	32139	9137	13717	
u113	20447			
u116	23703		3649	
u118	3901			
	11738			
u120	3962	8315	3292	
	12062	8315	3292	
u121	20553		6954	
u123	22015			
u125	3904			
	11741			
u126	3905			
	11742			
u127	3906			
	11743			
u129	19606			2221
u131	2963	358	853	
u135	22008			
u142	17640		7370	4032
v3	453			
	11495			
v24	16101			
v25	324		123	
v27	8085	30	38	
	92	30	38	
v28	16104			
x1	6855		1913	533
x2	6359		1715	
	6358		1715	
x3	6857		1915	
x4	20592			2373
x5	31476	7467	13160	
x6	645			
x7	31618	7605		
x10	18462			
x12	5608		1555	
x16	13753			
x24	4299			
x29	1039			
x30	1068			
y3	12279			
	12280			
y4	12281			

PHYSICAL CONSTANTS OF ORGANOMETALLIC COMPOUNDS

Metallic salts of organic acids will be found in Physical Constants of Organic Compounds table

No.	Name	Formula	Mol. wt.	Crystalline form, color and index of refraction	Sp. gr. or density	Melting point, °C	Boiling point, °C	Solubility in grams per 100 ml of: Cold water	Hot water	Alcohol, acids, etc.
	Aluminum									
1	Aluminum isopropoxide	$Al[OCH(CH_3)_2]_3$	204.25	col. liq.		118	145–150[5] solidifies			
2	Aluminum isopropylate	$Al(OC_3H_7)_3$	204.25	brittle wh. solid		118	125–130[4]	d.		s. bz.
3	Aluminum magnesium ethoxide	$Mg[Al(OC_2H_5)_4]_2$	438.77	iceblue cr.		129	225[4]			s. bz.
4	Diethylaluminum chloride	$(C_2H_5)_2AlCl$	120.56	col. liq.		−50	125–126[60]			
5	Diethylaluminum malonate	$Al(C_7H_{11}O_4)_3$	504.48	wh. need. or pr.	1.084[100]	98		i.		s. org. solv.
6	Dimethylaluminum chloride	$(CH_3)_2AlCl$	92.50	col. liq.		−50	83–84[200]			
7	Methylaluminum dichloride	CH_3AlCl_2	112.92		1.00[22]	72.7	97–100[100]			
8	Triethoxyaluminum	$Al(OC_2H_5)_3$	162.17	sol.	1.142[20]	150–160	d.			i. al., sl. s. bz., v. sl. s. eth.
9	Triethylaluminum	$Al(C_2H_5)_3$	114.17	col. liq., ign. in air, 1.480[6.5]	0.837	<−18 (−50.5)	194	exp.; d. to $Al(OH)_3 + C_2H_6$		
10	Triethylaluminum etherate	$4Al(C_2H_5)_3.3(C_2H_5)_2O$	679.04	col. liq.			112[16]	exp.		d. al; s. bz., eth.
11	Trimethylaluminum	$Al(CH_3)_3$	72.09	col. liq, ign. in air, 1.432[12]	0.752	0(15)	130 (125[26])	d. to $Al(OH)_3 + CH_4$		s. eth., al.
12	Triphenylaluminum	$Al(C_6H_5)_3$	258.30	wh. need		196–200		d.		d. al., chl., CCl_4; s. bz.
	Antimony									
1	Antimony ethoxide(ous) (triethyl antimonite)	$Sb(C_2H_5O)_3$	256.93	col. liq.	1.524[17]		95[11]	d.		s. org. liqs.
2	Pentamethylantimony	$Sb(CH_3)_5$	196.93	(exist?)		96–100		i.	i.	
3	Phenyldimethyl-antimony	$C_6H_5Sb(CH_3)_2$	228.93	col. oil, fumes in air			112[15–18]			
4	Phenylstibinic acid	$(C_6H_5SbO_2)_3H_2O]2H_2O$	746.61	wh. fine cr.				i.		s. chl, al., dil., a.; sl. s. acet.
5	Tetramethyldistibyl	$(C_6H_5)_2Sb.Sb(C_6H_5)_2$	551.93	col.		121–122 (in N_2)				
6	Tributylstibene	$(C_4H_9)_3Sb$	293.10	col. liq.			133–134[14]	i.	i.	s. org. solv
7	Triethylantimony	$Sb(C_2H_5)_3$	208.94	liq. 15 1.42	1.324[16]	<−29 (−98)	159.5	i.	i.	s. al., eth.,
8	Triethylantimony chloride	$(C_2H_5)_3SbCl_2$	279.84	col. liq.	1.540[17]		d.	i.		s. al., eth.; d. conc. H_2SO_4
9	Trimethylantimony	$Sb(CH_3)_3$	166.86	liq. n_D^{15}	1.523[15]		80.6	sl. s.	sl. s.	s. eth.; i. al.
10	Triphenylantimony	$Sb(C_6H_5)_3$	353.07	col. tricl. pl.	1.4343[25] (1.4998)	50, (46–53)	>220[1]; >360[760]	i.	i.	s. org. solv.; sl. s. al.
11	Triphenylantimony dichloride	$(C_6H_5)_3SbCl_2$	423.98	wh. sol.		143		i.	i.	s. bz., CS_2, hot al.
12	Triphenylantimony sulfide	$(C_6H_5)_3SbS$	385.13	wh. cr.		108–110		i.	i.	s. org. solv.
	Arsenic									
1	Acetylarsanilic acid	$CH_3CONHC_6H_4AsO(OH)_2$	259.09	cryst.		>200				s. $NaCO_3$; v. sl. s. HCl; i. eth.
2	3-Amino-4-hydroxy-phenylarsonic acid	$H_2O_3AsC_6H_3(OH)(NH_2)$	233.06	col. prism			d.290	sl. s.		s. alk.; min. a.; i. org. solv.
3	2-Aminophenylarsonic acid (Arsanilic acid)	$H_2NC_6H_4AsO(OH)_2$	217.06	need.		153		s.		s. al.; alk., ac.; sl. s. eth.
4	4-Aminophenylarsonic acid (*p*-Arsanilic acid)	$H_2NC_6H_4AsO(OH)_2$	217.06	wh. need.			d. 300	sl. s.	s.	s. alk., dil. a.; sl. s. al.; i. eth., acet.
5	Arsanilic acid (*p*) (*p*-aminophenylarsinic acid)	$H_2NC_6H_4AsO(OH)_2$	217.06	wh. need		232				s. eth., MeOH; sl. s. al., acet; i. bz., chl.
6	Arsenoacetic acid	$(AsCH_2COOH)_2$	267.93	sm. yel. need.		>260, d. 205		i.		s. pyr., alks., alk. carb; i. al., eth., chl.

PHYSICAL CONSTANTS OF ORGANOMETALLIC COMPOUNDS (Continued)

No.	Name	Formula	Mol. wt.	Crystalline form, color and index of refraction	Sp. gr. or density	Melting point, °C	Boiling point, °C	Solubility in grams per 100 ml of: Cold water	Hot water	Alcohol, acids, etc.
	Arsenic									
7	Arsenobenzene	$C_6H_5As{:}\ AsC_6H_5$	304.06	wh. need.		212		i.		s. bz., chl., CS_2; sl. s. al.; i. eth.
8	Arsenophenylglycinamide	$H_2NCOCH_2NHC_6H_4AsO(OH)_2$	274.11			280			v. s.	v. sl. s. al.
9	Benzophenone-4-arsonic acid	$H_2O_3AsC_6H_4COC_6H_5$	306.15	lustr. pl.		260		i.		s. al., alk; i. bz., eth.
10	Cacodyl oxide(dicacodyl oxide)	$[(CH_3)_2As]_2O$	225.98	col. liq.	1.486^{15}	−25	149–51	sl s.		s. al., eth.
11	Cacodyl sulfide (dicacodyl sulfide)	$[(CH_3)_2As]_2S$	242.05	oil			211	sl. s.		s. al., eth.
12	β-Chlorovinyl dichlorarsine	$C_2H_2Cl_3As$	207.32	liq.	1.8648^{20}_{4}	0.1	76.1^{10}			
13	Dimethylarsine (cacodyl hydride)	$(CH_3)_2AsH$	106.00	col. liq., ign. in air	1.213^{29}		35.6^{747}			s. al., eth., chl., bz., CS_2
14	Dimethylarsinic acid (cacodylic acid)	$As(CH_3)_2O.OH$	138.00	odorl., col. pr.		200		82.9^{22}	v. s.	s. al.; i. eth.
15	Dimethylbromarsine (cacodyl bromide)	$(CH_3)_2AsBr$	184.90	yel. oil			130			
16	Dimethylchlorarsine (cacodyl chloride)	$(CH_3)_2AsCl$	140.44	col. liq., infl.	>1	<−45	106.5–107, (109)	i.		v. s. al.; i. eth.
17	Dimethylcyanoarsine (Cacodylcyanide)	$(CH_3)_2AsCN$	131.01	lustr., col. pr., very pois.		33	140	sl. s.		s. al., eth.
18	Diphenylarsinic acid	$(C_6H_5)_2AsO(OH)$	262.14	wh. need.		174	subl. 190–200	s.		s. al., alk.; i. eth.; bz.
19	Diphenylchloroarsine	$(C_6H_5)_2AsCl$	264.59	pa. yel. liq.	1.4223^{26}		333 (in CO_2)	i.		s.abs.al.;eth.; bz.; NH_4OH.
20	Ethylarsinedisulfide	$C_2H_5AsS_2$	168.11	yel. oil	1.836^{24}			i.		s. bz.; chl.; i. al., eth.
21	Ethylarsonic acid	$C_2H_5AsO(OH)_2$	154.00	need.		99.9		70^{25}		39.4^{25} 95 % al.
22	*p*-Hydroxybenzenearsonic acid (Phenol-*p*-arsonic acid)	$HO(C_6H_4)AsO(OH)_2$	218.04	monocl. pr.		d. 174		s.	s.	s. al.; acet. a.; dil. a.; sl. s. acet. v. sl. s. eth.
23	Methylarsine	CH_3AsH_2	91.97	col. liq.			2	0.00085		s. al., eth.
24	Methyldichloroarsine	CH_3AsCl_2	160.86	col. liq.			133	d.		
25	Phenylarsine	$C_6H_5AsH_2$	154.04	col. oil	1.356^{25}_{25}		148			sl. a., eth., CS_2
26	Phenylarsonic acid (benzenearsonic acid)	$C_6H_5AsO_3H_2$	202.04	col. pr.	1.760	158–62 d.		3.36^{28}	31.6^{84}	18.4^{25} 95 %al
27	Phenylcyclotetramethylenearsine	$C_4H_8AsC_6H_5$	208.14	col. oil	1.2794^{20}_{4}		128.5^{15-16}	sl. s.		s. al., eth.
28	Phenyldimethylarsine	$(C_6H_5)(CH_3)_2As$	182.10	col. liq.			200	i.		s. al., bz.
29	Sodiumarsenophenylglycine	$(AsC_6H_5NHCH_2COONa)_2$	494.12	liq.		d.		s.		
30	Tetraethyldiarsine (ethyl cacodyl)	$[(C_2H_5)_2As]_2$	266.09	oil			185–90	i.		s. al, eth..
31	Tetramethylbiarsine (Tetramethylbiarsine)	$[As(CH_3)_2]_2$	209.98	col.-yel. oily liq., highly poisonous	1.447^{15}	−5	163	sl. s.		s. al., eth.
32	Tribenzylarsine	$(C_6H_5CH_2)_3As$	348.32	col. monocl.		104		i.		s. eth., bz.; sl. s. al.
33	Tri(β-chlorovinyl)arsine	$(CHClCH)_3As$	259.40	col. liq.	3–4	$151–158^{28}$	$151–158^{25}$	i.		i. al., dil. a.
34	Triethylarsine (arsenic triethyl)	$As(C_2H_5)_3$	162.11	col. liq. n^{20}_{D} 1.467	1.152		140^{736}	i.		
35	Trimethylarsine (arsenic trimethyl)	$As(CH_3)_3$	120.03	col. liq.	1.124		70	sl. s.		s. eth.
36	Triphenylarsine (arsenic triphenyl)	$As(C_6H_5)_3$	306.24	wh. need. or rhomb. pl., 1.6139^{48}	1.2225^{48}	60–60.5(57)	>360 (In CO_2)	i.		v. s. eth., bz.; sl. s. cold. al.
37	Triphenylarsinedihydroxide	$(C_6H_5)_3As(OH)_2$	340.26	col. hex. pr.		115–116		s.		s. al.; u. sl. s. eth.
38	Triphenylarsinesulfide	$(C_6H_5)_3AsS$	338.31	lustr. need.				i.		sl. s. hot al.; i. eth., a.

No.	Name	Formula	Mol. wt.	Crystalline form, color and index of refraction	Sp. gr. or density	Melting point, °C	Boiling point, °C	Solubility in grams per 100 ml of Cold water	Hot water	Alcohol, acids, etc.
	Beryllium									
1	Di-*n*-butylberyllium	$Be(C_4H_9)_2$	123.24	col. liq.			170[25]	d.	d.	
2	Diethylberyllium	$Be(C_2H_5)_2$	67.14	col. liq.		12	110[15]	d. to C_2H_6		
3	Dimethylberyllium	$Be(CH_3)_2$	39.09	wh. need.			subl. 200	d. to CH_4		
4	Dipropylberyllium	$Be(C_3H_7)_2$	95.19	liq.		<−17	245			
	Bismuth									
1	Bismuthethyl-camphorate	$C_{36}H_{57}BiO_{12}$	890.83	wh. amorph. solid		61–67		i.	i.	s. eth., chl.
2	Diphenylbismuthyl iodide	$(C_6H_5)_2BiI$	490.10	yel. flakes		132–134		i.		s. acet.
3	Methylbismuthine	CH_3BiH_2	226.03	liq. (exist ?)	2.30[18]		110	i.	i.	s. al., eth.
4	Triethylbismuthine (bismuth triethyl)	$Bi(C_2H_5)_3$	296.17	liq.	1.82		107[79]	i.		s. al., eth.
5	Trimethylbismuthine (bismuth trimethyl)	$Bi(CH_3)_3$	254.09	liq.	2.300[18]		110	i.	i.	s. al., eth.
6	Trinitrotriphenyl-bismuth dinitrate	p-$(NO_2C_6H_4)_3Bi(NO_3)_2$	699.30	pa. yel. cr.		d. 140–147				s. eth. ac., gla. ac. a.
7	Triphenylbismuth diacetate	$(C_6H_5)_3Bi(CH_3COO)_2$	558.39	wh. cr.		152–153				s. acet.
8	Triphenylbismuthine (bismuth triphenyl)	$Bi(C_6H_5)_3$	440.30	monocl., tan. cr.	1.585	78				v. s. chl.; s. eth., acet.; sl. s. al.
9	Tri-*n*-propylbismuth	$(C_3H_7)_3Bi$	338.25	col. liq.	1.621		86–87[8]			s. eth.
10	Tri-*m*-tolylbismuth dichloride	m-$(CH_3C_6H_4)_3BiCl_2$	553.29	wh. cr. (from acet.)		132–133				s. acet., al-chl.
	Boron									
1	Aminophenylboric acid (*m*)	$(NH_2C_6H_4)B(OH)_2$	136.95	wh. hex. pl.		d.	d.	sl. s.		s. al.; sl. s. eth.
2	Amylboric acid (*n*)	$(C_5H_{11})B(OH)_2$	115.97	col. fl.		93–4, d.	d.		s.	s. eth., dichloroethane
3	Anisylboric acids, *o,m,p* (methoxyphenyl-boric acids)	$CH_3OC_6H_4B(OH)_2$	151.96	wh. cr.		d.	d.	sl. s.		s. al., eth., bz.
4	Benzeneboronic acid	$C_6H_5B(OH)_2$	121.93			215–216		s.		s. bz., CCl_4; v. s. eth., MeOH
5	Borine carbonyl	BH_3CO	41.85	col. unst. gas		−137.0	−63	d.	d.	
6	Butylboric acid (*n*)	$C_4H_9B(OH)_2$	101.94	col. cr.		92–4	d.	s.	s.	v. s. al., eth., chl., acet., acet. a. and esters; sl. s. bz., CCl_4, pet. eth.
7	Butylboric acid (*tert*)	$C_4H_9B(OH)_2$	101.94	wh. cr.		105 d.	d.	s.		s. eth.
8	*n*-Butylphenylchloro-boronite	n-$C_4H_9O(C_6H_5)BCl$	196.49	n_D^{20} 1.4996	1.021_4^{20}	−32	65[0.4]			
9	Diethoxyboron chloride	$(C_2H_5O)_2BCl$	136.29	col. liq.			112.3	d.	d.	
10	Diisoamyloxyboron chloride	$(C_5H_{11}O)_2BCl$	220.55	col. liq.			110–15[14]	d.	d.	
11	Dimethoxyborine	$(CH_3O)_2BH$	73.89	col. liq., unst.		−130.6	25.9	d.	d.	
12	Dimethoxyboron chloride	$(CH_3O)_2BCl$	108.33	col. liq.		−87.5	74.7	d.	d.	
13	Dimethylboric acid (dimethylhydroxy-borine)	$(CH_3)_2BOH$	57.89	col. liq.			0[36]	v. s.		
14	Dimethylboric anhydride	$(CH_3)_2BOB(CH_3)_2$	97.76	col.		−37.3	43	hyd.	hyd.	
15	Dimethylborine trimethylamine	$(CH_3)_3NBH(CH_3)_2$	101.00	col. liq.		−18.0	d. 172	d.	d.	s. eth.
16	Dimethylboron bromide	$(CH_3)_2BBr$	120.79	col. liq. or gas		−123.4	22	d.		
17	Dimethylboron iodide	$(CH_3)_2BI$	167.79	col. liq.		−110.7	65	d.		
18	Dimethyldiborane(1,1) (unsym.)	$B_2H_4(CH_3)_2$	55.72	col. gas		−150.2	−2.6	d.	d.	
19	Dimethyldiborane(1,2) (sym.)	$B_2H_4(CH_3)_2$	55.72	col. unst. gas		−125	4.9	d.	d.	

PHYSICAL CONSTANTS OF ORGANOMETALLIC COMPOUNDS (Continued)

No.	Name	Formula	Mol. wt.	Crystalline form, color and index of refraction	Sp. gr. or density	Melting point, °C	Boiling point, °C	Solubility in grams per 100 ml of: Cold water	Hot water	Alcohol, acids, etc.
	Boron									
20	Dimethyltriborine triamine (*B*)	$(CH_3)_2B_3N_3H_4$	108.55	col. liq.		−48	107	hyd.	hyd.	
21	Dimethyltriborine triamine (*N*)	$(CH_3)_2B_3N_3H_4$	108.55	col. liq.			108	hyd.	hyd.	
22	Dimethyltriborine triamine (*N-B*)	$(CH_3)_2B_3N_3H_4$	108.55	col. liq.			124	hyd.	hyd.	
23	Diphenylboric acid (diphenylhydroxyborine)	$(C_6H_5)_2BOH$	182.03	col. radiating cr.		264–67	$215–35^{17}$	i.	i.	s. eth., al., pet. eth.
24	Diphenylboron bromide	$(C_6H_5)_2BBr$	244.93	col. visc. liq. or cr.		25	$150–60^{6}$	d.	d.	s. bz.
25	Diphenylboron chloride	$(C_6H_5)_2BCl$	200.48	col. visc. liq.			271	d.	d.	s. bz., pet. eth.
26	Di-*p*-tolylboric anhydride	$(C_7H_7)_2BOB(C_7H_7)_2$	402.16	wh. powd.		78		i.		s. al., eth., bz.
27	Ethoxyboron dichloride	$C_2H_5OBCl_2$	126.78	col. liq.			77.9	d.	d.	
28	Ethyl boric acid	$(C_2H_5)B(OH)_2$	73.89	wh. cr.		subl. 40		s.	s.	s. al., eth.
29	Furanylboric acid (*β*)	$(C_4H_3O)B(OH)_2$	111.89	wh. cr.		110 d.	d.	s.		v. s. eth., al., acet.; sl. s. bz., tol.
30	Hexylboric acid (*n*)	$C_6H_{13}B(OH)_2$	130.00	wh. cr.		88–90, d.	d.	sl. s.		s. eth.
31	Isobutylboric acid	$C_4H_9B(OH)_2$	101.94	col. cr.		106–12, d.	d.	s.		s. eth., dichloroethane
32	Isopropylamineborine	$(CH_3)_2CHNH_2BH_3$	72.95	wh. cr.	0.829^{25}_{4}	62	d. 75	sl. s.	sl. s.	s. al., eth.
33	Methoxyboron dichloride	CH_3OBCl_2	112.75	col. liq.		−15	58.0	d.	d.	
34	Methoxyboron difluoride	CH_3OBF_2	79.84	col. liq.	$1.417^{35.5}$; $1.354^{76.5}$	41.9	86	d.	d.	
35	Methylboric acid	$CH_3B(OH)_2$	59.86	wh. pl.		d.	d.	sl. s.		s. al., eth.
36	Methylborine trimethylammine	$(CH_3)_3NBH_2CH_3$	86.97	col. liq.		0.8	177	d.	d.	s. eth.
37	Methyldiborane	$B_2H_5CH_3$	41.70	col. very unst. gas			-80^{60}; d. appr. −20	d.	d.	
38	Methyltriborine triamine (*B*)	$CH_3B_3N_3H_5$	94.53	col. liq.		−59	87	hyd.	hyd.	
39	Methyltriborine triamine (*N*)	$CH_3B_3N_3H_5$	94.53	col. liq.			84	hyd.	hyd.	
40	Nitrophenylboric acids, *o*, *m*, *p*	$NO_2C_6H_4B(OH)_2$	166.93	yel. need. or pr.		d.	d.	sl. s.	s.	s. al., eth.
41	Phenylboron dibromide	$C_6H_5BBr_2$	247.74	col. cr.		34	100^{20}	d	d.	s. bz.
42	Phenylboron dichloride	$C_6H_5BCl_2$	158.82	col. liq. n_D^{20} 1.5385	1.194^{20}_{4}	7	175	d.	d.	s. bz.
43	Sodium tri-*α*-naphthyl boride	$Na_2B(C_{10}H_7)_3$	438.29	bl. cr. (purple in dil. soln.)			d.	d.		s. eth.; sl. s. lgr.
44	Tetramethoxydiborine	$(CH_3O)_4B_2$	145.76	col. liq.		−24	d. 93; 21^{44}	d.	d.	
45	Tetramethylammoniumborane	$(CH_3)_4NBH_4$	88.99	wh. cr.	0.813^{25}_{4}		>150	48^{20}	61^{40}	0.39; 1.04 abs. al. 95 % al.
46	Tetramethyldiborane (1,1,2,2)	$B_2H_2(CH_3)_4$	83.78	col. liq.		−72.5	68.6	d.	d.	
47	Tetramethyltriborine triamine (*N-B-B′-B″*)	$(CH_3)_4B_3N_3H_2$	136.61	col. liq.			158	hyd.	hyd.	
48	Thiophenylboric acid (*α*) ("thienylboric" acid)	$(C_4H_3S)B(OH)_2$	127.96	col. starformed need.		134	d.	s.		s. eth., al., acet., bz., CCl_4
49	Tribenzylborine	$B(C_6H_5CH_2)_3$	284.21	prismatic need. or col. oily liq.		47	230^{13}	i.		v. s. al., bz.; sl. s. eth.
50	Tri-*n*-butylborine	$B(C_4H_9)_3$	182.16	col. mobile liq.			$90–1^{9}$; $108–110^{20}$	i.	i.	v. s. eth., al.
51	Tri-*tert*-butylborine	$B(C_4H_9)_3$	182.16	col. mobile liq.		glass at low temp.	71^{12}	i.		s. eth.
52	Tri-*n*-butyltriborine trioxane (*n*-butyl boron oxide)	$(C_4H_9)_3B_3O_3$	251.78	col. liq.			154^{20}	hyd.	hyd.	v. s. eth.

No.	Name	Formula	Mol. wt.	Crystalline form, color and index of refraction	Sp. gr. or density	Melting point, °C	Boiling point, °C	Solubility in grams per 100 ml of: Cold water	Hot water	Alcohol, acids, etc.
	Boron									
53	Tri-*tert*-butyltriborine trioxane (*tert*-butyl boric oxide)	$(C_4H_9)_3B_3O_3$	251.78	col. liq.	...	20	66–8[5]	hyd.	hyd.	s. eth.
54	Trichloroborine dimethyletherate	$(CH_3)_2OBCl_3$	163.24	col. cr.	...	d. 76	d.	d.	...	...
55	Trichloroborine trimethylammine	$(CH_3)_3NBCl_3$	176.28	col. cr.	...	243	...	...	s.	s. al.
56	Tricyclohexylborine (boron tricyclohexyl)	$B(C_6H_{11})_3$	260.27	col. interlocking cr.	...	100	194[15]	i.	...	s. eth.
57	Triethyl borate (triethoxyborine)	$B(OC_2H_5)_3$	146.00	col. liq., 1.381	0.8746[10]; 0.864[26.5]	−84.8	117.4, (120)	d.	...	...
58	Triethylboron (triethylborine)	$B(C_2H_5)_3$	98.00	col. liq.	0.6961[23]	−92.9	0[12.5]	i.	i.	s. al., eth.
59	Tri-*n*-hexyltriborine trioxane (hexylboric oxide)	$(C_6H_{13})_3B_3O_3$	335.94	col. liq., 1.4323[20]	0.8876	...	178–82[24]	hyd.	hyd.	s. org. solv.
60	Triisoamyl borate (triisoamyloxyborine)	$B(OC_5H_{11})_3$	272.24	liq., 1.421	0.872[0]	...	255	...	...	...
61	Triisoamylborine	$B(C_5H_{11})_3$	224.24	col. mobile liq., 1.43207[22.6]	0.76	...	119[14]	i.	...	s. eth.
62	Tri-*p*-anisylborine	$B(CH_3OC_6H_4)_3$	332.21	wh. need.	...	128	...	i.	i.	s. al., eth., bz.
63	Triisobutyl borate (triisobutoxyborine)	$B(OC_4H_9)_3$	230.16	liq., 1.408	0.864[0]	...	212	...	...	...
64	Triisobutylborine	$B(C_4H_9)_3$	182.16	col., mobile liq., 1.41882[22.8]	0.74	...	188.86[20]	i.	i.	s. eth.
65	Trimethoxyboroxine	$(BO)_3(OCH_3)_3$	173.55	col. liq., 1.3986	1.216[25]	−30	(flashpt. 37)	...	...	v. s. bz., tol., CCl_4
66	Trimethylamminoborine	$(CH_3)_3NBH_3$	72.94	col. hex., columns or need.	0.792_4^{25}	94	172	i.	sl. s.	s. eth., al., bs., NH_3
67	Trimethyl borate (trimethoxyborine)	$B(OCH_3)_3$	103.92	col. liq. n_D^{20} 1.3610	0.915; 0.9205[24.3]	−29	68.7, (65)	d.	...	s. al., eth.
68	Trimethylboron (trimethylborine)	$B(CH_3)_3$	55.92	col. gas	1.9108 g/1, 0.625^{-100}	−161.5	−20.2	v. sl. s.	...	v. s. al., eth.
69	Trimethyldiborane (1,1,2)	$B_2H_3(CH_3)_3$	69.75	col. liq.	...	−123	45.5	d.	...	...
70	Trimethyltriborine triamine (*B*)	$(CH_3)_3B_3N_3H_3$	122.58	col. cr. or liq.	...	31.5	129	hyd.	hyd.	...
71	Trimethyltriborine triamine (*N*)	$(CH_3)_3B_3N_3H_3$	122.58	col. liq.	...	...	134	hyd.	hyd.	...
72	Trimethyltriborine triamine (*N-B-B′*)	$(CH_3)_3B_3N_3H_3$	122.58	col. liq.	...	...	139	hyd.	hyd.	...
73	Trimethyltriborine trioxane (methylboric anhydride)	$(CH_3)_3B_3O_3$	122.54	col. mobile liq.	...	−37vc,	79.3[755]	hyd.	hyd.	s. eth.
74	Tri-*β*-naphthyl borate	$B(C_{10}H_7O)_3$	440.31	col. leaflets	...	115	...	d.	...	s. bz.
75	Tri·*α*-naphthylborine	$B(C_{10}H_7)_3$	392.31	col. need.	...	203	d.	i.	...	sl. s. eth., al.; v. s. bz., CCl_4, chl., CS_2
76	Triphenyl borate (triphenoxyborine)	$(C_6H_5O)_3B$	280.13	col. cr.	...	*ca.* 35	>360	d.	...	s. eth., bz.
77	Triphenylborine ammine*	$(C_6H_5)_3BNH_3$	259.16	col. cr.	...	d. 216	d.	...	...	s. al.; sl. s. bz.
78	Triphenylboron	$B(C_6H_5)_3$	242.13	hex. need.	...	136	203[15]	d.	...	d. al.; s. bz.
79	Tripropyl borate (tripropoxyboron)	$B(OC_3H_7)_3$	188.08	liq.	liq. 0.867[16]	...	175	...	...	...
80	Tri-*n*-propylborine	$B(C_3H_7)_3$	140.08	col., mobile liq., 1.41352[22.8]	0.725	...	156, 60[20]	i.	...	s. eth.
81	Tri-*sec*-propylborine	$B(C_3H_7)_3$	140.08	col., mobile liq.	...	...	148–54, 33–5[12]	i.	...	s. eth.

* This compound is the prototype of numerous stable complex compounds formed from organic amines and tri-aryl-borines.

No.	Name	Formula	Mol. wt.	Crystalline form, color and index of refraction	Sp. gr. or density	Melting point, °C	Boiling point, °C	Solubility in grams per 100 ml of: Cold water	Hot water	Alcohol, acids, etc.
	Boron									
82	Tri-*p*-tolylborine	$B(CH_3C_6H_4)_3$	284.21	separate wh. cr.		175	233^{12}	i.		v. s. bz.; sl. s. eth.
83	Tri-*p*-xylylborine	$B(CH_3C_6H_3CH_3)_3$	326.29	col. bushed need.		147	221^{12}	i.		v. s. bz., chl., CCl_4; sl. s. eth.
	Cadmium									
1	Dibutylcadmium	$Cd(C_4H_9)_2$	226.63	oil n_D 1.5155	$1.3056^{19.5}$	−48	$103.5^{12.5}$	d.	d.	
2	Diethylcadmium	$Cd(C_2H_5)_2$	170.52	oil	$1.6564^{18.1}$	−21	64^{19}	d.	d.	v. s. eth.
3	Diisoamylcadmium	$Cd(C_5H_{11})_2$	254.69	oil n_D 1.5039	1.2210^{19}	−115	121.5^{15}			
4	Diisobutylcadmium	$Cd(C_4H_9)_2$	226.63	oil n_D 1.4997	1.2693^{18}	−37	90.5^{20}	d.	d.	
5	Dimethylcadmium	$Cd(CH_3)_2$	142.47	oil	$1.9846^{17.9}$	−4.5	105.5^{758}	d.	d.	
6	Dipropylcadmium	$Cd(C_3H_7)_2$	198.58	oil n_D 1.5291	$1.4201^{17.6}$	−83	$84^{21.5}$	d.	d.	
	Calcium									
1	Dianilinecalcium	$Ca(NHC_6H_5)_2$	224.32	wh. cr.		d.		d.		i. eth., bz., lgr.
2	Ethylcalcium iodide	C_2H_5CaI	196.05	amor. powd.				d.		sl. s.
3	Glycocollcalcium	$(CH_2NHCOO)Ca$	113.13	cr.				s.		
	Chromium									
1	Anilinechromium tricarbonyl	$H_2NC_6H_5Cr(CO)_3$	229.16	yel. cr.		173–5	subl. 110–130			s. org. solv. HCl
2	Anisolechromium tricarbonyl	$CH_3OC_6H_5Cr(CO)_3$	244.17	yel. cr.		80–92	subl. 70–80 (vac.)			
3	Benzoic acid chromium tricarbonyl	$HO_2CC_6H_5Cr(CO)_3$	258.15	or. cr.		201–2	subl. 150 (vac.)			
4	Benzenechromium tricarbonyl	$C_6H_6Cr(CO)_3$	214.14	yel. monocl.		165.5–166.5 (162–3)	subl. 60–90			s. bz., s. org. solv.
5	Bis-cyclopentadienyl-chromium	$(C_5H_5)_2Cr$	182.19	red cr.		170–2	subl. 75–90 (vac.)	d.		d. in CS_2, CCl_4
6	Bis-diphenyl chromium iodide ("Tetraphenyl-chromium" iodide)	$(C_6H_5C_6H_5)_2Cr^+I^-$	487.33	or. yel.		177–8 (178)		s.		s. chl.; i. pet. eth.
7	Chlorobenzene-chromium tricarbonyl	$ClC_6H_5Cr(CO)_3$	248.59	yel.		96–8; 150–5 d.	50–8 (vac.)			
8	Cyclopentadienyl-chromium dinitrosomethyl	$C_5H_5Cr(NO)_2CH_3$	192.14	grn.		83.0				s. org. solv.
9	Dimethylanilinechromium tricarbonyl	$(CH_3)_2NC_6H_5Cr(CO)_3$	257.21	cr.		146–146.5				s. org. solv.
10	Diphenylbenzene-chromium iodide ("Triphenyl chromium" iodide)	$C_6H_5C_6H_5Cr(I)(C_6H_5)$	410.22	or. yel. pl.		110–2 (146–8)		s.		sl. s. et., al., chl., *m*-dinitrobenzene.
11	Mesitylenechromium tricarbonyl	$(CH_3)_3C_6H_3Cr(CO)_3$	256.22	yel. cr.		177–8 (165, 170 d)	subl. 80–100			s. org. solv.
12	Methylbenzoate chromium tricarbonyl	$CH_3O_2CC_6H_5Cr(CO)_3$	272.18	red cr.		93–5	80–90 (vac.)			
13	Phenolchromium tricarbonyl	$HOC_6H_5Cr(CO)_3$	230.14	yel.		193–240				
14	Thiophenechromium tricarbonyl	$C_4H_4SCr(CO)_3$	220.17	red cr.			subl. 85–95 (vac.)			s. bz., eth, pet. eth.
15	Toluenechromium tricarbonyl	$CH_3C_6H_5Cr(CO)_3$	228.17	yel.		82–3 (80° d.)	subl.			s. org. solv.
16	*m*-Xylenechromium tricarbonyl	$(CH_3)_2C_6H_4Cr(CO)_3$	242.20	yel.		104–5	subl. 65			
17	*o*-Xylenechromium tricarbonyl	$(CH_3)_2C_6H_4Cr(CO)_3$	242.20	yel.		88–90	subl. 65			
18	*p*-Xylenechromium tricarbonyl	$(CH_3)_2C_6H_4Cr(CO)_3$	242.20	yel. cr.		97–8	subl. 65			
	Cesium									
1	Monocesiumacetylide	CsC_2H	157.94	col. cr.		300		d.		s. NH_3; i. bz.
	Cobalt									
1	Bis-dimethylglyoxime cobaltochloride	$HON{:}C(CH_3)C(CH_3){:}NOH.Co$ $HON{:}C(CH_3)C(CH_3){:}NOCl_2$	361.07	lt. grn. cr.				s.		s. al.

No.	Name	Formula	Mol. wt.	Crystalline form, color and index of refraction	Sp. gr. or density	Melting point, °C	Boiling point, °C	Solubility in grams per 100 ml of Cold water	Hot water	Alcohol, acids, etc.
	Cobalt									
2	Cobalt(ous) hexa-methylenetetramine	$CoCl_2C_6H_{12}N_4$	270.03	ultramarine blue				s.		
3	Cobalt(ous) hydroxy-quinone	$Co(C_{10}H_5O_3)_2$	405.23	ruby red		d. 210–15				
	Copper									
1	Cyclopentadienyl-triethylphosphine copper	$C_5H_5CuP(C_2H_5)_3$	246.80	wh. cr.		127–8		i.		s. dil. HCl, pet. eth., eth., bz, pyr.
2	Diazoamino-benzene(ous)	$CuN_3(C_6H_5)_2$	259.77	or. cr.		d. 270		i.		s. bz.; i. al., lgr.
	Gallium									
1	Dimethylgallium amide	$Ga(CH_3)_2NH_2$	115.81	wh. cr.			subl. 60 vac.			
2	Dimethylgallium chloride monammine	$Ga(CH_3)_2Cl.NH_3$	152.27	wh. cr.		54		d.	d.	v. s. NH_3; s. eth.
3	Dimethylgallium chloride diammine	$Ga(CH_3)_2Cl.2NH_3$	169.30	wh. cr.		112		d.	d.	v. s. NH_3; i. eth.
4	Methylgallium dichloride	$Ga(CH_3)Cl_2$	155.66	wh. cr.		75		d.		v. s. eth.
5	Methylgallium dichloride monammine	$Ga(CH_3)Cl_2.NH_3$	172.69	wh. cr.				d.		i. eth.
6	Methylgallium dichloride pentammine	$Ga(CH_3)Cl_2.5NH_3$	240.81	wh. cr.		d. >80		d.		i. NH_3
7	Triethylgallium	$Ga(C_2H_5)_3$	156.91	col. liq.	1.0576^{30}	−82.3	142.6	d.		s. eth.
8	Triethylgallium monammine	$Ga(C_2H_5)_3.NH_3$	173.94	col. liq.				d.		
9	Triethylgallium monoetherate	$Ga(C_2H_5)_3.(C_2H_5)_2O$	231.03	col. liq.				d.		s. eth.
10	Trimethylgallium	$Ga(CH_3)_3$	114.83	col. liq.		−19	55.7 ± 2^{762}	d.		s. eth., NH_3
11	Trimethylgallium monammine	$Ga(CH_3)_3.NH_3$	131.86	wh. cr.		31	subl. vac.	d.		s. eth., NH_3; i. pet. eth.
12	Trimethylgallium monoetherate	$Ga(CH_3)_3.(C_2H_5)_2O$	188.95	col. liq.		<−76	99	d.		s. NH_3, eth.
	Germanium									
1	Amyltriphenyl-germanium (*n*)	$Ge(C_5H_{11})(C_6H_5)_3$	375.05	col. pl.		42–3	i.	i.	i.	v. s. bz., pet. eth.; sl. s. me. al.
2	Benzyltriphenyl-germanium	$Ge(CH_2C_6H_5)(C_6H_5)_3$	395.04	col. pl.		82.5–3.5		i.	i.	v. s. bz., pet. eth., chl.; sl. s. iso-propyl al.; i. me. al.
3	Bis-acetylacetone germanium dibromide	$[CH(CCH_3O)_2]_2GeBr_2$	430.63	col. micr. cr.		226				sl. s. h. acetyl acet.; i. org. solv.
4	Bis-acetylacetone germanium dichloride	$[CH(CCH_3O)_2]_2GeCl_2$	341.72	col. pr.		240d.				sl. s. org. solv.
5	Bis(5-oxy-2,8-dithio-octane) germanium	$[(SCH_2CH_2)_2O]_2Ge$	354.06	col. cr.		159.0–.5				s. bz., abs. al.
6	Bis-propionylacetone germanium dichloride	$[CHC_2(C_2H_5)(CH_3)O_2]_2GeCl_2$	369.77	wh. cr. powd.		128–9				s. c. chl.; i. pet. eth.
7	Bis-tribenzyl germanyl sulfide	$[(C_6H_5CH_2)_3Ge]_2S$	724.05	col. cr.		124				s. al., me. al., bz.; i. alk.
8	Bis-tribiphenylyl germanyl sulfide	$[(C_6H_5C_6H_4)_3Ge]_2S$	1096.48	col. cr.		238				s. org. solv.; i. alk.
9	Bis-trichlorogermanyl methane	$CH_2(GeCl_3)_2$	371.93	col. liq.			110^{18}	hyd.	hyd.	s. org. solv.
10	Bis-tricyclohexyl-germanium disulfide	$[(C_6H_{11})_3Ge]_2S_2$	708.24	col. cr.		87–8	i.			s. abs. al.
11	Bis-triethylgermanyl sulfide	$[(C_2H_5)_3Ge]_2S$	351.62	col. oily liq.			$148–50^{12}$			s. org. solv.; i. alk.
12	Bis-triphenylgermanyl sulfide	$[(C_6H_5)_3Ge]_2S$	639.88	col. cr.		138				s. org. solv.; i. alk.
13	Bis-tritolylgermanyl sulfide	$[(C_6H_4CH_3)_3Ge]_2S$	724.05	col. cr.		156–7				s. org. solv.; i. alk.

No.	Name	Formula	Mol. wt.	Crystalline form, color and index of refraction	Sp. gr. or density	Melting point, °C	Boiling point, °C	Solubility in grams per 100 ml of		
								Cold water	Hot water	Alcohol, acids, etc.
	Germanium									
14	Butyltriphenyl-germanium (*n*)	$Ge(C_4H_9)(C_6H_5)_3$	361.01	col. need.		84.5–5.5		i.	i.	v. s. pet. eth., bz., chl., eth., sl. s. isopropyl al., i. me. al.
15	Cyclopentamethylene germanium dichloride	$(CH_2)_5GeCl_2$	213.63	col. liq.			55–60^{12}			
16	Diethylcyclopenta-methylenegermanium (1,1)	$(CH_2)_5Ge(C_2H_5)_2$	200.85	col. liq.			52^{13}			
17	Diethyldiphenyl-germanium	$Ge(C_2H_5)_2(C_6H_5)_2$	284.93	col. liq.			316	i.	i.	v. s. org. solv.
18	Diethyldinylgermanium	$(C_2H_5)_2Ge)CH{:}CH_2)_2$	184.81	n_D^{20} 1.4575	1.0193_4^{20}		149.8^{760}			
19	Diethylgermanium bromide	$(C_2H_5)_2GeBr_2$	290.53	col. liq.		<−33	202	d.	d.	d. liq. NH_3; s. org. solv.
20	Diethylgermanium chloride	$(C_2H_5)_2GeCl_2$	201.62	col. liq.		−39 to −37	175	d.	d.	d. liq. NH_3; s. org. solv.
21	Diethylgermanium imine	$(C_2H_5)_2GeNH$	145.73	col. liq.			$100^{0.01}$			
22	Diethylgermanium iodide	$(C_2H_5)_2GeI_2$	384.52	col. liq.		−2 to −1	252		d.	d. liq. NH_3; s. org. solv.
23	Diethylgermanium oxide(α)	$[(C_2H_5)_2GeO]_x$	146.71	stable wh. amor. sol.		175		i.	i.	i. org. solv., liq. NH_3
24	Diethylgermanium oxide(β)	$[(C_2H_5)_2GeO]_3$	440.14	unst. col. liq.		18		i.	i.	s. org. solv.; i. liq. NH_3
25	Diphenylgermanium	$[(C_6H_5)_2Ge]_4$	907.21	wh. cr.		294–5		i.	i.	sl. s. bz., tol., chl.; i. pet. eth.
26	Diphenylgermanium dibromide	$(C_6H_5)_2GeBr_2$	386.62	col. liq.			$120^{.007}$ 205–7^{512}	hyd.	hyd.	s. org. solv.
27	Diphenylgermanium dichloride	$(C_6H_5)_2GeCl_2$	297.71	col. liq.	71	9	223^{13}	hyd.	hyd.	s. org. solv.
28	Diphenylgermanium difluoride	$(C_6H_5)_2GeF_2$	264.80	col. liq.			$100^{0.007}$	hyd.	hyd.	s. org. solv.
29	Diphenyl-*sec*-propyl-germanium bromide	$(C_6H_5)_2(C_3H_7)GeBr$	349.80	col. liq.			215–50^{13}			
30	Di-*p*-tolylgermanium dibromide	$(CH_3C_6H_4)_2GeBr_2$	414.68	lt. yel. liq.			230–33^{13}	hyd.	hyd.	s. org. solv.
31	Di-*p*-tolylphenyl-germanium bromide	$(CH_3C_6H_4)_2(C_6H_5)GeBr$	411.87	col. pr.		119				
32	Di-triphenylgermanyl methane	$[(C_6H_5)_3Ge]_2CH_2$	621.85	lt. col. pr		132–33				v. s. bz., eth., pet. eth., chl.; i. liq. NH_3, al.
33	Ethyl-*tris*-*p*-biphenylyl-germanium	$Ge(C_2H_5)(C_6H_4C_6H_5)_3$	561.27	col. cr.		154–6				
34	Ethylgermanium oxide	$(C_2H_5GeO)_2O$	251.30	wh. powd.		>300	d.	s.	s.	s. HCl, al.; i. pet. eth.
35	Ethylgermanium tribromide bromide	$C_2H_5GeBr_3$	341.38	col. liq.		<−33	200^{763}	d.	d.	d. liq. NH_3; s. bz., eth.
36	Ethylgermanium trichloride	$C_2H_5GeCl_3$	208.01	col. liq.		<−33	144^{762}	d.	d.	d. liq. NH_3; s. bz., eth.
37	Ethylgermanium trifluoride	$C_2H_5GeF_3$	158.65	col. liq.		−16.5 to −15.5	112^{750}	d.	d.	d. liq. NH_3; s. bz., eth.
38	Ethylgermanium triiodide	$C_2H_5GeI_3$	482.37	yel. liq.		−2.5 to −1.5	281^{755} d. >350	d.	d.	d. liq. NH_3; s. bz., eth.
39	Ethylphenyldi-*p*-tolyl-germanium	$Ge(C_2H_5)(C_6H_5)(C_6H_4CH_3)_2$	361.03	wh. cr.		55				
40	Ethyl-*sec*-propyl-diphenylgermanium	$Ge(C_2H_5)(C_3H_7)(C_6H_5)_2$	298.95	liq.		175–90				
41	Ethyltribenzyl-germanium	$Ge(C_2H_5)(CH_2C_6H_5)_3$	375.05	col. cr.		56–7				s. meth. al.
42	Ethyltriphenyl-germanium	$Ge(C_2H_5)(C_6H_5)_3$	332.97	col. sld.		78.0–.5		i.	i.	s. eth., pet. eth., bz., chl., acet.; i. meth. al.
43	Hexabenzyldigermane	$[(C_6H_5CH_2)_3Ge]_2$	691.98	col. cr.		183–4				s. glac. acet. a.

PHYSICAL CONSTANTS OF ORGANOMETALLIC COMPOUNDS (Continued)

No.	Name	Formula	Mol. wt.	Crystalline form, color and index of refraction	Sp. gr. or density	Melting point, °C	Boiling point, °C	Solubility in grams per 100 ml of		
								Cold water	Hot water	Alcohol, acids, etc.
	Germanium									
44	Hexaethyldigermane	$[(C_2H_5)_3Ge]_2$	319.55	col. liq.		<−60	265^{758}	i.	i.	s. bz., eth.
45	Hexaphenyldigermane	$[(C_6H_5)_3Ge]_2$	607.82	wh. cr.		340		i.	i.	sl. s. h. bz., h. chl.; i. liq. NH_3, lgr.
46	Hexaphenyldigermane tribenzene	$[(C_6H_5)_3Ge]_2.3C_6H_6$	842.16	col. cr.		d. $-C_6H_6$		i.		s. bz.
47	Hexa-*p*-tolyldigermane	$[(CH_3C_6H_4)_3Ge]_2$	691.98	col. cr.		226–7				
48	Hexavinylgermanium	$(CH_2{:}CH)_6Ge_2$	307.44	n_D^{25} 1.5217	1.171_4^{25}		$55^{0.35}$			
49	Methyltriphenyl-germanium	$Ge(CH_3)(C_6H_5)_3$	318.95	trans. col. cr.		70.5–1.0		i.	i.	v. s. bz., eth., acet., chl., pet. eth.; i. c. meth. al., liq. NH_3
50	Octaphenyltrigermane	$(C_6H_5)_8Ge_3$	834.62	wh. cr.		247–8		i.	i.	s. h. bz., h. chl.
51	Phenylethyl-*sec*-propyl-germanium bromide	$(C_6H_5)(C_2H_5)[CH(CH_3)_2]GeBr$	301.76	col. oil; opt. act., *d.* & *l.* forms			$130–5^{13}$			
52	Phenylgermanium tribromide	$C_6H_5GeBr_3$	389.42	col. liq.			$120–2^{12}$	hyd.	hyd.	d. liq. NH_3; s. org. solv.
53	Phenylgermanium trichloride	$C_6H_5GeCl_3$	256.06	col. liq.			$105–6^{13}$	hyd.	hyd.	d. liq. NH_3; s. org. solv.
54	Phenylgermanium triiodide	$C_6H_5GeI_3$	530.41	wh. sol., dec. by light		55–6		hyd.	hyd.	d. liq. NH_3; s. glac. acet. a.; org. solv.
55	Phenyltri-*p*-tolyl-germanium	$Ge(C_6H_5)(C_6H_4CH_3)_3$	423.10	wh. pr.		191		i.	i.	s. org. solv.
56	Propyltriphenyl-germanium (*n*)	$Ge(C_3H_7)(C_6H_5)_3$	347.00	col. need.		86.0–.5		i.	i.	v. s. chl., bz., pet. eth.; sl. s. isopropyl al.; i. meth. al., liq. NH_3
57	Tetra-*i*-amylgermanium	$Ge(C_5H_{11})_4$	357.16	col. oily liq., $1.457^{17.5}$	0.9147_{20}^{20}		163–4			
58	Tetra-*n*-amylthio-germanium	$Ge[S(CH_2)_4CH_3]_4$	485.42	col. liq., 1.5336^{25}	1.0697^{25}		$240–1^{3-4}$			s. bz., abs. al.
59	Tetraanhydro-tetrakisdiphenyl-germanediol (*cyclo*)	$[Ge(C_6H_5)_2O]_4$	971.20	monocl. pr. & cubes		218				s. ethyl acetate, pet. eth., eth.
60	Tetrabenzylgermanium	$Ge(CH_2C_6H_5)_4$	437.13	col. sol.		107–8				
61	Tetra-*p*-biphenylyl-germanium	$Ge(C_6H_4C_6H_5)_4$	685.41	wh. need.		270–2		i.	i.	s. bz.
62	Tetra-*p*-bromophenyl-thiogermanium	$Ge(SC_6H_4Br)_4$	824.88	col. cr.		196.0–.5				s. bz., abs. al.
63	Tetra-*n*-butyl-germanium	$Ge(C_4H_9)_4$	301.06	col. oily liq.			$178–80^{733}$?			
64	Tetra-*p*-*tert*-butyl-phenylthiogermanium	$Ge[SC_6H_4C(CH_3)_3]_4$	733.71	col. tetrag.		155–6				sl. s. al., glac. acet. a.; s. pet. eth., eth., acet.; v. s. chl., bz.
65	Tetra-*n*-butylthio-germanium	$Ge[S(CH_2)_3CH_3]_4$	429.31	liq., 1.5439^{25}	1.1072^{25}		$222.5^{4.5}$			s. abs. al.
66	Tetra-*sec*-butylthio-germanium	$Ge[SCH(CH_3)(C_2H_5)]_4$	429.31	liq., 1.5497^{25}	1.1119^{25}		200.5^{4}			s. bz.
67	Tetra-*tert*-butylthio-germanium	$Ge[SC(CH_3)_3]_4$	429.31	tetrag. columns		172–73	subl. 170^{4}			s. abs. al.
68	Tetracetylthio-germanium	$Ge[SCH_2(CH_2)_{14}CH_3]_4$	1102.61	wh. cr.		50–1				v. s. chl., bz.; s. pet. eth., eth.; sl. s. acet. al., glac. acet. a.
69	Tetracyclohexylthio-germanium	$Ge(SC_6H_{11})_4$	533.46	2 cr. mod. α (stab.) tetrag	$\alpha 1.270^{15}$	84		i.		s. abs. al., pet. eth.
				(metastab.) monocl.	$\beta 1.259^{15}$	88		i.		s. abs. al., pet. eth.

No.	Name	Formula	Mol. wt.	Crystalline form, color and index of refraction	Sp. gr. or density	Melting point, °C	Boiling point, °C	Solubility in grams per 100 ml of: Cold water	Hot water	Alcohol, acids, etc.
	Germanium									
70	Tetraethoxylgermanium (tetraethyl germanate)	$Ge(OC_2H_5)_4$	252.84	col. liq.		−81	185–7			
71	Tetraethylgermanium	$Ge(C_2H_5)_4$	188.84	col. oil, $1.443^{17.5}$; 1.554^{0}; 1.439^{30}	1.198^{0}; $0.991^{24.5}_{24.5}$	−90	162.5–3.0	d.	d.	s. bz., eth., HCl
72	Tetraethylthiogermanium	$Ge(SC_2H_5)_4$	317.09	liq., 1.5886^{25}	1.2547^{25}		$164.5–5.0^{6}$			s. bz.
73	Tetra-iso-butylthiogermanium	$Ge[SCH_2CH(CH_3)_2]_4$	429.31	liq., 1.5381^{25}	1.0984^{25}		$199–200^{4-5}$			s. al.
74	Tetraisopropylthiogermanium	$Ge[SCH(CH_3)_2]_4$	373.20	liq., 1.5535^{25}	1.1478^{25}	15	$162–64^{4}$			s. abs. al.
75	Tetramethylgermanium	$Ge(CH_3)_4$	132.73	col. liq.	1.006^{0}	−88	43.4			s. al., eth., bz.
76	Tetramethylthiogermanium	$Ge(SCH_3)_4$	260.90	liq., 1.6379^{25}	1.4364^{25}	−3	$138–40^{4}$			s. al., bz.
77	Tetraphenoxygermanium	$Ge(OC_6H_5)_4$	445.01	col. oil			$210–20^{0.3}$			s. bz.
78	Tetraphenylgermanium	$Ge(C_6H_5)_4$	381.02	tetr., col.		235.7	>400	i.	i.	s. chl., bz., tol.; sl. s. eth., acet., lgr.
79	Tetra(2-phenylethyl)-germanium	$Ge(C_6H_5C_2H_4)_4$	493.23	col. cr.		56–7				s. eth., al.
80	Tetraphenylthiogermanium	$Ge(SC_6H_5)_4$	509.27	col., rhomb. cr., 1.7348, 1.782 (H green)		101.5				s. bz., abs. al., meth. al.
81	Tetra-*n*-propylgermanium	$Ge(C_3H_7)_4$	244.95	col. mob. liq., $1.451^{17.5}$	0.9539^{20}_{20}	−73	225^{746}			
82	Tetrapropylthiogermanium	$Ge(SC_3H_7)_4$	373.20	liq., 1.5612^{25}	1.1662^{25}		$191–92^{5}$			s. abs. al.
83	Tetra-*N*-pyrrylgermanium	$Ge(C_4H_4N)_4$	336.92	lt. yel. cr.		202				s. pet. eth., chl.
84	Tetra-α-thienylgermanium	$Ge(C_4H_3S)_4$	405.12	wh. need., doubly refract.		149–50		i.	i.	s. bz., tol., acet., chl., CCl_4; sl. s. al., meth. al.; i. pet. eth.
85	Tetra-*o*-tolylgermanium	$Ge(C_6H_4CH_3)_4$	437.13	wh. hex. cr.		175–6		i.	i.	s. CCl_4, bz., xylene; sl. s. h. al.; i. pet. eth., al.
86	Tetra-*m*-tolylgermanium	$Ge(C_6H_4CH_3)_4$	437.13	wh. need.		146		i.	i.	s. bz., tol., CCl_4; sl. s. meth. al.
87	Tetra-*p*-tolylgermanium	$Ge(C_6H_4CH_3)_4$	437.13	wh. rhbdr. tab.		227		i.	i.	s. bz.
88	Tetra-*p*-tolylthiogermanium	$Ge(SC_6H_4CH_3)_4$	565.38	col., rhomb. cr., 1.726, 1.771 (H green)		110–11				s. bz., abs. al.
89	Tetravinylgermanium	$(CH_2{:}CH)_4Ge$	180.77	n_D^{25} 1.4676	1.040^{25}_{4}		54^{27}			
90	Tolylgermanium tribromide(*p*)	$(CH_3C_6H_4)GeBr_3$	403.45	col. liq.		$155–6^{13}$		hyd.	hyd.	d. liq. NH_3; s. org. solv.
91	Tolylgermanium trichloride(*p*)	$(CH_3C_6H_4)GeCl_3$	270.08	col. liq.			$115–6^{12}$	hyd.	hyd.	d. liq. NH_3; s. org. solv.
92	Tolylgermanium triiodide(*p*)	$(CH_3C_6H_4)GeI_3$	544.44	col. cr., sensit. to light		72		hyd.	hyd.	d. liq. NH_3; s. org. solv.
93	Trianhydrotetrakis-diphenylgermanediol	$[HO\text{-}Ge(C_6H_5)_2\text{-}O\text{-}Ge(C_6H_5)_2]_2O$	989.23			149				s. eth. acetate
94	Tribenzylgermanium bromide	$(C_6H_5CH_2)_3GeBr$	425.90	col. cr.		145				
95	Tribenzylgermanium chloride	$(C_6H_5CH_2)_3GeCl$	381.44	col. cr.		155				

No.	Name	Formula	Mol. wt.	Crystalline form, color and index of refraction	Sp. gr. or density	Melting point, °C	Boiling point, °C	Solubility in grams per 100 ml of: Cold water	Hot water	Alcohol, acids, etc.
	Germanium									
96	Tribenzylgermanium fluoride	$(C_6H_5CH_2)_3GeF$	364.99	col. need.		96				
97	Tribenzylgermanium iodide	$(C_6H_5CH_2)_3GeI$	472.90	col. cr.		141				
98	Tribenzylgermanium oxide	$[(C_6H_5CH_2)_3Ge]_2O$	707.99			135				s. pet. eth.
99	Tri-*p*-biphenylyl-germanium bromide	$(C_6H_5C_6H_4)_3GeBr$	612.12	wh. cr.		242				s. bz.
100	Tri-*tert*-butylthio-germanium chloride	$Ge[SC(CH_3)_3]_3Cl$	375.58	col. cr.		66–7	$156–7^{3-4}$			s. al., org. solv.
101	Tri-*n*-butylvinyl-germanium	$(n\text{-}C_4H_9)_3Ge(CH{:}CH_2)$	270.99	n_D^{20} 1.4598	0.9479^{20}_4		$108–109^2$			
102	Tricyclohexyl-germanium bromide	$(C_6H_{11})_3GeBr$	401.96	col. cr.		110		hyd.	hyd.	s. al.
103	Tricyclohexyl-germanium chloride	$(C_6H_{11})_3GeCl$	357.51	col. cr.		102		hyd.	hyd.	s. meth. al.
104	Tricyclohexyl-germanium fluoride	$(C_6H_{11})_3GeF$	341.05	col. need.		92		hyd.	hyd.	s. meth. al.
105	Tricyclohexyl-germanium hydroxide	$(C_6H_{11})_3GeOH$	339.06			176–7				s. al., bz., pet. eth.
106	Tricyclohexyl-germanium iodide	$(C_6H_{11})_3GeI$	448.96	col. cr.		99–100		hyd.	hyd.	s. meth. al.
107	Triethylgermanium bromide	$(C_2H_5)_3GeBr$	239.69	col. liq.		−33	190.9	hyd.	hyd.	s. bz., eth., chl., CCl_4
108	Triethylgermanium chloride	$(C_2H_5)_3GeCl$	195.23	col. liq.		<−50	175.9	hyd.	hyd.	s. bz., eth., chl., CCl_4
109	Triethylgermanium fluoride	$(C_2H_5)_3GeF$	178.77	col. liq.			149.0^{751}	hyd.	hyd.	s. bz., eth., chl., CCl_4
110	Triethylgermanium hydride	$(C_2H_5)_3GeH$	160.78	col. liq.			124.4^{751}	i.	i.	s. bz., eth.; i. liq. NH_3
111	Triethylgermanium-imine	$[(C_2H_5)_3Ge]_2NH$	334.57	col. liq.			$100^{0.1}$	hyd.	hyd.	s. bz., eth., CCl_4, chl.; i. liq. NH_3
112	Triethylgermanium-iodide	$(C_2H_5)_3GeI$	286.68	col. liq.		<−50	212.3	hyd.	hyd.	s. bz., eth., chl., CCl_4; i. liq. NH_3
113	Triethylgermanium oxide	$[(C_2H_5)_3Ge]_2O$	335.55	col. liq.		<−50	253.9	i.	i.	s. $C_2H_5NH_2$, bz., eth. org. solv.; i. liq. NH_3
114	Triethylphenyl-germanium	$Ge(C_2H_5)_3(C_6H_5)$	236.88	col. liq.			$116–7^{13}$	i.		s. org. solv.
115	Triethyl-*p*-tolyl-germanium	$Ge(C_2H_5)_3CH_3C_6H_4$	250.91	col. liq.			$125–6^{12}$	i.		s. org. solv.
116	Triethyl-2,2,2,-tri-phenyldigermane-(1,1,1)	$(C_2H_5)_3GeGe(C_6H_5)_3$	463.69	rhomb. cr.		89.5–90.5		i.		v. s. bz., chl.; s. pet. eth., al.; sl. s. meth. al.
117	Triethylvinyl-germanium	$(C_2H_5)_3Ge(CH{:}CH_2)$	186.82	n_D^{20} 1.4501	1.0048^{20}_4		61^{28}			
118	Trimethylgermanium bromide	$Ge(CH_3)_3Br$	197.60	col. oily liq., 1.4705	1.544^{18}_{40}	−25	113.7	d.	d.	s. org. solv.
119	Trimethylphenyl-germanium	$Ge(CH_3)_3(C_6H_5)$	194.80	col. liq.			182–3	i.		s. org. solv.
120	Trimethylstannyl-tri-phenylgermanium	$(CH_3)_3SnGe(C_6H_5)_3$	467.71	wh. cr.		88		i.		s. pet. eth., CCl_4, chl., bz.; sl. s. al.; i. liq. NH_3
121	Triphenylanisyl-germanium	$Ge(C_6H_5)_3(CH_3OC_6H_4)$	411.04	wh. sld.		158–9				s. al., glac. acet. a.
122	Triphenyldimethyl-aminophenyl-germanium	$Ge(C_6H_5)_3C_6H_4N(CH_3)_2$	424.09	wh. need.		140–1				
123	Triphenylgermanium amide	$(C_6H_5)_3GeNH_2$	319.93	wh. ppt.		d. $-NH_3$		d.	d.	i. liq. NH_3

No.	Name	Formula	Mol. wt.	Crystalline form, color and index of refraction	Sp. gr. or density	Melting point, °C	Boiling point, °C	Solubility in grams per 100 ml of: Cold water	Hot water	Alcohol, acids, etc.
	Germanium									
124	Triphenylgermanium bromide	$(C_6H_5)_3GeBr$	383.82	hex. col.		138.7		i.	hyd.	s. bz., chl.; sl. s. lgr.
125	Triphenylgermanium chloride.	$(C_6H_5)_3GeCl$	339.36	wh. cr.		117–8	285^{12}	i.	hyd.	s. bz., eth.
126	Triphenylgermanium fluoride	$(C_6H_5)_3GeF$	322.91	wh. cr.		76.6		i.	hyd.	v. s. bz., eth., lgr., chl.; i. liq. NH_3
127	Triphenylgermanium hydride	$(C_6H_5)_3GeH$	304.92	wh. cr. (two forms)		α, 47; β, 27		i.	i.	v. s. bz., tol., eth., chl., CCl_4; sl. s. liq. NH_3
128	Triphenylgermanium hydroxide	$(C_6H_5)_3GeOH$	320.92	wh. cr.		134.2		i.	i.	s. bz., chl.; sl. s. lgr.
129	Triphenylgermanium iodide	$(C_6H_5)_3GeI$	430.81	wh. cr.		157		hyd.	hyd.	s. bz., eth.
130	Triphenylgermanium oxide	$[(C_6H_5)_3Ge]_2O$	623.82	col. pl.		183–4		i.	i.	s. bz., lgr., eth.
131	Triphenylgermanium-sodium	$Ge(C_6H_5)_3Na$	326.90	lt. yel.		v. high		d.	d.	v. s. liq. NH_3; sl. s. eth., bz.
132	Triphenylgermanium-sodiumoxide	$Ge(C_6H_5)_3ONa$	342.90	wh. sld.		high		d.	d.	i. liq. NH_3
133	Triphenylgermanium-sodiumtriammine	$Ge(C_6H_5)_3Na.3NH_3$	377.99	yel. sld.		d.		d.	d.	v. s. liq. NH_3
134	Triphenyl-*m*-tolyl germanium	$Ge(C_6H_5)_3(C_6H_4CH_3)$	395.04	wh. anisotropic need.		136.5–8.5		i.	i.	s. bz., h. pet. eth.; i. meth. al.
135	Triphenyl-*p*-tolyl germanium	$Ge(C_6H_5)_3(C_6H_4CH_3)$	395.04	wh. sld.		123–4				s. org. solv.
136	Tri-*o*-tolylgermanium bromide	$(CH_3C_6H_4)_3GeBr$	425.90	col. oil, (blue fluores.)			$205–10^{1}$			
137	Tri-*m*-tolylgermanium bromide	$(CH_3C_6H_4)_3GeBr$	425.90	wh. need., anisotropic		78.0–.9	$222–3^{1}$			s. CCl_4, bz., eth.
138	Tri-*p*-tolylgermanium bromide	$(CH_3C_6H_4)_3GeBr$	425.90	col. cr.		128–9				s. pet. eth.
139	Tri-*o*-tolylgermanium chloride	$(CH_3C_6H_4)_3GeCl$	381.44	col. oil			$216–22^{1}$			
140	Tri-*m*-tolylgermanium chloride	$(CH_3C_6H_4)_3GeCl$	381.44	sm. silky need.; opt. act.		84–5	$221–4^{1}$			s. pet. eth., bz.; i. c. meth. al.
141	Tri-*p*-tolylgermanium chloride	$(CH_3C_6H_4)_3GeCl$	381.44	wh. cr.		121				s. pet. eth.
142	Tri-*o*-tolylgermanium hydroxide	$(CH_3C_6H_4)_3GeOH$	363.00	amor. powd.			$212–4^{1}$			
143	Tri-*m*-tolylgermanium oxide	$[(CH_3C_6H_4)_3Ge]_2O$	707.98	wh. cr.		125.0–.2		i.	i.	s. al., bz., pet. eth.
144	Tri-*p*-tolylgermanium oxide	$[(CH_3C_6H_4)_3Ge]_2O$	707.98	wh. pr. anisotropic		148–50		i.	i.	s. h. lgr., c. bz.; sl. s. eth., al., meth. al.
145	Tri-*m*-tolyl-*p*-tolyl-germanium	$Ge(C_6H_4CH_3)_3(C_6H_4CH_3)$	437.13	wh. sld.		98.5–100.5				s. meth. al.
146	Tri-*p*-tolyl-*o*-tolyl-germanium	$Ge(C_6H_4CH_3)_3(C_6H_4CH_3)$	437.13	wh. cr.		164–6				
147	Tri-triphenylgermanium nitride	$[(C_6H_5)_3Ge]_3N$	925.74	col. need.		163–4		hyd.	hyd.	s. lgr., eth., bz.
148	Tris-acetylacetone-germanium cupribromide	$[(C_5H_7O_2)_3Ge]CuBr_3$	673.19	gr.-blk. cr.		139				i. chl.
149	Tris-acetylacetone-germanium cuprobromide	$[(C_5H_7O_2)_3Ge]CuBr_2$	593.28	col. rect. pr.		165–6				s. chl.
150	Tris-acetylacetone-germanium cuprochloride	$[(C_5H_7O_2)_3Ge]CuCl_2$	504.37	col. pr.		147–8				s. chl.
151	Tris-acetylacetone-germanium dicuprobromide	$[(C_5H_7O_2)_3Ge]Cu_2Br_3$	736.73	col. pr.		195d.				s. h. acet. acet.; i. chl.

No.	Name	Formula	Mol. wt.	Crystalline form, color and index of refraction	Sp. gr. or density	Melting point, °C	Boiling point, °C	Solubility in grams per 100 ml of: Cold water	Hot water	Alcohol, acids, etc.
	Gold									
1	Aminopyridinotribromogold	$H_2NC_5H_4NAuBr_3$	530.81	blk. pr.		160d.		s.		s. chl.
2	α-Auromercaptoacetanilide	C_8H_8AuNOS	363.19	gray-yel. powd.		238–241		i.		i. a., alk., eth., bz.
3	Diethylgold bromide	$[(C_2H_5)_2AuBr]_2$	670.00	col. need.		58; subl. vac.	d. 70, expl.			s. NH_4OH
4	Ethylenediaminodibutylgold bromide	$(CH_2NH_2)_2Au(C_4H_9)_2Br$	451.21	col. need.		190 d.		s.		s. al.
5	Ethylenediaminodipropylgold bromide	$(CH_2NH_2)_2Au(C_3H_7)_2Br$	423.15	col. cr.		volat. 130	d. 190			
6	Pyridinotribromogold	$C_5H_5NAuBr_3$	515.80	red need.		150 d.		s.		s. al., act.
7	Quinolinotribromogold	$C_9H_7NBr_3Au$	565.86	deep red lust. pr.		d. >200				s. chl.
	Indium									
1	Trimethylindium	$In(CH_3)_3$	159.93	col. cr.	1.568^{19}_{19}	89.0–.8	subl.	d.		s. eth.
	Iridium									
1	Cyclopentadienyliridium cyclopentadiene	$(C_5H_5)Ir(C_5H_6)$	323.40	yel. cr.		130–132	subl.			s. bz., acet., pet. eth.; sl. s. meth. alc.; alc.
	Iron									
1	(Acetylcyclopentadienyl) cyclopentadienyl iron (Monoacetylferrocene)	$(C_5H_4COCH_3)FeC_5H_5$	228.08	lng. or. (red) need.		85–6	86–7			sl. s. isooctane; s. H_2SO_4, HCl
2	*Di*(4-aminobutyl)-cyclopentadienyl iron	$[C_5H_4(CH_2)_4NH_2]_2Fe$	328.28	or. cr.		137–8				s. eth. al.; sl. s. *n*-heptane
3	(Aminocyclopentadienyl)cyclopentadienyliron (Aminoferrocene)	$(C_5H_4NH_2)FeC_5H_5$	201.05	yel. cr.		155				s. eth., al.
4	Mono(*p*-anisylisonitrile)tetracarbonyliron	$Fe(CO)_4(NCC_6H_4OCH_3)$	301.04	yel. pr. need.		39–40				s. eth., pet. eth., bz., chl.
5	Azulene-di-iron pentacarbonyl	$C_{10}H_8Fe_2(CO)_5$	379.92	sld.		100d.				
6	(Benzoylcyclopentadienyl)cyclopentadienyliron (Benzoyl ferrocene)	$(C_6H_5COC_5H_4)FeC_5H_5$	290.15	red need.		108.1–108.3 (108–9)				sl. s. meth. al.
7	(Benzylcyclopentadienyl)cyclopentadienyliron. (Phenylferrocenylmethane)	$(C_6H_5CH_2C_5H_4)FeC_5H_5$	276.16	yel. cr.		70–4 73.4				s. al.
8	Benzylcyclopentadienyliron dicarbonyl bromide	$C_6H_5CH_2C_5H_4Fe(CO)_2Br$	347.00	red cr.		82				s. eth.; i. lgr.
9	Bicycloheptadieneiron tricarbonyl. (Bicyclo-[2:2:1] hepta-2:5-dieneiron tricarbonyl)	$C_7H_8Fe(CO)_3$	232.02	or.-red liq. (yel. liq.)			$60.2^{0.2}$			s. cold H_2SO_4
10	Butadieneiron tricarbonyl	$C_4H_6Fe(CO)_3$	193.97	pa. yel.		19				sl. s. meth. al. ligr.
11	(Carboxyazidecyclopentadienyl)cyclopentadienyliron	$(C_5H_4CON_3)Fe(C_5H_5)$	255.06	cr.		74–5				s. eth.
12	(Carboxycyclopentadienyl)cyclopentadienyliron (Ferrocene monocarboxylic acid)	$(C_5H_4CO_2H)FeC_5H_5$	230.05	rdsh. br. need.		225–330 (208.5 d.) (219–225)				sl. s. chl.
13	(Carboxylamidecyclopentadienyl) cyclopentadienyliron	$(C_5H_4CONH_2)FeC_5H_5$	229.06	cr.		168–170				s. chl.

No.	Name	Formula	Mol. wt.	Crystalline form, color and index of refraction	Sp. gr. or density	Melting point, °C	Boiling point, °C	Solubility in grams per 100 ml of: Cold water	Hot water	Alcohol, acids, etc.
	Iron									
14	*Di*(3-Carboxypropionyl)cyclopentadienylron	$[C_5H_4CO(CH_2)_2CO_2H]_2Fe$	386.19	cr.		164–6d.				
15	Chloromercuriferrocene	$(C_5H_4HgCl)FeC_5H_5$	421.07	golden-yel. leaf.		193–4d. (194–6)				
16	(*p*-Chlorophenylcyclopentadienyl) cyclopentadienyliron	$(ClC_6H_4C_5H_4)FeC_5H_5$	296.58	yel. cr.		122				s. al., H_2SO_4; sl. s. ac. a.
17	(Cyanocyclopentadienyl)cyclopentadienyliron. (Ferrocenyl cyanide)	$(C_5H_4CN)FeC_5H_5$	211.05			107–8 (103–4)				s. CH_2Cl_2; sl.. s. *n*-heptane
18	*Di*[(3-Cyanopropionyl)-cyclopentadienyl]iron	$[C_5H_4CO(CH_2)_2CN]_2Fe$	348.19	or. cr.		133–4				sl. s. tol., al.
19	Cycloheptatrienyliron dicarbonyl	$C_7H_8Fe(CO)_2$	204.01	yel. liq.		$70^{0.4}$				
20	Cyclohexadieneiron tricarbonyl	$C_6H_8Fe(CO)_3$	220.01	yel. liq.			96^{12}			s. acet.
21	Cyclooctatetraenediiron hexacarbonyl	$C_8H_8Fe_2(CO)_6$	383.91	ye.-or.		d. <185				v. sl. s. common org. solv.; sl. s. bz.
22	Cyclooctatetraeneiron tricarbonyl	$C_8H_8Fe(CO)_3$	244.04	lng. red (deep red) need.		94–5(92) 155 d.				sl. s. hot hexane; s. common org. solv.
23	Cyclopentadienyliron dicarbonylbromide	$C_5H_5FeBr(CO)_2$	256.88			18–102 d.				s. chl.; i. lgr.
24	Cyclopentadienyliron dicarbonyliodide	$C_5H_5FeI(CO)_2$	303.88			117–8 d.				i. lgr.
25	(Cyclopentenylcyclopentadienyl)cyclopentadienyliron	$(C_5H_7C_5H_4)FeC_5H_5$	252.14	cr.		64–5				sl. s. meth. al., bz. gasoline
26	(Cyclopentylcyclopentadienyl)cyclopentadienyliron	$(C_5H_9C_5H_4)FeC_5H_5$	254.16	red liq.		16.3 frz.				s. bz.
27	1,1′-Di(acetylcyclopentadienyl)iron. (Diacetyl ferrocene)	$(C_5H_4COCH_3)_2Fe$	270.12	red cr.		130–1				
28	1-1′Di(benzhydryldicyclopentadienyl)-iron. (1,1′-Dibenzhydrylferrocene)	$[(C_6H_5)_2CHC_5H_4]_2Fe$	518.49	yel. need.		162–3				s. eth., cyclohexane; sl. s. acet.
29	1,1′-Di(benzoylcyclopentadienyl)iron. (1,1′-Dibenzoyl ferrocene)	$(C_6H_5COC_5H_4)_2Fe$	394.26	purple need.		106.5–106.7 (105–106)				s. CH_2Cl_2
30	1,1′-Di(benzylcyclopentadienyl)iron. (1,1′-Dibenzyl ferrocene)	$(C_6H_5CH_2C_5H_4)_2Fe$	366.29	yel. need.		102 (97–8) (105–6)				sl. s. meth. al.; s. *n*-butanol.
31	1,1′-Di(carboxycyclopentadienyl)iron (Ferrocene dicarboxylic acid)	$(C_5H_4CO_2H)_2Fe$	274.06				subl. <230			
32	(1,3-Dicarboxycyclopentadienyl)iron (Ferrocene 1,3-dicarboxylic acid)	$[C_5H_3(CO_2H)_2]FeC_5H_5$	274.06	red need.		206–206.5 d.				
33	1,1′-Di(chloromercuri)-ferrocene	$(C_5H_4HgCl)_2Fe$	656.11	sld. yel. powd.		no melt up to 300				
34	1,1′-Di(*p*-chlorophenylcyclopentadienyl)iron. (1,1′-Di′-*p*-chlorophenyl ferrocene).	$(ClC_6H_4C_5H_4)_2Fe$	407.13	cr.		192				s. me. al., H_2SO_4; sl. s. pet. eth.
35	Dicyclopentadienyl-diiron tetracarbonyl	$(C_5H_5)_2Fe_2(CO)_4$	353.92	cr.		192				s. pyr.

PHYSICAL CONSTANTS OF ORGANOMETALLIC COMPOUNDS (Continued)

No.	Name	Formula	Mol. wt.	Crystalline form, color and index of refraction	Sp. gr. or density	Melting point, °C	Boiling point, °C	Solubility in grams per 100 ml of: Cold water	Hot water	Alcohol, acids, etc.
	Iron									
36	Diferrocenyl mercury	$(C_5H_5FeC_5H_4)_2Hg$	570.65	or. cr.		235–6 d. 248–9 233–4				s. bz. sl. s. xylene
37	1,1′-Di(hydroxybenzyl-cyclopentadienyl)-iron. (Ferrocenyl-bis-phenyl methanol)	$[C_6H_5CH(OH)C_5H_4]_2Fe$	398.29	yel. leaf		136–7				sl. s. meth. al., al.
38	(Dimethylamino-methylcyclopenta-dienyl)cyclopenta-dienyl iron. (Di-methylaminomethyl-ferrocene).	$[(CH_3)_2NCH_2C_5H_4]Fe\,C_5H_5$	243.13	amber oil 1.5839^{25}_{4}			$91–2^{0.45}$			s. eth.
39	1,1′-Di(methylcarboxy-cyclopentadienyl)iron	$(C_5H_4CO_2CH_3)_2Fe$	302.11	red cr.			114–5			
40	(1,3-Diphenylcyclo-pentadienyl) cyclo-pentadienyliron. (1,3-Diphenylferrocene)	$[(C_6H_5)_2C_5H_3]Fe(C_5H_5)$	338.24	or. need.		107				s. lgr.
41	1,2-Diphenyl-1,2-di-ferrocenylethanediol	$[C(C_5H_5FeC_5H_4)(OH)(C_6H_5)]_2$	582.31	yel. cr		125–145 d.				s. eth., chl., dioxane,; sl. s. bz.
42	1,1′-Diphenyl-dicyclo-pentadieenyliron. (1,1′-Diphenyl-ferrocene)	$(C_6H_5C_5H_4)_2Fe$	338.24	or.-yel.-br. leaf		154 (140–4)				s. bz.; sl. s. lt. pet.
43	1,1′-Di(trimethylsilyl) ferrocene	$[(CH_3)_3Si\,C_5H_4]_2Fe$	330.40	1.5454^{25}_{D}		16	$87–88^{0.06}$			s. eth.
44	*bis*-(Ethylisonitrile)tri-carbonyl iron	$Fe(CO)_3(CNC_2H_5)_2$	250.04	yel. pl. need.		65.5–66				s. eth., bz., al.
45	Ferrocene. (Dicyclo-pentadienyliron)	$C_5H_5FeC_5H_5$	186.04	yel. need.		172.5–173	subl.	i.	i.	s. al., eth., bz., MeOH
46	Ferrocenyl-acetic acid	$(C_5H_4CH_2CO_2H)FeC_5H_5$	244.08	lt. yel. (yel.) need		150–2 (125–135)				s. eth.; i. pentane; sl. s. meth. al.
47	1,1′-Ferrocenyl-diacetic acid	$(C_5H_4CH_2CO_2H)_2Fe$	302.11	yel.		sintered 140–3(d)				s. chl., eth.; i. pentane
48	*β*-Ferrocenyl-ethyl alcohol	$(C_5H_4CH_2CH_2OH)Fe(C_5H_5)$	230.09			49–50				s. eth.; i. pentane
49	*β*-Ferrocenyl-propionic acid	$[C_5H_4CO(CH_2)_2CO_2H]FeC_5H_5$	286.11	or. cr.		166.5–167.5				s. CH_2Cl_2; sl. s. meth. al.
50	(Formylcyclopenta-dienyl)cyclopenta-dienyliron. (Ferrocene monoaldehyde)	$(C_5H_4CHO)FeC_5H_5$	214.05	sld. red (rdsh br.)cr.		120–2 (130–2)				s. eth., chl.; sl. s. al.
51	(Hydroxybenzylcyclo-pentadienyl)cyclo-pentadienyliron. (Phenyl-ferrocenyl methanol)	$[C_6H_5CH(OH)C_5H_4]FeC_5H_5$	292.16	yel. cr.		81–2 (80–80.5)				s. eth.; i. pet. eth.
52	(Hydroxymethylcyclo-pentadienyl)cyclo-pentadienyliron. (Hydroxymethyl-ferrocene)	$(C_5H_4CH_2OH)FeC_5H_5$	216.06	yel. pl. lng. need.		81–2 (74–6)				s. chl.
53	*p*-Hydroxyphenyl-cyclopentadienyl)-cyclopentadienyliron	$(HOC_6H_4C_5H_4)FeC_5H_5$	278.14	cr. yel. grn. fluorescence		165				s. al.
54	(Methylcarboxycyclo-pentadienyl)cyclo-pentadienyliron	$(C_5H_4CO_2CH_3)FeC_5H_5$	244.08			70–1				s. meth. al.
55	*bis*(Methylisonitrile)-tricarbonyliron	$Fe(CO)_3(CNCH_3)_2$	221.98			100–130 d.				s. al., eth., bz., ac. a.
56	Mono(ethylisonitrile)-tetracarbonyliron	$Fe(CO)_4(CNC_2H_5)$	222.97	yel.		−3		i. d.		s. org. solv.

PHYSICAL CONSTANTS OF ORGANOMETALLIC COMPOUNDS (Continued)

No.	Name	Formula	Mol. wt.	Crystalline form, color and index of refraction	Sp. gr. or density	Melting point, °C	Boiling point, °C	Solubility in grams per 100 ml of: Cold water	Hot water	Alcohol, acids, etc.
	Iron									
57	(*p*-Nitrophenylcyclopentadienyl)cyclopentadienyliron	$(NO_2C_6H_4C_5H_4)FeC_5H_5$	307.13	dk. purple cr.	....	163	....	....	....	s. al.
58	(Phenylcyclopentadienyl) cyclopentadienyliron (Phenylferrocene)	$(C_6H_5C_5H_5)FeC_5H_5$	263.14	yel. (or.) cr.	....	110–111 (109–110)	....	....	....	s. al.
59	Phenyl-diferrocenylmethane	$(C_5H_5FeC_5H_4)_2CHC_6H_5$	460.19	cr.	....	123–4	....	....	....	s. eth., al.
60	1,3,1′,3′-Tetraphenyldicyclopentadienyliron. (1,3,1′,3′-tetraphenylferrocene)	$[(C_6H_5)_2C_5H_3]_2Fe$	490.43	flat or. pr. (deep or. red)	....	219–220 (220–222)	....	....	....	sl. s. acet.
61	Thiopheniron dicarbonyl	$C_6H_4SFe(CO)_2$	220.03	pale red cr.	....	51	subl.	....	....	s. org. solv.
62	Trimethylsilylferrocene	$[C_5H_4Si(CH_3)_3]FeC_5H_5$	258.22	dk. or.-red liq. 1.5696_4^{25}	....	23	$64–5^{0.45}$	....	....	....
	Lanthanum									
1	Hexaantipyrineiodide	$La(COC_{10}H_{12}N_2)_6I_3$	1649.01	yel. cr.	....	268–9 d.	....	41.8^{20}	....	....
	Lead									
1	Hexaethyldilead (triethyllead)	$Pb_2(C_2H_5)_6$	588.75	liq.	1.471	....	d.	i.	....	....
2	Tetraethyllead	$Pb(C_2H_5)_4$	323.44	col. liq.; or. flame, grn. marg. 1.5195^{20}	1.659^{11}	−136.80	200d.; 91^{19}	i.	....	s. bz., pet., al., eth.
3	Tetra-iso-amyllead	$Pb(C_5H_{11})_4$	491.76	col. oily liq., 1.4946	n_D $1.2332_{4.5}^{20}$	....	....	i.	....	s. al., bz., eth
4	Tetraisobutyllead	$Pb[CH_2CH(CH_3)_2]_4$	435.66	col. liq., 1.5042	1.324	−23	....	....	....	....
5	Tetraisopropyllead	$Pb[CH(CH_3)_2]_4$	379.55	col. liq., dec. in air, 1.5223	1.4504	−53.5	120^{14}; $133–8^{27}$	i.	....	s. bz. pet. eth.
6	Tetramethyllead	$Pb(CH_3)_4$	267.33	liq. 1.5120^{20}.	1.995	−27.5	110	i.	....	s. bz., pet. eth., al.
7	Tetraphenyllead	$Pb(C_6H_5)_4$	515.62	wh. need.	1.530^{20}	227.7	d. 270	i.	....	s. bz.
8	Tetra-*n*-propyllead	$Pb(C_3H_7)_4$	379.55	col. liq., 1.5094	1.44	....	126^{13}	....	....	s. bz., pet. eth.; sl. s. al.
9	Tetravinyllead	$(CH_2{:}CH)_4Pb$	315.37	n_D^{20} 1.5462	1.7882_4^{20}	....	$69–70^{11}$	....	....	....
10	Tricyclohexyllead	$(C_6H_{11})_3Pb$	456.65	yel. hex. pl., decol. in light	....	d. 195	....	....	....	....
11	Triphenylbenzyllead	$(C_6H_5)_3PbCH_2C_6H_5$	529.64	col. pl.	....	93	d. 205–210	....	....	v. s. bz., eth., chl.
12	Triphenylethyllead	$(C_6H_5)_3Pb(C_2H_5)$	467.57	wh. cr. $n_{H\infty}^{61}$ 1.6263	1.5885_4^{61} (vac)	42	....	....	....	s. hot al.
13	Triphenylmethyllead	$(C_6H_5)_3PbCH_3$	453.55	col. rhomb. (from bz.)	....	60	d. 260	....	....	sl. sl al.; eth.; v. s. bz., chl.
14	Triphenylleadbromide	$(C_6H_5)_3PbBr$	518.42	need.	....	166	....	i.	....	s. al., bz., eth.
	Lithium									
1	*n*-Butyllithium	$n\text{-}C_4H_9Li$	64.06	inflamm. liq.	0.765^{25}	subl. vac. 80 100	d. 150	d.	d.	s. bz., eth.
2	Ethyllithium	LiC_2H_5	36.00	hex. transp. pl.	....	95 (in N_2)	subl. 95	d.	....	d. eth.; s. bz lgr.
	Magnesium									
1	Dimethylmagnesium	$(CH_3)_2Mg$(polymer?)	54.38	wh. sol., inflamm. in air	....	stable to 240	....	....	....	v. sl. s. eth.
2	Diphenylmagnesium	$(C_6H_5)_2Mg$	178.53	wh. sol.	....	d. *ca*, 280	....	....	....	i. bz., eth.
	Magnesiumaluminum ethoxide-*see* Aluminum(3)									

PHYSICAL CONSTANTS OF ORGANOMETALLIC COMPOUNDS (Continued)

No.	Name	Formula	Mol. wt.	Crystalline form, color and index of refraction	Sp. gr. or density	Melting point, °C	Boiling point, °C	Solubility in grams per 100 ml of: Cold water	Hot water	Alcohol, acids, etc.
	Manganese									
1	Acetylcyclopentadienyl manganese tricarbonyl	$CH_3COC_5H_4Mn(CO)_3$	246.10			41.5–42.5 (39.5–40.0)				s. pet. eth.
2	Acetylmanganese pentacarbonyl	$CH_3COMn(CO)_5$	238.04	wh. cr.		54–5				
3	(Benzoylcyclopentadienyl)manganese tricarbonyl	$C_5H_4COC_6H_5Mn(CO)_3$	308.17	yel. cr.		73.5–74.5				s. bz., hexane; i. pet. eth.
4	Benzoylmanganese pentacarbonyl	$C_6H_5COMn(CO)_5$	300.11	wh.		95.6				
5	Benzylmanganese pentacarbonyl	$C_6H_5CH_2Mn(CO)_5$	286.12	yel.		37.5–38.5				
6	(*tert*-Butylcyclopentadienyl) manganese tricarbonyl	$C_4H_9C_5H_4Mn(CO)_3$	260.17	1.5600^{20}_{D}			150^{28-29}			
7	(γ-Butyrocyclopentadienyl)manganese tricarbonyl	$[C_5H_4(CH_2)_3CO_2H]Mn(CO)_3$	290.16	yel. cr.		63–5				s. bz., eth., al.
8	(Carboxycyclopentadienyl)manganese tricarbonyl	$(C_5H_4CO_2H)Mn(CO)_3$	248.08	yel. cr.		195–6				s. al.; sl. s. bz.
9	(Di-*tert*-butylcyclopentadienyl) manganese tricarbonyl	$(C_4H_9)_2C_5H_3Mn(CO)_3$	316.28			70–1	$164–8^{10}$			
10	(Dicyclohexyl cyclopentadienyl) manganese tricarbonyl	$(C_6H_{11})_2C_5H_3Mn(CO)_3$	368.36	1.5804^{20}_{D}			$149–152^{2}$			
11	(Formylcyclopentadienyl)manganese tricarbonyl	$(C_5H_4CHO)Mn(CO)_3$	232.08	red oil.			$105^{0.15}$			s. org. solv.
12	(Hexylcyclopentadienyl)manganese tricarbonyl	$C_6H_{13}C_5H_4Mn(CO)_3$	288.23	1.5686^{20}_{D}			$110–114^{2}$			s. eth.
13	(Methylcyclopentadienyl)manganese (benzene)	$CH_3C_5H_4MnC_6H_6$	212.18	red cr.		116–8				
14	Methylmanganese pentacarbonyl	$CH_3Mn(CO)_5$	210.03	col. cr.		94.5–95				
15	Phenylmanganese pentacarbonyl	$C_6H_5Mn(CO)_5$	272.10	wh. cr.		52				
16	(Tri-*tert*-butylcyclopentadienyl) manganese tricarbonyl	$(C_4H_9)_3C_5H_2Mn(CO)_3$	372.39	1.5326^{20}_{D}			$160–3^{3.5}$			
17	(δ-Valerocyclopentadienyl)manganese tricarbonyl	$[C_5H_4(CH_2)_4CO_2H]Mn(CO)_3$	304.18	yel. cr.			69–71			s. bz., eth., al.
	Mercury									
1	Aminophenylmercuric acetate(*p*)	$C_6H_4(NH_2)HgO_2C_2H_3$	351.76	col. pr.		167		i.	i.	s. dil. a.; sl. s. chl., al.; i. eth.
2	Biphenylmercury	$Hg(C_6H_5C_6H_4)_2$	507.00	sm. scales		216		difficultly	soluble in	common solv.
3	*n*-Butylmercuric chloride	C_4H_9ClHg	293.16	wh. need.		130		14×10^{-4}, 18°	3.3×10^{-4}, 100°	0.5^{18} al.; 5.6^{18} $CHCl_3$
4	Chloromercuriphenol(*o*)	$C_6H_4OHHgCl$	329.15			152.5^{0}				s. NaOH
5	Di-*n*-amylmercury	$Hg(C_5H_{11})_2$	342.88	1.4998	1.6369		133^{10}			
6	Di-(*dl*)-amylmercury	$Hg(C_5H_{11})_2$	342.88	1.5014	1.6700		93^{1}			
7	Dibenzylmercury	$Hg(C_7H_7)_2$	382.86	long brittle col. need.						s. al., eth., chl., CS_2, ac. a., bz. eth. acet.; sl. s. lgr.

PHYSICAL CONSTANTS OF ORGANOMETALLIC COMPOUNDS (Continued)

No.	Name	Formula	Mol. wt.	Crystalline form, color and index of refraction	Sp. gr. or density	Melting point, °C	Boiling point, °C	Solubility in grams per 100 ml of		
								Cold water	Hot water	Alcohol, acids, etc.
	Mercury									
8	Di-*n*-butylmercury	$Hg(C_4H_9)_2$	314.82	liq. 1.5057	1.7779		105^{10}			
9	Diethylmercury	$Hg(C_2H_5)_2$	258.71	col. liq. of hazel odor 1.5399	liq. 2.444		159	i.	i.	v. s. eth.; sl. s. al.
	Diferrocenylmercury-*see* **Iron** (36)									
10	Di-*n*-hexylmercury	$Hg(C_6H_{13})_2$	370.94	1.4973	1.5361		158^{10}			
11	Diisoamylmercury	$Hg(C_5H_{11})_2$	342.89	1.4989	1.6397		125^{10}			
12	Diisobutylmercury	$Hg(C_4H_9)_2$	314.82	col. liq., 1.4965	1.835^{15}; 1.7678	volat. 100	205–7; 86^{10}	v. sl. s.		s. eth. al.
13	Diisopropylmercury	$Hg(C_3H_7)_2$	286.78	1.5263	2.0024		63^{10}			
14	Dimethylaminophenyl-mercuric acetate(*p*)	$C_6H_4N(CH_3)_2HgO_2C_2H_3$	379.82	long. col. need		165		i.	i.	s. bz., chl., al., dil. a.
15	Dimethylaniline-mercury(*p*)	$Hg[C_6H_4N(CH_3)_2]_2$	440.95	lust. need.		169				s. chl.; sl. s. al., eth., dil. HCl
16	Dimethylmercury	$Hg(CH_3)_2$	230.66	col. liq., sweet odor, 1.5327	3.069		96			s. al., eth.
17	Dinaphthylmercury(*α*)	$Hg(C_{10}H_7)_2$	454.92	wh. rhomb.	1.929	188 (243)	249	i.	sl. s.	s. h. CS_2, chl.; sl. s. bz., eth. v. sl. s. h. bz
18	Dinaphthylmercury(*β*)	$Hg(C_{10}H_7)_2$	454.92	cryst. from bz.		247–8		i.		sl. s. al., eth.
19	Diphenylmercury	$Hg(C_6H_5)_2$	354.81	wh. glassy need.	2.318	121.8, subl.	$204^{10.5}$ >306 d.	i.	i.	s. chl., CS_2, bz.; sl. s. eth., h. al.
20	Dipropylmercury	$Hg(C_3H_7)_2$	286.78	col. mobile liq., 1.5170	2.0208		189–91; 73^{10}	i.		v. s. eth.; s. al.
21	Ditolylmercury(*o*)	$Hg(C_7H_7)_2$	382.86	wh. tabl.		107	219^{14}			s. h. bz.
22	Ditolylmercury(*m*)	$Hg(C_7H_7)_2$	382.86	col. or. lt. yel. need.		102		i.		s. bz., chl., acet., eth. acet.
23	Ditolylmercury(*p*)	$Hg(C_7H_7)_2$	382.86	need.		238		i.		s. h. bz., chl., CS_2; sl. s. c. al.
24	Ethane hexamercarbide	$C_2Hg_6O_2(OH)_2$	1293.58	yelsh.-wh. powd.		exp. 230		i.	i.	i.
25	Ethylmercuric chloride	C_2H_5HgCl	265.10	silv. irid. leaf.	3.482	193	subl. 40	i.		v. s. h. al.; sl. s. eth.
26	Ethylmercuric hydroxide	C_2H_4HgOH	245.65	silv. irid. leaf.		37		i.	i.	v. s. h. al.; s. eth.; sl. s. c. al.
27	Ethylmercuric iodide	C_2H_5HgI	356.56	cr. from EtOH		186				s. al.
28	Mercury ethyl-mercaptide(ic)	$Hg(SC_2H_5)_2$	322.85	leaf.		76–7	d.	i.		d. a.
29	Mercuryphenyl-mercaptide(ic)	$Hg(SC_6H_5)_2$	418.93	yelsh. need.		153 d.	d.	i.		sl. s. h. al.; v. s. bz., pyr.
30	Methylmercuric chloride	CH_3HgCl	251.09	wh. cr., disg. odor	4.063	170	volat. 100			
31	Methylmercuric iodide	CH_3HgI	342.53	col. pearly leaf.		143		i.		v. s. meth. al.; s. eth., al.
32	Naphthylmercuric acetate(*α*)	$C_{10}H_7HgO_2C_2H_3$	386.81	fine need.		154		i.	i.	s. al., ac. a., bz., CS_2, fats; sl. s. eth.
33	Naphthylmercuric chloride(*α*)	$C_{10}H_7HgCl$	363.22	silk quad. tabl.		188–9		i.	i.	sl. s. bz., al.
34	Phenylmercuric acetate	$C_6H_5HgO_2C_2H_3$	336.75	rhomb. sm. wh. lust. pr.		149		sl. s.	sl. s.	s. glac. ac. a. ac. a., bz., al.
35	Phenylmercuric bromide	C_6H_5HgBr	357.61	rhomb. wh. lust. tabl.		276		i.	i.	s. al., bz., pyr.
36	Phenylmercuric chloride	C_6H_5HgCl	313.15	wh. satiny leaf.		251				sl. s. h. al., bz., pyr., eth.
37	Phenylmercuric cyanide	C_6H_5HgCN	303.73	rhomb. long pr.		204			sl. s.	s. h. al., bz.
38	Phenylmercuric iodide	C_6H_5HgI	404.60	rhomb. satiny tabl.		266		i.		s. chl., CS_2; sl. s. al., eth., bz.
39	Phenylmercuric nitrate	$C_6H_5HgNO_3$	339.70	rhomb. tabl.		176–86		i.	sl. s.	s. h. al., bz.

PHYSICAL CONSTANTS OF ORGANOMETALLIC COMPOUNDS (Continued)

No.	Name	Formula	Mol. wt.	Crystalline form, color and index of refraction	Sp. gr. or density	Melting point, °C	Boiling point, °C	Solubility in grams per 100 ml of: Cold water	Hot water	Alcohol, acids, etc.
	Mercury									
40	Phenylmercuric nitrate, basic	$C_6H_5HgOH.C_6H_5HgNO_3$	634.41	wh.-grayish powd.		d. 178–184		0.8		sl. s. al., glyc.
41	Tolylmercuric bromide(*p*)	C_7H_7HgBr	371.63	thin lust. gray sc.		228				s. chl., al., bz.; i. c. CS_2
42	Tolylmercuric chloride(*p*)	C_7H_7HgCl	327.19	rhomb. silky tabl.		233		i.	i.	sl. s. h. al., bz, chl., acet., pyr.; i. eth.
	Molybdenum									
1	Benzenemolybdenum tricarbonyl	$C_6H_6Mo(CO)_3$	258.09	yel. monocl.		120–5 d.				
2	Bicycloheptadiene-molybdenum tetracarbonyl	$C_7H_8Mo(CO)_4$	300.12	yel. pl.		76–7				sl. s. pet. eth.
3	Cycloheptatriene-molybdenum tricarbonyl	$C_7H_8Mo(CO)_3$	272.11	or.-red pl.		100.5–101.5				s. lt. pet.; chl., bz.
4	*bis*-Cyclopentadienyl-bimolybdenum pentacarbonyl (*bis*-Cyclopentadienyl-μ-pentacarbon monoxide bimolybdenum I)	$(C_5H_5)_2Mo_2(CO)_5$	462.12	purp.-red.		215–7 d.				s. chl.; sl. s. CCl_4 al., CS_2, bz.
5	Mesitylenemolybdenum tricarbonyl	$(CH_3)_3C_6H_3Mo(CO)_3$	300.17	yel.		130–40 d.	90 subl.			
6	Tropeniummolybdenum tricarbonyl fluoroborate	$C_7H_7Mo(CO)_3BF_4$	357.91	lt. or. need.		>270				
	Neodymium									
1	Neodymium hexaantipyrine iodide	$[Nd(COC_{10}H_{12}N_2)_6I_3]$	1654.34	rose cr.		270–2		12.7^{20}		
	Nickel									
1	Dicyclopentadienyl-nickel	$(C_5H_5)_2Ni$	188.90	drk. grn. cr.			130 subl.			sl. s. lgr.; s. al.
2	*bis*(Cyclopentadienyl-nickel carbonyl)	$(C_5H_5NiCO)_2$	303.63	purple red.		136		i.		s. bz., chl., pet. eth., eth., al.
3	Cyclopentadienylnickel carbonyl iodide	$C_5H_5Ni(CO)I$	278.72	blk. vlt.		20 d.				sl. s. org. solv.
4	Dicyclopentadiene-nickel	$(C_5H_6)_2Ni$	190.92	red need.		41–42 cl. tube				
	Niobium									
1	*bis*-Cyclopentadienylniobium monohydroxy-dibromide	$(C_5H_5)_2Nb(OH)Br_2$	399.92	or.-red.				s.		
2	*bis*-Cyclopentadienylniobium tribromide	$(C_5H_5)_2NbBr_3$	462.82	reddsh. br. cr.		260 d.		s.		s. chl., al.; i. pet. eth.; sl. s. bz.;
	Palladium									
1	Cycloocta-1,5-dienedibromopalladium	$C_8H_{12}PdBr_2$	374.40	or.-red. need.		213 d.		i.		sl. s. glac.; i. acet., ac. a.
2	Cycloocta-1,5-dienedichloropalladium	$C_8H_{12}PdCl_2$	285.49	pa. or. need.	d. 205–210					s. h. glac. ac. a.; sl. s. c. al., bx. sl. s. h. bz., chl., acet.
3	Cyclohexenepalladous chloride	$(C_6H_{10}PdCl_2)_2$	519.00	lt. br. cr.						v. i. in common org. solv.
4	Dicyclopentadienedichloropalladium	$(C_5H_6)_2PdCl_2$	309.51			165–70 d.		d.		s. bz., eth., chl.

No.	Name	Formula	Mol. wt.	Crystalline form, color and index of refraction	Sp. gr. or density	Melting point, °C	Boiling point, °C	Solubility in grams per 100 ml of Cold water	Hot water	Alcohol, acids, etc.
	Palladium									
5	*bis*(Dicyclopentadiene-methoxide)dichlorodipalladium	$C_{22}H_{30}O_2Pd_2Cl_2$	610.19	yel. pl.		166–70 d.		d.		i. meth. al.; s. boil. chl.; sl. s. c. chl., bz.
6	*bis*(Dicyclopentadiene-isopropoxide)dichlorodipalladium	$C_{26}H_{38}O_2Pd_2Cl_2$	666.30	yel. pl.		150–60 d.		i.		i. meth. al.; s. boil. chl.; sl. s. cold chl.; bz.
7	*bis*(Dicyclopentadiene-*n*-propoxide) dichlorodipalladium	$C_{26}H_{38}O_2Pd_2Cl_2$	666.30	yel. pl.		150–56 d.		i.		i. meth. al.; s. boil chl.; sl. s. cold chl.; bz.
8	Di(8-methoxycyclooct-4-enyl)dibromopalladium	$(C_8H_{12}OMe)_2Pd_2Br_2$	651.06	pa. yel. cr.		125–35 d.				
9	Di(8-methoxycyclooct-4-enyl)dichloropalladium	$(C_8H_{12}OMe)_2Pd_2Cl_2$	562.14	pa. yel. cr.		130–35 d.				sl. s. meth. al.,
10	Ethylenepalladous chloride	$C_4H_8Pd_2Cl_4$	410.72	lt. canary yel. need.						s. xylene; i. pet. eth.
11	1,5-Hexadienedichloropalladium. (Biallyldichloropalladium)	$C_6H_{10}PdCl_2$	259.46	brnsh. yel. cr.						sl. s. bz.; i. pet. eth.
12	Styrenepalladous chloride	$C_{16}H_{16}Pd_2Cl_4$	562.92	lt. red. br. need.						sl. s. bz.; i. common org. solv.
13	Trimethylethylenepalladous chloride	$C_5H_{10}PdCl_2$	247.45	red. pr.		85–90 d.		s. with d.		s. bz.; i. isopentane, gasoline
	Platinum									
1	Bromochlorothylene amine platinum	$C_2H_4PtNH_3BrCl$	355.55	pa. grn. pr.		d. 163		sl. s.		i. c. HCl
2	*cis*-2-Buteneplatinous chloride	$(C_4H_8PtCl_2)_2$	644.21			d. 165–75				s. liq. butene
3	*trans*-2-Buteneplatinous chloride	$(C_4H_8PtCl_2)_2$	644.21	pink cr.		110–115 (d. 125–135)				s. liq. butene
4	Cyclohexenedibromoplatinum (Cyclohexene-platinous bromide)	$(C_6H_{10}PtBr_2)_2$	874.11	lng. slend. or need.		150–1				s. chl.; v. sl. s. bz., al.; i. eth. ac. a.
5	Cyclohexenedichloroplatinum (Cyclohexeneplatinous chloride)	$(C_6H_{10}PtCl_2)_2$	696.29	lng. sl. or. need.		145–6				v. s. al., acet.; s. chl., eth.; sl. s. bz. i. ac. a.
6	Cycloocta-1,5-dienedichloroplatinum	$C_8H_{12}PtCl_2$	374.18	wh. need.		d. >220				s. boil. chl., ac. a., CH_2Cl_2
7	Cycloocta-1,5-dienediiodoplatinum	$C_8H_{12}PtI_2$	557.08	or.-yel. need.		d. 250				s. boil. CH_2Cl_2
8	Cyclooctatetraenedichloroplatinum	$C_8H_8PtCl_2$	370.15	or.						s. org. solv.
9	Cyclooctatetraenediiodoplatinum	$C_8H_8PtI_2$	553.05	or.-red. cr. powd.		Turns black (no change in cr. form)				s. boil. chl.
10	Cycloocta-1,5-dienedibromoplatinum	$C_8H_{12}PtBr_2$	463.09	pa. yel. need.		d. >200				s. h. ch., ac. ac. CH_2Cl_2
11	Cyclopropaneplatinous chloride	$C_3H_6PtCl_2$	308.08	br. powd.		d. >100		v. sl. s.		s. 10N HCl, al.; sl. s. eth., acet., chl., CCl_4
12	Diallyletherdichloroplatinum	$C_6H_{10}OPtCl_2$	364.14	wh. cr.		180 d.		i.	i.	s. h. chl.
13	*trans*-Dichloroethylenedichloroplatinum (*trans*-Dichloroethyleneplatinous chloride)	$[C_2H_2Cl_2PtCl_2]_2$	725.88	or. cr.		155–60				s. acet., chl., bz., nitrobz.
14	Dicyclopentadienedibromoplatinum	$C_{10}H_{12}PtBr_2$	487.11	yel. pr.		d. 200–225		i.		s. h. chl.; i. eth.

No.	Name	Formula	Mol. wt.	Crystalline form, color and index of refraction	Sp. gr. or density	Melting point, °C	Boiling point, °C	Solubility in grams per 100 ml of: Cold water	Hot water	Alcohol, acids, etc.
	Platinum									
15	Dicyclopentadienedichloroplatinum	$(C_5H_6)_2PtCl_2$	398.21	sm. ivory need.		218–220 d.				i. al.; s. h. chl.
16	Dicyclopentadienediiodoplatinum	$(C_5H_6)_2PtI_2$	581.11			209–211 d.		i.		s. acet.
17	Dicyclopentadienemethoxy chloroplatinum	$C_{11}H_{15}OPtCl$	393.78	wh. cr.						i. eth.; s. chl.
18	Dimethylplatinum diiodide	$(CH_3)_2PtI_2$	478.97	blk.				i.	i.	i. eth.; et. ac. acet.
19	Dipentenedichloroplatinum (Dipenteneplatinous chloride)	$[C_{10}H_{16}PtCl_2]_2$	804.47	pa. yel. biax. pr.		151–2				v. s. chl.; s. acet.; i. eth., al., ac. a., bz.
20	β-Dipenteneplatinous iodide	$C_{10}H_{16}PtI_2$	585.14	red cr.		122–4 d.				i. chl.; s. acet.
21	Dipyridinecyclopropanedichloroplatinum	$C_{13}H_{16}N_2Cl_2Pt$	466.28			d. >100		i.	d.	s. warm 10N HCl
22	Dipyridinotrimethyliodoplatinum	$C_{13}H_{19}N_2PtI$	525.30	pa. yel. cr.		168		sl. s.	s. s.	sl. s. al.; s. bz., chl., acet., h. al.
23	Dipyridinotrimethylplatinic iodide	$(CH_3)_3Pt(Py)_2I$	525.31	col. pr.		168				
24	Ethyl(trimethylplatinic) acetoacetate	$C_6H_9O_3Pt(CH_3)_3$	369.33	wh. hex. pl.		200 d.		i.		s. org. solv.
25	Ethylenedichloroplatinum (Ethyleneplatinous chloride)	$[C_2H_4PtCl_2]_2$	588.10	or. tabl.		170–180 d. 160–165		sl. s.		v. s. acet., al., eth.; s. bz., chl. sl. s. NaCl soln.
26	1,5-Hexadienedichloroplatinum (Biallyldichloroplatinum)	$C_6H_{10}PtCl_2$	348.15	or.-pa. yel.		172–3		i.	s. sl. with d.	s. chl.; v. sl. s. eth.; sl. s. bz.
27	1,5-Hexadienediiodoplatinum (Biallyldiiodoplatinum)	$C_6H_{10}PtI_2$	531.05	or. red cr.		d.		i.		s. chl.
28	Hexamethyldiplatinum	$(CH_3)_6Pt_2$	480.39	col. cr.		d.		i.	i.	v. s. bz., acet.; eth., sl. s. c. pet. eth.
29	Isobutylenedichloroplatinum (Isobutyleneplatinous chloride)	$[C_4H_8PtCl_2]_2$	644.21	or. rhomb. pl.		144–5		i.		v. s., chl., eth., bz., al. acet., NaCl soln.; sl. s. ac. a.; i. pet. eth.
30	8-Methoxycyclooct-4-enyl p-toluidinechloroplatinum)	$(C_8H_{12}OCH_3)(CH_3C_6H_4NH_2)PtCl$	476.92	wh. need.		140–2				s. bz., meth. al., nitrobz.; i. lgr.
31	Methoxydicyclopentadiene-p-toluidinechloroplatinum	$C_{18}H_{24}ONPtCl$	500.93	wh. need.		160–170				v. s. c. chl.; s. bz., meth. al., acet. eth.; i. nitrobz.
32	Methylplatinum pentaiodide	CH_3PtI_5	844.65	blk. cr.				sl. s.		sl. s. acet., et. ac., s. eth.
33	Methylplatinum triiodide	CH_3PtI_3	590.84	blk. amor.					i.	s. HCl
34	Tetramethyl platinum	$(CH_3)_4Pt$	255.23	lg. col. hex.		d.		i.	i.	v. s. bz., acet., eth., pet. eth.
35	Sesquimethylenediaminotrimethylplatiniciodide	$C_{12}H_{42}N_6I_2Pt_2$	914.50	col. cr.		d. 269		s.		
36	p-Toluidineethylenedichloroplatinum	$(C_2H_4)(CH_3C_6H_4NH_2)PtCl_2$	401.20	lem. yel. cr.		d. 125–135		i.		v. s. bz., acet., meth. al., eth., chl.; sl. s. cyclohexane; i. lt. pet.
37	Trimethylplatinicmonoethylenediamineiodide	$(CH_3)_3Pt(NH_2CH_2CH_2NH_2)I$	427.19			204 d.		s.		

No.	Name	Formula	Mol. wt.	Crystalline form, color and index of refraction	Sp. gr. or density	Melting point, °C	Boiling point, °C	Solubility in grams per 100 ml of: Cold water	Hot water	Alcohol, acids, etc.
	Platinum									
38	Trimethylplatinum benzoylacetone	$C_{13}H_{18}O_2Pt$	401.37	lng. wh. pr.						v. sl. s. lgr.; v. s. meth. al.
39	Trimethylplatinum chloride	$(CH_3)_3PtCl$	275.64	wh. cr.						sl. s. c. pet. eth.; s. h. pet. eth.
40	Trimethylplatinum dipropionyl methane	$C_7H_{11}O_2Pt(CH_3)_3$	367.36	wh. need.		d. 190		i.		s. org. solv.
41	Trimethylplatinum iodide	C_3H_9PtI	367.09	cr. yel. powd.		d. *ca* 250		i.	i.	sl. s. al. eth. acet., bz., s. warm bz.
42	Trimethylpyridinoplatinum iodide	$[(CH_3)_3Pt(C_5H_5N)I]_2$	892.40			d. >150				s. pyr., chl., bz.
43	Pinenedichloroplatinum (Pineneplatinous chloride)	$[C_{10}H_{16}PtCl_2]_2$	804.48	pr.		138–141				v. s. chl.; s. bz., acet.; sl. s. eth., et. ac., ac. a.
44	Stilbenedichloroplatinum (Stilbeneplatinous chloride)	$[C_{14}H_{12}PtCl_2]_2$	892.50	sm. hex. or. pr.		191–2				w. s. chl., acet., nitrobz.; v. sl. s. bz., al.; i. ac. a.
45	Styrenedichloroplatinum (Styreneplatinous chloride)	$[C_8H_8PtCl_2]_2$	704.30	or. hex. pr.		169–171		i.		s. chl., eth., acet.; v. sl. s. ac. a., NaCl soln.; sl. s. bz.
46	Styrenedibromoplatinum (Styreneplatinous bromide)	$[C_8H_8PtBr_2]_2$	918.12	pink hex. pr.		153–4				s. chl.; sl. s. bz., acet.; i. eth., al., ac. a.
	Potassium									
1	Potassium saccharate, acid(*d*)	$KHC_6H_8O_8$	248.23	rhomb. need.				1.1^{6}		
2	Potassium-*m*-nitrophenoxide	$KOC_6H_4NO_2.2H_2O$	213.23	flat or. need.	1.691^{20}	$-2H_2O$, 130	d.	16.3^{15}		s. al.
3	Potassium-*p*-nitrophenoxide	$KOC_6H_4NO_2.2H_2O$	213.23	yel. leaf	1.652^{20}	$-2H_2O$, 130	d.	7.5^{15}		sl. s. al.
	Rhenium									
1	Bis-2:2′-dipyridyl rhenichloride	$(C_{10}H_8N_2)_2ReCl_6$	713.31	pa. gr. cr.			sl. s.			
2	Dipyridyl *per*-rhenate(2:2′)	$(C_5H_4N)_2HReO_4$	407.39	col. need.			2.1			
3	Dipyridyl rhenichloride(2:2′)	$(C_5H_4N)_2ReCl_6$	555.12	yel. need.			sl. s.			
4	Trimethylrhenium	$Re(CH_3)_3$	231.31	col. oil		60				
5	Tripyridyl rhenichloride(2:2′:2″)	$(C_5H_4N)_3HReCl_6$	634.21	pa. grn. cr.			i.			
	Rubidium									
1	Monorubidium acetylide	$RbHC_2$	110.50	col. cr., very hygr.		d. 300				s. NH_3; i. eth.
	Silicon									
1	Aminotrisilane	$(H_3Si)_3N$	107.34	liq.	0.895^{-106}	−105.6	52	d.		s. eth.
2	Benzyltriethoxysilane	$C_6H_5CH_2Si(OC_2H_5)_3$	254.39	liq.	0.9864^{20}_{4}		253	i.	i.	s. al., acet., eth.
3	Butyltrifluorosilane	$C_4H_9SiF_3$	142.21	col. liq.	1.006^{25}_{4}	−96.6	52.4	d.		s. eth.
4	Carboxyethyl-dimethylsilane	$(CH_3)_2SiCH_2CH_2COOH$	131.23	n_D 1.4279		22				
5	Chloromethylsilicane	SiH_2ClCH_3	80.59		0.935^{-80}	−134.1	7			
6	Chlorotri-isocyanatesilane	$ClSi(NCO)_3$	189.59	solid; n^{20}_{D} 1.4507	1.437^{20}					s. bz., CS_2, CCl_4

No.	Name	Formula	Mol. wt.	Crystalline form, color and index of refraction	Sp. gr. or density	Melting point, °C	Boiling point, °C	Solubility in grams per 100 ml of		
								Cold water	Hot water	Alcohol, acids, etc.
	Silicon									
7	Di-*p*-aminoazobenzene fluosilicate	$(NH_2C_6H_4N_2C_6H_5)_2.H_2SiF_6$	538.57	long cinnamon br. need.	...	220 d.	...	...	...	187^{3}, 95% al.
8	Di-*p*-aminobenzoic acid fluosilicate	$(NH_2C_6H_4COOH)_2.H_2SiF_6$	418.37	pr. wh. long, narrow	...	242	...	...	...	0.91^{25}, 95% al.
9	Dianiline fluosilicate	$(C_6H_5NH_2)_2.H_2SiF_6$	330.33	irreg. wh. pl.	...	subl. 230	...	v. s.	...	s. h. al.
10	Di-*p*-biphenylyl diphenylsilane	$(C_6H_5)_2Si(C_6H_4C_6H_5)_2$	488.71	col. cr.	1.14^{20}	170	570	...	...	0.40^{20} al., 36.46^{20} bz., 12.91^{20} pyr.
11	Dichloromethylsilicane	$SiHCl_2CH_3$	115.02	...	0.93^{0}	−93	...	...	...	...
12	Didiphenylamine fluosilicate	$[(C_6H_5)_2NH]_2H_2SiF_6$	482.55	wh. rods forming rosettes	...	169	...	...	...	2.4492^{26}, 95% al.
13	Diethoxydibutoxysilane	$(C_2H_5O)_2Si(C_4H_9O)_2$	264.45	col. liq.; n_D^{20} 1.4010	0.909	...	128^{32}	d.	d.	s. org. solv.
14	Diethylaniline fluosilicate	$(C_6H_5NHC_2H_5)_2H_2SiF_6$	386.46	wh. pointed pr.	...	165.3	...	...	...	$.979^{26}$, 95% al.
15	Diethyldichlorosilane	$(C_2H_5)_2SiCl_2$	157.13	col. liq.; n_D^{20} 1.4809	1.0504_4^{20}	...	129	d.		s. eth.
16	Diethylsilanediol	$(C_2H_5)_2Si(OH)_2$	120.23	col. cr.	...	96	d. 140	s.	...	s. org. solv.
17	Dimethyoxdichlorosilane	$(CH_3O)_2SiCl_2$	161.07	col. liq.	1.2529_{15}^{0}	...	98–103	...	...	s. al., bz., eth.
18	Dimethylaniline fluosilicate	$(C_6H_5NHCH_3)_2H_2SiF_6$	358.38	monocl. wh. need.	...	...	...	...	...	s. h. al.; i. c. al.
19	Dimethyldi(β-chloroethoxy)silane	$(CH_3)_2Si(OC_2H_4Cl)_2$	217.17	col. liq. n_D^{20} 1.4420	1.135_4^{25}		213	i.	i.	s. al., acet., eth.
20	Dimethylfluorochlorosilane	$(CH_3)_2SiFCl$	112.61	col. liq.	1.18_4^{25}	−85.1	36.4	d.	...	s. eth.
21	Dimethylsilicane	$SiH_2(CH_3)_2$	60.17	...	0.68^{-80}	−150	−20.1	...	...	...
22	Di-α-naphthylamine fluosilicate	$(C_{10}H_7NH_2)_2.H_2SiF_6$	430.47	rosettes of wh. need.	...	218	...	...	...	$.1504^{25}$, 95% al.
23	Di-β-naphthylamine fluosilicate	$(C_{10}H_7NH_2)_2H_2SiF_6$	430.47	hex. wh. pl.	...	236.3	...	...	...	$.0816^{25}$; $.1248^{35}$, 95% al.
24	Di-*m*-nitraniline fluosilicate	$(C_6H_4NH_2NO_2)_2.H_2SiF_6$	420.50	rhomb. wh. pl.	...	200	...	...	...	$.121^{25}$; $.4736^{35}$, 95% al.
25	Dinitrosodiphenylamine	$[(C_6H_5)_2N=NO]_2.H_2SiF_6$	540.54	butterfly shaped indigo cr.	...	124.5	...	...	...	$.84^{25}$ 95% al.
26	Diphenylarsinophenylene triethylsilane	$(C_2H_5)_3SiC_6H_4As(C_6H_5)_2$	420.51	liq., n_D^{20} 1.61455	1.1661_4^{21}	...	$279–281^{17}$	...	...	...
27	Diphenyldichlorophenoxysilane	$(C_6H_5)_2Si(OC_6H_4Cl)_2$	437.40	col. liq., n_4^{20} 1.5510	1.2027_4^{20}	...	144	i.	i.	s. al., acet., eth.
28	Di-*o*-toluidine fluosilicate	$(C_6H_4NH_2CH_3)_2.H_2SiF_6$	358.41	wh. rhomb.	...	...	...	...	...	s. h. al.; i. c. al.
29	Di-*m*-toluidine fluosilicate	$(C_6H_4NH_2CH_3)_2.H_2SiF_6$	358.41	wh. rect. pr. pl.	...	...	...	...	...	s. h. al.; i. c. al.
30	Di-*p*-toluidine fluosilicate	$(C_6H_4NH_2CH_3)_2.H_2SiF_6$	358.41	wh. need., unst. irreg. outline	...	...	...	...	...	...
31	Docosamethyldecasiloxane	$(CH_3)_{22}O_9Si_{10}$	755.63	n_D^{20} 1.3988	0.925	...	183^{41}	i.	i.	s. bz.; sl. s. al.
32	Dodecamethylcyclohexasiloxane	$(CH_3)_{12}Si_6O_6$	444.93	oily liq., n_D^{20} 1.4015	0.9672	−3	128^{20}	...	...	...
33	Dodecamethylpentasiloxane	$(CH_3)_{12}O_4Si_5$	384.85	liq., n_D^{20} 1.3925	0.8755	*ca.* −80	229^{710}	i.	i.	s. bz.; sl. s. al.
34	Eicosamethylnonasiloxane	$(CH_3)_{20}O_7Si_9$	665.47	liq., n_D^{20} 1.3980	0.918	...	$173^{4.9}$	i.	i.	s. bz.; sl. s. al.

No.	Name	Formula	Mol. wt.	Crystalline form, color and index of refraction	Sp. gr. or density	Melting point, °C	Boiling point, °C	Solubility in grams per 100 ml of		
								Cold water	Hot water	Alcohol, acids, etc.
	Silicon									
35	Ethyldiethoxy-acetoxysilane	$C_2H_5Si(OC_2H_5)_2(OCOCH_3)$	206.31	liq., n_D^{20} 1.404	1.020_4^{20}		94^{16}	i.	i.	s. al., acet., eth.
36	Ethyldiethoxychloro-silane	$C_2H_5SiCl(OC_2H_5)_2$	182.72					i.	i.	s. al., acet., eth.
37	Ethylisocyanatesilane	$C_2H_5Si(NCO)_3$	183.20	sol., n_D^{20} 1.4468	1.2192^{20}	183.5		sl. d.		s. bz., CS_2, CCl_4
38	Ethyltriethoxysilane	$C_2H_5Si(OC_2H_5)_3$	192.33	liq., n_D^{20} 1.3853	0.9207_4^{20}		158.5	i.	i.	s. al., acet., eth.
39	Ethyltriphenylsilicane	$(C_2H_5)(C_6H_5)_3Si$	288.47	rhomboidal pr.		76		i.		s. chl., bz., eth., acet.; sl. s. al.
40	Hexadecamethylcyclo-octasiloxane	$(CH_3)_{16}Si_8O_8$	593.24	oily liq., n_D^{20} 1.4060		*ca.* 30	$175.^{20}$			
41	Hexamethylsilicane	$Si_2(CH_3)_6$	146.38	col. liq.	0.723_4^{20}	12.5–14	112.5	i.	i.	s. acet., bz., eth.; col.
42	Hexamethylmethyl-enedisilane	$(CH_3)_3SiCH_2Si(CH_3)_3$	160.41	liq., n_D^{20} 1.4172	0.7520_4^{20}		134^{760}			
43	Hydroxymethyltri-methylsilane	$OHCH_2Si(CH_3)_3$	104.23	liq., n_D^{25} 1.4169	0.8261_4^{20}		121.6			
44	Methylsilicane	SiH_3CH_3	46.14		0.62^{-67}	−156.4	31			
45	Methyltriethoxysilane	$CH_3Si(OC_2H_5)_3$	178.31	liq., n_D^{20} 1.3861	0.9383_4^{20}		151	i.	i.	s. al., acet., eth.
46	Methyltriphenyl-silicane	$(CH_3)(C_6H_5)_3Si$	274.44			67.3		i.		s. eth., chl., bz.; sl. s. al.
47	Octadecamethyl-octasiloxane	$(CH_3)_{18}O_7Si_8$	607.31	liq., n_D^{20} 1.3970	0.913		$153^{5.1}$			s. bz., sl. s. al.
48	Octamethylcyclo-tetrasiloxane	$(CH_3)_8Si_4O_4$	296.62	oily liq., n_D^{20} 1.3968	0.9558		175.0^{0}			
49	Octamethyltrisiloxane	$(CH_3)_8O_2Si_3$	236.54	liq., n_D^{20} 1.3848	0.8200	−80	153			s. bz.; sl. s. al.
50	Phenylenediamine fluosilicate(*m*)	$Si(C_6H_5CH_2)_4$	252.24	choc. br. need.-like pr.		243–4				0.065^{26}, 95 % al.
51	Phenylenediamine fluosilicate(*p*)	$C_6H_4(NH_2)_2H_2SiF_6$	252.24	pink irreg. six-sided pl.		d.				0.014^{26}, 95 % al.
52	Phenylisocyanatesilane	$C_6H_5Si(NCO)_3$	231.24	solid, n_D^{20} 1.5210	1.273^{20}	251.9		sl. d.		s. bz., CS_2, CCl_4
53	Phenyltrichlorosilicane	$C_6H_5SiCl_3$	211.55	liq.	1.326_4^{18}		201.5	d.		s. eth., chl.; d. al.
54	Silicobenzoic acid	C_6H_5SiOOH	138.20	col. flaky resin		40–50	215.6–			s. eth., bz., chl.; sl. s. al., ac. a.
55	Tetra-*m*-aminophenyl-silane	$(NH_2C_6H_4)_4Si$	396.57	solid		380				
56	Tetrabenzylsilicane	$Si(C_6H_5CH_2)_4$	392.62				127.5	i.		s. eth., chl., bz.; sl. s. al.
57	Tetra-*p*-Biphenylyl-silane	$Si(C_6H_4C_6H_5)_4$	640.91	col. cr.		283	600			0.049^{20} al.; 0.27^{20} bz.; 16.69^{20} pyr.
58	Tetradecamethylcyclo-heptasiloxane	$C_{14}H_{42}Si_7O_7$	519.09	oily liq., n_D^{20} 1.4040	0.9730	−26	154^{20}			
59	Tetradecamethyl-hexasiloxane	$C_{14}H_{42}O_5Si_6$	459.00	liq., n_D^{20} 1.3948	0.8910	−100	142^{20}			s. bz.; sl. s. al.
60	Tetraethylsilicane	$Si(C_2H_5)_4$	144.33	liq., 1.4246	0.762^{25}		152.8–3.2			

No.	Name	Formula	Mol. wt.	Crystalline form, color and index of refraction	Sp. gr. or density	Melting point, °C	Boiling point, °C	Solubility in grams per 100 ml of: Cold water	Hot water	Alcohol, acids, etc.
	Silicon									
61	Tetraethylthiosilane	$(C_2H_5S)_4Si$	272.59	liq., n_D^{25} 1.5638	1.0860^{25}_4	−6	170^{12}			
62	Tetrahexyloxysilane	$(C_6H_{13}O)_4Si$	432.77	col. liq., n_D^{20} 1.4300	0.876		$232–234^{13}$	d.	d.	s. org. solv.
63	Tetraisopropyl-mercaptane silicon	$C_{12}H_{28}S_4Si$	392.83	solid, n_D^{35} 1.5350	1.0099^{35}_4	335	$176–178^{13}$			
64	Tetramethoxysilane	$(OCH_3)_4Si$	152.22	col. liq., n_D^{20} 1.3681	1.0523		$121–122^{759}$	d.	d.	s. org. solv.
65	Tetramethyl-mercaptanesilicon	$(CH_3)_4S_4Si$	216.48	solid, n_D^{20} 1.5989	1.1888^{25}_4	31	$144–146^{12}$			
66	Tetramethylsilicane	$Si(CH_3)_4$	88.23	col. liq.	0.651^{15}		26.5			s. eth.; i. conc. H_2SO_4
67	Tetraphenoxysilane	$(C_6H_5O)_4Si$	400.51	col. cr.		47–48	$415–420^7$	d.	d.	s. org. solv.
68	Tetraphenylsilicane	$Si(C_6H_5)_4$	336.51	col. flocc. amor. part.						s. acet. anh. and chloro-sulphonic acid
69	Tetrapropoxysilane	$(OC_3H_7)_4Si$	264.44	col. liq., n_D^{20} 1.4015	0.918		$225–227^{769}$	d.	d.	s. org. solv.
70	Tetratriethyl-siloxysilane	$Si[OSi(CH_3)_2C_2H_5]_4$	440.96	col. liq., 7_D^{20} 1.4112	$0.895^{20}_{15\ 6}$	−45	102^2			
71	Thioisocyanatotri-ethylsilane	$(C_2H_5)_3SiNCS$	173.35	liq., n_D^{20} 1.4944	0.934^{20}_4		210.5			
72	Tolidine fluosilicate(*o*)	$(CH_3NH_2C_6H_3)_2H_2SiF_6$	356.39	tiny, micr. wh. pr.		268–9				$.013^{26}$, 0.41^{36} 95 % al.
73	Tri-*p*-Biphenylyl-phenylsilane	$(C_6H_5)_3SiC_6H_4C_6H_5$	412.61	col. cr.	1.10^{20}	174	580			0.238^{20} al.; 17.776^{20} bz.; 2.725^{20} pyr.
74	Trichloromethyl-triethyloxysilane	$CCl_3Si(OC_2H_5)_3$	281.64	col. liq., n_D^{20} 1.4320	1.8660^{20}_4		81^3	i.	i.	s. al., acet., eth.
75	Triethylbromosilane	$(C_2H_5)_3SiBr$	195.18	col. liq., n_D^{20} 1.4561	1.777^{20}_4		162	d.		s. eth.
76	Triethylchlorosilane	$(C_2H_5)_3SiCl$	150.73	col. liq., n_D^{20} 1.4314	0.8967^{20}_4		144^{735}	d.		s. eth.
77	Triethylfluorosilane	$(C_2H_5)_3SiF$	134.27	col. liq., n_D^{20} 1.3915	0.9354^{20}_4		$109–100^{745}$	d.		s. eth.
78	Triethylphenylsilicane	$(C_2H_5)_3(C_6H_5)Si$	192.38			149	230			
79	Trimethylchloro-methylsilane	$(CH_3)_3SiCH_2Cl$	122.67	col. liq., n_D^{20} 1.4180	0.8791^{20}_4		97.1^{734}			
80	Trimethylethoxysilane	$(CH_3)_3Si(OC_2H_5)$	118.25	liq., n_D^{20} 1.3741	0.7573^{20}_4		75^{745}	i.	i.	s. al., acet., eth.
81	Triphenylacetoxysilane	$(C_6H_5)_3SiOCOCH_3$	318.45	col. cr.		91–92				
82	Vinyltriphenoxysilane	$CH_2{:}CHSi(OC_6H_5)_3$	334.45	liq., n_D^{20} 1.5617	1.1300^{20}_4		210.2^7	i.	i.	s. al., acet., eth.
	Silver									
1	Acetylene silver nitrate (complex.)	$Ag_2C_2.AgNO_3$	409.64	fine. nd. and crosses		detonating pt. *ca.* 212				
2	Cycloocta-1,3-dienesilver nitrate	$C_8H_{12}.2AgNO_3$	447.94	col. need.		150 d.				s. h. meth. al.
3	Cycloocta-1,4-dienesilver nitrate	$C_8H_{12}.2AgNO_3$	447.94	lg. col. hex. tab.		110–1				sl. s. boil. meth. al.
4	Cycloocta-1,5-dienesilver nitrate	$C_8H_{12}.AgNO_3$	278.06	lng. col. need.		135–6 (128.5–131)				v. s. meth. al.

No.	Name	Formula	Mol. wt.	Crystalline form, color and index of refraction	Sp. gr. or density	Melting point, °C	Boiling point, °C	Solubility in grams per 100 ml of: Cold water	Hot water	Alcohol, acids, etc.
	Silver									
5	*cis-trans*-1,3-Cyclooctadienesilver nitrate	$C_8H_{12}.AgNO_3$	278.06	wh. cr.		126–127.5 d.				sl. s. abs. al.
6	Cyclooctatetraene dimer silver nitrate	$C_{16}H_{16}.AgNO_3$	378.18	pr.		153				s. h. al.
7	Cyclooctatetraenesilver nitrate	$C_8H_8.AgNO_3$	274.03	lg. cr.		173 d.		d.		s. aq. $AgNO_3$; s. al.
8	Cyclooctatetraenylphenylketonesilver nitrate	$C_{15}H_{12}O.AgNO_3$	378.13	pa. yel. need.		121.4–122 d.				sl. s. abs. al.
9	Cycloocta-1,3,6-trienesilver nitrate	$C_8H_{10}.3AgNO_3$	615.80	flat need		138–9				sl. s. meth. al.
10	1,3,5-Cyclooctatriene-silver nitrate	$C_8H_{10}.AgNO_3$	276.04	cr.		125–6				sl. s. abs. al.
11	*cis*-Cyclooctenesilver nitrate	$(C_8H_{14})_2.AgNO_3$	390.28	lg. cr.		51		s.		v. s. meth. al.
12	1,2-Dimethylcyclooctatetraenesilver nitrate	$C_{10}H_{12}.2AgNO_3$	471.95	cr.		142.5–144.5				s. h. abs. al.
13	Ethylcyclooctatetraene-silver nitrate	$C_{10}H_{12}.2AgNO_3$	471.95	gray wh. cr.		124–125.5				i. eth.; sl. s. abs. al.
14	Ethylsilver	C_2H_4Ag	135.93			d. > −40				
15	Methylcyclooctatetraenesilver nitrate	$2C_9H_{10}.3AgNO_3$	745.98	lt. yel. pr.		123–124.5				sl. s. abs. al.; i. eth.
16	Methylsilver	$AgCH_3$	122.90	yel.		d. > −40				meth. al. 0.04^{-80}; al. 0.04^{-80}; amyl al. 0.06^{-80}
17	Phenylcyclooctatetraenesilver nitrate	$C_6H_5.C_8H_7.AgNO_3$	350.12	yel. gr. cr.		144.5 d.				i. c. eth.; s. h. abs. al.
18	*n*-Propylcyclooctatetraenesilver nitrate	$C_{11}H_{14}.2AgNO_3$	485.98	pa. yel. cr.		141 d.				s. h. abs. al.
19	Silver acetylide silver nitrate complex	$Ag_2C_2.6AgNO_3$	1259.01	wh. rhomb.		d. >308		d.		i. acet.
	Sodium									
1	Bis(1-methylamyl)-sodium sulfosuccinate (Aerosol MA)	$C_{16}H_{29}NaO_7S$	388.46	wh. waxlike pellets				34.3^{25}	44.7^{70}	s. acet., bz., glyc., CCl_4
2	Hydroxydionsodium (Viadril)	$C_{25}H_{35}NaO_6$	454.54	wh. powd.		d. 193–203		s.	s.	s. acet., chl.
3	Sodium acetamide	$NaNHCOCH_3$	81.05	wh. tabl.		300–50 d.		d.		d. al.; sl. s. bz., liq. NH_3
4	Sodium acetylide (ethinylsodium)	$NaHC_2$	48.02	wh. cr.		d. >210		d.		d. a.; s. liq. NH_3
5	Sodium anilide	$NaNHC_6H_5$	115.11	wh. cr. v. hygr.		d.		d.		d. a., al.; s. liq. NH_3
6	Sodium anthraquinone-*β*-sulfonate ("Silver salt")	$NaC_{14}H_7O_5S.H_2O$	328.28	silvery leaf		d.		0.84	27	v. sl. s. al.
7	Sodium arsanilate (atoxyl, soamim)	$NaC_6H_7O_3NAs.6H_2O$	347.13	wh. cr. powd., monocl.				$16^{1.7}$		sl. s. al.; s. meth. al.
8	Sodium benzamide	$NaNHCOC_6H_5$	143.12	wh. powd.		d.		d.		d. al.; i. eth., bz., chl.
9	Sodium-*N*-chloro-*p*-toluenesulfonamide	$NaC_7H_7O_2NClS.3H_2O$	281.69	col. pr.		expl. 175–180		s.	v. s.	
10	Sodium ethoxide (sodium ethylate)	$NaOC_2H_5.2C_2H_5OH$	160.19	wh. powd. or need.		$-2C_2H_5OH$, 200	d.	d.	d.	v. s. al.; i. NH_3
11	Sodium, ethyl-	NaC_2H_5	52.05	wh. cr., d. air		d.		d.		d. al., eth.; s. diethylzinc; i. bz., lgr.
12	Sodium-*β*-naphthoxide	$NaOC_{10}H_7$	166.16	wh. powd., v. hygr.		d.		s.		v. s. al., eth.; i. lgr.
13	Sodium-*p*-nitrobenzene isodiazotate	$NaC_6H_4O_3N_3.2H_2O$	225.14	gold. leaf. or need.		$-H_2O$ over H_2SO_4	exp.	v. s.		

No.	Name	Formula	Mol. wt.	Crystalline form, color and index of refraction	Sp. gr. or density	Melting point, °C	Boiling point, °C	Solubility in grams per 100 ml of: Cold water	Hot water	Alcohol, acids, etc.
	Sodium									
14	Sodium-*p*-nitrophenoxide	$NaOC_6H_4NO_2.4H_2O$	233.15	yel. monocl. pr.		$-2H_2O$, 36; $-4H_2O$, 120	d.	5.97^{25}		sl. s. al.
15	Sodium-*o*-sulfo-benzoicimide (soluble saccharin)	$NaC_7H_4O_3NS.2H_2O$	241.20	wh. tabl.		$-H_2O$		v. s.		sl. s. h. al.
16	Triphenylborylsodium*	$NaB(C_6H_5)_3$	265.12	yel.-or. silky need.			d.	d.	d.	0.08^{18} eth.
17	Triphenylmethylsodium	$NaC(C_6H_5)_3$	266.32	red. cr.				d.		s. eth., bz., liq. NH_3
	Tantalum									
1	*bis*-Cyclopentadienyl-tantalum tribromide	$(C_5H_5)_2TaBr_3$	550.87	rust. cr.		280 d.		s.		sl. s. chl.;
2	Tantalumpenta-methoxide	$Ta(OCH_3)_5$	336.12	wh. solid		50	189^{10}	d.		
	Tellurium									
1	Di-*n*-butyl telluride	$(C_4H_9)_2Te$	241.83	yel. oil	1.334^{40}		132–5			
2	Diethyl telluride	$(C_2H_5)_2Te$	185.72				137–8	sl. s.		
3	Diethyltelluroketone	$C_2H_5.CTe.C_2H_5$	197.74	yelsh., oily liq. n_D^{20} 1.5480	0.8821^{15}_{4}		$69–72^{8}$	i.	i.	v. s. eth., 95% al. s.
4	Dimethyl telluride	$(CH_3)_2Te$	157.67	pa. yel. oil.; garlic-like odor		sld. in liq. air	82.97^{770}			
5	Dimethyltelluroketone (Telluroacetone)	CH_3CTeCH_3	169.68	col. oily liq., n_D^{25} 1.4883	0.8578^{18}_{4}		$55–58^{10}$	i.	i.	s. eth., sl. s. 95% al.
6	Dimethyltelluronium dibromide(α)	$C_2H_6Br_2Te$	317.49	or. leaf-like cr.		142 d.				s. al., eth.
7	Dimethyltelluronium dichloride(α)	$C_2H_6Cl_2Te$	228.58	leaf-like cr.		92		s.	s.	s. al., eth.
8	Dimethyltelluronium dichloride(β)	$C_2H_6Cl_2Te$	228.58	leaf-like cr.		134				s. al., eth.
9	Dimethyltelluronium diiodide(α)	$C_2H_6I_2Te$	411.48	red cr.		127 d.		i.	v. sl. s.	s. chl., bz.
10	Di-*p*-phenetyl ditelluride	$(C_2H_5OC_6H_4)_2Te_2$	497.52	or. brown need.	1.666	108	d. >108			s. lgr.
11	Ditelluromethane	CH_2Te_2	269.23	dk. red. amor. sld.		d. 124		i.	i.	i.
12	Ethylmethyl-tellurophetone	$C_2H_5CTeCH_3$	183.71	dark yel. oil, 1.5055^{25}	1.8711		63–6			
	Thallium									
1	Di-*n*-Butylthallium chloride	$(n\text{-}C_4H_9)_2TlCl$	354.06	col. pl. or flakes (from pyr.)		d. 240–250 (expl.)		sl. s.		s. hot bz.; 0.28^{30} al.; 2.98^{30} pyr; v. sl. s. eth.
2	Dicyclohexylthallium chloride	$(C_6H_{11})_2TlCl$	406.13	long col. need.		d. 210–230		v. sl. s.	i.	s. hot al., hot bz.; 1.34^{30} pyr.; i. eth.
3	Diethylthallium hydroxide	$Tl(C_2H_5)_2OH$	279.50			127–128		v. s.		v. s. al.
4	Di-*n*-hexylthallium chloride	$(n\text{-}C_6H_{13})_2TlCl$	410.16	col. pl.		d. 198		0.0117^{23}		s. pyr.; 0.123^{23} al.; 0.0304^{23} bz.; i. HNO_3
5	Di-*n*-hexylthallium nitrate	$(n\text{-}C_6H_{13})_2TlNO_3$	436.72	shiny flakes		d. 271		0.008^{23}		0.1856^{32} al.; v. sl. s. bz.
6	Diisoamylthallium chloride	$(i\text{-}C_5H_{11})_2TlCl$	382.11	col. cr.		d. 253		sl. s.		0.31^{30} al.; 1.74^{30} pyr.
7	Dipropylyhallium chloride	$Tl(C_3H_7)_2Cl$	326.00	silvery leaflets		d. 200				s. eth.; i. HCl
8	Triethylthallium	$Tl(C_2H_5)_3$	291.56	yel. liq.	1.957^{23}_{23}	−63	192			

* All of the tri-arylborines form analogous addition-salts of the alkali metals.

No.	Name	Formula	Mol. wt.	Crystalline form, color and index of refraction	Sp. gr. or density	Melting point, °C	Boiling point, °C	Solubility in grams per 100 ml of: Cold water	Hot water	Alcohol, acids, etc.
	Thallium									
9	Trimethylthallium	$Tl(CH_3)_3$	249.48	col. need.		38.5	147; may expl. 90	d.		v. s. bz., eth.
	Thorium									
1	Thoriumisotetra-propoxide	$Th(OC_3H_7)_4$	468.39	wh. cr.		$80^{0.1}$	subl. 200–$210^{0.1}$	d.		s. bz.
	Tin									
1	Amyltetrathioortho-stannate(*n*)	$Sn(SC_5H_{11})_4$	531.52				$162^{.004}$			
2	Amyltetrathioortho-stannate(*tert*)	$[CH_3CH_2C(CH_3)_2S]_4Sn$	531.52			44				
3	Bis(tributyltin) oxide	$[(C_4H_9)_3Sn]_2O$	596.07	yel. liq.	1.17^{25}		254^{50}		0.1 %	
4	Bis(triphenytin) sulfide	$[(C_6H_5)_3Sn]_2S$	732.07	solid		141–143				
5	Bromobenzenetetra-thioorthostannate(*p*)	$Sn(SC_6H_4Br)_4$	870.98			217				
6	Butylbenzenetetra-thioorthostannate(*p*)	$Sn(SC_6H_4C_4H_9)_4$	779.81			106				
7	Butyltetrathioortho-stannate(*n*)	$Sn(SC_4H_9)_4$	475.41					$136^{.001}$		
8	Butyltetrathioortho-stannate(*sec*)	$Sn(SC_4H_9)_4$	475.41				$111^{.001}$			
9	Butyltin trichloride	$C_4H_9SnCl_3$	282.17	col. liq., 1.5244^{20}	1.7^{20}	−63	98^{10}			s. org. solv.
10	*n*-Butylvinyltin dichloride	$n\text{-}C_4H_9(CH_2{:}CH)SnCl_2$	273.78	solid, n_D^{25} 1.5254	1.533_4^{25}	27–28	$99–101^{3}$			
11	Carbomethoxyphenyl-trichlorostannane(*o*)	$(CH_3OCOC_6H_4)SnCl_3$	360.19			164				
12	Chlorobenzenetetra-thioorthostannate(*p*)	$Sn(SC_6H_4Cl)_4$	693.15			189				
13	Cyclohexyltetrathio-orthostannate	$Sn(SC_6H_{11})_4$	579.56			53–4				
14	Diallylsdibutyltin	$(C_4H_9)_2(CH_2{:}CHCH_2)_2Sn$	315.07	liq., 1.4986	1.0999		$93^{0.1}$			
15	Diallyldiphenyltin	$(C_6H_5)_2(CH_2{:}CHCH_2)_2Sn$	355.05	liq., 1.6013^{25}	1.2688^{25}		$173–174^{5}$			
16	Di-*o*-anisyldichloro-stannane	$(CH_3OC_6H_4)_2SnCl_2$	403.86			113				
17	Dibenzyldiethyl-stannane	$(C_6H_5CH_2)_2Sn(C_2H_5)_2$	359.08	liq.	1.+	<20	$223–4^{20}$			s. org. solv.
18	Dibenzylethylpropyl-stannane	$(C_6H_5CH_2)_2(C_2H_5)(C_3H_7)Sn$	373.11	liq.		>0	$220–5^{15}$			misc. all org. solv.
19	Dibenzyltin acetate	$(C_6H_5CH_2)_2Sn(OCOCH_3)_2$	419.05	col. need. f. al.		136–7				s. acet. chl., bz.
20	Dibenzyltin dibromide	$(C_6H_5CH_2)_2SnBr_2$	460.78	col. need. f. pet.		130				s. acet., al., eth., chl., CCl_4
21	Dibenzyltin dichloride	$(C_6H_5CH_2)_2SnCl_2$	371.86	col. need. f. acet.-HCl		163–4				s. acet., al., eth., chl., CCl_4, h. ac. a.
22	Dibenzyltin diiodide	$(C_6H_5CH_2)_2SnI_2$	554.77	col. lng. silky yel. need f. pet. eth.		86–7				s. acet., al., eth., chl., CCl_4
23	Dibutyldiphenyltin	$(C_4H_9)_2(C_6H_5)_2Sn$	387.14	col. liq., 1.5602	1.1882		180^{2}			
24	Dibutyldivinyltin	$(C_4H_9)_2(CH_2{:}CH)_2Sn$	287.02	liq., 1.4824	1.1270		$78–80^{2}$			
25	Dibutyltin diacetate	$(C_4H_9)_2Sn(OOCH_3)_2$	326.99	col. liq., 1.482^{20}	1.32	10	$142–145^{10}$	i.	i.	s. org. solv.
26	Dibutyltin dibromide	$(C_4H_9)_2SnBr_2$	392.74	sm. need.		20	$118–170^{28}$	i.	i.	
27	Dibutyltin dichloride	$(C_4H_9)_2SnCl_2$	303.83	need. wh. 1.499^{50}	1.36^{50}	142^{10}	d. 113.6^{60}	d.	d.	s. eth., bz., al.
28	Dibutyltin dilaurate	$(C_4H_9)_2Sn(OOC_{11}H_{23})_2$	631.55	oily liq., 1.474^{14}	1.05	27		i.	i.	s. ac., bz., eth., CCl_4
29	Dibutyltin oxide	$(C_4H_9)_2SnO$	248.92	wh. amorph. powd.	1.58	d.		i.	i.	i. org. solv.

No.	Name	Formula	Mol. wt.	Crystalline form, color and index of refraction	Sp. gr. or density	Melting point, °C	Boiling point, °C	Solubility in grams per 100 ml of		
								Cold water	Hot water	Alcohol, acids, etc.
	Tin									
30	Dibutyltin sulfide	$(C_4H_9)_2SnS$	264.99	col.-yel. lip., 1.58	1.417					
31	Dibutylvinyltin bromide	$(C_4H_9)_2(CH_2{:}CH)SnBr$	339.88	col. liq., 1.4970	1.3913		$96^{0.55}$			
32	Dibutylvinyltin chloride	$(C_4H_9)_2(CH_2{:}CH)SnCl$	295.42	liq. 1.4987^{20}	1.2662^{20}		$112–114^{4}$			
33	Dichlorodi-*m*-tolyl stannane	$(CH_3C_6H_4)_2SnCl_2$	371.86			39–40				
34	Diethyldibromo-dipyridinetin	$(C_2H_5)_2SnBr_2(C_5H_5N)_2$	494.84			140				
35	Diethyldiisoamyltin	$(C_2H_5)_2Sn(C_5H_{11})_2$	319.10		1.0725^{19}		$131^{13.5}$			
36	Diethyldiisobutyltin	$(C_2H_5)_2Sn(C_4H_9)_2$	291.05		1.1030		108.2^{13}			
37	Diethyldiphenyltin	$(C_2H_5)_2(C_6H_5)_2Sn$	331.03				$154–6^{4}$			
38	Diethylisoamyltin bromide	$(C_2H_5)_2(C_5H_{11})SnBr$	327.87		1.4881^{17}		137.5^{17}			
39	Diethylisoamyltin chloride	$(C_2H_5)_2(C_5H_{11})SnCl$	283.41		$1.2994^{18.9}$		125.5^{13}			
40	Diethylisobutyltin bromide	$(C_2H_5)_2(C_4H_9)SnBr$	313.84		1.5108		122^{17}			
41	Diethyl-*n*-propyltin bromide	$(C_2H_5)_2(C_3H_7)SnBr$	299.81		1.5910^{21}		112.2^{16}			
42	Diethyl-*n*-propyltin chloride	$(C_2H_5)_2(C_3H_7)SnCl$	255.36		$1.3848^{15.7}$		108^{17}			
43	Diethyl-*n*-propyltin fluoride	$(C_2H_5)_2(C_3H_7)SnF$	238.90			271				6.93^{31}, meth. al.; 3.78^{12} al.; $.05^{31}$ bz.
44	Diethyltin	$Sn(C_2H_5)_2$	176.81	sl. yel. oily liq.	1.654	<−12	150 d.	i.	i.	s. bz., eth., lgr., chl., CCl_4
45	Diethyltin dibromide	$(C_2H_5)_2SnBr_2$	336.63	col. need.	2.068^{14}	63	232–3	s.	s.	s. eth., org. solv.
46	Diethyltin dichloride	$(C_2H_5)_2SnCl_2$	247.72	wh. need.		84–5	220	s.	s.	s. HCl, org. solv.
47	Diethyltin difluoride	$(C_2H_5)_2SnF_2$	214.81	sp. pl. or long rhomb. tab. f. meth. al.		229				$.45^{31}$, al.; 2.64^{31}, meth. al.; 0.47^{31}, bz.
48	Diethyltin diiodide	$(C_2H_5)_2SnI_2$	430.62	wh. need.		44.5–5.0	240–5 d.	v. sl. s.	sl. s.	s. org. solv.
49	Diethyltin oxide	$(C_2H_5)_2SnO$	192.81	wh. powd.		infus.		i.	i.	s. HCl, dil. al., conc. alk.; i. org. solv.
50	Diisoamyltin dibromide	$(C_5H_{11})_2SnBr_2$	420.79			−25 to −24				
51	Diisoamyltin dichloride	$(C_5H_{11})_2SnCl_2$	331.88			28				
52	Diisoamyltin diiodide	$(C_5H_{11})_2SnI_2$	514.79	oily liq.			$202–5^{8}$			
53	Diisobutyltin dichloride	$(CH_3)_2(CHCH_2)_2SnCl_2$	273.76	liq.	1.5012	9	135.5			
54	Diisobutyltin diiodide	$(C_4H_9)_2SnI_2$	486.73				290–5			
55	Diisopropyltin dibromide	$(C_3H_7)_2SnBr_2$	364.69	pa. yel. hyg. cr.		54		d.	d.	i. org. solv.
56	Diisopropyltin dichloride	$(C_3H_7)_2SnCl_2$	275.77	col. transp. cr.		80–4		s.	s.	s. al., h. bz., glac. ac. a.
57	Diisopropyltin oxide	$(C_3H_7)_2SnO$	220.87			d.		i.	i.	s. h. HCl; i. org. solv., alk.
58	Dimethyldibromo-dipyridinetin	$(CH_3)_2SnBr_2(C_5H_5N)_2$	466.78			172				
59	Dimethyldichloro-dipyridinetin	$(CH_3)_2SnCl_2(C_5H_5N)_2$	377.87			163				
60	Dimethyldiethyltin	$(CH_3)_2Sn(C_2H_5)_2$	206.88	col. liq.	1.2319^{19}	<−13	144–6	i.	i.	s. org. solv.
61	Dimethyldiisobutyltin	$(CH_3)_2Sn(C_4H_9)_2$	262.99		$1.1179^{20.1}$		$85^{16.5}$			
62	Dimethyldioctyltin	$(CH_3)_2(C_8H_{17})_2Sn$	375.21	liq., 1.4659	1.0168		$121–122^{0.2}$			
63	Dimethyldivinyltin	$(CH_3)_2Sn(CH{:}CH_2)_2$	202.85	n_D^{25} 1.4720	1.284_4^{25}		$120–121^{760}$			
64	Dimethylethylpropyltin	$(CH_3)_2(C_2H_5)C_3H_7Sn$	220.91		1.2014_{20}^{20}		149–51			

No.	Name	Formula	Mol. wt.	Crystalline form, color and index of refraction	Sp. gr. or density	Melting point, °C	Boiling point, °C	Solubility in grams per 100 ml of: Cold water	Hot water	Alcohol, acids, etc.
	Tin									
65	Dimethylethyltin iodide	$(CH_3)_2C_2H_5SnI$	304.73	1.5705^{18}	2.0264^{20}_{20}		$77–8^{11}$; $185–7^{718}$			
66	Dimethyltin	$[(CH_3)_2Sn]_x$	$(148.76)_x$	yel. sld.				i.	i.	i. org. solv.
67	Dimethyltin dibromide	$(CH_3)_2SnBr_2$	308.58	col. pr.		74–6	208–13	s.	s.	s. org. solv.
68	Dimethyltin dichloride	$(CH_3)_2SnCl_2$	219.67	col. cr.		90 (107)	188–90	s.		s. org. solv.
69	Dimethyltin difluoride	$(CH_3)_2SnF_2$	186.76	wh. fine pl.			d. <360	$4.66^{30.7}$		$.08^{31}$, al., 33^{31}, meth. al.
70	Dimethyltin diiodide	$(CH_3)_2SnI_2$	402.57	rhomb. wh.	2.872	43 (30)	228	sl. s.	s.	s. org. solv.
71	Dimethyltin oxide	$(CH_3)_2SnO$	164.76	wh. powd.	1.269	infus.	d.	i.	i.	s. a., NaOH; i. org. solv., NH_4OH
72	Dimethyltin sulfide	$(CH_3)_2SnS$	180.82			148				
73	Dimethylvinyltin bromide	$(CH_3)_2(CH_2{:}CH)SnBr$	255.72	solid n_D^{25} 1.5350	1.738^{66}	$59–61^{27}$				
74	Dimethylvinyltin iodide	$(CH_3)_2(CH{:}CH_2)SnI$	302.71	n_D^{25} 1.3762	2.033^{25}_{4}		$57.5–59^{5.2}$			
75	Di-α-naphthyltin	$Sn(C_{10}H_7)_2$	373.02			200	d. 225			
76	Diphenyldivinyltin	$(C_6H_5)_2Sn(CH{:}CH_2)_2$	327.00	n_D^{25} 1.5949	1.3195^{25}_{4}		$153–154^{5}$			
77	Diphenyldivinyltin	$(C_6H_5)_2Sn(CH{:}CH_2)_2$	326.98	liq., 1.5227^{20}	1.3049^{20}	$143–144^{5}$				
78	Di(phenylthiol)-diphenylstannane	$Sn(C_6H_5)_2(SC_6H_5)_2$	491.25			65–65.5				
79	Diphenyltin	$Sn(C_6H_5)_2$	272.90	yel. amor. powd.		225.7; (126–30)		i.	i.	s. chl., bz., eth.; i. abs. al.
80	Diphenyltin dibromide	$(C_6H_5)_2SnBr_2$	432.72	col. cr.		38	230^{42}			s. al., eth.
81	Diphenyltin dichloride	$(C_6H_5)_2SnCl_2$	343.81	col. cr.		42	333–7 d.	v. sl. s., d.		s. al., eth., lgr.
82	Diphenyltin difluoride	$(C_6H_5)_2SnF_2$	310.90			360				
83	Diphenyltin diiodide	$(C_6H_5)_2SnI_2$	526.71	col. cr.		71–72	$176–82^{2}$	i.	i.	s. org. solv.
84	Diphenyltin hydroxychloride	$(C_6H_5)_2Sn(OH)Cl$	325.36	amor. wh. powd.		187		i.	i.	s. conc. a.; i. org. solv.
85	Diphenyltin oxide	$(C_6H_5)_2SnO$	288.90	col. amor. powd.		infus.		i.	i.	s. conc. a.; i. org. solv.
86	Diphenyltin sulfide	$(C_6H_5)_2SnS$	304.97	wh. cr.		171–173				
87	Di-*n*-propyldibromodipyridinetin	$(C_3H_7)_2SnBr_2(C_5H_5N)_2$	522.89			128				
88	Dipropyltin dibromide	$(C_3H_7)_2SnBr_2$	364.69	col. need.		49		v. sl. s.		s. org. solv.
89	Dipropyltin dichloride	$(C_3H_7)_2SnCl_2$	275.77	col. cr.		81		v. sl. s.		s. org. solv.
90	Dipropyltin difluoride	$(C_3H_7)_2SnF_2$	242.87	leaf.		205		0.22^{32}		$.93^{32}$, al.. 1.91^{32}, meth. al.
91	Dipropyltin diiodide	$(C_3H_7)_2SnI_2$	458.68	col. oily liq.		<−15	270–3	i.	i.	s. org. solv.
92	Di-*m*-tolylstannane	$(CH_3C_6H_4)_2SnO$	316.96	wh. amor. infus.				i.	i.	s. min. a.; i. org. solv.
93	Di-*m*-tolyl thiostannane	$(CH_3C_6H_4)_2SnS$	333.02			121.5–2				v. s. chl., bz., eth. acetate, pyr., eth.; s. HCl
94	Di-*p*-tolyltin	$(CH_3C_6H_4)_2Sn$	300.96	or.-yel. amor. powd.		111.5	d. <245			s. bz.
95	Di-*o*-tolyltin dichloride	$(CH_3C_6H_4)_2SnCl_2$	371.86			49–50				
96	Di-*p*-tolyltin dichloride	$(CH_3C_6H_4)_2SnCl_2$	371.86			49–50				
97	Ditriphenylstannylmethane	$[(C_6H_5)_3Sn]_2CH_2$	714.05	wh. cr. sld.		104.5				v. s. bz. eth., chl.; s. h. pet. eth.
98	Divinylbutyltin chloride	$(C_4H_9)(CH_2{:}CH)_2SnCl$	265.35	liq., 1.4970	1.370		82.84^{3}			
99	Divinyltin dichloride	$(CH_2{:}CH)_2SnCl_2$	243.69	col. liq., 1.541	1.762		$54–56^{3}$	i.	i.	s. bz., MeOH
100	Di-*p*-xylyltin	$[(CH_3)_2C_6H_3]_2Sn$	329.01			157	d. 240			
101	Dodecyltetrathioorthostannate(*n*)	$Sn(SC_{12}H_{25})_4$	924.28			35.5				
102	Ethylchlorostannic acid	$H_2SnC_2H_5Cl_5$	327.03	col. deliq. pr.		d.		d.		
103	Ethyldiisoamyltin bromide	$(C_2H_5)(C_5H_{11})_2SnBr$	369.95		1.3650		$154–5^{16}$			

PHYSICAL CONSTANTS OF ORGANOMETALLIC COMPOUNDS (Continued)

No.	Name	Formula	Mol. wt.	Crystalline form, color and index of refraction	Sp. gr. or density	Melting point, °C	Boiling point, °C	Solubility in grams per 100 ml of: Cold water	Hot water	Alcohol, acids, etc.
	Tin									
104	Ethyldiisobutyltin bromide	$(C_2H_5)(C_4H_9)_2SnBr$	341.90		$1.4089^{19.5}$		130.6^{13}			
105	Ethylmethylpropyltin-iodide	$(CH_3)(C_2H_5)(C_3H_7)SnI$	332.78	1.5548^{17}	1.8182^{20}_{20}		$108–11^{11}$; $226–30^{720}$ sl. d.			
106	Ethyl-*n*-propyldi-isoamyltin	$(C_2H_5)(C_3H_7)Sn(C_5H_{11})_2$	333.13		$1.0654^{21.9}$		141^{17}			
107	Ethylpropyltin dichloride	$(C_2H_5)(C_3H_7)SnCl_2$	261.76	need. f. lt. pet.		57–8		s.		s. eth., al.
108	Ethyl stannic acid	C_2H_5SnOOH	180.77	wh. amor. gel. or powd.		d. below red	heat	i.	i.	s. dil. min. a., KOH; i. al., eth., chl., xylene
109	Ethyltetrathioortho-stannate	$Sn(SC_2H_5)_4$	363.19					$105^{.001}$		
110	Ethyltin tribromide	$C_2H_5SnBr_3$	387.48	col. feath. cr.		310		s.		s. al.
111	Ethyltin triiodide	$C_2H_5SnI_3$	528.47				$181–4^{19}$			
112	Ethyltri-*n*-butyltin	$C_2H_5(C_4H_9)_3Sn$	319.10	1.4732	1.0783		129^{10}			
113	Ethyltri-*n*-propyltin	$C_2H_5(C_3H_7)_3Sn$	277.02				101^{10}			
114	Hexabutylditin	$[(C_4H_9)_3Sn]_2$	580.08	col. liq.	1.1480	198^{10}		i.	i.	s. org. solv.
115	Hexadecyltetrathio-orthostannate(*n*)	$Sn(SC_{16}H_{33})_4$	1148.71			53–54				
116	Hexaethyl disitannane	$[Sn(C_2H_5)_3]_2$	411.75				160^{23}			
117	Hexaethylditin	$[(C_2H_5)_3Sn]_2$	411.76	liq.	1.412^{0}		d. 270			
118	Hexaphenylditin	$[(C_6H_5)_3Sn]_2$	700.02	wh. cr.		232.5	d. <280	i.	i.	.029 eth.; 18.08 chl.; 7.82 bz.
119	Hexa-*p*-tolylditin	$[(C_6H_4CH_3)_3Sn]_2$	784.18	flat tabl. f. bz.		143.5	d. 335			sl. s. bz., eth.; v. sl. s. abs. al.
120	Hexa-*p*-xylylditin	$[(CH_3)_2C_6H_3)_3Sn]_2$	868.36	flat rhomb. tabl. f. bz.-al.		192.5	d. 368			$21^{20.4}$ bz.
121	Isopropylstannic acid	C_3H_7SnOOH	194.79	wh. amor.		d.		i.		s. dil. min. a., KOH; i. org. solv.
122	Isopropyltetrathio-orthostannate	$Sn(SC_3H_7)_4$	419.30				$92^{.001}$			
123	Isopropyltin tribromide	$C_3H_7SnBr_3$	401.51	pa. yel. deliq. pr.		112				s. glac. ac. a.; sl. s. h. bz., chl.; i. dry eth.
124	Isopropyltin trichloride	$C_3H_7SnCl_3$	268.14				75^{16}			
125	Methylstannic acid	$(CH_3)SnOOH$	166.73	wh. amor. powd.		infus.		i.	i.	s. a., alk.; i. org. solv.
126	Methyltetrathio-orthostannate	$Sn(SCH_3)_4$	307.09			31	$81^{.001}$			
127	Methyltin tribromide	CH_3SnBr_3	373.45	wh. need.		53–5	$210–11^{746}$	s.		s. eth., al., bz., lgr., hyd. by alk.
128	Methyltin trichloride	CH_3SnCl_3	240.08	col. cr.		43	171	s.		hyd. by alk.; s. org. solv.
129	Methyltin triiodide	CH_3SnI_3	514.44	lt. yel. need.		86.5		s.	s.	s. eth., al., bz., chl., meth. al.
130	Methyltribromo-dipyridinetin	$CH_3SnBr_3(C_5H_5N)_2$	531.66			203				
131	Methyltri-*n*-butyltin	$CH_3(C_4H_9)_3Sn$	305.07	1.4735	1.0898^{20}_{4}		121^{10}			
132	Methyltri-*n*-propyltin	$CH_3(C_3H_7)_3Sn$	262.99				93^{10}			
133	Phenylbenzyltin dichloride	$(C_6H_5)(C_6H_5CH_2)SnCl_2$	357.84	col. need. f. dil. HCl		83–4	80–100			
134	Phenyltetrathio-orthostannate	$Sn(SC_6H_5)_4$	555.37			67				
135	Phenyltin tribomide	$C_6H_5SnBr_3$	435.52				$182–3^{29}$			
136	Phenyltin trichloride	$C_6H_5SnCl_3$	302.16	col. liq.			$142–3^{25}$	i.	i.	s. bz., MeOH

No.	Name	Formula	Mol. wt.	Crystalline form, color and index of refraction	Sp. gr. or density	Melting point, °C	Boiling point, °C	Solubility in grams per 100 ml of: Cold water	Hot water	Alcohol, acids, etc.
	Tin									
137	Phenyltribenzyltin	$(C_6H_5)Sn(C_6H_5CH_2)_3$	469.20	liq.			290^{5}			s. all ord. org. solv. except al.
138	Propyltetrathioorthostannate(*n*)	$Sn(SC_3H_7)_4$	419.30				$123^{.001}$			
139	Propyltin triiodide	$C_3H_7SnI_3$	542.49				d. 200^{16}			
140	Propyltri-*n*-amyltin(*n*)	$C_3H_7(C_5H_{11})_3Sn$	375.21	1.4732	1.0368		163^{10}			
141	Stannic bisacetylacetone dibromide	$(C_5H_7O_2)_2SnBr_2$	476.73	col. six-sided cr.			187			s. bz., chl., acet.; sl. s. eth., CCl_4
142	Stannic bisacetylacetone dichloride	$(C_5H_7O_2)_2SnCl_2$	387.82	col. six-sided cr.			202–3	s.		s. bz., acet.
143	Stannic bisbenzoylacetone dibromide	$(C_{10}H_{16}O_2)_2SnBr_2$	614.98	pa. yel. powd.			213–4			sl. s. org. solv.
144	Stannic bisdibenzoylmethane dibromide	$(C_{15}H_{10}O_2)_2SnBr_2$	723.00	sulfur-yel. cr.			276–8	i.		sl. s. org. solv.
145	Stannic bis-3-ethylacetylacetone dibromide	$(C_7H_{11}O_2)_2SnBr_2$	532.84	col. six-sided pr.		164–6				s. c. chl., bz.; sl. s. lt. pet.
146	Tetra-*dl*-amyltin	$Sn(C_5H_{11})_4$	403.26	1.4730	1.0222		174^{10}			
147	Tetra-*n*-amyltin	$Sn(C_5H_{11})_4$	403.26	col. stable liq., 1.4720	1.0206		181^{10}			
148	Tetraaquastannic bisacetylacetone stannibromide	$(C_5H_7O_2)_2Sn(OH_2)_4SnBr_6$	987.12	col. tab. pr.		105–7				s. bz.
149	Tetrabenzyltin	$Sn(C_6H_5CH_2)_4$	483.23	col. pr. f. lt. pet.		42–3		i.	i.	s. common org. solv.; sl. s. lt. pet.
150	Tetra-*n*-butyltin	$Sn(C_4H_9)_4$	347.16	col. stable liq., 1.4730	1.0572	<−70	145^{10}			
151	Tetracyclohexyltin	$Sn(C_6H_{11})_4$	451.31	wh. micr. grains		263–4		i.	i.	6.25^{30} bz.; 0.86^{30} al.; s. chl. CS_2
152	Tetraethyltin	$Sn(C_2H_5)_4$	234.94	col. liq. $n_D^{19.7}$ 1.4724	1.187^{23}	−112	181	i.	i.	s. org. solv.
153	Tetra-*n*-heptyltin	$Sn(C_7H_{15})_4$	515.48	1.4698	0.9748		239^{10}			
154	Tetra-*n*-hexyltin	$Sn(C_6H_{13})_4$	459.37	1.4706	0.9959		209^{10}			
155	Tetraisoamyltin	$Sn(C_5H_{11})_4$	403.26	liq.	$1.035^{19.6}$		188^{24}			
156	Tetraisobutyltin	$Sn(C_4H_9)_4$	347.16	col. liq.	1.054^{23}	−13	267; $143^{16.5}$	i.	i.	s. org. solv.
157	Tetralauryltin	$(C_{12}H_{25})_4Sn$	796.02	1.4736^{20}	0.895	15–16		i.	i.	
158	Tetramethyltin	$Sn(CH_3)_4$	178.83	col. liq. 1.4386	1.314^{0}	−54.8	78	i.	i.	s. org. solv.
159	Tetra-*n*-octyltin	$Sn(C_8H_{17})_4$	571.59	1.4691	0.9605		268^{10}			
160	Tetraphenyltin	$Sn(C_6H_5)_4$	427.12	tetr. col. f. xylene	1.490^{0}	226	>420	i.	i.	s. h. bz., pyr., CCl_4, chl., ac. a.; sl. s. al.
161	Tetrapropyltin	$Sn(C_3H_7)_4$	291.05	col. liq.	$1.1065^{20.2}$		222–5	i.	i.	s. org. solv.
162	Tetra-*o*-tolyltin	$Sn(C_6H_4CH_3)_4$	483.23	wh. cr. powd.		158–9 (215)		i.	i.	s. bz., eth.; i. al.
163	Tetra-*m*-tolyltin	$Sn(C_6H_4CH_3)_4$	483.23	col. need.		128.5		i.	i.	s. bz., h. eth., h. al.
164	Tetra-*p*-tolyltin	$Sn(C_6H_4CH_3)_4$	483.23	col. need.		230–3		i.	i.	s. bz., chl., CS_2, pyr.; sl. s. al., eth.
165	Tetravinyltin	$(CH{:}CH_2)_4Sn$	226.87	col. liq., n_D^{25} 1.4993	1.267_4^{25}		$55–57^{17}$			
166	Tetra-*m*-xylyltin	$[(CH_3)_2C_6H_3]_4Sn$	539.33	rhomb. need. f. bz.-al.		219.5	d. 360			$.314^{30}$ al.; 5.28^{30} eth.; 35.1^{30} bz.; 43.2^{30} chl.
167	Tetra-*p*-xylyltin	$[(CH_3)_2C_6H_3]_4Sn$	539.33	wh. quad. pr.		272–3	d. 360	i.	i.	0.015^{30} al.; 1.73^{30} bz.; 2.80^{30} chl. 0.29^{30} eth.; $.017^{30}$ meth. al.

PHYSICAL CONSTANTS OF ORGANOMETALLIC COMPOUNDS (Continued)

No.	Name	Formula	Mol. wt.	Crystalline form, color and index of refraction	Sp. gr. or density	Melting point, °C	Boiling point, °C	Solubility in grams per 100 ml of: Cold water	Hot water	Alcohol, acids, etc.
	Tin									
168	Tolylstannonic acid(*o*)	$CH_3C_6H_4SnO_2H$	242.83	amor. powd.	....	d. 295				
169	Tolylstannonic acid(*m*)	$CH_3C_6H_4SnO_2H$	242.83	....	....	....	d. 295	i.	i.	s. c. meth. al., al., eth., chl., eth. acet., pyr., a. and bases; i. pet. eth.
170	Tolyltetrathioortho-stannate(*p*)	$Sn(SCH_3C_6H_4)_4$	611.48	....	....	100	....	....	....	....
171	Tolyltin trichloride(*o*)	$CH_3C_6H_4SnCl_3$	316.18	....	1.7619	....	154–8^{20}	....	....	....
172	Tolyltin trichloride(*p*)	$CH_3C_6H_4SnCl_3$	316.18	....	1.7522	....	156–7^{23}	....	....	....
173	Tolyltrichloro-stannane(*m*)	$CH_3C_6H_4SnCl_3$	316.18	col. liq.	1.7516	<−20	150–1^{23}	sl. d.	....	....
174	Triallylbutyltin	$(CH_2{:}CHCH_3)_3(C_4H_9)Sn$	302.05	liq., 1.5162	1.3315	....	116–119^{10}	....	....	....
175	Tri-*n*-amyltin bromide	$(C_5H_{11})_3SnBr$	412.03	1.4963	1.2678	....	....	....	....	....
176	Tribenzylethyltin	$(C_6H_5CH_2)_3(C_2H_5)Sn$	421.15	col. tabl. f. al.-lt. pet.	....	31–2	....	....	....	s. eth. bz., chl.; sl. s. al.
177	Tribenzyltin chloride	$(C_6H_5CH_2)_3SnCl$	427.54	wh. need.	....	142–4	d.	i.	i.	s. ac. a., acet., bz., eth., chl. pyr.; i. al.
178	Tribenzyltin hydroxide	$(C_6H_5CH_2)_3SnOH$	409.10	rhomb., col. tabl.	....	117–21	....	....	....	s. h. al., CS_2, bz.; sl. s. eth., lgr.; i. KOH
179	Tribenzyltiniodide	$(C_6H_5CH_2)_3SnI$	519.00	need. like pr. f. glac. ac. a.	....	102–3	....	....	....	....
180	Tributyltin acetate	$(C_4H_9)_3Sn(OOCH_3)$	349.08	wh., waxy solid	1.27	80–83	....	i.	i.	s. bz., MeOH
181	Tri-*n*-butyltin bromide	$(C_4H_9)_3SnBr$	369.95	1.5000	1.3365	....	....	....	....	....
182	Tri-*n*-Butylvinyltin	$(n\text{-}C_4H_9)_3Sn(CH{:}CH_2)$	317.08	n_D^{25} 1.4761	1.085^{25}_{4}	....	114^{3}	....	....	....
183	Triethyl-*n*-amyltin	$(C_2H_5)_3C_5H_{11}Sn$	277.02	....	....	....	102^{10}	....	....	....
184	Triethyl(*p*-dimethyl-aminophenyl) stannane	$(C_2H_5)_3(CH_3)_2NC_6H_4Sn$	326.05	1.5610^{22}	1.2425	....	172–3^{3}	....	....	....
185	Triethyl-*o*-hydroxy-phenyl stannane	$(C_2H_5)_3OHC_6H_4Sn$	298.98	1.5377^{25}	1.3229^{25}	....	197–200^{3}	....	....	....
186	Triethylisoamyltin	$(C_2H_5)_3Sn(C_5H_{11})$	277.02	....	1.1203$^{20.1}$	....	111$^{18.5}$	....	....	....
187	Triethylisobutyltin	$(C_2H_5)_3Sn(C_4H_9)$	262.99	....	1.139$^{20.3}$	....	96.5^{17}	....	....	....
188	Triethylphenyltin	$(C_2H_5)_3Sn(C_6H_5)$	282.98	col. liq.	1.2639	....	254	i.	i.	s. al., eth., org. solv.
189	Triethyl-*n*-propyltin	$(C_2H_5)_3Sn(C_3H_7)$	248.97	....	1.1680$^{20.6}$	....	82^{13}	....	....	....
190	Triethyltin	$(C_2H_5)_3Sn$	205.88	col. liq.	1.3774	<−75	161^{23}	i.	i.	s. al., org. solv.
191	Triethyltin bromide	$(C_2H_5)_3SnBr$	285.79	col. liq.	1.630	−13.5	223–4	v. sl. s.	....	s. org. solv.
192	Triethyltin chloride	$(C_2H_5)_3SnCl$	241.33	col. liq. 1.5017$^{23.3}$	1.4288$^{23.3}$	15.5	208–10	i.	....	s. org. solv.
193	Triethyltin ethoxide	$(C_2H_5)_3Sn(OC_2H_5)$	250.94	col. liq.	1.2634		190	d.	....	s. org. solv.
194	Triethyltin hydroxide	$(C_2H_5)_3SnOH$	222.88	col. cr.	....	43	271	s.	s.	s. org. solv.
195	Triethyltin iodide	$(C_2H_5)_3SnI$	332.78	col. liq.	1.833	−34.5	225 (231)	v. sl. s.	....	s. org. solv.
196	Triisoamyltin bromide	$(C_5H_{11})_3SnBr$	412.03	....	1.2613$^{20.7}$	21	177^{15}	....	....	....
197	Triisoamyltin chloride	$(C_5H_{11})_3SnCl$	367.57	....	1.1290$^{34.2}$	−30.2	174^{13}	....	....	....
198	Triisoamyltin fluoride	$(C_5H_{11})_3SnF$	351.12	need.	....	288	....	....	....	1.03^{31} al.; .967^{31} bz.; 1.22^{31} meth. al.
199	Triisoamyltin iodide	$(C_5H_{11})_3SnI$	459.02	....	1.3777$^{26.5}$	−22	182^{13}	....	....	....
200	Triisobutylethyltin	$(C_4H_9)_3Sn(C_2H_5)$	319.10	....	1.0779^{21}	....	125^{16}	....	....	....
201	Triisobutylisoamyltin	$(C_4H_9)_3Sn(C_5H_{11})$	361.18	....	1.0356$^{26.8}$	....	152.9$^{16.5}$	....	....	....
202	Triisobutyltin bromide	$(C_4H_9)_3SnBr$	369.95	....	1.3523	−26.5	148^{13}	....	....	....
203	Triisobutyltin chloride	$(C_4H_9)_3SnCl$	325.49	....	1.1290$^{34.2}$	+30.2	174^{13}	....	....	....
204	Triisobutyltin fluoride	$(C_4H_9)_3SnF$	309.04	fine long pr.	....	244	....	....	....	.414^{32} al.; 0.614^{32} meth. al., .1^{32}bz.
205	Triisobutyltin iodide	$(C_4H_9)_3SnI$	416.94	col. liq.	1.378$^{26.5}$	−22	284–6	....	....	s. eth., org. solv.
206	Triisopropyltin bromide	$(C_3H_7)_3SnBr$	327.87	....	1.4263$^{25.2}$	−49	133^{12}	....	....	....
207	Triisopropyltin idodide	$(C_3H_7)_3SnI$	374.86	....	1.4378$^{22.2}$	....	151^{13}	....	....	s. org. solv.

PHYSICAL CONSTANTS OF ORGANOMETALLIC COMPOUNDS (Continued)

No.	Name	Formula	Mol. wt.	Crystalline form, color and index of refraction	Sp. gr. or density	Melting point, °C	Boiling point, °C	Solubility in grams per 100 ml of: Cold water	Hot water	Alcohol, acids, etc.
	Tin									
208	Trimethyldecyltin	$(CH_3)_3(C_{10}H_{21})Sn$	305.07	liq., 1.4602	1.0487		$67^{0.05}$			
209	Trimethyldodecyltin	$(CH_3)_3(C_{12}H_{25})Sn$	333.13	liq., 1.4610	1.0285		$93–98^{0.15}$			
210	Trimethylethyltin	$(CH_3)_3(C_2H_5)Sn$	192.86	col. liq.			108.2	i.	i.	s. org. solv.
211	Trimethyltin	$(CH_3)_3Sn$	163.80	col. liq.	1.570^{25}	23	182	i.	i.	s. org. solv.
212	Trimethyltin bromide	$(CH_3)_3SnBr$	243.70	col. cr. or liq.		27	165	s.	s.	s. org. solv.
213	Trimethyltin chloride	$(CH_3)_3SnCl$	199.24	col. cr.		37	154	s.	s.	s. org. solv.
214	Trimethyltin fluoride	$(CH_3)_3SnF$	182.79	wh. short. thick rect. pr.		360 seal. tube	d. <375			2.45^{31} meth. al.; 1.08^{31} al., 0.05^{31} bz.
215	Trimethyltin hydride	$(CH_3)_3SnH$	164.80	col. oily liq.			60	v. sl. s.		s. org. solv.
216	Trimethyltin hydroxide	$(CH_3)_3SnOH$	180.80	col. pr.		118 d.	subl. >80	s.	s.	s. a. al.,, bz., chl., CCl_4, alk.
217	Trimethyltin iodide	$(CH_3)_3SnI$	290.70	col. liq.	2.1432	3.4	170	v. sl. s.		s. bz., al., eth., acet.
218	Trimethyltin oxide	$[(CH_3)_3Sn]_2O$	343.59	wh. amor. powd.		d.		i.	i.	s. a., alk.; i. org. solv.
219	Trimethyltin sulfide	$[(CH_3)_3Sn]_2S$	359.65	lt. yel. oil	1.649^{25}	6	233.5	i.		s. org. solv., HNO_3
220	Triphenylallyltin	$(C_6H_5)_3(CH_2{:}CHCH_2)Sn$	391.08	wh. powd.		73–74		i.	i.	s. org. solv.
221	Triphenylbenzyltin	$(C_6H_5)_3Sn(C_6H_5CH_2)$	441.14	col. pl. f. al.		90	250^{3}			s. org. solv. except al.
222	Triphenylbutyltin	$(C_6H_5)_3(C_4H_9)Sn$	407.13	solid		59 60	222^{3}			sl. s. MeOH
223	Triphenylethyltin	$(C_6H_5)_3SnC_2H_5$	379.07	wh. pr. f. al.	1.2953^{62}	56				
224	Triphenylmethyltin	$(C_6H_5)_3SnCH_3$	365.05	col. tetr. f. eth.	$1.3113^{63.85}$	60–1				s. bz., chl., eth.
225	Triphenyl-α-naphthyltin	$(C_6H_5)_3Sn(C_{10}H_7)$	477.18	col. pr.		125				s. bz., chl., eth.
226	Triphenyltin	$(C_6H_5)_3Sn$	350.01	wh. powd.		232.5	d. 280	i.	i.	$.079^{30}$ al.; 7.82^{30} bz.; 0.92^{30} eth.; 18.1^{30} chl.
227	Triphenyltin bromide	$(C_6H_5)_3SnBr$	429.92	col. cr.		120.5	$249^{13.5}$	i.	i.	s. al., eth. org. solv.
228	Triphenyltin chloride	$(C_6H_5)_3SnCl$	385.46	col. cr.		106	$240^{13.5}$	i.	i.	s. org. solv.
229	Triphenyltin fluoride	$(C_6H_5)_3SnF$	369.01	fine pr.		357		sl. s.		sl. s. c. al., eth.
230	Triphenyltin hydroxide	$(C_6H_5)_3SnOH$	367.02			118				
231	Triphenyltin iodide	$(C_6H_5)_3SnI$	476.91	4-sided monocl. wh.		121	$253^{13.5}$	i.	i.	s. org. solv.
232	Triphenyl-*p*-tolyltin	$(C_6H_5)_3Sn(C_7H_7)$	441.14	need. f. eth.		124				s. bz., chl., eth.
233	Triphenyl-*p*-xylytin	$(C_6H_5)_3Sn[C_6H_3(CH_3)_2]_2$	560.33	col. lng.hex. sheets f. al.		100.5				s. bz., chl. eth.
234	Tri-*n*-propyl-*n*-butyltin	$(C_3H_7)_3SnC_4H_9$	305.07				121^{10}			
235	Tri-*n*-propylethyltin	$(C_3H_7)_3Sn(C_2H_5)$	277.02		$1.1225^{21.8}$		$117.5^{23.3}$			
236	Tri-*n*-propylisobutyltin	$(C_3H_7)_3Sn(C_4H_9)$	305.07		$1.0841^{24.1}$		128^{8}			
237	Tri-*n*-propyltin chloride	$(C_3H_7)_3SnCl$	283.41	col. liq.	1.2678^{28}	−23.5	123^{13}			s. org. solv.
238	Tri-*n*-propyltin fluoride	$(C_3H_7)_3SnF$	266.96	flat pr.		275				4.26^{31} meth. al.; 2.73^{31} al.; 0.118^{31} bz.
239	Tri-*n*-propyltin iodide	$(C_3H_7)_3SnI$	374.86	col. liq.	1.692^{13}	−53	260–2; 141^{13}			s. org. solv.
240	Tri-*o*-tolyltin bromide	$(C_6H_4CH_3)_3SnBr$	472.00	rhomb. tab f. al.		99.5				s. bz., eth.; sl. s. al.
241	Tri-*p*-tolyltin bromide	$(C_6H_4CH_3)_3SnBr$	472.00	rhbdr. f. al.		98.5				s. bz., eth.; sl. s. al.
242	Tri-*o*-tolyltin chloride	$(C_6H_4CH_3)_3SnCl$	427.54	short, thick pr. f. al.		99.5				s. bz., eth.; sl. s. al.
243	Tri-*m*-tolyltin chloride	$(C_6H_4CH_3)_3SnCl$	427.54			108				
244	Tri-*p*-tolytin chloride	$(C_6H_4CH_3)_3SnCl$	427.54	rhomb. pl. f. al.		97.5				sl. s. al. bz., eth.
245	Tri-*p*-tolyltin fluoride	$(C_6H_4CH_3)_3SnF$	411.09	hairlike felted need		305				s. al.
246	Tri-*p*-tolyltin hydroxide	$(C_6H_4CH_3)_3SnOH$	409.10			108–9				
247	Tri-*o*-tolytiniodide	$(C_6H_4CH_3)_3SnI$	519.00	rhomb. cr. fr. al.-eth		119.5				s. bz., eth.; sl. s. al.

PHYSICAL CONSTANTS OF ORGANOMETALLIC COMPOUNDS (Continued)

No.	Name	Formula	Mol. wt.	Crystalline form, color and index of refraction	Sp. gr. or density	Melting point, °C	Boiling point, °C	Solubility in grams per 100 ml of: Cold water	Hot water	Alcohol, acids, etc.
	Tin									
248	Tri-*p*-tolyltiniodide	$(C_6H_4CH_3)_3SnI$	519.00	rhomb. pl. fr. eth.-al.		120.5				s. bz.,-eth.; sl. s. al.
249	Tritriphenylstannylmethane	$[(C_6H_5)_3Sn]_3CH$	1063.05	wh. cr. sld.		128		i.	i.	v. s. bz., eth. chl.; s. h. pet. eth.; sl. s. c. pet. eth., al.
250	Trivinyldecyltin	$(CH_2{:}CH)_3(C_{10}H_{21})Sn$	341.11	1.4820	1.0672		90–94$^{0.06}$			
251	Trivinylhexyltin	$(CH_2{:}CH)_3(C_6H_{13})Sn$	285.00	1.4851	1.1266		57–91$^{0.03}$			
252	Trivinyloctylitin	$(CH_2{:}CH)_3(C_8H_{17})Sn$	313.05	1.4819	1.0865	90–93^{02}				
253	Trivinyltin chloride	$(CH_2{:}CH)_3SnCl$	235.28	1.5235	1.5139		59–60^{5}			
254	Tri-*p*-xylyltin bromide	$[(CH_3)_2C_6H_3]_3SnBr$	514.08	lng. hex. cr. f. al.		151				s. bz., chl., eth.; i. c. al.
255	Tri-*p*-xylyltin chloride	$[(CH_3)_2C_6H_3]SnCl$	469.63	6-cornered col. f. al.		141.5				s. bz., chl., eth.; i. c. al.
256	Tri-*m*-xylyltin fluoride	$[(CH_3)_2C_6H_3]_3SnF$	453.17	fine felted need.		205				s. bz., eth., al.
257	Tri-*p*-xylyltin fluoride	$[(CH_3)_2C_6H_3]_3SnF$	453.17	fine lng. need.		247				sl. s. bz., h. eth., al.
258	Tri-*p*-xylyltin iodide	$[(CH_3)_2C_6H_3]_3SnI$	561.08	hex. tabl. f. al.		159.5				s. bz., chl., eth.; i. c. al.
259	Vinyltin trichloride	$(CH_2{:}CH)SnCl_3$	252.10	1.5361	1.9981		63–65^{10}			
	Titanium									
1	*bis*-(cyclopentadienyl)-titanium diiodide	$(C_5H_5)_2TiI_2$	431.90	purp.		319 d.				sl. s. tol.; s. org. solv.
2	*bis*-(cyclopentadienyl)-titanium dichloride	$(C_5H_5)_2TiCl_2$	249.00	red-or.		289–291				sl. s. tol., chl., al., eth., bz., CS_2 CCl_4; i. pet. eth.
3	*bis*-(cyclopentadienyl)-titanium dibromide	$(C_5H_5)_2TiBr_2$	337.91	dk. red cr.	1.920	240–3 (314 ± 2)	subl.			s. tol., org. solv.
4	Di(*p*-dimethylaminophenol)[*bis*(cyclopentadienyl)] titanium	$[(CH_3)_2NC_6H_3OH]_2Ti(C_5H_5)_2$	450.44	maroon		unst.				
5	Diphenyl[*bis*(cyclopentadienyl)] titanium	$(C_6H_5)_2Ti(C_5H_5)_2$	332.30	or.-yel.		146–8 d.				s. CH_2Cl_2; i. pet. eth.
6	(Di-*m*-tolyl)[*bis*(cyclopentadienyl)] titanium	$(CH_3C_6H_4)_2Ti(C_5H_5)_2$	360.36	or.-yel.		135–40 d.				
7	(Di-*p*-tolyl)[*bis*(cyclopentadienyl)] titanium	$(CH_3C_6H_4)_2Ti(C_5H_5)_2$	360.36	or.-yel.						i. pet. eth.
8	Methylaminotitanium trichloride	$(CH_3N)_2TiCl_3$	213.34	bluish-grn. cr.	1.33^{25}	subl. 5–20	d. 75	s.		dil. HCl, dil. H_2SO_4
9	Monochlorotriethoxytitanium	$TiCl(OC_2H_5)_3$	218.54	pa. yel. liq.			170^{16}			
10	Phenyltitanium triisopropylate	$(C_6H_5)Ti[(OCH(CH_3)_2)]_3$	302.27	wh. cr.		100–120 d.				s. pet. eth.
11	Tetraethoxytitanium	$Ti(OC_2H_5)_4$	228.15	col. oily liq. supercooled, n_D^{25} 1.5082	1.1066^{25}		133–135^{5}			
12	Tetramethoxytitanium	$Ti(OCH_3)_4$	172.04	solid		210	243^{52}	i.	i.	i. bz.
13	Titanium isopropoxide	$Ti(OC_3H_7)_4$	284.25	liq.	0.9550	20	58^{1}			
14	Titaniummonochloro-tri-2-ethoxyethoxide	$Ti(OCH_2CH_2OCH_2CH_3)Cl$	350.70	col. liq., n_D^{25} 1.516	1.203^{20}		182^{1}			
15	Titanium tetra-*n*-hexoide	$Ti[O{:}(CH_2)_5{:}CH_3]_4$	452.58	liq., n_D^{25} 1.483.	0.950^{25}		336			
16	Titanium tetra-isobutoxide	$Ti[OCH_2CH(CH_3)_2]_4$	340.36	cr.; liq. n_D^{25} 1.475	liq., 0.960	30	269^{760}			

PHYSICAL CONSTANTS OF ORGANOMETALLIC COMPOUNDS (Continued)

No.	Name	Formula	Mol. wt.	Crystalline form, color and index of refraction	Sp. gr. or density	Melting point, °C	Boiling point, °C	Solubility in grams per 100 ml of: Cold water	Hot water	Alcohol, acids, etc.
	Titanium									
17	Titanium tetra-*n*-pentoxide	$Ti[O:(CH_2)_4:CH_3]_4$	396.47	liq. n_D^{25} 1.485	0.974^{25}		314			
	Tungsten									
1	*bis*-(Cyclopentadienyl)-hexacarbonmonoxide-bitungsten *bis*-(Cyclopentadienylbitungsten hexacarbonyl)	$(C_5H_5)_2W_2(CO)_6$	665.95	purp. red. cr.		240–2 d.				s. chl., CCl_4, CS_2; i. ligr.
2	Mesitylenetungsten tricarbonyl	$(CH_3)_3C_6H_3W(CO)_3$	388.08	yel.		160–165 d.	subl.			
	Uranium									
1	Uranium(IV)-dibenzoylmethane	$U(C_6H_5COCH:COC_6H_5)_4$	1131.04	vlt. blk. cr.		192–193	subl. vac.			v. s., bz.
2	Uranium(IV)-diethylamide	$U[N(C_2H_5)_2]_4$	526.55	grn. liq.		35.5–36.5		d.		s. bz., s. eth.
3	Uranium(V)ethoxide	$U(OC_2H_5)_5$	463.33	dk. brn. liq.	1.711^{25}		$160^{0.05}$			s. al., eth. bz.; d. acet.
4	Uranium isipropoxide	$U(OC_4H_9)_5$	603.61	brn. cr.		100–104	$192^{0.009}$			
5	Uranium(V)methoxide	$U(OCH_3)_5$	393.20	red, cr. solid		210	subl. 190–$210^{0.01}$	d.		s. eth., bz.
6	Uranium(V)2,2,2-trifluororethoxide	$U(OCH_2CF_3)_5$	733.19	grn.-brn. solid		$130^{0.008}$				s. eth., s. bz.
	Vanadium									
1	*bis*-Cyclopentadienyl-vanadium dichloride	$(C_5H_5)_2VCl_2$	252.04	pa. gr. cr.		1.60 g/ml d >250				s. chl., al.; sl. s. eth., CS_2, CCl_4, bz.; i. pet. eth.
2	*bis*-Cyclopentadienyl-vanadium dibromide	$(C_5H_5)_2VBr_2$	340.95	dk. gn.		unst.				s. chl., CCl_4, lgr.
	Zinc									
1	Di-*n*-butylzinc	$Zn(CH_2CH_2CH_2CH_3)_2$	179.60	liq.			$81–2^{9}$	d.		
2	Diethylzinc	$Zn(C_2H_5)_2$	123.49	col. liq. ign. in air or Cl	1.182^{18}		118	d.		
3	Dimethylzinc	$Zn(CH_3)_2$	95.44	col. liq.	$1.386^{10.5}$	−42.2	46	d.		d. al., a.; s. eth. xylene
4	Diphenylzinc	$Zn(C_6H_5)_2$	219.58	wh. cr.		107 (in H_2)	280–285 (in H_2)	d.		v. s. bz., eth., s. $CHCl_3$
5	Di-*n*-propylzinc	$Zn(CH_2CH_2CH_3)_2$	151.55	liq. n_4^{48} 1.4845			146; $39–40^{8}$	d.		
6	Di-*o*-tolylzinc	$Zn(C_6H_4CH_3)_2$	247.64	wh. cr.		207–10				s. xylene; v. sl. s. pet. eth.
	Zirconium									
1	*bis*-Cyclopentadienyl-zirconium dibromide	$(C_5H_5)_2ZrBr_2$	381.23	col. cr.		260 d.		s.		

SUBLIMATION DATA FOR ORGANIC COMPOUNDS

Compiled by Mansel Davies

The tables quote the parameters from what appear to be the best data inthe literature expressed in the form

$$\log_{10} p(\text{mm}) = A - B/T$$

and the temperature range for which they apply. The corresponding heats and entropies (taking the standard state of the vapor to be

$$\Delta H \text{ (sublimation)} = 2.303 \text{ R.B. cal/mol.}$$
$$\Delta S \text{ (sublimation)} = 2.303R(A - 2.881) \text{ cal/mole °K.}$$

Compound	Temp. range °C	A	B	Ref.
Acenaphthene	18 to 37	11.758	4290.5	1
Acetamide	25 to 77	11.8468	4050.1	2
Acetic acid	−35 to +10	8.502	2177.4	3
" , *m*-cresyl ester	2 to 44	9.759	3170	14
Acetophenone, 1-chloro-	5 to 50	13.779	4740	14
" , 1-chloro-*o*-nitro	23 to 54	14.24	5413	14
" , 1-chloro-*m*-nitro	26 to 70	14.080	5700	14
" , *p*-methoxy	3 to 27	11.367	4056	1
Acetone, benzoyl	5 to 26	12.317	4375	1
Adipic acid	86 to 133	15.463	6757	4
Anthracene	65 to 80	12.638	5320	5
"	105 to 125	12.002	5102	6
" , 9.10 diphenyl	208 to 229	16.058	8213	22
Anthraquinone	224 to 286	12.305	5747	3
Arachidic acid	63 to 73	25.453	10,424	23
Arsine, diphenylcyano	23 to 53	10.724	4420	14
Azobenzene (*cis*)	30 to 60	9.652	3914	7
" (*trans*)	30 to 60	9.721	3911	7
Behenic acid	71 to 79	23.604	10,100	23
Benzanthrone	—	13.416	6030	8
Benzene	−30 to 5	9.846	2309	3
"	−58 to −30	9.556	2241	3
" , *p*-chloroiodo	30 to 50	9.819	3200	15
" , *p*-dichloro	10 to 50	11.985	3570	17
" , α-hexachloro	51 to 71	11.950	4850	14
" , β-hexachloro	95 to 117	11.790	5375	14
" , γ-hexchloro	60 to 92	15.515	6022	14
" , λ-hexachloro	55 to 75	12.635	5100	14
" , -hexamethyl		11.070	4215	37
" , 1,2,3-trichloro	16 to 30	10.662	3440	6
" , 1,2,4-trichloro	6 to 25	10.445	3254	6
" , 1,3,5-trichloro	9 to 28	9.176	2956	6
Benzil	45 to 67	12.708	5140	1
Benzoic acid	70 to 114	12.870	4776	9
" , *p*-hydroxy	125 to 160	13.623	6063	9
" , *o*-methoxy	80 to 95	11.871	4746	9
Benzophenone (stable)	16 to 42	17.46	4966	10
" , (meta stable)	11 to 25	17.19	4818	10
Benzoquinone	—	10.00	3280	18
" , 2.6-dichloro	1 to 42	9.85	3670	18
" , trichloro	28 to 54	12.03	4630	18
" , tetrachloro	60 to 83	12.06	5170	18
" , *p*-xylo	0 to 20	11.53	4030	18
Bibenzyl	13 to 34	12.194	4386	1
Biphenyl	6 to 26	11.168	3959	1
Butyramide	25 to 68	12.739	4546	2
Butyramide	63 to 109	12.594	4513	2
Camphor	0 to 180	8.799	2797	3
Capramide	80 to 97	16.471	6577	2
" , *N*-methyl	30 to 52	14.594	5371	11
Capric acid	16 to 28	17.130	6119	23
Caproamide	65 to 95	13.328	4968	2
ε-Caprolactam	21 to 41	11.839	4339	34
Caprylamide	52 to 101	14.920	5783	2
Carbamic acid, *n*-butyl ester	19 to 43	14.582	4919	11
" , ethyl ester	19 to 43	14.090	4646	11
" , *n*-hexyl ester	18 to 41	14.748	5018	11
" , methyl ester	14 to 32	11.966	3883	11
Carbon tetrabromide (monoclinic)	22 to 46	9.3867	2841	12
(cubic)	48 to 56	8.5670	2579	12
Carbon tetrachloride	−64 to −48	9.089	2027	13
o-Cresol, 3,5-dinitro	17 to 51	14.140	5400	14
Cyclohexane	−5 to +5	8.594	1953	3
Cyclo-trimethylene-trinitramine	110 to 138	11.870	5850	16
Diphenylamine	25 to 51	12.434	4654	21
Dodecanedioic acid	102 to 123	17.728	8006	4
Eicosanedioic acid	107 to 122	18.185	8644	4
Enanthamide	72 to 93	13.617	5182	2
Ethane, 1.1:pp dichloro diphenyl tri-chloro	66 to 100	14.191	6160	14
" , hexachloro (cubic)	13 to 174	8.731	2677	26
" , hexachloro (triclinic)	13 to 174	9.890	3077	26
Ethylene dibromide	−21 to +8	9.884	2606	13
Ethylene, *trans* di-iodo	−8 to 20	5.86	2130	19
Fluorene	33 to 49	11.325	4324	5
Formic acid	−5 to +8	12.486	3160	36
2 Fuoric acid	44 to 55	14.62	5667	25
Hendecanoic acid	20 to 28	16.432	6037	24
Heneicosanoic acid	68 to 73	22.602	9642	24
Heptadecanoic acid	48 to 58	21.836	8769	24
Hydroquinone tetrachloro	77 to 86	10.08	4650	18
" , *p*-xylo	59 to 88	12.36	5280	18
Lauramide	76 to 95	19.169	7980	2
Lauric acid	22 to 41	19.897	7322	23
Methane	−194 to −184	7.651	5169	3
" , triphenyl	52 to 76	12.661	5228	1
Myristamide	85 to 100	20.940	8746	2
Myristic acid	38 to 52	18.740	7291	23
Naphthalene	6 to 21	11.597	3783	5
1-Naphthol	25 to 39	13.074	4873	1
	39 to 50	11.526	4389	1
2-Naphthol	25 to 39	13.356	5109	1
	39 to 58	11.660	4579	1
Nonadecanoic acid	58 to 64	35.916	13,815	24
n-Octadecane	15 to 25	22.83	7995	27
Oxalic acid, anhyd.	60 to 105	12.223	4727	29
" , anhyd. (α)	38 to 52	13.17	5130	28
" , anhyd. (β)	38 to 50	12.57	4875	28
Oxamic acid	82 to 90	12.58	5639	30
Oxamide	80 to 96	12.57	5893	30
Palmitamide	91 to 105	22.690	9489	2
Palmitic acid	46 to 60	20.217	8069	23
Pelargonamide	80 to 97	15.249	5997	2
Pentadecanoic acid	38 to 48	23.110	8813	24
Pentaerythritol (tetrag)	106 to 135	16.17	7528	28
" , tetranitrate	97 to 138	17.73	7750	16
Phenanthrene	37 to 50	11.388	4519	5
Phenol	5 to 32	11.421	3540	14
" , *p*-acetyl	47 to 75	12.216	5003	31
" , *p*-benzyl	40 to 62	12.600	5072	31
" , *p*-*tert* butyl	8 to 30	12.332	4402	31
" , 2-*tert* butyl-4-methyl	2 to 20	11.685	4036	31
" , 4-*tert* butyl-2-methyl	3 to 24	11.199	3952	31
Phenol, *p*-formyl	39 to 63	11.795	4762	31
" , *p*-methoxy	5 to 27	13.132	4624	31
" , *o*-phenyl	19 to 40	11.754	4331	31
" , *p*-phenyl	54 to 74	12.056	5068	31
" , 2:4:6-tri*tert* butyl	18 to 40	11.507	4383	31
Phthalic anhydride	30 to 60	12.249	4632	32
Propionamide	45 to 73	12.041	4139	2
Pyrene	72 to 85	11.270	4904	5
Pyrrole 2-carboxylic acid	77 to 81	16.60	6633	25
Rubeanic acid	87 to 105	12.713	5515	30
Salicylic acid	95 to 134	12.859	4969	9
Sebacic acid	102 to 130	18.911	8395	4
Stearamide	94 to 106	24.449	10,230	2
Steraric acid	57 to 67	21.180	8696	23
Suberic acid	106 to 134	16.937	7472	4
Succinic acid	99 to 128	14.068	6132	4
d-Tartaric acid, dimethyl ester	35 to 44	16.610	5903	20
dl- " , "	42 to 85	16.127	5941	20
Thapsic acid	104 to 125	17.165	7885	4
2-Thenoic acid	42 to 50	13.53	5065	25
Thymol	0 to 40	14.201	4766	14
Toluene, 2,4,6-trinitro	50 to 143	15.34	6180	33
Tridecanoic acid	31 to 39	20.939	7764	24
Valeramide	60 to 101	12.846	4666	2

References

1. Aihara, Bull. Chem. Soc., Japan.
2. Davies, Jones and Thomas, Trans. Faraday Soc., **55**, 1100 (1959).
3. This Handbook, 41st Edition, p. 2428 et seq.
4. Davies and Thomas, Trans. Faraday Soc., **56**, 185 (1960).
5. Bradley and Cleasby, J. Chem. Soc., 1690 (1953).
6. Sears and Hopke, J. Amer. Chem. Soc., **71**, 1632 (1949).
7. Bright et al., Research, **3**, 185 (1950).
8. Inokuchi et al., Bull. Chem. Soc., Japan, **25**, 299 (1952).
9. Davies and Jones, Trans. Farad. Soc., **50**, 1042 (1954).
10. Neumann and Volker, Z. Physik. Chem. **161A**, 33 (1932).
11. Davies and Jones, Trans. Farad. Soc., **55**, 1329 (1959).
12. Bradley and Drury, ibid., **55**, 1844 (1959).
13. Nitta and Seki, J. Chem. Soc., Japan, **69**, 85 (1948).
14. Balson, Trans. Faraday Soc., **43**, 54 (1947).
15. Ewald, ibid., **49**, 1401 (1953).
16. Edwards, ibid., **49**, 152 (1953).
17. Darkis et al., Ind. Eng. Chem., **32**, 946 (1940).
18. Coolidge and Coolidge, J. Amer. Chem. Soc., **49**, 100 (1927).
19. Broadway and Fraser, J. Chem. Soc., 429 (1933).
20. Crowell and Jones, J. Phys. Chem., **58**, 666 (1954).
21. Aihara, J. Chem. Soc., Japan, **74**, 437, 1953.
22. Stevens, J. Chem. Soc., 2973 (1953).
23. Davies, Malpass and Stenhagen, Arkiv for Kemi.
24. Thomas, M.Sc. thesis, Univ. of Wales, 1959.
25. Bradley and Care, J. Chem. Soc., 1688 (1953).
26. Ivin and Dainton, Trans. Farad. Soc., **43**, 32 (1947).
27. Bradley and Shellard, Proc. Roy. Soc., **198A**, 239 (1949).
28. Bradley and Cotson, J. Chem. Soc., 1684 (1953).
29. Noyes and Wobbe, J. Amer. Chem. Soc., **48**, 1882 (1926).
30. Bradley and Cleasby, J. Chem. Soc., 1681 (1953).
31. Aihara, Bull Chem. Soc., Japan.
32. Crooks and Feetham, J. Chem. Soc., 899 (1946).
33. Edwards, Trans. Faraday Soc., **46**, 423 (1950).
34. Aihara, J. Chem. Soc., Japan, **74**, 631 (1953).
35. Seki and Suzuki, Bull. Chem. Soc., Japan, **70**, 387 (1949).
36. Coolidge, J. Amer. Chem. Soc., **53**, 1874 (1930).
37. Nitta et al. J. Chem. Soc., Japan, **70**, 387 (1949).

HEAT OF FUSION OF SOME ORGANIC COMPOUNDS

Compiled by R. Loebel

Compounds in the following tables are listed in order of the increasing number of carbon and hydrogen atoms in the molecule. Melting points are listed in degrees Celcius and heats of fusion in calories per gram.

Formula	Name	M.P., °C	H_f, cal/g	Formula	Name	M.P., °C	H_f, cal/g
$CHCl_3$	Trichloromethane	−63.6	17.62	$C_4H_4O_3$	Succinic anhydride	119.0	48.74
CHN	Hydrogen cyanide	−13.4	74.38	C_4H_4S	Thiophene	−39.4	14.11
CH_2Cl_2	Dichloromethane	−95.14	16.89	C_4H_6	1,3-Butadiene	−108.9	35.28
CH_2N_2	Cyanamide	44.0	49.81	C_4H_6	2-Butyne	−32.26	40.80
CH_2O_2	Formic acid	8.3	66.05	$C_4H_6O_2$	Crotonic acid	72.0	25.32
CH_3D	Monodeuteromethane	−182.7	12.76	$C_4H_6O_2$	cis-Crotonic acid	71.2	34.90
CH_3Br	Methyl bromide	−93.7	15.05	$C_4H_6O_4$	Methyl oxalate	54.35	42.64
CH_4	Methane	−182.5	13.96	C_4H_8	Isobutene	−140.4	25.25
CD_4	Deuteromethane	−183.4	10.75	C_4H_8	cis-2-Isobutene	−138.9	32.30
CH_4O	Methanol	−97.8	23.70	$C_4H_8N_2S$	Thiosinamine	77.0	33.45
CH_4S	Methyl mercaptan	−121.0	29.35	$C_4H_8O_2$	Ethyl acetate	−83.6	28.43
CH_5N	Methylamine	−93.5	47.20	$C_4H_8O_2$	n-Butyric acid	−5.7	30.04
$CBrCl_3$	Bromotrichloromethane	−5.7	3.05	$C_4H_8O_2$	Dioxane	11.0	34.85
CCl_2O	Phosgene	−127.9	13.86	C_4H_9Br	sec-Butylbromide	−112.7	12.01
CCl_3NO_2	Chloropicrin	−64.0	48.16	C_4H_{10}	n-Butane	−138.3	19.18
CCl_4	Tetrachloromethane	−23.0	5.09	C_4H_{10}	Isobutane	−159.42	18.96
CS_2	Carbon disulfide	−111.5	13.80	$C_4H_{10}O$	n-Butanol	−89.8	29.93
$C_2HCl_3O_2$	Trichloroacetic acid	57.5	8.60	$C_4H_{10}O$	tert-Butyl alcohol	25.4	21.88
$C_2H_2Br_2Cl_2$	1,2-dibromo-1,1-dichloro-			$C_4H_{10}O$	Ethyl ether	−116.3	23.45
	methane	−66.9	7.73	$C_4H_{12}Si$	Tetramethyl silane	−99.04	18.64
$C_2H_2Cl_2O_2$	Dichloroacetic acid	10.8	14.21	C_5H_8	Cyclopentene	−135.1	11.80
C_2H_3Br	Bromoethane	−139.5	11.44	C_5H_8	Isoprene	−145.9	16.80
$C_2H_3Br_3$	1,1,2-Tribromoethane	−29.2	8.16	C_5H_8	1,4-Pentadiene	−148.8	21.55
C_2H_3Cl	Chloroethane	−153.8	18.14	$C_5H_8O_3$	Levulinic acid	33.0	18.97
$C_2H_3ClO_2$	α-Chloroacetic acid	61.2	31.06	C_5H_{10}	1-Pentene	−166.2	19.81
$C_2H_3ClO_2$	β-Chloroacetic acid	56.0	35.12	C_5H_{10}	cis-2-Pentene	−151.4	24.25
$C_2H_3Cl_3$	Methyl chloroform	−30.4	4.90	C_5H_{10}	trans-2-Pentene	−140.2	28.48
$C_2H_3Cl_3$	1,1,2-Trichloroethane	−36.6	20.68	C_5H_{10}	Cyclopentane	−93.8	2.07
$C_2H_3Cl_3$	Chloral hydrate	47.4	33.18	$C_5H_{10}O_2$	Valeric acid	−59.0	27.83
$C_2H_3F_3$	1,1,1-Trifluoroethane	−111.3	17.61	C_5H_{12}	n-Pentane	−129.7	27.89
$C_2H_4Br_2$	1,2-Dibromoethane	9.93	13.79	C_5H_{12}	Isopentane	−159.9	17.05
$C_2H_4Cl_2$	1,2-Dichloroethane	−35.5	21.12	C_5H_{12}	2,2-Dimethylpropane	−16.7	10.79
C_2H_4O	Ethylene oxide	−112.5	28.07	$C_5H_{12}O$	Amyl alcohol	−78.9	26.65
$C_2H_4O_2$	Acetic acid	16.6	45.91	$C_6H_3Br_3O$	2,4,6-Tribromophenol	93.0	13.38
C_2H_5Cl	Ethyl chloride	−138.3	16.49	C_6H_4BrI	o-Bromoiodobenzene	21.0	12.18
C_2H_6	Ethane	−183.3	22.73	C_6H_4BrI	m-Bromoiodobenzene	9.3	10.27
C_2H_6O	Dimethyl ether	−141.5	25.62	C_6H_4BrI	p-Bromoiodobenzene	90.1	16.16
C_2H_6O	Ethanol	−114.5	26.05	$C_6H_4Br_2$	o-Dibromobenzene	1.8	12.78
$C_2H_6O_2$	Ethyleneglycol	−11.5	43.26	$C_6H_4Br_2$	m-Dibromobenzene	−6.9	13.38
C_2H_6S	Dimethyl sulfide	−98.3	30.73	$C_6H_4Br_2$	p-Dibromobenzene	86.0	20.55
C_2H_6S	Ethyl mercaptan	−121.0	19.14	$C_6H_4Br_2O$	2,4-Dibromophenol	12.0	13.97
$C_2H_6S_2$	Methyl disulfide	−120.5	23.32	$C_6H_4ClNO_2$	m-Chloronitrobenzene	44.4	29.38
C_2H_7N	Dimethylamine	−92.2	31.51	$C_6H_4ClNO_2$	p-Chloronitrobenzene	83.5	31.51
$C_3H_4O_2$	Acrylic acid	12.3	37.03	$C_6H_4Cl_2$	o-Dichlorobenzene	−16.7	21.02
$C_3H_5Br_3$	1,2,3-Tribromopropane	16.19	20.24	$C_6H_4Cl_2$	m-Dichlorobenzene	−24.8	20.55
$C_3H_5N_3O_9$	Trinitroglycerol	12.3	23.02	$C_6H_4Cl_2$	p-Dichlorobenzene	53.13	29.07
C_3H_6	Cyclopropane	−127.4	30.92	$C_6H_4I_2$	o-Diiodobenzene	23.4	10.15
C_3H_6	Propene	−185.3	17.06	$C_6H_4I_2$	m-Diiodobenzene	34.2	11.54
C_3H_6BrCl	Trimethylenebromochloride	−58.9	14.0	$C_6H_4I_2$	p-Diiodobenzene	129.0	16.20
$C_3H_6Br_2$	1,3-Dibromopropane	−34.2	16.10	$C_6H_4N_2O_4$	o-Dinitrobenzene	116.93	32.25
$C_3H_6Cl_2$	1,2-Dichloropropane	−100.5	13.53	$C_6H_4N_2O_4$	m-Dinitrobenzene	89.7	24.70
C_3H_6O	Acetone	−94.8	23.42	$C_6H_4N_2O_4$	p-Dinitrobenzene	173.5	39.99
C_3H_7Cl	2-Chloropropane	−117.2	22.48	$C_6H_4O_2$	Benzoquinone	112.9	40.97
$C_3H_7NO_2$	Urethane	48.7	40.85	C_6H_5Br	Bromobenzene	−30.6	16.17
C_3H_8	Propane	−181.7	19.11	C_6H_5BrO	p-Bromophenol	63.5	20.50
C_3H_8O	Propanol	−126.1	20.66	C_6H_5Cl	Chlorobenzene	−45.2	20.40
C_3H_8O	Isopropanol	−89.5	21.37	C_6H_5I	Iodobenzene	−31.3	11.43
$C_3H_8O_3$	Glycerol	18.2	47.95	$C_6H_5NO_2$	Nitrobenzene	5.7	22.50
C_3H_9N	Trimethylamine	−117.1	26.47	$C_6H_5NO_3$	p-Nitrophenol	113.8	41.70
$C_4H_4N_2$	Succinonitrile	54.5	11.71	$C_6H_5NO_3$	o-Nitrophenol	45.13	26.76

HEAT OF FUSION OF SOME ORGANIC COMPOUNDS (Continued)

Formula	Name	M.P., °C	H_f, cal/g
C_6H_6	Benzene	5.53	30.45
$C_6H_6N_2O_2$	*o*-Nitroaniline	71.2	27.88
$C_6H_6N_2O_2$	*p*-Nitroaniline	147.0	36.50
$C_6H_6N_2O_2$	*m*-Nitroaniline	114.0	40.97
C_6H_6O	Phenol	40.9	28.67
$C_6H_6O_2$	Pyrocatechol	105.0	49.40
$C_6H_6O_2$	Quinol	172.3	58.84
$C_6H_6O_2$	Resorcinol	110.0	46.22
C_6H_6S	Thiophenol	−14.9	24.90
C_6H_7N	Aniline	−6.3	27.09
$C_6H_8N_2$	Phenylhydrazine	19.6	36.31
$C_6H_8O_4$	Glutaric acid	97.5	37.39
$C_6H_8O_4$	Methyl fumarate	102.0	57.93
C_6H_{10}	Cyclohexene	−103.5	9.58
$C_6H_{10}O_4$	Methyl succinate	19.5	35.72
$C_6H_{10}O_6$	*d*-Dimethyl tartrate	49.0	21.50
$C_6H_{10}O_6$	*dl*-Dimethyl tartrate	87.0	35.12
C_6H_{12}	Methylcyclopentane	−142.5	19.68
C_6H_{12}	Cyclohexane	6.6	7.47
C_6H_{12}	Tetramethylethylene	−74.6	15.51
$C_6H_{12}O$	Cyclohexanol	25.46	4.19
$C_6H_{12}O_3$	Paraldehyde	10.5	25.02
C_6H_{14}	2,2-Dimethylbutane	−99.0	1.61
C_6H_{14}	2,3-Dimethylbutane	−128.8	2.22
C_6H_{14}	*n*-Hexane	−95.3	36.27
C_6H_{14}	Isohexane	−153.7	17.41
C_6H_{14}	2-Methylpentane	−153.7	17.38
$C_6H_{14}O$	Isopropyl ether	−86.8	25.79
$C_6H_{14}O$	*n*-Propyl ether	−126.1	20.66
$C_7H_5ClO_2$	*o*-Chlorobenzoic acid	140.2	39.27
$C_7H_5ClO_2$	*m*-Chlorobenzoic acid	154.2	36.39
$C_7H_5ClO_2$	*p*-Chlorobenzoic acid	239.7	49.23
$C_7H_5NO_4$	*o*-Nitrobenzoic acid	145.8	40.06
$C_7H_5NO_4$	*m*-Nitrobenzoic acid	141.1	27.59
$C_7H_5NO_4$	*p*-Nitrobenzoic acid	239.20	52.80
$C_7H_5N_3O_6$	2,4,6-Trinitrotoluene	80.83	22.34
$C_7H_6N_2O_4$	2,4-Dinitrotoluene	70.14	26.40
$C_7H_6O_2$	Benzoic acid	122.4	33.89
C_7H_7Br	*p*-Bromotoluene	28.0	20.86
C_7H_7I	9-Iodotoluene	34.0	18.75
$C_7H_7NO_2$	*o*-Aminobenzoic acid	145.0	35.98
$C_7H_7NO_2$	*m*-Aminobenzoic acid	179.5	38.03
$C_7H_7NO_2$	*p*-Aminobenzoic acid	188.5	36.46
C_7H_8	Toluene	−94.99	17.17
C_7H_8O	Benzyl alcohol	−15.2	19.83
C_7H_8O	*p*-Cresol	34.6	26.28
$C_7H_8O_2$	Dimethylpyrone	132.0	56.14
C_7H_9O	*p*-Toluidine	43.3	39.90
C_7H_{14}	1-Heptene	−119.7	30.82
C_7H_{14}	Methylcyclohexane	−126.6	16.43
C_7H_{16}	*n*-Heptane	−90.6	33.78
C_7H_{16}	2-Methylhexane	−118.2	21.16
C_7H_{16}	2,2-Dimethylpentane	−123.8	13.98
C_7H_{16}	2,4-Dimethylpentane	−119.9	15.95
C_7H_{16}	3,3-Dimethylpentane	−134.9	16.86
C_7H_{16}	3-Ethylpentane	−118.6	22.78
C_7H_{16}	2,2,3-Trimethylbutane	−25.0	5.25
$C_8H_6Cl_4$	*o*-Tetrachloroxylene	86.0	21.02
$C_8H_6Cl_4$	*p*-Tetrachloroxylene	95.0	22.10
C_8H_8	Cyclooctatetraene	−4.7	25.87
$C_8H_8Br_2$	*o*-Xylenedibromide	95.0	24.25
$C_8H_8Br_2$	*m*-Xylenedibromide	77.0	21.45
$C_8H_8Cl_2$	*o*-Xylenedichloride	55.0	29.03
$C_8H_8Cl_2$	*m*-Xylenedichloride	34.0	26.64
$C_8H_8Cl_2$	*p*-Xylenedichloride	100.0	32.73
$C_8H_8O_2$	Phenylacetic acid	76.7	25.44
$C_8H_8O_2$	*o*-Toluic acid	103.7	35.40
$C_8H_8O_2$	*m*-Toluic acid	108.75	27.59
$C_8H_8O_2$	*p*-Toluic acid	179.6	39.90
$C_8H_9NO_2$	Hydroxyacetanilide	91.3	33.59
C_8H_{10}	*o*-Xylene	−25.2	30.64
C_8H_{10}	*m*-Xylene	−47.8	26.01
C_8H_{10}	*p*-Xylene	13.2	37.83
C_8H_{16}	Ethylcyclohexane	−111.3	17.75
C_8H_{16}	*trans*-1,1-dimethylcyclohexane	−33.3	4.38
C_8H_{16}	*cis*-1,2-dimethylcyclohexane	−49.9	3.50
C_8H_{16}	*trans*-1,2-dimethylcyclohexane	−88.2	22.35
C_8H_{16}	*cis*-1,3-dimethylcyclohexane	−75.6	23.05
C_8H_{16}	*trans*-1,3-dimethylcyclohexane	−90.1	21.01
C_8H_{16}	*cis*-1,4-dimethylcyclohexane	−87.4	19.82
C_8H_{16}	*trans*-1,4-dimethylcyclohexane	−36.9	26.27
$C_8H_{16}O_2$	Caprylic acid	16.3	35.40
C_8H_{18}	*n*-Octane	−56.8	43.21
C_8H_{18}	3-Methylheptane	−120.5	23.81
C_8H_{18}	4-Methylheptane	−121.0	22.68
C_8H_{18}	2,2,4-Trimethylpentane	−107.3	18.92
C_9H_7N	Quinoline	−15.6	19.98
$C_9H_8O_2$	Cinnamic acid	133.0	36.50
$C_9H_8O_2$	Allocinnamic acid	68.0	27.35
$C_9H_{10}O_2$	Hydrocinnamic acid	48.0	28.14
C_9H_{12}	Pseudocumene	−43.8	7.47
C_9H_{12}	Hemimellitine	−25.4	16.65
C_9H_{12}	Mesitylene	−44.7	18.90
$C_9H_{18}O_2$	Pelargonic acid	12.35	30.63
C_9H_{20}	*n*-Nonane	−53.5	28.83
$C_{10}H_7NO_2$	Nitronaphthalene	56.7	25.44
$C_{10}H_8$	Naphthalene	80.2	35.06
$C_{10}H_8O$	α-Naphthol	95.0	38.94
$C_{10}H_8O$	β-Naphthol	120.6	31.30
$C_{10}H_8O_2$	Methylphenyl propionate	18.0	22.86
$C_{10}H_9N$	α-Naphthylamine	50.0	22.34
$C_{10}H_{10}O_2$	Methyl cinnamate	36.0	26.53
$C_{10}H_{12}$	Camphene	51.0	57.0
$C_{10}H_{12}O$	Anethole	22.5	25.80
$C_{10}H_{14}$	Durene	79.3	37.40
$C_{10}H_{14}$	Prehnitene	−7.7	20.0
$C_{10}H_{14}$	*p*-Cymene	−68.9	17.10
$C_{10}H_{14}O$	Thymol	51.5	27.47
$C_{10}H_{15}NO$	*d*-Carvoxime	71.5	23.29
$C_{10}H_{15}NO$	1-Carvoxime	71.0	23.41
$C_{10}H_{15}NO$	*d*1-Carvoxime	91.0	23.61
$C_{10}H_{20}O$	1-Menthol	43.5	18.63
$C_{10}H_{20}O_2$	*n*-Capric acid	31.99	38.87
$C_{10}H_{22}$	*n*-Decane	−29.7	48.34
$C_{11}H_{10}$	2-Methylnaphthalene	−34.4	20.11
$C_{11}H_{22}O_2$	*n*-Undecilic acid	28.25	32.20
$C_{11}H_{24}$	Undecane	−25.6	34.12
$C_{12}H_9N$	Carbazole	243.0	42.05
$C_{12}H_{10}$	Diphenyl	165.5	30.40
$C_{12}H_{10}N_2$	Azobenzene	67.1	28.91
$C_{12}H_{10}N_2O$	Azoxybenzene	36.0	21.62
$C_{12}H_{11}N$	Diphenylamine	52.98	25.23
$C_{12}H_{12}N_2$	Hydrazobenzene	134.0	22.89
$C_{12}H_{14}O_4$	Apiol	29.5	25.80
$C_{12}H_{24}O_2$	*n*-Lauric acid	43.22	43.72
$C_{12}H_{26}$	*n*-Dodecane	−9.6	51.33
$C_{13}H_{10}O$	Benzophenone	47.85	23.53
$C_{13}H_{13}N$	Benzylaniline	32.37	21.86
$C_{14}H_8O_2$	Anthraquinone	284.8	37.48
$C_{14}H_{10}$	Anthracene	216.5	38.70

HEAT OF FUSION OF SOME ORGANIC COMPOUNDS (Continued)

Formula	Name	M.P., °C	H_f, cal/g
$C_{14}H_{10}$	Phenanthrene	96.3	25.0
$C_{14}H_{10}O_2$	Benzil	95.2	22.15
$C_{14}H_{12}$	Stilbene	124.0	40.0
$C_{14}H_{28}O_2$	Myristic acid	53.96	47.49
$C_{16}H_{32}O_2$	Palmitic acid	61.82	39.18
$C_{16}H_{34}O$	Cetyl alcohol	49.27	33.80
$C_{18}H_{14}O_3$	Cinnamic anhydride	48.0	28.14
$C_{18}H_{30}O_2$	Stearic acid	68.82	47.54
$C_{18}H_{34}O_2$	Elaidic acid	44.4	52.08
$C_{18}H_{38}$	*n*-Octadiene	28.2	57.65
$C_{19}H_{40}$	*n*-Nonadecane	32.1	40.78
$C_{20}H_{42}$	*n*-Eicosane	36.8	59.11
$C_{21}H_{44}$	*n*-Heneicosane	40.5	38.44
$C_{22}H_{46}$	*n*-Docosane	44.4	37.67
$C_{23}H_{48}$	*n*-Tricosane	47.6	30.74
$C_{24}H_{50}$	*n*-Tetracosane	50.9	38.74
$C_{25}H_{52}$	*n*-Pentacosane	53.7	39.13
$C_{27}H_{56}$	*n*-Heptacosane	59.0	37.93
$C_{28}H_{58}$	*n*-Octacosane	61.4	39.14
$C_{57}H_{110}O_6$	Tristearin	54.5	45.63

SOLUBILITY PARAMETERS OF ORGANIC COMPOUNDS

While the solubility of organic compounds in selected solvents may be estimated from the "like dissolves like" rule of thumb, more useful information is available under the "Solubility" heading of the table "Physical Constants of Organic Compounds" in Section C of this book. This table provides qualitative data on solubility in water (w), ethanol (al), ethyl ether (eth), acetone (ace), and benzene (bz).

More quantitative solubility data for nonpolar organic compounds may be calculated from the Hildebrand expression for the square root of the cohesive energy density which is defined as the solubility parameter (δ). As shown by the following expression, δ values may be calculated if information for ΔH_v, Kelvin temperature (T), molecular weight (M), and density (D) is available:

$$\delta = \left(\frac{\Delta E_V}{V}\right)^{1/2} = \left(\frac{D(\Delta H_V - RT)}{M}\right)^{1/2}$$

The value for ΔH_v may be found in the table "Heats of Vaporization of Organic Compounds" immediately following this table. Values for D and M are listed under physical constants in Section C. Thus, the δ value for heptane may be calculated as follows:

$$\delta = \left[\frac{0.684\ (8670 - 1.99(298))}{100}\right]^{1/2} = 7.4\ \text{H}$$

The dimensions for δ are $(\text{cal cm}^{-3})^{1/2}$ but the Hildebrand unit (H) is used for convenience. It is important to note that the law of mixtures applies for δ values of mixed nonpolar solvents. Thus, the δ value for an equimolar mixture of heptane and carbon disulfide (δ = 10.0 H) is 8.7 H.

When ΔH_v values are not available, Small's molar attraction constants shown in Table I may be used.[a] As illustrated in the following expression which is solved for heptane, the summation of these constants (ΣG) may be used to estimate δ values at 298 K:

$$\delta = \frac{D\Sigma G}{M} = \frac{(2 \times 214) + 5(133)0.684}{100} = 7.5\ \text{H}$$

Solvents such as acetone (δ = 9.9 H) and water (δ = 23.4 H) are completely miscible when a large difference in δ values exists. The critical δ range for solubility of solid solutes and liquid solutes is less than that for liquids, and the critical δ range for nonpolar polymers in nonpolar liquids is less than 2 H at temperatures below 50°C.

These δ values are most useful for nonpolar solvents that are listed in Section A of Table II. Some consideration must be given to the dipole-dipole interactions in more polar solvents that are listed in Section B of Table II. The values shown for hydrogen-bonded solvents that are shown in Section C of Table II must be used with more discretion because of the stronger intermolecular forces present. However, these values are much more useful than the "like dissolves like" rule of thumb and can be used to predict the solubility of most solutes in most solvents.

[a]Small, P. A., *J. Appl. Chem.*, 3, 71, 1953.

Table I
MOLAR-ATTRACTION CONSTANTS AT 298 K

Group	G	Group	G
$-CH_3$	214	Conjugation	20–30
$-CH_2-$ single bonded	133	H (variable)	80–100
–CH<	28	O ether	70
>C<	–93	CO ketones	275
$CH_2=$	190	COO esters	310
–CH= double bonded	111	CN	410
>C=	19	Cl (mean)	260
CH≡C–	285	Cl single	270
–C≡C–	222	Cl twinned as in $>CCl_2$	260
Phenyl	735	Cl triple as in $-CCl_3$	250
Phenylene (*o,m,p*)	658	Br single	340
Naphthyl	1146	I single	425
Ring, 5 membered	105–115	CF_2 n-fluorocarbons only	150
Ring, 6 membered	95–105	CF_3 n-fluorocarbons only	274

Table I (continued)
MOLAR-ATTRACTION CONSTANTS AT 298 K

Group	G	Group	G
S sulfides	225	NO_2 (aliphatic nitro-compounds)	~440
SH thiols	315	PO_4 (organic phosphates)	~500
ONO_2 nitrates	~440	Si (in silicones)	−38

Table compiled by R. B. Seymour.

Table II
SOLUBILITY PARAMETER VALUES

A. Nonpolar Solvents

Name	δ (H)	Name	δ (H)
Acetic acid nitrile (acetonitrile)	11.9	Ethene, (ethylene)	6.1
Anthracene	9.9	Ethene, tetrachloro (perchloroethylene)	9.3
Benzene	9.2	Ethene, trichloro	9.2
Benzene, chloro	9.5	Heptane	7.4
Benzene, 1,2-dichloro	10.0	Heptane, perfluoro	5.8
Benzene, ethyl	8.8	Hexane	7.3
Benzene, isopropyl (cumene)	8.5	Hexene-1	7.4
Benzene, 1-isopropyl-4-methyl (ρ-cymene)	8.2	Malonic acid dinitrile (malononitrile)	15.1
Benzene, nitro	10.0	Methane	5.4
Benzene, propyl	8.6	Methane, bromo	9.6
Benzene, 1,3,5-trimethyl (mesitylene)	8.8	Methane, dichloro (methylene chloride)	9.7
Benzoic acid nitrile (benzonitrile)	8.4	Methane, dichloro-difluoro (Freon 12®)	5.5
Biphenyl, perchloro	8.8	Methane, dichloro, manofluoro (Freon 21®)	8.3
1,3-Butadiene	7.1	Methane, nitro	12.7
1,3-Butadiene, 2-methyl (isoprene)	7.4	Methane, tetrachloro-difluoro (Freon 112®)	7.8
Butane	6.8	Methane, trichloro-monofluoro (Freon 11®)	7.6
Butanoic acid nitrile	10.5	Naphthalene	9.9
Carbon disulfide	10.0	Nonane	7.8
Carbon tetrachloride	8.6	Octane	7.6
Chloroform	9.3	Pentane	7.0
Cyclohexane	8.2	Pentane, 1-bromo	7.6
Cyclohexane, methyl	7.8	Pentane, 1-chloro	8.3
Cyclohexane, perfluoro	6.0	Pentanoic acid, nitrile (valeronitrile)	9.6
Cyclopentane	8.7	Pentene-1	6.9
Decalin	8.8	Phenanthrene	9.8
Decane	8.0	Propane	6.4
Dimethyl sulfide	9.4	Propane, 1-bromo	8.9
Ethane	6.0	Propane, 2,2-dimethyl (neopentane)	6.3
Ethane, bromo (ethyl bromide)	9.6	Propane, 1-nitro	16.3
Ethane, chloro (ethyl chloride)	9.2	Propane-2-nitro	9.9
Ethane, 1,2-dibromo	10.4	Propene (propylene)	6.5
Ethane, 1,1-dichloro (ethylidene chloride)	8.9	Propene, 2-methyl (isobutylene)	6.7
Ethane, difluoro-tetrachloro (Freon 112®)	7.8	Propenoic acid nitrile (acrylonitrile)	10.5
Ethane, nitro	11.1	Propionic acid nitrile	10.8
Ethane, pentachloro	9.4	Styrene	9.3
Ethane, 1,1,2,2-tetrachloro	9.7	Terphenyl, hydrogenated	9.0
Ethanethiol (ethyl mercaptan)	9.2	Tetralin	9.5
Ethane, 1,1,2-trichloro	9.6	Toluene	8.9
Ethane trichloro-trifluoro (Freon 113®)	7.3	Xylene, *m*-	8.8

Table II (continued)
SOLUBILITY PARAMETER VALUES

B. Moderately Polar Solvents

Name	δ (H)
Acetic acid, butyl ester	8.5
Acetic acid, ethyl ester	9.1
Acetic acid, methyl ester	9.6
Acetic acid, pentyl ester	8.0
Acetic acid, propyl ester	8.8
Acetic acid amide, *N,N*-diethyl	9.9
Acetic acid amide, *N,N*-dimethyl	10.8
Acrylic acid, butyl ester	8.4
Acrylic acid, ethyl ester	8.6
Acrylic acid, methyl ester	8.9
Adipic acid, dioctyl ester	8.7
Aniline, *N,N*-dimethyl	9.7
Benzene, 1-methoxy-4-propenyl (anethole)	8.4
Benzoic acid, ethyl ester	8.2
Benzoic acid, methyl ester	10.5
Butanal	9.0
Butane, 1-iodo	8.6
Butanoic acid, 4-hydroxylactone (butyrolactone)	12.6
2-Butanone	9.3
Carbonic acid, diethyl ester	8.8
Carbonic acid, dimethyl ester	9.9
Cyclohexanone	9.9
Cyclopentanone	10.4
2-Decanone	7.8
Diethylene glycol, monobutyl ether (butyl carbitol)	9.5
Diethylene glycol, monoethyl ether (ethyl carbitol)	10.2
Dimethyl sulfoxide	12.0
1,4-Dioxane	10.0
Ethene, chloro (vinyl chloride)	7.8
Ether, 1,1-dichloroethyl	10.0
Ether, diethyl	7.4
Ether, dimethyl	8.8
Ether, dipropyl	7.8
Ethylene glycol, monobutyl ether (butyl Cellosolve®)	9.5
Ethylene glycol, monoethyl ether (ethyl Cellosolve)	10.5
Ethylene glycol, monomethyl ether (methyl Cellosolve)	11.4
Formic acid amide, *N,N*-diethyl	10.6
Formic acid amide, *N,N*-dimethyl	12.1
Formic acid, ethyl ester	9.4
Formic acid, methyl ester	10.2
Formic acid, 2-methylbutyl ester	8.0
Formic acid, propyl ester	9.2
Furan	9.4
Furan, tetrahydro	9.1
Furfural	11.2
2-Heptanone	8.5
Hexanoic acid, 6-aminolactam (ϵ-caprolactam)	12.7
Hexanoic acid, 6-hydroxylactone (caprolactone)	10.1
Isophorone	9.1
Lactic acid, butyl ester	9.4
Lactic acid, ethyl ester	10.0
Methacrylic acid, butyl ester	8.3
Methacrylic acid, ethyl ester	8.5
Methacrylic acid, methyl ester	8.8
Oxalic acid, diethyl ester	8.6
Oxalic acid, dimethyl ester	11.0
Oxirane (ethylene oxide)	11.1
Pentane, 1-iodo	8.4
2-Pentanone	8.7
Pentanone-2,4-hydroxy,4-methyl (diacetone alcohol)	9.2
Pentanone-2,4-methyl (mesityl oxide)	9.0
Phosphoric acid, triphenyl ester	8.6
Phosphoric acid, tri-2-tolyl ester	8.4
Phthalic acid, dibutyl ester	9.3
Phthalic acid, diethyl ester	10.0
Phthalic acid, dihexyl ester	8.9
Phthalic acid, dimethyl ester	10.7
Phthalic acid, di-2-methylnonyl ester	7.2
Phthalic acid dioctyl ester	7.9
Phthalic acid, dipentyl ester	9.1
Phthalic acid, dipropyl ester	9.7
Propane, 1,2-epoxy (propylene oxide)	9.2
Propionic acid, ethyl ester	8.4
Propionic acid, methyl ester	8.9
4-Pyrone	13.4
2-Pyrrolidone, 1-methyl	11.3
Sebacic acid, dibutyl ester	9.2
Sebacic acid, dioctyl ester	8.6
Stearic acid, butyl ester	7.5
Sulfone, diethyl	12.4
Sulfone, dimethyl	14.5
Sulfone, dipropyl	11.3

Table II (continued)
SOLUBILITY PARAMETER VALUES

C. Hydrogen-bonded Solvents

Name	δ (H)	Name	δ (H)
Acetic acid	10.1	1-Hexanol	10.7
Acetic acid amide, N-ethyl	12.3	1-Hexanol-2-ethyl	9.5
Acetic acid, dichloro	11.0	Maleic acid anhydride	13.6
Acetic acid, anhydride	10.3	Methacrylic acid	11.2
Acrylic acid	12.0	Methacrylic acid amide, N-Methyl	14.6
Amine, diethyl	8.0	Methanol	14.5
Amine, ethyl	10.0	Methanol, 2-furil (furfuryl alcohol)	12.5
Amine, methyl	11.2	1-Nonanol	8.4
Ammonia	16.3	Pentane, 1-amino	8.7
Aniline	10.3	1,3-Pentanediol, 2-methyl	10.3
1,3-Butanediol	10.9	1-Pentanol	11.6
1,4-Butanediol	10.0	2-Pentanol	12.1
2,3-Butanediol	8.7	Piperidine	11.1
1-Butanol	13.6	2-Piperidone	11.4
2-Butanol	12.6	1,2-Propanediol	10.8
1-Butanol, 2-ethyl	11.9	1-Propanol	10.5
1-Butanol, 2-methyl	11.5	2-Propanol	10.0
Butyric acid	10.5	1-Propanol, 2-methyl	10.5
Cyclohexanol	10.6	2-Propanol, 2-methyl	11.4
Diethylene glycol	11.8	2-Propenol (allyl alcohol)	12.1
1-Dodecanol	9.9	Propionic acid	8.1
Ethanol	10.0	Propionic acid anhydride	12.7
Ethanol, 2-chloro (ethylene chlorohydrin)	12.6	1,2-Propanediol	12.2
Ethylene glycol	10.7	Pyridine	14.6
Formic acid	14.7	2-Pyrrolidone	12.1
Formic acid amide, N-ethyl	10.8	Quinoline	13.9
Formic acid amide, N-methyl	15.4	Succinic acid anhydride	16.1
Glycerol	9.9	Tetraethylene glycol	16.5
2,3-Hexanediol	10.2	Toluene, 3-hydroxy (meta cresol)	10.3
1,3-Hexanediol-2-ethyl	23.4	Water	9.4

Table compiled by R. B. Seymour.

HEATS OF VAPORIZATION OF ORGANIC COMPOUNDS

Numerical values in the following table are in the units of gram calories per gram mole. To convert to joules per gram mole, multiply the listed value by 4.184.

Formula	Name	Δ H_v
$CBrN$	Cyanogen bromide	10,882.8
CBr_4	Carbon tetrabromide	10,771.4
$CBrF_3$	Bromotrifluoromethane	–
CBr_2F_2	Dibromodifluoromethane	–
$CClF_3$	Chlorotrifluoromethane	3,996.3
$CClN$	Cyanogen chloride	5,243.4
CCl_2F_2	Dichlorodifluoromethane	8,363.1
CCl_2O	Phosgene	6,224.3
CCl_3F	Trichlorofluoromethane	6,424.1
CCl_3NO_2	Trichloronitromethane	9,109.7
CCl_4	Carbon tetrachloride	8,271.5
		7,628.8
CFN	Cyanogen fluoride	5,875.3
CF_4	Carbon tetrafluoride	3,016.5
$CHBr_3$	Tribromomethane	9,673.3
$CHClF_2$	Chlorodifluoromethane	5,212.9
$CHCl_2F$	Dichlorofluoromethane	6,286.8
$CHCl_3$	Chloroform	7,500.5
CHF_3	Trifluoromethane	–
CHN	Hydrogen cyanide	7,338.8
CH_2Br_2	Dibromomethane	8,722.0
CH_2Cl_2	Dichloromethane	7,572.3
CH_2O	Formaldehyde	5,917.9
CH_2O_2	Formic acid	9,896.5
CH_3ASCl_2	Dichloromethylarsine	9,636.8
CH_3BO	Borine carbonyl	4,867.6
CH_3Br	Methyl bromide	5,925.9
CH_3Cl	Methyl chloride	5,375.3
CH_3Cl_3Si	Trichloromethylsilane	7,450.0
CH_3F	Methyl fluoride	3,986.4
CH_3I	Methyl iodide	6,616.5
CH_3NO	Formamide	15,556.6
CH_3NO_2	Nitromethane	9,210.9
CH_4	Methane	2,128.8
CH_4Cl_2Si	Dichloromethylsilane	7,011.0
CH_4O	Methanol	9,377.2
		8,978.8
CH_4S	Methanethiol	6,331.9
CH_5ClSi	Chloromethylsilane	6,349.5
CH_5N	Methylamine	6,469.5
CH_6Si	Methylsilane	4,683.6
CH_9NSi_2	2-Methyldisilazane	7,185.6
CIN	Cyanogen iodide	14,065.4
CN_4O_8	Tetranitromethane	9,848.7
CO	Carbon monoxide	1,613.3
COS	Carbonyl sulfide	4,992.2
$COSe$	Carbonyl selenide	5,366.5
CO_2	Carbon dioxide	5,539.0
$CSSe$	Carbon selenosulfide	8,003.0
CS_2	Carbon disulfide	6,786.8
C_2BrCl_3O	Trichloroacetyl bromide	9,673.9
C_2HClF_2	1-Chloro-2,2-difluoroethylene	–
C_2ClF_3	1-Chloro-1,2,2-trifluoroethylene	5,421.5
$C_2Cl_2F_2$	1,2-Dichloro-1,2-difluoroethylene	7,185.6
C_2F_4	Tetrafluoroethylene	–
$C_2Cl_2F_4$	1,1-Dichloro-1,2,2,2-tetrafluoroethane	–
$C_2Cl_2F_4$	1,2-Dichloro-1,1,2,2-tetrafluoroethane	6,134.6
$C_2Cl_3F_3$	1,1,2-Trichloro-1,2,2-trifluoroethane	7,115.4
C_2Cl_4	Tetrachloroethylene	9,240.5
CCl_4F_2	Tetrachlorodifluoroethane	–
$C_2Cl_4F_2$	1,1,2,2-Tetrachloro-1,2-difluoroethane	8,746.2
$C_2Cl_3F_3$	1,1,2-Trifluoro-1,2,2-trichloroethane	–
C_2Cl_6	Hexachloroethane	11,711.3
C_2ClF_5	Chloropentafluoroethane	–
C_2F_6	Hexafluoroethane	–
C_2HBr_3O	Tribromoacetaldehyde	11,057.8
C_2HCl_3	Trichloroethylene	8,314.7
C_2HCl_3O	Trichloroacetaldehyde	8,469.2
$C_2HCl_3O_2$	Trichloroacetic acid	13,817.9
C_2HCl_5	Pentachloroethane	9,800.1
C_2H_2	Acetylene	4,665.8
$C_2H_2Br_4$	1,1,1,2-Tetrabromoethane	14.517.3
$C_2H_2Br_4$	1,1,2,2-Tetrabromoethane	12,911.5
$C_2H_2O_4$	Oxalic acid	21,630.6
$C_2H_2Cl_2$	*cis*-1,2-Dichloroethylene	7,420.6
$C_2H_2Cl_2$	*trans*-1,2-Dichloroethylene	7,243.1
$C_2H_2Cl_2$	1,1-Dichloroethylene	7,211.8
$C_2H_2F_2$	1,1-Difluoroethylene	–
$C_2H_2Cl_2O_2$	Dichloroacetic acid	12,952.9
$C_2H_2Cl_4$	1,1,1,2-Tetrachloroethane	9,296.5
		8,725.6
$C_2H_2Cl_4$	1,1,2,2-Tetrachloroethane	9,917.1
C_2H_3Br	1-Bromoethylene	6,076.9
$C_2H_3BrO_2$	Bromoacetic acid	13,537.8
$C_2H_3Br_3$	1,1,2-Tribromoethane	11,874.1
C_2H_3Cl	1-Chloroethylene	6,263.0
$C_2H_3ClO_2$	Chloroacetic acid	13,134.5
$C_2H_3Cl_3$	1,1,1-Trichloroethane	8,012.7
$C_2H_3Cl_3$	1,1,2-Trichloroethane	9,163.2
$C_2H_3ClF_2$	1-Chloro-1,1-difluoroethane	–
$C_2H_3F_3$	1,1,1-Trifluoroethane	–
$C_2H_3Cl_3O_2$	Trichloroacetaldehyde-hydrate	12,141.5
C_2H_3F	1-Fluoroethylene	4,198.1
C_2H_3N	Acetonitrile	8,173.2
C_2H_3NS	Methyl thiocyanate	9,424.1
C_2H_3NS	Methyl isothiocyanate	7,990.1
C_2H_4	Ethylene	3,453.7
C_2H_4BrCl	1-Bromo-2-chloroethane	9,314.9
C_2H_4BrCl	1-Bromo-2-chloroethane	8,995.6
$C_2H_4Br_2$	1,2-Dibromoethane	9,229.4
$C_2H_4Cl_2$	1,1-Dichloroethane	7,288.0
$C_2H_4Cl_2$	1,2-Dichloroethane	7,950.7
$C_2H_4F_2$	1,1-Difluoroethane	6,068.8
C_2H_4O	Acetaldehyde	7,267.8
		6,622.1
C_2H_4O	Ethylene oxide	6,823.3
$C_2H_4O_2$	Acetic acid	9,963.9
		9,486.6
$C_2H_4O_2$	Methyl formate	7,027.8
$C_2H_4O_2S$	Mercaptoacetic acid	13,790.7
C_2H_5Br	Ethyl bromide	6,843.1
C_2H_5Cl	Ethyl chloride	6,310.6
C_2H_5ClO	2-Chloroethanol	10,740.6
$C_2H_5Cl_3Si$	Trichloroethylsilane	9,457.8
$C_2H_5Cl_3OSi$	Trichloroethoxysilane	8,811.4
C_2H_5F	Ethyl fluoride	5,519.5
$C_2H_5F_3Si$	Ethyltrifluorosilane	6,945.7
C_2H_5I	Ethyl iodide	7,851.8
C_2H_5NO	Acetamide	14,025.3
C_2H_5NO	Acetaldoxime	11,317.8
$C_2H_5NO_2$	Nitroethane	9,531.1
$C_2H_5N_3O_2$	Di(nitrosomethyl)amine	10,326.7
C_2H_6	Ethane	3,739.5
$C_2H_6Cl_2Si$	Dichlorodimethylsilane	7,995.7
C_2H_6O	Ethyl alcohol	9,673.9
C_2H_6O	Dimethyl ether	5,409.8
$C_2H_6O_2$	1,2-Ethanediol	14,032.4
C_2H_6S	Dimethyl sulfide	6,742.3
C_2H_6S	Ethane thiol	6,728.7
C_2H_6Sb	Dimethylantimony	12,075.7
C_2H_7N	Ethylamine	6,845.1
C_2H_7N	Dimethylamine	6,660.0
$C_2H_8N_2$	1,2-Ethanediamine	10,510.5
C_2H_8Si	Dimethylsilane	5,497.8
$C_2H_{10}B_2$	Dimethyldiborane	5,696.7
$C_2H_{11}NSi_2$	2-Ethyldisilazane	7,348.3
C_2N_2	Cyanogen	6,597.3
C_3H_3N	Acrylonitrile	7,941.4
C_3H_4	Propadiene	5,141.2
C_3H_4	Propyne	5,632.4
$C_3H_4Br_2$	2,3-Dibromopropene	9,886.2
$C_3H_4Cl_2O_2$	Methyl dichloroacetate	10,820.5
C_3H_4O	2-Propenal	7,628.8
$C_3H_4O_2$	Acrylic acid	10,955.1
$C_3H_4O_3$	Pyruvic acid	11,815.7
$C_3H_5Br_3$	1,2,3-Tribromopropane	12,047.1
C_3H_5Cl	1-Chloropropene	6,594.3
C_3H_5Cl	Allyl chloride	7,386.8
C_3H_5ClO	Epichlorohydrine	9,815.4
$C_3H_5ClO_2$	Methyl chloroacetate	10,815.0
$C_3H_5Cl_3$	1,1,1-Trichloropropane	8,933.9
$C_3H_5Cl_3$	1,2,3-Trichloropropane	10,714.3
$C_3H_5Cl_3Si$	Allyltrichlorosilane	9,386.1
C_3H_5N	Propionitrile	8,769.0
C_3H_5NO	3-Hydroxypropionitrile	13,287.2
C_3H_5NS	Ethylisothiocyanate	9,574.7
$C_3H_5N_3O_9$	Nitroglycerine	13,753.1
C_3H_6	Propene	4,697.4
C_3H_6	Cyclopropane	5,897.7
C_3H_6BrNO	2-Bromo-2-nitrosopropane	9,619.6
$C_3H_6Br_2$	1,2-Dibromopropane	9,801.9
$C_3H_6Br_2$	1,3-Dibromopropane	10,374.4
$C_3H_6Br_2O$	2,3-Dibromo-1-propanol	13,190.0
$C_3H_6Cl_2$	1,2-Dichloropropane	8,428.5
$C_3H_6Cl_2O$	1,3-Dichloro-2-propanol	12,067.6
C_3H_6O	Acetone	7,641.5
C_3H_6O	Allyl alcohol	10,577.7
C_3H_6O	Propylene oxide	7,295.8
$C_3H_6O_2$	Propanoic acid	12,454.4
$C_3H_6O_2$	Methyl acetate	7,732.8
$C_3H_6O_2$	Ethyl formate	7,511.7
$C_3H_6O_3$	Methyl glycolate	11,105.0
$C_3H_6O_3$	Methoxyacetic acid	13,451.0
C_3H_7Br	*n*-Propyl bromide	8,029.8
C_3H_7Br	2-Bromopropane	7,591.7
C_3H_7Cl	*n*-Propyl chloride	7,485.7
		6,905.8
C_3H_7Cl	2-Chloropropane	6,855.2
$C_3H_7Cl_3Si$	Trichloroisopropylsilane	8,973.3
C_3H_7I	*n*-Propyl iodide	8,467.1
C_3H_7I	2-Iodopropane	8,243.4
	Propionamide	14,554.0
C_3H_7NO	1-Nitropropane	9,949.9
$C_3H_7NO_2$	2-Nitropropane	9,476.9
$C_3H_7NO_2$	Ethyl carbamate	13,078.6
$C_3H_7NO_2$	Propane	4,550.0
C_3H_8		4,811.8
C_3H_8O	*n*-Propanol	11,298.8
		10,421.1
C_3H_8O	Isopropanol	10,063.5
C_3H_8O	Ethyl methyl ether	6,388.3
$C_3H_8O_2$	1,2-Propanediol	13,575.2
$C_3H_8O_2$	1,3-Propanediol	13,782.3
$C_3H_8O_2$	2-Methoxyethanol	9,893.8
$C_3H_8O_2$	Glycerol	18,188.9
C_3H_8S	Methyl ethyl sulfide	–
C_3H_8S	Propanethiol	7,855.3
C_3H_9B	Trimethylborine	5,375.4
C_3H_9ClSi	Chlorotrimethylsilane	7,589.1
C_3H_9Ga	Trimethylgallium	7,758.8
C_3H_9N	*n*-Propylamine	7,408.0
C_3H_9N	Trimethylamine	6,361.7
$C_3H_9O_4P$	Trimethyl phosphate	11,019.7
$C_3H_{12}B_2$	Trimethyldiborane	6,981.8
C_3O_2	Carbon suboxide	6,446.3
C_3S_2	Carbon subsulfide	10,466.0
$C_4Cl_6O_3$	Trichloroacetic anhydride	12,929.0
C_3F_8	Octafluoropropane	–
C_4H_4O	Furan	–
C_4H_2	1,3-Butadiyne	7,761.0
$C_4H_2Br_2O_3$	α,β-Dibromomaleic anhydride	12,579.2
$C_4H_2Cl_2O_2$	*trans*-Fumaryl chloride	11,251.0
$C_4H_2O_3$	Maleic anhydride	12,122.3
$C_4H_3NO_2S$	2-Nitrothiophene	11,926.2
C_4H_4	Butenyne	6,677.2
$C_4H_4Cl_2O_2$	Succinyl chloride	12,466.1
$C_4H_4Cl_2O_3$	Chloroacetic anhydride	14,645.1
$C_4H_4O_3$	Succinic anhydride	14,726.0
$C_4H_4O_4$	1,4-Dioxane-2,6-dione	14,013.6
C_4H_4S	Thiophene	8,748.3
C_4H_4Se	Selenophene	7,766.1
$C_4H_5ClO_2$	α-Chlorocrotonic acid	15,440.1
$C_4H_5ClO_3$	Ethyl chloroglyoxylate	10,268.4
$C_4H_5Cl_3O_2$	Ethyl trichloroacetate	11,625.1
C_4H_5N	3-Butenenitrile	9,447.8
C_4H_5N	Methacrylonitrile	8,083.8
C_4H_5N	*cis*-Crotononitrile	8,905.4
C_4H_5N	*trans*-Crotononitrile	9,277.1
$C_4H_5NO_2$	Succinimide	16,422.0
C_4H_5NS	Allylisothiocyanate	9,967.8
C_4H_6	1,2-Butadiene	6,539.1
C_4H_6	1,3-Butadiene	5,688.2
C_4H_6	Cyclobutene	6,167.5
C_4H_6	1-Butyne	6,596.9
C_4H_6	2-Butyne	7,868.5
$C_4H_6Cl_2O_2$	Ethyl dichloroacetate	10,842.8
$C_4H_6Cl_2O_2$	2-Chloroethyl chloroacetate	12,588.7
$C_4H_6O_2$	*cis*-Crotonic acid	12,964.7
$C_4H_6O_2$	*trans*-Crotonic acid	13,252.2
$C_4H_6O_2$	Methyl acrylate	8,598.0
$C_4H_6O_2$	Methacrylic acid	12,526.6
$C_4H_6O_2$	Vinyl acetate	8,470.4
$C_4H_6O_2$	Acetic anhydride	10,930.4
$C_4H_6O_4$	Dimethyl oxalate	11,519.4
C_4H_7Br	*cis*-1-Bromo-1-butene	8,300.2
C_4H_7Br	*trans*-1-Bromo-1-butene	8,515.7
C_4H_7Br	2-Bromo-1-butene	8,389.7
C_4H_7Br	*cis*-2-Bromo-2-butene	8,486.3
C_4H_7Br	*trans*-2-Bromo-2-butene	8,238.1
C_4H_7BrO	1-Bromo-2-butanone	10,980.7
C_4H_7BrO	2-Methylpropionyl bromide	10,974.6
$C_4H_7Br_3$	1,1,2-Tribromobutane	11,936.5
$C_4H_7Br_3$	1,2,2-Tribromobutane	11,622.3

Formula	Name	Δ H_v
$C_4H_7Br_3$	1,2,2,-Tribromobutane	11.622.3
$C_4H_7Br_3$	2,2,3-Tribromobutane	11,664.2
$C_4H_7ClO_2$	Ethyl chloroacetate	10,522.6
$C_4H_8Cl_2$	1,1-Dichloro-2-methylpropane	9,111.1
$C_4H_7Cl_3$	1,2,3-Trichlorobutane	9,447.0
C_4H_7N	Butyronitrile	9,462.9
$C_4H_7NO_2$	Diacetamide	14,508.1
C_4H_8	1-Butene	5,996.7
C_4H_8	*cis*-2-Butene	6,401.0
C_4H_8	*trans*-2-Butene	6,221.6
C_4H_8	2-Methylpropene	5,742.9
C_4H_8	Cyclobutane	6,464.8
C_4H_8BrClO	2-Bromoethyl-2-chloroethyl ether	12,010.5
$C_4H_8Br_2$	1,2-Dibromobutane	10,182.1
$C_4H_8Br_2$	*dl*-2,3-Dibromobutane	10,136.1
$C_4H_8Br_2$	*meso*-2,3-Dibromobutane	9,966.9
$C_4H_8Br_2$	1,4-Dibromobutane	11,369.3
$C_4H_8Br_2O$	Di(2-bromoethyl)ether	12,454.4
$C_4H_8Cl_2$	1,2-Dichlorobutane	8;850.6
$C_4H_8Cl_2$	2,3-Dichlorobutane	8,975.3
$C_4H_8Cl_2$	1,1-Dichloro-2-methylpropane	8,795.6
$C_4H_8Cl_2$	1,2-Dichloro-2-methylpropane	9,260.1
$C_4H_8Cl_2$	1,3-Dichloro-2-methylpropane	10,519.7
$C_4H_8Cl_2O$	Di(chloroethyl)ether	11,376.8
C_4H_8O	1,2-Epoxy-2-methylpropane	7,066.6
C_4H_8O	Methyl ethyl ketone	8,149.5
$C_4H_8O_2$	Dioxane	8,546.2
$C_4H_8O_2$	*n*-Butyric acid	11,881.2
$C_4H_8O_2$	Isobutyric acid	11,182.8
$C_4H_8O_2$	Ethyl acetate	8,301.1
$C_4H_8O_2$	Methyl propanoate	8,356.2
$C_4H_8O_2$	*n*-Propyl formate	8,208.1
$C_4H_8O_2$	Isopropyl formate	8,230.2
$C_4H_8O_3$	α-Hydroxyisobutyric acid	15,967.0
$C_4H_8O_3$	Ethyl glycolate	11,318.1
C_4H_9Br	*n*-Butyl bromide	8,789.1
C_4H_9BrO	1-Bromo-2-butanol	13,473.7
C_4H_9Cl	*n*-Butyl chloride	8,144.8
C_4H_9Cl	*sec*-Butyl chloride	7,407.9
C_4H_9Cl	Isobutyl chloride	8,045.1
C_4H_9Cl	*tert*-Butyl chloride	6,876.0
$C_4H_9ClO_2$	2-(2-Chloroethoxy)ethanol	14,082.1
C_4H_9I	*n*-Butyl iodide	–
C_4H_9I	1-Iodo-2-methylpropane	9,650.7
$C_4H_9NO_2$	Ethyl methylcarbamate	12,161.2
$C_4H_9NO_2$	Propyl carbamate	14,071.8
$C_4H_9N_2O_2$	Di(nitrosoethyl)amine	10,894.8
C_4H_{10}	*n*-Butane	5,801.2
C_4H_{10}	2-Methylpropane	5,084.4
		5,416.2
$C_4H_{10}Cl_2Si$	Dichlorodiethylsilane	10,038.6
$C_4H_{10}F_2Si$	Diethyldifluorosilane	8,214.9
$C_4H_{10}O$	*n*-Butyl alcohol	10,970.5
$C_4H_{10}O$	*sec*-Butyl alcohol	10,712.3
$C_4H_{10}O$	Isobutyl alcohol	10,936.0
$C_4H_{10}O$	*tert*-Butyl alcohol	10,413.2
$C_4H_{10}O$	Diethyl ether	6,946.2
$C_4H_{10}O$	Methyl propyl ether	7,409.7
$C_4H_{10}O_2$	1,3-Butanediol	10,479.1
$C_4H_{10}O_2$	2,3-Butanediol	13,708.6
$C_4H_{10}O_2$	1,2-Dimethoxyethane	7,681.0
$C_4H_{10}O_2S$	2,2-Thiodiethanol	6,597.0
$C_4H_{10}O_3$	Diethylene glycol	16,146.7
$C_4H_{10}O_3$	1,2,3-Butanetriol	16,345.8
$C_4H_{10}O_3S$	Diethyl sulfite	10,783.0
$C_4H_{10}O_4S$	Diethyl sulfate	12,518.2
$C_4H_{10}S$	*n*-Butanethiol	–
$C_4H_{10}S$	Diethyl sulfide	8,210.8
$C_4H_{10}Se$	Diethyl selenide	9,274.7
$C_4H_{10}Zn$	Diethyl zinc	9,162.3
$C_4H_{11}N$	Diethyl amine	7.307.5
$C_4H_{11}N$	Isobutylamine	478.3
$C_4H_{12}Cl_2Si_2$	1,3-Diethoxytetramethyl-disiloxane	9,881.6
$C_4H_{12}Pb$	Tetramethyllead	8,843.8
$C_4H_{12}Si$	Tetramethylsilane	6,439.2
$C_4H_{12}Sn$	Tetramethyl tin	7,897.8
$C_4H_{14}B_2$	Tetramethyldiborane	7.517.1
C_4F_8	Octafluorocyclobutane	–
C_4F_{10}	Perfluoro-*n*-butane	–
C_5H_4BrN	3-Bromopyridine	10,863.7
C_5H_4Cln	2-Chloropyridine	10,614.5
$C_5H_4O_2$	2-Furaldehyde	11,614.6
$C_5H_4O_3$	Citraconic anhydride	12,307.8
C_5H_5N	Pyridine	9,649.4
$C_5H_6Cl_2O_2$	Glutaryl chloride	13,192.1
$C_5H_6N_2$	Glutaronitrile	13,767.5
$C_5H_6O_2$	Furfuryl alcohol	12,815.8
$C_5H_6O_3$	Glutaric anhydride	14,814.1
$C_5H_6O_3$	Pyrotartaric anhydride	13,251.2

Formula	Name	Δ H_v
C_5H_6S	2-Methylthiophene	8,884.2
C_5H_6S	3-Methylthiophene	9,084.1
$C_5H_7ClO_3$	Propyl chloroglyoxylate	11,430.0
C_5H_7N	Tiglonitrile	8,704.6
C_5H_7N	Angelonitrile	9,707.5
C_5H_7N	α-Ethylacrylonitrile	8,679.1
$C_5H_7NO_2$	Ethyl cyanoacetate	15,615.6
C_5H_8	Cyclopentene	–
C_5H_8	Isoprene	6,901.8
C_5H_8	1,3-Pentadiene	7,313.9
C_5H_8	1,4-Pentadiene	6,826.6
C_5H_8O	Tiglaldehyde	9,009.2
$C_5H_8O_2$	Levulinaldehyde	11,483.8
$C_5H_8O_2$	Tiglic acid	13,756.5
$C_5H_8O_2$	α-Valerolactone	11,537.0
$C_5H_8O_2$	α-Ethylacrylic acid	14,417.8
$C_5H_8O_2$	Ethyl acrylate	9,259.4
$C_5H_8O_2$	Methyl methacrylate	8,974.9
$C_5H_8O_3$	Levulinic acid	17,795.0
$C_5H_8O_4$	Glutaric acid	22,085.2
$C_5H_8O_4$	Dimethyl malonate	12,608.1
$C_5H_9ClO_2$	Ethyl α-chloropropionate	11,032.8
$C_5H_9ClO_2$	Isopropyl chloroacetate	10,575.7
C_5H_9N	Valeronitrile	9,931.3
C_5H_9NO	α-Hydroxybutyronitrile	13,577.0
C_5H_{10}	1-Pentene	6,931.2
C_5H_{10}	2-Pentene	–
C_5H_{10}	3-Methyl-2-butene	7,112.8
C_5H_{10}	2-Methyl-1-butene	6,474.6
C_5H_{10}	3-Methyl-1-butene	–
C_5H_{10}	Cyclopentane	7,411.1
C_5H_{10}	Methylcyclobutane	6,413.2
$C_5H_{10}Br_2$	1,2-Dibromopentane	11,130.0
$C_5H_{10}Br_2$	1,2-Dibromo-2-methylbutane	7,616.9
$C_5H_{10}Br_2$	1,3-Dibromo-3-methylbutane	10,639.6
$C_5H_{10}Cl_2Si$	Allyldichloroethylsilane	9,833.9
$C_5H_{10}O$	Diethyl ketone	11,183.0
$C_5H_{10}O$	Methyl *n*-propyl ketone	11.240.6
$C_5H_{10}O$	Methyl isopropyl ketone	11,073.2
$C_5H_{10}Cl_2O$	2-Chloroethyl 2-chloroisopropyl ether	11,420.8
$C_5H_{10}Cl_2O$	2-Chloroethyl 2-chloropropyl ether	11,316.9
	Di(2-chloroethoxy)methane	12,908.0
$C_5H_{10}Cl_2O_2$	4-Hydroxy-3-methyl-2-butanone	13,639.4
$C_5H_{10}O_2$	Valeric acid	13,370.3
$C_5H_{10}O_2$	Isovaleric acid	12,951.1
$C_5H_{10}O_2$	Ethyl propanoate	8,877.8
$C_5H_{10}O_2$	*n*-Propyl acetate	8,921.1
$C_5H_{10}O_2$	Isopropyl acetate	8,794.8
$C_5H_{10}O_2$	Methyl butyrate	8,886.0
$C_5H_{10}O_2$	Methyl isobutyrate	8,593.3
$C_5H_{10}O_2$	*n*-Butyl formate	9,285.9
$C_5H_{10}O_2$	Isobutyl formate	8,678.8
$C_5H_{10}O_2$	*sec*-Butyl formate	8,975.7
$C_5H_{10}O_2$	*tert*-Butyl formate	8,955.3
$C_5H_{10}O_2$	Diethyl carbonate	10,159.0
$C_5H_{10}O_3$	1-Bromopentane	–
$C_5H_{11}Br$	1-Bromo-3-methylbutane	9,282.7
$C_5H_{11}Br$	1-Chloropentane	–
$C_5H_{11}Br$	1-Iodopentane	–
$C_5H_{11}I$	1-Iodo-3-methylbutane	9,951.6
$C_5H_{11}I$	Piperidine	8,911.8
$C_5H_{11}N$	Pentanoic acid	–
$C_5H_{11}NO_2$	Isobutyl carbamate	13,897.1
$C_5H_{11}NO_3$	Isoamyl nitrate	10,817.2
C_5H_{12}	*n*-Pentane	6,595.1
C_5H_{12}	2-Methylbutane	6,470.8
C_5H_{12}	2,2-Dimethylpropane	5,648.6
$C_5H_{12}O$	Amyl alcohol	12,495.5
$C_5H_{12}O$	Isoamyl alcohol	12,497.9
$C_5H_{12}O$	2-Pentanol	12,086.2
$C_5H_{12}O$	*tert*-Amyl alcohol	11,239.2
$C_5H_{12}O$	Ethyl propyl ether	7,092.7
$C_5H_{12}O$	Methyl *n*-butyl ether	–
$C_5H_{12}O_3$	2,3,4-Pentanetriol	19,694.4
$C_5H_{12}S$	1-Pentanethiol	–
$C_5H_{14}OSi$	Ethoxytrimethylsilane	8,030.6
$C_5H_{14}Si$	Ethyltrimethylsilane	7,633.4
$C_5H_{14}Sn$	Ethyltrimethyltin	8,820.9
$C_6Cl_4O_2$	Chloranil	21,514.3
C_6Cl_6	Hexachlorobenzene	15,199.1
C_6HCl_5	Pentachlorobenzene	15,124.2
C_6HCl_5O	Pentachlorophenol	16,742.6
$C_6H_2BrCl_3O$	3-Bromo-2,4,6-trichlorophenol	15,231.9
$C_6H_2Cl_4$	1,2,3,4-Tetrachlorobenzene	12,872.5
$C_6H_2Cl_4$	1,2,3,5-Tetrachlorobenzene	11,982.1
$C_6H_2Cl_4$	1,2,4,5-Tetrachlorobenzene	12,828.8
$C_6H_2Cl_4O$	2,3,4,6-Tetrachlorophenol	15,362.7
$C_6H_3BrCl_2O$	2-Bromo-4,6-dichlorophenol	13,829.1
$C_6H_3Cl_3$	1,2,3-Trichlorobenzene	11,349.5

Formula	Name	Δ H_v
$C_6H_3Cl_3$	1,2,4-Trichlorobenzene	11,425.1
$C_6H_3Cl_3$	1,3,5-Trichlorobenzene	11,211.0
$C_6H_3Cl_3O$	2,4,5-Trichlorophenol	13,237.0
$C_6H_3Cl_3O$	2,4,6-Trichlorophenol	14,092.8
$C_6H_4Br_2$	1,4-Dibromobenzene	13,047.8
C_6H_4BrCl	1,4-Bromochlorobenzene	16,671.8
		11,451.1
$C_6H_4Cl_2$	1,2-Dichlorobenzene	10,943.0
$C_6H_4Cl_2$	1,3-Dichlorobenzene	10,446.8
$C_6H_4Cl_2$	1,4-Dichlorobenzene	17,260.5
		10,611.0
$C_6H_4Cl_2O$	2,4-Dichlorophenol	13,230.4
$C_6H_4Cl_2O$	2,6-Dichlorophenol	13,472.0
$C_6H_4Cl_3N$	2,4,6-Trichloroaniline	22,297.3
$C_6H_5AsCl_2$	Dichlorophenylarsine	12,229.5
C_6H_5Br	Bromobenzene	10,157.7
C_6H_5Cl	Chlorobenzene	10,098.0
		9,067.3
C_6H_5ClO	2-Chlorophenol	10,341.1
C_6H_5ClO	3-Chlorophenol	11,979.7
C_6H_5ClO	4-Chlorophenol	12,281.6
$C_6H_5ClO_2S$	Benzenesulfonylchloride	12,621.0
$C_6H_5Cl_2O_2P$	Phenyl dichlorophosphate	13,319.6
$C_6H_5Cl_3Si$	Trichlorophenylsilane	11,385.9
C_6H_5F	Fluorobenzene	7,980.4
$C_6H_5F_3Si$	Trifluorophenylsilane	9,171.6
C_6H_5I	Iodobenzene	10,277.2
		10,377.8
$C_6H_5NO_2$	Nitrobenzene	12,168.2
$C_6H_5NO_3$	2-Nitrophenol	12,497.3
C_6H_6	1,5-Hexadiene-3-yne	8,288.0
C_6H_6	Benzene	10,254.3
		8,146.5
C_6H_6ClN	2-Chloroaniline	12,441.0
C_6H_6ClN	3-Chloroaniline	13,385.6
C_6H_6ClN	4-Chloroaniline	12,832.8
C_6H_6ClO	4-Chlorophenol	12,964.7
$C_6H_6N_2O_2$	2-Nitroaniline	15,284.0
$C_6H_6N_2O_2$	3-Nitroaniline	15,996.3
$C_6H_6N_2O_2$	4-Nitroaniline	17,220.2
C_6H_6O	Phenol	11,891.5
$C_6H_6O_2$	Pyrocatechol	13,779.7
$C_6H_6O_2$	Resorcinol	16,400.8
$C_6H_6O_2$	Hydroquinone	18,734.0
$C_6H_6O_3$	Pyrogallol	15,731.8
C_6H_6S	Benzenethiol	11,320.1
C_6H_7N	Aniline	11,307.6
C_6H_7N	2-Picoline	9,933.2
C_6H_7N	3-Methylpyridine (β-picoline)	–
C_6H_8	1,3-Cyclohexadiene	–
$C_6H_8Cl_2O_4$	Ethylene-bis-chloroacetate	16,499.1
$C_6H_8N_2$	1,3-Phenylenediamine	14,761.1
$C_6H_8N_2$	Phenylhydrazine	13,711.9
$C_6H_8O_3$	α-Methylglutaric anhydride	14,204.9
$C_6H_8O_3$	α,α-Dimethylsuccinic anhydride	13,683.1
$C_6H_8O_4$	Dimethyl maleate	12,615.7
C_6H_{10}	Cyclohexene	–
C_6H_{10}	1,5-Hexadiene	–
$C_6H_{10}Cl_2O_2$	Isobutyl dichloroacetate	11,733.1
$C_6H_{10}Cl_2Si$	Diallyldichlorosilane	10,462.8
$C_6H_{10}O$	Cyclohexanone	10,037.6
$C_6H_{10}O$	Mesityl oxide	10,109.4
$C_6H_{10}O_2$	Isocaprolactone	11,685.0
$C_6H_{10}O_3$	Propionic anhydride	11,572.6
$C_6H_{10}O_3$	Ethyl acetoacetate	11,842.0
$C_6H_{10}O_3$	Methyl levulinate	12,249.8
$C_6H_{10}O_4$	Adipic acid	19,570.2
$C_6H_{10}O_4$	Diethyl oxalate	14,016.9
$C_6H_{10}O_4$	Glycol diacetate	12,496.1
$C_6H_{10}O_5$	Dimethyl-*l*-malate	14,127.6
$C_6H_{10}O_6$	Dimethyl-*d*-tartrate	15,372.6
$C_6H_{10}O_6$	Dimethyl-*dl*-tartrate	14,999.1
$C_6H_{10}S$	Diallyl sulfide	9,652.6
$C_6H_{11}BrO_2$	Ethyl α-bromoisobutyrate	10,635.8
$C_6H_{11}ClO_2$	*sec*-Butylchloroacetate	11,152.0
$C_6H_{11}N$	Capronitrile	10,492.3
C_6H_{12}	1-Hexene	7,787.6
C_6H_{12}	2-Hexene	–
C_6H_{12}	Cyclohexane	7,830.9
C_6H_{12}	Methylcyclopentane	7,940.0
$C_6H_{12}Cl_2O$	Dichlorodiisopropyl ether	11,881.1
$C_6H_{12}Cl_2O_2$	bis(2-Chloroethyl)acetal	13,497.1
$C_6H_{12}O$	2-Hexanone	12,358.3
$C_6H_{12}O$	4-Methyl-2-pentanone	11,669.6
$C_6H_{12}O$	Allyl propyl ether	8,621.5
$C_6H_{12}O$	Allyl isopropyl ether	8,637.5
$C_6H_{12}O$	Cyclohexanol	11,935.8
$C_6H_{12}O_2$	Caproic acid	16,189.4
$C_6H_{12}O_2$	Isocaproic acid	14,874.8
$C_6H_{12}O_2$	4-Hydroxy-4-methyl-2-pentanone	11,718.8

Formula	Name	Δ H_v
$C_6H_{12}O_2$	Methyl pentanoate	–
$C_6H_{12}O_2$	Methyl isovalerate	9,567.5
$C_6H_{12}O_2$	Ethyl *n*-butyrate	9,468.5
$C_6H_{12}O_2$	Ethyl isobutyrate	8,945.7
$C_6H_{12}O_2$	*n*-Propyl propanoate	9,857.2
$C_6H_{12}O_2$	*n*-Butyl acetate	–
$C_6H_{12}O_2$	Isobutyl acetate	9,300.8
$C_6H_{12}O_2$	*n*-Amyl formate	–
$C_6H_{12}O_2$	Isoamyl formate	9,438.2
$C_6H_{12}O_3$	Paraformaldehyde	10,348.2
C_6H_{14}	Hexane	7,627.2
C_6H_{14}	2-Methylpentane	7,676.6
C_6H_{14}	3-Methylpentane	7,743.9
C_6H_{14}	2,2-Dimethylbutane	7,271.0
C_6H_{14}	2,3-Dimethylbutane	7,120.0
$C_6H_{14}O$	1-Hexanol	12,708.5
$C_6H_{14}O$	2-Hexanol	12,386.5
$C_6H_{14}O$	3-Hexanol	11,157.9
$C_6H_{14}O$	2-Methyl-1-pentanol	12,036.6
$C_6H_{14}O$	2-Methyl-2-pentanol	11,132.0
$C_6H_{14}O$	2-Methyl-4-pentanol	10,985.5
$C_6H_{14}O$	Ethyl butyl ether	
$C_6H_{14}O$	Di-*n*-propyl ether	8,229.6
$C_6H_{14}O$	Diisopropyl ether	7,777.3
$C_6H_{14}O_2$	Acetal	9,853.9
$C_6H_{14}O_2$	1,2-Diethoxyethane	8,102.6
$C_6H_{14}O_3$	Di(2-methoxyethyl)ether	11,105.2
$C_6H_{14}O_3$	Diethyleneglycol-diethyl ether	12,669.0
$C_6H_{14}O_3$	Dipropyleneglycol	14,610.4
$C_6H_{14}O_4$	Triethyleneglycol	17,097.1
$C_6H_{15}N$	Dipropyl sulfide	–
$C_6H_{15}N$	Di-*n*-Propylamine	–
$C_6H_{15}N$	Triethylamine	–
$C_6H_{15}B$	Triethylboron	2,535.0
$C_6H_{15}ClSi$	Chlorotriethylsilane	9,806.9
$C_6H_{15}O_4P$	Triethyl phosphate	11,549.9
$C_6H_{15}Tl$	Triethylthallium	9,458.6
$C_6H_{16}O_2Si$	Diethoxydimethylsilane	9,758.2
$C_6H_{16}Si$	Trimethylpropylsilane	7,964.6
$C_6H_{16}Sn$	Trimethylpropyltin	9,659.6
$C_6H_{18}Cl_2O_2Si_3$	1,5-Dichlorohexamethyltrisiloxane	11,391.5
$C_6H_{18}O_3Si_3$	Hexamethylcyclotrisiloxane	10,503.3
$C_7H_3Cl_2F_3$	3,4-Dichloro-α,α,α-trifluorotoluene	10,253.5
$C_7H_4ClF_3$	2-Chloro-α,α,α-trifluorotoluene	10,016.9
$C_7H_4Cl_4$	2-α,α,α-Tetrachlorotoluene	12,501.3
C_7H_5BrO	Benzoyl bromide	12,070.8
C_7H_5ClO	Benzoyl chloride	11,438.0
$C_7H_5Cl_3$	α,α,α-Trichlorotoluene	12,168.6
$C_7H_5F_3$	α,α,α-Trifluorotoluene	8,869.7
C_7H_5N	Benzonitrile	11,341.0
C_7H_5N	Phenyl isocyanide	10,736.7
C_7H_5NO	Phenyl isocyanate	10,556.7
$C_7H_5NO_3$	2-Nitrobenzaldehyde	13,773.6
$C_7H_5NO_3$	3-Nitrobenzaldehyde	14,726.9
C_7H_5NS	Phenyl isothiocyanate	12,132.7
$C_7H_6Cl_2$	α,α-Dichlorotoluene	11,075.9
C_7H_6O	Benzaldehyde	11,657.8
$C_7H_6O_2$	Benzoic acid	15,253.3
		16,295.1
$C_7H_6O_2$	Salicylaldehyde	11,536.5
$C_7H_6O_2$	4-Hydroxybenzaldehyde	16,043.4
$C_7H_6O_3$	Salicylic acid	18,920.7
C_7H_7Br	α-Bromotoluene	11,360.4
C_7H_7Br	2-Bromotoluene	11,365.0
C_7H_7Br	3-Bromotoluene	10,537.1
C_7H_7Br	4-Bromotoluene	10,076.2
C_7H_7BrO	4-Bromoanisole	12,075.4
C_7H_7Cl	α-Chlorotoluene	11,158.7
C_7H_7Cl	2-Chlorotoluene	10,279.3
C_7H_7Cl	3-Chlorotoluene	10,081.1
C_7H_7Cl	4-Chlorotoluene	10,151.7
C_7H_7F	2-Fluorotoluene	9,164.8
C_7H_7F	3-Fluorotoluene	9,251.8
C_7H_7F	4-Fluorotoluene	9,281.0
C_7H_7I	2-Iodotoluene	11,380.7
$C_7H_7NO_2$	2-Nitrotoluene	12,239.1
$C_7H_7NO_2$	3-Nitrotoluene	11,831.1
$C_7H_7NO_2$	4-Nitrotoluene	11,915.0
C_7H_8	Toluene	9,368.5
		8,580.5
$C_7H_8Cl_2Si$	Benzyldichlorosilane	13,128.7
$C_7H_8Cl_2Si$	Dichloromethylphenylsilane	11,464.7
$C_7H_8Cl_2Si$	Dichloro-4-tolysilane	13,125.7
C_7H_8O	Anisole	10,440.9
C_7H_8O	Benzyl alcohol	14,093.2
C_7H_8O	*o*-Cresol	12,487.3
C_7H_8O	*m*-Cresol	13,483.8

Formula	Name	Δ H_v
C_7H_8O	*p*-Cresol	13,611.7
$C_7H_8O_2$	3,5-Dimethyl-1,2-pyrone	14,470.6
$C_7H_8O_2$	2-Methoxyphenol	13,425.8
$C_7H_8O_3$	Ethyl 2-furoate	12,144.0
C_7H_9N	2,6-Dimethylpyridine	–
C_7H_9N	Benzylamine	11,703.2
C_7H_9N	*N*-Methylaniline	11,982.3
C_7H_9N	2-Toluidine	12,663.4
C_7H_9N	3-Toluidine	12,104.1
$C_7H_{10}O_4$	Dimethyl citraconate	12,917.3
$C_7H_{10}O_4$	Dimethyl itaconate	15,613.7
$C_7H_{10}O_4$	*trans*-Dimethyl mesaconate	12,688.1
$C_7H_{11}NO_2$	2-Cyano-2-butyl acetate	12,720.8
$C_7H_{12}O_2$	Butyl acrylate	10,194.0
C_7H_9N	4-Toluidine	12,428.6
C_7H_9NO	2-Methoxyaniline	13,684.6
$C_7H_{10}N_2$	Toluene-2,4-diamine	15,928.1
$C_7H_{10}N_2$	4-Tolylhydrazine	15,063.1
$C_7H_{10}O_3$	Trimethylsuccinic anhydride	12,196.7
$C_7H_{12}O_3$	Ethyl levulinate	12,733.6
$C_7H_{12}O_4$	Pimelic acid	19,840.8
$C_7H_{12}O_4$	Diethyl malonate	12,227.7
$C_7H_{13}ClO$	Enanthyl chloride	15,242.7
$C_7H_{13}N$	Heptanonitrile	10,830.5
C_7H_{14}	Ethylcyclopentane	8,797.7
C_7H_{14}	2-Heptene	8,643.2
C_7H_{14}	Methylcyclohexane	8,549.2
$C_7H_{14}O$	Enanthaldehyde	11,413.4
$C_7H_{14}O$	2-Heptanone	12,478.9
$C_7H_{14}O$	4-Heptanone	13,451.9
$C_7H_{14}O$	2,5-Dimethyl-3-pentanone	12,266.9
$C_7H_{14}O_2$	Enanthic acid	15,893.8
$C_7H_{14}O_2$	Methyl caproate	10,676.8
$C_7H_{14}O_2$	Ethyl isovalerate	10,183.9
$C_7H_{14}O_2$	Propyl butyrate	10,283.7
$C_7H_{14}O_2$	Propyl isobutyrate	10,259.7
$C_7H_{14}O_2$	Isopropyl isobutyrate	9,717.6
$C_7H_{14}O_2$	Isobutyl propionate	10,495.8
$C_7H_{14}O_2$	Isoamyl acetate	10,494.9
C_7H_{16}	Perfluoro-*n*-Heptane	–
C_7H_{16}	*n*-Heptane	8,928.8
		8,409.6
C_7H_{16}	2-Methylhexane	8,538.7
C_7H_{16}	3-Methylhexane	8,596.3
C_7H_{16}	3-Ethylpentane	8,642.8
C_7H_{16}	2,2-Dimethylpentane	8,106.7
C_7H_{16}	2,3-Dimethylpentane	8,390.9
C_7H_{16}	2,4-Dimethylpentane	8,167.1
C_7H_{16}	3,3-Dimethylpentane	8,145.4
C_7H_{16}	2,2,3-Trimethylbutane	7,767.1
$C_7H_{16}O$	*n*-Heptanol	13,920.9
$C_7H_{16}O_3$	Triethyl orthoformate	10,935.0
$C_7H_{18}O_3Si$	Triethoxymethylsilane	10,306.7
$C_7H_{18}Si$	Butyltrimethylsilane	9,206.0
$C_7H_{18}Si$	Triethylmethylsilane	9,232.5
$C_8H_4Cl_2O_2$	Phthaloyl chloride	13,716.0
$C_8H_4O_3$	Phthalic anhydride	13,919.0
$C_8H_5Cl_2N$	α,α-Dichlorophenylacetonitrile	12,829.9
$C_8H_5Cl_5$	Pentachloroethylbenzene	13,728.7
$C_8H_6Cl_2$	2,3-Dichlorostyrene	12,827.2
$C_8H_6Cl_2$	2,4-Dichlorostyrene	12,511.7
$C_8H_6Cl_2$	2,5-Dichlorostyrene	12,592.5
$C_8H_6Cl_2$	2,6-Dichlorostyrene	12,186.0
$C_8H_6Cl_2$	3,4-Dichlorostyrene	12,626.5
$C_8H_6Cl_2$	3,5-Dichlorostyrene	12,511.7
$C_8H_6Cl_4$	3,4,5,6-Tetrachloro-1,2-xylene	14,763.1
$C_8H_6Cl_4$	1,2,3,5-Tetrachloro-4-ethylbenzene	12,980.3
$C_8H_6O_2$	Phenylglyoxal	13,731.6
$C_8H_6O_2$	Phthalide	14,021.6
$C_8H_6O_3$	Piperonal	14,425.5
C_8H_7Cl	3-Chlorostyrene	10,990.2
C_8H_7ClO	Phenylacetyl chloride	12,627.1
C_8H_7N	2-Tolunitrile	11,557.7
C_8H_7N	4-Tolunitrile	11,562.8
C_8H_7N	Phenylacetonitrile	12,796.2
C_8H_7N	2-Tolyl isocyanide	11,303.3
$C_8H_7NO_4$	2-Nitrophenyl acetate	16,875.3
C_8H_7NS	2-Methylbenzothiazole	14,492.3
C_8H_8	Styrene	9,634.7
$C_8H_8Br_2$	(1,2-Dibromoethyl)benzene	14,874.7
$C_8H_8Cl_2$	1,2-Dichloro-3-ethylbenzene	11,784.3
$C_8H_8Cl_2$	1,2-Dichloro-4-ethylbenzene	11,711.5
$C_8H_8Cl_2$	1,4-Dichloro-2-ethylbenzene	11,262.7
C_8H_8O	Acetophenone	11,731.5
$C_8H_8O_2$	Phenylacetate	12,174.9
$C_8H_8O_2$	Phenylacetic acid	15,568.7
$C_8H_8O_2$	Anisaldehyde	13,581.8
$C_8H_8O_2$	Methyl benzoate	12,077.2
$C_8H_8O_3$	Methyl salicylate	12,658.8
$C_8H_8O_3$	Vanillin	15,703.2

Formula	Name	Δ H_v
$C_8H_8O_4$	Dihydroacetic acid	14,663.8
C_8H_9Br	2-Bromo-1,4-xylene	11,603.7
C_8H_9Br	1-Bromo-4-ethylbenzene	10,170.0
C_8H_9Br	(2-Bromoethyl)benzene	12,152.5
C_8H_9Cl	1-Chloro-2-ethylbenzene	10,749.7
C_8H_9Cl	1-Chloro-3-ethylbenzene	10,724.1
C_8H_9Cl	1-Chloro-4-ethylbenzene	10,659.9
C_8H_9ClO	1-Chloro-2-ethoxybenzene	12,411.1
C_8H_9ClO	4-Chlorophenylethyl alcohol	14,298.5
$C_8H_{10}Cl_2Si$	Dichlorophenylethylsilane	11,895.1
C_8H_9NO	Acetanilide	15,474.1
$C_8H_9NO_2$	Methyl anthranilate	13,186.3
$C_8H_9NO_2$	4-Nitro-1,3-xylene	12,948.0
C_8H_{10}	Ethylbenzene	9,301.3
C_8H_{10}	*o*-Xylene	9,998.5
C_8H_{10}	*m*-Xylene	9,904.2
C_8H_{10}	*p*-Xylene	9,809.9
$C_8H_{10}Cl_2OSi$	Dichloroethoxyphenylsilane	12,516.5
$C_8H_{10}Cl_2Si$	Dichloroethylphenylsilane	11,721.2
$C_8H_{10}O$	2-Ethylphenol	12,516.7
$C_8H_{10}O$	3-Ethylphenol	13,856.4
$C_8H_{10}O$	4-Ethylphenol	13,437.9
$C_8H_{10}O$	Xylenol	–
$C_8H_{10}O$	2,3-Xylenol	13,106.9
$C_8H_{10}O$	2,4-Xylenol	13,130.2
$C_8H_{10}O$	2,5-Xylenol	13,130.2
$C_8H_{10}O$	3,4-Xylenol	13,991.0
$C_8H_{10}O$	3,5-Xylenol	13,767.7
$C_8H_{10}O$	Phenetole	11,075.8
$C_8H_{10}O$	α-Methyl benzyl alcohol	13,087.4
$C_8H_{10}O$	Phenylethylalcohol	13,307.4
$C_8H_{10}O_2$	4,6-Dimethylresorcinol	12,433.1
$C_8H_{10}O_2$	2-Phenoxyethanol	14,368.3
$C_8H_{10}O_6$	Diethyl dioxosuccinate	13,973.3
$C_8H_{11}ClSi$	Chlorodimethylphenylsilane	11,382.2
$C_8H_{11}N$	*N*-Ethylaniline	11,817.0
$C_8H_{11}N$	*N,N*-Dimethylaniline	11,320.4
$C_8H_{11}N$	4-Ethylaniline	12,679.9
$C_8H_{11}N$	2,4-Xylidine	13,099.2
$C_8H_{11}N$	2,6-Xylidine	11,742.6
$C_8H_{11}NO$	2-Phenetidine	13,877.8
$C_8H_{11}NO$	2-Anilinoethanol	15,643.2
$C_8H_{12}AsNO_2$	Dimethyl arsanilate	11,277.7
$C_8H_{12}Cl_2O_5$	Diethyleneglycol-bis-chloroacetate	19,830.5
$C_8H_{12}O_4$	Diethyl maleate	12,908.0
$C_8H_{12}O_4$	Diethyl fumarate	12,747.4
$C_8H_{12}Si$	Dimethylphenylsilane	10,274.2
$C_8H_{14}O_3$	Ethyl-α-ethylacetoacetate	12,344.2
$C_8H_{14}O_4$	Propyl levulinate	13,354.4
$C_8H_{14}O_4$	Isopropyl levulinate	12,689.6
$C_8H_{14}O_4$	Dipropyl oxalate	13,056.4
$C_8H_{14}O_4$	Diisopropyl oxalate	12,949.3
$C_8H_{14}O_4$	Diethyl succinate	13,076.1
$C_8H_{14}O_4$	Diethyl isosuccinate	12,087.6
$C_8H_{14}O_4$	Suberic acid	21,089.8
$C_8H_{14}O_5$	Diethyl malate	14,202.9
$C_8H_{14}O_6$	Diethyl-*dl*-tartrate	15,150.4
$C_8H_{14}O_6$	Diethyl-*d*-tartrate	15,517.8
$C_8H_{15}Br$	(2-Bromoethyl)cyclohexane	11,462.7
$C_8H_{15}N$	*n*-Caprylonitrile	12,221.8
$C_8H_{15}NO_3$	Ethyl-*N,N*-diethyloxamate	13,758.4
C_8H_{16}	1-Octene	–
C_8H_{16}	2-Octene	–
C_8H_{16}	2-Methyl-2-heptene	9,643.8
C_8H_{16}	1,1-Dimethylcyclohexane	8,949.1
C_8H_{16}	*cis*-1,2-Dimethylcyclohexane	9,364.9
C_8H_{16}	*trans*-1,2-Dimethylcyclohexane	9,097.1
C_8H_{16}	*cis*-1,3-Dimethylcyclohexane	9,232.6
C_8H_{16}	*trans*-1,3-Dimethylcyclohexane	9,080.3
C_8H_{16}	*cis*-1,4-Dimethylcyclohexane	9,188.9
C_8H_{16}	*trans*-1,4-Dimethylcyclohexane	8,951.2
C_8H_{16}	Ethylcyclohexane	9,441.2
$C_8H_{16}O$	Caprylaldehyde	21,201.0
$C_8H_{16}O$	Cyclohexaneethanol	13,152.4
$C_8H_{16}O$	6-Methyl-3-hepten-2-ol	13,864.1
$C_8H_{15}O$	6-Methyl-5-hepten-2-ol	13,999.1
$C_8H_{16}O$	2-Octanone	11,649.2
$C_8H_{16}O$	2,2,4-Trimethyl-3-pentanone	12,854.6
$C_8H_{16}O_2$	Caprylic acid	16,745.7
$C_8H_{16}O_2$	Ethyl isocaproate	10,826.7
$C_8H_{16}O_2$	Propyl isovalerate	10,715.7
$C_8H_{16}O_2$	Isobutyl butyrate	10,283.9
$C_8H_{16}O_2$	Isobutyl isobutyrate	10,706.3
$C_8H_{16}O_2$	Amylisopropionate	10,567.2
$C_8H_{16}ClO_4$	Tetraethyleneglycol-chlorohydrin	16,371.2
$C_8H_{17}I$	1-Iodooctane	11,625.1
$C_8H_{17}NO_2$	Ethyl-*l*-leucinate	11,383.5
C_8H_{18}	Octane	9,221.0
C_8H_{18}	2-Methylheptane	9,362.0

HEATS OF VAPORIZATION OF ORGANIC COMPOUNDS (continued)

Formula	Name	Δ H_v
C_8H_{18}	3-Methylheptane	9,432.0
C_8H_{18}	4-Methylheptane	9,404.8
C_8H_{18}	2,2-Dimethylhexane	8,927.8
C_8H_{18}	2,3-Dimethylhexane	9,224.9
C_8H_{18}	2,4-Dimethylhexane	9,086.6
C_8H_{18}	2,5-Dimethylhexane	9,110.2
C_8H_{18}	3,3-Dimethylhexane	9,065.2
C_8H_{18}	3,4-Dimethylhexane	9,239.4
C_8H_{18}	3-Ethylhexane	9,416.3
C_8H_{18}	2,2,3-Trimethylpentane	8,861.1
C_8H_{18}	2,2,4-Trimethylpentane	8,548.0
C_8H_{18}	2,3,3-Trimethylpentane	8,960.9
C_8H_{18}	2,3,4-Trimethylpentane	8,988.2
C_8H_{18}	2-Methyl-3-ethylpentane	9,134.3
C_8H_{18}	3-Methyl-3-ethylpentane	9,028.7
C_8H_{18}	2,2,3,3-Tetramethylbutane	10,351.5
$C_8H_{18}N_2$	Tetramethylpiperazine	11,187.5
C_8H_{18}	*n*-Octanol	14,262.4
C_8H_{18}	2-Octanol	12,468.4
$C_8H_{18}O$	Di *n*-butyl ether	–
$C_8H_{18}O$	Methyl heptyl ether	–
$C_8H_{18}O_2$	1,2-Dipropoxy ethane	6,370.7
$C_8H_{18}O_3$	Diethylene glycol butyl ether	14,127.0
$C_8H_{18}O_5$	Tetraethylene glycol	21,296.6
$C_8H_{18}S$	Di *n*-butyl sulfide	11,183.6
$C_8H_{18}S_2$	Dibutyl disulfide	8,254.1
$C_8H_{19}N$	Diisobutylamine	10,058.3
$C_8H_{20}O_4Si$	Tetraethoxysilane	10,968.6
$C_8H_{20}Pb$	Tetraethyllead	12,959.7
$C_8H_{20}Si$	Amyltrimethylsilane	9,659.6
$C_8H_{20}Si$	Tetraethylsilane	9,893.0
$C_8H_{20}Sb_2$	Tetraethylbistibine	12,975.4
$C_8H_{22}O_3Si_2$	1,3-Dichlorotetramethyldisiloxane	11,261.9
$C_8H_{24}Cl_2O_3Si_4$	1,7-Dichlorooctamethyltetrasiloxane	12,602.9
$C_8H_{24}O_2Si_3$	Octamethyltrisiloxane	10,956.0
$C_8H_{24}O_4Si_4$	Octamethylcyclotetra siloxane	11,515.0
$C_9H_6O_2$	Coumarin	15,202.7
C_9H_7N	Quinoline	12,575.4
C_9H_7N	Isoquinoline	12,847.6
C_9H_8	Indene	10,496.7
C_9H_8O	Cinnamylaldehyde	14,048.4
$C_9H_8O_2$	*trans*-Cinnamic acid	17,492.9
C_9H_9N	Skatole	15,232.7
$C_9H_9NO_4$	Ethyl 3-nitrobenzoate	15,056.1
C_9H_{10}	α-Methyl styrene	10,214.6
C_9H_{10}	β-Methyl styrene	10,701.3
C_9H_{10}	2-Methyl styrene	–
C_9H_{10}	3-Methyl styrene	–
C_9H_{10}	4-Methyl styrene	10,724.2
$C_9H_{10}O$	2,4-Xylaldehyde	13,618.4
$C_9H_{10}O$	Cinnamyl alcohol	13,421.6
$C_9H_{10}O$	Propiophenone	12,407.6
$C_9H_{10}O$	3-Vinylanisole	12,756.4
$C_9H_{10}O$	3-Vinylanisole	12,735.8
$C_9H_{10}O$	4-Vinylanisole	12,554.7
$C_9H_{10}O_2$	Benzyl acetate	12,107.2
$C_9H_{10}O_2$	Ethyl benzoate	11,981.5
$C_9H_{10}O_2$	Hydrocinnamic acid	15,411.9
$C_9H_{10}O_2$	Ethyl salicylate	13,030.1
$C_9H_{11}NO$	*N*-Methylacetanilide	13,235.2
$C_9H_{11}NO_2$	Ethyl carbanilate	19,791.8
C_9H_{12}	1,2,3-Trimethylbenzene	10,781.9
C_9H_{12}	1,2,4-Trimethylbenzene	10,710.2
C_9H_{12}	1,3,5-Trimethylbenzene	10,516.8
C_9H_{12}	*o*-Ethyl toluene	10,488.8
C_9H_{12}	*m*-Ethyl toluene	10,416.6
C_9H_{12}	*p*-Ethyl toluene	10,461.1
C_9H_{12}	Isopropylbenzene	10,335.3
C_9H_{12}	*N*-Propylbenzene	10,424.1
$C_9H_{12}O$	2-Ethylanisole	11,642.8
$C_9H_{12}O$	3-Ethylanisole	11,616.7
$C_9H_{12}O$	4-Ethylanisole	11,625.7
$C_9H_{12}O$	3-Phenyl-1-propanol	14,493.9
$C_9H_{12}O$	2-Isopropylphenol	13,402.3
$C_9H_{12}O$	3-Isopropylphenol	13,292.2
$C_9H_{12}O$	4-Isopropylphenol	13,878.7
$C_9H_{12}O$	Benzyl ethyl ether	11,315.5
$C_9H_{13}ClOSi$	Chloroethoxymethylphenylsilane	12,270.3
$C_9H_{13}N$	2,4,5-Trimethylaniline	13,975.0
$C_9H_{13}N$	*N,N*-Dimethyl-*O*-toluidine	11,648.3
$C_9H_{13}N$	*N,N*-Dimethyl-4-toluidine	12,738.4
$C_9H_{13}N$	4-Cumidine	13,127.9
$C_9H_{14}O$	Phorone	12,557.2
$C_9H_{14}O$	Isophorone	11,277.6
$C_9H_{14}O_4$	*cis*-Diethyl citraconate	12,913.2
$C_9H_{14}O_4$	Diethyl itaconate	12,075.8
$C_9H_{14}O_4$	Diethyl mesaconate	13,326.1
$C_9H_{14}O_7$	Trimethyl citrate	15,807.7

Formula	Name	Δ H_v
$C_9H_{16}O_3$	Isobutyl levulinate	13,571.2
$C_9H_{16}O_4$	Azelaic acid	20,944.2
$C_9H_{16}O_4$	Diethyl ethylmalonate	12,842.0
$C_9H_{16}O_4$	Diethyl glutarate	13,261.5
$C_9H_{18}O$	2-Nonanone	11,529.5
$C_9H_{18}O$	Di-isobutyl ketone	–
$C_9H_{18}O$	Azelaldehyde	12,143.4
$C_9H_{18}O_2$	Pelargonic acid	17,807.8
$C_9H_{18}O_2$	Methyl caprylate	11,914.9
$C_9H_{18}O_2$	Isobutyl isovalerate	10,999.7
$C_9H_{18}O_2$	Isoamyl butyrate	11,104.5
$C_9H_{18}O_2$	Isoamyl isobutyrate	10,870.6
$C_9H_{19}I$	1-Iodononane	14,853.0
C_9H_{20}	*n*-Nonane	10,456.9
C_9H_{20}	2,6-Dimethylheptane	–
C_9H_{20}	2-Methyloctane	–
C_9H_{20}	3-Methyloctane	–
$C_9H_{20}O$	1-Nonanol	13,849.2
$C_9H_{20}O$	Diisobutyl carbinol	–
$C_9H_{20}O_3$	Dipropyleneglycol isopropyl ether	12,583.8
$C_9H_{20}O_4$	Tripropyleneglycol	15,291.4
$C_9H_{22}Si$	Hexyltrimethylsilane	10,264.9
$C_9H_{22}Si$	Triethylpropylsilane	10,709.3
$C_{10}H_7Br$	1-Bromonaphthalene	13,274.9
$C_{10}H_7Cl$	1-Chloronaphthalene	13,570.5
$C_{10}H_{12}$	Dicyclopentadiene	10,165.9
$C_{10}H_8$	Naphthalene	17,065.2
		12,311.6
$C_{10}H_8Cl_2Si$	Dichloro-1-naphthysilane	16,325.3
$C_{10}H_8O$	1-Naphthol	14,205.6
$C_{10}H_8O$	2-Naphthol	14,138.5
$C_{10}H_9N$	1-Naphthylamine	14,529.5
$C_{10}H_9N$	2-Naphthylamine	14,679.6
$C_{10}H_9N$	2-Methylquinoline	14,154.0
$C_{10}H_{10}$	1,3-Divinylbenzene	11,384.7
$C_{10}H_{10}O$	4-Phenyl-3-buten-2-one	13,913.9
$C_{10}H_{10}O_2$	α-Methylcinnamic acid	18,149.4
$C_{10}H_{10}O_2$	Methyl cinnamate	13,325.5
$C_{10}H_{10}O_2$	Safrole	13,255.8
$C_{10}H_{10}O_4$	1,2-Phenylene diacetate	14,986.0
$C_{10}H_{10}O_4$	Dimethylphthalate	14,922.2
$C_{10}H_{12}$	2,4-Dimethylstyrene	11,454.0
$C_{10}H_{12}$	2,5-Dimethylstyrene	11,283.5
$C_{10}H_{12}$	3-Ethylstyrene	11,285.7
$C_{10}H_{12}$	4-Ethylstyrene	11,146.6
$C_{10}H_{12}$	Tetralin	11,613.0
$C_{10}H_{12}O$	Anethole	13,006.8
$C_{10}H_{12}O$	4-Methylpropiophenone	12,505.0
$C_{10}H_{12}O$	Estragole	12,879.3
$C_{10}H_{12}O$	Cuminal	12,668.0
$C_{10}H_{12}O$	4-Vinylphenetole	13,728.7
$C_{10}H_{12}O_2$	Eugenol	13,907.8
$C_{10}H_{12}O_2$	Isoeugenol	14,084.2
$C_{10}H_{12}O_2$	Chavibetol	14,527.7
$C_{10}H_{12}O_2$	Propyl benzoate	12,318.7
$C_{10}H_{12}O_3$	2-Phenoxyethyl acetate	14,070.3
$C_{10}H_{13}ClO$	2-Chloroethyl-α-methylbenzyl ether	12,969.2
$C_{10}H_{13}Cl_2O_2P$	4-*tert*-Butylphenyl dichlorophosphate	13,711.0
$C_{10}H_{14}$	1,2,3,4-Tetramethylbenzene	12,258.0
$C_{10}H_{14}$	1,2,3,5-Tetramethylbenzene	12,358.4
$C_{10}H_{14}$	1,2,4,5-Tetramethylbenzene	12,583.6
$C_{10}H_{14}$	4-Ethyl-1,3-xylene	11,070.4
$C_{10}H_{14}$	5-Ethyl-1,3-xylene	11,045.5
$C_{10}H_{14}$	2-Ethyl-1,4-xylene	11,144.6
$C_{10}H_{14}$	1,2-Diethylbenzene	11,695.5
$C_{10}H_{14}$	1,3-Diethylbenzene	10,993.9
$C_{10}H_{14}$	1,4-Diethylbenzene	10,746.3
$C_{10}H_{14}$	1-Methyl-2-isopropylbenzene	–
$C_{10}H_{14}$	1-Methyl-4-isopropylbenzene	11,038.7
$C_{10}H_{14}$	*N*-Butylbenzene	11,052.1
$C_{10}H_{14}$	Isobutylbenzene	8,567.8
$C_{10}H_{14}$	*sec*-Butylbenzene	11,069.3
$C_{10}H_{14}$	*tert*-Butylbenzene	10,705.5
$C_{10}H_{14}O$	Carvacrol	13,765.7
$C_{10}H_{14}O$	Carbone	12,796.2
$C_{10}H_{14}O$	Cuminyl alcohol	13,799.2
$C_{10}H_{14}O$	4-Ethylphenetole	12,766.4
$C_{10}H_{14}O$	2-Isopropyl-5-methylphenol	13,352.8
$C_{10}H_{14}O$	4-Isobutylphenol	14,053.5
$C_{10}H_{14}O$	4-*sec*-Butylphenol	13,690.2
$C_{10}H_{14}O$	2-*sec*-Butylphenol	12,781.3
$C_{10}H_{14}O$	2-*tert*-Butylphenol	13,112.3
$C_{10}H_{14}O$	4-*tert*-Butylphenol	13,787.7
$C_{10}H_{14}N_2$	Nicotine	12,337.1
$C_{10}H_{15}N$	*N*-Diethylaniline	12,539.2
$C_{10}H_{15}NO_2$	*N*-Phenyliminodiethanol	17,482.1
$C_{10}H_{16}$	Camphene	10,505.4
$C_{10}H_{16}$	Dipentene	10,538.3
$C_{10}H_{16}$	*d*-Limonene	10,508.4

Formula	Name	Δ H_v
$C_{10}H_{16}$	Myrcene	10,704.8
$C_{10}H_{16}$	α-Phellandrene	11,139.5
$C_{10}H_{16}$	α-Pinene	9,813.6
$C_{10}H_{16}$	β-Pinene	10,235.8
$C_{10}H_{16}$	Terpenoline	12,030.8
$C_{10}H_{16}AsNO_3$	Diethyl arsanilinate	12,973.9
$C_{10}H_{16}O$	*d*-Camphor	12,800.9
		11,978.0
$C_{10}H_{16}O$	*l*-dihydrocarvone	11,825.9
$C_{10}H_{16}O$	α-Citral	13,255.5
$C_{10}H_{16}O$	*d*-Fenchone	11,273.4
$C_{10}H_{16}O$	Pulegone	13,395.4
$C_{10}H_{16}O$	α-Thujone	11,950.8
$C_{10}H_{16}OSi$	Ethoxydimethylphenylsilane	11,718.6
$C_{10}H_{16}O_2$	Campholenic acid	16,324.1
$C_{10}H_{16}O_2$	Diosphenol	13,644.0
$C_{10}H_{16}O_2$	Fencholic acid	16,442.8
$C_{10}H_{18}$	*cis*-Decalin	10,515.4
$C_{10}H_{18}$	*trans*-Decalin	8,749.1
$C_{10}H_{18}O$	*d*-Citronellal	12,305.1
$C_{10}H_{18}O$	Cineol	10,570.8
$C_{10}H_{18}O$	Dihydrocarveol	13,698.5
$C_{10}H_{18}O$	*dl*-Fenchyl alcohol	12,955.9
$C_{10}H_{18}O$	Geraniol	14,060.7
$C_{10}H_{18}O$	*d*-Linalool	12,269.7
$C_{10}H_{18}O$	Nerol	13,366.1
$C_{10}H_{18}O$	α-Terpineol	12,754.5
$C_{10}H_{18}O_2$	Citronellic acid	16,455.4
$C_{10}H_{18}O_3$	Amyl levulinate	14,321.7
$C_{10}H_{18}O_3$	Isoamyl levulinate	13,867.9
$C_{10}H_{18}O_4$	Diethyl ethylmethylmalonate	12,345.6
$C_{10}H_{18}O_4$	Diethyl adipate	14,240.6
$C_{10}H_{18}O_4$	Diisobutyl oxalate	13,343.1
$C_{10}H_{18}O_4$	Dipropyl succinate	13,975.7
$C_{10}H_{18}O_4$	Sebacic acid	21,978.3
$C_{10}H_{18}O_6$	Dipropyl-*d*-tartrate	15,754.0
$C_{10}H_{18}O_6$	Diisopropyl-*d*-tartrate	15,836.6
$C_{10}H_{19}N$	Camphylamine	13,224.1
$C_{10}H_{20}$	Menthane	10,293.1
$C_{10}H_{20}$	1-Decene	10,233.3
$C_{10}H_{20}$	*n*-Decane	10,912.0
$C_{10}H_{20}Br_2$	1,2-Dibromodecane	16,407.7
$C_{10}H_{20}O$	Decanol	14,065.1
$C_{10}H_{20}O$	Citronellol	14,214.1
$C_{10}H_{20}O$	Capraldehyde	13,154.9
$C_{10}H_{20}O$	*l*-Menthol	13,475.3
$C_{10}H_{20}O$	Decan-2-one	12,114.7
$C_{10}H_{20}O_2$	Capric acid	19,372.6
$C_{10}H_{20}O_2$	Isoamyl isovalerate	11,040.8
$C_{10}H_{22}$	2,7-Dimethyloctane	10,339.3
$C_{10}H_{22}O$	Diisoamyl ether	11,072.2
$C_{10}H_{22}O_2$	2-Butyl-2-ethylbutane-1,3-diol	15,833.7
$C_{10}H_{22}O$	Dihydrocitronellol	16,769.8
$C_{10}H_{22}O_3$	Dipropylene glycol monobutyl ether	13,721.1
$C_{10}H_{22}S$	Diisoamyl sulfide	11,829.9
$C_{10}H_{24}Si$	Heptyltrimethylsilane	10,987.3
$C_{10}H_{24}Si$	Butyltriethylsilane	11,124.0
$C_{10}H_{28}O_4Si_3$	1,5-Diethoxyhexamethyl trisiloxane	12,586.4
$C_{10}H_{30}O_3Si_4$	Decamethyltetrasiloxane	11,981.2
$C_{10}H_{30}O_5Si_5$	Decamethylcyclopenta siloxane	12,272.1
$C_{11}H_8O_2$	1-Naphthoic acid	22,581.4
$C_{11}H_8O_2$	2-Naphthoic acid	22,630.8
$C_{11}H_{10}$	1-Methylnaphthalene	–
$C_{11}H_{12}O_2$	Ethyl-*trans*-cinnamate	13,639.9
$C_{11}H_{12}O_2$	1 Phenyl-1,3-pentanedione	15,033.9
$C_{11}H_{12}O_3$	Ethyl benzoylacetate	17,115.4
$C_{11}H_{12}O_3$	Myristicine	14,471.4
$C_{11}H_{14}$	1-Phenylpentane	–
$C_{11}H_{14}$	2,4,5-Trimethylstyrene	12,076.1
$C_{11}H_{14}$	2,4,6-Trimethylstyrene	11,588.8
$C_{11}H_{14}$	4-Isopropylstryene	11,471.0
$C_{11}H_{14}O$	Isobutyrophenone	12,878.8
$C_{11}H_{14}O$	Pivalophenone	13,221.3
$C_{11}H_{14}O$	2,3,5-Trimethylacetophenone	14,283.6
$C_{11}H_{14}O_2$	Isobutyl benzoate	13,105.8
$C_{11}H_{14}O_2$	4-Allylveratrole	15,027.1
$C_{11}H_{16}$	Pentamethylbenzene	–
$C_{11}H_{16}$	3,5-Diethyltoluene	11,167.4
$C_{11}H_{16}$	1,2,4-Trimethyl-5-ethylbenzene	12,145.3
$C_{11}H_{16}$	1,3,5-Trimethyl-2-ethylbenzene	11,677.3
$C_{11}H_{16}$	3-Ethylcumene	11,233.5
$C_{11}H_{16}$	4-Ethylcumene	11,425.6
$C_{11}H_{16}$	*sec*-Amylbenzene	11,886.0
$C_{11}H_{16}O$	4-*tert*-Butyl-2-cresol	13,798.1
$C_{11}H_{16}O$	2-*tert*-Butyl-4-cresol	14,037.9
$C_{11}H_{16}O$	4-*tert*-Amylphenol	13,154.3
$C_{11}H_{16}O_5$	Ethylcamphoronic anhydride	16,373.6

HEATS OF VAPORIZATION OF ORGANIC COMPOUNDS (continued)

Formula	Name	ΔH_v
$C_{11}H_{18}O_2$	Bornyl formate	12,276.0
$C_{11}H_{18}O_2$	Geranyl formate	13,189.7
$C_{11}H_{18}O_2$	Neryl formate	12,959.3
$C_{11}H_{18}O_2Si$	Diethoxymethylphenyl silane	13,267.3
$C_{11}H_{18}O_5$	Diethyl-gamma-oxoazelate	17,543.6
$C_{11}H_{20}O_2$	10-Hendecenoic acid	17,247.5
$C_{11}H_{20}O_2$	Menthyl formate	12,077.7
$C_{11}H_{20}O_2$	2-Ethylhexyl acrylate	12,522.5
$C_{11}H_{20}O_2$	Octyl acrylate	12,957.5
$C_{11}H_{20}O_3$	Hexyl levulinate	14,626.2
$C_{11}H_{22}O$	Hendecan-2-one	14,353.5
$C_{11}H_{22}O_2$	Methyl caprate	13.831.7
$C_{11}H_{22}O_2$	Hendecanoic acid	14,689.9
$C_{11}H_{24}$	Undecane	11,481.7
$C_{11}H_{24}O$	Hendecan-2-ol	14,216.2
$C_{11}H_{26}Si$	Trimethyloctylsilane	12,285.8
$C_{11}H_{26}Si$	Amyltriethylsilane	11,859.7
$C_{12}H_9Br$	4-Bromobiphenyl	13,493.4
$C_{12}H_9BrO$	2-Bromo-4-phenylphenol	13,589.9
$C_{12}H_9Cl$	2-Chlorobiphenyl	13,925.7
$C_{12}H_9Cl$	4-Chlorobiphenyl	14,017.4
$C_{12}H_9ClO$	2-Chloro-3-phenylphenol	15,258.0
$C_{12}H_9ClO$	2-Chloro-6-phenylphenol	15,508.4
$C_{12}H_9Cl_2PO$	2-Xenyl dichlorophosphate	17,127.6
$C_{12}H_9N$	Carbazole	15,421.6
$C_{12}H_{10}$	Acenaphthene	13,078.5
$C_{12}H_{10}$	Diphenyl	12,910.0
$C_{12}H_{10}ClPO_3$	Diphenyl chlorophosphate	13,191.2
$C_{12}H_{10}Cl_2Si$	Dichlorodiphenylsilane	14,968.5
$C_{12}H_{10}F_2Si$	Difluorodiphenylsilane	12,913.3
$C_{12}H_{10}N_2$	Azobenzene	14,786.7
$C_{12}H_{10}O$	1-Acetonaphthone	16,095.5
$C_{12}H_{10}O$	2-Acetonaphthone	16,496.7
$C_{12}H_{10}O$	Diphenyl ether	12,325.5
$C_{12}H_{10}O$	2-Phenylphenol	15,397.8
$C_{12}H_{10}O$	4-Phenylphenol	16,974.3
$C_{12}H_{10}S$	Diphenyl sulfide	13,974.8
$C_{12}H_{10}S$	Diphenyl disulfide	17,452.0
$C_{12}H_{10}Se$	Diphenyl selenide	14,603.4
$C_{12}H_{11}N$	Diphenylamine	14,920.3
$C_{12}H_{12}$	1-Ethylnaphthalene	12,751.3
$C_{12}H_{12}N_2$	1,1-Diphenylhydrazine	15,940.4
$C_{12}H_{14}N_2O_5$	2-Cyclohexyl-4,6-dinitrophenol	19,100.0
$C_{12}H_{14}O_3$	Eugenyl acetate	15,120.7
$C_{12}H_{14}O_4$	Apiole	16,881.7
$C_{12}H_{14}O_4$	Diethyl phthalate	15,383.0
$C_{12}H_{16}$	2,5-Diethylstyrene	12,150.3
$C_{12}H_{16}$	Phenylcyclohexane	13,345.6
$C_{12}H_{16}O_2$	Isoamyl benzoate	12,782.9
$C_{12}H_{18}$	Hexamethylbenzene	–
$C_{12}H_{18}$	1,2,4-Triethylbenzene	11,957.9
$C_{12}H_{18}$	1,3,4-Triethylbenzene	12,215.0
$C_{12}H_{18}$	1,3,5-Triethylbenzene	–
$C_{12}H_{18}$	1,2-Diisopropylbenzene	11,751.4
$C_{12}H_{18}$	1,3-Diisopropylbenzene	11,498.9
$C_{12}H_{18}O$	2-*tert*-Butyl-4-ethylphenol	13,994.0
$C_{12}H_{18}O$	4-*tert*-Butyl-2,5-xylenol	14,477.9
$C_{12}H_{18}O$	4-*tert*-Butyl-2,6-xylenol	14,142.5
$C_{12}H_{18}O$	6-*tert*-Butyl-2,4-xylenol	13,882.4
$C_{12}H_{18}O$	6-*tert*-Butyl-3,4-xylenol	14,848.3
$C_{12}H_{20}O_2$	*d*-Bornyl acetate	11,838.7
$C_{12}H_{20}O_2$	Geranyl acetate	13,879.9
$C_{12}H_{20}O_2$	Linalyl acetate	12,910.6
$C_{12}H_{20}O_3Si$	Triethoxyphenylsilane	14,117.9
$C_{12}H_{20}O_7$	Triethyl citrate	14,818.4
$C_{12}H_{21}PO_4$	Trimethallyl phosphate	12,566.1
$C_{12}H_{22}O_2$	Citronellyl acetate	15,781.3
$C_{12}H_{22}O_2$	Menthyl acetate	12,819.2
$C_{12}H_{22}O_4$	Dimethyl sebacate	14,861.3
$C_{12}H_{22}O_4$	Diisoamyl oxalate	14,123.7
$C_{12}H_{22}O_6$	Diisobutyl-*d*-tartrate	14,874.9
$C_{12}H_{24}$	1-Dodecene	12,587.8
$C_{12}H_{24}$	Triisobutylene	10,790.4
$C_{12}H_{24}O$	Dodecan-2-one	14,138.7
$C_{12}H_{24}O$	Lauraldehyde	13,644.2
$C_{12}H_{24}O_2$	Lauric acid	16,585.3
$C_{12}H_{26}$	*n*-Dodecane	11,857.7
$C_{12}H_{26}O$	Dodecyl alcohol	15,16[illegible]
$C_{12}H_{26}O_4$	Tripropylene glycol monoisopropyl ether	14,171.5
$C_{12}H_{27}N$	Triisobutylamine	12,390.6
$C_{12}H_{27}N$	Dodecylamine	14,836.4
$C_{12}H_{28}Si$	Triethylhexylsilane	12,119.4
$C_{12}H_{34}O_5Si_4$	1,7-Diethoxyoctamethyltetrasiloxane	14,095.9
$C_{12}H_{36}O_4Si_5$	Dodecamethylpentasiloxane	12,942.6
$C_{12}H_{36}O_6Si_6$	Dodecamethylcyclohexasiloxane	13,760.6
$C_{13}H_9N$	Acridine	15,174.6
$C_{13}H_{10}$	Fluorene	13,682.8
$C_{13}H_{10}O$	Benzophenone	14,725.4
$C_{13}H_{10}O_2$	Phenyl benzoate	14,181.7
$C_{13}H_{10}O_2$	Salol	15,441.6
$C_{13}H_{12}$	Diphenylmethane	13,089.4
$C_{13}H_{12}O$	Benzhydrol	15,220.2
$C_{13}H_{12}O$	Benzyl phenyl ether	14,156.7
$C_{13}H_{12}O$	1-Proprionaphthone	16,630.8
$C_{13}H_{13}ClSi$	Chloromethyldiphenylsilane	14,924.6
$C_{13}H_{14}Si$	Methyldiphenylsilane	15,396.8
$C_{13}H_{14}$	2-Isopropylnaphthalene	13,036.9
$C_{13}H_{18}O$	Enanthophenone	15,597.7
$C_{13}H_{20}$	Heptylbenzene	13,535.4
$C_{13}H_{20}O$	α-Ionone	14,253.4
$C_{13}H_{22}O_2$	Bornyl propionate	13,245.0
$C_{13}H_{26}O$	2-Tridecanone	14,416.1
$C_{13}H_{26}O_2$	Methyl laurate	14,853.5
$C_{13}H_{26}O_2$	Tridecanoic acid	19,214.8
$C_{13}H_{28}$	Tridecane	12,991.3
$C_{13}H_{28}O_4$	Tripropyleneglycol monobutyl ether	15,937.6
$C_{13}H_{30}Si$	Decyltrimethylsilane	13,311.1
$C_{13}H_{30}Si$	Triethylheptylsilane	13,298.3
$C_{14}H_8O_2$	Anthraquinone	21,163.1
$C_{14}H_8O_4$	1,4-Dihydroxyanthraquinone	17,677.9
$C_{14}H_{10}$	Anthracene	16,823.6
$C_{14}H_{10}$	Phenanthrene	14,184.0
$C_{14}H_{10}O_2$	Benzil	15,046.4
$C_{14}H_{10}O_3$	Benzoic anhydride	16,060.9
$C_{14}H_{12}$	1,1-Diphenylethylene	13,778.1
$C_{14}H_{12}$	*trans*-Diphenylethylene	15,010.1
$C_{14}H_{12}O$	Desoxybenzoin	15,642.1
$C_{14}H_{12}O$	Benzoin	15,952.5
$C_{14}H_{14}$	Dibenzyl	13,387.6
$C_{14}H_{14}O$	2-Isobutyronaphthone	17,133.8
$C_{14}H_{15}O$	Dibenzylamine	16,260.1
$C_{14}H_{15}N$	Ethyldiphenylamine	14,569.4
$C_{14}H_{20}Cl_2$	1,2-Dichlorotetraethylbenzene	14,629.0
$C_{14}H_{20}Cl_2$	1,4-Dichlorotetraethylbenzene	13,397.5
$C_{14}H_{20}O_3$	2-(4-*tert*-Butylphenoxy) ethyl acetate	16,017.6
$C_{14}H_{22}$	1,2,3,4-Tetraethylbenzene	12,763.5
$C_{14}H_{22}O$	2,4-Di-*tert*-butylphenol	14,237.7
$C_{14}H_{24}O_2$	Bornyl butyrate	13,746.1
$C_{14}H_{24}O_2$	Bornyl isobutyrate	13,501.8
$C_{14}H_{24}O_2$	Geranyl butyrate	16,086.4
$C_{14}H_{24}O_2$	Geranyl isobutyrate	15,699.5
$C_{14}H_{26}O_4$	Diethyl sebacate	16,819.6
$C_{14}H_{28}O$	2-Tetradecanone	15,102.7
$C_{14}H_{28}O$	Myristaldehyde	14,088.9
$C_{14}H_{28}O_2$	Myristic acid	18,380.1
$C_{14}H_{29}Cl$	1-Chlorotetradecane	14,083.5
$C_{14}H_{30}$	Tetradecane	13,750.0
$C_{14}H_{31}N$	Tetradecylamine	14,840.8
$C_{14}H_{32}Si$	Triethyloctysilane	12,954.8
$C_{14}H_{40}O_6Si_5$	1,9-Diethoxydeca methylpentasiloxane	15,296.9
$C_{14}H_{42}O_5Si_6$	Tetradecamethylhexasiloxane	13,800.0
$C_{14}H_{42}O_7Si_7$	Tetradecamethylcyclo heptasiloxane	14,263.8
$C_{15}H_{14}O$	1,3-Diphenyl-2-propanone	15,429.8
$C_{15}H_{14}O_2$	1-Biphenyloxy-2,3-epoxypropane	16,160.6
$C_{15}H_{16}O_2$	4,4-Isopropylidenebisphenol	23,254.0
$C_{15}H_{18}O$	Isocapronaphthone	17,360.3
$C_{15}H_{18}OSi$	Ethoxymethyldiphenylsilane	16,106.4
$C_{15}H_{20}O_2$	Helenin	26,532.7
$C_{15}H_{24}$	Cadinene	15,518.3
$C_{15}H_{24}O$	2,6-Di-*tert*-butyl-4-cresol	14,338.6
$C_{15}H_{24}O$	4,6-Di-*tert*-butyl-2-cresol	14,006.9
$C_{15}H_{24}O$	4,6-Di-*tert*-butyl-3-cresol	15,464.6
$C_{15}H_{26}O$	Champacol	14,655.9
$C_{15}H_{26}O_6$	Triethyl camphoronate	16,112.2
$C_{15}H_{30}O_2$	Methyl myristate	16,051.0
$C_{15}H_{32}$	Pentadecane	14,635.9
$C_{15}H_{32}O_5$	Tetrapropylene glycol monoisopropyl ether	16,494.6
$C_{15}H_{34}Si$	Dodecyltrimethylsilane	14,374.6
$C_{16}H_{14}O_2$	Benzyl cinnamate	20,840.6
$C_{16}H_{18}O$	Di(α-methylbenzyl)ether	14,628.1
$C_{16}H_{20}O_2Si$	Diethoxydiphenylsilane	15,828.8
$C_{16}H_{22}O_4$	Dibutyl phthalate	17,747.0
$C_{16}H_{25}Cl$	Pentaethylchlorobenzene	13,707.3
$C_{16}H_{26}$	Pentaethylbenzene	13,670.1
$C_{16}H_{26}O$	2,6-Di-*tert*-butyl-4-ethylphenol	14,438.0
$C_{16}H_{26}O$	4,6-Di-*tert*-butyl-3-ethylphenol	15,954.8
$C_{16}H_{30}O$	Muscone	14,722.5
$C_{16}H_{31}N$	Palmitonitrile	16,433.7
$C_{16}H_{32}$	1-Hexadecene	15,634.7
$C_{16}H_{32}$	Tetraisobutylene	12,937.2
$C_{16}H_{32}O$	2-Hexadecanone	15,194.4
$C_{16}H_{32}O$	Palmitaldehyde	15,454.2
$C_{16}H_{32}O_2$	Palmitic acid	17,603.6
$C_{16}H_{34}$	Hexadecane	15,405.5
$C_{16}H_{34}O$	Cetyl alcohol	14,483.4
$C_{16}H_{35}N$	Cetylamine	15,238.0
$C_{16}H_{36}Si$	Decyltriethylsilane	15,393.7
$C_{16}H_{46}O_7Si_6$	1,1,1-Diethoxydodeca methylhexasiloxane	15,945.3
$C_{16}H_{48}O_6Si_7$	Hexadecamethylheptasiloxane	14,841.5
$C_{16}H_{48}O_8Si_8$	Hexadecamethylcyclooctasiloxane	14,986.3
$C_{17}H_{10}O$	Benzanthrone	18,309.6
$C_{17}H_{18}O_3$	4-*tert*-Butylphenyl salicylate	16,455.6
$C_{17}H_{24}O_2$	Menthyl benzoate	16,804.5
$C_{17}H_{34}O$	2-Heptadecanone	16,559.8
$C_{17}H_{34}O_2$	Methyl palmitate	17,003.5
$C_{17}H_{36}$	Heptadecane	15,608.5
$C_{17}H_{38}Si$	Tetradecyltrimethylsilane	16,439.7
$C_{18}H_{12}Cl_3O_3PS$	Tri-2-chlorophenylthiophosphate	24,386.1
$C_{18}H_{15}O_4P$	Triphenyl phosphate	19,272.3
$C_{18}H_{30}$	Hexaethylbenzene	14,184.9
$C_{18}H_{30}O$	2,4,6-Tri-*tert*-butylphenol	14,703.7
$C_{18}H_{34}O_2$	Oleic acid	20,326.7
$C_{18}H_{34}O_2$	Elaidic acid	19,538.0
$C_{18}H_{36}O$	Stearaldehyde	16,555.6
$C_{18}H_{36}O_2$	Stearic acid	19,306.6
$C_{18}H_{38}$	Octadecane	15,447.0
$C_{18}H_{38}$	2-Methylheptadecane	16,095.9
$C_{18}H_{38}O$	1-Octadecanol	17,508.0
$C_{18}H_{39}N$	Ethylcetylamine	15,718.3
$C_{18}H_{52}O_8Si_7$	1,1,3-Diethoxytetradeca methylheptasiloxane	16,765.8
$C_{18}H_{54}O_7Si_8$	Octadecamethyloctasiloxane	15,270.3
$C_{19}H_{16}$	Triphenylmethane	34,470.8
$C_{19}H_{40}$	Nonadecane	16,497.3
$C_{20}H_{20}OSi$	Ethoxytriphenylsilane	20,214.2
$C_{20}H_{43}N$	Diethylhexadecylamine	15,871.3
$C_{20}H_{58}O_9Si_8$	1,1,5-Diethoxyhexadeca methyloctasiloxane	17,626.6
$C_{20}H_{60}O_8Si_9$	Eicosamethylnonasiloxane	19,522.9
$C_{21}H_{21}O_4P$	Tritolyl phosphate	20,835.9
$C_{21}H_{44}$	Heneicosane	17,702.2
$C_{22}H_{42}O_2$	Erucic acid	23,655.2
$C_{22}H_{42}O_2$	Brassidic acid	24,085.7
$C_{22}H_{46}$	Docosane	16,941.1
$C_{22}H_{66}O_9Si_{10}$	Docosamethyldecasiloxane	21,878.6
$C_{23}H_{48}$	Tricosane	19,082.1
$C_{24}H_{50}$	Tetracosane	19,642.5
$C_{24}H_{72}O_{10}Si_{11}$	Tetracosamethylhendecasiloxane	23,941.2
$C_{25}H_{52}$	Pentacosane	20,815.9
$C_{26}H_{54}$	Hexacosane	21,605.7
$C_{27}H_{33}O_4P$	Dicarvacryl-2-tolyl phosphate	24,233.3
$C_{27}H_{56}$	Heptacosane	21,958.1
$C_{28}H_{58}$	Octacosane	24,144.2
$C_{29}H_{60}$	Nonacosane	24,816.8
$C_{32}H_{34}ClO_4P$	Dicarvacryl-mono-(6-chloro-2-xenyl)-phosphate	25,299.5

MISCIBILITY OF ORGANIC SOLVENT PAIRS

Table A

Doctor J. S. Drury

Industrial and Engineering Chemistry Vol. 44, No. 11, Nov. 1952

(Reprinted by permission)

The classifications were made by shaking together 5 ml. of each of the solvents listed in a test tube for 1 minute, then allowing the mixture to settle. If no interfacial meniscus was observed, the solvent pair was considered miscible. If such a meniscus was present, the solvent pair was regarded as immiscible. The classification of immiscible is a qualitative one since solvent pairs may exhibit some degree of partial miscibility while existing as separate phases. Solvent pairs possessing a pronounced degree of partial miscibility are designated by the symbol Is.

Compound number	Compounds	Acetone	Acetyl acetone	2-Amino-2-methyl-1-propanol	Aniline	Benzaldehyde	Benzene	Benzin	Benzyl alcohol	Butyl acetate	Butyl alcohol	*n*-Butyl ether	Capryl alcohol	Carbon tetrachloride	Diacetone alcohol	Diethanolamine	Diethyl cellosolve	Diethyl ether	Dimethylaniline	Ethyl alcohol	Ethyl benzoate	Ethylene glycol	2-Ethylhexanol	Formamide	Furfuryl alcohol	Glycerol	Hydroxyethyl-ethylenediamine	Isoamyl alcohol	Methyl isobutyl ketone	Nitromethane	Dibutoxytetra-ethylene glycol	Pyridine	Triethanolamine	Trimethylene glycol
1	Acetone	..	M	M	..	M	M	M	M	M	M	M	M	M	M	M	M	M	M	M	M	M	M	M	M	I	M	M	M	M	M	M	..	M
2	Acetyl acetone	M	..	R	..	M	M	M	M	M	M	M	M	M	M	R	M	M	M	M	M	M	M	M	M	I	R	M	M	M	M	M	..	M
3	Adiponitrile	M	M	M	M	..	M	..	M	M	M	I	..	I	M	..	M	I	M	M	M	I	I	M	M	I	M	I	M	..	M	M	M	M
4	2-Amino-2-methyl-1-propanol	M	R	..	..	M	M	I	M	M	M	Is	M	M	R	M	M	M	M	M	M	M	M	M	M	M	M	M	M	M	M	M	..	M
5	Benzaldehyde	M	M	M	..	..	M	M	M	M	M	M	M	M	M	I	M	M	M	M	M	Is	M	M	M	Is	R	M	M	M	M	M	..	M
6	Benzene	M	M	M	..	M	..	M	M	M	M	M	M	M	Is	I	M	M	M	M	M	I	M	I	M	I	I	M	M	I	M	M	..	I
7	Benzin	M	M	I	..	M	M	..	I	M	M	M	M	M	I	I	M	M	M	M	M	M	M	I	I	I	Is	M	M	M	M	M	..	I
8	Benzonitrile	M	M	M	M	..	M	..	M	M	M	M	..	M	M	..	M	M	M	M	M	I	M	I	M	I	M	M	M	..	M	M	M	I
9	Benzothiazole	M	M	M	M	..	M	..	M	M	M	M	..	M	M	..	M	M	M	M	M	M	M	I	M	I	M	M	M	..	M	M	M	M
10	Benzyl alcohol	M	M	M	..	M	M	I	..	M	M	M	M	M	M	M	M	M	M	M	M	M	M	M	M	M	M	M	M	M	M	M	..	M
11	Benzyl mercaptan	M	M	I	M	..	M	..	M	M	M	M	..	M	M	..	M	M	M	M	M	I	M	I	M	I	M	M	M	..	M	M	R	I
12	Butyl acetate	M	M	M	..	M	M	M	M	..	M	M	M	M	M	I	M	M	M	M	M	Is	M	I	M	I	I	M	M	M	M	M	..	Is
13	Butyl alcohol	M	M	M	..	M	M	M	M	M	..	M	M	M	M	M	M	M	M	M	M	M	M	M	M	M	M	M	M	M	M	M	..	M
14	*n*-Butyl ether	M	M	Is	..	M	M	M	M	M	M	..	M	M	M	I	M	M	M	M	M	I	M	I	M	I	I	M	M	I	M	M	..	I
15	Capryl alcohol	M	M	M	..	M	M	M	M	M	M	M	..	M	M	M	M	M	M	M	M	M	M	I	M	I	M	M	M	Is	M	M	..	M
16	Carbon tetrachloride	M	M	M	..	M	M	M	M	M	M	M	M	..	Is	I	M	M	M	M	M	I	M	I	M	I	I	M	M	M	M	M	..	I
17	Diacetone alcohol	M	M	R	..	M	Is	I	M	M	M	M	M	Is	..	M	M	M	M	M	M	M	M	M	M	I	R	M	M	M	M	M	..	M
18	Diethanolamine	M	R	M	..	I	I	I	M	I	M	I	M	I	M	..	I	I	Is	M	M	M	M	M	M	M	M	M	I	I	I	M	..	M
19	Diethyl Cellosolve	M	M	M	..	M	M	M	M	M	M	M	M	M	M	I	..	M	M	M	M	M	M	M	M	I	I	M	M	M	M	M	..	M
20	Diethyl ether	M	M	M	..	M	M	M	M	M	M	M	M	M	M	I	M	..	M	M	M	I	M	I	M	I	I	M	M	M	M	M	..	I
21	Dimethylaniline	M	M	M	..	M	M	M	M	M	M	M	M	M	M	Is	M	M	..	M	M	I	M	I	M	I	I	M	M	M	M	M	..	I
22	Di-*N*-propylaniline	M	M	I	M	..	M	..	M	M	M	M	..	M	I	..	M	M	M	M	M	I	M	I	M	I	I	M	M	..	M	M	I	I
23	Ethyl alcohol	M	M	M	..	M	M	M	M	M	M	M	M	M	M	M	M	M	M	..	M	M	M	M	M	M	M	M	M	M	M	M	..	M
24	Ethyl benzoate	M	M	M	..	M	M	M	M	M	M	M	M	M	M	M	M	M	M	M	..	I	M	I	M	I	M	M	M	M	M	M	..	Is
25	Ethyl isothiocyanate	M	M	R	M	..	M	..	M	M	M	M	..	M	M	..	M	M	M	M	M	I	M	I	M	I	R	M	M	..	M	M	M	I
26	Ethyl thiocyanate	M	M	M	M	..	M	..	M	M	M	M	..	M	M	..	M	M	M	M	M	I	M	I	M	I	M	M	M	..	M	M	M	I
27	Ethylene glycol	M	M	M	..	Is	I	M	M	Is	M	I	M	I	M	M	M	I	I	M	I	..	M	M	M	M	M	M	I	I	M	M	..	M
28	2-Ethylhexanol	M	M	M	..	M	M	M	M	M	M	M	M	M	M	M	M	M	M	M	M	M	..	I	M	I	M	M	M	I	M	M	..	M
29	Formamide	M	M	M	..	M	I	I	M	I	M	I	I	I	M	M	M	I	I	M	I	M	I	..	M	M	M	M	Is	M	M	M	..	M
30	Furfuryl alcohol	M	M	M	..	M	M	I	M	M	M	M	M	M	M	M	M	M	M	M	M	M	M	M	..	M	M	M	M	M	M	M	..	M
31	Glycerol	I	I	M	..	I	I	I	M	I	M	I	I	I	I	M	I	I	I	M	I	M	I	M	M	..	M	I	I	I	I	M	..	M
32	Hydroxyethyl-ethylenediamine	M	R	M	..	R	I	Is	M	I	M	I	M	I	R	M	I	I	I	M	M	M	M	M	M	M	..	M	M	M	M	M	..	M
33	Isoamyl alcohol	M	M	M	..	M	M	M	M	M	M	M	M	M	M	M	M	M	M	M	M	M	M	M	M	I	M	..	M	M	M	M	..	M
34	Isoamyl sulfide	M	M	I	M	..	M	..	M	M	M	M	..	M	I	..	M	M	M	M	M	I	M	I	I	I	I	M	M	..	M	M	I	I
35	Isobutyl mercaptan	M	M	M	M	..	M	..	M	M	M	M	..	M	M	..	M	M	M	M	M	I	M	I	M	I	I	M	M	..	M	M	R	R
36	Methyl disulfide	M	M	M	M	..	M	..	M	M	M	M	..	M	M	..	M	M	M	M	M	I	M	I	M	I	I	M	M	..	M	M	I	R
37	Methyl isobutyl ketone	M	M	M	..	M	M	M	M	M	M	M	M	M	M	I	M	M	M	M	M	I	M	Is	M	I	M	M	..	M	M	M	..	I
38	Nitromethane	M	M	M	..	M	I	M	M	M	M	I	Is	M	M	I	M	M	M	M	M	I	I	M	M	I	M	M	M	..	M	M	..	I
39	Dibutoxytetra-ethylene glycol	M	M	M	..	M	M	M	M	M	M	M	M	M	M	I	M	M	M	M	M	M	M	M	M	I	M	M	M	M	..	M	..	M
40	Pyridine	M	M	M	..	M	M	M	M	M	M	M	M	M	M	M	M	M	M	M	M	M	M	M	M	M	M	M	M	M	M	..	..	M
41	Tri-*n*-butylamine	M	M	I	I	..	M	..	M	M	M	M	..	M	I	..	M	M	M	M	M	I	M	I	M	I	I	M	M	..	M	M	I	I
42	Trimethylene glycol	M	M	M	..	M	I	I	M	Is	M	I	M	I	M	M	M	I	I	M	Is	M	M	M	M	M	M	M	I	I	M	M	..	..

MISCIBILITY OF ORGANIC SOLVENT PAIRS (Continued)

Tables B and C

W. M. Jackson and J. S. Drury

The classifications were made at 20°C in the following manner. One-milliliter portions of each solvent comprising a pair were shaken together for approximately a minute. If no interfacial meniscus was observed after the contents of the tube were allowed to settle, the solvent pair was considered to be miscible, M. If a meniscus was observed without apparent change in the volume of either solvent, the pair was regarded as immiscible, I. This classification is a qualitative one, since solvent pairs may exhibit various degrees of partial miscibility while existing as separate phases. If an obvious change occurred in the volume of each solvent, but a meniscus was present, the pair was classified as partially miscible, S. The designation R indicates that the two solvents reacted.

Table B

Compound number	Compounds	Acetone	Isoamyl acetate	*n*-Amyl cyanide	Benzene	Benzyl ether	2-Bromoethyl acetate	Chloroform	Cinnamaldehyde	Di-*n*-amylamine	Di-*n*-butyl carbonate	Diethylacetic acid	Diethylenetriamine	Diethyl formamide	Diisobutyl ketone	Diisopropylamine	Di-*n*-propyl aniline	Ethyl alcohol	Ethyl benzoate	Ethyl ether	Ethyl phenylacetate	Heptadecanol[a]	3-Heptanol	*n*-Heptyl acetate	*n*-Hexyl ether	Methyl isopropyl ketone	4-Methyl-*n*-valeric acid	*o*-Phenetidine	Sulfuric acid (concd.)	Tetradecanol[a]	Tri-*n*-butyl phosphate	Triethylene glycol	Triethylenetetramine	2,6,8-Trimethyl 4-nonanone	Compound number
1	Acetone	..	M	M	M	M	M	M	M	M	M	M	M	M	M	M	M	M	M	M	M	M	M	M	M	M	M	M	R	M	M	M	M	M	1
2	Isoamyl acetate	M	..	M	M	M	M	M	M	M	M	M	M	M	M	M	M	M	M	M	M	M	M	M	M	M	M	M	R	M	M	I	M	M	2
3	*n*-Amyl cyanide	M	M	..	M	M	M	M	M	M	M	M	M	M	M	M	M	M	M	M	M	M	M	M	M	M	M	M	R	M	M	M	M	M	3
4	Benzene	M	M	M	..	M	S	M	M	M	M	M	M	M	M	M	M	M	M	M	M	M	M	M	M	M	M	M	I	M	M	S	M	M	4
5	Benzyl ether	M	M	M	M	..	M	M	M	M	M	M	M	M	M	M	M	M	M	M	M	M	M	M	M	M	M	M	R	M	M	I	M	M	5
6	2-Bromoethyl acetate	M	M	M	S	M	..	M	M	R	M	M	R	R	M	R	M	M	M	M	M	M	M	M	S	M	M	M	R	M	M	M	R	M	6
7	Chloroform	M	M	M	M	M	M	..	M	M	M	M	M	M	M	M	M	M	M	M	M	M	M	M	M	M	M	M	I	M	M	M	M	M	7
8	Cinnamaldehyde	M	M	M	M	M	M	M	..	M	M	M	R	M	M	M	M	M	M	M	M	M	M	M	M	M	M	R	R	M	M	M	R	M	8
9	Di-*n*-amylamine	M	M	M	M	M	R	M	M	..	M	R	I	R	M	M	M	M	M	M	M	M	M	M	M	M	M	M	R	M	M	S	I	M	9
10	Di-*n*-butyl carbonate	M	M	M	M	M	M	M	M	M	..	M	M	M	M	M	M	M	M	M	M	M	M	M	M	M	M	M	R	M	M	I	I	M	10
11	Diethylacetic acid	M	M	M	M	M	M	M	M	R	M	..	R	M	M	R	M	M	M	M	M	M	M	M	M	M	M	M	R	M	M	M	R	M	11
12	Diethylenetriamine	M	M	M	M	M	R	M	R	I	M	R	..	R	M	M	M	M	M	M	M	M	M	R	I	R	R	M	R	M	M	M	M	I	12
13	Diethyl formamide	M	M	M	M	M	R	M	M	R	M	M	R	..	M	R	M	M	M	M	M	M	M	M	I	M	M	M	R	M	M	M	R	M	13
14	Diisobutyl ketone	M	M	M	M	M	M	M	M	M	M	M	M	M	..	M	M	M	M	M	M	M	M	M	M	M	M	M	R	M	M	I	M	M	14
15	Diisopropylamine	M	M	M	M	M	R	M	M	M	M	R	M	R	M	..	M	M	M	M	M	M	M	M	M	M	R	M	R	M	M	M	M	M	15
16	Di-*n*-propylaniline	M	M	M	M	M	M	M	M	M	M	M	M	M	M	M	..	M	M	M	M	M	M	M	M	M	M	M	R	M	M	I	M	M	16
17	Ethyl alcohol	M	M	M	M	M	M	M	M	M	M	M	M	M	M	M	M	..	M	M	M	M	M	M	M	M	M	M	M	M	M	M	M	M	17
18	Ethyl benzoate	M	M	M	M	M	M	M	M	M	M	M	M	M	M	M	M	M	..	M	M	M	M	M	M	M	M	M	M	M	M	M	M	M	18
19	Ethyl ether	M	M	M	M	M	M	M	M	M	M	M	M	M	M	M	M	M	M	..	M	M	M	M	M	M	M	M	R	M	M	I	M	M	19
20	Ethyl phenylacetate	M	M	M	M	M	M	M	M	M	M	M	M	M	M	M	M	M	M	M	..	M	M	M	M	M	M	M	R	M	M	M	M	M	20
21	Heptadecanol[a]	M	M	M	M	M	M	M	M	M	M	M	M	M	M	M	M	M	M	M	M	..	M	M	M	M	M	M	R	M	M	I	M	M	21
22	3-Heptanol	M	M	M	M	M	M	M	M	M	M	M	M	M	M	M	M	M	M	M	M	M	..	M	M	M	M	M	R	M	M	M	M	M	22
23	*n*-Heptyl acetate	M	M	M	M	M	M	M	M	M	M	M	R	M	M	M	M	M	M	M	M	M	M	..	M	M	M	M	R	M	M	I	R	M	23
24	*n*-Hexyl ether	M	M	M	M	M	S	M	M	M	M	M	I	I	M	M	M	M	M	M	M	M	M	M	..	M	M	M	R	M	M	I	I	M	24
25	Methyl isopropyl ketone	M	M	M	M	M	M	M	M	M	M	M	R	M	M	M	M	M	M	M	M	M	M	M	M	..	M	M	R	M	M	M	R	M	25
26	4-Methyl-*n*-valeric acid	M	M	M	M	M	M	M	M	M	M	M	R	M	M	R	M	M	M	M	M	M	M	M	M	M	..	M	R	M	M	M	R	M	26
27	*o*-Phenetidine	M	M	M	M	M	M	M	R	M	M	M	M	M	M	M	M	M	M	M	M	M	M	M	M	M	M	..	R	M	M	M	M	M	27
28	Sulfuric acid (concd.)	R	R	R	I	R	R	I	R	R	R	R	R	R	R	R	R	M	M	R	R	R	R	R	R	R	R	R	..	R	R	R	R	R	28
29	Tetradecanol[a]	M	M	M	M	M	M	M	M	M	M	M	M	M	M	M	M	M	M	M	M	M	M	M	M	M	M	M	R	..	M	I	M	M	29
30	Tri-*n*-butyl phosphate	M	M	M	M	M	M	M	M	M	M	M	M	M	M	M	M	M	M	M	M	M	M	M	M	M	M	M	R	M	..	M	M	M	30
31	Triethylene glycol	M	I	M	S	I	M	M	M	S	I	M	M	M	I	M	I	M	M	I	M	I	M	I	I	M	M	M	R	I	M	..	M	I	31
32	Triethylenetetramine	M	M	M	M	M	R	M	R	I	I	R	M	R	M	M	M	M	M	M	M	M	M	R	I	R	R	M	R	M	M	M	..	I	32
33	2,6,8-Trimethyl 4-nonanone	M	M	M	M	M	M	M	M	M	M	M	I	M	M	M	M	M	M	M	M	M	M	M	M	M	M	M	R	M	M	I	I	..	33

[a] Union Carbide name.

MISCIBILITY OF ORGANIC SOLVENT PAIRS (Continued)

Table C

Compound number	Compounds	Acetone	Isoamyl acetate	*n*-Amyl cyanide	Anisaldehyde	Benzene	Benzyl ether	Chloroform	*o*-Cresol	Diisobutyl ketone	Diethylacetic acid	Diethyl formamide	Di-*n*-propyl aniline	Ethyl alcohol	Ethyl ether	3-Heptanol	*n*-Heptyl acetate	*n*-Hexyl ether	α-Methylbenzylamine	α-Methylbenzyldiethanolamine	α-Methylbenzyldimethylamine	α-Methylbenzylethanolamine	2-Methyl-5-ethylpyridine	Methyl isopropyl ketone	4-Methyl-*n*-valeric acid	*o*-Phenetidine	2-Phenylethylamine	Isopropanolamine	Pyridine	Salicylaldehyde	Tetradecanol[a]	Tri-*n*-butyl phosphate	Triethylenetetramine	2,6,8-Trimethyl 4-nonanone
1	1,3-Butylene glycol	M	I	M	I	I	I	M	M	I	M	M	I	M	S	M	I	I	M	M	M	M	M	M	M	M	M	M	M	M	M	M	M	I
2	2,3-Butylene glycol	M	M	M	M	S	I	M	M	M	M	M	I	M	M	M	M	I	M	M	M	M	M	M	M	M	M	M	M	M	M	M	M	I
3	2-Chloroethanol	M	M	M	M	M	M	M	M	M	M	M	M	M	M	M	M	M	R	M	M	M	R	M	M	M	R	R	M	M	M	M	M	M
4	3-Chloro-1,2-propanediol	M	M	M	M	I	M	M	M	M	M	M	I	M	M	M	M	I	R	M	M	M	M	M	M	M	R	R	M	M	S	M	R	S
5	Dibutyl hydrogen phosphite	M	M	M	M	M	M	M	M	M	M	M	M	M	M	M	M	M	M	M	M	M	M	M	M	M	M	M	M	M	M	M	M	M
6	Diethylene glycol dibutyl ether	M	M	M	M	M	M	M	M	M	M	M	M	M	M	M	M	M	R	M	S	M	M	M	M	M	R	R	M	M	M	M	R	M
7	Diethylene glycol diethyl ether	M	M	M	M	M	M	M	M	M	M	M	M	M	M	M	M	M	M	M	M	M	M	M	M	M	M	M	M	M	M	M	M	M
8	Diethylene glycol monobutyl ether	M	M	M	M	M	M	M	M	M	M	M	M	M	M	M	M	M	M	M	M	M	M	M	M	M	M	M	M	M	M	M	M	M
9	Diethylene glycol monoethyl ether	M	M	M	M	M	M	M	M	M	M	M	M	M	M	M	M	I	M	M	M	M	M	M	M	M	M	M	M	M	M	M	M	M
10	Diethylene glycol monomethyl ether	M	M	M	M	M	M	M	M	M	M	M	M	M	M	M	M	I	M	M	M	M	M	M	M	M	M	M	M	M	M	M	M	M
11	Dipropylene glycol	M	M	M	M	M	M	M	M	M	M	M	M	M	M	M	M	I	M	M	M	M	M	M	M	M	M	M	M	M	M	M	M	M
12	Ethylene diacetate	M	M	M	M	M	M	M	M	M	M	M	M	M	M	M	M	I	M	M	M	M	M	M	M	M	M	M	M	M	M	M	M	M
13	Ethylene glycol	M	I	I	I	I	I	S	M	I	M	M	I	M	I	M	I	I	M	M	M	M	M	I	M	M	M	M	M	I	I	S	M	I
14	Ethyl glycol ethylbutyl ether	M	M	M	M	M	M	M	M	M	M	M	M	M	M	M	M	M	M	M	M	M	M	M	M	M	M	M	M	M	M	M	M	M
15	Ethylene glycol monobutyl ether	M	M	M	M	M	M	M	M	M	M	M	M	M	M	M	M	M	M	M	M	M	M	M	M	M	M	M	M	M	M	M	M	M
16	Ethylene glycol monoethyl ether	M	M	M	M	M	M	M	M	M	M	M	M	M	M	M	M	M	M	M	M	M	M	M	M	M	M	M	M	M	M	M	M	M
17	Ethylene glycol monomethyl ether	M	M	M	M	M	M	M	M	M	M	M	M	M	M	M	M	M	M	M	M	M	M	M	M	M	M	M	M	M	M	M	M	M
18	Ethylene glycol monophenyl ether	M	M	M	M	M	M	M	M	M	M	M	M	M	M	M	M	M	M	M	M	M	M	M	M	M	M	M	M	M	M	M	M	M
19	Glycerol	I	I	I	I	I	I	I	M	I	I	M	I	M	I	I	I	I	M	M	I	M	M	I	I	I	M	M	M	I	I	I	M	I
20	1,2-Propanediol	M	M	M	M	I	I	M	M	I	M	M	I	M	S	M	I	I	M	M	M	M	M	M	M	M	M	M	M	M	M	M	M	I
21	1,3-Propanediol	M	I	I	I	I	I	M	M	I	M	M	I	M	I	M	I	I	M	M	M	M	M	M	M	M	M	M	M	I	S	M	M	I
22	Triethylene glycol	M	I	M	M	S	I	M	M	I	M	M	I	M	I	M	I	I	M	M	M	M	M	M	M	M	M	M	M	M	I	M	M	I
23	Triethyl phosphate	M	M	M	M	M	M	M	M	M	M	M	M	M	M	M	M	M	M	M	M	M	M	M	M	M	M	M	M	M	M	M	M	M
24	Trimethylene chlorohydrin	M	M	M	M	M	M	M	M	M	M	M	M	M	M	M	M	M	R	M	M	M	R	M	M	M	R	R	M	M	M	M	R	M

[a] Union Carbide name.

STEROID HORMONES AND OTHER STEROIDAL SYNTHETICS

Compiled by Erwin Di Cyan, Ph.D.

The field of steroids has expanded considerably and rapidly in degree and in kind, because synthetic steroids have been synthesized which though resembling the hormones in the body have no natural counterpart, but exert an effect comparable to those of the natural hormones.

In fact, the term *steroid hormone* thus becomes a misnomer when applied to the newer synthetically prepared steroids which do not have a counterpart in the body of man or other animals—as prednisone. (A hormone, by definition, is a material with certain functions and characteristics, *secreted by the ductless glands*. That part of the definition cannot be met by prednisone or by similar steroids as these are not secreted by the ductless, or endocrine glands.)

All the hormones as well as the synthetic analogues have in common the cyclopentanophenanthrene nucleus. Although chemically very similar, a comparatively slight structural change is in many instances productive of substances which have physiologically dissimilar effects, often acting upon different physiologic systems. But in many cases a small change in structure will result merely in an accentuation of certain effects.

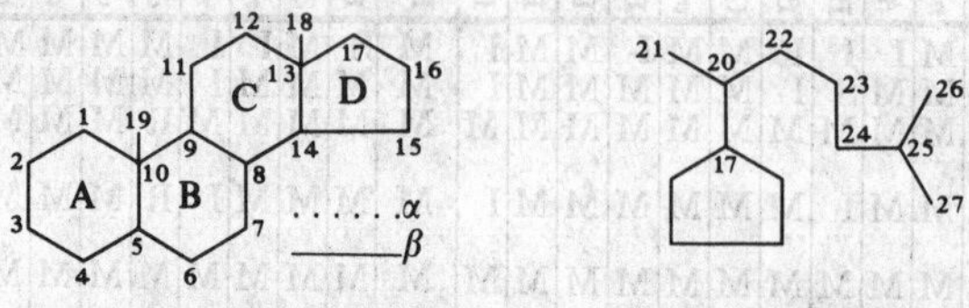

The Cyclopentanophenanthrene Nucleus

Classification. Classification becomes a bizarre problem by reason of the (a) overlapping uses to which these substances are put, and (b) the multiple purposes for which the hormones or synthetic substances are used. Indeed, the steroids may be classified by structure; that however would be uninformative to the student as to their use. Classification by origin, as adrenal, would also be unsuitable because, for example, a number of the adrenal corticosteroids are not found in the adrenal cortex at all, but merely resemble the natural hormones found in the adrenal cortex.

For those reasons the hormonal or hormonelike entries in the tables are classified by-and-large, by their predominant pharmacologic effects. Even that classification has its disparities as for example, the use of male sex hormones, i.e. the androgens, is neither limited to men, nor to uses which entail their effect upon male sex characteristics.

Uses. Originally, the use of steroid hormones was largely based upon one or more of the following predicates:

(a) To supplement the progressively declining secretion of a specific hormone due to natural biologic aging of the organism; in the menopause as an example of such declining secretion, a female sex hormone is used for such supplementation;

(b) To make available to the body a specific hormone, the natural secretion of which is inhibited because of a congenital or developmental anomaly; the underdevelopment of male secondary sex characteristics is an example of such an inhibited secretion, in which a male sex hormone is used—and correspondingly, female sex hormones in underdevelopment in females;

(c) To cause a reversal of hormonal balance in the treatment of diseases peculiar to a sex; for example, in the case of cancer of the female breast, a male sex hormone is administered, and in cancer of the prostate, a female sex hormone is used;

(d) To mimic a natural function, as menstruation, by the administration of estrogens—on withdrawal of which bleeding occurs; or by the alternate use of estrogenic and progestational—both female sex hormones.

(e) To delay a function, as ovulation, as in oral contraceptives, or *birth control pills*.

Since the finding that cortisone ameliorates the symptoms of rheumatoid arthritis (1949) the adrenal corticosteroid hormones and especially the synthetically prepared steroid analogues which have no natural counterpart in the body, have been successfully employed in the treatment of diseases not related to sex or sex function.

Androgens and Anabolic Agents. The agents listed in the tables under this classification have the effect of male sex hormones (androgens) i.e., to stimulate sexual maturation, in the "male climacteric," etc. But all androgens have in greater or lesser degree the ability to stimulate muscle development, i.e., an anabolic effect. Among the synthetically prepared agents which have no counterpart in the body (Methandrostenolone or Oxymetholone) are those which have a lessened androgenic, but a heightened anabolic effect. These qualities are determined by biological tests on animals but principally confirmed by clinical use in man. The anabolic effect includes remineralization of bone, which may be partially demineralized (osteoporosis) by age, or by certain drugs, as the adrenal corticosteroids (q.v.).

Anabolic agents are used for muscle and bone nutrition in men as well as women. The reason for the high interest in synthetic steroidal substances for anabolic use, is based on the need for materials, which within a given effective dose have a greater anabolic-to-androgenic ratio than such androgens as methyl testosterone. Otherwise, the administration of androgens to women produces manifestations of virilism, such as growth of hair on the face, a deepening of the voice, etc. Androgens are also used in the female in the suppression of excessive bleeding and in the treatment of cancer of the breast and cervix. (For other androgen-like agents, see also Progestogens and Progestins.)

Estrogens. Estrogenic agents hasten sexual maturation in the female. Therefore, they are used in underdevelopment in the female. The widest use of estrogens is in the treatment of the menopause, in which they supplement from without, the secretion of natural estrogens by the ovary, which begins to decline at about the 40th year. The menopause is usually a slow process, and the declining secretion gives rise to various symptoms during the time that the secretion declines, until adjustment to the new status takes place. The menopause, a period of physical and psychological stress, is made less precipitous by estrogens.

Frequently, a menopause must be quickly induced, as in cancer of the ovary or in uterine hemorrhage. This is done by radiation or by the removal of the uterus. Severe vasomotor symptoms occur when the menopause is thus suddenly induced. Estrogens —among other drugs—are used in the amelioration of these symptoms.

Estrogens (especially diethylstilbestrol which though not a hormone has an estrogenic effect) are also used in the control of cancer of the prostate in the male. Note the inverse correspondence to the use of male sex hormones in cancer of the breast in the female.

Progestogens and Progestins (Including 19-Norsteroid Compounds). The agents under that listing include progesterone, a female sex hormone, as well as progestins, i.e., synthetic progesterone-like compounds which have no natural counterpart in the body. Their use includes a variety of conditions: functional uterine bleeding, absence of menstruation (amenorrhea) used at times with estrogens, painful menstruation (dysmenorrhea), infertility, habitual abortion in order to maintain pregnancy, and in fact, to suppress ovulation hence their use as antifertility drugs. Certain progestins—as norethindrone combined with

an estrogen, are the principal components of birth control pills—suppressing ovulation, there is no egg to fertilize, hence conception does not take place.

Adrenal Corticosteroids, Including Antiinflammatory, Antiallergic and Antirheumatic Agents. The adrenal cortex secretes a large number of hormones. They usually differ from each other in the accentuation of some phases of their properties. Virtually all of the cortical hormones are catabolic, thus having an effect in this respect, diametrically opposed to the androgens which are anabolic. Nearly all the cortical hormones—differing in degree from each other—cause retention of sodium and water by the body and hasten the excretion of potassium. These effects are utilized in the treatment of adrenal insufficiency or Addison's disease, in which conversely, there is an undue excretion of sodium and a strong retention of potassium. Desoxycorticosterone is used in Addison's disease because it has a particularly strong sodium retaining and potassium excreting effect.

Since the finding in 1949 of the usefulness of cortisone in profoundly reducing the symptoms of rheumatoid arthritis, the adrenal corticosteroids, including hydrocortisone, a natural hormone secreted by the adrenal cortex, and particularly the synthetic analogues not found in the body, as prednisone, have been used in the treatment of a wide variety of inflammatory diseases—especially diseases of collagen tissue. The same antiinflammatory effect is also brought into use in the reduction of inflammations associated with diseases of the skin, allergy, asthma, and in such systematic diseases as disseminated lupus erythematosus, also a collagen disease.

The drawbacks of cortisone, also shared in lesser measure by hydrocortisone, gave the impetus to the synthesis of steroidal substances not native to the body but differing somewhat from cortisone and hydrocortisone, in order to reduce the drawbacks attendant to the use of the latter. The sideeffects—especially those of cortisone—are retention of water and sodium, excretion of potassium, loss of mineral from bone leading to osteoporosis and fractures, hypertension, at times diabetes, personality changes or gastric ulcer. Prednisone and prednisolone among others (see tables) are two such steroidal synthetics which have the effects of cortisone, but fewer or less severe sideeffects. Whereas the synthetic steroidal substances are superior to cortisone with respect to lessened sideeffects, it cannot be said that the sideeffects are absent—they vary in degree from substance to substance.

Diuretic, Antidiuretic and Local Anesthetic Agents. Aldosterone, a natural hormone of the adrenal cortex promotes retention in the body of sodium and water, and facilitates excretion of potassium. Hence its effect is almost diametrically opposed to diuretics—especially the thiazide diuretics. Aldosterone is much more active in this respect than desoxycorticosterone, and is used in the treatment of Addison's disease, a hypofunction of the adrenal glands.

Spironolactone is an antagonist to aldosterone—the latter when elaborated in the body in excessive amounts gives rise to a syndrome called aldosteronism. Spironolactone, a synthetically produced steroid does not have a natural counterpart in the body, is diuretic when mercurial or thiazide diuretics are ineffective; it prevents sodium retention and potassium excretion—effects opposite to aldosterone. Hence spironolactone is used in aldosteronism, against edema, in the treatment of congestive heart failure and in other conditions in which an accumulation of water, and water-retaining salt, is to be corrected.

Doses. The amount of substance which comprises a dose of steroid hormones, or of the steroidal synthetics varies from substance to substance—from 0.1 mg for an estradiol ester, to 50 mg for a 19-norsteroid compound. The dose is conditioned upon the order of activity of the substance, the purpose for which it is administered, as well as the patient's response. However, as additional steroids for hormonal use are synthesized—especially those with adrenocortical activity, their average dose is usually smaller than the previously available steroid. The smaller effective dose of the more recent steroid is cited as an advantage over the previously available steroid.

However, a smaller dose cannot be claimed as an inherent advantage of a new steroid in comparison with an existing one, unless the lower dosage exhibits either greater or more prolonged activity or lesser sideeffects. One cannot meaningfully compare a dose, milligram for milligram, without taking into consideration if a heightened effect of the smaller dose produces fewer sideeffects. For example, it does not make any difference if a given effect and the same accompanying sideeffects are produced by a 50 mg or a 5 mg dose.

ADRENAL CORTICOSTEROIDS, INCLUDING ANTIINFLAMMATORY, ANTIALLERGIC AND ANTIRHEUMATIC AGENTS

Names & synonyms:	BETAMETHASONE; 9α-fluoro-16β-methylprednisolone; 16β-methyl-11β,17α,21-trihydroxy-9α-fluoro-1,4-pregnadiene-3,20-dione.	BETAMETHASONE ACETATE; 9α-fluoro-16β-methylprednisolone-21-acetate.	BETAMETHASONE DISODIUM PHOSPHATE; 9α-fluoro-16β-methylprednisolone-21-disodium phosphate.
Formulae:	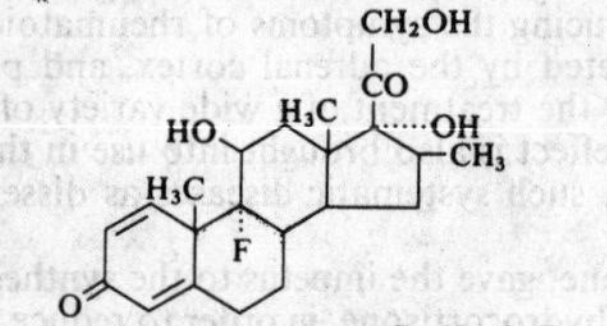 $C_{22}H_{29}O_5F$	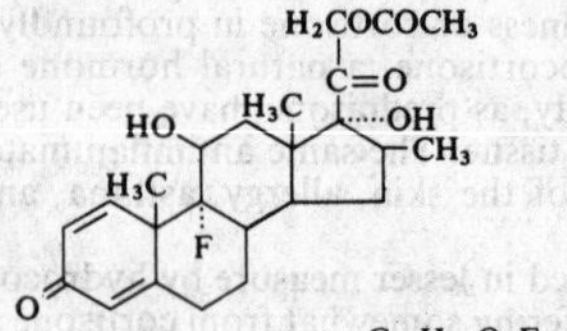 $C_{24}H_{31}O_6F$	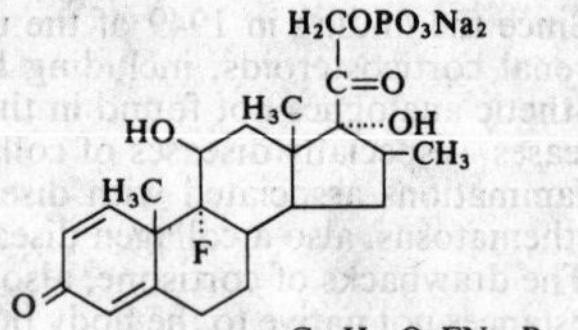 $C_{22}H_{28}O_8FNa_2P$
Molecular weight	392.5	434.5	516.4
Melting point (°C)	240 (dec.)	200 to 220 (dec.)	decomposes
Specific rotation	$(\alpha)\frac{25}{D}$ +112 to +120 (100 mg. in 10 ml. dioxane)	$(\alpha)\frac{25}{D}$ +120 to +128 (100 mg. in 10 ml. dioxane)	$(\alpha)\frac{25}{D}$ +99 to +105 (100 mg. in 10 ml. water)
Absorption max.	239 mμ, E(1 %, 1 cm) 390, methanol	239 mμ, methanol	241 mμ, water

Names & synonyms:	CHLOROPREDNISONE ACETATE; 6α-chloroprednisone acetate; 6α-chloro-$\Delta^{1,4}$-pregnadien-17β,21-diol-3,11,20-trione 21-acetate.	CORTICOSTERONE; 11,21-dihydroxyprogesterone; Δ^4-pregnene-11β,21-diol-3,20-dione; 11β,21-dihydroxy-4-pregnene-3,20-dione; Kendall compound B; Reichstein substance H.	CORTISONE; 17-hydroxy-11-dehydrocorticosterone; 17α,21-dihydroxy-4-pregnene-3,11,20-trione; Δ^4-pregnene-17α,21-diol-3,11,20-trione; Kendall compound E; Wintersteiner compound F.
Formulae:	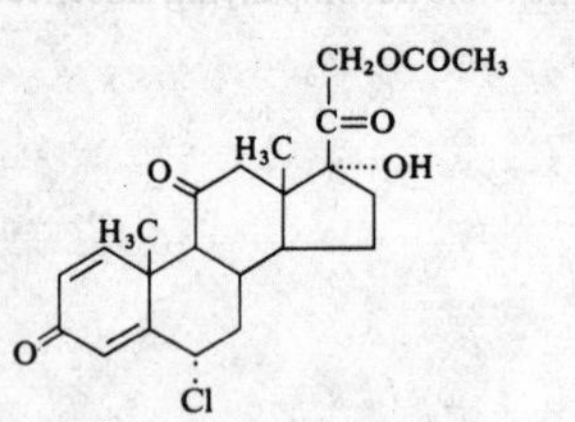	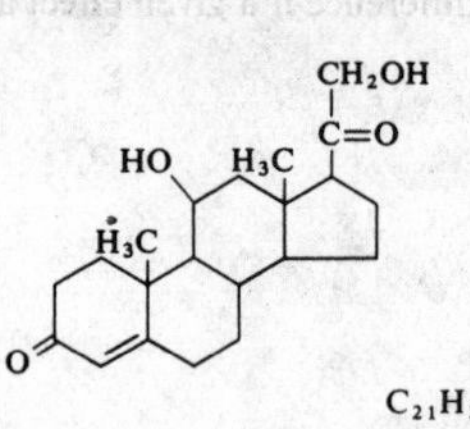 $C_{21}H_{30}O_4$	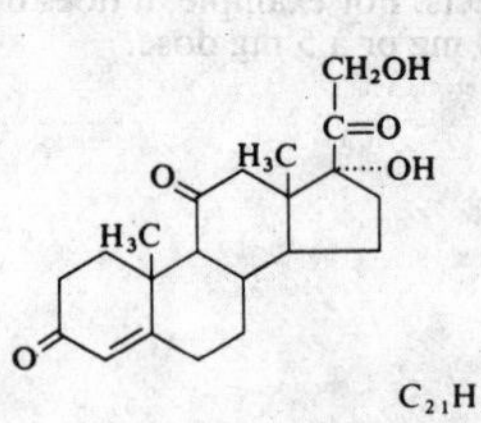 $C_{21}H_{28}O_5$
Molecular weight	436.6	346.40	360.4
Melting point (°C)	207–213	180–182	220–224
Specific rotation	$(\alpha)\frac{25}{D}$ +137 to +142 (100 mg. in 10 ml. chloroform)	$(\alpha)\frac{15}{D}$ +222 (110 mg. in 10 ml. alcohol)	$(\alpha)\frac{25}{D}$ +209 (120 mg. in 10 ml. alcohol)
Absorption max.		240 mμ	237 mμ

STEROID HORMONES AND OTHER STEROIDAL SYNTHETICS (Continued)

ADRENAL CORTICOSTEROIDS, INCLUDING ANTIINFLAMMATORY, ANTIALLERGIC AND ANTIRHEUMATIC AGENTS (Continued)

Names & synonyms:	DESOXYCORTICOSTERONE; deoxycorticosterone; 11-desoxycorticosterone; 21-hydroxyprogesterone; 4-pregnen-21-ol-3,20-dione; Kendall desoxy compound B; Reichstein substance Q.	DESOXYCORTICOSTERONE ACETATE; DCA; 11-desoxycorticosterone acetate.	DESOXYCORTICOSTERONE PIVALATE; desoxycorticosterone trimethylacetate; 21-hydroxy-4-pregnene-3,20-dione pivalate.
Formulae:	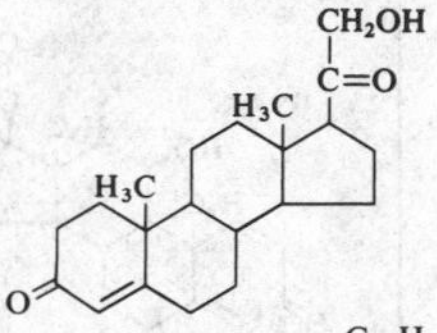 $C_{21}H_{30}O_3$	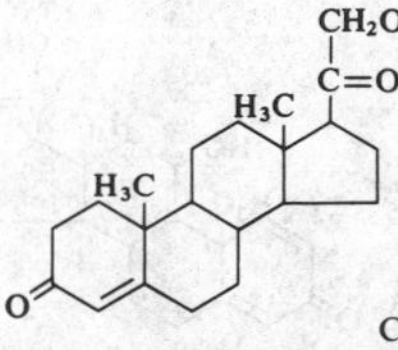 $C_{23}H_{32}O_4$	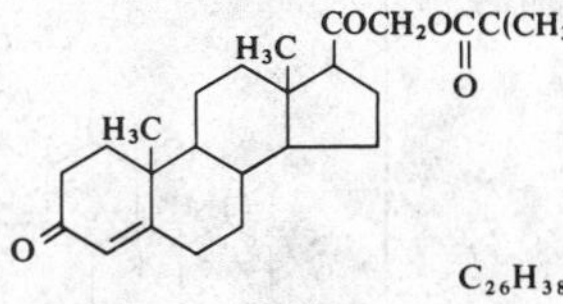 $C_{26}H_{38}O_4$
Molecular weight	330.2	372.4	414.6
Melting point (°C)	140–142	154–160	198–204
Specific rotation	$(\alpha)_D^{22}$ +176 – +178 (100 mg. in 10 ml. alcohol)	$(\alpha)_D^{20}$ +168 – +178 (100 mg. in 10 ml. dioxane)	$(\alpha)_D^{25}$ +157 ±4 (1 % in dioxane)
Absorption max.	240 mμ		240 mμ (in ethanol)

Names & synonyms:	DEXAMETHASONE; hexadecadrol; 9α-fluoro-16α-methyl prednisolone; 9α-fluoro-11β,17α-21-trihydroxy-16α-methyl-1,4-pregnadiene-3,20-dione; 16α-methyl-9α-fluoro-1,4-pregnadiene-11β,17α-21-triol-3,20-dione; 16α-methyl-9α-fluoro-Δ¹-hydrocortisone; 1-dehydro-16α-methyl-9α-fluorohydrocortisone.	DICHLORISONE ACETATE; 9α-11β-dichloro-1,4-pregnadiene-17α,21-diol-3,20-dione-21-acetate	FLUOCINOLONE ACETONIDE; 6α,9α-difluoro-16α hydroxyprednisolone-16, 17-acetonide; 6α,9α-difluoro-16α,17α-isopropylidenediosy-1,4-pregnadiene-3,20-dione.
Formulae:	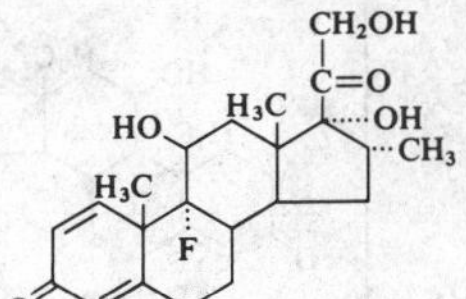 $C_{22}H_{29}FO_5$	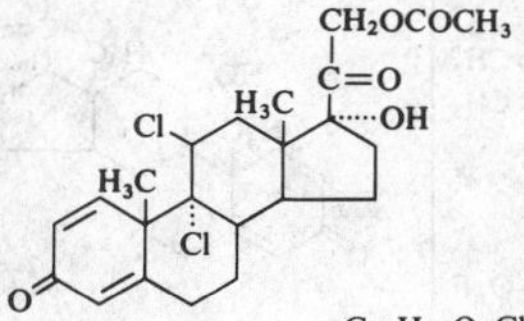 $C_{23}H_{28}O_5Cl$	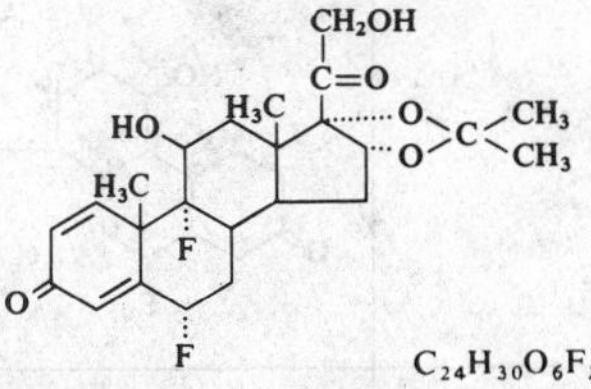 $C_{24}H_{30}O_6F_2$
Molecular weight	392.4	455.3	452.50
Melting point (°C)	262–264	235 (dec.)	255–266
Specific rotation	$(\alpha)_D^{25}$ +78 (100 mg. in 10 ml. dioxane)	$(\alpha)_D^{25}$ +160 – 168 (100 mg. in 10 ml. dioxane)	not less than +95° and not more than +105°C at 25°C.
Absorption max.		237 mμ – 316 – 337 (ε_1^1)	237 mμ ± 1 mμ

STEROID HORMONES AND OTHER STEROIDAL SYNTHETICS (Continued)

ADRENAL CORTICOSTEROIDS, INCLUDING ANTIINFLAMMATORY, ANTIALLERGIC AND ANTIRHEUMATIC AGENTS (Continued)

Names & synonyms:	FLUOROHYDROCORTISONE; fludrocortisone; 9α-fluorohydrocortisone; 9α-fluorocortisol; fluohydrisone; 9α-fluoro-11β,17α,21-trihydroxy-4-pregnene-3,20-dione; 9α-fluoro-17-hydroxycorticosterone.	FLUOROMETHOLONE; 9α-fluoro-11β,17α-dihydroxy-6α-methyl-1,4-pregnadiene-3,20-dione; 21-desoxy-9α-fluoro-6α-methyl-prednisolone.	FLUPREDNISOLONE; 6α-fluoroprednisolone; 6α-fluoro-1-dehydrohydrocortisone; 6α-fluoro-11β,17α,21-trihydroxy-1,4-pregnadiene-3,20-dione.
Formulae:	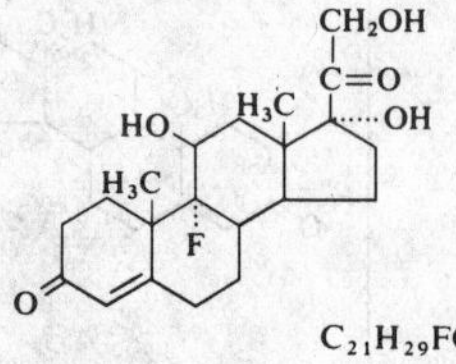 $C_{21}H_{29}FO_5$	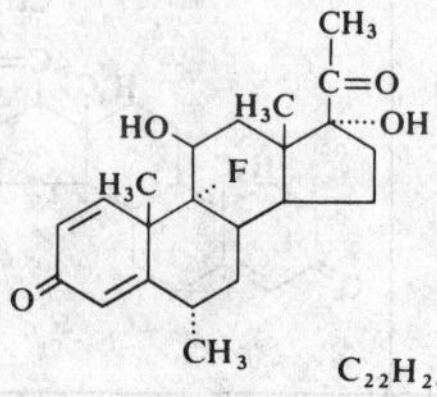 $C_{22}H_{24}FO_4$	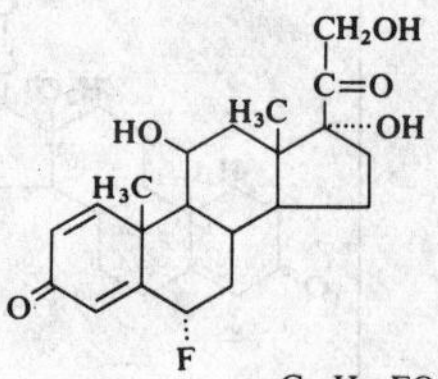 $C_{21}H_{27}FO_5$
Molecular weight	380.4	376.4	378.4
Melting point (°C)	260–262 (dec.)	290 (dec.)	205–210
Specific rotation	$(\alpha)\frac{23}{D}$ +139 (55 mg. in 10 ml. alcohol)	$(\alpha)\frac{25}{D}$ +56 (pyridine)	$(\alpha)_D$ +88 (dioxane)
Absorption max.		239 mu (a_M = 15,050) methanol	λ_{max} 241.5 mu (ε 16,000)

Names & synonyms:	FLURANDRENOLONE; 6-fluoro-16α-hydroxyhydrocortisone-16,17-acetonide; 6α-fluoro-11β,21-dehydroxy-16α,17α-isopropylidenedioxy-pregna-4-ene-3,20-dione.	HYDROCORTISONE; cortisol; 17-hydroxycorticosterone; hydrocortisone free alcohol; 11β,17α,21-trihydroxy-4-pregnene-3,20-dione; 4-pregnene-11β,17α,21-triol-3,20-dione; Kendall compound F; Reichstein substance M.	HYDROCORTISONE ACETATE; cortisol acetate; hydrocortisone-21-acetate; 17-hydroxycorticosterone-21-acetate.
Formulae:	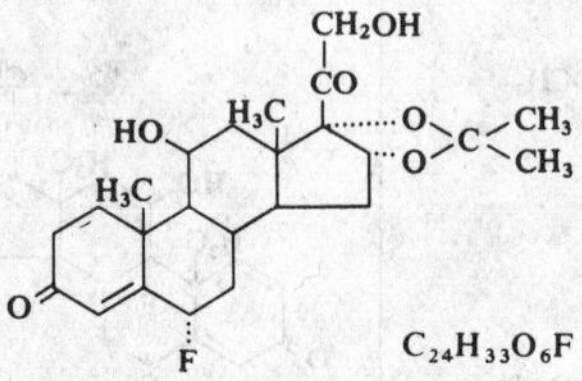 $C_{24}H_{33}O_6F$	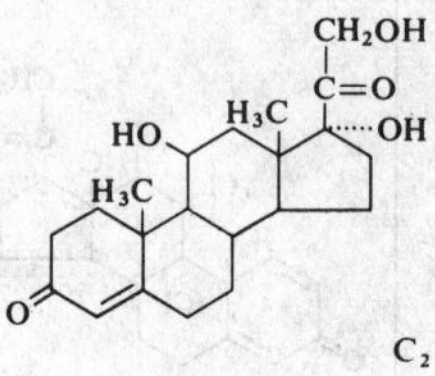 $C_{21}H_{30}O_5$	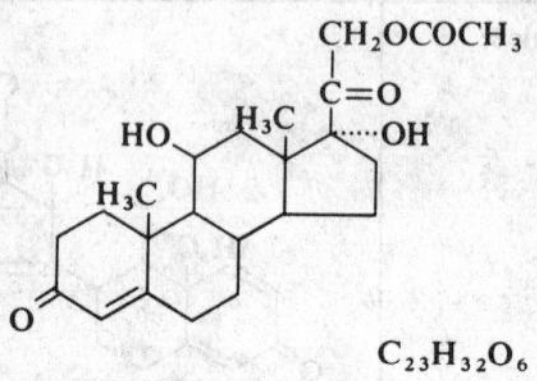 $C_{23}H_{32}O_6$
Molecular weight	436.5	362.5	404.5
Melting point (°C)	240–250	215–220 (dec.)	223 (dec.)
Specific rotation	$(\alpha)\frac{25}{D}$ = +145 (1% in $CHCl_3$)	$(\alpha)\frac{25}{D}$ +150 – +156 (100 mg. in 10 ml.) dioxane)	$(\alpha)\frac{25}{D}$ +158 – +165 (100 mg. in 10 ml. dioxane)
Absorption max.	236 mμ (methanol)	242 mμ	242 mμ (methanol)

ADRENAL CORTICOSTEROIDS, INCLUDING ANTIINFLAMMATORY, ANTIALLERGIC AND ANTIRHEUMATIC AGENTS (Continued)

Names & synonyms:	HYDROCORTISONE SODIUM SUCCINATE; 11β, 17α, 21-trihydroxy-4-pregnene-3, 20-dione, 21 hydrogen succinate, sodium salt; hydrocortisone, 21 hydrogen succinate, sodium salt.	METHYLPREDNISOLONE; 6α-methylprednisolone; Δ[1]-6α-methylhydrocortisone; 1-dehydro-6α-methylhydrocortisone; 11β, 17α,21-trihydroxy-6α-methyl-1,4-pregnadiene-3,20-dione.	METHYLPREDNISOLONE SODIUM SUCCINATE; 1-dehydro-6α-methylhydrocortisone, 21-hydrogen succinate, sodium salt; 6α-methylprednisolone 21-hydrogen succinate, sodium salt; 11β, 17α, 21-trihydroxy-6α-methyl-1, 4-pregnadiene-3, 20-dione, 21-hydrogen succinate, sodium salt.
Formulae:	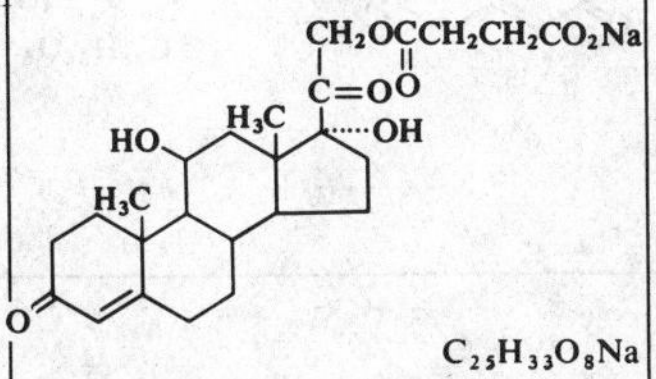 $C_{25}H_{33}O_8Na$	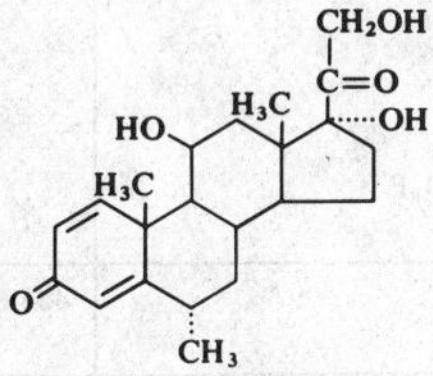 $C_{22}H_{30}O_5$	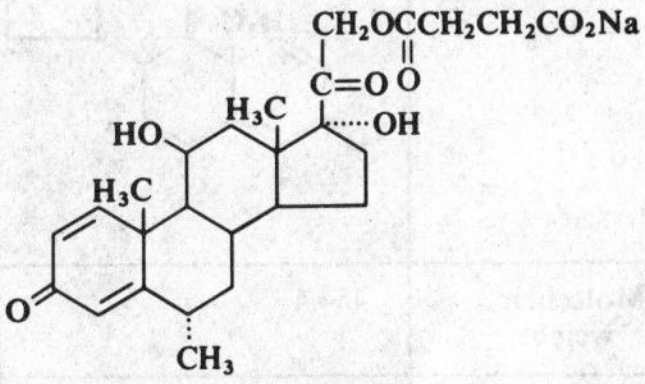 $C_{26}H_{33}O_8Na$
Molecular weight	484.5	374.5	496.5
Melting point (°C)	decomposes	230–240 (dec.)	decomposes
Specific rotation	$(\alpha)_D$ +140 ± 5 (alcohol)	$(\alpha)^{25}_{D}$ +85 (dioxane)	$(\alpha)_D$ +100 ± 4 (alcohol)
Absorption max.	λ 242 mμ (ε 15,700)	243 mμ	λ_{max} 242 mμ (ε 14,500)

Names & synonyms:	PARAMETHASONE; 6α-fluoro-16α-methylprednisolone; 6α-fluoro-11β-17α,21-trihydroxy-16α-methyl-1,4-pregnadiene-3,20-dione.	PARAMETHASONE ACETATE; 6α-fluoro-16α-methylprednisolone-21-acetate; 6α-fluoro-16α-methylpregna-1,4-diene-11β,21-diol-3,20-dione-21-acetate; 6α-fluoro-17β,17α,21-trihydroxy-16α-methyl-1,4-pregnadiene-3,20-dione-21-acetate.	PREDNISOLONE; metacortandralone; Δ[1]-dehydrocortisol; delta F; Δ[1]-hydrocortisone; Δ[1]-dehydrohydrocortisone; 1,4-pregnadiene-3,20-dione-11β,17α,21-triol; 11β,17α,21-trihydroxy-1,4-pregnadiene-3,20-dione.
Formulae:	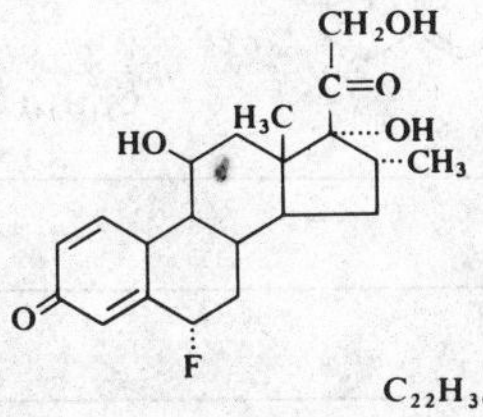 $C_{22}H_{30}O_5$	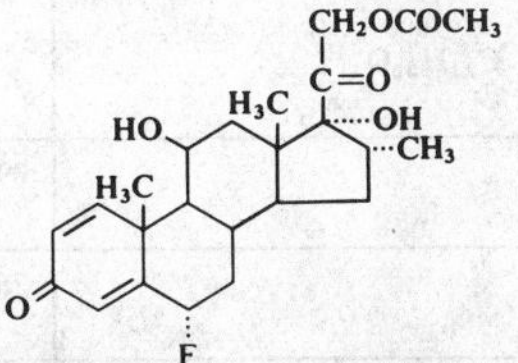 $C_{24}H_{31}O_6F$	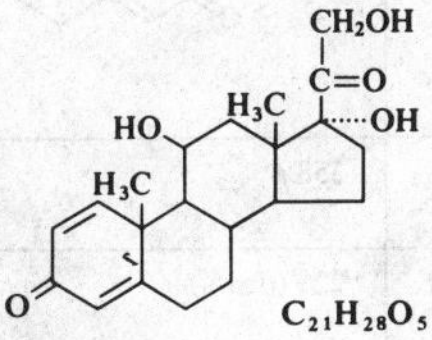 $C_{21}H_{28}O_5$
Molecular weight	392.45	434.5	360.4
Melting point (°C)	228–241	233–246	240 (dec.)
Specific rotation	+59 to +69 at 25°C	$(\alpha)^{25}_{D}$ +72 (1 % in $CHCl_3$)	$(\alpha)^{25}_{D}$ +97 – +103 (100 mg. in 10 ml. dioxane)
Absorption max.	242 mμ	242 mμ (methanol)	242 mμ (ε = 15,000) methanol

ADRENAL CORTICOSTEROIDS, INCLUDING ANTIINFLAMMATORY, ANTIALLERGIC AND ANTIRHEUMATIC AGENTS (Continued)

Names & synonyms:	PREDNISOLONE PHOSPHATE SODIUM; disodium prednisolone 21-phosphate.	PREDNISOLONE PIVALATE; prednisolone trimethylacetate; 11β,17α,21-trihydroxy-1,4-pregnadiene-3,20-dione 21-pivalate.
Formulae:	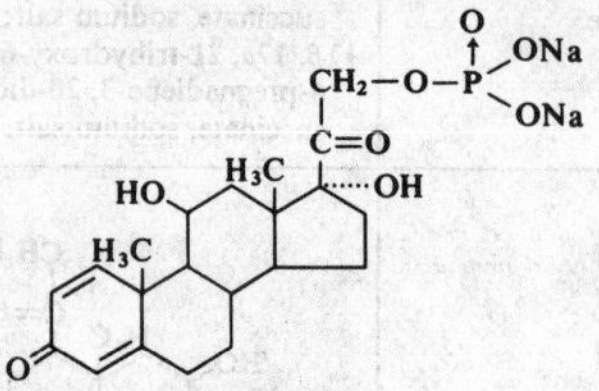 $C_{21}H_{27}Na_2O_8P$	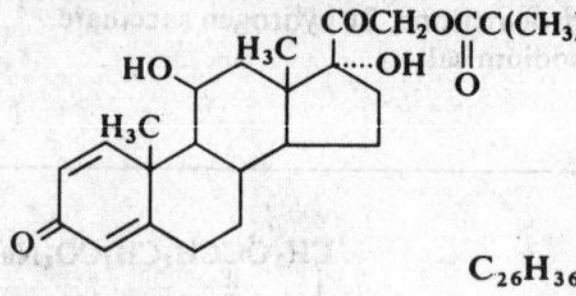 $C_{26}H_{36}O_6$
Molecular weight	484.4	444.6
Melting point (°C)		229
Specific rotation	$(\alpha)_D^{25}$ +102.5 (100 mg. in 10 ml. H_2O)	+108 ±4 (1 % in dioxane)
Absorption max.	243 mμ	240 and 263 mμ (in absolute ethanol)

Names & synonyms:	PREDNISONE; metacortandricin; Δ^1-dehydrocortisone; delta E; Δ^1-cortisone; 1,4-pregnadiene-17α,21-diol-3,11,20-trione; 17α,21-dihydroxy-1,4-pregnadiene-3,11,20-trione.	TRIAMCINOLONE; 9α-fluoro-16α-hydroxyprednisolone; 9α-fluoro-11β,16α,17α,21-tetrahydroxy-1,4-pregnadiene-3,20-dione.
Formulae:	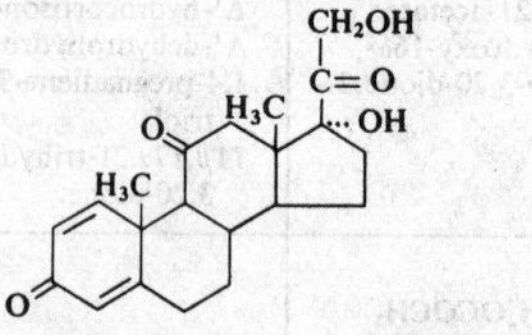 $C_{21}H_{26}O_5$	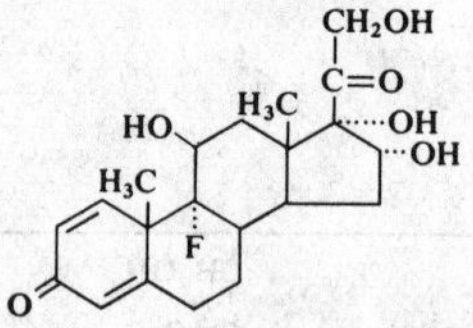 $C_{21}H_{27}FO_6$
Molecular weight	358.4	394.4
Melting point (°C)	225 (dec.)	260–262.5 (dec.)
Specific rotation	$(\alpha)_D^{25}$ +167 – +175 (100 mg. in 10 ml. dioxane)	$(\alpha)_D^{25}$ +75 (200 mg. in 100 ml. acetone)
Absorption max.	239 mμ (ε = 15,500) methanol	238 mμ (ε = 15,800)

ADRENAL CORTICOSTEROIDS, INCLUDING ANTIINFLAMMATORY, ANTIALLERGIC AND ANTIRHEUMATIC AGENTS (Continued)

Names & synonyms:	TRIAMCINOLONE ACETONIDE; 9α-fluoro-11β,21-dihydroxy-16α,17α-isopropylidenedioxy-1,4-pregnadiene-3,20-dione; 9α-fluoro-16α-hydroxyprednisolone 16,17-acetonide.	TRIAMCINOLONE DIACETATE; 16α,21-diacetoxy-9α-fluoro-11β,17α-dihydroxy-1,4-pregnadiene-3,20-dione; 9α-fluoro-16α-hydroxyprednisolone 16,21-diacetate.
Formulae:	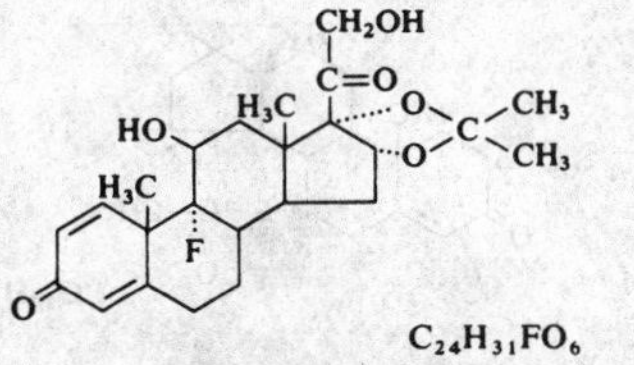 $C_{24}H_{31}FO_6$	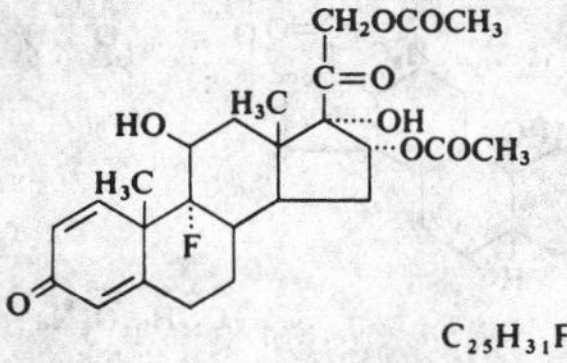 $C_{25}H_{31}FO_8$
Molecular weight	434.4	478.49
Melting point (°C)	274–278 (dec.); 292–294	variable: 158–235
Specific rotation	$(\alpha)^{25}_{D}$ +109 – +112 (53.7 mg. in 10 ml. chloroform)	$(\alpha)^{25}_{D}$ +22 (78.8 mg. in 10 ml. chloroform)
Absorption max.	238–239 mμ (ε = 14,600)	239 mμ (ε = 15,200)

ANDROGENS AND ANABOLIC AGENTS

Names & synonyms:	ANDROSTERONE; cis-androsterone; 3α-hydroxy-17-androstanone; androstane-3α-ol-17-one.	FLUOXYMESTERONE; 9α-fluoro-11β-hydroxy-17α-methyltestosterone 9α-fluoro-11β,17β-dihydroxy-17α-methyl-4-androsten-3-one.	ALDOSTERONE; electrocortin; 18-oxocorticosterone; 18-formyl-11β,21-dihydroxy-4-pregnene-3,20-dione.
Formulae:	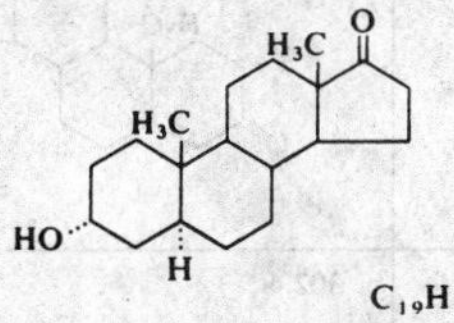 $C_{19}H_{30}O_2$	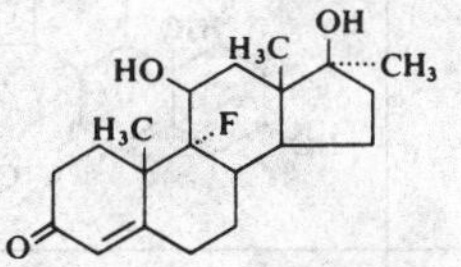 $C_{20}H_{29}FO_3$	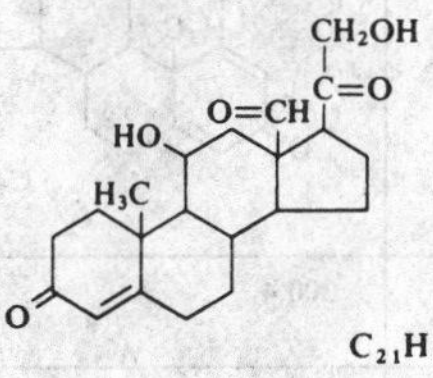 $C_{21}H_{28}O_5$
Molecular weight	290.4	336.4	360.4
Melting point (°C)	185–185.5	270 (dec.)	108–112 (hydrate); 164 (anhydrous)
Specific rotation	$(\alpha)^{15}_{D}$ +85 – +90 (150 mg. in 10 ml. dioxane)	$(\alpha)^{25}_{D}$ +107 – +109 (alcohol)	$(\alpha)^{25}_{D}$ +161 (10 mg. in 10 ml. chloroform)
Absorption max.		240 mμ (ε = 16,700) alcohol	240 mμ (log ε = 4.20 monohydr.; ε mol. 15,000 anhydr.)

STEROID HORMONES AND OTHER STEROIDAL SYNTHETICS (Continued)

ANDROGENS AND ANABOLIC AGENTS (Continued)

Names & synonyms:	HYDROXYDIONE SODIUM; 21-hydroxypregnane-3,20-dione-21-sodium hemisuccinate.	SPIRONOLACTONE: 3-(3-oxo-7α-acetylthio-17β-hydroxy-4-androsten-17α-yl)-propionic acid γ lactone.
Formulae:	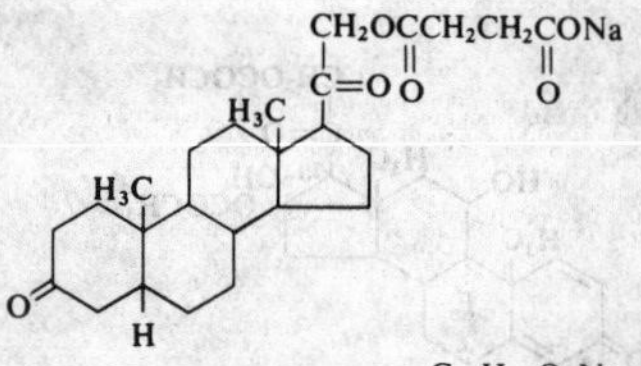 $C_{25}H_{35}O_6Na$	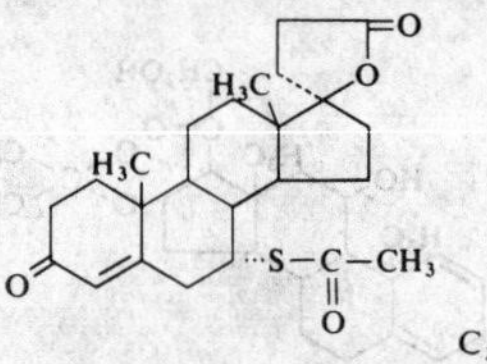 $C_{24}H_{32}O_4S$
Molecular weight	454.5	416.5
Melting point (°C)	193–203 (dec.)	135 (preliminary)–202 (dec.)
Specific rotation	$(\alpha)\frac{25}{D}+95$ (chloroform) for free acid.	$(\alpha)\frac{25}{D}-34$ (chloroform)
Absorption max.	280 mμ (ε = 93.2)	ε^{238} = 20,200

Names & synonyms:	METHANDROSTENOLONE; 17α-methyl-17β-hydroxy-1,4-androstadien-3-one.	METHYLANDROSTENEDIOL; MAD; methandriol; 17α-methyl-5-androsten-3β,17β-diol.	METHYL TESTOSTERONE; 17-methyl testosterone; 17α-methyl-Δ^4-androsten-17-β-ol-3-one; 17(β)-hydroxy-17(α-methyl-4-androsten-3-one.
Formulae:	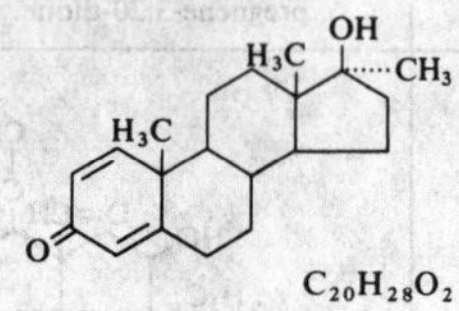 $C_{20}H_{28}O_2$	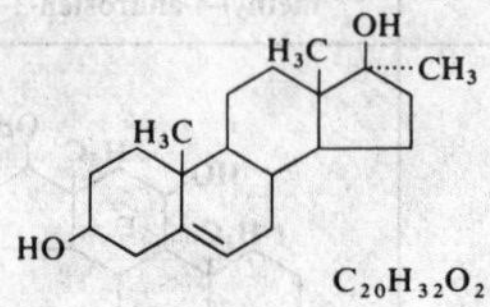 $C_{20}H_{32}O_2$	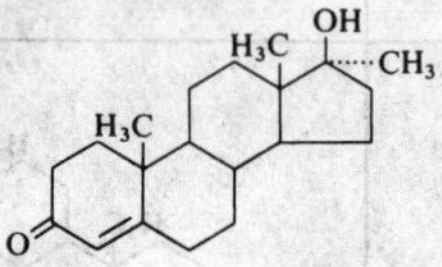 $C_{20}H_{30}O_2$
Molecular weight	300.4	304.4	302.4
Melting point (°C)	166–167	205–207	161–166
Specific rotation	$(\alpha)\frac{20}{D}+9-+17$ (100 mg. in 10 ml. alcohol)	$(\alpha)\frac{20}{D}-73$ (100 mg. in 10 ml. alcohol)	$(\alpha)\frac{25}{D}+69-+75$ (100 mg. in 10 ml. dioxane)
Absorption max.			

ANDROGENS AND ANABOLIC AGENTS (Continued)

Names & synonyms:	NORETHANDROLONE; 17α-ethyl-19-nortestosterone; 17α-ethyl-17-hydroxy-4-norandrosten-3-one; 17α-ethyl-17-hydroxy-19-norandrost-4-en-3-one.	OXANDROLONE; 17β-hydroxy-17α-methyl-2-oxa-5α-androstane-3-one.
Formulae:	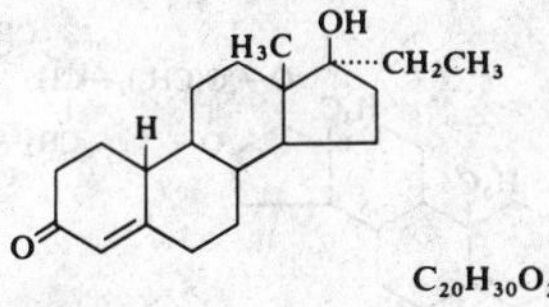 $C_{20}H_{30}O_2$	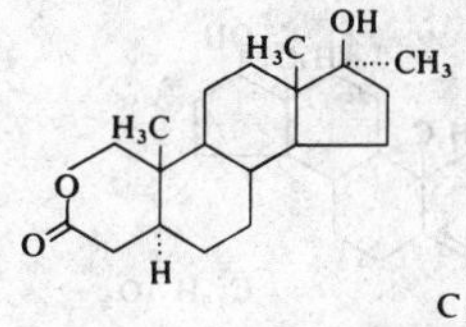 $C_{19}H_{30}O_3$
Molecular weight	302.4	306.4
Melting point (°C)	130–136	230–233
Specific rotation	$(\alpha)\frac{25}{D}$ +21 (dioxane)	$(\alpha)\frac{25}{D}$ −21 (1 % in chloroform)
Absorption max.	240 mμ (ε = 16,500)	None

Names & synonyms:	OXYMETHOLONE; 17β-hydroxy-2-hydroxymethylene-17α-methyl-3-androstanone; 2-hydroxymethylene-17-α-methyl dihydrotestosterone.	PROMETHOLONE; 2α-methyl-dihydro-testosterone propionate; 2α-methyl-5α-androstane-17β-ol-3-one-propionate.
Formulae:	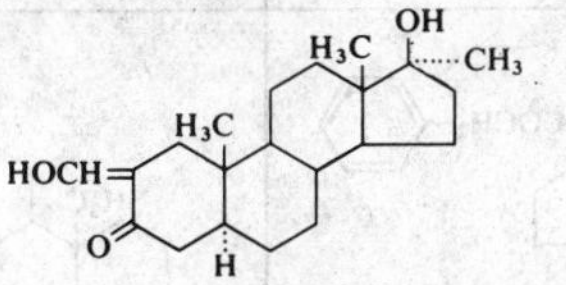 $C_{21}H_{32}O_3$	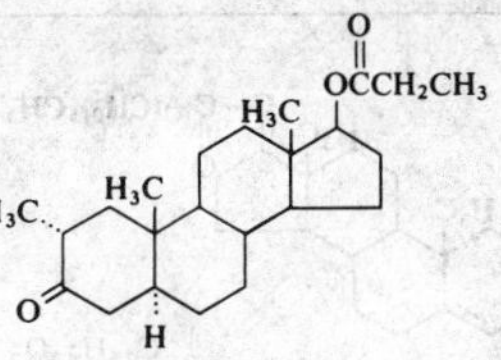
Molecular weight	332.4	360.5
Melting point (°C)	182	124–130
Specific rotation	$(\alpha)\frac{25}{D}$ = +36 (200 mg. in 10 ml. dioxane)	$(\alpha)\frac{25}{D}$ +22 – +29 (200 mg. in 10 ml. chloroform)
Absorption max.	E_1^1 = 547 at 315 mμ (in alkaline methanol made 0.01 N with NaOH)	without significant absorption from 220–300 mμ (methanol)

ANDROGENS AND ANABOLIC AGENTS (Continued)

Names & synonyms:	TESTOSTERONE; trans-testosterone; Δ^4-androsten-17-β-ol-3-one; 17β-hydroxy-4-androsten-3-one.	TESTOSTERONE CYPIONATE; testosterone cyclopentylpropionate; 17β-hydroxy-4-androsten-3-one, cyclopentanepropionate.
Formulae:	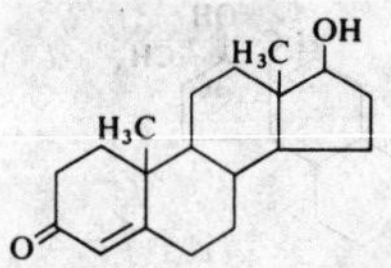 $C_{19}H_{28}O_2$	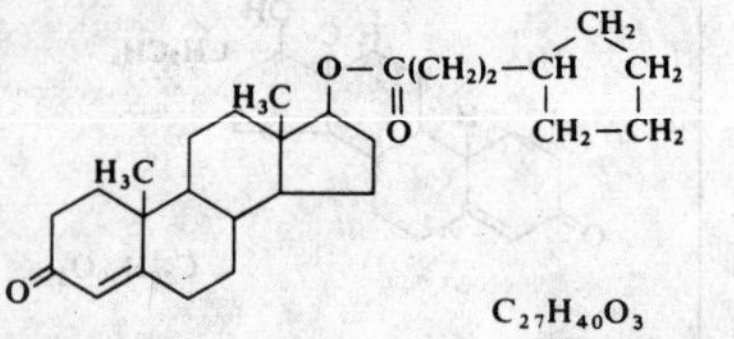 $C_{27}H_{40}O_3$
Molecular weight	288.4	412.6
Melting point (°C)	151–156	100–102
Specific rotation	$(\alpha)\frac{24}{D}$ +109 (400 mg. in 10 ml. alcohol)	$(\alpha)_D$ +88.5 ±3.5 ($CHCl_3$)
Absorption max.	238 mμ	λ_{max} 241 mμ (ε 16,125)

Names & synonyms:	TESTOSTERONE ENANTHATE; testosterone heptanoate; 17β-hydroxyandrost-4-en-3-one-17-enanthate.	TESTOSTERONE PHENYLACETATE; 17β-hydroxy-4-androsten-3-one phenyl-acetate; testosterone α-toluate.	TESTOSTERONE PROPIONATE; Δ^4-androstene-17-β-propionate-3-one.
Formulae:	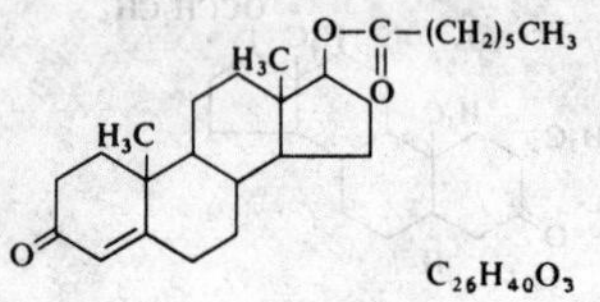 $C_{26}H_{40}O_3$	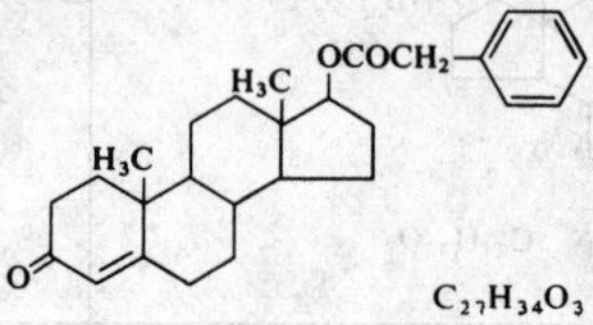 $C_{27}H_{34}O_3$	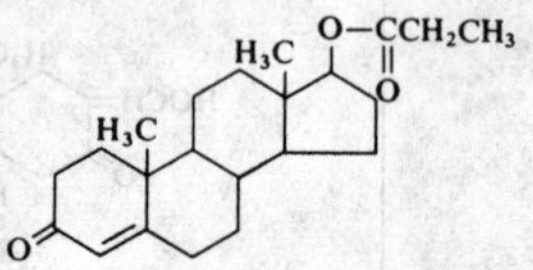 $C_{22}H_{32}O_3$
Molecular weight	400.6	406.5	344.4
Melting point (°C)	34–39	129–131	118–122
Specific rotation	$(\alpha)\frac{25}{D}$ +77 – +82 (2 % in dioxane)	$(\alpha)\frac{25}{D}$ +101 ±3 (1 % in chloroform)	$(\alpha)\frac{25}{D}$ +83 – +90 (100 mg. in 10 ml. dioxane)
Absorption max.	241 mμ (in ethanol)	241 mμ (in ethanol)	

STEROID HORMONES AND OTHER STEROIDAL SYNTHETICS (Continued)

ESTROGENS

Names & synonyms:	EQUILENIN; 3-hydroxy-17-keto-$\Delta^{1,3,5-10,6,8}$ estrapentaene; 1,3,5–10,6,8-estrapentaen-3-ol-17-one.	EQUILIN; 3-hydroxy-17-keto-$\Delta^{1,3,5-10,7}$ estratetraene; 1,3,5,7-estratetraen-3-ol-17-one.	ESTRADIOL (formerly called α-estradiol); β-estradiol; dihydrofolliculin; dihydroxyestrin; 1,3,5-estratriene-3,17β-diol; 3,17-dihydroxy-$\Delta^{1,3,5-10}$-estratriene; 3,17-epidihydroxyestratriene.
Formulae:	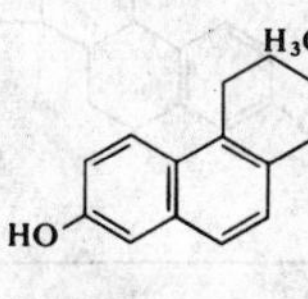 $C_{18}H_{18}O_2$	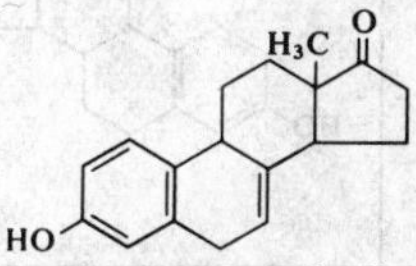 $C_{18}H_{20}O_2$	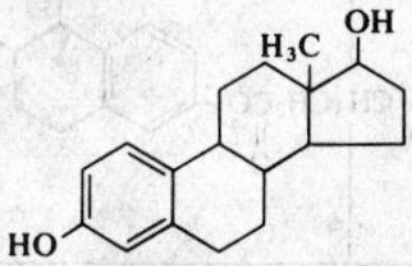 $C_{18}H_{24}O_2$
Molecular weight	266.3	268.3	272.3
Melting point (°C)	258–259	236–240	173–179
Specific rotation	$(\alpha)\frac{25}{D}+89$ (dioxane)	$(\alpha)\frac{25}{D}+308$ (200 mg. in 10 ml. dioxane); +325 (200 mg. in 10 ml. alcohol).	$(\alpha)\frac{25}{D}+76-+83$ (100 mg. in 10 ml. dioxane)
Absorption max.	231, 270, 282, 292, 325, 340 mμ	283–285 mμ	225, 280 mμ

Names & synonyms:	ESTRADIOL BENZOATE; β-estradiol-3-benzoate; estradiol monobenzoate.	ESTRADIOL CYPIONATE; estradiol cyclopentylpropionate; β-estradiol 17-cyclopentanepropionate; 1,3,5(10)-estratriene-3,17β-diol,17-cyclopentanepropionate.
Formulae:	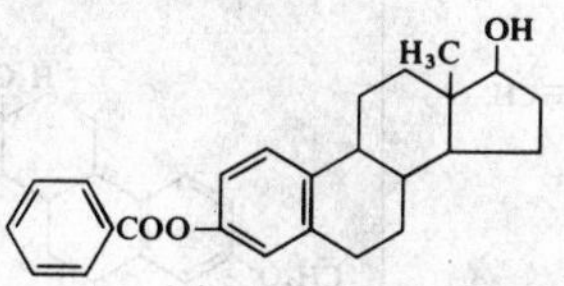 $C_{25}H_{28}O_3$	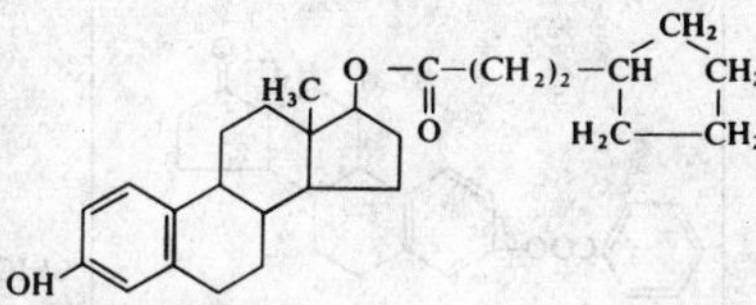 $C_{26}H_{36}O_3$
Molecular weight	376.4	396.6
Melting point (°C)	191–196	151–154
Specific rotation	$(\alpha)\frac{25}{D}+58-+63$ (200 mg. in 10 ml. dioxane)	$(\alpha)_D+41.5\pm3.5$ (dioxane)
Absorption max.		223 mμ

ESTROGENS (Continued)

Names & synonyms:	ESTRADIOL DIPROPIONATE; α-estradiol dipropionate; 17β-estradiol dipropionate.	ESTRIOL; trihydroxyestrin; $\Delta^{1,3,5-10}$-estratriene-3-16-cis-17-trans-diol; 1,3,5-estratriene-3,16α,17β-triol.	ESTRONE; folliculin; ketohydroxyestrin; 1,3,5-estratrien-3-ol-17-one.
Formulae:	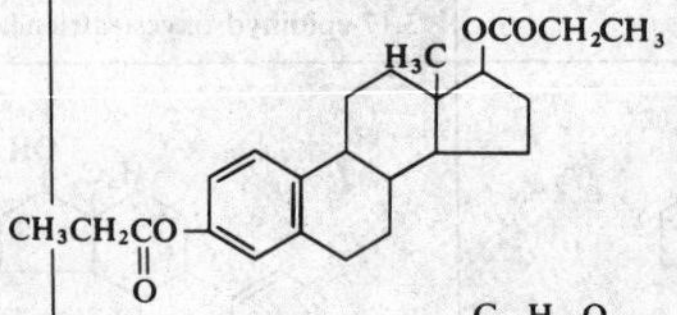 $C_{24}H_{32}O_4$	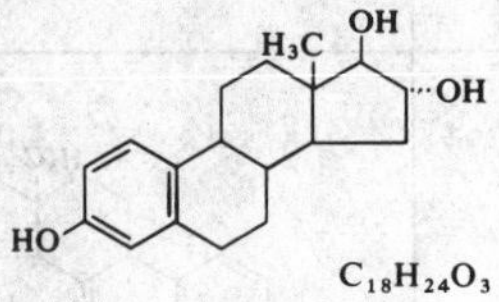 $C_{18}H_{24}O_3$	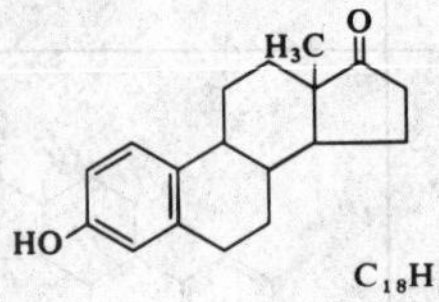 $C_{18}H_{22}O_2$
Molecular weight	384.5	288.3	270.3
Melting point (°C)	104–109	282	258–262
Specific rotation	$(\alpha)\frac{25}{D}$ +39 ±2 (1% in dioxane)	$(\alpha)\frac{25}{D}$ +53 – +63 (40 mg. in 1 ml. dioxane)	$(\alpha)\frac{25}{D}$ +158 – +168 (100 mg. in 10 ml. dioxane)
Absorption max.	268 mμ	280 mμ	283–285 mμ

Names & synonyms:	ESTRONE BENZOATE	ETHYNYL ESTRADIOL; 17-ethinyl estradiol; 17α-ethynyl-1,3,5-estratriene-3,17β-diol.	MESTRANOL; ethynylestradiol 3-methyl ether; 3-methoxy-17α-ethynyl-1,3,5(10)-estratriene-17β-ol; 17α-ethynyl-estradiol-3-methyl ether; 3-methoxy-19-nor-17α-pregna-1,3,5,trien-20-yn-17-ol.
Formulae:	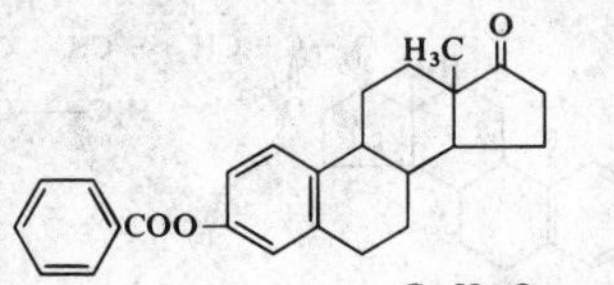 $C_{25}H_{26}O_3$	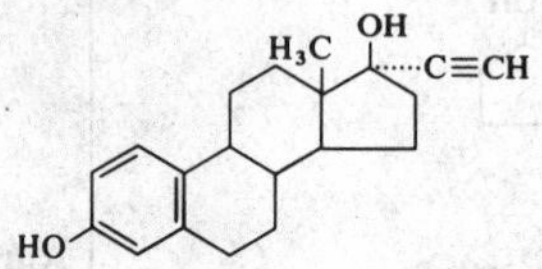 $C_{20}H_{24}O_2$	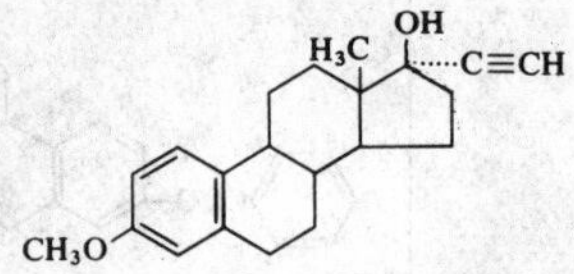 $C_{21}H_{26}O_2$
Molecular weight	374.4	296.4	310.4
Melting point (°C)	220	141–146	148–154
Specific rotation	$(\alpha)\frac{25}{D}$ +120 (dioxane)	$(\alpha)\frac{25}{D}$ +1 – +10 (100 mg. in 10 ml. dioxane)	$(\alpha)\frac{25}{D}$ +2 to +8 (200 mg. in 10 ml. dioxane)
Absorption max.		248 mμ	278 to 287 mμ (methanol)

PROGESTOGENS AND PROGESTINS (INCLUDING 19-NORSTEROID COMPOUNDS)

Names & synonyms:	ACETOXYPREGNENOLONE; 21-acetoxypregnenolone; prebediolone acetate; Δ^5-pregnene-3β,21-diol-20-one-21-monoacetate; 21-acetoxy-5-pregnene-3-ol-20-one; 3-hydroxy-21-acetoxy-5-pregnen-20-one.	ANAGESTONE ACETATE; 6α-methyl-4-pregnen-17α-ol-20-one acetate; 17α-acetoxy-6α-methylpregn-4-en-20-one; 17α-acetoxy-6α-methyl-4-pregnen-20-one.	CHLORMADINONE ACETATE; 6-chloro-Δ^6-dehydro-17α-acetoxyprogesterone; 6-chloro-$\Delta^{4,6}$-pregnadiene-17α-ol-3,20-dioneacetate.
Formulae:	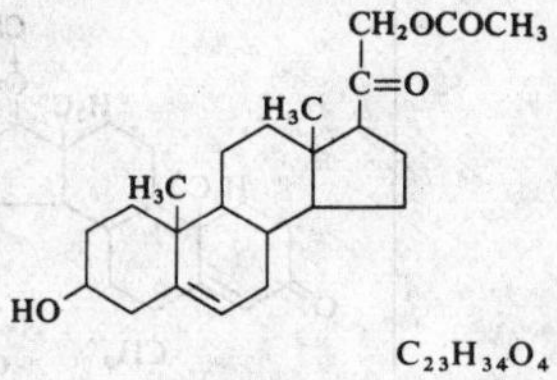 $C_{23}H_{34}O_4$	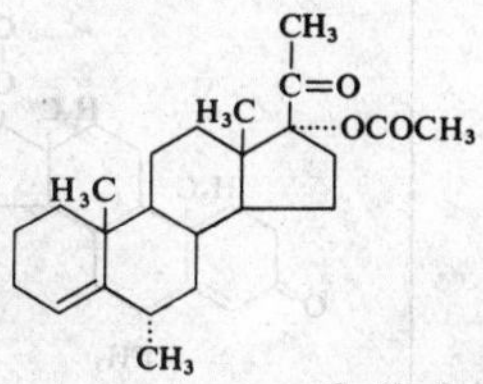 $C_{24}H_{36}O_3$	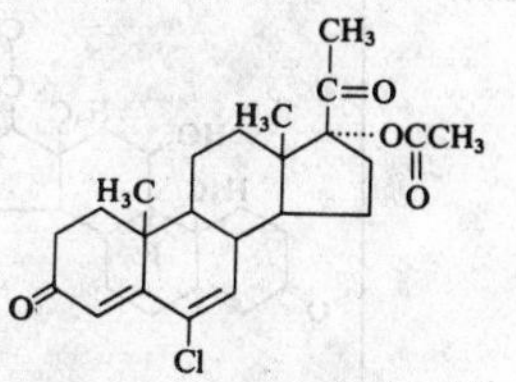 $C_{23}H_{29}ClO_4$
Molecular weight	374.5	372.6	404.9
Melting point (°C)	184–185	172–178	204–212
Specific rotation	$(\alpha)\frac{20}{D}$ +37 – +43 (dioxane)	$(\alpha)\frac{25}{D}$ +40 to +45 (10 mg. in 10 ml. chloroform)	$(\alpha)\frac{25}{D}$ 0 to −6 (200 mg. in 10 ml. chloroform)
Absorption max.			284 mμ (methanol) Log ε = 4.34 ±0.02

Names & synonyms:	DIMETHISTERONE; 6α,21-dimethylethisterone; 6α,21-dimethyl-17β-hydroxy-17α-pregn-4-en-20-yn-3-one; 6α-methyl-17α-propynylandrost-4-en-17β-ol-3-one; 17β-hydroxy-6α-methyl-17α-(prop-1-ynyl)-androst-4-ene-3-one.	ETHISTERONE; anhydrohydroxyprogesterone; ethinyl testosterone; pregneninolone; 17α ethynyl testosterone; 17α-ethynyl-17β-hydroxy-4-androsten-3-one.	ETHYNODIOL DIACETATE; 17α-ethynyl-4-estrene-3β,17β-diol-17-diacetate; 19-nor-17α-pregn-4-en-20-yne-3β,17-diol diacetate.
Formulae:	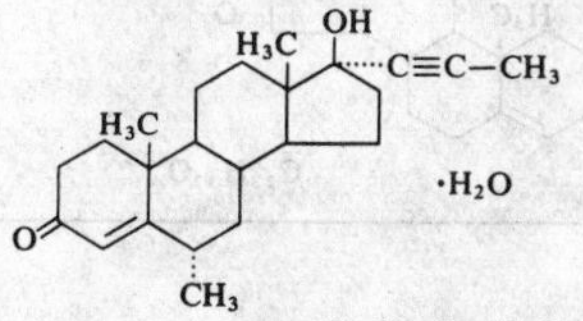 $C_{23}H_{32}O_2.H_2O$	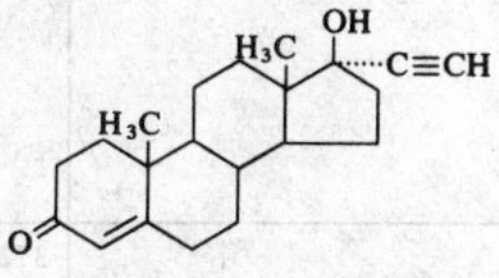 $C_{21}H_{28}O_2$	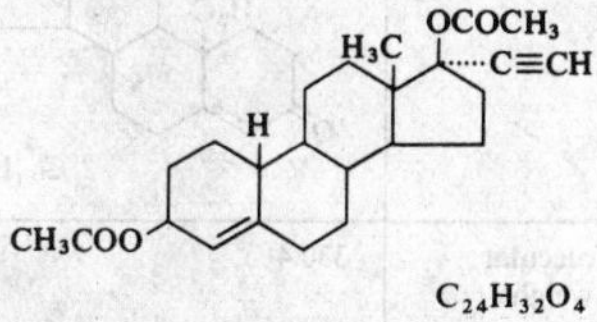 $C_{24}H_{32}O_4$
Molecular weight	358.5	312.4	384.5
Melting point (°C)	App. 100 (dec.)	266–273	126–132
Specific rotation	$(\alpha)\frac{20}{D}$ +16.5 to +18.5 (2 % solution in chloroform) (calculated to the anhydrous basis)	$(\alpha)\frac{25}{D}$ −32° (100 mg. in 10 ml. pyridine)	$(\alpha)\frac{25}{D}$ −74 (1 % in chloroform)
Absorption max.	App. 240 mμ (anhydrous ethanol) $E_{1\,cm}^{1\%}$ = 443	241 mμ (methanol)	None

PROGESTOGENS AND PROGESTINS (INCLUDING 19-NORSTEROID COMPOUNDS) (Continued)

Names & synonyms:	FLUROGESTONE ACETATE; 17α-acetoxy-9α-fluoro-11β-hydroxy-4-pregnene-3,20-dione.	HYDROXYMETHYLPROGESTERONE; medroxyprogesterone; 17α-hydroxy-6α-methylprogesterone; 17α-hydroxy-6α-methyl-4-pregnene-3,20-dione.	HYDROXYMETHYLPROGESTERONE ACETATE; medroxyprogesterone acetate; 17α-hydroxy-6α-methylprogesterone acetate; 17α-hydroxy-6α-methyl-4-pregnene-3,20-dione acetate.
Formulae:	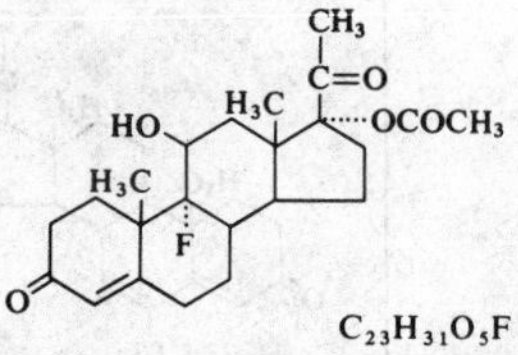 $C_{23}H_{31}O_5F$	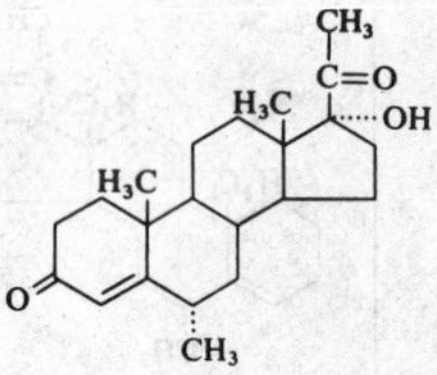 $C_{22}H_{32}O_3$	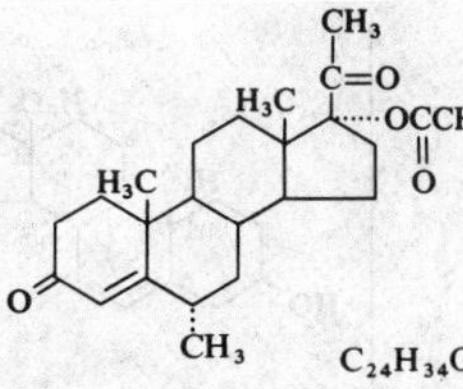 $C_{24}H_{34}O_4$
Molecular weight	406.5	344.5	386.5
Melting point (°C)	250–251	220–223.5	202–207
Specific rotation	$(\alpha)\frac{25}{D}+78$	$(\alpha)\frac{25}{D}+75$	$(\alpha)\frac{25}{D}+51$ (dioxane)
Absorption max.	238 mμ (ε = 17,100)	241 mμ (ε = 16,150)	241 mμ (α_M = 16,500) ethanol

Names & synonyms:	HYDROXYPROGESTERONE; 17α-hydroxyprogesterone; 17α-hydroxy-4-pregene-3,20 dione; 4-pregnen-17α-ol-3,20-dione.	HYDROXYPROGESTERONE ACETATE; 17α-acetoxyprogesterone; 17α-hydroxyprogesterone acetate; 17α-hydroxy-4-pregnene-3,20 dione acetate.
Formulae:	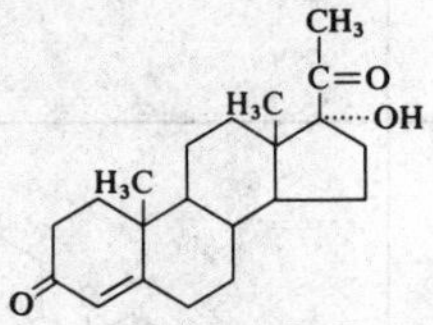 $C_{21}H_{30}O_3$	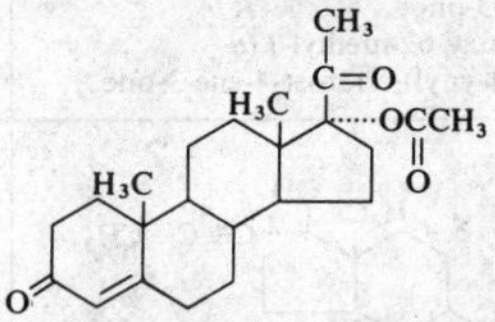 $C_{23}H_{32}O_4$
Molecular weight	330.4	372.5
Melting point (°C)	276	249–250
Specific rotation	$(\alpha)\frac{17}{D}+105$ (104 mg. in 10 ml. chloroform)	$(\alpha)\frac{25}{D}+72$ (chloroform)
Absorption max.		240 mμ (a_M = 16,875) ethanol

PROGESTOGENS AND PROGESTINS (INCLUDING 19-NORSTEROID COMPOUNDS) (Continued)

Names & synonyms:	HYDROXYPROGESTERONE CAPROATE; 17α-hydroxyprogesterone caproate; 17α-hydroxy-4-pregnene-3,20-dione caproate.	MELENGESTROL ACETATE: MGA; 17α-hydroxy-6-methyl-16-methylene-4,6-pregnadiene-3, 20-dione acetate; 6-dehydro-17-hydroxy-6-methyl-16-methylene-progesterone acetate.
Formulae:	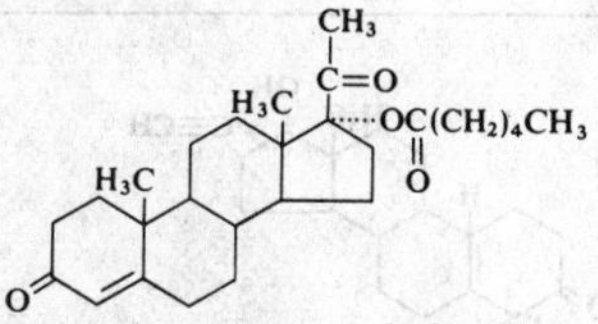 $C_{27}H_{40}O_4$	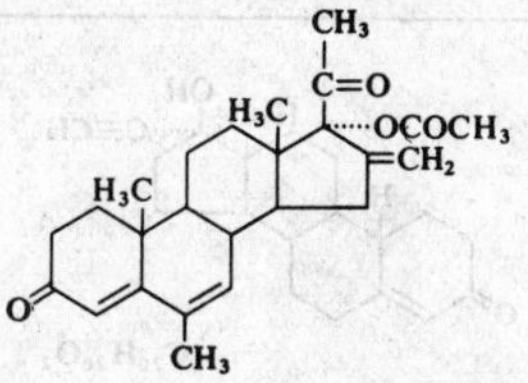 $C_{25}H_{32}O_4$
Molecular weight	428.6	396.51
Melting point (°C)	121–123	215–227
Specific rotation	$(\alpha)\frac{25}{D}+57$ (chloroform)	$(\alpha)_D -127$ to -135 (in $CHCl_3$)
Absorption max.		288 mμ ($\varepsilon_1^1 = 24{,}000$) (ethanol)

Names & synonyms:	NORETHINDRONE; Norethisterone; 17α-ethynyl-19-nortestosterone; 17α-ethynyl-17-hydroxy-19-nor-17α-4-en-20-yn-3-one.	NORETHINDRONE ACETATE; 17α-ethinyl-19-nortestosterone acetate.
Formulae:	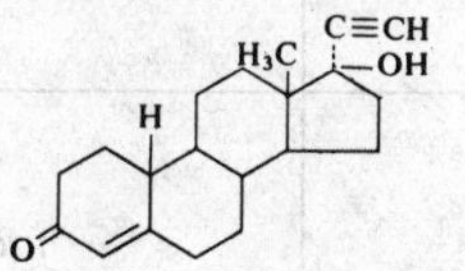 $C_{20}H_{26}O_2$	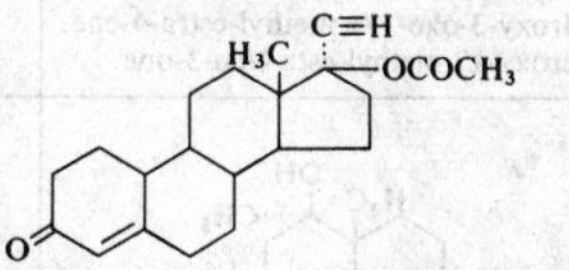 $C_{22}H_{28}O_3$
Molecular weight	298.4	340.4
Melting point (°C)	202 and 208	157–163
Specific rotation	$(\alpha)\frac{25}{D}-30 - -35$ (200 mg. in 10 ml. dioxane)	$(\alpha)\frac{25}{D}-32 - -35$ (200 mg. in 10 ml. dioxane)
Absorption max.	α (1 %, 1 cm) λ 240 = 535 ± 15	α (1 %, 1 cm.) λ 240 = 490 to 520 (505 ± 15) (ethanol)

PROGESTOGENS AND PROGESTINS (INCLUDING 19-NORSTEROID COMPOUNDS) (Continued)

Names & synonyms:	**NORETHISTERONE;** norethindrone; 19-norethisterone; 17α-ethynyl-19-nor-Δ^4-androstan-17β-ol-3-one; 17α-ethynyl-19-nor-testosterone; 17-hydroxy-3-oxo-19-nor-17α-pregn-4-ene-20-yne; 17-hydroxy-19-nor-17α-pregn-4-en-20-yn-3-one.	**NORETHYNODREL;** 17α-ethynyl-17β-hydroxy-5(10)-estren-3-one.
Formulae:	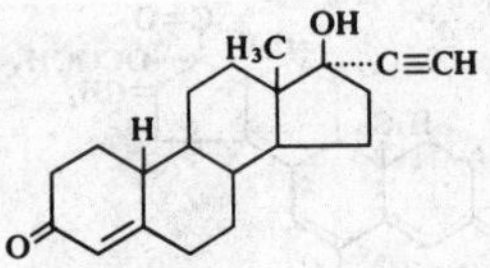 $C_{20}H_{26}O_2$	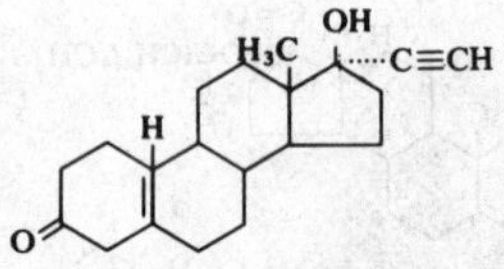 $C_{20}H_{26}O_2$
Molecular weight	298.4	298.4
Melting point (°C)	200–207	174–184
Specific rotation	$(\alpha)\frac{25}{D}$ −30 – −38 (200 mg. in 10 ml. dioxane)	$(\alpha)\frac{25}{D}$ +125 (dioxane)
Absorption max.	240 mμ (ε_1^1 = 576)	

Names & synonyms:	**NORMETHISTERONE;** 19-normethisterone; normethandrolone; metalutin; normetandrone; 17α-methyl-19-nor-Δ^4-androsten-17β-ol-3-one; 17α-methyl-19-nor-testosterone; 17β-hydroxy-3-oxo-17α-methyl-estra-4-ene; 17β-hydroxy-17-methyl-estr-4-en-3-one.	**PREGNENOLONE;** Δ^5-pregnenolone; Δ^5-pregnen-3β-ol-20-one; 17β(1-ketoethyl)-Δ^5-androstene-3β-ol.	**PROGESTERONE;** progestin; progestone; pregnendione; Δ^4-pregnene-3,20-dione.
Formulae:	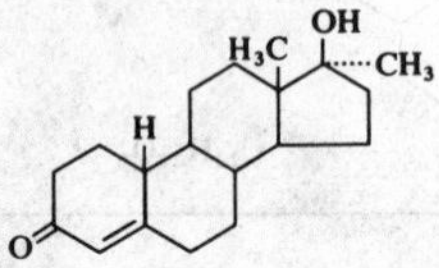 $C_{19}H_{28}O_2$	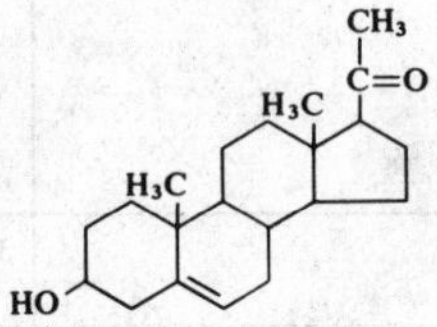 $C_{21}H_{32}O_2$	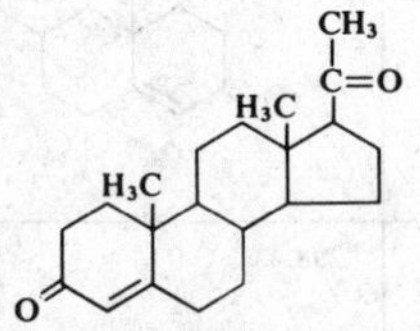 $C_{21}H_{30}O_2$
Molecular weight	288.4	308.4	314.4
Melting point (°C)	153–158	193	(β) isomer 121; (α) isomer 127–131
Specific rotation	$(\alpha)\frac{25}{D}$ +25 to +29 (200 mg. in 10 ml. chloroform)	$(\alpha)\frac{20}{D}$ +28 – +30 (alcohol)	$(\alpha)\frac{20}{D}$ +172 – +182 (200 mg. in 10 ml. dioxane)
Absorption max.	241 mμ – 565 ±15		240 mμ

DIURETIC, ANTIDIURETIC AND LOCAL ANESTHETIC AGENTS

Names & synonyms:	ALDOSTERONE; electrocortin; 18-oxocorticosterone; 18-formyl-11β,21-dihydroxy-4-pregnene-3,20-dione.	HYDROXYDIONE SODIUM; 21-hydroxypregnane-3,20-dione-21-sodium hemisuccinate.	SPIRONOLACTONE; 3-(3-oxo-7α-acetylthio-17β-hydroxy-4-androsten-17α-yl)-propionic acid γ lactone.
Formulae:	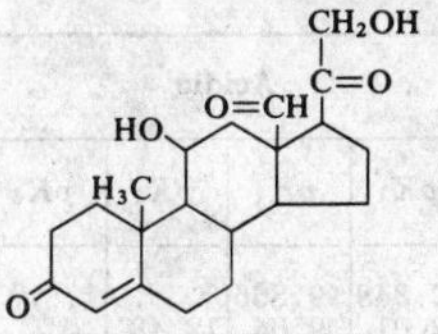 $C_{21}H_{28}O_5$	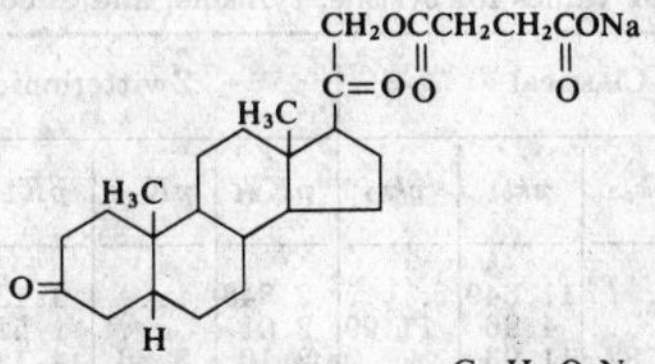 $C_{25}H_{35}O_6Na$	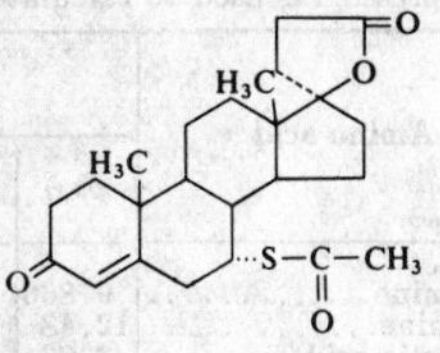 $C_{24}H_{32}O_4S$
Molecular weight	360.4	454.5	416.5
Melting point (°C)	108–112 (hydrate); 164 (anhydrous)	193–203 (dec.)	135; 202 (dec.)
Specific rotation	$(\alpha)\frac{25}{D}+161$ (10 mg. in 10 ml. chloroform)	$(\alpha)\frac{25}{D}+95$ (chloroform) for free acid.	$(\alpha)\frac{25}{D}-34$ (chloroform)
Absorption max.	240 mμ (log ε = 4.20 monohydr.; ε mol. 15,000 anhydr.)	280 mμ (ε = 93.2)	ε^{238} = 20,200

PROPERTIES OF THE AMINO ACIDS

IONIZATION CONSTANTS AND pH VALUES AT THE ISOELECTRIC POINTS OF THE AMINO ACIDS IN WATER AT 25°C

The majority of the recorded values are true thermodynamic constants calculated from electrometric force measurements of cells without liquid junctions. The values for the constants given in the table were derived from the classical, the zwitterionic (Bjerrum), and the acidic (Brönsted) formulations of ionization and the corresponding mass law expressions. pH values at the isoelectric points were calculated from the expression, $pI = \frac{1}{2}(pk_{a1} + pk_w - pk_{b1})$. The error is approximately 0.5 per cent when this expression is used to calculate pI values for cystine, tyrosine, and diiodotyrosine.

Amino acid	Classical				Zwitterionic				Acidic				pI	Ref. no.
	pk_{a1}	pk_{a2}	pk_{b1}	pk_{b2}	pK_{A1}	pK_{A2}	pK_{B1}	pP_{B2}	pK_1	pK_2	pK_3	pK_4		
DL-Alanine	9.866	.....	11.649	.....	2.348	....	4.131	....	2.348	9.866	.....	....	6.107	1
L-Arginine	12.48	.....	4.96	11.99	2.01	....	1.52	4.96	2.01	9.04	12.48	....	10.76	2
L-Aspartic acid	3.86	9.82	11.93	.....	2.10	3.86	4.18	....	2.10	3.86	9.82	....	2.98	3
L-Cystine	8.00	10.25	11.95	12.96	1.04	2.05	3.75	6.00	1.04	2.05	8.00	10.25	5.02	4
Diiodo-L-tyrosine	6.48	7.82	11.88	.....	2.12	6.48	6.18	....	2.12	6.48	7.82	....	4.29	5,6
L-Glutamic acid	4.07	9.47	11.90	.....	2.10	4.07	4.53	....	2.10	4.07	9.47	....	3.08	7
Glycine	9.778	.....	11.647	.....	2.350	....	4.219	....	2.350	9.778	.....	....	6.064	8
L-Histidine	9.18	.....	7.90	12.23	1.77	....	4.82	7.90	1.77	6.10	9.18	....	7.64	3
Hydroxy-L-proline	9.73	.....	12.08	.....	1.92	....	4.27	....	1.92	9.73	.....	....	5.82	9
DL-Isoleucine	9.758	.....	11.679	.....	2.318	....	4.239	....	2.318	9.758	.....	....	6.038	1
DL-Leucine	9.744	.....	11.669	.....	2.328	....	4.253	....	2.328	9.744	.....	....	6.036	1
L-Lysine	10.53	.....	5.05	11.82	2.18	....	3.47	5.05	2.18	8.95	10.53	....	9.47	2
DL-Methionine	9.21	.....	11.72	.....	2.28	....	4.79	....	2.28	9.21	.....	....	5.74	10
DL-Phenylalanine	9.24	.....	11.42	.....	2.58	....	4.76	....	2.58	9.24	.....	....	5.91	11
L-Proline	10.60	.....	12.0	.....	2.00	....	3.40	....	2.00	10.60	.....	....	6.3	12
DL-Serine	9.15	.....	11.79	.....	2.21	....	4.85	....	2.21	9.15	.....	....	5.68	9
L-Tryptophan	9.39	.....	11.62	.....	2.38	....	4.61	....	2.38	9.39	.....	....	5.88	13
L-Tyrosine	9.11	10.07	11.80	.....	2.20	9.11	3.93	....	2.20	9.11	10.07	....	5.63	6
DL-Valine	9.719	.....	11.711	.....	2.286	....	4.278	....	2.286	9.719	.....	....	6.002	1

References

1. Smith, P. K., Taylor, A. C., and Smith, E. R. B., J. Biol. Chem., **122**, 109 (1937–38).
2. Schmidt, C. L. A., Kirk, P. L., and Appleman, W. K., J. Biol. Chem., **88**, 285 (1930).
3. Greenstein, J. P., J. Biol. Chem., **93**, 479 (1931).
4. Borsook, H., Ellis, E. L., and Huffman, H. M., J. Biol. Chem., **117**, 281 (1937).
5. Dalton, J. B., Kirk, P. L., and Schmidt, C. L. A., J. Biol. Chem., **88**, 589 (1930).
6. Winnek, P. S., and Schmidt, C. L. A., J. Gen. Physiol., **18**, 889 (1935).
7. Simms, H. S., J. Gen. Physiol., **11**, 629 (1928); **12**, 231 (1928).
8. Owen, B. B., J. Am. Chem. Soc., **56**, 24 (1934).
9. Kirk, P. L., and Schmidt, C. L. A., J. Biol. Chem., **81**, 237 (1929).
10. Emerson, O. H., Kirk, P. L., and Schmidt, C. L. A., J. Biol. Chem., **92**, 449 (1931).
11. Miyamoto, S., and Schmidt, C. L. A., J. Biol. Chem., **90**, 165 (1931).
12. McCay, C. M., and Schmidt, C. L. A., J. Gen. Physiol., **9**, 333 (1926).
13. Schmidt, C. L. A., Appleman, W. K., and Kirk, P. L., J. Biol. Chem., **85**, 137 (1929–30).

Ionization Constants of the Amino Acids in Aqueous Ethanol Solutions

Amino acid	pK_1	pK_2	pK_3	Volume per cent ethanol	Temperature °C	Ref. No.
Alanine	3.55	10.02	.....	72	25	1
Arginine	3.34	9.40	14.1	72	25	1
Aspartic acid	2.85	5.20	10.51	72	25	1
Glutamic acid	3.16	5.63	10.75	72	25	2
Glycine	2.66	9.82	.....	10	19.5	2
	2.96	9.76	.....	40	19.5	2
	3.46	9.82	.....	72	25	1
	3.79	9.99	.....	90	19.5	2
Histidine	3.00	5.85	9.45	72	25	1
Isoleucine	3.69	9.81	.....	72	25	1
Lysine	2.75	8.95	10.53	48	25	1
	3.56	8.95	10.49	84	25	1
Proline	3.04	10.55	.....	72	25	1
Valine	3.60	9.73	.....	72	25	1

References

1. Jukes, T. H., and Schmidt, C. L. A., J. Biol. Chem., **105**, 359 (1934).
2. Michaelis, L., and Mizutani, M., Z. physik. Chem., **116**, 135 (1925).

Ionization Constants of the Amino Acids in Aqueous Formaldehyde Solution[a]

Amino acid	Mole per cent formaldehyde				
	0.99	3.95	5.60	10.0	17.9
DL-Alanine	8.36	7.42	6.96[b]	6.56	6.10
L-Arginine	....	3.45[c]	3.40[d]	....	....
L-Aspartic acid	....	......	7.21[d]	≷3.8[e] 6.85[f]	....
L-Glutamic acid	....	......	6.91[d]	≷4.2[e] 6.8[f]	....
Glycine	7.16	6.08	5.92[b]	5.34	5.04
L-Histidine	....	7.90[c]	7.90[d]	....	....
Hydroxy-L-Proline	....	......	7.19[d]	....	....
L-Leucine	8.44	7.50	6.92[d]	6.62	6.20
DL-Leucine	8.44	7.48	......	6.60	6.20
L-Lysine	....	7.35[c]	7.15[d]	....	....
L-Phenylalanine	....	......	6.62[d]	5.9[e]	....
DL-Phenylalanine	8.09	7.16	6.80[b]	6.35	6.13
L-Proline	....	......	7.78[d]	....	....
DL-Serine	6.66	5.74	5.63[b]	....	4.94
L-Tryptophan	....	......	6.88[d]	....	....
L-Tyrosine	....	......	7.50[d]	6.2[e] >9[f]	....
DL-Valine	8.52	7.65	7.47[b]	....	6.52

[a] Dunn and Weiner (1), pK_2 at 22°.
[b] Dunn and Loshakoff (2), pK_2 at 22°.
[c] Levy (3) pK_2 at 30° for arginine and pK_3 at 30° for histidine and lysine.
[d] Levy and Silberman (4), pK_2 at 30°, pK_3 at 30° for histidine and lysine.
[e] Harris (5), pK_2 at 25° for aspartic acid, glutamic acid, phenylalanine and tyrosine.
[f] Harris (5), pK_3 at 30° for aspartic acid, glutamic acid, and tyrosine.

References

1. Dunn, M. S., and Weiner, J. G., J. Biol. Chem., **117**, 381 (1937).
2. Dunn, M. S., and Loshakoff, A., J. Biol. Chem., **113**, 691 (1936).
3. Levy, M., J. Biol. Chem., **109**, 365 (1935).
4. Levy M., and Silberman, D. E., J. Biol. Chem., **118**, 723 (1937).
5. Harris, L. J., Proc. Roy. Soc. London, Series B, **95**, 440 (1923–24).

Specific Rotations of the Amino Acids Using Sodium Light (5893 Å)

Abbreviations

c—grams of solute per 100 ml. of solution.
d—density of the solution.
p—grams of solute per 100 grams of solution.
l—length of the tube in decimeters.
α—observed rotation in angular degrees.
[*α*]—specific rotation in angular degrees calculated from

$$[\alpha]^t_\lambda = \frac{\alpha \times 100}{c \times l} = \frac{\alpha \times 100}{p \times d \times l}$$

where *t* is temperature in °C and λ is wave length of the incident light in Ångstroms.

A—prepared from a protein or other naturally occurring material.
B—prepared by resolution of the inactive synthetic form.
C—prepared by resolution of the inactive racemized form.
D—prepared from the inactive synthetic form by a biological method.
E—prepared from the inactive racemized form by a biological method.
?—source not given.

Source	c	Solvent	d	p	Moles acid or base per mole amino acid	l	Temp. °C.	α	[α]	Ref. No.
				L-Alanine						
A	5.790	0.97 N HCl	1.033	5.605	1.5	2	15	+1.70	+14.7	1
A	10.3	Water	1.03	1.00	0	2	22	+0.55	+2.7	2
A	1.781	3 N NaOH	......	......	15	2	20	....	+3.0	3
				D-Alanine						
B	1.344	6 N HCl	......	......	39.4	2	30.4	−0.392	−14.6	4
				L-Arginine						
A	1.653	6.0 N HCl	......	......	63	4.001	23.4	+1.777	+26.9	5
A	3.48	Water	......	......	0	2	20	....	+12.5	6
A	0.87	0.50 N NaOH	......	......	10	2	20	....	+11.8	6
				L-Aspartic acid						
A	2.002	6.0 N HCl	......	......	39	4.001	24.0	+1.972	+24.6	7
A	1.3300	Water	......	......	0	3	18	....	+4.7	3
A	1.3300	3 N NaOH	......	......	30	3	18	....	−1.7	3
				D-Aspartic acid						
C	4.289	0.97 N HCl	1.032	4.156	3	1	20	−1.09	−25.5	8
				L-Cystine						
A	0.9974	1.02 N HCl	1.0181	0.9797	24.6	2	24.35	−4.277	−214.40	9
A	0.400	0.20 N NaOH	......	......	12	2	18.5	....	−70.0	3
				D-Cystine						
C	....	1 N HCl	......	1	24	.....	20	....	+223	10
				Diiodo-D-tyrosine						
A	5.08	1.1 N HCl	1.05	4.84	9.4	1	20	+0.15	+2.89	11
A	4.41	13.4 N NH_4OH	0.9779	4.51	132	1	20	+0.10	+2.27	11
				L-Glutamic acid						
A	1.002	6.0 N HCl	......	......	87	4.001	22.4	+1.25	+31.2	12
A	1.471	Water	......	......	0	2	18	....	+11.5	3
A	1.471	1 N NaOH	......	......	10	2	18	....	+10.96	3
				D-Glutamic acid						
C	5.425	0.37 N HCl	1.0233	5.3011	1	1	20	−1.63	−30.05	8
				L-Histidine						
A	1.480	6.0 N HCl	......	......	63	4.001	22.7	+0.766	+13.0	7
A	1.128	Water	1.0012	1.127	0	4	25.00	−1.714	−39.01	13
A	0.775	0.50 N NaOH	......	......	10	2	20	....	−10.9	6
				D-Histidine						
?	4.000	1.0 N HCl	......	......	4	1	20	−0.407	−10.2	14
B	2.66	Water	......	......	0	2	23	+2.11	+39.8	14
				Hydroxy-L-proline						
A	1.31	1.0 N HCl	......	......	10	2	20	....	−47.3	6
A	1.001	Water	......	......	0	4.001	22.5	−3.009	−75.2	7
A	0.655	0.50 N NaOH	......	......	10	2	20	....	−70.6	6
				Hydroxy-D-proline						
B	4.48	Water	1.03	4.35	0	1	21	+3.37	+75.2	16
				Allo-Hydroxy-L-proline						
B	2.617	Water	1.014	2.581	0	1	18	−1.52	−58.1	16
				Allo-Hydroxy-D-proline						
B	2.530	Water	1.013	2.998	0	1	17	+1.48	+58.5	16
				L-Isoleucine						
B	5.09	6.1 N HCl	1.098	4.64	15	1	20	+2.07	+40.61	17
B	3.10	Water	1.008	3.08	0	2	20	+0.70	+11.29	17
A	3.34	0.33 N NaOH	1.017	3.28	1.3	2	20	+0.74	+11.09	18
				D-Isoleucine						
B	4.53	6.1 N HCl	1.083	4.18	17	1	20	−1.85	−40.86	17
B	3.12	Water	1.006	3.10	0	2	20	−0.66	−10.55	17
				D-allo-Isoleucine						
D	5.14	6.0 N HCl	1.094	4.70	15.0	2	20	−3.80	−36.95	19
B	2.00	Water	......	......	0	1	20	−0.285	−14.2	20
				L-allo-Isoleucine						
B	3.97	6.0 N HCl	......	......	20	1	20	+1.50	+38.1	20
B	2.00	Water	......	......	0	1	20	+0.28	+14.0	20
				L-Leucine						
A	1.999	6.0 N HCl	......	......	38	4.001	25.9	+1.212	+15.1	5
A	2.001	Water	......	......	0	4.001	24.7	−0.863	−10.8	5
A	1.31	3.00 N NaOH	......	......	30	2	20	....	+7.6	3
				D-Leucine						
?	4.0	6.0 N HCl	1.1	3.664	19	2	20	+1.26	−15.6	21
?	....	Water	......	2.08	0	2	20	+0.43	+10.34	38
				L-Lysine						
A	2.00	6.0 N HCl	......	......	43	4	22.9	+1.652	+25.9	5
A	6.496	Water	......	......	0	2	20	+1.90	+14.6	22
				D-Lysine						
B	2.00	0.27 N HCl	......	......	2	2	20	−0.939	−23.48	23
				L-Methionine						
B	0.80	Water	......	......	0	2	25	−0.13	− 8.11	24
				D-Methionine						
B	0.80	0.2001 N HCl	......	......	4	2	25	−0.34	−21.18	24
B	0.80	Water	......	......	0	2	25	+0.13	+8.12	24
B	0.80	0.6 N $NaHCO_3$	......	......	11	2	25	−0.12	−7.47	24
				L-Phenylalanine						
B	1.936	Water	1.0040	1.928	0	2	20	−1.36	−35.14	27
				D-Phenylalanine						
B	3.814	5.4 N HCl	1.0895	3.501	23	2	20	+0.54	+7.07	28
B	2.043	Water	1.0045	2.034	0	2	20	+1.43	+35.0	27
				L-Proline						
A	0.575	0.50 N HCl	......	......	10	2	20	....	−52.6	6
A	1.001	Water	......	......	0	4.001	23.4	−3.402	−85.0	7
B	2.42	0.6 N KOH	1.031	2.35	3	1	20	−2.25	−93.0	29
				D-Proline						
B	3.90	Water	1.01	3.865	0	1	20	+3.18	+81.5	29
				L-Serine						
B	9.344	1 N HCl	1.0465	8.929	1	1	25	+1.35	+14.45	30
B	10.414	Water	1.0414	9.997	0	2	20	−1.42	−6.83	30
				D-Serine						
B	9.359	1 N HCl	1.0465	8.943	1	1	25	−1.34	−14.32	30
B	10.412	Water	1.0414	9.998	0	2	20	+1.43	+6.87	30
				D-Threonine						
B	....	Water	......	1.092	0	2	26	−0.625	−28.3	31
				L-Threonine						
B	....	Water	......	1.331	0	2	26	+0.780	+28.4	31
				D-allo-Threonine*						
B	....	Water	......	1.634	0	2	26	−0.302	−9.1	31
				L-allo-Threonine						
B	....	Water	......	1.643	0	2	26	+0.320	+9.6	31
				L-Thyroxine						
A	....	0.13 N NaOH in 70% EtOH by weight	......	3	3	1	......	−0.147	−4.4	32
				L-Tryptophan						
A	1.02	0.50 N HCl	......	......	10	2	20	....	+2.4	6
A	1.004	Water	......	......	0	4.001	22.7	−1.266	−31.5	7
A	2.426	0.5 N NaOH	1.0243	2.368	4.2	1	20	+0.15	+6.17	33
				D-Tryptophan						
C	0.5024	Water	......	......	0	2	25	+0.326	+32.45	34
				L-Tyrosine						
B	4.40	6.3 N HCl	1.116	3.94	28	2	20	−0.76	−8.64	35
A	0.906	3.0 N NaOH	......	......	60	3	18	....	−13.2	3
				D-Tyrosine						
B	5.1484	6.3 N HCl	1.1175	4.6071	24	2	20	+0.89	+8.64	35
				L-Valine						
B	3.4	6.0 N HCl	1.1	3.05	20	2	20	+1.93	+28.8	36
B	3.58	Water	1.007	3.56	0	2	20	+0.46	+6.42	36
				D-Valine						
B	3.2	6.0 N HCl	1.1	2.91	21	2	20	−1.86	−29.04	36
E	6.24	Water	1.00	6.24	0	1	20	−0.37	−6.06	37

* The levorotatory allothreonine probably belongs to the D family and its enantiomorph to the L family.

Specific Rotations of the Amino Acids Using Sodium Light (5893 Å) (Continued)

References

1. Clough G. W., J. Chem. Soc., **113**, 526 (1918).
2. Fischer, E., and Raske, K., Ber., **40**, 3717 (1907).
3. Lutz, O., and Jirgensons, B., Ber., **63**, 448 (1930).
4. Dunn, M. S., Butler, A. W., and Naiditch, M. J., unpublished data.
5. Dunn, M. S., and Courtney, G., unpublished data.
6. Lutz, O., and Jirgensons, B., Ber., **64**, 1221 (1931).
7. Dunn, M. S., and Stoddard, M. P., unpublished data.
8. Fischer, E., Ber., **32**, 2451 (1899).
9. Toennies, G., and Lavine, T. F., J. Biol. Chem., **89**, 153 (1930).
10. Loring, H. S., and du Vigneaud, V., J. Biol. Chem., **107**, 267 (1934).
11. Abderhalden, E., and Guggenheim, M., Ber., **41**, 1237 (1908).
12. Dunn, M. S., and Sexton, E. L., unpublished data.
13. Dunn, M. S., and Frieden, E. H., unpublished data.
14. Cox, G. J., and Berg, C. P., J. Biol. Chem., **107**, 497 (1934).
15. Dakin, H. D., Biochem. J., **13**, 398 (1919).
16. Leuchs, H., and Bormann, K., Ber., **52**, 2086 (1919).
17. Locquin, R., Bull. Soc. Chim., **(4)**, **1**, 601 (1907).
18. Ehrlich, F., Ber., **37**, 1809 (1904).
19. Ehrlich, F., Ber., **40**, 2538 (1907).
20. Abderhalden, E., and Zeisset, W., Z. physiol. Chem., **196**, 121 (1931).
21. Fischer, E., and Warburg, O., Ber., **38**, 3997 (1905).
22. Vickery, H. B., private communication, April, 1940.
23. Berg, C. P., J. Biol. Chem., **115**, 9 (1936); private communication, June, 1940.
24. Windus, W., and Marvel, C. S., J. Am. Chem. Soc., **53**, 3490 (1931).
27. Fischer, E., and Schoeller, W., Ann., **357**, 1 (1907).
28. Fischer, E., and Mouneyrat, A., Ber., **33**, 2383 (1900).
29. Fischer, E., and Zemplén, G., Ber., **42**, 2989 (1909).
30. Fischer, E., and Jacobs, W. A., Ber., **39**, 2942 (1906).
31. West, H. D., and Carter, H. E., J. Biol. Chem., **119**, 109 (1937); private communication from H. E. Carter, July, 1940.
32. Foster, G. L., Palmer, W. W., and Leland, J. P., J. Biol. Chem., **115**, 467 (1936).
33. Abderhalden, E., and Baumann, L., Z. physiol. Chem., **55**, 412 (1908).
34. Berg, C. P., J. Biol. Chem., **100**, 79 (1933); private communication, July, 1940.
35. Fischer, E., Ber., **32**, 3638 (1899).
36. Fischer, E., Ber., **39**, 2320 (1906).
37. Ehrlich, F., and Wendel, A., Biochem. Z. **8**, 399 (1908).
38. Ehrlich, F., Biochem. Z., **1**, 8 (1906).

Solubilities of the Amino Acids in Grams per 100 Grams of Water

Amino acid	Temperature, °C.					Ref. No.
	0°	25°	50°	75°	100°	
DL-Alanine	12.11	16.72	23.09	31.89	44.04	1
L-Alanine	12.73	16.65	21.79	28.51	37.30	1
DL-Aspartic acid	0.262	0.778	2.000	4.456	8.594	1
L-Aspartic acid	0.209	0.500	1.199	2.875	6.893	1
L-Cystine‡ × 10²	0.502	1.096	2.394	5.229	11.42	2
Diiodo-DL-tyrosine × 10	0.149	0.340	0.773	......	......	3
Diiodo-L-tyrosine × 10	0.204	0.617	1.862	5.62	17.00	1
DL-Glutamic acid	0.855	2.054	4.934	11.86	28.49	1
L-Glutamic acid	0.341	0.864	2.186	5.532	14.00	1
Glycine	14.18	24.99	39.10	54.39	67.17	1
L-Histidine	......	4.19	......	......	......	4
Hydroxy-L-Proline	28.86	36.11	45.18	51.67*	......	5
DL-Isoleucine	1.826	2.229	3.034	4.607	7.802	1
L-Isoleucine	3.791	4.117	4.818	6.076	8.255	2
DL-Leucine	0.797	0.991	1.406	2.276	4.206	1
L-Leucine	2.270	2.426†	2.887†	3.823	5.638	1
DL-Methionine	1.818	3.381	6.070	10.52	17.60	2
DL-Phenylalanine	0.997	1.411	2.187	3.708	6.886	1
L-Phenylalanine	1.983	2.965	4.431	6.624	9.900	2
L-Proline × 10⁻¹	12.74	16.23	20.67	23.90*	......	3
DL-Serine	2.204	5.023	10.34	19.21	32.24	2
L-Tryptophan	0.823	1.136	1.706	2.795	4.987	2
DL-Tyrosine × 10	0.147	0.351	0.836	......	......	3
L-Tyrosine × 10	0.196	0.453	1.052	2.438	5.650	1
D-Tyrosine × 10	0.196	0.453	1.052	......	......	3
DL-Valine	5.98	7.09	9.11	12.61	18.81	1
L-Valine	8.34	8.85	9.62	10.24*	......	6

* Value at 65°.

† Dunn and Stoddard (7) report 2.19 g. at 25° for L-Leucine rendered methionine-free by repeated recrystallization from 6 *N* HCl. Hlynka (8) found 2.20 g. at 25° and 2.66 g. at 50° for L-Leucine rendered methionine-free [by S. W. Fox (9)] by fractional crystallization of the formyl derivative and identical values for D-Leucine obtained by resolution of the DL form.

‡ The following values were found by Loring and du Vigneaud (10): DL-Cystine (0.0049 g), D-Cystine (0.0108 g), and *meso*-cystine (0.0056 g) at 25°.

References

1. Dalton, J. B., and Schmidt, C. L. A., J. Biol. Chem., **103**, 549 (1933).
2. Dalton, J. B., and Schmidt, C. L. A., J. Biol. Chem., **109**, 241 (1935).
3. Winnek, P. S., and Schmidt, C. L. A., J. Gen. Physiol., **18**, 889 (1934–35).
4. Dunn, M. S., Frieden, E. H., and Brown, H. V., unpublished data.
5. Tomiyama, T., and Schmidt, C. L. A., J. Gen. Physiol., **19**, 379 (1935–36).
6. Dalton, J. B., and Schmidt, C. L. A., J. Gen. Physiol., **19**, 767 (1935–36).
7. Dunn, M. S., and Stoddard, M. P., unpublished data.
8. Hlynka, I., Thesis (1939), California Institute of Technology, Pasadena, California.
9. Fox, S. W., Science, **84**, 163 (1936).
10. Loring, H. S., and du Vigneaud, V., J. Biol. Chem., **107**, 270 (1934).

Solubilities of the Amino Acids in Grams per 100 Grams of Water-Ethanol Mixtures

DL-Alanine

Per cent ethanol by volume	Temp. °C	Grams amino acid per 100 grams solvent	Ref. No.
24.93	0.00	3.84	1
50.10	0.00	1.16	1
74.50	0.00	0.305	1
95.14	0.00	0.0167	1
10	25	12.25	2
24.93	24.97	7.09	1
50.10	24.97	2.52	1
74.20	24.97	0.573	1
95.14	25.09	0.0329	1
25.28	45.16	10.6	1
50.10	44.96	4.25	1
74.20	44.98	0.949	1
95.14	45.19	0.0545	1
24.93	64.96	15.9	1
50.10	64.94	6.68	1
74.20	64.94	1.48	1
95.09	65.15	0.0851	1

DL-Aspartic acid

Per cent ethanol by volume	Temp. °C	Grams amino acid per 100 grams solvent	Ref. No.
24.93	0.03	0.0703	1
50.10	0.03	0.0267	1
74.20	0.02	0.0111	1
24.55	25.06	0.266	1
50.25	25.06	0.0992	1
74.28	25.14	0.0317	1
95.14	25.07	0.0020	1
24.74	45.25	0.680	1
50.18	45.25	0.255	1
74.28	45.27	0.0608	1
95.14	45.21	0.0042	1
24.93	64.91	1.53	1
50.10	64.91	0.588	1
74.20	65.07	0.132	1
95.14	65.00	0.0129	1

L-Aspartic acid

Per cent ethanol by volume	Temp. °C	Grams amino acid per 100 grams solvent	Ref. No.
20	25	0.204	3
50	25	0.0633	3
70	25	0.0224	3
90	25	0.0034	3

L-Glutamic acid

Per cent ethanol by volume	Temp. °C	Grams amino acid per 100 grams solvent	Ref. No.
24.74	0.01	0.0855	1
50.18	0.01	0.0371	1
74.28	0.03	0.0163	1
24.56	25.05	0.292	1
50.25	25.08	0.131	1
74.35	25.07	0.0370	1
95.14	25.04	0.0044	1
24.55	45.01	0.811	1
50.18	45.27	0.378	1
74.35	44.93	0.0885	1
95.14	45.20	0.0127	1

Glycine

Per cent ethanol by volume	Temp. °C	Grams amino acid per 100 grams solvent	Ref. No.
24.93	0.02	3.95	1
50.10	0.02	1.03	1
74.50	0.02	0.200	1
95.09	0.01	0.0080	1
10	25	17.13	2
24.93	24.97	8.72	1
50.10	24.97	2.47	1
74.20	24.97	0.448	1
95.14	25.09	0.0172	1
24.93	44.98	15.0	1
50.10	44.98	4.62	1
74.20	44.97	0.756	1
95.14	45.19	0.0294	1
24.93	65.11	24.5	1
50.10	65.10	8.03	1
74.20	65.07	1.23	1
95.14	65.00	0.0488	1

Solubilities of the Amino Acids in Grams per 100 Grams of Water-Ethanol Mixtures (Continued)

Per cent ethanol by volume	Temp. °C	Grams amino acid per 100 grams solvent	Ref No.
L-Isoleucine			
80	20	0.46	4
80	78–80	1.16	4
L-allo-Isoleucine			
80	20	0.81	4
80	78–80	1.97	4
DL-Leucine			
24.93	0.00	0.251	1
50.10	0.00	0.118	1
74.50	0.00	0.0693	1
95.14	0.00	0.0116	1
10	25	0.771	2
24.93	24.97	0.493	1
50.10	24.97	0.318	1
74.20	24.97	0.175	1
95.14	25.09	0.0258	1
24.93	45.24	0.853	1
50.10	45.24	0.633	1
74.50	45.18	0.323	1
95.14	45.18	0.0471	1
24.93	65.16	1.45	1
50.10	65.20	1.16	1
74.20	65.15	0.584	1
95.09	65.07	0.0844	1

Per cent ethanol by volume	Temp. °C	Grams amino acid per 100 grams solvent	Ref. No.
L-Leucine			
20	25	1.33	2
60	25	0.641	2
90	25	0.123	2
L-Proline			
100	19	1.5	5
DL-Serine			
24.93	0.00	0.1530	1
50.10	0.00	0.146	1
74.50	0.00	0.0304	1
95.14	0.00	0.0008	1
24.93	25.14	1.54	1
50.10	25.14	0.461	1
74.50	25.10	0.0840	1
95.14	25.09	0.0028	1
24.93	45.15	3.14	1
50.10	45.04	0.985	1
74.20	45.04	0.185	1
95.14	45.18	0.0058	1
24.93	65.26	5.99	1
50.10	65.25	1.88	1
74.50	65.24	0.318	1
95.14	65.01	0.0152	1

Per cent ethanol by volume	Temp. °C	Grams amino acid per 100 grams solvent	Ref. No.
DL-Threonine			
95	25	0.07*	6
DL-allo-Threonine			
95	25	0.03*	6
L-Tyrosine			
95	17	0.10	7
DL-Tyrosine			
95.09	0.00	0.0031	8
25.28	24.85	0.0285	8
50.99	24.75	0.0226	8
74.63	24.75	0.0117	8
95.09	25.24	0.0032	8
25.28	45.15	0.0630	8
50.99	45.16	0.0513	8
74.63	44.93	0.0230	8
95.09	44.98	0.0035	8
95.09	65.06	0.0067	8

Per cent ethanol by volume	Temp. °C	Grams amino acid per 100 grams solvent	Ref. No.
DL-Valine			
24.93	0.02	2.10	1
50.10	0.02	0.769	1
74.20	0.02	0.269	1
95.14	0.01	0.0277	1
10	25	5.50	2
25.28	24.85	3.30	1
50.99	24.85	1.53	1
74.35	24.93	0.570	1
95.14	25.04	0.0569	1
24.55	44.91	5.10	1
50.25	44.92	2.74	1
74.35	44.92	0.999	1
95.14	45.21	0.0979	1
24.55	65.07	7.44	1
50.10	64.94	4.49	1
74.20	64.34	1.62	1
95.09	65.15	0.167	1
L-Valine			
20	25	5.11	2
40	25	2.93	2
60	25	1.61	2
80	25	0.52	2

* Grams per 100 ml. of solution.

References

1. Dunn, M. S., and Ross, F. J., J. Biol. Chem., **125**, 309 (1938).
2. Cohn, E. J., McMeekin, T. L., Edsall, J. T., and Weare, J. H., J. Am. Chem. Soc., **56**, 2270 (1934).
3. McMeekin, T. L., Cohn, E. J., and Weare, J. H., J. Am. Chem. Soc., **57**, 626 (1935).
4. Abderhalden, E., and Zeisset, W., Z. physiol. Chem., **196**, 121 (1931).
5. Kapfhammer, J., and Eck, R., Z. physiol. Chem., **170**, 294 (1927).
6. West, H. D., and Carter, H. E., J. Biol. Chem., **119**, 109 (1937).
7. Stutzer, A., Z. anal. Chem., **31**, 501 (1892).
8. Dunn, M. S., and Ross, F. J., unpublished data.

Solubilities of the Amino Acids in Grams per 100 Grams of Organic Solvent

Solvent	Grams amino acid per 100 grams solvent	Temp. °C	Ref. No.
DL-Alanine			
Ethanol	0.0087	25	1
L-Aspartic acid			
Ethanol	0.000196	25	2
L-Glutamic acid			
Ethanol	0.000347	25	2
Ethanol	0.0056	44.93	3
Glycine			
Acetone	0.000291	25	4
Butanol	0.000892	25	4
Ethanol	0.0037	25	1
Formamide	0.558	25	4
Methanol	0.0407	25	4

Solvent	Grams amino acid per 100 grams solvent	Temp. °C	Ref. No.
L-Isoleucine			
Ethanol	0.09	20	5
Ethanol	0.13	78–80	5
L-allo-Isoleucine			
Ethanol	0.13	20	5
Ethanol	0.19	78–80	5
L-Leucine			
Ethanol	0.0217	25	1
L-Proline			
Ethanol	1.5	19	6
DL-Valine			
Ethanol	0.0136	0.03	3
Ethanol	0.019	25	1

References

1. Cohn, E. J., McMeekin, T. L., Edsall, J. T., and Weare, J. H., J. Am. Chem. Soc., **56**, 2270 (1934).
2. McMeekin, T. L., Cohn, E. J., and Weare, J. H., J. Am. Chem. Soc., **57**, 626 (1935).
3. Dunn, M. S., and Ross, F. J., J. Biol. Chem., **125**, 309 (1938).
4. McMeekin, T. L., Cohn, E. J., and Weare, J. H., J. Am. Chem. Soc., **58**, 2173 (1936).
5. Abderhalden, E., and Zeisset, W., Z. physiol. Chem., **196**, 121 (1931).
6. Kapfhammer, J., and Eck, R., Z. physiol. Chem., **170**, 294 (1927).

Densities of Crystalline Amino Acids

Amino acid	Density	Ref. No.	Amino acid	Density	Ref. No.
DL-Alanine	1.424	1	DL-Leucine	1.191	1
L-Alanine	1.401	2	L-Leucine	1.165	1
β-Alanine	1.404	1	DL-Methionine	1.340	5
DL-α-Amino-n-butyric acid	1.231	1	DL-Serine	1.537	5
α-Aminoisobutyric acid	1.278	1	L-Tyrosine	1.456	1
L-Arginine	1.1	3	DL-Valine	1.316	1
L-Aspartic acid	1.66	3	L-Valine	1.230	1
DL-Glutamic acid	1.460	4			
L-Glutamic acid	1.538	4			
Glycine*	1.601	3			
	1.607	1			

* The density of glycine at 50° is 1.5753 according to Houck (6) who concluded that the figure 1.1607, reported by Curtius (7) and reproduced in chemical handbooks, is a typographical error.

References

1. Cohn, E. J., McMeekin, T. L., Edsall, J. T., and Weare, J. H., J. Am. Chem. Soc., **56**, 2270 (1934).
2. Dalton, J. B., and Schmidt, C. L. A., J. Biol. Chem., **103**, 549 (1933).
3. Huffman, H. M., Ellis, E. L., and Fox, S. W., J. Am. Chem. Soc., **58**, 1728 (1936). Huffman, H. M., Fox, S. W., and Ellis, E. L., J. Am. Chem. Soc., **59**, 2144 (1937).
4. Schmidt, C. L. A., Chemistry of the Amino Acids and Proteins, C. C. Thomas, Springfield, Illinois, 1938, p. 900.
5. Albrecht, G., and Dunn, M. S., unpublished data.
6. Houck, R. C., J. Am. Chem. Soc., **52**, 2420 (1930).
7. Curtius, T., J. prakt. Chem., **26**, 145 (1882).

CARBOHYDRATES

These data for carbohydrates were compiled originally for the Biology Data Book by M. L. Wolfram, G. G. Maher and R. G. Pagnucco (1964). Data are reproduced here by permission of the copyright owners of the above publication, the Federation of American Societies for Experimental Biology, Washington, D.C. pp. 351–359.

All data are for crystalline substances, unless otherwise specified. Selection of substances was restricted to natural carbohydrates found free (or in chemical combination and released on hydrolysis) and to biological oxidation products of the natural carbohydrates. The nomenclature conforms with that of the British-American report as published in the *Journal of Organic Chemistry*, 28:281 (1963). Substances have been arranged alphabetically under the name of the parent sugar within groups formulated according to increasing carbon content (excluding carbon in substituents), with synonymous common names in parentheses. **Melting Point:** b.p. = boiling point; d. = decomposes; s. = sinters. **Specific Rotation** was determined in water at concentrations of 1–5 g per 100 ml. of solution and at 20°–25°C, unless otherwise specified; other temperatures or wavelengths are shown in brackets; *c* = grams solute per 100 ml of solution.

Part I. NATURAL MONOSACCHARIDES: ALDOSES AND KETOSES

	Substance (Synonym)	Chemical Formula	Melting Point °C	Specific Rotation $[\alpha]_D$
	(A)	(B)	(C)	(D)
	Aldoses			
1	D-Glyceraldehyde	$C_3H_6O_3$	....	+13.5 ± 0.5 (syrup)
2	D-Glyceraldehyde, 3-deoxy-3,3-*C*-bis-(hydroxymethyl)- (Cordycepose)	$C_5H_{10}O_4$	....	−26 (*c* 0.6, C_2H_5OH
3	D-Glyceraldehyde, 3,3-bis(*C*-hydroxy-methyl)- (Apiose)	$C_5H_{10}O_5$	....	+5.6 (*c* 10) [15°] syrup
4	β-D-Arabinose	$C_5H_{10}O_5$	155	−175 → −103
5	D-Arabinose, 2-*O*-methyl-	$C_6H_{12}O_5$	Syrup	−102
6	α-L-Arabinose	$C_5H_{10}O_5$	158 amorphous	+55.4 → +105
7	β-L-Arabinose	$C_5H_{10}O_5$	160	+190.6 → +104.5
8	DL-Arabinose	$C_5H_{10}O_5$	163.5–164.5	None
9	α-L-Lyxose	$C_5H_{10}O_5$	105	+5.8 → +13.5
10	L-Lyxose, 5-deoxy-3-*C*-formyl- (Streptose)	$C_6H_{10}O_5$	....	....
11	L-Lyxose, 3-*C*-formyl- (Hydroxy-streptose)	$C_6H_{10}O_6$	....	....
12	Pentose, 4,5-anhydro-5-deoxy-D-*erythro*-	$C_5H_8O_3$	....	....
13	Pentose, 2-deoxy-D-*erythro*-	$C_5H_{10}O_4$	96–98	−91 → −58
14	D-Ribose	$C_5H_{10}O_5$	87	−23.1 → −23.7
15	D-Ribose, 2-*C*-hydroxymethyl- (Hamamelose)	$C_6H_{12}O_6$	....	−7.1 [λ578]
16	α-D-Xylose	$C_5H_{10}O_5$	145	+93.6 → +18.8
17	D-Xylose, 5-deoxy-	$C_5H_{10}O_4$	....	+16
18	β-D-Xylose, 2-*O*-methyl-	$C_6H_{12}O_5$	137–138	−21 → +34
19	α-D-Xylose, 3-*O*-methyl-	$C_6H_{12}O_5$	95	+45 → +19
20	D-Allose, 6-deoxy-	$C_6H_{12}O_5$	140–143 146–148	+1.6 [18°] (*c* 0.6) −4.7 → 0
21	D-Allose, 6-deoxy-2,3-di-*O*-methyl- (Mycinose)	$C_8H_{16}O_5$	102–106	−46 → −29
22	Amicetose (a trideoxy hexose)	$C_6H_{12}O_3$	Oil, b.p. 65–70	+28.6 ($CHCl_3$)
23	Antiarose	$C_6H_{12}O_5$	....	Levo
24	α-D-Galactose	$C_6H_{12}O_6$	167	+150.7 → +80.2
25	β-D-Galactose	$C_6H_{12}O_6$	143–145	+52.8 → +80.2
26	D-Galactose, 3,6-anhydro-	$C_6H_{10}O_5$	....	+21.3 [10°]
27	α-D-Galactose, 6-deoxy- (D-Fucose; Rhodeose)	$C_6H_{12}O_5$	140–145	+127 → +76.3 (*c* 10)
28	D-Galactose, 6-deoxy-3-*O*-methyl- (Digitalose)	$C_7H_{14}O_5$	106[1], 119[2]	+106

CARBOHYDRATES (Continued)
Part I. NATURAL MONOSACCHARIDES: ALDOSES AND KETOSES (Continued)

	Substance (Synonym) (A)	Chemical Formula (B)	Melting Point °C (C)	Specific Rotation $[\alpha]_D$ (D)
	Aldoses (Con't)			
29	D-Galactose, 6-deoxy-4-*O*-methyl-	$C_7H_{14}O_5$	131–132	+82
30	D-Galactose, 6-deoxy-2,3-di-*O*-methyl-	$C_8H_{16}O_5$	....	+73
31	α-D-Galactose, 3-*O*-methyl-	$C_7H_{14}O_6$	144–147	+150.6 → +108.6
32	α-D-Galactose, 6-*O*-methyl-	$C_7H_{14}O_6$	122–123	+117 → +77.3
33	L-Galactose	$C_6H_{12}O_6$		*See* D-Galactose
34	α-L-Galactose, 3,6-anhydro-	$C_6H_{10}O_5$	....	−39.4 → −25.2
35	α-L-Galactose, 6-deoxy- (L-Fucose)	$C_6H_{12}O_5$	145	−124.1 → −76.4
36	L-Galactose, 6-deoxy-2-*O*-methyl-	$C_7H_{14}O_5$	149–150	−75 ± 4 (*c* 0.5)
37	L-Galactose, 6-sulfate	$C_6H_{12}O_9S$	....	−47 (*c* 0.2) (Na salt)
38	DL-Galactose	$C_6H_{12}O_6$	143–144, 163	None (racemic)
39	α-D-Glucose	$C_6H_{12}O_6$	146, 83 (H_2O)	+112 → +52.7
40	β-D-Glucose	$C_6H_{12}O_6$	148–150	+18.7 → +52.7
41	D-Glucose, 6-acetate	$C_7H_{14}O_7$	135	+48
42	D-Glucose, 2,3-di-*O*-methyl-	$C_8H_{16}O_6$	85–86, 121	+50
43	D-Glucose, 6-*O*-benzoyl- (Vaccinin)	$C_{13}H_{16}O_7$	Amorphous	+48 (C_2H_5OH)
44	α-D-Glucose, 6-deoxy- (Chinovose; Epirhamnose; Glucomethylose; Isorhamnose; Isorhodeose; Quinovose)	$C_6H_{12}O_5$	139–140	+73.3 → +29.7 (*c* 8)
45	α-D-Glucose, 6-deoxy-3-*O*-methyl- (D-Thevetose)	$C_7H_{14}O_5$	116	+84 → +33
46	D-Glucose, 6-sulfonic acid, 6-deoxy- (6-Sulfoquinovose)	$C_6H_{12}O_8S$	173–174	+87[3]
47	D-Glucose, 3-*O*-methyl-	$C_7H_{14}O_6$	162–167	+98 → +59.5
48	α-L-Glucose	$C_6H_{12}O_6$	141–143	−95.5 → −51.4
49	L-Glucose, 6-deoxy-3-*O*-methyl- (L-Thevetose)	$C_7H_{14}O_5$	126–129	−36.9 ± 2
50	D-Gulose, 6-deoxy-	$C_6H_{12}O_5$	....	....
51	Hexose, 2-deoxy-D-*arabino*-[4]	$C_6H_{12}O_5$	148	+46.6 [18°]
52	Hexose, 2,6-dideoxy-3-*O*-methyl-D-*arabino*- (D-Oleandrose)	$C_7H_{14}O_4$	....	−11
53	Hexose, 3,6-dideoxy-D-*arabino*- (Tyvelose)	$C_6H_{12}O_4$	....	+24 ± 2
54	Hexose, 2,6-dideoxy-3-*O*-methyl-L-*arabino*- (L-Oleandrose)	$C_7H_{14}O_4$	62–63	+11.9 ± 2.5
55	Hexose, 3,6-dideoxy-L-*arabino*- (Ascarylose)	$C_6H_{12}O_4$	....	−24 ± 2
56	Hexose, 2,6-dideoxy-3-*O*-methyl-D-*lyxo*- (Diginose)	$C_7H_{14}O_4$	90–92	+56 ± 4
57	Hexose, 2,6-dideoxy-L-*lyxo*- (L-Fucose, 2-deoxy-)	$C_6H_{12}O_4$	103–106	−61.6
58	Hexose, 2,6-dideoxy-3-*O*-methyl-L-*lyxo*-	$C_7H_{14}O_4$	78–85	−65
59	Hexose, 2,6-dideoxy-D-*ribo*- (Digitoxose; D-Altrose, 2,6-dideoxy-)	$C_6H_{12}O_4$	110	+46.4
60	Hexose, 2,6-dideoxy-3-*O*-methyl-D-*ribo*- (Cymarose)	$C_7H_{14}O_4$	93	+52
61	Hexose, 3,6-dideoxy-D-*ribo*- (Paratose)	$C_6H_{12}O_4$	....	+10 ± 2 (*c* 0.9)
62	Hexose, 4,6-dideoxy-3-*O*-methyl-D-*ribo*- (D-Gulose, 4,6-dideoxy-3-*O*-methyl-; Chalcose)	$C_7H_{14}O_4$	96–99	+120 → +76
63	Hexose, 2,6-dideoxy-D-*xylo*- (Boivinose)	$C_6H_{12}O_4$	96–98	−3.9 → +3.9

Part I. NATURAL MONOSACCHARIDES: ALDOSES AND KETOSES (Continued)

	Substance (Synonym) (A)	Chemical Formula (B)	Melting Point °C (C)	Specific Rotation $[\alpha]_D$ (D)
	Aldoses (Con't)			
64	Hexose, 2,6-dideoxy-3-*O*-methyl-D-*xylo*- (Sarmentose)	$C_7H_{14}O_4$	78–79	+12 → +15.8
65	Hexose, 3,6-dideoxy-D-*xylo*- (Abequose)	$C_6H_{12}O_4$	....	−3.2 ± 0.6
66	Hexose, 2,6-dideoxy-3-*C*-methyl-L-*xylo*-(Mycarose)	$C_7H_{14}O_4$	129–129	−31.1
67	Hexose, 2,6-dideoxy-3-*C*-methyl-3-*O*-methyl-L-*xylo*-(Cladinose)	$C_8H_{16}O_4$	oil, b.p. 120–132 (0.25 mm)	−23.1
68	Hexose, 3,6-dideoxy-L-*xylo*- (Colitose)	$C_6H_{12}O_4$	....	+4 (H_2O); −51 ± 2 (CH_3OH)
69	D-Idose[5]	$C_6H_{12}O_6$	....	....
70	L-Idose, 1,6-anhydro-	$C_6H_{10}O_5$	....	....
71	α-D-Mannose	$C_6H_{12}O_6$	133	+29.3 → +14.5
72	β-D-Mannose	$C_6H_{12}O_6$	132	−16.3 → +14.5
73	D-Mannose, 6-deoxy- (D-Rhamnose)	$C_6H_{12}O_5$	86–90	−7.0
74	α-L-Mannose, 6-deoxy-monohydrate (L-Rhamnose)	$C_6H_{14}O_6$	93–94	−8.6 → +8.2
75	β-L-Mannose, 6-deoxy-	$C_6H_{12}O_5$	123–125	+38.4 → +8.9
76	L-Mannose, 6-deoxy-2-*O*-methyl-	$C_7H_{14}O_5$	....	....
77	L-Mannose, 6-deoxy-3-*O*-methyl- (L-Acofriose)	$C_7H_{14}O_5$	114–115	+30 [18°]
78	L-Mannose, 6-deoxy-2,4-di-*O*-methyl-	$C_8H_{16}O_5$	82	−19 [16°]
79	L-Mannose, 6-deoxy-5-*C*-methyl-4-*O*-methyl-(Noviose)	$C_8H_{16}O_5$	128–130	+19.9 (50% C_2H_5OH)
80	Rhodinose (a 2,3,6-trideoxyhexose)	$C_6H_{12}O_3$	....	−11 ± 1.6
81	D-Talose	$C_6H_{12}O_6$	128–132	+16.9
82	D-Talose, 6-deoxy- (D-Talomethylose)	$C_6H_{12}O_5$	129–131	+20.6
83	L-Talose, 6-deoxy- (L-Talomethylose)	$C_6H_{12}O_5$	116–118	−19.5 ± 2 [18°]
84	L-Talose, 6-deoxy-2-*O*-methyl- (L-Acovenose)	$C_7H_{14}O_5$	....	−19.4
85	Heptose, D-*glycero*-D-*galacto*-	$C_7H_{14}O_7$	139–140	+47 → +64 (*c* 0.5)
86	Heptose, D-*glycero*-D-*manno*-	$C_7H_{14}O_7$	....	....
87	Heptose, D-*glycero*-L-*manno*-	$C_7H_{14}O_7$	....	....
	Ketoses			
88	Dihydroxyacetone	$C_3H_6O_3$	80 (dimer)	None
89	Tetrulose, L-*glycero*-[8] (L-Erythrulose; Ketoerythritol; L-Threulose)	$C_4H_8O_4$	Syrup	+12
90	Pentulose, D-*erythro*- (Adonose; D-Ribulose)	$C_5H_{10}O_5$	Syrup	+16.6 [27°]
91	Pentulose, L-*erythro*- (L-Ribulose)	$C_5H_{10}O_5$	....	−16.6
92	Pentulose, D-*threo*- (D-Xylulose)	$C_5H_{10}O_5$	....	−33
93	Pentulose, 5-deoxy-D-*threo*-	$C_5H_{10}O_4$	....	−5 ± 1 (CH_3OH)
94	Pentulose, L-*threo*- (L-Xylulose; L-Lyxulose; Xyloketose)	$C_5H_{10}O_5$	Syrup	+33.1
95	Hexulose, β-D-*arabino*-(β-D-Fructose; Levulose)	$C_6H_{12}O_6$	102–104[7]	−133.5 → −92
96	Hexulose, 6-deoxy-D-*arabino*- (D-Rhamnulose)	$C_6H_{12}O_5$	....	−13 ± 2
97	Hexulose, D-*lyxo*- (D-Tagatose)	$C_6H_{12}O_6$	131–132	+2.7 → −4, −5
98	5-Hexulose, D-*lyxo*	$C_6H_{12}O_6$	158	−86.6
99	Hexulose, 6-deoxy-L-*lyxo*- (L-Fuculose)	$C_6H_{12}O_5$	....	....

CARBOHYDRATES (Continued)

Part I. NATURAL MONOSACCHARIDES: ALDOSES AND KETOSES (Continued)

	Substance (Synonym)	Chemical Formula	Melting Point °C	Specific Rotation $[\alpha]_D$
	(A)	(B)	(C)	(D)
	Ketoses (Con't)			
100	Hexulose, D-*ribo*- (D-Psicose)	$C_6H_{12}O_6$	Amorphous	+4.7
101	Hexulose, L-*xylo*- (L-Sorbose)	$C_6H_{12}O_6$	159–161	−43.1
102	Hexulose, 6-deoxy-L-*xylo*-	$C_6H_{12}O_5$	88	−25 ± 2 (*c* 0.7)
103	Heptulose, D-*altro*- (Sedoheptulose; Sedoheptose)	$C_7H_{14}O_7$	Amorphous	+2.5 (*c* 10)
104	Heptulose·hemihydrate, L-*galacto*- (Perseulose)	$C_7H_{14}O_7 \cdot \frac{1}{2}H_2O$	110–115	−90 → −80
105	Heptulose, L-*gulo*-	$C_7H_{14}O_7$	....	−28
106	Heptulose, D-*ido*-	$C_7H_{14}O_7$	172	−34 ± 8 (*c* 0.3)
107	Heptulose, D-*manno*- (Mannoketoheptose; D-Mannotagatoheptose)	$C_7H_{14}O_7$	152	+29.4
108	Heptulose, D-*talo*-	$C_7H_{14}O_7$	....	....
109	Octulose, D-*glycero*-L-*galacto*-	$C_8H_{16}O_8$	....	−57, −43.4 → −13.4
110	Octulose, D-*glycero*-D-*manno*-	$C_8H_{16}O_8$	....	+20 (CH_3OH)

[1] Original melting point. [2] Melting point after four-months' storage. [3] As a methyl glycoside cyclohexylamine salt. [4] Included because of speculations concerning it in biological processes. [5] Either D-idose or L-altrose is in the polysaccharide varianose. [6] Early literature refers to this as D-erythrose. [7] The $\cdot\frac{1}{2}H_2O$ and $\cdot 2H_2O$ forms also exist.

Part II. NATURAL MONOSACCHARIDES: AMINO SUGARS

	Substance (Synonym)	Chemical Formula	Melting Point °C	Specific Rotation $[\alpha]_D$
	(A)	(B)	(C)	(D)
	Aldosamines			
1	D-Ribose, 3-amino-3-deoxy-	$C_5H_{11}NO_4$	158–158.5 d.	−24.6 (hydrochloride)
2	D-Galactose, 2-amino-2-deoxy- (Galactosamine; Chondrosamine)	$C_6H_{13}NO_5$	185	+121 → +80 (hydrochloride)
3	α-L-Galactose, 2-amino-2,6-dideoxy- (L-Fucosamine)	$C_6H_{13}NO_4$	192–193 d.	−119 → −92 [27°] (hydrochloride)
4	α-D-Glucose, 2-amino-2-deoxy- (Glucosamine; Chitosamine)	$C_6H_{13}NO_5$	88	+100 → +47.5
5	β-D-Glucose, 2-amino-2-deoxy-	$C_6H_{13}NO_5$	110–111	+28 → +47.5
6	D-Glucose, 3-amino-3-deoxy- (Kanosamine)	$C_6H_{13}NO_5$	128 d.	+19 [14°]
7	D-Glucose, 6-amino-6-deoxy-	$C_6H_{13}NO_5$	161–162 d.	+23 → +50.1 (hydrochloride)
8	D-Glucose, 2,6-diamino-2,6-dideoxy- (Neosamine C)	$C_6H_{14}N_2O_4$	>230	+61.5 (dihydrochloride)
9	D-Glucose, 3,6-dideoxy-3-dimethylamino- (Mycaminose)	$C_8H_{17}NO_4$	115–116	+31 (hydrochloride)
10	D-Glucose, 4,6-dideoxy-4-dimethylamino-	$C_8H_{17}NO_4$	192–193	+45.5 (hydrochloride)
11	L-Glucose, 2-deoxy-2-methylamino-	$C_7H_{15}NO_5$	130–132	−64
12	D-Gulose, 2-amino-1,6-anhydro-2-deoxy-	$C_6H_{11}NO_4$	250–260 d.	+41 ± 2 (hydrochloride)
13	D-Gulose, 2-amino-2-deoxy-	$C_6H_{13}NO_5$	152–162 d.	+5.6 → −18.7 (hydrochloride)

CARBOHYDRATES (Continued)

Part II. NATURAL MONOSACCHARIDES: AMINO SUGARS (Continued)

	Substance (Synonym)	Chemical Formula	Melting Point °C	Specific Rotation $[\alpha]_D$
	(A)	(B)	(C)	(D)
	Aldosamines (Con't)			
14	Hexose, 3,4,6-trideoxy-3-dimethyl-amino-D-*xylo*- (Desosamine; Picrocine)	$C_8H_{17}NO_3$	189–191 d.	+49.5 (*c* 10) (hydrochloride)
15	Hexose, a 4-acetamido-2-amino-2,4,6-trideoxy-	$C_8H_{16}N_2O_4$	216–219	+115 → +94 [26°] (*c* 0.05)
16	Hexose, an amino-deoxy-3-*O*-carboxy-ethyl-	$C_9H_{17}NO_7$	....	
17	Hexose, a 2,6-diamino-2,6-dideoxy- (Neosamine B; Paramose)	$C_6H_{14}N_2O_4$	135–150 d.	+17.5 (*c* 0.9 (hydrochloride)
18	Hexose, a 3-dimethylamino-2,3,6-trideoxy- (Rhodosamine)	$C_8H_{17}NO_3$	....	....
19	D-Mannose, 2-amino-2-deoxy- (Mannosamine)	$C_6H_{13}NO_5$	142 d.	−4.3 (*c* 9) (hydrochloride)
20	D-Mannose, 3-amino-3,6-dideoxy- (Mycosamine)	$C_6H_{13}NO_4$	162	−11.5 (hydrochloride)
21	D-Talose, 2-amino-2-deoxy- (Talosamine)	$C_6H_{13}NO_5$	151–153	+3.4 → −5.7 (*c* 0.9) (hydrochloride)
22	L-Talose, 2-amino-2,6-dideoxy- (Pneumosamine)	$C_6H_{13}NO_4$	162–163	+6.9 → +10.4 (hydrochloride)
	Ketosamines			
23	Pentulose, 1-(*o*-carboxyanilino)-1-deoxy-D-*erythro*-	$C_{12}H_{14}NO_6$	....	....
24	Hexulose, 1-(*o*-carboxyanilino)-1-deoxy-D-*arabino*-	$C_{13}H_{16}NO_7$	....	....
25	Hexulose, 5-amino-5-deoxy-L-*xylo*-	$C_6H_{13}NO_5$	174–176	−62
26	Hexulose, 6-deoxy-6-(*N*-methyl-acetamido)-L-*xylo*-	$C_9H_{17}NO_6$	....	....

Part III. NATURAL ALDITOLS AND INOSITOLS (with Inososes and Inosamines)

	Substance (Synonym)	Chemical Formula	Melting Point °C	Specific Rotation $[\alpha]_D$
	(A)	(B)	(C)	(D)
	Alditols			
1	Glycerol	$C_3H_8O_3$	20	None
2	Glycerol, 1-deoxy- (1,2-Propane-diol)[1]	$C_3H_8O_2$	Oil, b.p. 188–189	None (racemic)
3	Erythritol	$C_4H_{10}O_4$	118–120	None (meso)
4	Erythritol, 1,4-dideoxy- (2,3-Butylene-glycol)	$C_4H_{10}O_2$	25, 34	None (meso)
5	D-Threitol, 1,4-dideoxy-	$C_4H_{10}O_2$	19	−13.0
6	L-Threitol, 1,4-dideoxy-	$C_4H_{10}O_2$	....	+10.2
7	DL-Threitol, 1,4-dideoxy-	$C_4H_{10}O_2$	7.6	None (racemic)
8	D-Arabinitol	$C_5H_{12}O_5$	103	+7.82 (*c* 8, borax solution)
9	L-Arabinitol	$C_5H_{12}O_5$	101–102	−32 (*c* 0.4, 5% molybdate)
10	Ribitol (Adomitol)	$C_5H_{12}O_5$	102	None (meso)
11	Galactitol (Dulcitol)	$C_6H_{14}O_6$	186–188	None (meso)

CARBOHYDRATES (Continued)
Part III. NATURAL ALDITOLS AND INOSITOLS
(with Inososes and Inosamines) (Continued)

	Substance (Synonym)	Chemical Formula	Melting Point °C	Specific Rotation $[\alpha]_D$
	(A)	(B)	(C)	(D)
	Alditols (Con't)			
12	D-Glucitol (Sorbitol)	$C_6H_{14}O_6$	112	−1.8 [15°]
13	D-Glucitol, 1,5-anhydro- (Polygalitol)	$C_6H_{12}O_5$	140–141	+42.4
14	L-Iditol	$C_6H_{14}O_6$	73.5	−3.5 (*c* 10)
15	D-Mannitol	$C_6H_{14}O_6$	166	−0.21
16	D-Mannitol, 1,5-anhydro- (Styracitol)	$C_6H_{12}O_5$	157	−49.9
17	Heptitol, D-*glycero*-D-*galacto*- (Heptitol, L-*glycero*-D-*manno*-; Perseitol)	$C_7H_{16}O_7$	183–185, 188	−1.1
18	Heptitol, D-*glycero*-D-*gluco*- (Heptitol, L-*glycero*-D-*talo*-; β-Sedoheptitol)	$C_7H_{16}O_7$	131–132	+46 (5% NH_4 molybdate)
19	Heptitol, D-*glycero*-D-*manno*- (Heptitol, D-*glycero*-D-*talo*-; Volemitol)	$C_7H_{16}O_7$	153	+2.65
20	Octitol, D-*erythro*-D-*galacto*-	$C_8H_{18}O_8 \cdot H_2O$	169–170	−11 (5% NH_4 molybdate)
	Inositols			
21	Betitol (a dideoxy inositol)	$C_6H_{12}O_4$	224	....
22	Bioinosose (*scyllo*-Inosose; *myo*-Inosose-2; a deoxy keto inositol)	$C_6H_{10}O_6$	198–200	None (meso)
23	*h*-Bornesitol (a *myo*-inositol monomethyl ether)	$C_7H_{14}O_6$	200	+31.6
24	*l*-Bornesitol (a *myo*-inositol monomethyl ether)	$C_7H_{14}O_6$	205–206	−32.1
25	Conduritol (a 2,3-dehydro-2,3-dideoxyinositol)	$C_6H_{10}O_4$	142–143	None (meso)
26	Cordycepic acid (a tetrahydroxycyclohexanecarboxylic acid)[2]	$C_7H_{12}O_6$	....	....
27	Dambonitol (a *myo*-inositol dimethyl ether)	$C_8H_{16}O_6$	206	None (meso)
28	DL-Inositol	$C_6H_{12}O_6$	253	None (racemic)
29	*d*-Inositol	$C_6H_{12}O_6$	....	+60
30	*l*-Inositol	$C_6H_{12}O_6$	240	−65
31	Laminitol (a *C*-methyl *myo*-inositol)	$C_7H_{14}O_6$	266–269	−3
32	Liriodendritol (a *myo*-inositol dimethyl ether)	$C_8H_{16}O_6$	224	−25
33	*muco*-Inositol monomethyl ether	$C_7H_{14}O_6$	322–325	....
34	*myo*-Inositol (*meso*-Inositol)	$C_6H_{12}O_6$	217–218	None (meso)
35	*d*-*myo*-Inosose-1 (a deoxy keto inositol)	$C_6H_{10}O_6$	138–139	+19.6
36	Mytilitol (a *C*-methyl *scyllo*-inositol)	$C_7H_{14}O_6$	259	None (meso)
37	*neo*-Inosamine-2 (a deoxy amino inositol)	$C_6H_{13}O_5N$	239–241 d.	None (meso)
38	*d*-Ononitol (a *myo*-inositol monomethyl ether)	$C_7H_{14}O_6$	172	+6.6
39	*h*-Pinitol (a *dextro*-inositol monomethyl ether)	$C_7H_{14}O_6$	186	+65.5
40	*l*-Pinitol (a *levo*-inositol monomethyl ether)	$C_7H_{14}O_6$	186	−65
41	*l*-Quebrachitol (a *levo*-inositol monomethyl ether)	$C_7H_{14}O_6$	190–191	−80.2 [28°]
42	*d*-Quercitol (a deoxy *dextro*-inositol)	$C_6H_{12}O_5$	235	+24.2
43	*d*-Quinic acid (a trideoxy carboxy *dextro*-inositol)	$C_7H_{12}O_6$	164	+44 (*c* 10)

CARBOHYDRATES (Continued)
Part III. NATURAL ALDITOLS AND INOSITOLS
(with Inososes and Inosamines) (Continued)

	Substance (Synonym)	Chemical Formula	Melting Point °C	Specific Rotation $[\alpha]_D$
	(A)	(B)	(C)	(D)
	Inositols (Con't)			
44	*l*-Quinic acid (a trideoxy carboxy *levo*-inositol)	$C_7H_{12}O_6$	162	−42.1
45	Quinic acid, 5-dehydro-	$C_7H_{10}O_6$	140–142 (138 s.)	−82.4 [28°]
46	Scyllitol (*scyllo*-Inositol; Cocositol)	$C_6H_{12}O_6$	352–353	None (meso)
47	Sequoyitol (a *myo*-inositol monomethyl ether)	$C_7H_{14}O_6$	234–235	None (meso)
48	Shikimic acid (a 3,4-anhydro-quinic acid)	$C_7H_{10}O_5$	183–184	−200 [16°]
49	Shikimic acid, 5-dehydro-	$C_7H_8O_5$	150–152	−57.5 [28°] (EtOH)
50	Streptamine (2,4-diaminodideoxy-scyllitol)	$C_6H_{14}O_4N_2$	88, 210–250 d.	None (meso)
51	Streptamine, 2-deoxy-	$C_6H_{14}O_3N_2$	....	None (meso)
52	Streptadine (1,3-Dideoxy-1,3-diguanidino-scyllitol)	$C_8H_{18}N_6O_4$	....	None (meso)
53	Viburnitol (a deoxy *levo*-inositol)[3]	$C_6H_{12}O_5$	174	−73.9

[1] The 1-phosphate ester of this diol is said to occur in brain tissue and sea-urchin eggs. [2] Strong evidence that cordycepic acid is really D-mannitol. [3] Not an enantiomorph of *d*-quercitol; other isomeric relationship is involved.

Part IV. NATURAL ALDONIC, URONIC, AND ALDARIC ACIDS

	Substance (Synonym)	Chemical Formula	Melting Point °C	Specific Rotation $[\alpha]_D$
	(A)	(B)	(C)	(D)
	Aldonic Acids			
1	D-Glyceric acid	$C_3H_6O_4$	Gum	Dextro
2	L-Glyceric acid	$C_3H_6O_4$	Gum	Levo
3	D-Arabinonic acid	$C_5H_{10}O_6$	114–116	+10.5 (*c* 6)
4	L-Arabinonic acid	$C_5H_{10}O_6$	118–119	−9.6 → −41.7[1]
5	L-Arabinonic-1,4-lactone	$C_5H_8O_5$	97–99	−72
6	D-Ribonic acid	$C_5H_{10}O_6$	112–113	−17.0
7	D-Xylonic acid	$C_5H_{10}O_6$	....	−2.9 → +20.1[1]
8	L-Xylonic acid	$C_5H_{10}O_6$	....	−91.8[1]
9	D-Altronic acid	$C_6H_{12}O_7$	....	+11.5 → +24.8[1] (Ca salt, *N* HCl)
10	D-Galactonic acid	$C_6H_{12}O_7$	122	−11.2 → +57.6[1]
11	D-Gluconic acid	$C_6H_{12}O_7$	130–132 (110–112 s.)	−6.7 → +11.9[1]
12	L-Gulonic acid	$C_6H_{12}O_7$	Exists only in soln.	[ca. 0°]
13	Hexsonic acid, 2-deoxy-D-*arabino*-	$C_6H_{12}O_6$	93–95	+68 (lactone)
14	2-Hexulosonic acid, D-*arabino*-	$C_6H_{10}O_7$	....	−81.7 (Na salt)
15	2-Hexulosonic acid, 3-deoxy-D-*erythro*-	$C_6H_{10}O_6$	....	−29.2 (*c* 6, Ca salt)
16	2-Hexulosonic acid, D-*lyxo*-	$C_6H_{10}O_7$	169	−5
17	5-Hexulosonic acid, D-*arabino*-	$C_6H_{10}O_7$	108–109	....
18	5-Hexulosonic acid, D-*xylo*-	$C_6H_{10}O_7$	....	−14.5
19	D-Mannonic acid	$C_6H_{12}O_7$	....	−15.6
20	D-Gluconic acid, *O*-β-D-galactopyranosyl- (1 → 4)- (Lactobionic acid)	$C_{12}H_{22}O_{12}$	....	+25.1 (Ca salt)

CARBOHYDRATES (Continued)

Part IV. NATURAL ALDONIC, URONIC, AND ALDARIC ACIDS (Continued)

	Substance (Synonym) (A)	Chemical Formula (B)	Melting Point °C (C)	Specific Rotation $[\alpha]_D$ (D)
	Uronic Acids			
21	L-Lyxuronic acid	$C_5H_8O_6$	....	....
22	β-D-Galacturonic acid	$C_6H_{10}O_7$	160	+27 → +55.6
23	α-D-Galacturonic acid·monohydrate	$C_6H_{12}O_8$	159–160 (110–115 s.)	+97.9 → +50.9
24	D-Galacturonic acid, 2-amino-2-deoxy-	$C_6H_{11}O_6N$	160 d.	+84.5 (pH 2 HCl)
25	β-D-Glucuronic acid	$C_6H_{10}O_7$	156	+11.7 → +36.3
26	D-Glucuronic acid, 2-amino-2-deoxy-	$C_6H_{11}O_6N$	120–172 d.	+55
27	D-Glucuronic acid, 3-*O*-methyl-	$C_7H_{12}O_7$	Syrup	+6
28	L-Guluronic acid	$C_6H_{10}O_7$	....	....
29	L-Iduronic acid	$C_6H_{10}O_7$	....	+30
30	β-D-Mannuronic acid	$C_6H_{10}O_7$	165–167	−47.9 → −23.9
31	α-D-Mannuronic acid·monohydrate	$C_6H_{12}O_8$	110 s., 120–130 d.	+16 → −6.1 (*c* 6.8)
	Aldaric Acids			
32	D-Tartaric acid	$C_4H_6O_6$	170	−15
33	L-Tartaric acid	$C_4H_6O_6$	170	+15 [15°]
34	L-Malic acid	$C_4H_6O_5$	100	−2.3 (*c* 8.4)

[1] Equilibrates with the lactone.

WAXES

These data for waxes were compiled originally for the Biology Data Book by A. H. Warth. Data are reproduced here by permission of the copyright owners of the above publication, the American Societies for Experimental Biology, Washington, D.C. p. 382.

Specific Gravity (column C) was calculated at the specified temperature, degrees centigrade, and referred to water at the same temperature. **Density.** shown in parentheses (column C), and **Refractive Index** (column D) were measured at the specified temperature, degrees centigrade.

	Wax	Melting Point °C	Specific Gravity or (Density)	Refractive Index $n\frac{°C}{D}$	Iodine Value	Acid Value	Saponification Value
	(A)	(B)	(C)	(D)	(E)	(F)	(G)
1	Bamboo leaf	79–80	(0.961$^{25°}$)	....	7.8[1]	14.5	43.4
2	Bayberry (myrtle)	46.7–48.8	(0.985$^{15°}$)	1.436$^{80°}$	2.9[2]–3.9[3]	3.5	20.5–21.7
3	Beeswax, crude	62–66	(0.927–0.970$^{15°}$)	1.439–1.483$^{40°}$	6.8–16.4[2]	16.8–35.8	89.3–149.0
4	Beeswax, white, U.S.P.	61–69	(0.959–0.975$^{15°}$)	1.447–1.465$^{65°}$	7–11[3]	17–24	90–96
5	Beeswax, yellow	62–65	(0.960–0.964$^{15°}$)	1.443–1.449$^{65°}$	6–11	18–24	90–97
6	Candelilla, refined	67–69	(0.982–0.986$^{15°}$)	1.454–1.463$^{85°}$	14.4–20.4	12.7–18.1	35–86
7	Cape berry[4]	40.5–45.0	(1.004–1.007$^{15°}$)	1.450$^{45°}$	0.6–2.4	2.5–3.7	211–215
8	Carandá	79.7–84.5	(0.990$^{25°}$)	....	8.0–8.9	5.0–9.5	64.5–78.5
9	Carnauba	83–86	0.990–1.001$^{15°}$	1.467–1.472$^{40°}$	7.2–13.5	2.9–9.7	78–95
10	Castor oil, hydrogenated	83–88	(0.980–0.990$^{20°}$)	....	2.5–8.5	1.0–5.0	177–181
11	Chinese insect	81.5–84.0	0.950–0.970$^{15°}$	1.457$^{40°}$	1.4	0.2–1.5	73–93
12	Cotton	68–71	0.959$^{15°}$	....	24.5	32	70.6
13	Cranberry	207–218	(0.970–0.975$^{15°}$)	....	44.2–53.2[2]	42.2–59.1	131–134
14	Douglas-fir bark	59.0–72.8	(1.030$^{25°}$)	1.468$^{80°}$	25.8–62.5	58.6–80.7	112–200
15	Esparto	67.5–78.1	0.988$^{15°}$	....	22–23	22.7–23.9	69.8–79.3
16	Flax	61.5–69.8	0.908–0.985$^{15°}$	....	21.6–28.8	17.5–48.3	77.5–101.5
17	Ghedda, E. Indian beeswax	60.5–66.4	0.956–0.973$^{15°}$	1.440$^{50°}$	5.6–12.6	5.8–7.9	84.5–118.3
18	Indian corn	80–81	....	....	4.2[2]	1.9	120.3
19	Japan wax	48–53	0.975–0.993$^{15°}$	....	4.5–12.5	6–20	206.5–237.5
20	Jojoba	11.2–11.8	0.864–0.899$^{25°}$	1.465$^{25°}$	81.7–88.4[2]	0.2–0.6	92.2–95.0
21	Madagascar	88	....	....	3.2–5.3	17.7–28.0	140.0–159.6
22	Microcrystalline, amber	64–91	0.913–0.943$^{15°}$	1.424–1.452$^{80°}$	0	0	0
23	Microcrystalline, white	71–89	0.928–0.941$^{15°}$	1.441$^{80°}$	0	0	0
24	Montan, crude	76–86	(1.010–1.020$^{25°}$)	....	13.9–17.6	22.7–31.0	59.4–92.0
25	Montan, refined	77–84	(1.010–1.030$^{25°}$)	....	10–14	24–43	72–103
26	Orange peel	44.0–46.5	0.985$^{15°}$	1.502$^{20°}$	115.7[2]	48.3	120.9
27	Ouricury, refined	79.0–83.8	1.053$^{15°}$	....	6.9–7.8[2]	3.4–21.1	61.8–85.8
28	Ozocerite, refined	74.4–75.0	0.907–0.920$^{15°}$	....	0	0	0
29	Palm	74–86	(0.991–1.045$^{15°}$)	....	8.9–16.9[2]	5.0–10.6	64.5–104.0
30	Paraffin, American	49–63	0.896–0.925$^{15°}$	1.442–1.448$^{80°}$	0	0	0
31	Peat wax, natural	73–76	0.980$^{15°}$	....	16–40	60.0–73.3	73.9–136.0
32	Rice bran, refined	75.3–79.9	....	1.469$^{30°}$	11.1–19.4	15–17	56.9–104.4
33	Shellac wax	79–82	0.971–0.980$^{15°}$	....	6.0–8.8[3]	12.1–24.3	63.8–83.0
34	Sisal hemp	74–81	1.007–1.010$^{15°}$	....	28–29[2]	16–19[2]	56–58
35	Sorghum grain	77–82	....	....	15.7–20.9	10.1–16.2	16–44
36	Spanish moss	79–80	....	....	33.0	25.0	120.4
37	Spermaceti	42–50	0.905–0.945$^{15°}$	1.440$^{70°}$	4.8–5.9	2.0–5.2	108–134
38	Sugarcane, crude	52–67	0.988–0.998$^{25°}$	....	32–84	24–57	128–177
39	Sugarcane, double-refined	77–82	0.961–0.979$^{25°}$	1.510$^{25°}$	13–29	8–23	55–95
40	Wool wax, refined	36–43	0.932–0.945$^{15°}$	1.478–1.482$^{40°}$	15.0–46.9	5.6–22.0	80–127

[1] Wijs test. [2] Hanus test. [3] Hubl test. [4] *Myrica cordifolia.*

TRADE NAMES OF DYESTUFF INTERMEDIATES

Trade name	Chemical name
A acid	1,7-Hydroxynaphthalene-3,6-disulfonic acid
Acetyl H acid	N-Acetyl-1-amino-8-naphthol-3,6-disulfonic acid
Alen's acid	1-Naphthylamine-3,6-disulfonic acid (also Freund's ac.)
Alizarin	1,2-Dihydroxyanthraquinone
Amido acid	2-Amino-7-hydroxynaphthalene-5-sulfonic acid
Amido J acid	2-Naphthylamine-5,7-disulfonic acid
Amino G acid	7-Amino-1,3-naphthalene disulfonic acid 2-Naphthylamine-6,8-disulfonic acid
Amino R acid	3-Amino-2,7-naphthalene disulfonic acid 2-Naphthylamine-3,6-disulfonic acid
Aminophenolic acid V	1-Amino-3-oxybenzene-5-sulfonic acid
Aminophenol sulfonic acid III	1-Amino-3-oxybenzene-6-sulfonic acid
Andresen's acid	1-Naphthol-3,8-disulfonic acid
Anisidine	o-Aminophenol methylether
Anthrachrysone	1,3,5,7-Tetrahydroanthraquinone
Anthraflavic acid	2,6-Dihydroxyanthraquinone
Anthranilic acid	o-Aminobenzoic acid
Anthrarufin	1,5-Dihydroxyanthraquinone
Anthranol	9-Hydroxyanthracene
α Anthrol	1-Hydroxyanthracene
Armstrong's acid	Naphthalene-1,5-disulfonic acid
Armstrong & Wynne acid	1-Naphthol-3-sulfonic acid
Armstrong & Wynne acid II	2-Naphthylamino-5,7-disulfonic acid
B acid	8-Amino-1-naphthol-4,6-disulfonic acid
Badische acid	2-Naphthylamino-8-sulfonic acid
Bayer's acid	2-Naphthol-8-sulfonic acid
Benzidine	p,p′-Diaminodiphenyl
Bronner's acid	2-Naphthylamino-6-sulfonic acid
β acid	Anthraquinone-2-sulfonic acid
C acid CLT acid	6-Chloro-m-toluidine-4-sulfonic acid
Casella's acid	2-Naphthol-7-sulfonic acid (F acid)
Chicago acid	1-Amino-8-naphthyl-2,4-disulfonic acid
Chloro H acid	8-Chloro-1-naphthol-3,6-disulfonic acid
Chromogene I	4,5-Dihydroxy-2,7-naphthalene disulfonic acid
Chromotrope acid	1,8-Dihydronaphthalene-3,6-disulfonic acid
Chromotropic acid	4,5-Dihydroxy-2,7-naphthalene disulfonic acid
Chrysazine	1,8-Dihydroxyanthraquinone
Cleve's acid	1-Naphthylamine-3-sulfonic acid
Cleve's acid	1-Naphthylamine-5-sulfonic acid
Cleve's acid	1-Naphthylamine-6-sulfonic acid
Cleve's acid	1-Naphthylamine-7-sulfonic acid
α Coccinic acid	1-Methyl-5-oxybenzyl-2,4-dicarbonic acid
Cresidine	3-Amino-4-methoxytoluene
Cresotic acid	Cresol carboxylic acid
Croceine acid	2-Naphthol-1-sulfonic acid
DS	4,4′-Diamino-2,2′-stilbene disulfuric acid
DTS	Dihydrothio-p-toluidine sulfonic acid
Dahl's acid	2-Naphthylamine-5-sulfonic acid
Dahl's acid II	1-Naphthylamine-4,6-disulfonic acid
Dahl's acid III	1-Naphthylamine-4,7-disulfonic acid
Dimethyl-γ-acid	7-Dimethylamino-1-naphthol sulfonic acid
Dioxy G acid	1,7-Dihydroxynaphthalene-3-sulfonic acid
Dioxy J acid	1,6-Dihydroxynaphthalene-3-sulfonic acid
Dioxy S acid	1,8-Dihydroxynaphthalene-4-sulfonic acid
Diphenylblack base	p-Aminodiphenylamine
Disulfo acid S	1-Naphthylamine-4,8-disulfonic acid
δ acid	1-Naphthol-4,8-disulfonic acid 1-Naphthylamine-4,8-disulfonic acid

TRADE NAMES OF DYESTUFF INTERMEDIATES (Continued)

Trade name	Chemical name
Ebert & Merz acid	Naphthalene-2,7-disulfonic acid Naphthalene-2,6-disulfonic acid
Ethyl-γ-acid	7-Ethylamino-1-naphthol-3-sulfonic acid
Ethyl F acid	7-Ethylamino-2-naphthalene sulfonic acid
Ewer & Pick's acid	Naphthalene-1,6-disulfonic acid
ϵ acid	1-Naphthol-3,8-disulfonic acid 1-Naphthylamine-3,8-disulfonic acid
F acid	2-Naphthol-7-sulfonic acid (Casella's acid)
Fast Black B base	4,4′-Diamino diphenylamine
Fast Blue base	Dianisidine
Fast Blue Red O base	3-Nitro-p-phenetidine
Fast Bordeaux GP	3-Nitro-p-anisidine
Fast Orange GR	o-Nitroaniline
Fast Orange R	m-Nitroaniline
Fast Red base AL	α-Aminoanthraquinone
Fast Red B base	5-Nitro-o-anisidine
Fast Red GG base	p-Nitroaniline
Fast Red GL base	3-Nitro-p-toluidine
Fast Red 3 GL base	4-Chloro-2-nitroaniline
Fast Red RL base	5-Nitro-o-toluidine
Fast Scarlet G base	4-Nitro-α-toluidine
Fast Scarlet R base	4-Nitro-O-anisidine
Forsling's acid I	2-Naphthylamine-8-sulfonic acid
Forsling's acid II	2-Naphthylamine-5-sulfonic acid
Freund's acid	1-Naphthylamine-3,6-disulfonic acid
G acid	2-Naphthol-6,8-disulfonic acid
GR acid	α-Naphthol-3,6-disulfonic acid
Gallic acid	3,4,5-Trihydroxybenzoic acid
γ acid	2-Amino-8-naphthol-6-sulfonic acid
H acid	1-Amino-8-naphthol-3,6-disulfonic acid
Histazarin	2,3-Dihydroxyanthraquinone
Isoanthraflavic acid	2,7-Dihydroxyanthraquinone
J acid	2-Amino-5-naphthol-7-sulfonic acid
K acid	1-Amino-8-naphthol-4,6-disulfonic acid
Kalle's acid	1-Naphthylamine-2,7-disulfonic acid
Ketone base	Tetramethyl aminobenzophenone
Koch's acid	1-Naphthylamine-3,6,8-trisulfonic acid
L acid	1-Naphthol-5-sulfonic acid
Laurent's acid	1-Naphthylamine-5-sulfonic acid
Lepidine	4-Methylquinoline
Leucotrop	Phenyldimethyl benzylammonium chloride
M acid	1-Amino-5-naphthol-7-sulfonic acid
Mesidine	2,4,6-Thimethylaniline
Metanilic acid	Aniline-m-sulfonic acid
Methyl-γ-acid	7-Methyl-8-naphthol disulfonic acid
Michler's hydrol	Tetramethyl diaminobenzohydrol
Michler's ketone	Tetramethyl diaminobenzophenone
Myrbane oil	Nitrobenzene
Naphthacetol	1-Acetylamino-4-naphthol
Naphthazarin	5,8-Dihydroxy-1,4-naphthoquinone
Naphthionic acid	1-Naphthylamine-4-sulfonic acid
o-Naphthionic acid	1-Naphthylamine-2-sulfonic acid
Naphthol AS	Anilide of hydronaphthoic acid
Naphthoresorcine	1,3-Dihydroxynaphthalene
Nekal BX	Na-salt of 1,4-bis,sec-butylnaphthalene-6-sulfonic acid
Nevile and Winther's acid	1-Naphthol-4-sulfonic acid
Nigrotic acid	1,3,6,7-Dihydroxysulfonaphthoic acid
Nitron 1,2,4-acid	1-Amino-8-nitro-7-naphthol-4-sulfonic acid
Nitroso base	p-Nitrodimethyl aniline

TRADE NAMES OF DYESTUFF INTERMEDIATES (Continued)

Trade name	Chemical name
NW acid	Nevile and Winther s acid
Oxy L acid	1-Naphthol-5-sulfonic acid
Oxy Tobias acid	β-Naphthol-1-sulfonic acid
Peri acid	1-Naphthylamine-8-sulfonic acid
p-Phenetidine	p-Aminophenol ethylether
Phenyl gamma acid	2-Phenylamine-8-naphthol-6-sulfonic acid
Phenyl Peri acid	Phenyl-1-naphthylamine-8-sulfonic acid
Phosxgene	Carbonyl chloride
Phthalic acid	O-Benzenedicarbolic acid
Picramic acid	2-Amino-4,6-dinitrophenol
Picric acid	2,4,6-Trinitrophenol
Pirio's acid	4-Amino-1-naphthalene sulfonic acid
Primuline base	p-Toluidine heated with sulfur
Purpurine	1,2,4-Trihydroxyanthraquinone
Pyrogallol	1,2,3-Trihydroxybenzene
Quinaldine	2-Methylquinoline
Quinazarin	1,4-Dihydroxyanthraquinone
R acid	2-Naphthol-3,6-disulfonic acid
2 R acid	2-Amino-8-naphthol-3,6-disulfonic acid
Red acid	1,5-Dihydroxynaphthalene-3,7-disulfonic acid
RG acid	1-Naphthol-3,6-disulfonic acid
Resorcinol	1,3-Dihydroxybenzene
Rumpff acid	2-Naphthol-8-sulfonic acid (Croceine acid)
S acid	1-Amino-8-naphthol-4-sulfonic acid
2 S acid	1-Amino-8-naphthol-2,4-disulfonic acid
Salicylic acid	o-Hydroxybenzoic acid
Schäffer's acid	2-Naphthol-6-sulfonic acid
Schäffer and Baum acid	α-Naphthol-2-sulfonic acid
Schollkopf's acid	1-Naphthol-4,8-disulfonic acid 1-Naphthylamine-4,8-disulfonic acid 1-Naphthylamine-8-sulfonic acid
Sulfanilic acid	Aniline-p-sulfonic acid
Thiocarbanilide	Diphenylthiourea
Tobias acid	2-Naphthylamine-1-sulfonic acid
Tolidine	Di-p-aminoditolyl
Toluidine	Aminotoluene
Violet acid	α-Naphthol-3,6-disulfuric acid
Xylidine	Aminoxylene
Y acid	2-Naphthol-6,8-disulfuric acid
Yellow acid	1,3-Dihydroxynaphthalene-5,7-disulfuric acid
1:2:4 acid	1-Amino-2-naphthol-4-sulfonic acid

TRADE NAMES, COMPOSITION AND MANUFACTURERS OF SOME PLASTICS

Trade name	Composition	Manufacturer
A-100 SM	Acrylics	Resolite Corp.
Abson	Acrylonitrile-butadiene ABS polymers	B. F. Goodrich Chemical Co.
Acco	Polymers, copolymers of acrylics and methacrylate esters	Acco Polymers Company
Acetex	Synthetic resin latices	United States Rubber Co., Naugatuck Chemical Div.
Acrilan	Acrylic fiber	Monsanto Co.
Actol	Polyesters	Allied Chemical Corp.
Adipol BCA	Di-butoxyethyl adipate	Food Machinery and Chemical Corp.
Agile	Polyesters, polyethylenes, polypropylenes, polyvinyl chlorides	American Agile Corp.
Alathon	Polyethylene resins	E. I. duPont de Nemours & Co., Inc.
Alkor	Furane resin cement	Atlas Mineral Products
Alpex	Hydrocarbon resins	Alkydol Laboratories, Div. of Reichhold Chemicals, Inc.
Alpha	Nylons, polyethylenes, ABS-polymers, polypropylenes, polyvinyl chlorides	Alpha Plastics, Inc.
Ameripol	High density polyethylenes	Goodrich-Gulf Chemicals, Inc.
Ameripol CB	Polybutadienes	" " " "
Ameripol SBR	Butediene-styrene rubber	" " " "
Amerith	Cellulose nitrate	Celanese Corp. of America
Amphenol	Polystyrene	American Phenolic Co.
Amres	Phenolics, resorcinol, urea	American Marietta Co.
Anaconda	Polyethylenes, polyvinyl chlorides	Anaconda American Brass Co.
Aquaplex	Alkyd-resin emulsions	Rohm & Hass Company
Araldite	Epoxy resins	Ciba Products Corp.
Arcolite	Phenolic	Consolidated Molded Products Co.
Aritemp	Epoxy resins	Aries Laboratories, Inc.
Arociro	Polychlorinated polyphenyls	Monsanto Chemical Company
Arodures	Ureas	National Distillers & Chemical Corp.
Atlac	Polyester resins in solid state	Atlas Chemical Industries, Inc.
Avisco	Urea-formaldehyde compounds	American Viscose Corporation
AviSun	Polypropylene resins	AviSun Corporation
Bakelite	Acrylics, epoxies, phenolics, polyethylenes, copolymers	Union Carbide Corp., Plastics Div.
Bavick-11	Acrylics	J. T. Baker Chemical Co.
Beckacite	Maleic, fumaric, modified phenolic resins	Reichold Chemicals, Inc.
Beckamine	Alcohol-soluble urea-formaldehyde resins for laminates, adhesives, coating	" " "
Beckopol	High melt point modified phenolic resin	" " "
Blacar	Vinyl chloride resin; vinyl compound	Cary Chemicals, Inc.
Boltaflex	Supported and unsupported flexible vinyl sheeting	The General Tire & Rubber Co.
Boltaron	ABS polymers, polyethylenes, polypropylenes, polyvinyl chlorides	Bolta Products Company
Bondstrand	Epoxies, polyesters	Amercoat Corporation
Brea	Polyethylenes	" "
Budene	Polybutadiene	Goodyear Tire & Rubber Company
Butacite	Polyvinyl butyral resins	E. I. duPont de Nemours & Co.
Butakon	Butadiene-acrylonitriles	Imperial Chemical Ind. Ltd.
Buton	Thermosetting butadiene-styrene resins	Enjay Chemical Company, Div. Humble Oil & Refining Company
Butvar	Polyvinyl butyral resins	Shawinigan Resins Corporation

TRADE NAMES, COMPOSITION AND MANUFACTURERS OF SOME PLASTICS (Continued)

Trade name	Composition	Manufacturer
C-8 Epoxy	Epoxy resins	Union Carbide Corporation
Cadco	Acrylics, ABS polymers, nylons, phenolics, polyesters, polyethylenes, polypropylenes	Cadillac Plastic & Chemical Co.
Cadco Penton	Chlorinated polyethers	" " " "
Cadco Epoxy	Epoxies	" " " "
Cadco Teflon	Fluorocarbons	" " " "
Cadco PVC	Polyvinyl chlorides urethanes, elastomers	" " " "
Carbo-Korez	Phenolics	Atlas Mineral Products
Carbomix	Butadiene-styrene rubber	Copolymer Rubber & Chemical Corp.
Carbowax	Polyethylene glycols	Union Carbide Corporation
Cardolite	Phenol-aldehyde resins	Irvington Varnish Co.
Catabond	Phenolic resins	Catalin Corp. of America
Catalin	Acrylics, ABS polymers, nylons, phenolics, polyethylenes, polypropylenes, polystyrene	" " " "
Ceilcrete	Polyesters	The Ceilcote Co.
Centri-Cast	Epoxies, polyesters	Apex Fiberglass Products Co.
Chemigum	Butadiene-acrylonitrile rubber and latices	The Goodyear Tire & Rubber Co.
Chemline	Epoxies	A. O. Smith Corporation
Chempro	Fluorocarbons	Chemical & Power Products
Chemtite	Epoxies, phenolics	Johns-Manville Corp.
Cis-4	Polybutadienes	Phillips Petroleum Company
Cohrlastic	Silicon rubber	Connecticut Hardrubber Co.
Coltrock	Phenolics	Colt's Plastics Co.
Conolite	Polyester resins and laminates	Shellmar-Betner, Div. Continental Can Co.
Copo	Butadiene-styrene rubber	Copolymer Rubber & Chemical Corp.
Coroband	Furanes	The Ceilcote Company
Corocrete	Polyesters	" " "
Corvel	Epoxies, vinyls Fusion bond finishes	The Polymer Co.
Corvic	Polyvinyl chloride resins	Canadian Industries Limited
Cosmalite	Phenol-formaldehyde resins	Colton Chemical Co.
CR	Polyesters	Resolite Corp.
CR-39	Polycarbonates	Pittsburgh Plate Glass Co.
Crystalex	Acrylics	Rohm & Hass Company
Cumar	Paracoumarone-indene resins	Allied Chemical Corporation
Cycolac	ABS polymers Acrylonitrile-butadiene-styrene, copolymer	Borg-Warner Corporation
Cymel	Melamines molding compounds, adhesive and laminating resins	American Cyanamid Company
Dacovin	Polyvinyl chlorides	Diamond Alkali Company
Dapon	Diallyl phthalate	Food Machinery and Chemical Corp.
Delrin	Acetal resins	E. I. duPont de Nemours & Co., Inc.
D.E.R.	Epoxy resins	Dow Chemical Company
Devran	Epoxy resins	Devoe & Reynolds
Diamond PVS	Polyvinyl chloride resins; copolymer resins	Diamond Alkali Company
Dow Corning	Silicones	Dow Corning
Duradene	Butanediene-styrene rubber	Firestone Tire & Rubber Co.
Duragene 1203	Polybutadienes	The General Tire & Rubber Co.
Durcon	Epoxies	The Duriron Co.
Dur-X	Polyethylenes	Johns-Manville Company

TRADE NAMES, COMPOSITION AND MANUFACTURERS OF SOME PLASTICS (Continued)

Trade name	Composition	Manufacturer
Durez	Phenolic resins	Durez Plastic Division, Hooker Chemical Corporation
Dylan	Polyethylene	Koppers Company, Inc.
Dylene	Polystyrene	" " "
Dylite	Expandable polystyrene	" " "
Dyphene	Phenol-formaldehyde resins	The Sherwin-Williams Co.
Ebolene	Fluorocarbons	Chicago Gasket Co.
Enjay Butyl	Butylrubber	Enjay Chemical Company, Div. Humble Oil & Refining Co.
Enjay Butyl latex	Butylrubber in aqueous solution	Enjay Chemical Company, Div. Humble Oil & Refining Co.
Epikote	Epoxy resins	The Shell Chemical Company
Epiphen	Epoxide resins	The Borden Chemical Company
Epi-Rez	Epoxy resins	Jones-Dabney Company
Epi-Tex	Epoxy ester resins	" " "
Epolene	Low molecular weight polyethylene resins	Eastman Kodak Company
Epoxical	Epoxies	United States Gypsum Co.
Epon	Epoxy resins	The Shell Chemical Company
Escon	Polypropylene resins	Enjay Chemical Co., Div. Humble Oil & Refining Company
Estane	Polyurethane elastomers	B. F. Goodrich Chemical Company
Ethafoam	Polyethlene, low density	The Dow Chemical Company
eXtren 200	Polyesters	Universal Molded Fiberglass Co.
Fibercast	Epoxies, polyesters	Fibercast, Div. of Youngstown Sheet & Tube Co.
Fibestos	Cellulose acetate	Monsanto Chemical Co.
Fire-Snuf	Polyesters	Resolite Corp.
Flakoline	Polyesters	The Ceilcote Co.
Flamenol	Polyvinyl chlorides	General Electric Company
Flovic	Polyvinyl chlorides	Canadian Industries, Limited
Fluorogreen	Teflon with glass and ceramic fibers, fluorocarbons	John L. Dore Co.
Fluororay	Fluorocarbons	Raybestos-Manhattan, Inc.
Formica	Melamines	Formica Corp.
Formvar	Polyvinyl formal resins	Shawinigan Resins Corp.
Forticel	Cellulose propionate sheet films, molding powders	Celanese Polymer Co.
Fortiflex-A	Linear polyolefines	Celanese Polymer Company
Fortiflex-B	Linear copolymers	" "
Fortiflex-C	Medium density polyethylenes	" "
Fortiflex-D	Low-density polyethylenes	" "
Fosta-Tuf-Flex	Polystyrene, high impact	Foster-Grant, Inc.
FR-N	Nitrile rubber	The Firestone Tire & Rubber Co., Firestone Plastics Company
FR-S	Butadiene-styrene rubber	The Firestone Tire & Rubber Co., Firestone Plastics Company
Frostwhite	Polystyrene, medium impact	Sheffield Plastics, Inc.,
Furnane	Furanes	Atlas Mineral Products Co.
Gabrite	Urea-formaldehyde	Chemore Corporation
Gaco	Epoxies	Gates Engineering Division, Glidden Company
GE Methylon	Phenolics	General Electric Company, Chemical Materials Dept.
Gelva	Polyvinyl acetate resins " copolymers	Shawinigan Resins Corp.
Gelvatol	Polyvinyl alcohol resins	" " "

TRADE NAMES, COMPOSITION AND MANUFACTURERS OF SOME PLASTICS (Continued)

Trade name	Composition	Manufacturer
Genco	Acrylics, ABS polymers Polyethylenes, copolymers	General Plastics Corp.
GenEpoxy	Epoxy resins for adhesives, coatings, etc.	General Mills, Inc.
Genetron	Fluorinated hydrocarbons, monomers and polymers	Allied Chemical Corp., General Chemical Div.
Gen-Flo	Styrene-butadiene latex	General Tire & Rubber Co.
Genthanes	Urethane elastomers; prepolymers	" " " "
Geon	Polyvinyl chlorides	B. F. Goodrich Chemical Co.
Gering	Acrylics, ABS polymers, nylons, polyethylenes, polypropylenes polyvinyl chlorides, copolymers	Gering Plastics Company
Ger-Pak	Polyethylene film	Gering Plastics Company
Glykon	Polyester resins	The General Tire & Rubber Co.
GP excel	Polyvinyl chlorides	Glamorgan Plastics, Div. of Glamorgan Pipe & Foundry Co.
Grace	Polystyrenes	W. R. Grace & Company, Polymer Chemical Division
Grex	High density polyethylenes, polypropylenes	W. R. Grace & Company, Polymer Chemical Division
Halon	Fluorohalocarbon resins	Allied Chemical Corp.
Haveg	Phenolic resins, furanes	Haveg Industries, Inc.
Haysite	Reinforced polyester	Haysite Corp.
Hercose AP	Cellulose acetate-propionate	Hercules Powder Company
Herculoid	Cellulose nitrate	" " "
Hetrofoam	Polyurethane fire-retardant rigid foam	Hooker Chemical Corp., Durez Plastics Div.
Hetron	Fire retardant polyester resin	Hooker Chemical Corp., Durez Plastics Div.
Hi-flexible	Polyethylenes	Triangle Conduit & Cable Co.
Hycar	Butadiene-styrene, butadiene-acronitriles, acrylate emulsions	B. F. Goodrich Chemical Co.
Hystran	Epoxies, polyesters	Lamtex Industries of Koppers Co.
Iporca	Urea resin and foam	Badische Aniline & Soda Fabrik
Isothane	Urethane elastomers	Carborundum Company
J-M ABS	ABS polymers	Johns-Manville Company
J-M PVC	Polyvinyl chlorides	" " "
Kaylite	Polyvinyl chlorides	Kaykor Products Corp.
Kayrex	Polyvinyl chlorides	" " "
Kel-F	Chlorotrifluoroethylene, molding resins and dispersions	Minnesota Mining & Manufacturing Co.
Koroseal	Polyvinyl chlorides	B. F. Goodrich Industrial Products Co.
Kralac A-EP	High styrene resins	United States Rubber Co., Naugatuck Chemical Div.
Kralac Latex	Styrene-butadiene copolymers	United States Rubber Co., Naugatuck Chemical Div.
Kralastic	ABS Polymers, copolymers	United States Rubber Co., Naugatuck Chemical Div.
Kynar	Polyvinyldene fluoride	Pennsalt Chemical Corp.
La Favorite	Polyvinyl chlorides	La Favorita Rubber Mfg. Co.
Laminac	Polyester resins	American Cyanamid Company
Lemac	Polyvinyl acetate	Borden Chemical Company
Lexan	Polycarbonate resin	General Electric Company, Chemical Materials Dept.
Lucite	Acrylic resin	E. I. duPont de Nemours & Co.
Lustran	ABS polymers	Monsanto Chemical Company

TRADE NAMES, COMPOSITION AND MANUFACTURERS OF SOME PLASTICS (Continued)

Trade name	Composition	Manufacturer
Lustrex	Styrene molding and extrusion resins	Monsanto Chemical Company
Lytron RJ-100	Styrene molding and extrusion resins	" " "
Madurit	Melamine resins for plastic industry	Casella Farbwerke Mainkur, A.G.
Maplen	Polypropylenes	Chemore Corporation
Maraglas	Crystal clear epoxy resin	The Marblette Corporation
Marawood	Carvable epoxy resin	" " "
Marbon 8000	High styrene resins	Borg-Warner Corporation, Marbon Chemical Division
Marco	Polyester resins	Celanese Corp. of America
Marlex	Polyethylenes, polypropylenes, copolymers	Phillips Petroleum Company
Marvinol	Vinyl chloride resins and compounds	United States Rubber Company, Naugatuck Chemical Division
Melantine	Melamine resins	Ciba Limited
Melmac	Melamine resins	American Cyanamid Company
Melopas	Polyamide formaldehyde	Ciba Limited
Meltiplast 101	Polyethylenes	International Protected Metals
Meltiplast 301	Polyethylenes	" " "
Meltiplast 501	Chlorinated polyesters	" " "
Merlon	Polycarbonate resins	Mobay Chemical Co.
Methacrol	Acrylic emulsions	E. I. duPont de Nemours & Co., Inc.
Micarta	Melamines, phenolics, polyesters	Westinghouse Electric Co.
Microthene	Polyethylenes	U. S. Industrial Chemicals Co.
Monsanto	Polyethylene resins	Monsanto Chemical Company
MR Resin	Polyester resins	Celanese Corporation of America
Multrathane	Urethane elastomers	Mobay Chemical Company
Napcofoam	Polyurethane, flexible	Nopco Chemical Company
Natsyn	Polyisoprene	The Goodyear Tire & Rubber Co.
Naugatex 2700 series	Butadiene-styrene latices	United States Rubber Co., Naugatuck Chemical Division
Neville	Coumarone-indene resins	Neville Chemical Company
Niacet	Vinyl acetate, vinyl chloride	Union Carbide Corp.
Nopcofoam	Polyurethane plastics (flexible)	Nopco Chemical Co., Plastics Div.
Novodur	ABS copolymers	Farbenfabriken Bayer, A.G.
Nu-Klad	Epoxies	Amercoat Corporation
Nypene	Polystyrene resins	Neville Chemical Company
Opalon	Vinyl chloride resins and compounds	Monsanto Chemical Co.
Palmetto	Fluorocarbons	Greene, Tweed & Co.
Paracon	Polyester rubber	Bell Telephone Laboratories
Paracril	Nitrile rubber	United States Rubber Co., Naugatuck Chemical Div.
Paradene	Coumarone-indene resins	Neville Chemical Company
Paraplex	Polyester resins, acrylic modified polyester resin	Rohm & Haas Company
Pentacite	Pentaerythritol resins	Reichold Chemicals, Inc.
Permelite	Melamines	Melamine Plastics, Inc.
Petrothene	Polyethylene resin, polypropylene resin	U. S. Industrial Chemicals Co.
Pfaudlon 201	Epoxies	Pfaudler Co.
Pfaudlon 301	Chlorinated polyethers	" "
Philprene	Butadiene-styrene rubber	Phillips Petroleum Co.
Piccoflex	Styrene copolymer resins	Pennsylvania Industrial Chemical Corp.
Piccolastic	Styrene polymer resins	Pennsylvania Industrial Chemical Corp.
Plaskon	Nylons, melamines, phenolics polyesters	Allied Chemical Corp.
Pleogen	Polyester resins	American Petrochemical Corp.
Plexiglas	Acrylics	Rohm & Haas Company

TRADE NAMES, COMPOSITION AND MANUFACTURERS OF SOME PLASTICS (Continued)

Trade name	Composition	Manufacturer
Plioflex	Polyvinyl chlorides	The Goodyear Tire & Rubber Co.
Pliofoam	Expanded urea resins	" " " "
Pliolite	Styrene-butadiene resins	" " " "
Plio-Tuf series	Modified styrene resins	" " " "
Pliovic	Polyvinyl chlorides	" " " "
Plyophen	Phenol-formaldehyde resins	Reichhold Chemicals, Inc.
Poly-Eth	Polyethylene resins	Spencer Chemical Co.
Polylite	Polyester resins	Reichhold Chemicals, Inc.
Polypenco	Acrylics, chlorinated polyethers, fluorocarbons, nylons, polycarbonates,	Polymer Corp.
Red Thread	Epoxies, glassfiber filled	A. O. Smith Corp.
Resimene	Urea and melamine resins	Monsanto Chemical Co.
Resinox	Phenolic resins and compounds	" " "
Rezklad	Epoxies	Atlas Mineral Products, Div. Electric Storage Battery Co.
Rhonite	Urea resins	Rohm & Hass Company
Rock Island	Epoxies	Rock Island Fiberglass Pipe Co.
Root-Pruf	Butadiene-styrene	Triangle Conduit & Cable Co.
Roylar	Polyurethanes	United States Rubber Co., Naugatuck Chemical Division
RX	Epoxies, fluorocarbons, phenolics	Rogers Corp.
Ryercite	Phenolics	Joseph T. Ryerson & Son, Inc.
Ryertex	Phenolics	" " " "
Ryertex-Omicron	Polyvinyl chlorides	" " " "
S-4	Natural latex, liquid	Firestone Tire & Rubber Co.
Saran	Polyvinylidene chloride, vinylidene chloride copolymers	Saran Lined Pipe Co.
Sauereisen	Furanes	Sauereisen Cement Co.
Seilon ETH	Polyethylenes	Seiberling Rubber Co.
" PRO	Polypropylenes	" " "
" PVC	Polyvinyl chlorides	" " "
" S-3	ABS polymers	" " "
" UR	Urethane polymers	" " "
Semi-rigid	ABS polymers	Triangle Conduit & Cable Co.
Silastic®	Silicone rubber	Dow Corning
Solprene	Fluoro elastomers	Phillips Petroleum Company
S-polymers	Butadiene-styrene copolymers	Esso Laboratories
Stauffer	ABS polymers, polyethylenes, polypylenes, polyvinyl chlorides	Stauffer Chemical Company, Molded Products Div.
Stylplast	Urea-formaldehyde compounds	American Viscose Corporation
Styrex	Styrene-acrylonitrile copolymer	The Dow Chemical Company
Super Beckamine	Melamine-formaldehyde resins	Reichhold Chemicals, Inc.
Super Dylan	Polyethylene	Koppers Company, Inc.
Supreme	Polyethylenes	Johns-Manville Company
Sylkyd®	Silicone alkyd resins	Dow Corning
Sylplast	Urea formaldehyde	Food Machinery and Chemical Corp., Organic Chemicals Div.
Synthane	Epoxies, melamines, phenolics, silicones	Synthane Corporation
T/Na-100	Polyvinylfluorides	The Ruberoid Company
Teflon	Fluorocarbons, tetrafluoroethylene (TFE) fluorinated ethylpropylene resins (FEP)	E. I. duPont de Nemours & Co. Inc.
Tempron	Polybutadienes	Ace Molded Products Co.
Temp-R-Tape	Fluorocarbons	Connecticut Hardrubber Co.
Tenite	Cellulose acetate, cellulose-acetate-polyethylene, polypropylenes, urethane elastomers, copolymers	Eastman Chemical Products, Inc.

TRADE NAMES, COMPOSITION AND MANUFACTURERS OF SOME PLASTICS (Continued)

Trade name	Composition	Manufacturer
Tetran	Fluorocarbons	Pennsalt Chemicals Corp.
Texin	Urethane elastomers	Mobay Chemical Company
Thermo-Seal	Polyethylenes, polypropylenes	Cabot Piping Systems, Plastics Division
Thioment	Polyisoprenes	Atlas Mineral Products Co.
Trans-4	Rubberlike materials	Phillips Petroleum Company, Bartlesville, Okla.
TPC	Polyvinyl chlorides	Thermoplastics Corp.
ttP	ABS polymers, chlorinated polyesters, fluorocarbons, polyvinylchlorides	Cabot Piping Systems, Plastics Division
Tyril	Styrene-acrylonitrile, copolymer molding resins	The Dow Chemical Company
Ultrapas	Melamine resins	Dynamite A.G.
Ultrathene	Ethylene vinyl acetate	U. S. Industrial Chemicals Co.
Ultron	Polyvinyl chlorides	Monsanto Chemical Company
Unox	Epoxides	Union Carbide Corporation
Upalon	Polyvinyl chlorides	Monsanto Chemical Company
Vibrathane	Urethane elastomers	United States Rubber Co., Naugatuck Chemical Div.
Vibrin	Polyester resins	United States Rubber Co., Naugatuck Chemical Div.
Vinelle	Vinyl compounds	The General Tire & Rubber Co.
Vipla	Polyvinyl chloride, vinyl acetate, copolymers	Chemore Corp.
Vitalic	Chlorosulfonated polyethylenes, polybutadienes, polyisoprenes, ethylpropylenes, urethane elastomers	Continental Rubber Works
Vitel	Polyesters	Goodyear Chemical, Div., The Goodyear Tire & Rubber Co.
Viton	Synthetic rubbers	E. I. duPont de Nemours & Co., Inc.
Vitroplast	Polyester cement	Atlas Mineral Products Div.
Vygen	Polyvinyl chloride resins	The General Tire & Rubber Co.
Vyron	Polyvinyl chlorides	Monsanto Chemical Co.
X 2 B	Natural latex	The Firestone Tire & Rubber Co.
Yardley	ABS polymers, polyethylenes, polyvinyl chlorides	Celanese Plastics Co.
Zerlon	Methyl methacrylate styrene resins	The Dow Chemical Co.

Properties of Commercial Plastics

Of the many plastics commercially available in each chemical class only one or a very few individuals have been selected for this table as typical of the class. In some cases the range of properties has been expanded to include several grades or types. It is impractical to include a comprehensive list of materials or known properties of these materials in a table of convenient size. Properties vary widely with amount and kind of modifier such as filler and plasticizer. Within any type of thermoplastic resins molecular weight is an important variable. This property is controlled to afford the best physical properties available consistent with economical processing properties.

The information shown refers in all cases, except for "Forms available" and "Fabrication," to material in the fabricated form, which in the case of thermosetting materials means commercially cured. Physical and electrical properties will vary, to a greater or less degree with different materials, with humidity conditioning environment and with orientation. Strength values are quoted on the basis of short time tests at normal room temperature and are not suitable for engineering design purposes for load bearing applications. Maximum continuous service temperature refers to unloaded structures. The user of this table is referred to the specifications and test procedures of the American Society for Testing Materials.

PROPERTIES OF COMMERCIAL PLASTICS

Chemical Class	Cellulose Acetate	Cellulose Acetate	Cellulose Acetate Butyrate
Resin Type	Thermoplastic	Thermoplastic	Thermoplastic
Subclass or Modification	Soft	Hard	Soft
FORMS AVAILABLE Cs—castings, F—film, Fb—fibers, I—impregnants, L—laminations, Lq—lacquers, Mf—monofilaments, P—Powder, pellet, or granules, R—rods, tubes, or other extruded forms, S—sheets.	F, Lq, P, R, S	F, Lq, P, R, S	F, Lq, P, R, S
FABRICATION Cl—calendering, Cs—casting, E—extrusion, F—hot forming or drawing, I—impregnation, MB—blow molding, MC—compression molding, MI—injection molding, S—spreading.	Cs, E, F, MB, MC, MI, S	Cs, E, F, MB, MC, MI, S	Cs, E, F, MB, MC, MI, S
ELECTRICAL PROPERTIES			
D.C. Resistivity, ohm-cm	10^{10}–10^{13}	10^{10}–10^{13}	10^{10}–10^{12}
Dielectric constant, 60 cps	3.5–7.5	3.5–7.5	3.5–6.4
Dielectric constant, 10^6 cps	3.2–7.0	3.2–7.0	3.2–6.2
Dissipation factor, 60 cps	0.01–0.06	0.01–0.06	0.01–0.04
Dissipation factor, 10^6 cps	0.01–0.10	0.01–0.10	0.01–0.04
MECHANICAL PROPERTIES			
Modulus of elasticity, 10^3 psi	86–250	190–400	74–126
Tensile strength, psi	1,900–4,700	4,600–8,500	1,900–3,800
Ultimate elongation, %	32–50	6–40	60–74
Yield stress, psi	2,200–4,200	4,100–7,600	1,200–2,600
Yield strain, %			
Rockwell hardness	R 49–R 103	R 101–R 123	R 59–R 95
Notched Izod impact strength, ft.lb/in	2.0–5.2	0.4–2.7	2.5–5.4
Specific gravity	1.27–1.34	1.27–1.34	1.15–1.22
THERMAL PROPERTIES			
Burning rate	Medium	Medium	Medium
Heat distortion 264 psi, °C	44–57	60–113	49–58
Specific heat, cal/g	0.3–0.42	0.3–0.42	0.3–0.4
Linear thermal expansion coefficient, 10^{-5}, °C	8–16	8–16	11–17
Maximum continuous service temperature, °C			
CHEMICAL RESISTANCE			
Mineral acids, weak	Fair to good	Fair to good	Good
Mineral acids, strong	Poor	Poor	Fair to good
Oxidizing acids, concentrated	Very poor	Very poor	
Alkalies, weak	Poor	Poor	Good
Alkalies, strong	Very poor	Very poor	Poor
Alcohols	Poor	Poor	Poor
Ketones	Poor	Poor	Poor
Esters	Poor	Poor	Poor
Hydrocarbons, aliphatic	Fair to good	Fair to good	Fair to good
Hydrocarbons, aromatic	Poor to fair	Poor to fair	Poor
Oils, vegetable, animal, mineral	Fair to good	Fair to good	Good
MISCELLANEOUS PROPERTIES			
Clarity	Excellent	Excellent	Good to excellent
Color	Pale to colorless	Pale to colorless	Pale to colorless
Refractive index, n_D	1.46–1.50	1.46–1.50	1.46–1.49
Applicable ASTM specifications and test methods	D786, D706, D257, D150, D638, D785, D256, D792, D648, D696, D543, D542	D786, D706, D257, D150, D638, D785, D256, D792, D648, D696, D543, D542	D707, D257, D150, D638, D785, D256, D792, D648, D696

Chemical Class	Cellulose Acetate Butyrate	Nylon	Polycarbonates
Resin Type	Thermoplastic	Thermoplastic	Thermoplastic
Subclass or Modification	Hard	6/6	Unfilled
FORMS AVAILABLE Cs—castings, F—film, Fb—fibers, I—impregnants, L—laminations, Lq—lacquers, Mf—monofilaments, P—Powder, pellet, or granules, R—rods, tubes, or other extruded forms, S—sheets.	F, Lq, P, R, S	F, Fb, Mf, P, R, S	F, Fb, Mf, P, R, S
FABRICATION Cl—calendering, Cs—casting, E—extrusion, F—hot forming or drawing, I—impregnation, MB—blow molding, MC—compression molding, MI—injection molding, S—spreading.	Cs, E, F, MB, MC, MI, S	E, F, MB, MC, MI	Cs, E, F, MB, MC, MI
ELECTRICAL PROPERTIES			
D.C. Resistivity, ohm-cm	10^{10}–10^{12}		2 x 10^{16}
Dielectric constant, 60 cps	3.5–6.4	4.0–4.6	3.17
Dielectric constant, 10^6 cps	3.2–6.2	3.4–3.6	2.96
Dissipation factor, 60 cps	0.01–0.04	0.014–0.04	0.0009
Dissipation factor, 10^6 cps	0.01–0.04	0.04	0.01
MECHANICAL PROPERTIES			
Modulus of elasticity, 10^3 psi	150–200		290–325
Tensile strength, psi	5,000–6,800	9,000–12,000	8,000–9,500
Ultimate elongation, %	38–54	60–300	20–100
Yield stress, psi	3,600–6,100		8,000–10,000
Yield strain, %			
Rockwell hardness	R 108–R 117	R 108–R 120	M 70–M 180
Notched Izod impact strength, ft.lb/in	0.7–2.4	1.0–2.0	8–16
Specific gravity	1.19–1.25	1.13–1.15	1.2
THERMAL PROPERTIES			
Burning rate	Medium	Self extinguishing	Self extinguishing
Heat distortion 264 psi, °C	70–99		135–145
Specific heat, cal/g	0.3–0.4	0.4	0.3
Linear thermal expansion coefficient, 10^{-5}, °C	11–17	8.0	6.6
Maximum continuous service temperature, °C		80–150	138–143
CHEMICAL RESISTANCE			
Mineral acids, weak	Good	Very good	Excellent
Mineral acids, strong	Fair to good	Poor	Fair
Oxidizing acids, concentrated		Poor	
Alkalies, weak	Good	No effect	Poor
Alkalies, strong	Poor	No effect	Poor
Alcohols	Poor	Good	Poor
Ketones	Poor	Good	Poor
Esters	Poor	Good	Poor
Hydrocarbons, aliphatic	Fair to good	Very good	Poor
Hydrocarbons, aromatic	Poor	Fair to good	Poor
Oils, vegetable, animal, mineral	Good	Good	Poor
MISCELLANEOUS PROPERTIES			
Clarity	Good to excellent	Clear	Clear
Color	Pale to colorless	Pale amber to colorless	Colorless
Refractive index, n_D	1.46–1.49	1.53	1.60
Applicable ASTM specifications and test methods	D707, D257, D150, D638, D785, D256, D792, D648, D696, D543, D542	D257, D150, D638, D785, D256, D792, D648, D696, D543, D542	D257, D150, D638, D785, D256, D792, D648, D696, D543, D542

Chemical Class	Polyethylene	Polyethylene	Polyethylene
Resin Type	Thermoplastic	Thermoplastic	Thermoplastic
Subclass or Modification	Low density	Medium density	High density
FORMS AVAILABLE Cs—castings, F—film, Fb—fibers, I—impregnants, L—laminations, Lq—lacquers, Mf—monofilaments, P—Powder, pellet or granules, R—rods, tubes, or other extruded forms, S—sheets.	F, Mf, P, R, S	F, Mf, P, R, S	F, Fb, Mf, P, R, S
FABRICATION Cl—calendering, Cs—casting, E—extrusion, F—hot forming or drawing, I—impregnation, MB—blow molding, MC—compression molding, MI—injection molding, S—spreading.	Cl, E, F, MB, MC, MI	Cl, E, F, MB, MC, MI	Cl, E, F, MB, MC, MI
ELECTRICAL PROPERTIES			
D.C. Resistivity, ohm-cm	$>10^{15}$	$>10^{15}$	$>10^{15}$
Dielectric constant, 60 cps	2.3–2.35	2.3	2.3–2.35
Dielectric constant, 10^6 cps	2.3–2.35	2.3	2.3–2.35
Dissipation factor, 60 cps	<0.0005	<0.0005	<0.0005
Dissipation factor, 10^6 cps	<0.0005	<0.0005	<0.0005
MECHANICAL PROPERTIES			
Modulus of elasticity, 10^3 psi	14–38	35–90	85–160
Tensile strength, psi	1,000–1,400	1,200–3,500	3,100–5,500
Ultimate elongation, %	400–700	50–600	15–100
Yield stress, psi	1,100–1,700	1,500–2,600	2,400–5,000
Yield strain, %	20–40	10–20	5–10
Rockwell hardness			R 30–R 50
Notched Izod impact strength, ft.lb/in.	no break	0.5–>16	1.5–20
Specific gravity	0.91–0.925	0.926–0.941	0.941–0.965
THERMAL PROPERTIES			
Burning rate	Very slow	Slow	Slow
Heat distortion 264 psi, °C			
Specific heat, cal/g	0.55	0.55	0.55
Linear thermal expansion coefficient, 10^{-5}, °C	10–20	14–16	11–13
Maximum continuous service temperature, °C	60–77	71–93	92–200
CHEMICAL RESISTANCE			
Mineral acids, weak	Good	Excellent	Excellent
Mineral acids, strong	Good	Excellent	Excellent
Oxidizing acids, concentrated	Good to poor	Good to poor	Good to poor
Alkalies, weak	Good	Excellent	Excellent
Alkalies, strong	Good	Excellent	Excellent
Alcohols	Excellent to poor	Excellent to poor	Excellent to poor
Ketones	Excellent to poor	Excellent to poor	Excellent to poor
Esters	Excellent to poor	Excellent to poor	Excellent to poor
Hydrocarbons, aliphatic	Fair	Fair	Fair
Hydrocarbons, aromatic	Fair	Good	Fair
Oils, vegetable, animal, mineral	Good	Excellent	Good
MISCELLANEOUS PROPERTIES			
Clarity	Translucent	Translucent	Translucent
Color	Colorless	Colorless	Colorless
Refractive index, n_D	1.50–1.54	1.52–1.54	1.54
Applicable ASTM specifications and test methods	D702, D788, D257, D150, D412, D638, D696, D543, D542, D1248	D257, D150, D412, D638, D785, D256, D696, D543, D542, D1248	D257, D150, D412, D638, D785, D256, D696, D543, D542, D1248

PROPERTIES OF COMMERCIAL PLASTICS

PROPERTIES OF COMMERCIAL PLASTICS

Chemical Class	Methylmethacrylate	Polypropylene	Polypropylene
Resin Type	Thermoplastic	Thermoplastic	Thermoplastic
Subclass or Modification	Unmodified	Unmodified	Copolymer
FORMS AVAILABLE Cs—castings, F—film, Fb—fibers, I—impregnants, L—laminations, Lq—lacquers, Mf—monofilaments, P—Powder, pellet, or granules, R—rods, tubes, or other extruded forms, S—sheets.	Cs, P, R, S	F, Fb, Mf, P, R, S	F, Fb, Mf, P, R, S
FABRICATION Cl—calendering, Cs—casting, E—extrusion, F—hot forming or drawing, I—impregnation, MB—blow molding, MC—compression molding, MI—injection molding, S—spreading	Cs, E, F, Lq, MB, MC, MI	Cl, E, F, MB, MC, MI	Cl, E, F, MB, MC, MI
ELECTRICAL PROPERTIES			
D.C. Resistivity, ohm-cm	$>10^{14}$	$>10^{15}$	10^{17}
Dielectric constant, 60 cps	3.5–4.5	2.2–2.6	2.3
Dielectric constant, 10^6 cps	3.0–3.5	2.2–2.6	2.3
Dissipation factor, 60 cps	0.04–0.06	<0.0005	0.0001–0.0005
Dissipation factor, 10^6 cps	0.02–0.03	0.0005–0.002	0.0001–0.002
MECHANICAL PROPERTIES			
Modulus of elasticity, 10^3 psi	350–500	1.4–1.7	
Tensile strength, psi	7,000–11,000	4,300–5,500	2,900–4,500
Ultimate elongation, %	2.0–10	>220	200–700
Yield stress, psi		4,900	
Yield strain, %		15	
Rockwell hardness	M 80–M 105	93	R 50–R 96
Notched Izod impact strength, ft.lb/in	0.3–0.6	1.0	1.1–12
Specific gravity	1.18–1.20	0.90	0.90
THERMAL PROPERTIES			
Burning rate	Slow	Medium	Medium
Heat distortion 264 psi, °C	66–99		
Specific heat, cal/g	0.35	0.5	0.5
Linear thermal expansion coefficient, 10^{-5}, °C	5.0–9.0	5.8–10	8–10
Maximum continuous service temperature, °C	60–93		190–240
CHEMICAL RESISTANCE			
Mineral acids, weak	Good	Excellent	Excellent
Mineral acids, strong	Fair to poor	Excellent	Excellent
Oxidizing acids, concentrated	Attacked	Good to poor	Poor
Alkalies, weak	Good	Excellent to good	Excellent
Alkalies, strong	Poor	Excellent to good	Good
Alcohols		Excellent to good	Good below 80°C
Ketones	Dissolves	Excellent to good	Good below 80°C
Esters	Dissolves	Excellent to good	Good below 80°C
Hydrocarbons, aliphatic	Good	Good to fair	Good below 80°C
Hydrocarbons, aromatic	Softens	Good to fair	Good below 80°C
Oils, vegetable, animal, mineral	Good	Good	
MISCELLANEOUS PROPERTIES			
Clarity	Excellent	Transparent	Transparent
Color	Colorless	Colorless to sl. yellow	Colorless to sl. yellow
Refractive index, n_D	1.48–1.50	1.49	
Applicable ASTM specifications and test methods	D257, D150, D638, D785, D256, D792, D648, D696, D543, D542	D257, D150, D412, D638, D785, D256, D648, D543, D542	D257, D150, D412, D638, D785, D256, D648, D543, D542

PROPERTIES OF COMMERCIAL PLASTICS

PROPERTIES OF COMMERCIAL PLASTICS

Chemical Class	Polystyrene	Polystyrene-acrylonitrile	Polytetrafluoro ethylene
Resin Type	Thermoplastic	Thermoplastic	Thermoplastic
Subclass or Modification	Unmodified	Unmodified	Unmodified
FORMS AVAILABLE Cs—castings, F—film, Fb—fibers, I—impregnants, L—laminations, Lq—lacquers, Mf—monofilaments, P—Powder, pellet, or granules, R—rods, tubes, or other extruded forms, S—sheets.	F, Fb, Mf, P, R, S	F, Mf, P, R, S	F, L, P, R, S
FABRICATION Cl—calendering, Cs—casting, E—extrusion, F—hot forming or drawing, I—impregnation, MB—blow molding, MC—compression molding, MI—injection molding, S—spreading.	E, F, MB, MC, MI	Cl, E, F, MB, MC, MI	E, F, MC, MI
ELECTRICAL PROPERTIES			
D.C. Resistivity, ohm-cm	$>10^{16}$	10^{13}–10^{17}	10^{18}
Dielectric constant, 60 cps	2.5–2.65	2.6–3.4	2.
Dielectric constant, 10^6 cps	2.5–2.65	2.5–3.1	2.
Dissipation factor, 60 cps	0.0001–0.0003	0.006–0.008	0.0002
Dissipation factor, 10^6 cps	0.0001–0.0004	0.007–0.01	0.0002
MECHANICAL PROPERTIES			
Modulus of elasticity, 10^3 psi	400–600	$>10^{16}$	33–65
Tensile strength, psi	5,000–10,000	9,000–12,000	2,000–4,500
Ultimate elongation, %	1.0–2.5	1.0–2.5	200–400
Yield stress, psi			1,600–2,000
Yield strain, %			50–75
Rockwell hardness	M 65–M 85	M 75–M 90	D 50–D 65
Notched Izod impact strength, ft.lb/in	0.25–0.60	0.3–0.6	2.5–4.0
Specific gravity	1.04–1.08	1.05–1.1	2.1–2.3
THERMAL PROPERTIES			
Burning rate	Medium to slow	Slow	Self extinguishing
Heat distortion 264 psi, °C		91–104	60
Specific heat, cal/g	0.32–0.35	0.32–0.35	0.25
Linear thermal expansion coefficient, 10^{-5}, °C	6.0–8.0	3.6–3.8	10
Maximum continuous service temperature, °C	66–82	77–88	260
CHEMICAL RESISTANCE			
Mineral acids, weak	Excellent	Excellent	Excellent
Mineral acids, strong	Excellent	Good to excellent	Excellent
Oxidizing acids, concentrated	Poor	Poor	Excellent
Alkalies, weak	Excellent	Excellent	Excellent
Alkalies, strong	Excellent	Good to excellent	Excellent
Alcohols	Excellent	Good to excellent	Excellent
Ketones	Dissolves	Dissolves	Excellent
Esters	Poor	Dissolves	Excellent
Hydrocarbons, aliphatic	Poor	Good	Excellent
Hydrocarbons, aromatic	Dissolves	Fair to good	Excellent
Oils, vegetable, animal, mineral	Fair to poor	Good to excellent	Excellent
MISCELLANEOUS PROPERTIES			
Clarity	Transparent	Transparent	Translucent
Color	Colorless	Colorless to amber	Colorless to grey
Refractive index, n_D	1.59–1.60	1.56–1.57	1.30–1.40
Applicable ASTM specifications and test methods	D257, D150, D638, D785, D256, D792, D648, D696, D543, D542	D257, D150, D638, D785, D256, D792, D648, D696, D543, D542	

PROPERTIES OF COMMERCIAL PLASTICS

Chemical Class	Polytrifluorochloro ethylene	Polyvinylchloride and Vinylchloride acetate	Polyvinylchloride and Vinylchloride acetate
Resin Type	Thermoplastic	Thermoplastic	Thermoplastic
Subclass or Modification	Unmodified	Unmodified, rigid	Plasticized (non rigid)
FORMS AVAILABLE Cs—castings, F—film, Fb—fibers, I—impregnants, L—laminations, Lq—lacquers, Mf—monofilaments, P—Powder, pellet, or granules, R—rods, tubes, or other extruded forms, S—sheets.	F, Mf, P, R, S	F, Fb, I, Lq, Mf, P, R, S	F, L, P, R, S
FABRICATION Cl—calendering, Cs—casting, E—extrusion, F—hot forming or drawing, I—impregnation, MB—blow molding, MC—compression molding, MI—injection molding, S—spreading.	Cs, E, F, I, MC, MI, S	Cl, Cs, E, F, I, MB, MC, MI, S	Cl, Cs, E, MB, MC, MI, S
ELECTRICAL PROPERTIES			
D.C. Resistivity, ohm-cm	10^{18}	10^{12}–10^{16}	10^{11}–10^{14}
Dielectric constant, 60 cps	2.2–2.8	3.2–4.0	5.0–9.0
Dielectric constant, 10^6 cps	2.3–2.5	3.0–4.0	3.0–4.0
Dissipation factor, 60 cps	0.001	0.01–0.02	0.03–0.05
Dissipation factor, 10^6 cps	0.005	0.006–0.02	0.06–0.1
MECHANICAL PROPERTIES			
Modulus of elasticity, 10^3 psi	150	200–600	
Tensile strength, psi	4,500–6,000	5,000–9,000	1,500–3,000
Ultimate elongation, %	250	2.0–40	200–400
Yield stress, psi	4,200		
Yield strain, %	10	1.0–5.0	
Rockwell hardness	J 75–J 95	R 110–R 120	
Notched Izod impact strength, ft.lb/in.	2.5–4.0	0.4–2.0	
Specific gravity	2.1–2.3	1.36–1.4	1.15–1.35
THERMAL PROPERTIES			
Burning rate	Self extinguishing	Self extinguishing	Slow to self extinguishing
Heat distortion 264 psi, °C		60–80	
Specific heat, cal/g	0.22	0.2–0.28	0.36–0.5
Linear thermal expansion coefficient, 10^{-5}, °C	7.0	5.0–18	7.0–25
Maximum continuous service temperature, °C	200	70–74	80–105
CHEMICAL RESISTANCE			
Mineral acids, weak	Excellent	Excellent	Fair to good
Mineral acids, strong	Excellent	Good to excellent	Fair to good
Oxidizing acids, concentrated	Excellent	Fair to good	Poor to fair
Alkalies, weak	Excellent	Excellent	Fair to good
Alkalies, strong	Excellent	Good	Fair to good
Alcohols	Excellent	Excellent	Fair
Ketones	Excellent	Poor	Poor
Esters	Excellent	Poor	Poor
Hydrocarbons, aliphatic	Excellent	Excellent	Poor
Hydrocarbons, aromatic	Excellent	Poor	Poor
Oils, vegetable, animal, mineral	Excellent	Excellent	Poor
MISCELLANEOUS PROPERTIES			
Clarity	Transparent	Transparent	Transparent
Color	Colorless to pale	Colorless to amber	Colorless to amber
Refractive index, n_D	1.43	1.54	1.50–1.55
Applicable ASTM specifications and test methods	D1430, D257, D150, D638, D785, D256, D792, D648, D696, D543, D542	D708, D728, D257, D150, D638, D256, D792, D648, D696, D543, D542	D1432, D257, D150, D412, D792, D543, D542

Chemical Class	Epoxy	Melamine-Formaldehyde	Melamine-Formaldehyde
Resin Type	Thermosetting	Thermosetting	Thermosetting
Subclass or Modification	Unfilled	α-Cellulose filled	Mineral filled, (electrical)
FORMS AVAILABLE Cs—castings, F—film, Fb—fibers, I—impregnants, L—laminations, Lq—lacquers, Mf—monofilaments, P—Powder, pellet, or granules, R—rods, tubes, or other extruded forms, S—sheets.	Cs, Lq	P, R, S	P, R, S
FABRICATION Cl—calendering, Cs—casting, E—extrusion, F—hot forming or drawing, I—impregnation, MB—blow molding, MC—compression molding, MI—injection molding, S—spreading.	Cs, I, S	MC	MC
ELECTRICAL PROPERTIES			
D.C. Resistivity, ohm-cm	10^{12}–10^{14}	10^{12}–10^{14}	10^{13}–10^{14}
Dielectric constant, 60 cps	3.5–5.0	7.9–9.4	10.2
Dielectric constant, 10^6 cps	3.4–4.4	7.2–8.4	6.1
Dissipation factor, 60 cps	0.001–0.005	0.03–0.08	0.10
Dissipation factor, 10^6 cps	0.03–0.05	0.03–0.043	0.051
MECHANICAL PROPERTIES			
Modulus of elasticity, 10^3 psi	>300	1,300	1,950
Tensile strength, psi	4,000–13,000	7,000–13,000	5,500–6,500
Ultimate elongation, %	2.0–6.0	0.6–0.9	
Yield stress, psi			
Yield strain, %			
Rockwell hardness	M 75–M 110	M 110–M 124	E 90
Notched Izod impact strength, ft.lb/in	0.2–1.0	0.24–0.35	0.3–0.4
Specific gravity	1.115	1.47–1.52	1.78
THERMAL PROPERTIES			
Burning rate	Slow	Self extinguishing	Self extinguishing
Heat distortion 264 psi, °C	Up to 120	204	130
Specific heat, cal/g	0.25–0.4	0.4	
Linear thermal expansion coefficient, 10^{-5}, °C	4.5–9.0	2.0–5.7	2.1–4.3
Maximum continuous service temperature, °C	80	99.0	149
CHEMICAL RESISTANCE			
Mineral acids, weak	Excellent	Good	Fair
Mineral acids, strong	Fair to good	Poor	Poor
Oxidizing acids, concentrated	Poor	Poor	Poor
Alkalies, weak	Excellent	Good	Fair
Alkalies, strong	Excellent	Poor	Poor
Alcohols	Excellent	Good	Good
Ketones	Poor	Good	Good
Esters		Good	Good
Hydrocarbons, aliphatic	Excellent	Good	Good
Hydrocarbons, aromatic	Excellent	Good	Good
Oils, vegetable, animal, mineral	Excellent	Good	Good
MISCELLANEOUS PROPERTIES			
Clarity	Transparent	Translucent	Opaque
Color	Colorless	Colorless	Dark
Refractive index, n_D	1.58		
Applicable ASTM specifications and test methods	D257, D150, D651, D785, D256, D792, D648, D696, D543	D704, D257, D150, D638, D785, D256, D792, D648, D696, D543	D704, D257, D150, D638, D785, D256, D792, D648, D696, D543

Chemical Class	Phenol-Formaldehyde	Phenol-Formaldehyde	Phenol-Formaldehyde
Resin Type	Thermosetting	Thermosetting	Thermosetting
Subclass or Modification	Cordfilled	Cellulose filled	Unfilled Cast Phenolic, mechanical and chemical grade
FORMS AVAILABLE Cs—castings, F—film, Fb—fibers, I—impregnants, L—laminations, Lq—lacquers, Mf—monofilaments, P—Powder, pellet, or granules, R—rods, tubes, or other extruded forms, S—sheets.	L, P, S	L, P, S	Cs, R, S
FABRICATION Cl—calendering, Cs—casting, E—extrusion, F—hot forming or drawing, I—impregnation, MB—blow molding, MC—compression molding, MI—injection molding, S—spreading.	MC	MC	Cs, F
ELECTRICAL PROPERTIES			
D.C. Resistivity, ohm-cm	10^{11}–10^{12}	10^{11}–10^{13}	1.0-7.0 x 10^{12}
Dielectric constant, 60 cps	7.0–10.0	5.0–9.0	6.5–7.5
Dielectric constant, 10^6 cps	5.0–6.0	4.0–7.0	4.0–5.5
Dissipation factor, 60 cps	0.1–0.3	0.04–0.3	0.10–0.15
Dissipation factor, 10^6 cps	0.04–0.09	0.03–0.07	0.04–0.05
MECHANICAL PROPERTIES			
Modulus of elasticity, 10^3 psi	900–1,300	800–1,200	4.0–5.0
Tensile strength, psi	6,000–9,000	6,500–8,500	6,000–9,000
Ultimate elongation, %	0.5–1.0	0.6–1.0	1.5–2.0
Yield stress, psi			
Yield strain, %			
Rockwell hardness		M 110–M 120	M 93–M 120
Notched Izod impact strength, ft.lb/in	4.0–8.0	0.24–0.34	0.25–0.4
Specific gravity	1.36–1.43	1.32–1.55	1.307–1.318
THERMAL PROPERTIES			
Burning rate	Self extinguishing	Self extinguishing	Self extinguishing
Heat distortion 264 psi, °C	121–127	143–171	74–80
Specific heat, cal/g		0.35–0.40	
Linear thermal expansion coefficient, 10^{-5}, °C		3.0–4.5	6.0–8.0
Maximum continuous service temperature, °C	121	149–177	
CHEMICAL RESISTANCE			
Mineral acids, weak	Variable	Variable	Fair to good
Mineral acids, strong	Poor	Poor	Poor to good
Oxidizing acids, concentrated	Poor	Poor	Poor
Alkalies, weak	Variable	Variable	Poor to good
Alkalies, strong	Poor	Poor	Poor
Alcohols	Good	Good to excellent	Good to excellent
Ketones	Poor to fair	Fair	Fair
Esters	Fair to good	Fair to good	Fair to good
Hydrocarbons, aliphatic	Fair to good	Excellent	Good to excellent
Hydrocarbons, aromatic	Fair to good	Excellent	Good
Oils, vegetable, animal, mineral	Good	Excellent	Excellent
MISCELLANEOUS PROPERTIES			
Clarity	Opaque	Opaque	Clear
Color			Colorless to amber
Refractive index, n_D			
Applicable ASTM specifications and test methods	D700, D257, D150, D638, D651, D785, D256, D792, D648, D543	D700, D257, D150, D638, D651, D785, D256, D792, D648, D696, D543	D257, D150, D638, D785, D256, D792, D648, D696, D543

PROPERTIES OF COMMERCIAL PLASTICS

Chemical Class	Polyester (Styrene-Alkyd)	Silicones	Urea Formaldehyde
Resin Type	Thermosetting	Thermosetting	Thermosetting
Subclass or Modification	Glassfiber mat reinforced	Mineral filled	α-Cellulose filled
FORMS AVAILABLE Cs—castings, F—film, Fb—fibers, I—impregnants, L—laminations, Lq—lacquers, Mf—monofilaments, P—Powder, pellet, or granules, R—rods, tubes, or other extruded forms, S—sheets.	L, S	P	P, R, S
FABRICATION Cl—calendering, Cs—casting, E—extrusion, F—hot forming or drawing, I—impregnation, MB—blow molding, MC—compression molding, MI—injection molding, S—spreading.	I	MC	MC
ELECTRICAL PROPERTIES			
D.C. Resistivity, ohm-cm	10^{11}	$>10^{12}$	0.5–5.0
Dielectric constant, 60 cps	4.0–5.5	3.5–3.6	7.7–9.5
Dielectric constant, 10^6 cps	4.0–5.5	3.4–3.6	6.7–8.0
Dissipation factor, 60 cps	0.01–0.04	0.004	0.036–0.043
Dissipation factor, 10^6 cps	0.01–0.06	0.005–0.007	0.025–0.035
MECHANICAL PROPERTIES			
Modulus of elasticity, 10^3 psi	500–1,500		1,300–1,400
Tensile strength, psi	30,000–50,000	3,000–4,000	5,500–13,000
Ultimate elongation, %	0.5–1.5		0.6
Yield stress, psi			
Yield strain, %			
Rockwell hardness	M 80–M 120	M 85–M 95	E 94–E 97
Notched Izod impact strength, ft.lb/in	7.0–30	0.25–0.35	0.24–0.40
Specific gravity	1.5–2.1	1.8–2.8	1.47–1.52
THERMAL PROPERTIES			
Burning rate	Self extinguishing	Self extinguishing	Self extinguishing
Heat distortion 264 psi, °C	93–288	>260	130
Specific heat, cal/g	0.2–0.4	0.2–0.3	0.6
Linear thermal expansion coefficient, 10^{-5}, °C	1.8–3.0	2.0–4.0	2.2–3.6
Maximum continuous service temperature, °C	121–204	288	77
CHEMICAL RESISTANCE			
Mineral acids, weak	Good	Fair to good	Poor
Mineral acids, strong	Poor	Poor to good	Poor
Oxidizing acids, concentrated	Poor		Poor
Alkalies, weak	Good	Fair	Fair
Alkalies, strong	Poor	Poor	Poor
Alcohols	Good	Poor	Good
Ketones	Poor	Poor	Good
Esters	Good		Good
Hydrocarbons, aliphatic	Good	Fair to good	Good
Hydrocarbons, aromatic	Poor to fair	Poor	Good
Oils, vegetable, animal, mineral	Good	Good	
MISCELLANEOUS PROPERTIES			
Clarity	Translucent	Opaque	Translucent
Color	Colorless	Pale to dark	Colorless
Refractive index, n_D			1.54–1.56
Applicable ASTM specifications and test methods	D257, D150, D638, D785, D256, D792, D648, D696, D543	D257, D150, D785, D256, D792, D648, D696, D543	D705, D257, D150, D638, D785, D256, D792, D648

Chemical Class	Acrylonitrile-Butadiene-Styrene (ABS)	Acetal	Alkyd resins
Resin Type	Thermoplastic	Thermoplastic	Thermosetting
Subclass or Modification	High Heat resistant	Homopolymer	Synthetic fiber filled
FORMS AVAILABLE Cs—castings, F—film, Fb—fibers, I—impregnants, L—laminations, Lq—lacquers, Mf—monofilaments, P—Powder, pellet, or granules, R—rods, tubes, or other extruded forms, S—sheets.	P, S, L, R	C, R	P,
FABRICATION Cl—calendering, Cs—casting, E—extrusion, F—hot forming or drawing, I—impregnation, MB—blow molding, MC—compression molding, MI—injection molding, S—spreading.	Cl, E, MB, MI	MI, E	Cs, MC, MI
ELECTRICAL PROPERTIES			
D.C. Resistivity, ohm-cm			
Dielectric constant, 60 cps	2.4–5.0		3.8–5.0
Dielectric constant, 10^6 cps	2.4–3.8	3.7	3.6–4.7
Dissipation factor, 60 cps	0.003–0.008		0.012–0.026
Dissipation factor, 10^6 cps	0.007–0.015	0.004	0.01–0.016
MECHANICAL PROPERTIES			
Modulus of elasticity, 10^3 psi			
Tensile strength, psi	7,000–8,000	10,000–12,000	4,500–6,500
Ultimate elongation, %	1.0–20	15–75	
Yield stress, psi	4,000–9,000		10,000–13,000
Yield strain, %			
Rockwell hardness	R 110–115	M 94, R 120	E 76
Notched Izod impact strength, ft.lb/in.	2.0–4.0	1.4–2.3	0.50–4.5
Specific gravity	1.06–1.08	1.43	1.24–2.6
THERMAL PROPERTIES			
Burning rate	Slow	Slow	Self extinguishing
Heat distortion 264 psi, °C	115–118		
Specific heat, cal/g	0.3–0.4	0.35	
Linear thermal expansion coefficient, 10^{-5}, °C	6.0–6.5	8.1	4.0–5.5
Maximum continuous service temperature, °C	88–110	84	149–220
CHEMICAL RESISTANCE			
Mineral acids, weak	Good	Fair	Good
Mineral acids, strong	Good	Poor	Fair
Oxidizing acids, concentrated	Poor	Poor	
Alkalies, weak	Good	Poor	Good
Alkalies, strong	Good	Poor	Fair
Alcohols	Good	Good	Fair to good
Ketones	Poor	Good	Fair to good
Esters	Poor	Good	Fair to good
Hydrocarbons, aliphatic	Fair	Good	Fair to good
Hydrocarbons, aromatic	Fair	Good	Fair to good
Oils, vegetable, animal, mineral	Good	Good	
MISCELLANEOUS PROPERTIES			
Clarity	Translucent to opaque	Translucent to opaque	Opaque
Color	Colorless	Colorless	Colorless
Refractive index, n_D		1.48	
Applicable ASTM specifications and test methods	D638, D150, D792, D651, D648, D256, D758, D696, D543	D638, D150, D792, D651, D648, D256, D758, D696, D543	D638, D150, D792, D651, D648, D256, D758, D543

IONIC EXCHANGE RESINS

ANION EXCHANGE RESINS

The following table is divided into two parts; the first lists properties of some anionic resins and the second, properties of some cationic resins.

Character S = strong W = weak	Trade name	Manufacturer*	Active group	Matrix	Effective pH	Selectivity	Order of selectivity	Total exchange capacity; meq/ml	Total exchange capacity; meq/gm	Maximum thermal stability; °C	Physical form; s = sphere b = beads	Standard mesh range	Ionic form as shipped	Shipping density; lb./cu. ft.
S	Dowex 1	1	Trimethyl benzyl ammonium	Polystyrene	0–14	Cl/H approx. 25	I, NO_3, Br, Cl, Acetate, OH, F	1.33	3.5	OH^- 50 Cl^- 150	s	20–50 (wet)	Cl^-	44
S	Dowex 21 K	1	Trimethyl benzyl ammonium	Polystyrene	0–14	Cl/H approx. 15	I, NO_3, Br, Cl, Acetate, OH, F	1.25	4.5	OH^- 50 Cl^- 150	s	20–50 (wet)	Cl^-	43
S	Duolite A-101 D	2	Quaternary ammonium	Polystyrene	0–14	—	—	1.4	4.2	OH^- 60 Cl^- 100	b	16–50	Cl^-	—
S	Ionac A-540	3	Quaternary ammonium	Polystyrene	0–14	—	—	1.0	3.6	salt 100 OH^- 60	b	16–50	salt	43–66
S	Dowex 2	1	Dimethyl ethanol benzyl ammonium	Polystyrene	0–14	Cl/H approx. 1.5	I, NO_3, Br, Cl, Acetate, OH, F	1.33	3.5	OH^- 30 Cl^- 150	s	20–50 (wet)	Cl^-	44
S	Duolite A-102 D	2	Quaternary ammonium	Polystyrene	0–14	—	—	1.4	4.2	OH^- 40 Cl^- 100	b	16–50	Cl^-	—
S	Ionac A-550	3	Dimethyl ethanol benzyl ammonium	Polystyrene	0–14	—	—	1.3	3.5	salt 100 OH^- 40	b	16–50	salt	43–46
W	Duolite A-30 B	2	Tertiary amine; Quaternary ammonium	Epoxy polyamines	0–9	—	—	2.6	8.7	80	b	16–50	salt	—

IONIC EXCHANGE RESINS (Continued)

ANION EXCHANGE RESINS (Continued)

Character S = strong W = weak	Trade name	Manu-facturer*	Active group	Matrix	Effective pH	Selectivity	Order of selectivity	Total exchange capacity; meq/ml	Total exchange capacity; meq/gm	Maximum thermal stability; °C	Physical form; s = sphere b = beads	Standard mesh range	Ionic form as shipped	Shipping density; lb./cu. ft.
W	Ionac A-300	3	Tertiary amine; Quaternary ammonium	Epoxy amine	0–12	—	—	1.8	5.5	40	g	16–50	salt	19–21
W	Duolite A-6	2	Tertiary amine	Phenolic	0–5	—	—	2.4	7.6	60	g	16–50	salt	—
W	Duolite A-7	2	Secondary amine	Phenolic	0–4	—	—	2.4	9.1	40	g	16–50	salt	—

* 1. Dow
2. Diamond Shamrock
3. Ionac
4. Nalco

CATION EXCHANGE RESINS

Character S = strong W = weak	Trade name	Manu-facturer*	Active group	Matrix	Effective pH	Selectivity	Order of selectivity	Total exchange capacity; meq/ml	Total exchange capacity; meq/mg	Maximum thermal stability; °C	Physical form; s = sphere b = beads	Standard mesh range	Ionic form as shipped	Shipping density; lb./cu. ft.
S	Dowex 50	1	Nuclear sulfonic acid	Polystyrene	0–14	Na/H approx. 1.2	Ag, Cs, Rb, K, NH_4, Na, H, Li, Ba, Sr, Ca, Mg, Be	Na^+ 1.9 H^+ 1.7	Na^+ 4.8 H^+ 5.0	150	s	20–50 (wet)	H^+ or Na^+	H^+ 50 Na^+ 53
S	Dowex MPC-1	4	Nuclear sulfonic acid	Polystyrene	0–14	—	—	1.6–1.8 H^+ form	4.5–4.9 H^+ form	150	b	20–40 (wet)	Na^+	50
S	Duolite C-20	2	Nuclear sulfonic acid	Polystyrene	0–14	—	—	2.2	5.1	150	b	16–50	Na^+	—
S	Ionac 240	3	Nuclear sulfonic acid	Polystyrene	0–14	—	—	1.9	4.6	140 (Na^+) 130 (H^+)	b	16–50	Na^+	50–55

IONIC EXCHANGE RESINS (Continued)

CATION EXCHANGE RESINS (Continued)

Character S = strong W = weak	Trade name	Manu-facturer*	Active group	Matrix	Effective pH	Selectivity	Order of selectivity	Total exchange capacity; meq/ml	Total exchange capacity; meq/gm	Maximum thermal stability; °C	Physical form; s = sphere b = beads	Standard mesh range	Ionic form as shipped	Shipping density; lb./cu. ft.
S	Duolite C-3	2	Methylene sulfonic	Phenolic	0–9	—	—	1.1	2.9	60	g	16–50	H^+	—
W	Dowex CCR-1	4	Carboxylic	Phenolic	0–9	—	—	—	—	38	g	20–50 (wet)	H^+ (dry)	21
W	Duolite ES-63	2	Phosphonic	Polystyrene	4–14	—	—	3.3	6.5	100	b	16–50	H^+	—
W	Duolite ES-80	2	Aliphatic	Acrylic	6–14	—	—	3.5	10.2	100	b	16–50	H^+	—

* 1. Dow
2. Diamond Shamrock
3. Ionac
4. Nalco

AZEOTROPES

BINARY SYSTEMS

No.	Components: Compounds	Components: BP, °C	Azeotrope: BP, °C	Percent composition: In azeotrope	Percent composition: Upper layer	Percent composition: Lower layer	Relative volume of layers at 20°C	Specific gravity of layers or azeotrope
1	a. Acetal	102–4	78.2	84.5				
	b. Chloroform	61.2		15.5				
2	a. Acetaldehyde	20.9	−7.0	16.0				
	b. Butane	−0.5		84.0				
3	a. Acetaldehyde	20.9	20.5	76.0				0.762
	b. Ethyl ether	34.6		24.0				
4	a. Acetaldehyde } 1551 mm	55.0	52.6	77.5				0.763
	b. Ethyl ether } 1551 mm	70.0		22.5				
5	a. Acetamide	222.0	178.6	6.5				
	b. Benzaldehyde	179.5		93.5				
6	a. Acetamide	222.0	223.0	50.0				
	b. o-Bromophenol	195.0		50.0				
7	a. Acetamide	222.0	213.6	55.0				
	b. p-Chloronitrobenzene	242.0		45.0				
8	a. Acetamide	222.0	157.8	8.0				
	b. o-Chlorotoluene	159.0		92.0				
9	a. Acetamide	222.0	199.4	18.0				
	b. p-Dichlorobenzene	219.0		82.0				
10	a. Acetamide	222.0	190.7	5.0				
	b. Glycol monoacetate	182.0		95.0				
11	a. Acetamide	222.0	176.9	5.4				0.926
	b. 2-Methyl-5-ethyl pyridine	174.0		94.6				
12	a. Acetamide	222.0	206.5	32.5				
	b. o-Nitrotoluene	222.3		67.5				
13	a. Acetamide	222.0	202.0	24.0				
	b. Nitrobenzene	210.9		76.0				
14	a. Acetamide	222.0	194.5	9.5				
	b. Octyl alcohol	195.0		90.5				
15	a. Acetamide	222.0	198.6	12.0				
	b. o-Toluidine	199.8		88.0				
16	a. Acetamide	222.0	142.6	11.0				
	b. o-Xylene	144.1		89.0				
17	a. Acetic acid	118.1	80.1	2.0				0.882
	b. Benzene	80.1		98.0				
18	a. Acetic acid	118.1	118.4	95.0				
	b. Bromobenzene	156.0		5.0				
19	a. Acetic acid	118.1	116.7	81.0				0.983
	b. Butyl ether	142.0		19.0				

AZEOTROPES (Continued)

BINARY SYSTEMS (Continued)

No.	Components: Compounds	Components: BP, °C	Azeotrope: BP, °C	Percent composition: In azeotrope	Percent composition: Upper layer	Percent composition: Lower layer	Relative volume of layers at 20°C	Specific gravity of layers or azeotrope
20	a. Acetic acid	118.1	114.7	58.5				
	b. Chlorobenzene	132.0		41.5				
21	a. Acetic acid	118.1	79.7	2.0				
	b. Cyclohexane	81.4		98.0				
22	a. Acetic acid	118.1	81.8	6.5				
	b. Cyclohexene	83.0		93.5				
23	a. Acetic acid	118.1	103.7	25.0				
	b. 1,1-Dibromoethane	109.5		75.0				
24	a. Acetic acid	118.1	114.4	55.0				
	b. 1,2-Dibromoethane			45.0				
25	a. Acetic acid	118.1	94.8	16.0				
	b. Dibromomethane	98.2		84.0				
26	a. Acetic acid	118.1	159.0	26.0				
	b. Dimethyl formamide	153.0		74.0				1.004
27	a. Acetic acid	118.1	119.5	77.0				1.05
	b. 1,4-Dioxane	101.5		23.0				
28	a. Acetic acid	118.1	115.1	34.5				
	b. Epichlorohydrin	117.0		65.5				
29	a. Acetic acid	118.1	114.7	66.0				0.882
	b. Ethyl benzene	136.2		34.0				
30	a. Acetic acid	118.1	139.7	35.0				1.024
	b. Pyridine	115.3		65.0				
31	a. Acetic acid	118.1	107.4	38.5				
	b. Tetrachloroethylene	121.0		61.5				
32	a. Acetic acid	118.1	86.5	3.8				
	b. Trichloroethylene	87.0		96.2				
33	a. Acetic acid	118.1	105.4	28.0				0.905
	b. Toluene	110.6		72.0				
34	a. Acetic acid	118.1	163.0	69.0				1.023
	b. Triethyl amine	89.5		31.0				
35	a. Acetic acid	118.1	115.4	27.0				0.908
	b. m-Xylene	139.1		73.0				
37	a. Acetone	56.2	54.1	42.0				
	b. Bromopropane	59.6		58.0				
38	a. Acetone	56.2	39.3	33.0				1.04
	b. Carbon disulfide	46.3		67.0				
39	a. Acetone	56.2	56.1	88.5				
	b. Carbon tetrachloride	76.8		11.5				

AZEOTROPES (Continued)

BINARY SYSTEMS (Continued)

	Components		Azeotrope					
				Percent composition				
No.	Compounds	BP, °C	BP, °C	In azeo- trope	Upper layer	Lower layer	Relative volume of layers at 20°C	Specific gravity of layers or azeotrope
40	a. Acetone b. Chloroform	56.2 61.2	64.7	20.0 80.0				1.268
41	a. Acetone b. 1-Chloropropane	56.2 47.2	45.8	15.0 85.0				
42	a. Acetone b. Cyclohexane	56.2 81.4	53.0	67.0 33.0				
43	a. Acetone b. Cyclopentane	56.2 49.5	41.0	36.0 64.0				
44	a. Acetone b. Diethyl amine	56.2 55.5	51.3	38.2 61.8				0.732
45	a. Acetone b. Hexane	56.2 69.0	49.8	59.0 41.0				
46	a. Acetone b. Iodoethane	56.2 72.2	55.0	60.0 40.0				
47	a. Acetone b. Isobutyl amine	56.2 68.0	<56.0	<96.0 >4.0				
48	a. Acetone b. Isobutyl chloride	56.2 68.9	55.8	73.0 27.0				
49	a. Acetone b. Isopropyl ether	56.2 67.5	53.3	56.5 43.5				0.764
50	a. Acetone b. Isopropyl ether } 775.5 mm	78.3 91.9	75.0	55.0 45.0				0.769
51	a. Acetone b. Methanol	56.2 64.7	55.7	88.0 12.0				0.795
52	a. Acetone b. Methanol } 11.6 atm.	151.0	140.0	44.0 56.0				0.796
53	a. Acetone b. Methanol } 4.56 atm.	108.0 109.0	102.0	68.0 32.0				0.796
54	a. Acetone b. Methyl acetate	56.2 57.1	55.6	48.0 52.0				0.854
55	a. Acetone b. Pentane	56.2 36.0	32.0	21.0 79.0				
58	a. Acetone cyanohydrin b. Water	dec. 100.0	99.9	15.0 85.0				1.001
59	a. Acetonitrile b. Benzene	82.0 80.1	73.0	34.0 66.0				

BINARY SYSTEMS (Continued)

No.	Components: Compounds	Components: BP, °C	Azeotrope: BP, °C	Percent composition: In azeotrope	Percent composition: Upper layer	Percent composition: Lower layer	Relative volume of layers at 20°C	Specific gravity of layers or azeotrope
60	a. Acetonitrile b. Diethyl amine	82.0 55.5	54.5	8.0 92.0				0.705
61	a. Acetonitrile b. Ethanol	82.0 78.5	72.9	43.0 57.0				0.788
62	a. Acetonitrile b. Ethyl acetate	82.0 77.2	74.8	23.0 77.0				
63	a. Acetonitrile b. Isopropyl ether	82.0 67.5	61.7	17.0 83.0				0.743
64	a. Acetonitrile b. Methanol	82.0 64.7	63.5	81.0 19.0				
65	a. Acetonitrile } 1240 mm b. Pentane }	118.0 65.0	58.0	13.0 87.0	3.1 96.9	85.3 14.7	U 90.0 L 10	U 0.616 L 0.779
66	a. Acetonitrile } 152 mm b. Triethyl amine }	36.5 43.4	29.0	36.8 63.2				0.749
67	a. Acetonitrile b. Water	82.0 100.0	76.5	83.7 16.3				0.818
68	a. Acetophenone b. Octyl alcohol	202.3 195.0	195.0	12.5 87.5				
69	a. Acetylene b. Ethane	−83.6 −88.3	−94.5	40.7 59.3				
70	a. Acrolein b. Water	52.5 100.0	52.4	97.4 2.6				
71	a. Acrylonitrile b. Benzene	78.0 80.1	73.3	47.0 53.0				
72	a. Acrylonitrile b. Isopropanol	78.0 82.3	71.7	56.0 44.0				
73	a. Acrylonitrile } 175 mm b. Methanol }	37.0 31.3	29.0	53.0 47.0				
74	a. Acrylonitrile b. Water	78.0 100.0	70.6	85.7 14.3	96.8 3.2	7.3 92.7	U 89.5 L 10.5	U 0.813 L 0.99
75	a. Allyl acetate b. Water	104.1 100.0	83.0	83.3 16.7	98.4 1.6	3.0 97.0	U 84.0 L 16.0	U 0.930 L 0.998
76	a. Allyl acetone b. Water	129.5 100.0	92.1	64.7 35.3	97.75 2.25	2.2 97.8	U 69.0 L 31.0	U 0.849 L 0.997
77	a. Allyl alcohol b. Benzene	97.1 80.1	76.8	17.4 82.6				0.874
78	a. Allyl alcohol b. Carbon tetrachloride	97.1 76.8	72.3	11.5 88.5				1.450
79	a. Allyl alcohol b. Hexane	97.1 69.0	65.5	4.5 95.5				

No.	Components: Compounds	Components: BP, °C	Azeotrope: BP, °C	Percent composition: In azeotrope	Percent composition: Upper layer	Percent composition: Lower layer	Relative volume of layers at 20°C	Specific gravity of layers or azeotrope
80	a. Allyl alcohol	97.1	94.6	52.0				
	b. Propyl acetate	101.6		48.0				
81	a. Allyl alcohol	97.1	91.5	50.0				
	b. Toluene	110.6		50.0				
82	a. Allyl alcohol	97.1	81.0	16.0				1.313
	b. Trichloroethylene	87.0		84.0				
83	a. Allyl alcohol	97.1	66.6	5.0				
	b. Vinylallyl ether	67.4		95.0				
84	a. Allyl alcohol	97.1	88.2	72.9				0.905
	b. Water	100.0		27.1				
85	a. Allyl chloride	44.9	43.0	97.8				
	b. Water	100.0		2.2				
86	a. Allyl cyanide	118.9	89.4	66.0	97.5	3.6	U 70.0	U 0.837
	b. Water	100.0		34.0	2.5	96.4	L 30.0	L 0.994
87	a. 2-Aminoethanol	172.2	128.6	13.5				
	b. Chlorobenzene	132.0		86.5				
88	a. 2-Aminoethanol	172.2	146.5	26.0				
	b. o-Chlorotoluene	159.0		74.0				
89	a. Amyl acetate	148.7	147.6	94.0				
	b. Glycol(1,2-ethanediol)	197.2		6.0				
90	a. tert. Amyl alcohol	101.8	78.5	16.0				
	b. Cyclohexane	81.4		84.0				
91	a. tert. Amyl alcohol	101.8	100.5	56.0				
	b. Toluene	110.6		44.0				
92	a. Aniline	184.4	191.3	8.0				
	b. o-Cresol	191.5		92.0				
93	a. Aniline	184.4	180.6	76.0				
	b. Glycol(1,2-ethanediol)	197.2		24.0				
94	a. Aniline	184.4	184.0	83.0				
	b. Octyl alcohol	195.0		17.0				
95	a. Benzaldehyde	179.5	177.9	50.0				
	b. α-Chlorotoluene	179.0		50.0				
96	a. Benzaldehyde	179.5	185.6	49.0				
	b. Phenol	182.0		51.0				
97	a. Benzene	80.1	78.8	84.0				
	b. 2-Butanol	99.5		16.0				
98	a. Benzene	80.1	77.8	55.0				0.834
	b. Cyclohexane	81.4		45.0				
99	a. Benzene	80.1	67.8	67.6				0.848
	b. Ethanol	78.5		32.4				

AZEOTROPES (Continued)

BINARY SYSTEMS (Continued)

No.	Components: Compounds	Components: BP, °C	Azeotrope: BP, °C	Percent composition: In azeotrope	Percent composition: Upper layer	Percent composition: Lower layer	Relative volume of layers at 20°C	Specific gravity of layers or azeotrope
100	a. Benzene b. Ethyl nitrate	80.1 88.7	80.0	88.0 12.0				
101	a. Benzene b. Formic acid	80.1 100.7	71.1	69.0 31.0				
102	a. Benzene b. Heptane	80.1 98.4	80.1	99.3 0.7				
103	a. Benzene b. Isobutanol	80.1 108.3	79.8	90.7 9.3				0.870
104	a. Benzene b. Isopropanol	80.1 82.3	71.5	66.7 33.3				0.838
105	a. Benzene b. Methanol	80.1 64.7	58.3	60.5 39.5				0.844
106	a. Benzene b. Methylethyl ketone	80.1 79.6	78.4	62.5 37.5				0.853
107	a. Benzene b. 2-Methyl-2-propanol	80.1 82.8	74.0	63.4 36.6				0.842
108	a. Benzene b. Methyl propionate	80.1 79.6	79.5	48.0 52.0				
109	a. Benzene b. Nitromethane	80.1 101.0	79.2	86.0 14.0				
110	a. Benzene b. Propanol	80.1 97.2	77.1	83.1 16.9				0.865
111	a. Benzene b. Water	80.1 100.0	69.4	91.1 8.9	99.94 0.06	0.07 99.93	U 92.0 L 8.0	U 0.880 L 0.999
112	a. Benzonitrile b. p-Bromotoluene	190.7 185.0	184.3	15.0 85.0				
113	a. Benzonitrile b. o-Cresol	190.7 191.5	196.0	49.0 51.0				
114	a. Benzyl alcohol b. p-Cresol	205.2 202.5	206.8	62.0 38.0				
115	a. Benzyl alcohol b. Nitrobenzene	205.2 210.9	204.2	62.0 38.0				
116	a. Benzyl alcohol b. Water	205.2 100.0	99.9	9.0 91.0				
117	a. Bromobenzene b. Chloroacetic acid	156.0 189.0	154.3	89.0 11.0				
118	a. 1-Bromobutane b. Ethanol	101.6 78.5	75.0	57.0 43.0				
119	a. 1-Bromobutane b. Glycol	101.6 197.2	101.3	98.3 1.7				

BINARY SYSTEMS (Continued)

No.	Components: Compounds	Components: BP, °C	Azeotrope: BP, °C	Percent composition: In azeotrope	Percent composition: Upper layer	Percent composition: Lower layer	Relative volume of layers at 20°C	Specific gravity of layers or azeotrope
120	a. 1-Bromobutane b. Propyl acetate	101.6 101.6	99.9	52.0 48.0				
121	a. Bromodichloromethane b. Ethanol	90.2 78.5	75.5	72.0 28.0				
122	a. Bromoethane b. Ethanol	38.0 78.5	37.0	97.0 3.0				
123	a. o-Bromophenol b. o-Cresol	195.0 191.5	189.8	25.0 75.0				
124	a. o-Bromophenol b. Octyl alcohol	195.0 195.0	204.0	50.0 50.0				
125	a. 1-Bromopropane b. Hexane	70.9 69.0	67.2	50.0 50.0				
126	a. 2-Bromopropene b. Ethanol	48.4 78.5	46.2	94.0 6.0				
127	a. o-Bromotoluene b. Caproic acid	181.8 205.0	180.8	94.0 6.0				
128	a. o-Bromotoluene b. Chloroacetic acid	181.8 189.0	173.0	68.0 32.0				
129	a. o-Bromotoluene b. Ethylacetoacetate	181.8 180.0	174.7	49.0 51.0				
130	a. o-Bromotoluene b. Glycol	181.8 197.2	166.8	75.0 25.0				
131	a. m-Bromotoluene b. Octyl alcohol	183.7 195.0	184.1	91.0 9.0				
132	a. p-Bromotoluene b. Octyl alcohol	185.0 195.0	184.6	90.0 10.0				
133	a. Butane b. Ethylene oxide	−0.5 10.4	−6.5	78.0 22.0				
134	a. 2,3-Butanedione b. Ethanol	88.0 78.5	73.9	53.0 47.0				
135	a. 1-Butanol b. Butyl acetate	117.7 126.5	117.6	67.2 32.8				0.832
136	a. 1-Butanol b. Butyl ether	117.7 142.0	117.6	82.5 17.5				0.804
137	a. 1-Butanol b. Butyronitrile	117.7 118.0	113.0	50.0 50.0				
138	a. 1-Butanol b. Chlorobenzene	117.7 132.0	115.3	56.0 46.0				
139	a. 1-Butanol b. 1-Chloro-2-propanone	117.7 119.0	112.5	43.0 57.0				

AZEOTROPES (Continued)

BINARY SYSTEMS (Continued)

No.	Components: Compounds	Components: BP, °C	Azeotrope: BP, °C	Percent composition: In azeo-trope	Percent composition: Upper layer	Percent composition: Lower layer	Relative volume of layers at 20°C	Specific gravity of layers or azeotrope
140	a. 1-Butanol b. Cyclohexane	117.7 81.4	79.8	10.0 90.0				
141	a. 1-Butanol b. 1,1-Dibromoethane	117.7 109.5	104.5	20.0 80.0				
142	a. 1-Butanol b. 1,2-Dibromoethane	117.7 131.6	114.8	44.0 56.0				
143	a. 1-Butanol b. Epichlorohydrin	117.7 117.0	112.0	43.0 57.0				
144	a. 1-Butanol b. Ethylene diamine	117.7 116.9	124.7	64.3 35.7				0.849
145	a. 1-Butanol b. Ethyl nitrate	117.7 88.7	87.5	4.0 96.0				
146	a. 1-Butanol b. Heptane	117.7 98.4	93.3	18.0 82.0				0.701
147	a. 1-Butanol b. Hexaldehyde	117.7 128.5	116.8	77.1 22.9				0.829
148	a. 1-Butanol b. Hexane	117.7 69.0	67.0	3.0 97.0				0.69
149	a. 1-Butanol b. 1-Iodopropane	117.7 102.4	99.5	13.5 86.5				
150	a. 1-Butanol b. 3-Iodopropane	117.7 103.1	98.7	13.0 87.0				
151	a. 1-Butanol b. Methyl chloroacetate	117.7 130.0	116.3	74.0 26.0				
152	a. 1-Butanol b. Nitroethane	117.7 114.8	107.7	45.0 55.0				
153	a. 1-Butanol b. Nitromethane	117.7 101.9	97.8	30.0 70.0				
154	a. 1-Butanol b. Octane	117.7 125.8	110.2	50.0 50.0				
155	a. 1-Butanol b. Propylene diamine	117.7 120.9	126.5	51.0 49.0				0.843
156	a. 1-Butanol b. Pyridine	117.7 115.3	118.7	71.0 29.0				
157	a. 1-Butanol b. Tetrachloroethylene	117.7 121.2	110.0	32.0 68.0				
158	a. 1-Butanol b. Toluene	117.7 110.6	105.6	27.0 73.0				0.846
159	a. 1-Butanol b. Trichloroethylene	117.7 87.0	86.7	3.0 97.0				

AZEOTROPES (Continued)

BINARY SYSTEMS (Continued)

	Components		Azeotrope					
No.	Compounds	BP, °C	BP, °C	Percent composition: In azeotrope	Percent composition: Upper layer	Percent composition: Lower layer	Relative volume of layers at 20°C	Specific gravity of layers or azeotrope
160	a. 1-Butanol	117.7	93.3	7.8				
	b. Vinylbutyl ether	94.2		92.2				
161	a. 1-Butanol	117.7	93.0	55.5	79.9	7.7	U 71.5	U 0.849
	b. Water	100.0		44.5	20.1	92.3	L 28.5	L 0.990
162	a. 1-Butanol } 30 mm	48.0	29.0	47.6	79.9	7.7	U 59.0	U 0.849
	b. Water } 30 mm	29.0		52.4	20.1	92.3	L 41.0	L 0.989
163	a. 1-Butanol	117.7	116.8	75.0				
	b. o-Xylene	144.4		25.0				
164	a. 2-Butanol	99.5	76.0	18.0				
	b. Cyclohexane	81.4		82.0				
165	a. 2-Butanol	99.5	91.0	35.0				
	b. Diisobutylene	102.3		65.0				
166	a. 2-Butanol	99.5	95.3	55.0				
	b. Toluene	110.6		45.0				
167	a. 2-Butanol	99.5	88.5	68.0				0.863
	b. Water	100.0		32.0				
168	a. *tert.* Butanol	82.8	74.0	36.6				
	b. Benzene	80.1		63.4				
169	a. *tert.* Butanol	82.8	71.3	37.0				
	b. Cyclohexane	81.4		63.0				
170	a. *tert.* Butanol	82.8	63.7	22.0				
	b. Hexane	69.0		78.0				
171	a. *tert.* Butanol	82.8	79.9	88.2				
	b. Water	100.0		11.8				
172	a. 2-Butanone	79.6	71.8	40.0				
	b. Cyclohexane	81.4		60.0				
173	a. 2-Butanone	79.6	77.0	18.0				
	b. Ethyl acetate	77.2		82.0				
174	a. 1-Butenylethyl ether, cis	72.0	64.0	93.9	99.8	0.32	U 95.2	U 0.777
	b. Water	100.0		6.1	0.2	99.68	L 4.8	L 1.000
175	a. 1-Butenylmethyl ether, trans	76.7	67.0	92.8	99.8	0.01	U 94.2	U 0.787
	b. Water	100.0		7.2	0.2	99.99	L 5.8	L 1.000
176	a. 1-Butoxy-2-propanol	170.1	98.6	28.0	80.8	6.4	U 31.0	U 0.910
	b. Water	100.0		72.0	19.2	93.6	L 69.0	L 0.997
177	a. Butyl acetate	126.5	125.9	95.0				0.879
	b. Butyl ether	142.1		5.0				
178	a. Butyl acetate	126.5	125.6	69.0				
	b. 2-Chloroethanol	128.0		31.0				

BINARY SYSTEMS (Continued)

No.	Components: Compounds	Components: BP, °C	Azeotrope: BP, °C	Percent composition: In azeotrope	Percent composition: Upper layer	Percent composition: Lower layer	Relative volume of layers at 20°C	Specific gravity of layers or azeotrope
179	a. Butyl acetate	126.5	125.8	64.3				0.896
	b. 2-Ethoxy ethanol	135.6		35.7				
180	a. Butyl acetate	126.5	126.0	82.5				
	b. Isoamyl alcohol	130.5		17.5				
181	a. Butyl acetate	126.5	119.0	52.0				
	b. Octane	125.8		48.0				
182	a. Butyl acetate	126.5	90.7	72.9	98.8	0.68	U 75.8	U 0.882
	b. Water	100.0		27.1	1.2	99.32	L 24.2	L 0.998
183	a. Butyl acetoacetate	213.9	99.4	15.9	97.2	0.4	U 16.0	U 0.970
	b. Water	100.0		84.1	2.8	99.6	L 84.0	L 1.000
184	a. Butyl acrylate	118[2]	94.3	62.0	99.3	0.2	U 65.0	U 0.902
	b. Water	100.0		38.0	0.7	99.8	L 35	L 0.999
185	a. Butyl amine	77.1	76.5	60.0				
	b. Cyclohexane	81.4		40.0				
186	a. Butyl amine	77.1	82.2	51.0				0.766
	b. Ethanol	78.5		49.0				
187	a. Butyl amine	77.1	84.7	40.0				0.775
	b. Isopropanol	82.3		60.0				
188	a. Butyl amine } 575 mm	69.0	69.0	98.7				
	b. Water } 575 mm	92.4		1.3				
189	a. *n*-Butyl aniline	240.9	99.8	5.6	99.76	0.01	U 6.0	U 0.929
	b. Water	100.0		94.4	0.24	99.99	L 94.0	L 1.000
190	a. Butyl benzoate	250.2	99.9	5.0	0.01	99.68	U 95.0	U 1.000
	b. Water	100.0		95.0	99.99	0.32	L 5.0	L 1.007
191	a. Butyl butyrate	166.4	97.9	47.0	99.52	0.06	U 50.6	U 0.871
	b. Water	100.0		53.0	0.48	99.94	L 49.4	L 0.998
192	a. Butyl-2-ethoxyethanol	171.1	170.6	42.0				0.887
	b. Dibutyl acetal	189.4		58.0				
193	a. Butyl-2-ethoxyethanol	171.1	98.8	20.8				0.989
	b. Water	100.0		79.2				
194	a. Butyl chloride	78.0	68.0	93.0	99.92	0.11	U 94.0	U 0.887
	b. Water	100.0		7.0	0.08	99.89	L 6.0	L 1.000
195	a. Butyl ether	142.0	127.0	50.0				
	b. 2-Ethoxy ethanol	135.1		50.0				
196	a. Butyl ether	142.0	139.5	93.6	98.0	1.0	U 95.0	U 0.777
	b. Ethylene glycol	197.2		6.4	2.0	99.0	L 5.0	L 1.114
197	a. Butyl ether	142.0	126.7	27.0				0.924
	b. Morpholine	128.3		73.0				

AZEOTROPES (Continued)

BINARY SYSTEMS (Continued)

No.	Components: Compounds	Components: BP, °C	Azeotrope: BP, °C	Percent composition: In azeotrope	Percent composition: Upper layer	Percent composition: Lower layer	Relative volume of layers at 20°C	Specific gravity of layers or azeotrope
198	a. Butyl ether	142.0	136.0	55.0				
	b. Propionic acid	141.1		45.0				
199	a. Butyl ether	142.0	94.1	66.6	99.97	0.19	U 72.0	U 0.769
	b. Water	100.0		33.4	0.03	99.81	L 28.0	L 0.998
200	a. Butylisopropenyl ether	114.8	86.3	81.2	0.01	99.99	U 84.7	U 0.790
	b. Water	100.0		18.8	99.99	0.01	L 15.3	L 1.000
201	a. 2-Butyl octanol	253.4	99.9	2.5	98.9	0.01	U 3.0	U 0.85
	b. Water	100.0		97.5	1.1	99.99	L 97.0	L 1.00
202	a. Butyl salicylate	268.2	99.9	4.2	0.02	99.87	U 96.0	U 1.000
	b. Water	100.0		95.8	99.98	0.13	L 4.0	L 1.075
203	a. Butyraldehyde	75.7	70.7	39.4				0.835
	b. Ethanol	78.5		60.6				
204	a. Butyraldehyde	75.7	68.0	90.3	96.8	7.1	U 94.0	U 0.815
	b. Water	100.0		9.7	3.2	92.9	L 6.0	
205	a. Butyric acid	163.5	163.0	65.0				
	b. o-Dichlorobenzene	180.0		35.0				
206	a. Butyric acid	163.5	99.4	18.4				1.007
	b. Water	100.0		81.6				
207	a. Butyric acid	163.5	143.0	10.0				
	b. o-Xylene	144.4		90.0				
208	a. Butyric acid	163.5	138.1	6.6				
	b. m-Xylene	139.1		93.4				
209	a. Butyric acid	163.5	137.5	5.4				
	b. p-Xylene	138.4		94.6				
210	a. Butyronitrile	118.0	107.0	27.0				
	b. Toluene	110.6		73.0				
211	a. Butyronitrile	118.0	88.7	67.5	97.5	3.53	U 72.9	U 0.795
	b. Water	100.0		32.5	2.5	96.47	L 27.1	L 0.994
212	a. Caproic acid	205.0	201.9	13.0				
	b. m-Cresol	202.8		87.0				
213	a. Carbon disulfide	46.3	46.0	94.0				
	b. 1,1-Dichloroethane	57.2		6.0				
214	a. Carbon disulfide	46.3	42.4	91.0				1.197
	b. Ethanol	78.5		9.0				
215	a. Carbon disulfide	46.3	46.1	97.0				1.249
	b. Ethyl acetate	77.1		3.0				
216	a. Carbon disulfide	46.3	34.4	1.0				0.719
	b. Ethyl ether	34.6		99.0				
217	a. Carbon disulfide	46.3	39.4	63.0				
	b. Ethyl formate	54.3		37.0				

BINARY SYSTEMS (Continued)

No.	Components: Compounds	BP, °C	Azeotrope: BP, °C	Percent composition: In azeotrope	Percent composition: Upper layer	Percent composition: Lower layer	Relative volume of layers at 20°C	Specific gravity of layers or azeotrope
218	a. Carbon disulfide b. Isopropanol	46.3 82.3	44.6	92.0 8.0				1.202
219	a. Carbon disulfide b. Methanol	46.3 64.7	37.7	86.0 14.0	50.8 49.2	97.2 2.8	U 26.3 L 73.7	U 0.979 L 1.261
220	a. Carbon disulfide b. Methyl acetate	46.3 57.0	40.2	70.0 30.0				1.126
221	a. Carbon disulfide b. Methylethyl ketone	46.3 79.6	45.9	84.7 15.3				1.157
222	a. Carbon disulfide b. 2-Methyl-2-propanol	46.3 82.2	45.3	94.0 6.0				1.219
223	a. Carbon disulfide b. Propyl chloride	46.3 46.6	45.2	55.0 45.0				
224	a. Carbon disulfide b. Water	46.3 100.0	43.6	98.0 2.0	0.29 99.71	99.99 0.01	U 2.3 L 97.7	U 1.001 L 1.265
225	a. Carbon tetrachloride b. Ethanol	76.8 78.5	65.0	84.2 15.8				1.377
226	a. Carbon tetrachloride b. Ethyl acetate	76.8 77.1	74.8	57.0 43.0				1.202
227	a. Carbon tetrachloride } 286 mm b. Ethyl acetate } 286 mm	47.5 50.5	47.4	82.2 17.8				
228	a. Carbon tetrachloride b. Ethylene dichloride	76.8 84.0	75.3	78.0 22.0				1.500
229	a. Carbon tetrachloride b. Isobutanol	76.8 108.3	75.8	94.5 5.5				1.515
230	a. Carbon tetrachloride b. Isopropanol	76.8 82.3	67.0	82.0 18.0				1.344
231	a. Carbon tetrachloride b. Methanol	76.8 64.7	55.7	79.4 20.6				1.322
232	a. Carbon tetrachloride b. Methylethyl ketone	76.8 79.6	73.8	71.0 29.0				1.247
233	a. Carbon tetrachloride b. 2-Methyl-2-propanol	76.8 82.2	69.5	83.0 17.0				1.358
234	a. Carbon tetrachloride b. Propanol	76.8 97.2	72.8	88.5 11.5				1.437
235	a. Carbon tetrachloride b. Water	76.8 100.0	66.8	95.9 4.1	0.03 99.97	99.97 0.03	U 6.4 L 93.6	U 1.000 L 1.597
236	a. Chloral b. Heptane	98.0 98.4	93.0	53.0 47.0				

AZEOTROPES (Continued)

BINARY SYSTEMS (Continued)

No.	Components: Compounds	Components: BP, °C	Azeotrope: BP, °C	Percent composition: In azeotrope	Percent composition: Upper layer	Percent composition: Lower layer	Relative volume of layers at 20°C	Specific gravity of layers or azeotrope
237	a. Chloral	98.0	95.0	93.0				
	b. Water	100.0		7.0				
238	a. Chloroacetic acid	189.0	156.8	12.0				
	b. o-Chlorotoluene	159.0		88.0				
239	a. Chloroacetic acid	189.0	187.1	78.0				
	b. Naphthalene	217.9		22.0				
240	a. Chloroacetic acid	189.0	186.3	3.0				
	b. Valeric acid	187.0		97.0				
241	a. Chloroacetic acid	189.0	143.5	12.0				
	b. o-Xylene	144.4		88.0				
242	a. Chlorobenzene	132.0	120.0	58.0				
	b. 2-Chloroethanol	128.0		42.0				
243	a. Chlorobenzene	132.0	127.2	68.0				
	b. 2-Ethoxy ethanol	135.1		32.0				
244	a. Chlorobenzene	132.0	124.4	66.0				
	b. Isoamyl alcohol	120.5		34.0				
245	a. Chlorobenzene	132.0	107.1	37.0				
	b. Isobutyl alcohol	108.4		63.0				
246	a. Chlorobenzene	132.0	131.2	92.0				
	b. Isobutyric acid	154.4		8.0				
247	a. Chlorobenzene	132.0	96.5	20.0				
	b. Propanol	97.2		80.0				
248	a. Chlorobenzene	132.0	128.9	82.0				
	b. Propionic acid	141.1		18.0				
249	a. Chlorobenzene	132.0	124.5	57.0				
	b. Pyrrole	131.0		43.0				
250	a. Chlorobenzene	132.0	90.2	71.6				
	b. Water	100.0		28.4				
251	a. 1-Chlorobutane	78.0	65.7	79.7				
	b. Ethanol	78.5		20.3				
252	a. 1-Chlorobutane	78.0	70.8	77.0				
	b. Isopropyl alcohol	82.3		23.0				
253	a. 2-Chloroethanol	128.0	121.0	62.0				
	b. Ethyl benzene	136.2		38.0				
254	a. 2-Chloroethanol	128.8	127.8	75.0				
	b. Isoamyl alcohol	130.5		25.0				
255	a. 2-Chloroethyl ether	179.2	141.2	28.0				
	b. 3-Heptanol	157.0		72.0				
256	a. 2-Chloroethyl ether	179.2	98.0	34.5	1.07	99.72	U 70.4	U 1.002
	b. Water	100.0		65.5	98.03	0.28	L 29.6	L 1.220

No.	Components: Compounds	Components: BP, °C	Azeotrope: BP, °C	Percent composition: In azeotrope	Percent composition: Upper layer	Percent composition: Lower layer	Relative volume of layers at 20°C	Specific gravity of layers or azeotrope
257	a. Chloroform	61.2	59.4	93.0				1.403
	b. Ethanol	78.5		7.0				
258	a. Chloroform	61.2	60.0	72.0				1.101
	b. Hexane	68.7		28.0				
259	a. Chloroform	61.2	53.5	87.0				1.342
	b. Methanol	64.7		13.0				
260	a. Chloroform	61.2	79.9	17.0				0.877
	b. Methylethyl ketone	79.6		83.0				
261	a. Chloroform	61.2	56.3	97.0	0.8	99.8	U 4.4	U 1.004
	b. Water	100.0		3.0	99.2	0.2	L 95.6	L 1.491
262	a. Chloroisopropyl ether	187.0	98.5	37.4	0.17	99.86	U 65.0	U 1.00
	b. Water	100.0		62.6	99.83	0.14	L 35.0	L 1.11
263	a. o-Chloronitrobenzene	245.7	233.5	59.0				
	b. Diethylene glycol	244.5		41.0				
264	a. Chloronitromethane	122.5	115.2	45.0				
	b. Tetrachloroethylene	121.0		55.0				
265	a. o-Chlorophenol	122.5	115.2	48.0				
	b. o-Dichlorobenzene	180.0		52.0				
266	a. o-Chlorophenol	122.5	174.5	25.0				
	b. Phenol	182.0		75.0				
267	a. p-Chlorophenol	217.0	216.3	36.5				
	b. Naphthalene	217.9		63.5				
268	a. p-Chlorophenol	217.0	219.9	8.0				
	b. Nitrobenzene	210.9		92.0				
269	a. 1-Chloro-2-propanol	119.0	109.2	28.5				
	b. Toluene	110.6		71.5				
270	a. 1-Chloro-2-propanol	119.0	95.4	54.2				
	b. Water	100.0		45.8				
271	a. o-Chlorotoluene	159.0	155.5	62.0				
	b. Cyclohexanol	161.5		38.0				
272	a. o-Chlorotoluene	159.0	155.4	65.0				
	b. 2-Furaldehyde	161.7		35.0				
273	a. 1-Chlorotoluene	159.0	152.5	87.0				
	b. Glycol	197.2		13.0				
274	a. o-Chlorotoluene	159.0	153.5	56.0				
	b. Hexyl alcohol	157.2		44.0				
275	a. o-Chlorotoluene	159.0	157.5	88.0				
	b. Isovaleric acid	176.7		12.0				
276	a. o-Chlorotoluene	159.0	159.2	97.0				
	b. Phenol	182.0		3.0				

BINARY SYSTEMS (Continued)

No.	Components: Compounds	BP, °C	Azeotrope: BP, °C	Percent composition: In azeo-trope	Upper layer	Lower layer	Relative volume of layers at 20°C	Specific gravity of layers or azeotrope
277	a. o-Chlorotoluene b. Valeric acid	159.0 187.0	158.5	95.0 5.0				
278	a. o-Cresol b. Octyl alcohol	191.5 195.0	196.9	38.0 62.0				
279	a. m-Cresol b. Octyl alcohol	202.8 195.0	203.3	62.0 38.0				
280	a. p-Cresol b. Octyl alcohol	202.5 195.0	202.3	70.0 30.0				
281	a. Croton aldehyde b. Water	104.0 100.0	84.0	75.2 24.8	90.4 9.6	15.6 84.4	U 81.4 L 18.6	U 0.876 L 0.987
282	a. 1,3-Cyclohexadiene b. Cyclohexane	80.5 81.4	79.0	45.0 55.0				
283	a. Cyclohexane b. Ethanol	81.4 78.5	64.9	69.5 30.5				
284	a. Cyclohexane b. Ethyl acetate	81.4 77.2	72.8	46.0 54.0				
285	a. Cyclohexane b. Ethyl nitrate	81.4 88.7	74.5	64.0 36.0				
286	a. Cyclohexane b. Isobutanol	81.4 108.3	78.1	86.0 14.0				
287	a. Cyclohexane b. Isopropanol	81.4 82.3	68.6	67.0 33.0				0.777
288	a. Cyclohexane b. Isopropyl acetate	81.4 89.0	78.9	75.0 25.0				
289	a. Cyclohexane b. Methanol	81.4 64.7	45.2	63.0 37.0	97.0 3.0	39.0 61.0	U 43.0 L 57.0	
290	a. Cyclohexane b. Nitromethane	81.4 101.0	70.2	72.0 28.0				
291	a. Cyclohexane b. Propanol	81.4 97.2	74.2	80.0 20.0				
292	a. Cyclohexane b. Vinyl acetate	81.4 72.0	67.4	38.7 61.3				0.850
293	a. Cyclohexane b. Water	81.4 100.0	69.8	91.5 8.5	99.99 0.01	0.01 99.99	U 93.2 L 6.8	U 0.780 L 1.000
294	a. Cyclohexanol b. 2-Furaldehyde	161.5 161.7	156.5	94.5 5.5				
295	a. Cyclohexanol b. Phenol	161.5 182.0	183.0	13.0 87.0				
296	a. Cyclohexanol b. Propylchloroacetate	161.5 163.5	159.0	53.0 47.0				

No.	Components: Compounds	Components: BP, °C	Azeotrope: BP, °C	Percent composition: In azeotrope	Percent composition: Upper layer	Percent composition: Lower layer	Relative volume of layers at 20°C	Specific gravity of layers or azeotrope
297	a. Cyclohexanol	161.5	~97.8	~20.0				
	b. Water	100.0		~80.0				
298	a. Cyclohexanol	161.5	143.0	14.0				
	b. o-Xylene	144.4		86.0				
299	a. Cyclohexanone	155.4	159.0	55.0				
	b. 1,1,2,2-Tetrachloroethane	146.3		45.0				
300	a. Cyclohexanone	155.4	160.0	39.0				
	b. 1,2,3-Trichloropropane	156.3		61.0				
301	a. Cyclohexanone	155.4	95.0	38.4	92.0	2.3	U 41.5	U 0.953
	b. Water	100.0		61.6	8.0	97.7	L 58.5	L 1.000
302	a. Cyclohexylamine	134.0	96.4	44.2				
	b. Water	100.0		55.8				
303	a. Cyclopentanol	140.9	96.3	42.0				
	b. Water	100.0		58.0				
304	a. Cyclopentanol	140.9	132.8	40.0				
	b. m-Xylene	139.1		60.0				
305	a. Cyclopentanone	130.7	93.5	58.0	86.2	29.3	U 51.0	U 0.965
	b. Water	100.0		42.0	13.8	70.7	L 49.0	L 0.998
306	a. Diacetone alcohol	169.2	99.6	13.0				1.002
	b. Water	100.0		87.0				
307	a. Diallyl acetal	150.9	95.3	59.0	99.3	0.7	U 62.0	U 0.876
	b. Water	100.0		41.0	0.7	99.3	L 38.0	L 0.998
308	a. Diallyl amine	110.5	87.2	76.0	71.1	5.09		0.852
	b. Water	100.0		24.0	28.9	94.91		
309	a. p-Dibromobenzene	219.0	210.5	22.5				
	b. Nitrobenzene	210.9		77.5				
310	a. p-Dibromobenzene	219.0	218.0	73.0				
	b. o-Nitrotoluene	222.3		27.0				
311	a. 1,2-Dibromobutane	140.5	139.0	94.0				
	b. Glycol	197.2		6.0				
312	a. 1,2-Dibromethane	131.6	130.9	96.5				
	b. Glycol	197.2		3.5				
313	a. Dibromethane	98.2	76.0	62.0				
	b. Ethanol	78.5		38.0				
314	a. Dibutyl acetal	189.4	98.7	33.7	99.8	0.03	U 33.0	U 0.83
	b. Water	100.0		66.3	0.2	99.97	L 67.0	L 1.000
315	a. Dibutyl amine	159.6	97.0	49.5	93.8	0.47	U 58.0	U 0.780
	b. Water	100.0		50.5	6.2	99.53	L 42.0	L 0.990
316	a. Dibutyl ethanolamine	228.7	99.9	9.0	93.2	0.39	U 10.0	U 0.877
	b. Water	100.0		91.0	6.8	99.61	L 90.0	L 0.999

BINARY SYSTEMS (Continued)

No.	Components: Compounds	BP, °C	Azeotrope: BP, °C	Percent composition: In azeotrope	Percent composition: Upper layer	Percent composition: Lower layer	Relative volume of layers at 20°C	Specific gravity of layers or azeotrope
317	a. Dibutyl formal b. Water	181.8 100.0	98.2	38.0 62.0				
318	a. Dibutyl fumarate b. Water	285.1 100.0	99.9	1.5 98.5	99.62 0.38	0.1 99.9	U 1.5 L 98.5	U 0.987 L 1.000
319	a. Dibutyl maleate b. Water	280.6 100.0	99.9	1.6 98.4	99.36 0.64	0.01 99.99	U 1.7 L 98.3	U 0.996 L 1.000
320	a. o-Dichlorobenzene b. 2-Furaldehyde	180.0 161.7	161.0	22.0 78.0				
321	a. o-Dichlorobenzene b. Glycol	180.0 197.2	165.8	80.0 20.0				
322	a. o-Dichlorobenzene b. Phenol	180.0 182.0	173.7	65.0 35.0				
323	a. o-Dichlorobenzene b. Valeric acid	180.0 187.0	175.8	78.0 22.0				
324	a. p-Dichlorobenzene b. Furfuryl alcohol	173.4 171.0	172.5	30.0 70.0				
325	a. p-Dichlorobenzene b. Phenol	173.4 182.0	171.1	74.8 25.2				
326	a. 1,2-Dichloroethane b. Ethanol	83.5 78.5	70.5	63.0 27.0				
327	a. 1,2-Dichloroethane b. Isobutyl alcohol	83.5 108.4	83.5	93.5 6.5				
328	a. 1,1-Dichloroethane b. Methanol	57.2 64.5	49.1	88.5 11.5				
329	a. 1,2-Dichloroethane b. Water	83.5 100.0	72.0	80.5 19.5				
330	a. Di(2-chloroethyl) formal b. Water	218.1 100.0	99.4	13.2 86.8	0.78 99.22	99.56 0.44	U 89.0 L 11.0	U 1.002 L 1.233
331	a. Dichloromethane b. Ethanol	40.1 78.5	<39.9	>95.0 <5.0				
332	a. Dichloromethane b. Glycol	40.1 197.2	168.7	86.0 14.0				
333	a. 2,2-Dichloropropane b. Isopropyl alcohol	69.7 82.3	66.8	83.0 17.0				
334	a. 2,3-Dichloropropanol b. Water	182.0 100.0	99.4	13.0 87.0	11.3 88.7	89.7 10.3	U 99.0 L 1.0	U 1.04 L 1.33
335	a. Dicyclopentadiene b. Water	170.0 100.0	98.0	32.3 67.7	99.99 0.01	0.05 99.95	U 33.0 L 67.0	U 0.978 L 1.000
336	a. Diethyl acetal b. Water	102.1 100.0	82.6	85.7 14.3	98.8 1.2	5.5 94.5	U 88.0 L 12.0	U 0.826 L 0.99

AZEOTROPES (Continued)

BINARY SYSTEMS (Continued)

No.	Components: Compounds	Components: BP, °C	Azeotrope: BP, °C	Percent composition: In azeotrope	Percent composition: Upper layer	Percent composition: Lower layer	Relative volume of layers at 20°C	Specific gravity of layers or azeotrope
337	a. Diethyl amine (740 mm)	54.7	66.2	40.0				
	b. Methanol (740 mm)	63.8		60.0				
338	a. Diethylaminoethyl amine	144.9	99.8	20.5				0.988
	b. Water	100.0		79.5				
339	a. Diethyl butyral	146.3	94.2	65.5	99.52	0.46	U 30.0	U 0.830
	b. Water	100.0		34.5	0.48	99.54	L 70.0	L 0.999
340	a. Diethylene glycol (50 mm)	163.0	129.0	15.5	0.2	99.6	U 89.0	U 0.795
	b. Hexyl ether (50 mm)	137.0		84.5	99.8	0.4	L 11.0	L 1.117
341	a. Diethylene glycol	244.5	212.6	22.0				
	b. Naphthalene	217.9		78.0				
342	a. Diethylene glycol	244.5	210.0	10.0				
	b. Nitrobenzene	210.9		90.0				
343	a. Diethylene glycol	244.5	218.2	17.5				
	b. o-Nitrotoluene	222.3		82.5				
344	a. Diethyleneglycol dibutyl ether	254.6	99.8	5.3	98.6	0.3	U 6.0	U 0.887
	b. Water	100.0		94.7	1.4	99.7	L 94.0	L 1.000
345	a. Diethyl ethanolamine	162.1	98.9	25.6				0.993
	b. Water	100.0		74.4				
346	a. Diethyl formal	87.5	74.2	43.0				
	b. Ethanol	78.5		57.0				
347	a. Diethyl fumarate	218.1	99.5	12.5	0.4	98.87	U 88.0	U 1.000
	b. Water	100.0		87.5	99.6	1.13	L 12.0	L 1.052
348	a. Di(2-ethylhexyl) acetate	dec.	99.9	1.0	99.9	0.1	U 1.2	U 0.845
	b. Water	100.0		99.0	0.1	99.9	L 98.8	L 1.000
349	a. Diethylisopropanol amine	150.5	97.2	45.0				0.962
	b. Water	100.0		55.0				
350	a. Diethyl ketone	102.7	94.9	43.0				
	b. Propanol	97.2		57.0				
351	a. Diethyl maleate	225.0	99.6	11.9	1.50	98.60	U 89.0	U 1.001
	b. Water	100.0		88.1	98.50	1.40	L 11.0	L 1.069
352	a. Diethyl phthalate	296.0	99.9	1.6	0.02	99.5	U 98.7	U 1.000
	b. Water	100.0		98.4	99.98	0.5	L 1.3	L 1.118
353	a. Diethyl pimelate	268.1	100.0	1.7	98.7	0.14	U 1.7	U 0.995
	b. Water	100.0		98.3	1.3	99.86	L 98.3	L 1.000
354	a. Diethyl succinate	217.7	99.9	9.0	2.05	98.2	U 93.0	U 1.005
	b. Water	100.0		91.0	97.95	1.8	L 7.0	L 1.04
355	a. Dihexyl amine	239.8	99.8	7.2	97.5	1.1	U 7.7	U 0.801
	b. Water	100.0		92.8	2.5	98.9	L 92.3	L 0.999

AZEOTROPES (Continued)

BINARY SYSTEMS (Continued)

No.	Components: Compounds	Components: BP, °C	Azeotrope: BP, °C	Percent composition: In azeotrope	Percent composition: Upper layer	Percent composition: Lower layer	Relative volume of layers at 20°C	Specific gravity of layers or azeotrope
356	a. Diisobutylene b. Isopropanol	102.6 82.3	77.8	45.5 54.5				
357	a. Diisobutylene b. Water	102.6 100.0	82.0	88.0 12.0	99.99 0.01	0.01 99.99	U 91.0 L 9.0	U 0.721 L 1.000
358	a. Diisobutyl ketone b. Morpholine	168.0 128.3	128.0	2.0 98.0				0.996
359	a. Diisobutyl ketone b. Water	168.0 100.0	97.0	48.1 51.9	99.25 0.75	0.05 99.95	U 53.4 L 46.6	U 0.810 L 1.000
360	a. Diisopropyl amine b. Isopropanol	84.1 82.3	79.7	60.0 40.0				0.755
361	a. Diisopropyl amine b. Water	84.1 100.0	74.1	91.0 9.0				
362	a. Diisopropylethanol amine b. Water	190.0 100.0	99.2	15.0 85.0	81.8 18.2	1.28 98.72	U 18.0 L 82.0	U 0.90 L 0.998
363	a. Diisopropyl maleate b. Water	228.7 100.0	99.9	7.0 93.0	0.23 99.77	99.19 0.81	U 93.0 L 7.0	U 1.000 L 1.012
364	a. Dimethyl acetal b. Methanol	64.5 64.7	57.5	75.8 24.2				0.841
365	a. Dimethyl acetal b. Water	64.5 100.0	61.3	96.4 3.6				
366	a. 1,3-Dimethylbutyl amine b. Water	108.5 100.0	89.5	71.4 28.6				0.83
367	a. Dimethyl butyral b. Water	114.0 100.0	87.3	79.7 20.3	99.1 0.9	2.9 97.1	U 82.3 L 17.7	U 0.851 L 0.997
368	a. Dimethylethanol amine b. Water } 540 mm	123.4 90.7	91.0	4.8 95.2				
369	a. Dimethyl formal b. Methanol	42.3 64.7	41.8	92.1 7.9				0.860
370	a. 2,5-Dimethyl furan b. Methanol	93.3 64.7	61.5	49.0 51.0				0.841
371	a. 2,5-Dimethyl furan b. Water	93.3 100.0	77.0	88.3 11.7	99.87 0.13	0.13 99.87	U 89.0 L 11.0	U 0.902 L 1.000
372	a. 2,6-Dimethyl-4-heptanol b. Water	178.1 100.0	98.5	29.6 70.4	99.01 0.99	0.06 99.94	U 34.4 L 65.6	L 0.814 L 1.000
373	a. Dimethyl isobutyral b. Water	104.7 100.0	83.9	85.7 14.3	99.23 0.77	3.48 96.52	U 87.7 L 12.3	U 0.847 L 0.995
374	a. 2,6-Dimethyl morpholine b. Water	146.6 100.0	99.6	30.0 70.0				

BINARY SYSTEMS (Continued)

Components			Azeotrope					
				Percent composition				
No.	Compounds	BP, °C	BP, °C	In azeo-trope	Upper layer	Lower layer	Relative volume of layers at 20°C	Specific gravity of layers or azeotrope
375	a. Dimethyl phthalate	282.9	100.0	1.1	0.43	98.5	U 99.0	U 1.001
	b. Water	100.0		98.9	99.57	1.5	L 1.0	L 1.191
376	a. Dimethyl pimelate	248.9	99.9	3.2	0.99	98.0	U 97.8	U 1.005
	b. Water	100.0		96.8	99.01	2.0	L 2.2	L 1.04
377	a. 1,4-Dioxane	101.3	87.8	81.6				1.04
	b. Water	100.0		18.4				
378	a. 1,3-Dioxolane	75.6	72.3	81.0				0.957
	b. Heptane	98.4		19.0				
379	a. 1,3-Dioxolane	75.6	71.9	93.0				1.068
	b. Water	100.0		7.0				
380	a. Dipropyl acetal	146.6	95.0	63.4	99.66	0.41	U 67.5	U 0.837
	b. Water	100.0		36.6	0.34	99.59	L 32.5	L 0.999
381	a. Dipropyl ketone	144.0	94.3	59.5	99.1	0.4	U 64.6	U 0.820
	b. Water	100.0		40.5	0.9	99.6	L 35.4	L 1.000
382	a. Epichlorohydrin	117.0	96.0	23.0				
	b. Propanol	97.2		77.0				
383	a. Epichlorohydrin	117.0	108.3	26.0				
	b. Toluene	110.6		74.0				
384	a. Epichlorohydrin	117.0	88.5	74.0	5.9	98.8	U 30.0	U 1.010
	b. Water	100.0		26.0	94.1	1.2	L 70.0	L 1.175
385	a. Ethanol	78.5	71.8	31.0				0.863
	b. Ethyl acetate	77.1		69.0				
386	a. Ethanol	78.5	77.5	72.7				
	b. Ethyl acrylate	99.3		27.3				
387	a. Ethanol	78.5	73.8	49.3				0.776
	b. Ethylbutyl ether	92.2		50.7				
388	a. Ethanol	78.5	71.0	33.5				1.049
	b. Ethylene dichloride	83.5		66.5				
389	a. Ethanol	78.5	71.9	44.0				
	b. Ethyl nitrate	88.7		56.0				
390	a. Ethanol	78.5	72.0	48.0				0.729
	b. Heptane	98.4		52.0				
391	a. Ethanol	78.5	58.7	21.0				0.687
	b. Hexane	69.0		79.0				
392	a. Ethanol	78.5	63.0	14.0				
	b. Iodoethane	72.2		86.0				
393	a. Ethanol	78.5	41.2	3.2				
	b. Iodomethane	42.5		96.8				
394	a. Ethanol	78.5	61.3	16.3				
	b. Isobutyl chloride	68.9		83.7				

AZEOTROPES (Continued)

BINARY SYSTEMS (Continued)

No.	Components: Compounds	Components: BP, °C	Azeotrope: BP, °C	Percent composition: In azeotrope	Percent composition: Upper layer	Percent composition: Lower layer	Relative volume of layers at 20°C	Specific gravity of layers or azeotrope
395	a. Ethanol b. Isoprene	78.5 34.0	32.7	3.0 97.0				
396	a. Ethanol b. Isopropyl acetate	78.5 89.0	76.8	53.0 47.0				
397	a. Ethanol b. Isopropyl ether	78.5 67.5	64.0	17.1 82.9				0.741
398	a. Ethanol b. Methyl acrylate	78.5 80.9	73.5	42.4 57.6				
399	a. Ethanol b. Methylbutyl ether	78.5 70.3	65.5	20.0 80.0				
400	a. Ethanol b. Methylethyl ketone	78.5 79.6	74.8	34.0 66.0				0.802
401	a. Ethanol b. Nitromethane	78.5 101.0	76.0	73.2 26.8				
402	a. Ethanol b. Tetrachloroethylene	78.5 121.0	76.8	~63.0 ~37.0				
403	a. Ethanol b. Thiophene	78.5 84.1	70.0	45.0 55.0				
404	a. Ethanol b. Toluene	78.5 110.6	76.7	68.0 32.0				0.815
405	a. Ethanol b. Trichloroethylene	78.5 87.1	70.9	27.0 73.0				1.197
406	a. Ethanol b. 1,1,2-Trichloro-trifluoroethane	78.5 47.7	43.8	3.8 96.2				1.517
407	a. Ethanol b. Triethyl amine	78.5 89.5	76.9	51.0 49.0				0.775
408	a. Ethanol b. Vinylbutyl ether	78.5 94.2	73.0	48.0 52.0				0.786
409	a. Ethanol b. Vinylisobutyl ether	78.5 83.4	69.2	33.0 67.0				0.778
410	a. Ethanol b. Vinylpropyl ether	78.5 65.1	60.0	18.4 81.6				
411	a. Ethanol b. Water	78.5 100.0	78.2	95.6 4.4				0.804
412	a. Ethanol } 3 atm. b. Water }	109.0 134.0	109.0	95.2 4.8				0.805
413	a. Ethanol } 95 mm b. Water }	33.5 51.0	33.4	99.5 0.5				0.792

BINARY SYSTEMS (Continued)

No.	Components: Compounds	Components: BP, °C	Azeotrope: BP, °C	Percent composition: In azeotrope	Percent composition: Upper layer	Percent composition: Lower layer	Relative volume of layers at 20°C	Specific gravity of layers or azeotrope
414	a. Ethoxy diglycol b. Ethylene glycol	202.8 197.5	192.0	54.5 45.5				1.050
415	a. 2-Ethoxy ethanol b. Toluene	135.1 110.6	110.2	10.8 89.2				0.874
416	a. 2-Ethoxy ethanol b. Water	135.1 100.0	99.4	28.8 71.2				1.003
417	a. 2-Ethoxy ethanol } 200 mm b. Water	96.5 66.3	66.4	15.0 85.0				
418	a. 2-Ethoxy ethanol b. o-Xylene	135.1 144.4	130.8	55.0 45.0				
419	a. 1-Ethoxy-2-propanol b. Water	132.2 100.0	97.3	49.9 50.1				0.98
420	a. Ethyl acetate b. Iodoethane	77.2 72.2	70.0	22.0 78.0				
421	a. Ethyl acetate b. Isopropanol	77.2 82.3	74.8	77.0 23.0				0.869
422	a. Ethyl acetate b. Methanol	77.2 64.7	62.1	51.4 48.6				0.846
423	a. Ethyl acetate b. Water	77.2 100.0	70.4	91.9 8.1	96.7 3.3	8.7 91.3	U 95.0 L 5.0	U 0.907 L 0.999
424	a. Ethyl acrylate b. Water	99.8 100.0	81.0	84.9 15.1	98.5 1.5	2.0 98.0	U 87.0 L 13.0	U 0.910 L 0.998
425	a. n-Ethyl aniline b. Water	204.8 100.0	99.2	16.1 83.9	99.3 0.7	0.2 99.8	U 16.6 L 83.4	U 0.963 L 1.000
426	a. Ethyl benzene b. Propionic acid	136.2 141.6	131.1	72.0 28.0				
427	a. Ethyl benzene b. Water	136.2 100.0	92.0	67.0 33.0	99.95 0.05	0.02 99.98	U 70.0 L 30.0	U 0.870 L 1.000
428	a. 2-Ethyl butanol } 50 mm b. 2-Ethylhexyl chloride	77.0 89.0	77.0	61.0 39.0				
429	a. 2-Ethyl butanol b. Water	146.0 100.0	96.7	42.0 58.0	95.44 4.56	0.43 99.57	U 48.0 L 52.0	U 0.841 L 0.999
430	a. 2-Ethylbutyl acetate b. Water	162.3 100.0	97.0	47.6 52.4	99.43 0.57	0.06 99.94	U 51.3 L 48.7	U 0.871 L 1.000
431	a. 2-Ethylbutyl acetate b. Hexanol	162.3 157.1	154.4	27.5 72.5				0.838
432	a. 2-Ethylbutyl butyrate b. Water	199.6 100.0	98.6	25.1 74.9	99.5 0.5	0.1 99.9	U 27.8 L 72.2	U 0.873 L 0.999

BINARY SYSTEMS (Continued)

No.	Components: Compounds	Components: BP, °C	Azeotrope: BP, °C	Percent composition: In azeotrope	Percent composition: Upper layer	Percent composition: Lower layer	Relative volume of layers at 20°C	Specific gravity of layers or azeotrope
433	a. Ethylbutyl ether b. Methanol	92.2 64.7	62.6	44.0 56.0				0.770
434	a. Ethylbutyl ether b. Water	92.2 100.0	76.6	88.1 11.9	99.6 0.4	0.44 99.56	U 91.0 L 9.0	U 0.753 L 0.998
435	a. Ethylbutyl ketone b. Water	148.5 100.0	94.6	57.8 42.2	99.2 0.8	1.4 98.6	U 62.5 L 37.5	U 0.822 L 0.997
436	a. 2-Ethylbutyraldehyde b. Water	116.9 100.0	87.5	76.3 23.7	99.2 0.8	0.4 99.6	U 80.2 L 19.8	U 0.816 L 0.997
437	a. 2-Ethylbutyric acid b. Water	194.2 100.0	99.7	13.0 87.0	96.7 3.3	1.55 98.45	U 12.0 L 88.0	U 0.929 L 1.000
438	a. Ethyl carbamate b. Octyl alcohol	180.0 195.0	183.5	72.5 27.5				
439	a. Ethyl carbamate b. Phenol	180.0 182.0	190.8	53.5 46.5				
440	a. Ethyl crotonate b. Water	137.8 100.0	93.5	62.0 38.0	98.48 1.52	0.63 99.37	U 65.0 L 35.0	U 0.921 L 1.000
441	a. n-Ethylcyclohexyl amine b. Water	164.0 100.0	97.1	42.0 58.0	77.0 23.0	1.5 98.5	U 56.0 L 44.0	U 0.895 L 0.998
442	a. Ethylene chlorohydrin b. Vinyl-2-chloroethyl ether	128.7 109.1	108.8	3.0 97.0				1.050
443	a. Ethylene chlorohydrin b. Water	128.7 100.0	97.8	42.3 57.7				1.093
444	a. Ethylene diamine b. Isobutanol	116.5 108.3	120.5	50.0 50.0				0.856
445	a. Ethylene diamine b. Toluene	116.5 110.6	103.0	30.0 70.0				
446	a. Ethylene diamine b. Water	116.5 100.0	119.0	81.6 18.4				0.953
447	a. Ethylene diamine } 300 mm b. Water	90.5 75.9	94.5	77.1 22.9				0.963
448	a. Ethylene diamine } 50 mm b. Water	49.0 38.0	55.0	71.0 29.0				
449	a. Ethylene dichloride b. Isopropanol	84.0 82.3	72.7	60.8 39.2				1.012
450	a. Ethylene dichloride b. Methanol	84.0 64.7	59.5	65.0 35.0				1.045
451	a. Ethylene dichloride b. Trichloroethylene	84.0 87.1	82.3	59.1 40.9				1.330

BINARY SYSTEMS (Continued)

No.	Components: Compounds	Components: BP, °C	Azeotrope: BP, °C	Percent composition: In azeotrope	Percent composition: Upper layer	Percent composition: Lower layer	Relative volume of layers at 20°C	Specific gravity of layers or azeotrope
452	a. Ethylene dichloride b. Water	84.0 100.0	71.6	91.8 8.2	0.8 99.2	99.8 0.2	U 10.0 L 90.0	U 1.002 L 1.254
453	a. Ethylene glycol b. 2-Ethylhexyl ether } 10 mm	87.0 135.0					U 50.0 L 50.0	
454	a. Ethylene glycol b. Hexyl ether } 50 mm	123.0 137.0	112.8	35.6 64.4	0.1 99.9	99.9 0.1	U 71.8 L 28.2	U 0.795 L 1.115
455	a. Ethylene glycol b. Phenyl ether	197.5 257.4	192.3	64.5 35.5	0.22 99.78	98.28 1.72	U 35.3 L 64.7	U 1.068 L 1.108
456	a. Ethyl ether b. Isoprene	34.6 34.0	33.2	48.0 52.0				
457	a. Ethyl ether b. Methyl formate	34.6 31.5	28.2	44.0 56.0				
458	a. Ethyl ether b. Methyl sulfide	34.6 37.5	34.0	80.0 20.0				
459	a. Ethyl ether b. Water	34.6 100.0	34.2	98.8 1.2				0.720
460	a. Ethyl-3-ethoxy propionate b. Water	170.1 100.0	97.0	37.0 63.0	98.1 1.9	1.6 98.5	U 38.0 L 62.0	U 0.94 L 0.99
461	a. Ethyl formate b. Methanol	54.2 64.7	51.0	84.0 16.0				
462	a. Ethyl formate b. Water	54.2 100.0	52.6	95.0 5.0	95.5 4.5	13.6 86.4	U >99.5 L <0.5	U 0.920 L 0.995
463	a. 2-Ethylhexanoic acid b. Water	228.0 100.0	99.9	3.6 96.4	98.77 1.23	0.25 99.75	U 3.7 L 96.3	U 0.906 L 1.000
464	a. 2-Ethyl hexanol b. Phenol } 25 mm	95.6 89.6	95.0	66.0 34.0				0.900
465	a. 2-Ethyl hexanol b. Water	185.0 100.0	99.1	20.0 80.0	97.4 2.6	0.10 99.90	U 23.0 L 77.0	U 0.838 L 1.000
466	a. 2-Ethylhexyl acetate b. Water	199.0 100.0	99.0	26.5 73.5	99.45 0.55	0.03 99.97	U 25.0 L 75.0	U 0.873 L 1.000
467	a. 2-Ethylhexyl amine b. Water	169.1 100.0	98.2	36.0 64.0	74.7 25.3	0.25 99.75	U 52.0 L 48.0	U 0.848 L 1.000
468	a. n-Ethylhexyl aniline b. Water	dec. 100.0	100.0	0.7 99.3	99.9 0.1	0.01 99.99	U 0.7 L 99.3	U 0.848 L 1.000
469	a. 2-Ethylhexyl chloride b. Water	173.0 100.0	97.3	45.0 55.0	99.9 0.1	0.1 99.9	U 48.0 L 52.0	U 0.883 L 1.000
470	a. 2-Ethylhexyl crotonate b. Water	241.2 100.0	99.9	6.6 93.4	99.55 0.45	0.01 99.99	U 7.2 L 92.8	U 0.889 L 0.999

BINARY SYSTEMS (Continued)

No.	Components: Compounds	BP, °C	Azeotrope: BP, °C	Percent composition: In azeotrope	Percent composition: Upper layer	Percent composition: Lower layer	Relative volume of layers at 20°C	Specific gravity of layers or azeotrope
471	a. 2-Ethylhexyl ether b. Water	269.8 100.0	99.8	3.6 96.4	99.97 0.03	0.01 99.99	U 4.0 L 96.0	U 0.911 L 0.998
472	a. 2-Ethylhexyl hexanoate b. Water	267.2 100.0	99.9	3.6 96.4	99.81 0.19	0.01 99.99	U 4.1 L 95.9	U 0.865 L 1.000
473	a. Ethylidene acetone b. Water	123.5 100.0	92.0	71.4 28.6	82.8 17.2	38.0 62.0	U 76.3 L 23.7	U 0.892 L 0.975
474	a. N-Ethyl morpholine b. Water	138.3 100.0	96.7	53.8 46.2				1.000
475	a. 4-Ethyl octanol b. Water	220.5 100.0	99.9	6.0 94.0	97.57 2.43	0.01 99.99	U 7.2 L 92.8	U 0.842 L 1.000
476	a. Ethyl propionate b. Propanol	99.0 97.2	93.4	49.0 51.0				
477	a. Ethyl propionate } 350 mm b. Water	76.0 79.7	61.0	86.7 13.3	98.6 1.4	1.7 98.3	U 89.0 L 11.0	U 0.886 L 0.998
478	a. Formic acid b. Toluene	100.7 110.6	85.8	50.0 50.0				
479	a. Formic acid b. Water	100.7 100.0	107.1	77.5 22.5				
480	a. 2-Furaldehyde b. o-Xylene	161.7 144.4	140.5	13.0 87.0				
481	a. Glycol b. Naphthalene	197.2 217.9	183.9	51.0 49.0				
482	a. Glycol b. Nitrobenzene	197.2 210.9	185.9	59.0 41.0				
483	a. Glycol b. Octyl alcohol	197.2 195.0	184.4	36.5 63.5				
484	a. Glycol b. Quinoline	197.2 237.7	196.4	79.5 20.5				
485	a. Glycol b. Styrene	197.2 146.0	139.5	16.5 83.5				
486	a. Glycol b. Toluene	197.2 110.6	110.2	6.5 93.5				
487	a. Glycol b. o-Toluidine	197.2 199.8	186.5	42.5 57.5				
488	a. Glycol b. o-Xylene	197.2 144.4	139.6	16.0 84.0				
489	a. Glycol diacetate b. Water	190.8 100.0	99.7	15.4 84.6				1.024

AZEOTROPES (Continued)

BINARY SYSTEMS (Continued)

	Components		Azeotrope					
				Percent composition				
No.	Compounds	BP, °C	BP, °C	In azeotrope	Upper layer	Lower layer	Relative volume of layers at 20°C	Specific gravity of layers or azeotrope
490	a. Heptane b. Methanol	98.4 64.7	59.1	48.5 51.5				
491	a. Heptane b. Vinyl acetate	98.4 72.7	72.0	16.5 83.5				0.880
492	a. Heptane b. Water	98.4 100.0	79.2	87.1 12.9	99.98 0.02	0.01 99.99	U 90.8 L 9.2	U 0.685 L 1.000
493	a. 2-Heptyl acetate b. Water	176.4 100.0	97.8	41.1 58.9	99.49 0.51	0.03 99.67	U 45.0 L 55.0	U 0.864 L 1.000
494	a. 3-Heptyl acetate b. Water	173.8 100.0	97.5	42.4 57.6	99.56 0.44	0.03 99.97	U 46.0 L 54.0	U 0.863 L 1.000
495	a. Hexaldehyde b. Water	128.5 100.0	91.0	68.7 31.3	98.8 1.2	0.6 99.4	U 73.4 L 26.6	U 0.817 L 0.997
496	a. Hexane b. Isobutyl chloride	69.0 68.9	66.3	45.0 55.0				
497	a. Hexane b. Isopropanol	69.0 82.3	61.0	78.0 22.0				0.686
498	a. Hexane b. Isopropyl ether	69.0 67.5	67.5	47.0 53.0				
499	a. Hexane b. Methanol	69.0 64.7	50.0	73.1 26.9	85.0 15.0	42.0 58.0	U 67.8 L 32.2	U 0.675 L 0.724
500	a. Hexane b. Methylethyl ketone	69.0 79.6	64.3	71.7 28.3				0.698
501	a. Hexane b. 2-Methyl-2-propanol	69.0 82.6	63.7	75.0 25.0				0.691
502	a. Hexane b. Nitromethane	69.0 101.0	62.0	79.0 21.0				
503	a. Hexane b. Propanol	69.0 97.2	65.7	96.0 4.0				0.67
504	a. Hexane b. Propionitrile	69.0 97.1	63.5	91.0 9.0				
505	a. Hexane b. Water	69.0 100.0	61.6	94.4 5.6			U 96.2 L 3.8	U 0.660 L 1.000
506	a. Hexanoic acid b. Water	205.0 100.0	99.9	7.9 92.1	94.6 5.4	1.1 98.9	U 8.5 L 91.5	U 0.934 L 0.999
507	a. Hexanol b. Water	158.0 100.0	97.8	32.8 67.2	92.8 7.2	0.58 99.42	U 39.0 L 61.0	U 0.835 L 0.999
508	a. 2-Hexenal b. Water	149.0 100.0	95.1	51.4 48.6	98.3 1.7	0.2 99.8	U 56.3 L 43.7	U 0.848 L 0.999
509	a. Hexyl acetate b. Water	169.2 100.0	97.4	39.0 61.0	99.44 0.56	0.05 99.95	U 42.3 L 57.7	U 0.873 L 1.000

BINARY SYSTEMS (Continued)

No.	Components: Compounds	BP, °C	Azeotrope: BP, °C	Percent composition: In azeotrope	Percent composition: Upper layer	Percent composition: Lower layer	Relative volume of layers at 20°C	Specific gravity of layers or azeotrope
510	a. Hexyl amine b. Water	132.7 100.0	95.5	51.0 49.0				0.881
511	a. Hexyl chloride b. Water	133.9 100.0	91.8	70.3 29.7	99.83 0.17	0.01 99.99	U 73.0 L 27.0	U 0.873 L 1.000
512	a. Hexyl-2-ethyl butyrate b. Water	230.3 100.0	99.7	11.2 88.8	99.76 0.24	0.01 99.99	U 13.0 L 87.0	U 0.861 L 1.000
513	a. Hexyl hexanoate b. Water	245.2 100.0	99.8	6.7 93.3	99.70 0.30	0.01 99.99	U 7.7 L 92.3	U 0.864 L 1.000
514	a. Hydrogen bromide b. Water	−67.0 100.0	126.0	47.5 52.5				1.481
515	a. Hydrogen chloride b. Water	−83.7 100.0	108.6	20.2 79.8				1.102
516	a. Hydrogen chloride } 6.8 atm b. Water	−31.0 169.0	177.0	14.8 85.2				
517	a. Hydrogen fluoride b. Water	19.4 100.0	111.4	35.6 64.4				
518	a. Hydrogen iodide b. Water	−35.5 100.0	127.0	57.0 43.0				
519	a. Hydrogen nitrate b. Water	86.0 100.0	121.0	68.5 31.5				1.405
520	a. Iodoethane b. Propanol	72.2 97.2	70.0	93.0 7.0				
521	a. Iodomethane b. Isopropyl alcohol	42.5 82.3	42.4	98.2 1.8				
522	a. Iodomethane b. Methanol	42.5 64.7	37.8	95.5 4.5				
523	a. 1-Iodopropane b. Isopropyl alcohol	102.4 82.3	79.8	58.0 42.0				
524	a. Isoamyl acetate b. Isoamyl alcohol	142.5 130.5	129.1	2.6 97.4				
525	a. Isoamyl alcohol b. Water	130.5 100.0	95.2	50.4 49.6				
526	a. Isoamyl alcohol b. o-Xylene	130.5 144.4	127.0	>52.0 <48.0				
527	a. Isobutyl acetate b. Isobutyl alcohol	116.5 108.4	107.4	45.0 55.0				
528	a. Isobutyl alcohol b. Propyl acetate	108.4 101.6	101.0	17.0 83.0				

BINARY SYSTEMS (Continued)

No.	Components: Compounds	Components: BP, °C	Azeotrope: BP, °C	Percent composition: In azeotrope	Percent composition: Upper layer	Percent composition: Lower layer	Relative volume of layers at 20°C	Specific gravity of layers or azeotrope
529	a. Isobutyl alcohol	108.4	123.0	35.0				
	b. Propylene diamine	120.9		65.0				
530	a. Isobutyl alcohol	108.4	101.2	44.5				0.836
	b. Toluene	110.6		55.5				
531	a. Isobutyl alcohol	108.4	85.4	9.0				1.368
	b. Trichloroethylene	87.1		91.0				
532	a. Isobutyl alcohol	108.4	89.7	70.0	85.0	8.7	U 82.3	U 0.839
	b. Water	100.0		30.0	15.0	91.3	L 17.7	L 0.988
533	a. Isobutyl chloride	68.9	63.8	81.0				
	b. Isopropanol	82.3		19.0				
534	a. Isobutyl chloride	68.9	53.1	79.4				
	b. Methanol	64.7		20.6				
535	a. Isophorone	215.2	99.5	16.1	95.7	1.2	U 16.0	U 0.929
	b. Water	100.0		83.9	4.3	98.8	L 84.0	L 0.999
536	a. Isoprene	34.0	34.0	86.0				
	b. 2-Methyl-2-butene	38.4		14.0				
537	a. Isopropyl acetate	89.0	64.0	29.8				0.816
	b. Methanol	64.7		70.2				
538	a. Isopropyl acetate	89.0	75.9	88.9	98.2	2.9	U 91.4	U 0.870
	b. Water	100.0		11.1	1.8	97.1	L 8.6	L 0.995
539	a. Isopropyl alcohol	82.3	80.1	52.6				0.822
	b. Isopropyl acetate	89.0		47.4				
540	a. Isopropyl alcohol	82.3	77.3	30.0				0.800
	b. Methylethyl ketone	79.6		70.0				
541	a. Isopropyl alcohol	82.3	77.0	28.0				
	b. Methyl propionate	79.6		72.0				
542	a. Isopropyl alcohol	82.3	79.3	71.8				
	b. Nitromethane	101.0		28.2				
543	a. Isopropyl alcohol	82.3	81.7	81.0				
	b. Tetrachloroethylene	121.2		19.0				
544	a. Isopropyl alcohol	82.3	80.6	58.0				
	b. Toluene	110.6		42.0				
545	a. Isopropyl alcohol	82.3	74.0	28.0				1.182
	b. Trichloroethylene	87.1		72.0				
546	a. Isopropyl alcohol	82.3	70.8	22.4				0.899
	b. Vinyl acetate	72.7		77.6				
547	a. Isopropyl alcohol	82.3	80.4	87.8				0.818
	b. Water	100.0		12.2				
548	a. Isopropyl benzene	152.4	95.0	56.2	99.95	0.01	U 60.0	U 0.884
	b. Water	100.0		43.8	0.05	99.99	L 40.0	L 1.000

AZEOTROPES (Continued)

BINARY SYSTEMS (Continued)

No.	Components: Compounds	Components: BP, °C	Azeotrope: BP, °C	Percent composition: In azeotrope	Percent composition: Upper layer	Percent composition: Lower layer	Relative volume of layers at 20°C	Specific gravity of layers or azeotrope
549	a. Isopropyl chloride	36.5	35.0	99.0	99.67	0.31	U 99.0	U 0.862
	b. Water	100.0		1.0	0.33	99.69	L 1.0	L 0.999
550	a. Isopropyl ether	67.5	62.2	95.4	99.43	0.90	U 97.0	U 0.727
	b. Water	100.0		4.6	0.57	99.10	L 3.0	L 0.998
551	a. Isopropyl ether } 481 mm	54.8	50.0	96.0	99.4	0.90	U 97.4	U 0.727
	b. Water	87.7		4.0	0.6	99.10	L 2.6	L 0.998
552	a. Isoveralaldehyde	92.5	77.0	88.0				
	b. Water	100.0		12.0				
553	a. Mesityl oxide	128.7	91.8	65.3	96.6	2.8	U 69.8	U 0.860
	b. Water	100.0		34.7	3.4	97.2	L 30.2	L 0.995
554	a. Methacrylaldehyde	67.5	63.6	92.3	98.02	2.65	U 94.9	U 0.845
	b. Water	100.0		7.7	1.98	97.35	L 5.1	L 0.996
555	a. Methanol	64.7	54.0	18.7				0.908
	b. Methyl acetate	57.0		81.3				
556	a. Methanol } 4.4 atm.	107.0	99.0	29.0				0.890
	b. Methyl acetate	107.0		71.0				
557	a. Methanol	64.7	41.8	92.2				
	b. Methylal	42.3		7.8				
558	a. Methanol	64.7	62.5	54.0				
	b. Methyl acrylate	80.5		46.0				
559	a. Methanol	64.7	64.5	92.0				
	b. Nitromethane	101.0		8.0				
560	a. Methanol	64.7	63.0	72.0				
	b. Octane	125.8		28.0				
561	a. Methanol	64.7	63.7	72.4				0.813
	b. Toluene	110.6		27.6				
562	a. Methanol	64.7	60.2	36.0				1.126
	b. Trichloroethylene	87.1		64.0				
563	a. Methanol	64.7	39.9	6.0				1.476
	b. 1,1,2-Trichloro-trifluoroethane	47.7		94.0				
564	a. Methanol	64.7	54.0	27.0				0.892
	b. Trimethyl borate	65.0		73.0				
565	a. Methanol	64.7	58.5	36.6				0.880
	b. Vinyl acetate	72.7		63.4				
566	a. Methanol	64.7	62.0	57.5				
	b. 1-Methoxy-1,3-butadiene	90.9		42.5				
567	a. Methanol	64.7	62.0	52.0				
	b. Vinylbutyl ether	94.2		48.0				

BINARY SYSTEMS (Continued)

No.	Components: Compounds	Components: BP, °C	Azeotrope: BP, °C	Percent composition: In azeotrope	Percent composition: Upper layer	Percent composition: Lower layer	Relative volume of layers at 20°C	Specific gravity of layers or azeotrope
568	a. 1-Methoxy-1,3-butadiene	90.9	76.2	87.3	99.74	0.31	U 89.4	U 0.832
	b. Water	100.0		12.7	0.26	99.69	L 10.6	L 0.999
569	a. 3-Methoxybutyl acetate	171.3	96.5	34.6	95.9	6.2	U 33.0	U 0.960
	b. Water	100.0		65.4	4.1	93.8	L 67.0	L 1.005
570	a. 2-Methoxy ethanol	124.6	105.9	25.0				0.887
	b. Toluene	110.6		75.0				
571	a. 2-Methoxy ethanol	124.6	99.9	15.3				1.005
	b. Water	100.0		84.7				
572	a. 1-Methoxy-2-propanol	118.5	97.5	51.0				0.994
	b. Water	100.0		49.0				
573	a. Methyl acetate	57.0	56.1	95.0				0.940
	b. Water	100.0		5.0				
574	a. Methylal	42.3	42.1	98.6				
	b. Water	100.0		1.4				
575	a. Methylamyl ketone	150.5	95.2	54.6	98.5	0.43	U 59.5	U 0.819
	b. Water	100.0		45.4	1.5	99.57	L 40.5	L 0.999
576	a. α-Methylbenzyl amine	187.4	99.4	16.2	52.0	4.8	U 26.0	U 0.994
	b. Water	100.0		83.8	48.0	95.2	L 74.0	L 1.000
577	a. α-Methylbenzyl ether	286.7	100.0	1.3	0.01	99.78	U 98.7	U 1.000
	b. Water	100.0		98.7	99.99	0.22	L 1.3	L 1.003
578	a. n-Methylbutyl amine	91.1	82.7	85.0				0.802
	b. Water	100.0		15.0				
579	a. n-Methyldibutyl amine	162.9	96.5	52.0	99.6	0.07	U 58.9	U 0.761
	b. Water	100.0		48.0	0.4	99.93	L 41.1	L 1.000
580	a. Methylene chloride	40.0	38.8	99.0	2.0	99.9	U 1.6	U 1.009
	b. Water	100.0		1.0	98.0	<0.1	L 98.4	L 1.328
581	a. Methylethyl ketone	79.6	73.4	88.0				0.834
	b. Water	100.0		12.0				
582	a. 2-Methyl-5-ethyl pyridine	177.8	98.4	28.0	80.6	1.22	U 35.0	U 0.960
	b. Water	100.0		72.0	19.4	98.78	L 65.0	L 1.002
583	a. 5-Methyl-2-hexanone	144.0	94.7	56.0	98.6	0.55	U 63.0	U 0.814
	b. Water	100.0		44.0	1.4	99.45	L 37.0	L 1.004
584	a. Methylisobutyl ketone	115.1	87.9	76.0	98.4	2.0	U 80.4	U 0.806
	b. Water	100.0		24.0	1.6	98.0	L 19.6	L 0.999
585	a. Methylisopropenyl ketone	97.9	81.5	81.6	97 0	0.91	U 96.0	U 0.853
	b. Water	100.0		18.4	3.0	99.09	L 4.0	L 0.997
586	a. N-Methyl morpholine	115.0	94.2	76.0				
	b. Water	100.0		24.0				

AZEOTROPES (Continued)

BINARY SYSTEMS (Continued)

No.	Components: Compounds	Components: BP, °C	Azeotrope: BP, °C	Percent composition: In azeo-trope	Percent composition: Upper layer	Percent composition: Lower layer	Relative volume of layers at 20°C	Specific gravity of layers or azeotrope
587	a. 2-Methyl pentanal b. Propanol	118.3 97.2	95.0	14.0 86.0				0.814
588	a. 2-Methyl pentanal b. Water	118.3 100.0	88.5	77.0 23.0	99.17 0.83	0.42 99.58	U 81.0 L 19.0	U 0.811 L 0.999
589	a. 2-Methylpentanoic acid b. Water	196.4 100.0	99.4	12.1 87.9	97.1 2.9	1.3 98.7	U 13.0 L 87.0	U 0.924 L 1.000
590	a. 2-Methyl pentanol b. Water	148.0 100.0	97.2	40.0 60.0	94.6 5.4	0.3 99.7	U 47.0 L 53.0	U 0.826 L 1.003
591	a. 4-Methyl-2-pentanol b. Water	131.0 100.0	94.3	56.7 43.3	94.2 5.8	1.7 98.3	U 64.0 L 36.0	U 0.820 L 0.999
592	a. 4-Methyl-2-pentanol b. Vinyl-2-chloryl ether	131.0 109.1	109.0	1.0 99.0				1.004
593	a. 4-Methyl-2-pentanone b. Water	116.9 100.0	87.9	75.7 24.3				
594	a. 4-Methyl-2-pentene b. Water	56.7 100.0	53.3	96.5 3.5	99.91 0.09	0.15 99.85	U 97.7 L 2.3	U 0.670 L 0.999
595	a. 4-Methyl-2-pentyl acetate b. Water	146.1 100.0	94.8	63.3 36.7	99.42 0.58	0.13 99.87	U 67.0 L 33.0	U 0.858 L 0.998
596	a. 4-Methyl-2-pentyl butyrate b. Water	182.6 100.0	98.2	39.2 60.8	97.2 2.8	0.81 99.19	U 43.0 L 57.0	U 0.859 L 0.999
597	a. Methylphenyl carbinol b. Water	205.0 100.0	99.7	11.0 89.0	2.3 97.7	94.1 5.9	U 91.0 L 9.0	U 1.000 L 1.010
598	a. Methylphenyl ketone b. Water	202.3 100.0	99.1	18.5 81.5	0.55 99.45	98.35 1.65	U 91.0 L 9.0	U 1.00 L 1.03
599	a. 2-Methyl-2-propanol b. Water	82.6 100.0	79.9	88.3 11.7				0.814
600	a. 2-Methylpropyl acetate b. Water	117.3 100.0	88.4	78.0 22.0	98.98 1.02	0.63 99.37	U 81.0 L 9.0	U 0.874 L 1.000
601	a. Methylpropyl ketone b. Water	101.7 100.0	83.8	80.4 19.6	96.7 3.3	4.3 95.7	U 85.0 L 15.0	U 0.812 L 0.99
602	a. Methylvinyl chloride (cis) b. Water	32.8 100.0	33.0	99.1 0.9	99.9 0.1	0.2 99.8	U 99.3 L 0.7	U 0.927 L 1.000
603	a. Methylvinyl ketone b. Water	80.0 100.0	75.8	85.0 15.0				
604	a. Nitroethane b. Toluene	114.8 110.6	106.2	25.0 75.0				
605	a. Nitromethane b. Toluene	101.0 110.6	96.5	55.0 45.0				

	Components		Azeotrope					
				Percent composition				
No.	Compounds	BP, °C	BP, °C	In azeotrope	Upper layer	Lower layer	Relative volume of layers at 20°C	Specific gravity of layers or azeotrope
606	a. Nitromethane b. Trichloroethylene	101.0 87.0	81.4	20.0 80.0				
607	a. o-Nitrophenol b. Propionamide	217.3 213.0	211.2	75.2 24.8				
608	a. Nonane b. Water	150.8 100.0	95.0	60.2 39.8	100.0	 100.0	U 68.0 L 32.0	U 0.719 L 1.000
609	a. 1-Octanol b. Phenol	195.0 182.0	195.4	87.0 13.0				
610	a. 1-Octanol b. Water	195.0 100.0	99.4	10.0 90.0				
611	a. Paraldehyde b. Water	124.5 100.0	90.8	74.8 25.2	98.9 1.1	10,5 89.5	U 73.0 L 27.0	U 0.998 L 1.01
612	a. Pentachloroethane b. Phenol	162.0 182.0	160.9	90.5 9.5				
613	a. Pentane b. Water	36.1 100.0	34.6	98.6 1.4	99.95 0.05	0.04 99.96	U 99.1 L 0.9	U 0.627 L 1.000
614	a. 2,4-Pentanedione b. Water	140.6 100.0	94.4	59.0 41.0	95.5 4.5	16.6 83.4	U 55.0 L 45.0	U 0.981 L 1.011
615	a. 3-Pentanol b. Pyridine	115.6 115.5	117.4	55.0 45.0				
616	a. Pentanol b. Water	138.0 100.0	95.4	45.0 55.0				
617	a. 3-Pentanol b. Water	115.6 100.0	91.5	65.0 35.0	90.1 9.9	5.5 94.5	U 74.0 L 26.0	U 0.833 L 0.992
618	a. 2-Pentanone b. Water	102.0 100.0	83.3	80.5 19.5				
619	a. 4-Pentenal b. Water	106.2 100.0	84.3	79.0 21.0				
620	a. Phenol b. Water	182.0 100.0	99.5	9.21 90.79				
621	a. Phenyl ether b. Water	259.0 100.0	99.8	4.3 95.7	0.02 99.98	99.97 0.03	U 96.0 L 4.0	U 0.997 L 1.067
622	a. γ-Picoline b. Water	144.6 100.0	97.4	36.5 63.5				0.996
623	a. Propadiene b. Propane	−32.0 −42.1	−42.0	11.6 88.4	by volume by volume			
624	a. Propane b. Propyne	−42.1 −23.2	−42.0	88.3 11.7	by volume by volume			
625	a. Propanol b. Propyl acetate	97.2 101.6	94.0	63.0 37.0				0.833

	Components		Azeotrope					
No.	Compounds	BP, °C	BP, °C	Percent composition In azeo-trope	Percent composition Upper layer	Percent composition Lower layer	Relative volume of layers at 20°C	Specific gravity of layers or azeotrope
626	a. Propanol b. Tetrachloroethylene	97.2 121.0	94.1	48.0 52.0				
627	a. Propanol b. Toluene	97.2 110.6	92.6	49.0 51.0				0.836
628	a. Propanol b. Trichloroethylene	97.2 87.1	81.8	17.0 83.0				1.287
629	a. Propanol b. Water	97.2 100.0	88.1	71.8 28.2				0.866
630	a. Propanol b. m-Xylene	97.2 139.1	97.1	94.0 6.0				
631	a. Propionaldehyde b. Water	48.8 100.0	47.5	98.0 2.0				0.81
632	a. Propionaldehyde } 250 mm b. Water	21.0 71.6	19.0	99.5 0.5				
633	a. Propionamide b. o-Xylene	213.0 144.4	144.0	2.0 98.0				
634	a. Propionic acid b. Water	141.6 100.0	99.9	17.7 82.3				1.016
635	a. Propionic acid b. o-Xylene	141.6 144.4	135.4	43.0 57.0				
636	a. Propionic acid b. p-Xylene	141.6 138.4	132.0	36.0 64.0				
637	a. Propionitrile b. Water	97.2 100.0	82.2	76.0 24.0	95.0 5.0	9.4 90.6	U 83.3 L 16.7	U 0.793 L 0.983
638	a. Propyl acetate b. Water	101.6 100.0	82.4	86.0 14.0				
639	a. Propyl chloride b. Water	46.6 100.0	44.0	97.8 2.2	99.7 0.3	0.3 99.7	U 98.0 L 2.0	U 0.892 L 1.000
640	a. Propylene chlorohydrin b. Water	127.0 100.0	95.4	54.2 45.8				
641	a. Propylene diamine b. Toluene	120.9 110.6	105.0	32.0 68.0				0.865
642	a. Propylene dichloride b. Water	96.8 100.0	78.4	89.4 10.6	0.26 99.74	99.94 0.06	U 12.0 L 88.0	U 1.000 L 1.159
643	a. Propylene oxide } 60 psig b. Water	88.0 155.0	86.5	99.8 0.2				
644	a. Pyridine b. 3-Pentanol	115.5 115.6	117.4	45.0 55.0				
645	a. Pyridine b. Toluene	115.5 110.6	110.2	22.0 78.0				

AZEOTROPES (Continued)

BINARY SYSTEMS (Continued)

No.	Components: Compounds	Components: BP, °C	Azeotrope: BP, °C	Percent composition: In azeotrope	Percent composition: Upper layer	Percent composition: Lower layer	Relative volume of layers at 20°C	Specific gravity of layers or azeotrope
646	a. Pyridine	115.5	92.6	57.0				1.010
	b. Water	100.0		43.0				
647	a. Styrene	145.2	93.9	59.1	99.95	0.03	U 61.0	U 0.91
	b. Water	100.0		40.9	0.05	99.97	L 39.0	L 1.00
648	a. Styrene oxide	194.2	99.2	22.4	0.3	99.5	U 79.1	U 1.000
	b. Water	100.0		77.6	99.7	0.5	L 20.9	L 1.054
649	a. Tetrachloroethylene	121.0	88.5	82.8	0.02	99.99	U 25.0	U 1.000
	b. Water	100.0		17.2	99.98	<0.01	L 75.0	L 1.625
650	a. 1,2,3,6-Tetrahydrobenzaldehyde	164.2	96.9	40.0	98.98	0.51	U 41.0	U 0.972
	b. Water	100.0		60.0	1.02	99.49	L 59.0	L 1.000
651	a. Tetrahydrobenzonitrile	195.1	98.8	21.7	99.46	0.63	U 22.3	U 0.958
	b. Water	100.0		78.3	0.54	99.37	L 77.7	L 0.999
652	a. 1,4-Thioxane	149.2	95.6	52.0	6.85	98.38	U 53.0	U 1.014
	b. Water	100.0		48.0	93.15	1.62	L 47.0	L 1.120
653	a. Toluene	110.6	85.0	79.8	99.95	0.06	U 82.0	U 0.868
	b. Water	100.0		20.2	0.05	99.94	L 18.0	L 1.000
654	a. Triallyl amine	151.1	95.0	62.0	99.57	0.13	U 67.0	U 0.802
	b. Water	100.0		38.0	0.43	99.87	L 33.0	L 1.000
655	a. Tributyl amine	213.9	99.8	18.0	99.7	0.01	U 22.0	U 0.781
	b. Water	100.0		82.0	0.3	99.99	L 78.0	L 1.000
656	a. 1,1,2-Trichloroethane	113.7	86.0	83.6	0.45	99.95	U 22.0	U 1.000
	b. Water	100.0		16.4	99.55	0.05	L 78.0	L 1.443
657	a. 1,1,2-Trichloroethylene	87.1	73.1	93.7	0.2	99.98	U 9.0	U 1.003
	b. Water	100.0		6.3	99.8	0.02	L 91.0	L 1.466
658	a. 1,1,2-Trichlorotrifluoroethane	47.7	44.5	99.0	1.0	99.9	U 2.0	U 1.00
	b. Water	100.0		1.0	99.0	0.1	L 98.0	L 1.57
659	a. Tridecanol	244.0	100.0	2.2	98.74	0.01	U 2.6	U 0.843
	b. Water	100.0		97.8	1.26	99.99	L 97.4	L 1.000
660	a. Tridecyl acrylate	dec.	100.0	1.2	99.8	0.01	U 1.3	U 0.882
	b. Water	100.0		98.8	0.2	99.99	L 98.7	L 1.000
661	a. 1,1,3-Triethoxyethane	dec.	99.6	15.0	99.41	0.01	U 17.0	U 0.875
	b. Water	100.0		85.0	0.59	99.99	L 83.0	L 1.000
662	a. 1,1,3-Triethoxypropane		99.0	30.0				
	b. Water	100.0		70.0				
663	a. Triethyl amine	89.5	75.8	90.0				0.769
	b. Water	100.0		10.0				
664	a. Triglycol dichloride	240.9	99.7	6.0	1.9	99.2	U 96.5	U 1.003
	b. Water	100.0		94.0	98.1	0.8	L 3.5	L 1.188

BINARY SYSTEMS (Continued)

No.	Components: Compounds	BP, °C	Azeotrope: BP, °C	Percent composition: In azeotrope	Percent composition: Upper layer	Percent composition: Lower layer	Relative volume of layers at 20°C	Specific gravity of layers or azeotrope
665	a. Trimethyltetrahydro-benzaldehyde	204.5	99.0	23.0	99.72	0.01	U 24.5	U 0.919
	b. Water	100.0		77.0	0.28	99.99	L 75.5	L 1.000
666	a. Valeraldehyde	103.3	83.0	81.0	98.7	1.35	U 85.0	U 0.810
	b. Water	100.0		19.0	1.3	98.65	L 15.0	L 0.998
667	a. Valeric acid	186.2	99.8	11.0	87.0	2.4	U 10.0	U 0.957
	b. Water	100.0		89.0	13.0	97.6	L 90.0	L 1.000
668	a. Vinyl acetate	72.7	66.0	92.7	98.97	2.0	U 94.0	U 0.933
	b. Water	100.0		7.3	1.03	98.0	L 6.0	L 0.999
669	a. Vinylallyl ether	67.4	60.0	94.6	99.73	0.4	U 95.8	U 0.806
	b. Water	100.0		5.4	0.27	99.6	L 4.2	L 0.999
670	a. Vinyl benzoate	dec.	99.3	17.4	0.01	99.68	U 83.5	U 1.000
	b. Water	100.0		82.6	99.99	0.32	L 16.5	L 1.070
671	a. Vinylbutyl ether	94.2	77.5	88.4	99.91	0.3	U 91.0	U 0.780
	b. Water	100.0		11.6	0.09	99.7	L 9.0	L 1.000
672	a. Vinyl butyrate	116.7	87.2	79.6	99.7	0.3	U 81.0	U 0.902
	b. Water	100.0		20.4	0.3	99.7	L 19.0	L 0.999
673	a. Vinyl-2-chloroethyl ether	109.1	84.0	83.0	0.61	99.63	U 17.0	U 1.000
	b. Water	100.0		17.0	99.39	0.37	L 83.0	L 1.049
674	a. Vinyl crotonate	133.9	92.0	69.0	98.9	0.3	U 71.0	U 0.944
	b. Water	100.0		31.0	1.1	99.7	L 29.0	L 1.000
675	a. Vinylethyl ether	35.5	34.6	98.5	99.8	0.9	U 99.3	U 0.754
	b. Water	100.0		1.5	0.2	99.1	L 0.7	L 0.999
676	a. Vinyl-2-ethyl hexanoate	185.2	98.6	32.0	99.8	0.01	U 35.0	U 0.875
	b. Water	100.0		68.0	0.2	99.99	L 65.0	L 1.000
677	a. Vinyl-2-ethylhexyl ether	177.7	97.8	40.9	99.95	0.01	U 46.0	U 0.810
	b. Water	100.0		59.1	0.05	99.99	L 54.0	L 1.000
678	a. 2-Vinyl-5-ethyl pyridine	dec.	99.4	15.0	94.64	0.01	U 16.0	U 0.95
	b. Water	100.0		85.0	5.36	99.99	L 84.0	L 1.00
679	a. Vinylisobutyl ether	83.4	70.5	92.2	99.92	0.2	U 16.0	U 0.771
	b. Water	100.0		7.8	0.08	99.8	L 84.0	L 1.000
680	a. Vinyl isobutyrate	105.4	83.5	83.0	99.7	0.36	U 85.0	U 0.891
	b. Water	100.0		17.0	0.3	99.64	L 15.0	L 1.000
681	a. Vinyl isopropyl ether	55.7	51.8	97.3	99.81	0.64	U 98.0	U 0.760
	b. Water	100.0		2.7	0.19	99.36	L 2.0	L 0.991
682	a. Vinyl-2-methyl pentanoate	148.8	95.0	62.0	99.81	0.03	U 65.0	U 0.881
	b. Water	100.0		38.0	0.19	99.97	L 35.0	L 1.000

BINARY SYSTEMS (Continued)

No.	Components: Compounds	Components: BP, °C	Azeotrope: BP, °C	Percent composition: In azeotrope	Percent composition: Upper layer	Percent composition: Lower layer	Relative volume of layers at 20°C	Specific gravity of layers or azeotrope
683	a. Vinyl propionate b. Water	94.9 100.0	79.0	87.0 13.0	99.40 0.60	0.82 99.18	U 88.0 L 12.0	U 0.918 L 0.999
684	a. Vinylpropyl ether b. Water	65.1 100.0	59.0	95.0 5.0	99.8 0.2	0.4 99.6	U 96.0 L 4.0	U 0.770 L 0.999
685	a. m-Xylene b. Water	139.1 100.0	94.5	60.0 40.0	99.95 0.05	0.05 99.95	U 63.4 L 36.6	U 0.868 L 1.000

AZEOTROPES

TERNARY SYSTEMS

No.	Components: Compounds	Components: BP, °C	Azeotrope: BP, °C	Percent composition: In azeotrope	Percent composition: Upper layer	Percent composition: Lower layer	Relative volume of layers at 20°C	Specific gravity of layers or azeotrope
1	a. Acetal b. Ethanol c. Water	103.6 78.5 100.0	77.8	61.0 27.6 11.4				
2	a. Acetone b. Carbon disulfide c. Water	56.2 46.3 100.0	38.04	23.98 75.21 0.81				
3	a. Acetone b. Chloroform c. Methanol	56.2 61.2 64.7	57.5	30.0 47.0 23.0				
4	a. Acetone b. Chloroform c. Water	56.2 61.2 100.0	60.4	38.4 57.6 4.0				
5	a. Acetone b. Cyclohexane c. Methanol	56.2 81.4 64.7	51.5	43.5 40.5 16.0				
6	a. Acetone b. Isoprene c. Water	56.2 34.0 100.0	32.5	7.6 92.0 0.4				
7	a. Acetone, 775.5 mm b. Isopropyl ether c. Water	78.5 91.9 120.9	75.0	49.0 48.0 3.0				0.769
8	a. Acetone b. Methanol c. Methyl acetate	56.2 64.7 57.0	53.7	5.8 17.4 76.8				0.898
9	a. Acetonitrile b. Benzene c. Water	82.0 80.1 100.0	66.0	23.3 68.5 8.2				
10	a. Acetonitrile b. Ethanol c. Triethyl amine	82.0 78.5 89.5	70.1	34.0 8.0 58.0				0.757

AZEOTROPES (Continued)

TERNARY SYSTEMS (Continued)

No.	Components: Compounds	Components: BP, °C	Azeotrope: BP, °C	Percent composition: In azeotrope	Percent composition: Upper layer	Percent composition: Lower layer	Relative volume of layers at 20°C	Specific gravity of layers or azeotrope
11	a. Acetonitrile	82.0	72.9	44.0				
	b. Ethanol	78.5		55.0				
	c. Water	100.0		1.0				
12	a. Acetonitrile	82.0	59.0	13.0	13.0	13.0	U 97.0	U 0.742
	b. Isopropyl ether	68.3		82.0	85.5	1.0	L 3.0	L 0.976
	c. Water	100.0		5.0	1.5	86.0		
13	a. Acetonitrile	82.0	67.0	20.5				
	b. Trichloroethylene	87.0		73.1				
	c. Water	100.0		6.4				
14	a. Acetonitrile	82.0	68.6	31.0				0.77
	b. Triethyl amine	89.5		63.0				
	c. Water	100.0		6.0				
15	a. Acrylonitrile	78.0	69.5	71.0				
	b. Ethanol	78.5		20.3				
	c. Water	100.0		8.7				
16	a. Allyl alcohol	97.0	77.8	8.7				
	b. Allyl ether	94.3		12.4				
	c. Water	100.0		77.8				
17	a. Allyl alcohol	97.0	68.2	9.2	8.7	17.7	U 91.2	U 0.877
	b. Benzene	80.1		82.2	90.7	0.4	L 8.8	L 0.985
	c. Water	100.0		8.6	0.6	80.9		
18	a. Allyl alcohol	97.0	65.2	11.0	25.6	10.1	U 8.6	U 0.777
	b. Carbon tetrachloride	76.8		84.0	2.7	89.1	L 91.4	L 1.464
	c. Water	100.0		5.0	71.7	0.8		
19	a. Allyl alcohol	97.0	66.2	81.0				
	b. Cyclohexane	81.4		11.0				
	c. Water	100.0		8.0				
20	a. Allyl alcohol	97.0	68.0	11.0				
	b. Cyclohexene	83.0		80.5				
	c. Water	100.0		8.5				
21	a. Allyl alcohol	97.0	59.7	5.0	3.6	34.8	U 98.2	U 0.668
	b. Hexane	69.0		90.0	95.9	0.8	L 1.8	L 0.964
	c. Water	100.0		5.0	0.5	64.4		
22	a. Allyl alcohol	97.0	80.6	31.4	32.4	27.4	U 83.0	U 0.856
	b. Toluene	110.6		53.4	64.9	2.2	L 17.0	L 0.95
	c. Water	100.0		15.2	2.7	70.4		
23	a. Allyl alcohol	97.0	71.4	12.5	21.6	12.5	U 11.8	U 0.983
	b. Trichloro ethylene	87.0		80.0	4.7	86.5	L 88.2	L 1.354
	c. Water	100.0		7.5	73.7	1.0		
24	a. Benzene	80.1	64.6	74.1	86.0	4.8	U 85.5	U 0.866
	b. Ethanol	78.5		18.5	12.7	52.1	L 14.2	L 0.892
	c. Water	100.0		7.4	1.3	43.1		
25	a. Benzene	80.1	65.7	72.0	77.5	0.5	U 93.6	U 0.855
	b. Isopropanol	82.3		19.8	20.2	14.4	L 6.4	L 0.966
	c. Water	100.0		8.2	2.3	85.1		

AZEOTROPES (Continued)

TERNARY SYSTEMS (Continued)

	Components		Azeotrope					
No.	Compounds	BP, °C	BP, °C	Percent composition: In azeotrope	Percent composition: Upper layer	Percent composition: Lower layer	Relative volume of layers at 20°C	Specific gravity of layers or azeotrope
26	a. Benzene	80.1	68.2	65.1	71.3	0.1	U 92.5	U 0.858
	b. 2-Butanone (Methylethyl ketone)	79.6		26.1	28.1	5.2	L 7.5	L 0.992
	c. Water	100.0		8.8	0.6	94.7		
27	a. Benzene	80.1	67.3	70.5	76.7	0.1	U 92.4	U 0.857
	b. 2-Methyl-2-propanol	82.6		21.4	21.6	3.2	L 7.6	L 0.979
	c. Water	100.0		8.1	1.7	96.7		
28	a. Benzene	80.1	68.5	82.4	92.2	0.1	U 92.1	U 0.873
	b. Propanol	97.2		9.0	7.7	15.3	L 7.9	L 0.979
	c. Water	100.0		8.6	0.1	84.6		
29	a. Benzene	80.1	67.0	82.3				
	b. Propyl alcohol	97.2		10.1				
	c. Water	100.0		7.6				
30	a. Bromodichloromethane	90.2	72.0	>70.0				
	b. Ethanol	78.5		<22.5				
	c. Water	100.0		7.5				
31	a. 1-Bromopropane	70.8	60.0	83.0				
	b. Ethanol	78.5		12.0				
	c. Water	100.0		5.0				
32	a. 1-Butanol	117.7	90.7	8.0	11.0	2.0	U 75.5	U 0.874
	b. Butyl acetate	126.5		63.0	86.0	1.0	L 24.5	L 0.997
	c. Water	100.0		29.0	3.0	97.0		
33	a. 1-Butanol, 100 mm	69.8	46.0	26.0	41.0	2.6	U 65.0	U 0.862
	b. Butyl acrylate			33.0	53.0	0.4	L 35.0	L 0.990
	c. Water	51.6		41.0	6.0	97.0		
34	a. 1-Butanol	117.7	90.6	34.6	46.0	4.9	U 76.5	U 0.796
	b. Butyl ether	142.0		34.5	49.0	0.3	L 23.5	L 0.995
	c. Water	100.0		29.9	5.0	94.8		
35	a. 1-Butanol	117.7	77.4	2.0	2.0	2.0	U 92.0	U 0.78
	b. Vinylbutyl ether	94.2		88.0	98.0	0.3	L 8.0	L 1.00
	c. Water	100.0		10.0	0.1	98.0		
36	a. 2-Butanol	99.5	85.5	27.4	31.7	4.6	U 86.0	U 0.858
	b. 2-Butyl acetate	112.2		52.4	62.3	0.6	L 14.0	L 0.994
	c. Water	100.0		20.2	6.0	94.8		
37	a. 2-Butanol	99.5	86.6	56.1	65.0	10.0	U 86.0	U 0.816
	b. Butyl ether	142.0		19.2	23.0	0.2	L 14.0	L 0.981
	c. Water	100.0		24.7	12.0	89.8		
38	a. 2-Butanol	99.5	67.0					
	b. Cyclohexane	81.0						
	c. Water	100.0						
39	a. 2-Butanol	99.5	77.5	19.0	20.0	9.0±1	U 92.0	U 0.736
	b. Diisobutylene	102.6		70.0	78.8	0.5	L 8.0	L 0.987
	c. Water	100.0		11.0	1.2	91.0±1		

TERNARY SYSTEMS (Continued)

No.	Components: Compounds	Components: BP, °C	Azeotrope: BP, °C	Percent composition: In azeotrope	Percent composition: Upper layer	Percent composition: Lower layer	Relative volume of layers at 20°C	Specific gravity of layers or azeotrope
40	a. 2-Butanol	99.5	61.1					
	b. Hexane	69.0						
	c. Water	100.0						
41	a. 2-Butanone	79.6	65.7	22.2				
	b. Carbon tetrachloride	76.8		74.8				
	c. Water	100.0		3.0				
42	a. 2-Butanone	79.6	58.5	22.0				
	b. Hexane	68.0		74.0				
	c. Water	100.0		4.0				
43	a. Butenylmethyl ether		61.4	78.9	84.8	2.1	U 94.0	U 0.787
	b. Ethanol	78.5		14.3	12.6	37.1	L 6.0	L 0.996
	c. Water	100.0		6.8	2.6	60.8		
44	a. tert. Butyl alcohol	82.2	64.7	11.9				
	b. Carbon tetrachloride	76.8		85.0				
	c. Water	100.0		3.1				
45	a. Butyl amine	77.1	81.8	50.0				0.795
	b. Ethanol	78.5		42.5				
	c. Water	100.0		7.5				
46	a. Butyl amine	77.1	83.0	47.0				
	b. Isopropanol	82.3		40.5				
	c. Water	100.0		12.5				
47	a. Butyraldehyde	75.7	67.2	80.0	82.0		U 97.8	
	b. Ethanol	78.5		11.0	11.0		L 2.2	L 0.838
	c. Water	100.0		9.0	7.0			
48	a. Carbon disulfide	46.2	41.3	93.4				
	b. Ethanol	78.5		5.0				
	c. Water	100.0		1.6				
49	a. Carbon disulfide	46.2	37.0					
	b. Methanol	64.7						
	c. Methyl acetate	57.0						
50	a. Carbon disulfide	46.2	35.6	55.0				
	b. Methanol	64.7		7.0				
	c. Methylal	44.0		38.0				
51	a. Carbon tetrachloride	76.8	61.8	86.3	7.0	94.8	U 15.2	U 0.935
	b. Ethanol	78.5		10.3	48.5	5.2	L 84.4	L 1.519
	c. Water	100.0		3.4	44.5	<0.1		
52	a. Carbon tetrachloride	76.8	65.7	74.8	0.1	77.3	U 3.9	U 0.993
	b. 2-Butanone	79.6		22.2	5.5	22.6	L 96.1	L 1.313
	c. Water	100.0		3.0	94.4	0.1		
53	a. Carbon tetrachloride	76.8	65.4	84.0	0.1	88.0	U 7.0	U 0.979
	b. Propanol	97.2		11.0	15.0	11.0	L 93.0	L 1.436
	c. Water	100.0		5.0	84.9	1.0		
54	a. Chloroform	61.2	55.5	92.5	1.0	95.8	U 6.2	U 0.976
	b. Ethanol	78.5		4.0	18.2	3.7	L 93.8	L 1.441
	c. Water	100.0		3.5	80.8	0.5		

TERNARY SYSTEMS (Continued)

No.	Components: Compounds	Components: BP, °C	Azeotrope: BP, °C	Percent composition: In azeotrope	Percent composition: Upper layer	Percent composition: Lower layer	Relative volume of layers at 20°C	Specific gravity of layers or azeotrope
55	a. Chloroform	61.2	52.6	81.0	32.0	83.0	U 3.0	U 1.022
	b. Methanol	64.7		15.0	41.0	14.0	L 97.0	L 1.399
	c. Water	100.0		4.0	27.0	3.0		
56	a. 1-Chloro-2-methylpropane	68.9	58.6	82.5				
	b. Ethanol	78.5		13.0				
	c. Water	100.0		4.5				
57	a. Crotonaldehyde	104.0	78.0	7.3				0.810
	b. Ethanol	78.5		87.9				
	c. Water	100.0		4.8				
58	a. 1,3-Cyclohexadiene	80.5	67.8	79.0				
	b. n-Propanol	97.1		12.0				
	c. Water	100.0		9.0				
59	a. Cyclohexane	81.0	62.1	76.0				
	b. Ethanol	78.5		17.0				
	c. Water	100.0		7.0				
60	a. Cyclohexane	81.0	68.3					
	b. Ethyl acetate	77.1						
	c. Isopropanol	82.3						
61	a. Cyclohexane	81.0	50.8	33.6				
	b. Methanol	64.7		17.8				
	c. Methyl acetate	57.1		48.6				
62	a. Cyclohexane	81.0	63.6	35.0	62.4	0.1	U 94.5	U 0.769
	b. 2-Butanone	79.6		60.0	37.0	10.0	L 5.5	L 0.98
	c. Water	100.0		5.0	0.6	89.9		
63	a. Cyclohexane	81.0	65.0	71.0				
	b. 2-Methyl-2-propanol	82.6		21.0				
	c. Water	100.0		8.0				
64	a. Cyclohexane	81.0	66.6	81.5				
	b. Propanol	97.2		10.0				
	c. Water	100.0		8.5				
65	a. Cyclohexane	81.0	66.1	71.0				
	b. Isopropanol	82.3		21.5				
	c. Water	100.0		7.5				
66	a. Cyclohexene	82.9	64.1	73.0				
	b. Ethanol	78.5		20.0				
	c. Water	100.0		7.0				
67	a. 1,2-Dichloroethane	83.7	66.7	78.0				
	b. Ethanol	78.5		17.0				
	c. Water	100.0		5.0				
68	a. cis-1,2-Dichloroethylene	60.3	53.8	90.5				
	b. Ethanol	78.5		6.65				
	c. Water	100.0		2.85				

TERNARY SYSTEMS (Continued)

No.	Components: Compounds	Components: BP, °C	Azeotrope: BP, °C	Percent composition: In azeotrope	Percent composition: Upper layer	Percent composition: Lower layer	Relative volume of layers at 20°C	Specific gravity of layers or azeotrope
69	a. trans-1,2-Dichloroethylene	47.5	44.4	94.5				
	b. Ethanol	78.5		4.4				
	c. Water	100.0		1.1				
70	a. Diethoxymethane	87.5	73.2	69.5				
	b. Ethanol	78.5		18.4				
	c. Water	100.0		12.8				
71	a. Diethyl formal	87.5	73.2	69.5				
	b. Ethanol	78.5		18.4				
	c. Water	100.0		12.1				
72	a. Diethyl ketone	102.7	81.2	60.0				
	b. Propanol	97.2		20.0				
	c. Water	100.0		20.0				
73	a. Diisobutylene	102.6	72.3	59.1	70.0	5.4	U 83.0	
	b. Isopropanol	82.3		31.6	26.9	55.2	L 17.0	
	c. Water	100.0		9.3	3.1	39.4		
74	a. Dipropyl acetal	146.6	87.6	21.0				
	b. Propanol	97.2		51.6				
	c. Water	100.0		27.4				
75	a. Dipropyl formal	137.4	86.4	47.2				
	b. Propanol	97.2		44.5				
	c. Water	100.0		8.0				
76	a. Ethanol	78.5	70.2	8.4				0.901
	b. Ethyl acetate	77.1		82.6				
	c. Water	100.0		9.0				
77	a. Ethanol } 1446 mm		88.9	12.1				
	b. Ethyl acetate } 1446 mm			77.6				
	c. Water } 1446 mm			10.3				
78	a. Ethanol	78.5	77.1	48.3				0.867
	b. Ethyl acrylate	99.8		41.6				
	c. Water	100.0		10.1				
79	a. Ethanol	78.5	71.6	4.2	2.6	33.4	U 95.5	U 0.774
	b. Ethylbutyl ether	92.2		86.5	91.2	1.4	L 4.5	L 0.941
	c. Water	100.0		9.3	6.2	65.2		
80	a. Ethanol	78.5	81.4	61.7				
	b. Ethyl chloroacetate	143.5		20.8				
	c. Water	100.0		17.5				
81	a. Ethanol	78.5	67.8	15.7	41.8	12.5	U 13.3	U 0.941
	b. Ethylene dichloride	84.0		77.1	11.6	85.2	L 86.7	L 1.167
	c. Water	100.0		7.2	46.6	2.3		
82	a. Ethanol	78.5	66.0	15.8	14.6	33.0	U 94.5	U 0.747
	b. Ethylisobutyl ether	81.0		77.7	82.6	7.8	L 5.5	L 0.90
	c. Water	100.0		6.5	2.8	59.2		

AZEOTROPES (Continued)

TERNARY SYSTEMS (Continued)

	Components		Azeotrope					
				Percent composition				
No.	Compounds	BP, °C	BP, °C	In azeo-trope	Upper layer	Lower layer	Relative volume of layers at 20°C	Specific gravity of layers or azeotrope
83	a. Ethanol	78.5	68.8	33.0	5.0	75.9	U 64.6	U 0.686
	b. Heptane	98.4		60.9	94.8	9.1	L 35.4	L 0.801
	c. Water	100.0		6.1	0.2	15.0		
84	a. Ethanol	78.5	56.0	12.0	3.0	75.0	U 90.0	U 0.672
	b. Hexane	69.0		85.0	96.5	6.0	L 10.0	L 0.833
	c. Water	100.0		3.0	0.5	19.0		
85	a. Ethanol	78.5	58.6	13.0				
	b. Isobutyl chloride	68.9		82.5				
	c. Water	100.0		4.5				
86	a. Ethanol	78.5	74.8	19.4				0.874
	b. Isopropyl acetate	93.0		70.8				
	c. Water	100.0		9.8				
87	a. Ethanol	78.5	61.0	6.5	5.9	20.2	U 97.1	U 0.737
	b. Isopropyl ether	67.5		89.5	92.9	1.8	L 2.9	L 0.967
	c. Water	100.0		4.0	1.2	78.0		
88	a. Ethanol	78.5	62.0	8.6	8.1	17.0	U 96.0	U 0.743
	b. Methylbutyl ether	71.0		85.1	89.6	2.2	L 4.0	L 0.95
	c. Water	100.0		6.3	2.3	80.8		
89	a. Ethanol	78.5	73.2	14.0				0.832
	b. Methylethyl ketone	79.6		75.0				
	c. Water	100.0		11.0				
90	a. Ethanol	78.5	66.0	14.7	13.7	30.5	U 95.0	U 0.745
	b. Propylisopropyl ether	83.0		78.3	83.3	0.7	L 5.0	L 0.925
	c. Water	100.0		7.0	3.0	68.3		
91	a. Ethanol	78.5	74.4	37.0	15.6	54.8	U 46.5	U 0.849
	b. Toluene	110.6		51.0	81.3	24.5	L 53.5	L 0.855
	c. Water	100.0		12.0	3.1	20.7		
92	a. Ethanol	78.5	74.7	15.0				0.774
	b. Triethyl amine	89.5		75.0				
	c. Water	100.0		10.0				
93	a. Ethanol	78.5	60.0	22.0	20.3	38.7	U 91.0	U 0.777
	b. Vinylisobutyl ether	83.4		70.0	77.6	3.6	L 9.0	L 0.90
	c. Water	100.0		8.0	2.1	57.7		
94	a. Ethanol	78.5	57.0	21.2	20.6	32.0	U 95.5	U 0.722
	b. Vinylpropyl ether	65.1		73.7	77.8	0.2	L 4.5	L 0.923
	c. Water	100.0		5.1	1.6	67.8		
95	a. 2-Ethoxyethanol	135.6	97.7	11.0	0.5	17.0	U 43.0	U 0.81
	b. Vinyl-2-ethylhexyl ether	177.7		38.0	99.4	0.1	L 57.0	L 1.00
	c. Water	100.0		51.0	0.1	82.9		
96	a. Ethyl acetate	77.1	68.3					
	b. Isopropanol	82.3						
	c. Cyclohexane	81.0						
97	a. Ethylbutyl ether	92.2	73.4	67.7	73.1	0.5	U 94.0	U 0.762
	b. Isopropanol	82.3		21.9	22.5	14.5	L 6.0	L 0.97
	c. Water	100.0		10.4	4.4	85.0		

No.	Components: Compounds	Components: BP, °C	Azeotrope: BP, °C	Percent composition: In azeotrope	Percent composition: Upper layer	Percent composition: Lower layer	Relative volume of layers at 20°C	Specific gravity of layers or azeotrope
98	a. Ethylene dichloride	84.0	69.7	75.3	0.9	78.1	U 7.0	U 0.968
	b. Isopropanol	82.3		19.0	20.7	18.8	L 93.0	L 1.117
	c. Water	100.0		7.7	78.4	3.1		
99	a. Ethyl ether	34.6	20.4	8.0				
	b. Methyl formate	31.5		40.0				
	c. Pentane	36.2		52.0				
100	a. Fluosilicic acid	d	116.1	36.0				
	b. Hydrofluoric acid	19.5		10.0				
	c. Water	100.0		54.0				
101	a. Formic acid	100.7	97.5	40.4				
	b. m-Xylene	139.0		49.0				
	c. Water	100.0		10.6				
102	a. Hexane	69.0	45.0	59.0				0.73
	b. Methanol	64.7		14.0				
	c. Methyl acetate	57.0		27.0				
103	a. Isoamyl acetate	142.5	93.6	24.0				
	b. Isoamyl alcohol	130.5		31.2				
	c. Water	100.0		44.8				
104	a. Isoamyl alcohol	130.5	89.8	19.6				
	b. Isoamyl formate	123.5		48.0				
	c. Water	100.0		32.4				
105	a. Isoamyl formate	123.5	117.6	25.0				
	b. Paraldehyde	124.0		30.0				
	c. Tetrachloroethylene	121.0		45.0				
106	a. Isobutyl acetate	116.5	86.8	46.5				
	b. Isobutyl alcohol	108.4		23.1				
	c. Water	100.0		30.4				
107	a. Isopropanol	82.3	75.5	13.0	13.0	11.5	U 94.0	U 0.870
	b. Isopropyl acetate	89.0		76.0	81.4	2.9	L 6.0	L 0.981
	c. Water	100.0		11.0	5.6	85.6		
108	a. Isopropanol	82.3	61.8	4.0	4.0	5.0	U 97.2	U 0.732
	b. Isopropyl ether	67.5		91.0	94.7	1.0	L 2.8	L 0.990
	c. Water	100.0		5.0	1.3	94.0		
109	a. Isopropanol } 776 mm	100.7	81.0	7.0	7.0	9.0	U 96.1	U 0.737
	b. Isopropyl ether } 776 mm	91.9		87.0	91.5	1.0	L 3.9	L 0.984
	c. Water } 776 mm	120.9		6.0	1.5	90.0		
110	a. Isopropanol	82.3	73.4	1.0				0.834
	b. Methylethyl ketone	79.6		88.0				
	c. Water	100.0		11.0				
111	a. Isopropanol	82.3	76.3	38.2	38.2	38.0	U 92.0	U 0.845
	b. Toluene	110.6		48.7	53.3	1.0	L 8.0	L 0.930
	c. Water	100.0		13.1	8.5	61.0		
112	a. Isopropyl ether	67.5	61.0	92.4	97.4	3.0	U 97.0	U 0.724
	b. Propanol	97.2		2.0	2.0	3.0	L 3.0	L 0.990
	c. Water	100.0		5.6	0.6	94.0		

AZEOTROPES (Continued)

TERNARY SYSTEMS (Continued)

No.	Components: Compounds	BP, °C	Azeotrope: BP, °C	Percent composition: In azeotrope	Percent composition: Upper layer	Percent composition: Lower layer	Relative volume of layers at 20°C	Specific gravity of layers or azeotrope
113	a. Methanol	64.7	67.9	81.20				
	b. Methyl chloroacetate	131.5		13.54				
	c. Water	100.0		5.26				
114	a. 2-Methoxyethanol	124.6	97.7	4.0	0.2	6.0	U 45.0	U 0.81
	b. Vinyl-2-ethylhexyl ether	177.7		39.0	99.7	0.1	L 55.0	L 1.00
	c. Water	100.0		57.0	0.1	93.9		
115	a. 3-Pentanone	102.2	~81.2	~60.0				
	b. n-Propyl alcohol	97.2		~20.0				
	c. Water	100.0		~20.0				
116	a. Propanol	97.2	82.2	19.5				
	b. Propyl acetate	101.6		59.5				
	c. Water	100.0		21.0				
117	a. n-Propanol	97.2	70.8	5.0				
	b. Propyl formate	81.3		82.0				
	c. Water	100.0		13.0				
118	a. Propanol	97.2	71.6	12.0				
	b. Trichloroethylene	87.1		81.0				
	c. Water	100.0		7.0				
119	a. Propanol	97.2	74.8	20.2				
	b. Propyl ether	91.0		68.1				
	c. Water	100.0		11.7				

HEAT OF FORMATION OF INORGANIC OXIDES

The ΔH_o values are given in gram calories per mole. The *a*, *b*, and *I* values listed here make it possible for one to calculate the ΔF and ΔS values by use of the following equations:

$$\Delta F_t = \Delta H_o + 2.303aT \log T + b \times 10^{-3}T^2 + c \times 10^5 T^{-1} + IT$$
$$\Delta S_t = -a - 2.303a \log T - 2b \times 10^{-3}T + c \times 10^5 T^{-2} - I$$

Ref: Bulletin 542, U. S. Bureau of Mines, 1954.

Coefficients in Free-Energy Equations

Reaction and temperature range of validity	ΔH_o	2.303*a*	*b*	*c*	*I*
$2\,Ac(c) + 3/2\,O_2(g) = Ac_2O_3(c)$ .... (298.16°–1,000° K.)	−446,090	−16.12	.....	.....	+109.89
$2\,Al(c) + 1/2\,O_2(g) = Al_2O(g)$ .... (298.16°–931.7° K.)	−31,660	+14.97	.....	.....	−72.74
$2\,Al(l) + 1/2\,O_2(g) = Al_2O(g)$ .... (931.7°–2,000° K.)	−38,670	+10.36	.....	.....	−51.53
$Al(c) + 1/2\,O_2(g) = AlO(g)$ .... (298.16°–931.7° K.)	+10,740	+5.76	.....	.....	−37.61
$Al(l) + 1/2\,O_2(g) = AlO(g)$ .... (931.7°–2,000° K.)	+8,170	+5.76	.....	.....	−34.85
$2\,Al(c) + 3/2\,O_2(g) = Al_2O_3$ (*corundum*). (298.16°–931.7° K.)	−404,080	−15.68	+2.18	+3.935	+123.64
$2\,Al(l) + 3/2\,O_2(g) = Al_2O_3$ (*corundum*). (931.7°–2,000° K.)	−407,950	−6.19	−.78	+3.935	+102.37
$2\,Am(c) + 3/2\,O_2(g) = Am_2O_3(c)$ .... (298.16°–1,000° K.)	−422,090	−16.12	.....	.....	+107.89
$Am(c) + O_2(g) = AmO_2(c)$ .... (298.16°–1,000° K.)	−240,600	−4.61	.....	.....	+55.91
$2\,Sb(c) + 3/2\,O_2(g) = Sb_2O_3$ (*cubic*) ..... (298.16°–842° K.)	−169,450	+6.12	−6.01	−.30	+52.21
$2\,Sb(c) + 3/2\,O_2(g) = Sb_2O_3$ (*orthorhombic*). (298.16°–903° K.)	−168,060	+6.12	−6.01	−.30	+50.56
$2\,Sb(l) + 3/2\,O_2(g) = Sb_2O_3$ (*orthorhombic*). (903°–928° K.)	−175,370	+15.29	−7.75	−.30	+33.12
$2\,Sb(l) + 3/2\,O_2(g) = Sb_2O_3(l)$ .... (928°–1,698° K.)	−173,940	−32.84	+.75	−.30	+166.52
$2\,Sb(l) + 3/2\,O_2(g) = 1/2\,Sb_4O_6(g)$ ..... (1,698°–1,713° K.)	−132,760	+10.91	+.75	−.30	+.96
$2\,Sb(g) + 3/2\,O_2(g) = 1/2\,Sb_4O_6(g)$ ..... (1,713°–2,000° K.)	−234,760	−.74	+.75	−.30	+98.17
$2\,Sb(c) + 2\,O_2(g) = Sb_2O_4(c)$ .... (298.16°–903° K.)	−208,310	+6.31	−5.36	−.40	+73.02
$2\,Sb(l) + 2\,O_2(g) = Sb_2O_4(c)$ .... (903°–1,500° K.)	−215,610	+15.47	−7.10	−.40	+55.61
$6\,Sb(c) + 13/2\,O_2(g) = Sb_6O_{13}(c)$ .... (298.16°–903° K.)	−649,160	+38.46	−25.13	−1.30	+192.54
$6\,Sb(c) + 13/2\,O_2(g) = Sb_6O_{13}(c)$ .... (903°–1,500° K.)	−691,370	+14.13	−30.35	−1.30	+315.93
$2\,Sb(c) + 5/2\,O_2(g) = Sb_2O_5(c)$ .... (298.16°–903° K.)	−226,060	+37.12	−22.66	−.50	+18.61
$2\,Sb(l) + 5/2\,O_2(g) = Sb_2O_5(c)$ .... (903°–1,500° K.)	−240,130	+29.01	−24.40	−.50	+59.74
$2\,As(c) + 3/2\,O_2(g) = As_2O_3$ (*orthorhombic*). (298.16°–542° K.)	−154,870	+29.54	−21.33	−.30	−8.83
$2\,As(c) + 3/2\,O_2(g) = As_2O_3$ (*monoclinic*). (298.16°–586° K.)	−150,760	+29.54	−21.33	−.30	−16.95
$2\,As(c) + 3/2\,O_2(g) = As_2O_3(l)$ .... (542°–730.3° K.)	−156,260	−43.29	+2.97	−.30	+180.95
$2\,As(c) + 3/2\,O_2(g) = 1/2\,As_4O_6(g)$ ..... (730.3°–883° K.)	−135,930	+.46	+2.97	−.30	+26.88
$1/2\,As_4(g) + 3/2\,O_2(g) = 1/2\,As_4O_6(g)$ .. (883°–2,000° K.)	−154,450	−2.90	+.75	−.30	+59.71
$2\,As(c) + 2\,O_2(g) = As_2O_4(c)$ .... (298.16°–883° K.)	−173,690	+21.52	−13.42	−.40	+34.38
$1/2\,As_4(g) + 2\,O_2(g) = As_2O_4(c)$ .... (883°–1,500° K.)	−192,210	+18.15	−15.64	−.40	+67.22
$2\,As(c) + 5/2\,O_2(g) = As_2O_5(c)$ .... (298.16°–883° K.)	−217,080	+12.32	−4.65	−.50	+80.50
$1/2\,As_4(g) + 5/2\,O_2(g) = As_2O_5(c)$ ..... (883°–2,000° K.)	−235,600	+8.96	−6.87	−.50	+113.33
$Ba(\alpha) + 1/2\,O_2(g) = BaO(c)$ .... (298.16°–648° K.)	−134,590	−7.60	+.87	+.42	+45.76
$Ba(\beta) + 1/2\,O_2(g) = BaO(c)$ .... (648°–977° K.)	−134,140	−3.34	−.56	+.42	+34.01
$Ba(l) + 1/2\,O_2(g) = BaO(c)$ .... (977°–1,911° K.)	−135,900	−2.19	−.56	+.42	+32.37
$Ba(g) + 1/2\,O_2(g) = BaO(c)$ .... (1,911°–2,000° K.)	−176,400	−8.01	−0.56	+0.42	+72.66
$Ba(\alpha) + O_2(g) = BaO_2(c)$ .... (298.16°–648° K.)	−154,830	−11.05	+.87	+.42	+74.48
$Ba(\beta) + O_2(g) = BaO_2(c)$ .... (648°–977° K.)	−154,380	−6.79	−.56	+.42	+62.73
$Ba(l) + O_2(g) = BaO_2(c)$ .... (977°–1,500° K.)	−156,140	−5.64	−.56	+.42	+61.09
$Be(c) + 1/2\,O_2(g) = BeO(c)$ .... (298.16°–1,556° K.)	−144,220	−1.91	−.46	+1.24	+30.64
$Be(l) + 1/2\,O_2(g) = BeO(c)$ .... (1,556°–2,000° K.)	−144,300	+6.06	−1.75	+1.485	+7.25
$Bi(c) + 1/2\,O_2(g) = BiO(c)$ .... (298.16°–544° K.)	−50,450	−4.61	.....	.....	+35.51
$Bi(l) + 1/2\,O_2(g) = BiO(c)$ .... (544°–1,600° K.)	−52,920	−4.61	.....	.....	+40.05
$2\,Bi(c) + 3/2\,O_2(g) = Bi_2O_3(c)$ .... (298.16°–544° K.)	−139,000	−11.56	+2.15	−.30	+96.52
$2\,Bi(l) + 3/2\,O_2(g) = Bi_2O_3(c)$ .... (544°–1,090° K.)	−142,270	+2.30	−3.25	−.30	+67.55
$2\,Bi(l) + 3/2\,O_2(g) = Bi_2O_3(l)$ .... (1,090°–1,600° K.)	−147,350	−32.84	+.75	−.30	+174.59
$2\,B(c) + 3/2\,O_2(g) = B_2O(c)$ .... (298.16°–723° K.)	−304,690	+11.72	−7.55	+.355	+34.25
$2\,B(c) + 3/2\,O_2(g) = B_2O_3(gl)$ .... (298.16°–723° K.)	−298,670	+26.57	−15.90	−.30	−10.40
$2\,B(c) + 3/2\,O_2(g) = B_2O_3(l)$ .... (723°–2,000° K.)	−308,100	−38.41	+5.15	−.30	+173.24
$Cd(c) + 1/2\,O_2(g) = CdO(c)$ .... (298.16°–594° K.)	−62,330	−2.05	+.71	−.10	+29.17
$Cd(l) + 1/2\,O_2(g) = CdO(c)$ .... (594°–1,038° K.)	−63,240	+2.07	−.76	−.10	+20.14
$Cd(g) + 1/2\,O_2(g) = CdO(c)$ .... (1,038°–2,000° K.)	−89,320	−2.83	−.76	−.10	+60.05
$Ca(\alpha) + 1/2\,O_2(g) = CaO(c)$ .... (298.16°–673° K.)	−151,850	−6.56	+1.46	+.68	+43.93
$Ca(\beta) + 1/2\,O_2(g) = CaO(c)$ .... (673°–1,124° K.)	−151,730	−4.14	+.41	+.68	+37.63
$Ca(l) + 1/2\,O_2(g) = CaO(c)$ .... (1,124°–1,760° K.)	−153,480	−1.36	−.29	+.68	+31.49
$Ca(g) + 1/2\,O_2(g) = CaO(c)$ .... (1,760°–2,000° K.)	−194,670	−7.18	−.29	+.68	+73.84
$Ca(\alpha) + O_2(g) = CaO_2(c)$ .... (298.16°–500° K.)	−158,230	−12.32	+1.46	+.68	+78.28
$C(graphite) + 1/2\,O_2(g) = CO(g)$ .... (298.16°–2,000° K.)	−25,400	+2.05	+.27	−1.095	−28.79

HEAT OF FORMATION OF INORGANIC OXIDES (Continued)
Coefficients in Free-Energy Equations (Continued)

Reaction and temperature range of validity	ΔH_o	2.303*a*	*b*	*c*	*I*
$C(graphite) + O_2(g) = CO_2(g)$ (298.16°–2,000° K.)	−93,690	+1.63	−.07	−.23	−5.64
$2\ Ce(c) + 3/2\ O_2(g) = Ce_2O_3(c)$ (298.16°–1,048° K.)	−435,600	−4.60	.....	.....	+92.84
$2\ Ce(l) + 3/2\ O_2(g) = Ce_2O_3(c)$ (1,048°–1,900° K.)	−440,400	−4.60	.....	.....	+97.42
$Ce(c) + O_2(g) = CeO_2(c)$ (298.16°–1,048° K.)	−245,490	−6.42	+2.34	−.20	+67.79
$Ce(l) + O_2(g) = CeO_2(c)$ (1,048°–2,000° K.)	−247,930	+.71	−.66	−.20	+51.73
$2\ Cs(c) + 1/2\ O_2(g) = Cs_2O(c)$ (298.16°–301.5°K.)	−75,900	.....	.....	.....	+36.60
$2\ Cs(l) + 1/2\ O_2(g) = Cs_2O(c)$ (301.5°–763° K.)	−76,900	.....	.....	.....	+39.92
$2\ Cs(l) + 1/2\ O_2(g) = Cs_2O(l)$ (763°–963° K.)	−75,370	−9.21	.....	.....	+64.47
$2\ Cs(g) + 1/2\ O_2(g) = Cs_2O(l)$ (963°–1,500° K.)	−113,790	−23.03	.....	.....	+145.60
$2\ Cs(c) + O_2(g) = Cs_2O_2(c)$ (298.16°–301.5° K.)	−96,500	−2.30	.....	.....	+62.30
$2\ Cs(l) + O_2(g) = Cs_2O_2(c)$ (301.5°–870° K.)	−97,800	−4.61	.....	.....	+72.34
$2\ Cs(l) + O_2(g) = Cs_2O_2(l)$ (870°–963° K.)	−96,060	−18.42	.....	.....	+110.94
$2\ Cs(g) + O_2(g) = Cs_2O_2(l)$ (963°–1,500° K.)	−134,000	−31.08	.....	.....	+188.11
$2\ Cs(c) + 3/2\ O_2(g) = Cs_2O_3(c)$ (298.16°–301.5° K.)	−112,690	−11.51	.....	.....	+110.10
$2\ Cs(l) + 3/2\ O_2(g) = Cs_2O_3(c)$ (301.5°–775° K.)	−113,840	−12.66	.....	.....	+116.77
$2\ Cs(l) + 3/2\ O_2(g) = Cs_2O_3(l)$ (775°–963° K.)	−110,740	−26.48	.....	.....	+152.70
$2\ Cs(g) + 3/2\ O_2(g) = Cs_2O_3(l)$ (963°–1,500° K.)	−148,680	−39.14	.....	.....	+229.87
$Cs(c) + O_2(g) = CsO_2(c)$ (298.16°–301.5° K.)	−63,590	−11.51	.....	.....	+72.29
$Cs(l) + O_2(g) = CsO_2(c)$ (301.5°–705° K.)	−64,240	−12.66	.....	.....	+77.30
$Cs(l) + O_2(g) = CsO_2(l)$ (705°–963° K.)	−61,770	−18.42	.....	.....	+90.20
$Cs(g) + O_2(g) = CsO_2(l)$ (963°–1,500° K.)	−80,500	−24.18	.....	.....	+126.83
$Cl_2(g) + 1/2\ O_2(g) = Cl_2O(g)$ (298.16°–2,000° K.)	+17,770	−.71	−.12	+.49	+16.81
$1/2\ Cl_2(g) + 1/2\ O_2(g) = ClO(g)$ (298.16°–1,000° K.)	+33,000	.....	.....	.....	−.24
$1/2\ Cl_2(g) + O_2(g) = ClO_2(g)$ (298.16°–2,000° K.)	+24,150	−.76	−.105	−.665	+19.08
$1/2\ Cl_2(g) + 3/2\ O_2(g) = ClO_3(g)$ (298.16°–500° K.)	+37,740	+5.76	.....	.....	+21.42
$Cl_2(g) + 7/2\ O_2(g) = Cl_2O_7(g)$ (298.16°–500° K.)	+65,040	+12.66	.....	.....	+78.01
$2\ Cr(c) + 3/2\ O_2(g) = Cr_2O_3(\beta)$ (298.16°–1,823° K.)	−274,670	−14.07	+2.01	+.69	+105.65
$2\ Cr(l) + 3/2\ O_2(g) = Cr_2O_3(\beta)$ (1,823°–2,000° K.)	−278,030	+2.33	−.35	+1.57	+58.29
$Cr(c) + O_2(g) = CrO_2(c)$ (298.16°–1,000° K.)	−142,500	.....	.....	.....	+42.00
$Cr(c) + 3/2\ O_2(g) = CrO_3(c)$ (298.16°–471° K.)	−141,590	−13.82	.....	.....	+103.90
$Cr(c) + 3/2\ O_2(g) = CrO_3(l)$ (471°–600° K.)	−141,580	−32.24	.....	.....	+153.14
$Co(\alpha, \beta) + 1/2\ O_2(g) = CoO(c)$ (298.16°–1,400° K.)	−56,910	+.69	.....	.....	+16.03
$Co(\gamma) + 1/2\ O_2(g) = CoO(c)$ (1,400°–1,763° K.)	−58,160	−1.15	.....	.....	+22.71
$Co(l) + 1/2\ O_2(g) = CoO(c)$ (1,763°–2,000° K.)	−65,680	−6.22	.....	.....	+43.43
$3\ Co(\alpha, \beta, \gamma) + 2\ O_2(g) = Co_3O_4(c)$ (298.16°–1,500° K.)	−207,300	−2.30	.....	.....	+90.56
$2\ Cu(c) + 1/2\ O_2(g) = Cu_2O(c)$ (298.16°–1,357° K.)	−40,550	−1.15	−1.10	−.10	+21.92
$2\ Cu(l) + 1/2\ O_2(g) = Cu_2O(c)$ (1,357°–1,502° K.)	−43,880	+8.47	−2.60	−.10	−3.72
$2\ Cu(l) + 1/2\ O_2(g) = Cu_2O(l)$ (1,502°–2,000° K.)	−37,710	−12.48	+.25	−.10	+54.44
$Cu(c) + 1/2\ O_2(g) = CuO(c)$ (298.16°–1,357° K.)	−37,740	−.64	−1.40	−.10	+24.87
$Cu(l) + 1/2\ O_2(g) = CuO(c)$ (1,357°–1,720° K.)	−39,410	+4.17	−2.15	−.10	+12.05
$Cu(l) + 1/2\ O_2(g) = CuO(l)$ (1,720°–2,000° K.)	−41,060	−11.35	+.25	−.10	+59.09
$F_2(g) + 1/2\ O_2(g) = F_2O(g)$ (298.16°–2,000° K.)	+5,070	−.41	−.15	+.535	+16.04
$2\ Ga(c) + 1/2\ O_2(g) = Ga_2O(c)$ (298.16°–302.7° K.)	−81,110	+10.32	−5.75	−.10	−3.66
$2\ Ga(l) + 1/2\ O_2(g) = Ga_2O(c)$ (302.7°–1,000° K.)	−83,360	+13.49	−5.75	−.10	−4.08
$2\ Ga(c) + 3/2\ O_2(g) = Ga_2O_3(c)$ (298.16°–302.7° K.)	−256,240	+14.64	−3.75	−.30	+32.23
$2\ Ga(l) + 3/2\ O_2(g) = Ga_2O_3(c)$ (302.7°–2,000° K.)	−258,490	+17.82	−3.75	−.30	+31.79
$Ge(c) + 1/2\ O_2(g) = GeO(c)$ (298.16°–1,200° K.)	−60,900	+1.27	−1.49	−.10	+17.19
$Ge(c) + 1/2\ O_2(g) = GeO(g)$ (298.16°–1,200° K.)	−21,870	+6.72	−.075	−.10	−41.25
$Ge(c) + O_2(g) = GeO_2(gl)$ (298.16°–1,200° K.)	−127,830	+4.28	−2.52	−0.20	+30.54
$2\ Au(c) + 3/2\ O_2(g) = Au_2O_3(c)$ (298.16°–500° K.)	−2,160	−10.36	.....	.....	+95.14
$Hf(c) + O_2(g) = HfO_2(monocl.)$ (298.16°–2,000° K.)	−268,380	−9.74	−.28	+1.54	+78.16
$H_2(g) + 1/2\ O_2(g) = H_2O(l)$ (298.16°–373.16° K.)	−70,600	−18.26	+.64	−.04	+91.67
$H_2(g) + 1/2\ O_2(g) = H_2O(g)$ (298.16°–2,000° K.)	−56,930	+6.75	−.64	−.08	−8.74
$D_2(g) + 1/2\ O_2(g) = D_2O(l)$ (298.16°–374.5° K.)	−72,760	−18.10	.....	.....	+93.59
$D_2(g) + 1/2\ O_2(g) = D_2O(g)$ (298.16°–2,000° K.)	−58,970	+5.50	−.75	+.085	−3.74
$1/2\ H_2(g) + 1/2\ O_2(g) = OH(g)$ (298.16°–2,000° K.)	+10,350	+.90	+.005	−.26	−6.69
$H_2(g) + O_2(g) = H_2O_2(l)$ (298.16°–425° K.)	−47,140	−13.52	−7.13	.....	+99.39
$H_2(g) + O_2(g) = H_2O_2(g)$ (298.16°–1,500° K.)	−32,570	+4.77	−.96	+.97	+13.84
$2\ In(c) + 3/2\ O_2(g) = In_2O_3(c)$ (298.16°–429.6° K.)	−220,410	+5.43	−.50	−.30	+59.49
$2\ In(l) + 3/2\ O_2(g) = In_2O_3(c)$ (429.6°–2,000° K.)	−220,970	+13.22	−3.00	−.30	+41.36
$I_2(c) + 5/2\ O_2(g) = I_2O_5(c)$ (298.16°–386.8° K.)	−42,040	+2.30	.....	.....	+113.71
$I_2(l) + 5/2\ O_2(g) = I_2O_5(c)$ (386.8°–456° K.)	−43,490	+16.12	.....	.....	+81.70
$I_2(g) + 5/2\ O_2(g) = I_2O_5(c)$ (456°–500° K.)	−58,020	−6.91	.....	.....	+174.79
$Ir(c) + O_2(g) = IrO_2(c)$ (298.16°–1,300° K.)	−39,480	+8.17	−6.39	−.20	+20.33
$0.947\ Fe(\alpha) + 1/2\ O_2(g) = Fe_{0.947}O(c)$ (298.16°–1,033° K.)	−65,320	−11.26	+2.61	+0.44	+48.60
$0.947\ Fe(\beta) + 1/2\ O_2(g) = Fe_{0.947}O(c)$ (1,033°–1,179° K.)	−62,380	+4.08	−.75	+.235	+3.00
$0.947\ Fe(\gamma) + 1/2\ O_2(g) = Fe_{0.947}O(c)$ (1,179°–1,650° K.)	−66,750	−8.04	+.67	−.10	+42.28
$0.947\ Fe(\gamma) + 1/2\ O_2(g) = Fe_{0.947}O(l)$ (1,650°–1,674° K.)	−64,200	−18.72	+1.67	−.10	+73.45
$0.947\ Fe(\delta) + 1/2\ O_2(g) = Fe_{0.947}O(l)$ (1,674°–1,803° K.)	−59,650	−6.84	+.25	−.10	+34.81

HEAT OF FORMATION OF INORGANIC OXIDES (Continued)
Coefficients in Free-Energy Equations (Continued)

Reaction and temperature range of validity	ΔH_0	2.303a	b	c	I
$0.947\,Fe(l) + 1/2\,O_2(g) = Fe_{0.947}O(l)$ (1,803°–2,000° K.)	−63,660	−7.48	+.25	−.10	+39.12
$3\,Fe(\alpha) + 2\,O_2(g) = Fe_3O_4$ (*magnetite*) (298.16°–900° K.)	−268,310	+5.87	−12.45	+.245	+73.11
$3\,Fe(\alpha) + 2\,O_2(g) = Fe_3O_4(\beta)$ (900°–1,033° K.)	−272,300	−54.27	+11.65	+.245	+233.52
$3\,Fe(\beta) + 2\,O_2(g) = Fe_3O_4(\beta)$ (1,033°–1,179° K.)	−262,990	−5.71	+1.00	−.40	+89.19
$3\,Fe(\gamma) + 2\,O_2(g) = Fe_3O_4(\beta)$ (1,179°–1,674° K.)	−276,990	−44.05	+5.50	−.40	+213.52
$3\,Fe(\delta) + 2\,O_2(g) = Fe_3O_4(\beta)$ (1,674°–1,803° K.)	−262,560	−6.40	+1.00	−.40	+91.05
$3\,Fe(l) + 2\,O_2(g) = Fe_3O_4(\beta)$ (1,803°–1,874° K.)	−275,280	−8.74	+1.00	−.40	+104.84
$3\,Fe(l) + 2\,O_2(g) = Fe_3O_4(l)$ (1,874°–2,000° K.)	−257,240	−26.89	+1.00	−.40	+155.46
$2\,Fe(\alpha) + 3/2\,O_2(g) = Fe_2O_3$ (*hematite*) (298.16°–950° K.)	−200,000	−13.84	−1.45	+1.905	+108.26
$2\,Fe(\alpha) + 3/2\,O_2(g) = Fe_2O_3(\beta)$ (950°–1,033° K.)	−202,960	−42.64	+7.85	+.13	+188.48
$2\,Fe(\beta) + 3/2\,O_2(g) = Fe_2O_3(\beta)$ (1,033°–1,050° K.)	−196,740	−10.27	+.75	−.30	+92.26
$2\,Fe(\beta) + 3/2\,O_2(g) = Fe_2O_3(\gamma)$ (1,050°–1,179° K.)	−193,200	−.39	−.13	−.30	+59.96
$2\,Fe(\gamma) + 3/2\,O_2(g) = Fe_2O_3(\gamma)$ (1,179°–1,674° K.)	−202,540	−25.95	+2.87	−.30	+142.85
$2\,Fe(\delta) + 3/2\,O_2(g) = Fe_2O_3(\gamma)$ (1,674°–1,800° K.)	−192,920	−.85	−.13	−.30	+61.21
$2\,La(c) + 3/2\,O_2(g) = La_2O_3(c)$ (298.16°–1,153° K.)	−431,120	−13.31	+.80	+1.34	+112.36
$2\,La(l) + 3/2\,O_2(g) = La_2O_3(c)$ (1,153°–2,000° K.)	−434,330	−4.88	−.80	+1.34	+91.17
$Pb(c) + 1/2\,O_2(g) = PbO$ (*red*) (298.16°–600.5° K.)	−52,800	−2.76	−.80	−.10	+32.49
$Pb(l) + 1/2\,O_2(g) = PbO$ (*red*) (600.5°–762° K.)	−53,780	−.51	−1.75	−.10	+28.44
$Pb(c) + 1/2\,O_2(g) = PbO$ (*yellow*) (298.16°–600.5° K.)	−52,040	+.81	−2.00	−.10	+22.13
$Pb(l) + 1/2\,O_2(g) = PbO$ (*yellow*) (600.5°–1,159° K.)	−53,020	+3.06	−2.95	−.10	+18.08
$Pb(l) + 1/2\,O_2(g) = PbO(l)$ (1,159°–1,745° K.)	−53,980	−12.94	+.25	−.10	+64.22
$Pb(c) + 1/2\,O_2(g) = PbO(g)$ (298.16°–600.5° K.)	+10,270	+1.91	+1.08	+.295	−23.21
$Pb(l) + 1/2\,O_2(g) = PbO(g)$ (600.5°–2,000° K.)	+9,300	+4.17	+.13	+.295	−27.29
$3\,Pb(c) + 2\,O_2(g) = Pb_3O_4(c)$ (298.16°–600.5° K.)	−174,920	+8.82	−8.20	−.40	+72.78
$3\,Pb(l) + 2\,O_2(g) = Pb_3O_4(c)$ (600.5°–1,000° K.)	−177,860	+15.59	−11.05	−.40	+60.57
$Pb(c) + O_2(g) = PbO_2(c)$ (298.16°–600.5° K.)	−66,120	+.64	−2.45	−.20	+45.58
$Pb(l) + O_2(g) = PbO_2(c)$ (600.5°–1,100° K.)	−67,100	+2.90	−3.40	−.20	+41.50
$2\,Li(c) + 1/2\,O_2(g) = Li_2O(c)$ (298.16°–452° K.)	−142,220	−3.06	+5.77	−.10	+34.19
$2\,Li(l) + 1/2\,O_2(g) = Li_2O(c)$ (452°–1,600° K.)	−141,380	+16.97	−2.63	−.10	−17.05
$2\,Li(c) + O_2(g) = Li_2O_2(c)$ (298.16°–452° K.)	−151,880	−1.38	......	.....	+54.83
$2\,Li(l) + O_2(g) = Li_2O_2(c)$ (452°–500° K.)	−153,260	−1.38	......	.....	+57.88
$Mg(c) + 1/2\,O_2(g) = MgO$ (*periclase*) (298.16°–923° K.)	−144,090	−1.06	+.13	+.25	+29.16
$Mg(l) + 1/2\,O_2(g) = MgO$ (*periclase*) (923°–1,393° K.)	−145,810	+1.84	−.62	+.64	+23.07
$Mg(g) + 1/2\,O_2(g) = MgO$ (*periclase*) (1,393°–2,000° K.)	−180,700	−3.75	−.62	+.64	+65.69
$Mg(c) + O_2(g) = MgO_2(c)$ (298.16°–500° K.)	−150,230	−9.12	+.13	+.25	+70.84
$Mn(\alpha) + 1/2\,O_2(g) = MnO(c)$ (298.16°–1,000° K.)	−92,600	−4.21	+.97	+.155	+29.66
$Mn(\beta) + 1/2\,O_2(g) = MnO(c)$ (1,000°–1,374° K.)	−91,900	+1.84	−.39	+.34	+12.15
$Mn(\gamma) + 1/2\,O_2(g) = MnO(c)$ (1,374°–1,410° K.)	−89,810	+7.30	−.72	+.34	−6.05
$Mn(\delta) + 1/2\,O_2(g) = MnO(c)$ (1,410°–1,517° K.)	−89,390	+8.68	−.72	+.34	−10.70
$Mn(l) + 1/2\,O_2(g) = MnO(c)$ (1,517°–2,000° K.)	−93,350	+7.99	−.72	+.34	−5.90
$3\,Mn(\alpha) + 2\,O_2(g) = Mn_3O_4(\alpha)$ (298.16°–1,000° K.)	−332,400	−7.41	+.66	+.145	+106.62
$3\,Mn(\beta) + 2\,O_2(g) = Mn_3O_4(\alpha)$ (1,000°–1,374° K.)	−330,310	+10.75	−3.42	+.70	+54.07
$3\,Mn(\gamma) + 2\,O_2(g) = Mn_3O_4(\alpha)$ (1,374°–1,410° K.)	−324,050	+27.12	−4.41	+.70	−.50
$3\,Mn(\delta) + 2\,O_2(g) = Mn_3O_4(\alpha)$ (1,410°–1,445° K.)	−322,800	+31.27	−4.41	+.70	−14.46
$3\,Mn(\delta) + 2\,O_2(g) = Mn_3O_4(\beta)$ (1,445°–1,517° K.)	−328,870	−4.56	+1.00	−.40	+95.20
$3\,Mn(l) + 2\,O_2(g) = Mn_3O_4(\beta)$ (1,517°–1,800° K.)	−340,730	−6.63	+1.00	−.40	+109.60
$2\,Mn(\alpha) + 3/2\,O_2(g) = Mn_2O_3(c)$ (298.16°–1,000° K.)	−230,610	−5.96	−.06	+.945	+80.74
$2\,Mn(\beta) + 3/2\,O_2(g) = Mn_2O_3(c)$ (1,000°–1,374° K.)	−229,210	+6.15	−2.78	+1.315	+45.70
$2\,Mn(\gamma) + 3/2\,O_2(g) = Mn_2O_3(c)$ (1,374°–1,410° K.)	−225,030	+17.06	−3.44	+1.315	+9.33
$2\,Mn(\delta) + 3/2\,O_2(g) = Mn_2O_3(c)$ (1,410°–1,517° K.)	−224,200	+19.82	−3.44	+1.315	+0.05
$2\,Mn(l) + 3/2\,O_2(g) = Mn_2O_3(c)$ (1,517°–1,700° K.)	−232,110	+18.44	−3 44	+1.315	+9.65
$Mn(\alpha) + O_2(g) = MnO_2(c)$ (298.16°–1,000° K.)	−126,400	−8.61	+.97	+1.555	+70.14
$2\,Hg(l) + 1/2\,O_2(g) = Hg_2O(c)$ (298.16°–629.88° K.)	−22,400	−4.61	......	.....	+43.29
$2\,Hg(g) + 1/2\,O_2(g) = Hg_2O(c)$ (629.88°–1,000° K.)	−53,800	−16.12	......	.....	+125.36
$Hg(l) + 1/2\,O_2(g) = HgO$ (*red*) (298.16°–629.88° K.)	−21,760	+.85	−2.47	−.10	+24.81
$Hg(g) + 1/2\,O_2(g) = HgO$ (*red*) (629.88°–1,500° K.)	−36,920	−2.92	−2.47	−.10	+59.42
$Mo(c) + O_2(g) = MoO_2(c)$ (298.16°–2,000° K.)	−132,910	−3.91	......	.....	+47.42
$Mo(c) + 3/2\,O_2(g) = MoO_3(c)$ (298.16°–1,068° K.)	−182,650	−8.86	−1.55	+1.54	+90.07
$Mo(c) + 3/2\,O_2(g) = MoO_3(l)$ (1,068°–1,500° K.)	−179,770	−36.34	+1.40	−.30	+167.61
$2\,Nd(c) + 3/2\,O_2(g) = Nd_2O_3$ (*hexagonal*). (298.16°–1,113° K.)	−435,150	−16.19	+3.21	+1.78	+125.68
$2\,Nd(l) + 3/2\,O_2(g) = Nd_2O_3$ (*hexagonal*). (1,113°–1,500° K.)	−437,090	+4.03	−2.13	+1.78	+71.77
$Np(c) + O_2(g) = NpO_2(c)$ (298.16°–913° K.)	−246,450	−3.45	......	.....	+52.44
$Np(l) + O_2(g) = NpO_2(c)$ (913°–1,500° K.)	−249,010	−4.61	......	.....	+58.68
$Ni(\alpha) + 1/2\,O_2(g) = NiO(c)$ (298.16°–633° K.)	−57,640	−4.61	+2.16	−.10	+34.41
$Ni(\beta) + 1/2\,O_2(g) = NiO(c)$ (633°–1,725° K.)	−57,460	−.14	− 46	−.10	+23.27
$Ni(l) + 1/2\,O_2(g) = NiO(c)$ (1,725°–2,000° K.)	−58,830	+7.23	−1.36	−.10	+1.76
$2\,Nb(c) + 2\,O_2(g) = Nb_2O_4(c)$ (298.16°–2,000° K.)	−382,050	−9.67	......	.....	+116.23

HEAT OF FORMATION OF INORGANIC OXIDES (Continued)
Coefficients in Free-Energy Equations (Continued)

Reaction and temperature range of validity	ΔH_0	2.303*a*	*b*	*c*	*I*
$2\,Nb(c) + 5/2\,O_2(g) = Nb_2O_5(c)$ (298.16°–1,785° K.)	−458,640	−16.14	−.56	+1.94	+157.66
$2\,Nb(c) + 5/2\,O_2(g) = Nb_2O_5(l)$ (1,785°–2,000° K.)	−463,630	−66.04	+2.21	−.50	+317.84
$N_2(g) + 1/2\,O_2(g) = N_2O(g)$ (298.16°–2,000° K.)	+18,650	−1.57	−.27	+.92	+23.47
$1/2\,N_2(g) + 1/2\,O_2(g) = NO(g)$ (298.16°–2,000° K.)	+21,590	−.28	+.45	−.03	−2.20
$N_2(g) + 3/2\,O_2(g) = N_2O_3(g)$ (298.16°–500° K.)	+17,390	−.35	.....	.....	+54.30
$1/2\,N_2(g) + O_2(g) = NO_2(g)$ (298.16°–2,000° K.)	+7,730	+.53	−.265	+.605	+13.74
$N_2(g) + 2\,O_2(g) = N_2O_4(g)$ (298.16°–1,000° K.)	+1,370	+2.14	−3.24	+1.38	+68.34
$N_2(g) + 5/2\,O_2(g) = N_2O_5(c)$ (298.16°–305° K.)	−10,200	.....	.....	.....	+131.70
$N_2(g) + 5/2\,O_2(g) = N_2O_5(g)$ (298.16°–500° K.)	+3,600	.....	.....	.....	+86.50
$Os(c) + 2\,O_2(g) = OsO_4$ (*yellow*) (298.16°–329° K.)	−92,260	+14.97	.....	.....	+32.45
$Os(c) + 2\,O_2(g) = OsO_4$ (*white*) (298.16°–315° K.)	−90,560	+14.97	.....	.....	+27.60
$Os(c) + 2\,O_2(g) = OsO_4(l)$ (315°–403° K.)	−88,970	+9.67	.....	.....	+35.78
$Os(c) + 2\,O_2(g) = OsO_4(g)$ (298.16°–1,000° K.)	−81,200	+8.17	−2.86	+1.94	+20.34
$3/2\,O_2(g) = O_3(g)$ (298.16°–2,000° K.)	+33,980	+2.03	−.48	+.36	+11.45
P (*white*) $+ 1/2\,O_2(g) = PO(g)$ (298.16°–317.4° K.)	−9,370	+2.53	.....	.....	−25.40
$P(l) + 1/2\,O_2(g) = PO(g)$ (317.4°–553° K.)	−9,390	+3.45	.....	.....	−27.63
$1/4\,P_4(g) + 1/2\,O_2(g) = PO(g)$ (553°–1,500° K.)	−12,640	+2.30	.....	.....	−18.61
$4\,P$ (*white*) $+ 5\,O_2(g) = P_4O_{10}$ (*hexagonal*). (298.16°–317.4° K.)	−711,520	+95.67	−51.50	−1.00	−28.24
$4\,P(l) + 5\,O_2(g) = P_4O_{10}$ (*hex.*) (317.4°–553° K.)	−711,800	+97.98	−51.50	−1.00	−33.13
$P_4(g) + 5\,O_2(g) = P_4O_{10}$ (*hex.*) (553°–631° K.)	−725,560	+87.45	−51.07	−2.405	+20.87
$P_4(g) + 5\,O_2(g) = P_4O_{10}(g)$ (631°–1,500° K.)	−722,330	−43.45	+2.93	−2.405	+348.20
$Pu(c) + O_2(g) = PuO_2(c)$ (298.16°–1,500° K.)	−246,450	−3.45	.....	.....	+52.48
$Po(c) + O_2(g) = PoO_2(c)$ (298.16°–900° K.)	−61,510	−9.21	.....	.....	+72.80
$2\,K(c) + 1/2\,O_2(g) = K_2O(c)$ (298.16°–336.4° K.)	−86,400	.....	.....	.....	+33.90
$2\,K(l) + 1/2\,O_2(g) = K_2O(c)$ (336.4°–1,049° K.)	−87,380	+1.15	.....	.....	+33.90
$2\,K(g) + 1/2\,O_2(g) = K_2O(c)$ (1,049°–1,500° K.)	−133,090	−16.12	.....	.....	+129.64
$2\,K(c) + O_2(g) = K_2O_2(c)$ (298.16°–336.4° K.)	−118,300	−2.30	.....	.....	+59.60
$2\,K(l) + O_2(g) = K_2O_2(c)$ (336.4°–763° K.)	−119,780	−4.61	.....	.....	+69.85
$2\,K(l) + O_2(g) = K_2O_2(l)$ (763°–1,049° K.)	−118,250	−18.42	.....	.....	+107.66
$2\,K(g) + O_2(g) = K_2O_2(l)$ (1,049°–1,500° K.)	−161,870	−31.08	.....	.....	+187.49
$2\,K(c) + 3/2\,O_2(g) = K_2O_3(c)$ (298.16°–336.4° K.)	−126,640	−12.66	.....	.....	+111.75
$2\,K(l) + 3/2\,O_2(g) = K_2O_3(c)$ (336.4°–703° K.)	−127,790	−12.66	.....	.....	+115.16
$2\,K(l) + 3/2\,O_2(g) = K_2O_3(l)$ (703°–1,000° K.)	−125,330	−27.63	.....	.....	+154.28
$K(c) + O_2(g) = KO_2(c)$ (298.16°–336.4° K.)	−68,940	−10.36	.....	.....	+66.45
$K(l) + O_2(g) = KO_2(c)$ (336.4°–653° K.)	−69,510	−10.36	.....	.....	+68.15
$K(l) + O_2(g) = KO_2(l)$ (653°–1,000° K.)	−67,880	−18.42	.....	.....	+88.34
$K(c) + 3/2\,O_2(g) = KO_3(c)$ (298.16°–336.4° K.)	−63,340	−10.36	.....	.....	+85.85
$K(l) + 3/2\,O_2(g) = KO_3(c)$ (336.4°–500° K.)	−63,910	−10.36	.....	.....	+87.55
$2\,Pr(c) + 3/2\,O_2(g) = Pr_2O_3$(*c, C-type*) (298.16°–1,205° K.)	−440,600	−4.60	.....	.....	+78.38
$2\,Pr(l) + 3/2\,O_2(g) = Pr_2O_3$(*c, C-type*) (1,205°–2,000° K.)	−446,100	−4.60	.....	.....	+82.94
$6\,Pr(c) + 11/2\,O_2(g) = Pr_6O_{11}(c)$ (298.16°–1,205° K.)	−1,374,000	.....	.....	.....	+241.04
$6\,Pr(l) + 11/2\,O_2(g) = Pr_6O_{11}(c)$ (1,205°–1,500° K.)	−1,390,500	.....	.....	.....	+254.73
$Pr(c) + O_2(g) = PrO_2(c)$ (298.16°–1,200° K.)	−230,990	−6.42	+2.34	−.20	+61.07
$Ra(c) + 1/2\,O_2(g) = RaO(c)$ (298.16°–1,000° K.)	−130,000	.....	.....	.....	+23.50
$Re(c) + 3/2\,O_2(g) = ReO_3(c)$ (298.16°–433° K.)	−149,090	−16.12	.....	.....	+110.49
$Re(c) + 3/2\,O_2(g) = ReO_3(l)$ (433°–1,000° K.)	−146,750	−31.32	.....	.....	+145.16
$2Re(c) + 7/2\,O_7(g) = Re_2O_7(c)$ (298.16°–569° K.)	−301,470	−34.54	.....	.....	+250.57
$2\,Re(c) + 7/2\,O_7(g) = Re_2O_7(l)$ (569°–635.5° K.)	−295,810	−73.68	.....	.....	+348.45
$2\,Re(c) + 7/2\,O_2(g) = Re_2O_7(g)$ (635.5°–1,500° K.)	−256,460	+3.45	.....	.....	+70.33
$2\,Re(c) + 4\,O_2(g) = Re_2O_8(c)$ (298.16°–420° K.)	−313,870	−41.45	.....	.....	+293.57
$2\,Re(c) + 4\,O_2(g) = Re_2O_8(l)$ (420°–600° K.)	−318,470	−87.50	.....	.....	+425.32
$2\,Rh(c) + 1/2\,O_2(g) = Rh_2O(c)$ (298.16°–2,000° K.)	−23,740	−8.06	.....	.....	+35.64
$Rh(c) + 1/2\,O_2(g) = RhO(c)$ (298.16°–1,500° K.)	−22,650	−7.37	.....	.....	+40.54
$2\,Rh(c) + 3/2\,O_2(g) = Rh_2O_3(c)$ (298.16°–1,500° K.)	−70,060	−13.58	.....	.....	+101.72
$2\,Rb(c) + 1/2\,O_2(g) = Rb_2O(c)$ (298.16°–312.2° K.)	−78,900	.....	.....	.....	+32.20
$2\,Rb(l) + 1/2\,O_2(g) = Rb_2O(c)$ (312.2°–750° K.)	−79,950	.....	.....	.....	+35.56
$2\,Rb(l) + 1/2\,O_2(g) = Rb_2O(l)$ (750°–952° K.)	−78,830	−10.36	.....	.....	+63.85
$2\,Rb(g) + 1/2\,O_2(g) = Rb_2O(l)$ (952°–1,500° K.)	−120,290	−23.03	.....	.....	+145.14
$2\,Rb(c) + O_2(g) = Rb_2O_2(c)$ (298.16°–312.2° K.)	−102,000	−2.30	.....	.....	+57.40
$2\,Rb(l) + O_2(g) = Rb_2O_2(c)$ (312.2°–840° K.)	−103,360	−4.61	.....	.....	+67.52
$2\,Rb(l) + O_2(g) = Rb_2O_2(l)$ (840°–952° K.)	−101,680	−18.42	.....	.....	+105.91
$2\,Rb(g) + O_2(g) = Rb_2O_2(l)$ (952°–1,500 K.)	−143,130	−31.08	.....	.....	+187.16
$2\,Rb(c) + 3/2\,O_2(g) = Rb_2O_3(c)$ (298.16°–312.2° K.)	−118,190	−11.51	.....	.....	+104.70
$2\,Rb(l) + 3/2\,O_2(g) = Rb_2O_3(c)$ (312.2°–760° K.)	−119,400	−12.66	.....	.....	+111.43
$2\,Rb(l) + 3/2\,O_2(g) = Rb_2O_3(l)$ (760°–952° K.)	−116,740	−27.63	.....	.....	+151.06
$2\,Rb(g) + 3/2\,O_2(g) = Rb_2O_3(l)$ (952°–1,500° K.)	−157,720	−39.14	.....	.....	+228.39
$Rb(c) + O_2(g) = RbO_2(c)$ (298.16°–312.2° K.)	−52,330	−10.36	.....	.....	+66.25
$Rb(l) + O_2(g) = RbO_2(c)$ (312.2°–685° K.)	−65,120	−11.51	.....	.....	+71.30

HEAT OF FORMATION OF INORGANIC OXIDES (Continued)
Coefficients in Free-Energy Equations (Continued)

Reaction and temperature range of validity	ΔH_o	2.303*a*	*b*	*c*	*I*
$Rb(l) + O_2(g) = RbO_2(l)$ (685°–952° K.)	−63,070	−18.42	.....	.....	+87.89
$Rb(g) + O_2(g) = RbO_2(l)$ (952°–1,500° K.)	−83,560	−24.18	.....	.....	+126.57
$Ru(\alpha, \beta, \gamma) + O_2(g) = RuO_2(c)$ (298.16°–1,500° K.)	−57,290	−6.91	.....	.....	+62.01
$2\ Sm(c) + 3/2\ O_2(g) = Sm_2O_3(c)$ (298.16°–1,623° K.)	−430,600	−4.60	.....	.....	+78.38
$2\ Sm(l) + 3/2\ O_2(g) = Sm_2O_3(c)$ (1,623°–2,000° K.)	−438,000	−4.60	.....	.....	+82.94
$2\ Sc(c) + 3/2\ O_2(g) = Sc_2O_3(c)$ (298.16°–1,673° K.)	−409,960	+7.78	−1.84	−0.30	+52.73
$2\ Sc(l) + 3/2\ O_2(g) = Sc_2O_3(c)$ (1,673°–2,000° K.)	−412,950	+18.47	−2.93	−.30	+21.88
$Se(c) + 1/2\ O_2(g) = SeO(g)$ (298.16°–490° K.)	+9,280	−3.04	+4.40	+.30	−14.78
$Se(l) + 1/2\ O_2(g) = SeO(g)$ (490°–1,027° K.)	+9,420	+8.70	.....	+.30	−44.50
$1/2\ Se_2(g) + 1/2\ O_2(g) = SeO(g)$ (1,027°–2,000° K.)	−7,400	−.37	.....	+.19	−.80
$Se(c) + O_2(g) = SeO_2(c)$ (298.16°–490° K.)	−53,770	+14.94	−9.41	−.20	+6.94
$Se(l) + O_2(g) = SeO_2(c)$ (490°–595° K.)	−53,640	+27.59	−14.31	.....	−25.05
$Se(l) + O_2(g) = SeO_2(g)$ (595°–1,027° K.)	−32,840	+6.79	.....	.....	−10.80
$1/2\ Se_2(g) + O_2(g) = SeO_2(g)$ (1,027°–2,000° K.)	−49,000	−.74	.....	.....	+27.61
$Si(c) + 1/2\ O_2(g) = SiO(g)$ (298.16°–1,683° K.)	−21,090	+3.84	+.16	−.295	−33.14
$Si(l) + 1/2\ O_2(g) = SiO(g)$ (1,683°–2,000° K.)	−30,170	−7.78	−.12	+.25	−40.01
$Si(c) + O_2(g) = SiO_2$(*α-quartz*) (298.16°–848° K.)	−210,070	+3.98	−3.32	+.605	+34.59
$Si(c) + O_2(g) = SiO_2$(*β-quartz*) (848°–1,683° K.)	−209,920	−3.36	−.19	−.745	+53.44
$Si(l) + O_2(g) = SiO_2$(*β-quartz*) (1,683°–1,883° K.)	−219,000	+.58	−.47	−.20	+46.58
$Si(l) + O_2(g) = SiO_2(l)$ (1,883°–2,000° K.)	−228,590	−15.66	.....	.....	+103.97
$Si(c) + O_2(g) = SiO_2$(*α-crist.*) (298.16°–523° K.)	−207,330	+19.96	−9.75	−.745	−9.78
$Si(c) + O_2(g) = SiO_2$(*β-crist.*) (523°–1,683° K.)	−209,820	−3.34	−.24	−.745	+53.35
$Si(l) + O_2(g) = SiO_2$(*β-crist.*) (1,683°–2,000° K.)	−218,900	+.60	−.52	−.20	+46.49
$Si(c) + O_2(g) = SiO_2$(*α-trid.*) (298.16°–390° K.)	−207,030	+22.29	−11.62	−.745	−15.64
$Si(c) + O_2(g) = SiO_2$(*β-trid.*) (390°–1,683° K.)	−209,350	−1.59	−.54	−.745	+47.86
$Si(l) + O_2(g) = SiO_2$(*β-trid.*) (1,683°–1,953° K.)	−218,430	+2.35	−.82	−.20	+41.00
$2\ Ag(c) + 1/2\ O_2(g) = Ag_2O(c)$ (298.16°–1,000° K.)	−7,740	−4.14	.....	.....	+27.84
$2\ Ag(c) + O_2(g) = Ag_2O_2(c)$ (298.16°–500° K.)	−6,620	−3.22	.....	.....	+52.17
$2\ Na(c) + 1/2\ O_2(g) = Na_2O(c)$ (298.16°–371° K.)	−99,820	−7.51	+5.47	−.10	+50.43
$2\ Na(l) + 1/2\ O_2(g) = Na_2O(c)$ (371°–1,187° K.)	−100,150	+4.97	−2.45	−.10	+22.19
$2\ Na(g) + 1/2\ O_2(g) = Na_2O(c)$ (1,187°–1,190° K.)	−156,200	−20.72	.....	.....	+145.48
$2\ Na(g) + 1/2\ O_2(g) = Na_2O(l)$ (1,190°–2,000° K.)	−150,250	−23.03	.....	.....	+147.58
$2\ Na(c) + O_2(g) = Na_2O_2(c)$ (298.16°–371° K.)	−122,500	−2.30	.....	.....	+57.51
$2\ Na(l) + O_2(g) = Na_2O_2(c)$ (371°–733° K.)	−124,320	−5.76	.....	.....	+71.30
$2\ Na(l) + O_2(g) = Na_2O_2(l)$ (733°–1,187° K.)	−123,220	−20.72	.....	.....	+112.66
$2\ Na(g) + O_2(g) = Na_2O_2(l)$ (1,187°–1,500° K.)	−174,800	−31.08	.....	.....	+187.97
$Na(c) + O_2(g) = NaO_2(c)$ (298.16°–371° K.)	−63,040	−8.06	.....	.....	+56.98
$Na(l) + O_2(g) = NaO_2(c)$ (371°–1,000° K.)	−64,220	−11.51	.....	.....	+69.04
$Sr(c) + 1/2\ O_2(g) = SrO(c)$ (298.16°–1,043° K.)	−142,410	−6.79	+.305	+.675	+44.33
$Sr(l) + 1/2\ O_2(g) = SrO(c)$ (1,043°–1,657° K.)	−143,370	−2.42	−.38	+.675	+32.77
$Sr(g) + 1/2\ O_2(g) = SrO(c)$ (1,657°–2,000° K.)	−181,180	−8.24	−.38	+.675	+74.32
$Sr(c) + O_2(g) = SrO_2(c)$ (298.16°–1,000° K.)	−155,510	−11.40	+.305	+.675	+75.44
$S(rh) + 1/2\ O_2(g) = SO(g)$ (298.16°–368.6° K.)	+19,250	−1.24	+2.95	+.225	−18.84
$S(mon) + 1/2\ O_2(g) = SO(g)$ (368.6°–392° K.)	+19,200	−1.29	+3.31	+.225	−18.72
$S(l\lambda,\mu) + 1/2\ O_2(g) = SO(g)$ (392°–718° K.)	+20,320	+10.22	−.17	+.225	−50.05
$1/2\ S_2(g) + 1/2\ O_2(g) = SO(g)$ (298.16°–2,000° K.)	+3,890	+.07	.....	.....	−1.50
$S(rh) + O_2(g) = SO_2(g)$ (298.16°–368.6° K.)	−70,980	+.83	+2.35	+.51	−5.85
$S(mon) + O_2(g) = SO_2(g)$ (368.6°–392° K.)	−71,020	+.78	+2.71	+.51	−5.74
$S(l\lambda,\mu) + O_2(g) = SO_2(g)$ (392°–718° K.)	−69,900	+12.30	−.77	+.51	−37.10
$1/2\ S_2(g) + O_2(g) = SO_2(g)$ (298.16°–2,000° K.)	−86,330	+2.42	−.70	+.31	+10.71
$S(rh) + 3/2\ O_2(g) = SO_3(c—I)$ (298.16°–335.4° K.)	−111,370	−6.45	.....	.....	+88.32
$S(rh) + 3/2\ O_2(g) = SO_3(c—II)$ (298.16°–305.7° K.)	−108,680	−11.97	.....	.....	+94.95
$S(rh) + 3/2\ O_2(g) = SO_3(l)$ (298.16°–335.4° K.)	−107,430	−21.18	.....	.....	+113.76
$S(rh) + 3/2\ O_3(g) = SO_3(g)$ (298.16°–368.6° K.)	−95,070	+1.43	+0.66	+1.26	+16.81
$S(mon) + 3/2\ O_2(g) = SO_3(g)$ (368.6°–392° K.)	−95,120	+1.38	+1.02	+1.26	+16.93
$S(l\lambda,\mu) + 3/2\ O_2(g) = SO_3(g)$ (392°–718° K.)	−94,010	+12.89	−2.46	+1.26	−14.40
$1/2\ S_2(g) + 3/2\ O_2(g) = SO_3(g)$ (298.16°–1,500° K.)	−110,420	+3.02	−2.39	+1.06	+33.41
$2\ Ta(c) + 5/2\ O_2(g) = Ta_2O_5(c)$ (298.16°–2,000° K.)	−492,790	−17.18	−1.25	+2.46	+161.68
$Tc(c) + O_2(g) = TcO_2(c)$ (298.16°–500° K.)	−103,400	.....	.....	.....	+41.00
$Tc(c) + 3/2\ O_2(g) = TcO_3(c)$ (298.16°–500° K.)	−129,000	.....	.....	.....	+64.50
$2\ Tc(c) + 7/2\ O_2(g) = Tc_2O_7(c)$ (298.16°–392.7° K.)	−266,000	.....	.....	.....	+147.00
$2\ Tc(c) + 7/2\ O_2(g) = Tc_2O_7(l)$ (392.7°–500° K.)	−258,930	.....	.....	.....	+129.00
$Te(c) + 1/2\ O_2(g) = TeO(g)$ (298.16°–723° K.)	+43,110	+1.91	+.84	+.315	−27.22
$Te(l) + 1/2\ O_2(g) = TeO(g)$ (723°–1,360° K.)	+39,750	+6.08	+.09	+.315	−33.94
$1/2\ Te_2(g) + 1/2\ O_2(g) = TeO(g)$ (1,360°–2,000° K.)	+23,730	−.90	+.09	+.315	−.29
$Te(c) + O_2(g) = TeO_2(c)$ (298.16°–723° K.)	−78,090	−2.10	−2.35	−.20	+51.27
$Te(l) + O_2(g) = TeO_2(c)$ (723°–1,006° K.)	−81,530	+1.84	−3.10	−.20	+45.30
$Te(l) + O_2(g) = TeO_2(l)$ (1,006°–1,300° K.)	−82,090	−21.74	+.50	−.20	+113.04
$2\ Tl(\alpha) + O_2(g) = Tl_2O(c)$ (298.16°–505.5° K.)	−44,110	−6.91	.....	.....	+42.30

HEAT OF FORMATION OF INORGANIC OXIDES (Continued)
Coefficients in Free-Energy Equations (Continued)

Reaction and temperature range of validity		ΔH_0	2.303*a*	*b*	*c*	*I*
$2\ Tl(\beta) + O_2(g) = Tl_2O(c)$	(505.5°–573° K.)	−44,260	−6.91	.....	.....	+42.60
$2\ Tl(\beta) + 1/2\ O_2(g) = Tl_2O(l)$	(573°–576° K.)	−40,880	−13.82	.....	.....	+55.76
$2\ Tl(l) + 1/2\ O_2(g) = Tl_2O(l)$	(576°–773° K.)	−42,320	−11.51	.....	.....	+51.89
$2\ Tl(l) + 1/2\ O_2(g) = Tl_2O(g)$	(773°–1,730° K.)	−18,400	+11.51	.....	.....	−45.55
$2\ Tl(g) + 1/2\ O_2(g) = Tl_2O(g)$	(1,730°–2,000° K.)	−104,670	.....	.....	.....	+41.59
$2\ Tl(\alpha) + 3/2\ O_2(g) = Tl_2O_3(c)$	(298.16°–505.5° K.)	−99,410	−16.12	.....	.....	+119.09
$2\ Tl(\beta) + 3/2\ O_2(g) = Tl_2O_3(c)$	(505.5°–576° K.)	−99,560	−16.12	.....	.....	+119.39
$2\ Tl(l) + 3/2\ O_2(g) = Tl_2O_3(c)$	(576°–990° K.)	−101,010	−13.82	.....	.....	+115.55
$2\ Tl(l) + 3/2\ O_2(g) = Tl_2O_3(l)$	(990°–1,500° K.)	−94,550	−27.63	.....	.....	+150.39
$2\ Tl(\alpha) + 2\ O_2(g) = Tl_2O_4(c)$	(298.16°–505.5° K.)	−117,680	−23.03	.....	.....	+161.19
$2\ Tl(\beta) + 2\ O_2(g) = Tl_2O_4(c)$	(505.5°–576° K.)	−117,830	−23.03	.....	.....	+161.49
$2\ Tl(l) + 2\ O_2(g) = Tl_2O_4(c)$	(576°–1,000° K.)	−119,270	−20.72	.....	.....	+157.63
$Th(c) + O_2(g) = ThO_2(c)$	(298.16°–2,000° K.)	−294,350	−5.25	+.59	+.775	+62.81
$Sn(c) + 1/2\ O_2(g) = SnO(c)$	(298.16°–505° K.)	−68,600	−3.57	+1.65	−.10	+32.59
$Sn(l) + 1/2\ O_2(g) = SnO(c)$	(505°–1,300° K.)	−69,670	+3.06	−1.50	−.10	+18.39
$Sn(c) + 1/2\ O_2(g) = SnO(g)$	(298.16°–505° K.)	−1,000	−.97	+3.24	+.32	−17.41
$Sn(l) + 1/2\ O_2(g) = SnO(g)$	(505°–2,000° K.)	−2,070	+5.66	+.09	+.32	−31.62
$Sn(c) + O_2(g) = SnO_2(c)$	(298.16°–505° K.)	−142,010	−14.00	+2.45	+2.38	+90.74
$Sn(l) + O_2(g) = SnO_2(c)$	(505°–1,898° K.)	−143,080	−7.37	−.70	+2.38	+76.53
$Sn(l) + O_2(g) = SnO_2(l)$	(1,898°–2,000° K.)	−139,130	−21.97	+.50	−.20	+120.11
$Ti(\alpha) + 1/2\ O_2(g) = TiO(\alpha)$	(298.16°–1,150° K.)	−125,010	−4.01	−.29	+.83	+36.28
$Ti(\beta) + 1/2\ O_2(g) = TiO(\alpha)$	(1,150°–1,264° K.)	−125,040	+1.17	−1.55	+.83	+21.90
$Ti(\beta) + 1/2\ O_2(g) = TiO(\beta)$	(1,264°–2,000° K.)	−125,210	−1.77	−1.25	−.10	+30.83
$Ti(\alpha) + 1/2\ O_2(g) = TiO(g)$	(298.16°–1,150° K.)	+11,710	+3.71	+1.07	−.10	−35.50
$Ti(\beta) + 1/2\ O_2(g) = TiO(g)$	(1,150°–2,000° K.)	+11,680	+8.89	−.19	−.10	−49.88
$2\ Ti(\alpha) + 3/2\ O_2(g) = Ti_2O_3(\alpha)$	(298.16°–473° K.)	−360,660	+32.08	−23.49	−.30	−10.66
$2\ Ti(\alpha) + 3/2\ O_2(g) = Ti_2O_3(\beta)$	(473°–1,150° K.)	−369,710	−30.95	+2.62	+4.80	+162.79
$2\ Ti(\beta) + 3/2\ O_2(g) = Ti_2O_3(\beta)$	(1,150°–2,000° K.)	−369,760	−20.59	+.10	+4.80	+134.03
$3\ Ti(\alpha) + 5/2\ O_2(g) = Ti_3O_5(\alpha)$	(298.16°–450° K.)	−587,980	−4.19	−9.72	−.50	+131.05
$3\ Ti(\alpha) + 5/2\ O_2(g) = Ti_3O_5(\beta)$	(450°–1,150° K.)	−586,330	−18.31	+1.03	−.50	+159.98
$3\ Ti(\beta) + 5/2\ O_2(g) = Ti_3O_5(\beta)$	(1,150°–2,000° K.)	−586,420	−2.76	−2.75	−.50	+116.81
$Ti(\alpha) + O_2(g) = TiO_2$ (*rutile*)	(298.16°–1,150° K.)	−228,360	−12.80	+1.62	+1.975	+82.81
$Ti(\beta) + O_2(g) = TiO_2$ (*rutile*)	(1,150°–2,000° K.)	−228,380	−7.62	+.36	+1.975	+68.43
$W(c) + O_2(g) = WO_2(c)$	(298.16°–1,500° K.)	−137,180	−1.38	.....	.....	+45.56
$4\ W(c) + 11/2\ O_2(g) = W_4O_{11}(c)$	(298,16°–1,700° K.)	−745,730	−32.70	.....	.....	+321.84
$W(c) + 3/2\ O_2(g) = WO_3(c)$	(298.16°–1,743° K.)	−201,180	−2.92	−1.81	−.30	+70.89
$W(c) + 3/2\ O_2(g) = WO_3(l)$	(1,743°–2,000° K.)	−203,140	−35.74	+1.13	−.30	+173.27
$U(\alpha) + O_2(g) = UO_2(c)$	(298.16°–935° K.)	−262,880	−19.92	+3.70	+2.13	+100.54
$U(\beta) + O_2(g) = UO_2(c)$	(935°–1,045° K.)	−260,660	−4.28	−.31	+1.78	+55.50
$U(\gamma) + O_2(g) = UO_2(c)$	(1,045°–1,405° K.)	−262,830	−6.54	−.31	+1.78	+64.41
$U(l) + O_2(g) = UO_2(c)$	(1,405°–1,500° K.)	−264,790	−5.92			+63.50
$3\ U(\alpha) + 4\ O_2(g) = U_3O_8(c)$	(298.16°–935° K.)	−863,370	−56.57	+10.68	+5.20	+330.19
$3\ U(\beta) + 4\ O_2(g) = U_3O_8(c)$	(935°–1,045° K.)	−856,720	−9.67	−1.35	+4.15	+195.12
$3\ U(\gamma) + 4\ O_2(g) = U_3O_8(c)$	(1,045°–1,405° K.)	−863,230	−16.44	−1.35	+4.15	+221.79
$3\ U(l) + 4\ O_2(g) = U_3O_8(c)$	(1,405°–1,500° K.)	−869,460	−10.91	−1.35	+4.15	+208.82
$U(\alpha) + 3/2\ O_2(g) = UO_3$ (*hex*)	(298.16°–935° K.)	−294,090	−18.33	+3.49	+1.535	+114.94
$U(\beta) + 3/2\ O_2(g) = UO_3$ (*hex*)	(935°–1,045° K.)	−291,870	−2.69	−.52	+1.185	+69.90
$U(\gamma) + 3/2\ O_2(g) = UO_3$ (*hex*)	(1,045°–1,400° K.)	−294,040	−4.95	−.52	+1.185	+78.80
$V(c) + 1/2\ O_2(g) = VO(c)$	(298.16°–2,000° K.)	−101,090	−5.39	−.36	+.53	+38.69
$V(c) + 1/2\ O_2(g) = VO(g)$	(298.16°–2,000° K.)	+52,090	+1.80	+1.04	+.35	−28.42
$2\ V(c) + 3/2\ O_2(g) = V_2O_3(c)$	(298.16°–2,000° K.)	−299,910	−17.98	+.37	+2.41	+118.83
$2\ V(c) + 2\ O_2(g) = V_2O_4(\alpha)$	(298.16°–345° K.)	−342,890	−11.03	+3.00	−.40	+117.38
$2\ V(c) + 2\ O_2(g) = V_2O_4(\beta)$	(345°–1,818° K.)	−345,330	−24.36	+1.30	+3.545	+155.55
$2\ V(c) + 2\ O_2(g) = V_2O_4(l)$	(1,818°–2,000° K.)	−339,880	−59.59	+3.00	−0.40	+264.42
$6\ V(c) + 13/2\ O_2(g) = V_6O_{13}(c)$	(298.16°–1,000° K.)	−1,076,340	−95.33			+557.61
$2\ V(c) + 5/2\ O_2(g) = V_2O_5(c)$	(298.16°–943° K.)	−381,960	−41.08	+5.20	+6.11	+228.50
$2\ V(c) + 5/2\ O_2(g) = V_2O_5(l)$	(943°–2,000° K.)	−365,840	−38.91	+3.25	−.50	+207.54
$2\ Y(c) + 3/2\ O_2(g) = Y_2O_3(c)$	(298.16°–1,773° K.)	−419,600	+2.76	−1.73	−.30	+66.36
$2\ Y(l) + 3/2\ O_2(g) = Y_2O_3(c)$	(1,773°–2,000° K.)	−422,850	+13.36	−2.75	−.30	+35.56
$Zn(c) + 1/2\ O_2(g) = ZnO(c)$	(298.16°–692.7° K.)	−84,070	−6.40	+.84	+.99	+43.25
$Zn(l) + 1/2\ O_2(g) = ZnO(c)$	(692.7°–1,180° K.)	−85,520	−1.45	−.36	+.99	+31.25
$Zn(g) + 1/2\ O_2(g) = ZnO(c)$	(1,180°–2,000° K.)	−115,940	−7.28	−.36	+.99	+74.94
$Zr(\alpha) + O_2(g) = ZrO_2(\alpha)$	(298.16°–1,135° K.)	−262,980	−6.10	+.16	+1.045	+65.00
$Zr(\beta) + O_2(g) = ZrO_2(\alpha)$	(1,135°–1,478° K.)	−264,190	−5.09	−.40	+1.48	+63.58
$Zr(\beta) + O_2(g) = ZrO_2(\beta)$	(1,478°–2,000° K.)	−262,290	−7.76	+.50	−.20	+69.50

REFRACTORY MATERIALS

Borides

Name	Formula	Molecular weight	Melting point, °C	Crystalline form	Lattice parameter, Å	X-ray density, g/cm³
Chromium boride	CrB_2	73.65	1,850[29]a	Hexagonal $A1B_2$ type [C 32]	a = 2.969 c = 3.066[39]	5.16
Hafnium boride	HfB_2	200.14	3,100[31]	Hexagonal $A1B_2$ type [C 32]	a = 3.14[6] c = 3.47	10.5
Molybdenum boride	MoB	106.77	2,180[40]	Tetragonal	a = 3.110[41] c = 16.95	8.77
	MoB_2	117.59	2,100[40]	Hexagonal $A1B_2$ type [C 32]	a = 3.05 c = 3.113[42]	7.78
	Mo_2B	202.72	2,000[40] (decomposes)	Tetragonal $CuAl_2$ type [C 16]	a = 5.543[41] c = 4.735	9.31
Niobium boride	NbB	103.73	>0.2000[3]	Orthorhombic	a = 3.298 b = 8.724 c = 3.137[35]	
	NbB_2	114.55	2,900[36] (decomposes)	Hexagonal AlB_2 type [C 32]	a = 3.086[28] c = 3.306	7.21
Tantalum boride	TaB	191.77	>0.2000[3]	Orthorhombic	a = 3.276 b = 8.669 c = 3.157[37]	14.29
	TaB_2	202.59	3,000[6]	Hexagonal AlB_2 type [C 32]	a = 3.088[28] c = 3.241	12.60
Thorium boride	ThB_4	275.53	>0.2500[6]	Tetragonal D_{4k}^5-P4/mbm	a = 7.256[43] c = 4.113	8.45
Titanium boride	TiB_2	69.54	2,980[6]	Hexagonal AlB_2 type [C 32]	a = 3.028[28] c = 3.228	4.52
Tungsten boride	WB	194.68	2,860[29]	Tetragonal	a = 3.115[6] c = 16.92	16.0
	W_2B	378.54	2,770[29]	Tetragonal $CuAl_2$ type [C 16]	a = 5.564[41] c = 4.740	16.72
Uranium boride	UB_{12}	367.91	>0.1500[6] (decomposes)	Face-centered cubic	a = 7.473[44]	5.82
Vanadium boride	VB_2	72.59	2,100[29]	Hexagonal AlB_2 type [C 32]	a = 2.998[28] c = 3.057	5.10
Zirconium boride	ZrB_2	112.86	3,040 ± 50[30]	Hexagonal AlB_2 type [C 32]	a = 3.169[28] c = 3.530	6.09
	ZrB_{12}	221.06	2,680[30]	Face-centered cubic	a = 7.408[33]	

Note: See end of table for footnotes.

REFRACTORY MATERIALS (Continued)

Borides (Continued)

Name	Thermal conductivity, cal-sec^{-1}-cm^{-2}-cm-°C^{-1}	Electrical resistivity, microhm-cm	Ductility relative scale[b]	Resistance to oxidation[c]	Hardness[d]
Chromium boride		21 at room temperature[38]			1,800 kg/mm^2 [29]
Hafnium boride		10[31]	3[3]		
Molybdenum boride		α-MoB = 45 at room temperature[40] β-MoB = 25 at room temperature 45 at room temperature[40] 40 at room temperature[40]	3[3]	3[3]	8 Mohs[20] 1,570 kg/mm^2 [29] 1,280 kg/mm^2 [29] 8–9 Mohs[20]
Niobium boride	0.040 at 20°C[36]	6.45 at room temperature[36] 32.0 at room temperature[36]			>0.8 Mohs[36]
Tantalum boride	0.026 at 20°C[6]	100 at room temperature[38] 68 at room temperature[38]	3[3]	3[3]	
Thorium boride					
Titanium boride	0.0624 at 200°C[28] (15% porosity)	28.4[28]	3[3]	2–3[3]	3,400 kg/mm^2 [29]
Tungsten boride			3[3]	3[3]	9 Mohs[6]
Uranium boride					
Vanadium boride	16 at 20°C[34]				8–9 Mohs[6]
Zirconium boride	0.0550 at 200°C[28] 0.029 at room temperature[33]	9.2 at 20°C[31] 60–80 at room temperature[30]	2[3]	2–3[3]	8 Mohs[32]

Carbides

Name	Formula	Molecular weight	Melting point, °C	Crystalline form	Lattice parameter, Å	X-ray density, g/cm^3
Boron carbide	B_4C	55.29	2,450	Hexagonal	a_0 = 5.60[6] c_0 = 12.12	2.52
Chromium carbide	Cr_3C_2	180.05	1,895[15]	Orthorhombic ($D5_{10}$)	a = 2.82 b = 5.53 c = 11.47[16]	6.7
	Cr_7Cr_3	400.01	1,780[15]	Hexagonal [$C_{3v}{}^4$]	a = 14.01 c = 4.532[17]	6.92
Graphite	C	12.01	3,700 ± 100[3]	Hexagonal	Orthohexagonal axes a_0 = 2.46 b_0 = 4.28 c_0 = 6.71	2.25
Hafnium carbide	HfC	190.51	3,890[9]	Cubic NaCl type (B1)	a = 4.46[10]	12.7

REFRACTORY MATERIALS (Continued)

Carbides (Continued)

Name	Formula	Molecular weight	Melting point, °C	Crystalline form	Lattice parameter, Å	X-ray density, g/cm³
	MoC	107.96	2,695[9]	Face-centered cubic	a = 4.28[19]	
Molybdenum carbide	Mo_2C	203.91	2,690[6]	Hexagonal (L′3)	a = 3.002 c = 4.724[18]	9.2
Niobium carbide	NbC	104.92	3,500[9]	Cubic NaCl type (B1)	a = 4.461	7.85
Silicon carbide						
β	SiC	40.07	Trans. to α at 2,100	Face-centered cubic	a_0 = 4.3590[6]	3.22
α	SiC		2,700[6]	Hexagonal (Wurtzite)	a_0 = 3.081[6] c_0 = 5.0394	
Tantalum carbide	TaC	192.96	3,880[9], 4,730[4]	Cubic NaCl type (B1)	a = 4.455[14]	14.5
Thorium carbide	ThC	244.06	2,625[23]	Cubic NaCl type (B1)	a = 5.34[9]	10.67
	ThC_2	256.07	2,655[29]	Monoclinic [C 2/e]	a = 6.53[24] b = 4.24	9.6
Titanium carbide	TiC	59.91	3,160[4]	Cubic NaCl type (B1)	a = 4.32[6]	4.938
Tungsten carbide	W_2C	379.73	2,730[21]	Hexagonal (L′3)	a = 2.98[6] c = 4.71	17.34
	WC	195.87	2,630[21] (decomposes)	Hexagonal (L′3)	a = 2.900[22] c = 2.831	15.77
Uranium carbide	UC	250.08	2,450–2,500[25]	Cubic NaCl type (B1)	a = 4.955[26]	13.6
	UC_2	262.09	2,350–2,400[25]	Body-centered tetragonal $CaCl_2$ type	a = 3.517[27] c = 5.987	11.68
Vanadium carbide	VC	62.96	2,830[1]	Cubic NaCl type (B1)	a = 4.16[6]	5.8
Zirconium carbide	ZrC	103.23	3,030[4]	Cubic NaCl type (B1)	a = 4.689[5]	6.44

Name	Thermal conductivity, cal-sec^{-1}-cm^{-2}-cm-°C^{-1}	Electrical resistivity, microhm-cm	Ductility relative scale[b]	Resistance to oxidation[c]	Hardness[d]
Boron carbide	0.05 at 20–425°C[1]	0.30–0.80[1]	3[3]	3[3]	9.3 Mohs
Chromium carbide			3[3]	3[3]	1,300 kg/mm²[8]
			3[3]	3[3]	
Graphite	0.268–0.451 at room temperature	65 at room temperature[71]	3[3]	4[3]	
Hafnium carbide		109 at room temperature[1]	3[3]	3[3]	
Molybdenum carbide		97[6]	3[3]	5[3]	1,800 kg/mm²[6]
			3[3]	5[3]	>0.7 Mohs[20]
Niobium carbide	0.034 at room temperature[7]	74[6]	2–3[3]	3[3]	2,470 kg/mm²[13]
Silicon carbide					
β	0.10 at 20–425°C[1]	107–200 ohm cm[1] at room temperature	3[3]	2[3]	9.2 Mohs
Tantalum carbide	0.053 at room temperature[6]	30[6]	2–3[3]	3[3]	1,800 kg/mm²[8]
Thorium carbide			3[3]	3[3]	

REFRACTORY MATERIALS (Continued)

Carbides (Continued)

Name	Thermal conductivity, cal-sec $^{-1}$-cm $^{-2}$-cm-°C $^{-1}$	Electrical resistivity, microhm-cm	Ductility relative scale[b]	Resistance to oxidation[c]	Hardness[d]
Titanium carbide	0.049 at room temperature[6]	180–250[4]	3[3]	3[4]	3,200 kg/mm^2 [8]
Tungsten carbide		80[6]	3[3]	5[3]	3,000 kg/mm^2 [6]
		53[6]	3[3]	5[3]	2,400 kg/mm^2 [6]
Uranium carbide			3[3]		
				5[3]	
Vanadium carbide		150[6]	3[3]		2,800 kg/mm^2 [11]
					2,100 kg/mm^2 [12]
Zirconium carbide	0.049 at room temperature[6]	70 at room temperature[4]	3[3]	3[3]	2,600 kg/mm^2 [6]

Nitrides

Name	Formula	Molecular weight	Melting point, °C	Crystalline form	Lattice parameter, Å	X-ray density, g/cm^3
Boron nitride	BN	24.83	2,730[6]	Hexagonal (B 12)	a = 2.51 ± .02 c = 6.70 ± .04[52]	2.25
Chromium nitride	CrN	66.02	1,500 (decomposes)	Cubic NaCl type (B1)	a = 4.140[6]	6.14
Hafnium nitride	HfN	192.60	3,310[6]			
Niobium nitride	NbN	106.92	2,050[46]	Cubic NaCl type (B1)	a = 4.41–4.375[6]	7.28
Tantalum nitride	Ta_2N	375.77	3,090[46]	Hexagonal	a = 3.05[48] c = 4.95[53]	14.1
Thorium nitride	ThN	246.13	2,630[48]	Cubic NaCl type (B1)	a = 5.2[48]	
Titanium nitride	TiN	61.91	2,930 N_2 liberated on melting[46]	Cubic NaCl type (B1)	a = 4.23[53]	5.43
Uranium nitride	UN		2,650	Cubic NaCl type (B1)	a = 4.880[58]	14.32
Vanadium nitride	VN	64.96	2,050[46]	Cubic NaCl type (B1)	a = 4.129[6]	6.102
Zirconium nitride	ZrN	105.22	2,980[47]	Cubic NaCl type (B1)	a = 4.567[53]	7.349

REFRACTORY MATERIALS (Continued)

Nitrides (Continued)

Name	Thermal conductivity, cal-sec^{-1}-cm^{-2}-cm-°C^{-1}	Electrical resistivity, microhm-cm	Ductility relative scale[b]	Resistance to oxidation[c]	Hardness[d]
Boron nitride		1,900 at 2,000°C[1]	3[3]	2[3]	2.0 Mohs
Chromium nitride					
Hafnium nitride					
Niobium nitride		200 at room temperature[46]	5[3]	3[3]	+8 Mohs[46]
Tantalum nitride		135 at room temperature[47]	3[3]	5[3]	+8
Thorium nitride					
Titanium nitride		21.7 at room temperature	3[3]	3[3]	Between 9 and 10 Mohs[46]
Uranium nitride					
Vanadium nitride		200 at room temperature[46]	3[3]		
Zirconium nitride		13.6 at room temperature[46]	3[3]	3[3]	+8 Mohs[6]

Oxides

Name	Formula	Molecular weight	Melting point, °C	Crystalline form	Lattice parameter, Å	X-ray density, g/cm^3
Aluminum oxide	Al_2O_3	101.92	2,015[1]	Rhombohedral	a = 5.13 axial angle = 55° 6′	3.965
Beryllium oxide	BeO	25.02	2,550[1]	Hexagonal	a = 2.70 c = 4.39	3.03
Cerium oxide	CeO_2	172.3	2,600[1]	Face-centered cubic	a = 5.41	7.13
Chromic oxide	Cr_2O_3	152.02	2,265[1]	Rhombohedral	a = 5.38 axial angle = 54° 50′	5.21
Hafnium oxide	HfO_2	210.6	2,777[1]	Face-centered cubic	a = 5.11	9.68
Magnesium oxide	MgO	40.32	2,800[1]	Face-centered cubic	a = 4.20	3.58
Silicon oxide	SiO_2	60.06	1,728[1]	Hexagonal	a = 4.90 c = 5.39	2.32 (low cristobalite)
Thorium oxide	ThO_2	264.12	3,300[1]	Face-centered cubic	a = 5.59	9.69
Titanium oxide	TiO_2	79.90	1,840[1]	Tetragonal	a = 4.58 c = 2.98	4.24
Uranium oxide	UO_2	270.07	2,280[1]	Face-centered cubic	a = 5.47	10.96
Zirconium oxide	ZrO_2	123.22	2,677[1]	Monoclinic	a = 5.21 c = 5.37	5.56

REFRACTORY MATERIALS (Continued)

Oxides (Continued)

Name	Thermal conductivity, cal-sec^{-1}-cm^{-2}-cm-°C^{-1}	Electrical resistivity, microhm-cm	Ductility relative scale[b]	Resistance to oxidation[c]	Hardness[d]
Aluminum oxide	0.0723 at 100°C[2]	1×10^{22} at 14°C[1] 3×10^{19} at 300°C[1] 3.5×10^{14} at 800°C[1]	3[3]	1[3]	9 Mohs[1]
Beryllium oxide	0.525 at 100°C[2]	4×10^{14} at 600°C[1] 5×10^{12} at 1,100°C[1] 8×10^{8} at 2,100°C[1]			9 Mohs[1]
Cerium oxide		6.5×10^{10} at 800°C[1] 3.4×10^{8} at 1,200°C[1]			6 Mohs[1]
Chromium oxide		1.3×10^{9} at 350°C[1] 2.3×10^{7} at 1,200°C[1] 6.8×10^{3} at 600°C[1] 4.5×10^{3} at 1,100°C[1]	3[3]	1[3]	
Hafnium oxide		5×10^{15} at 400°C[1] 1×10^{3} at 1,500°C[1]			
Magnesium oxide	0.0860 at 100°C[2]	2×10^{14} at 850°C[1] 3×10^{13} at 980°C[1] 4.5×10^{8} at 2,100°C[1]			6 Mohs[1]
Silicon oxide		1×10^{21} at 20°C[1] 7×10^{12} at 600°C[1] 4×10^{5} at 1,300°C[1] (vitreous)	3[4] (vitreous)	1[4] (vitreous)	6–7 Mohs[1] (cristobalite)
Thorium oxide	0.0245 at 100°C[2]	2.6×10^{13} at 550°C[1] 8×10^{11} at 800°C[1] 1.5×10^{10} at 1,200°C[1]			6.5 Mohs[1]
Titanium oxide	0.0156 at 100°C[2]	1.2×10^{10} at 800°C[1] 8.5×10^{6} at 1,200°C[1]			5.5–6.0 Mohs[1]
Uranium oxide	0.0234 at 100°C[2]	3.8×10^{10} at 20°C[1] 5×10^{8} at 500°C[1]			
Zirconium oxide	0.00466 at 100°C[2]	1×10^{12} at 385°C[1] 2.2×10^{10} at 700°C[1] 3.6×10^{8} at 1,200°C[1]	3[3]	1[3]	6.5 Mohs[1]

REFRACTORY MATERIALS (Continued)

Silicides

Name	Formula	Molecular weight	Melting point, °C	Crystalline form	Lattice parameter, Å	X-ray density g/cm³
Chromium silicide	$CrSi_2$	108.13	1,570[51]	Hexagonal	a = 4.42 c = 6.35[51]	
	CrSi	80.07	1,870 (decomposes in presence of C)	Tetragonal (C 11b)	a = 3.20 c = 7.86[6]	
Molybdenum silicide	$MoSi_2$	152.07	1,870 (decomposes in presence of C)	Tetragonal (C 11b)	a = 3.20 c = 7.86[64][51]	6.24
Niobium silicide	$NbSi_2$		1,950[50]	Hexagonal $CrSi_2$ type	a = 4.785 c = 6.576[51]	5.29
Tantalum silicide	$TaSi_2$	237.00	2,400[50]	Hexagonal $CrSi_2$ type	a = 4.773 c = 6.552[51]	8.83
Titanium silicide	$TiSi_2$	104.02	1,540[49]	Orthorhombic	a = 8.24 b = 4.79 c = 8.52[59]	4.13
	Ti_5Si_3	323.68	2,120[49]	Hexagonal	a = 7.465 c = 5.162[60]	4.32
Tungsten silicide	WSi_2	240.04	2,050[50]	Tetragonal $MoSi_2$ Structure (C 11b)	a = 3.21 c = 7.83[66]	9.3
Uranium silicide	βUSi_2	294.19	1,700[6]	Hexagonal	a = 3.85[6] c = 4.06	9.25
	U_3Si_2	770.33	1,665[6]	Tetragonal	a = 7.3151[6] c = 3.8925	12.20
Vanadium silicide	VSi_2	107.07	1,750[50]	Hexagonal $CrSi_2$ type	a = 4.562 c = 6.359[51]	4.71

REFRACTORY MATERIALS (Continued)

Silicides (Continued)

Name	Thermal conductivity, cal-sec^{-1}-cm^{-2}-cm-°C^{-1}	Electrical resistivity, microhm-cm	Ductility relative scale[b]	Resistance to oxidation[c]	Hardness[d]
Chromium silicide					1,150 kg/mm^2 (50)
Molybdenum silicide	0.075 at room temperature to 200°C(6)	21.5 at room temperature 18.9 at –80°C(69)	2(3)	1(3)	1,290 kg/mm^2 (50)
Niobium silicide		6.3(68)	2(3)	4(3)	1,050 kg/mm^2 (50)
Tantalum silicide		8.5(68)	2(3)	3(3)	1,560 kg/mm^2 (50)
Titanium silicide		123 (hot pressed)(1)	3(3)	4(3)	870 kg/mm^2 (70)
			3(3)	4(3)	986 kg/mm^2 (49)
Tungsten silicide		33.4(68)		1–2(3)	1,310 kg/mm^2 (68)
					1,090 kg/mm^2 (68)
Uranium silicide					
Vanadium silicide		9.5(68)			1,090 kg/mm^2 (50)

[a]Numbers in parentheses refer to references at end of table.
[b]Ductility – 1: capable of being severely drawn, rolled, or otherwise worked without failure; 2: capable of withstanding slight deformation, or consisting of individually ductile crystals fragilely bound together; 3: incapable of being worked, of glasslike brittleness.
[c]Resistance to oxidation – classed according to the temperature range in which the rate of attack by air would cause severe erosion or failure of the coated specimen within a few hours. 1: above 1,700°C; 2: 1,400 to 1,700°C; 1,100 to 1,400°C; 4: 800 to 1,000°C; 5: 500 to 800°C.
[d]Microhardness values taken with 100-g load.

REFRACTORY MATERIALS (Continued)

REFERENCES

1. **Campbell, I. E.,** *High Temperature Technology,* Electrochemical Society, John Wiley & Sons New York, 1956.
2. **Kingery, W. D., Franch, J., Coble, R. L., and Vasilos, T.,** Thermal conductivity: X, data for several pure oxide materials corrected to zero porosity, *J. Am. Ceram. Soc.,* 37, 2, 107, 1954.
3. **Powel, C. F., Campbell, I. E., and Gonser, B. W.,** *Vapor Plating,* John Wiley & Sons, New York, 1955.
4. **Friederich, E. and Sittig, L.,** *Z. Anorg. Allg. Chem.,* 144, 169, 1925.
5. **Norton, J. T. and Morory, A. L.,** *Trans. Am. Inst. Min. Metall. Pet. Eng.,* 185, 133, 1949.
6. **Schwarzkopf, P. and Kieffer, R.,** *Refractory Hard Metals,* Macmillan, New York, 1953.
7. **Schwarzkopf, P. and Sindeband, S. J.,** Electrochemical Society, 97th Meeting, Cleveland, Ohio, 1950.
8. **Kieffer, R. and Kölbl, F.,** *Powder Metall. Bull.,* 4, 4, 1949.
9. **Agte, C. and Alterthum, H.,** *Z. Tech. Phys.,* 11, 182, 1930.
10. **Becker, K. and Ebert, F.,** *Z. Phys.,* 31, 268, 1925.
11. **Ruff, O. and Martin, W.,** *Z. Angew. Chem.,* 25, 49, 1912.
12. **Hinnuber, J.,** *Z. VDI,* 92, 111, 1950.
13. **Foster, L. S., Forbes, L. W., Jr., Friar, L. B., Moody, L. S., and Smith, W. H.,** *J. Am. Ceram. Soc.,* 33, 27, 1950.
14. **Ellinger, F. H.,** *Trans. Am. Soc. Met.,* 31, 89, 1943.
15. **Bloom, D. S. and Grant, N. J.,** *Trans. Am. Inst. Min. Metall. Pet. Eng.,* 188, 41, 1950.
16. **Hellstrom, K. and Westgren, A.,** *Kem. Tidskr.,* 45, 141, 1933.
17. **Westgren, A.,** *Jernkontorets Ann.,* 119, 231, 1935.
18. **Kuo, K. and Hägg, G.,** *Nature,* 170, 245, 1952.
19. **Nowotny, H. and Kieffer, R.,** *Z. Anorg. Allg. Chem.,* 267, 261, 1952.
20. **Weiss, G.,** *Ann. Chem.,* 1, 446, 1946.
21. **Brewer, L., Bromely, L. A., Gilles, P. W., and Lofgren, N. L.,** *The Chemistry and Metallurgy of Miscellaneous Materials: Thermodynamics,* McGraw-Hill, New York, 1950.
22. **Becker, K.,** *Z. Phys.,* 51, 481, 1928.
23. **Wilhelm, H. A. and Chiotti, P.,** *Trans. Am. Soc. Met.,* 42, 1295, 1950.
24. **Hunt, E. B. and Rundle, R. E.,** *J. Am. Chem. Soc.,* 73, 4777, 1951.
25. **Mallet, W., Gerds, A. F., and Nelson, H. R.,** *J. Electrochem. Soc.,* 99, 197, 1952.
26. **Litz, L. M., Gurrett, A. B., and Croxton, F. C.,** *J. Am. Chem. Soc.,* 70, 1718, 1948.
27. **Rundle, R. E., Baenziger, N. C., Wilson, A. S., and McDonald, R. A.,** *J. Am. Chem. Soc.,* 70, 99, 1948.
28. **Norton, J. T., Blumenthal, H., and Sindeband, S. J.,** *Trans. Am. Inst. Min. Metall. Pet. Eng.,* 185, 749, 1949.
29. **Honak, E. R.,** Thesis, Tech. Hochsch. Graz, 1951.
30. **Glaser, F. W. and Post, B.,** *J. Met.,* 1953.
31. **Moers, K.,** *Z. Anorg. Allg. Chem.,* 198, 262, 1931.
32. **Andrieux, L.,** *Rev. Mét.,* 45, 49, 1948.
33. **Post, B. and Glaser, F. W.,** *J. Met.,* 4, 631, 1952.
34. **Moers, K.,** *Z. Anorg. Allg. Chem.,* 198, 243, 1931.
35. **Anderson, L. H. and Kiessling, R.,** *Acta Chem. Scand.,* 4, 160, 209, 1950.
36. **Glaser, F. W.,** *J. Met.,* 4, 391, 1952.
37. **Kiessling, R.,** *Acta Chem. Scand.,* 3, 603, 1949.
38. **Moers, K.,** *Z. Anorg. Allg. Chem.,* 198, 262, 1931.

REFRACTORY MATERIALS (Continued)

REFERENCES (Continued)

39. **Kiessling, R.,** *Acta Chem. Scand.,* 3, 595, 1949.
40. **Steintz, R.,** *J. Met.,* 4, 148, 1952.
41. **Kiessling, R.,** *Acta Chem. Scand.,* 1, 893, 1947.
42. **Bertaut, F. and Blum, P.,** *Acta Crystallogr.,* 4, 72, 1951.
43. **Zalkin, A. and Templeton, D. H.,** *J. Chem. Phys.,* 18, 391, 1950.
44. **Bertaut, F. and Blum, P.,** *Comptes rendus,* 229, 666, 1949.
45. **Baumann, H. N., Jr.,** Electrochemical Society, 99th Meeting, Washington, D.C., April 1951.
46. **Friederich, E. and Sittig, L.,** *Z. Anorg. Allg. Chem.,* 143, 293, 1925.
47. **Agte, C. and Moers, K.,** *Z. Anorg. Chem.,* 198, 233, 1931.
48. **Chiotti, P.,** *J. Am. Ceram. Soc.,* 35, 123, 1952.
49. **Hansen, M., Klasler, H. D., and McPherson, D. J.,** *Trans. Am. Soc. Met.,* 44, 518, 1952.
50. **Cerwenka, E.,** Thesis, Tech. Hochsch. Graz, 1951.
51. **Wallbaum, H. J.,** *Z. Metallkd.,* 33, 378, 1941.
52. **Pease, R. S.,** *Acta Crystallogr.,* 5, 356, 1952.
53. **van Arkel, A. E.,** *Physica,* 4, 296, 1924.
54. **Horn, F. H. and Ziegler, W. T.,** *J. Am. Chem. Soc.,* 69, 2762, 1947.
55. **Pauling, L., Killeffer, D. H., and Linz, A.,** *Molybdenum Compounds,* Interscience, New York, 1952.
56. **Hägg, G.,** *Z. Phys. Chem. Abt. B,* 7, 339, 1930.
57. **Kiessling, R. and Lier, Y. H.,** *J. Met.,* 3, 639, 1951.
58. **Rundle, R. E., Baenziger, N. C., Wilson, A. S., and McDonald, R. A.,** *J. Am. Chem. Soc.,* 70, 99, 1941.
59. **Laues, F. and Wallbaum, H. J.,** *Z. Kristallogr. A,* 101, 78, 1939.
60. **Pietrokowsky, P. and Duwez, P.,** *J. Met.,* 3, 772, 1951.
61. **Naray Szako, S. V.,** *Z. Kristallogr. A,* 97, 223, 1937.
62. **Lundin, C. E., McPherson, D. J., and Hansen, M.,** American Society for Metals, 34th Ann. Convention, Preprint No. 41, 1952.
63. **Wallbaum, H. J.,** *Z. Metallkd.,* 31, 362, 1939.
64. **Zachariasen, W. H.,** *Z. Phys. Chem.,* 128, 39, 1927.
65. **Templeton, D. H. and Dauben, C. H.,** *Acta Crystallogr.,* 3, 261, 1950.
66. **Nowotny, H., Kieffer, R., and Schachner, H.,** *Mh. Chem.,* 83, 1243, 1952.
67. **Brauer, G. and Mitices, A.,** *Z. Anorg. Allg. Chem.,* 249, 325, 1942.
68. **Gallistl, E.,** Thesis, Tech. Hochsch. Graz, 1951.
69. **Glaser, F. W.,** *J. Appl. Phys.,* 22, 103, 1951.
70. **Cerwenka, E.,** Thesis, Tech. Hochsch. Graz, 1951.
71. **Currie, L. M., Hamister, V. C., and MacPherson, H. G.,** paper presented at the United Nations International Conference on The Peaceful Uses of Atomic Energy, Geneva, 1955.

THERMODYNAMIC PROPERTIES OF ELEMENTS AND OXIDES

Thermodynamic calculations over a wide range of temperatures are generally made with the aid of algebraic equations representing the characteristic properties of the substances being considered. The necessary integrations and differentiations, or other mathematical manipulations, are then most easily effected.

The most convenient starting point in making such calculations for a given substance is the heat capacity at constant pressure. From this quantity and a knowledge of the properties of any phase transitions, the other thermodynamic properties may be computed by the well-known equations given in standard texts on thermodynamics.

Users of the following equations and tables are cautioned that the units for a, b, c, and d are cal/g mole, whereas those for A are Kcal/g mole. The necessary adjustment must be made when the data are substituted into the equations.

Empirical heat capacity equations are generally of the form of a power series with the absolute temperature T as the independent variable:

$$C_p = a' + (b' \times 10^{-3})T + (c' \times 10^{-6})T^2$$

or

$$C_p = a'' + (b'' \times 10^{-3})T + \frac{d \times 10^5}{T^2}.$$

Since both forms are used in the ensuing, let

$$(1) \quad C_p = a + (b \times 10^{-3})T + (c \times 10^{-6})T^2 + \frac{d \times 10^5}{T^2}.$$

The constants a, b, c, and d are to be determined either experimentally or by some theoretical or semi-empirical approach.

The heat content or enthalpy H is determined from the heat capacity by a simple integration over the range of temperatures for which (1) is applicable. Thus, if 298° K is taken as a reference temperature,

$$(2) \quad H_T - H_{298} = \int_{298}^{T} C_p dT$$

$$= a(T - 298) + \tfrac{1}{2}(b \times 10^{-3})(T^2 - 298^2) + \frac{1}{3}(c \times 10^{-6})(T^3 - 298^3) - (d \times 10^5)\left(\frac{1}{T} - \frac{1}{298}\right)$$

$$= aT + \tfrac{1}{2}(b \times 10^{-3})T^2 + \frac{1}{3}(c \times 10^{-6})T^3 - \frac{d \times 10^5}{T} - A,$$

where all the constants on the right hand side of the equation have been incorporated in the term −A.

In general, the enthalpy is given by a sum of terms such as (2) for each phase of the substance involved in the temperature range considered plus terms which represent the heats of transitions:

$$H_T - H_{298} = \sum \int_{T_1}^{T_2} C_p dT + \sum \Delta H_{tr.}$$

In a similar manner, the entropy S is obtained from (1) by performing the integration

$$(3) \quad S_T - S_{298} = \int_{298}^{T} (C_p/T)dt$$

$$= a\ln(T/298) + (b \times 10^{-3})(T - 298) + \tfrac{1}{2}(c \times 10^{-6})(T^2 - 298^2) - \tfrac{1}{2}(d \times 10^5)\left(\frac{1}{T^2} - \frac{1}{298^2}\right)$$

$$= a\ln T + (b \times 10^{-3})T + \tfrac{1}{2}(c \times 10^{-6})T^2 - \left(\frac{\tfrac{1}{2}(d \times 10^5)}{T^2}\right) - B'$$

or

$$(4) \quad S_T = 2.303\, a \log T + (b \times 10^{-3})T + \tfrac{1}{2}(c \times 10^{-6})T^2 - \frac{\tfrac{1}{2}(d \times 10^5)}{T^2} - B$$

where

$$(5) \quad B = B' - S_{298}.$$

From the definition of free energy F:

$$F = H - TS$$

the quantity

$$F_T - H_{298} = (H_T - H_{298}) - TS_T$$

is obtained from (2) and (4):

$$(6) \quad F_T - H_{298} = -2.303aT \log T - \tfrac{1}{2}(b \times 10^{-3})T^2 - \frac{1}{6}(c \times 10^{-6})T^3 - \frac{\tfrac{1}{2}(d \times 10^5)}{T} + (B + a)T - A$$

and also the free energy function

$$(7) \quad \frac{F_T - H_{298}}{T} = -2.303a \log T - \tfrac{1}{2}(b \times 10^{-3})T - \frac{1}{6}(c \times 10^{-6})T^2 - \frac{\tfrac{1}{2}(d \times 10^5)}{T^2} + (B + a) - \frac{A}{T}$$

In the following two tables there has been collected values of the constants. The first column lists the element or the oxide. The second column gives the phase to which they are applicable. The third, fourth, and fifth columns specify the thermodynamic properties for the transition to the succeding phase. In column 6, the value of the entropy at 298.15° K, the reference temperature, is given. The remaining columns, except for the last, give the values of the constants a, b, c, d, A, and B required in the thermodynamic equations.

All values throughout the table which represent estimates have been enclosed in parentheses.

The heat capacities at temperatures beyond the range of experimental determination were estimated by extrapolation. Where no experimental values were found, use of analogy with compounds of neighboring elements in the Periodic Table was employed.

THERMODYNAMIC PROPERTIES OF THE ELEMENTS

Ref.: U. S. Atomic Energy Commission Report ANL-5750.

Element	Phase	Temperature of Transition (°K)	Heat of Transition (kcal/g mole)	Entropy of Transition (e.u.)	Entropy at 298°K (e.u.)	Element	cal/g mole a	cal/g mole b	cal/g mole c	cal/g mole d	A (kcal/g mole)	B (e.u.)
Ac	solid	(1090)	(2.5)	(2.3)	(13)	Ac	(5.4)	(3.0)	—	—	(1.743)	(18.7)
	liquid	(2750)	(70)	(25)	—		(8)	—	—	—	(0.295)	(31.3)
Ag	solid	1234	2.855	2.313	10.20	Ag	5.09	1.02	—	0.36	1.488	19.21
	liquid	2485	60.72	24.43	—		7.30	—	—	—	0.164	30.12
	gas	—	—	—	—		(4.97)	—	—	—	(−66.34)	(−12.52)
Al	solid	931.7	2.57	2.76	6.769	Al	4.94	2.96	—	—	1.604	22.26
	liquid	2600	67.9	26	—		7.0	—	—	—	0.33	30.83
Am	solid	(1200)	(2.4)	(2.0)	(13)	Am	(4.9)	(4.4)	—	—	(1.657)	(16.2)
	liquid	2733	51.7	18.9	—		(8.5)	—	—	—	(0.409)	(34.5)
As	solid	883	31/4	35.1/4	8.4	As	5.17	2.34	—	—	1.646	21.8
Au	solid	1336.16	3.03	2.27	11.32	Au	6.14	−0.175	0.92	—	1.831	23.65
	liquid	2933	74.21	25.30	—		7.00	—	—	—	−0.631	26.99
B	solid	2313	(3.8)	(1.6)	1.42	B	1.54	4.40	—	—	0.655	8.67
	liquid	2800	75	27	—		(6.0)	—	—	—	(−4.599)	(31.4)
Ba	solid, α	648	0.14	0.22	16	Ba	5.55	1.50	—	—	1.722	16.1
	solid, β	977	1.83	1.87	—		5.55	1.50	—	—	1.582	15.9
	liquid	1911	35.665	18.63	—		(7.4)	—	—	—	(0.843)	(25.3)
	gas	—	—	—	—		(4.97)	—	—	—	(−39.65)	(−11.7)
Be	solid	1556	2.919	1.501	2.28	Be	5.07	1.21	—	−1.15	1.951	27.62
	liquid	—	—	—	—		5.27	—	—	—	−1.611	25.68
Bi	solid	544.2	2.63	4.83	13.6	Bi	5.38	2.60	—	—	1.720	17.8
	liquid	1900	41.1	21.6	—		7.60	—	—	—	−0.087	25.6
	gas	—	—	—	—		(4.97)	—	—	—	(−46.19)	(−15.9)
C	solid	—	—	—	1.3609	C	4.10	1.02	—	−2.10	1.972	23.484
Ca	solid, α	723	0.24	0.33	9.95	Ca	5.24	3.50	—	—	1.718	20.95
	solid, β	1123	2.2	1.96	—		6.29	1.40	—	—	1.689	26.01
	liquid	1755	38.6	22.0	—		7.4	—	—	—	−0.147	30.28
	gas	—	—	—	—		(4.97)	—	—	—	(−43.015)	(−9.88)
Cd	solid	594.1	1.46	2.46	12.3	Cd	5.31	2.94	—	—	1.714	18.8
	liquid	1040	23.86	22.94	—		7.10	—	—	—	0.798	26.1
	gas	—	—	—	—		(4.97)	—	—	—	(−25.28)	(−11.7)
Ce	solid	1048	2.1	2.0	13.8	Ce	4.40	6.0	—	—	1.579	13.1
	liquid	2800	73	26	—		(7.9)	—	—	—	(−0.148)	(29.1)
Cl_2	gas	—	—	—	53.286	Cl_2	8.76	0.27	—	−0.65	2.845	−2.929
Co	solid, α	723	0.005	0.007	6.8	Co	4.72	4.30	—	—	1.598	21.4
	solid, β	1398	0.095	0.068	—		3.30	5.86	—	—	0.974	13.1
	solid, γ	1766	3.7	2.1	—		9.60	—	—	—	3.961	50.5
	liquid	3370	93	28	—		8.30	—	—	—	−2.034	38.7
Cr	solid	2173	3.5	1.6	5.68	Cr	5.35	2.36	—	−0.44	1.848	25.75
	liquid	2495	72.97	29.25	—		9.40	—	—	—	1.556	50.13
	gas	—	—	—	—		(4.97)	—	—	—	(−82.47)	(−13.8)
Cs	solid	301.9	0.50	1.7	19.8	Cs	7.42	—	—	—	2.212	22.5
	liquid	963	16.32	17.0	—		8.00	—	—	—	1.887	24.1
	gas	—	—	—	—		(4.97)	—	—	—	(−17.35)	(−13.6)
Cu	solid	1356.2	3.11	2.29	7.97	Cu	5.41	1.50	—	—	1.680	23.30
	liquid	2868	72.8	25.4	—		7.50	—	—	—	0.024	34.05
F_2	gas	—	—	—	48.58	F_2	8.29	0.44	—	−0.80	2.760	−0.76
Fe	solid, α	1033	0.410	0.397	6.491	Fe	3.37	7.10	—	0.43	1.176	14.59
	solid, β	1180	0.217	0.184	—		10.40	—	—	—	4.281	55.66
	solid, γ	1673	0.15	0.084	—		4.85	3.00	—	—	0.396	19.76
	solid, δ	1808	3.86	2.14	—		10.30	—	—	—	4.382	55.11
	liquid	3008	84.62	28.1	—		10.00	—	—	—	−0.021	50.73
Ga	solid	302.94	1.335	4.407	9.82	Ga	5.237	3.33	—	—	1.710	21.01
	liquid	2700	—	—	—		(6.645)	—	—	—	(0.648)	(23.64)
Ge	solid	1232	8.3	6.7	10.1	Ge	5.90	1.13	—	—	1.764	23.8
	liquid	2980	68	23	—		(7.3)	—	—	—	(−5.668)	(25.7)
H_2	gas	—	—	—	31.211	H_2	6.62	0.81	—	—	2.010	6.75
Hf	solid	(2600)	(6.0)	(2.3)	13.1	Hf	(6.00)	(0.52)	—	—	(1.812)	(21.2)
Hg	liquid	629.73	13.985	22.208	18.46	Hg	6.61	—	—	—	1.971	19.20
	gas	—	—	—	—		4.969	—	—	—	−13.048	−13.54
In	solid	430	0.775	1.80	13.88	In	5.81	2.50	—	—	1.844	19.97
	liquid	2440	53.8	22.0	—		7.50	—	—	—	1.564	27.34
	gas	—	—	—	—		(4.97)	—	—	—	(−58.42)	(−14.46)
Ir	solid	2727	6.6	2.4	8.7	Ir	5.56	1.42	—	—	1.721	23.4
K	solid	336.4	0.5575	1.657	15.2	K	1.3264	19.405	—	—	1.258	−1.86
	liquid	1052	18.88	17.95	—		8,8825	−4.565	2.9369	—	1.923	32.55
	gas	—	—	—	—		(4.97)	—	—	—	(−19.689)	(−9.46)
La	solid	1153	(2.3)	(2.0)	13.7	La	6.17	1.60	—	—	1.911	21.9
	liquid	3000	80	27	—		(7.3)	—	—	—	(−0.15)	(26.0)
Li	solid	459	0.69	1.5	6.70	Li	3.05	8.60	—	—	1.292	12.92
	liquid	1640	32.48	19.81	—		7.0	—	—	—	1.509	32.00
	gas	—	—	—	—		(4.97)	—	—	—	(−34.30)	(−2.84)
Mg	solid	923	2.2	2.4	7.77	Mg	5.33	2.45	—	−0.103	1.733	23.39
	liquid	1393	31.5	22.6	—		(8.0)	—	—	—	0.942	36.97
	gas	—	—	—	—		(4.97)	—	—	—	(−34.78)	(−7.60)
Mn	solid, α	1000	0.535	0.535	7.59	Mn	5.70	3.38	—	−0.37	1.974	26.11
	solid, β	1374	0.545	0.397	—		8.33	0.66	—	—	2.672	41.02
	solid, γ	1410	0.430	0.305	—		10.70	—	—	—	4.760	56.84
	solid, δ	1517	3.5	2.31	—		11.30	—	—	—	5.176	60.88
	liquid	2368	53.7	22.7	—		11.00	—	—	—	1.221	56.38
	gas	—	—	—	—		6.26	—	—	—	−63.704	−3.13
Mo	solid	2883	(5.8)	(2.0)	6.83	Mo	5.48	1.30	—	—	1.692	24.78
N_2	gas	—	—	—	45.767	N_2	6.76	0.606	0.13	—	2.044	−7.064
Na	solid	371	0.63	1.7	12.31	Na	5.657	3.252	0.5785	—	1.836	20.92
	liquid	1187	23.4	20.1	—		8.954	−4.577	2.540	—	1.924	36.0
	gas	—	—	—	—		(4.97)	—	—	—	(−24.40)	(−8.7)
Nb	solid	2760	(5.8)	(2.1)	8.3	Nb	5.66	0.96	—	—	1.730	24.24
Nd	solid	1297	(2.55)	(1.97)	13.9	Nd	5.61	5.34	—	—	1.910	19.7
	liquid	(2750)	(61)	(22)	—		(9.1)	—	—	—	−0.606	35.8
Ni	solid, α	626	0.092	0.15	7.137	Ni	4.06	7.04	—	—	1.523	18.095
	solid, β	1728	4.21	2.44	—		6.00	1.80	—	—	1.619	27.16
	liquid	3110	90.48	29.0	—		9.20	—	—	—	0.251	45.47

THERMODYNAMIC PROPERTIES OF THE ELEMENTS

Element	Phase	Temperature of Transition (°K)	Heat of Transition (kcal/g mole)	Entropy of Transition (e.u.)	Entropy at 298°K (e.u.)	Element	cal/g mole a	cal/g mole b	cal/g mole c	cal/g mole d	A (kcal/g mole)	B (e.u.)
Np	solid	913	(2.3)	(2.5)	(14)	Np	(5.3)	(3.4)	—	—	(1.731)	(17.9)
	liquid	(2525)	(55)	(22)	—		(9.0)	—	—	—	(1.392)	(37.5)
O_2	gas	—	—	—	49.003	O_2	8.27	0.258	—	−1.877	3.007	−0.750
Os	solid	2970	(6.4)	(2.2)	7.8	Os	5.69	0.88	—	—	1.736	24.9
P_4	solid, white	317.4	0.601	1.89	42.4	P_4	13.62	28.72	—	—	5.338	43.8
	liquid	553	11.9	21.3	—		19.23	0.51	—	−2.98	6.035	66.7
	gas	—	—	—	—		(19.5)	(−0.4)	(1.3)	—	(−6.32)	(46.1)
Pa	solid	(1825)	(4.0)	(2.2)	(13.5)	Pa	(5.2)	(4.0)	—	—	(1.728)	(17.3)
	liquid	(4500)	(115)	(26)	—		(8.0)	—	—	—	(−3.823)	(28.8)
Pb	solid	600.6	1.141	1.900	15.49	Pb	5.64	2.30	—	—	1.784	17.33
	liquid	2023	42.5	21.0	—		7.75	−0.73	—	—	1.362	27.11
	gas	—	—	—	—		(4.97)	—	—	—	(−45.25)	(−13.6)
Pd	solid	1828	4.12	2.25	8.9	Pd	5.80	1.38	—	—	1.791	24.6
	liquid	3440	89	26	—		(9.0)	—	—	—	(1.215)	(43.8)
Po	solid	525	(2.4)	(4.6)	13	Po	(5.2)	(3.2)	—	—	(1.693)	(17.6)
	liquid	(1235)	(24.6)	(19.9)	—		(9.0)	—	—	—	(0.847)	(35.2)
	gas	—	—	—	—		(4.97)	—	—	—	(−28.73)	(−13.5)
Pr	solid	1205	(2.5)	(2.1)	(13.5)	Pr	(5.0)	(4.6)	—	—	(1.705)	(16.4)
	liquid	3563	—	—	—		(8.0)	—	—	—	(−0.519)	(30.0)
Pt	solid	2042.5	5.2	2.5	10.0	Pt	5.74	1.34	—	0.10	1.737	23.0
	liquid	4100	122	29.8	—		(9.0)	—	—	—	(0.406)	(42.6)
Pu	solid	913	(2.26)	(2.48)	(13.0)	Pu	(5.2)	(3.6)	—	—	(1.710)	(17.7)
	liquid	—	—	—	—		(8.0)	—	—	—	(0.506)	(31.0)
Ra	solid	1233	(2.3)	(1.9)	(17)	Ra	(5.8)	(1.2)	—	—	(1.783)	(16.4)
	liquid	(1700)	(35)	(21)	—		(8.0)	—	—	—	(1.284)	(28.6)
	gas	—	—	—	—		(4.97)	—	—	—	(−38.87)	(−14.5)
Rb	solid	312.0	0.525	1.68	16.6	Rb	3.27	13.1	—	—	1.557	5.9
	liquid	952	18.11	19.0	—		7.85	—	—	—	1.814	26.5
	gas	—	—	—	—		(4.97)	—	—	—	(−19.04)	(−12.3)
Re	solid	3440	(7.9)	(2.3)	(8.89)	Re	(5.85)	(0.8)	—	—	(1.780)	(24.7)
Rh	solid	2240	(5.2)	(2.3)	7.6	Rh	5.40	2.19	—	—	1.707	23.8
	liquid	4150	127	30.7	—		(9.0)	—	—	—	(−0.923)	(44.4)
Ru	solid, α	1308	0.034	0.026	6.9	Ru	5.25	1.50	—	—	1.632	23.5
	solid, β	1473	0	0	—		7.20	—	—	—	2.867	35.5
	solid, γ	1773	0.23	0.13	—		7.20	—	—	—	2.867	35.5
	solid, δ	2700	(6.1)	(2.3)	—		7.50	—	—	—	3.169	37.6
S	solid, α	368.6	0.088	0.24	7.62	S	3.58	6.24	—	—	1.345	14.64
	solid, β	392	0.293	0.747	—		3.56	6.95	—	—	1.298	14.54
	liquid	717.76	2.5	3.5	—		5.4	5.0	—	—	1.576	24.02
$\frac{1}{2}S_2$	gas	—	—	—	—	$\frac{1}{2}S_2$	(4.25)	(0.15)	—	(−1.0)	(−2.859)	(9.57)
Sb	solid (α,β,γ	903.7	4.8	5.3	10.5	Sb	5.51	1.74	—	—	1.720	21.4
	liquid	1713	46.665	27.3	—		7.50	—	—	—	−1.992	28.1
$\frac{1}{2}Sb_2$	gas	—	—	—	—	$\frac{1}{2}Sb_2$	4.47	—	—	−0.11	−53.876	−21.7
Sc	solid	1670	(4.0)	(2.4)	(9.0)	Sc	(5.13)	(3.0)	—	—	1.663	21.1
	liquid	3000	80	27	—		(7.50)	—	—	—	(−2.563)	31.3
Se	solid	490.6	1.25	2.55	10.144	Se	3.30	8.80	—	—	1.375	11.28
	liquid	1000	14.27	14.27	—		7.0	—	—	—	0.881	27.34
Si	solid	1683	11.1	6.60	4.50	Si	5.70	1.02	—	−1.06	2.100	28.88
	liquid	2750	71	26	—		7.4	—	—	—	−7.646	33.17
Sm	solid	1623	3.7	2.3	(15)	Sm	(6.7)	(3.4)	—	—	(2.149)	(24.2)
	liquid	(2800)	(70)	(25)	—		(9.0)	—	—	—	(−2.296)	(33.4)
Sn	solid (α,β)	505.1	1.69	3.35	12.3	Sn	4.42	6.30	—	—	1.598	14.8
	liquid	2473	(55)	(22)	—		7.30	—	—	—	0.559	26.2
	gas	—	—	—	—		(4.97)	—	—	—	(−60.21)	(−14.3)
Sr	solid	1043	2.2	2.1	13.0	Sr	(5.60)	(1.37)	—	—	(1.731)	(19.3)
	liquid	1657	33.61	20.28	—		(7.7)	—	—	—	(0.976)	(30.4)
	gas	—	—	—	—		(4.97)	—	—	—	(−37.16)	(−10.2)
Ta	solid	3250	7.5	2.3	9.9	Ta	5.82	0.78	—	—	1.770	23.4
Tc	solid	(2400)	(5.5)	(2.3)	(8.0)	Tc	(5.6)	(2.0)	—	—	(1.759)	(24.5)
	liquid	(3800)	(120)	(32)	—		(11)	—	—	—	(3.459)	(59.4)
Te	solid, α	621	0.13	0.21	11.88	Te	4.58	5.25	—	—	1.599	15.78
	solid, β	723	4.28	5.92	—		4.58	5.25	—	—	1.469	15.57
	liquid	1360	11.9	8.75	—		9.0	—	—	—	−0.988	34.96
$\frac{1}{2}Te_2$	gas	—	—	—	—	$\frac{1}{2}Te_2$	4.47	—	—	−0.10	−19.048	−6.47
Th	solid	2173	(4.6)	(2.1)	12.76	Th	8.2	−0.77	2.04	—	2.591	33.64
	liquid	4500	(130)	(29)	—		(8.0)	—	—	—	(−7.602)	(26.84)
Ti	solid, α	1155	0.950	0.822	7.334	Ti	5.25	2.52	—	—	1.677	23.33
	solid, β	2000	(4.6)	(2.3)	—		7.50	—	—	—	1.645	35.46
	liquid	3550	(101)	(28)	—		(7.8)	—	—	—	(−2.355)	(35.45)
Tl	solid, α	508.3	0.082	0.16	15.4	Tl	5.26	3.46	—	—	1.722	15.6
	solid, β	576.8	1.03	1.79	—		7.30	—	—	—	2.230	26.4
	liquid	1730	38.81	22.4	—		7.50	—	—	—	1.315	25.9
	gas	—	—	—	—		(4.97)	—	—	—	(−41.88)	(−15.4)
U	solid, α	938	0.665	0.709	12.02	U	3.25	8.15	—	0.80	1.063	8.47
	solid, β	1049	1.165	1.111	—		10.28	—	—	—	3.493	48.27
	solid, γ	1405	(3.0)	(2.1)	—		9.12	—	—	—	1.110	39.09
	liquid	3800	—	—	—		(8.99)	—	—	—	(−2.073)	36.01
V	solid	2003	(4.0)	(2.0)	7.05	V	5.57	0.97	—	—	1.704	24.97
	liquid	3800	—	—	—		(8.6)	—	—	—	1.827	44.06
W	solid	3650	8.42	2.3	8.0	W	5.74	0.76	—	—	1.745	24.9
Y	solid	1750	(4.0)	(2.3)	(11)	Y	(5.6)	(2.2)	—	—	(1.767)	(21.6)
	liquid	3500	(90)	(26)	—		(7.5)	—	—	—	(−2.277)	(29.6)
Zn	solid	692.7	1.595	2.303	9.95	Zn	5.35	2.40	—	—	1.702	21.25
	liquid	1180	27.43	23.24	—		7.50	—	—	—	1.020	31.35
	gas	—	—	—	—		(4.97)	—	—	—	(−29.407)	(−9.81)
Zr	solid, α	1135	0.920	0.811	9.29	Zr	6.83	1.12	—	−0.87	2.378	30.45
	solid, β	2125	(4.9)	(2.3)	—		7.27	—	—	—	1.159	31.43
	liquid	(3900)	(100)	(26)	—		(8.0)	—	—	—	(−2.190)	(34.7)

THERMODYNAMIC PROPERTIES OF THE OXIDES

For description of headings and symbols see preceding table.
Ref.: U. S. Atomic Energy Report ANL-5750.

Oxide	Phase	Temperature of Transition (°K)	Heat of Transition kcal/mole	Entropy of Transition (e.u.)	Entropy at 298°K (e.u.)
Ac_2O_3	solid	(2250)	(20)	(8.9)	(36.5)
	liquid	—			
Ag_2O	solid	dec. 460	—	—	29.09
Ag_2O_2	solid	dec	—	—	(20.4)
Al_2O_3	solid	2300	26	11	12.186
	liquid	dec.	—	—	
Am_2O_3	solid	(2225)	(17)	(7.6)	(37)
	liquid	(3400)	(85)	(25)	—
AmO_2	solid	dec.	—	—	(20)
As_2O_3	solid, α	503	4.1	8.2	25.6
	solid, β	586	4.4	7.5	—
	liquid	730	7.15	9.79	—
	gas	—	—	—	—
AsO_2	solid	(1200)	(9.0)	(7.5)	(13)
	liquid	(dec.)	—	—	—
As_2O_5	solid	dec. >1100	—	—	25.2
Au_2O_3	solid	dec.	—	—	30
B_2O_3	solid	723	5.27	7.29	12.91
	liquid	2520	(55)	(22)	—
Ba_2O	solid	(880)	(5.2)	(5.9)	(23.5)
	liquid	(1040)	(20)	(19)	—
	gas	—	—	—	—
BaO	solid	2196	13.8	6.28	16.8
	liquid	3000	(62)	(21)	—
BaO_2	solid	723	(5.7)	(7.9)	(18.5)
	liquid	dec. 1110	—	—	—
BeO	solid	dec.	—	—	3.37
BiO	solid	(1175)	(3.7)	(3.1)	(15)
	liquid	(1920)	(54)	(28)	—
	gas	—	—	—	—
Bi_2O_3	solid	1090	6.8	6.2	36.2
	liquid	(dec.)	—	—	—
CO	gas	—	—	—	47.30
CO_2	gas	—	—	—	51.06
CaO	solid	2860	(18)	(6.3)	9.5
CdO	solid	dec.	—	—	13.1
Ce_2O_3	solid	1960	(20)	(10)	(33.5)
	liquid	(3500)	(80)	(23)	—
CeO_2	solid	3000	(19)	(6.3)	17.7
CoO	solid	2078	(12)	(5.8)	10.5
	liquid	(2900)	(61)	(21)	—
Co_3O_4	solid	dec., 1240	—	—	(35.5)
Cr_2O_3	solid	2538	(25)	(10)	19.4
CrO_2	solid	dec. 700	—	—	(11.5)
CrO_3	solid	460	(6.1)	(13)	(17.5)
	liquid	(1000)	(25)	(25)	—
	gas	—	—	—	—
Cs_2O	solid	763	(4.58)	(6.0)	(23)
	liquid	dec.	—	—	—
Cs_2O_2	solid	867	(5.5)	(6.3)	(40)
	liquid	dec.	—	—	—
Cs_2O_3	solid	775	(7.75)	(10)	(47)
	liquid	dec.	—	—	—
Cu_2O	solid	1503	13.4	8.92	22.44
	liquid	dec.	—	—	—
CuO	solid	1609	(8.9)	(5.5)	10.4
	liquid	dec.	—	—	—
FeO	solid	1641	7.5	4.6	12.9
	liquid	(2700)	(55)	(20)	—
Fe_3O_4	solid, α	900	(0)	(0)	35 0
	solid, β	dec.	—	—	—
Fe_2O_3	solid, α	950	0.16	0.17	21.5
	solid, β	1050	0	0	—
	solid, γ	dec.	—	—	—
Ga_2O	solid	(925)	(8.5)	(9.2)	(22.5)
	liquid	(1000)	(20)	(20)	—
	gas	—	—	—	—
Ga_2O_3	solid	2013	(22)	(11)	20.23
	liquid	(2900)	(75)	(26)	—
GeO	solid	983	(50)	(51)	(12.5)
	gas	—	—	—	—
GeO_2	solid (α,β)	1389	10.5	7.56	(12.5)
	liquid	(2625)	(61)	(23)	—
H_2O	liquid	373.16	9.770	26.18	16.716
	gas	—	—	—	—
HfO_2	solid	3063	(17)	(5.6)	14.18
Hg_2O	solid	dec.	—	—	(30)
HgO	solid	dec.	—	—	16.839
In_2O	solid	(600)	(4.5)	(7.5)	(28)
	liquid	(800)	(16)	(20)	—
	gas	—	—	—	—
InO	solid	(1325)	(4.0)	(3.0)	(14.5)
	liquid	(2000)	(60)	(30)	—
	gas	—	—	—	—
In_2O_3	solid	(2000)	(20)	(10)	30.1
	liquid	(3600)	(85)	(24)	—

Oxide	cal/g mole a	cal/g mole b	cal/g mole c	cal/g mole d	A (kcal/mole)	B (e.u.)
Ac_2O_3	(20.0)	(20.4)	—	—	(6.870)	(80.9)
	(40)	—	—	—	(−19.767)	(180.5)
Ag_2O	13.26	7.04	—	—	4.266	48.56
Ag_2O_2	(16.4)	(12.2)	—	—	(5.432)	(76.7)
Al_2O_3	26.12	4.388	—	−7.269	10.422	142.03
	(33)	—	—	—	(−11.655)	(174.1)
Am_2O_3	(20.0)	(15.6)	—	—	(6.657)	(81.6)
	(38.5)	—	—	—	(−7.796)	(181.8)
AmO_2	(14.0)	(6.8)	—	—	(4.477)	(61.8)
As_2O_3	8.37	48.6	—	—	4.656	36.6
	8.37	48.6	—	—	0.556	28.4
	(39)	—	—	—	(5.760)	(187.6)
	(21.5)	—	—	—	(−14.164)	(62.5)
AsO_2	(8.5)	(9.4)	—	—	(2.952)	(38.2)
	(21)	—	—	—	(2.184)	(108.0)
As_2O_5	(31.1)	(16.4)	—	(−5.4)	(11.813)	(159.9)
Au_2O_3	(23.5)	(4.8)	—	—	(7.220)	(105.3)
B_2O_3	8.73	25.40	—	−1.31	4.171	45.04
	30.50	—	—	—	7.822	161.59
Ba_2O	(20.0)	(2.2)	—	—	(6.061)	(91.1)
	(22)	—	—	—	(1.769)	(96.8)
	(15)	—	—	—	(−25.51)	(29.0)
BaO	12.74	1.040	—	−1.984	4.510	57.2
	(13.9)	—	—	—	(− 9.341)	(57.5)
BaO_2	(13.6)	(2.0)	—	—	(4.144)	(59.6)
	(21)	—	—	—	(3.241)	(99.0)
BeO	8.69	3.65	—	−3.13	3.803	48.99
BiO	(9.7)	(3.0)	—	—	(3.025)	(41.2)
	(14)	—	—	—	(2.306)	(64.9)
	(8.9)	—	—	—	(−61.49)	(−1.8)
Bi_2O_3	23.27	11.05	—	—	7.429	99.7
	(35.7)	—	—	—	(7.614)	(168.3)
CO	6.60	1.2	—	—	2.021	−9.34
CO_2	7.70	5.3	−0.83	—	2.490	−5.64
CaO	10.00	4.84	—	−1.08	3.559	49.5
CdO	9.65	2.08	—	—	2.970	42.5
Ce_2O_3	(23.0)	(9.0)	—	—	(7.258)	(100.2)
	(37)	—	—	—	(−2.591)	(178.5)
CeO_2	15.0	2.5	—	—	4.579	68.5
CoO	(9.8)	(2.2)	—	—	(3.020)	(46.0)
	(15.5)	—	—	—	(−1.886)	(79.2)
Co_3O_4	(29.5)	(17.0)	—	—	(9.551)	(137.6)
Cr_2O_3	28.53	2.20	—	−3.736	9.857	145.9
CrO_2	(16.1)	(3.0)	—	(−3.0)	(5.946)	(82.8)
CrO_3	(18.1)	(4.0)	—	(−2.0)	(6.245)	(87.9)
	(27)	—	—	—	(3.381)	(127.0)
	(20)	—	—	—	(−28.62)	(53.6)
Cs_2O	(16.5)	(5.4)	—	—	(5.160)	(72.6)
	(22)	—	—	—	(3.205)	(99.0)
Cs_2O_2	(21.4)	(11.4)	—	—	(6.887)	(85.3)
	(29.5)	—	—	—	(4.125)	(123.8)
Cs_2O_3	(24.0)	(22.6)	—	—	(8.160)	(96.5)
	(35)	—	—	—	(2.148)	(142.2)
Cu_2O	(13.4)	(8.6)	—	—	(4.378)	(54.9)
	(21.5)	—	—	—	(3.721)	(96.0)
CuO	14.34	6.2	—	—	4.551	61.11
	(22)	—	—	—	(−4.339)	(98.91)
FeO	9.27	4.80	—	—	(2.977)	(43.8)
	(14.5)	—	—	—	(−3.721)	(69.2)
Fe_3O_4	12.38	1.62	—	−0.38	3.826	58.3
	(14.5)	—	—	—	(−2.399)	(66.7)
Fe_2O_3	21.88	48.20	—	—	8.666	104.0
	48.00	—	—	—	12.652	238.3
	23.49	18.6	—	−3.55	9.021	119.9
Ga_2O	36.00	—	—	—	11.979	187.6
	31.71	1.8	—	—	8.467	159.7
	(13.8)	(8.6)	—	—	(4.497)	(58.7)
Ga_2O_3	(21.5)	—	—	—	(−0.559)	(94.1)
	(14)	—	—	—	(−28.06)	(22.3)
GeO	11.77	25.2	—	—	(4.630)	(54.35)
	(35.5)	—	—	—	(−20.66)	(173.2)
GeO_2	(10.4)	(2.6)	—	(−0.5)	(3.384)	(47.8)
	(8.2)	(0.4)	—	(−0.2)	(−49.67)	(−20.2)
H_2O	11.2	7.17	—	—	(3.658)	(53.5)
	(21.7)	—	—	—	(1.149)	(111.9)
HfO_2	18.03	—	—	—	5.376	86.01
Hg_2O	7.17	2.56	—	−0.08	−8.290	−3.56
HgO	17.39	2.08	—	−3.48	6.445	87.48
	(15.6)	(28)	—	—	(4.776)	(59.7)
	(8.5)	(7.0)	—	—	(2.846)	(33.7)
In_2O	(14.7)	(7.8)	—	—	(4.730)	(58.1)
	(22)	—	—	—	(3.206)	(92.6)
	(15)	—	—	—	(−18.39)	(25.8)
InO	(10.0)	(3.2)	—	—	(3.124)	(43.4)
	(14)	—	—	—	(1.615)	(64.9)
	(9.0)	—	—	—	(−68.38)	(−3.1)
In_2O_3	(22.6)	(6.0)	—	—	(7.005)	(100.5)
	(35)	—	—	—	(−0.195)	(172.8)

THERMODYNAMIC PROPERTIES OF THE OXIDES

Oxide	Phase	Temperature of Transition (°K)	Heat of Transition kcal/mole	Entropy of Transition (e.u.)	Entropy at 298°K (e.u.)	Oxide	cal/g mole a	cal/g mole b	cal/g mole c	cal/g mole d	A (kcal/mole)	B (e.u.)
Ir_2O_3	solid	(1450)	(10)	(6.8)	(26.5)	Ir_2O_3	(21.8)	(14.4)	—	—	(7.140)	(102.0)
	liquid	(2250)	(50)	(22)	—		(35)	—	—	—	(0.706)	(170.3)
	gas	—	—	—			(20)	(10)	—	—	(−57.73)	(54.8)
IrO_2	solid	dec. 1373	—	—	(15.9)	IrO_2	9.17	15.20	—	—	3.410	40.9
K_2O	solid	(980)	(6.8)	(6.9)	(23)	K_2O	(15.9)	(6.4)	—	—	(5.025)	(69.5)
	liquid	dec.	—	—	—		(22)	—	—	—	(1.130)	(98.3)
K_2O_2	solid	763	(7.0)	(9.2)	(27)	K_2O_2	(20.8)	(5.4)	—	—	(6.442)	(93.1)
	liquid	(1800)	(45)	(25)	—		(29)	—	—	—	(4.127)	(134.2)
	gas	—	—	—			(20)	—	—	—	(−57.07)	(41.7)
K_2O_3	solid	703	(6.1)	(8.7)	(33.5)	K_2O_3	(19.1)	(23.2)	—	—	(6.750)	(82.2)
	liquid	(975)	(25)	(26)	—		(35.5)	—	—	—	(6.447)	(164.7)
	gas	—	—	—			(20)	(5.0)	—	—	(−31.29)	(37.3)
KO_2	solid	653	(4.9)	(7.5)	27.9	KO_2	(15.0)	(12.0)	—	—	(5.006)	(61.1)
	liquid	dec.	—	—	—		(24)	—	—	—	(3.424)	(105.5)
La_2O_3	solid	2590	(18)	(7)	(36.5)	La_2O_3	28.86	3.076	—	−3.275	9.840	(130.7)
Li_2O	solid	2000	(14)	(7)	9.06	Li_2O	(11.4)	(5.4)	—	—	(3.639)	(57.5)
	liquid	2600	(56)	(22)	—		(21)	—	—	—	(−1.961)	(112.7)
Li_2O_2	solid	dec. 470	—	—	(16.5)	Li_2O_2	(17.0)	(5.4)	—	—	(5.309)	(82.0)
MgO	solid	3075	18.5	5.8	6.4	MgO	10.86	1.197	—	−2.087	3.991	57.0
MgO_2	solid	dec. 361	—	—	(20.5)	MgO_2	(12.1)	(2.4)	—	—	(3.714)	(49.2)
MnO	solid	2058	13.0	6.32	14.27	MnO	11.11	1.94	—	−0.88	3.689	50.10
	liquid	dec.	—	—	—		(13.5)	—	—	—	(−8.543)	(58.02)
Mn_3O_4	solid, α	1445	4.97	3.44	35.5	Mn_3O_4	34.64	10.82	—	−2.20	11.312	166.3
	solid, β	1863	(33)	(18)	—		50.20	—	—	—	17.376	260.4
	liquid	(2900)	(75)	(26)	—		(49)	—	—	—	(−17.86)	(233.4)
Mn_2O_3	solid	dec. 1620	—	—	26.4	Mn_2O_3	24.73	8.38	—	−3.23	8.829	118.8
MnO_2	solid	dec. 1120	—	—	12.7	MnO_2	16.60	2.44	—	−3.88	6.359	84.8
MoO_2	solid	(2200)	(16)	(7.3)	(14.5)	MoO_2	(16.2)	(3.0)	—	(−3.0)	(5.973)	(80.4)
	liquid	dec. 2250	—	—	—		(23)	—	—	—	(−2.463)	(118.4)
MoO_3	solid	1068	12.54	11.74	18.68	MoO_3	13.6	13.5	—	—	4.655	62.83
	liquid	1530	33	22	—		(28.4)	—	—	—	(0.222)	(139.88)
	gas	—	—	—			(18.1)	—	—	—	(−48.54)	(42.8)
N_2O	gas	—	—	—	52.58	N_2O	10.92	2.06	—	−2.04	4.032	11.40
Na_2O	solid	1193	(7.1)	(6.0)	17.4	Na_2O	15.70	5.40	—	—	4.921	73.7
	liquid	dec.	—	—	—		(22)	—	—	—	(1.494)	(105.9)
Na_2O_2	solid	dec. 919	—	—	22.6	Na_2O_2	(20.2)	(3.8)	—	—	(6.192)	(93.6)
NaO_2	solid	(825)	(6.2)	(7.5)	27.7	NaO_2	(16.2)	(3.6)	—	—	(4.990)	(65.7)
	liquid	(1300)	(28)	(22)	—		(23)	—	—	—	(3.175)	(100.9)
	gas	—	—	—			(15)	—	—	—	(−35.22)	(22.0)
NbO	solid	(2650)	(16)	(6.0)	(12)	NbO	(9.6)	(4.4)	—	—	(3.058)	(44.0)
NbO_2	solid	(2275)	(16)	(7.0)	(12.7)	NbO_2	(17.1)	(1.6)	—	(−2.8)	(6.109)	(84.6)
	liquid	(3800)	(85)	(22)	—		(24)	—	—	—	(1.033)	(127.2)
Nb_2O_5	solid	1733	(28)	(16)	32.8	Nb_2O_5	21.88	28.2	—	—	7.776	100.3
	liquid	(3200)	(80)	(25)	—		(44.2)	—	—	—	(−24.09)	(201.6)
Nd_2O_3	solid	2545	(22)	(8.8)	(35.3)	Nd_2O_3	28.99	5.760	—	(−4.159)	10.295	(133.9)
NiO	solid	2230	(12.1)	(5.43)	9.22	NiO	13.69	0.83	—	−2.915	5.097	70.67
	liquid	dec.	—	—	—		(14.3)	—	—	—	(−7.861)	(67.91)
NpO_2	solid	(2600)	(15)	(5.7)	19.19	NpO_2	(17.7)	(3.2)	—	(−2.6)	(6.292)	(84.08)
Np_2O_5	solid	dec. 800–900°K	—	—	(43)	Np_2O_5	(32.4)	(12.6)	—	—	(10.22)	(145.4)
OsO_2	solid	dec. 923	—	—	(14.5)	OsO_2	(11.5)	(6.0)	—	—	(3.696)	(52.8)
OsO_4	solid	313.3	3.41	10.9	34.7	OsO_4	(16.4)	(23.1)	—	(−2.4)	(6.726)	(67.0)
	liquid	403	9.45	23.4	—		(33)	—	—	—	(6.612)	(143.0)
	gas	—	—	—	—		16.46	8.60	—	−4.6	(−7.644)	(25.3)
P_2O_3	liquid	448.5	4.5	10	(34)	P_2O_3	(34.5)	—	—	—	(10.287)	(162.6)
	gas	—	—	—	—		(15)	(10)	—	—	(−1.953)	(38.0)
PO_2	solid	(350)	(2.7)	(7.7)	(11.5)	PO_2	(11.3)	(5.0)	—	—	(3.591)	(54.4)
	liquid	(dec)	—	—	—		(20)	—	—	—	(3.640)	(95.9)
P_2O_5	solid	631	8.8	13.9	33.5	P_2O_5	8.375	5.40	—	—	4.897	30.3
	gas	—	—	—			36.80	—	—	—	3.284	165.6
PaO_2	solid	(2560)	(20)	(7.8)	(17.8)	PaO_2	(14.4)	(2.6)	—	—	(4.409)	(65.0)
Pa_2O_5	solid	(2050)	(26)	(13)	(37.5)	Pa_2O_5	(28.4)	(11.4)	—	—	(8.975)	(127.7)
	liquid	(3350)	(95)	(28)	—		(48)	—	—	—	(−0.800)	(241.1)
PbO	solid, red	762	(0.4)	(0.5)	16.2	PbO	10.60	4.00	—	—	3.338	45.4
	solid, yellow	1159	2.8	2.4	—		9.05	6.40	—	—	2.454	36.4
	liquid	1745	51	29	—		(14.6)	—	—	—	1.788	65.7
	gas	—	—	—			(8.1)	(0.4)	—	—	(−59.94)	(−11.0
Pb_3O_4	solid	dec.	—	—	50.5	Pb_3O_4	(31.1)	(17.6)	—	—	(10.055)	(132.0)
PbO_2	solid	dec.	—	—	18.3	PbO_2	12.7	7.80	—	—	4.133	56.4
PdO	solid	dec. 1150	—	—	(9.1)	PdO	3.30	14.2	—	—	1.615	(13.9)
PoO_2	solid	(825)	(5.5)	(6.7)	(17)	PoO_2	(14.3)	(5.6)	—	—	(4.513)	(66.1)
	liquid	(dec.)	—	—	—		(22)	—	—	—	(3.460)	(106.5)
Pr_2O_3	solid	(2200)	(22)	(10)	(35.5)	Pr_2O_3	(29.0)	(4.0)	—	(−4.0)	(10.166)	(133.2)
	liquid	(4000)	(90)	(23)	—		(36)	—	—	—	(−6.298)	(168.3)
PrO_2	solid	dec. 700	—	—	(17)	PrO_2	(17.6)	(3.4)	—	(−2.8)	(6.338)	(85.9)
PtO	solid	dec. 780	—	—	(13.5)	PtO	(9.0)	(6.4)	—	—	(2.968)	(39.7)
Pt_3O_4	solid	(dec.)	—	—	(41)	Pt_3O_4	(30.8)	(17.4)	—	—	(9.957)	(139.7)
PtO_2	solid	723	(4.6)	(6.4)	(16.5)	PtO_2	(11.1)	(9.6)	—	—	(3.736)	(49.6)
	liquid	dec. 750	—	—	—		(21)	—	—	—	(3.785)	(101.5)
PuO	solid	(1290)	(7.2)	(5.6)	(20)	PuO	(12.0)	(2.4)	—	—	(3.685)	(49.1)
	liquid	(2325)	(47)	(20)	—		(14.5)	—	—	—	(−2.287)	(58.3)
	gas	—	—	—			(8.9)	—	—	—	(−62.307)	(−5.3)
Pu_2O_3	solid	(1880)	(16)	(8.5)	(38)	Pu_2O_3	(21.2)	(18.2)	—	—	(7.130)	(88.2)
	liquid	(3250)	(75)	(23)	—		(40)	—	—	—	(−5.691)	(187.2)
PuO_2	solid	(2400)	(15)	(6.2)	(19.7)	PuO_2	(17.1)	(3.4)	—	(−2.6)	(6.122)	(80.2)
	liquid	(3500)	(90)	(26)	—		(20.5)	—	—	—	(−10.62)	(92.2)
RaO	solid	(>2500)	—	—	(17)	RaO	(10.5)	(2.0)	—	—	(3.220)	(43.4)
Rb_2O	solid	(910)	(5.7)	(6.3)	(27)	Rb_2O	(15.4)	(5.8)	—	—	(4.850)	(62.5)
	liquid	dec.	—	—	—		(22)	—	—	—	(2.754)	(95.9)
Rb_2O_2	solid	843	(7.3)	(8.7)	(27.5)	Rb_2O_2	(20.9)	(8.0)	—	—	(6.587)	(94.0)
	liquid	(dec.)	—	—	—		(29)	—	—	—	(3.273)	(133.2)
Rb_2O_3	solid	762	(7.6)	(10)	(32.5)	Rb_2O_3	(20.5)	(13.0)	—	—	(6.690)	(88.2)
	liquid	dec.	—	—	—		(34)	—	—	—	(5.603)	(157.8)
RbO_2	solid	685	(4.1)	(6.0)	(21.5)	RbO_2	(13.8)	(6.4)	—	—	(4.399)	(59.0)
	liquid	dec.	—	—	—		(21)	—	—	—	(3.720)	(95.7)
ReO_2	solid	(1475)	(12)	(8.1)	(15)	ReO_2	(10.8)	(9.8)	—	—	(3.656)	(49.5)

THERMODYNAMIC PROPERTIES OF THE OXIDES

Oxide	Phase	Temperature of Transition (°K)	Heat of Transition kcal/mole	Entropy of Transition (e.u.)	Entropy at 298°K (e.u.)	Oxide	cal/g mole a	cal/g mole b	cal/g mole c	cal/g mole d	A (kcal/mole)	B (e.u.)
ReO_2	liquid	(3250)	(80)	(25)	—	ReO_2	(24.5)	—	—	—	(1.204)	(127.0)
ReO_3	solid	433	5.2	12	19.8	ReO_3	(18.0)	(5.8)	—	—	(5.625)	(84.5)
	liquid	dec.	—	—	—		29	—	—	—	(4.644)	(136.8)
Re_2O_7	solid	569	15.8	27.8	44	Re_2O_7	(41.8)	(14.8)	—	(−3.0)	(14.127)	(200.3)
	liquid	635.5	17.7	27.9	—		(65.7)	—	—	—	(9.203)	(314.7)
	gas	—	—	—	—		(38.2)	—	—	—	(−25.97)	(109.3)
ReO_4	solid	420	(4.2)	(10)	(34.5)	ReO_4	(21.4)	(10.8)	—	(−2.0)	(7.531)	(91.8)
	liquid	(460)	(9.3)	(20)	—		(33)	—	—	—	(6.775)	(146.7)
	gas	—	—	—	—		(16.5)	(8.6)	—	(−5.0)	(−8.118)	(30.6)
Rh_2O	solid	dec. 1400	—	—	(25.5)	Rh_2O	15.59	6.47	—	—	4.936	(65.3)
RhO	solid	dec. 1394	—	—	(12)	RhO	(9.84)	(5.53)	—	—	(3.179)	(45.7)
Rh_2O_3	solid	dec. 1388	—	—	(23)	Rh_2O_3	20.73	13.80	—	—	6.794	(99.2)
RuO_2	solid	dec. 1400	—	—	(12.5)	RuO_2	(11.4)	(6.0)	—	—	3.666	(54.2)
RuO_4	solid	300	(3.2)	(11)	(32.5)	RuO_4	(20)	—	—	—	(5.963)	(81.5)
	liquid	dec.	—	—	—		(33)	—	—	—	(6.663)	(144.9)
SO_2	gas	—	—	—	59.40	SO_2	11.4	1.414	—	−2.045	4.148	7.12
Sb_2O_3	solid	928	14.74	15.88	29.4	Sb_2O_3	19.10	17.1	—	—	6.455	84.5
	liquid	1698	8.92	5.25	—		(36)	—	—	—	(0.035)	(168.2)
	gas	—	—	—	—		(20.8)	—	—	—	(−34.70)	(49.9)
SbO_2	solid	dec.	—	—	15.2	SbO_2	11.30	8.1	—	—	3.725	51.6
Sb_2O_5	solid	dec.	—	—	29.9	Sb_2O_5	(22.4)	(23.6)	—	—	(7.728)	(104.8)
Sc_2O_3	solid	(2500)	(23)	(9.3)	24.8	Sc_2O_3	23.17	5.64	—	—	7.159	108.9
SeO	solid	(1375)	(7.6)	(5.5)	(11)	SeO	(9.1)	(3.8)	—	—	(2.882)	(42.0)
	liquid	(2075)	(45)	(22)	—		(15.5)	—	—	—	(0.490)	(77.5)
	gas	—	—	—	—		8.20	0.50	—	−0.80	(−58.54)	(0.7)
SeO_2	solid	603	(24.5)	(40.6)	(15)	SeO_2	(12.8)	(6.1)	—	(−0.2)	(4.150)	(59.9)
	gas	—	—	—	—		(14.5)	—	—	—	(−20.45)	(26.4)
SiO	solid	(2550)	(12)	(4.7)	(6.5)	SiO	(7.3)	(2.4)	—	—	(2.283)	(35.8)
SiO_2	solid, β	856	0.15	0.18	10.06	SiO_2	11.22	8.20	—	−2.70	4.615	57.83
	solid, α	1883	2.04	1.08	—		14.41	1.94	—	—	4.602	73.67
	liquid	dec. 2250	—	—	—		(20)	—	—	—	(9.649)	(111.08)
Sm_2O_3	solid	(2150)	(20)	(9.3)	(36.5)	Sm_2O_3	(25.9)	(7.0)	—	—	(8.033)	(113.2)
	liquid	(3800)	(80)	(21)	—		(36)	—	—	—	(−6.431)	(166.3)
SnO	solid	(1315)	(6.4)	(4.9)	13.5	SnO	9.40	3.62	—	—	2.964	41.1
	liquid	(1800)	(60)	(33)	—		(14.5)	—	—	—	(0.141)	(68.1)
	gas	—	—	—	—		(9.0)	—	—	—	(−69.76)	(−6.4
SnO_2	solid	1898	(11.39)	(5.95)	12.5	SnO_2	17.66	2.40	—	−5.16	7.103	91.7
	liquid	(3200)	(75)	(23)	—		(22.5)	—	—	—	(0.304)	(117.7)
SrO	solid	2703	16.7	6.2	13.0	SrO	12.34	1.120	—	−1.806	4.335	58.7
SrO_2	solid	dec. 488	—	—	(14.8)	SrO_2	(16.8)	(2.2)	—	(−3.0)	(6.113)	(83.3)
Ta_2O_5	solid	2150	(16)	(7.4)	34.2	Ta_2O_5	29.2	10.0	—	—	9.151	135.2
	liquid	—	—	—	—		(46)	—	—	—	(6.158)	(235.1)
TcO_2	solid	(2400)	(18)	(7.5)	(13.5)	TcO_2	(10.4)	(9.2)	—	—	(3.510)	(48.6)
	liquid	(4000)	(105)	(26)	—		(25)	—	—	—	(−5.946)	(132.7)
TcO_3	solid	(dec. <1200)	—	—	(19.5)	TcO_3	(19.4)	(5.2)	—	(−2.0)	(6.686)	(93.7)
Tc_2O_7	solid	392.7	(11)	(28)	(42.5)	Tc_2O_7	(39.1)	(18.6)	—	(−2.4)	(13.29)	(187.2)
	liquid	583.8	(14)	(24)	—		(64)	—	—	—	(10.02)	(299.8)
	gas	—	—	—	—		(25)	(28)	—	—	(−21.98)	(43.8)
TeO	solid	(1020)	(7.1)	(7.0)	(13)	TeO	(8.6)	(6.2)	—	—	(2.840)	(37.8)
	liquid	(1775)	(50)	(28)	—		(15.5)	—	—	—	(−0.448)	(72.3)
	gas	—	—	—	—		(8.9)	—	—	—	(−62.16)	(−5.2)
TeO_2	solid	1006	3.2	3.2	16.99	TeO_2	13.85	6.87	—	—	4.435	63.97
	liquid	dec.	—	—	—		(20)	—	—	—	(3.940)	(96.4)
ThO	solid	(2150)	(13)	(6.0)	(16)	ThO	(11.0)	(2.4)	—	—	(3.386)	(47.4)
	liquid	(3250)	(65)	(20)	—		(15)	—	—	—	(−6.561)	(66.9)
ThO_2	solid	3225	(18)	(5.6)	15.59	ThO_2	16.45	2.346	—	−2.124	5.721	80.03
TiO	solid, α	1264	0.82	0.65	8.31	TiO	10.57	3.60	—	−1.86	3.935	54.03
	solid, β	dec. 2010	—	—	—		11.85	3.00	—	—	4.108	61.71
Ti_2O_3	solid, α	473	0.215	0.455	18.83	Ti_2O_3	7.31	53.52	—	—	4.559	38.78
	solid, β	2400	(24)	(10)	—		34.68	1.30	—	−10.20	13.605	184.48
	liquid	3300	—	—	—		(37.5)	—	—	—	(−7.796)	(193.2)
Ti_3O_5	solid, α	450	2.24	4.98	30.92	Ti_3O_5	35.47	29.50	—	—	11.887	179.98
	solid, β	(2450)	(50)	(20)	—		41.60	8.00	—	—	10.230	202.80
	liquid	(3600)	(85)	(24)	—		(60)	—	—	—	(−18.701)	(306.4)
TiO_2	solid	2128	(16)	(7.5)	12.01	TiO_2	17.97	0.28	—	−4.35	6.829	92.92
	liquid	dec. 3200	—	—	—		(21.4)	—	—	—	(−2.610)	(111.08)
Tl_2O	solid	573	(5.0)	(8.7)	23.8	Tl_2O	(15.8)	(6.0)	—	(−0.3)	(5.078)	(68.2)
	liquid	773	(17)	(22)	—		(22.1)	—	—	—	(2.651)	(96.0)
	gas	—	—	—	—		(13.7)	—	—	—	(−20.94)	(18.0)
Tl_2O_3	solid	990	(12.4)	(13)	(33.5)	Tl_2O_3	(23.0)	(5.0)	—	—	(7.080)	(99.0)
	liquid	(dec)	—	—	—		(35.5)	—	—	—	(4.604)	(167.8)
UO	solid	(2750)	(14)	(5.1)	(16)	UO	(10.6)	(2.0)	—	—	(3.249)	(45.0)
UO_2	solid	3000	—	—	18.63	UO_2	19.20	1.62	—	−3.957	7.124	93.37
U_3O_8	solid	dec.	—	—	(66)	U_3O_8	(65)	(7.5)	—	(−10.9)	(23.37)	(312.7)
UO_3	solid	dec. 925	—	—	23.57	UO_3	22.09	2.54	—	−2.973	7.696	104.72
VO	solid	(2350)	(15)	(6.4)	9.3	VO	11.32	1.61	—	−1.26	3.869	56.4
	liquid	(3400)	(70)	(21)	—		(14.5)	—	—	—	(−8.157)	(70.9)
V_2O_3	solid	2240	(24)	(11)	23.58	V_2O_3	29.35	4.76	—	−5.42	10.780	148.12
	liquid	dec. 3300	—	—	—		(38)	—	—	—	(−6.028)	(193.4)
V_3O_5	solid	(2100)	(42)	(20)	(32)	V_3O_5	(36)	(30)	—	—	(12.07)	(182.1)
	liquid	(dec.)	—	—	—		(55.6)	—	—	—	(−54.72)	(249.1)
VO_2	solid, α	345	1.02	2.96	12.32	VO_2	14.96	—	—	—	4.460	72.92
	solid, β	1818	13.60	7.48	—		17.85	1.70	—	−3.94	5.680	89.09
	liquid	dec. 3300	—	—	—		25.50	—	—	—	2.962	135.87
V_2O_5	solid	943	15.56	16.50	313	V_2O_5	46.54	−3.90	—	−13.22	18.136	240.2
	liquid	(2325)	(63)	(27)	—		45.60	—	—	—	2.122	220.1
	gas	—	—	—	—		(40)	—	—	—	(−73.90)	(149.6)
WO_2	solid	(1543)	(11.5)	(7.45)	(15)	WO_2	(17.6)	(4.2)	—	(−4.0)	(6.772)	(88.8)
	liquid	dec. 2125	—	—	—		(24)	—	—	—	(−0.112)	(121.8)
WO_3	solid	1743	(17)	(9.8)	19.90	WO_3	17.33	7.74	—	—	5.511	81.15
	liquid	(2100)	(43)	(20)	—		(30)	—	—	—	(−1.162)	(152.5)
	gas	—	—	—	—		(18)	—	—	—	(−69.36)	(40.2)
Y_2O_3	solid	(2500)	(25)	(10)	(29.5)	Y_2O_3	(26.0)	(8.2)	—	(−2.2)	(8.846)	(122.3)
ZnO	solid	dec.	—	—	10.4	ZnO	11.71	1.22	—	−2.18	4.277	57.88
ZrO_2	solid, α	1478	1.420	0.961	12.03	ZrO_2	16.64	1.80	—	−3.36	6.168	85.21
	solid, β	2950	20.8	7.0	—		17.80	—	—	—	4.270	89.96

VALUES OF CHEMICAL THERMODYNAMIC PROPERTIES

All values of energy in these tables are expressed, insofar as possible, in terms of the thermochemical calorie, now defined in terms of the absolute joule. 1 thermochemical calorie = 1 calorie = 4.1840 absolute joule = 4.1833 international joule. The notations used in these tables are as follows:

ΔHf°	=	the standard heat of formation of a given substance from its elements at 25°C., kilo -cal/g mole.
ΔFf°	=	the standard free energy of formation of a given substance from its elements at 25°C., kilo -cal/g mole.
$\log_{10}$ Kf	=	the logarithm of the equilibrium constant for the reaction for forming a given substance from its elements at 25°C.
S°	=	the entropy of the given substance in its thermodynamic reference state at the reference temperature at 25°C., cal/deg. mole.
c	=	crystalline; in certain cases where a substance exists in more than one crystalline form there is an indication as to which form is concerned.
g	=	gaseous.
am	=	amorphous.
aq	=	aqueous; unless otherwise indicated the aqueous solution is taken as the hypothetical ideal state of unit molality.
gls	=	glass.
lq	=	liquid.
ppi	=	precipitate.

The values in these tables were taken from Circular of the National Bureau of Standards 500. "Selected Values of Chemical Thermodynamic Properties, issued February 1, 1952.

RELATIONSHIP TO SI UNITS

The symbols cal. mole^{-1} deg^{-1} and gibbs/mol are identical and refer to units of calories per degree-mole. These units can be converted to SI units of joules per degree-mole by multiplying the tabulated values by 4.184. Similarly values in kilocalories per mole can be converted to joules per mole by multiplying with the factor 4184. For further discussions of the SI system and for conversions from other units the reader should consult Pure and Applied Chemistry, *21,* 1 (1970).

Substance	State	ΔHf°	ΔFf°	Log₁₀ Kf	S°
Aluminum					
Al	*g*	75.00	65.3	−47.86	39.303
	c	0.00	0.00	0.000	6.77
Al^{+++}	*aq*	1307.44	....	....	....
$AlBr_3$	*c*	−125.8	−120.7	88.47	44
	aq	−211.9	....	....	....
Al_4C_3	*c*	−30.9	−29.0	21.26	25
$Al(CH_3)_3$	*lq*	−26.9	....	....	....
Al_2Cl_6	*g*	−303.6	....	....	....
$AlCl_3$	*c*	−166.2	−152.2	111.56	40
	aq, 600	−245.5	....	....	....
$AlCl_3 \cdot 6H_2O$	*c*	−641.1	−542 4	397.57	90
AlF_3	*c*	−311	−294	215.5	23
$AlF_3 \cdot 3H_2O$	*c*	−549.1	−490.4	359.46	50
	aq	−361.4	....	....	....
$AlF_3 \cdot \frac{1}{2}H_2O$	*c*	−357.4	−333.6	244.52	28
AlI_3	*c*	−75.2	−75.0	54.97	48
	aq	−165.8	....	....	....
AlN	*c*	−57.7	−50.1	36.72	5
$Al(NO_2)_3$	*aq*	−273.65	....	....	....
$Al(NO_3)_3 \cdot 6H_2O$	*c*	−680.65	−525.82	385.419	111.8
$Al(NO_3)_3 \cdot 9H_2O$	*c*	−897.34	−700.2	513.24	136
$Al_2 \cdot O_3(\alpha)$*	*c*	−399.09	−376.77	276.167	12.19
(γ)	*c*	−384.84	....	....	....
$Al_2O_3 \cdot H_2O$	*c*	−471	−435	318.8	23.15
$Al_2O_3 \cdot 3H_2O$†	*c*	−613.7	−547.9	401.60	33.51
$Al(OH)_3$	*am*	−304.2	....	....	....
Al_2S_3	*c*	−121.6	−117.7	86.27	23
$Al_2(SO_4)_3$	*c*	−820.98	−738.99	541.670	57.2
	aq	−897.1	....	....	....
$Al_2(SO_4)_3 \cdot 6H_2O$	*c*	−1268.14	−1105.14	810.054	112.1
$Al_2(SO_4)_3 18H_2O$	*c*	−2118.5	....	....	....
$AlNH_4(SO_4)_2$	*c*	−561.24	−485.95	356.195	51.7
	aq	−591.74	....	....	....
$AlNH_4(SO_4)_2 \cdot 12H_2O$	*c*	−1419.40	−1179.02	864.207	166.6
Ammonium					
NH_3	*g*	−11.04	−3.976	2.914	46.01
	aq	−19.32	−6.37	−4.669	26.3
	aq ∞	−19.32	....	....	....
NH_4^+	*aq*	−31.74	−19.00	13 027	26.97
NH_4OH	*aq*	−87.64	....	....	....
$NH_4H_2AsO_4$	*c*	−251.47	−197.24	144.574	41.12
	aq	247.9	....	....	....
$(NH_4)_2HAsO_4$	*c*	−280.24	....	....	....
	aq	−278.3	....	....	....
$(NH_4)_3AsO_4$	*c*	−306.11	....	....	....
	aq	−303	....	....	....
$(NH_4)_3AsO_4 \cdot 3H_2O$	*c*	−516.6	....	....	....
NH_4BO_2	*aq*	−215.2	....	....	....
NH_4BO_3	*aq*	−193.7	....	....	....
$NH_4BO_3 \cdot H_2O$	*c*	−270.8	....	....	....
$(NH_4)_2HBO_3$	*aq*	−305.5	....	....	....
NH_4Br	*c*	−64.61	....	....	....
	aq, ∞	−60.74	....	....	....
$(NH_4)_2CO_3$	*aq*	−225.11	−164.22	120.371	41.2
NH_4HCO_3	*c*	−203.7	....	....	....
	aq	−196.92	−159.31	116.772	49.7
$NH_4CO_2NH_2$	*c*	−154.21	−109.47	80.240	39.70
	aq	−150.4	....	....	....
NH_4CN	*c*	0.0	....	....	....
	aq	4.6	20.4	....	....
NH_4CNO	*c*	−74.7	....	....	....
	aq	−68.5	....	....	....
NH_4CNS	*c*	−20.0	....	....	....
	aq	−14.5	4.4	....	....
$NH_4C_2H_3O_2$	*c*	−147.8	....	....	....
	aq, 400	−148.1	....	....	....
$(NH_4)_2C_2O_4$	*c*	−268.72	....	....	....
	aq, 2100	−260.6	....	....	....

Substance	State	ΔHf°	ΔFf°	Log₁₀ Kf	S°
Ammonium					
$(NH_4)_2C_2O_4 \cdot H_2O$	*c*	−340.62	....	....	....
$NH_4HC_2O_4$	*aq*	−227.02	....	....	....
NH_4Cl	*c*	−75.38	−48.73	35.718	22.6
	aq, ∞	−71.76	....	....	....
NH_4ClO_4	*c*	−69.42	....	....	....
NH_4F	*c*	−111.6	....	....	....
	aq, ∞	−110.40	....	....	....
NH_4I	*c*	−48.30	....	....	....
	aq, ∞	−45.11	....	....	....
NH_4NO_2	*c*	−63.1	....	....	....
	aq	−57.1	....	....	....
NH_4NO_3	*c*	−87.27	....	....	....
	aq	−81.11	....	....	....
$NH_4H_2PO_4$	*c*	−346.75	−290.46	212.903	36.32
	aq, 500	−342.91	....	....	....
$(NH_4)_2HPO_4$	*c*	−376.12	....	....	....
	aq, 500	−373.04	....	....	....
$(NH_4)_3PO_4$	*c*	−401 8	....	....	....
	aq, 660	−394.0	....	....	....
$(NH_4)_3PO_4 \cdot 3H_2O$	*c*	−612.8	....	....	....
$(NH_4)_2PtCl_4$	*c*	−195.3	....	....	....
	aq	−186.9	....	....	....
$(NH_4)_2S$	*aq*	−54.5	....	....	....
NH_4HS	*c*	−38.10	....	....	....
	aq	−35.1	....	....	....
NH_4S_4	*c*	−34.0	....	....	....
	aq	−29 7	....	....	....
$(NH_4)_2S_5$	*c*	−69.4	....	....	....
	aq	−65.4	....	....	....
$(NH_4)_2SO_3$	*c*	−212.0	....	....	....
	aq	−211.3	....	....	....
$(NH_4)_2SO_3 \cdot H_2O$	*c*	−284.22	....	....	....
NH_4HSO_3	*c*	−183.8	....	....	....
	aq	−181.5	....	....	....
$(NH_4)_2SO_4$	*c*	−281.86	−215.19	157.732	52.65
	aq, ∞	−280.38	....	....	....
NH_4HSO_4	*c*	−244.83	....	....	....
	aq	−245.6	....	....	....
$(NH_4)_2S_2O_8$	*c*	−396.4	....	....	....
	aq	−387.8	....	....	....
$(NH_4)_2Se$	*aq*	−26.2	....	....	....
NH_4HSe	*aq*	−5.6	....	....	....
$(NH_4)_2SiF_6$	*c*	−629.7	....	....	....
	aq	−622.0	....	....	....
NH_4VO_3	*c*	−251.2	−211.8	155.25	33.6
NH_2OH	*c*	−25.5	....	....	....
	aq	−21.7	....	....	....
$NH_2OH \cdot HCl$	*c*	−74.0	....	....	....
	aq	−70.7	....	....	....
$NH_2OH \cdot HNO_3$	*c*	−86.3	....	....	....
	aq	−80.5	....	....	....
$(NH_2OH)_2 \cdot H_2SO_4$	*c*	−282.5	....	....	....
	aq	−276.7	....	....	....
$NH_2OH \cdot H_2SO_4$	*c*	−245.1	....	....	....
	aq	−244.4	....	....	....
N_2H_4	*lq*	12.05	....	....	....
	aq, 300	8.16	....	....	....
$N_2H_4 \cdot H_2O$	*lq*	−57.95	....	....	....
$N_2H_4 \cdot HCl$	*c*	−46.93	....	....	....
	aq	−41.7	....	....	....
$N_2H_4 \cdot 2HCl$	*c*	−90.0	....	....	....
	aq	−84.0	....	....	....
$(N_2H_4)_2 \cdot H_2SO_4$	*aq*	−217.8	....	....	....
$N_2H_4 \cdot H_2SO_4$	*c*	−231.6	....	....	....
	aq	−223.4	....	....	....
HN_3	*g*	70.3	78.5	−57.54	56.74
	aq, 100	61.51	....	....	....

* Corundum.
† Hydrargillite.

VALUES OF CHEMICAL THERMODYNAMIC PROPERTIES (Continued)

Substance	State	ΔHf°	ΔFf°	Log_{10} Kf	S°
Antimony					
Sb	*g*	60.8	51.1	−37.46	43.06
	c, III	0.00	0.00	0.000	10.5
SbO^+	*aq*	....	−42.0	30.79	....
Sb_2	*g*	52.	40	−29.32	60.9
$SbBr_3$	*c*	−62.1	....	....	....
$SbCl_3$	*g*	−75.2	−72.3	52.99	80.8
	c	−91.34	−77.62	56.894	44.5
$SbCl_5$	*g*	−93.9	....	....	....
	lq	−104.8	....	....	....
$SbOCl$	*c*	−90.8	....	....	....
SbF_3	*c*	−217.2	....	....	....
	aq	−216.1	....	....	....
SbI_3	*c*	−23.0	....	....	....
	aq	−22.6	....	....	....
Sb_2O_3	*aq*	−166.5	....	....	....
Sb_2O_4	*c*	−214	−188	137.8	30.3
Sb_2O_5	*c*	−234.4	−200.5	146.96	29.9
	aq	−226.4	....	....	....
Sb_2O_6	*c*	−336.8	−298.0	218.43	59.8
Sb_2S_3 (black)	*c*	−43.5	....	....	....
(orange)	*am*	−36.0	....	....	....
$Sb_2(SO_4)_3$	*c*	575.3	....	....	....
H_3SbO_4	*aq*	−215.7	....	....	....
Argon					
A	*g*	0.00	0.00	0.000	36.983
$A \cdot 5H_2O$	*c*	−357.2	....	....	....
Arsenic					
As (gray)	*c*	0.00	0.00	0.000	8.4
(β)	*am*	1.0	....	....	....
(yellow)	*c*	3.53	....	....	....
AsO^+	*aq*	....	−39.1	28.69	....
As_2	*g*	29.6	17.5	−12.83	57.3
As_4	*g*	35.7	25.2	−18.47	69
$AsBr_3$	*c*	−46.61	....	....	....
$AsCl_3$	*g*	−71.5	−68.5	50.21	78.2
	lq	−80.2	−70.5	51.68	55.8
AsF_3	*g*	−218.3	−214.7	157.37	69.08
	lg	−226.8	−215.5	157.96	43.31
AsH_3	*g*	41.0	....	....	....
$AsH_3 \cdot 6H_2O$	*c*	−386.7	....	....	....
AsI_3	*c*	−13.7	....	....	....
As_2O_5	*c*	−218.6	−184.6	135.41	25.2
	aq	−224.6	....	....	....
$As_2O_5 \cdot 4H_2O$	*c*	−500.3	....	....	....
$3As_2O_5 \cdot 5H_2O$	*c*	−1007.5	....	....	....
$As_2O_3 \cdot As_2O_5$	*c*	−351.1	....	....	....
As_4O_6 (oct)	*c*	−313.94	−275.36	201.835	51.2
(mon)	*c*	−312.8	....	....	....
	aq	−299.4	....	....	....
As_2S_3	*c*	−35	....	....	....
H_3AsO_3	*aq*	−177.3	−152.9	112	47.0
H_3AsO_4	*c*	−215.2	....	....	....
	aq	−214.8	−183.8	134:72	49.3
Barium					
Ba	*g*	41.96	34.60	−25.361	40.699
	c	0.00	0.00	0.000	16
Ba^{++}	*aq*	−128.67	−134.0	98.22	3
$Ba_3(AsO_4)_2$	*c*	−817.8	....	....	....
$BaHAsO_4 \cdot H_2O$	*c*	−411.5	....	....	....
$Ba(H_2AsO_4)_2 \cdot 2H_2O$	*c*	−694.7	....	....	....
$Ba(C_2H_3O_2)_2$	*c*	−355.1	....	....	....
$Ba(C_2H_3O_2)_2 3H_2O$	*c*	−567.3	....	....	....
$BaBr_2$	*c*	−180.4	....	....	....
	aq	−186.47	−183.1	134.21	42
$BaBr_2 \cdot H_2O$	*c*	−254.9	....	....	....
$BaBr_2 \cdot 2H_2O$	*c*	−326.3	....	....	....
$BaOBr_2$	*aq*	−181.6	....	....	....
$Ba(HCO_2)_2$	*c*	−326.5	....	....	....
$Ba(CN)_2$	*c*	−47.9	....	....	....
	aq	−50.7	....	....	....
$Ba(CN)_2 \cdot H_2O$	*c*	−120.1	....	....	....
$Ba(CN)_2 \cdot 2H_2O$	*c*	−191.1	....	....	....
$BaCN_2$	*c*	−63.8	....	....	....
$Ba(CNO)_2$	*c*	−209.6	....	....	....
$BaCO_3$	*c*	−291.3	−272.2	199.52	26.8
	aq	−290.30	−260.2	190.72	−10
$Ba(HCO_3)_2$	*aq*	−459.0	−414.6	303.90	48
$BaC_2O_4 \cdot \frac{1}{2}H_2O$	*c*	−363.7	....	....	....
$BaC_2O_4 \cdot 2H_2O$	*c*	−470.1	....	....	....
$BaC_2O_4 \cdot 3\frac{1}{2}H_2O$	*c*	−575.3	....	....	....
$BaCl_2$	*c*	−205.56	−193.8	142.05	30
	aq	−208.72	−196.7	144.18	29
$BaCl_2 \cdot H_2O$	*c*	−278.4	−253.1	185.52	40
$BaCl_2 \cdot 2H_2O$	*c*	−349.35	−309.7	227.01	48.5
$BaOCl_2$	*aq*	−194.0	....	....	....
$Ba(OCl)_2$	*aq*	−176.1	....	....	....
$Ba(ClO_2)_2$	*c*	−158.2	....	....	....
$Ba(ClO_3)_2$	*c*	−181.7	....	....	....
	aq	−175.6	....	....	....
Barium					
$Ba(ClO_3)_2 \cdot H_2O$	*c*	−254.9	....	....	....
$Ba(ClO_4)_2$	*c*	−192.8	....	....	....
$Ba(ClO_4)_2 \cdot 3H_2O$	*c*	−405.4	....	....	....
$BaPtCl_6$	*c*	−286.8	....	....	....
	aq	−296.1	....	....	....
$BaPtCl_6 \cdot 6H_2O$	*c*	−707.2	....	....	....
$BaCrO_4$	*c*	−341.3	....	....	....
BaF_2	*c*	−286.9	....	....	....
	aq	−286.0	−265.3	....	....
$BaSiF_6$	*c*	−691.6	....	....	....
BaH	*g*	52	46	−33.7	52.97
BaH_2	*c*	−40.9	....	....	....
BaI_2	*c*	−144.0	....	....	....
	aq	−155.41	−158.7	116.32	55
$BaI_2 \cdot H_2O$	*c*	−219.8	....	....	....
$BaI_2 \cdot 2H_2O$	*c*	−290.9	....	....	....
$BaI_2 \cdot 2\frac{1}{2}H_2O$	*c*	−326.0	....	....	....
$BaI_2 \cdot 7H_2O$	*c*	−640.1	....	....	....
$Ba(IO_3)_2$	*aq*	−238.4	....	....	....
$Ba(IO_3)_2 \cdot H_2O$	*c*	−319.6	....	....	....
$BaMoO_4$	*c*	−373.8	....	....	....
$Ba(N_3)_2$	*c*	−8.0	....	....	....
	aq	0.2	....	....	....
Ba_3N_2	*c*	−86.9	....	....	....
$Ba(NO_2)_2$	*c*	−174.0	....	....	....
$Ba(NO_2)_2 \cdot H_2O$	*c*	−254.5	....	....	....
$Ba(NO_3)_2$	*c*	−237.06	−190.0	139.27	51.1
	aq	−227.41	−186.8	136.92	73
BaO	*c*	−133.4	−126.3	92.58	16.8
BaO_2	*c*	−150.5	....	....	....
$BaO_2 \cdot H_2O$	*c*	−223.5	....	....	....
$BaO_2 \cdot 8H_2O$	*c*	−719.3	....	....	....
$Ba(OH)_2$	*c*	−226.2	....	....	....
	aq	−238.58	−209.2	153.34	−2
$Ba(OH)_2 \cdot H_2O$	*c*	−299.0	....	....	....
$Ba(OH)_2 \cdot 8H_2O$	*c*	−799.5	....	....	....
$Ba_3(PO_4)_2$	*c*	−998.0	....	....	....
$BaHPO_4$	*c*	−465.8	....	....	....
$Ba(H_2PO_4)_2$	*c*	−749.6	....	....	....
BaS	*g*	41	....	....	....
	c	−106.0	....	....	....
	aq	−118.4	....	....	....
$Ba(HS)_2$	*aq*	−134.8	....	....	....
$Ba(HSO_3)_2$	*aq*	−430.7	....	....	....
$BaSO_3$	*c*	−282.6	....	....	....
$BaSO_4$	*c*	−350.2	−323.4	237.05	31.6
	aq, ∞	−345.57	−311.3	228.18	7
BaS_2O_6	*aq*	−409.3	....	....	....
$BaS_2O_6 \cdot 2H_2O$	*c*	−552.5	....	....	....
BaS_2O_8	*aq*	−454.0	....	....	....
$BaS_2O_8 \cdot 4H_2O$	*c*	−738.7	....	....	....
BaS_4O_6	*aq*	−401.3	....	....	....
$BaS_4O_6 \cdot 2H_2O$	*c*	−554.5	....	....	....
BaSe	*c*	−142.0	....	....	....
$BaSeO_4$	*c*	−280.0	....	....	....
$BaSiO_3$	*c*	−359.5	....	....	....
Ba_2SiO_4	*c*	−496.8	....	....	....
$BaWO_4$	*c*	−407.7	....	....	....
Beryllium					
Be	*g*	76.63	67.60	−49.550	32.55
	c	0.00	0.00	0.000	2.28
Be^{++} (in acid solution)	*aq*	....	−93	....	....
$BeBr_2$	*c*	−88.4	....	....	....
$BeBr_2$ (in HCl)	*aq*	151	....	....	....
$BeCl_2$	*c*	−122.3	....	....	....
$BeCl_2$ (in HCl)	*aq*	173.4	....	....	....
$BeCl_2 \cdot 4H_2O$	*c*	−436.8	....	....	....
BeF_2	*aq*	−251.4	....	....	....
BeH	*g*	78.1	71.3	−52.26	40.84
BeI_2	*c*	−50.6	....	....	....
BeI_2 (in HCl)	*aq*	−120	....	....	....
$BeMoO_4$	*c*	−330	....	....	....
Be_3N_2	*c*	−135.7	−122.4	....	....
$Be(NO_3)_2$	*aq*	−188.3	....	....	....
BeO	*g*	11.8	5.7	−4.18	47.18
	c	−146.0	−139.0	101.88	3.37
$Be(OH)_2$	*c* (α)	−216.8	....	....	....
	c (β)	−216.1	....	....	....
BeS	*c*	−55.9	....	....	....
$BeSO_4$	*c*	−286.0	....	....	....
$BeSO_4 \cdot H_2O$	*c*	−361	....	....	....
$BeSO_4 \cdot 2H_2O$	*c*	−433.2	....	....	....
$BeSO_4 \cdot 4H_2O$	*c*	−576.3	....	....	....
$BeSO_4 \cdot 4BeO$	*c*	−871.4	....	....	....
Bismuth					
Bi	*g*	49.7	40.4	−29.61	44.67
	c	0.00	0.00	0.000	13.6
BiO^+	*aq*	....	−35.54	25.317	....
$BiCl_3$	*g*	−64.7	−62.2	45.59	85.3
	c	−90.61	−76.23	55.876	45.3

Substance	State	ΔHf°	ΔFf°	Log_{10} Kf	S°
Bismuth					
$BiCl_3$ (in HCl)	*aq*	−101.6	....	....	....
BiOCl	*c*	−87.3	−77.0	56.44	20.6
Bi_2O_3	*c*	−137.9	−118.7	87.01	36.2
$Bi(OH)_3$	*c*	−169.6	....	....	....
Bi_2S_3	*c*	−43.8	−39.4	28.88	35.3
Boron					
B	*g*	97.2	86.7	−63.55	36.649
	c	0.00	0.00	0.000	1.56
B	*am*	0.4	....	....	....
BBr_3	*g*	−44.6	−51.0	37.38	77.49
	lq	−52.8	−52.4	38.41	54.7
BCl_3	*g*	−94.5	−90.9	66.63	69.29
	lq	−100.0	−90.6	66.41	50.0
B_4C	*c*	....	....	....	6.47
$B(CH_3)_3$	*lq*	−31.4	....	....	....
BF_3	*g*	−265.4	−261.3	191.53	60.70
	aq	−289.8	....	....	....
BF_4^-	*aq*	−365	−347	251.4	40
HBF_4	*aq*	−365	....	....	....
BH	*g*	73.8	67.1	−49.18	39.62
B_2H_6	*g*	7.5	19.8	−14.51	55.66
B_5H_9	*g*	15.0	39.6	−29.03	65.88
$B_{10}H_{14}$	*c*	8	....	....	....
BN	*g*	90.6	....	....	....
	c	−32.1	−27.2	19 94	8
BO	*g*	−5.3	−11.6	8 50	47.22
B_2O_3	*c*	−302.0	−283.0	207.44	12.91
	gls	−297.6	−280.4	205.53	18.8
B_2S_3	*c*	−57.0	....	....	....
HBO_2	*c*	−186.9	−170.5	124.97	11
	aq	−186.9	....	....	....
HBO_3	*aq*	−251.8	−217.63	159 520	7.3
$H_2B_4O_7$	*c*	−676.5	....	....	....
H_3BO_3	*c*	−260.2	−230.2	168.73	21.41
	aq	−255.2	−230.24	168.762	38.2
Bromine					
Br_2	*g*	7.34	0.75	−0.551	58.64
	lq	0.00	0.00	0.000	36.4
	aq	−1.1	....	....	....
$Br_2 \cdot 10H_2O$	*c*	−700	....	....	....
BrCl	*g*	3 51	−0.21	0 154	57.34
HBr	*g*	−8.66	−12.72	9.327	47.44
	aq, 1	−18.56	....	....	....
	aq, 3	−24.36	....	....	....
	aq	−28.90	−24.57	18.012	19.29
$HBrO_3$	*aq*	−11.63	5.00	....	....
Cadmium					
Cd	*g*	26.97	18.69	−13.700	40.07
(α)	*c*	0.00	0.00	0.000	12.3
(γ)	*c*	....	0.14	−0.103	....
$CdBr_2$	*c*	−75.15	−70.14	51.412	31.9
	aq	−75.822	−67.6	....	....
$CdBr_2 \cdot 4H_2O$	*c*	−356.32	−297.64	218 166	74.7
$CdCl_2$	*c*	−93.00	−81.88	60.017	28.3
	aq	−97.39	−81.28	59.577	11.6
$CdCl_2 \cdot H_2O$	*c*	−164 13	−140.13	102.713	40.8
$CdCl_2 \cdot 2\frac{1}{2}H_2O$	*c*	−269.97	−225.47	165.266	55.6
$Cd(CH_3)_2$	*lq*	16.2	....	....	....
$Cd(C_2H_5)_2$	*lq*	14.7	....	....	....
$Cd(CN)_2$	*c*	39.0	....	....	....
	aq	30.3	....	....	....
$Cd(ONC)_2$	*c*	37.7	....	....	....
$CdCO_3$	*c*	−178.7	−160.2	117.42	25.2
CdF_2	*c*	−164.9	−154.8	113.47	27
	aq	−173.6	....	....	....
CdH	*g*	62.54	55.73	−40.849	50.76
CdI_2	*c*	−48.0	−48.00	35.183	40.2
	aq, ∞	−44.66	....	....	....
$Cd(N_3)_2$	*c*	107.8	....	....	....
Cd_3N_2	*c*	38 6	....	....	....
$Cd(NO_3)_2$	*c*	−107 98	....	....	....
	aq	−116.04	−71.41	52.342	20.4
$Cd(NO_3)_2 \cdot 2H_2O$	*c*	−251 19	....	....	....
$Cd(NO_3)_2 \cdot 4H_2O$	*c*	−394.02	....	....	....
CdO	*c*	−60 86	−53.79	39.427	13.1
$Cd(OH)_2$	*c*	−133.26	−112.46	82.432	22.8
CdS	*c*	−34.5	−33.6	24.63	17
$CdSO_4$	*c*	−221.36	−195.99	143.657	32.8
	aq	−234.20	−195 92	143.607	−10.47
$CdSO_4 \cdot H_2O$	*c*	−294.37	−254.84	186.794	41.1
Cd_3Sb_2	*c*	7.83	1.59	−1.165	78.8
CdSe	*g*	1.6	....	....	....
CdTe	*c*	−24 30	−23.82	17.460	22.6
Calcium					
Ca	*g*	46.04	37.98	−27.839	36.99
	c	0.00	0.00	0.000	9.95
Ca^{++}	*aq*	−129.77	−132.18	96.886	−13.2
$2CaO \cdot Al_2O_3$	*c*	−704	....	....	....
$2CaOAl_2O_3 \cdot 5H_2O$	*c*	−1078	....	....	....
$3CaO \cdot Al_2O_3$	*c*	−861	....	....	....
$3CaO \cdot Al_2O_3 \cdot 6H_2O$	*c*	−1329	....	....	....
$4CaO \cdot Al_2O_3$	*c*	−1026	....	....	....
$Ca_3(AsO_4)_2$	*c*	−796	....	....	....
$CaHAsO_4$	*aq*	−344.1	....	....	....
$CaHAsO_4 \cdot H_2O$	*c*	−410	−363	266.1	35

Substance	State	ΔHf°	ΔFf°	Log_{10} Kf	S°
Calcium					
$Ca(H_2AsO_4)_2$	*aq*	−560.8	....	....	....
$CaBr_2$	*c*	−161.3	−156.8	114.93	31
	aq	−187.57	−181.33	132.912	25.4
$CaBr_2 \cdot 6H_2O$	*c*	−597.2	....	....	....
$CaO \cdot B_2O_3$	*c*	−483.3	−457.7	335.49	25.1
$CaO \cdot 2B_2O_3$	*c*	−798.8	−752.4	551.50	32.2
$2CaO \cdot B_2O_3$	*c*	−651.6	−618.6	453.42	34.7
$3CaO \cdot B_2O_3$	*c*	−817.7	−777.1	569.60	43.9
CaC_2	*c*	−15.0	−16.2	11.87	16.8
$CaCO_3$ (calcite)	*c*	−288.45	−269.78	197.745	22.2
(aragonite)	*c*	−288.49	−269.53	197 562	21.2
$Ca(HCO_3)_2$	*aq*	−460.13	−412.80	302.577	32.2
$Ca(HCO_2)_2$	*c*	−323.5	....	....	....
CaC_2O_4	*c*	−332.2	....	....	....
$CaC_2O_4 \cdot H_2O$	*ppt*	−399.1	−360.6	264.31	37.28
$CaC_2O_4 \cdot 2H_2O$	*c*	−469.1	−416 9	305.58	47
$Ca(C_2H_3O_2)_2$	*c*	−355.0	....	....	....
	aq	−362.5	....	....	....
$Ca(C_2H_3O_2)_2H_2O$	*c*	−425.1	....	....	....
$Ca(CN)_2$	*c*	−44.2	....	....	....
	aq	....	−54	....	....
$CaCN_2$	*c*	−84.0	....	....	....
$CaCl_2$	*c*	−190.0	−179.3	131.42	27.2
$CaCl_2$	*aq*	−209.82	−194.88	142.884	13.1
$CaCl_2 \cdot H_2O$	*c*	−265.1	....	....	....
$CaCl_2 \cdot 2H_2O$	*c*	−335.5	....	....	....
$CaCl_2 \cdot 4H_2O$	*c*	−480.2	....	....	....
$CaCl_2 \cdot 6H_2O$	*c*	−623.15	....	....	....
$CaCl_2 \cdot 2CaO$	*c*	−505	....	....	....
$CaCl_2 \cdot 3CaO$	*c*	−654	....	....	....
$CaCl_2 \cdot 3CaO \cdot 3H_2O$	*c*	−910.6	....	....	....
$CaCl_2 \cdot 3CaO \cdot 16H_2O$	*c*	−1833	....	....	....
$CaCl_2 \cdot 3C_2H_5OH$	*c*	−399.8	....	....	....
$CaCl_2 \cdot 4C_2H_5OH$	*c*	−467.4	....	....	....
$CaOCl_2$	*c*	−178.6	....	....	....
	aq	−189.1	....	....	....
$CaOCl_2 \cdot H_2O$	*c*	−249.2	....	....	....
$Ca(OCl)_2$	*aq*	−180.0	....	....	....
$CaCrO_4$	*c*	−329.6	−305.3	223.78	32
	aq	−336.0	....	....	....
CaF_2	*c*	−290.3	−277.7	203.55	16.46
	aq	−287.09	−264.34	193.758	−17.8
CaH	*g*	58.7	....	....	....
CaH_2	*c*	−45.1	−35.8	26.24	10
CaI_2	*c*	−127.8	−126.6	92.80	34
	aq	−156.51	−156.88	114.991	39.1
$CaI_2 \cdot 8H_2O$	*c*	−700.7	....	....	....
CaN_6	*c*	75.8	....	....	....
Ca_3N_2	*c*	−103.2	−88.1	64.57	25
$Ca(NO_2)_2$	*c*	−178.3	....	....	....
	aq	−180.5	....	....	....
$Ca(NO_3)_2$	*c*	−224.0	−177.34	129.988	46.2
	aq	−228.51	−185.00	135.602	56.8
$Ca(NO_3)_2 \cdot 2H_2O$	*c*	−368.00	−293.51	215.139	64.3
$Ca(NO_3)_2 \cdot 3H_2O$	*c*	−437.18	−349.0	255.81	74
$Ca(NO_3)_2 \cdot 4H_2O$	*c*	−509.37	−406.5	297.96	81
CaO	*c*	−151.9	−144.4	105.84	9.5
CaO_2	*c*	−157.5	....	....	....
$Ca(OH)_2$	*c*	−235.80	−214.33	157.101	18.2
	aq	−239.68	−207.37	152.000	−18.2
Ca_3P_2	*c*	−120.5	....	....	....
$Ca_3(PO_4)_2$	*c* (α)	−986.2	−929.7	681.46	57.6
	c (β)	−988.9	−932.0	683.14	56.4
$CaHPO_4$	*c*	−435.2	−401.5	294.29	21
$CaHPO_4 \cdot 2H_2O$	*c*	−576.0	−514.6	377.19	40
$Ca(H_2PO_4)_2$	*ppt*	−744.4	....	....	....
CaS	*c*	−115.3	−114.1	83.63	13.5
	aq	−119.8	....	....	....
$CaSO_3 \cdot 2H_2O$	*c*	−421.2	−374.1	274.21	44
$CaSO_4$ (anhydrite)	*c*	−342.42	−315.56	231.301	25.5
(soluble α)	*c*	−340.27	−313.52	229.806	25.9
(soluble β)	*c*	−339.21	−312.46	229.029	25.9
	aq	−346.67	−309.52	226.874	−9.1
$CaSO_4 \cdot \frac{1}{2}H_2O$	*c* (α)	−376.47	−343.02	251.429	31.2
	c (β)	−375.97	−342.78	251.253	32.1
$CaSO_4 \cdot 2H_2O$	*c*	−483.06	−429.19	314.590	46.36
CaS_2O_3	*aq*, 1000	−283 4	....	....	....
$CaS_2O_3 \cdot 6H_2O$	*c*	....	−602.2	441.40	....
CaSe	*c*	−74 7	−73.5	53.87	16
$CaSi_2$	*c*	−36	....	....	....
Ca_2Si	*c*	−50	....	....	....
Ca_2Si_2	*c*	−72	....	....	....
$CaSiO_3$*	*c* (α)	−377.4	−357.4	261.97	20.9
	c (β)	−378.6	−358.2	262.55	19.6
Ca_2SiO_4	*c* (β)	−538.0	....	....	....
	c (γ)	−539.0	....	....	....
Ca_3SiO_5	*c*	−688.4	....	....	....
$CaWO_4$	*c*	−392.5	....	....	....
Carbon**					
C	*g*	171.70	160.85	−117.897	37.76
C (diamond)	*c*	0.45	0.69	−0.502	0.58
(graphite)	*c*	0.00	0.00	0.000	1.36
CO_2	*g*	−94 05	−94.26	69.092	51.06
	aq	−98.69	−92.31	67.662	29.0
CO	*g*	−26.42	−32.81	24.048	47.30

* (α) Pseudowollastonite; (β) wollast

** For the values of organic compounds see table following zirconium compounds.

VALUES OF CHEMICAL THERMODYNAMIC PROPERTIES (Continued)

Substance	State	ΔHf°	ΔFf°	Log₁₀ Kf	S°
Cerium					
Ce	c	0.00	0.00	0.000	13.8
Ce^{+++}	aq	−173.7	−170.5	124.97	−44
$CeCl_3$	c (α)	−260.3	....	....	....
	aq	−293.9	−264 5	193.87	−5
Ce_3H_8	c	−170	....	....	....
CeI_3	c (α)	−164.4	....	....	....
	aq	−213.9	−207.5	152.09	34
CeO_2	c	−233	....	....	....
$CeO_2{\cdot}2H_2O$	c	−389	....	....	....
CeS_2	c	−153.9	....	....	....
Ce_2S_3	c	−298.7	....	....	....
$Ce(SO_4)_2$	c	−560	....	....	....
$Ce_2(SO_4)_3$	aq	−998.3	−873.0	639.90	−76
$Ce_2(SO_4)_3{\cdot}5H_2O$	c	−1308	....	....	....
$Ce_2(SO_4)_3{\cdot}8H_2O$	c	....	−1340.2	982.35	....
$Ce_2(SO_4)_3{\cdot}9H_2O$	c	....	−1396.8	1023.83	....
Cesium					
Cs	c	0.00	0.00	0.000	19.8
Cs^{+}	aq	−59.2	−67.41	49.111	31.8
CsBr	c	−94.3	−91.6	67.142	29
	aq	−88.1	−91.98	67.420	51
Cs_2CO_3	c	−267.4	....	....	....
$CsHCO_3$	c	−228.4	....	....	....
	aq	−224.4	....	....	....
CsCl	c	−103.5	....	....	....
	aq	−99.2	−88.76	65.060	45.0
$CsClO_4$	c	−103.86	−73.28	53.713	41.89
	aq	−90.6	−69.98	51.294	75.3
CsF	c	−126.9	....	....	....
	aq	−135.9	−133.49	97.846	29.5
Cs_2SiF_6	c	−669.5	....	....	....
CsH	g	29.0	24.3	−17.81	51.25
CsI	c	−80.5	−79.7	58.419	31
	aq	−72.6	−79.76	58.463	57.9
$CsNH_2$	c	−25.4	....	....	....
$CsNO_3$	c	−118.11	....	....	....
	aq	−108.6	−93.82	68.769	86.8
Cs_2O	c	−75.9	....	....	....
Cs_2O_2	c	−96.2	....	....	....
CsOH	c	−97.2	....	....	....
	aq	−114.2	−105.00	76.964	29.3
$CsOH{\cdot}H_2O$	c	−186.9	....	....	....
$CsReO_4$	c	−257.2	....	....	....
	aq	−249.5	....	....	....
Cs_2S	c	−81.1	....	....	....
	aq	−108.4	....	....	....
CsHS	c	−62.9	....	....	....
	aq	−63.2	....	....	....
Cs_2SO_4	c	−339.38	....	....	....
	aq	−335.3	−312.16	228.809	67.7
$CsHSO_4$	c	−274.0	....	....	....
CsHSe	c	−36.7	....	....	....
	aq	−34.6	....	....	....
Chlorine					
Cl_2	g	0.00	0.00	0.000	53.29
Cl^{-}	aq	−40.023	−31.350	22.98	13.17
ClF	g	−13.3	−13.6	9.97	52.05
ClO	g	33	....	....	....
ClO_2	g	24.7	29.5	−21.623	59.6
ClO_3	g	37.0	....	....	....
Cl_2O	g	18.20	22.40	−16.419	63.70
Cl_2O_7	g	63.4	....	....	....
HCl	g	−22.06	−22.77	16.690	44.62
	aq	−40.02	−31.35	22.979	13.17
HClO	aq, 400	−28.18	....	....	....
	aq, 1000	−27.83	....	....	....
$HClO_2$	aq	−14.0	....	....	....
$HClO_3$	aq	−23.50	....	....	....
$HClO_4$	lq (lq)	−11.1	....	....	....
	aq	−31.41	....	....	....
$HClO_4{\cdot}H_2O$	c	−92.1	....	....	....
$HClO{\cdot}2H_2O$	lq	−162.8	....	....	....
Chromium					
Cr	g	80.5	69.8	−51.16	41.64
	c	0.00	0.00	0.000	5.68
Cr^{+++}	aq	1310	....	....	....
Cr_3C_2	c	−21.0	−21.2	15.54	20.4
Cr_4C	c	−16.4	−16.8	12.31	25.3
Cr_7C_3	c	−42.5	−43.8	32.10	48.0
$CrCl_2$	c	−94.56	−85.15	62.414	27.4
	aq	−113.2	....	....	....
$CrCl_2{\cdot}2H_2O$	c	−237	....	....	....
$CrCl_2{\cdot}3H_2O$	c	−309.0	....	....	....
$CrCl_2{\cdot}4H_2O$	c	−384.5	....	....	....
$CrCl_3$	c	−134.6	−118.0	86.492	30.0
$CrCl_4$	g	−104.	....	....	....
CrO_2Cl_2	lq	−135.7	....	....	....
CrF_2	c	−181.0	....	....	....
CrF_3	c	−265.2	....	....	....
Cr_7H_2	c	−3.7	....	....	....
CrI_2	c	−54 2	....	....	....

Substance	State	ΔHf°	ΔFf°	Log₁₀ Kf	S°
Chromium					
	aq	−59.9	....	....	....
CrN	c	−29.8	....	....	....
Cr_2N	c	−23.4	....	....	....
Cr_2O_3	c	−269.7	−250.2	183.39	19.4
$Cr_2O_3{\cdot}H_2O$	c	−358	....	....	....
$Cr_2O_3{\cdot}2H_2O$	c	−439	....	....	....
$Cr_2O_3{\cdot}3H_2O$	c	−517	....	....	....
$Cr(OH)_3$	c	−247.1	....	....	....
H_2CrO_4	aq	−213.3	−178.5	130.84	17.5
Cobalt					
Co	g	105	94	−68.9	42.88
	c	0.00	0.00	0.000	6.8
Co^{++}	aq	−16.1	−12.3	9.016	−37.1
$CoBr_2$	c	−55.5	....	....	....
$CoBr_2{\cdot}6H_2O$	aq	−73 9	....	....	....
	c	−485.1	....	....	....
Co_3C	c	9.5	7.1	−5.20	29.8
$CoCO_3$	c	−172.7	−155.36	....	....
$CoCl_2$	c	−77.8	−67.5	49.48	25.4
	aq	−96.1	....	....	....
$CoCl_2{\cdot}2H_2O$	c	−222 9	....	....	....
$CoCl_2{\cdot}4H_2O$	c	−367 2	....	....	....
$CoCl_2{\cdot}6H_2O$	c	−508.9	....	....	....
CoF_2	c	−159	....	....	....
	aq	−173.6	....	....	....
$CoF_2{\cdot}4H_2O$	c	....	−380.9	279.19	....
CoF_3	c	−187	....	....	....
CoH	c	−4.1	....	....	....
CoH_2	c	−10.2	....	....	....
CoI_2	c	−24.4	....	....	....
	aq	−42.8	....	....	....
$Co(IO_3)_2$	c	−124 3	....	....	....
	aq	−125.9	....	....	....
$Co(IO_3)_2{\cdot}2H_2O$	c	−264.9	....	....	....
$Co(IO_3)_2{\cdot}4H_2O$	c	−401 9	....	....	....
$Co(NO_3)_2$	c	−102.9	....	....	....
	aq	−114 8	....	....	....
$Co(NO_3)_2{\cdot}6H_2O$	c	−529.7	....	....	....
CoO	c	−57.2	−51.0	37.38	10.5
Co_3O_4	c	−210	....	....	....
$Co(OH)_2$	c	−131 2	−108.9	....	....
$Co(OH)_3$	c	−176 6	−142.0	....	....
CoP	c	−35	....	....	....
CoP_3	c	−65	....	....	....
Co_2P	c	−47 3	....	....	....
CoS	c	−20 2	−19.8	....	....
	ppt	−21 4	....	....	....
Co_2S_3	c	−51	....	....	....
$CoSO_4$	c	−207.5	−182.1	133.48	27.1
	aq	−232 0	....	....	....
$CoSO_4{\cdot}6H_2O$	c	−643.2	....	....	....
$CoSO_4{\cdot}7H_2O$	c	−713.8	....	....	....
Columbium (See Niobium)					
Copper					
Cu	g	81.52	72.04	−52.804	39.74
	c	0.00	0.00	0.000	7.96
Cu^{+}	aq	12.4	12.0	−8.796	−6.3
Cu^{++}	aq	15.39	15.53	−11.383	−23.6
CuBr	g	38	28	−20.5	59.22
	c	−25.1	−23.81	17.452	21.9
$CuBr_2$	c	−33.8	....	....	....
$CuBr_2{\cdot}4H_2O$	c	316.4	....	....	....
$CuCO_3$	c	−142.2	−123.8	90.744	21
$Cu(C_2H_3O_2)_2$	c	−213.2	....	....	....
$Cu(C_2H_3O_2)_2{\cdot}H_2O$	c	−284.2	....	....	....
CuONC	c	26.3	....	....	....
CuCl	g	32	25	−18.3	56.50
	c	−32.5	−28.2	20.67	20.2
$CuCl_2$	c	−49.2	....	....	....
CuF	g	44	....	....	....
CuF_2	c	−126.9	....	....	....
$CuF_2{\cdot}2H_2O$	c	−274.5	−235.2	172.40	36.2
CuH	g	71	64	−46.9	46.89
CuI	g	62	50	−36.6	61.06
	c	−16.2	−16.62	12.182	23.1
CuI_2	c	−1.7	....	....	....
CuN_3	c	60.5	....	....	....
Cu_3N	c	17.8	....	....	....
$Cu(NO_3)_2$	c	−73.4	....	....	....
CuO	g	35	....	....	....
	c	−37.1	−30.4	22.28	10.4
Cu_2O	c	−39.84	−34.98	25.640	24.1
$Cu(OH)_2$	c	−107.2	....	....	....
CuS	c	−11.6	−11.7	8.576	15.9
Cu_2S	c	−19.0	−20.6	15.10	28 9
$CuSO_4$	c	−184.00	--158 2	115.96	27.1
	aq	−201.51	−161 81	118.604	19.5
$CuSO_4{\cdot}H_2O$	c	−259.00	−219.2	160.67	35.8
$CuSO_4{\cdot}3H_2O$	c	−402.27	−334 6	245.26	53.8
$CuSO_4{\cdot}5H_2O$	c	−544 45	−449.3	329 33	73.0
Cu_2SO_4	c	−179.2	....	....	....
	aq	−190.8	....	....	....
CuSe	c	−15.1	....	....	....

* (α) Pseudowollastonite; (β) wollastonite.
** For the values of organic compounds see table following zirconium compounds.

VALUES OF CHEMICAL THERMODYNAMIC PROPERTIES (Continued)

Substance	State	ΔHf°	ΔFf°	Log₁₀ Kf	S°
Dysprosium					
Dy^{+++}	aq	−166.0	−161.2	118.16	....
$DyCl_3$	c (β)	−237.8	....	....	....
	c (γ)	−234.8	....	....	....
	aq	−286.1	−255.2	187.06	....
DyI_3	c (β)	−144.5	....	....	....
	aq	−206 1	−198.2	145.28	....
$Dy(OH)_3$	c	....	−305.8	224.15	....
$Dy_2(SO_4)_3$	aq	−982 7	−854.4	626.26	....
$Dy_2(SO_4)_3 \cdot 8H_2O$	c	....	−1322.0	969.01	....
Erbium					
Er^{+++}	aq	−162.3	−157.5	115.45	....
$ErCl_3$	c (γ)	−231.8	....	....	....
	aq	−282 4	−251.5	184.35	....
ErI_3	c (β)	−140.0	....	....	....
ErI_3	aq	−202.4	−194.5	142.57	....
$Er(OH)_3$	c	−340.5	....	....	....
$Er_2(SO_4)_3$	aq	−975.3	−847.0	620.84	....
$Er_2(SO_4)_3 \cdot 8H_2O$	c	....	−1313.9	963.07	....
Europium					
Eu^{+++}	aq	−169.3	−165.1	121.02	....
$EuCl_3$	c (α)	−247.1	....	....	....
	aq	−289.4	−259.2	189.99	....
$Eu_2(SO_4)_3$	aq	−989.3	−862.2	631.98	....
$Eu_2(SO_4)_3 \cdot 8H_2O$	c	....	−1331.0	975.60	....
Fluorine					
F_2	g	0.00	0.00	0.000	48.6
F^-	aq	−78.66	66.08	48.435	−2.3
F_2O	g	5.5	9.7	−7.126	58.95
HF	g	−64.2	−64.7	47.402	41.47
	aq	−78.66	−66.08	48.435	−2.3
Gadolinium					
Gd	g	87	77	−56.4	46.41
Gd^{+++}	aq	−168.8	−164.6	120.65	−47.1
$GdCl_3$	c (α)	−245.5	....	....	....
	aq	−288.9	−258.6	189.55	−7.6
GdI_3	c (β)	−147.6	....	....	....
	aq	−208.9	−201.6	147.77	−31.3
$Gd(OH)_3$	c	....	−308.1	225.83	....
$Gd_2(SO_4)_3$	aq	−988.3	−861.2	631.25	−81.9
$Gd_2(SO_4)_3 \cdot 8H_2O$	c	−1518.9	−1329.8	974.72	155.8
Gallium					
Ga	g	66.0	57.0	−41.78	40.38
Ga^{+++}	aq	50.4	36.6	26.83	−83
$GaBr_3$	c	−92.4	....	....	....
$GaCl_3$	c	−125.4	....	....	....
	aq	−170.5	−130.7	95.801	−43
GaI_3	c	−51.2	....	....	....
GaN	c	−25	....	....	....
GaO	g	66	....	....	....
Ga_2O	c	−82	....	....	....
Ga_2O_3	c	−258	....	....	....
$Ga(OH)_3$	c	....	−199	145.9	....
Germanium					
Ge	g	78.44	69.50	−50.943	40.11
	c	0.00	0.00	0.000	10.14
GeBr	g	34.8	....	....	....
GeCl	g	32.6	32.45	....	....
$GeCl_4$	lq	−130	....	....	....
GeH_4	g	....	....	....	51.21
Ge_3N_4	c	−14.8	....	....	....
GeO	g	−22.8	−28.2	20.63	52.56
GeO_2 (gls)	am	−128.3	....	....	....
(tetr.)	c	−128.3	....	....	....
GeS	g	1.35	....	....	....
H_2GeO_3	aq	−199.3	....	....	....
Gold					
Au	g	82.29	72.83	−53.383	43.12
AuO_3^{---}	aq	....	−5.8	4.25	....
$AuBr_3$	c	−13.0	....	....	....
	aq	−9.2	....	....	....
$HAuBr_4$	aq	−45.5	−38.1	27.93	75
$HAuBr_4 \cdot 5H_2O$	c	−398.5	....	....	....
$AuCl_3$	c	−28.3	....	....	....
	aq	−32.8	....	....	....
$AuCl_3 \cdot 2H_2O$	c	−167.7	....	....	....
$HAuCl_4$	aq	−77.8	−56.2	41.19	61
$HAuCl_4 \cdot 3H_2O$	c	−279.2	....	....	....
$HAuCl_4 \cdot 4H_2O$	c	−356.9	....	....	....
AuI	c	0.2	....	....	....
Au_2O_3	c	19.3	39.0	−28.6	30
$Au(OH)_3$	c	−100.0	−69.3	50.80	29
Hafnium					
Hf	g	....	....	....	44.65
	c	0.00	0.00	0.000	13 1
HfO_2	c	−271.5	....	....	....
Helium					
He	g	0.00	0.00	0.000	30.13
Holmium					
Ho^{+++}	aq	−163.7	−159.2	116.69	....
$HoCl_3$	c (γ)	−232.8	....	....	....
	aq	−283.8	−253.2	185.59	....
HoI_3	c (β)	−141.7	....	....	....
	aq	−203.8	−196.2	143.81	....
$Ho_2(SO_4)_3$	aq	−978.1	−850.4	623.33	....
$Ho_2(SO_4)_3 \cdot 8H_2O$	c	....	−1318.0	966.08	....

Substance	State	ΔHf°	ΔFf°	Log₁₀ Kf	S°
Hydrogen					
H_2 (mass 1)*	g	0.00	0.00	0.000	31.21
(mass 2)**	g	0.00	0.00	0.000	34.602
H*	g	52.09	48.58	−35.60	27.39
H**	g	52.98	49.36	−36.18	29.46
H_2O*	g	−57.80	−54.64	40.047	45.11
	lq	−68.32	−56.69	41.553	16.72
H_2O**	g	−59.56	−56.07	41.096	47.38
	lq	−70.41	−58.21	42.664	18.16
H*H**O	g	−58.74	−55.83	40.921	47.66
	lq	−69.39	−57.93	42.459	18.85
H_2O_2	g	−31.83	....	....	....
	lq	−44.84	−28.2	....	....
	aq	−45.68	....	....	....
Indium					
In	g	58.2	49.6	−36.36	41.51
In^{+++}	aq	27.3	−32.0	23.46	−62
$InBr_3$	c	−96.5	....	....	....
InCl	g	−18	−23	16.9	59.3
	c	−44.5	....	....	....
$InCl_2$	c	−86.8	....	....	....
$InCl_3$	c	−128.4	....	....	....
InH	g	51	45	−33.0	49.6
InI	g	20	9	−6.6	63.8
InI_3	c	−55.0	....	....	....
InN	c	−4.8	....	....	....
InO	g	91	....	....	....
In_2O_3	c	−222.5	....	....	....
$In(OH)_3$	c	−214	−182	133.4	25
$In_2(SO_4)_3$	c	−695	....	....	....
Iodine					
I_2	g	14.88	4.63	−3.394	62.28
	c	0.00	0.00	0.000	27.9
I^-	aq	−13.37	−12.35	9.252	26.14
IBr	g	9.75	0.91	0.667	61.80
ICl	g	4.20	−1.32	0.968	59.12
ICl_3	c	−21.1	−5.36	3.929	41.1
HI	g	6.20	0.31	−0.227	49.31
	aq	−13.37	−12.35	9.052	26.14
I_2O_5	c	−42.34	....	....	....
HIO_3	c	−57.03	....	....	....
Iridium					
Ir	g	165	154	−112.9	46.25
	c	0.00	0.00	0.000	8.7
IrCl	c	−22.3	....	....	....
$IrCl_2$	c	−42.8	....	....	....
$IrCl_3$	c	−61.5	....	....	....
IrF_6	lq	−130	....	....	....
IrO_2	c	−40.1	....	....	....
IrS_2	c	−30	....	....	....
Ir_2S_3	c	−51	....	....	....
Iron					
Fe	g	96.68	85.76	−62.861	43.11
	c	0.00	0.00	0.000	6.49
Fe^{++}	aq	−21.0	−20.30	14.88	−27.1
Fe^{+++}	aq	−11.4	−2.52	1.847	−70.1
$FeBr_2$	c	−60.02	....	....	....
$FeBr_3$	aq	−98.1	....	....	....
Fe_3C (cementite)	c	5.0	3.5	−2.56	25.7
$Fe(CO)_5$	lq	−187.8	....	....	....
$FeCC_3$ (siderite)	c	−178.70	−161.06	118.055	22.2
$H_3Fe(CN)_6$	aq	153.	....	....	....
$H_4Fe(CN)_6$	c	127.8	....	....	....
	aq	127.4	....	....	....
$FeCl_2$	c	−81.5	−72.2	52.92	28.6
	aq	−101.0	....	....	....
$FeCl_2 \cdot 2H_2O$	c	−228.2	....	....	....
$FeCl_2 \cdot 4H_2O$	c	−370.7	....	....	....
$FeCl_3$	c	−96.8	....	....	....
	aq	−127.9	−96.5	....	....
$FeCl_3 \cdot 6H_2O$	c	−532.0	....	....	....
$FeCr_2O_4$	c	−341.9	−317.7	232.87	34.9
FeF_2	aq	−177.8	....	....	....
FeF_3	aq	−243.1	....	....	....
FeI_2	c	−29.98	....	....	....
	aq	−49.03	....	....	....
Fe_2N	c	−0.9	2.6	−1.91	24.2
Fe_4N	c	−2.55	0.89	−0.652	37.3
$Fe(NO_3)_2$	aq	−118.9	....	....	....
$Fe_{0.95}O$ "FeO" (wüstite)	c	−63.7	−58.4	42.81	12.9
Fe_2O_3 (hematite)	c	−196.5	−177.1	129.81	21.5
Fe_3O_4 (magnetite)	c	−267 0	−242.4	177 68	35.0
$Fe(OH)_2$	c	−135.8	−115.57	84.711	19
$Fe(OH)_3$	c	−197.0	....	....	....
FeP	c	−28	....	....	....
FeP_2	c	−42	....	....	....
Fe_2P	c	−36	....	....	....
Fe_3P	c	−40	....	....	....
$FePO_4$	c	−299.6	....	....	....
$FePO_4 \cdot 2H_2O$	c	−440.8	....	....	....
$FePO_4 \cdot 4H_2O$	c	−578.8	....	....	....
FeS	c (α)	−22.72	−23.32	17.093	16.1
	c (β)	−21.35	....	....	....
FeS_2 (pyrites)	c	−42.52	−39.84	29.202	12.7
(markasite)	c	−36.88	....	....	....

* Atomic weight 1.0078.
** Atomic weight 2.0142.

Substance	State	ΔHf°	ΔFf°	Log₁₀ Kf	S°
Krypton					
Kr	g	0.00	0.00	0.000	39.19
$Kr\cdot 5H_2O$	c	−357.1	....	....	....
Lanthanum					
La	g	88	79	−57.9	43.57
	c	0.00	0.00	0.000	13.7
La^{+++}	aq	−176.2	−172.9	126.73	−44
$La_2(CN_2)_3$	c	−229	....	....	....
$LaCl_3$	c (α)	−263.6	....	....	....
	aq	−296.3	−266.9	195.63	−5
LaI_3	c (α)	−167.4	....	....	....
	aq	−216.3	−209.9	153.85	34
LaN	c	−72.1	....	....	....
La_2O_3	c	−458.	....	....	....
$La(OH)_3$	c	....	−312.8	229.28	....
La_2S_3	c	−306.8	....	....	....
$La_2(SO_4)_3$	aq	−1003.1	−877.8	643.41	−76
Lanthanum					
$La_2(SO_4)_3\cdot 9H_2O$	c	....	−1403.1	1028.45	....
Lead					
pb	g	46.34	38.47	−28.198	41.89
	c	0.00	0.00	0.000	15.51
Pb^{++}	aq	0.39	−5.81	4.259	5.1
$PbBr_2$	c	−66.21	−62.24	45.621	38.6
	aq	−57.41	−54.95	40.278	43.7
$PbCO_3$	c	−167.3	−149.7	109.73	31.3
$PbCO_3\cdot PbO$	c	−220.0	−195.6	143.37	48.5
$PbCO_3\cdot 2PbO$	c	−273	−242	177.4	65
PbC_2O_4	c	−205.1	....	....	....
$Pb(C_2H_3O_2)_2$	c	230.5	....	....	....
	aq	−232.6	....	....	....
$Pb(C_2H_3O_2)_2\cdot 3H_2O$	c	−443.1	....	....	....
$Pb(C_2H_5)_4$	lq	52	....	....	....
$PbCl_2$	c	−85.85	−75.04	55.003	32.6
	aq	−79.65	−68.51	50.217	31.4
PbF_2	c	−158.5	−148.1	108.55	29
PbF_4	c	−222.3	....	....	....
PbI_2	c	−41.85	−41.53	30.441	42.3
	aq	−26.35	−30.51	22.363	57.8
$Pb(N_3)_2$	c	104.3	....	....	....
$Pb(NO_3)_2$	c	−107.35	....	....	....
	aq	−98.35	−58.64	42.982	75.1
PbO	g	....	....	....	57.4
(red)	c	−52.40	−45.25	33.168	16.2
(yellow)	c	−52.07	−45.05	33.021	16.6
PbO_2	c	−66.12	−52.34	38.364	18.3
Pb_2O	c	−51.2	....	....	....
Pb_3O_4	c	−175.6	−147 6	108.19	50.5
$Pb(OH)_2$	c	−123.0	−100.6	73.738	21
$Pb_3(PO_4)_2$	c	−620.3	−581.4	426.16	84.45
PbS	c	−22.54	−22.15	16.236	21.8
$PbSO_4$	c	−219 50	−193.89	142.119	35.2
PbS_2O_3	c	−150.1	....	....	....
PbSe	c	−18	....	....	....
$PbSeO_4$	c	−148	....	....	....
$PbSiO_3$	c	−258 8	−239.0	175.18	27
Pb_2SiO_4	c	−312.7	−285.7	209.41	43
Lithium					
Li	g	37.07	29.19	−21.396	33.14
	c	0.00	0.00	0 000	6.70
Li^+	aq	−66.54	−70.22	51.470	3.4
LiBr	g	−41	−50	36.6	53.78
	c	−83.72	....	....	....
	aq	−95.45	−94.69	69.483	22.7
$LiBr\cdot H_2O$	c	−158.34	....	....	....
$LiBr\cdot 2H_2O$	c	−229.94	....	....	....
$LiBr\cdot 3H_2O$	c	−301.9	....	....	....
Li_2CO_3	c	−290.54	−270.66	198.390	21.60
	aq	−294.74	−266.66	195.458	−5.9
$LiHCO_3$	aq	−231.73	−210.53	154.316	29.5
LiCl	g	−53	−58	42.5	51.01
	c	−97.70	....	....	....
	aq	−106.58	−101.57	74.449	16.6
$LiCl\cdot H_2O$	c	−170.31	−151.2	110.83	24.8
$LiCl\cdot 2H_2O$	c	−242.1	....	....	....
$LiCl\cdot 3H_2O$	c	−313.5	....	....	....
LiF	c	−146.3	−139.5	102.32	8.57
	aq	−145.21	−136.30	99.906	1.1
LiH	g	30.7	25.2	−18.47	40.77
	c	−21.61	−16.72	12.255	5.9
LiI	g	−16	−26	19.1	55.68
	c	−64.79	....	....	....
	aq	−79.92	−82.57	−60.523	29.5
$LiI\cdot \frac{1}{2}H_2O$	c	−103.8	....	....	....
$LiI\cdot H_2O$	c	−141.16	....	....	....
$LiI\cdot 2H_2O$	c	−213.03	....	....	....
$LiI\cdot 3H_2O$	c	−285.02	....	....	....
Li_3N	c	−47.2	....	....	....
$LiNO_2$	c	−96.6	....	....	....
$LiNO_3$	c	−115.28	....	....	....
	aq	−115.93	−96.63	70.828	38.4
$LiNO_3\cdot 3H_2O$	c	−328.6	....	....	....
Li_2O	c	−142.4	....	....	....
Li_2O_2	c	−151.7	....	....	....
	aq	−159.0	....	....	....
Lithium					
LiOH	c	−116.45	−106.1	77.77	12
	aq	−121.51	−107.82	79.031	0.9
$LiOH\cdot H_2O$	c	−188.77	−164.8	120.80	22
Li_2SO_4	c	−342.83	....	....	....
	aq	−350.01	−317.78	232.928	10.9
$Li_2SO_4\cdot H_2O$	c	−414.20	....	....	....
Lutetium					
Lu	g	87	....	....	44.14
	c	0.00	0.00	0.000	....
Lu^{++}	aq	−160.1	−155.0	113.61	....
$LuCl_3$	c (γ)	−227.9	....	....	....
	aq	−280.2	−249.0	182.51	....
LuI_3	c (β)	−133.2	....	....	....
	aq	−200.2	−192.0	140.73	....
$Lu_2(SO_4)_3$	aq, ∞	−970.9	−842.0	617.17	....
$Lu_2(SO_4)_3\cdot 8H_2O$	c	....	−1308.1	958.82	....
Magnesium					
Mg	g	35.9	27.6	−20.23	35.50
	c	0.00	0.00	0.000	7.77
Mg^{++}	aq	−110.41	−108.99	79.89	−28.2
$Mg_3(AsO_4)_2$	c	−731.3	....	....	....
$MgHAsO_4$	aq	−349.2	....	....	....
$Mg(H_2AsO_4)_2$	aq	−541.0	....	....	....
$Mg(NH_4)AsO_4\cdot 6H_2O$	c	−800.7	....	....	....
Mg_3Bi_2	c	−36.5	....	....	....
$MgBr_2$	c	−123.7	....	....	....
	aq	−168.21	−158.14	115.914	10.4
$MgBr_2\cdot 6H_2O$	c	−575.4	−491.0	359.90	95
$Mg(CN)_2$	aq	−31.9	....	....	....
$MgCN_2$	c	−60.3	....	....	....
$MgCO_3$	c	−266	−246	180.3	15.7
$MgCl_2$	c	−153.40	−141.57	103.769	21.4
	aq	−190.46	−171.69	125.846	−1.9
$MgCl_2\cdot H_2O$	c	−231.15	−206.11	151.076	32.8
$MgCl_2\cdot 2H_2O$	c	−305.99	−267.32	195.942	43.0
$MgCl_2\cdot 4H_2O$	c	−454.00	−390.49	286.224	63.1
$MgCl_2\cdot 6H_2O$	c	−597.42	−305.65	370.635	87.5
$Mg(OH)Cl$	c	−191.3	−175.0	128.27	19.8
$Mg(ClO_4)_2$	c	−140.6	....	....	....
$Mg(ClO_4)_2\cdot 2H_2O$	c	−290.7	....	....	....
$Mg(ClO_4)_2\cdot 4H_2O$	c	−438.6	....	....	....
$Mg(ClO_4)_2\cdot 6H_2O$	c	−583.2	....	....	....
$MgCrO_4$	c	−318.3	....	....	....
	aq	−321.2	....	....	....
MgF_2	c	−263.5	−250.8	183.83	13.68
MgH	g	41	34	−24.9	47.61
MgI_2	c	-86.0	....	....	....
	aq	−137.15	−133.69	97.993	24.1
Mg_3N_2	c	−110.24	....	....	....
$Mg(NO_3)_2$	c	−188.72	−140.63	103.080	39.2
	aq	−209.15	−161.81	118.605	41.8
$Mg(NO_3)_2\cdot 6H_2O$	c	−624.36	....	....	....
MgO	c	−143.84	−136.13	99.781	6.4
MgO_2	c	−148.9	....	....	....
$Mg(OH)_2$	c	−221.00	−199.27	146.062	15.09
$Mg_3(PO_4)_2$	c	−961.5	....	....	....
$Mg(NH_4)PO_4\cdot 6H_2O$	c	−881.0	....	....	....
MgS	c	−83.0	....	....	....
$MgSO_4$	c	−305.5	−280.5	152.83	21.9
	aq	−327.31	−286.33	209.876	−24.0
$MgSO_4\cdot 2H_2O$	c	−381.9	....	....	....
$MgSO_4\cdot 4H_2O$	c	−595.5	....	....	....
$MgSO_4\cdot 6H_2O$	c	−736.6	....	....	....
$MgSO_4\cdot 7H_2O$	c	−808.7	....	....	....
Mg_3Sb_2	c	−68.1	....	....	....
$MgSiO_3$	c	−357.9	−337.2	247.16	16.2
Mg_2SiO_4 (forsterite)	c	−488.2	−459.8	337.03	22.7
Mg_2Sn	c	−17.0	....	....	....
$MgWO_4$	c	−345.2	....	....	....
Manganese					
Mn	g	68.34	58.23	−42.682	41.49
	c (α)	0.00	0.00	0.000	7.59
	c (γ)	0.37	0.33	−0.242	7.72
Mn^{++}	aq	−52.3	−53.4	39.14	−20
$MnBr_2$	c	−90.7	....	....	....
	aq	−110.1	....	....	....
$MnBr_2\cdot H_2O$	c	−164.4	....	....	....
$MnBr_2\cdot 4H_2O$	c	−367.5	....	....	....
$MnBr_3$	aq	−111	....	....	....
Mn_3C	c	−1	−1	0.73	23.6
$MnCO_3$	c	−213.9	−195.4	143.23	20.5
	ppt	−211.0	....	....	....
	aq	−213.9	−179.6	131.64	−32.7
MnC_2O_4	c	−258.2	....	....	....
$MnC_2O_4\cdot 2H_2O$	c	−388.6	....	....	....
$MnC_2O_4\cdot 3H_2O$	c	−455.4	....	....	....
$Mn(C_2H_3O_2)_2$	c	−273.0	....	....	....
	aq	−285.2	....	....	....
$Mn(C_2H_3O_2)_2\cdot 4H_2O$	c	−556.9	....	....	....
$MnCl_2$	c	−115.3	−105.5	77.330	28.0
$MnCl_2\cdot H_2O$	c	−188.5	....	....	....
$MnCl_2\cdot 2H_2O$	c	−276.7	....	....	....

VALUES OF CHEMICAL THERMODYNAMIC PROPERTIES (Continued)

Substance	State	ΔHf°	ΔFf°	Log₁₀ Kf	S°
Manganese					
$MnCl_2 \cdot 4H_2O$	*c*	−407.0	......	......	......
MnF_2	*c*	−189	−179	131.2	22.2
	aq	−209.2	......	......	......
MnI_2	*c*	−59.3	......	......	......
MnI_2	*aq*	−79.0	......	......	......
$MnI_2 \cdot H_2O$	*c*	−127.9	......	......	......
$MnI_2 \cdot 2H_2O$	*c*	−194.5	......	......	......
$MnI_2 \cdot 4H_2O$	*c*	−327.5	......	......	......
$MnI_2 \cdot 6H_2O$	*c*	−451	......	......	......
$Mn(N_3)_2$	*c*	92.2	......	......	......
Mn_5N_2	*c*	−57.8	......	......	......
Mn_3N_2	*c*	−81	......	......	......
$Mn(NO_3)_2$	*c*	−166.32	......	......	......
$Mn(NO_3)_2 \cdot 3H_2O$	*c*	−355.1	......	......	......
$Mn(NO_3)_2 \cdot 6H_2O$	*c*	−566.50	......	......	......
MnO	*g*	34.6	......	......	......
	c	−92.0	−86.8	63.62	14.4
MnO_2	*c*	−124.5	−111.4	81.655	12.7
Mn_2O_3	*c*	−232.1	......	......	......
Mn_3O_4	*c*	−331.4	−306.0	224.29	35.5
$Mn(OH)_2$	*am*	−165.8	−145.9	106.94	21.1
$Mn(OH)_3$	*am*	−212	......	......	......
$Mn_3(PO_4)_2$	*c*	−771.	......	......	......
MnS (green)	*c*	−48.8	−49.9	36.58	18.7
(red)	*c*	−47.6	......	......	......
$MnSO_4$	*c*	−254.24	−228.48	167.473	26.8
$MnSO_4$	*aq*	−269.2	−230.7	169.10	−16
$MnSO_4 \cdot H_2O$	*c* (I)	−328.5	......	......	......
	c (II)	−322.4	......	......	......
$MnSO_4 \cdot 4H_2O$	*c*	−539.3	......	......	......
$MnSO_4 \cdot 5H_2O$	*c*	−609.6	......	......	......
$MnSO_4 \cdot 7H_2O$	*c*	−750.0	......	......	......
$Mn_2(SO_4)_3$	*c*	−666.9	......	......	......
	aq	−699	......	......	......
$MnSiO_3$	*c*	−302.5	−283.3	207.65	21.3
	gls	−294.0	......	......	......
Mercury					
Hg	*g*	14.54	7.59	−5.563	41.8
	lq	0.00	0.00	0.00	18.5
Hg^{++}	*aq*	......	39.38	−28.865	......
Hg_2^{++}	*aq*	......	36.79	−26.967	......
HgBr	*g*	23	18	−13.2	65.0
Hg_2Br_2	*c*	−49.42	−42.714	31.309	50.9
$HgBr_2$	*c*	−40.5	−38.8	......	......
$Hg(CN)_2$	*c*	62.5	......	......	......
$Hg(ONC)_2$	*c*	64	......	......	......
$Hg(CNS)_2$	*c*	48.0	......	......	......
HgC_2O_4	*c*	−161.8	......	......	......
$Hg(C_2H_3O_2)_2$	*c*	−199.4	......	......	......
$Hg_2(C_2H_3O_2)_2$	*c*	−201.1	......	......	......
$Hg(CH_3)_2$	*lq*	18.	......	......	......
$Hg(C_2H_5)_2$	*lq*	15	......	......	......
HgCl	*g*	19	14	−10.3	62.2
Hg_2Cl_2	*c*	−63.32	−50.35	36.906	46.8
$HgCl_2$	*c*	−55.0	−42.2	......	......
HgF	*g*	14	......	......	......
HgH	*g*	58.06	52.60	−38.555	52.42
HgI	*g*	33	23	−16.9	67.1
Hg_2I_2 (yellow)	*c*	−28.91	−26.60	19.497	57.2
HgI_2 (red)	*c*	−25.2	......	......	......
(yellow)	*c*	−24.55	......	......	......
$Hg_2(N_3)_2$	*c*	133	......	......	......
$Hg(NO_3)_2 \cdot \frac{1}{2}H_2O$	*c*	−93.0	......	......	......
$Hg_2(NO_3)_2 \cdot 2H_2O$	*c*	−206.9	......	......	......
HgO (red)	*c*	−21.68	−13.990	10.255	17.2
(yellow)	*c*	−21.56	−13.959	10.232	17.5
Hg_2O	*c*	−21.8	−12.80	......	......
$Hg(OH)_2$	*aq*	......	−66.0	43.38	......
HgS (red)	*c*	−13.90	−11.67	8.554	18.6
(black)	*c*	−12.90	−11.05	8.100	19.9
$HgSO_4$	*c*	−168.3	......	......	......
Hg_2SO_4	*c*	−177.34	−149.12	109.303	47.98
Molybdenum					
Mo	*g*	155.5	144.2	−105.70	43.46
	c	0.00	0.00	0.000	6.83
$MoBr_2$	*c*	−29	......	......	......
$MoBr_3$	*c*	−41	......	......	......
$MoBr_4$	*c*	−45	......	......	......
$MoBr_5$	*c*	−51	......	......	......
Mo_2C	*c*	4.3	2.9	−2.13	19.7
$MoCl_2$	*c*	−44	......	......	......
$MoCl_3$	*c*	−65	......	......	......
$MoCl_4$	*c*	−79	......	......	......
$MoCl_5$	*c*	−90.8	......	......	......
$MoCl_6$	*c*	−90	......	......	......
MoI_2	*c*	−12	......	......	......
MoI_3	*c*	−15	......	......	......
MoI_4	*c*	−18	......	......	......
MoI_5	*c*	−18	......	......	......
Mo_2N	*c*	−8 3	......	......	......
MoO_2	*c*	−130	......	......	......
MoO_3	*c*	−180.33	−161.95	118.707	18.68
	aq	−188.1	......	......	......
Molybdenum					
MoO_4	*aq*	−173.5	......	......	......
MoO_5	*aq*	−155.1	......	......	......
MoS_2	*c*	−55.5	−53.8	39.43	15.1
MoS_3	*c*	−61.2	......	......	......
H_2MoO_4 (white)	*c*	−256.9	......	......	......
$H_2MoO_4 \cdot H_2O$ (yellow)	*c*	−331.4	......	......	......
Neodymium					
Nd	*g*	87	......	......	......
	c	0.00	0.00	0.000	......
Nd^{+++}	*aq*	−171.2	−167.6	122.85	......
$NdCl_3$	*c* (α)	−254.3	......	......	......
	aq	−291.3	−261.6	191.75	......
$NdCl_3 \cdot 6H_2O$	*c*	−692.3	......	......	......
$Nd(OH)_3$	*c*	......	−309.6	226.93	......
NdI_3	*c* (α)	−158.9	......	......	......
	aq	−211.3	−204.6	149.97	......
Nd_2O_3	*c*	−442.0	......	......	......
Nd_2S_3	*c*	−281.8	......	......	......
$Nd_2(SO_4)_3$	*c*	−948.1	......	......	......
	aq	−993.1	−867.2	635.65	......
$Nd_2(SO_4)_3 \cdot 5H_2O$	*c*	−1318.4	......	......	......
$Nd_2(SO_4)_3 \cdot 8H_2O$	*c*	−1524.7	−1334.5	978.17	......
Neon					
Ne	*g*	0.00	0.00	0.000	34.95
Neptunium					
Np	*c*	0.00	0.00	0.000	......
$NpBr_3$	*c*	−174	......	......	......
$NpBr_4$	*c*	−183	......	......	......
$NpCl_3$	*c*	−216	......	......	......
$NpCl_4$	*c*	−237	......	......	......
$NpCl_5$	*c*	−246	......	......	......
NpF_3	*c*	−360	......	......	......
NpF_4	*c*	−428	......	......	......
NpI_3	*c*	−120	......	......	......
Nickel					
Ni	*g*	101.61	90.77	−66.533	43.59
	c	0.00	0.00	0.000	7.20
Ni^{++}	*aq*	−15.3	−11.1	8.136	−38.1
$NiBr_2$	*c*	−54.2	......	......	......
$NiBr_2 \cdot 3H_2O$	*c*	−277.8	......	......	......
Ni_3C	*c*	11.0	......	......	......
$Ni(CN)_2$	*c*	27.1	......	......	......
$NiCO_3$	*c*	......	−146.7	107.53	......
$NiCl_2$	*c*	−75.5	−65.1	47.72	25.6
$NiCl_2 \cdot 2H_2O$	*c*	−220.8	......	......	......
$NiCl_2 \cdot 4H_2O$	*c*	−364.7	......	......	......
$NiCl_2 \cdot 6H_2O$	*c*	−505.8	−410.5	300.89	75.2
NiF_2	*c*	−159.5	......	......	......
$NiF_2 \cdot 4H_2O$	*c*	......	−379.9	278.46	......
NiH	*g*	93	......	......	......
	c	−2.7	......	......	......
NiH_2	*c*	−6.2	......	......	......
NiI_2	*c*	−20.5	......	......	......
$Ni(IO_3)_2$	*c*	−124.5	......	......	......
$Ni(IO_3)_2 \cdot 2H_2O$	*c* (I)	−264.9	......	......	......
	c (II)	−263.9	......	......	......
$Ni(IO_3)_2 \cdot 4H_2O$	*c*	−401.7	......	......	......
$Ni(NO_3)_2$	*c*	−102.2	......	......	......
$Ni(NO_3)_2 \cdot 6H_2O$	*c*	−531.4	......	......	......
NiO	*g*	59.3	51.8	−37.97	57
	c	−58.4	−51.7	37.90	9.22
$Ni(OH)_2$	*c*	−128.6	−108.3	79.382	19
$Ni(OH)_3$	*c*	−162.1	......	......	......
NiS	*c*	−17.5	......	......	......
$NiSO_4$	*c*	−213.0	−184.9	135.53	18.6
	aq	−232.2	−188.4	138.09	−34.0
$NiSO_4 \cdot 6H_2O$ (green)	*c*	−644.98	......	......	......
(blue)	*c*	−642.5	−531.0	389.2	73.1
$NiSO_4 \cdot 7H_2O$	*c*	−712.9	......	......	......
Niobium					
Nb	*g*	184.5	173.7	−127.32	44.49
	c	0.00	0.00	0.000	8.3
Nb_2O_4	*c*	−387.8	......	......	......
Nb_2O_5	*c*	−463.2	......	......	......
Nitrogen (NH_4OH; NH_2OH; N_2H_4; and N_3H; see under ammonia)					
N_2	*g*	0.00	0.00	0.000	45.77
NOBr	*g*	19.56	19.70	−14.44	65.16
NOCl	*g*	12.57	15.86	−11.625	63.0
NF_3	*g*	−27.2	......	......	......
NO	*g*	21.60	20.72	−15.187	50.34
NO_2	*g*	8.09	12.39	−9.082	57.47
N_2O_4	*g*	2.31	23.49	−17.219	72.73
N_2O	*g*	19.49	24.76	−18.149	52.58
N_2O_5	*g*	3.6	......	......	......
	c	−10.0	......	......	......
HNO_3	*lq*	−41.40	−19.10	14.000	37.19
	aq	−49.37	−26.41	19.36	35.0
$HNO_3 \cdot H_2O$	*lq*	−112.96	−78.41	57.474	51.83
$HNO_3 \cdot 3H_2O$	*lq*	−252.20	−193.70	141.980	82.92
Osmium					
Os	*g*	174	163	−119.5	45.97
	c	0.00	0.00	0.000	7.8
OsO_4	*g*	−79.9	−67.9	49.77	65.6

VALUES OF CHEMICAL THERMODYNAMIC PROPERTIES (Continued)

Substance	State	ΔHf°	ΔFf°	Log₁₀ Kf	S°
Osmium					
OsO_4	*g*	−79.9	−67.9	49.77	65.6
(white)	*c*	−91.7	−70.5	51.68	34.7
(yellow)	*c*	−93.4	−70.7	51.82	29.7
	aq	....	−68.59	50.276	....
OsS_2	*c*	−35	....	....	....
H_2OsO_5	*aq*	....	−125.28	91.829	....
Oxygen					
O_2	*g*	0.00	0.00	0.000	49.003
O_3	*g*	34.0	39.06	−28.631	56.8
OH^-	*aq*	−54.957	−37.595	27.56	−2.519
Palladium					
Pd	*g*	93	84	−61.6	39.91
	c	0.00	0.00	0.000	8.9
$PdBr_2$	*c*	−24.9	....	....	....
	aq	−205.6	....	....	....
$Pd(CN)_2$	*c*	52.1	....	....	....
$PdCl_2$	*c*	−45.4	....	....	....
$PdCl_4^{--}$	*aq*	−128.3	−96.7	70.88	41.
H_2PdCl_4	*aq*	−129.3	....	....	....
Pd_2H	*c*	−8.9	....	....	....
PdO	*c*	−20.4	....	....	....
$Pd(OH)_2$	*c*	−92.1	....	....	....
$Pd(OH)_4$	*c*	−169.4	....	....	....
Phosphorus					
P	*g*	75.18	66.71	−48.897	38.98
(white)	*c*	0.00	0.00	0.000	10.6
(red)	*c*	−4.4	....	....	....
(black)	*c*	−10.3	....	....	....
P_2	*g*	33.82	24.60	−18.031	52.13
P_4	*g*	13.12	5.82	−4.266	66.90
PBr_3	*g*	−35.9	−41.2	30.19	83.11
PBr_5	*c*	−66	....	....	....
$POBr_3$	*c*	−114.6	....	....	....
PCl_3	*g*	−73.22	−68.42	50.151	74.49
PCl_5	*g*	−95.35	−77.59	56.872	84.3
$POCl_3$	*g*	−141.5	−130.3	95.508	77.59
PH_3	*g*	2.21	4.36	−3.196	50.2
PI_3	*c*	−10.9	....	....	....
PN	*g*	−20.2	−25.3	18.54	50.45
HPO_3	*c*	−228.2	....	....	....
	aq	−234.9	....	....	....
H_3PO_2	*c*	−145.5	....	....	....
	aq	−145.6	....	....	....
H_3PO_3	*c*	−232.2	....	....	....
	aq	−232.2	....	....	....
H_3PO_4	*c*	−306.2	....	....	....
$H_4P_2O_7$	*c*	−538.0	....	....	....
Platinum					
Pt	*g*	121.6	110.9	−81.288	45.96
	c	0.00	0.00	0.000	10.0
$PtBr_4$	*c*	−41.3	....	....	....
$H_2PtBr_6 \cdot 9H_2O$	*c*	−734.7	....	....	....
$PtCl_2$	*c*	−35.5	....	....	....
$PtCl_4$	*c*	−62.9	....	....	....
$PtCl_4^{--}$	*aq*	−123.4	−91.9	67.36	42.
$PtCl_4 \cdot 5H_2O$	*c*	−425.8	....	....	....
$PtCl_6^{--}$	*aq*	−167.4	123.1	90.231	52.6
H_2PtCl_6	*aq*	−167.3	....	....	....
$H_2PtCl_6 \cdot 6H_2O$	*c*	−572.9	....	....	....
PtI_4	*c*	−21.6	....	....	....
$Pt(OH)_2$	*c*	−87.2	−68.2	49.99	26.5
PtS	*c*	−20.8	....	....	....
PtS_2	*c*	−27.8	....	....	....
Plutonium					
$PuBr_3$	*c*	−187.2	....	....	....
$PuCl_3$	*c*	−230.0	....	....	....
PuOCl	*c*	−222.8	....	....	....
PuF_3	*aq*	−374.6	....	....	....
PuH_2	*c*	−31.7	....	....	....
PuI_3	*c*	−133	....	....	....
PuO_2	*c*	−251	....	....	....
Polonium					
PoO_2	*c*	....	−46.3	33.7	....
Potassium					
K	*g*	21.51	14.62	−10.716	38.30
	c	0.00	0.00	0.000	15.2
K^+	*aq*	−60.04	−67.466	49.452	24.5
$KAl(SO_4)_7$	*c*	−589.24	−534.29	391.627	48.9
$KAl(SO_4)_2 \cdot 12H_2O$	*c*	−1447.74	−1227.8	899.96	164.3
KH_2AsO_4	*c*	−271.5	−237.0	173.72	37.08
KBr	*c*	−93.73	−90.63	66.430	23.05
	aq	−88.94	−92.04	67.464	43.8
KBrO	*aq*	−82.0	....	....	....
$KBrO_3$	*c*	−79.4	−58.2	42.66	35.65
	aq	69.6	−56.6	41.49	63.4
K_2CO_3	*c*	−273.93	....	....	....
$K_2CO_3 \cdot \frac{1}{2}H_2O$	*c*	−210.43	....	....	....
$K_2CO_3 \cdot 1\frac{1}{2}H_2O$	*c*	−283.40	....	....	....
$KHCO_3$	*c*	−229.3	....	....	....
$KC_2H_3O_2$	*c*	−173.2	....	....	....
$K_2C_2O_4$	*c*	−320.8	....	....	....
$K_2C_2O_4 \cdot H_2O$	*c*	−392.17	....	....	....
KCN	*c*	−26.90	....	....	....
KCNO	*c*	−98.5	....	....	....

Substance	State	ΔHf°	ΔFf°	Log₁₀ Kf	S°
Potassium					
KCNS	*c*	−48.62	....	....	....
KCl	*g*	−51.6	−56.2	41.19	57.24
	c	−104.18	−97.592	71.534	19.76
	aq	−100.06	−98.82	72.431	37.7
KClO	*aq*	−85.4	....	....	....
$KClO_3$	*e*	−93.50	−69.29	50.789	34.17
	aq, 100	−84.09	....	....	....
	aq	−83.54	−68.09	49.909	63.5
$KClO_4$	*c*	−103.6	−72.7	53.29	36.1
	aq	−91.45	−70.04	51.338	68.0
K_2CrO_4	*c*	−330.49	....	....	....
$K_2Cr_2O_7$	*c*	−485.90	....	....	....
KF	*c*	−134.46	−127.42	93.397	15.91
$KF \cdot 2H_2O$	*c*	−277.0	−242.7	177.90	36
$KF \cdot 4H_2O$	*c*	−418.0	....	....	....
KHF_2	*c*	−219.98	−203.73	149.331	24.92
KI	*c*	−78.31	−77.03	56.462	24.94
	aq	−73.41	−79.82	58.504	50.6
KIO_3	*c*	−121.5	−101.7	74.54	36.20
	aq	−115.0	−99.9	73.22	52.2
KIO_4	*aq*, 1200	−97.6	....	....	....
$KMnO_4$	*c*	−194.4	−170.6	125.05	41.04
KNH_2	*c*	−28.3	....	....	....
KNO_2	*c*	−88.5	....	....	....
KNO_3	*c*	−117.76	−93.96	68.871	31.77
	aq	−109.41	−93.88	68.813	69.5
K_2O	*c*	−86.4	....	....	....
K_2O_2	*c*	−118	....	....	....
K_2O_4	*c*	−134	....	....	....
KOH	*c*	−101.78	....	....	....
	aq	−115.00	−105.06	77.008	22.0
K_2S	*c*	−100	....	....	....
K_2S_4	*c*	−113.0	....	....	....
K_2SO_3	*c*	−266.9	....	....	....
K_2SO_4	*c*	−342.66	−314.62	230.612	42.0
$KHSO_4$	*c*	−276.8	....	....	....
$K_2S_2O_6$	*c*	−413.6	....	....	....
$K_2S_2O_8$	*c*	−458.3	....	....	....
$K_2S_4O_6$	*c*	−422	....	....	....
K_2Se	*c*	−79.3	....	....	....
Praseodymium					
Pr	*g*	87	....	....	....
	c	0.00	0.00	0.000	....
Pr^{+++}	*aq*	−172.7	−169.1	123.95	....
$PrCl_3$	*c* (α)	−257.8	....	....	....
	aq	−292.8	−263.1	192.85	....
$PrCl_3 \cdot H_2O$	*c*	−330.8	....	....	....
$PrCl_3 \cdot 7H_2O$	*c*	−764.5	....	....	....
PrI_3	*c* (α)	−162.0	....	....	....
	aq	−212.8	−206.1	151.07	....
PrO_2	*c*	−234.0	....	....	....
Pr_2O_3	*c*	−444.5	....	....	....
$Pr(OH)_3$	*c*	....	−310.7	227.74	....
$Pr_2(SO_4)_3$	*aq*	−996.1	−870.2	637.84	....
$Pr_2(SO_4)_3 \cdot 8H_2O$	*c*	....	−1337.0	980.00	....
Promethium					
Pm^{+++}	*aq*	−170.4	....	....	....
$PmCl_3$	*c*	−251.9	....	....	....
	aq	−290.5	....	....	....
Radium					
Ra	*g*	31	23	−16.8	42.15
	c	0.00	0.00	0.000	17
Ra^{++}	*aq*	−126.	−134.5	98.58	13
$RaCl_2 \cdot 2H_2O$	*c*	−351	−311.7	228.47	50
$Ra(NO_3)_2$	*c*	−237	−190.3	139.49	52
RaO	*c*	−125	....	....	....
$RaSO_4$	*c*	−352	−326.0	238.95	34
Radon					
Rn	*g*	0.00	0.00	0.000	42.10
Rhenium					
Re	*g*	189	179	−131.2	45.13
	c	0.00	0.00	0.000	10
ReF_6	*g*	−273	....	....	....
Re_2O_7	*c*	−297.5	....	....	....
Re_2O_8	*c*	−148.3	....	....	....
ReS_2	*c*	−44.3	....	....	....
ReO_4^-	*aq*	−190.3	....	....	....
Rhodium					
Rh	*g*	138	127	−93.09	44.39
	c	0.00	0.00	0.000	7.6
RhCl	*c*	−16	....	....	....
$RhCl_2$	*c*	−36	....	....	....
$RhCl_3$	*c*	−56	....	....	....
RhO	*c*	−21.7	....	....	....
RH_2O	*c*	−22.7	....	....	....
Rh_2O_3	*c*	−68.3	....	....	....
Rubidium					
Rb	*g*	20.51	13.35	−9.785	40.63
	c	0.00	0.00	0.000	16.6
Rb^{+++}	*aq*	−58.9	−67.45	47.974	29.7
RbBr	*c*	−93.03	−90.38	66.247	25.88
	aq	−87.8	−92.02	67.449	49.0
Rb_2CO_3	*c*	−269.6	....	....	....
$Rb_2CO_3 \cdot H_2O$	*c*	−344.2	....	....	....

VALUES OF CHEMICAL THERMODYNAMIC PROPERTIES (Continued)

Substance	State	ΔHf°	ΔFf°	Log_{10} Kf	S°
Rubidium					
$RbHCO_3$	*c*	−228.5	...	...	...
RbCNS	*c*	−54	...	...	...
	aq	−41.7	...	...	...
RbCl	*c*	−102.9	−98.48	...	...
	aq	−98.9	−98.80	72.419	42.9
$RbClO_3$	*c*	−93.8	−69.8	51.16	36.3
	aq	−82.4	−68.07	49.894	68.7
$RbClO_4$	*c*	−103.87	−73.19	53.647	38.4
	aq	−90.3	−70.02	51.324	73.2
RbF	*c*	−131.28	...	...	...
	aq	−137.6	−133.53	97.876	27.4
$RbHF_2$	*c*	−217.3	...	...	...
RbI	*c*	−78.5	−77.8	57.03	28.21
	aq	−72.3	−79.80	58.492	55.[illegible]
$RbNO_3$	*c*	−117.04	...	...	...
	aq	−108.5	−93.86	68.798	64.7
$RbNH_2$	*c*	−25.7	...	...	...
Rb_2O	*c*	−78.9	...	...	...
Rb_2O_2	*c*	−101.7	...	...	...
RbOH	*c*	−98.9	...	...	...
	aq	−113.9	−105.05	77.000	27.2
$RbOH \cdot H_2O$	*c*	−177.8	...	...	...
$RbOH \cdot 2H_2O$	*c*	−250.8	...	...	...
Rb_2S	*c*	−83.2	...	...	...
RbHS	*c*	−62.4	...	...	...
$RbHSO_4$	*c*	−273.7	...	...	...
Rb_2SO_4	*c*	−340.50	...	...	...
	aq	−334.7	−312.24	228.868	33.8
Ruthenium					
Ru	*g*	160	149	−109.2	44.57
	c	0.00	0.00	0.000	6.9
$RuCl_3$	*c*	−63	...	...	...
RuO_2	*c*	−52.5	...	...	...
RuS_2	*c*	−48.1	−44.1	...	...
Samarium					
Sm	*g*	87	...	...	43.74
Sm^{+++}	*aq*	−169.8	−165.9	121.60	...
$SmCl_3$	*c* (α)	−249.8	...	...	...
	aq	−289.9	−259.9	190.50	...
SmI_3	*c* (β)	−153.4	...	...	...
	aq	−209.9	−202.9	148.72	...
$Sm(OH)_3$	*c*	...	−308.8	226.35	...
$Sm_2(SO_4)_3$	*aq*	−990.3	−863.8	633.15	...
$Sm_2(SO_4)_3 \cdot 8H_2O$	*c*	...	−1332.6	976.78	...
Scandium					
Sc	*g*	93	...	...	41.76
Sc^{+++}	*aq*	−148.8	−143.7	105.33	...
$ScBr_3$	*c*	−179.4	...	...	...
	aq	−235.5	−217.4	159.35	...
$ScCl_3$	*c*	−220.8	...	...	...
	aq	−268.9	−237.7	174.23	...
Selenium					
Se	*g*	48.37	38.77	−28.42	42.21
Se (gray, hex.)	*c*	0.00	0.00	0.000	10.0
(red, mon.)	*c*	0.2	...	...	...
	gls	1.05	...	...	...
	ppt	1.05	...	...	...
H_2Se	*g*	20.5	17	−12.468	52.9
	aq	18.1	18.4	−13.49	39.9
SeO_2	*c*	−55.00	...	...	...
H_2SeO_3	*c*	−126.5	...	...	...
	aq	−122.39	−101.8	74.622	45.7
H_2SeO_4	*c*	−128.6	...	...	...
	aq	−145.3	−105.42	77.276	5.7
Silicon					
Si	*g*	88.04	77.41	−56.740	40.12
	c	0.00	0.00	0.000	4.47
$SiBr_4$	*lq*	−95.1	...	...	...
SiC	*c*	−26.7	−26.1	19.13	3.94
$SiCl_4$	*g*	−145.7	−136.2	99.83	79.2
	lq	−153.0	−136.9	100.34	57.2
SiF_4	*g*	−370	−360	263.9	68.0
SiH_4	*g*	−14.8	−9.4	6.89	48.7
SiI_4	*c*	−31.6	...	...	...
Si_3N_4	*c*	−179.3	−154.7	113.39	22.4
SiO_2 (quartz)	*c*	−205.4	−192.4	141.03	10.00
(cristobalite)	*c*	−205.0	−192.1	140.81	10.19
(tridymite)	*c*	−204.8	−191.9	140.66	10.36
	gls	−202.5	−190.9	139.92	11.2
	aq	−201.0	...	...	...
SiS_2 (white)	*c*	−34.7	...	...	...
H_2SiF_6	*aq*	−557.2	...	...	...
H_2SiO_3	*c*	−270.7	...	...	...
H_2SiO_4	*c*	−340.6	...	...	...
H_2SiO_5	*c*	−472.4	...	...	...
$H_6Si_2O_7$	*c*	−611.2	...	...	...
Silver					
Ag	*g*	69.12	59.84	−43.862	41.32
	c	0.00	0.00	0.000	10.21
Ag^+	*aq*	25.31	18.43	−13.51	17.67
AgBr	*c*	−23.78	−22.39	16.807	25.60
Ag_2CO_3	*c*	−120.97	−104.48	76.582	40.0
$AgC_2H_3O_2$	*c*	−93.41	...	...	...
AgCN	*c*	34.94	39.20	−28.733	20.0

Substance	State	ΔHf°	ΔFf°	Log_{10} Kf	S°
Silver					
AgCNO	*c*	−21.1	...	...	...
AgONC	*c*	43.2	...	...	...
AgCNS	*c*	21.0	...	...	...
AgCl	*g*	23.23	−16.79	−12.307	58.5
	c	−30.36	−26.22	19.222	22.97
$AgClO_2$	*c*	0.00	16.0	−11.73	32.16
$AgClO_3$	*c*	−5.73	...	...	...
$AgClO_4$	*c*	−7.75	...	...	...
Ag_2CrO_4	*c*	−170.15	−148.57	108.90	51.8
AgF	*c*	−48.5	−44.2	32.40	20
	aq	−53.35	−47.65	34.927	15.4
AgH	*g*	67.7	60.8	−44.56	48.86
AgI	*c*	−14.91	−15.85	11.618	27.3
$AgIO_3$	*c*	−42.02	−24.08	...	35.7
AgN_3	*c*	66.8	...	...	...
$AgNO_2$	*c*	−10.61	4.74	−3.477	30.62
	aq	−0.09	9.79	−7.176	49.0
$AgNO_3$	*c*	−29.43	−7.69	5.637	33.68
	aq	−24.06	−7.98	5.849	52.67
Ag_2O	*c*	−7.31	−2.59	1.896	29.09
Ag_2O_2	*c*	−6.3	...	...	...
Ag_2S	*c* (α)	−7.60	−9.62	7.051	34.8
	c (β)	−7.01	−9.36	6.860	35.9
Ag_2SO_4	*c*	−170.50	−147.17	107.873	47.8
Sodium					
Na	*g*	25.98	18.67	−13.685	36.72
	c	0.00	0.00	0.000	12.2
Na^+	*aq*	−57.28	−62.59	45.88	14.4
Na_3AsO_4	*c*	−365	...	...	...
$Na_3AsO_4 \cdot 12H_2O$	*c*	−1213.9	...	...	...
$NaBO_2$	*c*	−253	...	...	...
$NaBO_3 \cdot 4H_2O$	*c*	−504.8	...	...	...
$Na_2B_4O_7$	*c*	−777.7	...	...	...
$Na_2B_4O_7 \cdot 4H_2O$	*c*	−1072.9	...	...	...
$Na_2B_4O_7 \cdot 5H_2O$	*c*	−1143.5	...	...	...
$Na_2B_4O_7 \cdot 10H_2O$	*c*	−1497.2	...	...	...
Na_3BiO_4	*c*	−288.	...	...	...
NaBr	*c*	−86.03	...	...	...
	aq	−86.18	−87.16	63.889	33.7
$NaBr \cdot 2H_2O$	*c*	−227.25	...	...	...
NaCN	*c*	−21.46	...	...	...
$NaCN \cdot \frac{1}{2}H_2O$	*c*	−56.19	...	...	...
$NaCN \cdot 2H_2O$	*c*	−162.25	...	...	...
NaCNO	*c*	−95.6	...	...	...
NaCNS	*c*	−41.73	...	...	...
	aq	−40.1	...	...	...
Na_2CO_3	*c*	−270.3	−250.4	183.53	32.5
$Na_2CO_3 \cdot H_2O$	*c*	−341.8	...	...	...
$Na_2CO_3 \cdot 7H_2O$	*c*	−765.1	...	...	...
$Na_2CO_3 \cdot 10H_2O$	*c*	−975.6	...	...	...
$NaHCO_3$	*c*	−226.5	−203.6	149.24	24.4
$NaC_2H_3O_2$	*c*	−169.8	...	...	...
$NaC_2H_3O_2 \cdot 3H_2O$	*c*	−383.50	...	...	...
$Na_2C_2O_4$	*c*	−314.3	...	...	...
NaCl	*g*	−43.50	...	...	...
	c	−98.23	−91.79	67.277	17.30
	aq	−97.302	−93.94	68.856	27.6
NaClO	*aq*	−82.7	...	...	...
$NaClO_2$	*c*	−72.65	...	...	...
$NaClO_3$	*c*	−85.73	...	...	...
	aq, 400	−80.74	−62.8	...	...
	aq, 1000	−80.73	...	...	...
	aq, 5000	−80.74	...	...	...
	aq	−80.78	−63.21	46.332	53.4
$NaClO_4$	*c*	−92.18	...	...	...
	aq	−88.69	−65.16	47.688	57.9
Na_2CrO_4	*c*	−317.6	...	...	...
$Na_2CrO_4 \cdot 4H_2O$	*c*	−601.3	...	...	...
$Na_2Cr_2O_7$	*aq*, 600	−463.4	...	...	...
NaF	*c*	−136.0	−129.3	94.78	14.0
	aq	−135.94	−128.67	94.313	12.1
$NaHF_2$	*c*	−216.6	...	...	...
NaH	*g*	29.88	24.78	−18.163	44.93
	c	−13.7	...	...	...
NaI	*g*	−20.94	...	...	...
	c	−68.84	...	...	...
	aq	−70.65	−70.94	51.998	40.5
$NaI \cdot 2H_2O$	*c*	−211.05	...	...	...
	aq, ∞	−112.3	...	...	...
$NaNH_2$	*c*	−28.4	...	...	...
$NaNO_2$	*c*	−85.9	...	...	...
$NaNO_3$	*c*	−101.54	−87.45	64.100	27.8
	aq	−106.65	−89.00	65.236	49.4
Na_2O	*c*	−99.4	−90.0	65.97	17.4
Na_2O_2	*c*	−120.6	...	...	...
NaOH	*c*	−101.99	...	...	...
	aq	−112.24	−100.184	73.434	11.9
$NaOH \cdot H_2O$	*c*	−175.17	−149.00	109.215	20.2
$NaPO_3$	*c*	−288.6	...	...	...
NaH_2PO_3	*c*	−289.4	...	...	...
$NaH_2PO_3 \cdot 2\frac{1}{2}H_2O$	*c*	−454.8	...	...	...
Na_2HPO_3	*c*	−338.0	...	...	...
$Na_2HPO_3 \cdot 5H_2O$	*c*	−684.2	...	...	...
NaH_2PO_4	*aq*, 300	−367.7	...	...	...

VALUES OF CHEMICAL THERMODYNAMIC PROPERTIES (Continued)

Substance	State	ΔHf°	ΔFf°	Log₁₀ Kf	S°
Sodium					
Na_2HPO_4	*c*	−417.4	....	....	....
$Na_2HPO_4 \cdot 2H_2O$	*c*	−560.2	....	....	....
$Na_2HPO_4 \cdot 7H_2O$	*c*	−913.3	....	....	....
$Na_2HPO_4 \cdot 12H_2O$	*c*	−1266.4	....	....	....
Na_3PO_4	*c*	−460	....	....	....
$Na_3PO_4 \cdot 12H_2O$	*c*	−1309.0	....	....	....
$NaNH_4HPO_4 \cdot 4H_2O$	*c*	−682.7	....	....	....
$NaH_3P_2O_7$	*c*	−602.7	....	....	....
$NaH_3P_2 \cdot O_7H_2O$	*c*	−670.6	....	....	....
$Na_2H_2P_2O_7$	*c*	−663.4	....	....	....
$Na_2H_2P_2O_7 \cdot 6H_2O$	*c*	−1085.5	....	....	....
$Na_3HP_2O_7$	*c*	−711.4	....	....	....
$Na_3HP_2O_7 \cdot H_2O$	*c*	−788.2	....	....	....
$Na_3HP_2O_7 \cdot 6H_2O$	*c*	−1135.7	....	....	....
$Na_4P_2O_7$	*c*	−760.8	....	....	....
$Na_4P_2O_7 \cdot 10H_2O$	*c*	−1468.2	....	....	....
Na_2PbO_3	*c*	−205.	....	....	....
Na_2PtCl_6	*c*	−273.6	....	....	....
$Na_2PtCl_6 \cdot 2H_2O$	*c*	−418.5	....	....	....
$Na_2PtCl_6 \cdot 6H_2O$	*c*	−702.5	....	....	....
Na_2S	*c*	−89.2	....	....	....
$Na_2S \cdot 4\frac{1}{2}H_2O$	*c*	−416.9	....	....	....
$Na_2S \cdot 5H_2O$	*c*	−452.7	....	....	....
$Na_2S \cdot 9H_2O$	*c*	−736.7	....	....	....
Na_2S_4	*c*	−98.4	....	....	....
Na_2SO_3	*c*	−260.6	−239.5	175.55	34.9
$Na_2SO_3 \cdot 7H_2O$	*c*	−753.4	....	....	....
Na_2SO_4	*c*	−330.90	−302.78	221.934	35.73
	aq	−331.46	−302.52	221.743	32.9
$Na_2SO_4 \cdot 10H_2O$	*c*	−1033.48	−870.93	638.380	141.7
$NaHSO_4$	*c*	−269.2	....	....	....
$NaHSO_4 \cdot H_2O$	*c*	−339.2	....	....	....
$Na_2S_2O_3$	*c*	−267.0	....	....	....
$Na_2S_2O_3 \cdot 5H_2O$	*c* (I)	−621.89	....	....	....
	c (II)	−620.60	....	....	....
$Na_2S_2O_5$	*c*	−349.1	....	....	....
$Na_2S_2O_6$	*c*	−399.9	....	....	....
$Na_2S_2O_6 \cdot 2H_2O$	*c*	−542.5	....	....	....
$Na_2S_3O_6 \cdot 3H_2O$	*c*	−623.0	....	....	....
$Na_2S_4O_6 \cdot 2H_2O$	*c*	−550.0	....	....	....
Na_2Se	*c*	−63.0	....	....	....
$Na_2Se \cdot 4\frac{1}{2}H_2O$	*c*	−398.2	....	....	....
$Na_2Se \cdot 9H_2O$	*c*	−709.1	....	....	....
$Na_2Se \cdot 16H_2O$	*c*	−1199.4	....	....	....
$NaHSe$	*c*	−27.8	....	....	....
Na_2SeO_4	*c*	−258.	....	....	....
Na_2SiO_3	*c*	−363	−341	249.9	27.2
$Na_2SiO_3 \cdot 5H_2O$	*c*	−720.0	....	....	....
$Na_2SiO_3 \cdot 9H_2O$	*c*	−1002.0	....	....	....
Na_2SiF_6	*c*	−677	....	....	....
Na_2SnO_3	*c*	−276	....	....	....
Na_4SnO_4	*aq*, 1200	−455.5	....	....	....
Na_2UO_4	*c*	−501	....	....	....
$Na_2U_2O_7 \cdot 1\frac{1}{2}H_2O$	*c*	−880	....	....	....
Na_3VO_4	*c*	−420	....	....	....
Na_2WO_4	*c*	−395	....	....	....
Na_2ZnO_2	*c*	−188	....	....	....
Strontium					
Sr	*g*	39.2	26.3	−19.28	39.33
	c	0.00	0.00	0.000	13.0
Sr^{++}	*aq*	−130.38	−133.2	97.49	−9.4
$Sr_3(AsO_4)_2$	*c*	−800.7	....	....	....
$SrBr_2$	*c*	−171.1	....	....	....
	aq	−188.2	−182.1	133.48	29.2
$SrBr_2 \cdot H_2O$	*c*	−246.2	....	....	....
$SrBr_2 \cdot 6H_2O$	*c*	−604.4	....	....	....
$SrCO_3$	*c*	−291.2	−271 9	199.30	23.2
$Sr(HCO_3)_2$	*aq*	−460.7	−413.8	303.31	36.0
$Sr(C_2H_3O_2)_2$	*c*	−356.7	....	....	....
	aq	−362.7	−311.8	....	....
$Sr(C_2H_3O_2)_2 \cdot \frac{1}{2}H_2O$	*c*	−391.2	....	....	....
SrC_2O_4	*aq*	−327.4	−294.3	215.72	3
$SrC_2O_4 \cdot H_2O$	*c*	......	360.8	264.46	....
$SrC_2O_4 \cdot 2\frac{1}{2}H_2O$	*c*	−504.2	....	....	....
$Sr(CN)_2 \cdot 4H_2O$	*c*	−335.2	....	....	....
$SrCl_2$	*c*	−198.0	−186.7	136.85	28
	aq	−210.43	−195.7	143.45	16.9
$SrCl_2 \cdot H_2O$	*c*	−271.7	....	....	....
$SrCl_2 \cdot 2H_2O$	*c*	−343.7	....	....	....
$SrCl_2 \cdot 6H_2O$	*c*	−627.1	....	....	....
SrF_2	*c*	−290.3	....	....	....
SrH	*g*	52.4	45.8	−33.57	49.43
SrH_2	*c*	−42.3	....	....	....
SrI_2	*c*	−135.5	....	....	....
	aq	−157.1	−157.7	115.59	42.9
$SrI_2 \cdot H_2O$	*c*	−212.2	....	....	....
$SrI_2 \cdot 2H_2O$	*c*	−282.8	....	....	....
$SrI_2 \cdot 6H_2O$	*c*	−571.2	....	....	....
$Sr(N_3)_2$	*c*	48.9	....	....	....
Sr_3N_2	*c*	−93.4	....	....	....
$Sr(NO_2)_2$	*c*	−179.3	....	....	....
$Sr(NO_3)_2$	*c*	−233.25	....	....	....
	aq	−229.02	−185.8	136.19	60.06
$Sr(NO_3)_2 \cdot 4H_2O$	*c*	−514.5	....	....	....
Strontium					
SrO	*c*	−141 1	−133.8	98.07	13.0
SrO_2	*c*	−153.6	....	....	....
$SrO_2 \cdot 8H_2O$	*c*	−722.6	....	....	....
$Sr(OH)_2$	*c*	−229.3	....	....	....
	aq	−240.29	−208.2	152.61	−14.4
$Sr(OH)_2 \cdot H_2O$	*c*	−302.3	....	....	....
$Sr(OH)_2 \cdot 8H_2O$	*c*	−801.2	....	....	....
$Sr_3(PO_4)_2$	*c*	−987.3	....	....	....
$SrHPO_4$	*c*	−431.3	....	....	....
$Sr(H_2PO_4)_2 \cdot H_2O$	*c*	−819.4	....	....	....
SrS	*g*	19	....	....	....
	c	−108.1	....	....	....
$SrSO_4$	*c*	−345.3	−318.9	233.75	29.1
	aq	−347.38	−310.3	227.45	−5.3
Sulfur					
S	*g*	53.25	43.57	−31.9362	40.085
(rhb)	*c*	0.00	0.00	0.000	7.62
(mon)	*c*	0.071	0.023	−0.017	7.78
S_2Br_2	*lq*	−3.6	....	....	....
S_2Cl_2	*lq*	−14.4	....	....	....
S_2Cl_4	*lq*	−24.0	....	....	....
$SOCl_2$	*lq*	−49.2	....	....	....
SO_2Cl_2	*lq*	−93 0	−75	....	....
SF_6	*g*	−262	−237	173.7	69.5
H_2S	*g*	−4.815	−7.892	5.785	49.15
	aq	−9.4	−6.54	4.797	29.2
$H_2S \cdot 6H_2O$	*c*	−431.2	....	....	....
SO_2	*g*	−70.96	−71.79	52.621	59.40
SO_3	*g*	−94.45	−88.52	64.884	61.24
	lq	−104.67	....	....	....
S_2O_7	*c*	−194.3	....	....	....
H_2SO_3	*aq*, 200	−146.82	−128.54	....	....
	aq, 500	−147.17	....	....	....
	aq, 1000	−147.59	....	....	....
	aq, 5000	−148.73	....	....	....
H_2SO_4	*lq*	−193.91	....	....	....
	aq	−216.90	−177.34	129.99	4.1
S^{--}	*aq*	10	20	−14.660	5.3
SO_3^{--}	*aq*	−149.2	−118.8	87.08	10.4
SO_4^{--}	*aq*	−216.90	−177.34	129.99	4.1
$S_2O_3^{--}$	*aq*	−154.	−127.2	93.24	29.
HS^-	*aq*	−4.22	3.01	−2.206	14.6
HSO_3^-	*aq*	−150.09	−126.03	92.378	31.64
HSO_4^-	*aq*	−211.70	−179.94	131.893	30.32
Tantalum					
Ta	*g*	185	175	−128.3	44.24
	c	0.00	0.00	0.000	9.9
TaN	*c*	−58.2	....	....	....
Ta_2O_5	*c*	−499.9	−470.6	344.94	34.2
Technetium					
Tc	*c*	0.00	0.00	0.000	9
Tellurium					
Te	*g*	47.6	38.1	−27.93	43.64
	c	0.00	0.00	0.000	11.88
	(*am. ppt*)	2.7	....	....	....
Te_2	*g*	41.0	29.0	−21.26	64.07
$TeBr_4$	*c*	−49.8	....	....	....
$TeCl_4$	*c*	−77.2	....	....	....
TeF_6	*g*	−315	−292	214.0	80.67
H_2Te	*g*	36.9	33.1	−24.26	56
TeO_2	*c*	−77.69	−64.60	47.353	16.99
H_2TeO_3	*c*	−144.7	−115.7	84.81	47.7
	aq	−144.7	....	....	....
H_2TeO_4	*aq*	−166.7	....	....	....
$H_2TeO_4 \cdot 2H_2O$	*c*	−306.6	−245.3	179.80	47
Terbium					
Tb	*g*	87	....	....	....
	c	0.00	0.00	0.000	....
Tb^{+++}	*aq*	−168.4	−163.9	120.14	....
$TbCl_3$	*c* (β)	−241.6	....	....	....
	aq	−288.5	−257.9	189.04	....
$Tb_2(SO_4)_3$	*aq*	−987.5	−859.8	630.22	....
$Tb_2(SO_4)_3 \cdot 8H_2O$	*c*	....	−1328.2	973.55	....
Thallium					
Tl	*g*	44.5	36.2	−26.53	43.23
	c	0.00	0.00	0.000	15.4
Tl^+	*aq*	1.38	−7.755	5.68	30.4
Tl^{+++}	*aq*	27.7	50.0	−36.65	−106
$TlBr$	*g*	−5.	−14	10.3	63.8
	c	−41.2	−39.7	29.10	26.6
	aq	−27.5	−32.3	23.68	49.7
$TlBr_3$	*c*	−59.0	....	....	....
$TlBr_3 \cdot 4H_2O$	*c*	−334.6	....	....	....
$TlBrO_3$	*c*	−24.0	....	....	....
	aq	−10.2	1.3	−0.953	68.6
$TlCl$	*g*	−16.	−22	16.1	61.1
	c	−48.99	−44.19	32.391	25.9
	aq	−38.64	−39.105	28.663	43.6
$TlCl_3$	*c*	−83.9	....	....	....
$TlCl_3 \cdot 4H_2O$	*c*	−367.7	....	....	....
TlF	*g*	−33	....	....	....
TlH	*g*	48	42	−30.8	51.39
TlI	*c* (II)	−29.7	−29.7	21.77	29.4
	aq	−12.0	−19.9	14.59	56.5

Substance	State	ΔHf°	ΔFf°	Log_{10} Kf	S°
Thallium					
$TlIO_3$	c	....	−47.6	34.89	......
	aq	−53.5	−40.1	29.39	58.1
$TlNO_3$	c	−58.01	−36.07	26.439	38.2
	aq	−47.99	−34.22	25.083	65.6
Tl_2O	c	−41.9	−32.5	23.82	23.8
$TlOH$	c	−56.9	−45.5	33.35	17.3
$Tl(OH)_3$	c	−122.6	....	....	......
Tl_2S	c	−20.8	....	....	......
Tl_2SO_4	c	−221.7	....	....	......
	aq	−214.14	−192.85	141.356	65.1
Thorium					
Th	c	0.00	0.00	0.000	13.6
Th^{++++}	aq	−183.0	....	....	......
$ThBr_4$	c	−227.1	....	....	......
ThC_2	c	−45	....	....	......
$ThCl_4$	c	−285	....	....	......
ThF_4	c	−477	....	....	......
ThH_4	c	−43	....	....	......
ThI_4	c	−131	....	....	......
$Th(NO_3)_4$	aq, 100	−380.48	....	....	......
ThO_2	c	−292	....	....	......
$Th(OH)_4$	c*	−421.5	....	....	......
Th_2S_3	c	−262.0	....	....	......
$Th(SO_4)_2$	c	−602	....	....	......
Thulium					
Tm	c	0.00	0.00	0.000	......
Tm^{+++}	aq	−161.3	−156.5	114.71	......
$TmCl_3$	c (γ)	−229.5	....	....	......
	aq	−281.4	−250.5	183.61	......
TmI_3	c (β)	−137.8	....	....	......
	aq	−201.4	−193.5	141.83	......
Tin					
Sn	g	72	64	−47.6	40.25
(gray), cubic	c (III)	0.6	1.1	−0.806	10.7
(tet), white	c (II)	0.00	0.00	0.000	12.3
$SnBr_2$	c	−63.6	....	....	......
$SnBr_4$	c (I)	−97.1	....	....	......
$SnBr_4 \cdot 8H_2O$	lq	−656.6	....	....	......
$Sn(C_2H_5)_4$	lq	−49	....	....	......
$SnCl_2$	c	−83.6	....	....	......
$SnCl_2 \cdot 2H_2O$	c	−225.9	....	....	......
$SnCl_4$	lq	−130.3	−113.3	83.047	61.8
$SnOCl_2$	aq	−147.5	....	....	......
SnI_2	c	−34.4	....	....	......
SnO	c	−68.4	−61.5	45.08	13.5
SnO_2	c	−138.8	−124.2	91.037	12.5
$Sn(OH)_2$	c	−138.3	−117.6	86.199	23.1
$Sn(OH)_4$	c	−270.5	....	....	......
SnS	c	−18.6	−19.7	14.44	23.6
$Sn(SO_4)_2$	c	−393.4	....	....	......
Titanium					
Ti	g	112	101	−74	43.07
	c	0.00	0.00	0.000	7.24
$TiBr_2$	c	−95	....	....	......
$TiBr_3$	c	−132	....	....	......
$TiBr_4$	c	−155	....	....	......
TiC	c	−54	−53	38.8	5.8
$TiCl$	g	122	....	....	......
$TiCl_2$	c	−114	....	....	......
$TiCl_3$	c	−165	....	....	......
$TiCl_4$	g	....	....	....	84.4
	lq	−179.3	−161.2	118.16	60.4
TiF_2	c	−198	....	....	......
TiF_3	c	−315	....	....	......
TiF_4	c	−370	....	....	......
H_2TiF_6	lq	−555.1	....	....	......
TiI_2	c	−61	....	....	......
TiI_3	c	−80	....	....	......
TiI_4	c	−102	....	....	......
TiN	c	−73.0	−66.1	48.45	7.2
TiO	g	43	....	....	......
	c (II)	....	....	....	8.31
TiO_2 (rutile)	c (III)	−218.0	−203.8	149.38	12.01
(hydrated)	(am. ppt)	−207.	....	....	......
Ti_2O_3	c (II)	−367	−346	253.6	18.83
Ti_3O_5	c	−584	−550	403.1	30.92
Tungsten					
W	g	201.6	191.6	−140.44	41.55
	c	0.00	0.00	0.000	8.0
WC	c	−9.09	....	....	......
WO_2	c	−136.3	....	....	......
WO_3 (yellow)	c	−200.84	−182.47	133.748	19.90
$WO_3 \cdot H_2O$ (H_2WO_4)	c	−279.6	....	....	......
W_2O_5	c	−337.9	....	....	......
WS_2	c	−46.3	−46.2	33.86	23
Uranium					
U	g	125	....	....	......
	c (III)	0.00	0.00	0.000	12.03
U^{+++}	aq	−128.0	−124.4	91.18	−30
U^{++++}	aq	−146.7	−138.4	101.44	−78
UBr_3	c	−170.1	−164.7	120.72	49
UBr_4	c	−196.6	−188.5	138.17	58
UC_2	c	−42	−42	30.8	14
UCl_3	c	−213.0	−196.9	144.32	37.99
UCl_4	c	−251.2	−230.0	168.59	47.4
Uranium					
UCl_5	c	−262.1	−237.4	174.01	62
UCl_6	c	−272.4	−241.5	177.02	68.3
UF_3	c	−357	−339	248.5	26
UF_4	c	−443	−421	308.6	36.1
UF_5	c	−488	−461	337.9	43
UF_6	g	−505	−485	355.5	90.76
UH_3	c	−30.4	....	....	......
UI_3	c	−114.7	−115.3	84.51	56
UI_4	c	−127.0	−126.1	92.43	65
$UIBr_3$	c	−177.1	....	....	......
$UICl_3$	c	−219.9	−204.4	149.82	54
UN	c	−80	−75	55	18
U_2N_3	c	−213	−194	142.2	29
UO_2	c	−270	−257	188.4	18.6
UO_2^+	aq	−247.4	−237.6	174.16	12
UO_2^{++}	aq	−250.4	−236.4	173.28	−17
UO_3	c	−302	−283	207.4	23.57
$UO_3 \cdot H_2O$	c	−375.4	....	....	......
$UO_3 \cdot 2H_2O$	c	−446.2	....	....	......
$UO_4 2H_2O$	c	−436	....	....	......
U_3O_8	c	−898	....	....	......
$U(OH)_3^{+++}$	aq	−204.1	−193.5	141.83	−30
$U(SO_4)_2$	c	−563	....	....	......
UO_2Br_2	aq (dil)	−308.2	....	....	......
$UO_2(C_2H_3O_2)_2$	aq (dil)	−484.0	....	....	......
$UO_2(C_2H_3O_2)_2 \cdot 2H_2O$	c	−624.9	....	....	......
$UO_2(C_2H_3O_2)_2 \cdot NH_4C_2H_3O_2 \cdot 6H_2O$	c	−1045.8	....	....	......
UO_2Cl_2	aq (dil)	−331	....	....	......
UO_2CrO_4	aq (dil)	−456.7	....	....	......
$UO_2CrO_4 \cdot 5\frac{1}{2}H_2O$	c	−838.8	....	....	......
$UO_2(NO_3)_2$	c	−329.2	−273.1	200.18	66
	aq	−349.1	−289.2	211.98	53
$UO_2(NO_3)_2 \cdot H_2O$	c	−404.8	−335.3	245.77	76
$UO_2(NO_3)_2 \cdot 2H_2O$	c	−480.0	−396.6	290.70	85
$UO_2(NO_3)_2 \cdot 3H_2O$	c	−552.2	−454.7	333.29	94
$UO_2(NO_3) \cdot 6H_2O$	c	−764.3	−625.0	458.12	120.85
UO_2SO_4	aq	−467.3	−413.7	303.24	−13
$UO_2SO_4 \cdot 3H_2O$	c	−666.8	−586.0	429.53	63
Vanadium					
V	g	120	109	−79.90	43.55
	c	0.00	0.00	0.000	7.05
VCl_2	c	−108	−97	71.1	23.2
VCl_3	c	−137	−120	87.96	31.3
VCl_4	lq	−138	....	....	......
VN	c	−41	−35	25.7	8.91
VO	g	52	....	....	......
V_2O_2	c	−200	....	....	......
V_2O_3	c	−290	−271	198.6	23.58
V_2O_4	c (II)	−344	−318	233.1	24.65
V_2O_5	c	−373	−344	252.1	31.3
$VOCl_3$	c	−172	....	....	......
$VOSO_4$	c	−312.5	....	....	......
Wolfram	(See Tungsten)				
Xenon					
Xe	g	0.00	0.00	0.000	40.53
$Xe \cdot 6H_2O$	c	−428	....	....	......
Ytterbium					
Yb	g	87	....	....	41.30
	c	0.00	0.00	0.000	......
Yb^{++}	aq	....	−129.0	94.56	......
Yb^{+++}	aq	−160.6	−155.5	113.98	......
$YbCl_3$	c (γ)	−228.7	....	....	......
	aq	−280.7	−249.5	182.88	......
$Yb_2(SO_4)_3$	aq	−971.9	−843.0	617.91	......
$Yb_2(SO_4)_3 \cdot 8H_2O$	c	....	−1308.8	959.33	......
Yttrium					
Y	g	103	....	....	42.87
	c	0.00	0.00	0.000	......
Y^{+++}	aq	−168.0	−164.1	120.28	......
YCl_3	c (γ)	−234.8	....	....	......
	aq	−288.1	−258.1	189.18	......
YI_3	c (β)	−143.2	....	....	......
	aq	−208.1	−201.1	147.40	......
$Y(OH)_3$	c	−337.6	−308.3	225.98	......
$Y_2(SO_4)_3$	aq	−986.7	−860.2	630.51	......
$Y_2(SO_4)_3 \cdot 8H_2O$	c	−1512.6	−1327.3	972.89	......
Zinc					
Zn	g	31.19	22.69	−16.631	38.45
	c	0.00	0.00	0.000	9.95
Zn^{++}	aq	−36.43	−35.184	25.79	−25.45
$ZnBr_2$	c	−78.17	−74.142	54.345	32.84
	aq	−94.23	−84.33	61.814	13.13
$ZnBr_2 \cdot 2H_2O$	c	−220.9	−190.6	139.71	56.2
$Zn(CN)_2$	c	18.4	....	....	......
$ZnCO_3$	c	−194.2	−174.8	128.13	19.7
$Zn(C_2H_3O_2)_2$	c	−258.1	....	....	......
$Zn(C_2H_3O_2)_2 \cdot H_2O$	c	−329.2	....	....	......
$Zn(C_2H_3O_2)_2 \cdot 2H_2O$	c	−398.8	....	....	......
$ZnC_2O_4 \cdot 2H_2O$	c	−373.8	....	....	......
$Zn(CH_3)_2$	lq	6.0	....	....	......
$Zn(C_2H_5)_2$	lq	5.0	....	....	......
$ZnCl_2$	c	−99.40	−88.26	64.690	25.9
	aq	−116.48	−97.88	71.748	0.89

* Soluble

VALUES OF CHEMICAL THERMODYNAMIC PROPERTIES (Continued)

Substance	State	ΔHf°	ΔFf°	Log₁₀ Kf	S°
Zinc					
ZnF_2	*aq*	−187.9	−166	......	......
ZnH	*g*	54.4	47.5	−34.82	48.70
ZnI_2	*c*	−49.98	−50.01	36.657	38.0
	aq	−63.17	−59.88	43.894	26.83
$Zn(NO_3)_2$	*c*	−115.12	......	......	......
$Zn(NO_3)_2 \cdot H_2O$	*c*	−191.99	......	......	......
$Zn(NO_3)_2 \cdot 2H_2O$	*c*	−264.94	......	......	......
$Zn(NO_3)_2 \cdot 4H_2O$	*c*	−405.76			
$Zn(NO_3)_2 \cdot 6H_2O$	*c*	−550.92	......	......	......
ZnO	*c*	−83.17	−76.05	55.744	10.5
$Zn(OH)_2$ (stable)	*c* (I)	−153.5	......	......	......
(unstable)	*c* (II)	−153.1	......	......	......
ZnS	*g*	−14.0	......	......	......
(sphalerite)	*c* (II)	−48.5	−47.4	34.74	13.8
(würtzite)	*c* (I)	−45.3	......	......	......
$ZnSO_4$	*c*	−233.88	−208.31	152.688	29.8
	aq	−253.33	−212.52	155 774	−21.3
$ZnSO_4 \cdot H_2O$	*c*	−310.6	−269.9	197 83	34.9
$ZnSO_4 \cdot 6H_2O$	*c*	−663.3	−555.0	406.81	86.8
$ZnSO_4 \cdot 7H_2O$	*c*	−735.1	−611.9	448.51	92.4
Zirconium					
Zr	*g*	125	115	−84.29	43.31
	c	0.00	0.00	0.000	9.18
$ZrBr_2$	*c*	−120	......	......	......
$ZrBr_3$	*c*	−174	......	......	......
$ZrBr_4$	*c*	−192	......	......	......
ZrC	*c*	−45	......	......	......
$ZrCl_2$	*c*	−145	......	......	......
$ZrCl_3$	*c*	−208	......	......	......
$ZrCl_4$	*c*	−230	−209	153.2	44.5
ZrF_2	*c*	−230	......	......	......
ZrF_3	*c*	−350	......	......	......
ZrF_4	*c*	−445	......	......	......
ZrI_2	*c*	−90	......	......	......
ZrI_3	*c*	−128	......	......	......
ZrI_4	*c*	−130	......	......	......
ZrN	*c*	−82.2	−75.4	55.27	9.23
$ZrO(NO_3)_2 \cdot 2H_2O$	*c*	−456.3	......	......	......
$ZrO(NO_3)_2 \cdot 3H_2O$	*c*	−527.3	......	......	......
$ZrO(NO_3)_2 \cdot 3\frac{1}{2}H_2O$	*c*	−562.8	......	......	......
$ZrO(NO_3)_2 \cdot 6H_2O$	*c*	−737.6	......	......	......
ZrO_2	*c*	−258.2	−244.4	179.14	12.03
$Zr(OH)_4$	*c*	−411.2	......	......	......
$Zr(OH)_4 \cdot H_2O$	*c*	−482.9	......	......	......
$Zr(OH)_4 \cdot 2H_2O$	*c*	−554.2	......	......	......
$Zr(SO_4)_2$	*c*	−597.4	......	......	......
$Zr(SO_4)_2 \cdot H_2O$	*c*	−677.9	......	......	......
$Zr(SO_4)_2 \cdot 4H_2O$	*c*	−893.1	......	......	......

Formula	Name	State	ΔHf°	ΔFf°	Log₁₀ Kf	S°
C	Carbon	*g*	171.698	160.845	−117.897	37.761
	(diamond)	*c*	0.4532	0.6850	−0.5021	0.5829
	(graphite)	*c*	0.000	0.000	0.000	1.3609
CO	Carbon monoxide	*g*	−26.4157	−32.8079	24.048	47.301
CO_2	Carbon dioxide	*g*	−94.052	−94.260	69.091	51.061
		aq	−98.69	−92.31	67.662	29.0
CO_3^{--}	Carbonate ion	*aq*	−161.63	−126.22	92.517	−12.7
CH_3	Methyl	*g*	32.0	......	......	......
CH_4	Methane	*g*	−17.889	−12.140	8.8985	44.50
$HCOO^-$	Formate ion	*aq*	−98.0	−80.0	58.64	21.9
HCO_3^-	Bicarbonate ion	*aq*	−165.18	−140.31	102.845	22.7
CH_2O	Formaldehyde	*g*	−27.7	−26.3	19.28	52.26
CH_2O_2	Formic acid, monomer	*g*	−86.67	−80.24	58.815	60.0
	Formic acid	*lq*	−97.8	−82.7	60.62	30.82
		aq	−98.0	−85.1	62.38	39.1
H_2CO_3	Carbonic acid	*aq*	−167.0	−149.00	109.215	45.7
CH_3OH	Methanol	*g*	−48.08	−38.69	28.359	56.8
		lq	−57.02	−39.73	29.122	30.3
CF_4	Tetrafluoromethane	*g*	−162.5	−151.8	111.27	62.7
CCl_4	Tetrachloromethane	*g*	−25.5	−15.3	11.21	73.95
		lq	−33.3	−16.4	12.02	51.25
$COCl_2$	Carbonyl chloride	*g*	−53.30	−50.31	36.877	69.13
CH_3Cl	Chloromethane	*g*	−19.6	−14.0	10.26	55.97
CH_2Cl_2	Dichloromethane	*g*	−21.	−14.	10.3	64.68
		lq	−28	−15.1	11.07	42.7
$CHCl_3$	Trichloromethane	*g*	−24	−16	11.7	70.86
		lq	−31.5	−17.1	12.53	48.5
CBr_4	Tetrabromomethane	*g*	12.	8.6	−6.30	85.6
CH_3Br	Bromomethane	*g*	−8.5	−6.2	4.54	58.74
CH_2Br_2	Dibromomethane	*g*	−1.	−1.4	1.03	70.16
$CHBr_3$	Tribromomethane	*g*	6.	3.8	−2.78	79.18
		lq	−4.8	0.7	−0.51	53.0
CH_3I	Iodomethane	*g*	4.9	5.3	−3.88	60.85
		lq	−2.0	4.9	−3.59	38.9

VALUES OF CHEMICAL THERMODYNAMIC PROPERTIES (Continued)

Formula	Name	State	ΔHf°	ΔFf°	Log₁₀ Kf	S°
CH_2I_2	Diodomethane	*lq*	15.9	....	....	....
CHI_3	Triodomethane	*c*	33.6	....	....	....
CS_2	Carbon disulfide	*g*	27.55	15.55	−11.398	56.84
		lq	21.0	15.2	−11.14	36.10
COS	Carbon oxysulfide	*g*	−32.80	−40.45	29.649	55.34
CN^-	Cyanide ion	*aq*	36.1	39.6	−29.03	28.2
CNO^-	Cyanate ion	*aq*	−33.5	−23.6	17.30	31.1
$C(NO_2)_4$	Tetranitromethane	*lq*	8.8	....	....	....
HCN	Hydrogen cyanide	*g*	31.2	28.7	−21.04	48.23
		lq	25.2	29.0	−21.26	26.97
	Hydrocyanic acid	*aq*	25.2	26.8	−19.64	30.8
CH_5N	Methyl amine	*g*	−6.7	6.6	−4.84	57.73
CH_2N_2	Cyanamide	*c*	9.2	....	....	....
CH_5N_3	Guanidine	*c*	−17.0	....	....	....
$HCNO$	Cyanic acid	*aq*	−35.1	−28.9	21.18	43.6
CH_3O_2N	Nitromethane	*lq*	−21.28	2.26	−1.656	41.1
CH_4ON_2	Urea	*c*	−79.634	−47.120	34.538	25.00
		aq	−76.30	−48.72	35.711	41.55
$CH_6O_2N_2$	Ammonium carbamate	*c*	−154.21	−109.47	80.240	39.70
$CNCl$	Cyanogen chloride	*g*	34.5	32.9	−24.12	56.31
CH_6NCl	Methylamine hydrochloride	*c*	−68.31	−35.09	25.720	33.13
CNI	Cyanogen iodide	*g*	54.6	47.7	−34.96	61.26
		c	40.4	42.6	−31.23	30.8
		aq	43.2	43.3	−31.74	37.9
CNS^-	Thiocyanate ion	*aq* (very dilute)	17.2	....	....	....
CH_4N_2S	Thiourea	*c*	−22.1	....	....	....
$C_2O_4^{--}$	Oxalate ion	*aq*	−197.0	−161.3	118.23	12.2
C_2H_2	Ethyne (acetylene)	*g*	54.194	50.000	−36.649	47.997
$C_2H_2 \cdot 6H_2O$	Ethyne hexahydrate	*c*	−371.1	....	....	....
C_2H_4	Ethene (ethylene)	*g*	12.496	16.282	−11.935	52.45
C_2H_6	Ethane	*g*	−20.236	−7.860	5.7613	54.85
$HC_2O_4^-$	Bioxalate ion	*aq*	−195.5	−167.1	122.48	36.7
C_2H_2O	Ketene	*g*	−14.6	....	....	....
$C_2H_2O_2$	Glyoxal	*c*	−83.7	....	....	....
$C_2H_2O_4$	Oxalic acid	*c*	−197.6	−166.8	122.28	28.7
$C_2H_3O_2^-$	Acetate ion	*aq*	−116.843	....	....	....
C_2H_4O	Ethylene oxide	*g*	−12.19	−2.79	2.045	58.1
C_2H_4O	Acetaldehyde	*g*	−39.76	−31.96	23.426	63.5
$C_2H_4O_2$	Acetic acid	*lq*	−116.4	−93.8	68.75	38.2
$C_2H_4O_4$	Formic acid dimer	*g*	−187.7	−163.8	120.09	83.1
C_2H_6O	Ethanol	*g*	−56.24	−40.30	29.536	67.4
		lq	−66.356	−41.77	30.617	38.4
	Dimethyl ether	*g*	−44.3	−27.3	19.98	63.72
$C_2H_6O_2$	1-2 Ethanediol	*lq*	−108.58	−77.12	56.528	39.9
C_2H_5Cl	Chloroethane	*g*	−25.1	−12.7	9.331	65.90
$C_2H_4Cl_2$	1-2 Dichloroethane	*lq*	−39.7	−19.2	14.11	49.84
C_2HOCl_3	Chloral trichloroacetaldehyde	*lq*	−51.1	....	....	....
$C_2HO_2Cl_3$	Trichloroacetic acid	*c*	−122.8	....	....	....
C_2H_5Br	Bromoethane	*g*	−13.0	....	....	....
		lq	−20.4	....	....	....
$C_2H_4Br_2$	1-2 Dibromoethane	*lq*	−19.30	−4.94	3.621	53.37
C_2H_5I	Iodoethane	*lq*	−7.4	....	....	....
$C_2H_4I_2$	1-2 Diiodoethane	*c*	0.1	....	....	....
C_2H_6S	Ethanethiol	*lq*	−16.	....	....	....
	Dimethyl sulfide	*g*	−6.9	3.7	−2.74	68.28
		lq	−13.6	3.4	−2.50	46.94
C_2N_2	Cyanogen	*g*	73.60	70.81	−51.903	57.86
C_2H_3N	Acetonitrile	*g*	21.0	25.2	−18.47	58.18
		lq	12.7	24.0	−17.59	34.5
	Methyl isocyanide	*g*	35.9	40.0	−29.32	58.78
C_2H_7N	Ethylamine	*g*	−11.6	....	....	....
$C_2H_5O_2N$	Aminoacetic acid (glycine)	*c*	−126.33	−88.61	64.950	26.1
	Nitroethane	*lq*	−30.	....	....	....
C_2H_3SN	Methyl thiocyanate	*lq*	25.	....	....	....
	Methyl isothiocyanate	*c*	18.	....	....	....
$C_2H_7O_3SN$	2-Aminoethylsulfonic acid (taurine)	*c*	−181.66	....	....	....

THERMODYNAMIC FUNCTIONS OF COPPER, SILVER AND GOLD

1 cal = 4.1840J H°_0 is the enthalpy of the solid at 0°K and 1 atm pressure

T °K	C°_P J/deg-mol			$H^\circ_T - H^\circ_0$ J/mol			$(H^\circ_T - H^\circ_0)/T$ J/deg-mol		
	Cu	Ag	Au	Cu	Ag	Au	Cu	Ag	Au
1.00	0.000743	0.000818	0.00118	0.000359	0.000367	0.000478	0.000359	0.000367	0.000478
2.00	0.00177	0.00265	0.00504	0.00158	0.00197	0.00326	0.000790	0.000987	0.00163
3.00	0.00337	0.00650	0.0141	0.00409	0.00633	0.0123	0.00136	0.00211	0.00410
4.00	0.00582	0.0134	0.0306	0.00860	0.0160	0.0340	0.00215	0.00399	0.00849
5.00	0.00943	0.0243	0.0570	0.0161	0.0344	0.0768	0.00322	0.00689	0.0154
6.00	0.0145	0.0403	0.0955	0.0279	0.0663	0.152	0.00466	0.0110	0.0253
7.00	0.0213	0.0626	0.149	0.0456	0.117	0.273	0.00652	0.0167	0.0390
8.00	0.0301	0.0927	0.220	0.0712	0.194	0.456	0.00889	0.0243	0.0570
9.00	0.0414	0.132	0.313	0.107	0.306	0.720	0.0119	0.0340	0.0800
10.00	0.0555	0.183	0.431	0.155	0.462	1.090	0.0155	0.0462	0.109
11.00	0.0727	0.247	0.577	0.219	0.676	1.592	0.0199	0.0614	0.145
12.00	0.0936	0.325	0.755	0.302	0.961	2.255	0.0251	0.0801	0.188
13.00	0.119	0.421	0.963	0.407	1.332	3.112	0.0313	0.102	0.239
14.00	0.149	0.535	1.203	0.541	1.809	4.193	0.0386	0.129	0.299
15.00	0.184	0.670	1.474	0.706	2.409	5.529	0.0471	0.161	0.369
16.00	0.225	0.826	1.772	0.910	3.155	7.149	0.0569	0.197	0.447
17.00	0.273	1.002	2.096	1.158	4.067	9.081	0.0681	0.239	0.534
18.00	0.328	1.199	2.442	1.458	5.166	11.35	0.0810	0.287	0.630
19.00	0.390	1.414	2.807	1.816	6.471	13.97	0.0956	0.341	0.735
20.00	0.462	1.647	3.187	2.242	8.001	16.97	0.112	0.400	0.848
25.00	0.963	3.066	5.245	5.703	19.62	37.97	0.228	0.785	1.519
30.00	1.693	4.774	7.375	12.25	39.14	69.53	0.408	1.305	2.318
35.00	2.638	6.612	9.395	22.99	67.58	111.5	0.657	1.931	3.186
40.00	3.740	8.419	11.22	38.89	105.2	163.2	0.972	2.630	4.079
45.00	4.928	10.11	12.86	60.54	151.6	223.4	1.345	3.368	4.965
50.00	6.154	11.66	14.29	88.23	206.1	291.4	1.765	4.121	5.828
55.00	7.385	13.04	15.52	122.1	267.9	366.0	2.220	4.871	6.654
60.00	8.595	14.27	16.59	162.0	336.2	446.3	2.701	5.604	7.438
65.00	9.759	15.35	17.51	208.0	410.4	531.6	3.199	6.313	8.179
70.00	10.86	16.30	18.31	259.5	489.5	621.2	3.708	6.993	8.874
75.00	11.89	17.14	19.01	316.4	573.2	714.6	4.219	7.642	9.528
80.00	12.85	17.87	19.63	378.4	660.7	811.2	4.729	8.259	10.14
85.00	13.74	18.53	20.17	444.9	751.8	910.7	5.234	8.844	10.71
90.00	14.56	19.11	20.64	515.7	845.9	1013.	5.730	9.399	11.25
95.00	15.31	19.63	21.06	590.4	942.8	1117.	6.215	9.924	11.76
100.00	16.01	20.10	21.44	668.7	1042.	1223.	6.687	10.42	12.23
105.00	16.64	20.52	21.77	750.3	1144.	1331.	7.146	10.89	12.68
110.00	17.22	20.89	22.06	835.0	1247.	1441.	7.591	11.34	13.10
115.00	17.76	21.23	22.33	922.5	1353.	1552.	8.021	11.76	13.49
120.00	18.25	21.54	22.56	1013.	1460.	1664.	8.438	12.16	13.87
125.00	18.70	21.82	22.78	1105.	1568.	1777.	8.839	12.54	14.22
130.00	19.12	22.07	22.97	1199.	1678.	1892.	9.227	12.91	14.55
135.00	19.51	22.31	23.15	1296.	1789.	2007.	9.601	13.25	14.87
140.00	19.87	22.52	23.31	1395.	1901.	2123.	9.961	13.58	15.17
145.00	20.20	22.72	23.45	1495.	2014.	2240.	10.31	13.89	15.45
150.00	20.51	22.90	23.59	1597.	2128.	2358.	10.64	14.19	15.72
155.00	20.79	23.07	23.70	1700.	2243.	2476.	10.97	14.47	15.97
160.00	21.05	23.22	23.81	1804.	2358.	2595.	11.28	14.74	16.22
165.00	21.30	23.37	23.91	1910.	2475.	2714.	11.58	15.00	16.45
170.00	21.53	23.50	24.00	2017.	2592.	2834.	11.87	15.25	16.67
175.00	21.74	23.63	24.08	2125.	2710.	2954.	12.15	15.49	16.88
180.00	21.94	23.75	24.15	2235.	2828.	3075.	12.42	15.71	17.08
185.00	22.13	23.86	24.22	2345.	2947.	3196.	12.68	15.93	17.27
190.00	22.31	23.96	24.29	2456.	3067.	3317.	12.93	16.14	17.46
195.00	22.47	24.06	24.35	2568.	3187.	3438.	13.17	16.34	17.63
200.00	22.63	24.16	24.41	2681.	3308.	3650.	13.40	16.54	17.80
205.00	22.77	24.24	24.48	2794.	3429.	3683.	13.63	16.72	17.96
210.00	22.91	24.33	24.54	2908.	3550.	3805.	13.85	16.90	18.12
215.00	23.04	24.41	24.60	3023.	3672.	3928.	14.06	17.08	18.27
220.00	23.17	24.49	24.65	3139.	3794.	4051.	14.27	17.25	18.41
225.00	23.28	24.56	24.71	3255.	3917.	4174.	14.47	17.41	18.55
230.00	23.39	24.63	24.76	3372.	4040.	4298.	14.66	17.56	18.69
235.00	23.50	24.69	24.82	3489.	4163.	4422.	14.85	17.71	18.82
240.00	23.60	24.76	24.87	3607.	4287.	4546.	15.03	17.86	18.94
245.00	23.69	24.82	24.92	3725.	4411.	4671.	15.20	18.00	19.06
250.00	23.78	24.88	24.97	3844.	4535.	4796.	15.37	18.14	19.18
255.00	23.86	24.93	25.02	3963.	4659.	4921.	15.54	18.27	19.30
260.00	23.94	24.99	25.07	4082.	4784.	5046.	15.70	18.40	19.41
265.00	24.02	25.04	25.12	4202.	4909.	5171.	15.86	18.53	19.51
270.00	24.09	25.09	25.17	4322.	5035.	5297.	16.01	18.65	19.62
273.15	24.13	25.12	25.20	4398.	5114.	5376.	16.10	18.72	19.68
275.00	24.15	25.14	25.21	4443.	5160.	5423.	16.16	18.76	19.72
280.00	24.22	25.19	25.26	4564.	5286.	5549.	16.30	18.88	19.82
285.00	24.28	25.24	25.31	4685.	5412.	5676.	16.44	18.99	19.91
290.00	24.34	25.28	25.35	4807.	5538.	5802.	16.57	19.10	20.01
295.00	24.40	25.32	25.39	4929.	5665.	5929.	16.71	19.20	20.10
298.15	24.44	25.35	25.42	5005.	5745.	6009.	16.79	19.27	20.15
300.00	24.46	25.37	25.43	5051.	5792.	6056.	16.84	19.31	20.19

THERMODYNAMIC FUNCTIONS OF COPPER, SILVER AND GOLD (Continued)

From NSRDS-NBS George T. Furukawa, William G. Saba and Martin L. Reilly

T °K	S°_T J/deg-mol			$-(G^\circ_T - H^\circ_0)$ J/mol			$-(G^\circ_T - H^\circ_0)/T$ J/deg-mol		
	Cu	Ag	Au	Cu	Ag	Au	Cu	Ag	Au
1.00	0.000711	0.000706	0.000880	0.000351	0.000339	0.000402	0.000351	0.000339	0.000402
2.00	0.00152	0.00175	0.00266	0.00145	0.00152	0.00206	0.000727	0.000762	0.00103
3.00	0.00251	0.00347	0.00620	0.00345	0.00406	0.00631	0.00115	0.00135	0.00210
4.00	0.00379	0.00619	0.0123	0.00657	0.00879	0.0153	0.00164	0.00220	0.00383
5.00	0.00546	0.0103	0.0218	0.0112	0.0169	0.0321	0.00223	0.00338	0.00641
6.00	0.00760	0.0160	0.0354	0.0176	0.0299	0.0603	0.00294	0.00498	0.0100
7.00	0.0103	0.0238	0.0539	0.0265	0.0496	0.104	0.00379	0.00709	0.0149
8.00	0.0137	0.0341	0.0782	0.0385	0.0783	0.170	0.00481	0.00979	0.0212
9.00	0.0179	0.0472	0.109	0.0542	0.119	0.263	0.00602	0.0132	0.0292
10.00	0.0229	0.0636	0.148	0.0746	0.174	0.391	0.00746	0.0174	0.0391
11.00	0.0290	0.0839	0.196	0.100	0.247	0.562	0.00913	0.0225	0.0511
12.00	0.0362	0.109	0.253	0.133	0.343	0.786	0.0111	0.0286	0.0655
13.00	0.0447	0.138	0.322	0.173	0.466	1.073	0.0133	0.0359	0.0825
14.00	0.0545	0.174	0.402	0.223	0.622	1.434	0.0159	0.0444	0.102
15.00	0.0660	0.215	0.494	0.283	0.815	1.880	0.0189	0.0544	0.125
16.00	0.0791	0.263	0.598	0.355	1.054	2.426	0.0222	0.0659	0.152
17.00	0.0941	0.318	0.715	0.442	1.344	3.081	0.0260	0.0790	0.181
18.00	0.111	0.381	0.845	0.544	1.693	3.861	0.0302	0.0940	0.214
19.00	0.131	0.452	0.987	0.665	2.109	4.775	0.0350	0.111	0.251
20.00	0.152	0.530	1.140	0.806	2.599	5.838	0.0403	0.130	0.292
25.00	0.305	1.043	2.069	1.917	6.446	13.76	0.0767	0.258	0.550
30.00	0.541	1.750	3.214	3.995	13.35	26.89	0.133	0.445	0.896
35.00	0.871	2.623	4.505	7.487	24.22	46.14	0.214	0.692	1.318
40.00	1.294	3.625	5.881	12.86	39.79	72.08	0.322	0.995	1.802
45.00	1.802	4.715	7.299	20.57	60.61	105.0	0.457	1.347	2.334
50.00	2.385	5.862	8.729	31.01	87.04	145.1	0.620	1.741	2.902
55.00	3.029	7.040	10.15	44.52	119.3	192.3	0.809	2.169	3.496
60.00	3.724	8.228	11.55	61.38	157.5	246.5	1.023	2.624	4.109
65.00	4.458	9.414	12.91	81.82	201.6	307.7	1.259	3.101	4.734
70.00	5.222	10.59	14.24	106.0	251.6	375.6	1.514	3.594	5.366
75.00	6.007	11.74	15.53	134.1	307.4	450.0	1.788	4.099	6.001
80.00	6.806	12.87	16.78	166.1	368.9	530.8	2.076	4.612	6.635
85.00	7.612	13.97	17.98	202.1	436.1	617.7	2.378	5.130	7.267
90.00	8.421	15.05	19.15	242.2	508.6	710.6	2.691	5.652	7.895
95.00	9.229	16.10	20.28	286.4	586.5	809.1	3.014	6.174	8.517
100.00	10.03	17.12	21.37	334.5	669.6	913.3	3.345	6.696	9.133
105.00	10.83	18.11	22.42	386.7	757.7	1023.	3.683	7.216	9.740
110.00	11.62	19.07	23.44	442.8	850.6	1137.	4.025	7.733	10.34
115.00	12.39	20.01	24.43	502.8	948.3	1257.	4.372	8.246	10.93
120.00	13.16	20.92	25.38	566.7	1051.	1382.	4.723	8.755	11.51
125.00	13.91	21.80	26.31	634.4	1157.	1511.	5.075	9.260	12.09
130.00	14.66	22.66	27.20	705.8	1269.	1645.	5.429	9.759	12.65
135.00	15.39	23.50	28.07	780.9	1384.	1783.	5.785	10.25	13.21
140.00	16.10	24.32	28.92	859.7	1504.	1925.	6.140	10.74	13.75
145.00	16.80	25.11	29.74	941.9	1627.	2072.	6.496	11.22	14.29
150.00	17.49	25.88	30.54	1028.	1755.	2223.	6.851	11.70	14.82
155.00	18.17	26.64	31.31	1117.	1886.	2377.	7.206	12.17	15.34
160.00	18.84	27.37	32.07	1209.	2021.	2536.	7.559	12.63	15.85
165.00	19.49	28.09	32.80	1305.	2160.	2698.	7.910	13.09	16.35
170.00	20.13	28.79	33.52	1404.	2302.	2864.	8.260	13.54	16.85
175.00	20.75	29.47	34.21	1506.	2448.	3033.	8.608	13.99	17.33
180.00	21.37	30.14	34.89	1612.	2597.	3206.	8.954	14.43	17.81
185.00	21.97	30.79	35.55	1720.	2749.	3382.	9.298	14.86	18.28
190.00	22.57	31.43	36.20	1831.	2904.	3561.	9.639	15.29	18.74
195.00	23.15	32.05	36.83	1946.	3063.	3744.	9.978	15.71	19.20
200.00	23.72	32.66	37.45	2063.	3225.	3930.	10.31	16.12	19.65
205.00	24.28	33.26	38.05	2183.	3390.	4118.	10.65	16.54	20.09
210.00	24.83	33.85	38.64	2306.	3558.	4310.	10.98	16.94	20.52
215.00	25.37	34.42	39.22	2431.	3728.	4505.	11.31	17.34	20.95
220.00	25.90	34.98	39.79	2559.	3902.	4702.	11.63	17.74	21.37
225.00	26.42	35.53	40.34	2690.	4078.	4903.	11.96	18.12	21.79
230.00	26.94	36.07	40.89	2824.	4257.	5106.	12.28	18.51	22.20
235.00	27.44	36.60	41.42	2960.	4439.	5312.	12.59	18.89	22.60
240.00	27.94	37.12	41.94	3098.	4623.	5520.	12.91	19.26	23.00
245.00	28.42	37.63	42.46	3239.	4810.	5731.	13.22	19.63	23.39
250.00	28.90	38.14	42.96	3382.	4999.	5945.	13.53	20.00	23.78
255.00	29.37	38.63	43.46	3528.	5191.	6161.	13.83	20.36	24.16
260.00	29.84	39.11	43.94	3676.	5386.	6379.	14.14	20.71	24.53
265.00	30.30	39.59	44.42	3826.	5582.	6600.	14.44	21.07	24.91
270.00	30.75	40.06	44.89	3979.	5782.	6823.	14.74	21.41	25.27
273.15	31.02	40.35	45.18	4076.	5908.	6965.	14.92	21.63	25.50
275.00	31.19	40.52	45.35	4134.	5983.	7049.	15.03	21.76	25.63
280.00	31.62	40.97	45.81	4291.	6187.	7277.	15.32	22.10	25.99
285.00	32.05	41.42	46.25	4450.	6393.	7507.	15.61	22.43	26.34
290.00	32.48	41.86	46.69	4611.	6601.	7739.	15.90	22.76	26.69
295.00	32.89	42.29	47.13	4775.	6811.	7974.	16.19	23.09	27.03
298.15	33.15	42.56	47.40	4879.	6945.	8123.	16.36	23.29	27.24
300.00	33.30	42.72	47.56	4940.	7024.	8211.	16.47	23.41	27.37

VALUES OF CHEMICAL THERMODYNAMIC PROPERTIES OF HYDROCARBONS

The values in this table are for the ideal gas state at 298.15 K. The units for ΔHf°, ΔFf°, and $Log_{10}Kf$ are Kcal/g mol. The units for absolute entropy, S°, are cal/°K g mol.

It is frequently possible to calculate values for compounds not listed since the following increments are known for an addition of a methylene group, CH_2, to the following types of compounds:

Normal alkyl cyclohexanes
Normal alkyl benzenes
Normal alkyl cyclopentanes
Normal monoolefins (1-alkenes)
Normal acetylenes (1-alkynes)

For each of the above types of compounds the increments per CH_2 group are

ΔHf°:	-4.926 kcal/g mol
ΔFf°:	-2.048 kcal/g mol
$Log_{10}Kf$:	-1.5012 kcal/g mol
S°:	-9.183 cal/deg g mol

Relationships to SI units – The symbols cal $mole^{-1}$ deg^{-1} and gibbs/mol are identical and refer to units of calories per degree-mole. These units can be converted to SI units of joules per degree-mole by multiplying the tabulated values by 4.184. Similarly, values in kilocalories per mole can be converted to joules per mole by multiplying with the factor 4184. For further discussions of the SI system and for conversions from other units, see *Pure and Applied Chemistry*, 21, 1, 1970.

Formula	Compound	ΔHf°	ΔFf°	$Log_{10}Kf$	S°
CH_4	Methane	-17.889	-12.140	8.8985	44.50
C_2H_2	Ethyne (acetylene)	54.194	50.000	-36.6490	47.997
C_2H_4	Ethene (ethylene)	12.496	16.282	-11.9345	52.54
C_2H_6	Ethane	-20.236	-7.860	5.7613	54.85
C_3H_4	Propadiene (allene)	45.92	48.37	-35.4519	58.30
C_3H_4	Propyne (methyl-acetylene)	44.319	46.313	-33.9469	59.30
C_3H_6	Propene (propylene)	4.879	14.990	-10.9875	63.80
C_3H_8	Propane	-24.820	-5.614	4.1150	64.51
C_4H_6	1,2-Butadiene	39.55	48.21	-35.3377	70.03
C_4H_6	1,3-Butadiene	26.75	36.43	-26.7004	66.62
C_4H_6	1-Butyne (ethyl acetylene)	39.70	48.52	-35.5616	69.51
C_4H_6	2-Butyne (dimethylacetylene)	35.374	44.725	-32.7823	67.71
C_4H_8	1-Butene	0.280	17.217	-12.6199	73.48
C_4H_8	cis-2-Butene	-1.362	16.046	-11.7618	71.90
C_4H_8	trans-2-Butene	-2.405	15.315	-11.2255	70.86
C_4H_8	2-Methylpropane (isobutene)	-3.343	14.582	-10.6888	70.17
C_4H_{10}	*n*-Butane	-29.812	-3.754	2.7516	74.10
C_4H_{10}	2-Methylpropane (isobutane)	-31.452	-4.296	3.1489	70.42
C_5H_8	1-Pentyne	34.50	50.17	-36.7712	79.10
C_5H_8	2-Pentyne	30.80	46.41	-34.0177	79.30
C_5H_8	3-Methyl-1-butyne	32.60	49.12	-36.0061	76.23
C_5H_8	1,2-Pentadiene	34.80	50.29	-36.861	79.7
C_5H_8	cis-1,3-Pentadiene (cis-piperylene)	18.70	34.88	-25.563	77.5
C_5H_8	trans-1,3-Pentadiene (trans-piperylene)	18.60	35.07	-25.707	76.4
C_5H_8	1,4-Pentadiene	25.20	40.69	-29.824	79.7
C_5H_8	2,3-Pentadiene	33.10	49.22	-36.074	77.6
C_5H_8	3-Methyl-1,2-butadiene	31.00	47.47	-34.657	76.4
C_5H_8	2-Methyl-1,3-butadiene (isoprene)	18.10	34.87	-25.560	75.44
C_5H_{10}	1-Pentene	-5.000	18.787	-13.7704	83.08
C_5H_{10}	cis-2-Pentene	-6.710	17.173	-12.5874	82.76

VALUES OF CHEMICAL THERMODYNAMIC PROPERTIES OF HYDROCARBONS (Continued)

Formula	Compound	ΔHf°	ΔFf°	Log_{10} Kf	S°
C_5H_{10}	trans-2-Pentene	– 7.590	16.575	–12.1495	81.81
C_5H_{10}	2-Methyl-1-butene	– 8.680	15.509	–11.3680	81.73
C_5H_{10}	3-Methyl-1-butene	– 6.920	17.874	–13.1017	79.70
C_5H_{10}	2-Methyl-2-butene	–10.170	14.267	–10.4572	80.90
C_5H_{10}	Cyclopentane	–18.46	9.23	– 6.7643	70.00
C_5H_{12}	*n*-Pentane	–35.00	– 1.96	1.4366	83.27
C_6H_6	Benzene	19.820	30.989	–22.7143	64.34
C_6H_{10}	1-Hexyne	29.55	52.19	–38.258	88.27
C_6H_{12}	1-Hexene	– 9.96	20.80	–15.2491	92.25
C_6H_{12}	cis-2-Hexene	–11.56	19.18	–14.0549	92.35
C_6H_{12}	trans-2-Hexene	–12.56	18.46	–13.5291	91.40
C_6H_{12}	cis-3-Hexene	–11.56	19.66	–14.4094	90.73
C_6H_{12}	trans-3-Hexene	–12.56	18.86	–13.8262	90.04
C_6H_{12}	2-Methyl-1-pentene	–13.56	17.48	–12.8135	91.32
C_6H_{12}	3-Methyl-1-pentene	–11.02	20.28	–14.8655	90.45
C_6H_{12}	4-Methyl-1-pentene	–11.66	19.90	–14.5865	89.58
C_6H_{12}	2-Methyl-2-pentene	–14.96	16.34	–11.9780	90.45
C_6H_{12}	cis-3-Methyl-2-pentene	–14.32	16.98	–12.4471	90.45
C_6H_{12}	trans-3-Methyl-2-pentene	–14.32	16.74	–12.2697	91.26
C_6H_{12}	cis-4-Methyl-2-pentene	–13.26	18.40	–13.4903	89.23
C_6H_{12}	trans-4-Methyl-2-pentene	–14.26	17.77	–13.0216	88.02
C_6H_{12}	2-Ethyl-1-butene	–12.92	18.51	–13.5690	90.01
C_6H_{12}	2,3-Dimethyl-1-butene	–14.78	17.43	–12.7782	89.39
C_6H_{12}	3,3-Dimethyl-1-butene	–14.25	19.04	–13.9578	83.79
C_6H_{12}	2,3-Dimethyl-2-butene	–15.91	16.52	–12.1073	86.67
C_6H_{12}	Methylcyclopentane	–25.50	8.55	– 6.2649	81.24
C_6H_{12}	Cyclohexane	–29.43	7.59	– 5.5605	71.28
C_6H_{14}	*n*-Hexane	–39.96	0.05	0.037	92.45
C_7H_8	Methylbenzene (toluene)	11.950	29.228	–21.4236	76.42
C_7H_{12}	1-Heptyne	24.62	54.24	–39.759	97.25
C_7H_{12}	1-Heptene	–14.85	22.84	–16.742	101.43
C_7H_{14}	Ethylcyclopentane	–30.37	10.59	– 7.7632	90.62
C_7H_{14}	1,1-Dimethylcyclopentane	–33.05	9.33	– 6.8372	85.87
C_7H_{14}	1,cis-2-Dimethylcyclopentane	–30.96	10.93	– 8.0107	87.51
C_7H_{14}	1,trans-2-Dimethylcyclopentane	–32.67	9.17	– 6.7224	87.67
C_7H_{14}	1,cis-3-Dimethylcyclopentane	–31.93	9.91	– 7.2648	87.67
C_7H_{14}	1,trans-3-Dimethylcyclopentane	–32.47	9.37	– 6.8690	87.67
C_7H_{14}	Methylcyclohexane	–36.99	6.52	– 4.7819	82.06
C_7H_{16}	*n*-Heptane	–44.89	2.09	– 1.532	101.64
C_8H_8	Ethenylbenzene (styrene)	35.32	51.10	–37.4532	82.48
C_8H_{10}	Ethylbenzene	7.120	31.208	–22.8750	86.15
C_8H_{10}	1,2-Dimethylbenzene (*o*-xylene)	4.540	29.177	–21.3860	84.31
C_8H_{10}	1,3-Dimethylbenzene (*m*-xylene)	4.120	28.405	–20.8202	85.49
C_8H_{10}	1,4-Dimethylbenzene (*p*-xylene)	4.290	28.952	–21.2214	84.23
C_8H_{14}	1-Octyne	19.70	56.29	–41.260	106.63
C_8H_{16}	1-Octene	–19.82	24.89	–18.244	110.61
C_8H_{16}	*n*-Propylcyclopentane	–35.39	12.54	– 9.195	99.80
C_8H_{16}	1,1-Dimethylcyclohexane	–43.26	8.42	– 6.174	87.24
C_8H_{16}	cis-1,2-Dimethylcyclohexane	–41.15	9.85	– 7.225	89.51
C_8H_{16}	trans-1,2-Dimethylcyclohexane	–43.02	8.24	– 6.038	88.65

Formula	Compound	ΔHf°	ΔFf°	$Log_{10}Kf$	S°
C_8H_{16}	cis-1,3-Dimethylcyclohexane	–44.16	7.13	– 5.228	88.54
C_8H_{16}	trans-1,3-Dimethylcyclohexane	–42.20	8.68	– 6.363	89.92
C_8H_{16}	cis-1,4-Dimethylcyclohexane	–42.22	9.07	– 6.650	88.54
C_8H_{16}	trans-1,4-Dimethylcyclohexane	–44.12	7.58	– 5.552	87.19
C_8H_{18}	*n*-Octane	–49.82	4.14	– 3.035	110.82
C_8H_{18}	2-Methylheptane	-51.50	3.06	– 2.243	108.81
C_8H_{18}	3-Methylheptane	-50.82	3.29	– 2.412	110.32
C_8H_{18}	4-Methylheptane	-50.69	4.00	– 2.932	108.35
C_8H_{18}	3-Ethylhexane	-50.40	3.95	– 2.895	109.51
C_8H_{18}	2,2-Dimethylhexane	-53.71	2.56	– 1.876	103.06
C_8H_{18}	2,3-Dimethylhexane	-51.13	4.23	– 3.101	106.11
C_8H_{18}	2,4-Dimethylhexane	-52.44	2.80	– 2.052	106.51
C_8H_{18}	2,5-Dimethylhexane	-53.21	2.50	– 1.832	104.93
C_8H_{18}	3,3-Dimethylhexane	-52.61	3.17	– 2.324	104.70
C_8H_{18}	3,4-Dimethylhexane	-50.91	4.14	– 3.035	107.15
C_8H_{18}	2-Methyl-3-ethylpentane	-50.48	5.08	– 3.724	105.43
C_8H_{18}	3-Methyl-3-ethylpentane	-51.38	4.76	– 3.489	103.48
C_8H_{18}	2,2,3-Trimethylpentane	-52.61	4.09	– 2.998	101.62
C_8H_{18}	2,2,4-Trimethylpentane	-53.57	3.13	– 2.294	101.62
C_8H_{18}	2,3,3-Trimethylpentane	-51.73	4.52	– 3.313	103.14
C_8H_{18}	2,3,4-Trimethylpentane	-51.97	4.32	– 3.167	102.99
C_8H_{18}	2,2,3,3-Tetramethylbutane	-53.99	4.88	– 3.577	94.34
C_9H_{10}	Isopropenylbenzene (α-methylstyrene; 2-phenyl-1-propene)	27.00	49.84	-36.531	91.70
C_9H_{10}	1-Methyl-2-ethenylbenzene (*o*-Methylstyrene)	28.30	51.14	-37.484	91.70
C_9H_{10}	1-Methyl-3-ethenylbenzene (*m*-methylstyrene)	27.60	50.02	-36.665	93.1
C_9H_{10}	1-Methyl-4-ethenylbenzene (*p*-methylstyrene)	27.40	50.24	-36.825	91.7
C_9H_{12}	*n*-Propylbenzene	1.870	32.810	-24.049	95.74
C_9H_{12}	Isopropylbenzene (Cumene)	0.940	32.738	-23.996	92.87
C_9H_{12}	1,3,5-Trimethylbenzene (Mesitylene)	– 3.840	28.172	-20.6497	92.15
C_9H_{16}	1-Nonyne	14.77	58.34	-42.761	115.82
C_9H_{18}	1-Nonene	-24.74	26.94	-19.747	119.80
C_9H_{18}	*n*-Butylcyclopentane	-40.22	14.69	-10.768	108.99
C_9H_{20}	*n*-Nonane	-54.74	6.18	– 4.536	120.00
$C_{10}H_{14}$	*n*-Butylbenzene	– 3.30	34.62	-25.374	104.91
$C_{10}H_{18}$	1-Decyne	9.85	60.39	-44.262	125.00
$C_{10}H_{10}$	1-Decene	-29.67	28.99	-21.249	128.98
$C_{10}H_{22}$	*n*-Decane	-59.67	8.23	6.037	129.19
$C_{11}H_{22}$	1-Undecene	-34.60	31.03	-22.745	138.16
$C_{11}H_{24}$	*n*-Undecane	-64.60	10.28	– 7.539	138.37
$C_{12}H_{24}$	1-Dodecene	-39.52	33.08	-24.297	147.34

From Rossini, F. D., Pitzer, K. S., Arnett, R. L., Braun, R. M., and Pimentel, G. C., *Selected Values of Physical and Thermodynamic Properties of Hydrocarbons and Related Compounds,* Carnegie Press, Pittsburgh, 1953.

KEY VALUES FOR THERMODYNAMICS

In this report, the CODATA Task Group on Key Values for Thermodynamics presents recommended values for the quantities ΔH°_f(298.15 K), S°(298.15 K), and H°(298.15 K) – H°(0). The recommended values in the table below were given the approval of IUPAC Commission I.2 on Thermodynamics and Thermochemistry at the XXVIIth IUPAC Conference in Munich, August 1973. The values depend on the following values of the fundamental constants: gas constant, R, (8.31433 ± 0.00120) J K^{-1} mol^{-1}; constant relating wave number and energy (0.1196256 ± 0.0000107) J m mol^{-1}. Relative atomic masses were taken from the recommendations of the IUPAC Commission on Atomic Weights (1970). Data in the table relate to the natural mixture of isotopic species. Nuclear spin contributions were ignored. The uncertainties in the table represent the Task Group's estimates of the overall uncertainties at a 95% confidence level.

In the table, the standard state for crystalline solids (symbolized as c) and for liquids (symbolized as l) is that of the pure substance under a pressure of 101,325 Pa (l standard atmosphere). For gases (symbolized as g) the standard state is that of the ideal gas at a pressure of 101,325 Pa. For species in aqueous solution (symbolized as aq) the standard state is the hypothetical ideal solution at unit activity; the properties of ideal aqueous ionic solutions are taken equal to the sum of the properties of the individual ions.

The data listed in joules are the primary data. The values in calories (thermochemical) are derived from the corresponding values in joules by dividing by 4.184.

Recommended Key Values for Thermodynamics

Substance	State	ΔH°_f(298.15 K) kJ mol^{-1}	ΔH°_f(298.15 K) kcal mol^{-1}	S°(298.15 K) J K^{-1} mol^{-1}	S°(298.15 K) cal K^{-1} mol^{-1}	H°(298.15 K) – H°(0) kJ mol^{-1}	H°(298.15 K) – H°(0) kcal mol^{-1}
O	g	249.17 ± 0.10	59.553 ± 0.024	160.946 ± 0.020	38.467 ± 0.005	6.728 ± 0.003	1.608 ± 0.001
O_2	g	0	0	205.037 ± 0.033	49.005 ± 0.008	8.682 ± 0.004	2.075 ± 0.001
H	g	217.997 ± 0.006	52.103 ± 0.001	114.604 ± 0.015	27.391 ± 0.004	6.197 ± 0.002	1.481 ± 0.001
H+	aq	0	0	0	0	—	—
H_2	g	0	0	130.570 ± 0.033	31.207 ± 0.008	8.468 ± 0.003	2.024 ± 0.001
OH^-	aq	–230.025 ± 0.045	–54.977 ± 0.011	–10.71 ± 0.20	–2.56 ± 0.05	—	—
H_2O	l	–285.830 ± 0.042	–68.315 ± 0.010	69.950 ± 0.080	16.718 ± 0.019	13.293 ± 0.021	3.177 ± 0.005
H_2O	g	–241.814 ± 0.042	–57.795 ± 0.010	188.724 ± 0.040	45.106 ± 0.010	9.908 ± 0.008	2.368 ± 0.002
He	g	0	0	126.039 ± 0.012	30.124 ± 0.003	6.197 ± 0.002	1.481 ± 0.001
Ne	g	0	0	146.214 ± 0.016	34.946 ± 0.004	6.197 ± 0.002	1.481 ± 0.001
Ar	g	0	0	154.732 ± 0.020	36.982 ± 0.005	6.197 ± 0.002	1.481 ± 0.001
Kr	g	0	0	163.971 ± 0.020	39.190 ± 0.005	6.197 ± 0.002	1.481 ± 0.001
Xe	g	0	0	169.573 ± 0.020	40.529 ± 0.005	6.197 ± 0.002	1.481 ± 0.001
Cl	g	121.290 ± 0.008	28.989 ± 0.002	165.076 ± 0.020	39.454 ± 0.005	6.272 ± 0.003	1.499 ± 0.001
Cl^-	aq	–167.080 ± 0.088	–39.933 ± 0.021	56.73 ± 0.16	13.56 ± 0.04	—	—
Cl^2	g	0	0	222.965 ± 0.040	53.290 ± 0.010	9.180 ± 0.008	2.194 ± 0.002
HCl	g	–92.31 ± 0.13	–22.063 ± 0.031	186.786 ± 0.033	44.643 ± 0.008	8.640 ± 0.004	2.065 ± 0.001
Br	g	111.86 ± 0.12	26.735 ± 0.029	174.904 ± 0.020	41.803 ± 0.005	6.197 ± 0.002	1.481 ± 0.001
Br^-	aq	–121.50 ± 0.15	– 29.039 ± 0.035	82.84 ± 0.20	19.80 ± 0.05	—	—
Br_2	l	0	0	152.210 ± 0.040	36.379 ± 0.010	24.52 ± 0.13	5.860 ± 0.031
Br_2	g	30.91 ± 0.11	7.388 ± 0.026	245.350 ± 0.054	58.640 ± 0.013	9.724 ± 0.012	2.324 ± 0.003
HBr	g	–36.38 ± 0.17	–8.695 ± 0.041	198.585 ± 0.033	47.463 ± 0.008	8.648 ± 0.004	2.067 ± 0.001
I	g	106.762 ± 0.040	25.517 ± 0.010	180.673 ± 0.020	43.182 ± 0.008	6.197 ± 0.002	1.481 ± 0.001

Recommended Key Values for Thermodynamics (Continued)

Substance	State	ΔH°_f(298.15 K) kJ mol^{-1}	ΔH°_f(298.15 K) kcal mol^{-1}	S°(298.15 K) J K^{-1} mol^{-1}	S°(298.15 K) cal K^{-1} mol^{-1}	H°(298.15 K) − H°(0) kJ mol^{-1}	H°(298.15 K) − H°(0) kcal mol^{-1}
I^-	aq	−56.90 ± 0.84	−13.60 ± 0.20	106.70 ± 0.20	25.50 ± 0.05	—	—
I_2	c	0	0	116.139 ± 0.080	27.758 ± 0.019	13.196 ± 0.040	3.154 ± 0.010
I_2	g	62.421 ± 0.080	14.919 ± 0.019	260.567 ± 0.063	62.277 ± 0.015	10.117 ± 0.021	2.418 ± 0.005
HI	g	26.36 ± 0.80	6.30 ± 0.19	206.480 ± 0.040	49.350 ± 0.010	8.657 ± 0.006	2.069 ± 0.001
S	g	276.98 ± 0.25	66.20 ± 0.06	167.715 ± 0.035	40.085 ± 0.008	6.657 ± 0.004	1.591 ± 0.001
S_2	g	128.49 ± 0.30	30.71 ± 0.07	228.055 ± 0.050	54.506 ± 0.012	9.088 ± 0.008	2.172 ± 0.002
SO_2	g	−296.81 ± 0.20	−70.94 ± 0.05	248.11 ± 0.06	59.300 ± 0.014	10.548 ± 0.013	2.521 ± 0.003
N	g	472.68 ± 0.40	112.97 ± 0.10	153.189 ± 0.020	36.613 ± 0.005	6.197 ± 0.002	1.481 ± 0.001
N_2	g	0	0	191.502 ± 0.025	45.770 ± 0.006	8.669 ± 0.003	2.072 ± 0.001
NO_3^-	aq	—	—	146.94 ± 0.85	35.12 ± 0.20	—	—
NH_3	g	−45.94 ± 0.35	−10.980 ± 0.08	192.67 ± 0.08	46.049 ± 0.019	10.046 ± 0.008	2.401 ± 0.002
NH_4^+	aq	−133.26 ± 0.25	−31.850 ± 0.06	111.17 ± 0.75	26.57 ± 0.18	—	—
C	c	0	0	5.74 ± 0.12	1.372 ± 0.029	1.050 ± 0.020	0.251 ± 0.005
C	g	716.67 ± 0.44	171.29 ± 0.11	157.988 ± 0.020	37.760 ± 0.03	6.535 ± 0.006	1.562 ± 0.002
CO	g	−110.53 ± 0.17	−26.417 ± 0.041	197.556 ± 0.032	47.217 ± 0.008	8.673 ± 0.008	2.073 ± 0.002
CO_2	g	−393.51 ± 0.13	−94.051 ± 0.031	213.677 ± 0.040	51.070 ± 0.010	9.364 ± 0.008	2.238 ± 0.002
Zn	c	0	0	41.63 ± 0.13	9.950 ± 0.031	5.657 ± 0.020	1.352 ± 0.005
Zn	g	130.42 ± 0.20	31.17 ± 0.05	160.875 ± 0.025	38.450 ± 0.006	6.197 ± 0.004	1.481 ± 0.001
Ag	c	0	0	42.55 ± 0.21	10.17 ± 0.03	5.745 ± 0.020	1.373 ± 0.003
Ag	g	284.9 ± 0.08	68.09 ± 0.19	172.883 ± 0.025	41.320 ± 0.006	6.197 ± 0.004	1.481 ± 0.005
Ag^+	aq	105.750 ± 0.085	25.275 ± 0.020	73.38 ± 0.40	17.54 ± 0.10	—	—
AgCl	c	−127.070 ± 0.085	−30.370 ± 0.020	96.23 ± 0.20	23.00 ± 0.05	12.033 ± 0.040	2.876 ± 0.010
Li	c	0	0	29.12 ± 0.20	6.96 ± 0.05	4.632 ± 0.040	1.107 ± 0.010
Li^+	aq	−278.455 ± 0.090	−66.552 ± 0.022	11.30 ± 0.35	2.70 ± 0.08	—	—
Na	c	0	0	51.30 ± 0.20	12.26 ± 0.05	6.460 ± 0.020	1.544 ± 0.005
Na^+	aq	−240.300 ± 0.065	−57.433 ± 0.015	58.41 ± 0.20	13.96 ± 0.05	—	—
K	c	0	0	64.68 ± 0.20	15.46 ± 0.05	7.088 ± 0.020	1.694 ± 0.005
K^+	aq	−252.17 ± 0.10	−60.271 ± 0.025	101.04 ± 0.25	24.15 ± 0.06	—	—
Rb	c	0	0	76.78 ± 0.30	18.35 ± 0.07	7.489 ± 0.020	1.790 ± 0.005
Rb^+	aq	−251.12 ± 0.13	−60.018 ± 0.030	120.46 ± 0.40	28.79 ± 0.10	—	—
Cs	c	0	0	85.23 ± 0.40	20.37 ± 0.10	7.711 ± 0.020	1.843 ± 0.005
Cs^+	aq	−258.04 ± 0.13	−61.673 ± 0.030	132.84 ± 0.40	31.75 ± 0.10	—	—

Data from CODATA Bulletin 10, December 1973. For information and references upon which these values are based, consult this bulletin, available at no cost from ICSU CODATA Central Office, 19 Westendstrasse, 6 Frankfurt/Main, Germany Federal Republic.

KEY VALUES FOR THERMODYNAMICS

This list of recommended values of $\Delta H^\circ f$(298.15 K), S°(298.15 K) and H°(298.15 K)$-H^\circ$ (0) for 20 chemical species augments the list published in *CODATA Bulletin* 17, January 1976.

The usual definitions of reference and standard states have been adopted. The reference state for each element at 298.15 K is the thermodynamically stable standard state except for phosphorus, for which the "white" crystal modification has been selected, as it is the most reproducible. For crystalline solids (c) and liquids (l) the standard state is that of the pure substance under a pressure of 101325 Pa (one standard atmosphere). For gases (g) the standard state is that of the ideal gas at a pressure of 101325 Pa. For species in aqueous solution (aq) the standard state is the hypothetical ideal solution at unit activity. The properties of ideal aqueous ionic solutions are taken equal to the sum of the properties of the individual ions. For substances that are solids or liquids at 298.15 K, the values of H° (298.15 K) $-$ H° (0) are given relative to the stable crystalline form at zero temperature (unless otherwise noted); for gases the values are relative to the hypothetical ideal gas at zero temperature.

The following values of the fundamental constants were employed in the calculations: gas constant, R = (8.31433 ± 0.00080) J/K/mol; Faraday constant, F= (96487.0 ± 1.0) J/mol/V, and the constant relating wave number and energy, $N_A hc$ = (0.1196256 ± 0.0000026)J/m/mol. These values, which were also used in the earlier work of the Task Group on Key Values for Thermodynamics, differ very slightly from the values recommended by CODATA in 1973 (*CODATA Bulletin* 11, December 1973). However, adoption of the 1973 values of the fundamental constants would change the values in the present tables by far less than their assigned uncertainties, so the present values may be said to be consistent with the 1973 CODATA set of fundamental constants.

Substance	State	$\Delta H^\circ f$(298.15 K) kJ/mol	$\Delta H^\circ f$(298.15 K) kcal/mol	S°(298.15 K) J/K/mol	S° (298.15 K) cal/K/mol	H° (298.15 K) − H° (0) kJ/mol	H° (298.15 K) − H° (0) kcal/mol
F	g	79.39 ± 0.30	18.97 ± 0.07	158.640 ± 0.020	37.916 ± 0.005	6.518 ± 0.004	1.558 ± 0.001
SO_4^{2}	aq	− 909.60 ± 0.40	− 217.40± 0.10	18.83 ± 0.50	4.50 ± 0.12	—	—
P	c, white	0	0	41.09 ± 0.25	9.82 ± 0.06	5.360 ± 0.015	1.281 ± 0.004
AlF_3	c	− 1510.4 ± 1.3	− 361.0 ± 0.3	66.5 ± 0.4	15.89 ± 0.10	11.62 ± 0.04	2.777 ± 0.010
Cu^{+2}	aq	65.69 ± 0.80	15.70 ± 0.19	− 97.1 ± 1.2	− 23.2 ± 0.3	—	—
$CuSO_4$	c	− 771.1 ± 1.2	− 184.3 ± 0.3	109.2 ± 0.4	26.1 ± 0.1	16.86 ± 0.08	4.03 ± 0.02
Th	c	0	0	53.39 ± 0.40	12.76 ± 0.10	6.510 ± 0.020	1.556 ± 0.005
Th	g	598 ± 6	142.9 ± 1.4	190.06 ± 0.04	45.425 ± 0.010	6.197 ± 0.004	1.481 ± 0.001
ThO_2	c	− 1226.4 ± 3.5	− 293.12 ± 0.84	65.23 ± 0.20	15.59 ± 0.05	10.560 ± 0.020	2.524 ± 0.005
U	c	0	0	50.20 ± 0.20	12.00 ± 0.05	6.364 ± 0.020	1.521 ± 0.005
Be	c	0	0	9.50 ± 0.08	2.27 ± 0.02	1.950 ± 0.020	0.466 ± 0.005
Be	g	324 ± 5	77.4 ± 1.2	136.165 ± 0.020	32.544 ± 0.005	6.197 ± 0.004	1.481 ± 0.001
BeO	c	− 609.4 ± 2.5	− 145.6 ± 0.6	13.77 ± 0.04	3.29 ± 0.01	2.837 ± 0.008	0.678 ± 0.002
Mg	c	0	0	32.68 ± 0.10	7.81 ± 0.02	5.000 ± 0.030	1.195 ± 0.007
Mg	g	147.10 ± 0.80	35.16 ± 0.19	148.535 ± 0.020	35.501 ± 0.005	6.197 ± 0.004	1.481 ± 0.001
MgO	c	− 601.5 ± 0.3	− 143.76 ± 0.07	26.95 ± 0.15	6.44 ± 0.04	5.160 ± 0.020	1.233 ± 0.005
MgF_2	c	− 1124.2 ± 1.2	− 268.7 ± 0.3	57.2 ± 0.4	13.7 ± 0.1	9.92 ± 0.06	2.37 ± 0.01
Ca	c	0	0	41.6 ± 0.4	9.94 ± 0.10	5.73 ± 0.04	1.37 ± 0.01
Ca	g	177.8 ± 0.8	42.5 ± 0.2	154.775 ± 0.020	36.992 ± 0.005	6.197 ± 0.004	1.481 ± 0.001
CaO	c	− 635.09 ± 0.90	− 151.79 ± 0.22	38.1 ± 0.4	9.1 ± 0.1	6.75 ± 0.06	1.61 ± 0.01

LATTICE ENERGIES

H. D. B. Jenkins

Table 1 contains calculated values of lattice energies, U_{Pot}, of crystalline salts M_aX_b. U_{Pot} is expressed in the units of kilojoules per mole, $KJmol^{-1}$. M and X can be complex or simple ions. Also cited is the lattice energy obtained from the Born-Fajans-Haber Cycle, $U^{BFHC}{}_{Pot}$ using thermochemical data published in U.S. Government publications plus certain other data which are located at the end of this table. The values quoted are of variable reliability and a full discussion of the values is to appear in a review by Jenkins and Waddington currently (1978) nearing completion.

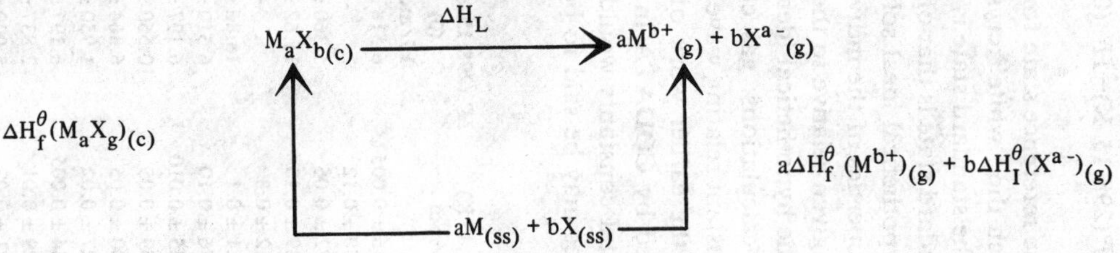

where, (ss) is the standard state of the ion or element

$$\Delta H_L = U_{Pot}\,(M_aX_b) + \left[a\left(\frac{n_{M^{b+}}}{2} - 2\right) + b\left(\frac{n_{X^{a-}}}{2} - 2\right)\right]RT$$

$$\Delta H_L = a\Delta H_f^\theta\,(M^{b+})\,(g) + b\Delta H_f^\theta\,(X^{a-})\,(g) - \Delta H_f^\theta\,(M_aX_b)(c)$$

Where $n_{M^{b}+}$, $n_{X^{a}-}$ is equal to 3 for monatomic ions, 5 for linear polyatomic ions, and 6 for polyatomic nonlinear ions.

The data listed in Table 2 were employed in the calculation of the Born-Haber Cycle values in Table 1 and are not listed in Technical Notes 270 of the National Bureau of Standards.

Table 1
LATTICE ENERGIES

Salt	Calculated lattice energy ($kJmol^{-1}$)	Literature source	Thermochemical cycle lattice energy ($kJmol^{-1}$)
Acetates			
$Li(CH_3COO)$	—	—	881
$Na\,(CH_3COO)$	761	Morris (1959)	763
$K(CH_3COO)$	686	Morris (1959)	682
$Rb\,(CH_3COO)$	715	Yatsimirskii (1956)	656
$Cs\,(CH_3COO)$	682	Yatsimirskii (1956)	—
$NH_4\,(CH_3COO)$	725	Yatsimirskii (1956)	695
$Ag\,(CH_3COO)$	—	—	863
$Tl\,(CH_3COO)$	—	—	750
$Ca\,(CH_3COO)_2$	2431	Yatsimirskii (1956)	2294
$Sr\,(CH_3COO)_2$	2280	Yatsimirskii (1956)	2166
$Ba\,(CH_3COO)_2$	2180	Yatsimirskii (1956)	2033
$Mn\,(CH_3COO)_2$	2548	Yatsimirskii (1956)	2616

Table 1 (continued)
LATTICE ENERGIES

Salt	Calculated lattice energy ($kJmol^{-1}$)	Literature source	Thermochemical cycle lattice energy ($kJmol^{-1}$)
CU $(CH_3COO)_2$	—	—	2835
Zn $(CH_3COO)_2$	2615	Yatsimirskii (1956)	2750
Hg $(CH_3COO)_2$	2368	Yatsimirskii (1961)	2595
Pb $(CH_3COO)_2$	2247	Yatsimirskii (1961)	2225
Acetylides			
CaC_2	2911	Vinek, Neckel, Nowatny (1967)	2904
SrC_2	2788	Vinek, Neckel, Nowatny (1967)	2784
BaC_2	2647	Vinek, Neckel, Nowatny (1967)	2654
Ammonium salts			
NH_4 (CH_3COO)	725	Yatsimirskii (1956)	656
NH_4HF_2	705	Jenkins, Pratt (1977)	705
NH_4HCO_3	—	—	741
NH_4HSO_4	640	Yatsimirskii (1956)	(645)
NH_4BF_4	582	Ladd, Lee (1961)	—
NH_4NCO	724	Waddington (1959)	—
NH_4CN	617	Ladd (1970)	670
NH_4HCO_2	715	Morris (1959)	—
NH_4 $(NH_2CH_2CO_3)$	650	Bernard, Businot, Decker (1974)	—
NH_4IO_3	—	—	808
$NH_4IO_2F_2$	678	Finch, Gates, Jenkinson (1972)	685
NH_4HS	661	Yatsimirskii (1956)	666
NH_4NO_3	661	Morris (1958)	676
NH_4ClO_4	583	Jenkins, Pratt (1978)	580
$(NH_4)_2S$	2008	Karapet'yanto (1954)	(2026)
$(NH_4)_2SO_4$	1766	Jenkins, Smith (1975)	1777
$(NH_4)_2GeF_6$	1657	Jenkins, Pratt (1978)	—
$(NH_4)_2IrCl_6$	1442	Jenkins, Pratt (1978)	1440
$(NH_4)_2OsCl_6$	1433	Jenkins, Pratt (1978)	—
$(NH_4)_2PdCl_6$	1481	Jenkins, Pratt (1978)	—
$(NH_4)_2PbCl_6$	1355	Jenkins, Pratt (1978)	—
$(NH_4)_2PbCl_6$	1468	Jenkins, Pratt (1978)	—
$(NH_4)_2ReCl_6$	1402	Jenkins, Pratt (1978)	1390
$(NH_4)_2SiF_6$	1657	Jenkins, Pratt (1978)	1727
$(NH_4)_2SnCl_6$	1370	Jenkins, Pratt (1978)	1334
$(NH_4)_2SnBr_6$	1319	Jenkins, Pratt (1978)	—
$(NH_4)_2TeCl_6$	1318	Jenkins, Pratt (1978)	—
$(NH_4)_2TeBr_6$	1294	Jenkins, Pratt (1978)	—
$(NH_4)_2TiCl_6$	1413	Jenkins, Pratt (1978)	—
NH_4CNS	605	Gill, Singla, Paul, Narula (1972)	611
$(NH_4)_2SeCl_6$	1420	Jenkins, Pratt (1978)	—
$(NH_4)_2SeBr_6$	1380	Jenkins, Pratt (1978)	—
Arsenates			
Mg_3 $(AsO_4)_2$	10669	Grekenschikov (1967)	10716
Ca_3 $(AsO_4)_2$	9749	Grekenschikov (1967)	9653
Sr_3 $(AsO_4)_2$	9330	Grekenschikov (1967)	9266
Ba_3 $(AsO_4)_2$	8870	Grekenschikov (1967)	8985
$AlAsO_4$	7255	Grekenschikov (1967)	—
$GaAsO_4$	7243	Grekenschikov (1967)	—
Astatides			
LiAt	720	Ladd, Lee (1961)	—
NaAt	657	Ladd, Lee (1961)	—
KAt	615	Ladd, Lee (1961)	—
RbAt	594	Ladd, Lee (1961)	—

Table 1 (continued)
LATTICE ENERGIES

Salt	Calculated lattice energy (kJmol⁻¹)	Literature source	Thermochemical cycle lattice energy (kJmol⁻¹)
CsAt	586	Ladd, Lee (1961)	—
FrAt	573	Ladd, Lee (1961)	—
Azides			
LiN_3	812	Jenkins, Pratt (1977)	818
NaN_3	732	Jenkins, Pratt (1977)	731
KN_3	659	Jenkins, Pratt (1977)	658
RbN_3	637	Jenkins, Pratt (1977)	632
CsN_3	612	Jenkins, Pratt (1977)	604
AgN_3	854	Jenkins, Pratt (1977)	—
TlN_3	689	Jenkins, Pratt (1977)	686
$Ca(N_3)_2$	2186	Jenkins, Pratt (1977)	2162
$Sr(N_3)_2$	2056	Jenkins, Pratt (1977)	2066
$Ba(N_3)_2$	2021	Gora (1971)	1965
$Mn(N_3)_2$	2408	Jenkins, Pratt (1977)	2416
$Cu(N_3)_2$	2730	Jenkins, Pratt (1977)	2738
$Zn(N_3)_2$	2840	Jenkins, Pratt (1977)	2848
$Cd(N_3)_2$	2446	Jenkins, Pratt (1977)	2454
$Pb(N_3)_2$	—	Jenkins, Pratt (1977)	2173
Bihalide salts			
$LiHF_2$	893	Jenkins, Pratt (1977)	866
$NaHF_2$	788	Jenkins, Pratt (1977)	788
KHF_2	703	Jenkins, Pratt (1977)	698
$RbHF_2$	674	Jenkins, Pratt (1977)	676
$CSHF_2$	646	Jenkins, Pratt (1977)	628
NH_4HF_2	705	Jenkins, Pratt (1977)	705
$CsHCl_2$	601	Thomson, Clark, Waddington, Jenkins (1975)	—
Me_4NHCl_2	427	Thompson, Clark, Waddington, Jenkins (1975)	—
Eb_4NHCl_2	346	Thompson, Clark, Waddington, Jenkins (1975)	—
Bu_4NHCl_2	290	Thompson, Clark, Waddington, Jenkins (1975)	—
Bicarbonates			
$NaHCO_3$	820	Yatsimirskii (1956)	818
$KHCO_3$	741	Yatsimirskii (1956)	736
$RbHCO_3$	707	Yatsimirskii (1956)	714
$CsHCO_3$	678	Yatsimirskii (1956)	709
NH_4HCO_3	—	—	741
$Ca(HCO_3)_2$	2402	Yatsimirskii (1956)	(2403)
$Sr(HCO_3)_2$	2255	Yatsimirskii (1956)	(2272)
$Ba(HCO_3)_2$	2155	Yatsimirskii (1956)	(2159)
Bisulphates			
NH_4HSO_4	640	Yatsimirskii (1956)	(645)
Borides			
CaB_6	5146	Samsanov, Shulishova (1956)	—
SrB_6	5104	Samsanov, Shulishova (1956)	—
BaB_6	5021	Samsanov, Shulishova (1956)	—
YB_6	7447	Samsanov, Shulishova (1956)	—
LaB_6	7406	Samsanov, Shulishova (1956)	—
CeB_6	10083	Samsanov, Shulishova (1956)	—
PrB_6	7447	Samsanov, Shulishova (1956)	—
NdB_6	7447	Samsanov, Shulishova (1956)	—
PmB_6	7406	Samsanov, Shulishova (1956)	—

Table 1 (continued)
LATTICE ENERGIES

Salt	Calculated lattice energy (kJmol⁻¹)	Literature source	Thermochemical cycle lattice energy (kJmol⁻¹)
SmB_6	7447	Samsanov, Shulishova (1956)	—
EuB_6	5104	Samsanov, Shulishova (1956)	—
GdB_6	7489	Samsonov, Shulishova (1956)	—
TbB_6	7489	Samsanov, Shulishova (1956)	—
DyB_6	7489	Samsanov, Shulishova (1956)	—
HoB_6	7489	Samsanov, Shulishova (1956)	—
ErB_6	7489	Samsanov, Shulishova (1956)	—
TmB_6	7489	Samsanov, Shulishova (1956)	—
YbB_6	5146	Samsanov, Shulishova (1956)	—
LuB_6	7489	Samsanov, Shulishova (1956)	—
ThB_6	10167	Samsanov, Shulishova (1956)	—
Borohydrides			
$LiBH_4$	778	Altschuller (1955)	—
$NaBH_4$	703	Altschuller (1955)	—
KBH_4	665	Altschuller (1955)	—
$RbBH_4$	648	Altschuller (1955)	—
$CsBH_4$	628	Altschuller (1955)	—
Borohalides			
$LiBF_4$	699	Kapustinskii, Yatsimirskii (1949)	—
$NaBF_4$	657	Kapustinskii, Yatsimirskii (1949)	619
KBF_4	611	Ladd, Lee (1961)	631
$RbBF_4$	577	Kapustinskii, Yatsimirskii (1949)	605
$CsBF_4$	556	Kapustinskii, Yatsimirskii (1949)	—
NH_4BF_4	582	Ladd, Lee (1961)	—
$Co(BF_4)_2$	2127	Bhattacharya, Das Gupta (1970)	—
$Ni(BF_4)_2$	2136	Bhattacharya, Das Gupta (1970)	—
$Zn(BF_4)_2$	2063	Bhattacharya, Das Gupta (1970)	—
$Cd(BF_4)_2$	1937	Bhattacharya, Das Gupta (1970)	—
$KBCl_4$	506	Krivtsov, Titova, Rosolovskii (1973)	—
$RbBCl_4$	489	Krivtsov, Titova, Rosolovskii (1973)	—
$CsBCl_4$	473	Krivtsov, Titova, Rosolovskii (1973)	—
Carbonates			
Li_2CO_3	2523	Yatsimirskii (1956)	2269
Na_2CO_3	2301	Yatsimirskii (1956)	2030
K_2CO_3	2084	Yatsimirskii (1956)	1858
Rb_2CO_3	2000	Yatsimirskii (1956)	1795
Cs_2CO_3	1920	Yatsimirskii (1956)	1702
$MgCO_3$	3180	Waddington (1959)	3122
$CaCO_3$	2804	Jenkins, Waddington, Pratt (1976)	2810
$SrCO_3$	2720	Waddington (1959)	2688
$BaCO_3$	2615	Waddington (1959)	2554
$MnCO_3$	3046	Waddington (1959)	3151
$FeCO_3$	3121	Waddington (1959)	3171
$CoCO_3$	3443	Yatsimirskii (1958)	3232
$NiCO_3$	—	—	3297
$CuCO_3$	3494	Yatsimirskii (1958)	3327
$ZnCO_3$	3121	Waddington (1959)	3273
$CdCO_3$	2929	Waddington (1959)	3052
$SnCO_3$	2904	Yatsimirskii (1961)	(2853)
$PbCO_3$	2728	Yatsimirskii (1961)	2750
Cyanates			
LiNCO	770	Waddington (1959)	—
NaNCO	736	Waddington (1959)	—
KNCO	653	Waddington (1959)	—

Table 1 (continued)
LATTICE ENERGIES

Salt	Calculated lattice energy (kJmol⁻¹)	Literature source	Thermochemical cycle lattice energy (kJmol⁻¹)
RbNCO	615	Waddington (1959)	—
CsNCO	586	Waddington (1959)	—
NH_4NCO	724	Waddington (1959)	—
Cyanides			
LiCN	—	—	849
NaCN	738	Jenkins, Pratt (1977)	739
KCN	674	Jenkins, Pratt (1977)	669
RbCN	646	Jenkins, Pratt (1977)	(640)
CsCN	612	Jenkins, Pratt (1977)	602
NH_4CN	617	Ladd (1970)	670
$Ca(CN)_2$	2268	Yatsimirskii (1956)	2191
$Sr(CN)_2$	2138	Yatsimirskii (1956)	(2076)
$Ba(CN)_2$	2046	Yatsimirskii (1956)	1960
CuCN	—	—	1035
AgCN	741	Yatsimirskii (1961)	914
$Zn(CN)_2$	2431	Ladd, Lee (1960)	2768
$Cd(CN)_2$	2284	Ladd, Lee (1960)	2542
Formates			
$Li(HCO_2)$	865	Morris (1959)	872
$Na(HCO_2)$	791	Morris (1959)	811
$K(HCO_2)$	713	Morris (1959)	729
$Rb(HCO_2)$	685	Morris (1959)	682
$Cs(HCO_2)$	651	Morris (1959)	644
NH_4HCO_2	715	Morris (1959)	—
$Mg(HCO_2)_2$	2674	Morris (1959)	—
$Ca(HCO_2)_2$	2360	Morris (1959)	2390
$Sr(HCO_2)_2$	2221	Morris (1959)	2261
$Ba(HCO_2)_2$	2092	Morris (1959)	2134
$Mn(HCO_2)_2$	2598	Morris (1959)	2701
$Co(HCO_2)_2$	—	—	2792
$Ni(HCO_2)_2$	—	—	2880
$Cu(HCO_2)_2$	2870	Morris (1959)	2913
$Zn(HCO_2)_2$	2791	Morris (1959)	2847
$Cd(HCO_2)_2$	2556	Morris (1959)	—
$Pb(HCO_2)_2$	2276	Morris (1959)	2330
Germanates			
Mg_2GeO_4	7991	Grekenschikov (1967)	—
Ca_2GeO_4	7301	Grekenschikov (1967)	7306
Sr_2GeO_4	6987	Grekenschikov (1967)	—
Ba_2GeO_4	6653	Grekenschikov (1967)	6643
Glycinates			
$Na(NH_2CH_2COO_2)$	739	Bernard, Businot, Decker (1974)	—
$K(NH_2CH_2COO_2)$	668	Bernard, Businot, Decker (1974)	—
$Rb(NH_2CH_2COO_2)$	648	Bernard, Businot, Decker (1974)	—
$NH_4(NH_2CH_2COO_2)$	650	Bernard, Businot, Decker (1974)	—
$Cu(NH_2CH_2CO_2)_2$	2694	Bernard, Businot, Decker (1974)	—
Halates			
$LiBrO_3$	883	Finch, Gardner (1965)	897
$NaBrO_3$	803	Finch, Gardner (1965)	814
$KBrO_3$	740	Finch, Gardner (1965)	745
$RbBrO_3$	720	Finch, Gardner (1965)	742
$CsBrO_3$	694	Finch, Gardner (1965)	681
$NaClO_3$	770	Morris (1958)	770
$KClO_3$	711	Morris (1958)	706

Table 1 (continued)
LATTICE ENERGIES

Salt	Calculated lattice energy (kJmol⁻¹)	Literature source	Thermochemical cycle lattice energy (kJmol⁻¹)
$RbClO_3$	690	Morris (1958)	687
$CsClO_3$	—	—	647
$LiIO_3$	975	Finch, Gardner (1965)	—
$NaIO_3$	883	Finch, Gardner (1965)	882
KIO_3	820	Finch, Gardner (1965)	806
$RbIO_3$	791	Finch, Gardner (1965)	—
$CsIO_3$	761	Finch, Gardner (1965)	769
NH_4IO_3	—	Finch, Gardner (1965)	808
$Mg(ClO_3)_2$	2535	Finch, Gardner (1965)	(2475)
$Ca(ClO_3)_2$	2259	Finch, Gardner (1965)	2286
$Sr(ClO_3)_2$	2138	Finch, Gardner (1965)	2155
$Ba(ClO_3)_2$	2021	Finch, Gardner (1965)	2027
Halides			
LiF	1030	Jenkins, Pratt (1977)	1036
LiCL	834	Jenkins, Pratt (1977)	853
LiBr	788	Jenkins, Pratt (1977)	807
LiI	730	Jenkins, Pratt (1977)	757
NaF	910	Jenkins, Pratt (1977)	923
NaCl	769	Jenkins, Pratt (1977)	786
NaBr	732	Jenkins, Pratt (1977)	747
NaI	682	Jenkins, Pratt (1977)	704
KF	808	Jenkins, Pratt (1977)	821
KCl	701	Jenkins, Pratt (1977)	715
KBr	671	Jenkins, Pratt (1977)	682
KI	632	Jenkins, Pratt (1977)	649
RbF	774	Jenkins, Pratt (1977)	785
RbCl	680	Jenkins, Pratt (1977)	689
RbBr	651	Jenkins, Pratt (1977)	660
RbI	617	Jenkins, Pratt (1977)	630
CsF	744	Jenkins, Pratt (1977)	740
CsCl	657	Jenkins, Pratt (1977)	659
CsBr	632	Jenkins, Pratt (1977)	631
CsI	600	Jenkins, Pratt (1977)	604
FrF	715	Ladd, Lee (1961)	—
FrCL	632	Ladd, Lee (1961)	—
FrBr	611	Ladd, Lee (1961)	—
FrI	582	Ladd, Lee (1961)	—
CuCl	921	Sharma, Madan (1964)	996
CuBr	879	Sharma, Madan (1964)	979
CuI	835	Sharma, Madan (1964)	966
AgF	953	Sharma, Madan (1964)	967
AgCl	864	Sharma, Madan (1964)	915
AgBr	830	Sharma, Madan (1964)	904
AgI	808	Sharma, Madan (1964)	889
AuCl	1013	Mamulov (1961)	1066
AuBr	1015	Mamulov (1961)	1061
AuI	1015	Mamulov (1961)	1070
InCl	—	—	763
InBr	—	—	767
InI	—	—	732
TLF	—	—	845
TLCl	782	Jenkins, Alcock (1975)	751
TlBr	713	Sharma, Madan (1964)	735
TlI	687	Sharma, Madan (1964)	709

Table 1 (continued)
LATTICE ENERGIES

Salt	Calculated lattice energy (kJmol⁻¹)	Literature source	Thermochemical cycle lattice energy (kJmol⁻¹)
BeF_2	3150	Krasnov (1961)	3505
$BeCl_2$	3004	Krasnov (1961)	3020
$BeBr_2$	2950	Brackett, Brackett (1965)	2914
BeI_2	2653	Krasnov (1961)	2800
MgF_2	2913	Brackett, Brackett (1965)	2957
$MgCl_2$	2326	Brackett, Brackett (1965)	2526
$MgBr_2$	2097	Brackett, Brackett (1965)	2440
MgI_2	1944	Brackett, Brackett (1965)	2327
CaF_2	2609	Brackett, Brackett (1965)	2630
$CaCl_2$	2223	Brackett, Brackett (1965)	2258
$CaBr_2$	2132	Brackett, Brackett (1965)	2176
CaI_2	1905	Brackett, Brackett (1965)	2074
SrF_2	2476	Brackett, Brackett (1965)	2492
$SrCl_2$	2127	Brackett, Brackett (1965)	2156
$SrBr_2$	2008	Brackett, Brackett (1965)	2075
SrI_2	1937	Karapet'yants (1954)	1963
BaF_2	2341	Brackett, Brackett (1965)	2352
$BaCl_2$	2033	Brackett, Brackett (1965)	2056
$BaBr_2$	1950	Brackett, Brackett (1965)	1985
BaI_2	1831	Brackett, Brackett (1965)	1877
RaF_2	2284	Yatsimirskii, Krestov (1960)	—
$RaCl_2$	2004	Yatsimirskii, Krestov (1960)	—
$RaBr_2$	1929	Yatsimirskii, Krestov (1960)	—
RaI_2	1803	Yatsimirskii, Krestov (1960)	—
$ScCl_2$	2380	Nelson, Sharpe (1966)	—
$ScBr_2$	2291	Nelson, Sharpe (1966)	—
ScI_2	2201	Nelson, Sharpe (1966)	—
TiF_2	2724	Mamulov (1961)	—
$TiCl_2$	2431	Mamulov (1961)	2501
$TiBr_2$	2360	Mamulov (1961)	2419
TiI_2	2259	Mamulov (1961)	2329
VCl_2	2607	Mamulov (1961)	2579
VBr_2	—	Mamulov (1961)	2523
VI_2	—	Mamulov (1961)	2456
CrF_2	2778	Mamulov (1961)	2917
$CrCl_2$	2455	Mamulov (1961)	2586
$CrBr_2$	2377	Mamulov (1961)	2523
CrI_2	2269	Mamulov (1961)	2425
$MoCl_2$	2485	Mamulov (1961)	2733
$MoBr_2$	2448	Mamulov (1961)	2742
MoI_2	2422	Mamulov (1961)	2630
MnF_2	2644	Mamulov (1961)	—
$MnCl_2$	2368	Mamulov (1961)	2537
$MnBr_2$	2304	Mamulov (1961)	2471
MnI_2	2212	Mamulov (1961)	—
FeF_2	2769	Mamulov (1961)	—
$FeCl_2$	2525	Mamulov (1961)	2631
$FeBr_2$	2464	Mamulov (1961)	2569
FeI_2	2382	Mamulov (1961)	2480
CoF_2	2878	Morris (1957)	3018
$CoCl_2$	2709	Mamulov (1961)	2691
$CoBr_2$	2648	Yatsimirskii (1958)	2629
CoI_2	2569	Yatsimirskii (1958)	2545
NiF_2	2845	VonBaur (1961)	3066

Table 1 (continued)
LATTICE ENERGIES

Salt	Calculated lattice energy (kJmol⁻¹)	Literature source	Thermochemical cycle lattice energy (kJmol⁻¹)
$NiCl_2$	2753	Yatsimirskii (1958)	2772
$NiBr_2$	2699	Yatsimirskii (1958)	2709
NiI_2	2607	Yatsimirskii (1958)	2623
$PdCl_2$	2766	Yatsimirskii (1958)	2778
$PdBr_2$	—	—	2741
PdI_2	—	—	2748
CuF_2	3046	Perret (1961)	3082
$CuCl_2$	2774	Perret (1961)	2811
$CuBr_2$	2711	Perret (1961)	2763
CuI_2	2640	Perret (1961)	—
AgF_2	2919	Mamulov (1961)	2942
ZnF_2	2930	Mamulov (1961)	3032
$ZnCl_2$	2690	Mamulov (1961)	2734
$ZnBr_2$	2632	Mamulov (1961)	2678
ZnI_2	2549	Mamulov (1961)	2605
CdF_2	2740	Karapet'yants (1954)	2809
$CdCl_2$	2226	Krasnov (1961)	2552
$CdBr_2$	2468	Karapet'yants (1954)	2507
CdI_2	2406	Karapet'yants (1954)	2441
HgF_2	2757	Karapet'yants (1954)	(2798)
$HgCl_2$	2569	Yatsimirskii (1961)	2651
$HgBr_2$	2598	Karapet'yants (1954)	2628
HgI_2	2569	Karapet'yants (1954)	2610
SnF_2	2551	Mamulov (1961)	—
$SnCl_2$	2276	Mamulov (1961)	2297
$SnBr_2$	2211	Mamulov (1961)	2245
SnI_2	2123	Mamulov (1961)	2193
PbF_2	2460	Mamulov (1961)	2522
$PbCl_2$	2229	Mamulov (1961)	2269
$PbBr_2$	2169	Mamulov (1961)	2219
PbI_2	2086	Mamulov (1961)	2163
ScF_3	5096	Mamulov (1961)	5492
$ScCl_3$	4874	Kapustinskii, Yatsimirskii (1956)	4866
$ScBr_3$	4711	Kapustinskii, Yatsimirskii (1956)	4729
ScI_3	4640	Mamulov (1961)	—
YF_3	4983	Mamulov (1961)	—
YCl_3	4447	Mamulov (1961)	4506
YBr_3	4410	Wen, Sho (1975)	—
YI_3	4125	Mamulov (1961)	4240
TiF_3	5644	Mamulov (1961)	—
$TiCl_3$	5134	Mamulov (1961)	5134
$TiBr_3$	5012	Mamulov (1961)	5007
TiI_3	4845	Mamulov (1961)	—
ZrF_3	—	—	(5400)
$ZrCl_3$	—	—	4791
$ZrBr_3$	—	—	4758
ZrI_3	—	—	(4591)
VF_3	5895	Cavell, Clark (1965)	—
VCl_3	5322	Mamulov (1961)	5315
VBr_3	5192	Nelson, Sharpe (1966)	5214
VI_3	5058	Nelson, Sharpe (1966)	5121
$NbCl_3$	5062	Jenkins, Waddington[a]	—
$NbBr_3$	4980	Jenkins, Waddington[a]	—
NbI_3	4860	Jenkins, Waddington[a]	—

Table 1 (continued)
LATTICE ENERGIES

Salt	Calculated lattice energy (kJmol⁻¹)	Literature source	Thermochemical cycle lattice energy (kJmol⁻¹)
CrF_3	5958	Mamulov (1961)	6033
$CrCl_3$	5473	Mamulov (1961)	5509
$CrBr_3$	5355	Mamulov (1961)	—
CrI_3	5201	Mamulov (1961)	5274
MoF_3	(6459)	Van Gool, Picken (1964)	5230
$MoCl_3$	—	Van Gool, Picken (1964)	5230
$MoBr_3$	—	Van Gool, Picken (1964)	5156
MoI_3	—	Van Gool, Picken (1964)	5073
MnF_3	6017	Cavell, Clark (1965)	—
$MnCl_3$	5544	Nelson, Sharpe (1966)	—
$MnBr_3$	5448	Nelson, Sharpe (1966)	—
MnI_3	5330	Nelson, Sharpe (1966)	—
$TcCl_3$	5270	Jenkins, Waddington[a]	—
$TcBr_3$	5215	Jenkins, Waddington[a]	—
TcI_3	5188	Jenkins, Waddington[a]	—
FeF_3	5870	Perret (1961)	—
$FeCl_3$	5364	Perret (1961)	5359
$FeBr_3$	5268	Perret (1961)	5333
FeI_3	5117	Perret (1961)	—
$RuCl_3$	—	—	5245
$RuBr_3$	—	—	5223
RuI_3	—	—	5222
CoF_3	5991	Cavell, Clark (1965)	6118
$RhCl_3$	—	—	5641
IrF_3	(6112)	Hoppe (1970)	—
$IrBr_3$	(4794)	Hoppe (1970)	—
NiF_3	(6111)	Van Gool, Picken (1969)	—
AuF_3	(5777)	Hoppe (1970)	—
$AuCl_3$	(4605)	Hoppe (1970)	—
$ZnCl_3$	5832	Nelson, Sharpe (1963)	—
$ZnBr_3$	5732	Nelson, Sharpe (1963)	—
ZnI_3	5636	Nelson, Sharpe (1963)	—
AlF_3	5924	Mamulov (1961)	5215
$AlCl_3$	5376	Mamulov (1961)	5492
$AlBr_3$	5247	Mamulov (1961)	5361
AlI_3	5070	Mamulov (1961)	5218
GaF_3	—	—	6205
$GaCl_3$	5217	Mamulov (1961)	5645
$GaBr_3$	4966	Mamulov (1961)	5552
GaI_3	4611	Mamulov (1961)	5476
$InCl_3$	4736	Yatsimirskii (1961)	5187
$InBr_3$	4535	Yatsimirskii (1961)	5124
InI_3	4234	Yatsimirskii (1961)	5005
TlF_3	5493	Mamulov (1961)	—
$TlCl_3$	5252	Mamulov (1961)	5258
$TlBr_3$	5171	Mamulov (1961)	—
TlI_3	5088	Mamulov (1961)	—
$AsBr_3$	—	—	5497
AsI_3	(3758)	Van Gool, Picken (1969)	4824
SbF_3	—	—	5295
$SbCl_3$	—	—	5032
$SbBr_3$	—	—	4954
SbI_3	—	—	4867
$BiCl_3$	—	—	4689

Table 1 (continued)
LATTICE ENERGIES

Salt	Calculated lattice energy (kJmol⁻¹)	Literature source	Thermochemical cycle lattice energy (kJmol⁻¹)
$BiBr_3$	—	—	—
BiI_3	(3774)	Van Gool, Picken (1969)	—
LaF_3	4682	Mamulov (1961)	—
$LaCl_3$	4343	Yatsimirskii (1961)	4242
$LaBr_3$	4209	Wen, Sho (1975)	—
LaI_3	3916	Yatsimirskii (1961)	3986
$CeCl_3$	4297	Ladd, Lee (1961)	4284
CeI_3	—	—	4029
$PrCl_3$	4322	Ladd, Lee (1961)	4326
PrI_3	—	—	4071
$NdCl_3$	4343	Ladd, Lee (1961)	—
$SmCl_3$	4376	Ladd, Lee (1961)	
$EuCl_3$	4393	Ladd, Lee (1961)	—
$GdCl_3$	4406	Ladd, Lee (1961)	—
$DyCl_3$	4481	Ladd, Lee (1961)	—
$HoCl_3$	4501	Ladd, Lee (1961)	—
$ErCl_3$	4527	Ladd, Lee (1961)	—
$TmCl_3$	4548	Ladd, Lee (1961)	4550
TmI_3	—	—	4314
$YbCl_3$	—	—	4546
$AcCl_3$	4096	Ladd, Lee (1961)	—
UCl_3	4243	Ladd, Lee (1961)	—
$NpCl_3$	4268	Ladd, Lee (1961)	—
$PuCl_3$	4289	Ladd, Lee (1961)	—
$PuBr_3$	(3959)	Van Gool, Picken (1969)	—
$AmCl_3$	4293	Van Gool, Picken (1969)	—
TiF_4	10012	Mamulov (1961)	9908
$TiCl_4$	9431	Mamulov (1961)	—
$TiBr_4$	9288	Manulov (1961)	9039
TiI_4	9108	Mamulov (1961)	8893
ZrF_4	8853	Mamulov (1961)	8971
$ZrCl_4$	8096	Mamulov (1961)	8144
$ZrBr_4$	7916	Mamulov (1961)	7984
ZrI_4	7661	Mamulov (1961)	7801
MoF_4	8795	Mamulov (1961)	—
$MoCl_4$	8556	Mamulov (1961)	9573
$MoBr_4$	8510	Mamulov (1961)	9475
MoI_4	8427	Mamulov (1961)	—
$SnCl_4$	8355	Mamulov (1961)	—
$SnBr_4$	7970	Mamulov (1961)	8833
PbF_4	—	—	9461
CrF_2Cl	5795	Perret (1961)	—
CrF_2Br	5753	Perret (1961)	—
CrF_2I	5669	Perret (1961)	—
$CrCl_2Br$	5448	Perret (1961)	—
$CrCl_2I$	5381	Perret (1961)	—
$CrBr_2I$	5330	Perret (1961)	—
CuFCl	2891	Perret (1961)	—
CuFBr	2853	Perret (1961)	—
CuFI	2803	Perret (1961)	—
CuClBr	2753	Perret (1961)	—
CuClI	2694	Perret (1961)	—
CuBrI	2669	Perret (1961)	—
FeF_2Cl	5711	Perret (1961)	—
FeF_2Br	5653	Perret (1961)	—

Table 1 (continued)
LATTICE ENERGIES

Salt	Calculated lattice energy (kJmol⁻¹)	Literature source	Thermochemical cycle lattice energy (kJmol⁻¹)
FeF_2I	5569	Perret (1961)	—
$FeCl_2Br$	5339	Perret (1961)	—
$FeCl_2I$	5272	Perret (1961)	—
FBr_2I	5209	Perret (1961)	—
$LiIO_2F_2$	845	Finch, Gates, Jenkins (1972)	—
$NaIo_2F_2$	766	Finch, Gates, Jenkins (1972)	764
KIo_2F_2	699	Finch, Gates, Jenkins (1972)	697
$RbIo_2F2$	674	Finch, Gates, Jenkins (1972)	671
$CsIo_2F_2$	636	Finch, Gates, Jenkins (1972)	—
$NH_4IO_2F_2$	678	Finch, Gates, Jenkins (1972)	—
$AgIO_2F_2$	736	Finch, Gates, Jenkins (1972)	685
Hydrides			
LiH	858	Tsai (1964)	920
NaH	782	Tsai (1964)	808
KH	699	Tsai (1964)	714
RbH	674	Tsai (1964)	685
CsH	648	Tsai (1964)	644
TiH	—	—	1407
ZrH	—	—	1590
VH	—	—	(1344)
NbH	—	—	(1633)
PdH	—	—	(1368)
CuH	—	—	1254
TiH	996	Gibb (1962)	1407
ZrH	916	Gibb (1962)	1590
HfH	904	Gibb (1962)	—
LaH	828	Gibb (1962)	—
VH	1184	Gibb (1962)	(1344)
NbH	1163	Gibb (1962)	(1633)
TaH	1021	Gibb (1962)	—
CrH	1050	Gibb (1962)	—
NiH	929	Gibb (1962)	—
PdH	979	Gibb (1962)	1368
PtH	937	Gibb (1962)	—
CuH	828	Gibb (1962)	1254
AgH	941	Karapet'yants (1954)	—
AuH	1033	Karapet'yants (1954)	—
TlH	745	Karapet'yants (1954)	—
GeH	950	Gibb (1962)	—
PbH	778	Gibb (1962)	—
BeH_2	3205	Gibb (1962)	3295
MgH_2	2791	Gibb (1962)	2706
CaH_2	2410	Karapet'yants (1954)	2394
SrH_2	2250	Karapet'yants (1954)	2253
BaH_2	2121	Karapet'yants (1954)	2121
ScH_2	2711	Gibb (1962)	2659
YH_2	(2598)	Gibb (1962)	2670
LaH_2	2380	Gibb (1962)	2522
CeH_2	2414	Gibb (1962)	2484
PrH_2	2448	Gibb (1962)	2398
NdH_2	2464	Gibb (1962)	2367
PmH_2	2519	Gibb (1962)	—
SmH_2	2510	Gibb (1962)	2389
GdH_2	2494	Gibb (1962)	2706
AcH_2	2372	Gibb (1962)	—

Table 1 (continued)
LATTICE ENERGIES

Salt	Calculated lattice energy (kJmol⁻¹)	Literature source	Thermochemical cycle lattice energy (kJmol⁻¹)
ThH_2	2711	Gibb (1962)	—
PuH_2	2519	Gibb (1962)	—
AmH_2	2544	Gibb (1962)	—
TiH_2	2866	Gibb (1962)	2845
ZrH_2	2711	Gibb (1962)	2999
CuH_2	2941	Karapet'yants (1954)	—
ZnH_2	2870	Karapet'yants (1954)	—
HgH_2	2707	Karapet'yants (1954)	—
AlH_3	5924	Gibb (1962)	—
FeH_3	5724	Gibb (1962)	—
ScH_3	5439	Gibb (1962)	—
YH_3	5063	Gibb (1962)	—
LaH_3	4895	Gibb (1962)	4493
FeH_3	5724	Gibb (1962)	—
GaH_3	5690	Gibb (1962)	—
InH_3	5092	Gibb (1962)	—
TlH_3	5092	Gibb (1962)	—
Hydroselenides			
NaHse	703	Yatsimirskii (1956)	—
KHSe	644	Yatsimirskii (1956)	—
RbHSe	623	Yatsimirskii (1956)	—
CsHSe	598	Yatsimirskii (1956)	—
Hydrosulphides			
LiHS	759	Waddington (1959)	821
NaHS	704	Waddington (1959)	747
KHS	650	Waddington (1959)	659
RbHS	623	Waddington (1959)	637
CsHS	582	Waddington (1959)	595
NH_4HS	661	Yatsimirskii (1956)	666
$Ca(HS)_2$	2184	Yatsimirskii (1956)	(2171)
$Sr(HS)_2$	2063	Yatsimirskii (1956)	—
$Ba(HS)_2$	1979	Yatsimirskii (1956)	(1956)
Hydroxides			
LiOH	1021	Saloman (1970)	1039
NaOH	887	Saloman (1970)	900
KOH	789	Saloman (1970)	804
RbOH	766	Yatsimirskii (1956)	773
CsOH	721	Yatsimirskii (1956)	724
$Be(OH)_2$	3477	Finch, Gardner (1965)	3629
$Mg(OH)_2$	2870	Finch, Gardner (1965)	3006
$Ca(OH)_2$	2506	Finch, Gardner (1965)	2645
$Sr(OH)_2$	2330	Finch, Gardner (1965)	2483
$Ba(OH)_2$	2141	Finch, Gardner (1965)	2339
$Ti(OH)_2$	—	—	2962
$Mn(OH)_2$	2909	Wen, Sho (1975)	3008
$Fe(OH)_2$	2653	Karapet'yants (1954)	3055
$Co(OH)_2$	2786	Karapet'yants (1954)	3115
$Ni(OH)_2$	2832	Karapet'yants (1954)	3193
$Pd(OH)_2$	—	—	3175
$Cu(OH)_2$	2870	Karapet'yants (1954)	3237
CuOH	1006	Karapet'yants (1954)	—
AgOH	918	Karapet'yants (1954)	—
AuOH	1033	Karapet'yants (1954)	—
TlOH	705	Karapet'yants (1954)	—
$Zn(OH)_2$	2795	Yatsimirskii (1961)	3158

Table 1 (continued)
LATTICE ENERGIES

Salt	Calculated lattice energy (kJmol⁻¹)	Literature source	Thermochemical cycle lattice energy (kJmol⁻¹)
$Cd(OH)_2$	2607	Yatsimirskii (1961)	2918
$Hg(OH)_2$	2669	Karapet'yants (1954)	—
$Sn(OH)_2$	2489	Yatsimirskii (1961)	2729
$Pb(OH)_2$	2376	Yatsimirskii (1961)	2623
$Sc(OH)_3$	5063	Karapet'yants (1954)	—
$Y(OH)_3$	4707	Karapet'yants (1954)	—
$La(OH)_3$	4443	Karapet'yants (1954)	—
$Cr(OH)_3$	5556	Karapet'yants (1954)	—
$Mn(OH)_3$	6213	Wen,Sho (1975)	—
$Al(OH)_3$	5627	Karapet'yants (1954)	—
$Ga(OH)_3$	5732	Karapet'yants (1954)	—
$In(OH)_3$	5280	Karapet'yants (1954)	—
$Tl(OH)_3$	5314	Karapet'yants (1954)	—
$Ti(OH)_4$	9456	Karapet'yants (1954)	—
$Zr(OH)_4$	8619	Karapet'yants (1954)	—
$Mn(OH)_4$	10933	Brenet, Goelfier, Cabano (1963)	—
$Sn(OH)_4$	9188	Karapet'yants (1954)	—
Imides			
CaNH	3293	Altschuller (1955)	—
SrNH	3146	Altschuller (1955)	—
BaNH	2975	Altschuller (1955)	—
Metaniobates			
$NaNbO_3$	789	Lapitskii, Nebyhtsyn (1961)	—
$Ca(NbO_3)_2$	2315	Lapitskii, Nebyhtsyn (1961)	—
$Fe(NBO_3)_2$	2502	Lapitskii, Nebyhtsyn (1961)	—
Metatantalates			
$NaTaO_3$	789	Lapitskii, Nebyhtsyn (1961)	—
$Ca(TaO_3)_2$	2315	Lapitskii, Nebyhtsyn (1961)	—
$Fe(TaO_3)_2$	2502	Lapitskii, Nebyhtsyn (1961)	—
Metavanadates			
Li_3VO_4	3945	Golvkin, Fotiev (1970)	—
Na_3VO_4	3766	Golvkin, Fotiev (1970)	—
K_3VO_4	3376	Golvkin, Fotiev (1970)	—
Rb_3VO_4	3243	Golvkin, Fotiev (1970)	—
Cs_3VO_4	3137	Golvkin, Fotiev (1970)	—
Nitrates			
$LiNO_3$	848	Jenkins, Moms (1977)	848
$NaNO_3$	755	Jenkins, Moms (1977)	756
KNO_3	685	Jenkins, Moms (1977)	687
$RbNO_3$	662	Jenkins, Moms (1977)	658
$CSNO_3$	648	Jenkins, Moms (1977)	625
NH_4NO_3	661	Moms (1958)	676
$AgNO_3$	820	Moms (1958)	822
$TlNO_3$	690	Moms (1958)	700
$Mg(NO_3)_2$	2468	Finch, Gardner (1965)	2503
$Ca(NO_3)_2$	2209	Finch, Gardner (1965)	2228
$Sr(NO_3)_2$	2092	Finch, Gardner (1965)	2132
$Ba(NO_3)_2$	1975	Finch, Gardner (1965)	2016
$Mn(NO_3)_2$	2318	Yatsimirskii (1961)	2519
$Fe(NO_3)_2$	—	—	(2563)
$CO(NO_3)_2$	2560	Wen,Sho (1975)	2626
$Ni(NO_3)_2$	—	—	2709
$Cu(NO_3)_2$	—	—	2720
$Zn(NO_3)_2$	2376	Yatsimirskii (1961)	2628
$Cd(NO_3)_2$	2238	Yatsimirskii (1961)	2443

Table 1 (continued)
LATTICE ENERGIES

Salt	Calculated lattice energy (kJmol⁻¹)	Literature source	Thermochemical cycle lattice energy (kJmol⁻¹)
$Ha(NO_3)_2$	2255	Yatsimirskii (1961)	—
$Sn(NO_3)_2$	2155	Yatsimirskii (1961)	2254
$Pb(NO_3)_2$	2067	Yatsimirskii (1961)	2189
Nitrides			
ScN	7547	Baughan (1959)	7506
LaN	6876	Baughan (1959)	6793
TiN	8130	Baughan (1959)	8033
ZrN	7633	Baughan (1959)	7723
VN	8283	Baughan (1959)	8233
NbN	7939	Baughan (1959)	8022
CrN	8269	Baughan (1959)	8358
Nitrites			
$NaNO_2$	774	Morris (1958)	748
KNO_2	660	Jenkins (1977)	664
$RbNO_2$	638	Jenkins (1977)	765
$CsNO_2$	598	Jenkins (1977)	—
$Ca(NO_2)_2$	2460	Yatsimirskii (1956)	2225
$Sr(NO_2)_2$	2305	Yatsimirskii (1956)	2111
$Ba(NO_2)_2$	2205	Yatsimirskii (1956)	1987
Oxides			
Li_2O	2799	Baughan (1959)	
Na_2O	2481	Baughan (1959)	
K_2O	2238	Baughan (1959)	
Rb_2O	2163	Baughan (1959)	
Cu_2O	3273	Mamulov (1961)	
Ag_2O	3002	Mamulov (1961)	
Tl_2O	2659	Mamulov (1961)	
LiO_2	(878)	D'Orazio, Wood (1965)	(872)
NaO_2	799	Yatsimirskii (1959)	796
KO_2	741	D'Orazio, Wood (1965)	725
RbO_2	706	D'Orazio, Wood (1965)	695
CsO_2	679	D'Orazio, Wood (1965)	668
Li_2O_2	2592	Wood, D'Orazio (1965)	256
Na_2O_2	2309	Wood, D'Orazio (1965)	2305
K_2O_2	2114	Wood, D'Orazio (1965)	2078
Rb_2O_2	2025	Wood, D'Orazio (1965)	2006
Cs_2O_2	1948	Wood, D'Orazio (1965)	1861
MgO_2	3356	Wood, D'Orazio (1965)	3526
CaO_2	3144	Wood, D'Orazio (1965)	3133
SrO_2	3037	Wood, D'Orazio (1965)	2849
KO_3	697	Wood, D'Orazio (1965)	—
BEO	4293	Huggins, Sakamoto (1957)	4443
MgO	3795	Huggins, Sakomoto (1957)	3791
CaO	3414	Huggins, Sakamoto (1957)	3401
SrO	3217	Huggins, Sakamoto (1957)	3223
BaO	3029	Huggins, Sakamoto (1957)	3054
TiO	3832	Huggins, Sakamoto (1957)	3811
VO	3932	Ladd, Lee (1961)	3863
MnO	3724	Ladd, Lee (1961)	3745
FeO	3795	Ladd, Lee (1961)	3865
CoO	3837	Ladd, Lee (1961)	3910
NiO	3908	Ladd, Lee (1961)	4010
PdO	3736	Ladd, Lee (1961)	—
CuO	4135	Mamulov (1961)	4050
ZnO	4142	Ladd, Lee (1961)	3971

Table 1 (continued)
LATTICE ENERGIES

Salt	Calculated lattice energy (kJmol^{-1})	Literature source	Thermochemical cycle lattice energy (kJmol^{-1})
CdO	3806	Ladd, Lee (1961)	—
HgO	3907	Ladd, Lee (1961)	—
GeO	3919	Ladd, Lee (1961)	—
SnO	3652	Ladd, Lee (1961)	—
PbO	3520	Ladd, Lee (1961)	—
Sc_2O_3	13557	Gasharov, Sovers (1970)	13708
Y_2O_3	12705	Gasharov, Sovers (1970)	—
La_2O_3	12452	Johnson (1969)	—
Ce_2O_3	12661	Johnson (1969)	—
Pr_2O_3	12703	Johnson (1969)	—
Nd_2O_3	12736	Johnson (1969)	—
Pm_2O_3	12811	Johnson (1969)	—
Sm_2O_3	12878	Johnson (1969)	—
Eu_2O_3	12945	Johnson (1969)	—
Gd_2O_3	12996	Johnson (1969)	—
Tb_2O_3	13071	Johnson (1969)	—
Dy_2O_3	13138	Johnson (1969)	—
Ho_2O_3	13180	Johnson (1969)	—
Er_2O_3	13263	Johnson (1969)	—
Tm_2O_3	13322	Johnson (1969)	—
Yb_2O_3	13380	Johnson (1969)	—
Lu_2O_3	13665	Ladd, Lee (1961)	—
Ac_2O_3	12573	Krestov, Krestova (1969)	—
Ti_2O_3	—	—	14149
V_2O_3	15096	Mamulov (1961)	14520
Cr_2O_3	15276	Mamulov (1961)	114957
Mn_2O_3	15146	Mamulov (1961)	15035
Fe_2O_3	14309	Mamulov (1961)	14774
Al_2O_3	15916	Yatsimirskii (1961)	—
Ga_2O_3	15590	Yatsimirskii (1961)	15220
In_2O_3	13928	Yatsimirskii (1961)	—
Ti_2O_3	14702	Mamulov (1961)	—
Pb_2O_3	(14841)	Van Gool, Picken (1969)	—
CeO_2	9627	VanBaur (1961)	—
ThO_2	10397	Ladd, Lee (1961)	—
PaO_2	10573	Ladd, Lee (1961)	—
VO_2	10644	Ladd, Lee (1961)	—
NpO_2	10707	Ladd, Lee (1961)	—
PuO_2	10786	Ladd, Lee (1961)	—
AmO_2	10799	Ladd, Lee (1961)	—
CmO_2	10832	Ladd, Lee (1961)	—
TiO_2	12150	Ladd, Lee (1961)	—
ZrO_2	11188	Ladd, Lee (1961)	—
MoO_2	11648	Ladd, Lee (1961)	—
MnO_2	12970	Ladd, Lee (1961)	—
SiO_2	13125	Ladd, Lee (1961)	—
GeO_2	12828	Ladd, Lee (1961)	—
SnO_2	11807	Ladd, Lee (1961)	—
PbO_2	11217	—	
Perchlorates			
$LiClO_4$	709	Gill, Singla, Paul, Narula (1972)	723
$NaClO_4$	643	Jenkins, Pratt (1978)	648
$KClO_4$	599	Jenkins, Pratt (1978)	602
$RbClO_4$	564	Gill, Singla, Paul, Narula (1972)	582
$CsClO_4$	636	Morris (1958)	(542)

Table 1 (continued)
LATTICE ENERGIES

Salt	Calculated lattice energy (kJmol^{-1})	Literature source	Thermochemical cycle lattice energy (kJmol^{-1})
NH_4ClO_4	583	Jenkins, Pratt (1978)	580
$Ca(ClO_4)_2$	1958	Yatsimirskii (1956)	1971
$Sr(ClO_4)_2$	1862	Yatsimirskii (1956)	1862
$Ba(ClO_4)_2$	1795	Yatsimirskii (1958)	1769
$NaMnO_4$	661	Yatsimirskii (1956)	—
$KMnO_4$	607	Yatsimirskii (1956)	—
$RbMnO_4$	586	Yatsimirskii (1956)	—
$CsMnO_4$	565	Yatsimirskii (1956)	—
$Ca(MnO_4)_2$	1937	Yatsimirskii (1956)	—
$Sr(MnO_4)_2$	1845	Yatsimirskii (1956)	—
$Ba(MnO_4)_2$	1778	Yatsimirskii (1956)	—
Phosphates			
$Mg_3(PO_4)_2$	11632	Grekenschikov (1967)	11407
$Ca_3(PO_4)_2$	10602	Grekenschikov (1967)	10479
$Sr_3(PO_4)_2$	10125	Grekenschikov (1967)	10075
$Ba_3(PO_4)_2$	9652	Grekenschikov (1967)	9654
$MnPO_4$	7397	Grekenschikov (1967)	—
$FePO_4$	7251	Grekenschikov (1967)	7303
BPO_4	8201	Grekenschikov (1967)	—
$AlPO_4$	7427	Grekenschikov (1967)	7509
$GaPO_4$	7381	Grekenschikov (1967)	—
Phosphonium salts			
PH_4Br	616	Waddington (1965)	—
PH_4I	590	Waddington (1965)	—
Selenides			
Li_2Se	2364	Bevan, Moms (1960)	—
Na_2Se	2130	Bevan, Moms (1960)	—
K_2Se	1933	Bevan, Moms (1960)	—
Rb_2Se	1837	Bevan, Moms (1960)	—
CS_2Se	1745	Bevan, Moms (1960)	—
Ag_2Se	2686	Mamulov (1961)	—
Tl_2Se	2209	Mamulov (1961)	—
BeSe	3431	Huggins, Sakamoto (1957)	—
MgSe	3071	Huggins, Sakamoto (1957)	—
CaSe	2858	Huggins, Sakamoto (1957)	2862
SrSe	2736	Huggins, Sakamoto (1957)	—
BaSe	2611	Huggins, Sakamoto (1957)	—
MnSe	3176	Mamulov (1961)	3194
FeSe	3499	Mamulov (1961)	3396
CoSe	3554	Mamulov (1961)	3471
NiSe	3658	Mamulov (1961)	3558
CuSe	3736	Mamulov (1961)	3662
ZnSe	3502	Lamberdi (1969)	3514
CdSe	3330	Mamulov (1961)	—
HgSe	3501	Mamulov (1961)	—
SnSe	3058	Mamulov (1961)	—
PbSe	3050	Mamulov (1961)	—
Selenites			
Li_2SeO_3	2171	Klushina, Selivanova (1972)	—
Na_2SeO_3	1950	Klushina, Selivanova (1972)	1931
K_2SeO_3	1774	Klushina, Selivanova (1972)	—
Rb_2SeO_3	1715	Klushina, Selivanova (1972)	1675
Cs_2SeO_3	1640	Klushina, Selivanova (1972)	—
Tl_2,SeO_3	1879	Klushina, Selivanova (1972)	—
Ag_2SeO_3	2113	Klushina, Selivanova (1972)	—
$BeSeO_3$	3322	Klushina, Selivanova (1972)	—

Table 1 (continued)
LATTICE ENERGIES

Salt	Calculated lattice energy (kJmol⁻¹)	Literature source	Thermochemical cycle lattice energy (kJmol⁻¹)
$MgSeO_3$	3012	Klushina, Selivanova (1972)	2996
$CaSeO_3$	2732	Klushina, Selivanova (1972)	
$SrSeO_3$	2586	Klushina, Selivanova (1972)	2586
$BaSeO_3$	2460	Klushina, Selivanova (1972)	2448
$RaSeO_3$	2456	Klushina, Selivanova (1972)	—
$MnSeO_3$	2975	Klushina, Selivanova (1972)	—
$FeSeO_3$	2895	Klushina, Selivanova (1972)	—
$CoSeO_3$	3155	Klushina, Selivanova (1972)	—
$NiSeO_3$	2945	Klushina, Selivanova (1972)	—
$CuSeO_3$	3209	Klushina, Selivanova (1972)	—
$ZnSeO_3$	3167	Klushina, Selivanova (1972)	—
$CdSeO_3$	2962	Klushina, Selivanova (1972)	—
$PbSeO_3$	2669	Klushina, Selivanova (1972)	—
Selenates			
Li_2SeO_4	2054	Selivanova, Karapet'yants (1963)	—
Na_2SeO_4	1879	Selivanova, Karapet'yants (1963)	—
K_2SeO_4	1732	Selivanova, Katapet'yants (1963)	—
Rb_2SeO_4	1686	Selivanova, Karapet'yants (1963)	—
Cs_2SeO_4	1615	Selivanova, Karapet'yants (1963)	—
Cu_2SeO_4	2201	Selivanova, Karapet'yants (1963)	—
Ag_2SeO_4	2033	Selivanova, Karapet'yants (1963)	—
Tl_2SeO_4	1766	Selivanova, Karapet'yants (1963)	—
Hg_2SeO_4	2163	Selivanova, Karapet'yants (1963)	—
$BeSeO_4$	3448	Selivanova, Karapet'yants (1963)	—
$MgSeO_4$	2895	Selivanova, Karapet'yants (1963)	—
$CaSeO_4$	2632	Selivanova, Karapet'yants (1963)	—
$SrSeO_4$	2489	Selivanova, Karapet'yants (1963)	—
$BaSeO_4$	2385	Selivanova, Karapet'yants (1963)	—
$RaSeO_4$	2364	Selivanova, Karapet'yants (1963)	—
$GdSeO_4$	2753	Selivanova, Karapet'yants (1963)	—
$MnSeO_4$	2837	Selivanova, Karapet'yants (1963)	—
$FeSeO_4$	3008	Selivanova, Karapet'yants (1963)	—
$CoSeO_4$	2912	Selivanova, Karapet'yants (1963)	—
$NiSeO_4$	3079	Selivanova, Karapet'yants (1963)	—
$CuSeO_4$	3104	Selivanova, Karapet'yants (1963)	—
$ZnSeO_4$	3021	Selivanova, Karapet'yants (1963)	—
$CdSeO_4$	2833	Selivanova, Karapet'yants (1963)	—
$HgSeO_4$	2845	Selivanova, Karapet'yants (1963)	—
$PbSeO_4$	2561	Selivanova, Karapet'yants (1963)	—
Sulphides			
Li_2S	2464	Morris (1958)	2472
Na_2S	2192	Morris (1958)	2203
K_2S	1979	Morris (1958)	(2052)
Rb_2S	1929	Morris (1958)	1949
Cs_2S	1892	Karapet'yants (1954)	1850
$(NH_4)_2S$	2008	Karapet'yants (1954)	(2026)
Cu_2S	2786	Karapet'yants (1954)	2865
Ag_2S	2606	Karapet'yants (1954)	2677
Au_2S	2908	Karapet'yants (1954)	—
Tl_2S	2298	Mamulov (1961)	2258
Sulphates			
Li_2SO_4	2229	Jenkins (1975)	2142
Na_2SO_4	1827	Jenkins (1975)	1938
K_2SO_4	1700	Jenkins (1975)	1796
Rb_2SO_4	1636	Jenkins (1975)	1748

Table 1 (continued)
LATTICE ENERGIES

Salt	Calculated lattice energy (kJmol⁻¹)	Literature source	Thermochemical cycle lattice energy (kJmol⁻¹)
Cs_2SO_4	1596	Jenkins (1975)	1658
$(NH_4)_2SO_4$	1766	Jenkins, Smith (1975)	1777
Cu_2SO_4	2276	Selivanova, Karapet'yants (1963)	2166
Ag_2SO_4	2104	Selivanova, Karapet'yants (1963)	1989
Tl_2SO_4	1828	Selivanova, Karapet'yants (1963)	1722
Hg_2SO_4	—	Selivanova, Karapet'yants (1963)	2127
$CaSo_4$	2489	Ladd, Lee (1968)	2480
$SrSO_4$	2577	Selivanova, Karapet'yants (1963)	2484
$BaSO_4$	2469	Selivanova, Karapet'yants (1963)	2374
$MnSO_4$	2920	Selivanova, Karapet'yants (1963)	2825
$FeSO_4$	2983	Selivanova, Karapet'yants (1963)	2921
$CoSo_4$	3088	Selivanova, Karapet'yants (1963)	2917
$NiSO_4$	3167	Selivanova, Karapet'yants (1963)	3044
$CuSO_4$	3167	Selivanova, Karapet'yants (1963)	3066
$ZnSO_4$	3100	Selivanova, Karapet'yants (1963)	3006
$CdSO_4$	2891		
$PbSO_4$	2635	Selivanova, Karapet'yants (1963)	2534
$Sn(SO_4)_2$	—	—	9616
Ternary salts			
CS_2CuCl_4	1393	Blake, Colton (19#)	—
Rb_2ZnCl_4	1529	Paoletti (1965)	—
Cs_2ZnCl_4	1492	Paoletti (1965)	—
Rb_2ZnBr_4	1498	Paoletti (1965)	—
Cs_2ZnBr_4	1454	Paoletti (1965)	—
$(Me_4N)_2ZnBr_4$	1364	Paoletti (1965)	—
Cs_2ZnI_4	1386	Paoletti (1965)	—
$CsGaCl_4$	494	Jenkins (1976)	—
$NaAlCl_4$	556	Jenkins (1976)	—
$CsAlCl_4$	486	Jenkins (1976)	—
$NaFeCl_4$	492	Blake, Colton (1963)	—
Rb_2CoCl_4	1447	Lister, Nyburg, Poyntz (1974)	—
Cs_2CoCl_4	1391	Lister, Nyburg, Poyntz (1974)	—
K_2PtCl_4	1574	Hartley (1972)	1550
Cs_2GeF_6	1573	Jenkins, Pratt (1978)	—
$(NH_4)_2GeF_6$	1657	Jenkins, Pratt (1978)	—
Cs_2GeCl_6	1375	Jenkins, Pratt (1078)	1375
K_2HfCl_6	1345	Jenkins, Pratt (1978)	1461
K_2IrCl_6	1442	Jenkins, Pratt (1978)	1440
$(NH_4)_2IrCl_6$	1442	Jenkins, Pratt (1978)	—
Na_2MoCl_6	1526	Jenkins, Pratt (1978)	—
K_2MoCl_6	1418	Jenkins, Pratt (1978)	1433
Rb_2MoCl_6	1399	Jenkins, Pratt (1978)	1399
Cs_2MoCl_6	1347	Jenkins, Pratt (1978)	1347
K_2NbCl_6	1375	Jenkins, Pratt (1978)	1398
Rb_2NbCl_6	1371	Jenkins, Pratt (1978)	1385
Cs_2NbCl_6	1381	Jenkins, Pratt (1978)	1344
K_2OsCl_6	1447	Jenkins, Pratt (1978)	1447
Cs_2OsCl_6	1409	Jenkins, Pratt (1978)	—
$(NH_4)_2OsCl_6$	1433	Jenkins, Pratt (1978)	—
K_2OsBr_6	1396	Jenkins, Pratt (1978)	—
K_2PdCl_6	1450	Jenkins, Pratt (1978)	1466
Rb_2PdCl_6	1449	Jenkins, Pratt (1978)	—
Cs_2PdCl_6	1426	Jenkins, Pratt (1978)	—
$(NH_4)_2PdCl_6$	1481	Jenkins, Pratt (1978)	—
Rb_2PbCl_6	1343	Jenkins, Pratt (1978)	1343

Table 1 (continued)
LATTICE ENERGIES

Salt	Calculated lattice energy (kJmol⁻¹)	Literature source	Thermochemical cycle lattice energy (kJmol⁻¹)
Cs_2PbCl_6	1344	Jenkins, Pratt (1978)	—
$(NH_4)_2PbCl_6$	1355	Jenkins, Pratt (1978)	—
K_2PtCl_6	1468	Jenkins, Pratt (1978)	—
Rb_2PtCl_6	1464	Jenkins, Pratt (1978)	—
Cs_2PtCl_6	1444	Jenkins, Pratt (1978)	—
$(NH_4)_2PtCl_6$	1468	Jenkins, Pratt (1978)	—
Tl_2PtCl_6	1546	Jenkins, Pratt (1978)	—
Ag_2PtCl_6	1773	Jenkins, Pratt (1978)	—
$BaPtCl_6$	2047	Jenkins, Pratt (1978)	—
K_2PtBr_6	1423	Jenkins, Pratt (1978)	—
Ag_2PtBr_6	1791	Jenkins, Pratt (1978)	—
K_2PtI_6	1421	Lister, Nyburg, Poyntz (1974)	—
K_2ReCl_6	1416	Jenkins, Pratt (1978)	1442
Rb_2ReCl_6	1414	Jenkins, Pratt (1978)	—
Cs_2ReCl_6	1398	Jenkins, Pratt (1978)	—
$(NH_4)_2ReCl_6$	1402	Jenkins, Pratt (1978)	1390
K_2ReBr_6	1375	Jenkins, Pratt (1978)	1375
K_2SiF_6	1670	Jenkins, Pratt (1978)	1628
Rb_2SiF_6	1639	Jenkins, Pratt (1978)	1621
Cs_2SiF_6	1604	Jenkins, Pratt (1978)	1498
Tl_2SiF_6	1675	Jenkins, Pratt (1978)	—
$(NH_4)_2SiF_6$	1657	Jenkins, Pratt (1978)	1727
K_2SnCl_6	1363	Jenkins, Pratt (1978)	1390
Rb_2SnCl_6	1361	Jenkins, Pratt (1978)	1363
Cs_2SnCl_6	1358	Jenkins, Pratt (1978)	—
Tl_2SnCl_6	1437	Jenkins, Pratt (1978)	—
$(NH_4)_2SnCl_6$	1370	Jenkins, Pratt (1978)	1334
Rb_2SnBr_6	1309	Jenkins, Pratt (1978)	—
Cl_2SnBr_6	1306	Jenkins, Pratt (1978)	—
$(NH_4)_2SnBr_6$	1319	Jenkins, Pratt (1978)	—
Rb_2SnI_6	1226	Jenkins, Pratt (1978)	—
Cs_2SnI_6	1243	Jenkins, Pratt (1978)	—
K_2TeCl_6	1318	Jenkins, Pratt (1978)	1320
Rb_2TeCl_6	1321	Jenkins, Pratt (1978)	—
Cs_2TeCl_6	1323	Jenkins, Pratt (1978)	—
Tl_2TeCl_6	1392	Jenkins, Pratt (1978)	—
$(NH_4)_2TeCl_6$	1318	Jenkins, Pratt (1978)	—
K_2RuCl_6	1451	Jenkins, Pratt (1978)	—
Rb_2CoF_6	1688	Jenkins, Pratt (1978)	—
Cs_2CoF_6	1632	Jenkins, Pratt (1978)	—
K_2NiF_6	1721	Jenkins, Pratt (1978)	—
Rb_2NiF_6	1688	Jenkins, Pratt (1978)	—
Rb_2SbCl_6	1357	Jenkins, Pratt (1978)	—
Rb_2SeCl_6	1409	Jenkins, Pratt (1978)	—
Cs_2SeCl_6	1397	Jenkins, Pratt (1978)	—
$(NH_4)_2SeCl_6$	1420	Jenkins, Pratt (1978)	—
K_2SeBr_6	1379	Jenkins, Pratt (1978)	—
$(NH_4)_2SeBr_6$	1380	Jenkins, Pratt (1978)	—
$(NH_4)_2PoCl_6$	1338	Jenkins, Pratt (1978)	—
Cs_2PoBr_6	1286	Jenkins, Pratt (1978)	—
$(NH_4)_2PoBr_6$	1292	Jenkins, Pratt (1978)	—
Cs_2CrF_6	1603	Jenkins, Pratt (1978)	—
Rb_2MnF_6	1688	Jenkins, Pratt (1978)	—
Cs_2MnF_6	1620	Jenkins, Pratt (1978)	—
K_2MnCl_6	1462	Jenkins, Pratt (1978)	—
Rb_2MnCl_6	1451	Jenkins, Pratt (1978)	—

Table 1 (continued)
LATTICE ENERGIES

Salt	Calculated lattice energy (kJmol⁻¹)	Literature source	Thermochemical cycle lattice energy (kJmol⁻¹)
$(NH_4)_2MnCl_6$	1464	Jenkins, Pratt (1978)	—
Cs_2TeBr_6	1306	Jenkins, Pratt (1978)	—
$(NH_4)_2TeBr_6$	1294	Jenkins, Pratt (1978)	—
Cs_2TeI_6	1246	Jenkins, Pratt (1978)	—
K_2TiCl_6	1412	Jenkins, Pratt (1978)	1443
Rb_2TiCl_6	1415	Jenkins, Pratt (1978)	1425
Cs_2TiCl_6	1402	Jenkins, Pratt (1978)	1370
Tl_2TiCl_6	1560	Jenkins, Pratt (1978)	1553
$(NH_4)_2TiCl_6$	1413	Jenkins, Pratt (1978)	—
K_2TiBr_6	1379	Jenkins, Pratt (1978)	1379
Rb_2TiBr_6	1341	Jenkins, Pratt (1978)	1365
Cs_2TiBr_6	1339	Jenkins, Pratt (1978)	1316
Na_2UBr_6	1504	Vdorenko, Suglobova, Chirkst (1974)	—
K_2UBr_6	1484	Vdorenko, Suglobova, Chirkst (1974)	—
Rb_2UBr_6	1473	Vdorenko, Suglobova, Chirkst (1974)	—
Cs_2UBr_6	1459	Vdorenko, Suglobova, Chirkst (1974)	—
K_2WCl_6	1398	Jenkins, Pratt (1978)	1423
Rb_2WCl_6	1397	Jenkins, Pratt (1978)	1434
Cs_2WCl_6	1392	Jenkins, Pratt (1978)	1366
K_2WBr_6	1408	Jenkins, Pratt (1978)	1408
Rb_2WBr_6	1361	Jenkins, Pratt (1978)	1391
Cs_2WBr_6	1362	Jenkins, Pratt (1978)	1332
K_2ZrCl_6	1339	Jenkins, Pratt (1978)	1371
Rb_2ZrCl_6	1341	Jenkins, Pratt (1978)	—
Cs_2ZrCl_6	1339	Jenkins, Pratt (1978)	1308
Tetraalkyl ammonium salts			
Me_4NCl	566	Wilson (1976)	—
Me_4NBr	553	Wilson (1976)	—
Me_4NI	544	Wilson (1976)	—
Tellurides			
Li_2Te	2212	Beván, Moms (1960)	—
Na_2Te	1997	Bevan, Moms (1960)	2095
K_2Te	1830	Bevans, Moms (1960)	—
Rb_2Te	1837	Bevan, Moms (1960)	—
Cs_2Te	1745	Bevan, Moms (1960)	—
Cu_2Te	2706	Mamulov (1961)	2683
Ag_2Te	2607	Mamulov (1961)	2600
Tl_2Te	2084	Mamulov (1961)	2172
BeTe	3319	Das, Keer, Rao (1963)	—
MgTe	2878	Das, Keer, Rao (1963)	3081
CaTe	2721	Das, Keer, Rao (1963)	—
SrTe	2599	Das, Keer, Rao (1963)	—
BaTe	2473	Das, Keer, Rao (1963)	—
RaTe	2481	Yatsimirskii, Krestov (1960)	—
MnTe	3041	Mamulov (1961)	—
FeTe	3399	Mamulov (1961)	—
CoTe	3429	Mamulov (1961)	—
NiTe	3534	Mamulov (1961)	—
CuTe	3639	Mamulov (1961)	—
ZnTe	3416	Mamulov (1961)	—
CdTe	3212	Mamulov (1961)	—
PbTe	2930	Mamulov (1961)	—
Thiocyanates			
LiCNS	764	Gill, Singla, Paul, Narula (1972)	(765)
NaCNS	682	Morris (1958)	682

Table 1 (continued)
LATTICE ENERGIES

Salt	Calculated lattice energy. (kJmol⁻¹)	Literature source	Thermochemical cycle lattice energy (kJmol⁻¹)
KCNS	623	Morris (1958)	616
RbCNS	623	Morris (1958)	619
CsCNS	623	Yatsimirskii (1956)	568
NH_4CNS	605	Gill, Singla, Paul, Narula (1972)	611
$Ca(CNS)_2$	2184	Yatsimirskii (1956)	2118
$Sr(CNS)_2$	2063	Yatsimirskii (1956)	1957
$Ba(CNS)_2$	1979	Yatsimirskii (1956)	1852
$Mn(CNS)_2$	2280	Yatsimirskii (1961)	2351
$Zn(CNS)_2$	2335	Yatsimirskii (1961)	2560
$Cd(CNS)_2$	2201	Yatsimirskii (1961)	2374
$Hg(CNS)_2$	2146	Yatsimirskii (1961)	2492
$Sn(CNS)_2$	2117	Yatsimirskii (1961)	2142
$Pb(CNS)_2$	2058	Morris (1958)	—
Vanadates			
$LiVO_3$	810	Golovkin, Fotiev (1970)	—
$NaVO_3$	761	Golovkin, Fotiev (1970)	—
KVO_3	686	Golovkin, Fotiev (1970)	—
$RbVO_3$	657	Golovkin, Fotiev (1970)	—
$CsVO_3$	628	Golovkin, Fotiev (1970)	—

Jenkins and Waddington refers to a forthcoming review article: "The Use of Lattice Energy in Inorganic Chemistry"; publication details can be obtained by writing to H. D. B. Jenkins, Department of Chemistry, University of Warwick, Coventry CV 4 7AL, Warwickshire.

Table 2
ANCILLARY THERMODYNAMIC DATA

Salt or ion	State	Source	ΔH_f^θ (kJmol⁻¹)
Li^+	g	JANAF	687.16
Na^+	g	JANAF	609.84
K^+	g	JANAF	514.19
Cs^+	g	JANAF	452.3
O_2^-	g	Jenkins, Waddington[a]	−74
CN^-	g	Jenkins, Pratt, Waddington (1977)	41
NH_4^+	g	Jenkins, Morris (1977)	630.2
C_2^{-2}	g	Jenkins, Waddington[a]	918
O_2^{-2}	g	Jenkins, Waddington[a]	553
NO_3^-	g	Jenkins, Moms (1977)	−320.1
$IO_2F_2^-$	g	Jenkins, Waddington[a]	−693
IO_3^-	g	Jenkins, Waddington[a]	−208
ClO_3^-	g	Jenkins, Waddington[a]	−200
BrO_3^-	g	Jenkins, Waddington[a]	−145
HSO_4^-	g	Jenkins, Waddington[a]	(−1012)
HCO_3^-	g	Jenkins, Waddington[a]	−738
$NH_2CH_2CO_2^-$	g	Bernard	−564
N^{-3}	g	Jenkins, Waddington[a]	(2588)
CO_3^{-2}	g	Jenkins, Waddington, Pratt	−321
PO_4^{-3}	g	Jenkins, Waddington[a]	(291)
AsO_4^{-3}	g	Jenkins, Waddington[a]	(289)

Table 2 (continued)
ANCILLARY THERMODYNAMIC DATA

Salt or ion	State	Source	ΔH_f^θ (kJmol^{-1})
GeO_4^{-4}	g	Jenkins, Waddington[a]	(1460)
CH_3COO^-	g	Jenkins, Waddington[a]	−554
ClO_4^-	g	Jenkins, Pratt (1978)	−344
NO_2^-	g	Jenkins (1977)	−219
SeO_3^{-2}	g	Jenkins, Waddington[a]	(−249)
HCO_2^-	g	Jenkins, Waddington[a]	−463
BH_4^-	g	Waddington (1959)	−96
HS^-	g	Jenkins, Waddington[a]	−120
$TiCl_6^{-2}$	g	Jenkins, Pratt (1978)	−1330
$TiBr_6^{-2}$	g	Jenkins, Pratt (1978)	−1142
$ZrCl_6^{-2}$	g	Jenkins, Pratt (1978)	−1526
$HfCl_6^{-2}$	g	Jenkins, Pratt (1978)	−1640
$NbCl_6^{-2}$	g	Jenkins, Pratt (1978)	−1224
$TaCl_6^{-2}$	g	Jenkins, Pratt (1978)	−1275
$MoCl_6^{-2}$	g	Jenkins, Pratt (1978)	−1070
WCl_6^{-2}	g	Jenkins, Pratt (1978)	−985
WBr_6^{-2}	g	Jenkins, Pratt (1978)	−705
$ReCl_6^{-2}$	g	Jenkins, Pratt (1978)	−919
$ReBr_6^{-2}$	g	Jenkins, Pratt (1978)	−689
$OsCl_6^{-2}$	g	Jenkins, Pratt (1978)	−752
$IrCl_6^{-2}$	g	Jenkins, Pratt (1978)	−785
$PdCl_6^{-2}$	g	Jenkins, Pratt (1978)	−749
$PtCl_6^{-2}$	g	Jenkins, Pratt (1978)	−774
$PtBr_6^{-2}$	g	Jenkins, Pratt (1978)	−645
SiF_6^{-2}	g	Jenkins, Pratt (1978)	−2207
$GeCl_6^{-2}$	g	Jenkins, Pratt (1978)	−981
$SuCl_6^{-2}$	g	Jenkins, Pratt (1978)	−1156
$PbCl_6^{-2}$	g	Jenkins, Pratt (1978)	−940
$TeCl_6^{-2}$	g	Jenkins, Pratt (1978)	−902
NO_2^-	g	Jenkins, Pratt (1978)	−202
LiO_2	c	D'Orazio, Wood (1965)	−259.4
NaO_2	c	JANAF	−260.7
KO_2	c	JANAF 1974 Suppl.	−284.5
CsO_2	c	D'Orazio, Wood (1965)	−289.5
Li_2O_2	c	JANAF	−632.6
Na_2O_2	c	JANAF	−531.2
K_2O_2	c	JANAF	−495.8
Cs_2O_2	c	NBS Circular 500	−402.5
ThO_2	c	Smirnov, Ivanovskii (1957)	−1230.5
MgO_2	c	NBS Circular 500	−623.0
$NaIO_2F_2$	c	Finch, Gates, Jenkinson	−848.7
KIO_2F_2	c	Finch, Gates, Jenkinson	−876.6
$RbIO_2F_2$	c	Finch, Gates, Jenkinson	−875.3
$NH_4IO_2F_2$	c	Finch, Gates, Jenkinson	−747.7
$LiBrO_3$	c	Boyd, Vaslow (1962)	−356.4
$CsBrO_3$	c	Boyd, Vaslow (1962)	−374.5
$CsHCO_3$	c	NBS Circular 500	−995.6
$Ca(HCO_3)_2$	c	Wilcox, Bromley (1963)	(−1950)
$Sr(HCO_3)_2$	c	Wilcox, Bromley (1963)	(−1954)
$Ba(HCO_3)_2$	c	Wilcox, Bromley (1963)	(−1971)
CH_3COONH_4	c	NBS Circular 500	−618.4
CH_3COOLi	c	Rudnitskii (1961)	−748.9
CH_3COORb	c	Morris (1959)	−720.9
$LiClO_4$	c	JANAF	−380.7
$CsClO_4$	c	Wilcox, Bromley (1963)	(−435.1)
LiH	c	JANAF	−90.62
NaH	c	JANAF	−56.4

Table 2 (continued)
ANCILLARY THERMODYNAMIC DATA

Salt or ion	State	Source	ΔH_f^θ (kJmol^{-1})
KH	c	JANAF	−57.8
CsH	c	Smith, Bass (1963)	−49.9
LiN_3	c	Gray, Waddington (1956)	10.8
NaN_3	c	Gray, Waddington (1956)	21.25
KN_3	c	Gray, Waddington (1956)	−1.38
RbN_3	c	Gray, Waddington (1956)	−0.29
CsN_3	c	Gray, Waddington (1956)	−9.92
TlN_3	c	Gray, Waddington (1956)	233.4
AgN_3	c	Gray, Waddington (1956)	310.3
CsHS	c	NBS Circular 500	−263.2
$Ca(HS)_2$	c	Wilcox, Brawley (1963)	(−481)
$Ba(HS)_2$	c	Wilcox, Brawley (1956)	(−531.4)
LiHS	c	Juza, Laurer (1953)	−254.4
K_2TiCl_6	c	Jenkins, Pratt (1978)[b]	−1747
Rb_2TiCl_6	c	Jenkins, Pratt (1978)[b]	−1767
Cs_2TiCl_6	c	Jenkins, Pratt (1978)[b]	−1797
Tl_2TiCl_6	c	Jenkins, Pratt (1978)[b]	−1330
K_2TiBr_6	c	Jenkins, Pratt (1978)[b]	−1493
Rb_2TiBr_6	c	Jenkins, Pratt (1978)[b]	−1517
Cs_2TiBr_6	c	Jenkins, Pratt (1978)[b]	−1553
K_2ZrCl_6	c	Jenkins, Pratt (1978)[b]	−1932
Cs_2ZrCl_6	c	Jenkins, Pratt (1978)[b]	−1992
K_2HfCl_6	c	Jenkins, Pratt (1978)[b]	−1957
K_2NbCl_6	c	Jenkins, Pratt (1978)[b]	−1594
Rb_2NbCl_6	c	Jenkins, Pratt (1978)[b]	−1619
Cs_2NbCl_6	c	Jenkins, Pratt (1978)[b]	−1663
K_2ReCl_6	c	Jenkins, Pratt (1978)[b]	−1333
$(NH_4)_2ReCl_6$	c	Jenkins, Pratt (1978)[b]	−1056
K_2ReBr_6	c	Jenkins, Pratt (1978)[b]	−1036
K_2OsCl_6	c	Jenkins, Pratt (1978)[b]	−1171
K_2IrCl_6	c	Jenkins, Pratt (1978)[b]	−1197
K_2SiF_6	c	Jenkins, Pratt (1978)[b]	−2807
Rb_2SiF_6	c	Jenkins, Pratt (1978)[b]	−2838
Cs_2SiF_6	c	Jenkins, Pratt (1978)[b]	−2801
$(NH_4)_2SiF_6$	c	Jenkins, Pratt (1978)[b]	−2681
Rb_2PbCl_6	c	Jenkins, Pratt (1978)[b]	−1293
$Mn(CN)_2$	c	Karapet'yants, Karapet'yants	60.7
RbCN	c	Pritchard (1953)	(−108.8)
CsCN	c	Pritchard (1953)	(−108.7)
LiCN	c	Jenkins, Pratt, Waddington (1977)	−121
$Sr(CN)_2$	c	Wilcox, Bromley (1963)	(−205)
$LiNo_3$	c	NBS Circular 500	−482.3
$RbNO_3$	c	NBS Circular 500	−489.7
$CsNO_3$	c	NBS Circular 500	−494.2
$Fe(NO_3)_2$	c	Wilcox, Bromley (1963)	(−447.7)
$Sn(NO_3)_2$	c	Yatsimirskii (1961)	−456.1
$NaBF_4$	c	Altschuller (1955)	−1774
$CsIO_3$	c	Shedlovskii, Voskresenskii (1966)	−526.3
$CsClO_3$	c	Volodina, Shidlovskii, Voskresenskii (1966)	−395.8
$Mg(ClO_3)_2$	c	Wilcox, Bromley (1963)	(−523)
$Ca(ClO_3)_2$	c	Rudnitskii (1961)	−757
$Sr(ClO_3)_2$	c	Rudnitskii (1961)	−766
ScN	c	Neuman, Kroger, Kunz	−284.5
Li_2CO_3	c	JANAF	−1216.0

Table 2 (continued)
ANCILLARY THERMODYNAMIC DATA

Salt or ion	State	Source	ΔH_f^θ (kJmol⁻¹)
Cs_2CO_3	c	NBS Circular 500	−1118.8
$NiCO_3$	c	Karapet'yants (1955)	−689.1
$CuCO_3$	c	NBS Circular 500	−595.0
$SnCO_3$	c	Wilcox, Bromley (1963)	(−740.6)
HCO_2Li	c	Moms (1959)	−649.3
HCO_2Rb	c	Moms (1959)	−656.0
TiH	c	Gibb (1962)	−130.5
ZrH	c	Gibb (1962)	−173.5
VH	c	Gibb (1962)	(−32)
NbH	c	Gibb (1962)	(−100)
PdH	c	Gibb (1962)	(−37)
CuH	c	Gibb (1962)	(−21.4)
LaH_2	c	Gibb (1962)	−230.1
CeH_2	c	Gibb (1962)	−141.8
PrH_2	c	Gibb (1962)	−200.0
NdH_2	c	Gibb (1962)	−187.4
SmH_2	c	Gibb (1962)	−187.4
GdH_2	c	Gibb (1962)	−196.2
CsOH	c	NBS Circular 500	−406.7
$Ti(OH)_2$	c	Wilcox, Bromley (1963)	(−778.0)
LiOH	c	NBS Circular 500	−487.2
K_2TaCl_6	c	Jenkins, Pratt (1978)[b]	−1648
Rb_2TaCl_6	c	Jenkins, Pratt (1978)[b]	−1669
Cs_2TaCl_6	c	Jenkins, Pratt (1978)[b]	−1711
Na_2MoCl_6	c	Jenkins, Pratt (1978)[b]	−1376
K_2MoCl_6	c	Jenkins, Pratt (1978)[b]	−1475
Rb_2MoCl_6	c	Jenkins, Pratt (1978)[b]	−1479
Cs_2MoCl_6	c	Jenkins, Pratt (1978)[b]	−1512
K_2WCl_6	c	Jenkins, Pratt (1978)[b]	−1380
Rb_2WCl_6	c	Jenkins, Pratt (1978)[b]	−1429
Cs_2WCl_6	c	Jenkins, Pratt (1978)[b]	−1446
K_2WBr_6	c	Jenkins, Pratt (1978)[b]	−1085
Rb_2WBr_6	c	Jenkins, Pratt (1978)[b]	−1106
Cs_2WBr_6	c	Jenkins, Pratt (1978)[b]	−1132
K_2PdCl_6	c	Jenkins, Pratt (1978)[b]	−1187
Ag_2PtCl_6	c	Jenkins, Pratt (1978)[b]	−527
$BaPtCl_6$	c	Jenkins, Pratt (1978)[b]	−1180
K_2PtBr_6	c	Jenkins, Pratt (1978)[b]	−1040
Ag_2PtBr_6	c	Jenkins, Pratt (1978)[b]	−398
Cs_2GeCl_6	c	Jenkins, Pratt (1978)[b]	−1451
K_2SnCl_6	c	Jenkins, Pratt (1978)[b]	−1518
Rb_2SnCl_6	c	Jenkins, Pratt (1978)[b]	−1529
$(NH_4)_2SnCl_6$	c	Jenkins, Pratt (1978)[b]	−1237
Rb_2TeCl_6	c	Jenkins, Pratt (1978)[b]	−1237

[a] Jenkins and Waddington refers to a forthcoming review article: "The Use of Lattice Energy in Inorganic Chemistry"; publication details can be obtained by writing to H. D. B. Jenkins, Department of Chemistry, University of Warwick, Coventry CV 4 7AL, Warwickshire.

[b] For sources of data used in this ancillary table, refer to *Adv. Inorg. Chem. Radiochem.*, 22, 1, 1978.

ATOMIC TRANSITION PROBABILITIES

Compiled by W. L. Wiese and G. A. Martin

These tables were prepared under the auspices of the Committee on Line Spectra of the Elements of the National Academy of Sciences — National Research Council. They contain critically evaluated atomic transition probabilities A for about 4000 selected lines of elements from hydrogen to nickel. The material is largely for neutral and singly ionized spectra, but includes a number of prominent lines of more highly charged ions.

Most of the data are obtained from comprehensive compilations of the National Bureau of Standards Data Center on Atomic Transition Probabilities. Specifically, data have been taken from critical compilations on Sc,[1] Ti,[1] V,[2] Cr,[2] Mn,[2] and Fe[3] without changes. Material from earlier compilations for the elements H through Ne[4] and for Na through Ca[5] was supplemented by more recent material taken directly from the original literature. For the higher ions, most data were derived from studies of the systematic behavior of transition probabilities.[6-8] The original literature is cited in a very recent bibliography on transition probabilities.[9]

The wavelength range for the neutral species has normally been restricted to the visible or shorter wavelengths; only the most prominent near infrared lines are included. For the higher ions, most prominent lines are located in the extreme uv. The tabulation is limited to electric dipole — including intercombination — lines and comprises essentially the fairly strong transitions (usually $gf > 10^{-3}$) with estimated uncertainties of 50% or less. With the exception of hydrogen and helium, most transitions are between states with fairly low principal quantum numbers, i.e., not exceeding the quantum number of the ground state by more than two.

Whenever the wavelengths of individual lines within a multiplet are extremely close, an average wavelength for the multiplet is given, followed by an asterisk (*). This has also been done when the transition probability for an entire multiplet has been taken from the literature and values for individual lines cannot be determined because of insufficient knowledge of the coupling of electrons. The statistical weights, g_i and g_k, for the lower (i) and upper (k) atomic states are listed since the product g_kA (or g_if) is needed in many applications. A is given in units of $10^8\ s^{-1}$and is listed to as many digits as is consistent with the indicated accuracy. A number in parentheses following the tabulated value of the transition probability indicates the power of ten by which this value has to be multiplied. The uncertainties of the transition probabilities are indicated by code letters as follows: AA for uncertainties within 1%; A for uncertainties within 3%; B for uncertainties within 10%; C for uncertainties within 25%; D for uncertainties within 50%. The listed uncertainties are used with the connotation of "extent of possible error."

In the table for hydrogen, the uncertainty column has been eliminated since the transition probabilities for this element are known very precisely (< 1% error). Due to the hydrogen degeneracy, a "transition" is actually the sum of all transitions between the principal quantum numbers listed in the transition column, and the tabulation represents the properly weighted total A values.

In addition to the transition probability A, the atomic oscillator-strength f and the line-strength S are often used in the literature. The conversion factors between these quantities are:

$$g_if = 1.499 \times 10^{-8} \lambda^2 g_kA = 303.8 \lambda^{-1}S$$

where λ is in Angströms, A is in $10^8\ s^{-1}$, and S is in atomic units, which are $a_o^2e^2 = 7.188 \times 10^{-59} m^2C^2$ for electric dipole transitions.

We acknowledge the very valuable preparatory work by D. Trahan and W. Croom in arranging and compiling the numerical data.

Table 1a
TRANSITION PROBABILITIES FOR ALLOWED LINES OF HYDROGEN

Wavelength (λ{Å})	Transition	Statistical weights g_i	g_k	Average transition probability (A {10^8 s^{-1}})
914.039	1—20	2	800	3.928(−5)
914.286	1—19	2	722	5.077(−5)
914.576	1—18	2	648	6.654(−5)
914.919	1—17	2	578	8.858(−5)
915.329	1—16	2	512	1.200(−4)
915.824	1—15	2	450	1.657(−4)
916.429	1—14	2	392	2.341(−4)
917.181	1—13	2	338	3.393(−4)
918.129	1—12	2	288	5.066(−4)
919.352	1—11	2	242	7.834(−4)
920.963	1—10	2	200	1.263(−3)
923.150	1—9	2	162	2.143(−3)
926.226	1—8	2	128	3.869(−3)
930.748	1—7	2	98	7.568(−3)
937.803	1—6(Lε)	2	72	1.644(−2)
949.743	1—5(Lδ)	2	50	4.125(−2)
972.537	1—4(Lγ)	2	32	1.278(−1)
1025.72	1—3(Lβ)	2	18	5.575(−1)
1215.67	1—2(Lα)	2	8	4.699
3682.81	2—20	8	800	2.172(−5)
3686.83	2—19	8	722	2.809(−5)
3691.55	2—18	8	648	3.685(−5)
3697.15	2—17	8	578	4.910(−5)
3703.85	2—16	8	512	6.658(−5)
3711.97	2—15	8	450	9.210(−5)
3721.94	2—14	8	392	1.303(−4)
3734.37	2—13	8	338	1.893(−4)
3750.15	2—12	8	288	2.834(−4)
3770.63	2—11	8	242	4.397(−4)
3797.90	2—10	8	200	7.122(−4)
3835.38	2—9	8	162	1.216(−3)
3889.05	2—8	8	128	2.215(−3)
3970.07	2—7(Hε)	8	98	4.389(−3)
4101.73	2—6(Hδ)	8	72	9.732(−3)
4340.46	2—5(Hγ)	8	50	2.530(−2)
4861.32	2—4(Hβ)	8	32	8.419(−2)
6562.80	2—3(Hα)	8	18	4.410(−1)
8392.40	3—20	18	800	1.517(−5)
8413.32	3—19	18	722	1.964(−5)
8437.96	3—18	18	648	2.580(−5)
8467.26	3—17	18	578	3.444(−5)
8502.49	3—16	18	512	4.680(−5)
8545.39	3—15	18	450	6.490(−5)
8598.39	3—14	18	392	9.211(−5)
8665.02	3—13	18	338	1.343(−4)
8750.47	3—12	18	288	2.021(−4)
8862.79	3—11	18	242	3.156(−4)
9014.91	3—10	18	200	5.156(−4)
9229.02	3—9	18	162	8.905(−4)
9545.98	3—8(Pε)	18	128	1.651(−3)
10049.4	3—7(Pδ)	18	98	3.358(−3)
10938.1	3—6(Pγ)	18	72	7.783(−3)
12818.1	3—5(Pβ)	18	50	2.201(−2)
16407.2	4—12	32	288	1.620(−4)
16806.5	4—11	32	242	2.556(−4)
17362.1	4—10	32	200	4.235(−4)
18174.1	4—9	32	162	7.459(−4)
18751.0	3—4(Pα)	18	32	8.986(−2)
19445.6	4—8	32	128	1.424(−3)
21655.0	4—7	32	98	3.041(−3)
26252.0	4—6	32	72	7.711(−3)
27575	5—12	50	288	1.402(−4)
28722	5—11	50	242	2.246(−4)
30384	5—10	50	200	3.800(−4)
32961	5—9	50	162	6.908(−4)
37395	5—8	50	128	1.388(−3)
40512.0	4—5	32	50	2.699(−2)
43753	6—12	72	288	1.288(−4)
46525	5—7	50	98	3.253(−3)
46712	6—11	72	242	2.110(−4)
51273	6—10	72	200	3.688(−4)
59066	6—9	72	162	7.065(−4)
74578	5—6	50	72	1.025(−2)
75005	6—8	72	128	1.561(−3)
123680	6—7	72	98	4.561(−3)

For hydrogen-like ions of nuclear charge Z, the following scaling laws hold:

$A_Z = Z^4 A_{Hydrogen}$ $f_Z = f_H$; $S_Z = Z^{-2} S_H$
(For wavelengths $\lambda_Z = Z^{-2}\lambda_H$).

For lines which, due to the scaling, shift from the air to the vacuum region, the (tabulated) experimental wavelengths in air and corresponding A-values have to be converted first to vacuum values.[10] For highly charged ions, relativistic effects need to be taken into account.[11]

Table 1b
TRANSITION PROBABILITIES FOR SELECTED ATOMIC AND IONIC SPECIES

Spectrum	Wavelength (λ[Å])	Statistical weights g_i	g_k	Transition probability (A [10^8 s^{-1}])	Uncertainty
Aluminum					
Al I	2263.5	2	4	0.782	C
	2269.1	4	6	0.932	C
	2269.2	4	4	0.16	C
	2367.0	2	4.	0.721	C
	2373.1	4	6	0.859	C
	2373.4	4	4	0.14	C
	2568.0	2	4	0.228	C
	2575.1	4	6	0.271	C
	2575.4	4	4	0.045	C
	2652.5	2	2	0.133	C
	2660.4	4	2	0.264	C
	3082.2	2	4	0.630	C
	3092.7	4	6	0.748	C
	3092.8	4	4	0.12	C
	3944.0	2	2	0.493	C
	3961.5	4	2	0.98	C
	6696.0	2	4	0.0169	C
	6698.7	2	2	0.0169	C
	7835.3	4	6	0.057	D
	7836.1	6	8	0.062	D
Al II	1047.9	1	3	0.36	D
	1048.6	3	5	0.48	D
	1539.7	3	5	8.8	D
	1670.8	1	3	14.6	B
	1719.5	1	3	6.79	B
	1764.0	5	5	9.8	C
	1856.0	1	3	0.832	B
	1858.0	3	3	2.48	B
	1862.3	5	3	4.12	B
	1910.9	5	5	5.8	D
	1911.0	5	3	3.4	D
	1931.0	3	1	10.8	C
	1958.4	1	3	7.0	D
	1958.9	*15	9	6.8	D
	1965.3	5	7	13.	D
	1989.9	3	5	14.7	C
	2015.4	*15	15	4.1	C
	2816.2	3	1	3.83	C
	4663.1	5	3	0.53	C
	5593.2	3	5	1.1	D
	6226.2	1	3	0.62	D
	6231.8	3	5	0.84	D
	6243.4	5	7	1.1	D
	6335.7	5	3	0.14	D
	6823.5	3	3	0.34	D
	6837.1	5	3	0.57	D
	6920.6	3	1	0.96	D
	7042.1	3	5	0.59	C
	7056.6	3	3	0.58	C
	7471.4	5	7	0.94	D
Al III	560.39	*2	6	0.41	D
	695.82	2	4	0.73	C
	696.21	2	2	0.73	C
	1352.8	*10	14	4.40	C
	1379.6	2	2	4.64	C
	1384.2	4	2	9.18	C
	1605.7	2	4	12.2	B
	1611.9	4	6	14.5	B
	1611.9	4	4	2.42	B
	1854.7	2	4	5.39	B
	1862.8	2	2	5.32	B
	1935.9	*10	14	12.2	C
	3601.6	6	4	1.34	C
	3601.9	4	4	0.150	C
	3612.4	4	2	1.48	C
Al X	36.925	1	3	2590.	C
	51.979	1	3	4780.	C
	55.227	1	3	5300.	D
	55.272	3	5	7180.	D
	55.376	5	7	9510.	D
	59.107	3	5	4580.	C
	332.78	1	3	56.	C
	394.83	3	1	834.	C
	395.36	3	5	13.0	C
	397.76	1	3	17.0	C
	400.43	3	3	12.5	C
	401.12	5	5	37.3	C
	403.55	3	1	48.8	C
	406.31	5	3	19.9	C
	670.06	3	5	9.81	C
	2529.6	1	3	0.38	D
Al XI	36.675	*2	6	1520.	C
	39.091	2	4	2620.	C
	39.180	4	6	3130.	C
	39.530	2	2	190.	D
	39.623	4	2	370.	D
	48.298	2	4	3090.	B
	48.338	2	2	3080.	B
	52.299	2	4	8200.	C
	52.446	4	6	9700.	C
	52.461	4	4	1620.	C
	54.217	2	2	483.	C
	54.388	4	2	960.	C
	99.1	*2	6	224.	C
	103.7	2	4	422.	C
	103.8	4	6	510.	C
	141.6	*2	6	407.	C
	150.31	2	4	840.	C
	150.61	4	6	1010.	C
	157.0	2	2	43.3	C
	157.5	4	2	172.	C
	204.9	*2	6	63.	C
	308.6	*2	6	98.5	C
	341.3	*2	6	14.5	C
	550.05	2	4	8.45	B
	568.12	2	2	7.92	B
	1997.6	2	4	1.06	C
	2069.3	2	2	0.99	C
	4738.	*2	6	0.258	C
	5204.2	2	4	0.0373	C
	5547.9	4	6	0.0394	C
	5725.8	4	4	0.0062	D
Argon					
AR I	3554.3	5	5	0.0029	D
	3567.7	5	7	0.0012	D
	3606.5	3	1	0.0081	D
	3649.8	3	1	0.0085	D
	3834.7	3	1	0.0080	D
	3949.0	5	3	0.00467	C
	4044.4	3	5	0.00346	C
	4158.6	5	5	0.0145	C
	4164.2	5	3	0.00295	C
	4181.9	1	3	0.0058	C
	4190.7	5	5	0.00254	C
	4191.0	1	3	0.0056	C
	4198.3	3	1	0.0276	C
	4200.7	5	7	0.0103	C
	4259.4	3	1	0.0415	C
	4266.3	3	5	0.00333	C
	4272.2	3	3	0.0084	C
	4300.1	3	5	0.00394	C
	4333.6	3	5	0.0060	C
	4335.3	3	3	0.00387	C
	4345.2	3	3	0.00313	C
	4510.7	3	1	0.0123	C
	4768.7	3	5	0.0090	D
	4836.7	3	5	0.00106	C
	4876.3	3	5	0.0081	D
	4888.0	3	3	0.014	D
	4894.7	3	1	0.019	D
	5048.8	3	5	0.0048	D
	5054.2	3	3	0.0047	D
	5056.5	3	1	7.5 (−4)	D
	5118.2	5	7	0.0028	D
	5151.4	3	1	0.0249	C
	5152.3	3	5	0.0011	D
	5162.3	3	3	0.0198	C
	5177.5	7	5	0.0025	D
	5187.8	3	5	0.0138	C
	5194.0	3	1	0.0081	D
	5210.5	7	7	0.0011	D
	5214.8	5	3	0.0022	D

Table 1b (continued)
TRANSITION PROBABILITIES FOR SELECTED ATOMIC AND IONIC SPECIES

Spectrum	Wavelength (λ(Å))	Statistical weights g_i	g_k	Transition probability (A {10^8 s^{-1}})	Uncertainty
	5216.3	5	3	0.0014	D
	5221.3	7	9	0.0092	D
	5241.1	5	5	0.0014	D
	5252.8	5	7	0.0056	D
	5373.5	3	5	0.0028	D
	5393.3	5	5	0.0010	D
	5410.5	5	7	0.0021	D
	5421.4	7	5	0.0062	D
	5440.0	3	3	0.0020	D
	5442.2	7	7	9.7 (−4)	D
	5451.7	3	5	0.0049	D
	5457.4	5	3	0.0037	D
	5467.2	5	5	7.9 (−4)	D
	5473.5	5	3	0.0021	D
	5490.1	5	5	8.9 (−4)	D
	5495.9	7	9	0.0176	C
	5506.1	5	7	0.0037	D
	5525.0	7	7	0.0018	D
	5534.5	5	3	0.0028	D
	5558.7	3	5	0.0148	C
	5559.7	3	5	0.0023	D
	5572.5	5	7	0.0069	C
	5588.7	5	5	0.0016	D
	5606.7	3	3	0.0229	C
	5618.0	3	3	0.0022	D
	5620.9	3	1	0.0038	D
	5623.8	5	5	0.0015	D
	5635.6	3	5	0.0010	D
	5639.1	1	3	0.0022	D
	5641.4	3	5	9.1 (−4)	D
	5648.7	5	3	0.0013	D
	5650.7	3	1	0.0333	C
	5659.1	5	5	0.0027	D
	5681.9	5	7	0.0021	D
	5683.7	5	5	0.0021	D
	5700.9	5	7	0.0061	D
	5739.5	3	5	0.0091	D
	5772.1	5	7	0.0021	D
	5774.0	5	5	0.0011	D
	5783.5	3	5	8.4 (−4)	D
	5802.1	5	3	0.0044	D
	5834.3	5	5	0.0052	C
	5860.3	3	3	0.00285	C
	5882.6	3	1	0.0128	C
	5888.6	7	5	0.0134	C
	5912.1	3	3	0.0105	C
	5928.8	5	3	0.011	D
	5940.9	1	3	0.0012	D
	5942.7	5	5	0.0019	D
	5949.3	3	3	0.0016	D
	5968.3	3	3	0.0019	D
	5971.6	3	1	0.011	D
	5987.3	7	7	0.0013	D
	5988.1	3	5	6.4 (−4)	D
	5999.0	5	5	0.0015	D
	6005.7	5	3	0.0015	D
	6013.7	7	5	0.0015	D
	6025.2	5	3	0.0094	D
	6032.1	7	9	0.0246	C
	6043.2	5	7	0.0153	C
	6052.7	3	5	0.0020	D
	6059.4	3	5	0.00423	C
	6064.8	5	7	6.0 (−4)	D
	6090.8	1	3	0.0031	D
	6098.8	3	3	0.0054	D
	6101.2	3	3	0.0034	D
	6104.6	3	1	0.0035	D
	6105.6	3	5	0.0126	C
	6127.4	5	3	0.0011	D
	6128.7	3	5	9.0 (−4)	D
	6145.4	5	7	0.0079	C
	6155.2	5	3	0.0053	D
	6165.1	5	5	0.00103	C
	6170.2	5	5	0.0052	C
	6173.1	3	5	0.0070	C
	6212.5	5	7	0.0041	D
	6215.9	5	5	0.0059	D
	6248.4	3	5	7.1 (−4)	D
	6296.9	3	5	0.0094	D
	6307.7	5	5	0.0063	D
	6364.9	3	1	0.0058	D
	6369.6	5	3	0.0044	D
	6384.7	3	3	0.00439	C
	6416.3	3	5	0.0121	C
	6466.6	1	3	0.0016	D
	6481.1	1	3	9.8 (−4)	D
	6538.1	7	7	0.0011	D
	6604.0	7	5	0.0029	D
	6660.7	3	1	0.0081	D
	6664.1	5	5	0.0016	D
	6677.3	3	1	0.00241	C
	6698.9	5	3	0.0017	D
	6719.2	1	3	0.0025	D
	6752.8	3	5	0.0201	C
	6754.4	3	3	0.0022	D
	6756.1	5	5	0.0038	D
	6766.6	5	3	0.0042	D
	6779.9	1	3	0.00126	C
	6827.3	5	3	0.0025	D
	6851.9	3	5	7.0 (−4)	D
	6871.3	3	3	0.0290	C
	6879.6	3	5	0.0019	D
	6887.1	5	7	0.0014	D
	6888.2	3	5	0.0026	D
	6925.0	3	3	0.0012	D
	6937.7	3	1	0.0077	C
	6951.5	5	5	0.0023	D
	6960.2	5	5	0.0025	D
	6965.4	5	3	0.067	C
	6992.2	3	1	0.0078	D
	7030.3	7	5	0.0278	C
	7067.2	5	5	0.0395	C
	7068.7	5	3	0.021	D
	7086.7	1	3	0.0016	D
	7107.5	5	5	0.0047	D
	7125.8	3	3	0.0063	D
	7147.0	5	3	0.0065	C
	7158.8	3	1	0.022	D
	7207.0	5	3	0.0258	C
	7229.9	5	5	6.9 (−4)	D
	7265.2	3	3	0.0018	D
	7270.7	7	7	0.0011	D
	7272.9	3	3	0.0200	C
	7285.4	5	3	0.0013	D
	7311.7	5	3	0.018	D
	7316.0	3	3	0.010	D
	7350.8	3	1	0.012	D
	7353.3	5	7	0.010	D
	7372.1	7	9	0.020	D
	7384.0	3	5	0.087	C
	7393.0	5	3	0.0075	D
	7412.3	3	5	0.0041	D
	7422.3	3	5	6.9 (−4)	D
	7425.3	5	7	0.0032	D
	7435.3	5	5	0.0094	D
	7436.3	7	5	0.0028	D
	7484.2	3	5	0.0035	D
	7503.9	3	1	0.472	C
	7510.4	5	5	0.0047	D
	7514.7	3	1	0.430	C
	7618.3	3	5	0.0030	D
	7628.9	3	5	0.0030	D
	7635.1	5	5	0.274	C
	7670.0	5	3	0.0029	D
	7704.8	5	7	6.6 (−4)	D
	7723.8	5	3	0.057	C
	7724.2	1	3	0.127	C
	7798.6	3	5	9.1 (−4)	D
	7868.2	1	3	0.00365	C
	7891.1	5	5	0.0099	C
	7916.5	3	3	0.0013	D
	7948.2	1	3	0.196	C

Table 1b (continued)
TRANSITION PROBABILITIES FOR SELECTED ATOMIC AND IONIC SPECIES

Spectrum	Wavelength (λ(Å))	Statistical weights g_i	g_k	Transition probability (A {10^8 s^{-1}})	Uncertainty
Ar II	3000.4	4	4	1.5	D
	3028.9	2	4	2.3	D
	3093.4	4	6	4.4	D
	3139.0	6	6	1.0	D
	3161.4	2	4	1.8	D
	3169.7	4	6	0.82	D
	3181.0	6	4	0.63	D
	3236.8	2	4	0.52	D
	3243.7	4	2	2.0	D
	3249.8	2	4	1.0	D
	3293.6	4	4	1.7	D
	3307.2	2	2	3.4	D
	3350.9	6	6	1.5	D
	3376.4	8	8	1.5	D
	3388.5	2	4	1.9	D
	3454.1	6	4	0.45	D
	3464.1	6	6	0.37	D
	3476.8	6	6	1.34	C
	3491.2	4	4	2.2	D
	3509.8	2	2	2.5	D
	3514.4	4	6	1.23	C
	3520.0	6	6	0.80	D
	3521.3	8	8	0.23	D
	3535.3	2	4	0.82	D
	3545.6	4	6	3.4	D
	3545.8	6	8	3.9	D
	3548.5	4	4	1.1	D
	3559.5	6	8	3.9	D
	3565.0	2	4	1.1	D
	3576.6	6	8	2.77	C
	3581.6	2	4	1.8	D
	3582.4	4	6	3.72	C
	3588.5	8	10	3.39	C
	3600.2	4	4	2.2	D
	3622.1	4	2	0.64	D
	3639.8	4	6	1.4	D
	3655.3	4	6	0.23	D
	3671.0	4	2	0.71	D
	3680.1	2	4	1.2	D
	3718.2	4	6	2.0	D
	3724.5	6	6	0.34	D
	3729.3	6	4	0.60	D
	3737.9	6	8	2.3	D
	3765.3	6	6	0.98	D
	3770.5	2	4	0.41	D
	3780.8	8	8	0.94	D
	3796.6	4	6	0.25	D
	3803.2	6	6	1.5	D
	3809.5	4	6	0.44	D
	3825.7	6	4	0.76	D
	3850.6	4	4	0.47	D
	3868.5	4	6	1.9	D
	3925.7	6	4	1.4	D
	3928.6	2	4	0.30	D
	3932.6	4	4	1.1	D
	3946.1	8	6	1.4	D
	3952.7	4	4	0.35	D
	3979.4	4	2	1.3	D
	4033.8	4	2	0.98	D
	4042.9	4	4	1.4	D
	4072.0	6	6	0.57	C
	4076.6	2	2	0.80	D
	4076.9	4	2	0.99	D
	4079.6	6	4	0.26	D
	4131.7	4	2	1.4	D
	4156.1	4	4	0.39	D
	4179.3	6	6	0.13	D
	4218.7	4	4	0.36	D
	4222.6	4	2	0.69	D
	4227.0	4	6	0.41	D
	4266.5	6	6	0.156	C
	4275.2	2	4	0.26	D
	4277.5	6	4	1.0	D
	4331.2	4	4	0.56	C
	4337.1	2	4	0.34	D
	4348.1	6	8	1.24	C
	4370.8	4	4	0.65	C
	4371.3	6	4	0.233	C
	4379.7	2	2	1.04	C
	4401.0	8	6	0.322	C
	4426.0	4	6	0.83	C
	4430.2	2	4	0.53	C
	4448.9	6	6	0.65	D
	4481.8	6	6	0.494	C
	4545.1	4	4	0.413	B
	4547.8	4	4	0.077	D
	4589.9	4	6	0.82	C
	4609.6	6	8	0.91	C
	4637.3	6	6	0.090	D
	4657.9	4	2	0.81	C
	4726.9	4	4	0.50	C
	4735.9	6	4	0.58	C
	4764.9	2	4	0.575	B
	4806.0	6	6	0.79	C
	4847.8	4	2	0.85	C
	4865.9	4	6	0.15	D
	4879.9	4	6	0.78	C
	4965.1	2	4	0.347	C
	5009.3	4	6	0.147	C
	5017.2	4	6	0.231	C
	5062.0	2	4	0.221	C
	5141.8	6	8	0.095	C
	6638.2	6	4	0.129	C
	6643.7	10	8	0.167	C
	6684.3	8	6	0.113	C
Ar III	769.1	5	3	6.0	D
	871.10	5	3	1.58	C
	875.53	3	1	3.74	C
	878.73	5	5	2.78	C
	879.62	3	3	0.92	C
	883.18	1	3	1.22	C
	887.40	3	5	0.90	C
	3024.0	5	7	2.6	D
	3027.1	5	5	0.64	D
	3054.8	3	5	1.9	D
	3064.8	3	3	1.0	D
	3078.2	1	3	1.4	D
	3285.8	5	7	2.0	D
	3301.9	5	5	2.0	D
	3311.2	5	3	2.0	D
	3336.1	7	9	2.0	D
	3344.7	5	7	1.8	D
	3352.2	7	7	0.22	D
	3358.5	3	5	1.6	D
	3361.3	5	5	0.30	D
	3472.6	5	7	0.20	D
	3480.6	7	7	1.6	D
	3499.7	3	3	1.3	D
	3500.5	3	5	0.26	D
	3502.7	5	3	0.43	D
	3503.6	5	5	1.2	D
	3511.7	7	5	0.26	D
Ar IV	840.03	4	2	2.73	C
	843.77	4	4	2.70	C
	850.60	4	6	2.63	C
Ar VI	292.15	2	2	69.	C
	294.05	4	2	136.	C
Ar VII	250.41	*9	3	278.	C
	477.55	*9	15	99.2	B
	585.75	1	3	78.3	B
	637.30	*9	9	67.	C
Ar VIII	157.17	2	2	116.	C
	158.92	2	4	112.	C
	229.44	2	2	112.	C
	230.88	4	2	220.	C
	337.09	4	4	12.	D
	337.26	6	4	100.	C
	338.22	4	2	110.	C
	519.43	2	4	96.	C
	526.45	4	6	72.	C
	526.87	4	4	12.	D
	700.24	2	4	25.3	C

Table 1b (continued)
TRANSITION PROBABILITIES FOR SELECTED ATOMIC AND IONIC SPECIES

Spectrum	Wavelength (λ{Å})	Statistical weights g_i	g_k	Transition probability (A {10^8 s^{-1}})	Uncertainty
	713.81	2	2	23.9	C
Ar IX	48.73	1	3	1300.	D
Ar XIII	162.96	5	3	340.	C
	163.08	*9	3	530.	C
	184.90	5	5	156.	C
	186.38	1	3	90.	C
	207.88	*9	9	95.	C
	245.04	*9	15	37.	C
Ar XIV	180.29	2	4	44.6	C
	183.41	2	2	169.	C
	187.95	4	4	197.	C
	191.35	2	2	46.	C
	191.35	4	2	74.5	C
	194.39	4	2	88.	C
Ar XV	25.05	1	3	1.7 (+4)	B
	221.10	1	3	105.	B
	239.8	*9	9	98.	C
Ar XVI	23.52	*2	6	1.43 (+4)	B
	24.96	*6	10	4.4 (+4)	B
	353.88	2	4	15.4	B
	389.11	2	2	11.4	B
	1268.	2	4	1.8	B
	1401.	2	2	1.5	B
	2975.	2	4	0.081	C
	3514.	4	6	0.070	C
Boron					
B I	1378.6	2	4	3.49	C
	1379.2	4	2	7.0	C
	1465.5	2	4	3.34	C
	1465.7	4	4	6.7	C
	1465.8	6	4	10.0	C
	1825.9	2	4	6.5	C
	1826.4	4	6	7.7	C
	2088.8	2	4	0.28	D
	2089.6	4	6	0.33	D
	2496.8	2	2	0.85	C
	2497.7	4	2	1.70	C
Beryllium					
Be I	1491.8	1	3	0.013	D
	1661.5	1	3	0.20	D
	2348.6	1	3	0.540	B
	2494.6	*9	15	1.03	C
	2650.6	*9	9	4.30	C
	3455.2	3	1	2.35	C
	4572.7	3	5	0.79	C
	7209.3	3	5	0.0010	D
Be II	1197.1	2	2	0.47	D
	1197.2	4	2	0.93	D
	1512.3	2	4	9.5	C
	1512.4	4	6	11.4	C
	1776.1	2	2	1.41	C
	1776.3	4	2	2.81	C
	2453.8	*2	6	0.071	C
	3046.5	2	4	0.482	C
	3046.7	4	6	0.58	C
	3130.4	2	4	1.15	B
	3131.1	2	2	1.15	B
	3241.6	2	2	0.141	C
	3241.8	4	2	0.282	C
	3274.6	2	4	0.185	C
	3274.7	2	2	0.185	C
	4360.7	2	4	0.93	C
	4361.0	4	6	1.12	C
	5255.9	*2	6	0.0256	C
	5270.3	2	2	0.329	C
	5270.8	4	2	0.66	C
	6279.4	2	4	0.119	C
	6279.7	4	6	0.143	C
	6756.7	2	2	0.0171	C
	6757.1	2	4	0.0171	C
	7401.2	2	4	0.0296	C
	7401.4	2	2	0.0296	C
Carbon					
C I	941.19	1	3	9.6	D
	945.34	3	3	28.	D
	945.58	5	3	47.	D
	1260.7	1	3	0.40	D
	1260.9	3	1	1.2	D
	1261.0	3	3	0.31	D
	1261.1	3	5	0.30	D
	1261.4	5	3	0.50	D
	1261.6	5	5	0.93	D
	1274.1	5	7	0.0068	D
	1277.2	1	3	0.88	D
	1277.3	3	5	1.2	D
	1277.5	3	3	0.65	D
	1277.6	5	7	1.5	D
	1277.8	5	5	0.39	D
	1277.9	5	3	0.042	D
	1279.2	5	7	0.11	D
	1279.9	3	5	0.21	D
	1280.1	1	3	0.27	D
	1280.3	5	5	0.62	D
	1280.4	3	3	0.20	D
	1280.6	3	1	0.81	D
	1280.8	5	3	0.35	D
	1328.8	1	3	0.49	D
	1364.2	5	5	0.047	D
	1431.6	5	7	1.5	D
	1432.1	5	5	1.4	D
	1432.5	5	3	1.3	D
	1459.0	5	3	0.37	D
	1463.3	5	7	2.1	D
	1467.4	5	3	0.46	D
	1468.4	5	3	0.019	D
	1470.1	5	7	0.0088	D
	1472.2	5	3	0.0051	D
	1481.8	5	5	0.33	D
	1560.3	1	3	0.84	D
	1561.3	5	5	0.37	D
	1561.4	5	7	1.5	D
	1656.3	3	5	0.79	D
	1656.9	1	3	1.1	D
	1657.0	5	5	2.4	D
	1657.4	3	3	0.79	D
	1657.9	3	1	3.2	D
	1658.1	5	3	1.3	D
	1751.8	1	3	0.57	D
	1763.9	1	3	0.022	D
	1765.4	3	5	0.0071	D
	1930.9	5	3	3.7	D
	2478.6	1	3	0.54	D
	2902.3	1	3	0.0066	D
	2903.3	3	3	0.017	D
	2905.0	5	3	0.022	D
	4269.0	3	5	0.0032	D
	4371.4	3	3	0.0097	D
	4762.3	1	3	0.0052	D
	4762.5	3	5	0.0038	D
	4766.7	3	3	0.0039	D
	4770.0	3	1	0.015	D
	4771.7	5	5	0.012	D
	4775.9	5	3	0.0062	D
	4812.9	1	3	9.7 (−4)	D
	4817.4	3	3	0.0028	D
	4826.8	5	3	0.0047	D
	4932.1	3	1	0.046	D
	5052.2	3	5	0.017	D
	5380.3	3	3	0.016	D
	5793.1	7	5	0.0033	D
	5794.5	5	5	5.8 (−4)	D
	5800.2	3	3	9.7 (−4)	D
	5800.6	5	3	0.0029	D
	5805.2	3	1	0.0039	D
	6587.6	3	3	0.024	D
C II	687.35	4	6	27.2	C
	858.09	2	2	0.362	C
	858.56	4	2	1.09	C
	903.62	2	4	6.6	C
	903.96	2	2	26.3	C
	904.14	4	4	33.0	C
	904.48	4	2	13.2	C

Table 1b (continued)
TRANSITION PROBABILITIES FOR SELECTED ATOMIC AND IONIC SPECIES

Spectrum	Wavelength (λ(Å))	Statistical weights g_i	g_k	Transition probability (A $\{10^8\ s^{-1}\}$)	Uncertainty
	1009.9	2	4	5.8	C
	1010.1	4	4	11.5	C
	1010.4	6	4	17.2	C
	1036.3	2	2	8.0	C
	1037.0	4	2	15.9	C
	1323.9	4	4	4.56	C
	1324.0	6	6	4.72	C
	1334.5	2	4	2.42	C
	1335.7	4	6	2.89	C
	2509.1	2	4	0.64	C
	2511.7	4	4	0.125	C
	2512.1	4	6	0.76	C
	6578.1	2	4	0.370	C
	6582.9	2	2	0.369	C
	7231.3	2	4	0.364	C
	7236.4	4	6	0.436	C
	7237.2	4	4	0.073	C
C III	310.17	1	3	18.0	C
	386.20	1	3	32.2	C
	459.46	1	3	59.	C
	459.52	3	5	79.	C
	459.63	5	7	105.	C
	574.28	3	5	64.	C
	977.02	1	3	17.8	B
	1174.9	3	5	3.34	C
	1175.3	1	3	4.44	C
	1175.6	3	3	3.33	C
	1175.7	5	5	10.	C
	1176.0	3	1	13.3	C
	1176.4	5	3	5.5	C
	1247.4	3	1	21.2	C
	2296.9	3	5	1.38	C
	4647.4	3	5	0.73	C
	4650.3	3	3	0.73	C
	4651.5	3	1	0.73	C
C IV	312.43	*2	6	44.9	B
	384.12	*6	10	180.	C
	1548.2	2	4	2.66	B
	1550.8	2	2	2.64	B
	5801.3	2	4	0.318	B
	5812.0	2	2	0.316	B
C V	34.973	1	3	2554.	AA
	40.268	1	3	8873.	AA
	227.19	*3	9	136.3	AA
	247.31	1	3	127.9	AA
	248.70	*9	15	425.	A
	260.19	*9	3	66.83	AA
	267.27	3	5	396.	A
	2273.9	*3	9	0.5650	AA
	3526.7	1	3	0.1663	AA
	8432.2	*3	9	0.06870	AA
Calcium					
Ca I	2275.5	1	3	0.301	C
	2995.0	1	3	0.367	C
	2997.3	3	5	0.241	C
	2999.6	3	3	0.279	C
	3000.9	3	1	1.58	C
	3006.9	5	5	0.75	C
	3009.2	5	3	0.430	C
	3344.5	1	3	0.151	C
	3350.2	3	5	0.178	C
	3361.9	5	7	0.223	C
	3624.1	1	3	0.212	C
	3630.8	3	5	0.297	C
	3631.0	3	3	0.153	C
	3644.4	5	7	0.355	C
	3644.8	5	5	0.094	C
	3870.5	3	5	0.072	D
	3957.0	3	3	0.098	C
	3973.7	5	3	0.175	C
	4092.6	3	5	0.11	D
	4094.9	5	7	0.12	D
	4098.5	7	9	0.13	D
	4108.5	5	7	0.90	D
	4226.7	1	3	2.18	B
	4283.0	3	5	0.434	C
	4289.4	1	3	0.60	C
	4299.0	3	3	0.466	C
	4302.5	5	5	1.36	C
	4307.7	3	1	1.99	C
	4318.6	5	3	0.74	C
	4355.1	5	7	0.19	D
	4425.4	1	3	0.498	C
	4435.0	3	5	0.667	C
	4435.7	3	3	0.342	C
	4454.8	5	7	0.872	C
	4455.9	5	5	0.204	C
	4526.9	5	3	0.41	D
	4578.6	3	5	0.176	C
	4581.4	5	7	0.209	C
	4585.9	7	9	0.229	C
	4685.3	3	5	0.080	D
	4878.1	5	7	0.188	C
	5041.6	5	3	0.33	D
	5188.8	3	5	0.40	D
	5261.7	3	3	0.15	D
	5262.2	3	1	0.60	D
	5264.2	5	5	0.091	D
	5265.6	5	3	0.44	D
	5270.3	7	5	0.50	D
	5582.0	5	7	0.060	D
	5588.8	7	7	0.49	D
	5590.1	3	5	0.083	D
	5594.5	5	5	0.38	D
	5598.5	3	3	0.43	D
	5601.3	7	5	0.086	D
	5602.8	5	3	0.14	D
	5857.4	3	5	0.66	D
	6102.7	1	3	0.0096	C
	6122.2	3	3	0.286	C
	6161.3	5	5	0.033	D
	6162.2	5	3	0.477	C
	6163.8	3	3	0.056	D
	6166.4	3	1	0.22	D
	6169.1	5	3	0.17	D
	6169.6	7	5	0.19	D
	6439.1	7	9	0.53	D
	6449.8	3	5	0.090	D
	6462.6	5	7	0.47	D
	6471.7	7	7	0.059	D
	6493.8	3	5	0.44	D
	6499.6	5	5	0.081	D
Ca II	1341.9	2	4	0.015	D
	1342.5	2	2	0.015	D
	1649.9	2	4	0.0032	D
	1652.0	2	2	0.0032	D
	1673.8	2	4	0.224	C
	1680.0	4	6	0.265	C
	1680.1	4	4	0.0441	C
	1807.3	2	4	0.354	C
	1814.6	4	4	0.070	C
	1815.0	4	6	0.420	C
	1843.6	2	2	0.156	C
	1851.1	4	2	0.308	C
	2103.2	2	4	0.824	C
	2112.8	4	6	0.976	C
	2113.2	4	4	0.163	C
	2197.8	2	2	0.315	C
	2208.6	4	2	0.62	C
	3158.9	2	4	3.02	C
	3179.3	4	6	3.55	C
	3181.3	4	4	0.51	C
	3706.0	2	2	0.88	C
	3736.9	4	2	1.71	C
	3933.7	2	4	1.48	C
	3968.5	2	2	1.45	C
Ca III	357.97	1	3	2600.	D
	439.69	1	3	0.56	D
	490.55	1	3	0.049	D
Ca V	558.60	5	3	22.	D
	637.93	5	3	4.1	D
	643.12	3	1	9.5	D

Table 1b (continued)
TRANSITION PROBABILITIES FOR SELECTED ATOMIC AND IONIC SPECIES

Spectrum	Wavelength (λ(Å))	Statistical weights g_i	Statistical weights g_k	Transition probability (A (10⁸ s⁻¹))	Uncertainty
	646.56	5	5	7.0	D
	647.87	3	3	2.3	D
	651.55	1	3	3.1	D
	656.77	3	5	2.2	D
Ca VII	550.19	5	5	18.	D
	624.38	1	3	3.2	D
	630.51	3	5	4.3	D
	630.78	3	3	2.4	D
	639.15	5	7	5.4	D
	640.38	5	5	1.4	D
Ca VIII	182.71	2	2	160.	C
	184.16	4	2	320.	C
Ca IX	163.23	5	3	376.	C
	371.90	1	3	88.	C
	373.80	3	5	116.	C
	378.09	5	7	150.	C
	398.7	3	5	220.	D
	466.23	1	3	112.	B
	498.00	3	5	24.9	C
	506.16	5	5	72.	C
	515.57	5	3	37.5	C
Ca X	110.96	2	4	295.	C
	111.20	2	2	293.	C
	151.83	2	2	232.	C
	153.01	4	2	454.	C
	206.57	4	4	29.	D
	206.75	6	4	260.	C
	207.39	4	2	280.	C
	411.69	2	4	83.	C
	419.76	4	6	95.	C
	420.49	4	4	16.	D
	557.74	2	4	35.2	C
	574.01	2	2	32.3	C
Ca XI	30.448	1	3	6200.	D
	30.867	1	3	4.9 (+4)	D
	35.213	1	3	2000.	D
Ca XII	140.05	4	2	2300.	C
	147.27	2	2	980.	C
Ca XV	141.66	5	3	410.	C
	142.22	*9	3	630.	C
	160.96	5	5	190.	C
Ca XVII	19.638	1	3	2.5 (+4)	C
	21.212	3	5	4.8 (+4)	C
	22.113	3	1	3300.	C
	192.90	1	3	121.	C
	218.82	3	5	27.6	C
	223.02	1	3	34.4	C
	228.72	3	3	23.7	C
	232.83	5	5	64.8	C
	244.06	5	3	32.8	C
Ca XVIII	18.70	*2	6	2.28 (+4)	B
	19.74	*6	10	7.0 (+4)	B
	302.19	2	4	19.7	B
	344.76	2	2	13.5	B
Chlorine					
Cl I	1188.8	4	4	0.271	C
	1188.8	4	6	2.33	C
	1201.4	2	4	2.39	C
	1335.7	4	2	1.74	C
	1347.2	4	4	4.19	C
	1351.7	2	2	3.23	C
	1363.4	2	4	0.75	C
	4323.4	4	4	0.011	D
	4363.3	4	6	0.068	D
	4379.9	4	4	0.014	D
	4389.8	6	8	0.014	D
	4526.2	4	4	0.051	C
	4601.0	2	2	0.042	C
	4661.2	2	4	0.012	D
	7256.6	6	4	0.15	D
	7414.1	6	4	0.047	D
	7547.1	4	4	0.12	D
	7717.6	4	4	0.030	D
	7744.9	2	4	0.063	D
	7769.2	6	6	0.060	D
	7821.4	6	8	0.098	D
	7830.8	4	4	0.097	D
	7878.2	6	6	0.018	D
	7899.3	4	6	0.051	D
	7924.6	2	4	0.021	D
	7935.0	6	8	0.039	D
	7997.8	4	4	0.021	D
Cl II	3329.1	5	7	1.5	D
	3522.1	7	7	1.4	D
	3798.8	5	7	1.6	D
	3805.2	7	9	1.8	D
	3809.5	3	5	1.5	D
	3851.0	5	7	1.8	D
	3851.4	5	5	1.6	D
	3854.8	3	5	2.2	D
	3860.8	7	9	2.7	D
	3862.0	5	7	2.4	D
	3868.6	7	9	2.7	D
	3913.9	9	9	0.82	D
	3990.2	5	7	0.84	D
	4132.5	5	5	1.6	D
	4208.0	5	5	1.1	D
	4276.5	9	7	0.76	D
	4768.7	3	5	0.77	D
	4771.7	3	5	1.0	D
	4781.3	5	7	1.0	D
	4794.5	5	7	1.01	C
	4810.1	5	5	0.99	C
	4819.5	5	3	1.00	C
	4904.8	5	7	0.81	D
	4917.7	3	5	0.75	D
	5078.2	7	7	0.77	D
	5220.6	*3	9	0.86	C
	5392.1	5	7	1.0	D
Cl III	2298.5	4	4	4.2	D
	2340.6	6	6	4.2	D
	2370.4	8	6	2.8	D
	2531.8	2	4	4.4	D
	2532.5	4	6	5.3	D
	2577.1	4	6	4.3	D
	2580.7	6	8	4.7	D
	2601.2	2	4	4.6	D
	2603.6	4	6	5.0	D
	2609.5	6	8	5.7	D
	2617.0	8	10	6.6	D
	2661.6	4	6	3.4	D
	2665.5	6	8	4.8	D
	2691.5	4	4	3.5	D
	2710.4	4	6	3.5	D
	3340.4	6	6	1.5	D
	3392.9	4	4	1.9	D
	3393.4	6	6	1.9	D
	3530.0	6	8	1.8	D
	3560.7	4	6	1.7	D
	3602.1	6	8	1.7	D
	3612.8	4	6	1.2	D
	3720.4	4	6	1.7	D
Cobalt					
Co I	2987.2	10	8	0.065	D
	2989.6	10	10	0.051	D
	3013.6	10	8	0.019	D
	3017.5	8	6	0.092	D
	3042.5	8	6	0.024	D
	3044.0	10	10	0.31	D
	3048.9	6	4	0.11	D
	3061.8	8	8	0.20	D
	3064.4	8	10	0.0090	D
	3082.6	10	12	0.034	D
	3086.8	4	4	0.25	D
	3089.6	8	8	0.032	D
	3098.2	6	6	0.030	D
	3139.9	8	6	0.035	D
	3147.1	6	8	0.058	D
	3149.3	6	4	0.036	D
	3395.4	6	8	0.39	D
	3405.1	10	10	2.4	D
	3409.2	8	8	0.60	D
	3412.3	8	10	1.0	D
	3412.6	10	8	0.16	D

Table 1b (continued)
TRANSITION PROBABIILITIES FOR SELECTED ATOMIC AND IONIC SPECIES

Spectrum	Wavelength (λ(Å))	Statistical weights g_i	g_k	Transition probability (A {10^8 s^{-1}})	Uncertainty
	3417.2	6	6	0.54	D
	3431.6	8	6	0.15	D
	3433.0	4	4	1.5	D
	3442.9	6	4	0.16	D
	3443.6	8	8	1.01	D
	3449.2	6	6	0.55	D
	3449.4	10	10	0.23	D
	3455.2	4	2	0.25	D
	3462.8	4	6	1.3	D
	3465.8	10	12	0.13	D
	3483.4	8	10	0.025	D
	3489.4	8	6	2.5	D
	3491.3	4	4	0.080	D
	3495.7	4	6	0.76	D
	3496.7	8	8	0.078	D
	3518.3	6	4	2.8	D
	3520.1	8	6	0.054	D
	3521.6	10	8	0.19	D
	3523.4	4	2	2.4	D
	3526.8	10	10	0.16	D
	3529.0	6	8	0.13	D
	3529.8	8	10	0.69	D
	3533.4	4	6	0.12	D
	3543.5	10	12	2.1	D
	3550.6	6	4	0.073	D
	3560.9	4	4	0.24	D
	3564.9	6	8	0.12	D
	3569.4	8	8	2.3	D
	3575.0	6	6	0.23	D
	3575.4	8	8	0.12	D
	3585.1	8	8	0.13	D
	3587.2	6	6	2.4	D
	3594.9	6	6	0.11	D
	3602.1	4	4	0.13	D
	3605.4	8	8	0.048	D
	3627.8	8	10	0.066	D
	3631.4	8	10	0.0076	D
	3647.7	4	6	0.0013	D
	3652.5	6	8	0.011	D
	3704.1	6	8	0.18	D
	3745.5	8	8	0.15	D
	3842.0	8	6	0.24	D
	3845.5	8	10	0.81	D
	3873.1	10	8	0.19	D
	3874.0	8	6	0.15	D
	3876.9	10	10	0.0068	D
	3881.9	6	4	0.14	D
	3894.0	6	8	0.96	D
	3895.0	4	2	0.16	D
	3909.9	10	12	0.0021	D
	3936.0	8	10	0.12	D
	3995.3	8	10	0.52	D
	3997.9	6	8	0.13	D
	4020.9	10	10	0.0067	D
	4092.4	8	8	0.079	D
	4110.5	6	6	0.072	D
	4118.8	6	8	0.30	D
	4121.3	8	10	0.31	D
	4230.0	6	4	0.0037	D
Chromium					
Cr I	2000.0	9	9	1.4	D
	2383.3	9	11	0.43	D
	2385.7	9	9	0.17	D
	2389.2	3	5	0.23	D
	2408.6	9	7	0.29	D
	2408.7	7	5	0.29	D
	2492.6	3	5	0.45	C
	2496.3	5	7	0.56	C
	2502.6	7	9	0.22	D
	2504.3	7	9	0.45	C
	2549.6	3	3	0.48	D
	2560.7	5	5	0.43	D
	2577.7	7	7	0.26	D
	2591.8	9	7	0.65	C
	2622.9	9	9	0.13	D
	2702.0	9	11	0.21	C
	2726.5	5	7	0.75	C
	2731.9	5	5	0.78	C
	2736.5	5	3	0.75	D
	2752.9	3	3	0.87	D
	2757.1	5	5	0.68	C
	2761.7	5	3	0.68	D
	2764.4	7	7	0.37	D
	2769.9	7	5	1.1	C
	2780.7	9	7	1.4	C
	2871.6	7	9	0.12	D
	2874.2	1	3	0.12	D
	2879.3	5	7	0.21	D
	2887.0	3	5	0.27	D
	2889.3	9	9	0.66	C
	2899.2	3	3	0.15	D
	2909.1	5	3	0.61	C
	2967.6	7	9	0.39	D
	2971.1	5	7	0.71	C
	2975.5	3	5	0.89	C
	2980.8	1	3	0.34	D
	2988.6	5	7	0.52	C
	2991.9	3	1	3.0	D
	2995.1	5	5	0.43	D
	2996.6	5	3	2.0	C
	2998.8	5	3	0.59	D
	3000.9	7	5	1.6	C
	3005.1	9	7	0.92	C
	3013.7	3	5	0.83	C
	3020.7	3	3	1.5	D
	3021.6	9	11	3.2	C
	3024.4	5	5	2.3	C
	3030.3	7	7	1.1	C
	3034.2	7	7	0.35	D
	3037.0	9	9	0.54	C
	3040.9	7	5	0.74	D
	3053.9	9	7	1.2	C
	3148.5	9	11	0.59	C
	3155.2	11	13	0.54	C
	3163.8	13	15	0.52	C
	3237.7	9	9	1.3	D
	3238.1	11	11	0.20	D
	3578.7	7	9	1.48	B
	3593.5	7	7	1.50	B
	3605.3	7	5	1.62	B
	3639.8	13	11	1.8	D
	3768.7	7	5	0.22	D
	3804.8	9	9	0.69	D
	3879.2	3	5	0.56	D
	3963.6	13	15	1.3	D
	3969.8	11	13	1.2	D
	3981.2	3	5	0.11	D
	3983.9	7	9	1.05	C
	4001.4	9	11	0.65	D
	4031.1	3	5	0.79	D
	4039.1	15	15	0.68	C
	4048.8	13	13	0.65	D
	4058.8	11	11	0.69	D
	4161.4	13	15	0.80	D
	4165.5	11	13	0.75	D
	4211.4	7	9	0.085	D
	4224.5	9	9	0.067	D
	4239.0	9	9	0.071	D
	4254.4	7	9	0.315	B
	4274.8	7	7	0.306	B
	4280.4	13	15	0.47	D
	4289.7	7	5	0.313	B
	4297.7	11	13	0.49	D
	4344.5	7	9	0.11	C
	4351.8	9	11	0.12	C
	4374.2	13	11	0.10	C
	4375.3	11	9	0.072	D
	4381.1	5	3	0.10	D
	4387.5	13	11	0.066	D
	4413.9	7	5	0.41	D
	4432.2	1	3	0.17	D
	4458.5	9	11	0.13	D
	4482.9	3	3	0.30	D
	4488.1	7	7	0.63	D

Table 1b (continued)
TRANSITION PROBABILITIES FOR SELECTED ATOMIC AND IONIC SPECIES

Spectrum	Wavelength (λ(Å))	Statistical weights		Transition probability (A (10^8 s^{-1}))	Uncertainty
		g_i	g_k		
	4506.9	13	11	0.27	D
	4511.9	9	9	0.13	D
	4526.5	13	13	0.20	D
	4530.8	11	11	0.20	D
	4544.6	5	5	0.26	D
	4595.6	13	13	0.47	D
	4632.2	7	7	0.071	D
	4639.5	5	7	0.095	D
	4646.2	9	7	0.087	C
	4689.4	7	5	0.23	D
	4708.0	11	9	0.37	D
	4718.4	13	11	0.42	D
	4723.1	7	7	0.093	D
	4724.4	9	9	0.063	D
	4727.2	13	13	0.051	D
	4729.7	5	3	0.17	D
	4730.7	7	5	0.28	D
	4737.4	9	7	0.24	D
	4789.4	13	11	0.076	D
	4792.5	7	5	0.26	D
	4801.0	9	7	0.23	D
	4870.8	7	9	0.35	D
	4887.0	9	11	0.32	D
	4922.3	11	13	0.40	D
	4936.3	7	9	0.14	D
	4954.8	9	11	0.12	D
	5139.7	7	7	0.13	D
	5177.4	9	11	0.061	D
	5184.6	7	9	0.11	D
	5192.0	5	7	0.14	D
	5196.6	11	9	0.12	D
	5200.2	3	5	0.16	D
	5204.5	5	3	0.55	D
	5206.0	5	5	0.53	D
	5208.4	5	7	0.51	D
	5243.4	5	3	0.20	D
	5261.8	7	9	0.13	D
	5272.0	7	9	0.11	D
	5287.2	5	7	0.078	D
	5304.2	9	9	0.066	D
	5312.9	7	7	0.11	D
	5318.8	5	5	0.13	D
	5328.3	9	11	0.60	D
	5340.4	5	3	0.16	D
	5400.6	5	5	0.16	D
	5409.8	9	7	0.062	D
Cr II	2653.6	4	6	0.35	D
	2658.6	2	4	0.58	D
	2666.0	6	8	0.59	D
	2668.7	4	2	1.4	D
	2671.8	6	4	1.0	D
	2672.8	8	6	0.55	D
	2693.5	10	8	1.4	D
	2727.3	10	8	1.7	D
	2740.1	6	8	0.11	D
	2745.0	4	6	0.85	D
	2768.6	6	8	2.8	D
	2774.4	8	8	1.7	D
	2778.1	10	10	3.2	D
	2782.4	6	4	1.6	D
	2785.7	10	8	2.1	D
	2787.6	6	6	1.5	D
	2792.2	12	10	2.3	D
	2800.8	12	14	2.2	D
	2818.4	8	10	2.2	D
	2822.0	6	8	2.3	D
	2832.5	12	10	1.3	D
	2838.8	8	8	2.7	D
	2840.0	10	12	2.7	D
	2843.2	8	10	0.64	D
	2849.8	6	8	0.92	D
	2851.4	8	10	2.2	D
	2856.8	4	6	0.43	D
	2857.4	6	8	0.28	D
	2860.9	2	4	0.69	D
	2862.6	8	8	0.63	D
	2866.7	4	4	1.2	D
	2867.1	4	4	1.1	D
	2867.7	2	2	1.1	D
	2870.4	6	6	1.3	D
	2873.8	4	2	0.88	D
	2878.5	10	8	0.074	D
	2880.9	6	4	0.79	D
	2888.7	10	12	0.88	D
	2898.5	10	12	1.2	D
	2921.8	8	10	0.90	D
	2927.1	10	10	2.8	D
	2930.8	2	4	1.1	D
	2935.1	6	8	1.8	D
	2953.3	2	2	1.8	D
	2953.7	10	10	0.92	D
	2966.0	10	8	0.54	D
	2971.9	14	14	2.0	D
	2979.7	12	12	1.8	D
	2985.3	10	10	2.2	D
	2989.2	8	8	2.2	D
	3040.9	10	12	4.8	D
	3041.7	10	10	3.1	D
	3050.1	12	14	1.8	D
	3093.5	10	12	0.67	D
	3096.1	10	8	0.75	D
	3107.6	8	10	0.62	D
	3118.6	2	4	1.7	D
	3120.4	4	6	1.5	D
	3122.6	12	12	0.44	D
	3128.7	4	4	0.81	D
	3136.7	6	6	0.64	D
	3152.2	4	4	1.8	D
	3180.7	12	10	0.70	D
	3183.3	8	6	0.87	D
	3209.2	8	6	0.68	D
	3217.4	6	4	0.77	D
	3234.1	10	8	0.92	D
	3238.8	12	10	0.54	D
	3295.4	12	14	0.32	D
	3336.3	2	2	0.42	D
	3339.8	4	4	0.49	D
	3342.6	6	6	0.39	D
	3347.8	4	2	0.52	D
	3358.5	6	4	1.1	D
	3360.3	8	8	1.3	D
	3368.0	8	6	1.4	D
	3378.4	8	6	0.41	D
	3379.4	4	6	0.48	D
	3382.7	6	6	0.45	D
	3391.4	2	4	0.19	D
	3393.0	2	4	0.46	D
	3393.9	4	4	0.66	D
	3394.3	6	4	0.75	D
	3402.4	2	2	0.80	D
	3408.8	8	6	0.95	D
	3421.2	2	2	1.7	D
	3422.7	6	4	1.4	D
	3433.3	4	2	1.3	D
	3511.8	8	6	0.079	D
	4242.4	10	8	0.12	D
Cr XXI	149.89	1	3	160.	C
Cr XXII	8.51	*2	6	1.2(+4)	C
	9.493	*2	6	2.5(+4)	C
	9.809	2	4	4.1(+4)	B
	9.865	4	6	4.9(+4)	B
	12.620	2	4	5.13(+4)	B
	12.662	2	2	5.28(+4)	B
	13.147	2	4	1.3(+5)	B
	13.294	4	6	1.6(+5)	B
	13.306	4	4	2.6(+4)	C
	25.2	*2	6	3750.	C
	37.00	*2	6	7000.	C
	37.52	*6	10	1.7(+4)	B
	223.00	2	4	33.	B
	279.69	2	2	17.	B
Cr XXIII	2.20	1	3	3.3(+6)	B
Fluorine					
F I	806.96	4	6	3.3	C

Table 1b (continued)
TRANSITION PROBABILITIES FOR SELECTED ATOMIC AND IONIC SPECIES

Spectrum	Wavelength (λ(Å))	Statistical weights g_i	g_k	Transition probability (A {10^8 s^{-1}})	Uncertainty
	809.60	2	4	2.8	C
	951.87	4	2	2.7	C
	954.83	4	4	6.4	C
	955.55	2	2	5.0	C
	958.52	2	4	1.2	C
	6239.6	6	4	0.25	D
	6348.5	4	4	0.18	D
	6413.7	2	4	0.11	D
	6708.3	6	4	0.014	D
	6774.0	6	6	0.10	D
	6795.5	4	2	0.052	D
	6834.3	4	4	0.21	D
	6856.0	6	8	0.42	D
	6870.2	2	2	0.38	D
	6902.5	4	6	0.32	D
	6909.8	2	4	0.22	D
	6966.4	4	2	0.11	D
	7037.5	4	4	0.30	D
	7127.9	2	2	0.38	D
	7309.0	6	8	0.47	D
	7311.0	4	2	0.39	D
	7314.3	4	6	0.48	D
	7332.0	6	4	0.31	D
	7398.7	6	6	0.31	D
	7425.6	4	2	0.34	D
	7482.7	4	4	0.056	D
	7489.1	2	2	0.11	D
	7514.9	2	2	0.052	D
	7552.2	4	6	0.078	D
	7573.4	2	4	0.10	D
	7607.2	4	4	0.070	D
	7754.7	4	6	0.30	D
	7800.2	2	4	0.21	D
Iron					
Fe I	1934.5	9	7	0.24	C
	1937.3	9	7	0.21	C
	1940.7	7	5	0.25	C
	2084.1	9	7	0.36	C
	2102.4	7	7	0.084	C
	2113.0	1	3	0.18	C
	2132.0	9	9	0.073	C
	2166.8	9	7	2.6	C
	2191.8	5	5	1.1	C
	2196.0	3	3	1.1	C
	2200.4	1	3	0.86	C
	2200.7	3	5	0.27	C
	2259.5	9	11	0.067	C
	2276.0	9	7	0.16	C
	2277.1	7	5	0.19	C
	2287.3	5	3	0.32	C
	2294.4	3	1	0.58	C
	2300.1	5	7	0.077	C
	2309.0	3	5	0.14	C
	2313.1	5	7	0.13	C
	2320.4	7	9	0.12	C
	2373.6	7	7	0.064	C
	2374.5	1	3	0.28	C
	2462.2	7	5	0.15	C
	2462.7	9	9	0.56	C
	2479.8	5	5	1.8	C
	2483.3	9	11	4.7	C
	2484.2	3	3	2.5	C
	2488.2	7	9	4.5	C
	2490.6	5	7	3.6	C
	2491.2	3	5	2.9	C
	2501.1	9	7	0.65	C
	2510.8	7	5	1.3	C
	2518.1	5	3	1.8	C
	2522.9	9	9	2.8	C
	2524.3	3	1	3.2	C
	2527.4	7	7	1.8	C
	2529.1	5	5	0.93	C
	2535.6	1	3	0.93	C
	2541.0	3	5	0.88	C
	2546.0	5	7	0.64	C
	2549.6	7	9	0.34	C
	2584.5	11	13	0.39	C
	2719.0	9	7	1.3	C
	2720.9	7	5	1.0	C
	2723.6	5	3	0.61	C
	2737.3	3	3	0.76	C
	2742.4	5	5	0.60	C
	2744.1	1	3	0.34	C
	2750.1	7	7	0.37	C
	2756.3	3	5	0.19	C
	2788.1	11	13	0.60	C
	2894.5	5	5	0.68	C
	2899.4	5	3	0.65	C
	2923.3	11	11	1.8	C
	2925.4	7	9	0.20	C
	2936.9	9	9	0.14	C
	2954.7	5	7	0.13	C
	2966.9	9	11	0.27	C
	2980.5	7	7	0.24	C
	2983.6	9	7	0.29	C
	2990.4	9	11	0.43	C
	2996.4	3	5	0.20	C
	2999.5	11	11	0.25	C
	3000.9	5	3	0.70	C
	3008.1	3	1	1.2	C
	3009.1	13	11	0.084	C
	3009.6	9	9	0.19	C
	3011.5	7	9	0.52	C
	3019.0	7	7	0.16	C
	3037.4	3	5	0.35	C
	3042.7	5	7	0.071	C
	3047.6	5	7	0.30	C
	3053.1	3	5	0.19	C
	3057.4	11	9	0.48	C
	3059.1	7	9	0.18	C
	3067.2	9	7	0.38	C
	3075.7	7	5	0.32	C
	3083.7	5	3	0.38	C
	3098.2	11	11	0.12	C
	3100.7	7	7	0.17	C
	3119.5	11	9	0.10	C
	3120.4	9	7	0.11	C
	3160.7	9	9	0.20	C
	3161.9	11	13	0.13	C
	3166.4	9	7	0.15	C
	3175.4	11	11	0.14	C
	3199.5	9	9	0.29	C
	3205.4	3	3	1.3	C
	3215.9	5	5	0.87	C
	3217.4	11	9	0.24	C
	3225.8	11	13	0.96	C
	3231.0	7	5	0.42	C
	3233.1	13	15	0.54	C
	3234.0	9	9	0.22	C
	3239.4	9	9	0.54	C
	3248.2	7	7	0.24	C
	3253.6	7	9	0.20	C
	3254.4	11	13	0.58	C
	3265.6	7	5	0.41	C
	3271.0	5	3	0.72	C
	3280.3	9	11	0.59	C
	3282.9	3	5	0.33	C
	3292.0	7	9	0.67	C
	3292.6	3	3	0.33	C
	3306.0	5	7	0.51	C
	3307.2	13	13	0.22	C
	3314.7	5	7	0.76	C
	3323.7	5	5	0.33	C
	3328.9	11	11	0.29	C
	3337.7	11	9	0.071	C
	3355.2	9	9	0.35	C
	3369.5	9	9	0.27	C
	3370.8	11	11	0.36	C
	3380.1	7	7	0.26	C
	3384.0	7	7	0.12	C
	3392.7	7	7	0.28	C
	3399.3	5	5	0.42	C
	3402.3	13	13	0.31	C
	3406.4	3	5	0.32	C

Table 1b (continued)
TRANSITION PROBABILITIES FOR SELECTED ATOMIC AND IONIC SPECIES

Spectrum	Wavelength (λ(Å))	Statistical weights g_i	Statistical weights g_k	Transition probability (A {10^8 s^{-1}})	Uncertainty
	3407.5	7	9	0.64	C
	3410.2	3	5	0.51	C
	3411.4	9	9	0.070	C
	3413.1	5	7	0.39	C
	3417.8	3	3	0.56	C
	3418.5	3	1	1.4	C
	3424.3	7	7	0.22	C
	3425.0	9	7	0.31	C
	3427.1	7	9	0.60	C
	3428.2	5	5	0.23	C
	3445.2	5	7	0.31	C
	3447.3	5	5	0.11	C
	3450.3	3	3	0.26	C
	3495.3	9	7	0.14	C
	3497.1	7	7	0.16	C
	3521.8	3	5	0.12	C
	3524.1	7	5	0.11	C
	3527.8	9	9	0.24	C
	3529.8	3	3	0.83	C
	3536.6	5	7	0.85	C
	3540.1	7	9	0.13	C
	3541.1	9	11	0.68	C
	3542.1	7	9	0.82	C
	3552.8	5	5	0.19	C
	3553.7	11	9	0.89	C
	3556.9	9	11	0.48	C
	3559.5	3	3	0.25	C
	3560.7	7	9	0.088	C
	3565.4	7	9	0.41	C
	3570.1	9	11	0.69	C
	3572.0	11	11	0.27	C
	3581.2	11	13	0.96	C
	3582.2	13	11	0.27	C
	3587.0	5	5	0.16	C
	3594.6	9	9	0.30	C
	3599.6	11	9	0.20	C
	3603.2	11	11	0.29	C
	3605.5	9	9	0.70	C
	3606.7	11	13	0.90	C
	3608.9	3	5	0.83	C
	3610.2	13	13	0.53	C
	3612.1	11	13	0.082	C
	3617.8	5	7	0.71	C
	3618.8	5	7	0.73	C
	3621.5	9	11	0.56	C
	3622.0	7	7	0.56	C
	3623.2	13	13	0.082	C
	3631.5	7	9	0.52	C
	3632.0	3	5	0.53	C
	3638.3	7	9	0.29	C
	3640.4	9	11	0.42	C
	3645.8	1	3	0.62	C
	3647.8	9	11	0.32	C
	3649.5	11	9	0.46	C
	3651.5	7	9	0.68	C
	3655.5	5	5	0.14	C
	3659.5	9	9	0.073	C
	3669.5	9	7	0.32	C
	3670.1	11	13	0.083	C
	3676.3	9	11	0.061	C
	3677.6	7	5	0.88	C
	3682.2	5	5	1.9	C
	3684.1	9	7	0.39	C
	3686.0	9	11	0.28	C
	3687.5	11	9	0.10	C
	3690.7	11	11	0.30	C
	3694.0	5	7	0.75	C
	3697.4	7	7	0.23	C
	3701.1	7	9	0.53	C
	3704.5	11	9	0.15	C
	3709.2	9	7	0.19	C
	3719.9	9	11	0.16	C
	3724.4	5	7	0.14	C
	3727.6	5	5	0.29	C
	3730.4	9	11	0.14	C
	3732.4	5	5	0.30	C
	3734.9	11	11	0.89	C
	3737.1	7	9	0.14	C
	3738.3	11	13	0.41	C
	3740.3	7	7	0.19	C
	3742.6	9	9	0.12	C
	3744.1	5	3	0.41	C
	3745.6	5	7	0.12	C
	3749.5	9	9	0.78	C
	3753.6	7	5	0.12	C
	3756.9	11	11	0.27	C
	3758.2	7	7	0.67	C
	3760.1	13	15	0.064	C
	3763.8	5	5	0.61	C
	3765.5	13	15	1.1	C
	3767.2	3	3	0.72	C
	3787.2	5	5	0.12	C
	3787.9	3	5	0.15	C
	3794.3	9	11	0.046	C
	3795.0	5	7	0.14	C
	3798.5	9	11	0.042	C
	3799.6	7	9	0.093	C
	3804.0	11	9	0.055	C
	3805.3	9	11	1.1	C
	3806.2	3	3	0.30	C
	3806.7	11	11	0.59	C
	3807.5	3	5	0.10	C
	3810.8	5	3	0.27	C
	3813.0	7	5	0.13	C
	3813.9	13	11	0.10	C
	3815.8	9	7	1.2	C
	3820.4	11	9	0.68	C
	3821.2	11	13	0.65	C
	3821.8	5	5	0.098	C
	3825.9	9	7	0.65	C
	3827.8	7	5	1.2	C
	3833.3	9	9	0.068	C
	3834.2	7	5	0.49	C
	3836.3	5	5	0.49	C
	3839.3	9	9	0.31	C
	3840.4	5	3	0.52	C
	3841.1	5	3	1.5	C
	3843.3	9	7	0.51	C
	3846.8	7	7	0.72	C
	3850.0	3	1	0.73	C
	3859.2	13	11	0.093	C
	3859.9	9	9	0.087	C
	3865.5	3	3	0.20	C
	3867.2	5	5	0.37	C
	3871.8	11	11	0.074	C
	3872.5	5	5	0.067	C
	3873.8	11	9	0.088	C
	3878.0	7	7	0.096	C
	3883.3	7	7	0.18	C
	3884.4	11	9	0.045	C
	3888.5	5	5	0.29	C
	3891.9	3	3	0.45	C
	3893.4	11	11	0.16	C
	3900.5	7	7	0.093	C
	3902.9	7	7	0.25	C
	3903.9	9	9	0.10	C
	3916.7	13	11	0.13	C
	3919.1	9	9	0.049	C
	3933.6	3	5	0.086	C
	3935.8	5	5	0.16	C
	3942.4	3	5	0.11	C
	3951.2	3	5	0.38	C
	3952.6	11	11	0.050	C
	3953.2	7	9	0.050	C
	3963.1	3	5	0.17	C
	3967.4	9	7	0.25	C
	3969.3	9	7	0.25	C
	3971.3	11	9	0.066	C
	3973.7	5	7	0.083	C
	3977.7	5	5	0.088	C
	3981.8	9	9	0.049	C
	3984.0	9	7	0.095	C
	3985.4	5	5	0.078	C
	3986.2	7	9	0.071	C

Table 1b (continued)
TRANSITION PROBABILITIES FOR SELECTED ATOMIC AND IONIC SPECIES

Spectrum	Wavelength (λ{Å})	Statistical weights g_i	Statistical weights g_k	Transition probability (A {10^8 s^{-1}})	Uncertainty
	3997.0	9	9	0.072	C
	3997.4	9	11	0.17	C
	3998.1	11	9	0.081	C
	4005.2	7	5	0.24	C
	4014.5	11	11	0.26	C
	4017.2	9	11	0.057	C
	4021.9	7	9	0.011	C
	4032.0	3	5	0.088	C
	4045.8	9	9	0.81	C
	4062.4	3	3	0.25	C
	4063.6	7	7	0.74	C
	4068.0	9	9	0.18	C
	4070.8	7	5	0.13	C
	4071.7	5	5	0.86	C
	4073.8	5	3	0.19	C
	4074.8	9	9	0.059	C
	4076.6	9	9	0.21	C
	4084.5	11	9	0.13	C
	4085.3	7	7	0.11	C
	4098.2	7	7	0.084	C
	4107.5	5	3	0.27	C
	4109.8	3	3	0.20	C
	4113.0	11	13	0.14	C
	4127.6	1	3	0.17	C
	4132.9	3	5	0.12	C
	4134.7	5	7	0.20	C
	4137.0	3	5	0.25	C
	4143.9	7	9	0.17	C
	4149.4	11	13	0.045	C
	4153.9	7	9	0.25	C
	4154.8	9	11	0.16	C
	4156.8	5	5	0.20	C
	4170.9	5	5	0.078	C
	4172.1	7	5	0.12	C
	4175.6	3	5	0.18	C
	4176.6	9	11	0.089	C
	4184.9	5	5	0.13	C
	4187.0	7	5	0.25	C
	4187.8	9	7	0.18	C
	4196.2	7	7	0.092	C
	4210.4	3	3	0.21	C
	4217.6	3	5	0.25	C
	4219.4	11	13	0.39	C
	4222.2	7	7	0.067	C
	4224.2	9	11	0.14	C
	4225.5	5	7	0.19	C
	4227.4	11	13	0.52	C
	4233.6	3	5	0.22	C
	4238.8	7	9	0.24	C
	4246.1	7	5	0.074	C
	4250.1	5	7	0.24	C
	4282.4	7	5	0.15	C
	4299.2	9	11	0.13	C
	4307.9	7	9	0.37	C
	4315.1	5	5	0.097	C
	4325.8	5	7	0.55	C
	4327.1	5	5	0.092	C
	4369.8	9	9	0.079	C
	4383.6	9	11	0.49	C
	4388.4	7	7	0.14	C
	4401.3	7	7	0.074	C
	4404.8	7	9	0.27	C
	4415.1	5	7	0.14	C
	4443.2	1	3	0.14	C
	4466.6	5	7	0.13	C
	4469.4	5	7	0.29	C
	4476.0	3	5	0.19	C
	4525.1	7	5	0.32	C
	4528.6	7	9	0.067	C
	4547.9	5	7	0.084	C
	4611.3	5	5	0.14	C
	4736.8	9	11	0.051	C
	4789.7	5	5	0.090	C
	4859.8	5	3	0.18	C
	4871.3	7	5	0.24	C
	4872.1	3	3	0.26	C
	4878.2	1	3	0.11	C
	4890.8	5	5	0.23	C
	4891.5	9	7	0.32	C
	4903.3	3	5	0.064	C
	4919.0	7	7	0.18	C
	4920.5	11	9	0.39	C
	4966.1	11	11	0.040	C
	4973.1	3	3	0.13	C
	4989.0	7	7	0.062	C
	5001.9	9	7	0.43	C
	5015.0	7	5	0.33	C
	5022.2	5	3	0.29	C
	5074.8	9	11	0.16	C
	5090.8	7	5	0.22	C
	5121.6	5	5	0.095	C
	5137.4	11	9	0.12	C
	5208.6	7	5	0.068	C
	5233.0	9	11	0.17	C
	5242.5	13	11	0.032	C
	5263.3	5	5	0.065	C
	5266.6	7	9	0.095	C
	5281.8	5	7	0.041	C
	5302.3	3	5	0.079	C
	5324.2	9	9	0.17	C
	5339.9	5	7	0.077	C
	5367.5	7	9	0.63	C
	5370.0	9	11	0.52	C
	5383.4	11	13	0.66	C
	5393.2	7	9	0.039	C
	5404.1	11	11	0.77	C
	5410.9	7	9	0.51	C
	5415.2	11	13	0.59	C
	5463.3	9	9	0.35	C
	5473.9	7	7	0.061	C
	5569.6	5	3	0.23	C
	5572.9	7	5	0.23	C
	5586.8	9	7	0.22	C
	5615.7	11	9	0.19	C
	5624.6	5	5	0.067	C
	5658.8	7	7	0.045	C
	5753.1	3	5	0.077	C
	5763.0	5	7	0.11	C
Fe II	2029.2	10	8	0.076	D
	2040.7	10	10	0.46	D
	2051.0	8	8	0.42	D
	2296.7	10	8	0.037	D
	2303.3	12	10	0.054	D
	2369.2	10	10	0.026	D
	2379.0	8	10	0.064	D
	2388.4	10	12	0.14	D
	2414.1	14	12	0.0094	D
	2433.5	10	12	0.091	D
	2555.0	8	8	0.019	D
	2559.8	6	8	0.22	D
	2561.6	10	10	0.0081	D
	2573.2	8	10	0.11	D
	2591	5	6.6	0.51	D
	2592.8	14	16	2.25	D
	2598.0	10	10	0.020	D
	2598.4	8	6	1.3	D
	2599.4	10	10	2.2	D
	2623.1	14	14	0.092	D
	2625.5	12	14	2.0	D
	2625.7	8	10	0.34	D
	2645.0	8	8	0.062	D
	2664.7	8	10	1.5	D
	2666.6	6	8	1.6	D
	2684.9	12	12	0.0043	D
	2712.4	10	12	0.11	D
	2753.3	10	12	1.7	D
	2879.2	10	8	0.029	D
	2902.5	10	10	0.038	D
	2910.8	8	8	0.0055	D
	2934.5	8	10	0.013	D
	2997.7	8	10	0.0048	D
	3002.3	6	8	0.018	D
	3044.8	8	10	0.011	D
	3131.7	12	10	0.012	D

Table 1b (continued)
TRANSITION PROBABILITIES FOR SELECTED ATOMIC AND IONIC SPECIES

Spectrum	Wavelength (λ[Å])	Statistical weights		Transition probability (A [10^8 s^{-1}])	Uncertainty
		g_i	g_k		
	3162.8	8	8	0.042	D
	3186.7	4	4	0.039	D
	3187.3	10	10	0.028	D
	3213.3	4	6	0.064	D
	3277.3	8	10	0.0023	D
	3360.1	12	12	0.0084	D
	4515.3	6	6	0.0018	D
	4520.2	10	8	0.0010	D
	4583.8	10	8	0.0063	D
	4629.3	10	10	0.0013	D
	4923.9	6	4	0.030	D
	4954.0	6	8	0.0016	D
	5018.4	6	6	0.026	D
	5019.5	8	10	0.0015	D
Fe XVI	50.35	2	4	1860.	C
	50.56	2	2	1970.	C
	62.88	2	4	367.	C
	63.72	2	2	352.	C
	335.41	2	4	80.5	C
	360.80	2	2	64.3	C
Fe XVIII	93.94	4	2	4200.	C
	103.96	2	2	1600.	C
Fe XXI	98.37	5	3	680.	D
	99.42	*9	3	110.	D
	113.31	5	5	290.	D
Fe XXIII	11.164	1	3	6.1 (+4)	D
	132.83	1	3	204.	B
	222.16	3	5	46.	D
Fe XXIV	10.62	*2	6	7.37 (+4)	B
	11.12	*6	10	2.2 (+5)	B
	192.04	2	4	43.4	B
	255.10	2	2	18.4	B
Helium					
He I	510.00	1	3	0.482	B
	512.10	1	3	0.763	B
	515.62	1	3	1.30	A
	522.21	1	3	2.46	A
	537.03	1	3	5.66	A
	584.33	1	3	17.99	AA
	2696.1	*3	9	0.00550	B
	2723.2	*3	9	0.007 80	B
	2763.8	*3	9	0.0111	B
	2829.1	*3	9	0.017	B
	2945.1	*3	9	0.0320	A
	3187.7	*3	9	0.05639	AA
	3354.6	1	3	0.0130	B
	3447.6	1	3	0.0232	A
	3554.4	*9	15	0.0131	A
	3587.3	*9	15	0.0205	C
	3613.6	1	3	0.0390	A
	3634.2	*9	15	0.0261	A
	3705.0	*9	15	0.0444	C
	3819.6	*9	15	0.0636	A
	3833.6	3	5	0.00971	B
	3867.5	*9	3	0.025	B
	3871.8	3	5	0.0126	C
	3888.7	*3	9	0.09478	AA
	3926.5	3	5	0.0195	B
	3964.7	1	3	0.0719	A
	4009.3	3	5	0.0279	C
	4026.2	*9	15	0.116	A
	4120.8	*9	3	0.0444	A
	4143.8	3	5	0.0485	A
	4387.9	3	5	0.0894	A
	4437.6	3	1	0.033	B
	4471.5	*9	15	0.246	A
	4713.2	*9	3	0.0955	A
	4921.9	3	5	0.198	A
	5015.7	1	3	0.1338	AA
	5047.7	3	1	0.0675	A
	5875.7	*9	15	0.7053	AA
	6678.2	3	5	0.6339	AA
	7065.3	*9	3	0.2786	AA
	7281.4	3	1	0.1829	AA
	8361.8	*3	9	0.00334	A
	9463.6	*3	9	0.00501	A
	9603.4	1	3	0.00610	A
	9702.7	*9	3	0.00858	B
	10311.	*9	15	0.0201	A
	10668	*9	3	0.0152	A
	10830	*3	9	0.1022	AA
	10913.	*15	21	0.0212	B
	10917.	5	7	0.0212	B
	10997.	*15 .	9	0.0013	B
	11013.	1	3	0.0100	A
	11045.	3	5	0.0185	A
	11226.	1	3	0.0108	A
	11969.	*9	15	0.0358	A
	12528.	*3	9	0.00710	A
	12756.	5	3	0.0012	B
	12785.	*15	21	0.0462	B
	12790.	5	7	0.0461	B
	12846.	*9	3	0.0289	A
	12968.	3	5	0.0343	A
	12985.	*15	9	0.0025	B
Potassium					
K I	4044.2	2	4	0.0124	C
	4047.2	2	2	0.0124	C
	5084.3	2	2	0.00350	C
	5099.2	4	2	0.0070	C
	5323.4	2	2	0.0063	C
	5339.8	4	2	0.0126	C
	5343.1	2	4	0.0040	D
	5359.7	4	6	0.0046	D
	5782.4	2	2	0.0123	C
	5801.8	4	2	0.0246	C
	5812.2	2	4	0.0028	D
	5831.9	4	6	0.0032	D
	6911.1	2	2	0.0272	C
	6938.8	4	2	0.054	C
	7664.9	2	4	0.387	B
	7699.0	2	2	0.382	B
K II	550.30	1	3	0.46	D
	607.92	1	3	0.039	D
K III	2550.0	6	4	2.0	D
	2635.1	4	4	1.2	D
	2992.2	6	8	2.5	D
	3052.1	4	6	1.7	D
	3202.0	4	4	1.8	D
	3289.1	4	6	2.0	D
	3322.4	6	6	1.3	D
	3421.8	2	4	1.5	D
K XVI	211.81	1	3	89.	C
KXVII	22.020	2	4	4.68 (+4)	C
	22.163	4	6	5.5 (+4)	C
	22.18	4	4	9200.	C
	22.60	2	2	800.	D
	22.76	4	2	3160.	C
Lithium					
Li I	2741.2	*2	6	0.013	D
	3232.6	*2	6	0.0055	B
	4602.8	2	4	0.197	C
	4602.9	4	6	0.236	C
	6103.5	2	4	0.60	C
	6103.6	4	6	0.72	C
	6103.7	4	4	0.119	C
	6707.8	2	4	0.372	B
	6707.9	2	2	0.372	B
Magnesium					
Mg I	2025.8	1	3	0.98	D
	2779.9	*9	9	5.2	C
	2850.0	*9	15	0.23	C
	3095.0	*9	15	0.53	C
	3329.9	1	3	0.032	C
	3332.2	3	3	0.10	C
	3336.7	5	3	0.16	C
	3835.3	*9	15	1.68	B
	4703.0	3	5	0.25	C
	5167.3	1	3	0.116	B
	5172.7	3	3	0.346	B
	5183.6	5	3	0.575	B
	5528.4	3	5	0.20	C
Mg II	1239.9	2	4	14.	C
	1240.4	2	2	14.	C

Table 1b (continued)
TRANSITION PROBABIILITIES FOR SELECTED ATOMIC AND IONIC SPECIES

Spectrum	Wavelength (λ(Å))	Statistical weights g_i	g_k	Transition probability (A {10^8 s^{-1}})	Uncertainty
	2680.8	*10	14	0.38	D
	2790.8	2	4	3.88	C
	2795.5	2	4	26.0	C
	2798.0	4	4	0.779	C
	2802.7	2	2	25.8	C
	2928.6	2	2	1.16	C
	2936.5	4	2	2.30	C
	3104.8	*10	14	0.811	C
	3848.2	4	4	0.0030	D
	3848.2	6	4	0.0270	C
	3850.4	4	2	0.0300	C
	4481.2	*10	14	2.23	C
	9218.2	2	4	0.364	C
	9244.3	2	2	0.362	C
Mg IV	321.00	4	2	120.	D
	323.31	2	2	59.	D
	1245.2	6	6	5.9	D
	1375.4	4	4	4.5	D
	1459.6	6	4	4.6	D
	1525.2	4	4	6.7	D
	1548.1	4	6	6.4	D
	1683.0	6	8	5.8	D
	1698.9	4	6	3.9	D
	1893.9	6	6	2.8	D
Mg VI	269.92	*10	6	310.	D
	292.53	*6	6	90.	D
	314.64	*6	2	180.	D
	349.15	*10	10	61.	D
	387.94	*6	10	13.	D
	399.29	4	2	28.	D
	400.68	4	4	28.	D
	403.32	4	6	27.	D
Mg VII	277.01	3	3	94.	C
	278.40	5	3	150.	C
	280.74	5	3	200.	D
	319.02	5	5	89.	C
	366.42	*9	9	44.	C
	433.04	*9	15	16.	C
	1334.3	5	5	5.32	C
	1410.0	5	5	2.60	C
	1487.0	3	5	3.04	C
	1487.9	5	7	3.66	C
Mg VIII	74.98	*6	10	4300.	D
	315.02	4	4	130.	C
	342.29	*10	6	63.	D
	353.84	4	4	41.	C
	356.60	6	4	60.	C
	428.52	*10	10	34.	C
	434.62	*6	10	16.	C
	489.33	*6	6	39.	D
	686.92	*6	10	9.4	D
Mg IX	62.75	1	3	2870.	B
	67.19	*9	15	6000.	C
	71.96	*9	3	1200.	C
	73.31	3	5	3900.	C
	77.74	3	1	400.	C
	368.07	1	3	52.	B
	438.69	3	1	78.	C
	443.74	*9	9	41.9	B
	751.56	3	5	8.24	C
	1639.8	3	5	2.1	D
	2814.2	1	3	0.34	C
Mg X	57.88	2	4	21.00	B
	57.92	2	2	2090.	B
	63.15	2	4	5600.	B
	63.30	4	6	6700.	B
	609.79	2	4	7.47	B
	624.94	2	2	7.12	B
	2212.	2	4	0.95	B
	2279.	2	2	0.90	B
	5919.	2	4	0.031	C
	6230.	4	6	0.033	C
Mg XI	7.31	1	3	1.15 (+5)	B
	7.47	1	3	2.27 (+4)	B
	7.85	1	3	5.50 (+4)	B
	9.17	1	3	1.97 (+5)	B
Manganese					
Mn I	2794.8	6	8	3.7	C
	2798.3	6	6	3.6	C
	2801.1	6	4	3.7	C
	3007.7	6	6	1.8	D
	3011.4	8	10	0.31	D
	3016.5	10	12	0.29	D
	3043.4	8	8	0.58	D
	3044.6	10	8	0.57	D
	3045.6	10	10	0.67	D
	3045.8	8	10	0.17	D
	3046.6	10	12	0.13	D
	3047.0	12	12	0.61	D
	3054.4	8	6	0.46	D
	3066.0	8	8	0.16	D
	3073.2	4	4	0.37	D
	3082.7	14	14	0.29	D
	3110.7	6	8	0.27	D
	3113.8	12	10	0.26	D
	3122.9	10	10	0.19	D
	3126.9	8	6	0.23	D
	3132.3	10	10	0.21	D
	3132.8	8	8	0.27	D
	3160.2	10	12	0.14	D
	3175.6	8	10	0.18	D
	3175.7	10	12	0.12	D
	3190.0	6	8	0.16	D
	3201.1	4	6	0.22	D
	3212.9	10	10	0.16	D
	3228.1	10	12	0.64	D
	3230.2	10	12	0.19	D
	3230.7	8	8	0.35	D
	3238.7	8	10	0.12	D
	3243.8	6	6	0.53	D
	3256.1	4	6	0.50	D
	3258.4	2	2	0.97	D
	3260.2	2	4	0.38	D
	3264.7	8	10	0.14	D
	3267.8	14	14	0.35	D
	3268.7	6	8	0.33	D
	3270.4	12	12	0.26	D
	3273.0	10	10	0.27	D
	3278.6	8	8	0.0091	D
	3298.2	6	4	0.28	D
	3420.8	14	14	0.12	D
	3463.7	8	8	0.32	C
	3470.0	6	8	0.24	C
	3511.8	12	12	0.27	C
	3535.3	10	10	0.17	C
	3559.8	6	6	0.21	C
	3577.9	10	8	0.94	C
	3601.3	12	10	0.23	C
	3607.5	8	8	0.23	C
	3608.5	6	6	0.36	C
	3610.3	4	4	0.42	C
	3635.7	10	8	0.21	C
	3660.4	12	14	0.91	C
	3677.0	10	12	0.73	D
	3680.2	12	10	0.19	C
	3682.1	8	10	0.76	D
	3684.9	6	8	0.26	C
	3706.1	12	14	1.4	C
	3718.9	10	12	0.96	C
	3729.5	10	12	0.066	D
	3731.9	8	10	1.0	C
	3746.6	12	12	0.16	C
	3756.6	10	10	0.14	C
	3767.7	8	8	0.14	C
	3768.2	10	12	0.071	C
	3771.4	14	14	0.19	C
	3773.9	12	12	0.25	C
	3800.6	6	8	0.27	C
	3801.9	12	12	0.38	C
	3806.7	10	12	0.38	C
	3809.6	8	8	0.20	C
	3823.5	8	10	0.44	C
	3823.9	6	6	0.36	C
	3833.9	4	4	0.52	C
	3834.4	6	8	0.52	D
	3839.8	2	2	0.58	C

Table 1b (continued)
TRANSITION PROBABILITIES FOR SELECTED ATOMIC AND IONIC SPECIES

Spectrum	Wavelength (λ(Å))	Statistical weights g_i	g_k	Transition probability (A {10^8 s^{-1}})	Uncertainty
	3844.0	2	4	0.29	C
	3872.1	10	12	0.077	C
	3873.2	8	10	0.11	C
	3889.5	12	14	0.31	C
	3898.4	6	8	0.17	C
	3899.3	4	6	0.24	C
	3911.1	6	6	0.13	D
	3919.3	8	8	0.088	D
	3923.3	12	10	0.13	C
	3924.1	2	4	0.94	D
	3926.5	6	8	0.54	C
	3929.7	10	8	0.092	C
	3931.5	10	10	0.082	C
	3936.8	8	6	0.12	C
	3952.8	6	6	0.41	D
	3975.9	2	4	0.18	D
	3977.1	4	6	0.16	D
	3980.1	10	8	0.13	D
	3982.2	4	2	0.35	D
	3982.6	6	4	0.23	D
	3982.9	6	4	0.55	D
	3986.8	12	10	0.11	D
	3987.1	10	8	0.10	D
	4003.3	12	10	0.11	D
	4011.9	8	8	0.23	D
	4018.1	10	8	0.33	C
	4026.4	12	14	0.089	D
	4030.8	6	8	0.19	C
	4031.8	10	12	0.073	D
	4033.1	6	6	0.18	C
	4034.5	6	4	0.18	C
	4041.4	10	10	1.0	C
	4048.8	6	4	0.75	C
	4052.5	6	8	0.38	D
	4055.6	8	8	0.61	C
	4058.9	4	2	1.0	C
	4061.7	8	6	0.19	D
	4063.5	6	6	0.22	C
	4065.1	12	14	0.25	D
	4066.2	10	8	0.22	D
	4079.4	2	4	0.38	C
	4083.0	4	6	0.37	C
	4083.6	6	8	0.28	C
	4089.9	8	10	0.17	D
	4092.4	8	8	0.14	D
	4099.4	8	8	0.11	D
	4105.4	10	8	0.17	D
	4107.9	12	10	0.097	D
	4114.4	8	8	0.15	D
	4116.6	6	6	0.12	D
	4125.8	10	10	0.070	D
	4132.3	8	10	0.15	D
	4135.0	12	12	0.30	D
	4141.1	10	10	0.26	D
	4147.5	6	6	0.066	D
	4148.8	8	8	0.23	D
	4176.6	14	12	0.21	D
	4182.3	12	14	0.092	D
	4190.0	12	10	0.20	D
	4201.8	10	8	0.23	D
	4220.6	6	4	0.16	D
	4239.7	4	2	0.39	D
	4257.7	2	2	0.37	D
	4261.3	12	12	0.081	D
	4265.9	4	4	0.35	D
	4278.7	10	10	0.068	D
	4281.1	6	6	0.23	D
	4300.2	12	10	0.087	D
	4381.7	8	10	0.14	D
	4411.9	12	10	0.26	D
	4414.9	8	6	0.18	D
	4419.8	10	8	0.21	D
	4451.6	8	8	0.71	D
	4452.5	14	14	0.059	D
	4455.8	4	6	0.17	D
	4457.0	6	4	0.20	D
	4457.6	6	6	0.38	D
	4458.3	6	8	0.28	D
	4461.1	8	8	0.17	D
	4462.0	8	10	0.43	D
	4464.7	6	6	0.26	D
	4479.4	8	10	0.34	D
	4498.9	4	6	0.11	D
	4502.2	6	8	0.078	D
	4503.9	12	14	0.083	D
	4605.4	10	12	0.36	D
	4626.5	12	14	0.36	D
	4709.7	8	8	0.077	D
	4727.5	6	6	0.084	D
	4754.1	6	8	0.38	D
	4761.5	2	4	0.28	D
	4762.4	8	10	0.57	D
	4765.9	4	6	0.28	D
	4766.4	6	8	0.45	D
	4783.4	8	8	0.39	D
	4823.5	10	8	0.45	D
Mn II	2933.1	5	3	1.7	C
	2939.3	5	5	1.8	C
	2949.2	5	7	1.7	C
	3439.0	5	7	0.0041	D
	3442.0	9	7	0.43	C
	3460.3	7	5	0.32	C
	3474.0	7	7	0.079	C
	3474.1	5	3	0.15	C
	3482.9	5	5	0.20	C
	3488.7	3	3	0.25	C
	3495.8	1	3	0.11	C
	3496.8	5	7	0.016	C
	3497.5	3	5	0.051	C
Mn XXIII	12.03	2	4	1.57 (+5)	C
	12.158	4	6	1.84 (+5)	C
Nitrogen					
N I	1134.2	4	2	1.5	D
	1134.4	4	4	1.5	D
	1135.0	4	6	1.5	D
	1163.9	6	6	0.43	D
	1164.0	4	6	0.032	D
	1164.2	6	4	0.048	D
	1164.3	4	4	0.43	D
	1167.5	6	8	1.1	D
	1168.4	6	6	0.095	D
	1168.5	4	6	1.3	D
	1169.7	6	8	0.030	D
	1176.5	6	4	0.95	D
	1176.6	4	4	0.11	D
	1177.7	4	2	1.3	D
	1199.6	4	6	5.5	D
	1200.2	4	4	5.3	D
	1200.7	4	2	5.5	D
	1310.5	4	6	1.3	D
	1316.3	4	6	0.025	D
	1492.6	6	4	5.3	D
	1492.8	4	4	0.58	D
	1494.7	4	2	5.0	D
	4100.0	2	4	0.034	D
	4110.0	4	6	0.040	D
	4114.0	4	4	0.0068	D
	4137.6	2	4	0.0039	D
	4143.4	4	4	0.0078	D
	4151.6	6	4	0.013	D
	4214.8	4	6	0.022	D
	4216.1	2	4	0.031	D
	4218.9	2	2	0.012	D
	4222.1	4	4	0.0098	D
	4223.0	6	6	0.051	D
	4224.9	4	2	0.061	D
	4230.5	6	4	0.033	D
	4385.5	2	2	0.0052	C
	4392.4	4	2	0.0102	C
	4914.9	2	2	0.00759	B
	4935.1	4	2	0.0158	B
	5169.6	6	4	0.00209	C
	5181.4	4	4	0.00144	C

Table 1b (continued)
TRANSITION PROBABILITIES FOR SELECTED ATOMIC AND IONIC SPECIES

Spectrum	Wavelength (λ(Å))	Statistical weights g_i	Statistical weights g_k	Transition probability (A {10^8 s^{-1}})	Uncertainty
	5186.6	2	4	7.3 (−4)	C
	5197.9	2	2	0.023	D
	5200.3	2	4	0.023	D
	5281.2	6	6	0.00282	C
	5292.7	6	4	0.00167	C
	5293.5	4	6	0.00113	C
	5309.4	4	2	0.00273	C
	5310.6	2	4	0.00137	C
	5344.0	6	6	6.2 (−4)	C
	5356.6	4	6	0.00189	C
	5367.0	4	4	0.00118	C
	5372.6	2	4	0.00107	C
	5378.3	2	2	0.00210	C
	5816.5	4	6	0.00278	C
	5829.5	6	6	0.0064	C
	5834.6	2	4	0.00383	C
	5840.9	4	4	0.00122	C
	5849.7	2	2	0.00152	C
	5854.0	6	4	0.00409	C
	5856.0	4	2	0.0076	C
	6606.2	4	6	7.9 (−4)	C
	6622.5	6	6	0.0071	C
	6627.0	2	4	0.00197	C
	6636.9	4	4	0.0125	C
	6645.0	8	6	0.0311	C
	6646.5	2	2	0.0194	C
	6653.5	6	4	0.0244	C
	6656.5	4	2	0.0193	C
	6926.7	4	6	0.0064	C
	6945.2	6	6	0.0149	C
	6951.6	2	4	0.0088	C
	6960.5	4	4	0.00281	C
	6973.1	2	2	0.00350	C
	6979.2	6	4	0.0094	C
	6982.0	4	2	0.0174	C
	7423.6	2	4	0.052	C
	7442.3	4	4	0.106	C
	7468.3	6	4	0.161	C
N II	474.89	5	5	4.4	D
	475.65	1	3	15.	D
	475.70	3	5	20.	D
	475.76	3	3	11.	D
	475.80	5	7	26.	D
	475.88	5	5	6.5	D
	508.70	5	5	2.7	D
	510.76	5	7	28.	D
	513.85	5	5	6.9	D
	529.36	1	3	6.5	D
	529.41	3	1	20.	D
	529.49	3	3	4.9	D
	529.64	3	5	4.9	D
	529.72	5	3	8.1	D
	529.87	5	5	15.	D
	533.51	1	3	20.	D
	533.58	3	5	27.	D
	533.65	3	3	15.	D
	533.73	5	7	36.	D
	533.82	5	5	9.1	D
	547.82	5	3	5.1	D
	555.76	1	3	12.	D
	574.65	5	7	35.	D
	582.16	5	5	13.	D
	635.20	1	3	18.	D
	644.63	1	3	12.	C
	644.84	3	3	35.	C
	645.18	5	3	58.	C
	660.29	5	3	40.	C
	671.02	3	5	3.0	D
	671.39	5	5	8.9	D
	671.41	1	3	3.5	D
	671.63	3	3	2.6	D
	671.77	3	1	12.	D
	672.00	5	3	4.4	D
	745.84	1	3	10.	C
	746.98	5	3	40.	C
	748.37	5	3	2.1	D
	775.97	5	5	35.	C
	915.61	1	3	3.6	C
	916.96	3	1	11.	C
	1084.0	1	3	1.9	C
	1085.5	5	5	0.85	C
	1085.7	5	7	3.4	C
	2139.5	3	3	0.30	D
	3593.6	3	5	0.23	D
	3609.1	3	3	0.23	D
	3615.9	3	1	0.23	D
	3829.8	3	5	0.15	D
	3838.4	5	5	0.58	D
	3842.2	1	3	0.19	D
	3847.4	3	3	0.15	D
	3855.1	3	1	0.58	D
	3856.1	5	3	0.24	D
	3919.0	3	3	1.0	C
	3995.0	3	5	1.3	D
	4114.4	3	3	0.00185	D
	4447.0	3	5	1.3	C
	4477.7	5	3	0.035	D
	4507.6	7	5	0.038	D
	4601.5	3	5	0.27	C
	4607.2	1	3	0.34	C
	4613.9	3	3	0.20	C
	4621.4	3	1	0.90	C
	4630.5	5	5	0.84	C
	4643.1	5	3	0.47	C
	4774.2	3	5	0.054	C
	4779.7	3	3	0.27	C
	4781.2	5	7	0.040	C
	4788.1	5	5	0.25	C
	4793.7	5	3	0.089	C
	4803.3	7	7	0.313	C
	4810.3	7	5	0.055	C
	4987.4	3	1	0.63	C
	4994.4	3	3	0.74	C
	5001.1	3	5	1.0	C
	5001.5	5	7	1.1	C
	5002.7	1	3	0.085	C
	5005.1	7	9	1.2	C
	5007.3	3	5	0.77	C
	5010.6	3	3	0.27	C
	5025.7	7	7	0.13	C
	5040.7	7	5	0.0053	C
	5045.1	5	3	0.41	C
	5452.1	1	3	0.18	D
	5454.3	3	1	0.060	D
	5462.6	3	3	0.045	D
	5478.1	3	5	0.075	D
	5480.1	5	3	0.045	D
	5495.7	5	5	0.14	D
	5666.6	3	5	0.42	C
	5676.0	1	3	0.31	C
	5679.6	5	7	0.56	C
	5686.2	3	3	0.23	C
	5710.8	5	5	0.14	C
	5927.8	1	3	0.32	C
	5931.8	3	5	0.43	C
	5940.2	3	3	0.24	C
	5941.7	5	7	0.56	C
	5952.4	5	5	0.14	C
	6482.1	3	3	0.38	D
	6610.6	5	7	0.52	D
N III	374.20	2	4	101.	C
	451.87	2	2	8.8	C
	452.23	4	2	35.9	C
	685.00	2	4	11.5	C
	685.51	2	2	45.7	C
	685.82	4	4	17.3	C
	686.34	4	2	23.0	C
	763.34	2	2	9.6	C
	764.36	4	2	18.7	C
	771.54	2	4	8.2	C
	771.90	4	4	16.5	C
	772.39	6	4	24.7	C
	772.89	6	4	20.3	C
	772.98	4	2	22.8	C

Table 1b (continued)
TRANSITION PROBABILITIES FOR SELECTED ATOMIC AND IONIC SPECIES

Spectrum	Wavelength (λ(Å))	Statistical weights g_i	Statistical weights g_k	Transition probability (A {10^8 s^{-1}})	Uncertainty
	979.84	4	4	8.9	C
	979.92	6	6	9.3	C
	989.79	2	4	4.19	C
	991.51	4	4	0.81	C
	991.58	4	6	4.97	C
	1747.9	2	4	1.32	C
	1751.2	4	4	0.261	C
	1751.7	4	6	1.57	C
	4097.3	2	4	0.85	C
	4103.4	2	2	0.84	C
	4634.1	2	4	0.65	C
	4640.6	4	6	0.78	C
	4641.9	4	4	0.130	C
N IV	247.20	1	3	114.	B
	283.53	*9	15	290.	C
	322.65	*9	3	84.	C
	355.05	3	5	160.	C
	387.35	3	1	28.	D
	765.14	1	3	24.	B
	923.15	*9	9	18.	B
	955.34	3	1	30.	C
	1718.5	3	5	2.37	C
	3480.8	*3	9	1.1	C
	4057.8	3	5	0.68	C
	6380.8	1	3	0.14	B
	7117.0	*9	15	0.12	C
N V	209.29	*2	6	118.	B
	247.66	*6	10	430.	C
	1238.8	2	4	3.41	B
	1242.8	2	2	3.38	B
	4603.7	2	4	0.41	B
	4620.0	2	2	0.41	B
N VI	24.898	1	3	5158.	AA
	28.787	1	3	180.9 (+2)	AA
	161.22	*3	9	285.9	AA
	173.34	1	3	270.	A
	173.92	*9	15	876.	A
	185.09	3	5	824.	A
	1901.5	*3	9	0.6777	AA
	2896.4	1	3	0.2080	AA
Sodium					
Na I	3302.4	2	4	0.0262	C
	3303.0	2	2	0.0262	C
	4390.0	2	4	0.0077	D
	4393.3	4	6	0.0092	D
	4393.3	4	4	0.0015	D
	4494.2	2	4	0.0126	D
	4497.7	4	6	0.0145	C
	4497.7	4	4	0.0026	D
	4658.6	4	6	0.0248	C
	4658.6	4	4	0.0041	D
	4664.8	2	4	0.0207	C
	4747.9	2	2	0.0063	D
	4751.8	4	2	0.0127	C
	4978.5	2	4	0.041	C
	4982.8	4	4	0.0082	D
	4982.8	4	6	0.049	C
	5148.8	2	2	0.0117	C
	5153.4	4	2	0.0253	C
	5682.6	2	4	0.102	C
	5688.2	4	4	0.0204	C
	5688.2	4	6	0.123	C
	5890.0	2	4	0.615	C
	5895.9	2	2	0.613	C
	6154.2	2	2	0.0257	C
	6160.7	4	2	0.0512	C
	8183.4	2	4	0.455	C
	8194.8	4	6	0.540	C
	8194.8	4	4	0.090	C
	11382.	2	2	0.0879	C
	11404.	4	2	0.175	C
Na II	300.15	1	3	30.	D
	301.43	1	3	49.	D
	372.07	1	3	34.	D
Na III	378.14	4	2	76.	C
	380.11	2	2	37.	C
	1976.4	4	6	8.3	D
	2004.8	2	4	4.6	D
	2011.9	6	8	8.4	D
	2151.2	2	4	4.4	D
	2174.5	4	6	5.3	D
	2180.8	4	6	3.6	D
	2194.8	4	4	3.7	D
	2230.3	6	8	3.7	D
	2232.2	4	4	3.3	D
	2246.7	4	6	2.4	D
	2459.4	4	6	3.0	D
	2468.9	2	4	2.4	D
	2497.0	6	6	1.7	D
Na V	333.46	*6	6	78.	D
	369.01	*10	6	120.	D
	370.89	*10	6	270.	D
	400.72	*10	10	50.	D
	445.14	*6	10	11.	D
	459.90	4	2	23.	D
	461.05	4	4	23.	D
	463.26	4	6	23.	D
	510.09	2	2	56.	D
	511.21	4	4	68.	D
Na VI	313.74	5	3	130.	C
	361.25	5	5	77.	D
	416.53	*9	9	37.	C
	492.80	*9	15	14.	C
	1550.6	5	5	4.36	C
	1567.8	5	3	2.00	C
	1608.5	3	1	2.64	C
	1649.4	5	5	2.06	C
	1741.5	3	5	2.59	C
	1747.5	5	7	3.07	C
Na VII	94.41	*6	10	2700.	C
	105.27	*6	2	450.	C
	353.29	4	4	98.	C
	381.30	4	2	40.	C
	397.52	4	4	36.	C
	399.21	6	4	52.	C
	483.28	*10	10	29.	C
	486.74	2	4	11.	C
	492.60	4	6	13.	C
	555.80	4	4	28.	D
	786.65	4	6	7.6	D
Na VIII	83.34	*9	15	3940.	C
	89.88	*9	3	809.	C
	90.54	3	5	2860.	C
	411.15	1	3	44.2	C
	1239.4	3	3	3.02	C
	1802.7	3	1	2.70	C
	1867.7	3	5	1.72	C
	2059.1	3	5	1.80	C
	2558.2	5	3	0.0226	C
	2772.0	3	5	0.419	C
	3021.0	5	7	0.490	C
	3108.9	1	3	0.258	C
	3182.3	1	3	0.292	C
Na IX	70.62	2	4	1360.	B
	70.64	2	2	1350.	B
	77.76	2	4	3700.	B
	77.91	4	6	4400.	B
	681.72	2	4	6.60	B
	694.17	2	2	6.37	B
	2487.7	2	4	0.41	B
	2535.8	2	2	1.6	B
	6841.8	2	4	0.025	C
	7103.4	4	6	0.028	C
Neon					
Ne I	615.62	1	3	0.38	C
	618.67	1	3	0.93	C
	619.09	1	3	0.33	C
	626.82	1	3	0.74	C
	629.73	1	3	0.48	C
	735.89	1	3	6.11	B
	743.70	1	3	0.484	B
	3369.8	5	5	0.0010	D
	3369.9	5	3	0.0076	D

Table 1b (continued)
TRANSITION PROBABILITIES FOR SELECTED ATOMIC AND IONIC SPECIES

Spectrum	Wavelength (λ(Å))	Statistical weights g_i	Statistical weights g_k	Transition probability (A {10^8 s^{-1}})	Uncertainty
	3375.6	5	3	0.0022	D
	3417.9	3	5	0.0092	D
	3418.0	3	3	0.0022	D
	3423.9	3	3	0.0010	D
	3447.7	5	5	0.021	D
	3450.8	5	3	0.0049	D
	3454.2	3	1	0.037	D
	3460.5	1	3	0.0070	D
	3464.3	5	5	0.0067	D
	3466.6	1	3	0.013	D
	3472.6	5	7	0.017	D
	3498.1	3	5	0.0051	D
	3501.2	3	3	0.012	D
	3510.7	5	3	0.0022	D
	3515.2	3	5	0.0069	D
	3520.5	3	1	0.0093	D
	3593.5	3	5	0.0099	D
	3593.6	3	3	0.0066	D
	3600.2	3	3	0.0043	D
	3633.6	3	1	0.011	D
	3682.2	3	5	0.0016	D
	3685.7	3	3	0.0039	D
	3701.2	3	5	0.0022	D
	4536.3	3	3	0.0050	D
	4702.5	3	3	0.0021	D
	4708.6	3	3	0.042	D
	4955.4	3	3	0.0033	D
	5113.7	3	3	0.010	D
	5120.5	3	3	0.0056	D
	5154.4	3	3	0.019	D
	5191.3	3	3	0.013	D
	5326.4	3	3	0.0068	D
	5333.3	3	3	0.0053	D
	5341.1	3	3	0.11	D
	5400.6	3	1	0.0090	B
	5418.6	3	3	0.0052	D
	5433.7	3	3	0.00283	B
	5652.6	3	3	0.0089	D
	5662.5	3	3	0.0069	D
	5852.5	3	1	0.682	B
	5868.4	3	3	0.014	D
	5881.9	5	3	0.115	B
	5913.6	3	3	0.048	D
	5939.3	5	3	0.00200	B
	5944.8	5	5	0.113	B
	5961.6	3	3	0.033	D
	5975.5	5	3	0.0351	B
	6030.0	3	3	0.0561	B
	6046.1	3	3	0.00226	B
	6074.3	3	1	0.603	B
	6096.2	3	5	0.181	B
	6118.0	5	3	0.00609	B
	6128.5	3	3	0.0067	B
	6143.1	5	5	0.282	B
	6150.3	3	3	0.015	D
	6163.6	1	3	0.146	B
	6217.3	5	3	0.0637	B
	6266.5	1	3	0.249	B
	6273.0	3	3	0.0097	D
	6293.7	3	3	0.00639	B
	6304.8	3	5	0.0416	B
	6328.2	5	3	0.0339	B
	6330.9	3	3	0.023	D
	6334.4	5	5	0.161	B
	6351.9	1	3	0.00345	B
	6383.0	3	3	0.321	B
	6401.1	3	3	0.0139	B
	6402.3	5	7	0.0514	B
	6506.5	3	5	0.300	B
	6532.9	1	3	0.108	B
	6599.0	3	3	0.232	B
	6602.9	3	3	0.0059	D
	6652.1	3	1	0.0029	B
	6678.3	3	5	0.233	B
	6717.0	3	3	0.217	B
	6721.1	3	3	4.9 (−4)	D
	6929.5	3	5	0.174	B
	7024.1	3	3	0.0189	B
	7032.4	5	3	0.253	B
	7051.3	3	3	0.030	D
	7059.1	3	5	0.068	C
	7173.9	3	5	0.0287	B
	7245.2	3	3	0.0935	B
	7304.8	1	3	0.00255	B
	7438.9	1	3	0.0231	B
	7472.4	3	3	0.040	D
	7535.8	3	3	0.43	D
	7937.0	5	5	0.0078	C
	8082.5	3	3	0.0012	B
	8118.5	3	3	0.049	D
	8128.9	3	5	0.0072	C
	8259.4	5	5	0.0203	C
	8571.4	3	3	0.055	D
	8582.9	3	5	0.0100	C
	8647.0	5	5	0.0391	C
	8681.9	3	3	0.21	D
	8767.6	3	3	0.0011	D
	8771.7	3	3	0.16	D
	8783.7	3	5	0.313	C
	8865.8	3	3	0.0094	D
	9201.8	3	3	0.091	D
	9432.9	3	3	0.0011	D
	9486.7	3	3	0.025	D
	9534.2	3	3	0.063	D
	10621.	3	3	0.0024	D
	11409.	3	3	0.042	D
	11525.	3	3	0.084	D
	11767.	3	3	0.069	D
	12459.	3	3	0.015	D
Ne II	357.03	*6	10	38.	C
	361.77	*6	2	16.	C
	406.28	*6	10	18.	C
	446.37	*6	6	40.7	C
	460.73	4	2	46.2	B
	462.39	2	2	22.9	B
	1907.5	4	2	0.28	D
	1916.1	4	4	0.69	D
	1930.0	2	2	0.57	D
	1938.8	2	4	0.13	D
	2858.0	6	6	0.79	D
	2870.0	6	6	0.17	D
	2873.0	6	4	0.38	D
	2876.3	4	6	0.78	D
	2876.5	6	4	0.33	D
	2878.1	2	2	0.069	D
	2888.4	4	6	0.070	D
	2891.5	4	4	0.061	D
	2897.0	6	8	0.052	D
	2906.8	2	4	0.55	D
	2910.1	4	2	1.7	D
	2910.4	2	4	0.59	D
	2916.2	6	4	0.096	D
	2925.6	2	2	0.56	D
	2933.7	6	6	0.069	D
	2955.7	6	4	1.2	D
	3001.7	4	4	0.87	D
	3017.3	6	4	0.35	D
	3027.0	6	6	1.4	D
	3028.7	4	2	0.85	D
	3028.9	2	4	0.47	D
	3034.5	6	8	3.1	D
	3037.7	4	4	2.1	D
	3045.6	2	2	2.5	D
	3047.6	4	6	1.8	D
	3054.7	2	4	0.94	D
	3092.9	6	6	1.3	D
	3097.1	8	8	1.3	D
	3118.0	8	6	0.042	D
	3134.1	6	4	0.26	D
	3140.4	8	6	0.24	D
	3151.1	6	6	0.048	D
	3154.8	8	6	0.018	D

Table 1b (continued)
TRANSITION PROBABILITIES FOR SELECTED ATOMIC AND IONIC SPECIES

Spectrum	Wavelength (λ(Å))	Statistical weights g_i	Statistical weights g_k	Transition probability (A {10^8 s^{-1}})	Uncertainty
	3164.4	8	8	0.16	D
	3165.6	6	6	0.12	D
	3173.6	6	4	0.045	D
	3176.1	4	6	0.060	D
	3187.6	4	6	0.014	D
	3188.7	6	6	0.39	D
	3190.9	4	6	0.15	D
	3194.6	4	4	0.52	D
	3198.6	6	8	1.7	D
	3198.9	4	4	0.23	D
	3209.0	8	8	0.16	D
	3209.4	2	4	0.60	D
	3213.7	2	4	1.7	D
	3214.3	4	6	2.2	D
	3218.2	8	10	3.6	D
	3224.8	6	8	3.5	D
	3229.5	8	8	0.13	D
	3229.6	8	10	3.6	D
	3230.1	6	6	1.8	D
	3230.4	4	6	0.14	D
	3232.0	6	4	0.27	D
	3232.4	4	4	1.6	D
	3243.4	6	6	0.23	D
	3244.1	6	8	1.5	D
	3248.1	4	4	0.24	D
	3255.4	6	4	0.038	D
	3263.4	2	4	0.39	D
	3269.9	4	6	0.51	D
	3270.8	6	4	0.057	D
	3297.7	6	6	0.43	D
	3309.7	4	2	0.31	D
	3310.5	4	4	0.069	D
	3311.3	4	2	0.26	D
	3314.7	6	6	0.044	D
	3319.7	4	2	1.6	D
	3320.2	8	6	0.21	D
	3323.7	4	4	1.6	D
	3327.2	4	4	0.91	D
	3329.2	8	8	0.88	D
	3330.7	6	6	0.039	D
	3334.8	6	8	1.8	D
	3336.1	4	6	1.1	D
	3344.4	2	2	1.5	D
	3345.5	6	4	1.4	D
	3345.8	4	4	0.22	D
	3353.6	4	2	0.12	D
	3355.0	4	6	1.3	D
	3356.3	6	6	0.20	D
	3357.8	6	6	0.50	D
	3360.6	2	4	0.82	D
	3360.8	2	4	0.86	D
	3362.9	4	2	0.35	D
	3371.8	4	6	0.22	D
	3374.1	4	4	0.30	D
	3378.2	2	2	1.7	D
	3379.3	2	2	0.30	D
	3386.2	4	6	0.055	D
	3388.4	4	6	2.2	D
	3390.6	2	4	0.077	D
	3392.8	2	4	0.44	D
	3404.8	4	6	1.9	D
	3406.9	6	8	2.3	D
	3411.4	4	2	0.61	D
	3413.1	4	4	1.8	D
	3414.9	4	6	0.018	D
	3416.9	6	6	0.64	D
	3417.7	6	8	1.6	D
	3438.9	2	2	1.4	D
	3440.7	2	4	0.35	D
	3453.1	4	4	0.46	D
	3454.8	4	4	1.6	D
	3456.6	2	4	0.96	D
	3457.1	4	6	0.099	D
	3459.3	6	6	1.6	D
	3475.2	4	4	0.012	D
	3477.6	4	6	0.43	D
	3481.9	4	2	1.4	D
	3503.6	2	2	2.0	D
	3522.7	4	2	0.023	D
	3538.0	4	2	0.76	D
	3539.9	4	4	0.036	D
	3542.2	6	4	0.60	D
	3542.9	4	6	1.2	D
	3546.2	2	4	0.063	D
	3551.6	2	4	0.037	D
	3557.8	2	2	0.19	D
	3561.2	4	6	0.21	D
	3565.8	4	4	0.62	D
	3568.5	6	8	1.4	D
	3571.2	4	4	0.63	D
	3574.2	6	6	0.10	D
	3574.6	4	6	1.3	D
	3590.4	4	6	0.036	D
	3594.2	4	2	1.3	D
	3612.3	2	4	0.26	D
	3628.0	4	4	0.60	D
	3632.7	4	4	0.13	D
	3643.9	4	4	0.32	D
	3644.9	2	4	0.99	D
	3659.9	4	6	0.067	D
	3664.1	6	4	0.70	D
	3679.8	4	2	0.32	D
	3694.2	6	6	1.0	D
	3697.1	2	2	0.28	D
	3701.8	4	6	0.27	D
	3709.6	4	2	1.1	D
	3713.1	4	6	1.3	D
	3721.8	4	6	0.20	D
	3726.9	4	4	0.12	D
	3727.1	2	4	0.98	D
	3734.9	4	4	0.19	D
	3744.6	2	4	0.26	D
	3751.2	2	2	0.18	D
	3753.8	4	6	0.45	D
	3766.3	4	6	0.29	D
	3777.1	2	4	0.42	D
	3800.0	4	4	0.37	D
	3818.4	2	4	0.61	D
	3829.7	4	6	0.84	D
	3942.3	4	6	0.010	D
Ne V	142.61	*9	9	670.	C
	143.32	*9	15	1200.	C
	147.13	5	7	1500.	C
	151.42	5	5	337.	C
	156.61	1	3	675.	C
	167.70	*9	9	146.	C
	358.93	*9	3	21.0	C
	365.59	5	3	135.	C
	482.15	*9	9	30.1	C
	571.04	*9	15	10.4	C
	2259.6	3	5	1.7	D
	2265.7	5	7	2.2	D
Ne VII	97.50	1	3	1070.	B
	115.52	*9	3	480.	C
	116.69	3	5	1600.	C
	127.66	3	1	190.	C
	465.22	1	3	40.6	B
	558.6	3	5	8.11	B
	559.9	1	3	10.7	B
	561.4	3	3	7.99	B
	561.7	5	5	23.9	B
	563.0	3	1	31.7	B
	564.5	5	3	13.1	B
Ne VIII	88.09	*2	6	839.	B
	98.22	*6	10	2800.	B
	770.41	2	4	5.90	B
	780.32	2	2	5.73	B
	2820.7	2	4	0.72	B
	2860.1	2	2	0.70	B
Nickel					
Ni I	3200.4	7	5	8.9 (−4)	D
	3271.1	5	5	0.0051	D

Table 1b (continued)
TRANSITION PROBABIILITIES FOR SELECTED ATOMIC AND IONIC SPECIES

Spectrum	Wavelength (λ[Å])	Statistical weights g_i	Statistical weights g_k	Transition probability (A {10^8 s^{-1}})	Uncertainty
	3286.9	7	5	0.036	D
	3320.3	7	5	0.041	D
	3361.6	5	5	0.044	D
	3362.8	3	5	0.0011	D
	3409.6	9	7	0.0032	D
	3413.5	7	5	0.047	D
	3420.7	5	5	0.011	D
	3433.6	7	7	0.15	D
	3458.5	3	5	0.56	D
	3493.0	5	3	0.87	D
	3515.1	5	7	0.44	D
	3519.8	5	5	0.042	D
	3524.5	7	5	0.88	D
	3566.4	5	5	0.55	D
	3571.9	7	7	0.16	D
	3597.7	3	3	0.16	D
	3610.5	5	5	0.10	D
	3664.1	5	3	0.015	D
	3670.4	7	5	0.0084	D
	3674.2	5	5	0.026	D
	3688.4	5	7	0.0038	D
	3722.5	3	5	0.0059	D
	3793.6	5	5	0.0015	D
	3831.7	5	3	0.012	D
	3858.3	5	7	0.069	D
	3973.6	5	5	0.0031	D
	4401.5	9	11	0.33	D
	4714.4	13	11	0.40	D
	4786.5	11	11	0.16	D
	4901.0	9	11	0.045	D
	5017.6	11	11	0.18	D
	5158.0	9	11	0.0052	D
	5265.7	9	11	0.0033	D
	5578.7	5	5	5.5 (−4)	D
	5847.0	5	5	1.3 (−4)	D
	6256.4	5	3	0.0019	D
	6314.7	5	5	0.031	D
	6327.6	5	7	4.2 (−4)	D
	6364.6	3	5	3.3 (−4)	D
	6643.6	5	5	0.0014	D
	6767.8	1	3	0.0031	D
	7197.1	5	3	8.4 (−4)	D
	7261.9	3	3	7.9 (−4)	D
	7291.5	5	7	2.6 (−4)	D
	7414.5	1	3	9.9 (−4)	D
	7714.3	5	5	0.0013	D
	7789.0	3	5	8.4 (−4)	D
Ni II	2033.4	8	6	0.038	D
	2053.3	10	8	0.042	D
	2080.8	4	4	0.13	D
	2090.1	4	6	0.11	D
	2093.6	8	8	0.11	D
	2125.1	8	8	0.010	D
	2125.9	10	8	0.074	D
	2128.6	6	8	0.41	D
	2138.6	8	6	0.028	D
	2158.7	6	4	0.057	D
	2161.2	6	8	0.32	D
	2165.6	10	10	3.9	D
	2169.1	8	8	2.3	D
	2174.7	8	10	2.4	D
	2175.2	6	6	2.8	D
	2184.6	4	4	4.7	D
	2188.1	8	6	0.091	D
	2201.4	4	6	2.1	D
	2206.7	6	8	2.5	D
	2210.4	8	10	0.63	D
	2216.5	10	12	5.5	D
	2220.4	6	8	3.7	D
	2223.0	10	10	1.6	D
	2224.4	8	6	0.51	D
	2224.9	8	8	2.6	D
	2226.3	6	6	2.0	D
	2253.9	4	6	3.2	D
	2264.5	6	8	2.4	D
	2270.2	8	10	2.5	D
	2278.8	8	6	4.4	D
	2287.1	6	4	4.5	D
	2296.6	8	8	3.2	D
	2297.1	6	4	4.6	D
	2297.5	4	2	5.3	D
	2298.3	6	6	4.5	D
	2303.0	8	6	4.7	D
	2316.0	10	8	4.9	D
	2326.4	4	4	0.99	D
	2334.6	8	8	1.3	D
	2356.4	6	6	0.44	D
	2367.4	8	8	0.13	D
	2375.4	6	8	1.1	D
	2387.8	8	8	0.23	D
	2394.5	8	10	2.9	D
	2410.7	8	6	0.018	D
	2412.3	6	8	0.0039	D
	2413.0	6	4	0.13	D
	2416.1	6	8	3.3	D
	2433.6	6	6	0.11	D
	2437.9	8	10	0.88	D
	2510.9	8	10	0.94	D
	2545.9	6	8	0.26	D
	2630.3	8	8	0.012	D
Oxygen					
O I	1028.2	1	3	0.20	D
	1152.2	5	5	5.5	D
	1217.6	1	3	1.8	C
	1302.2	5	3	3.3	C
	1304.9	3	3	2.0	C
	1306.0	1	3	0.66	C
	5435.2	3	5	0.0061	C
	5435.8	5	5	0.0102	C
	5436.9	7	5	0.0142	C
	6453.6	3	5	0.0142	B
	6454.4	5	5	0.0237	B
	6456.0	7	5	0.0331	B
	6653.8	3	1	0.600	B
	7156.7	5	5	0.473	B
	7471.4	5	3	0.0114	B
	7473.2	5	5	0.102	B
	7477.2	3	3	0.170	B
	7479.1	3	5	0.306	B
	7480.7	1	3	0.226	B
	7676.4	5	7	0.408	B
	7772.0	5	7	0.340	B
	7774.2	5	5	0.340	B
	7775.4	5	3	0.340	B
	7886.3	3	5	0.370	B
	7939.5	7	5	0.00165	C
	7943.2	7	7	0.0417	C
	7947.2	5	5	0.058	C
	7947.6	7	9	0.373	C
	7950.8	5	7	0.331	C
	7952.2	3	5	0.313	C
	7981.9	3	3	0.12	D
	7982.4	1	3	0.16	D
	7987.0	3	5	0.21	D
	7987.3	5	5	0.072	D
	7995.1	5	7	0.29	D
O II	429.92	4	2	39.	D
	430.04	4	4	39.	D
	430.18	4	6	39.	D
	483.75	4	2	0.84	D
	483.98	6	4	0.76	D
	484.03	4	4	0.084	D
	485.09	6	8	25.	D
	485.47	6	6	1.6	D
	485.52	4	6	23.	D
	3007.1	8	10	0.84	C
	3007.7	6	8	0.72	C
	3013.4	6	8	0.74	C
	3032.1	8	10	0.85	C
	3032.5	6	8	0.82	C
	3134.8	8	6	1.23	C
	3273.5	8	6	1.14	C
	3377.2	2	2	1.88	C
	3390.3	2	4	1.86	C

Table 1b (continued)
TRANSITION PROBABILITIES FOR SELECTED ATOMIC AND IONIC SPECIES

Spectrum	Wavelength (λ(Å))	Statistical weights g_i	g_k	Transition probability (A (10⁸ s⁻¹))	Uncertainty
	3407.4	6	6	0.75	C
	3749.5	6	4	0.90	C
	3882.2	8	8	0.493	C
	3912.0	6	4	1.27	C
	3919.3	4	2	1.40	C
	3973.3	4	4	1.27	C
	4069.6	2	4	1.39	C
	4069.9	4	6	1.49	C
	4072.2	6	8	1.70	C
	4075.9	8	10	1.98	C
	4085.1	6	6	0.478	C
	4087.2	4	6	2.24	C
	4089.3	10	12	2.62	C
	4095.6	6	8	2.23	C
	4097.3	8	10	2.37	C
	4104.7	4	6	1.04	C
	4105.0	4	4	0.80	C
	4108.8	8	8	0.349	C
	4119.2	6	8	1.48	C
	4120.3	6	6	0.443	C
	4132.8	2	4	0.84	C
	4153.3	4	6	0.77	C
	4276.7	6	8	1.82	C
	4277.4	2	4	1.49	C
	4277.9	8	8	0.302	C
	4281.4	6	6	0.60	C
	4282.8	4	4	1.06	C
	4283.0	4	6	1.58	C
	4283.1	6	6	0.51	C
	4283.8	4	4	0.59	C
	4294.8	4	6	1.39	C
	4303.8	6	8	1.97	C
	4328.6	4	2	1.21	C
	4340.4	6	8	2.23	C
	4347.4	4	4	0.94	C
	4349.4	6	6	0.74	C
	4351.3	6	6	0.97	C
	4396.0	6	6	0.398	C
	4414.9	4	6	1.15	C
	4417.0	2	4	0.95	C
	4443.1	6	6	0.57	C
	4448.2	8	8	0.57	C
	4465.4	6	8	0.92	C
	4467.9	6	6	0.92	C
	4469.3	6	4	0.92	C
	4489.5	2	4	1.51	C
	4491.3	4	6	1.81	C
	4596.2	4	6	1.03	C
	4602.1	4	6	1.70	C
	4609.4	6	8	1.82	C
	4641.8	4	6	0.79	C
	4661.6	4	4	0.52	C
	4701.2	4	4	0.87	C
	4703.2	4	6	0.82	C
	4705.4	6	8	1.38	C
	4871.6	4	6	0.435	C
	4906.9	4	4	0.68	C
	4924.6	4	6	0.67	C
	4941.1	2	4	0.83	C
	4943.1	4	6	1.06	C
	5206.7	4	4	0.391	C
	6627.6	4	4	0.089	C
	6669.9	4	2	0.0349	C
	6678.2	2	4	0.0173	C
	6718.1	2	2	0.068	C
	6721.4	4	2	0.189	C
	6810.6	6	8	0.00180	C
	6844.1	4	6	0.00325	C
	6847.0	8	8	0.0347	C
	6869.7	6	6	0.059	C
	6885.1	4	4	0.067	C
	6895.3	10	8	0.298	C
	6906.5	8	6	0.272	C
	6908.1	4	2	0.332	C
	6910.8	6	4	0.267	C
O III	262.98	5	5	39.	D
	263.69	1	3	51.	D
	263.73	3	5	69.	D
	263.77	3	3	38.	D
	263.82	5	7	92.	D
	263.86	5	5	23.	D
	277.38	5	7	110.	D
	279.79	5	5	2.6	D
	295.94	1	3	49.	D
	303.41	1	3	34.	D
	303.46	3	1	100.	D
	303.52	3	3	26.	D
	303.62	3	5	25.	D
	303.69	5	3	42.	D
	303.80	5	5	76.	D
	305.60	1	3	100.	D
	305.66	3	5	140.	D
	305.70	3	3	76.	D
	305.77	5	7	180.	D
	305.84	5	5	46.	D
	320.98	5	7	188.	D
	328.45	5	5	61.	D
	345.31	1	3	99.	D
	374.08	5	5	26.	D
	395.56	5	3	49.	D
	507.39	1	3	16.	C
	507.68	3	3	47.	C
	508.18	5	3	79.	C
	525.80	5	3	88.	C
	597.82	1	3	18.	C
	599.60	5	5	55.	C
	702.33	1	3	5.7	C
	702.82	3	1	17.	C
	832.93	1	3	3.2	C
	835.10	5	5	2.2	C
	835.29	5	7	5.7	C
	1109.5	3	3	2.7	D
	1679.2	3	5	0.92	D
	1686.9	3	3	0.90	D
	1760.5	3	5	0.59	D
	1764.6	5	5	1.8	D
	1766.5	1	3	0.78	D
	1772.4	3	1	2.3	D
	1773.1	5	3	0.97	D
	2390.4	3	3	7.1	D
	2959.7	3	5	2.1	D
	2996.5	3	3	0.51	D
	3004.4	5	5	0.47	D
	3017.6	7	7	0.59	D
	3115.7	3	1	1.5	D
	3121.7	3	3	1.5	D
	3132.9	3	5	1.5	D
	3261.0	5	7	1.8	D
	3265.5	7	9	2.1	D
	3267.3	3	5	1.7	D
	3282.0	5	5	0.32	D
	3284.6	7	7	0.23	D
	3405.7	1	3	0.27	D
	3408.1	3	1	0.81	D
	3415.3	3	3	0.20	D
	3428.7	3	5	0.20	D
	3430.6	5	3	0.33	D
	3444.1	5	5	0.59	D
	3702.7	1	3	0.62	D
	3707.2	3	5	0.83	D
	3714.0	3	3	0.46	D
	3715.0	5	7	1.1	D
	3725.2	5	5	0.27	D
	3961.6	5	7	1.2	D
	5268.1	1	3	0.28	D
	5500.1	5	5	0.11	D
	5592.4	3	3	0.36	D
O IV	238.36	2	4	288.	C
	238.57	4	6	345.	C
	238.58	4	4	58.	C
	279.63	2	2	23.9	C
	279.93	4	2	47.7	C

Table 1b (continued)
TRANSITION PROBABILITIES FOR SELECTED ATOMIC AND IONIC SPECIES

Spectrum	Wavelength (λ(Å))	Statistical weights g_i	g_k	Transition probability (A {10^8 s^{-1}})	Uncertainty
	553.33	2	4	12.1	C
	554.08	2	2	47.4	C
	554.51	4	4	60.	C
	555.26	4	2	24.9	C
	608.40	2	2	12.5	C
	609.83	4	2	23.6	C
	616.95	6	4	14.3	C
	617.01	4	4	2.94	C
	617.04	4	2	51.	C
	624.62	2	4	10.8	C
	625.13	4	4	21.5	C
	625.85	6	4	32.2	C
	779.73	6	4	1.47	C
	779.82	4	4	13.2	C
	779.91	6	6	13.7	C
	780.00	4	6	0.98	C
	787.71	2	4	5.9	C
	790.11	4	4	1.15	C
	790.20	4	6	7.1	C
	921.30	2	4	2.27	C
	921.37	2	2	9.1	C
	923.37	4	4	11.3	C
	923.43	4	2	4.51	C
	1338.6	2	4	2.23	C
	1343.0	4	4	0.425	C
	1343.5	4	6	2.64	C
O V	172.17	1	3	280.	B
	192.85	*9	15	690.	B
	215.18	*9	3	170.	C
	220.35	3	5	440.	C
	248.46	3	1	65.	D
	629.73	1	3	29.5	B
	758.7	3	5	5.68	B
	759.4	1	3	7.55	B
	760.2	3	3	5.64	B
	760.4	5	5	16.9	B
	761.1	3	1	22.5	B
	762.0	5	3	9.34	B
	774.52	3	1	35.4	C
	1371.3	3	5	3.29	C
	2784.0	*3	9	1.6	D
	3144.7	3	5	0.93	C
	5114.1	1	3	0.17	B
	5590.0	*9	15	0.15	C
O VI	150.13	*2	6	253.	B
	173.03	*6	10	880.	B
	1031.9	2	4	4.15	B
	1037.6	2	2	4.11	B
	3811.4	2	4	0.51	B
	3834.2	2	2	0.51	B
O VII	18.627	1	3	9360.	A
	21.601	1	3	330.9 (+2)	AA
	120.35	*3	9	533.3	AA
	128.25	1	3	505.	A
	128.46	*9	15	1620.	A
	135.77	3	5	1530.	A
	1630.2	*3	9	0.7935	AA
	2450.0	1	3	0.2514	AA
Phosphorus					
P I	1671.7	4	2	0.40	D
	1674.6	4	4	0.40	D
	1679.7	4	6	0.40	D
	1775.0	4	6	2.17	C
	1782.9	4	4	2.14	C
	1787.7	4	2	2.13	C
	2135.5	4	4	0.211	C
	2136.2	6	4	2.83	C
	2149.1	4	2	3.18	C
	2152.9	2	4	0.485	C
	2154.1	4	4	0.173	C
	2154.1	4	6	0.58	C
	2534.0	2	4	0.200	C
	2535.6	4	4	0.95	C
	2553.2	2	2	0.71	C
	2554.9	4	2	0.300	C
P II	1301.9	1	3	0.50	C
	1304.5	3	1	1.50	C
	1304.7	3	3	0.38	C
	1305.5	3	5	0.37	C
	1309.9	5	3	0.62	C
	1310.7	5	5	1.11	C
	4475.3	5	7	1.3	D
	4499.2	5	7	1.4	D
	4530.8	3	5	1.0	D
	4554.8	3	5	0.96	D
	4588.0	5	7	1.7	D
	4589.9	3	5	1.6	D
	4602.1	7	9	1.9	D
	4943.5	7	5	0.63	D
	5253.5	3	5	1.0	D
	5425.9	5	5	0.69	D
	6024.2	3	5	0.51	D
	6043.1	5	7	0.68	D
P III	1334.9	2	4	0.33	D
	1344.3	4	6	0.64	D
	1344.9	4	4	0.064	D
	4057.4	4	4	0.10	D
	4059.3	6	4	0.90	D
	4080.0	4	2	0.99	D
Sulfur					
S I	1295.7	5	5	4.9	D
	1296.2	5	3	2.7	D
	1302.3	3	5	1.8	D
	1302.9	3	3	1.6	D
	1303.1	3	1	6.6	D
	1303.4	5	3	1.9	D
	1305.9	1	3	2.4	D
	1401.5	5	3	0.91	D
	1409.4	3	3	0.50	D
	1412.9	1	3	0.16	D
	1425.1	5	7	4.5	D
	1425.2	5	5	1.2	D
	1433.3	3	3	1.9	D
	1433.3	3	5	3.3	D
	1437.0	1	3	2.4	D
	1448.2	5	3	7.3	D
	1472.9	5	7	0.42	D
	1474.0	5	7	1.6	D
	1474.4	5	5	0.50	D
	1474.6	5	3	0.11	D
	1481.7	3	5	0.17	D
	1483.0	3	5	1.2	D
	1483.2	3	3	0.75	D
	1487.2	1	3	0.87	D
	1666.7	5	5	6.3	C
	1687.5	1	3	0.94	D
	1782.3	1	3	1.9	D
	1807.3	5	3	3.8	C
	1820.4	3	3	2.2	C
	1826.3	1	3	0.72	C
	4694.1	5	7	0.0067	D
	4695.4	5	5	0.0067	D
	4696.2	5	3	0.0067	D
	6403.6	3	5	0.0057	D
	6408.1	5	5	0.095	D
	6415.5	7	5	0.013	D
	6751.2	*15	25	0.079	D
	7679.6	3	5	0.012	D
	7686.1	5	5	0.020	D
	7696.7	7	5	0.028	D
S II	1124.4	2	4	1.0	D
	1125.0	4	4	4.6	D
	1131.0	2	2	3.4	D
	1131.6	4	2	1.4	D
	1250.5	4	2	0.46	C
	1253.8	4	4	0.42	C
	1259.5	4	6	0.34	C
	4463.6	8	6	0.53	D
	4483.4	6	4	0.31	D
	4486.7	4	2	0.66	D
	4524.7	4	4	0.093	D
	4525.0	6	4	1.2	D

Table 1b (continued)
TRANSITION PROBABILITIES FOR SELECTED ATOMIC AND IONIC SPECIES

Spectrum	Wavelength (λ(Å))	Statistical weights g_i	Statistical weights g_k	Transition probability (A {10^8 s^{-1}})	Uncertainty
	4552.4	4	2	1.2	D
	4656.7	2	4	0.09	D
	4716.2	4	4	0.29	D
	4815.5	6	4	0.88	D
	4885.6	2	4	0.17	D
	4917.2	2	2	0.66	D
	4924.1	4	6	0.22	D
	4925.3	2	4	0.24	D
	4942.5	2	2	0.15	D
	4991.9	4	4	0.15	D
	5009.5	4	2	0.70	D
	5014.0	4	4	0.84	D
	5027.2	4	2	0.26	D
	5032.4	6	6	0.81	D
	5047.3	4	2	0.36	D
	5103.3	6	4	0.50	D
	5142.3	2	2	0.19	D
	5201.0	4	4	0.75	D
	5201.3	6	4	0.065	D
	5212.6	4	6	0.098	D
	5212.6	6	6	0.85	D
	5320.7	6	8	0.92	D
	5345.7	6	6	0.11	D
	5345.7	4	6	0.88	D
	5428.6	2	4	0.42	D
	5432.8	4	6	0.68	D
	5453.8	6	8	0.85	D
	5473.6	2	2	0.73	D
	5509.7	4	4	0.40	D
	5526.2	8	8	0.081	D
	5536.8	4	6	0.066	D
	5556.0	4	2	0.11	D
	5564.9	6	6	0.17	D
	5578.8	6	6	0.11	D
	5606.1	10	8	0.54	D
	5616.6	4	4	0.12	D
	5640.0	4	6	0.66	D
	5645.6	6	4	0.018	D
	5647.0	2	4	0.57	D
	5660.0	6	4	0.46	D
	5664.7	4	2	0.58	D
	5819.2	4	4	0.085	D
	6305.5	8	6	0.18	D
	6312.7	6	4	0.30	D
S III	2496.2	7	5	2.5	D
	2508.2	5	3	2.3	D
	2636.9	3	5	0.45	D
	2665.4	5	5	1.4	D
	2680.5	1	3	0.62	D
	2691.7	3	3	0.46	D
	2702.8	3	1	1.9	D
	2718.9	3	3	1.2	D
	2721.4	5	3	0.77	D
	2726.8	3	5	0.60	D
	2731.1	5	5	1.1	D
	2756.9	7	7	1.4	D
	2785.5	3	3	0.61	D
	2856.0	5	7	5.1	D
	2863.5	7	9	5.7	D
	2872.0	3	5	4.7	D
	2950.2	3	5	3.0	D
	2964.8	5	7	4.0	D
	3662.0	3	3	0.64	D
	3717.8	5	3	1.0	D
	3778.9	3	5	0.44	D
	3831.8	1	3	0.56	D
	3837.8	3	3	0.42	D
	3838.3	5	5	1.3	D
	3860.6	3	1	1.6	D
	3899.1	5	3	0.67	D
	4253.6	5	7	1.2	D
	4285.0	3	5	0.90	D
S IV	551.17	2	2	20.6	C
	554.07	4	2	40.8	C
	3097.5	2	4	2.6	D
	3117.8	2	2	2.5	D
S V	437.37	1	3	11.2	C
	438.19	3	3	33.3	C
	439.65	5	3	55.	C
	661.52	*9	15	64.4	B
	679.01	*9	15	86.	D
	690.75	*9	9	50.	D
	786.48	1	3	52.5	B
	854.85	*9	9	41.8	C
S VI	248.98	2	4	30.6	C
	249.27	2	2	30.5	C
	388.94	2	2	44.2	C
	390.86	4	2	87.4	C
	706.48	2	4	41.7	C
	712.68	4	6	48.5	C
	712.84	4	4	8.1	D
	933.38	2	4	16.4	C
	944.52	2	2	15.9	C
S VII	60.16	1	3	980.	D
	60.80	1	3	8400.	D
	72.03	1	3	730.	D
S VIII	198.57	4	2	250.	C
	202.65	2	2	120.	C
S XI	189.90	*9	3	440.	C
	190.37	5	3	280.	C
	215.95	5	5	130.	C
	217.63	1	3	71.	C
	239.81	1	3	27.	D
	242.57	3	5	20.	D
	242.82	3	3	19.	D
	246.90	5	5	56.	D
	247.12	5	3	31.	D
	288.65	*9	15	29.	C
S XII	212.14	2	4	36.	C
	215.18	2	2	140.	C
	218.20	4	4	167.	C
	221.44	4	2	64.	C
	227.5	2	2	12.	C
	234.48	2	4	11.	C
S XIII	32.41	1	3	1.09 (+4)	B
	37.6	3	1	1300.	C
	256.66	1	3	86.5	C
	299.89	3	5	17.8	C
	303.37	1	3	22.8	C
	307.36	3	3	16.4	C
	308.91	5	5	48.2	C
	312.68	3	1	62.5	C
	316.84	5	3	25.0	C
	500.42	3	5	14.3	C
S XIV	30.44	*2	6	8280.	B
	32.52	*6	10	2.6 (+4)	B
	417.67	2	4	12.2	B
	445.71	2	2	10.4	B
	1550.	2	4	1.4	B
	1663.	2	2	1.4	B
	3967.	2	4	0.053	C
	4153.	4	6	0.058	C
Scandium					
Sc I	2974.0	4	4	2.3	D
	2980.7	6	6	2.3	D
	3015.4	4	6	2.3	D
	3019.3	6	8	2.5	D
	3269.9	4	2	3.1	C
	3273.6	6	4	2.7	C
	3907.5	4	6	1.35	C
	3911.8	6	8	1.58	C
	4020.4	4	4	1.35	C
	4023.7	6	6	1.47	C
	4737.7	6	4	1.1	D
	4741.0	8	6	1.1	D
	4743.8	10	8	1.3	D
	5081.5	10	10	1.0	D
	5083.7	8	8	0.84	D
	5085.5	6	6	0.76	D
	5086.9	4	4	0.88	D
	5349.3	6	4	0.54	D
	5356.1	8	6	0.51	D

Table 1b (continued)
TRANSITION PROBABILITIES FOR SELECTED ATOMIC AND IONIC SPECIES

Spectrum	Wavelength (λ(Å))	Statistical weights g_i	g_k	Transition probability (A {10^8 s^{-1}})	Uncertainty
	5482.0	8	8	0.45	D
	5484.6	6	6	0.44	D
	5514.2	6	8	0.55	D
	5520.5	8	10	0.57	D
	5671.8	10	12	0.76	D
	5686.8	8	10	0.69	D
	5700.1	6	8	0.65	D
	5711.7	4	6	0.64	D
Sc II	2273.1	1	3	7.7	C
	2552.4	7	5	2.3	C
	3353.7	5	7	1.98	C
	3372.1	7	5	1.16	C
	3535.7	5	3	0.82	D
	3572.5	7	7	1.86	C
	3576.3	5	5	1.45	C
	3613.8	7	9	1.88	C
	3630.7	5	7	1.58	C
	3642.8	3	5	1.50	C
	4246.8	5	5	1.55	C
	4314.1	9	7	0.46	D
	4374.5	9	9	0.16	D
	4400.4	7	7	0.15	D
	4415.6	5	5	0.16	D
	4670.4	5	7	0.183	C
	5031.0	5	3	0.490	C
	5239.8	1	3	0.14	D
	5526.8	9	7	0.41	D
	5657.9	5	5	0.12	D
	6245.6	5	7	0.026	D
Silicon					
Si I	1977.6	1	3	0.18	D
	1979.2	3	1	0.51	D
	1980.6	3	3	0.13	D
	1983.2	3	5	0.14	D
	1986.4	5	3	0.21	D
	1989.0	5	5	0.41	D
	2208.0	1	3	0.311	C
	2210.9	3	5	0.416	C
	2211.7	3	3	0.232	C
	2216.7	5	7	0.55	C
	2218.1	5	5	0.138	C
	2506.9	3	5	0.465	C
	2514.3	1	3	0.612	C
	2516.1	5	5	1.21	C
	2519.2	3	3	0.452	C
	2524.1	3	1	1.82	C
	2528.5	5	3	0.765	C
	2532.4	1	3	0.26	D
	2631.3	1	3	0.97	D
	2881.6	5	3	1.89	C
	3905.5	1	3	0.118	C
	4738.8	3	3	0.010	D
	4783.0	5	3	0.017	D
	4792.3	5	5	0.017	D
	4818.1	5	7	0.011	D
	4821.2	3	5	0.0080	D
	4947.6	3	1	0.042	D
	5006.1	3	5	0.028	D
	5622.2	3	3	0.016	D
	5690.3	3	3	0.012	D
	5708.4	5	5	0.014	D
	5754.2	5	3	0.015	D
	5772.2	3	1	0.036	D
	5948.6	3	5	0.022	D
	7226.2	3	5	0.0079	D
	7405.8	3	5	0.039	D
	7409.1	5	7	0.023	D
	7680.3	3	5	0.046	D
	7918.4	3	5	0.052	D
	7932.4	5	7	0.051	D
	7944.0	7	9	0.058	D
	7970.3	5	5	0.0071	D
Si II	989.87	2	4	6.7	D
	992.68	4	6	8.0	D
	1020.7	2	2	1.3	D
	1190.4	2	4	6.9	D
	1193.3	2	2	27.	D
	1194.5	4	4	34.	D
	1197.4	4	2	14.	D
	1248.4	4	4	13.	D
	1251.2	6	4	19.	D
	1260.4	2	4	21.	D
	1264.7	4	6	25.	D
	1304.4	2	2	3.6	D
	1309.3	4	2	7.0	D
	1526.7	2	2	3.73	C
	1533.4	4	2	7.4	C
	1808.0	2	4	0.041	D
	2904.3	4	6	0.67	D
	2905.7	6	8	0.71	D
	3210.0	4	6	0.46	D
	4128.1	4	6	1.32	C
	4130.9	6	8	1.42	C
	5041.0	2	4	0.98	D
	5056.0	4	6	1.2	D
	5957.6	2	2	0.42	D
	5978.9	4	2	0.81	D
	6347.1	2	4	0.70	C
	6371.4	2	2	0.69	C
	7848.8	4	6	0.39	D
	7849.7	6	8	0.42	D
Si III	883.40	5	7	63.	D
	994.79	3	3	7.89	B
	997.39	5	3	13.1	B
	1141.6	3	5	30.	D
	1144.3	5	7	39.	D
	1161.6	5	5	16.	D
	1206.5	1	3	25.9	B
	1206.5	3	5	48.9	B
	1207.5	5	5	19.	D
	1294.5	1	3	5.05	B
	1296.7	3	5	6.70	B
	1298.9	3	3	5.00	B
	1299.0	5	5	15.0	B
	1301.2	3	1	19.9	B
	1303.3	5	3	8.25	B
	1328.8	1	3	27.	D
	1417.2	3	1	26.0	C
	1435.8	5	7	21.	D
	1589.0	5	3	11.	D
	1778.7	7	9	4.4	D
	1783.2	5	7	3.8	D
	3241.6	5	3	2.3	D
	3486.9	*15	21	1.8	D
	3590.5	3	5	3.9	D
	4552.6	3	5	1.26	C
	4554.0	5	3	0.76	D
	4567.8	3	3	1.25	C
	4683.0	5	5	0.95	D
	4716.6	5	7	2.0	D
	5451.5	3	5	0.60	D
	5473.0	5	7	0.79	D
	5716.3	9	7	0.19	D
	5739.7	1	3	0.47	D
	7462.6	5	3	0.49	D
	7466.3	7	5	0.54	D
	7612.4	3	5	1.1	D
Si IV	457.82	2	4	3.6	D
	458.16	2	2	3.6	D
	515.12	2	2	4.1	D
	516.35	4	2	8.2	D
	560.50	*6	10	1.0	D
	749.94	*10	14	14.5	C
	815.05	2	2	12.3	C
	818.13	4	2	24.4	C
	860.74	*10	6	1.8	D
	1066.6	*10	14	39.1	C
	1126.4	*6	10	25.0	C
	1393.8	2	4	7.73	B
	1402.8	2	2	7.58	B
	1724.1	*10	6	5.5	C

Table 1b (continued)
TRANSITION PROBABILITIES FOR SELECTED ATOMIC AND IONIC SPECIES

Spectrum	Wavelength (λ(Å))	Statistical weights g_i	g_k	Transition probability (A {10^8 s^{-1}})	Uncertainty
Si V	96.439	1	3	480.	D
	97.143	1	3	2000.	D
	117.86	1	3	300.	D
Si VI	246.00	4	2	170.	C
	249.13	2	2	85.	C
Si VII	217.83	5	3	430.	D
	272.64	5	3	51.	C
	274.18	3	1	120.	C
	275.35	5	5	89.	C
	275.66	3	3	30.	C
	276.84	1	3	39.	C
	278.44	3	5	29.	C
Si VIII	214.75	4	2	410.	D
	216.92	6	4	360.	D
	232.85	2	2	80.	D
	235.56	4	4	97.	D
	250.60	2	2	77.	D
	250.97	4	2	160.	D
	314.31	4	2	52.	D
	316.20	4	4	50.	D
	319.83	4	6	49.	D
Si IX	223.72	1	3	41.	C
	225.03	3	3	122.	C
	227.00	5	3	198.	C
	227.35	5	3	250.	D
	258.10	5	5	104.	C
	294.44	*9	9	59.	C
	347.4	*9	15	22.	C
Si X	253.81	2	4	29.4	C
	256.58	2	2	114.	C
	258.39	4	4	140.	C
	261.27	4	2	54.0	C
	272.00	2	2	19.	C
	277.27	4	2	37.	C
	287.25	2	4	26.	C
	289.28	4	4	51.	C
	292.31	6	4	74.	C
	348.89	*10	10	44.	D
	353.14	*6	10	21.	C
Si XI	43.76	1	3	6110.	B
	49.12	*9	3	2450.	C
	49.22	3	5	8900.	C
	52.30	3	1	7540.	C
	303.58	1	3	64.2	B
	358.29	3	1	103.	C
	358.54	3	5	13.8	C
	361.31	1	3	18.0	C
	364.50	3	3	13.2	C
	365.42	5	5	39.0	C
	368.38	3	1	51.0	C
	371.61	5	3	20.7	C
	609.76	3	5	11.2	C
	2295.0	1	3	0.435	C
Si XII	40.92	*2	6	4420.	B
	44.12	*6	10	1.4 (+4)	B
	499.43	2	4	9.45	B
	520.76	2	2	8.68	B
	1862.	2	4	1.1	B
	1949.	2	2	1.0	B
	4620.	2	4	0.044	C
	4942.	4	6	0.046	C
Titanium					
Ti I	2912.1	5	7	1.3	D
	3199.9	9	11	0.96	D
	3314.4	5	7	1.3	D
	3314.5	3	3	0.73	D
	3371.4	9	11	0.13	D
	3653.5	9	11	0.56	D
	3724.6	9	9	1.3	D
	3858.1	9	11	0.55	D
	3926.3	11	9	0.78	D
	4030.5	11	13	0.49	D
	4137.3	9	7	0.65	D
	4237.9	5	5	1.2	D
	4256.0	9	9	0.63	D
	4263.1	11	9	0.74	D
	4282.7	7	5	0.89	D
	4300.6	7	5	1.1	D
	4301.1	9	7	0.93	D
	4305.9	11	9	0.89	D
	4318.6	13	11	0.76	D
	4321.7	9	7	0.61	D
	4325.1	11	9	0.72	D
	4427.1	9	11	0.53	D
	4449.1	11	11	0.86	D
	4455.3	7	7	0.66	D
	4457.4	9	9	0.69	D
	4533.2	11	11	1.0	D
	4534.8	9	9	0.88	D
	4535.6	7	7	0.62	D
	4617.3	7	9	0.83	D
	4623.1	5	7	0.62	D
	4742.8	9	9	0.56	D
	4758.1	11	11	0.77	D
	4759.3	13	13	0.82	D
	4808.5	9	11	0.33	D
	4856.0	13	15	0.59	D
	4868.3	9	11	0.41	D
	4870.1	11	13	0.48	D
	4885.1	11	13	0.53	D
	4899.9	9	11	0.46	D
	4913.6	7	9	0.46	D
	4975.3	5	7	0.62	D
	4981.7	11	13	0.51	D
	4991.1	9	11	0.38	D
	4999.5	7	9	0.45	D
	5007.2	5	7	0.50	D
	5014.3	3	5	0.60	D
	5035.9	9	11	0.43	D
	5036.5	7	9	0.41	D
	5038.4	5	7	0.45	D
	5120.4	11	13	0.47	D
	5206.1	3	5	5.8	D
Ti II	2635.6	4	4	1.9	D
	2638.7	6	6	1.7	D
	2642.2	8	8	1.8	D
	2646.1	10	10	2.7	D
	2746.7	6	8	2.6	D
	2751.7	8	10	3.7	D
	2752.8	8	10	1.1	D
	2800.6	10	8	1.8	D
	2805.0	6	8	4.6	D
	2810.3	8	10	5.1	D
	2817.8	10	12	3.8	D
	2827.2	8	10	1.0	D
	2828.2	12	14	4.4	D
	2828.9	10	10	0.91	D
	2834.1	10	12	0.79	D
	2836.6	8	8	1.2	D
	2839.7	12	12	0.83	D
	2846.1	10	10	1.2	D
	2856.2	12	12	1.5	D
	2931.3	6	6	3.2	D
	2936.2	4	6	2.7	D
	2938.7	6	8	2.4	D
	2942.0	8	10	1.7	D
	2943.1	8	8	1.1	D
	2945.5	10	12	2.7	D
	2954.8	10	12	4.0	D
	2959.0	10	10	4.0	D
	3022.8	10	10	1.2	D
	3023.9	8	8	1.0	D
	3081.6	10	8	1.1	D
	3088.0	10	8	1.3	C
	3089.4	8	6	1.3	C
	3103.8	10	8	1.1	C
	3127.9	6	6	1.7	D
	3128.6	8	8	1.3	D
	3190.9	6	8	1.3	C
	3202.5	4	6	1.1	C
	3224.2	12	10	0.70	C
	3234.5	10	10	1.3	C

Table 1b (continued)
TRANSITION PROBABILITIES FOR SELECTED ATOMIC AND IONIC SPECIES

Spectrum	Wavelength (λ[Å])	Statistical weights g_i	g_k	Transition probability (A_{ki}) (10^8 s^{-1})	Uncertainty
	3236.6	8	8	1.2	C
	3287.7	8	10	1.4	C
	3341.9	6	8	0.96	D
	3349.0	8	10	1.0	D
	3349.4	10	12	1.3	D
	3361.2	8	10	1.2	D
	3372.8	6	8	1.1	C
	3383.8	4	6	1.1	C
	3483.8	10	8	0.97	D
	3492.4	8	6	0.98	D
	3504.9	10	10	0.80	C
	3510.8	8	8	0.91	C
	3759.3	8	8	0.98	C
	3761.3	6	6	1.0	C
Ti III	2375.0	5	3	4.0	D
	2414.0	5	7	3.8	D
	2516.0	7	9	3.4	D
	2527.8	5	7	2.2	D
	2540.0	3	5	2.0	D
	2563.4	7	7	2.1	D
	2565.4	5	5	1.6	D
	2567.5	3	3	2.3	D
	2576.4	5	3	0.92	D
	2984.8	5	5	1.9	D
Ti XIX	179.31	1	3	115.	C
Ti XX	11.452	*2	6	1.68 (+4)	C
	11.872	2	4	2.84 (+4)	C
	11.958	4	6	3.36 (+4)	C
	15.914	2	4	9.0 (+4)	C
	16.059	4	6	1.06 (+5)	C
	16.067	4	4	1.76 (+4)	C
Vanadium					
V I	3050.4	10	8	0.47	D
	3053.7	4	4	1.1	D
	3056.3	6	6	1.0	D
	3060.5	8	8	1.1	D
	3066.4	10	10	1.6	D
	3088.1	4	6	0.38	D
	3089.1	4	4	0.45	D
	3093.8	6	6	0.36	D
	3112.9	4	2	0.43	D
	3183.4	6	8	1.3	D
	3185.4	10	12	1.4	D
	3198.0	6	6	0.31	D
	3202.4	8	8	0.29	D
	3205.6	8	10	1.1	D
	3212.4	10	12	1.2	D
	3218.9	8	6	0.31	D
	3233.2	10	8	0.28	D
	3273.0	8	8	0.24	D
	3284.4	10	10	0.24	D
	3309.2	4	4	0.28	D
	3329.9	6	4	0.69	D
	3356.4	4	6	0.27	D
	3365.6	2	4	0.41	D
	3376.1	4	4	0.28	D
	3377.4	4	2	0.80	D
	3377.6	6	6	0.53	D
	3400.4	8	8	0.22	D
	3529.7	4	6	0.36	D
	3533.7	2	4	0.32	D
	3533.8	2	4	0.32	D
	3543.5	2	2	0.58	D
	3545.3	4	4	0.32	D
	3663.6	4	6	0.82	D
	3667.7	6	8	0.64	D
	3671.2	8	10	0.18	D
	3672.4	12	12	0.79	D
	3673.4	8	10	2.4	D
	3676.7	14	14	1.1	D
	3680.1	10	12	1.9	D
	3686.3	10	12	0.20	D
	3687.5	12	14	2.6	D
	3688.1	8	8	0.28	D
	3690.3	2	4	0.37	D
	3692.2	6	6	0.46	D
	3695.3	14	16	2.5	D
	3695.9	4	4	0.54	D
	3703.6	10	8	0.79	D
	3704.7	8	6	0.58	D
	3705.0	6	4	0.31	D
	3706.0	10	10	0.46	D
	3708.7	12	12	0.39	D
	3790.5	10	8	0.20	D
	3795.0	10	10	0.21	D
	3806.8	10	10	0.22	D
	3818.2	4	2	0.56	C
	3828.6	6	4	0.431	C
	3840.1	8	8	0.18	D
	3840.8	8	6	0.46	D
	3855.4	4	4	0.28	D
	3855.8	10	8	0.451	C
	3863.9	8	6	0.27	D
	3864.3	8	6	0.17	D
	3864.9	6	6	0.208	C
	3871.1	10	8	0.24	D
	3875.1	8	8	0.17	D
	3886.6	10	8	0.14	D
	3902.3	10	10	0.217	C
	3921.9	4	2	0.23	D
	3922.4	6	6	0.23	D
	3930.0	10	10	0.29	D
	3934.0	10	10	0.29	D
	3992.8	12	10	1.1	D
	3998.7	14	12	0.92	D
	4051.0	10	10	1.2	D
	4051.4	12	12	1.2	D
	4090.6	8	10	0.77	D
	4092.7	8	10	0.21	D
	4095.5	6	8	0.54	D
	4099.8	6	8	0.39	D
	4102.2	4	6	0.50	D
	4104.8	10	8	1.9	D
	4105.2	4	6	0.42	D
	4109.8	2	4	0.47	D
	4111.8	10	10	0.91	D
	4113.5	6	8	0.15	D
	4115.2	8	8	0.59	D
	4116.5	6	6	0.24	D
	4123.6	4	2	0.94	D
	4128.1	6	4	0.70	D
	4132.0	8	6	0.52	D
	4134.5	10	8	0.27	D
	4232.5	10	10	0.86	D
	4233.0	8	8	0.67	D
	4268.6	14	14	1.0	D
	4271.6	12	12	0.84	D
	4277.0	10	10	0.84	D
	4284.1	8	8	1.0	D
	4291.8	12	14	0.78	D
	4296.1	10	12	0.69	D
	4297.7	8	10	0.61	D
	4298.0	6	8	0.70	D
	4342.8	10	10	0.13	D
	4355.0	12	12	0.11	D
	4379.2	10	12	1.2	D
	4384.7	8	10	0.97	D
	4390.0	6	8	0.70	D
	4395.2	4	6	0.48	D
	4400.6	2	4	0.33	D
	4406.6	10	10	0.19	D
	4407.6	8	8	0.38	D
	4408.2	6	6	0.51	D
	4416.5	4	2	0.21	D
	4452.0	4	6	0.80	D
	4457.8	10	12	0.24	D
	4460.3	10	8	0.26	D
	4462.4	12	14	0.66	D
	4468.0	8	10	0.20	D
	4469.7	12	14	0.66	D
	4474.1	10	8	0.41	D
	4490.8	10	12	0.10	D

Table 1b (continued)
TRANSITION PROBABILITIES FOR SELECTED ATOMIC AND IONIC SPECIES

Spectrum	Wavelength (λ(Å))	Statistical weights g_i	g_k	Transition probability (A (10⁸ s⁻¹))	Uncertainty
	4496.1	8	6	0.35	D
	4514.2	6	4	0.29	D
	4524.2	12	10	0.26	D
	4525.2	4	2	0.36	D
	4529.6	10	8	0.21	D
	4545.4	10	12	0.67	D
	4560.7	8	10	0.62	D
	4571.8	6	8	0.53	D
	4578.7	4	6	0.59	D
	4579.2	8	8	0.13	D
	4706.2	6	4	0.21	D
	4706.6	12	14	0.15	D
	4710.6	10	12	0.17	D
	4746.6	4	4	0.17	D
	4751.0	8	8	0.15	D
	4754.0	10	10	0.13	D
	4757.5	4	2	0.65	D
	4766.6	6	4	0.48	D
	4776.4	8	6	0.43	D
	4786.5	10	8	0.40	D
	4796.9	12	10	0.42	D
	4807.5	14	12	0.51	D
	5193.0	12	12	0.35	D
	5195.4	8	8	0.21	D
	5234.1	10	10	0.41	D
	5240.9	12	12	0.39	D
	5415.3	12	14	0.27	D
	5487.9	12	10	0.25	D
	5507.8	10	8	0.30	D
	5559.9	6	2	0.14	D
	5698.5	6	8	0.28	D
	5703.6	4	6	0.19	D
	5707.0	2	4	0.19	D
	5725.6	8	10	0.18	D
	5727.0	8	10	0.18	D
	6090.2	8	6	0.13	D
V II	2503.0	5	7	0.23	C
	2506.2	7	9	0.23	C
	2514.6	9	11	0.24	C
	2672.0	5	7	0.18	D
	2677.8	3	5	0.29	D
	2679.3	7	7	0.26	D
	2688.0	9	9	0.60	D
	2690.8	5	3	0.43	D
	2700.9	9	11	0.29	C
	2702.2	7	7	0.20	D
	2706.2	7	9	0.29	C
	2728.6	3	5	0.20	C
	2768.6	11	9	0.82	C
	2774.3	9	7	0.88	C
	2799.5	5	5	0.60	C
	2802.8	7	7	0.46	C
	2803.5	9	9	0.58	C
	2836.5	9	11	0.25	C
	2841.0	7	9	0.27	C
	2877.7	7	5	0.67	C
	2880.0	7	7	0.15	D
	2882.5	5	5	0.27	D
	2888.2	11	9	0.54	C
	2889.6	3	1	1.4	D
	2891.6	5	3	0.92	D
	2893.3	9	7	0.70	D
	2903.1	3	5	0.21	D
	2906.5	7	7	0.53	D
	2907.5	9	11	0.19	C
	2908.8	11	9	1.1	D
	2910.0	5	5	0.72	D
	2910.4	3	3	0.87	D
	2911.1	7	9	0.30	C
	2924.6	9	9	0.91	C
	2930.1	7	7	0.27	C
	2944.6	9	7	0.65	C
	2950.3	3	3	0.34	C
	2952.1	7	5	0.58	C
	2957.5	5	3	0.44	C
	3048.9	9	7	0.54	C
	3093.1	11	13	1.8	D
	3100.9	7	7	1.0	C
	3102.3	9	11	1.6	D
	3110.7	7	9	1.5	D
	3118.4	5	7	1.5	D
	3121.1	11	11	0.22	D
	3125.3	3	5	1.6	D
	3126.2	9	9	0.41	D
	3130.3	7	7	0.50	D
	3133.3	5	5	0.48	D
	3187.7	5	5	0.85	D
	3188.5	7	7	0.80	D
	3190.7	9	9	0.88	D
	3208.4	7	5	0.17	D
	3232.0	3	5	0.23	C
	3271.1	7	9	1.1	C
	3276.1	9	11	1.1	C
	3287.7	5	7	1.2	C
	3321.5	9	7	0.15	C
	3517.3	9	7	0.14	D
	3530.8	5	3	0.19	D
	3545.2	7	5	0.18	D
	3556.8	9	7	0.20	D
	3589.8	5	3	0.37	D
	3592.0	7	5	0.23	D
	3715.5	13	11	0.16	C
	3727.4	9	9	0.22	C
	3732.8	11	9	0.16	C
	3745.8	9	7	0.17	C
	3750.9	7	7	0.20	C
	3771.0	5	5	0.22	C
V XX	14.38	1	3	6.9 (+4)	C
	159.36	1	3	150.	C
V XXI	8.843	*2	6	6030.	C
	8.882	4	6	6410.	C
	9.111	2	4	8910.	C
	9.175	4	6	1.04 (+4)	C
	9.352	*2	6	1.0 (+4)	C
	9.633	2	4	1.63 (+4)	C
	9.704	4	6	1.91 (+4)	C
	10.412	*2	6	2.0 (+4)	C
	10.770	2	4	3.5 (+4)	B
	10.853	4	6	4.2 (+4)	B
	13.823	2	4	4.26 (+4)	B
	13.865	2	2	4.37 (+4)	B
	14.430	2	4	1.1 (+5)	B
	14.572	4	6	1.3 (+5)	B
	14.592	4	4	2.1 (+4)	C
	28.18	*2	6	3000.	C
	28.510	2	4	5690.	C
	28.689	4	6	6700.	C
	40.612	*2	6	5800.	C
	41.394	2	4	1.15 (+4)	B
	41.715	4	6	1.35 (+4)	B
	41.773	4	4	2250.	B
	48.18	*6	10	1060.	C
	58.29	*6	10	1660.	C
	89.63	*6	10	2900.	C
	239.5	2	4	29.	B
	291.5	2	2	16.	B
V XXII	2.3823	1	3	2.9 (+6)	B

REFERENCES

1. **Wiese, W. L. and Fuhr, J. R.,** *J. Phys. Chem. Ref. Data,* 4, 263, 1975.
2. **Younger, S. M., Fuhr, J. R., Martin, G. A., and Wiese, W. L.,** *J. Phys. Chem. Ref. Data,* to be published, 1978.
3. **Fuhr, J. R., Younger, S. M., Martin, G. A., and Wiese, W. L.,** to be published.
4. **Weise, W. L., Smith, M. W., and Glennon, B. M.,** Atomic Transition Probabilities (H through Ne — A Critical Data Compilation), National Standards Reference Data Series, National Bureau of Standards 4, Vol. I, U.S. Government Printing Office, Washington, D. C., 1966.
5. **Wiese, W. L., Smith, M. W., and Miles, B. M.,** Atomic Transition Probabilities (Na through Ca — Critical Data Compilation), National Standards Reference Data Series National Bureau of Standards 22, Vol. 11, U.S. Government Printing Office, Washington, D.C, 1969.
6. **Wiese, W. L. and Weiss, A. W.,** *Phys. Rev.,* 175, 50, 1968.
7. **Smith, M. W. and Wiese, W. L.,** *Astrophys. J. Suppl. Ser.,* 23, 196, 103, 1971.
8. **Martin, G. A. and Wiese, W. L.,** *J. Phys. Chem. Ref. Data,* 5, 537, 1976.
9. **Fuhr, J. R., Miller, B. J., and Martin, G. A.,** Bibliography on Atomic Transition Probabilities (1914 through October 1977), National Bureau of Standards Special Publication, 505, 1978.
10. **Coleman, C. D., Bozman, W. R., and Meggers, W. F.,** Table of Wavenumbers, *Natl. Bur. Stand. (U.S.) Monogr.,* 3, 1960.
11. **Younger, S. M. and Weiss, A. W.,** *J. Res. Natl. Bur. Stand.,* 79A, 629, 1975.

THERMODYNAMIC PROPERTIES OF ALKANE HYDROCARBONS[a]

Symbols and Units

$(G° - H°_0)/T$	=	Gibbs energy function, cal K^{-1} mol^{-1}
$(H° - H°_0)/T$	=	enthalpy function, cal K^{-1} mol^{-1}
S°	=	entropy, cal K^{-1} mol^{-1}
C_p	=	heat capacity at constant pressure, cal K^{-1} mol^{-1}
ΔHf°	=	enthalpy of formation, kcal mol^{-1}
ΔGf°	=	Gibbs energy of formation, kcal mol^{-1}
log_{10} Kf	=	common logarithm of the equilibrium constant of formation
T	=	temperature, kelvins
$H° - H°_0$	=	enthalpy, kcal mol^{-1}

Conversion Factors

The tabulated values may be converted to SI units by use of the relations:

calories X 4.184 = joules
kilocalories X 4184 = joules

T, K	$(G° - H°_0)/T$, cal K^{-1} mol^{-1}	$(H° - H°_0)/T$, cal K^{-1} mol^{-1}	$H° - H°_0$, kcal mol^{-1}	S°, cal K^{-1} mol^{-1}	$C_p°$, cal K^{-1} mol^{-1}	$\Delta H_f°$, kcal mol^{-1}	$\Delta G_f°$, kcal mol^{-1}	log_{10} K_f
				Methane (CH_4)				
0	0	0	0	0	0	-15.9_5	-15.9_5	∞
200	-33.3_1	7.9_4	1.58_8	41.2_5	8.0_1	-17.1_8	-13.8_8	15.1_7
273.15	-35.7_9	7.9_9	2.18_4	43.7_8	8.3_3	-17.6_8	-12.6_0	10.0_8
298.15	-36.4_9	8.0_3	2.39	44.5_2	8.5_3	-17.8_6	-12.1	8.8_8
400	-38.8_8	8.3_0	3.32	47.1_8	9.7_1	-18.6_0	-10.0_4	5.4_9
600	-42.4	9.2	5.5_4	51.6	12.5	-19.9	-5.5	1.9_9
800	-45.2	10.4	8.3	55.6	15.2	-20.8	-0.5	0.1_4
1,000	-47.7	11.6	11.6	59.3	$17._4$	-21.4	4.6	-1.0_0
1,200	$-49._9$	$12._7$	$15._3$	$62._6$	$19._2$	$-21._7$	$9._8$	-1.7_9
1,500	$-52._9$	$14._2$	$21._3$	$67._1$	21.	$-21._8$	$17._7$	-2.5_9

[a]The original source of the data in this table contains the thermodynamic properties of all the alkane hydrocarbons through decane. Including the *meso*, *(d, l)*, *cis*, and *trans* isomers, there are 182 compounds in the original source.

THERMODYNAMIC PROPERTIES OF ALKANE HYDROCARBONS (Continued)

T, K	$(G° - H°_0)/T$, cal K^{-1} mol^{-1}	$(H° - H°_0)/T$, cal K^{-1} mol^{-1}	$H° - H°_0$, kcal mol^{-1}	S°, cal K^{-1} mol^{-1}	$C_p°$, cal K^{-1} mol^{-1}	$\Delta H_f°$, kcal mol^{-1}	$\Delta G_f°$, kcal mol^{-1}	$\log_{10} K_f$
				Ethane (C_2H_6)				
0	0	0	0	0	0	-16.2_6	-16.2_6	∞
200	-41.6_3	8.6_7	1.73_5	50.3_0	10.1_0	-18.8_0	-11.4_6	12.5_2
273.15	-44.4_2	9.2_8	2.53_4	53.7_0	11.8_5	-19.7_0	-8.6_3	6.9_1
298.15	-45.2_4	9.5_2	2.84	54.7_6	12.5_7	-20.0_0	-7.6_0	5.5_7
400	-48.2_0	10.7_1	4.28	58.9_1	15.8_3	-21.1_8	-3.1_8	1.7_4
600	-53.1	13.5	8.0_9	66.6	22.2	-23.0	6.2	-2.2_7
800	-57.3	16.7	13.3	74.0	27.4	-23.7	16.1	-4.4_1
1,000	-61.3	19.0	19.0	80.3	$31._7$	-24.3	26.2	-5.7_3
1,200	$-65._0$	$21._4$	$25._7$	$86._4$	$35._1$	$-24._1$	$36._3$	-6.6_1
1,500	$-70._1$	$24._6$	$36._9$	$94._7$	39.	$-23._1$	$51._3$	-7.4_7
				Propane (C_3H_8)				
0	0	0	0	0	0	-19.7_4	-19.7_4	∞
200	-48.3_6	10.2_1	2.04_1	58.5_7	12.6_4	-23.4_2	-11.8_8	12.9_8
273.15	-51.7_0	11.3_3	3.09_6	63.0_3	16.2_2	-24.6_6	-7.4_7	5.9_7
298.15	-52.7_1	11.8_0	3.52	64.5_1	17.4_4	-25.0_7	-5.8_7	4.3_0
400	-56.4_6	13.8_8	5.55	70.3_4	22.4_6	-26.6_1	0.9_4	-0.5_1
600	-62.9	18.2	10.9_5	81.1	31.2	-28.9	15.2	-5.5_5
800	-68.7	22.4	17.9	91.1	38.0	-30.2	30.2	-8.2_5
1,000	-74.1	26.1	26.1	100.2	$43._4$	-30.8	45.4	-9.9_2
1,200	$-79._2$	$29._3$	$35._2$	$108._5$	$47._6$	$-30._7$	$60._6$	-11.0_4
1,500	$-86._2$	$33._5$	$50._2$	$119._7$	52.	$-29._7$	$83._3$	-12.1_4
				n-Butane (C_4H_{10})				
0	0	0	0	0	0	-23.6_3	-23.6_3	∞
200	-52.5_7	13.4_1	2.68_1	65.9_8	18.3_6	-28.1_3	-12.2_3	13.3_6
273.15	-57.0_1	15.1_8	4.14_5	72.1_9	21.8_9	-29.5_7	-6.1_9	4.9_5
298.15	-58.3_7	15.8_0	4.71	74.1_7	23.3_0	-30.0_5	-4.0_2	2.9_5
400	-63.3_8	18.4_9	7.40	81.8_7	29.4_5	-31.9_0	5.1_7	-2.8_2
600	-71.9	24.0	14.4_2	95.9	40.4	-34.7	24.3	-8.8_7
800	-79.6	29.2	23.4	108.8	48.8	-36.3	44.3	-12.1_0
1,000	-86.6	33.8	33.8	120.4	$55._2$	-37.0	64.5	-14.1_0
1,200	$-93._1$	$37._8$	$45._4$	$130._9$	$60._1$	$-37._0$	$84._8$	-15.4_5
1,500	$-102._1$	$42._8$	$64._3$	$144._9$	66.	$-36._1$	$115._2$	-16.7_8

THERMODYNAMIC PROPERTIES OF ALKANE HYDROCARBONS (Continued)

T, K	$(G° - H°_0)/T$, cal K^{-1} mol^{-1}	$(H° - H°_0)/T$, cal K^{-1} mol^{-1}	$H° - H°_0$, kcal mol^{-1}	S°, cal K^{-1} mol^{-1}	$C_p°$, cal K^{-1} mol^{-1}	$\Delta H_f°$, kcal mol^{-1}	$\Delta G_f°$, kcal mol^{-1}	$\log_{10} K_f$
				2-Methylpropane (C_4H_{10})				
0	0	0	0	0	0	−25.3$_4$	−25.3$_4$	∞
200	−50.7$_9$	11.7$_2$	2.34$_4$	62.5$_1$	16.8$_5$	−30.1$_7$	−13.5$_8$	14.8$_4$
273.15	−54.7$_3$	13.7$_1$	3.74$_4$	68.4$_4$	21.5$_2$	−31.6$_7$	−7.2$_7$	5.8$_2$
298.15	−55.9$_7$	14.4$_4$	4.30	70.4$_1$	23.1$_3$	−32.1$_6$	−5.0$_0$	3.6$_7$
400	−60.6$_3$	17.4$_9$	6.99	78.1$_2$	29.6$_2$	−34.0$_0$	4.5$_6$	−2.4$_9$
600	−68.9	23.4	14.0$_7$	92.3	40.8	−36.7	24.5	−8.9$_2$
800	−76.4	28.9	23.1	105.3	49.5	−38.2	45.1	−12.3$_3$
1,000	−83.3	33.8	33.8	117.1	56.$_4$	−38.7	66.1	−14.4$_4$
1,200	−89.$_9$	38.$_0$	45.$_6$	127.$_9$	61.$_7$	−38.$_4$	87.$_0$	−15.8$_4$
1,500	−99.$_0$	43.$_3$	65.$_0$	142.$_3$	68.	−37.$_0$	118.$_2$	−17.2$_2$
				n-Pentane (C_5H_{12})				
0	0	0	0	0	0	−27.3$_5$	−27.3$_5$	∞
200	−57.0$_3$	16.4$_6$	3.29$_1$	73.4$_9$	22.3$_6$	−32.7$_0$	−12.4$_5$	13.6$_0$
273.15	−62.4$_7$	18.6$_1$	5.08$_3$	81.0$_8$	26.9$_0$	−34.4$_1$	−4.7$_7$	3.8$_2$
298.15	−64.1$_4$	19.3$_8$	5.78	83.5$_2$	28.6$_9$	−34.9$_8$	−2.0$_3$	1.4$_9$
400	−70.2$_9$	22.7$_4$	9.10	93.0$_3$	36.4$_6$	−37.1$_6$	9.5$_8$	−5.2$_3$
600	−80.8	29.7	17.7$_9$	110.5	49.9	−40.3	33.7	−12.2$_7$
800	−90.3	36.0	28.8	126.3	59.9	−42.2	58.7	−16.0$_3$
1,000	−98.9	41.6	41.6	140.5	67.$_3$	−43.0	84.0	−18.3$_6$
1,200	−106.$_9$	46.$_3$	55.$_6$	153.$_2$	72.$_8$	−43.$_0$	109.$_4$	−19.9$_2$
1,500	−117.$_9$	52.$_3$	78.$_4$	170.$_2$	79.	−42.$_1$	147.$_4$	−21.4$_8$
				2-Methybutane (C_5H_{12})				
0	0	0	0	0	0	−28.4$_4$	−28.4$_4$	∞
200	−58.1$_6$	14.3$_4$	2.86$_7$	72.5$_0$	20.3$_0$	−34.2$_0$	−13.7$_6$	15.0$_3$
273.15	−62.9$_8$	16.7$_4$	4.57$_3$	79.7$_2$	26.3$_8$	−36.0$_0$	−6.0$_0$	4.8$_0$
298.15	−64.4$_9$	17.6$_4$	5.26	82.1$_3$	28.4$_1$	−36.5$_8$	−3.2$_1$	2.3$_6$
400	−70.2$_0$	21.4$_3$	8.57	91.6$_3$	36.5$_4$	−38.7$_6$	8.5$_3$	−4.6$_6$
600	−80.3	28.8	17.3$_0$	109.1	50.2	−41.9	32.9	−11.9$_9$
800	−89.5	35.5	28.4	125.0	60.5	−43.7	58.2	−15.8$_9$
1,000	−98.1	41.3	41.3	139.4	68.$_4$	−44.3	83.7	−18.3$_0$
1,200	−106.$_1$	46.$_3$	55.$_6$	152.$_4$	74.$_4$	−44.$_1$	109.$_3$	−19.9$_1$
1,500	−117.$_1$	52.$_7$	79.$_0$	169.$_8$	81.	−42.$_6$	147.$_5$	−21.4$_9$

THERMODYNAMIC PROPERTIES OF ALKANE HYDROCARBONS (Continued)

T, K	$(G° - H°_0)/T$, cal K^{-1} mol^{-1}	$(H° - H°_0)/T$, cal K^{-1} mol^{-1}	$H° - H°_0$, kcal mol^{-1}	S°, cal K^{-1} mol^{-1}	$C_p°$, cal K^{-1} mol^{-1}	$\Delta H_f°$, kcal mol^{-1}	$\Delta G_f°$, kcal mol^{-1}	$\log_{10} K_f$
2,2-Dimethylpropane (C_5H_{12})								
0	0	0	0	0	0	-32.3_8	-32.3_8	∞
200	-48.3_2	13.4_3	2.68_5	61.7_5	19.2_5	-38.3_3	-15.7_4	17.2_0
273.15	-53.0_1	17.4_6	4.77_0	70.4_7	26.6_8	-39.7_6	-7.2_2	5.7_8
298.15	-54.5_3	18.5_8	5.54	73.1_1	28.8_8	-40.2_5	-4.1_9	3.0_7
400	-60.3_3	22.7_2	9.09	83.0_5	37.2_8	-42.2_0	8.5_3	-4.6_6
600	-70.7	30.2	18.1_1	100.9	51.3	-45.0	34.7	-12.6_5
800	-80.2	36.9	29.6	117.1	62.4	-46.5	61.7	-16.8_6
1,000	-88.9	43.0	43.0	131.9	$71._2$	-46.6	88.9	-19.4_3
1,200	$-97._2$	$48._3$	$57._9$	$145._5$	$78._2$	$-45._7$	$116._0$	-21.1_3
1,500	$-108._7$	$55._1$	$82._7$	$163._8$	86.	$-42._9$	$156._2$	-22.7_6
***n*-Hexane (C_6H_{14})**								
0	0	0	0	0	0	-31.0_9	-31.0_9	∞
200	-61.4_5	19.5_6	3.91_1	81.0_1	26.4_3	-37.2_6	-12.6_8	13.8_5
273.15	-67.9_2	22.1_0	6.03_5	90.0_2	31.9_2	-39.2_5	-3.3_7	2.7_0
298.15	-69.9_0	23.0_1	6.86	92.9_1	34.0_8	-39.9_1	-0.0_5	0.0_3
400	-77.2_1	27.0_1	10.80	104.2_2	43.3_9	-42.4_2	13.9_7	-7.6_3
600	-89.7	35.3	21.1_5	125.0	59.3	-46.0	43.0	-15.6_7
800	-100.9	42.8	34.2	143.7	70.8	-48.1	73.1	-19.9_6
1,000	-111.2	49.3	49.3	160.5	$79._2$	-49.0	103.5	-22.6_2
1,200	$-120._7$	$54._8$	$65._8$	$175._5$	$85._4$	$-49._1$	$134._0$	-24.4_0
1,500	$-133._7$	$61._7$	$92._5$	$195._4$	93.	$-48._2$	$179._7$	-26.1_8
2-Methylpentane (C_6H_{14})								
0	0	0	0	0	0	-32.1_6	-32.1_6	∞
200	-62.3_5	17.1_8	3.43_6	79.5_3	24.2_2	-38.8_1	-13.9_3	15.2_2
273.15	-68.1_2	20.0_4	5.47_3	88.1_6	31.5_2	-40.8_9	-4.5_0	3.6_0
298.15	-69.9_2	21.1_1	6.29	91.0_3	33.9_9	-41.5_5	-1.1_3	0.8_3
400	-76.7_5	25.6_7	10.27	102.4_2	43.8_6	-44.0_3	13.0_8	-7.1_5
600	-88.9	34.5	20.7_3	123.4	60.0	-47.5	42.5	-15.4_7
800	-99.9	42.5	34.0	142.4	71.8	-49.5	72.8	-19.8_8
1,000	-110.1	49.3	49.3	159.4	$80._6$	-50.1	103.5	-22.6_1
1,200	$-119._6$	$55._1$	$66._1$	$174._7$	$87._2$	$-49._9$	$134._2$	-24.4_3
1,500	$-132._7$	$62._3$	$93._4$	$195._0$	95.	$-48._4$	$180._0$	-26.2_3

T, K	$(G° - H°_0)/T$, cal K^{-1} mol^{-1}	$(H° - H°_0)/T$, cal K^{-1} mol^{-1}	$H° - H°_0$, kcal mol^{-1}	S°, cal K^{-1} mol^{-1}	$C_p°$, cal K^{-1} mol^{-1}	$\Delta H_f°$, kcal mol^{-1}	$\Delta G_f°$, kcal mol^{-1}	$\log_{10} K_f$
				3-Methylpentane (C_6H_{14})				
0	0	0	0	0	0	-31.4_8	-31.4_8	∞
200	-63.1_3	17.1_0	3.41_9	80.2_3	23.6_4	-38.1_4	-13.4_0	14.6_4
273.15	-68.8_5	19.8_4	5.41_9	88.6_9	31.0_3	-40.2_6	-4.0_1	3.2_1
298.15	-70.6_3	20.8_9	6.23	91.5_2	33.4_9	-40.9_3	-0.6_5	0.4_8
400	-77.3_9	25.3_6	10.14	102.7_5	43.3_0	-43.4_6	13.5_1	-7.3_8
600	-89.3	34.2	20.5_0	123.5	59.5	-47.1	42.9	-15.6_1
800	-100.3	42.1	33.6	142.4	71.4	-49.1	73.2	-19.9_9
1,000	-110.4	48.9	48.9	159.3	$80._3$	-49.8	103.9	-22.7_0
1,200	$-119._9$	$54._7$	$65._6$	$174._6$	$87._0$	$-49._7$	$134._6$	-24.5_1
1,500	$-132._9$	$61._9$	$92._9$	$194._8$	95.	$-48._2$	$180._5$	-26.3_0
				2,2-Dimethylbutane (C_6H_{14})				
0	0	0	0	0	0	-34.2_8	-34.2_8	∞
200	-58.3_5	15.8_2	3.16_3	74.1_7	24.2_5	-41.2_0	-15.2_4	16.6_6
273.15	-63.7_5	19.0_0	5.19_0	82.7_5	31.3_3	-43.2_8	-5.4_2	4.3_3
298.15	-65.4_7	20.1_5	6.01	85.6_2	33.8_1	-43.9_5	-1.9_1	1.4_0
400	-72.0_5	24.9_1	9.96	96.9_6	43.7_7	-46.4_4	12.8_5	-7.0_2
600	-83.9	34.1	20.4_6	118.0	60.5	-49.9	43.3	-15.7_8
800	-94.9	42.3	33.9	137.2	73.3	-51.7	74.7	-20.4_1
1,000	-105.1	49.6	49.6	154.7	$83._2$	-51.9	106.4	-23.2_5
1,200	$-114._7$	$55._8$	$67._0$	$170._5$	$90._9$	$-51._1$	$137._9$	-25.1_2
1,500	$-128._1$	$63._8$	$95._7$	$191._9$	100.	$-48._3$	$184._9$	-26.9_9
				2,3-Dimethylbutane (C_6H_{14})				
0	0	0	0	0	0	-32.2_9	-32.2_9	∞
200	-60.8_6	15.5_6	3.11_1	76.4_2	22.2_6	-39.2_7	-13.7_6	15.0_4
273.15	-66.1_2	18.5_0	5.05_1	84.6_2	30.6_7	-41.4_4	-4.0_8	3.2_7
298.15	-67.8_0	19.6_3	5.85	87.4_3	33.3_2	-42.1_2	-0.6_2	0.4_6
400	-74.2_3	24.4_3	9.77	98.6_6	43.4_3	-44.6_5	13.9_6	-7.6_3
600	-85.9	33.6	20.1_7	119.5	59.8	-48.2	44.1	-16.0_7
800	-96.7	41.8	33.4	138.5	72.1	-50.1	75.2	-20.5_5
1,000	-106.8	48.8	48.8	155.6	$81._4$	-50.7	106.7	-23.3_1
1,200	$-116._3$	$54._8$	$65._8$	$171._1$	$88._6$	$-50._3$	$138._1$	-25.1_5
1,500	$-129._4$	$62._4$	$93._7$	$191._8$	97.	$-48._3$	$185._1$	-26.9_5

THERMODYNAMIC PROPERTIES OF ALKANE HYDROCARBONS (Continued)

T, K	$(G° - H°_0)/T$, cal K^{-1} mol^{-1}	$(H° - H°_0)/T$, cal K^{-1} mol^{-1}	$H° - H°_0$, kcal mol^{-1}	S°, cal K^{-1} mol^{-1}	$C_p°$, cal K^{-1} mol^{-1}	$\Delta H_f°$, kcal mol^{-1}	$\Delta G_f°$, kcal mol^{-1}	$\log_{10} K_f$
				n-Heptane (C_7H_{16})				
0	0	0	0	0	0	-34.8_3	-34.8_3	∞
200	-65.8_6	22.6_7	4.53_4	88.5_3	30.5_1	-41.8_3	-12.9_0	14.1_0
273.15	-73.3_5	25.5_8	6.98_8	98.9_4	36.9_6	-44.0_9	-1.9_6	1.5_7
298.15	-75.6_4	26.6_5	7.94	102.2_9	39.4_8	-44.8_4	1.9_4	-1.4_2
400	-84.1_1	31.3_0	12.52	115.4_1	50.3_5	-47.6_7	18.3_7	-10.0_4
600	-98.6	40.9	24.5_2	139.5	68.7	-51.7	52.4	-19.0_7
800	-111.6	49.6	39.6	161.2	81.8	-54.1	87.4	-23.8_8
1,000	-123.5	57.0	57.0	180.5	$91._2$	-55.1	123.0	-26.8_7
1,200	$-134._4$	$63._3$	$76._0$	$197._7$	$98._1$	$-55._2$	$158._6$	-28.8_8
1,500	$-149._4$	$71._1$	$106._6$	$220._5$	106.	$-54._4$	$212._0$	-30.8_8
				2-Methylhexane (C_7H_{16})				
0	0	0	0	0	0	-35.9_1	-35.9_1	∞
200	-66.7_7	20.3_5	4.07_0	87.1_2	28.3_4	-43.3_8	-14.1_7	15.4_8
273.15	-73.5_8	23.5_8	6.43_9	97.1_6	36.5_1	-45.7_3	-3.1_1	2.4_9
298.15	-75.7_0	24.7_8	7.39	100.4_8	39.3_2	-46.4_8	0.8_4	-0.6_1
400	-83.6_9	29.9_4	11.98	113.6_3	50.6_6	-49.3_0	17.4_5	-9.5_4
600	-97.8	40.1	24.0_6	137.9	69.2	-53.3	51.8	-18.8_6
800	-110.6	49.1	39.3	159.7	82.6	-55.5	87.2	-23.8_1
1,000	-122.4	56.9	56.9	179.3	$92._4$	-56.3	123.0	-26.8_7
1,200	$-133._4$	$63._4$	$76._1$	$196._8$	$99._8$	$-56._2$	$158._8$	-28.9_2
1,500	$-148._4$	$71._6$	$107._4$	$220._0$	108.	$-54._7$	$212._4$	-30.9_5
				3-Methylhexane (C_7H_{16})				
0	-1.38	0	0	1.38	0	-35.2_2	-35.2_2	∞
200	-68.6_8	19.9_5	3.98_9	88.6_3	27.5_8	-42.7_7	-13.8_5	15.1_4
273.15	-75.3_6	23.1_4	6.32_2	98.5_0	36.2_0	-45.1_5	-2.9_0	2.3_2
298.15	-77.4_4	24.3_6	7.26	101.8_0	39.1_0	-45.9_1	1.0_1	-0.7_4
400	-85.3_2	29.6_0	11.84	114.9_2	50.6_0	-48.7_4	17.5_0	-9.5_6
600	-99.3	39.9	23.9_2	139.2	69.2	-52.7	51.6	-18.7_8
800	-112.0	49.0	39.2	161.0	82.6	-54.9	86.7	-23.6_8
1,000	-123.8	56.7	56.7	180.5	$92._4$	-55.8	122.2	-26.7_1
1,200	$-134._8$	$63._3$	$76._0$	$198._1$	$99._8$	$-55._6$	$157._8$	-28.7_4
1,500	$-149._8$	$71._5$	$107._2$	$221._3$	108.	$-54._1$	$211._0$	-30.7_5

THERMODYNAMIC PROPERTIES OF ALKANE HYDROCARBONS (Continued)

T, K	$(G° - H°_0)/T$, cal K^{-1} mol^{-1}	$(H° - H°_0)/T$, cal K^{-1} mol^{-1}	$H° - H°_0$, kcal mol^{-1}	S°, cal K^{-1} mol^{-1}	$C_p°$, cal K^{-1} mol^{-1}	$\Delta H_f°$, kcal mol^{-1}	$\Delta G_f°$, kcal mol^{-1}	$\log_{10} K_f$
				3-Ethylpentane (C_7H_{16})				
0	0	0	0	0	0	-34.8_4	-34.8_4	∞
200	-64.1_4	20.5_0	4.09_9	84.6_4	29.9_6	-42.2_8	-12.5_6	13.7_3
273.15	-71.0_4	23.9_5	6.54_1	94.9_9	37.0_6	-44.5_5	-1.3_4	1.0_7
298.15	-73.2_0	25.1_6	7.50	98.3_6	39.6_7	-45.2_9	2.6_6	-1.9_5
400	-81.2_9	30.2_5	12.10	111.5_4	50.5_6	-48.1_0	19.4_9	-10.6_5
600	−95.5	40.2	24.1_3	135.7	68.9	−52.1	54.2	-19.7_6
800	−108.3	49.1	39.3	157.4	82.2	−54.4	90.1	-24.6_1
1,000	−120.1	56.8	56.8	176.9	$92._1$	−55.3	126.3	-27.6_1
1,200	$-131._0$	$63._3$	$76._0$	$194._3$	$99._5$	$-55._2$	$162._6$	-29.6_2
1,500	$-146._1$	$71._4$	$107._2$	$217._5$	108.	$-53._8$	$217._0$	-31.6_2
				2,2-Dimethylpentane (C_7H_{16})				
0	0	0	0	0	0	-38.0_2	-38.0_2	∞
200	-62.1_9	18.2_8	3.65_5	80.4_7	27.9_3	-45.9_0	-15.3_6	16.7_8
273.15	-68.4_3	22.0_3	6.01_7	90.4_6	36.7_9	-48.2_6	-3.8_1	3.0_5
298.15	-70.4_3	23.4_0	6.98	93.8_3	39.8_4	-49.0_0	0.3_0	-0.2_2
400	-78.1_0	29.1_4	11.66	107.2_4	51.8_1	-51.7_3	17.5_8	-9.6_1
600	−92.0	40.0	24.0_2	132.0	70.9	−55.4	53.1	-19.3_5
800	−104.9	49.6	39.7	154.5	84.9	−57.2	89.6	-24.4_8
1,000	−116.8	57.8	57.8	174.6	$95._6$	−57.5	126.4	-27.6_3
1,200	$-128._0$	$64._8$	$77._8$	$192._8$	$103._9$	$-56._6$	$163._1$	-29.7_1
1,500	$-143._5$	$73._6$	$110._4$	$217._1$	114.	$-53._8$	$217._8$	-31.7_3
				2,3-Dimethylpentane (C_7H_{16})				
0	−1.38	0	0	1.38	0	-35.2_0	-35.2_0	∞
200	-68.3_2	18.1_6	3.63_1	86.4_8	25.0_9	-43.1_1	-13.7_7	15.0_5
273.15	-74.4_4	21.4_0	5.84_6	95.8_4	35.2_8	-45.6_1	-2.6_4	2.1_1
298.15	-76.3_7	22.7_1	6.77	99.0_8	38.4_4	-46.3_9	1.3_4	-0.9_8
400	-83.8_1	28.2_8	11.31	112.0_9	50.4_4	-49.2_6	18.1_1	-9.9_0
600	−97.3	39.0	23.3_8	136.3	69.3	−53.3	52.7	-19.2_1
800	−109.9	48.4	38.7	158.3	83.1	−55.4	88.4	-24.1_6
1,000	−121.5	56.4	56.4	177.9	$93._5$	−56.1	124.5	-27.2_1
1,200	$-132._4$	$63._3$	$75._9$	$195._7$	$101._3$	$-55._7$	$160._6$	-29.2_4
1,500	$-147._5$	$71._8$	$107._7$	$219._3$	110.	$-53._6$	$214._4$	-31.2_4

T, K	$(G° - H°_0)/T$, cal K^{-1} mol^{-1}	$(H° - H°_0)/T$, cal K^{-1} mol^{-1}	$H° - H°_0$, kcal mol^{-1}	S°, cal K^{-1} mol^{-1}	$C_p°$, cal K^{-1} mol^{-1}	$\Delta H_f°$, kcal mol^{-1}	$\Delta G_f°$, kcal mol^{-1}	$log_{10} K_f$
				2,4-Dimethylpentane (C_7H_{16})				
0	0	0	0	0	0	-37.2_1	-37.2_1	∞
200	-63.1_8	18.1_2	3.62_3	81.3_0	27.5_9	-45.1_3	-14.7_5	16.1_2
273.15	-69.3_9	22.0_2	6.01_4	91.4_1	37.6_6	-47.4_5	-3.2_7	2.6_1
298.15	-71.3_9	23.4_7	7.00	94.8_6	40.8_1	-48.1_7	0.8_2	-0.6_0
400	-79.1_1	29.4_3	11.77	108.5_4	52.6_0	-50.8_0	17.9_8	-9.8_3
600	-93.2	40.3	24.2_1	133.5	70.8	-54.4	53.2	-19.3_9
800	-106.1	49.7	39.8	155.8	84.2	-56.3	89.5	-24.4_4
1,000	-118.0	57.7	57.7	175.7	$94._2$	-56.8	126.0	-27.5_4
1,200	$-129._2$	$64._4$	$77._3$	$193._6$	$101._9$	$-56._3$	$162._5$	-29.6_0
1,500	$-144._5$	$72._8$	$109._2$	$217._3$	111.	$-54._1$	$217._0$	-31.6_1
				3,3-Dimethylpentane (C_7H_{16})				
0	0	0	0	0	0	-36.6_4	-36.6_4	∞
200	-63.7_0	18.2_4	3.64_8	81.9_4	27.1_8	-44.5_3	-14.2_8	15.6_0
273.15	-69.9_1	21.9_0	5.98_2	91.8_1	36.5_6	-46.9_1	-2.8_4	2.2_7
298.15	-71.8_9	23.2_7	6.94	95.1_6	39.6_2	-47.6_6	1.2_4	-0.9_1
400	-79.5_1	28.9_7	11.59	108.4_8	51.4_7	-50.4_2	18.3_9	-10.0_5
600	-93.3	39.8	23.8_9	133.1	70.7	-54.2	53.7	-19.5_6
800	-106.1	49.4	39.5	155.5	84.9	-56.0	90.0	-24.5_8
1,000	-118.1	57.6	57.6	175.7	$95._7$	-56.3	126.6	-27.6_6
1,200	$-129._2$	$64._7$	$77._6$	$193._9$	$104._1$	$-55._4$	$163._0$	-29.6_9
1,500	$-144._7$	$73._6$	$110._4$	$218._3$	114.	$-52._4$	$217._3$	-31.6_6
				2,2,3-Trimethylbutane (C_7H_{16})				
0	0	0	0	0	0	-37.4_2	-37.4_2	∞
200	-61.3_4	17.3_0	3.45_9	78.6_4	26.6_4	-45.5_0	-14.5_9	15.9_4
273.15	-67.2_7	21.0_5	5.74_9	88.3_2	35.9_6	-47.9_3	-2.9_0	2.3_2
298.15	-69.1_8	22.4_4	6.69	91.6_2	39.0_2	-48.6_9	1.2_7	-0.9_3
400	-76.5_7	28.2_0	11.28	104.7_7	50.8_7	-51.5_1	18.7_9	-10.2_7
600	-90.1	39.1	23.4_7	129.2	70.2	-55.4	54.9	-19.9_8
800	-102.7	48.8	39.0	151.5	84.9	-57.3	91.9	-25.1_2
1,000	-114.5	57.2	57.2	171.7	$96._3$	-57.5	129.3	-28.2_6
1,200	$-125._6$	$64._5$	$77._4$	$190._1$	$105._2$	$-56._4$	$166._6$	-30.3_4
1,500	$-141._0$	$73._7$	$110._6$	$214._7$	115.	$-53._0$	$222._0$	-32.3_4

T, K	$(G° - H°_0)/T$, cal K^{-1} mol^{-1}	$(H° - H°_0)/T$, cal K^{-1} mol^{-1}	$H° - H°_0$, kcal mol^{-1}	S°, cal K^{-1} mol^{-1}	$C_p°$, cal K^{-1} mol^{-1}	$\Delta H_f°$, kcal mol^{-1}	$\Delta G_f°$, kcal mol^{-1}	$\log_{10} K_f$
				n-Octane (C_8H_{18})				
0	0	0	0	0	0	-38.5_7	-38.5_7	∞
200	-70.2_6	25.7_9	5.15_8	96.0_5	34.6_0	-46.4_0	-13.1_2	14.3_4
273.15	-78.7_7	29.0_9	7.94_5	107.8_6	41.9_9	-48.9_4	-0.5_6	0.4_5
298.15	-81.3_8	30.2_9	9.03	111.6_7	44.8_8	-49.7_7	3.9_2	-2.8_7
400	-91.0_0	35.5_9	14.24	126.5_9	57.3_0	-52.9_3	22.7_7	-12.4_4
600	−107.5	46.5	27.8_8	154.0	78.1	−57.4	61.7	-22.4_7
800	−122.3	56.3	45.1	178.6	92.8	−60.0	101.8	-27.8_2
1,000	−135.7	64.7	64.7	200.4	$103._1$	−61.2	142.4	-31.1_3
1,200	$-148._2$	$71._8$	$86._1$	$220._0$	$110._7$	$-61._4$	$183._2$	-33.3_6
1,500	$-165._2$	$80._5$	$120._7$	$245._7$	119.	$-60._5$	$244._3$	-35.5_9
				2-Methylheptane (C_8H_{18})				
0	0	0	0	0	0	-39.6_5	-39.6_5	∞
200	-71.1_5	23.4_7	4.69_3	94.6_2	32.4_2	-47.9_5	-14.3_9	15.7_2
273.15	-78.9_9	27.0_7	7.39_4	106.0_6	41.5_6	-50.5_7	-1.7_0	1.3_6
298.15	-81.4_2	28.4_2	8.47	109.8_4	44.7_5	-51.4_1	2.8_2	-2.0_7
400	-90.5_7	34.2_4	13.70	124.8_1	57.6_6	-54.5_5	21.8_6	-11.9_4
600	−106.7	45.7	27.4_4	152.4	78.7	−59.0	61.1	-22.2_6
800	−121.3	55.9	44.8	177.2	93.6	−61.4	101.5	-27.7_4
1,000	−134.7	64.6	64.6	199.3	$104._4$	−62.4	142.4	-31.1_3
1,200	$-147._1$	$72._0$	$86._3$	$219._1$	$112._5$	$-62._2$	$183._4$	-33.4_0
1,500	$-164._2$	$81._0$	$121._5$	$245._2$	122.	$-60._8$	$244._6$	-35.6_4
				3-Methylheptane (C_8H_{18})				
0	−1.38	0	0	1.38	0	-38.9_7	-38.9_7	∞
200	-73.1_0	23.1_3	4.62_5	96.2_3	31.7_1	-47.3_4	-14.0_9	15.4_0
273.15	-80.8_2	26.6_9	7.28_9	107.5_1	41.1_8	-50.0_0	-1.5_2	1.2_1
298.15	-83.2_2	28.0_4	8.36	111.2_6	44.4_1	-50.8_4	2.9_7	-2.1_8
400	-92.2_6	33.8_8	13.55	126.1_4	57.4_0	-54.0_1	21.8_6	-11.9_5
600	−108.2	45.4	27.2_5	153.6	78.4	−58.5	60.9	-22.1_7
800	−122.7	55.7	44.5	178.4	93.5	−60.9	101.0	-27.6_1
1,000	−136.1	64.4	64.4	200.5	$104._3$	−61.9	141.7	-30.9_7
1,200	$-148._5$	$71._7$	$86._1$	$220._2$	$112._4$	$-61._8$	$182._4$	-33.2_2
1,500	$-165._5$	$80._8$	$121._2$	$246._3$	122.	$-60._4$	$243._4$	-35.4_6

T, K	$(G^\circ - H^\circ_0)/T$, cal K^{-1} mol^{-1}	$(H^\circ - H^\circ_0)/T$, cal K^{-1} mol^{-1}	$H^\circ - H^\circ_0$, kcal mol^{-1}	S°, cal K^{-1} mol^{-1}	C_p°, cal K^{-1} mol^{-1}	ΔH_f°, kcal mol^{-1}	ΔG_f°, kcal mol^{-1}	$\log_{10} K_f$
			4-Methylheptane (C_8H_{18})					
0	0	0	0	0	0	-38.9_6	-38.9_6	∞
200	-71.4_7	22.8_0	4.56_0	94.2_7	31.5_2	-47.3_9	-13.7_6	15.0_4
273.15	-79.1_0	26.4_5	7.22_5	105.5_5	41.3_7	-50.0_5	-1.0_4	0.8_3
298.15	-81.4_8	27.8_4	8.30	109.3_2	44.7_0	-50.8_9	3.5_0	-2.5_7
400	-90.4_8	33.8_5	13.54	124.3_3	57.9_2	-54.0_2	22.5_8	-12.3_4
600	−106.4	45.6	27.3_4	152.0	79.0	−58.4	61.9	-22.5_5
800	−121.0	55.9	44.7	176.9	93.9	−60.8	102.4	-27.9_8
1,000	−134.5	64.6	64.6	199.1	$104._6$	−61.7	143.4	-31.3_3
1,200	$-146._9$	$72._0$	$86._4$	$218._9$	$112._6$	$-61._5$	$184._3$	-33.5_7
1,500	$-164._0$	$81._0$	$121._6$	$245._0$	122.	$-60._0$	$245._7$	-35.7_9
			2,2-Dimethylhexane (C_8H_{18})					
0	0	0	0	0	0	-41.7_7	-41.7_7	∞
200	-66.6_3	21.5_8	4.31_5	88.2_1	32.2_0	-50.4_5	-15.6_1	17.0_5
273.15	-73.9_5	25.6_7	7.01_0	99.6_2	41.6_8	-53.0_8	-2.4_4	1.9_6
298.15	-76.2_7	27.1_5	8.10	103.4_2	45.0_0	-53.9_1	2.2_4	-1.6_4
400	-85.1_1	33.4_2	13.37	118.5_3	58.2_8	-57.0_0	21.9_2	-11.9_8
600	−101.0	45.5	27.2_8	146.5	79.8	−61.2	62.4	-22.7_3
800	−115.6	56.1	44.9	171.7	95.6	−63.4	104.0	-28.4_0
1,000	−129.1	65.3	65.3	194.4	$107._4$	−63.8	145.9	-31.8_9
1,200	$-141._7$	$73._1$	$87._7$	$214._8$	$116._4$	$-63._0$	$187._8$	-34.2_0
1,500	$-159._1$	$82._8$	$124._2$	$241._9$	127.	$-60._2$	$250._2$	-36.4_5
			3,3-Dimethylhexane (C_8H_{18})					
0	0	0	0	0	0	-40.3_8	-40.3_8	∞
200	-68.9_2	20.7_0	4.14_0	89.6_2	30.8_6	-49.2_4	-14.6_7	16.0_3
273.15	-75.9_7	24.9_3	6.80_7	100.9_0	42.0_0	-51.8_9	-1.6_1	1.2_8
298.15	-78.2_3	26.5_1	7.90	104.7_4	45.6_2	-52.7_1	3.0_5	-2.2_3
400	-86.9_4	33.1_8	13.27	120.1_2	59.4_4	-55.7_1	22.5_8	-12.3_4
600	−102.8	45.7	27.4_3	143.5	81.0	−59.7	62.7	-22.8_3
800	−117.5	56.6	45.2	174.1	96.5	−61.6	103.8	-28.3_6
1,000	−131.1	65.8	65.8	196.9	$108._1$	−61.9	145.3	-31.7_5
1,200	$-143._8$	$73._6$	$88._3$	$217._4$	$117._0$	$-61._0$	$186._6$	-33.9_8
1,500	$-161._3$	$83._4$	$125._0$	$244._7$	127.	$-58._0$	$248._2$	-36.1_6

T, K	$(G° - H°_0)/T$, cal K^{-1} mol^{-1}	$(H° - H°_0)/T$, cal K^{-1} mol^{-1}	$H° - H°_0$, kcal mol^{-1}	S°, cal K^{-1} mol^{-1}	$C_p°$, cal K^{-1} mol^{-1}	$\Delta H_f°$, kcal mol^{-1}	$\Delta G_f°$, kcal mol^{-1}	log_{10} K_f
				2,2,4-Trimethylpentane (Isooctane) (C_8H_{18})				
0	0	0	0	0	0	-41.2_3	-41.2_3	∞
200	-66.3_0	19.7_3	3.94_5	86.0_3	31.2_1	-50.2_8	-15.0_0	16.3_9
273.15	-73.1_0	24.1_8	6.60_6	97.2_8	41.5_8	-52.9_4	-1.6_7	1.3_3
298.15	-75.2_9	25.8_0	7.69	101.0_9	45.0_3	-53.7_7	3.0_7	-2.2_5
400	-83.7_9	32.4_4	12.98	116.2_3	58.4_6	-56.8_5	22.9_9	-12.5_6
600	−99.3	44.9	26.9_4	144.2	80.2	−61.0	63.9	-23.2_8
800	−113.8	55.8	44.7	169.6	96.4	−63.1	105.9	-28.9_4
1,000	−127.3	65.2	65.2	192.5	$108._7$	−63.3	148.2	-32.4_0
1,200	$-139._9$	$73._3$	$88._0$	$213._2$	$118._2$	$-62._2$	$190._4$	-34.6_8
1,500	$-157._4$	$83._4$	$125._1$	$240._8$	129.	$-58._7$	$253._2$	-36.9_0
				n-Nonane (C_9H_{20})				
0	0	0	0	0	0	-42.3_2	-42.3_2	∞
200	-74.6_4	28.9_2	5.78_3	103.5_6	38.7_0	-50.9_8	-13.3_6	14.6_0
273.15	-84.2_0	32.5_9	8.90_2	116.7_9	47.0_3	-53.7_9	0.8_4	-0.6_7
298.15	-87.1_0	33.9_4	10.12	121.0_4	50.2_9	-54.7_1	5.9_0	-4.3_2
400	-97.8_9	39.8_8	15.95	137.7_7	64.2_5	-58.2_0	27.1_6	-14.8_4
600	−116.4	52.1	31.2_5	168.5	87.5	−63.2	71.0	-25.8_6
800	−132.9	63.1	50.5	196.0	103.7	−66.0	116.2	-31.7_4
1,000	−148.0	72.4	72.4	220.4	$115._1$	−67.3	161.9	-35.3_9
1,200	$-161._9$	$80._3$	$96._3$	$242._2$	$123._4$	$-67._5$	$207._8$	-37.8_4
1,500	$-180._9$	$89._9$	$134._8$	$270._8$	133.	$-66._7$	$276._5$	-40.2_9
				2-Methyloctane (C_9H_{20})				
0	0	0	0	0	0	-43.4_0	-43.4_0	∞
200	-75.5_4	26.5_9	5.31_8	102.1_3	36.5_2	-52.5_3	-14.6_2	15.9_8
273.15	-84.4_0	30.5_8	8.35_2	114.9_8	46.6_0	-55.4_2	-0.3_0	0.2_4
298.15	-87.1_5	32.0_7	9.56	119.2_2	50.1_5	-56.3_5	4.8_0	-3.5_2
400	-97.4_5	38.5_4	15.41	135.9_9	64.6_1	-59.8_2	26.2_5	-14.3_4
600	−115.5	51.4	30.8_1	166.9	88.1	−64.7	70.4	-25.6_6
800	−131.9	62.7	50.2	194.6	104.6	−67.4	115.9	-31.6_7
1,000	−146.9	72.4	72.4	219.3	$116._4$	−68.4	161.9	-35.3_8
1,200	$-160._9$	$80._4$	$96._5$	$241._3$	$125._1$	$-68._4$	$208._0$	-37.8_7
1,500	$-180._0$	$90._4$	$135._6$	$270._4$	135.	$-66._9$	$276._9$	-40.3_5

THERMODYNAMIC PROPERTIES OF ALKANE HYDROCARBONS (Continued)

T, K	$(G° - H°_0)/T$, cal K^{-1} mol^{-1}	$(H° - H°_0)/T$, cal K^{-1} mol^{-1}	$H° - H°_0$, kcal mol^{-1}	S°, cal K^{-1} mol^{-1}	$C_p°$, cal K^{-1} mol^{-1}	$\Delta H_f°$, kcal mol^{-1}	$\Delta G_f°$, kcal mol^{-1}	$\log_{10} K_f$
				3-Methyloctane (C_9H_{20})				
0	−1.38	0	0	1.38	0	-42.7_1	-42.7_1	∞
200	-77.4_8	26.2_4	5.24_8	103.7_2	35.7_9	-51.9_1	-14.3_2	15.6_4
273.15	-86.2_3	30.1_8	8.24_4	116.4_1	46.2_4	-54.8_3	-0.1_0	0.0_8
298.15	-88.9_3	31.6_9	9.45	120.6_2	49.8_5	-55.7_7	4.9_6	-3.6_4
400	-99.1_4	38.1_8	5.00	137.3_2	64.4_0	-59.2_6	26.2_7	-14.3_5
600	−117.1	51.0	30.6_3	168.1	87.9	−64.2	70.2	-25.5_7
800	−133.4	62.4	50.0	195.8	104.5	−66.9	115.4	-31.5_4
1,000	−148.4	72.1	72.1	220.5	$116._3$	−68.0	161.1	-35.2_2
1,200	$-162._3$	$80._2$	$96._3$	$242._5$	$125._0$	$-67._9$	$207._0$	-37.7_0
1,500	$-181._3$	$90._2$	$135._4$	$271._5$	135.	$-66._5$	$275._6$	-40.1_6
				4-Methyloctane (C_9H_{20})				
0	−1.38	0	0	1.38	0	-42.7_0	-42.7_0	∞
200	-77.2_7	25.9_7	5.19_4	103.2_4	35.6_4	-51.9_5	-14.2_6	15.5_9
273.15	-85.9_4	29.9_9	8.19_0	115.9_3	46.3_5	-54.8_8	-0.0_2	0.0_1
298.15	-88.6_3	31.5_2	9.40	120.1_5	50.0_2	-55.8_1	5.0_6	-3.7_1
400	-98.8_0	38.1_2	15.25	136.9_2	64.7_1	-59.2_8	26.4_2	-14.4_3
600	−116.8	51.1	30.6_7	167.9	88.2	−64.1	70.4	-25.6_5
800	−133.1	62.5	50.1	195.6	104.7	−66.8	115.7	-31.6_0
1,000	−148.1	72.2	72.2	220.3	$116._5$	−67.8	161.5	-35.2_9
1,200	$-162._0$	$80._4$	$96._4$	$242._4$	$125._2$	$-67._8$	$207._3$	-37.7_6
1,500	$-181._0$	$90._4$	$135._5$	$271._4$	135.	$-66._3$	$276._0$	-40.2_1
				3-Ethylheptane (C_9H_{20})				
0	0	0	0	0	0	-42.3_4	-42.3_4	∞
200	-74.9_0	26.5_0	5.30_0	101.4_0	38.0_7	-51.4_8	-13.4_3	14.6_7
273.15	-83.8_0	30.7_9	8.41_0	114.5_9	47.2_7	-54.3_0	0.9_3	-0.7_4
298.15	-86.5_6	32.3_2	9.64	118.8_7	50.6_4	-55.2_1	6.0_4	-4.4_3
400	-96.9_4	38.7_8	15.51	135.7_2	64.6_2	-58.6_6	27.5_2	-15.0_4
600	−115.1	51.4	30.8_6	166.5	87.7	−63.6	71.8	-26.1_4
800	−131.5	62.6	50.1	194.1	104.2	−66.3	117.3	-32.0_5
1,000	−146.5	72.2	72.2	218.7	$116._0$	−67.5	163.4	-35.7_2
1,200	$-160._4$	$80._3$	$96._3$	$240._7$	$124._8$	$-67._5$	$209._6$	-38.1_7
1,500	$-179._5$	$90._2$	$135._4$	$269._7$	135.	$-66._1$	$278._8$	-40.6_1

THERMODYNAMIC PROPERTIES OF ALKANE HYDROCARBONS (Continued)

T, K	$(G° - H°_0)/T$, cal K^{-1} mol^{-1}	$(H° - H°_0)/T$, cal K^{-1} mol^{-1}	$H° - H°_0$, kcal mol^{-1}	S°, cal K^{-1} mol^{-1}	$C_p°$, cal K^{-1} mol^{-1}	$\Delta H_f°$, kcal mol^{-1}	$\Delta G_f°$, kcal mol^{-1}	$\log_{10} K_f$
			4-Ethylheptane (C_9H_{20})					
0	0	0	0	0	0	-42.3_4	-42.3_4	∞
200	-74.6_3	26.1_7	5.23_3	100.8_0	37.9_3	-51.5_6	-13.3_8	14.6_2
273.15	-83.4_3	30.5_7	8.34_9	114.0_0	47.5_2	-54.3_6	1.0_2	-0.8_2
298.15	-86.1_8	32.1_3	9.58	118.3_1	50.9_8	-55.2_7	6.1_5	-4.5_1
400	-96.5_3	38.7_5	15.50	135.2_8	65.0_9	-58.6_7	27.6_8	-15.1_2
600	−114.7	51.6	30.9_4	166.3	88.1	−63.5	72.0	-26.2_2
800	−131.1	62.9	50.3	194.0	104.5	−66.2	117.6	-32.1_3
1,000	−146.2	72.4	72.4	218.6	$116._2$	−67.3	163.7	-35.7_8
1,200	$-160._1$	$80._5$	$96._6$	$240._6$	$124._9$	$-67._3$	$209._9$	-38.2_3
1,500	$-179._2$	$90._4$	$135._6$	$269._6$	135.	$-65._9$	$279._1$	-40.6_6
			n-Decane					
0	0	0	0	0	0	-46.0_6	-46.0_6	∞
200	-79.0_2	32.0_5	6.40_8	111.0_7	42.8_0	-55.5_6	-13.5_8	14.8_4
273.15	-89.6_1	36.1_0	9.86_1	125.7_1	52.0_8	-58.6_3	2.2_5	-1.8_0
298.15	-92.8_3	37.5_9	11.21	130.4_2	55.7_0	-59.6_4	7.8_8	-5.7_8
400	-104.7_7	44.1_8	17.67	148.9_5	71.2_2	-63.4_5	31.5_6	-17.2_4
600	−125.3	57.7	34.6_3	183.0	97.0	−68.8	80.3	-29.2_6
800	−143.6	69.9	55.9	213.5	114.7	−71.9	130.6	-35.6_7
1,000	−160.3	80.2	80.2	240.5	$127._1$	−73.3	181.4	-39.6_4
1,200	$-175._7$	$88._8$	$106._5$	$264._5$	$136._1$	$-73._6$	$232._4$	-42.3_2
1,500	$-196._7$	$99._3$	$149._0$	$296._0$	146.	$-72._8$	$308._8$	-44.9_9
			2-Methylnonane ($C_{10}H_{22}$)					
0	0	0	0	0	0	-47.1_5	-47.1_5	∞
200	-79.9_2	29.7_2	5.94_4	109.6_4	40.6_2	-57.1_0	-14.8_5	16.2_2
273.15	-89.8_1	34.0_9	9.31_0	123.9_0	51.6_5	-60.2_6	1.1_0	-0.8_8
298.15	-92.8_7	35.7_2	10.65	128.5_9	55.5_6	-61.2_8	6.7_8	-4.9_7
400	-104.3_4	42.8_4	17.13	147.1_8	71.5_8	-65.0_8	30.6_4	-16.7_4
600	−124.4	57.0	34.1_9	181.4	97.5	−70.4	79.8	-29.0_5
800	−142.6	69.5	55.6	212.1	115.6	−73.3	130.3	-35.6_0
1,000	−159.2	80.1	80.1	239.3	$128._4$	−74.5	181.4	-39.6_4
1,200	$-174._7$	$88._9$	$106._8$	$263._6$	$137._8$	$-74._5$	$232._5$	-42.3_5
1,500	$-195._7$	$99._9$	$149._8$	$295._6$	149.	$-73._0$	$309._2$	-45.0_4

THERMODYNAMIC PROPERTIES OF ALKANE HYDROCARBONS (Continued)

T, K	$(G^\circ - H^\circ_0)/T$, cal K^{-1} mol^{-1}	$(H^\circ - H^\circ_0)/T$, cal K^{-1} mol^{-1}	$H^\circ - H^\circ_0$, kcal mol^{-1}	S°, cal K^{-1} mol^{-1}	C_p°, cal K^{-1} mol^{-1}	ΔH_f°, kcal mol^{-1}	ΔG_f°, kcal mol^{-1}	$\log_{10} K_f$
			3-Methylnonane ($C_{10}H_{22}$)					
0	−1.38	0	0	1.38	0	-46.4_5	-46.4_5	∞
200	-81.8_6	29.3_7	5.87_3	111.2_3	39.8_8	-56.4_8	-14.5_4	15.8_9
273.15	-91.6_4	33.6_9	9.20_2	125.3_3	51.2_8	-59.6_7	1.3_0	-1.0_4
298.15	-94.6_6	35.3_3	10.54	129.9_9	55.2_5	-60.7_0	6.9_5	-5.0_9
400	-106.0_2	42.4_8	16.99	148.5_0	71.3_5	-64.5_2	30.6_7	-16.7_6
600	−125.9	56.7	34.0_0	182.6	97.3	−69.9	79.5	-28.9_7
800	−144.1	69.2	55.4	213.3	115.4	−72.8	129.8	-35.4_7
1,000	−160.7	79.8	79.8	240.5	$128._3$	−74.0	180.7	-39.4_8
1,200	$-176._0$	$88._7$	$106._5$	$264._7$	$137._7$	$-74._0$	$231._6$	-42.1_8
1,500	$-197._1$	$99._6$	$149._5$	$296._7$	149.	$-72._6$	$307._9$	-44.8_6
			4-Methylnonane ($C_{10}H_{22}$)					
0	−1.38	0	0	1.38	0	-46.4_5	-46.4_5	∞
200	-81.6_5	29.0_9	5.81_7	110.7_4	39.7_2	-56.5_3	-14.5_0	15.8_4
273.15	-91.3_5	33.4_8	9.14_6	124.8_3	51.4_1	-59.7_3	1.3_8	-1.1_1
298.15	-94.3_5	35.1_6	10.48	129.5_1	55.4_5	-60.7_5	7.0_4	-5.1_6
400	-105.6_8	42.4_2	16.97	148.1_0	71.7_1	-64.5_4	30.8_1	-16.8_3
600	−125.6	56.8	34.0_5	182.4	97.7	−69.8	79.7	-29.0_4
800	−143.7	69.4	55.5	213.1	115.7	−72.7	130.1	-35.5_3
1,000	−160.4	80.0	80.0	240.4	$128._5$	−73.9	180.9	-39.5_4
1,200	$-175._8$	$88._9$	$106._7$	$264._7$	$137._8$	$-73._8$	$231._9$	-42.2_3
1,500	$-196._8$	$99._8$	$149._7$	$296._6$	149.	$-72._4$	$308._2$	-44.9_0
			5-Methylnonane ($C_{10}H_{22}$)					
0	0	0	0	0	0	-46.4_3	-46.4_3	∞
200	-80.3_3	29.1_5	5.82_9	109.4_8	39.7_5	-56.5_1	-14.2_2	15.5_4
273.15	-90.0_4	33.5_2	9.15_5	123.5_6	51.3_1	-59.7_0	1.7_6	-1.4_0
298.15	-93.0_5	35.1_8	10.49	128.2_3	55.3_2	-60.7_3	7.4_4	-5.4_6
400	-104.3_8	42.3_9	16.96	146.7_7	71.5_2	-64.5_4	31.3_4	-17.1_3
600	−124.3	56.7	34.0_0	181.0	97.5	−69.8	80.5	-29.3_4
800	−142.4	69.3	55.4	211.7	115.6	−72.8	131.2	-35.8_3
1,000	−159.0	79.9	79.9	238.9	$128._4$	−73.9	182.3	-39.8_4
1,200	$-174._4$	$88._8$	$106._6$	$263._2$	$137._8$	$-73._9$	$233._6$	-42.5_4
1,500	$-195._4$	$99._7$	$149._6$	$295._1$	149.	$-72._5$	$310._3$	-45.2_1

Data from Scott, D. W., U.S. Bureau of Mines Bulletin 666 (stock number 2404-01547), U.S. Government Printing Office, Washington, D.C., 1974.

HEAT OF DILUTION OF ACIDS

From National Standards Reference Data Systems NSRDS-NBS 2
Vivian B. Parker

ΔH_{diln}, the integral heat of dilution, is the change in enthalpy, per mole of solute, when a solution of concentration m_1 is diluted to a final finite concentration m_2. When the dilution is carried out by addition of an infinite amount of solvent, so the final solution is infinitely dilute, the enthalpy change is the integral heat of dilution to infinite dilution. Since Φ_L, the relative apparent molal enthalpy, is equal to and opposite in sign to this, only Φ_L is referred to here.

Φ_L, cal/mole, at 25°C

n	m	HF	HCl	$HClO_4$	HBr	HI	HNO_3	CH_2O_2	$C_2H_4O_2$
∞	0.00	0	0	0	0	0	0	0	0
500,000	.000111	300	5	5	5	5	5	9	40
100,000	.000555	900	10	10	9	9	11	13	50
50,000	.00111	1,300	16	14	13	12	15	20	53
20,000	.00278	1,800	25	22	22	20	23	23	55
10,000	.00555	2,130	34	30	31	29	31	25	58
7,000	.00793	2,250	40	35	37	34	36	26	59
5,000	.01110	2,360	47	40	44	41	42	26	61
4,000	.01388	2,450	54	43	49	46	46	27	62
3,000	.01850	2,550	60	47	56	52	51	28	62
2,000	.02775	2,700	74	54	68	63	59	28	63
1,500	.03700	2,812	85	58	77	71	65	29	64
1,110	.05000	2,927	97	62	89	81	73	29	65
1,000	.05551	2,969	102	62	92	84	76	29	65
900	.0617	2,989	107	63	97	88	78	30	66
800	.0694	3,015	113	64	102	92	81	31	67
700	.0793	3,037	120	65	108	96	84	32	68
600	.0925	3,057	129	65	115	102	88	32	68
555.1	.1000	3,060	133	65	119	105	89	32	69
500	.1110	3,077	140	65	124	108	92	32	70
400	.1388	3,097	156	64	135	116	97	33	72
300	.1850	3,126	176	61	150	125	103	34	76
277.5	.2000	3,129	182	59	155	128	105	35	79
200	.2775	3,142	212	50	176	140	117	36	82
150	.3700	3,148	242	36	197	154	118	39	88
111.0	.5000	3,156	280	18	225	170	119	42	97
100	.5551	3,160	295	+12	235	176	120	44	101
75	.7401	3,167	343	−14	270	194	121	49	113
55.51	1.0000	3,179	405	−48	314	223	121	54	130
50	1.1101	3,184	431	−61	331	234	121	56	147
40	1.3877	3,192	493	−91	379	260	121	60	155
37.00	1.5000	3,194	518	−103	398	269	121	62	162
30	1.8502	3,200	595	−138	455	301	124	65	183
27.75	2.0000	3,203	627	−149	477	315	126	66	192
25	2.2202	3,208	674	−162	510	336	130	67	204
22.20	2.5000	3,211	732	−173	550	365	139	68	218
20	2.7753	3,214	792	−182	590	396	149	69	233
18.50	3.0000	3,216	838	−187	624	427	159	69	245
15.86	3.500	3,221	946	−196	709	503	189	69	268
15	3.7004	3,227	988	−195	743	536	203	69	277
13.88	4.0000	3,234	1,052	−188	796	588	229	69	291
12.33	4.5000	3,246	1,171	−175	887	676	265	69	313
12	4.6255	3,249	1,190	−170	911	700	277	69	318
11.10	5.0000	3,256	1,271	−150	983	764	313	69	333
10	5.5506	3,265	1,396	−117	1,097	855	368	68	353
9.5	5.8427	3,269	1,462	−97	1,156	920	400	68	363
9.251	6.0000	3,272	1,498	−84	1,196	950	418	67	368
9.0	6.1674	3,274	1,535	−72	1,230	980	437	67	373
8.5	6.5301	3,278	1,618	−40	1,313	1,050	480	66	383
8.0	6.9383	3,282	1,710	+4	1,401	1,115	530	65	392
7.929	7.0000	3,283	1,725	11	1,416	1,130	538	65	394

HEAT OF DILUTION OF ACIDS (Continued)

n	m	HF	HCl	$HClO_4$	HBr	HI	HNO_3	CH_2O_2	$C_2H_4O_2$
7.5	7.4008	3,286	1,820	61	1,497	1,210	595	63	402
7.0	7.9295	3,290	1,942	135	1,608	1,325	661	61	411
6.938	8.0000	3,291	1,960	146	1,622	1,340	667	61	412
6.5	8.5394	3,296	2,090	229	1,738	1,450	745	58	420
6.167	9.0000	3,302	2,202	306	1,845	1,570	805	55	426
6.0	9.2510	3,305	2,265	348	1,903	1,630	840	53	429
5.551	10.0000	3,316	2,447	481	2,078	1,820	940	49	436
5.5	10.0920	3,317	2,472	499	2,102	1,850	950	49	437
5.0	11.1012	3,335	2,721	730	2,344	2,100	1,098	43	445
4.5	12.3346	3,362	3,025	1,144	2,655	2,460	1,270	37	453
4.0	13.8765	3,400	3,404	1,574	3,089	2,960	1,495	29	462
3.700	15.0000	3,428	3,680	1,893	3,415	3,350	1,645	26	469
3.5	15.8589	3,450	3,882	2,150	3,668	3,660	1,770	21	473
3.25	17.0788	3,483	4,160	2,460	4,005	4,110	1,920	17	481
3.0	18.5020	3,520	4,460	2,880	4,370	4,630	2,101	13	488
2.775	20.0000	3,557	4,750	3,300	4,760	5,190	2,270	9	496
2.5	22.2024	3,607	5,180	4,000	5,300	6,000	2,520	+4	506
2.0	27.7530	3,712	6,260	5,500	6,650	.....	3,060	−5	528
1.5	37.0040	.....	8,240	.....	8,530	.....	3,770	−13	532
1.0	55.506	.....	10,900	.....	11,670	.....	4,715	+11	518
0.5	111.012	.....	.....	.....	.....	.....	.....	77	495
0.25	222.02	.....	.....	.....	.....	.....	.....	129	.....

HEATS OF SOLUTION

From National Standards Reference Data Systems NSRDS-NBS 2

Vivian B. Parker

ΔH°_∞ 25°C for uni-univalent electrolytes in H_2O

Substance	State	ΔH°_∞	Substance	State	ΔH°_∞	Substance	State	ΔH°_∞
		cal/mole			*cal/mole*			*cal/mole*
HF	g	−14,700	$LiBr \cdot 2H_2O$	c	−2,250	KCl	c	4,115
HCl	g	−17,888	$LiBrO_3$	c	340	$KClO_3$	c	9,890
$HClO_4$	l	−21,215	LiI	c	−15,130	$KClO_4$	c	12,200
$HClO_4 \cdot H_2O$	c	−7,875	$LiI \cdot H_2O$	c	−7,090	KBr	c	4,750
HBr	g	−20,350	$LiI \cdot 2H_2O$	c	−3,530	$KBrO_3$	c	9,830
HI	g	−19,520	$LiI \cdot 3H_2O$	c	140	KI	c	4,860
HIO_3	c	2,100	$LiNO_2$	c	−2,630	KIO_3	c	6,630
HNO_3	l	−7,954	$LiNO_2 \cdot H_2O$	c	1,680	KNO_2	c	3,190
HCOOH	l	−205	$LiNO_3$	c	−600	KNO_3	c	8,340
CH_3COOH	l	−360				$KC_2H_3O_2$	c	−3,665
			NaOH	c	−10,637	KCN	c	2,800
NH_3	g	−7,290	$NaOH \cdot H_2O$	c	−5,118	KCNO	c	4,840
NH_4Cl	c	3,533	NaF	c	218	KCNS	c	5,790
NH_4ClO_4	c	8,000	NaCl	c	928	$KMnO_4$	c	10,410
NH_4Br	c	4,010	$NaClO_2$	c	80			
NH_4I	c	3,280	$NaClO_2 \cdot 3H_2O$	c	6,830	RbOH	c	−14,900
NH_4IO_3	c	7,600	$NaClO_3$	c	5,191	$RbOH \cdot H_2O$	c	−4,310
NH_4NO_2	c	4,600	$NaClO_4$	c	3,317	$RbOH \cdot 2H_2O$	c	210
NH_4NO_3	c	6,140	$NaClO_4 \cdot H_2O$	c	5,380	RbF	c	−6,240
$NH_4C_2H_3O_2$	c	−570	NaBr	c	−144	$RbF \cdot H_2O$	c	−100
NH_4CN	c	4,200	$NaBr \cdot 2H_2O$	c	4,454	$RbF \cdot 1\frac{1}{2}H_2O$	c	320
NH_4CNS	c	5,400	$NaBrO_3$	c	6,430	RbCl	c	4,130
CH_3NH_3Cl	c	1,378	NaI	c	−1,800	$RbClO_3$	c	11,410
$(CH_3)_3NHCl$	c	350	$NaI \cdot 2H_2O$	c	3,855	$RbClO_4$	c	13,560
$N(CH_3)_4Cl$	c	975	$NaIO_3$	c	4,850	RbBr	c	5,230
$N(CH_3)_4Br$	c	5,800	$NaNO_2$	c	3,320	$RbBrO_3$	c	11,700
$N(CH_3)_4I$	c	10,055	$NaNO_3$	c	4,900	RbI	c	6,000
			$NaC_2H_3O_2$	c	−4,140	$RbNO_3$	c	8,720
$AgClO_4$	c	1,760	$NaC_2H_3O_2 \cdot 3H_2O$	c	4,700			
$AgNO_2$	c	8,830	NaCN	c	290	CsOH	c	−17,100
$AgNO_3$	c	5,400	$NaCN \cdot \frac{1}{2}H_2O$	c	790	$CsOH \cdot H_2O$	c	−4,900
			$NaCN \cdot 2H_2O$	c	4,440	CsF	c	−8,810
LiOH	c	−5,632	NaCNO	c	4,590	$CsF \cdot H_2O$	c	−2,500
$LiOH \cdot H_2O$	c	−1,600	NaCNS	c	1,632	$CsF \cdot 1\frac{1}{2}H_2O$	c	−1,300
LiF	c	1,130				CsCl	c	4,250
LiCl	c	−8,850	KOH	c	−13,769	$CsClO_4$	c	13,250
$LiCl \cdot H_2O$	c	−4,560	$KOH \cdot H_2O$	c	−3,500	CsBr	c	6,210
$LiClO_4$	c	−6,345	$KOH \cdot 1\frac{1}{2}H_2O$	c	−2,500	$CsBrO_3$	c	12,060
$LiClO_4 \cdot 3H_2O$	c	7,795	KF	c	−4,238	CsI	c	7,970
LiBr	c	−11,670	$KF \cdot 2H_2O$	c	1,666	$CsNO_3$	c	9,560
$LiBr \cdot H_2O$	c	−5,560						

HEAT CAPACITY OF AQUEOUS SOLUTIONS OF VARIOUS ACIDS

From National Standards Reference Data Systems NSRDS-NBS 2

Vivian B. Parker

Φ_C is the apparent molal heat capacity of the solute, equal to $[(1000 + mM_2)C - 1000C°]/m$ where C and C° are the specific heats (per unit mass) of the solution and pure solvent, respectively, m is the molality, and M_2 is the molecular weight of the solute.

Φ_C, cal/deg mole, at 25°C

n	*m*	HF	HCl	HBr	HI	HIO_3	HNO_3	CH_2O_2	$C_2H_4O_2$	$C_3H_6O_2$
∞	0.00	−25.5	−32.6	−33.9	−34.0	−29.6	−20.7	−21.0	−1.5	+26.7
500,000	.000111	−23.0	.....	.....	.....	.....	.....	−9.8	+25.8	38.0
100,000	.000555	−18.8	−32.4	−33.8	−33.9	−29.4	−20.6	−1.2	32.6	45.1
50,000	.00111	−16.6	−32.4	−33.7	−33.8	−29.3	−20.5	+3.1	34.0	48.3
20,000	.00278	−12.4	−32.3	−33.6	−33.7	−29.1	−20.4	10.2	35.7	52.2
10,000	.00555	−8.6	−32.2	−33.4	−33.5	−28.5	−20.3	12.7	36.9	54.1
7,000	.00793	−6.7	−32.1	−33.4	−33.4	−28.4	−20.2	13.7	37.4	54.8
5,000	.01110	−4.9	−32.0	−33.3	−33.3	−28.1	−20.1	14.3	37.8	55.3
4,000	.01388	−3.6	−31.9	−33.2	−33.2	−27.7	−20.1	14.7	37.9	55.6
3,000	.01850	−2.2	−31.8	−33.1	−33.1	−27.2	−20.0	15.1	38.1	56.1
2,000	.02775	−0.5	−31.6	−32.9	−32.9	−25.8	−19.8	15.8	38.5	56.7
1,500	.03700	+0.4	−31.4	−32.7	−32.7	−24.9	−19.6	16.4	38.7	57.1
1,000	.05551	1.6	−31.2	−32.5	−32.5	−23.0	−19.3	16.8	39.0	57.7
900	.0617	1.8	−31.2	−32.4	−32.4	−22.3	−19.2	17.0	39.0	57.8
800	.0694	2.0	−31.1	−32.3	−32.3	−21.5	−19.1	17.2	39.1	57.9
700	.0793	2.2	−30.9	−32.2	−32.1	−20.2	−19.0	17.3	39.2	58.0
600	.0925	2.4	−30.8	−32.1	−32.0	−18.7	−18.9	17.5	39.3	58.3
500	.1110	2.7	−30.6	−31.9	−31.7	−16.7	−18.7	17.7	39.4	58.4
400	.1388	2.8	−30.3	−31.6	−31.4	−13.9	−18.4	18.0	39.4	58.6
300	.1850	3.2	−30.0	−31.2	−30.9	−10.1	−17.9	18.3	39.4	58.8
200	.2775	3.3	−29.3	−30.6	−30.2	−4.7	−17.2	18.7	39.3	58.9
150	.3700	3.5	−28.8	−30.1	−29.6	−0.7	−16.4	18.9	39.2	58.8
100	.5551	3.8	−27.8	−29.1	−28.4	+4.8	−15.0	19.0	39.1	58.7
75	.7401	4.1	−27.0	−28.5	−27.5	9.2	−13.7	19.2	38.9	58.6
50	1.1101	4.5	−25.6	−26.8	−25.9	15.7	−11.5	19.8	38.6	58.3
40	1.3877	4.9	−24.8	−25.9	−24.8	19.0	−9.9	19.8	38.4	58.0
30	1.8502	5.3	−23.6	−24.5	−23.3	23.1	−7.3	20.0	38.0	57.1
25	2.2202	5.6	−22.7	−23.6	−22.2	25.7	−5.4	20.1	37.7	56.2
20	2.7753	5.7	−21.5	−22.2	−20.8	.....	−2.7	20.2	37.1	55.0
15	3.7004	6.0	−19.8	−20.3	.....	.....	+1.4	20.4	36.3	53.1
12	4.6255	6.2	−18.2	−18.5	.....	.....	5.0	20.4	35.4	51.3
10	5.5506	6.3	−16.8	−16.8	.....	.....	8.5	20.6	34.8	49.7
9.5	5.8427	6.4	−16.4	−16.3	.....	.....	9.2	20.6	34.7	.....
9.0	6.1674	6.4	−15.8	−15.7	.....	.....	10.3	20.6	34.5	.....
8.5	6.5301	6.5	−15.4	−15.1	.....	.....	11.4	20.7	34.4	.....
8.0	6.9383	6.6	−14.8	−14.4	.....	.....	12.5	20.7	34.2	.....
7.5	7.4008	6.7	−14.2	−13.2	.....	.....	13.7	20.8	34.0	.....
7.0	7.9295	6.9	−13.5	−12.7	.....	.....	14.9	20.8	33.8	.....
6.5	8.5394	7.0	−12.7	−11.9	.....	.....	16.1	20.8	33.5	.....
6.0	9.2510	7.1	−11.8	−10.8	.....	.....	17.1	20.9	33.3	.....
5.5	10.0920	7.2	−10.8	−9.6	.....	.....	18.3	20.9	33.0	.....
5.0	11.1012	7.3	−9.6	−8.2	.....	.....	19.3	21.0	32.8	.....
4.5	12.3346	7.4	−8.7	−6.8	.....	.....	20.4	21.1	32.5	.....
4.0	13.8765	7.5	−6.6	−5.5	.....	.....	21.3	21.2	32.2	.....
3.5	15.8589	7.6	−4.7	−4.0	.....	.....	22.1	21.3	31.8	.....
3.25	17.0788	7.6	.....	−3.2	.....	.....	22.6	21.4	31.7	.....
3.0	18.5020	7.7	.....	−2.3	.....	.....	23.0	21.5	31.5	.....
2.5	22.2024	7.8	.....	−0.4	.....	.....	23.8	21.6	31.2	.....
2.0	27.7530	7.9	.....	.....	.....	.....	24.6	21.7	30.8	.....
1.5	37.0040	.....	.....	.....	.....	.....	25.2	21.8	30.4	.....
1.0	55.506	.....	.....	.....	.....	.....	25.7	22.0	30.1	.....

THERMODYNAMIC FORMULAS

Compiled by Doctor E. A. Coomes

Legend:
p = Pressure
V = Volume
T = Temperature
n = Number of mols
S = Entropy
U = Internal energy (some books use E)
C_p = Molal specific heat at constant pressure
β = Coefficient volume expansion
K = Compressibility
H = U + pV = Total heat or enthalpy
A = U − TS = Helmholtz free energy
G = H − TS = Gibbs' free energy (some books use F).

Use of Table.—Partial derivatives of the first order for the eight fundamental thermodynamic variables, namely, p, V, T, U, S, H, A, G, may be obtained in terms of $(\partial V/\partial T)_p$, $(\partial V/\partial p)_T$, and $(\partial H/\partial T)_p$; the latter three are connected to measurable quantities as follows:

$$\frac{1}{V}\left(\frac{\partial V}{\partial T}\right)_p = \beta; \; -\frac{1}{V}\left(\frac{\partial V}{\partial p}\right)_T = K; \; \left(\frac{\partial H}{\partial T}\right)_p = nC_p$$

Computation by the Table.—The method of using the table will become apparent in working several examples. Suppose it is desired to know $(\partial H/\partial p)_S$ in terms of p, V, and T. Under caption "Constant" move horizontally to column marked "S"; across from "H" beside caption "differential" find "$-VnC_p/T$." Across from "p" in column "S" find "$-nC_p/T$"; $(\partial H/\partial p)_S$ is found by taking the ratio of the two:

$$(\partial H/\partial p)_S = (-VnC_p/T)/(-nC_p/T) = V$$

To find $(\partial S/\partial V)_T$ in terms of p, V, T, move to column "T" under "constant." Opposite "S" beside "differential" find "$(\partial V/\partial T)_p$"; opposite "V" find "$-(\partial V/\partial p)_T$."

Taking the ratio:

$$(\partial S/\partial V)_T = (\partial V/\partial T)_p/-(\partial V/\partial p)_T = (\partial p/\partial T)_V.$$

Differential \ Constant	T	p	V	S
T	0	1	$\left(\frac{\partial V}{\partial p}\right)_T$	$-\left(\frac{\partial V}{\partial T}\right)_p$
p	−1	0	$-\left(\frac{\partial V}{\partial T}\right)_p$	$-\frac{nC_p}{T}$
V	$-\left(\frac{\partial V}{\partial p}\right)_T$	$\left(\frac{\partial V}{\partial T}\right)_p$	0	$\left(-\frac{1}{T}\right)\left[nC_p\left(\frac{\partial V}{\partial p}\right)_T + T\left(\frac{\partial V}{\partial T}\right)_p^2\right]$
S	$\left(\frac{\partial V}{\partial T}\right)_p$	$\frac{nC_p}{T}$	$\left(\frac{1}{T}\right)\left[nC_p\left(\frac{\partial V}{\partial p}\right)_T + T\left(\frac{\partial V}{\partial T}\right)_p^2\right]$	0
U	$T\left(\frac{\partial V}{\partial T}\right)_p + p\left(\frac{\partial V}{\partial p}\right)_T$	$nC_p - p\left(\frac{\partial V}{\partial T}\right)_p$	$nC_p\left(\frac{\partial V}{\partial p}\right)_T + T\left(\frac{\partial V}{\partial T}\right)_p^2$	$\left(\frac{p}{T}\right)\left[nC_p\left(\frac{\partial V}{\partial p}\right)_T + T\left(\frac{\partial V}{\partial T}\right)_p^2\right]$
H	$-V + T\left(\frac{\partial V}{\partial T}\right)_p$	nC_p	$nC_p\left(\frac{\partial V}{\partial p}\right)_T + T\left(\frac{\partial V}{\partial T}\right)_p^2 - V\left(\frac{\partial V}{\partial T}\right)$	$-\frac{VnC_p}{T}$
A	$p\left(\frac{\partial V}{\partial p}\right)_T$	$-S - p\left(\frac{\partial V}{\partial T}\right)_p$	$-S\left(\frac{\partial V}{\partial p}\right)_T$	$\left(\frac{1}{T}\right)\left[pnC_p\left(\frac{\partial V}{\partial p}\right)_T + pT\left(\frac{\partial V}{\partial T}\right)_p^2 + TS\left(\frac{\partial V}{\partial T}\right)_p\right]$
G	−V	−S	$-V\left(\frac{\partial V}{\partial T}\right)_p - S\left(\frac{\partial V}{\partial p}\right)_T$	$\left(-\frac{1}{T}\right)\left[nC_pV - TS\left(\frac{\partial V}{\partial T}\right)_p\right]$

Differential \ Constant	U	H	A	G
T	$-T\left(\frac{\partial V}{\partial T}\right)_p - p\left(\frac{\partial V}{\partial p}\right)_T$	$V - T\left(\frac{\partial V}{\partial T}\right)_p$	$-p\left(\frac{\partial V}{\partial p}\right)_T$	V
p	$-nC_p + p\left(\frac{\partial V}{\partial T}\right)_p$	$-nC_p$	$S + p\left(\frac{\partial V}{\partial T}\right)_p$	S
V	$-nC_p\left(\frac{\partial V}{\partial p}\right)_T - T\left(\frac{\partial V}{\partial T}\right)_p^2$	$-nC_p\left(\frac{\partial V}{\partial p}\right)_T - T\left(\frac{\partial V}{\partial T}\right)_p^2 + V\left(\frac{\partial V}{\partial T}\right)_p$	$S\left(\frac{\partial V}{\partial p}\right)_T$	$V\left(\frac{\partial V}{\partial T}\right)_p + S\left(\frac{\partial V}{\partial p}\right)_T$
S	$\left(-\frac{p}{T}\right)\left[nC_p\left(\frac{\partial V}{\partial p}\right)_T + T\left(\frac{\partial V}{\partial T}\right)_p^2\right]$	$\frac{VnC_p}{T}$	$\left(-\frac{1}{T}\right)\left[pnC_p\left(\frac{\partial V}{\partial p}\right)_T + pT\left(\frac{\partial V}{\partial T}\right)_p^2 + TS\left(\frac{\partial V}{\partial T}\right)_p\right]$	$\left(\frac{1}{T}\right)\left[nC_pV - TS\left(\frac{\partial V}{\partial T}\right)_p\right]$
U	0	$V\left[nC_p - p\left(\frac{\partial V}{\partial T}\right)_p\right] + p\left[nC_p\left(\frac{\partial V}{\partial p}\right)_T + T\left(\frac{\partial V}{\partial T}\right)_p^2\right]$	$-p\left[nC_p\left(\frac{\partial V}{\partial p}\right)_T + T\left(\frac{\partial V}{\partial T}\right)_p^2\right] - S\left[T\left(\frac{\partial V}{\partial T}\right)_p + p\left(\frac{\partial V}{\partial p}\right)_T\right]$	$V\left[nC_p - p\left(\frac{\partial V}{\partial T}\right)_p\right] - S\left[T\left(\frac{\partial V}{\partial T}\right)_p + p\left(\frac{\partial V}{\partial p}\right)_T\right]$
H	$-V\left[nC_p - p\left(\frac{\partial V}{\partial T}\right)_p\right] - p\left[nC_p\left(\frac{\partial V}{\partial p}\right)_T + T\left(\frac{\partial V}{\partial T}\right)_p^2\right]$	0	$\left[S + p\left(\frac{\partial V}{\partial T}\right)_p\right] \times \left[V - T\left(\frac{\partial V}{\partial T}\right)_p - pnC_p\left(\frac{\partial V}{\partial p}\right)_T\right]$	$VnC_p + VS - TS\left(\frac{\partial V}{\partial T}\right)_p$
A	$p\left[nC_p\left(\frac{\partial V}{\partial p}\right)_T + T\left(\frac{\partial V}{\partial T}\right)_p^2\right] + S\left[T\left(\frac{\partial V}{\partial T}\right)_p + p\left(\frac{\partial V}{\partial p}\right)_T\right]$	$-\left[S + p\left(\frac{\partial V}{\partial T}\right)_p\right] \times \left[V - T\left(\frac{\partial V}{\partial T}\right)_p - pnC_p\left(\frac{\partial V}{\partial p}\right)_T\right]$	0	$-S\left[V + p\left(\frac{\partial V}{\partial p}\right)_T\right] + pV\left(\frac{\partial V}{\partial T}\right)_p$
G	$-V\left[nC_p - p\left(\frac{\partial V}{\partial T}\right)_p\right] + S\left[T\left(\frac{\partial V}{\partial T}\right)_p + p\left(\frac{\partial V}{\partial p}\right)_T\right]$	$-VnC_p - VS + TS\left(\frac{\partial V}{\partial T}\right)_p$	$S\left[V + p\left(\frac{\partial V}{\partial p}\right)_T\right] + pV\left(\frac{\partial V}{\partial T}\right)_p$	0

LIMITS OF INFLAMMABILITY

Reprinted from "Combustion Flame and Explosions of Gases," B. Lewis and G. von Elbe, authors, Academic Press (1951), publishers, by special permission.

The limits of inflammability given in the following tables were all determined at atmospheric pressure and room* temperature for upward propagation in a tube or bomb 2 inches or more in diameter. Values are on a percentage-by-volume basis.

LIMITS OF INFLAMMABILITY OF GASES AND VAPORS IN AIR

Compound	Empirical formula	Limits of inflammability	
		Lower	Upper
Paraffin hydrocarbons			
Methane	CH_4	5.00	15.00
Ethane	C_2H_6	3.00	12.50
Propane	C_3H_8	2.12	9.35
Butane	C_4H_{10}	1.86	8.41
Isobutane	C_4H_{10}	1.80	8.44
Pentane	C_5H_{12}	1.40	7.80
Isopentane	C_5H_{12}	1.32	.....
2,2-Dimethylpropane	C_5H_{12}	1.38	7.50
Hexane	C_6H_{14}	1.18	7.40
Heptane	C_7H_{16}	1.10	6.70
2,3-Dimethylpentane	C_7H_{16}	1.12	6.75
Octane	C_8H_{18}	0.95	.....
Nonane	C_9H_{20}	0.83	.....
Decane	$C_{10}H_{22}$	0.77	5.35
Olefins			
Ethylene	C_2H_4	2.75	28.60
Propylene	C_3H_6	2.00	11.10
Butene-1	C_4H_8	1.65	9.95
Butene-2	C_4H_8	1.75	9.70
Amylene	C_5H_{10}	1.42	8.70
Acetylenes			
Acetylene	C_2H_2	2.50	80.00
Aromatics			
Benzene	C_6H_6	1.40	7.10
Toluene	C_7H_8	1.27	6.75
o-Xylene	C_8H_{10}	1.00	6.00
Cyclic hydrocarbons			
Cyclopropane	C_3H_6	2.40	10.40
Cyclohexane	C_6H_{12}	1.26	7.75
Methylcyclohexane	C_7H_{14}	1.15	.....
Terpenes			
Turpentine	$C_{10}H_{16}$	0.80	.....
Alcohols			
Methyl alcohol	CH_4O	6.72	36.50
Ethyl alcohol	C_2H_6O	3.28	18.95
Allyl alcohol	C_3H_6O	2.50	18.00
n-Propyl alcohol	C_3H_8O	2.15	13.50
Isopropyl alcohol	C_3H_8O	2.02	11.80
n-Butyl alcohol	$C_4H_{10}O$	1.45	11.25
Isobutyl alcohol	$C_4H_{10}O$	1.68	.....
n-Amyl alcohol	$C_5H_{12}O$	1.19	.....
Isoamyl alcohol	$C_5H_{12}O$	1.20	.....
Aldehydes			
Acetaldehyde	C_2H_4O	3.97	57.00
Crotonic aldehyde	C_4H_6O	2.12	15.50
Furfural	$C_5H_4O_2$	2.10	.....
Paraldehyde	$C_6H_{12}O_3$	1.30	.....
Ethers			
Methylethyl ether	C_3H_8O	2.00	10.00
Diethyl ether	$C_4H_{10}O$	1.85	36.50
Divinyl ether	C_4H_6O	1.70	27.00
Ketones			
Acetone	C_3H_6O	2.55	12.80
Methylethyl ketone	C_4H_8O	1.81	9.50
Methylpropyl ketone	$C_5H_{10}O$	1.55	8.15
Methylbutyl ketone	$C_6H_{12}O$	1.35	7.60
Acids			
Acetic acid	$C_2H_4O_2$	5.40	.....
Hydrocyanic acid	HCN	5.60	40.00
Esters			
Methyl formate	$C_2H_4O_2$	5.05	22.70
Ethyl formate	$C_3H_6O_2$	2.75	16.40
Methyl acetate	$C_3H_6O_2$	3.15	15.60
Ethyl acetate	$C_4H_8O_2$	2.18	11.40
Propyl acetate	$C_5H_{10}O_2$	1.77	8.00
Isopropyl acetate	$C_5H_{10}O_2$	1.78	7.80
Butyl acetate	$C_6H_{12}O_2$	1.39	7.55
Amyl acetate	$C_7H_{14}O_2$	1.10	.....
Hydrogen			
Hydrogen	H_2	4.00	74.20
Nitrogen compounds			
Ammonia	NH_3	15.50	27.00
Cyanogen	C_2N_2	6.60	42.60
Pyridine	C_5H_5N	1.81	12.40
Ethyl nitrate	$C_2H_5NO_3$	3.80	.....
Ethyl nitrite	$C_2H_5NO_2$	3.01	50.00
Oxides			
Carbon monoxide	CO	12.50	74.20
Ethylene oxide	C_2H_4O	3.00	80.00
Propylene oxide	C_3H_6O	2.00	22.00
Dioxan	$C_4H_8O_2$	1.97	22.25
Diethyl peroxide	$C_4H_{10}O_2$	2.34	.....
Sulfides			
Carbon disulfide	CS_2	1.25	50.00
Hydrogen sulfide	H_2S	4.30	45.50
Carbon oxysulfide	COS	11.90	28.50
Chlorides			
Methyl chloride	CH_3Cl	8.25	18.70
Ethyl chloride	C_2H_5Cl	4.00	14.80
Propyl chloride	C_3H_7Cl	2.60	11.10
Butyl chloride	C_4H_9Cl	1.85	10.10
Isobutyl chloride	C_4H_9Cl	2.05	8.75
Allyl chloride	C_3H_5Cl	3.28	11.15
Amyl chloride	$C_5H_{11}Cl$	1.60	8.63
Vinyl chloride	C_2H_3Cl	4.00	21.70
Ethylene dichloride	$C_2H_4Cl_2$	6.20	15.90
Propylene dichloride	$C_3H_6Cl_2$	3.40	14.50
Bromides			
Methyl bromide	CH_3Br	13.50	14.50
Ethyl bromide	C_2H_5Br	6.75	11.25
Allyl bromide	C_3H_5Br	4.36	7.25
Amines			
Methyl amine	CH_5N	4.95	20.75
Ethyl amine	C_2H_7N	3.55	13.95
Dimethyl amine	C_2H_7N	2.80	14.40
Propyl amine	C_3H_9N	2.01	10.35
Diethyl amine	$C_4H_{11}N$	1.77	10.10
Trimethyl amine	C_3H_9N	2.00	11.60
Triethyl amine	$C_6H_{15}N$	1.25	7.90

* The upper limits of some vapors were determined at somewhat higher temperatures because of their low vapor pressures.

LIMITS OF INFLAMMABILITY OF GASES AND VAPORS IN OXYGEN

Compound	Formula	Limits of inflammability	
		Lower	Upper
Hydrogen	H_2	4.65	93.9
Deuterium	D_2	5.00	95.0
Carbon monoxide	CO	15.50	93.9
Methane	CH_4	5.40	59.2
Ethane	C_2H_6	4.10	50.5
Ethylene	C_2H_4	2.90	79.9
Propylene	C_3H_6	2.10	52.8
Cyclopropane	C_3H_6	2.45	63.1
Ammonia	NH_3	13.50	79.0
Diethyl ether	$C_4H_{10}O$	2.10	82.0
Divinyl ether	C_4H_6O	1.85	85.5

LIMITS FOR HUMAN EXPOSURE TO AIR CONTAMINANTS

From United States Federal Register
Volume 36, Number 105

Exposures by inhalation, ingestion, skin absorption, or contact to any material or substance (1) at a concentration above those specified in the "Threshold Limit Values of Airborne Contaminants for 1970" of the American Conference of Governmental Industrial Hygienists, listed in Table 1, except for the American National Standards listed in Table 2 of this section and except for values of mineral dusts listed in Table 3 of this section, and (2) concentrations above those specified in Table 1, 2, and 3 of this section, shall be avoided, or protective equipment shall be provided and used.

Table 1

Substance	ppm[a]	mg/m³ [b]
Abate		15
Acetaldehyde	200	360
Acetic acid	10	25
Acetic anhydride	5	20
Acetone	1,000	2,400
Acetonitrile	40	70
Acetylene dichloride, see 1,2-Dichloroethylene		
Acetylene tetrabromide	1	14
Acrolein	0.1	0.25
Acrylamide–Skin		0.3
Acrylonitrile–Skin	20	45
Aldrin–Skin		0.25
Allyl alcohol–Skin	2	5
Allyl chloride	1	3
**C Allyl glycidyl ether (AGE)	10	45
Allyl propyl disulfide	2	12
2-Aminoethanol, see Ethanolamine		
2-Aminopyridine	0.5	2
**Ammonia	50	35
Ammonium sulfamate (Ammate)		15
n-Amyl acetate	100	525
sec-Amyl acetate	125	650
Aniline–Skin	5	19
Anisidine (*o,p*-isomers)–Skin		0.5
Antimony and compounds (as Sb)		0.5
ANTU (alpha naphthyl thiourea)		0.3
Arsenic and compounds (as As)		0.5
Arsine	0.05	0.2
Azinphos-methyl–Skin		0.2
Barium (soluble compounds)		0.5
p-Benzoquinone, see Quinone		
Benzoyl peroxide		5
Benzyl chloride	1	5
Biphenyl, see Diphenyl		
Bisphenol A, see Diglycidyl ether		
Boron oxide		15
Boron tribromide	1	10
C Boron trifluoride	1	3
Bromine	0.1	0.7
*Bromine pentafluoride	0.1	0.7
Bromoform–Skin	0.5	5
Butadiene (1,3-butadiene)	1,000	2,200
Butanethiol, see Butyl mercaptan		
2-Butanone	200	590
2-Butoxy ethanol (Butyl Cellosolve)–Skin	50	240
Butyl acetate (*n*-butyl acetate)	150	710
sec-Butyl acetate	200	950
tert-Butyl acetate	200	950
Butyl alcohol	100	300
sec-Butyl alcohol	150	450
tert-Butyl alcohol	100	300
C Butylamine–Skin	5	15
C *tert*-Butyl chromate (as CrO_3)–Skin		0.1
n-Butyl glycidyl ether (BGE)	50	270
*Butyl mercaptan	0.5	1.5
p-tert-Butyltoluene	10	60
Calcium arsenate		1
Calcium oxide		5
**Camphor (Synthetic)	2	
Carbaryl (Sevin®)		5
Carbon black		3.5
Carbon dioxide	5,000	9,000
Carbon monoxide	50	55
Chlordane–Skin		0.5
Chlorinated camphene–Skin		0.5
Chlorinated diphenyl oxide		0.5
*Chlorine	1	3
Chlorine dioxide	0.1	0.3
C Chlorine trifluoride	0.1	0.4
C Chloroacetaldehyde	1	3
α-Chloroacetophenone (phenacylchloride)	0.05	0.3
Chlorobenzene (monochlorobenzene)	75	350
o-Chlorobenzylidene malononitrile (OCBM)	0.05	0.4
Chlorobromomethane	200	1,050
2-Chloro-1,3-butadiene, see Chloroprene		
Chlorodiphenyl (42 percent Chlorine)–Skin		1
Chlorodiphenyl (54 percent Chlorine)–Skin		0.5
1-Chloro-2,3-epoxypropane, see Epichlorhydrin		

LIMITS FOR HUMAN EXPOSURE TO AIR CONTAMINANTS (*Continued*)

Table 1 (continued)

Substance	ppm[a]	mg/m³ [b]
2-Chloroethanol, see Ethylene chlorohydrin		
Chloroethylene, see Vinyl chloride		
C Chloroform (trichloromethane)	50	240
1-Chloro-1-nitropropane	20	100
Chloropicrin	0.1	0.7
Chloroprene (2-chloro-1,3-butadiene)–Skin	25	90
Chromium, sol. chromic, chromous salts as Cr		0.5
Metal and insol. salts		1
Coal tar pitch volatiles (benzene soluble fraction) anthracene, BaP, phenanthrene, acridine, chrysene, pyrene		0.2
Cobalt, metal fume and dust		0.1
Copper fume		0.1
Dusts and Mists		1
Cotton dust (raw)		1
Crag® herbicide		15
Cresol (all isomers)–Skin	5	22
Crotonaldehyde	2	6
Cumene–Skin	50	245
Cyanide (as CN)–Skin		5
*Cyanogen	100	
Cyclohexane	300	1,050
Cyclohexanol	50	200
Cyclohexanone	50	200
Cyclohexene	300	1,015
Cyclopentadiene	75	200
2,4-D		10
DDT–Skin		1
DDVP, see Dichlorvos		
Decaborane–Skin	0.05	0.3
Demeton®–Skin		0.1
Diacetone alcohol (4-hydroxy-4-methyl-2-pentanone)	50	240
1,2-Diaminoethane, see Ethylenediamine		
Diazomethane	0.2	0.4
Diborane	0.1	0.1
Dibutyl phosphate	1	5
Dibutylphthalate		5
*C Dichloroacetylene	0.1	0.4
C *o*-Dichlorobenzene	50	300
p-Dichlorobenzene	75	450
Dichlorodifluoromethane	1,000	4,950
1,3-Dichloro-5,5-dimethyl hydantoin		0.2
1,1-Dichloroethane	100	400
1,2-Dichloroethylene	200	790
C Dichloroethyl ether–Skin	15	90
Dichloromethane, see Methylenechloride		
Dichloromonofluoromethane	1,000	4,200
C 1,1-Dichloro-1-nitroethane	10	60
1,2-Dichloropropane, see Propylenedichloride		
Dichlorotetrafluoroethane	1,000	7,000
Dichlorvos (DDVP)–Skin		1

Substance	ppm[a]	mg/m³ [b]
Dieldrin–Skin		0.25
Diethylamine	25	75
Diethylamino ethanol–Skin	10	50
**C Diethylene triamine–Skin	10	42
Diethylether, see Ethyl ether		
Difluorodibromomethane	100	860
C Diglycidyl ether (DGE)	0.5	2.8
Dihydroxybenzene, see Hydroquinone		
Diisobutyl ketone	50	290
Diisopropylamine–Skin	5	20
Dimethoxymethane, see Methylal		
Dimethyl acetamide–Skin	10	35
Dimethylamine	10	18
Dimethylaminobenzene, see Xylidene		
Dimethylaniline (*N*-dimethylaniline)–Skin	5	25
Dimethylbenzene, see Xylene		
Dimethyl 1,2-dibromo-2,2-dichloroethyl phosphate, (Dibrom)		3
Dimethylformamide–Skin	10	30
2,6-Dimethylheptanone, see Diisobutyl ketone		
1,1-Dimethylhydrazine–Skin	0.5	1
Dimethylphthalate		5
Dimethylsulfate–Skin	1	5
Dinitrobenzene (all isomers)–Skin		1
Dinitro-*o*-cresol–Skin		0.2
Dinitrotoluene–Skin		1.5
Dioxane (Diethylene dioxide)–Skin	100	360
Diphenyl	0.2	1
Diphenylamine		10
Diphenylmethane diisocyanate (see Methylene bisphenyl isocyanate (MDI)		
Dipropylene glycol methyl ether–Skin	100	600
Di-*sec*, octyl phthalate (Di-2-ethylhexylphthalate)		5
*Endosulfan (Thiodan®)–Skin		0.1
Endrin–Skin		0.1
Epichlorhydrin–Skin	5	19
EPN–Skin		0.5
1,2-Epoxypropane, see Propyleneoxide		
2,3-Epoxy-1-propanol, see Glycidol		
Ethanethiol, see Ethylmercaptan		
Ethanolamine	3	6
2-Ethoxyethanol–Skin	200	740
2-Ethoxyethylacetate (Cellosolve acetate)–Skin	100	540
Ethyl acetate	400	1,400
Ethyl acrylate–Skin	25	100
Ethyl alcohol (ethanol)	1,000	1,900
Ethylamine	10	18

LIMITS FOR HUMAN EXPOSURE TO AIR CONTAMINANTS (*Continued*)

Table 1 (continued)

Substance	ppm[a]	mg/m³ [b]
Ethyl *sec*-amyl ketone (5-methyl-3-heptanone)	25	130
Ethyl benzene	100	435
Ethyl bromide	200	890
Ethyl butyl ketone (3-Heptanone)	50	230
Ethyl chloride	1,000	2,600
Ethyl ether	400	1,200
Ethyl formate	100	300
Ethyl mercaptan	0.5	1
Ethyl silicate	100	850
Ethylene chlorohydrin–Skin	5	16
Ethylenediamine	10	25
Ethylene dibromide, see 1,2-Dibromoethane		
Ethylene dichloride, see 1,2-Dichloroethane		
C Ethylene glycol dinitrate and/or Nitroglycerin–Skin	[d]0.2	
Ethylene glycol monomethyl ether acetate, see Methyl cellosolve acetate		
Ethylene imine–Skin	0.5	1
Ethylene oxide	50	90
Ethylidine chloride, see 1,1-Dichloroethane		
N-Ethylmorpholine–Skin	20	94
Ferbam		15
Ferrovanadium dust		1
Fluoride (as F)		2.5
Fluorine	0.1	0.2
Fluorotrichloromethane	1,000	5,600
Formic acid	5	9
Furfural–Skin	5	20
Furfuryl alcohol	50	200
Glycidol (2,3-Epoxy-1-propanol)	50	150
Glycol monoethyl ether, see 2-Ethoxyethanol		
Guthion®, see Azinphosmethyl		
Hafnium		0.5
Heptachlor–Skin		0.5
Heptane (*n*-heptane)	500	2,000
Hexachloroethane–Skin	1	10
Hexachloronaphthalene–Skin		0.2
Hexane (*n*-hexane)	500	1,800
2-Hexanone	100	410
Hexone (Methyl isobutyl ketone)	100	410
sec-Hexyl acetate	50	300
Hydrazine–Skin	1	1.3
Hydrogen bromide	3	10
C Hydrogen chloride	5	7
Hydrogen cyanide–Skin	10	11
Hydrogen peroxide	1	1.4
Hydrogen selenide	0.05	0.2
Hydroquinone		2
*Indene	10	45
Indium and compounds, as In		0.1
C Iodine	0.1	1
Iron oxide fume		10
Iron salts, soluble, as Fe		1
Isoamyl acetate	100	525
Isoamyl alcohol	100	360
Isobutyl acetate	150	700
Isobutyl alcohol	100	300
Isophorone	25	140
Isopropyl acetate	250	950
Isopropyl alcohol	400	980
Isopropylamine	5	12
Isopropylether	500	2,100
Isopropyl glycidyl ether (IGE)	50	240
Ketene	0.5	0.9
Lead arsenate		0.15
Lindane–Skin		0.5
Lithium hydride		0.025
L.P.G. (liquefied petroleum gas)	1,000	1,800
Magnesium oxide fume		15
Malathion–Skin		15
Maleic anhydride	0.25	1
C Manganese and compounds, as Mn		5
Mesityl oxide	25	100
Methanethiol, see Methyl mercaptan		
Methoxychlor		15
2-Methoxyethanol, see Methyl cellosolve		
Methyl acetate	200	610
Methyl acetylene (propyne)	1,000	1,650
Methyl acetylene-propadiene mixture (MAPP)	1,000	1,800
Methyl acrylate–Skin	10	35
Methylal (dimethoxymethane)	1,000	3,100
Methyl alcohol (methanol)	200	260
Methylamine	10	12
Methyl amyl alcohol, see Methyl isobutyl carbinol		
*Methyl isoamyl ketone	100	475
Methyl (*n*-amyl) ketone (2-Heptanone)	100	465
C Methyl bromide–Skin	20	80
Methyl butyl ketone, see 2-Hexanone		
Methyl cellosolve–Skin	25	80
Methyl cellosolve acetate–Skin	25	120
Methyl chloroform	350	1,900
Methylcyclohexane	500	2,000
Methylcyclohexanol	100	470
o-Methylcyclohexanone–Skin	100	460
Methyl ethyl ketone (MEK), see 2-Butanone		
Methyl formate	100	250
Methyl iodide–Skin	5	28
Methyl isobutyl carbinol–Skin	25	100
Methyl isobutyl ketone, see Hexone		
Methyl isocyanate–Skin	0.02	0.05
*Methyl mercaptan	0.5	1
Methyl methacrylate	100	410
Methyl propyl ketone, see 2-Pentanone		
C Methyl silicate	5	30

Table 1 (continued)

Substance	ppm[a]	mg/m^3 [b]
C α-Methyl styrene	100	480
C Methylene bisphenyl isocyanate (MDI)	0.02	0.2
Molybdenum:		
Soluble compounds		5
Insoluble compounds		15
Monomethyl aniline–Skin	2	9
C Monomethyl hydrazine–Skin	0.2	0.35
Morpholine–Skin	20	70
Naphtha (coaltar)	100	400
Naphthalene	10	50
Nickel carbonyl	0.001	0.007
Nickel, metal and soluble cmpds, as Ni		1
Nicotine–Skin		0.5
Nitric acid	2	5
Nitric oxide	25	30
p-Nitroaniline–Skin	1	6
Nitrobenzene–Skin	1	5
p-Nitrochlorobenzene–Skin		1
Nitroethane	100	310
Nitrogen dioxide	5	9
Nitrogen trifluoride	10	29
Nitroglycerin–Skin	0.2	2
Nitromethane	100	250
1-Nitropropane	25	90
2-Nitropropane	25	90
Nitrotoluene–Skin	5	30
Nitrotrichloromethane, see Chloropicrin		
Octachloronaphthalene–Skin		0.1
*Octane	400	1,900
*Oil mist, particulate		5
Osmium tetroxide		0.002
Oxalic acid		1
Oxygen difluoride	0.05	0.1
Ozone	0.1	0.2
Paraquat–Skin		0.5
Silver, metal and soluble compounds		0.01
Sodium fluoroacetate (1080)–Skin		0.05
Sodium hydroxide		2
Stibine	0.1	0.5
*Stoddard solvent	200	1,150
Strychnine		0.15
Sulfur dioxide	5	13
Sulfur hexafluoride	1,000	6,000
Sulfuric acid		1
Sulfur monochloride	1	6
Sulfur pentafluoride	0.025	0.25
Sulfuryl fluoride	5	20
Systox, see Demeton®		
2,4,5T		10
Tantalum		5
TEDP–Skin		0.2
Tellurium		0.1
Tellurium hexafluoride	0.02	0.2
TEPP–Skin		0.05
C Terphenyls	1	9
1,1,1,2-Tetrachloro-2,2-difluoroethane	500	4,170
1,1,2,2-Tetrachloro-1,2-difluoroethane	500	4,170
1,1,2,2-Tetrachloroethane–Skin	5	35
Tetrachloroethylene, see Perchloroethylene		
Tetrachloromethane, see Carbon tetrachloride		
Tetrachloronaphthalene–Skin		2
Tetraethyl lead (as Pb)–Skin		[f]0.100
Tetrahydrofuran	200	590
Tetramethyl lead (as Pb)–Skin		[f]0.150
Tetramethyl succinonitrile–Skin	0.5	3
Tetranitromethane	1	8
Tetryl (2,4,6-trinitrophenylmethylnitramine)–Skin		1.5
Thallium (soluble compounds)–Skin as Tl		0.1
Thiram		5
Tin (inorganic cmpds, except SnH_4 and SnO_2)		2
Tin (organic cmpds)		0.1
C Toluene-2,4-diisocyanate	0.02	0.14
o-Toluidine–Skin	5	22
Toxaphene, see Chlorinated camphene		
Tributyl phosphate		5
1,1,1-Trichloroethane (see Methyl chloroform		
1,1,2-Trichloroethane–Skin	10	45
Parathion–Skin		0.1
Pentaborane	0.005	0.01
Pentachloronaphthalene–Skin		0.5
Pentachlorophenol–Skin		0.5
*Pentane	500	1,500
2-Pentanone	200	700
Perchloromethyl mercaptan	0.1	0.8
Perchloryl fluoride	3	13.5
Phenol–Skin	5	19
p-Phenylene diamine–Skin		0.1
Phenyl ether (vapor)	1	7
Phenyl ether-biphenyl mixture (vapor)	1	7
Phenylethylene, see Styrene		
Phenyl glycidyl ether (PGE)	10	60
Phenylhydrazine–Skin	5	22
Phosdrin (Mevinphos®)–Skin		0.1
Phosgene (carbonyl chloride)	0.1	0.4
Phosphine	0.3	0.4
Phosphoric acid		1
Phosphorus (yellow)		0.1
Phosphorus pentachloride		1
Phosphorus pentasulfide		1
Phosphorus trichloride	0.5	3
Phthalic anhydride	2	12

LIMITS FOR HUMAN EXPOSURE TO AIR CONTAMINANTS (*Continued*)

Table 1 (continued)

Substance	ppm[a]	mg/m^3 [b]
Picric acid–Skin		0.1
Pival® (2-Pivalyl-1,3-indandione)		0.1
Platinum (Soluble Salts) as Pt		0.002
Propargyl alcohol–Skin	1	
n-Propyl acetate	200	840
Propyl alcohol	200	500
n-Propyl nitrate	25	110
Propylene dichloride	75	350
Propylene imine–Skin	2	5
Propylene oxide	100	240
Propyne, see Methylacetylene		
Pyrethrum		5
Pyridine	5	15
Quinone	0.1	0.4
RDX–Skin		1.5
Rhodium, Metal fume and dusts, as Rh		0.1
Soluble salts		0.001
Ronnel		10
Rotenone (commercial)		5
Selenium compounds (as Se)		0.2
Selenium hexafluoride	0.05	0.4
Trichloromethane, see Chloroform		
Trichloronaphthalene–Skin		5
1,2,3-Trichloropropane	50	300
1,1,2-Trichloro 1,2,2-trifluoroethane	1,000	7,600
Triethylamine	25	100
Trifluoromonobromomethane	1,000	6,100
*Trimethyl benzene	25	120
2,4,6-Trinitrophenol, see Picric acid		
2,4,6-Trinitrophenylmethylnitramine, see Tetryl		
Trinitrotoluene–Skin		1.5
Triorthocresyl phosphate		0.1
Triphenyl phosphate		3
Tungsten and compounds, as W:		
Soluble		1
Insoluble		5
Turpentine	100	560
Uranium (natural) sol. and insol. compounds as U		0.2
C Vanadium:		
V_2O_5 dust		0.5
V_2O_5 fume		0.1
Vinyl benzene, see Styrene		
**C Vinyl chloride	500	1,300
Vinylcyanide, see Acrylonitrile		
Vinyl toluene	100	480
Warfarin		0.1
Xylene (xylol)	100	435
Xylidine–Skin	5	25
Yttrium		1
Zinc chloride fume		1
Zinc oxide fume		5
Zirconium compounds (as Zr)		5

*1970 Addition.

[a]Parts of vapor or gas per million parts of contaminated air by volume at 25°C and 760 mm Hg pressure.
[b]Approximate milligrams of particulate per cubic meter of air.

(No footnote "c" is used to avoid confusion with ceiling value notations.)

[d]An atmospheric concentration of not more than 0.02 ppm, or personal protection may be necessary to avoid headache.

[e]As sampled method that does not collect vapor.
[f]For control of general room air, biologic monitoring is essential for personnel control.

Table 2

	8-hour time weighted average
Benzene (Z37.4–1969)	10 ppm
Beryllium and beryllium compounds (Z37.29–1970)	0.002 mg/m^3
Cadmium dust (as Cd) (Z37.5–1970)	0.2 mg/m^3
Cadmium fume (as Cd) (Z37.5–1970)	0.1 mg/m^3
Carbon disulfide (Z37.3–1968)	20 ppm
Carbon tetrachloride (Z37.17–1967)	10 ppm
Ethylene dibromide (Z37.31–1970)	20 ppm
Ethylene dichloride (Z37.21–1969)	50 ppm
Formaldehyde (Z37.16–1967)	3 ppm
Hydrogen fluoride (Z37.28–1969)	3 ppm
Fluoride as dust (Z37.28–1966)	2.5 mg/m^3
Lead and its inorganic compounds (Z37.11–1969)	0.2 mg/m^3
Methyl chloride (Z37.18–1969)	100 ppm
Methylene chloride (Z37.23–1969)	500 ppm
Organo (alkyl) mercury (Z37.30–1969)	0.01 mg/m^3
Styrene (Z37.12–1969)	100 ppm
Tetrachloroethylene (Z37.22–1967)	100 ppm
Toluene (Z37.12–1967)	200 ppm
	Acceptable ceiling concentration
Hydrogen sulfide (Z37.2–1966)	20 ppm
Chromic acid and chromates (Z37.3–1971)	1 mg/10 m^3
Mercury (Z37.8–1971)	1 mg/10 m^3

Table 3

Substance	Mppcf[a]	mg/m³
Silica:		
Crystalline:		
Quartz (respirable)	$\frac{250^{f}}{\%SiO_2+5}$	$\frac{10\ mg/m^3\ {}^{r}}{\%SiO_2+2}$
Quartz (total dust)		$\frac{30\ mg/m^3}{\%SiO_2+2}$
Cristobalite: Use ½ the value calculated from the count or mass formulae for quartz.		
Trioymite: Use ½ the value calculated from the formulae for quartz.		
Amorphous, including natural diatomaceous earth	20	$\frac{80\ mg/m^3}{\%SiO_2}$
Tremolite	5	$\frac{20\ mg/m^3}{\%SiO_2}$
Silicates (less than 1% crystalline silica):		
Asbestos – 12 fibers per milliliter greater than 5 microns in length,[i] or	2	
Mica	20	
Soapstone	20	
Talc	20	
Portland cement	50	
Graphite (natural)	15	
Coal dust (respirable fraction less than 5% SiO_2)		2.4 mg/m³ or
For more than 5% SiO_2		$\frac{10\ mg/m^3}{\%SiO_2+2}$

Substance	Mppcf[a]	mg/m³
Inert or Nuisance Dust:		
Respirable fraction	15	5 mg/m³
Total dust	50	15 mg/m³

NOTE: Conversion factors–
mppcf x 35.3 = million particles per cubic meter
= particles per cc

[a]Millions of particles per cubic foot of air, based on impinger samples counted by light-field technics.

[f]The percentage of crystalline silica in the formula is the amount determined from air-borne samples, except in those instances in which other methods have been shown to be applicable.

[i]As determined by the membrane filter method at 430 x phase contrast magnification.

[r]Both concentration and percent quartz for the application of this limit are to be determined from the fraction passing a size-selector with the following characteristics:

Aerodynamic diameter (unit density sphere)	Percent passing selector
2	90
2.5	75
3.5	50
5.0	25
10	0

The measurements under this note refer to the use of an AEC instrument. If the respirable fraction of coal dust is determined with a MRE the figure corresponding to that of 2.4 mg/m³ in the table for coal dust is 4.5 mg/m³

FLAME AND BEAD TESTS

Flame Colorations

Violet

Potassium compounds. Purple red through blue glass. Easily obscured by sodium flame. Bluish green through green glass. Rubidium and Caesium compounds impart same flame as potassium compounds

Blues

Azure.—Copper chloride. Copper bromide gives azure blue followed by green. Other copper compounds give same coloration when moistened with hydrochloric acid.

Light Blue.—Lead, Arsenic, Selenium.

Greens

Emerald.—Copper compounds except the halides, and when not moistened with hydrochloric acid.

Pure Green.—Compounds of thallium and tellurium.

Yellowish.—Barium compounds. Some molybdenum compounds. Borates, especially when treated with sulphuric acid or when burned with alcohol.

Bluish.—Phosphates with sulphuric acid.

Feeble.—Antimony compounds. Ammonium compounds.

Whitish.—Zinc.

Reds

Carmine.—Lithium compounds. Violet through blue glass. Invisible through green glass. Masked by barium flame.

Scarlet.—Strontium compounds. Violet through blue glass. Yellowish through green glass. Masked by barium flame.

Yellowish.—Calcium compounds. Greenish through blue glass. Green through green glass. Masked by barium flame.

Yellow

Yellow.—All sodium compounds. Invisible with blue glass.

Borax Beads

Abbreviations employed: s., saturated; s.s., supersaturated; n.s., not saturated; h., hot; c., cold.

Substance	Oxidizing flame	Reducing flame
Aluminum	Colorless (h.c., n.s.); opaque (s.s.)	Colorless; opaque (s.)
Antimony	Colorless; yellow or brownish (h., s.s.)	Gray and opaque
Barium	Colorless (n.s.)	....
Bismuth	Colorless; yellow or brownish (h., s.s.)	Gray and opaque
Cadmium	Colorless	Gray and opaque
Calcium	Colorless (n.s.)	....
Cerium	Red (h.)	Colorless (h.c.)
Chromium	Green (c.)	Green
Cobalt	Blue (h.c.)	Blue (h.c.)
Copper	Green (h.); blue (c.)	Red (c.): opaque (s.s.): colorless (h.)
Iron	Yellow or brownish red (h., n.s.)	Green (s.s.)
Lead	Colorless; yellow or brownish (h., s.s.)	Gray and opaque
Magnesium	Colorless (n.s.)	....
Manganese	Violet (h.c.)	Colorless (h.c.)
Molybdenum	Colorless	Yellow or brown (h.)
Nickel	Brown; red (c.)	Gray and opaque
Silicon	Colorless (h.c.); opaque (s.s.)	Colorless; opaque (s.)
Silver	Colorless (n.s.)	Gray and opaque
Strontium	Colorless (n.s.)	....
Tin	Colorless (h.c.); opaque (s.s.)	Colorless; opaque (s.)
Titanium	Colorless	Yellow (h.); violet (c.)
Tungsten	Colorless	Brown
Uranium	Yellow or brownish (h., n.s.)	Green
Vanadium	Colorless	Green

Beads of Microcosmic Salt
$NaNH_4HPO_4$

Substance	Oxidizing flame	Reducing flame
Aluminum	Colorless; opaque (s.)	Colorless; not clear (s.s.)
Antimony	Colorless (n.s.)	Gray and opaque
Barium	Colorless; opaque (s.)	Colorless; not clear (s.s.)
Bismuth	Colorless (n.s.)	Gray and opaque
Cadmium	Colorless (n.s.)	Gray and opaque
Calcium	Colorless; opaque (s.)	Colorless; not clear (s.s.)
Cerium	Yellow or brownish red (h., s.)	Colorless
Chromium	Red (h., s.); green (c.)	Green (c.)
Cobalt	Blue (h.c.)	Blue (h.c.)
Copper	Blue (c.); green (h.)	Red and opaque (c.)
Iron	Yellow or brown (h., s.)	Colorless; yellow or brownish (h.)
Lead	Colorless (n.s.)	Gray and opaque
Magnesium	Colorless; opaque (s.)	Colorless; not clear (s.s.)
Manganese	Violet (h.c.)	Colorless
Molybdenum	Colorless; green (h.)	Green (h.)
Nickel	Yellow (c.); red (h., s.)	Yellow (c.); red (h.); gray and opaque
Silicon	(Swims undissolved)	(Swims undissolved)
Silver	....	Gray and opaque
Strontium	Colorless; opaque (s.)	Colorless; not clear (s.s.)
Tin	Colorless; opaque (s.)	Colorless
Titanium	Colorless (n.s.)	Violet (c.); yellow or brownish (h.)
Uranium	Green; yellow or brownish (h., s.)	Green (h.)
Vanadium	Yellow	Green
Zinc	Colorless (n.s.)	Gray and opaque

Sodium Carbonate Bead

Substance	Oxidizing flame	Reducing flame
Manganese	Green	Colorless

PREPARATION OF REAGENTS

The following pages present directions for the preparation of various reagents. The collection has been prepared with the active collaboration of W. D. Bonner, R. K. Carleton, L. L. Carrick, Giles B. Cooke, E. J. Cragoe, Thos. De Vries, James L. Kassner, Thos. W. Mason, F. C. Mathers, M. G. Mellon, W. C. Pierce, J. H. Reedy, Arthur A. Vernon and S. R. Wood. Many others have contributed valuable suggestions.

Volumes have been stated in milliliters (ml) and liters (l). One milliliter is equivalent to one cubic centimeter (cm^3 or cc.). Masses are indicated in grams (g).

The relation to molar solution (*M*) or normal solution (*N*) is indicated in many cases.

Distilled water should be used.

LABORATORY REAGENTS FOR GENERAL USE

DILUTE ACIDS, 3 molar. Use the amount of concentrated acid indicated and dilute to one liter.

Acetic acid, 3 *N*. Use 172 ml of 17.4 *M* acid (99–100 %).

Hydrochloric acid, 3 *N*. Use 258 ml of 11.6 *M* acid (36 % HCl).

Nitric acid, 3 *N*. Use 195 ml of 15.4 *M* acid (69% HNO_3).

Phosphoric acid, 9 *N*. Use 205 ml of 14.6 *M* acid (85% H_3PO_4).

Sulfuric acid, 6 *N*. Use 168 ml of 17.8 *M* acid (95% H_2SO_4).

DILUTE BASES.

Ammonium hydroxide, 3 *M*, 3 *N*. Dilute 200 ml of concentrated solution (14.8 *M*, 28% NH_3) to 1 liter.

Barium hydroxide, 0.2 *M*, 0.4 *N*. Saturated solution, 63 g per liter of $Ba(OH)_2.8H_2O$. Use some excess, filter off $BaCO_3$ and protect from CO_2 of the air with soda lime or ascarite in a guard tube.

Calcium hydroxide, 0.02 *M*, 0.04 *N*. Saturated solution, 1.5 g per liter of $Ca(OH)_2$. Use some excess, filter off $CaCO_3$ and protect from CO_2 of the air.

Potassium hydroxide, 3 *M*, 3 *N*. Dissolve 176 g of the sticks (95 %) in water and dilute to 1 liter.

Sodium hydroxide, 3 *M*, 3 *N*. Dissolve 126 g of the sticks (95 %) in water and dilute to 1 liter.

GENERAL REAGENTS. (See also *Decinormal Solutions of Salts and Other Reagents.*)

Aluminum chloride, 0.167 *M*, 0.5 *N*. Dissolve 22 g of $AlCl_3$ in 1 liter of water.

Aluminum nitrate, 0.167 *M*, 0.5 *N*. Dissolve 58 g of $Al(NO_3)_3.7.5H_2O$ in 1 liter of water.

Aluminum sulfate, 0.083 *M*, 0.5 *N*. Dissolve 56 g of $Al_2(SO_4)_3.18H_2O$ in 1 liter of water.

Ammonium acetate, 3 *M*, 3 *N*. Dissolve 230 g of $NH_4C_2H_3O_2$ in water and dilute to 1 liter.

Ammonium carbonate, 1.5 *M*. Dissolve 144 g of the commercial salt (mixture of $(NH_4)_2CO_3.H_2O$ and $NH_4CO_2NH_2$) in 500 ml of 3 *N* NH_4OH and dilute to 1 liter.

Ammonium chloride, 3 *M*, 3 *N*. Dissolve 160 g of NH_4Cl in water. Dilute to 1 liter.

Ammonium molybdate.

1. 0.5 *M*, 1 *N*. Mix well 72 g of pure MoO_3 (or 81 g of H_2MoO_4) with 200 ml of water, and add 60 ml of conc. ammonium hydroxide. When solution is complete, filter and pour filtrate, *very slowly* and with *rapid stirring*, into a mixture of 270 ml of conc. HNO_3 and 400 ml of water. Allow to stand over night, filter and dilute to 1 liter.

2. The reagent is prepared as two solutions which are mixed as needed, thus always providing fresh reagent of proper strength and composition. Since ammonium molybdate is an expensive reagent, and since an acid solution of this reagent as usually prepared keeps for only a few days, the method proposed will avoid loss of reagent and provide more certain results for quantitative work.

Solution 1. Dissolve 100 g of ammonium molybdate (C.P. grade) in 400 ml of water and 80 ml of 15 *M* NH_4OH. Filter if necessary, though this seldom has to be done.

Solution 2. Mix 400 ml of 16 *M* nitric acid with 600 ml of water.

For use, mix the calculated amount of solution 1 with twice its volume of solution 2, adding solution 1 to solution 2 slowly with vigorous stirring. Thus, for amounts of phosphorus up to 20 mg, 10 ml of solution 1 to 20 ml of solution 2 is adequate. Increase amount as needed.

Ammonium nitrate, 1 *M*, 1 *N*. Dissolve 80 g of NH_4NO_3 in 1 liter of water.

Ammonium oxalate, 0.25 *M*, 0.5 *N*. Dissolve 35.5 g of $(NH_4)_2C_2O_4.H_2O$ in water. Dilute to 1 liter.

Ammonium sulfate, 0.25 *M*, 0.5 *N*. Dissolve 33 g of $(NH_4)_2SO_4$ in 1 liter of water.

Ammonium sulfide, colorless.

1. 3 *M*. Treat 200 ml of conc. NH_4OH with H_2S until saturated, keeping the solution cold. Add 200 ml of conc. NH_4OH and dilute of 1 liter.

2. 6 *N*. Saturate 6 *N* ammonium hydroxide (40 ml conc. ammonia solution + 60 ml H_2O) with washed H_2S gas. The ammonium hydroxide bottle must be completely full and must be kept surrounded by ice while being saturated (about 48 hours for two liters). The reagent is best preserved in brown, completely filled, glass-stoppered bottles.

LABORATORY REAGENTS (Continued)

Ammonium sulfide, yellow. Treat 150 ml of conc. NH_4OH with H_2S until saturated, keeping the solution cool. Add 250 ml of conc. NH_4OH and 10 g of powdered sulfur. Shake the mixture until the sulfur is dissolved and dilute to 1 liter with water. In the solution the concentration of $(NH_4)_2S_2$, $(NH_4)_2S$ and NH_4OH are 0.625, 0.4 and 1.5 normal respectively. On standing, the concentration of $(NH_4)_2S_2$ increases and that of $(NH_4)_2S$ and NH_4OH decreases.

Antimony pentachloride, 0.1 *M*, 0.5 *N*. Dissolve 30 g of $SbCl_5$ in 1 liter of water.

Antimony trichloride, 0.167 *M*, 0.5 *N*. Dissolve 38 g of $SbCl_3$ in 1 liter of water.

Aqua regia. Mix 1 part concentrated HNO_3 with 3 parts of concentrated HCl. This formula should include one volume of water if the aqua regia is to be stored for any length of time. Without water, objectionable quantities of chlorine and other gases are evolved.

Barium chloride, 0.25 *M*, 0.5 *N*. Dissolve 61 g of $BaCl_2.2H_2O$ in water. Dilute to 1 liter.

Barium hydroxide, 0.1 *M*, about 0.2 *N*. Dissolve 32 g of $Ba(OH)_2.8H_2O$ in 1 liter of water.

Barium nitrate, 0.25 *M*, 0.5 *N*. Dissolve 65 g of $Ba(NO_3)_2$ in 1 liter of water.

Bismuth chloride, 0.167 *M*, 0.5 *N*. Dissolve 53 g of $BiCl_3$ in 1 liter of dilute HCl. Use 1 part HCl to 5 parts water.

Bismuth nitrate, 0.083 *M*, 0.25 *N*. Dissolve 40 g of $Bi(NO_3)_3.5H_2O$ in 1 liter of dilute HNO_3. Use 1 part of HNO_3 to 5 parts of water.

Cadmium chloride, 0.25 *M*, 0.5 *N*. Dissolve 46 g of $CdCl_2$ in 1 liter of water.

Cadmium nitrate, 0.25 *M*, 0.5 *N*. Dissolve 77 g of $Cd(NO_3)_2.4H_2O$ in 1 liter of water.

Cadmium sulfate, 0.25 *M*, 0.5 *N*. Dissolve 70 g of $CdSO_4.4H_2O$ in 1 liter of water.

Calcium chloride, 0.25 *M*, 0.5 *N*. Dissolve 55 g of $CaCl_2.6H_2O$ in water. Dilute to 1 liter.

Calcium nitrate, 0.25 *M*, 0.5 *N*. Dissolve 41 g of $Ca(NO_3)_2$ in 1 liter of water.

Chloroplatinic acid.

1. 0.0512 *M*, 0.102 *N*. Dissolve 26.53 g of $H_2PtCl_6.6H_2O$ in water. Dilute to 100 ml. Contains 0.100 g Pt per ml.

2. Make a 10% solution by dissolving 1 g of $H_2PtCl_6.6H_2O$ in 9 ml of water. Shake thoroughly to insure complete mixing. Keep in a dropping bottle.

Chromic chloride, 0.167 *M*, 0.5 *N*. Dissolve 26 g of $CrCl_3$ in 1 liter of water.

Chromic nitrate, 0.167 *M*, 0.5 *N*. Dissolve 40 g of $Cr(NO_3)_3$ in 1 liter of water.

Chromic sulfate, 0.083 *M*, 0.5 *N*. Dissolve 60 g of $Cr_2(SO_4)_3.18H_2O$ in 1 liter of water.

Cobaltous nitrate, 0.25 *M*, 0.5 *N*. Dissolve 73 g of $Co(NO_3)_2.6H_2O$ in 1 liter of water.

Cobaltous sulfate, 0.25 *M*, 0.5 *N*. Dissolve 70 g of $CoSO_4.7H_2O$ in 1 liter of water.

Cupric chloride, 0.25 *M*, 0.5 *N*. Dissolve 43 g of $CuCl_2.2H_2O$ in 1 liter of water.

Cupric nitrate, 0.25 *M*, 0.5 *N*. Dissolve 74 g of $Cu(NO_3)_2.6H_2O$ in 1 liter of water.

Cupric sulfate, 0.5 *M*, 1 *N*. Dissolve 124.8 g of $CuSO_4.5H_2O$ in water to which 5 ml of H_2SO_4 has been added. Dilute to 1 liter.

Ferric chloride, 0.5 *M*, 1.5 *N*. Dissolve 135.2 g of $FeCl_3.6H_2O$ in water containing 20 ml of conc. HCl. Dilute to 1 liter.

Ferric nitrate, 0.167 *M*, 0.5 *N*. Dissolve 67 g of $Fe(NO_3)_3.9H_2O$ in 1 liter of water.

Ferric sulfate, 0.25 *M*, 0.5 *N*. Dissolve 140.5 g of $Fe_2(SO_4)_3.9H_2O$ in water containing 100 ml of conc. H_2SO_4. Dilute to 1 liter.

Ferrous ammonium sulfate, 0.5 *M*, 1 *N*. Dissolve 196 g of $Fe(NH_4SO_4)_2.6H_2O$ in water containing 10 ml of conc. H_2SO_4. Dilute to 1 liter. Prepare fresh solutions for best results.

Ferrous sulfate, 0.5 *M*, 1 *N*. Dissolve 139 g of $FeSO_4.7H_2O$ in water containing 10 ml of conc. H_2SO_4. Dilute to 1 liter. Solution does not keep well.

Lead acetate, 0.5 *M*, 1 *N*. Dissolve 190 g of $Pb(C_2H_3O_2)_2.3H_2O$ in water. Dilute to 1 liter.

Lead nitrate, 0.25 *M*, 0.5 *N*. Dissolve 83 g of $Pb(NO_3)_2$ in water. Dilute to one liter.

Lime water. See *Calcium hydroxide.*

Magnesium chloride, 0.25 *M*, 0.5 *N*. Dissolve 51 g of $MgCl_2.6H_2O$ in 1 liter of water.

Magnesium chloride reagent. Dissolve 50 g of $MgCl_2.6H_2O$ and 100 g of NH_4Cl in 500 ml of water. Add 10 ml of conc. NH_4OH, allow to stand over night and filter if a precipitate has formed. Make acid to methyl red with dilute HCl. Dilute to 1 liter. Solution contains 0.25 *M* $MgCl_2$ and 2 *M* NH_4Cl. Solution may also be diluted with 133 ml of conc. NH_4OH and water to make 1 liter. Such a solution will contain 2 *M* NH_4OH.

Magnesium nitrate, 0.25 *M*, 0.5 *N*. Dissolve 64 g of $Mg(NO_3)_2.6H_2O$ in 1 liter of water.

Magnesium sulfate, 0.25 *M*, 0.5 *N*. Dissolve 62 g of $MgSO_4.7H_2O$ in 1 liter of water.

Manganous chloride, 0.25 *M*, 0.5 *N*. Dissolve 50 g of $MnCl_2.4H_2O$ in 1 liter of water.

Manganous nitrate, 0.25 *M*, 0.5 *N*. Dissolve 72 g of $Mn(NO_3)_2.6H_2O$ in 1 liter of water.

Manganous sulfate, 0.25 *M*, 0.5 *N*. Dissolve 69 g of $MnSO_4.7H_2O$ in 1 liter of water.

Mercuric chloride, 0.25 *M*, 0.5 *N*. Dissolve 68 g of $HgCl_2$ in water. Dilute to 1 liter.

LABORATORY REAGENTS (Continued)

Mercuric nitrate, 0.25 *M*, 0.5 *N*. Dissolve 81 g of $Hg(NO_3)_2$ in 1 liter of water.

Mercuric sulfate, 0.25 *M*, 0.5 *N*. Dissolve 74 g of $HgSO_4$ in 1 liter of water.

Mercurous nitrate. Use 1 part $HgNO_3$, 20 parts water and 1 part HNO_3.

Nickel chloride, 0.25 *M*, 0.5 *N*. Dissolve 59 g of $NiCl_2.6H_2O$ in 1 liter of water.

Nickel nitrate, 0.25 *M*, 0.5 *N*. Dissolve 73 g of $Ni(NO_3)_2.6H_2O$ in 1 liter of water.

Nickel sulfate, 0.25 *M*, 0.5 *N*. Dissolve 66 g of $NiSO_4.6H_2O$ in 1 liter of water.

Potassium bromide, 0.5 *M*, 0.5 *N*. Dissolve 60 g of KBr in 1 liter of water.

Potassium carbonate, 1.5 *M*, 3 *N*. Dissolve 207 g of K_2CO_3 in 1 liter of water.

Potassium chloride, 0.5 *M*, 0.5 *N*. Dissolve 37 g of KCl in 1 liter of water.

Potassium chromate, 0.25 *M*, 0.5 *N*. Dissolve 49 g of K_2CrO_4 in 1 liter of water.

Potassium cyanide, 0.5 *M*, 0.5 *N*. Dissolve 33 g of KCN in 1 liter of water.

Potassium dichromate, 0.125 *M*. Dissolve 37 g of $K_2Cr_2O_7$ in 1 liter of water.

Potassium ferricyanide, 0.167 *M*, 0.5 *N*. Dissolve 55 g of $K_3Fe(CN)_6$ in 1 liter of water.

Potassium ferrocyanide, 0.5 *M*, 2 *N*. Dissolve 211 g of $K_4Fe(CN)_6.3H_2O$ in water. Dilute to 1 liter.

Potassium iodide, 0.5 *M*, 0.5 *N*. Dissolve 83 g of KI in 1 liter of water.

Potassium nitrate, 0.5 *M*, 0.5 *N*. Dissolve 51 g of KNO_3 in 1 liter of water.

Potassium sulfate, 0.25 *M*, 0.5 *N*. Dissolve 44 g of K_2SO_4 in 1 liter of water.

Silver nitrate, 0.5 *M*, 0.5 *N*. Dissolve 85 g of $AgNO_3$ in water. Dilute to 1 liter.

Sodium acetate, 3 *M*, 3 *N*. Dissolve 408 g of $NaC_2H_3O_2.3H_2O$ in water. Dilute to 1 liter.

Sodium carbonate, 1.5 *M*, 3 *N*. Dissolve 159 g of Na_2CO_3, or 430 g of $Na_2CO_3.10H_2O$ in water. Dilute to 1 liter.

Sodium chloride, 0.5 *M*, 0.5 *N*. Dissolve 29 g of NaCl in 1 liter of water.

Sodium cobaltinitrite, 0.08 *M* (reagent for potassium). Dissolve 25 g of $NaNO_2$ in 75 ml of water, add 2 ml of glacial acetic acid and then 2.5 g of $Co(NO_3)_2.6H_2O$. Allow to stand for several days, filter and dilute to 100 ml. Reagent is somewhat unstable.

Sodium hydrogen phosphate, 0.167 *M*, 0.5 *N*. Dissolve 60 g of $Na_2HPO_4.12H_2O$ in 1 liter of water.

Sodium nitrate, 0.5 *M*, 0.5 *N*. Dissolve 43 g of $NaNO_3$ in 1 liter of water.

SPECIAL SOLUTIONS AND REAGENTS

Sodium sulfate, 0.25 *M*, 0.5 *N*. Dissolve 36 g of Na_2SO_4 in 1 liter of water.

Sodium sulfide, 0.5 *M*, 1 *N*. Dissolve 120 g of $Na_2S.9H_2O$ in water and dilute to 1 liter. Or, saturate 500 ml of 1 *M* NaOH (21 g of 95% NaOH sticks) with H_2S, keeping the solution cool, and dilute with 500 ml of 1 *M* NaOH.

Stannic chloride, 0.125 *M*, 0.5 *N*. Dissolve 33 g of $SnCl_4$ in 1 liter of water.

Stannous chloride, 0.5 *M*, 1 *N*. Dissolve 113 g of $SnCl_2.2H_2O$ in 170 ml of conc. HCl, using heat if necessary. Dilute with water to 1 liter. Add a few pieces of tin foil. Prepare solution fresh at frequent intervals.

Stannous chloride (for Bettendorf test). Dissolve 113 g of $SnCl_2.2H_2O$ in 75 ml of conc. HCl. Add a few pieces of tin foil.

Strontium chloride, 0.25 *M*, 0.5 *N*. Dissolve 67 g of $SrCl_2.6H_2O$ in 1 liter of water.

Zinc nitrate, 0.25 *M*, 0.5 *N*. Dissolve 74 g of $Zn(NO_3)_2.6H_2O$ in 1 liter of water.

Zinc sulfate, 0.25 *M*, 0.5 *N*. Dissolve 72 g of $ZnSO_4.7H_2O$ in 1 liter of water.

SPECIAL SOLUTIONS AND REAGENTS

Aluminon (qualitative test for aluminum). Aluminon is a trade name for the ammonium salt of aurin tricarboxylic acid. Dissolve 1 g of the salt in 1 liter of distilled water. Shake the solution well to insure thorough mixing.

Bang's reagent (for glucose estimation). Dissolve 100 g of K_2CO_3, 66 g of KCl and 160 g of $KHCO_3$ in the order given in about 700 ml of water at 30° C. Add 4.4 g of $CuSO_4$ and dilute to 1 liter after the CO_2 is evolved. This solution should be shaken only in such a manner as not to allow entry of air. After 24 hours 300 ml are diluted to 1 liter with saturated KCl solution, shaken gently and used after 24 hours; 50 ml equivalent to 10 mg glucose.

Barfoed's reagent (test for glucose). See *Cupric acetate.*

Baudisch's reagent. See *Cupferron.*

Benedict's solution (qualitative reagent for glucose). With the aid of heat, dissolve 173 g of sodium citrate and 100 g of Na_2CO_3 in 800 ml of water. Filter, if necessary, and dilute to 850 ml. Dissolve 17.3 g of $CuSO_4.5H_2O$ in 100 ml of water. Pour the latter solution, with constant stirring, into the arbonate-citrate solution, and make up to 1 liter.

Benzidine hydrochloride solution (for sulfate determination). Make a paste of 8 g of benzidine hydrochloride ($C_{12}H_8(NH_2)_2.2HCl$) and 20 ml of water, add 20 ml of HCl (sp. gr. 1.12) and dilute to 1 liter with water. Each ml of this solution is equivalent to 0.00357 g of H_2SO_4.

Bertrand's reagent (glucose estimation). Consists of the following solutions:

(a) Dissolve 200 g of Rochelle salts and 150 g of NaOH in sufficient water to make 1 liter of solution.

(b) Dissolve 40 g of $CuSO_4$ in enough water to make 1 liter of solution.

(c) Dissolve 50 g of $Fe_2(SO_4)_3$ and 200 g of H_2SO_4 (sp. gr. 1.84) in sufficient water to make 1 liter of solution.

(d) Dissolve 5 g of $KMnO_4$ in sufficient water to make 1 liter of solution.

Bial's reagent (for pentose). Dissolve 1 g of orcinol ($CH_3.C_6H_3(OH)_2$) in 500 ml of 30 % HCl to which 30 drops of a 10 % solution of $FeCl_3$ has been added.

Boutron-Boudet soap solution.

(a) Dissolve 100 g of pure castile soap in about 2500 ml of 56 % ethyl alcohol.

(b) Dissolve 0.59 g of $Ba(NO_3)_2$ in 1 liter of water.

Adjust the castile soap solution so that 2.4 ml of it will give a permanent lather with 40 ml of solution (b). When adjusted, 2.4 ml of soap solution is equivalent to 220 parts per million of hardness (as $CaCO_3$) for a 40 ml sample.

See also *Soap solution.*

Brucke's reagent (protein precipitation). See *Potassium iodide-mercuric iodide.*

Clarke's soap solution (or A.P.H.A. standard method). Estimation of hardness in water.

(a) Dissolve 100 g of pure powdered castile soap in 1 liter of 80 % ethyl alcohol and allow to stand over night.

(b) Prepare a standard solution of $CaCl_2$ by dissolving 0.5 g of $CaCO_3$ in HCl (sp. gr. 1.19), neutralize with NH_4OH and make slightly alkaline to litmus, and dilute to 500 ml. One ml is equivalent to 1 mg of $CaCO_3$.

Titrate (a) against (b) and dilute (a) with 80 % ethyl alcohol until 1 ml of the resulting solution is equivalent to 1 ml of (b) after making allowance for the lather factor (the amount of standard soap solution required to produce a permanent lather in 50 ml of distilled water). One ml of the adjusted solution after subtracting the lather factor is equivalent to 1 mg of $CaCO_3$.

See also *Soap solution.*

Cobalticyanide paper (Rinnmann's test for Zn). Dissolve 4 g of $K_3Co(CN)_6$ and 1 g of $KClO_3$ in 100 ml of water. Soak filter paper in solution and dry at 100° C. Apply drop of zinc solution and burn in an evaporating dish. A green disk is obtained if zinc is present.

Cochineal. Extract 1 g of cochineal for four days with 20 ml of alcohol and 60 ml of distilled water. Filter.

Congo red. Dissolve 0.5 g of congo red in 90 ml of distilled water and 10 ml of alcohol.

Cupferron (Baudisch's reagent for iron analysis). Dissolve 6 g of the ammonium salt of nitroso-phenyl-hydroxylamine (cupferron) in 100 ml of H_2O. Reagent good for one week only and must be kept in the dark.

Cupric acetate (Barfoed's reagent for reducing monosaccharides). Dissolve 66 g of cupric acetate and 10 ml of glacial acetic acid in water and dilute to 1 liter.

Cupric oxide, ammoniacal; Schweitzer's reagent (dissolves cotton, linen and silk, but not wool).

1. Dissolve 5 g of cupric sulfate in 100 ml of boiling water, and add sodium hydroxide until precipitation is complete. Wash the precipitate well, and dissolve it in a minimum quantity of ammonium hydroxide.

2. Bubble a slow stream of air through 300 ml of strong ammonium hydroxide containing 50 g of fine copper turnings. Continue for one hour.

Cupric sulfate in glycerin-potassium hydroxide (reagent for silk). Dissolve 10 g of cupric sulfate, $CuSO_4.5H_2O$, in 100 ml of water and add 5 g of glycerin. Add KOH solution slowly until a deep blue solution is obtained.

Cupron (benzoin oxime). Dissolve 5 g in 100 ml of 95 % alcohol.

Cuprous chloride, acidic (reagent for CO in gas analysis).

1. Cover the bottom of a two-liter flask with a layer of cupric oxide about one-half inch deep, suspend a bunch of copper wire so as to reach from the bottom to the top of the solution, and fill the flask with hydrochloric acid (sp. gr. 1.10). Shake occasionally. When the solution becomes nearly colorless, transfer to reagent bottles, which should also contain copper wire. The stock bottle may be refilled with dilute hydrochloric acid until either the cupric oxide or the copper wire is used up.

Copper sulfate may be substituted for copper oxide in the above procedure.

2. Dissolve 340 g of $CuCl_2.2H_2O$ in 600 ml of conc. HCl and reduce the cupric chloride by adding 190 ml of a saturated solution of stannous chloride or until the solution is colorless. The stannous chloride is prepared by treating 300 g of metallic tin in a 500 ml flask with conc. HCl until no more tin goes into solution.

3. (Winkler method). Add a mixture of 86 g of CuO and 17 g of finely divided metallic Cu, made by the reduction of CuO with hydrogen, to a solution of HCl, made by diluting 650 ml of conc. HCl with 325 ml of water. After the mixture has been added slowly and with frequent stirring, a spiral of copper wire is suspended in the bottle, reaching all the way to the bottom. Shake occasionally, and when the solution becomes colorless, it is ready for use.

Cuprous chloride, ammoniacal (reagent for CO in gas analysis).

1. The acid solution of cuprous chloride as prepared above is neutralized with ammonium hydroxide until an ammonia odor persists. An excess of metallic copper must be kept in the solution.

2. Pour 800 ml of acidic cuprous chloride, prepared by the Winkler method, into about 4 liters of water. Transfer the

SPECIAL SOLUTIONS AND REAGENTS (Continued)

precipitate to a 250 ml graduate. After several hours, siphon off the liquid above the 50 ml mark and refill with 7.5% NH_4OH solution which may be prepared by diluting 50 ml of conc. NH_4OH with 150 ml of water. The solution is well shaken and allowed to stand for several hours. It should have a faint odor of ammonia.

Dichlorofluorescein indicator. Dissolve 1 g in 1 liter of 70% alcohol or 1 g of the sodium salt in 1 liter of water.

Dimethylglyoxime (diacetyl dioxime), 0.01 *N*. Dissolve 0.6 g of dimethylglyoxime, $(CH_3CNOH)_2$, in 500 ml of 95% ethyl alcohol. This is an especially sensitive test for nickel, a very definite crimson color being produced.

Diphenylamine (reagent for rayon). Dissolve 0.2 g in 100 ml of concentrated sulfuric acid.

Diphenylamine sulfonate (for titration of iron with $K_2Cr_2O_7$). Dissolve 0.32 g of the barium salt of diphenylamine sulfonic acid in 100 ml of water, add 0.5 g of sodium sulfate and filter off the precipitate of $BaSO_4$.

Diphenylcarbazide. Dissolve 0.2 g of diphenylcarbazide in 10 ml of glacial acetic acid and dilute to 100 ml with 95% ethyl alcohol.

Esbach's reagent (estimation of protein). To a water solution of 10 g of picric acid and 20 g of citric acid, add sufficient water to make one liter of solution.

Eschka's compound. Two parts of calcined ("light") magnesia are thoroughly mixed with one part of anhydrous sodium carbonate.

Fehling's solution (reagent for reducing sugars).

(a) Copper sulfate solution. Dissolve 34.66 g of $CuSO_4 \cdot 5H_2O$ in water and dilute to 500 ml.

(b) Alkaline tartrate solution. Dissolve 173 g of potassium sodium tartrate (Rochelle salts, $KNaC_4H_4O_6 \cdot 4H_2O$) and 50 g of NaOH in water and dilute when cold to 500 ml.

For use, mix equal volumes of the two solutions at the time of using.

Ferric-alum indicator. Dissolve 140 g of ferric-ammonium sulfate crystals in 400 ml of hot water. When cool, filter, and make up to a volume of 500 ml with dilute (6 *N*) nitric acid.

Folin's mixture (for uric acid). To 650 ml of water add 500 g of $(NH_4)_2SO_4$, 5 g of uranium acetate and 6 g of glacial acetic acid. Dilute to 1 liter.

Formaldehyde-sulfuric acid (Marquis' reagent for alkaloids). Add 10 ml of formaldehyde solution to 50 ml of sulfuric acid.

Froehde's reagent. See *Sulfomolybdic acid.*

Fuchsin (reagent for linen). Dissolve 1 g of fuchsin in 100 ml of alcohol.

Fuchsin-sulfurous acid (Schiff's reagent for aldehydes). Dissolve 0.5 g of fuchsin and 9 g of sodium bisulfite in 500 ml of water, and add 10 ml of HCl. Keep in well-stoppered bottles and protect from light.

Gunzberg's reagent (detection of HCl in gastric juice). Prepare as needed a solution containing 4 g of phloroglucinol and 2 g of vanillin in 100 ml of absolute ethyl alcohol.

Hager's reagent. See *Picric acid.*

Hanus solution (for iodine number). Dissolve 13.2 g of resublimed iodine in one liter of glacial acetic acid which will pass the dichromate test for reducible matter. Add sufficient bromine to double the halogen content, determined by titration (3 ml is about the proper amount). The iodine may be dissolved by the aid of heat, but the solution should be cold when the bromine is added.

Iodine, tincture of. To 50 ml of water add 70 g of I_2 and 50 g of KI. Dilute to 1 liter with alcohol.

Iodo-potassium iodide (Wagner's reagent for alkaloids). Dissolve 2 g of iodine and 6 g of KI in 100 ml of water.

Litmus (indicator). Extract litmus powder three times with boiling alcohol, each treatment consuming an hour. Reject the alcoholic extract. Treat residue with an equal weight of cold water and filter; then exhaust with five times its weight of boiling water, cool and filter. Combine the aqueous extracts.

Magnesia mixture (reagent for phosphates and arsenates). Dissolve 55 g of magnesium chloride and 105 g of ammonium chloride in water, barely acidify with hydrochloric acid, and dilute to 1 liter. The ammonium hydroxide may be omitted until just previous to use. The reagent, if completely mixed and stored for any period of time, becomes turbid.

Magnesium reagent. See *S and O reagent.*

Magnesium uranyl acetate. Dissolve 100 g of $UO_2(C_2H_3O_2)_2 \cdot 2H_2O$ in 60 ml of glacial acetic acid and dilute to 500 ml. Dissolve 330 g of $Mg(C_2H_3O_2)_2 \cdot 4H_2O$ in 60 ml of glacial acetic acid and dilute to 200 ml. Heat solutions to the boiling point until clear, pour the magnesium solution into the uranyl solution, cool and dilute to 1 liter. Let stand over night and filter if necessary.

Marme's reagent. See *Potassium-cadmium iodide.*

Marquis' reagent. See *Formaldehyde-sulfuric acid.*

Mayer's reagent (white precipitate with most alkaloids in slightly acid solutions). Dissolve 1.358 g of $HgCl_2$ in 60 ml of water and pour into a solution of 5 g of KI in 10 ml of H_2O. Add sufficient water to make 100 ml.

Methyl orange indicator. Dissolve 1 g of methyl orange in 1 liter of water. Filter, if necessary.

Methyl orange, modified. Dissolve 2 g of methyl orange and 2.8 g of xylene cyanole FF in 1 liter of 50% alcohol.

Methyl red indicator. Dissolve 1 g of methyl red in 600 ml of alcohol and dilute with 400 ml of water.

Methyl red, modified. Dissolve 0.50 g of methyl red and 1.25 g of xylene cyanole FF in 1 liter of 90% alcohol. Or, dissolve 1.25 g of methyl red and 0.825 g of methylene blue in 1 liter of 90% alcohol.

SPECIAL SOLUTIONS AND REAGENTS (Continued)

Millon's reagent (for albumins and phenols). Dissolve 1 part of mercury in 1 part of cold fuming nitric acid. Dilute with twice the volume of water and decant the clear solution after several hours.

Mixed indicator. Prepared by adding about 1.4 g of xylene cyanole FF to 1 g of methyl orange. The dye is seldom pure enough for these proportions to be satisfactory. Each new lot of dye should be tested by adding additional amounts of the dye until a test portion gives the proper color change. The acid color of this indicator is like that of permanganate; the neutral color is gray; and the alkaline color is green. Described by Hickman and Linstead, J. Chem. Soc. (Lon.), **121,** 2502 (1922).

Molisch's reagent. See *α-Naphthol.*

α-Naphthol (Molisch's reagent for wool). Dissolve 15 g of α-naphthol in 100 ml of alcohol or chloroform.

Nessler's reagent (for ammonia). Dissolve 50 g of KI in the smallest possible quantity of cold water (50 ml). Add a saturated solution of mercuric chloride (about 22 g in 350 ml of water will be needed) until an excess is indicated by the formation of a precipitate. Then add 200 ml of 5 *N* NaOH and dilute to 1 liter. Let settle, and draw off the clear liquid.

Nickel oxide, ammoniacal (reagent for silk). Dissolve 5 g of nickel sulfate in 100 ml of water, and add sodium hydroxide solution until nickel hydroxide is completely precipitated. Wash the precipitate well and dissolve in 25 ml of concentrated ammonium hydroxide and 25 ml of water.

***p*-Nitrobenzene-azo-resorcinol** (reagent for magnesium). Dissolve 1 g of the dye in 10 ml of *N* NaOH and dilute to 1 liter.

Nitron (detection of nitrate radical). Dissolve 10 g of nitron ($C_{20}H_{16}N_4$, 4, 5-dihydro-1, 4-diphenyl-3, 5-phenylimino-1, 2, 4-triazole) in 5 ml of glacial acetic acid and 95 ml of water. The solution may be filtered with slight suction through an alundum crucible and kept in a dark bottle.

α-Nitroso-β-naphthol. Make a saturated solution in 50% acetic acid (1 part of glacial acetic acid with 1 part of water). Does not keep well.

Nylander's solution (carbohydrates). Dissolve 20 g of bismuth subnitrate and 40 g of Rochelle salts in 1 liter of 8% NaOH solution. Cool and filter.

Obermayer's reagent (for indoxyl in urine). Dissolve 4 g of $FeCl_3$ in one liter of HCl (sp. gr. 1.19).

Oxine. Dissolve 14 g of HC_9H_6ON in 30 ml of glacial acetic acid. Warm slightly, if necessary. Dilute to 1 liter.

Oxygen absorbent. Dissolve 300 g of ammonium chloride in one liter of water and add one liter of concentrated ammonium hydroxide solution. Shake the solution thoroughly. For use as an oxygen absorbent, a bottle half full of copper turnings is filled nearly full with the NH_4Cl-NH_4OH solution and the gas passed through.

Pasteur's salt solution. To one liter of distilled water add 2.5 g of potassium phosphate, 0.25 g of calcium phosphate, 0.25 g of magnesium sulfate and 12.00 g of ammonium tartrate.

Pavy's solution (glucose reagent). To 120 ml of Fehling's solution, add 300 ml of NH_4OH (sp. gr. 0.88) and dilute to 1 liter with water.

Phenanthroline ferrous ion indicator. Dissolve 1.485 g of phenanthroline monohydrate in 100 ml of 0.025 *M* ferrous sulfate solution.

Phenolphthalein. Dissolve 1 g of phenolphthalein in 50 ml of alcohol and add 50 ml of water.

Phenolsulfonic acid (determination of nitrogen as nitrate). Dissolve 25 g of phenol in 150 ml of conc. H_2SO_4, add 75 ml of fuming H_2SO_4 (15% SO_3), stir well and heat for two hours at 100° C.

Phloroglucinol solution (pentosans). Make a 3% phloroglucinol solution in alcohol. Keep in a dark bottle.

Phosphomolybdic acid (Sonnenschein's reagent for alkaloids).

1. Prepare ammonium phosphomolybdate and after washing with water, boil with nitric acid and expel NH_3; evaporate to dryness and dissolve in 2 *N* nitric acid.

2. Dissolve ammonium molybdate in HNO_3 and treat with phosphoric acid. Filter, wash the precipitate, and boil with aqua regia until the ammonium salt is decomposed. Evaporate to dryness. The residue dissolved in 10% HNO_3 constitutes Sonnenschein's reagent.

Phosphoric acid—sulfuric acid mixture. Dilute 150 ml of conc. H_2SO_4 and 100 ml of conc. H_3PO_4 (85%) with water to a volume of 1 liter.

Phosphotungstic acid (Scheibler's reagent for alkaloids).

1. Dissolve 20 g of sodium tungstate and 15 g of sodium phosphate in 100 ml of water containing a little nitric acid.

2. The reagent is a 10% solution of phosphotungstic acid in water. The phosphotungstic acid is prepared by evaporating a mixture of 10 g of sodium tungstate dissolved in 5 g of phosphoric acid (sp. gr. 1.13) and enough boiling water to effect solution. Crystals of phosphotungstic acid separate.

Picric acid (Hager's reagent for alkaloids, wool and silk). Dissolve 1 g of picric acid in 100 ml of water.

Potassium antimonate (reagent for sodium). Boil 22 g of potassium antimonate with 1 liter of water until nearly all of the salt has dissolved, cool quickly, and add 35 ml of 10% potassium hydroxide. Filter after standing over night.

Potassium-cadmium iodide (Marme's reagent for alkaloids). Add 2 g of CdI_2 to a boiling solution of 4 g of KI in 12 ml of water, and then mix with 12 ml of saturated KI solution.

Potassium hydroxide (for CO_2 absorption). Dissolve 360 g of KOH in water and dilute to 1 liter.

Potassium iodide-mercuric iodide (Brucke's reagent for proteins). Dissolve 50 g of KI in 500 ml of water, and saturate with mercuric iodide (about 120 g). Dilute to 1 liter.

Potassium pyrogallate (for oxygen absorption). For mixtures of gases containing less than 28% oxygen, add 100 ml of KOH solution (50 g of KOH to 100 ml of water) to 5 g of pyrogallol. For mixtures containing more than 28% oxygen the KOH solution should contain 120 g of KOH to 100 ml of water.

Pyrogallol, alkaline.

(a) Dissolve 75 g of pyrogallic acid in 75 ml of water.

(b) Dissolve 500 g of KOH in 250 ml of water. When cool, adjust until sp. gr. is 1.55.

For use, add 270 ml of solution (b) to 30 ml of solution (a).

Rosolic acid (indicator). Dissolve 1 g of rosolic acid in 10 ml of alcohol and add 100 ml of water.

S and O reagent (Suitsu and Okuma's test for Mg). Dissolve 0.5 g of the dye (*o-p*-dihydroxy-monazo-*p*-nitrobenzene) in 100 ml of 0.25 *N* NaOH.

Scheibler's reagent. See *Phosphotungstic acid.*

Schiff's reagent. See *Fuchsin-sulfurous acid.*

Schweitzer's reagent. See *Cupric oxide, ammoniacal.*

Soap solution (reagent for hardness in water). Dissolve 100 g of dry castile soap in 1 liter of 80% alcohol (5 parts alcohol to 1 part water). Allow to stand several days and dilute with 70% to 80% alcohol until 6.4 ml produces a permanent lather with 20 ml of standard calcium solution. The latter solution is made by dissolving 0.2 g of $CaCO_3$ in a small amount of dilute HCl, evaporating to dryness and making up to 1 liter.

Sodium bismuthate (oxidation of manganese). Heat 20 parts of NaOH nearly to redness in an iron or nickel crucible and add slowly 10 parts of basic bismuth nitrate which has been previously dried. Add two parts of sodium peroxide, and pour the brownish-yellow fused mass on an iron plate to cool. When cold, break up in a mortar, extract with water, and collect on an asbestos filter.

Sodium hydroxide (for CO_2 absorption). Dissolve 330 g of NaOH in water and dilute to 1 liter.

Sodium nitroprusside (reagent for hydrogen sulfide and wool). Use a freshly prepared solution of 1 g of sodium nitroprusside in 10 ml of water.

Sodium oxalate, according to Sörensen (primary standard). Dissolve 30 g of the commercial salt in 1 liter of water, make slightly alkaline with sodium hydroxide, and let stand until perfectly clear. Filter and evaporate the filtrate to 100 ml. Cool and filter. Pulverize the residue and wash it several times with small volumes of water. The procedure is repeated until the mother liquor is free from sulfate and is neutral to phenolphthalein.

Sodium plumbite (reagent for wool). Dissolve 5 g of sodium hydroxide in 100 ml of water. Add 5 g of litharge and boil until dissolved.

Sodium polysulfide. Dissolve 480 g of $Na_2S.9H_2O$ in 500 ml of water, add 40 g of NaOH and 18 g of sulfur. Stir thoroughly and dilute to 1 liter with water.

Sonnenschein's reagent. See *Phosphomolybdic acid.*

Starch solution.

1. Make a paste with 2 g of soluble starch and 0.01 g of HgI_2 with a small amount of water. Add the mixture slowly to 1 liter of boiling water and boil for a few minutes. Keep in a glass stoppered bottle. If other than soluble starch is used, the solution will not clear on boiling; it should be allowed to stand and the clear liquid decanted.

2. A solution of starch which keeps indefinitely is made as follows: Mix 500 ml of saturated NaCl solution (filtered), 80 ml of glacial acetic acid, 20 ml of water and 3 g of starch. Bring slowly to a boil and boil for two minutes.

3. Make a paste with 1 g of soluble starch and 5 mg of HgI_2, using as little cold water as possible. Then pour about 200 ml of boiling water on the paste and stir immediately. This will give a clear solution if the paste is prepared correctly and the water actually boiling. Cool and add 4 g of KI. Starch solution decomposes on standing due to bacterial action, but this solution will keep a long time if stored under a layer of toluene.

Stoke's reagent. Dissolve 30 g of $FeSO_4$ and 20 g of tartaric acid in water and dilute to 1 liter. Just before using, add concentrated NH_4OH until the precipitate first formed is redissolved.

Sulfanilic acid (reagent for nitrites). Dissolve 0.5 g of sulfanilic acid in a mixture of 15 ml of glacial acetic acid and 135 ml of recently boiled water.

Sulfomolybdic acid (Froehde's reagent for alkaloids and glucosides). Dissolve 10 g of molybdic acid or sodium molybdate in 100 ml of conc. H_2SO_4.

Tannic acid (reagent for albumen, alkaloids and gelatin). Dissolve 10 g of tannic acid in 10 ml of alcohol and dilute with water to 100 ml.

Titration mixture. See *Zimmermann-Reinhardt reagent.*

***o*-Tolidine solution** (residual chlorine in water analysis). Prepare 1 liter of dilute HCl (100 ml of HCl (sp. gr. 1.19) in sufficient water to make 1 liter). Dissolve 1 g of *o*-tolidine in 100 ml of the dilute HCl and dilute to 1 liter with dilute HCl solution.

Trinitrophenol solution. See *Picric acid.*

Turmeric paper. Impregnate white, unsized paper with the tincture, and dry.

Turmeric tincture (reagent for borates). Digest ground turmeric root with several quantities of water which are discarded. Dry the residue and digest it several days with six times its weight of alcohol. Filter.

Uffelmann's reagent (turns yellow in presence of a lactic acid). To a 2% solution of pure phenol in water, add a water solution of $FeCl_3$ until the phenol solution becomes violet in color.

SPECIAL SOLUTIONS AND REAGENTS (Continued)

Wagner's reagent. See *Iodo-potassium iodide.*

Wagner's solution (used in phosphate rock analysis to prevent precipitation of iron and aluminum). Dissolve 25 g of citric acid and 1 g of salicylic acid in water and dilute to 1 liter. Use 50 ml of the reagent.

Wij's iodine monochloride solution (for iodine number). Dissolve 13 g of resublimed iodine in 1 liter of glacial acetic acid which will pass the dichromate test for reducible matter. Set aside 25 ml of this solution. Pass into the remainder of the solution dry chlorine gas (dried and washed by passing through H_2SO_4 (sp. gr. 1.84)) until the characteristic color of free iodine has been discharged. Now add the iodine solution which was reserved, until all free chlorine has been destroyed. A slight excess of iodine does little or no harm, but an excess of chlorine must be avoided. Preserve in well stoppered, amber colored bottles. Avoid use of solutions which have been prepared for more than 30 days.

Wij's special solution (for iodine number—Analyst **58**, 523–7, 1933). To 200 ml of glacial acetic acid that will pass the dichromate test for reducible matter, add 12 g of dichloroamine T (paratoluene-sulfonedichloroamide), and 16.6 g of dry KI (in small quantities with continual shaking until all the KI has dissolved). Make up to 1 liter with the same quality of acetic acid used above and preserve in a dark colored bottle.

Zimmermann-Reinhardt reagent (determination of iron). Dissolve 70 g of $MnSO_4.4H_2O$ in 500 ml of water, add 125 ml of conc. H_2SO_4 and 125 ml of 85% H_3PO_4, and dilute to 1 liter.

Zinc chloride solution, basic (reagent for silk). Dissolve 1000 g of zinc chloride in 850 ml of water, and add 40 g of zinc oxide. Heat until solution is complete.

Zinc uranyl acetate (reagent for sodium). Dissolve 10 g of $UO_2(C_2H_3O_2)_2.2H_2O$ in 6 g of 30% acetic acid with heat, if necessary, and dilute to 50 ml. Dissolve 30 g of $Zn(C_2H_3O_2)_2.2H_2O$ in 3 g of 30% acetic acid and dilute to 50 ml. Mix the two solutions, add 50 mg of NaCl, allow to stand over night and filter.

DECI-NORMAL SOLUTIONS OF SALTS AND OTHER REAGENTS

Atomic and molecular weights in the following table are based upon the 1965 atomic weight scale and the isotope C-12. The weight in grams of the compound in 1 cc of the following deci-normal solutions is found by dividing the H equivalent in the last column by 1000.

Name	Formula	Atomic or molecular weight	Hydrogen equivalent	0.1 Hydrogen equivalent in g
Acetic acid	$HC_2H_3O_2$	60.0530	$HC_2H_3O_2$	6.0053
Ammonia	NH_3	17.0306	NH_3	1.7031
Ammonium ion	NH_4^+	18.0386	NH_4	1.8039
Ammonium chloride	NH_4Cl	53.4916	NH_4Cl	5.3492
Ammonium sulfate	$(NH_4)_2SO_4$	132.1388	$\frac{1}{2}(NH_4)_2SO_4$	6.6069
Ammonium thiocyanate	NH_4CNS	76.1204	NH_4CNS	7.6120
Barium	Ba	137.34	$\frac{1}{2}Ba$	6.867
Barium carbonate	$BaCO_3$	197.3494	$\frac{1}{2}BaCO_3$	9.8675
Barium chloride hydrate	$BaCl_2 \cdot 2H_2O$	244.2767	$\frac{1}{2}BaCl_2 \cdot 2H_2O$	12.2138
Barium hydroxide	$Ba(OH)_2$	171.3547	$\frac{1}{2}Ba(OH)_2$	8.5677
Barium oxide	BaO	153.3394	$\frac{1}{2}BaO$	7.6670
Bromine	Br	79.909	Br	7.9909
Calcium	Ca	40.08	$\frac{1}{2}Ca$	2.004
Calcium carbonate	$CaCO_3$	100.0894	$\frac{1}{2}CaCO_3$	5.0045
Calcium chloride	$CaCl_2$	110.9860	$\frac{1}{2}CaCl_2$	5.5493
Calcium chloride hydrate	$CaCl_2.6H_2O$	219.0150	$\frac{1}{2}CaCl_2.6H_2O$	10.9508
Calcium hydroxide	$Ca(OH)_2$	74.0947	$\frac{1}{2}Ca(OH)_2$	3.7047
Calcium oxide	CaO	56.0794	$\frac{1}{2}CaO$	2.8040
Chlorine	Cl	35.453	Cl	3.5453
Citric acid	$C_6H_8O_7 \cdot H_2O$	210.1418	$\frac{1}{3}C_6H_8O_7 \cdot H_2O$	7.0047
Cobalt	Co	58.9332	$\frac{1}{2}Co$	2.9466
Copper	Cu	63.54	$\frac{1}{2}Cu$	3.177
Copper oxide (cupric)	CuO	79.5394	$\frac{1}{2}CuO$	3.9770
Copper sulfate hydrate	$CuSO_4 \cdot 5H_2O$	249.6783	$\frac{1}{2}CuSO_4 \cdot 5H_2O$	12.4839
Cyanogen	$(CN)_2$	26.0179	CN	2.6018
Hydrochloric acid	HCl	36.4610	HCl	3.6461
Hydrocyanic acid	HCN	27.0258	HCN	2.7026
Iodine	I	126.9044	I	12.6904
Lactic acid	$C_3H_6O_3$	90.0795	$C_3H_6O_3$	9.0080
Malic acid	$C_4H_6O_5$	134.0894	$\frac{1}{2}C_4H_6O_5$	6.7045
Magnesium	Mg	24.312	$\frac{1}{2}Mg$	1.2156
Magnesium carbonate	$MgCO_3$	84.3214	$\frac{1}{2}MgCO_3$	4.2161
Magnesium chloride	$MgCl_2$	95.2180	$\frac{1}{2}MgCl_2$	4.7609
Magnesium chloride hydrate	$MgCl_2 \cdot 6H_2O$	203.2370	$\frac{1}{2}MgCl_2 \cdot 6H_2O$	10.1623
Magnesium oxide	MgO	40.3114	$\frac{1}{2}MgO$	2.0156
Manganese	Mn	54.938	$\frac{1}{2}Mn$	2.7469
Manganese sulfate	$MnSO_4$	150.9996	$\frac{1}{2}MnSO_4$	7.5500
Mercuric chloride	$HgCl_2$	271.4960	$\frac{1}{2}HgCl_2$	13.5748
Nickel	Ni	58.71	$\frac{1}{2}Ni$	2.9356
Nitric acid	HNO_3	63.0129	HNO_3	6.3013
Nitrogen	N	14.0067	N	1.4007
Nitrogen pentoxide	N_2O_5	108.0104	$\frac{1}{2}N_2O_5$	5.4005
Oxalic acid	$H_2C_2O_4$	90.0358	$\frac{1}{2}H_2C_2O_4$	4.5018
Oxalic acid hydrate	$H_2C_2O_4 \cdot 2H_2O$	126.0665	$\frac{1}{2}H_2C_2O_4 \cdot 2H_2O$	6.3033
Oxalic acid anhydride	C_2O_3	72.0205	$\frac{1}{2}C_2O_3$	3.6010
Phosphoric acid	H_3PO_4	97.9953	$\frac{1}{3}H_3PO_4$	3.2665
Potassium	K	39.102	K	3.9102
Potassium bicarbonate	$KHCO_3$	100.1193	$KHCO_3$	10.0119
Potassium carbonate	K_2CO_3	138.2134	$\frac{1}{2}K_2CO_3$	6.9106
Potassium chloride	KCl	74.5550	KCl	7.4555
Potassium cyanide	KCN	65.1199	KCN	6.5120
Potassium hydroxide	KOH	56.1094	KOH	5.6109
Potassium oxide	K_2O	94.2034	$\frac{1}{2}K_2O$	4.7102
Potassium permanganate for Co estimation	$KMnO_4$	158.0376	$\frac{1}{6}KMnO_4$	2.6339
Potassium permanganate for Mn estimation	$KMnO_4$	158.0376	$\frac{1}{3}KMnO_4$	5.2678
Potassium tartrate	$K_2H_4C_4O_6$	226.2769	$\frac{1}{2}K_2H_4C_4O_6$	11.3139
Silver	Ag	107.87	Ag	10.787
Silver nitrate	$AgNO_3$	169.8749	$AgNO_3$	16.9875
Sodium	Na	22.9898	Na	2.2990
Sodium bicarbonate	$NaHCO_3$	84.0071	$NaHCO_3$	8.4007
Sodium carbonate	Na_2CO_3	105.9890	$\frac{1}{2}Na_2CO_3$	5.2995
Sodium chloride	NaCl	58.4428	NaCl	5.8443
Sodium hydroxide	NaOH	39.9972	NaOH	3.9997
Sodium oxide	Na_2O	61.9790	$\frac{1}{2}Na_2O$	3.0990
Sodium sulfide	Na_2S	78.0436	$\frac{1}{2}Na_2S$	3.9022
Succinic acid	$H_2C_4H_4O_4$	118.0900	$\frac{1}{2}H_2C_4H_4O_4$	5.9045
Sulfuric acid	H_2SO_4	98.0775	$\frac{1}{2}H_2SO_4$	4.9039
Sulfur trioxide	SO_3	80.0622	$\frac{1}{2}SO_3$	4.0031
Tartaric acid	$C_4H_6O_6$	150.0888	$\frac{1}{2}C_4H_6O_6$	7.5044
Zinc	Zn	65.37	$\frac{1}{2}Zn$	3.269
Zinc sulfate	$ZnSO_4 \cdot 7H_2O$	287.5390	$\frac{1}{2}ZnSO_4 7 \cdot H_2O$	14.3769

DECI-NORMAL SOLUTIONS OF OXIDATION AND REDUCTION REAGENTS

Atomic and molecular weights in the following table are based upon the 1965 atomic weight scale and the isotope C-12. The weight in grams of the compound in 1 cc of the following deci-normal solutions is found by dividing the H equivalent in the last column by 1000.

Name	Formula	Atomic or molecular weight	Hydrogen equivalent	0.1 Hydrogen equivalent in g
Antimony	Sb	121.75	$\frac{1}{2}Sb$	6.0875
Arsenic	As	74.9216	$\frac{1}{2}As$	3.7461
Arsenic trisulfide	As_2S_3	246.0352	$\frac{1}{4}As_2S_3$	6.1509
Arsenous oxide	As_2O_3	197.8414	$\frac{1}{4}As_2O_3$	4.9460
Barium peroxide	BaO_2	169.3388	$\frac{1}{2}BaO_2$	8.4669
Barium peroxide hydrate	$BaO_2{\cdot}8H_2O$	313.4615	$\frac{1}{2}BaO_2{\cdot}8H_2O$	15.6730
Calcium	Ca	40.08	$\frac{1}{2}Ca$	2.004
Calcium carbonate	$CaCO_3$	100.0894	$\frac{1}{2}CaCO_3$	5.0045
Calcium hypochlorite	$Ca(OCl)_2$	142.9848	$\frac{1}{4}Ca(OCl)_2$	3.5746
Calcium oxide	CaO	56.0794	$\frac{1}{2}CaO$	2.8040
Chlorine	Cl	35.453	Cl	3.5453
Chromium trioxide	CrO_3	99.9942	$\frac{1}{3}CrO_3$	3.3331
Ferrous ammonium sulfate	$FeSO_4(NH_4)SO_4{\cdot}6H_2O$	392.0764	$FeSO_4(NH_4)_2SO_4{\cdot}6H_2O$	39.2076
Hydroferrocyanic acid	$H_4Fe(CN)_6$	215.9860	$H_4Fe(CN)_6$	21.5986
Hydrogen peroxide	H_2O_2	34.0147	$\frac{1}{2}H_2O_2$	1.7007
Hydrogen sulfide	H_2S	34.0799	$\frac{1}{2}H_2S$	1.7040
Iodine	I	126.9044	I	12.6904
Iron	Fe	55.847	Fe	5.5847
Iron oxide (ferrous)	FeO	71.8464	FeO	7.1846
Iron oxide (ferric)	Fe_2O_3	159.6922	$\frac{1}{2}Fe_2O_3$	7.9846
Lead peroxide	PbO_2	239.1888	$\frac{1}{2}PbO_2$	11.9594
Manganese dioxide	MnO_2	86.9368	$\frac{1}{2}MnO_2$	4.3468
Nitric acid	HNO_3	63.0129	$\frac{1}{3}HNO_3$	2.1004
Nitrogen trioxide	N_2O_3	76.0116	$\frac{1}{4}N_2O_3$	1.9002
Nitrogen pentoxide	N_2O_5	108.0104	$\frac{1}{6}N_2O_5$	1.8001
Oxalic acid	$C_2H_2O_4$	90.0358	$\frac{1}{2}C_2H_2O_4$	4.5018
Oxalic acid hydrate	$C_2H_2O_4{\cdot}2H_2O$	126.0665	$\frac{1}{2}C_2H_2O_4{\cdot}2H_2O$	6.3033
Oxygen	O	15.9994	$\frac{1}{2}O$	0.8000
Potassium dichromate	$K_2Cr_2O_7$	294.1918	$\frac{1}{6}K_2Cr_2O_7$	4.9032
Potassium chlorate	$KClO_3$	122.5532	$\frac{1}{6}KClO_3$	2.0425
Potassium chromate	K_2CrO_4	194.1076	$\frac{1}{3}K_2CrO_4$	6.4733
Potassium ferrocyanide	$K_4Fe(CN)_6$	368.3621	$K_4Fe(CN)_6$	36.8362
Potassium ferrocyanide	$K_4Fe(CN)_6{\cdot}3H_2O$	422.4081	$K_4Fe(CN)_6{\cdot}3H_2O$	42.2408
Potassium iodide	KI	166.0064	KI	16.6006
Potassium nitrate	KNO_3	101.1069	$\frac{1}{3}KNO_3$	3.3702
Potassium perchlorate	$KClO_4$	138.5526	$\frac{1}{8}KClO_4$	1.7319
Potassium permanganate	$KMnO_4$	158.0376	$\frac{1}{5}KMnO_4$	3.1608
Sodium chlorate	$NaClO_3$	106.4410	$\frac{1}{6}NaClO_3$	1.7740
Sodium nitrate	$NaNO_3$	84.9947	$\frac{1}{3}NaNO_3$	2.8332
Sodium thiosulfate	$Na_2S_2O_3{\cdot}5H_2O$	248.1825	$Na_2S_2O_3{\cdot}5H_2O$	24.8183
Stannous chloride	$SnCl_2$	189.5960	$\frac{1}{2}SnCl_2$	9.4798
Stannous oxide	SnO	134.6894	$\frac{1}{2}SnO$	6.7345
Sulfur dioxide	SO_2	64.0628	$\frac{1}{2}SO_2$	3.2031
Tin	Sn	118.69	$\frac{1}{2}Sn$	5.935

ORGANIC ANALYTICAL REAGENTS

Compiled by John H. Yoe

Determination	Reagent	Reference
Acetate	*o*-Nitrobenzaldehyde	Feigl, p. 342 (3)
Aldehydes	Dimethyl-dihydroresorcin (Dimedon)	Ind. Eng. Chem., Anal. Ed. **3**, 365 (1931)
Aluminum	Alizarin S	J. Am. Chem. Soc. **50**, 748 (1928)
	Ammonium salt of aurin tricarboxylic acid ("Aluminon")	J. Am. Chem. Soc. **49**, 2395 (1927) J. Am. Chem. Soc. **55**, 2437 (1933)
	Ammonium salt of nitrosophenyl hydroxylamine ("Cupferron")	Bull. soc. chim. Belg. **36**, 288 (1927)
	Eriochrome cyanine	Z. anal. Chem. **96**, 91 (1934)
	Hematoxylin	Ind. Eng. Chem. **16**, 233 (1924)
	8-Hydroxyquinoline	J. Am. Chem. Soc. **50**, 1900 (1928)
	Morin	Feigl, p. 182 (2) Ind. Eng. Chem., Anal. Ed. **12**, 229 (1940)
	Quinalizarine	J. Am. Pharm. Assoc. **17**, 260 (1928)
	Sodium or zinc salt of 4-sulfo-2 hydroxy-α-naphthalene-azo-β-naphthol (Pontachrome Blue Black R)	Sandell, p. 241 (6)
	Urea	Ind. Eng. Chem., Anal. Ed. **9**, 357 (1937)
Ammonia	Hematoxylin	Helv. Chim. Acta **12**, 730 (1929)
	p-Nitrobenzenediazonium chloride	Feigl, p. 235 (2)
	Phenol and sodium hypochloride	Snell, Vol. II, p. 818 (8)
	Tannin—$AgNO_3$	Snell, Vol. II, p. 819 (8)
	Thymol	J. Biol. Chem. **131**, 309 (1939)
	Zinc oxinate	Feigl, p. 674 (9)
Antimony	Hexamethylenetetramine	Z. anal. Chem. **67**, 298 (1925)
	9-Methyl-2,3,7-tryhydroxyfluorone	Helv. Chim. Acta **20**, 1427 (1937)
	Phenylthiohydantoic acid	Compt. rend. **176**, 1221 (1923)
	Pyridine	Analyst **53**, 373 (1928)
	Pyrogallol	Z. anal. Chem. **64**, 44 (1924)
	Rhodamine B	Z. anal. Chem. **70**, 400 (1927)
Arsenic	Cocaine-molybdate	Biochem. Z. **185**, 14 (1927)
	N-Ethyl-8-hydroxytetrahydroquinoline hydrochloride	Z. anal. Chem. **99**, 180 (1934)
	Quinine arsenomolybdate	Analyst **47**, 317 (1922)
	Strychnine-molybdate	Ann. Chim. applicata **23**, 517 (1933)
Barium	Sodium rhodizonate	Feigl, p. 216 (2)
	Tetrahydroxyquinone	Welcher, Vol. I, p. 225 (9)
Beryllium	1-Amino-4-hydroxyanthraquinone	Ind. Eng. Chem., Anal. **13**, 809 (1941)
	Aurin trycarboxylic	J. Am. Chem. Soc. **48**, 2125 (1926); ibid., **50**, 353 (1928)
	Curcumin	J. Am. Chem. Soc. **50**, 393 (1928)
	1,4-Dihydroxyanthraquinone (Quinizarin)	Ind. Eng. Chem. Anal. Ed., **18**, 179 (1946)
	1.4-Dihydroxyanthraquinone-2-sulfonic acid (quinizarine-2-sulfonic acid)	Sandell, p. 318 (6)
	8-Hydroxyquinoline	Bur. Standards J. Research **3**, 91 (1929)
	Morin	Ind. Eng. Chem., Anal. Ed. **12**, 674, 762 (1940)
	Naphthochrome Azurine 2B	Sandell, p. 318 (6)
	Naphthochrome Green G	Sandell, p. 318 (6)
	p-Nitrobenzeneazo-orcinol	Mikrochemie **14**, 315 (1934)
	1,2,5,8-Tetrahydroxyanthraquinone (Quinalizarin)	Siemens-Konzerns, Beryllium, p. 25 (1932)
Bismuth	Caffeine (as sulfate or nitrate)	Ind. Eng. Chem., Anal. Ed. **14**, 43 (1942)
	Cinchonine	Scott, p. 158 (7)
	Diethyldithiocarbamate	Sandell, p. 338 (6)
	Dimercaptothiodiazole	Z. anal. Chem. **98**, 184 (1934); ibid., **100**, 408 (1935)
	Dimethylglyoxime	Z. anal. Chem. **72**, 11 (1927)
	Diphenylthiocarbazone	Z. Angew. Chem. **47**, 685 (1934); Ind. Eng. Chem., Anal. Ed. **7**, 285 (1935)
	8-Hydroxyquinoline	Z. anal. Chem. **72**, 177 (1927)
	Phenyldithiobiazolonethiol	J. Indian Chem. Soc., **21**, 240, 347 (1944)
	Pyrogallol	Z. anal. Chem. **65**, 448 (1925)
	Thiourea	Z. anal. Chem. **94**, 161 (1933)
Boron	Curcumin	Chem. News **87**, 27 (1903)
	1,1-Dianthramide	Boltz, p. 346 (1)
	Diaminochrysazin	Anal. Chem. **29**, 1251 (1957)
	Diaminoanthrarufin	Ibid.
	Tribromoanthrarufin	Ibid.
	Mannitol	Scott, p. 168 (7)
	Methyl alcohol	J. Am. Chem. Soc. **50**, 1385 (1928)
	p-Nitrobenzeneazochromotropic acid (Chromotrope 2B)	Feigl, p. 341 (2)
	Quinalizarin	Ind. Eng. Chem., Anal. Ed. **11**, 540 (1939)
	Titan Yellow (Clayton Yellow)	Compt. rend. **138**, 1046 (1904)
	Turmeric	Ind. Eng. Chem., Anal. Ed. **4**, 180 (1932)
Bromine	Fluorescein	Snell, Vol. II, p. 725 (8)
	Fuchsin	Snell, Vol. II, p. 724 (8)
	Phenol red	Snell, Vol. II, p. 725 (8)
Cadmium	Allythiourea	Helvetica Chim. Acta. **12**, 718 (1929)
	Di-β-naphthylthiocarbazone	Ind. Eng. Chem., Anal. Ed., **16**, 333, (1944)
	Dinitrodiphenylcarbazide	Feigl, p. 96 (2)
	Diphenylcarbazide	Feigl, p. 99 (2)
	Diphenylthiocarbazone	Z. angew. Chem. **47**, 685 (1934); Ind. Eng. Chem., Anal. Ed. **11**, 364 (1939)
	Ethylenediamine	Z. anal. Chem. **77**, 340 (1929)
	Hexamethylenetetramine alliodide	C. A., **24**, 311 (1930)
	β-Naphthoquinoline	Analyst, **58**, 667 (1933)
	Nitrophenolarsinic acid	Mikrochemie 8, 277 (1930)
	4-Nitrophthalene-diazo-aminobenzene-4-azobenzene (Cadion 2B)	C. A. **32**, 2871 (1938)
	Phenyl-trimethyl-ammonium iodide	Analyst **58**, 667 (1933)
	Pyridine	Z. Anal. Chem. **73**, 279 (1928)
Calcium	Alizarin	Biochem. J. **16**, 494 (1922); Yoe, Vol. I, p. 139 (2)
	1-amino-2-naphthol-4-sulfonic acid	J. Biol. Chem. **81**, 1 (1929)
	Ammonium oxalate	Snell, Vol. II, p. 592 (8)
	Ammonium stearate	J. Biol. Chem. **29**, 169 (1917); Yoe, Vol. II, p. 119 (3)
	2,5-Dichloro-3,6-dihydroxyquinone (Chloranilic acid)	Anal. Chem., **20**, 76 (1948)
	Dihydroxytartaric acid osazone (sodium salt)	Feigl, p. 221 (2)
	Picrolonic acid	Biochem. Z. **265**, 85 (1933)
	Potassium oleate	Sandell, p. 377 (6)
	Pyrogallol carboxylic acid	Sandell, p. 380 (6)
	Sodium sulforicinate	Biochem. Z. **137**, 157 (1923); Yoe, Vol. II, p. 125 (3)
Cerium	Benzidine	Feigl, p. 211 (2)
	Brucine	Sandell, p. 386 (6)
	Gallic acid	Snell, Vol. II, p. 608 (8)
	8-Hydroxyquinoline	Analyst, **75**, 275 (1948)
	Malachite Green	C. A. **30**, 5143 (1936), **31**, 626 (1937)
	Morphine	Sandell, p. 386 (6)
	Sulfanilic acid	Sandell, p. 386 (6)
Cesium	Dipicrylamine (Hexanitrodiphenylamine)	Mikrochemie 18, 175 (1935)
Chlorate	Aniline hydrochloride	Snell, Vol. II, p. 717 (8)
Chlorine	Benzidine hydrochloride	Ind. Eng. Chem., Anal. Ed. **4**, 2 (1932)
	Dimethyl-*p*-phenylene-diamine	Chem. Weekblad. **23**, 203 (1926)
	Oleic acid	J. Soc. Chem. Ind. **42**, 427A (1923)
	Sodium sulforicinate	Biochem. Z., **137**, 157 (1923); Yoe, Vol. II, p. 125 (3)
	Thymolphthalein	Ind. Eng. Chem. **19**, 112 (1927)
	o-Tolidine	Yoe, Vol. I, p. 157 (2)
Columbium	Benzidine	Feigl, p. 171 (2)
	1,8-Dihydroxynaphthalene-3,6-Disulfonate	Ind. Eng. Chem. **5**, 298 (1913)
	s-Diphenylcarbazide	J. Am. Chem. Soc. **50**, 2363 (1928)
	Pyrogallol dimethyl ether	C. A. **4**, 3178 (1910)
	Serichrome Blue R	Ind. Eng. Chem., Amal. Ed., **4**, 245 (1932)

ORGANIC ANALYTICAL REAGENTS (Continued)

Determination	Reagent	Reference
Cobalt	Anthranilic acid (o-aminobenzoic acid)	Z. anal. Chem. **93**, 241 (1933)
	Cysteine hydrochloride	J. Biol. Chem. **83**, 367 (1929)
	Dimethylglyoxime	J. Am. Chem. Soc. **43**, 482 (1921)
	3,5-Dimethylpyrazole	Ind. Eng. Chem., Anal. Ed. **2**, 38 (1930)
	Dinitrosoresorcinol	J. Am. Chem. Soc. **45**, 1439 (1923)
	Formaldoxime reagent	C. A. **27**, 927 (1933)
	o-Nitrosocresol	Sandell, p. 422 (6)
	α-Nitroso-β-naphthol	Chem. Zeit. **46**, 430 (1922)
	Nitrose-R-salt	J. Am. Chem. Soc. **43**, 746 (1921)
	β-Nitroso-α-naphthol	Ind. Eng. Chem., Anal. Ed. **12**, 405 (1940)
	o-Nitrosoresorcinol	Ind. Eng. Chem., Anal. Ed. **15**, 310 (1943)
	o-Nitrosophenol	Sandell, p. 422 (6)
	Phenylthiohydantoic acid	J. Am. Chem. Soc. **44**, 2219 (1922)
	Rubeanic acid (Dithio-oxamide)	Anal. Chim. Acta **20**, 332, 435 (1959)
	2,2′,2″-Terpyridyl	Ind. Eng. Chem., Anal. Ed., **15**, 74 (1943)
	Niobium	
Columbium	See **Nobium**	
Copper	Ammonium salt of nitrosophenyl-hydroxylamine ("Cupferron")	Ind. Eng. Chem. **3**, 629 (1911)
	m-Benzamino-semicarbazide	Snell, Vol. II, p. 129 (6)
	Benzidine	Z. anal. Chem. **67**, 31 (1925)
	α-Benzionoxime (Cupron)	Ber. **56**, 2083 (1923)
	Benzotriazole	Ind. Eng. Chem., Anal. Ed. **13**, 349 (1941)
	2-Carboxy-a′-hydroxy-5′ sulfoformazylbenzene (Zincon)	Anal. Chem. **26**, 1345 (1954)
	Diacetyl-dioxime	Analyst **54**, 333 (1929)
	Dibenzyldithiocarbamate	Sandell, p. 443 (6)
	Dihydroxyethyldithiocarbamic acid	Sandell, p. 443 (6)
	p-Dimethylaminobenzalrhodanine	J. Am. Chem. Soc. **52**, 2222 (1930); J. Am. Chem. Soc. **47**, 1268 (1925)
	s-Diphenylcarbazide	Chem. Weekblad. **21**, 20 (1924)
	Diphenylthiocarbazone (Dithizone)	J. Assoc. Official Agr. Chem. **18**, 192 (1935)
	Hydroquinone	Bull. soc. chim. **31**, 1176 (1922)
	Isatin	Rec trav chim. **42**, 199 (1923)
	Mercaptobenzothiazole	Z. anal. Chem. **102**, 24, 108 (1935)
	α-Naphthol	Bull. soc. chim. **31**, 1176 (1922)
	β-Naphthol	Am. J. Pharm. **105**, 62 (1933)
	Phenolphthalein	Compt. rend. **173**, 1082 (1921)
	Phenylthiohydantoic acid	J. Am. Chem. Soc. **44**, 225 (1922)
	Piperidinium piperidyl-dithioformate	Analyst **56**, 736 (1931)
	Potassium ethyl xanthate	Yoe, Vol. I, p. 184 (2)
	Pyridine	Z. anal. Chem. **67**, 27 (1925)
	Rubeanic acid (Dithio-oxamide)	Anal. Chim. Acta **20**, 332, 435 (1959)
	Salicylaldoxime	J. Chem. Soc. (1933) 314; Feigl, p. 40 (8)
	Salicylic acid	Yoe, Vol. I, p. 183 (2)
	Sodium diethyldithiocarbamate	Analyst **54**, 650 (1929)
	o-Toluidine	Z. anal. Chem. **67**, 31 (1925)
	Urobilin	Chem. Weekblad. **27**, 552 (1930)
Cyanide	Benzidine	Feigl, p. 276 (2)
	Phenolphthalin	Analyst **60**, 294 (1935)
	Picric acid	Helv. Chim. Acta **12**, 713 (1929); J. Am. Chem. Soc. **51**, 1171 (1929)
Fluoride	Acetylacetone	Ind. Eng. Chem., Anal. Ed. **5**, 300 (1933)
	Alizarin sodium sulfonate—$Zr(NO_3)_4$	Ind. Eng. Chem., Anal. Ed. **7**, 23 (1935)
	p-Dimethylaminoazophenylarsonic acid	Feigl, p. 271 (2)
	Triphenyltin chloride	J. Am. Chem. Soc. **54**, 4625 (1932)
	Quinalizarine—$Zr(NO_3)_4$	Ind. Eng. Chem., Anal. Ed. **6**, 61 (1934)
Gallium	Camphoric acid	Welcher, Vol. II, p. 31 (9)
	8-Hydroxyquinoline	Ind. Eng. Chem., Anal. Ed. **13**, 844 (1941)
	Morin	Mikrochemie **20**, 194 (1936)
	Quinalizarin	J. Am. Chem. Soc. **59**, 40 (1937)
Germanium	Benzidine	Mikrochemie **18**, 66 (1935)
	Quinalizarin	Mikrochemie **18**, 48 (1935)
	Tannin	Welcher, Vol. II, p. 160 (9)
Gold	Benzidine	Bull. Chim. Farm. **52**, 461 (1912)
	o-Dianisidine	Sandell, p. 507 (6)
	Dimethylaminobenzylidene rhodanine	Feigl, p. 127 (2)
	Formaldehyde	Bull. soc. chim. **31**, 717 (1922)
	p-Fuchsine	Ind. Eng. Chem., Anal. Ed., **18**, 400 (1946)
	Malachite Green	Sandell, p. 506 (6)
	m-Phenylenediamine sulfate	Chem. Zeit. **36**, 934 (1912)
	Phenylhydrazine	Ann. chim. anal. **12**, 90 (1907)
	o-Toluidine	Analyst **44**, 94 (1919)
Hafnium	2-Hydroxy-5-methylazobenzene-4-sulfonic acid	Feigl, p. 288 (4)
Hydrogen sulfide	p-phenylenedimethyldiamine sulfate	Yoe, Vol. I, p. 375 (2)
Indium	Diphenylthiocarbazone (Dithizone)	Sandell, p. 516 (6)
	8-Hydroxyquinoline	Z. anorg. allgem. Chem. **209**, 129 (1932)
	Morin	Mikrochim. Acta **2**, 287 (1937)
Iodine	Starch	Snell, Vol. II, p. 740 (8)
	o-Toluidine	J. Am. Chem. Soc. **47**, 1000 (1925)
Iridium	Benzidine	Snell, Vol. II, p. 526 (8)
	Malachite Green, leuco base	C. A. **24**, 2689 (1930)
Iron	Acetylacetone	J. Am. Chem. Soc. **26**, 967 (1904)
	Alloxantin	Compt. rend. **180**, 519 (1925)
	Ammonium salt of nitrosophenyl hydroxylamine ("Cupferron")	Ind. Eng. Chem. **3**, 629 (1911)
	Bis-p-chlorophenylphosphoric acid	Feigl, p. 294 (4)
	Cysteine	Biochem. Z. **187**, 255 (1927)
	Dimethyl glyoxime	Z. anorg. Chemie **89**, 401 (1914)
	Dinitrosoresorcinol	J. Am. Chem. Soc. **47**, 1268 (1925)
	Diisonitrosoacetone	Chem. Listy **23**, 496 (1929); C. A. **24**, 801 (1930)
	Dioximes (various)	Anal. Chem. **19**, 1017 (1947)
	Diphenylamine	J. Am. Chem. Soc. **46**, 263 (1924)
	2,2-Bipyridyl	Snell, Vol. II, p. 316 (8)
	Disodium-1,2-dihydroxybenzene-3,5-disulfonate ("Tiron")	Ind. Eng. Chem., Anal. Ed. **16**, 111 (1944)
	Hexamethylenetetramine	Bull soc. chim. Rom. **2**, 89 (1921)
	4 Hydroxybiphenyl-3-carboxylic acid	J. Am. Chem. Soc. **70**, 648 (1948)
	8-Hydroxyquinoline	Bull. soc. chim. biol., **17**, 432 (1935)
	4-Hydroxybiphenyl-3-carboxylic acid	J. Am. Chem. Soc. **70**, 648 (1948)
	7-Iodo-8-hydroxyquinoline-5-sulfonic acid (Ferron)	J. Am. Chem. Soc. **59**, 872 (1937)
	Isonitrosoacetophenone	Ber. **60**, 527 (1927)
	Isonitrosodimethyldihydroresorcinol	Anal. Chem., **20**, 1205 (1948)
	Kojic acid	Ind. Eng. Chem., Anal. Ed., **13**, 612 (1941)
	α-Nitrose-β-naphthol	Bull. soc. chim. **35**, 641 (1924)
	Nitroso-R salt	Ind. Chem., Anal. Ed., **14**, 756 (1942); **16**, 276 (1944)
	o-Phenanthroline	Ind. Eng. Chem., Anal. Ed. **9**, 67 (1937)
	Protacatechvic acid	J. Biol. Chem., **137**, 417 (1941)
	Pyramidone	Pharm.Weekblad, **63**, 1121 (1926)
	Pyrocatechol	Helv. chim. Acta **9**, 835 (1926)
	Salicylaldoxime	Ind Eng. Chem., Anal. Ed., **12**, 448 (1940)
	Salicylic acid	J. Chem. Soc. **93**, 93 (1908)
	Salicylsulfonic acid	Snell, Vol. II, p. 321 (8)
	Sulfosalicylic acid	Biochem. Z. **181**, 391 (1927)
	Thioglycollic acid	J. Am. Chem. Soc. **49**, 1916 (1927)
Lanthanum	8-Hydroxyquinoline	Z. anal. Chem. **107**, 191 (1936)
	Sodium alizarinesulfonate	J. Am. Chem. Soc., **53**, 1217 (1931)
Lead	Ammonium thiocyanate and pyridine	Z. anal. Chem. **72**, 289 (1927)
	Aniline	Ind. Eng. Chem. **11**, 1055 (1919); Yoe, Vol. I, p. 257 (2)
	Anthranilic acid	Z. anal. Chem. **101**, 85 (1935)
	Carminic acid	Mikrochemie **7**, 301 (1929)
	s-Diphenylcarbazide	Yoe, Vol. I, p. 255 (2)
	Diphenylthiocarbazone (Dithizone)	Snell, Vol. IIA, p. 10 (8)
	Hematein	Yoe, Vol. I, p. 257 (2)
	Salicylaldoxime	Ind. Eng. Chem., Anal. Ed. **14**, 359 (1942)
	Tetramethyldiamidodiphenylmethane	Snell, Vol. II, p. 43 (8)
	Thiourea	Z. anorg. Chem., **234**, 224 (1937)

ORGANIC ANALYTICAL REAGENTS (Continued)

Determination	Reagent	Reference
Lithium......	Ammonium stearate	J. Am. Chem. Soc. **52,** 2754 (1930)
	Hexamethylenetetramine	Welcher, Vol. III, p. 131 (9)
	Pyridine	Welcher, Vol. III, p. 33 (9)
Magnesium..	Brilliant yellow	Sandell, p. 598 (6)
	Curcumin	Ind. Eng. Chem., Anal. Ed. **4, 426** (1932)
	Dimethylamine	Z. anorg. Chem. **26,** 347 (1901)
	Hydroquinone	Yoe, Vol. I, p. 264 (2)
	8-Hydroxyquinoline	Z. anal. Chem. **71,** 122 (1927)
	p-Nitrobenzeneazo-α-naphthol	Feigl, p. 225 (2)
	p-Nitrobenzeneazo-resorcinol	J. Am. Chem. Soc. **51,** 1456 (1929)
	Oleic acid	Yoe, Vol. I, p. 270 (2)
	Quinalizarin	Feigl, p. 224 (2)
	Sodium 1-azo-2-hydroxy-3-(2,4-dimethylcarboxaniliodonaphthalene-1'(2-hydroxybenzene-5-sulfonate)	Anal. Chem. **28,** 202 (1956); cf. Anal. Chim. Acta **16,** 155 (1957)
	Thiazole yellow	Sandell, p. 598 (6)
	Titan yellow	Sandell, p. 591 (6)
	Tropeoline OO	Rec. trav. chim., **61,** 849 (1942)
Manganese..	Benzidine	Snell, Vol. II, p. 397 (8)
	Formaldoxime	Ind. Eng. Chem., Anal. Ed. **9,** 445 (1937); **12,** 307 (1940)
	Tetramethyldiaminodiphenylmethane	Feigl, p. 174 (2); Snell, Vol. IIA, p. 311 (8)
	4,4-Tetramethyldiaminotriphenylmethane	J. Biol. Chem., **168,** 537 (1947)
Mercury....	Anthranilic acid	Z. anal. Chem. **101,** 88 (1935)
	p-Dimethylaminobenzalrhodamine	J. Am. Chem. Soc. **52,** 2222 (1930)
	Di-β-naphthylthiocarbazone	Sandell, p. 630 (6)
	s-Diphenylcarbazide	Z. angew. Chem. **39,** 791 (1926)
	Diphenylthiocarbazone (Dithizone)	Z. anal. Chem. **103,** 241 (1935)
	Potassium *s*-diphenylcarbazone	Snell, Vol. II 78 (8)
	Strychnine	" II 76 (8)
Molybdenum	α-Benzoin-oxime (Cupron)	B. S. J. Research **9,** 1 (1932)
	Disodium-1,2-dihydroxy-benzene-3,5-disulfonate (Tiron)	Anal. Chim. Acta **8,** 546 (1953)
	4-Methyl-1,2-dimercaptobenzene (Dithiol)	J. Am. Pharm. Assoc., **37,** 255 (1948)
	Phenylhydrazine	Ber. **36,** 512 (1903)
	Potassium ethyl xanthate and chloroform	J. Am. Chem. Soc. **44,** 1462 (1922)
	Tannic acid	Chem. Eng. Mining Rev. **11,** 258 (1919)
Nickel......	α-Benzil-dioxime	Analyst **38,** 316 (1913)
	Cyclohexanedione-dioxime	Ann. **437,** 148 (1925)
	Dicyandiamidine sulfate	Chem. Zeit. **31,** 335, 911 (1907)
	Diethyldithiocarbamate	Ind. Eng. Chem., Anal. Ed., **18,** 206 (1946)
	Dimethylglyoxime	Chem. Weekblad. **21,** 358 (1924)
	Formaldoxime	Snell, Vol. IIA, p. 265 (8)
	α-Furildioxime	J. Am. Chem. Soc. **47,** 918 (1925)
	Potassium dithiooxalate	J. Am. Chem. Soc. **54,** 1866 (1932)
	Rubianic acid (Dithio-oxamide)	Anal. Chim. Acta **20,** 332 435 (1959)
Niobium (Columbium)	Ammonium salt of nitrosophenyl hydroxylamine ("Cupferron")	Hillebrand et al., p. 120 (5)
	8 Hydroxyquinolin	Wechler, Vol. I, p. 302 (2)
Nitrate....	Brucine	Yoe, Vol. I, p. 318 (2)
	Diphenylamine sulfonic acid	J. Am. Chem. Soc. **55,** 1448 (1933)
	Diphenylbenzidine	Yoe, Vol. I, p. 316 (2)
	Diphenyl-endo-anilo-hydrotriazole ("Nitron")	Fales and Kenny, Inorg. Quant. Anal., p. 347 (1938)
	Phenoldisulfonic acid	Yoe, Vol. I, p. 313 (2)
	Pyrogallol	" p. 319 (2)
	Strychnine sulfate	" p. 320 (2)
	2:4-Xylenol	J. Assoc. Off. Agri. Chem. **18,** 459 (1935)
Nitrite.....	Antipyrin	Yoe, Vol. I, p. 311 (2)
	Dimethylaniline	" p. 311 (2)
	Dimethyl-α-Naphthylamine	Ind. Eng. Chem., Anal. Ed. **1,** 28 (1929)
	Diphenylamine sulfate	Yoe, Vol. I, p. 654 (2)
	α-Naphthylamine and β-Naphthylamine-6,8-Disulfonic acid	J. Pharmacol. **51,** 398 (1934)
	α-Naphthylamine hydrochloride	Yoe, Vol. I, p. 309 (2)
	m-Phenylenediamine	Yoe, Vol. I, p. 310 (2)
	Sulfanilic acid and α-naphthylamine	Yoe, Vol. I, p. 308 (2)
Osmium....	*s*-Diphenylthiourea	Yoe and Sarver, p. 155 (12)
	1-Naphthylamine-4,6,8-trisulfonic acid	Anal. Chim. Acta **20,** 205 (1959)
	Strychnine	Welcher, Vol. IV, p. 272 (9)
	Thiourea	Compt. rend. **167,** 235 (1918)
Oxygen.....	Indigo carmine	Snell, Vol. IIA, p. 726 (8)
	Pyrogallol	Dennis, Gas Analysis, p. 174 (1929)
Palladium..	Dimethylglyoxime	J. Am. Chem. Soc. **57,** 2565 (1935)
	p-Fuchsine	Ind. Eng. Chem., Anal. Ed., **18,** 400 (1946)
	β-Furfuraldoxime	Ind. Eng. Chem., Anal. Ed. **14,** 491 (1942)
	6-Nitroquinoline	J. Am. Chem. Soc. **50,** 3018 (1928)
	p-Nitrosodiphenylamine	J. Am. Chem. Soc. **61,** 2058 (1939)
	p-Nitrosodimethylaniline	J. Am. Chem. Soc. **63,** 3224 (1941)
	p-Nitrosodiethylaniline	Ibid.
	Thiomalic acid	Talanta **2,** 223 (1959)
Phosphate..	1,2,4-Aminonaphthosulfonic acid	Yoe, Vol. I, p. 348 (2)
	Hydroquinone	Yoe, Vol. I, pp. 346 and 353 (2)
	Quinine-molybdate	Yoe, Vol. I, p. 343 (2)
	Strychnine-molybdate	Yoe, Vol. II, p. 142 (3)
Phosphorus.	Hydrazine sulfate	Yoe, Vol. I, p. 341 (2)
Platinum...	*p*-Nitrosodimethylaniline	Anal. Chem. **26,** 1335, 1340 (1954)
Potassium..	6-Chloro-5-nitrotoluene-3-sulfonic acid	Mikrochem. **14,** 368 (1934)
	Dipicrylamine	Z. angew. Chem. **49,** 827 (1936)
	Picric acid	J. Am. Chem. Soc. **53,** 539 (1931)
Rhodium...	Thiomalic acid	Talanta **2,** 239 (1959)
Ruthenium.	*s*-Diphenylthiourea	Yoe and Sarver, p. 155 (12)
	5-Hydroxyquinoline-8-carboxylic acid	Canadian J. Research B25, **49,** (1945)
	1-Naphthylamine-3,5,7-trisulfonic acid	Anal. Chim Acta **20,** 211 (1959)
	Rubianic acid (Dithio-oxamide)	Mikrochemie **15,** 295 (1934)
	Thioglycolyl-β-amidonaphthalide (Thionalid)	Ind. Eng. Chem., Anal. Ed. **12,** 5611 (1940)
	Thiourea	Sandell, p. 781 (6)
Scandium..	Morin	Mikrochem. Acta **2,** 9, 287 (1937)
Selenium...	Codeine phosphate	Arch Pharm. **252,** 161 (1914)
	Hydrazine	Boltz, p. 321 (1)
	Hydroquinone	Am. J. Sci. **15,** 253 (1928)
	Hydroxylamine hydrochloride	J. Am. Chem. Soc. **47,** 2456 (1925)
	Pyrrol	Snell, Vol. II, p. 779 (8)
	Thiourea	Ann. chim. applicata **17,** 357 (1927); Feigl, p. 231 (8)
Silver......	Chromotropic acid	Helvetica Chim. Acta **12,** 714 (1929)
	Dichlorofluorescein	J. Am. Chem. Soc. **51,** 3273 (1929)
	p-Dimethylaminobenzalrhodamine	J. Am. Chem. Soc. **52,** 2222 (1930)
	Diphenylthiocarbazone (Dithizone)	Z. anal. Chem. **101,** 1 (1935)
	Formazylcarboxylic acid	J. Anal. Chem. Russ., **2,** 131 (1934); abs. in Analyst, **73,** 352 (1948)
	Methylamine	Mikrochemie **7,** 233 (1929)
	2-Thio-5-keto-4-carbethoxy-1,3-dihydropyrimidine	Ind. Eng. Chem., Anal. Ed. **14,** 148 (1942)
Sodium....	6,8-Dichlorobenzoylurea	J. Org. Chem. **3,** 414 (1938)
	Dihydroxy-tartaric acid	J. Russ. Phys. Chem. Soc. **60,** 661 (1928)
	Uranyl zinc acetate	J. Am. Chem. Soc. **51,** 1664 (1929)
Strontium.	Sodium rhodizonate	Mikrochemie **2,** 187 (1924)
Sulfide.....	*p*-Aminodimethylaniline	Snell, Vol. IIA, p. 658 (8)
Sulfur.....	*p*-Phenylenedimethyl-diamine-hydrochloride	Yoe, Vol. I, p. 373 (2)

ORGANIC ANALYTICAL REAGENTS

Determination	Reagent	Reference
Tantalum	Ammonium salt of nitrosophenyl hydroxylamine ("Cupferron")	Hillebrand et. al., p. 120 (5)
	Pyrogallol	Sandell, p. 695 (6)
Tellurium	Hydrazine hydrochloride	J. Am. Chem. Soc. **47**, 2456 (1925)
	Hydroquinone	Am. J. Sci. **15**, 253 (1928)
	Thiourea	Ann. chim. applicata **17**, 359 (1927)
Thallium	Diphenylthiocarbazone (Dithizone)	Analyst **60**, 394 (1935)
	Thionalid (Thioglycolyl-β-amido-naphthalide)	Z. angew. Chem. **48**, 430, 597 (1935)
Thorium	1-Amino-4-hydroxy-anthraquinone	Ind. Eng. Chem., Anal. Ed. **13**, 809 (1941)
	Cupferron (Ammonium nitrosophenyl-hydroxylamine)	Chem. Ztg. **33**, 1298 (1908)
	8-Hydroxyquinoline	Z. anal. Chem. **100**, 98 (1935)
	Phenylarsonic acid	J. Am. Chem. Soc. **48**, 895 (1926)
Tin	Ammonium salt of nitrosophenyl hydroxylamine ("Cupferron")	Hillebrand et al., p. 120 (5)
	Cacotheline	Ind. Eng. Chem., Anal. Ed. **7**, 26 (1935)
	4-Chloro-1,2-dimercaptobenzene (1-chloro-benzene-3,4-dithiol)	J. Chem. Soc. **149**, 175 (1936)
	Hematoxylin	Sandell, p. 866 (6)
	Quinalizarin	Sandell, p. 866 (6)
	Toluene-3,4-dithiol (4-methyl-1,2-dimercaptobenzene)	Analyst **61**, 242 (1936); Ibid., **62**, 661 (1937)
Titanium	Ammonium salt of nitrosophenyl hydroxylamine ("Cupferron")	Hillebrand et al., p. 119 (5); Z. anal. Chem. **83**, 345 (1931)
	Chromotropic acid	Feigl, p. 197 (2)
	5,7-Dibromo-8-hydroxyquinoline	Z. anorg. Chem. **204**, 215 (1932)
	Dihydroxymaleic acid	Snell, Vol. II, p. 445 (8)
	Disodium 1,2-dihydroxybenzene-3,5-disulfonate ("Tiron")	Ind. Eng. Chem., Anal. Ed. **19**, 100 (1947)
	Gallic acid	Snell, Vol. II, p. 444 (8)
	p-Hydroxyphenyl-arsonic acid	Ind. Eng. Chem., Anal. Ed. **10**, 642 (1938)
	8-Hydroxyquinoline	Z. anal. Chem. **81**, 1 (1930)
	Resoflavine (Color Index 1015)	Anal. Chim. Acta, **1**, 244 (1947)
	Tannic acid	Analyst **55**, 605 (1930)
	Thymol	Yoe, Vol. I, p. 381 (2)
Tungsten (Wolfram)	Anti-1,5-di-(p-methoxyphenyl)-1-hydroxylamino-3-oximino-4-pentene ("Wolfron")	Ind. Eng. Chem., Anal. Ed. **16**, 45 (1944)
	Benzidine	Ber. **38**, 783 (1905)
	α-Benzoinoxime ("Cupron")	Bur. Std. J. Research **9**, 1 (1932)
	Cinchonine	Hillebrand et al., p. 689 (5)
	Hydroquinone	Z. angew. Chem. **44**, 237 (1931)
	Phenylhydrazine	Bull. soc. chim. Belg. **38**, 385 (1929)
	Rhodamine B.	Snell, Vol. II, p. 469 (8)
Tungsten (Wolfram) (Cont.)	Toluene-3,4-dithiol	Analyst, **69**, 109 (1944); J. Am. Pharm. Assoc., **37**, 255 (1948)
	Uric acid	Ann. chim. anal. **9**, 371 (1904)
	α-Benzoinoxime ("Cupron")	Bur. Std. J. Research **9**, 1 (1932)
	Cinchonine	Hillebrand et al., p. 689 (5)
	Hydroquinone	Z. angew. Chem. **44**, 237 (1931)
	Phenylhydrazine	Bull. soc. chim. Belg. **38**, 385 (1929)
	Rhodamine B.	Snell, Vol. II, p. 469 (8)
	Toluene-3,4-dithiol	Analyst, **69**, 109 (1944); J. Am. Pharm. Assoc., **37**, 255 (1948)
	Uric acid	Ann. chim. anal. **9**, 371 (1904)
Uranium	Dibenzoylmethane	Anal. Chem. **25**, 1200 (1953)
	o-Hydroxybenzoic acid	Snell, Vol. II, p. 492 (8)
	α-Quinaldinic acid	Z. anal. Chem. **95**, 400 (1933)
	Sodium diethyldithiocarbamate	Sandell, p. 921 (6)
	Sodium salicylate	Chem. Zeit. **43**, 739 (1919)
	Thioglycolic acid (Mercapto acetic acid)	Anl. Chem., **21**, 1093 (1949)!
Urea	Xanthydrol	Mikrochem. **14**, 132 (1934)
Vanadium	Ammonium benzoate	C. A. **29**, 2880 (1935)
	Aniline	C. A. **24**, 567 (1930)
	Benzidine	J. Applied Chem. (U.S.S.R.) **17**, 83 (1944); C. A. **39**, 1115 (1945)
	Diphenylamine	Yoe, Vol. I, p. 715 (2)
	Diphenylbenzidine	Ind. Eng. Chem. **20**, 764 (1928)
	8-Hydroxyquinoline	Feigl, p. 125 (2)
	Safranine	Vol. Anal., Vol. II, p. 326 (6)
	Strychnine	Yoe, Vol. I, p. 393 (2)
Wolfram	See Tungsten	
Zinc	Anthranilic acid	Z. anal. Chem. **91**, 332 (1933)
	2 Carboxy-2′hydroxy-5′-sulfoformazybenzene (Zincon)	Anal. Chem. **26**, 1345 (1954)
	Di-β-napthylthiocarbazone	Sandell, pp. 945, 962 (6)
	Diphenylamine	J. Am. Chem. Soc. **49**, 2214 (1927)
	Diphenylbenzidine	J. Am. Chem. Soc. **49**, 356 (1927)
	Diphenylthiocarbazone (Dithizone)	Ind. Eng. Chem., Anal. Ed. **9**, 127 (1937)
	8-Hydroxyquinoline	Z. anal. Chem. **71**, 171 (1927)
	5-Nitroquinaldic acid	Ind. Eng. Chem., Anal. Ed. **10**, 335 (1938)
	Pyridine	Z. anal. Chem. **73**, 356 (1928)
	α-Quinaldinic acid	Z. anal. Chem. **100**, 324 (1935)
	Resorcinol	Yoe, Vol. I, p. 396 (2)
	Urobilin	J. Ind. Hyg. **7**, 273 (1925)
Zirconium	Ammonium salt of nitrosophenyl hydroxylamine ("Cupferron")	Hillebrand et al., p. 572 (5)
	5-Chlorobromamine acid	Ind. Eng. Chem., Anal. Ed. **15**, 73 (1943)
	p-Dimethylaminoazophenylarsenic acid	Ind. Eng. Chem., Anal. Ed., **13**, 603 (1941)
	2-Hydroxy-5-methylazobenzene-4-sulfonic acid	Feigl, p. 288 (4)
	Morin	Sandell, p. 971 (6)
	Phenylarsenic acid	J. Am. Chem. Soc. **48**, 895 (1926)
	Sodium alizarin sulfonate (Alizarin S)	Chem. Weekblad **20**, 404 (1924); Feigl, p. 200 (2)

(1) Colorimetric Determination of Nonmetals, Boltz, Edition 1958.
(2) Feigl, Spot Tests in Inorganic Analysis, 5th English Ed., translated by Oesper, 1958.
(3) Feigl, Spot Tests in Organic Analysis, 5th English Ed., translated by Oesper, 1956.
(4) Feigl, Chemistry of Specific, Selective and Sensitive Reactions translated by Oesper, 1949.
(5) Hillebrand, Lundell, Bright and Hoffman, Applied Inorganic Analysis, 2nd Edition, 1953.
(6) Sandell, Colorimetric Determination of Traces of Metals, 3rd Edition, 1959.
(7) Scott's Standard Methods of Chemical Analysis, Furman, 5th Edition 1939.
(8) Snell, Snell and Snell, Colorimetric Methods of Analysis, Volume II (1949), II A (1959).
(9) Welcher, Organic Analytical Reagents, Vols. I–IV, 1947–1948.
(10) Yoe, Photometric Chemical Analysis, Vol. I, Colorimetry, 1928.
(11) Yoe, Photometric Chemical Analysis Vol. II, Nephelometry, 1929
(12) Yoe and Sarver, Organic Analytical Reagents, 1941

CALIBRATION OF VOLUMETRIC GLASSWARE FROM THE WEIGHT OF THE CONTAINED WATER OR MERCURY WHEN WEIGHED IN AIR WITH BRASS WEIGHTS

D. F. SWINEHART

A borosilicate glass vessel containing g_t grams of water at a temperature of t°C has, at the same temperature, a volume $V_t = W_t \times g_t$ cubic centimeters. Similarly when filled with G_t grams of mercury at a temperature of t°C the volume at the same temperature is given by $V_t = M_t \times G_t$ cubic centimeters.

When filled with g_t grams of water at a temperature of t°C the volume of the vessel at 18°C is given by $V_{18} = W_{18} \times g_t$ cubic centimeters and the true volume at 25°C is given by $V_{25} = W_{25} \times g_t$. The volumes at 18°C and 25°C are given similarly when using mercury by using the values under M_{18} and M_{25}, respectively.

The data on water are adapted from the data of G. S. Kell, *Journal of Chemical and Engineering Data*, 12, 67–68 (1967) (Table on p. F5, 52nd Edition, this handbook) and the data on mercury are adapted from *Smithsonian Tables*, Ninth Revised Edition, Volume 120, Publication No. 4169. The coefficient of linear expansion for borosilicate glass used here is 32.5×10^{-7} deg^{-1} and the volume coefficient of expansion is 97.5×10^{-7} deg^{-1}.

t°C	W_t	W_{18}	W_{25}	M_t	M_{18}	M_{25}
0	1.001 220	1.001 396	1.001 466	0.0735 519	0.0735 648	0.0735 698
1	1.001 161	1.001 327	1.001 395	0.0735 653	0.0735 775	0.0735 825
2	1.001 120	1.001 276	1.001 345	0.0735 787	0.0735 902	0.0735 952
3	1.001 096	1.001 242	1.001 311	0.0735 920	0.0736 028	0.0736 078
4	1.001 088	1.001 225	1.001 293	0.0736 054	0.0736 154	0.0736 205
5	1.001 096	1.001 223	1.001 291	0.0736 188	0.0736 281	0.0736 332
6	1.001 120	1.001 237	1.001 306	0.0736 322	0.0736 408	0.0736 458
7	1.001 158	1.001 265	1.001 334	0.0736 456	0.0736 535	0.0736 585
8	1.001 211	1.001 309	1.001 377	0.0736 590	0.0736 662	0.0736 712
9	1.001 279	1.001 367	1.001 435	0.0736 724	0.0736 789	0.0736 839
10	1.001 360	1.001 438	1.001 506	0.0736 858	0.0736 915	0.0736 966
11	1.001 455	1.001 523	1.001 592	0.0736 992	0.0737 042	0.0737 093
12	1.001 563	1.001 622	1.001 690	0.0737 125	0.0737 168	0.0737 218
13	1.001 684	1.001 733	1.001 801	0.0737 259	0.0737 295	0.0737 345
14	1.001 816	1.001 855	1.001 923	0.0737 393	0.0737 422	0.0737 472
15	1.001 961	1.001 990	1.002 059	0.0737 526	0.0737 548	0.0737 598
16	1.002 118	1.002 138	1.002 206	0.0737 660	0.0737 674	0.0737 725
17	1.002 286	1.002 296	1.002 364	0.0737 794	0.0737 801	0.0737 852
18	1.002 466	1.002 466	1.002 534	0.0737 928	0.0737 928	0.0737 978
19	1.002 658	1.002 648	1.002 717	0.0738 062	0.0738 055	0.0738 105
20	1.002 859	1.002 839	1.002 908	0.0738 196	0.0738 182	0.0738 232
21	1.003 072	1.003 043	1.003 111	0.0738 330	0.0738 308	0.0738 359
22	1.003 294	1.003 255	1.003 323	0.0738 463	0.0738 434	0.0738 485
23	1.003 528	1.003 479	1.003 548	0.0738 597	0.0738 561	0.0738 611
24	1.003 771	1.003 712	1.003 781	0.0738 731	0.0738 688	0.0738 738
25	1.004 024	1.003 955	1.004 024	0.0738 864	0.0738 814	0.0738 864
26	1.004 287	1.004 209	1.004 277	0.0738 998	0.0738 940	0.0738 991
27	1.004 560	1.004 472	1.004 540	0.0739 132	0.0739 067	0.0739 118
28	1.004 842	1.004 744	1.004 813	0.0739 266	0.0739 194	0.0739 244
29	1.005 133	1.005 025	1.005 094	0.0739 400	0.0739 321	0.0739 371
30	1.005 434	1.005 316	1.005 385	0.0739 534	0.0739 447	0.0739 498
31	1.005 743	1.005 615	1.005 684	0.0739 669	0.0739 575	0.0739 626
32	1.006 060	1.005 923	1.005 991	0.0739 801	0.0739 700	0.0739 750
33	1.006 388	1.006 241	1.006 310	0.0739 934	0.0739 826	0.0739 876
34	1.006 723	1.006 566	1.006 635	0.0740 068	0.0739 953	0.0740 003
35	1.007 066	1.006 899	1.006 968	0.0740 202	0.0740 079	0.0740 130
36	1.007 418	1.007 242	1.007 311	0.0740 335	0.0740 205	0.0740 256
37	1.007 780	1.007 593	1.007 669	0.0740 469	0.0740 332	0.0740 382
38	1.008 149	1.007 952	1.008 021	0.0740 603	0.0740 459	0.0740 509
39	1.008 525	1.008 318	1.008 387	0.0740 737	0.0740 585	0.0740 636
40	1.008 910	1.008 694	1.008 762	0.0740 871	0.0740 712	0.0740 763
41	1.009 303	1.009 077	1.009 146	0.0741 007	0.0740 841	0.0740 891
42	1.009 703	1.009 467	1.009 536	0.0741 139	0.0740 966	0.0741 016
43	1.010 112	1.009 866	1.009 935	0.0741 273	0.0741 092	0.0741 143
44	1.010 528	1.010 272	1.010 341	0.0741 407	0.0741 219	0.0741 270
45	1.010 951	1.010 685	1.010 754	0.0741 541	0.0741 346	0.0741 396
46	1.011 382	1.011 106	1.011 175	0.0741 675	0.0741 473	0.0741 523
47	1.011 820	1.011 534	1.011 603	0.0741 810	0.0741 600	0.0741 651
48	1.012 266	1.011 970	1.012 039	0.0741 944	0.0741 727	0.0741 778
49	1.012 719	1.012 413	1.012 482	0.0742 078	0.0741 854	0.0741 904
50	1.013 180	1.012 864	1.012 933	0.0742 213	0.0741 981	0.0742 031

SOLUBILITY CHART

Abbreviations: W, soluble in water; A, insoluble in water but soluble in acids; w, sparingly soluble in water but soluble in acids; a, insoluble in water and only sparingly soluble in acids; I, insoluble in both water and acids; d, decomposes in water. * Certain salts occur in two modifications.

No.		Al	NH_4	Sb	Ba	Bi	Cd	Ca	Cr	Co	Cu	Au′	Au‴	H	Fe″	Fe‴
1	Acetates $-(C_2H_3O_2)$	W $Al(-)_3$	W $NH_4(-)$	....	W $Ba(-)_2$	W $Bi(-)_3$	W $Cd(-)_2$	W $Ca(-)_2$	W $Cr(-)_3$	W $Co(-)_2$	W $Cu(-)_2$	....	....	W $C_2H_4O_2$	W $Fe(-)_2$	W $Fe_2(-)_6$
2	Arsenate $-(AsO_4)$	a $Al(-)$	W $(NH_4)_3(-)$	A $Sb(-)$	w $Ba_3(-)_2$	A $Bi(-)$	A $Cd_3(-)_2$	w $Ca_3(-)_2$	....	A $Co_3(-)_2$	A $Cu_3(-)_2$	....	....	W H_3AsO_4	A $Fe_3(-)_2$	A $Fe(-)$
3	Arsenite $-(AsO_3)$	....	W NH_4AsO_2	A $Sb(-)$	....	....	....	w $Ca_3(-)_2$	....	A $Co_3H_6(-)_4$	A $CuH(-)$	....	....	....	....	....
4	Benzoate $-(C_7H_5O_2)$	....	W $NH_4(-)$	....	W $Ba(-)_2$	A $Bi(-)_3$	W $Cd(-)_2$	W $Ca(-)_2$	....	W $Co(-)_2$	w $Cu(-)_2$	....	....	W $C_7H_6O_2$	W $Fe(-)_2$	A $Fe_2(-)_6$
5	Bromide	W $AlBr_3$	W NH_4B1	d $SbBr_3$	W $BaBr_2$	d $BiBr_3$	W $CdBr_2$	W $CaBr_2$	W(I)* $CrBr_3$	W $CoBr_2$	W $CuBr_2$	w $AuBr$	W $AuBr_3$	W HBr	W $FeBr_2$	W $FeBr_3$
6	Carbonate	....	W $(NH_4)_2CO_3$	....	w $BaCO_3$	....	A $CdCO_3$	w $CaCO_3$	W $CrCO_3$	A $CoCO_3$	....	....	....	....	w $FeCO_3$	....
7	Chlorate $-(ClO_3)$	W $Al(-)_3$	W $NH_4(-)$	....	W $Ba(-)_2$	W $Bi(-)_3$	W $Cd(-)_2$	W $Ca(-)_2$	....	W $Co(-)_2$	W $Cu(-)_2$	....	....	W $HClO_3$	W $Fe(-)_2$	W $Fe(-)_3$
8	Chloride	W $AlCl_3$	W NH_4Cl	W $SbCl_3$	W $BaCl_2$	d $BiCl_3$	W $CdCl_2$	W $CaCl_2$	I $CrCl_3$	W $CoCl_2$	W $CuCl_2$	w $AuCl$	W $AuCl_3$	W HCl	W $FeCl_2$	W $FeCl_3$
9	Chromate $-(CrO_4)$	....	W $(NH_4)_2(-)$	....	A $Ba(-)$	....	A $Cd(-)$	W $Ca(-)$	....	A $Co(-)$	....	....	....	....	....	A $Fe_2(-)_3$
10	Citrate $-(C_6H_5O_7)$	W $Al(-)$	W $(NH_4)_3(-)$	....	w $Ba_3(-)_2$	A $Bi(-)$	A $Cd_3(-)_2$	w $Ca_3(-)_2$	....	w $Co_3(-)_2$	....	....	....	W $C_6H_8O_7$	....	W $Fe(-)$
11	Cyanide	....	W NH_4CN	....	W $Ba(CN)_2$	w $Bi(CN)_3$	W $Cd(CN)_2$	W $Ca(CN)_2$	A $Cr(CN)_2$	A $Co(CN)_2$	A $Cu(CN)_2$	w $AuCN$	W $Au(CN)_3$	W HCN	a $Fe(CN)_2$	....
12	Ferricy'de $-(Fe(CN)_6)$	....	W $(NH_4)_3(-)$	....	w $Ba_3(-)_2$	....	A $Cd_3(-)_2$	W $Ca_3(-)_2$	....	I $Co_3(-)_2$	I $Cu_3(-)_2$	....	....	W $H_3(-)$	I $Fe_3(-)_2$	....
13	Ferrocy'de $-(Fe(CN)_6)$	w $Al_4(-)_3$	W $(NH_4)_4(-)$	....	W $Ba_2(-)$	....	A $Cd_2(-)$	W $Ca_2(-)$	....	I $Co_2(-)$	I $Cu_2(-)$	....	....	W $H_4(-)$	I $Fe_2(-)$	a $Fe_4(-)_3$
14	Fluoride	W AlF_3	W NH_4F	W SbF_3	w BaF_2	W BiF_3	W CdF_2	w CaF_2	W(a)* CrF_3	W CoF_2	w CuF_2	....	....	W HF	w FeF_2	w FeF_3
15	Formate $-(CHO_2)$	W $Al(-)_3$	W $NH_4(-)$	....	W $Ba(-)_2$	W $Bi(-)_3$	W $Cd(-)_2$	W $Ca(-)_2$	....	W $Co(-)_2$	W $Cu(-)_2$	....	....	W CH_2O_2	W $Fe(-)_2$	W $Fe(-)_3$
16	Hydroxide	A $Al(OH)_3$	W NH_4OH	....	W $Ba(OH)_2$	A $Bi(OH)_3$	A $Cd(OH)_2$	W $Ca(OH)_2$	A $Cr(OH)_3$	A $Co(OH)_2$	A $Cu(OH)_2$	W $AuOH$	A $Au(OH)_3$	....	A $Fe(OH)_2$	A $Fe(OH)_3$
17	Iodide	W AlI_3	W NH_4I	d SbI_3	W BaI_2	A BiI_3	W CdI_2	W CaI_2	W CrI_2	W CoI_2	a CuI	a AuI	a AuI_3	W HI	W FeI_2	W FeI_3
18	Nitrate	W $Al(NO_3)_3$	W NH_4NO_3	....	W $Ba(NO_3)_2$	d $Bi(NO_3)_3$	W $Cd(NO_3)_2$	W $Ca(NO_3)_2$	W $Cr(NO_3)_3$	W $Co(NO_3)_2$	W $Cu(NO_3)_2$	....	....	W HNO_3	W $Fe(NO_3)_2$	W $Fe(NO_3)$
19	Oxalate $-(C_2O_4)$	A $Al_2(-)_3$	W $(NH_4)_2(-)$	....	w $Ba(-)$	A $Bi_2(-)_3$	w $Cd(-)$	A $Ca(-)$	W $Cr(-)$	A $Co(-)$	A $Cu(-)$	....	....	W $C_2H_2O_4$	A $Fe(-)$	W $Fe_2(-)_3$
20	Oxide	a Al_2O_3	....	w Sb_2O_3	W BaO	A Bi_2O_3	A CdO	w CaO	a Cr_2O_3	A CoO	A CuO	Au_2O	A Au_2O_3	W H_2O_2	A FeO	A Fe_2O_3
21	Phosphate	A $AlPO_4$	W $NH_4H_2PO_4$	....	A $Ba_3(PO_4)_2$	A $BiPO_4$	A $Cd_3(PO_4)_2$	w $Ca_3(PO_4)$	w $Cr_2(PO_4)_2$	A $Co_3(PO_4)_2$	A $Cu_3(PO_4)_2$	....	....	W H_3PO_4	A $Fe_3(PO_4)_2$	w $FePO_4$
22	Silicate, $-(SiO_3)$	I $Al_2(-)_3$	....	....	W $Ba(-)$	....	A $Cd(-)$	w $Ca(-)$	....	A Co_2SiO_4	A $Cu(-)$	....	....	I H_2SiO_3	....	....
23	Sulfate	W $Al_2(SO_4)_3$	W $(NH_4)_2SO_4$	A $Sb_2(SO_4)_3$	a $BaSO_4$	d $Bi_2(SO_4)_3$	W $CdSO_4$	w $CaSO_4$	W(I)* $Cr_2(SO_4)_3$	W $CoSO_4$	W $CuSO_4$	....	....	W H_2SO_4	W $FeSO_4$	w $Fe_2(SO_4)_3$
24	Sulfide	d Al_2S_3	W $(NH_4)_2S$	A Sb_2S_3	d BaS	A Bi_2S_3	A CdS	w CaS	d Cr_2S_3	A CoS	A CuS	I Au_2S	I Au_2S_3	W H_2S	A FeS	d Fe_2S_3
25	Tartrate $-(C_4H_4O_6)$	w $Al_2(-)_3$	W $(NH_4)_2(-)$	W $Sb_2(-)_3$	w $Ba(-)$	A $Bi_2(-)_3$	A $Cd(-)$	w $Ca(-)$	....	w $Co(-)$	w $Cu(-)$	....	....	W $C_4H_6O_6$	w $Fe(-)$	W $Fe_2(-)_3$
26	Thiocy'te	....	W NH_4CNS	....	W $Ba(CNS)_2$	....	....	W $Ca(CNS)$	....	W $Co(CNS)_2$	d $CuCNS$	....	....	W $CNSH$	W $Fe(CNS)_2$	W $Fe(CNS)_3$

No.		Pb	Mg	Mn	Hg′	Hg″	Ni	K	Ag	Na	Sn''''	Sn″	Sr	Zn	Pt
1	Acetate $-(C_2H_3O_2)$	W $Pb(-)_2$	W $Mg(-)_2$	W $Mn(-)_2$	w $Hg(-)$	W $Hg(-)_2$	W $Ni(-)_2$	W $K(-)$	w $Ag(-)$	W $Na(-)$	W $Sn(-)_4$	d $Sn(-)_2$	W $Sr(-)_2$	W $Zn(-)_2$	....
2	Arsenate $-(AsO_4)$	A $PbH(-)$	A $Mg_3(-)$	w $MnH(-)$	A $Hg_3(-)$	w $Hg_3(-)_2$	A $Ni_3(-)_2$	W $K_3(-)$	A $Ag_3(-)$	W $Na_3(-)$	....	....	w $SrH(-)$	A $Zn_3(-)_2$	....
3	Arsenite $-(AsO_3)$	....	W $Mg_3(-)_2$	A $Mn_3H_6(-)_4$	A $Hg_3(-)$	A $Hg_3(-)$	A $Ni_3H_6(-)_4$	W K_3AsO_3	A $Ag_3(-)$	W $Na_2H(-)$	....	A $Sn_3(-)_2$	w $Sr_3(-)_2$	....	....
4	Benzoate $-(C_7H_5O_2)$	w $Pb(-)_2$	W $Mg(-)_2$	W $Mn(-)_2$	A $Hg_2(-)_2$	w $Hg(-)_2$	w $Ni(-)_2$	W $K(-)$	w $Ag(-)$	W $Na(-)$	....	....	....	W $Zn(-)_2$	....
5	Bromide	W $PbBr_2$	W $MgBr_2$	W $MnBr_2$	A $HgBr$	W $HgBr_2$	W $NiBr_2$	W KBr	a $AgBr$	W $NaBr$	W $SnBr_4$	W $SnBr_2$	W $SrBr_2$	W $ZnBr_2$	w $PtBr_4$
6	Carbonate	A $PbCO_3$	w $MgCO_3$	w $MnCO_3$	A Hg_2CO_3	....	w $NiCO_3$	W K_2CO_3	A Ag_2CO_3	W Na_2CO_3	....	....	w $SrCO_3$	w $ZnCO_3$	....
7	Chlorate $-(ClO_3)$	W $Pb(-)_2$	W $Mg(-)_2$	W $Mn(-)_2$	W $Hg(-)$	W $Hg(-)_2$	W $Ni(-)_2$	W $K(-)$	W $Ag(-)$	W $Na(-)$	....	W $Sn(-)_2$	W $Sr(-)_2$	W $Zn(-)_2$	....
8	Chloride	W $PbCl_2$	W $MgCl_2$	W $MnCl_2$	a $HgCl$	W $HgCl_2$	W $NiCl_2$	W KCl	a $AgCl$	W $NaCl$	W $SnCl_4$	W $SnCl_2$	W $SrCl_2$	W $ZnCl_2$	W $PtCl_4$
9	Chromate $-(CrO_4)$	A $Pb(-)$	W $Mg(-)$	....	w $Hg_2(-)$	w $Hg(-)$	A $Ni(-)$	W $K_2(-)$	w $Ag_2(-)$	W $Na_2(-)$	W $Sn(-)_2$	A $Sn(-)$	w $Sr(-)$	w $Zn(-)$	....
10	Citrate $-(C_6H_5O_7)$	W $Pb_3(-)_2$	W $Mg_3(-)_2$	w $MnH(-)$	w $Hg_3(-)$	....	W $Ni_3(-)_2$	W $K_3(-)$	w $Ag_3(-)$	W $Na_3(-)$	....	....	A $SrH(-)$	w $Zn_3(-)_2$	....

SOLUBILITY CHART (Continued)

No.		Pb	Mg	Mn	Hg′	Hg″	Ni	K	Ag	Na	Sn''''	Sn″	Sr	Zn	Pt
11	Cyanide	w $Pb(CN)_2$	W $Mg(CN)_2$	....	A HgCN	W $Hg(CN)_2$	a $Ni(CN)_2$	W KCN	a AgCN	W NaCN	....	....	W $Sr(CN)_2$	A $Zn(CN)_2$	I $Pt(CN)_2$
12	Ferricy'de $-Fe(CN)_6$	w $Pb_3(-)_2$	W $Mg_3(-)_2$	....	......	A $Hg_3(-)_2$	I $Ni_3(-)_2$	W $K_3(-)$	I $Ag_3(-)$	W $Na_3(-)$	....	A $Sn_3(-)_2$	W $Sr_3(-)_2$	A $Zn_3(-)_2$	....
13	Ferrocy'de $-Fe(CN)_6$	a $Pb_2(-)$	W $Mg_2(-)$	A $Mn_2(-)$	......	I $Hg_2(-)$	I $Ni_2(-)$	W $K_4(-)$	I $Ag_4(-)$	W $Na_4(-)$	....	a $Sn_2(-)$	W $Sr_2(-)$	I $Zn_2(-)$	....
14	Fluoride	w PbF_2	w MgF_2	A MnF_2	d HgF	d HgF_2	w NiF_2	W KF	W AgF	W NaF	W SnF_4	W SnF_2	w SrF_2	w ZnF_2	W PtF_4
15	Formate $-(CHO_2)$	W $Pb(-)_2$	W $Mg(-)_2$	W $Mn(-)_2$	w $Hg(-)$	W $Hg(-)_2$	W $Ni(-)_2$	W $K(-)$	W $Ag(-)$	W $Na(-)$	....	....	W $Sr(-)_2$	W $Zn(-)_2$	....
16	Hydroxide	w $Pb(OH)_2$	A $Mg(OH)_2$	A $Mn(OH)_2$	......	A $Hg(OH)_2$	w $Ni(OH)_2$	W KOH	....	W NaOH	w $Sn(OH)_4$	A $Sn(OH)_2$	W $Sr(OH)_2$	A $Zn(OH)_2$	A $Pt(OH)_4$
17	Iodide	w PbI_2	W MgI_2	W MnI_2	A HgI	w HgI_2	W NiI_2	W KI	I AgI	W NaI	d SnI_4	W SnI_2	W SrI_2	W ZnI_2	I PtI_2
18	Nitrate	W $Pb(NO_3)_2$	W $Mg(NO_3)_2$	W $Mn(NO_3)_2$	W $HgNO_3$	W $Hg(NO_3)_2$	W $Ni(NO_3)_2$	W KNO_3	W $AgNO_3$	W $NaNO_3$	....	d $Sn(NO_3)_2$	W $Sr(NO_3)_2$	W $Zn(NO_3)_2$	W $Pt(NO_3)_4$
19	Oxalate $-(C_2O_4)$	A $Pb(-)$	w $Mg(-)$	w $Mn(-)$	a $Hg_2(-)$	A $Hg(-)$	A $Ni(-)$	W $K_2(-)$	a $Ag_2(-)$	W $Na_2(-)$	....	A $Sn(-)$	w $Sr(-)$	A $Zn(-)$	....
20	Oxide	w PbO	A MgO	A MnO	A Hg_2O	w HgO	A NiO	W K_2O	w Ag_2O	d Na_2O	A SnO_2	A SnO	W SrO	w ZnO	A PtO
21	Phosphate	A $Pb_3(PO_4)_2$	w $Mg_3(PO_4)_2$	w $Mn_3(PO_4)_2$	A Hg_3PO_4	A $Hg_3(PO_4)_2$	A $Ni_3(PO_4)_2$	W K_3PO_4	A Ag_3PO_4	W Na_3PO_4	....	A $Sn_3(PO_4)_2$	A $Sr_3(PO_4)_2$	A $Zn_3(PO_4)_2$	....
22	Silicate $--(SiO_3)$	A $Pb(-)$	A $Mg(-)$	I $Mn(-)$	......	....	....	W $K_2(-)$	....	W $Na_2(-)$	....	....	A $Sr(-)$	A $Zn(-)$	....
23	Sulfate	w $PbSO_4$	W $MgSO_4$	W $MnSO_4$	w Hg_2SO_4	d $HgSO_4$	W $NiSO_4$	W K_2SO_4	w Ag_2SO_4	W Na_2SO_4	W $Sn(SO_4)_2$	W $SnSO_4$	w $SrSO_4$	W $ZnSO_4$	W $Pt(SO_4)_2$
24	Sulfide	A PbS	d MgS	A MnS	I Hg_2S	I HgS	A NiS	W K_2S	A Ag_2S	W Na_2S	A SnS_2	A SnS	W SrS	A ZnS	I PtS
25	Tartrate $-(C_4H_4O_6)$	A $Pb(-)$	w $Mg(-)$	w $Mn(-)$	I $Hg_2(-)$	....	A $Ni(-)$	W $K_2(-)$	w $Ag_2(-)$	W $Na_2(-)$	....	W $Sn(-)$	w $Sr(-)$	w $Zn(-)$	....
26	Thiocy'te	w $Pb(CNS)_2$	W $Mg(CNS)_2$	W $Mn(CNS)_2$	A HgCNS	w $Hg(CNS)_2$	....	W KCNS	I AgCNS	W NaCNS	....	....	W $Sr(CNS)_2$	W $Zn(CNS)_2$	....

REDUCTIONS OF WEIGHINGS IN AIR TO VACUO

When the weight M in grams of a body is determined in air, a correction is necessary for the buoyancy of the air. The following table is computed for an air density of 0.0012. The corrected weight $= M + kM/1000$, values of k being found in the table.

Density of body weighed	Correction factor, k.		
	Pt Ir weights	Brass weights	Quartz or Al weights
.5	+2.34	+2.26	+1.95
.6	+1.94	+1.86	+1.55
.7	+1.66	+1.57	+1.26
.75	+1.55	+1.46	+1.15
.80	+1.44	+1.36	+1.05
.85	+1.36	+1.27	+0.96
.90	+1.28	+1.19	+ .88
.95	+1.21	+1.12	+ .81
1.00	+1.14	+1.06	+ .75
1.1	+1.04	+0.95	+ .64
1.2	+0.94	+ .86	+ .55
1.3	+ .87	+ .78	+ .47
1.4	+ .80	+ .71	+ .40
1.5	+ .75	+ .66	+ .35

Density of body weighed	Correction factor, k.		
	Pt Ir weights	Brass weights	Quartz or Al weights
1.6	+ .69	+ .61	+ .30
1.7	+ .65	+ .56	+ .25
1.8	+ .62	+ .52	+ .21
1.9	+ .58	+ .49	+ .18
2.0	+ .54	+ .46	+ .15
2.5	+ .43	+ .34	+ .03
3.0	+ .34	+ .26	— .05
4.0	+ .24	+ .16	— .15
6.0	+ .14	+ .06	— .25
8.0	+ .09	+ .01	— .30
10.0	+ .06	— .02	— .33
15.0	+ .03	— .06	— .37
20.0	+ .004	— .08	— .39
22.0	— .001	— .09	— .40

OPERATIONAL DEFINITIONS OF pH

Prepared by R. A. Robinson

The operational definition of pH is:

$$pH = pH(s) + E/k$$

where E is the e.m.f. of the cell:

$$H_2|\text{Solution, pH}|\text{Saturated KCl}|\text{Solution, pH(s)}|H_2$$

the half cell on the left containing the solution whose pH is being measured and that on the right a standard buffer mixture of known pH; k = 2.303RT/F, where R is the gas constant, T the temperature in degrees Kelvin and F the value of the faraday.

Alternatively, the cell:

Glass electrode|Solution, pH|Saturated calomel electrode

can be used, the glass electrode being calibrated using a standard buffer mixture or, if possible, two standard buffer mixtures whose pH values lie on either side of that of the solution which is being measured. Suitable standard buffer mixtures are:

0.05 M potassium hydrogen phthalate (pH = 4.008 at 25°C)
0.025 M potassium dihydrogen phosphate
0.025 M disodium hydrogen phosphate (pH = 6.865 at 25°C)
0.01 M borax (pH = 9.180 at 25°C)

For most purposes pH can be equated to $-\log_{10} \gamma_{H^+} m_{H^+}$, i.e., to the negative logarithm of the hydrogen ion activity. There is a small difference between those two quantities if pH > 9.2 or pH < 4.0, given by:

$$-\log \gamma_{H^+} m_{H^+} = pH + 0.014(pH - 9.2) \text{ for } pH > 9.2$$
$$= pH + 0.009(4.0 - pH) \text{ for } pH < 4.0$$

It should be noted that in the table titled "Solutions giving Round Values of pH at 25°C" it is $-\log \gamma_{H^+} m_{H^+}$ and not pH which is quoted when there is a difference between them.

References:

R. G. Bates, "Electrometric pH Determinations: Theory and Practice" Wiley, New York, 1954.
R. A. Robinson and R. H. Stokes, "Electrolyte Solutions," 2nd edition, Butterworths, London; Academic Press, Inc. New York, 1959. R. C. Bates, J. Res. of N.B.S. 66 A, 179 (1962).

National Bureau of Standards
R. G. Bates and S. F. Acree, Res. **34, 373** (1945); W. J. Hammer, C. D. Pinching and S. F. Acree, ibid. **36,** 47 (1946); G. G. Manor, N. J. DeLollis, P. W. Lindwall and S. F. Acree, ibid., **36,** 543 (1946); R. G. Bates, ibid., **39,** 411 (1947); R. G. Bates, V. E. Bower, R. G. Miller and E. R. Smith, ibid., **47,** 433 (1951); V. E. Bower, R. G. Bates and E. R. Smith, ibid., **51,** 189 (1953); V. E. Bower and R. G. Bates, ibid., **55,** 197 (1955); R. G. Bates, V. E. Bower and E. R. Smith, ibid., **56,** 305 (1956); V. E. Bower and R. G. Bates, ibid., **59,** 261; R. G. Bates and V. E. Bower, Anal. Chem., **28,** 1322 (1956).

PROPERTIES OF STANDARD AQUEOUS BUFFER SOLUTIONS AT 25°C

Solution	Buffer substance	Molality m	Weight of salt in air per liter solution	Density g/ml	Molarity M	Dilution value $\Delta pH_{\frac{1}{2}}$	ΔpH_s^a	Buffer value, equiv. per pH	Temp coeff., dpH_s/dt. Units per °C
Tetroxalate	$KH_3(C_2O_4)_2 \cdot 2H_2O$	0.05	12.61	1.0032	0.04962	+0.186	−0.0028	0.070	+0.001
Tartrate	$KHC_4H_4O_6$, sat. sol'n. at 25°C	0.0341	......	1.0036	0.034	+0.049	−0.0003	0.027	−0.0014
Phthalate	$KHC_8O_4H_4$	0.05	10.12	1.0017	0.04958	+0.052	−0.0009	0.016	+0.0012
Phosphate	KH_2PO_4 + Na_2HPO_4	0.025[b]	3.39 3.53	1.0028	0.0249[b]	+0.080	−0.0006	0.029	−0.0028
Phosphate	KH_2PO_4 + Na_2HPO_4	0.008695[c] 0.03043[d]	1.179 4.30	1.0020	0.008665[c] 0.03032[d]	+0.07[e]	−0.0005	0.016	−0.0028
Borax	$Na_2B_4O_7 \cdot 10H_2O$	0.01	3.80	0.9996	0.009971	+0.01	−0.0001	0.020	−0.0082
Calcium hydroxide	$Ca(OH)_2$, sat. sol'n. at 25°C	0.0203	.....	0.9991	0.02025	−0.28	+0.0014	0.09	−0.033

[a] ΔpH_s = pHs (M Molar solution) − pH_s (m molal solution).
[b] Concentration of each phosphate salt.
[c] KH_2PO_4.
[d] Na_2HPO_4.
[e] Calculated value.

BUFFER SOLUTIONS (Continued)

SOLUTIONS GIVING ROUND VALUES OF pH AT 25°C

Reproduced from "Electrolyte Solutions" by permission from Robinson and Stokes, authors, and Butterworth's Scientific Publications.

A*		B*		C*		D*		E*	
pH	*x*	pH	*x*	pH	*x*	pH	*x*	pH	*x*
1.00	67.0	2.20	49.5	4.10	1.3	5.80	3.6	7.00	46.6
1.10	52.8	2.30	45.8	4.20	3.0	5.90	4.6	7.10	45.7
1.20	42.5	2.40	42.2	4.30	4.7	6.00	5.6	7.20	44.7
1.30	33.6	2.50	38.8	4.40	6.6	6.10	6.8	7.30	43.4
1.40	26.6	2.60	35.4	4.50	8.7	6.20	8.1	7.40	42.0
1.50	20.7	2.70	32.1	4.60	11.1	6.30	9.7	7.50	40.3
1.60	16.2	2.80	28.9	4.70	13.6	6.40	11.6	7.60	38.5
1.70	13.0	2.90	25.7	4.80	16.5	6.50	13.9	7.70	36.6
1.80	10.2	3.00	22.3	4.90	19.4	6.60	16.4	7.80	34.5
1.90	8.1	3.10	18.8	5.00	22.6	6.70	19.3	7.90	32.0
2.00	6.5	3.20	15.7	5.10	25.5	6.80	22.4	8.00	29.2
2.10	5.1	3.30	12.9	5.20	28.8	6.90	25.9	8.10	26.2
2.20	3.9	3.40	10.4	5.30	31.6	7.00	29.1	8.20	22.9
		3.50	8.2	5.40	34.1	7.10	32.1	8.30	19.9
		3.60	6.3	5.50	36.6	7.20	34.7	8.40	17.2
		3.70	4.5	5.60	38.8	7.30	37.0	8.50	14.7
		3.80	2.9	5.70	40.6	7.40	39.1	8.60	12.2
		3.90	1.4	5.80	42.3	7.50	41.1	8.70	10.3
		4.00	0.1	5.90	43.7	7.60	42.8	8.80	8.5
						7.70	44.2	8.90	7.0
						7.80	45.3	9.00	5.7
						7.90	46.1		
						8.00	46.7		

F*		G*		H*		I*		J*	
pH	*x*	pH	*x*	pH	*x*	pH	*x*	pH	*x*
8.00	20.5	9.20	0.9	9.60	5.0	10.90	3.3	12.00	6.0
8.10	19.7	9.30	3.6	9.70	6.2	11.00	4.1	12.10	8.0
8.20	18.8	9.40	6.2	9.80	7.6	11.10	5.1	12.20	10.2
8.30	17.7	9.50	8.8	9.90	9.1	11.20	6.3	12.30	12.8
8.40	16.6	9.60	11.1	10.00	10.7	11.30	7.6	12.40	16.2
8.50	15.2	9.70	13.1	10.10	12.2	11.40	9.1	12.50	20.4
8.60	13.5	9.80	15.0	10.20	13.8	11.50	11.1	12.60	25.6
8.70	11.6	9.90	16.7	10.30	15.2	11.60	13.5	12.70	32.2
8.80	9.6	10.00	18.3	10.40	16.5	11.70	16.2	12.80	41.2
8.90	7.1	10.10	19.5	10.50	17.8	11.80	19.4	12.90	53.0
9.00	4.6	10.20	20.5	10.60	19.1	11.90	23.0	13.00	66.0
9.10	2.0	10.30	21.3	10.70	20.2	12.00	26.9		
		10.40	22.1	10.80	21.2				
		10.50	22.7	10.90	22.0				
		10.60	23.3	11.00	22.7				
		10.70	23.8						
		10.80	24.25						

*A. 25 ml of 0.2 molar KCl + *x* ml of 0.2 molar HCl.
*B. 50 ml of 0.1 molar potassium hydrogen phthalate + *x* ml of 0.1 molar HCl.
*C. 50 ml of 0.1 molar potassium hydrogen phthalate + *x* ml of 0.1 molar NaOH.
*D. 50 ml of 0.1 molar potassium dihydrogen phosphate + *x* ml 0.1 molar NaOH.
*E. 50 ml of 0.1 molar tris(hydroxymethyl) aminomethane + *x* ml of 0.1 M HCl.
*F. 50 ml of 0.025 molar borax + *x* ml of 0.1 molar HCl.
*G. 50 ml of 0.025 molar borax + *x* ml of 0.1 molar NaOH.
*H. 50 ml of 0.05 molar sodium bicarbonate + *x* ml of 0.1 molar NaOH.
*I. 50 ml of 0.05 molar disodium hydrogen phosphate + *x* ml of 0.1 molar NaOH.
*J. 25 ml of 0.2 molar KCl + *x* ml of 0.2 molar NaOH.
Final Volume of Mixtures = 100 ml

BUFFER SOLUTIONS (Continued)

STANDARD VALUES OF pH AT TEMPERATURE 0–95°C

Temperature	Tetroxalate 0.05 molal	Tartrate 0.0341 molal (sat'd at 25°C)	Phthalate 0.05 molal	Phosphate[a]	Phosphate[b]	Borax 0.01 molal	Calcium hydroxide (sat'd at 25°C)
0	1.666	.....	4.003	6.984	7.534	9.464	13.423
5	1.668	.....	3.999	6.951	7.500	9.395	13.207
10	1.670	.....	3.998	6.923	7.472	9.332	13.003
15	1.672	.....	3.999	6.900	7.448	9.276	12.810
20	1.675	.....	4.002	6.881	7.429	9.225	12.627
25	1.679	3.557	4.008	6.865	7.413	9.180	12.454
30	1.683	3.552	4.015	6.853	7.400	9.139	12.289
35	1.688	3.549	4.024	6.844	7.389	9.102	12.133
38	1.691	3.548	4.030	6.840	7.384	9.081	12.043
40	1.694	3.547	4.035	6.838	7.380	9.068	11.984
45	1.700	3.547	4.047	6.834	7.373	9.038	11.841
50	1.707	3.549	4.060	6.833	7.367	9.011	11.705
55	1.715	3.554	4.075	6.834	.....	8.985	11.574
60	1.723	3.560	4.091	6.836	.....	8.962	11.449
70	1.743	3.580	4.126	6.845	.....	8.921	.....
80	1.766	3.609	4.164	6.859	.....	8.885	.....
90	1.792	3.650	4.205	6.877	.....	8.850	.....
95	1.806	3.674	4.227	6.886	.....	8.833	.....

[a] Solution 0.025 m KH_2PO_4 and 0.025 m Na_2HPO_4.

[b] Solution 0.008695 m KH_2PO_4 and 0.03043 m Na_2HPO_4.

APPROXIMATE pH VALUES

The following tables give approximate pH values for a number of substances such as acids, bases, foods, biological fluids, etc. All values are rounded off to the nearest tenth and are based on measurements made at 25° C. A few buffer systems with their pH values are also given.

From Modern pH and Chlorine Control, W. A. Taylor & Co., by permission

ACIDS

Hydrochloric, N 0.1
Hydrochloric, 0.1N 1.1
Hydrochloric, 0.01N 2.0
Sulfuric, N 0.3
Sulfuric, 0.1N 1.2
Sulfuric, 0.01N 2.1
Orthophosphoric, 0.1N 1.5
Sulfurous, 0.1N 1.5
Oxalic, 0.1N 1.6
Tartaric, 0.1N 2.2
Malic, 0.1N 2.2
Citric, 0.1N 2.2
Formic, 0.1N 2.3
Lactic, 0.1N 2.4
Acetic, N 2.4
Acetic, 0.1N 2.9
Acetic, 0.01N 3.4
Benzoic, 0.01N 3.1
Alum, 0.1N 3.2
Carbonic (saturated) 3.8
Hydrogen sulfide, 0.1N 4.1
Arsenious (saturated) 5.0
Hydrocyanic, 0.1N 5.1
Boric, 0.1N 5.2

BASES

Sodium hydroxide, N 14.0
Sodium hydroxide, 0.1N 13.0
Sodium hydroxide, 0.01N 12.0
Potassium hydroxide, N 14.0
Potassium hydroxide, 0.1N 13.0
Potassium hydroxide, 0.01N 12.0
Sodium metasilicate, 0.1N 12.6
Lime (saturated) 12.4
Trisodium phosphate, 0.1N 12.0
Sodium carbonate, 0.1N 11.6
Ammonia, N 11.6
Ammonia, 0.1N 11.1
Ammonia, 0.01N 10.6
Potassium cyanide, 0.1N 11.0
Magnesia (saturated) 10.5
Sodium sesquicarbonate, 0.1M 10.1
Ferrous hydroxide (saturated) 9.5
Calcium carbonate (saturated) 9.4
Borax, 0.1N 9.2
Sodium bicarbonate, 0.1N 8.4

BIOLOGIC MATERIALS

Blood, plasma, human 7.3–7.5
Spinal fluid, human 7.3–7.5
Blood, whole, dog 6.9–7.2
Saliva, human 6.5–7.5
Gastric contents, human 1.0–3.0
Duodenal contents, human 4.8–8.2
Feces, human 4.6–8.4
Urine, human 4.8–8.4
Milk, human 6.6–7.6
Bile, human 6.8–7.0

FOODS

Apples 2.9–3.3
Apricots 3.6–4.0
Asparagus 5.4–5.8
Bananas 4.5–4.7
Beans 5.0–6.0
Beers 4.0–5.0
Beets 4.9–5.5
Blackberries 3.2–3.6
Bread, white 5.0–6.0
Butter 6.1–6.4
Cabbage 5.2–5.4
Carrots 4.9–5.3
Cheese 4.8–6.4
Cherries 3.2–4.0
Cider 2.9–3.3
Corn 6.0–6.5
Crackers 6.5–8.5
Dates 6.2–6.4
Eggs, fresh white 7.6–8.0
Flour, wheat 5.5–6.5
Gooseberries 2.8–3.0
Grapefruit 3.0–3.3
Grapes 3.5–4.5
Hominy (lye) 6.8–8.0
Jams, fruit 3.5–4.0
Jellies, fruit 2.8–3.4
Lemons 2.2–2.4
Limes 1.8–2.0
Maple syrup 6.5–7.0
Milk, cows 6.3–6.6
Olives 3.6–3.8
Oranges 3.0–4.0
Oysters 6.1–6.6
Peaches 3.4–3.6
Pears 3.6–4.0
Peas 5.8–6.4
Pickles, dill 3.2–3.6
Pickles, sour 3.0–3.4
Pimento 4.6–5.2
Plums 2.8–3.0
Potatoes 5.6–6.0
Pumpkin 4.8–5.2
Raspberries 3.2–3.6
Rhubarb 3.1–3.2
Salmon 6.1–6.3
Sauerkraut 3.4–3.6
Shrimp 6.8–7.0
Soft drinks 2.0–4.0
Spinach 5.1–5.7
Squash 5.0–5.4
Strawberries 3.0–3.5
Sweet potatoes 5.3–5.6
Tomatoes 4.0–4.4
Tuna 5.9–6.1
Turnips 5.2–5.6
Vinegar 2.4–3.4
Water, drinking 6.5–8.0
Wines 2.8–3.8

ACID BASE INDICATORS

Indicator	Approximate pH range	Color-change	Preparation
Methyl Violet	0.0–1.6	yel to bl	0.01–0.05% in water
Crystal Violet	0.0–1.8	yel to bl	0.02% in water
Ethyl Violet	0.0–2.4	yel to bl	0.1 g in 50 ml of MeOH + 50 ml of water
Malachite Green	0.2–1.8	yel to bl grn	water
Methyl Green	0.2–1.8	yel to bl	0.1% in water
2-(p-dimethylaminophenylazo)pyridine	0.2–1.8 4.4–5.6	yel to bl red to yel	0.1% in EtOH
o-Cresolsulfonephthalein (Cresol Red)	0.4–1.8 7.0–8.8	yel to red yel to red	0.1 g in 26.2 ml 0.01N NaOH + 223.8 ml water
Quinaldine Red	1.0–2.2	col to red	1% in EtOH
p-(p-dimethylaminophenylazo)-benzoic acid, Na-salt (Paramethyl Red)	1.0–3.0	red to yel	EtOH
m-(p-anilnophenylazo)benzene sulfonic acid, Na-salt (Metanil Yellow)	1.2–2.4	red to yel	0.01% in water
4-Phenylazodiphenylamine	1.2–2.6	red to yel	0.01 g in 1 ml 1N HCl + 50 ml EtOH + 49 ml water
Thymolsulfonephthalein (Thymol Blue)	1.2–2.8 8.0–9.6	red to yel yel to bl	0.1 g in 21.5 ml 0.01N NaOH + 229.5 ml water
m-Cresolsulfonephthalein (Metacresol Purple)	1.2–2.8 7.4–9.0	red to yel yel to purp	0.1 g in 26.2 ml 0.01N NaOH + 223.8 ml water
p-(p-anilinophenylazo)benzenesulfonic acid, Na-salt (Orange IV)	1.4–2.8	red to yel	0.01% in water
4-o-Tolylazo-o-toluidine	1.4–2.8	or to yel	water
Erythrosine, disodium salt	2.2–3.6	or to red	0.1% in water
Benzopurpurine 48	2.2–4.2	vt to red	0.1% in water
N,N-dimethyl-p-(m-tolylazo)aniline	2.6–4.8	red to yel	0.1% in water
4,4′-Bix(2-amino-1-naphthylazo)2,2′-stilbenedisulfonic acid	3.0–4.0	purp to red	0.1 g in 5.9 ml 0.05N NaOH + 94.1 ml water
Tetrabromophenolphthaleinethyl ester, K-salt	3.0–4.2	yel to bl	0.1% in EtOH
3′,3″,5′,5″-tetrabromophenolsulfonephthalein (Bromophenol Blue)	3.0–4.6	yel to bl	0.1 g in 14.9 ml 0.01N NaOH + 235.1 ml water
2,4-Dinitrophenol	2.8–4.0	col to yel	saturated water solution
N,N-Dimethyl-p-phenylazoaniline (p-Dimethylaminoazobenzene)	2.8–4.4	red to yel	0.1 g in 90 ml in EtOH + 10 ml water
Congo Red	3.0–5.0	blue to red	0.1% in water
Methyl Orange-Xylene Cyanole solution	3.2–4.2	purp to grn	ready solution
Methyl Orange	3.2–4.4	red to yel	0.01% in water
Ethyl Orange	3.4–4.8	red to yel	0.05–0.2% in water or aqueous EtOH
4-(4-Dimethylamino-1-naphthylazo)-3-methoxybenzenesulfonic acid	3.5–4.8	vt to yel	0.1% in 60% EtOH
3′,3″,5′,5″-Tetrabromo-m-cresolsulfonephthalein (Bromocresol Green)	3.8–5.4	yel to blue	0.1 g in 14.3 ml 0.01N NaOH + 235.7 ml water
Resazurin	3.8–6.4	or to vt	water
4-Phenylazo-1-naphthylamine	4.0–5.6	red to yel	0.1% in EtOH
Ethyl Red	4.0–5.8	col to red	0.1 g in 50 ml MeOH + 50 ml water
2-(p-Dimethylaminophenylazo)-pyridine	0.2–1.8 4.4–5.6	yel to red red to yel	0.1% in EtOH
4-(p-ethoxyphenylazo)-m-phenylenediamine monohydrochloride	4.4–5.8	or to yel	0.1% in water
Lacmoid	4.4–6.2	red to bl	0.2% in EtOH
Alizarin Red S	4.6–6.0	yel to red	dilute solution in water
Methyl Red	4.8–6.0	red to yel	0.02 g in 60 ml EtOH + 40 ml water

Indicator	Approximate pH range	Color-change	Preparation
Propyl Red	4.8–6.6	red to yel	EtOH
5′,5″-Dibromo-o-cresolsulfone-phthalein (Bromocresol Purple)	5.2–6.8	yel to purp	0.1 g in 18.5 ml 0.01N NaOH + 231.5 ml water
3′,3″-Dichlorophenolsulfonephthalein (Chlorophenol Red)	5.2–6.8	yel to red	0 1 g in 23.6 ml 0.01N NaOH + 226.4 ml water
p-Nitrophenol	5.4–6.6	col to yel	0.1% in water
Alizarin	5.6–7.2	yel to red	0.1% in MeOH
	11.0–12.4	red to purp	
2-(2,4-Dinitrophenylazo)-1-naphthol-3, 6-disulfonic acid, di-Na salt	6.0–7.0	yel to bl	0.1% in water
3′,3″-Dibromothymolsulfonephthalein (Bromothymol Blue)	6.0–7.6	yel to bl	0.1 g in 16 ml 0.01N NaOH + 234 ml water
6,8-Dinitro-2,4-(1H)quinazolinedione (m-Dinitrobenzoylene urea)	6.4–8.0	col to yel	25 g in 115 ml M NaOH + 50 ml boiling water
			0.292 g of NaCl in 100 ml water
Brilliant Yellow	6.6–7.8	yel to or	1% in water
Phenolsulfonephthalein (Phenol Red)	6.6–8.0	yel to red	0.1 g in 28.2 ml 0.01N NaOH + 221.8 ml water
Neutral Red	6.8–8.0	red to amb	0.01 g in 50 ml EtOH + 50 ml water
m-Nitrophenol	6.8–8.6	col to yel	0.3% in water
o-Cresolsulfonephthalein (Cresol Red)	0.0–1.0	red to yel	0.1 g in 26.2 ml 0.01N
	7.0–8.8	yel to red	NaOH + 223.8 ml water
Curcumin	7.4–8.6	yel to red	EtOH
	10.2–11.8		
m-Cresolsulfonephthalein (Metacresol Purple)	1.2–2.8	red to yel	0.1 g in 26.2 ml 0.01N
	7.4–9.0	yel to purp	NaOH + 223.8 ml water
4,4′-Bis(4-amino-1-naphthylazo) 2,2′stilbene disulfonic acid	8.0–9.0	bl to red	0.1 g in 5.9 ml 0.05N NaOH + 94.1 ml water
Thymolsulfonephthalein (Thymol Blue)	1.2–2.8	red to yel	0.1 g in 21.5 ml 0.01N
	8.0–9.6		NaOH + 228.5 ml water
o-Cresolphthalein	8.2–9.8	col to red	0.04% in EtOH
p-Naphtholbenzene	8.2–10.0	or to bl	1% in dil. alkali
Phenolphthalein	8.2–10.0	col to pink	0.05 g in 50 ml EtOH + 50 ml water
Ethyl-bis(2,4-dimethylphenyl)acetate	8.4–9.6	col to bl	saturated solution in 50% acetone alcohol
Thymolphthalein	9.4–10.6	col to bl	0.04 g in 50 ml EtOH + 50 ml water
5-(p-Nitrophenylazo)salicylic acid, Na-salt (Alizarin Yellow R)	10.1–12.0	yel to red	0.01% in water
p-(2,4-Dihydroxyphenylazo)benzene-sulfonic acid, Na-salt	11.4–12.6	yel to or	0.1% in water
5,5′-Indigodisulfonic acid, di-Na-salt	11.4–13.0	bl to yel	water
2,4,6-Trinitrotoluene	11.5–13.0	col to or	0.1–0.5% in EtOH
1,3,5-Trinitrobenzene	12.0–14.0	col to or	0.1–0.5% in EtOH
Clayton Yellow	12.2–13.2	yel to amb	0.1% in water

FLUORESCENT INDICATORS

Jack DeMent

Fluorescent indicators are substances which show definite changes in fluorescence with change in pH. Some fluorescent materials are not suitable for indicators since their change in fluorescence is too gradual. Fluorescent indicators find greatest utility in the titration of opaque, highly turbid or deeply colored solutions. A long wavelength ultraviolet ("black light") lamp in a dimly lighted room provides the best environment for titrations involving fluorescent indicators, although bright daylight is sometimes sufficient to evoke a response in the bright green, yellow and orange fluorescent indicators. Titrations are carried out in non-fluorescent glassware. One should check the glassware prior to use to make certain that it does not fluoresce due to the wavelengths of light involved in the titration. The meniscus of the liquid in the burette can be followed when a few particles of an insoluble fluorescent solid are dropped onto its surface.

In this table the indicators are arranged by approximate pH range covered. In the case of some of the dyestuffs the end point may vary slightly with the source or manufacturer.

Indicator	C.I.	From pH	To pH
pH 0 to 2			
Benzoflavine	—	0.3, yellow fl.	1.7, green fl.
3,6-Dioxyphthalimide	—	0, blue fl.	2.4, green fl.
Eosine YS	768	0, yellow colored	3.0, yellow fl.
Erythrosine	772	0, yellow colored	3.6, yellow fl.
Esculin	—	1.5, colorless	2, blue fl.
4-Ethoxyacridone	—	1.2, green fl.	3.2, blue fl.
3,6-Tetramethyldiaminooxanthone	—	1.2, green fl.	3.4, blue fl.
pH 2 to 4			
Chromotropic acid		3.5, colorless	4.5, blue fl.
Fluorescein	766	4, colorless	4.5, green fl.
Magdala Red	—	3.0, purple colored	4.0, fl.
α-Naphthylamine	—	3.4, colorless	4.8, blue fl.
β-Naphthylamine	—	2.8, colorless	4.4, violet fl.
Phloxine	774	3.4, colorless	5.0, bright yellow fl.
Salicylic acid	—	2.5, colorless	3.5, blue fl.
pH 4 to 6			
Acridine	788	4.9, green fl.	5.1, violet colored
Dichlorofluorescein	—	4.0, colorless	5.0, green fl.
3,6-Dioxyxanthone	—	5.4, colorless	7.6, blue-violet fl.
Erythrosine	772	4.0, colorless	4.5, yellow-green fl.
β-Methylesculetin	—	4.0, colorless	6.2, blue fl.
Neville-Winther acid	—	6.0, colorless	6.5, blue fl.
Resorufin	—	4.4, yellow fl.	6.4, weak orange fl.
Quininic acid	—	4.0, yellow colored	5.0, blue fl.
Quinine [first end point]	—	5.0, blue fl.	6.1, violet fl.
pH 6 to 8			
Acid R Phosphine	—	(claimed for range pH 6.0–7.0)	
Brilliant Diazol Yellow	—	6.5, colorless	7.5, violet fl.
Cleves acid	—	6.5, colorless	7.5, green fl.
Coumaric acid	—	7.2, colorless	9.0, green fl.
3,6-Dioxyphthalic dinitrile	—	5.8, blue fl.	8.2, green fl.
Magnesium 8-hydroxyquinolinate	—	6.5, colorless	7.5, golden fl.
β-Methylumbelliferone	—	7.0, colorless	7.5, blue fl.
1-Naphthol-4-sulfonic acid	—	6.0, colorless	6.5, blue fl.
Orcinaurine	—	6.5, colorless	8.0, green fl.
Patent Phosphine	789	(for the range pH 6.0–7.0, green-yellow fl.)	
Thioflavine	816	(for the region pH 6.5–7.0, yellow fl.)	
Umbelliferone	—	6.5, colorless	7.6, blue fl.

pH 8 to 10

Indicator	C.I.	From pH	To pH
Acridine Orange	788	8.4, orange colored	10.4, green fl.
Ethoxyphenylnaphthostilbazonium chloride	—	9, green fl.	11, non-fl.
G Salt	—	9.0, dull blue fl.	9.5, bright blue fl.
Naphthazol derivatives	—	8.2, colorless	10.0, yellow or green fl.
α-Naphthionic acid	—	9, blue fl.	11, green fl.
2-Naphthol-3,6-disulfonic acid	—	9.5, dark blue fl.	Light blue fl. at higher pH
β-Naphthol	—	8.6, colorless	Blue fl. at higher pH
α-Naphtholsulfonic acid	—	8.0, dark blue fl.	9.0, bright violet fl.
1,4-Naphtholsulfonic acid	—	8.2, dark blue fl.	Light blue fl. at higher pH
Orcinsulfonphthalein	—	8.6, yellow colored	10.0 fl.
Quinine [second end point]	—	9.5, violet fl.	10.0, colorless
R-Salt	—	9.0, dull blue fl.	9.5, bright blue fl.
Sodium 1-naphthol-2-sulfonate	—	9.0, dark blue fl.	10.0, bright violet fl.

pH 10 to 12

Coumarin	—	9.8, deep green fl.	12, light green fl.
Eosine BN	771	10.5, colorless	14.0, yellow fl.
Papaverine (permanganate oxidized)	—	9.5, yellow fl.	11.0, blue fl.
Schaffers Salt	—	5.0, violet fl.	11.0, green-blue fl.
SS-Acid (sodium salt)	—	10.0, violet fl.	12.0, yellow colored

pH 12 to 14

Cotarnine	—	12.0, yellow fl.	13.0, white fl.
α-Naphthionic acid	—	12, blue fl.	13, green fl.
β-Naphthionic acid	—	12, blue fl.	13, violet fl.

FORMULAS FOR CALCULATING TITRATION DATA, *p*H VS. ML. OF REAGENT

A Substance Titrated V_0 ml. of solution M_0 its molarity	B Initial $[H^+]$ or $[OH^-]$	C Intermediate Points 10, 50, 99, etc., per cent neutralized, V_1 ml. of reagent of M_1 molarity added	D Equivalence Point	E Excess of Reagent V_1 volume reagent M_1 its molarity V_T total volume
(1) Strong Acid......................	$[H^+] = M_0$	$[H^+] = \frac{V_0M_0 - V_1M_1}{V_0 + V_1}$	$\sqrt{K_w}$	$[OH^-] = \frac{V_1M_1}{V_T}$
(2) Strong Base......................	$[OH^-] = M_0$	$[OH^-] = \frac{V_0M_0 - V_1M_1}{V_0 + V_1}$	$\sqrt{K_w}$	$[H^+] = \frac{V_1M_1}{V_T}$
(3) Weak Acid ($K_a = 10^{-5}$ to 10^{-8}).......	$[H^+] = \sqrt{M_0K_a}$	$[H^+] = \frac{[Acid]}{[Salt]} K_a$	$[OH^-] = \sqrt{\frac{K_w}{K_a} c}$	$[OH^-] = \frac{V_1M_1}{V_T}$ (Value in column D to be added)
(4) Weak Base ($K_b = 10^{-5}$ to 10^{-8}).......	$[OH^-] = \sqrt{M_0K_b}$	$[OH^-] = \frac{[Base]}{[Salt]} K_b$	$[H^+] = \sqrt{\frac{K_w}{K_b} c}$	$[H^+] = \frac{V_1M_1}{V_T}$ (Value in column D to be added)
(5) Salt of a Very Weak Acid (e.g. KCN).	$[OH^-] = \sqrt{\frac{K_w}{K_a} c}$	$[H^+] = \frac{[Acid]}{[Salt]} K_a$	$[H^+] = \sqrt{[Acid]K_a}$	$[H^+] = \frac{V_1M_1}{V_T}$ (Correct for value in column D)
(6) Salt of a Very Weak Base............	$[H^+] = \sqrt{\frac{K_w}{K_b} c}$	$[OH^-] = \frac{[Base]}{[Salt]} K_b$	$[OH^-] = \sqrt{[Base]K_b}$	$[OH^-] = \frac{V_1M_1}{V_T}$ (Add to $[OH^-]$ found in column D)

CONVERSION FORMULAE FOR SOLUTIONS HAVING CONCENTRATIONS EXPRESSED IN VARIOUS WAYS

A = Weight per cent of solute
B = Molecular weight of solvent
E = Molecular weight of solute
F = Grams of solute per liter of solution
G = Molality
M = Molarity
N = Mole fraction
R = Density of solution grams per cc

Concentration of solute—SOUGHT	Concentration of solute—GIVEN A	N	G	M	F
A	—	$\frac{100N \times E}{N \times E + (1 - N)B}$	$\frac{100G \times E}{1000 + G \times E}$	$\frac{M \times E}{10R}$	$\frac{F}{10R}$
N	$\frac{\frac{A}{E}}{\frac{A}{E} + \frac{100 - A}{B}}$	—	$\frac{B \times G}{B \times G + 1000}$	$\frac{B \times M}{M(B - E) + 1000R}$	$\frac{B \times F}{F(B - E) + 1000R \times E}$
G	$\frac{1000A}{E(100 - A)}$	$\frac{1000N}{B - N \times B}$	—	$\frac{1000M}{1000R - (M \times E)}$	$\frac{1000F}{E(1000R - F)}$
M	$\frac{10R \times A}{E}$	$\frac{1000R \times N}{N \times E + (1 - N)B}$	$\frac{1000R \times G}{1000 + E \times G}$	—	$\frac{F}{E}$
F	10AR	$\frac{1000R \times N \times E}{N \times E + (1 - N)B}$	$\frac{1000R \times G \times E}{1000 + G \times E}$	$M \times E$	—

ELECTROCHEMICAL SERIES

Table I
Alphabetical listing
Compiled by J. F. Hunsberger
Values listed are Standard Reduction Potentials

Reaction	Potential, volts
$Ag^{+} + e^{-} \rightarrow Ag$	0.7996
$Ag^{+2} + e^{-} \rightarrow Ag^{+1}$(4f $HClO_4$)	1.987
$AgAc + e^{-} \rightarrow Ag + Ac^{-}$	0.64
$AgBr + e^{-} \rightarrow Ag + Br^{-}$	0.0713
$AgBrO_3 + e^{-} \rightarrow Ag + BrO_3^{-}$	0.680
$AgC_2O_4 + 2e^{-} \rightarrow Ag + C_2O_4^{-2}$	0.4776
$AgCl + e^{-} \rightarrow Ag + Cl^{-}$	0.2223
$AgCN + e^{-} \rightarrow Ag + CN^{-}$	−0.02
$Ag_2CO_3 + 2e^{-} \rightarrow 2Ag + CO_3^{-2}$	0.4769
$Ag_2CrO_4 + 2e^{-} \rightarrow 2Ag + CrO_4^{-2}$	0.4463
$Ag_4Fe(CN)_6 + 4e^{-} \rightarrow 4Ag + Fe(CN)_6^{-4}$	0.1943
$AgI + e^{-} \rightarrow Ag + I^{-}$	−0.1519
$AgIO_3 + e^{-} \rightarrow Ag + IO_3^{-}$	0.3551
$Ag_2MoO_4 + 2e^{-} \rightarrow 2Ag + MoO_4^{-2}$	0.49
$AgNO_2 + e^{-} \rightarrow Ag + NO_2^{-}$	0.59
$Ag_2O + H_2O + 2e \rightarrow 2Ag + 2OH^{-}$	0.342
$Ag_2O_3 + H_2O + 2e^{-} \rightarrow 2AgO + 2OH^{-}$	0.74
$2AgO + H_2O + 2e^{-} \rightarrow Ag_2O + 2OH^{-}$	0.599
$AgOCN + e^{-} \rightarrow Ag + OCN^{-}$	0.41
$Ag_2S + 2e^{-} \rightarrow 2Ag + S^{-2}$	−0.7051
$Ag_2S + 2H^{+} + 2e^{-} \rightarrow 2Ag + H_2S$	−0.0366
$AgSCN + e^{-} \rightarrow Ag + SCN^{-}$	0.0895
$Ag_2SeO_3 + 2e \rightarrow 2Ag + SeO_3^{-2}$	0.3629
$Ag_2SO_4 + 2e^{-} \rightarrow 2Ag + SO_4^{-2}$	0.653
$Ag_2(WO_4) + 2e^{-} \rightarrow 2Ag + WO_4^{-2}$	0.466
$Al^{+3} + 3e^{-} \rightarrow Al$ (0.1f NaOH)	−1.706
$H_2AlO_3^{-} + H_2O + 3e^{-} \rightarrow Al + 4OH^{-}$	−2.35
$As + 3H^{+} + 3e^{-} \rightarrow AsH_3$	−0.54
$As_2O_3 + 6H^{+} + 6e^{-} \rightarrow 2As + 3H_2O$	0.234
$HAsO_2 + 3H^{+} + 3e^{-} \rightarrow As + 2H_2O$	0.2475
$AsO_2^{-} + 2H_2O + 3e^{-} \rightarrow As + 4OH^{-}$	−0.68
$H_3AsO_4 + 2H^{+} + 2e^{-} \rightarrow HAsO_2 + 2H_2O$(1f HCl)	0.58
$AsO_4^{-3} + 2H_2O + 2e^{-} \rightarrow AsO_2^{-} + 4OH^{-}$	−0.71
$AsO_4^{-3} + 2H_2O + 2e^{-} \rightarrow AsO_2^{-} + 4OH^{-}$ (1f NaOH)	−0.08
$Au^{+} + e^{-} \rightarrow Au$	1.68
$Au^{+3} + 2e^{-} \rightarrow Au^{+1}$	1.29
$Au^{+3} + 3e^{-} \rightarrow Au$	1.42
$AuBr_2^{-} + e^{-} \rightarrow Au + 2Br^{-}$	0.963
$AuBr_4^{-} + 3e^{-} \rightarrow Au + 4Br^{-}$	0.858
$AuCl_4^{-} + 3e^{-} \rightarrow Au + 4Cl^{-}$	0.994
$Au(OH)_3 + 3H^{+} + 3e^{-} \rightarrow Au + 3H_2O$	1.45
$H_2BO_3^{-} + 5H_2O + 8e^{-} \rightarrow BH_4^{-} + 8OH^{-}$	−1.24
$H_2BO_3^{-} + H_2O + 3e^{-} \rightarrow B + 4OH^{-}$	−2.5
$H_3BO_3 + 3H^{+} + 3e^{-} \rightarrow B + 3H_2O$	−0.73
$Ba^{+2} + 2e^{-} \rightarrow Ba$	−2.90
$Ba^{+2} + 2e^{-} \rightarrow Ba(Hg)$	−1.570
$Ba(OH)_2 \cdot 8H_2O + 2e^{-} \rightarrow Ba + 2OH^{-} + 8H_2O$	−2.97
$Be^{+2} + 2e^{-} \rightarrow Be$	−1.70
	(−1.85)
$Be_2O_3^{-2} + 3H_2O + 4e^{-} \rightarrow 2Be + 6OH^{-}$	−2.28
$Bi(Cl)_4^{-} + 3e^{-} \rightarrow Bi + 4Cl^{-}$	0.168
$Bi_2O_3 + 3H_2O + 6e^{-} \rightarrow 2Bi + 6OH^{-}$	−0.46
$Bi_2O_4 + 4H^{+} + 2e^{-} \rightarrow 2BiO^{+} + 2H_2O$	1.59
$BiO^{+} + 2H^{+} + 3e^{-} \rightarrow Bi + H_2O$	0.32
$BiOCl + 2H^{+} + 3e \rightarrow Bi + Cl^{-} + H_2O$	0.1583
$BiOOH + H_2O + 3e^{-} \rightarrow Bi + 3OH^{-}$	−0.46
$Br_2(aq) + 2e^{-} \rightarrow 2Br^{-}$	1.087
$Br_2(l) + 2e^{-} \rightarrow 2Br^{-}$	1.065
$HBrO + H^{+} + e^{-} \rightarrow 1/2Br_2 + H_2O$	1.59
$HBrO + H^{+} + 2e^{-} \rightarrow Br^{-} + H_2O$	1.33
$2HBrO + 2H^{+} + 2e^{-} \rightarrow Br_2(l) + 2H_2O$	1.6
$BrO^{-} + H_2O + 2e^{-} \rightarrow Br^{-} + 2OH^{-}$(1f NaOH)	0.70
$BrO_3^{-} + 6H^{+} + 5e^{-} \rightarrow 1/2Br_2 + 3H_2O$	1.52
$BrO_3^{-} + 6H^{+} + 6e^{-} \rightarrow Br^{-} + 3H_2O$	1.44
$BrO_3^{-} + 3H_2O + 6e^{-} \rightarrow Br^{-} + 6OH^{-}$	0.61
$C_6H_4O_2 + 2H^{+} + 2e \rightarrow C_6H_4(OH)_2$	0.6992
$Ca^{+} + e^{-} \rightarrow Ca$	−3.02
$Ca^{+2} + 2e^{-} \rightarrow Ca$	−2.76
Calomel Electrode, Molal KCl	0.2800
*Calomel Electrode, N KCl	0.2807
*Calomel Electrode 0.1 N KCl	0.3337
*Calomel Electrode, Sat'd. KCl	0.2415
*Calomel Electrode, Sat'd NaCl	0.2360
$Ca(OH)_2 + 2e^{-} \rightarrow Ca + 2OH^{-}$	−3.02
$Cd^{+2} + 2e^{-} \rightarrow Cd$	−0.4026
$Cd^{+2} + 2e^{-} \rightarrow Cd(Hg)$	−0.3521
$Cd(OH)_2 + 2e^{-} \rightarrow Cd(Hg) + 2OH^{-}$	−0.761
	(−0.81)
$CdSO_4 \cdot 8/3H_2O + 2e^{-} \rightarrow Cd(Hg) + CdSO_4$(sat'd aq)	−0.4346
$Ce^{+3} + 3e^{-} \rightarrow Ce$	−2.335

Reaction	Potential, volts
$Ce^{+3} + 3e^{-} \rightarrow Ce(Hg)$	−1.4373
$Ce^{+4} + e^{-} \rightarrow Ce^{+3}$	1.4430
	(1.61)
$Ce^{+4} + e^{-} \rightarrow Ce^{+3}$(0.5f H_2SO_4)	1.4587
$CeOH^{+3} + H^{+} + e^{-} \rightarrow Ce^{+3} + H_2O$	1.7134
$Cl_2(g) + 2e^{-} \rightarrow 2Cl^{-}$	1.3583
$HClO + H^{+} + e^{-} \rightarrow 1/2Cl_2 + H_2O$	1.63
$HClO + H^{+} + 2e^{-} \rightarrow Cl^{-} + H_2O$	1.49
$ClO^{-} + H_2O + 2e^{-} \rightarrow Cl^{-} + 2OH^{-}$	0.90
$ClO_2 + e^{-} \rightarrow ClO_2^{-}$	1.15
$ClO_2 + H^{+} + e^{-} \rightarrow HClO_2$	1.27
$HClO_2 + 2H^{+} + 2e^{-} \rightarrow HClO + H_2O$	1.64
$HClO_2 + 3H^{+} + 3e^{-} \rightarrow 1/2Cl_2 + 2H_2O$	1.63
$HClO_2 + 3H^{+} + 4e^{-} \rightarrow Cl^{-} + 2H_2O$	1.56
$ClO_2^{-} + H_2O + 2e^{-} \rightarrow ClO^{-} + 2OH^{-}$	0.59
$ClO_2^{-} + 2H_2O + 4e^{-} \rightarrow Cl^{-} + 4OH^{-}$	0.76
$ClO_2(aq) + e^{-} \rightarrow ClO_2^{-}$	0.954
$ClO_3^{-} + 2H^{+} + e^{-} \rightarrow ClO_2 + H_2O$	1.15
$ClO_3^{-} + 3H^{+} + 2e \rightarrow HClO_2 + H_2O$	1.21
	(1.23)
$ClO_3^{-} + 6H^{+} + 5e^{-} \rightarrow \frac{1}{2}Cl_2 + 3H_2O$	1.47
$ClO_3^{-} + 6H^{+} + 6e^{-} \rightarrow Cl^{-} + 3H_2O$	1.45
$ClO_3^{-} + H_2O + 2e^{-} \rightarrow ClO_2^{-} + 2OH^{-}$	0.35
$ClO_3^{-} + 3H_2O + 6e^{-} \rightarrow Cl^{-} + 6OH^{-}$	0.62
$ClO_4^{-} + 2H^{+} + 2e^{-} \rightarrow ClO_3^{-} + H_2O$	1.19
$ClO_4^{-} + 8H^{+} + 7e^{-} \rightarrow \frac{1}{2}Cl_2 + 4H_2O$	1.34
$ClO_4^{-} + 8H^{+} + 8e- \rightarrow Cl^{-} + 4H_2O$	1.37
$ClO_4^{-} + H_2O + 2e^{-} \rightarrow ClO_3^{-} + 2OH^{-}$	0.17
$(CN)_2 + 2H^{+} + 2e^{-} \rightarrow 2HCN$	0.37
$2HCNO + 2H^{+} + 2e \rightarrow (CN)_2 + 2H_2O$	0.33
$(CNS)_2 + 2e^{-} \rightarrow 2CNS^{-}$	0.77
$Co^{+2} + 2e^{-} \rightarrow Co$	−0.28
$Co^{+3} + e^{-} \rightarrow Co^{+2}$(3f HNO_3)	1.842
$CO_2 + 2H^{+} + 2e^{-} \rightarrow HCOOH$	−0.2
$2CO_2 + 2H^{+} + 2e^{-} \rightarrow H_2C_2O_4$	−0.49
$Co(NH_3)_6^{+3} + e^{-} \rightarrow Co(NH_3)_6^{+2}$	0.1
$Co(OH)_2 + 2e^{-} \rightarrow Co + 2OH^{-}$	−0.73
$Co(OH)_3 + e^{-} \rightarrow Co(OH)_2 + OH^{-}$	0.2 (0.17)
$Cr^{+2} + 2e \rightarrow Cr$	−0.557
$Cr^{+3} + e^{-} \rightarrow Cr^{+2}$	−0.41
$Cr^{+3} + 3e^{-} \rightarrow Cr$	−0.74
$Cr^{+6} + 3e^{-} \rightarrow Cr^{+3}$(2f H_2SO_4)	1.10
$Cr^{+6} + 3e^{-} \rightarrow Cr^{+3}$(1f NaOH)	−0.12
$Cr_2O_7^{-2} + 14H^{+} + 6e^{-} \rightarrow 2Cr^{+3} + 7H_2O$	1.33
$CrO_2^{-} + 2H_2O + 3e^{-} \rightarrow Cr + 4OH^{-}$	−1.2
$HCrO_4^{-} + 7H^{+} + 3e^{-} \rightarrow Cr^{+3} + 4H_2O$	1.195
$CrO_4^{-2} + 4H_2O + 3e^{-} \rightarrow Cr(OH)_3 + 5OH^{-}$	−0.12
$Cr(OH)_3 + 3e^{-} \rightarrow Cr + 3OH^{-}$	−1.3
$Cs^{+} + e^{-} \rightarrow Cs$	−2.923
$Cu^{+} + e^{-} \rightarrow Cu$	0.522
$Cu^{+2} + 2CN^{-} + e^{-} \rightarrow Cu(CN)_2^{-}$	1.12
$Cu^{+2} + e \rightarrow Cu^{+}$	0.158
	(0.167)
$Cu^{+2} + 2e^{-} \rightarrow Cu$	0.3402
$Cu^{+2} + 2e \rightarrow Cu(Hg)$	0.345
$CuI_2^{-} + e^{-} \rightarrow Cu + 2I^{-}$	0.00
$Cu_2O + H_2O + 2e^{-} \rightarrow 2Cu + 2OH^{-}$	−0.361
$Cu(OH)_2 + 2e^{-} \rightarrow Cu + 2OH^{-}$	−0.224
$2Cu(OH)_2 + 2e^{-} \rightarrow Cu_2O + 2OH^{-} + H_2O$	−0.09
$D^{+} + e^{-} \rightarrow 1/2D_2$	−0.0034
$2D^{+} + 2e^{-} \rightarrow D_2$	−0.044
$Eu^{+3} + e^{-} \rightarrow Eu^{+2}$	−0.43
$1/2F_2 + e^{-} \rightarrow F^{-}$	2.85
$1/2F_2 + H^{+} + e^{-} \rightarrow HF$	3.03
$F_2 + 2e^{-} \rightarrow 2F^{-}$	2.87
$F_2O + 2H^{+} + 4e^{-} \rightarrow H_2O + 2F^{-}$	2.1
$Fe^{+2} + 2e^{-} \rightarrow Fe$	−0.409
$Fe^{+3} + 3e^{-} \rightarrow Fe$	−0.036
$Fe^{+3} + e^{-} \rightarrow Fe^{+2}$	0.770
$Fe^{+3} + e^{-} \rightarrow Fe^{+2}$(1f HCl)	0.770
$Fe^{+3} + e^{-} \rightarrow Fe^{+2}$(1f $HClO_4$)	0.747
$Fe^{+3} + e^{-} \rightarrow Fe^{+2}$(1f H_3PO_4)	0.438
$Fe^{+3} + e^{-} \rightarrow Fe^{+2}$(0.5f H_2SO_4)	0.679
$Fe(CN)_6^{-3} + e^{-} \rightarrow Fe(CN)_6^{-4}$(0.01f NaOH)	0.46
$Fe(CN)_6^{-3} + e^{-} \rightarrow Fe(CN)_6^{-4}$(1f H_2SO_4)	0.69
$FeO_4^{-2} + 8H^{+} + 3e^{-} \rightarrow Fe^{+3} + 4H_2O$	1.9
$Fe(OH)_3 + e^{-} \rightarrow Fe(OH)_2 + OH^{-}$	−0.56
Fe (phenanthroline)$_3^{+3} + e^{-} \rightarrow Fe(ph)_3^{+2}$	1.14
Fe (phenanthroline)$_3^{+3} + e^{-} \rightarrow Fe(ph)_3^{+2}$(2f H_2SO_4)	1.056
$Ga^{+3} + 3e^{-} \rightarrow Ga$	−0.560
$H_2GaO_3^{-} + H_2O + 3e^{-} \rightarrow Ga + 4OH^{-}$	−1.22

Reaction	Potential, volts
$GeO_2 + 2H^+ + 2e^- \rightarrow GeO + H_2O$	−0.12
$H_2GeO_3 + 4H^+ + 4e^- \rightarrow Ge + 3H_2O$	−0.13
$2H^+ + 2e^- \rightarrow H_2$	0.0000
$1/2H_2 + e^- \rightarrow H^-$	−2.23
$2H_2O + 2e^- \rightarrow H_2 + 2OH^-$	−0.8277
$H_2O_2 + 2H^+ + 2e^- \rightarrow 2H_2O$	1.776
$HfO^{+2} + 2H^+ + 4e^- \rightarrow Hf + H_2O$	−1.68
$HfO_2 + 4H^+ + 4e^- \rightarrow Hf + 2H_2O$	−1.57
$HfO(OH)_2 + H_2O + 4e^- \rightarrow Hf + 4OH^-$	−2.60
$Hg^{+2} + 2e^- \rightarrow Hg$	0.851
$2Hg^{+2} + 2e^- \rightarrow Hg_2^{+2}$	0.905
$1/2Hg_2^{+2} + e^- \rightarrow Hg$	0.7986
$Hg_2^{+2} + 2e^- \rightarrow 2Hg$	0.7961
$Hg_2(AcO)_2 + 2e^- \rightarrow 2Hg + 2AcO^-$	0.5113
$Hg_2Br_2 + 2e^- \rightarrow 2Hg + 2Br^-$	0.1396
$Hg_2Cl_2 + 2e^- \rightarrow 2Hg + 2Cl^-$	0.2682
$Hg_2Cl_2 + 2e \rightarrow 2Hg + 2Cl^-$(0.1f NaOH)	0.3419
	(0.268)
$Hg_2HPO_4 + H^+ + 2e^- \rightarrow 2Hg + H_2PO_4^-$	0.639
$Hg_2I_2 + 2e^- \rightarrow 2Hg + 2I^-$	−0.0405
$Hg_2O + H_2O + 2e^- \rightarrow 2Hg + 2OH^-$	0.123
$HgO + H_2O + 2e^- \rightarrow Hg + 2OH^-$	0.0984
$Hg_2SO_4 + 2e^- \rightarrow 2Hg + SO_4^{-2}$	0.6158
$HO_2 + H^+ + e^- \rightarrow H_2O_2$	1.5
$I_2 + 2e^- \rightarrow 2I^-$	0.535
$I_3^- + 2e^- \rightarrow 3I^-$	0.5338
$In^{+2} + e^- \rightarrow In^{+1}$	−0.40
$In^{+3} + e^- \rightarrow In^{+2}$	−0.49
$In^{+3} + 2e^- \rightarrow In^{+1}$	−0.40
$In^{+3} + 3e^- \rightarrow In$	−0.338
$H_3IO_6^{-2} + 2e^- \rightarrow IO_3^- + 3OH^-$	~0.70
$H_5IO_6 + H^+ + 2e^- \rightarrow IO_3^- + 3H_2O$	~1.7
$HIO + H^+ + e^- \rightarrow 1/2I_2 + H_2O$	1.45
$HIO + H^+ + 2e^- \rightarrow I^- + H_2O$	0.99
$IO^- + H_2O + 2e^- \rightarrow I^- + 2OH^-$	0.49
$IO_3^- + 6H^+ + 5e^- \rightarrow 1/2I_2 + 3H_2O$	1.195
$IO_3^- + 6H^+ + 6e^- \rightarrow I^- + 3H_2O$	1.085
$2IO_3^- + 12H^+ + 10e^- \rightarrow I_2 + 6H_2O$	1.19
$IO_3^- + 2H_2O + 4e^- \rightarrow IO^- + 4OH^-$	0.56
$IO_3^- + 3H_2O + 6e^- \rightarrow I^- + 6OH^-$	0.26
$IrCl_6^{-2} + e^- \rightarrow IrCl_6^{-3}$	1.02
$IrCl_6^{-3} + 3e^- \rightarrow Ir + 6Cl^-$	0.77
$Ir_2O_3 + 3H_2O + 6e^- \rightarrow 2Ir + 6OH^-$	0.1
$K^+ + e^- \rightarrow K$	−2.924
	(−2.923)
$La^{+3} + 3e^- \rightarrow La$	−2.37
$La(OH)_3 + 3e \rightarrow La + 3OH^-$	−2.76
$Li^+ + e^- \rightarrow Li$	−3.045
	(−3.02)
$Mg^{++} + 2e^- \rightarrow Mg$	−2.375
$Mg(OH)_2 + 2e^- \rightarrow Mg + 2OH^-$	−2.67
$Mn^{+2} + 2e^- \rightarrow Mn$	−1.029
$Mn^{+3} + e^- \rightarrow Mn^{+2}$	1.51
$MnO_2 + 4H^+ + 2e^- \rightarrow Mn^{+2} + 2H_2O$	1.208
$MnO_4^- + e^- \rightarrow MnO_4^{-2}$	0.564
$MnO_4^- + 4H^+ + 3e^- \rightarrow MnO_2 + 2H_2O$	+1.679
$MnO_4^- + 8H^+ + 5e^- \rightarrow Mn^{+2} + 4H_2O$	1.491
$MnO_4^- + 2H_2O + 3e^- \rightarrow MnO_2 + 4OH^-$	0.588
$MnO_4^{-1} + 2H_2O + 3e^- \rightarrow MnO_2 + 4OH^-$	0.58
$Mn(OH)_2 \rightarrow Mn + 2OH^-$	−1.47
$Mn(OH)_3 + e^- \rightarrow Mn(OH)_2 + OH^-$	−0.40
$H_2MoO_4 + 6H^+ + 6e^- \rightarrow Mo + 4H_2O$	0.0
$N_2 + 2H_2O + 4H^+ + 2e^- \rightarrow 2NH_3OH^+$	−1.87
$3N_2 + 2H^+ + 2e^- \rightarrow 2HN_3$	−3.1
$N_2H_5^+ + 3H^+ + 2e^- \rightarrow 2NH_4^+$	1.27
$N_2O + 2H^+ + 2e^- \rightarrow N_2 + H_2O$	1.77
$H_2N_2O_2 + 2H^+ + 2e^- \rightarrow N_2 + 2H_2O$	2.65
$N_2O_4 + 2e^- \rightarrow 2NO_2^-$	0.88
$N_2O_4 + 2H^+ + 2e^- \rightarrow 2HNO_2$	1.07
$N_2O_4 + 4H^+ + 4e^- \rightarrow 2NO + 2H_2O$	1.03
$Na^+ + e^- \rightarrow Na$	−2.7109
$Nb_2O_5 + 10H^+ + 10e^- \rightleftharpoons 2Nb + 5H_2O$	−0.62
$Nb^{+5} + 2e \rightarrow Nb^{+3}$(2fHCl)	0.344
$Nd^{+3} + 3e^- \rightarrow Nd$	−2.246
$2NH_3OH^+ + H^+ + 2e^- \rightarrow N_2H_5^+ + 2H_2O$	1.42
$Ni^{+2} + 2e^- \rightarrow Ni$	−0.23
$Ni(OH)_2 + 2e^- \rightarrow Ni + 2OH^-$	−0.66
$NiO_2 + 4H^+ + 2e^- \rightarrow Ni^{+2} + 2H_2O$	1.93
$NiO_2 + 2H_2O + 2e^- \rightarrow Ni(OH)_2 + 2OH^-$	0.49
$2NO + 2e^- \rightarrow N_2O_2^{-2}$	0.10
$2NO + 2H^+ + 2e^- \rightarrow N_2O + H_2O$	1.59
$2NO + H_2O + 2e^- \rightarrow N_2O + 2OH^-$	0.76
$HNO_2 + H^+ + e^- \rightarrow NO + H_2O$	0.99
$2HNO_2 + 4H^+ + 4e^- \rightarrow H_2N_2O_2 + 2H_2O$	0.80
$2HNO_2 + 4H^+ + 4e^- \rightarrow N_2O + 3H_2O$	1.27
	(1.29)

Reaction	Potential, volts
$NO_2^- + H_2O + e^- \rightarrow NO + 2OH^-$	−0.46
$2NO_2^- + 2H_2O + 4e^- \rightarrow N_2O_2^{-2} + 4OH^-$	−0.18
$2NO_2^- + 3H_2O + 4e^- \rightarrow N_2O + 6OH^-$	0.15
$NO_3^- + 3H^+ + 2e^- \rightarrow HNO_2 + H_2O$	0.94
$NO_3^- + 4H^+ + 3e^- \rightarrow NO + 2H_2O$	0.96
$2NO_3^- + 4H^+ + 2e^- \rightarrow N_2O_4 + 2H_2O$	0.81
$NO_3^- + H_2O + 2e^- \rightarrow NO_2^- + 2OH^-$	0.01
$2NO_3^- + 2H_2O + 2e^- \rightarrow N_2O_4 + 4OH^-$	−0.85
$Np^{+3} + 3e^- \rightarrow Np$	−1.9
$Np^{+4} + e^- \rightarrow Np^{+3}$(1fHClO$_4$)	0.155
$Np^{+5} + e^- \rightarrow Np^{+4}$(1fHClO$_4$)	0.739
$Np^{+6} + e^- \rightarrow Np^{+5}$(1fHClO$_4$)	1.137
$1/2O_2 + 2H^+(10^{-7}M) + 2e^- \rightarrow H_2O$	0.815
$O_2 + 2H^+ + 2e^- \rightarrow H_2O_2$	0.682
$O_2 + 4H^+ + 4e^- \rightarrow 2H_2O$	1.229
$O_2 + H_2O + 2e^- \rightarrow HO_2^- + OH^-$	−0.076
$O_2 + 2H_2O + 2e^- \rightarrow H_2O_2 + 2OH^-$	−0.146
$O_2 + 2H_2O + 4e^- \rightarrow 4OH^-$	0.401
$O_3 + 2H^+ + 2e^- \rightarrow O_2 + H_2O$	2.07
$O_3 + H_2O + 2e^- \rightarrow O_2 + 2OH^-$	1.24
$O_{(g)} + 2H^+ + 2e^- \rightarrow H_2O$	2.42
$OH + e^- \rightarrow OH^-$	1.4
$HO_2^- + H_2O + 2e^- \rightarrow 3OH^-$	0.87
$OsO_4 + 8H^+ + 8e^- \rightarrow Os + 4H_2O$	0.85
$P + 3H^+ + 3e^- \rightarrow PH_3(g)$	−0.04
$P + 3H_2O + 3e^- \rightarrow PH_3(g) + 3OH^-$	−0.87
$Pb^{+2} + 2e^- \rightarrow Pb$	−0.1263
	(−0.126)
$Pb^{+2} + 2e^- \rightarrow Pb(Hg)$	−0.1205
$PbBr_2 + 2e^- \rightarrow Pb(Hg) + 2Br^-$	−0.275
$PbCl_2 + 2e^- \rightarrow Pb(Hg) + 2Cl^-$	−0.262
$PbF_2 + 2e^- \rightarrow Pb(Hg) + 2F^-$	−0.3444
$PbHPO_4 + H^+ + 2e^- \rightarrow Pb(Hg) + HPO_4^=$	−0.2448
$PbI_2 + 2e^- \rightarrow Pb(Hg) + 2I^-$	−0.358
$PbO + H_2O + 2e^- \rightarrow Pb + 2OH^-$	−0.576
$PbO_2 + 4H^+ + 2e^- \rightarrow Pb^{+2} + 2H_2O$	1.46
$HPbO_2^- + H_2O + 2e^- \rightarrow Pb + 3OH^-$	−0.54
$PbO_2 + H_2O + 2e^- \rightarrow PbO + 2OH^-$	0.28
$PbO_2 + SO_4^{-2} + 4H^+ + 2e^- \rightarrow PbSO_4 + 2H_2O$	1.685
$PbSO_4 + 2e^- \rightarrow Pb + SO_4^{-2}$	−0.356
$PbSO_4 + 2e^- \rightarrow Pb(Hg) + SO_4^{-2}$	−0.3505
$Pd^{+2} + 2e^- \rightarrow Pd$	0.83
$Pd^{+2} + 2e^- \rightarrow Pd$(1fHCl)	0.623
$Pd^{+2} + 2e^- \rightarrow Pd$(4f HClO$_4$)	0.987
$PdCl_4^{-2} + 2e^- \rightarrow Pd + 4Cl^-$	0.623
$PdCl_6^{-2} + 2e^- \rightarrow PdCl_4^{-2} + 2Cl^-$	1.29
$Pd(OH)_2 + 2e^- \rightarrow Pd + 2OH^-$	0.1
$H_2PO_2^- + e^- \rightarrow P + 2OH^-$	−1.82
$H_3PO_2 + H^+ + e^- \rightarrow P + 2H_2O$	−0.51
$H_3PO_3 + 2H^+ + 2e^- \rightarrow H_3PO_2 + H_2O$	−0.50
	(−0.59)
$H_3PO_3 + 3H^+ + 3e^- \rightarrow P + 3H_2O$	−0.49
$HPO_3^{-2} + 2H_2O + 2e^- \rightarrow H_2PO_2^- + 3OH^-$	−1.65
$HPO_3^{-2} + 2H_2O + 3e^- \rightarrow P + 5OH^-$	−1.71
$H_3PO_4 + 2H^+ + 2e^- \rightarrow H_3PO_3 + H_2O$	−0.276
	(−0.2)
$PO_4^{-3} + 2H_2O + 2e^- \rightarrow HPO_3^{-2} + 3OH^-$	−1.05
$Pt^{+2} + 2e^- \rightarrow Pt$	~1.2
$PtCl_4^{-2} + 2e^- \rightarrow Pt + 4Cl^-$	0.73
$PtCl_6^{-2} + 2e^- \rightarrow PtCl_4^{-2} + 2Cl^-$	0.74
$Pt(OH)_2 + 2e^- \rightarrow Pt + 2OH^-$	0.16
$Pu^{+4} + e^- \rightarrow Pu^{+3}$(1f HClO$_4$)	0.982
$Pu^{+5} + e^- \rightarrow Pu^{+4}$(0.5f HCl)	1.099
$Pu^{+6} + e^- \rightarrow Pu^{+5}$(1f HClO$_4$)	0.9184
$Pu^{+6} + 2e^- \rightarrow Pu^{+4}$(1f HCl)	1.052
Quinhydrone Elec. H^+, a = 1	0.6995
$Rb^+ + e^- \rightarrow Rb$	−2.925
	(−2.99)
$Re^{+3} + 3e^- \rightarrow Re$	0.3 ~
$ReO_4^- + 4H^+ + 3e^- \rightarrow ReO_2 + 2H_2O$	0.51
$ReO_2 + 4H^+ + 4e^- \rightarrow Re + 2H_2O$	0.26
$ReO_4^- + 2H^+ + e^- \rightarrow ReO_{3(cc)} + 2H_2O$	0.768
$ReO_4^- + 4H_2O + 7e^- \rightarrow Re + 8OH^-$	−0.81
$ReO_4^- + 8H^+ + 7e^- \rightarrow Re + 4H_2O$	0.367
$Rh^{+4} + e^- \rightarrow Rh^{+3}$	1.43
$Rh(Cl)_6^{-3} + 3e^- \rightarrow Rh + 6Cl^-$	0.44
$Ru^{+3} + e^- \rightarrow Ru^{+2}$(0.1f HClO$_4$)	−0.11
$Ru^{+3} + e^- \rightarrow Ru^{+2}$(1–6f HCl)	−0.084
$Ru^{+4} + e^- \rightarrow Ru^{+3}$(0.1f HClO$_4$)	0.49
$Ru^{+4} + e^- \rightarrow Ru^{+3}$(2f HCl)	0.858
$RuO_2 + 4H^+ + 4e^- \rightarrow Ru + 2H_2O$	−0.8
$RuO_4^- + e^- \rightarrow RuO_4^{-2}$	0.59
$RuO_{4(c)} + e^- \rightarrow RuO_4^-$	1.00
$S + 2e^- \rightarrow S^{-2}$	−0.508
$S + 2H^+ + 2e^- \rightarrow H_2S_{(aq)}$	0.141
$S + H_2O + 2e^- \rightarrow HS^- + OH^-$	−0.478

ELECTROCHEMICAL SERIES—TABLE I (Continued)

Reaction	Potential, volts
$S_2O_6^{-2} + 4H^+ + 2e^- \rightarrow 2H_2SO_3$	0.6
$S_2O_8^{-2} + 2e^- \rightarrow 2SO_4^{-2}$	2.0
	(2.05)
$S_4O_6^{=} + 2e^- \rightarrow 2S_2O_3^{-2}$	0.09
	(0.10)
$Sb + 3H^+ + 3e^- \rightarrow H_3Sb$	−0.51
$Sb^{+5} + 2e^- \rightarrow Sb^{+3}$(3.5f HCl)	0.75
$Sb_2O_3 + 6H^+ + 6e^- \rightarrow 2Sb + 3H_2O$	0.1445
	(0.152)
$Sb_2O_5 + 4H^+ + 4e^- \rightarrow Sb_2O_3 + 2H_2O$	0.69
$Sb_2O_{5(s)} + 6H^+ + 4e^- \rightarrow 2SbO^+ + 3H_2O$	0.64
$SbO^+ + 2H^+ + 3e^- \rightarrow Sb + 2H_2O$	0.212
$SbO_2^- + 2H_2O + 3e^- \rightarrow Sb + 4OH^-$	−0.66
$SbO_3^- + H_2O + 2e^- \rightarrow SbO_2^- + 2OH^-$	−0.59
$Sc^{+3} + 3e^- \rightarrow Sc$	−2.08
$Se + 2e^- \rightarrow Se^{-2}$	−0.78
$Se + 2H^+ + 2e^- \rightarrow H_2Se(aq)$	−0.36
$H_2SeO_3 + 4H^+ + 4e^- \rightarrow Se + 3H_2O$	0.74
$SeO_3^{-2} + 3H_2O + 4e^- \rightarrow Se + 6OH^-$	−0.35
$SeO_4^{-2} + 4H^+ + 2e^- \rightarrow H_2SeO_3 + H_2O$	1.15
$SeO_4^{-2} + H_2O + 2e^- \rightarrow SeO_3^{-2} + 2OH^-$	0.03
$SiF_6^{-2} + 4e^- \rightarrow Si + 6F^-$	−1.2
$SiO_2 + 4H^+ + 4e^- \rightarrow Si + 2H_2O$	−0.84
$SiO_3^{-2} + 3H_2O + 4e^- \rightarrow Si + 6OH^-$	−1.73
$Sn^{+2} + 2e^- \rightarrow Sn$	−0.1364
$Sn^{+4} + 2e^- \rightarrow Sn^{+2}$	0.15
$Sn^{+4} + 2e^- \rightarrow Sn^{+2}$(0.1f HCl)	0.070
$Sn^{+4} + 2e^- \rightarrow Sn^{+2}$(1f HCl)	0.139
$HSnO_2^- + H_2O + 2e^- \rightarrow Sn + 3OH^-$	−0.79
$Sn(OH)_6^{-2} + 2e^- \rightarrow HSnO_2^- + 3OH^- + H_2O$	−0.96
$2H_2SO_3 + H^+ + 2e^- \rightarrow HS_2O_4^- + 2H_2O$	−0.08
$H_2SO_3 + 4H^+ + 4e^- \rightarrow S + 3H_2O$	0.45
$2SO_3^{-2} + 2H_2O + 2e^- \rightarrow S_2O_4^{-2} + 4OH^-$	−1.12
$2SO_3^{-2} + 3H_2O + 4e^- \rightarrow S_2O_3^{-2} + 6OH^-$	−0.58
$SO_4^{-2} + 4H^+ + 2e^- \rightarrow H_2SO_3 + H_2O$	0.20
$2SO_4^{-2} + 4H^+ + 2e^- \rightarrow S_2O_6^{-2} + H_2O$	−0.2
$SO_4^{-2} + H_2O + 2e^- \rightarrow SO_3^{-2} + 2OH^-$	−0.92
$Sr^{+2} + 2e^- \rightarrow Sr$	−2.89
$Sr^{+2} + 2e^- \rightarrow Sr(Hg)$	−1.793
$Sr(OH)_2 \cdot 8H_2O + 2e^- \rightarrow Sr + 2OH^- + 8H_2O$	−2.99
$Ta_2O_5 + 10H^+ + 10e^- \rightarrow 2Ta + 5H_2O$	−0.71
$TcO_4^- + 4H^+ + 3e^- \rightarrow TcO_{2(c)} + 2H_2O$	0.738
$Te + 2e^- \rightarrow Te^{-2}$	−0.92
$Te + 2H^+ + 2e^- \rightarrow H_2Te(Ag)$	−0.69
	(−0.72)
$Te^{+4} + 4e^- \rightarrow Te$(2.5f HCl)	0.63
$TeO_2 + 4H_2 + 4e^- \rightarrow Te + 2H_2O$	0.593
$TeO_3^{-2} + 3H_2O + 4e^- \rightarrow Te + 6OH^-$	−0.02
$TeO_4^- + 8H^+ + 7e^- \rightarrow Te + 4H_2O$	0.472
$H_6TeO_{6(s)} + 2H^+ + 2e^- \rightarrow TeO_{2(s)} + 4H_2O$	1.02
$Th^{+4} + 4e^- \rightarrow Th$	−1.90
$ThO_2 + 4H^+ + 4e^- \rightarrow Th + 2H_2O$	−1.80
$ThO_2 + 2H_2O + 4e^- \rightarrow Th + 4OH^-$	−2.64
$Ti^{+2} + 2e^- \rightarrow Ti$	−1.63
$Ti^{+3} + e^- \rightarrow Ti^{+2}$	−2.0
$TiO_2 + 4H^+ + 4e^- \rightarrow Ti + 2H_2O$	−0.86
$Ti(OH)^{+3} + H^+ + e^- \rightarrow Ti^{+3} + H_2O$	0.06
$Tl^+ + e^- \rightarrow Tl$	−0.3363
$Tl^+ + e^- \rightarrow Tl(Hg)$	−0.3338
$Tl^{+3} + e^- \rightarrow Tl^{+2}$	−0.37
$Tl^{+3} + 2e^- \rightarrow Tl^{+1}$	1.247
$Tl^{+3} + 2e^- \rightarrow Tl^{+1}$(1f HCl)	0.783
$TlBr + e^- \rightarrow Tl(Hg) + Br^-$	−0.606
$TlCl + e^- \rightarrow Tl(Hg) + Cl^-$	−0.555
$TlI + e^- \rightarrow Tl(Hg) + I^-$	−0.769
$Tl_2O_3 + 3H_2O + 4e^- \rightarrow 2Tl^+ + 6OH^-$	0.02
$TlOH + e^- \rightarrow Tl + OH^-$	−0.3445
$Tl(OH)_3 + 2e^- \rightarrow TlOH + 2OH^-$	−0.05
$Tl_2SO_4 + 2e^- \rightarrow Tl(Hg) + SO_4^{-2}$	−0.4360
$U^{+3} + 3e^- \rightarrow U$	−1.8
$U^{+4} + e^- \rightarrow U^{+3}$	−0.61
$U^{+4} + e^- \rightarrow U^{+3}$(1f $HClO_4$)	−0.631
$U^{+5} + e^- \rightarrow U^{+4}$(1f HCl)	1.02
$U^{+6} + e^- \rightarrow U^{+5}$(1f $HClO_4$)	0.063
$UO_2^+ + 4H^+ + e^- \rightarrow U^{+4} + 2H_2O$	0.62
$UO_2^{+2} + e^- \rightarrow UO_2^+$	0.062
$UO_2^{+2} + 4H^+ + 2e^- \rightarrow U^{+4} + 2H_2O$	0.334
$UO_2^{+2} + 4H^+ + 6e^- \rightarrow U + 2H_2O$	−0.82
$V^{+2} + 2e^- \rightarrow V$	−1.2
$V^{+3} + e^- \rightarrow V^{+2}$	−0.255
$V^{+5} + e^- \rightarrow V^{+4}$(1f NaOH)	−0.74
$VO^{+2} + 2H^+ + e^- \rightarrow V^{+3} + H_2O$	0.337
$VO_2^+ + 2H^+ + e^- \rightarrow VO^{+2} + H_2O$	1.00
$V(OH)_4^+ + 2H^+ + e^- \rightarrow VO^{+2} + 3H_2O$	1.00
$V(OH)_4^+ + 4H^+ + 5e^- \rightarrow V + 4H_2O$	−0.25
$W_2O_5 + 2H^+ + 2e^- \rightarrow 2WO_2 + H_2O$	−0.04
$WO_2 + 4H^+ + 4e^- \rightarrow W + 2H_2O$	−0.12
$WO_3 + 6H^+ + 6e^- \rightarrow W + 3H_2O$	−0.09
$2WO_3 + 2H^+ + 2e^- \rightarrow W_2O_5 + H_2O$	−0.03
$Y^{+3} + 3e^- \rightarrow Y$	−2.37
$Zn^{+2} + 2e^- \rightarrow Zn$	−0.7628
$Zn^{+2} + 2e^- \rightarrow Zn(Hg)$	−0.7628
$ZnO_2^{-2} + 2H_2O + 2e^- \rightarrow Zn + 4OH^-$	−1.216
$ZnSO \cdot 7H_2O + 2e^- \rightarrow Zn(Hg) + SO_4^{-2}$(Sat'd $ZnSO_4$)	−0.7993
$ZrO_2 + 4H^+ + 4e \rightarrow Zr + 2H_2O$	−1.43
$ZrO(OH)_2 + H_2O + 4e^- \rightarrow Zr + 4OH^-$	−2.32

ELECTROCHEMICAL SERIES

Table II
Compiled by J. F. Hunsberger
Values listed are Standard Reduction Potentials

This table is divided into two parts. The first part lists reduction reactions which are positive with respect to the potential of the Standard Hydrogen Electrode. In this first part of Table II the reduction reactions are written in increasing order of the positive potential, beginning with 0.00 and ending with +3.03 volts. The second part of this Table II lists reduction reactions having reduction potentials more negative than that of the Standard Hydrogen Electrode. This second part of the table lists reactions in their order of increasing negative potential, beginning with 0.00 and ending with −3.1 volts.

PART 1

Reaction	Potential, volts
$2H^+ + 2e^- \rightarrow H_2$	0.0000
$CuI_2^- + e^- \rightarrow Cu + 2I^-$	0.00
$H_2MoO_4 + 6H^+ + 6e^- \rightarrow Mo + 4H_2O$	0.0
$NO_3^- + H_2O + 2e^- \rightarrow NO_2^- + 2OH^-$	0.01
$Tl_2O_3 + 3H_2O + 4e^- \rightarrow 2Tl^+ + 6OH^-$	0.02
$SeO_4^{-2} + H_2O + 2e^- \rightarrow SeO_3^{-2} + 2OH^-$	0.03
$Ti(OH)^{+3} + H^+ + e^- \rightarrow Ti^{+3} + H_2O$	0.06
$UO_2^{+2} + e^- \rightarrow UO_2^+$	0.062
$U^{+6} + e^- \rightarrow U^{+5}$(1f $HClO_4$)	0.063
$Sn^{+4} + 2e^- \rightarrow Sn^{+2}$(0.1f HCl)	0.070
$AgBr + e^- \rightarrow Ag + Br^-$	0.0713
$AgSCN + e^- \rightarrow Ag + SCN^-$	0.0895
$S_4O_6^{=} + 2e^- \rightarrow 2S_2O_3^{-2}$	0.09
	(0.10)
$HgO + H_2O + 2e^- \rightarrow Hg + 2OH^-$	0.0984
$2NO + 2e^- \rightarrow N_2O_2^{-2}$	0.10
$Ir_2O_3 + 3H_2O + 6e^- \rightarrow 2Ir + 6OH^-$	0.1
$Pd(OH)_2 + 2e^- \rightarrow Pd + 2OH^-$	0.1
$Co(NH_3)_6^{+3} + e^- \rightarrow Co(NH_3)_6^{+2}$	0.1
$Hg_2O + H_2O + 2e^- \rightarrow 2Hg + 2OH^-$	0.123
$Sn^{+4} + 2e^- \rightarrow Sn^{+2}$(1f HCl)	0.139
$Hg_2Br_2 + 2e^- \rightarrow 2Hg + 2Br^-$	0.1396
$S + 2H^+ + 2e^- \rightarrow H_2S_{(aq)}$	0.141
$Sb_2O_3 + 6H^+ + 6e^- \rightarrow 2Sb + 3H_2O$	0.1445
	(0.152)
$2NO_2^- + 3H_2O + 4e^- \rightarrow N_2O + 6OH^-$	0.15
$ReO_4^- + 8H^+ + 7e^- \rightarrow Re + 4H_2O$	0.15
$Sn^{+4} + 2e^- \rightarrow Sn^{+2}$	0.15
$Np^{+4} + e^- \rightarrow Np^{+3}$(1f $HClO_4$)	0.155
$Cu^{+2} + e \rightarrow Cu^+$	0.158
	(0.167)
$BiOCl + 2H^+ + 3e \rightarrow Bi + Cl^- + H_2O$	0.1583
$Pt(OH)_2 + 2e^- \rightarrow Pt + 2OH^-$	0.16
$Bi(Cl)_4^- + 3e^- \rightarrow Bi + 4Cl^-$	0.168

Reaction	Potential, volts
$ClO_4^- + H_2O + 2e^- \rightarrow ClO_3^- + 2OH^-$	0.17
$Ag_4Fe(CN)_6 + 4e^- \rightarrow 4Ag + Fe(CN)_6^{-4}$	0.1943
$SO_4^{-2} + 4H^+ + 2e^- \rightarrow H_2SO_3 + H_2O$	0.20
$Co(OH)_3 + e^- \rightarrow Co(OH)_2 + OH^-$	0.2
	(0.17)
$SbO^+ + 2H^+ + 3e^- \rightarrow Sb + 2H_2O$	0.212
$AgCl + e^- \rightarrow Ag + Cl^-$	0.2223
*Calomel Electrode, Sat'd NaCl	0.2360
$As_2O_3 + 6H^+ + 6e^- \rightarrow 2As + 3H_2O$	0.234
*Calomel Electrode, Sat'd. KCl	0.2415
$HAsO_2 + 3H^+ + 3e^- \rightarrow As + 2H_2O$	0.2475
$IO_3^- + 3H_2O + 6e^- \rightarrow I^- + 6OH^-$	0.26
$ReO_2 + 4H^+ + 4e^- \rightarrow Re + 2H_2O$	0.26
$Hg_2Cl_2 + 2e^- \rightarrow 2Hg + 2Cl^-$	0.2682
Calomel Electrode, Molal KCl	0.2800
$PbO_2 + H_2O + 2e^- \rightarrow PbO + 2OH^-$	0.28
*Calomel Electrode, N KCl	0.2807
$Re^{+3} + 3e^- \rightarrow Re$	0.3~
$BiO^+ + 2H^+ + 3e^- \rightarrow Bi + H_2O$	0.32
$2HCNO + 2H^+ + 2e \rightarrow (CN)_2 + 2H_2O$	0.33
*Calomel Electrode 0.1 N KCl	0.3337
$UO_2^{+2} + 4H^+ + 2e^- \rightarrow U^{+4} + 2H_2O$	0.334
$VO^{+2} + 2H^+ + e^- \rightarrow V^{+3} + H_2O$	0.337
$Cu^{+2} + 2e^- \rightarrow Cu$	0.3402
$Hg_2Cl_2 + 2e^- \rightarrow 2Hg + 2Cl^-(0.1f NaOH)$	0.3419
	(0.268)
$Ag_2O + H_2O + 2e^- \rightarrow 2Ag + 2OH^-$	0.342
$Nb^{+5} + 2e \rightarrow Nb^{+3}(2f HCl)$	0.344
$Cu^{+2} + 2e \rightarrow Cu(Hg)$	0.345
$ClO_3^- + H_2O + 2e^- \rightarrow ClO_2^- + 2OH^-$	0.35
$AgIO_3 + e^- \rightarrow Ag + IO_3$	0.3551
$Ag_2SeO_3 + 2e \rightarrow 2Ag + SeO_3^{-2}$	0.3629
$ReO_4^- + 8H^+ + 7e^- \rightarrow Re + 4H_2O$	0.367
$(CN)_2 + 2H^+ + 2e^- \rightarrow 2HCN$	0.37
$O_2 + 2H_2O + 4e^- \rightarrow 4OH^-$	0.401
$AgOCN + e^- \rightarrow Ag + OCN^-$	0.41
$Fe^{+3} + e^- \rightarrow Fe^{+2}(1f H_3PO_4)$	0.438
$Rh(Cl)_6^{-3} + 3e^- \rightarrow Rh + 6Cl^-$	0.44
$Ag_2CrO_4 + 2e^- \rightarrow 2Ag + CrO_4^{-2}$	0.4463
$H_2SO_3 + 4H^+ + 4e^- \rightarrow S + 3H_2O$	0.45
$Fe(CN)_6^{-3} + e^- \rightarrow Fe(CN)_6^{-4}(0.01f NaOH)$	0.46
$Ag_2(WO_4) + 2e^- \rightarrow 2Ag + WO_4^{-2}$	0.466
$TeO_4^- + 8H^+ + 7e^- \rightarrow Te + 4H_2O$	0.472
$Ag_2CO_3 + 2e^- \rightarrow 2Ag + CO_3^{-2}$	0.4769
$AgC_2O_4 + 2e^- \rightarrow Ag + C_2O_4^{-2}$	0.4776
$Ag_2MoO_4 + 2e^- \rightarrow 2Ag + MoO_4^{-2}$	0.49
$IO^- + H_2O + 2e^- \rightarrow I^- + 2OH^-$	0.49
$NiO_2 + 2H_2O + 2e^- \rightarrow Ni(OH)_2 + 2OH^-$	0.49
$Ru^{+4} + e^- \rightarrow Ru^{+3}(0.1f HClO_4)$	0.49
$ReO_4^- + 4H^+ + 3e^- \rightarrow ReO_2 + 2H_2O$	0.51
$Hg_2(AcO)_2 + 2e^- \rightarrow 2Hg + 2AcO^-$	0.5113
$Cu^+ + e^- \rightarrow Cu$	0.522
$I_3^- + 2e^- \rightarrow 3I^-$	0.5338
$I_2 + 2e^- \rightarrow 2I^-$	0.535
$IO_3^- + 2H_2O + 4e^- \rightarrow IO^- + 4OH^-$	0.56
$MnO_4^- + e^- \rightarrow MnO_4^{-2}$	0.564
$MnO_4^{-1} + 2H_2O + 3e^- \rightarrow MnO_2 + 4OH^-$	0.58
$H_3AsO_4 + 2H^+ + 2e^- \rightarrow HAsO_2 + 2H_2O(1f HCl)$	0.58
$MnO_4^- + 2H_2O + 3e^- \rightarrow MnO_2 + 4OH^-$	0.588
$AgNO_2 + e^- \rightarrow Ag + NO_2^-$	0.59
$ClO_2^- + H_2O + 2e^- \rightarrow ClO^- + 2OH^-$	0.59
$RuO_4^- + e^- \rightarrow RuO_4^{-2}$	0.59
$TeO_2 + 4H_2 + 4e^- \rightarrow Te + 2H_2O$	0.593
$2AgO + H_2O + 2e^- \rightarrow Ag_2O + 2OH^-$	0.599
$S_2O_6^{-2} + 4H^+ + 2e^- \rightarrow 2H_2SO_3$	0.6
$BrO_3^- + 3H_2O + 6e^- \rightarrow Br^- + 6OH^-$	0.61
$Hg_2SO_4 + 2e^- \rightarrow 2Hg + SO_4^{-2}$	0.6158
$ClO_3^- + 3H_2O + 6e^- \rightarrow Cl^- + 6OH^-$	0.62
$UO_2^+ + 4H^+ + e^- \rightarrow U^{+4} + 2H_2O$	0.62
$Pd^{+2} + 2e^- \rightarrow Pd(1f HCl)$	0.623
$PdCl_4^{-2} + 2e^- \rightarrow Pd + 4Cl^-$	0.623
$Te^{+4} + 4e^- \rightarrow Te(2.5f HCl)$	0.63
$Hg_2HPO_4 + H^+ + 2e^- \rightarrow 2Hg + H_2PO_4^-$	0.639
$AgAc + e^- \rightarrow Ag + Ac^-$	0.64
$Sb_2O_{5(s)} + 6H^+ + 4e^- \rightarrow 2SbO^+ + 3H_2O$	0.64
$Ag_2SO_4 + 2e^- \rightarrow 2Ag + SO_4^{-2}$	0.653
$Fe^{+3} + e^- \rightarrow Fe^{+2}(0.5f H_2SO_4)$	0.679
$AgBrO_3 + e^- \rightarrow Ag + BrO_3^-$	0.680
$O_2 + 2H^+ + 2e^- \rightarrow H_2O_2$	0.682
$Fe(CN)_6^{-3} + e^- \rightarrow Fe(CN)_6^{-4}(1f H_2SO_4)$	0.69
$Sb_2O_5 + 4H^+ + 4e^- \rightarrow Sb_2O_3 + 2H_2O$	0.69
$C_6H_4O_2 + 2H^+ + 2e \rightarrow C_6H_4(OH)_2$	0.6992
Quinhydrone Elec. H^+, a = 1	0.6995
$BrO^- + H_2O + 2e^- \rightarrow Br^- + 2OH^-(1f NaOH)$	0.70
$H_3IO_6^{-2} + 2e^- \rightarrow IO_3^- + 3OH^-$	~0.70
$PtCl_4^{-2} + 2e^- \rightarrow Pt + 4Cl^-$	0.73
$TcO_4^- + 4H^+ + 3e^- \rightarrow TcO_{2(c)} + 2H_2O$	0.738
$Np^{+5} + e^- \rightarrow Np^{+4}(1f HClO_4)$	0.739
$Ag_2O_3 + H_2O + 2e^- \rightarrow 2AgO + 2OH^-$	0.74
$H_2SeO_3 + 4H^+ + 4e^- \rightarrow Se + 3H_2O$	0.74
$PtCl_6^{-2} + 2e^- \rightarrow PtCl_4^{-2} + 2Cl^-$	0.74
$Fe^{+3} + e^- \rightarrow Fe^{+2}(1f HClO_4)$	0.747
$Sb^{+5} + 2e^- \rightarrow Sb^{+3}(3.5f HCl)$	0.75
$ClO_2^- + 2H_2O + 4e^- \rightarrow Cl^- + 4OH^-$	0.76
$NiO_2 + 2H_2O + 2e^- \rightarrow Ni(OH)_2 + 2OH^-$	0.76
$2NO + H_2O + 2e^- \rightarrow N_2O + 2OH^-$	0.76
$ReO_4^- + 2H^+ + e^- \rightarrow ReO_{3(cc)} + 2H_2O$	0.768
$(CNS)_2 + 2e^- \rightarrow 2CNS^-$	0.77
$Fe^{+3} + e^- \rightarrow Fe^{+2}$	0.770
$Fe^{+3} + e^- \rightarrow Fe^{+2}(1f HCl)$	0.770
$IrCl_6^{-3} + 3e^- \rightarrow Ir + 6Cl^-$	0.77
$Tl^{+3} + 2e^- \rightarrow Tl^{+1}(1f HCl)$	0.783
$Hg_2^{+2} + 2e^- \rightarrow 2Hg$	0.7961
$1/2Hg_2^{+2} + e^- \rightarrow Hg$	0.7986
$Ag^+ + e^- \rightarrow Ag$	0.7996
$2HNO_2 + 4H^+ + 4e^- \rightarrow H_2N_2O_2 + 2H_2O$	0.80
$2NO_3^- + 4H^+ + 2e^- \rightarrow N_2O_4 + 2H_2O$	0.81
$1/2O_2 + 2H^+(10^{-7}M) + 2e^- \rightarrow H_2O$	0.815
$Pd^{+2} + 2e^- \rightarrow Pd$	0.83
$OsO_4 + 8H^+ + 8e^- \rightarrow Os + 4H_2O$	0.85
$Hg^{+2} + 2e^- \rightarrow Hg$	0.851
$AuBr_4^- + 3e^- \rightarrow Au + 4Br^-$	0.858
$Ru^{+4} + e^- \rightarrow Ru^{+3}(2f HCl)$	0.858
$TiO_2 + 4H^+ + 4e^- \rightarrow Ti + 2H_2O$	0.86
$HO_2^- + H_2O + 2e^- \rightarrow 3OH^-$	0.87
$N_2O_4 + 2e^- \rightarrow 2NO_2^-$	0.88
$ClO^- + H_2O + 2e^- \rightarrow Cl^- + 2OH^-$	0.90
$2Hg^{+2} + 2e^- \rightarrow Hg_2^{+2}$	0.905
$Pu^{+6} + e^- \rightarrow Pu^{+5}(1f HClO_4)$	0.9184
$NO_3^- + 3H^+ + 2e^- \rightarrow HNO_2 + H_2O$	0.94
$ClO_2(aq) + e^- \rightarrow ClO_2^-$	0.954
$NO_3^- + 4H^+ + 3e^- \rightarrow NO + 2H_2O$	0.96
$AuBr_2^- + e^- \rightarrow Au + 2Br^-$	0.963
$Pu^{+4} + e^- \rightarrow Pu^{+3}(1f HClO_4)$	0.982
$Pd^{+2} + 2e^- \rightarrow Pd(4f HClO_4)$	0.987
$HIO + H^+ + 2e^- \rightarrow I^- + H_2O$	0.99
$HNO_2 + H^+ + e^- \rightarrow NO + H_2O$	0.99
$AuCl_4^- + 3e^- \rightarrow Au + 4Cl^-$	0.994
$RuO_{4(c)} + e^- \rightarrow RuO_4^-$	1.00
$VO_2^+ + 2H^+ + e^- \rightarrow VO^{+2} + H_2O$	1.00
$V(OH)_4^+ + 2H^+ + e^- \rightarrow VO^{+2} + 3H_2O$	1.00
$H_6TeO_{6(s)} + 2H^+ + 2e^- \rightarrow TeO_{2(s)} + 4H_2O$	1.02
$IrCl_6^{-2} + e^- \rightarrow IrCl_6^{-3}$	1.02
$U^{+5} + e^- \rightarrow U^{+4}(1f HCl)$	1.02
$N_2O_4 + 4H^+ + 4e^- \rightarrow 2NO + 2H_2O$	1.03
$Pu^{+6} + 2e^- \rightarrow Pu^{+4}(1f HCl)$	1.052
$Fe(phenanthroline)_3^{+3} + e^- \rightarrow Fe(ph)_3^{+2}(2f H_2SO_4)$	1.056
$Br_{2(l)} + 2e^- \rightarrow 2Br^-$	1.065
$N_2O_4 + 2H^+ + 2e^- \rightarrow 2HNO_2$	1.07
$IO_3^- + 6H^+ + 6e^- \rightarrow I^- + 3H_2O$	1.085
$Br_{2(aq)} + 2e^- \rightarrow 2Br^-$	1.087
$Pu^{+5} + e^- \rightarrow Pu^{+4}(0.5f HCl)$	1.099
$Cr^{+6} + 3e^- \rightarrow Cr^{+3}(2f H_2SO_4)$	1.10
$Cu^{+2} + 2CN^- + e^- \rightarrow Cu(CN)_2^-$	1.12
$Np^{+6} + e^- \rightarrow Np^{+5}(1f HClO_4)$	1.137
$Fe(phenanthroline)_3^{+3} + e^- \rightarrow Fe(ph)_3^{+2}$	1.14
$SeO_4^{-2}\ 4H^+ + 2e^- \rightarrow H_2SeO_3 + H_2O$	1.15
$ClO_2 + e^- \rightarrow ClO_2^-$	1.15
$ClO_3^- + 2H^+ + e^- \rightarrow ClO_2 + H_2O$	1.15
$ClO_4^- + 2H^+ + 2e^- \rightarrow ClO_3^- + H_2O$	1.19
$2IO_3^- + 12H^+ + 10e^- \rightarrow I_2 + 6H_2O$	1.19
$HCrO_4^- + 7H^+ + 3e^- \rightarrow Cr^{+3} + 4H_2O$	1.195
$IO_3^- + 6H^+ + 5e^- \rightarrow 1/2I_2 + 3H_2O$	1.195
$Pt^{+2} + 2e^- \rightarrow Pt$	~1.2
$MnO_2 + 4H^+ + 2e^- \rightarrow Mn^{+2} + 2H_2O$	1.208
$ClO_3^- + 3H^+ + 2e \rightarrow HClO_2 + H_2O$	1.21
	(1.23)
$O_2 + 4H^+ + 4e^- \rightarrow 2H_2O$	1.229
$O_3 + H_2O + 2e^- \rightarrow O_2 + 2OH^-$	1.24
$Tl^{+3} + 2e^- \rightarrow Tl^{+1}$	1.247
$ClO_2 + H^+ + e^- \rightarrow HClO_2$	1.27
$2HNO_2 + 4H^+ + 4e^- \rightarrow N_2O + 3H_2O$	1.27
	(1.29)
$N_2H_5^+ + 3H^+ + 2e^- \rightarrow 2NH_4^+$	1.27
$Au^{+3} + 2e^- \rightarrow Au^{+1}$	~1.29
$PdCl_6^{-2} + 2e^- \rightarrow PdCl_4^{-2} + 2Cl^-$	1.29
$HBrO + H^+ + 2e^- \rightarrow Br^- + H_2O$	1.33
$Cr_2O_7^{-2} + 14H^+ + 6e^- \rightarrow 2Cr^{+3} + 7H_2O$	1.33
$ClO_4^- + 8H^+ + 7e^- \rightarrow 1/2Cl_2 + 4H_2O$	1.34
$Cl_{2(g)} + 2e^- \rightarrow 2Cl^-$	1.3583
$ClO_4^- + 8H^+ + 8e^- \rightarrow Cl^- + 4H_2O$	1.37

Reaction	Potential, volts
$OH + e^- \rightarrow OH^-$	1.4
$Au^{+3} + 3e^- \rightarrow Au$	1.42
$2NH_3OH^+ + H^+ + 2e^- \rightarrow N_2H_5^+ + 2H_2O$	1.42
$Rh^{+4} + e^- \rightarrow Rh^{+3}$	1.43
$BrO_3^- + 6H^+ + 6e^- \rightarrow Br^- + 3H_2O$	1.44
$Ce^{+4} + e^- \rightarrow Ce^{+3}$	1.4430
	(1.61)
$Au(OH)_3 + 3H^+ + 3e^- \rightarrow Au + 3H_2O$	1.45
$ClO_3^- + 6H^+ + 6e^- \rightarrow Cl^- + 3H_2O$	1.45
$HIO + H^+ + e^- \rightarrow 1/2I_2 + H_2O$	1.45
$Ce^{+4} + e^- \rightarrow Ce^{+3}$(0.5f H_2SO_4)	1.4587
$PbO_2 + 4H^+ + 2e^- \rightarrow Pb^{+2} + 2H_2O$	1.46
$ClO_3^- + 6H^+ + 5e^- \rightarrow 1/2Cl_2 + 3H_2O$	1.47
$HClO + H^+ + 2e^- \rightarrow Cl^- + H_2O$	1.49
$MnO_4^- + 8H^+ + 5e^- \rightarrow Mn^{+2} + 4H_2O$	1.491
$HO_2 + H^+ + e^- \rightarrow H_2O_2$	1.5
$Mn^{+3} + e^- \rightarrow Mn^{+2}$	1.51
$BrO_3^- + 6H^+ + 5e^- \rightarrow 1/2Br_2 + 3H_2O$	1.52
$HClO_2 + 3H^+ + 4e^- \rightarrow Cl^- + 2H_2O$	1.56
$Bi_2O_4 + 4H^+ + 2e^- \rightarrow 2BiO^+ + 2H_2O$	1.59
$HBrO + H^+ + e^- \rightarrow 1/2Br_2 + H_2O$	1.59
$2NO + 2H^+ + 2e^- \rightarrow N_2O + H_2O$	1.59
$2HBrO + 2H^+ + 2e^- \rightarrow Br_{2(l)} + 2H_2O$	1.6

Reaction	Potential, volts
$HClO_2 + 3H^+ + 3e^- \rightarrow 1/2Cl_2 + 2H_2O$	1.63
$HClO + H^+ + e^- \rightarrow 1/2Cl_2 + H_2O$	1.63
$HClO_2 + 2H^+ + 2e^- \rightarrow HClO + H_2O$	1.64
$MnO_4^- + 4H^+ + 3e^- \rightarrow MnO_2 + 2H_2O$	1.679
$Au^+ + e^- \rightarrow Au$	1.68
$PbO_2 + SO_4^{-2} + 4H^+ + 2e^- \rightarrow PbSO_4 + 2H_2O$	1.685
$H_5IO_6 + H^+ + 2e^- \rightarrow IO_3^- + 3H_2O$	~1.7
$CeOH^{+3} + H^+ + e^- \rightarrow Ce^{+3} + H_2O$	1.7134
$N_2O + 2H^+ + 2e^- \rightarrow N_2 + H_2O$	1.77
$H_2O_2 + 2H^+ + 2e^- \rightarrow 2H_2O$	1.776
$Co^{+3} + e^- \rightarrow Co^{+2}$(3f HNO_3)	1.842
$FeO_4^{-2} + 8H^+ + 3e^- \rightarrow Fe^{+3} + 4H_2O$	1.9
$NiO_2 + 4H^+ + 2e^- \rightarrow Ni^{+2} + 2H_2O$	1.93
$Ag^{+2} + e^- \rightarrow Ag^{+1}$(4f $HClO_4$)	1.987
$S_2O_8^{-2} + 2e^- \rightarrow 2SO_4^{-2}$	2.0
	(2.05)
$O_3 + 2H^+ + 2e^- \rightarrow O_2 + H_2O$	2.07
$F_2O + 2H^+ + 4e^- \rightarrow H_2O + 2F^-$	2.1
$O_{(g)} + 2H^+ + 2e^- \rightarrow H_2O$	2.42
$H_2N_2O_2 + 2H^+ + 2e^- \rightarrow N_2 + 2H_2O$	2.65
$1/2F_2 + e^- \rightarrow F^-$	2.85
$F_2 + 2e^- \rightarrow 2F^-$	2.87
$1/2F_2 + H^+ + e^- \rightarrow HF$	3.03

PART 2

Reaction	Potential, volts
$2H^+ + 2e^- \rightarrow H_2$	0.0000
$D^+ + e^- \rightarrow 1/2D_2$	-0.0034
$AgCN + e^- \rightarrow Ag + CN^-$	-0.02
$TeO_3^{-2} + 3H_2O + 4e^- \rightarrow Te + 6OH^-$	-0.02
$2WO_3 + 2H^+ + 2e^- \rightarrow W_2O_5 + H_2O$	-0.03
$Fe^{+3} + 3e^- \rightarrow Fe$	-0.036
$Ag_2S + 2H^+ + 2e^- \rightarrow 2Ag + H_2S$	-0.0366
$P + 3H^+ + 3e^- \rightarrow PH_3(g)$	-0.04
$W_2O_5 + 2H^+ + 2e^- \rightarrow 2WO_2 + H_2O$	-0.04
$Hg_2I_2 + 2e^- \rightarrow 2Hg + 2I^-$	-0.0405
$2D^+ + 2e^- \rightarrow D_2$	-0.044
$Tl(OH)_3 + 2e^- \rightarrow TlOH + 2OH^-$	-0.05
$O_2 + H_2O + 2e^- \rightarrow HO_2^- + OH^-$	-0.076
$AsO_4^{-3} + 2H_2O + 2e^- \rightarrow AsO_2^- + 4OH^-$(1f NaOH)	-0.08
$2H_2SO_3 + H^+ + 2e^- \rightarrow HS_2O_4^- + 2H_2O$	-0.08
$Ru^{+3} + e^- \rightarrow Ru^{+2}$(1–6f HCL)	-0.084
$2Cu(OH)_2 + 2e^- \rightarrow Cu_2O + 2OH^- + H_2O$	-0.09
$WO_3 + 6H^+ + 6e^- \rightarrow W + 3H_2O$	-0.09
$Ru^{+3} + e^- \rightarrow Ru^{+2}$(0.1f $HClO_4$)	-0.11
$Cr^{+6} + 3e^- \rightarrow Cr^{+3}$(1f NaOH)	-0.12
$CrO_4^{-2} + 4H_2O + 3e^- \rightarrow Cr(OH)_3 + 5OH^-$	-0.12
$GeO_2 + 2H^+ + 2e^- \rightarrow GeO + H_2O$	-0.12
$WO_2 + 4H^+ + 4e^- \rightarrow W + 2H_2O$	-0.12
$Pb^{+2} + 2e^- \rightarrow Pb(Hg)$	-0.1205
$Pb^{+2} + 2e^- \rightarrow Pb$	-0.1263
	(0.126)
$H_2GeO_3 + 4H^+ + 4e^- \rightarrow Ge + 3H_2O$	-0.13
$Sn^{+2} + 2e^- \rightarrow Sn$	-0.1364
$O_2 + 2H_2O + 2e^- \rightarrow H_2O_2 \rightarrow 2OH^-$	-0.146
$AgI + e^- \rightarrow Ag + I^-$	-0.1519
$2NO_2^- + 2H_2O + 4e^- \rightarrow N_2O_2^{-2} + 4OH^-$	-0.18
$CO_2 + 2H^+ + 2e^- \rightarrow HCOOH$	-0.2
$2SO_4^{-2} + 4H^+ + 2e^- \rightarrow S_2O_6^{-2} + 2H_2O$	-0.2
$Cu(OH)_2 + 2e^- \rightarrow Cu + 2OH^-$	-0.224
$Ni^{+2} + 2e^- \rightarrow Ni$	-0.23
$PbHPO_4 + H^+ + 2e^- \rightarrow Pb(Hg) + HPO_4^=$	-0.2448
$V(OH)_4^+ + 4H^+ + 5e^- \rightarrow V + 4H_2O$	-0.25
$V^{+3} + e^- \rightarrow V^{+2}$	-0.255
$PbCl_2 + 2e^- \rightarrow Pb(Hg) + 2Cl^-$	-0.262
$PbBr_2 + 2e^- \rightarrow Pb(Hg) + Br^-$	-0.275
$H_3PO_4 + 2H^+ + 2e^- \rightarrow H_3PO_3 + H_2O$	-0.276
	(-0.2)
$Co^{+2} + 2e^- \rightarrow Co$	-0.28
$Tl^+ + e^- \rightarrow Tl(Hg)$	-0.3338
$Tl^+ + e^- \rightarrow Tl$	-0.3363
$In^{+3} + 3e^- \rightarrow In$	-0.338
$PbF_2 + 2e^- \rightarrow Pb(Hg) + 2F^-$	-0.3444
$TlOH + e^- \rightarrow Tl + OH^-$	-0.3445
$SeO_3^{-2} + 3H_2O + 4e^- \rightarrow Se + 6OH^-$	-0.35
$PbSO_4 + 2e^- \rightarrow Pb(Hg) + SO_4^{-2}$	-0.3505
$Cd^{+2} + 2e^- \rightarrow Cd(Hg)$	-0.3521
$PbSO_4 + 2e^- \rightarrow Pb + SO_4^{-2}$	-0.356
$PbI_2 + 2e^- \rightarrow Pb(Hg) + 2I^-$	-0.358
$Se + 2H^+ + 2e^- \rightarrow H_2Se(aq)$	-0.36
$Cu_2O + H_2O + 2e^- \rightarrow 2Cu + 2OH^-$	-0.361
$Tl^{+3} + e^- \rightarrow Tl^{+2}$	-0.37

Reaction	Potential, volts
$In^{+2} + e^- \rightarrow In^{+1}$	-0.40
$In^{+3} + 2e^- \rightarrow In^{+1}$	-0.40
$Mn(OH)_3 + e^- \rightarrow Mn(OH)_2 + OH^-$	-0.40
$Cd^{+2} + 2e^- \rightarrow Cd$	-0.4026
$Fe^{+2} + 2e^- \rightarrow Fe$	-0.409
$Cr^{+3} + e^- \rightarrow Cr^{+2}$	-0.41
$Eu^{+3} + e^- \rightarrow Eu^{+2}$	-0.43
$CdSO_4 \cdot 8/3H_2O + 2e^- \rightarrow Cd(Hg) + CdSO_4$ (sat'd aq)	-0.4346
$Tl_2SO_4 + 2e^- \rightarrow Tl(Hg) + SO_4^{-2}$	-0.4360
$Bi_2O_3 + 3H_2O + 6e^- \rightarrow 2Bi + 6OH^-$	-0.46
$BiOOH + H_2O + 3e^- \rightarrow Bi + 3OH^-$	-0.46
$NO_2^- + H_2O + e^- \rightarrow NO + 2OH^-$	-0.46
$S + H_2O + 2e^- \rightarrow HS^- + OH^-$	-0.478
$H_3PO_3 + 3H^+ + 3e^- \rightarrow P + 3H_2O$	-0.49
$In^{+3} + e^- \rightarrow In^{+2}$	-0.49
$2CO_2 + 2H^+ + 2e^- \rightarrow H_2C_2O_4$	-0.49
$H_3PO_3 + 2H^+ + 2e^- \rightarrow H_3PO_2 + H_2O$	-0.50
	(-0.59)
$S + 2e^- \rightarrow S^{-2}$	-0.508
$H_3PO_2 + H^+ + e^- \rightarrow P + 2H_2O$	-0.51
$Sb + 3H^+ + 3e^- \rightarrow H_3Sb$	-0.51
$As + 3H^+ + 3e^- \rightarrow AsH_3$	-0.54
$HPbO_2^- + H_2O + 2e^- \rightarrow Pb + 3OH^-$	-0.54
$TlCl + e^- \rightarrow Tl(Hg) + Cl^-$	-0.555
$Cr^{+2} + 2e \rightarrow Cr$	-0.557
$Ga^{+3} + 3e^- \rightarrow Ga$	-0.560
$Fe(OH)_3 + e^- \rightarrow Fe(OH)_2 + OH^-$	-0.56
$PbO + H_2O + 2e^- \rightarrow Pb + 2OH^-$	-0.576
$2SO_3^{-2} + 3H_2O + 4e^- \rightarrow S_2O_3^{-2} + 6OH^-$	-0.58
$SbO_3^- + H_2O + 2e^- \rightarrow SbO_2^- + 2OH^-$	-0.59
$TlBr + e^- \rightarrow Tl(Hg) + Br^-$	-0.606
$U^{+4} + e^- \rightarrow U^{+3}$	-0.61
$Cb_2O_5 + 10H^+ + 10e^- \rightarrow 2Cb + 5H_2O$	-0.62
$U^{+4} + e^- \rightarrow U^{+3}$(1 f$HCLO_4$)	-0.631
$Ni(OH)_2 + 2e^- \rightarrow Ni + 2OH^-$	-0.66
$SbO_2^- + 2H_2O + 3e^- \rightarrow Sb + 4OH^-$	-0.66
$AsO_2^- + 2H_2O + 3e^- \rightarrow As + 4OH^-$	-0.68
*$Te + 2H^+ + 2e^- \rightarrow H_2Te(Ag)$	-0.69
	(-0.72)
$Ag_2S + 2e^- \rightarrow 2Ag + S^{-2}$	-0.7051
$Ta_2O_5 + 10H^+ + 10e^- \rightarrow 2Ta + 5H_2O$	-0.71
$AsO_4^{-3} + 2H_2O + 2e^- \rightarrow AsO_2^- + 4OH^-$	-0.71
$Co(OH)_2 + 2e^- \rightarrow Co + 2OH^-$	-0.73
$H_3BO_3 + 3H^+ + 3e^- \rightarrow B + 3H_2O$	-0.73
$Cr^{+3} + 3e^- \rightarrow Cr$	-0.74
$V^{+5} + e^- \rightarrow V^{+4}$(1f NaOH)	-0.74
$Cd(OH)_2 + 2e^- \rightarrow Cd(Hg) + 2OH^-$	-0.761
	(-0.81)
$Zn^{+2} + 2e^- \rightarrow Zn$	-0.7628
$Zn^{+2} + 2e^- \rightarrow Zn(Hg)$	-0.7628
$TlI + e^- \rightarrow Tl(Hg) + I^-$	-0.769
$Se + 2e^- \rightarrow Se^{-2}$	-0.78
$HSnO_2^- + H_2O + 2e^- \rightarrow Sn + 3OH^-$	-0.79
$ZnSO_4 \cdot 7H_2O + 2e^- \rightarrow Zn(Hg) + SO_4^{-2}$ (Sat'd $ZnSO_4$)	-0.7993
$RuO_2 + 4H^+ + 4e^- \rightarrow Ru + 2H_2O$	-0.8
$ReO_4^- + 4H_2O + 7e^- \rightarrow Re + 8OH^-$	-0.81

ELECTROCHEMICAL SERIES—TABLE II (Continued)

Reaction	Potential, volts
$UO_2^{+2} + 4H^+ + 6e^- \rightarrow U + 2H_2O$	−0.82
$2H_2O + 2e^- \rightarrow H_2 + 2OH^-$	−0.8277
$SiO_2 + 4H^+ + 4e^- \rightarrow Si + 2H_2O$	−0.84
$2NO_3^- + 2H_2O + 2e^- \rightarrow N_2O_4 + 4OH^-$	−0.85
$TiO_2 + 4H^+ + 4e^- \rightarrow Ti + 2H_2O$	−0.86
$P + 3H_2O + 3e^- \rightarrow PH_3(g) + 3OH^-$	−0.87
$SO_4^{-2} + H_2O + 2e^- \rightarrow SO_3^{-2} + 2OH^-$	−0.92
$Te + 2e^- \rightarrow Te^{-2}$	−0.92
$Sn(OH)_6^{-2} + 2e^- \rightarrow HSnO_2^- + 3OH^- + H_2O$	−0.96
$Mn^{+2} + 2e^- \rightarrow Mn$	−1.029
$PO_4^{-3} + 2H_2O + 2e^- \rightarrow HPO_3^{-2} + 3OH^-$	−1.05
$2SO_3^{-2} + 2H_2O + 2e^- \rightarrow S_2O_4^{-2} + 4OH^-$	−1.12
$CrO_2^- + 2H_2O + 3e^- \rightarrow Cr + 4OH^-$	−1.2
$SiF_6^{-2} + 4e^- \rightarrow Si + 6F^-$	−1.2
$V^{+2} + 2e^- \rightarrow V$	−1.2
$ZnO_2^{-2} + 2H_2O + 2e^- \rightleftharpoons Zn + 4OH^-$	−1.216
$H_2GaO_3^- + H_2O + 3e^- \rightarrow Ga + 4OH^-$	−1.22
$H_2BO_3^- + 5H_2O + 8e^- \rightarrow BH_4^- + 8OH^-$	−1.24
$Cr(OH)_3 + 3e^- \rightarrow Cr + 3OH^-$	−1.3
$ZrO_2 + 4H^+ + 4e^- \rightarrow Zr + 2H_2O$	−1.43
$Ce^{+3} + 3e^- \rightarrow Ce(Hg)$	−1.4373
$Mn(OH)_2 \rightarrow Mn + 2OH^-$	−1.47
$Ba^{+2} + 2e^- \rightarrow Ba(Hg)$	−1.570
$HfO_2 + 4H^+ + 4e^- \rightarrow Hf + 2H_2O$	−1.57
$Ti^{+2} + 2e^- \rightarrow Ti$	−1.63
$HPO_3^{-2} + 2H_2O + 2e^- \rightarrow H_2PO_2^- + 3OH^-$	−1.65
$HfO^{+2} + 2H^+ + 4e^- \rightarrow Hf + H_2O$	−1.68
$Be^{+2} + 2e^- \rightarrow Be$	−1.70
	(−1.85)
$Al^{+3} + 3e^- \rightarrow Al(0.1f\ NaOH)$	−1.706
$HPO_3^{-2} + 2H_2O + 3e^- \rightarrow P + 5OH^-$	−1.71
$SiO_3^{-2} + 3H_2O + 4e^- \rightarrow Si + 6OH^-$	−1.73
$Sr^{+2} + 2e^- \rightarrow Sr(Hg)$	−1.793
$ThO_2 + 4H^+ + 4e^- \rightarrow Th + 2H_2O$	−1.80
$U^{+3} + 3e^- \rightarrow U$	−1.8
$H_2PO_2^- + e^- \rightarrow P + 2OH^-$	−1.82
$N_2 + 2H_2O + 4H^+ + 2e^- \rightarrow 2NH_3OH^+$	−1.87
$Th^{+4} + 4e^- \rightarrow Th$	−1.90
$Np^{+3} + 3e^- \rightarrow Np$	−1.9
$Ti^{+3} + e^- \rightarrow Ti^{+2}$	−2.0
$Sc^{+3} + 3e^- \rightarrow Sc$	−2.08
$1/2H_2 + e^- \rightarrow H^-$	−2.23
$Nd^{+3} + 3e^- \rightarrow Nd$	−2.246
$Be_2O_3^{-2} + 3H_2O + 4e^- \rightarrow 2Be + 6OH^-$	−2.28
$ZrO(OH)_2 + H_2O + 4e^- \rightarrow Zr + 4OH^-$	−2.32
$Ce^{+3} + 3e^- \rightarrow Ce$	−2.335
$H_2AlO_3^- + H_2O + 3e^- \rightarrow Al + 4OH^-$	−2.35
$Y^{+3} + 3e^- \rightarrow Y$	−2.37
$La^{+3} + 3e^- \rightarrow La$	−2.37
$Mg^{++} + 2e^- \rightarrow Mg$	−2.375
$H_2BO_3^- + H_2O + 3e^- \rightarrow B + 4OH^-$	−2.5
$HfO(OH)_2 + H_2O + 4e^- \rightarrow Hf + 4OH^-$	−2.60
$ThO_2 + 2H_2O + 4e^- \rightarrow Th + 4OH^-$	−2.64
$Mg(OH)_2 + 2e^- \rightarrow Mg + 2OH^-$	−2.67
$Na^+ + e^- \rightarrow Na$	−2.7109
	(−2.712)
$Ca^{+2} + 2e^- \rightarrow Ca$	−2.76
$La(OH)_3 + 3e \rightarrow La + 3OH^-$	−2.76
$Sr^{+2} + 2e^- \rightarrow Sr$	−2.89
$Ba^{+2} + 2e^- \rightarrow Ba$	−2.90
$Cs^+ + e^- \rightarrow Cs$	−2.923
$K^+ + e^- \rightarrow K$	−2.924
	(−2.923)
$Rb^+ + e^- \rightarrow Rb$	−2.925
	(−2.99)
$Ba(OH)_2 \cdot 8H_2O + 2e^- \rightarrow Ba + 2OH^- + 8H_2O$	−2.97
$Sr(OH)_2 \cdot 8H_2O + 2e^- \rightarrow Sr + 2OH^- + 8H_2O$	−2.99
$Ca(OH)_2 + 2e^- \rightarrow Ca + 2OH^-$	−3.02
$Ca^+ + e^- \rightarrow Ca$	−3.02
$Li^+ + e^- \rightarrow Li$	−3.045
	(−3.02)
$3N_2 + 2H^+ + 2e^- \rightarrow 2HN_3$	−3.1

DISSOCIATION CONSTANTS OF ORGANIC BASES IN AQUEOUS SOLUTION

No.	Compound	Temp. °C	Step	pK_a	K_a
a1	Acetamide	25		0.63	2.34×10^{-1}
a2	Acridine	20		5.58	2.63×10^{-6}
a3	α-Alanine	25		2.345	4.52×10^{-3}
a4	Alanine, glycyl-	25		3.153	7.03×10^{-4}
a5	Alanine, methoxy- (DL)	25		2.037	9.18×10^{-3}
a6	Alanine, phenyl	25	2	9.18	6.61×10^{-10}
a7	Allothreonine	25	1	2.108	7.80×10^{-3}
	Allothreonine	25	2	9.096	8.02×10^{-10}
a8	*n*-Amylamine	25		10.63	2.34×10^{-11}
a9	Aniline	25		4.63	2.34×10^{-5}
a10	Aniline, *n*-allyl	25		4.17	6.76×10^{-5}
a11	Aniline, 4-(p-aminobenzoyl)	25	1	2.932	1.17×10^{-3}
a12	Aniline, 4-benzyl	25		2.17	6.76×10^{-3}
a13	Aniline, 2-bromo	25		2.53	2.95×10^{-3}
a14	Aniline, 3-bromo	25		3.58	2.63×10^{-4}
a15	Aniline, 4-bromo	25		3.86	1.38×10^{-4}
a16	Aniline, 4-bromo-N,N-dimethyl	25		4.232	5.86×10^{-5}
a17	Aniline, o-chloro	25		2.65	2.24×10^{-3}
a18	Aniline, m-chloro	25		3.46	3.47×10^{-4}
a19	Aniline, p-chloro	25		4.15	7.08×10^{-5}
a20	Aniline, 3-chloro-N,N-dimethyl	20		3.837	1.46×10^{-4}
a21	Aniline, 4-chloro-N,N-dimethyl	20		4.395	4.03×10^{-5}
a22	Aniline, 3,5-dibromo	25		2.34	4.57×10^{-3}
a23	Aniline, 2,4-dichloro	22		2.05	8.91×10^{-3}
a24	Aniline, N,N-diethyl	22		6.61	2.46×10^{-7}
a25	Aniline, N,N-dimethyl	25		5.15	7.08×10^{-6}
a26	Aniline, N,N-dimethyl-3-nitro	25		2.626	2.37×10^{-3}
a27	Aniline, N-ethyl	24		5.12	7.59×10^{-6}
a28	Aniline, 2-fluoro	25		3.20	6.31×10^{-4}
a29	Aniline, 3-fluoro	25		3.50	3.16×10^{-4}
a30	Aniline, 4-fluoro	25		4.65	2.24×10^{-5}
a31	Aniline, 2-iodo	25		2.60	2.51×10^{-3}
a32	Aniline, N-methyl	25		4.848	1.41×10^{-5}
a33	Aniline, 4-methylthio	25		4.35	4.46×10^{-5}
a34	Aniline, 3-nitro	25		2.466	3.42×10^{-3}
a35	Aniline, 4-nitro	25		1.0	1.00×10^{-1}
a36	Aniline, 2-sulfonic acid	25	2	2.459	3.47×10^{-3}
a37	Aniline, 3-sulfonic acid	25	2	3.738	1.82×10^{-4}
a38	Aniline, 4-sulfonic acid	25	2	3.227	5.92×10^{-4}
a39	*o*-Anisidine	25		4.52	3.02×10^{-5}
a40	*m*-Anisidine	25		4.23	5.89×10^{-5}
a41	*p*-Anisidine	25		5.34	4.57×10^{-6}
a42	Arginine	25	1	1.8217	1.51×10^{-2}
		25	2	8.9936	1.01×10^{-9}
a43	Asparagine	20	1	2.213	6.12×10^{-3}
		20	2	8.85	1.41×10^{-9}
a44	Asparagine, glycyl	25	1	2.942	1.14×10^{-3}
		18	2	8.44	3.63×10^{-9}
a45	DL-Aspartic acid	1	1	2.122	7.55×10^{-3}
		1	2	4.006	1.00×10^{-4}
a46	Azetidine (Trimethylimidine)	25		11.29	5.12×10^{-12}
a47	Aziridine	25		8.01	9.77×10^{-9}
b1	Benzene, 4-aminoazo	25		2.82	1.51×10^{-3}
b2	Benzene, 2-aminoethyl (β-Phenylamine)	25		9.84	1.45×10^{-10}
b3	Benzene, 4-dimethylaminoazo	25		3.226	5.94×10^{-4}
b4	Benzidine	30	1	4.66	2.19×10^{-5}
		30	2	3.57	2.69×10^{-4}
b5	Benzimidazole	25		5.532	2.94×10^{-6}
b6	Benzimidazole, 2-ethyl	25		6.18	6.61×10^{-7}
b7	Benzimidazole, 2-methyl	25		6.19	6.46×10^{-7}
b8	Benzimidazole, 2-phenyl	25	1	5.23	5.89×10^{-6}
		25	2	11.91	1.23×10^{-12}
b9	Benzoic acid, 2-amino (Anthranilic acid)	25	1	2.108	7.80×10^{-3}
		25	2	4.946	1.13×10^{-5}
b10	Benzoic acid, 4-amino	25	1	2.501	3.15×10^{-3}
		25	2	4.874	1.33×10^{-5}
b11	Benzylamine	25		9.33	4.67×10^{-10}
b12	Betaine	0		1.83	1.48×10^{-2}
b13	Biphenyl, 2-amino	22		3.82	1.51×10^{-4}
b14	Bornylamine(trans-)	25		10.17	6.76×10^{-11}
b15	Brucine	25	1	8.28	5.24×10^{-9}
b16	Butane, 1-amino-3-methyl	25		10.60	2.51×10^{-11}
b17	Butane, 2-amino-2-methyl	19		10.85	1.41×10^{-11}
b18	Butane, 1,4-diamino (Putrescine)	10	1	11.15	7.08×10^{-12}
		10	2	9.71	1.95×10^{-10}
b19	*n*-Butylamine	20		10.77	1.69×10^{-11}
b20	*t*-Butylamine	18		10.83	1.48×10^{-11}
b21	Butyric acid, 4-amino	25	1	4.0312	9.31×10^{-5}
		25	2	10.5557	2.78×10^{-11}
b22	*n*-Butyric acid, glycyl-2-amino	25	1	3.1546	7.01×10^{-4}
c1	Cacodylic acid	25	1	1.57	2.69×10^{-2}
			2	6.27	5.37×10^{-7}
c2	β-Chlortriethyl-ammonium	25		8.80	1.59×10^{-9}
c3	Cinnoline	20		2.37	4.27×10^{-3}
c4	Codeine	25		8.21	6.15×10^{-9}
c5	Cyclohexaneamine, *n*-butyl	25		11.23	5.89×10^{-12}
c6	Cyclohexylamine	24		10.66	2.19×10^{-11}
c7	Cystine	30	1	1.90	1.25×10^{-2}
		30	2	8.24	5.76×10^{-9}
d1	*n*-Decylamine	25		10.64	2.29×10^{-11}
d2	Diethylamine	40		10.489	3.24×10^{-11}
d3	Diisobutylamine	21		10.91	1.23×10^{-11}
d4	Diisopropylamine	28.5		10.96	1.09×10^{-11}
d5	Dimethylamine	25		10.732	1.85×10^{-11}
d6	*n*-Diphenylamine	25		0.79	1.62×10^{-1}
d7	*n*-Dodecaneamine (Laurylamine)	25		10.63	2.35×10^{-11}
e1	*d*-Ephedrine	10		10.139	7.26×10^{-11}
e2	*l*-Ephedrine	10		9.958	1.10×10^{-10}
e3	Ethane, 1-amino-3-methoxy	10		9.89	1.29×10^{-10}
e4	Ethane, 1,2-bismethylamino	25	1	10.40	3.98×10^{-11}
		25	2	8.26	5.50×10^{-9}
e5	Ethanol, 2-amino	25		9.50	3.16×10^{-10}
e6	Ethylamine	20		10.807	1.56×10^{-11}
e7	Ethylenediamine	0	1	10.712	1.94×10^{-11}
		0	2	7.564	2.73×10^{-8}
g1	*l*-Glutamic acid	25	1	2.13	7.41×10^{-3}
		25	2	4.31	4.90×10^{-5}
e2	Glutamic acid, α-monoethyl	25	1	3.846	1.42×10^{-4}
		25	2	7.838	1.45×10^{-8}
e3	*l*-Glutamine	—		9.28	5.25×10^{-10}
e4	*l*-Glutathione	25	2	3.59	2.57×10^{-4}
e5	Glycine	25	1	2.3503	4.46×10^{-3}
		25	2	9.7796	1.68×10^{-10}
e6	Glycine, *n*-acetyl	25		3.6698	2.14×10^{-4}
e7	Glycine, dimethyl	5		10.3371	4.60×10^{-11}
e8	Glycine, glycyl	25		3.1397	7.25×10^{-4}
e9	Glycine, glycylglycyl	25	1	3.225	5.96×10^{-4}
		25	2	8.090	8.13×10^{-9}
e10	Glycine, leucyl	25	1	3.25	5.62×10^{-4}
		25	2	8.28	5.25×10^{-9}
e11	Glycine, methyl (Sarcosine)	25	1	2.21	6.16×10^{-3}
		25	2	10.12	7.58×10^{-11}
g12	Glycine, phenyl	25	1	1.83	1.48×10^{-2}
		25	2	4.39	4.07×10^{-5}
g13	Glycine, N,*n*-propyl	25	1	2.35	4.46×10^{-3}
		25	2	10.19	6.46×10^{-11}
g14	Glycine, tetraglycyl	20	1	3.10	7.94×10^{-4}
		20	2	8.02	9.55×10^{-9}
g15	Glycylserine	25	1	2.9808	1.04×10^{-3}
		25	2	8.38	4.17×10^{-9}
h1	Hexadecaneamine	25		10.63	2.35×10^{-11}
h2	Heptane, 1-amino	25		10.66	2.19×10^{-11}
h3	Heptane, 2-amino	19		10.88	1.58×10^{-11}
h4	Heptane, 2-methylamino	17		10.99	1.02×10^{-11}
h5	Hexadecaneamine	25		10.61	2.46×10^{-11}

DISSOCIATION CONSTANTS OF ORGANIC BASES IN AQUEOUS SOLUTION (*Continued*)

No.	Compound	Temp. °C	Step	pK_a	K_a
h6	Hexamethylene-diamine	0	1	11.857	1.39×10^{-12}
		0	2	10.762	1.73×10^{-11}
h7	Hexanoic acid, 6-amino.	25	1	4.373	4.23×10^{-5}
		25	2	10.804	1.57×10^{-11}
h8	*n*-Hexylamine	25		10.56	2.75×10^{-11}
h9	*dl*-Histidine	25	1	1.80	1.58×10^{-2}
		25	2	6.04	9.12×10^{-7}
		25	3	9.33	4.67×10^{-10}
h10	Histidine, β-alanyl (Carnosine)	20	1	2.73	1.86×10^{-3}
		20	2	6.87	1.35×10^{-7}
		20	3	9.73	1.48×10^{-10}
i1	Imidazol	25		6.953	1.11×10^{-7}
i2	Imidazol, 2,4-dimethyl.	25		8.359	5.50×10^{-9}
i3	Imidazol, 1-methyl (Oxalmethyline)	25		6.95	1.12×10^{-7}
i4	Indane, 1-amino (*d*-1-Hydrindamine	22.5		9.21	6.17×10^{-10}
i5	Isobutyric acid, 2-amino	25	1	2.357	4.30×10^{-3}
		25	2	10.205	6.23×10^{-11}
i6	Isoleucine	25	1	2.318	4.81×10^{-3}
		25	2	9.758	1.74×10^{-10}
i7	Isoquinoline (Leucoline)	20		5.42	3.80×10^{-6}
i8	Isoquinoline, 1-amino	20		7.59	2.57×10^{-8}
i9	Isoquinoline, 7-hydroxy	20	1	5.68	2.09×10^{-6}
		20	2	8.90	1.26×10^{-9}
l1	*L*-Leucine	25	1	2.328	4.70×10^{-3}
		25	2	9.744	1.80×10^{-10}
l2	Leucine, glycyl	25		3.18	6.61×10^{-4}
m1	Methionine	25	1	2.22	6.02×10^{-3}
		25	2	9.27	5.37×10^{-10}
m2	Methylamine	25		10.657	2.70×10^{-11}
m3	Morphine	25		8.21	6.16×10^{-9}
m4	Morpholine	25		8.33	4.67×10^{-9}
n1	Naphthalene, 1-amino-6-hydroxy	25		3.97	1.07×10^{-4}
n2	Naphthalene, dimethylamino	25		4.566	2.72×10^{-5}
n3	α-Naphthylamine	25		3.92	1.20×10^{-4}
n4	β-Naphthylamine	25		4.16	6.92×10^{-5}
n5	α-Naphthylamine, *n*-methyl	27		3.67	2.13×10^{-4}
n6	Neobornylamine(cis-)	25		10.01	9.77×10^{-11}
n7	Nicotine	25	1	8.02	9.55×10^{-9}
		25	2	3.12	7.59×10^{-4}
n8	*n*-Nonylamine	25		10.64	2.29×10^{-11}
n9	Norleucine	25		2.335	4.62×10^{-3}
o1	Octadecaneamine	25		10.60	2.51×10^{-11}
o2	Octylamine	25		10.65	2.24×10^{-11}
o3	Ornithine	25	1	1.705	1.97×10^{-2}
		25	2	8.690	2.04×10^{-9}
p1	Papaverine	25		6.40	3.98×10^{-7}
p2	Pentane, 3-amino	17		10.59	2.57×10^{-11}
p3	Pentane, 3-amino-3-methyl	16		11.01	9.77×10^{-12}
p4	*n*-Pentadecylamine	25		10.61	2.46×10^{-11}
p5	Pentanoic acid, 5-amino(Valeric acid)	25	1	4.270	5.37×10^{-5}
		25	2	10.766	1.71×10^{-11}
p6	Perimidine	20		6.35	4.47×10^{-7}
p7	Phenanthridine	20		5.58	2.63×10^{-6}
p8	1,10-Phenanthroline	25		4.84	1.44×10^{-5}
p9	*o*-Phenetidine (2-Ethoxyaniline)	28		4.43	3.72×10^{-5}
p10	*m*-Phenetidine (3-Ethoxyaniline)	25		4.18	6.60×10^{-5}
p11	*p*-Phenetidine (4-Ethoxyaniline)	28		5.20	6.31×10^{-6}
p12	α-Picoline	20		5.97	1.07×10^{-6}
p13	β-Picoline	20		5.68	2.09×10^{-6}
p14	γ-Picoline	20		6.02	9.55×10^{-7}
p15	Pilocarpine	30		6.87	1.35×10^{-7}
p16	Piperazine	23.5	1	9.83	1.48×10^{-10}
		23.5	2	5.56	2.76×10^{-6}
p17	Piperazine, 2,5-dimethyl(trans-)	25	1	9.66	2.19×10^{-10}
		25	2	5.20	6.31×10^{-6}
p18	Piperidine	25		11.123	7.53×10^{-12}
p19	Piperidine, 3-acetyl	25		3.18	6.61×10^{-4}
p20	Piperidine, 1-*n*-butyl	23		10.47	3.39×10^{-11}
p21	Piperidine, 1,2-dimethyl	25		10.22	6.03×10^{-11}
p22	Piperidine, 1-ethyl	23		10.45	3.55×10^{-11}
p23	Piperidine, 1-methyl	25		10.08	8.32×10^{-11}
p24	Piperidine, 2,2,6,6-tetramethyl	25		11.07	8.51×10^{-12}
p25	Piperidine, 2,2,4-trimethyl	30		11.04	9.12×10^{-12}
p26	Proline	25	1	1.952	1.11×10^{-2}
		25	2	10.640	2.29×10^{-11}
p27	Proline, hydroxy	25	1	1.818	1.52×10^{-2}
		25	2	9.662	2.18×10^{-10}
p28	Propane, 1-amino-2,2-dimethyl	25		10.15	7.08×10^{-11}
p29	Propane, 1,2-diamino	25	1	9.82	1.52×10^{-10}
		25	2	6.61	2.46×10^{-7}
p30	Propane, 1,3-diamino	10	1	10.94	1.15×10^{-11}
		10	2	9.03	9.33×10^{-10}
p31	Propane, 1,2,3-triamino	20	1	9.59	2.57×10^{-10}
		20	2	7.95	1.12×10^{-8}
p32	Propanoic acid, 3-amino (β-Alanine)	25	1	3.551	2.81×10^{-4}
		25	2	10.238	5.78×10^{-11}
p33	Propylamine	20		10.708	1.96×10^{-11}
p34	Pteridine	20		4.05	8.91×10^{-5}
p35	Pteridine, 2-amino-4,6-dihydroxy	20	2	6.59	2.57×10^{-7}
		20	3	9.31	4.90×10^{-10}
p36	Pteridine, 2-amino-4-hydroxy	20	1	2.27	5.37×10^{-3}
		20	2	7.96	1.10×10^{-8}
p37	Pteridine, 6-chloro	20		3.68	2.09×10^{-4}
p38	Pteridine, 6-hydroxy-4-methyl	20	1	4.08	8.32×10^{-5}
			2	6.41	3.89×10^{-7}
p39	Purine	20	1	2.30	5.01×10^{-3}
		20	2	8.96	1.10×10^{-9}
p40	Purine, 6-amino (Adenine)	25	1	4.12	7.59×10^{-5}
		25	2	9.83	1.48×10^{-10}
p41	Purine, 2-dimethylamino	20	1	4.00	1.00×10^{-4}
		20	2	10.24	5.75×10^{-11}
p42	Purine, 8-hydroxy	20	1	2.56	2.75×10^{-3}
		20	2	8.26	9.49×10^{-9}
p43	Pyrazine	27		0.65	2.24×10^{-1}
p44	Pyrazine, 2-methyl	27		1.45	3.54×10^{-2}
p45	Pyrazine, methylamino.	25		3.39	4.07×10^{-4}
p46	Pyridazine	20		2.24	5.76×10^{-3}
p47	Pyrimidine, 2-amino	20		3.45	3.54×10^{-4}
p48	Pyrimidine, 2-amino-4,6-dimethyl	20		4.82	1.51×10^{-5}
p49	Pyrimidine, 2-amino-5-nitro	20		0.35	4.46×10^{-1}
p50	Pyridine	25		5.25	5.62×10^{-6}
p51	Pyridine, 2-aldoxime	20	1	3.59	2.57×10^{-4}
		20	2	10.18	6.61×10^{-11}
p52	Pyridine, 2-amino	20		6.82	1.51×10^{-7}
p53	Pyridine, 4-amino	25		9.1141	7.69×10^{-10}
p54	Pyridine, 2-benzyl	25		5.13	7.41×10^{-6}
p55	Pyridine, 3-bromo	25		2.84	1.45×10^{-3}
p56	Pyridine, 3-chloro	25		2.84	1.45×10^{-3}
p57	Pyridine, 2,5-diamino	20		6.48	3.31×10^{-7}
p58	Pyridine, 2,3-dimethyl (2,3-Lutidine)	25		6.57	2.69×10^{-7}
p59	Pyridine, 2,4-dimethyl (2,4-Lutidine)	25		6.99	1.02×10^{-7}
p60	Pyridine, 3,5-dimethyl (3,5-Lutidine)	25		6.15	7.08×10^{-7}
p61	Pyridine, 2-ethyl	25		5.89	1.28×10^{-6}
p62	Pyridine, 2-formyl	20		3.80	1.59×10^{-4}
p63	Pyridine, 2-hydroxy (2-Pyridol)	20	1	0.75	9.82×10^{-1}
		20	2	11.65	2.24×10^{-12}

DISSOCIATION CONSTANTS OF ORGANIC BASES IN AQUEOUS SOLUTION (*Continued*)

No.	Compound	Temp. °C	Step	pK_a	K_a
p64	Pyridine, 4-hydroxy	20	1	3.20	6.31×10^{-4}
		20	2	11.12	7.59×10^{-12}
p65	Pyridine, methoxy	25		6.47	3.30×10^{-7}
p66	Pyridine, 4-methylamino	20		9.65	2.24×10^{-10}
p67	Pyridine, 2,4,6-trimethyl	25		7.43	3.72×10^{-8}
p68	Pyrrolidine	25		11.27	5.37×10^{-12}
p69	Pyrrolidine, 1,2-dimethyl	26		10.20	6.31×10^{-11}
p70	Pyrrolidine, *n*-methyl	25		10.32	4.79×10^{-11}
q1	Quinazoline	20		3.43	3.72×10^{-4}
q2	Quinazoline, 5-hydroxy	20	1	3.62	2.40×10^{-4}
		20	2	7.41	3.89×10^{-8}
q3	Quinine	25	1	8.52	3.02×10^{-9}
		25	2	4.13	7.41×10^{-5}
q4	Quinoline	20		4.90	1.25×10^{-5}
q5	Quinoline, 3-amino	20		4.91	1.23×10^{-5}
q6	Quinoline, 3-bromo	25		2.69	2.04×10^{-3}
q7	Quinoline, 8-carboxy	25		1.82	1.51×10^{-2}
q8	Quinoline, 3-hydroxy (3-Quinolinol)	20	1	4.28	5.25×10^{-5}
		20	2	8.08	8.32×10^{-9}
q9	Quinoline, 8-hydroxy (8-Quinolinol)	20	1	5.017	1.21×10^{-6}
		25	2	9.812	1.54×10^{-10}
q10	Quinoline, 8-hydroxy-5-sulfo	25	1	4.112	7.73×10^{-5}
		25	2	8.757	1.75×10^{-9}
q11	Quinoline, 6-methoxy	20		5.03	9.33×10^{-6}
q12	Quinoline, 2-methyl (Quinaldine)	20		5.83	1.48×10^{-6}
q13	Quinoline, 4-methyl (Lepidine)	20		5.67	2.14×10^{-6}
q14	Quinoline, 5-methyl	20		5.20	6.31×10^{-6}
q15	Quinoxaline (Quinazine)	20		0.56	3.63×10^{-1}
s1	Serine (2-amino-3-hydroxypropanoic acid)	25	1	2.186	5.49×10^{-3}
		25	2	9.208	6.19×10^{-10}
s2	Strychnine	25		8.26	5.49×10^{-9}
t1	Taurine (2-Aminoethane sulfonic acid)	25	2	9.0614	8.69×10^{-10}
t2	Tetradecaneamine (Myristilamine)	25		10.62	2.40×10^{-11}
t3	Thiazole	20		2.44	3.63×10^{-3}
t4	Thiazole, 2-amino	20		5.36	4.36×10^{-6}
t5	Threonine	25	1	2.088	8.16×10^{-3}
		25	2	9.10	7.94×10^{-10}
t6	*o*-Toluidine	25		4.44	3.63×10^{-5}
t7	*m*-Toluidine	25		4.73	1.86×10^{-5}
t8	*p*-Toluidine	25		5.08	8.32×10^{-6}
t9	1,3,5-Triazine, 2,4,6-triamino	25		5.00	1.00×10^{-5}
t10	Tridecaneamine	25		10.63	2.35×10^{-11}
t11	Triethylamine	18		11.01	9.77×10^{-12}
t12	Trimethylamine	25		9.81	1.55×10^{-10}
t13	Tryptophan	25	1	2.43	3.72×10^{-3}
		25	2	9.44	3.63×10^{-10}
t14	Tyrosine	25	2	9.11	7.76×10^{-10}
		25	3	10.13	7.41×10^{-11}
t15	Tyrosineamide	25		7.33	4.68×10^{-8}
u1	Urea	21		0.10	7.94×10^{-1}
v1	Valine	25	1	2.286	5.17×10^{-3}
		25	2	9.719	1.91×10^{-10}

DISSOCIATION CONSTANTS OF INORGANIC BASES IN AQUEOUS SOLUTIONS

(Approximately 0.1–0.01 N)

Compound	*T*°C	Step	K_b	pK_b
Ammonium hydroxide	25		1.79×10^{-5}	4.75
Arsenous oxide	25		1.1×10^{-4}	3.96
Beryllium hydroxide	25	2	5×10^{-11}	10.30
Calcium hydroxide	25	1	3.74×10^{-3}	2.43
Calcium hydroxide	30	2	4.0×10^{-2}	1.40
Deuteroammonium hydroxide	25		1.1×10^{-5}	4.96
Hydrazine	20		1.7×10^{-6}	5.77
Hydroxylamine	20		1.07×10^{-8}	7.97
Lead Hydroxide	25		9.6×10^{-4}	3.02
Silver Hydroxide	25		1.1×10^{-4}	3.96
Zinc Hydroxide	25		9.6×10^{-4}	3.02

DISSOCIATION CONSTANTS OF ORGANIC ACIDS IN AQUEOUS SOLUTIONS

Compound	T°C	Step	K	pK
Acetic	25		1.76×10^{-5}	4.75
Acetoacetic	18		2.62×10^{-4}	3.58
Acrylic	25		5.6×10^{-5}	4.25
Adipamic	25		2.35×10^{-5}	4.63
Adipic	25	1	3.71×10^{-5}	4.43
Adipic	25	2	3.87×10^{-5}	4.41
d-Alanine	25		1.35×10^{-10}	9.87
Allantoin	25		1.10×10^{-9}	8.96
Alloxanic	25		2.3×10^{-7}	6.64
α-Aminoacetic (glycine)	25		1.67×10^{-10}	9.78
o-Aminobenzoic	25		1.07×10^{-7}	6.97
m-Aminobenzoic	25		1.67×10^{-5}	4.78
p-Aminobenzoic	25		1.2×10^{-5}	4.92
o-Aminobenzosulfonic	25		3.3×10^{-3}	2.48
m-Aminobenzosulfonic	25		1.85×10^{-4}	3.73
p-Aminobenzosulfonic	25		5.81×10^{-4}	3.24
Anisic	25		3.38×10^{-5}	4.47
o-β-Anisylpropionic	25		1.59×10^{-5}	4.80
m-β-Anisylpropionic	25		2.24×10^{-5}	4.65
p-β-Anisylpropionic	25		2.04×10^{-5}	4.69
Ascorbic	24	1	7.94×10^{-5}	4.10
Ascorbic	16	2	1.62×10^{-12}	11.79
DL-Aspartic	25	1	1.38×10^{-4}	3.86
DL-Aspartic	25	2	1.51×10^{-10}	9.82
Barbituric	25		9.8×10^{-5}	4.01
Benzoic	25		6.46×10^{-5}	4.19
Benzosulfonic	25		2×10^{-1}	0.70
Bromoacetic	25		2.05×10^{-3}	2.69
o-Bromobenzoic	25		1.45×10^{-3}	2.84
m-Bromobenzoic	25		1.37×10^{-4}	3.86
n-Butyric	20		1.54×10^{-5}	4.81
iso-Butyric	18		1.44×10^{-5}	4.84
Cacodylic	25		6.4×10^{-7}	6.19
n-Caproic	18		1.43×10^{-5}	4.83
iso-Caproic	18		1.46×10^{-5}	4.84
Chloroacetic	25		1.40×10^{-3}	2.85
o-Chlorobenzoic	25		1.20×10^{-3}	2.92
m-Chlorobenzoic	25		1.51×10^{-4}	3.82
p-Chlorobenzoic	25		1.04×10^{-4}	3.98
α-Chlorobutyric	R.T.		1.39×10^{-3}	2.86
β-Chlorobutyric	R.T.		8.9×10^{-5}	4.05
γ-Chlorobutyric	R.T.		3.0×10^{-5}	4.52
o-Chlorocinnamic	25		5.89×10^{-5}	4.23
m-Chlorocinnamic	25		5.13×10^{-5}	4.29
p-Chlorocinnamic	25		3.89×10^{-5}	4.41
o-Chlorophenoxyacetic	25		8.91×10^{-4}	3.05
m-Chlorophenoxyacetic	25		8.51×10^{-4}	3.07
p-Chlorophenoxyacetic	25		7.94×10^{-4}	3.10
o-Chlorophenylacetic	25		1.18×10^{-5}	4.07
m-Chlorophenylacetic	25		7.25×10^{-5}	4.14
p-Chlorophenylacetic	25		6.46×10^{-5}	4.19
β-(*o*-Chlorophenyl) propionic	25		2.63×10^{-4}	4.58
β-(*m*-Chlorophenyl) propionic	25		2.57×10^{-5}	4.59
β-(*p*-Chlorophenyl) propionic	25		2.46×10^{-5}	4.61
α-Chloropropionic	25		1.47×10^{-3}	2.83
β-Chloropropionic	25		1.04×10^{-4}	3.98
cis-Cinnamic	25		1.3×10^{-4}	3.89
trans-Cinnamic	25		3.65×10^{-5}	4.44
Citric	18	1	7.10×10^{-4}	3.14
Citric	18	2	1.68×10^{-5}	4.77
Citric	18	3	6.4×10^{-6}	6.39
o-Cresol	25		6.3×10^{-11}	10.20
m-Cresol	25		9.8×10^{-11}	10.01
p-Cresol	25		6.7×10^{-11}	10.17
Crotonic (*trans*-)	25		2.03×10^{-5}	4.69
Cyanoacetic	25		3.65×10^{-3}	2.45
γ-Cyanobutyric	25		3.80×10^{-3}	2.42
o-Cyanophenoxyacetic	25		1.05×10^{-3}	2.98
m-Cyanophenoxyacetic	25		9.33×10^{-4}	3.03
p-Cyanophenoxyacetic	25		1.18×10^{-3}	2.93
Cyanopropionic	25		3.6×10^{-3}	2.44
Cyclohexane-1:1-dicarboxylic	25	1	3.55×10^{-4}	3.45
Cyclohexane-1:1-dicarboxylic	25	2	7.76×10^{-7}	6.11
Cyclopropane-1:1-dicarboxylic	25	1	1.51×10^{-2}	1.82
Cyclopropane-1:1-dicarboxylic	25	2	3.72×10^{-8}	7.43
DL-Cysteine	30	1	7.25×10^{-9}	8.14
DL-Cysteine	30	2	4.6×10^{-11}	10.34
L-Cystine	25	1	1.4×10^{-8}	7.85
L-Cystine	25	2	1.4×10^{-10}	9.85
Deuteroacetic (in D_2O)	25		5.5×10^{-6}	5.25
Dichloroacetic	25		3.32×10^{-2}	1.48
Dichloroacetylacetic	?		7.8×10^{-3}	2.11
Dichlorophenol (2,3-)	25		3.6×10^{-8}	7.44
Dihydroxybenzoic (2,2-)	25		1.14×10^{-3}	2.94
Dihydroxybenzoic (2,5-)	25		1.08×10^{-3}	2.97
Dihydroxybenzoic (3,4-)	25		3.3×10^{-5}	4.48
Dihydroxybenzoic (3,5-)	25		9.1×10^{-5}	4.04
Dihydroxymalic	25		1.2×10^{-2}	1.92
Dihydroxytartaric	25		1.2×10^{-2}	1.92
Dimethylglycine	25		1.3×10^{-10}	9.89
Dimethylmalic	25	1	6.83×10^{-4}	3.17
Dimethylmalic	25	2	8.72×10^{-7}	6.06
Dimethylmalonic	25		7.08×10^{-4}	3.15
Dinicotinic	25		1.6×10^{-3}	2.80
Dinitrophenol (2,4-)	15		1.1×10^{-4}	3.96
Dinitrophenol (3,6-)	15		7.1×10^{-6}	5.15
Diphenylacetic	25		1.15×10^{-4}	3.94
Ethylbenzoic	25		4.47×10^{-5}	4.35
Ethylphenylacetic	25		4.27×10^{-5}	4.37
Fluorobenzoic	17		1.25×10^{-3}	2.90
Formic	20		1.77×10^{-4}	3.75
Fumaric (*trans*-)	18	1	9.30×10^{-4}	3.03
Fumaric (*trans*-)	18	2	3.62×10^{-5}	4.44
Furancarboxylic	25		7.1×10^{-4}	3.15
Furoic	25		6.76×10^{-4}	3.17
Gallic	25		3.9×10^{-5}	4.41
Glutaramic	25		3.98×10^{-5}	4.60
Glutaric	25	1	4.58×10^{-5}	4.31
Glutaric	25	2	3.89×10^{-6}	5.41
Glycerol	25		7×10^{-15}	14.15
Glycine	25		1.67×10^{-10}	9.78
Glycol	25		6×10^{-15}	14.22
Glycolic	25		1.48×10^{-4}	3.83
Heptanoic	25		1.28×10^{-5}	4.89
Hexahydrobenzoic	25		1.26×10^{-5}	4.90
Hexanoic	25		1.31×10^{-5}	4.88
Hippuric	25		1.57×10^{-4}	3.80
Histidine	25		6.7×10^{-10}	9.17
Hydroquinone	20		4.5×10^{-11}	10.35
o-Hydroxybenzoic	19	1	1.07×10^{-3}	2.97
o-Hydroxybenzoic	18	2	4×10^{-14}	13.40
m-Hydroxybenzoic	19	1	8.7×10^{-5}	4.06
m-Hydroxybenzoic	19	2	1.2×10^{-10}	9.92
p-Hydroxybenzoic	19	1	3.3×10^{-5}	4.48
p-Hydroxybenzoic	19	2	4.8×10^{-10}	9.32
β-Hydroxybutyric	25		2×10^{-5}	4.70
γ-Hydroxybutyric	25		1.9×10^{-5}	4.72
β-Hydroxypropionic	25		3.1×10^{-5}	4.51
γ-Hydroxyquinoline	20		3.1×10^{-10}	9.51
Iodoacetic	25		7.5×10^{-4}	3.12
o-Iodobenzoic	25		1.4×10^{-3}	2.85
m-Iodobenzoic	25		1.6×10^{-4}	3.80
Itaconic	25	1	1.40×10^{-4}	3.85
Itaconic	25	2	3.56×10^{-6}	5.45
Lactic	100		8.4×10^{-4}	3.08
Lutidinic	25		7.0×10^{-3}	2.15
Lysine	25		2.95×10^{-11}	10.53
Maleic	25	1	1.42×10^{-2}	1.83
Maleic	25	2	8.57×10^{-7}	6.07
Malic	25	1	3.9×10^{-4}	3.40
Malic	25	2	7.8×10^{-6}	5.11
Malonic	25	1	1.49×10^{-3}	2.83
Malonic	25	2	2.03×10^{-6}	5.69
DL-Mandelic	25		1.4×10^{-4}	3.85
Mesaconic	25	1	8.22×10^{-4}	3.09
Mesaconic	25	2	1.78×10^{-5}	4.75
Mesitylenic	25		4.8×10^{-5}	4.32
Methyl-*o*-aminobenzoic	25		4.6×10^{-6}	5.34
Methyl-*m*-aminobenzoic	25		8×10^{-6}	5.10
Methyl-*p*-aminobenzoic	25		9.2×10^{-6}	5.04
o-Methylcinnamic	25		3.16×10^{-5}	4.50
m-Methylcinnamic	25		3.63×10^{-5}	4.44
p-Methylcinnamic	25		2.76×10^{-5}	4.56
β-Methylglutaric	25		5.75×10^{-5}	4.24
n-Methylglycine	18		1.2×10^{-10}	9.92
Methylmalonic	25		1.17×10^{-4}	3.07
Methylsuccinic	25	1	7.4×10^{-5}	4.13
Methylsuccinic	25	2	2.3×10^{-6}	5.64
o-Monochlorophenol	25		3.2×10^{-9}	8.49
m-Monochlorophenol	25		1.4×10^{-9}	8.85
p-Monochlorophenol	25		6.6×10^{-10}	9.18
Naphthalenesulfonic	25		2.7×10^{-1}	0.57
α-Naphthoic	25		2×10^{-4}	3.70
β-Naphthoic	25		6.8×10^{-5}	4.17
α-Naphthol	25		4.6×10^{-10}	9.34
β-Naphthol	25		3.1×10^{-10}	9.51
Nitrobenzene	0		1.05×10^{-4}	3.98
o-Nitrobenzoic	18		6.95×10^{-3}	2.16
m-Nitrobenzoic	25		3.4×10^{-4}	3.47
p-Nitrobenzoic	25		3.93×10^{-4}	3.41
o-Nitrophenol	25		6.8×10^{-8}	7.17
m-Nitrophenol	25		5.3×10^{-9}	8.28
p-Nitrophenol	25		7×10^{-8}	7.15
o-Nitrophenylacetic	25		1.00×10^{-4}	4.00
m-Nitrophenylacetic	25		1.07×10^{-4}	3.97
p-Nitrophenylacetic	25		1.41×10^{-4}	3.85
o-β-Nitrophenylpropionic	25		3.16×10^{-5}	4.50
p-β-Nitrophenylpropionic	25		3.39×10^{-5}	4.47
Nonanic	25		1.09×10^{-5}	4.96
Octanoic	25		1.28×10^{-5}	4.89
Oxalic	25	1	5.90×10^{-2}	1.23
Oxalic	25	2	6.40×10^{-5}	4.19

DISSOCIATION CONSTANTS OF ORGANIC ACIDS IN AQUEOUS SOLUTIONS (Continued)

Compound	T°C	Step	K	pK
Phenol	20		1.28×10^{-10}	9.89
Phenylacetic	18		5.2×10^{-5}	4.28
o-Phenylbenzoic	25		3.47×10^{-4}	3.46
γ-Phenylbutyric	25		1.74×10^{-5}	4.76
α-Phenylpropionic	25		2.27×10^{-5}	4.64
β-Phenylpropionic	25		4.25×10^{-5}	4.37
o-Phthalic	25	1	1.3×10^{-3}	2.89
o-Phthalic	25	2	3.9×10^{-6}	5.51
m-Phthalic	25	1	2.9×10^{-4}	3.54
m-Phthalic	18	2	2.5×10^{-5}	4.60
p-Phthalic	25	1	3.1×10^{-4}	3.51
p-Phthalic	16	2	1.5×10^{-5}	4.82
Picric	25		4.2×10^{-1}	0.38
Pimelic	25		3.09×10^{-5}	4.71
Propionic	25		1.34×10^{-5}	4.87
iso-Propylbenzoic	25		3.98×10^{-5}	4.40
2-Pyridinecarboxylic	25		3×10^{-6}	5.52
3-Pyridinecarboxylic	25		1.4×10^{-5}	4.85
4-Pyridinecarboxylic	25		1.1×10^{-5}	4.96
Pyrocatechol	20		1.4×10^{-10}	9.85
Quinolinic	25		3×10^{-3}	2.52
Resorcinol	25		1.55×10^{-10}	9.81
Saccharin	18		2.1×10^{-12}	11.68
Suberic	25		2.99×10^{-5}	4.52
Succinic	25	1	6.89×10^{-5}	4.16
Succinic	25	2	2.47×10^{-6}	5.61
Sulfanilic	25		5.9×10^{-4}	3.23
α-Tartaric	25	1	1.04×10^{-3}	2.98
α-Tartaric	25	2	4.55×10^{-5}	4.34
meso-Tartaric	25	1	$6 \times 10^{\ 4}$	3.22
meso-Tartaric	25	2	1.53×10^{-5}	4.82
Theobromine	18		1.3×10^{-8}	7.89
Terephthalic	25		3.1×10^{-4}	3.51
Thioacetic	25		4.7×10^{-4}	3.33
Thiophenecarboxylic	25		3.3×10^{-4}	3.48
o-Toluic	25		1.22×10^{-4}	3.91
m-Toluic	25		5.32×10^{-5}	4.27
p-Toluic	25		4.33×10^{-5}	4.36
Trichloroacetic	25		2×10^{-1}	0.70
Trichlorophenol	25		1×10^{-6}	6.00
Trihydroxybenzoic (2,4,6-)	25		2.1×10^{-2}	1.68
Trimethylacetic	18		9.4×10^{-6}	5.03
Trinitrophenol (2,4,6-)	25		4.2×10^{-1}	0.38
Tryptophan	25		4.2×10^{-10}	9.38
Tyrosine	17		3.98×10^{-9}	8.40
Uric	12		1.3×10^{-4}	3.89
n-Valeric	18		1.51×10^{-5}	4.82
iso-Valeric	25		1.7×10^{-5}	4.77
Veronal	25		3.7×10^{-8}	7.43
Vinylacetic	25		4.57×10^{-5}	4.34
Xanthine	40		1.24×10^{-10}	9.91

DISSOCIATION CONSTANTS OF INORGANIC ACIDS IN AQUEOUS SOLUTIONS

(Approximately 0.1–0.01 N)

Compound	T°C	Step	K	pK
Arsenic	18	1	5.62×10^{-3}	2.25
Arsenic	18	2	1.70×10^{-7}	6.77
Arsenic	18	3	3.95×10^{-12}	11.60
Arsenious	25		6×10^{-10}	9.23
o-Boric	20	1	7.3×10^{-10}	9.14
o-Boric	20	2	1.8×10^{-13}	12.74
o-Boric	20	3	1.6×10^{-14}	13.80
Carbonic	25	1	4.30×10^{-7}	6.37
Carbonic	25	2	5.61×10^{-11}	10.25
Chromic	25	1	1.8×10^{-1}	0.74
Chromic	25	2	3.20×10^{-7}	6.49
Germanic	25	1	2.6×10^{-9}	8.59
Germanic	25	2	1.9×10^{-13}	12.72
Hydrocyanic	25		4.93×10^{-10}	9.31
Hydrofluoric	25		3.53×10^{-4}	3.45
Hydrogen sulfide	18	1	9.1×10^{-8}	7.04
Hydrogen sulfide	18	2	1.1×10^{-12}	11.96
Hydrogen peroxide	25		2.4×10^{-12}	11.62
Hypobromous	25		2.06×10^{-9}	8.69
Hypochlorous	18		2.95×10^{-8}	7.53
Hypoiodous	25		2.3×10^{-11}	10.64
Iodic	25		1.69×10^{-1}	0.77
Nitrous	12.5		4.6×10^{-4}	3.37
Periodic	25		2.3×10^{-2}	1.64
o-Phosphoric	25	1	7.52×10^{-3}	2.12
o-Phosphoric	25	2	6.23×10^{-8}	7.21
o-Phosphoric	18	3	2.2×10^{-13}	12.67
Phosphorous	18	1	1.0×10^{-2}	2.00
Phosphorous	18	2	2.6×10^{-7}	6.59
Pyrophosphoric	18	1	1.4×10^{-1}	0.85
Pyrophosphoric	18	2	3.2×10^{-2}	1.49
Pyrophosphoric	18	3	1.7×10^{-6}	5.77
Pyrophosphoric	18	4	6×10^{-9}	8.22
Selenic	25	2	1.2×10^{-2}	1.92
Selenious	25	1	3.5×10^{-3}	2.46
Selenious	25	2	5×10^{-8}	7.31
m-Silicic	R.T.	1	2×10^{-10}	9.70
m-Silicic	R.T.	2	1×10^{-12}	12.00
o-Silicic	30	1	2.2×10^{-10}	9.66
o-Silicic	30	2	2×10^{-12}	11.70
o-Silicic	30	3	1×10^{-12}	12.00
o-Silicic	30	4	1×10^{-12}	12.00
Sulfuric	25	2	1.20×10^{-2}	1.92
Sulfurous	18	1	1.54×10^{-2}	1.81
Sulfurous	18	2	1.02×10^{-7}	6.91
Telluric	18	1	2.09×10^{-8}	7.68
Telluric	18	2	6.46×10^{-12}	11.29
Tellurous	25	1	3×10^{-3}	2.48
Tellurous	25	2	2×10^{-8}	7.70
Tetraboric	25	1	$\sim 10^{-4}$	~4.00
Tetraboric	25	2	$\sim 10^{-9}$	~9.00

DISSOCIATION CONSTANTS (K_b) OF AQUEOUS AMMONIA FROM 0 TO 50°C.

Temperature °C.	pK_b	K_b
0	4.862	1.374×10^{-5}
5	4.830	1.479×10^{-5}
10	4.804	1.570×10^{-5}
15	4.782	1.652×10^{-5}
20	4.767	1.710×10^{-5}
25	4.751	1.774×10^{-5}
30	4.740	1.820×10^{-5}
35	4.733	1.849×10^{-5}
40	4.730	1.862×10^{-5}
45	4.726	1.879×10^{-5}
50	4.723	1.892×10^{-5}

Values of K_b accurate to ±0.005; determined by e.m.f. method by: R. G. Bates and G. D. Pinching, J. Am. Chem. Soc., 1950, 72, 1393.

IONIZATION CONSTANT FOR WATER (K_w)

$-\log_{10} K_w$	Temperature °C.	$-\log_{10} K_w$	Temperature °C.
14.9435	0	13.8330	30
14.7338	5	13.6801	35
14.5346	10	13.5348	40
14.3463	15	13.3960	45
14.1669	20	13.2617	50
14.0000	24	13.1369	55
13.9965	25	13.0171	60

IONIZATION CONSTANTS OF ACIDS IN WATER AT VARIOUS TEMPERATURES

Acids		0°	5°	10°	15°	20°	25°	30°	35°	40°	45°	50°
Formic	$K_A \cdot 10^4$	1.638	1.691	1.728	1.749	1.765	1.772	1.768	1.747	1.716	1.685	1.650
Acetic	$K_A \cdot 10^5$	1.657	1.700	1.729	1.745	1.753	1.754	1.750	1.728	1.703	1.670	1.633
Propionic	$K_A \cdot 10^5$	1.274	1.305	1.326	1.336	1.338	1.336	1.326	1.310	1.280	1.257	1.229
n-Butyric	$K_A \cdot 10^5$	1.563	1.574	1.576	1.569	1.542	1.515	1.484	1.439	1.395	1.347	1.302
Chloracetic	$K_A \cdot 10^3$	1.528	....	1.488	....	....	1.379	....	....	1.230	....	....
Lactic	$K_A \cdot 10^4$	1.287	....	....	....	....	1.374	....	....	....	....	1.270
Glycollic	$K_A \cdot 10^4$	1.334	....	....	....	....	1.475	....	....	....	....	1.415
Oxalic	$K_{2A} \cdot 10^5$	5.91	5.82	5.70	5.55	5.40	5.18	4.92	4.67	4.41	4.09	3.83
Malonic	$K_{2A} \cdot 10^6$	2.140	2.165	2.152	2.124	2.076	2.014	1.948	1.863	1.768	1.670	1.575
Phosphoric	$K_A \cdot 10^3$	8.968	....	....	....	....	7.516	....	....	....	....	5.495
Phosphoric	$K_{2A} \cdot 10^8$	4.85	5.24	5.57	5.89	6.12	6.34	6.46	6.53	6.58	6.59	6.55
Boric	$K_A \cdot 10^{10}$	....	3.63	4.17	4.72	5.26	5.79	6.34	6.86	7.38	....	8.32
Carbonic	$K_{1A} \cdot 10^7$	2.64	3.04	3.44	3.81	4.16	4.45	4.71	4.90	5.04	5.13	5.19
Phenol-sulfonic	$K_{2A} \cdot 10^{10}$	4.45	5.20	6.03	6.92	7.85	8.85	9.89	10.94	12.00	13.09	14.16
Glycine	$K_{1A} \cdot 10^7$	....	3.82	3.99	4.17	4.32	4.46	4.57	4.66	4.73	4.77	4.79
Citric	$K_{1A} \cdot 10^4$	6.03	6.31	6.69	6.92	7.21	7.45	7.66	7.78	7.96	7.99	8.04
	$K_{2A} \cdot 10^5$	1.45	1.54	1.60	1.65	1.70	1.73	1.76	1.77	1.78	1.76	1.75
	$K_{3A} \cdot 10^7$	4.05	4.11	4.14	4.13	4.09	4.02	3.99	3.78	3.69	3.45	3.28

Reproducibility between various workers is about ± (0.01 − 0.02).10^5.
All values are on the m-scale.

COMPOSITION OF SOME INORGANIC ACIDS AND BASES

The following acids and bases are frequently supplied as concentrated aqueous solutions. This table presents certain data concerning these solutions.

	Formula weight	Molarity	Specific gravity	Weight percent	
Acetic acid	60.05	17.5	1.05	99–100	CH_3COOH
Ammonium hydroxide	35.05	14.8	0.90	28–30	NH_3
Hydriodic acid	127.91	5.5	1.5	47–47.5	HI
Hydrobromic acid	80.93	9.0	1.5	47–49	HBr
Hydrochloric acid	36.46	12.0	1.18	36.5–38	HCl
Hydrofluoric acid	20.01	28.9	1.17	48–51	HF
Phosphoric acid	98.00	14.7	1.7	85	H_3PO_4
Sulfuric acid	98.08	18.0	1.84	95–98	H_2SO_4

IONIZATION CONSTANTS FOR DEUTERIUM OXIDE FROM 10 TO 50°C.

From NBS Technical Note 400

The subscript *m* indicates values on the molal scale, whereas the subscript *c* indicates values on the molar scale.

t, °C	pK_m	pK_c
10	15.526	15.439
20	15.136	15.049
25	14.955	14.869
30	14.784	14.699
40	14.468	14.385
50	14.182	14.103

EQUIVALENT CONDUCTANCES OF SOME ELECTROLYTES IN AQUEOUS SOLUTIONS AT 25°C

Compound	Concentration in gram equivalents per 1000 cm^3							
	Infinite dilution	0.0005	0.001	0.005	0.01	0.02	0.05	0.1
$AgNO_3$	133.36	131.36	130.51	127.20	124.76	121.41	115.24	109.14
$BaCl_2$	139.98	135.96	134.34	128.02	123.94	119.09	111.48	105.19
$CaCl_2$	135.84	131.93	130.36	124.25	120.36	115.65	108.47	102.4
$Ca(OH)_2$	257.9			232.9	225.9	213.9		
$CuSO_4$	133.6	121.6	115.26	94.07	83.12	72.20	59.05	50.58
HCl	426.16	422.74	421.36	415.80	412.00	407.24	399.09	391.32
KBr	151.9			146.09	143.43	140.48	135.68	131.39
KCl	149.86	147.81	146.95	143.35	141.27	138.34	133.37	128.96
$KClO_4$	140.04	138.76	137.87	134.16	131.46	127.92	121.62	115.20
$K_3Fe(CN)_6$	174.5	166.4	163.1	150.7				
$K_4Fe(CN)_6$	184.5		167.24	146.09	134.83	122.82	107.70	97.87
$KHCO_3$	118.0	116.10	115.34	112.24	110.08	107.22		
KI	150.38			144.37	142.18	139.45	134.97	131.11
KIO_4	127.92	125.80	124.94	121.24	118.51	114.14	106.72	98.12
KNO_3	144.96	142.77	141.84	138.48	132.82	132.41	126.31	120.40
$KReO_4$	128.20	126.03	125.12	121.31	118.49	114.49	106.40	97.40
$LaCl_3$	145.8	139.6	137.0	127.5	121.8	115.3	106.2	99.1
LiCl	115.03	113.15	112.40	109.40	107.40	104.65	100.11	95.86
$LiClO_4$	105.98	104.18	103.44	100.57	98.61	96.18	92.20	88.56
$MgCl_2$	129.40	125.61	124.11	118.31	114.55	110.04	103.08	97.10
NH_4Cl	149.7		146.8	143.5	141.28	138.33	133.29	128.75
NaCl	126.45	124.50	123.74	120.65	118.51	115.51	111.06	106.74
$NaClO_4$	117.48	115.64	114.87	111.75	109.59	106.96	102.40	98.43
NaI	126.94	125.36	124.25	121.25	119.24	116.70	112.79	108.78
$NaOOCCH_3$	91.0	89.2	88.5	85.72	83.76	81.24	76.92	72.80
$NaOOCC_2H_5$	85.9		83.5	80.9	79.1	76.6		
$NaOOCC_3H_7$	82.70	81.04	80.31	77.58	75.76	73.39	69.32	65.27
NaOH	247.8	245.6	244.7	240.8	238.0			
Na_2SO_4	129.9	125.74	124.15	117.15	112.44	106.78	97.75	89.98
$SrCl_2$	135.80	131.90	130.33	124.24	120.24	115.54	108.25	102.19
$ZnSO_4$	132.8	121.4	114.53	95.49	84.91	74.24	61.20	52.64

THE EQUIVALENT CONDUCTANCE OF THE SEPARATE IONS

(From Smithsonian Physical Tables)

Ion.	0°	18°	25°	50°	75°	100°	128°	156°
K	40.4	64.6	74.5	115	159	206	263	317
Na	26.	43.5	50.9	82	116	155	203	249
NH_4	40.2	64.5	74.5	115	159	207	264	319
Ag	32.9	54.3	63.5	101	143	188	245	299
$\frac{1}{2}$Ba	33.	55.	65.	104	149	200	262	322
$\frac{1}{2}$Ca	30.	51.	60.	98	142	191	252	312
$\frac{1}{3}$La	35.	61.	72.	119	173	235	312	388
Cl	41.1	65.5	75.5	116	160	207	264	318
NO_3	40.4	61.7	70.6	104	140	178	222	263
$C_2H_3O_2$	20.3	34.6	40.8	67	96	130	171	211
$\frac{1}{2}SO_4$	41.	68.	79.	125	177	234	303	370
$\frac{1}{2}C_2O_4$	39.	63.	73.	115	163	213	27	336
$\frac{1}{3}C_6H_5O_7$	36.	60.	70.	113	161	214		
$\frac{1}{4}Fe(CN)_6$	58.	95.	111.	173	244	321		
H	240.	314.	350.	465	565	644	722	777
OH	105.	172.	192.	284	360	439	525	592

ACTIVITY COEFFICIENTS OF ACIDS, BASES AND SALTS

The following coefficients are valid only at 25°C. The concentrations are expressed as molalities.

Name	0.1	0.2	0.3	0.4	0.5	0.6	0.7	0.8	0.9	1.0
$AgNO_3$	0.734	0.657	0.606	0.567	0.536	0.509	0.485	0.464	0.446	0.429
$AlCl_3$	(0.337)	0.305	0.302	0.313	0.331	0.356	0.388	0.429	0.479	0.539
$Al_2(SO_4)_3$	(0.0350)	0.0225	0.0176	0.0153	0.0143	0.0140	0.0142	0.0149	0.0159	0.0175
$CdSO_4$	(0.150)	0.102	0.082	0.069	0.061	0.055	0.050	0.046	0.043	0.041
$CrCl_3$	(0.331)	0.298	0.294	0.300	0.314	0.335	0.362	0.397	0.436	0.481
$Cr(NO_3)_3$	(0.319)	0.285	0.279	0.281	0.291	0.304	0.322	0.344	0.371	0.401
$Cr_2(SO_4)_3$	(0.0458)	0.0300	0.0238	0.0207	0.0190	0.0182	0.0181	0.0185	0.0194	0.0208
CsBr	0.754	0.694	0.654	0.626	0.603	0.586	0.571	0.558	0.547	0.538
CsCl	0.756	0.694	0.656	0.628	0.606	0.589	0.575	0.563	0.553	0.544
CsI	0.754	0.692	0.651	0.621	0.599	0.581	0.567	0.554	0.543	0.533
$CsNO_3$	0.733	0.655	0.602	0.561	0.528	0.501	0.478	0.458	0.439	0.422
CsOAc	0.799	0.771	0.761	0.759	0.762	0.768	0.776	0.783	0.792	0.802
$CuSO_4$	(0.150)	0.104	0.083	0.071	0.062	0.056	0.052	0.048	0.045	0.043
HBr	0.805	0.782	0.777	0.781	0.789	0.801	0.815	0.832	0.850	0.871
HCl	0.796	0.767	0.756	0.755	0.757	0.763	0.772	0.783	0.795	0.809
$HClO_4$	0.803	0.778	0.768	0.766	0.769	0.776	0.785	0.795	0.808	0.823
HI	0.818	0.807	0.811	0.823	0.839	0.860	0.883	0.908	0.935	0.963
HNO_3	0.791	0.754	0.735	0.725	0.720	0.717	0.717	0.718	0.721	0.724
KBr	0.772	0.722	0.693	0.673	0.657	0.646	0.636	0.629	0.622	0.617
KCl	0.770	0.718	0.688	0.666	0.649	0.637	0.626	0.618	0.610	0.604
KCNS	0.769	0.716	0.685	0.663	0.646	0.633	0.623	0.614	0.606	0.599
KF	0.775	0.727	0.700	0.682	0.670	0.661	0.654	0.650	0.646	0.645
KI	0.778	0.733	0.707	0.689	0.676	0.667	0.660	0.654	0.649	0.645
KNO_3	0.739	0.663	0.614	0.576	0.545	0.519	0.496	0.476	0.459	0.443
KOAc	0.796	0.766	0.754	0.750	0.751	0.754	0.759	0.766	0.774	0.783
KOH	0.798	0.760	0.742	0.734	0.732	0.733	0.736	0.742	0.749	0.756
LiBr	0.796	0.766	0.756	0.752	0.753	0.758	0.767	0.777	0.789	0.803
LiCl	0.790	0.757	0.744	0.740	0.739	0.743	0.748	0.755	0.764	0.774
$LiClO_4$	0.812	0.794	0.792	0.798	0.808	0.820	0.834	0.852	0.869	0.887
LiI	0.815	0.802	0.804	0.813	0.824	0.838	0.852	0.870	0.888	0.910
$LiNO_3$	0.788	0.752	0.736	0.728	0.726	0.727	0.729	0.733	0.737	0.743
LiOAc	0.784	0.742	0.721	0.709	0.700	0.691	0.689	0.688	0.688	0.689
$MgSO_4$	(0.150)	0.108	0.088	0.076	0.068	0.062	0.057	0.054	0.051	0.049
$MnSO_4$	(0.150)	0.106	0.085	0.073	0.064	0.058	0.053	0.049	0.046	0.044
NaBr	0.782	0.741	0.719	0.704	0.697	0.692	0.689	0.687	0.687	0.687
NaCl	0.778	0.735	0.710	0.693	0.681	0.673	0.667	0.662	0.659	0.657
$NaClO_4$	0.775	0.729	0.701	0.683	0.668	0.656	0.648	0.641	0.635	0.629
NaCNS	0.787	0.750	0.731	0.720	0.715	0.712	0.710	0.710	0.711	0.712
NaF	0.765	0.710	0.676	0.651	0.632	0.616	0.603	0.592	0.582	0.573
NaH_2PO_4	0.744	0.675	0.629	0.593	0.563	0.539	0.517	0.499	0.483	0.468
NaI	0.787	0.751	0.735	0.727	0.723	0.723	0.724	0.727	0.731	0.736
$NaNO_3$	0.762	0.703	0.666	0.638	0.617	0.599	0.583	0.570	0.558	0.548
NaOAc	0.791	0.757	0.744	0.737	0.735	0.736	0.740	0.745	0.752	0.757
NaOH	0.766	0.727	0.708	0.697	0.690	0.685	0.681	0.679	0.678	0.678
$NiSO_4$	(0.150)	0.105	0.084	0.071	0.063	0.056	0.052	0.047	0.044	0.042
RbBr	0.763	0.706	0.673	0.650	0.632	0.617	0.605	0.595	0.586	0.578
RbCl	0.764	0.709	0.675	0.652	0.634	0.620	0.608	0.599	0.590	0.583
RbI	0.762	0.705	0.671	0.647	0.629	0.614	0.602	0.591	0.583	0.575
$RbNO_3$	0.734	0.658	0.606	0.565	0.534	0.508	0.485	0.465	0.446	0.430
RbOAc	0.796	0.767	0.756	0.753	0.755	0.759	0.766	0.773	0.782	0.792
$TlNO_3$	0.702	0.606	0.545	0.500	....	....	....	....	....	....
$ZnSO_4$	(0.150)	0.104	0.083	0.071	0.063	0.057	0.052	0.048	0.046	0.043

EQUIVALENT CONDUCTANCES OF AQUEOUS SOLUTIONS OF HYDROHALOGEN ACIDS

From NSRDS-NBS 33
Walter J. Hamer and Harold J. DeWane

One may wish to refer to the above document which defines terms relating to conductance of electrolytic solutions and also discusses some general considerations of the migration of ions under applied potential gradients. In NSRDS-NBS 33 conductance equations are given and some treatment of the Debye–Hückel–Onsager–Fuoss theories is presented.

Equivalent conductances (Ω^{-1} cm^2 equiv^{-1} of aqueous solutions of HF at 0, 16, 18, 20, and 25°C

c	0°C	16°C	18°C	20°C	25°C
mol l^{-1}					
0.004	106.7	—	—	—	140.5
0.005	97.7	—	—	—	128.1
0.006	90.9	112.9	114.6	116.3	118.8
0.007	85.5	105.7	107.3	108.9	111.4
0.008	81.1	99.9	101.3	102.9	105.4
0.009	77.3	95.0	96.4	97.8	100.4
0.01	74.2	90.8	92.1	93.5	96.1
0.02	56.2	67.7	68.8	69.8	72.2
0.05	39.3	46.8	47.6	48.3	50.1
0.07	34.7	41.3	41.9	42.6	44.3
0.10	30.7	36.4	37.0	37.7	39.1
0.20	24.8	29.6	30.1	30.7	31.7
0.50	20.4	—	—	—	26.3
0.70	19.5	—	—	—	25.1
1.0	18.8	—	—	—	24.3

EQUIVALENT CONDUCTANCES OF AQUEOUS SOLUTIONS OF HYDROHALOGEN ACIDS (Continued)

Equivalent conductances (Ω^{-1} cm^2 equiv^{-1}) of aqueous solutions of HCl from −20 to 65°C

c	−20°C	−10°C	0°C	5°C	10°C	15°C	20°C	25°C	30°C	35°C	40°C	45°C	50°C	55°C	65°C
mol l^{-1}															
0.0001	—	—	—	296.4	—	360.8	—	424.5	—	487.0	—	547.9	577.7	606.6	662.9
0.0005	—	—	—	295.2	—	359.2	—	422.6	—	484.7	—	545.2	575.1	603.5	660.0
0.001	—	—	—	294.3	—	358.0	—	421.2	—	483.1	—	543.2	573.1	601.3	657.8
0.005	—	—	—	291.0	—	353.5	—	415.7	—	476.7	—	535.5	564.4	592.6	647.3
0.01	—	—	—	288.6	—	350.3	—	411.9	—	472.2	—	530.3	558.7	586.5	641.2
0.05	—	—	—	280.3	—	339.9	—	398.9	—	456.7	—	512.4	—	565.6	616.9
0.1	—	—	—	275.0	—	333.3	—	391.1	—	446.8	—	501.1	—	552.8	602.8
0.5	—	—	228.7	254.8	283.0	308.1	336.4	360.7	386.8	411.9	436.9	461.1	482.4	508.0	552.3
1.0	—	—	211.7	235.2	261.6	283.9	312.2	332.2	359.0	379.4	402.9	424.8	445.3	468.1	509.3
1.5	—	—	196.2	216.9	241.5	261.5	287.5	305.8	331.1	349.4	371.6	391.5	410.8	431.7	469.9
2.0	—	—	182.0	199.9	222.7	240.7	262.9	281.4	303.3	321.6	342.4	360.6	378.2	398.0	433.6
2.5	—	131.7	168.5	184.3	205.1	221.4	239.8	258.9	277.0	295.8	315.2	332.0	347.6	366.7	399.9
3.0	—	120.8	154.6	169.5	188.5	203.4	219.3	237.6	253.3	271.5	289.3	304.8	319.0	336.9	368.0
3.5	85.5	111.3	139.6	155.6	172.2	186.5	201.6	218.3	232.9	248.6	263.9	279.4	292.1	308.6	337.2
4.0	79.3	102.7	129.2	143.4	158.1	171.5	185.6	200.0	214.2	228.4	242.2	256.6	268.2	283.6	310.1
4.5	73.7	94.9	119.5	132.0	145.4	157.4	170.6	183.1	196.6	209.5	222.5	235.2	246.7	260.2	284.7
5.0	68.5	87.8	110.3	121.3	133.5	144.4	156.6	167.4	180.2	191.9	204.1	215.4	226.5	238.4	261.0
5.5	63.6	81.1	101.7	111.4	122.5	132.3	143.6	152.9	165.0	175.6	187.1	197.1	207.7	218.3	239.1
6.0	58.9	74.9	93.7	102.2	112.3	121.2	131.5	139.7	151.0	160.6	171.3	180.2	190.3	199.7	218.9
6.5	54.4	69.1	86.2	93.8	103.0	111.0	120.4	127.7	138.2	146.8	156.9	164.8	174.3	182.7	200.4
7.0	50.2	63.7	79.3	86.0	94.4	101.7	110.2	116.9	126.4	134.3	143.3	150.8	159.7	167.2	183.5
7.5	46.3	58.6	73.0	78.9	86.5	93.3	100.9	107.0	115.7	122.9	131.6	138.1	146.2	153.1	168.1
8.0	42.7	54.0	67.1	72.4	79.4	85.6	92.4	98.2	106.1	112.6	120.6	126.7	134.0	140.4	154.1
8.5	39.4	49.8	61.7	66.5	72.9	78.7	84.7	90.3	97.3	103.2	110.7	116.4	123.0	128.8	141.5
9.0	36.4	45.9	56.8	61.2	67.1	72.5	77.8	83.1	89.4	94.8	101.7	107.1	112.9	118.3	130.1
9.5	33.6	42.3	52.3	56.4	61.8	66.8	71.5	76.6	82.3	87.3	93.6	98.7	103.9	108.8	119.7
10.0	31.2	39.1	48.2	52.0	57.0	61.6	65.8	70.7	75.9	80.5	86.3	91.1	95.7	—	—
10.5	28.9	36.1	44.5	48.0	52.7	56.9	60.7	65.3	70.1	74.3	79.6	84.1	88.4	—	—
11.0	26.8	33.4	41.1	44.4	48.8	52.6	56.1	60.2	64.9	68.7	73.6	77.7	81.7	—	—
11.5	24.9	31.0	38.0	41.1	45.3	48.5	51.9	55.3	60.1	63.6	68.0	71.7	75.6	—	—
12.0	23.1	28.7	35.3	38.0	42.0	—	48.0	—	55.6	—	62.8	—	70.0	—	—
12.5	21.4	26.7	32.7	—	39.0	—	44.4	—	51.4	—	57.9	—	64.8	—	—
13.0	20.0	24.9	—	—	—	—	—	—	—	—	—	—	—	—	—

Equivalent conductances (Ω^{-1} cm^2 equiv^{-1}) of aqueous solutions of HBr at 25°C

c	Λ	c	Λ	c	Λ	c	Λ	c	Λ	c	Λ
mol l^{-1}		mol l^{-1}		mol l^{-1}		mol l^{-1}		mol l^{-1}		mol l^{-1}	
0.0002	425.5	0.003	419.7	0.04	402.7	0.30	374.0	0.80	345.3	4.0	199.4
0.0003	425.0	0.004	418.5	0.05	400.4	0.35	370.8	0.85	342.6	4.5	182.4
0.0004	424.6	0.005	417.6	0.06	398.4	0.40	367.7	0.90	339.9	5.0	166.5
0.0005	424.3	0.006	416.7	0.07	396.5	0.45	364.7	0.95	337.2	5.5	151.8
0.0006	424.0	0.007	415.8	0.08	394.9	0.50	361.9	1.0	334.5	6.0	138.2
0.0007	423.7	0.008	415.1	0.09	393.4	0.55	359.0	1.5	307.6	6.5	125.7
0.0008	423.4	0.009	414.4	0.10	391.9	0.60	356.2	2.0	281.7	7.0	114.2
0.0009	423.2	0.01	413.7	0.15	386.0	0.65	353.5	2.5	257.8	7.5	103.8
0.001	422.9	0.02	408.9	0.20	381.4	0.70	350.8	3.0	236.8	8.0	94.4
0.002	421.1	0.03	405.4	0.25	377.5	0.75	348.0	3.5	217.5	8.5	85.8

EQUIVALENT CONDUCTANCES OF AQUEOUS SOLUTIONS OF HYDROHALOGEN ACIDS (Continued)

Equivalent conductances (Ω^{-1} cm^2 equiv^{-1}) of aqueous solutions of HBr from −20 to 50°C

c	−20°C	−10°C	0°C	10°C	20°C	30°C	40°C	50°C
mol l^{-1}								
0.50	—	—	240.9	295.9	347.0	398.9	453.6	496.8
0.75	—	—	234.7	284.9	339.0	387.2	433.8	480.6
1.00	—	—	229.6	276.0	329.0	380.4	418.6	465.2
1.25	—	—	221.7	265.8	314.9	362.8	401.8	442.9
1.50	—	—	209.5	254.9	298.9	340.6	381.8	421.4
1.75	—	—	198.3	243.1	284.6	327.3	366.2	404.8
2.00	—	150.8	188.6	231.3	271.8	314.1	350.5	387.4
2.25	—	143.4	180.1	219.6	258.3	296.9	332.3	367.0
2.50	—	136.8	171.7	208.3	244.8	281.7	316.0	349.1
2.75	—	131.1	164.1	198.3	232.9	267.6	301.2	333.2
3.00	—	125.7	157.2	189.5	222.2	255.0	287.8	318.6
3.25	—	120.6	150.8	181.7	212.4	244.3	275.4	304.9
3.50	—	116.1	144.1	174.6	203.2	234.4	263.7	291.9
3.75	87.1	112.0	137.6	167.4	194.7	224.2	252.2	279.4
4.00	84.0	107.5	132.3	160.2	186.8	214.2	239.7	266.9
4.25	80.9	103.1	127.7	153.2	179.0	204.5	228.8	254.6
4.50	78.0	99.0	123.0	146.4	171.2	195.1	218.8	242.6
4.75	75.1	95.1	117.7	139.9	163.1	186.1	209.3	231.3
5.00	72.3	91.4	112.6	134.0	155.7	178.2	199.6	221.3
5.25	69.6	87.8	107.8	128.3	148.7	170.5	190.0	211.4
5.50	67.0	84.2	103.1	122.7	142.1	162.8	181.4	201.8
5.75	64.4	80.6	98.6	117.3	135.7	155.3	173.2	192.4
6.00	61.8	77.2	94.3	112.0	129.6	148.0	165.4	183.4
6.25	59.3	73.8	90.1	106.9	123.7	140.9	157.8	174.7
6.50	56.8	70.7	86.0	102.0	118.0	134.1	150.5	166.3
6.75	54.4	67.7	82.1	97.2	112.5	127.6	143.3	158.4
7.00	51.9	64.6	78.4	92.6	107.1	121.4	136.3	150.8

Equivalent conductances (Ω^{-1} cm^2 equiv^{-1}) of aqueous solutions of HI at 25°C

c	Λ	*c*	Λ	*c*	Λ	*c*	Λ	*c*	Λ
mol l^{-1}		mol l^{-1}		mol l^{-1}		mol l^{-1}		mol l^{-1}	
0.00045	423.2								
0.00050	423.0	0.0035	417.9	0.025	406.6	0.10	394.0	0.90	349.2
0.00055	422.9	0.0040	417.3	0.030	405.2	0.15	389.5	0.95	346.6
0.00060	422.7	0.0045	416.8	0.035	403.9	0.20	385.9	1.0	343.9
0.00065	422.6	0.0050	416.4	0.040	402.8	0.25	382.9	1.5	316.4
0.00070	422.4	0.0055	415.9	0.045	401.7	0.30	380.1	2.0	288.9
0.00075	422.3	0.0060	415.5	0.050	400.8	0.40	374.9	2.5	262.5
0.00080	422.2	0.0065	415.1	0.055	399.9	0.45	372.3	3.0	237.9
0.00085	422.0	0.0070	414.8	0.060	399.1	0.50	369.8	3.5	215.4
0.00090	421.9	0.0075	414.4	0.065	398.3	0.55	367.3	4.0	195.1
0.00095	421.8	0.0080	414.1	0.070	397.6	0.60	364.8	4.5	176.8
0.0010	421.7	0.0090	413.4	0.075	396.9	0.65	362.2	5.0	160.4
0.0015	420.7	0.0095	413.1	0.080	396.3	0.70	359.7	5.5	145.5
0.0020	419.8	0.010	412.8	0.085	395.6	0.75	357.1	6.0	131.7
0.0025	419.1	0.015	410.3	0.090	395.1	0.80	354.5	6.5	118.6
0.0030	418.5	0.020	408.3	0.095	394.5	0.85	351.9	7.0	105.7

EQUIVALENT CONDUCTANCES OF AQUEOUS SOLUTIONS OF HYDROHALOGEN ACIDS (Continued)

Equivalent conductances (Ω^{-1} cm^2 $equiv^{-1}$) of aqueous solutions of HI from −20to 50°C

c	−20°C	−10°C	0°C	10°C	20°C	30°C	40°C	50°C
mol l^{-1}								
0.4	—	—	253.9	300.7	354.5	411.9	453.6	501.8
0.6	—	—	249.1	293.6	344.6	401.1	441.5	488.4
0.8	—	—	242.4	285.5	333.9	389.1	429.3	474.8
1.0	—	—	234.8	276.9	322.7	376.3	416.5	460.8
1.2	—	—	226.6	267.9	311.4	363.1	403.4	446.3
1.4	—	—	218.2	258.8	300.2	349.8	390.0	431.4
1.6	—	—	209.9	249.7	289.1	336.6	376.5	416.3
1.8	—	—	201.8	240.7	278.4	323.7	363.0	401.2
2.0	—	—	194.0	231.9	268.1	311.1	349.6	386.1
2.2	—	147.7	186.6	223.4	258.2	299.0	336.3	371.2
2.4	—	143.5	179.5	215.1	248.8	287.3	323.3	356.5
2.6	—	138.8	172.8	207.2	239.7	276.1	310.5	342.2
2.8	—	133.9	166.5	199.5	231.0	265.4	298.1	328.2
3.0	99.9	129.0	160.5	192.0	222.7	255.1	286.1	314.7
3.2	97.4	124.1	154.7	184.8	214.6	245.2	274.5	301.7
3.4	94.2	119.5	149.1	177.8	206.7	235.7	263.4	289.2
3.6	90.6	115.0	143.7	171.0	198.9	226.4	252.6	277.3
3.8	86.9	110.9	138.3	164.3	191.2	217.4	242.3	266.0
4.0	83.6	106.9	132.8	157.6	183.5	208.4	232.5	255.3
4.2	80.9	103.0	127.3	151.0	175.6	199.6	223.1	245.3
4.4	79.3	99.2	121.6	144.3	167.5	190.7	214.1	235.9

SPECIFIC HEAT OF WATER

Heat Capacity of Air-free Water 0°–100°C at 1 Atmosphere Pressure

The heat capacity of air-free water is given in international steam table calories per gram and in absolute joules per gram. (1 absolute joule—0.238846 I.T. Cal.).

The enthalpy or heat content is given for air-free water in I.T. Cal. per gram and in absolute joules per gram.

From Osborne, Stimson and Ginnings; B. of S. Jour. Res. **23**. 238, 1939.

Temp. °C.	Thermal Capacity		Enthalpy		Temp. °C	Thermal Capacity		Enthalpy	
	Cal./g/°C	Joules/g/°C	Cal./g	Joules/g		Cal./g/°C	Joules/g/°C	Cal/g	Joules/g
0	1.00738	4.2177	0.0245	0.1026	50	.99854	4.1807	50.0079	209.3729
1	1.00652	4.2141	1.0314	4.3184	51	.99862	4.1810	51.0065	213.5538
2	1.00571	4.2107	2.0376	8.5308	52	.99871	4.1814	52.0051	217.7350
3	1.00499	4.2077	3.0429	12.7400	53	.99878	4.1817	53.0039	221.9166
4	1.00430	4.2048	4.0475	16.9462	54	.99885	4.1820	54.0027	226.0984
5	1.00368	4.2022	5.0515	21.1498	55	.99895	4.1824	55.0016	230.2806
6	1.00313	4.1999	6.0549	25.3508	56	.99905	4.1828	56.0006	234.4632
7	1.00260	4.1977	7.0578	29.5496	57	.99914	4.1832	56.9997	238.6462
8	1.00213	4.1957	8.0602	33.7463	58	.99924	4.1836	57.9989	242.8296
9	1.00170	4.1939	9.0621	37.9410	59	.99933	4.1840	58.9982	247.0134
10	1.00129	4.1922	10.0636	42.1341	60	.99943	4.1844	59.9975	251.1976
11	1.00093	4.1907	11.0647	46.3255	61	.99955	4.1849	60.9970	255.3822
12	1.00060	4.1893	12.0654	50.5155	62	.99964	4.1853	61.9966	259.5673
13	1.00029	4.1880	13.0659	54.7041	63	.99976	4.1858	62.9963	263.7529
14	1.00002	4.1869	14.0660	58.8916	64	.99988	4.1863	63.9962	267.9390
15	.99976	4.1858	15.0659	63 0779	65	1.00000	4.1868	64.9961	272.1256
16	.99955	4.1849	16.0655	67.2632	66	1.00014	4.1874	65.9962	276.3127
17	.99933	4.1840	17.0650	71.4476	67	1.00026	4.1879	66.9964	280.5003
18	.99914	4.1832	18.0642	75.6312	68	1.00041	4.1885	67.9967	284.6885
19	.99897	4.1825	19.0633	79.8141	69	1.00053	4.1890	68.9972	288.8772
20	.99883	4.1819	20.0622	83.9963	70	1.00067	4.1896	69.9977	293.0665
21	.99869	4.1813	21.0609	88.1778	71	1.00081	4.1902	70.9985	297.2564
22	.99857	4.1808	22.0596	92.3589	72	1.00096	4.1908	71.9994	301.4469
23	.99847	4.1804	23.0581	96.5395	73	1.00112	4.1915	73.0004	305.6381
24	.99838	4.1800	24.0565	100.7196	74	1.00127	4.1921	74.0016	309.8299
25	.99828	4.1796	25.0548	104.8994	75	1.00143	4.1928	75.0030	314.0224
26	.99821	4.1793	26.0530	109.0788	76	1.00160	4.1935	76.0045	318.2155
27	.99814	4.1790	27.0512	113.2580	77	1.00177	4.1942	77.0062	322.4094
28	.99809	4.1788	28.0493	117.4369	78	1.00194	4.1949	78.0080	326.6039
29	.99804	4.1786	29.0474	121.6157	79	1.00213	4.1957	79.0101	330.7992
30	.99802	4 1785	30.0455	125 7943	80	1.00229	4.1964	80.0123	334.9952
31	.99799	4.1784	31.0435	129.9727	81	1.00248	4.1972	81.0147	339.1920
32	.99797	4.1783	32.0414	134.1510	82	1.00268	4.1980	82.0172	343.3897
33	.99797	4.1783	33.0394	138.3293	83	1.00287	4.1988	83.0200	347.5881
34	.99795	4.1782	34.0374	142.5076	84	1.00308	4.1997	84.0230	351.7873
35	.99795	4.1782	35.0353	146.6858	85	1.00327	4.2005	85.0262	355.9874
36	.99797	4.1783	36.0333	150.8641	86	1.00349	4.2014	86.0295	360.1883
37	.99797	4.1783	37.0312	155.0423	87	1.00370	4.2023	87.0331	364.3902
38	.99799	4.1784	38.0292	159.2207	88	1.00392	4.2032	88.0369	368.5929
39	.99802	4.1785	39.0272	163.3991	89	1.00416	4.2042	89.0410	372.7966
40	.99804	4.1786	40.0253	167.5777	90	1.00437	4.2051	90.0452	377.0012
41	.99807	4.1787	41.0233	171.7563	91	1.00461	4.2061	91.0497	381.2068
42	.99811	4.1789	42.0214	175.9351	92	1.00485	4.2071	92.0545	385.4135
43	.99816	4.1791	43.0195	180.1141	93	1.00509	4.2081	93.0594	389.6211
44	.99819	4.1792	44.0177	184.2933	94	1.00535	4.2092	94.0647	393.8297
45	.99826	4.1795	45.0159	188.4726	95	1.00561	4.2103	95.0701	398.0395
46	.99830	4.1797	46.0142	192.6522	96	1.00588	4.2114	96.0759	402.2503
47	.99835	4.1799	47.0125	196.8320	97	1.00614	4.2125	97.0819	406.4622
48	.99842	4.1802	48.0109	201.0120	98	1.00640	4.2136	98.0882	410.6753
49	.99847	4.1804	49.0094	205.1923	99	1.00669	4.2148	99.0947	414.8895
					100	1.00697	4.2160	100.1015	419.1049

SPECIFIC HEAT OF WATER (Continued)

Enthalpy of Air-saturated Water
1 Atmosphere Pressure 0–100°C

Temp. °C	Enthalpy Cal/g	Enthalpy Joules/g	Temp. °C	Enthalpy Cal/g	Enthalpy Joules/g
0	0	0	75	74.9907	313.9712
5	5.0276	21.0496	80	80.0019	334.9519
10	10.0402	42.0363	85	85.0180	355.9532
15	15.0431	62.9826	90	90.0395	376.9773
20	20.0400	83.9034	95	95.0671	398.0270
25	25.0332	104.8089	100	100.1016	419.1053
30	30.0244	125.7063			
35	35.0149	146.6003			
40	40.0055	167.4949			
45	44.9968	188.3928			
50	49.9896	209.2964			
55	54.9842	230.2077			
60	59.9811	251.1289			
65	64.9808	272.0619			
70	69.9839	293.0087			

Specific Heat of Water Above 100°C

Mean specific heat of water in 15°C calories between 0°C and the temperature stated.

Heat content (Enthalpy) in joules per gram between 0°C and the temperature stated.

From data by Osborne, Stimson and Fiock, B of S Jour. Res. 5, 411, 1930.

Temp. °C	Specific heat mean 0–t°C	Heat content 0–t joules/g	Temp. °C	Specific heat mean 0–t°C	Heat content 0–t joules/g
100	1.0008	418.75	190	1.0153	807.15
110	1.0015	460.97	200	1.0181	852.02
120	1.0025	503.36	210	1.0212	897.35
130	1.0037	545.93	220	1.0247	943.24
140	1.0050	588.71	230	1.0285	989.75
150	1.0067	631.75	240	1.0326	1036.97
160	1.0083	675.06	250	1.0376	1084.97
170	1.0103	718.66	260	1.0423	1133.87
180	1.0127	762.72	270	1.0483	1184.32

Specific Heat of Ice—Cal./g/°C

Temp. °C	Specific heat	Observer	Temp. °C.	Specific heat	Observer
−252 to −188	.146	Dieterici, 1903	−31.8	.4454	Dickinson-Osborne, 1915
−250	.0361		−23.7	.4599	Dickinson-Osborne, 1915
−200	.162	Mean	−24.5	.4605	Dickinson-Osborne, 1915
−188 to −78	.285	Dieterici, 1903	−20.8	.4668	Dickinson-Osborne, 1915
−180	.199	Nernst, 1910	−14.8	.4782	Dickinson-Osborne, 1915
−160	.230	Nernst, 1910	−14.6	.4779	Dickinson-Osborne, 1915
−150	.246		−11.0	.4861	Dickinson-Osborne, 1915
−140	.262	Nernst, 1910	− 8.1	.4896	Dickinson-Osborne, 1915
−100	.329	Mean	− 4.3	.4989	Dickinson-Osborne, 1915
− 78 to −18	.463	Dieterici, 1903	− 4.5	.4984	Dickinson-Osborne, 1915
− 60	.392		− 4.9	.4932	Dickinson-Osborne, 1915
− 38.3	.4346	Dickinson-Osborne, 1915	− 2.6	.5003	Dickinson-Osborne, 1915
− 34.3	.4411	Dickinson-Osborne, 1915	− 2.2	.5018	Dickinson-Osborne, 1915
− 30.6	.4488	Dickinson-Osborne, 1915			
		Water Below 0°C			
− 6	1.0119	Martinetti, 1890	− 3	1.0102	Martinetti, 1890
− 5	1.0155	Barnes, 1902	− 2	1.0097	Martinetti, 1890
− 5	1.0113	Martinetti, 1890	− 1	1.0092	Martinetti, 1890
− 4	1.0105	Martinetti, 1890			

Specific Heat of Super-heated Steam

Specific heat of steam under constant pressure given in atmospheres and at temperatures above saturation in Cal./g/°C

Temp. °C	Pressure in atmospheres 1	2	4	6	8	10	12
110	0.481						
120	0.477	0.498					
130	0.475	0.494					
140	0.473	0.489					
150	0.472	0.486	0.519				
160	0.471	0.483	0.512	0.549			
170	0.470	0.481	0.507	0.538			
180	0.469	0.479	0.502	0.528	0.561	0.602	
190	0.469	0.478	0.498	0.522	0.549	0.583	0.625
200.	0.469	0.478	0.495	0.515	0.539	0.567	0.601
210	0.470	0.477	0.493	0.510	0.531	0.555	0.584
220	0.470	0.477	0.491	0.506	0.524	0.545	0.569
230	0.471	0.477	0.489	0.504	0.519	0.537	0.557
240	0.472	0.477	0.488	0.501	0.515	0.530	0.548
250	0.473	0.477	0.488	0.499	0.512	0.525	0.540
260	0.474	0.478	0.487	0.498	0.509	0.521	0.534
270	0.474	0.478	0.487	0.497	0.507	0.518	0.529
280	0.475	0.479	0.487	0.496	0.505	0.515	0.525
290	0.476	0.480	0.487	0.495	0.504	0.513	0.523
300	0.477	0.481	0.488	0.495	0.503	0.511	0.519
310	0.478	0.482	0.488	0.495	0.502	0.510	0.518
320	0.480	0.483	0.489	0.496	0.502	0.509	0.516
330	0.482	0.484	0.490	0.496	0.502	0.508	0.515
340	0.483	0.485	0.491	0.496	0.502	0.507	0.513
350	0.484	0.486	0.492	0.497	0.502	0.507	0.512
360	0.485	0.487	0.492	0.497	0.502	0.507	0.511
370	0.486	0.488	0.493	0.498	0.503	0.507	0.511
380	0.488	0.490	0.494	0.498	0.503	0.507	0.511
390	0.489	0.491	0.495	0.499	0.503		
400	0.490	0.492	0.496	0.500	0.504		
410	0.492	0.494	0.497	0.501	0.505		
420	0.494	0.496	0.498	0.502	0.506		
430	0.495	0.497	0.500	0.504	0.507		
440	0.497	0.499	0.501	0.505	0.508		
450	0.498	0.500	0.503	0.506	0.509		
460	0.500	0.501	0.505	0.507	0.510		
470	0.502	0.503	0.506	0.508	0.512		
480	0.504	0.505	0.507	0.509	0.513		
490	0.505	0.506	0.509	0.511	0.514		
500	0.506	0.508	0.510	0.512	0.515		

HEAT CAPACITY OF MERCURY

The specific heat of solid mercury is given in relation to water at 15°C The values are from Carpenter and Stoodley, Phil. Mag. **10,** 249, 1930. Heat capacity is given in calories per gram and in calories per gram atom (1 cal = 4.1840 absolute joules and the atomic weight of mercury 200.61). Values for the liquid and vapor are from Douglas, Ball and Ginnings, Jour of Res. Bureau of Standards **46,** 334, 1951.

Temp. °C	Specific heat	Heat capacity cal/g-atom
	Solid	Solid
−75.6	.0319	6.3995
−72.9	.0324	6.4998
−65.4	.0324	6.4998
−59.5	.0324	6.4998
−44.9	.0336	6.7405
−42.2	.0336	6.7405
−40.0	.0337	6.7606
Temp. °C	**Heat capacity cal/g**	**Heat capacity cal/g-atom**
	Liquid	Liquid
−38.88	.033686	6.7578
−20	.033534	6.7272
0	.033382	6.6967
20	.033240	6.6683
25	.033206	6.6615
40	.033109	6.6419
60	.032987	6.6176
80	.032877	6.5954
100	.032776	6.5752
120	.032686	6.5571
140	.032606	6.5410
160	.032535	6.5270
180	.032476	6.5150

Temp. °C	Heat capacity cal/g	Heat capacity cal/g-atom
	Liquid	Liquid
200	.032426	6.5050
220	.032386	6.4970
240	.032356	6.4910
260	.032336	6.4869
280	.032325	6.4847
300	.032322	6.4843
320	.032330	6.4858
340	.032346	6.4890
356.58	.032366	6.4930
360	.032371	6.4940
380	.032404	6.5005
400	.032445	6.5087
420	.032494	6.5186
440	.032550	6.5298
460	.032614	6.5426
480	.032684	6.5567
500	.032762	6.5723
	Vapor	Vapor
0	.02476	4.968
100	.02476	4.968
200	.02477	4.969
300	.02480	4.975
400	.02489	4.993
500	.02507	5.030

HEAT CAPACITY (C_p) OF ORGANIC LIQUIDS AND VAPORS AT 25C

The values in this table are expressed as cal/(g mole) (°K) at 1 atm pressure. Calories, g × 4.184 = joules.

Robert Shaw

	C_p(l)	C_p(g)
Alkanes		
2-Methyl butane	39.4	28.5
n-Hexane	46.6	34.2
2-Methyl pentane	46.2	34.5
3-Methyl pentane	45.6	34.2
2,3-Dimethyl butane	45.0	33.6
2,2-Dimethyl butane	45.0	33.9
n-Heptane	53.7	39.7
2-Methyl hexane	53.1	
3-Ethyl pentane	52.5	
2,2-Dimethyl pentane	52.8	
2,3-Dimethyl pentane	51.8	
2,4-Dimethyl pentane	53.5	
3,3-Dimethyl pentane	51.4	
2,2,3-Trimethyl butane	51.0	39.3
n-Octane	60.7	45.1
2,2,4-Trimethyl pentane	56.5	
2,2,3,3-Tetramethyl butane	55.9	46.0
n-Nonane	68.0	50.6
n-Decane	75.2	56.1
2-Methyl nonane	75.6	
n-Undecane	82.5	61.5
n-Dodecane	89.7	67.0
n-Tridecane	97.2	72.5
n-Tetradecane	104.8	77.9
n-Pentadecane	112.2	83.4
n-Hexadecane	119.9	88.9
Olefins		
Pentene-1	37.1	26.2
cis-Pentene-2	36.3	24.3
trans-Pentene-2	37.6	25.9
2-Methyl butene-1	37.5	26.7
3-Methyl butene-1	37.3	28.4
2-Methyl butene-2	36.4	25.1
Hexene-1	43.8	31.6
2,3-Dimethyl butene-2	41.7	30.5
Heptene-1	50.5	37.1
Octene-1	57.7	42.6
Decene-1	71.8	53.5
Undecene-1	78.9	59.0
Dodecene-1	86.2	64.4
Hexadecene-1	115.9	86.3
Butadiene-1,3	29.5	19.0
Pentadiene-1,4	34.8	25.1
2-Methyl butadiene-1,3	36.7	25.0
Butyne-2	29.4	18.6
Aromatics		
Benzene	32.4	19.5
Toluene	37.3	24.8
Ethyl benzene	44.5	30.7
p-Xylene	43.6	30.3
1,2,3-Trimethyl benzene	51.8	36.9
1,2,4-Trimethyl benzene	51.6	37.1
1,3,5-Trimethyl benzene	50.1	35.9
Oxygen-Containing		
Methanol	19.5	10.5
Ethanol	27.0	15.7
Propanol-1	34.3	20.9
Propanol-2	37.1	21.3
2-Methyl propanol-2	53.2	27.2
estimated alternative	42.2	
Diethyl ether	41.1	25.8
Acetone	30.2	18.0
Formic acid	23.6	10.8
Acetic acid	29.4	16.0
Ethyl acetate	40.6	
Benzyl alcohol	52.1	
Diphenyl ether	64.2	
Furan	27.4	
Nitrogen-Containing		
Hydrogen cyanide	16.9	8.6
Hydrazine	22.2	12.2
Unsym. dimethyl hydrazine	39.2	
Trimethyl hydrazine	44.5	
Aniline	46.2	26.1
Quinoline	47.6	
Acrylonitrile	26.5	14.9
Perfluoro piperidene	71.0	
Methyl nitrate	37.5	
Tetrahydropyrole	37.4	
Nitromethane	25.4	13.8
Ring Compounds		
Cyclopentane	30.4	19.8
Methyl cyclopentane	38.0	26.2
Ethyl cyclopentane	44.5	31.5
1,1-Dimethyl cyclopentane	44.6	31.9
cis-1,2-Dimethyl cyclopentane	45.1	32.1
trans-1,2-Dimethyl cyclopentane	45.1	32.1
trans-1,3-Dimethyl cyclopentane	45.1	32.1
Cyclopentane	29.4	18.0
Cyclohexane	36.4	25.4
Methyl cyclohexane	44.2	32.3
Ethyl cyclohexane	50.0	38.0
1,2-Dimethyl cyclohexane	50.4	37.4
cis-1,3-Dimethyl cyclohexane	50.0	37.6
trans-1,3-Dimethyl cyclohexane	50.8	37.6
cis-1,4-Dimethyl cyclohexane	50.5	37.6
trans-1,4-Dimethyl cyclohexane	50.3	37.7
Cyclohexane	35.6	25.1
Cyclooctatetraene	44	
Tetralin $C_6H_4(CH_2)_4$	52	
cis-Decalin	55.4	
trans-Decalin	54.5	
Halogen-Containing		
Chloroform	27.6	15.7
Carbon tetrachloride	31.7	19.9
1,2-Dibromoethane	32.2	18.2
1,1,1-Trichloro ethane	34.4	22.4
1,1-Dichloro ethylene	26.6	16.1
Fluoro benzene	35.1	22.7
Chloro benzene	34.9	23.6
Bromo benzene	37.1	24.2
Sulfur-Containing		
Carbon disulfide[d]	18.2	10.9
Dimethyl sulfide	28.3	17.8
Methylethyl sulfide	34.6	22.8
Methyl *n*-propyl sulfide	41.0	28.2
Methyl *n*-butyl sulfide	48.0	33.8
Diethyl sulfide	40.9	28.1
Ethyl *n*-propyl sulfide	47.4	33.4
Di-*n*-propyl sulfide	54.0	38.7

	C_p(l)	C_p(g)
Sulfur-Containing (Continued)		
Di-*n*-butyl sulfide	68.0	49.7
Methyl isopropyl sulfide	41.2	28.8
Ethyl mercaptan	28.2	17.4
N-butyl mercaptan	41.3	28.4
Iso-butyl mercaptan	41.1	28.4
Amyl mercaptan	48.2	33.9
Iso-propyl mercaptan	34.7	23.1
Sec-butyl mercaptan	40.9	28.6
Cyclo-pentyl mercaptan	39.5	25.9
Tert-butyl mercaptan	41.8	29.0
Dimethyl disulfide	34.9	22.6
Diethyl disulfide	48.7	33.9
$(CH_2)_3S$	27.1	16.7
$(CH_2)_4S$	33.5	21.9
$(CH_2)_5S$	39.0	26.0
Phenyl mercaptan	39.1	25.2
Silicon-Containing		
Tetramethylsilane	49.8	33.5
Hexamethyldisiloxane	74.4	

HEAT CAPACITY (C_p) OF ORGANIC GASES AT 300 and 800 K

These data are a portion of those contained in "Additivity Rules for the Estimation of Thermochemical Properties", authored by S. W. Benson, F. R. Cruickshank, D. M. Golden, G. R. Haugen, H. E. O'Neal, A. S. Rodgers, R. Shaw and R. Walsh in *Chemical Reviews*, *69*, 279 (1969). These data are reproduced here with permission of the copyright owner, the American Chemical Society.

	C_p	
	300 K	800 K
Alkanes		
Ethane	12.65	25.83
Propane	17.66	37.08
n-Butane	23.40	48.23
n-Hexane	34.37	70.36
n-Octane	43.35	92.50
n-Decane	(56.34)	(114.63)
n-Dodecane	(67.33)	(136.76)
2-Methylpropane	23.25	48.49
2-Methylbutane	28.54	59.71
2,2-Dimethylpropane	29.21	60.78
2,3-Dimethylbutane	33.76	70.7
2,2,3-Trimethylbutane	39.54	82.73
2,2,3,3-Tetramethylbutane	46.29	96.18
Alkenes		
Ethene	10.45	20.20
Propene	15.34	30.68
But-1-ene	20.57	41.80
cis-But-2-ene	18.96	40.87
trans-But-2-ene	21.08	41.50
2-Methylpropene	21.39	41.86
Pent-1-ene	26.31	52.95
cis-Pent-2-ene	(24.45)	(52.29)
trans-Pent-2-ene	(26.04)	(52.45)
2-Methylbut-1-ene	26.41	53.15
3-Methylbut-1-ene	(28.47)	(53.85)
2-Methylbut-2-ene	25.22	52.05
Allenic Dienes		
Allene	14.16	25.42
1,2-Butadiene	19.23	36.01
1,2-Pentadiene	(25.3)	(47.7)
2,3-Pentadiene	(24.3)	(46.6)
3-Methyl-1,2-butadiene	(25.3)	(47.2)
Nonallenic Dienes		
1,3-Butadiene	19.11	36.84
cis-1,3-Pentadiene	(22.7)	(47.0)
trans-1,3-Pentadiene	(24.9)	(47.7)
1,4-Pentadiene	(25.2)	(47.6)
2-Methyl-1,3-butadiene (isoprene)	(25.2)	(48.0)
Alkynes		
Ethyne	10.53	14.93
Propyne	14.55	25.14
But-1-yne	19.54	35.95
But-2-yne	18.70	35.14
Pent-1-yne	(25.65)	(47.1)
Pent-2-yne	(23.69)	(45.9)
Aromatics		
Benzene	19.65	45.06
Toluene	24.94	56.61
Ethylbenzene	30.88	67.15
1,2-Dimethylbenzene	32.10	66.50
1,3-Dimethylbenzene	30.66	66.41
1,4-Dimethylbenzene	30.49	66.14
n-Propylbenzene	(36.99)	(78.30)
n-Butylbenzene	(42.09)	(89.37)
1-Propylbenzene	(36.47)	(78.6)
1-Methyl-2-ethylbenzene	(37.94)	(78.1)
1-Methyl-3-ethylbenzene	(36.59)	(77.8)
1-Methyl-4-ethylbenzene	(36.42)	(77.6)
1,2,3-Trimethylbenzene	37.82	77.56

	C_p	
	300 K	800 K
1,2,4-Trimethylbenzene	36.99	76.93
1,3,5-Trimethylbenzene	36.10	76.84
1,2,3,4-Tetramethylbenzene	45.50	89.42
1,2,3,5-Tetramethylbenzene	44.57	88.79
1,2,4,5-Tetramethylbenzene	44.77	88.41
Pentamethylbenzene	51.99	101.29
Hexamethylbenzene	59.73	113.51
Unsaturated Benzenes		
Styrene	29.35	61.40
S-Methylstyrene	(34.9)	(71.8)
cis-β-Methylstyrene	(34.9)	(71.8)
trans-β-Methylstyrene	(35.1)	(72.2)
o-Methylstyrene	(34.9)	(71.8)
m-Methylstyrene	(34.9)	(71.8)
p-Methylstyrene	(34.9)	(71.8)
Ethynylbenzene	27.63	55.79
Polyaromatics		
Biphenyl	39.05	86.92
Nonaromatic Rings: Ring Corrections		
Cyclopropane	13.44	31.44
Cyclobutane	17.37	42.42
Cyclobutene	16.03	32.26
Cyclopentane	19.98	52.44
Cyclopentene	18.08	45.78
Methylcyclopentane	26.43	63.8
1-Methylcyclopentene	(24.3)	(57.0)
1,1-Dimethylcyclopentane	(32.16)	(76.18)
cis-1,2-Dimethylcyclopentane	(32.34)	74.98
trans-1,2-Dimethylcyclopentane	(32.44)	(75.84)
cis-1,3-Dimethylcyclopentane	32.72	75.68
trans-1,3-Dimethylcyclopentane	(32.72)	(75.68)
Cyclohexane	25.58	66.76
Cyclohexene	25.28	59.49
Methylcyclohexane	32.51	78.74
Ethylcyclohexane	38.23	90.1
1,1-Dimethylcyclohexane	(37.2)	(90.7)
cis-1,2-Dimethylcyclohexane	(37.7)	(90.1)
trans-1,2-Dimethylcyclohexane	(38.3)	(90.5)
cis-1,3-Dimethylcyclohexane	(37.9)	(90.5)
trans-1,3-Dimethylcyclohexane	(37.9)	(89.8)
cis-1,4-Dimethylcyclohexane	(37.9)	(89.8)
trans-1,4-Dimethylcyclohexane	(38.0)	(90.6)
Spiropentane	21.19	47.91
Alcohols and Phenols		
Methanol, CH_3OH	10.5	19.0
Ethanol, C_2H_5OH	15.7	30.3
1-Propanol, C_3H_7OH	20.9	41.0
2-Propanol $(CH_3)_2CHOH$	21.3	42.1
2-Butanol $(CH_3CH_2)(CH_3)CHOH$	27.2	52.7
2-Methyl-2-propanol, $(CH_3)_3COH$	27.2	53.3
Phenol, C_6H_5OH	24.8	50.7
o-Cresol, $CH_3C_6H_4OH$	31.2	61.6
m-Cresol, $CH_3C_6H_4OH$	29.8	61.3
p-Cresol, $CH_3C_6H_4OH$	29.8	61.1
Ethers		
Dimethyl ether, CH_3OCH_3	15.8	30.4

	C_p 300 K	C_p 800 K
Peroxides and Hydroperoxides		
Hydrogen peroxide, HOOH	10.3	14.3
Cyclics		
Ethylene oxide	11.5	24.6
Propylene oxide	17.4	35.7
Trimethylene oxide	14.3	35.1
Aldehydes and Ketones		
Acetaldehyde, CH_3CHO	13.2	24.2
Propionaldehyde, C_2H_5CHO	18.8	35.2
Acetone, CH_3COCH_3	18.0	34.9
Methyl ethyl ketone, $CH_3CH_2COCH_3$	24.7	46.1
Acids, Esters, and Anhydrides		
Formic acid, HCOOH	10.8	18.3
Acetic acid, CH_3COOH	16.0	29.1
Aliphatic and Aromatic Amines		
CH_3NH_2	11.91	22.43
$(CH_3)_2NH$	16.58	33.94
$(CH_3)_3N$	22.05	45.62
$C_6H_5NH_2$	26.07	53.79
Cyanides		
C_2H_5CN	17.30	32.7
C_6H_5CN	26.2	52.1
c-C_3H_5CN	18.95	37.68
$CH_2{=}CHCN$	15.30	26.43
trans-$CH_3CH{=}CHCN$	19.71	36.97
$C_2(CN)_2$	20.58	27.26
Nitrites		
CH_3ONO	15.36	26.27
Hydrazines, Azo Compounds, and Tetrazines		
N_2H_4	12.19	21.08
CH_3NHNH_2	17.1	31.3
$NH{=}NH$	8.75	13.55
Amides		
$HCONH_2$	11.10	21.12
CH_3CONH_2	15.63	32.60
Heterocyclic N-Atom-Containing Ring Compounds in the Ideal Gas State		
Pyrrole	17.16	...
Pyridine	18.80	...
Picolines, α	24.05	...
β	23.94	...
Unique Groups		
NH_3	8.53	12.23
HCN	8.61	11.45
CH_3CN	12.55	21.32
$(CN)_2$	13.59	17.42
$CH_2(CN)_2$	17.38	27.72
HNC	9.33	11.49
CH_3NC	12.83	21.34
HN_3	10.46	15.10
CH_2N_2	12.58	18.73
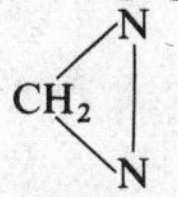	10.23	18.35
HCNO	10.57	15.19
HNO	8.29	10.77

	C_p 300 K	C_p 800 K
$HONO_2$	12.80	20.33
CH_2NO_2	13.76	25.56
N_2	6.96	7.51
NO	7.13	7.83
NO_2	8.85	11.88
NO_3 sym	11.26	17.51
N_2O	9.25	12.49
N_2O_3	15.72	21.39
N_2O_4	18.52	27.11
N_2O_5	23.09	32.74
$CH_2{=}CHNO_2$	17.43	31.7
Haloalkanes		
CF_3CCl_3	28.1	39.6
CF_2ClCF_2Cl	28.9	...
CF_3CF_3	25.5	38.3
CCl_3CCl_3	32.7	41.3
CBr_3CBr_3	33.4	39.9
$CF_3CHClBr$	25.1	37.2
CF_3CHCl_2	24.4	37.3
CF_3CH_2Cl	21.4	34.8
CF_3CH_2Br	21.8	35.1
CF_3CH_2I	21.9	35.1
CF_3CH_2F	20.9	34.5
CF_3CH_3	18.9	...
$CHCl_2CH_2Cl$	21.3	33.3
CH_3CCl_3	22.4	33.3
CH_3CHF_2	16.1	...
CH_3CHCl_2	18.2	30.9
CH_3CH_2F	14.1	...
CH_3CH_2Cl	15.1	28.3
CH_3CH_2Br	15.5	28.6
$CF_2ClCF_2CF_2Cl$	37.4	...
$CF_3CF_2CH_3$	28.8	...
$CH_3CH_2CH_2Cl$	20.3	39.2
$CH_3CHClCH_3$	21.3	...
$CH_3CH_2CH_2Br$	20.7	39.4
$(CH_3)_3CCl$	27.9	...
$CH_2ICH_{22}I$	19.2	
CH_3CHICH_2I	(24.1)	
Haloalkenes, Alkynes, and Arenes		
$CFCl{=}CCl_2$	21.8	28.8
$CF_2{=}CClBr$	21.8	28.5
$CFCl{=}CFCl$	21.0	28.4
$CF_2{=}CBr_2$	22.2	...
$CF_2{=}CFCl$	20.0	28.0
$CF_2{=}CFBr$	20.5	28.1
$CF_2{=}CF_2$	19.3	27.6
$CCl_2{=}CCl_2$	22.7	29.2
$CBr_2{=}CBr_2$	24.5	29.7
$CHF{=}CCl$	18.3	26.5
$CF_2{=}CHBr$	17.6	26.2
$CF_2{=}CHF$	16.5	25.6
$CHCl{=}CCl_2$	19.3	26.9
$CHBr{=}CBr_2$	20.5	...
$CH_2{=}CFCl$	15.3	24.4
$CH_2{=}CCl_2$	16.1	24.7
$CHCl{=}CHCl$ (av)	15.8	24.6
$CHBr{=}CHBr$ *cis*	16.5	...
$CH_2{=}CHCl$	12.9	22.4
$CH_2{=}CHBr$	13.4	22.5
$CH_2{=}CHI$	13.9	22.7
$CH_3C{\equiv}CCl$	17.2	27.0
$CH_3C{\equiv}CBr$	17.6	27.1

	C_p 300 K	C_p 800 K
Haloalkenes, Alkynes, and Arenes (Continued)		
$CH_3C{\equiv}CI$	17.8	27.2
C_6H_5F	22.7	47.6
o-$C_6H_4F_2$	25.6	49.5
m-$C_6H_4F_2$	25.5	49.7
p-$C_6H_4F_2$	25.7	49.7
p-$FC_6H_4CH_3$	27.9	58.6
$C_6H_5CF_3$	31.4	62.8
C_6H_5Cl	23.6	47.9
C_6H_5Br	24.0	48.0
C_6H_5I	24.2	48.0
Organosulfur Compounds		
CH_3SH	12.05	20.32
CH_3CH_2SH	17.44	31.83
$CH_3(CH_2)_2SH$	22.75	43.60
$CH_3(CH_2)_3SH$	28.37	55.68
$CH_3(CH_2)_4SH$	33.91	66.78
$((CH_3)_2CH)SH$	23.04	43.26
$(CH_3)_2CHCH_2SH$	28.41	53.77
$(CH_3)(CH_3CH_2)CHSH$	28.64	54.29
$(CH_3)_3CSH$	29.04	55.53
$(CH_3)_2(CH_3CH_2)CSH$	34.46	66.28
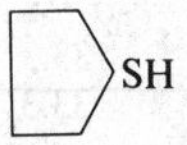	25.94	58.61
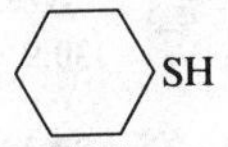	32.13	75.19
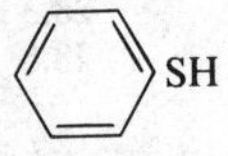	25.22	51.59
$(CH_3)_2S$	17.77	31.59
$CH_3CH_2S_3CH$	22.82	42.93
$CH_3S(CH_2)_2CH_3$	28.17	54.45
$(CH_3)_2CHSCH_3$	28.82	54.89
$CH_3S(CH_2)_3CH_3$	33.79	66.53
$CH_3SC(CH_3)_3$	34.66	66.74
$CH_3CH_2SCH_2CH_3$	28.09	54.91
$CH_3CH_2SCH_2CH_2CH_3$	33.40	66.68
$CH_3(CH_2)_2S(CH_2)_2CH_3$	38.71	78.45
$(CH_3)_2CHSCH(CH_3)_2$	40.65	77.12
$CH_3(CH_2)_3S(CH_2)_3CH_3$	49.70	100.58
CH_3SSCH_3	22.61	37.66
$CH_3CH_2SSCH_2CH_3$	33.91	60.19
$CH_3(CH_2)_2SS(CH_2)_2CH_3$	44.50	83.70
$CH_3(SO)CH_3$	21.26	37.02
$CH_3(SO_2)CH_3$	23.9	
NH_2CSNH_2	17.72	29.06
CH_3NCS	15.70	25.87
S_8	37.25	42.62

	C_p 300 K	C_p 800 K
	12.90	25.61
	16.66	36.40
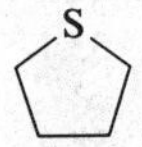	21.85	47.66
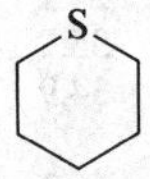	26.03	64.00
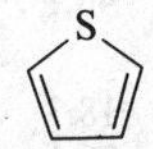	17.52	36.01
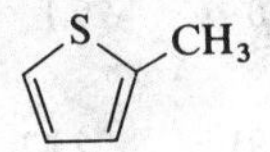	22.92	46.43
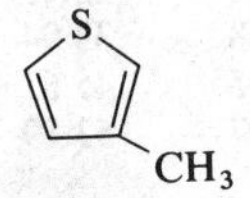	22.80	45.95
Compounds Containing Unique Groups		
SOF	9.97	12.79
$SOCl_2$	15.94	18.88
$SOCl$	10.88	13.03
S_2F_2	15.33	18.94
SF_2	10.44	13.14
SF	7.57	8.68
S_2Cl_2	17.44	19.46
SCl_2	12.19	13.61
SCl	8.20	8.95
$CSCl_2$	15.34	18.78
NSF_3	17.23	23.56
BO_2	10.08	13.45
H_3SiBr	12.69	20.02
H_3SiI	12.86	
H_2SiBr_2	15.68	21.73
$SiBr_4$	23.21	25.39
$ClSiBr_3$	22.78	25.31
Cl_2SiBr_2	22.39	25.23
Cl_3SiBr	21.72	25.13
$ISiBr_3$	23.39	25.42
$Si(CH_3)_4$	33.52	60.55
$(CH_3)_3SiCl$	31.36	52.15
$(CH_3)_2SiCl_2$	26.17	42.54
$(CH_3)_3SiF$	29.42	51.30
$(CH_3)_2SiF_2$	25.26	42.16
CH_3SiF_3	22.73	
$[(CH_3)_3Si]_2O$	57.22	99.95
$(CH_3)_3SiH$	28.36	50.50
$(CH_3)_2SiH_2$	22.03	39.79

SPECIFIC HEAT OF THE ELEMENTS AT 25°C

C_p = cal g^{-1} °K^{-1}

Element	Kelly: Bureau of Mines Bulletin 592 (1961)	Hultgren: Selected values of Thermodynamic properties of Metals and Alloys (1963)	N.B.S. Circular #500 Part 1 (1952)
Aluminum	0.215	0.215	0.2154
Antimony	0.049	0.0495	0.0501
Argon	0.124	......	0.124
Arsenic	0.0785	......	0.0796
Barium	0.046	0.0362	0.0458
Beryllium	0.436	0.436	0.4733
Bismuth	0.0296	0.0238	0.0292
Boron	0.245	......	0.2463
Bromine (Br_2)	0.113	0.0537	
Cadmium	0.0555	0.0552	0.0554
Calcium	0.156	0.155	0.1566
Carbon (Diamond)	0.124	......	0.120
" (Graphite)	0.170	......	0.172
Cerium	0.049	0.0459	0.0442
Cesium	0.057	0.0575	0.0558
Chlorine (Cl_2)	0.114	......	0.114
Chromium	......	0.107	0.1073
Cobalt	0.109	0.107	0.1037
Columbium	See Niobium		
Copper	0.092	0,0924	0.0920
Dysprosium	0.0414	0.0414	
Erbium	0.0401	0.0401	
Europium	0.0421	0.0326	
Fluorine (F_2)	0.197	0.197	
Gadolinium	0.055	0.056	
Gallium	0.089	0.088	0.0911
Germanium	0.077		
Gold	0.0308	0.0308	0.0305
Hafnium	0.035	0.028	0.0344
Helium	1.24	......	1.242
Holmium	0.0393	0.0394	
Hydrogen (H_2)	3.41	......	3.42
Indium	0.056	0.0556	0.0570
Iodine (I_2)	0.102	......	0.034
Iridium	0.0317	0.0312	0.0305
Iron (α)	0.106	0.1075	0.1078
Krypton	0.059	......	0.059
Lanthanum	0.047	0.0479	0.0475
Lead	0.038	0.0305	0.0308
Lithium	0.85	0.834	0.814
Lutetium	0.037	0.0285	
Magnesium	0.243	0.245	0.235
Manganese (α)	0.114	0.114	0.1147
" (β)	0.119	......	0.1120
Mercury	0.0331	0.0333	0.0331
Molybdenum	0.0599	0.0597	0.0584
Neodymium	0.049	0.0453	0.0499
Neon	0.246	......	0.246
Nickel	0.106	0.1061	0.1057
Niobium	0.064	0.0633	
Nitrogen (N_2)	0.249	......	0.249
Osmium	0.03127	0.0310	0.0310
Oxygen (O_2)	0.219	0.219	
Palladium	0.0584	0.0583	0.0590

SPECIFIC HEAT OF THE ELEMENTS AT 25°C (Continued)

Element	Kelly: Bureau of Mines Bulletin 592 (1961)	Hultgren: Selected values of Thermodynamic properties of Metals and Alloys (1963)	N.B.S. Circular #500 Part 1 (1952)
Phosphorus, white	0.181	......	0.178
" red, triclinic	0.160		
Platinum	0.0317	0.0317	0.0325
Polonium	0.030		
Potassium	0.180	0.180	0.1787
Praseodymium	0.046	(0.0467)	0.0482
Promethium	0.0442		
Protactinium	0.029		
Radium	0.0288		
Radon	0.0224	......	0.0224
Rhenium	0.0329	0.0330	0.0327
Rhodium	0.0583	0.0580	0.0592
Rubidium	0.0861	0.0860	0.0850
Ruthenium	0.057	0.0569	
Samarium	0.043	0.0469	
Scandium	0.133	0.1173	
Selenium (Se_2)	0.0767	......	0.0535
Silicon	0.168	......	0.169
Silver	0.0566	0.0562	0.0564
Sodium	0.293	0.292	0.2952
Strontium	0.0719	0.0719	0.0684
Sulfur, yellow	0.175	......	0.177
Tantalum	0.0334	0.0334	0.0335
Technetium	0.058		
Tellurium	0.0481	......	0.482
Terbium	0.0437	0.0435	
Thallium	0.0307	0.0307	0.0310
Thorium	0.0271	0.0281	0.0331
Thulium	0.0382		
Tin (α)	0.0510	0.0519	0.0518
Tin (β)	0.0530	0.0543	0.0530
Titanium	0.125	0.1248	0.1231
Tungsten	0.0317	0.0322	0.0324
Uranium	0.0276	0.0278	0.0276
Vanadium	0.116	0.116	0.1147
Xenon	0.0378	......	0.0379
Ytterbium	0.0346	0.0287	
Yttrium	0.068	0.0713	
Zinc	0.0928	0.0922	0.0916
Zirconium	0.0671	0.0660	

SPECIFIC HEAT AND ENTHALPY OF SOME SOLIDS AT LOW TEMPERATURES

R. J. Corruccini and J. J. Gniewek

For a more extensive listing of data one is referred to N.B.S. Monograph 21 (1960)

Joules/gm × 453.6 = joules/lb × 0.239 = cal/gm × 0.4299 = Btu/lb

METALS

T	Aluminum		Beryllium		Bismuth		Cadmium	
	C_p	$H - H_0$	C_p	$H - H_0$	C_p	$H - H_0$	C_p	$H - H_0$
°K	$jg^{-1}\ deg^{-1}\ K$	jg^{-1}	$jg^{-1}\ deg^{-1}\ K$	jg^{-1}	$jg^{-1}\ deg^{-1}\ K$	jg^{-1}	$jg^{-1}\ deg^{-1}\ K$	jg^{-1}
1	0.000 10[a]	....	....	....			....	....
1	.000 051	0.000 025	0.000 025	0.000 013	0.000 00598	0.000 00158	0.000 008	0.000 003
2	.000 108	.000 105	.000 051	.000 051	.000 0461	.000 0233	.000 033	.000 022
3	.000 176	.000 246	.000 079	.000 116	.000 170	.000 123	.000 090	.000 082
4	.000 261	.000 463	.000 109	.000 209	.000 493	.000 432	.000 21	.000 22
6	.000 50	.001 21	.000 180	.000 496	.002 14	.002 88	.001 30	.001 5
8	.000 88	.002 6	.000 271	.000 944	.005 47	.010 2	.004 3	.007 0
10	.001 4	.004 9	.000 389	.001 60	.010 4	.025 9	.008 0	.019
15	.004 0	.018	.000 842	.004 57	.023 8	.111	.025	.102
20	.008 9	.048	.001 61	.010 5	.036 3	.262	.046	.28
25	.017 5	.112	.002 79	.021 2	.047 7	.472	.066	.56
30	.031 5	.232	.004 50	.039 2	.057 2	.734	.086	.94
35	.051 5	.436	....	....	....	....	....	....
40	.077 5	.755	.009 96	.109	.072 7	1.38	.117	1.96
50	.142	1.85	.019 2	.253	.084 6	2.17	.141	3.26
60	.214	3.64	.034 1	.523	.093 5	3.06	.159	4.76
70	.287	6.15	.056 2	.971	.100	4.03	.172	6.43
80	.357	9.37	.090 6	1.69	.105	5.05	.182	8.20
90	.422	13.25	.139	2.82	.108	6.12	.190	10.1
100	.481	17.76	.199	4.51	.111	7.21	.196	12.0

T	Chromium		Copper		Germanium[b]		Gold	
	C_v	$H - H_0$	C_p	$H - H_0$	C_p	$H - H_0$	C_p	$H - H_0$
°K	$jg^{-1}\ deg^{-1}\ K$	jg^{-1}	$jg^{-1}\ deg^{-1}\ K$	jg^{-1}	$g^{-1}\ deg^{-1}\ K$	jg^{-1}	$jg^{-1}\ deg^{-1}\ K$	jg^{-1}
1	0.000 0285	0.000 0142	0.000 012	0.000 006	0.000 000 528	0.000 000 132	0.000 006	0.000 002
2	.000 058	.000 0573	.000 028	.000 025	.000 004 23	.000 002 11	.000 025	.000 016
3	.000 089	.000 131	.000 053	.000 064	.000 014 4	.000 010 7	.000 070	.000 061
4	.000 124	.000 237	.000 091	.000 13	.000 034 4	.000 034 3	.000 16	.000 17
6	.000 206	.000 567	.000 23	.000 44	.000 125	.000 179	.000 50	.000 78
8	.000 312	.001 07	.000 47	.001 12	.000 335	.000 612	.001 2	.002 4
10	.000 451	.001 82	.000 86	.002 4	.000 813	.001 69	.002 2	.005 6
15	.001 02	.005 28	.002 7	.010 7	.004 45	.013 6	.007 4	.028
20	.002 10	.012 8	.007 7	.034	.012 5	.054 0	.015 9	.086
25	.003 92	.027 4	.016	.090	.024 0	.145	.026 3	.191
30	.006 83	.053 2	.027	.195	.036 6	.296	.037 1	.349
40	.017 1	.163	.060	.61	.061 7	.786	.057 2	.821
50	.035 8	.421	.099	1.40	.085 8	1.52	.072 6	1.47
60	.062 1	.904	.137	2.58	.108	2.50	.084 2	2.25
70	.093	1.68	.173	4.13	.131	3.70	.092 8	3.14
80	.127	2.77	.205	6.02	.153	5.12	.099 2	4.10
90	.161	4.21	.232	8.22	.173	6.74	.104 3	5.12
100	.193	5.98	.254	10.6	.191	8.55	.108 3	6.18

T	Indium		α-Iron[c]		γ-Iron[d]		Lead	
	C_p	$H - H_0$	C_p	$H - H_0$	C_p	$H - H_0$	C_p	$H - H_0$
°K	$jg^{-1}\ deg^{-1}\ K$	jg^{-1}	$jg^{-1}\ deg^{-1}\ K$	jg^{-1}	$jg^{-1}\ deg^{-1}\ K$	jg^{-1}	$jg^{-1}\ deg^{-1}\ K$	jg^{-1}
1	0.000 029	0.000 011	0.000 090	0.000 045	....	....	0.000 026	0.000 010
1	[a].000 019	[a].000 006	....	....	....	....	[a].000 012	[a].000 003
2	.000 138	.000 085	.000 183	.000 181	....	....	.000 12	.000 07
2	[a].000 141	[a].000 073	....	....	....	....	[a].000 09	[a].000 05
								.000 28
3	.000 410	.000 341	.000 279	.000 412	....	....	.000 33	[a].000 23
3	[a].000 464	[a].000 357	....	....	....	....	[a].000 31	
[e]3.40	.000 584	.000 537	....	....	....	....	....	....
3.40	[a].000 669	[a].000 581	....	....	....	....	....	....
4	.000 95	.000 99	.000 382	.000 742	....	....	.000 7	.000 8
4	....	....	....	....	....	....	[a].000 7	[a].000 7
5	....	....	....	....	....	....	.001 5	.001 8
5	....	....	....	....	....	....	[a].001 5	[a].001 8
6	.003 59	.005 20	.000 615	.001 73	....	....	.002 9	.003 9
6	....	....	....	....	....	....	[a].003 0	[a].004 0
7	....	....	....	....	....	....	.004 8	.008
7	....	....	....	....	....	....	[a].005 0	[a].008
8	.008 55	.017 0	.000 90	.003 23	....	....	.007 3	.014
10	.015 5	.040 8	.001 24	.005 37	....	....	.013 7	.034
15	.036 7	.170	.002 49	.014 5	....	....	.033 5	.150
20	.060 8	.413	.004 5	.031 6	0.007	0	.053 1	.368
25	.085 7	.778	.007 5	.061	....	....	.068 1	.672
30	.108	1.265	.012 4	.110	.016	.11	.079 6	1.042
40	.141	2.52	.029	.31	.041	.39	.094 4	1.920
50	.162	4.04	.055	.73	.090	1.0_2	.103	2.91
60	.176	5.73	.087	1.43	$.13_7$	2.1_6	.108	3.97
70	.186	7.53	.121	2.46	$.18_0$	3.7_5	.112	5.07
80	.193	9.42	.154	3.84	$.21_8$	5.7_4	.114	6.20
90	.198	11.38	.186	5.55	$.25_5$	8.1_1	.116	7.35
100	.203	13.39	.216	7.56	$.28_8$	$10._8$	.118	8.53

METALS (*Continued*)

T	Molybdenum		Nickel		Palladium	
	C_p	$H - H_0$	C_p	$H - H_0$	C_p	$H - H_0$
°K	$jg^{-1}\ deg^{-1}\ K$	jg^{-1}	$jg^{-1}\ deg^{-1}\ K$	jg^{-1}	$jg^{-1}\ deg^{-1}\ K$	jg^{-1}
1	0.000 0229	0.000 0105	0.000 120	0.000 060	0.000 099	0.000 0493
2	.000 0472	.000 0445	.000 242	.000 241	.000 203	.000 200
2	....	....	....	....	....	....
3	.000 0745	.000 105	.000 369	.000 546	.000 318	.000 459
3	....	....	....	....	....	....
4	.000 106	.000 194	.000 503	.000 98	.000 447	.000 840
4	....	....	....	....	....	....
5	....	....	....	....	....	....
5	....	....	....	....	....	....
6	.000 191	.000 484	.000 82	.002 28	.000 891	.002 31
6	....	....	....	....	....	....
7	....	....	....	....	....	....
7	....	....	....	....	....	....
8	.000 317	.000 981	.001 19	.004 28	.001 41	.004 60
8	....	....	....	....	....	....
9	....	....	....	....	....	....
9	....	....	....	....	....	....
10	.000 498	.001 78	.001 62	.007 1	.002 10	.008 07
15	.001 31	.006 10	.003 1	.018 5	.004 71	.024 5
20	.002 87	.016 1	.005 8	.041	.009 22	.058 6
25	.005 77	.037 4	.010 1	.079	.016 0	.120
30	.009 60	.072 9	.016 7	.145	.025 8	.223
40	.023 6	.232	.038 1	.413	.050 7	.600
50	.041 0	.554	.068 2	.937	.077 7	1.24
60	.061 9	1.07	.103	1.79	.101	2.14
70	.083 8	1.80	.139	3.00	.122	3.26
80	.104	2.74	.173	4.56	.139	4.56
90	.123	3.88	.204	6.45	.154	6.03
100	.139	5.20	.232	8.63	.167	7.63

T	Platinum		Rhodium		Silicon[l]		Silver	
	C_p	$H - H_0$	C_p	$H - H_0$	C_p	$H - H_0$	C_p	$H - H_0$
°K	$jg^{-1}\ deg^{-1}\ K$	jg^{-1}	$jg^{-1}\ deg^{-1}\ K$	jg^{-1}	$jg^{-1}\ deg^{-1}\ K$	jg^{-1}	$jg^{-1}\ deg^{-1}\ K$	jg^{-1}
1	0.000 035	0.000 0175	0.000 048	0.000 024	0.000 000 263	0.000 000 0658	0.000 0072	0.000 0032
2	.000 074	.000 071	.000 097	.000 096	.000 002 10	.000 001 05	.000 0239	.000 0176
3	.000 122	.000 168	.000 147	.000 218	.000 007 09	.000 005 32	.000 0595	.000 0574
4	.000 186	.000 320	.000 201	.000 392	.000 016 8	.000 016 8	.000 124	.000 146
6	.000 37	.000 85	.000 32	.000 91	.000 059 6	.000 085 3	.000 39	.000 62
8	.000 67	.001 88	.000 47	.001 70	.000 140	.000 279	.000 91	.001 87
10	.001 12	.003 65	.000 65	.002 81	.000 275	.000 679	.001 8	.004 52
15	.003 3	.013 5	.001 35	.007 65	.001 09	.003 74	.006 4	.023 3
20	.007 4	.039 5	.002 71	.017 4	.003 37	.013 8	.015 5	.076
25	.013 7	.092	.005 61	.037 3	.008 49	.042 3	.028 7	.185
30	.021 2	.182	.010 6	.077 1	.017 1	.105	.044 2	.368
40	.038	.48	.026 6	.256	.044 0	.400	.078	.979
50	.055	.95	.048 9	.633	.078 5	1.00	.108	1.91
60	.068	1.56	.072 4	1.238	.115	1.97	.133	3.12
70	.079	2.29	.094	2.07	.152	3.31	.151	4.54
80	.088	3.12	.114	3.11	.188	5.01	.166	6.13
90	.094	4.02	.132	4.34	.224	7.06	.177	7.85
100	.100	5.01	.147	5.74	.259	9.47	.187	9.67

T	Sodium[m]		Tantalum		Tin (white)		Titanium	
	C_p	$H - H_0$	C_p	$H - H_0$	C_p	$H - H_0$	C_p	$H - H_0$
°K	$jg^{-1}\ deg^{-1}\ K$	jg^{-1}	$jg^{-1}\ deg^{-1}\ K$	jg^{-1}	$jg^{-1}\ deg^{-1}\ K$	jg^{-1}	$jg^{-1}\ deg^{-1}\ K$	jg^{-1}
1	0.000 081	0.000 035	0.000 032	0.000 016	0.000 0170	0.000 0079	0.000 071	0.000 035
1	....	....	[a].000 0063	[a].000 0021	[a].000 0041	[a].000 0009	....	....
2	.000 289	.000 204	.000 068	.000 065	.000 047	.000 0383	.000 146	.000 143
2	....	....	[a].000 054	[a].000 026	[a].000 048	[a].000 0228	....	....
3	.000 76	.000 70	.000 112	.000 155	.000 109	.000 113	.000 226	.000 329
3	....	....	[a].000 178	[a].000 138	[a].000 151	[a].000 116	....	....
[n]3.72	....	....	....	....	.000 198	.000 221	....	....
3.72	....	....	....	....	[a].000 285	[a].000 270	....	....
4	.001 60	.001 84	.000 171	.000 295	.000 245	.000 283	.000 317	.000 599
4	....	....	[a].000 352	[a].000 400	....	....	....	....
[o]4.39	....	....	.000 201	.000 368	....	....	....	....
4.39	....	....	[a].000 433	[a].000 553	....	....	....	....
5	.002 98	.004 08	....	....	.000 54	.000 65	....	....
6	.005 1	.008 1	.000 333	.000 776	.001 27	.001 51	.000 54	.001 45
8	.012 2	.024 7	.000 648	.001 73	.004 2	.006 8	.000 84	.002 81
10	.023 8	.060 2	.001 17	.003 52	.008 1	.019 0	.001 26	.004 89
12	.039 7	.123	....	....	....	....	....	....
14	.063	.225	....	....	....	....	....	....
15	....	....	.003 60	.014 5	.022 6	.093	.003 3	.015 6
16	.093	.380	....	....	....	....	....	....
18	.124	.597	....	....	....	....	....	....
20	.155	.875	.008 23	.043 2	.040	.251	.007 0	.040
25	.259	1.90	.015 3	.102	.058	.498	.013 4	.090
30	.364	3.45	.024 0	.202	.076	.834	.024 5	.182
40	.544	8.03	.043 0	.540	.106	1.75	.057 1	.581
50	.695	14.2	.060 4	1.06	.130	2.93	.099 2	1.358
60	.793	21.7	.075 4	1.74	.148	4.33	.146 7	2.592
70	.86	30.0	.087 9	2.56	.162	5.88	.189	4.27
80	.91	38.9	.097 6	3.49	.173	7.55	.230	6.37
90	.95	48.2	.105	4.50	.182	9.33	.267	8.86
100	.98	57.9	.111	5.58	.189	11.18	.300	11.69

METALS (*Continued*)

T	Tungsten		Zinc	
	C_p	$H - H_0$	C_p	$H - H_0$
°K	$jg^{-1}\ deg^{-1}\ K$	jg^{-1}	$jg^{-1}\ deg^{-1}\ K$	jg^{-1}
1	0.000 0074	0.000 0037	0.000 011	0.000 005
2	.000 0158	.000 0152	.000 028	.000 023
3	.000 0262	.000 0360	.000 058	.000 065
4	.000 0393	.000 0685	.000 11	.000 14
6	.000 0783	.000 182	.000 29	.000 53
8	.000 141	.000 396	.000 96	.001 6
10	.000 234	.000 765	.002 5	.005 0
15	.000 725	.002 97	.011	.034
20	.001 89	.009 27	.026	.125
25	.004 21	.023 7	.049	.31
30	.007 83	.053 4	.076	.62
40	.018 4	.181	.125	1.62
50	.033 2	.436	.171	3.11
60	.048 3	.843	.208	5.01
70	.060 5	1.39	.236	7.23
80	.071 5	2.05	.258	9.70
90	.081 0	2.81	.277	12.38
100	.088 8	3.66	.293	15.24

[a] Superconducting.

[b] In germanium the electronic specific heat is markedly dependent on impurities. The values given are for pure germanium (negligible electronic specific heat).

[c] α-Iron is the form that is thermodynamically stable at low temperatures. It has the body-centered cubic lartice which is the basis of the ferritic steels.

[d] γ-Iron is stable between 910° and 1,400°C. It has the face-centered cubic structure which is the basis of the austenitic steels. Since pure γ-iron is not stable at low temperatures the above values were calculated by application of the Kopp-Neumann rule to experimental data on two austenitic Fe-Mn alloys and are of uncertain accuracy.

[e] Superconducting transition temperature.

[i] Superconducting transition temperature of mercury.

[j] Melting temperature of mercury.

[l] In silicon the electronic specific heat, γT, is markedly dependent on impurities. Values of the coefficient, γ, from zero to $2.4 \times 10^{-6}\ jg^{-1}\ deg^{-2}\ K$ have been reported. The values in the above table are for pure silicon ($\gamma = 0$).

[m] It has been shown (Barrett 1956, Hull & Rosenberg 1959) that sodium partially transforms at low temperatures from the normal body-centered cubic structure to close-packed hexagonal. The transformation is of the martensitic type and is promoted by cold-working at the low temperatures. Inasmuch as none of the calorimetric measurements on sodium were accompanied by crystallographic analysis, the tabulated data below 100°K are to some degree ambiguous.

[n] Superconducting transition temperature of tin.

[o] Superconducting transition temperature of tantalum.

CONSTANTS OF DEBYE-SOMMERFELD EQUATION

$C_v = \gamma T + \alpha T^3$; $\alpha = 12\pi^4 R/5\theta_0^3$; $0 \lesseqgtr T \lesseqgtr T_{max}$; T_{max} = maximum temperature to which the equation can be used with the limiting value of θ.

Substance	$10^6\gamma$	γ	$10^6\alpha$	θ_0	T_{max}
Metals:	$jg^{-1}\ deg^{-2}\ K$	$mjg\text{-}atom^{-1}\ deg^{-2}\ K$	$jg^{-1}\ deg^{-4}\ K$	deg K	deg K
Aluminum	50.4	1.36	0.93	426	4
Beryllium	25	0.226	.138	1160	20
Bismuth	0.32	.067	5.66	118	2
Cadmium	5.6	.63	2.69	186	3
Chromium	28.3	1.47	0.165	610	4
Copper	10.81	0.687	.746	344.5	10
Germanium	(a)	(a)	.528	370	2
Gold	3.75	0.74	2.19	165	15
Indium	15.8	1.81	13.1	109	2
α-Iron	90	5.0	0.349	464	10
Lead	15.1	3.1	10.6	96	4
Magnesium	54	1.32	1.19	406	4
α-Manganese	251	13.8	0.328	476	12
Molybdenum	23	2.18	.238	440	4
Nickel	120	7.0	.39	440	4
Niobium	85	7.9	.64	320	1
Palladium	98	10.5	.89	274	4
Platinum	34.1	6.7	.72	240	3
Rhodium	48	4.9	.173	478	4
Silicon	(a)	(a)	.263	640	4
Silver	5.65	0.610	1.58	225	4
Sodium[a]	60	1.37	21.4	158	4
Tantalum	31.5	5.7	0.69	250	4
Tin (white)	14.7	1.75	2.21	195	2
Titanium	71	3.4	0.54	420	10
Tungsten	7	1.3	.16	405	4
Zinc	9.6	0.63	1.10	300	4
Alloys:					
Constantan[a]	113	6.9	0.56	384	15
Monel[a]	108	6.5	.62	374	20
Other inorg. subs.:					
Diamond	....	....	0.0152	2200	50
Ice	....	....	15.2	192	10
Pyrex	....	....	3.14	......	5
Organic subs.:					
Glyptal	....	....	27	......	4
Lucite	....	....	35	......	4
Polystyrene	....	....	63	......	4

(a) Superconducting.

BOILING POINT OF WATER
(Hydrogen Scale)

Pressure mm	Tenths of millimeters .0	.1	.2	.3	.4	.5	.6	.7	.8	.9
700	97.714	718	722	725	729	733	737	741	745	749
701	753	757	761	765	769	773	777	781	785	789
702	792	796	800	804	808	812	816	820	824	828
703	832	836	840	844	847	851	855	859	863	867
704	871	875	879	883	887	891	895	899	902	906
705	97.910	914	918	922	926	930	934	938	942	946
706	949	953	957	961	965	969	973	977	981	985
707	989	993	996	*000	*004	*008	*012	*016	*020	*024
708	98.028	032	036	040	043	047	051	055	059	063
709	067	071	075	079	082	086	090	094	098	102
710	98.106	110	114	118	121	125	129	133	137	141
711	145	149	153	157	160	164	168	172	176	180
712	184	188	192	195	199	203	207	211	215	219
713	223	227	230	234	238	242	246	250	254	258
714	261	265	269	273	277	281	285	289	292	296
715	98.300	304	308	312	316	320	323	327	331	335
716	339	343	347	351	355	358	362	366	370	374
717	378	382	385	389	393	397	401	405	409	412
718	416	420	424	428	432	436	440	443	447	451
719	455	459	463	467	470	474	478	482	486	490
720	98.493	497	501	505	509	513	517	520	524	528
721	532	536	540	544	547	551	555	559	563	567
722	570	574	578	582	586	590	593	597	601	605
723	609	613	617	620	624	628	632	636	640	643
724	647	651	655	659	662	666	670	674	678	682
725	98.686	689	693	697	701	705	709	712	716	720
726	724	728	732	735	739	743	747	751	755	758
727	762	766	770	774	777	781	785	789	793	797
728	800	804	808	812	816	819	823	827	831	835
729	838	842	846	850	854	858	861	865	869	873
730	98.877	880	884	888	892	896	899	903	907	911
731	915	918	922	926	930	934	937	941	945	949
732	953	956	960	964	968	972	975	979	983	987
733	991	994	998	*002	*006	*010	*013	*017	*021	*025
734	99.029	032	036	040	044	048	051	055	059	063
735	99.067	070	074	078	082	085	089	093	097	101
736	104	108	112	116	119	123	127	131	135	138
737	142	146	150	153	157	161	165	169	172	176
738	180	184	187	191	195	199	203	206	210	214
739	218	221	225	229	233	236	240	244	248	252
740	99.255	259	263	267	270	274	278	282	285	289
741	293	297	300	304	308	312	316	319	323	327
742	331	334	338	342	346	349	353	357	361	364
743	368	372	376	379	383	387	391	394	398	402
744	406	409	413	417	421	424	428	432	436	439
745	99.443	447	451	454	458	462	466	469	473	477
746	481	484	488	492	495	499	503	507	510	514
747	518	522	525	529	533	537	540	544	548	551
748	555	559	563	566	570	574	578	581	585	589
749	592	596	600	604	607	611	615	619	622	626

Pressure mm	Tenths of millimeters .0	.1	.2	.3	.4	.5	.6	.7	.8	.9
750	99.630	633	637	641	645	648	652	656	659	663
751	667	671	674	678	682	686	689	693	697	700
752	704	708	712	715	719	723	726	730	734	738
753	741	745	749	752	756	760	764	767	771	775
754	778	782	786	790	793	797	801	804	808	812
755	99.815	819	823	827	830	834	838	841	845	849
756	852	856	860	863	867	871	875	878	882	886
757	889	893	897	900	904	908	911	915	919	923
758	926	930	934	937	941	945	948	952	956	959
759	963	967	970	974	978	982	985	989	993	996
760	100.000	004	007	011	015	018	022	026	029	033
761	037	040	044	048	052	055	059	063	066	070
762	074	077	081	085	088	092	096	099	103	107
763	110	114	118	121	125	129	132	136	140	143
764	147	151	154	158	162	165	169	173	176	180
765	100.184	187	191	195	198	202	206	209	213	216
766	220	224	227	231	235	238	242	246	249	253
767	257	260	264	268	271	275	279	283	286	290
768	293	297	300	304	308	311	315	319	322	326
769	330	333	337	341	344	348	352	355	359	363
770	100.366	370	373	377	381	384	388	392	395	399
771	403	406	410	414	417	421	424	428	432	435
772	439	442	446	450	453	457	461	464	468	472
773	475	479	483	486	490	493	497	501	504	508
774	511	515	519	522	526	530	533	537	540	544
775	100.548	551	555	559	562	566	569	573	577	580
776	584	588	591	595	598	602	606	609	613	616
777	620	624	627	631	634	638	642	645	649	653
778	656	660	663	667	671	674	678	681	685	689
779	692	696	689	703	707	710	714	718	721	725
780	100.728	732	735	739	743	746	750	753	757	761
781	764	768	772	775	779	782	786	789	793	797
782	800	804	807	811	815	818	822	825	829	833
783	836	840	843	847	851	854	858	861	865	869
784	872	876	879	883	886	890	894	897	901	904
785	100.908	912	915	919	922	926	929	933	937	940
786	944	947	951	954	958	962	965	969	972	976
787	979	983	987	990	994	997	*001	*005	*008	*012
788	101.015	019	022	026	029	033	037	040	044	047
789	051	054	058	062	065	069	072	076	079	083
790	101.087	090	094	097	101	104	108	112	115	119
791	122	126	129	133	136	140	144	147	151	154
792	158	161	165	168	172	176	179	183	186	190
793	193	197	200	204	207	211	215	218	222	225
794	229	232	236	239	243	246	250	254	257	261
795	101.264	268	271	275	278	282	286	289	293	296
796	300	303	307	310	314	317	321	324	328	332
797	335	339	342	346	349	353	356	360	363	367
798	370	374	377	381	385	388	392	395	399	402
799	406	409	413	416	420	423	427	430	434	437
800	101.441	...	...	...	.....	...	...	...	...	...

For lower pressures see under Vapor Tension of Water

MELTING POINTS OF MIXTURES OF METALS

(Smithsonian Physical Tables)

Melting-points, °C.

Metals	Percentage of metal in second column. 0%*	10%	20%	30%	40%	50%	60%	70%	80%	90%	100%
Pb. Sn.	326	295	276	262	240	220	190	185	200	216	232
Bi.	322	290	...	...	179	145	126	168	205	...	268
Te.	322	710	790	880	917	760	600	480	410	425	446
Ag.	328	460	545	590	620	650	705	775	840	905	959
Na.	...	360	420	400	370	330	290	250	200	130	96
Cu.	326	870	920	925	945	950	955	985	1005	1020	1084
Sb.	326	250	275	330	395	440	490	525	560	600	632
Al. Sb.	650	750	840	925	945	950	970	1000	1040	1010	632
Cu.	650	630	600	560	540	580	610	755	930	1055	1084
Au.	655	675	740	800	855	915	970	1025	1055	675	1062
Ag.	650	625	615	600	590	580	575	570	650	750	954
Zn.	654	640	620	600	580	560	530	510	475	425	419
Fe.	653	860	1015	1110	1145	1145	1220	1315	1425	1500	1515
Sn.	650	645	635	625	620	605	590	570	560	540	232
Sb. Bi.	632	610	590	575	555	540	520	470	405	330	268
Ag.	630	595	570	545	520	500	505	545	680	850	959
Sn.	622	600	570	525	480	430	395	350	310	255	232
Zn.	632	555	510	540	570	565	540	525	510	470	419

Metals	Percentage of metal in second column. 0%*	10%	20%	30%	40%	50%	60%	70%	80%	90%	100%
Ni. Sn.	1455	1380	1290	1200	1235	1290	1305	1230	1060	800	232
Na. Bi.	96	425	520	590	645	690	720	730	715	570	268
Cd.	96	125	185	245	285	325	330	340	360	390	322
Cd. Ag.	322	420	520	610	700	760	805	850	895	940	954
Tl.	321	300	285	270	262	258	245	230	210	235	302
Zn.	322	280	270	295	313	327	340	355	370	390	419
Au. Cu.	1063	910	890	895	905	925	975	1000	1025	1060	1084
Ag.	1064	1062	1061	1058	1054	1049	1039	1025	1006	982	963
Pt.	1075	1125	1190	1250	1320	1380	1455	1530	1610	1685	1775
K. Na.	62	17.5	−10	−3.5	5	11	26	41	58	77	97.5
Hg.	...	...	...	...	...	90	110	135	162	265	...
Tl.	62.5	133	165	188	205	215	220	240	280	305	301
Cu. Ni.	1080	1180	1240	1290	1320	1355	1380	1410	1430	1440	1455
Ag.	1082	1035	990	945	910	870	830	788	814	875	960
Sn.	1084	1005	890	755	725	680	630	580	530	440	232
Zn.	1084	1040	995	930	900	880	820	780	700	580	419
Ag. Zn.	959	850	755	705	690	660	630	610	570	505	419
Sn.	959	870	750	630	550	495	450	420	375	300	232
Na. Hg.	96.5	90	80	70	60	45	22	55	95	215	...

* The data in this table are compiled from various sources,— hence variations in the melting point of the metals as shown in this column.

The triple point of water, 0.01°C (273.16°K) is the thermodynamic point for temperature measurements.

COMMERCIAL METALS AND ALLOYS

Miscellaneous Properties

Properties (Typical Only)

Common name and classification	Thermal conductivity J/sec cm °K	Thermal conductivity Btu/hr ft °F	Thermal conductivity kcal/sec cm °C	Specific gravity	Coeff. of linear expansion, μ in./in. °F	Electrical resistivity, microhm-cm	Modulus of elasticity, millions of psi	Approximate melting point °F	Approximate melting point °C
Ingot iron (included for comparison)	1.3	77	0.32	7.86	6.8	9	30	2800	1538
Plain carbon steel AISI–SAE 1020	1.0	56	0.23	7.86	6.7	10	30	2760	1515
Stainless steel type 304	0.3	19	0.08	8.02	9.6	72	28	2600	1427
Cast gray iron ASTM A48–48. Class 25	0.8	48	0.20	7.2	6.7	67	13	2150	1177
Malleable iron ASTM A47				7.32	6.6	30	25	2250	1232
Ductile cast iron ASTM A339, A395	0.6	34	0.14	7.2	7.5	60	25	2100	1149
Ni-resist cast iron, type 2	0.7	41	0.17	7.3	9.6	170	15.6	2250	1232
Cast 28-7 alloy (IID) ASTM A297–63T	0.04	2	0.01	7.6	9.2	41	27	2700	1482
Hastelloy C	0.2	10	0.04	3.94	6.3	139	30	2350	1288
Inconel X, annealed	0.3	17	0.07	8.25	6.7	122	31	2550	1399
Haynes Stellite alloy 25 (L605)	0.2	10	0.04	9.15	7.61	88	34	2500	1371
Aluminum alloy 3003, rolled ASTM B221	2.8	164	0.68	2.73	12.9	4	10	1200	649
Aluminum alloy 2017, annealed ASTM B221	3.0	174	0.72	2.8	12.7	4	10.5	1185	641
Aluminum alloy 380 ASTM SC84B	1.8	102	0.42	2.7	11.6	7.5	10.3	1050	566
Copper ASTM B152, B124, B133, B1, B2, B3	7.1	411	1.70	8.91	9.3	1.7	17	1980	1082
Yellow brass (high brass) ASTM B36, B134, B135	2.2	126	0.52	8.47	10.5	7	15	1710	932
Aluminum bronze ASTM B169, alloy A; ASTM B124, B150	1.3	75	0.31	7.8	9.2	12	17	1900	1038
Beryllium copper 25 ASTM B194	0.2	12	0.05	8.25	9.3	–	19	1700	927

COMMERCIAL METALS AND ALLOYS (continued)

Common name and classification	Thermal conductivity J/sec cm °K	Thermal conductivity Btu/hr ft °F	Thermal conductivity kcal/sec cm °C	Specific gravity	Coeff. of linear expansion, μ in./in. °F	Electrical resistivity, microhm-cm	Modulus of elasticity, millions of psi	Approximate melting point °F	Approximate melting point °C
Nickel silver 18% alloy A (wrought) ASTM B122, No. 2	0.6	34	0.14	8.8	9.0	29	18	2030	1110
Cupronickel 30%	0.5	31	0.13	8.95	8.5	35	22	2240	1227
Red brass (cast) ASTM B30, No. 4A	1.3	77	0.32	8.7	10	11	13	1825	996
Chemical lead	0.6	36	0.15	11.35	16.4	21	2	621	327
Antimonial lead (hard lead)	0.5	31	0.13	10.9	15.1	23	3	554	290
Solder 50–50	0.8	48	0.20	8.89	13.1	15	–	420	216
Magnesium alloy AZ31B	1.4	82	0.34	1.77	14.5	9	6.5	1160	627
K Monel	0.3	19	0.08	8.47	7.4	58	26	2430	1332
Nickel ASTM B160, B161, B162	1.1	63	0.26	8.89	6.6	10	30	2625	1441
Cupronickel 55–45 (Constantan)	0.4	24	0.10	8.9	8.1	49	24	2300	1260
Commercial titanium	0.3	19	0.08	5	4.9	80	16.5	3300	1816
Zinc ASTM B69	2.0	114	0.47	7.14	18	6	–	785	418
Zirconium, commercial	0.3	19	0.08	6.5	2.9	41	12	3350	1843

THERMAL PROPERTIES OF PURE METALS

From Handbook of Tables for Applied Engineering Science by R. E. Bolz and G. L. Tuve, The Chemical Rubber Co., 1970

	At Atmospheric Pressure						Liquid Metal			
	At 100°K			At 25°C (77°F)				Vapor pressure		
								10^{-3} atm	10^{-6} atm	10^{-9} atm
Metal	Latent heat of fusion, cal/g	Thermal conductivity, watts cm°C	Specific heat, cal/g°C	Specific heat, cal/g°C	Coeff. of linear expansion, ($\times 10^6$) (°C)$^{-1}$	Thermal conductivity, watts/cm°C*	Specific heat (liquid) at 2000°K cal/g°C	Boiling point temperatures, °K		
Aluminum	95	3.00	0.115	0.215	25	2.37	0.26	1,782	1,333	1,063
Antimony	38.5	—	0.040	0.050	9	0.185	0.062	1,007	741	612
Beryllium	324	—	0.049	0.436	12	2.18	0.78	1,793	1,347	1,085
Bismuth	12.4	—	0.026	0.030	13	0.084	0.036	1,155	851	677
Cadmium	13.2	1.03	0.047	0.055	30	0.93	0.063	655	486	388
Chromium	79	1.58	0.046	0.110	6	0.91	0.224	1,992	1,530	1,247
Cobalt	66	—	0.057	0.10	12	0.69	0.164	2,167	1,652	1,345
Copper	49	4.83	0.061	0.092	16.6	3.98	0.118	1,862	1,391	1,120
Gold	15	3.45	0.026	0.031	14.2	3.15	0.0355	2,023	1,510	1,211
Iridium	33	—	0.022	0.031	6	1.47	0.0434	3,253	2,515	2,062
Iron	65	1.32	0.052	0.108	12	0.803	0.197	2,093	1,594	1,297
Lead	5.5	0.396	0.028	0.031	29	0.346	0.033	1,230	889	698
Magnesium	88.0	1.69	0.016	0.243	25	1.59	0.32	857	638	509
Manganese	64	—	0.064	0.114	22	—	0.20	1,495	1,131	913
Mercury	2.7	—	0.029	0.033	—	0.0839	—	393	287	227
Molybdenum	69	1.79	0.033	0.060	5	1.4	0.089	3,344	2,558	2,079
Nickel	71	1.58	0.055	0.106	13	0.899	0.175	2,156	1,646	1,343
Niobium (Columbium)	68	0.552	0.045	0.064	7	0.52	0.083	3,523	2,721	2,232
Osmium	34	—	—	0.031	5	0.61	0.039	—	—	—
Platinum	24	0.79	0.024	0.032	9	0.73	0.043	2,817	2,155	1,757
Plutonium	3	—	0.019	0.032	54	0.08	0.041	2,200	1,596	1,252
Potassium	14.5	—	0.150	0.180	83	0.99	—	606	430	335
Rhodium	50	—	—	0.058	8	1.50	0.092	—	—	—
Selenium	16	—	—	0.077	37	0.005	—	—	—	—
Silicon	430	—	0.062	0.17	3	0.835	0.217	2,340	1,749	1,427
Silver	26.5	4.50	0.045	0.057	19	4.27	0.068	1,582	1,179	952
Sodium	27	—	0.234	0.293	70	1.34	—	701	504	394
Tantalum	41	0.592	0.026	0.034	6.5	0.54	0.040	3,959	3,052	2,495
Thorium	17	—	0.024	0.03	12	0.41	0.047	3,251	2,407	1,919
Tin	14.1	0.85	0.039	0.054	20	0.64	0.058	1,857	1,366	1,080
Titanium	100	0.312	0.072	0.125	8.5	0.2	0.188	2,405	1,827	1,484
Tungsten	46	2.35	0.021	0.032	4.5	1.78	0.040	4,139	3,228	2,656
Uranium	12	—	0.022	0.028	13.4	0.25	0.048	2,861	2,128	1,699
Vanadium	98	—	0.061	0.116	8	0.60	0.207	2,525	1,948	1,591
Zinc	27	1.32	0.063	0.093	35	1.15	—	752	559	449

* (watts/cm°C) × 860.421 = Cal(gm)hr^{-1}cm^{-1}°C^{-1}
(watts/cm°C) × 57.818 = Btu hr^{-1} ft^{-1} °F.

MELTING AND BOILING POINTS OF THE ELEMENTS

Element	Melting Point °C	Boiling Point °C	Element	Melting Point °C	Boiling Point °C
Actinium	1050	3200 ± 300	Manganese	1244 ± 3	1962
Aluminum	660.37	2467	Mendelevium	—	—
Americium	994 ± 4	2607	Mercury	−38.87	356.58
Antimony	630.74	1750	Molybdenum	2617	4612
Argon	−189.2	−185.7	Neodymium	1021	3068
Arsenic (gray)	817 (28 atm.)	613 (sub.)	Neon	−248.67	−246.048
Astatine	302	337	Neptunium	640 ± 1	3902
Barium	725	1640	Nickel	1453	2732
Berkelium	—	—	Niobium	2468 ± 10	4742
Beryllium	1278 ± 5	2970 (5 mm)	Nitrogen	−209.86	−195.8
Bismuth	271.3	1560 ± 5	Nobelium	—	—
Boron	2300	2550 (sub.)	Osmium	3045 ± 30	5027 ± 100
Bromine	−7.2	58.78	Oxygen	−218.4	−182.962
Cadmium	320.9	765	Ozone	−192.7 ± 2	−111.9
Calcium	839 ± 2	1484	Palladium	1552	3140
Californium	—	—	Phosphorus (white)	44.1	280
Carbon	~3550	4827	Platinum	1772	3827 ± 100
Cerium	799 ± 3	3426	Plutonium	641	3232
Cesium	28.40 ± 0.01	678.4	Polonium	254	962
Chlorine	−100.98	−34.6	Potassium	63.65	774
Chromium	1857 ± 20	2672	Praseodymium	931	3512
Cobalt	1495	2870	Promethium	~1080	2460 (?)
Columbium (See *Niobium*)			Protactinium	<1600	—
			Radium	700	1140
Copper	1083.4 ± 0.2	2567	Radon	−71	−61.8
Curium	1340 ± 40	—	Rhenium	3180	5627 (est.)
Deuterium (See *Hydrogen*)			Rhodium	1966 ± 3	3727 ± 100
			Rubidium	38.89	688
Dysprosium	1412	2562	Ruthenium	2310	3900
Einsteinium	—	—	Samarium	1077 ± 5	1791
Element 104	—	—	Scandium	1541	2831
Element 105	—	—	Selenium (gray)	217	684.9 ± 1.0
Erbium	1529	2863	Silicon	1410	2355
Europium	822	1597	Silver	961.93	2212
Fermium	—	—	Sodium	97.81 ± 0.03	882.9
Fluorine	−219.62	−188.14	Strontium	769	1384
Francium	(27)	(677)	Sulfur (rhombic)	112.8	444.674
Gadolinium	1313 ± 1	3266	(monoclinic)	119.0	444.674
Gallium	29.78	2403	Tantalum	2996	5425 ± 100
Germanium	937.4	2830	Technetium	2172	4877
Gold (Aurum)	1064.43	2807	Tellurium	449.5 ± 0.3	989.8 ± 3.8
Hafnium	2227 ± 20	4602	Terbium	1356 ± 4	3123
Helium	−272.2 (26 atm.)	−268.934	Thallium	303.5	1457 ± 10
Holmium	1474	2695	Thorium	1750	~4790
Hydrogen	−259.14	−252.87	Thulium	1545	1947
Indium	156.61	2080	Tin	231.9681	2270
Iodine	113.5	184.35	Titanium	1660 ± 10	3287
Iridium	2410	4130	Tungsten	3410 ± 20	5660
Iron (ferrum)	1535	2750	Uranium	1132.3 ± 0.8	3818
Krypton	−156.6	−152.30 ± 0.10	Vanadium	1890 ± 10	3380
Lanthanum	921 ± 5	3457	Wolfram (See *Tungsten*)		
Lawrencium	—	—	Xenon	−111.9	−107.1 ± 3
Lead	327.502	1740	Ytterbium	819 ± 5	1194
Lithium	180.54	1347	Yttrium	1522 ± 8	3338
Lutetium	1663 ± 5	3395	Zinc	419.58	907
Magnesium	648.8 ± 0.5	1090	Zirconium	1852 ± 2	4377

BOILING POINTS AND TRIPLE POINTS OF SOME LOW BOILING ELEMENTS

Element	Boiling point °K	Triple point °K	λ point °K
He	4.216		2.174
p-H_2	20.27	13.81	
n-H_2	20.379	13.95	
N_2	77.35		
O_2	90.188	54.34	
A	87.45		

MOLECULAR ELEVATION OF THE BOILING POINT

(Most values from Hoyt, C.S. and Fink, C.K., Journal of Physical Chemistry, Vol. 41. No. 3., March, 1937.)

Molecular elevation of the boiling point showing the elevation of the boiling point in degrees C due to the addition of one gram molecular weight of the dissolved substance to 1000 grams of any one of the solvents below. The correction in the last column gives the number of degrees to be subtracted for each mm. of difference between the barometric reading and 760 mm.

Solvent	K_B	Barometric Correction per mm.
Acetic acid	3.07	0.0008
Acetone	1.71	0.0004
Aniline	3.52	0.0009
Benzene	2.53	0.0007
Bromobenzene	6.26	0.0016
Carbon bisulfide	2.34	0.0006
Carbon tetrachloride	5.03	0.0013
Chloroform	3.63	0.0009
Cyclohexane	2.79	0.0007
Ethanol (ethyl alcohol)	1.22	0.0003
Ethyl acetate	2.77	0.0007
Ethyl ether	2.02	0.0005
n-Hexane	2.75	0.0007
Methanol (methyl alcohol)	0.83	0.0002
Methyl acetate	2.15	0.0005
Nitrobenzene	5.24	0.0013
n-Octane	4.02	0.0010
Phenol	3.56	0.0009
Toluene	3.33	0.0008
Water	0.512	0.0001

MOLECULAR DEPRESSION OF THE FREEZING POINT

Showing the depression of the freezing point due to the addition of one gram molecular weight of dissolved substance, for various solvents.

Solvent	Depression for one gram molecular weight dissolved in 100 gms. °C
Acetic acid	39.0
Benzene	49.0
Benzophenone	98.0
Diphenyl	80.0
Diphenylamine	86.0
Ethylene dibromide	118.0
Formic acid	27.7
Naphthalene	68–69
Nitrobenzene	70.0
Phenol	74.0
Stearic acid	45.0
Triphenyl methane	124.5
Urethane	51.4
Water	18.5–18.7

CORRECTION OF BOILING POINTS TO STANDARD PRESSURE

By H. B. Hass and R. F. Newton

This correction may be made by using the equation:

$$\Delta t = \frac{(273.1 + t)(2.8808 - \log p)}{\phi + .15(2.8808 - \log p)} \quad (1)$$

where Δt = degrees C to be added to the observed boiling point.
t = the observed boiling point.
$\log p$ = the logarithm of the observed pressure in millimeters of mercury.
ϕ = the entropy of vaporization at 760 mm.

The value of ϕ may be estimated from the graph and the table. Substances not included in the table may be classified by grouping them with compounds which bear a close physical or structural resemblance to them.

Example 1. Benzene boils at 20°C. at 75 mm pressure. What is its normal boiling point? We do not find benzene in the table but we find hydrocarbons in group 2, and a group 2 compound with a boiling point of 20° has a ϕ of 4.6.

Substituting in the equation

$$\Delta t = \frac{(273.1 + 20)(2.8808 - 1.8751)}{4.60 + .15(2.8808 - 1.8751)} = 62°$$

Adding this to 20° gives 82° as a first approximation.

The graph shows that the ϕ for a compound of group 2 boiling at 82° is 4.72 instead of 4.60 which we originally used. Since ϕ is in the denominator, this increase will lower our Δt by the ratio, 4.60/4.72, or the corrected Δt is $62 \times 4.60/4.72 = 60.4$. Adding Δt to t, gives 80.4° as a second approximation.

The formula can best be used in a slightly different form when the reverse calculation is desired, *i.e.*, when one calculates the vapor pressure at a given temperature, lower than the normal boiling point.

$$2.8808 - \log p = \frac{\phi \Delta t}{273.1 + t - .15\Delta t} \quad (2)$$

Example 2. Alcohol boils at 78.4°C. What is its vapor pressure at 20°C.? Substituting in equation 2:

$$2.8808 - \log p = \frac{6.06 \times 58.4}{293.1 - (.15 \times 58.4)} = 1.245$$

$$\log p = 2.8808 - 1.245 = 1.6358$$

$$p = 43.2 \text{ mm.}$$

Here no second approximation is necessary, since the correct value of ϕ was taken immediately, the normal boiling point having been known.

Compound	Group	Compound	Group
Acetaldehyde	3	Amines	3
Acetic acid	4	*n*-Amyl alcohol	8
Acetic anhydride	6	Anthracene	1
Acetone	3	Anthraquinone	1
Acetophenone	4	Benzaldehyde	2

CORRECTION OF BOILING POINTS TO STANDARD PRESSURE (Continued)

Compound	Group	Compound	Group
Benzoic acid	5	Hydrogen cyanide	3
Benzonitrile	2	Isoamyl alcohol	7
Benzophenone	2	Isobutyl alcohol	8
Benzyl alcohol	5	Isobutyric acid	6
Butylethylene	1	Isocaproic acid	7
Butyric acid	7	Methane	1
Camphor	2	Methanol	7
Carbon monoxide	1	Methyl amine	5
Carbon oxysulfide	2	Methyl benzoate	3
Carbon suboxide	2	Methyl ether	3
Carbon sulfoselenide	2	Methyl ethyl ether	3
m.p. Chloroanilines	3	Methyl ethyl ketone	2
Chlorinated derivatives	Same group as though Cl were H	Methyl fluoride	3
		Methyl formate	4
o.m.p. Cresols	4	Methyl salicylate	2
Cyanogen	4	Methyl silicane	1
Cyanogen chloride	3	α, β Naphthols	3
Dibenzyl ketone	2	Nitrobenzene	3
Dimethyl amine	4	Nitromethane	3
Dimethyl oxalate	4	o.m.p. Nitrotoluenes	2
Dimethyl silicane	2	o.m.p. Nitrotoluidines	2
Esters	3	Phenanthrene	1
Ethanol	8	Phenol	5
Ethers	2	Phosgene	2
Ethylamine	4	Phthalic anhydride	2
Ethylene glycol	7	Propionic acid	5
Ethylene oxide	3	*n*-Propyl alcohol	8
Formic acid	3	Quinoline	2
Glycol diacetate	4	Sulfides	2
Halogen derivatives	Same group as though halogen were hydrogen.	Tetranitromethane	3
		Trichloroethylene	1
		Valeric acid	7
Heptylic acid	7	Water	6
Hydrocarbons	2		

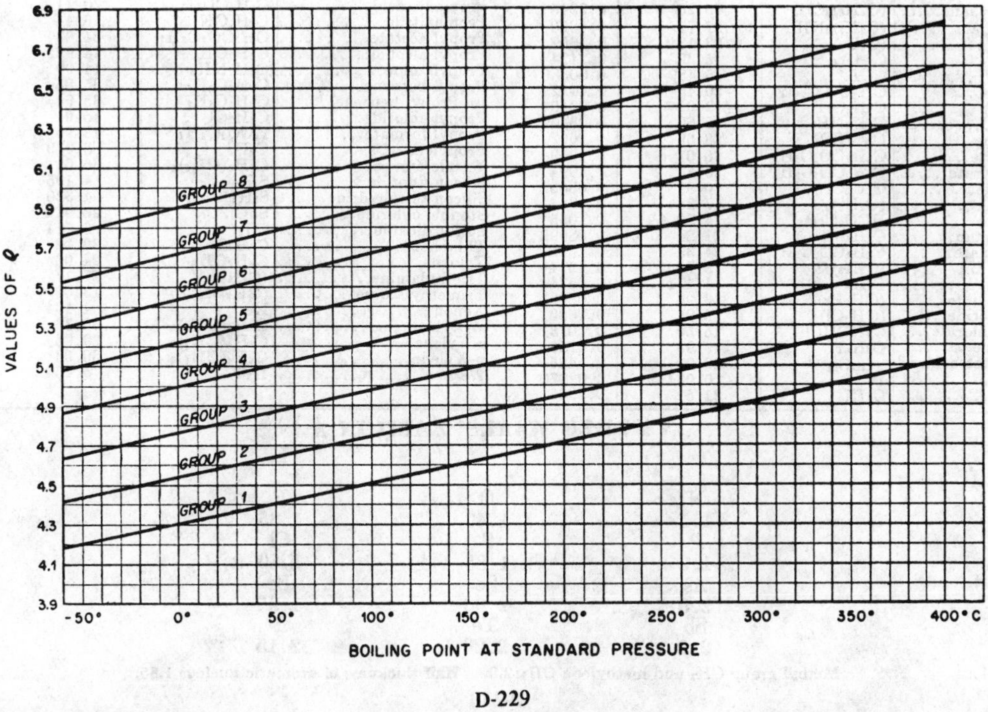

VAN DER WAALS' CONSTANTS FOR GASES

(Calculated from Amagat units in Landolt-Bornstein Physical Chemical Tables)

Van der Waals' equation is an equation of state for real gases. It may be written

$$\left(P + \frac{a}{V^2}\right)(V - b) = RT \text{ for one mole.} \qquad \text{or} \qquad \left(P + \frac{n^2a}{V^2}\right)(V - nb) = nRT \text{ for } n \text{ moles.}$$

The term a is a measure of the attractive force between the molecules. The term b is due to the finite volume of the molecules and to their general incompressibility. It is known that a and b vary to some extent with temperature.

The values for a and b in the following table are those to be used when the pressure is in atmospheres and the volume is in liters. Thus R in the above equation will be 0.08206 liter atmospheres per mole per degree. T is degrees Kelvin.

Name	Formula	a (liters)² × atm. / (mole)²	b liters / mole
Acetic acid	CH_3CO_2H	17.59	0.1068
Acetic anhydride	$(CH_3CO)_2O$	19.90	0.1263
Acetone	$(CH_3)_2CO$	13.91	0.0994
Acetonitrile	CH_3CN	17.58	0.1168
Acetylene	C_2H_2	4.390	0.05136
Ammonia	NH_3	4.170	0.03707
Amyl formate	$HCO_2C_5H_{11}$	27.58	0.1730
Amylene	C_5H_{10}	15.90	0.1207
*Iso*amylene	C_5H_{10}	18.08	0.1405
Aniline	$C_6H_5NH_2$	26.50	0.1369
Argon	A	1.345	0.03219
Benzene	C_6H_6	18.00	0.1154
Benzonitrile	C_6H_5CN	33.39	0.1724
Bromobenzene	C_6H_5Br	28.56	0.1539
n-Butane	C_4H_{10}	14.47	0.1226
iso-Butane	C_4H_{10}	12.87	0.1142
iso-Butyl acetate	$CH_3CO_2C_4H_9$	28.50	0.1833
iso-Butyl alcohol	C_4H_9OH	17.03	0.1143
iso-Butyl benzene	$C_6H_5C_4H_9$	38.59	0.2144
iso-Butyl formate	$HCO_2C_4H_9$	22.54	0.1476
Butyronitrile	C_3H_7CN	25.72	0.1596
Capronitrile	$C_5H_{11}CN$	34.16	0.1984
Carbon dioxide	CO_2	3.592	0.04267
Carbon disulfide	CS_2	11.62	0.07685
Carbon monoxide	CO	1.485	0.03985
Carbon oxysulfide	COS	3.933	0.05817
Carbon tetrachloride	CCl_4	20.39	0.1383
Chlorine	Cl_2	6.493	0.05622
Chlorobenzene	C_6H_5Cl	25.43	0.1453
Chloroform	$CHCl_3$	15.17	0.1022
m-Cresol	$CH_3C_6H_4OH$	31.38	0.1607
Cyanogen	C_2N_2	7.667	0.06901
Cyclohexane	C_6H_{12}	22.81	0.1424
Cymene	$C_{10}H_{14}$	42.16	0.2336
Decane	$C_{10}H_{12}$	48.55	0.2905
Di-isobutyl	C_8H_{18}	34.97	0.2296
Diethylamine	$(C_2H_5)_2NH$	19.15	0.1392
Dimethylamine	$(CH_3)_2NH$	10.38	0.08570
Dimethylaniline	$C_6H_5N(CH_3)_2$	37.49	0.1970
Diphenyl	$(C_6H_5)_2$	52.79	0.2480
Diphenyl methane	$(C_6H_5)_2CH_2$	38.20	0.2240
Dipropylamine	$(C_2H_7)_2NH$	27.72	0.1820
Di-isopropyl	$(C_3H_7)_2$	23.13	0.1669
Durene	$C_{10}H_{14}$	45.32	0.2424
Ethane	C_2H_6	5.489	0.06380
Ethyl acetate	$CH_3CO_2C_2H_5$	20.45	0.1412
Ethyl alcohol	C_2H_5OH	12.02	0.08407
Ethylamine	$C_2H_5NH_2$	10.60	0.08409
Ethyl benzene	$C_2H_5C_6H_5$	28.60	0.1667
Ethyl butyrate	$C_3H_7CO_2C_2H_5$	30.07	0.1919
Ethyl isobutyrate	$C_3H_7CO_2C_2H_5$	28.87	0.1994
Ethyl chloride	C_2H_5Cl	10.91	0.08651
Ethyl ether	$(C_2H_5)_2O$	17.38	0.1344
Ethyl formate	$HCO_2C_2H_5$	14.80	0.1056
Ethyl mercaptan	C_2H_5SH	11.24	0.08098
Ethyl propionate	$C_2H_5CO_2C_2H_5$	24.39	0.1615
Ethyl sulfide	$(C_2H_5)_2S$	18.75	0.1214
Ethylene	C_2H_4	4.471	0.05714
Ethylene bromide	$(CH_2Br)_2$	13.98	0.08664
Ethylene chloride	$(CH_2Cl)_2$	16.91	0.1086
Ethylidene chloride	CH_3CHCl_2	15.50	0.1073
Fluorobenzene	C_6H_5F	19.93	0.1286
Germanium tetrachloride	$GeCl_4$	22.60	0.1485
Helium	He	0.03412	0.02370
n-Heptane	C_7H_{16}	31.51	0.2065

Name	Formula	a (liters)² × atm. / (mole)²	b liters / mole
n-Hexane	C_6H_{14}	24.39	0.1735
Hydrogen	H_2	0.2444	0.02661
Hydrogen bromide	HBr	4.451	0.04431
Hydrogen chloride	HCl	3.667	0.04081
Hydrogen selenide	H_2Se	5.268	0.04637
Hydrogen sulfide	H_2S	4.431	0.04287
Iodobenzene	C_6H_5I	33.08	0.1656
Krypton	Kr	2.318	0.03978
Mercury	Hg	8.093	0.01696
Mesitylene	$(CH_3)_3C_6H_3$	34.32	0.1979
Methane	CH_4	2.253	0.04278
Methyl acetate	$CH_3CO_2CH_3$	15.29	0.1091
Methyl alcohol	CH_3OH	9.523	0.06702
Methylamine	CH_3NH_2	7.130	0.05992
Methyl butyrate	$C_3H_7CO_2CH_3$	23.94	0.1569
Methyl isobutyrate	$C_3H_7CO_2CH_3$	24.50	0.1637
Methyl chloride	CH_3Cl	7.471	0.06483
Methyl ether	$(CH_3)_2O$	8.073	0.07246
Methyl ethyl ether	$CH_3OC_2H_5$	11.95	0.09775
Methyl ethyl sulfide	$CH_3SC_2H_5$	19.23	0.1304
Methyl fluoride	CH_3F	4.631	0.05264
Methyl formate	HCO_2CH_3	10.84	0.08068
Methyl propionate	$C_2H_5CO_2CH_3$	19.91	0.1360
Methyl sulfide	$(CH_3)_2S$	12.87	0.09213
Methyl valerate	$C_4H_9CO_2CH_3$	28.96	0.1845
Naphthalene	$C_{10}H_8$	39.74	0.1937
Neon	Ne	0.2107	0.01709
Nitric oxide	NO	1.340	0.02789
Nitrogen	N_2	1.390	0.03913
Nitrogen dioxide	NO_2	5.284	0.04424
Nitrous oxide	N_2O	3.782	0.04415
n-Octane	C_8H_{18}	37.32	0.2368
Oxygen	O_2	1.360	0.03183
n-Pentane	C_5H_{12}	19.01	0.1460
iso-Pentane	C_5H_{12}	18.05	0.1417
Phenetole	$C_6H_5OC_2H_5$	35.16	0.1963
Phosphine	PH_3	4.631	0.05156
Phosphonium chloride	PH_4Cl	4.054	0.04545
Phosphorus	P	52.94	0.1566
Propane	C_3H_8	8.664	0.08445
Propionic acid	$C_2H_5CO_2H$	20.11	0.1187
Propionitrile	C_2H_5CN	16.44	0.1064
Propyl acetate	$CH_3CO_2C_3H_7$	24.63	0.1619
Propyl alcohol	C_3H_7OH	14.92	0.1019
Propylamine	$C_3H_7NH_2$	14.99	0.1090
Propyl benzene	$C_6H_5C_3H_7$	35.85	0.2028
iso-Propyl benzene	$C_6H_5C_3H_7$	35.64	0.2025
Propyl chloride	C_3H_7Cl	15.91	0.1141
Propyl formate	$HCO_2C_3H_7$	18.95	0.1280
Propylene	C_3H_6	8.379	0.08272
Pseudo-cumene	$C_6H_3(CH_3)_3$	36.61	0.2021
Silicon fluoride	SiF_4	4.195	0.05571
Silicon tetrahydride	SiH_4	4.320	0.05786
Stannic chloride	$SnCl_4$	26.91	0.1642
Sulfur dioxide	SO_2	6.714	0.05636
Thiophene	C_4H_4S	20.72	0.1270
Toluene	$C_6H_5CH_3$	24.06	0.1463
Triethylamine	$(C_2H_5)_3N$	27.17	0.1831
Trimethylamine	$(CH_3)_3N$	13.02	0.1084
Xenon	Xe	4.194	0.05105
m-Xylene	$C_6H_4(CH_3)_2$	30.36	0.1772
o-Xylene	$C_6H_4(CH_3)_2$	29.98	0.1755
p-Xylene	$C_6H_4(CH_3)_2$	30.93	0.1809
Water	H_2O	5.464	0.03049

VAN DER WAALS' RADII IN Å

		H
		1.2
N	O	F
1.5	1.40	1.35
P	S	Cl
1.9	1.85	1.80
As	Se	Br
2.0	2.00	1.95
Sb	Te	I
2.2	2.20	2.15

Methyl group CH_3 and methylene CH_2: 2.0. Half thickness of aromatic nucleus 1.85.

EMERGENT STEM CORRECTION FOR LIQUID-IN-GLASS THERMOMETERS

Accurate thermometers are calibrated with the entire stem immersed in the bath which determines the temperature of the thermometer bulb. However, for reasons of convenience it is common practice when using a thermometer to permit its stem to extend out of the apparatus. Under these conditions both the stem and the mercury in the exposed stem are at a temperature different from that of the bulb. This introduces an error into the observed temperature. Since the coefficient of thermal expansion of glass is less than that of mercury, the observed temperature will be less than the true temperature if the bulb is hotter than the stem and greater than the true temperature, providing the thermal gradient is reversed. For exact work the magnitude of this error can only be determined by experiment. However, for most purposes it is sufficiently accurate to apply the following equation which takes into account the difference of the thermal expansion of glass and mercury:

$$T_c = T_o + F \times L(T_o - T_m)$$

Where

T_c = corrected temperature

T_o = observed temperature

T_m = mean temperature of exposed stem. The mean temperature of the exposed stem may be determined by fastening the bulb of a second thermometer against the midpoint of the exposed liquid column.

L = the length of the exposed column in degrees above the surface of the substance whose temperature is being determined.

F = correction factor. For approximate work and when the liquid in the thermometer is mercury a value for F of 0.00016 is generally used. For more accurate work with mercury filled thermometers values as given in the following table are used. For thermometers filled with organic liquids it is customary to use 0.001 for the value of F.

Values of F for various glasses

Tm°C.	Corning 0041	Corning 8800	Corning 8810	Jena 16 III	Jena 59 III
50	0.000157	0.000166	0.000156	0.000158	0.000164
150	0.000159	0.000167	0.000157	0.000158	0.000165
250	0.000163	0.000168	0.000161	0.000161	0.000170
350	0.000168	0.000173	0.000166	....	0.000177

PRESSURE OF AQUEOUS VAPOR

Vapor Pressure of Ice

Pressure of aqueous vapor over ice in mm of Hg for temperatures from −98 to 0°C.

Temp. °C	0	2	4	6	8
−90	.000070	.000048	.000033	.000022	.000015
−80	.00040	.00029	.00020	.00014	.00010
−70	.00194	.00143	.00105	.00077	.00056
−60	.00808	.00614	.00464	.00349	.00261
−50	.02955	.0230	.0178	.0138	.0106
−40	.0966	.0768	.0609	.0481	.0378
−30	.2859	.2318	.1873	.1507	.1209

Temp. °C	0.0	0.2	0.4	0.6	0.8
−29	0.317	0.311	0.304	0.298	0.292
−28	0.351	0.344	0.337	0.330	0.324
−27	0.389	0.381	0.374	0.366	0.359
−26	0.430	0.422	0.414	0.405	0.397
−25	0.476	0.467	0.457	0.448	0.439
−24	0.526	0.515	0.505	0.495	0.486
−23	0.580	0.569	0.558	0.547	0.536
−22	0.640	0.627	0.615	0.603	0.592
−21	0.705	0.691	0.678	0.665	0.652
−20	0.776	0.761	0.747	0.733	0.719

Temp. °C	0.0	0.2	0.4	0.6	0.8
−19	0.854	0.838	0.822	0.806	0.791
−18	0.939	0.921	0.904	0.887	0.870
−17	1.031	1.012	0.993	0.975	0.956
−16	1.132	1.111	1.091	1.070	1.051
−15	1.241	1.219	1.196	1.175	1.153
−14	1.361	1.336	1.312	1.288	1.264
−13	1.490	1.464	1.437	1.411	1.386
−12	1.632	1.602	1.574	1.546	1.518
−11	1.785	1.753	1.722	1.691	1.661
−10	1.950	1.916	1.883	1.849	1.817
− 9	2.131	2.093	2.057	2.021	1.985
− 8	2.326	2.285	2.246	2.207	2.168
− 7	2.537	2.493	2.450	2.408	2.367
− 6	2.765	2.718	2.672	2.626	2.581
− 5	3.013	2.962	2.912	2.862	2.813
− 4	3.280	3.225	3.171	3.117	3.065
− 3	3.568	3.509	3.451	3.393	3.336
− 2	3.880	3.816	3.753	3.691	3.630
− 1	4.217	4.147	4.079	4.012	3.946
− 0	4.579	4.504	4.431	4.359	4.287

VAPOR PRESSURE OF WATER BELOW 100°C

Pressure of aqueous vapor over water in mm of Hg for temperatures from −15.8 to 100°C. Values for fractional degrees between 50 and 89 were obtained by interpolation.

Temp. °C	0.0	0.2	0.4	0.6	0.8
−15	1.436	1.414	1.390	1.368	1.345
−14	1.560	1.534	1.511	1.485	1.460
−13	1.691	1.665	1.637	1.611	1.585
−12	1.834	1.804	1.776	1.748	1.720
−11	1.987	1.955	1.924	1.893	1.863
−10	2.149	2.116	2.084	2.050	2.018
− 9	2.326	2.289	2.254	2.219	2.184
− 8	2.514	2.475	2.437	2.399	2.362
− 7	2.715	2.674	2.633	2.593	2.553
− 6	2.931	2.887	2.843	2.800	2.757
− 5	3.163	3.115	3.069	3.022	2.976
− 4	3.410	3.359	3.309	3.259	3.211
− 3	3.673	3.620	3.567	3.514	3.461
− 2	3.956	3.898	3.841	3.785	3.730
− 1	4.258	4.196	4.135	4.075	4.016
− 0	4.579	4.513	4.448	4.385	4.320
0	4.579	4.647	4.715	4.785	4.855
1	4.926	4.998	5.070	5.144	5.219
2	5.294	5.370	5.447	5.525	5.605
3	5.685	5.766	5.848	5.931	6.015
4	6.101	6.187	6.274	6.363	6.453
5	6.543	6.635	6.728	6.822	6.917
6	7.013	7.111	7.209	7.309	7.411
7	7.513	7.617	7.722	7.828	7.936
8	8.045	8.155	8.267	8.380	8.494
9	8.609	8.727	8.845	8.965	9.086
10	9.209	9.333	9.458	9.585	9.714
11	9.844	9.976	10.109	10.244	10.380
12	10.518	10.658	10.799	10.941	11.085
13	11.231	11.379	11.528	11.680	11.833
14	11.987	12.144	12.302	12.462	12.624
15	12.788	12.953	13.121	13.290	13.461
16	13.634	13.809	13.987	14.166	14.347
17	14.530	14.715	14.903	15.092	15.284
18	15.477	15.673	15.871	16.071	16.272
19	16.477	16.685	16.894	17.105	17.319
20	17.535	17.753	17.974	18.197	18.422
21	18.650	18.880	19.113	19.349	19.587
22	19.827	20.070	20.316	20.565	20.815
23	21.068	21.324	21.583	21.845	22.110
24	22.377	22.648	22.922	23.198	23.476
25	23.756	24.039	24.326	24.617	24.912
26	25.209	25.509	25.812	26.117	26.426
27	26.739	27.055	27.374	27.696	28.021
28	28.349	28.680	29.015	29.354	29.697
29	30.043	30.392	30.745	31.102	31.461
30	31.824	32.191	32.561	32.934	33.312
31	33.695	34.082	34.471	34.864	35.261
32	35.663	36.068	36.477	36.891	37.308
33	37.729	38.155	38.584	39.018	39.457
34	39.898	40.344	40.796	41.251	41.710
35	42.175	42.644	43.117	43.595	44.078
36	44.563	45.054	45.549	46.050	46.556
37	47.067	47.582	48.102	48.627	49.157
38	49.692	50.231	50.774	51.323	51.879
39	52.442	53.009	53.580	54.156	54.737
40	55.324	55.91	56.51	57.11	57.72
41	58.34	58.96	59.58	60.22	60.86

Temp. °C	0.0	0.2	0.4	0.6	0.8
42	61.50	62.14	62.80	63.46	64.12
43	64.80	65.48	66.16	66.86	67.56
44	68.26	68.97	69.69	70.41	71.14
45	71.88	72.62	73.36	74.12	74.88
46	75.65	76.43	77.21	78.00	78.80
47	79.60	80.41	81.23	82.05	82.87
48	83.71	84.56	85.42	86.28	87.14
49	88.02	88.90	89.79	90.69	91.59
50	92.51	93.5	94.4	95.3	96.3
51	97.20	98.2	99.1	100.1	101.1
52	102.09	103.1	104.1	105.1	106.2
53	107.20	108.2	109.3	110.4	111.4
54	112.51	113.6	114.7	115.8	116.9
55	118.04	119.1	120.3	121.5	122.6
56	123.80	125.0	126.2	127.4	128.6
57	129.82	131.0	132.3	133.5	134.7
58	136.08	137.3	138.5	139.9	141.2
59	142.60	143.9	145.2	146.6	148.0
60	149.38	150.7	152.1	153.5	155.0
61	156.43	157.8	159.3	160.8	162.3
62	163.77	165.2	166.8	168.3	169.8
63	171.38	172.9	174.5	176.1	177.7
64	179.31	180.9	182.5	184.2	185.8
65	187.54	189.2	190.9	192.6	194.3
66	196.09	197.8	199.5	201.3	203.1
67	204.96	206.8	208.6	210.5	212.3
68	214.17	216.0	218.0	219.9	221.8
69	223.73	225.7	227.7	229.7	231.7
70	233.7	235.7	237.7	239.7	241.8
71	243.9	246.0	248.2	250.3	252.4
72	254.6	256.8	259.0	261.2	263.4
73	265.7	268.0	270.2	272.6	274.8
74	277.2	279.4	281.8	284.2	286.6
75	289.1	291.5	294.0	296.4	298.8
76	301.4	303.8	306.4	308.9	311.4
77	314.1	316.6	319.2	322.0	324.6
78	327.3	330.0	332.8	335.6	338.2
79	341.0	343.8	346.6	349.4	352.2
80	355.1	358.0	361.0	363.8	366.8
81	369.7	372.6	375.6	378.8	381.8
82	384.9	388.0	391.2	394.4	397.4
83	400.6	403.8	407.0	410.2	413.6
84	416.8	420.2	423.6	426.8	430.2
85	433.6	437.0	440.4	444.0	447.5
86	450.9	454.4	458.0	461.6	465.2
87	468.7	472.4	476.0	479.8	483.4
88	487.1	491.0	494.7	498.5	502.2
89	506.1	510.0	513.9	517.8	521.8
90	525.76	529.77	533.80	537.86	541.95
91	546.05	550.18	554.35	558.53	562.75
92	566.99	571.26	575.55	579.87	584.22
93	588.60	593.00	597.43	601.89	606.38
94	610.90	615.44	620.01	624.61	629.24
95	633.90	638.59	643.30	648.05	652.82
96	657.62	662.45	667.31	672.20	677.12
97	682.07	687.04	692.05	697.10	702.17
98	707.27	712.40	717.56	722.75	727.98
99	733.24	738.53	743.85	749.20	754.58
100	760.00	765.45	770.93	776.44	782.00
101	787.57	793.18	798.82	804.50	810.21

VAPOR PRESSURE OF WATER ABOVE 100° C.

Based on values given by Keyes in the International Critical Tables.

Temp. °C	Pressure mm	Pressure Pounds per sq. in.	Temp. °F	Temp. °C	Pressure mm	Pressure Pounds per sq. in.	Temp. °F	Temp °C	Pressure mm	Pressure Pounds per sq. in.	Temp. °F	Temp. °C	Pressure mm	Pressure Pounds per sq. in.	Temp. °F
100	760.	14.696	212.0	170	5940.92	114.879	338.0	240	25100.52	485.365	464.0	310	74024.00	1431.390	590.0
101	787.51	15.228	213.8	171	6085.32	117.671	339.8	241	25543.60	493.933	465.8	311	75042.40	1451.083	591.8
102	815.86	15.776	215.6	172	6233.52	120.537	341.6	242	25994.28	502.647	467.6	312	76076.00	1471.070	593.6
103	845.12	16.342	217.4	173	6383.24	123.432	343.4	243	26449.52	511.450	469.4	313	77117.20	1491.203	595.4
104	875.06	16.921	219.2	174	6538.28	126.430	345.2	244	26912.36	520.400	471.2	314	78166.00	1511.484	597.2
105	906.07	17.521	221.0	175	6694.08	129.442	347.0	245	27381.28	529.467	473.0	315	79230.00	1532.058	599.0
106	937.92	18.136	222.8	176	6852.92	132.514	348.8	246	27855.52	538.638	474.8	316	80294.00	1552.632	600.8
107	970.60	18.768	224.6	177	7015.56	135.659	350.6	247	28335.84	547.926	476.6	317	81373.20	1573.501	602.6
108	1004.42	19.422	226.4	178	7180.48	138.848	352.4	248	28823.76	557.360	478.4	318	82467.60	1594.663	604.4
109	1038.92	20.089	228.2	179	7349.20	142.110	354.2	249	29317.00	566.898	480.2	319	83569.60	1615.972	606.2
110	1074.56	20.779	230.0	180	7520.20	145.417	356.0	250	29817.84	576.583	482.0	320	84686.80	1637.575	608.0
111	1111.20	21.487	231.8	181	7694.24	148.782	357.8	251	30324.00	586.370	483.8	321	85819.20	1659.472	609.8
112	1148.74	22.213	233.6	182	7872.08	152.221	359.6	252	30837.76	596.305	485.6	322	86959.20	1681.516	611.6
113	1187.42	22.961	235.4	183	8052.96	155.719	361.4	253	31356.84	606.342	487.4	323	88114.40	1703.854	613.4
114	1227.25	23.731	237.2	184	8236.88	159.275	363.2	254	31885.04	616.556	489.2	324	89277.20	1726.339	615.2
115	1267.98	24.519	239.0	185	8423.84	162.890	365.0	255	32417.80	626.858	491.0	325	90447.60	1748.971	617.0
116	1309.94	25.330	240.8	186	8616.12	166.609	366.8	256	32957.40	637.292	492.8	326	91633.20	1771.897	618.8
117	1352.95	26.162	242.6	187	8809.92	170.356	368.6	257	33505.36	647.888	494.6	327	92826.40	1794.969	620.6
118	1397.18	27.017	244.4	188	9007.52	174.177	370.4	258	34059.40	658.601	496.4	328	94042.40	1818.483	622.4
119	1442.63	27.896	246.2	189	9208.16	178.057	372.2	259	34618.76	669.417	498.2	329	95273.60	1842.291	624.2
120	1489.14	28.795	248.0	190	9413.36	182.025	374.0	260	35188.00	680.425	500.0	330	96512.40	1866.245	626.0
121	1536.80	29.717	249.8	191	9620.08	186.022	375.8	261	35761.80	691.520	501.8	331	97758.80	1890.346	627.8
122	1586.04	30.669	251.6	192	9831.36	190.107	377.6	262	36343.20	702.763	503.6	332	99020.40	1914.742	629.6
123	1636.36	31.642	253.4	193	10047.20	194.281	379.4	263	36932.20	714.152	505.4	333	100297.20	1939.431	631.4
124	1687.81	32.637	255.2	194	10265.32	198.499	381.2	264	37529.56	725.703	507.2	334	101581.60	1964.267	633.2
125	1740.93	33.664	257.0	195	10488.76	202.819	383.0	265	38133.00	737.372	509.0	335	102881.20	1989.398	635.0
126	1795.12	34.712	258.8	196	10715.24	207.199	384.8	266	38742.52	749.158	510.8	336	104196.00	2014.822	636.8
127	1850.83	35.789	260.6	197	10944.76	211.637	386.6	267	39361.92	761.135	512.6	337	105526.00	2040.540	638.6
128	1907.83	36.891	262.4	198	11179.60	216.178	388.4	268	39986.64	773.215	514.4	338	106871.20	2066.552	640.4
129	1966.35	38.023	264.2	199	11417.48	220.778	390.2	269	40619.72	785.457	516.2	339	108224.00	2092.710	642.2
130	2026.16	39.180	266.0	200	11659.16	225.451	392.0	270	41261.16	797.861	518.0	340	109592.00	2119.163	644.0
131	2087.42	40.364	267.8	201	11905.40	230.213	393.8	271	41910.20	810.411	519.8	341	110967.60	2145.763	645.8
132	2150.42	41.582	269.6	202	12155.44	235.048	395.6	272	42566.08	823.094	521.6	342	112358.40	2172.657	647.6
133	2214.64	42.824	271.4	203	12408.52	239.942	397.4	273	43229.56	835.923	523.4	343	113749.20	2199.550	649.4
134	2280.76	44.103	273.2	204	12666.16	244.924	399.2	274	43902.16	848.929	525.2	344	115178.00	2227.179	651.2
135	2347.26	45.389	275.0	205	12929.12	250.008	401.0	275	44580.84	862.053	527.0	345	116614.40	2254.954	653.0
136	2416.34	46.724	276.8	206	13197.40	255.196	402.8	276	45269.40	875.367	528.8	346	118073.60	2283.171	654.8
137	2488.16	48.113	278.6	207	13467.96	260.428	404.6	277	45964.04	888.799	530.6	347	119532.80	2311.387	656.6
138	2560.67	49.515	280.4	208	13742.32	265.733	406.4	278	46669.32	902.437	532.4	348	121014.80	2340.044	658.4
139	2634.84	50.950	282.2	209	14022.76	271.156	408.2	279	47382.20	916.222	534.2	349	122504.40	2368.848	660.2
140	2710.92	52.421	284.0	210	14305.48	276.623	410.0	280	48104.20	930.183	536.0	350	124001.60	2397.799	662.0
141	2788.44	53.920	285.8	211	14595.04	282.222	411.8	281	48833.80	944.291	537.8	351	125521.60	2427.191	663.8
142	2867.48	55.448	287.6	212	14888.40	287.895	413.6	282	49570.24	958.532	539.6	352	127049.20	2456.730	665.6
143	2948.80	57.020	289.4	213	15184.80	293.626	415.4	283	50316.56	972.963	541.4	353	128599.60	2486.710	667.4
144	3031.64	58.622	291.2	214	15488.04	299.490	417.2	284	51072.76	987.586	543.2	354	130157.60	2516.837	669.2
145	3116.76	60.268	293.0	215	15792.80	305.383	419.0	285	51838.08	1002.385	545.0	355	131730.80	2547.258	671.0
146	3203.40	61.944	294.8	216	16104.40	311.408	420.8	286	52611.76	1017.345	546.8	356	133326.80	2578.119	672.8
147	3292.32	63.663	296.6	217	16420.56	317.522	422.6	287	53395.32	1032.497	548.6	357	134945.60	2609.422	674.6
148	3382.76	65.412	298.4	218	16742.04	323.738	424.4	288	54187.24	1047.810	550.4	358	136579.60	2641.018	676.4
149	3476.24	67.220	300.2	219	17067.32	330.028	426.2	289	54989.04	1063.314	552.2	359	138228.80	2672.908	678.2
150	3570.48	69.042	302.0	220	17395.64	336.377	428.0	290	55799.20	1078.980	554.0	360	139893.20	2705.093	680.0
151	3667.00	70.908	303.8	221	17731.56	342.872	429.8	291	56612.40	1094.705	555.8	361	141572.80	2737.571	681.8
152	3766.56	72.833	305.6	222	18072.80	349.471	431.6	292	57448.40	1110.871	557.6	362	143275.20	2770.490	683.6
153	3866.88	74.773	307.4	223	18417.84	356.143	433.4	293	58284.40	1127.036	559.4	363	144992.80	2803.703	685.4
154	3970.24	76.772	309.2	224	18766.68	362.888	435.2	294	59135.60	1143.496	561.2	364	146733.20	2837.357	687.2
155	4075.88	78.815	311.0	225	19123.12	369.781	437.0	295	59994.40	1160.102	563.0	365	148519.20	2871.892	689.0
156	4183.80	80.901	312.8	226	19482.60	376.732	438.8	296	60860.80	1176.856	564.8	366	150320.40	2906.722	690.8
157	4293.24	83.018	314.6	227	19848.92	383.815	440.6	297	61742.40	1193.903	566.6	367	152129.20	2941.698	692.6
158	4404.96	85.178	316.4	228	20219.80	390.987	442.4	298	62624.00	1210.950	568.4	368	153960.80	2977.116	694.4
159	4519.72	87.397	318.2	229	20596.76	398.276	444.2	299	63528.40	1228.439	570.2	369	155815.20	3012.974	696.2
160	4636.00	89.646	320.0	230	20978.28	405.654	446.0	300	64432.80	1245.927	572.0	370	157692.40	3049.273	698.0
161	4755.32	91.953	321.8	231	21365.12	413.134	447.8	301	65352.40	1263.709	573.8	371	159584.80	3085.866	699.8
162	4876.92	94.304	323.6	232	21757.28	420.717	449.6	302	66279.60	1281.638	575.6	372	161507.60	3123.047	701.6
163	5000.04	96.685	325.4	233	22154.00	428.388	451.4	303	67214.40	1299.714	577.4	373	163468.40	3160.963	703.4
164	5126.96	99.139	327.2	234	22558.32	436.207	453.2	304	68156.80	1317.937	579.2	374	165467.20	3199.613	705.2
165	5256.16	101.638	329.0	235	22967.96	444.128	455.0	305	69114.40	1336.454	581.0				
166	5386.88	104.165	330.8	236	23382.92	452.152	456.8	306	70072.00	1354.971	582.8				
167	5521.40	106.766	332.6	237	23802.44	460.264	458.6	307	71052.40	1373.929	584.6				
168	5658.20	109.412	334.4	238	24229.56	468.523	460.4	308	72048.00	1393.181	586.4				
169	5798.04	112.116	336.2	239	24661.24	476.871	462.2	309	73028.40	1412.139	588.2				

VAPOR PRESSURE OF MERCURY

Vapor pressure of mercury in mm. of Hg for temperatures from −38 to 400°C. Note that the values for the first four lines only, are to be multiplied by 10^{-6}

Temp. °C	0	2	4	6	8
	10^{-6}	10^{-6}	10^{-6}	10^{-6}	10^{-6}
−30	4.78	3.59	2.66	1.97	1.45
−20	18.1	14.0	10.8	8.28	6.30
−10	60.6	48.1	38.0	29.8	23.2
− 0	185.	149.	119.	95.4	76.2
+ 0	.000185	.000228	.000276	.000335	.000406
+10	.000490	.000588	.000706	.000846	.001009
20	.001201	.001426	.001691	.002000	.002359
30	.002777	.003261	.003823	.004471	.005219
40	.006079	.007067	.008200	.009497	.01098
50	.01267	.01459	.01677	.01925	.02206
60	.02524	.02883	.03287	.03740	.04251
70	.04825	.05469	.06189	.06993	.07889
80	.08880	.1000	.1124	.1261	.1413
90	.1582	.1769	.1976	.2202	.2453
100	.2729	.3032	.3366	.3731	.4132
110	.4572	.5052	.5576	.6150	.6776
120	.7457	.8198	.9004	.9882	1.084
130	1.186	1.298	1.419	1.551	1.692
140	1.845	2.010	2.188	2.379	2.585
150	2.807	3.046	3.303	3.578	3.873
160	4.189	4.528	4.890	5.277	5.689
170	6.128	6.596	7.095	7.626	8.193
180	8.796	9.436	10.116	10.839	11.607
190	12.423	13.287	14.203	15.173	16.200

Temp. °C	0	2	4	6	8
200	17.287	18.437	19.652	20.936	22.292
210	23.723	25.233	26.826	28.504	30.271
220	32.133	34.092	36.153	38.318	40.595
230	42.989	45.503	48.141	50.909	53.812
240	56.855	60.044	63.384	66.882	70.543
250	74.375	78.381	82.568	86.944	91.518
260	96.296	101.28	106.48	111.91	117.57
270	123.47	129.62	136.02	142.69	149.64
280	156.87	164.39	172.21	180.34	188.79
290	197.57	206.70	216.17	226.00	236.21
300	246.80	257.78	269.17	280.98	293.21
310	305.89	319.02	332.62	346.70	361.26
320	376.33	391.92	408.04	424.71	441.94
330	459.74	478.13	497.12	516.74	537.00
340	557.90	579.45	601.69	624.64	648.30
350	672.69	697.83	723.73	750.43	777.92
360	806.23	835.38	865.36	896.23	928.02
370	960.66	994.34	1028.9	1064.4	1100.9
380	1138.4	1177.0	1216.6	1257.3	1299.1
390	1341.9	1386.1	1431.3	1477.7	1525.2
400	1574.1	.....	.....	.....	.....

VAPOR PRESSURES OF SOME OF THE ELEMENTS* AT VARIOUS TEMPERATURES °C

Element	Melting point (°C)	Pressure (mm Hg)					
		10^{-5}	10^{-4}	10^{-3}	10^{-2}	10^{-1}	1
Ag	961	767	848	936	1047	1184	1353
Al	660	724	808	889	996	1123	1279
Au	1063	1083	1190	1316	1465	1646	1867
Ba	717	418	476	546	629	730	858
Be	1284	942	1029	1130	1246	1395	1582
Bi	271	474	536	609	698	802	934
C	....	2129	2288	2471	2681	2926	3214
Cd	321	148	180	220	264	321	....
Co	1478	1249	1362	1494	1649	1833	2056
Cr	1900	907	992	1090	1205	1342	1504
Cu	1083	946	1035	1141	1273	1432	1628
Fe	1535	1094	1195	1310	1447	1602	1783
Hg	−38.9	-23.9	−5.5	18.0	48.0	82.0	126
In	157	667	746	840	952	1088	1260
Ir	2454	1993	2154	2340	2556	2811	3118
Mg	651	287	331	383	443	515	605
Mn	1244	717	791	878	980	1103	1251
Mo	2622	1923	2095	2295	2533	....	....
Ni	1455	1157	1257	1371	1510	1679	1884
Os	2697	2101	2264	2451	2667	2920	3221
Pb	328	483	548	625	718	832	975
Pd	1555	1156	1271	1405	1566	1759	2000
Pt	1774	1606	1744	1904	2090	2313	2582
Sb	630	466	525	595	678	779	904
Si	1410	1024	1116	1223	1343	1485	1670
Sn	232	823	922	1042	1189	1373	1609
Ta	2996	2407	2599	2820	....	....	....
W	3382	2554	2767	3016	3309	....	....
Zn	419	211	248	292	343	405	....
Zr	2127	1527	1660	1816	2001	2212	2459

Reprinted with permission from Saul Dushman, "Scientific Foundations of Vacuum Technique," 1949, John Wiley and Sons, Inc., New York.
* The values given in this table are from a variety of sources, not all of which are in agreement. The table is to be used, therefore, only as a general guide. For a detailed discussion of vapor pressure data, see Dr. Dushman's book, pages 752–754.

VAPOR PRESSURE OF CARBON DIOXIDE

Solid

From Bureau of Standards Journal of Research
(Mercury column, density = 13.5951 g/cm³, g = 980.665)

Pressure in microns of mercury

°C	0	1	2	3	4	5	6	7	8	9
−180	0.013	0.008	0.006	0.004	0.003	0.0017	0.0011	0.0007	0.0005	0.0003
−170	.37	.27	.20	.14	.10	.074	.052	.037	.026	.018
−160	5.9	4.6	3.6	2.7	2.1	1.58	1.19	.90	.67	.50
−150	60.5	48.8	39.2	31.4	25.1	19.9	15.8	12.4	9.8	7.6
−140	431	359	298	247	204	168	138	113	92	75

Pressure in mm of mercury

°C	0	1	2	3	4	5	6	7	8	9
−130	2.31	1.97	1.68	1.43	1.22	1.03	0.87	0.73	0.61	0.51
−120	9.81	8.57	7.46	6.49	5.63	4.88	4.22	3.64	3.13	2.69
−110	34.63	30.76	27.27	24.14	21.34	18.83	16.58	14.58	12.80	11.22
−100	104.81	94.40	84.91	76.27	68.43	61.30	54.84	48.99	43.71	38.94
− 90	279.5	254.7	231.8	210.8	191.4	173.6	157.3	142.4	128.7	116.2
− 80	672.2	618.3	568.2	521.7	478.5	438.6	401.6	367.4	335.7	306.5
− 70	1486.1	1377.3	1275.6	1180.5	1091.7	1008.9	931.7	859.7	792.7	730.3
− 60	3073.1	2865.1	2669.7	2486.3	2314.2	2152.8	2001.5	1859.7	1726.9	1602.5
− 50	......	......	......	......	......	......	......	3780.9	3530.2	3294.6

Liquid

°C	0	1	2	3	4	5	6	7	8	9
−50	5127.8	4922.7	4723.9	4531.1	4344.3	4163.2	3987.9	3818.2*	3653.9*	3495.0*
−40	7545	7271	7005	6746	6494	6250	6012	5781	5557	5339
−30	10718	10363	10017	9679	9350	9029	8716	8412	8115	7826
−20	14781	14331	13891	13461	13040	12630	12229	11838	11455	11082
−10	19872	19312	18764	18228	17703	17189	16686	16194	15712	15241
− 0	26142	25457	24786	24127	23482	22849	22229	21622	21026	20443
0	26142	26840	27552	28277	29017	29771	30539	31323	32121	32934
10	33763	34607	35467	36343	37236	38146	39073	40017	40980	41960
20	42959	43977	45014	46072	47150	48250	49370	50514	51680	52871
30	54086	55327	......	......	......	......	......	......	......	......

* Undercooled liquid.
Critical temperature = 31.0°C. Triple point, −56.602 ± 0.005°C; 3885.2 ± 0.4 mm.

VAPOR PRESSURE

The following table is an abridged form of the very extensive compilation by Daniel R. Stull and published in Industrial and Engineering Chemistry **39,** 517 (1947).

The table gives the temperatures in degrees centigrade at which the vapor of the compound listed at the left has the pressure indicated at the top of the column. Organic compounds are listed in the order of their empirical formulae and inorganic compounds in the alphabetic order of their names. Pressures greater than one atmosphere are listed in separate tables.

Abbreviations:
- d = decomposes
- *d* = dextrorotatory
- *dl* = inactive (50% *d* and 50% *l*)
- e = explodes
- *l* = levorotatory
- M.P. = melting point
- p = polymerizes
- s = solid

INORGANIC COMPOUNDS
Pressures Less than One Atmosphere

Name	Formula	Temperature, °C						M.P.
		1 mm	10 mm	40 mm	100 mm	400 mm	760 mm	
Aluminum	Al	1284	1487	1635	1749	1947	2056	660
Aluminum borohydride	AlB_3H_{12}	s	-42.9	-20.9	-3.9	+28.1	45.9	-64.5
Aluminum bromide	$AlBr_3$	81.3_s	118.0	150.6	176.1	227.0	256.3	97.5
Aluminum chloride	$AlCl_3$	100.0_s	123.8_s	139.9_s	152.0_s	171.6_s	180.2_s	192.4
Aluminum fluoride	AlF_3	1238	1324	1378	1422	1496	1537	1040
Aluminum iodide	AlI_3	178.0_s	225.8	265.0	294.5	354.0	385.5	......
Aluminum oxide	Al_2O_3	2148	2385	2549	2665	2874	2977	2050
Ammonia	NH_3	-109.1_s	-91.9_s	-79.2_s	-68.4	-45.4	-33.6	-77.7
Deutero ammonia	ND_3	s	s	s	-67.4	-45.4	-33.4	-74.0
Ammonium azide	NH_4N_3	29.2_s	59.2_s	80.1_s	95.2_s	120.4_s	133.8_s	......
Ammonium bromide	NH_4Br	198.3_s	252.0_s	290.0_s	320.0_s	370.9_s	396.0_s	......
Ammonium carbamate	$NH_4CO_2NH_2$	-26.1_s	-2.9_s	$+14.0_s$	26.7_s	48.0_s	58.3_s	......
Ammonium chloride	NH_4Cl	160.4_s	209.8_s	245.0_s	271.5_s	316.5_s	337.8_s	520
Ammonium hydrogen sulfide	NH_4HS	-51.1	-28.7	-12.3	0.0	+21.8	33.3	......
Ammonium iodide	NH_4I	210.9_s	263.5_s	302.8_s	331.8_s	381.0_s	404.9_s	......
Ammonium cyanide	NH_4CN	-50.6_s	-28.6_s	-12.6_s	-0.5_s	$+20.5_s$	31.7_s	36
Antimony	Sb	886	1033	1141	1223	1364	1440	630.5
Antimony tribromide	$SbBr_3$	93.9	142.7	177.4	203.5	250.2	275.0	96.6
Antimony trichloride	$SbCl_3$	49.2_s	85.2	117.8	143.3	192.2	219.0	73.4
Antimony pentachloride	$SbCl_5$	22.7	61.8	91.0	114.1	......	......	2.8
Antimony triiodide	SbI_3	163.6_s	223.5	267.8	303.5	368.5	401.0	167
Antimony trioxide	Sb_2O_3	574_s	666	812	957	1242	1425	656
Argon	A	-218.2_s	-210.9_s	-204.9_s	-200.5_s	-190.6_s	-185.6	-189.2
Arsenic (metallic)	As	372_s	437_s	483_s	518_s	579_s	610_s	814
Arsenic tribromide	$AsBr_3$	41.8	85.2	118.7	145.2	193.6	220.0	......
Arsenic trichloride	$AsCl_3$	-11.4	+23.5	50.0	70.9	109.7	130.4	-18
Arsenic trifluoride	AsF_3	s	s	-2.5	+13.2	41.5	56.3	-5.9
Arsenic pentafluoride	AsF_5	-117.9_s	-103.1_s	-92.4_s	-84.3_s	-64.0	-52.8	-79.8
Arsenic hydride (arsine)	AsH_3	-142.6_s	-124.7_s	-110.2	-98.0	-75.2	-62.1	-116.3
Arsenic trioxide	As_2O_3	212.5_s	259.7_s	299.2_s	332.5	412.2	457.2	312.8
Barium	Ba	s	1049	1195	1301	1518	1638	850
Beryllium borohydride	BeB_2H_8	$+1.0_s$	28.1_s	46.2_s	58.6_s	79.7_s	90.0_s	123
Beryllium bromide	$BeBr_2$	289_s	342_s	379_s	405_s	451_s	474_s	490
Beryllium chloride	$BeCl_2$	291_s	346_s	384_s	411	461	487	405
Beryllium iodide	BeI_2	283_s	341_s	382_s	411_s	461_s	487_s	488
Bismuth	Bi	1021	1136	1217	1271	1370	1420	271
Bismuth tribromide	$BiBr_3$	s	282	327	360	425	461	218
Bismuth trichloride	$BiCl_3$	s	264	311	343	405	441	230
Borine carbonyl	BH_3CO	-139.2	-121.1	-106.6	-95.3	-74.8	-64.0	-137.0
Boron tribromide	BBr_6	-41.4	-10.1	+14.0	33.5	70.0	91.7	-45
Boron trichloride	BCl_3	-91.5	-66.9	-47.8	-32.4	-3.6	+12.7	-107
Boron trifluoride	BF_3	-154.6	-141.3_s	-131.0_s	-123.0	-108.3	-110.7	-126.8
Dihydrodiborane	B_2H_6	-159.7	-144.3	-131.6	-120.9	-99.6	-86.5	-169
Diborane hydrobromide	B_2BrH_5	-93.3	-66.3	-45.4	-29.0	0.0	+16.3	-104.2
Triborine triamine	$B_3H_6N_3$	-63.0_s	-35.3	-13.2	+4.0	34.3	50.6	-58.2
Tetrahydrotetraborane	B_4H_{10}	-90.9	-64.3	-44.3	-28.1	+0.8	16.1	-119.9
Dihydropentaborane	B_5H_9	s	-30.7	-8.0	+9.6	40.8	58.1	-47.0
Tetrahydropentaborane	B_5H_{11}	-50.2	-19.9	+2.7	20.1	51.2	67.0	......
Dihydrodecaborane	$B_{10}H_{14}$	60.0_s	90.2_s	117.4	142.3	d	......	99.6
Bromine	Br_2	-48.7_s	-25.0_s	-8.0_s	+9.3	41.0	58.2	-7.3
Bromine pentafluoride	BrF	-69.3_s	-41.9	-21.0	-4.5	+25.7	40.0	-61.4
Cadmium	Cd	394	484	553	611	711	765	320.9

VAPOR PRESSURE (Continued)

INORGANIC COMPOUNDS (Continued)

Pressures Less than One Atmosphere (Continued)

Name	Formula	Temperature, °C 1 mm	10 mm	40 mm	100 mm	400 mm	760 mm	M.P.
Cadmium chloride	$CdCl_2$	s	656	736	797	908	967	568
Cadmium fluoride	CdF_2	1112	1286	1400	1486	1651	1751	520
Cadmium iodide	CdI_2	416	512	584	640	742	796	385
Cadmium oxide	CdO	1000_s	1149_s	1257_s	1341_s	1484_s	1559_s	......
Calcium	Ca	s	983	1111	1207	1388	1487	851
Carbon	C	3586_s	3946_s	4196_s	4373_s	4660_s	4827_s	.....
Carbon tetrabromide	CBr_4	s	s	96.3	119.7	163.5	189.5	90.1
Carbon tetrachloride	CCl_4	-50.0_s	−19.6	+4.3	23.0	57.8	76.7	−22.6
Carbon tetrafluoride	CF_4	-184.6_s	−169.3	−158.8	−150.7	−135.5	−127.7	−183.7
Carbon dioxide	CO_2	-134.3_s	-119.5_s	-108.6_s	-100.2_s	-85.7_s	-78.2_s	−57.5
Carbon suboxide	C_3O_2	−94.8	−71.0	−52.0	−36.9	−8.9	+6.3	−107
Carbon disulfide	CS_2	−73.8	−44.7	−22.5	−5.1	+28.0	46.5	−110.8
Carbon subsulfide	C_3S_2	14.0	54.9	85.6	109.9	p	....	+0.4
Carbon selenosulfide	$CSSe$	−47.3	−16.0	+8.6	28.3	65.2	85.6	−75.2
Carbon monoxide	CO	-222.0_s	-215.0_s	-210.0_s	-205.7_s	−196.3	−191.3	−205.0
Carbonyl chloride	$COCl_2$	−92.9	−69.3	−50.3	−35.6	−7.6	+8.3	−104
Carbonyl selenide	$COSe$	−117.1	−95.0	−76.4	−61.7	−35.6	−21.9	....
Carbonyl sulfide	COS	−132.4	−113.3	−98.3	−85.9	−62.7	−49.9	−138.8
Chloropicrin	CCl_3NO_2	−25.5	+7.8	33.8	53.8	91.8	111.9	−64
Chlorotrifluoromethane	$CClF_3$	−149.5	−134.1	−121.9	−111.7	−92.7	−81.2	....
Cyanogen	C_2N_2	-95.8_s	-76.8_s	-62.7_s	-51.8_s	−33.0	−21.0	−34.4
Cyanogen bromide	$CBrN$	-35.7_s	-10.0_s	$+8.6_s$	22.6_s	46.0_s	61.5	58
Cyanogen chloride	$CClN$	-76.7_s	-53.8_s	-37.5_s	-24.9_s	−2.3	+13.1	−6.5
Cyanogen fluoride	CFN	-134.4_s	-118.5_s	-106.4_s	-97.0_s	-80.5_s	-72.6_s	....
Cyanogen iodide	CIN	25.2_s	57.7_s	80.3_s	97.6_s	126.1_s	141.1_s	....
Deuterocyanic acid	CDN	-68.9_s	-46.7_s	-30.1_s	-17.5_s	+10.0	26.2	−12
Dichlorodifluoromethane	CCl_2F_2	−118.5	−97.8	−81.6	−68.6	−43.9	−29.8	....
Dichlorofluoromethane	$CHCl_2F$	−91.3	−67.5	−48.8	−33.9	−6.2	+8.9	−135
Chlorodifluoromethane	$CHClF_2$	−122.8	−103.7	−88.6	−76.4	−53.6	−40.8	−160
Trichlorofluoromethane	CCl_3F	−84.3	−59.0	−39.0	−23.0	+6.8	23.7	....
Cesium	Cs	279	375	449	509	624	690	28.5
Cesium bromide	$CsBr$	748	887	993	1072	1221	1300	636
Cesium chloride	$CsCl$	744	884	989	1069	1217	1300	646
Cesium fluoride	CsF	712	844	947	1025	1170	1251	683
Cesium iodide	CsI	738	873	976	1055	1200	1280	621
Chlorine	Cl_2	-118.0_s	-101.6_s	−84.5	−71.7	−47.3	−33.8	−100.7
Chlorine fluoride	ClF	s	−139.0	−128.8	−120.8	−107.0	−100.5	−145
Chlorine trifluoride	ClF_3	s	−71.8	−51.3	−34.7	−4.9	+11.5	−83
Chlorine monoxide	Cl_2O	−98.5	−73.1	−54.3	−39.4	−12.5	+2.2	−116
Chlorine dioxide	ClO_2	s	−59.0	−42.8	−29.4	−4.0	+11.1	−59
Dichlorine hexoxide	Cl_2O_6	+7.5	42.0	68.0	87.7	123.8	142.0	3.5
Chlorine heptoxide	Cl_2O_7	−45.3	−13.2	+10.3	29.1	62.2	78.8	−91
Chlorosulfonic acid	HSO_3Cl	32.0	64.0	87.6	105.3	136.1	151.0_d	−80
Chromium	Cr	1616	1845	2013	2139	2361	2482	1615
Chromium carbonyl	$Cr(CO)_6$	36.0	68.3	91.2	108.0	137.2	151.0	....
Chromyl chloride	CrO_2Cl_2	−18.4	+13.8	38.5	58.0	95.2	117.1	....
Cobaltous chloride	$COCl_2$	s	s	770	843	974	1050	735
Cobalt nitrosyl tricarbonyl	$Co(CO)_2NO$	s	s	+11.0	29.0	62.0	80.0	−11
Columbium pentafluoride	CbF_5	s	86.3	121.5	148.5	198.0	225.0	75.5
Copper	Cu	1628	1879	2067	2207	2465	2595	1083
Cuprous bromide	Cu_2Br_2	572	718	844	951	1189	1355	504
Cuprous chloride	Cu_2Cl_2	546	702	838	960	1249	1490	422
Cuprous iodide	Cu_2I_2	s	656	786	907	1158	1336	605
Ferric chloride	$FeCl_2$	194.0_s	235.5_s	256.8_s	272.5_s	298.0_s	319.0	304
Ferrous chloride	$FeCl_2$	....	700	779	842	961	1026	....
Fluorine	F_2	−223.0	−214.1	−207.7	−202.7	−193.2	−187.9	−223
Fluorine monoxide	F_2O	−196.1	−182.3	−173.0	−165.8	−151.9	−144.6	−223.9
Gallium	Ga	1349	1541	1680	1784	1974	2071	30
Gallium trichloride	$GaCl_3$	48.0_s	76.5_s	107.5	132.0	176.3	200.0	77.0
Germanium hydride	GeH_4	−163.0	−145.3	−131.6	−120.3	−100.2	−88.9	−165
Germanium bromide	$GeBr_4$	s	56.8	88.1	113.2	161.6	189.0	26.1
Germanium chloride	$GeCl_4$	−45.0	−15.0	+8.0	27.5	63.8	84.0	−49.5
Trichlorogermane	$GeHCl_3$	−41.3	−13.0	+8.8	26.5	58.3_d	75.0_d	−71.1
Tetramethylgermanium	$Ge(CH_3)_4$	−73.2	−45.2	−23.4	−6.3	+26.0	44.0	−88
Digermane	Ge_2H_6	−88.7	−60.1	−38.2	−20.3	+13.3	31.5	−109
Trigermane	Ge_3H_8	−36.9	−0.9	+26.3	47.9	88.6	110.8	−105.6

VAPOR PRESSURE (Continued)

INORGANIC COMPOUNDS (Continued)
Pressures Less than One Atmosphere (Continued)

Name	Formula	Temperature, °C						M.P.
		1 mm	10 mm	40 mm	100 mm	400 mm	760 mm	
Gold	Au	1869	2154	2363	2521	2807	2966	1063
Helium	He	−271.7	−271.3	−270 7	−270.3	−269.3	−268.6	......
Hydrogen	H_2	-263.3_s	-261.3_s	-259.6_s	−257.9	−254.5	−252.5	−259.1
Hydrogen deuteride	HD	....	−259.8	−258 2	−256.6	−253.0	−251.0	......
Hydrogen bromide	HBr	-138.8_s	-121.8_s	-108.3_s	-97.7_s	− 78.0	− 66.5	− 87.0
Hydrogen chloride	HCl	-150.8_s	-135.6_s	-123.8_s	−114.0	− 95.3	− 84.8	−114.3
Hydrogen cyanide	HCN	-71.0_s	-47.7_s	-30.9_s	-17.8_s	+ 10.2	25.9	− 13.2
Hydrogen fluoride	HF	s	− 65.8	− 45.0	− 28.2	+ 2.5	19.7	− 83.7
Hydrogen iodide	HI	-123.3_s	-102.3_s	$-85\ 6_s$	-72.1_s	− 48.3	− 35.1	− 50.9
Hydrogen peroxide	H_2O_2	15.3	50.4	77.0	97.9	137.4_d	158.0_d	− 0.9
Hydrogen selenide	H_2Se	-115.3_s	-97.9_s	-84.7_s	-74.2_s	− 53.6	− 41.1	− 64
Hydrogen sulfide	H_2S	-134.3_s	-116.3_s	-102.3_s	-91.6_s	− 71.8	− 60.4	− 85.5
Hydrogen disulfide	H_2S_2	− 43.2	− 15.2	+ 6.0	22.0	49.6	64.0	− 89.7
Hydrogen teluride	H_2Te	-96.4_s	-75.4_s	-59.1_s	− 45.7	− 17.2	− 2.0	− 49.0
Hydroxylamine	NH_2OH	s	47.2	64.6	77.5	99.2	110.0	34.0
Iodine	I_2	$38\ 7_s$	73.2_s	97.5_s	116.5	159.8	183.0	112.9
Iodine pentafluoride	IF_5	− 15.2	+ 8.5	32.2	50.0	81.2	97.0	8.0
Iodine heptafluoride	IF_7	-87.0_s	-63.0_s	-45.3_s	-31.9_s	$-8\ 3_s$	$+4.0_s$	5.5
Iron	Fe	1787	2039	2224	2360	2605	2735	1535
Iron pentacarbonyl	$Fe(CO)_5$	....	+ 4.6	30.3	50.3	86.1	105.0	− 21
Krypton	Kr	-199.3_s	-187.2_s	-178.4_s	-171.8_s	-159.0_s	−152.0	−156.7
Lead	Pb	973	1162	1309	1421	1630	1744	327.5
Lead bromide	$PbBr_2$	513	610	686	745	856	914	373
Lead chloride	$PbCl_2$	547	648	725	784	893	954	501
Lead fluoride	PbF_2	s	904	1003	1080	1219	1293	855
Lead iodide	PbI	479	571	644	701	807	872	402
Lead oxide	PbO	943	1085	1189	1265	1402	1472	890
Lead sulfide	PbS	852_s	975_s	1048_s	1108_s	1221	1281	1114
Lithium	Li	723	881	1003	1097	1273	1372	186
Lithium bromide	LiBr	748	888	994	1076	1226	1310	547
Lithium chloride	LiCl	783	932	1045	1129	1290	1382	614
Lithium fluoride	LiF	1047	1211	1333	1425	1591	1681	870
Lithium iodide	LiI	723	841	927	993	1110	1171	446
Magnesium	Mg	621_s	743	838	909	1034	1107	651
Magnesium chloride	$MgCl_2$	778	930	1050	1142	1316	1418	712
Manganese	Mn	1292	1505	1666	1792	2029	2151	1260
Manganous chloride	$MnCl_2$	s	778	879	960	1108	1190	650
Mercury	Hg	126.2	184.0	228.8	261.7	323.0	357.0	− 38.9
Mercuric bromide	$HgBr_2$	136.5_s	179.8_s	211.5_s	237.8	290.0	319.0	237
Mercuric chloride	$HgCl_2$	136.2_s	180.2_s	212.5_s	237.0_s	275.5_s	304.0	277
Mercuric iodide	HgI_2	157.5_s	204.5_s	238.2_s	261.8	324.2	354.0	259
Molybdenum	Mo	3102	3535	3859	4109	4553	5560	2622
Molybdenum hexafluoride	MoF_6	-65.5_s	-40.8_s	-22.1_s	-8.0_s	+ 17.2	36.0	17
Molybdenum trioxide	MoO_3	734_s	814	892	955	1082	1151	795
Neon	Ne	-257.3_s	-254.6_s	-252.6_s	-251.0_s	−248.1	−246.0	−248.7
Nickel	Ni	1810	2057	2234	2364	2603	2732	1452
Nickel chloride	$NiCl_2$	671_s	759_s	821_s	866_s	945_s	987_s	1001

INORGANIC COMPOUNDS (Continued)
Pressures Less than One Atmosphere (Continued)

Name	Formula	Temperature, °C						M.P.
		1 mm	10 mm	40 mm	100 mm	400 mm	760 mm	
Nickel carbonyl	$Ni(CO)_4$	s	s	−23.0	−6.0	+25.8	42.5	−25
Nitrogen	N_2	-226.1_s	-219.1_s	-214.0_s	−209.7	−200.9	−195.8	−210.0
Nitrogen trifluoride	NF_3	s	−170.7	−160.2	−152.3	−137.4	−129.0	−183.7
Nitric oxide	NO	-184.5_s	-178.2_s	-171.7_s	-166.0_s	-156.8_s	−151.7	−161
Nitrous oxide	N_2O	-143.4_s	-128.7_s	-118.3_s	-110.3_s	-96.2_s	−88.5	−90.9
Nitrogen tetroxide	N_2O_4	-55.6_s	-36.7_s	-23.9_s	-14.7_s	+8.0	21.0	−9.3
Nitrogen pentoxide	N_2O_5	-36.8_s	-16.7_s	-2.9_s	$+7.4_s$	24.4_s	32.4	30
Nitrosyl chloride	$NOCl$	s	s	−60.2	−46.3	−20.3	−6.4	−64.5
Nitrosyl fluoride	NOF	−132.0	−114.3	−100.3	−88.8	−68.2	−56.0	−134
Nitroxyl fluoride	NO_2F	-143.7_s	−126.2	−112.8	−102.3	−83.2	−72.0	−139
Osmium tetroxide (white)	OsO_4	-5.6_s	$+26.0_s$	50.5	71.5	109.3	130.0	42
Osmium tetroxide (yellow)	OsO_4	3.2_s	31.3_s	51.7_s	71.5	109.3	130.0	56
Oxygen	O_2	-219.1_s	−210.6	−204.1	−198.8	−188.8	−182.96	−218.4
Ozone	O_3	−180.4	−163.2	−150.7	−141.0	−122.5	−111.9	−192.1 ±.1
Phosphorus (yellow)	P	76.6	128.0	166.7	197.3	251.0	280.0	44.1
Phosphorus (violet)	P	237_s	287_s	323_s	349_s	391_s	417_s	590
Phosphorus (black)	P	290_s	338_s	371_s	393_s	432_s	453_s	......
Phosphorus tribromide	PBr_3	7.8	47.8	79.0	103.6	149.7	175.3	−40
Phosphorus trichloride	PCl_3	−51.6	−21.3	+2.3	21.0	56.9	74.2	−111.8
Phosphorus pentachloride	PCl_5	55.5_s	83.2_s	102.5_s	117.0_s	147.2_s	162.0_s	......
Phosphorus hydride (Phosphene)	PH_3	s	s	−129.4	−118.8	−98.3	−87.5	−132.5
Phosphonium bromide	PH_4Br	-43.7_s	-21.2_s	-5.0_s	$+7.4_s$	28.0_s	38.3_d	......
Phosphonium chloride	PH_4Cl	-91.0_s	-74.0_s	-61.5_s	-52.0_s	-35.4_s	-27.0_s	−28.5
Phosphonium iodide	PH_4I	-25.2_s	-1.1_s	16.1_s	29.3_s	51.6_s	62.3_s	......
Phosphorus trioxide	P_2O_3	s	53.0	84.0	108.3	150.3	173.1	22.5
Phosphorus oxychloride	$POCl_3$	s	2.0	27.3	47.4	84.3	105.1	2
Phosphorus pentoxide (stable form)	P_2O_5	384_s	442_s	481_s	510_s	556_s	591	569
Phosphorus pentoxide (Metastable) form)	P_2O_5	189	236	270	294	336	358	......
Phosphorus thiobromide	$PSBr_3$	50.0	83.6	108.0	126.3	157.8	175.0_d	38
Phosphorus thiochloride	$PSCl_3$	−18.3	+16.1	42.7	63.8	102.3	124.0	−36.2
Platinum	Pt	2730	3146	3469	3714	4169	4407	1755
Potassium	K	341	443	524	586	708	774	62.3
Potassium bromide	KBr	795	982	1050	1137	1297	1383	730
Potassium chloride	KCl	821	968	1078	1164	1322	1407	790
Potassium fluoride	KF	885	1039	1156	1245	1411	1502	880
Potassium hydroxide	KOH	719	863	976	1064	1233	1327	380
Potassium iodide	KI	745	887	995	1080	1238	1324	723
Radon	Rn	-144.2_s	-126.3_s	-111.3_s	-99.0_s	-75.0_s	−61.8	−71
Rhenium heptoxide	Re_2O_7	212.5_s	248.0_s	272.0_s	289.0_s	336.0	362.4	296
Rubidium	Rb	297	389	459	514	620	679	38.5
Rubidium bromide	RbBr	781	923	1031	1114	1267	1352	682
Rubidium chloride	RbCl	792	937	1047	1133	1294	1381	715
Rubidium fluoride	RbF	921	1016	1096	1168	1322	1408	760
Rubidium iodide	RbI	748	884	991	1072	1223	1304	642
Selenium	Se	356	442	506	554	637	680	217
Selenium dioxide	SeO_2	157.0_s	202.5_s	234.1_s	258.0_s	297.7_s	317.0_s	340
Selenium hexafluoride	SeF_6	-118.6_s	-98.9_s	-84.7_s	-73.9_s	-55.2_s	-45.8_s	−34.7
Selenium oxychloride	$SeOCl_2$	34.8	71.9	98.0	118.0	151.7	168.0	8.5
Selenium tetrachloride	$SeCl_4$	74.0_s	107.4_s	130.1_s	147.5_s	176.4_s	191.5_d	......
Silane	SiH_4	−179.3	−163.0	−150.3	−140.5	−122.0	−111.5	−185
Silicon	Si	1724	1888	2000	2083	2220	2287	1420
Silicon dioxide	SiO_2	s	1732	1867	1969	2141	2227	1710
Silicon tetrachloride	$SiCl_4$	−63.4	−34.4	−12.1	+5.4	38.4	56.8	−68.8
Silicon tetrafluoride	SiF_4	-144.0_s	-130.4_s	-120.8_s	-113.3_s	-100.7_s	-94.8_s	−90
Bromosilane	SiH_3Br	s	−77.3	−57.8	−42.3	−13.3	+2.4	−93.9
Chlorosilane	SiH_3Cl	−117.8	−97.7	−81.8	−68.5	−44.5	−30.4	......
Fluorosilane	SiH_3F	−153.0	−141.2	−130.8	−122.4	−106.8	−98.0	......
Iodosilane	SiH_3I	s	−43.7	−21.8	−4.4	+27.9	45.4	−57.0

INORGANIC COMPOUNDS (Continued)
Pressures Less than One Atmosphere (Continued)

Name	Formula	Temperature, °C						M.P.
		1 mm	10 mm	40 mm	100 mm	400 mm	760 mm	
Bromodichlorofluorosilane	$SiBrCl_2F$	−86.5	−59.0	−37.0	−19.5	+15.4	35.4	−112.3
Bromotrifluorosilane	$SiBrF_3$	s	s	s	s	−55.9	−41.7	−70.5
Chlorotrifluorosilane	$SiClF_3$	−144.0$_s$	−127.0	−112.8	−101.7	−81.0	−70.0	−142
Dibromochlorofluorosilane	$SiBr_2ClF$	−65.2	−35.6	−12.0	+6.3	43.0	59.5	−99.3
Dibromodifluorosilane	$SiBr_2F_2$	s	−66.8	−47.4	−31.9	−2.6	+13.7	−66.9
Dibromosilane	SiH_2Br_2	−60.9	−29.4	−5.2	+14.1	50.7	70.5	−70.2
Dichlorodifluorosilane	$SiCl_2F_2$	−124.7	−102.9	−85.0	−70.3	−45.0	−31.8	−139.7
Difluorosilane	SiH_2F_2	−146.7	−130.4	−117.6	−107.3	−87.6	−77.8	......
Diiodosilane	SiH_2I_2	s	18.0	52.6	79.4	125.5	149.5	−1.0
Disilane	Si_2H_6	−114.8	−91.4	−72.8	−57.5	−29.0	−14.3	−132.6
Disiloxane	$(SiH_3)_2O$	−112.5	−88.2	−70.4	−55.9	−29.3	−15.4	−144.2
Fluorotrichlorosilane	$SiCl_3F$	−92.6	−68.3	−48.8	−33.2	−4.0	+12.2	−120.8
Hexachlorodisilane	Si_2Cl_6	+4.0	38.8	65.3	85.4	120.6	139.0	−1.2
Hexachlorodisiloxane	$(SiCl_3)_2O$	−5.0	+29.4	55.2	75.4	113.6	135.6	−33.2
Hexafluorodisilane	Si_2F_6	−81.0$_s$	−63.1$_s$	−50.6$_s$	−41.7$_s$	−26.4$_s$	−18.9$_s$	−18.6
Octachlorotrisilane	Si_3Cl_8	46.3	89.3	121.5	146.0	189.5	211.4	......
Tetrasilane	Si_4H_{10}	−27.7	+4.3	28.4	47.4	81.7	100.0	−93.6
Tribromofluorosilane	$SiBr_3F$	−46.1	−15.1	+9.2	28.6	64.6	83.8	−82.5
Tribromosilane	$SiHBr_3$	−30.5	+3.4	30.0	51.6	90.2	111.8	−73.5
Trichlorosilane	$SiHCl_3$	−80.7	−53.4	−32.9	−16.4	+14.5	31.8	−126.6
Trifluorosilane	$SiHF_3$	−152.0$_s$	−138.2$_s$	−127.3	−118.7	−102.8	−95.0	−131.4
Trisilane	Si_3H_8	−68.9	−40.0	−16.9	+1.6	35.5	53.1	−117.2
Disilazane	$(SiH_3)_3N$	−68.7	−40.4	−18.5	−1.1	+31.0	48.7	−105.7
Silver	Ag	1357	1575	1743	1865	2090	2212	960.5
Silver chloride	AgCl	912	1074	1200	1297	1467	1564	455
Silver iodide	AgI	820	983	1111	1210	1400	1506	552
Sodium	Na	439	549	633	701	823	892	97.5
Sodium bromide	NaBr	806	952	1063	1148	1304	1392	755
Sodium chloride	NaCl	865	1017	1131	1220	1379	1465	800
Sodium cyanide	NaCN	817	983	1115	1214	1401	1497	564
Sodium fluoride	NaF	1077	1240	1363	1455	1617	1704	992
Sodium hydroxide	NaOH	739	897	1017	1111	1286	1378	318
Sodium iodide	NaI	767	903	1005	1083	1225	1304	651
Stannic bromide	$SnBr_4$	s	72.7	105.5	131.0	177.7	204.7	31.0
Stannic chloride	$SnCl_4$	−22.7	+10.0	35.2	54.7	92.1	113.0	−30.2
Stannic hydride	SnH_4	−140.0	−118.5	−102.3	−89.2	−65.2	−52.3	−149.9
Stannic iodide	SnI_4	s	175.8	218.8	254.2	315.5	348.0	144.5
Stannous chloride	$SnCl_2$	316	391	450	493	577	623	246.8
Strontium	Sr	s	898	1018	1111	1285	1384	800
Strontium oxide	SrO	2068$_s$	2262$_s$	2410$_s$	....	....	....	2430
Sulfur	S	183.8	243.8	288.3	327.2	399.6	444.6	112.8
Sulfur hexafluoride	SF_6	−132.7$_s$	−114.7$_s$	−101.5$_s$	−90.9$_s$	−72.6$_s$	−63.5$_s$	−50.2
Sulfur dioxide	SO_2	−95.5$_s$	−76.8$_s$	−60.5	−46.9	−23.0	−10.0	−73.2
Sulfur monochloride	S_2Cl_2	−7.4	+27.5	54.1	75.3	115.4	138.0	−80
Sulfuryl chloride	SO_2Cl_2	s	−24.8	−1.0	+17.8	51.3	69.2	−54.1
Sulfur trioxide (α)	SO_3	−39.0$_s$	−16.5$_s$	−1.0$_s$	+10.5$_s$	32.6	44.8	16.8
Sulfur trioxide (β)	SO_3	−34.0$_s$	−12.3$_s$	+3.2$_s$	14.3$_s$	32.6	44.8	32.3
Sulfur trioxide (γ)	SO_3	−15.3$_s$	+4.3$_s$	17.9$_s$	28.0$_s$	44.0$_s$	51.6$_s$	62.1
Sulfuric acid	H_2SO_4	145.8	194.2	229.7	257.0	305.0	330.0$_d$	10.5
Thionyl bromide	$SOBr_2$	−6.7	+31.0	58.8	80.6	119.2	139.5	−52.2
Thionyl chloride	$SOCl_2$	−52.9	−21.9	+2.2	21.4	56.5	75.4	−104.5
Tantalum pentafluoride	TaF_5	s	s	s	130.0	194.0	230.0	96.8
Tellurium	Te	520	650	753	838	997	1087	452
Tellurium tetrachloride	$TeCl_4$	s	233	273	304	360	392	224
Tellurium hexafluoride	TeF_6	−111.3$_s$	−92.4$_s$	−78.4$_s$	−67.9$_s$	−48.2$_s$	−38.6$_s$	−37.8
Thallium	Tl	825	983	1103	1196	1364	1457	303.5
Thallium bromide	TlBr	s	522	598	653	759	819	460
Thallium chloride	TlCl	s	517	589	645	748	807	430
Thallium iodide	TlI	440	531	607	663	763	823	440
Tin	Sn	1492	1703	1855	1968	2169	2270	231.9
Titanium tetrachloride	$TiCl_4$	−13.9	+21.3	48.4	71.0	112.7	136.0	−30
Tungsten	W	3990	4507	4886	5168	5666	5927	3370
Tungsten hexafluoride	WF_6	−71.4$_s$	−49.2$_s$	−33.0$_s$	−20.3$_s$	+1.2	17.3	−0.5
Uranium hexafluoride	UF_6	−38.8$_s$	−13.8$_s$	+4.4$_s$	18.2$_s$	42.7$_s$	55.7$_s$	69.2
Vanadyl trichloride	$VOCl_3$	−23.2	+12.2	40.0	62.5	103.5	127.2	......

INORGANIC COMPOUNDS (Continued)
Pressures Less than One Atmosphere (Continued)

Name	Formula	Temperature, °C						M.P.
		1 mm	10 mm	40 mm	100 mm	400 mm	760 mm	
Water	H_2O	− 17.3$_s$	+ 11.3	34.1	51.6	83.0	100.0	0.0
Xenon	Xe	−168.5$_s$	−152.8$_s$	−141.2$_s$	−132.8$_s$	−117.1$_s$	−108.0	−111.6
Zinc	Zn	487	593	673	736	844	907	419.4
Zinc chloride	$ZnCl_2$	428	508	566	610	689	732	365
Zinc fluoride	ZnF_2	970	1086	1175	1254	1417	1497	872
Zirconium tetrabromide	$ZrBr_4$	207$_s$	250$_s$	281$_s$	301$_s$	337$_s$	357$_s$	450
Zirconium tetrachloride	$ZrCl_4$	190$_s$	230$_s$	259$_s$	279$_s$	312$_s$	331$_s$	437
Zirconium tetraiodide	ZrI_4	264$_s$	311$_s$	344$_s$	369$_s$	409$_s$	431$_s$	499

Pressures Greater than One Atmosphere

Name	Formula	Temperature, °C						
		1 atm.	2 atm.	5 atm.	10 atm.	20 atm.	40 atm.	60 atm.
Ammonia	NH_3	− 33.6	− 18.7	+ 4.7	25.7	50.1	78.9	98.3
Argon	A	−185.6	−179.0	−166.7	−154.9	−141.3	−124.9	......
Boron trichloride	BCl_3	12.7	33.2	66.0	96.7	135.4	......	......
Boron trifluoride	BF_3	−100.7	− 89.4	− 72.6	− 57.7	− 40.0	− 19.0	......
Bromine	Br_2	58.2	78.8	110.3	139.8	174.0	216.0	243.5
Carbontetrachloride	CCl_4	76.7	102.0	141.7	178.0	222.0	276.0	......
Carbon dioxide	CO_2	− 78.2$_s$	− 69.1$_s$	− 56.7	− 39.5	− 18.9	+ 5.9	22.4
Carbon disulfide	CS_2	46.5	69.1	104.8	136.3	175.5	222.8	256.0
Carbon monoxide	CO	−191.3	−183.5	−170.7	−161.0	−149.7	......	......
Carbonyl chloride	$COCl_2$	+ 8.3	27.3	57.2	85.0	119.0	159.8	......
Chlorotrifluoromethane	$CClF_3$	− 81.2	− 66.7	− 42.7	− 18.5	+ 12.0	52.8	......
Cyanogen	C_2N_2	− 21.0	− 4.4	+ 21.4	44.6	72.6	106.5	......
Dichlorodifluoromethane	CCl_2F_2	− 29.8	− 12.2	+ 16.1	42.4	74.0	......	......
Dichlorofluoromethane	$CHCl_2F$	8.9	28.4	59.0	87.0	121.2	162.6	......
Chlorodifluoromethane	$CHClF_2$	− 40.8	− 24.7	+ 0.3	24.0	52.0	85.3	......
Trichlorofluoromethane	CCl_3F	23.7	44.1	77.3	108.2	146.7	194.0	......
Chlorine	Cl_2	− 33.8	− 16.9	+ 10.3	35.6	65.0	101.6	127.1
Helium	He	−268.6	−268.0	......	......	......	......	......
Hydrogen	H_2	−252.5	−250.2	−246.0	−241.8	......	......	......
Hydrogen bromide	HBr	− 66.5	− 51.5	− 29.1	− 8.4	+ 16.8	48.1	70.6
Hydrogen chloride	HCl	− 84.8	− 71.4	− 50.5	− 31.7	− 8.8	+ 17.8	36.2
Hydrogen cyanide	HCN	25.9	45.8	75.8	102.7	135.0	169.9	......
Hydrogen iodide	HI	− 35.1	− 18.9	+ 7.3	32.0	62.2	100.7	127.5
Hydrogen sulfide	H_2S	− 60.4	− 45.9	− 22.3	− 0.4	+ 25.5	55.8	76.3
Hydrogen selenide	H_2Se	− 41.1	− 25.2	0.0	+ 23.4	50.8	84.6	108.7
Krypton	Kr	−152.0	−143.5	−130.0	−118.0	−101.7	− 78.4	......
Neon	Ne	−246.0	−243.8	−239.9	−236.0	−230.8	......	......
Nitrogen	N_2	−195.8	−189.2	−179.1	−169.8	−157.6	......	......
Nitric oxide	NO	−151.7	−145.1	−135.7	−127.3	−116.8	−103.2	− 94.8
Nitrous oxide	N_2O	− 88.5	− 76.8	− 58.0	− 40.7	− 18.8	+ 8.0	27.4
Nitrogen tetroxide	N_2O_4	21.0	37.3	59.8	79.4	100.3	121.4	132.2
Oxygen	O_2	−183.1	−176.0	−164.5	−153.2	−140.0	−124.1	......
Silicon tetrafluoride	SiF_4	− 94.8$_s$	− 84.4	− 67.9	− 52.6	− 33.4	......	......
Chlorotrifluorosilane	$SiClF_3$	− 70.0	− 57.3	− 37.2	− 18.6	+ 4.1	......	......
Dichlorodifluorosilane	$SiCl_2F_2$	− 31.8	− 15.1	+ 11.6	36.6	66.2	......	......
Fluorotrichlorosilane	$SiCl_3F$	12.2	32.4	64.6	94.2	131.8	......	......
Stannic chloride	$SnCl_4$	113.0	141.3	184.3	223.0	270.0	......	......
Sulfur dioxide	SO_2	− 10.0	+ 6.3	32.1	55.5	83.8	118.0	141.7
Sulfur trioxide	SO_3	44.8	60.0	82.5	104.0	138.0	175.0	198.0
Water	H_2O	100.0	120.1	152.4	180.5	213.1	251.1	276.5

Name	Formula	Temperature °C						M.P.
		1 mm	10 mm	40 mm	100 mm	400 mm	760 mm	
Cyanogen bromide	$CBrN$	-35.7$_s$	-10.0$_s$	+8.6$_s$	22.6$_s$	46.0$_s$	61.5	58
Carbon tetrabromide	CBr_4	s	s	96.3	119.7	163.5	189.5	90.1
Chlorotrifluoromethane	$CClF_3$	-149.5	-134.1	-121.9	-111.7	-92.7	-81.2	......
Cyanogen chloride	$CClN$	-76.7$_s$	-53.8$_s$	-37.5$_s$	-24.9$_s$	-2.3	+13.1	-6.5
Dichlorodifluoromethane	CCl_2F_2	-118.5	-97.8	-81.6	-68.6	-43.9	-29.8	......
Carbonyl chloride	CCl_2O	-92.9	-69.3	-50.3	-35.6	-7.6	+8.3	-104
Trichlorofluoromethane	CCl_3F	-84.3	-59.0	-39.0	-23.0	+6.8	23.7	......
Trichloronitromethane	CCl_3NO_2	-25.5	+7.8	33.8	53.8	91.8	111.9	-64
Carbontetrachloride	CCl_4	-50.0$_s$	-19.6	+4.3	23.0	57.8	76.7	-22.6
Cyanogen fluoride	CFN	-134.4$_s$	-118.5$_s$	-106.4$_s$	-97.0$_s$	-80.5$_s$	-72.6$_s$	......
Carbontetrafluoride	CF_4	-184.6$_s$	-169.3	-158.8	-150.7	-135.5	-127.7	......
Tribromomethane	$CHBr_3$	s	34.0	63.6	85.9	127.9	150.5	8.5
Chlorodifluoromethane	$CHClF_2$	-122.8	-103.7	-88.6	-76.4	-53.6	-40.8	-160
Dichlorofluoromethane	$CHCl_2F$	-91.3	-67.5	-48.8	-33.9	-6.2	+8.9	-135
Trichloromethane	$CHCl_3$	-58.0	-29.7	-7.1	+10.4	42.7	61.3	-63.5
Hydrocyanic acid	CHN	-70.8$_s$	-48.2$_s$	-31.3$_s$	-18.8$_s$	+9.8	25.8	-14
Dibromomethane	CH_2Br_2	-35.1	-2.4	+23.3	42.3	79.0	98.6	-52.8
Dichloromethane	CH_2Cl_2	-70.0	-43.3	-22.3	-6.3	24.1	40.7	-96.7
Formaldehyde	CH_2O	s	-88.0	-70.6	-57.3	-33.0	-19.5	-92
Formic acid	CH_2O_2	-20.0$_s$	+2.1$_s$	24.0	43.8	80.3	100.6	8.2
Dichloromethylarsine	CH_3ASCl_2	-11.1	+24.3	51.5	73.0	112.7	134.5	-59
Borine carbonyl	CH_3BO	-139.2	-121.1	-106.6	-95.3	-74.8	-64.0	......
Methyl bromide	CH_3Br	-96.3$_s$	-72.8	-54.2	-39.4	-11.9	+3.6	-93
Methyl chloride	CH_3Cl	s	-92.4	-76.0	-63.0	-38.0	-24.0	-97.7
Trichloromethylsilane	CH_3Cl_3Si	-60.8	-30.7	-7.0	+12.1	47.0	66.4	-90
Methyl fluoride	CH_3F	-147.3	-131.6	-119.1	-109.0	-89.5	78.2	......
Methyl iodide	CH_3I	s	-45.8	-24.2	-7.0	+25.3	42.4	-64.4
Formamide	CH_3NO	70.5	109.5	137.5	157.5	193.5	210.5$_d$	......
Nitromethane	CH_3NO_2	-29.0	+2.8	27.5	46.6	82.0	101.2	-29
Methane	CH_4	-205.9$_s$	-195.5$_s$	-187.7$_s$	-181.4	-168.8	-161.5	-182.5
Dichloromethylsilane	CH_4Cl_2Si	-75.0	-47.8	-26.2	-9.0	+23.7	41.9	......
Methanol	CH_4O	-44.0	-16.2	+5.0	21.2	49.9	64.7	-97.8
Methanethiol	CH_4S	-90.7	-67.5	-49.2	-34.8	-7.9	+6.8	-121
Chloromethylsilane	CH_5ClSi	-95.0	-71.0	-51.7	-36.4	-7.8	+8.7	......
Methylamine	CH_5N	-95.8$_s$	-73.8	-56.9	-43.7	-19.7	-6.3	-93.5
Methylsilane	CH_6Si	-138.5	-120.0	-104.8	-93.0	-70.3	-56.9	......
2-Methyldisilazane	CH_9NSi_2	-76.3	-50.1	-29.6	-13.1	+17.2	34.0	......
Cyanogen iodide	CIN	25.2$_s$	57.7$_s$	80.3$_s$	97.6$_s$	126.1$_s$	141.1$_s$	......
Tetranitromethane	CN_4O_8	s	22.7	48.4	68.9	105.9	125.7$_d$	13
Carbon monoxide	CO	-222.0$_s$	-215.0$_s$	-210.0$_s$	-205.7$_s$	-196.3	-191.3	-205.0
Carbonyl sulfide	COS	-132.4	-113.3	-98.3	-85.9	-62.7	-49.9	-138.8
Carbonyl selenide	$COSe$	-117.1	-95.0	-76.4	-61.7	-35.6	-21.9	......
Carbon dioxide	CO_2	-134.3$_s$	-119.5$_s$	-108.6$_s$	-100.2$_s$	-85.7$_s$	-78.2$_s$	-57.5
Carbon Selenosulfide	$CSSe$	-47.3	-16.0	+8.6	28.3	65.2	85.6	-75.2
Carbon disulfide	CS_2	-73.8	-44.7	-22.5	-5.1	+28.0	46.5	-110.8
Trichloroacetyl bromide	C_2BrCl_3O	-7.4	+29.3	57.2	79.5	120.2	143.0	......
1-Chloro-1,2,2-trifluoroethylene	C_2ClF_3	-116.0	-95.9	-79.7	-66.7	-41.7	-27.9	-157.5
1,2-Dichloro-1,2-difluoroethylene	$C_2Cl_2F_2$	-82.0	-57.3	-38.2	-23.0	+5.0	20.9	-112
1,2-Dichloro-1,1,2,2-tetrafluoroethane	$C_2Cl_2F_4$	-95.4	-72.3	-53.7	-39.1	-12.0	+3.5	-94
1,1,2-Trichloro-1,2,2-trifluoroethane	$C_2Cl_3F_3$	-68.0$_s$	-40.3$_s$	-18.5	-1.7	+30.2	47.6	-35
Tetrachloroethylene	C_2Cl_4	-20.6$_s$	+13.8	40.1	61.3	100.0	120.8	-19.0
1,1,2,2-Tetrachloro-1,2-difluoroethane	$C_2Cl_4F_2$	-37.5$_s$	-5.0$_s$	+19.8$_s$	38.6	73.1	92.0	26.5
Hexachloroethane	C_2Cl_6	32.7$_s$	73.5$_s$	102.3$_s$	124.2$_s$	163.8$_s$	185.6$_s$	186.6
Tribromoacetaldehyde	C_2HBr_3O	18.5	58.0	87.8	110.2	151.6	174.0$_d$	......
Trichloroethylene	C_2HCl_3	-43.8	-12.4	+11.9	31.4	67.0	86.7	-73
Trichloroacetaldehyde	C_2HCl_3O	-37.8	-5.0	20.2	40.2	77.5	97.7	-57
Trichloroacetic acid	$C_2HCl_3O_2$	51.0$_s$	88.2	116.3	137.8	175.2	195.6	57
Pentachloroethane	C_2HCl_5	+1.0	39.8	69.9	93.5	137.2	160.5	-22
Acetylene	C_2H_2	-142.9$_s$	-128.2$_s$	-116.7$_s$	-107.9$_s$	-92.0$_s$	-84.0$_s$	-81.5
1,1,1,2-Tetrabromoethane	$C_2H_2Br_4$	58.0	95.7	123.2	144.0	181.0	200.0$_d$	......
1,1,2,2-Tetrabromoethane	$C_2H_2Br_4$	65.0	110.0	144.0	170.0	217.5	243.5	......
cis-1,2-Dichloroethylene	$C_2H_2Cl_2$	-58.4	-29.9	-7.9	+9.5	41.0	59.0	-80.5
trans-1,2-Dichloroethylene	$C_2H_2Cl_2$	-65.4$_s$	-38.0	-17.0	-0.2	+30.8	47.8	-50.0
1,1-Dichloroethene	$C_2H_2Cl_2$	-77.2	-51.2	-31.1	-15.0	+14.8	31.7	-122.5
Dichloroacetic acid	$C_2H_2Cl_2O_2$	44.0	82.6	111.8	134.0	173.7	194.4	9.7
1,1,1,2-Tetrachloroethane	$C_2H_2Cl_4$	-16.3	+19.3	46.7	68.0	108.2	130.5	-68.7
1,1,2,2-Tetrachloroethane	$C_2H_2Cl_4$	-3.8	+33.0	60.8	83 2	124.0	145.9	-36
1-Bromoethylene	C_2H_3Br	-95.4	-68.8	-48.1	-31.9	-1.1	+15.8	-138
Bromoacetic acid	$C_2H_3BrO_2$	54 7	94.1	124.0	146.3	186.7	208.0	49.5

ORGANIC COMPOUNDS (Continued)

Pressures Less than One Atmosphere (Continued)

Name	Formula	Temperature °C						M.P.
		1 mm	10 mm	40 mm	100 mm	400 mm	760 mm	
1,1,2-Tribromoethane	$C_2H_3Br_3$	32.6	70.6	100.0	123.5	165.4	188.4	−26
1-Chloroethylene	C_2H_3Cl	−105.6	−83.7	−66.8	−53.2	−28.0	−13.8	−153.7
Chloroacetic acid	$C_2H_3ClO_2$	43.0$_s$	81.0	109.2	130.7	169.0	189.5	61.2
1,1,1-Trichloroethane	$C_2H_3Cl_3$	−52.0	−21.9	+1.6	20.0	54.6	74.1	−30.6
1,1,2-Trichloroethane	$C_2H_3Cl_3$	−24.0	+8.3	35.2	55.7	93.0	113.9	−36.7
Trichloroacetaldehyde hydrate	$C_2H_3Cl_3O_2$	−9.8$_s$	+19.5$_s$	39.7$_s$	55.0	82.1	96.2$_d$	51.7
1-Fluoroethylene	C_2H_3F	−149.3	−132.2	−118.0	−106.2	−84.0	−72.2	−160.5
Acetonitrile	C_2H_3N	−47.0$_s$	−16.3	+7.7	27.0	62.5	81.8	−41
Methyl thiocyanate	C_2H_3NS	−14.0	+21.6	49.0	70.4	110.8	132.9	−51
Methyl isothiocyanate	C_2H_3NS	−34.7$_s$	+5.4$_s$	38.2	59.3	97.8	119.0	35.5
Ethylene	C_2H_4	−168.3	−153.2	−141.3	−131.8	−113.9	−103.7	−169
1-Bromo-1-chloroethane	C_2H_4BrCl	−36.0$_s$	−9.4$_s$	+10.4$_s$	28.0	63.4	82.7	16.6
1-Bromo-2-chloroethane	C_2H_4BrCl	−28.8$_s$	+4.1	29.7	49.5	86.0	106.7	−16.6
1,2-Dibromoethane	$C_2H_4Br_2$	−27.0$_s$	+18.6	48.0	70.4	110.1	131.5	10
1,1-Dichloroethane	$C_2H_4Cl_2$	−60.7	−32.3	−10.2	+7.2	39.8	57.4	−96.7
1,2-Dichloroethane	$C_2H_4Cl_2$	−44.5$_s$	−13.6	+10.0	29.4	64.0	82.4	−35.3
1,1-Difluoroethane	$C_2H_4F_2$	−112.5	−91.7	−75.8	−63.2	−39.5	−26.5	−117
Acetaldehyde	C_2H_4O	−81.5	−56.8	−37.8	−22.6	+4.9	20.2	−123.5
Ethylene oxide	C_2H_4O	−89.7	−65.7	−46.9	−32.1	−4.9	+10.7	−111.3
Acetic acid	$C_2H_4O_2$	−17.2$_s$	+17.5	43.0	63.0	99.0	118.1	16.7
Methyl formate	$C_2H_4O_2$	−74.2	−48.6	−28.7	−12.9	16.0	32.0	−99.8
Mercaptoacetic acid	$C_2H_4O_2S$	60.0	101.5	131.8	154.0$_d$	......	......	−16.5
Ethyl bromide	C_2H_5Br	−74.3	−47.5	−26.7	−10.0	+21.0	38.4	−117.8
Ethyl chloride	C_2H_5Cl	−89.8	−65.8	−47.0	−32.0	−3.9	+12.3	−139
2-Chloroethanol	C_2H_5ClO	−4.0	+30.3	56.0	75.0	110.0	128.8	−69
Trichloroethylsilane	$C_2H_5Cl_3Si$	−27.9	+3.6	27.9	46.3	80.3	99.5	−40
Trichloroethoxysilane	$C_2H_5Cl_3OSi$	−32.4	0.0	+25.3	45.2	82.2	102.4	......
Ethyl fluoride	C_2H_5F	−117.0	−97.7	−81.8	−69.3	−45.5	−32.0	......
Ethyltrifluorosilane	$C_2H_5F_3Si$	−95.4	−73.7	−56.8	−43.6	−19.1	−5.4	......
Ethyl Iodide	C_2H_5I	−54.4	−24.3	−0.9	+18.0	52.3	72.4	−105
Acetamide	C_2H_5NO	65.0$_s$	105.0	135.8	158.0	200.0	222.0	81
Acetaldoxime	C_2H_5NO	−5.8$_s$	+25.8	48.6	66.2	98.0	115.0	47
Nitroethane	$C_2H_5NO_2$	−21.0	+12.5	38.0	57.8	94.0	114.0	−90
Di(nitrosomethyl)amine	$C_2H_5N_3O_2$	+3.2	40.0	68.2	90.3	131.3	153.0	......
Ethane	C_2H_6	−159.5	−142.9	−129.8	−119.3	−99.7	−88.6	−183.2
Dichlorodimethylsilane	$C_2H_6Cl_2Si$	−53.5	−23.8	−0.4	+17.5	51.9	70.3	−86.0
Ethanol	C_2H_6O	−31.3	−2.3	+19.0	34.9	63.5	78.4	−112
Dimethyl ether	C_2H_6O	−115.7	−93.3	−76.2	−62.7	−37.8	−23.7	−138.5
1,2-Ethanediol	$C_2H_6O_2$	53.0	92.1	120.0	141.8	178.5	197.3	−15.6
Dimethyl sulfide	C_2H_6S	−75.6	−49.2	−28.4	−12.0	+18.7	36.0	−83.2
Ethanethiol	C_2H_6S	−76.7	−50.2	−29.8	−13.0	+17.7	35.5	−121
Dimethylantimony	C_2H_6Sb	44.0	86.0	118.3	143.5	187.2	211.0	......
Ethylamine	C_2H_7N	−82.3$_s$	−58.3	−39.8	−25.1	+2.0	16.6	−80.6
Dimethylamine	C_2H_7N	−87.7	−64.6	−46.7	−32.6	−7.1	+7.4	−96
1,2-Ethanediamine	$C_2H_8N_2$	−11.0$_s$	+21.5	45.8	62.5	99.0	117.2	8.5
Dimethylsilane	C_2H_8Si	−115.0	−93.1	−75.7	−61.4	−35.0	−20.1	......
Dimethyldiborane	$C_2H_{10}B_2$	−106.5	−82.1	−62.4	−47.0	−18.8	−2.6	−150.2
2-Ethyldisilazane	$C_2H_{11}NSi_2$	−62.0	−32.2	−8.3	+10.4	45.9	65.9	−127
Cyanogen	C_2N_2	−95.8$_s$	−76.8$_s$	−62.7$_s$	−51.8$_s$	−33.0	−21.0	−34.4
Acrylonitrile	C_3H_3N	−51.0	−20.3	+3.8	22.8	58.3	78.5	−82
Propadiene	C_3H_4	−120.6	−101.0	−85.2	−72.5	−48.5	−35.0	−136
Propyne	C_3H_4	−111.0$_s$	−90.5	−74.3	−61.3	−37.2	−23.3	−102.7
2,3-Dibromopropene	$C_3H_4Br_2$	−6.0	+30.0	57.8	79.5	119.5	141.2	......
Methyl dichloroacetate	$C_3H_4Cl_2O_2$	3.2	38.1	64.7	85.4	122.6	143.0	......
2-Propenal	C_3H_4O	−64.5	−36.7	−15.0	+2.5	34.5	52.5	−87.7
Acrylic acid	$C_3H_4O_2$	+3.5$_s$	39.0	66.2	86.1	122.0	141.0	14
Pyruvic acid	$C_3H_4O_3$	21.4	57.9	85.3	106.5	144.7	165.0$_d$	13.6
1,2,3-Tribromopropene	$C_3H_5Br_3$	47.5	90.0	122.8	148.0	195.0	220.0	16.5
1-Chloropropene	C_3H_5Cl	−81.3	−54.1	−32.7	−15.1	+18.0	37.0	−99.0
3-Chloropropene	C_3H_5Cl	−70.0	−42.9	−21.2	−4.5	+27.5	44.6	−136.4
Epichlorohydrine	C_3H_5ClO	−16.5	+16.6	42.0	62.0	98.0	117.9	−25.6
Methyl chloroacetate	$C_3H_5ClO_2$	−2.9	+30.0	54.5	73.5	109.5	130.3	−31.9
1,1,1-Trichloropropane	$C_3H_5Cl_3$	−28.8	+4.2	29.9	50.0	87.5	108.2	−77.7
1,2,3-Trichloropropane	$C_3H_5Cl_3$	+9.0	46.0	74.0	96.1	137.0	158.0	−14.7
Allyltrichlorosilane	$C_3H_5Cl_3Si$	−20.7	+13.2	39.2	59.3	97.1	118.0	......
Propionitrile	C_3H_5N	−35.0	−3.0	+22.0	41.4	77.7	97.1	−91.9
3-Hydroxypropionitrile	C_3H_5NO	58.7	102.0	134.1	157.7	200.0	221.0	......
Ethylisothiocyanate	C_3H_5NS	−13.2$_s$	+22.8	50.8	71.9	110.1	131.0	−5.9
Nitroglycerine	$C_3H_5N_3O_9$	127	188	235	e	......	......	11
Propylene	C_3H_6	−131.9	−112.1	−96.5	−84.1	−60.9	−47.7	−185
Cyclopropane	C_3H_6	−116.8	−97.5	−82.3	−70.0	−46.9	−33.5	−126.6
2-Bromo-2-nitrosopropane	C_3H_6BrNO	−33.5	−4.3	+17.9	35.2	66.2	83.0	......

VAPOR PRESSURE (Continued)
ORGANIC COMPOUNDS (Continued)
Pressures Less than One Atmosphere (Continued)

Name	Formula	Temperature °C						M.P.
		1 mm	10 mm	40 mm	100 mm	400 mm	760 mm	
1,2-Dibromopropane	$C_3H_6Br_2$	−7.0	+29.4	57.2	78.7	118.5	141.6	−55.5
1,3-Dibromopropane	$C_3H_6Br_2$	+9.7	48.0	77.8	101.3	144.1	167.5	−34.4
2,3-Dibromo-1-propanol	$C_3H_6Br_2O$	57.0	98.2	129.8	153.0	196.0	219.0	......
1,2-Dichloropropane	$C_3H_6Cl_2$	−38.5	−6.1	+19.4	39.4	76.0	96.8	......
1,3-Dichloro-2-propanol	$C_3H_6Cl_2O$	28.0	64.7	93.0	114.8	153.5	174.3	......
Acetone	C_3H_6O	−59.4	−31.1	−9.4	+7.7	39.5	56.5	−94.6
Allyl alcohol	C_3H_6O	−20.0	+10.5	33.4	50.0	80.2	96.6	−129
Propylene oxide	C_3H_6O	−75.0	−49.0	−28.4	−12.0	+17.8	34.5	−112.1
Propionic acid	$C_3H_6O_2$	4.6	39.7	65.8	85.8	122.0	141.1	−22
Methyl acetate	$C_3H_6O_2$	−57.2	−29.3	−7.9	+9.4	40.0	57.8	−98.7
Ethyl formate	$C_3H_6O_2$	−60.5	−33.0	−11.5	+5.4	37.1	54.3	−79
Methyl glycolate	$C_3H_6O_3$	+9.6	45.3	72.3	93.7	131.7	151.5	......
Methoxyacetic acid	$C_3H_6O_3$	52.5	92.0	122.0	144.5	184.2	204.0	......
1-Bromopropane	C_3H_7Br	−53.0	−23.3	−0.3	+18.0	52 0	71.0	−109.9
2-Bromopropane	C_3H_7Br	−61.8	−32.8	−10.1	+8 0	41.5	60 0	−89 0
1-Chloropropane	C_3H_7Cl	−68.3	−41.0	−19.5	−2.5	+29.4	46 4	−122.8
2-Chloropropane	C_3H_7Cl	−78.8	−52 0	−31 0	−13.7	+18.0	36.5	−117
Trichloroisopropylsilane	$C_3H_7Cl_3Si$	−24.3	+9.9	36.5	57.8	96.8	118.5	....
1-Iodopropane	C_3H_7I	−36.0	−2.4	+23.6	43.8	81 8	102.5	−98 8
2-Iodopropane	C_3H_7I	−43.3	−11.7	+13.2	32.8	69 5	89.5	−90
Propionamide	C_3H_7NO	65.0_s	105.0	134.8	156.0	194.0	213.0	79
1-Nitropropane	$C_3H_7NO_2$	−9.6	+25.3	51.8	72.3	110.6	131 6	−108
2-Nitropropane	$C_3H_7NO_2$	−18.8	+15.8	41 8	62.0	99.8	120 3	−93
Ethyl Carbamate	$C_3H_7NO_2$	s	77.8	105.6	126 2	164.0	184 0	49
Propane	C_3H_8	−128.9	−108.5	−92.4	−79.6	−55.6	−42.1	−187.1
Dichloroethoxymethylsilane	$C_3H_8Cl_2OSi$	−33.8	−1.3	+24.4	44.1	80.3	100 6	......
1-Propanol	C_3H_8O	−15.0	+14.7	36.4	52.8	82.0	97.8	−127
2-Propanol	C_3H_8O	−26.1	+2.4	23 8	39.5	67.8	82 5	−85.8
Ethyl methyl ether	C_3H_8O	−91.0	−67 8	−49.4	−34.8	−7.8	+7.5	......
1,2-Propanediol	$C_3H_8O_2$	45.5	83.2	111.2	132.0	168.1	188.2	......
1,3-Propanediol	$C_3H_8O_2$	59.4	100.6	131.0	153.4	193.8	214 2	....
2-Methoxyethanol	$C_3H_8O_2$	−13.5	+22.0	47.8	68.0	104.3	124 4	....
Glycerol	$C_3H_8O_3$	125.5	167.2	198.0	220.1	263.0	290.0	17 9
1-Propanethiol	C_3H_8S	−56.0	−26.3	−3.2	+15 3	49.2	67.4	−112
Trimethylborine	C_3H_9B	−118.0	−92.4	−74.7	−60.8	−34.7	−20 1	......
Chlorotrimethylsilane	C_3H_9ClSi	−62.8	−34.0	−11.4	+6.0	39 4	57.9	......
Trimethylgallium	C_3H_9Ga	-62.3_s	-31.7_s	−9.0	+8.0	39.0	55.6	−19
Propylamine	C_3H_9N	−64.4	−37.2	−16.0	+0.5	31.5	48.5	−83
Trimethylamine	C_3H_9N	−97.1	−73.8	−55.2	−40.3	−12.5	+2.9	−117.1
Trimethyl phosphate	$C_3H_9O_4P$	26.0	67.8	100.0	124.0	167.8	192.7	......
Trimethyldiborane	$C_3H_{12}B_2$	−74.0	−44.8	−22.0	−4.4	+27.8	45.5	−122.9
Carbon suboxide	C_3O_2	−94.8	−71.0	−52.0	−36.9	−8.9	+6.3	−107
Carbon subsulfide	C_3S_2	14.0	54.9	85.6	109 9	......	......	+0.4
Trichloroacetic anhydride	$C_4Cl_6O_3$	56.2	99.6	131.2	155 2	199.8	223.0	......
1,3-Butadiyne	C_4H_2	-82.5_s	-61.2_s	-45.9_s	−34.0	−6.1	+9.7	−34.9
α,β-Dibromomaleic anhydride	$C_4H_2Br_2O_3$	50.0	92.0	123.5	147.7	192.0	215.0	......
trans-Fumaryl chloride	$C_4H_2Cl_2O_2$	+15.0	51.8	79.5	101.0	140.0	160.0	......
Maleic anhydride	$C_4H_2O_3$	$44\ 0_s$	78 7	111.8	135.8	179.5	202.0	58
2-Nitrothiophene	$C_4H_3NO_2S$	48.2	92.0	125.8	151.5	199.6	224.5	46
Butenyne	C_4H_4	−93.2	−70.0	−51.7	−37.1	−10.1	+5.3	......
Succinyl chloride	$C_4H_4Cl_2O_2$	39 0	78 0	107.5	130.0	170.0	192.5	17
Chloroacetic anhydride	$C_4H_4Cl_2O_3$	67.2	108.0	138.2	159.8	197.0	217.0	46
Succinic anhydride	$C_4H_4O_3$	92.0_s	128 2	163.0	189.0	237.0	261.0	119.6
1,4-Dioxane-2,6-dione	$C_4H_4O_4$	s	116 6	148.6	173.2	217 0	240.0	97
Thiophene	C_4H_4S	-40.7_s	−10.9	+12 5	30.5	64.7	84 4	−38.3
Selenophene	C_4H_4Se	−39.0	−4.0	+24.1	47 0	89 8	114.3	......
α-Chlorocrotonic acid	$C_4H_5ClO_2$	70.0	108.0	135.6	155.9	193.2	212.0	......
Ethyl chloroglyoxylate	$C_4H_5ClO_3$	−5.1	+29.9	56.0	76.6	114.7	135.0	......
Ethyl trichloroacetate	$C_4H_5Cl_3O_2$	20.7	57.7	85.5	107 4	146.0	167.0	......
3-Butenenitrile	C_4H_5N	−19.6	+14.1	40.0	60.2	98.0	119.0	......
Methacrylonitrile	C_4H_5N	−44.5	−12.5	+12.8	32.8	32.8	90 3	......
cis-Crotononitrile	C_4H_5N	−29.0	+4.0	30.0	50.1	88.0	108.0	......
trans-Crotononitrile	C_4H_5N	−19.5	+15 0	41.8	62.8	101.5	122.8	......
Succinimide	$C_4H_5NO_2$	115.0_s	157.0	192.0	217.4	263.5	287.5	125.5
Allylisothiocyanate	C_4H_5NS	−2.0	+38.3	67.4	89.5	129.8	150.7	−80
1,2-Butadiene	C_4H_6	−89.0	−64.2	−44.3	−28.3	+1.8	18.5	......
1,3-Butadiene	C_4H_6	−102.8	−79.7	−61.3	−46.8	−19.3	−4.5	−108.9
Cyclobutene	C_4H_6	−99.1	−75.4	−56.4	−41.2	−12.2	+2.4	......
1 Butyne	C_4H_6	−92.5	−68.7	−50.0	−34.9	−6.9	+8.7	−130
2-Butyne	C_4H_6	-73.0_s	-50.5_s	-33.9_s	−18.8	+10.6	27.2	−32.5
Ethyl dichloroacetate	$C_4H_6Cl_2O_2$	9.6	46.3	74.0	96.1	135.9	156.5	......
2-Chloroethyl chloroacetate	$C_4H_6Cl_2O_2$	46.0	86.0	116.0	140.0	182.2	205.0	......
cis-Crotonic acid	$C_4H_6O_2$	33.5	69.0	96 0	116.3	152.2	171.9_d	15.5
trans-Crotonic acid	$C_4H_6O_2$	s	80.0	107.8	128.0	165 5	185.0	72
Methyl acrylate	$C_4H_6O_2$	−43.7	−13.5	+9.2	28.0	61.8	80.2_p	......
Methacrylic acid	$C_4H_6O_2$	25.5	60.0	86.4	106.6	142.5	161.0	15
Vinyl acetate	$C_4H_6O_2$	−48.0	−18.0	+5.3	23 3	55.5	72.5	......
Acetic anhydride	$C_4H_6O_3$	1.7	36.0	62.1	82.2	119 8	139.6	−73
Dimethyl oxalate	$C_4H_6O_4$	20.0	56.0	83 6	104.8	143 3	163.3	......
cis-1 Bromo-1-butene	C_4H_7Br	−44.0	−12.8	+11.5	30 8	66 8	86 2	......
trans-1-Bromo-1-butene	C_4H_7Br	−38.4	−6 4	+18.4	38 1	75 0	94.7	−100.3
2-Bromo-1-butene	C_4H_7Br	−47.3	−16.8	+7.2	26.3	61.9	81 0	−133 4
cis-2-Bromo-2-butene	C_4H_7Br	−39.0	−7.2	+17 7	37.5	74.0	93.0	−111.2
trans-2-Bromo-2-butene	C_4H_7Br	−45.0	−13.8	+10.5	29.9	66.0	85.5	−114.6

VAPOR PRESSURE (Continued)
ORGANIC COMPOUNDS (Continued)
Pressures Less than One Atmosphere (Continued)

Name	Formula	Temperature °C						M.P
		1 mm	10 mm	40 mm	100 mm	400 mm	760 mm	
1-Bromo-2-Butanone	C_4H_7BrO	+6.2	41.8	68.2	89.2	126.3	147.0	......
2-Methylpropionyl bromide	C_4H_7BrO	13.5	50.6	79.4	101.6	141.7	163.0	......
1,1,2-Tribromobutane	$C_4H_7Br_3$	45.0	87.8	120.2	146.0	192.0	216.2	......
1,2,2-Tribromobutane	$C_4H_7Br_3$	41.0	83.2	116.0	141.8	188.0	213.8	......
2,2,3-Tribromobutane	$C_4H_7Br_3$	38.2	79.8	111.8	136.3	182.2	206.5	......
Ethyl chloroacetate	$C_4H_7ClO_2$	+1.0	37.5	65.2	86.0	123.8	144.2	-26
1,2,3-Trichlorobutane	$C_4H_7Cl_3$	+0.5	40.0	71.5	96.2	143.0	169.0	......
Butyronitrile	C_4H_7N	-20.0	+13.4	38.4	59.0	96.8	117.5	......
Diacetamide	$C_4H_7NO_2$	70.0$_s$	108.0	138.2	160.6	202.0	223.0	78.5
1-Butene	C_4H_8	-104.8	-81.6	-63.4	-48.9	-21.7	-6.3	-130
cis-2-Butene	C_4H_8	-96.4	-73.4	-54.7	-39.8	-12.0	+3.7	-138.9
trans-2-Butene	C_4H_8	-99.4	-76.3	-57.6	-42.7	-14.8	+0.9	-105.4
2-Methylpropene	C_4H_8	-105.1	-81.9	-63.8	-49.3	-22.2	-6.9	-140.3
Cyclobutane	C_4H_8	-92.0$_s$	-67.9$_s$	-48.4	-32.8	-3.4	+12.9	-50
Methylcyclopropane	C_4H_8	-96.0	-72.8	-54.2	-39.3	-11.3	+4.5	......
2-Bromoethyl 2-chloroethyl ether	C_4H_8BrClO	36.5	76.3	106.6	129.8	172.3	195.8	......
1,2-Dibromobutane	$C_4H_8Br_2$	7.5	46.1	76.0	99.8	143.5	166.3	-64.5
dl-2,3-Dibromobutane	$C_4H_8Br_2$	+5.0	41.6	72.0	95.3	138.0	160.5	......
meso-2,3-Dibromobutane	$C_4H_8Br_2$	+1.5	39.3	68.2	91.7	134.2	157.3	-34.5
1,4-Dibromobutane	$C_4H_8Br_2$	32.0	72.4	104.0	128.7	173.8	197.5	-20
1,2-Dibromo-2-methylpropane	$C_4H_8Br_2$	-28.8	+10.5	42.3	68.8	119.8	149.0	-70.3
1,3-Dibromo-2-methylpropane	$C_4H_8Br_2$	14.0	53.0	83.5	107.4	150.6	174.6	......
Di(2-bromoethyl) ether	$C_4H_8Br_2O$	47.7	88.5	119.8	144.0	188.0	212.5	......
1,2-Dichlorobutane	$C_4H_8Cl_2$	-23.6	+11.5	37.7	60.2	100.8	123.5	......
2,3-Dichlorobutane	$C_4H_8Cl_2$	-25.2	+8.5	35.0	56.0	94.2	116.0	-80.4
1,1-Dichloro-2-methylpropane	$C_4H_8Cl_2$	-31.0	+2.6	28.2	48.2	85.4	-106.0$_d$	......
1,2-Dichloro-2-methylpropane	$C_4H_8Cl_2$	-25.8	+6.7	32.0	51.7	87.8	108.0	......
1,3-Dichloro-2-methylpropane	$C_4H_8Cl_2$	-3.0	+32.0	58.6	78.8	115.4	135.0	......
di(2-chloroethyl) ether	$C_4H_8Cl_2O$	23.5	62.0	91.5	114.5	155.4	178.5	......
1,2-Epoxy-2-methylpropane	C_4H_8O	-69.0	-40.3	-17.3	+1.2	36.0	55.5	......
2-Butanone	C_4H_8O	-48.3	-17.7	+6.0	25.0	60.0	79.6	-85.9
1,4-Dioxane	$C_4H_8O_2$	-35.8$_s$	-1.2$_s$	+25.2	45.1	81.8	101.1	10
Butyric acid	$C_4H_8O_2$	25.5	61.5	88.0	108.0	144.5	163.5	-4.7
Isobutyric acid	$C_4H_8O_2$	14.7	51.2	77.8	98.0	134.5	154.5	-47
Ethyl acetate	$C_4H_8O_2$	-43.4	-13.5	+9.1	27.0	59.3	77.1	-82.4
Methyl propionate	$C_4H_8O_2$	-42.0	-11.8	+11.0	29.0	61.8	79.8	-87.5
Propyl formate	$C_4H_8O_2$	-43.0	-12.6	+10.8	29.5	62.6	81.3	-92.9
Isopropyl formate	$C_4H_8O_2$	-52.0	-22.7	-0.2	+17.8	50.5	68.3	......
α-Hydroxyisobutyric acid	$C_4H_8O_3$	73.5$_s$	110.5	138.0	157.7	193.8	212.0	79
Ethyl glycolate	$C_4H_8O_3$	14.3	50.5	78.1	99.8	138.0	158.2	......
1-Bromobutane	C_4H_9Br	-33.0	-0.3	+24.8	44.7	81.7	101.6	-112.4
1-Bromo-2-butanol	C_4H_9BrO	23.7	55.8	79.5	97.6	128.3	145.0	......
1-Chlorobutane	C_4H_9Cl	-49.0	-18.6	+5.0	24.0	58.8	77.8	-123.1
sec-Butyl chloride	C_4H_9Cl	-60.2	-29.2	-5.0	+14.2	50.0	68.0	-131.3
Isobutyl chloride	C_4H_9Cl	-53.8	-24.5	-1.9	+16.0	50.0	68.9	-131.2
tert-Butyl chloride	C_4H_9Cl	s	s	-19.0	-1.0	+32.6	51.0	-26.5
2-(2-Chloroethoxy) ethanol	$C_4H_9ClO_2$	53.0	90.7	118.4	139.5	176.5	196.0	......
1-Iodo-2-methylpropane	C_4H_9I	-17.0	+17.0	42.8	63.5	100.3	120.4	-90.7
Ethyl methylcarbamate	$C_4H_9NO_2$	26.5	63.2	91.0	112.0	149.8	170.0	......
Propyl carbamate	$C_4H_9NO_2$	52.4	90.0	117.7	138.3	175.8	195.0	......
Di(nitrosoethyl)amine	$C_4H_9N_2O_2$	18.5	57.7	87.6	111.0	153.5	176.9	......
Butane	C_4H_{10}	-101.5	-77.8	-59.1	-44.2	-16.3	-0.5	-135
2-Methylpropane	C_4H_{10}	-109.2	-86.4	-68.4	-54.1	-27.1	-11.7	-145
Dichlorodiethylsilane	$C_4H_{10}Cl_2Si$	-9.2	+25.4	51.6	71.8	110.0	130.4	......
Diethyldifluorosilane	$C_4H_{10}F_2Si$	-56.8	-28.8	-7.3	+9.8	40.5	58.0	......
Butyl alcohol	$C_4H_{10}O$	-1.2	+30.2	53.4	70.1	100.8	117.5	-79.9
sec-Butyl alcohol	$C_4H_{10}O$	-12.2	+16.9	38.1	54.1	83.9	99.5	-114.7
Isobutyl alcohol	$C_4H_{10}O$	-9.0	+21.7	44.1	61.5	91.4	108.0	-108
tert-Butyl alcohol	$C_4H_{10}O$	-20.4$_s$	+5.5$_s$	24.5$_s$	39.8	68.0	82.9	25.3
Diethyl ether	$C_4H_{10}O$	-74.3	-48.1	-27.7	-11.5	+17.9	34.6	-116.3
Methyl propyl ether	$C_4H_{10}O$	-72.2	-45.4	-24.3	-8.1	+22.5	39.1	......
1,3-Butanediol	$C_4H_{10}O_2$	22.2$_s$	85.3	117.4	141.2	183.8	206.5	77
2,3-Butanediol	$C_4H_{10}O_2$	44.0	80.3	107.8	127.8	164.0	182.0	22.5
1,2-Dimethoxyethane	$C_4H_{10}O_2$	-48.0	-15.3	+10.7	31.8	70.8	93.0	......
2,2′-Thiodiethanol	$C_4H_{10}O_2S$	42.0	128.0	210.0$_d$	285$_d$	......	......	......
Diethylene glycol	$C_4H_{10}O_3$	91.8	133.8	164.3	187.5	226.5	244.8	......
1,2,3-Butanetriol	$C_4H_{10}O_3$	102.0	146.0	178.0	202.5	243.5	264.0	......
Diethyl sulfite	$C_4H_{10}O_3S$	10.0	46.4	74.2	96.3	137.0	159.0	......
Diethyl sulfate	$C_4H_{10}O_4S$	47.0	87.7	118.0	142.5	185.5	209.5$_d$	-25.0
Diethyl sulfide	$C_4H_{10}S$	-39.6	-8.0	+16.1	35.0	69.7	88.0	-99.5
Diethyl selenide	$C_4H_{10}Se$	-25.7	+7.0	31.2	51.8	88.0	108.0	......
Diethylzinc	$C_4H_{10}Zn$	-22.4	+11.7	38.0	59.1	97.3	118.0	-28
Diethylamine	$C_4H_{11}N$	s	-33.0	-11.3	+6.0	38.0	55.5	-38.9
Isobutylamine	$C_4H_{11}N$	-50.0	-21.0	+1.3	18.8	50.7	68.6	-85.0
1,3-Dichlorotetramethyldisiloxane	$C_4H_{12}Cl_2Si_2$	-7.4	+28.3	55.7	76.9	116.3	138.0	-37
Tetramethyllead	$C_4H_{12}Pb$	-29.0$_s$	+4.4	30.3	50.8	89.0	110.0	-27.5
Tetramethylsilane	$C_4H_{12}Si$	-83.8	-58.0	-37.4	-20.9	+10.0	27.0	-102.1
Tetramethyltin	$C_4H_{12}Sn$	-51.3	-20.6	+3.5	22.8	58.5	78.0	......
Tetramethyldiborane	$C_4H_{14}B_2$	-59.6	-27.4	-3.4	+15.3	49.8	68.6	-72.5
3-Bromopyridine	C_5H_4BrN	16.8	55.2	84.1	107.8	150.0	173.4	......
2-Chloropyridine	C_5H_4ClN	13.3	51.7	81.7	104.6	147.7	170.2	......

VAPOR PRESSURE (Continued)
ORGANIC COMPOUNDS (Continued)
Pressures Less than One Atmosphere (Continued)

Name	Formula	Temperature °C						M.P.
		1 mm	10 mm	40 mm	100 mm	400 mm	760 mm	
2-Furaldehyde	$C_5H_4O_2$	18.5	54.8	82.1	103.4	141.8	161.8	− 36.5
Citraconic anhydride	$C_5H_4O_3$	47.1	88.9	120.3	145.4	189.8	213.5	....
Pyridine	C_5H_5N	− 18.9	+ 13.2	38.0	57.8	95.6	115.4	− 42
Glutaryl chloride	$C_5H_6Cl_2O_2$	56.1	97.8	128.3	151.8	195.3	217.0	....
Glutaronitrile	$C_5H_6N_2$	91.3	140.0	176.4	205.5	257.3	286.2	....
Furfuryl alcohol	$C_5H_6O_2$	31.8	68.0	95.7	115.9	151.8	170.0	....
Glutaric anhydride	$C_5H_6O_3$	100.8	149.5	185.5	212.5	261.0	287.0	....
Pyrotartaric anhydride	$C_5H_6O_3$	69.7	114.2	147.8	173.8	221.0	247.4	....
2-Methylthiophene	C_5H_6S	− 27.4	+ 6.0	32.3	53.1	91.8	112.5	− 63.5
3-Methylthiophene	C_5H_6S	− 24.5	+ 9.1	35.4	55.8	93.8	115.4	− 68.9
Propyl chloroglyoxylate	$C_5H_7ClO_3$	9.7	43.5	68.8	88.0	123.0	150.0	....
Tiglonitrile	C_5H_7N	− 25.5	+ 9.2	36.7	58.2	99.7	122.0	....
Angelonitrile	C_5H_7N	− 8.0	+ 28.0	55.8	77.5	117.7	140.0	....
α-Ethylacrylonitrile	C_5H_7N	− 29.0	+ 5.0	31.8	53.0	92.2	114.0	....
Ethyl cyanoacetate	$C_5H_7NO_2$	67.8	106.0	133.8	152.8	187.8	206.0	....
Isoprene	C_5H_8	− 79.8	− 53.3	− 32.6	− 16.0	+ 15.4	32.6	−146.7
1,3-Pentadiene	C_5H_8	− 71.8	− 45.0	− 23.4	− 6.7	+ 24.7	42.1	....
1,4-Pentadiene	C_5H_8	− 83.5	− 57.1	− 37.0	− 20.6	+ 8.3	26.1	....
Tiglaldehyde	C_5H_8O	− 25.0	+ 10.0	37.0	57.7	95.5	116.4	....
Levulinaldehyde	$C_5H_8O_2$	28.1	68.0	98.3	121.8	164.0	187.0	....
Tiglic acid	$C_5H_8O_2$	52.0_s	90.2	119.0	140.5	179.2	198.5	64.5
α-Valerolactone	$C_5H_8O_2$	37.5	79.8	101.9	136.5	182.3	207.5	....
α-Ethylacrylic acid	$C_5H_8O_2$	47.0	82.0	108.1	127.5	160.7	179.2	....
Ethyl acrylate	$C_5H_8O_2$	− 29.5	+ 2.0	26.0	44.5	80.0	99.5	− 71.2
Methyl methacrylate	$C_5H_8O_2$	− 30.5	+ 1.0	25.5	47.0	82.0	101.0	....
Levulinic acid	$C_5H_8O_3$	102.0	141.8	169.5	190.2	227.4	245.8_d	33.5
Glutaric acid	$C_5H_8O_4$	155.5	196.0	226.3	247.0	283.5	303.0	97.5
Dimethyl malonate	$C_5H_8O_4$	35.0	72.0	100.0	121.9	159.8	180.7	− 62
Ethyl α-chloropropionate	$C_5H_9ClO_2$	+ 6.6	41.9	68.2	89.3	126.2	146.5	....
Isopropyl chloroacetate	$C_5H_9ClO_2$	+ 3.8	40.2	68.7	90.3	128.0	148.6	....
Valeronitrile	C_5H_9N	− 6.0	+ 30.0	57.8	78.6	118.7	140.8	....
α-Hydroxybutyronitrile	C_5H_9NO	41.0	77.8	104.8	125.0	159.8	178.8	....
1-Pentene	C_5H_{10}	− 80.4	− 54.5	− 34.1	− 17.7	+ 12.8	30.1	....
2-Methyl-2-butene	C_5H_{10}	− 75.4	− 47.9	− 26.7	− 9.9	+ 21.6	38.5	−133
2-Methyl-1-butene	C_5H_{10}	− 89.1	− 64.3	− 44.1	− 28.0	+ 2.5	20.2	−135
Cyclopentane	C_5H_{10}	− 68.0	− 40.4	− 18.6	− 1.3	+ 31.0	49.3	− 93.7
1,2-Dibromopentane	$C_5H_{10}Br_2$	19.8	58.0	87.4	110.1	151.8	175.0	....
2-Chloroethyl 2-chloroisopropyl ether	$C_5H_{10}Cl_2O$	24.7	63.0	92.4	115.8	156.5	180.0	....
2-Chloroethyl 2-chloropropyl ether	$C_5H_{10}Cl_2O$	29.8	70.0	101.5	125.6	169.8	194.1	....
Di(2-chloroethoxy)methane	$C_5H_{10}Cl_2O_2$	53.0	94.0	125.5	149.6	192.0	215.0	....
Allyldichloroethylsilane	$C_5H_{10}Cl_2Si$	− 3.0	34.2	62.7	85.2	127.0	150.3	....
3-Pentanone	$C_5H_{10}O$	− 12.7	+ 17.2	39.4	56.2	86.3	102.7	− 42
2-Pentanone	$C_5H_{10}O$	− 12.0	+ 17.9	39.8	56.8	86.8	103.3	− 77.8
3-Methyl-2-butanone	$C_5H_{10}O$	− 19.9	+ 8.3	29.6	45.5	73.8	88.9	− 92
4-Hydroxy-3-methyl-2-butanone	$C_5H_{10}O_2$	44.6	81.0	108.2	129.0	165.5	185.0	....
Valeric acid	$C_5H_{10}O_2$	42.2	79.8	107.8	128.3	165.0	184.4	− 34.5
Isovaleric acid	$C_5H_{10}O_2$	34.5	71.3	98.0	118.9	155.2	175.1	− 37.6
Ethyl propionate	$C_5H_{10}O_2$	− 28.0	+ 3.4	27.2	45.2	79.8	99.1	− 72.6
Propyl acetate	$C_5H_{10}O_2$	− 26.7	+ 5.0	28.8	47.8	82.0	101.8	− 92.5
Isopropyl acetate	$C_5H_{10}O_2$	− 38.3	− 7.2	+ 17.0	35.7	69.8	89.0	....
Methyl butyrate	$C_5H_{10}O_2$	− 26.8	+ 5.0	29.6	48.0	83.1	102.3	....
Methyl isobutyrate	$C_5H_{10}O_2$	− 34.1	− 2.9	+ 21.0	39.6	73.6	92.6	− 84.7
Butyl formate	$C_5H_{10}O_2$	− 26.4	+ 6.1	31.6	51.0	86.2	106.0	....
Isobutyl formate	$C_5H_{10}O_2$	− 32.7	− 0.8	+ 24.1	43.4	79.0	98.2	− 95.3
sec-Butyl formate	$C_5H_{10}O_2$	− 34.4	− 3.1	+ 21.3	40.2	75.2	93.6	....
Diethyl carbonate	$C_5H_{10}O_3$	− 10.1	+ 23.8	49.5	69.7	105.8	125.8	− 43
1-Bromo-3-methylbutane	$C_5H_{11}Br$	− 20.4	+ 13.6	39.8	60.4	99.4	120.4	....
1-Iodo-3-methylbutane	$C_5H_{11}I$	− 2.5	+ 34.1	62.3	84.4	125.8	148.2	....
Piperidine	$C_5H_{11}N$	s	+ 3.9	29.2	49.0	85.7	106.0	− 9
Isobutyl carbamate	$C_5H_{11}NO_2$	s	96.4	125.3	147.2	186.0	206.5	65
Isoamyl nitrate	$C_5H_{11}NO_3$	+ 5.2	40.3	67.6	88.6	126.5	147.5	....
Pentane	C_5H_{12}	− 76.6	− 50.1	− 29.2	− 12.6	+ 18.5	36.1	−129.7
2-Methylbutane	C_5H_{12}	− 82.9	− 57.0	− 36.5	− 20.2	+ 10.5	27.8	−159.7
2,2-Dimethylpropane	C_5H_{12}	−102.0	− 76.7_s	− 56.1_s	− 39.1_s	− 7.1	+ 9.5	− 16.6
Amyl alcohol	$C_5H_{12}O$	+ 13.6	44.9	68.0	85.8	119.8	137.8	....
Isoamyl alcohol	$C_5H_{12}O$	+ 10.0	40.8	63.4	80.7	113.7	130.6	−117.2
2-Pentanol	$C_5H_{12}O$	+ 1.5	32.2	54.1	70.7	102.3	119.7	....
tert-Amyl alcohol	$C_5H_{12}O$	− 12.9_s	+ 17.2	38.8	55.3	85.7	101.7	− 11.9
Ethyl propyl ether	$C_5H_{12}O$	− 64.3	− 35.0	− 12.0	+ 6.8	41.6	61.7	....
2,3,4-Pentanetriol	$C_5H_{12}O_3$	155.0	204.5	239.6	263.5	307.0	327.2	....
Ethoxytrimethylsilane	$C_5H_{14}OSi$	− 50.9	− 20.7	+ 3.7	22.1	56.3	75.7	....
Ethyltrimethylsilane	$C_5H_{14}Si$	− 60.6	− 31.8	− 9.0	+ 9.2	42.8	62.0	....
Ethyltrimethyltin	$C_5H_{14}Sn$	− 30.0	+ 3.8	30.0	50.0	87.6	108.8	....
Chloranil	$C_6Cl_4O_2$	70.7_s	97.8_s	116.1_s	129.5_s	151.3_s	162.6_s	290
Hexachlorobenzene	C_6Cl_6	114.4_s	166.4_s	206.0_s	235.5	283.5	309.4	230
Pentachlorobenzene	C_6HCl_5	98.6	144.3	178.5	205.5	251.6	276.0	85.5
Pentachlorophenol	C_6HCl_5O	s	s	211.2	239.6	285.0	309.3_d	188.5
3-Bromo-2,4,6-trichlorophenol	$C_6H_2BrCl_3O$	112.4	163.2	200.5	229.3	278.0	305.8	....
1,2,3,4-Tetrachlorobenzene	$C_6H_2Cl_4$	68.5	114.7	149.2	175.7	225.5	254.0	46.5
1,2,3,5-Tetrachlorobenzene	$C_6H_2Cl_4$	58.2	104.1	140.0	168.0	220.0	246.0	54.5
1,2,4,5-Tetrachlorobenzene	$C_6H_2Cl_4$	s	s	146.0	173.5	220.5	245.0	139
2,3,4,6-Tetrachlorophenol	$C_6H_2Cl_4O$	100.0	145.3	179.1	205.2	250.4	275.0	69.5
2-Bromo-4,6-dichlorophenol	$C_6H_3BrCl_2O$	84.0	130.8	165.8	193.2	242.0	268.0	68
1,2,3-Trichlorobenzene	$C_6H_3Cl_3$	40.0_s	85.6	119.8	146.0	193.5	218.5	52.5

VAPOR PRESSURE (Continued)
ORGANIC COMPOUNDS (Continued)
Pressures Less than One Atmosphere (Continued)

Name	Formula	Temperature °C						M.P.
		1 mm	10 mm	40 mm	100 mm	400 mm	760 mm	
1,2,4-Trichlorobenzene	$C_6H_3Cl_3$	38.4	81.7	114.8	140.0	187.7	213.0	17
1,3,5-Trichlorobenzene	$C_6H_3Cl_3$	s	78.0	110.8	136.0	183.0	208.4	63.5
2,4,5-Trichlorophenol	$C_6H_3Cl_3O$	72.0	117.3	151.5	178.0	226.5	251.8	62
2,4,6-Trichlorophenol	$C_6H_3Cl_3O$	76.5	120.2	152.2	177.8	222.5	246.0	68.5
1,4-Dibromobenzene	$C_6H_4Br_2$	61.0_s	87.7	120.8	146.5	192.5	218.6	87.5
1,4-Bromochlorobenzene	C_6H_4BrCl	32.0	72.7	103.8	128.0	172.6	196.9	......
1,2-Dichlorobenzene	$C_6H_4Cl_2$	20.0	59.1	89.4	112.9	155.8	179.0	− 17.6
1,3-Dichlorobenzene	$C_6H_4Cl_2$	12.1	52.0	82.0	105.0	149.0	173.0	− 24.2
1,4-Dichlorobenzene	$C_6H_4Cl_2$	s	54.8	84.8	108.4	150.2	173.9	53.0
2,4-Dichlorophenol	$C_6H_4Cl_2O$	53.0	92.8	123.4	146.0	187.5	210.0	45.0
2,6-Dichlorophenol	$C_6H_4Cl_2O$	59.5	101.0	131.6	154.6	197.7	220.0	......
2,4,6-Trichloroaniline	$C_6H_4Cl_3N$	134.0	170.0	195.8	214.6	246.4	262.0	78
Dichlorophenylarsine	$C_6H_5AsCl_2$	61.8	116.0	151.0	178.9	228.8	256.5	......
Bromobenzene	C_6H_5Br	+ 2.9	40.0	68.6	90.8	132.3	156.2	− 30.7
Chlorobenzene	C_6H_5Cl	− 13.0	+ 22.2	49.7	70.7	110.0	132.2	− 45.2
2-Chlorophenol	C_6H_5ClO	12.1	51.2	82.0	106.0	149.8	174.5	7.0
3-Chlorophenol	C_6H_5ClO	44.2	86.1	118.0	143.0	188.7	214.0	32.5
4-Chlorophenol	C_6H_5ClO	49.8	92.2	125.0	150.0	196.0	220.0	42
Benzenesulfonylchloride	$C_6H_5ClO_2S$	65.9	112.0	147.7	174.5	224.0	251.5_d	14.5
Phenyl dichlorophosphate	$C_6H_5Cl_2O_2P$	66.7	110.0	143.4	168.0	213.0	239.5	......
Trichlorophenylsilane	$C_6H_5Cl_3Si$	33.0	74.2	105.8	130.5	175.7	201.0	......
Fluorobenzene	C_6H_5F	− 43.4_s	− 12.4	+ 11.5	30.4	65.7	84.7	− 42.1
Trifluorophenylsilane	$C_6H_5F_3Si$	− 31.0	+ 0.8	25.4	44.2	78.7	98.3	......
Iodobenzene	C_6H_5I	24.1	64.0	94.4	118.3	163.9	188.6	− 28.5
Nitrobenzene	$C_6H_5NO_2$	44.4	84.9	115.4	139.9	185.8	210.6	+ 5.7
2-Nitrophenol	$C_6H_5NO_3$	49.3	90.4	122.1	146.4	191.0	214.5	45
1,5-Hexadiene-3-yne	C_6H_6	− 45.1	− 14.0	+ 10.0	29.5	64.4	84.0	......
Benzene	C_6H_6	− 36.7	− 11.5_s	+ 7.6	26.1	60.6	80.1	+ 5.5
2-Chloroaniline	C_6H_6ClN	46.3	84.8	115.6	139.5	183.7	208.8	......
3-Chloroaniline	C_6H_6ClN	63.5	102.0	133.6	158.0	203.5	228.5	− 10.4
4-Chloroaniline	C_6H_6ClN	59.3_s	102.1	135.0	159.9	206.6	230.5	70.5
4-Chlorophenol	C_6H_6ClO	54.3	95.8	126.8	150 6	194.3	217.0	42
2-Nitroaniline	$C_6H_6N_2O_2$	104.0	150.4	186.0	213.0	260.0	284.5_d	71.5
3-Nitroaniline	$C_6H_6N_2O_2$	119.3	167.8	204.2	232.1	280.2	305.7_d	114
4-Nitroaniline	$C_6H_6N_2O_2$	142.4_s	194.4	234.2	261.8	310.2_d	336.0_d	146.5
Phenol	C_6H_6O	40.1_s	73.8	100.1	121.4	160.0	181.9	40.6
Pyrocatechol	$C_6H_6O_2$	s	118.3	150.6	176.0	221.5	245.5	105
Resorcinol	$C_6H_6O_2$	108.4_s	152.1	185.3	209.8	253.4	276.5	110.7
Hydroquinone	$C_6H_6O_2$	132.4_s	163.5_s	192.0	216.5	262.5	286.2	170.3
Pyrogallol	$C_6H_6O_3$	s	167.7	204 2	232.0	281.5	309.0_d	133
Benzenethiol	C_6H_6S	18.6	56.0	84.2	106.6	146.7	168.0	......
Aniline	C_6H_7N	34.8	69.4	96.7	119.9	161.9	184.4	− 6.2
2-Picoline	C_6H_7N	− 11.1	+ 24.4	51.2	71.4	108.4	128.8	− 70
Ethylene-bis-(chloroacetate)	$C_6H_8Cl_2O_4$	112.0	158.0	191.0	215.0	259.5	283.5	......
1,3-Phenylenediamine	$C_6H_8N_2$	99.8	147.0	182.5	209.9	259.0	285.5	62.8
Phenylhydrazine	$C_6H_8N_2$	71.8	115.8	148.2	173.5	218.2	243.5_d	19.5
α-Methylglutaric anhydride	C_6H[illegible]O_3	93.8	141.8	177.5	205.0	255.5	282.5	......
α,α-Dimethylsuccinic anhydride	C_6H[illegible]O_3	61.4	102.0	132.3	155.3	197.5	219.5	......
Dimethyl maleate	$C_6H_8O_4$	45.7	86.4	117.2	140.3	182.2	205.0	......
Isobutyl dichloroacetate	$C_6H_{10}Cl_2O_2$	28.6	67.5	96.7	119.8	160.0	183.0	......
Diallyldichlorosilane	$C_6H_{10}Cl_2Si$	+ 9.5	47.4	76.4	99.7	142.0	165.3	......
Cyclohexanone	$C_6H_{10}O$	+ 1.4	38.7	67.8	90.4	132.5	155.6	− 45.0
Mesityl oxide	$C_6H_{10}O$	− 8.7	+ 26.0	51.7	72.1	109.8	130.0	− 59
Isocaprolactone	$C_6H_{10}O_2$	38.3	80.3	112 3	137.2	182 1	207.0	......
Propionic anhydride	$C_6H_{10}O_3$	20.6	57.7	85.6	107.2	146.0	167.0	− 45
Ethyl acetoacetate	$C_6H_{10}O_3$	28.5	67.3	96.2	118.5	158.2	180.8	− 45
Methyl levulinate	$C_6H_{10}O_3$	39 8	79.7	109.5	133 0	175 8	197.7	......
Adipic acid	$C_6H_{10}O_4$	159.5	205.5	240 5	265.0	312.5	337.5	152
Diethyl oxalate	$C_6H_{10}O_4$	47.4	83 8	110.6	130 8	166.2	185.7	− 40.6
Glycol diacetate	$C_6H_{10}O_4$	38.3	77.1	106.1	128.0	168.3	190.5	− 31
Dimethyl-*l*-malate	$C_6H_{10}O_5$	75.4	118.3	150.1	175.1	219.5	242.6	......
Dimethyl-*d*-tartrate	$C_6H_{10}O_6$	102.1	148.2	182.4	208.8	255.0	280.0	61.5
Dimethyl-*dl*-tartrate	$C_6H_{10}O_6$	100.4	147.5	182.4	209.5	257.4	282.0	89
Diallyl sulfide	$C_6H_{10}S$	− 9.5	+ 26.6	54.2	75.8	116.1	138.6	− 83
Ethyl α-bromoisobutyrate	$C_6H_{11}BrO_2$	10.6	48.0	77.0	99.8	141.2	163.6	......
sec-Butyl chloroacetate	$C_6H_{11}ClO_2$	17.0	54.6	83.6	105.5	146.0	167.8	......
Capronitrile	$C_6H_{11}N$	+ 9.2	47.5	76.9	99.8	141.0	163.7	......
1-Hexene	C_6H_{12}	− 57.5	− 28.1	− 5.0	+ 13.0	46.8	66.0	− 98.5
Cyclohexane	C_6H_{12}	− 45.3_s	− 15.9_s	+ 6.7	25.5	60.8	80.7	+ 6.6
Methylcyclopentane	C_6H_{12}	− 53.7	− 23.7	− 0.6	+ 17.9	52.3	71.8	−142.4
Dichlorodiisopropyl ether	$C_6H_{12}Cl_2O$	29.6	68.2	97.3	119.7	159.8	182.7	......
Bis-2-chloroethyl acetal	$C_6H_{12}Cl_2O_2$	56.2	97.6	127.8	150.7	190.5	212.6	......
2-Hexanone	$C_6H_{12}O$	+ 7.7	38.8	62.0	79.8	111.0	127.5	− 56.9
4-Methyl-2-pentanone	$C_6H_{12}O$	− 1.4	+ 30.0	52.8	70.4	102.0	119.0	− 84.7
Allyl propyl ether	$C_6H_{12}O$	− 39.0	− 7.9	+ 16.4	35.8	71.4	90.5	......
Allyl isopropyl ether	$C_6H_{12}O$	− 43.7	− 12.9	+ 10.9	29.0	61.7	79.5	......
Cyclohexanol	$C_6H_{12}O$	21.0_s	56.0	83.0	103.7	141.4	161.0	23.9
Caproic acid	$C_6H_{12}O_2$	71.4	99.5	125.0	144.0	181.0	202.0	1.5
Isocaproic acid	$C_6H_{12}O_2$	66.2	94.0	120.4	141.4	181.0	207.7	− 35
4-Hydroxy-4-methyl-2-pentanone	$C_6H_{12}O_2$	22.0	58.8	86.7	108.2	147.5	167.9	− 47
Methyl isovalerate	$C_6H_{12}O_2$	− 19.2	+ 14.0	39.8	59.8	96.7	116.7	......
Ethyl butyrate	$C_6H_{12}O_2$	− 18.4	+ 15.3	41.5	62.0	100.0	121.0	− 93.3
Ethyl isobutyrate	$C_6H_{12}O_2$	− 24.3	+ 8.4	33.8	53.5	90.0	110.1	− 88.2
Propyl propionate	$C_6H_{12}O_2$	− 14.2	+ 19.4	45.0	65.2	102.0	122.4	− 76
Isobutyl acetate	$C_6H_{12}O_2$	− 21.2	+ 12.8	39.2	59.7	97.5	118.0	− 98.9

VAPOR PRESSURE (Continued)
ORGANIC COMPOUNDS (Continued)
Pressures Less than One Atmosphere (Continued)

Name	Formula	Temperature °C						M.P.
		1 mm	10 mm	40 mm	100 mm	400 mm	760 mm	
Isoamyl formate	$C_6H_{12}O_2$	−17.5	+17.1	44.0	65.4	102.7	123.3	......
Paraformaldehyde	$C_6H_{12}O_3$	−9.4$_s$	+24.1$_s$	49.5$_s$	69.0$_s$	104.3$_s$	124.0$_s$	155 ± 5
sec-Butyl glycolate	$C_6H_{12}O_3$	28.3	66.0	94.2	116.4	155.6	177.5	......
Hexane	C_6H_{14}	−53.9	−25.0	−2.3	+15.8	49.6	68.7	−95.3
2-Methylpentane	C_6H_{14}	−60.9	−32.1	−9.7	+8.1	41.6	60.3	−154
3-Methylpentane	C_6H_{14}	−59.0	−30.1	−7.3	+10.5	44.2	63.3	−118
2,2-Dimethylbutane	C_6H_{14}	−69.3	−41.5	−19.5	−2.0	+31.0	49.7	−99.8
2,3-Dimethylbutane	C_6H_{14}	−63.6	−34.9	−12.4	+5.4	39.0	58.0	−128.2
1-Hexanol	$C_6H_{14}O$	24.4	58.2	83.7	102.8	138.8	157.0	−51 6
2-Hexanol	$C_6H_{14}O$	14.6	45.0	67.9	87.3	121.8	139.9	......
3-Hexanol	$C_6H_{14}O$	+2.5	36.7	62.2	81.8	117.0	135.5	......
2-Methyl-1-pentanol	$C_6H_{14}O$	15.4	49 6	74.7	94.2	129.8	147.9	......
2-Methyl-2-pentanol	$C_6H_{14}O$	−4.5	+27.6	51.3	69.2	102.6	121 1	−103
2-Methyl-4-pentanol	$C_6H_{14}O$	−0.3	+33.3	58.2	78.0	113.5	131.7	......
Dipropyl ether	$C_6H_{14}O$	−43.3	−11.8	+13.2	33.0	69.5	89.5	−122
Diisopropyl ether	$C_6H_{14}O$	−57.0	−27.4	−4.5	+13.7	48.2	67.5	−60
Acetal	$C_6H_{14}O_2$	−23.0	+8.0	31.9	50.1	84.0	102.0	......
1,2-Diethoxyethane	$C_6H_{14}O_2$	−33.5	+1.6	29.7	51.8	94.1	119.5	......
Di(2-methoxyethyl) ether	$C_6H_{14}O_3$	13.0	50.0	77.5	99.5	138.5	159.8	......
Diethylene glycol, ethyl ether	$C_6H_{14}O_3$	45.3	85.8	116.7	140.3	180.3	201.9	......
Dipropyleneglycol	$C_6H_{14}O_3$	73.8	116.2	147.4	169.9	210.5	231.8	......
Triethyleneglycol	$C_6H_{14}O_4$	114.0	158.1	191.3	214.6	256.6	278.3	......
Triethylboron	$C_6H_{15}B$	......	−148.0	−131.4	−116.0	−81.0	−56.2	......
Chlorotriethylsilane	$C_6H_{15}ClSi$	−4.9	+32.0	60.2	82.3	123.6	146.3	......
Triethyl phosphate	$C_6H_{15}O_4P$	39.6	82.1	115.7	141.6	187.0	211.0	......
Triethylthallium	$C_6H_{15}Tl$	+9.3	51.7	85.4	112.1	163.5	192.1$_d$	−63.0
Diethoxydimethylsilane	$C_6H_{16}O_2Si$	−19.1	+13.3	38.0	57.6	93.2	113.5	......
Trimethylpropylsilane	$C_6H_{16}Si$	−46.0	−13.9	+11.3	31.6	69.2	90.0	......
Trimethylpropyltin	$C_6H_{16}Sn$	−12.0	+21.8	48.5	69.8	109.6	131.7	......
1,5-Dichlorohexamethyltrisiloxane	$C_6H_{18}Cl_2O_2Si_3$	26.0	65.1	94.8	118.2	160.2	184.0	−53
Hexamethylcyclotrisiloxane	$C_6H_{18}O_3Si_3$	s	s	s	78.7	114.7	134.0	64
Hexamethyldisiloxane	$C_6H_{18}OSi_2$	−29.0	+2.8	26.7	45.6	80.0	99.2	......
3,4-Dichloro-α,α,α-trifluorotoluene	$C_7H_3Cl_2F_3$	11.0	52.2	84.0	109.2	150.5	172.8	−12.1
2-Chloro-α,α,α-trifluorotoluene	$C_7H_4ClF_3$	0.0	37.1	65.9	88.3	130.0	152.2	−6.0
2-α,α,α-tetrachlorotoluene	$C_7H_4Cl_4$	69.0	117.9	155.0	185.0	233.0	262.1	28.7
Benzoyl bromide	C_7H_5BrO	47.0	89.8	122.6	147.7	193.7	218.5	0
Benzoyl chloride	C_7H_5ClO	32.1	73.0	103.8	128.0	172.8	197.2	−0.5
α,α,α-Trichlorotoluene	$C_7H_5Cl_3$	45.8	87.6	119.8	144.3	189.2	213.5	−21.2
α,α,α-Trifluorotoluene	$C_7H_5F_3$	−32.0$_s$	+0.4	25.7	45.3	82.0	102.2	−29.3
Benzonitrile	C_7H_5N	28.2	69.2	99.6	123.5	166.7	190.6	−12.9
Phenyl isocyanide	C_7H_5N	12.0	49.7	78.3	101.0	142.3	165.0$_d$	......
Phenyl isocyanate	C_7H_5NO	10.6	48.5	77.7	100.6	142.7	165.6	......
2-Nitrobenzaldehyde	$C_7H_5NO_3$	85.8	133.4	168.8	196.2	246.8	273.5	40.9
3-Nitrobenzaldehyde	$C_7H_5NO_3$	96.2	142.8	177.7	204.3	252.1	278.3	58
Phenyl isothiocyanate	C_7H_5NS	47.2	89.8	122.5	147.7	194.0	218.5	−21.0
α,α-dichlorotoluene	$C_7H_6Cl_2$	35.4	78.7	112.1	138.3	187.0	214.0	−16.1
Benzaldehyde	C_7H_6O	26.2	62.0	90.1	112.5	154.1	179.0	−26
Benzoic acid	$C_7H_6O_2$	96.0$_s$	132.1	162.6	186.2	227.0	249.2	121.7
Salicylaldehyde	$C_7H_6O_2$	33.0	73.8	105.2	129.4	173.7	196.5	−7
4-Hydroxybenzaldehyde	$C_7H_6O_2$	121.2	169.7	206.0	233.5	282.6	310.0	115.5
Salicylic acid	$C_7H_6O_3$	113.7$_s$	146.2$_s$	172.2	193.4	230.5	256.0	159
α-Bromotoluene	C_7H_7Br	32.2	73.4	104.8	129.8	175.2	198.5	−4
2-Bromotoluene	C_7H_7Br	24.4	62.3	91.0	112.0	157.3	181.8	−28
3-Bromotoluene	C_7H_7Br	14.8	64.0	93.9	117.8	160.0	183.7	−39.8
4-Bromotoluene	C_7H_7Br	10.3	61.1	91.8	116.4	160.2	184.5	28.5
4-Bromoanisole	C_7H_7BrO	48.8	91.9	125.0	150.1	197.5	223.0	12.5
α-Chlorotoluene	C_7H_7Cl	22.0	60.8	90.7	114.2	155.8	179.4	−39
2-Chlorotoluene	C_7H_7Cl	+5.4	43.2	72.0	94.7	137.1	159.3	......
3-Chlorotoluene	C_7H_7Cl	+4.8	43.2	73.0	96.3	139.7	162.3	......
4-Chlorotoluene	C_7H_7Cl	+5.5	43.8	73.5	96.6	139.8	162.3	+7.3
2-Fluorotoluene	C_7H_7F	−24.2	+8.9	34.7	55.3	92.8	114.0	−80
3-Fluorotoluene	C_7H_7F	−22.4	+11.0	37.0	57.5	95.4	116.0	−110.8
4-Fluorotoluene	C_7H_7F	−21.8	+11.8	37.8	58.1	96.1	117.0	......
2-Iodotoluene	C_7H_7I	37.2	79.8	112.4	138.1	185.7	211.0	......
2-Nitrotoluene	$C_7H_7NO_2$	50.0	93.8	126.3	151.5	197.7	222.3	−4.1
3-Nitrotoluene	$C_7H_7NO_2$	50.2	96.0	130.7	156.9	206.8	231.9	15.1
4-Nitrotoluene	$C_7H_7NO_2$	53.7	100.5	136.0	163.0	212.5	238.3	51.9
Toluene	C_7H_8	−26.7	+6.4	31.8	51.9	89.5	110.6	−95.0
Benzyldichlorosilane	$C_7H_8Cl_2Si$	45.3	83.2	111.8	133.5	173.0	194.3	......
Dichloromethylphenylsilane	$C_7H_8Cl_2Si$	35.7	77.4	109.5	134.2	180.2	205.5	......
Dichloro-4-tolysilane	$C_7H_8Cl_2Si$	46.2	84.2	113.2	135.5	175.2	196.3	......
Anisole	C_7H_8O	+5.4	42.2	70.7	93.0	133.8	155.5	−37.3
Benzyl alcohol	C_7H_8O	58.0	92.6	119.8	141 7	183.0	204.7	−15.3
2-Cresol	C_7H_8O	38.2	76.7	105.8	127.4	168.4	190 8	30.8
3-Cresol	C_7H_8O	52.0	87.8	116 0	138.0	179.0	202.8	10.9
4-Cresol	C_7H_8O	53.0	88.6	117.7	140.0	179.4	201.8	35.5
3,5-Dimethyl-1,2-pyrone	$C_7H_8O_2$	78.6	122.0	152.7	177.5	221.0	245.0	51.5
2-Methoxyphenol	$C_7H_8O_2$	52.4	92.0	121.6	144.0	184.1	205.0	28.3
Ethyl 2-furoate	$C_7H_8O_3$	37.6	77.1	107.5	130.4	172.5	195.0	34

Name	Formula	Temperature °C						M.P.
		1 mm	10 mm	40 mm	100 mm	400 mm	760 mm	
Benzylamine	C_7H_9N	29.0	67.7	97.3	120.0	161.3	184.5	......
N-Methylaniline	C_7H_9N	36.0	76.2	106.0	129.8	172.0	195.5	− 57
2-Toluidine	C_7H_9N	44.0	81.4	110.0	133.0	176.2	199.7	− 16.3
3-Toluidine	C_7H_9N	41.0	82.0	113.5	136.7	180.6	203.3	− 31.5
4-Toluidine	C_7H_9N	42.0	81.8	111.5	133.7	176.9	200.4	44.5
2-Methoxyaniline	C_7H_9NO	61.0	101.7	132.0	155.2	197.3	218.5	5.2
Toluene-2,4-diamine	$C_7H_{10}N_2$	106.5	151.7	185.7	211.5	256.0	280.0	99
4-Tolylhydrazine	$C_7H_{10}N_2$	82.0	123.8	154.1	178.0	219.5	242.0_d	65.5
Trimethysuccinic anhydride	$C_7H_{10}O_3$	53.5	97.4	131.0	156.5	205.5	231.0	......
Dimethyl citraconate	$C_7H_{10}O_4$	50.8	91.8	122.6	145.8	188.0	210.5	......
Dimethyl itaconate	$C_7H_{10}O_4$	69.3	106.6	133.7	153.7	189.8	208.0	38
trans-Dimethyl mesaconate	$C_7H_{10}O_4$	46.8	87.8	118.0	141.5	183.5	206.0	......
2-Cyano-2-butyl acetate	$C_7H_{11}NO_2$	42.0	82.0	111.8	133.8	173.4	195.2	......
Butyl acrylate	$C_7H_{12}O_2$	− 0.5	+ 35.5	63.4	85.1	125.2	147.4	− 64.6
Ethyl levulinate	$C_7H_{12}O_3$	47.3	87.3	117.7	141.3	183.0	206.2	......
Pimelic acid	$C_7H_{12}O_4$	163.4	212.0	247.0	272.0	318.5	342.1	103
Diethyl malonate	$C_7H_{12}O_4$	40.0	81.3	113.3	136.2	176.8	198.9	− 49.8
Enanthyl chloride	$C_7H_{13}ClO$	34.2	64.6	86.4	102.7	130.7	145.0	......
Enanthonitrile	$C_7H_{13}N$	21.0	61.6	92.6	116.8	160.0	184.6	......
Ethylcyclopentane	C_7H_{14}	− 32.2	− 0.1	+ 25.0	45.0	82.3	103.4	−138.6
Methylcyclohexane	C_7H_{14}	− 35.9	− 3.2	+ 22.0	42.1	79.6	100.9	−126.4
2-Heptene	C_7H_{14}	− 35.8	− 3.5	+ 21.5	41.3	78.1	98.5	......
Enanthaldehyde	$C_7H_{14}O$	12.0	43.0	66.3	84.0	125.5	155.0	− 42
2-Heptanone	$C_7H_{14}O$	19.3	55.5	81.2	100.0	133.2	150.2	......
4-Heptanone	$C_7H_{14}O$	23.0	55.0	78.1	96.0	127.3	143.7	− 32.6
2,4-Dimethyl-3-pentanone	$C_7H_{14}O$	+ 5.2	36.7	59.6	77.0	108.0	123.7	......
Enanthic acid	$C_7H_{14}O_2$	78.0	113.2	139.5	160.0	199.6	221.5	− 10
Methyl caproate	$C_7H_{14}O_2$	+ 5.0	42.0	70.0	91.4	129.8	150.0_d	......
Ethyl isovalerate	$C_7H_{14}O_2$	− 6.1	+ 28.7	55.2	75.9	114.0	134.3	− 99.3
Propyl butyrate	$C_7H_{14}O_2$	− 1.6	+ 34.0	61.5	82.6	121.7	142.7	− 95.2
Propyl isobutyrate	$C_7H_{14}O_2$	− 6.2	+ 28.3	54.3	73.9	112.0	133.9	......
Isopropyl isobutyrate	$C_7H_{14}O_2$	− 16.3	+ 17.0	42.4	62.3	100.0	120.5	......
Isobutyl propionate	$C_7H_{14}O_2$	− 2.3	+ 32.3	58.5	79.5	116.4	136.8	− 71
Isoamyl acetate	$C_7H_{14}O_2$	0.0	+ 35.2	62.1	83.2	121.5	142.0	......
Heptane	C_7H_{16}	− 34.0	− 2.1	+ 22.3	41.8	78.0	98.4	− 90.6
2-Methylhexane	C_7H_{16}	− 40.4	− 9.1	+ 14.9	34.1	69.8	90.0	−118.2
3-Methylhexane	C_7H_{16}	− 39.0	− 7.8	+ 16.4	35.6	71.6	91.9	−119.4
3-Ethylpentane	C_7H_{16}	− 37.8	− 6.8	+ 17.5	36.9	73.0	93.5	−118.6
2,2-Dimethylpentane	C_7H_{16}	− 49.0	− 18.7	+ 5.0	23.9	59.2	79.2	−123.7
2,3-Dimethylpentane	C_7H_{16}	− 42.0	− 10.3	+ 13.9	33.3	69.4	89.8	−135
2,4-Dimethylpentane	C_7H_{16}	− 48.0	− 17.1	+ 6.5	25.4	60.6	80.5	−119.5
3,3-Dimethylpentane	C_7H_{16}	− 45.9	− 14.4	+ 9.9	29.3	65.5	86.1	−135.0
2,2,3-Trimethylbutane	C_7H_{16}	s	− 18.8	+ 5.2	24.4	60.4	80.9	− 25.0
1-Heptanol	$C_7H_{16}O$	42.4	74.7	99.8	119.5	155.6	175.8	34.6
Triethyl orthoformate	$C_7H_{16}O_3$	+ 5.5	40.5	67.5	88.0	125.7	146.0	......
Triethoxymethylsilane	$C_7H_{18}O_3Si$	− 1.5	+ 34.6	61.7	82.7	121.8	143.5	......
Butyltrimethylsilane	$C_7H_{18}Si$	− 23.4	+ 9.9	35.9	56.3	93.8	115.0	......
Triethylmethylsilane	$C_7H_{18}Si$	− 18.2	+ 16.6	44.0	65.6	105.3	127.0	......
Phthaloyl chloride	$C_8H_4Cl_2O_2$	86.3_s	134.2	170.0	197.8	248.3	275.8	88.5
Phthalic anhydride	$C_8H_4O_3$	96.5_s	134.0	172.0	202.3	256.8	284.5	130.8
α,α-Dichlorophenylacetonitrile	$C_8H_5Cl_2N$	56.0	98.1	130.0	154.5	199.5	223.5	......
Pentachloroethylbenzene	$C_8H_5Cl_5$	96.2	148.0	186.2	216.0	269.3	299.0	......
Phenylglyoxylonitrile	C_8H_5NO	44.5	85.5	116.6	141.0	185.0	208.0	33.5
2,3-Dichlorostyrene	$C_8H_6Cl_2$	61.0	104.6	137.8	163.5	210.0_p	235.0_p	......
2,4-Dichlorostyrene	$C_8H_6Cl_2$	53.5	97.4	129.2	153.8	200.0_p	225.0_p	......
2,5-Dichlorostyrene	$C_8H_6Cl_2$	55.5	98.2	131.0	155.8	202.5_p	227.0_p	......
2,6-Dichlorostyrene	$C_8H_6Cl_2$	47.8	90.0	122.4	147.6	193.5_p	217.0_p	......
3,4-Dichlorostyrene	$C_8H_6Cl_2$	57.2	100.4	133.7	158.2	205.7_p	230.0_p	......
3,5-Dichlorostyrene	$C_8H_6Cl_2$	53.5	97.4	129.2	153.8	200.0_p	225.0_p	......
3,4,5,6-Tetrachloro-1,2-xylene	$C_8H_6Cl_4$	94.4	140.3	174.2	200.5	248.3	273.5	......
1,2,3,5-Tetrachloro-4-ethylbenzene	$C_8H_6Cl_4$	77.0	126.0	162.1	191.6	243.0	270.0	......
Phenylglyoxal	$C_8H_6O_2$	s	87.8	115.5	136.2	173.5	193.5	73
Phthalide	$C_8H_6O_2$	95.5	144.0	181.0	210.0	261.8	290.0	73
Piperonal	$C_8H_6O_3$	87.0	132.0	165.7	191.7	238.5	263.0	37
3-Chlorostyrene	C_8H_7Cl	25.3	65.2	96.5	121.2	165.7_p	190.0_p	......
4-Chlorostyrene	C_8H_7Cl	28.0	67.5	98.0	122.0	166.0_p	191.0_p	− 15.0
Phenylacetyl chloride	C_8H_7ClO	48.0	89.0	119.8	143.5	186.0	210.0	......
2-Tolunitrile	C_8H_7N	36.7	77.9	110.0	135.0	180.0	205.2	− 13
4-Tolunitrile	C_8H_7N	42.5	85.8	109.5	145.2	193.0	217.6	29.5
Phenylacetonitrile	C_8H_7N	60.0	103.5	136.3	161.8	208.5	233.5	− 23.8
2-Tolyl isocyanide	C_8H_7N	25.2	64.0	94.0	117.7	159.9	183.5	......
2-Nitrophenyl acetate	$C_8H_7NO_4$	100.0	142.0	172.8	194.1	233.5	253.0_d	......
2-Methylbenzothiazole	C_8H_7NS	70.0	111.2	141.2	163.9	204.5	225.5	15.4
Benzylisothiocyanate	C_8H_7NS	79.5	121.8	153.0	177.7	220.4	243.0	......
Styrene	C_8H_8	− 7.0	+ 30.8	59.8	82.0_p	122.5_p	145.2_p	− 30.6
(1,2-Dibromoethyl) benzene	$C_8H_8Br_2$	86.0	129.8	161.8	186.3	230.0	254.0	......
1,2-Dichloro-3-ethylbenzene	$C_8H_8Cl_2$	46.0	90.0	123.8	149.8	197.0	222.1	− 40.8
1,2-Dichloro-4-ethylbenzene	$C_8H_8Cl_2$	47.0	92.3	127.5	153.3	201.7	226.6	− 76.4
1,4-Dichloro-2-ethylbenzene	$C_8H_8Cl_2$	38.5	83.2	118.0	144.0	191.5	216.3	− 61.2
Acetophenone	C_8H_8O	37.1	78.0	109.4	133.6	178.0	202.4	20.5
Phenyl acetate	$C_8H_8O_2$	38.2	78.0	108.1	131.6	173.5	195.9	......
Phenylacetic acid	$C_8H_8O_2$	97.0	141.3	173.6	198.2	243.0	265.5	76.5
Anisaldehyde	$C_8H_8O_2$	73.2	117.8	150.5	176.7	223.0	248.0	2.5
Methyl benzoate	$C_8H_8O_2$	39.0	77.3	107.8	130.8	174.7	199.5	− 12.5
Methyl salicylate	$C_8H_8O_3$	54.0	95.3	126.2	150.0	197.5	223.2	− 8.3
Vanillin	$C_8H_8O_3$	107.0	154.0	188.7	214.5	260.0	285.0	81.5
Dehydroacetic acid	$C_8H_8O_4$	91.7	137.3	171.0	197.5	244.5	269.0	......
2-Bromo-1,4-xylene	C_8H_9Br	37.5	78.8	110.6	135.7	181.0	206.7	+ 9.5

VAPOR PRESSURE (Continued)

ORGANIC COMPOUNDS (Continued)

Pressures Less than One Atmosphere (Continued)

Name	Formula	Temperature °C						M.P.
		1 mm	10 mm	40 mm	100 mm	400 mm	760 mm	
1-Bromo-4-ethylbenzene	C_8H_9Br	30.4	74.0	108.5	135.5	182.0	206.0	− 45.0
(2-Bromoethyl) benzene	C_8H_9Br	48.0	90.5	123.2	148.2	194.0	219.0	......
1-Chloro-2-ethylbenzene	C_8H_9Cl	17.2	56.1	86.2	110.0	152.2	177.6	− 80.2
1-Chloro-3-ethylbenzene	C_8H_9Cl	18.6	58.2	89.2	113.6	156.7	181.1	− 53.3
1-Chloro-4-ethylbenzene	C_8H_9Cl	19.2	60.0	91.8	116.0	159.8	184.3	− 62.6
1-Chloro-2-ethoxybenzene	C_8H_9ClO	45.8	86.5	117.8	141.8	185.5	208.0	......
4-Chlorophenethyl alcohol	C_8H_9ClO	84.0	129.0	162.0	188.1	234.5	259.3	......
Acetanilide	C_8H_9NO	114.0	162.0	199.6	227.2	277.0	303.8	113.5
Methyl anthranilate	$C_8H_9NO_2$	77.6	124.2	159.7	187.8	238.5	266.5	24
4-Nitro-1,3-xylene	$C_8H_9NO_2$	65.6	109.8	143.3	168.5	217.5	244.0	+ 2
Ethylbenzene	C_8H_{10}	− 9.8	+ 25.9	52.8	74.1	113.8	136.2	− 94.9
2-Xylene	C_8H_{10}	− 3.8	+ 32.1	59.5	81.3	121.7	144.4	− 25.2
3-Xylene	C_8H_{10}	− 6.9	+ 28.3	55.3	76.8	116.7	139.1	− 47.9
4-Xylene	C_8H_{10}	− 8.1	+ 27.3	54.4	75.9	115.9	138.3	+ 13.3
Dichloroethoxyphenylsilane	$C_8H_{10}Cl_2OSi$	52.4	94.6	126.2	151.4	197.2	222.2	......
Dichloroethylphenylsilane	$C_8H_{10}Cl_2Si$	48.5	92.4	126.7	153.3	203.5	230.0	......
2-Ethylphenol	$C_8H_{10}O$	46.2	87.0	117.9	141.8	184.5	207.5	− 45
3-Ethylphenol	$C_8H_{10}O$	60.0	100.2	130.0	152.0	193.3	214.0	− 4
4-Ethylphenol	$C_8H_{10}O$	59.3	100.2	131.3	154.2	197.4	219.0	46.5
2,3-Xylenol	$C_8H_{10}O$	56.0_8	97.6	129.2	152.2	196.0	218.0	75
2,4-Xylenol	$C_8H_{10}O$	51.8	91.3	121.5	143.0	184.2	211.5	25.5
2,5-Xylenol	$C_8H_{10}O$	51.8_8	91.3	121.5	143.0	184.2	211.5	74.5
3,4-Xylenol	$C_8H_{10}O$	66.2	107.7	138.0	161.0	203.6	225.2	62.5
3,5-Xylenol	$C_8H_{10}O$	62.0_8	102.4	133.3	156.0	197.8	219.5	68
Phenetole	$C_8H_{10}O$	18.1	56.4	86.6	108.4	149.8	172.0	− 30.2
α-Methyl benzyl alcohol	$C_8H_{10}O$	49.0	88.0	117.8	140.3	180.7	204.0	......
Phenethylalcohol	$C_8H_{10}O$	58.2	100.0	130.5	154.0	197.5	219.5	......
4,6-Dimethylresorcinol	$C_8H_{10}O_2$	49.0	90.7	122.5	147.3	192.0	215.0	......
2-Phenoxyethanol	$C_8H_{10}O_2$	78.0	121.2	152.2	176.5	221.0	245.3	11.6
Diethyl dioxosuccinate	$C_8H_{10}O_6$	70.0	112.0	143.8	167.7	210.8	233.5	......
Chlorodimethylphenylsilane	$C_8H_{11}ClSi$	29.8	70.0	101.2	124.7	168.6	193.5	......
N-Ethylaniline	$C_8H_{11}N$	38.5	80.6	113.2	137.3	180.8	204.0	− 63.5
N,N-Dimethylaniline	$C_8H_{11}N$	29.5	70.0	101.6	125.8	169.2	193.1	+ 2.5
4-Ethylaniline	$C_8H_{11}N$	52.0	93.8	125.7	149.8	194.2	217.4	− 4
2,4-Xylidine	$C_8H_{11}N$	52.6	93.0	123.8	146.8	188.3	211.5	......
2,6-Xylidine	$C_8H_{11}N$	44.0	87.0	120.2	146.0	193.7	217.9	......
2-Phenetidine	$C_8H_{11}NO$	67.0	108.6	139.9	163.5	207.0	228.0	......
2-Anilinoethanol	$C_8H_{11}NO$	104.0	149.6	183.7	209.5	254.5	279.6	......
Dimethyl arsanilate	$C_8H_{12}AsNO_3$	15.0	51.8	79.7	101.0	140.3	160.5	......
Diethyleneglycol-bis-chloroacetate	$C_8H_{12}Cl_2O_5$	148.3	195.8	229.0	252.0	291.8	313.0	......
Diethyl maleate	$C_8H_{12}O_4$	57.3	100.0	131.8	156.0	201.7	225.0	......
Diethyl fumarate	$C_8H_{12}O_4$	53.2	95.3	126.7	151.1	195.8	218.5	+ 0.6
Dimethylphenylsilane	$C_8H_{12}Si$	+ 5.3	42.6	71.4	94.2	136.4	159.3	......
Ethyl-α-ethylacetoacetate	$C_8H_{14}O_3$	40.5	80.2	110.3	133.8	175.6	198.0	......
Propyl levulinate	$C_8H_{14}O_3$	59.7	99.9	130.1	154.0	198.0	221.2	......
Isopropyl levulinate	$C_8H_{14}O_3$	48.0	88.0	118.1	141.8	185.2	208.2	......
Dipropyl oxalate	$C_8H_{14}O_4$	53.4	93.9	124.6	148.1	190.3	213.5	......
Diisopropyl oxalate	$C_8H_{14}O_4$	43.2	81.9	110.5	132.6	171.8	193.5	......
Diethyl succinate	$C_8H_{14}O_4$	54.6	96.6	127.8	151.1	193.8	216.5	− 20.8
Diethyl isosuccinate	$C_8H_{14}O_4$	39.8	80.0	111.0	134.8	177.7	201.3	......
Suberic acid	$C_8H_{14}O_4$	172.8	219.5	254.6	279.0	322.8	345.5	142
Diethyl malate	$C_8H_{14}O_5$	80.7	125.6	157.8	183.9	229.5	253.4	......
Diethyl-*dl*-tartrate	$C_8H_{14}O_6$	100.0	147.2	181.7	208.0	254.3	280.0	......
Diethyl-*d*-tartrate	$C_8H_{14}O_6$	102.0	148.0	182.3	208.5	254.8	280.0	......
(2-Bromoethyl) cyclohexane	$C_8H_{15}Br$	38.7	80.5	113.0	138.0	186.2	213.0	......
Caprylonitrile	$C_8H_{15}N$	43.0	80.4	110.6	134.8	179.5	204.5	......
Ethyl *N,N*-diethyloxamate	$C_8H_{15}NO_3$	76.0	121.7	154.4	180.3	226.5	252.0	......
2-Methyl-2-heptene	C_8H_{16}	− 16.1	+ 17.8	44.0	64.6	102.2	122.5	......
1,1-Dimethylcyclohexane	C_8H_{16}	− 24.4	+ 10.3	37.3	57.9	97.2	119.5	− 34
cis-1,2-Dimethylcyclohexane	C_8H_{16}	− 15.9	+ 18.4	45.3	66.8	107.0	129.7	− 50.0
trans-1,2-Dimethylcyclohexane	C_8H_{16}	− 21.1	+ 13.0	39.7	61.0	100.9	123.4	− 88.0
cis-1,3-Dimethylcyclohexane	C_8H_{16}	− 22.7	+ 11.2	37.5	58.5	97.8	120.1	− 76.2
trans-1,3-Dimethylcyclohexane	C_8H_{16}	− 19.4	+ 14.9	41.4	62.5	102.1	124.4	− 92.0
cis-1,4-Dimethylcyclohexane	C_8H_{16}	− 20.0	+ 14.5	41.1	62.3	101.9	124.3	− 87.4
trans-1,4-Dimethylcyclohexane	C_8H_{16}	− 24.3	+ 10.1	36.5	57.6	97.0	119.3	− 36.9
Ethylcyclohexane	C_8H_{16}	− 14.5	+ 20.6	47.6	69.0	109.1	131.8	−111.3
Caprylaldehyde	$C_8H_{16}O$	73.4	101.2	120.0	133.9	156.5	168.5	......
Cyclohexaneethanol	$C_8H_{16}O$	50.4	90.0	119.8	142.7	183.5	205.4	......
6-Methyl-3-hepten-2-ol	$C_8H_{16}O$	41.6	76.7	102.7	122.6	156.6	175.5	......
6-Methyl-5-hepten-2-ol	$C_8H_{16}O$	41.9	77.8	104.0	123.8	156.6	174.3	......
2-Octanone	$C_8H_{16}O$	23.6	60.9	89.8	111.7	151.0	172.9	− 16
2,2,4-Trimethyl-3-pentanone	$C_8H_{16}O$	14.7	46.4	69.8	87.6	118.4	135.0	......
Caprylic acid	$C_8H_{16}O_2$	92.3	124.0	150.6	172.2	213.9	237.5	16
Ethyl isocaproate	$C_8H_{16}O_2$	11.0	48.0	76.3	98.4	139.2	160.4	......
Propyl isovalerate	$C_8H_{16}O_2$	+ 8.0	45.1	72.8	95.0	135.0	155.9	......
Isobutyl butyrate	$C_8H_{16}O$	+ 4.6	42.2	71.7	94.0	135.7	156.9	......
Isobutyl isobutyrate	$C_8H_{16}O_2$	+ 4.1	39.9	67.2	88.0	126.3	147.5	− 80.7
Amyl propionate	$C_8H_{16}O_2$	+ 8.5	46.3	75.5	97.6	138.4	160.2	......
Tetraethyleneglycol chlorohydrin	$C_8H_{16}ClO_4$	110.1	156.1	190.0	214.7	258.2	281.5	......
1-Iodoöctane	$C_8H_{17}I$	45.8	90.0	123.8	150.0	199.3	225.5	− 45.9
Ethyl-*l*-leucinate	$C_8H_{17}NO_2$	27.8	72.1	106.0	131.8	167.3	184.0	......
Octane	C_8H_{18}	− 14.0	+ 19.2	45.1	65.7	104.0	125.6	− 56.8

VAPOR PRESSURE (Continued)
ORGANIC COMPOUNDS (Continued)
Pressures Less than One Atmosphere Continued

Name	Formula	Temperature °C 1 mm	10 mm	40 mm	100 mm	400 mm	760 mm	M.P.
2-Methylheptane	C_8H_{18}	−21.0	+12.3	37.9	58.3	96.2	117.6	−109.5
3-Methylheptane	C_8H_{18}	−19.8	+13.3	38.9	59.4	97.4	118.9	−120.8
4-Methylheptane	C_8H_{18}	−20.4	+12.4	38.0	58.3	96.3	117.7	−121.1
2,2-Dimethylhexane	C_8H_{18}	−29.7	+3.1	28.2	48.2	85.6	106.8	......
2,3-Dimethylhexane	C_8H_{18}	−23.0	+9.9	35.6	56.0	94.1	115.6	......
2,4-Dimethylhexane	C_8H_{18}	−26.9	+5.2	30.5	50.6	88.2	109.4	......
2,5-Dimethylhexane	C_8H_{18}	−26.7	+5.3	30.4	50.5	87.9	109.1	−90.7
3,3-Dimethylhexane	C_8H_{18}	−25.8	+6.1	31.7	52.5	90.4	112.0	......
3,4-Dimethylhexane	C_8H_{18}	−22.1	+11.3	37.1	57.7	96.0	117.7	......
3-Ethylhexane	C_8H_{18}	−20.0	+12.8	38.5	58.9	97.0	118.5	......
2,2,3-Trimethylpentane	C_8H_{18}	−29.0	+3.9	29.5	49.9	88.2	109.8	−112.3
2,2,4-Trimethylpentane	C_8H_{18}	−36.5	−4.3	+20.7	40.7	78.0	99.2	−107.3
2,3,3-Trimethylpentane	C_8H_{18}	−25.8	+6.9	33.0	53.8	92.7	114.8	−101.5
2,3,4-Trimethylpentane	C_8H_{18}	−26.3	+7.1	32.9	53.4	91.8	113.5	−109.2
2-Methyl-3-ethylpentane	C_8H_{18}	−24.0	+9.5	35.2	55.7	94.0	115.6	−114.5
3-Methyl-3-ethylpentane	C_8H_{18}	−23.9	+9.9	36.2	57.1	96.2	118.3	−90
2,2,3,3-Tetramethylbutane	C_8H_{18}	−17.4	+13.5	36.8	54.8	87.4	106.3	+100.7
Tetramethylpiperazine	$C_8H_{18}N_2$	23.7	61.7	90.0	113.8	157.8	183.5	......
1-Octanol	$C_8H_{18}O$	54.0	88.3	115.2	135.2	173.8	195.2	−15.4
2-Octanol	$C_8H_{18}O$	32.8	70.0	98.0	119.8	157.5	178.5	−38.6
1,2-Dipropoxy ethane	$C_8H_{18}O_2$	−38.8	+5.0	42.3	74.2	140.0	180.0	......
Diethylene glycol butyl ether	$C_8H_{18}O_3$	70.0	107.8	135.5	159.8	205.0	231.2	......
Tetraethylene glycol	$C_8H_{18}O_5$	153.9	197.1	228.0	250.0	228.0	307.8	......
Dibutyl sulfide	$C_8H_{18}S$	+21.7	66.4	96.0	118.6	159.0	182.0	−79.7
Dibutyl disulfide	$C_8H_{18}S_2$	+34.6	94.0	145.1	188.0	275.5	330.5	......
Diisobutylamine	$C_8H_{19}N$	−5.1	+30.6	57.8	79.2	118.0	139.5	−70
Tetraethoxysilane	$C_8H_{20}O_4Si$	16.0	52.6	81.1	103.6	146.2	168.5	......
Tetraethyllead	$C_8H_{20}Pb$	38.4	74.8	102.4	123.8	161.8	183.0	−136
Amyltrimethylsilane	$C_8H_{20}Si$	−9.2	+26.7	54.4	76.2	116.6	139.0	......
Tetraethylsilane	$C_8H_{20}Si$	−1.0	+36.3	65.3	88.0	130.2	153.0	......
Tetraethyl bistibine	$C_8H_{20}Sb_2$	97.0	151.2	193.2	225.6	286.2	320.3	......
1,3-Diethoxytetramethyldisiloxane	$C_8H_{22}O_3Si_2$	14.8	51.2	78.7	100.3	139.8	160.7	......
1,7-Dichlorooctamethyltetra-siloxane	$C_8H_{24}Cl_2O_3Si_4$	53.3	95.8	127.8	152.7	197.8	222.0	−62
Octamethyltrisiloxane	$C_8H_{24}O_2Si_3$	7.4	43.1	70.0	91.1	129.4	150.2	......
Octamethylcyclotetrasiloxane	$C_8H_{24}O_4Si_4$	21.7	59.0	87.4	110.0	149.6	171.2	17.4
Coumarin	$C_9H_6O_2$	106.0	153.4	189.0	216.5	264.7	291.0	70
Quinoline	C_9H_7N	59.7	103.8	136.7	163.2	212.3	237.7	−15
Isoquinoline	C_9H_7N	63.5	107.8	141.6	167.6	214.5	240.5	24.6
Indene	C_9H_8	16.4	58.5	90.7	114.7	157.8	181.6	−2
Cinnamylaldehyde	C_9H_8O	76.1	120.0	152.2	177.7	222.4	246.0	−7.5
trans-Cinnamic acid	$C_9H_8O_2$	127.5_s	173.0	207.1	232.4	276.7	300.0	133
Skatole	C_9H_9N	95.0	139.6	171.9	197.4	242.5	266.2	95
Ethyl 3-nitrobenzoate	$C_9H_9NO_4$	108.1	155.0	192.6	220.3	270.6	298.0	47
α-Methyl styrene	C_9H_{10}	7.4	47.1	77.8	102.2	143.0	165.4_p	−23.2
β-Methyl styrene	C_9H_{10}	17.5	57.0	87.7	111.7	154.7	179.0	−30.1
4-Methyl styrene	C_9H_{10}	16.0	55.1	85.0	108.6	151.2	175.0_p	......
Propenylbenzene	C_9H_{10}	17.5	57.0	87.7	111.7	154.7	179.0	−30.1
2,4-Xylaldehyde	$C_9H_{10}O$	59.0_s	99.0	129.7	152.2	194.1	215.5	75
Cinnamyl alcohol	$C_9H_{10}O$	72.6	117.8	151.0	177.8	224.6	250.0	33
Propiophenone	$C_9H_{10}O$	50.0	92.2	124.3	149.3	194.2	218.0	21
2-Vinylanisole	$C_9H_{10}O$	41.9	81.0	110.0	132.3	172.1	194.0_p	......
3-Vinylanisole	$C_9H_{10}O$	43.4	83.0	112.5	135.3	175.8	197.5_p	......
4-Vinylanisole	$C_9H_{10}O$	45.2	85.7	116.0	139.7	182.0	204.5_p	......
Benzyl acetate	$C_9H_{10}O_2$	45.0	87.6	119.6	144.0	189.0	213.5	−51.5
Ethyl Benzoate	$C_9H_{10}O_2$	44.0	86.0	118.2	143.2	188.4	213.4	−34.6
Hydrocinnamic acid	$C_9H_{10}O_2$	102.2	148.7	183.3	209.0	255.0	279.8	48.5
Ethyl salicylate	$C_9H_{10}O_3$	61.2	104.2	136.7	161.5	207.0	231.5	1.3
N-Methylacetanilide	$C_9H_{11}NO$	$_s$	118.6	152.2	179.8	227.4	253.0	102
Ethyl carbanilate	$C_9H_{11}NO_2$	107.8	143.7	168.8	187.9	220.0	237.0	52.5
1,2,3-Trimethylbenzene	C_9H_{12}	16.8	55.9	85.4	108.8	152.0	176.1	−25.5
1,2,4-Trimethylbenzene	C_9H_{12}	13.6	50.7	79.8	102.8	145.4	169.2	−44.1
1,3,5-Trimethylbenzene	C_9H_{12}	9.6	47.4	76.1	98.9	141.0	164.7	−44.8
2-Ethyltoluene	C_9H_{12}	9.4	47.6	76.4	99.0	141.4	165.1	......
3-Ethyltoluene	C_9H_{12}	7.2	44.7	73.3	95.9	137.8	161.3	−95.5
4-Ethyltoluene	C_9H_{12}	7.6	44.9	73.6	96.3	136.4	162.0	......
Cumene	C_9H_{12}	2.9	38.3	66.1	88.1	129.2	152.4	−96.0
Propylbenzene	C_9H_{12}	6.3	43.4	71.6	94.0	135.7	159.2	−99.5
2-Ethylanisole	$C_9H_{12}O$	29.7	69.0	98.8	122.3	164.2	187.1	......
3-Ethylanisole	$C_9H_{12}O$	33.7	73.9	104.8	129.2	172.8	196.5	......
4-Ethylanisole	$C_9H_{12}O$	33.5	73.9	104.7	128.4	172.3	196.5	......
3-Phenyl-1-propanol	$C_9H_{12}O$	74.7	116.0	147.4	170.3	212.8	235.0	......
2-Isopropylphenol	$C_9H_{12}O$	56.6	97.0	127.5	150.3	192.6	214.5	15.5
3-Isopropylphenol	$C_9H_{12}O$	62.0	104.1	136.2	160.2	205.0	228.0	26
4-Isopropylphenol	$C_9H_{12}O$	67.0	108.0	139.8	163.3	206.1	228.2	61
Benzyl ethyl ether	$C_9H_{12}O$	26.0	65.0	95.4	118.9	161.5	185.0	......
Chloroethoxymethylphenylsilane	$C_9H_{13}ClOSi$	44.8	94.6	117.8	142.6	187.7	212.0	......
2,4,5-Trimethylaniline	$C_9H_{13}N$	68.4	109.0	139.8	162.0	203.7	234.5	67
N,*N*-Dimethyl-2-toluidine	$C_9H_{13}N$	28.8	66.2	95.0	118.1	161.5	184.8	−61
N,*N*-Dimethyl-4-toluidine	$C_9H_{13}N$	50.1	86.7	116.3	140.3	185.4	209.5	......
4-Cumidine	$C_9H_{13}N$	60.0	102.2	134.2	158.0	203.2	227.0	......
Phorone	$C_9H_{14}O$	42.0	81.5	111.3	134.0	175.3	197.2	28

Name	Formula	Temperature °C						M.P.
		1 mm	10 mm	40 mm	100 mm	400 mm	760 mm	
Isophorone	$C_9H_{14}O$	38.0	81.2	114.5	140.6	188.7	215.2	......
cis-Diethyl citraconate	$C_9H_{14}O_4$	59.8	103.0	135.7	160.0	206.5	230.3	......
Diethyl itaconate	$C_9H_{14}O_4$	51.3	95.2	128.2	154.3	203.1	227.9	......
Diethyl mesaconate	$C_9H_{14}O_4$	62.8	105.3	137.3	161.6	205.8	229.0	......
Trimethyl citrate	$C_9H_{14}O_7$	106.2	160.4	194.2	219.6	264.2	287.0d	78.5
Isobutyl levulinate	$C_9H_{16}O_3$	65.0	105.9	136.2	160.2	205.5	229.9	......
Azelaic acid	$C_9H_{16}O_4$	178.3	225.5	260.0	286.5	332.8	356.5d	106.5
Diethyl ethylmalonate	$C_9H_{16}O_4$	50.8	91.6	122.4	146.0	188.7	211.5	......
Diethyl glutarate	$C_9H_{16}O_4$	65.6	109.7	142.8	167.8	212.8	237.0	......
2-Nonanone	$C_9H_{18}O$	32.1	72.3	103.4	127.4	171.2	195.0	− 19
Azelaldehyde	$C_9H_{18}O$	33.3	71.6	100.2	123.0	163.4	185.0	......
Pelargonic acid	$C_9H_{18}O_2$	108.2	137.4	163.7	184.4	227.5	253.5	12.5
Methyl caprylate	$C_9H_{18}O_2$	34.2	74.9	105.3	128.0	170.0	193.0d	− 40
Isobutyl isovalerate	$C_9H_{18}O_2$	16.0	53.8	82.7	105.2	146.4	168.7	......
Isoamyl butyrate	$C_9H_{18}O_2$	21.2	59.9	90.0	113.1	155.3	178.6	......
Isoamyl isobutyrate	$C_9H_{18}O_2$	14.8	52.8	81.8	104.4	146.0	168.8	......
Iodononane	$C_9H_{19}I$	70.0	109.0	138.1	159.8	199.3	219.5	......
Nonane	C_9H_{20}	+ 1.4	38.0	66.0	88.1	128.2	150.8	− 53.7
1-Nonanol	$C_9H_{20}O$	59.5	99.7	129.0	151.3	192.1	213.5	− 5
Dipropyleneglycol, isopropyl ether	$C_9H_{20}O_3$	46.0	86.2	117.0	140.3	183.1	205.6	......
Tripropyleneglycol	$C_9H_{20}O_4$	96.0	140.5	173.7	199.0	244.3	267.2	......
Hexyltrimethylsilane	$C_9H_{22}Si$	+ 6.7	44.8	74.0	97.2	139.9	163.0	......
Triethylpropylsilane	$C_9H_{22}Si$	15.2	54.0	83.7	107.4	149.8	173.0	......
1-Bromonaphthalene	$C_{10}H_7Br$	84.2	133.6	170.2	198.8	252.0	281.1	5.5
1-Chloronaphthalene	$C_{10}H_7Cl$	80.6	118.6	153.2	180.4	230.8	259.3	− 20
Dicyclopentadiene	$C_{10}H_{12}$	s	47.6	77.9	101.7	144.2	166.6d	32.9
Naphthalene	$C_{10}H_8$	52.6s	85.8	119.3	145.5	193.2	217.9	80.2
Dichloro-1-naphthylsilane	$C_{10}H_8Cl_2Si$	106.2	149.2	181.7	205.9	249.7	273.3	......
1-Naphthol	$C_{10}H_8O$	94.0s	142.0	177.8	206.0	255.8	282.5	96
2-Naphthol	$C_{10}H_8O$	s	145.5	181.7	209.8	260.6	288.0	122.5
1-Naphthylamine	$C_{10}H_9N$	104.3	153.8	191.5	220.0	272.2	300.8	50
2-Naphthylamine	$C_{10}H_9N$	108.0	157.6	195.7	224.3	277.4	306.1	111.5
2-Methylquinoline	$C_{10}H_9N$	75.3	119.0	150.8	176.2	211.7	246.5	− 1
1,3-Divinylbenzene	$C_{10}H_{10}$	32.7	73.8	105.5	130.0	175.2	199.5p	− 66.9
4-Phenyl-3-buten-2-one	$C_{10}H_{10}O$	81.7	127.4	161.3	187.8	235.4	261.0	41.5
α-Methylcinnamic acid	$C_{10}H_{10}O_2$	125.7	169.8	201.8	224.8	266.8	288.0	......
Methyl cinnamate	$C_{10}H_{10}O_2$	77.4	123.0	157.9	185.8	235.0	263.0	33.4
Safrole	$C_{10}H_{10}O_2$	63.8	107.6	140.1	165.1	210.0	233.0	11.2
1,2-Phenylene diacetate	$C_{10}H_{10}O_4$	98.0	145.7	179.8	206.5	253.3	278.0	......
Dimethyl phthalate	$C_{10}H_{10}O_4$	100.3	147.6	182.8	210.0	257.8	283.7	......
2,4-Dimethylstyrene	$C_{10}H_{12}$	34.2	75.8	107.7	132.3	177.5	202.0p	......
2,5-Dimethylstyrene	$C_{10}H_{12}$	29.0	69.0	100.2	124.7	168.7	193.0p	......
3-Ethylstyrene	$C_{10}H_{12}$	28.3	68.3	99.2	123.2	167.2	191.5p	......
4-Ethylstyrene	$C_{10}H_{12}$	26.0	66.3	97.3	121.5	165.0	189.0p	......
Tetralin	$C_{10}H_{12}$	38.0	79.0	110.4	135.3	181.8	207.2	− 31.0
Anethole	$C_{10}H_{12}O$	62.6	106.0	139.3	164.2	210.5	235.3	22.5
4-Methylpropiophenone	$C_{10}H_{12}O$	59.6	103.8	138.0	164.2	212.7	238.5	......
Estragole	$C_{10}H_{12}O$	52.6	93.7	124.6	148.5	192.0	215.0	......
Cuminal	$C_{10}H_{12}O$	58.0	102.0	135.2	160.0	206.7	232.0	......
4-Vinylphenetole	$C_{10}H_{12}O$	64.0	105.6	136.3	159.8	202.8	225.0p	......
Eugenol	$C_{10}H_{12}O_2$	78.4	123.0	155.8	182.2	228.3	253.5	......
Isoeugenol	$C_{10}H_{12}O_2$	86.3	132.4	167.0	194.0	242.3	267.5	− 10
Chavibetol	$C_{10}H_{12}O_2$	83.6	127.0	159.8	185.5	229.8	254.0	......
Propyl benzoate	$C_{10}H_{12}O_2$	54.6	98.0	131.8	157.4	205.2	231.0	− 51.6
2-Phenoxyethyl acetate	$C_{10}H_{12}O_3$	82.6	128.0	162.3	189.2	235.0	259.7	− 6.7
2-Chloroethyl α-methylbenzyl ether	$C_{10}H_{13}ClO$	62.3	106.0	139.6	164.8	210.8	235.0	......
4-*tert*-Butylphenyl dichlorophosphate	$C_{10}H_{13}Cl_2O_2P$	96.0	146.0	184.3	214.3	268.2	299.0	......
1,2,3,4-Tetramethylbenzene	$C_{10}H_{14}$	42.6	81.8	111.5	135.7	180.0	204.4	− 6.2
1,2,3,5-Tetramethylbenzene	$C_{10}H_{14}$	40.6	77.8	105.8	128.3	173.7	197.9	− 24.0
1,2,4,5-Tetramethylbenzene	$C_{10}H_{14}$	45.0s	74.6s	104.2	128.1	172.1	195.9	79.5
4-Ethyl-1,3-xylene	$C_{10}H_{14}$	26.3	66.4	97.2	121.2	164.4	188.4	......
5-Ethyl-1,3-xylene	$C_{10}H_{14}$	22.1	62.1	92.6	116.5	159.6	183.7	......
2-Ethyl-1,4-xylene	$C_{10}H_{14}$	25.7	65.6	96.0	120.0	163.1	186.9	......
1,2-Diethylbenzene	$C_{10}H_{14}$	22.3	62.0	92.5	116.2	159.0	183.5	− 31.4
1,3-Diethylbenzene	$C_{10}H_{14}$	20.7	59.9	90.4	114.4	156.9	181.1	− 83.9
1,4-Diethylbenzene	$C_{10}H_{14}$	20.7	60.3	91.1	115.3	159.0	183.8	− 43.2
Cymene	$C_{10}H_{14}$	17.3	57.0	87.0	110.8	153.5	177.2	− 68.2
Butylbenzene	$C_{10}H_{14}$	22.7	62.0	92.4	116.2	159.2	183.1	− 88.0
Isobutylbenzene	$C_{10}H_{14}$	14.1	53.7	83.3	107.0	149.6	172.8	− 51.5
sec-Butylbenzene	$C_{10}H_{14}$	18.6	57.0	86.2	109.5	150.3	173.5	− 75.5
tert-Butylbenzene	$C_{10}H_{14}$	13.0	51.7	80.8	103.8	145.8	168.5	− 58
Carvacrol	$C_{10}H_{14}O$	70.0	113.2	145.2	169.7	213.8	237.0	+ 0.5
Carbone	$C_{10}H_{14}O$	57.4	100.4	133.0	157.3	203.5	227.5	......
Cuminyl alcohol	$C_{10}H_{14}O$	74.2	118.0	150.3	176.2	221.7	246.6	......
4-Ethylphenetole	$C_{10}H_{14}O$	48.5	89.5	119.8	143.5	185.7	208.0	......
Thymol	$C_{10}H_{14}O$	64.3	107.4	139.8	164.1	209.2	231.8	51.5
4-Isobutylphenol	$C_{10}H_{14}O$	72.1	115.5	147.2	171.2	214.7	237.0	......
4-*sec*-Butylphenol	$C_{10}H_{14}O$	71.4	114.8	147.8	172.4	217.6	242.1	......
2-*sec*-Butylphenol	$C_{10}H_{14}O$	57.4	100.8	133.4	157.3	203.8	228.0	......
2-*tert*-Butylphenol	$C_{10}H_{14}O$	56.6	98.1	129.2	153.5	196.3	219.5	......

ORGANIC COMPOUNDS (Continued)
Pressures Less than One Atmosphere (Continued)

Name	Formula	Temperature °C						M.P.
		1 mm	10 mm	40 mm	100 mm	400 mm	760 mm	
4-*tert*-Butylphenol	$C_{10}H_{14}O$	70.0	114.0	146.0	170.2	214.0	238.0	99
Nicotine	$C_{10}H_{14}N_2$	61.8	107.2	142.1	169.5	219.8	247.3	......
N-Diethylaniline	$C_{10}N_{15}N$	49.7	91.9	123.6	147.3	192.4	215.5	− 34.4
N-Phenyliminodiethanol	$C_{10}H_{15}NO_2$	145.0	195.8	233.0	260.6	311.3	337.8	......
Camphene	$C_{10}H_{16}$	$_s$	47.2$_s$	75.7	97.9	138.7	160.5	50
Dipentene	$C_{10}H_{16}$	14.0	53.8	84.3	108.3	150.5	174.6	......
d-Limonene	$C_{10}H_{16}$	14.0	53.8	84.3	108.3	151.4	175.0	− 96.9
Myrcene	$C_{10}H_{16}$	14.5	53.2	82.6	106.0	148.3	171.5	......
α-Phellandrene	$C_{10}H_{16}$	20.0	58.0	87.8	110.6	152.0	175.0	......
α-Pinene	$C_{10}H_{16}$	− 1.0	+ 37.3	66.8	90.1	132.3	155.0	− 55
β-Pinene	$C_{10}H_{16}$	+ 4.2	42.3	71.5	94.0	136.1	158.3	......
Terpenoline	$C_{10}H_{16}$	32.3	70.6	100.0	122.7	163.5	185.0	......
Diethyl arsanilate	$C_{10}H_{16}AsNO_3$	38.0	74.8	102.6	123.8	161.0	181.0	......
d-Camphor	$C_{10}H_{16}O$	41.5$_s$	82.3$_s$	114.0$_s$	138.0$_s$	182.0	209.2	178.5
l-Dihydrocarvone	$C_{10}H_{16}O$	46.6	90.0	123.7	149.7	197.0	223.0	......
α-Citral	$C_{10}H_{16}O$	61.7	103.9	135.9	160.0	205.0	228.0$_d$	......
d-Fenchone	$C_{10}H_{16}O$	28.0	68.3	99.5	123.6	166.8	191.0	5
Pulegone	$C_{10}H_{16}O$	58.3	94.0	121.7	143.1	189.8	221.0	......
α-Thyjone	$C_{10}H_{16}O$	38.3	79.3	110.0	134.0	177.8	201.	......
Ethoxydimethylphenylsilane	$C_{10}H_{16}OSi$	36.3	76.2	107.2	131.4	175.0	199.5	......
Campholenic acid	$C_{10}H_{16}O_2$	97.6	139.8	170.0	193.7	234.0	256.0	......
Diosphenol	$C_{10}H_{16}O_2$	66.7	109.0	141.2	165.6	209.5	232.0	......
Fencholic acid	$C_{10}H_{16}O_2$	101.7	142.3	171.8	194.0	237.8	264.1	19
cis-Decalin	$C_{10}H_{18}$	22.5	64.2	97.2	123.2	169.9	194.6	− 43.3
trans-Decalin	$C_{10}H_{18}$	− 0.8	+ 47.2	85.7	114.6	160.1	186.7	− 30.7
d-Citronellal	$C_{10}H_{18}O$	44.0	84.8	116.1	140.1	183.8	206.5	......
Cineol	$C_{10}H_{18}O$	15.0	54.1	84.2	108.2	151.6	176.0	− 1
Dihydrocarveol	$C_{10}H_{18}O$	63.9	105.0	136.1	159.8	202.8	225.0	......
dl-Fenchyl alcohol	$C_{10}H_{18}O$	45.8	82.1	110.8	132.3	173.2	201.0	35
Geraniol	$C_{10}H_{18}O$	69.2	110.0	141.8	165.3	207.8	230.0	......
d-Linalool	$C_{10}H_{18}O$	40.0	79.8	109.9	133.8	175.6	198.0	......
Nerol	$C_{10}H_{18}O$	61.7	104.0	136.1	159.8	203.5	226.0	......
α-Terpineol	$C_{10}H_{18}O$	52.8	94.3	126.0	150.1	194.3	217.5	35
Citronellic acid	$C_{10}H_{18}O_2$	99.5	141.4	171.9	195.4	236.6	257.0	......
Amyl levulinate	$C_{10}H_{18}O_3$	81.3	124.0	155.8	180.5	227.4	253.2	......
Isoamyl levulinate	$C_{10}H_{18}O_3$	75.6	118.8	151.7	177.0	222.7	247.9	......
Diethyl ethylmethylmalonate	$C_{10}H_{18}O_4$	44.7	85.7	116.7	140.8	134.1	207.5	......
Diethyl adipate	$C_{10}H_{18}O_4$	74.0	123.0	154.6	179.0	219.1	240.0	− 21
Diisobutyl oxalate	$C_{10}H_{18}O_4$	63.2	105.3	137.5	161.8	205.8	229.5	......
Dipropyl succinate	$C_{10}H_{18}O_4$	77.5	122.2	154.8	180.3	226.5	250.8	......
Sebacic acid	$C_{10}H_{18}O_4$	183.0	232.0	268.2	294.5	332.8	352.3$_d$	134.5
Dipropyl-*d*-tartrate	$C_{10}H_{18}O_6$	115.6	163.5	199.7	227.0	275.6	303.0	......
Diisopropyl-*d*-tartrate	$C_{10}H_{18}O_6$	103.7	148.2	181.8	207.3	251.8	275.0	......
Camphylamine	$C_{10}H_{19}N$	45.3	83.7	112.5	134.6	173.8	195.0	......
Menthane	$C_{10}H_{20}$	+ 9.7	48.3	78.3	102.1	146.0	169.5	......
1-Decene	$C_{10}H_{20}$	14.7	53.7	83.3	106.5	149.2	172.0	......
1,2-Dibromodecane	$C_{10}H_{20}Br_2$	95.7	137.3	167.4	190.2	229.8	250.4	......
Citronellol	$C_{10}H_{20}O$	66.4	107.0	137.2	159.8	201.0	221.5	......
Capraldehyde	$C_{10}H_{20}O$	51.9	92.0	122.2	145.3	186.3	208.5	......
***l*-Menthol**	$C_{10}H_{20}O$	56.0	96.0	126.1	149.4	190.2	212.0	42.5
Decan-2-one	$C_{10}H_{20}O$	44.2	85.8	117.1	142.0	186.7	211.0	+ 3.5
Capric acid	$C_{10}H_{20}O_2$	125.0	152.2	179.9	200.0	240.3	268.4	31.5
Isoamyl isovalerate	$C_{10}H_{20}O_2$	27.0	68.6	100.6	125.1	169.5	194.0	......
Decane	$C_{10}H_{22}$	16.5	55.7	85.5	108.6	150.6	174.1	− 29.7
2,7-Dimethyloctane	$C_{10}H_{22}$	+ 6.3	42.3	71 2	93.9	136.0	159.7	− 52.8
Decyl alcohol	$C_{10}H_{22}O$	69.5	111.3	142.1	165.8	208.8	231.0	+ 7
Diisoamyl ether	$C_{10}H_{22}O$	18.6	57.0	86.3	109.6	150.3	173.4	......
2-Butyl-2-ethylbutane-1,3-diol	$C_{10}H_{22}O_2$	94.1	136.8	167.8	191.9	233.5	255.0	......
Dihydrocitronellol	$C_{10}H_{22}O$	68.0	103.0	127.6	145.9	176.8	193.5	......
Dipropylene glycol monobutyl ether	$C_{10}H_{22}O_3$	64.7	106.0	136.3	159.8	203.8	227.0	......
Diisoamyl sulfide	$C_{10}H_{22}S$	43.0	87.6	120.0	145.3	191.0	216.0	......
Heptyltrimethylsilane	$C_{10}H_{24}Si$	22.3	62.1	92.4	116.5	159.8	184.0	......
Butyltriethylsilane	$C_{10}H_{24}Si$	27.1	67.5	98.3	123.2	167.5	192.0	......
1,5-Diethoxyhexamethyltrisiloxane	$C_{10}H_{28}O_4Si_3$	41.8	80.7	110.0	133.2	174.0	196.6	......
Decamethyltetrasiloxane	$C_{10}H_{30}O_3Si_4$	35.3	74.3	104.0	127.3	169.8	193.5	......
Decamethylcyclopentasiloxane	$C_{10}H_{30}O_5Si_5$	45.2	86.2	117.7	142.0	186.0	210.0	− 38.0
1-Naphthoic acid	$C_{11}H_8O_2$	156.0	196.8	225.0	245.8	281.4	300.0	160.5
2-Naphthoic acid	$C_{11}H_8O_2$	160.8	202.8	231.5	252.7	289.5	308.5	184
Ethyl-*trans*-cinnamate	$C_{11}H_{12}O_2$	87.6	134.0	169.2	196.0	245.0	271.0	12
1-Phenyl-1,3-pentanedione	$C_{11}H_{12}O_2$	98.0	144.0	178.0	204.5	251.2	276.5	......
Ethyl benzoylacetate	$C_{11}H_{12}O_3$	107.6	150.3	181.8	205.0	244.7	265.0$_d$	......
Myristicine	$C_{11}H_{12}O_3$	95.2	142.0	177.7	205.0	253.5	280.0	......
2,4,5-Trimethylstyrene	$C_{11}H_{14}$	48.1	91.6	124.2	149.8	196.1	221.2$_p$	......
2,4,6-Trimethylstyrene	$C_{11}H_{14}$	37.5	79.7	111.8	136.8	182.3	207.0$_p$	......
4-Isopropylstyrene	$C_{11}H_{14}$	34.7	76.0	108.0	132.8	178.0	202.5$_p$	......
Isovalerophenone	$C_{11}H_{14}O$	58.3	101.4	133.8	158.0	204.2	228.0	......
Pivalophenone	$C_{11}H_{14}O$	57.8	99.0	130.4	154.0	197.7	220.0	......
2,3,5-Trimethylacetophenone	$C_{11}H_{14}O$	79.0	122.3	154.2	179.7	224.3	247.5	......
Isobutyl benzoate	$C_{11}H_{14}O_2$	64.0	108.6	141.8	166.4	212.8	237.0	......
4-Allylveratrole	$C_{11}H_{14}O_2$	85.0	127.0	158.3	183.7	226 2	248.0	......
3,5-Diethyltoluene	$C_{11}H_{16}$	34.0	75.3	107.0	131.7	176.5	200.7	......
1,2,4-Trimethyl-5-ethylbenzene	$C_{11}H_{16}$	43.7	84.6	106.0	140 3	184.5	208.1	......
1,3,5-Trimethyl-2-ethylbenzene	$C_{11}H_{16}$	38.8	80 5	113.2	137.9	183.5	208.0	......
3-Ethylcumene	$C_{11}H_{16}$	28.3	68.8	99 9	124.3	168.2	193 0	......
4-Ethylcumene	$C_{11}H_{16}$	31.5	72.0	103.3	127.2	171.8	195.8	......

VAPOR PRESSURE (Continued)

ORGANIC COMPOUNDS (Continued)

Pressures Less than One Atmosphere (Continued)

Name	Formula	Temperature °C						M.P.
		1 mm	10 mm	40 mm	100 mm	400 mm	760 mm	
sec-Amylbenzene	$C_{11}H_{16}$	29.0	69.2	100.0	124.1	168.0	193.0	......
4-*tert*-Butyl-2-cresol	$C_{11}H_{16}O$	74.3	118.0	150.8	176.2	221.8	247.0	......
2-*tert*-Butyl-4-cresol	$C_{11}H_{16}O$	70.0	112.0	143.9	167.0	210.0	232.6	......
4-*tert*-Amylphenol	$C_{11}H_{16}O$	s	125.5	160.3	189.0	239.5	266.0	93
Ethylcamphoronic anhydride	$C_{11}H_{16}O_5$	118.2	165.0	199.8	226.6	272.8	298.0	......
Bornyl formate	$C_{11}H_{18}O_2$	47.0	89.3	121.2	145.8	190.2	214.0	......
Geranyl formate	$C_{11}H_{18}O_2$	61.8	104.3	136.2	160.7	205.8	230.0	......
Neryl formate	$C_{11}H_{18}O_2$	57.3	99.7	131.5	155.6	200.0	224.5	......
Diethoxymethylphenylsilane	$C_{11}H_{18}O_2Si$	56.5	97.2	127.5	151.2	193.8	216.5	......
Diethyl-γ-oxoazelate	$C_{11}H_{18}O_5$	121.0	165.7	197.7	221.6	264.5	286.0	......
10-Hendecenoic acid	$C_{11}H_{20}O_2$	114.0	156.3	188.7	213.5	254.0	275.0	24.5
Menthyl formate	$C_{11}H_{20}O_2$	47.3	90.0	123.0	148.0	194.2	219.0	......
2-Ethylhexyl acrylate	$C_{11}H_{20}O_2$	50.0	91.8	123.7	147.9	192.2	216.0	......
Octyl acrylate	$C_{11}H_{20}O_2$	58.5	102.0	135.6	159.1	204.0	227.0	......
Hexyl levulinate	$C_{11}H_{20}O_3$	90.0	134.7	167.8	193.6	241.0	266.8	......
Hendecan-2-one	$C_{11}H_{22}O$	68.2	108.9	139.0	161.0	202.3	224.0	15
Methyl caprate	$C_{11}H_{22}O_2$	63.7	108.0	139.0	161.5	202.9	224.0_d	− 18
Hendecanoic acid	$C_{11}H_{22}O_2$	101.4	149.0	185.6	212.5	262.8	290.0	29.5
Hendecane	$C_{11}H_{24}$	32.7	73.9	104.4	128.1	171.9	195.8	− 25.6
Hendecan-2-ol	$C_{11}H_{24}O$	71.1	112.8	143.7	167.2	209.8	232.0	......
Trimethyloctylsilane	$C_{11}H_{26}Si$	41.8	82.3	113.0	136.5	179.5	202.0	......
Amyltriethylsilane	$C_{11}H_{26}Si$	41.8	83.8	116.0	141.2	186.3	211.0	......
1-Bromobiphenyl	$C_{12}H_9Br$	98.0	150.6	190.8	221.8	277.7	310.0	90.5
2-Bromo-4-Phenylphenol	$C_{12}H_9BrO$	100.0	152.3	193.8	224.5	280.2	311.0	95
2-Chlorobiphenyl	$C_{12}H_9Cl$	89.3	134.7	169.9	197.0	243.8	267.5	34
4-Chlorobiphenyl	$C_{12}H_9Cl$	96.4	146.0	183.8	212.5	264.5	292.9	75.5
2-Chloro-3-phenylphenol	$C_{12}H_9ClO$	118.0	169.7	207.4	237.0	289.4	317.5	+ 6
2-Chloro-6-phenylphenol	$C_{12}H_9ClO$	119.8	170.7	208.2	237.1	289.5	317.0	......
2-Xenyl dichlorophosphate	$C_{12}H_9Cl_2PO$	138.2	187.0	223.8	251.5	301.5	328.5	......
Carbazole	$C_{12}H_9N$	s	s	s	265.0	323.0	354.8	244.8
Acenaphthene	$C_{12}H_{10}$	s	131.2	168.2	197.5	250.0	277.5	95
Biphenyl	$C_{12}H_{10}$	70.6	117.0	152.5	180.7	229.4	254.9	69.5
Diphenyl chlorophosphate	$C_{12}H_{10}ClPO_3$	121.5	182.0	227.9	265.0	337.2	378.0	......
Dichlorodiphenylsilane	$C_{12}H_{10}Cl_2Si$	109.6	158.0	195.5	223.8	275.5	304.0	......
Difluorodiphenylsilane	$C_{12}H_{10}F_2Si$	68.4	115.5	149.8	176.3	225.4	252.5	......
Azobenzene	$C_{12}H_{10}N_2$	103.5	151.5	187.9	216.0	266.1	293.0	68
1-Acetonaphthone	$C_{12}H_{10}O$	115.6	161.5	196.8	223.8	270.5	295.5	......
2-Acetonaphthone	$C_{12}H_{10}O$	120.2	168.5	203.8	229.8	275.8	301.0	55.5
Diphenyl ether	$C_{12}H_{10}O$	66.1	114.0	150.0	178.8	230.7	258.5	27
2-Phenylphenol	$C_{12}H_{10}O$	100.0	146.2	180.3	205.9	251.8	275.0	56.5
4-Phenylphenol	$C_{12}H_{10}O$	s	176.2	213.0	240.9	285.5	308.0	164.5
Diphenyl sulfide	$C_{12}H_{10}S$	96.1	145.0	182.8	211.8	263.9	292.5	......
Diphenyl disulfide	$C_{12}H_{10}S$	131.6	180.0	214.8	241.3	285.8	310.0	61
Diphenyl selenide	$C_{12}H_{10}Se$	105.7	154.4	192.2	220.8	273.2	301.5	+ 2.5
Diphenylamine	$C_{12}H_{11}N$	108.3	157.0	194.3	222.8	274.1	302.0	52.9
1-Ethylnaphthalene	$C_{12}H_{12}$	70.0	116.8	152.0	180.0	230.8	258.1_d	− 27
1,1-Diphenylhydrazine	$C_{12}H_{12}N_2$	126.0	176.1	213.5	242.5	294.0	322.2	44
2-Cyclohexyl-4,6-dinitrophenol	$C_{12}H_{14}N_2O_5$	132.8	175.9	206.7	229.0	269.8	291.5	......
Eugenyl acetate	$C_{12}H_{14}O_3$	101.6	148.0	183.0	209.7	257.4	282.0	29.5
Apiole	$C_{12}H_{14}O_4$	116.0	160.2	193.7	218.0	262.1	285.0	30
Diethyl phthalate	$C_{12}H_{14}O_4$	108.8	156.0	192.1	219.5	267.5	294.0	......
2,5-Diethylstyrene	$C_{12}H_{16}$	49.7	92.6	125.8	151.0	198.0	223.0_p	......
Phenylcyclohexane	$C_{12}H_{16}$	67.5	111.3	144.0	169.3	214.6	240.0	+ 7.5
Isoamyl benzoate	$C_{12}H_{16}O_2$	72.0	121.6	158.3	186.8	235.8	262.0	......
1,2,4-Triethylbenzene	$C_{12}H_{18}$	46.0	88.[illegible]	121.7	146.8	193.7	218.0	......
1,4-Diisopropylbenzene	$C_{12}H_{18}$	40.0	81.8	114.0	138.7	184.3	209.0	......
1,3-Diisopropylbenzene	$C_{12}H_{18}$	34.7	76.0	107.9	132.3	177.6	202.0	−105
2-*tert*-Butyl-4-ethylphenol	$C_{12}H_{18}O$	76.3	121.0	154.0	179.0	223.8	247.8	......
4-*tert*-Butyl-2,5-xylenol	$C_{12}H_{18}O$	88.2	135.0	169.8	195.0	241.3	265.3	......
4-*tert*-Butyl-2,6-xylenol	$C_{12}H_{18}O$	74.0	119.0	152.2	176.0	217.8	239.8	......
6-*tert*-Butyl-2,4-xylenol	$C_{12}H_{18}O$	70.3	115.0	148.5	172.0	214.2	236.5	......
6-*tert*-Butyl-3,4-xylenol	$C_{12}H_{18}O$	83.9	127.0	159.7	184.0	226.7	249.5	......
d-Bornyl acetate	$C_{12}H_{20}O_2$	46.9	90.2	123.7	149.8	197.5	223.0	29
Geranyl acetate	$C_{12}H_{20}O_2$	73.5	117.9	150.0	175.2	219.8	243.3_d	......
Linalyl acetate	$C_{12}H_{20}O_2$	55.4	96.0	127.7	151.8	196.2	220.0_d	......
Triethoxyphenylsilane	$C_{12}H_{20}O_3Si$	71.0	112.6	143.5	167.5	210.5	233.5	......
Triethyl citrate	$C_{12}H_{20}O_7$	107.0	144.0	190.4	217.8	267.5	294.0_d	......
Trimethallyl phosphate	$C_{12}H_{21}PO_4$	93.7	149.8	192.0	225.7	288.5	324.0	......
Citronellyl acetate	$C_{12}H_{22}O_2$	74.7	113.0	140.5	161.0	197.8	217.0	......
Menthyl acetate	$C_{12}H_{22}O_2$	57.4	100.0	132.1	156.7	202.8	227.0	......
Dimethyl sebacate	$C_{12}H_{22}O_4$	104.0	156.2	196.0	222.6	269.6	293.5	38
Diisoamyl oxalate	$C_{12}H_{22}O_4$	85.4	131.4	165.7	192.2	240.0	265.0	......
Diisobutyl-*d*-tartrate	$C_{12}H_{22}O_6$	117.8	169.0	208.5	239.5	294.0	324.0	73.5
1-Dodecene	$C_{12}H_{24}$	47.2	87.8	118.6	142.3	185.5	208.0	− 31.5
Triisobutylene	$C_{12}H_{24}$	18.0	56.5	86.7	110.0	153.0	179.0	......
Dodecan-2-one	$C_{12}H_{24}O$	77.1	120.4	152.4	177.5	222.5	246.5	......
Lauraldehyde	$C_{12}H_{24}O$	77.7	123.7	157.8	184.5	231.8	257.0	44.5
Lauric acid	$C_{12}H_{24}O_2$	121.0	166.0	201.4	227.5	273.8	299.2	48
Dodecane	$C_{12}H_{26}$	47.8	90.0	121.7	146.2	191.0	216.2	− 9.6
Dodecyl alcohol	$C_{12}H_{26}O$	91.0	134.7	167.2	192.0	235.7	259.0	24
Tripropylene glycol monoisopropyl ether	$C_{12}H_{26}O_4$	82.4	127.3	161.4	187.8	232.8	256.6	......
Triisobutylamine	$C_{12}H_{27}N$	32.3	69.8	97.8	119.7	157.8	179.0	− 22
Dodecylamine	$C_{12}H_{27}N$	82.8	127.8	157.4	182.1	225.0	248.0	......
Triethylhexylsilane	$C_{12}H_{28}Si$	52.4	96.4	130.0	156.0	204.6	230.0	......

VAPOR PRESSURE (Continued)

ORGANIC COMPOUNDS (Continued)

Pressures Less than One Atmosphere (Continued)

Name	Formula	Temperature °C 1 mm	10 mm	40 mm	100 mm	400 mm	760 mm	M.P.
1,7-Diethoxyoctamethyltetrasiloxane	$C_{12}H_{34}O_5Si_4$	67.7	108.6	319.0	162.0	204.0	227.5	......
Dodecamethylpentasiloxane	$C_{12}H_{36}O_4Si_5$	56.6	98.0	128.8	162.8	196.5	220.5	......
Dodecamethylcyclohexasiloxane	$C_{12}H_{36}O_6Si_6$	67.3	110.0	141.8	166.3	210.6	236.0	− 3.0
Acridine	$C_{13}H_9N$	129.4	184.0	224.2	256.0	314.3	346.0	110.5
Fluorene	$C_{13}H_{10}$	s	146.0	185.2	214.7	268.6	295.0	113
Benzophenone	$C_{13}H_{10}O$	108.2	157.6	195.7	224.4	276.8	305.4	48.5
Phenyl benzoate	$C_{13}H_{10}O_2$	106.8	157.8	197.6	227.8	283.5	314.0	70.5
Salol	$C_{13}H_{10}O_3$	117.8	167.0	205.0	233.8	284.8	313.0	42.5
Diphenylmethane	$C_{13}H_{12}$	76.0	122.8	157.8	186.3	237.5	264.5	26.5
Benzhydrol	$C_{13}H_{12}O$	110.0	162.0	200.0	227.5	275.6	301.0	68.5
Benzyl phenyl ether	$C_{13}H_{12}O$	95.4	144.0	180.1	209.2	259.8	287.0	......
1-Propionaphthone	$C_{13}H_{12}O$	124.0	171.0	206.9	233.5	280.2	306.0	......
Chloromethyldiphenylsilane	$C_{13}H_{13}ClSi$	105.0	152.7	189.2	216.0	266.5	295.5	......
Methyldiphenylamine	$C_{13}H_{13}N$	103.5	149.7	184.0	210.1	257.0	282.0	− 7.6
2-Isopropylnaphthalene	$C_{13}H_{14}$	76.0	123.4	159.0	187.6	238.5	266.0	......
Methyldiphenylsilane	$C_{13}H_{14}Si$	88.0	132.8	166.4	193.7	241.5	266.8	......
Enanthophenone	$C_{13}H_{18}O$	100.0	145.5	178.9	204.2	248.3	271.3	......
Heptylbenzene	$C_{13}H_{20}$	64.0	110.0	144.0	170.2	217.8	244.0	......
α-Ionone	$C_{13}H_{20}O$	79.5	123.0	155.6	181.2	225.2	250.0	......
Bornyl propionate	$C_{13}H_{22}O_2$	64.6	108.0	140.4	165.7	211.2	235.0	......
2-Tridecanone	$C_{13}H_{26}O$	86.8	131.8	165.7	191.5	238.3	262.5	28.5
Methyl laurate	$C_{13}H_{26}O_2$	87.8	133.2	166.0	190.8	d	d	5
Tridecanoic acid	$C_{13}H_{26}O_2$	137.8	181.0	212.4	236.0	276.5	299.0	41
Tridecane	$C_{13}H_{28}$	59.4	104.0	137.7	162.5	209.4	234.0	− 6.2
Tripropyleneglycol, monobutyl ether	$C_{13}H_{28}O_4$	101.5	147.0	179.8	204.4	247.0	269.5	......
Decyltrimethylsilane	$C_{13}H_{30}Si$	67.4	111.0	144.0	169.5	215.5	240.0	......
Triethylheptylsilane	$C_{13}H_{30}Si$	70.0	114.6	148.0	174.0	221.0	247.0	......
Anthraquinone	$C_{14}H_8O_2$	190.0$_s$	234.2$_s$	264.3$_s$	285.0$_s$	346.2	379.9	286
1,4-Dihydroxyanthraquinone	$C_{14}H_8O_4$	196.7	259.8	307.4	344.5	413.0$_d$	450.0$_d$	194
Anthracene	$C_{14}H_{10}$	145.0$_s$	187.2$_s$	217.5$_s$	250.0	310.2	342.0	217.5
Phenanthrene	$C_{14}H_{10}$	118.2	173.0	215.8	249.0	308.0	340.2	99.5
Benzil	$C_{14}H_{10}O_2$	128.4	183.0	224.5	255.8	314.3	347.0	95
Benzoic anhydride	$C_{14}H_{10}O_3$	143.8	198 0	239.8	270.4	328.8	360.0	42
1,1-Diphenylethylene	$C_{14}H_{12}$	87.4	135.0	170.8	198.6	249.8	277.0	......
trans-Diphenylethylene	$C_{14}H_{12}$	113.2	161.0	199.0	227.4	287.3	306.5	124
Desoxybenzoin	$C_{14}H_{12}O$	123.3	173.5	212.0	241.3	293.0	321.0	60
Benzoin	$C_{14}H_{12}O$	135.6	188.0	227.6	258.0	313.5	343.0	132
Dibenzyl	$C_{14}H_{14}$	86.8	136.0	173.7	202.8	255.0	284.0	51.5
2-Isobutyronaphthone	$C_{14}H_{14}O$	133.2	181.0	215.6	242.3	288.2	313.0	......
Dibenzylamine	$C_{14}H_{15}O$	118.3	165.6	200.2	227.3	274.3	300.0	− 26
Ethyldiphenylamine	$C_{14}H_{15}N$	98.3	146.0	182.0	209.8	258.8	286.0	......
1,2-Dichlorotetraethylbenzene	$C_{14}H_{20}Cl_2$	105.6	155.0	192.2	220.7	272.8	302.0	......
1,4-Dichlorotetraethylbenzene	$C_{14}H_{20}Cl_2$	91.7	143.8	183.2	212.0	265.8	296.5	......
2-(4-*tert*-Butylphenoxy) ethyl acetate	$C_{14}H_{20}O_3$	118.0	165.8	201.5	228.0	277.6	304.4	......
1,2,4,5-Tetraethylbenzene	$C_{14}H_{22}$	65.7	111.6	145.8	172.4	221.4	248.0	11.6
2,4-Di-*tert*-butylphenol	$C_{14}H_{22}O$	84.5	130.0	164.3	190.0	237.0	260.8	......
Bornyl butyrate	$C_{14}H_{24}O_2$	74.0	118.0	150.7	176.4	222.2	247.0	......
Bornyl isobutyrate	$C_{14}H_{24}O_2$	70.0	114.0	147.2	172.2	218.2	243.0	......
Geranyl butyrate	$C_{14}H_{24}O_2$	96.8	139.0	170.1	193.8	235.0	257.4	......
Geranyl isobutyrate	$C_{14}H_{24}O_2$	90.7	133.0	164.0	187.7	228.5	251.0	......
Diethyl sebacate	$C_{14}H_{26}O_4$	125.3	172.1	207.5	234.4	280.3	305.5	1.3
2-Tetradecanone	$C_{14}H_{28}O$	99.3	145.5	179.8	206.0	253.0	278.0	......
Myristaldehyde	$C_{14}H_{28}O$	99.0	148.3	186.0	214.5	267.9	297.8	23.5
Myristic acid	$C_{14}H_{28}O_2$	142.0	190.8	223.5	250.5	294.6	318.0	57.5
1-Chlorotetradecane	$C_{14}H_{29}Cl$	98.5	148.2	187.0	215.5	267.5	296.0	+ 0.9
Tetradecane	$C_{14}H_{30}$	76.4	120.7	152.7	178.5	226.8	252.5	5.5
Tetradecylamine	$C_{14}H_{31}N$	102.6	152.0	189.0	215.7	264.6	291.2	......
Triethyloctylsilane	$C_{14}H_{32}Si$	73.7	120.6	155.7	184.3	235.0	262.0	......
1,9-Diethoxydecamethylpentasiloxane	$C_{14}H_{40}O_6Si_5$	89.0	131.5	162.2	187.0	230.0	253.3	......
Tetradecamethylhexasiloxane	$C_{14}H_{42}O_5Si_6$	73.7	117.6	149.8	175.2	220.5	245.5	......
Tetradecamethylcycloheptasiloxane	$C_{14}H_{42}O_7Si_7$	86.3	131.5	165.3	191.8	239.2	264.0	− 32
1,3-Diphenyl-2-propanone	$C_{15}H_{14}O$	125.5	177.6	216.6	246.6	301.7	330.5	34.5
1-Biphenyloxy-2,3-epoxypropane	$C_{15}H_{14}O_2$	135.3	187.2	226.3	255.0	309.8	340.0	......
1-Isovaleronaphthone	$C_{15}H_{16}O$	136.0	184.0	219.7	246.7	294.0	320.0	......
4,4-Isopropylidenebisphenol	$C_{15}H_{16}O_2$	193.0	240.8	273.0	297.0	339.0	360.5	......
Ethoxymethyldiphenylsilane	$C_{15}H_{18}OSi$	109.0	152.7	186.0	211.8	256.8	282.0	......
Helenin	$C_{15}H_{20}O_2$	157.7	192.1	215.2	232.6	260.6	275.0	76
Cadinene	$C_{15}H_{24}$	101.3	146.0	179.8	205.6	250.7	275.0	......
2,6-Di-*tert*-butyl-4-cresol	$C_{15}H_{24}O$	85.8	131.0	164.1	190.0	237.6	262.5	......
4,6-Di-*tert*-butyl-2-cresol	$C_{15}H_{24}O$	86.2	132.4	167.4	194.0	243.4	269.3	......
4,6-Di-*tert*-butyl-3-cresol	$C_{15}H_{24}O$	103.7	150.0	185.3	211.0	257.1	282.0	......
Champacol	$C_{15}H_{26}O$	100.0	148.0	184.0	211.9	261.2	288.0	91
Triethyl camphoronate	$C_{15}H_{26}O_6$	s	166.0	201.8	228.6	276.0	301.0	135
Methyl myristate	$C_{15}H_{30}O_2$	115.0	160.8	195.8	222.6	269.8	295.8	18.5
Pentadecane	$C_{15}H_{32}$	91.6	135.4	167.7	194.0	242.8	270.5	10
Tetrapropylene glycol monoisopropyl ether	$C_{15}H_{32}O_5$	116.6	163.0	197.7	223.3	268.3	292.7	......
Dodecyltrimethylsilane	$C_{15}H_{34}Si$	91.2	137.7	172.1	199.5	248.0	273.0	......
Benzyl cinnamate	$C_{16}H_{14}O_2$	173.8	221.5	255.8	281.5	326.7	350.0	39
Di(α-methylbenzyl) ether	$C_{16}H_{18}O$	96.7	144.0	179.6	206.8	254.8	281.0	......
Diethoxydiphenylsilane	$C_{16}H_{20}O_2Si$	111.5	157.6	193.2	220.0	259.7	296.0	......
Dibutyl phthalate	$C_{16}H_{22}O_4$	148.2	198.2	235.8	263.7	313.5	340.0	......
Pentaethylchlorobenzene	$C_{16}H_{25}Cl$	90.0	140.7	178.2	208.0	257.2	285.0	......

ORGANIC COMPOUNDS (Continued)

Pressures Less than One Atmosphere (Continued)

Name	Formula	Temperature °C						M.P.
		1 mm	10 mm	40 mm	100 mm	400 mm	760 mm	
Ethylcetylamine	$C_{18}H_{39}N$	133.2	186.0	226.5	256.8	313.0	342.0d	......
1,1,3-Diethoxytetradecamethylheptasiloxane	$C_{18}H_{52}O_8Si_7$	119.0	163.5	197.0	223.2	268.3	293.5	......
Octadecamethyloctasiloxane	$C_{18}H_{54}O_7Si_8$	105.8	152.3	187.5	214.5	263.5	290.0	......
Triphenylmethane	$C_{19}H_{16}$	169.7	197.0	215.5	228.4	249.8	259.2	93.4
Diphenyl-2-tolyl thiophosphate	$C_{19}H_{17}O_3PS$	159.7	201.6	230.6	252.5	290.0	310.0	......
Nonadecane	$C_{19}H_{40}$	133.2	183.5	220.0	248.0	299.8	330.0	32
Ethoxytriphenylsilane	$C_{20}H_{20}OSi$	167.0	213.5	247.0	273.5	319.5	344.0	......
Diethylhexadecylamine	$C_{20}H_{43}N$	139.8	194.0	235.0	265.5	324.6	355.0	......
1,1,5-Diethoxyhexadecamethyloctasiloxane	$C_{20}H_{58}O_9Si_8$	133.7	179.7	213.8	240.0	286.0	311.5	......
Eicosamethylnonasiloxane	$C_{20}H_{60}O_8Si_9$	144.0	189.0	220.5	244.3	286.0	307.5	......
Tritolyl phosphate	$C_{21}H_{21}O_4P$	154.6	198.0	229.7	252.2	292.7	313.0	......
Heneicosane	$C_{21}H_{44}$	152.6	205.4	243.4	272.0	323.8	350.5	40.4
Erucic acid	$C_{22}H_{42}O_2$	206.7	254.5	289.1	314.4	358.8	381.5d	33.5
Brassidic acid	$C_{22}H_{42}O_2$	209.6	256.0	290.0	316.2	359.6	382.5d	61.5
Docosane	$C_{22}H_{46}$	157.8	213.0	254.5	286.0	343.5	376.0	44.5
Docosamethyldecasiloxane	$C_{22}H_{66}O_9Si_{10}$	160.3	202.8	233.8	255.0	293.8	314.0	......
Tricosane	$C_{23}H_{48}$	170.0	223.0	261.3	289.8	339.8	366.5	47.7
Tetracosane	$C_{24}H_{50}$	183.8	237.6	276.3	305.2	358.0	386.4	51.1
Tetracosamethylhendecasiloxane	$C_{24}H_{72}O_{10}Si_{11}$	175.2	216.7	246.2	266.3	303.7	322.8	......
Pentacosane	$C_{25}H_{52}$	194.2	248.2	285.6	314.0	365.4	390.3	53.3
Hexacosane	$C_{26}H_{54}$	204.0	257.4	295.2	323.2	374.6	399.8	56.6
Dicarvacryl-2-tolyl phosphate	$C_{27}H_{33}O_4P$	180.2	221.8	251.5	272.5	309.8	330.0	......
Heptacosane	$C_{27}H_{56}$	211.7	266.8	305.7	333.5	385.0	410.6	59.5
Octacosane	$C_{28}H_{58}$	226.5	277.4	314.2	341.8	388.9	412.5	61.6
Nonacosane	$C_{29}H_{60}$	234.2	286.4	323.2	350.0	397.2	421.8	63.8
Dicarvacryl-mono-(6-chloro-2-xenyl) phosphate	$C_{32}H_{34}ClO_4P$	204.2	249.3	280.5	304.9	342.0	361.0	......
Pentaethylbenzene	$C_{16}H_{26}$	86.0	135.8	171.9	200.0	250.2	277.0	......
2,6-Di-*tert*-butyl-4-ethylphenol	$C_{16}H_{26}O$	89.1	137.0	172.1	198.0	244.0	268.6	......
4,6-Di-*tert*-butyl-3-ethylphenol	$C_{16}H_{26}O$	111.5	157.4	192.3	218.0	264.6	290.0	......
Muscone	$C_{16}H_{30}O$	118.0	170.0	210.0	241.5	297.2	328.0	......
Palmitonitrile	$C_{16}H_{31}N$	134.3	185.8	223.8	251.5	304.5	332.0	31
1-Hexadecene	$C_{16}H_{32}$	101.6	146.2	178.8	205.3	250.0	274.0	4
Tetraisobutylene	$C_{16}H_{32}$	63.8	108.5	142.2	167.5	214.6	240.0	......
2-Hexadecanone	$C_{16}H_{32}O$	109.8	167.3	203.7	230.5	279.8	307.0	......
Palmitaldehyde	$C_{16}H_{32}O$	121.6	171.8	210.0	239.5	292.3	321.0	34
Palmitic acid	$C_{16}H_{32}O_2$	153.6	205.8	244.4	271.5	326.0	353.8	64.0
Hexadecane	$C_{16}H_{34}$	105.3	149.8	181.3	208.5	258.3	287.5	18.5
Cetyl alcohol	$C_{16}H_{34}O$	122.7	177.8	219.8	251.7	312.7	344.0	49.3
Cetylamine	$C_{16}H_{36}N$	123.6	176.0	215.7	245.8	300.4	330.0	......
Decyltriethylsilane	$C_{16}H_{36}Si$	108.5	155.6	191.7	218.3	267.5	293.0	......
1,1,1-Diethoxydodecamethylhexasiloxane	$C_{16}H_{46}O_7Si_6$	103.6	147.5	180.0	205.5	250.0	273.5	......
Hexadecamethylheptasiloxane	$C_{16}H_{48}O_6Si_7$	93.2	138.5	171.8	198.0	244.7	270.0	......
Hexadecamethylcyclooctasiloxane	$C_{16}H_{48}O_8Si_8$	103.5	150.5	186.3	213.8	263.0	290.0	31.5
Benzanthrone	$C_{17}H_{10}O$	225.0	297.2	350.0	390.0	......	......	174
4-*tert*-Butylphenyl salicylate	$C_{17}H_{18}O_3$	166.2	225.0	270.7	305.8	370.6	404.0d	......
Menthyl benzoate	$C_{17}H_{24}O_2$	123.2	170.0	204.3	230.4	277.1	301.0	54.5
2-Heptadecanone	$C_{17}H_{34}O$	129.6	178.0	214.3	242.0	291.7	319.5	......
Methyl palmitate	$C_{17}H_{34}O_2$	134.3	184.3	d	......	......	......	30
Heptadecane	$C_{17}H_{36}$	115.0	160.0	195.8	223.0	274.5	303.0	22.5
Tetradecyltrimethylsilane	$C_{17}H_{38}Si$	120.0	166.2	201.5	227.8	275.0	300.0	......
Tri-2-chlorophenylthiophosphate	$C_{18}H_{12}Cl_3O_3PS$	188.2	231.2	261.7	283.8	322.0	341.3	......
Triphenyl phosphate	$C_{18}H_{15}O_4P$	193.5	249.8	290.3	322.5	379.2	413.5	49.4
Hexaethylbenzene	$C_{18}H_{30}$	s	150.3	187.7	216.0	268.5	298.3	130
2,4,6-Tri-*tert*-butylphenol	$C_{18}H_{30}O$	95.2	142.0	177.4	203.0	250.6	276.3	......
Oleic Acid	$C_{18}H_{34}O_2$	176.5	223.0	257.2	286.0	334.7	360.0d	14
Elaidic acid	$C_{18}H_{34}O_2$	171.3	223.5	260.8	288.0	337.0	362.0	51.5
Stearaldehyde	$C_{18}H_{36}O$	140.0	192.1	230.8	260.0	313.8	342.5	63.5
Stearic acid	$C_{18}H_{36}O_2$	173.7	225.0	263.3	291.0	343.0	370.0d	69.3
Octadecane	$C_{18}H_{38}$	119.6	169.6	207.4	236.0	288.0	317.0	28
2-Methylheptadecane	$C_{18}H_{38}$	119.8	168.7	204.8	231.5	279.8	306.5	......
1-Octadecanol	$C_{18}H_{38}O$	150.3	202.0	240.4	269.4	320.3	349.5	58.5

Pressures Greater than One Atmosphere

Name	Formula	Temperature °C						
		1 atm.	2 atm.	5 atm.	10 atm.	20 atm.	40 atm.	60 atm.
Chlorotrifluoromethane	$CClF_3$	− 81.2	− 66.7	− 42.7	− 18.5	+ 12.0	52.8	......
Dichlorodifluoromethane	CCl_2F_2	− 29.8	− 12.2	+ 16.1	42.4	74.0	......	......

Name	Formula	Temperature °C						
		1 atm.	2 atm.	5 atm.	10 atm.	20 atm.	40 atm.	60 atm.
Carbonyl chloride	CCl_2O	8.3	27.3	57.2	85.0	119.0	159.8	.....
Trichlorofluoromethane	CCl_3F	23.7	44.1	77.3	108.2	146.7	194.0	.....
Carbontetrachloride	CCl_4	76.7	102.0	141.7	178.0	222.0	276.0	.....
Chlorodifluoromethane	$CHClF_2$	−40.8	−24.7	+0.3	24.0	52.0	85.3	.....
Dichlorofluoromethane	$CHCl_2F$	8.9	28.4	59.0	87.0	121.2	162.6	.....
Trichloromethane	$CHCl_3$	61.3	83.9	120.0	152.3	191.8	237.5	.....
Hydrocyanic acid	CHN	25.8	45.5	75.5	103.5	134.2	170.2	.....
Methyl bromide	CH_3Br	3.6	23.3	54.8	84.0	121.7	170.2	.....
Methyl chloride	CH_3Cl	−24.0	−6.4	+22.0	47.3	77.3	113.8	137.5
Methyl fluoride	CH_3F	−78.2	−64.5	−42.0	−21.0	+2.6	26.5	43.5
Methyl iodide	CH_3I	42.4	65.5	101.8	138.0	176.5	228.5	.....
Methane	CH_4	−161.5	−152.3	−138.3	−124.8	−108.5	−86.3	.....
Methanol	CH_4O	64.7	84.0	112.5	138.0	167.8	203.5	224.0
Methanethiol	CH_4S	6.8	26.1	55.9	83.4	117.5	157.7	185.0
Methylamine	CH_5N	−6.3	+10.1	36.0	59.5	87.8	121.8	144.6
Carbon monoxide	CO	−191.3	−183.5	−170.7	−161.0	−149.7	......	.....
Carbon dioxide	CO_2	-78.2_8	-69.1_8	−56.7	−39.5	−18.9	+5.9	22.4
Carbon disulfide	CS_2	46.5	69.1	104.8	136.3	175.5	222.8	256.0
1-Chloro-1,2,2-trifluoroethylene	C_2ClF_3	−27.9	−11.1	+15.5	40.0	71.1	......	.....
1,2-Dichloro-1,1,2,2-tetrafluoroethane	$C_2Cl_2F_4$	3.5	22.8	54.0	82.3	117.5	......	.....
1,1,2-Trichloro-1,2,2-trifluoroethane	$C_2Cl_3F_3$	47.6	70.0	105.5	138.0	177.7	......	.....
Acetylene	C_2H_2	-84.0_8	−71.6	−50.2	−32.7	−10.0	+16.8	34.8
cis-1,2-Dichloroethylene	$C_2H_2Cl_2$	59.0	82.1	119.3	152.3	194.0	244.5	.....
trans-1,2-Dichloroethylene	$C_2H_2Cl_2$	47.8	69.8	104.0	135.7	174.0	220.0	.....
Ethylene	C_2H_4	−103.7	−90.8	−71.1	−52.8	−29.1	−1.5	.....
1,2-Dibromoethane	$C_2H_4Br_2$	131.5	157.7	200.0	237.0	269.0	295.0	304.5
1,1-Dichloroethane	$C_2H_4Cl_2$	57.3	80.2	117.3	150.3	192.7	243.0	.....
1,2-Dichloroethane	$C_2H_4Cl_2$	83.7	108.1	147.8	183.5	226.5	272.0	.....
Acetic acid	$C_2H_4O_2$	118.1	143.5	180.3	214.0	252.0	297.0	.....
Methyl formate	$C_2H_4O_2$	32.0	51.9	83.5	112.0	147.2	188.5	.....
Ethyl bromide	C_2H_5Br	38.4	60.2	95.0	126.8	164.3	206.5	229.5
Ethyl chloride	C_2H_5Cl	12.3	32.5	64.0	92.6	127.3	167.0	.....
Ethyl fluoride	C_2H_5F	−32.0	−16.7	+7.7	30.2	57.5	90.0	.....
Ethane	C_2H_6	−88.6	−75.0	−52.8	−32.0	−6.4	+23.6	.....
Ethanol	C_2H_6O	78.4	97.5	126.0	151.8	183.0	218.0	242.0
Dimethyl ether	C_2H_6O	−23.7	−6.4	+20.8	45.5	75.7	112.1	.....
Ethanethiol	C_2H_6S	35.0	56.6	90.7	121.9	159.5	204.7	.....
Dimethyl sulfide	C_2H_6S	36.0	57.8	92.3	124.5	163.8	209.0	.....
Ethylamine	C_2H_7N	16.6	35 7	65.3	91.8	124.0	163.0	.....
Dimethylamine	C_2H_7N	7.4	25.0	53.9	80.0	111.7	149.8	.....
Cyanogen	C_2N_2	−21.0	−4.4	+21.4	44.6	72.6	106.5	.....
Propadiene	C_3H_4	−35.0	−18.4	+8.0	33.2	64.5	103.5	.....
Propyne	C_3H_4	−23.3	−7.1	+19.5	43.8	74.0	111.5	.....
Propylene	C_3H_6	−47.7	−31.4	−4.8	+19.8	49.5	85.0	.....
Acetone	C_3H_6O	56.5	78.6	113.0	144.5	181.0	214.5	.....
Propionic acid	$C_3H_6O_2$	141.1	160.0	186.0	203.5	220.0	233.0	.....
Methyl acetate	$C_3H_6O_2$	57.8	79.5	113.1	144.2	181.0	225.0	.....
Ethyl formate	$C_3H_6O_2$	54.3	76.0	110.5	142.2	180.0	225.0	.....
Propane	C_3H_8	−42.1	−25.6	+1.4	26.9	58.1	94.8	.....
1-Propanol	C_3H_8O	97.8	117.0	149.0	177.0	210.8	250.0	.....
2-Propanol	C_3H_8O	82.5	101.3	130.2	155.7	186.0	220.2	.....
Ethyl methyl ether	C_3H_8O	7.5	26.5	56.4	84.0	108.0	160.0	.....
Propylamine	C_3H_9N	48.5	69.8	102.8	133 4	170.0	214.5	.....
1,3-Butadiene	C_4H_6	−4 5	+15.3	47.0	76.0	114.0	158.0	.....
Acetic anhydride	$C_4H_6O_3$	139.6	162.0	194.0	221.5	253.0	288.5	.....
Dimethyl oxalate	$C_4H_6O_4$	163 3	189.6	228.7	......	......	......	.....
Butyric acid	$C_4H_8O_2$	163.5	188.3	225.0	257.0	295.0	338.0	.....
Isobutyric acid	$C_4H_8O_2$	154.5	179.8	217.0	250.0	289.0	336.0	.....
Ethyl acetate	$C_4H_8O_2$	77.1	100.6	136.6	169.7	209.5	......	.....
Methyl propionate	$C_4H_8O_2$	79.8	103.0	139.8	172.6	212.5	......	.....
Propyl formate	$C_4H_8O_2$	81.3	104.3	142.0	176.4	217.5	......	.....
Butane	C_4H_{10}	−0.5	+18.8	50.0	79.5	116.0	......	.....
2-Methylpropane	C_4H_{10}	−11.7	+7.5	39.0	66.8	99.5	......	.....
Butyl alcohol	$C_4H_{10}O$	117.5	139.8	172.5	203.0	237.0	277.0	.....
sec-Butyl alcohol	$C_4H_{10}O$	99.5	118.2	147.5	172.0	204.0	251.0	.....
Isobutyl alcohol	$C_4H_{10}O$	108.0	127.3	156.2	182.0	212.5	251.0	.....
tert-Butyl alcohol	$C_4H_{10}O$	82.9	102.0	130.0	154.2	184.5	222.5	.....
Diethyl ether	$C_4H_{10}O$	34.6	56.0	90.0	122.0	159.0	......	.....
Diethyl sulfide	$C_4H_{10}S$	88.0	112.0	153.8	190.2	234.0	......	.....
Diethylamine	$C_4H_{11}N$	55.5	77.8	113.0	145.3	184.5	......	.....
Tetramethylsilane	$C_4H_{12}Si$	27.0	48.0	82.0	113.0	152.0	......	.....
Ethyl propionate	$C_5H_{10}O_2$	99.1	123.8	162.7	197.8	240.0	......	.....
Propyl acetate	$C_5H_{10}O_2$	101.8	126.8	165.7	200.5	242.8	......	.....
Isobutyl formate	$C_5H_{10}O_2$	98.2	121.8	157.8	192.4	234.0	......	.....
Methyl butyrate	$C_5H_{10}O_2$	102.3	127.5	166.7	203.0	244.5	......	.....

VAPOR PRESSURE (Continued)
ORGANIC COMPOUNDS (Continued)
Pressures Greater than One Atmosphere (Continued)

Name	Formula	Temperature °C						
		1 atm.	2 atm.	5 atm.	10 atm.	20 atm.	40 atm.	60 atm.
Methyl isobutyrate	$C_5H_{10}O_2$	92.6	116.7	155.2	190.2	232.0	.....	.....
Pentane	C_5H_{12}	36.1	58.0	92.4	124.7	164.3	.....	.....
2-Methylbutane	C_5H_{12}	27.8	48.8	82.8	114.5	154.0	.....	.....
2,2-Dimethylpropane	C_5H_{12}	+ 9.5	29.5	61.1	90.7	127.6	.....	.....
Ethyl propyl ether	$C_5H_{12}O$	61.7	85.3	123.1	156.2	197.2	.....	.....
Bromobenzene	C_6H_5Br	156.2	186.2	232.5	274.5	327.0	387.5	.....
Chlorobenzene	C_6H_5Cl	132.2	160.2	205.0	245.3	292.8	349.8	.....
Fluorobenzene	C_6H_5F	84.7	109.9	148.5	184.4	227.6	279.3	.....
Iodobenzene	C_6H_5I	188.6	220.0	270.0	315.7	371.5	437.2	.....
Benzene	C_6H_6	80.1	103.8	142.5	178.8	221.5	272.3	.....
Phenol	C_6H_6O	181.9	208.0	248.2	283.8	328.7	382.1	418.7
Aniline	C_6H_7N	184.4	212.8	254.8	292.7	342.0	400.0	.....
Cyclohexane	C_6H_{12}	80.7	106.0	146.4	184.0	228.4	.....	.....
Ethyl isobutyrate	$C_6H_{12}O_2$	110.1	135.5	174.2	210.0	253.0	.....	.....
Hexane	C_6H_{14}	68.7	93.0	131.7	166.6	209.4	.....	.....
2,3-Dimethylbutane	C_6H_{14}	58.0	82.0	120.3	155.7	198.7	.....	.....
Toluene	C_7H_8	110.6	136.5	178.0	215.8	262.5	319.0	.....
Heptane	C_7H_{16}	98.4	124.8	165.7	202.8	247.5	.....	.....
Ethylbenzene	C_8H_{10}	136.2	163.5	207.5	246.3	294.5	.....	.....
Octane	C_8H_{18}	125.6	152.7	196.2	235.8	281.4	.....	.....
Dodecane	$C_{12}H_{26}$	216.2	249.2	300.0	345.8	.....	.....	.....

VAPOR PRESSURE OF THE ELEMENTS

Rudolf Loebel

This table lists the temperature in degrees Celsius at which an element has a vapor pressure indicated by the headings of the columns. For pressures of one atmosphere and lower, the pressures are given in millimeters of mercury; for pressures above one atmosphere, the pressures are given in atmospheres.

Element		mm Hg					atm.				
		1	10	100	400	760	2	5	10	20	40
Aluminum	Al	1540	1780	2080	2320	2467	2610	2850	3050	3270	3530
Antimony	Sb		960	1280	1570	1750	1960	2490			
Arsenic	As	380	440	510	580	610					
Barium	Ba	860	1050	1300	1520	1640	1790	2030	2230		
Beryllium	Be	1520	1860	2300	2770	2970	3240	3730	4110	4720	5610
Bismuth	Bi		1060	1280	1450	1560	1660	1850	2000	2180	
Boron	B	2660	3030	3460	3810	4000					
Bromine	Br	−60	−30	+9	39	59	78	110			
Cadmium	Cd	393	486	610	710	765	830	930	1030	1120	1240
Calcium	Ca	800	970	1200	1390	1490	1630	1850	2020	2290	
Cesium	Cs		373	513	624	690					
Chlorine	Cl	−123	−101	−71	−46	−34	−17	+9	30	55	97
Chromium	Cr	1610	1840	2140	2360	2480	2630	2850	3010	3180	
Cobalt	Co	1910	2170	2500	2760	2870	3040	3270			
Copper	Cu		1870	2190	2440	2600	2760	3010	3500	3460	3740
Fluorine	F			−203	−193	−188	−180.7	−169.1	−159.6		
Gallium	Ga	1350	1570	1850	2060	2180	2320	2560	2730		
Germanium	Ge		2080	2440	2710	2830	2970	3200	3430		
Gold	Au	1880	2160	2520	2800	2940	3120	3490	3630	3890	
Indium	In				1960	2080	2230	2440	2600		
Iridium	Ir	2830	3170	3630	3960	4130	4310	4650			
Iron	Fe	1780	2040	2370	2620	2750	2900	3150	3360	3570	
Iodine	I	40	72	115	160	185	216	265			
Lanthanum	La				3230	3420	3620	3960	4270		
Lead	Pb	970	1160	1420	1630	1740	1880	2140	2320	2620	
Lithium	Li	750	890	1080	1240	1310	1420	1518			
Magnesium	Mg	620	740	900	1040	1110	1190	1330	1430	1560	
Manganese	Mn		1510	1810	2050	2100	2360	2580	2850		
Mercury	Hg			260	330	356.9	398	465	517	581	657
Molybdenum	Mo	3300	3770	4200	4580	4830	5050	5340	5680	5980	
Neodymium	Nd				2870	3100	3300	3680	3990		
Nickel	Ni	1800	2090	2370	2620	2730	2880	3120	3300	3310	
Palladium	Pd	1470	2290	2670	2950	3140	3270	3560	3840		
Phosphorus	P		127	199	253	283	319				
Platinum	Pt	2600	2940	3360	3650	3830	4000	4310	4570	4860	
Polonium	Po	472	587	752	890	960	1060	1200	1340		
Potassium	K			590	710	770	850	950	1110	1240	1420
Rhodium	Rh	2530	2850	3260	3590	3760	3930	4230	4440		
Rubidium	Rb		390	527	640	700					
Selenium	Se		429	547	640	685	750	850	920	1010	1120
Silver	Ag	1310	1540	1850	2060	2210	2360	2600	2850	3050	3300
Sodium	Na	440	546	700	830	890	980	1120	1230	1370	
Strontium	Sr	740	900	1100	1280	1380	1480	1670	1850	2030	
Sulfur	S		246	333	407	445	493	574	640	720	
Tellurium	Te	520	633	792	900	962	1030	1160	1250		
Thallium	Tl		1000	1210	1370	1470	1560	1750	1900	2050	2260
Tin	Sn	1610	1890	2270	2580	2750	2950	3270	3540	3890	
Titanium	Ti	2180	2480	2860	3100	3260	3400	3650	3800		
Tungsten	W	3980	4490	5160	5470	5940	6260	6670	7250	7670	
Uranium	U	2450	2800	3270	3620	3800	4040	4420			
Vanadium	V	2290	2570	2950	3220	3380	3540	3800			
Zinc	Zn		590	730	840	907	970	1090	1180	1290	

VAPOR PRESSURE OF ELEMENTS THAT ARE GASEOUS AT STANDARD CONDITIONS

The following tables contain vapor pressure data for helium, hydrogen, neon, nitrogen, and oxygen.

Vapor Pressure (atm) vs. Temperature (K) for Helium-4

Temperature, K	Pressure atm
2.177	0.04969
2.20	0.05256
2.25	0.05916
2.30	0.06629
2.35	0.07399
2.40	0.08228
2.45	0.09120
2.50	0.1008
2.55	0.1110
2.60	0.1219
2.65	0.1336
2.70	0.1460
2.75	0.1591
2.80	0.1730
2.85	0.1878
2.90	0.2033
2.95	0.2198
3.00	0.2371
3.05	0.2553
3.10	0.2744
3.15	0.2945
3.20	0.3156
3.25	0.3376
3.30	0.3607
3.35	0.3848
3.40	0.4100
3.45	0.4363
3.50	0.4637
3.55	0.4923
3.60	0.5220
3.65	0.5528
3.70	0.5849
3.75	0.6182
3.80	0.6528
3.85	0.6886
3.90	0.7257
3.95	0.7642
4.00	0.8040
4.05	0.8452
4.10	0.8878
4.15	0.9318
4.20	0.9772
4.224	1.000
4.25	1.024
4.30	1.073
4.35	1.123
4.40	1.174
4.45	1.227
4.50	1.282
4.55	1.339
4.60	1.397
4.65	1.457
4.70	1.519
4.75	1.582
4.80	1.648
4.85	1.715
4.90	1.784
4.95	1.856
5.00	1.929
5.05	2.004
5.10	2.082
5.201	2.245

Vapor Pressure (atm) vs. Temperature (K) for Equilibrium Hydrogen

Temperature, K	Pressure, atm
13.803	0.069_5
14	0.077_8
15	0.133
16	0.213
17	0.325
18	0.476
19	0.673
20	0.923
20.268	1.000
21	1.233
22	1.613
23	2.069
24	2.611
25	3.245
26	3.982
27	4.829
28	5.794
29	6.887
30	8.118
31	9.501
32	11.051
32.976	12.759

Vapor Pressure (atm) vs. Temperature (K) for Neon

Temperature, K	Pressure, atm
25	0.50366
26	0.70902
27	0.97255
28	1.3037
29	1.7124
30	2.2088
31	2.8031
32	3.5061
33	4.3286
34	5.2818
35	6.3773
36	7.6271
37	9.0439
38	10.641
39	12.432
40	14.434
41	16.661
42	19.133
43	21.867
44	24.887
44.4	26.19

Vapor Pressure (atm) vs. Temperature (K) for Nitrogen

Temperature, K	Pressure, atm
63.148	0.1237
64	0.1443
66	0.2037
68	0.2813
70	0.3807
72	0.5059
74	0.6610
76	0.8506
77.347	1.0000
78	1.0793
80	1.3520
82	1.6739
84	2.0503
86	2.4865
88	2.9882
90	3.5607
92	4.2099
94	4.9415
98	6.6748
100	7.6885
102	8.8083
104	10.041
106	11.392
108	12.870
110	14.481
112	16.233
114	18.133
116	20.190
118	22.411
120	24.806
122	27.386
124	30.174
126	33.227
126.200	33.555

Vapor Pressure (atm) vs. Temperature (K) for Oxygen

Temperature, K	Pressure, atm
54.351	0.001
56	0.002
58	0.004
60	0.007
62	0.012
64	0.018
66	0.028
68	0.042
70	0.061
72	0.087
74	0.122
76	0.167
78	0.224
80	0.297
82	0.387
84	0.497
86	0.631
88	0.791
90	0.981
90.180	1.000
92	1.205
94	1.466
96	1.768
98	2.114
100	2.509
102	2.957
104	3.462
106	4.029
108	4.661
110	5.363

VAPOR PRESSURE OF ELEMENTS THAT ARE GASEOUS AT STANDARD CONDITIONS (Continued)

Vapor Pressure (atm) vs. Temperature (K) for Oxygen (Continued)

Temperature, K	Pressure atm	Temperature, K	Pressure, atm
112	6.139	136	22.986
114	6.995	138	25.170
116	7.934	140	27.501
118	8.961		
120	10.082	142	29.986
		144	32.631
122	11.300	146	35.448
124	12.621	148	38.446
126	14.049	150	41.638
128	15.591		
130	17.249	152	45.041
		154	48.675
132	19.031	154.576	49.767
134	20.942		

Data from Sparks, L. L., ASRDI Oxygen Technology Survey, Vol. IV: Low Temperature Measurement, NASA-3073, 1974.

VAPOR PRESSURE OF NITRIC ACID

Temperature °C	Vapor Pressure, mm. of Hg 100% HNO_3	90% of HNO_3
0	14.4	5.5
10	26.6	11.
20	47.9	20.
30	81.3	37.3
40	133.	64.4
50	208.	107.
70	467.	242.
80	670.	352.
90	937.	504.
100	1282.	710.

LOW TEMPERATURE LIQUID BATHS

Liquid thermostat baths suitable for many physical measurements can be produced by using a stirred solid–liquid mixture at its melting point. Dry-ice or liquid air can be used to produce the solid. A Dewar flask is preferable as a container and, with adequate insulation or immersion in another somewhat colder bath, good temperature constancy can be maintained over several hours. Such baths are especially useful over the temperature range between dry-ice (−78C) and liquid air (−190). The following table gives the melting and normal boiling points of some readily available organic liquids suitable for this purpose. The compounds are listed in order of their increasing melting points. Temperatures are in degrees Celcius.

Compound	M.P.	B.P.	Compound	M.P.	B.P.
Isopentane (2-Methyl butane)	−159.9	27.85	Ethyl acetate	− 84	77
Methyl cyclopentane	−142.4	71.8	(Dry-ice + acetone)	− 78	——
Allyl chloride	−134.5	45	*p*-Cymene	− 67.9	177.1
n-Pentane	−129.7	36.1	Chloroform	− 63.5	61.7
Allyl alcohol	−129	97	N-Methyl aniline	− 57	196
Ethyl alcohol	−117.3	78.5	Chlorobenzene	− 45.6	132
Carbon disulfide	−110.8	46.3	Anisole	− 37.5	155
Isobutyl alcohol	−108	108.1	Bromobenzene	− 30.8	156
Acetone	− 95.4	56.2	Carbon tetrachloride	− 23	76.5
Toluene	− 95	110.6	Benzonitrile	− 13	205

FATS AND OILS

These data for fats and oils were compiled originally for the Biology Data Book by H. J. Harwood, and R. P. Geyer. 1964. Data are reproduced here by permission of the copyright owners of the above publication, the Federation of American Societies for Experimental Biology, Washington, D.C. pp. 380–382.

Values are typical rather than average, and frequently were derived from specific analyses for particular samples (especially the constituent fatty acids). Extreme variations may occur, depending on a number

	Fat or Oil	Source	Constants				
			Melting (or Solidification) Point, °C	Specific Gravity (or Density)	Refractive Index $n_D^{40°C}$	Iodine Value	Saponification Value
	(A)	(B)	(C)	(D)	(E)	(F)	(G)
	Land Animals						
1	Butterfat	*Bos taurus*	32.2	$0.911^{40°/15°}$	1.4548	36.1	227
2	Depot fat	*Homo sapiens*	(15)	$0.918^{15°}$	1.4602	67.6	196.2
3	Lard oil	*Sus scrofa*	(30.5)	$0.919^{15°}$	1.4615	58.6	194.6
4	Neat's-foot oil	*B. taurus*	....	$0.910^{25°}$	$1.464^{25°}$	69–76	190–199
5	Tallow, beef	*B. taurus*	....	....	....	49.5	197
6	Tallow, mutton	*Ovis aries*	(42.0)	$0.945^{15°}$	1.4565	40	194
	Marine Animals						
7	Cod-liver oil	*Gadus morhua*	....	0.925^{25}	$1.481^{25°}$	165	186
8	Herring oil	*Clupea harengus*	....	$0.900^{60°}$	$1.4610^{60°}$	140	192
9	Menhaden oil	*Brevoortia tyrannus*	....	$0.903^{60°}$	$1.4645^{60°}$	170	191
10	Sardine oil	*Sardinops caerulea*	....	$0.905^{60°}$	$1.4660^{60°}$	185	191
11	Sperm oil, body	*Physeter macrocephalus*	....	....	....	76–88	122–130
12	Sperm oil, head	*P. macrocephalus*	....	....	....	70	140–144
13	Whale oil	*Balaena mysticetus*	....	0.892^{60}	$1.460^{60°}$	120	195
	Plants						
14	Babassu oil	*Attalea funifera*	22–26	$(0.893^{60°})$	$1.443^{60°}$	15.5	247
15	Castor oil	*Ricinus communis*	(−18.0)	$0.961^{15°}$	1.4770	85.5	180.3
16	Cocoa butter	*Theobroma cacao*	34.1	$0.964^{15°}$	1.4568	36.5	193.8
17	Coconut oil	*Cocos nucifera*	25.1	$0.924^{15°}$	1.4493	10.4	268
18	Corn oil	*Zea mays*	(−20.0)	$0.922^{15°}$	1.4734	122.6	192.0
19	Cotton seed oil	*Gossypium hirsutum*	(−1.0)	$0.917^{25°}$	1.4735	105.7	194.3
20	Linseed oil	*Linum usitatissimum*	(−24.0)	$0.938^{15°}$	$1.4782^{25°}$	178.7	190.3
21	Mustard oil	*Brassica hirta*	....	$0.9145^{15°}$	1.475	102	174
22	Neem oil	*Melia azadirachta*	−3	$0.917^{15°}$	1.4615	71	194.5
23	Niger-seed oil	*Guizotia abyssinica*	....	$0.925^{15°}$	1.471	128.5	190
24	Oiticica oil	*Licania rigida*	....	$0.974^{25°}$	....	140–180	....
25	Olive oil	*Olea europaea sativa*	(−6.0)	$0.918^{15°}$	1.4679	81.1	189.7
26	Palm oil	*Elaeis guineensis*	35.0	$0.915^{15°}$	1.4578	54.2	199.1
27	Palm-kernel oil	*E. guineensis*	24.1	$0.923^{15°}$	1.4569	37.0	219.9
28	Peanut oil	*Arachis hypogaea*	(3.0)	$0.914^{15°}$	1.4691	93.4	192.1
29	Perilla oil	*Perilla frutescens*	....	$(0.935^{15°})$	$1.481^{25°}$	195	192
30	Poppy-seed oil	*Papaver somniferum*	(−15)	$0.925^{15°}$	1.4685	135	194
31	Rapeseed oil	*Brassica campestris*	(−10)	$0.915^{15°}$	1.4706	98.6	174.7
32	Safflower oil	*Carthamus tinctorius*	....	$(0.900^{60°})$	$1.462^{60°}$	145	192
33	Sesame oil	*Sesamum indicum*	(−6.0)	$0.919^{25°}$	1.4646	106.6	187.9
34	Soybean oil	*Glycine soja*	(−16.0)	$0.927^{15°}$	1.4729	130.0	190.6
35	Sunflower-seed oil	*Helianthus annuus*	(−17.0)	$0.923^{15°}$	1.4694	125.5	188.7
36	Tung oil	*Aleurites fordi*	(−2.5)	$0.934^{15°}$	$1.5174^{25°}$	168.2	193.1
37	Wheat-germ oil	*Triticum aestivum*	....	....	....	125	....

[1] Caproic. [2] Capryli. [3] Capric. [4] Butyric. [5] Decenoic. [6] C_{12} monoethenoic. [7] C_{14} monoethenoic. [8] Gadoleic plus erucic. [9] C_{12} n-pentadecanoic. [10] C_{17} margaric. [11] 12-Methyl tetradecanoic. [12] C_{20} polyethenoic.

of variables such as source, treatment, and age of a fat or oil. **Specific Gravity** (column D) was calculated at the specified temperature (degrees centigrade) and referred to water at the same temperature, unless otherwise specified. **Density,** shown in parentheses (column D), was measured at the specified temperature (degrees centigrade). **Refractive Index** (column E) was measured at 50°C, unless otherwise specified.

Constituent Fatty Acids, g/100 g total fatty acids

	Saturated						Unsaturated				
	Lauric	Myristic	Palmitic	Stearic	Arachidic	Other	Palmitoleic	Oleic	Linoleic	Linolenic	Other
	(H)	(I)	(J)	(K)	(L)	(M)	(N)	(O)	(P)	(Q)	(R)
1	2.5	11.1	29.0	9.2	2.4	2.0[1]; 0.5[2]; 2.3[3]	4.6	26.7	3.6	....	3.6[4]; 0.1[5]; 0.1[6]; 0.9[7] 1.4[8]; 1.0[9]; 1.0[10]; 0.4[11]
2	....	2.7	24.0	8.4	....	....	5	46.9	10.2	....	2.5[8]
3	....	1.3	28.3	11.9	....	....	2.7	47.5	6	....	0.2[7]; 2.1[8]
4	....	.....	17–18	2–3	....	....	....	74–76	....	....	....
5	....	6.3	27.4	14.1	....	....	....	49.6	2.5	....	....
6	....	4.6	24.6	30.5	....	....	....	36.0	4.3	....	....
7	....	5.8	8.4	0.6	....	....	20.0	← 29.1 →		....	25.4[12]; 9.6[13]
8	....	7.3	13.0	Trace	....	....	4.9	....	....	20.7	30.1[12]; 23.2[13]
9	....	5.9	16.3	0.6	0.6	....	15.5	....	....	29.6	19.0[12]; 11.7[13]; 0.8[14]
10	....	5.1	14.6	3.2	....	....	11.8	← 17.8 →		....	18.1[12]; 14.0[13]; trace[7]; 15.4[15]
11	1	5	6.5	.....	....	....	26.5	37	19	....	1[13]; 4[7]; 19[16]
12	16	14	8	2	....	3.5[3]	15	17	6.5	....	4[6]; 14[7]; 6.5[16]
13	0.2	9.3	15.6	2.8	....	....	14.4	35.2	....	....	13.6[12]; 5.9[13]; 2.5[7]; 0.2[17]
14	44.1	15.4	8.5	2.7	0.2	0.2[1]; 4.8[2]; 6.6[3]	....	16.1	1.4	....	....
15	← 2.4 →				....	....	....	7.4	3.1	....	87[18]
16	....	.....	24.4	35.4	....	....	....	38.1	2.1	....	....
17	45.4	18.0	10.5	2.3	0.4[19]	0.8[1]; 5.4[2]; 8.4[3]	0.4	7.5	Trace	....	....
18	....	1.4	10.2	3.0	....	....	1.5	49.6	34.3	....	....
19	....	1.4	23.4	1.1	1.3	....	2.0	22.9	47.8	....	....
20	....	.....	6.3	2.5	0.5	....	....	19.0	24.1	47.4	0.2[14]
21	....	1.3[20]	.....	.....	....	....	....	27.2[20]	16.6[20]	1.8[20]	1.1[14]; 1.0[21]; 51.0[22]
22	....	2.6[20]	14.1[20]	24.0[20]	0.8[20]	....	....	58.5[20]	....	....	....
23	....	3.3[20]	8.2[20]	4.8[20]	0.5[20]	....	....	30.3[20]	57.3[20]	....	....
24	← 11.3[23] →					....	....	6.2	....	....	82.5[24]
25	....	Trace	6.9	2.3	0.1	....	....	84.4	4.6	....	....
26	....	1.4	40.1	5.5	....	....	....	42.7	10.3	....	....
27	46.9	14.1	8.8	1.3	....	2.7[2]; 7.0[3]	....	18.5	0.7	....	....
28	....	.....	8.3	3.1	2.4	....	......	56.0	26.0	....	3.1[14]; 1.1[21]
29	← 9.6[23] →					....	....	17.8	....	17.5	....
30	....	.....	4.8[20]	2.9[20]	....	....	....	30.1[20]	62.2[20]	....	....
31	....	.....	1	.....	....	....	....	32	15	1	50[22]
32	← 6.8[23] →					....	....	18.6	70.1	3.4	....
33	....	.....	9.1	4.3	0.8	....	....	45.4	40.4	....	....
34	0.2	0.1	9.8	2.4	0.9	....	0.4	28.9	50.7	6.5	0.1[7]
35	....	.....	5.6	2.2	0.9	....	....	25.1	66.2	....	....
36	← 4.6[23] →					....	....	4.1	0.6	....	90.7[25]
37	← 16.0[23] →					....	....	28.1	52.3	3.6	....

[13] C_{22} polyethenoic. [14] Behenic. [15] C_{14} polyethenoic. [16] Gadoleic. [17] C_{24} polyethenoic. [18] Ricinoleic. [19] Includes behenic and lignoceric. [20] Percent by weight. [21] Lignoceric. [22] Erucic. [23] Includes behenic. [24] Licanic. [25] Eleostearic.

CONCENTRATIVE PROPERTIES OF AQUEOUS SOLUTIONS: CONVERSION TABLES

A. V. Wolf, Morden G. Brown and Phoebe G. Prentiss

The table columns are:

A% = anhydrous solute weight per cent, g solute/100 g solution.
H% = hydrated solute weight per cent, g solute/100 g solution.
ρ or D_4^{20} = relative density at 20°C, kg/l.
D_{20}^{20} = specific gravity at 20°C.
C_s = anhydrous solute concentration, g/l.
M = molar concentration, g-mol/l.
C_w = total water concentration, g/l.
$(C_0 - C_w)$ = water displaced by anhydrous solute, g/l.
$(n - n_0) \times 10^4$ = index of refraction increment above index of refraction of pure water $\times 10^4$; refraction at 20°C.
n = index of refraction at 20°C relative to air for sodium yellow light.
Δ = freezing point depression, °C.
O = osmolality, Os/kg water.
S = osmosity, molar concentration of NaCl solution having same freezing point or osmotic pressure as given solution, g-mol/l.
η/η_0 = relative viscosity, ratio of the absolute viscosity of a solution at 20°C to the absolute viscosity of water at 20°C.
η/ρ = kinematic viscosity, ratio of absolute viscosity at 20°C (centipoise, cP) to relative density, (centistokes, cS).
ϕ = fluidity, reciprocal of absolute viscosity at 20°C, (poise^{-1}), rhe.
γ = specific conductance (electrical) at 20°C, mmho/cm.
T = condosity, molar concentration of NaCl solution having same specific conductance (electrical) at 20°C as given solution, g-mol/l.

Some related measures are:

Specific gravity, $D_{20}^{20} = D_4^{20}/.99823$
Relative density, $D_4^{20} = (C_s + C_w)/1000$
Molality, g-mol/kg water = $1000 \times M/C_w$
Osmolality, $O = \Delta/1.86$
Absolute viscosity (cP) = kinematic viscosity (cS) $\times D_4^{20}$
Specific viscosity = $\eta/\eta_0 - 1$
Specific refraction = $(n - 1)/\rho$
Specific refractive increment = $1000(n - n_0)/C_w$
Molar conductance (ohm^{-1} cm^2 g-mol^{-1}), $\Lambda_M = \lambda/M$
Relative salinity, $\Sigma = T/S$
Ratios such as S/M, T/M, r/A%, $(D_{20}^{20} - 1)/S$, etc.
G water displaced/g solute = $(C_0 - C_w)/C_s$
Ml water displaced/g solute = $(C_0 - C_w)/(.99823 \times C_s)$
G water displaced/mol solute = $(C_0 - C_w)/M$
Ml water displaced/mol solute = $(C_0 - C_w)/(.99823 \times M)$
Mols water displaced/mol solute = $(C_0 - C_w)/(18.015 \times M)$
Concentration of solute relative to water, g/kg water = $1000 \times C_s/C_w$

Relative specific refractivity = contribution of unit weight of solute relative to unit weight of water to the imaginary concentration of water with same refractive index as solution, equivalent concentration of water (ECW) = $3694.1788 \times n - 300 \times n^2 - 3394.1788$. The refractive index for this equation is relative to a vacuum whereas the tabulated values are relative to air. The above equation for the equivalent concentration of water provides better constancy of relative specific refractivity than Lorentz and Lorenz relation, ECW = $4848.3431 - 14545.029/(n^2 + 2)$.

Tables are based upon *Handbook of Chemistry and Physics,* 51st Edition (1970); G. F. Hewitt, Tables of the resistivity of aqueous chloride solutions, Chem. Eng. Div., U.K.A.E.A. Res. Group, Harwell, Gr. Brit. Oct. 1960, HL 60/5450 (S.C. 2), AERE-R 3497; A. V. Wolf, *Aqueous Solutions and Body Fluids,* Hoeber, 1966; and new data.

LIST OF TABLES

1 ACETIC ACID, CH_3COOH

MOLECULAR WEIGHT = 60.05
RELATIVE SPECIFIC REFRACTIVITY = 1.065

0.00 % by wt. data are the same for all compounds.
For Values of 0.00 wt. % solutions see Table 1, Acetic Acid.

A% by wt.	ρ D_4^{20}	D_{20}^{20}	C_s g/l	M g-mol/l	C_w g/l	$(C_o - C_w)$ g/l	$(n - n_o)$ × 10^4	n	Δ °C	O Os/kg	S g-mol/l	η/η_o	η/ρ cS	ϕ rhe	γ mmho/cm	T g-mol/l
0.00	0.9982	1.0000	0.0	0.000	998.2	0.0	0	1.3330	0.000	0.000	0.000	1.000	1.004	99.80	0.0	0.000
0.50	0.9989	1.0007	5.0	0.083	993.9	4.3	4	1.3334	0.159	0.086	0.045	1.010	1.013	98.81	0.3	0.003
1.00	0.9996	1.0014	10.0	0.166	989.6	8.6	7	1.3337	0.317	0.170	0.091	1.020	1.022	97.84	0.6	0.006
1.50	1.0003	1.0021	15.0	0.250	985.3	12.9	11	1.3341	0.474	0.255	0.137	1.030	1.032	96.90	0.7	0.007
2.00	1.0011	1.0028	20.0	0.333	981.0	17.2	15	1.3345	0.630	0.339	0.183	1.040	1.041	95.96	0.8	0.008
2.50	1.0018	1.0035	25.0	0.417	976.7	21.5	18	1.3348	0.786	0.423	0.229	1.050	1.051	95.02	0.9	0.009
3.00	1.0025	1.0042	30.1	0.501	972.4	25.9	22	1.3352	0.942	0.507	0.275	1.061	1.060	94.07	1.0	0.010
3.50	1.0031	1.0049	35.1	0.585	968.0	30.2	26	1.3355	1.099	0.591	0.321	1.072	1.070	93.14	1.1	0.010
4.00	1.0038	1.0056	40.2	0.669	963.7	34.5	29	1.3359	1.256	0.675	0.367	1.082	1.080	92.24	1.1	0.011
4.50	1.0045	1.0063	45.2	0.753	959.3	38.9	33	1.3363	1.415	0.761	0.414	1.092	1.090	91.36	1.2	0.011
5.00	1.0052	1.0070	50.3	0.837	955.0	43.3	36	1.3366	1.576	0.847	0.461	1.103	1.099	90.50	1.2	0.012
5.50	1.0059	1.0077	55.3	0.921	950.6	47.6	40	1.3370	1.737	0.934	0.508	1.113	1.109	89.66	1.3	0.012
6.00	1.0066	1.0084	60.4	1.006	946.2	52.0	44	1.3373	1.899	1.021	0.555	1.123	1.118	88.87	1.3	0.013
6.50	1.0073	1.0091	65.5	1.090	941.8	56.4	47	1.3377	2.062	1.108	0.602	1.132	1.126	88.14	1.4	0.013
7.00	1.0080	1.0098	70.6	1.175	937.4	60.8	51	1.3381	2.225	1.196	0.650	1.141	1.134	87.46	1.4	0.013
7.50	1.0087	1.0105	75.7	1.260	933.0	65.2	54	1.3384	2.390	1.285	0.697	1.150	1.142	86.78	1.4	0.014
8.00	1.0093	1.0111	80.7	1.345	928.6	69.6	58	1.3388	2.555	1.374	0.745	1.160	1.152	86.03	1.4	0.014
8.50	1.0100	1.0118	85.9	1.430	924.2	74.1	62	1.3391	2.722	1.463	0.792	1.171	1.162	85.20	1.5	0.014
9.00	1.0107	1.0125	91.0	1.515	919.7	78.5	65	1.3395	2.889	1.553	0.840	1.184	1.173	84.31	1.5	0.014
9.50	1.0114	1.0132	96.1	1.600	915.3	82.9	69	1.3399	3.057	1.644	0.888	1.196	1.185	83.43	1.5	0.015
10.00	1.0121	1.0138	101.2	1.685	910.8	87.4	72	1.3402	3.226	1.734	0.935	1.208	1.196	82.62	1.5	0.015
11.00	1.0134	1.0152	111.5	1.856	901.9	96.3	79	1.3409	3.567	1.918	1.031	1.229	1.215	81.19	1.6	0.015
12.00	1.0147	1.0165	121.8	2.028	893.0	105.3	86	1.3416	3.911	2.103	1.127	1.250	1.234	79.84	1.6	0.016
13.00	1.0161	1.0178	132.1	2.200	884.0	114.3	93	1.3423	4.259	2.290	1.223	1.272	1.255	78.43	1.6	0.016
14.00	1.0174	1.0192	142.4	2.372	874.9	123.3	100	1.3430	4.611	2.479	1.320	1.295	1.275	77.07	1.7	0.016
15.00	1.0187	1.0205	152.8	2.545	865.9	132.4	107	1.3437	4.967	2.670	1.415	1.317	1.295	75.79	1.7	0.016
16.00	1.0200	1.0218	163.2	2.718	856.8	141.5	114	1.3444	5.33	2.86	1.511	1.338	1.314	74.59	1.7	0.016
17.00	1.0213	1.0231	173.6	2.891	847.6	150.6	121	1.3451	5.69	3.06	1.607	1.357	1.331	73.54	1.7	0.017
18.00	1.0225	1.0243	184.1	3.065	838.5	159.8	128	1.3458	6.06	3.26	1.703	1.377	1.349	72.48	1.7	0.017
19.00	1.0238	1.0256	194.5	3.239	829.3	169.0	135	1.3465	6.43	3.46	1.798	1.402	1.372	71.18	1.7	0.017
20.00	1.0250	1.0269	205.0	3.414	820.0	178.2	142	1.3472	6.81	3.66	1.894	1.428	1.396	69.89	1.7	0.017
22.00	1.0275	1.0293	226.1	3.764	801.5	196.8	155	1.3485	7.57	4.07	2.084	1.475	1.438	67.66	1.7	0.017
24.00	1.0299	1.0318	247.2	4.116	782.8	215.5	169	1.3498	8.36	4.49	2.273	1.522	1.481	65.57	1.7	0.016
26.00	1.0323	1.0341	268.4	4.470	763.9	234.3	182	1.3512	9.17	4.93	2.462	1.569	1.523	63.61	1.6	0.015
28.00	1.0346	1.0365	289.7	4.824	744.9	253.3	195	1.3525	10.00	5.38	2.650	1.610	1.559	61.99	1.5	0.015
30.00	1.0369	1.0388	311.1	5.180	725.8	272.4	207	1.3537	10.84	5.83	2.835	1.666	1.610	59.90	1.4	0.014
32.00	1.0391	1.0410	332.5	5.537	706.6	291.6	220	1.3550	11.70	6.29	3.017	1.712	1.651	58.29	1.4	0.013
34.00	1.0413	1.0431	354.0	5.896	687.3	311.0	232	1.3562	12.55	6.75	3.192	1.758	1.692	56.77	1.3	0.012
36.00	1.0434	1.0452	375.6	6.255	667.8	330.5	245	1.3574	13.38	7.20	3.359	1.808	1.736	55.20	1.2	0.011
38.00	1.0454	1.0473	397.3	6.615	648.2	350.1	257	1.3586				1.848	1.771	54.00	1.1	0.011
40.00	1.0474	1.0492	419.0	6.977	628.4	369.8	268	1.3598				1.908	1.825	52.31	1.1	0.010
42.00	1.0493	1.0511	440.7	7.339	608.6	389.7	280	1.3610				1.956	1.868	51.02	1.0	0.010
44.00	1.0510	1.0529	462.5	7.701	588.6	409.7	291	1.3621				2.003	1.910	49.83	1.0	0.009
46.00	1.0528	1.0547	484.3	8.065	568.5	429.7	302	1.3632				2.048	1.949	48.73	0.9	0.009
48.00	1.0545	1.0564	506.2	8.429	548.4	449.9	312	1.3642				2.106	2.001	47.39	0.8	0.008
50.00	1.0562	1.0581	528.1	8.794	528.1	470.1	323	1.3653				2.154	2.043	46.33	0.8	0.007
52.00	1.0577	1.0596	550.0	9.159	507.7	490.5	333	1.3663				2.208	2.092	45.20	0.7	0.007
54.00	1.0592	1.0611	572.0	9.525	487.2	511.0	343	1.3673				2.260	2.138	44.16	0.7	0.006
56.00	1.0605	1.0624	593.9	9.890	466.6	531.6	352	1.3682				2.303	2.176	43.33	0.5	0.005
58.00	1.0618	1.0636	615.8	10.255	445.9	552.3	361	1.3691				2.355	2.222	42.38	0.3	0.003
60.00	1.0629	1.0648	637.7	10.620	425.2	573.1	370	1.3700				2.404	2.266	41.51	0.4	0.004
62.00	1.0640	1.0659	659.7	10.985	404.3	593.9	378	1.3708				2.451	2.308	40.72		
64.00	1.0650	1.0668	681.6	11.350	383.4	614.8	386	1.3716				2.497	2.349	39.97		
66.00	1.0659	1.0678	703.5	11.715	362.4	635.8	394	1.3724				2.548	2.395	39.17		
68.00	1.0668	1.0687	725.4	12.080	341.4	656.9	402	1.3732				2.589	2.432	38.55		
70.00	1.0673	1.0692	747.1	12.441	320.2	678.0	409	1.3738				2.624	2.463	38.03		
72.00	1.0676	1.0695	768.7	12.800	298.9	699.3	415	1.3745				2.657	2.494	37.56		
74.00	1.0678	1.0697	790.2	13.158	277.6	720.6	421	1.3751				2.682	2.517	37.21		
76.00	1.0680	1.0699	811.7	13.516	256.3	741.9	427	1.3757				2.709	2.542	36.84		
78.00	1.0681	1.0700	833.1	13.956	235.0	763.2	432	1.3762				2.715	2.547	36.76		
80.00	1.0680	1.0699	854.4	14.227	213.6	784.6	437	1.3767				2.715	2.547	36.76		
82.00	1.0677	1.0696	875.5	14.579	192.2	806.0	441	1.3770				2.691	2.525	37.09		
84.00	1.0673	1.0692	896.5	14.928	170.8	827.4	443	1.3773				2.653	2.491	37.62		
86.00	1.0666	1.0685	917.3	15.275	149.3	848.9	444	1.3774				2.591	2.434	38.52		
88.00	1.0658	1.0677	937.9	15.618	127.9	870.3	444	1.3774				2.506	2.356	39.82		
90.00	1.0644	1.0663	958.0	15.953	106.4	891.8	441	1.3771				2.381	2.241	41.92		
92.00	1.0629	1.0648	977.9	16.284	85.0	913.2	436	1.3766				2.236	2.108	44.63		
94.00	1.0606	1.0625	997.0	16.602	63.6	934.6	429	1.3759				2.032	1.920	49.11		
96.00	1.0578	1.0597	1015.5	16.912	42.3	955.9	418	1.3748				1.809	1.714	55.17		
98.00	1.0538	1.0557	1032.7	17.196	21.1	977.1	404	1.3734				1.532	1.457	65.14		
100.00	1.0477	1.0496	1047.7	17.446	0.0	998.2	386	1.3716				1.221	1.168	81.73		

2 ACETONE, CH_3COCH_3

MOLECULAR WEIGHT = 58.05
RELATIVE SPECIFIC REFRACTIVITY = 1.349

0.00 % by wt. data are the same for all compounds.
For Values of 0.00 wt. % solutions see Table 1, Acetic Acid.

A% by wt.	ρ D_4^{20}	D_{20}^{20}	C_s g/l	M g-mol/l	C_w g/l	$(C_o - C_w)$ g/l	$(n - n_o)$ × 10^4	n	Δ °C	O Os/kg	S g-mol/l	η/η_o	η/ρ cS	ϕ rhe	γ mmho/cm	T g-mol/l
0.50	0.9975	0.9993	5.0	0.086	992.5	5.7	4	1.3334	0.160	0.086	0.045	1.011	1.015	98.73		
1.00	0.9968	0.9985	10.0	0.172	986.8	11.4	7	1.3337	0.321	0.173	0.092	1.022	1.027	97.65		
1.50	0.9961	0.9978	14.9	0.257	981.1	17.1	11	1.3341	0.483	0.260	0.140	1.033	1.039	96.58		

2 ACETONE, CH_3COCH_3—(Continued)

A% by wt.	ρ D_4^{20}	D_{20}^{20}	C_s g/l	M g-mol/l	C_w g/l	$(C_o - C_w)$ g/l	$(n - n_o)$ × 10^4	n	Δ °C	O Os/kg	S g-mol/l	η/η_o	η/ρ cS	φ rhe	γ mmho/cm	T g-mol/l
2.00	0.9954	0.9971	19.9	0.343	975.5	22.8	15	1.3344	0.645	0.347	0.187	1.045	1.052	95.50		
2.50	0.9947	0.9964	24.9	0.428	969.8	28.4	18	1.3348	0.807	0.434	0.235	1.057	1.065	94.39		
3.00	0.9940	0.9957	29.8	0.514	964.1	34.1	22	1.3352	0.970	0.521	0.283	1.070	1.079	93.25		
3.50	0.9933	0.9950	34.8	0.599	958.5	39.7	25	1.3355	1.133	0.609	0.331	1.083	1.093	92.11		
4.00	0.9926	0.9943	39.7	0.684	952.9	45.4	29	1.3359	1.296	0.697	0.379	1.097	1.107	90.99		
4.50	0.9919	0.9937	44.6	0.769	947.3	51.0	33	1.3363	1.460	0.785	0.427	1.110	1.121	89.90		
5.00	0.9912	0.9930	49.6	0.854	941.7	56.6	36	1.3366	1.625	0.874	0.475	1.123	1.135	88.87		
5.50	0.9906	0.9923	54.5	0.939	936.1	62.1	40	1.3370	1.790	0.962	0.523	1.136	1.149	87.89		
6.00	0.9899	0.9917	59.4	1.023	930.5	67.7	44	1.3373	1.955	1.051	0.571	1.148	1.162	86.94		
6.50	0.9893	0.9910	64.3	1.108	925.0	73.3	47	1.3377	2.120	1.140	0.619	1.160	1.175	86.03		
7.00	0.9886	0.9904	69.2	1.192	919.4	78.8	51	1.3381	2.286	1.229	0.667	1.172	1.188	85.14		
7.50	0.9880	0.9897	74.1	1.276	913.9	84.4	54	1.3384	2.452	1.318	0.715	1.184	1.201	84.28		
8.00	0.9874	0.9891	79.0	1.361	908.4	89.9	58	1.3388	2.619	1.408	0.763	1.196	1.214	83.44		
8.50	0.9867	0.9885	83.9	1.445	902.9	95.4	62	1.3392	2.785	1.498	0.811	1.208	1.226	82.64		
9.00	0.9861	0.9879	88.8	1.529	897.4	100.9	65	1.3395	2.952	1.587	0.858	1.219	1.239	81.85		
9.50	0.9855	0.9873	93.6	1.613	891.9	106.3	69	1.3399	3.119	1.677	0.905	1.231	1.251	81.09		
10.00	0.9849	0.9867	98.5	1.697	886.4	111.8	73	1.3402	3.287	1.767	0.953	1.242	1.264	80.35		

3 AMMONIA, NH_3, AND AMMONIUM HYDROXIDE, NH_4OH

MOLECULAR WEIGHT, NH_3 = 17.03 FORMULA WEIGHT, NH_4OH = 35.05
RELATIVE SPECIFIC REFRACTIVITY = 1.582

0.00 % by wt. data are the same for all compounds.
For Values of 0.00 wt. % solutions see Table 1, Acetic Acid.

NH_3 % by wt.	NH_4OH % by wt.	ρ D_4^{20}	D_{20}^{20}	C_s (NH_3) g/l	M g-mol/l	C_w g/l	$(C_o - C_w)$ g/l	$(n - n_o)$ × 10^4	n	Δ °C	O Os/kg	S g-mol/l	η/η_o	η/ρ cS	φ rhe	γ mmho/cm	T g-mol/l
0.50	1.03	0.9960	0.9978	5.0	0.292	991.0	7.2	2	1.3332	0.550	0.296	0.160	1.007	1.013	99.14	0.5	0.005
1.00	2.06	0.9938	0.9956	9.9	0.584	983.9	14.3	5	1.3335	1.135	0.610	0.332	1.013	1.022	98.48	0.7	0.007
1.50	3.09	0.9917	0.9934	14.9	0.873	976.8	21.5	7	1.3337	1.725	0.927	0.504	1.020	1.031	97.82	0.9	0.008
2.00	4.12	0.9895	0.9913	19.8	1.162	969.7	28.5	9	1.3339	2.319	1.247	0.677	1.027	1.040	97.17	1.0	0.009
2.50	5.15	0.9874	0.9891	24.7	1.449	962.7	35.5	12	1.3342	2.921	1.570	0.849	1.034	1.049	96.52	1.0	0.010
3.00	6.17	0.9853	0.9870	29.6	1.736	955.7	42.5	14	1.3344	3.531	1.898	1.021	1.041	1.059	95.88	1.1	0.011
3.50	7.20	0.9832	0.9849	34.4	2.021	948.8	49.5	17	1.3347	4.150	2.231	1.193	1.048	1.068	95.24	1.1	0.011
4.00	8.23	0.9811	0.9828	39.2	2.304	941.8	56.4	19	1.3349	4.781	2.570	1.365	1.055	1.077	94.60	1.1	0.011
4.50	9.26	0.9790	0.9808	44.1	2.587	935.0	63.3	22	1.3352	5.42	2.92	1.537	1.062	1.087	93.98	1.1	0.011
5.00	10.29	0.9770	0.9787	48.8	2.868	928.1	70.1	24	1.3354	6.08	3.27	1.708	1.069	1.096	93.36	1.1	0.011
5.50	11.32	0.9750	0.9767	53.6	3.149	921.3	76.9	27	1.3357	6.75	3.63	1.879	1.076	1.106	92.75	1.1	0.011
6.00	12.35	0.9730	0.9747	58.4	3.428	914.6	83.6	30	1.3359	7.43	4.00	2.049	1.083	1.115	92.15	1.1	0.011
6.50	13.38	0.9710	0.9727	63.1	3.706	907.9	90.4	32	1.3362	8.13	4.37	2.219	1.090	1.125	91.55	1.1	0.011
7.00	14.41	0.9690	0.9707	67.8	3.983	901.2	97.1	35	1.3365	8.85	4.76	2.388	1.097	1.134	90.96	1.1	0.011
7.50	15.44	0.9671	0.9688	72.5	4.259	894.5	103.7	38	1.3367	9.58	5.15	2.557	1.104	1.144	90.38	1.1	0.011
8.00	16.47	0.9651	0.9668	77.2	4.534	887.9	110.3	40	1.3370	10.34	5.56	2.725	1.111	1.154	89.81	1.1	0.011
8.50	17.49	0.9632	0.9649	81.9	4.807	881.3	116.9	43	1.3373	11.11	5.97	2.892	1.118	1.163	89.25	1.1	0.010
9.00	18.52	0.9613	0.9630	86.5	5.080	874.8	123.5	46	1.3376	11.90	6.40	3.059	1.125	1.173	88.70	1.1	0.010
9.50	19.55	0.9594	0.9611	91.1	5.352	868.3	130.0	48	1.3378	12.71	6.84	3.225	1.132	1.182	88.16	1.0	0.010
10.00	20.58	0.9575	0.9592	95.8	5.623	861.8	136.5	51	1.3381	13.55	7.28	3.391	1.139	1.192	87.63	1.0	0.010
11.00	22.64	0.9538	0.9555	104.9	6.161	848.9	149.3	57	1.3387	15.29	8.22	3.719	1.153	1.211	86.57	1.0	0.009
12.00	24.70	0.9502	0.9519	114.0	6.695	836.2	162.1	63	1.3393	17.13	9.21	4.044	1.167	1.230	85.53	0.9	0.009
13.00	26.76	0.9466	0.9483	123.1	7.226	823.5	174.7	68	1.3398	19.07	10.25	4.365	1.180	1.250	84.55	0.8	0.008
14.00	28.81	0.9431	0.9447	132.0	7.753	811.0	187.2	74	1.3404	21.13	11.36		1.193	1.268	83.64	0.8	0.007
15.00	30.87	0.9396	0.9412	140.9	8.276	798.6	199.6	80	1.3410	23.32	12.54		1.205	1.285	82.84	0.7	0.007
16.00	32.93	0.9361	0.9378	149.8	8.795	786.4	211.9	86	1.3416	25.63	13.78		1.216	1.301	82.10	0.7	0.006
17.00	34.99	0.9327	0.9344	158.6	9.311	774.2	224.1	92	1.3422	28.09	15.10		1.226	1.317	81.43	0.6	0.006
18.00	37.05	0.9294	0.9310	167.3	9.823	762.1	236.1	98	1.3428	30.70	16.51		1.235	1.331	80.82	0.6	0.005
19.00	39.10	0.9261	0.9277	176.0	10.332	750.1	248.1	104	1.3434	33.47	18.00		1.243	1.345	80.26	0.5	0.005
20.00	41.16	0.9228	0.9245	184.6	10.838	738.3	260.0	110	1.3440	36.42	19.58		1.251	1.359	79.75	0.5	0.005
22.00	45.28	0.9164	0.9181	201.6	11.839	714.8	283.4	123	1.3453	43.36	23.31		1.265	1.383	78.88	0.4	0.004
24.00	49.40	0.9102	0.9118	218.4	12.827	691.7	306.5	135	1.3465	51.38	27.62		1.277	1.405	78.18	0.4	0.004
26.00	53.51	0.9040	0.9056	235.0	13.802	669.0	329.3	148	1.3477	60.77	32.67		1.285	1.425	77.64	0.4	0.004
28.00	57.63	0.8980	0.8996	251.4	14.764	646.5	351.7	160	1.3490	71.66	38.53						
30.00	61.74	0.8920	0.8936	267.6	15.713	624.4	373.8	172	1.3502	84.06	45.19						

4 AMMONIUM CHLORIDE, NH_4Cl

MOLECULAR WEIGHT = 53.50
RELATIVE SPECIFIC REFRACTIVITY = 1.241

0.00 % by wt. data are the same for all compounds.
For Values of 0.00 wt. % solutions see Table 1, Acetic Acid.

A% by wt.	ρ D_4^{20}	D_{20}^{20}	C_s g/l	M g-mol/l	C_w g/l	$(C_o - C_w)$ g/l	$(n - n_o)$ × 10^4	n	Δ °C	O Os/kg	S g-mol/l	η/η_o	η/ρ cS	φ rhe	γ mmho/cm	T g-mol/l
0.50	0.9998	1.0016	5.0	0.093	994.8	3.4	10	1.3340	0.322	0.173	0.092	0.997	0.999	100.13	10.5	0.110
1.00	1.0014	1.0032	10.0	0.187	991.4	6.9	19	1.3349	0.637	0.343	0.185	0.994	0.995	100.40	20.4	0.225
1.50	1.0030	1.0047	15.0	0.281	987.9	10.3	29	1.3359	0.953	0.512	0.278	0.992	0.991	100.62	30.3	0.347
2.00	1.0045	1.0063	20.1	0.376	984.4	13.8	39	1.3369	1.270	0.683	0.371	0.990	0.988	100.81	40.3	0.474
2.50	1.0061	1.0079	25.2	0.470	980.9	17.3	48	1.3378	1.590	0.855	0.465	0.988	0.984	101.00	49.9	0.601
3.00	1.0076	1.0094	30.2	0.565	977.4	20.8	58	1.3388	1.913	1.029	0.559	0.986	0.981	101.18	59.2	0.728
3.50	1.0092	1.0110	35.3	0.660	973.8	24.4	68	1.3397	2.240	1.205	0.654	0.985	0.978	101.35	68.3	0.858
4.00	1.0107	1.0125	40.4	0.756	970.3	28.0	77	1.3407	2.571	1.382	0.749	0.983	0.975	101.52	77.3	0.992
4.50	1.0122	1.0140	45.6	0.851	966.7	31.6	87	1.3417	2.907	1.563	0.845	0.982	0.972	101.68	86.3	1.13
5.00	1.0138	1.0155	50.7	0.947	963.1	35.2	96	1.3426	3.246	1.745	0.941	0.980	0.969	101.84	95.3	1.27
5.50	1.0153	1.0171	55.8	1.044	959.4	38.8	106	1.3436	3.591	1.931	1.038	0.979	0.966	101.99	104.	1.42
6.00	1.0168	1.0186	61.0	1.140	955.8	42.5	115	1.3445	3.941	2.119	1.135	0.977	0.963	102.15	113.	1.58
6.50	1.0183	1.0201	66.2	1.237	952.1	46.1	125	1.3455	4.296	2.310	1.233	0.976	0.960	102.30	122.	1.74

4 AMMONIUM CHLORIDE, NH_4Cl—(Continued)

A% by wt.	ρ D_4^{20}	D_{20}^{20}	C_s g/l	M g-mol/l	C_w g/l	$(C_o - C_w)$ g/l	$(n - n_o)$ × 10^4	n	Δ °C	O Os/kg	S g-mol/l	η/η_o	η/ρ cS	φ rhe	γ mmho/cm	T g-mol/l
7.00	1.0198	1.0216	71.4	1.334	948.4	49.8	135	1.3464	4.657	2.504	1.332	0.974	0.957	102.44	131.	1.91
7.50	1.0213	1.0231	76.6	1.432	944.7	53.6	144	1.3474	5.02	2.70	1.431	0.973	0.955	102.58	139.	2.09
8.00	1.0227	1.0246	81.8	1.529	940.9	57.3	154	1.3483	5.40	2.90	1.530	0.972	0.952	102.71	147.	2.27
8.50	1.0242	1.0260	87.1	1.627	937.2	61.1	163	1.3493	5.77	3.10	1.629	0.971	0.950	102.83	156.	2.46
9.00	1.0257	1.0275	92.3	1.725	933.4	64.9	172	1.3502	6.16	3.31	1.728	0.970	0.947	102.93	164.	2.66
9.50	1.0272	1.0290	97.6	1.824	929.6	68.7	182	1.3512	6.55	3.52	1.828	0.969	0.945	103.02	172.	2.86
10.00	1.0286	1.0304	102.9	1.923	925.8	72.5	191	1.3521	6.95	3.73	1.928	0.968	0.943	103.10	180.	3.07
11.00	1.0315	1.0333	113.5	2.121	918.1	80.2	210	1.3540	7.76	4.17	2.129	0.967	0.939	103.20	200.	3.71
12.00	1.0344	1.0362	124.1	2.320	910.3	88.0	229	1.3559	8.60	4.62	2.330	0.967	0.936	103.23	220.	4.84
13.00	1.0373	1.0391	134.8	2.520	902.4	95.8	248	1.3578	9.47	5.09	2.531	0.967	0.934	103.22		
14.00	1.0401	1.0419	145.6	2.722	894.5	103.8	267	1.3596				0.967	0.932	103.17		
15.00	1.0429	1.0447	156.4	2.924	886.5	111.8	285	1.3615				0.968	0.930	103.10		
16.00	1.0457	1.0475	167.3	3.127	878.4	119.8	304	1.3634				0.969	0.928	103.01		
17.00	1.0485	1.0503	178.2	3.332	870.2	128.0	322	1.3652				0.970	0.927	102.90		
18.00	1.0512	1.0531	189.2	3.537	862.0	136.2	341	1.3671				0.971	0.926	102.75		
19.00	1.0540	1.0558	200.3	3.743	853.7	144.5	360	1.3689				0.973	0.925	102.54		
20.00	1.0567	1.0586	211.3	3.950	845.3	152.9	378	1.3708				0.976	0.925	102.25		
22.00	1.0621	1.0640	233.7	4.367	828.4	169.8	415	1.3745				0.984	0.928	101.46		
24.00	1.0674	1.0693	256.2	4.788	811.2	187.0	452	1.3782				0.994	0.933	100.40		

5 AMMONIUM SULFATE, $(NH_4)_2SO_4$

MOLECULAR WEIGHT = 132.14
RELATIVE SPECIFIC REFRACTIVITY = 0.886

0.00 % by wt. data are the same for all compounds.
For Values of 0.00 wt. % solutions see Table 1, Acetic Acid.

A% by wt.	ρ D_4^{20}	D_{20}^{20}	C_s g/l	M g-mol/l	C_w g/l	$(C_o - C_w)$ g/l	$(n - n_o)$ × 10^4	n	Δ °C	O Os/kg	S g-mol/l	η/η_o	η/ρ cS	φ rhe	γ mmho/cm	T g-mol/l
0.50	1.0012	1.0030	5.0	0.038	996.2	2.1	8	1.3338	0.173	0.093	0.049	1.006	1.007	99.22	7.4	0.077
1.00	1.0042	1.0059	10.0	0.076	994.1	4.1	16	1.3346	0.331	0.178	0.095	1.012	1.010	98.62	14.2	0.151
2.00	1.0101	1.0119	20.2	0.153	989.9	8.3	33	1.3363	0.631	0.339	0.183	1.025	1.017	97.37	25.7	0.290
3.00	1.0160	1.0178	30.5	0.231	985.6	12.7	49	1.3379	0.921	0.495	0.269	1.039	1.025	96.03	36.7	0.428
4.00	1.0220	1.0238	40.9	0.309	981.1	17.1	65	1.3395	1.208	0.649	0.353	1.055	1.034	94.63	47.2	0.565
5.00	1.0279	1.0297	51.4	0.389	976.5	21.7	82	1.3411	1.489	0.800	0.435	1.071	1.044	93.18	57.4	0.704
6.00	1.0338	1.0356	62.0	0.469	971.8	26.5	98	1.3428	1.768	0.951	0.517	1.088	1.055	91.71	67.4	0.845
7.00	1.0397	1.0416	72.8	0.551	966.9	31.3	114	1.3444	2.047	1.101	0.598	1.106	1.066	90.21	77.0	0.987
8.00	1.0456	1.0475	83.6	0.633	962.0	36.3	130	1.3460	2.327	1.251	0.679	1.125	1.079	88.67	86.3	1.13
9.00	1.0515	1.0534	94.6	0.716	956.9	41.4	146	1.3476	2.608	1.402	0.760	1.145	1.091	87.13	95.5	1.28
10.00	1.0574	1.0593	105.7	0.800	951.6	46.6	162	1.3492	2.892	1.555	0.841	1.166	1.105	85.59	105.	1.43
12.00	1.0691	1.0710	128.3	0.971	940.8	57.4	193	1.3523	3.473	1.867	1.005	1.208	1.133	82.58	122.	1.75
14.00	1.0808	1.0827	151.3	1.145	929.5	68.8	225	1.3555	4.073	2.190	1.171	1.253	1.162	79.63	139.	2.09
16.00	1.0924	1.0943	174.8	1.323	917.6	80.6	256	1.3586	4.689	2.521	1.341	1.302	1.194	76.66	156.	2.44
18.00	1.1039	1.1059	198.7	1.504	905.2	93.0	287	1.3616				1.356	1.231	73.59	171.	2.82
20.00	1.1154	1.1174	223.1	1.688	892.3	105.9	317	1.3647				1.418	1.274	70.38	185.	3.21
22.00	1.1269	1.1289	247.9	1.876	879.0	119.3	347	1.3677				1.487	1.323	67.10	198.	3.67
24.00	1.1383	1.1403	273.2	2.067	865.1	133.2	377	1.3707				1.563	1.376	63.84	210.	4.18
26.00	1.1496	1.1516	298.9	2.262	850.7	147.5	407	1.3737				1.647	1.435	60.60	220.	4.84
28.00	1.1609	1.1629	325.0	2.460	835.8	162.4	436	1.3766				1.740	1.502	57.37		
30.00	1.1721	1.1742	351.6	2.661	820.5	177.7	465	1.3795				1.843	1.576	54.15		
32.00	1.1833	1.1854	378.7	2.866	804.7	193.6	494	1.3824				1.957	1.657	50.99		
34.00	1.1945	1.1966	406.1	3.073	788.3	209.9	523	1.3853				2.082	1.747	47.93		
36.00	1.2056	1.2077	434.0	3.284	771.6	226.7	551	1.3881				2.218	1.844	44.99		
38.00	1.2166	1.2188	462.3	3.499	754.3	243.9	580	1.3909				2.366	1.948	42.19		
40.00	1.2277	1.2298	491.1	3.716	736.6	261.6	608	1.3938				2.525	2.061	39.52		

6 BARIUM CHLORIDE, $BaCl_2 \cdot 2H_2O$

MOLECULAR WEIGHT = 208.27 FORMULA WEIGHT, HYDRATE = 244.31
RELATIVE SPECIFIC REFRACTIVITY = 0.552

0.00 % by wt. data are the same for all compounds.
For Values of 0.00 wt. % solutions see Table 1, Acetic Acid.

A% by wt.	H% by wt.	ρ D_4^{20}	D_{20}^{20}	C_s g/l	M g-mol/l	C_w g/l	$(C_o - C_w)$ g/l	$(n - n_o)$ × 10^4	n	Δ °C	O Os/kg	S g-mol/l	η/η_o	η/ρ cS	φ rhe	γ mmho/cm	T g-mol/l
0.50	0.59	1.0026	1.0044	5.0	0.024	997.6	0.6	7	1.3337	0.119	0.064	0.033	1.007	1.007	99.06	4.7	0.047
1.00	1.17	1.0070	1.0088	10.1	0.048	996.9	1.3	15	1.3345	0.233	0.125	0.066	1.014	1.009	98.42	9.1	0.094
2.00	2.35	1.0159	1.0177	20.3	0.098	995.6	2.6	30	1.3360	0.461	0.248	0.133	1.024	1.010	97.46	17.4	0.189
3.00	3.52	1.0249	1.0267	30.7	0.148	994.2	4.0	45	1.3375	0.691	0.372	0.201	1.035	1.012	96.41	25.2	0.284
4.00	4.69	1.0341	1.0359	41.4	0.199	992.7	5.5	61	1.3391	0.928	0.499	0.271	1.047	1.015	95.28	32.8	0.379
5.00	5.87	1.0434	1.0452	52.2	0.250	991.2	7.0	76	1.3406	1.177	0.633	0.344	1.060	1.018	94.15	40.4	0.476
6.00	7.04	1.0528	1.0547	63.2	0.303	989.6	8.6	92	1.3422	1.435	0.772	0.420	1.073	1.021	93.05	47.9	0.575
7.00	8.21	1.0624	1.0643	74.4	0.357	988.0	10.2	108	1.3438	1.704	0.916	0.498	1.085	1.024	91.95	55.3	0.675
8.00	9.38	1.0721	1.0740	85.8	0.412	986.3	11.9	124	1.3454	1.984	1.067	0.580	1.099	1.027	90.85	62.6	0.777
9.00	10.56	1.0820	1.0839	97.4	0.468	984.6	13.6	140	1.3470	2.274	1.222	0.664	1.112	1.030	89.72	69.8	0.880
10.00	11.73	1.0921	1.0940	109.2	0.524	982.9	15.4	157	1.3487	2.575	1.385	0.751	1.127	1.034	88.55	76.7	0.983
12.00	14.08	1.1128	1.1148	133.5	0.641	979.3	19.0	191	1.3520	3.218	1.730	0.933	1.159	1.043	86.14	90.1	1.19
14.00	16.42	1.1342	1.1362	158.8	0.762	975.4	22.8	225	1.3555	3.921	2.108	1.129	1.193	1.054	83.63	103.0	1.40
16.00	18.77	1.1564	1.1584	185.0	0.888	971.4	26.9	261	1.3591	4.689	2.521	1.341	1.232	1.067	81.03	115.0	1.61
18.00	21.11	1.1793	1.1814	212.3	1.019	967.0	31.2	297	1.3627				1.274	1.083	78.32	126.0	1.82
20.00	23.46	1.2031	1.2052	240.6	1.155	962.5	35.8	334	1.3664				1.322	1.101	75.49	137.0	2.04
22.00	25.81	1.2277	1.2299	270.1	1.297	957.6	40.6	373	1.3703				1.375	1.122	72.57	147.0	2.26
24.00	28.15	1.2531	1.2553	300.7	1.444	952.4	45.9	411	1.3741				1.434	1.147	69.57	157.0	2.48
26.00	30.50	1.2793	1.2816	332.6	1.597	946.7	51.6	451	1.3781				1.500	1.175	66.53	166.0	2.70

7 CADMIUM CHLORIDE, $CdCl_2$

MOLECULAR WEIGHT = 183.32

A% by wt.	ρ D_4^{20}	D_{20}^{20}	C_s g/l	M g-mol/l	C_w g/l	$(C_o - C_w)$ g/l	$(n - n_o)$ × 10^4	n	Δ °C	O Os/kg	S g-mol/l	η/η_o	η/ρ cS	φ rhe	γ mmho/cm	T g-mol/l
1.00	1.0062	1.0080	10.1	0.055	996.1	2.1	16.	1.3346	0.186	0.100	0.053	1.012	1.008	98.60	5.3	0.053
2.00	1.0148	1.0166	20.3	0.111	994.5	3.7	31.	1.3361	0.368	0.198	0.106	1.028	1.015	97.09	9.0	0.093
3.00	1.0237	1.0255	30.7	0.168	993.0	5.2	47.	1.3377	0.555	0.299	0.161	1.044	1.022	95.59	11.8	0.124
4.00	1.0325	1.0344	41.3	0.225	991.2	7.0	62.	1.3392	0.715	0.384	0.208	1.059	1.028	94.21	14.1	0.150
5.00	1.0415	1.0434	52.1	0.284	989.4	8.8	78.	1.3408	0.878	0.472	0.256	1.076	1.035	92.77	16.0	0.172
6.00	1.0508	1.0527	63.0	0.344	987.8	10.4	94.	1.3424	1.035	0.557	0.302	1.092	1.041	91.42	17.5	0.190
7.00	1.0600	1.0619	74.2	0.405	985.8	12.4	110.	1.3440	1.191	0.640	0.348	1.109	1.048	90.02	18.9	0.207
8.00	1.0692	1.0711	85.5	0.467	983.7	14.5	127.	1.3457	1.351	0.726	0.395	1.125	1.054	88.74	20.2	0.222
9.00	1.0794	1.0813	97.1	0.530	982.3	15.9	143.	1.3473	1.491	0.802	0.436	1.146	1.064	87.08	21.2	0.235
10.00	1.0897	1.0917	109.0	0.594	980.7	17.5	159.	1.3489	1.634	0.878	0.478	1.166	1.072	85.60	22.1	0.246
12.00	1.1103	1.1123	133.2	0.727	977.1	21.1	193.	1.3523	1.936	1.041	0.566	1.208	1.090	82.63	23.7	0.265
14.00	1.1308	1.1328	158.3	0.864	972.5	25.7	228.	1.3558	2.236	1.203	0.653	1.264	1.120	78.96	24.7	0.278
16.00	1.1531	1.1552	184.5	1.006	968.6	29.6	264.	1.3594	2.564	1.378	0.747	1.322	1.149	75.48	25.5	0.287
18.00	1.1749	1.1770	211.5	1.154	963.4	34.8	301.	1.3631	2.889	1.553	0.840	1.386	1.182	72.01	25.9	0.293
20.00	1.1988	1.2010	239.8	1.308	959.0	39.2	338.	1.3668	3.246	1.745	0.941	1.455	1.216	68.60	26.3	0.297
24.00	1.2476	1.2498	299.4	1.633	948.2	50.0	420.	1.3750	3.984	2.142	1.147	1.596	1.282	62.52	25.8	0.291
28.00	1.3003	1.3026	364.1	1.986	936.2	62.0	506.	1.3836	4.750	2.554	1.357	1.832	1.412	54.47	24.5	0.275
32.00	1.3558	1.3582	433.8	2.367	922.0	76.2	597.	1.3927				2.135	1.578	46.74	22.5	0.251
36.00	1.4178	1.4204	510.4	2.784	907.4	90.8	696.	1.4026				2.556	1.806	39.05	20.2	0.222
40.00	1.4843	1.4870	593.7	3.239	890.6	107.6	803.	1.4133				3.259	2.200	30.62	17.2	0.187
44.00	1.5563	1.5591	684.8	3.735	871.5	126.7	918.	1.4248				4.155	2.675	24.02	14.3	0.152
48.00	1.6345	1.6374	784.5	4.280	850.0	148.2	1044.	1.4374				5.740	3.519	17.39	11.3	0.119
52.00	1.7206	1.7237	894.7	4.881	825.9	172.3	1180.	1.4510				8.811	5.131	11.33	8.4	0.087
56.00	1.8131	1.8164	1015.4	5.539	797.7	200.5	1326.	1.4656				14.67	8.109	6.802	5.9	0.060

8 CADMIUM SULFATE, $CdSO_4$

MOLECULAR WEIGHT = 208.46

A% by wt.	ρ D_4^{20}	D_{20}^{20}	C_s g/l	M g-mol/l	C_w g/l	$(C_o - C_w)$ g/l	$(n - n_o)$ × 10^4	n	Δ °C	O Os/kg	S g-mol/l	η/η_o	η/ρ cS	φ rhe	γ mmho/cm	T g-mol/l
1.00	1.0078	1.0096	10.1	0.048	997.7	0.5	14.	1.3344	0.135	0.053	0.028	1.029	1.023	97.00	4.2	0.042
2.00	1.0175	1.0193	20.3	0.098	997.2	1.0	28.	1.3358	0.194	0.104	0.055	1.059	1.043	94.23	7.4	0.075
3.00	1.0271	1.0290	30.8	0.148	996.3	1.9	42.	1.3372	0.290	0.156	0.083	1.089	1.062	91.67	10.2	0.105
4.00	1.0375	1.0394	41.5	0.199	996.0	2.2	56.	1.3386	0.382	0.205	0.110	1.125	1.086	88.75	12.6	0.133
5.00	1.0478	1.0497	52.4	0.251	995.4	2.8	69.	1.3399	0.477	0.256	0.138	1.162	1.111	85.90	14.9	0.160
6.00	1.0579	1.0598	63.5	0.304	994.4	3.8	84.	1.3414	0.569	0.306	0.165	1.197	1.134	83.36	17.1	0.186
7.00	1.0684	1.0703	74.8	0.359	993.6	4.6	98.	1.3428	0.657	0.354	0.191	1.238	1.161	80.62	20.1	0.211
8.00	1.0789	1.0808	86.3	0.414	992.6	5.6	114.	1.3444	0.742	0.399	0.216	1.280	1.189	77.96	21.2	0.235
9.00	1.0894	1.0914	98.0	0.470	991.4	6.8	129.	1.3459	0.831	0.447	0.242	1.325	1.219	75.30	23.2	0.259
10.00	1.1004	1.1024	110.0	0.528	990.4	7.8	143.	1.3473	0.933	0.502	0.272	1.375	1.252	72.58	25.1	0.282
12.00	1.1233	1.1253	134.8	0.647	988.5	9.7	173.	1.3503	1.137	0.611	0.332	1.480	1.320	67.44	28.7	0.327
14.00	1.1470	1.1491	160.6	0.770	986.4	11.8	203.	1.3533	1.354	0.728	0.396	1.603	1.400	62.27	32.0	0.369
16.00	1.1715	1.1736	187.4	0.899	984.1	14.1	234.	1.3564	1.604	0.862	0.469	1.742	1.490	57.29	35.6	0.405
18.00	1.1967	1.1989	215.4	1.033	981.3	16.9	266.	1.3596	1.854	0.997	0.542	1.909	1.598	52.29	37.5	0.439
20.00	1.2229	1.2251	244.6	1.173	978.3	19.9	299.	1.3629	2.140	1.150	0.625	2.081	1.705	47.96	39.8	0.469
24.00	1.2784	1.2807	306.8	1.472	971.6	26.6	369.	1.3699	2.864	1.540	0.833	2.515	1.971	39.69	43.2	0.513
28.00	1.3385	1.3409	374.8	1.798	963.7	34.5	445.	1.3775	3.959	2.129	1.140	3.150	2.358	31.68	45.2	0.539
32.00	1.3996	1.4021	447.9	2.148	951.7	46.5	520.	1.3850				4.083	2.923	24.44	45.5	0.542
36.00	1.4655	1.4681	527.6	2.531	937.9	60.3	597.	1.3927				5.334	3.647	18.71	44.1	0.524
40.00	1.5456	1.5484	618.2	2.966	927.4	70.8	676.	1.4006				7.381	4.785	13.52	40.3	0.475

9 CALCIUM CHLORIDE, $CaCl_2 \cdot 2H_2O$

MOLECULAR WEIGHT = 110.99 FORMULA WEIGHT, HYDRATE = 147.03
RELATIVE SPECIFIC REFRACTIVITY = 0.863

0.00 % by wt. data are the same for all compounds.
For Values of 0.00 wt. % solutions see Table 1, Acetic Acid.

A% by wt.	H% by wt.	ρ D_4^{20}	D_{20}^{20}	C_s g/l	M g-mol/l	C_w g/l	$(C_o - C_w)$ g/l	$(n - n_o) \times 10^4$	n	Δ °C	O Os/kg	S g-mol/l	η/η_o	η/ρ cS	φ rhe	γ mmho/cm	T g-mol/l
0.50	0.66	1.0024	1.0041	5.0	0.045	997.3	0.9	12	1.3342	0.222	0.119	0.063	1.013	1.013	98.47	8.1	0.083
1.00	1.32	1.0065	1.0083	10.1	0.091	996.4	1.8	24	1.3354	0.440	0.237	0.127	1.026	1.021	97.27	15.7	0.169
1.50	1.99	1.0106	1.0124	15.2	0.137	995.5	2.7	36	1.3366	0.661	0.355	0.192	1.037	1.028	96.24	22.7	0.254
2.00	2.65	1.0148	1.0166	20.3	0.183	994.5	3.7	48	1.3378	0.880	0.473	0.257	1.048	1.035	95.23	29.4	0.337
2.50	3.31	1.0190	1.0208	25.5	0.230	993.5	4.7	60	1.3390	1.102	0.593	0.322	1.061	1.043	94.06	36.1	0.421
3.00	3.97	1.0232	1.0250	30.7	0.277	992.5	5.8	72	1.3402	1.330	0.715	0.389	1.076	1.053	92.79	42.6	0.504
3.50	4.64	1.0274	1.0292	36.0	0.324	991.4	6.8	84	1.3414	1.567	0.843	0.458	1.091	1.064	91.45	48.9	0.588
4.00	5.30	1.0316	1.0334	41.3	0.372	990.3	7.9	96	1.3426	1.815	0.976	0.531	1.108	1.076	90.09	55.1	0.672
4.50	5.96	1.0358	1.0377	46.6	0.420	989.2	9.0	109	1.3438	2.074	1.115	0.606	1.125	1.088	88.75	61.1	0.756
5.00	6.62	1.0401	1.0419	52.0	0.469	988.1	10.2	121	1.3451	2.345	1.261	0.684	1.141	1.099	87.47	67.0	0.840
5.50	7.29	1.0443	1.0462	57.4	0.518	986.9	11.3	133	1.3463	2.630	1.414	0.766	1.157	1.110	86.25	72.8	0.924
6.00	7.95	1.0486	1.0505	62.9	0.567	985.7	12.5	145	1.3475	2.930	1.575	0.852	1.173	1.121	85.07	78.3	1.01
6.50	8.61	1.0529	1.0548	68.4	0.617	984.5	13.8	158	1.3487	3.244	1.744	0.941	1.189	1.132	83.91	83.6	1.09
7.00	9.27	1.0572	1.0591	74.0	0.667	983.2	15.0	170	1.3500	3.573	1.921	1.033	1.206	1.143	82.77	88.7	1.17
7.50	9.94	1.0615	1.0634	79.6	0.717	981.9	16.3	182	1.3512	3.917	2.106	1.128	1.222	1.154	81.64	93.6	1.25
8.00	10.60	1.0659	1.0678	85.3	0.768	980.6	17.6	195	1.3525	4.275	2.299	1.227	1.240	1.165	80.51	98.4	1.33
8.50	11.26	1.0703	1.0722	91.0	0.820	979.3	18.9	207	1.3537	4.649	2.499	1.330	1.257	1.177	79.38	103.	1.41
9.00	11.92	1.0747	1.0766	96.7	0.871	977.9	20.3	220	1.3549	5.04	2.71	1.434	1.276	1.190	78.22	108.	1.49
9.50	12.58	1.0791	1.0810	102.5	0.924	976.6	21.7	232	1.3562	5.44	2.92	1.541	1.295	1.203	77.04	112.	1.57
10.00	13.25	1.0835	1.0854	108.3	0.976	975.1	23.1	245	1.3575	5.86	3.15	1.651	1.316	1.217	75.84	117.	1.65
11.00	14.57	1.0923	1.0943	120.2	1.083	972.2	26.1	270	1.3600	6.74	3.63	1.877	1.359	1.247	73.41	125.	1.81
12.00	15.90	1.1014	1.1033	132.2	1.191	969.2	29.0	295	1.3625	7.70	4.14	2.113	1.405	1.278	71.03	133.	1.97
13.00	17.22	1.1105	1.1125	144.4	1.301	966.2	32.1	321	1.3651	8.72	4.69	2.358	1.454	1.311	68.66	141.	2.13
14.00	18.55	1.1198	1.1218	156.8	1.412	963.0	35.2	347	1.3677	9.83	5.28	2.611	1.505	1.347	66.30	148.	2.28
15.00	19.87	1.1292	1.1312	169.4	1.526	959.8	38.4	374	1.3704	11.01	5.92	2.872	1.561	1.385	63.93	154.	2.42
16.00	21.20	1.1386	1.1407	182.2	1.641	956.5	41.8	400	1.3730	12.28	6.60	3.138	1.622	1.427	61.54	160.	2.55
17.00	22.52	1.1482	1.1502	195.2	1.759	953.0	45.2	427	1.3757	13.65	7.34	3.410	1.688	1.473	59.14	165.	2.67
18.00	23.84	1.1579	1.1599	208.4	1.878	949.5	48.8	454	1.3784	15.11	8.12	3.686	1.760	1.523	56.72	169.	2.78
19.00	25.17	1.1677	1.1697	221.9	1.999	945.8	52.4	482	1.3812	16.7	8.98	3.97	1.839	1.578	54.27	173.	2.88
20.00	26.49	1.1775	1.1796	235.5	2.122	942.0	56.2	509	1.3839	18.3	9.84	4.24	1.926	1.639	51.82	177.	2.97
22.00	29.14	1.1976	1.1997	263.5	2.374	934.1	64.1	565	1.3895	21.7	11.67		2.123	1.777	47.00	182.	3.11
24.00	31.79	1.2180	1.2201	292.3	2.634	925.7	72.6	621	1.3951	25.3	13.60		2.351	1.934	42.45	183.	3.16
26.00	34.44	1.2388	1.2410	322.1	2.902	916.7	81.5	678	1.4008	29.7	15.97		2.640	2.135	32.81	182.	3.12
28.00	37.09	1.2600	1.2622	352.8	3.179	907.2	91.0	736	1.4066	34.7	18.66		2.994	2.381	33.33	179.	3.03
30.00	39.74	1.2816	1.2838	384.5	3.464	897.1	101.1	794	1.4124	41.0	22.04		3.460	2.705	28.85	172.	2.85
32.00	42.39	1.3036	1.3059	417.1	3.758	886.4	111.8	853	1.4183	49.7	26.72		4.027	3.095	24.79	162.	2.61
34.00	45.04	1.3260	1.3283	450.8	4.062	875.1	123.1	912	1.4242				4.810	3.635	20.75	150.	2.34
36.00	47.69	1.3488	1.3512	485.6	4.375	863.2	135.0	971	1.4301				5.795	4.305	17.22	137.	2.05
38.00	50.34	1.3720	1.3745	521.4	4.697	850.7	147.6	1031	1.4361				7.306	5.336	13.66	123.	1.76
40.00	52.99	1.3957	1.3982	558.3	5.030	837.4	160.8	1090	1.4420				8.979	6.446	11.12	106.	1.45

10 CESIUM CHLORIDE, CsCl

MOLECULAR WEIGHT = 168.37
RELATIVE SPECIFIC REFRACTIVITY = 0.464

0.00 % by wt. data are the same for all compounds.
For Values of 0.00 wt. % solutions see Table 1, Acetic Acid.

A% by wt.	ρ D_4^{20}	D_{20}^{20}	C_s g/l	M g-mol/l	C_w g/l	$(C_o - C_w)$ g/l	$(n - n_o) \times 10^4$	n	Δ °C	O Os/kg	S g-mol/l	η/η_o	η/ρ cS	φ rhe	γ mmho/cm	T g-mol/l
0.50	1.0020	1.0038	5.0	0.030	997.0	1.2	4	1.3334	0.100	0.054	0.028	0.998	0.997	100.05	3.8	0.038
1.00	1.0058	1.0076	10.1	0.060	995.8	2.5	8	1.3337	0.201	0.108	0.057	0.995	0.991	100.30	7.4	0.075
1.50	1.0097	1.0114	15.1	0.090	994.5	3.7	11	1.3341	0.302	0.162	0.087	0.992	0.985	100.56	10.6	0.111
2.00	1.0135	1.0153	20.3	0.120	993.3	5.0	15	1.3345	0.403	0.217	0.116	0.990	0.979	100.81	13.8	0.147
2.50	1.0174	1.0192	25.4	0.151	992.0	6.2	19	1.3349	0.505	0.272	0.146	0.988	0.973	101.04	17.0	0.184
3.00	1.0214	1.0232	30.6	0.182	990.7	7.5	23	1.3353	0.607	0.326	0.176	0.986	0.967	101.26	20.2	0.222
3.50	1.0253	1.0271	35.9	0.213	989.4	8.8	27	1.3357	0.709	0.381	0.206	0.984	0.961	101.46	23.3	0.260
4.00	1.0293	1.0311	41.2	0.245	988.2	10.1	31	1.3361	0.812	0.436	0.236	0.982	0.956	101.65	26.5	0.300
4.50	1.0334	1.0352	46.5	0.276	986.9	11.4	35	1.3365	0.915	0.492	0.267	0.980	0.950	101.84	29.7	0.340
5.00	1.0374	1.0392	51.9	0.308	985.5	12.7	39	1.3369	1.018	0.547	0.297	0.978	0.945	102.02	32.9	0.380
5.50	1.0415	1.0433	57.3	0.340	984.2	14.0	43	1.3373	1.121	0.602	0.327	0.976	0.939	102.20	36.1	0.421
6.00	1.0456	1.0475	62.7	0.373	982.9	15.3	47	1.3377	1.223	0.658	0.358	0.975	0.934	102.37	39.3	0.462
6.50	1.0498	1.0516	68.2	0.405	981.6	16.7	52	1.3382	1.326	0.713	0.388	0.973	0.929	102.53	42.5	0.504
7.00	1.0540	1.0558	73.8	0.438	980.2	18.0	56	1.3386	1.428	0.768	0.418	0.972	0.924	102.67	45.8	0.546
7.50	1.0582	1.0601	79.4	0.471	978.8	19.4	60	1.3390	1.531	0.823	0.448	0.971	0.919	102.82	49.0	0.589
8.00	1.0625	1.0643	85.0	0.505	977.5	20.8	64	1.3394	1.635	0.879	0.478	0.969	0.914	102.95	52.3	0.633
8.50	1.0668	1.0686	90.7	0.539	976.1	22.2	69	1.3399	1.739	0.935	0.509	0.968	0.909	103.09	55.6	0.678
9.00	1.0711	1.0730	96.4	0.573	974.7	23.5	73	1.3403	1.845	0.992	0.539	0.967	0.904	103.22	59.0	0.724
9.50	1.0754	1.0773	102.2	0.607	973.3	25.0	77	1.3407	1.952	1.049	0.571	0.966	0.900	103.36	62.4	0.772
10.00	1.0798	1.0818	108.0	0.641	971.9	26.4	82	1.3412	2.060	1.108	0.602	0.964	0.895	103.51	65.8	0.822
11.00	1.0887	1.0907	119.8	0.711	969.0	29.3	91	1.3421	2.281	1.227	0.666	0.961	0.885	103.80	72.8	0.926
12.00	1.0978	1.0997	131.7	0.782	966.1	32.2	100	1.3430	2.507	1.348	0.731	0.959	0.875	104.09	80.0	1.03
13.00	1.1070	1.1089	143.9	0.855	963.1	35.2	109	1.3439	2.737	1.472	0.797	0.956	0.865	104.38	87.3	1.15
14.00	1.1163	1.1183	156.3	0.928	960.0	38.2	118	1.3448	2.972	1.598	0.864	0.953	0.856	104.67	94.8	1.27
15.00	1.1258	1.1278	168.9	1.003	956.9	41.3	128	1.3458	3.212	1.727	0.932	0.951	0.846	104.96	102.	1.40
16.00	1.1355	1.1375	181.7	1.079	953.8	44.5	138	1.3468	3.457	1.859	1.000	0.948	0.837	105.27	110.	1.54
17.00	1.1453	1.1473	194.7	1.156	950.6	47.7	147	1.3477	3.708	1.993	1.070	0.945	0.827	105.58	118.	1.68
18.00	1.1552	1.1573	207.9	1.235	947.3	51.0	157	1.3487	3.963	2.131	1.141	0.943	0.817	105.89	126.	1.83
19.00	1.1653	1.1674	221.4	1.315	943.9	54.3	167	1.3497	4.224	2.271	1.213	0.940	0.808	106.19	134.	1.99
20.00	1.1756	1.1777	235.1	1.396	940.5	57.7	178	1.3507	4.491	2.415	1.287	0.937	0.799	106.49	142.	2.17
22.00	1.1967	1.1989	263.3	1.564	933.5	64.8	198	1.3528				0.932	0.781	107.05	159.	2.53
24.00	1.2185	1.2207	292.4	1.737	926.1	72.1	220	1.3550				0.928	0.763	107.56	176.	2.95
26.00	1.2411	1.2433	322.7	1.916	918.4	79.8	242	1.3572				0.924	0.746	107.97	192.	3.48

10 CESIUM CHLORIDE, CsCl—(Continued)

A % by wt.	ρ D_4^{20}	D_{20}^{20}	C_s g/l	M g-mol/l	C_w g/l	$(C_o - C_w)$ g/l	$(n - n_o)$ × 10^4	n	Δ °C	O Os/kg	S g-mol/l	η/η_o	η/ρ cS	ϕ rhe	γ mmho/cm	T g-mol/l
28.00	1.2644	1.2666	354.0	2.103	910.4	87.9	264	1.3594				0.922	0.730	108.27	207.	4.15
30.00	1.2885	1.2908	386.6	2.296	902.0	96.3	288	1.3617				0.920	0.716	108.44	221.	4.95
32.00	1.3135	1.3158	420.3	2.496	893.1	105.1	311	1.3641				0.920	0.702	108.43		
34.00	1.3393	1.3417	455.4	2.705	883.9	114.3	336	1.3666				0.922	0.690	108.25		
36.00	1.3661	1.3685	491.8	2.921	874.3	124.0	361	1.3691				0.924	0.678	107.97		
38.00	1.3938	1.3963	529.6	3.146	864.2	134.1	387	1.3717				0.928	0.667	107.56		
40.00	1.4226	1.4251	569.0	3.380	853.5	144.7	414	1.3744				0.932	0.657	107.02		
42.00	1.4525	1.4550	610.0	3.623	842.4	155.8	442	1.3771				0.938	0.647	106.40		
44.00	1.4835	1.4861	652.7	3.877	830.8	167.5	470	1.3800				0.945	0.638	105.63		
46.00	1.5158	1.5185	697.3	4.141	818.5	179.7	500	1.3829				0.954	0.630	104.64		
48.00	1.5495	1.5522	743.7	4.417	805.7	192.5	530	1.3860				0.965	0.624	103.45		
50.00	1.5846	1.5874	792.3	4.706	792.3	206.0	562	1.3892				0.979	0.619	101.94		
52.00	1.6212	1.6241	843.0	5.007	778.2	220.0	595	1.3925				0.998	0.617	100.02		
54.00	1.6596	1.6625	896.2	5.323	763.4	234.8	630	1.3960				1.021	0.616	97.76		
56.00	1.6999	1.7029	951.9	5.654	747.9	250.3	666	1.3996				1.048	0.618	95.25		
58.00	1.7422	1.7453	1010.5	6.001	731.7	266.5	705	1.4035				1.078	0.620	92.54		
60.00	1.7868	1.7900	1072.1	6.367	714.7	283.5	746	1.4076				1.118	0.627	89.27		
62.00	1.8340	1.8373	1137.1	6.754	696.9	301.3	790	1.4120				1.170	0.639	85.28		
64.00	1.8842	1.8875	1205.9	7.162	678.3	319.9	837	1.4167				1.236	0.657	80.74		

11 CITRIC ACID, $(COOH)CH_2C(OH)(COOH)CH_2COOH \cdot 1H_2O$

MOLECULAR WEIGHT = 192.12 FORMULA WEIGHT, HYDRATE = 210.14
RELATIVE SPECIFIC REFRACTIVITY = 0.951

0.00 % by wt. data are the same for all compounds.
For Values of 0.00 wt. % solutions see Table 1, Acetic Acid.

A % by wt.	H % by wt.	ρ D_4^{20}	D_{20}^{20}	C_s g/l	M g-mol/l	C_w g/l	$(C_o - C_w)$ g/l	$(n - n_o)$ × 10^4	n	Δ °C	O Os/kg	S g-mol/l	η/η_o	η/ρ cS	ϕ rhe	γ mmho/cm	T g-mol/l
0.50	0.55	1.0002	1.0020	5.0	0.026	995.2	3.0	7	1.3336	0.052	0.028	0.014	1.011	1.013	98.74	1.2	0.012
1.00	1.09	1.0022	1.0040	10.0	0.052	992.2	6.0	13	1.3343	0.105	0.056	0.029	1.022	1.022	97.65	2.1	0.021
2.00	2.19	1.0063	1.0081	20.1	0.105	986.2	12.0	26	1.3356	0.211	0.113	0.060	1.046	1.041	95.41	3.0	0.030
3.00	3.28	1.0105	1.0123	30.3	0.158	980.2	18.1	38	1.3368	0.318	0.171	0.091	1.071	1.062	93.21	3.7	0.037
4.00	4.38	1.0147	1.0165	40.6	0.211	974.1	24.1	51	1.3381	0.428	0.230	0.123	1.096	1.082	91.05	4.2	0.042
5.00	5.47	1.0189	1.0207	50.9	0.265	968.0	30.3	64	1.3394	0.539	0.290	0.156	1.123	1.104	88.87	4.7	0.047
6.00	6.56	1.0232	1.0250	61.4	0.320	961.8	36.5	77	1.3407	0.651	0.350	0.189	1.151	1.128	86.68	5.1	0.051
7.00	7.66	1.0274	1.0292	71.9	0.374	955.5	42.8	90	1.3420	0.764	0.411	0.222	1.181	1.152	84.50	5.5	0.055
8.00	8.75	1.0316	1.0335	82.5	0.430	949.1	49.1	103	1.3433	0.879	0.473	0.256	1.212	1.177	82.32	5.7	0.058
9.00	9.84	1.0359	1.0377	93.2	0.485	942.7	55.6	116	1.3446	0.997	0.536	0.291	1.245	1.204	80.15	6.0	0.060
10.00	10.94	1.0402	1.0420	104.0	0.541	936.1	62.1	129	1.3459	1.120	0.602	0.327	1.280	1.233	77.97	6.2	0.063
12.00	13.13	1.0490	1.0509	125.9	0.655	923.2	75.1	156	1.3486	1.379	0.741	0.403	1.354	1.293	73.72	6.6	0.067
14.00	15.31	1.0580	1.0599	148.1	0.771	909.9	88.3	184	1.3514	1.656	0.890	0.484	1.433	1.358	69.62	6.9	0.070
16.00	17.50	1.0672	1.0690	170.7	0.889	896.4	101.8	211	1.3541	1.950	1.048	0.570	1.522	1.429	65.58	7.1	0.072
18.00	19.69	1.0764	1.0783	193.8	1.009	882.7	115.6	239	1.3569	2.255	1.213	0.658	1.622	1.510	61.53	7.1	0.072
20.00	21.88	1.0858	1.0877	217.2	1.130	868.6	129.6	268	1.3598	2.567	1.380	0.748	1.737	1.603	57.46	7.2	0.073
22.00	24.06	1.0953	1.0972	241.0	1.254	854.3	143.9	296	1.3626	2.884	1.550	0.839	1.868	1.709	53.43	7.2	0.073
24.00	26.25	1.1049	1.1069	265.2	1.380	839.8	158.5	325	1.3655	3.210	1.726	0.931	2.013	1.826	49.57	7.2	0.073
26.00	28.44	1.1147	1.1167	289.8	1.509	824.9	173.4	355	1.3684	3.547	1.907	1.025	2.174	1.954	45.91	7.1	0.072
28.00	30.63	1.1246	1.1266	314.9	1.639	809.7	188.5	384	1.3714	3.893	2.093	1.122	2.351	2.094	42.46	6.8	0.069
30.00	32.81	1.1346	1.1366	340.4	1.772	794.2	204.0	414	1.3744	4.250	2.285	1.220	2.544	2.247	39.23	6.5	0.066

12 COBALTOUS CHLORIDE, $CoCl_2 \cdot 6H_2O$

MOLECULAR WEIGHT = 129.85 FORMULA WEIGHT, HYDRATE = 237.95
RELATIVE SPECIFIC REFRACTIVITY = 0.733

0.00 % by wt. data are the same for all compounds.
For Values of 0.00 wt. % solutions see Table 1, Acetic Acid.

A % by wt.	H % by wt.	ρ D_4^{20}	D_{20}^{20}	C_s g/l	M g-mol/l	C_w g/l	$(C_o - C_w)$ g/l	$(n - n_o)$ × 10^4	n	Δ °C	O Os/kg	S g-mol/l	η/η_o	η/ρ cS	ϕ rhe	γ mmho/cm	T g-mol/l
0.50	0.92	1.0027	1.0045	5.0	0.039	997.7	0.5	11	1.3341	0.191	0.103	0.054	1.017	1.017	98.09	6.7	0.068
1.00	1.83	1.0073	1.0090	10.1	0.078	997.2	1.1	23	1.3352	0.380	0.204	0.110	1.034	1.029	96.52	12.7	0.134
2.00	3.66	1.0164	1.0182	20.3	0.157	996.0	2.2	45	1.3375	0.771	0.415	0.224	1.064	1.049	93.80	23.4	0.262
3.00	5.50	1.0256	1.0274	30.8	0.237	994.8	3.4	68	1.3398	1.187	0.638	0.347	1.096	1.071	91.05	33.6	0.389
4.00	7.33	1.0349	1.0368	41.4	0.319	993.5	4.7	91	1.3421	1.627	0.875	0.476	1.130	1.094	88.30	43.3	0.514
5.00	9.16	1.0444	1.0462	52.2	0.402	992.2	6.1	114	1.3444	2.099	1.128	0.613	1.166	1.119	85.59	52.5	0.636
6.00	10.99	1.0540	1.0559	63.2	0.487	990.8	7.5	137	1.3467	2.611	1.404	0.761	1.203	1.143	82.97	61.0	0.754
7.00	12.83	1.0637	1.0656	74.5	0.573	989.3	9.0	161	1.3491	3.164	1.701	0.918	1.241	1.169	80.43	69.0	0.868
8.00	14.66	1.0736	1.0755	85.9	0.661	987.7	10.5	185	1.3515	3.761	2.022	1.085	1.281	1.195	77.91	76.4	0.979
9.00	16.49	1.0837	1.0856	97.5	0.751	986.2	12.1	209	1.3539				1.324	1.224	75.40	83.4	1.08
10.00	18.32	1.0939	1.0958	109.4	0.842	984.5	13.7	234	1.3563				1.370	1.255	72.85	89.9	1.18
12.00	21.99	1.1148	1.1168	133.8	1.030	981.1	17.2	283	1.3613				1.474	1.324	67.73	102.	1.38
14.00	25.65	1.1364	1.1384	159.1	1.225	977.3	20.9	334	1.3664				1.592	1.403	62.71	112.	1.55
16.00	29.32	1.1587	1.1607	185.4	1.428	973.3	24.9	386	1.3716				1.725	1.492	57.85	120.	1.70
18.00	32.98	1.1816	1.1836	212.7	1.638	968.9	29.4	439	1.3769				1.876	1.591	53.19	126.	1.82
20.00	36.65	1.2050	1.2071	241.0	1.856	964.0	34.3	492	1.3822				2.046	1.701	48.78	130.	1.90

13 CREATININE, $CH_3NC(:NH)NHCOCH_2$

MOLECULAR WEIGHT = 113.12
RELATIVE SPECIFIC REFRACTIVITY = 1.301

0.00 % by wt. data are the same for all compounds.
For Values of 0.00 wt. % solutions see Table 1, Acetic Acid.

A % by wt.	D_4^{20}	D_{20}^{20}	C_s g/l	M g-mol/l	C_w g/l	$(C_o - C_w)$ g/l	$(n - n_o)$ × 10^4	n	Δ °C	O Os/kg	S g-mol/l	η/η_o	η/ρ cS	ϕ rhe	γ mmho/cm	T g-mol/l
0.50	0.9995	1.0012	5.0	0.044	994.5	3.8	9	1.3339	0.080	0.043	0.022	1.009	1.011	98.95		
1.00	1.0007	1.0025	10.0	0.088	990.7	7.5	19	1.3349	0.150	0.081	0.042	1.019	1.020	97.97		
1.50	1.0019	1.0037	15.0	0.133	986.9	11.3	28	1.3358	0.220	0.118	0.063	1.030	1.030	96.89		
2.00	1.0032	1.0050	20.1	0.177	983.1	15.1	38	1.3368	0.300	0.161	0.086	1.043	1.041	95.73		
2.50	1.0045	1.0063	25.1	0.222	979.4	18.9	48	1.3378	0.380	0.204	0.110	1.056	1.053	94.52		
3.00	1.0058	1.0075	30.2	0.267	975.6	22.6	57	1.3387	0.460	0.247	0.133	1.070	1.066	93.29		
3.50	1.0071	1.0088	35.2	0.312	971.8	26.4	67	1.3397				1.084	1.079	92.05		
4.00	1.0084	1.0102	40.3	0.357	968.0	30.2	77	1.3407				1.099	1.092	90.83		
4.50	1.0097	1.0115	45.4	0.402	964.3	34.0	87	1.3417				1.113	1.105	89.65		
5.00	1.0110	1.0128	50.6	0.447	960.5	37.7	97	1.3427				1.128	1.117	88.51		
5.50	1.0124	1.0142	55.7	0.492	956.7	41.5	107	1.3437				1.141	1.130	87.45		
6.00	1.0138	1.0156	60.8	0.538	952.9	45.3	117	1.3447				1.154	1.141	86.46		
6.50	1.0151	1.0169	66.0	0.583	949.2	49.1	127	1.3457				1.166	1.151	85.57		
7.00	1.0165	1.0183	71.2	0.629	945.4	52.8	137	1.3467				1.177	1.160	84.78		
7.50	1.0180	1.0198	76.3	0.675	941.6	56.6	148	1.3478				1.186	1.168	84.11		
8.00	1.0194	1.0212	81.6	0.721	937.8	60.4	158	1.3488				1.194	1.174	83.58		

14 CUPRIC SULFATE, $CuSO_4 \cdot 5H_2O$

MOLECULAR WEIGHT = 159.61   FORMULA WEIGHT, HYDRATE = 249.69
RELATIVE SPECIFIC REFRACTIVITY = 0.517

0.00 % by wt. data are the same for all compounds.
For Values of 0.00 wt. % solutions see Table 1, Acetic Acid.

A % by wt.	H % by wt.	ρ D_4^{20}	D_{20}^{20}	C_s g/l	M g-mol/l	C_w g/l	$(C_o - C_w)$ g/l	$(n - n_o)$ × 10^4	n	Δ °C	O Os/kg	S g-mol/l	η/η_o	η/ρ cS	ϕ rhe	γ mmho/cm	T g-mol/l
0.50	0.78	1.0033	1.0051	5.0	0.031	998.3	−0.1	9	1.3339	0.075	0.040	0.021	1.015	1.014	98.30	2.9	0.028
1.00	1.56	1.0085	1.0103	10.1	0.063	998.4	−0.2	18	1.3348	0.140	0.075	0.039	1.034	1.027	96.52	5.4	0.054
1.50	2.35	1.0137	1.0155	15.2	0.095	998.5	−0.3	28	1.3358	0.200	0.108	0.057	1.057	1.045	94.40	7.4	0.076
2.00	3.13	1.0190	1.0208	20.4	0.128	998.6	−0.4	37	1.3367	0.259	0.139	0.074	1.082	1.064	92.24	9.3	0.096
2.50	3.91	1.0243	1.0261	25.6	0.160	998.7	−0.4	47	1.3377	0.316	0.170	0.091	1.105	1.081	90.31	11.1	0.116
3.00	4.69	1.0296	1.0314	30.9	0.194	998.7	−0.5	56	1.3386	0.372	0.200	0.107	1.127	1.097	88.54	12.8	0.135
3.50	5.48	1.0349	1.0368	36.2	0.227	998.7	−0.5	66	1.3396	0.428	0.230	0.124	1.149	1.112	86.86	14.4	0.154
4.00	6.26	1.0403	1.0421	41.6	0.261	998.7	−0.5	75	1.3405	0.484	0.260	0.140	1.171	1.128	85.23	16.0	0.172
4.50	7.04	1.0457	1.0475	47.1	0.295	998.6	−0.4	85	1.3414	0.539	0.290	0.156	1.194	1.144	83.58	17.5	0.190
5.00	7.82	1.0511	1.0530	52.6	0.329	998.6	−0.3	94	1.3424	0.594	0.319	0.172	1.219	1.162	81.87	19.0	0.208
5.50	8.60	1.0565	1.0584	58.1	0.364	998.4	−0.2	104	1.3433	0.648	0.349	0.188	1.246	1.181	80.12	20.5	0.226
6.00	9.39	1.0620	1.0639	63.7	0.399	998.3	0.0	113	1.3443	0.703	0.378	0.205	1.273	1.201	78.37	21.9	0.243
6.50	10.17	1.0675	1.0694	69.4	0.435	998.1	0.1	123	1.3452	0.759	0.408	0.221	1.302	1.222	76.63	23.3	0.260
7.00	10.95	1.0730	1.0749	75.1	0.471	997.9	0.3	132	1.3462	0.816	0.439	0.238	1.333	1.244	74.90	24.6	0.277
7.50	11.73	1.0786	1.0805	80.9	0.507	997.7	0.5	142	1.3472	0.874	0.470	0.255	1.364	1.267	73.17	26.0	0.293
8.00	12.52	1.0842	1.0861	86.7	0.543	997.5	0.8	151	1.3481	0.933	0.501	0.272	1.397	1.291	71.46	27.2	0.309
8.50	13.30	1.0898	1.0918	92.6	0.580	997.2	1.0	161	1.3491	0.992	0.533	0.289	1.431	1.315	69.77	28.5	0.325
9.00	14.08	1.0955	1.0975	98.6	0.618	996.9	1.3	171	1.3501	1.052	0.566	0.307	1.466	1.341	68.09	29.7	0.341
9.50	14.86	1.1012	1.1032	104.6	0.655	996.6	1.6	181	1.3511	1.113	0.599	0.325	1.502	1.367	66.44	31.0	0.356
10.00	15.64	1.1070	1.1090	110.7	0.694	996.3	1.9	191	1.3520	1.177	0.633	0.344	1.540	1.394	64.81	32.2	0.371
11.00	17.21	1.1186	1.1206	123.0	0.771	995.6	2.7	210	1.3540	1.308	0.703	0.382	1.617	1.449	61.70	34.4	0.400
12.00	18.77	1.1304	1.1324	135.6	0.850	994.7	3.5	230	1.3560	1.447	0.778	0.423	1.698	1.505	58.77	36.6	0.427
13.00	20.34	1.1424	1.1444	148.5	0.930	993.8	4.4	251	1.3581	1.596	0.858	0.467	1.786	1.566	55.88	38.6	0.453
14.00	21.90	1.1545	1.1566	161.6	1.013	992.9	5.3	271	1.3601	1.753	0.942	0.513	1.885	1.636	52.95	40.5	0.478
15.00	23.47	1.1669	1.1690	175.0	1.097	991.9	6.4	292	1.3622				2.000	1.717	49.90	42.3	0.501
16.00	25.03	1.1796	1.1817	188.7	1.182	990.9	7.4	314	1.3644				2.132	1.811	46.81	44.0	0.523
17.00	26.59	1.1926	1.1947	202.7	1.270	989.8	8.4	336	1.3666				2.280	1.916	43.77	45.6	0.543
18.00	28.16	1.2059	1.2080	217.1	1.360	988.8	9.4	359	1.3689				2.444	2.031	40.83	47.0	0.562

15 DEXTRAN, $(C_6H_{10}O_5)_x$

MOLECULAR WEIGHT, AVERAGE = 72,000
RELATIVE SPECIFIC REFRACTIVITY = 1.045

0.00 % by wt. data are the same for all compounds.
For Values of 0.00 wt. % solutions see Table 1, Acetic Acid.

A % by wt.	ρ D_4^{20}	d D_{20}^{20}	C_s g/l	M g-mol/l	C_w g/l	$(C_o - C_w)$ g/l	$(n - n_o)$ × 10^4	n	Δ °C	O Os/kg	S g-mol/l	η/η_o	η/ρ cS	ϕ rhe	γ mmho/cm	T g-mol/l
0.50	1.0001	1.0019	5.0		995.1	3.1	7	1.3337				1.129	1.131	88.38		
1.00	1.0020	1.0037	10.0		992.0	6.3	15	1.3345				1.280	1.280	77.97		
1.50	1.0039	1.0056	15.1		988.8	9.4	22	1.3352				1.453	1.450	68.69		
2.00	1.0057	1.0075	20.1		985.6	12.6	29	1.3359				1.647	1.641	60.60		
2.50	1.0076	1.0094	25.2		982.5	15.8	37	1.3367				1.858	1.848	53.71		
3.00	1.0096	1.0113	30.3		979.3	19.0	44	1.3374				2.086	2.071	47.83		
3.50	1.0115	1.0133	35.4		976.1	22.2	52	1.3382				2.336	2.314	42.73		
4.00	1.0134	1.0152	40.5		972.9	25.4	59	1.3389				2.611	2.581	38.23		
4.50	1.0153	1.0171	45.7		969.6	28.6	67	1.3397				2.915	2.877	34.24		
5.00	1.0173	1.0191	50.9		966.4	31.8	74	1.3404				3.254	3.205	30.67		
5.50	1.0192	1.0210	56.1		963.2	35.1	82	1.3412				3.627	3.566	27.52		
6.00	1.0212	1.0230	61.3		959.9	38.3	89	1.3419				4.031	3.955	24.76		
6.50	1.0232	1.0250	66.5		956.7	41.6	97	1.3427				4.466	4.373	22.35		
7.00	1.0251	1.0270	71.8		953.4	44.8	105	1.3435				4.933	4.821	20.23		
7.50	1.0271	1.0290	77.0		950.1	48.1	112	1.3442				5.431	5.299	18.37		
8.00	1.0291	1.0310	82.3		946.8	51.4	120	1.3450				5.963	5.805	16.74		
8.50	1.0312	1.0330	87.6		943.5	54.7	128	1.3458				6.527	6.342	15.29		
9.00	1.0332	1.0350	93.0		940.2	58.0	135	1.3465				7.124	6.909	14.01		
9.50	1.0352	1.0371	98.3		936.9	61.4	143	1.3473				7.754	7.506	12.87		
10.00	1.0373	1.0391	103.7		933.5	64.7	151	1.3481				8.419	8.133	11.85		

16 ETHANOL, CH_3CH_2OH

MOLECULAR WEIGHT = 46.07
RELATIVE SPECIFIC REFRACTIVITY = 1.363

0.00 % by wt. data are the same for all compounds.
For Values of 0.00 wt. % solutions see Table 1, Acetic Acid.

A% by wt.	ρ D_4^{20}	D_{20}^{20}	C_s g/l	M g-mol/l	C_w g/l	$(C_o - C_w)$ g/l	$(n - n_o)$ × 10^4	n	Δ °C	O Os/kg	S g-mol/l	η/η_o	η/ρ cS	φ rhe	γ mmho/cm	T g-mol/l
0.50	0.9973	0.9991	5.0	0.108	992.3	5.9	3	1.3333	0.20	0.11	0.057	1.021	1.026	97.71		
1.00	0.9963	0.9981	10.0	0.216	986.4	11.8	6	1.3336	0.40	0.22	0.116	1.044	1.050	95.59		
1.50	0.9954	0.9972	14.9	0.324	980.5	17.8	9	1.3339	0.60	0.32	0.175	1.068	1.075	93.42		
2.00	0.9945	0.9962	19.9	0.432	974.6	23.6	12	1.3342	0.81	0.44	0.236	1.093	1.101	91.31		
2.50	0.9936	0.9953	24.8	0.539	968.7	29.5	15	1.3345	1.02	0.55	0.297	1.116	1.126	89.42		
3.00	0.9927	0.9945	29.8	0.646	962.9	35.3	18	1.3348	1.23	0.66	0.359	1.138	1.149	87.70		
3.50	0.9918	0.9936	34.7	0.754	957.1	41.1	21	1.3351	1.44	0.77	0.421	1.159	1.171	86.08		
4.00	0.9910	0.9927	39.6	0.860	951.4	46.9	24	1.3354	1.65	0.89	0.484	1.181	1.194	84.52		
4.50	0.9902	0.9919	44.6	0.967	945.6	52.6	27	1.3357	1.87	1.01	0.547	1.203	1.217	82.98		
5.00	0.9893	0.9911	49.5	1.074	939.9	58.4	31	1.3360	2.09	1.12	0.611	1.226	1.242	81.40		
5.50	0.9885	0.9903	54.4	1.180	934.2	64.1	34	1.3364	2.31	1.24	0.675	1.250	1.267	79.81		
6.00	0.9878	0.9895	59.3	1.286	928.5	69.7	37	1.3367	2.54	1.36	0.739	1.276	1.294	78.24		
6.50	0.9870	0.9887	64.2	1.393	922.8	75.4	41	1.3370	2.76	1.49	0.804	1.301	1.321	76.69		
7.00	0.9862	0.9880	69.0	1.498	917.2	81.0	44	1.3374	2.99	1.61	0.870	1.328	1.349	75.16		
7.50	0.9855	0.9872	73.9	1.604	911.6	86.7	48	1.3377	3.23	1.74	0.936	1.355	1.378	73.66		
8.00	0.9847	0.9865	78.8	1.710	906.0	92.3	51	1.3381	3.47	1.86	1.003	1.382	1.407	72.19		
8.50	0.9840	0.9857	83.6	1.816	900.4	97.9	55	1.3384	3.71	2.00	1.071	1.411	1.436	70.75		
9.00	0.9833	0.9850	88.5	1.921	894.8	103.4	58	1.3388	3.96	2.13	1.140	1.439	1.467	69.34		
9.50	0.9826	0.9843	93.3	2.026	889.2	109.0	62	1.3392	4.21	2.27	1.210	1.468	1.497	67.97		
10.00	0.9819	0.9836	98.2	2.131	883.7	114.5	65	1.3395	4.47	2.40	1.282	1.498	1.529	66.62		
11.00	0.9805	0.9822	107.9	2.341	872.6	125.6	73	1.3403	5.00	2.69	1.426	1.560	1.594	63.99		
12.00	0.9792	0.9809	117.5	2.550	861.7	136.6	80	1.3410	5.56	2.99	1.572	1.624	1.662	61.44		
13.00	0.9778	0.9796	127.1	2.759	850.7	147.5	87	1.3417	6.13	3.30	1.722	1.691	1.732	59.03		
14.00	0.9765	0.9782	136.7	2.967	839.8	158.4	95	1.3425	6.73	3.62	1.874	1.757	1.803	56.80		
15.00	0.9752	0.9769	146.3	3.175	828.9	169.3	102	1.3432	7.36	3.96	2.030	1.822	1.872	54.78		
16.00	0.9739	0.9756	155.8	3.382	818.1	180.2	110	1.3440	8.01	4.31	2.189	1.886	1.941	52.91		
17.00	0.9726	0.9743	165.3	3.589	807.3	191.0	117	1.3447	8.69	4.67	2.351	1.951	2.010	51.16		
18.00	0.9713	0.9730	174.8	3.795	796.5	201.8	125	1.3455	9.40	5.05	2.516	2.015	2.078	49.54		
19.00	0.9700	0.9717	184.3	4.000	785.7	212.5	132	1.3462	10.14	5.45	2.683	2.077	2.146	48.04		
20.00	0.9687	0.9704	193.7	4.205	774.9	223.3	140	1.3469	10.92	5.87	2.853	2.138	2.212	46.68		
22.00	0.9660	0.9677	212.5	4.613	753.5	244.8	154	1.3484	12.60	6.78	3.203	2.254	2.338	44.27		
24.00	0.9632	0.9649	231.2	5.018	732.0	266.2	168	1.3498	14.47	7.78	3.568	2.365	2.460	42.20		
26.00	0.9602	0.9619	249.7	5.419	710.6	287.7	181	1.3511	16.41	8.82	3.920	2.471	2.579	40.39		
28.00	0.9571	0.9588	268.0	5.817	689.1	309.1	194	1.3524	18.43	9.91	4.262	2.576	2.696	38.75		
30.00	0.9539	0.9556	286.2	6.211	667.7	330.5	205	1.3535	20.47	11.01	4.583	2.662	2.796	37.49		
32.00	0.9504	0.9521	304.1	6.601	646.3	352.0	216	1.3546	22.44	12.07		2.721	2.869	36.68		
34.00	0.9468	0.9485	321.9	6.988	624.9	373.3	227	1.3557	24.27	13.05		2.762	2.923	36.13		
36.00	0.9431	0.9447	339.5	7.369	603.6	394.7	236	1.3566	25.98	13.97		2.797	2.971	35.69		
38.00	0.9392	0.9408	356.9	7.747	582.3	415.9	245	1.3575	27.62	14.85		2.823	3.012	35.35		
40.00	0.9352	0.9369	374.1	8.120	561.1	437.1	253	1.3583	29.26	15.73		2.840	3.043	35.14		
42.00	0.9311	0.9328	391.1	8.488	540.0	458.2	260	1.3590	30.98	16.66		2.846	3.063	35.06		
44.00	0.9269	0.9286	407.8	8.853	519.1	479.2	268	1.3598	32.68	17.57		2.844	3.074	35.09		
46.00	0.9227	0.9243	424.4	9.213	498.2	500.0	274	1.3604	34.36	18.47		2.837	3.081	35.18		
48.00	0.9183	0.9199	440.8	9.568	477.5	520.7	280	1.3610	36.04	19.38		2.826	3.083	35.32		
50.00	0.9139	0.9155	457.0	9.919	457.0	541.3	286	1.3616	37.67	20.25		2.807	3.078	35.55		
52.00	0.9095	0.9111	472.9	10.265	436.5	561.7	291	1.3621	39.20	21.08		2.783	3.066	35.87		
54.00	0.9049	0.9065	488.7	10.607	416.3	582.0	296	1.3626	40.65	21.85		2.749	3.044	36.31		
56.00	0.9004	0.9020	504.2	10.944	396.2	602.1	300	1.3630	42.06	22.61		2.696	3.000	37.02		
58.00	0.8958	0.8974	519.5	11.277	376.2	622.0	304	1.3634	43.49	23.38		2.627	2.938	38.00		
60.00	0.8911	0.8927	534.7	11.606	356.5	641.8	308	1.3638	44.93	24.15		2.542	2.858	39.26		
62.00	0.8865	0.8880	549.6	11.930	336.9	661.4	311	1.3641	46.28	24.88		2.474	2.796	40.34		
64.00	0.8818	0.8833	564.3	12.250	317.4	680.8	315	1.3644	47.52	25.55		2.410	2.739	41.41		
66.00	0.8771	0.8786	578.9	12.565	298.2	700.0	317	1.3647	48.64	26.15		2.342	2.676	42.61		
68.00	0.8724	0.8739	593.2	12.876	279.2	719.1	320	1.3650	49.52	26.62		2.276	2.614	43.85		
70.00	0.8676	0.8692	607.3	13.183	260.3	737.9	322	1.3652				2.210	2.552	45.17		
72.00	0.8629	0.8644	621.3	13.486	241.6	756.6	324	1.3654				2.144	2.490	46.55		
74.00	0.8581	0.8596	635.0	13.784	223.1	775.1	325	1.3655				2.078	2.426	48.03		
76.00	0.8533	0.8549	648.5	14.077	204.8	793.4	327	1.3657				2.011	2.361	49.63		
78.00	0.8485	0.8500	661.8	14.366	186.7	811.6	327	1.3657				1.944	2.296	51.33		
80.00	0.8436	0.8451	674.9	14.650	168.7	829.5	328	1.3658				1.877	2.229	53.15		
82.00	0.8387	0.8401	687.7	14.927	151.0	847.3	328	1.3657				1.804	2.155	55.31		
84.00	0.8335	0.8350	700.2	15.198	133.4	864.9	326	1.3656				1.738	2.089	57.42		
86.00	0.8284	0.8299	712.4	15.464	116.0	882.3	325	1.3655				1.671	2.021	59.74		
88.00	0.8232	0.8247	724.4	15.725	98.8	899.4	323	1.3653				1.603	1.951	62.26		
90.00	0.8180	0.8194	736.2	15.979	81.8	916.4	320	1.3650				1.539	1.885	64.85		
92.00	0.8125	0.8140	747.5	16.226	65.0	933.2	316	1.3646				1.472	1.815	67.80		
94.00	0.8070	0.8084	758.6	16.466	48.4	949.8	312	1.3642				1.404	1.743	71.07		
96.00	0.8013	0.8027	769.2	16.697	32.1	966.2	306	1.3636				1.339	1.674	74.53		
98.00	0.7954	0.7968	779.5	16.920	15.9	982.3	300	1.3630				1.270	1.600	78.59		
100.00	0.7893	0.7907	789.3	17.133	0.0	998.2	284	1.3614				1.201	1.525	83.10		

17 (ETHYLENEDINITRILO)TETRAACETIC ACID DISODIUM SALT, EDTA DISODIUM*, $Na_2C_{10}H_{14}O_8N_2 \cdot 2H_2O$

MOLECULAR WEIGHT = 336.21 FORMULA WEIGHT, HYDRATE = 372.24
RELATIVE SPECIFIC REFRACTIVITY = 0.977

0.00 % by wt. data are the same for all compounds.
For Values of 0.00 wt. % solutions see Table 1, Acetic Acid.

A% by wt.	H% by wt.	ρ D_4^{20}	D_{20}^{20}	C_s g/l	M g-mol/l	C_w g/l	$(C_o - C_w)$ g/l	$(n - n_o) \times 10^4$	n	Δ °C	O Os/kg	S g-mol/l	η/η_o	η/ρ cS	φ rhe	γ mmho/cm	T g-mol/l
0.50	0.55	1.0009	1.0027	5.0	0.015	995.9	2.3	9	1.3339	0.07	0.04	0.020	1.015	1.016	98.35	1.9	0.018
1.00	1.11	1.0036	1.0053	10.0	0.030	993.5	4.7	18	1.3348	0.14	0.07	0.039	1.030	1.028	96.94	3.5	0.034
1.50	1.66	1.0062	1.0080	15.1	0.045	991.1	7.1	26	1.3356	0.21	0.11	0.059	1.044	1.040	95.55	5.0	0.049
2.00	2.21	1.0089	1.0107	20.2	0.060	988.7	9.5	35	1.3365	0.27	0.15	0.078	1.060	1.052	94.19	6.3	0.064
2.50	2.77	1.0115	1.0133	25.3	0.075	986.2	12.0	44	1.3374	0.33	0.18	0.096	1.075	1.065	92.84	7.6	0.077
3.00	3.32	1.0142	1.0160	30.4	0.090	983.8	14.5	53	1.3383	0.40	0.21	0.115	1.091	1.078	91.51	8.8	0.091
3.50	3.88	1.0169	1.0187	35.6	0.106	981.3	16.9	62	1.3392	0.46	0.25	0.133	1.107	1.091	90.18	10.1	0.105
4.00	4.43	1.0196	1.0214	40.8	0.121	978.8	19.5	70	1.3400	0.52	0.28	0.151	1.123	1.104	88.85	11.3	0.119
4.50	4.98	1.0223	1.0241	46.0	0.137	976.3	22.0	79	1.3409	0.58	0.31	0.169	1.140	1.118	87.53	12.5	0.133
5.00	5.54	1.0250	1.0268	51.2	0.152	973.7	24.5	88	1.3418	0.65	0.35	0.187	1.158	1.132	86.20	13.7	0.145
5.50	6.09	1.0277	1.0295	56.5	0.168	971.2	27.0	97	1.3427	0.71	0.38	0.206	1.176	1.147	84.86	14.6	0.156
6.00	6.64	1.0305	1.0323	61.8	0.184	968.6	29.6	106	1.3436	0.77	0.42	0.225	1.195	1.162	83.51	15.3	0.164

* Sodium(Di)Ethylenediamine Tetraacetate.

18 ETHYLENE GLYCOL, CH_2OHCH_2OH

MOLECULAR WEIGHT = 62.07
RELATIVE SPECIFIC REFRACTIVITY = 1.147

0.00 % by wt. data are the same for all compounds.
For Values of 0.00 wt. % solutions see Table 1, Acetic Acid.

A% by wt.	ρ D_4^{20}	D_{20}^{20}	C_s g/l	M g-mol/l	C_w g/l	$(C_o - C_w)$ g/l	$(n - n_o) \times 10^4$	n	Δ °C	O Os/kg	S g-mol/l	η/η_o	η/ρ cS	φ rhe	γ mmho/cm	T g-mol/l
0.50	0.9988	1.0006	5.0	0.080	993.8	4.4	5	1.3335	0.15	0.08	0.042	1.008	1.011	99.03		
1.00	0.9995	1.0012	10.0	0.161	989.5	8.8	9	1.3339	0.30	0.16	0.086	1.018	1.021	98.04		
2.00	1.0007	1.0025	20.0	0.322	980.7	17.6	19	1.3348	0.61	0.33	0.176	1.046	1.047	95.41		
3.00	1.0019	1.0037	30.1	0.484	971.9	26.4	28	1.3358	0.92	0.50	0.269	1.072	1.072	93.09		
4.00	1.0032	1.0049	40.1	0.646	963.0	35.2	37	1.3367	1.24	0.67	0.364	1.097	1.096	90.98		
5.00	1.0044	1.0062	50.2	0.809	954.2	44.0	47	1.3377	1.58	0.85	0.461	1.123	1.120	88.87		
6.00	1.0057	1.0075	60.3	0.972	945.3	52.9	56	1.3386	1.91	1.03	0.560	1.151	1.147	86.72		
7.00	1.0070	1.0087	70.5	1.136	936.5	61.8	66	1.3396	2.26	1.22	0.660	1.180	1.174	84.60		
8.00	1.0082	1.0100	80.7	1.299	927.6	70.6	75	1.3405	2.62	1.41	0.763	1.210	1.202	82.49		
9.00	1.0095	1.0113	90.9	1.464	918.7	79.6	85	1.3415	2.99	1.61	0.868	1.241	1.232	80.41		
10.00	1.0108	1.0126	101.1	1.629	909.8	88.5	95	1.3425	3.37	1.81	0.975	1.274	1.263	78.34		
12.00	1.0134	1.0152	121.6	1.959	891.8	106.4	114	1.3444	4.16	2.24	1.196	1.345	1.330	74.21		
14.00	1.0161	1.0179	142.3	2.292	873.8	124.4	134	1.3464	5.01	2.69	1.426	1.421	1.401	70.25		
16.00	1.0188	1.0206	163.0	2.626	855.8	142.5	154	1.3484	5.91	3.18	1.665	1.497	1.473	66.65		
18.00	1.0214	1.0232	183.9	2.962	837.6	160.7	174	1.3503	6.89	3.70	1.913	1.575	1.545	63.35		
20.00	1.0241	1.0259	204.8	3.300	819.3	178.9	194	1.3523	7.93	4.27	2.171	1.658	1.622	60.19		
24.00	1.0296	1.0314	247.1	3.981	782.5	215.8	234	1.3564	10.28	5.53	2.714	1.839	1.790	54.26		
28.00	1.0350	1.0369	289.8	4.669	745.2	253.0	275	1.3605	13.03	7.01	3.289	2.043	1.978	48.84		
32.00	1.0405	1.0424	333.0	5.364	707.6	290.7	316	1.3646	16.23	8.73	3.889	2.275	2.191	43.87		
36.00	1.0460	1.0478	376.6	6.067	669.4	328.8	357	1.3687	19.82	10.66	4.483	2.532	2.425	39.42		
40.00	1.0514	1.0532	420.6	6.775	630.8	367.4	398	1.3728	23.84	12.82		2.826	2.693	35.32		
44.00	1.0567	1.0586	465.0	7.491	591.8	406.5	440	1.3769	28.32	15.23		3.160	2.996	31.59		
48.00	1.0619	1.0638	509.7	8.212	552.2	446.0	481	1.3811	33.30	17.90		3.537	3.337	28.22		
52.00	1.0670	1.0689	554.8	8.939	512.2	486.1	521	1.3851	38.81	20.87		3.973	3.731	25.12		
56.00	1.0719	1.0738	600.3	9.671	471.6	526.6	562	1.3892	44.83	24.10		4.466	4.175	22.35		
60.00	1.0765	1.0784	645.9	10.406	430.6	567.6	602	1.3931	51.23	27.54		5.016	4.669	19.90		

19 FERRIC CHLORIDE, $FeCl_3 \cdot 6H_2O$

MOLECULAR WEIGHT = 162.22 FORMULA WEIGHT, HYDRATE = 270.32
RELATIVE SPECIFIC REFRACTIVITY = 0.946

0.00 % by wt. data are the same for all compounds.
For Values of 0.00 wt. % solutions see Table 1, Acetic Acid.

A% by wt.	H% by wt.	ρ D_4^{20}	D_{20}^{20}	C_s g/l	M g-mol/l	C_w g/l	$(C_o - C_w)$ g/l	$(n - n_o) \times 10^4$	n	Δ °C	O Os/kg	S g-mol/l	η/η_o	η/ρ cS	φ rhe	γ mmho/cm	T g-mol/l
0.50	0.83	1.0025	1.0043	5.0	0.031	997.5	0.7	14	1.3344	0.206	0.111	0.059	1.022	1.022	97.61		
1.00	1.67	1.0068	1.0086	10.1	0.062	996.7	1.5	28	1.3358	0.385	0.207	0.111	1.045	1.040	95.49		
2.00	3.33	1.0153	1.0171	20.3	0.125	995.0	3.3	56	1.3386	0.753	0.405	0.219	1.091	1.077	91.48		
3.00	5.00	1.0238	1.0256	30.7	0.189	993.1	5.2	83	1.3413	1.146	0.616	0.335	1.137	1.113	87.79		
4.00	6.67	1.0323	1.0341	41.3	0.255	991.0	7.3	111	1.3441	1.562	0.840	0.457	1.185	1.150	84.22		
5.00	8.33	1.0408	1.0426	52.0	0.321	988.7	9.5	138	1.3468	2.002	1.076	0.585	1.236	1.190	80.76		
6.00	10.00	1.0493	1.0512	63.0	0.388	986.4	11.9	166	1.3496	2.475	1.331	0.722	1.289	1.231	77.41		
7.00	11.66	1.0580	1.0599	74.1	0.457	984.0	14.3	194	1.3524	2.994	1.609	0.870	1.347	1.275	74.11		
8.00	13.33	1.0668	1.0687	85.3	0.526	981.5	16.7	222	1.3552	3.566	1.917	1.031	1.409	1.324	70.82		
9.00	15.00	1.0760	1.0779	96.8	0.597	979.2	19.0	251	1.3581	4.185	2.250	1.202	1.477	1.376	67.55		
10.00	16.66	1.0853	1.0872	108.5	0.669	976.8	21.5	281	1.3611	4.854	2.610	1.385	1.550	1.431	64.39		
12.00	20.00	1.1040	1.1059	132.5	0.817	971.5	26.8	340	1.3670	6.38	3.43	1.785	1.704	1.547	58.57		
14.00	23.33	1.1228	1.1248	157.2	0.969	965.6	32.6	400	1.3730	8.22	4.42	2.240	1.875	1.673	53.22		
16.00	26.66	1.1420	1.1440	182.7	1.126	959.3	39.0			10.45	5.62	2.750	2.076	1.822	48.07		
18.00	29.99	1.1615	1.1636	209.1	1.289	952.5	45.8			13.08	7.03	3.299	2.306	1.990	43.27		
20.00	33.33	1.1816	1.1837	236.3	1.457	945.3	53.0			16.14	8.68	3.873	2.565	2.175	38.91		
24.00	39.99	1.2234	1.2256	293.6	1.810	929.8	68.4			23.79	12.79		3.172	2.598	31.47		
28.00	46.66	1.2679	1.2702	355.0	2.189	912.9	85.3			33.61	18.07		4.030	3.185	24.76		
32.00	53.32	1.3153	1.3176	420.9	2.595	894.4	103.8			49.16	26.43		5.263	4.010	18.96		
36.00	59.99	1.3654	1.3678	491.5	3.030	873.9	124.4						7.116	5.222	14.03		
40.00	66.66	1.4176	1.4201	567.0	3.496	850.6	147.7						9.655	6.824	10.34		

20 FORMIC ACID, HCOOH

MOLECULAR WEIGHT = 46.03
RELATIVE SPECIFIC REFRACTIVITY = 0.916

0.00 % by wt. data are the same for all compounds.
For Values of 0.00 wt. % solutions see Table 1, Acetic Acid.

A% by wt.	ρ D^{20}_4	D^{20}_{20}	C_s g/l	M g-mol/l	C_w g/l	$(C_o - C_w)$ g/l	$(n - n_o)$ × 10^4	n	Δ °C	O Os/kg	S g-mol/l	η/η_o	η/ρ cS	φ rhe	γ mmho/cm	T g-mol/l
0.50	0.9994	1.0012	5.0	0.109	994.4	3.8	3	1.3333	0.210	0.113	0.060	1.004	1.007	99.36	1.4	0.013
1.00	1.0006	1.0023	10.0	0.217	990.6	7.7	6	1.3336	0.418	0.225	0.121	1.009	1.010	98.96	2.4	0.024
2.00	1.0029	1.0047	20.1	0.436	982.9	15.3	12	1.3342	0.824	0.443	0.240	1.015	1.015	98.28	3.5	0.034
3.00	1.0053	1.0071	30.2	0.655	975.2	23.1	18	1.3348	1.243	0.668	0.363	1.023	1.019	97.59	4.3	0.043
4.00	1.0077	1.0095	40.3	0.876	967.4	30.8	24	1.3354	1.672	0.899	0.489	1.030	1.024	96.90	5.0	0.050
5.00	1.0102	1.0119	50.5	1.097	959.6	38.6	29	1.3359	2.103	1.131	0.615	1.037	1.029	96.24	5.6	0.056
6.00	1.0126	1.0144	60.8	1.320	951.8	46.4	35	1.3365	2.534	1.363	0.739	1.044	1.033	95.63	6.1	0.062
7.00	1.0150	1.0168	71.1	1.544	944.0	54.3	41	1.3371	2.967	1.595	0.862	1.050	1.036	95.06	6.6	0.067
8.00	1.0175	1.0193	81.4	1.768	936.1	62.2	46	1.3376	3.402	1.829	0.985	1.056	1.040	94.51	7.0	0.071
9.00	1.0199	1.0217	91.8	1.994	928.1	70.1	52	1.3382	3.838	2.063	1.106	1.062	1.043	93.98	7.4	0.075
10.00	1.0224	1.0242	102.2	2.221	920.1	78.1	57	1.3387	4.267	2.294	1.225	1.068	1.047	93.45	7.8	0.079
12.00	1.0273	1.0291	123.3	2.678	904.0	94.2	67	1.3397	5.19	2.79	1.475	1.080	1.053	92.41	8.4	0.086
14.00	1.0322	1.0340	144.5	3.139	887.7	110.5	78	1.3408	6.11	3.28	1.715	1.092	1.060	91.40	8.8	0.091
16.00	1.0371	1.0389	165.9	3.605	871.1	127.1	88	1.3418	7.06	3.80	1.957	1.104	1.067	90.38	9.3	0.096
18.00	1.0419	1.0438	187.5	4.074	854.4	143.9	98	1.3428	8.08	4.35	2.207	1.117	1.074	89.35	9.6	0.100
20.00	1.0467	1.0486	209.3	4.548	837.4	160.9	107	1.3437	9.11	4.90	2.448	1.130	1.081	88.36	9.9	0.103
24.00	1.0562	1.0580	253.5	5.507	802.7	195.5	126	1.3456	11.10	5.97	2.891	1.154	1.094	86.52	10.3	0.108
28.00	1.0654	1.0673	298.3	6.481	767.1	231.1	145	1.3475	13.10	7.04	3.303	1.177	1.107	84.79	10.5	0.109
32.00	1.0746	1.0765	343.9	7.471	730.7	267.5	163	1.3493	15.28	8.21	3.717	1.201	1.120	83.09	10.5	0.109
36.00	1.0839	1.0858	390.2	8.477	693.7	304.6	181	1.3511	17.65	9.49	4.132	1.225	1.133	81.44	10.3	0.107
40.00	1.0935	1.0955	437.4	9.503	656.1	342.1	199	1.3529	20.18	10.85	4.539	1.251	1.146	79.78	9.9	0.103
44.00	1.1015	1.1035	484.7	10.529	616.9	381.4	217	1.3547	22.93	12.33		1.278	1.162	78.11	9.5	0.098
48.00	1.1097	1.1117	532.7	11.572	577.1	421.2	235	1.3565	26.06	14.01		1.306	1.179	76.42	8.9	0.092
52.00	1.1183	1.1203	581.5	12.634	536.8	461.4	251	1.3581	29.69	15.96		1.337	1.198	74.65	8.3	0.085
56.00	1.1273	1.1293	631.3	13.714	496.0	502.2	267	1.3597	33.81	18.18		1.371	1.218	72.81	7.6	0.078
60.00	1.1364	1.1385	681.9	14.813	454.6	543.7	282	1.3612	38.26	20.57		1.407	1.241	70.93	6.9	0.070
64.00	1.1456	1.1476	733.2	15.928	412.4	585.8	296	1.3626	43.02	23.13		1.446	1.265	69.03	6.2	0.062
68.00	1.1544	1.1565	785.0	17.055	369.4	628.8	311	1.3641				1.487	1.291	67.11	5.4	0.054
70.00	1.1586	1.1607	811.1	17.620	347.6	650.6	318	1.3648				1.508	1.305	66.16	5.1	0.051

21 D-FRUCTOSE (LEVULOSE), $C_6H_{12}O_6$

MOLECULAR WEIGHT = 180.16
RELATIVE SPECIFIC REFRACTIVITY = 1.021

0.00 % by wt. data are the same for all compounds.
For Values of 0.00 wt. % solutions see Table 1, Acetic Acid.

A% by wt.	ρ D^{20}_4	D^{20}_{20}	C_s g/l	M g-mol/l	C_w g/l	$(C_o - C_w)$ g/l	$(n - n_o)$ × 10^4	n	Δ °C	O Os/kg	S g-mol/l	η/η_o	η/ρ cS	φ rhe	γ mmho/cm	T g-mol/l
0.50	1.0002	1.0020	5.0	0.028	995.2	3.0	7	1.3337	0.052	0.028	0.014	1.013	1.015	98.53		
1.00	1.0021	1.0039	10.0	0.056	992.1	6.1	14	1.3344	0.104	0.056	0.029	1.026	1.026	97.28		
1.50	1.0041	1.0059	15.1	0.084	989.1	9.2	21	1.3351	0.157	0.085	0.045	1.039	1.037	96.06		
2.00	1.0061	1.0079	20.1	0.112	986.0	12.3	28	1.3358	0.211	0.113	0.060	1.052	1.048	94.87		
2.50	1.0081	1.0098	25.2	0.140	982.9	15.4	36	1.3366	0.265	0.142	0.076	1.065	1.059	93.71		
3.00	1.0101	1.0118	30.3	0.168	979.7	18.5	43	1.3373	0.319	0.172	0.092	1.078	1.069	92.59		
3.50	1.0120	1.0138	35.4	0.197	976.6	21.6	50	1.3380	0.374	0.201	0.108	1.091	1.080	91.50		
4.00	1.0140	1.0158	40.6	0.225	973.5	24.8	57	1.3387	0.430	0.231	0.124	1.104	1.091	90.40		
4.50	1.0160	1.0178	45.7	0.254	970.3	27.9	65	1.3395	0.487	0.262	0.141	1.118	1.102	89.30		
5.00	1.0181	1.0199	50.9	0.283	967.2	31.1	72	1.3402	0.544	0.292	0.158	1.132	1.114	88.16		
5.50	1.0201	1.0219	56.1	0.311	964.0	34.3	79	1.3409	0.602	0.324	0.175	1.147	1.127	87.00		
6.00	1.0221	1.0239	61.3	0.340	960.8	37.5	87	1.3417	0.660	0.355	0.192	1.163	1.140	85.83		
6.50	1.0241	1.0259	66.6	0.369	957.6	40.7	94	1.3424	0.720	0.387	0.209	1.179	1.153	84.66		
7.00	1.0262	1.0280	71.8	0.399	954.3	43.9	101	1.3431	0.780	0.419	0.227	1.196	1.167	83.48		
7.50	1.0282	1.0300	77.1	0.428	951.1	47.1	109	1.3439	0.841	0.452	0.245	1.213	1.182	82.29		
8.00	1.0303	1.0321	82.4	0.457	947.8	50.4	116	1.3446	0.902	0.485	0.263	1.230	1.197	81.11		
8.50	1.0323	1.0342	87.7	0.487	944.6	53.7	124	1.3454	0.965	0.519	0.282	1.249	1.212	79.93		
9.00	1.0344	1.0362	93.1	0.517	941.3	56.9	131	1.3461	1.028	0.553	0.300	1.267	1.228	78.75		
9.50	1.0365	1.0383	98.5	0.547	938.0	60.2	139	1.3469	1.092	0.587	0.319	1.286	1.244	77.58		
10.00	1.0385	1.0404	103.9	0.576	934.7	63.5	147	1.3476	1.158	0.622	0.338	1.306	1.260	76.42		
11.00	1.0427	1.0446	114.7	0.637	928.0	70.2	162	1.3492	1.290	0.694	0.377	1.346	1.294	74.13		
12.00	1.0469	1.0488	125.6	0.697	921.3	76.9	177	1.3507	1.427	0.767	0.417	1.388	1.329	71.89		
13.00	1.0512	1.0530	136.7	0.758	914.5	83.7	193	1.3522	1.567	0.842	0.458	1.432	1.365	69.67		
14.00	1.0554	1.0573	147.8	0.820	907.7	90.6	208	1.3538	1.710	0.920	0.500	1.480	1.405	67.45		
15.00	1.0597	1.0616	159.0	0.882	900.8	97.5	224	1.3554	1.858	0.999	0.543	1.530	1.447	65.23		
16.00	1.0640	1.0659	170.2	0.945	893.8	104.5	240	1.3569	2.009	1.080	0.587	1.584	1.491	63.02		
17.00	1.0684	1.0703	181.6	1.008	886.8	111.5	255	1.3585	2.162	1.163	0.632	1.640	1.538	60.84		
18.00	1.0728	1.0747	193.1	1.072	879.7	118.6	272	1.3601	2.319	1.247	0.677	1.700	1.588	58.69		
19.00	1.0772	1.0791	204.7	1.136	872.5	125.7	288	1.3618	2.479	1.333	0.723	1.764	1.641	56.56		
20.00	1.0816	1.0835	216.3	1.201	865.3	133.0	304	1.3634	2.640	1.419	0.769	1.833	1.698	54.45		
22.00	1.0906	1.0925	239.9	1.332	850.6	147.6	337	1.3667	3.054	1.642	0.887	1.982	1.821	50.35		
24.00	1.0996	1.1016	263.9	1.465	835.7	162.5	370	1.3700	3.431	1.845	0.993	2.150	1.959	46.41		
26.00	1.1089	1.1108	288.3	1.600	820.5	177.7	404	1.3734	3.815	2.051	1.100	2.343	2.117	42.60		
28.00	1.1182	1.1202	313.1	1.738	805.1	193.1	438	1.3768	4.196	2.256	1.205	2.557	2.291	39.03		
30.00	1.1276	1.1296	338.3	1.878	789.3	208.9	473	1.3803				2.811	2.498	35.50		
32.00	1.1372	1.1392	363.9	2.020	773.3	224.9	509	1.3839				3.106	2.737	32.13		
34.00	1.1469	1.1490	390.0	2.165	757.0	241.3	545	1.3874				3.455	3.018	28.89		
36.00	1.1568	1.1588	416.4	2.312	740.3	257.9	581	1.3911				3.891	3.370	25.65		
38.00	1.1668	1.1688	443.4	2.461	723.4	274.8	618	1.3948				4.409	3.786	22.64		
40.00	1.1769	1.1790	470.8	2.613	706.1	292.1	655	1.3985				5.036	4.288	19.82		
42.00	1.1871	1.1892	498.6	2.767	688.5	309.7	694	1.4023				5.761	4.863	17.32		
44.00	1.1975	1.1996	526.9	2.925	670.6	327.6	732	1.4062				6.631	5.548	15.05		
46.00	1.2080	1.2102	555.7	3.084	652.3	345.9	771	1.4101				7.738	6.419	12.90		
48.00	1.2187	1.2208	585.0	3.247	633.7	364.5	811	1.4141				9.042	7.434	11.04		

21 D-FRUCTOSE (LEVULOSE), $C_6H_{12}O_6$—(Continued)

A% by wt.	ρ D_4^{20}	D_{20}^{20}	C_s g/l	M g-mol/l	C_w g/l	$(C_o - C_w)$ g/l	$(n - n_o)$ × 10^4	n	Δ °C	O Os/kg	S g-mol/l	η/η_o	η/ρ cS	φ rhe	γ mmho/cm	T g-mol/l
50.00	1.2295	1.2316	614.7	3.412	614.7	383.5	851	1.4181				10.80	8.80	9.24		
52.00	1.2404	1.2426	645.0	3.580	595.4	402.9	892	1.4222				12.99	10.49	7.69		
54.00	1.2514	1.2536	675.8	3.751	575.7	422.6	934	1.4264				15.84	12.68	6.30		
56.00	1.2626	1.2649	707.1	3.925	555.6	442.7	976	1.4306				20.05	15.91	4.98		
58.00	1.2739	1.2762	738.9	4.101	535.1	463.2	1019	1.4348				25.60	20.13	3.90		
60.00	1.2854	1.2877	771.2	4.281	514.2	484.1	1062	1.4392				32.47	25.31	3.07		
62.00	1.2970	1.2992	804.1	4.463	492.8	505.4	1105	1.4435				42.97	33.20	2.32		
64.00	1.3086	1.3110	837.5	4.649	471.1	527.1	1150	1.4480				58.17	44.54	1.72		
66.00	1.3204	1.3228	871.5	4.837	448.9	549.3	1194	1.4524				82.19	62.37	1.21		
68.00	1.3323	1.3347	906.0	5.029	426.3	571.9	1240	1.4570				118.4	89.08	0.84		
70.00	1.3443	1.3467	941.0	5.223	403.3	594.9	1285	1.4615				177.8	132.5	0.56		

22 D-GLUCOSE (DEXTROSE), $C_6H_{12}O_6 \cdot 1H_2O$

MOLECULAR WEIGHT = 180.16 FORMULA WEIGHT, HYDRATE = 198.17
RELATIVE SPECIFIC REFRACTIVITY = 1.031

0.00 % by wt. data are the same for all compounds.
For Values of 0.00 wt. % solutions see Table 1, Acetic Acid.

A% by wt.	H% by wt.	ρ D_4^{20}	D_{20}^{20}	C_s g/l	M g-mol/l	C_w g/l	$(C_o - C_w)$ g/l	$(n - n_o)$ × 10^4	n	Δ °C	O Os/kg	S g-mol/l	η/η_o	η/ρ cS	φ rhe	γ mmho/cm	T g-mol/l
0.50	0.55	1.0001	1.0019	5.0	0.028	995.1	3.1	7	1.3337	0.047	0.025	0.013	1.008	1.010	99.01		
1.00	1.10	1.0020	1.0038	10.0	0.056	992.0	6.2	14	1.3344	0.107	0.057	0.030	1.019	1.019	97.94		
1.50	1.65	1.0039	1.0057	15.1	0.084	988.9	9.4	21	1.3351	0.158	0.085	0.045	1.034	1.032	96.54		
2.00	2.20	1.0058	1.0076	20.1	0.112	985.7	12.5	28	1.3358	0.214	0.115	0.061	1.050	1.046	95.05		
2.50	2.75	1.0078	1.0095	25.2	0.140	982.6	15.7	36	1.3366	0.270	0.145	0.077	1.065	1.059	93.68		
3.00	3.30	1.0097	1.0115	30.3	0.168	979.4	18.8	43	1.3373	0.323	0.174	0.093	1.081	1.072	92.36		
3.50	3.85	1.0116	1.0134	35.4	0.197	976.2	22.0	50	1.3380	0.377	0.202	0.108	1.096	1.085	91.08		
4.00	4.40	1.0136	1.0154	40.5	0.225	973.0	25.2	57	1.3387	0.433	0.233	0.125	1.111	1.098	89.82		
4.50	4.95	1.0155	1.0173	45.7	0.254	969.8	28.4	65	1.3395	0.490	0.264	0.142	1.127	1.112	88.57		
5.00	5.50	1.0175	1.0193	50.9	0.282	966.6	31.6	72	1.3402	0.549	0.295	0.159	1.143	1.126	87.31		
5.50	6.05	1.0194	1.0212	56.1	0.311	963.4	34.9	79	1.3409	0.608	0.327	0.176	1.160	1.140	86.06		
6.00	6.60	1.0214	1.0232	61.3	0.340	960.1	38.1	87	1.3417	0.668	0.359	0.194	1.177	1.154	84.82		
6.50	7.15	1.0234	1.0252	66.5	0.369	956.9	41.4	94	1.3424	0.729	0.392	0.212	1.194	1.169	83.59		
7.00	7.70	1.0254	1.0272	71.8	0.398	953.6	44.6	102	1.3432	0.790	0.425	0.230	1.212	1.184	82.37		
7.50	8.25	1.0274	1.0292	77.1	0.428	950.3	47.9	109	1.3439	0.851	0.458	0.248	1.230	1.199	81.17		
8.00	8.80	1.0294	1.0312	82.4	0.457	947.0	51.2	117	1.3447	0.913	0.491	0.266	1.248	1.215	79.97		
8.50	9.35	1.0314	1.0332	87.7	0.487	943.7	54.5	124	1.3454	0.975	0.524	0.284	1.267	1.231	78.77		
9.00	9.90	1.0334	1.0352	93.0	0.516	940.4	57.8	132	1.3462	1.038	0.558	0.303	1.286	1.247	77.58		
9.50	10.45	1.0354	1.0373	98.4	0.546	937.1	61.2	139	1.3469	1.102	0.592	0.322	1.306	1.264	76.39		
10.00	11.00	1.0375	1.0393	103.7	0.576	933.7	64.5	147	1.3477	1.167	0.627	0.341	1.327	1.282	75.21		
11.00	12.10	1.0416	1.0434	114.6	0.636	927.0	71.2	162	1.3492	1.303	0.700	0.381	1.369	1.317	72.89		
12.00	13.20	1.0457	1.0475	125.5	0.697	920.2	78.0	178	1.3508	1.443	0.776	0.422	1.413	1.354	70.63		
13.00	14.30	1.0498	1.0517	136.5	0.758	913.4	84.9	193	1.3523	1.586	0.853	0.464	1.459	1.393	68.40		
14.00	15.40	1.0540	1.0559	147.6	0.819	906.4	91.8	209	1.3539	1.731	0.931	0.506	1.509	1.434	66.15		
15.00	16.50	1.0582	1.0601	158.7	0.881	899.5	98.8	225	1.3555	1.880	1.011	0.550	1.563	1.480	63.85		
16.00	17.60	1.0624	1.0643	170.0	0.944	892.5	105.8	241	1.3571	2.033	1.093	0.594	1.622	1.530	61.53		
17.00	18.70	1.0667	1.0686	181.3	1.007	885.4	112.9	257	1.3587	2.190	1.178	0.640	1.685	1.583	59.21		
18.00	19.80	1.0710	1.0729	192.8	1.070	878.2	120.0	273	1.3603	2.353	1.265	0.687	1.753	1.640	56.93		
19.00	20.90	1.0753	1.0772	204.3	1.134	871.0	127.2	289	1.3619	2.521	1.355	0.735	1.825	1.700	54.70		
20.00	22.00	1.0797	1.0816	215.9	1.199	863.7	134.5	306	1.3635	2.696	1.449	0.785	1.900	1.763	52.53		
22.00	24.20	1.0884	1.0904	239.5	1.329	849.0	149.3	339	1.3668	3.073	1.652	0.892	2.059	1.896	48.47		
24.00	26.40	1.0973	1.0993	263.4	1.462	834.0	164.3	372	1.3702	3.480	1.871	1.007	2.238	2.044	44.59		
26.00	28.60	1.1063	1.1083	287.6	1.597	818.7	179.6	406	1.3736	3.900	2.097	1.124	2.453	2.222	40.69		
28.00	30.80	1.1154	1.1174	312.3	1.734	803.1	195.1	440	1.3770	4.337	2.332	1.244	2.702	2.427	36.94		
30.00	33.00	1.1246	1.1266	337.4	1.873	787.2	211.0	475	1.3805	4.794	2.577	1.369	2.992	2.666	33.36		
32.00	35.20	1.1340	1.1360	362.9	2.014	771.1	227.1	511	1.3840				3.317	2.931	30.09		
34.00	37.40	1.1434	1.1454	388.8	2.158	754.6	243.6	546	1.3876				3.697	3.240	26.99		
36.00	39.60	1.1529	1.1550	415.1	2.304	737.9	260.3	582	1.3912				4.185	3.637	23.85		
38.00	41.80	1.1626	1.1647	441.8	2.452	720.8	277.4	619	1.3949				4.776	4.116	20.90		
40.00	44.00	1.1724	1.1745	469.0	2.603	703.5	294.8	656	1.3986				5.482	4.685	18.21		
42.00	46.20	1.1823	1.1844	496.6	2.756	685.8	312.5	694	1.4024				6.275	5.318	15.90		
44.00	48.40	1.1924	1.1945	524.7	2.912	667.7	330.5	732	1.4062				7.221	6.068	13.82		
46.00	50.60	1.2026	1.2047	553.2	3.071	649.4	348.8	771	1.4101				8.437	7.029	11.83		
48.00	52.80	1.2130	1.2151	582.2	3.232	630.7	367.5	811	1.4141				9.863	8.148	10.12		
50.00	55.00	1.2235	1.2257	611.7	3.396	611.7	386.5	851	1.4181				11.86	9.713	8.41		
52.00	57.20	1.2342	1.2364	641.8	3.562	592.4	405.8	892	1.4222				14.46	11.74	6.90		
54.00	59.40	1.2451	1.2473	672.4	3.732	572.8	425.5	933	1.4263				17.88	14.39	5.58		
56.00	61.60	1.2562	1.2585	703.5	3.905	552.7	445.5	976	1.4306				22.84	18.22	4.37		
58.00	63.80	1.2676	1.2699	735.2	4.081	532.4	465.8	1020	1.4349				29.33	23.18	3.40		
60.00	66.00	1.2793	1.2816	767.6	4.261	511.7	486.5	1064	1.4394				37.37	29.27	2.67		

23 GLYCEROL, $CH_2OHCHOHCH_2OH$

MOLECULAR WEIGHT = 92.09
RELATIVE SPECIFIC REFRACTIVITY = 1.109

0.00 % by wt. data are the same for all compounds.
For Values of 0.00 wt. % solutions see Table 1, Acetic Acid.

A% by wt.	ρ D_4^{20}	D_{20}^{20}	C_s g/l	M g-mol/l	C_w g/l	$(C_o - C_w)$ g/l	$(n - n_o)$ × 10^4	n	Δ °C	O Os/kg	S g-mol/l	η/η_o	η/ρ cS	φ rhe	γ mmho/cm	T g-mol/l
0.50	0.9994	1.0011	5.0	0.054	994.4	3.9	6	1.3336	0.072	0.039	0.020	1.009	1.012	98.89		
1.00	1.0005	1.0023	10.0	0.109	990.5	7.7	12	1.3342	0.180	0.097	0.051	1.020	1.022	97.84		
2.00	1.0028	1.0046	20.1	0.218	982.7	15.5	23	1.3353	0.411	0.221	0.119	1.046	1.045	95.41		
3.00	1.0051	1.0069	30.2	0.327	974.9	23.3	35	1.3365	0.627	0.337	0.182	1.072	1.068	93.12		
4.00	1.0074	1.0092	40.3	0.438	967.1	31.2	46	1.3376	0.849	0.456	0.247	1.098	1.092	90.93		

A% by wt.	ρ D_4^{20}	D_{20}^{20}	C_s g/l	M g-mol/l	C_w g/l	$(C_o - C_w)$ g/l	$(n - n_o) \times 10^4$	n	Δ °C	O Os/kg	S g-mol/l	η/η_o	η/ρ cS	φ rhe	γ mmho/cm	T g-mol/l
5.00	1.0097	1.0115	50.5	0.548	959.2	39.0	58	1.3388	1.078	0.580	0.315	1.125	1.116	88.71		
6.00	1.0120	1.0138	60.7	0.659	951.3	46.9	70	1.3400	1.316	0.708	0.385	1.155	1.143	86.44		
7.00	1.0144	1.0162	71.0	0.771	943.4	54.9	82	1.3412	1.561	0.839	0.457	1.186	1.171	84.17		
8.00	1.0167	1.0185	81.3	0.883	935.4	62.9	94	1.3424	1.811	0.974	0.530	1.218	1.201	81.90		
9.00	1.0191	1.0209	91.7	0.996	927.4	70.9	106	1.3436	2.064	1.110	0.603	1.253	1.232	79.67		
10.00	1.0215	1.0233	102.1	1.109	919.3	78.9	118	1.3448	2.323	1.249	0.678	1.288	1.263	77.48		
12.00	1.0262	1.0281	123.1	1.337	903.1	95.1	142	1.3472	2.880	1.548	0.837	1.362	1.330	73.28		
14.00	1.0311	1.0329	144.4	1.568	886.7	111.5	167	1.3496	3.469	1.865	1.004	1.442	1.401	69.22		
16.00	1.0360	1.0378	165.8	1.800	870.2	128.0	191	1.3521	4.094	2.201	1.177	1.530	1.480	65.22		
18.00	1.0409	1.0428	187.4	2.035	853.6	144.7	217	1.3547	4.756	2.557	1.359	1.627	1.566	61.34		
20.00	1.0459	1.0478	209.2	2.272	836.8	161.5	242	1.3572	5.46	2.93	1.546	1.734	1.661	57.56		
24.00	1.0561	1.0580	253.5	2.752	802.6	195.6	294	1.3624	7.01	3.77	1.944	1.984	1.882	50.31		
28.00	1.0664	1.0683	298.6	3.243	767.8	230.4	347	1.3676	8.77	4.71	2.370	2.274	2.136	43.89		
32.00	1.0770	1.0789	344.6	3.742	732.3	265.9	400	1.3730	10.74	5.78	2.814	2.632	2.449	37.91		
36.00	1.0876	1.0896	391.5	4.252	696.1	302.2	455	1.3785	12.96	6.97	3.276	3.082	2.839	32.38		
40.00	1.0984	1.1003	439.4	4.771	659.0	339.2	511	1.3841	15.50	8.33	3.757	3.646	3.326	27.37		
44.00	1.1092	1.1112	488.1	5.300	621.2	377.1	567	1.3897				4.434	4.005	22.51		
48.00	1.1200	1.1220	537.6	5.838	582.4	415.8	624	1.3954				5.402	4.833	18.47		
52.00	1.1308	1.1328	588.0	6.385	542.8	455.4	681	1.4011				6.653	5.895	15.00		
56.00	1.1419	1.1439	639.4	6.944	502.4	495.8	739	1.4069				8.332	7.311	11.98		
60.00	1.1530	1.1551	691.8	7.513	461.2	537.0	799	1.4129				10.66	9.264	9.36		
64.00	1.1643	1.1663	745.1	8.091	419.1	579.1	859	1.4189				13.63	11.73	7.32		
68.00	1.1755	1.1775	799.3	8.680	376.1	622.1	919	1.4249				18.42	15.70	5.42		
72.00	1.1866	1.1887	854.3	9.277	332.2	666.0	980	1.4310				27.57	23.28	3.62		
76.00	1.1976	1.1997	910.2	9.883	287.4	710.8	1040	1.4370				40.49	33.88	2.46		
80.00	1.2085	1.2106	966.8	10.498	241.7	756.5	1101	1.4431				59.78	49.57	1.67		
84.00	1.2192	1.2214	1024.2	11.121	195.1	803.2	1162	1.4492				84.17	69.18	1.19		
88.00	1.2299	1.2320	1082.3	11.752	147.6	850.7	1223	1.4553				147.2	119.9	0.68		
92.00	1.2404	1.2426	1141.1	12.392	99.2	899.0	1284	1.4613				383.7	310.0	0.26		
96.00	1.2508	1.2530	1200.7	13.039	50.0	948.2	1344	1.4674				778.9	624.0	0.13		
100.00	1.2611	1.2633	1261.1	13.694	0.0	998.2	1405	1.4735				1759.6	1398.1	0.06		

24 HYDROCHLORIC ACID, HCl

MOLECULAR WEIGHT = 36.47
RELATIVE SPECIFIC REFRACTIVITY = 1.152

0.00 % by wt. data are the same for all compounds.
For Values of 0.00 wt. % solutions see Table 1, Acetic Acid.

A% by wt.	ρ D_4^{20}	D_{20}^{20}	C_s g/l	M g-mol/l	C_w g/l	$(C_o - C_w)$ g/l	$(n - n_o) \times 10^4$	n	Δ °C	O Os/kg	S g-mol/l	η/η_o	η/ρ cS	φ rhe	γ mmho/cm	T g-mol/l
0.50	1.0007	1.0025	5.0	0.137	995.7	2.5	12	1.3341	0.486	0.261	0.141	1.006	1.008	99.16	45.1	0.537
1.00	1.0031	1.0049	10.0	0.275	993.1	5.1	23	1.3353	0.989	0.532	0.289	1.013	1.012	98.50	92.9	1.23
1.50	1.0056	1.0074	15.1	0.414	990.5	7.7	35	1.3365	1.519	0.817	0.444	1.020	1.016	97.84	140.	2.10
2.00	1.0081	1.0098	20.2	0.553	987.9	10.3	46	1.3376	2.076	1.116	0.607	1.027	1.021	97.17	183.	3.14
2.50	1.0105	1.0123	25.3	0.693	985.3	13.0	58	1.3388	2.662	1.431	0.775	1.034	1.026	96.49	220.	4.85
3.00	1.0130	1.0148	30.4	0.833	982.6	15.6	69	1.3399	3.276	1.761	0.949	1.042	1.030	95.81		
3.50	1.0154	1.0172	35.5	0.975	979.9	18.3	81	1.3411	3.916	2.105	1.128	1.049	1.035	95.12		
4.00	1.0179	1.0197	40.7	1.116	977.2	21.0	92	1.3422	4.579	2.462	1.311	1.057	1.040	94.43		
4.50	1.0204	1.0222	45.9	1.259	974.4	23.8	104	1.3434	5.27	2.83	1.496	1.065	1.046	93.73		
5.00	1.0228	1.0246	51.1	1.402	971.7	26.5	115	1.3445	5.98	3.22	1.683	1.073	1.051	93.04		
5.50	1.0253	1.0271	56.4	1.546	968.9	29.3	127	1.3457	6.73	3.62	1.874	1.081	1.056	92.34		
6.00	1.0278	1.0296	61.7	1.691	966.1	32.1	138	1.3468	7.52	4.04	2.070	1.089	1.062	91.64		
6.50	1.0302	1.0321	67.0	1.836	963.3	35.0	150	1.3480	8.34	4.49	2.269	1.097	1.067	90.94		
7.00	1.0327	1.0345	72.3	1.982	960.4	37.8	162	1.3491	9.22	4.96	2.474	1.106	1.073	90.24		
7.50	1.0352	1.0370	77.6	2.129	957.5	40.7	173	1.3503	10.14	5.45	2.681	1.115	1.079	89.54		
8.00	1.0377	1.0395	83.0	2.276	954.6	43.6	185	1.3515	11.10	5.97	2.890	1.123	1.085	88.85		
8.50	1.0401	1.0420	88.4	2.424	951.7	46.5	196	1.3526	12.10	6.51	3.101	1.132	1.091	88.15		
9.00	1.0426	1.0445	93.8	2.573	948.8	49.5	208	1.3538	13.15	7.07	3.313	1.141	1.097	87.46		
9.50	1.0451	1.0469	99.3	2.722	945.8	52.4	219	1.3549	14.25	7.66	3.526	1.150	1.103	86.77		
10.00	1.0476	1.0494	104.8	2.872	942.8	55.4	231	1.3561	15.40	8.28	3.740	1.159	1.109	86.08		
11.00	1.0526	1.0544	115.8	3.175	936.8	61.4	254	1.3584	17.85	9.60	4.166	1.178	1.122	84.71		
12.00	1.0576	1.0594	126.9	3.480	930.7	67.6	277	1.3607	20.51	11.03	4.590	1.197	1.134	83.36		
13.00	1.0626	1.0645	138.1	3.788	924.4	73.8	300	1.3630				1.217	1.148	82.01		
14.00	1.0676	1.0695	149.5	4.098	918.1	80.1	323	1.3653				1.237	1.161	80.68		
15.00	1.0726	1.0745	160.9	4.412	911.8	86.5	347	1.3676				1.258	1.175	79.35		
16.00	1.0777	1.0796	172.4	4.728	905.3	93.0	370	1.3700				1.279	1.189	78.03		
17.00	1.0828	1.0847	184.1	5.047	898.7	99.5	393	1.3723				1.301	1.204	76.72		
18.00	1.0878	1.0898	195.8	5.369	892.0	106.2	416	1.3746				1.323	1.219	75.42		
19.00	1.0929	1.0949	207.7	5.694	885.3	113.0	439	1.3769				1.347	1.235	74.11		
20.00	1.0980	1.1000	219.6	6.022	878.4	119.8	462	1.3792				1.371	1.251	72.80		
22.00	1.1083	1.1102	243.8	6.686	864.5	133.8	509	1.3838				1.423	1.286	70.15		
24.00	1.1185	1.1205	268.4	7.361	850.1	148.1	555	1.3884				1.480	1.326	67.44		
26.00	1.1288	1.1308	293.5	8.047	835.3	162.9	600	1.3930				1.544	1.371	64.63		
28.00	1.1391	1.1411	318.9	8.745	820.1	178.1	646	1.3976				1.617	1.423	61.71		
30.00	1.1492	1.1513	344.8	9.454	804.5	193.8	691	1.4020				1.702	1.484	58.64		
32.00	1.1594	1.1614	371.0	10.173	788.4	209.9	736	1,4066				1.795	1.551	55.61		
34.00	1.1693	1.1714	397.6	10.901	771.8	226.5	782	1.4112				1.896	1.625	52.63		
36.00	1.1791	1.1812	424.5	11.639	754.6	243.6	828	1.4158				1.998	1.698	49.95		
38.00	1.1886	1.1907	451.7	12.385	736.9	261.3	874	1.4204				2.101	1.771	47.51		
40.00	1.1977	1.1999	479.1	13.137	718.6	279.6	920	1.4250								

25 INULIN, $(C_6H_{10}O_5)_x$*

MOLECULAR WEIGHT, AVERAGE = 5200
RELATIVE SPECIFIC REFRACTIVITY = 1.034

0.00 % by wt. data are the same for all compounds.
For Values of 0.00 wt. % solutions see Table 1, Acetic Acid.

A% by wt.	ρ D_4^{20}	D_{20}^{20}	C_s g/l	M g-mol/l	C_w g/l	$(C_o - C_w)$ g/l	$(n - n_o) \times 10^4$	n	Δ °C	O Os/kg	S g-mol/l	η/η_o	η/ρ cS	φ rhe	γ mmho/cm	T g-mol/l
0.50	1.0001	1.0019	5.0	0.001	995.1	3.1	7	1.3337	0.003	0.001	0.001	1.030	1.032	96.93		
1.00	1.0020	1.0038	10.0	0.002	992.0	6.3	14	1.3344	0.005	0.003	0.001	1.059	1.059	94.24		
1.50	1.0038	1.0056	15.1	0.003	988.8	9.5	22	1.3351	0.007	0.004	0.002	1.088	1.086	91.74		
2.00	1.0057	1.0074	20.1	0.004	985.6	12.7	29	1.3359	0.010	0.005	0.003	1.117	1.113	89.35		
2.50	1.0075	1.0093	25.2	0.005	982.3	15.9	36	1.3366	0.012	0.007	0.003	1.147	1.140	87.03		
3.00	1.0093	1.0111	30.3	0.006	979.0	19.2	43	1.3373	0.015	0.008	0.004	1.177	1.168	84.81		
3.50	1.0112	1.0129	35.4	0.007	975.8	22.5	51	1.3381	0.017	0.009	0.005	1.208	1.197	82.65		
4.00	1.0130	1.0148	40.5	0.008	972.5	25.8	58	1.3388	0.019	0.010	0.005	1.240	1.226	80.51		
4.50	1.0148	1.0166	45.7	0.009	969.2	29.1	65	1.3395	0.022	0.012	0.006	1.274	1.258	78.36		
5.00	1.0167	1.0185	50.8	0.010	965.9	32.4	73	1.3403	0.024	0.013	0.007	1.310	1.291	76.18		
5.50	1.0186	1.0204	56.0	0.011	962.6	35.7	80	1.3410	0.027	0.015	0.007	1.349	1.327	74.00		
6.00	1.0205	1.0223	61.2	0.012	959.3	38.9	88	1.3418	0.030	0.016	0.008	1.389	1.364	71.83		
6.50	1.0225	1.0243	66.5	0.013	956.0	42.2	95	1.3425	0.033	0.018	0.009	1.432	1.403	69.69		
7.00	1.0245	1.0263	71.7	0.014	952.7	45.5	103	1.3433	0.036	0.019	0.010	1.477	1.444	67.59		
7.50	1.0265	1.0283	77.0	0.015	949.5	48.7	110	1.3440	0.039	0.021	0.011	1.523	1.487	65.52		
8.00	1.0286	1.0304	82.3	0.016	946.3	52.0	118	1.3448	0.043	0.023	0.012	1.572	1.531	63.49		
8.50	1.0307	1.0325	87.6	0.017	943.1	55.2	125	1.3455	0.047	0.025	0.013	1.623	1.578	61.50		
9.00	1.0329	1.0347	93.0	0.018	939.9	58.3	133	1.3463	0.051	0.027	0.014	1.676	1.626	59.56		
9.50	1.0351	1.0370	98.3	0.019	936.8	61.4	141	1.3471	0.055	0.030	0.015	1.731	1.675	57.67		
10.00	1.0374	1.0393	103.7	0.020	933.7	64.5	148	1.3478	0.060	0.032	0.017	1.788	1.727	55.82		

* Supersaturated solutions.

26 LACTIC ACID, $CH_3CHOHCOOH$, (D-PROPANOIC ACID, 2-HYDROXY)

MOLECULAR WEIGHT = 90.08
RELATIVE SPECIFIC REFRACTIVITY = 1.058

0.00 % by wt. data are the same for all compounds.
For Values of 0.00 wt. % solutions see Table 1, Acetic Acid.

A% by wt.	ρ D_4^{20}	D_{20}^{20}	C_s g/l	M g-mol/l	C_w g/l	$(C_o - C_w)$ g/l	$(n - n_o) \times 10^4$	n	Δ °C	O Os/kg	S g-mol/l	η/η_o	η/ρ cS	φ rhe	γ mmho/cm	T g-mol/l
0.50	0.9992	1.0010	05.0	0.055	994.2	4.0	5	1.3335	0.095	0.051	0.027	1.012	1.015	98.62	1.0	0.010
1.00	1.0002	1.0020	10.0	0.111	990.2	8.0	10	1.3340	0.190	0.102	0.054	1.025	1.027	97.37	1.8	0.018
2.00	1.0023	1.0040	20.0	0.223	982.2	16.0	20	1.3350	0.378	0.203	0.109	1.054	1.053	94.69	2.6	0.026
3.00	1.0043	1.0061	30.1	0.334	974.2	24.0	30	1.3360	0.567	0.305	0.165	1.082	1.078	92.27	3.2	0.031
4.00	1.0065	1.0083	40.3	0.447	966.2	32.0	40	1.3370	0.758	0.407	0.221	1.108	1.102	90.05	3.5	0.035
5.00	1.0086	1.0104	50.4	0.560	958.2	40.0	50	1.3380	0.953	0.512	0.278	1.136	1.127	87.85	3.8	0.038
6.00	1.0108	1.0126	60.6	0.673	950.2	48.1	60	1.3390	1.155	0.621	0.337	1.165	1.154	85.64	4.1	0.041
7.00	1.0131	1.0149	70.9	0.787	942.1	56.1	70	1.3400	1.361	0.732	0.398	1.196	1.181	83.47	4.4	0.043
8.00	1.0153	1.0171	81.2	0.902	934.1	64.1	80	1.3410	1.573	0.846	0.460	1.227	1.210	81.34	4.6	0.045
9.00	1.0176	1.0194	91.6	1.017	926.0	72.2	90	1.3420	1.791	0.963	0.524	1.259	1.239	79.24	4.7	0.047
10.00	1.0199	1.0217	102.0	1.132	917.9	80.3	100	1.3430	2.015	1.084	0.589	1.293	1.269	77.19	4.9	0.049
12.00	1.0246	1.0264	123.0	1.365	901.6	96.6	120	1.3450	2.491	1.339	0.726	1.363	1.332	73.20	5.1	0.051
14.00	1.0294	1.0312	144.1	1.600	885.2	113.0	140	1.3470	2.988	1.607	0.868	1.438	1.399	69.38	5.2	0.052
16.00	1.0342	1.0360	165.5	1.837	868.7	129.5	161	1.3491	3.478	1.870	1.006	1.519	1.471	65.71	5.2	0.052
18.00	1.0390	1.0408	187.0	2.076	852.0	146.3	181	1.3511	3.962	2.130	1.141	1.604	1.546	62.22	5.2	0.052
20.00	1.0439	1.0457	208.8	2.318	835.1	163.1	202	1.3532	4.440	2.387	1.273	1.696	1.628	58.84	5.1	0.051
24.00	1.0536	1.0554	252.9	2.807	800.7	197.5	243	1.3573				1.898	1.806	52.57	4.9	0.048
28.00	1.0632	1.0651	297.7	3.305	765.5	232.7	285	1.3615				2.132	2.010	46.80	4.5	0.045
32.00	1.0728	1.0747	343.3	3.811	729.5	268.8	327	1.3657				2.409	2.251	41.43	4.1	0.041
36.00	1.0822	1.0841	389.6	4.325	692.6	305.7	370	1.3700				2.725	2.524	36.63	3.7	0.036
40.00	1.0915	1.0934	436.6	4.847	654.9	343.4	413	1.3743				3.108	2.853	32.11	3.2	0.032
44.00	1.1008	1.1028	484.4	5.377	616.5	381.8	456	1.3786				3.559	3.239	28.04	2.8	0.027
48.00	1.1105	1.1125	533.0	5.917	577.5	420.8	498	1.3828				4.098	3.699	24.35	2.3	0.022
52.00	1.1201	1.1221	582.4	6.466	537.6	460.6	541	1.3871				4.779	4.276	20.89	1.8	0.018
56.00	1.1297	1.1317	632.6	7.023	497.0	501.2	584	1.3914				5.568	4.940	17.93	1.5	0.014
60.00	1.1392	1.1412	683.5	7.588	455.7	542.6	628	1.3958				6.666	5.865	14.97	1.1	0.011
64.00	1.1486	1.1506	735.1	8.160	413.5	584.7	671	1.4001				8.008	6.988	12.46	0.8	0.008
68.00	1.1579	1.1599	787.4	8.741	370.5	627.7	715	1.4045				9.843	8.519	10.14	0.5	0.005
72.00	1.1670	1.1691	840.3	9.328	326.8	671.5	758	1.4088				12.84	11.02	7.77	0.3	0.003
76.00	1.1760	1.1781	893.8	9.922	282.2	716.0	801	1.4131				16.94	14.43	5.89	0.1	0.001
80.00	1.1848	1.1869	947.9	10.523	237.0	761.3	843	1.4173				22.12	18.70	4.51	0.1	0.001

27 LACTOSE, $C_{12}H_{22}O_{11} \cdot 1H_2O$

MOLECULAR WEIGHT = 342.30 FORMULA WEIGHT, HYDRATE = 360.31
RELATIVE SPECIFIC REFRACTIVITY = 1.036

0.00 % by wt. data are the same for all compounds.
For Values of 0.00 wt. % solutions see Table 1, Acetic Acid.

A% by wt.	H% by wt.	ρ D_4^{20}	D_{20}^{20}	C_s g/l	M g-mol/l	C_w g/l	$(C_o - C_w)$ g/l	$(n - n_o) \times 10^4$	n	Δ °C	O Os/kg	S g-mol/l	η/η_o	η/ρ cS	φ rhe	γ mmho/cm	T g-mol/l
0.50	0.53	1.0002	1.0019	5.0	0.015	995.2	3.1	7	1.3337	0.027	0.015	0.007	1.011	1.013	98.71		
1.00	1.05	1.0021	1.0039	10.0	0.029	992.1	6.1	15	1.3345	0.055	0.030	0.015	1.024	1.024	97.46		
1.50	1.58	1.0041	1.0059	15.1	0.044	989.0	9.2	22	1.3352	0.083	0.045	0.023	1.039	1.037	96.01		
2.00	2.11	1.0061	1.0079	20.1	0.059	986.0	12.3	30	1.3359	0.112	0.060	0.031	1.056	1.052	94.51		
2.50	2.63	1.0081	1.0099	25.2	0.074	982.9	15.3	37	1.3367	0.140	0.075	0.040	1.072	1.065	93.12		
3.00	3.16	1.0102	1.0119	30.3	0.089	979.8	18.4	45	1.3375	0.169	0.091	0.048	1.087	1.078	91.80		
3.50	3.68	1.0122	1.0140	35.4	0.103	976.8	21.5	53	1.3382	0.198	0.107	0.056	1.103	1.091	90.52		
4.00	4.21	1.0143	1.0160	40.6	0.119	973.7	24.6	60	1.3390	0.228	0.122	0.065	1.118	1.105	89.25		
4.50	4.74	1.0163	1.0181	45.7	0.134	970.6	27.7	68	1.3398	0.258	0.139	0.074	1.135	1.119	87.96		
5.00	5.26	1.0184	1.0202	50.9	0.149	967.5	30.8	76	1.3406	0.288	0.155	0.083	1.152	1.133	86.63		
5.50	5.79	1.0204	1.0223	56.1	0.164	964.3	33.9	84	1.3413	0.319	0.172	0.092	1.170	1.149	85.27		

27 LACTOSE, $C_{12}H_{22}O_{11} \cdot 1H_2O$—(Continued)

A% by wt.	H% by wt.	ρ D_4^{20}	D_{20}^{20}	C_s g/l	M g-mol/l	C_w g/l	$(C_o - C_w)$ g/l	$(n - n_o)$ × 10^4	n	Δ °C	O Os/kg	S g-mol/l	η/η_o	η/ρ cS	φ rhe	γ mmho/cm	T g-mol/l
6.00	6.32	1.0225	1.0243	61.4	0.179	961.2	37.1	91	1.3421	0.351	0.189	0.101	1.189	1.166	83.91		
6.50	6.84	1.0246	1.0264	66.6	0.195	958.0	40.2	99	1.3429	0.385	0.207	0.111	1.209	1.183	82.54		
7.00	7.37	1.0267	1.0285	71.9	0.210	954.8	43.4	107	1.3437	0.420	0.226	0.121	1.230	1.200	81.16		
7.50	7.89	1.0287	1.0305	77.2	0.225	951.6	46.7	115	1.3445	0.456	0.245	0.132	1.251	1.218	79.79		
8.00	8.42	1.0308	1.0326	82.5	0.241	948.3	49.9	123	1.3453	0.495	0.266	0.143	1.273	1.237	78.42		
8.50	8.95	1.0329	1.0347	87.8	0.256	945.1	53.2	130	1.3460				1.295	1.257	77.05		
9.00	9.47	1.0349	1.0367	93.1	0.272	941.8	56.5	138	1.3468				1.318	1.277	75.69		
9.50	10.00	1.0370	1.0388	98.5	0.288	938.5	59.8	146	1.3476				1.342	1.297	74.34		
10.00	10.53	1.0390	1.0409	103.9	0.304	935.1	63.1	154	1.3484				1.367	1.318	73.01		
11.00	11.58	1.0432	1.0450	114.7	0.335	928.4	69.8	170	1.3500				1.418	1.362	70.37		
12.00	12.63	1.0473	1.0492	125.7	0.367	921.7	76.6	185	1.3515				1.473	1.409	67.77		
13.00	13.68	1.0515	1.0534	136.7	0.399	914.8	83.4	201	1.3531				1.530	1.458	65.24		
14.00	14.74	1.0558	1.0577	147.8	0.432	908.0	90.2	218	1.3548				1.590	1.509	62.76		
15.00	15.79	1.0602	1.0621	159.0	0.465	901.2	97.0	234	1.3564				1.654	1.563	60.35		
16.00	16.84	1.0648	1.0667	170.4	0.498	894.4	103.8	252	1.3582				1.721	1.619	58.01		
17.00	17.89	1.0696	1.0714	181.8	0.531	887.7	110.5	270	1.3600				1.791	1.678	55.73		
18.00	18.95	1.0746	1.0765	193.4	0.565	881.2	117.1	289	1.3619				1.865	1.739	53.51		

28 LANTHANUM NITRATE, $La(NO_3)_3 \cdot 6H_2O$

MOLECULAR WEIGHT = 324.93 FORMULA WEIGHT, HYDRATE = 433.02

A% by wt.	H% by wt.	ρ D_4^{20}	D_{20}^{20}	C_s g/l	M g-mol/l	C_w g/l	$(C_o - C_w)$ g/l	$(n - n_o)$ × 10^4	n	Δ °C	O Os/kg	S g-mol/l	η/η_o	η/ρ cS	φ rhe	γ mmho/cm	T g-mol/l
1.00	1.33	1.0062	1.0080	10.1	0.031	996.1	2.1	14.	1.3344	0.180	0.097	0.051	1.013	1.009	98.50	7.8	0.080
2.00	2.67	1.0142	1.0160	20.3	0.062	993.9	4.3	28.	1.3358	0.358	0.192	0.103	1.025	1.013	97.33	14.2	0.151
3.00	4.00	1.0226	1.0244	30.7	0.094	991.9	6.3	42.	1.3372	0.525	0.283	0.152	1.040	1.019	95.97	20.1	0.221
4.00	5.33	1.0309	1.0328	41.2	0.127	989.7	8.5	57.	1.3387	0.698	0.375	0.203	1.056	1.026	94.54	25.6	0.289
5.00	6.66	1.0394	1.0413	52.0	0.160	987.4	10.8	71.	1.3401	0.889	0.478	0.259	1.074	1.035	92.96	31.1	0.358
6.00	8.00	1.0483	1.0502	62.9	0.194	985.4	12.8	86.	1.3416	1.079	0.580	0.315	1.091	1.043	91.46	36.0	0.420
7.00	9.33	1.0575	1.0594	74.0	0.228	983.5	14.7	101.	1.3431	1.266	0.681	0.370	1.110	1.052	89.89	40.6	0.479
8.00	10.66	1.0667	1.0686	85.3	0.263	981.4	16.8	115.	1.3445	1.463	0.787	0.428	1.130	1.061	88.36	45.0	0.535
9.00	11.99	1.0761	1.0780	96.8	0.298	979.3	18.9	130.	1.3460	1.676	0.901	0.490	1.149	1.070	86.85	49.5	0.596
10.00	13.33	1.0853	1.0873	108.5	0.334	976.8	21.4	145.	1.3475	1.874	1.007	0.548	1.170	1.080	85.32	53.4	0.648
12.00	15.99	1.1055	1.1075	132.7	0.408	972.8	25.4	176.	1.3506	2.340	1.258	0.683	1.214	1.100	82.23	60.7	0.750
14.00	18.66	1.1251	1.1271	157.5	0.485	967.6	30.6	208.	1.3538	2.822	1.517	0.821	1.262	1.124	79.08	67.4	0.845
16.00	21.32	1.1456	1.1477	183.3	0.564	962.3	35.9	241.	1.3571	3.331	1.791	0.965	1.322	1.156	75.51	73.2	0.930
18.00	23.99	1.1667	1.1688	210.0	0.646	956.7	41.5	274.	1.3604	3.869	2.080	1.115	1.390	1.194	71.79	77.7	0.999
20.00	26.65	1.1892	1.1914	237.9	0.732	951.3	46.9	308.	1.3638	4.418	2.375	1.267	1.463	1.233	68.20	81.4	1.055
24.00	31.98	1.2337	1.2359	296.1	0.911	937.6	60.6	379.	1.3709				1.641	1.333	60.81	86.7	1.137
28.00	37.31	1.2829	1.2852	359.2	1.105	923.7	74.5	453.	1.3783				1.854	1.448	53.83	86.8	1.138
32.00	42.64	1.3319	1.3343	426.2	1.312	905.7	92.5	530.	1.3860				2.156	1.622	46.29	84.0	1.095
36.00	47.98	1.3874	1.3899	499.5	1.537	887.9	110.3	613.	1.3943				2.580	1.863	38.69	81.4	1.055
40.00	53.31	1.4477	1.4503	579.1	1.782	868.6	129.6	700.	1.4030				3.205	2.218	31.14	71.9	0.911
44.00	58.64	1.5084	1.5111	663.7	2.043	844.7	153.5	788.	1.4118				4.105	2.727	24.31	61.9	0.767

29 LEAD NITRATE, $Pb(NO_3)_2$

MOLECULAR WEIGHT = 331.23
RELATIVE SPECIFIC REFRACTIVITY = 0.479

0.00 % by wt. data are the same for all compounds.
For Values of 0.00 wt. % solutions see Table 1, Acetic Acid.

A% by wt.	ρ D_4^{20}	D_{20}^{20}	C_s g/l	M g-mol/l	C_w g/l	$(C_o - C_w)$ g/l	$(n - n_o)$ × 10^4	n	Δ °C	O Os/kg	S g-mol/l	η/η_o	η/ρ cS	φ rhe	γ mmho/cm	T g-mol/l
0.50	1.0025	1.0043	5.0	0.015	997.5	0.8	6	1.3336	0.072	0.039	0.020	1.003	1.003	99.49	2.8	0.028
1.00	1.0068	1.0086	10.1	0.030	996.7	1.5	12	1.3342	0.137	0.074	0.039	1.006	1.001	99.21	5.4	0.054
2.00	1.0155	1.0173	20.3	0.061	995.2	3.1	24	1.3354	0.258	0.139	0.074	1.011	0.998	98.71	9.6	0.099
3.00	1.0243	1.0261	30.7	0.093	993.6	4.6	36	1.3366	0.372	0.200	0.107	1.016	0.993	98.27	13.3	0.141
4.00	1.0333	1.0352	41.3	0.125	992.0	6.2	49	1.3379	0.482	0.259	0.140	1.020	0.989	97.85	16.7	0.181
5.00	1.0425	1.0444	52.1	0.157	990.4	7.8	61	1.3391	0.587	0.316	0.170	1.025	0.985	97.37	19.9	0.219
6.00	1.0519	1.0537	63.1	0.191	988.7	9.5	73	1.3403	0.690	0.371	0.201	1.031	0.982	96.78	23.0	0.256
7.00	1.0614	1.0632	74.3	0.224	987.1	11.2	86	1.3415	0.792	0.426	0.231	1.038	0.980	96.12	25.9	0.292
8.00	1.0710	1.0729	85.7	0.259	985.4	12.9	98	1.3428	0.891	0.479	0.260	1.046	0.979	95.40	28.6	0.326
9.00	1.0809	1.0828	97.3	0.294	983.6	14.6	110	1.3440	0.987	0.531	0.288	1.054	0.978	94.64	31.3	0.360
10.00	1.0909	1.0929	109.1	0.329	981.8	16.4	124	1.3454	1.081	0.581	0.316	1.063	0.976	93.89	33.9	0.393
12.00	1.1115	1.1135	133.4	0.403	978.1	20.1	151	1.3481	1.265	0.680	0.370	1.081	0.974	92.33	38.9	0.457
14.00	1.1329	1.1349	158.6	0.479	974.3	24.0	180	1.3510	1.450	0.780	0.424	1.101	0.973	90.69	43.4	0.515
16.00	1.1550	1.1571	184.8	0.558	970.2	28.0	209	1.3539	1.639	0.881	0.479	1.122	0.973	88.94	47.5	0.570
18.00	1.1779	1.1800	212.0	0.640	965.9	32.3	239	1.3569	1.825	0.981	0.534	1.146	0.975	87.10	51.4	0.621
20.00	1.2017	1.2038	240.3	0.726	961.4	36.9	271	1.3601	2.008	1.080	0.587	1.172	0.977	85.15	55.2	0.673
24.00	1.2519	1.2541	300.5	0.907	951.5	46.8	337	1.3667	2.361	1.269	0.689	1.233	0.987	80.97	62.5	0.772
28.00	1.3059	1.3082	365.7	1.104	940.3	58.0	407	1.3737	2.714	1.459	0.790	1.306	1.002	76.43	68.4	0.858
32.00	1.3640	1.3664	436.5	1.318	927.5	70.7	481	1.3811				1.398	1.027	71.39	72.5	0.920
34.00	1.3947	1.3972	474.2	1.432	920.5	77.7	520	1.3850				1.459	1.048	68.42	74.8	0.954

30 LITHIUM CHLORIDE, LiCl

MOLECULAR WEIGHT = 42.40
RELATIVE SPECIFIC REFRACTIVITY = 1.023

0.00 % by wt. data are the same for all compounds.
For Values of 0.00 wt. "„ solutions see Table 1, Acetic Acid.

A% by wt.	ρ D_4^{20}	D_{20}^{20}	C_s g/l	M g-mol/l	C_w g/l	$(C_o - C_w)$ g/l	$(n - n_o)$ × 10^4	n	Δ °C	O Os/kg	S g-mol/l	η/η_o	η/ρ cS	φ rhe	γ mmho/cm	T g-mol/l
0.50	1.0012	1.0029	5.0	0.118	996.2	2.1	11	1.3341	0.415	0.223	0.120	1.017	1.018	98.08	10.1	0.105
1.00	1.0041	1.0059	10.0	0.237	994.0	4.2	21	1.3351	0.836	0.450	0.244	1.035	1.033	96.43	19.0	0.208
2.00	1.0099	1.0117	20.2	0.476	989.7	8.5	43	1.3373	1.724	0.927	0.504	1.070	1.062	93.27	34.9	0.406
3.00	1.0157	1.0175	30.5	0.719	985.2	13.0	64	1.3394	2.684	1.443	0.782	1.106	1.091	90.22	49.8	0.600
4.00	1.0215	1.0233	40.9	0.964	980.6	17.6	85	1.3415	3.727	2.004	1.075	1.144	1.122	87.26	63.6	0.791
5.00	1.0272	1.0290	51.4	1.211	975.9	22.4	106	1.3436	4.859	2.612	1.386	1.183	1.154	84.36	76.4	0.978
6.00	1.0330	1.0348	62.0	1.462	971.0	27.2	127	1.3457	6.14	3.30	1.723	1.224	1.187	81.55	88.3	1.16
7.00	1.0387	1.0405	72.7	1.715	966.0	32.2	148	1.3478	7.56	4.07	2.081	1.266	1.221	78.83	99.5	1.34
8.00	1.0444	1.0463	83.6	1.971	960.9	37.3	169	1.3499	9.11	4.90	2.449	1.310	1.257	76.16	110.	1.52
9.00	1.0502	1.0521	94.5	2.229	955.7	42.6	190	1.3520	10.79	5.80	2.825	1.357	1.295	73.52	119.	1.69
10.00	1.0560	1.0578	105.6	2.490	950.4	47.9	211	1.3541	12.61	6.78	3.205	1.408	1.336	70.88	127.	1.84
12.00	1.0675	1.0694	128.1	3.021	939.4	58.8	253	1.3583	16.59	8.92	3.951	1.519	1.426	65.68	140.	2.12
14.00	1.0792	1.0811	151.1	3.563	928.1	70.1	295	1.3625	21.04	11.31	4.670	1.644	1.527	60.70	151.	2.34
16.00	1.0910	1.0929	174.6	4.117	916.4	81.8	338	1.3668				1.783	1.637	55.99	160.	2.56
18.00	1.1029	1.1048	198.5	4.682	904.4	93.9	381	1.3711				1.938	1.761	51.50	167.	2.73
20.00	1.1150	1.1170	223.0	5.260	892.0	106.2	425	1.3755				2.124	1.909	46.99	170.	2.80
22.00	1.1274	1.1294	248.0	5.850	879.3	118.9	469	1.3799				2.336	2.076	42.72	170.	2.79
24.00	1.1399	1.1419	273.6	6.452	866.3	131.9	514	1.3844				2.595	2.281	38.46	167.	2.73
26.00	1.1527	1.1548	299.7	7.069	853.0	145.2	560	1.3890				2.919	2.537	34.19	163.	2.63
28.00	1.1658	1.1678	326.4	7.699	839.4	158.9	606	1.3936				3.311	2.846	30.14	156.	2.47
30.00	1.1791	1.1812	353.7	8.343	825.4	172.9	653	1.3983				3.777	3.210	26.42	146.	2.25

31 MAGNESIUM CHLORIDE, $MgCl_2 \cdot 6H_2O$

MOLECULAR WEIGHT = 95.23 FORMULA WEIGHT, HYDRATE = 203.33
RELATIVE SPECIFIC REFRACTIVITY = 0.893

0.00 % by wt. data are the same for all compounds.
For Values of 0.00 wt. % solutions see Table 1, Acetic Acid.

A% by wt.	H% by wt.	ρ D_4^{20}	D_{20}^{20}	C_s g/l	M g-mol/l	C_w g/l	$(C_o - C_w)$ g/l	$(n - n_o) \times 10^4$	n	Δ °C	O Os/kg	S g-mol/l	η/η_o	η/ρ cS	φ rhe	γ mmho/cm	T g-mol/l
0.50	1.07	1.0022	1.0040	5.0	0.053	997.2	1.0	13	1.3343	0.255	0.137	0.073	1.022	1.022	97.66	8.6	0.092
1.00	2.14	1.0062	1.0080	10.1	0.106	996.2	2.0	26	1.3356	0.517	0.278	0.150	1.044	1.040	95.59	16.6	0.183
2.00	4.27	1.0144	1.0162	20.3	0.213	994.1	4.1	51	1.3381	1.063	0.572	0.311	1.089	1.076	91.60	31.2	0.356
3.00	6.41	1.0226	1.0244	30.7	0.322	991.9	6.3	76	1.3406	1.653	0.889	0.483	1.137	1.114	87.80	44.5	0.530
4.00	8.54	1.0309	1.0328	41.2	0.433	989.7	8.5	102	1.3432	2.297	1.235	0.671	1.186	1.153	84.13	56.4	0.690
5.00	10.68	1.0394	1.0412	52.0	0.546	987.4	10.8	127	1.3457	3.012	1.619	0.875	1.239	1.194	80.57	66.9	0.840
6.00	12.81	1.0479	1.0497	62.9	0.660	985.0	13.2	153	1.3483				1.295	1.238	77.09	77.2	0.990
7.00	14.95	1.0564	1.0583	74.0	0.777	982.5	15.7	178	1.3508				1.355	1.285	73.67	86.4	1.13
8.00	17.08	1.0651	1.0670	85.2	0.895	979.9	18.3	204	1.3534				1.420	1.336	70.30	94.5	1.26
9.00	19.22	1.0738	1.0757	96.6	1.015	977.2	21.1	231	1.3560				1.490	1.390	66.98	102.0	1.38
10.00	21.35	1.0826	1.0845	108.3	1.137	974.4	23.9	257	1.3587				1.567	1.450	63.70	108.0	1.49
12.00	25.62	1.1005	1.1025	132.1	1.387	968.5	29.8	311	1.3641				1.742	1.586	57.30	119.0	1.69
14.00	29.89	1.1189	1.1209	156.6	1.645	962.3	36.0	365	1.3695				1.952	1.748	51.13	127.0	1.84
16.00	34.16	1.1372	1.1392	181.9	1.911	955.2	43.0	419	1.3749				2.203	1.941	45.30	132.0	1.94
18.00	38.43	1.1553	1.1574	208.0	2.184	947.4	50.8	474	1.3804				2.502	2.170	39.89	134.0	1.99
20.00	42.70	1.1742	1.1763	234.8	2.466	939.4	58.9	529	1.3859				2.861	2.441	34.89	134.0	1.99
22.00	46.97	1.1938	1.1959	262.6	2.758	931.2	67.1	585	1.3915				3.316	2.783	30.10	131.0	1.93
24.00	51.24	1.2140	1.2162	291.4	3.060	922.6	75.6	642	1.3972				3.909	3.226	25.53	126.0	1.81
26.00	55.51	1.2346	1.2368	321.0	3.371	913.6	84.6	700	1.4030				4.685	3.802	21.30	118.0	1.67
28.00	59.78	1.2555	1.2577	351.5	3.691	903.9	94.3	759	1.4089				5.698	4.547	17.52	109.0	1.50
30.00	64.05	1.2763	1.2786	382.9	4.021	893.4	104.8	818	1.4148				7.003	5.498	14.25	97.9	1.32

32 MAGNESIUM SULFATE, $MgSO_4 \cdot 7H_2O$

MOLECULAR WEIGHT = 120.37 FORMULA WEIGHT, HYDRATE = 246.48
RELATIVE SPECIFIC REFRACTIVITY = 0.572

0.00 % by wt. data are the same for all compounds.
For Values of 0.00 wt. % solutions see Table 1, Acetic Acid.

A% by wt.	H% by wt.	ρ D_4^{20}	D_{20}^{20}	C_s g/l	M g-mol/l	C_w g/l	$(C_o - C_w)$ g/l	$(n - n_o) \times 10^4$	n	Δ °C	O Os/kg	S g-mol/l	η/η_o	η/ρ cS	φ rhe	γ mmho/cm	T g-mol/l
0.50	1.02	1.0033	1.0051	5.0	0.042	998.3	0.0	10	1.3340	0.103	0.055	0.029	1.025	1.024	97.35	4.1	0.041
1.00	2.05	1.0084	1.0102	10.1	0.084	998.3	−0.1	20	1.3350	0.192	0.103	0.055	1.052	1.045	94.88	7.6	0.078
2.00	4.10	1.0186	1.0204	20.4	0.169	998.2	0.0	41	1.3371	0.361	0.194	0.104	1.110	1.092	89.91	13.3	0.141
3.00	6.15	1.0289	1.0307	30.9	0.256	998.0	0.2	61	1.3391	0.522	0.281	0.151	1.175	1.144	84.94	18.4	0.201
4.00	8.19	1.0392	1.0411	41.6	0.345	997.6	0.6	81	1.3411	0.692	0.372	0.201	1.247	1.202	80.05	23.1	0.258
5.00	10.24	1.0497	1.0515	52.5	0.436	997.2	1.1	101	1.3431	0.868	0.467	0.253	1.325	1.265	75.32	27.4	0.311
6.00	12.29	1.0602	1.0621	63.6	0.528	996.6	1.6	121	1.3451	1.048	0.564	0.306	1.408	1.330	70.90	31.1	0.358
7.00	14.34	1.0708	1.0727	75.0	0.623	995.9	2.3	142	1.3471	1.235	0.664	0.361	1.495	1.399	66.77	34.4	0.399
8.00	16.39	1.0816	1.0835	86.5	0.719	995.1	3.2	162	1.3492	1.431	0.770	0.419	1.590	1.473	62.75	37.3	0.436
9.00	18.44	1.0924	1.0944	98.3	0.817	994.1	4.1	182	1.3512	1.635	0.879	0.478	1.699	1.558	58.75	40.0	0.471
10.00	20.48	1.1034	1.1053	110.3	0.917	993.0	5.2	202	1.3532	1.849	0.994	0.541	1.825	1.657	54.69	42.7	0.506
12.00	24.57	1.1257	1.1276	135.1	1.122	990.6	7.7	242	1.3572	2.313	1.243	0.675	2.100	1.869	47.53	48.4	0.581
14.00	28.67	1.1484	1.1504	160.8	1.335	987.6	10.6	283	1.3613	2.861	1.538	0.832	2.407	2.100	41.46	53.3	0.647
16.00	32.76	1.1717	1.1737	187.5	1.557	984.2	14.0	324	1.3654	3.673	1.974	1.061	2.803	2.397	35.61	55.2	0.673
18.00	36.86	1.1955	1.1976	215.2	1.787	980.3	17.9	364	1.3694				3.353	2.810	29.77	53.7	0.652
20.00	40.95	1.2198	1.2220	244.0	2.026	975.9	22.4	405	1.3735				4.139	3.400	24.11	51.1	0.617
22.00	45.05	1.2447	1.2469	273.8	2.275	970.9	27.4	446	1.3776				5.189	4.177	19.23	48.8	0.586
24.00	49.14	1.2701	1.2724	304.8	2.532	965.3	32.9	487	1.3817				6.485	5.116	15.39	45.9	0.547
26.00	53.24	1.2961	1.2984	337.0	2.799	959.1	39.1	528	1.3858				8.050	6.223	12.40	42.3	0.500

33 MALTOSE, $C_{12}H_{22}O_{11} \cdot 1H_2O$

MOLECULAR WEIGHT = 342.29 FORMULA WEIGHT, HYDRATE = 360.31
RELATIVE SPECIFIC REFRACTIVITY = 1.038

0.00 % by wt. data are the same for all compounds.
For Values of 0.00 wt. % solutions see Table 1, Acetic Acid.

A% by wt.	H% by wt.	ρ D_4^{20}	D_{20}^{20}	C_s g/l	M g-mol/l	C_w g/l	$(C_o - C_w)$ g/l	$(n - n_o) \times 10^4$	n	Δ °C	O Os/kg	S g-mol/l	η/η_o	η/ρ cS	φ rhe	γ mmho/cm	T g-mol/l
0.50	0.53	1.0003	1.0020	5.0	0.015	995.3	3.0	7	1.3337	0.027	0.015	0.007	1.014	1.016	98.45		
1.00	1.05	1.0023	1.0041	10.0	0.029	992.3	6.0	15	1.3345	0.055	0.030	0.015	1.028	1.028	97.08		
1.50	1.58	1.0043	1.0061	15.1	0.044	989.2	9.0	22	1.3352	0.083	0.045	0.023	1.043	1.040	95.71		
2.00	2.11	1.0063	1.0081	20.1	0.059	986.2	12.0	29	1.3359	0.112	0.060	0.031	1.058	1.053	94.33		
2.50	2.63	1.0083	1.0101	25.2	0.074	983.1	15.1	37	1.3367	0.140	0.075	0.040	1.074	1.067	92.95		
3.00	3.16	1.0104	1.0121	30.3	0.089	980.0	18.2	44	1.3374	0.169	0.091	0.048	1.090	1.081	91.56		
3.50	3.68	1.0124	1.0142	35.4	0.104	976.9	21.3	52	1.3382	0.199	0.107	0.057	1.107	1.095	90.17		
4.00	4.21	1.0144	1.0162	40.6	0.119	973.8	24.4	59	1.3389	0.229	0.123	0.065	1.124	1.110	88.78		
4.50	4.74	1.0164	1.0182	45.7	0.134	970.7	27.6	67	1.3397	0.259	0.139	0.074	1.142	1.126	87.41		
5.00	5.26	1.0184	1.0202	50.9	0.149	967.5	30.7	74	1.3404	0.290	0.156	0.083	1.160	1.141	86.03		
5.50	5.79	1.0204	1.0222	56.1	0.164	964.3	33.9	82	1.3412	0.321	0.172	0.092	1.179	1.157	84.68		
6.00	6.32	1.0224	1.0243	61.3	0.179	961.1	37.1	90	1.3420	0.352	0.189	0.101	1.198	1.174	83.34		
6.50	6.84	1.0245	1.0263	66.6	0.195	957.9	40.4	97	1.3427	0.384	0.206	0.111	1.217	1.190	82.01		
7.00	7.37	1.0265	1.0283	71.9	0.210	954.6	43.6	105	1.3435	0.416	0.224	0.120	1.237	1.207	80.69		
7.50	7.89	1.0285	1.0303	77.1	0.225	951.3	46.9	113	1.3443	0.449	0.241	0.130	1.257	1.225	79.38		
8.00	8.42	1.0305	1.0323	82.4	0.241	948.0	50.2	120	1.3450	0.482	0.259	0.139	1.278	1.243	78.07		
8.50	8.95	1.0325	1.0343	87.8	0.256	944.7	53.5	128	1.3458	0.516	0.277	0.149	1.300	1.262	76.77		
9.00	9.47	1.0345	1.0363	93.1	0.272	941.4	56.8	136	1.3466	0.550	0.296	0.159	1.322	1.281	75.48		
9.50	10.00	1.0365	1.0383	98.5	0.288	938.0	60.2	144	1.3474	0.584	0.314	0.169	1.345	1.300	74.19		
10.00	10.53	1.0385	1.0403	103.8	0.303	934.6	63.6	152	1.3482	0.619	0.333	0.180	1.369	1.321	72.90		
11.00	11.58	1.0425	1.0443	114.7	0.335	927.8	70.4	167	1.3497	0.691	0.371	0.201	1.419	1.364	70.34		
12.00	12.63	1.0465	1.0483	125.6	0.367	920.9	77.3	183	1.3513	0.765	0.411	0.223	1.471	1.409	67.83		

33 MALTOSE, $C_{12}H_{22}O_{11} \cdot 1H_2O$—(Continued)

A% by wt.	H% by wt.	ρ D_4^{20}	D_{20}^{20}	C_s g/l	M g-mol/l	C_w g/l	$(C_o - C_w)$ g/l	$(n - n_o)$ × 10^4	n	Δ °C	O Os/kg	S g-mol/l	η/η_o	η/ρ cS	φ rhe	γ mmho/cm	T g-mol/l
13.00	13.68	1.0505	1.0523	136.6	0.399	913.9	84.3	199	1.3529	0.841	0.452	0.245	1.527	1.456	65.37		
14.00	14.74	1.0545	1.0563	147.6	0.431	906.8	91.4	216	1.3546	0.919	0.494	0.268	1.585	1.506	62.97		
15.00	15.79	1.0585	1.0603	158.8	0.464	899.7	98.6	232	1.3562	1.000	0.538	0.292	1.646	1.558	60.63		
16.00	16.84	1.0629	1.0648	170.1	0.497	892.8	105.4	248	1.3578	1.080	0.581	0.315	1.712	1.614	58.31		
17.00	17.89	1.0672	1.0691	181.4	0.530	885.8	112.4	265	1.3595	1.165	0.626	0.340	1.780	1.672	56.05		
18.00	18.95	1.0716	1.0735	192.9	0.563	878.7	119.5	282	1.3612	1.250	0.672	0.365	1.855	1.735	53.79		
19.00	20.00	1.0759	1.0778	204.4	0.597	871.5	126.7	298	1.3628	1.338	0.719	0.391	1.936	1.803	51.55		
20.00	21.05	1.0801	1.0820	216.0	0.631	864.1	134.1	314	1.3644	1.430	0.769	0.418	2.017	1.871	49.48		
22.00	23.16	1.0894	1.0913	239.7	0.700	849.7	148.5	348	1.3678	1.635	0.879	0.478	2.212	2.034	45.13		
24.00	25.26	1.0984	1.1004	263.6	0.770	834.8	163.4	384	1.3714	1.848	0.993	0.540	2.458	2.242	40.60		
26.00	27.37	1.1080	1.1100	288.1	0.841	819.9	178.3	419	1.3749	2.080	1.118	0.608	2.748	2.485	36.32		
28.00	29.47	1.1171	1.1191	312.8	0.913	804.3	193.9	455	1.3785	2.335	1.255	0.681	3.060	2.745	32.62		
30.00	31.58	1.1269	1.1289	338.1	0.987	788.8	209.4	491	1.3821	2.615	1.406	0.762	3.420	3.041	29.18		
32.00	33.68	1.1367	1.1387	363.7	1.062	773.0	225.2	528	1.3858	2.925	1.572	0.850	3.910	3.447	25.52		
34.00	35.79	1.1463	1.1483	389.7	1.138	756.6	241.6	566	1.3896	3.252	1.748	0.943	4.438	3.879	22.49		
36.00	37.89	1.1561	1.1582	416.2	1.215	739.9	258.3	605	1.3935	3.600	1.935	1.040	5.040	4.368	19.80		
38.00	40.00	1.1663	1.1684	443.2	1.294	723.1	275.1	644	1.3974	3.990	2.145	1.148	5.820	5.000	17.15		
40.00	42.10	1.1769	1.1790	470.8	1.375	706.1	292.1	683	1.4013	4.408	2.370	1.264	6.912	5.885	14.44		
42.00	44.21	1.1878	1.1899	498.9	1.457	688.9	309.3	721	1.4051	4.878	2.622	1.390	8.175	6.896	12.21		
44.00	46.31	1.1979	1.2000	527.1	1.539	670.8	327.4	764	1.4094	5.352	2.877	1.518	9.630	8.055	10.36		
46.00	48.42	1.2084	1.2105	555.9	1.623	652.5	345.7	806	1.4136				11.45	9.492	8.72		
48.00	50.52	1.2194	1.2216	585.3	1.709	634.1	364.1	847	1.4177				14.12	11.60	7.07		
50.00	52.63	1.2304	1.2326	615.2	1.796	615.2	383.0	887	1.4217				17.75	14.46	5.62		
52.00	54.74	1.2416	1.2438	645.6	1.885	596.0	402.2	930	1.4260				21.99	17.74	4.54		
54.00	56.84	1.2528	1.2550	676.5	1.975	576.3	421.9	978	1.4308				28.70	22.96	3.48		
56.00	58.94	1.2638	1.2660	707.7	2.066	556.1	442.1	1020	1.4350				38.15	30.25	2.62		
58.00	61.05	1.2746	1.2769	739.3	2.159	535.3	462.9	1064	1.4394				49.20	38.68	2.03		
60.00	63.16	1.2855	1.2878	771.3	2.252	514.2	484.0	1110	1.4440								

34 MANGANOUS SULFATE, $MnSO_4 \cdot 1H_2O$

MOLECULAR WEIGHT = 151.00 FORMULA WEIGHT, HYDRATE = 169.01
RELATIVE SPECIFIC REFRACTIVITY = 0.541

0.00 % by wt. data are the same for all compounds.
For Values of 0.00 wt. % solutions see Table 1, Acetic Acid.

A% by wt.	H% by wt.	ρ D_4^{20}	D_{20}^{20}	C_s g/l	M g-mol/l	C_w g/l	$(C_o - C_w)$ g/l	$(n - n_o)$ × 10^4	n	Δ °C	O Os/kg	S g-mol/l	η/η_o	η/ρ cS	φ rhe	γ mmho/cm	T g-mol/l
1.00	1.12	1.0080	1.0098	10.1	0.067	997.9	0.3	18	1.3348	0.160	0.086	0.045	1.044	1.038	95.59	6.2	0.063
2.00	2.24	1.0178	1.0196	20.4	0.135	997.5	0.8	36	1.3366	0.306	0.164	0.088	1.088	1.071	91.73	10.6	0.111
3.00	3.36	1.0277	1.0296	30.8	0.204	996.9	1.3	54	1.3384	0.437	0.235	0.126	1.135	1.107	87.93	14.6	0.157
4.00	4.48	1.0378	1.0396	41.5	0.275	996.3	2.0	72	1.3402	0.569	0.306	0.165	1.185	1.145	84.19	18.3	0.199
5.00	5.60	1.0480	1.0498	52.4	0.347	995.6	2.7	90	1.3420	0.701	0.377	0.204	1.240	1.186	80.48	21.6	0.240
6.00	6.72	1.0583	1.0602	63.5	0.421	994.8	3.4	108	1.3438	0.838	0.451	0.244	1.298	1.229	76.87	24.7	0.278
7.00	7.83	1.0688	1.0707	74.8	0.495	994.0	4.3	127	1.3457	0.978	0.526	0.285	1.360	1.275	73.36	27.4	0.312
8.00	8.95	1.0794	1.0813	86.4	0.572	993.1	5.2	145	1.3475	1.122	0.603	0.328	1.428	1.325	69.90	29.9	0.343
9.00	10.07	1.0902	1.0921	98.1	0.650	992.1	6.1	164	1.3494	1.275	0.685	0.373	1.502	1.380	66.46	32.3	0.372
10.00	11.19	1.1012	1.1031	110.1	0.729	991.1	7.2	183	1.3513	1.439	0.773	0.421	1.584	1.441	63.01	34.5	0.401
11.00	12.31	1.1123	1.1143	122.4	0.810	990.0	8.3	202	1.3532	1.612	0.867	0.472	1.675	1.509	59.57	36.7	0.428
12.00	13.43	1.1236	1.1256	134.8	0.893	988.8	9.5	221	1.3551	1.797	0.966	0.526	1.775	1.583	56.23	38.7	0.454
13.00	14.55	1.1351	1.1371	147.6	0.977	987.5	10.7	240	1.3570	1.995	1.073	0.583	1.883	1.663	52.99	40.5	0.478
14.00	15.67	1.1467	1.1487	160.5	1.063	986.2	12.1	259	1.3589	2.207	1.187	0.645	2.001	1.749	49.87	42.2	0.499
15.00	16.79	1.1585	1.1606	173.8	1.151	984.7	13.5	279	1.3609	2.431	1.307	0.709	2.129	1.841	46.88	43.7	0.518
16.00	17.91	1.1705	1.1726	187.3	1.240	983.2	15.0	299	1.3629	2.669	1.435	0.777	2.267	1.940	44.03	44.9	0.534
17.00	19.03	1.1827	1.1847	201.1	1.331	981.6	16.6	318	1.3648	2.922	1.571	0.850	2.415	2.046	41.32	45.9	0.548
18.00	20.15	1.1950	1.1971	215.1	1.425	979.9	18.3	338	1.3668	3.194	1.717	0.926	2.575	2.160	38.75	46.7	0.559
19.00	21.27	1.2075	1.2097	229.4	1.519	978.1	20.1	358	1.3688	3.486	1.874	1.008	2.747	2.280	36.32	47.3	0.566
20.00	22.39	1.2203	1.2224	244.1	1.616	976.2	22.0	378	1.3708	3.801	2.043	1.096	2.932	2.408	34.04	47.6	0.570

35 D-MANNITOL, $CH_2OH(CHOH)_4CH_2OH$

MOLECULAR WEIGHT = 182.17
RELATIVE SPECIFIC REFRACTIVITY = 1.063

0.00 % by wt. data are the same for all compounds.
For Values of 0.00 wt. % solutions see Table 1, Acetic Acid.

A% by wt.	ρ D_4^{20}	D_{20}^{20}	C_s g/l	M g-mol/l	C_w g/l	$(C_o - C_w)$ g/l	$(n - n_o)$ × 10^4	n	Δ °C	O Os/kg	S g-mol/l	η/η_o	η/ρ cS	φ rhe	γ mmho/cm	T g-mol/l
0.50	1.0000	1.0018	5.0	0.027	995.0	3.2	7	1.3337	0.051	0.028	0.014	1.015	1.017	98.30		
1.00	1.0017	1.0035	10.0	0.055	991.7	6.5	15	1.3345	0.103	0.055	0.029	1.030	1.030	96.89		
1.50	1.0035	1.0053	15.1	0.083	988.5	9.8	22	1.3352	0.156	0.084	0.044	1.044	1.042	95.59		
2.00	1.0053	1.0071	20.1	0.110	985.2	13.1	29	1.3359	0.209	0.112	0.059	1.058	1.055	94.33		
2.50	1.0070	1.0088	25.2	0.138	981.9	16.4	37	1.3367	0.262	0.141	0.075	1.072	1.067	93.07		
3.00	1.0088	1.0106	30.3	0.166	978.5	19.7	44	1.3374	0.317	0.170	0.091	1.087	1.079	91.84		
3.50	1.0106	1.0124	35.4	0.194	975.2	23.0	52	1.3381	0.371	0.200	0.107	1.101	1.092	90.61		
4.00	1.0124	1.0141	40.5	0.222	971.9	26.4	59	1.3389	0.427	0.230	0.123	1.117	1.105	89.38		
4.50	1.0141	1.0159	45.6	0.251	968.5	29.7	66	1.3396	0.483	0.260	0.140	1.132	1.119	88.13		
5.00	1.0159	1.0177	50.8	0.279	965.1	33.1	73	1.3403	0.540	0.290	0.156	1.149	1.133	86.86		
5.50	1.0177	1.0195	56.0	0.307	961.7	36.5	81	1.3411	0.598	0.321	0.173	1.166	1.148	85.56		
6.00	1.0195	1.0213	61.2	0.336	958.3	39.9	88	1.3418	0.656	0.353	0.191	1.185	1.164	84.24		
6.50	1.0212	1.0230	66.4	0.364	954.9	43.4	95	1.3425	0.715	0.384	0.208	1.204	1.181	82.92		
7.00	1.0230	1.0248	71.6	0.393	951.4	46.8	103	1.3433	0.774	0.416	0.225	1.223	1.198	81.59		
7.50	1.0248	1.0266	76.9	0.422	947.9	50.3	110	1.3440	0.835	0.449	0.243	1.244	1.216	80.26		
8.00	1.0266	1.0284	82.1	0.451	944.5	53.8	117	1.3447	0.896	0.482	0.261	1.264	1.234	78.93		
8.50	1.0284	1.0302	87.4	0.480	941.0	57.2	125	1.3455	0.958	0.515	0.279	1.286	1.253	77.62		
9.00	1.0302	1.0320	92.7	0.509	937.5	60.7	132	1.3462	1.020	0.549	0.298	1.308	1.272	76.31		

35 D-MANNITOL, $CH_2OH(CHOH)_4CH_2OH$—(Continued)

A% by wt.	ρ D_4^{20}	D_{20}^{20}	C_s g/l	M g-mol/l	C_w g/l	$(C_o - C_w)$ g/l	$(n - n_o)$ × 10^4	n	Δ °C	O Os/kg	S g-mol/l	η/η_o	η/ρ cS	φ rhe	γ mmho/cm	T g-mol/l
9.50	1.0320	1.0338	98.0	0.538	934.0	64.3	139	1.3469	1.084	0.583	0.316	1.330	1.291	75.03		
10.00	1.0338	1.0357	103.4	0.568	930.5	67.8	147	1.3477	1.148	0.617	0.335	1.353	1.311	73.76		
11.00	1.0375	1.0393	114.1	0.626	923.4	74.8	161	1.3491	1.279	0.688	0.374	1.400	1.352	71.28		
12.00	1.0412	1.0431	124.9	0.686	916.3	81.9	176	1.3506	1.413	0.760	0.413	1.450	1.395	68.85		
13.00	1.0450	1.0469	135.9	0.746	909.2	89.0	191	1.3521	1.551	0.834	0.454	1.501	1.440	66.47		
14.00	1.0489	1.0508	146.9	0.806	902.1	96.1	207	1.3536	1.692	0.910	0.495	1.555	1.486	64.16		
15.00	1.0529	1.0548	157.9	0.867	895.0	103.2	222	1.3552	1.837	0.988	0.537	1.612	1.534	61.91		

36 METHANOL, CH_3OH

MOLECULAR WEIGHT = 32.03
RELATIVE SPECIFIC REFRACTIVITY = 1.247

0.00 % by wt. data are the same for all compounds.
For Values of 0.00 wt. % solutions see Table 1, Acetic Acid.

A% by wt.	ρ D_4^{20}	D_{20}^{20}	C_s g/l	M g-mol/l	C_w g/l	$(C_o - C_w)$ g/l	$(n - n_o)$ × 10^4	n	Δ °C	O Os/kg	S g-mol/l	η/η_o	η/ρ cS	φ rhe	γ mmho/cm	T g-mol/l
0.50	0.9973	0.9991	5.0	0.156	992.4	5.9	1	1.3331	0.278	0.149	0.080	1.020	1.025	97.84		
1.00	0.9964	0.9982	10.0	0.311	986.5	11.7	2	1.3332	0.560	0.301	0.162	1.038	1.044	96.15		
1.50	0.9956	0.9973	14.9	0.466	980.6	17.6	3	1.3333	0.847	0.455	0.247	1.053	1.060	94.74		
2.00	0.9947	0.9965	19.9	0.621	974.8	23.4	4	1.3334	1.140	0.613	0.333	1.068	1.076	93.45		
2.50	0.9938	0.9956	24.8	0.776	969.0	29.2	5	1.3335	1.442	0.776	0.422	1.083	1.092	92.14		
3.00	0.9930	0.9947	29.8	0.930	963.2	35.0	6	1.3336	1.750	0.941	0.512	1.098	1.108	90.85		
3.50	0.9921	0.9939	34.7	1.084	957.4	40.8	8	1.3337	2.058	1.106	0.601	1.114	1.125	89.59		
4.00	0.9913	0.9930	39.7	1.238	951.6	46.6	9	1.3339	2.370	1.274	0.692	1.129	1.142	88.36		
4.50	0.9904	0.9922	44.6	1.392	945.9	52.4	10	1.3340	2.691	1.447	0.784	1.145	1.159	87.15		
5.00	0.9896	0.9914	49.5	1.545	940.1	58.1	11	1.3341	3.020	1.624	0.877	1.161	1.176	85.96		
5.50	0.9888	0.9905	54.4	1.698	934.4	63.8	12	1.3342	3.362	1.807	0.974	1.177	1.193	84.79		
6.00	0.9880	0.9897	59.3	1.851	928.7	69.5	14	1.3343	3.710	1.995	1.071	1.194	1.210	83.62		
6.50	0.9872	0.9889	64.2	2.003	923.0	75.2	15	1.3345	4.058	2.182	1.167	1.210	1.228	82.46		
7.00	0.9864	0.9881	69.0	2.156	917.3	80.9	16	1.3346	4.410	2.371	1.265	1.227	1.247	81.33		
7.50	0.9855	0.9873	73.9	2.308	911.6	86.6	17	1.3347	4.769	2.564	1.362	1.244	1.265	80.22		
8.00	0.9848	0.9865	78.8	2.460	906.0	92.3	19	1.3348	5.13	2.76	1.459	1.261	1.283	79.15		
8.50	0.9840	0.9857	83.6	2.611	900.3	97.9	20	1.3350	5.49	2.95	1.554	1.278	1.301	78.11		
9.00	0.9832	0.9849	88.5	2.763	894.7	103.5	21	1.3351	5.85	3.15	1.649	1.294	1.319	77.12		
9.50	0.9824	0.9841	93.3	2.914	889.1	109.2	23	1.3352	6.22	3.35	1.745	1.310	1.336	76.17		
10.00	0.9816	0.9833	98.2	3.065	883.5	114.8	24	1.3354	6.60	3.55	1.841	1.326	1.354	75.26		
11.00	0.9801	0.9818	107.8	3.366	872.3	126.0	27	1.3356	7.36	3.96	2.032	1.357	1.387	73.57		
12.00	0.9785	0.9803	117.4	3.666	861.1	137.1	29	1.3359	8.14	4.38	2.221	1.386	1.419	72.00		
13.00	0.9770	0.9787	127.0	3.965	850.0	148.2	32	1.3362	8.93	4.80	2.407	1.415	1.451	70.53		
14.00	0.9755	0.9772	136.6	4.264	838.9	159.3	35	1.3365	9.72	5.23	2.588	1.443	1.482	69.15		
15.00	0.9740	0.9757	146.1	4.561	827.9	170.3	38	1.3367	10.53	5.66	2.767	1.471	1.513	67.85		
16.00	0.9725	0.9742	155.6	4.858	816.9	181.3	40	1.3370	11.36	6.11	2.946	1.498	1.544	66.61		
17.00	0.9710	0.9727	165.1	5.154	805.9	192.3	43	1.3373	12.23	6.57	3.126	1.525	1.574	65.45		
18.00	0.9695	0.9712	174.5	5.449	795.0	203.2	46	1.3376	13.13	7.06	3.308	1.551	1.603	64.35		
19.00	0.9680	0.9698	183.9	5.742	784.1	214.1	49	1.3379	14.06	7.56	3.490	1.576	1.632	63.31		
20.00	0.9666	0.9683	193.3	6.035	773.2	225.0	51	1.3381	15.02	8.07	3.669	1.601	1.660	62.34		
22.00	0.9636	0.9653	212.0	6.618	751.6	246.6	57	1.3387	16.98	9.13	4.019	1.649	1.715	60.52		
24.00	0.9606	0.9623	230.5	7.198	730.1	268.2	62	1.3392	19.04	10.23	4.359	1.694	1.767	58.91		
26.00	0.9576	0.9593	249.0	7.773	708.6	289.6	67	1.3397	21.23	11.42		1.732	1.813	57.61		
28.00	0.9545	0.9562	267.3	8.344	687.2	311.0	72	1.3402	23.59	12.68		1.765	1.853	56.55		
30.00	0.9514	0.9531	285.4	8.911	666.0	332.3	77	1.3407	25.91	13.93		1.791	1.886	55.72		
32.00	0.9482	0.9499	303.4	9.473	644.8	353.5	81	1.3411	28.15	15.13		1.810	1.913	55.13		
34.00	0.9450	0.9466	321.3	10.031	623.7	374.6	85	1.3415	30.48	16.39		1.823	1.933	54.74		
36.00	0.9416	0.9433	339.0	10.583	602.6	395.6	89	1.3419	32.97	17.73		1.831	1.949	54.49		
38.00	0.9382	0.9399	356.5	11.131	581.7	416.6	92	1.3422	35.60	19.14		1.835	1.959	54.40		
40.00	0.9347	0.9363	373.9	11.672	560.8	437.4	95	1.3425	38.6	20.8		1.833	1.965	54.45		
42.00	0.9311	0.9327	391.0	12.209	540.0	458.2	97	1.3427	41.5	22.3		1.827	1.966	54.62		
44.00	0.9273	0.9290	408.0	12.739	519.3	478.9	99	1.3429	44.5	23.9		1.817	1.963	54.93		
46.00	0.9235	0.9251	424.8	13.263	498.7	499.5	100	1.3430	47.8	25.7		1.801	1.955	55.40		
48.00	0.9196	0.9212	441.4	13.781	478.2	520.1	101	1.3431	51.2	27.5		1.781	1.940	56.05		
50.00	0.9156	0.9172	457.8	14.292	457.8	540.5	101	1.3431	54.5	29.3		1.757	1.923	56.80		
52.00	0.9114	0.9130	473.9	14.797	437.5	560.7	101	1.3431	58.1	31.2		1.732	1.905	57.61		
54.00	0.9072	0.9088	489.9	15.295	417.3	580.9	100	1.3430	62.0	33.3		1.705	1.883	58.52		
56.00	0.9030	0.9046	505.7	15.787	397.3	600.9	99	1.3429	66.0	35.5		1.673	1.857	59.64		
58.00	0.8987	0.9003	521.2	16.274	377.5	620.8	98	1.3428	70.0	37.6		1.638	1.826	60.94		
60.00	0.8944	0.8960	536.6	16.754	357.8	640.5	96	1.3426	74.5	40.0		1.597	1.789	62.49		
62.00	0.8901	0.8917	551.9	17.230	338.3	660.0	95	1.3425	79.3	42.6		1.550	1.744	64.40		
64.00	0.8856	0.8872	566.8	17.695	318.8	679.4	92	1.3422	84.4	45.4		1.500	1.697	66.54		
66.00	0.8810	0.8825	581.4	18.153	299.5	698.7	89	1.3419	89.6	48.2		1.453	1.653	68.66		
68.00	0.8763	0.8778	595.9	18.603	280.4	717.8	85	1.3415	96.3	51.8		1.410	1.612	70.78		
70.00	0.8715	0.8730	610.0	19.046	261.4	736.8	81	1.3411				1.365	1.569	73.11		
72.00	0.8667	0.8682	624.0	19.482	242.7	755.6	77	1.3407				1.315	1.521	75.87		
74.00	0.8618	0.8633	637.7	19.910	224.1	774.2	72	1.3402				1.264	1.470	78.93		
76.00	0.8568	0.8583	651.2	20.331	205.6	792.6	67	1.3397				1.216	1.422	82.05		
78.00	0.8518	0.8533	664.4	20.744	187.4	810.8	61	1.3391				1.170	1.377	85.28		
80.00	0.8468	0.8483	677.4	21.149	169.4	828.9	55	1.3385				1.126	1.332	88.63		
82.00	0.8416	0.8431	690.2	21.547	151.5	846.7	49	1.3379				1.072	1.276	93.11		
84.00	0.8365	0.8380	702.6	21.937	133.8	864.4	42	1.3372				1.021	1.223	97.75		
86.00	0.8312	0.8327	714.9	22.318	116.4	881.9	35	1.3365				0.968	1.167	103.1		
88.00	0.8259	0.8274	726.8	22.691	99.1	899.1	27	1.3357				0.914	1.109	109.2		
90.00	0.8204	0.8219	738.4	23.053	82.0	916.2	19	1.3348				0.859	1.049	116.2		
92.00	0.8148	0.8163	749.6	23.404	65.2	933.0	9	1.3339				0.804	0.989	124.1		
94.00	0.8089	0.8103	760.3	23.738	48.5	949.7	−2	1.3328				0.748	0.927	133.4		
96.00	0.8034	0.8048	771.3	24.081	32.1	966.1	−14	1.3316				0.694	0.866	143.8		
98.00	0.7976	0.7990	781.6	24.402	16.0	982.2	−26	1.3304				0.638	0.801	156.4		
100.0	0.7917	0.7931	791.7	24.717	0.0	998.2	−40	1.3290				0.585	0.740	170.6		

37 NICKEL SULFATE, $NiSO_4 \cdot 6H_2O$

MOLECULAR WEIGHT = 154.75 FORMULA WEIGHT, HYDRATE = 262.85
RELATIVE SPECIFIC REFRACTIVITY = 0.524

0.00 % by wt. data are the same for all compounds.
For Values of 0.00 wt. % solutions see Table 1, Acetic Acid.

A% by wt.	H% by wt.	D_4^{20}	D_{20}^{20}	C_s g/l	M g-mol/l	C_w g/l	$(C_o - C_w)$ g/l	$(n - n_o) \times 10^4$	n	Δ °C	O Os/kg	S g-mol/l	η/η_o	η/ρ cS	ϕ rhe	γ mmho/cm	T g-mol/l
0.50	0.85	1.0035	1.0053	5.0	0.032	998.5	− 0.3	10	1.3340	0.081	0.044	0.023	1.022	1.020	97.66	3.1	0.031
1.00	1.70	1.0089	1.0107	10.1	0.065	998.8	− 0.6	20	1.3349	0.150	0.081	0.043	1.044	1.037	95.59	5.8	0.058
1.50	2.55	1.0142	1.0160	15.2	0.098	999.0	− 0.8	29	1.3359	0.217	0.117	0.062	1.066	1.053	93.60	8.0	0.081
2.00	3.40	1.0196	1.0214	20.4	0.132	999.2	− 1.0	39	1.3369	0.282	0.152	0.081	1.089	1.070	91.64	9.9	0.103
2.50	4.25	1.0250	1.0268	25.6	0.166	999.4	− 1.2	49	1.3379	0.346	0.186	0.099	1.113	1.088	89.67	11.9	0.125
3.00	5.10	1.0304	1.0323	30.9	0.200	999.5	− 1.3	60	1.3389	0.410	0.220	0.118	1.138	1.107	87.70	13.7	0.146
3.50	5.94	1.0359	1.0377	36.3	0.234	999.6	− 1.4	70	1.3400	0.475	0.255	0.137	1.164	1.126	85.75	15.5	0.167
4.00	6.79	1.0413	1.0431	41.7	0.269	999.6	− 1.4	80	1.3410	0.540	0.290	0.156	1.191	1.146	83.80	17.3	0.187
4.50	7.64	1.0467	1.0486	47.1	0.304	999.6	− 1.4	90	1.3420	0.605	0.325	0.176	1.219	1.167	81.87	18.9	0.207
5.00	8.49	1.0521	1.0540	52.6	0.340	999.5	− 1.3	100	1.3430	0.671	0.361	0.195	1.248	1.189	79.95	20.5	0.226
5.50	9.34	1.0575	1.0594	58.2	0.376	999.4	− 1.1	110	1.3440	0.736	0.396	0.214	1.279	1.211	78.06	22.0	0.245
6.00	10.19	1.0629	1.0648	63.8	0.412	999.1	− 0.9	120	1.3450	0.803	0.432	0.234	1.310	1.235	76.18	23.5	0.263

38 NITRIC ACID, HNO_3

MOLECULAR WEIGHT = 63.02
RELATIVE SPECIFIC REFRACTIVITY = 0.818

0.00 % by wt. data are the same for all compounds.
For Values of 0.00 wt. % solutions see Table 1, Acetic Acid.

A% by wt.	ρ D_4^{20}	d D_{20}^{20}	C_s g/l	M g-mol/l	C_w g/l	$(C_o - C_w)$ g/l	$(n - n_o) \times 10^4$	n	Δ °C	O Os/kg	S g-mol/l	η/η_o	η/ρ cS	ϕ rhe	γ mmho/cm	T g-mol/l
0.50	1.0009	1.0027	5.0	0.079	995.9	2.3	6	1.3336	0.281	0.151	0.080	1.002	1.003	99.64	28.4	0.323
1.00	1.0037	1.0054	10.0	0.159	993.6	4.6	13	1.3343	0.558	0.300	0.162	1.003	1.001	99.50	56.1	0.686
1.50	1.0064	1.0082	15.1	0.240	991.3	6.9	19	1.3349	0.837	0.450	0.244	1.004	1.000	99.40	84.7	1.10
2.00	1.0091	1.0109	20.2	0.320	988.9	9.3	26	1.3356	1.120	0.602	0.327	1.005	0.998	99.30	108.	1.50
2.50	1.0119	1.0137	25.3	0.401	986.6	11.7	32	1.3362	1.408	0.757	0.412	1.006	0.997	99.17	138.	1.97
3.00	1.0146	1.0164	30.4	0.483	984.2	14.0	39	1.3368	1.704	0.916	0.498	1.008	0.995	99.01	160.	2.57
3.50	1.0174	1.0192	35.6	0.565	981.8	16.5	45	1.3375	2.006	1.078	0.586	1.010	0.995	98.83	184.	3.18
4.00	1.0202	1.0220	40.8	0.648	979.4	18.9	51	1.3381	2.315	1.245	0.676	1.012	0.994	98.64	213.	4.31
4.50	1.0230	1.0248	46.0	0.730	976.9	21.3	58	1.3388	2.632	1.415	0.767	1.014	0.993	98.43		
5.00	1.0257	1.0276	51.3	0.814	974.5	23.8	64	1.3394	2.958	1.590	0.859	1.016	0.992	98.23		
5.50	1.0286	1.0304	56.6	0.898	972.0	26.3	71	1.3401	3.290	1.769	0.953	1.018	0.992	98.02		
6.00	1.0314	1.0332	61.9	0.982	969.5	28.7	78	1.3407	3.629	1.951	1.048	1.020	0.991	97.81		
6.50	1.0342	1.0360	67.2	1.067	967.0	31.3	84	1.3414	3.974	2.137	1.144	1.023	0.991	97.59		
7.00	1.0370	1.0389	72.6	1.152	964.4	33.8	91	1.3421	4.327	2.326	1.241	1.025	0.990	97.36		
7.50	1.0399	1.0417	78.0	1.238	961.9	36.3	97	1.3427	4.687	2.520	1.340	1.028	0.990	97.12		
8.00	1.0427	1.0446	83.4	1.324	959.3	38.9	104	1.3434	5.05	2.72	1.439	1.030	0.990	96.88		
8.50	1.0456	1.0475	88.9	1.410	956.7	41.5	110	1.3440	5.43	2.92	1.538	1.033	0.990	96.62		
9.00	1.0485	1.0504	94.4	1.497	954.1	44.1	117	1.3447	5.81	3.12	1.639	1.036	0.990	96.35		
9.50	1.0514	1.0533	99.9	1.585	951.5	46.7	124	1.3454	6.20	3.33	1.740	1.039	0.990	96.07		
10.00	1.0543	1.0562	105.4	1.673	948.9	49.4	130	1.3460	6.60	3.55	1.841	1.042	0.990	95.78		
11.00	1.0602	1.0620	116.6	1.850	943.5	54.7	144	1.3474	7.42	3.99	2.045	1.049	0.991	95.15		
12.00	1.0660	1.0679	127.9	2.030	938.1	60.1	157	1.3487	8.27	4.45	2.251	1.056	0.993	94.48		
13.00	1.0720	1.0739	139.4	2.211	932.6	65.6	170	1.3500	9.15	4.92	2.459	1.064	0.995	93.76		
14.00	1.0780	1.0799	150.9	2.395	927.1	71.2	184	1.3514	10.08	5.42	2.667	1.073	0.997	93.00		
15.00	1.0840	1.0859	162.6	2.580	921.4	76.8	198	1.3527	11.04	5.93	2.877	1.082	1.001	92.20		
16.00	1.0901	1.0921	174.4	2.768	915.7	82.5	211	1.3541	12.04	6.47	3.087	1.092	1.004	91.35		
17.00	1.0963	1.0982	186.4	2.957	909.9	88.3	225	1.3555	13.08	7.03	3.298	1.103	1.008	90.47		
18.00	1.1025	1.1044	198.4	3.149	904.0	94.2	239	1.3569	14.16	7.61	3.509	1.114	1.013	89.55		
19.00	1.1087	1.1107	210.7	3.343	898.0	100.2	253	1.3582	15.30	8.22	3.720	1.126	1.018	88.60		
20.00	1.1150	1.1170	223.0	3.538	892.0	106.2	266	1.3596				1.139	1.024	87.62		
22.00	1.1277	1.1297	248.1	3.937	879.6	118.6	294	1.3624				1.167	1.037	85.55		
24.00	1.1406	1.1426	273.7	4.344	866.8	131.4	322	1.3652				1.197	1.052	83.36		
26.00	1.1536	1.1557	299.9	4.759	853.7	144.6	350	1.3680				1.231	1.069	81.06		
28.00	1.1668	1.1688	326.7	5.184	840.1	158.1	378	1.3708				1.268	1.089	78.70		
30.00	1.1801	1.1822	354.0	5.618	826.0	172.2	406	1.3736				1.308	1.111	76.30		
32.00	1.1934	1.1955	381.9	6.060	811.5	186.7	433	1.3763				1.351	1.134	73.87		
34.00	1.2068	1.2090	410.3	6.511	796.5	201.7	460	1.3790				1.397	1.160	71.42		
36.00	1.2202	1.2224	439.3	6.970	780.9	217.3	487	1.3817				1.447	1.188	68.96		
38.00	1.2335	1.2357	468.7	7.438	764.8	233.5	513	1.3842				1.501	1.219	66.50		
40.00	1.2466	1.2489	498.7	7.913	748.0	250.2	537	1.3867				1.558	1.252	64.06		

39 OXALIC ACID, HO_2CCO_2H

MOLECULAR WEIGHT = 90.04

A% by wt.	ρ D_4^{20}	D_{20}^{20}	C_s g/l	M g-mol/l	C_w g/l	$(C_0 - C_w)$ g/l	$(n - n_0)$ × 10^4	n	Δ °C	O Os/kg	S g-mol/l	η/η_0	η/ρ cS	φ rhe	γ mmho/cm	T g-mol/l
0.50	1.0006	1.0024	5.0	0.056	995.6	2.6	6.	1.3336	0.159	0.086	0.045	1.011	1.012	98.76	14.0	0.149
1.00	1.0030	1.0048	10.0	0.111	993.0	5.2	12.	1.3342	0.304	0.163	0.087	1.021	1.020	97.75	21.8	0.242
1.50	1.0054	1.0072	15.1	0.167	990.3	7.9	17.	1.3347	0.440	0.237	0.127	1.031	1.028	96.75	29.3	0.335
2.00	1.0079	1.0097	20.2	0.224	987.7	10.5	23.	1.3353	0.572	0.308	0.166	1.042	1.036	95.77	35.3	0.411
2.50	1.0103	1.0121	25.3	0.281	985.0	13.2	29.	1.3359	0.705	0.379	0.205	1.053	1.044	94.81	41.2	0.486
3.00	1.0126	1.0144	30.4	0.337	982.2	16.0	34.	1.3364	0.838	0.450	0.244	1.063	1.052	93.87	46.9	0.561
3.50	1.0150	1.0168	35.5	0.395	979.5	18.7	40.	1.3370	0.967	0.520	0.282	1.074	1.060	92.95	51.7	0.626
4.00	1.0174	1.0192	40.7	0.452	976.7	21.5	45.	1.3375	1.093	0.587	0.319	1.084	1.068	92.03	57.1	0.700
4.50	1.0197	1.0215	45.9	0.510	973.8	24.4	51.	1.3381				1.095	1.076	91.14	61.2	0.757
5.00	1.0220	1.0238	51.1	0.568	970.9	27.3	56.	1.3386				1.106	1.084	90.27	65.6	0.819
5.50	1.0244	1.0262	56.3	0.626	968.1	30.1	62.	1.3392				1.116	1.092	89.39	69.7	0.879
6.00	1.0265	1.0284	61.6	0.684	964.9	33.3	67.	1.3397				1.127	1.100	88.56	73.7	0.937
6.50	1.0288	1.0307	66.9	0.743	961.9	36.3	72.	1.3402				1.138	1.108	87.73	77.2	0.990
7.00	1.0310	1.0329	72.2	0.802	958.8	39.4	77.	1.3407				1.148	1.116	86.91	80.4	1.038
7.50	1.0332	1.0351	77.5	0.861	955.7	42.5	83.	1.3413				1.159	1.124	86.11	83.4	1.085
8.00	1.0355	1.0374	82.8	0.920	952.7	45.5	88.	1.3418				1.170	1.132	85.31	86.1	1.128

40 PHOSPHORIC ACID, H_3PO_4

MOLECULAR WEIGHT = 98.00
RELATIVE SPECIFIC REFRACTIVITY = 0.727

0.00 % by wt. data are the same for all compounds.
For Values of 0.00 wt. % solutions see Table 1, Acetic Acid.

A % by wt.	ρ D_4^{20}	D_{20}^{20}	C_s g/l	M g-mol/l	C_w g/l	$(C_0 - C_w)$ g/l	$(n - n_0)$ × 10^4	n	Δ °C	O Os/kg	S g-mol/l	η/η_0	η/ρ cS	φ rhe	γ mmho/cm	T g-mol/l
0.50	1.0010	1.0028	5.0	0.051	996.0	2.2	5	1.3335	0.124	0.067	0.035	1.008	1.009	99.06	5.5	0.057
1.00	1.0038	1.0056	10.0	0.102	993.7	4.5	10	1.3340	0.243	0.131	0.069	1.018	1.016	98.04	10.1	0.105
1.50	1.0065	1.0083	15.1	0.154	991.4	6.8	15	1.3345	0.354	0.190	0.102	1.032	1.028	96.68	13.4	0.142
2.00	1.0092	1.0110	20.2	0.206	989.0	9.2	19	1.3349	0.464	0.249	0.134	1.048	1.041	95.23	16.2	0.175
2.50	1.0119	1.0137	25.3	0.258	986.6	11.6	24	1.3354	0.578	0.311	0.168	1.063	1.052	93.91	19.0	0.208
3.00	1.0146	1.0164	30.4	0.311	984.2	14.1	28	1.3358	0.694	0.373	0.202	1.077	1.064	92.64	21.6	0.239
3.50	1.0173	1.0191	35.6	0.363	981.7	16.5	33	1.3363	0.811	0.436	0.236	1.092	1.075	91.41	24.1	0.270
4.00	1.0200	1.0218	40.8	0.416	979.2	19.0	37	1.3367	0.929	0.499	0.271	1.106	1.087	90.22	26.5	0.300
4.50	1.0227	1.0245	46.0	0.470	976.7	21.6	42	1.3372	1.045	0.562	0.305	1.121	1.098	89.03	29.0	0.331
5.00	1.0254	1.0272	51.3	0.523	974.1	24.1	46	1.3376	1.160	0.624	0.339	1.136	1.110	87.85	31.5	0.363
5.50	1.0281	1.0299	56.5	0.577	971.6	26.7	51	1.3381	1.271	0.683	0.372	1.151	1.122	86.69	34.2	0.397

40 PHOSPHORIC ACID, H_3PO_4—(Continued)

A% by wt.	ρ D_4^{20}	D_{20}^{20}	C_s g/l	M g-mol/l	C_w g/l	$(C_o - C_w)$ g/l	$(n - n_o)$ × 10^4	n	Δ °C	O Os/kg	S g-mol/l	η/η_o	η/ρ cS	φ rhe	γ mmho/cm	T g-mol/l
6.00	1.0309	1.0327	61.9	0.631	969.0	29.2	55	1.3385	1.381	0.742	0.404	1.167	1.134	85.54	36.9	0.431
6.50	1.0336	1.0354	67.2	0.686	966.4	31.9	60	1.3390	1.496	0.804	0.438	1.182	1.146	84.42	39.6	0.466
7.00	1.0363	1.0381	72.5	0.740	963.8	34.5	64	1.3394	1.619	0.871	0.474	1.198	1.158	83.32	42.4	0.502
7.50	1.0391	1.0409	77.9	0.795	961.1	37.1	69	1.3399	1.748	0.940	0.511	1.214	1.170	82.23	45.2	0.539
8.00	1.0418	1.0437	83.3	0.850	958.5	39.7	73	1.3403	1.882	1.012	0.550	1.230	1.183	81.14	48.0	0.576
8.50	1.0446	1.0465	88.8	0.906	955.8	42.4	78	1.3408	2.020	1.086	0.590	1.247	1.196	80.05	50.8	0.614
9.00	1.0474	1.0493	94.3	0.962	953.2	45.1	83	1.3413	2.160	1.161	0.631	1.264	1.209	78.96	53.7	0.653
9.50	1.0503	1.0521	99.8	1.018	950.5	47.7	87	1.3417	2.305	1.239	0.673	1.282	1.223	77.87	56.5	0.692
10.00	1.0531	1.0550	105.3	1.075	947.8	50.4	92	1.3422	2.450	1.317	0.715	1.300	1.237	76.77	59.4	0.731
11.00	1.0589	1.0607	116.5	1.189	942.4	55.9	101	1.3431	2.721	1.463	0.792	1.338	1.267	74.56	65.1	0.812
12.00	1.0647	1.0665	127.8	1.304	936.9	61.3	111	1.3441	3.010	1.618	0.874	1.379	1.298	72.38	70.9	0.896
13.00	1.0705	1.0724	139.2	1.420	931.4	66.9	120	1.3450	3.382	1.818	0.979	1.421	1.330	70.22	76.7	0.983
14.00	1.0765	1.0784	150.7	1.538	925.8	72.5	130	1.3460	3.760	2.022	1.085	1.466	1.364	68.08	82.5	1.07
15.00	1.0825	1.0844	162.4	1.657	920.1	78.1	140	1.3470	4.075	2.191	1.172	1.513	1.400	65.98	88.4	1.17
16.00	1.0885	1.0905	174.2	1.777	914.4	83.9	150	1.3480	4.450	2.392	1.276	1.562	1.438	63.90	94.3	1.26
17.00	1.0947	1.0966	186.1	1.899	908.6	89.7	160	1.3489	4.820	2.591	1.376	1.613	1.477	61.86	100.	1.36
18.00	1.1009	1.1028	198.2	2.022	902.7	95.5	170	1.3500	5.25	2.82	1.491	1.668	1.518	59.85	106.	1.46
19.00	1.1071	1.1091	210.4	2.146	896.8	101.5	180	1.3510	5.72	3.07	1.614	1.724	1.561	57.88	112.	1.56
20.00	1.1135	1.1154	222.7	2.272	890.8	107.5	190	1.3520	6.23	3.35	1.747	1.784	1.605	55.94	118.	1.67
22.00	1.1263	1.1283	247.8	2.528	878.5	119.7	211	1.3540	7.38	3.97	2.037	1.910	1.699	52.26	129.	1.89
24.00	1.1395	1.1415	273.5	2.790	866.0	132.2	231	1.3561	8.69	4.67	2.350	2.045	1.798	48.80	141.	2.12
26.00	1.1528	1.1549	299.7	3.059	853.1	145.1	253	1.3582	10.12	5.44	2.677	2.194	1.907	45.50	152.	2.37
28.00	1.1665	1.1685	326.6	3.333	839.9	158.4	274	1.3604	11.64	6.26	3.004	2.360	2.027	42.29	163.	2.63
30.00	1.1804	1.1825	354.1	3.613	826.3	172.0	295	1.3625	13.23	7.11	3.328	2.548	2.163	39.17	173.	2.88
32.00	1.1945	1.1966	382.2	3.900	812.3	186.0	317	1.3647	14.94	8.03	3.656	2.760	2.315	36.16	183.	3.13
34.00	1.2089	1.2111	411.0	4.194	797.9	200.3	339	1.3669	16.81	9.04	3.989	2.995	2.482	33.33	191.	3.38
36.00	1.2236	1.2257	440.5	4.495	783.1	215.1	361	1.3691	18.85	10.13	4.329	3.253	2.664	30.68	199.	3.63
38.00	1.2385	1.2407	470.6	4.802	767.8	230.4	383	1.3713	21.09	11.34	4.677	3.537	2.862	28.21	205.	3.88
40.00	1.2536	1.2558	501.4	5.117	752.2	246.1	405	1.3735	23.58	12.68		3.848	3.076	25.94	209.	4.13

41 POTASSIUM BICARBONATE, $KHCO_3$

MOLECULAR WEIGHT = 100.12
RELATIVE SPECIFIC REFRACTIVITY = 0.667

0.00 % by wt. data are the same for all compounds.
For Values of 0.00 wt. % solutions see Table 1, Acetic Acid.

A% by wt.	ρ D_4^{20}	D_{20}^{20}	C_s g/l	M g-mol/l	C_w g/l	$(C_o - C_w)$ g/l	$(n - n_o)$ × 10^4	n	Δ °C	O Os/kg	S g-mol/l	η/η_o	η/ρ cS	φ rhe	γ mmho/cm	T g-mol/l
0.50	1.0014	1.0031	5.0	0.050	996.4	1.9	5	1.3335	0.175	0.094	0.050	1.007	1.007	99.14	4.6	0.046
1.00	1.0046	1.0064	10.0	0.100	994.6	3.7	11	1.3341	0.344	0.185	0.099	1.013	1.010	98.52	8.9	0.092
2.00	1.0114	1.0132	20.2	0.202	991.1	7.1	23	1.3353	0.668	0.359	0.194	1.025	1.016	97.37	17.0	0.184
3.00	1.0181	1.0199	30.5	0.305	987.6	10.6	35	1.3365	0.983	0.528	0.287	1.038	1.021	96.17	24.6	0.275
4.00	1.0247	1.0265	41.0	0.409	983.7	14.5	46	1.3376	1.291	0.694	0.378	1.051	1.028	94.95	31.7	0.365
5.00	1.0310	1.0329	51.6	0.515	979.5	18.8	56	1.3386	1.600	0.860	0.468	1.065	1.035	93.71	38.8	0.455
6.00	1.0379	1.0397	62.3	0.622	975.6	22.6	68	1.3397	1.911	1.028	0.559	1.079	1.042	92.46	45.8	0.547
7.00	1.0446	1.0465	73.1	0.730	971.5	26.7	79	1.3409	2.222	1.195	0.649	1.094	1.049	91.21	52.7	0.639
8.00	1.0514	1.0532	84.1	0.840	967.3	31.0	90	1.3419	2.533	1.362	0.738	1.110	1.057	89.95	59.4	0.732
9.00	1.0581	1.0600	95.2	0.951	962.9	35.3	100	1.3430	2.844	1.529	0.827	1.126	1.066	88.65	65.9	0.825
10.00	1.0650	1.0668	106.5	1.064	958.5	39.8	111	1.3441	3.157	1.697	0.916	1.143	1.075	87.31	72.4	0.918
12.00	1.0788	1.0807	129.5	1.293	949.3	48.9	132	1.3462	3.785	2.035	1.092	1.181	1.097	84.54	84.5	1.10
14.00	1.0929	1.0948	153.0	1.528	939.9	58.4	154	1.3484	4.414	2.373	1.266	1.222	1.120	81.68	95.9	1.29
16.00	1.1073	1.1093	177.2	1.770	930.1	68.1	176	1.3506				1.267	1.147	78.77	107.0	1.47
18.00	1.1221	1.1241	202.0	2.017	920.1	78.1	198	1.3528				1.316	1.175	75.83	118.0	1.67
20.00	1.1372	1.1392	227.4	2.272	909.7	88.5	220	1.3550				1.370	1.207	72.85	128.0	1.86
22.00	1.1527	1.1547	253.6	2.533	899.1	99.2	242	1.3572				1.429	1.242	69.83	137.0	2.05
24.00	1.1685	1.1706	280.4	2.801	888.1	110.2	265	1.3595				1.494	1.281	66.80	145.0	2.23

42 POTASSIUM BIPHTHALATE, $KHC_8H_4O_4$

MOLECULAR WEIGHT = 204.23
RELATIVE SPECIFIC REFRACTIVITY = 1.097

0.00 % by wt. data are the same for all compounds.
For Values of 0.00 wt. % solutions see Table 1, Acetic Acid.

A% by wt.	ρ D_4^{20}	D_{20}^{20}	C_s g/l	M g-mol/l	C_w g/l	$(C_o - C_w)$ g/l	$(n - n_o)$ × 10^4	n	Δ °C	O Os/kg	S g-mol/l	η/η_o	η/ρ cS	φ rhe	γ mmho/cm	T g-mol/l
0.50	1.0004	1.0021	5.0	0.024	995.4	2.9	9	1.3339	0.085	0.046	0.024	1.012	1.013	98.67	2.0	0.020
1.00	1.0025	1.0043	10.0	0.049	992.5	5.7	19	1.3349	0.169	0.091	0.048	1.023	1.022	97.56	4.0	0.040
1.50	1.0047	1.0065	15.1	0.074	989.6	8.6	28	1.3358	0.254	0.136	0.073	1.034	1.032	96.48	5.9	0.059
2.00	1.0069	1.0087	20.1	0.099	986.8	11.5	38	1.3368	0.337	0.181	0.097	1.046	1.041	95.41	7.7	0.079
2.50	1.0092	1.0109	25.2	0.124	983.9	14.3	47	1.3377	0.421	0.226	0.121	1.058	1.050	94.35	9.4	0.098
3.00	1.0114	1.0132	30.3	0.149	981.1	17.2	57	1.3387	0.503	0.271	0.146	1.070	1.060	93.29	11.0	0.115
3.50	1.0137	1.0155	35.5	0.174	978.2	20.0	66	1.3396	0.585	0.315	0.170	1.082	1.070	92.24	12.5	0.132
4.00	1.0160	1.0178	40.6	0.199	975.3	22.9	76	1.3406	0.667	0.359	0.194	1.094	1.079	91.19	14.0	0.149
4.50	1.0183	1.0201	45.8	0.224	972.5	25.8	86	1.3416	0.749	0.403	0.218	1.107	1.089	90.15	15.3	0.165
5.00	1.0206	1.0224	51.0	0.250	969.6	28.7	95	1.3425	0.831	0.447	0.242	1.120	1.100	89.11	16.7	0.181
5.50	1.0229	1.0247	56.3	0.275	966.7	31.6	105	1.3435	0.913	0.491	0.266	1.133	1.110	88.07	18.1	0.197
6.00	1.0252	1.0270	61.5	0.301	963.7	34.5	115	1.3444	0.995	0.535	0.290	1.147	1.121	87.04	19.4	0.213
6.50	1.0275	1.0294	66.8	0.327	960.8	37.5	124	1.3454	1.076	0.579	0.314	1.160	1.132	86.01	20.7	0.229
7.00	1.0298	1.0317	72.1	0.353	957.8	40.5	133	1.3463	1.158	0.622	0.338	1.174	1.143	84.99	21.9	0.244
7.50	1.0321	1.0339	77.4	0.379	954.7	43.5	143	1.3473	1.239	0.666	0.362	1.189	1.154	83.97	23.1	0.258
8.00	1.0344	1.0362	82.7	0.405	951.6	46.6	152	1.3482	1.320	0.710	0.386	1.203	1.165	82.96	24.2	0.272

43 POTASSIUM BROMIDE, KBr

MOLECULAR WEIGHT = 119.01
RELATIVE SPECIFIC REFRACTIVITY = 0.627

0.00 % by wt. data are the same for all compounds.
For Values of 0.00 wt. % solutions see Table 1, Acetic Acid.

A% by wt.	ρ D_4^{20}	D_{20}^{20}	C_s g/l	M g-mol/l	C_w g/l	$(C_o - C_w)$ g/l	$(n - n_o)$ × 10^4	n	Δ °C	O Os/kg	S g-mol/l	η/η_o	η/ρ cS	φ rhe	γ mmho/cm	T g-mol/l
0.50	1.0018	1.0036	5.0	0.042	996.8	1.4	6	1.3336	0.148	0.080	0.042	0.998	0.998	100.00	5.2	0.053
1.00	1.0054	1.0072	10.1	0.084	995.4	2.9	12	1.3342	0.294	0.158	0.084	0.996	0.993	100.20	10.2	0.106
1.50	1.0090	1.0108	15.1	0.127	993.9	4.3	18	1.3348	0.439	0.236	0.127	0.994	0.987	100.40	14.9	0.160
2.00	1.0127	1.0145	20.3	0.170	992.4	5.8	24	1.3354	0.585	0.315	0.170	0.992	0.982	100.61	19.5	0.214
2.50	1.0164	1.0182	25.4	0.214	990.9	7.3	30	1.3360	0.731	0.393	0.213	0.990	0.976	100.82	24.2	0.270
3.00	1.0200	1.0218	30.6	0.257	989.4	8.8	36	1.3366	0.878	0.472	0.256	0.988	0.970	101.04	28.8	0.328
3.50	1.0238	1.0256	35.8	0.301	987.9	10.3	42	1.3372	1.026	0.552	0.300	0.985	0.965	101.27	33.5	0.387
4.00	1.0275	1.0293	41.1	0.345	986.4	11.9	49	1.3379	1.175	0.632	0.343	0.983	0.959	101.50	38.3	0.447
4.50	1.0312	1.0331	46.4	0.390	984.8	13.4	55	1.3385	1.325	0.712	0.387	0.981	0.953	101.73	43.0	0.509
5.00	1.0350	1.0368	51.8	0.435	983.3	15.0	61	1.3391	1.476	0.794	0.432	0.979	0.948	101.94	47.7	0.572
5.50	1.0388	1.0406	57.1	0.480	981.7	16.6	67	1.3397	1.629	0.876	0.477	0.977	0.942	102.15	52.5	0.636
6.00	1.0426	1.0445	62.6	0.526	980.1	18.2	74	1.3403	1.784	0.959	0.522	0.975	0.937	102.34	57.2	0.701
6.50	1.0465	1.0483	68.0	0.572	978.4	19.8	80	1.3410	1.939	1.043	0.567	0.973	0.932	102.54	62.0	0.767
7.00	1.0503	1.0522	73.5	0.618	976.8	21.4	86	1.3416	2.097	1.127	0.613	0.972	0.927	102.72	66.7	0.834
7.50	1.0542	1.0561	79.1	0.664	975.1	23.1	92	1.3422	2.255	1.213	0.658	0.970	0.922	102.90	71.5	0.903
8.00	1.0581	1.0600	84.6	0.711	973.5	24.8	99	1.3429	2.416	1.299	0.705	0.968	0.917	103.08	76.3	0.974
8.50	1.0620	1.0639	90.3	0.759	971.8	26.5	105	1.3435	2.578	1.386	0.751	0.967	0.912	103.25	81.1	1.05
9.00	1.0660	1.0679	95.9	0.806	970.1	28.2	112	1.3441	2.741	1.474	0.798	0.965	0.907	103.41	85.9	1.12
9.50	1.0700	1.0719	101.6	0.854	968.3	29.9	118	1.3448	2.907	1.563	0.845	0.964	0.902	103.58	90.7	1.20
10.00	1.0740	1.0759	107.4	0.902	966.6	31.6	124	1.3454	3.074	1.653	0.893	0.962	0.898	103.74	95.6	1.28
11.00	1.0821	1.0840	119.0	1.000	963.1	35.2	137	1.3467	3.415	1.836	0.989	0.959	0.888	104.06	105.	1.45
12.00	1.0903	1.0922	130.8	1.099	959.5	38.8	151	1.3481	3.764	2.024	1.086	0.956	0.879	104.36	115.	1.63
13.00	1.0986	1.1005	142.8	1.200	955.8	42.5	164	1.3494	4.122	2.216	1.185	0.954	0.870	104.64	125.	1.82
14.00	1.1070	1.1090	155.0	1.302	952.0	46.2	177	1.3507	4.487	2.412	1.286	0.951	0.861	104.90	134.	2.02
15.00	1.1155	1.1175	167.3	1.406	948.2	50.0	191	1.3521	4.862	2.614	1.387	0.949	0.853	105.13	144.	2.23
16.00	1.1242	1.1262	179.9	1.511	944.3	53.9	205	1.3535	5.25	2.82	1.490	0.947	0.844	105.35	154.	2.46
17.00	1.1330	1.1350	192.6	1.618	940.4	57.9	218	1.3548	5.64	3.03	1.593	0.946	0.836	105.54	164.	2.69
18.00	1.1419	1.1439	205.5	1.727	936.4	61.9	232	1.3562	6.04	3.25	1.698	0.944	0.829	105.70	174.	2.95
19.00	1.1509	1.1530	218.7	1.837	932.2	66.0	247	1.3577	6.46	3.47	1.804	0.943	0.821	105.84	184.	3.22
20.00	1.1601	1.1621	232.0	1.950	928.1	70.2	261	1.3591	6.88	3.70	1.912	0.942	0.814	105.95	194.	3.50
22.00	1.1788	1.1809	259.3	2.179	919.4	78.8	290	1.3620	7.76	4.17	2.130	0.941	0.800	106.07	216.	4.55
24.00	1.1980	1.2002	287.5	2.416	910.5	87.7	320	1.3650	8.70	4.68	2.352	0.941	0.787	106.07		
26.00	1.2179	1.2200	316.6	2.661	901.2	97.0	350	1.3680	9.68	5.20	2.579	0.942	0.775	105.92		
28.00	1.2383	1.2405	346.7	2.913	891.5	106.7	382	1.3711	10.72	5.76	2.809	0.945	0.765	105.59		
30.00	1.2593	1.2615	377.8	3.174	881.5	116.7	413	1.3743	11.82	6.35	3.042	0.950	0.756	105.05		
32.00	1.2810	1.2832	409.9	3.444	871.0	127.2	446	1.3776	12.98	6.98	3.279	0.957	0.748	104.30		
34.00	1.3033	1.3056	443.1	3.723	860.2	138.1	479	1.3809				0.966	0.743	103.33		
36.00	1.3263	1.3287	477.5	4.012	848.8	149.4	513	1.3843				0.977	0.738	102.13		
38.00	1.3501	1.3525	513.0	4.311	837.0	161.2	548	1.3878				0.991	0.736	100.69		
40.00	1.3746	1.3770	549.8	4.620	824.8	173.5	584	1.3914				1.008	0.735	99.01		

44 POTASSIUM CARBONATE, $K_2CO_3 \cdot 1\frac{1}{2}H_2O$

MOLECULAR WEIGHT = 138.20 FORMULA WEIGHT, HYDRATE = 165.23
RELATIVE SPECIFIC REFRACTIVITY = 0.608

0.00 % by wt. data are the same for all compounds.
For Values of 0.00 wt. % solutions see Table 1, Acetic Acid.

A% by wt.	H% by wt.	ρ D_4^{20}	D_{20}^{20}	C_s g/l	M g-mol/l	C_w g/l	$(C_o - C_w)$ g/l	$(n - n_o)$ × 10^4	n	Δ °C	O Os/kg	S g-mol/l	η/η_o	η/ρ cS	φ rhe	γ mmho/cm	T g-mol/l
0.50	0.60	1.0027	1.0045	5.0	0.036	997.7	0.5	9	1.3339	0.176	0.095	0.050	1.011	1.011	98.68	7.0	0.072
1.00	1.20	1.0072	1.0090	10.1	0.073	997.2	1.1	17	1.3347	0.339	0.182	0.097	1.023	1.018	97.57	13.6	0.144
2.00	2.39	1.0163	1.0181	20.3	0.147	995.9	2.3	35	1.3365	0.661	0.355	0.192	1.046	1.031	95.41	25.4	0.287
3.00	3.59	1.0254	1.0272	30.8	0.223	994.6	3.6	52	1.3382	0.986	0.530	0.288	1.069	1.045	93.36	36.7	0.428
4.00	4.78	1.0345	1.0363	41.4	0.299	993.1	5.1	69	1.3399	1.322	0.711	0.387	1.092	1.058	91.37	47.4	0.568
5.00	5.98	1.0437	1.0455	52.2	0.378	991.5	6.7	86	1.3416	1.668	0.897	0.488	1.117	1.072	89.35	58.0	0.712
6.00	7.17	1.0529	1.0548	63.2	0.457	989.7	8.5	103	1.3433	2.025	1.089	0.592	1.144	1.088	87.27	68.5	0.862
7.00	8.37	1.0622	1.0640	74.4	0.538	987.8	10.4	120	1.3450	2.395	1.287	0.699	1.172	1.105	85.17	78.8	1.02
8.00	9.56	1.0715	1.0734	85.7	0.620	985.8	12.5	137	1.3467	2.774	1.492	0.807	1.202	1.124	83.06	88.9	1.17
9.00	10.76	1.0809	1.0828	97.3	0.704	983.6	14.6	154	1.3484	3.167	1.702	0.919	1.233	1.143	80.94	98.8	1.34
10.00	11.96	1.0904	1.0923	109.0	0.789	981.3	16.9	171	1.3501	3.574	1.921	1.033	1.266	1.163	78.83	109.	1.50
12.00	14.35	1.1095	1.1115	133.1	0.963	976.4	21.8	205	1.3535	4.445	2.390	1.274	1.336	1.206	74.73	127.	1.83
14.00	16.74	1.1291	1.1311	158.1	1.144	971.0	27.2	239	1.3569	5.39	2.90	1.529	1.411	1.252	70.75	144.	2.17
16.00	19.13	1.1490	1.1510	183.8	1.330	965.2	33.1	273	1.3603	6.42	3.45	1.794	1.494	1.303	66.79	160.	2.53
18.00	21.52	1.1692	1.1713	210.5	1.523	958.8	39.5	307	1.3637	7.55	4.06	2.077	1.591	1.363	62.74	175.	2.90
20.00	23.91	1.1898	1.1919	238.0	1.722	951.9	46.4	341	1.3671	8.82	4.74	2.381	1.704	1.435	58.57	188.	3.31
24.00	28.69	1.2320	1.2342	295.7	2.140	936.3	61.9	409	1.3739	11.96	6.43	3.071	1.974	1.605	50.57	211.	4.24
28.00	33.48	1.2755	1.2778	357.1	2.584	918.4	79.9	477	1.3807	16.01	8.61	3.849	2.326	1.828	42.90		
32.00	38.26	1.3204	1.3227	422.5	3.057	897.8	100.4	544	1.3874	21.46	11.54		2.828	2.146	35.29		
36.00	43.04	1.3665	1.3690	492.0	3.560	874.6	123.6	610	1.3940	28.58	15.37		3.496	2.563	28.55		
40.00	47.82	1.4142	1.4167	565.7	4.093	848.5	149.7	676	1.4006	37.55	20.19		4.351	3.083	22.94		
44.00	52.61	1.4633	1.4659	643.9	4.659	819.5	178.8	741	1.4071				5.709	3.909	17.48		
48.00	57.39	1.5142	1.5169	726.8	5.259	787.4	210.8	806	1.4136				7.749	5.128	12.88		
50.00	59.78	1.5404	1.5431	770.2	5.573	770.2	228.0	838	1.4168				9.350	6.082	10.67		

45 POTASSIUM CHLORIDE, KCl

MOLECULAR WEIGHT = 74.55
RELATIVE SPECIFIC REFRACTIVITY = 0.758

0.00 % by wt. data are the same for all compounds.
For Values of 0.00 wt. % solutions see Table 1, Acetic Acid.

A% by wt.	D_4^{20}	D_{20}^{20}	C_s g/l	M g-mol/l	C_w g/l	$(C_o - C_w)$ g/l	$(n - n_o)$ × 10^4	n	Δ °C	O Os/kg	S g-mol/l	η/η_o	η/ρ cS	φ rhe	γ mmho/cm	T g-mol/l
0.50	1.0014	1.0032	5.0	0.067	996.4	1.8	7	1.3337	0.234	0.126	0.067	0.998	0.999	99.97	8.2	0.084
1.00	1.0046	1.0064	10.0	0.135	994.6	3.7	14	1.3343	0.463	0.249	0.134	0.997	0.995	100.08	15.7	0.169
1.50	1.0078	1.0096	15.1	0.203	992.7	5.5	20	1.3350	0.691	0.372	0.201	0.997	0.991	100.11	22.7	0.253
2.00	1.0110	1.0128	20.2	0.271	990.8	7.4	27	1.3357	0.920	0.495	0.268	0.997	0.988	100.12	29.5	0.338
2.50	1.0142	1.0160	25.4	0.340	988.9	9.4	34	1.3364	1.149	0.618	0.336	0.996	0.984	100.16	36.5	0.426
3.00	1.0174	1.0192	30.5	0.409	986.9	11.3	41	1.3371	1.380	0.742	0.404	0.996	0.981	100.20	43.6	0.517
3.50	1.0207	1.0225	35.7	0.479	984.9	13.3	47	1.3377	1.613	0.867	0.472	0.995	0.977	100.26	50.6	0.610
4.00	1.0239	1.0257	41.0	0.549	982.9	15.3	54	1.3384	1.847	0.993	0.540	0.995	0.974	100.32	57.6	0.707
4.50	1.0271	1.0289	46.2	0.620	980.9	17.3	61	1.3391	2.083	1.120	0.609	0.994	0.970	100.38	64.8	0.807
5.00	1.0304	1.0322	51.5	0.691	978.8	19.4	68	1.3398	2.322	1.248	0.678	0.994	0.966	100.44	71.9	0.911
5.50	1.0336	1.0354	56.8	0.763	976.8	21.5	74	1.3404	2.561	1.377	0.747	0.993	0.963	100.51	79.1	1.01
6.00	1.0369	1.0387	62.2	0.835	974.7	23.6	81	1.3411	2.803	1.507	0.816	0.992	0.959	100.60	86.2	1.12
6.50	1.0402	1.0420	67.6	0.907	972.5	25.7	88	1.3418	3.047	1.638	0.885	0.991	0.955	100.69	93.3	1.24
7.00	1.0434	1.0453	73.0	0.980	970.4	27.8	95	1.3425	3.294	1.771	0.955	0.990	0.951	100.79	100.	1.36
7.50	1.0467	1.0486	78.5	1.053	968.2	30.0	102	1.3431	3.544	1.905	1.025	0.989	0.947	100.89	108.	1.48
8.00	1.0500	1.0519	84.0	1.127	966.0	32.2	108	1.3438	3.797	2.041	1.095	0.988	0.943	100.99	115.	1.61
8.50	1.0533	1.0552	89.5	1.201	963.8	34.4	115	1.3445	4.050	2.177	1.165	0.987	0.939	101.07	122.	1.74
9.00	1.0566	1.0585	95.1	1.276	961.6	36.7	122	1.3452	4.300	2.312	1.234	0.987	0.936	101.14	129.	1.88
9.50	1.0600	1.0619	100.7	1.351	959.3	39.0	129	1.3459	4.550	2.446	1.303	0.986	0.932	101.18	136.	2.02
10.00	1.0633	1.0652	106.3	1.426	957.0	41.2	136	1.3466	4.805	2.583	1.372	0.986	0.929	101.20	143.	2.17
11.00	1.0700	1.0719	117.7	1.579	952.3	45.9	149	1.3479	5.33	2.87	1.513	0.987	0.924	101.15	157.	2.49
12.00	1.0768	1.0787	129.2	1.733	947.6	50.7	163	1.3493	5.88	3.16	1.657	0.988	0.919	101.01	172.	2.84
13.00	1.0836	1.0855	140.9	1.890	942.7	55.5	177	1.3507	6.45	3.47	1.804	0.990	0.915	100.82	185.	3.21
14.00	1.0905	1.0924	152.7	2.048	937.8	60.4	191	1.3521				0.992	0.912	100.58	197.	3.61
15.00	1.0974	1.0993	164.6	2.208	932.8	65.5	205	1.3535				0.995	0.908	100.34	208.	4.05
16.00	1.1043	1.1063	176.7	2.370	927.7	70.6	219	1.3549				0.997	0.905	100.11	220.	4.85
17.00	1.1114	1.1133	188.9	2.534	922.4	75.8	233	1.3563				0.999	0.901	99.87		
18.00	1.1185	1.1204	201.3	2.701	917.1	81.1	247	1.3577				1.002	0.898	99.60		
19.00	1.1256	1.1276	213.9	2.869	911.7	86.5	262	1.3592				1.005	0.895	99.26		
20.00	1.1328	1.1348	226.6	3.039	906.2	92.0	276	1.3606				1.010	0.893	98.81		
22.00	1.1474	1.1494	252.4	3.386	895.0	103.2	305	1.3635				1.022	0.893	97.64		
24.00	1.1623	1.1643	278.9	3.742	883.3	114.9	335	1.3665				1.038	0.895	96.15		

46 POTASSIUM CHROMATE, K_2CrO_4

MOLECULAR WEIGHT = 194.20
RELATIVE SPECIFIC REFRACTIVITY = 0.820

0.00 % by wt. data are the same for all compounds.
For Values of 0.00 wt. % solutions see Table 1, Acetic Acid.

A% by wt.	ρ D_4^{20}	D_{20}^{20}	C_s g/l	M g-mol/l	C_w g/l	$(C_o - C_w)$ g/l	$(n - n_o)$ × 10^4	n	Δ °C	O Os/kg	S g-mol/l	η/η_o	η/ρ cS	φ rhe	γ mmho/cm	T g-mol/l
0.50	1.0023	1.0040	5.0	0.026	997.3	1.0	11	1.3341	0.118	0.063	0.033	1.004	1.004	99.35	5.6	0.056
1.00	1.0063	1.0081	10.1	0.052	996.2	2.0	22	1.3352	0.235	0.126	0.067	1.009	1.005	98.91	10.8	0.113
2.00	1.0144	1.0162	20.3	0.104	994.1	4.2	44	1.3374	0.464	0.249	0.134	1.018	1.006	98.04	20.5	0.226
3.00	1.0224	1.0242	30.7	0.158	991.7	6.5	66	1.3396	0.682	0.367	0.198	1.027	1.006	97.21	29.8	0.340
4.00	1.0305	1.0323	41.2	0.212	989.2	9.0	87	1.3417	0.892	0.480	0.260	1.035	1.007	96.41	38.8	0.455
5.00	1.0386	1.0404	51.9	0.267	986.6	11.6	109	1.3439	1.103	0.593	0.322	1.044	1.007	95.59	47.8	0.573
6.00	1.0467	1.0486	62.8	0.323	983.9	14.3	131	1.3461	1.314	0.707	0.384	1.053	1.008	94.73	56.7	0.695
7.00	1.0550	1.0568	73.8	0.380	981.1	17.1	153	1.3483	1.524	0.819	0.446	1.063	1.010	93.85	65.6	0.819
8.00	1.0633	1.0652	85.1	0.438	978.2	20.0	175	1.3505	1.734	0.932	0.507	1.074	1.012	92.95	74.3	0.947
9.00	1.0718	1.0736	96.5	0.497	975.3	22.9	197	1.3527	1.946	1.047	0.569	1.085	1.014	92.02	83.0	1.08
10.00	1.0803	1.0823	108.0	0.556	972.3	25.9	220	1.3550	2.164	1.163	0.632	1.096	1.017	91.06	91.6	1.22
12.00	1.0980	1.0999	131.8	0.678	966.2	32.0	267	1.3597	2.613	1.405	0.761	1.120	1.022	89.07	109.	1.50
14.00	1.1163	1.1183	156.3	0.805	960.0	38.2	315	1.3645	3.076	1.654	0.893	1.147	1.030	87.00	125.	1.81
16.00	1.1344	1.1364	181.5	0.935	952.9	45.3	363	1.3693	3.555	1.911	1.028	1.176	1.039	84.84	141.	2.14
18.00	1.1527	1.1547	207.5	1.068	945.2	53.0	410	1.3740	4.050	2.177	1.165	1.209	1.051	82.56	157.	2.49
20.00	1.1720	1.1740	234.4	1.207	937.6	60.7	459	1.3789	4.561	2.452	1.306	1.245	1.064	80.16	172.	2.86
22.00	1.1920	1.1941	262.2	1.350	929.7	68.5	509	1.3839				1.285	1.081	77.64	189.	3.32
24.00	1.2125	1.2146	291.0	1.498	921.5	76.7	561	1.3891				1.330	1.099	75.04	201.	3.78
26.00	1.2335	1.2356	320.7	1.651	912.8	85.5	614	1.3944				1.379	1.120	72.37	214.	4.40
28.00	1.2547	1.2570	351.3	1.809	903.4	94.8	668	1.3998				1.433	1.144	69.64		
30.00	1.2763	1.2785	382.9	1.972	893.4	104.8	724	1.4054				1.492	1.171	66.89		

47 POTASSIUM DICHROMATE, $K_2Cr_2O_7$

MOLECULAR WEIGHT = 294.21
RELATIVE SPECIFIC REFRACTIVITY = 0.835

0.00 % by wt. data are the same for all compounds.
For Values of 0.00 wt. % solutions see Table 1, Acetic Acid.

A% by wt.	ρ D_4^{20}	D_{20}^{20}	C_s g/l	M g-mol/l	C_w g/l	$(C_o - C_w)$ g/l	$(n - n_o)$ × 10^4	n	Δ °C	O Os/kg	S g-mol/l	η/η_o	η/ρ cS	φ rhe	γ mmho/cm	T g-mol/l
0.50	1.0017	1.0035	5.0	0.017	996.7	1.5	9	1.3339	0.101	0.054	0.028	1.000	1.001	99.75	3.7	0.037
1.00	1.0052	1.0070	10.1	0.034	995.1	3.1	18	1.3348	0.192	0.103	0.055	1.001	0.998	99.70	7.3	0.074
1.50	1.0087	1.0105	15.1	0.051	993.6	4.7	27	1.3357	0.278	0.149	0.080	1.001	0.995	99.65	10.6	0.111
2.00	1.0122	1.0140	20.2	0.069	992.0	6.3	36	1.3366	0.359	0.193	0.103	1.002	0.992	99.60	13.8	0.147
2.50	1.0157	1.0175	25.4	0.086	990.3	7.9	46	1.3375	0.435	0.234	0.126	1.003	0.989	99.55	17.0	0.184
3.00	1.0193	1.0211	30.6	0.104	988.7	9.5	55	1.3385				1.003	0.986	99.50	20.1	0.221
3.50	1.0228	1.0246	35.8	0.122	987.0	11.2	64	1.3394				1.003	0.983	99.45	23.1	0.258
4.00	1.0264	1.0282	41.1	0.140	985.3	12.9	73	1.3403				1.004	0.980	99.40	26.2	0.296
4.50	1.0300	1.0318	46.3	0.158	983.6	14.6	83	1.3413				1.004	0.977	99.35	29.2	0.334
5.00	1.0336	1.0354	51.7	0.176	981.9	16.3	92	1.3422				1.005	0.974	99.30	32.2	0.372

47 POTASSIUM DICHROMATE, $K_2Cr_2O_7$—(Continued)

A% by wt.	ρ D_4^{20}	D_{20}^{20}	C_s g/l	M g-mol/l	C_w g/l	$(C_o - C_w)$ g/l	$(n - n_o)$ × 10^4	n	Δ °C	O Os/kg	S g-mol/l	η/η_o	η/ρ cS	φ rhe	γ mmho/cm	T g-mol/l
5.50	1.0372	1.0390	57.0	0.194	980.1	18.1	102	1.3431				1.006	0.971	99.25	35.3	0.410
6.00	1.0408	1.0426	62.4	0.212	978.3	19.9	111	1.3441				1.006	0.969	99.21	38.3	0.449
6.50	1.0444	1.0463	67.9	0.231	976.5	21.7	121	1.3450				1.007	0.966	99.16	41.4	0.488
7.00	1.0481	1.0499	73.4	0.249	974.7	23.5	130	1.3460				1.007	0.963	99.11	44.4	0.528
7.50	1.0517	1.0536	78.9	0.268	972.8	25.4	140	1.3470				1.008	0.960	99.06	47.4	0.567
8.00	1.0554	1.0573	84.4	0.287	971.0	27.3	149	1.3479				1.008	0.957	99.01	50.3	0.607
8.50	1.0591	1.0610	90.0	0.306	969.1	29.2	159	1.3489				1.009	0.954	98.96	53.3	0.647
9.00	1.0628	1.0647	95.7	0.325	967.2	31.1	169	1.3499				1.009	0.951	98.91	56.3	0.688
9.50	1.0665	1.0684	101.3	0.344	965.2	33.0	179	1.3509				1.010	0.948	98.86	59.2	0.729
10.00	1.0703	1.0722	107.0	0.364	963.3	35.0	189	1.3519				1.010	0.946	98.81	62.1	0.770

48 POTASSIUM FERRICYANIDE, $K_3Fe(CN)_6$

MOLECULAR WEIGHT = 329.26
RELATIVE SPECIFIC REFRACTIVITY = 0.929

0.00 % by wt. data are the same for all compounds.
For Values of 0.00 wt. % solutions see Table 1, Acetic Acid.

A% by wt.	ρ D_4^{20}	D_{20}^{20}	C_s g/l	M g-mol/l	C_w g/l	$(C_o - C_w)$ g/l	$(n - n_o)$ × 10^4	n	Δ °C	O Os/kg	S g-mol/l	η/η_o	η/ρ cS	φ rhe	γ mmho/cm	T g-mol/l
0.50	1.0008	1.0026	5.0	0.015	995.8	2.4	8	1.3338	0.093	0.050	0.026	1.003	1.004	99.49	5.0	0.050
1.00	1.0034	1.0052	10.0	0.030	993.4	4.8	17	1.3347	0.180	0.097	0.051	1.006	1.005	99.21	9.7	0.100
1.50	1.0061	1.0079	15.1	0.046	991.0	7.3	25	1.3355	0.258	0.139	0.074	1.009	1.005	98.95	14.0	0.150
2.00	1.0087	1.0105	20.2	0.061	988.5	9.7	34	1.3364	0.334	0.180	0.096	1.011	1.004	98.71	18.2	0.199
2.50	1.0114	1.0132	25.3	0.077	986.1	12.1	42	1.3372	0.412	0.222	0.119	1.013	1.004	98.51	22.4	0.249
3.00	1.0140	1.0158	30.4	0.092	983.6	14.6	50	1.3380	0.491	0.264	0.142	1.015	1.003	98.33	26.5	0.298
3.50	1.0167	1.0185	35.6	0.108	981.1	17.1	59	1.3389	0.570	0.306	0.165	1.017	1.002	98.17	30.5	0.348
4.00	1.0194	1.0212	40.8	0.124	978.6	19.6	67	1.3397	0.649	0.349	0.189	1.018	1.001	98.02	34.5	0.399
4.50	1.0221	1.0239	46.0	0.140	976.1	22.1	75	1.3405	0.728	0.392	0.212	1.020	1.000	97.85	38.4	0.450
5.00	1.0249	1.0267	51.2	0.156	973.6	24.6	84	1.3414	0.807	0.434	0.235	1.022	0.999	97.65	42.4	0.502
5.50	1.0276	1.0294	56.5	0.172	971.1	27.2	92	1.3422	0.885	0.476	0.258	1.024	0.999	97.43	46.4	0.555
6.00	1.0304	1.0322	61.8	0.188	968.5	29.7	100	1.3430	0.962	0.517	0.281	1.027	0.999	97.19	50.5	0.609
6.50	1.0331	1.0350	67.2	0.204	966.0	32.3	109	1.3438	1.039	0.559	0.303	1.030	0.999	96.94	54.5	0.663
7.00	1.0359	1.0377	72.5	0.220	963.4	34.8	117	1.3447	1.115	0.600	0.326	1.032	0.999	96.67	58.5	0.719
7.50	1.0387	1.0406	77.9	0.237	960.8	37.4	125	1.3455	1.192	0.641	0.348	1.035	0.999	96.39	62.5	0.775
8.00	1.0415	1.0434	83.3	0.253	958.2	40.0	134	1.3464	1.268	0.682	0.371	1.039	0.999	96.10	66.5	0.832
8.50	1.0444	1.0462	88.8	0.270	955.6	42.6	142	1.3472	1.345	0.723	0.393	1.042	0.999	95.80	70.4	0.890
9.00	1.0472	1.0491	94.2	0.286	953.0	45.3	150	1.3480	1.422	0.764	0.416	1.045	1.000	95.49	74.4	0.949
9.50	1.0501	1.0519	99.8	0.303	950.3	47.9	159	1.3489	1.500	0.807	0.439	1.049	1.000	95.18	78.4	1.01
10.00	1.0530	1.0548	105.3	0.320	947.7	50.6	167	1.3497	1.580	0.849	0.462	1.052	1.001	94.87	82.4	1.07
11.00	1.0588	1.0607	116.5	0.354	942.3	55.9	184	1.3514	1.740	0.935	0.509	1.059	1.002	94.22	90.4	1.20
12.00	1,0647	1.0665	127.8	0.388	936.9	61.3	201	1.3531	1.901	1.022	0.556	1.067	1.004	93.55	98.4	1.33
13.00	1.0706	1.0725	139.2	0.423	931.4	66.8	219	1.3549	2.066	1.110	0.604	1.075	1.006	92.85	106.0	1.46
14.00	1.0766	1.0785	150.7	0.458	925.9	72.4	236	1.3566	2.235	1.202	0.653	1.084	1.008	92.11	114.0	1.60
15.00	1.0826	1.0845	162.4	0.493	920.2	78.0	254	1.3583	2.413	1.298	0.704	1.093	1.012	91.31	122.0	1.74
16.00	1.0887	1.0907	174.2	0.529	914.5	83.7	271	1.3601	2.600	1.398	0.758	1.103	1.015	90.46	129.0	1.88
17.00	1.0949	1.0968	186.1	0.565	908.8	89.5	289	1.3619	2.795	1.503	0.813	1.114	1.020	89.58	136.0	2.03
18.00	1.1011	1.1031	198.2	0.602	902.9	95.3	307	1.3637	2.997	1.611	0.871	1.126	1.024	88.65	143.0	2.17
19.00	1.1074	1.1093	210.4	0.639	897.0	101.3	325	1.3655	3.208	1.725	0.930	1.138	1.030	87.67	150.0	2.33
20.00	1.1137	1.1156	222.7	0.676	890.9	107.3	344	1.3674	3.427	1.843	0.992	1.152	1.036	86.63	157.0	2.50
22.00	1.1263	1.1283	247.8	0.753	878.6	119.7	381	1.3711				1.182	1.052	84.41	171.0	2.88
24.00	1.1391	1.1411	273.4	0.830	865.7	132.5	419	1.3749				1.217	1.070	82.02	185.0	3.31
26.00	1.1519	1.1540	299.5	0.910	852.4	145.8	457	1.3787				1.256	1.092	79.49	198.0	3.79
28.00	1.1647	1.1668	326.1	0.990	838.6	159.6	495	1.3825				1.299	1.117	76.83	210.0	4.34
30.00	1.1773	1.1794	353.2	1.073	824.1	174.1	533	1.3863				1.347	1.146	74.09	222.0	4.96

49 POTASSIUM FERROCYANIDE, $K_4Fe(CN)_6 \cdot 3H_2O$

MOLECULAR WEIGHT = 368.36 FORMULA WEIGHT, HYDRATE = 422.41
RELATIVE SPECIFIC REFRACTIVITY = 0.921

0.00 % by wt. data are the same for all compounds.
For Values of 0.00 wt. % solutions see Table 1, Acetic Acid.

A% by wt.	H% by wt.	ρ D_4^{20}	D_{20}^{20}	C_s g/l	M g-mol/l	C_w g/l	$(C_o - C_w)$ g/l	$(n - n_o)$ × 10^4	n	Δ °C	O Os/kg	S g-mol/l	η/η_o	η/ρ cS	φ rhe	γ mmho/cm	T g-mol/l
0.50	0.57	1.0016	1.0034	5.0	0.014	996.6	1.6	10	1.3340	0.083	0.044	0.023	1.007	1.008	99.07	4.8	0.048
1.00	1.15	1.0049	1.0067	10.0	0.027	994.9	3.3	21	1.3351	0.162	0.087	0.046	1.014	1.011	98.42	9.3	0.096
1.50	1.72	1.0083	1.0101	15.1	0.041	993.1	5.1	31	1.3361	0.239	0.129	0.068	1.020	1.013	97.87	13.4	0.142
2.00	2.29	1.0116	1.0134	20.2	0.055	991.4	6.9	41	1.3371	0.314	0.169	0.090	1.025	1.015	97.37	17.3	0.188
2.50	2.87	1.0149	1.0167	25.4	0.069	989.5	8.7	51	1.3381	0.385	0.207	0.111	1.031	1.017	96.84	21.3	0.235
3.00	3.44	1.0182	1.0200	30.5	0.083	987.7	10.6	62	1.3391	0.453	0.243	0.131	1.036	1.020	96.33	25.2	0.282
3.50	4.01	1.0215	1.0233	35.8	0.097	985.8	12.5	72	1.3402	0.518	0.278	0.150	1.042	1.022	95.81	29.0	0.331
4.00	4.59	1.0248	1.0266	41.0	0.111	983.8	14.4	82	1.3412	0.582	0.313	0.169	1.047	1.024	95.30	32.9	0.379
4.50	5.16	1.0281	1.0300	46.3	0.126	981.9	16.4	92	1.3422	0.645	0.347	0.187	1.053	1.026	94.77	36.7	0.428
5.00	5.73	1.0315	1.0333	51.6	0.140	979.9	18.4	102	1.3432	0.709	0.381	0.206	1.059	1.029	94.24	40.6	0.478
5.50	6.31	1.0348	1.0366	56.9	0.155	977.9	20.4	112	1.3442	0.773	0.415	0.225	1.065	1.031	93.70	44.4	0.528
6.00	6.88	1.0381	1.0400	62.3	0.169	975.8	22.4	122	1.3452	0.836	0.450	0.244	1.071	1.034	93.15	48.2	0.578
6.50	7.45	1.0415	1.0433	67.7	0.184	973.8	24.5	132	1.3462	0.899	0.483	0.262	1.078	1.037	92.60	51.9	0.629
7.00	8.03	1.0449	1.0467	73.1	0.199	971.7	26.5	143	1.3473	0.961	0.517	0.281	1.084	1.040	92.04	55.7	0.680
7.50	8.60	1.0483	1.0501	78.6	0.213	969.6	28.6	153	1.3483	1.023	0.550	0.299	1.091	1.043	91.47	59.4	0.731
8.00	9.17	1.0517	1.0535	84.1	0.228	967.5	30.7	163	1.3493	1.084	0.583	0.317	1.098	1.046	90.90	63.1	0.784
8.50	9.75	1.0553	1.0572	89.7	0.244	965.6	32.6	175	1.3505	1.144	0.615	0.334	1.105	1.049	90.32	66.8	0.837
9.00	10.32	1.0588	1.0607	95.3	0.259	963.5	34.7	185	1.3515	1.204	0.647	0.352	1.112	1.052	89.74	70.6	0.891
9.50	10.89	1.0623	1.0641	100.9	0.274	961.3	36.9	196	1.3525	1.262	0.679	0.369	1.119	1.056	89.15	74.3	0.947
10.00	11.47	1.0657	1.0676	106.6	0.289	959.2	39.1	206	1.3536	1.320	0.710	0.386	1.127	1.060	88.55	78.1	1.00
11.00	12.61	1.0727	1.0746	118.0	0.320	954.7	43.5	227	1.3557	1.433	0.771	0.419	1.143	1.067	87.35	85.8	1.12

49 POTASSIUM FERROCYANIDE, $K_4Fe(CN)_6 \cdot 3H_2O$—(Continued)

A% by wt.	H% by wt.	ρ D_4^{20}	D_{20}^{20}	C_s g/l	M g-mol/l	C_w g/l	$(C_o - C_w)$ g/l	$(n - n_o)$ × 10^4	n	Δ °C	O Os/kg	S g-mol/l	η/η_o	η/ρ cS	φ rhe	γ mmho/cm	T g-mol/l
12.00	13.76	1.0798	1.0817	129.6	0.352	950.2	48.0	248	1.3578				1.159	1.075	86.13	93.7	1.25
13.00	14.91	1.0869	1.0889	141.3	0.384	945.6	52.6	270	1.3600				1.176	1.084	84.89	101.	1.38
14.00	16.05	1.0941	1.0961	153.2	0.416	941.0	57.3	291	1.3621				1.193	1.093	83.63	109.	1.51
15.00	17.20	1.1014	1.1034	165.2	0.449	936.2	62.0	313	1.3643				1.212	1.103	82.34	117.	1.64
16.00	18.35	1.1088	1.1107	177.4	0.482	931.4	66.9	335	1.3665				1.232	1.113	81.03	124.	1.78
17.00	19.49	1.1162	1.1182	189.8	0.515	926.4	71.8	357	1.3687				1.252	1.124	79.70	131.	1.91
18.00	20.64	1.1237	1.1257	202.3	0.549	921.4	76.8	379	1.3709				1.274	1.136	78.35	137.	2.05
19.00	21.79	1.1312	1.1332	214.9	0.583	916.3	81.9	401	1.3731				1.296	1.148	76.98	143.	2.18
20.00	22.93	1.1389	1.1409	227.8	0.618	911.1	87.1	424	1.3754				1.320	1.161	75.61	149.	2.32

50 POTASSIUM HYDROXIDE, KOH

MOLECULAR WEIGHT = 56.11
RELATIVE SPECIFIC REFRACTIVITY = 0.680

0.00 % by wt. data are the same for all compounds.
For Values of 0.00 wt. % solutions see Table 1, Acetic Acid.

A% by wt.	ρ D_4^{20}	D_{20}^{20}	C_s g/l	M g-mol/l	C_w g/l	$(C_o - C_w)$ g/l	$(n - n_o)$ × 10^4	n	Δ °C	O Os/kg	S g-mol/l	η/η_o	η/ρ cS	φ rhe	γ mmho/cm	T g-mol/l
0.50	1.0025	1.0043	5.0	0.089	997.5	0.7	10	1.3340	0.299	0.161	0.086	1.008	1.008	98.98	20.0	0.220
1.00	1.0068	1.0086	10.1	0.179	996.7	1.5	20	1.3350	0.609	0.327	0.177	1.017	1.012	98.13	38.5	0.452
1.50	1.0111	1.0129	15.2	0.270	995.9	2.3	29	1.3359	0.924	0.497	0.269	1.026	1.017	97.25	56.9	0.697
2.00	1.0155	1.0172	20.3	0.362	995.1	3.1	39	1.3369	1.242	0.668	0.363	1.036	1.022	96.36	75.0	0.957
2.50	1.0198	1.0216	25.5	0.454	994.3	3.9	49	1.3379	1.562	0.840	0.457	1.046	1.027	95.45	92.8	1.23
3.00	1.0242	1.0260	30.7	0.548	993.5	4.8	58	1.3388	1.886	1.014	0.551	1.056	1.033	94.52	110.	1.53
3.50	1.0286	1.0304	36.0	0.642	992.6	5.6	68	1.3398	2.217	1.192	0.648	1.066	1.039	93.58	128.	1.85
4.00	1.0330	1.0348	41.3	0.736	991.7	6.5	78	1.3408	2.565	1.379	0.748	1.077	1.045	92.63	144.	2.20
4.50	1.0374	1.0393	46.7	0.832	990.8	7.5	87	1.3417	2.940	1.581	0.854	1.089	1.051	91.68	161.	2.58
5.00	1.0419	1.0437	52.1	0.928	989.8	8.4	97	1.3427	3.356	1.804	0.972	1.100	1.058	90.71	178.	3.00
5.50	1.0464	1.0482	57.6	1.026	988.8	9.4	106	1.3436	3.747	2.014	1.081	1.112	1.065	89.75	193.	3.47
6.00	1.0509	1.0527	63.1	1.124	987.8	10.4	116	1.3445	4.144	2.228	1.191	1.124	1.072	88.78	206.	3.97
6.50	1.0554	1.0572	68.6	1.223	986.8	11.5	125	1.3455	4.521	2.431	1.295	1.137	1.079	87.81	216.	4.52
7.00	1.0599	1.0618	74.2	1.322	985.7	12.5	134	1.3464	4.921	2.649	1.403	1.149	1.086	86.84		
7.50	1.0644	1.0663	79.8	1.423	984.6	13.6	144	1.3474				1.162	1.094	85.87		
8.00	1.0690	1.0709	85.5	1.524	983.5	14.8	153	1.3483				1.175	1.102	84.91		
8.50	1.0736	1.0755	91.3	1.626	982.3	15.9	162	1.3492				1.189	1.110	83.95		
9.00	1.0781	1.0801	97.0	1.729	981.1	17.1	172	1.3502				1.203	1.118	82.98		
9.50	1.0827	1.0847	102.9	1.833	979.9	18.4	181	1.3511				1.217	1.126	82.03		
10.00	1.0873	1.0893	108.7	1.938	978.6	19.6	190	1.3520				1.231	1.134	81.07		
11.00	1.0966	1.0985	120.6	2.150	976.0	22.3	209	1.3539				1.261	1.152	79.17		
12.00	1.1059	1.1079	132.7	2.365	973.2	25.0	228	1.3558				1.291	1.170	77.28		
13.00	1.1153	1.1172	145.0	2.584	970.3	28.0	246	1.3576				1.324	1.189	75.39		
14.00	1.1246	1.1266	157.5	2.806	967.2	31.0	265	1.3595				1.358	1.210	73.50		
15.00	1.1341	1.1361	170.1	3.032	964.0	34.3	284	1.3614				1.394	1.232	71.59		
16.00	1.1435	1.1456	183.0	3.261	960.6	37.7	302	1.3632				1.433	1.255	69.66		
17.00	1.1531	1.1551	196.0	3.493	957.0	41.2	321	1.3651				1.474	1.281	67.69		
18.00	1.1626	1.1647	209.3	3.730	953.3	44.9	340	1.3670				1.518	1.308	65.75		
19.00	1.1722	1.1743	222.7	3.969	949.5	48.8	358	1.3688				1.565	1.338	63.77		
20.00	1.1818	1.1839	236.4	4.212	945.4	52.8	377	1.3707				1.616	1.370	61.76		
22.00	1.2014	1.2035	264.3	4.710	937.1	61.2	414	1.3744				1.729	1.442	57.72		
24.00	1.2210	1.2231	293.0	5.223	927.9	70.3	451	1.3781				1.857	1.524	53.74		
26.00	1.2408	1.2430	322.6	5.750	918.2	80.0	488	1.3818				2.002	1.617	49.85		
28.00	1.2609	1.2632	353.1	6.292	907.9	90.4	524	1.3854				2.166	1.721	46.09		
30.00	1.2813	1.2836	384.4	6.851	896.9	101.3	559	1.3889				2.352	1.839	42.44		
32.00	1.3020	1.3043	416.6	7.425	885.4	112.9	593	1.3923				2.565	1.974	38.90		
34.00	1.3230	1.3254	449.8	8.017	873.2	125.0	628	1.3957				2.814	2.131	35.47		
36.00	1.3444	1.3468	484.0	8.626	860.4	137.8	664	1.3993				3.105	2.315	32.14		
38.00	1.3661	1.3685	519.1	9.252	847.0	151.3	700	1.4030				3.453	2.533	28.90		
40.00	1.3881	1.3906	555.2	9.896	832.9	165.4	738	1.4068				3.871	2.795	25.78		
42.00	1.4104	1.4129	592.4	10.558	818.1	180.2	776	1.4106				4.380	3.112	22.78		
44.00	1.4331	1.4356	630.6	11.238	802.5	195.7	813	1.4143				5.003	3.498	19.95		
46.00	1.4560	1.4586	669.8	11.936	786.2	212.0	849	1.4179				5.769	3.970	17.30		
48.00	1.4791	1.4817	710.0	12.653	769.1	229.1	884	1.4214				6.713	4.547	14.87		
50.00	1.5024	1.5050	751.2	13.388	751.2	247.0	917	1.4247				7.876	5.253	12.67		

51 POTASSIUM IODIDE, KI

MOLECULAR WEIGHT = 166.03
RELATIVE SPECIFIC REFRACTIVITY = 0.647

0.00 % by wt. data are the same for all compounds.
For Values of 0.00 wt. % solutions see Table 1, Acetic Acid.

A% by wt.	ρ D_4^{20}	D_{20}^{20}	C_s g/l	M g-mol/l	C_w g/l	$(C_o - C_w)$ g/l	$(n - n_o)$ × 10^4	n	Δ °C	O Os/kg	S g-mol/l	η/η_o	η/ρ cS	φ rhe	γ mmho/cm	T g-mol/l
0.50	1.0019	1.0037	5.0	0.030	996.9	1.3	7	1.3337	0.111	0.060	0.031	0.9972	0.9973	100.1	3.8	0.038
1.00	1.0056	1.0074	10.1	0.061	995.5	2.7	13	1.3343	0.220	0.118	0.063	0.9945	0.9910	100.3	7.5	0.076
2.00	1.0131	1.0149	20.3	0.122	992.8	5.4	27	1.3357	0.431	0.232	0.124	0.9892	0.9784	100.9	14.2	0.152
3.00	1.0206	1.0224	30.6	0.184	990.0	8.2	40	1.3370	0.642	0.345	0.187	0.9840	0.9661	101.4	21.1	0.233
4.00	1.0282	1.0301	41.1	0.248	987.1	11.1	54	1.3384	0.858	0.461	0.250	0.9788	0.9538	102.0	28.1	0.320
5.00	1.0360	1.0378	51.8	0.312	984.2	14.1	67	1.3397	1.079	0.580	0.315	0.9736	0.9417	102.5	35.2	0.409
6.00	1.0438	1.0457	62.6	0.377	981.2	17.1	81	1.3411	1.304	0.701	0.381	0.9670	0.9283	103.2	42.3	0.501
7.00	1.0517	1.0536	73.6	0.443	978.1	20.1	95	1.3425	1.534	0.825	0.449	0.9608	0.9153	103.9	49.6	0.597
8.00	1.0598	1.0617	84.8	0.511	975.0	23.2	110	1.3440	1.769	0.951	0.517	0.9549	0.9028	104.5	56.9	0.697
9.00	1.0679	1.0698	96.1	0.579	971.8	26.4	124	1.3454	2.010	1.081	0.588	0.9494	0.8908	105.1	64.3	0.801
10.00	1.0762	1.0781	107.6	0.648	968.6	29.6	139	1.3469	2.257	1.213	0.659	0.9442	0.8791	105.7	71.8	0.909
12.00	1.0931	1.0950	131.2	0.790	961.9	36.3	168	1.3498	2.767	1.488	0.805	0.9348	0.8569	106.8	86.8	1.14

51 POTASSIUM IODIDE, KI—(Continued)

A% by wt.	ρ D_4^{20}	D_{20}^{20}	C_s g/l	M g-mol/l	C_w g/l	$(C_o - C_w)$ g/l	$(n - n_o)$ × 10^4	n	Δ °C	O Os/kg	S g-mol/l	η/η_o	η/ρ cS	φ rhe	γ mmho/cm	T g-mol/l
14.00	1.1105	1.1125	155.5	0.936	955.0	43.2	199	1.3529	3.304	1.776	0.957	0.9265	0.8360	107.7	102.	1.39
16.00	1.1284	1.1304	180.5	1.087	947.9	50.4	231	1.3560	3.869	2.080	1.115	0.9193	0.8163	108.6	118.	1.66
18.00	1.1469	1.1489	206.4	1.243	940.4	57.8	263	1.3593	4.464	2.400	1.279	0.9131	0.7977	109.3	133.	1.96
20.00	1.1659	1.1680	233.2	1.405	932.8	65.5	296	1.3626	5.09	2.74	1.449	0.9077	0.7801	109.9	149.	2.30
22.00	1.1856	1.1877	260.8	1.571	924.8	73.4	331	1.3661	5.76	3.09	1.624	0.9031	0.7632	110.5	164.	2.66
24.00	1.2060	1.2081	289.4	1.743	916.5	81.7	366	1.3696	6.46	3.47	1.805	0.8991	0.7470	111.0	180.	3.05
26.00	1.2270	1.2291	319.0	1.921	908.0	90.3	403	1.3733	7.21	3.87	1.993	0.8956	0.7314	111.4	195.	3.53
28.00	1.2487	1.2509	349.6	2.106	899.1	99.2	441	1.3771	8.01	4.30	2.188	0.8928	0.7164	111.8	210.	4.16
30.00	1.2712	1.2734	381.4	2.297	889.8	108.4	480	1.3810	8.86	4.76	2.390	0.8904	0.7019	112.1	224.	5.20
32.00	1.2944	1.2967	414.2	2.495	880.2	118.0	521	1.3851	9.76	5.25	2.597	0.8888	0.6880	112.3		
34.00	1.3185	1.3208	448.3	2.700	870.2	128.0	563	1.3893	10.72	5.76	2.808	0.8880	0.6748	112.4		
36.00	1.3434	1.3458	483.6	2.913	859.8	138.5	606	1.3936	11.73	6.31	3.024	0.8884	0.6626	112.3		
38.00	1.3692	1.3716	520.3	3.134	848.9	149.3	651	1.3981	12.81	6.89	3.246	0.8905	0.6517	112.1		
40.00	1.3959	1.3984	558.4	3.363	837.6	160.7	697	1.4027	13.97	7.51	3.472	0.8952	0.6426	111.5		

52 POTASSIUM NITRATE, KNO_3

MOLECULAR WEIGHT = 101.10
RELATIVE SPECIFIC REFRACTIVITY = 0.651

0.00 % by wt. data are the same for all compounds.
For Values of 0.00 wt. % solutions see Table 1, Acetic Acid.

A% by wt.	ρ D_4^{20}	D_{20}^{20}	C_s g/l	M g-mol/l	C_w g/l	$(C_o - C_w)$ g/l	$(n - n_o)$ × 10^4	n	Δ °C	O Os/kg	S g-mol/l	η/η_o	η/ρ cS	φ rhe	γ mmho/cm	T g-mol/l
0.50	1.0014	1.0031	5.0	0.050	996.4	1.9	5	1.3335	0.171	0.092	0.049	0.997	0.998	100.10	5.5	0.056
1.00	1.0045	1.0063	10.0	0.099	994.5	3.8	10	1.3339	0.333	0.179	0.096	0.994	0.992	100.40	10.7	0.112
2.00	1.0108	1.0126	20.2	0.200	990.6	7.7	19	1.3349	0.642	0.345	0.187	0.988	0.979	101.01	20.1	0.222
3.00	1.0171	1.0189	30.5	0.302	986.6	11.7	28	1.3358	0.938	0.504	0.274	0.984	0.969	101.46	29.3	0.334
4.00	1.0234	1.0252	40.9	0.405	982.5	15.7	38	1.3368	1.223	0.657	0.357	0.981	0.960	101.77	38.3	0.448
5.00	1.0298	1.0317	51.5	0.509	978.3	19.9	47	1.3377	1.498	0.805	0.438	0.978	0.952	102.05	47.0	0.563
6.00	1.0363	1.0381	62.2	0.615	974.1	24.1	57	1.3386	1.763	0.948	0.516	0.975	0.943	102.33	55.5	0.678
7.00	1.0428	1.0447	73.0	0.722	969.8	28.4	66	1.3396	2.022	1.087	0.591	0.973	0.935	102.59	63.7	0.793
8.00	1.0494	1.0512	84.0	0.830	965.4	32.8	76	1.3405	2.273	1.222	0.664	0.971	0.927	102.81	71.7	0.908
9.00	1.0560	1.0579	95.0	0.940	961.0	37.3	85	1.3415	2.517	1.353	0.734	0.969	0.920	102.98	79.5	1.03
10.00	1.0627	1.0646	106.3	1.051	956.4	41.8	95	1.3425	2.754	1.481	0.802	0.968	0.913	103.10	87.3	1.15
12.00	1.0762	1.0781	129.1	1.277	947.0	51.2	114	1.3444				0.968	0.901	103.13	103.	1.40
14.00	1.0899	1.0918	152.6	1.509	937.3	60.9	133	1.3463				0.970	0.892	102.90	117.	1.66
16.00	1.1039	1.1058	176.6	1.747	927.3	71.0	152	1.3482				0.974	0.884	102.44	131.	1.94
18.00	1.1181	1.1201	201.3	1.991	916.9	81.4	172	1.3502				0.980	0.879	101.79	145.	2.22
20.00	1.1326	1.1346	226.5	2.241	906.1	92.2	191	1.3521				0.988	0.874	101.01	157.	2.49
22.00	1.1473	1.1493	252.4	2.497	894.9	103.3	211	1.3541				0.997	0.871	100.10	168.	2.75
24.00	1.1623	1.1644	279.0	2.759	883.3	114.9	231	1.3561				1.008	0.869	99.01	178.	3.00

53 POTASSIUM OXALATE, $K_2C_2O_4 \cdot 1H_2O$

MOLECULAR WEIGHT = 166.21 FORMULA WEIGHT, HYDRATE = 184.23
RELATIVE SPECIFIC REFRACTIVITY = 0.676

0.00 % by wt. data are the same for all compounds.
For Values of 0.00 wt. % solutions see Table 1, Acetic Acid.

A% by wt.	H% by wt.	ρ D_4^{20}	D_{20}^{20}	C_s g/l	M g-mol/l	C_w g/l	$(C_o - C_w)$ g/l	$(n - n_o)$ × 10^4	n	Δ °C	O Os/kg	S g-mol/l	η/η_o	η/ρ cS	φ rhe	γ mmho/cm	T g-mol/l
0.50	0.55	1.0018	1.0035	5.0	0.030	996.8	1.5	8	1.3337	0.137	0.074	0.039	1.005	1.005	99.33	5.8	0.059
1.00	1.11	1.0053	1.0071	10.1	0.060	995.3	3.0	15	1.3345	0.273	0.147	0.078	1.010	1.007	98.81	11.4	0.119
1.50	1.66	1.0089	1.0107	15.1	0.091	993.8	4.5	22	1.3352	0.407	0.219	0.118	1.016	1.009	98.25	16.6	0.180
2.00	2.22	1.0125	1.0143	20.3	0.122	992.3	6.0	29	1.3359	0.540	0.290	0.156	1.022	1.011	97.65	21.7	0.241
2.50	2.77	1.0161	1.0179	25.4	0.153	990.7	7.5	36	1.3366	0.669	0.360	0.194	1.028	1.014	97.04	26.8	0.302
3.00	3.33	1.0198	1.0216	30.6	0.184	989.2	9.0	43	1.3373	0.797	0.429	0.232	1.035	1.017	96.41	31.7	0.364
3.50	3.88	1.0235	1.0253	35.8	0.216	987.6	10.6	50	1.3380	0.925	0.497	0.270	1.042	1.020	95.77	36.6	0.426
4.00	4.43	1.0272	1.0290	41.1	0.247	986.1	12.2	56	1.3386	1.055	0.567	0.308	1.049	1.024	95.11	41.4	0.489
4.50	4.99	1.0309	1.0327	46.4	0.279	984.5	13.8	63	1.3393	1.186	0.638	0.347	1.057	1.027	94.45	46.2	0.552
5.00	5.54	1.0346	1.0364	51.7	0.311	982.9	15.4	70	1.3400	1.318	0.709	0.385	1.064	1.030	93.80	50.9	0.615
5.50	6.10	1.0383	1.0402	57.1	0.344	981.2	17.0	77	1.3407	1.451	0.780	0.425	1.071	1.034	93.14	55.6	0.679
6.00	6.65	1.0421	1.0439	62.5	0.376	979.6	18.7	84	1.3414	1.585	0.852	0.464	1.079	1.038	92.48	60.3	0.743
6.50	7.20	1.0459	1.0477	68.0	0.409	977.9	20.3	90	1.3420	1.720	0.925	0.503	1.087	1.041	91.82	64.8	0.807
7.00	7.76	1.0497	1.0516	73.5	0.442	976.2	22.0	97	1.3427	1.856	0.998	0.543	1.095	1.045	91.15	69.3	0.872
7.50	8.31	1.0535	1.0554	79.0	0.475	974.5	23.7	104	1.3434	1.994	1.072	0.583	1.103	1.049	90.48	73.8	0.937
8.00	8.87	1.0574	1.0592	84.6	0.509	972.8	25.5	111	1.3441	2.132	1.146	0.623	1.111	1.053	89.81	78.1	1.00
8.50	9.42	1.0612	1.0631	90.2	0.543	971.0	27.2	118	1.3448	2.273	1.222	0.663	1.120	1.057	89.13	82.4	1.07
9.00	9.98	1.0651	1.0670	95.9	0.577	969.3	29.0	126	1.3456	2.414	1.298	0.704	1.128	1.061	88.45	86.6	1.13
9.50	10.53	1.0691	1.0709	101.6	0.611	967.5	30.7	133	1.3463	2.557	1.375	0.745	1.137	1.066	87.77	90.8	1.20
10.00	11.08	1.0730	1.0749	107.3	0.646	965.7	32.5	140	1.3470	2.701	1.452	0.786	1.146	1.070	87.09	94.8	1.27
11.00	12.19	1.0810	1.0829	118.9	0.715	962.1	36.2	154	1.3484	2.993	1.609	0.870	1.166	1.081	85.58	104.	1.42
12.00	13.30	1.0891	1.0910	130.7	0.786	958.4	39.8	168	1.3498	3.291	1.769	0.954	1.188	1.093	84.01	112.	1.57
13.00	14.41	1.0973	1.0993	142.7	0.858	954.7	43.6	182	1.3512	3.594	1.932	1.039	1.211	1.106	82.40	121.	1.72
14.00	15.52	1.1057	1.1077	154.8	0.931	950.9	47.3	196	1.3526	3.903	2.098	1.125	1.234	1.118	80.89	129.	1.87

54 POTASSIUM PERMANGANATE, $KMnO_4$

MOLECULAR WEIGHT = 158.04
RELATIVE SPECIFIC REFRACTIVITY = 0.811

0.00 % by wt. data are the same for all compounds.
For Values of 0.00 wt. % solutions see Table 1, Acetic Acid.

A % by wt.	ρ D_4^{20}	D_{20}^{20}	C_s g/l	M g-mol/l	C_w g/l	$(C_o - C_w)$ g/l	$(n - n_o) \times 10^4$	n	Δ °C	O Os/kg	S g-mol/l	η/η_o	η/ρ cS	φ rhe	γ mmho/cm	T g-mol/l
0.50	1.0017	1.0034	5.0	0.032	996.7	1.6			0.110	0.059	0.031	0.999	0.999	99.93	3.5	0.035
1.00	1.0051	1.0068	10.1	0.064	995.0	3.2			0.217	0.117	0.062	0.998	0.995	100.04	6.9	0.070
1.50	1.0085	1.0102	15.1	0.096	993.3	4.9			0.323	0.174	0.093	0.997	0.990	100.14	10.0	0.104
2.00	1.0118	1.0136	20.2	0.128	991.6	6.6			0.426	0.229	0.123	0.996	0.986	100.24	13.0	0.138
2.50	1.0152	1.0170	25.4	0.161	989.9	8.4						0.994	0.981	100.36	16.0	0.173
3.00	1.0186	1.0204	30.6	0.193	988.1	10.2						0.993	0.977	100.49	19.0	0.209
3.50	1.0220	1.0238	35.8	0.226	986.3	12.0						0.992	0.972	100.63	22.0	0.244
4.00	1.0254	1.0272	41.0	0.260	984.4	13.8						0.990	0.968	100.78	24.9	0.280
4.50	1.0288	1.0306	46.3	0.293	982.5	15.7						0.989	0.963	100.94	27.7	0.315
5.00	1.0322	1.0340	51.6	0.327	980.6	17.7						0.987	0.958	101.12	30.5	0.350
5.50	1.0356	1.0374	57.0	0.360	978.6	19.6						0.985	0.953	101.30	33.3	0.385
6.00	1.0390	1.0408	62.3	0.394	976.6	21.6						0.983	0.948	101.51	36.0	0.420

55 POTASSIUM PHOSPHATE, DIHYDROGEN (MONOBASIC), KH_2PO_4

MOLECULAR WEIGHT = 136.13
RELATIVE SPECIFIC REFRACTIVITY = 0.632

0.00 % by wt. data are the same for all compounds.
For Values of 0.00 wt. % solutions see Table 1, Acetic Acid.

A % by wt.	ρ D_4^{20}	D_{20}^{20}	C_s g/l	M g-mol/l	C_w g/l	$(C_o - C_w)$ g/l	$(n - n_o) \times 10^4$	n	Δ °C	O Os/kg	S g-mol/l	η/η_o	η/ρ cS	φ rhe	γ mmho/cm	T g-mol/l
0.50	1.0018	1.0035	5.0	0.037	996.8	1.5	6	1.3336	0.130	0.070	0.037	1.008	1.008	98.98	3.0	0.030
1.00	1.0053	1.0071	10.1	0.074	995.3	3.0	12	1.3342	0.252	0.136	0.072	1.017	1.014	98.13	5.9	0.059
1.50	1.0089	1.0107	15.1	0.111	993.8	4.5	18	1.3348	0.371	0.200	0.107	1.026	1.019	97.25	8.5	0.087
2.00	1.0125	1.0143	20.2	0.149	992.2	6.0	24	1.3354	0.488	0.262	0.141	1.036	1.025	96.33	11.0	0.115
2.50	1.0161	1.0179	25.4	0.187	990.7	7.5	29	1.3359	0.603	0.324	0.175	1.046	1.032	95.37	13.5	0.143
3.00	1.0197	1.0215	30.6	0.225	989.1	9.1	35	1.3365	0.720	0.387	0.209	1.058	1.039	94.37	15.9	0.171
3.50	1.0233	1.0251	35.8	0.263	987.5	10.7	41	1.3371	0.838	0.451	0.244	1.069	1.047	93.34	18.3	0.199
4.00	1.0269	1.0288	41.1	0.302	985.9	12.4	47	1.3377	0.957	0.515	0.279	1.081	1.055	92.30	20.6	0.227
4.50	1.0306	1.0324	46.4	0.341	984.2	14.0	52	1.3382	1.075	0.578	0.314	1.094	1.063	91.26	22.8	0.255
5.00	1.0342	1.0360	51.7	0.380	982.5	15.8	58	1.3388	1.190	0.640	0.348	1.106	1.072	90.24	25.0	0.282
5.50	1.0378	1.0396	57.1	0.419	980.7	17.5	64	1.3394	1.303	0.700	0.381	1.119	1.080	89.23	27.2	0.309
6.00	1.0414	1.0432	62.5	0.459	978.9	19.3	70	1.3400	1.414	0.760	0.414	1.131	1.088	88.22	29.4	0.335
6.50	1.0450	1.0469	67.9	0.499	977.1	21.2	75	1.3405	1.523	0.819	0.446	1.144	1.097	87.21	31.5	0.361
7.00	1.0486	1.0505	73.4	0.539	975.2	23.0	81	1.3411	1.631	0.877	0.477	1.158	1.106	86.21	33.5	0.387
7.50	1.0522	1.0541	78.9	0.580	973.3	24.9	87	1.3417	1.736	0.933	0.508	1.171	1.115	85.21	35.5	0.412
8.00	1.0558	1.0577	84.5	0.620	971.4	26.9	92	1.3422	1.839	0.989	0.538	1.185	1.125	84.22	37.5	0.437
8.50	1.0594	1.0613	90.1	0.662	969.4	28.9	98	1.3428	1.941	1.043	0.567	1.199	1.134	83.23	39.3	0.461
9.00	1.0630	1.0649	95.7	0.703	967.4	30.9	104	1.3434	2.039	1.096	0.596	1.213	1.144	82.25	41.2	0.485
9.50	1.0667	1.0686	101.3	0.744	965.3	32.9	109	1.3439	2.136	1.148	0.624	1.228	1.154	81.27	42.9	0.508
10.00	1.0703	1.0722	107.0	0.786	963.3	35.0	115	1.3445	2.230	1.199	0.651	1.243	1.164	80.29	44.6	0.531

56 POTASSIUM PHOSPHATE, MONOHYDROGEN (DIBASIC), $K_2HPO_4 \cdot 3H_2O$

MOLECULAR WEIGHT = 174.18 FORMULA WEIGHT, HYDRATE = 228.23
RELATIVE SPECIFIC REFRACTIVITY = 0.589

0.00 % by wt. data are the same for all compounds.
For Values of 0.00 wt. % solutions see Table 1, Acetic Acid.

A % by wt.	H % by wt.	ρ D_4^{20}	D_{20}^{20}	C_s g/l	M g.mol/l	C_w g/l	$(C_o - C_w)$ g/l	$(n - n_o) \times 10^4$	n	Δ °C	O Os/kg	S g-mol/l	η/η_o	η/ρ cS	φ rhe	γ mmho/cm	T g-mol/l
0.50	0.66	1.0025	1.0043	5.0	0.029	997.5	0.7	8.	1.3338	0.126	0.068	0.036	1.011	1.010	98.76	5.2	0.052
1.00	1.31	1.0068	1.0086	10.1	0.058	996.7	1.5	15.	1.3345	0.250	0.134	0.071	1.021	1.016	97.76	9.9	0.102
1.50	1.97	1.0110	1.0128	15.2	0.087	995.8	2.4	23.	1.3353	0.371	0.199	0.107	1.032	1.023	96.69	14.2	0.151
2.00	2.62	1.0153	1.0171	20.3	0.117	994.9	3.3	31.	1.3361	0.490	0.263	0.142	1.044	1.030	95.62	18.3	0.200
2.50	3.28	1.0195	1.0213	25.5	0.146	994.0	4.2	38.	1.3368	0.613	0.329	0.178	1.055	1.037	94.59	23.0	0.248
3.00	3.93	1.0238	1.0256	30.7	0.176	993.1	5.2	46.	1.3376	0.732	0.394	0.213	1.067	1.044	93.56	26.8	0.294
3.50	4.59	1.0281	1.0299	36.0	0.207	992.1	6.1	54.	1.3384	0.855	0.457	0.249	1.079	1.052	92.46	29.7	0.340
4.00	5.24	1.0324	1.0342	41.3	0.237	991.1	7.1	62.	1.3392	0.974	0.523	0.284	1.092	1.060	92.18	33.3	0.385
4.50	5.90	1.0368	1.0386	46.7	0.268	990.1	8.1	69.	1.3399	1.096	0.589	0.320	1.105	1.068	90.31	36.8	0.430
5.00	6.55	1.0412	1.0430	52.1	0.299	989.1	9.1	77.	1.3407	1.218	0.655	0.356	1.118	1.076	89.26	40.3	0.475
5.50	7.21	1.0456	1.0474	57.5	0.330	988.1	10.2	85.	1.3415	1.337	0.719	0.391	1.131	1.084	88.23	43.8	0.520
6.00	7.86	1.0500	1.0519	63.0	0.362	987.0	11.2	92.	1.3422	1.460	0.785	0.427	1.145	1.093	87.13	47.2	0.565
6.50	8.52	1.0545	1.0564	68.5	0.394	986.0	12.3	100.	1.3430	1.580	0.850	0.462	1.160	1.102	86.05	50.6	0.610
7.00	9.17	1.0590	1.0609	74.1	0.426	984.9	13.3	108.	1.3438	1.703	0.915	0.498	1.175	1.112	84.92	53.8	0.654
7.50	9.83	1.0635	1.0654	79.8	0.458	983.8	14.5	115.	1.3445	1.823	0.980	0.533	1.191	1.122	83.80	57.0	0.698
8.00	10.48	1.0680	1.0699	85.4	0.491	982.6	15.7	123.	1.3453	1.947	1.046	0.569	1.207	1.132	82.71	60.0	0.740

57 POTASSIUM SULFATE, K_2SO_4

MOLECULAR WEIGHT = 174.26
RELATIVE SPECIFIC REFRACTIVITY = 0.565

0.00 % by wt. data are the same for all compounds.
For Values of 0.00 wt. % solutions see Table 1, Acetic Acid.

A % by wt.	D_4^{20}	D_{20}^{20}	C_s g/l	M g-mol/l	C_w g/l	$(C_o - C_w)$ g/l	$(n - n_o) \times 10^4$	n	Δ °C	O Os/kg	S g-mol/l	η/η_o	η/ρ cS	φ rhe	γ mmho/cm	T g-mol/l
0.50	1.0022	1.0040	5.0	0.029	997.2	1.0	6	1.3336	0.136	0.073	0.039	1.004	1.004	99.37	5.8	0.059
1.00	1.0062	1.0080	10.1	0.058	996.2	2.1	13	1.3343	0.262	0.141	0.075	1.009	1.005	98.91	11.2	0.117
1.50	1.0102	1.0120	15.2	0.087	995.1	3.1	19	1.3349	0.382	0.205	0.110	1.014	1.006	98.42	16.2	0.175
2.00	1.0143	1.0161	20.3	0.116	994.0	4.2	25	1.3355	0.499	0.268	0.144	1.019	1.007	97.92	21.0	0.232
2.50	1.0183	1.0201	25.5	0.146	992.9	5.4	32	1.3362	0.614	0.330	0.178	1.025	1.008	97.39	25.6	0.289
3.00	1.0224	1.0242	30.7	0.176	991.7	6.5	38	1.3368	0.726	0.390	0.211	1.031	1.010	96.84	30.2	0.347

57 POTASSIUM SULFATE, K_2SO_4—(Continued)

A% by wt.	D_4^{20}	D_{20}^{20}	C_s g/l	M g-mol/l	C_w g/l	$(C_o - C_w)$ g/l	$(n - n_o) \times 10^4$	n	Δ °C	O Os/kg	S g-mol/l	η/η_o	η/ρ cS	φ rhe	γ mmho/cm	T g-mol/l
3.50	1.0265	1.0283	35.9	0.206	990.6	7.7	44	1.3374	0.839	0.451	0.244	1.037	1.012	96.27	34.7	0.404
4.00	1.0306	1.0324	41.2	0.237	989.4	8.9	50	1.3380	0.950	0.511	0.277	1.043	1.014	95.69	39.2	0.461
4.50	1.0347	1.0365	46.6	0.267	988.1	10.1	57	1.3386	1.061	0.571	0.310	1.049	1.016	95.10	43.7	0.518
5.00	1.0388	1.0406	51.9	0.298	986.8	11.4	63	1.3393	1.172	0.630	0.342	1.056	1.019	94.50	48.0	0.576
5.50	1.0429	1.0447	57.4	0.329	985.5	12.7	69	1.3399				1.063	1.021	93.89	52.3	0.634
6.00	1.0470	1.0489	62.8	0.361	984.2	14.0	75	1.3405				1.070	1.024	93.27	56.6	0.692
6.50	1.0512	1.0530	68.3	0.392	982.8	15.4	81	1.3411				1.077	1.027	92.64	60.8	0.751
7.00	1.0553	1.0572	73.9	0.424	981.4	16.8	87	1.3417				1.085	1.030	92.02	64.9	0.810
7.50	1.0595	1.0614	79.5	0.456	980.0	18.2	93	1.3422				1.092	1.033	91.39	69.0	0.869
8.00	1.0637	1.0655	85.1	0.488	978.6	19.7	98	1.3428				1.100	1.036	90.77	73.1	0.928
8.50	1.0679	1.0697	90.8	0.521	977.1	21.1	104	1.3434				1.107	1.039	90.14	77.0	0.988
9.00	1.0721	1.0740	96.5	0.554	975.6	22.6	110	1.3440				1.115	1.042	89.53	80.9	1.04
9.50	1.0763	1.0782	102.3	0.587	974.1	24.2	116	1.3446				1.122	1.045	88.92	84.8	1.10
10.00	1.0806	1.0825	108.1	0.620	972.5	25.7	122	1.3452				1.130	1.048	88.32	88.6	1.16

58 POTASSIUM THIOCYANATE, KSCN

MOLECULAR WEIGHT = 97.18
RELATIVE SPECIFIC REFRACTIVITY = 1.025

0.00 % by wt. data are the same for all compounds.
For Values of 0.00 wt. % solutions see Table 1, Acetic Acid.

A% by wt.	ρ D_4^{20}	D_{20}^{20}	C_s g/l	M g-mol/l	C_w g/l	$(C_o - C_w)$ g/l	$(n - n_o) \times 10^4$	n	Δ °C	O Os/kg	S g-mol/l	η/η_o	η/ρ cS	φ rhe	γ mmho/cm	T g-mol/l
0.50	1.0007	1.0025	5.0	0.051	995.7	2.5	9	1.3339	0.177	0.095	0.050	0.9974	0.9987	100.1	5.8	0.058
1.00	1.0032	1.0049	10.0	0.103	993.1	5.1	19	1.3349	0.355	0.191	0.102	0.9947	0.9936	100.3	11.4	0.119
2.00	1.0081	1.0099	20.2	0.207	987.9	10.3	37	1.3367	0.709	0.381	0.206	0.9887	0.9827	100.9	21.9	0.243
3.00	1.0130	1.0148	30.4	0.313	982.6	15.6	55	1.3385	1.061	0.570	0.310	0.9833	0.9726	101.5	32.1	0.370
4.00	1.0180	1.0198	40.7	0.419	977.3	21.0	74	1.3404	1.414	0.760	0.413	0.9785	0.9632	102.0	42.1	0.499
5.00	1.0229	1.0248	51.1	0.526	971.8	26.4	93	1.3423	1.771	0.952	0.518	0.9741	0.9542	102.5	52.2	0.632
6.00	1.0279	1.0298	61.7	0.635	966.3	32.0	111	1.3441	2.136	1.148	0.624	0.9700	0.9456	102.9	62.4	0.773
7.00	1.0330	1.0348	72.3	0.744	960.7	37.6	130	1.3460	2.506	1.347	0.731	0.9664	0.9374	103.3	72.7	0.922
8.00	1.0380	1.0398	83.0	0.854	955.0	43.3	149	1.3479	2.882	1.549	0.838	0.9632	0.9298	103.6	82.9	1.08
9.00	1.0431	1.0449	93.9	0.966	949.2	49.0	168	1.3498	3.264	1.755	0.946	0.9604	0.9225	103.9	93.	1.24
10.00	1.0482	1.0500	104.8	1.079	943.3	54.9	187	1.3517	3.653	1.964	1.055	0.9579	0.9157	104.2	103.	1.40
12.00	1.0585	1.0603	127.0	1.307	931.4	66.8	226	1.3556	4.451	2.393	1.276	0.9541	0.9032	104.6	121.	1.72
14.00	1.0689	1.0708	149.6	1.540	919.2	79.0	265	1.3595	5.28	2.84	1.498	0.9521	0.8925	104.8	138.	2.06
16.00	1.0795	1.0814	172.7	1.777	906.8	91.5	304	1.3634				0.9523	0.8840	104.8	156.	2.47
18.00	1.0902	1.0922	196.2	2.019	894.0	104.3	345	1.3674				0.9550	0.8777	104.5	176.	2.95
20.00	1.1011	1.1031	220.2	2.266	880.9	117.3	385	1.3715				0.9597	0.8733	104.0	197.	3.60
24.00	1.1236	1.1255	269.7	2.775	853.9	144.3	469	1.3798				0.9755	0.8700	102.3		
28.00	1.1467	1.1488	321.1	3.304	825.6	172.6	554	1.3884				1.001	0.8747	99.69		
32.00	1.1706	1.1727	374.6	3.855	796.0	202.2	643	1.3972				1.037	0.8874	96.27		
36.00	1.1949	1.1970	430.2	4.427	764.7	233.5	732	1.4062				1.081	0.9069	92.28		
40.00	1.2199	1.2220	488.0	5.021	731.9	266.3	824	1.4154				1.150	0.9446	86.78		
44.00	1.2455	1.2477	548.0	5.639	697.5	300.7	918	1.4248				1.253	1.008	79.65		
48.00	1.2719	1.2741	610.5	6.282	661.4	336.9	1014	1.4343				1.389	1.094	71.86		
52.00	1.2990	1.3013	675.5	6.951	623.5	374.7	1112	1.4442				1.552	1.197	64.31		
56.00	1.3269	1.3292	743.1	7.646	583.8	414.4	1212	1.4542				1.735	1.310	57.52		
60.00	1.3556	1.3580	813.4	8.370	542.2	456.0	1315	1.4645				2.013	1.488	49.58		
64.00	1.3853	1.3877	886.6	9.123	498.7	499.5	1420	1.4750				2.434	1.761	41.00		

59 PROCAINE HYDROCHLORIDE, $C_6H_4[COOCH_2CH_2N(C_2H_5)_2][NH_2] \cdot HCl$-1,4

MOLECULAR WEIGHT = 272.78
RELATIVE SPECIFIC REFRACTIVITY = 1.460

0.00 % by wt. data are the same for all compounds.
For Values of 0.00 wt. % solutions see Table 1, Acetic Acid.

A% by wt.	ρ D_4^{20}	D_{20}^{20}	C_s g/l	M g-mol/l	C_w g/l	$(C_o - C_w)$ g/l	$(n - n_o) \times 10^4$	n	Δ °C	O Os/kg	S g-mol/l	η/η_o	η/ρ cS	φ rhe	γ mmho/cm	T g-mol/l
0.50	0.9991	1.0008	5.0	0.018	994.1	4.2	11	1.3341	0.062	0.033	0.017	1.014	1.017	98.42	1.4	0.014
1.00	0.9999	1.0017	10.0	0.037	989.9	8.3	22	1.3352	0.126	0.068	0.035	1.028	1.030	97.08	2.7	0.027
2.00	1.0016	1.0034	20.0	0.073	981.6	16.7	44	1.3374	0.237	0.127	0.068	1.056	1.056	94.51	5.2	0.052
3.00	1.0033	1.0051	30.1	0.110	973.2	25.0	66	1.3396	0.343	0.184	0.099	1.083	1.082	92.16	7.4	0.076
4.00	1.0050	1.0068	40.2	0.147	964.8	33.4	89	1.3419	0.433	0.233	0.125	1.110	1.106	89.95	9.4	0.098
5.00	1.0067	1.0085	50.3	0.185	956.4	41.8	111	1.3441	0.519	0.279	0.150	1.138	1.133	87.70	11.4	0.119
6.00	1.0085	1.0103	60.5	0.222	948.0	50.2	133	1.3463	0.593	0.319	0.172	1.169	1.161	85.38	13.1	0.139
7.00	1.0103	1.0120	70.7	0.259	939.5	58.7	156	1.3486	0.675	0.363	0.196	1.202	1.192	83.06	14.7	0.157
8.00	1.0120	1.0138	81.0	0.297	931.1	67.2	179	1.3508	0.756	0.406	0.220	1.236	1.224	80.73	16.2	0.174
9.00	1.0138	1.0156	91.2	0.334	922.6	75.6	201	1.3531	0.824	0.443	0.240	1.273	1.258	78.40	17.5	0.191
10.00	1.0156	1.0174	101.6	0.372	914.1	84.2	224	1.3554	0.889	0.478	0.259	1.312	1.294	76.07	18.8	0.206
12.00	1.0193	1.0211	122.3	0.448	897.0	101.3	270	1.3600	1.026	0.552	0.300	1.397	1.373	71.45	21.1	0.233
14.00	1.0230	1.0248	143.2	0.525	879.8	118.5	316	1.3646	1.164	0.626	0.340	1.491	1.460	66.94	23.0	0.256
16.00	1.0267	1.0285	164.3	0.602	862.4	135.8	362	1.3692	1.297	0.698	0.379	1.595	1.557	62.57	24.6	0.277
18.00	1.0305	1.0323	185.5	0.680	845.0	153.2	409	1.3739	1.425	0.766	0.417	1.708	1.661	58.44	26.0	0.294
20.00	1.0343	1.0361	206.9	0.758	827.4	170.8	456	1.3786	1.545	0.831	0.452	1.838	1.781	54.30	27.2	0.309
24.00	1.0420	1.0439	250.1	0.917	791.9	206.3	552	1.3881				2.143	2.061	46.57	29.0	0.331
28.00	1.0499	1.0517	294.0	1.078	755.9	242.3	649	1.3978				2.542	2.426	39.26	29.8	0.342
32.00	1.0578	1.0596	338.5	1.241	719.3	278.9	747	1.4077				3.089	2.927	32.30	30.2	0.346
36.00	1.0658	1.0676	383.7	1.407	682.1	316.2	848	1.4178				3.746	3.522	26.64	29.5	0.338
40.00	1.0738	1.0757	429.5	1.575	644.3	354.0	951	1.4280				4.757	4.439	20.98	28.3	0.323
44.00	1.0819	1.0838	476.0	1.745	605.9	392.4	1055	1.4385				6.028	5.583	16.56	26.8	0.304
48.00	1.0900	1.0920	523.2	1.918	566.8	431.4	1162	1.4491				7.888	7.251	12.65	24.7	0.278
52.00	1.0982	1.1002	571.1	2.094	527.2	471.1	1270	1.4600				11.22	10.24	8.89	22.3	0.248
56.00	1.1066	1.1085	619.7	2.272	486.9	511.3	1381	1.4711				15.97	14.46	6.25	19.5	0.214
60.00	1.1151	1.1171	669.1	2.453	446.0	552.2	1493	1.4823				22.12	19.88	4.51	16.1	0.174

61 2-PROPANOL, $CH_3CHOHCH_3$

MOLECULAR WEIGHT = 60.09

A% by wt.	ρ D_4^{20}	D_{20}^{20}	C_s g/l	M g-mol/l	C_w g/l	$(C_0 - C_w)$ g/l	$(n - n_0)$ × 10^4	n	Δ °C	O Os/kg	S g-mol/l	η/η_0	η/ρ cS	φ rhe	γ mmho/cm	T g-mol/l
1.00	0.9960	0.9978	10.0	0.166	986.0	12.2	8.	1.3338	0.304	0.163	0.087	1.054	1.060	94.72	0.0	0.000
2.00	0.9939	0.9957	19.9	0.331	974.0	24.2	16.	1.3346	0.603	0.324	0.175	1.110	1.119	89.91		
3.00	0.9920	0.9938	29.8	0.495	962.2	36.0	25.	1.3355	0.926	0.498	0.270	1.164	1.176	85.72		
4.00	0.9902	0.9920	39.6	0.659	950.6	47.6	34.	1.3364	1.259	0.677	0.368	1.223	1.238	81.57		
5.00	0.9884	0.9902	49.4	0.822	939.0	59.2	43.	1.3373	1.610	0.866	0.471	1.284	1.302	77.71		
6.00	0.9871	0.9889	59.2	0.986	927.9	70.3	52.	1.3382	1.960	1.054	0.573	1.349	1.369	74.00		
7.00	0.9855	0.9873	69.0	1.148	916.5	81.7	62.	1.3392	2.320	1.247	0.677	1.414	1.438	70.56		
8.00	0.9843	0.9861	78.7	1.310	905.6	92.6	71.	1.3401	2.679	1.440	0.780	1.482	1.509	67.33		
9.00	0.9831	0.9849	88.5	1.472	894.6	103.6	80.	1.3410	3.062	1.646	0.889	1.550	1.580	64.38		
10.00	0.9816	0.9834	98.2	1.634	883.4	114.8	90.	1.3420	3.481	1.871	1.007	1.626	1.660	61.37		
12.00	0.9793	0.9811	117.5	1.956	861.8	136.4	109.	1.3439	4.433	2.383	1.271	1.790	1.831	55.77		
14.00	0.9772	0.9790	136.8	2.277	840.4	157.8	129.	1.3459	5.29	2.85	1.503	1.966	2.016	50.76		
16.00	0.9751	0.9769	156.0	2.596	819.1	179.1	148.	1.3478	6.36	3.42	1.78	2.156	2.215	46.30		
18.00	0.9725	0.9743	175.1	2.913	797.4	200.8	166.	1.3496	7.40	3.98	2.04	2.347	2.418	42.53		
20.00	0.9696	0.9713	193.9	3.227	775.7	222.5	184.	1.3514	8.52	4.58	2.31	2.545	2.630	39.21		
24.00	0.9630	0.9647	231.1	3.846	731.9	266.3	217.	1.3547	10.91	5.87	2.85	2.851	2.966	35.01		
28.00	0.9555	0.9572	267.5	4.452	688.0	310.2	245.	1.3575				3.115	3.267	32.03		
32.00	0.9478	0.9495	303.3	5.047	644.5	353.7	270.	1.3600				3.330	3.520	29.97		
36.00	0.9393	0.9410	338.2	5.627	601.1	397.1	292.	1.3622				3.499	3.733	28.52		
40.00	0.9302	0.9319	372.1	6.192	558.1	440.1	312.	1.3642				3.630	3.910	27.49		
44.00	0.9207	0.9224	405.1	6.742	515.6	482.6	330.	1.3660				3.732	4.061	26.75		
48.00	0.9113	0.9129	437.4	7.279	473.9	524.3	346.	1.3676				3.772	4.147	26.46		
52.00	0.9019	0.9035	469.0	7.805	432.9	565.3	362.	1.3692				3.789	4.210	26.34		
56.00	0.8920	0.8936	499.5	8.313	392.5	605.7	375.	1.3705				3.781	4.247	26.40		
60.00	0.8824	0.8840	529.4	8.811	353.0	645.2	387.	1.3717				3.719	4.223	26.84		
64.00	0.8729	0.8745	558.7	9.297	314.2	684.0	400.	1.3730				3.636	4.174	27.45		
68.00	0.8632	0.8648	587.0	9.769	276.2	722.0	411.	1.3741				3.528	4.095	28.29		
72.00	0.8535	0.8550	614.5	10.226	239.0	759.2	421.	1.3751				3.391	3.981	29.43		
76.00	0.8438	0.8453	641.3	10.672	202.5	795.7	428.	1.3758				3.235	3.841	30.85		
80.00	0.8341	0.8356	667.3	11.105	166.8	831.4	435.	1.3765				3.072	3.690	32.49		
84.00	0.8243	0.8258	692.4	11.523	131.9	866.3	440.	1.3770				2.886	3.508	34.58		
88.00	0.8145	0.8160	716.8	11.929	97.7	900.5	445.	1.3775				2.684	3.302	37.18		
92.00	0.8046	0.8061	740.3	12.319	64.3	933.9	447.	1.3777				2.530	3.151	39.44		
96.00	0.7949	0.7963	763.1	12.699	31.8	966.4	446.	1.3776				2.424	3.055	41.18		
100.00	0.7848	0.7862	784.8	13.060	0.0	998.2	442.	1.3742				2.428	3.100	41.10		

60 1-PROPANOL, $CH_3CH_2CH_2OH$

MOLECULAR WEIGHT = 60.09

A% by wt.	ρ D_4^{20}	D_{20}^{20}	C_s g/l	M g-mol/l	C_w g/l	$(C_0 - C_w)$ g/l	$(n - n_0)$ × 10^4	n	Δ °C	O Os/kg	S g-mol/l	η/η_0	η/ρ cS	φ rhe	γ mmho/cm	T g-mol/l
1.00	0.9963	0.9981	10.0	0.166	986.3	11.9	9.	1.3339	0.307	0.165	0.088	1.049	1.055	95.14	0.0	0.000
2.00	0.9946	0.9964	19.9	0.331	974.7	23.5	18.	1.3348	0.613	0.329	0.178	1.098	1.106	90.91		
3.00	0.9928	0.9946	29.8	0.496	963.0	35.2	27.	1.3357	0.933	0.502	0.272	1.150	1.161	86.76		
4.00	0.9911	0.9929	39.6	0.660	951.5	46.7	36.	1.3366	1.239	0.666	0.362	1.206	1.219	82.77		
5.00	0.9896	0.9914	49.5	0.823	940.1	58.1	46.	1.3376	1.573	0.846	0.460	1.264	1.280	78.94		
6.00	0.9882	0.9900	59.3	0.987	928.9	69.3	55.	1.3385	1.906	1.024	0.557	1.322	1.340	75.52		
7.00	0.9868	0.9886	69.1	1.150	917.7	80.5	64.	1.3394	2.258	1.214	0.659	1.384	1.405	72.12		
8.00	0.9855	0.9873	78.8	1.312	906.7	91.5	74.	1.3404	2.609	1.403	0.760	1.446	1.470	69.03		
9.00	0.9842	0.9860	88.6	1.474	895.6	102.6	84.	1.3414	2.988	1.606	0.868	1.511	1.538	66.06		
10.00	0.9829	0.9847	98.3	1.636	884.6	113.6	93.	1.3423	3.356	1.805	0.972	1.574	1.605	63.39		
12.00	0.9804	0.9822	117.6	1.958	862.8	135.4	112.	1.3442	3.477	2.014	1.181	1.707	1.745	58.45		
14.00	0.9779	0.9797	136.9	2.278	841.0	157.2	130.	1.3460	4.914	2.642	1.401	1.845	1.890	54.10		
16.00	0.9749	0.9767	156.0	2.596	818.9	179.3	147.	1.3477	5.776	3.108	1.629	1.982	2.037	50.35		
18.00	0.9719	0.9737	174.9	2.911	797.0	201.2	164.	1.3494	6.667	3.588	1.859	2.102	2.167	47.48		
20.00	0.9686	0.9703	193.7	3.224	774.9	223.3	180.	1.3510	7.555	4.062	2.079	2.214	2.290	45.09		
24.00	0.9612	0.9629	230.7	3.839	730.5	267.7	209.	1.3539	9.115	4.901	2.450	2.427	2.530	41.12		
28.00	0.9533	0.9550	266.9	4.442	686.4	311.8	236.	1.3566	10.17	5.47	2.688	2.607	2.740	38.28		
32.00	0.9452	0.9469	302.5	5.034	642.7	355.5	262.	1.3592	10.66	5.73	2.797	2.759	2.925	36.17		
36.00	0.9370	0.9387	337.3	5.614	599.7	398.5	284.	1.3614				2.894	3.095	34.48		
40.00	0.9288	0.9305	371.5	6.183	557.3	440.9	305.	1.3635				3.004	3.241	33.22		
44.00	0.9206	0.9223	405.1	6.741	515.5	482.7	328.	1.3658				3.096	3.370	32.23		
48.00	0.9127	0.9143	438.1	7.291	483.6	514.6	348.	1.3678				3.162	3.472	31.56		
52.00	0.9043	0.9059	470.2	7.826	434.1	564.1	367.	1.3697				3.195	3.540	31.24		
56.00	0.8959	0.8975	501.7	8.349	394.2	604.0	385.	1.3715				3.197	3.576	31.21		
60.00	0.8875	0.8891	532.5	8.862	355.0	643.2	404.	1.3734				3.180	3.590	31.39		
64.00	0.8790	0.8806	562.6	9.362	316.4	681.8	422.	1.3752				3.135	3.574	31.83		
68.00	0.8706	0.8722	592.0	9.852	278.6	719.6	437.	1.3767				3.075	3.539	32.46		
72.00	0.8623	0.8639	620.9	10.332	241.4	756.8	453.	1.3783				2.997	3.482	33.30		
76.00	0.8549	0.8564	649.7	10.813	205.2	793.0	467.	1.3797				2.909	3.410	34.30		
80.00	0.8470	0.8485	677.6	11.276	169.4	828.8	482.	1.3812				2.816	3.332	35.43		
84.00	0.8390	0.8405	704.8	11.728	134.2	864.0	495.	1.3825				2.720	3.249	36.69		
88.00	0.8306	0.8321	730.9	12.164	99.7	898.5	505.	1.3835				2.590	3.125	38.53		
92.00	0.8218	0.8233	756.1	12.582	65.7	932.5	513.	1.3843				2.467	3.008	40.45		
96.00	0.8130	0.8145	780.5	12.989	32.5	965.7	518.	1.3848				2.345	2.890	42.56		
100.00	0.8034	0.8048	803.4	13.370	0.0	998.2	522.	1.3852				2.223	2.773	44.89		

62 PROPYLENE GLYCOL, $CH_2OHCHOHCH_3$

MOLECULAR WEIGHT = 76.09
RELATIVE SPECIFIC REFRACTIVITY = 1.236

0.00 % by wt. data are the same for all compounds.
For Values of 0.00 wt. % solutions see Table 1, Acetic Acid.

A % by wt.	ρ D_4^{20}	D_{20}^{20}	C g/l	M g-mol/l	C_w g/l	$(C_o - C_w)$ g/l	$(n - n_o)$ × 10^4	n	Δ °C	O Os/kg	S g-mol/l	η/η_o	η/ρ cS	φ rhe	γ mmho/cm	T g-mol/l
0.50	0.9985	1.0003	5.0	0.066	993.5	4.7	5	1.3335	0.131	0.071	0.037	1.013	1.017	98.52		
1.00	0.9988	1.0006	10.0	0.131	988.8	9.4	10	1.3340	0.259	0.139	0.074	1.028	1.031	97.08		
2.00	0.9994	1.0012	20.0	0.263	979.4	18.8	20	1.3350	0.501	0.269	0.145	1.064	1.067	93.80		
3.00	1.0001	1.0019	30.0	0.394	970.1	28.2	31	1.3361	0.758	0.407	0.221	1.101	1.104	90.61		
4.00	1.0008	1.0025	40.0	0.526	960.7	37.5	41	1.3371	1.027	0.552	0.300	1.140	1.142	87.52		
5.00	1.0015	1.0032	50.1	0.658	951.4	46.8	52	1.3382	1.303	0.701	0.381	1.181	1.182	84.50		
6.00	1.0022	1.0040	60.1	0.790	942.1	56.2	63	1.3393	1.583	0.851	0.463	1.223	1.223	81.59		
7.00	1.0030	1.0047	70.2	0.923	932.8	65.5	74	1.3403	1.870	1.005	0.547	1.267	1.266	78.77		
8.00	1.0037	1.0055	80.3	1.055	923.4	74.8	84	1.3414	2.164	1.163	0.632	1.313	1.310	76.03		
9.00	1.0045	1.0063	90.4	1.188	914.1	84.1	95	1.3425	2.465	1.325	0.719	1.361	1.357	73.34		
10.00	1.0054	1.0071	100.5	1.321	904.8	93.4	107	1.3436	2.773	1.491	0.807	1.412	1.407	70.68		
12.00	1.0070	1.0088	120.8	1.588	886.2	112.0	129	1.3459	3.534	1.900	1.022	1.524	1.516	65.50		
14.00	1.0088	1.0106	141.2	1.856	867.6	130.7	152	1.3482	4.393	2.362	1.260	1.646	1.635	60.65		
16.00	1.0106	1.0124	161.7	2.125	848.9	149.4	175	1.3504	5.44	2.92	1.540	1.776	1.761	56.19		
18.00	1.0124	1.0142	182.2	2.395	830.2	168.1	198	1.3528				1.915	1.895	52.12		
20.00	1.0142	1.0160	202.8	2.666	811.4	186.9	221	1.3551				2.066	2.041	48.31		
24.00	1.0178	1.0196	244.3	3.210	773.6	224.7	267	1.3597				2.410	2.372	41.42		
28.00	1.0214	1.0232	286.0	3.759	735.4	262.8	314	1.3644				2.811	2.758	35.50		
32.00	1.0248	1.0266	327.9	4.310	696.8	301.4	360	1.3690				3.280	3.207	30.43		
36.00	1.0279	1.0297	370.1	4.863	657.9	340.4	406	1.3736				3.811	3.715	26.19		
40.00	1.0308	1.0326	412.3	5.419	618.5	379.8	451	1.3780				4.432	4.308	22.52		
44.00	1.0333	1.0352	454.7	5.975	578.7	419.6	494	1.3824				5.144	4.988	19.40		
48.00	1.0356	1.0374	497.1	6.533	538.5	459.7	537	1.3867				5.963	5.769	16.74		
52.00	1.0377	1.0395	539.6	7.091	498.1	500.2	580	1.3910				6.939	6.700	14.38		
56.00	1.0396	1.0415	582.2	7.651	457.4	540.8	622	1.3952				8.067	7.775	12.37		
60.00	1.0418	1.0436	625.1	8.215	416.7	581.5	665	1.3995				9.348	8.991	10.68		

63 SEA WATER*

RELATIVE SPECIFIC REFRACTIVITY = 0.789

0.00 % by wt. data are the same for all compounds.
For Values of 0.00 wt. % solutions see Table 1, Acetic Acid.

A % by wt.	Salinity ‰	Chlorinity ‰	ρ D_4^{20}	D_{20}^{20}	C_s g/l	C_w g/l	$(C_o - C_w)$ g/l	$(n - n_o)$ × 10^4	n	Δ °C	O Os/kg	S g-mol/l	η/η_o	η/ρ cS	φ rhe	γ mmho/cm	T g-mol/l
0.50	4.94	2.72	1.0019	1.0037	5.0	996.9	1.3	9	1.3339	0.269	0.144	0.077	1.008	1.008	99.02	7.8	0.082
1.00	9.92	5.48	1.0057	1.0075	10.1	995.5	2.7	18	1.3348	0.537	0.289	0.156	1.016	1.012	98.23	15.2	0.163
1.50	14.91	8.24	1.0094	1.0112	15.1	994.1	4.1	27	1.3357	0.806	0.433	0.235	1.024	1.017	97.45	22.0	0.244
2.00	19.89	11.00	1.0132	1.0150	20.3	992.7	5.6	36	1.3366	1.075	0.578	0.314	1.033	1.022	96.61	28.4	0.324
2.50	24.87	13.76	1.0170	1.0188	25.4	991.2	7.0	45	1.3375	1.346	0.724	0.394	1.044	1.029	95.64	34.5	0.401
3.00	29.86	16.53	1.0207	1.0225	30.6	989.7	8.5	54	1.3384	1.616	0.869	0.473	1.055	1.036	94.58	40.4	0.476
3.50	34.84	19.29	1.0245	1.0263	35.8	988.1	10.1	63	1.3393	1.877	1.009	0.549	1.067	1.044	93.55	46.3	0.552
4.00	39.82	22.05	1.0283	1.0301	41.1	986.6	11.7	72	1.3402	2.129	1.145	0.622	1.077	1.050	92.67	52.4	0.635
4.50	44.81	24.81	1.0320	1.0339	46.4	985.0	13.3	81	1.3411	2.374	1.277	0.693	1.087	1.056	91.83	58.6	0.721
5.00	49.79	27.57	1.0358	1.0376	51.8	983.3	14.9	90	1.3420	2.612	1.404	0.761	1.100	1.065	90.75	64.2	0.799
5.50	54.78	30.33	1.0396	1.0414	57.1	981.6	16.6	99	1.3429	2.842	1.528	0.827	1.117	1.077	89.38	68.7	0.865
6.00	59.76	33.09	1.0434	1.0452	62.5	979.9	18.3	108	1.3438	3.064	1.647	0.890	1.137	1.093	87.80	72.6	0.922
6.50	64.74	35.85	1.0472	1.0490	68.0	978.1	20.1	118	1.3448								
7.00	69.73	38.61	1.0509	1.0528	73.5	976.3	21.9	127	1.3457								
7.50	74.71	41.37	1.0547	1.0566	79.0	974.6	23.6	136	1.3466								
8.00	79.69	44.13	1.0585	1.0604	84.6	972.9	25.3	146	1.3476								
8.50	84.68	46.90	1.0623	1.0642	90.2	971.2	27.1	155	1.3485								
9.00	89.66	49.66	1.0662	1.0680	95.9	969.3	28.9	164	1.3494								
9.50	94.64	52.42	1.0700	1.0719	101.6	967.5	30.8	173	1.3503								
10.00	99.63	55.18	1.0738	1.0757	107.3	965.6	32.6	182	1.3512								
11.00	109.6	60.70	1.0814	1.0833	118.9	961.7	36.5	200	1.3530								
12.00	119.6	66.22	1.0891	1.0910	130.6	957.8	40.5	219	1.3548								
13.00	129.5	71.74	1.0968	1.0987	142.5	953.7	44.5	237	1.3567								
14.00	139.5	77.27	1.1045	1.1065	154.6	949.5	48.7	255	1.3585								
15.00	149.5	82.79	1.1122	1.1142	166.8	945.2	53.0	273	1.3602								

* Values corresponding to A % from 0.00 6.00 are based upon measurements of artificial sea water as formulated by Lyman and Fleming, *J. Marine Res.* **3**, 134, 194 Values corresponding to A % above 6.00 are based upon natural sea water from the Gulf of Mexico at Aransas Pass, Texas, concentrated by evaporation, according Behrens, ibid. **23**: 165, 1965.

64 SERUM OR PLASMA, HUMAN

0.00 % by wt. data are the same for all compounds.
For Values of 0.00 wt. % solutions see Table 1, Acetic Acid.

A % by wt.	ρ D_4^{20}	D_{20}^{20}	C_s g/l	C_{Pr} g 100 ml	C_{NPr} g 100 ml	C_w g/l	$(C_o - C_w)$ g/l	$(n - n_o)$ × 10^4	n
1.00	1.0029	1.0047	10.0	0.0	1.00	993.	5.0	16	1.3346
1.50	1.0051	1.0069	15.0	0.0	1.50	990.	8.0	24	1.3354
2.00	1.0069	1.0087	20.0	0.3	1.69	987.	11.0	33	1.3363
2.50	1.0082	1.0100	25.0	0.8	1.68	983.	15.0	42	1.3372
3.00	1.0096	1.0114	30.0	1.4	1.68	979.	19.0	52	1.3382
3.50	1.0110	1.0128	35.0	1.9	1.67	976.	22.0	61	1.3391
4.00	1.0123	1.0141	40.0	2.4	1.66	972.	26.0	71	1.3401
4.50	1.0137	1.0155	46.0	2.9	1.66	968.	30.0	80	1.3410
5.00	1.0150	1.0168	51.0	3.4	1.65	964.	34.0	90	1.3420
5.50	1.0164	1.0182	56.0	3.9	1.64	960.	38.0	99	1.3429

64 SERUM OR PLASMA, HUMAN—(Continued)

A % by wt.	ρ D_4^{20}	D_{20}^{20}	C_s g/l	C_{Pr} g/100 ml	C_{NPr} g/100 ml	C_w g/l	$(C_o - C_w)$ g/l	$(n - n_o) \times 10^4$	n
6.00	1.0177	1.0195	61.	4.5	1.64	957.	41.	108	1.3438
6.50	1.0190	1.0208	66.	5.0	1.63	953.	45.	118	1.3448
7.00	1.0203	1.0221	71.	5.5	1.62	949.	49.	127	1.3457
7.50	1.0216	1.0234	77.	6.0	1.62	945.	53.	137	1.3467
8.00	1.0229	1.0247	82.	6.6	1.61	941.	57.	146	1.3476
8.50	1.0242	1.0260	87.	7.1	1.60	937.	61.	156	1.3486
9.00	1.0255	1.0273	92.	7.6	1.60	933.	65.	165	1.3495
9.50	1.0267	1.0286	98.	8.2	1.59	929.	69.	175	1.3505
10.00	1.0279	1.0298	103.	8.7	1.58	925.	73.	184	1.3514
11.00	1.0305	1.0324	113.	9.8	1.57	917.	81.	203	1.3533
12.00	1.0329	1.0348	124.	10.8	1.55	909.	89.	222	1.3552
13.00	1.0354	1.0373	135.	11.9	1.54	901.	97.	241	1.3571
14.00	1.0378	1.0397	145.	13.0	1.53	893.	105.	260	1.3590
15.00	1.0401	1.0420	156.	14.1	1.51	884.	114.	279	1.3609

Notes: C_{Pr} = g protein/100 ml; C_{NPr} = g nonprotein solute/100 ml.

65 SERUM OR PLASMA, RABBIT OR GUINEA PIG

0.00 % by wt. data are the same for all compounds.
For Values of 0.00 wt. % solutions see Table 1, Acetic Acid.

A % by wt.	ρ D_4^{20}	D_{20}^{20}	C_s g/l	C_{Pr} g/100 ml	C_w g/l	$(C_o - C_w)$ g/l	$(n - n_o) \times 10^4$	n
4.0	1.0125	1.0143	41.	3.3	972.	28.	71	1.3401
4.5	1.0139	1.0156	46.	3.8	968.	30.	80	1.3410
5.0	1.0152	1.0170	51.	4.2	964.	34.	89	1.3419
5.5	1.0165	1.0183	56.	4.6	961.	38.	98	1.3428
6.0	1.0179	1.0197	61.	5.0	957.	41.	107	1.3437
6.5	1.0192	1.0210	66.	5.4	953.	45.	116	1.3446
7.0	1.0206	1.0224	71.	5.8	949.	49.	125	1.3455
7.5	1.0219	1.0238	77.	6.2	945.	53.	134	1.3464
8.0	1.0233	1.0251	82.	6.6	941.	57.	143	1.3473
8.5	1.0247	1.0265	87.	7.0	938.	61.	152	1.3482
9.0	1.0260	1.0278	92.	7.4	934.	65.	161	1.3491
9.5	1.0274	1.0292	98.	7.8	930.	68.	170	1.3500
10.0	1.0287	1.0306	103.	8.2	926.	72.	179	1.3509
11.0	1.0315	1.0333	113.	9.0	918.	80.	197	1.3527
12.0	1.0342	1.0360	124.	9.8	910.	88.	215	1.3545

Note: C_{Pr} = g protein/100 ml.

66 SILVER NITRATE, $AgNO_3$

MOLECULAR WEIGHT = 169.89
RELATIVE SPECIFIC REFRACTIVITY = 0.475

0.00 % by wt. data are the same for all compounds.
For Values of 0.00 wt. % solutions see Table 1, Acetic Acid.

A % by wt.	ρ D_4^{20}	D_{20}^{20}	C_s g/l	M g-mol/l	C_w g/l	$(C_o - C_w)$ g/l	$(n - n_o) \times 10^4$	n	Δ °C	O Os/kg	S g-mol/l	η/η_o	η/ρ cS	φ rhe	γ mmho/cm	T g-mol/l
0.50	1.0027	1.0045	5.0	0.030	997.7	0.5	6	1.3336	0.103	0.056	0.029	1.001	1.001	99.66	3.1	0.031
1.00	1.0070	1.0088	10.1	0.059	996.9	1.3	12	1.3342	0.202	0.109	0.058	1.003	0.998	99.50	6.1	0.062
2.00	1.0154	1.0172	20.3	0.120	995.1	3.1	22	1.3352	0.395	0.213	0.114	1.007	0.994	99.11	12.0	0.126
3.00	1.0239	1.0257	30.7	0.181	993.2	5.1	33	1.3363	0.586	0.315	0.170	1.011	0.989	98.74	17.2	0.187
4.00	1.0327	1.0345	41.3	0.243	991.4	6.8	44	1.3374	0.775	0.416	0.226	1.014	0.984	98.39	22.0	0.245
5.00	1.0417	1.0436	52.1	0.307	989.6	8.6	55	1.3385	0.961	0.517	0.281	1.018	0.979	98.04	26.7	0.303
6.00	1.0506	1.0524	63.0	0.371	987.5	10.7	66	1.3396	1.146	0.616	0.335	1.022	0.974	97.69	31.4	0.362
7.00	1.0597	1.0616	74.2	0.437	985.5	12.7	78	1.3407	1.329	0.715	0.389	1.025	0.969	97.34	36.1	0.421
8.00	1.0690	1.0709	85.5	0.503	983.5	14.7	89	1.3419	1.511	0.812	0.442	1.029	0.964	97.00	40.7	0.480
9.00	1.0785	1.0804	97.1	0.571	981.4	16.8	101	1.3431	1.690	0.909	0.494	1.033	0.959	96.63	45.3	0.540
10.00	1.0882	1.0901	108.8	0.641	979.3	18.9	113	1.3443	1.868	1.004	0.546	1.037	0.955	96.24	49.8	0.600
12.00	1.1079	1.1099	133.0	0.783	975.0	23.2	138	1.3467	2.213	1.190	0.646	1.047	0.947	95.35	58.8	0.723
14.00	1.1284	1.1304	158.0	0.930	970.4	27.8	163	1.3493	2.546	1.369	0.742	1.058	0.939	94.35	67.6	0.849
16.00	1.1496	1.1516	183.9	1.083	965.7	32.6	189	1.3519	2.862	1.539	0.832	1.070	0.933	93.25	76.3	0.976
18.00	1.1715	1.1736	210.9	1.241	960.6	37.6	216	1.3546				1.084	0.927	92.07	84.7	1.10
20.00	1.1942	1.1963	238.8	1.406	955.4	42.9	244	1.3574				1.099	0.922	90.81	92.8	1.23
22.00	1.2177	1.2198	267.9	1.577	949.8	48.5	273	1.3602				1.115	0.918	89.49	101.0	1.36
24.00	1.2420	1.2442	298.1	1.755	943.9	54.3	302	1.3632				1.133	0.914	88.09	108.0	1.49
26.00	1.2672	1.2694	329.5	1.939	937.7	60.5	332	1.3662				1.152	0.911	86.60	115.0	1.61
28.00	1.2933	1.2956	362.1	2.132	931.2	67.1	364	1.3694				1.174	0.910	85.01	122.0	1.74
30.00	1.3204	1.3228	396.1	2.332	924.3	73.9	397	1.3726				1.198	0.909	83.31	129.0	1.88
32.00	1.3487	1.3510	431.6	2.540	917.1	81.1	430	1.3760				1.225	0.910	81.49	136.0	2.02
34.00	1.3780	1.3805	468.5	2.758	909.5	88.7	465	1.3795				1.254	0.912	79.58	142.0	2.16
36.00	1.4087	1.4112	507.1	2.985	901.6	96.7	502	1.3832				1.287	0.915	77.56	149.0	2.31
38.00	1.4407	1.4433	547.5	3.223	893.2	105.0	541	1.3871				1.323	0.920	75.44	156.0	2.46
40.00	1.4743	1.4769	589.7	3.471	884.6	113.7	581	1.3911				1.363	0.926	73.22	162.0	2.61

67 SODIUM ACETATE, CH_3COONa

MOLECULAR WEIGHT = 82.04
RELATIVE SPECIFIC REFRACTIVITY = 0.884

0.00 % by wt. data are the same for all compounds.
For Values of 0.00 wt. % solutions see Table 1, Acetic Acid.

A% by wt.	ρ D_4^{20}	D_{20}^{20}	C_s g/l	M g-mol/l	C_w g/l	$(C_o - C_w)$ g/l	$(n - n_o)$ × 10^4	n	Δ °C	O Os/kg	S g-mol/l	η/η_o	η/ρ cS	φ rhe	γ mmho/cm	T g-mol/l
0.50	1.0008	1.0026	5.0	0.061	995.8	2.4	7	1.3337	0.217	0.116	0.062	1.019	1.020	97.96	3.9	0.039
1.00	1.0034	1.0052	10.0	0.122	993.4	4.9	14	1.3344	0.433	0.233	0.125	1.038	1.037	96.15	7.6	0.078
2.00	1.0085	1.0103	20.2	0.246	988.3	9.9	28	1.3358	0.877	0.471	0.256	1.078	1.071	92.58	14.4	0.154
3.00	1.0135	1.0153	30.4	0.371	983.1	15.2	42	1.3372	1.339	0.720	0.391	1.122	1.109	88.98	20.4	0.225
4.00	1.0184	1.0202	40.7	0.497	977.7	20.6	56	1.3386	1.820	0.978	0.532	1.169	1.150	85.38	25.7	0.291
5.00	1.0234	1.0252	51.2	0.624	972.2	26.0	70	1.3400	2.323	1.249	0.678	1.220	1.195	81.80	30.9	0.355
6.00	1.0283	1.0302	61.7	0.752	966.6	31.6	85	1.3414	2.848	1.531	0.828	1.275	1.242	78.27	35.9	0.419
7.00	1.0334	1.0352	72.3	0.882	961.1	37.2	99	1.3428	3.398	1.827	0.984	1.334	1.294	74.79	40.8	0.482
8.00	1.0386	1.0404	83.1	1.013	955.5	42.7	112	1.3442	3.975	2.137	1.144	1.398	1.348	71.41	45.4	0.542
9.00	1.0440	1.0458	94.0	1.145	950.0	48.2	126	1.3456	4.566	2.455	1.307	1.465	1.406	68.13	49.6	0.598
10.00	1.0495	1.0514	105.0	1.279	944.6	53.6	140	1.3470				1.536	1.466	64.97	53.4	0.648
12.00	1.0607	1.0625	127.3	1.551	933.4	64.9	168	1.3498				1.685	1.592	59.22	59.0	0.725
14.00	1.0718	1.0737	150.1	1.829	921.8	76.5	196	1.3526				1.851	1.731	53.90	62.7	0.777
16.00	1.0830	1.0849	173.3	2.112	909.7	88.5	224	1.3554				2.050	1.897	48.68	65.5	0.818
18.00	1.0940	1.0960	196.9	2.400	897.1	101.1	253	1.3583				2.279	2.087	43.79	67.8	0.852
20.00	1.1050	1.1070	221.0	2.694	884.0	114.2	281	1.3611				2.562	2.323	38.95	69.3	0.873
22.00	1.1159	1.1179	245.5	2.993	870.4	127.8	309	1.3639				2.942	2.642	33.92	69.9	0.882
24.00	1.1268	1.1288	270.4	3.296	856.4	141.9	336	1.3666				3.393	3.017	29.41	69.7	0.878
26.00	1.1377	1.1397	295.8	3.606	841.9	156.3	363	1.3693				3.869	3.408	25.79	68.8	0.865
28.00	1.1488	1.1508	321.7	3.921	827.1	171.1	390	1.3720				4.379	3.819	22.79	67.0	0.840
30.00	1.1602	1.1623	348.1	4.243	812.2	186.1	418	1.3748				4.930	4.258	20.24	64.3	0.800

68 SODIUM BICARBONATE, $NaHCO_3$

MOLECULAR WEIGHT = 84.01
RELATIVE SPECIFIC REFRACTIVITY = 0.687

0.00 % by wt. data are the same for all compounds.
For Values of 0.00 wt. % solutions see Table 1, Acetic Acid.

A% by wt.	ρ D_4^{20}	D_{20}^{20}	C_s g/l	M g-mol/l	C_w g/l	$(C_o - C_w)$ g/l	$(n - n_o)$ × 10^4	n	Δ °C	O Os/kg	S g-mol/l	η/η_o	η/ρ cS	φ rhe	γ mmho/cm	T g-mol/l
0.50	1.0018	1.0036	5.0	0.060	996.8	1.4	7	1.3337	0.199	0.107	0.057	1.013	1.013	98.52	4.2	0.041
1.00	1.0054	1.0072	10.1	0.120	995.3	2.9	14	1.3344	0.396	0.213	0.114	1.026	1.023	97.23	8.2	0.083
1.50	1.0089	1.0107	15.1	0.180	993.8	4.4	21	1.3351	0.591	0.318	0.172	1.040	1.033	95.94	11.7	0.123
2.00	1.0125	1.0143	20.2	0.241	992.2	6.0	27	1.3357	0.784	0.422	0.228	1.055	1.044	94.64	15.0	0.161
2.50	1.0160	1.0178	25.4	0.302	990.6	7.6	34	1.3364	0.975	0.524	0.284	1.069	1.055	93.33	18.0	0.197
3.00	1.0196	1.0214	30.6	0.364	989.0	9.3	41	1.3370	1.164	0.626	0.340	1.084	1.066	92.03	20.8	0.230
3.50	1.0231	1.0249	35.8	0.426	987.3	11.0	47	1.3377	1.350	0.726	0.395	1.100	1.077	90.73	23.5	0.263
4.00	1.0266	1.0284	41.1	0.489	985.5	12.7	54	1.3383	1.535	0.825	0.449	1.116	1.089	89.44	26.1	0.295
4.50	1.0301	1.0320	46.4	0.552	983.8	14.5	60	1.3390	1.718	0.924	0.502	1.132	1.101	88.16	28.7	0.327
5.00	1.0337	1.0355	51.7	0.615	982.0	16.2	67	1.3396	1.899	1.021	0.555	1.149	1.114	86.88	31.4	0.361
5.50	1.0372	1.0391	57.0	0.679	980.2	18.1	73	1.3403	2.078	1.117	0.607	1.166	1.126	85.61	34.2	0.397
6.00	1.0408	1.0426	62.4	0.743	978.4	19.9	79	1.3409	2.256	1.213	0.659	1.183	1.139	84.36	37.3	0.437

69 SODIUM BROMIDE, NaBr

MOLECULAR WEIGHT = 102.91
RELATIVE SPECIFIC REFRACTIVITY = 0.626

0.00 % by wt. data are the same for all compounds.
For Values of 0.00 wt. % solutions see Table 1, Acetic Acid.

A% by wt.	ρ D_4^{20}	D_{20}^{20}	C_s g/l	M g-mol/l	C_w g/l	$(C_o - C_w)$ g/l	$(n - n_o)$ × 10^4	n	Δ °C	O Os/kg	S g-mol/l	η/η_o	η/ρ cS	φ rhe	γ mmho/cm	T g-mol/l
0.50	1.0021	1.0039	5.0	0.049	997.1	1.1	7	1.3337	0.173	0.093	0.049	1.002	1.002	99.55	5.0	0.050
1.00	1.0060	1.0078	10.1	0.098	995.9	2.3	14	1.3344	0.344	0.185	0.099	1.005	1.001	99.30	9.7	0.100
1.50	1.0099	1.0117	15.1	0.147	994.8	3.5	21	1.3351	0.515	0.277	0.149	1.007	1.000	99.06	14.1	0.150
2.00	1.0139	1.0157	20.3	0.197	993.6	4.7	28	1.3358	0.687	0.369	0.200	1.010	0.998	98.81	18.4	0.201
2.50	1.0178	1.0196	25.4	0.247	992.4	5.9	35	1.3365	0.860	0.463	0.251	1.013	0.997	98.56	22.7	0.253
3.00	1.0218	1.0236	30.7	0.298	991.1	7.1	42	1.3372	1.036	0.557	0.303	1.015	0.995	98.31	26.9	0.305
3.50	1.0258	1.0276	35.9	0.349	989.9	8.3	49	1.3379	1.214	0.653	0.355	1.018	0.994	98.06	31.2	0.359
4.00	1.0298	1.0317	41.2	0.400	988.6	9.6	57	1.3386	1.394	0.750	0.408	1.020	0.993	97.80	35.5	0.413
4.50	1.0339	1.0357	46.5	0.452	987.4	10.9	64	1.3394	1.577	0.848	0.461	1.023	0.992	97.54	39.8	0.468
5.00	1.0380	1.0398	51.9	0.504	986.1	12.2	71	1.3401	1.761	0.947	0.515	1.026	0.990	97.27	44.0	0.523
5.50	1.0421	1.0439	57.3	0.557	984.8	13.5	78	1.3408	1.949	1.048	0.570	1.029	0.989	97.00	48.2	0.578
6.00	1.0462	1.0481	62.8	0.610	983.5	14.8	86	1.3415	2.139	1.150	0.625	1.032	0.988	96.73	52.4	0.634
6.50	1.0504	1.0522	68.3	0.663	982.1	16.1	93	1.3423	2.332	1.254	0.681	1.035	0.987	96.46	56.5	0.691
7.00	1.0546	1.0564	73.8	0.717	980.8	17.5	100	1.3430	2.528	1.359	0.737	1.038	0.986	96.18	60.6	0.748
7.50	1.0588	1.0607	79.4	0.772	979.4	18.8	108	1.3438	2.728	1.466	0.794	1.041	0.985	95.90	64.7	0.806
8.00	1.0630	1.0649	85.0	0.826	978.0	20.2	115	1.3445	2.929	1.575	0.851	1.044	0.984	95.61	68.7	0.864
8.50	1.0673	1.0692	90.7	0.882	976.6	21.6	123	1.3453	3.133	1.685	0.909	1.047	0.983	95.31	72.7	0.923
9.00	1.0716	1.0735	96.4	0.937	975.2	23.1	130	1.3460	3.342	1.797	0.968	1.051	0.982	95.00	76.7	0.983
9.50	1.0760	1.0779	102.2	0.993	973.7	24.5	138	1.3468	3.554	1.911	1.027	1.054	0.982	94.67	80.6	1.04
10.00	1.0803	1.0822	108.0	1.050	972.3	25.9	145	1.3475	3.768	2.026	1.087	1.058	0.981	94.33	84.6	1.10
11.00	1.0892	1.0911	119.8	1.164	969.3	28.9	161	1.3491	4.206	2.261	1.208	1.066	0.981	93.61	92.4	1.22
12.00	1.0981	1.1000	131.8	1.280	966.3	31.9	176	1.3506	4.665	2.508	1.334	1.075	0.981	92.85	100.	1.35
13.00	1.1072	1.1091	143.9	1.399	963.2	35.0	192	1.3522	5.150	2.770	1.463	1.084	0.981	92.05	108.	1.48
14.00	1.1164	1.1184	156.3	1.519	960.1	38.1	208	1.3538	5.650	3.040	1.597	1.094	0.982	91.21	115.	1.61
15.00	1.1257	1.1277	168.9	1.641	956.9	41.4	224	1.3554	6.180	3.320	1.734	1.105	0.984	90.32	122.	1.75
16.00	1.1352	1.1372	181.6	1.765	953.6	44.7	241	1.3570	6.740	3.620	1.875	1.117	0.986	89.38	129.	1.89
17.00	1.1448	1.1469	194.6	1.891	950.2	48.0	257	1.3587	7.320	3.930	2.021	1.129	0.988	88.40	136.	2.03
18.00	1.1546	1.1566	207.8	2.019	946.8	51.5	274	1.3604				1.142	0.991	87.37	143.	2.18
19.00	1.1645	1.1665	221.3	2.150	943.2	55.0	291	1.3621				1.157	0.995	86.29	150.	2.33
20.00	1.1745	1.1766	234.9	2.283	939.6	58.6	308	1.3638				1.172	1.000	85.15	157.	2.50
22.00	1.1951	1.1972	262.9	2.555	932.2	66.1	343	1.3673				1.205	1.010	82.81	172.	2.84

69 SODIUM BROMIDE, NaBr—(Continued)

A% by wt.	ρ D_4^{20}	D_{20}^{20}	C_s g/l	M g-mol/l	C_w g/l	$(C_o - C_w)$ g/l	$(n - n_o)$ × 10^4	n	Δ °C	O Os/kg	S g-mol/l	η/η_o	η/ρ cS	φ rhe	γ mmho/cm	T g-mol/l
24.00	1.2163	1.2184	291.9	2.837	924.4	73.9	378	1.3708				1.242	1.023	80.37	185.0	3.21
26.00	1.2382	1.2404	321.9	3.128	916.2	82.0	415	1.3745				1.284	1.039	77.75	197.0	3.60
28.00	1.2608	1.2630	353.0	3.430	907.8	90.5	453	1.3783				1.333	1.059	74.88	207.0	4.03
30.00	1.2842	1.2864	385.2	3.744	889.9	99.3	492	1.3822				1.392	1.086	71.70	216.0	4.50
32.00	1.3083	1.3106	418.7	4.068	889.7	108.6	532	1.3862				1.462	1.120	68.26		
34.00	1.3333	1.3357	453.3	4.405	880.0	118.2	573	1.3903				1.543	1.160	64.68		
36.00	1.3592	1.3616	489.3	4.755	869.9	128.3	616	1.3946				1.636	1.206	61.01		
38.00	1.3860	1.3885	526.7	5.118	859.3	138.9	660	1.3990				1.742	1.259	57.30		
40.00	1.4138	1.4163	565.5	5.495	848.3	150.0	705	1.4035				1.862	1.320	53.60		

70 SODIUM CARBONATE, $Na_2CO_3 \cdot 10H_2O$

MOLECULAR WEIGHT = 106.00 FORMULA WEIGHT, HYDRATE = 286.16
RELATIVE SPECIFIC REFRACTIVITY = 0.619

0.00 % by wt. data are the same for all compounds.
For Values of 0.00 wt. % solutions see Table 1, Acetic Acid.

A% by wt.	H% by wt.	ρ D_4^{20}	D_{20}^{20}	C_s g/l	M g-mol/l	C_w g/l	$(C_o - C_w)$ g/l	$(n - n_o)$ × 10^4	n	Δ °C	O Os/kg	S g-mol/l	η/η_o	η/ρ cS	φ rhe	γ mmho/cm	T g-mol/l
0.50	1.35	1.0034	1.0052	5.0	0.047	998.4	−0.2	11	1.3341	0.222	0.119	0.063	1.023	1.021	97.58	7.0	0.072
1.00	2.70	1.0086	1.0104	10.1	0.095	998.5	−0.3	23	1.3352	0.425	0.229	0.123	1.047	1.040	95.32	13.1	0.139
1.50	4.05	1.0138	1.0156	15.2	0.143	998.6	−0.4	34	1.3364	0.598	0.322	0.174	1.073	1.060	93.03	18.4	0.201
2.00	5.40	1.0190	1.0208	20.4	0.192	998.6	−0.4	45	1.3375	0.751	0.404	0.219	1.100	1.082	90.73	23.3	0.260
2.50	6.75	1.0242	1.0260	25.6	0.242	998.6	−0.4	56	1.3386	0.915	0.492	0.267	1.128	1.104	88.46	27.8	0.316
3.00	8.10	1.0294	1.0312	30.9	0.291	998.5	−0.3	67	1.3397	1.082	0.582	0.316	1.157	1.127	86.23	32.0	0.369
3.50	9.45	1.0346	1.0364	36.2	0.342	998.4	−0.2	78	1.3408	1.250	0.672	0.365	1.188	1.150	84.01	36.0	0.420
4.00	10.80	1.0398	1.0416	41.6	0.392	998.2	0.0	89	1.3419	1.421	0.764	0.416	1.220	1.176	81.81	39.8	0.468
4.50	12.15	1.0450	1.0468	47.0	0.444	998.0	0.3	100	1.3430	1.593	0.857	0.466	1.253	1.202	79.62	43.5	0.515
5.00	13.50	1.0502	1.0521	52.5	0.495	997.7	0.5	110	1.3440	1.768	0.951	0.517	1.289	1.230	77.42	47.0	0.562
5.50	14.85	1.0554	1.0573	58.0	0.548	997.3	0.9	121	1.3451	1.945	1.046	0.569	1.326	1.259	75.26	50.4	0.608
6.00	16.20	1.0606	1.0625	63.6	0.600	997.0	1.3	132	1.3462	2.125	1.142	0.621	1.364	1.289	73.16	53.6	0.652
6.50	17.55	1.0658	1.0677	69.3	0.654	996.5	1.7	142	1.3472				1.404	1.320	71.09	56.7	0.694
7.00	18.90	1.0711	1.0730	75.0	0.707	996.1	2.1	153	1.3483				1.445	1.352	69.05	59.7	0.735
7.50	20.25	1.0763	1.0782	80.7	0.762	995.6	2.6	164	1.3494				1.489	1.386	67.04	62.4	0.774
8.00	21.60	1.0816	1.0835	86.5	0.816	995.1	3.1	174	1.3504				1.535	1.422	65.03	65.1	0.812
8.50	22.95	1.0869	1.0888	92.4	0.872	994.5	3.7	185	1.3515				1.583	1.460	63.03	67.6	0.848
9.00	24.30	1.0922	1.0942	98.3	0.927	993.9	4.3	196	1.3525				1.635	1.500	61.03	70.0	0.883
9.50	25.65	1.0975	1.0995	104.3	0.984	993.3	5.0	206	1.3536				1.691	1.543	59.03	72.2	0.916
10.00	27.00	1.1029	1.1048	110.3	1.040	992.6	5.6	217	1.3547				1.750	1.590	57.03	74.4	0.948
11.00	29.70	1.1136	1.1156	122.5	1.156	991.1	7.1	238	1.3568				1.880	1.691	53.09	78.3	1.00
12.00	32.40	1.1244	1.1264	134.9	1.273	989.5	8.8	259	1.3589				2.024	1.803	49.32	81.7	1.05
13.00	35.10	1.1353	1.1373	147.6	1.392	987.7	10.5	280	1.3610				2.182	1.926	45.74	84.6	1.10
14.00	37.79	1.1463	1.1483	160.5	1.514	985.8	12.4	301	1.3631				2.356	2.059	42.36	86.9	1.14
15.00	40.49	1.1574	1.1595	173.6	1.638	983.8	14.4	322	1.3652				2.546	2.204	39.20	88.6	1.16

71 SODIUM CHLORIDE, NaCl

MOLECULAR WEIGHT = 58.44
RELATIVE SPECIFIC REFRACTIVITY = 0.794

0.00 % by wt. data are the same for all compounds.
For Values of 0.00 wt. % solutions see Table 1, Acetic Acid.

A% by wt.	ρ D_4^{20}	D_{20}^{20}	C_s g/l	M g-mol/l	C_w g/l	$(C_o - C_w)$ g/l	$(n - n_o)$ × 10^4	n	Δ °C	O Os/kg	S g-mol/l	η/η_o	η/ρ cS	φ rhe	γ mmho/cm	T g-mol/l
0.10	0.9989	1.0007	1.0	0.017	997.9	0.3	2	1.3332	0.062	0.033	0.017	1.002	1.005	99.61	1.7	0.017
0.20	0.9997	1.0014	2.0	0.034	997.7	0.6	4	1.3333	0.121	0.065	0.034	1.004	1.006	99.43	3.3	0.034
0.30	1.0004	1.0021	3.0	0.051	997.4	0.9	5	1.3335	0.181	0.097	0.051	1.006	1.007	99.24	5.0	0.051
0.40	1.0011	1.0028	4.0	0.069	997.1	1.2	7	1.3337	0.240	0.129	0.069	1.007	1.008	99.06	6.6	0.069
0.50	1.0018	1.0036	5.0	0.086	996.8	1.5	9	1.3339	0.299	0.161	0.086	1.009	1.009	98.89	8.2	0.086
0.60	1.0025	1.0043	6.0	0.103	996.5	1.8	11	1.3340	0.358	0.192	0.103	1.011	1.011	98.71	9.8	0.103
0.70	1.0032	1.0050	7.0	0.120	996.2	2.1	12	1.3342	0.417	0.224	0.120	1.013	1.012	98.54	11.4	0.120
0.80	1.0039	1.0057	8.0	0.137	995.9	2.4	14	1.3344	0.475	0.256	0.137	1.015	1.013	98.37	12.9	0.137
0.90	1.0046	1.0064	9.0	0.155	995.6	2.7	16	1.3346	0.534	0.287	0.155	1.016	1.014	98.20	14.4	0.155
1.00	1.0053	1.0071	10.1	0.172	995.3	3.0	18	1.3347	0.593	0.319	0.172	1.018	1.015	98.04	16.0	0.172
1.10	1.0060	1.0078	11.1	0.189	995.0	3.3	19	1.3349	0.652	0.351	0.189	1.020	1.016	97.87	17.5	0.189
1.20	1.0068	1.0085	12.1	0.207	994.7	3.6	21	1.3351	0.711	0.382	0.207	1.021	1.017	97.72	18.9	0.207
1.30	1.0075	1.0093	13.1	0.224	994.4	3.9	23	1.3353	0.770	0.414	0.224	1.023	1.017	97.56	20.4	0.224
1.40	1.0082	1.0100	14.1	0.241	994.1	4.2	25	1.3354	0.829	0.446	0.241	1.025	1.018	97.41	21.8	0.241
1.50	1.0089	1.0107	15.1	0.259	993.8	4.5	26	1.3356	0.888	0.478	0.259	1.026	1.019	97.26	23.2	0.259
1.60	1.0096	1.0114	16.2	0.276	993.5	4.8	28	1.3358	0.948	0.509	0.276	1.028	1.020	97.11	24.6	0.276
1.70	1.0103	1.0121	17.2	0.294	993.1	5.1	30	1.3360	1.007	0.541	0.294	1.029	1.021	96.96	26.0	0.294
1.80	1.0110	1.0128	18.2	0.311	992.8	5.4	32	1.3362	1.067	0.573	0.311	1.031	1.022	96.82	27.4	0.311
1.90	1.0117	1.0135	19.2	0.329	992.5	5.7	33	1.3363	1.126	0.605	0.329	1.032	1.022	96.67	28.8	0.329
2.00	1.0125	1.0143	20.2	0.346	992.2	6.0	35	1.3365	1.186	0.638	0.346	1.034	1.023	96.52	30.2	0.346
2.10	1.0132	1.0150	21.3	0.364	991.9	6.3	37	1.3367	1.246	0.670	0.364	1.036	1.024	96.37	31.6	0.364
2.20	1.0139	1.0157	22.3	0.382	991.6	6.6	39	1.3369	1.306	0.702	0.382	1.037	1.025	96.22	33.0	0.382
2.30	1.0146	1.0164	23.3	0.399	991.3	7.0	40	1.3370	1.366	0.734	0.399	1.039	1.026	96.07	34.3	0.399
2.40	1.0153	1.0171	24.4	0.418	991.0	7.3	42	1.3372	1.426	0.767	0.418	1.040	1.027	95.93	35.7	0.418
2.50	1.0160	1.0178	25.4	0.435	990.6	7.6	44	1.3374	1.486	0.799	0.435	1.042	1.028	95.78	37.1	0.435
2.60	1.0168	1.0185	26.4	0.452	990.3	7.9	46	1.3376	1.547	0.832	0.452	1.044	1.028	95.64	38.5	0.452
2.70	1.0175	1.0193	27.5	0.470	990.0	8.2	47	1.3377	1.607	0.864	0.470	1.045	1.029	95.49	39.9	0.470
2.80	1.0182	1.0200	28.5	0.488	989.7	8.6	49	1.3379	1.668	0.897	0.488	1.047	1.030	95.35	41.2	0.488
2.90	1.0189	1.0207	29.5	0.505	989.4	8.9	51	1.3381	1.729	0.930	0.505	1.048	1.031	95.21	42.6	0.505
3.00	1.0196	1.0214	30.6	0.523	989.0	9.2	53	1.3383	1.790	0.962	0.523	1.050	1.032	95.06	44.0	0.523
3.10	1.0203	1.0221	31.6	0.541	988.7	9.5	54	1.3384	1.851	0.995	0.541	1.051	1.033	94.92	45.3	0.541
3.20	1.0211	1.0229	32.7	0.559	988.4	9.9	56	1.3386	1.913	1.028	0.559	1.053	1.033	94.78	46.7	0.559

71 SODIUM CHLORIDE, NaCl—(Continued)

A% by wt.	ρ D_4^{20}	D_{20}^{20}	C_s g/l	M g-mol/l	C_w g/l	$(C_o - C_w)$ g/l	$(n - n_o) \times 10^4$	n	Δ °C	O Os/kg	S g-mol/l	η/η_o	η/ρ cS	ϕ rhe	γ mmho/cm	T g-mol/l
3.30	1.0218	1.0236	33.7	0.577	988.0	10.2	58	1.3388	1.974	1.061	0.577	1.055	1.034	94.64	48.0	0.577
3.40	1.0225	1.0243	34.8	0.595	987.7	10.5	60	1.3390	2.036	1.094	0.595	1.056	1.035	94.50	49.4	0.595
3.50	1.0232	1.0250	35.8	0.613	987.4	10.8	61	1.3391	2.098	1.128	0.613	1.058	1.036	94.35	50.7	0.613
3.60	1.0239	1.0257	36.9	0.631	987.1	11.2	63	1.3393	2.160	1.161	0.631	1.059	1.037	94.21	52.0	0.631
3.70	1.0246	1.0265	37.9	0.649	986.7	11.5	65	1.3395	2.222	1.194	0.649	1.061	1.037	94.07	53.3	0.649
3.80	1.0254	1.0272	39.0	0.667	986.4	11.8	67	1.3397	2.284	1.228	0.667	1.063	1.038	93.93	54.7	0.667
3.90	1.0261	1.0279	40.0	0.685	986.1	12.2	68	1.3398	2.347	1.262	0.685	1.064	1.039	93.78	56.0	0.685
4.00	1.0268	1.0286	41.1	0.703	985.7	12.5	70	1.3400	2.409	1.295	0.703	1.066	1.040	93.64	57.3	0.703
4.10	1.0275	1.0293	42.1	0.721	985.4	12.8	72	1.3402	2.472	1.329	0.721	1.067	1.041	93.49	58.6	0.721
4.20	1.0282	1.0301	43.2	0.739	985.1	13.2	74	1.3404	2.535	1.363	0.739	1.069	1.042	93.35	59.9	0.739
4.30	1.0290	1.0308	44.2	0.757	984.7	13.5	75	1.3405	2.598	1.397	0.757	1.071	1.043	93.20	61.2	0.757
4.40	1.0297	1.0315	45.3	0.775	984.4	13.9	77	1.3407	2.662	1.431	0.775	1.072	1.044	93.06	62.5	0.775
4.50	1.0304	1.0322	46.4	0.794	984.0	14.2	79	1.3409	2.725	1.465	0.794	1.074	1.045	92.91	63.8	0.794
4.60	1.0311	1.0330	47.4	0.812	983.7	14.5	81	1.3411	2.789	1.500	0.812	1.076	1.046	92.76	65.1	0.812
4.70	1.0318	1.0337	48.5	0.830	983.4	14.9	82	1.3412	2.853	1.534	0.830	1.078	1.046	92.61	66.3	0.830
4.80	1.0326	1.0344	49.6	0.848	983.0	15.2	84	1.3414	2.917	1.569	0.848	1.079	1.047	92.46	67.6	0.848
4.90	1.0333	1.0351	50.6	0.866	982.7	15.6	86	1.3416	2.982	1.603	0.866	1.081	1.048	92.31	68.9	0.866
5.00	1.0340	1.0358	51.7	0.885	982.3	15.9	88	1.3418	3.046	1.638	0.885	1.083	1.049	92.15	70.1	0.885
5.20	1.0355	1.0373	53.8	0.921	981.6	16.6	91	1.3421	3.176	1.708	0.921	1.087	1.052	91.84	72.6	0.921
5.40	1.0369	1.0387	56.0	0.958	980.9	17.3	95	1.3425	3.307	1.778	0.958	1.090	1.054	91.53	75.1	0.958
5.60	1.0384	1.0402	58.1	0.995	980.2	18.0	98	1.3428	3.438	1.848	0.995	1.094	1.056	91.22	77.5	0.995
5.80	1.0398	1.0417	60.3	1.032	979.5	18.7	102	1.3432	3.570	1.919	1.032	1.098	1.058	90.90	80.0	1.032
6.00	1.0413	1.0431	62.5	1.069	978.8	19.4	105	1.3435	3.703	1.991	1.069	1.102	1.060	90.59	82.4	1.069
6.20	1.0427	1.0446	64.6	1.106	978.1	20.2	109	1.3439	3.837	2.063	1.106	1.106	1.062	90.27	84.7	1.106
6.40	1.0442	1.0460	66.8	1.144	977.4	20.9	112	1.3442	3.972	2.135	1.144	1.109	1.065	89.95	87.1	1.144
6.60	1.0456	1.0475	69.0	1.181	976.6	21.6	116	1.3446	4.107	2.208	1.181	1.113	1.067	89.63	89.4	1.181
6.80	1.0471	1.0490	71.2	1.218	975.9	22.3	119	1.3449	4.244	2.282	1.218	1.117	1.069	89.31	91.8	1.218
7.00	1.0486	1.0504	73.4	1.256	975.2	23.1	123	1.3453	4.378	2.354	1.256	1.122	1.072	88.99	94.1	1.256
7.20	1.0500	1.0519	75.6	1.294	974.4	23.8	126	1.3456	4.516	2.428	1.294	1.126	1.074	88.66	96.4	1.294
7.40	1.0515	1.0533	77.8	1.331	973.7	24.6	130	1.3460	4.655	2.503	1.331	1.130	1.077	88.33	98.7	1.331
7.60	1.0530	1.0548	80.0	1.369	972.9	25.3	133	1.3463	4.795	2.578	1.369	1.134	1.079	88.00	101.	1.369
7.80	1.0544	1.0563	82.2	1.407	972.2	26.1	137	1.3467	4.937	2.654	1.407	1.138	1.082	87.67	103.	1.407
8.00	1.0559	1.0578	84.5	1.445	971.4	26.8	140	1.3470	5.079	2.731	1.445	1.143	1.084	87.33	105.	1.445
8.20	1.0574	1.0592	86.7	1.484	970.7	27.6	144	1.3474	5.222	2,808	1.484	1.147	1.087	86.99	107.	1.484
8.40	1.0588	1.0607	88.9	1.522	969.9	28.3	147	1.3477	5.367	2.885	1.522	1.152	1.090	86.65	109.	1.522
8.60	1.0603	1.0622	91.2	1.560	969.1	29.1	151	1.3481	5.512	2.964	1.560	1.156	1.093	86.30	112.	1.560
8.80	1.0618	1.0637	93.4	1.599	968.3	29.9	154	1.3484	5.659	3.043	1.599	1.161	1.096	85.95	114.	1.599
9.00	1.0633	1.0651	95.7	1.637	967.6	30.7	158	1.3488	5.807	3.122	1.637	1.166	1.099	85.60	116.	1.637
9.20	1.0647	1.0666	98.0	1.676	966.8	31.4	161	1.3491	5.956	3.202	1.676	1.171	1.102	85.25	118.	1.676
9.40	1.0662	1.0681	100.2	1.715	966.0	32.2	165	1.3495	6.106	3.283	1.715	1.176	1.105	84.89	120.	1.715
9.60	1.0677	1.0696	102.5	1.754	965.2	33.0	168	1.3498	6.258	3.364	1.754	1.181	1.108	84.53	122.	1.754
9.80	1.0692	1.0711	104.8	1.793	964.4	33.8	172	1.3502	6.410	3.446	1.793	1.186	1.111	84.16	124.	1.793
10.00	1.0707	1.0726	107.1	1.832	963.6	34.6	175	1.3505	6.564	3.529	1.832	1.191	1.115	83.80	126.	1.832
10.50	1.0744	1.0763	112.8	1.930	961.6	36.6	184	1.3514	6.954	3.739	1.930	1.204	1.123	82.86	131.	1.930
11.00	1.0781	1.0801	118.6	2.029	959.5	38.7	193	1.3523	7.353	3.953	2.029	1.218	1.132	81.90	136.	2.029
11.50	1.0819	1.0838	124.4	2.129	957.5	40.8	202	1.3532	7.760	4.172	2.129	1.233	1.142	80.94	140.	2.129
12.00	1.0857	1.0876	130.3	2.229	955.4	42.9	211	1.3541	8.176	4.396	2.229	1.248	1.152	79.95	145.	2.229
12.50	1.0894	1.0914	136.2	2.330	953.3	45.0	220	1.3549	8.602	4.625	2.330	1.264	1.162	78.96	149.	2.330
13.00	1.0932	1.0952	142.1	2.432	951.1	47.1	228	1.3558	9.038	4.859	2.432	1.280	1.173	77.97	154.	2.432
13.50	1.0970	1.0990	148.1	2.534	948.9	49.3	237	1.3567	9.484	5.099	2.534	1.297	1.184	76.97	158.	2.534
14.00	1.1008	1.1028	154.1	2.637	946.7	51.5	246	1.3576	9.940	5.344	2.637	1.314	1.196	75.97	163.	2.637
14.50	1.1047	1.1066	160.2	2.741	944.5	53.7	255	1.3585	10.408	5.596	2.741	1.331	1.207	74.97	167.	2.741
15.00	1.1085	1.1105	166.3	2.845	942.2	56.0	264	1.3594	10.888	5.854	2.845	1.349	1.219	73.98	171.	2.845
16.00	1.1162	1.1182	178.6	3.056	937.6	60.6	282	1.3612	11.885	6.390	3.056	1.385	1.243	72.07	179.	3.056
17.00	1.1240	1.1260	191.1	3.270	932.9	65.3	300	1.3630	12.935	6.954	3.270	1.421	1.267	70.21	186.	3.270
18.00	1.1319	1.1339	203.7	3.486	928.1	70.1	318	1.3648	14.044	7.550	3.486	1.460	1.293	68.34	193.	3.486
19.00	1.1398	1.1418	216.6	3.706	923.2	75.0	336	1.3666	15.216	8.181	3.706	1.504	1.322	66.36	199.	3.705
20.00	1.1478	1.1498	229.6	3.928	918.2	80.0	354	1.3684	16.458	8.849	3.928	1.554	1.357	64.22	204.	3.928
21.00	1.1558	1.1579	242.7	4.153	913.1	85.1	372	1.3702	17.776	9.557	4.153	1.611	1.396	61.95	209.	4.153
22.00	1.1640	1.1660	256.1	4.382	907.9	90.4	391	1.3721	19.176	10.310	4.382	1.673	1.441	59.64	213.	4.382
23.00	1.1721	1.1742	269.6	4.613	902.6	95.7	409	1.3739	20.667	11.111	4.613	1.742	1.489	57.29	217.	4.613
24.00	1.1804	1.1825	283.3	4.848	897.1	101.1	428	1.3757			4.848	1.817	1.542	54.93	220.	4.848
25.00	1.1887	1.1909	297.2	5.085	891.6	106.7	446	1.3776			5.085	1.898	1.600	52.58	222.	5.085
26.00	1.1972	1.1993	311.3	5.326	885.9	112.3	465	1.3795			5.326	1.986	1.662	50.26	225.	5.326

72 SODIUM CITRATE, $Na_3C_6H_5O_7 \cdot 2H_2O$

MOLECULAR WEIGHT = 258.068 FORMULA WEIGHT, HYDRATE = 294.10

A% by wt.	H% by wt.	ρ D_4^{20}	D_{20}^{20}	C_S g/l	M g-mol/l	C_W g/l	$(C_0 - C_W)$ g/l	$(n - n_0)$ × 10^4	n	Δ °C	O Os/kg	S g-mol/l	η/η_0	η/ρ cS	φ rhe	γ mmho/cm	T g-mol/l
1.00	1.14	1.0049	1.0067	10.0	0.039	994.9	3.3	18.	1.3348	0.197	0.106	0.056	1.041	1.038	95.87	7.4	0.075
2.00	2.28	1.0120	1.0138	20.2	0.078	991.8	6.4	36.	1.3366	0.389	0.209	0.112	1.079	1.068	92.52	12.8	0.136
3.00	3.42	1.0186	1.0204	30.6	0.118	988.0	10.2	53.	1.3383	0.586	0.315	0.170	1.120	1.102	89.09	17.9	0.195
4.00	4.56	1.0260	1.0278	41.0	0.159	985.0	13.2	71.	1.3401	0.787	0.423	0.229	1.164	1.137	85.72	22.1	0.246
5.00	5.70	1.0331	1.0350	51.7	0.200	981.4	16.8	89.	1.3419	0.974	0.523	0.284	1.208	1.172	82.59	26.2	0.296
6.00	6.84	1.0405	1.0424	62.4	0.242	978.1	20.1	107.	1.3437	1.171	0.630	0.342	1.260	1.213	79.23	29.9	0.343
7.00	7.98	1.0482	1.0501	73.4	0.284	974.8	23.4	125.	1.3455	1.361	0.732	0.398	1.311	1.253	76.14	33.2	0.384
8.00	9.12	1.0557	1.0576	84.5	0.327	971.2	27.0	143.	1.3473	1.566	0.842	0.458	1.368	1.298	72.98	36.5	0.426
9.00	10.26	1.0632	1.0651	95.7	0.371	967.5	30.7	161.	1.3491	1.768	0.951	0.517	1.424	1.342	70.09	39.4	0.463
10.00	11.40	1.0708	1.0727	107.1	0.415	963.7	34.5	179.	1.3509	1.957	1.052	0.572	1.496	1.400	66.71	42.1	0.497
12.00	13.68	1.0861	1.0881	130.3	0.505	955.8	42.4	216.	1.3546	2.376	1.277	0.693	1.646	1.519	60.61	48.9	0.588
14.00	15.95	1.1019	1.1039	154.3	0.598	947.6	50.6	253.	1.3583	2.819	1.515	0.820	1.828	1.662	54.60	50.4	0.608
16.00	18.23	1.1173	1.1193	178.8	0.693	938.5	59.7	288.	1.3618	3.274	1.760	0.949	2.041	1.830	48.91	53.5	0.650
18.00	20.51	1.1327	1.1347	203.9	0.790	928.8	69.4	326.	1.3656	3.815	2.051	1.100	2.285	2.021	43.68	55.7	0.680
20.00	22.79	1.1492	1.1513	229.8	0.891	919.4	78.8	363.	1.3693	4.393	2.362	1.260	2.591	2.259	38.52	57.1	0.700
24.00	27.35	1.1813	1.1834	283.5	1.099	897.8	100.4	437.	1.3767				3.402	2.886	29.33	57.7	0.708
28.00	31.91	1.2151	1.2173	340.2	1.318	874.9	123.3	515.	1.3845				4.577	3.774	21.81	55.9	0.683
32.00	36.47	1.2487	1.2510	399.6	1.548	849.1	149.1	593.	1.3923				6.528	5.238	15.29	51.0	0.616
36.00	41.03	1.2843	1.2866	462.3	1.792	822.0	176.2	671.	1.4001				9.768	7.621	10.22	44.5	0.529

73 SODIUM DIATRIZOATE, $(CH_3CONH)_2C_6I_3COONa$, HYPAQUE®

MOLECULAR WEIGHT = 635.92
RELATIVE SPECIFIC REFRACTIVITY = 0.823

0.00 % by wt. data are the same for all compounds.
For Values of 0.00 wt. % solutions see Table 1, Acetic Acid.

A% by wt.	ρ D_4^{20}	D_{20}^{20}	C_s g/l	M g-mol/l	C_w g/l	$(C_o - C_w)$ g/l	$(n - n_o)$ × 10^4	n	Δ °C	O Os/kg	S g-mol/l	η/η_o	η/ρ cS	φ rhe	γ mmho/cm	T g-mol/l
1.00	1.0041	1.0059	10.0	0.016	994.1	4.1	16	1.3346	0.058	0.031	0.016	1.021	1.019	97.75	0.8	0.008
2.00	1.0102	1.0120	20.2	0.032	990.0	8.2	31	1.3361	0.110	0.059	0.031	1.041	1.033	95.87	1.6	0.016
3.00	1.0164	1.0182	30.5	0.048	985.9	12.3	48	1.3377	0.164	0.088	0.047	1.061	1.046	94.05	2.4	0.024
4.00	1.0227	1.0245	40.9	0.064	981.8	16.4	64	1.3394	0.220	0.118	0.063	1.082	1.060	92.26	3.1	0.031
5.00	1.0292	1.0310	51.5	0.081	977.7	20.5	80	1.3410	0.276	0.148	0.079	1.104	1.075	90.40	3.8	0.038
6.00	1.0357	1.0376	62.1	0.098	973.6	24.6	97	1.3427	0.332	0.178	0.095	1.128	1.092	88.44	4.5	0.045
7.00	1.0424	1.0443	73.0	0.115	969.5	28.8	114	1.3444	0.389	0.209	0.112	1.155	1.110	86.42	5.2	0.052
8.00	1.0493	1.0511	83.9	0.132	965.3	32.9	131	1.3461	0.446	0.240	0.129	1.183	1.130	84.38	5.9	0.059
9.00	1.0562	1.0581	95.1	0.150	961.1	37.1	148	1.3478	0.504	0.271	0.146	1.212	1.150	82.32	6.5	0.066
10.00	1.0632	1.0651	106.3	0.167	956.9	41.3	165	1.3495	0.562	0.302	0.163	1.243	1.171	80.29	7.2	0.073

® Registered trademark, sodium 3.5-diacetamido-2,4,6 tri-iodo benzoate.

73 SODIUM DIATRIZOATE, $(CH_3CONH)_2C_6I_3COONa$, HYPAQUE®—(Continued)

A% by wt.	ρ D_4^{20}	D_{20}^{20}	C_s g/l	M g-mol/l	C_w g/l	$(C_o - C_w)$ g/l	$(n - n_o)$ × 10^4	n	Δ °C	O Os/kg	S g-mol/l	η/η_o	η/ρ cS	φ rhe	γ mmho/cm	T g-mol/l
12.00	1.0776	1.0796	129.3	0.203	948.3	49.9	201	1.3530	0.683	0.367	0.199	1.305	1.213	76.50	8.4	0.086
14.00	1.0925	1.0944	152.9	0.240	939.5	58.7	237	1.3567	0.808	0.434	0.235	1.375	1.261	72.60	9.6	0.099
16.00	1.1077	1.1097	177.2	0.279	930.5	67.7	274	1.3603	0.936	0.503	0.273	1.470	1.329	67.91	10.6	0.110
18.00	1.1234	1.1254	202.2	0.318	921.2	77.0	313	1.3643	1.067	0.574	0.312	1.592	1.420	62.70	11.5	0.120
20.00	1.1395	1.1415	227.9	0.358	911.6	86.6	353	1.3683	1.205	0.648	0.352	1.724	1.516	57.89	12.2	0.129
22.00	1.1560	1.1581	254.3	0.400	901.7	96.5	394	1.3724	1.351	0.726	0.395	1.847	1.601	54.03	13.0	0.138
24.00	1.1730	1.1751	281.5	0.443	891.5	106.7	436	1.3766	1.506	0.810	0.441	1.965	1.678	50.79	13.7	0.146
26.00	1.1906	1.1927	309.5	0.487	881.0	117.2	480	1.3810	1.671	0.898	0.489	2.094	1.762	47.66	14.4	0.153
28.00	1.2086	1.2108	338.4	0.532	870.2	128.0	525	1.3854	1.846	0.993	0.540	2.254	1.869	44.27	15.0	0.160
30.00	1.2273	1.2295	368.2	0.579	859.1	139.1	571	1.3901	2.033	1.093	0.594	2.469	2.016	40.42	15.5	0.167
32.00	1.2467	1.2489	399.0	0.627	847.8	150.5	619	1.3949	2.232	1.200	0.652	2.742	2.204	36.39	16.2	0.174
34.00	1.2670	1.2692	430.8	0.677	836.2	162.0	669	1.3999	2.444	1.314	0.713	3.066	2.425	32.55	16.8	0.182
36.00	1.2882	1.2904	463.7	0.729	824.4	173.8	722	1.4052	2.672	1.437	0.778	3.449	2.683	28.93	17.2	0.187
38.00	1.3105	1.3128	498.0	0.783	812.5	185.8	778	1.4108	2.917	1.568	0.848	3.902	2.984	25.58	17.2	0.187
40.00	1.3286	1.3310	531.5	0.836	797.2	201.0	834	1.4164				4.400	3.318	22.68	16.6	0.180

®Registered trademark, sodium 3,5-diacetamido-2,4,6 tri-iodo benzoate.

74 SODIUM DICHROMATE, $Na_2Cr_2O_7 \cdot 2H_2O$

MOLECULAR WEIGHT = 261.97 FORMULA WEIGHT, HYDRATE = 298.00
RELATIVE SPECIFIC REFRACTIVITY = 0.859

0.00 % by wt. data are the same for all compounds.
For Values of 0.00 wt. % solutions see Table 1, Acetic Acid.

A% by wt.	H% by wt.	ρ D_4^{20}	D_{20}^{20}	C_s g/l	M g-mol/l	C_w g/l	$(C_o - C_w)$ g/l	$(n - n_o)$ × 10^4	n	Δ °C	O Os/kg	S g-mol/l	η/η_o	η/ρ cS	φ rhe	γ mmho/cm	T g-mol/l
0.50	0.57	1.0019	1.0037	5.0	0.019	996.9	1.4	10	1.3340	0.103	0.056	0.029	1.005	1.005	99.33	3.5	0.035
1.00	1.14	1.0056	1.0073	10.1	0.038	995.5	2.7	21	1.3351	0.204	0.110	0.058	1.009	1.005	98.91	6.9	0.070
2.00	2.28	1.0130	1.0148	20.3	0.077	992.7	5.5	42	1.3371	0.396	0.213	0.114	1.016	1.005	98.23	13.1	0.139
3.00	3.41	1.0204	1.0222	30.6	0.117	989.8	8.4	63	1.3392	0.581	0.312	0.169	1.024	1.005	97.47	19.1	0.209
4.00	4.55	1.0280	1.0298	41.1	0.157	986.9	11.4	84	1.3414	0.762	0.410	0.222	1.033	1.007	96.64	24.9	0.280
5.00	5.69	1.0356	1.0374	51.8	0.198	983.8	14.4	105	1.3435	0.946	0.509	0.276	1.042	1.008	95.78	30.5	0.350
6.00	6.83	1.0433	1.0452	62.6	0.239	980.7	17.5	127	1.3457	1.135	0.610	0.331	1.052	1.010	94.91	36.0	0.419
7.00	7.96	1.0511	1.0530	73.6	0.281	977.5	20.7	149	1.3479	1.325	0.712	0.387	1.061	1.012	94.02	41.2	0.486
8.00	9.10	1.0590	1.0609	84.7	0.323	974.3	23.9	171	1.3501	1.519	0.817	0.444	1.072	1.014	93.10	46.3	0.553
9.00	10.24	1.0670	1.0689	96.0	0.367	971.0	27.2	193	1.3523	1.720	0.924	0.503	1.083	1.017	92.15	51.5	0.622
10.00	11.38	1.0751	1.0770	107.5	0.410	967.6	30.6	216	1.3546	1.929	1.037	0.564	1.095	1.021	91.14	56.7	0.694
12.00	13.65	1.0916	1.0936	131.0	0.500	960.6	37.6	262	1.3592	2.388	1.284	0.697	1.121	1.029	89.01	67.8	0.851
14.00	15.93	1.1086	1.1105	155.2	0.592	953.4	44.8	310	1.3640	2.876	1.546	0.836	1.151	1.040	86.72	78.9	1.02
16.00	18.20	1.1260	1.1280	180.2	0.688	945.9	52.4	358	1.3688	3.356	1.804	0.972	1.184	1.054	84.28	89.2	1.18
18.00	20.48	1.1439	1.1460	205.9	0.786	938.0	60.2	408	1.3738	3.830	2.059	1.104	1.221	1.070	81.71	98.6	1.33
20.00	22.75	1.1624	1.1644	232.5	0.887	929.9	68.3	460	1.3790	4.301	2.312	1.234	1.264	1.090	78.96	107.	1.48
24.00	27.30	1.2009	1.2031	288.2	1.100	912.7	85.5	567	1.3897				1.364	1.138	73.18	121.	1.73
28.00	31.85	1.2418	1.2440	347.7	1.327	894.1	104.1	681	1.4011				1.494	1.206	66.79	131.	1.93
32.00	36.40	1.2851	1.2874	411.2	1.570	873.9	124.4	801	1.4131				1.677	1.308	59.51	140.	2.10
36.00	40.95	1.3302	1.3326	478.9	1.828	851.4	146.9	925	1.4255				1.914	1.442	52.13	145.	2.22
40.00	45.50	1.3786	1.3810	551.4	2.105	827.1	171.1	1058	1.4388				2.246	1.632	44.43	148.	2.29
44.00	50.05	1.4304	1.4330	629.4	2.403	801.0	197.2	1200	1.4530				2.646	1.853	37.72	147.	2.26
48.00	54.60	1.4860	1.4886	713.3	2.723	772.7	225.5	1353	1.4683				3.225	2.174	30.95	141.	2.12
52.00	59.15	1.5454	1.5482	803.6	3.068	741.8	256.4						4.297	2.786	23.23	131.	1.94
56.00	63.70	1.6088	1.6117	900.9	3.439	707.9	290.3						5.933	3.695	16.82	117.	1.67
60.00	68.25	1.6763	1.6793	1005.8	3.839	670.5	327.7						8.221	4.914	12.14	93.9	1.25

75 SODIUM FERROCYANIDE, $Na_4Fe(CN)_6 \cdot 10H_2O$

MOLECULAR WEIGHT = 303.91 FORMULA WEIGHT, HYDRATE = 484.07
RELATIVE SPECIFIC REFRACTIVITY = 0.991

0.00 % by wt. data are the same for all compounds.
For Values of 0.00 wt. % solutions see Table 1, Acetic Acid.

A% by wt.	H% by wt.	ρ D_4^{20}	D_{20}^{20}	C_s g/l	M g-mol/l	C_w g/l	$(C_o - C_w)$ g/l	$(n - n_o)$ × 10^4	n	Δ °C	O Os/kg	S g-mol/l	η/η_o	η/ρ cS	φ rhe	γ mmho/cm	T g-mol/l
0.50	0.80	1.0018	1.0036	5.0	0.016	996.8	1.4	13	1.3342	0.091	0.049	0.026	1.016	1.016	98.23	5.0	0.050
1.00	1.59	1.0055	1.0073	10.1	0.033	995.4	2.8	25	1.3355	0.192	0.103	0.055	1.032	1.028	96.73	9.8	0.102
1.50	2.39	1.0092	1.0110	15.1	0.050	994.0	4.2	38	1.3368	0.302	0.162	0.087	1.047	1.040	95.28	14.3	0.153
2.00	3.19	1.0129	1.0147	20.3	0.067	992.6	5.6	51	1.3381	0.394	0.212	0.114	1.063	1.052	93.87	18.7	0.204
2.50	3.98	1.0166	1.0184	25.4	0.084	991.2	7.0	64	1.3394	0.486	0.261	0.141	1.079	1.063	92.50	22.8	0.254
3.00	4.78	1.0204	1.0222	30.6	0.101	989.8	8.5	77	1.3407	0.576	0.310	0.167	1.095	1.075	91.16	26.7	0.303
3.50	5.57	1.0239	1.0258	35.8	0.118	988.1	10.1	89	1.3419	0.666	0.358	0.194	1.111	1.087	89.83	30.5	0.350
4.00	6.37	1.0276	1.0294	41.1	0.135	986.5	11.7	102	1.3432	0.755	0.406	0.220	1.127	1.099	88.52	34.1	0.396
4.50	7.17	1.0313	1.0331	46.4	0.153	984.9	13.3	115	1.3445	0.842	0.453	0.245	1.144	1.112	87.22	37.7	0.441
5.00	7.96	1.0350	1.0368	51.8	0.170	983.3	15.0	128	1.3457	0.929	0.499	0.271	1.161	1.124	85.93	41.1	0.485
5.50	8.76	1.0387	1.0406	57.1	0.188	981.6	16.6	140	1.3470	1.014	0.545	0.296	1.179	1.138	84.63	44.5	0.529
6.00	9.56	1.0425	1.0443	62.5	0.206	979.9	18.3	153	1.3483	1.098	0.590	0.321	1.198	1.151	83.33	47.7	0.572
6.50	10.35	1.0462	1.0481	68.0	0.224	978.2	20.0	166	1.3496	1.181	0.635	0.345	1.217	1.165	82.02	51.0	0.616
7.00	11.15	1.0500	1.0519	73.5	0.242	976.5	21.7	179	1.3509	1.262	0.679	0.369	1.237	1.180	80.70	54.2	0.659
7.50	11.95	1.0539	1.0557	79.0	0.260	974.8	23.4	192	1.3522	1.342	0.722	0.393	1.257	1.195	79.38	57.3	0.702
8.00	12.74	1.0577	1.0596	84.6	0.278	973.1	25.1	205	1.3535	1.421	0.764	0.416	1.279	1.211	78.04	60.5	0.746
8.50	13.54	1.0616	1.0634	90.2	0.297	971.3	26.9	218	1.3548	1.498	0.806	0.438	1.301	1.228	76.70	63.6	0.790
9.00	14.34	1.0655	1.0673	95.9	0.316	969.6	28.7	232	1.3562	1.574	0.846	0.460	1.325	1.246	75.34	66.7	0.835
9.50	15.13	1.0694	1.0713	101.6	0.334	967.8	30.4	245	1.3575	1.648	0.886	0.482	1.349	1.264	73.98	69.7	0.879
10.00	15.93	1.0733	1.0752	107.3	0.353	966.0	32.2	258	1.3588	1.720	0.925	0.503	1.375	1.283	72.61	72.8	0.925
11.00	17.52	1.0813	1.0832	118.9	0.391	962.4	35.9	285	1.3615				1.429	1.324	69.85	78.8	1.02
12.00	19.11	1.0894	1.0913	130.7	0.430	958.6	39.6	312	1.3642				1.488	1.368	67.09	84.7	1.10
13.00	20.71	1.0976	1.0995	142.7	0.469	954.9	43.4	340	1.3670				1.551	1.416	64.34	90.0	1.19
14.00	22.30	1.1058	1.1078	154.8	0.509	951.0	47.2	367	1.3697				1.619	1.467	61.63	94.7	1.27
15.00	23.89	1.1142	1.1162	167.1	0.550	947.1	51.1	395	1.3725				1.692	1.522	58.98	98.1	1.32

76 SODIUM HYDROXIDE, NaOH

MOLECULAR WEIGHT = 40.01
RELATIVE SPECIFIC REFRACTIVITY = 0.692

0.00 % by wt. data are the same for all compounds.
For Values of 0.00 wt. % solutions see Table 1, Acetic Acid.

A % by wt.	ρ D_4^{20}	D_{20}^{20}	C_s g/l	M g-mol/l	C_w g/l	$(C_o - C_w)$ g/l	$(n - n_o)$ × 10^4	n	Δ °C	O Os/kg	S g-mol/l	η/η_o	η/ρ cS	φ rhe	γ mmho/cm	T g-mol/l
0.50	1.0039	1.0057	5.0	0.125	998.9	− 0.6	14	1.3344	0.429	0.231	0.124	1.025	1.023	97.34	24.8	0.283
1.00	1.0095	1.0113	10.1	0.252	999.4	− 1.2	29	1.3358	0.860	0.462	0.251	1.052	1.044	94.87	48.6	0.584
1.50	1.0151	1.0169	15.2	0.381	999.9	− 1.7	43	1.3373	1.294	0.695	0.378	1.080	1.066	92.39	71.3	0.900
2.00	1.0207	1.0225	20.4	0.510	1000.3	− 2.0	56	1.3386	1.735	0.933	0.507	1.110	1.090	89.91	93.1	1.24
2.50	1.0262	1.0281	25.7	0.641	1000.6	− 2.4	70	1.3400	2.184	1.174	0.638	1.141	1.114	87.44	114.	1.61
3.00	1.0318	1.0336	31.0	0.774	1000.8	− 2.6	84	1.3414	2.642	1.420	0.770	1.174	1.140	84.98	134.	2.02
3.50	1.0373	1.0391	36.3	0.907	1001.0	− 2.8	97	1.3427	3.109	1.671	0.902	1.209	1.168	82.54	153.	2.45
4.00	1.0428	1.0446	41.7	1.043	1001.1	− 2.9	111	1.3441	3.587	1.929	1.037	1.246	1.197	80.11	171.	2.92
4.50	1.0483	1.0502	47.2	1.179	1001.1	− 2.9	124	1.3454	4.074	2.190	1.172	1.285	1.228	77.69	189.	3.42
5.00	1.0538	1.0557	52.7	1.317	1001.1	− 2.9	138	1.3467	4.569	2.457	1.308	1.326	1.261	75.26	206.	3.96
5.50	1.0593	1.0612	58.3	1.456	1001.0	− 2.8	151	1.3481	5.08	2.73	1.445	1.369	1.295	72.90	222.	4.54
6.00	1.0648	1.0667	63.9	1.597	1000.9	− 2.7	164	1.3494	5.60	3.01	1.582	1.413	1.330	70.62		
6.50	1.0703	1.0722	69.6	1.739	1000.7	− 2.5	177	1.3507	6.13	3.30	1.722	1.459	1.366	68.40		
7.00	1.0758	1.0777	75.3	1.882	1000.5	− 2.3	190	1.3520	6.69	3.60	1.864	1.507	1.404	66.21		
7.50	1.0813	1.0833	81.1	2.027	1000.2	− 2.0	203	1.3533	7.27	3.91	2.008	1.558	1.444	64.05		
8.00	1.0869	1.0888	86.9	2.173	999.9	− 1.7	216	1.3546	7.87	4.23	2.155	1.613	1.487	61.89		
8.50	1.0924	1.0943	92.9	2.321	999.5	− 1.3	229	1.3559	8.48	4.56	2.303	1.671	1.533	59.72		
9.00	1.0979	1.0998	98.8	2.470	999.1	− 0.9	242	1.3572	9.12	4.91	2.452	1.734	1.583	57.54		
9.50	1.1034	1.1054	104.8	2.620	998.6	− 0.3	255	1.3585	9.78	5.26	2.602	1.803	1.637	55.35		
10.00	1.1089	1.1109	110.9	2.772	998.0	0.2	268	1.3597	10.47	5.63	2.753	1.878	1.697	53.14		
11.00	1.1199	1.1219	123.2	3.079	996.7	1.5	293	1.3623	11.89	6.39	3.058	2.035	1.821	49.04		
12.00	1.1309	1.1329	135.7	3.392	995.2	3.0	318	1.3648	13.42	7.21	3.365	2.197	1.947	45.42		
13.00	1.1419	1.1440	148.5	3.710	993.5	4.8	343	1.3673	15.04	8.08	3.673	2.371	2.080	42.09		
14.00	1.1530	1.1550	161.4	4.034	991.6	6.7	367	1.3697	16.76	9.01	3.981	2.563	2.228	38.93		
15.00	1.1640	1.1661	174.6	4.364	989.4	8.8	392	1.3722				2.783	2.395	35.87		
16.00	1.1751	1.1771	188.0	4.699	987.0	11.2	416	1.3746				3.037	2.590	32.86		
17.00	1.1861	1.1882	201.6	5.040	984.5	13.8	440	1.3770				3.337	2.819	29.91		
18.00	1.1971	1.1993	215.5	5.386	981.7	16.6	463	1.3793				3.691	3.089	27.04		
19.00	1.2082	1.2103	229.6	5.737	978.6	19.6	487	1.3817				4.111	3.410	24.27		
20.00	1.2192	1.2214	243.8	6.094	975.4	22.9	510	1.3840				4.610	3.789	21.65		
22.00	1.2412	1.2434	273.1	6.825	968.1	30.1	555	1.3885				5.753	4.644	17.35		
24.00	1.2631	1.2653	303.1	7.576	959.9	38.3	599	1.3929				7.086	5.621	14.08		
26.00	1.2848	1.2871	334.0	8.349	950.8	47.5	641	1.3971				8.727	6.806	11.44		
28.00	1.3064	1.3087	365.8	9.142	940.6	57.6	682	1.4012				10.81	8.29	9.23		
30.00	1.3277	1.3301	398.3	9.956	929.4	68.8	721	1.4051				13.49	10.18	7.40		
32.00	1.3488	1.3512	431.6	10.788	917.2	81.0	758	1.4088				16.81	12.49	5.94		
34.00	1.3697	1.3721	465.7	11.639	904.0	94.3	793	1.4123				20.71	15.15	4.82		
36.00	1.3901	1.3926	500.5	12.508	889.7	108.5	826	1.4156				25.24	18.19	3.95		
38.00	1.4102	1.4127	535.9	13.394	874.3	123.9	856	1.4186				30.40	21.60	3.28		
40.00	1.4299	1.4324	571.9	14.295	857.9	140.3	885	1.4215				36.24	25.40	2.75		

77 SODIUM MOLYBDATE, $Na_2MoO_4 \cdot 2H_2O$

MOLECULAR WEIGHT = 205.94 FORMULA WEIGHT, HYDRATE = 241.98
RELATIVE SPECIFIC REFRACTIVITY = 0.642

0.00 % by wt. data are the same for all compounds.
For Values of 0.00 wt. % solutions see Table 1, Acetic Acid.

A % by wt.	H % by wt.	ρ D_4^{20}	D_{20}^{20}	C_s g/l	M g-mol/l	C_w g/l	$(C_o - C_w)$ g/l	$(n - n_o)$ × 10^4	n	Δ °C	O Os/kg	S g-mol/l	η/η_o	η/ρ cS	φ rhe	γ mmho/cm	T g-mol/l
0.50	0.59	1.0024	1.0042	5.0	0.024	997.4	0.8	8	1.3338	0.096	0.052	0.027	1.013	1.012	98.56	4.1	0.041
1.00	1.18	1.0067	1.0084	10.1	0.049	996.6	1.6	17	1.3347	0.203	0.109	0.058	1.025	1.020	97.37	7.8	0.080
1.50	1.76	1.0110	1.0128	15.2	0.074	995.8	2.4	25	1.3355	0.322	0.173	0.092	1.037	1.028	96.24	11.2	0.118
2.00	2.35	1.0153	1.0171	20.3	0.099	995.0	3.2	34	1.3364	0.440	0.237	0.127	1.049	1.035	95.14	14.4	0.154
2.50	2.94	1.0197	1.0215	25.5	0.124	994.2	4.0	43	1.3373	0.558	0.300	0.162	1.061	1.043	94.05	17.5	0.190
3.00	3.53	1.0242	1.0260	30.7	0.149	993.4	4.8	52	1.3382	0.670	0.360	0.195	1.073	1.050	92.98	20.5	0.226
3.50	4.11	1.0286	1.0305	36.0	0.175	992.6	5.6	61	1.3391	0.782	0.420	0.228	1.086	1.058	91.91	23.3	0.261
4.00	4.70	1.0332	1.0350	41.3	0.201	991.8	6.4	70	1.3400	0.900	0.484	0.263	1.098	1.065	90.85	26.1	0.296
4.50	5.29	1.0377	1.0395	46.7	0.227	991.0	7.2	79	1.3409	1.015	0.546	0.296	1.112	1.073	89.79	28.9	0.330
5.00	5.88	1.0423	1.0441	52.1	0.253	990.1	8.1	88	1.3418	1.126	0.606	0.329	1.125	1.082	88.71	31.6	0.364
5.50	6.46	1.0468	1.0487	57.6	0.280	989.3	9.0	97	1.3427	1.238	0.666	0.362	1.139	1.090	87.62	34.3	0.398
6.00	7.05	1.0514	1.0533	63.1	0.306	988.4	9.9	106	1.3436	1.350	0.726	0.395	1.153	1.099	86.53	36.9	0.431
6.50	7.64	1.0561	1.0579	68.6	0.333	987.4	10.8	115	1.3445				1.168	1.108	85.44	39.5	0.465
7.00	8.23	1.0607	1.0626	74.2	0.361	986.4	11.8	124	1.3454				1.183	1.118	84.35	42.1	0.497
7.50	8.81	1.0653	1.0672	79.9	0.388	985.4	12.8	133	1.3463				1.199	1.127	83.25	44.5	0.530
8.00	9.40	1.0700	1.0719	85.6	0.416	984.4	13.8	142	1.3472				1.215	1.138	82.16	46.9	0.562
8.50	9.99	1.0747	1.0766	91.3	0.444	983.3	14.9	151	1.3481				1.231	1.148	81.06	49.3	0.593
9.00	10.58	1.0794	1.0813	97.1	0.472	982.2	16.0	160	1.3490				1.248	1.159	79.97	51.6	0.624

78 SODIUM NITRATE, $NaNO_3$

MOLECULAR WEIGHT = 85.01
RELATIVE SPECIFIC REFRACTIVITY = 0.657

0.00 % by wt. data are the same for all compounds.
For Values of 0.00 wt. % solutions see Table 1, Acetic Acid.

A % by wt.	ρ D_4^{20}	D_{20}^{20}	C_s g/l	M g-mol/l	C_w g/l	$(C_o - C_w)$ g/l	$(n - n_o)$ × 10^4	n	Δ °C	O Os/kg	S g-mol/l	η/η_o	η/ρ cS	φ rhe	γ mmho/cm	T g-mol/l
0.50	1.0016	1.0034	5.0	0.059	996.6	1.6	6	1.3336	0.204	0.110	0.058	1.002	1.003	99.55	5.4	0.055
1.00	1.0050	1.0067	10.0	0.118	994.9	3.3	11	1.3341	0.403	0.216	0.116	1.005	1.002	99.30	10.6	0.111
2.00	1.0117	1.0135	20.2	0.238	991.5	6.7	23	1.3353	0.793	0.426	0.231	1.010	1.000	98.81	20.4	0.225
3.00	1.0185	1.0203	30.6	0.359	988.0	10.3	34	1.3364	1.177	0.633	0.344	1.016	0.999	98.24	29.5	0.336
4.00	1.0254	1.0272	41.0	0.482	984.4	13.9	46	1.3375	1.561	0.839	0.456	1.023	0.999	97.60	38.0	0.445
5.00	1.0322	1.0341	51.6	0.607	980.6	17.6	57	1.3387	1.942	1.044	0.568	1.030	1.000	96.89	46.2	0.552
6.00	1.0392	1.0410	62.4	0.733	976.8	21.4	68	1.3398	2.323	1.249	0.678	1.038	1.001	96.14	54.2	0.659
7.00	1.0462	1.0480	73.2	0.861	972.9	25.3	79	1.3409	2.703	1.453	0.787	1.047	1.003	95.32	61.7	0.764

78 SODIUM NITRATE, $NaNO_3$—(Continued)

A% by wt.	ρ D_4^{20}	D_{20}^{20}	C_s g/l	M g-mol/l	C_w g/l	$(C_o - C_w)$ g/l	$(n - n_o)$ × 10^4	n	Δ °C	O Os/kg	S g-mol/l	η/η₀	η/ρ cS	φ rhe	γ mmho/cm	T g-mol/l
8.00	1.0532	1.0550	84.3	0.991	968.9	29.3	91	1.3421	3.083	1.657	0.895	1.057	1.005	94.45	68.9	0.868
9.00	1.0603	1.0622	95.4	1.123	964.8	33.4	102	1.3432	3.462	1.862	1.002	1.067	1.009	93.51	75.8	0.971
10.00	1.0674	1.0693	106.7	1.256	960.7	37.6	113	1.3443	3.841	2.065	1.107	1.079	1.013	92.49	82.6	1.07
12.00	1.0819	1.0838	129.8	1.527	952.1	46.1	136	1.3466	4.599	2.472	1.316	1.105	1.024	90.28	95.3	1.27
14.00	1.0967	1.0986	153.5	1.806	943.2	55.1	159	1.3489	5.37	2.89	1.523	1.136	1.038	87.85	106.	1.46
18.00	1.1272	1.1292	202.9	2.387	924.3	73.9	206	1.3536	6.98	3.75	1.937	1.213	1.078	82.30	125.	1.82
20.00	1.1429	1.1449	228.6	2.689	914.3	83.9	230	1.3559	7.81	4.20	2.142	1.260	1.105	79.21	134.	1.99
24.00	1.1752	1.1772	282.0	3.318	893.1	105.1	277	1.3607	9.52	5.12	2.542	1.374	1.172	72.64	149.	2.30
28.00	1.2085	1.2106	338.4	3.981	870.1	128.1	325	1.3654	11.28	6.07	2.929	1.519	1.259	65.71	160.	2.56
30.00	1.2256	1.2278	367.7	4.325	858.0	140.3	348	1.3678				1.606	1.313	62.14	165.	2.68
34.00	1.2610	1.2632	428.7	5.043	832.2	166.0	397	1.3726				1.814	1.442	55.01	173.	2.87
40.00	1.3175	1.3198	527.0	6.199	790.5	207.7	472	1.3802				2.222	1.690	44.91	178.	3.00

79 SODIUM PHOSPHATE (TRIBASIC), $Na_3PO_4 \cdot 12H_2O$

MOLECULAR WEIGHT = 163.96 FORMULA WEIGHT, HYDRATE = 380.14
RELATIVE SPECIFIC REFRACTIVITY = 0.555

0.00 % by wt. data are the same for all compounds.
For Values of 0.00 wt. % solutions see Table 1, Acetic Acid.

A% by wt.	H% by wt.	ρ D_4^{20}	D_{20}^{20}	C_s g/l	M g-mol/l	C_w g/l	$(C_o - C_w)$ g/l	$(n - n_o)$ × 10^4	n	Δ °C	O Os/kg	S g-mol/l	η/η₀	η/ρ cS	φ rhe	γ mmho/cm	T g-mol/l
0.50	1.16	1.0042	1.0059	5.0	0.031	999.1	−0.9	13	1.3343	0.192	0.103	0.055	1.031	1.029	96.80	7.3	0.075
1.00	2.32	1.0100	1.0118	10.1	0.062	999.9	−1.7	26	1.3356	0.368	0.198	0.106	1.062	1.054	93.97	14.1	0.150
1.50	3.48	1.0158	1.0176	15.2	0.093	1000.6	−2.3	39	1.3369	0.527	0.283	0.153	1.092	1.078	91.36	18.8	0.206
2.00	4.64	1.0216	1.0234	20.4	0.125	1001.2	−2.9	51	1.3381	0.668	0.359	0.194	1.124	1.102	88.79	22.7	0.253
2.50	5.80	1.0275	1.0293	25.7	0.157	1001.8	−3.6	64	1.3394	0.790	0.425	0.230	1.159	1.130	86.13	26.7	0.302
3.00	6.96	1.0335	1.0353	31.0	0.189	1002.5	−4.2	76	1.3406				1.196	1.160	83.45	30.4	0.349
3.50	8.11	1.0395	1.0413	36.4	0.222	1003.1	−4.9	89	1.3419				1.236	1.191	80.76	34.0	0.394
4.00	9.27	1.0456	1.0474	41.8	0.255	1003.8	−5.5	102	1.3432				1.278	1.225	78.06	37.3	0.436
4.50	10.43	1.0517	1.0536	47.3	0.289	1004.4	−6.2	114	1.3444				1.324	1.261	75.39	40.5	0.477
5.00	11.59	1.0579	1.0598	52.9	0.323	1005.0	−6.8	127	1.3457				1.372	1.299	72.74	43.5	0.516
5.50	12.75	1.0642	1.0661	58.5	0.357	1005.7	−7.4	140	1.3470				1.423	1.340	70.13	46.4	0.554
6.00	13.91	1.0705	1.0724	64.2	0.392	1006.3	−8.0	152	1.3482				1.477	1.383	67.56	49.0	0.589
6.50	15.07	1.0768	1.0787	70.0	0.427	1006.8	−8.6	165	1.3495				1.535	1.428	65.03	51.5	0.622
7.00	16.23	1.0832	1.0851	75.8	0.462	1007.4	−9.1	177	1.3507				1.595	1.475	62.57	53.7	0.653
7.50	17.39	1.0896	1.0915	81.7	0.498	1007.9	−9.7	189	1.3519				1.659	1.525	60.16	55.8	0.681
8.00	18.55	1.0961	1.0980	87.7	0.535	1008.4	−10.1	202	1.3532				1.726	1.578	57.82	57.6	0.706

80 SODIUM PHOSPHATE, DIHYDROGEN (MONOBASIC), $NaH_2PO_4 \cdot 1H_2O$

MOLECULAR WEIGHT = 119.97 FORMULA WEIGHT, HYDRATE = 137.99
RELATIVE SPECIFIC REFRACTIVITY = 0.637

0.00 % by wt. data are the same for all compounds.
For Values of 0.00 wt. % solutions see Table 1, Acetic Acid.

A% by wt.	H% by wt.	ρ D_4^{20}	D_{20}^{20}	C_s g/l	M g-mol/l	C_w g/l	$(C_o - C_w)$ g/l	$(n - n_o)$ × 10^4	n	Δ °C	O Os/kg	S g-mol/l	η/η₀	η/ρ cS	φ rhe	γ mmho/cm	T g-mol/l
0.50	0.58	1.0019	1.0037	5.0	0.042	996.9	1.3	7	1.3336	0.140	0.075	0.040	1.016	1.017	98.18	2.2	0.021
1.00	1.15	1.0056	1.0074	10.1	0.084	995.6	2.7	13	1.3343	0.279	0.150	0.080	1.033	1.029	96.61	4.4	0.044
1.50	1.73	1.0094	1.0111	15.1	0.126	994.2	4.0	20	1.3349	0.418	0.225	0.120	1.049	1.042	95.10	6.8	0.068
2.00	2.30	1.0131	1.0149	20.3	0.169	992.8	5.4	26	1.3356	0.556	0.299	0.161	1.066	1.054	93.62	9.1	0.094
2.50	2.88	1.0168	1.0186	25.4	0.212	991.4	6.8	33	1.3362	0.695	0.374	0.202	1.083	1.067	92.13	11.3	0.119
3.00	3.45	1.0206	1.0224	30.6	0.255	990.0	8.3	39	1.3369	0.835	0.449	0.243	1.101	1.081	90.65	13.4	0.143
3.50	4.03	1.0244	1.0262	35.9	0.299	988.5	9.7	45	1.3375	0.976	0.525	0.285	1.119	1.095	89.17	15.4	0.166
4.00	4.60	1.0281	1.0300	41.1	0.343	987.0	11.2	52	1.3382	1.116	0.600	0.326	1.138	1.109	87.69	17.4	0.189
4.50	5.18	1.0319	1.0338	46.4	0.387	985.5	12.7	58	1.3388	1.254	0.674	0.367	1.158	1.124	86.21	19.2	0.211
5.00	5.75	1.0358	1.0376	51.8	0.432	984.0	14.3	65	1.3395	1.388	0.746	0.406	1.178	1.140	84.72	21.0	0.232
5.50	6.33	1.0396	1.0414	57.2	0.477	982.4	15.8	71	1.3401	1.519	0.817	0.444	1.199	1.156	83.23	22.6	0.252
6.00	6.90	1.0434	1.0453	62.6	0.522	980.8	17.4	78	1.3408	1.645	0.884	0.481	1.221	1.172	81.74	24.1	0.270
6.50	7.48	1.0473	1.0491	68.1	0.567	979.2	19.0	84	1.3414	1.767	0.950	0.517	1.243	1.190	80.26	25.5	0.287
7.00	8.05	1.0511	1.0530	73.6	0.613	977.6	20.7	91	1.3421	1.887	1.014	0.552	1.267	1.208	78.78	26.7	0.303
7.50	8.63	1.0550	1.0569	79.1	0.660	975.9	22.3	97	1.3427	2.004	1.077	0.586	1.291	1.226	77.31	27.9	0.317
8.00	9.20	1.0589	1.0608	84.7	0.706	974.2	24.0	104	1.3434	2.119	1.139	0.619	1.316	1.245	75.85	29.0	0.330
8.50	9.78	1.0628	1.0647	90.3	0.753	972.5	25.7	110	1.3440	2.234	1.201	0.652	1.342	1.265	74.39	30.0	0.344
9.00	10.35	1.0668	1.0686	96.0	0.800	970.8	27.5	117	1.3447	2.350	1.263	0.686	1.368	1.285	72.93	31.0	0.357
9.50	10.93	1.0707	1.0726	101.7	0.848	969.0	29.2	123	1.3453	2.466	1.326	0.719	1.396	1.307	71.48	32.1	0.370
10.00	11.50	1.0747	1.0766	107.5	0.896	967.2	31.0	130	1.3460	2.584	1.389	0.753	1.425	1.329	70.04	33.2	0.384
11.00	12.65	1.0826	1.0846	119.1	0.993	963.6	34.7	143	1.3473	2.821	1.517	0.821	1.485	1.374	67.20	35.4	0.412
12.00	13.80	1.0907	1.0926	130.9	1.091	959.8	38.4	156	1.3486	3.056	1.643	0.887	1.549	1.423	64.44	37.5	0.439
13.00	14.95	1.0988	1.1007	142.8	1.191	955.9	42.3	169	1.3499	3.291	1.769	0.954	1.617	1.474	61.73	39.6	0.465
14.00	16.10	1.1070	1.1089	155.0	1.292	952.0	46.2	182	1.3512	3.529	1.897	1.021	1.691	1.530	59.04	41.5	0.490
15.00	17.25	1.1152	1.1172	167.3	1.394	947.9	50.3	195	1.3525	3.775	2.029	1.089	1.771	1.591	56.35	43.3	0.513
16.00	18.40	1.1236	1.1255	179.8	1.498	943.8	54.4	209	1.3538	4.028	2.166	1.159	1.857	1.656	53.74	44.8	0.534
17.00	19.55	1.1320	1.1340	192.4	1.604	939.5	58.7	222	1.3552	4.287	2.305	1.230	1.948	1.724	51.24	46.2	0.552
18.00	20.70	1.1404	1.1425	205.3	1.711	935.2	63.1	235	1.3565	4.551	2.447	1.303	2.046	1.798	48.77	47.4	0.568
19.00	21.85	1.1490	1.1510	218.3	1.820	930.7	67.5	248	1.3578	4.820	2.592	1.376	2.155	1.880	46.30	48.5	0.583
20.00	23.00	1.1576	1.1597	231.5	1.930	926.1	72.1	262	1.3592	5.10	2.74	1.450	2.278	1.972	43.81	49.6	0.597
22.00	25.30	1.1752	1.1773	258.5	2.155	916.6	81.6	289	1.3618				2.545	2.170	39.21	51.4	0.622
24.00	27.60	1.1931	1.1952	286.3	2.387	906.7	91.5	316	1.3646				2.844	2.389	35.09	52.7	0.639
26.00	29.91	1.2113	1.2134	314.9	2.625	896.3	101.9	343	1.3673				3.208	2.654	31.11	53.5	0.650
28.00	32.21	1.2299	1.2320	344.4	2.870	885.5	112.7	370	1.3700				3.675	2.994	27.16	53.9	0.655
30.00	34.51	1.2488	1.2510	374.6	3.123	874.2	124.1	398	1.3728				4.291	3.443	23.26	54.0	0.656
32.00	36.81	1.2682	1.2704	405.8	3.383	862.4	135.9	426	1.3756				5.069	4.005	19.69	53.7	0.653
34.00	39.11	1.2879	1.2902	437.9	3.650	850.0	148.2	454	1.3784				5.996	4.665	16.64	52.9	0.642
36.00	41.41	1.3080	1.3103	470.9	3.925	837.1	161.1	482	1.3812				7.084	5.427	14.09	51.5	0.622
38.00	43.71	1.3285	1.3308	504.8	4.208	823.6	174.6	511	1.3840				8.346	6.295	11.96	49.3	0.593
40.00	46.01	1.3493	1.3517	539.7	4.499	809.6	188.6	539	1.3869				9.794	7.273	10.19	46.1	0.551

81 SODIUM PHOSPHATE, MONOHYDROGEN (DIBASIC), $Na_2HPO_4 \cdot 7H_2O$

MOLECULAR WEIGHT = 141.97 FORMULA WEIGHT, HYDRATE = 268.09
RELATIVE SPECIFIC REFRACTIVITY = 0.576

0.00 % by wt. data are the same for all compounds.
For Values of 0.00 wt. % solutions see Table 1, Acetic Acid.

A% by wt.	H% by wt.	ρ D_4^{20}	D_{20}^{20}	C_s g/l	M g-mol/l	C_w g/l	$(C_o - C_w)$ g/l	$(n - n_o) \times 10^4$	n	Δ °C	O Os/kg	S g-mol/l	η/η_o	η/ρ cS	φ rhe	γ mmho/cm	T g-mol/l
0.50	0.94	1.0032	1.0050	5.0	0.035	998.2	0.0	10	1.3340	0.167	0.090	0.047	1.019	1.018	97.91	4.6	0.046
1.00	1.89	1.0082	1.0100	10.1	0.071	998.1	0.1	19	1.3349	0.319	0.172	0.092	1.040	1.034	95.96	8.7	0.090
1.50	2.83	1.0131	1.0149	15.2	0.107	997.9	0.3	28	1.3358	0.462	0.248	0.133	1.062	1.051	93.94	12.3	0.130
2.00	3.78	1.0180	1.0198	20.4	0.143	997.7	0.6	38	1.3368				1.086	1.069	91.90	15.6	0.168
2.50	4.72	1.0229	1.0247	25.6	0.180	997.4	0.9	47	1.3377				1.111	1.088	89.86	18.7	0.205
3.00	5.67	1.0279	1.0297	30.8	0.217	997.0	1.2	56	1.3386				1.136	1.108	87.82	21.6	0.240
3.50	6.61	1.0328	1.0346	36.1	0.255	996.7	1.6	66	1.3396				1.163	1.129	85.80	24.3	0.273
4.00	7.55	1.0378	1.0396	41.5	0.292	996.3	1.9	75	1.3405				1.191	1.150	83.78	26.8	0.304
4.50	8.50	1.0428	1.0446	46.9	0.331	995.9	2.4	84	1.3414				1.221	1.173	81.77	29.2	0.334
5.00	9.44	1.0478	1.0496	52.4	0.369	995.4	2.8	94	1.3424				1.251	1.196	79.78	31.4	0.362
5.50	10.39	1.0528	1.0546	57.9	0.408	994.9	3.4	103	1.3433				1.283	1.221	77.80	33.5	0.388

82 SODIUM SULFATE, $Na_2SO_4 \cdot 10H_2O$

MOLECULAR WEIGHT = 142.06 FORMULA WEIGHT, HYDRATE = 322.22
RELATIVE SPECIFIC REFRACTIVITY = 0.548

0.00 % by wt. data are the same for all compounds.
For Values of 0.00 wt. % solutions see Table 1, Acetic Acid.

A% by wt.	H% by wt.	ρ D_4^{20}	D_{20}^{20}	C_s g/l	M g-mol/l	C_w g/l	$(C_o - C_w)$ g/l	$(n - n_o) \times 10^4$	n	Δ °C	O Os/kg	S g-mol/l	η/η_o	η/ρ cS	φ rhe	γ mmho/cm	T g-mol/l
0.50	1.13	1.0027	1.0044	5.0	0.035	997.7	0.6	8	1.3338	0.165	0.089	0.047	1.011	1.010	98.71	5.9	0.060
1.00	2.27	1.0071	1.0089	10.1	0.071	997.1	1.2	15	1.3345	0.320	0.172	0.092	1.024	1.019	97.46	11.2	0.117
1.50	3.40	1.0116	1.0134	15.2	0.107	996.4	1.8	23	1.3353	0.466	0.251	0.135	1.039	1.030	96.02	15.7	0.169
2.00	4.54	1.0161	1.0179	20.3	0.143	995.8	2.4	30	1.3360	0.606	0.326	0.176	1.056	1.041	94.51	19.8	0.218
2.50	5.67	1.0206	1.0225	25.5	0.180	995.1	3.1	38	1.3368	0.742	0.399	0.216	1.072	1.053	93.06	23.9	0.268
3.00	6.80	1.0252	1.0270	30.8	0.216	994.4	3.8	46	1.3376	0.873	0.469	0.254	1.089	1.065	91.63	27.9	0.317
3.50	7.94	1.0298	1.0316	36.0	0.254	993.7	4.5	53	1.3383	1.001	0.538	0.292	1.106	1.077	90.21	31.8	0.365
4.00	9.07	1.0343	1.0362	41.4	0.291	993.0	5.3	61	1.3391	1.125	0.605	0.329	1.124	1.089	88.80	35.5	0.413
4.50	10.21	1.0389	1.0408	46.8	0.329	992.2	6.0	68	1.3398	1.245	0.669	0.364	1.142	1.101	87.38	39.2	0.460
5.00	11.34	1.0436	1.0454	52.2	0.367	991.4	6.8	76	1.3406	1.359	0.731	0.397	1.161	1.115	85.96	42.7	0.506
5.50	12.48	1.0481	1.0499	57.6	0.406	990.4	7.8	83	1.3413	1.465	0.788	0.429	1.180	1.129	84.54	46.1	0.551
6.00	13.61	1.0526	1.0545	63.2	0.445	989.5	8.8	90	1.3420	1.560	0.839	0.456	1.200	1.143	83.14	49.4	0.594
6.50	14.74	1.0572	1.0591	68.7	0.484	988.5	9.7	97	1.3427				1.221	1.157	81.75	52.5	0.636
7.00	15.88	1.0619	1.0638	74.3	0.523	987.5	10.7	105	1.3435				1.242	1.172	80.36	55.5	0.676
7.50	17.01	1.0666	1.0684	80.0	0.563	986.6	11.7	112	1.3442				1.264	1.187	78.98	58.3	0.716
8.00	18.15	1.0713	1.0732	85.7	0.603	985.6	12.7	120	1.3449				1.286	1.203	77.59	61.1	0.755
8.50	19.28	1.0760	1.0779	91.5	0.644	984.6	13.7	127	1.3457				1.310	1.220	76.20	63.7	0.793
9.00	20.41	1.0808	1.0827	97.3	0.685	983.5	14.7	134	1.3464				1.334	1.237	74.80	66.3	0.830
9.50	21.55	1.0856	1.0875	103.1	0.726	982.5	15.7	142	1.3472				1.360	1.255	73.38	68.8	0.866
10.00	22.68	1.0905	1.0924	109.0	0.768	981.4	16.8	149	1.3479				1.387	1.274	71.95	71.3	0.902
11.00	24.95	1.1002	1.1022	121.0	0.852	979.2	19.0	164	1.3494				1.444	1.315	69.11	75.9	0.970
12.00	27.22	1.1101	1.1121	133.2	0.938	976.9	21.3	179	1.3509				1.505	1.359	66.30	80.1	1.03
13.00	29.49	1.1201	1.1221	145.6	1.025	974.5	23.8	194	1.3524				1.571	1.406	63.52	83.9	1.09
14.00	31.75	1.1301	1.1321	158.2	1.114	971.9	26.3	209	1.3539				1.643	1.457	60.74	87.5	1.15
15.00	34.02	1.1402	1.1422	171.0	1.204	969.2	29.1	223	1.3553				1.722	1.513	57.96	91.1	1.21
16.00	36.29	1.1503	1.1523	184.0	1.296	966.2	32.0	237	1.3567				1.808	1.575	55.20	94.9	1.27
17.00	38.56	1.1604	1.1625	197.3	1.389	963.1	35.1	251	1.3581				1.901	1.641	52.50	98.5	1.33
18.00	40.83	1.1705	1.1726	210.7	1.483	959.8	38.4	265	1.3595				2.001	1.713	49.88	102.0	1.39
19.00	43.10	1.1806	1.1827	224.3	1.579	956.3	41.9	278	1.3608				2.108	1.789	47.34	105.0	1.44
20.00	45.36	1.1907	1.1928	238.1	1.676	952.6	45.7	290	1.3620				2.223	1.871	44.89	109.0	1.50
22.00	49.90	1.2106	1.2127	266.3	1.875	944.2	54.0	313	1.3643				2.476	2.050	40.30	114.0	1.61

83 SODIUM TARTRATE, $NaOOC(CHOH)_2COONa \cdot 2H_2O$

MOLECULAR WEIGHT = 194.06 FORMULA WEIGHT, HYDRATE = 230.10
RELATIVE SPECIFIC REFRACTIVITY = 0.781

0.00 % by wt. data are the same for all compounds.
For Values of 0.00 wt. % solutions see Table 1, Acetic Acid.

A% by wt.	H% by wt.	ρ D_4^{20}	D_{20}^{20}	C_s g/l	M g-mol/l	C_w g/l	$(C_o - C_w)$ g/l	$(n - n_o) \times 10^4$	n	Δ °C	O Os/kg	S g-mol/l	η/η_o	η/ρ cS	φ rhe	γ mmho/cm	T g-mol/l
0.50	0.59	1.0017	1.0035	5.0	0.026	996.7	1.5	8	1.3338	0.111	0.060	0.031	1.015	1.016	98.28	3.8	0.037
1.00	1.19	1.0052	1.0070	10.1	0.052	995.2	3.1	17	1.3347	0.224	0.121	0.064	1.031	1.028	96.80	7.1	0.072
2.00	2.37	1.0123	1.0141	20.2	0.104	992.0	6.2	33	1.3363	0.457	0.246	0.132	1.062	1.051	93.97	12.6	0.133
3.00	3.56	1.0194	1.0212	30.6	0.158	988.8	9.4	50	1.3380	0.678	0.365	0.197	1.094	1.075	91.21	17.5	0.191
4.00	4.74	1.0266	1.0284	41.1	0.212	985.5	12.7	67	1.3397	0.891	0.479	0.260	1.128	1.101	88.49	22.1	0.245
5.00	5.93	1.0338	1.0356	51.7	0.266	982.1	16.1	84	1.3414	1.103	0.593	0.322	1.164	1.128	85.74	26.2	0.297
6.00	7.11	1.0410	1.0428	62.5	0.322	978.5	19.7	101	1.3431	1.317	0.708	0.385	1.203	1.157	82.99	30.1	0.346
7.00	8.30	1.0482	1.0501	73.4	0.378	974.9	23.4	118	1.3448	1.529	0.822	0.447	1.243	1.188	80.28	33.7	0.390
8.00	9.49	1.0555	1.0574	84.4	0.435	971.1	27.2	135	1.3465	1.742	0.937	0.509	1.287	1.222	77.56	37.0	0.432
9.00	10.67	1.0628	1.0647	95.7	0.493	967.2	31.1	152	1.3482	1.955	1.051	0.572	1.334	1.258	74.81	40.0	0.471
10.00	11.86	1.0702	1.0721	107.0	0.551	963.2	35.1	169	1.3499	2.171	1.167	0.634	1.386	1.298	72.01	43.0	0.509
12.00	14.23	1.0851	1.0870	130.2	0.671	954.9	43.4	203	1.3533	2.607	1.402	0.760	1.499	1.384	66.58	48.1	0.577
14.00	16.60	1.1002	1.1022	154.0	0.794	946.2	52.0	237	1.3567	3.048	1.639	0.885	1.625	1.480	61.43	52.2	0.632
15.00	17.79	1.1079	1.1098	166.2	0.856	941.7	56.5	254	1.3584	3.269	1.758	0.948	1.695	1.533	58.88	53.9	0.655
18.00	21.34	1.1313	1.1333	203.6	1.049	927.6	70.6	307	1.3637	3.938	2.117	1.134	1.951	1.728	51.14	57.9	0.710
20.00	23.71	1.1471	1.1492	229.4	1.182	917.7	80.5	342	1.3672	4.385	2.358	1.258	2.174	1.899	45.91	59.9	0.738
22.00	26.09	1.1633	1.1653	255.9	1.319	907.4	90.9	378	1.3708				2.444	2.105	40.84	61.4	0.759
24.00	28.46	1.1797	1.1818	283.1	1.459	896.6	101.7	414	1.3744				2.757	2.342	36.20	62.0	0.768
26.00	30.83	1.1963	1.1984	311.0	1.603	885.3	113.0	450	1.3780				3.117	2.611	32.01	61.6	0.762
28.00	33.20	1.2132	1.2153	339.7	1.750	873.5	124.7	487	1.3817				3.528	2.914	28.29	60.1	0.741

84 SODIUM THIOCYANATE, NaCNS

MOLECULAR WEIGHT = 81.07

A% by wt.	ρ D_4^{20}	D_{20}^{20}	C_s g/l	M g-mol/l	C_w g/l	$(C_0 - C_w)$ g/l	$(n - n_0)$ × 10^4	n	Δ °C	O Os/kg	S g-mol/l	η/η_0	η/ρ cS	φ rhe	γ mmho/cm	T g-mol/l
1.00	1.0030	1.0048	10.0	0.124	993.0	5.2	22.	1.3352	0.426	0.229	0.123	1.006	1.005	99.21	10.0	0.113
2.00	1.0078	1.0096	20.2	0.249	987.6	10.6	44.	1.3374	0.848	0.456	0.247	1.012	1.006	98.64	20.5	0.226
3.00	1.0127	1.0145	30.4	0.375	982.3	15.9	66.	1.3396	1.280	0.688	0.374	1.019	1.008	97.96	29.8	0.341
4.00	1.0176	1.0194	40.7	0.502	976.9	21.3	88.	1.3418	1.734	0.932	0.507	1.025	1.009	97.40	39.0	0.458
5.00	1.0226	1.0244	51.1	0.631	971.5	26.7	110.	1.3440	2.202	1.184	0.643	1.032	1.011	96.73	48.0	0.576
6.00	1.0275	1.0294	61.7	0.760	965.8	32.4	132.	1.3462	2.679	1.440	0.780	1.039	1.013	96.07	56.7	0.694
7.00	1.0327	1.0346	72.3	0.892	960.4	37.8	155.	1.3485	3.168	1.703	0.919	1.047	1.016	95.31	65.1	0.812
8.00	1.0381	1.0400	83.1	1.024	955.0	43.2	178.	1.3508	3.671	1.974	1.060	1.057	1.020	94.44	73.3	0.932
9.00	1.0435	1.0454	93.9	1.158	949.6	48.6	201.	1.3531	4.184	2.250	1.202	1.067	1.025	93.49	81.2	1.052
10.00	1.0491	1.0510	104.9	1.294	944.2	54.0	224.	1.3554	4.706	2.530	1.345	1.079	1.031	92.45	89.0	1.174
12.00	1.0603	1.0622	127.2	1.569	9.33.1	65.1	270.	1.3600				1.106	1.045	90.25	104.	1.418
14.00	1.0715	1.0734	150.0	1.850	921.5	76.7	318.	1.3648				1.138	1.064	87.72	118.	1.664
16.00	1.0828	1.0848	173.3	2.137	909.5	88.7	366.	1.3696				1.175	1.087	84.96	130.	1.912
18.00	1.0942	1.0962	197.0	2.430	897.2	101.0	414.	1.3744				1.217	1.114	82.04	142.	2.162
20.00	1.1056	1.1076	221.1	2.728	884.5	113.7	464.	1.3794				1.268	1.149	78.72	154.	2.412
24.00	1.1284	1.1304	270.8	3.340	857.6	140.6	564.	1.3894				1.396	1.240	71.47	170.	2.786
28.00	1.1515	1.1536	322.4	3.977	829.1	169.1	666.	1.3996				1.546	1.345	64.57	178.	3.016
32.00	1.1763	1.1784	376.4	4.643	799.9	198.3	774.	1.4104				1.764	1.503	56.56	184.	3.194
36.00	1.2018	1.2040	432.7	5.337	769.1	229.1	882.	1.4212				2.101	1.752	47.49	185.	3.227
40.00	1.2282	1.2304	491.3	6.060	736.9	261.3	990.	1.4320				2.483	2.026	40.19	177.	2.986
44.00	1.2549	1.2572	552.2	6.811	702.7	295.5	1105.	1.4435				3.151	2.516	31.67	165.	2.678
48.00	1.2821	1.2844	615.4	7.591	666.7	331.5	1223.	1.4553				3.936	3.076	25.36	148.	2.290
52.00	1.3100	1.3124	681.2	8.403	628.8	369.4	1342.	1.4672				5.279	4.038	18.90	130.	1.900
56.00	1.3388	1.3412	749.7	9.248	589.1	409.1	1461.	1.4791				7.554	5.654	13.21	108.	1.493

85 SODIUM THIOSULFATE, $Na_2S_2O_3 \cdot 5H_2O$

MOLECULAR WEIGHT = 158.13 FORMULA WEIGHT, HYDRATE = 248.21
RELATIVE SPECIFIC REFRACTIVITY = 0.773

0.00 % by wt. data are the same for all compounds.
For Values of 0.00 wt. % solutions see Table 1, Acetic Acid.

A% by wt.	H% by wt.	ρ D_4^{20}	D_{20}^{20}	C_s g/l	M g-mol/l	C_w g/l	$(C_0 - C_w)$ g/l	$(n - n_0)$ × 10^4	n	Δ °C	O Os/kg	S g-mol/l	η/η_0	η/ρ cS	φ rhe	γ mmho/cm	T g-mol/l
0.50	0.78	1.0024	1.0041	5.0	0.032	997.4	0.9	10	1.3340	0.139	0.075	0.039	1.010	1.010	98.76	5.7	0.057
1.00	1.57	1.0065	1.0083	10.1	0.064	996.4	1.8	21	1.3351	0.279	0.150	0.080	1.021	1.016	97.75	10.7	0.112
2.00	3.14	1.0148	1.0166	20.3	0.128	994.5	3.7	42	1.3371	0.566	0.304	0.164	1.042	1.029	95.78	19.5	0.214
3.00	4.71	1.0231	1.0249	30.7	0.194	992.4	5.8	62	1.3392	0.835	0.449	0.243	1.064	1.042	93.77	27.7	0.315
4.00	6.28	1.0315	1.0333	41.3	0.261	990.2	8.0	83	1.3413	1.090	0.586	0.318	1.088	1.057	91.74	35.6	0.415
5.00	7.85	1.0399	1.0417	52.0	0.329	987.9	10.3	104	1.3434	1.341	0.721	0.392	1.113	1.072	89.67	43.3	0.513
6.00	9.42	1.0483	1.0502	62.9	0.398	985.4	12.8	124	1.3454	1.588	0.854	0.464	1.139	1.089	87.59	50.6	0.610
7.00	10.99	1.0568	1.0587	74.0	0.468	982.8	15.4	145	1.3475	1.826	0.982	0.534	1.167	1.107	85.50	57.5	0.705
8.00	12.56	1.0654	1.0673	85.2	0.539	980.1	18.1	166	1.3496	2.062	1.108	0.602	1.197	1.126	83.38	64.2	0.799
9.00	14.13	1.0740	1.0759	96.7	0.611	977.3	20.9	187	1.3517	2.299	1.236	0.671	1.229	1.147	81.21	70.5	0.891
10.00	15.70	1.0827	1.0846	108.3	0.685	974.4	23.8	208	1.3538	2.546	1.369	0.742	1.264	1.170	78.96	76.7	0.982
12.00	18.84	1.1003	1.1023	132.0	0.835	968.3	29.9	251	1.3581	3.064	1.647	0.890	1.342	1.222	74.35	88.2	1.16
14.00	21.98	1.1182	1.1202	156.6	0.990	961.7	36.5	294	1.3624	3.604	1.938	1.041	1.432	1.283	69.71	98.8	1.33
16.00	25.11	1.1365	1.1385	181.8	1.150	954.6	43.6	337	1.3667	4.168	2.241	1.197	1.534	1.353	65.04	108.	1.49
18.00	28.25	1.1551	1.1571	207.9	1.315	947.1	51.1	381	1.3711	4.758	2.558	1.359	1.654	1.435	60.34	117.	1.64
20.00	31.39	1.1740	1.1760	234.8	1.485	939.2	59.1	426	1.3756	5.37	2.89	1.524	1.794	1.531	55.63	123.	1.77
22.00	34.53	1.1932	1.1953	262.5	1.660	930.7	67.5	471	1.3801				1.954	1.641	51.08	129.	1.87
24.00	37.67	1.2128	1.2150	291.1	1.841	921.8	76.5	517	1.3847				2.137	1.765	46.70	132.	1.94
26.00	40.81	1.2328	1.2350	320.5	2.027	912.3	85.9	563	1.3893				2.351	1.911	42.45	135.	2.00
28.00	43.95	1.2532	1.2554	350.9	2.219	902.3	95.9	610	1.3940				2.591	2.072	38.52	136.	2.02
30.00	47.09	1.2739	1.2762	382.2	2.417	891.7	106.5	657	1.3987				2.897	2.279	34.45	136.	2.03
32.00	50.23	1.2950	1.2973	414.4	2.621	880.6	117.6	705	1.4035				3.298	2.552	30.26	135.	1.99
34.00	53.37	1.3164	1.3188	447.6	2.831	868.8	129.4	754	1.4084				3.784	2.881	26.37	132.	1.93
36.00	56.51	1.3382	1.3406	481.8	3.047	856.5	141.8	802	1.4132				4.350	3.257	22.95	128.	1.86
38.00	59.65	1.3603	1.3627	516.9	3.269	843.4	154.8	851	1.4181				5.001	3.684	19.96	124.	1.77
40.00	62.79	1.3827	1.3852	553.1	3.498	829.6	168.6	899	1.4229				5.747	4.165	17.37	118.	1.66

86 SODIUM TUNGSTATE, $Na_2WO_4 \cdot 2H_2O$

MOLECULAR WEIGHT = 293.91   FORMULA WEIGHT, HYDRATE = 329.95
RELATIVE SPECIFIC REFRACTIVITY = 0.414

0.00 % by wt. data are the same for all compounds.
For Values of 0.00 wt. % solutions see Table 1, Acetic Acid.

A% by wt.	H% by wt.	ρ D_4^{20}	D_{20}^{20}	C_s g/l	M g-mol/l	C_w g/l	$(C_o - C_w)$ g/l	$(n - n_o) \times 10^4$	n	Δ °C	O Os/kg	S g-mol/l	η/η_o	η/ρ cS	φ rhe	γ mmho/cm	T g-mol/l
0.50	0.56	1.0028	1.0046	5.0	0.017	997.8	0.4	5	1.3335	0.081	0.044	0.023	1.007	1.006	99.13	2.8	0.028
1.00	1.12	1.0074	1.0091	10.1	0.034	997.3	0.9	11	1.3341	0.161	0.086	0.046	1.014	1.009	98.42	5.5	0.055
1.50	1.68	1.0119	1.0137	15.2	0.052	996.8	1.5	16	1.3346	0.239	0.129	0.068	1.022	1.012	97.67	7.9	0.081
2.00	2.25	1.0166	1.0184	20.3	0.069	996.2	2.0	21	1.3351	0.317	0.171	0.091	1.030	1.015	96.89	10.3	0.107
2.50	2.81	1.0212	1.0230	25.5	0.087	995.7	2.5	27	1.3357	0.396	0.213	0.114	1.039	1.019	96.08	12.5	0.132
3.00	3.37	1.0259	1.0278	30.8	0.105	995.2	3.1	33	1.3362	0.474	0.255	0.137	1.048	1.024	95.23	14.7	0.157
3.50	3.93	1.0307	1.0326	36.1	0.123	994.7	3.6	38	1.3368	0.553	0.297	0.160	1.058	1.028	94.35	16.8	0.182
4.00	4.49	1.0356	1.0374	41.4	0.141	994.1	4.1	44	1.3374	0.633	0.340	0.184	1.068	1.033	93.46	18.8	0.206
4.50	5.05	1.0405	1.0423	46.8	0.159	993.6	4.6	50	1.3380	0.714	0.384	0.208	1.078	1.038	92.58	20.8	0.230
5.00	5.61	1.0453	1.0472	52.3	0.178	993.1	5.2	56	1.3386	0.796	0.428	0.232	1.088	1.043	91.73	22.8	0.254
5.50	6.17	1.0502	1.0520	57.8	0.197	992.4	5.8	62	1.3392	0.879	0.472	0.256	1.098	1.048	90.89	24.7	0.278
6.00	6.74	1.0550	1.0569	63.3	0.215	991.7	6.5	67	1.3397	0.962	0.517	0.281	1.108	1.052	90.06	26.7	0.302
6.50	7.30	1.0599	1.0618	68.9	0.234	991.0	7.2	73	1.3403	1.046	0.562	0.305	1.118	1.057	89.24	28.6	0.326
7.00	7.86	1.0648	1.0667	74.5	0.254	990.3	8.0	79	1.3409	1.130	0.607	0.330	1.129	1.062	88.43	30.5	0.350
7.50	8.42	1.0698	1.0717	80.2	0.273	989.5	8.7	85	1.3414	1.212	0.652	0.354	1.139	1.067	87.62	32.4	0.374
8.00	8.98	1.0748	1.0767	86.0	0.293	988.8	9.4	91	1.3420	1.293	0.695	0.378	1.150	1.072	86.81	34.3	0.398
8.50	9.54	1.0799	1.0818	91.8	0.312	988.1	10.1	97	1.3427	1.371	0.737	0.401	1.160	1.077	86.02	36.1	0.422
9.00	10.10	1.0852	1.0871	97.7	0.332	987.5	10.7	103	1.3433	1.445	0.777	0.423	1.171	1.081	85.23	38.0	0.445

87 STRONTIUM CHLORIDE, $SrCl_2 \cdot 6H_2O$

MOLECULAR WEIGHT = 158.54   FORMULA WEIGHT, HYDRATE = 266.64
RELATIVE SPECIFIC REFRACTIVITY = 0.635

0.00 % by wt. data are the same for all compounds.
For Values of 0.00 wt. % solutions see Table 1, Acetic Acid.

A% by wt.	H% by wt.	ρ D_4^{20}	D_{20}^{20}	C_s g/l	M g-mol/l	C_w g/l	$(C_o - C_w)$ g/l	$(n - n_o) \times 10^4$	n	Δ °C	O Os/kg	S g-mol/l	η/η_o	η/ρ cS	φ rhe	γ mmho/cm	T g-mol/l
0.50	0.84	1.0027	1.0044	5.0	0.032	997.6	0.6	9	1.3339	0.158	0.085	0.045	1.010	1.009	98.85	5.9	0.060
1.00	1.68	1.0071	1.0089	10.1	0.064	997.0	1.2	18	1.3348	0.310	0.167	0.089	1.019	1.014	97.94	11.4	0.120
2.00	3.36	1.0161	1.0179	20.3	0.128	995.8	2.5	36	1.3366	0.619	0.333	0.180	1.037	1.023	96.24	22.0	0.244
3.00	5.05	1.0252	1.0270	30.8	0.194	994.4	3.8	54	1.3384	0.934	0.502	0.272	1.055	1.032	94.56	31.4	0.362
4.00	6.73	1.0344	1.0362	41.4	0.261	993.0	5.2	73	1.3402	1.261	0.678	0.369	1.074	1.041	92.89	40.4	0.475
5.00	8.41	1.0437	1.0456	52.2	0.329	991.5	6.7	91	1.3421	1.607	0.864	0.470	1.094	1.050	91.23	49.1	0.591
6.00	10.09	1.0532	1.0551	63.2	0.399	990.0	8.2	110	1.3440	1.978	1.064	0.578	1.114	1.060	89.59	58.0	0.712
7.00	11.77	1.0628	1.0647	74.4	0.469	988.4	9.8	129	1.3459	2.377	1.278	0.693	1.134	1.069	87.98	66.8	0.836
8.00	13.45	1.0726	1.0745	85.8	0.541	986.8	11.4	148	1.3478	2.800	1.505	0.815	1.155	1.079	86.37	75.3	0.962
9.00	15.14	1.0825	1.0844	97.4	0.614	985.0	13.2	168	1.3498	3.252	1.748	0.943	1.178	1.090	84.73	83.6	1.08
10.00	16.82	1.0925	1.0944	109.3	0.689	983.3	15.0	188	1.3518	3.736	2.009	1.078	1.202	1.102	83.03	91.5	1.21
12.00	20.18	1.1131	1.1150	133.6	0.842	979.5	18.7	228	1.3558	4.811	2.587	1.374	1.255	1.130	79.53	107.	1.46
14.00	23.55	1.1342	1.1362	158.8	1.002	975.4	22.9	269	1.3599	6.03	3.24	1.696	1.314	1.160	75.98	120.	1.71
16.00	26.91	1.1558	1.1579	184.9	1.166	970.9	27.3	311	1.3641	7.41	3.99	2.044	1.380	1.196	72.32	133.	1.96
18.00	30.27	1.1780	1.1801	212.0	1.337	966.0	32.3	354	1.3684	8.98	4.83	2.418	1.457	1.239	68.52	144.	2.19
20.00	33.64	1.2008	1.2029	240.2	1.515	960.6	37.6	398	1.3728	10.74	5.78	2.814	1.546	1.290	64.55	153.	2.40
22.00	37.00	1.2241	1.2263	269.3	1.699	954.8	43.4	442	1.3772	12.74	6.85	3.230	1.647	1.348	60.58	160.	2.56
24.00	40.36	1.2481	1.2503	299.5	1.889	948.6	49.7	487	1.3817	14.99	8.06	3.664	1.761	1.414	56.67	165.	2.68
26.00	43.73	1.2728	1.2751	330.9	2.087	941.9	56.3	534	1.3864				1.893	1.491	52.71	170.	2.79
28.00	47.09	1.2983	1.3006	363.5	2.293	934.8	63.4	582	1.3911				2.052	1.584	48.64	174.	2.90

A% by wt.	H% by wt.	ρ D_4^{20}	D_{20}^{20}	C_s g/l	M g-mol/l	C_w g/l	$(C_o - C_w)$ g/l	$(n - n_o) \times 10^4$	n	Δ °C	O Os/kg	S g-mol/l	η/η_o	η/ρ cS	φ rhe	γ mmho/cm	T g-mol/l
30.00	50.46	1.3248	1.3271	397.4	2.507	927.3	70.9	631	1.3961				2.241	1.695	44.53	178.0	3.02
32.00	53.82	1.3523	1.3547	432.7	2.729	919.6	78.7	683	1.4013				2.522	1.869	39.57	178.0	3.02
34.00	57.18	1.3811	1.3835	469.6	2.962	911.5	86.7	737	1.4067				2.840	2.060	35.14	175.0	2.92
36.00	60.55	1.4114	1.4139	508.1	3.205	903.3	94.9	794	1.4124				3.200	2.272	31.19	166.0	2.70

88 SUCROSE, $C_{12}H_{22}O_{11}$

MOLECULAR WEIGHT = 342.30
RELATIVE SPECIFIC REFRACTIVITY = 1.031

0.00% by wt. data are the same for all compounds.
For Values of 0.00 wt. % solutions see Table 1, Acetic Acid.

A% by wt.	ρ D_4^{20}	D_{20}^{20}	C_s g/l	M g-mol/l	C_w g/l	$(C_o - C_w)$ g/l	$(n - n_o) \times 10^4$	n	Δ °C	O Os/kg	S g-mol/l	η/η_o	η/ρ cS	φ rhe	γ mmho/cm	T g-mol/l
0.50	1.0002	1.0019	5.0	0.015	995.2	3.1	7	1.3337	0.027	0.015	0.007	1.013	1.015	98.53		
1.00	1.0021	1.0039	10.0	0.029	992.1	6.2	14	1.3344	0.055	0.030	0.015	1.026	1.026	97.27		
1.50	1.0040	1.0058	15.1	0.044	989.0	9.3	22	1.3351	0.083	0.045	0.023	1.039	1.037	96.03		
2.00	1.0060	1.0078	20.1	0.059	985.9	12.4	29	1.3359	0.112	0.060	0.031	1.053	1.049	94.78		
2.50	1.0079	1.0097	25.2	0.074	982.7	15.5	36	1.3366	0.140	0.076	0.040	1.067	1.061	93.51		
3.00	1.0099	1.0117	30.3	0.089	979.6	18.6	43	1.3373	0.170	0.091	0.048	1.082	1.074	92.24		
3.50	1.0119	1.0137	35.4	0.103	976.5	21.8	51	1.3381	0.199	0.107	0.057	1.097	1.086	90.99		
4.00	1.0139	1.0156	40.6	0.118	973.3	24.9	58	1.3388	0.229	0.123	0.066	1.112	1.099	89.75		
4.50	1.0158	1.0176	45.7	0.134	970.1	28.1	65	1.3395	0.260	0.140	0.074	1.128	1.112	88.49		
5.00	1.0178	1.0196	50.9	0.149	966.9	31.3	73	1.3403	0.291	0.156	0.083	1.144	1.126	87.24		
5.50	1.0198	1.0216	56.1	0.164	963.7	34.5	80	1.3410	0.322	0.173	0.093	1.160	1.140	86.02		
6.00	1.0218	1.0236	61.3	0.179	960.5	37.7	88	1.3418	0.354	0.190	0.102	1.177	1.154	84.79		
6.50	1.0238	1.0257	66.5	0.194	957.3	40.9	95	1.3425	0.386	0.208	0.111	1.195	1.169	83.54		
7.00	1.0259	1.0277	71.8	0.210	954.1	44.2	103	1.3433	0.419	0.225	0.121	1.213	1.185	82.28		
7.50	1.0279	1.0297	77.1	0.225	950.8	47.4	110	1.3440	0.452	0.243	0.131	1.232	1.201	81.02		
8.00	1.0299	1.0317	82.4	0.241	947.5	50.7	118	1.3448	0.485	0.261	0.140	1.251	1.217	79.78		
8.50	1.0320	1.0338	87.7	0.256	944.2	54.0	125	1.3455	0.520	0.279	0.150	1.271	1.234	78.54		
9.00	1.0340	1.0358	93.1	0.272	940.9	57.3	133	1.3463	0.554	0.298	0.161	1.291	1.251	77.30		
9.50	1.0361	1.0379	98.4	0.288	937.6	60.6	141	1.3471	0.589	0.317	0.171	1.312	1.269	76.09		
10.00	1.0381	1.0400	103.8	0.303	934.3	63.9	148	1.3478	0.625	0.336	0.181	1.333	1.287	74.87		
11.00	1.0423	1.0441	114.7	0.335	927.6	70.6	164	1.3494	0.698	0.375	0.203	1.378	1.325	72.42		
12.00	1.0465	1.0483	125.6	0.367	920.9	77.4	180	1.3509	0.773	0.415	0.225	1.426	1.365	69.99		
13.00	1.0507	1.0525	136.6	0.399	914.1	84.2	195	1.3525	0.850	0.457	0.248	1.477	1.409	67.57		
14.00	1.0549	1.0568	147.7	0.431	907.2	91.0	211	1.3541	0.930	0.500	0.271	1.531	1.454	65.19		
15.00	1.0592	1.0610	158.9	0.464	900.3	97.9	227	1.3557	1.012	0.544	0.295	1.589	1.503	62.81		
16.00	1.0635	1.0653	170.2	0.497	893.3	104.9	243	1.3573	1.097	0.590	0.320	1.650	1.555	60.49		
17.00	1.0678	1.0697	181.5	0.530	886.3	112.0	259	1.3589	1.185	0.637	0.346	1.716	1.610	58.16		
18.00	1.0722	1.0741	193.0	0.564	879.2	119.1	276	1.3606	1.275	0.686	0.373	1.786	1.669	55.88		
19.00	1.0766	1.0785	204.5	0.598	872.0	126.2	292	1.3622	1.369	0.736	0.400	1.861	1.732	53.63		
20.00	1.0810	1.0829	216.2	0.632	864.8	133.5	309	1.3639	1.465	0.788	0.429	1.941	1.799	51.42		
22.00	1.0899	1.0918	239.8	0.701	850.1	148.1	342	1.3672	1.668	0.897	0.488	2.120	1.949	47.08		
24.00	1.0990	1.1009	263.8	0.771	835.2	163.0	376	1.3706	1.886	1.014	0.551	2.326	2.121	42.91		
26.00	1.1082	1.1102	288.1	0.842	820.1	178.2	411	1.3741	2.120	1.140	0.619	2.568	2.322	38.86		
28.00	1.1175	1.1195	312.9	0.914	804.6	193.6	446	1.3776	2.371	1.275	0.692	2.849	2.554	35.03		
30.00	1.1270	1.1290	338.1	0.988	788.9	209.3	482	1.3812	2.644	1.421	0.770	3.181	2.828	31.37		
32.00	1.1366	1.1386	363.7	1.063	772.9	225.3	518	1.3848	2.942	1.582	0.855	3.754	3.309	26.59		
34.00	1.1464	1.1484	389.8	1.139	756.6	241.6	555	1.3885	3.268	1.757	0.947	4.044	3.535	24.68		
36.00	1.1562	1.1583	416.2	1.216	740.0	258.2	592	1.3922	3.625	1.949	1.047	4.612	3.997	21.64		
38.00	1.1663	1.1683	443.2	1.295	723.1	275.1	630	1.3960	4.018	2.160	1.156	5.304	4.557	18.82		
40.00	1.1765	1.1785	470.6	1.375	705.9	292.4	669	1.3999	4.452	2.394	1.276	6.150	5.238	16.23		
42.00	1.1868	1.1889	498.4	1.456	688.3	309.9	708	1.4038	4.932	2.652	1.406	7.220	6.096	13.82		
44.00	1.1972	1.1994	526.8	1.539	670.5	327.8	748	1.4078				8.579	7.180	11.63		
46.00	1.2079	1.2100	555.6	1.623	652.2	346.0	788	1.4118				10.28	8.53	9.71		
48.00	1.2186	1.2208	584.9	1.709	633.7	364.6	829	1.4159				12.49	10.27	7.99		
50.00	1.2295	1.2317	614.8	1.796	614.8	383.5	871	1.4201				15.40	12.55	6.48		
52.00	1.2406	1.2428	645.1	1.885	595.5	402.7	913	1.4243				19.30	15.59	5.17		
54.00	1.2518	1.2540	676.0	1.975	575.8	422.4	956	1.4286				24.63	19.71	4.05		
56.00	1.2632	1.2654	707.4	2.067	555.8	442.4	1000	1.4330				32.06	25.43	3.11		
58.00	1.2747	1.2770	739.3	2.160	535.4	462.8	1044	1.4374				42.69	33.56	2.34		
60.00	1.2864	1.2887	771.9	2.255	514.6	483.7	1089	1.4419				58.37	45.46	1.71		
62.00	1.2983	1.3006	804.9	2.352	493.3	504.9	1135	1.4465				82.26	63.49	1.21		
64.00	1.3103	1.3126	838.6	2.450	471.7	526.5	1181	1.4511				119.9	91.69	0.83		
66.00	1.3224	1.3248	872.8	2.550	449.6	548.6	1228	1.4558				181.7	137.6	0.55		
68.00	1.3348	1.3371	907.6	2.652	427.1	571.1	1276	1.4606				287.9	216.1	0.35		
70.00	1.3472	1.3496	943.1	2.755	404.2	594.1	1324	1.4654				480.6	357.4	0.21		
72.00	1.3599	1.3623	979.1	2.860	380.8	617.5	1373	1.4703				853.2	628.6	0.12		
74.00	1.3726	1.3751	1015.7	2.967	356.9	641.4	1423	1.4753				1628.	1188.	0.06		
76.00	1.3855	1.3880	1053.0	3.076	332.5	665.7	1473	1.4803								
78.00	1.3986	1.4011	1090.9	3.187	307.7	690.5	1524	1.4854								
80.00	1.4117	1.4142	1129.4	3.299	282.3	715.9	1576	1.4906								
82.00	1.4250	1.4275	1168.5	3.414	256.5	741.7	1628	1.4958								
84.00	1.4383	1.4409	1208.2	3.530	230.1	768.1	1681	1.5010								

89 SULFURIC ACID, H_2SO_4

MOLECULAR WEIGHT = 98.08
RELATIVE SPECIFIC REFRACTIVITY = 0.685

0.00 % by wt. data are the same for all compounds.
For Values of 0.00 wt. % solutions see Table 1, Acetic Acid.

A % by wt.	ρ D_4^{20}	D_{20}^{20}	C_s g/l	M g-mol/l	C_w g/l	$(C_o - C_w)$ g/l	$(n - n_o)$ $\times 10^4$	n	Δ °C	O Os/kg	S g-mol/l	η/η_o	η/ρ cS	ϕ rhe	γ mmho/cm	T g-mol/l
0.50	1.0016	1.0034	5.0	0.051	996.6	1.7	6	1.3336	0.210	0.113	0.060	1.008	1.009	98.96	24.3	0.277
1.00	1.0049	1.0067	10.0	0.102	994.9	3.3	13	1.3342	0.423	0.227	0.122	1.017	1.014	98.13	47.8	0.573
1.50	1.0083	1.0101	15.1	0.154	993.2	5.1	19	1.3349	0.662	0.356	0.192	1.025	1.019	97.34	70.3	0.886
2.00	1.0116	1.0134	20.2	0.206	991.4	6.8	25	1.3355	0.796	0.428	0.232	1.034	1.024	96.52	92.	1.22
2.50	1.0150	1.0168	25.4	0.259	989.6	8.6	31	1.3361	1.004	0.540	0.293	1.044	1.031	95.55	113.	1.58
3.00	1.0183	1.0201	30.6	0.311	987.8	10.4	37	1.3367	1.172	0.630	0.343	1.057	1.040	94.45	134.	1.98
3.50	1.0217	1.0235	35.8	0.365	985.9	12.3	43	1.3373	1.354	0.728	0.396	1.070	1.049	93.29	155.	2.42
4.00	1.0250	1.0269	41.0	0.418	984.0	14.2	49	1.3379	1.599	0.860	0.468	1.083	1.059	92.11	175.	2.93
4.50	1.0284	1.0302	46.3	0.472	982.1	16.1	55	1.3385	1.855	0.998	0.543	1.097	1.069	90.97	194.	3.57
5.00	1.0318	1.0336	51.6	0.526	980.2	18.0	61	1.3391	2.047	1.101	0.598	1.110	1.078	89.91	211.	4.25
5.50	1.0352	1.0370	56.9	0.580	978.2	20.0	67	1.3397	2.259	1.214	0.659	1.122	1.086	88.93		
6.00	1.0385	1.0404	62.3	0.635	976.2	22.0	73	1.3403	2.495	1.341	0.727	1.134	1.094	88.00		
6.50	1.0419	1.0438	67.7	0.691	974.2	24.0	79	1.3409	2.730	1.468	0.795	1.146	1.102	87.10		
7.00	1.0453	1.0472	73.2	0.746	972.2	26.1	85	1.3415	2.952	1.587	0.858	1.157	1.109	86.24		
7.50	1.0488	1.0506	78.7	0.802	970.1	28.1	91	1.3421	3.197	1.719	0.927	1.169	1.117	85.39		
8.00	1.0522	1.0541	84.2	0.858	968.0	30.2	97	1.3427	3.493	1.878	1.010	1.180	1.124	84.56		
8.50	1.0556	1.0575	89.7	0.915	965.9	32.3	103	1.3433	3.801	2.043	1.096	1.192	1.131	83.74		
9.00	1.0591	1.0610	95.3	0.972	963.8	34.4	109	1.3439	4.083	2.195	1.174	1.204	1.139	82.92		
9.50	1.0626	1.0645	100.9	1.029	961.6	36.6	115	1.3445	4.360	2.344	1.250	1.216	1.146	82.10		
10.00	1.0661	1.0680	106.6	1.087	959.5	38.8	121	1.3451	4.644	2.497	1.328	1.228	1.154	81.27		
11.00	1.0731	1.0750	118.0	1.204	955.1	43.2	133	1.3463	5.25	2.82	1.490	1.253	1.170	79.63		
12.00	1.0802	1.0821	129.6	1.322	950.6	47.6	145	1.3475	5.93	3.19	1.669	1.279	1.187	78.02		
13.00	1.0874	1.0893	141.4	1.441	946.0	52.2	158	1.3488	6.67	3.59	1.859	1.306	1.203	76.43		
14.00	1.0947	1.0966	153.3	1.563	941.4	56.8	170	1.3500	7.49	4.03	2.063	1.334	1.221	74.82		
15.00	1.1020	1.1039	165.3	1.685	936.7	61.5	183	1.3513	8.35	4.49	2.270	1.364	1.240	73.17		
16.00	1.1094	1.1114	177.5	1.810	931.9	66.3	195	1.3525	9.26	4.98	2.483	1.396	1.261	71.47		
17.00	1.1169	1.1189	189.9	1.936	927.0	71.2	208	1.3538	10.23	5.50	2.702	1.431	1.284	69.74		
18.00	1.1245	1.1265	202.4	2.064	922.1	76.2	221	1.3551	11.29	6.07	2.932	1.467	1.308	68.01		
19.00	1.1321	1.1341	215.1	2.193	917.0	81.2	233	1.3563	12.43	6.68	3.169	1.505	1.332	66.32		
20.00	1.1398	1.1418	228.0	2.324	911.9	86.4	246	1.3576	13.64	7.33	3.409	1.543	1.356	64.68		
22.00	1.1554	1.1575	254.2	2.592	901.2	97.0	272	1.3602	16.48	8.86	3.932	1.621	1.405	61.58		
24.00	1.1714	1.1735	281.1	2.866	890.3	108.0	298	1.3628	19.85	10.67	4.488	1.703	1.457	58.60		
26.00	1.1872	1.1893	308.7	3.147	878.5	119.7	323	1.3653	24.29	13.06		1.793	1.513	55.67		
28.00	1.2031	1.2052	336.9	3.435	866.2	132.0	347	1.3677	29.65	15.94		1.890	1.574	52.81		
30.00	1.2191	1.2213	365.7	3.729	853.4	144.9	371	1.3701	36.21	19.47		1.997	1.641	49.98		
32.00	1.2353	1.2375	395.3	4.030	840.0	158.2	395	1.3725	44.76	24.07		2.118	1.718	47.12		
34.00	1.2518	1.2540	425.6	4.339	826.2	172.1	419	1.3749	55.28	29.72		2.250	1.801	44.36		
36.00	1.2685	1.2707	456.7	4.656	811.8	186.4	443	1.3773				2.387	1.885	41.82		
38.00	1.2855	1.2878	488.5	4.981	797.0	201.2	467	1.3797				2.528	1.970	39.48		
40.00	1.3028	1.3051	521.1	5.313	781.7	216.5	491	1.3821				2.685	2.065	37.17		
42.00	1.3205	1.3229	554.6	5.655	765.9	232.3	516	1.3846				2.866	2.174	34.83		
44.00	1.3386	1.3410	589.0	6.005	749.6	248.6	540	1.3870				3.067	2.296	32.53		
46.00	1.3570	1.3594	624.2	6.365	732.8	265.4	565	1.3895				3.292	2.431	30.32		
48.00	1.3759	1.3783	660.4	6.734	715.5	282.8	590	1.3920				3.539	2.577	28.20		
50.00	1.3952	1.3977	697.6	7.113	697.6	300.6	616	1.3945				3.818	2.742	26.14		
52.00	1.4149	1.4174	735.8	7.502	679.2	319.1	641	1.3971				4.134	2.927	24.14		
54.00	1.4351	1.4377	775.0	7.901	660.2	338.1	667	1.3997				4.490	3.135	22.23		
56.00	1.4558	1.4584	815.3	8.312	640.6	357.7	694	1.4024				4.896	3.370	20.38		
58.00	1.4770	1.4796	856.7	8.734	620.3	377.9	720	1.4050				5.343	3.625	18.68		
60.00	1.4987	1.5013	899.2	9.168	599.5	398.8	747	1.4077				5.905	3.948	16.90		
62.00	1.5200	1.5227	942.4	9.608	577.6	420.6										
64.00	1.5421	1.5448	986.9	10.062	555.2	443.0										
66.00	1.5646	1.5674	1032.6	10.528	532.0	466.2										
68.00	1.5874	1.5902	1079.4	11.005	508.0	490.2										
70.00	1.6105	1.6134	1127.4	11.495	483.1	515.1										
72.00	1.6338	1.6367	1176.3	11.993	457.5	540.7										
74.00	1.6574	1.6603	1226.5	12.505	430.9	567.3										
76.00	1.6810	1.6840	1277.6	13.026	403.4	594.8										
78.00	1.7043	1.7073	1329.4	13.554	374.9	623.3										
80.00	1.7272	1.7303	1381.8	14.088	345.4	652.8										
82.00	1.7491	1.7522	1434.3	14.624	314.8	683.4										
84.00	1.7693	1.7724	1486.2	15.153	283.1	715.1										
86.00	1.7872	1.7904	1537.0	15.671	250.2	748.0										
88.00	1.8022	1.8054	1585.9	16.169	216.3	781.9										
90.00	1.8144	1.8176	1633.0	16.650	181.4	816.8										
92.00	1.8240	1.8272	1678.1	17.110	145.9	852.3										
94.00	1.8312	1.8344	1721.3	17.550	109.9	888.3										
96.00	1.8355	1.8388	1762.1	17.966	73.4	924.8										
98.00	1.8361	1.8394	1799.4	18.346	36.7	961.5										
100.00	1.8305	1.8337	1830.5	18.663	0.0	998.2										

90 TARTARIC ACID, $HO_2C(CHOH)_2CO_2H$

MOLECULAR WEIGHT = 150.09

A% by wt.	ρ D_4^{20}	D_{20}^{20}	C_s g/l	M g-mol/l	C_w g/l	$(C_0 - C_w)$ g/l	$(n - n_0)$ × 10^4	n	Δ °C	O Os/kg	S g-mol/l	η/η_0	η/ρ cS	ϕ rhe	γ mmho/cm	T g-mol/l
1.00	1.0029	1.0047	10.0	0.067	992.9	5.3	12.	1.3342	0.132	0.070	0.037	1.021	1.020	97.76	2.5	0.025
2.00	1.0072	1.0090	20.1	0.134	987.1	11.1	25.	1.3355	0.262	0.141	0.075	1.044	1.039	95.56	3.9	0.039
3.00	1.0121	1.0139	30.4	0.202	981.7	16.5	37.	1.3367	0.402	0.216	0.116	1.069	1.058	93.39	4.9	0.049
4.00	1.0163	1.0181	40.7	0.271	975.6	22.6	49.	1.3379	0.538	0.290	0.156	1.092	1.077	91.36	5.5	0.056
5.00	1.0208	1.0226	51.0	0.340	969.8	28.4	62.	1.3392	0.681	0.366	0.198	1.119	1.098	89.22	6.2	0.063
6.00	1.0255	1.0273	61.5	0.410	964.0	34.2	73.	1.3403	0.824	0.443	0.240	1.147	1.121	86.99	6.9	0.070
7.00	1.0310	1.0329	72.2	0.481	958.8	39.4	88.	1.3418	0.974	0.523	0.284	1.176	1.143	84.86	7.3	0.075
8.00	1.0348	1.0367	82.8	0.552	952.0	46.2	99.	1.3429	1.130	0.607	0.330	1.207	1.169	82.67	7.8	0.080
9.00	1.0393	1.0412	93.5	0.623	945.8	52.4	111.	1.3441	1.283	0.690	0.375	1.236	1.192	80.72	8.2	0.084
10.00	1.0443	1.0462	104.4	0.696	939.9	58.3	124.	1.3454	1.443	0.776	0.422	1.272	1.220	78.49	8.5	0.088
12.00	1.0540	1.0559	126.5	0.843	927.5	70.7	151.	1.3481	1.768	0.951	0.517	1.343	1.277	74.30	9.2	0.095
14.00	1.0639	1.0658	148.9	0.992	915.0	83.2	178.	1.3508	2.130	1.145	0.622	1.413	1.331	70.62	9.7	0.100
16.00	1.0738	1.0757	171.8	1.145	902.0	96.2	205.	1.3535	2.504	1.346	0.730	1.498	1.398	66.61	10.1	0.104
18.00	1.0839	1.0859	195.1	1.300	888.8	109.4	231.	1.3561	2.917	1.568	0.848	1.591	1.471	62.72	10.3	0.107
20.00	1.0942	1.0962	218.8	1.458	875.4	122.8	260.	1.3590	3.360	1.807	0.973	1.701	1.558	58.66	10.5	0.109
24.00	1.1152	1.1172	267.6	1.783	847.6	150.6	316.	1.3646	4.309	2.316	1.236	1.957	1.758	51.01	10.6	0.110
28.00	1.1366	1.1387	318.3	2.120	818.3	179.9	375.	1.3705	5.46	2.93	1.546	2.296	2.024	43.47	10.4	0.108
32.00	1.1588	1.1609	370.8	2.471	788.0	210.2	436.	1.3766	6.73	3.62	1.874	2.695	2.330	37.04	10.1	0.104
36.00	1.1813	1.1834	425.3	2.833	756.0	242.2	499.	1.3829	8.30	4.47	2.259	3.242	2.750	30.78	9.3	0.096
40.00	1.2049	1.2071	482.0	3.211	722.9	275.3	561.	1.3891	10.20	5.48	2.695	4.002	3.328	24.94	8.2	0.085
44.00	1.2290	1.2312	540.8	3.603	688.2	310.0	627.	1.3957				5.088	4.148	19.62	7.3	0.074
48.00	1.2535	1.2558	601.7	4.009	651.8	346.4	692.	1.4022				6.605	5.280	15.11	6.2	0.063
52.00	1.2787	1.2810	664.9	4.430	613.8	384.4	762.	1.4092				8.961	7.022	11.14	5.0	0.050
56.00	1.3058	1.3082	731.3	4.872	574.5	423.7	834.	1.4164				12.60	9.670	7.92	3.7	0.037
58.00	1.3196	1.3220	765.4	5.099	554.2	444.0	869.	1.4199				15.64	11.87	6.38	3.0	0.030

91 TETRACAINE HYDROCHLORIDE, $C_{15}H_{24}N_2O_2 \cdot HCl$

MOLECULAR WEIGHT = 300.84

A% by wt.	ρ D_4^{20}	D_{20}^{20}	C_s g/l	M g-mol/l	C_w g/l	$(C_0 - C_w)$ g/l	$(n - n_0)$ × 10^4	n	Δ °C	O Os/kg	S g-mol/l	η/η_0	η/ρ cS	ϕ rhe	γ mmho/cm	T g-mol/l
0.50	0.9988	1.0006	5.0	0.017	993.8	4.4	11.	1.3341	0.058	0.031	0.016	1.013	1.016	98.54	1.2	0.012
1.00	0.9995	1.0013	10.0	0.033	989.5	8.7	23.	1.3353	0.110	0.059	0.031	1.029	1.032	96.95	2.4	0.024
1.50	1.0001	1.0019	15.0	0.050	985.1	13.1	34.	1.3364	0.159	0.086	0.045	1.045	1.047	95.50	3.6	0.036
2.00	1.0008	1.0026	20.0	0.067	980.8	17.4	46.	1.3376	0.204	0.109	0.058	1.063	1.064	93.91	4.7	0.047
2.50	1.0014	1.0032	25.0	0.083	976.4	21.8	57.	1.3387				1.081	1.082	92.29	5.6	0.057
3.00	1.0021	1.0039	30.1	0.100	972.0	26.2	69.	1.3399				1.100	1.100	90.72	6.6	0.067
3.50	1.0027	1.0045	35.1	0.117	967.6	30.6	80.	1.3410				1.121	1.120	89.05	7.4	0.076
4.00	1.0034	1.0052	40.1	0.133	963.3	34.9	92.	1.3422				1.142	1.140	87.42	8.2	0.084
4.50	1.0040	1.0058	45.2	0.150	958.8	39.4	103.	1.3433				1.163	1.161	85.79	8.9	0.092
5.00	1.0046	1.0064	50.2	0.167	954.4	43.8	115.	1.3445				1.187	1.184	84.07	9.6	0.099
5.50	1.0053	1.0071	55.3	0.184	950.0	48.2	126.	1.3456				1.211	1.207	82.41	10.2	0.106
6.00	1.0059	1.0077	60.4	0.201	945.5	52.7	138.	1.3468				1.237	1.232	80.69	10.8	0.113
6.50	1.0066	1.0084	65.4	0.217	941.2	57.0	149.	1.3479				1.263	1.257	79.03	11.4	0.120
7.00	1.0072	1.0090	70.5	0.234	936.7	61.5	160.	1.3490				1.291	1.284	77.33	11.9	0.126
7.50	1.0079	1.0097	75.6	0.251	932.3	65.9	172.	1.3502				1.319	1.311	75.68	12.5	0.132
8.00	1.0085	1.0103	80.7	0.268	927.8	70.4	183.	1.3513				1.348	1.339	74.05	13.0	0.138
8.50	1.0092	1.0110	85.8	0.285	923.4	74.8	195.	1.3525				1.379	1.369	72.38	13.6	0.144
9.00	1.0098	1.0116	90.9	0.302	918.9	79.3	206.	1.3536				1.410	1.399	70.79	14.0	0.149
9.50	1.0105	1.0123	96.0	0.319	914.5	83.7	218.	1.3548				1.441	1.429	69.25	14.4	0.154
10.00	1.0111	1.0129	101.1	0.336	910.0	88.2	229.	1.3559				1.477	1.464	67.56	14.7	0.158
11.00	1.0124	1.0142	111.4	0.370	901.0	97.2	252.	1.3582				1.548	1.532	64.48	15.8	0.170
12.00	1.0137	1.0155	121.6	0.404	892.1	106.1	275.	1.3605				1.623	1.604	61.50	16.8	0.182

92 TRICHLOROACETIC ACID, CCl_3COOH

MOLECULAR WEIGHT = 163.38
RELATIVE SPECIFIC REFRACTIVITY = 0.865

0.00 % by wt. data are the same for all compounds.
For Values of 0.00 wt. % solutions see Table 1, Acetic Acid.

A% by wt.	ρ D_4^{20}	D_{20}^{20}	C_s g/l	M g-mol/l	C_w g/l	$(C_0 - C_w)$ g/l	$(n - n_0)$ × 10^4	n	Δ °C	O Os/kg	S g-mol/l	$(n - n_0)$ η/η_0	η/ρ cS	ϕ rhe	γ mmho/cm	T g-mol/l
0.50	1.0008	1.0026	5.0	0.031	995.8	2.4	7	1.3337	0.105	0.057	0.030	1.009	1.010	98.91	10.3	0.107
1.00	1.0034	1.0051	10.0	0.061	993.3	4.9	13	1.3343	0.211	0.113	0.060	1.019	1.018	97.94	19.6	0.215
2.00	1.0083	1.0101	20.2	0.123	988.2	10.1	26	1.3356	0.423	0.227	0.122	1.042	1.035	95.78	37.2	0.435
3.00	1.0133	1.0151	30.4	0.186	982.9	15.4	39	1.3369	0.638	0.343	0.185	1.067	1.055	93.53	54.0	0.656

92 TRICHLOROACETIC ACID, CCl_3COOH—(Continued)

A% by wt.	ρ D_4^{20}	D_{20}^{20}	C_s g/l	M g-mol/l	C_w g/l	$(C_o - C_w)$ g/l	$(n - n_o)$ $\times 10^4$	n	Δ °C	O Os/kg	S g-mol/l	η/η_o	η/ρ cS	ϕ rhe	γ mmho/cm	T g-mol/l
4.00	1.0182	1.0200	40.7	0.249	977.4	20.8	52	1.3381	0.857	0.461	0.250	1.094	1.077	91.24	69.8	0.880
5.00	1.0230	1.0248	51.2	0.313	971.9	26.4	64	1.3394	1.079	0.580	0.315	1.121	1.098	89.03	84.7	1.10
6.00	1.0279	1.0297	61.7	0.377	966.2	32.0	76	1.3406	1.303	0.700	0.381	1.148	1.119	86.95	99.	1.33
7.00	1.0328	1.0346	72.3	0.443	960.5	37.7	89	1.3418	1.529	0.822	0.447	1.175	1.140	84.95	113.	1.57
8.00	1.0378	1.0396	83.0	0.508	954.8	43.5	101	1.3431	1.758	0.945	0.514	1.202	1.161	83.01	125.	1.80
9.00	1.0428	1.0446	93.9	0.574	948.9	49.3	114	1.3444	1.992	1.071	0.582	1.231	1.182	81.10	137.	2.04
10.00	1.0479	1.0497	104.8	0.641	943.1	55.1	126	1.3456	2.233	1.201	0.652	1.260	1.205	79.21	148.	2.28
12.00	1.0583	1.0602	127.0	0.777	931.3	66.9	153	1.3483	2.734	1.470	0.796	1.323	1.252	75.44	169.	2.76
14.00	1.0692	1.0710	149.7	0.916	919.5	78.8	180	1.3510	3.261	1.753	0.945	1.390	1.302	71.81	186.	3.24
16.00	1.0806	1.0825	172.9	1.058	907.7	90.6	209	1.3539	3.815	2.051	1.100	1.459	1.353	68.41	200.	3.73
18.00	1.0921	1.0941	196.6	1.203	895.5	102.7	239	1.3568				1.530	1.404	65.21	211.	4.22
20.00	1.1035	1.1055	220.7	1.351	882.8	115.4	267	1.3597				1.605	1.457	62.18	221.	4.90
24.00	1.1260	1.1280	270.2	1.654	855.8	142.5	322	1.3652				1.764	1.570	56.58		
28.00	1.1485	1.1505	321.6	1.968	826.9	171.3	376	1.3705				1.931	1.685	51.68		
32.00	1.1713	1.1734	374.8	2.294	796.5	201.8	429	1.3759				2.114	1.808	47.22		
36.00	1.1947	1.1968	430.1	2.633	764.6	233.6	483	1.3813				2.315	1.942	43.11		
40.00	1.2188	1.2210	487.5	2.984	731.3	266.9	538	1.3868				2.540	2.088	39.29		
44.00	1.2435	1.2457	547.1	3.349	696.3	301.9	593	1.3923				2.791	2.249	35.75		
48.00	1.2682	1.2704	608.7	3.726	659.5	338.8	647	1.3977				3.070	2.425	32.51		
50.00	1.2803	1.2826	640.2	3.918	640.2	358.1	673	1.4003				3.219	2.519	31.00		

93 TRIS(HYDROXYMETHYL)AMINOMETHANE,* THAM,® $H_2NC(CH_2OH)_3$

MOLECULAR WEIGHT = 121.14
RELATIVE SPECIFIC REFRACTIVITY = 1.169

0.00 % by wt. data are the same for all compounds.
For Values of 0.00 wt. % solutions see Table 1, Acetic Acid.

A% by wt.	ρ D_4^{20}	D_{20}^{20}	C_s g/l	M g-mol/l	C_w g/l	$(C_o - C_w)$ g/l	$(n - n_o)$ $\times 10^4$	n	Δ °C	O Os/kg	S g-mol/l	η/η_o	η/ρ cS	ϕ rhe	γ mmho/cm	T g-mol/l
0.50	0.9994	1.0012	5.0	0.041	994.4	3.8	7	1.3337	0.080	0.043	0.022	1.012	1.015	98.59		
1.00	1.0006	1.0024	10.0	0.083	990.6	7.6	15	1.3344	0.159	0.085	0.045	1.025	1.026	97.37		
2.00	1.0030	1.0048	20.1	0.166	982.9	15.3	29	1.3359	0.314	0.169	0.090	1.052	1.051	94.87		
3.00	1.0054	1.0072	30.2	0.249	975.2	23.0	44	1.3374	0.473	0.254	0.137	1.081	1.078	92.29		
4.00	1.0078	1.0096	40.3	0.333	967.5	30.7	58	1.3388	0.636	0.342	0.185	1.113	1.106	89.67		
5.00	1.0103	1.0121	50.5	0.417	959.8	38.4	73	1.3403	0.804	0.432	0.234	1.146	1.137	87.09		
6.00	1.0128	1.0146	60.8	0.502	952.0	46.2	88	1.3418	0.974	0.524	0.284	1.180	1.168	84.57		
7.00	1.0153	1.0171	71.1	0.587	944.3	54.0	103	1.3433	1.149	0.618	0.336	1.216	1.200	82.10		
8.00	1.0179	1.0197	81.4	0.672	936.4	61.8	118	1.3448	1.328	0.714	0.388	1.253	1.233	79.66		
9.00	1.0204	1.0222	91.8	0.758	928.6	69.6	133	1.3463	1.510	0.812	0.442	1.292	1.269	77.24		
10.00	1.0230	1.0248	102.3	0.844	920.7	77.5	148	1.3478	1.696	0.912	0.496	1.334	1.307	74.81		
12.00	1.0282	1.0301	123.4	1.019	904.8	93.4	178	1.3508	2.076	1.116	0.606	1.424	1.388	70.09		
14.00	1.0335	1.0354	144.7	1.194	888.8	109.4	209	1.3539	2.473	1.329	0.721	1.524	1.478	65.48		
16.00	1.0389	1.0407	166.2	1.372	872.7	125.6	240	1.3570	2.899	1.559	0.843	1.639	1.581	60.88		
18.00	1.0443	1.0462	188.0	1.552	856.4	141.9	271	1.3601	3.357	1.805	0.972	1.768	1.696	56.45		
20.00	1.0498	1.0517	210.0	1.733	839.9	158.4	303	1.3633	3.847	2.068	1.109	1.916	1.829	52.09		
22.00	1.0554	1.0572	232.2	1.917	823.2	175.0	335	1.3665				2.079	1.974	48.00		
24.00	1.0610	1.0628	254.6	2.102	806.3	191.9	367	1.3697				2.256	2.131	44.23		
26.00	1.0666	1.0685	277.3	2.289	789.3	208.9	400	1.3730				2.459	2.310	40.59		
28.00	1.0723	1.0742	300.2	2.479	772.1	226.2	433	1.3763				2.700	2.523	36.97		
30.00	1.0781	1.0800	323.4	2.670	754.6	243.6	467	1.3797				2.992	2.781	33.36		
32.00	1.0839	1.0858	346.8	2.863	737.0	261.2	501	1.3831				3.337	3.085	29.91		
34.00	1.0897	1.0916	370.5	3.058	719.2	279.0	535	1.3865				3.729	3.429	26.76		
36.00	1.0956	1.0976	394.4	3.256	701.2	297.0	570	1.3900				4.169	3.812	23.94		
38.00	1.1016	1.1035	418.6	3.456	683.0	315.2	605	1.3935				4.658	4.237	21.43		
40.00	1.1076	1.1096	443.1	3.657	664.6	333.6	640	1.3970				5.198	4.702	19.20		

* 1,2-Propanediol, 2-amino-2(hydroxymethyl). ® Registered trademark.

94 UREA, NH_2CONH_2

MOLECULAR WEIGHT = 60.06
RELATIVE SPECIFIC REFRACTIVITY = 1.147

0.00 % by wt. data are the same for all compounds.
For Values of 0.00 wt. % solutions see Table 1, Acetic Acid.

A% by wt.	ρ D_4^{20}	D_{20}^{20}	C_s g/l	M g-mol/l	C_w g/l	$(C_o - C_w)$ g/l	$(n - n_o)$ $\times 10^4$	n	Δ °C	O Os/kg	S g-mol/l	η/η_o	η/ρ cS	ϕ rhe	γ mmho/cm	T g-mol/l
0.50	0.9995	1.0012	5.0	0.083	994.5	3.8	7	1.3337	0.155	0.083	0.044	1.005	1.007	99.33		
1.00	1.0007	1.0025	10.0	0.167	990.7	7.5	14	1.3344	0.310	0.167	0.089	1.008	1.009	99.01		
1.50	1.0020	1.0038	15.0	0.250	987.0	11.3	21	1.3351	0.465	0.250	0.134	1.009	1.009	98.89		
2.00	1.0033	1.0050	20.1	0.334	983.2	15.0	28	1.3358	0.620	0.333	0.180	1.010	1.009	98.81		
2.50	1.0045	1.0063	25.1	0.418	979.4	18.8	35	1.3365	0.775	0.417	0.226	1.012	1.010	98.60		
3.00	1.0058	1.0076	30.2	0.502	975.7	22.6	42	1.3372	0.928	0.499	0.271	1.015	1.011	98.30		
3.50	1.0071	1.0089	35.3	0.587	971.9	26.3	50	1.3379	1.082	0.582	0.316	1.019	1.014	97.95		
4.00	1.0085	1.0102	40.3	0.672	968.1	30.1	57	1.3387	1.237	0.665	0.362	1.023	1.016	97.56		
4.50	1.0098	1.0116	45.4	0.757	964.3	33.9	64	1.3394	1.393	0.749	0.407	1.027	1.019	97.17		
5.00	1.0111	1.0129	50.6	0.842	960.6	37.7	71	1.3401	1.552	0.835	0.454	1.031	1.022	96.80		
5.50	1.0125	1.0142	55.7	0.927	956.8	41.5	79	1.3409	1.715	0.922	0.501	1.035	1.024	96.45		
6.00	1.0138	1.0156	60.8	1.013	953.0	45.3	86	1.3416	1.880	1.011	0.550	1.039	1.027	96.09		
6.50	1.0151	1.0169	66.0	1.099	949.2	49.1	93	1.3423	2.048	1.101	0.598	1.043	1.029	95.72		
7.00	1.0165	1.0183	71.2	1.185	945.4	52.9	101	1.3431	2.218	1.192	0.648	1.047	1.032	95.35		
7.50	1.0179	1.0197	76.3	1.271	941.5	56.7	108	1.3438	2.389	1.285	0.697	1.051	1.034	94.98		
8.00	1.0192	1.0210	81.5	1.358	937.7	60.5	116	1.3446	2.562	1.378	0.747	1.055	1.037	94.60		
8.50	1.0206	1.0224	86.8	1.444	933.9	64.4	123	1.3453	2.737	1.471	0.797	1.059	1.040	94.23		

94 UREA, NH_2CONH_2—(Continued)

A% by wt.	ρ D^{20}_4	D^{20}_{20}	C_s g/l	M g-mol/l	C_w g/l	$(C_o - C_w)$ g/l	$(n - n_o)$ × 10^4	n	Δ °C	O Os/kg	S g-mol/l	η/η_o	η/ρ cS	φ rhe	γ mmho/cm	T g-mol/l
9.00	1.0220	1.0238	92.0	1.531	930.0	68.2	131	1.3461	2.911	1.565	0.846	1.063	1.043	93.85		
9.50	1.0234	1.0252	97.2	1.619	926.2	72.1	138	1.3468	3.086	1.659	0.896	1.068	1.045	93.47		
10.00	1.0248	1.0266	102.5	1.706	922.3	75.9	146	1.3476	3.260	1.753	0.945	1.072	1.048	93.10		
11.00	1.0276	1.0294	113.0	1.882	914.5	83.7	161	1.3491	3.606	1.939	1.042	1.081	1.054	92.36		
12.00	1.0304	1.0322	123.6	2.059	906.7	91.5	176	1.3506	3.952	2.125	1.138	1.089	1.059	91.64		
13.00	1.0332	1.0350	134.3	2.236	898.9	99.4	192	1.3521	4.301	2.312	1.234	1.098	1.065	90.91		
14.00	1.0360	1.0378	145.0	2.415	891.0	107.3	207	1.3537	4.656	2.503	1.332	1.107	1.071	90.15		
15.00	1.0388	1.0407	155.8	2.594	883.0	115.2	222	1.3552	5.02	2.70	1.430	1.117	1.077	89.35		
16.00	1.0417	1.0435	166.7	2.775	875.0	123.2	238	1.3568	5.40	2.90	1.531	1.128	1.085	88.50		
17.00	1.0445	1.0463	177.6	2.956	866.9	131.3	253	1.3583	5.79	3.11	1.633	1.139	1.093	87.62		
18.00	1.0473	1.0492	188.5	3.139	858.8	139.4	269	1.3599	6.19	3.33	1.736	1.151	1.101	86.72		
19.00	1.0502	1.0520	199.5	3.322	850.6	147.6	284	1.3614	6.59	3.54	1.839	1.163	1.110	85.80		
20.00	1.0530	1.0549	210.6	3.506	842.4	155.8	300	1.3629	7.00	3.76	1.941	1.176	1.119	84.86		
22.00	1.0586	1.0605	232.9	3.878	825.7	172.5	331	1.3661	7.81	4.20	2.141	1.203	1.139	82.93		
24.00	1.0643	1.0662	255.4	4.253	808.9	189.4	362	1.3692	8.64	4.65	2.339	1.233	1.161	80.95		
26.00	1.0699	1.0718	278.2	4.632	791.7	206.5	393	1.3723	9.52	5.12	2.543	1.263	1.183	79.00		
28.00	1.0756	1.0775	301.2	5.014	774.4	223.8	424	1.3754	10.45	5.62	2.751	1.295	1.206	77.09		
30.00	1.0812	1.0831	324.4	5.401	756.8	241.4	455	1.3785	11.40	6.13	2.955	1.329	1.232	75.09		
32.00	1.0869	1.0888	347.8	5.791	739.1	259.2	487	1.3817	12.34	6.63	3.150	1.368	1.261	72.97		
34.00	1.0926	1.0945	371.5	6.185	721.1	277.1	518	1.3848	13.27	7.14	3.337	1.410	1.293	70.77		
36.00	1.0984	1.1004	395.4	6.584	703.0	295.2	551	1.3881	14.20	7.63	3.516	1.456	1.329	68.52		
38.00	1.1044	1.1064	419.7	6.988	684.7	313.5	583	1.3913	15.11	8.12	3.686	1.506	1.367	66.26		
40.00	1.1106	1.1126	444.2	7.397	666.4	331.9	617	1.3947	15.99	8.60	3.845	1.562	1.409	63.89		
42.00	1.1171	1.1191	469.2	7.812	647.9	350.3	652	1.3982	16.83	9.05	3.993	1.626	1.458	61.40		
44.00	1.1239	1.1259	494.5	8.234	629.4	368.8	688	1.4018	17.62	9.47	4.127	1.697	1.512	58.83		
46.00	1.1313	1.1333	520.4	8.665	610.9	387.3	726	1.4056				1.776	1.573	56.21		

95 URINE, CAT

RELATIVE SPECIFIC REFRACTIVITY = 1.08

0.00 % by wt. data are the same for all compounds.
For Values of 0.00 wt. % solutions see Table 1, Acetic Acid.

A% by wt.	ρ D^{20}_4	D^{20}_{20}	C_s g/l	C_w g/l	$(C_o - C_w)$ g/l	$(n - n_o)$ × 10^4	n
0.50	1.000	1.002	5	995	3	8	1.3338
1.00	1.002	1.004	10	992	6	16	1.3346
1.50	1.004	1.006	15	989	9	24	1.3354
2.00	1.006	1.008	20	986	12	33	1.3363
2.50	1.008	1.010	25	983	15	41	1.3371
3.00	1.010	1.012	30	979	19	49	1.3379
3.50	1.011	1.013	35	976	22	57	1.3387
4.00	1.013	1.015	41	973	25	65	1.3395
4.50	1.015	1.017	46	970	28	73	1.3403
5.00	1.017	1.019	51	966	32	81	1.3411
5.50	1.019	1.021	56	963	35	89	1.3419
6.00	1.021	1.023	61	960	38	97	1.3427
6.50	1.023	1.025	67	956	42	105	1.3435
7.00	1.025	1.027	72	953	45	113	1.3443
7.50	1.026	1.028	77	950	48	121	1.3451
8.00	1.028	1.030	82	946	52	129	1.3459
8.50	1.030	1.032	88	943	55	136	1.3466
9.00	1.032	1.034	93	939	59	144	1.3474
9.50	1.034	1.036	98	936	62	152	1.3482
10.00	1.036	1.038	104	932	66	160	1.3490
11.00	1.039	1.041	114	925	73	175	1.3505
12.00	1.043	1.045	125	918	80	191	1.3521
13.00	1.047	1.049	136	911	87	206	1.3536
14.00	1.050	1.052	147	903	95	222	1.3552
15.00	1.054	1.056	158	896	102	237	1.3567
16.00	1.058	1.060	169	888	110	252	1.3582
17.00	1.061	1.063	180	881	117	267	1.3597
18.00	1.065	1.067	192	873	125	282	1.3612
19.00	1.068	1.070	203	865	133	296	1.3626
20.00	1.072	1.074	214	857	141	311	1.3641

96 URINE, GUINEA PIG

RELATIVE SPECIFIC REFRACTIVITIES: CLEAR* = 0.984; TURBID = 0.955

0.00 % by wt. data are the same for all compounds.
For Values of 0.00 wt. % solutions see Table 1, Acetic Acid.

CLEAR								TURBID					
A% by wt.	ρ D^{20}_4	D^{20}_{20}	C_s g/l	C_w g/l	$(C_o - C_w)$ g/l	$(n - n_o)$ × 10^4	n	A% by wt.	ρ D^{20}_4	D^{20}_{20}	C_s g/l	C_w g/l	$(C_o - C_w)$ g/l
0.5	1.001	1.002	5	996	2	8	1.3338	0.5	1.001	1.002	5	996	2
1.0	1.003	1.005	10	993	5	16	1.3346	1.0	1.003	1.005	10	993	5
1.5	1.005	1.007	15	990	8	23	1.3353	1.5	1.005	1.007	15	990	8
2.0	1.008	1.009	20	988	10	31	1.3361	2.1	1.008	1.010	21	987	11
2.5	1.010	1.012	25	985	13	39	1.3369	2.6	1.010	1.012	26	984	14
3.0	1.012	1.014	30	982	16	47	1.3377	3.1	1.013	1.015	31	982	16

* Cleared by centrifuging 15 minutes at 25,000 G.

96 URINE, GUINEA PIG—(Continued)

CLEAR								TURBID					
A% by wt.	ρ D_4^{20}	D_{20}^{20}	C_s g/l	C_w g/l	$(C_o - C_w)$ g/l	$(n - n_o)$ × 10^4	n	A% by wt.	ρ D_4^{20}	D_{20}^{20}	C_s g/l	C_w g/l	$(C_o - C_w)$ g/l
3.5	1.015	1.016	36	979	19	55	1.3385	3.6	1.015	1.017	37	978	20
4.0	1.017	1.019	41	976	22	63	1.3393	4.1	1.018	1.020	42	976	22
4.5	1.019	1.021	46	973	25	70	1.3400	4.6	1.020	1.022	47	973	25
5.0	1.022	1.024	51	971	27	78	1.3408	5.2	1.023	1.025	53	970	28
5.5	1.024	1.026	56	968	30	86	1.3416	5.7	1.026	1.027	58	968	30
6.0	1.026	1.028	62	964	34	94	1.3424	6.2	1.029	1.030	64	965	33
6.5	1.029	1.031	67	962	36	102	1.3432	6.7	1.031	1.032	69	962	36
7.0	1.031	1.033	72	959	39	110	1.3440	7.2	1.034	1.035	74	960	38
7.5	1.033	1.035	77	956	42	117	1.3447	7.8	1.036	1.038	81	955	43
8.0	1.036	1.038	83	953	45	125	1.3455	8.3	1.038	1.040	86	952	46
8.5	1.038	1.040	88	950	48	133	1.3463	8.8	1.040	1.042	92	948	50
9.0	1.040	1.042	94	946	52	141	1.3471	9.3	1.043	1.045	97	946	52
9.5	1.043	1.045	99	944	54	149	1.3479	9.8	1.046	1.047	103	943	55
10.0	1.045	1.047	104	941	57	157	1.3487	10.3	1.049	1.050	108	941	57
11.0	1.050	1.052	116	934	64	172	1.3502	11.4	1.054	1.055	120	934	64
12.0	1.054	1.056	126	928	70	188	1.3518	12.4	1.058	1.060	131	927	71
13.0	1.059	1.061	138	921	77	204	1.3534	13.4	1.064	1.066	143	921	77
14.0	1.064	1.066	149	915	83	220	1.3550	14.5	1.069	1.071	155	914	84
15.0	1.069	1.070	160	909	89	235	1.3565	15.5	1.074	1.076	166	908	90
16.0	1.073	1.075	172	901	97	251	1.3581	16.5	1.078	1.080	178	900	98
17.0	1.078	1.080	183	895	103	266	1.3596	17.6	1.084	1.086	191	893	105
18.0	1.083	1.085	195	888	110	282	1.3612	18.6	1.089	1.091	203	886	112
19.0	1.087	1.089	207	880	118	298	1.3628	19.6	1.094	1.096	214	880	118
20.0	1.092	1.094	218	874	124	314	1.3644	20.7	1.099	1.101	227	872	126

97 URINE, HUMAN

RELATIVE SPECIFIC REFRACTIVITY = 1.02

0.00 % by wt. data are the same for all compounds.
For Values of 0.00 wt. % solutions see Table 1, Acetic Acid.

A% by wt.	ρ D_4^{20}	D_{20}^{20}	C_s g/l	C_w g/l	$(C_o - C_w)$ g/l	$(n - n_o)$ × 10^4	n
0.50	1.000	1.002	5	996	2	8	1.3338
1.00	1.003	1.005	10	993	5	16	1.3346
1.50	1.005	1.007	15	990	8	24	1.3354
2.00	1.007	1.009	20	987	11	32	1.3362
2.50	1.009	1.011	24	984	14	40	1.3370
3.00	1.012	1.014	31	981	17	47	1.3377
3.50	1.014	1.016	36	978	20	55	1.3385
4.00	1.016	1.018	41	975	23	63	1.3393
4.50	1.018	1.020	46	972	26	71	1.3401
5.00	1.020	1.022	51	969	29	78	1.3408
5.50	1.022	1.024	56	966	32	86	1.3416
6.00	1.024	1.026	61	962	36	94	1.3424
6.50	1.026	1.028	67	959	39	101	1.3431
7.00	1.027	1.029	72	956	42	109	1.3439
7.50	1.029	1.031	77	952	46	116	1.3446
8.00	1.031	1.033	83	949	49	123	1.3453
8.50	1.033	1.035	88	945	53	131	1.3461
9.00	1.034	1.036	93	941	57	138	1.3468
9.50	1.036	1.038	98	937	61	145	1.3475
10.00	1.037	1.039	104	934	64	152	1.3482

98 URINE, RABBIT

RELATIVE SPECIFIC REFRACTIVITIES: CLEAR* = 0.997; TURBID = 0.958

0.00 % by wt. data are the same for all compounds.
For Values of 0.00 wt. % solutions see Table 1, Acetic Acid.

CLEAR								TURBID					
A% by wt.	ρ D_4^{20}	D_{20}^{20}	C_s g/l	C_w g/l	$(C_o - C_w)$ g/l	$(n - n_o)$ × 10^4	n	A% by wt.	ρ D_4^{20}	D_{20}^{20}	C_s g/l	C_w g/l	$(C_o - C_w)$ g/l
0.5	1.000	1.002	5	995	3	7	1.3337	0.6	1.002	1.003	6	996	2
1.0	1.003	1.004	10	993	5	15	1.3345	1.1	1.003	1.005	11	992	6
1.5	1.005	1.007	15	990	8	22	1.3352	1.7	1.006	1.008	17	989	9
2.0	1.007	1.009	20	987	11	30	1.3360	2.2	1.008	1.010	22	986	12
2.5	1.009	1.011	25	984	14	38	1.3368	2.7	1.011	1.012	27	984	14
3.0	1.011	1.013	30	981	17	45	1.3375	3.2	1.013	1.014	32	981	17
3.5	1.014	1.015	35	979	19	53	1.3383	3.8	1.016	1.017	39	977	21
4.0	1.016	1.018	41	975	23	60	1.3390	4.3	1.018	1.019	44	974	24
4.5	1.018	1.020	46	972	26	68	1.3398	4.9	1.020	1.021	50	970	28
5.0	1.020	1.022	51	969	29	76	1.3406	5.4	1.022	1.024	55	967	31
5.5	1.022	1.024	56	966	32	83	1.3413	5.9	1.024	1.026	60	964	34
6.0	1.025	1.026	62	963	35	91	1.3421	6.5	1.027	1.029	67	960	38
6.5	1.027	1.029	67	960	38	98	1.3428	7.0	1.029	1.031	72	957	41

* Cleared by centrifuging 15 minutes at 25,000 G.

CLEAR								TURBID					
A% by wt.	ρ D_4^{20}	D_{20}^{20}	C_s g/l	C_w g/l	$(C_o - C_w)$ g/l	$(n - n_o)$ × 10^4	n	A% by wt.	ρ D_4^{20}	D_{20}^{20}	C_s g/l	C_w g/l	$(C_o - C_w)$ g/l
7.0	1.029	1.031	72	957	41	106	1.3436	7.5	1.032	1.034	77	955	43
7.5	1.031	1.033	77	954	44	113	1.3443	8.0	1.034	1.036	83	951	47
8.0	1.033	1.035	83	950	48	121	1.3451	8.6	1.037	1.039	89	948	50
8.5	1.036	1.037	88	948	50	128	1.3458	9.1	1.039	1.041	95	944	54
9.0	1.038	1.040	93	945	53	136	1.3466	9.7	1.042	1.044	101	941	57
9.5	1.040	1.042	99	941	57	143	1.3473	10.2	1.044	1.046	106	938	60
10.0	1.042	1.044	104	938	60	151	1.3481	10.8	1.047	1.049	113	934	64
11.0	1.046	1.048	115	931	67	166	1.3496	11.8	1.051	1.053	124	927	71
12.0	1.051	1.053	126	925	73	181	1.3511	12.9	1.056	1.058	136	920	78
13.0	1.055	1.057	137	918	80	196	1.3526	14.0	1.061	1.063	149	912	86
14.0	1.060	1.062	148	912	86	211	1.3541	15.1	1.066	1.068	161	905	93
15.0	1.064	1.066	160	904	94	226	1.3556	16.2	1.071	1.073	174	897	101
16.0	1.068	1.070	171	897	101	242	1.3572	17.2	1.076	1.077	185	891	107
17.0	1.073	1.075	182	891	107	257	1.3587	18.3	1.080	1.082	198	882	116

99 ZINC SULFATE, $ZnSO_4 \cdot 7H_2O$

MOLECULAR WEIGHT = 161.44 FORMULA WEIGHT, HYDRATE = 287.56
RELATIVE SPECIFIC REFRACTIVITY = 0.493

0.00 % by wt. data are the same for all compounds.
For Values of 0.00 wt. % solutions see Table 1, Acetic Acid.

A% by wt.	H% by wt.	ρ D_4^{20}	D_{20}^{20}	C_s g/l	M g-mol/l	C_w g/l	$(C_o - C_w)$ g/l	$(n - n_o)$ × 10^4	n	Δ °C	O Os/kg	S g-mol/l	η/η_o	η/ρ cS	φ rhe	γ mmho/cm	T g-mol/l
0.50	0.89	1.0034	1.0051	5.0	0.031	998.3	−0.1	9	1.3339	0.079	0.043	0.022	1.019	1.017	97.97	2.8	0.028
1.00	1.78	1.0085	1.0103	10.1	0.062	998.4	−0.2	18	1.3348	0.150	0.081	0.042	1.038	1.031	96.15	5.4	0.054
1.50	2.67	1.0137	1.0155	15.2	0.094	998.5	−0.3	27	1.3357	0.217	0.117	0.062	1.058	1.046	94.32	7.8	0.080
2.00	3.56	1.0190	1.0208	20.4	0.126	998.6	−0.4	36	1.3366	0.282	0.152	0.081	1.079	1.061	92.49	10.0	0.104
2.50	4.45	1.0243	1.0261	25.6	0.159	998.7	−0.4	45	1.3375	0.346	0.186	0.099	1.101	1.077	90.65	12.1	0.127
3.00	5.34	1.0296	1.0314	30.9	0.191	998.7	−0.5	54	1.3384	0.408	0.219	0.118	1.124	1.094	88.80	13.9	0.148
3.50	6.23	1.0349	1.0368	36.2	0.224	998.7	−0.5	64	1.3393	0.470	0.253	0.136	1.148	1.111	86.94	15.7	0.169
4.00	7.12	1.0403	1.0421	41.6	0.258	998.7	−0.5	73	1.3403	0.531	0.286	0.154	1.173	1.130	85.10	17.3	0.188
4.50	8.02	1.0457	1.0475	47.1	0.291	998.6	−0.4	82	1.3412	0.591	0.318	0.172	1.198	1.148	83.27	18.9	0.207
5.00	8.91	1.0511	1.0530	52.6	0.326	998.5	−0.3	91	1.3421	0.651	0.350	0.189	1.225	1.168	81.47	20.5	0.226
5.50	9.80	1.0565	1.0584	58.1	0.360	998.4	−0.2	100	1.3430	0.711	0.382	0.207	1.252	1.188	79.70	22.0	0.245
6.00	10.69	1.0620	1.0639	63.7	0.395	998.3	0.0	109	1.3439	0.772	0.415	0.225	1.280	1.208	77.98	23.5	0.263
6.50	11.58	1.0675	1.0694	69.4	0.430	998.1	0.1	118	1.3448	0.833	0.448	0.243	1.308	1.228	76.28	24.9	0.280
7.00	12.47	1.0730	1.0749	75.1	0.465	997.9	0.3	127	1.3457	0.893	0.480	0.260	1.338	1.249	74.60	26.3	0.297
7.50	13.36	1.0786	1.0805	80.9	0.501	997.7	0.5	136	1.3466	0.953	0.512	0.278	1.368	1.271	72.95	27.6	0.314
8.00	14.25	1.0842	1.0861	86.7	0.537	997.5	0.8	146	1.3475	1.013	0.544	0.296	1.400	1.294	71.30	28.9	0.330
8.50	15.14	1.0899	1.0918	92.6	0.574	997.2	1.0	155	1.3485	1.073	0.577	0.313	1.433	1.317	69.65	30.1	0.345
9.00	16.03	1.0956	1.0975	98.6	0.611	997.0	1.3	164	1.3494	1.136	0.611	0.332	1.467	1.342	68.01	31.3	0.360
9.50	16.92	1.1013	1.1032	104.6	0.648	996.7	1.6	173	1.3503	1.200	0.645	0.351	1.504	1.368	66.37	32.5	0.375
10.00	17.81	1.1071	1.1091	110.7	0.686	996.4	1.8	183	1.3513	1.267	0.681	0.370	1.542	1.396	64.72	33.7	0.390
11.00	19.59	1.1188	1.1208	123.1	0.762	995.8	2.5	202	1.3532	1.407	0.756	0.411	1.624	1.454	61.45	35.9	0.418
12.00	21.37	1.1308	1.1328	135.7	0.841	995.1	3.1	221	1.3551	1.553	0.835	0.454	1.713	1.518	58.25	38.0	0.445
13.00	23.16	1.1429	1.1450	148.6	0.920	994.4	3.9	241	1.3570	1.711	0.920	0.500	1.810	1.587	55.14	39.9	0.470
14.00	24.94	1.1553	1.1573	161.7	1.002	993.6	4.7	260	1.3590	1.889	1.015	0.552	1.914	1.660	52.14	41.7	0.492
15.00	26.72	1.1679	1.1699	175.2	1.085	992.7	5.6	280	1.3610	2.087	1.122	0.610	2.027	1.739	49.24	43.3	0.513
16.00	28.50	1.1806	1.1827	188.9	1.170	991.7	6.5	300	1.3630	2.306	1.240	0.673	2.148	1.823	46.46	44.6	0.531

OSMOTIC PARAMETERS AND ELECTRICAL CONDUCTIVITIES OF AQUEOUS SOLUTIONS

For any aqueous solutions within the range, the two right hand columns provide values for either condosity (T) or specific conductance (γ) where the other value is given; the three left hand columns provide values for freezing point depression (Δ), osmolality (O) or osmosity (S) where any one of these values is given. Values in all four columns are interconvertible *only* for solutions of sodium chloride, in which case, by definition, osmosity, molarity (M) and condosity are always equal. Other concentrative properties of specific solutions may be found in conversion tables on preceding pages.

Table columns are:

Δ = freezing point depression, °C
O = osmolality (= Δ/1.86, Os/kg
S = osmosity, molar concentration of NaCl solution having same freezing point or osmotic pressure as given solution, g-mol/l
M = molarity of NaCl solution, g-mol/l
T = condosity, molar concentration of NaCl solution having same specific conductance at 20°C as given solution, g-mol/l
γ = specific conductance (electrical) at 20°C, mmho/cm.

Δ °C	O Os/Kg	S,M,T, g-mol/l	γ mmho/cm	Δ °C	O Os/Kg	S,M,T, g-mol/l	γ mmho/cm
0.000	0.000	0.000	0.0	1.881	1.011	0.550	46.1
0.037	0.020	0.010	1.0	1.916	1.030	0.560	46.8
0.072	0.039	0.020	2.0	1.950	1.048	0.570	47.6
0.107	0.057	0.030	3.0	1.985	1.067	0.580	48.3
0.142	0.076	0.040	4.0	2.019	1.086	0.590	49.1
0.176	0.095	0.050	5.0	2.054	1.104	0.600	49.8
0.211	0.113	0.060	5.9	2.088	1.123	0.610	50.6
0.245	0.132	0.070	6.9	2.123	1.141	0.620	51.3
0.279	0.150	0.080	7.8	2.157	1.160	0.630	52.0
0.314	0.169	0.090	8.7	2.192	1.178	0.640	52.8
0.348	0.187	0.100	9.7	2.226	1.197	0.650	53.5
0.382	0.205	0.110	10.6	2.261	1.216	0.660	54.2
0.416	0.224	0.120	11.4	2.296	1.234	0.670	55.0
0.450	0.242	0.130	12.3	2.330	1.253	0.680	55.7
0.484	0.260	0.140	13.2	2.365	1.271	0.690	56.4
0.518	0.279	0.150	14.1	2.400	1.290	0.700	57.1
0.552	0.297	0.160	14.9	2.434	1.309	0.710	57.9
0.586	0.315	0.170	15.8	2.469	1.328	0.720	58.6
0.620	0.333	0.180	16.6	2.504	1.346	0.730	59.3
0.654	0.352	0.190	17.5	2.539	1.365	0.740	60.0
0.688	0.370	0.200	18.3	2.574	1.384	0.750	60.7
0.722	0.388	0.210	19.2	2.609	1.403	0.760	61.4
0.756	0.407	0.220	20.0	2.644	1.421	0.770	62.1
0.790	0.425	0.230	20.8	2.679	1.440	0.780	62.9
0.824	0.443	0.240	21.6	2.714	1.459	0.790	63.6
0.858	0.461	0.250	22.4	2.749	1.478	0.800	64.3
0.892	0.480	0.260	23.3	2.784	1.497	0.810	65.0
0.926	0.498	0.270	24.1	2.819	1.515	0.820	65.7
0.960	0.516	0.280	24.9	2.854	1.534	0.830	66.3
0.994	0.534	0.290	25.7	2.889	1.553	0.840	67.0
1.028	0.553	0.300	26.5	2.924	1.572	0.850	67.7
1.062	0.571	0.310	27.3	2.959	1.591	0.860	68.4
1.096	0.589	0.320	28.1	2.995	1.610	0.870	69.1
1.130	0.607	0.330	28.9	3.030	1.629	0.880	69.8
1.164	0.626	0.340	29.7	3.065	1.648	0.890	70.5
1.198	0.644	0.350	30.5	3.101	1.667	0.900	71.2
1.232	0.662	0.360	31.3	3.136	1.686	0.910	71.8
1.266	0.681	0.370	32.1	3.171	1.705	0.920	72.5
1.300	0.699	0.380	32.9	3.207	1.724	0.930	73.2
1.334	0.717	0.390	33.7	3.242	1.743	0.940	73.9
1.368	0.736	0.400	34.4	3.278	1.762	0.950	74.5
1.402	0.754	0.410	35.2	3.313	1.781	0.960	75.2
1.436	0.772	0.420	36.0	3.349	1.801	0.970	75.9
1.470	0.791	0.430	36.8	3.385	1.820	0.980	76.5
1.505	0.809	0.440	37.6	3.420	1.839	0.990	77.2
1.539	0.827	0.450	38.4	3.456	1.858	1.000	77.8
1.573	0.846	0.460	39.2	3.492	1.877	1.010	78.5
1.607	0.864	0.470	39.9	3.527	1.896	1.020	79.2
1.641	0.882	0.480	40.7	3.563	1.916	1.030	79.8
1.676	0.901	0.490	41.5	3.599	1.935	1.040	80.5
1.710	0.919	0.500	42.3	3.635	1.954	1.050	81.1
1.744	0.938	0.510	43.0	3.671	1.974	1.060	81.8
1.778	0.956	0.520	43.8	3.707	1.993	1.070	82.4
1.813	0.975	0.530	44.6	3.743	2.012	1.080	83.1
1.847	0.993	0.540	45.3	3.779	2.032	1.090	83.7

Δ °C	O Os/Kg	S,M,T, g-mol/l	γ mmho/cm	Δ °C	O Os/Kg	S,M,T, g-mol/l	γ mmho/cm
3.815	2.051	1.100	84.3	5.86	3.15	1.650	117
3.851	2.070	1.110	85.0	5.89	3.17	1.660	117
3.887	2.090	1.120	85.6	5.93	3.19	1.670	118
3.923	2.109	1.130	86.2	5.97	3.21	1.680	119
3.959	2.129	1.140	86.9	6.01	3.23	1.690	119
3.995	2.148	1.150	87.5	6.05	3.25	1.700	120
4.032	2.168	1.160	88.1	6.09	3.27	1.710	120
4.068	2.187	1.170	88.8	6.13	3.29	1.720	121
4.104	2.207	1.180	89.4	6.16	3.31	1.730	121
4.141	2.226	1.190	90.0	6.20	3.34	1.740	122
4.177	2.246	1.200	90.6	6.24	3.36	1.750	122
4.214	2.265	1.210	91.3	6.28	3.38	1.760	123
4.250	2.285	1.220	91.9	6.32	3.40	1.770	123
4.287	2.305	1.230	92.5	6.36	3.42	1.780	124
4.323	2.324	1.240	93.1	6.40	3.44	1.790	125
4.360	2.344	1.250	93.7	6.44	3.46	1.800	125
4.393	2.362	1.260	94.3	6.48	3.48	1.810	126
4.429	2.381	1.270	94.9	6.52	3.50	1.820	126
4.466	2.401	1.280	95.6	6.56	3.52	1.830	127
4.503	2.421	1.290	96.2	6.60	3.55	1.840	127
4.539	2.441	1.300	96.8	6.64	3.57	1.850	128
4.576	2.460	1.310	97.4	6.67	3.59	1.860	128
4.613	2.480	1.320	98.0	6.71	3.61	1.870	129
4.650	2.500	1.330	98.6	6.75	3.63	1.880	129
4.687	2.520	1.340	99.2	6.79	3.65	1.890	129
4.724	2.540	1.350	99.8	6.83	3.67	1.900	130
4.761	2.560	1.360	100	6.87	3.70	1.910	130
4.798	2.580	1.370	101	6.91	3.72	1.920	131
4.835	2.600	1.380	102	6.95	3.74	1.930	131
4.872	2.620	1.390	102	6.99	3.76	1.940	132
4.910	2.640	1.400	103	7.03	3.78	1.950	132
4.947	2.660	1.410	103	7.07	3.80	1.960	133
4.984	2.680	1.420	104	7.11	3.82	1.970	133
5.02	2.70	1.430	104	7.15	3.85	1.980	134
5.06	2.72	1.440	105	7.19	3.87	1.990	134
5.10	2.74	1.450	106	7.23	3.89	2.000	135
5.13	2.76	1.460	106	7.27	3.91	2.010	135
5.17	2.78	1.470	107	7.32	3.93	2.020	136
5.21	2.80	1.480	107	7.36	3.95	2.030	136
5.25	2.82	1.490	108	7.40	3.98	2.040	137
5.28	2.84	1.500	109	7.44	4.00	2.050	137
5.32	2.86	1.510	109	7.48	4.02	2.060	137
5.36	2.88	1.520	110	7.52	4.04	2.070	138
5.40	2.90	1.530	110	7.56	4.06	2.080	138
5.44	2.92	1.540	111	7.60	4.09	2.090	139
5.47	2.94	1.550	111	7.64	4.11	2.100	139
5.51	2.96	1.560	112	7.68	4.13	2.110	140
5.55	2.98	1.570	112	7.72	4.15	2.120	140
5.59	3.00	1.580	113	7.76	4.17	2.130	141
5.63	3.02	1.590	114	7.81	4.20	2.140	141
5.66	3.05	1.600	114	7.85	4.22	2.150	142
5.70	3.07	1.610	115	7.89	4.24	2.160	142
5.74	3.09	1.620	115	7.93	4.26	2.170	142
5.78	3.11	1.630	116	7.97	4.29	2.180	143
5.82	3.13	1.640	116	8.01	4.31	2.190	143

Δ °C	O Os/Kg	S,M,T, g-mol/l	γ mmho/cm	Δ °C	O Os/Kg	S,M,T, g-mol/l	γ mmho/cm
8.05	4.33	2.200	144	10.45	5.62	2.750	168
8.10	4.35	2.210	144	10.50	5.64	2.760	168
8.14	4.38	2.220	145	10.54	5.67	2.770	169
8.18	4.40	2.230	145	10.59	5.69	2.780	169
8.22	4.42	2.240	146	10.63	5.72	2.790	170
8.26	4.44	2.250	146	10.68	5.74	2.800	170
8.31	4.47	2.260	147	10.73	5.77	2.810	170
8.35	4.49	2.270	147	10.77	5.79	2.820	171
8.39	4.51	2.280	148	10.82	5.82	2.830	171
8.43	4.53	2.290	148	10.86	5.84	2.840	172
8.47	4.56	2.300	148	10.91	5.87	2.850	172
8.52	4.58	2.310	149	10.96	5.89	2.860	172
8.56	4.60	2.320	149	11.00	5.92	2.870	173
8.60	4.62	2.330	150	11.05	5.94	2.880	173
8.64	4.65	2.340	150	11.10	5.97	2.890	173
8.69	4.67	2.350	151	11.14	5.99	2.900	174
8.73	4.69	2.360	151	11.19	6.02	2.910	174
8.77	4.72	2.370	152	11.24	6.04	2.920	175
8.82	4.74	2.380	152	11.28	6.07	2.930	175
8.86	4.76	2.390	153	11.33	6.09	2.940	175
8.90	4.79	2.400	153	11.38	6.12	2.950	176
8.94	4.81	2.410	153	11.43	6.14	2.960	176
8.99	4.83	2.420	154	11.47	6.17	2.970	176
9.03	4.85	2.430	154	11.52	6.19	2.980	177
9.07	4.88	2.440	155	11.57	6.22	2.990	177
9.12	4.90	2.450	155	11.62	6.25	3.000	178
9.16	4.92	2.460	156	11.71	6.30	3.020	178
9.20	4.95	2.470	156	11.81	6.35	3.040	179
9.25	4.97	2.480	156	11.90	6.40	3.060	180
9.29	5.00	2.490	157	12.00	6.45	3.080	181
9.33	5.02	2.500	157	12.10	6.50	3.100	181
9.38	5.04	2.510	158	12.20	6.56	3.120	182
9.42	5.07	2.520	158	12.29	6.61	3.140	183
9.47	5.09	2.530	159	12.39	6.66	3.160	183
9.51	5.11	2.540	159	12.49	6.71	3.180	184
9.55	5.14	2.550	160	12.59	6.77	3.200	185
9.60	5.16	2.560	160	12.69	6.82	3.220	185
9.64	5.18	2.570	160	12.79	6.87	3.240	186
9.69	5.21	2.580	161	12.89	6.93	3.260	187
9.73	5.23	2.590	161	12.99	6.98	3.280	187
9.78	5.26	2.600	162	13.09	7.04	3.300	188
9.82	5.28	2.610	162	13.19	7.09	3.320	189
9.86	5.30	2.620	163	13.29	7.15	3.340	189
9.91	5.33	2.630	163	13.39	7.20	3.360	190
9.95	5.35	2.640	163	13.49	7.26	3.380	190
10.00	5.38	2.650	164	13.60	7.31	3.400	191
10.04	5.40	2.660	164	13.70	7.37	3.420	192
10.09	5.42	2.670	165	13.80	7.42	3.440	192
10.13	5.45	2.680	165	13.91	7.48	3.460	193
10.18	5.47	2.690	165	14.01	7.53	3.480	193
10.22	5.50	2.700	166	14.12	7.59	3.500	194
10.27	5.52	2.710	166	14.22	7.65	3.520	195
10.31	5.55	2.720	167	14.33	7.70	3.540	195
10.36	5.57	2.730	167	14.43	7.76	3.560	196
10.41	5.59	2.740	168	14.54	7.82	3.580	196

Δ °C	O Os/Kg	S,M,T, g-mol/l	γ mmho/cm	Δ °C	O Os/Kg	S,M,T, g-mol/l	γ mmho/cm
14.65	7.87	3.600	197	18.67	10.04	4.300	212
14.75	7.93	3.620	197	18.79	10.10	4.320	213
14.86	7.99	3.640	198	18.92	10.17	4.340	213
14.97	8.05	3.660	198	19.04	10.24	4.360	213
15.08	8.11	3.680	199	19.17	10.30	4.380	214
15.19	8.16	3.700	199	19.29	10.37	4.400	214
15.30	8.22	3.720	200	19.42	10.44	4.420	214
15.41	8.28	3.740	200	19.55	10.51	4.440	215
15.52	8.34	3.760	201	19.67	10.58	4.460	215
15.63	8.40	3.780	201	19.80	10.65	4.480	215
15.74	8.46	3.800	202	19.93	10.72	4.500	216
15.85	8.52	3.820	202	20.06	10.78	4.520	216
15.96	8.58	3.840	203	20.19	10.85	4.540	216
16.07	8.64	3.860	203	20.32	10.92	4.560	216
16.19	8.70	3.880	204	20.45	10.99	4.580	217
16.30	8.76	3.900	204	20.58	11.07	4.600	217
16.41	8.82	3.920	205	20.71	11.14	4.620	217
16.53	8.89	3.940	205	20.85	11.21	4.640	218
16.64	8.95	3.960	206	20.98	11.28	4.660	218
16.76	9.01	3.980	206	21.11	11.35	4.680	218
16.87	9.07	4.000	206			4.700	218
16.99	9.13	4.020	207			4.750	219
17.11	9.20	4.040	207			4.800	220
17.22	9.26	4.060	208			4.850	220
17.34	9.32	4.080	208			4.900	221
17.46	9.39	4.100	209			4.950	221
17.58	9.45	4.120	209			5.000	222
17.70	9.51	4.140	209			5.050	222
17.82	9.58	4.160	210			5.100	223
17.94	9.64	4.180	210			5.150	223
18.06	9.71	4.200	210			5.200	224
18.18	9.77	4.220	211			5.250	224
18.30	9.84	4.240	211			5.300	225
18.42	9.90	4.260	212				
18.55	9.97	4.280	212				

SPECIFIC GRAVITY OF AQUEOUS INVERT SUGAR SOLUTIONS

t = 20 C

Invert sugar is the equimolar mixture of fructose and glucose obtained by hydrolyzing sucrose.

Wt % (Vacuum)	Specific Gravity	G solute per liter of solution	lb solute per cu ft of solution	lb solute per gallon (U.S.) of solution
1	1.00211	10.021	0.62561	0.08363
2	1.00601	20.120	1.25609	0.16790
3	1.00994	30.298	1.89150	0.25284
4	1.01389	40.556	2.53191	0.33844
5	1.01786	50.893	3.17725	0.42470
6	1.02186	61.312	3.82771	0.51165
7	1.02589	71.812	4.48322	0.59927
8	1.02995	82.396	5.14398	0.68759
9	1.03403	93.063	5.80992	0.77661
10	1.03814	103.814	6.48111	0.86633
11	1.04227	114.649	7.15754	0.95675
12	1.04644	125.573	7.83952	1.04791
13	1.05063	136.582	8.52681	1.13978
14	1.05484	147.678	9.21954	1.23237
15	1.05909	158.864	9.91788	1.32572
16	1.06337	170.139	10.62178	1.41981
17	1.06767	181.504	11.33129	1.51465
18	1.07200	192.960	12.04649	1.61025
19	1.07637	204.510	12.76756	1.70664
20	1.08076	216.152	13.49437	1.80379
21	1.08518	227.888	14.22705	1.90173
22	1.08963	239.719	14.96566	2.00046
23	1.09411	251.645	15.71020	2.09998
24	1.09862	263.669	16.46086	2.20032
25	1.10316	275.790	17.21757	2.30147
26	1.10773	288.010	17.98046	2.40344
27	1.11233	300.329	18.74954	2.50625
28	1.11697	312.752	19.52511	2.60992
29	1.12163	325.273	20.30679	2.71440
30	1.12633	337.899	21.09503	2.81977
31	1.13105	350.626	21.88958	2.92597
32	1.13581	363.459	22.69075	3.03307
33	1.14060	376.398	23.49853	3.14104
34	1.14542	389.443	24.31293	3.24990
35	1.15027	402.595	25.13401	3.35966
36	1.15516	415.858	25.96201	3.47034
37	1.16007	429.226	26.79658	3.58199
38	1.16502	442.708	27.63826	3.69440
39	1.17000	456.300	28.48681	3.80728
40	1.17502	470.008	29.34260	3.92222
41	1.18006	483.825	30.20519	4.03752
42	1.18514	497.759	31.07509	4.15380
43	1.19025	511.808	31.95217	4.27104
44	1.19539	525.972	32.83643	4.38924
45	1.20057	540.257	33.72824	4.50844
46	1.20577	554.654	34.62705	4.62859
47	1.21101	569.175	35.53360	4.74977
48	1.21629	583.819	36.44782	4.87197
49	1.22159	598.579	37.36929	4.99514
50	1.22693	613.465	38.29862	5.11937
51	1.23230	628.473	39.23557	5.24461
52	1.23771	643.609	40.18051	5.37092
53	1.24315	658.870	41.13325	5.49827
54	1.24861	674.249	42.09337	5.62661
55	1.25412	689.766	43.06209	5.75610
56	1.25965	705.404	44.03837	5.88660
57	1.26522	721.175	45.02296	6.01821
58	1.27082	737.076	46.01565	6.15090
59	1.27646	753.111	47.01672	6.28471
60	1.28112	768.672	47.98819	6.41457
61	1.28782	785.570	49.04310	6.55558
62	1.29355	802.001	50.06892	6.69270
63	1.29932	818.572	51.10345	6.83098
64	1.30511	835.270	52.14591	6.97033
65	1.31094	852.111	53.19729	7.11087
66	1.31680	869.088	54.25716	7.25254
67	1.32269	886.202	55.32559	7.39536
68	1.32862	903.462	56.40313	7.53939
69	1.33458	920.860	57.48929	7.68458
70	1.34057	938.399	58.58425	7.83094
71	1.34659	956.079	59.68801	7.79848
72	1.35264	973.901	60.80064	8.12720
73	1.35872	991.866	61.92219	8.27712
74	1.36484	1009.982	63.05318	8.42830
75	1.37099	1028.243	64.19321	8.58069
76	1.37717	1046.649	65.34230	8.73429
77	1.38337	1065.195	66.50012	8.88905
78	1.38962	1083.904	67.66813	9.04518
79	1.39589	1102.753	68.84487	9.20247
80	1.40219	1121.752	70.03098	9.36102
81	1.40852	1140.901	71.22645	9.52082
82	1.41488	1160.202	72.43141	9.68189
83	1.42128	1179.662	73.64630	9.84428
84	1.42770	1199.268	74.87030	10.00790
85	1.43415	1219.028	76.10392	10.17279

HEATS OF COMBUSTION

For Organic Compounds

The heat of combustion is given in kilogram calories per gram molecular weight of the substance when combustion takes place at atmospheric pressure and at either 20°C or 25°C. If the data are for 20°C there is no asterisk for the numerical value of the heat of combustion. If the numerical value is for 25°C there is an asterisk marking the numerical value of the heat of combustion. The final products of combustion are gaseous carbon dioxide, liquid water and nitrogen gas for C, H, N compounds. For method of computing heats of formation see statement following this table.

Name	Formula	Physical state	Heat of combustion, kg. calories
Acetaldehyde	CH_3CHO	liquid	278.77*
Acetamide	CH_3CONH_2	solid	282.6
Acetanilide	$CH_3CONHC_6H_5$	solid	1,010.4
Acetic acid	CH_3CO_2H	liquid	209.02*
Acetic anhydride	$(CH_3CO)_2O$	liquid	431.70*
Acetone	$(CH_3)_2CO$	liquid	427.92*
Acetonitrile	CH_3CN	liquid	302.4
Acetophenone	$C_6H_5COCH_3$	liquid	991.60*
Acetylacetone	$CH_3COCH_2COCH_3$	liquid	615.9
Acetylene	$(CH)_2$	gas	310.61*
Acrolein	$CH_2{:}CHCHO$	liquid	389.6
Acrylic acid	$CH_2{:}CHCO_2H$	liquid	327.0*
Adipic acid	$(CH_2)_4(CO_2H)_2$	solid	668.29*
Alanine	$CH_3CH(NH_2)CO_2H$	solid	387.7
Aldol, *see β-hydroxybutyr-aldehyde*			
Alizarin, *see Dihydroxyanthraquinone*			
Allyl alcohol	$CH_2{:}CHCH_2OH$	liquid	442.4
Allylene	$CH_3C{:}CH$	gas	465.1
p-Aminoazobenzene	$H_2NC_6H_4N_2C_6H_5$	solid	1,574.0
p-Aminophenol	$HOC_6H_4NH_2$	solid	760.0
Amygdalin	$C_{20}H_{27}O_{11}N$	solid	2,348.4
Amyl acetate	$C_4H_9CO_2C_2H_5$	liquid	1,042.5
Amyl alcohol	$(CH_3)_2CHCH_2CH_2OH$	liquid	793.7
Amylene	C_5H_{10}	liquid	803.4
Anethole	$C_{10}H_{12}O$	solid	1,324.4
Aniline	$C_6H_5NH_2$	liquid	811.7
p-Anisidine	$CH_3OC_6H_4NH_2$	solid	924.0
Anisole	$C_6H_5OCH_3$	liquid	905.1
Anthracene	$C_{14}H_{10}$	solid	1,712.0*
Anthraquinone	$C_{14}H_8O_2$	solid	1,544.5
Arabinose	$C_5H_{10}O_5$	solid	559.9
Arabitol	$C_5H_{12}O_5$	solid	661.2
Arachidic acid	$C_{20}H_{40}O_2$	solid	3,025.9
Azelaic acid	$(CH_2)_7(CO_2H)_2$	solid	1,141.7
Azobenzene	$(C_6H_5N)_2$	solid	1,545.9
Azoxybenzene	$(C_6H_5N)_2O$	solid	1,534.5
Behenic acid	$C_{22}H_{44}O_2$	solid	3,338.4
Benzalacetone	$C_6H_5CH{:}CHCOCH_3$	solid	1,257.4
Benzaldehyde	C_6H_5CHO	liquid	843.2*
Benzamide	$C_6H_5CONH_2$	solid	847.6
Benzanilide	$C_6H_5CONHC_6H_5$	solid	1,575.5
Benzene	C_6H_6	liquid	780.96*
Benzenediazonium nitrate	$C_6H_5N_2NO_3$	solid	782.6
Benzidine	$(C_6H_4NH_2)_2$	solid	1,560.9
Benzil	$(C_6H_5CO)_2$	solid	1,624.6
* Benzoic acid	$C_6H_5CO_2H$	solid	771.24*
Benzoic anhydride	$(C_6H_5CO)_2O$	solid	1,555.1
Benzoin	$C_6H_5.CHOH.COC_6H_5$	solid	1,671.4
Benzonitrile	C_6H_5CN	liquid	865.5
Benzophenone	$(C_6H_5)_2CO$	solid	1,556.5
Benzoyl chloride	C_6H_5COCl	liquid	782.8
Benzoyl peroxide	$(C_6H_5CO)_2O_2$	solid	1,551.7
Benzyl alcohol	$C_6H_5CH_2OH$	liquid	894.3
Benzylamine	$C_6H_5CH_2NH_2$	liquid	969.4
Benzyl carbylamine	$C_6H_5CH_2NC$	liquid	1,046.5
Benzyl chloride	$C_6H_5CH_2Cl$	liquid	886.4
Benzyl cyanide	$C_6H_5CH_2CN$	liquid	1,023.5
Borneol	$C_{10}H_{18}O$	liquid	1,469.6
Brucine	$C_{23}H_{26}O_4N_2$	gas	687.68*
n-Butyl alcohol	C_4H_9OH	liquid	639.53*
tert.-Butyl alcohol, see *Trimethyl carbinol*			
n-Butylamine	$C_4H_9NH_2$	liquid	710.6
sec.-Butylamine	$(CH_3)(C_2H_5){:}CHNH_2$	liquid	713.0
tert.-Butylamine	$(CH_3)_3CNH_2$	liquid	716.0
tert.-Butylbenzene	$C_6H_5C(CH_3)_3$	liquid	1,400.4
n-Butyramide	$C_3H_7CONH_2$	solid	596.0
n-Butyric acid	$C_3H_7CO_2H$	liquid	521.87*
n-Butyronitrile	C_3H_7CN	liquid	613.3
Caffeine	$C_8H_{10}O_2N_4$	solid	1,014.2
Camphene	$C_{10}H_{16}$	solid	1,468.8
Camphor	$C_{10}H_{16}O$	solid	1,411.0
Cane sugar, *see Sucrose*			
Capric acid	$C_9H_{18}O_2$	solid	1,453.07*
Caproic acid	$C_5H_{11}CO_2H$	liquid	834.49*
Carbon disulfide	CS_2	liquid	246.6
Carbon subnitride	$(C.CN)_2$	solid	514.8
Carbon tetrachloride	CCl_4	liquid	37.3
Carbonyl sulfide	COS	gas	130.5
Carvacrol	$C_{10}H_{14}O$	liquid	1,354.5
Cetyl alcohol	$C_{16}H_{34}O$	solid	2,504.4
Cetyl palmitate	$C_{32}H_{64}O$	solid	4,872.8
Chloracetic acid	$ClCH_2CO_2H$	solid	171.0
o-Chlorobenzoic acid	$ClC_6H_4CO_2H$	solid	734.5
Chloroform	$CHCl_3$	liquid	89.2
Chrysene	$C_{18}H_{12}$	solid	2,139.1
Cinnamic acid (*trans*)	$C_6H_5CH{:}CHCO_2H$	solid	1,040.2
Cinnamic aldehyde	$C_6H_5CH{:}CHCHO$	liquid	1,112.3

*25 C

HEAT OF COMBUSTION (Continued)

For Organic Compounds

Name	Formula	Physical state	Heat of combustion, kg. calories
Cinnamic anhydride	$C_{18}H_{14}O_3$	solid	2,091.3
d-Citrene	$C_{10}H_{16}$	liquid	1,473.0
Citric acid (anhydr.)	$C_6H_8O_7$	solid	468.6*
Codeine	$C_{18}H_{21}O_3N.H_2O$	solid	2,327.6
Coniine	$C_8H_{17}N$	liquid	1,275.5
Creatine (anhydr.)	$C_4H_9O_2N_3$	solid	559.8
Creatinine	$C_4H_7ON_3$	solid	563.4
o-Cresol	$CH_3C_6H_4OH$	liquid	882.6
o-Cresol	$CH_3C_6H_4OH$	solid	882.72*
m-Cresol	$CH_3C_6H_4OH$	liquid	880.5
p-Cresol	$CH_3C_6H_4OH$	liquid	882.5
p-Cresol	$CH_3C_6H_4OH$	solid	883.99*
m-Cresolmethyl ether	$CH_3C_6H_4OCH_3$	liquid	1,057.0
Crotonaldehyde	C_3H_5CHO	liquid	542.1
Cyanoacetic acid	$NC\ CH_2CO_2H$	solid	298.8
Cyanogen	$(CN)_2$	gas	258.3
Cyclobutane	C_4H_8	liquid	650.22*
Cycloheptane	$(CH_2)_7$	liquid	1,087.3
Cycloheptanol	$CH_2(CH_2)_5CHOH$	liquid	1,050.2
Cycloheptene	C_7H_{12}	liquid	1,099.09*
Cyclohexane	$(CH_2)_6$	liquid	936.87*
Cyclohexanol	$CH_2(CH_2)_4CHOH$	liquid	890.7
Cyclohexene, *see Tetrahydrobenzene*			
Cyclopentane	$(CH_2)_5$	liquid	786.55*
Cyclopropane, *see Trimethylene*			
Cymene	$C_6H_4(CH_3)(CH_3CHCH_3)$—(1, 4)	liquid	1,402.8
Decahydronaphthalene (*cis*)	$C_{10}H_{18}$	liquid	1,502.5
Decahydronaphthalene (*trons*)	$C_{10}H_{18}$	liquid	1,499.5
Decane	$C_{10}H_{22}$	liquid	1,610.2
Dextrose, *see Glucose*			
Diallyl	$(CH_2:CHCH_2)_2$	vapor	903.4
Diamyl ether	$(C_5H_{11})_2O$	liquid	1,609.3
Diamylene	$C_{10}H_{20}$	liquid	1,582.2
Dibenzyl	$(C_6H_5CH_2)_2$	solid	1,810.6
Dibenzyl amine	$(C_6H_5CH_2)_2NH$	solid	1,853.0
o-Dichlorobenzene	$C_6H_4Cl_2$	liquid	671.8
Diethylacetic acid	$(C_2H_5)_2CHCO_2H$	liquid	830.8
Diethyl amine	$(C_2H_5)_2NH$	liquid	716.9
Diethylaniline	$C_6H_5N(C_2H_5)_2$	liquid	1,451.6
Diethyl carbonate	$CO(OC_2H_5)_2$	liquid	647.9
Diethyl ether	$(C_2H_5)_2O$	liquid	657.52*
Diethyl ketone	$(C_2H_5)_2CO$	liquid	735.6
Diethyl malonate	$CH_2(CO_2C_2H_5)_2$	liquid	860.4
Diethyl oxalate	$(CO_2C_2H_5)_2$	liquid	716.0
Diethyl succinate	$(CH_2CO_2C_2H_5)_2$	liquid	1,007.3
Dihydrobenzene	C_6H_8	liquid	847.8
Δ_1-Dihydronaphthalene	$C_{10}H_{10}$	liquid	1,296.3
Δ_1-Dihydronaphthalene	$C_{10}H_{10}$	solid	1,298.3
Dihydroxyanthraquinone	$C_{14}H_6O_2(OH)_2$—(1, 2)	solid	1,448.9
Diisoamyl	$[(CH_3)_2CHCH_2CH_2]_2$	liquid	1,615.8
Diisobutylene	$[(CH_3)_2CHCH_2]_2$	liquid	1,252.4
Diisopropyl	$[(CH_3)_2CH]_2$	vapor	993.9
Diisopropyl ketone	$[(CH_3)_2CH]_2CO$	liquid	1,045.5
Dimethyl amine	$(CH_3)_2NH$	liquid	416.7
Dimethylaniline	$C_6H_5N(CH_3)_2$	liquid	1,142.7
Dimethyl carbonate	$CO(OCH_3)_2$	liquid	340.8
Dimethyl ether	$(CH_3)_2O$	gas	
Dimethylethyl carbinol	$C_2H_5(CH_3)_2CHOH$	liquid	784.6
Dimethyl fumarate	$(CHCO_2CH_3)_2$	solid	663.3*
2, 5-Dimethylhexane	$(CH_3)_2CH.C_2H_4CH(CH_3)_2$	liquid	1,303.3
3, 4-Dimethylhexane	$[(C_2H_5)(CH_3)CH]_2$	liquid	1,303.7
Dimethyl maleate	$(CHCO_2CH_3)_2$	liquid	669.4*
Dimethyl oxalate	$(CO_2CH_3)_2$	liquid	400.2*
2, 2-Dimethylpentane	$(CH_3)_3C.C_3H_7$	liquid	1,148.9
2, 3-Dimethylpentane	$(CH_3)_2CHCH(CH_3)C_2H_5$	liquid	1,148.9
2, 4-Dimethylpentane	$(CH_3)_2CHCH_2CH(CH_3)_2$	liquid	1,148.9
3, 3-Dimethylpentane	$(CH_3)_2C(C_2H_5)_2$	liquid	1,147.9
Dimethyl phthalate	$C_6H_4(CO_2CH_3)_2$	liquid	1,119.7
Dimethyl succinate	$(CH_2CO_2CH_3)_2$	solid	706.3*
m-Dinitrobenzene	$C_6H_4(NO_2)_2$	solid	696.8
Dinitrophenol	$C_6H_3(OH)(NO_2)_2$—(1, 2, 4)	solid	648.0
Dinitrotoluene	$C_6H_3(CH_3)(NO_2)_2$—(1, 2, 4)	solid	852.8
Diphenyl	$(C_6H_5)_2$	solid	1,493.6
Diphenyl amine	$(C_6H_5)_2NH$	solid	1,536.2
Diphenyl carbinol	$(C_6H_5)_2CHOH$	solid	1,615.4
Diphenylmethane	$(C_6H_5)_2CH_2$	solid	1,655.0
Diphenylnitrosamine	$(C_6H_5)_2N.NO$	solid	1,532.6
Dipropargyl	$(CH:C.CH_2)_2$	vapor	882.9
Dipropyl ketone	$(C_3H_7)_2CO$	liquid	1,050.5
Dulcitol	$C_6H_{14}O_6$	solid	729.1
Durene	$C_6H_2(CH_3)_4$—(1, 2, 4, 5)	solid	1,393.6
Eicosane	$C_{20}H_{42}$	solid	3,183.1
Erythritol	$C_4H_{10}O_4$	solid	504.1
Ethane	C_2H_6	gas	372.81*
Ethine, see *Acetylene*			
Ethyl acetate	$CH_3CO_2C_2H_5$	liquid	536.9
Ethyl acetoacetate	$CH_3COCH_2CO_2C_2H_5$	liquid	690.8
Ethyl alcohol	C_2H_5OH	liquid	326.68*
Ethyl amine	$C_2H_5NH_2$	liquid	409.5*
Ethylaniline	$C_6H_5NHC_2H_5$	liquid	1,121.5
Ethylbenzene	$C_2H_5C_6H_5$	liquid	1,091.2

*25 C

HEAT OF COMBUSTION (Continued)

For Organic Compounds

Name	Formula	Physical state	Heat of combustion, kg. calories
Ethyl benzoate	$C_6H_5CO_2C_2H_5$	liquid	1,098.7
Ethyl bromide	C_2H_5Br	vapor	340.5
Ethyl *n*-butyrate	$C_3H_7CO_2C_2H_5$	liquid	851.2
Ethyl carbylamine	C_2H_5NC	liquid	477.1
Ethyl chloride	C_2H_5Cl	vapor	316.7
Ethylcycloheptane	$C_2H_5C_7H_{13}$	liquid	1,406.8
Ethyl formate	$HCO_2C_2H_5$	liquid	391.7
3-Ethylhexane	$(C_2H_5)_2CH.C_3H_7$	liquid	1,302.3
Ethyl iodide	C_2H_5I	liquid	356.0
Ethyl isobutyrate	$(CH_3)_2CHCH_2CO_2C_2H_5$	liquid	845.7
Ethyl isocyanate	C_2H_5NCO	liquid	424.5
Ethyl nitrate	$C_2H_5ONO_2$	vapor	322.4
Ethyl nitrite	C_2H_5ONO	vapor	332.6
3-Ethylpentane	$(C_2H_5)_3CH$	liquid	1,149.9
Ethyl propionate	$C_2H_5CO_2C_2H_5$	liquid	690.8
Ethyl salicylate	$HOC_6H_4CO_2C_2H_5$	liquid	1,051.2
Ethyl valerate	$C_4H_9CO_2C_2H_5$	liquid	1,017.5
Ethylene	$CH_2{:}CH_2$	gas	337.23*
Ethylene chloride	$(CH_2Cl)_2$	vapor	271.0
Ethylene diamine	$(CH_2NH_2)_2$	liquid	452.6
Ethylene glycol	$(CH_2OH)_2$	liquid	281.9
Ethylene iodide	$(CH_2I)_2$	solid	324.8
Ethylene oxide	CH_2CH_2O	liquid	302.1
Ethylidene chloride	CH_3CHCl_2	liquid	267.1
Eugenol	$C_{10}H_{12}O_2$	liquid	1,286.6
Fenchane	$C_{10}H_{18}$	liquid	1,502.6
Fluorene	$(C_6H_4)_2{:}CH_2$	solid	1,584.9
Fluorobenzene	C_6H_5F	liquid	747.2
Formaldehyde	CH_2O	gas	136.42*
Formamide	$HCONH_2$	solid	134.9
Formic acid	HCO_2H	liquid	60.86*
β-D-Fructose	$C_6H_{12}O_6$	solid	672.0*
Fumaric acid (*trans*)	$(CHCO_2H)_2$	solid	318.99*
Furfural	C_4H_3OCHO	liquid	559.5
α-D-Galactose	$C_6H_{12}O_6$	solid	670.1*
Gallic acid	$C_6H_2(OH)_3CO_2H$—(1, 3, 5, 6)	solid	633.7
α-D-Glucose	$C_6H_{12}O_6$	solid	669.94*
Glutaric acid	$(CH_2)_3(CO_2H)_2$	solid	514.08*
Glycerol	$(CH_2OH)_2CHOH$	liquid	397.0
Glyceryl tributyrate	$C_{15}H_{26}O_6$	liquid	1,941.1
Glycine	$H_2NCH_2CO_2H$	solid	232.67*
Glycogen	$(C_6H_{10}O_5)x$ per kg	solid	4,186.8
Glycollic acid	CH_2OHCO_2H	solid	166.1*
Glycylglycine	$C_4H_8O_3N_2$	solid	470.7
n-Heptaldehyde	$CH_3(CH_2)_5CHO$	liquid	1,062.2*
n-Heptane	C_7H_{16}	liquid	1,149.9
Heptine-1	$CH{:}C(CH_2)_4CH_3$	liquid	1,091.2
n-Heptyl alcohol	$CH_3(CH_2)_5CH_2OH$	liquid	1,104.9
Heptyl amine	$C_7H_{15}NH_2$	liquid	1,178.9
Heptylic acid	$C_7H_{14}O_2$	liquid	986.1
n-Hexane	C_6H_{14}	liquid	995.01*
Hexachlorbenzene	C_6Cl_6	solid	509.0
Hexachlorethane	C_2Cl_6	solid	110.0
Hexadecane	$C_{16}H_{34}$	solid	2,559.1
Hexahydronaphthalene	$C_{10}H_{14}$	liquid	1,419.3
Hexamethylbenzene	$C_6(CH_3)_6$	solid	1,711.9
Hexamethylenetetramine	$(CH_2)_6N_4$	solid	1,006.7
Hexamethylethane	$[(CH_3)_3C]_2$	solid	1,301.8
Hexyl amine	$C_6H_{13}NH_2$	liquid	1,022.2
Hexylene	C_6H_{12}	liquid	952.6
Hippuric acid	$C_6H_5CONHCH_2CO_2H$	solid	1,012.4
Hydantoic acid	$C_3H_6O_3N_2$	solid	308.6
Hydrazobenzene	$(C_6H_5NH)_2$	solid	1,597.3
Hydroquinol	$C_6H_4(OH)_2$	solid	681.78*
Hydroquinoldimethyl ether	$(CH_3O)_2C_6H_4$	solid	1,014.7
p-Hydroxyazobenzene	$HOC_6H_4N_2C_6H_5$	solid	1,502.0
o-Hydroxybenzaldehyde	$C_6H_4(OH)CHO$	liquid	796.4*
m-Hydroxybenzaldehyde	$C_6H_4(OH)CHO$	solid	788.7
p-Hydroxybenzaldehyde	$C_6H_4(OH)CHO$	solid	792.7
m-Hydroxybenzoic acid	$HOC_6H_4CO_2H$	solid	726.1
p-Hydroxybenzoic acid	$HOC_6H_4CO_2H$	solid	725.4
β-Hydroxybutyraldehyde	$CH_3CHOHCH_2CHO$	liquid	546.6
Indigo	$C_{16}H_{10}O_2N_2$	solid	1,815.0
Indole	C_8H_7N	solid	1,022.2
Inositol	$C_6H_{12}O_6$	solid	662.1
Iodoform	CHI_3	solid	161.9
Isoamyl amine	$(CH_3)_2CHC_2H_4NH_2$	liquid	866.8
Isobutane	$(CH_3)_3CH$	gas	683.4
Isobutyl alcohol	$(CH_3)_2CHCH_2OH$	liquid	638.2
Isobutyl amine	$C_4H_9NH_2$	liquid	713.6
Isobutylene	$(CH_3)_2C{:}CH_2$	gas	647.2
Isobutyraldehyde	$(CH_3)_2CHCHO$	vapor	596.8
Isobutyramide	$(CH_3)_2CHCONH_2$	solid	595.9
Isobutyric acid	$(CH_3)_2CHCO_2H$	liquid	517.4
Isoeugenol	$C_{10}H_{12}O_2$	liquid	1,277.6
Isopentane	C_5H_{12}	gas	843.5(?)
Isopentane	C_5H_{12}	liquid	838.3(?)
Isophthalic acid	$C_6H_4(CO_2H)_2$	solid	768.3
Isopropyl alcohol	$(CH_3)_2CHOH$	liquid	474.8
Isopropylbenzene	$(CH_3)_2CHC_6H_5$	liquid	1,247.3
Isopropyltoluene	$C_6H_4(CH_3)(CH_3CHC_3)$—(1, 3)	liquid	1,409.5

Isopropyltoluene, *see Cymene*

*25 C

FOR ORGANIC COMPOUNDS

Name	Formula	Physical state	Heat of combustion, kg. calories
Isosafrole	$C_{10}H_{10}O_2$	liquid	1,233.9
Lactic acid, DL	$CH_3CHOHCO_2H$	liquid	326.8*
Lactose (anhydr.)	$C_{12}H_{22}O_{11}$	solid	1,350.0*
Lauric acid	$C_{12}H_{24}O_2$	solid	1,763.25*
Leucine	$C_6H_{13}O_2N$	solid	855.6
d-Limonene	$C_{10}H_{16}$	liquid	1,471.2
Maleic acid (*cis*)	$(CHCO_2H)_2$	solid	323.89*
Maleic anhydride	$(CHCO)_2O$	solid	332.10*
l-Malic acid	$(CHOHCH_2):(CO_2H)_2$	solid	317.37*
Malonic acid	$CH_2(CO_2H)_2$	solid	205.82*
Maltose	$C_{12}H_{22}O_{11}$	solid	1,349.3*
Mandelic acid	$C_6H_5CHOHCO_2H$	solid	890.3
d-Mannitol	$C_6H_{14}O_6$	solid	727.6
Menthene	$C_{10}H_{18}$	liquid	1,523.2
Menthol	$C_{10}H_{20}O$	solid	1,508.8
Mesitylene	$(CH_3)_3C_6H_3$—(1, 3, 5)	liquid	1,243.6
Mesityl oxide	$(CH_3)_2C:CHCOCH_3$	liquid	846.7
Mesotartaric acid	$(CHOH)_2(CO_2H)_2$	solid	276.0
Methane	CH_4	gas	212.79*
Methyl acetate	$CH_3CO_2CH_3$	liquid	381.2
Methyl alcohol	CH_3OH	liquid	173.64*
Methyl amine	CH_3NH_2	liquid	253.5*
Methylaniline	$C_6H_5NHCH_3$	liquid	973.5
Methyl benzoate	$C_6H_5CO_2CH_3$	liquid	945.9*
Methyl bromide	CH_3Br	vapor	184.0
Methyl butyl ketone	$CH_3COC_4H_9$	liquid	895.2
Methyl *tert*-butyl ketone, *see Pinacoline*			
Methyl butyrate	$C_3H_7COOCH_3$	liquid	692.8
Methyl carbylamine	CH_3NC	liquid	320.1
Methyl chloride	CH_3Cl	gas	164.2
Methyl cinnamate	$C_{10}H_{10}O_2$	solid	1,213.0
Methylcyclobutane	$CH_3CHCH_2CH_2CH_2$	liquid	784.2
Methylcycloheptane	$CH_3C_7H_{13}$	liquid	1,244.5
Methylcyclohexane	$CH_3C_6H_{11}$	liquid	1,091.8
Methylcyclopentane	$CH_3CH.C_3H_6CH_2$	liquid	937.9
Methyldiethyl carbinol	$CH_3(C_2H_5)_2CHOH$	liquid	927.0
Methylene chloride	CH_2Cl_2	vapor	106.8
Methylene iodide	CH_2I_2	liquid	178.4
Methylethyl ether	$CH_3OC_2H_5$	vapor	503.69*
Methylethyl ketone	$CH_3COC_2H_5$	liquid	584.17*
Methyl formate	HCO_2CH_3	liquid	234.1*
2-Methylheptane	$(CH_3)_2CH.C_5H_{11}$	liquid	1,306.1
2-Methylhexane	$(CH_3)_2CHC_4H_9$	liquid	1,148.9
3-Methylhexane	$(C_2H_5)(CH_3)CHC_3H_7$	liquid	1,148.9
Methylhexyl ketone	$CH_3COC_6H_{13}$	liquid	1,205.1
Methyl iodide	CH_3I	liquid	194.7
Methyl isobutyrate	$(CH_3)_2CHCO_2CH_3$	liquid	694.2
Methyl isocyanate	CH_3NCO	liquid	269.4
Methylisopropyl ketone	$CH_3COCH(CH_3)_2$	liquid	733.9
Methyl lactate	$CH_3CHOHCO_2CH_3$	liquid	497.2
Methyl propionate	$C_2H_5CO_2CH_3$	vapor	552.3
Methylpropyl ketone	$CH_3COC_3H_7$	liquid	740.78*
Methyl salicylate	$HOC_6H_4CO_2CH_3$	liquid	898.6*
Milk sugar, *see Lactose*			
Morphine	$C_{17}H_{19}O_3N.H_2O$	solid	2,146.3
Mucic acid	$C_6H_{10}O_8$	solid	483.0*
Myristic acid	$C_{14}H_{28}O_2$	solid	2,073.91*
Naphthalene	$C_{10}H_8$	solid	1,231.8*
α-Naphthoic acid	$C_{10}H_7CO_2H$	solid	1,231.8
β-Naphthoic acid	$C_{10}H_7CO_2H$	solid	1,227.6
α-Naphthol	$C_{10}H_7OH$	solid	1,185.4
β-Naphthol	$C_{10}H_7OH$	solid	1,187.2
α-Naphthonitrile	$C_{10}H_7CN$	solid	1,326.2
β-Naphthonitrile	$C_{10}H_7CN$	solid	1,321.0
α-Naphthoquinone	$C_{10}H_6O_2$	solid	1,100.8
β-Naphthoquinone	$C_{10}H_6O_2$	solid	1,106.4
α-Naphthyl amine	$C_{10}H_7NH_2$	solid	1,263.5
β-Naphthyl amine	$C_{10}H_7NH_2$	solid	1,261.0
Narceine	$C_{23}H_{27}O_8N.2H_2O$	solid	2,802.9
Narcotine	$C_{22}H_{23}O_7N$	solid	2,644.5
Nicotine	$C_{10}H_{14}N_2$	liquid	1,427.7
o-Nitraniline	$C_6H_4(NH_2)(NO_2)$	solid	765.8
m-Nitraniline	$C_6H_4(NH_2)(NO_2)$	solid	765.2
p-Nitraniline	$C_6H_4(NH_2)(NO_2)$	solid	761.0
m-Nitrobenzaldehyde	$O_2NC_6H_4CHO$	solid	800.4
Nitrobenzene	$C_6H_5NO_2$	liquid	739.2
m-Nitrobenzoic acid	$O_2NC_6H_4CO_2H$	solid	729.1
Nitroethane	$C_2H_5NO_2$	liquid	322.2
Nitroglycerine, *see Trinitroglycerol*			
Nitromethane	CH_3NO_2	liquid	169.4
o-Nitrophenol	$HOC_6H_6NO_2$	solid	689.1
m-Nitrophenol	$HOC_5H_4NO_2$	solid	684.4
p-Nitrophenol	$HOC_6H_4NO_2$	solid	688.8
Nitropropane	$C_3H_7NO_2$	liquid	477.9
o-Nitrotoluene	$CH_3C_6H_4NO_2$	liquid	897.0
p-Nitrotoluene	$CH_3C_6H_6NO_2$	solid	888.6
Octahydronaphthalene	$C_{10}H_{16}$	liquid	1,461.7
n-Octane	C_8H_{18}	liquid	1,302.7
Octyl alcohol	$C_8H_{18}O$	liquid	1,262.0
Oleic acid	$C_{18}H_{34}O_2$	liquid	2,657.4*
Oxalic acid, *a*	$(CO_2H)_2$	solid	58.7*
Oxamide	$(CONH_2)_2$	solid	203.2
Palmitic acid	$C_{16}H_{32}O_2$	solid	2,384.76*

*25C

HEAT OF COMBUSTION (Continued)

For Organic Compounds

Name	Formula	Physical state	Heat of combustion, kg. calories
Papaverine	$C_{20}H_{21}O_4N$	solid	2,478.1
Pentamethylbenzene	$C_6H(CH_3)_5$	solid	1,554.0
n-Pentane	C_5H_{12}	gas	845.16*
n-Pentane	C_5H_{12}	liquid	838.78*
Phenacetin	$C_{10}H_{13}O_2N$	solid	1,285.2
Phenanthraquinone	$C_{14}H_8O_2$	solid	1,544.0
Phenanthrene	$C_{14}H_{10}$	solid	1,685.6*
Phenetole	$C_6H_5OC_2H_5$	liquid	1,060.3
Phenol	C_6H_5OH	solid	729.80*
Phenylacetic acid	$C_6H_5CH_2CO_2H$	solid	930.4*
Phenylacetylene	$C_6H_5C{:}CH$	liquid	1,024.2
Phenylalanine	$C_9H_{11}O_2N$	solid	1,111.3
p-Phenylenediamine	$C_6H_4(NH_2)_2$	solid	843.4
Phenylethylene, *see Styrene*			
Phenylglycine	$C_2H_5NHCH_2CO_2H$	solid	955.1
Phenylhydrazine	$C_6H_5N_2H_3$	solid	875.4
Phenylhydroxylamine	C_6H_5NHOH	liquid	803.7
Phenyl iodide	C_6H_5I	liquid	770.7
Phloroglucinol	$C_6H_3(OH)_3$	solid	635.7
o-Phthalic acid	$C_6H_4(CO_2H)_2$	solid	770.44*
Phthalic anhydride	$C_6H_4(CO)_2O$	solid	783.4
Phthalimide	$C_8H_5O_2N$	solid	849.5
Picric acid	$C_6H_2(OH)(NO_2)_3$—(1, 2, 4, 6)	solid	611.8
Pinacoline	$CH_3COC(CH_3)_3$	solid	891.8
Piperidine	$C_5H_{11}N$	liquid	826.6
Piperonal	$C_8H_6O_3$	solid	870.7
Propane	C_3H_8	gas	530.57*
Propine, *see Allylene*			
Propionaldehyde	C_2H_5CHO	liquid	434.1*
Propionamide	$C_2H_5CONH_2$	solid	439.9
Propionic acid	$C_2H_5CO_2H$	liquid	365.03*
Propionic anhydride	$(C_2H_5CO)_2O$	liquid	746.6
Propionitrile	C_2H_5CN	liquid	456.4
n-Propyl alcohol	C_3H_7OH	liquid	482.75*
Propyl amine	$C_3H_7NH_2$	liquid	565.31*
n-Propylbenzene	$C_3H_7C_6H_5$	liquid	1,246.4
Propyl bromide	C_3H_7Br	vapor	497.3
Propyl carbylamine	C_3H_7NC	liquid	639.6
Propyl chloride	C_3H_7Cl	vapor	478.3
Propylene	$CH_3CH{:}CH_2$	gas	490.2
Propylene glycol	$CH_3CHOHCH_2OH$	liquid	431.0
n-Propyl iodide	C_3H_7I	liquid	514.3
n-Propyltoluene	$C_6H_4(CH_3)(C_3H_7)$—(1, 3)	liquid	1,405.4
Pseudocumene	$C_6H_3(CH_3)_3$—(1, 2, 4)	liquid	1,241.7
Pyridine	C_5H_5N	liquid	665.0*
Pyrocatechol	$C_6H_4(OH)_2$	solid	683.0*
Pyrogallol	$C_6H_3(OH)_3$	solid	638.7
Pyrrole	C_4H_5N	liquid	567.7
Quercitol	$C_6H_{12}O_5$	solid	704.2
Quinoline	C_9H_7N	liquid	1,123.5
Quinone	$O{:}C_6H_4{:}O$	solid	656.6
Raffinose	$C_{18}H_{32}O_{16}$	solid	2,025.5
Retene	$C_{18}H_{18}$	solid	2,306.8
Resorcinol	$C_6H_4(OH)_2$	solid	681.30*
Resorcinoldimethyl ether	$(CH_3O)_2C_6H_4$	liquid	1,022.6
Rhamnose	$C_6H_{12}O_6$	solid	718.3
Safrole	$C_{10}H_{10}O_2$	liquid	1,244.1
Salicylaldehyde, *see o-Hydroxybenzaldehyde*			
*Salicylic acid	$HOC_6H_4CO_2H$—(1, 2)	solid	722.4**
Sarcosine	$CH_3NHCH_2CO_2H$	solid	401.1
Sebacic acid	$(CH_2)_8(CO_2H)_2$	solid	1,297.3
Skatole	C_9H_9N	liquid	1,170.5
d-Sorbose	$C_6H_{12}O_6$	solid	668.3
Starch	$(C_6H_{10}O_5)x$ per kg	solid	4,178.8
Stearic acid	$C_{18}H_{36}O_2$	solid	2,696.12*
Strychnine	$C_{21}H_{22}O_2N_2$	solid	2,685.7
Styrene	$C_6H_5CH{:}CH_2$	liquid	1,047.1
Suberic acid	$(CH_2)_6(CO_2H)_2$	solid	985.2
Succinic acid	$(CH_2CO_2H)_2$	solid	356.36*
Succinic acid nitrile	$(CH_2CN)_2$	liquid	545.7
Succinic anhydride	$(CH_2CO)_2O$	solid	369.0*
Succinimide	$C_4H_5O_2N$	solid	437.9
Sucrose	$C_{12}H_{22}O_{11}$	solid	1,348.2*
Sylvestrene	$C_{10}H_{16}$	liquid	1,464.7
l-Tartaric acid	$(CHOH)_2(CO_2H)_2$	solid	274.7*
d, l-Tartaric acid (anhydr.)	$(CHOH)_2(CO_2H)_2$	solid	272.6*
Terephthalic acid	$C_6H_4(CO_2H)_2$	solid	770.4
Terpin hydrate	$C_{10}H_{22}O_3$	solid	1,451.0
Terpineol	$C_{10}H_{18}O$	solid	1,469.5
Tetrahydrobenzene	C_6H_{10}	liquid	891.9
Tetrahydronaphthalene	$C_{10}H_{12}$	liquid	1,352.4
Tetramethylmethane	$(CH_3)_4C$	gas	842.6
Tetraphenylmethane	$(C_6H_5)_4C$	solid	3,102.4
Tetryl	$C_7H_5N_5O_8$	solid	842.3
Thebaine	$C_{19}H_{21}O_3N$	solid	2,441.3
Thiophene	C_4H_4S	liquid	670.5
Thujane	$C_{10}H_{18}$	liquid	1,506.4
Thymol	$C_{10}H_{14}O$	liquid	1,353.4
Thymol	$C_{10}H_{14}O$	solid	1,349.7
Thymoquinone	$C_{10}H_{12}O_2$	solid	1,271.3
Toluene	$CH_3C_6H_5$	liquid	934.2
o-Toluic acid	$CH_3C_6H_4CO_2H$	solid	928.9

*25 C

**Recommended as a secondary thermochemical standard.

HEAT OF COMBUSTION (Continued)

FOR ORGANIC COMPOUNDS

Name	Formula	Physical state	Heat of combustion, kg. calories
m-Toluic acid	$CH_3C_6H_4CO_2H$	solid	928.6
p-Toluic acid	$CH_3C_6H_4CO_2H$	solid	926.9
o-Toluidine	$CH_3C_6H_4NH_2$	liquid	964.3
m-Toluidine	$CH_3C_6H_4NH_2$	liquid	965.3
p-Toluidine	$CH_3C_6H_4NH_2$	solid	958.4
o-Tolunitrile	$CH_3C_6H_4CN$	liquid	1,030.3
Toluquinone	$C_7H_6O_2$	solid	803.2
Triaminotriphenyl carbinol	$(C_6H_4NH_2)_3COH$	solid	2,483.5
Tribenzyl amine	$(C_6H_5CH_2)_3N$	solid	2,762.1
Trichloracetic acid	$Cl_3C.CO_2H$	solid	92.8
Triethyl amine	$(C_2H_5)_3N$	liquid	1,036.8
Triethyl carbinol	$(C_2H_5)_3CHOH$	liquid	1,080.0
Triisoamyl amine	$[(CH_3)_2CHCH_2CH_2]_3N$	liquid	2,459.3
Triisobutyl amine	$[(CH_3)_2CHCH_2]_3N$	liquid	1,973.6
Trimethyl amine	$(CH_3)_3N$	liquid	578.6
2, 2, 3-Trimethylbutane	$(CH_3)_3C.CH(CH_3)_2$	liquid	1,147.9
Trimethyl carbinol	$(CH_3)_3COH$	liquid	629.3
Trimethylene	$CH_2CH_2CH_2$ (ring)	gas	499.89*
Trimethylethylene	$(CH_3)_2C:CHCH_3$	liquid	796.0
Trimethylethylene	$(CH_3)_2C:CHCH_3$	vapor	803.6
2, 2, 4-Trimethylpentane	$(CH_3)_3C.CH_2CH(CH_3)_2$	liquid	1,303.9
Trinitrobenzene	$C_6H_3(NO_2)_3$—(1, 3, 5)	solid	663.7
Trinitroglycerol	$C_3H_5(NO_3)_3$	liquid	368.4
Trinitrotoluene	$C_6H_2(CH_3)(NO_2)_3$—(1, 2, 4, 6)	solid	820.7
Triphenyl amine	$(C_6H_5)_3N$	solid	2,267.8
Triphenylbenzene	$C_6H_3(C_6H_5)_3$—(1, 3, 5)	solid	2,936.7
Triphenyl carbinol	$(C_6H_5)_3CHOH$	solid	2,340.8
Triphenylmethane	$(C_6H_5)_3CH$	solid	2,388.7
Triphenyl methyl	$(C_6H_5)_3C$	solid	2,378.5
Tyrosine	$C_9H_{11}O_3N$	solid	1,070.2
Undecylic acid	$C_{11}H_{22}O_2$	solid	1,615.9
Urea	$(NH_2)_2CO$	solid	150.97*
Urethane	$NH_2CO_2C_2H_5$	solid	397.2
Uric acid	$C_5H_4O_3N_4$	solid	460.2
n-Valeric acid	$C_4H_9CO_2H$	liquid	678.12*
Vanillin	$C_6H_3(OH)(OCH_3)CHO$—(1, 2, 4)	solid	914.1
o-Xylene	$(CH_3)_2C_6H_4$	liquid	1,091.7
m-Xylene	$(CH_3)_2C_6H_4$	liquid	1,088.4
p-Xylene	$(CH_3)_2C_6H_4$	liquid	1,089.1
Xylose	$C_5H_{10}O_5$	solid	559.0*

*25 C

HEAT OF FORMATION

FOR ORGANIC COMPOUNDS

The heat of formation of a compound "A" is equal to the sum of the heats of formation of the products of combustion minus the heat of combustion (see preceding table) of the compound "A." The heat of formation of:

Free elements	0	kg-cal
CO_2 (gas)	94.38	"
$\frac{1}{2}H_2O$ (liquid from 1 H)	34.19	"
HF (dilute aqueous solution)	75.6	"
SO_2 (gas)	69.3	"
HBr (aqueous solution)	28.54	"
HCl (aqueous solution)	39.46	"
HNO_3 (aqueous solution)	49.80	"
H_2SO_4 (aqueous solution)	207.5	"

Example I

To calculate the heat of formation of methane (CH_4) where

Heat of combustion of methane = 210.8
Heat of formation of CO_2 = 94.38
Heat of formation of $\frac{1}{2}H_2O$ = 34.19

and where the combustion occurs according to the equation:

$$CH_4 + 2O_2 = CO_2\ (gas) + 2H_2O\ (liquid)$$

Then the heat of formation of $CH_4 = 94.38 + 4(34.19) - 210.8 = +20.34$ kg-cal. per gram molecular weight.

Example II

To calculate the heat of formation of ethylene (C_2H_4) where

Heat of combustion of ethylene = 337.23

and the combustion occurs according to the equation:

$$C_2H_4 + 3O_2 = 2CO_2 + 2H_2O$$

The heat of formation of $C_2H_4 = 2(94.38) + 4(34.19) - 337.23 = -11.71$ kg-cal. per gram molecular weight.

Example III

To calculate the heat of formation of ethylamine ($C_2H_5NH_2$) where

Heat of combustion of ethylamine = 409.5

and the combustion occurs according to the equation:

$$C_2H_5NH_2 + 3.75O_2 = 2CO_2 + 3.5(H_2O) + 0.5N_2$$

The heat of formation of $C_2H_5NH_2 = 2(94.38) + 7(34.19) + O(N_2) - 409.5 = +18.59$ kg-cal. per gram molecular weight.

LOWERING OF VAPOR PRESSURE BY SALTS IN AQUEOUS SOLUTIONS

The table gives the reduction of the vapor pressure in millimeters due to the presence of the number of grammolecules of salt per liter of water given at the head of the columns, at the temperature 100° C, at which temperature the vapor pressure of pure water is 760 millimeters.

(From Smithsonian Tables.)

Substance	0.5	1.0	2.0	3.0	4.0	5.0	6.0	8.0	10.0
$Al_2(SO_4)_3$	12.8	36.5							
$AlCl_3$	22.5	61.0	179.0	318.0					
BaS_2O_6	6.6	15.4	34.4						
$Ba(OH)_2$	12.3	22.5	39.0						
$Ba(NO_3)_2$	13.5	27.0							
$Ba(ClO_3)_2$	15.8	33.3	70.5	108.2					
$BaCl_2$	16.4	36.7	77.6						
$BaBr_2$	16.8	38.8	91.4	150.0	204.7				
CaS_2O_6	9.9	23.0	56.0	106.0					
$Ca(NO_3)_2$	16.4	34.8	74.6	139.3	161.7	205.4			
$CaCl_2$	17.0	39.8	95.3	166.6	241.5	319.5			
$CaBr_2$	17.7	44.2	105.8	191.0	283.3	368.5			
$CdSO_4$	4.1	8.9	18.1						
CdI_2	7.6	14.8	33.5	52.7					
$CdBr_2$	8.6	17.8	36.7	55.7	80.0				
$CdCl_2$	9.6	18.8	36.7	57.0	77.3	99.0			
$Cd(NO_3)_2$	15.9	36.1	78.0	122.2					
$Cd(ClO_3)_2$	17.5								
$CoSO_4$	5.5	10.7	22.9	45.5					
$CoCl_2$	15.0	34.8	83.0	136.0	186.4				
$Co(NO_3)_2$	17.3	39.2	89.0	152.0	218.7	282.0	332.0		
$FeSO_4$	5.8	10.7	24.0	42.4					
H_3BO_3	6.0	12.3	25.1	38.0	51.0				
H_3PO_4	6.6	14.0	28.6	45.2	62.0	81.5	103.0	146.9	189.5
H_3AsO_4	7.3	15.0	30.2	46.4	64.9				
H_2SO_4	12.9	26.5	62.8	104.0	148.0	198.4	247.0	343.2	
KH_2PO_4	10.2	19.5	33.3	47.8	60.5	73.1	85.2		
KNO_3	10.3	21.1	40.1	57.6	74.5	88.2	102.1	126.3	148.0
$KClO_3$	10.6	21.6	42.8	62.1	80.0				
$KBrO_3$	10.9	22.4	45.0						
$KHSO_4$	10.9	21.9	43.3	65.3	85.5	107.8	129.9	170.0	
KNO_2	11.1	22.8	44.8	67.0	90.0	110.5	130.7	167.0	198.8
$KClO_4$	11.5	22.3							
KCl	12.2	24.4	48.8	74.1	100.9	128.5	152.2		
$KHCO_3$	11.6	23.6	59.0	77.6	104.2	132.0	160.0	210.0	255.0
KI	12.5	25.3	52.2	82.6	112.2	141.5	171.8	225.5	278.5
$K_2C_2O_4$	13.9	28.3	59.8	94.2	131.0				
K_2WO_4	13.9	33.0	75.0	123.8	175.4	226.4			
K_2CO_3	14.4	31.0	68.3	105.5	152.0	209.0	258.5	350.0	
KOH	15.0	29.5	64.0	99.2	140.0	181.8	223.0	309.5	387.8
K_2CrO_4	16.2	29.5	60.0						
$LiNO_3$	12.2	25.9	55.7	88.9	122.2	155.1	188.0	253.4	309.2
$LiCl$	12.1	25.5	57.1	95.0	132.5	175.5	219.5	311.5	393.5
$LiBr$	12.2	26.2	60.0	97.0	140.0	186.3	241.5	341.5	438.0
Li_2SO_4	13.3	28.1	56.8	89.0					
$LiHSO_4$	12.8	27.0	57.0	93.0	130.0	168.0			
LiI	13.6	28.6	64.7	105.2	154.5	206.0	264.0	357.0	445.0
Li_2SiF_6	15.4	34.0	70.0	106.0					

Substance	0.5	1.0	2.0	3.0	4.0	5.0	6.0	8.0	10.0
$LiOH$	15.9	37.4	78.1						
Li_2CrO_4	16.4	32.6	74.0	120.0	171.0				
$MgSO_4$	6.5	12.0	24.5	47.5					
$MgCl_2$	16.8	39.0	100.5	183.3	277.0	377.0			
$Mg(NO_3)_2$	17.6	42.0	101.0	174.8					
$MgBr_2$	17.9	44.0	115.8	205.3	298.5				
$MgH_2(SO_4)_2$	18.3	46.0	116.0						
$MnSO_4$	6.0	10.5	21.0						
$MnCl_2$	15.0	34.0	76.0	122.3	167.0	209.0			
NaH_2PO_4	10.5	20.0	36.5	51.7	66.8	82.0	96.5	126.7	157.1
$NaHSO_4$	10.9	22.1	47.3	75.0	100.2	126.1	148.5	189.7	231.4
$NaNO_3$	10.6	22.5	46.2	68.1	90.3	111.5	131.7	167.8	198.8
$NaClO_3$	10.5	23.0	48.4	73.5	98.5	123.3	147.5	196.5	223.5
$(NaPO_3)_6$	11.6								
$NaOH$	11.8	22.8	48.2	77.3	107.5	139.1	172.5	243.3	314.0
$NaNO_2$	11.6	24.4	50.0	75.0	98.2	122.5	146.5	189.0	226.2
Na_2HPO_4	12.1	23.5	43.0	60.0	78.7	99.8	122.1		
$NaHCO_3$	12.9	24.1	48.2						
Na_2SO_4	12.6	25.0	48.9	74.2					
$NaCl$	12.3	25.2	52.1	80.0	111.0	143.0	176.5		
$NaBrO_3$	12.1	25.0	54.1	81.3	108.8	136.0			
$NaBr$	12.6	25.9	57.0	89.2	124.2	159.5	197.5	268.0	
NaI	12.1	25.6	60.2	99.5	136.7	177.5	221.0	301.5	370.0
$Na_4P_2O_7$	13.2	22.0							
Na_2CO_3	14.3	27.3	53.5	80.2	111.0				
$Na_2C_2O_4$	14.5	30.0	65.8	105.8	146.0				
Na_2WO_4	14.8	33.6	71.6	115.7	162.6				
Na_3PO_4	16.5	30.0	52.5						
$(NaPO_3)_3$	17.1	36.5							
NH_4NO_3	12.8	22.0	42.1	62.7	82.9	103.8	121.0	152.2	180.0
$(NH_4)_2SiF_6$	11.5	25.0	44.5						
NH_4Cl	12.0	23.7	45.1	69.3	94.2	118.5	138.2	179.0	213.8
NH_4HSO_4	11.5	22.0	46.8	71.0	94.5	118.	139.0	181.2	218.0
$(NH_4)_2SO_4$	11.0	24.0	46.5	69.5	93.0	117.0	141.8		
NH_4Br	11.9	23.9	48.8	74.1	99.4	121.5	145.5	190.2	228.5
NH_4I	12.9	25.1	49.8	78.5	104.5	132.3	156.0	200.0	243.5
$NiSO_4$	5.0	10.2	21.5						
$NiCl_2$	16.1	37.0	86.7	147.0	212.8				
$Ni(NO_3)_2$	16.1	37.3	91.3	156.2	235.0				
$Pb(NO_3)_2$	12.3	23.5	45.0	63.0					
$Sr(SO_3)_2$	7.2	20.3	47.0						
$Sr(NO_3)_2$	15.8	31.0	64.0	97.4	131.4				
$SrCl_2$	16.8	38.8	91.4	156.8	223.3	281.5			
$SrBr_2$	17.8	42.0	101.1	179.0	267.0				
$ZnSO_4$	4.9	10.4	21.5	42.1	66.2				
$ZnCl_2$	9.2	18.7	46.2	75.0	107.0	153.0	195.0		
$Zn(NO_3)_2$	16.6	39.0	93.5	157.5	223.8				

THERMAL CONDUCTIVITY OF GASES

The values in this table are given as cal/(sec)(cm²)(°C/cm) × 10⁻⁶. To convert these values to Btu/(hr)(ft²)(°F/ft) × 10⁻⁶ multiply by 241.909.

Gas	°F °C	−400 −240	−300 −184.4	−200 −128.9	−100 −73.3	−40 −40	−20 −28.9	0 −17.8	20 −6.7	40 4.4	60 15.6	80 26.7	100 37.8	120 48.9	200 93.3
Acetylene		...	...	...	28.10	34.71	37.19	39.67	42.15	45.04	47.94	50.83	53.72	56.62	69.43
Air		...	...	...	...	50.09	52.15	54.22	56.24	58.31	60.34	62.20	64.22	66.04	...
Ammonia		...	...	...	...	43.39	45.87	48.35	50.83	53.31	55.79	58.68	61.58	64.47	...
Argon		...	...	...	...	34.30	35.95	37.19	38.85	40.09	41.33	42.57	44.22	45.46	...
Bromine		...	...	...	...	...	...	9.09	...	...	...	...	11.57	...	...
n-Butane		...	...	...	...	...	...	...	30.99	33.06	35.54	38.02	40.91	43.39	54.14
i-Butane		...	...	...	...	...	...	...	32.65	33.89	36.37	38.85	41.74	44.22	55.79
Carbon dioxide		...	...	...	...	27.90	29.75	31.70	33.68	35.62	37.61	39.67	41.74	43.81	...
Carbon disulfide		...	...	...	...	...	...	14.05	15.29	16.53	17.77	19.01	19.84	...	...
Carbon monoxide		...	...	...	...	47.94	50.00	51.95	53.85	55.87	57.86	59.92	61.99	63.89	...
Chlorine		...	...	...	...	15.29	16.53	17.36	18.18	19.01	20.25	21.08	21.90	23.14	...
Deuterium		...	...	...	...	274.82	285.15	295.07	305.81	309.95	322.34	334.74	343.01	355.40	...
Ethane		...	...	...	23.97	32.65	35.54	38.43	41.33	44.63	47.94	51.24	54.55	58.27	74.39
Ethanol		...	...	...	...	...	...	...	29.34	30.99	32.65	34.71	36.78	...	...
Ethylamine		...	...	...	...	...	...	...	31.41	33.47	35.54	37.61	39.67	42.15	...
Ethylene		...	...	...	26.86	33.06	35.54	38.02	40.50	43.39	46.29	49.18	52.07	54.96	68.19
Fluorine		...	18.18	30.58	43.39	50.83	52.90	55.38	57.86	59.92	61.99	64.06	66.12	68.19	76.04
Helium		84.31	163.24	221.51	274.8	304.99	314.49	324.00	333.50	343.42	352.10	360.36	368.63	376.07	...
Hydrogen		59.92	142.57	227.29	308.7	357.47	371.93	388.46	405.00	417.39	433.92	446.32	458.72	471.11	...
Hydrogen bromide		...	...	...	...	15.29	16.11	16.49	17.77	18.60	19.84	20.66	21.49	...	...
Hydrogen chloride		...	...	...	...	25.62	26.86	28.51	29.75	30.99	32.23	33.89	35.12	...	...
Hydrogen cyanide		...	...	...	...	...	...	23.97	25.62	26.86	28.10	29.75	30.99	32.65	...
Hydrogen sulfide		...	...	...	...	...	...	28.10	29.75	31.41	33.47	...	36.78		...
Krypton		...	...	...	...	...	...	19.84	...	...	...	...	23.56	...	...
Methane		...	22.32	36.86	52.07	61.37	64.55	67.86	71.08	74.39	78.11	81.83	85.54	89.26	106.62
Neon		...	...	...	...	97.94	100.84	104.14	107.03	109.93	112.82	115.71	118.19	121.09	...
Nitric oxide		...	...	30.91	42.40	49.01	51.24	53.39	55.54	57.65	59.76	61.99	64.06	66.12	74.39
Nitrogen		...	20.25	33.06	44.22	50.42	52.48	54.55	56.20	58.27	60.34	62.40	64.06	65.71	...
Nitrous oxide		...	...	...	...	28.93	30.91	32.90	35.04	37.15	39.30	41.45	43.81	46.08	...
Oxygen		...	18.84	31.66	43.72	50.54	52.81	54.96	57.24	59.43	61.58	63.64	65.91	68.19	76.87
n-Propane		...	...	...	...	27.69	29.75	32.23	34.71	37.19	39.67	42.47	45.46	48.35	60.75
R-11(CCl_3F)		...	...	...	...	...	12.81	13.64	14.88	15.70	16.53	17.77	18.60	...	...
R-12(CCl_2F_2)		...	...	...	...	...	17.36	18.60	19.42	20.66	21.49	22.73	23.56	...	...
R-21($CHCl_2F$)		...	...	...	...	...	...	21.90	22.32	22.73	23.14	23.56	23.97	...	...
R-22($CHClF_2$)		...	...	...	...	...	...	24.80	25.62	26.45	27.28	28.10	28.93	...	...
Water		...	...	...	...	...	...	34.71	36.78	38.85	40.50	42.57	44.63	46.70	54.96

THERMAL CONDUCTIVITY OF GASEOUS HELIUM, NITROGEN AND WATER

From NSRDS-NBS 8

R. W. Powell, C. Y. Ho, and P. E. Liley

The thermal conductivity, k, is given in the units Milliwatt cm^{-1} °K^{-1}. To convert to Cal(*gm*) hr^{-1} cm^{-1} °K^{-1} multiply the values listed in the table by 0.860421

T (K)	k He	k N_2
0.08	0.00044	
0.09	0.00053	
0.10	0.00064	
0.15	0.00130	
0.20	0.00231	
0.25	0.0039	
0.30	0.0062	
0.35	0.0089	
0.40	0.0120	
0.45	0.0154	
0.5	0.0187	
0.6	0.0231	
0.7	0.0252	
0.8	0.0262	
0.9	0.0266	
1.0	0.0269	
1.25	0.0281	
1.5	0.0306	
2.0	0.0393	
2.5	0.0502	
3.0	0.0607	
3.5	0.0732	
4.0	0.0803	
4.5	0.0879	
5.0	0.0962	
6	0.1113	
7	0.1247	
8	0.1393	
9	0.1523	
10	0.1640	
12	0.1866	
14	0.2067	
16	0.2259	
18	0.2435	
20	0.2582	
25	0.2962	
30	0.3330	
35	0.3669	
40	0.4000	
45	0.4314	
50	0.4623	(0.0485)*
60	0.521	(0.0578)*
70	0.578	(0.0670)*
80	0.631	0.0762
90	0.679	0.0852
100	0.730	0.0941
110	0.776	0.1030
120	0.819	0.1119
130	0.863	0.1208
140	0.907	0.1296

T (K)	k He	k N_2	k H_2O
150	0.950	0.1385	
160	0.992	0.1474	
170	1.033	0.1562	
180	1.072	0.1651	
190	1.112	0.1739	
200	1.151	0.1826	
210	1.190	0.1908	
220	1.228	0.1989	
230	1.266	0.2067	
240	1.304	0.2145	
250	1.338	0.2222	(0.140)*
260	1.372	0.2298	(0.148)*
270	1.405	0.2374	(0.156)*
280	1.437	0.2449	0.164
290	1.468	0.2524	0.172
300	1.499	0.2598	0.181
310	1.530	0.2671	0.189
320	1.560	0.2741	0.197
330	1.590	0.2808	0.205
340	1.619	0.2874	0.214
350	1.649	0.2939	0.222
360	1.678	0.3002	0.231
370	1.708	0.3065	0.239
380	1.737	0.3127	0.248
390	1.766	0.3189	0.256
400	1.795	0.3252	0.264
410	1.824	0.3314	0.273
420	1.853	0.3376	0.282
430	1.882	0.3438	0.291
440	1.914	0.3501	0.300
450	1.947	0.3564	0.307
460	1.980	0.3626	0.317
470	2.013	0.3688	0.327
480	2.046	0.3749	0.337
490	2.080	0.3808	0.347
500	2.114	0.3864	0.357
510	2.15	0.392	0.368
520	2.18	0.398	0.378
530	2.22	0.403	0.389
540	2.25	0.408	0.400
550	2.29	0.414	0.411
560	2.33	0.420	0.422
570	2.36	0.425	0.432
580	2.40	0.431	0.443
590	2.43	0.436	0.454
600	2.47	0.441	0.464
610	2.51	0.446	0.475
620	2.54	0.452	0.486
630	2.58	0.457	0.497
640	2.61	0.462	0.508

T (K)	k He	k N_2	k H_2O
650	2.64	0.467	0.518
660	2.67	0.472	0.529
670	2.69	0.478	0.540
680	2.72	0.483	0.551
690	2.75	0.488	0.562
700	2.78	0.493	0.572
710	2.81	0.498	0.58
720	2.84	0.503	0.59
730	2.87	0.508	0.60
740	2.90	0.513	0.62
750	2.92	0.517	0.63
760	2.95	0.522	0.64
770	2.98	0.526	0.65
780	3.01	0.531	0.66
790	3.04	0.536	0.67
800	3.07	0.541	0.68
810	3.09	0.546	0.69
820	3.12	0.551	0.70
830	3.15	0.555	0.71
840	3.18	0.559	0.72
850	3.21	0.564	0.73
860	3.23	0.569	0.74
870	3.26	0.574	0.75
880	3.29	0.578	0.76
890	3.32	0.583	0.77
900	3.35	0.587	0.78
910	3.37	0.592	
920	3.40	0.596	
930	3.43	0.600	
940	3.46	0.605	
950	3.49	0.609	
960	3.52	0.613	
970	3.54	0.618	
980	3.57	0.622	
990	3.60	0.626	
1000	3.63	0.631	
1050	3.76	0.651	
1100	3.89	0.672	
1150	4.03	0.693	
1200	4.16	0.713	
1250	4.29	0.733	
1300	4.43	0.754	
1350	4.55	0.775	
1400	4.69	0.797	
1450	4.82	0.819	
1500	4.94	0.842	
1550	5.07	0.867	
1600	5.21	0.893	
1650	5.33	0.921	
1700	5.45	0.950	

T (K)	k He	k N_2
1750	5.57	0.981
1800	5.70	1.013
1850	5.83	1.046
1900	5.96	1.080
1950	6.08	1.113
2000	6.20	1.146
2100	6.44	1.207
2200	6.69	1.263
2300	6.93	1.314
2400	7.16	1.361
2500	7.39	1.406
2600	7.62	1.449
2700	7.85	1.494
2800	8.07	1.542
2900	8.29	1.590
3000	8.51	1.640
3100	8.72	1.691
3200	8.95	1.743
3300	9.16	1.795
3400	9.37	1.853
3500	9.58	1.915
3600	9.79	
3700	10.00	
3800	10.22	
3900	10.43	
4000	10.64	
4100	10.85	
4200	11.06	
4300	11.27	
4400	11.48	
4500	11.69	
4600	11.90	
4700	12.11	
4800	12.31	
4900	12.51	
5000	12.71	

THERMAL CONDUCTIVITY

THERMAL CONDUCTIVITY OF DIELECTRIC CRYSTALS

Name	Remarks	Conductivity mw/cm deg K 83° K	273° K
Marble	Small crystals, 99.9 % $CaCO_3$	42	33
Do	99.99 % $CaCO_3$	54	38
Do	Large crystals	50	33
Calcite	Main crystal axis perpendicular to rod axis.	180	46
Do	Main crystal axis parallel to rod axis.	293	54
Sylvite	Natural crystal	159	75
KCl	Pressed at 8,000 atm	314	88
KCl	From a melt	402	92
NaCl	do	343	92
NaCl	Pressed at 8,000 atm	251	71
Rock salt	do	180	63
Sylvite	do	343	84
KCl	Pressed at 1,250 atm	243	75
KCl	Pressed at 2,500 atm	368	92
KCl	Pressed at 8,900 atm	402	96
KBr	Pressed at 8,000 atm	92	38
NaBr	do	50	25
KI	do	121	29
KF	do	234	71
NaF	do	519	105
RbI	do	59	33
RbCl	do	29	21
90 % KBr, 10 % KCl	do	50	29
75 % KBr, 25 % KCl	do	29	21
50 % KBr, 50 % KCl	do	25	25
25 % KBr, 75 % KCl	Pressed at 8,000 atm	46	33
10 % KBr, 90 % KCl	do	80	50
50 % KCl, 50 % NaCl	do	188	71
KNO_3	do	17	21
Mercuric chloride	do	17	13
NH_4Cl	do	109	25
NH_4Br	do	67	25
$Ba(NO_3)_2$	do	33	13
Copper sulfate		29	21
Magnesium sulfate		25	25
$K_4Fe(CN_6)$		17	17
Chrom alum		13	21
Potassium alum		13	21
Potassium bichromate	Main crystal axis perpendicular to rod axis.	17	21
Do	Main crystal axis parallel to rod axis.	17	17
Topaz	Mineral		234
Zincblend	do	63	264
Beryll	do	88	84
Tourmaline	do	38	46

THERMAL CONDUCTIVITY OF ORGANIC COMPOUNDS

The values in this table are given as cal/(sec)(cm²)(°C/cm). To convert these values to Btu/(hr)(ft²)(°F/ft) multiply by 242.08.

Substance	k	t, °C	t, °F
Acetaldehyde	0.0004089	21	69.8
Acetic acid	0.0004109	20	68
Acetic anhydride	0.0005286	21	69.8
Acetone	0.0004750	−80	−112
	0.0004543	16	61
	0.0004031	75	167
Allyl alcohol	0.0004295	30	86
Amyl acetate (n)	0.0003085	20	68
(iso)	0.000310	20	68
Amyl alcohol (n)	0.0003874	30–100	86–212
(iso)	0.0003531	30	86
Amyl bromide (n)	0.0002350	18	64.4
Aniline	0.0004237	16.5	61.5
Benzene	0.0003780	22.5	72.5
	0.0003275	50	122
	0.0003630	60	140
	0.0002870	140	284
Bromobenzene	0.0002664	20	68
Butyl acetate (n)	0.000327	20	68
Butyl alcohol (n)	0.0003663	20	68
Carbon tetrachloride	0.0002470	20	68
	0.0002333	50	122
Chlorobenzene	0.0003457	30–100	86–212
Chlorotoluene (p)	0.000310	20	68
Chloroform	0.0002891	16	61
	0.000246	20	68
Cresol (m)	0.0003581	20	68
(p)	0.000345	20.1	68.2
Cumene	0.000298	20	68
Cymene (p)	0.0003217	30	86
Decane	0.0003349	30	86

THERMAL CONDUCTIVITY OF ORGANIC COMPOUNDS (Continued)

Substance	k	t, °C	t, °F
Diethyl ether	0.0003283	30	86
Dichloroethane, 1–2	0.000302	20	68
Di-isopropyl ether	0.000262	20	68
Ethyl acetate	0.0003560	16	60.8
Ethyl alcohol	0.0003995	20	68
Ethyl benzene	0.0003160	20	68
Ethyl bromide	0.0002862	30	86
Ethyl ether	0.0003283	30	86
Ethyl iodide	0.0002651	30	86
Ethylene glycol	0.0006236	20	68
	0.0006323	15	122
	0.0006443	80	176
Freon-12 (CCl_2F_2)	0.0002310	0–75	32–167
Freon-21 ($CHCl_2F$)	0.0003180	0–75	32–167
Freon-22 ($CHClF_2$)	0.0002309	40	104
Freon-113 (CCl_2FCCl_2F)	0.0002379	0–80	32–176
Freon-114 ($C_2H_2F_4$)	0.0002127	0–75	32–167
Glycerol	0.000703	20	68
Heptane (n)	0.0003354	30	86
Heptyl alcohol	0.0003882	70–100	86–212
Hexane (n)	0.0003287	30–100	86–212
Hexyl alcohol (n)	0.0003857	30–100	86–212
Iodobenzene	0.0002874	30–100	86–212
Mesitylene	0.0003246	20	68
Methyl alcohol	0.0004832	20	68
Methyl aniline	0.0004419	21.5	70.5
Methyl chloride	0.0004597	−15 (−) +30	5–86
Methyl cyclohexane	0.0003052	30	86
Methylene chloride	0.0002908	0	32
Nitrobenzene	0.0003907	30–100	86–212
Nitromethane	0.0005142	30	86
Nonane (n)	0.0003374	30–100	86–212
Nonyl alcohol (n)	0.0004014	30–100	86–212
Octane (n)	0.0003469	30	86
Octyl alcohol (n)	0.0003973	30–100	86–212
Oleic acid	0.0005514	26.5	79.7
Palmitic acid	0.0004097	72.5	162.5
Pentachloroethane	0.0002994	20	68
Pentane (n)	0.0003221	30	86
Phenetole	0.0003577	−20	−4
Phenyl hydrazine	0.0004121	25	69.8
Propyl acetate (iso)	0.000321	20	68
Propyl alcohol (iso)	0.0003362	20	68
Propylene chloride	0.0002994	20–50	68–122
Propylene glycol (1–2)	0.0004799	20–80	68–176
Stearic acid	0.0003824	72.5	162.5
Tetrachloroethane (sym)	0.000272	20	68
Tetrachloroethylene	0.0003866	20	68
Toluene	0.0003804	−80	−112
	0.0003221	20	68
	0.0002808	80	176
Trichloroethylene	0.0003246	−60	−76
	0.0002775	20	68
Triethylamine	0.0003498	−80	−112
	0.0002891	20	68
	0.0002664	44.4	112
Xylene (o)	0.0003411	−20 (−) +80	(−4)–176
Xylene (m)	0.0003767	25	77

THERMAL CONDUCTIVITY OF INORGANIC COMPOUNDS

Ammonia	0.0001198	−15(−) +30	5–86
Argon	0.0002895	−183	−297
	0.0001677	−133	−207
	0.0000553	−105	−157.5
	0.0000409	−75	−102.5
Carbon dioxide	0.0002040	−50	−58
	0.0002412	−40	−40
	0.0002664	−30	−22
	0.0002746	−20	−4
	0.0002495	0	32
	0.0001677	30	86
Nitrogen	0.0003400	−196	−321.5
	0.0002961	−189	−308
	0.0002028	−158	−253
	0.0000640	−105	−155
Oxygen	0.0000500	(−207)–(−191)	(−340)–(−312)
	0.0000504	(−178)–(−182)	(−288)–(−295)
Water	0.001326	−3	27
	0.001372	+7	45
	0.001456	27	81
	0.001522	47	117
	0.001575	67	153
	0.001625	97	207
	0.00163 5	107	225
	0.001628	157	315
	0.001580	197	387
	0.001463	247	441
	0.001288	297	567
	0.001004	347	657

Thermal Conductivity of Miscellaneous Substances

Chlorinated diphenyl 1242	0.0002936	30–100	86–212
Chlorinated diphenyl 1248	0.0002808	30–100	86–212
Kerosene	0.0003572	30	86
Light heat transfer oil	0.0003159	30–100	86–212
Petroleum ether	0.0003118	30	86
Red oil	0.0003366	30	86
Transformer oil	0.0004242	70–100	86–212

THERMAL CONDUCTIVITY OF MATERIALS

(Bureau of Standards Letter Circular No. 227)

D = Density in pounds per cubic foot.
K = Thermal conductivity in B.T.U. per hour, square foot, and temperature gradient of 1 degree Fahrenheit per inch thickness. The lower the conductivity, the greater the insulating values.

Soft Flexible Materials in Sheet Form

		D	K
Dry Zero	Kapok between burlap or paper	1.0	0.24
		2.0	0.25
Cabots Quilt	Eel grass between kraft paper	3.4	0.25
		4.6	0.26
Hair Felt	Felted cattle hair	11.0	0.26
		13.0	0.26
Balsam Wool	Chemically treated wood fibre	2.2	0.27
Hairinsul	75% hair 25% jute	6.3	0.27
	50% hair 50% jute	6.1	0.26
Linofelt	Flax fibres between paper	4.9	0.28
Thermofelt	Jute and asbestos fibres, felted	10.0	0.37
	Hair and asbestos fibres. felted	7.8	0.28

Loose Materials

		D	K
Rock Wool	Fibrous material made from rock, also made in sheet form, felted and confined with wire netting	6.0	0.26
		10.0	0.27
		14.0	0.28
		18.0	0.29
Glass Wool	Pyrex glass, curled	4.0	0.29
		10.0	0.29
Sil-O-Cel	Powdered diatomaceous earth	10.6	0.31
Regranulated Cork	Fine particles	9.4	0.30
	about 3/16 inch particles	8.1	0.31
Thermofill	Gypsum in powdered form	26.	0.52
		34.	0.60
Sawdust	Various	12.0	0.41
	redwood	10.9	0.42
Shavings	Various, from planer	8.8	0.41
Charcoal	From maple, beech and birch, coarse	13.2	0.36
	6 mesh	15.2	0.37
	20 mesh	19.2	0.39

Semi-flexible Materials in Sheet Form

		D	K
Flaxlinum	Flax fibre	13.0	0.31
Fibrofelt	Flax and rye fibre	13.6	0.32

Semi-rigid Materials in Board Form

		D	K
Corkboard	No added binder; very low density	5.4	0.25
Corkboard	No added binder; low density	7.0	0.27
Corkboard	No added binder; medium density	10.6	0.30
Corkboard	No added binder; high density	14.0	0.34
Eureka	Corkboard with asphaltic binder	14.5	0.32
Rock Cork	Rock wool block with binder. Also called "Tucork"	14.5	0.326
Lith	Board containing rock wool, flax and straw pulp	14.3	0.40

Stiff Fibrous Materials in Sheet Form

		D	K
Insulite	Wood pulp	16.2	0.34
		16.9	0.34
Celotex	Sugar cane fibre	13.2	0.34
		14.8	0.34

	K
*Masonite	0.33
*Inso-board	0.33
*Maizewood	0.33 to 0.39
*Cornstalk Pith Board	0.24 to 0.30
*Maftex	0.34

THERMAL CONDUCTIVITY OF MATERIALS (Continued)

Cellular Gypsum

	D	K
Insulex or Pyrocell	8	0.35
	12	0.44
	18	0.59
	24	0.77
	30	1.00

Woods (Across Grain)

	D	K
Balsa	7.3	0.33
	8.8	0.38
	20	0.58
Cypress	29	0.67
White pine	32	0.78
Mahogany	34	0.90
Virginia pine	34	0.98
Oak	38	1.02
Maple	44	1.10

Miscellaneous Building Materials

(Data taken from various sources)

	K		K
Cinder concrete	1.7 to 7	Limestone	4 to 9
Building gypsum	About 3	Concrete	9 to 25
Plaster	2 to 5	Sandstone	8 to 16
Building brick	3 to 6	Marble	14 to 20
Glass	5 to 6	Granite	13 to 28

* From various commercial laboratories and the work of O. R. Sweeney at Iowa State College.

THERMAL CONDUCTIVITY DATA ON CERAMIC MATERIALS

Description[a]	Class.[b]	Water Abs. %	Bulk Density gms/cc	Thermal Conductivity[c] 100°F	200°F	300°F
Single Crystals						
Silicon carbide	5	—	—	52.0	50.0	49.0
Periclase	5	—	—	26.7	22.5	19.5
Sapphire, c-axis	5	—	—	20.2	16.0	14.0
Sapphire, a-axis	5	—	—	18.7	15.0	12.9
Topaz, a-axis	5	—	—	10.8	9.4	7.9
Kyanite, c-axis	5	—	—	10.00	8.6	7.4
Kyanite, b-axis	5	—	—	9.6	8.3	7.1
Spinel, $MgO \cdot Al_2O_3$	5	—	—	6.80	6.20	5.50
Quartz, c-axis	4	—	—	6.40	5.40	5.02
Quartz, a-axis	4	—	—	3.40	3.00	2.60
Rutile, c-axis	5	—	—	5.60	4.80	4.40
Rutile, a-axis	5	—	—	3.20	3.20	3.20
Fluorite	5	—	—	5.30	4.37	3.45
Beryl, aquamarine, c-axis	4	—	—	3.18	3.15	3.12
Beryl, aquamarine, a-axis	4	—	—	2.52	2.52	2.52
Zircon, a-axis	4	—	—	2.45	2.45	2.45
Zircon, c-axis	4	—	—	2.34	2.34	2.35
Polycrystalline Single Oxide Ceramics						
Pure BeO, hot pressed	2	0.03	2.97	125.0	104.0	92.0
MgO (spec. pure)	1	0.83	3.21	21.2	18.4	16.0
SnO_2 98 %	1	0.03	6.62	17.5	15.0	12.7
ZnO (yellow)	1	0.00	5.28	16.8	14.6	12.5
ZnO (gray)	1	0.03	5.20	13.6	11.8	10.2
CuO (100 %)	1	0.04	6.76	10.2	9.00	7.80
ThO_2, hot pressed	2	—	9.58	8.00	7.02	6.50
CeO_2	1	0.00	6.20	6.63	6.29	5.20
Mn_3O_4	1	0.02	4.21	4.18	3.80	3.41
PbO (100 %)	1	0.38	7.98	1.6	1.25	0.98

[a] Composition: 90 % MgO, 10 % Al_2O_3 designates weight percent. $Li_2O:4B_2O_3$ designates mole composition, does not indicate compound formation.

[b] Classification: 1 = research body; 2 = industrial research body; 3 = commercial body; 4 = natural mineral; 5 = synthetic mineral.

[c] Thermal conductivity: Units in Btu/(hr) (sq ft)(°F/ft); to convert to cal/(sec) (sq cm)(°C/cm) multiply by 0.00413. (I) = determination made with high vacuum apparatus, inconel thermodes. No letters following value, determination made with high vacuum thermal conductivity apparatus, copper thermodes.

By permission from Engineering Research Bulletin No. 40 Rutgers University (1958).

THERMAL CONDUCTIVITIES OF GLASSES BETWEEN −150 AND +100°C

E. H. Ratcliffe

Type of glass	Approximate silica contents (wt. %)	Approximate contents other oxides normally present in quantity (wt. %)		Estimated approximate thermal conductivity at various temperatures: Temperature (°C)	Thermal conductivity $\left(\frac{\text{cal cm}}{\text{cm}^2\text{ s deg C}}\right) \times 10^4$
(a) Vitreous silica	100	....		−150	20.0
				−100	25.0
				− 50	28.8
				0	31.5
				50	33.7
				100	35.4
(b) 'Vycor' glass	96	B_2O_3	3	−100	24
				0	30
				100	34
(c) General information					
'Crown' glasses	50–75	Various		−100	12–20.5
				30	19–26
				100	21–29
'Flint' glasses	20–55	Various		−100	9–15
				30	13–21
				100	15–23
(d) Pyrex type chemically-resistant borosilicate glasses	80–81	B_2O_3	12–13	−100	21
		Na_2O	4	0	26
		Al	2	100	30
(e) Borosilicate crown glasses	60–65	B_2O_3	15–20	−100	16–17.5
				0	21–22.5
				100	24–25.5
	65–70	B_2O_3	10–15	−100	17.5–19
				0	22.5–24
				100	25.5–27
	70–75	B_2O_3	5–10	−100	19–20.5
				0	24.5–26
				100	27.5–29
(f) (i) Zinc crown glasses	55–65	ZnO	5–15	−100	21–22
		Remainder		0	26–27
		B_2O_3, Al_2O_3		100	28–30
		ZnO	5–15	−100	14–17
		Remainder		0	17–21
		Na_2O, K_2O		100	20–23
		ZnO	15–25	−100	21–22
		Remainder		0	26–27
		B_2O_3, Al_2O_3		100	27–29
		ZnO	15–25	−100	16–19
		Remainder		0	20–23
		Na_2O, K_2O		100	22–25
(f) (ii) Zinc crown glasses	65–75	ZnO	5–15	−100	21–22
		Remainder		0	27–28
		B_2O_3, Al_2O_3		100	29–31
		ZnO	5–15	−100	17–20
		Remainder		0	21–25
		Na_2O, K_2O		100	24–27
		ZnO	15–25	−100	21–23
		Remainder		0	27–28
		B_2O_3, Al_2O_3		100	29–30
		ZnO	15–25	−100	16–20
		Remainder		0	20–24
		Na_2O, K_2O		100	25–29

THERMAL CONDUCTIVITIES OF GLASSES BETWEEN −150 AND +100°C (Continued)

Type of glass	Approximate silica contents (wt. %)	Approximate contents other oxides normally present in quantity (wt. %)		Estimated approximate thermal conductivity at various temperatures: Temperature (°C)	Thermal conductivity $\left(\frac{\text{cal cm}}{\text{cm}^2\text{ s deg C}}\right) \times 10^4$
(g) Barium crown glasses	31	B_2O_3	12	−100	13
		Al_2O_3	8	0	17
		BaO	48	100	19
	41	B_2O_3	6	−100	14
		Al_2O_3	2	0	18
		ZnO	8	100	20
		BaO	43		
	47	B_2O_3	4	−100	15
		Na_2O	1	0	18
		K_2O	7	100	21
		ZnO	8		
		BaO	32		
	65	B_2O_3	2	−100	17
		Na_2O	5	0	21
		K_2O	15	100	24
		ZnO	2		
		BaO	10		
(h) Borate glasses					
Borate flint glass	9	B_2O_3	36	−100	13
		Na_2O	1	0	16
		K_2O	2	100	19
		PbO	36		
		Al_2O_3	10		
		ZnO	6		
Borate flint glass	.....	B_2O_3	56	−100	12
		Al_2O_3	12	0	16
		PbO	32	100	20
Borate flint glass	.....	B_2O_3	43	−100	9
		Al_2O_3	5	0	13
		PbO	52	100	17
Borate glass	4	B_2O_3	55	−100	15
		Al_2O_3	14	0	19
		PbO	11	100	21
		K_2O	4		
		ZnO	12		
Borate crown glass	.....	B_2O_3	64	−100	12
		Na_2O	8	0	16
		K_2O	3	100	20
		BaO	4		
		PbO	3		
		Al_2O_3	18		
Light borate crown glass	.....	B_2O_3	69	−100	13
		Na_2O	8	0	17
		BaO	5	100	21
		Al_2O_3	18		
Zinc borate glass	.....	B_2O_3	40	−100	16
		ZnO	60	0	18
				100	20
(i) Phosphate crown glasses					
Potash phosphate glass	.....	P_2O_5	70	0	18
		B_2O_3	3	100	20
		K_2O	12		

Type of glass	Approximate silica contents (wt. %)	Approximate contents other oxides normally present in quantity (wt. %)		Estimated approximate thermal conductivity at various temperatures: Temperature (°C)	Thermal conductivity $\left(\frac{\text{cal cm}}{\text{cm}^2\text{ s deg C}}\right) \times 10^4$
(i) Potash phosphate glass (*Continued*)	.....	Al_2O_3	10		
		MgO	4		
Baryta phosphate glass	.....	P_2O_5	60	45	18
		B_2O_3	3		
		Al_2O_3	8		
		BaO	28		
(j) Soda-lime glasses	75	Na_2O	17	−100	18
		CaO	8	0	23
				100	26
	75	Na_2O	12	−100	21
		CaO	13	0	26
				100	28
	72	Na_2O	15	−100	19
		CaO	11	0	24
		Al_2O_3	2	100	27
	65	Na_2O	25	−100	16
		CaO	10	0	20
				100	23
	65	Na_2O	15	−100	20
		CaO	20	0	24
				100	26
	60	Na_2O	20	−100	18
		CaO	20	0	22
				100	24
(k) Other crown glasses					
Crown glass	75	Na_2O	9	−100	19
		K_2O	11	0	24
		CaO	5	100	26
High dispersion crown glass	68	Na_2O	16	−100	16
		ZnO	3	0	20
		PbO	13	100	24
(l) Miscellaneous flint glasses					
(i) Silicate flint glasses					
Light flint glasses	65	PbO	25	−100	16–17
		Others	10	0	21–22
				100	24–25
	55	PbO	35	−100	14–16
		Others	10	0	18–20
				100	21–22
Ordinary flint glass	45	PbO	45	−100	12–14
		Others	10	0	16–18
				100	19–20
Heavy flint glass	35	PbO	60	−100	11–12
		Others	5	0	14–15
				100	17–18
Very heavy flint glasses	25	PbO	73	−100	10–11
		Others	2	0	13–14
				100	15–16
	20	PbO	80	−100	10
				0	12
				100	14

THERMAL CONDUCTIVITIES OF GLASSES BETWEEN −150 AND +100°C (Continued)

Type of glass	Approximate silica contents (wt. %)	Approximate contents other oxides normally present in quantity (wt. %)		Estimated approximate thermal conductivity at various temperatures: Temperature (°C)	Thermal conductivity $\left(\frac{\text{cal cm}}{\text{cm}^2\text{ s deg C}}\right) \times 10^4$
(ii) Borosilicate flint glass	33	B_2O_3	31	−100	15
		PbO	25	0	20
		Al_2O_3	7	100	23
		K_2O	3		
		Na_2O	1		
(iii) Barium flint glass	50	BaO	24	−100	14
		PbO	6	0	17
		K_2O	8	100	20
		Na_2O	3		
		ZnO	8		
		Sb_2O_3	1		
(m) Other glasses					
(i) Potassium glass	59	K_2O	33	50	21–22
		CaO	8		
(ii) Iron glasses	63	Fe_2O_3	10	−100	19
		Na_2O	17	0	23
		MgO	4	100	25
		CaO	3		
		Al_2O_3	2		
	67	Fe_2O_3	15	0	21–22
		Na_2O_3	18	100	24–25
	62	Fe_2O_3	20	0	20.5–21.5
		Na_2O	18	100	23–24
(ii) Rock glasses					
Obsidian				0	32
				100	35
Artificial diabase				0	27
				100	30

THERMAL CONDUCTIVITY OF CERTAIN METALS

From NSRDS-NBS 8

R. W. Powell, C. Y. Ho, and P. E. Liley

The thermal conductivity, k, is given in the units Watt cm^{-1} °K^{-1}.
To convert to Cal(*gm*) hr^{-1} cm^{-1} °C^{-1} multiply the values listed in the tables by 860.421
To convert to Btu hr^{-1} ft^{-1} °F^{-1} multiply the values listed in the tables by 57.818.
ρ_0 is the residual electrical resistivity and the value of ρ at 4.2°K is used approximately as ρ_0.

T,K	Aluminum	Copper	Gold	Iron	Manganin	Platinum	Silver	Tungsten
	99.996+%	99.999+%	99.999+%	99.998+%		99.999%	99.999+%	99.99+%
	ρ_0 = 0.00315	ρ_0 = 0.000851	ρ_0 = 0.0055	ρ_0 = 0.0327		ρ_0 = 0.0106	ρ_0 = 0.00062	ρ_0 = 0.0017
	μohm cm	μohm cm	μohm cm	μohm cm		μohm cm	μohm cm	μohm cm
0	0	0	0	0	0	0	0	0
1	7.8	28.7	4.4	0.75	0.0007	2.31	39.4	14.4
2	15.5	57.3	8.9	1.49	0.0018	4.60	78.3	28.7
3	23.2	85.5	13.1	2.24	0.0031	6.79	115	42.6
4	30.8	113	17.1	2.97	0.0046	8.8	147	55.6
5	38.1	138	20.7	3.71	0.0062	10.5	172	67.1
6	45.1	159	23.7	4.42	0.0078	11.8	187	76.2
7	51.5	177	26.0	5.13	0.0095	12.6	193	82.4
8	57.3	189	27.5	5.80	0.0111	12.9	190	85.3
9	62.2	195	28.2	6.45	0.0128	12.8	181	85.1
10	66.1	196	28.2	7.05	0.0145	12.3	168	82.4
11	69.0	193	27.7	7.62	0.0162	11.7	154	77.9
12	70.8	185	26.7	8.13	0.0180	10.9	139	72.4
13	71.5	176	25.5	8.58	0.0197	10.1	124	66.4
14	71.3	166	24.1	8.97	0.0215	9.3	109	60.4
15	70.2	156	22.6	9.30	0.0232	8.4	96	54.8
16	68.4	145	20.9	9.56	0.0250	7.6	85	49.3
18	63.5	124	17.7	9.88	0.0285	6.1	66	40.0
20	56.5	105	15.0	9.97	0.0322	4.9	51	32.6
25	40.0	68	10.2	9.36	0.0410	3.15	29.5	20.4
30	28.5	43	7.6	8.14	0.0497	2.28	19.3	13.1
35	21.0	29	6.1	6.81	0.0583	1.80	13.7	8.9
40	16.0	20.5	5.2	5.55	0.067	1.51	10.5	6.5
45	12.5	15.3	4.6	4.50	0.075	1.32	8.4	5.07
50	10.0	12.2	4.2	3.72	0.082	1.18	7.0	4.17
60	6.7	8.5	3.8	2.65	0.097	1.01	5.5	3.18
70	5.0	6.7	3.58	2.04	0.110	0.90	4.97	2.76
80	4.0	5.7	3.52	1.68	0.120	0.84	4.71	2.56
90	3.4	5.14	3.48	1.46	0.127	0.81	4.60	2.44
100	3.0	4.83	3.45	1.32	0.133	0.79	4.50	2.35
150	2.47	4.28	3.35	1.04	0.156	0.762	4.32	2.10
200	2.37	4.13	3.27	0.94	0.172	0.748	4.30	1.97
250	2.35	4.04	3.20	0.865	0.193	0.737	4.28	1.86
273	2.36	4.01	3.18	0.835	0.206	0.734	4.28	1.82
300	2.37	3.98	3.15	0.803	0.222	0.730	4.27	1.78
350	2.40	3.94	3.13	0.744	0.250	0.726	4.24	1.70
400	2.40	3.92	3.12	0.694	(0.279)	0.722	4.20	1.62
500	2.37	3.88	3.09	0.613	(0.338)	0.719	4.13	1.49
600	2.32	3.83	3.04	0.547	(0.397)	0.720	4.05	1.39
700	2.26	3.77	2.98	0.487		0.723	3.97	1.33
800	2.20	3.71	2.92	0.433		0.729	3.89	1.28
900	2.13	3.64	2.85	0.380		0.737	3.82	1.24
1000	[0.93]**	3.57	(2.78)	0.326		0.748	(3.74)	1.21
1100	[0.96]	3.50	(2.71)	0.297		0.760	(3.66)	1.18
1200	[0.99]	3.42	(2.62)	0.282		0.775	(3.58)	1.15
1300	[1.02]	(3.34)†	(2.51)	0.299		0.791		1.13
1400				0.309		0.807		1.11
1500				0.318		0.824		1.09
1600				(0.327)		0.842		1.07
						0.860		1.05
						0.877		1.03
						(0.895)		1.02
						(0.913)		1.00
								0.98
								0.96
								0.94
								0.925
								0.915
								0.905
								0.900
								(0.895)

* In the table the third significant figure is given only for the purpose of comparison and for smoothness and is not indicative of the degree of accuracy.

** Values in square brackets are for liquid state.

† Values in parentheses are extrapolated.

‡ Estimated.

THERMAL CONDUCTIVITY OF CERTAIN LIQUIDS

From NSRDS-NBS 8
R. W. Powell, C. Y. Ho, and P. E. Liley

The thermal conductivity, k, is given in the units Milliwatt cm^{-1} $°K^{-1}$. To convert to Cal(*gm*) hr^{-1} cm^{-1} $°K^{-1}$ multiply the values listed in the table by 0.860421

T (K)	Helium	Nitrogen	Argon	Carbon tetra-chloride	Diphenyl	m-Terphenyl	Toluene	Water
2.4	0.192							
2.6	0.193							
2.8	0.197							
3.0	0.204							
3.2	0.214							
3.4	0.227							
3.6	0.241							
3.8	0.260							
4.0	0.282							
4.2	0.307							
4.4	(0.335)‡							
4.6	(0.366)‡							
4.8	(0.400)‡							
5.0	(0.437)‡							
5.2	(0.477)‡							
60		1.692†						
65		1.598						
70		1.504						
75		1.411						
80		1.320‡	1.315†					
85		1.229‡	1.258					
90		1.140‡	1.200‡					
95		1.051‡	1.141‡					
100		0.965‡	1.082‡					
105		0.879‡	1.023‡					
110		0.794‡	0.963‡					
115		0.710‡	0.903‡					
120		0.627‡	0.842‡					
125		0.544‡	0.780‡					
130			0.717‡					
135			0.654‡					
140			0.591‡					
145			0.527‡					
150			0.463‡				(1.719)†	
160							(1.694)†	
170							(1.669)†	
180							1.644	
190							1.619	
200							1.594	
210							1.569	
220							1.543	
230				(1.169)†			1.518	
240				(1.150)†			1.492	
250				1.131			1.467	5.22†
260				1.112			1.442	5.39†
270				1.093			1.416	5.55†
280				1.074			1.391	5.74
290				1.055			1.365	5.92
300				1.036			1.340	6.09
310				1.017			1.315	6.23
320				0.997			1.289	6.37
330				0.978	(1.402)†		1.264	6.48
340				0.959	(1.387)†		1.238	6.59
350				0.940	1.373	(1.361)†	1.213	6.68
360				(0.921)	1.359	(1.356)†	1.188	6.75
370				(0.902)	1.345	1.351	1.162	6.80
380				(0.882)	1.331	1.346	1.137	6.84‡
390				(0.863)	1.316	1.341	(1.112)‡	6.86‡
400				(0.844)	1.302	1.335	(1.086)‡	6.86‡
410				(0.825)	1.288	1.329	(1.061)‡	6.86‡
420				(0.806)	1.274	1.323	(1.036)‡	6.84‡
430				(0.787)	1.259	1.317	(1.013)‡	6.81‡
440				(0.768)	1.245	1.310	(0.985)‡	6.78‡
450				(0.749)	1.231	1.304	(0.959)‡	6.73‡
460					1.217	1.297	(0.933)‡	6.67‡
470					1.202	1.290	(0.908)‡	6.61‡
480					1.188	1.283	(0.885)‡	6.53‡
490					1.174	1.276	(0.862)‡	6.45‡
500					1.160	1.268	(0.839)‡	6.35‡
510					1.146	1.261		6.24‡
520					1.131	1.254		6.12‡
530					1.117‡	1.246		5.99‡
540					1.103‡	1.238		5.86‡
550					1.089‡	1.230		5.71‡
560					1.074‡	1.222		5.55‡
570					1.060‡	1.213		5.39‡
580					1.046‡	1.205		5.20‡
590					1.032‡	1.197		5.01‡
600					1.018‡	1.188		4.81‡
610						1.180		4.60‡
620						1.172		4.40‡
630						1.163		(4.20)‡
640						1.155‡		(4.01)‡
650						1.146‡		

† Extrapolated for the supercooled liquid. [Approximate n.m.p. in K: N_2, 63; A, 84; CCl_4, 250; $C_{12}H_{10}$, 342; m-$C_{18}H_{14}$, 361; p-$C_{18}H_{14}$, 486; C_7H_{10}, 178; H_2O, 273.1].

‡Under saturation vapor pressure [Approximate n.b.p. in K: He, 4.3; N_2, 77; A, 88; CCl_4, 350; $C_{12}H_{10}$, 528; m-$C_{18}H_{14}$, 637; p-$C_{18}H_{14}$, 658; C_7H_{10}, 384; H_2O, 373].

THERMAL CONDUCTIVITY OF THE ELEMENTS

Data contained in the following table were extracted from the extensive compilation prepared by C. Y. Ho, R. W. Powell, and P. E. Liley under the National Standard Reference Data System (NSRDS) of the National Bureau of Standards project at the Thermophysical Properties Research Center (TPRC) at Purdue University and published in the Journal of Physical and Chemical Reference Data, *1*, 279-421 (1972). The data in the table below are used with the permission of the authors and the copyright owners, the American Institute of Physics and the American Chemical Society. Users are referred to their more extensive compilation for conductivities at temperatures other than those listed in the table below, and also to obtain an understanding of the basis of selection of recommended and provisional values. Temperatures are in kelvins (K) and conductivities, k, in watt per centimeter kelvin (W cm^{-1} K^{-1}), except as noted. If the numerical value of k has a superscript m, the units of k are milliwatt per centimeter kelvin (mW cm^{-1} K^{-1}). To convert the listed conductivities to units other than those in the tables, one should make use of the conversion factors listed in the table "Conversion Factors for Units of Thermal Conductivity", which follows this table. Conductivity values listed with an asterisk*, are provisional values.

		Conductivity at		
Element	**State or Condition**	**273.2K**	**298.2K**	**373.2K**
Aluminum	Solid	2.36	2.37	2.40
Antimony	Polycrystalline	0.255	0.244	0.219
Argon	Gas at 1 atm.	0.1619^m (270K)	0.1772^m (300K)	0.2103^m (370K)
Arsenic	Solid, Gray, Polycrystalline	0.539*	0.502*	0.427*
Barium	Solid	0.185*	0.184* (295K)	
Beryllium	Polycrystalline	2.18*	2.01	1.68
Bismuth	Solid			
	// to triagonal axis	0.0554	0.0530	0.0481
	⊥ to triagonal axis	0.0953	0.0919	0.0844
	Polycrystalline	0.0822	0.0792	0.0722
Boron	Solid	0.318	0.274	0.188
Bromine	Saturated liquid	1.30^m* (270K)	1.22^m* (300K)	1.06^m* (370K)
	Saturated vapor		0.048^m* (300K)	
	Gas	0.042^m* (270K)	0.048^m* (300K)	0.057^m* (350K)
Cadmium	Solid			
	// to c-axis	0.835	0.830	0.816
	⊥ to c-axis	1.04	1.04	1.02
	Polycrystalline	0.975	0.969	0.953
Calcium	Solid	2.06*	2.01*	1.92*
Carbon	Solid, Amorphous	0.0150	0.0159	0.0182
	Solid, Type I	9.94	9.90	7.03*
	Solid, Type IIa	26.2	23.2	17.0*
	Solid, Type IIb	15.2	13.6	10.2*
	Solid, Acheson graphite			
	// to axis of extrusion	1.69	1.65	1.50
	⊥ to axis of extrusion	1.21	1.19	1.11
	Solid, AGOT graphite			
	// to axis of extrusion	2.28	2.21	1.95
	⊥ to axis of extrusion	1.41	1.38	1.22
	Solid, ATJ graphite			
	// to molding pressure	0.984	0.982	0.933
	⊥ to molding pressure	1.31	1.29	1.21
	Solid, AWG graphite			
	// to molding pressure	0.807	0.796	0.733
	⊥ to molding pressure	1.32	1.28	1.16
	Solid, Pyrolytic graphite			
	// to layer planes	21.3	19.6	15.1
	⊥ to layer planes	0.0636	0.0573	0.0442

THERMAL CONDUCTIVITY OF THE ELEMENTS (*Continued*)

Element	State or Condition	Conductivity at 273.2K	298.2K	373.2K
	Solid, 875S graphite			
	∥ to axis of extrusion	1.97*	1.92*	1.75*
	⊥ to axis of extrusion	1.49*	1.46*	1.34*
	Solid, 890S graphite			
	∥ to axis of extrusion	1.87*	1.83*	1.66*
	⊥ to axis of extrusion	1.51*	1.48*	1.36*
Cerium	Solid, Polycrystalline	0.108*	0.113	0.128*
Cesium	Solid	0.361*	0.359 (301.9K)	
	Liquid		0.197 (301.9K)	0.201
Chlorine	Saturated liquid	1.49^m* (270K)	1.34^m* (300K)	0.95^m* (370K)
	Saturated vapor	0.082^m* (270K)	0.097^m* (300K)	0.155^m* (370K)
	Gas, 1 atm.	0.078^m (270K)	0.089^m (300K)	0.114^m (370K)
Chromium	Solid, Polycrystalline	0.965	0.939	0.921
Cobalt	Solid, Polycrystalline	1.05	1.00	0.890
Copper	Solid	4.03	4.01	3.95
Dysprosium	Solid			
	∥ to c-axis	0.114*	0.117*	
	⊥ to c-axis	0.101*	0.103*	
	Polycrystalline	0.105*	0.107*	0.108*
Erbium	Solid			
	∥ to c-axis	0.187*	0.184*	
	⊥ to c-axis	0.127*	0.126*	
	Polycrystalline	0.147*	0.145*	0.140*
Europium	Solid	0.140*	0.139*	
Fluorine	Gas, 1 atm.	0.251^m (270K)	0.279^m (300K)	0.344^m (370K)
Gadolinium	Solid			
	∥ to c-axis	0.104*	0.108*	
	⊥ to c-axis	0.103*	0.103*	
	Polycrystalline	0.103*	0.105*	
Gallium	Solid			
	∥ to a-axis	0.410	0.408	
	∥ to b-axis	0.884	0.883	
	∥ to c-axis	0.160	0.159	
	Liquid		0.281 (302.93K)	0.328
Germanium	Solid	0.667	0.602	0.465
Gold	Solid	3.19	3.18	3.13
Hafnium	Solid, Polycrystalline	0.233*	0.230	0.224
Helium	Solid, ^{3}He	0.033 (0.9K)	0.020 (1K)	0.0021 (2K)
	Solid, ^{4}He	0.650 (0.9K)	0.245 (1K)	0.0018 (2K)
	Liquid, saturated; He-I	0.191^m (2.5K)	0.232^m (3.5K)	0.434^m (5K)
	Gas, 1 atm.	1.411^m	1.520^m	1.766^m
Holmium	Solid			
	∥ to c-axis	0.215*	0.222*	
	⊥ to c-axis	0.136*	0.138*	
	Polycrystalline	0.159*	0.162*	0.170*
Hydrogen	Solid, Normal Hydrogen	2.30 (4K)	0.0158 (10K)	0.0090 (15K)
	Liquid, saturated; Normal Hydrogen	1.022^m (15K)	1.269^m (25K)	0.60^m* (33K)

Element	State or Condition	Conductivity at 273.2K	298.2K	373.2K
	Gas, 1 atm. Normal Hydrogen	1.665^{m}	1.815^{m}	2.106^{m}
	Liquid, saturated; para-Hydrogen	0.824^{m} (14K)	0.998^{m} (25K)	0.58^{m} (32K)
	Vapor, saturated; para-Hydrogen	0.081^{m}* (10K)	0.242^{m}* (25K)	0.58^{m}* (32K)
	Gas, 1 atm.; para-Hydrogen	1.768^{m}*	1.880^{m}*	2.126^{m}*
	Deuterium:			
	Liquid, saturated	1.26^{m} (20K)	1.37^{m} (30K)	0.83^{m}* (38K)
	Vapor, saturated	0.084^{m}* (20K)	0.26^{m}* (30K)	0.83^{m}* (38K)
	Gas, 1 atm.	1.294^{m}* (270K)	1.406^{m}* (300K)	1.66^{m}* (370K)
	Tritium:			
	Liquid, saturated	1.25 (21K)	1.34 (30K)	0.68 (44K)
Indium	Solid, Polycrystalline	0.837	0.818	0.762
Iodine	Solid	4.81^{m}*	4.49^{m}* (300K)	3.75^{m}* (386.8K)
	Liquid, saturated			1.16^{m}* (386.8K)
Iridium	Solid	1.48	1.47	1.45
Iron	Solid	0.865	0.804	0.720
	Armco Iron	0.747	0.728	0.676
Krypton	Solid	0.4^{m}* (1K)	17^{m} (10K)	2.5^{m} (116K)
	Liquid, saturated			0.931^{m} (116K)
	Vapor, saturated	0.0406^{m}* (120K)	0.0554^{m}* (150K)	0.21^{m}* (210K)
	Gas	0.0860^{m} (270K)	0.0949^{m} (300K)	0.1145^{m} (370K)
Lanthanum	Solid, Polycrystalline	0.131	0.134	0.145
Lead	Solid	0.356	0.353	0.344
Lithium	Solid	0.859	0.848	0.818
Lutetium	Solid			
	// to c-axis	0.236*	0.232*	
	⊥ to c-axis	0.140*	0.138*	
	Polycrystalline	0.167*	0.164*	
Magnesium	Solid, Polycrystalline	1.57	1.56	1.54
Manganese	Solid	0.0768*	0.0781*	
Mercury	Liquid	0.0782	0.0830	0.0947
Molybdenum	Solid	1.39*	1.38*	1.35*
Neodymium	Solid, Polycrystalline	0.165*	0.165*	0.167*
Neon	Gas	0.461^{m}* (270K)	0.493^{m}* (300K)	0.563^{m}* (370K)
Neptunium	Solid		0.063* (300K)	
Nickel	Solid	0.941	0.909	0.827
Niobium	Solid	0.533	0.537	0.548
Nitrogen	Solid	56^{m} (4K)	17^{m} (10K)	3.2^{m} (25K)
	Liquid, saturated	1.60^{m} (65K)	0.966^{m} (100K)	0.37^{m}* (126K)
	Vapor, saturated	0.061^{m}* (65K)	0.111^{m}* (100K)	0.37^{m}* (126K)
	Gas, 1 atm.	0.2374^{m} (270K)	0.2598^{m} (300K)	0.3065^{m} (370K)

THERMAL CONDUCTIVITY OF THE ELEMENTS (*Continued*)

Element	State or Condition	Conductivity at 273.2K	298.2K	373.2K
Cesium	Solid			
	∥ to c-axis	2.93	14.3	15.4
		(2K)	(10K)	(30K)
	⊥ to c-axis	1.76	8.65	11.1
		(2K)	(10K)	(30K)
	Polycrystalline	2.09	10.2	12.4
		(2K)	(10K)	(30K)
	Polycrystalline	0.880*	0.876*	0.870*
Oxygen	Liquid, saturated	1.501^m	1.023^m	0.41^m
		(90K)	(125K)	(155K)
	Vapor, saturated	0.081^{m}*	0.135^{m}*	0.41^{m}*
		(90K)	(125K)	(155K)
	Gas, 1 atm.	0.2424	0.2674	0.3204
		(270K)	(300K)	(370K)
Palladium	Solid	0.716*	0.718	0.730
Phosphorus	Solid			
	Black (Polycrystalline)	0.132	0.121	
	White	0.00250*	0.00236*	
	Liquid, White			0.00181
Platinum	Solid	0.717	0.716	0.717
Plutonium	Solid, polycrystalline	0.0616*	0.0670*	0.0790*
				(350K)
Potassium	Solid	1.036*	1.025	
	Liquid			0.532
Praseodymium	Solid, polycrystalline	0.120	0.125	0.134
Promethium	Solid, polycrystalline		0.179*	0.184*
Radium	Solid		0.186	
			(293.2K)	
Radon	Gas, 1 atm.	0.0327^{m}*	0.0364^{m}*	0.0445^{m}*
		(270K)	(300K)	(370K)
Rhenium	Solid, polycrystalline	0.486	0.480	0.466
Rhodium	Solid	1.51	1.50	1.47
Rubidium	Solid	0.583*	0.582	0.581
				(312.04K)
	Liquid			0.333
				(312.04K)
Ruthenium	Solid, polycrystalline	1.17	1.17	1.15
Samarium	Solid, polycrystalline	0.133*	0.133*	0.133*
Scandium	Solid, polycrystalline	0.157	0.158	
Selenium	Solid			
	∥ to c-axis	0.0481	0.0452	0.0483
	⊥ to c-axis	0.0137	0.0131	0.0139
	Amorphous	0.00428	0.00519	0.00818
				(323.2K)
Silicon	Solid	1.68	1.49	1.08
Silver	Solid	4.29		4.26
Sodium	Solid	1.42	1.42	1.32
				(371K)
Strontium	Solid, polycrystalline	0.364*	0.354*	0.325*
Sulfur	Solid, polycrystalline	0.00287	0.00270	0.00154
	Solid, amorphous	0.00200	0.00205	0.00216*
				(350K)
	Liquid			0.00129
				(392.2K)
Tantalum	Solid	0.574	0.575	0.577
Technetium	Solid, polycrystalline	0.509*	0.506	0.501
Tellurium	Solid			
	∥ to c-axis	0.0360	0.0338	0.0292
	⊥ to c-axis	0.0208	0.197	0.173

THERMAL CONDUCTIVITY OF THE ELEMENTS (*Continued*)

Element	State or Condition	Conductivity at 273.2K	298.2K	373.2K
Terbium	Solid			
	// to c-axis	0.138*	0.147*	
	⊥ to c-axis	0.0900*	0.0956*	
	Polycrystalline	0.104*	0.111*	
Thallium	Solid, polycrystalline	0.469	0.461	0.443
Thorium	Solid	0.540*	0.540*	0.543*
Thulium	Solid			
	// to c-axis	0.242*	0.242*	
	⊥ to c-axis	0.140*	0.141*	
	Polycrystalline	0.168*	0.169*	
Tin	Solid			
	// to c-axis	0.527	0.516	0.489
	⊥ to c-axis	0.759	0.743	0.704
	Polycrystalline	0.682	0.668	0.632
Titanium	Solid, polycrystalline	0.224	0.219	0.207
Tungsten	Solid	1.77	1.73	1.63
Uranium	Solid, polycrystalline	0.270	0.275	0.291
Vanadium	Solid	0.307*	0.307	0.310
Xenon	Liquid, saturated	0.31[m] (270K)	0.16[m]* (290K)	
	Vapor, saturated	0.084[m]*	0.16[m]*	
	Gas, 1 atm.	0.0514[m] (270K)	0.0569[m] (300K)	0.0695[m] (370K)
Ytterbium	Solid	0.354*	0.349*	0.343*
Yttrium	Solid, polycrystalline	0.170*	0.172*	0.177*
Zinc	Solid, polycrystalline	1.17	1.16	1.12
Zirconium	Solid, polycrystalline	0.232*	0.227 (300K)	0.218

THERMAL CONDUCTIVITY OF ROCKS

Rock	Temperature, °C	Conductivity, cal m^{-1} hr^{-1} deg^{-1}	Heat Capacity, cal g^{-1} deg^{-1}
Granite	0	3.02	0.192
	50	2.81	–
	100	2.59	
	200	2.34	0.228
	300	2.12	–
	400	–	0.258
Marble	118	1.44	0.21
	196	1.29	0.24
	245	1.19	–
	360	0.95	0.271
Dolomitic limestone	130	1.41	–
	181	1.37	–
	268	1.29	–
	377	1.15	–
Shale	0	1.65	0.17
	100	1.51	–
	120	1.33	–
	188	1.41	0.24
	304	1.26	0.245
Sandstone (quartzitic)	0	4.9	–
	100	3.82	0.26
	200	3.24	

CONVERSION FACTORS FOR UNITS OF THERMAL CONDUCTIVITY

MULTIPLY

by appropriate factor to / OBTAIN→	$Btu_{IT}\ h^{-1}\ ft^{-1}\ F^{-1}$	Btu_{IT} in. $h^{-1}\ ft^{-2}\ F^{-1}$	$Btu_{th}\ h^{-1}\ ft^{-1}\ F^{-1}$	Btu_{th} in. $h^{-1}\ ft^{-2}\ F^{-1}$	$cal_{IT}\ s^{-1}\ cm^{-1}\ C^{-1}$	$cal_{th}\ s^{-1}\ cm^{-1}\ C^{-1}$
$Btu_{IT}\ h^{-1}\ ft^{-1}\ F^{-1}$	1	12	1.00067	12.0080	4.13379×10^{-3}	4.13656×10^{-3}
Btu_{IT} in. $h^{-1}\ ft^{-2}\ F^{-1}$	8.33333×10^{-2}	1	8.33891×10^{-2}	1.00067	3.44482×10^{-4}	3.44713×10^{-4}
$Btu_{th}\ h^{-1}\ ft^{-1}\ F^{-1}$	0.999331	11.9920	1	12	4.13102×10^{-3}	4.13379×10^{-3}
Btu_{th} in. $h^{-1}\ ft^{-2}\ F^{-1}$	8.32776×10^{-2}	0.999331	8.33333×10^{-2}	1	3.44252×10^{-4}	3.44482×10^{-4}
$cal_{IT}\ s^{-1}\ cm^{-1}\ C^{-1}$	2.41909×10^{2}	2.90291×10^{3}	2.42071×10^{2}	2.90485×10^{3}	1	1.00067
$cal_{th}\ s^{-1}\ cm^{-1}\ C^{-1}$	2.41747×10^{2}	2.90096×10^{3}	2.41909×10^{2}	2.90291×10^{3}	0.999331	1
$kcal_{th}\ h^{-1}\ m^{-1}\ C^{-1}$	0.671520	8.05824	0.671969	8.06363	2.77592×10^{-3}	2.77778×10^{-3}
$J\ s^{-1}\ cm^{-1}\ K^{-1}$	57.7789	6.93347×10^{2}	57.8176	6.93811×10^{2}	0.238846	0.239006
$W\ cm^{-1}\ K^{-1}$	57.7789	6.93347×10^{2}	57.8176	6.93811×10^{2}	0.238846	0.239006
$W\ m^{-1}\ K^{-1}$	0.577789	6.93347	0.578176	6.93811	2.38846×10^{-3}	2.39006×10^{-3}
$mW\ cm^{-1}\ K^{-1}$	5.77789×10^{-2}	0.693347	5.78176×10^{-2}	0.693811	2.38846×10^{-4}	2.39006×10^{-4}

MULTIPLY

by appropriate factor to / OBTAIN→	$kcal_{th}\ h^{-1}\ m^{-1}\ C^{-1}$	$J\ s^{-1}\ cm^{-1}\ K^{-1}$	$W\ cm^{-1}\ K^{-1}$	$W\ m^{-1}\ K^{-1}$	$mW\ cm^{-1}\ K^{-1}$
$Btu_{IT}\ h^{-1}\ ft^{-1}\ F^{-1}$	1.48916	1.73073×10^{-2}	1.73073×10^{-2}	1.73073	17.3073
Btu_{IT} in. $h^{-1}\ ft^{-2}\ F^{-1}$	0.124097	1.44228×10^{-3}	1.44228×10^{-3}	0.144228	1.44228
$Btu_{th}\ h^{-1}\ ft^{-1}\ F^{-1}$	1.48816	1.72958×10^{-2}	1.72958×10^{-2}	1.72958	17.2958
Btu_{th} in. $h^{-1}\ ft^{-2}\ F^{-1}$	0.124014	1.44131×10^{-3}	1.44131×10^{-3}	0.144131	1.44131
$cal_{IT}\ s^{-1}\ cm^{-1}\ C^{-1}$	3.60241×10^{2}	4.1868	4.1868	4.1868×10^{2}	4.1868×10^{-3}
$cal_{th}\ s^{-1}\ cm^{-1}\ C^{-1}$	3.6×10^{2}	4.184	4.184	4.184×10^{2}	4.184×10^{-3}
$kcal_{th}\ h^{-1}\ m^{-1}\ C^{-1}$	1	1.16222×10^{-2}	1.16222×10^{-2}	1.16222	11.6222
$J\ s^{-1}\ cm^{-1}\ K^{-1}$	86.0421	1	1	1×10^{2}	1×10^{3}
$W\ cm^{-1}\ K^{-1}$	86.0421	1	1	1×10^{2}	1×10^{3}
$W\ m^{-1}\ K^{-1}$	0.860421	1×10^{-2}	1×10^{-2}	1	10
$mW\ cm^{-1}\ K^{-1}$	8.60421×10^{-2}	1×10^{-3}	1×10^{-3}	0.1	1

Reproduced by permission of the publishers and copyright owners of the 1967 ASME Steam Tables. Further data and information on the thermodynamic and transport properties of steam and water are contained in the above ASME publication. It is obtainable from The American Society of Mechanical Engineers, United Engineering Center, 345 East 47th Street, New York, New York 10017.

Properties of Saturated Steam and Saturated Water

Temp. F	Press. psia	Volume, ft³/lbm Water v_f	Volume Evap. v_{fg}	Volume Steam v_g	Enthalpy, Btu/lbm Water h_f	Enthalpy Evap. h_{fg}	Enthalpy Steam h_g	Entropy, Btu/lbm ×F Water s_f	Entropy Evap. s_{fg}	Entropy Steam s_g	Temp. F
705.47	3208.2	0.05078	0.00000	0.05078	906.0	0.0	906.0	1.0612	0.0000	1.0612	705.47
705.0	3198.3	0.04427	0.01304	0.05730	873.0	61.4	934.4	1.0329	0.0527	1.0856	705.0
704.5	3187.8	0.04233	0.01822	0.06055	861.9	85.3	947.2	1.0234	0.0732	1.0967	704.5
704.0	3177.2	0.04108	0.02192	0.06300	854.2	102.0	956.2	1.0169	0.0876	1.1046	704.0
703.5	3166.8	0.04015	0.02489	0.06504	848.2	115.2	963.5	1.0118	0.0991	1.1109	703.5
703.0	3156.3	0.03940	0.02744	0.06684	843.2	126.4	969.6	1.0076	0.1087	1.1163	703.0
702.5	3145.9	0.03878	0.02969	0.06847	838.9	136.1	974.9	1.0039	0.1171	1.1210	702.5
702.0	3135.5	0.03824	0.03173	0.06997	835.0	144.7	979.7	1.0006	0.1246	1.1252	702.0
701.5	3125.2	0.03777	0.03361	0.07138	831.5	152.6	984.0	0.9977	0.1314	1.1291	701.5
701.0	3114.9	0.03735	0.03536	0.07271	828.2	159.8	988.0	0.9949	0.1377	1.1326	701.0
700.5	3104.6	0.03697	0.03701	0.07397	825.2	166.5	991.7	0.9924	0.1435	1.1359	700.5
700.0	3094.3	0.03662	0.03857	0.07519	822.4	172.7	995.2	0.9901	0.1490	1.1390	700.0
699.0	3073.9	0.03600	0.04149	0.07749	817.3	184.2	1001.5	0.9858	0.1590	1.1447	699.0
698.0	3053.6	0.03546	0.04420	0.07966	812.6	194.6	1007.2	0.9818	0.1681	1.1499	698.0
697.0	3033.5	0.03498	0.04674	0.08172	808.4	204.0	1012.4	0.9783	0.1764	1.1547	697.0
696.0	3013.4	0.03455	0.04916	0.08371	804.4	212.8	1017.2	0.9749	0.1841	1.1591	696.0
695.0	2993.5	0.03415	0.05147	0.08563	800.6	221.0	1021.7	0.9718	0.1914	1.1632	695.0
694.0	2973.7	0.03379	0.05370	0.08749	797.1	228.8	1025.9	0.9689	0.1983	1.1671	694.0
693.0	2954.0	0.03345	0.05587	0.08931	793.8	236.1	1029.9	0.9660	0.2048	1.1708	693.0
692.0	2934.5	0.03313	0.05797	0.09110	790.5	243.1	1033.6	0.9634	0.2110	1.1744	692.0
690.0	2895.7	0.03256	0.06203	0.09459	784.5	256.1	1040.6	0.9583	0.2227	1.1810	690.0
688.0	2857.4	0.03204	0.06595	0.09799	778.8	268.2	1047.0	0.9535	0.2337	1.1872	688.0
686.0	2819.5	0.03157	0.06976	0.10133	773.4	279.5	1052.9	0.9490	0.2439	1.1930	686.0
684.0	2782.1	0.03114	0.07349	0.10463	768.2	290.2	1058.4	0.9447	0.2537	1.1984	684.0
682.0	2745.1	0.03074	0.07716	0.10790	763.3	300.4	1063.6	0.9406	0.2631	1.2036	682.0
680.0	2708.6	0.03037	0.08080	0.11117	758.5	310.1	1068.5	0.9365	0.2720	1.2086	680.0
678.0	2672.5	0.03002	0.08440	0.11442	753.8	319.4	1073.2	0.9326	0.2807	1.2133	678.0
676.0	2636.8	0.02970	0.08799	0.11769	749.2	328.5	1077.6	0.9287	0.2892	1.2179	676.0
674.0	2601.5	0.02939	0.09156	0.12096	744.7	337.2	1081.9	0.9249	0.2974	1.2223	674.0
672.0	2566.6	0.02911	0.09514	0.12424	740.2	345.7	1085.9	0.9212	0.3054	1.2266	672.0
670.0	2532.2	0.02884	0.09871	0.12755	735.8	354.0	1089.8	0.9174	0.3133	1.2307	670.0
668.0	2498.1	0.02858	0.10229	0.13087	731.5	362.1	1093.5	0.9137	0.3210	1.2347	668.0
666.0	2464.4	0.02834	0.10588	0.13421	727.1	370.0	1097.1	0.9100	0.3286	1.2387	666.0
664.0	2431.1	0.02811	0.10947	0.13757	722.9	377.7	1100.6	0.9064	0.3361	1.2425	664.0
662.0	2398.2	0.02789	0.11306	0.14095	718.8	385.1	1103.9	0.9028	0.3434	1.2462	662.0
660.0	2365.7	0.02768	0.11663	0.14431	714.9	392.1	1107.0	0.8995	0.3502	1.2498	660.0
658.0	2333.5	0.02748	0.12023	0.14771	711.1	399.0	1110.1	0.8963	0.3570	1.2533	658.0
656.0	2301.7	0.02728	0.12387	0.15115	707.4	405.7	1113.1	0.8931	0.3637	1.2567	656.0
654.0	2270.3	0.02709	0.12754	0.15463	703.7	412.2	1115.9	0.8899	0.3702	1.2601	654.0
652.0	2239.2	0.02691	0.13124	0.15816	700.0	418.7	1118.7	0.8868	0.3767	1.2634	652.0
650.0	2208.4	0.02674	0.13499	0.16173	696.4	425.0	1121.4	0.8837	0.3830	1.2667	650.0
648.0	2178.1	0.02657	0.13876	0.16534	692.9	431.1	1124.0	0.8806	0.3893	1.2699	648.0
646.0	2148.0	0.02641	0.14258	0.16899	689.4	437.2	1126.6	0.8776	0.3954	1.2730	646.0
644.0	2118.3	0.02625	0.14644	0.17269	685.9	443.1	1129.0	0.8746	0.4015	1.2761	644.0
642.0	2088.9	0.02610	0.15033	0.17643	682.5	448.9	1131.4	0.8716	0.4075	1.2791	642.0
640.0	2059.9	0.02595	0.15427	0.18021	679.1	454.6	1133.7	0.8686	0.4134	1.2821	640.0
638.0	2031.2	0.02580	0.15824	0.18405	675.8	460.2	1136.0	0.8657	0.4193	1.2850	638.0
636.0	2002.8	0.02566	0.16226	0.18792	672.4	465.7	1138.1	0.8628	0.4251	1.2879	636.0
634.0	1974.7	0.02553	0.16633	0.19185	669.1	471.1	1140.2	0.8599	0.4307	1.2907	634.0
632.0	1947.0	0.02539	0.17044	0.19583	665.9	476.4	1142.2	0.8571	0.4364	1.2934	632.0
630.0	1919.5	0.02526	0.17459	0.19986	662.7	481.6	1144.2	0.8542	0.4419	1.2962	630.0
628.0	1892.4	0.02514	0.17880	0.20394	659.5	486.7	1146.1	0.8514	0.4474	1.2988	628.0
626.0	1865.6	0.02501	0.18306	0.20807	656.3	491.7	1148.0	0.8486	0.4529	1.3015	626.0
624.0	1839.0	0.02489	0.18737	0.21226	653.1	496.6	1149.8	0.8458	0.4583	1.3041	624.0
622.0	1812.8	0.02477	0.19173	0.21650	650.0	501.5	1151.5	0.8430	0.4636	1.3066	622.0
620.0	1786.9	0.02466	0.19615	0.22081	646.9	506.3	1153.2	0.8403	0.4689	1.3092	620.0
618.0	1761.2	0.02455	0.20063	0.22517	643.8	511.0	1154.8	0.8375	0.4742	1.3117	618.0
616.0	1735.9	0.02444	0.20516	0.22960	640.8	515.6	1156.4	0.8348	0.4794	1.3141	616.0
614.0	1710.8	0.02433	0.20976	0.23409	637.8	520.2	1158.0	0.8321	0.4845	1.3166	614.0
612.0	1686.1	0.02422	0.21442	0.23865	634.8	524.7	1159.5	0.8294	0.4896	1.3190	612.0
610.0	1661.6	0.02412	0.21915	0.24327	631.8	529.2	1160.9	0.8267	0.4947	1.3214	610.0
608.0	1637.3	0.02402	0.22394	0.24796	628.8	533.6	1162.4	0.8240	0.4997	1.3238	608.0
606.0	1613.4	0.02392	0.22881	0.25273	625.9	537.9	1163.8	0.8214	0.5048	1.3261	606.0
604.0	1589.7	0.02382	0.23374	0.25757	622.9	542.2	1165.1	0.8187	0.5097	1.3284	604.0
602.0	1566.3	0.02373	0.23875	0.26248	620.0	546.4	1166.4	0.8161	0.5147	1.3307	602.0
600.0	1543.2	0.02364	0.24384	0.26747	617.1	550.6	1167.7	0.8134	0.5196	1.3330	600.0
598.0	1520.4	0.02354	0.24900	0.27255	614.3	554.7	1169.0	0.8108	0.5245	1.3353	598.0
596.0	1497.8	0.02345	0.25425	0.27770	611.4	558.8	1170.2	0.8082	0.5293	1.3375	596.0
594.0	1475.4	0.02337	0.25958	0.28294	608.6	562.8	1171.4	0.8056	0.5342	1.3398	594.0
592.0	1453.3	0.02328	0.26499	0.28827	605.7	566.8	1172.6	0.8030	0.5390	1.3420	592.0
590.0	1431.5	0.02319	0.27049	0.29368	602.9	570.8	1173.7	0.8004	0.5437	1.3442	590.0
588.0	1410.0	0.02311	0.27608	0.29919	600.1	574.7	1174.8	0.7978	0.5485	1.3464	588.0
586.0	1388.6	0.02303	0.28176	0.30478	597.3	578.5	1175.9	0.7953	0.5532	1.3485	586.0
584.0	1367.6	0.02295	0.28753	0.31048	594.6	582.4	1176.9	0.7927	0.5580	1.3507	584.0
582.0	1346.7	0.02287	0.29340	0.31627	591.8	586.1	1178.0	0.7902	0.5627	1.3528	582.0
580.0	1326.2	0.02279	0.29937	0.32216	589.1	589.9	1179.0	0.7876	0.5673	1.3550	580.0

Quantities for saturated liquid v_f h_f s_f

Quantities for saturated vapor v_g h_g s_g

Increment for evaporation v_{fg} h_{fg} s_{fg}

STEAM TABLES (Continued)

Properties of Saturated Steam and Saturated Water

Temp. F	Press. psia	Volume, ft³/lbm Water v_f	Evap. v_{fg}	Steam v_g	Enthalpy, Btu/lbm Water h_f	Evap. h_{fg}	Steam h_g	Entropy, Btu/lbm ×F Water s_f	Evap. s_{fg}	Steam s_g	Temp. F
580.0	1326.17	0.02279	0.29937	0.32216	589.1	589.9	1179.0	0.7876	0.5673	1.3550	580.0
578.0	1305.84	0.02271	0.30544	0.32816	586.4	593.6	1179.9	0.7851	0.5720	1.3571	578.0
576.0	1285.74	0.02264	0.31162	0.33426	583.7	597.2	1180.9	0.7825	0.5766	1.3592	576.0
574.0	1265.89	0.02256	0.31790	0.34046	581.0	600.9	1181.8	0.7800	0.5813	1.3613	574.0
572.0	1246.26	0.02249	0.32429	0.34678	578.3	604.5	1182.7	0.7775	0.5859	1.3634	572.0
570.0	1226.88	0.02242	0.33079	0.35321	575.6	608.0	1183.6	0.7750	0.5905	1.3654	570.0
568.0	1207.72	0.02235	0.33741	0.35975	572.9	611.5	1184.5	0.7725	0.5950	1.3675	568.0
566.0	1188.80	0.02228	0.34414	0.36642	570.3	615.0	1185.3	0.7699	0.5996	1.3696	566.0
564.0	1170.10	0.02221	0.35099	0.37320	567.6	618.5	1186.1	0.7674	0.6041	1.3716	564.0
562.0	1151.63	0.02214	0.35797	0.38011	565.0	621.9	1186.9	0.7650	0.6087	1.3736	562.0
560.0	1133.38	0.02207	0.36507	0.38714	562.4	625.3	1187.7	0.7625	0.6132	1.3757	560.0
558.0	1115.36	0.02201	0.37230	0.39431	559.8	628.6	1188.4	0.7600	0.6177	1.3777	558.0
556.0	1097.55	0.02194	0.37966	0.40160	557.2	632.0	1189.2	0.7575	0.6222	1.3797	556.0
554.0	1079.96	0.02188	0.38715	0.40903	554.6	635.3	1189.9	0.7550	0.6267	1.3817	554.0
552.0	1062.59	0.02182	0.39479	0.41660	552.0	638.5	1190.6	0.7525	0.6311	1.3837	552.0
550.0	1045.43	0.02176	0.40256	0.42432	549.5	641.8	1191.2	0.7501	0.6356	1.3856	550.0
548.0	1028.49	0.02169	0.41048	0.43217	546.9	645.0	1191.9	0.7476	0.6400	1.3876	548.0
546.0	1011.75	0.02163	0.41855	0.44018	544.4	648.1	1192.5	0.7451	0.6445	1.3896	546.0
544.0	995.22	0.02157	0.42677	0.44834	541.8	651.3	1193.1	0.7427	0.6489	1.3915	544.0
542.0	978.90	0.02151	0.43514	0.45665	539.3	654.4	1193.7	0.7402	0.6533	1.3935	542.0
540.0	962.79	0.02146	0.44367	0.46513	536.8	657.5	1194.3	0.7378	0.6577	1.3954	540.0
538.0	946.88	0.02140	0.45237	0.47377	534.2	660.6	1194.8	0.7353	0.6621	1.3974	538.0
536.0	931.17	0.02134	0.46123	0.48257	531.7	663.6	1195.4	0.7329	0.6665	1.3993	536.0
534.0	915.66	0.02129	0.47026	0.49155	529.2	666.6	1195.9	0.7304	0.6708	1.4012	534.0
532.0	900.34	0.02123	0.47947	0.50070	526.8	669.6	1196.4	0.7280	0.6752	1.4032	532.0
530.0	885.23	0.02118	0.48886	0.51004	524.3	672.6	1196.9	0.7255	0.6796	1.4051	530.0
528.0	870.31	0.02112	0.49843	0.51955	521.8	675.5	1197.3	0.7231	0.6839	1.4070	528.0
526.0	855.58	0.02107	0.50819	0.52926	519.3	678.4	1197.8	0.7206	0.6883	1.4089	526.0
524.0	841.04	0.02102	0.51814	0.53916	516.9	681.3	1198.2	0.7182	0.6926	1.4108	524.0
522.0	826.69	0.02097	0.52829	0.54926	514.4	684.2	1198.6	0.7158	0.6969	1.4127	522.0
520.0	812.53	0.02091	0.53864	0.55956	512.0	687.0	1199.0	0.7133	0.7013	1.4146	520.0
518.0	798.55	0.02086	0.54920	0.57006	509.6	689.9	1199.4	0.7109	0.7056	1.4165	518.0
516.0	784.76	0.02081	0.55997	0.58079	507.1	692.7	1199.8	0.7085	0.7099	1.4183	516.0
514.0	771.15	0.02076	0.57096	0.59173	504.7	695.4	1200.2	0.7060	0.7142	1.4202	514.0
512.0	757.72	0.02072	0.58218	0.60289	502.3	698.2	1200.5	0.7036	0.7185	1.4221	512.0
510.0	744.47	0.02067	0.59362	0.61429	499.9	700.9	1200.8	0.7012	0.7228	1.4240	510.0
508.0	731.40	0.02062	0.60530	0.62592	497.5	703.7	1201.1	0.6987	0.7271	1.4258	508.0
506.0	718.50	0.02057	0.61722	0.63779	495.1	706.3	1201.4	0.6963	0.7314	1.4277	506.0
504.0	705.78	0.02053	0.62938	0.64991	492.7	709.0	1201.7	0.6939	0.7357	1.4296	504.0
502.0	693.23	0.02048	0.64180	0.66228	490.3	711.7	1202.0	0.6915	0.7400	1.4314	502.0
500.0	680.86	0.02043	0.65448	0.67492	487.9	714.3	1202.2	0.6890	0.7443	1.4333	500.0
498.0	668.65	0.02039	0.66743	0.68782	485.6	716.9	1202.5	0.6866	0.7486	1.4352	498.0
496.0	656.61	0.02034	0.68065	0.70100	483.2	719.5	1202.7	0.6842	0.7528	1.4370	496.0
494.0	644.73	0.02030	0.69415	0.71445	480.8	722.1	1202.9	0.6818	0.7571	1.4389	494.0
492.0	633.03	0.02026	0.70794	0.72820	478.5	724.6	1203.1	0.6793	0.7614	1.4407	492.0
490.0	621.48	0.02021	0.72203	0.74224	476.1	727.2	1203.3	0.6769	0.7657	1.4426	490.0
488.0	610.10	0.02017	0.73641	0.75658	473.8	729.7	1203.5	0.6745	0.7700	1.4444	488.0
486.0	598.87	0.02013	0.75111	0.77124	471.5	732.2	1203.7	0.6721	0.7742	1.4463	486.0
484.0	587.81	0.02009	0.76613	0.78622	469.1	734.7	1203.8	0.6696	0.7785	1.4481	484.0
482.0	576.90	0.02004	0.78148	0.80152	466.8	737.2	1204.0	0.6672	0.7828	1.4500	482.0
480.0	566.15	0.02000	0.79716	0.81717	464.5	739.6	1204.1	0.6648	0.7871	1.4518	480.0
478.0	555.55	0.01996	0.81319	0.83315	462.2	742.1	1204.2	0.6624	0.7913	1.4537	478.0
476.0	545.11	0.01992	0.82958	0.84950	459.9	744.5	1204.3	0.6599	0.7956	1.4555	476.0
474.0	534.81	0.01988	0.84632	0.86621	457.5	746.9	1204.4	0.6575	0.7999	1.4574	474.0
472.0	524.67	0.01984	0.86345	0.88329	455.2	749.3	1204.5	0.6551	0.8042	1.4592	472.0
470.0	514.67	0.01980	0.88095	0.90076	452.9	751.6	1204.6	0.6527	0.8084	1.4611	470.0
468.0	504.83	0.01976	0.89885	0.91862	450.7	754.0	1204.6	0.6502	0.8127	1.4629	468.0
466.0	495.12	0.01973	0.91716	0.93689	448.4	756.3	1204.7	0.6478	0.8170	1.4648	466.0
464.0	485.56	0.01969	0.93588	0.95557	446.1	758.6	1204.7	0.6454	0.8213	1.4667	464.0
462.0	476.14	0.01965	0.95504	0.97469	443.8	761.0	1204.8	0.6429	0.8256	1.4685	462.0
460.0	466.87	0.01961	0.97463	0.99424	441.5	763.2	1204.8	0.6405	0.8299	1.4704	460.0
458.0	457.73	0.01958	0.99467	1.01425	439.3	765.5	1204.8	0.6381	0.8342	1.4722	458.0
456.0	448.73	0.01954	1.01518	1.03472	437.0	767.8	1204.8	0.6356	0.8385	1.4741	456.0
454.0	439.87	0.01950	1.03616	1.05567	434.7	770.0	1204.8	0.6332	0.8428	1.4759	454.0
452.0	431.14	0.01947	1.05764	1.07711	432.5	772.3	1204.8	0.6308	0.8471	1.4778	452.0
450.0	422.55	0.01943	1.07962	1.09905	430.2	774.5	1204.7	0.6283	0.8514	1.4797	450.0
448.0	414.09	0.01940	1.10212	1.12152	428.0	776.7	1204.7	0.6259	0.8557	1.4815	448.0
446.0	405.76	0.01936	1.12515	1.14452	425.7	778.9	1204.6	0.6234	0.8600	1.4834	446.0
444.0	397.56	0.01933	1.14874	1.16806	423.5	781.1	1204.6	0.6210	0.8643	1.4853	444.0
442.0	389.49	0.01929	1.17288	1.19217	421.3	783.2	1204.5	0.6185	0.8686	1.4872	442.0
440.0	381.54	0.01926	1.19761	1.21687	419.0	785.4	1204.4	0.6161	0.8729	1.4890	440.0
438.0	373.72	0.01923	1.22293	1.24216	416.8	787.5	1204.3	0.6136	0.8773	1.4909	438.0
436.0	366.03	0.01919	1.24887	1.26806	414.6	789.7	1204.2	0.6112	0.8816	1.4928	436.0
434.0	358.46	0.01916	1.27544	1.29460	412.4	791.8	1204.1	0.6087	0.8859	1.4947	434.0
432.0	351.00	0.01913	1.30266	1.32179	410.1	793.9	1204.0	0.6063	0.8903	1.4966	432.0
430.0	343.67	0.01909	1.33055	1.34965	407.9	796.0	1203.9	0.6038	0.8946	1.4985	430.0

STEAM TABLES (Continued)

Properties of Saturated Steam and Saturated Water

Temp. F	Press. psia	Volume, ft³/lbm Water v_f	Evap. v_{fg}	Steam v_g	Enthalpy, Btu/lbm Water h_f	Evap. h_{fg}	Steam h_g	Entropy, Btu/lbm ×F Water s_f	Evap. s_{fg}	Steam s_g	Temp. F
430.0	343.674	0.01909	1.3306	1.3496	407.9	796.0	1203.9	0.6038	0.8946	1.4985	430.0
428.0	336.463	0.01906	1.3591	1.3782	405.7	798.0	1203.7	0.6014	0.8990	1.5004	428.0
426.0	329.369	0.01903	1.3884	1.4075	403.5	800.1	1203.6	0.5989	0.9034	1.5023	426.0
424.0	322.391	0.01900	1.4184	1.4374	401.3	802.2	1203.5	0.5964	0.9077	1.5042	424.0
422.0	315.529	0.01897	1.4492	1.4682	399.1	804.2	1203.3	0.5940	0.9121	1.5061	422.0
420.0	308.780	0.01894	1.4808	1.4997	396.9	806.2	1203.1	0.5915	0.9165	1.5080	420.0
418.0	302.143	0.01890	1.5131	1.5320	394.7	808.2	1202.9	0.5890	0.9209	1.5099	418.0
416.0	295.617	0.01887	1.5463	1.5651	392.5	810.2	1202.8	0.5866	0.9253	1.5118	416.0
414.0	289.201	0.01884	1.5803	1.5991	390.3	812.2	1202.6	0.5841	0.9297	1.5137	414.0
412.0	282.894	0.01881	1.6152	1.6340	388.1	814.2	1202.4	0.5816	0.9341	1.5157	412.0
410.0	276.694	0.01878	1.6510	1.6697	386.0	816.2	1202.1	0.5791	0.9385	1.5176	410.0
408.0	270.600	0.01875	1.6877	1.7064	383.8	818.2	1201.9	0.5766	0.9429	1.5195	408.0
406.0	264.611	0.01872	1.7253	1.7441	381.6	820.1	1201.7	0.5742	0.9473	1.5215	406.0
404.0	258.725	0.01870	1.7640	1.7827	379.4	822.0	1201.5	0.5717	0.9518	1.5234	404.0
402.0	252.942	0.01867	1.8037	1.8223	377.3	824.0	1201.2	0.5692	0.9562	1.5254	402.0
400.0	247.259	0.01864	1.8444	1.8630	375.1	825.9	1201.0	0.5667	0.9607	1.5274	400.0
398.0	241.677	0.01861	1.8862	1.9048	372.9	827.8	1200.7	0.5642	0.9651	1.5293	398.0
396.0	236.193	0.01858	1.9291	1.9477	370.8	829.7	1200.4	0.5617	0.9696	1.5313	396.0
394.0	230.807	0.01855	1.9731	1.9917	368.6	831.6	1200.2	0.5592	0.9741	1.5333	394.0
392.0	225.516	0.01853	2.0184	2.0369	366.5	833.4	1199.9	0.5567	0.9786	1.5352	392.0
390.0	220.321	0.01850	2.0649	2.0833	364.3	835.3	1199.6	0.5542	0.9831	1.5372	390.0
388.0	215.220	0.01847	2.1126	2.1311	362.2	837.2	1199.3	0.5516	0.9876	1.5392	388.0
386.0	210.211	0.01844	2.1616	2.1801	360.0	839.0	1199.0	0.5491	0.9921	1.5412	386.0
384.0	205.294	0.01842	2.2120	2.2304	357.9	840.8	1198.7	0.5466	0.9966	1.5432	384.0
382.0	200.467	0.01839	2.2638	2.2821	355.7	842.7	1198.4	0.5441	1.0012	1.5452	382.0
380.0	195.729	0.01836	2.3170	2.3353	353.6	844.5	1198.0	0.5416	1.0057	1.5473	380.0
378.0	191.080	0.01834	2.3716	2.3900	351.4	846.3	1197.7	0.5390	1.0103	1.5493	378.0
376.0	186.517	0.01831	2.4279	2.4462	349.3	848.1	1197.4	0.5365	1.0148	1.5513	376.0
374.0	182.040	0.01829	2.4857	2.5039	347.2	849.8	1197.0	0.5340	1.0194	1.5534	374.0
372.0	177.648	0.01826	2.5451	2.5633	345.0	851.6	1196.7	0.5314	1.0240	1.5554	372.0
370.0	173.339	0.01823	2.6062	2.6244	342.9	853.4	1196.3	0.5289	1.0286	1.5575	370.0
368.0	169.113	0.01821	2.6691	2.6873	340.8	855.1	1195.9	0.5263	1.0332	1.5595	368.0
366.0	164.968	0.01818	2.7337	2.7519	338.7	856.9	1195.6	0.5238	1.0378	1.5616	366.0
364.0	160.903	0.01816	2.8002	2.8184	336.5	858.6	1195.2	0.5212	1.0424	1.5637	364.0
362.0	156.917	0.01813	2.8687	2.8868	334.4	860.4	1194.8	0.5187	1.0471	1.5658	362.0
360.0	153.010	0.01811	2.9392	2.9573	332.3	862.1	1194.4	0.5161	1.0517	1.5678	360.0
358.0	149.179	0.01809	3.0117	3.0298	330.2	863.8	1194.0	0.5135	1.0564	1.5699	358.0
356.0	145.424	0.01806	3.0863	3.1044	328.1	865.5	1193.6	0.5110	1.0611	1.5721	356.0
354.0	141.744	0.01804	3.1632	3.1812	326.0	867.2	1193.2	0.5084	1.0658	1.5742	354.0
352.0	138.138	0.01801	3.2423	3.2603	323.9	868.9	1192.7	0.5058	1.0705	1.5763	352.0
350.0	134.604	0.01799	3.3238	3.3418	321.8	870.6	1192.3	0.5032	1.0752	1.5784	350.0
348.0	131.142	0.01797	3.4078	3.4258	319.7	872.2	1191.9	0.5006	1.0799	1.5806	348.0
346.0	127.751	0.01794	3.4943	3.5122	317.6	873.9	1191.4	0.4980	1.0847	1.5827	346.0
344.0	124.430	0.01792	3.5834	3.6013	315.5	875.5	1191.0	0.4954	1.0894	1.5849	344.0
342.0	121.177	0.01790	3.6752	3.6931	313.4	877.2	1190.5	0.4928	1.0942	1.5871	342.0
340.0	117.992	0.01787	3.7699	3.7878	311.3	878.8	1190.1	0.4902	1.0990	1.5892	340.0
338.0	114.873	0.01785	3.8675	3.8853	309.2	880.5	1189.6	0.4876	1.1038	1.5914	338.0
336.0	111.820	0.01783	3.9681	3.9859	307.1	882.1	1189.1	0.4850	1.1086	1.5936	336.0
334.0	108.832	0.01781	4.0718	4.0896	305.0	883.7	1188.7	0.4824	1.1134	1.5958	334.0
332.0	105.907	0.01779	4.1788	4.1966	302.9	885.3	1188.2	0.4798	1.1183	1.5981	332.0
330.0	103.045	0.01776	4.2892	4.3069	300.8	886.9	1187.7	0.4772	1.1231	1.6003	330.0
328.0	100.245	0.01774	4.4030	4.4208	298.7	888.5	1187.2	0.4745	1.1280	1.6025	328.0
326.0	97.506	0.01772	4.5205	4.5382	296.6	890.1	1186.7	0.4719	1.1329	1.6048	326.0
324.0	94.826	0.01770	4.6418	4.6595	294.6	891.6	1186.2	0.4692	1.1378	1.6071	324.0
322.0	92.205	0.01768	4.7669	4.7846	292.5	893.2	1185.7	0.4666	1.1427	1.6093	322.0
320.0	89.643	0.01766	4.8961	4.9138	290.4	894.8	1185.2	0.4640	1.1477	1.6116	320.0
318.0	87.137	0.01764	5.0295	5.0471	288.3	896.3	1184.7	0.4613	1.1526	1.6139	318.0
316.0	84.688	0.01761	5.1673	5.1849	286.3	897.9	1184.1	0.4586	1.1576	1.6162	316.0
314.0	82.293	0.01759	5.3096	5.3272	284.2	899.4	1183.6	0.4560	1.1626	1.6185	314.0
312.0	79.953	0.01757	5.4566	5.4742	282.1	901.0	1183.1	0.4533	1.1676	1.6209	312.0
310.0	77.667	0.01755	5.6085	5.6260	280.0	902.5	1182.5	0.4506	1.1726	1.6232	310.0
308.0	75.433	0.01753	5.7655	5.7830	278.0	904.0	1182.0	0.4479	1.1776	1.6256	308.0
306.0	73.251	0.01751	5.9277	5.9452	275.9	905.5	1181.4	0.4453	1.1827	1.6279	306.0
304.0	71.119	0.01749	6.0955	6.1130	273.8	907.0	1180.9	0.4426	1.1877	1.6303	304.0
302.0	69.038	0.01747	6.2689	6.2864	271.8	908.5	1180.3	0.4399	1.1928	1.6327	302.0
300.0	67.005	0.01745	6.4483	6.4658	269.7	910.0	1179.7	0.4372	1.1979	1.6351	300.0
298.0	65.021	0.01743	6.6339	6.6513	267.7	911.5	1179.2	0.4345	1.2031	1.6375	298.0
296.0	63.084	0.01741	6.8259	6.8433	265.6	913.0	1178.6	0.4317	1.2082	1.6400	296.0
294.0	61.194	0.01739	7.0245	7.0419	263.5	914.5	1178.0	0.4290	1.2134	1.6424	294.0
292.0	59.350	0.01738	7.2301	7.2475	261.5	915.9	1177.4	0.4263	1.2186	1.6449	292.0
290.0	57.550	0.01736	7.4430	7.4603	259.4	917.4	1176.8	0.4236	1.2238	1.6473	290.0
288.0	55.795	0.01734	7.6634	7.6807	257.4	918.8	1176.2	0.4208	1.2290	1.6498	288.0
286.0	54.083	0.01732	7.8916	7.9089	255.3	920.3	1175.6	0.4181	1.2342	1.6523	286.0
284.0	52.414	0.01730	8.1280	8.1453	253.3	921.7	1175.0	0.4154	1.2395	1.6548	284.0
282.0	50.786	0.01728	8.3729	8.3902	251.2	923.2	1174.4	0.4126	1.2448	1.6574	282.0
280.0	49.200	0.01726	8.6267	8.6439	249.2	924.6	1173.8	0.4098	1.2501	1.6599	280.0

STEAM TABLES (Continued)

Properties of Saturated Steam and Saturated Water

Temp. F	Press. psia	Volume, ft³/lbm Water v_f	Evap. v_{fg}	Steam v_g	Enthalpy, Btu/lbm Water h_f	Evap. h_{fg}	Steam h_g	Entropy, Btu/lbm × F Water s_f	Evap. s_{fg}	Steam s_g	Temp. F
280.0	49.200	0.017264	8.627	8.644	249.17	924.6	1173.8	0.4098	1.2501	1.6599	280.0
278.0	47.653	0.017246	8.890	8.907	247.13	926.0	1173.2	0.4071	1.2554	1.6625	278.0
276.0	46.147	0.017228	9.162	9.180	245.08	927.5	1172.5	0.4043	1.2607	1.6650	276.0
274.0	44.678	0.017210	9.445	9.462	243.03	928.9	1171.9	0.4015	1.2661	1.6676	274.0
272.0	43.249	0.017193	9.738	9.755	240.99	930.3	1171.3	0.3987	1.2715	1.6702	272.0
270.0	41.856	0.017175	10.042	10.060	238.95	931.7	1170.6	0.3960	1.2769	1.6729	270.0
268.0	40.500	0.017157	10.358	10.375	236.91	933.1	1170.0	0.3932	1.2823	1.6755	268.0
266.0	39.179	0.017140	10.685	10.703	234.87	934.5	1169.3	0.3904	1.2878	1.6781	266.0
264.0	37.894	0.017123	11.025	11.042	232.83	935.9	1168.7	0.3876	1.2933	1.6808	264.0
262.0	36.644	0.017106	11.378	11.395	230.79	937.3	1168.0	0.3847	1.2988	1.6835	262.0
260.0	35.427	0.017089	11.745	11.762	228.76	938.6	1167.4	0.3819	1.3043	1.6862	260.0
258.0	34.243	0.017072	12.125	12.142	226.72	940.0	1166.7	0.3791	1.3098	1.6889	258.0
256.0	33.091	0.017055	12.520	12.538	224.69	941.4	1166.1	0.3763	1.3154	1.6917	256.0
254.0	31.972	0.017039	12.931	12.948	222.65	942.7	1165.4	0.3734	1.3210	1.6944	254.0
252.0	30.883	0.017022	13.358	13.375	220.62	944.1	1164.7	0.3706	1.3266	1.6972	252.0
250.0	29.825	0.017006	13.802	13.819	218.59	945.4	1164.0	0.3677	1.3323	1.7000	250.0
248.0	28.796	0.016990	14.264	14.281	216.56	946.8	1163.4	0.3649	1.3379	1.7028	248.0
246.0	27.797	0.016974	14.744	14.761	214.53	948.1	1162.7	0.3620	1.3436	1.7056	246.0
244.0	26.826	0.016958	15.243	15.260	212.50	949.5	1162.0	0.3591	1.3494	1.7085	244.0
242.0	25.883	0.016942	15.763	15.780	210.48	950.8	1161.3	0.3562	1.3551	1.7113	242.0
240.0	24.968	0.016926	16.304	16.321	208.45	952.1	1160.6	0.3533	1.3609	1.7142	240.0
238.0	24.079	0.016910	16.867	16.884	206.42	953.5	1159.9	0.3505	1.3667	1.7171	238.0
236.0	23.216	0.016895	17.454	17.471	204.40	954.8	1159.2	0.3476	1.3725	1.7201	236.0
234.0	22.379	0.016880	18.065	18.082	202.38	956.1	1158.5	0.3446	1.3784	1.7230	234.0
232.0	21.567	0.016864	18.701	18.718	200.35	957.4	1157.8	0.3417	1.3842	1.7260	232.0
230.0	20.779	0.016849	19.364	19.381	198.33	958.7	1157.1	0.3388	1.3902	1.7290	230.0
229.0	20.394	0.016842	19.707	19.723	197.32	959.4	1156.7	0.3373	1.3931	1.7305	229.0
228.0	20.015	0.016834	20.056	20.073	196.31	960.0	1156.3	0.3359	1.3961	1.7320	228.0
227.0	19.642	0.016827	20.413	20.429	195.30	960.7	1156.0	0.3344	1.3991	1.7335	227.0
226.0	19.274	0.016819	20.777	20.794	194.29	961.3	1155.6	0.3329	1.4021	1.7350	226.0
225.0	18.912	0.016812	21.149	21.166	193.28	962.0	1155.3	0.3315	1.4051	1.7365	225.0
224.0	18.556	0.016805	21.529	21.545	192.27	962.6	1154.9	0.3300	1.4081	1.7380	224.0
223.0	18.206	0.016797	21.917	21.933	191.26	963.3	1154.5	0.3285	1.4111	1.7396	223.0
222.0	17.860	0.016790	22.313	22.330	190.25	963.9	1154.2	0.3270	1.4141	1.7411	222.0
221.0	17.521	0.016783	22.718	22.735	189.24	964.6	1153.8	0.3255	1.4171	1.7427	221.0
220.0	17.186	0.016775	23.131	23.148	188.23	965.2	1153.4	0.3241	1.4201	1.7442	220.0
219.0	16.857	0.016768	23.554	23.571	187.22	965.8	1153.1	0.3226	1.4232	1.7458	219.0
218.0	16.533	0.016761	23.986	24.002	186.21	966.5	1152.7	0.3211	1.4262	1.7473	218.0
217.0	16.214	0.016754	24.427	24.444	185.21	967.1	1152.3	0.3196	1.4293	1.7489	217.0
216.0	15.901	0.016747	24.878	24.894	184.20	967.8	1152.0	0.3181	1.4323	1.7505	216.0
215.0	15.592	0.016740	25.338	25.355	183.19	968.4	1151.6	0.3166	1.4354	1.7520	215.0
214.0	15.289	0.016733	25.809	25.826	182.18	969.0	1151.2	0.3151	1.4385	1.7536	214.0
213.0	14.990	0.016726	26.290	26.307	181.17	969.7	1150.8	0.3136	1.4416	1.7552	213.0
212.0	14.696	0.016719	26.782	26.799	180.17	970.3	1150.5	0.3121	1.4447	1.7568	212.0
211.0	14.407	0.016712	27.285	27.302	179.16	970.9	1150.1	0.3106	1.4478	1.7584	211.0
210.0	14.123	0.016705	27.799	27.816	178.15	971.6	1149.7	0.3091	1.4509	1.7600	210.0
209.0	13.843	0.016698	28.324	28.341	177.14	972.2	1149.4	0.3076	1.4540	1.7616	209.0
208.0	13.568	0.016691	28.862	28.878	176.14	972.8	1149.0	0.3061	1.4571	1.7632	208.0
207.0	13.297	0.016684	29.411	29.428	175.13	973.5	1148.6	0.3046	1.4602	1.7649	207.0
206.0	13.031	0.016677	29.973	29.989	174.12	974.1	1148.2	0.3031	1.4634	1.7665	206.0
205.0	12.770	0.016670	30.547	30.564	173.12	974.7	1147.9	0.3016	1.4665	1.7681	205.0
204.0	12.512	0.016664	31.135	31.151	172.11	975.4	1147.5	0.3001	1.4697	1.7698	204.0
203.0	12.259	0.016657	31.736	31.752	171.10	976.0	1147.1	0.2986	1.4728	1.7714	203.0
202.0	12.011	0.016650	32.350	32.367	170.10	976.6	1146.7	0.2971	1.4760	1.7731	202.0
201.0	11.766	0.016643	32.979	32.996	169.09	977.2	1146.3	0.2955	1.4792	1.7747	201.0
200.0	11.526	0.016637	33.622	33.639	168.09	977.9	1146.0	0.2940	1.4824	1.7764	200.0
199.0	11.290	0.016630	34.280	34.297	167.08	978.5	1145.6	0.2925	1.4856	1.7781	199.0
198.0	11.058	0.016624	34.954	34.970	166.08	979.1	1145.2	0.2910	1.4888	1.7798	198.0
197.0	10.830	0.016617	35.643	35.659	165.07	979.7	1144.8	0.2894	1.4920	1.7814	197.0
196.0	10.605	0.016611	36.348	36.364	164.06	980.4	1144.4	0.2879	1.4952	1.7831	196.0
195.0	10.385	0.016604	37.069	37.086	163.06	981.0	1144.0	0.2864	1.4985	1.7848	195.0
194.0	10.168	0.016598	37.808	37.824	162.05	981.6	1143.7	0.2848	1.5017	1.7865	194.0
193.0	9.956	0.016591	38.564	38.580	161.05	982.2	1143.3	0.2833	1.5050	1.7882	193.0
192.0	9.747	0.016585	39.337	39.354	160.05	982.8	1142.9	0.2818	1.5082	1.7900	192.0
191.0	9.541	0.016578	40.130	40.146	159.04	983.5	1142.5	0.2802	1.5115	1.7917	191.0
190.0	9.340	0.016572	40.941	40.957	158.04	984.1	1142.1	0.2787	1.5148	1.7934	190.0
189.0	9.141	0.016566	41.771	41.787	157.03	984.7	1141.7	0.2771	1.5180	1.7952	189.0
188.0	8.947	0.016559	42.621	42.638	156.03	985.3	1141.3	0.2756	1.5213	1.7969	188.0
187.0	8.756	0.016553	43.492	43.508	155.02	985.9	1140.9	0.2740	1.5246	1.7987	187.0
186.0	8.568	0.016547	44.383	44.400	154.02	986.5	1140.5	0.2725	1.5279	1.8004	186.0
185.0	8.384	0.016541	45.297	45.313	153.02	987.1	1140.2	0.2709	1.5313	1.8022	185.0
184.0	8.203	0.016534	46.232	46.249	152.01	987.8	1139.8	0.2694	1.5346	1.8040	184.0
183.0	8.025	0.016528	47.190	47.207	151.01	988.4	1139.4	0.2678	1.5379	1.8057	183.0
182.0	7.850	0.016522	48.172	48.189	150.01	989.0	1139.0	0.2662	1.5413	1.8075	182.0
181.0	7.679	0.016516	49.178	49.194	149.00	989.6	1138.6	0.2647	1.5446	1.8093	181.0
180.0	7.511	0.016510	50.208	50.225	148.00	990.2	1138.2	0.2631	1.5480	1.8111	180.0

Properties of Saturated Steam and Saturated Water

Temp. F	Press. psia	Volume, ft³/lbm Water v_f	Evap. v_{fg}	Steam v_g	Enthalpy, Btu/lbm Water h_f	Evap. h_{fg}	Steam h_g	Entropy, Btu/lbm×F Water s_f	Evap. s_{fg}	Steam s_g	Temp. F
180.0	7.5110	0.016510	50.21	50.22	148.00	990.2	1138.2	0.2631	1.5480	1.8111	180.0
179.0	7.3460	0.016504	51.26	51.28	147.00	990.8	1137.8	0.2615	1.5514	1.8129	179.0
178.0	7.1840	0.016498	52.35	52.36	145.99	991.4	1137.4	0.2600	1.5548	1.8147	178.0
177.0	7.0250	0.016492	53.46	53.47	144.99	992.0	1137.0	0.2584	1.5582	1.8166	177.0
176.0	6.8690	0.016486	54.59	54.61	143.99	992.6	1136.6	0.2568	1.5616	1.8184	176.0
175.0	6.7159	0.016480	55.76	55.77	142.99	993.2	1136.2	0.2552	1.5650	1.8202	175.0
174.0	6.5656	0.016474	56.95	56.97	141.98	993.8	1135.8	0.2537	1.5684	1.8221	174.0
173.0	6.4182	0.016468	58.18	58.19	140.98	994.4	1135.4	0.2521	1.5718	1.8239	173.0
172.0	6.2736	0.016463	59.43	59.45	139.98	995.0	1135.0	0.2505	1.5753	1.8258	172.0
171.0	6.1318	0.016457	60.72	60.74	138.98	995.6	1134.6	0.2489	1.5787	1.8276	171.0
170.0	5.9926	0.016451	62.04	62.06	137.97	996.2	1134.2	0.2473	1.5822	1.8295	170.0
169.0	5.8562	0.016445	63.39	63.41	136.97	996.8	1133.8	0.2457	1.5857	1.8314	169.0
168.0	5.7223	0.016440	64.78	64.80	135.97	997.4	1133.4	0.2441	1.5892	1.8333	168.0
167.0	5.5911	0.016434	66.21	66.22	134.97	998.0	1133.0	0.2425	1.5926	1.8352	167.0
166.0	5.4623	0.016428	67.67	67.68	133.97	998.6	1132.6	0.2409	1.5961	1.8371	166.0
165.0	5.3361	0.016423	69.17	69.18	132.96	999.2	1132.2	0.2393	1.5997	1.8390	165.0
164.0	5.2124	0.016417	70.70	70.72	131.96	999.8	1131.8	0.2377	1.6032	1.8409	164.0
163.0	5.0911	0.016412	72.28	72.30	130.96	1000.4	1131.4	0.2361	1.6067	1.8428	163.0
162.0	4.9722	0.016406	73.90	73.92	129.96	1001.0	1131.0	0.2345	1.6103	1.8448	162.0
161.0	4.8556	0.016401	75.56	75.58	128.96	1001.6	1130.6	0.2329	1.6138	1.8467	161.0
160.0	4.7414	0.016395	77.27	77.29	127.96	1002.2	1130.2	0.2313	1.6174	1.8487	160.0
159.0	4.6294	0.016390	79.02	79.04	126.96	1002.8	1129.8	0.2297	1.6210	1.8506	159.0
158.0	4.5197	0.016384	80.82	80.83	125.96	1003.4	1129.4	0.2281	1.6245	1.8526	158.0
157.0	4.4122	0.016379	82.66	82.68	124.95	1004.0	1129.0	0.2264	1.6281	1.8546	157.0
156.0	4.3068	0.016374	84.56	84.57	123.95	1004.6	1128.6	0.2248	1.6318	1.8566	156.0
155.0	4.2036	0.016369	86.50	86.52	122.95	1005.2	1128.2	0.2232	1.6354	1.8586	155.0
154.0	4.1025	0.016363	88.50	88.52	121.95	1005.8	1127.7	0.2216	1.6390	1.8606	154.0
153.0	4.0035	0.016358	90.55	90.57	120.95	1006.4	1127.3	0.2199	1.6426	1.8626	153.0
152.0	3.9065	0.016353	92.66	92.68	119.95	1007.0	1126.9	0.2183	1.6463	1.8646	152.0
151.0	3.8114	0.016348	94.83	94.84	118.95	1007.6	1126.5	0.2167	1.6500	1.8666	151.0
150.0	3.7184	0.016343	97.05	97.07	117.95	1008.2	1126.1	0.2150	1.6536	1.8686	150.0
149.0	3.6273	0.016337	99.33	99.35	116.95	1008.7	1125.7	0.2134	1.6573	1.8707	149.0
148.0	3.5381	0.016332	101.68	101.70	115.95	1009.3	1125.3	0.2117	1.6610	1.8727	148.0
147.0	3.4508	0.016327	104.10	104.11	114.95	1009.9	1124.9	0.2101	1.6647	1.8748	147.0
146.0	3.3653	0.016322	106.58	106.59	113.95	1010.5	1124.5	0.2084	1.6684	1.8769	146.0
145.0	3.2816	0.016317	109.12	109.14	112.95	1011.1	1124.0	0.2068	1.6722	1.8789	145.0
144.0	3.1997	0.016312	111.74	111.76	111.95	1011.7	1123.6	0.2051	1.6759	1.8810	144.0
143.0	3.1195	0.016308	114.44	114.45	110.95	1012.3	1123.2	0.2035	1.6797	1.8831	143.0
142.0	3.0411	0.016303	117.21	117.22	109.95	1012.9	1122.8	0.2018	1.6834	1.8852	142.0
141.0	2.9643	0.016298	120.05	120.07	108.95	1013.4	1122.4	0.2001	1.6872	1.8873	141.0
140.0	2.8892	0.016293	122.98	123.00	107.95	1014.0	1122.0	0.1985	1.6910	1.8895	140.0
139.0	2.8157	0.016288	125.99	126.01	106.95	1014.6	1121.6	0.1968	1.6948	1.8916	139.0
138.0	2.7438	0.016284	129.09	129.11	105.95	1015.2	1121.1	0.1951	1.6986	1.8937	138.0
137.0	2.6735	0.016279	132.28	132.29	104.95	1015.8	1120.7	0.1935	1.7024	1.8959	137.0
136.0	2.6047	0.016274	135.55	135.57	103.95	1016.4	1120.3	0.1918	1.7063	1.8980	136.0
135.0	2.5375	0.016270	138.93	138.94	102.95	1016.9	1119.9	0.1901	1.7101	1.9002	135.0
134.0	2.4717	0.016265	142.40	142.41	101.95	1017.5	1119.5	0.1884	1.7140	1.9024	134.0
133.0	2.4074	0.016260	145.97	145.98	100.95	1018.1	1119.1	0.1867	1.7178	1.9046	133.0
132.0	2.3445	0.016256	149.64	149.66	99.95	1018.7	1118.6	0.1851	1.7217	1.9068	132.0
131.0	2.2830	0.016251	153.42	153.44	98.95	1019.3	1118.2	0.1834	1.7256	1.9090	131.0
130.0	2.2230	0.016247	157.32	157.33	97.96	1019.8	1117.8	0.1817	1.7295	1.9112	130.0
129.0	2.1642	0.016243	161.32	161.34	96.96	1020.4	1117.4	0.1800	1.7335	1.9134	129.0
128.0	2.1068	0.016238	165.45	165.47	95.96	1021.0	1117.0	0.1783	1.7374	1.9157	128.0
127.0	2.0507	0.016234	169.70	169.72	94.96	1021.6	1116.5	0.1766	1.7413	1.9179	127.0
126.0	1.9959	0.016229	174.08	174.09	93.96	1022.2	1116.1	0.1749	1.7453	1.9202	126.0
125.0	1.9424	0.016225	178.58	178.60	92.96	1022.7	1115.7	0.1732	1.7493	1.9224	125.0
124.0	1.8901	0.016221	183.23	183.24	91.96	1023.3	1115.3	0.1715	1.7533	1.9247	124.0
123.0	1.8390	0.016217	188.01	188.03	90.96	1023.9	1114.9	0.1697	1.7573	1.9270	123.0
122.0	1.7891	0.016213	192.94	192.95	89.96	1024.5	1114.4	0.1680	1.7613	1.9293	122.0
121.0	1.7403	0.016208	198.01	198.03	88.96	1025.0	1114.0	0.1663	1.7653	1.9316	121.0
120.0	1.6927	0.016204	203.25	203.26	87.97	1025.6	1113.6	0.1646	1.7693	1.9339	120.0
119.0	1.6463	0.016200	208.64	208.66	86.97	1026.2	1113.2	0.1629	1.7734	1.9362	119.0
118.0	1.6009	0.016196	214.20	214.21	85.97	1026.8	1112.7	0.1611	1.7774	1.9386	118.0
117.0	1.5566	0.016192	219.93	219.94	84.97	1027.3	1112.3	0.1594	1.7815	1.9409	117.0
116.0	1.5133	0.016188	225.84	225.85	83.97	1027.9	1111.9	0.1577	1.7856	1.9433	116.0
115.0	1.4711	0.016184	231.93	231.94	82.97	1028.5	1111.5	0.1559	1.7897	1.9457	115.0
114.0	1.4299	0.016180	238.21	238.22	81.97	1029.1	1111.0	0.1542	1.7938	1.9480	114.0
113.0	1.3898	0.016177	244.69	244.70	80.98	1029.6	1110.6	0.1525	1.7980	1.9504	113.0
112.0	1.3505	0.016173	251.37	251.38	79.98	1030.2	1110.2	0.1507	1.8021	1.9528	112.0
111.0	1.3123	0.016169	258.26	258.28	78.98	1030.8	1109.8	0.1490	1.8063	1.9552	111.0
110.0	1.2750	0.016165	265.37	265.39	77.98	1031.4	1109.3	0.1472	1.8105	1.9577	110.0
109.0	1.2385	0.016162	272.71	272.72	76.98	1031.9	1108.9	0.1455	1.8146	1.9601	109.0
108.0	1.2030	0.016158	280.28	280.30	75.98	1032.5	1108.5	0.1437	1.8188	1.9626	108.0
107.0	1.1684	0.016154	288.09	288.11	74.99	1033.1	1108.1	0.1419	1.8231	1.9650	107.0
106.0	1.1347	0.016151	296.16	296.18	73.99	1033.6	1107.6	0.1402	1.8273	1.9675	106.0
105.0	1.1017	0.016147	304.49	304.50	72.99	1034.2	1107.2	0.1384	1.8315	1.9700	105.0

STEAM TABLES (Continued)
Properties of Saturated Steam and Saturated Water

Temp. F	Press. psia	Volume, ft³/lbm Water v_f	Evap. v_{fg}	Steam v_g	Enthalpy, Btu/lbm Water h_f	Evap. h_{fg}	Steam h_g	Entropy, Btu/lbm×F Water s_f	Evap. s_{fg}	Steam s_g	Temp. F
105.0	1.10174	0.016147	304.5	304.5	72.990	1034.2	1107.2	0.1384	1.8315	1.9700	105.0
104.0	1.06965	0.016144	313.1	313.1	71.992	1034.8	1106.8	0.1366	1.8358	1.9725	104.0
103.0	1.03838	0.016140	322.0	322.0	70.993	1035.4	1106.3	0.1349	1.8401	1.9750	103.0
102.0	1.00789	0.016137	331.1	331.1	69.995	1035.9	1105.9	0.1331	1.8444	1.9775	102.0
101.0	0.97818	0.016133	340.6	340.6	68.997	1036.5	1105.5	0.1313	1.8487	1.9800	101.0
100.0	0.94924	0.016130	350.4	350.4	67.999	1037.1	1105.1	0.1295	1.8530	1.9825	100.0
99.0	0.92103	0.016127	360.5	360.5	67.001	1037.6	1104.6	0.1278	1.8573	1.9851	99.0
98.0	0.89356	0.016123	370.9	370.9	66.003	1038.2	1104.2	0.1260	1.8617	1.9876	98.0
97.0	0.86679	0.016120	381.7	381.7	65.005	1038.8	1103.8	0.1242	1.8660	1.9902	97.0
96.0	0.84072	0.016117	392.8	392.9	64.006	1039.3	1103.3	0.1224	1.8704	1.9928	96.0
95.0	0.81534	0.016114	404.4	404.4	63.008	1039.9	1102.9	0.1206	1.8748	1.9954	95.0
94.0	0.79062	0.016111	416.3	416.3	62.010	1040.5	1102.5	0.1188	1.8792	1.9980	94.0
93.0	0.76655	0.016108	428.6	428.6	61.012	1041.0	1102.1	0.1170	1.8837	2.0006	93.0
92.0	0.74313	0.016105	441.3	441.3	60.014	1041.6	1101.6	0.1152	1.8881	2.0033	92.0
91.0	0.72032	0.016102	454.5	454.5	59.016	1042.2	1101.2	0.1134	1.8926	2.0059	91.0
90.0	0.69813	0.016099	468.1	468.1	58.018	1042.7	1100.8	0.1115	1.8970	2.0086	90.0
89.0	0.67653	0.016096	482.2	482.2	57.020	1043.3	1100.3	0.1097	1.9015	2.0112	89.0
88.0	0.65551	0.016093	496.8	496.8	56.022	1043.9	1099.9	0.1079	1.9060	2.0139	88.0
87.0	0.63507	0.016090	511.9	511.9	55.024	1044.4	1099.5	0.1061	1.9105	2.0166	87.0
86.0	0.61518	0.016087	527.5	527.5	54.026	1045.0	1099.0	0.1043	1.9151	2.0193	86.0
85.0	0.59583	0.016085	543.6	543.6	53.027	1045.6	1098.6	0.1024	1.9196	2.0221	85.0
84.0	0.57702	0.016082	560.3	560.3	52.029	1046.1	1098.2	0.1006	1.9242	2.0248	84.0
83.0	0.55872	0.016079	577.6	577.6	51.031	1046.7	1097.7	0.0988	1.9288	2.0275	83.0
82.0	0.54093	0.016077	595.5	595.6	50.033	1047.3	1097.3	0.0969	1.9334	2.0303	82.0
81.0	0.52364	0.016074	614.1	614.1	49.035	1047.8	1096.9	0.0951	1.9380	2.0331	81.0
80.0	0.050683	0.016072	633.3	633.3	48.037	1048.4	1096.4	0.0932	1.9426	2.0359	80.0
79.0	0.49049	0.016070	653.2	653.2	47.038	1049.0	1096.0	0.0914	1.9473	2.0387	79.0
78.0	0.47461	0.016067	673.8	673.9	46.040	1049.5	1095.6	0.0895	1.9520	2.0415	78.0
77.0	0.45919	0.016065	695.2	695.2	45.042	1050.1	1095.1	0.0877	1.9567	2.0443	77.0
76.0	0.44420	0.016063	717.4	717.4	44.043	1050.7	1094.7	0.0858	1.9614	2.0472	76.0
75.0	0.42964	0.016060	740.3	740.3	43.045	1051.2	1094.3	0.0839	1.9661	2.0500	75.0
74.0	0.41550	0.016058	764.1	764.1	42.046	1051.8	1093.8	0.0821	1.9708	2.0529	74.0
73.0	0.40177	0.016056	788.8	788.8	41.048	1052.4	1093.4	0.0802	1.9756	2.0558	73.0
72.0	0.38844	0.016054	814.3	814.3	40.049	1052.9	1093.0	0.0783	1.9804	2.0587	72.0
71.0	0.37549	0.016052	840.8	840.9	39.050	1053.5	1092.5	0.0764	1.9852	2.0616	71.0
70.0	0.36292	0.016050	868.3	868.4	38.052	1054.0	1092.1	0.0745	1.9900	2.0645	70.0
69.0	0.35073	0.016048	896.9	896.9	37.053	1054.6	1091.7	0.0727	1.9948	2.0675	69.0
68.0	0.33889	0.016046	926.5	926.5	36.054	1055.2	1091.2	0.0708	1.9996	2.0704	68.0
67.0	0.32740	0.016044	957.2	957.2	35.055	1055.7	1090.8	0.0689	2.0045	2.0734	67.0
66.0	0.31626	0.016043	989.0	989.1	34.056	1056.3	1090.4	0.0670	2.0094	2.0764	66.0
65.0	0.30545	0.016041	1022.1	1022.1	33.057	1056.9	1089.9	0.0651	2.0143	2.0794	65.0
64.0	0.29497	0.016039	1056.5	1056.5	32.058	1057.4	1089.5	0.0632	2.0192	2.0824	64.0
63.0	0.28480	0.016038	1092.1	1092.1	31.058	1058.0	1089.0	0.0613	2.0242	2.0854	63.0
62.0	0.27494	0.016036	1129.2	1129.2	30.059	1058.5	1088.6	0.0593	2.0291	2.0885	62.0
61.0	0.26538	0.016035	1167.6	1167.6	29.059	1059.1	1088.2	0.0574	2.0341	2.0915	61.0
60.0	0.25611	0.016033	1207.6	1207.6	28.060	1059.7	1087.7	0.0555	2.0391	2.0946	60.0
59.0	0.24713	0.016032	1249.1	1249.1	27.060	1060.2	1087.3	0.0536	2.0441	2.0977	59.0
58.0	0.23843	0.016031	1292.2	1292.2	26.060	1060.8	1086.9	0.0516	2.0491	2.1008	58.0
57.0	0.23000	0.016029	1337.0	1337.0	25.060	1061.4	1086.4	0.0497	2.0542	2.1039	57.0
56.0	0.22183	0.016028	1383.6	1383.6	24.059	1061.9	1086.0	0.0478	2.0593	2.1070	56.0
55.0	0.21392	0.016027	1432.0	1432.0	23.059	1062.5	1085.6	0.0458	2.0644	2.1102	55.0
54.0	0.20625	0.016026	1482.4	1482.4	22.058	1063.1	1085.1	0.0439	2.0695	2.1134	54.0
53.0	0.19883	0.016025	1534.7	1534.8	21.058	1063.6	1084.7	0.0419	2.0746	2.1165	53.0
52.0	0.19165	0.016024	1589.2	1589.2	20.057	1064.2	1084.2	0.0400	2.0798	2.1197	52.0
51.0	0.18469	0.016023	1645.9	1645.9	19.056	1064.7	1083.8	0.0380	2.0849	2.1230	51.0
50.0	0.17796	0.016023	1704.8	1704.8	18.054	1065.3	1083.4	0.0361	2.0901	2.1262	50.0
49.0	0.17144	0.016022	1766.2	1766.2	17.053	1065.9	1082.9	0.0341	2.0953	2.1294	49.0
48.0	0.16514	0.016021	1830.0	1830.0	16.051	1066.4	1082.5	0.0321	2.1006	2.1327	48.0
47.0	0.15904	0.016021	1896.5	1896.5	15.049	1067.0	1082.1	0.0301	2.1058	2.1360	47.0
46.0	0.15314	0.016020	1965.7	1965.7	14.047	1067.6	1081.6	0.0282	2.1111	2.1393	46.0
45.0	0.14744	0.016020	2037.7	2037.8	13.044	1068.1	1081.2	0.0262	2.1164	2.1426	45.0
44.0	0.14192	0.016019	2112.8	2112.8	12.041	1068.7	1080.7	0.0242	2.1217	2.1459	44.0
43.0	0.13659	0.016019	2191.0	2191.0	11.038	1069.3	1080.3	0.0222	2.1271	2.1493	43.0
42.0	0.13143	0.016019	2272.4	2272.4	10.035	1069.8	1079.9	0.0202	2.1325	2.1527	42.0
41.0	0.12645	0.016019	2357.3	2357.3	9.031	1070.4	1079.4	0.0182	2.1378	2.1560	41.0
40.0	0.12163	0.016019	2445.8	2445.8	8.027	1071.0	1079.0	0.0162	2.1432	2.1594	40.0
39.0	0.11698	0.016019	2538.0	2538.0	7.023	1071.5	1078.5	0.0142	2.1487	2.1629	39.0
38.0	0.11249	0.016019	2634.1	2634.2	6.018	1072.1	1078.1	0.0122	2.1541	2.1663	38.0
37.0	0.10815	0.016019	2734.4	2734.4	5.013	1072.7	1077.7	0.0101	2.1596	2.1697	37.0
36.0	0.10395	0.016020	2839.0	2839.0	4.008	1073.2	1077.2	0.0081	2.1651	2.1732	36.0
35.0	0.09991	0.016020	2948.1	2948.1	3.002	1073.8	1076.8	0.0061	2.1706	2.1767	35.0
34.0	0.09600	0.016021	3061.9	3061.9	1.996	1074.4	1076.4	0.0041	2.1762	2.1802	34.0
33.0	0.09223	0.016021	3180.7	3180.7	0.989	1074.9	1075.9	0.0020	2.1817	2.1837	33.0
32.018	0.08865	0.016022	3302.4	3302.4	0.0003	1075.5	1075.5	0.0000	2.1872	2.1872	32.018
*32.0	0.08859	0.016022	3304.7	3304.7	−0.0179	1075.5	1075.5	−0.0000	2.1873	2.1873	32.0

*The states here shown are metastable

STEAM TABLES (Continued)

Specific Heat at constant pressure of Steam and of Water

Temp. F	c_p, Btu/lbm × F															Temp. F
Press., psia	1	1.5	2	3	4	6	8	10	15	20	30	40	60	80	100	Press., psia
Sat. Water	0.998	0.998	0.999	1.000	1.000	1.002	1.003	1.004	1.007	1.010	1.014	1.019	1.026	1.033	1.039	Sat. Water
Sat. Steam	0.450	0.452	0.454	0.458	0.461	0.466	0.471	0.475	0.485	0.493	0.508	0.521	0.543	0.564	0.582	Sat. Steam
1500	0.559	0.559	0.559	0.559	0.559	0.559	0.559	0.559	0.559	0.559	0.560	0.560	0.560	0.561	0.561	1500
1480	0.557	0.557	0.557	0.557	0.557	0.557	0.557	0.558	0.558	0.558	0.558	0.558	0.559	0.559	0.559	1480
1460	0.556	0.556	0.556	0.556	0.556	0.556	0.556	0.556	0.556	0.556	0.556	0.557	0.557	0.557	0.558	1460
1440	0.554	0.554	0.554	0.554	0.554	0.554	0.554	0.554	0.554	0.554	0.555	0.555	0.555	0.556	0.556	1440
1420	0.552	0.552	0.552	0.552	0.552	0.552	0.553	0.553	0.553	0.553	0.553	0.553	0.554	0.554	0.555	1420
1400	0.551	0.551	0.551	0.551	0.551	0.551	0.551	0.551	0.551	0.551	0.551	0.552	0.552	0.553	0.553	1400
1380	0.549	0.549	0.549	0.549	0.549	0.549	0.549	0.549	0.549	0.549	0.550	0.550	0.550	0.551	0.551	1380
1360	0.547	0.547	0.547	0.547	0.547	0.547	0.547	0.547	0.548	0.548	0.548	0.548	0.549	0.549	0.550	1360
1340	0.546	0.546	0.546	0.546	0.546	0.546	0.546	0.546	0.546	0.546	0.546	0.547	0.547	0.548	0.548	1340
1320	0.544	0.544	0.544	0.544	0.544	0.544	0.544	0.544	0.544	0.544	0.545	0.545	0.545	0.546	0.546	1320
1300	0.542	0.542	0.542	0.542	0.542	0.542	0.542	0.542	0.542	0.543	0.543	0.543	0.544	0.544	0.545	1300
1280	0.540	0.540	0.540	0.540	0.540	0.540	0.540	0.541	0.541	0.541	0.541	0.541	0.542	0.543	0.543	1280
1260	0.538	0.539	0.539	0.539	0.539	0.539	0.539	0.539	0.539	0.539	0.539	0.540	0.540	0.541	0.541	1260
1240	0.537	0.537	0.537	0.537	0.537	0.537	0.537	0.537	0.537	0.537	0.538	0.538	0.539	0.539	0.540	1240
1220	0.535	0.535	0.535	0.535	0.535	0.535	0.535	0.535	0.535	0.536	0.536	0.536	0.537	0.537	0.538	1220
1200	0.533	0.533	0.533	0.533	0.533	0.533	0.533	0.533	0.534	0.534	0.534	0.534	0.535	0.536	0.536	1200
1180	0.531	0.531	0.531	0.531	0.531	0.531	0.532	0.532	0.532	0.532	0.532	0.533	0.533	0.534	0.535	1180
1160	0.529	0.529	0.530	0.530	0.530	0.530	0.530	0.530	0.530	0.530	0.530	0.531	0.532	0.532	0.533	1160
1140	0.528	0.528	0.528	0.528	0.528	0.528	0.528	0.528	0.528	0.528	0.529	0.529	0.530	0.531	0.531	1140
1120	0.526	0.526	0.526	0.526	0.526	0.526	0.526	0.526	0.526	0.527	0.527	0.527	0.528	0.529	0.530	1120
1100	0.524	0.524	0.524	0.524	0.524	0.524	0.524	0.524	0.525	0.525	0.525	0.526	0.526	0.527	0.528	1100
1080	0.522	0.522	0.522	0.522	0.522	0.522	0.522	0.523	0.523	0.523	0.523	0.524	0.525	0.525	0.526	1080
1060	0.520	0.520	0.520	0.520	0.520	0.521	0.521	0.521	0.521	0.521	0.522	0.522	0.523	0.524	0.524	1060
1040	0.518	0.519	0.519	0.519	0.519	0.519	0.519	0.519	0.519	0.519	0.520	0.520	0.521	0.522	0.523	1040
1020	0.517	0.517	0.517	0.517	0.517	0.517	0.517	0.517	0.517	0.518	0.518	0.518	0.519	0.520	0.521	1020
1000	0.515	0.515	0.515	0.515	0.515	0.515	0.515	0.515	0.515	0.516	0.516	0.517	0.518	0.519	0.519	1000
980	0.513	0.513	0.513	0.513	0.513	0.513	0.513	0.513	0.514	0.514	0.514	0.515	0.516	0.517	0.518	980
960	0.511	0.511	0.511	0.511	0.511	0.511	0.512	0.512	0.512	0.512	0.513	0.513	0.514	0.515	0.516	960
940	0.509	0.509	0.509	0.509	0.509	0.510	0.510	0.510	0.510	0.510	0.511	0.511	0.512	0.514	0.515	940
920	0.507	0.508	0.508	0.508	0.508	0.508	0.508	0.508	0.508	0.509	0.509	0.510	0.511	0.512	0.513	920
900	0.506	0.506	0.506	0.506	0.506	0.506	0.506	0.506	0.506	0.507	0.507	0.508	0.509	0.510	0.512	900
880	0.504	0.504	0.504	0.504	0.504	0.504	0.504	0.504	0.505	0.505	0.506	0.506	0.508	0.509	0.510	880
860	0.502	0.502	0.502	0.502	0.502	0.502	0.503	0.503	0.503	0.503	0.504	0.505	0.506	0.507	0.509	860
840	0.500	0.500	0.500	0.500	0.500	0.501	0.501	0.501	0.501	0.502	0.502	0.503	0.504	0.506	0.507	840
820	0.498	0.498	0.499	0.499	0.499	0.499	0.499	0.499	0.499	0.500	0.501	0.501	0.503	0.504	0.506	820
800	0.497	0.497	0.497	0.497	0.497	0.497	0.497	0.497	0.498	0.498	0.499	0.500	0.501	0.503	0.505	800
780	0.495	0.495	0.495	0.495	0.495	0.495	0.495	0.496	0.496	0.496	0.497	0.498	0.500	0.502	0.503	780
760	0.493	0.493	0.493	0.493	0.493	0.494	0.494	0.494	0.494	0.495	0.496	0.497	0.499	0.500	0.502	760
740	0.491	0.491	0.491	0.492	0.492	0.492	0.492	0.492	0.493	0.493	0.494	0.495	0.497	0.499	0.501	740
720	0.490	0.490	0.490	0.490	0.490	0.490	0.490	0.491	0.491	0.492	0.493	0.494	0.496	0.498	0.500	720
700	0.488	0.488	0.488	0.488	0.488	0.488	0.489	0.489	0.490	0.490	0.491	0.492	0.495	0.497	0.500	700
680	0.486	0.486	0.486	0.486	0.487	0.487	0.487	0.487	0.488	0.489	0.490	0.491	0.494	0.496	0.499	680
660	0.484	0.485	0.485	0.485	0.485	0.485	0.485	0.486	0.486	0.487	0.489	0.490	0.493	0.496	0.499	660
640	0.483	0.483	0.483	0.483	0.483	0.484	0.484	0.484	0.485	0.486	0.487	0.489	0.492	0.495	0.499	640
620	0.481	0.481	0.481	0.481	0.482	0.482	0.482	0.483	0.483	0.484	0.486	0.488	0.491	0.495	0.499	620
600	0.479	0.480	0.480	0.480	0.480	0.480	0.481	0.481	0.482	0.483	0.485	0.487	0.491	0.495	0.499	600
580	0.478	0.478	0.478	0.478	0.478	0.479	0.479	0.480	0.481	0.482	0.484	0.486	0.491	0.495	0.500	580
560	0.476	0.476	0.476	0.477	0.477	0.477	0.478	0.478	0.479	0.481	0.483	0.485	0.490	0.496	0.501	560
540	0.475	0.475	0.475	0.475	0.475	0.476	0.476	0.477	0.478	0.480	0.482	0.485	0.491	0.497	0.503	540
520	0.473	0.473	0.473	0.474	0.474	0.475	0.475	0.476	0.477	0.479	0.482	0.485	0.491	0.498	0.505	520
500	0.472	0.472	0.472	0.472	0.473	0.473	0.474	0.475	0.476	0.478	0.481	0.485	0.492	0.500	0.508	500
480	0.470	0.470	0.470	0.471	0.471	0.472	0.473	0.473	0.475	0.477	0.481	0.485	0.493	0.502	0.511	480
460	0.469	0.469	0.469	0.469	0.470	0.471	0.472	0.472	0.475	0.477	0.481	0.486	0.495	0.505	0.516	460
440	0.467	0.467	0.468	0.468	0.469	0.470	0.470	0.471	0.474	0.476	0.481	0.487	0.498	0.509	0.522	440
420	0.466	0.466	0.466	0.467	0.467	0.468	0.470	0.471	0.473	0.476	0.482	0.488	0.501	0.514	0.528	420
400	0.464	0.465	0.465	0.466	0.466	0.467	0.469	0.470	0.473	0.476	0.483	0.490	0.504	0.520	0.536	400
380	0.463	0.463	0.464	0.464	0.465	0.466	0.468	0.469	0.473	0.477	0.484	0.492	0.509	0.527	0.546	380
360	0.462	0.462	0.462	0.463	0.464	0.466	0.467	0.469	0.473	0.477	0.486	0.495	0.515	0.536	0.558	360
340	0.460	0.461	0.461	0.462	0.463	0.465	0.467	0.469	0.473	0.478	0.488	0.499	0.521	0.546	0.572	340
320	0.459	0.460	0.460	0.461	0.462	0.464	0.467	0.469	0.474	0.480	0.491	0.504	0.530	0.558	1.036	320
300	0.458	0.459	0.459	0.460	0.462	0.464	0.466	0.469	0.475	0.482	0.495	0.509	0.539	1.029	1.029	300
280	0.457	0.458	0.458	0.460	0.461	0.464	0.467	0.469	0.477	0.484	0.500	0.516	1.022	1.022	1.022	280
260	0.456	0.457	0.457	0.459	0.461	0.464	0.467	0.470	0.478	0.487	0.505	1.017	1.017	1.017	1.016	260
240	0.455	0.456	0.457	0.458	0.460	0.464	0.468	0.471	0.481	0.491	1.012	1.012	1.012	1.012	1.012	240
220	0.454	0.455	0.456	0.458	0.460	0.464	0.468	0.473	0.484	1.008	1.008	1.008	1.008	1.008	1.008	220
200	0.453	0.454	0.455	0.458	0.460	0.465	0.470	0.475	1.005	1.005	1.005	1.005	1.005	1.005	1.005	200
180	0.452	0.454	0.455	0.458	0.460	0.466	1.003	1.003	1.003	1.003	1.003	1.003	1.003	1.002	1.002	180
160	0.451	0.453	0.455	0.458	0.461	1.001	1.001	1.001	1.001	1.001	1.001	1.001	1.001	1.001	1.001	160
140	0.451	0.453	0.454	1.000	1.000	1.000	1.000	1.000	0.999	0.999	0.999	0.999	0.999	0.999	0.999	140
120	0.450	0.452	0.999	0.999	0.999	0.999	0.999	0.999	0.999	0.999	0.998	0.998	0.998	0.998	0.998	120
100	0.998	0.998	0.998	0.998	0.998	0.998	0.998	0.998	0.998	0.998	0.998	0.998	0.998	0.998	0.998	100
80	0.998	0.998	0.998	0.998	0.998	0.998	0.998	0.998	0.998	0.998	0.998	0.998	0.998	0.998	0.998	80
60	1.000	1.000	1.000	1.000	1.000	1.000	1.000	1.000	1.000	1.000	1.000	1.000	0.999	0.999	0.999	60
40	1.004	1.004	1.004	1.004	1.004	1.004	1.004	1.004	1.004	1.004	1.004	1.004	1.004	1.004	1.003	40
32	1.007	1.007	1.007	1.007	1.007	1.007	1.007	1.007	1.007	1.007	1.007	1.007	1.007	1.007	1.006	32

STEAM TABLES (Continued)

Specific Heat at constant pressure of Steam and of Water

Temp. F / Press., psia	150	200	300	400	600	800	1000	1500	2000	3000	4000	6000	8000	10000	15000	Temp. F / Press., psia
	c_p, Btu/lbm × F															
Sat. Water	1.054	1.067	1.093	1.118	1.168	1.224	1.286	1.492	1.841	7.646	—	—	—	—	—	Sat. Water
Sat. Steam	0.624	0.661	0.729	0.792	0.915	1.046	1.191	1.667	2.557	13.66	—	—	—	—	—	Sat. Steam
1500	0.562	0.563	0.565	0.567	0.571	0.576	0.580	0.590	0.601	0.623	0.645	0.691	0.737	0.780	0.868	1500
1480	0.561	0.562	0.564	0.566	0.570	0.575	0.579	0.590	0.601	0.623	0.647	0.694	0.742	0.786	0.878	1480
1460	0.559	0.560	0.562	0.565	0.569	0.573	0.578	0.589	0.601	0.624	0.648	0.698	0.747	0.793	0.888	1460
1440	0.557	0.559	0.561	0.563	0.568	0.572	0.577	0.589	0.600	0.625	0.650	0.701	0.753	0.800	0.900	1440
1420	0.556	0.557	0.559	0.562	0.566	0.571	0.576	0.588	0.600	0.625	0.651	0.705	0.759	0.808	0.909	1420
1400	0.554	0.555	0.558	0.560	0.565	0.570	0.575	0.587	0.600	0.626	0.653	0.709	0.765	0.817	0.926	1400
1380	0.553	0.554	0.556	0.559	0.564	0.569	0.574	0.587	0.600	0.627	0.655	0.714	0.773	0.827	0.939	1380
1360	0.551	0.552	0.555	0.558	0.563	0.568	0.573	0.586	0.600	0.628	0.657	0.719	0.781	0.838	0.953	1360
1340	0.549	0.551	0.553	0.556	0.561	0.567	0.572	0.586	0.600	0.629	0.660	0.725	0.790	0.850	0.968	1340
1320	0.548	0.549	0.552	0.555	0.560	0.566	0.571	0.585	0.600	0.630	0.663	0.731	0.800	0.864	0.983	1320
1300	0.546	0.548	0.550	0.553	0.559	0.565	0.570	0.585	0.600	0.632	0.666	0.738	0.811	0.879	0.998	1300
1280	0.545	0.546	0.549	0.552	0.558	0.564	0.570	0.585	0.600	0.634	0.669	0.746	0.824	0.897	1.014	1280
1260	0.543	0.544	0.547	0.550	0.556	0.563	0.569	0.585	0.601	0.636	0.673	0.755	0.838	0.918	1.033	1260
1240	0.541	0.543	0.546	0.549	0.555	0.562	0.568	0.584	0.601	0.638	0.678	0.765	0.855	0.942	1.053	1240
1220	0.540	0.541	0.544	0.548	0.554	0.561	0.567	0.584	0.602	0.641	0.683	0.777	0.875	0.969	1.072	1220
1200	0.538	0.540	0.543	0.546	0.553	0.560	0.567	0.584	0.603	0.644	0.689	0.790	0.897	1.000	1.095	1200
1180	0.536	0.538	0.541	0.545	0.552	0.559	0.566	0.584	0.604	0.647	0.696	0.805	0.922	1.033	1.117	1180
1160	0.535	0.536	0.540	0.544	0.551	0.558	0.565	0.585	0.606	0.652	0.704	0.823	0.952	1.070	1.143	1160
1140	0.533	0.535	0.539	0.542	0.550	0.557	0.565	0.585	0.607	0.656	0.713	0.843	0.986	1.107	1.167	1140
1120	0.531	0.533	0.537	0.541	0.549	0.557	0.565	0.586	0.609	0.662	0.723	0.866	1.025	1.149	1.190	1120
1100	0.530	0.532	0.536	0.540	0.548	0.556	0.564	0.587	0.612	0.668	0.735	0.893	1.070	1.193	1.220	1100
1080	0.528	0.530	0.534	0.538	0.547	0.555	0.564	0.588	0.615	0.676	0.749	0.924	1.120	1.242	1.240	1080
1060	0.527	0.529	0.533	0.537	0.546	0.555	0.564	0.590	0.618	0.685	0.765	0.960	1.176	1.295	1.260	1060
1040	0.525	0.527	0.532	0.536	0.545	0.555	0.565	0.592	0.622	0.695	0.783	1.002	1.238	1.351	1.282	1040
1020	0.523	0.526	0.530	0.535	0.545	0.555	0.565	0.594	0.627	0.707	0.804	1.051	1.306	1.399	1.298	1020
1000	0.522	0.524	0.529	0.534	0.544	0.555	0.566	0.597	0.633	0.721	0.829	1.110	1.382	1.471	1.306	1000
980	0.520	0.523	0.528	0.533	0.544	0.555	0.567	0.601	0.640	0.737	0.858	1.180	1.475	1.531	1.312	980
960	0.519	0.521	0.527	0.532	0.543	0.556	0.568	0.605	0.648	0.756	0.893	1.267	1.598	1.595	1.310	960
940	0.517	0.520	0.526	0.531	0.543	0.556	0.570	0.610	0.658	0.778	0.934	1.376	1.708	1.639	1.299	940
920	0.516	0.519	0.525	0.531	0.544	0.558	0.573	0.617	0.669	0.803	0.984	1.520	1.819	1.667	1.281	920
900	0.515	0.518	0.524	0.530	0.544	0.559	0.576	0.624	0.683	0.834	1.048	1.716	1.932	1.660	1.259	900
880	0.513	0.516	0.523	0.530	0.545	0.561	0.580	0.633	0.699	0.872	1.130	1.993	2.000	1.633	1.232	880
860	0.512	0.515	0.523	0.530	0.546	0.564	0.584	0.644	0.718	0.918	1.240	2.316	2.019	1.593	1.212	860
840	0.511	0.514	0.522	0.530	0.548	0.568	0.590	0.657	0.740	0.977	1.395	2.653	1.978	1.547	1.192	840
820	0.510	0.514	0.522	0.531	0.550	0.572	0.597	0.672	0.767	1.054	1.620	2.886	1.888	1.503	1.175	820
800	0.509	0.513	0.522	0.532	0.553	0.577	0.605	0.690	0.800	1.160	1.967	2.872	1.768	1.459	1.157	800
780	0.508	0.513	0.522	0.533	0.557	0.584	0.615	0.712	0.840	1.312	2.550	2.547	1.670	1.416	1.142	780
760	0.507	0.512	0.523	0.535	0.561	0.592	0.628	0.738	0.892	1.542	4.462	2.156	1.576	1.370	1.126	760
740	0.507	0.512	0.524	0.537	0.567	0.602	0.642	0.770	0.960	1.913	8.119	1.886	1.493	1.332	1.114	740
720	0.506	0.512	0.525	0.540	0.574	0.613	0.660	0.811	1.052	2.584	3.458	1.696	1.421	1.290	1.100	720
700	0.506	0.513	0.528	0.544	0.582	0.627	0.681	0.861	1.181	6.145⊙	2.237	1.557	1.358	1.250	1.089	700
680	0.506	0.514	0.530	0.549	0.592	0.644	0.707	0.927	1.365	2.469	1.789	1.450	1.303	1.217	1.079	680
660	0.507	0.515	0.534	0.555	0.604	0.665	0.738	1.015	1.639	1.851	1.587	1.369	1.256	1.187	1.071	660
640	0.507	0.517	0.538	0.562	0.619	0.690	0.777	1.135	2.219	1.601	1.454	1.303	1.216	1.157	1.063	640
620	0.509	0.519	0.543	0.571	0.637	0.720	0.826	1.308	1.614	1.455	1.362	1.252	1.184	1.136	1.056	620
600	0.510	0.522	0.550	0.582	0.659	0.757	0.888	1.586	1.453	1.358	1.295	1.211	1.157	1.118	1.052	600
580	0.513	0.526	0.558	0.595	0.685	0.804	0.969	1.393	1.351	1.289	1.243	1.178	1.134	1.102	1.046	580
560	0.516	0.531	0.568	0.611	0.717	0.862	1.079	1.309	1.281	1.237	1.202	1.151	1.115	1.087	1.039	560
540	0.519	0.538	0.580	0.630	0.756	0.937	1.272	1.249	1.229	1.196	1.169	1.128	1.098	1.074	1.031	540
520	0.524	0.545	0.594	0.653	0.804	1.035	1.221	1.204	1.189	1.164	1.142	1.109	1.083	1.062	1.024	520
500	0.530	0.554	0.611	0.680	0.865	1.187	1.181	1.169	1.157	1.137	1.120	1.092	1.069	1.051	1.017	500
480	0.537	0.565	0.632	0.714	1.159	1.154	1.150	1.140	1.131	1.115	1.101	1.077	1.057	1.041	1.010	480
460	0.545	0.578	0.657	0.755	1.132	1.128	1.125	1.117	1.110	1.096	1.084	1.064	1.047	1.033	1.004	460
440	0.556	0.594	0.687	1.113	1.110	1.107	1.104	1.098	1.092	1.080	1.070	1.052	1.038	1.025	0.999	440
420	0.568	0.614	0.724	1.094	1.091	1.089	1.087	1.081	1.076	1.067	1.058	1.042	1.029	1.018	0.994	420
400	0.583	0.636	1.079	1.078	1.076	1.074	1.072	1.067	1.063	1.055	1.047	1.034	1.022	1.011	0.990	400
380	0.601	1.066	1.065	1.065	1.063	1.061	1.059	1.056	1.052	1.044	1.038	1.026	1.015	1.006	0.986	380
360	0.622	1.054	1.054	1.053	1.052	1.050	1.049	1.045	1.042	1.036	1.030	1.019	1.009	1.001	0.982	360
340	1.045	1.044	1.044	1.043	1.042	1.040	1.039	1.036	1.033	1.028	1.022	1.013	1.004	0.996	0.979	340
320	1.036	1.036	1.035	1.034	1.033	1.032	1.031	1.028	1.026	1.021	1.016	1.007	0.999	0.992	0.976	320
300	1.028	1.028	1.028	1.027	1.026	1.025	1.024	1.022	1.019	1.015	1.010	1.002	0.995	0.988	0.973	300
280	1.022	1.022	1.021	1.021	1.020	1.019	1.018	1.016	1.014	1.009	1.005	0.998	0.991	0.985	0.971	280
260	1.016	1.016	1.016	1.015	1.014	1.013	1.013	1.011	1.009	1.005	1.001	0.994	0.988	0.982	0.968	260
240	1.012	1.011	1.011	1.011	1.010	1.009	1.008	1.006	1.004	1.001	0.997	0.991	0.985	0.979	0.966	240
220	1.008	1.008	1.007	1.007	1.006	1.005	1.005	1.003	1.001	0.998	0.994	0.988	0.982	0.977	0.964	220
200	1.005	1.004	1.004	1.004	1.003	1.002	1.002	1.000	0.998	0.995	0.992	0.986	0.980	0.975	0.963	200
180	1.002	1.002	1.002	1.001	1.001	1.000	0.999	0.998	0.996	0.993	0.989	0.983	0.978	0.973	0.961	180
160	1.000	1.000	1.000	0.999	0.999	0.998	0.997	0.996	0.994	0.991	0.987	0.981	0.976	0.971	0.959	160
140	0.999	0.999	0.998	0.998	0.997	0.997	0.996	0.994	0.992	0.989	0.986	0.980	0.974	0.969	0.958	140
120	0.998	0.998	0.997	0.997	0.996	0.996	0.995	0.993	0.991	0.988	0.984	0.978	0.972	0.967	0.957	120
100	0.997	0.997	0.997	0.996	0.996	0.995	0.994	0.992	0.990	0.986	0.983	0.976	0.970	0.965	0.955	100
80	0.998	0.997	0.997	0.996	0.995	0.994	0.994	0.991	0.989	0.985	0.981	0.974	0.968	0.962	0.951	80
60	0.999	0.999	0.998	0.997	0.996	0.995	0.994	0.991	0.989	0.984	0.979	0.970	0.963	0.956	0.942	60
40	1.003	1.003	1.002	1.001	1.000	0.998	0.997	0.993	0.989	0.983	0.976	0.965	0.954	0.945	0.920	40
32	1.006	1.006	1.005	1.004	1.002	1.000	0.999	0.994	0.990	0.983	0.975	0.962	0.949	0.937	0.904	32

⊙Critical point.

STEAM TABLES (Continued)

Thermal Conductivity of Steam and Water

Temp. F	k, (Btu/hr × ft. × F) × 10^3												
Press., psia	1	2	5	10	20	50	100	200	500	1000	2000	5000	7500
Sat. Water	364.0	373.1	383.8	390.4	395.2	397.4	394.7	386.2	361.7	327.6	271.8	—	—
Sat. Steam	11.6	12.2	13.0	13.8	14.8	16.6	18.4	21.1	27.2	36.5	61.3	—	—
1500	63.7	63.7	63.7	63.7	63.7	63.8	64.0	64.3	65.4	67.1	70.7	82.0	92.2
1450	61.4	61.4	61.5	61.5	61.5	61.6	61.8	62.1	63.2	64.9	68.5	80.1	90.6
1400	59.2	59.2	59.2	59.2	59.3	59.4	59.6	59.9	60.9	62.7	66.3	78.2	89.2
1350	57.0	57.0	57.0	57.0	57.1	57.2	57.3	57.7	58.7	60.5	64.2	76.3	87.9
1300	54.8	54.8	54.8	54.8	54.8	54.9	55.1	55.5	56.5	58.3	62.0	74.6	86.9
1250	52.6	52.6	52.6	52.6	52.6	52.7	52.9	53.2	54.3	56.1	59.9	73.0	86.3
1200	50.4	50.4	50.4	50.4	50.4	50.5	50.7	51.0	52.1	53.9	57.8	71.6	86.2
1150	48.2	48.2	48.2	48.2	48.2	48.3	48.5	48.9	49.9	51.8	55.7	70.5	87.0
1100	46.0	46.0	46.0	46.0	46.1	46.2	46.3	46.7	47.8	49.6	53.7	69.8	89.0
1050	43.9	43.9	43.9	43.9	43.9	44.0	44.2	44.6	45.6	47.5	51.8	69.7	93.4
1000	41.7	41.7	41.8	41.8	41.8	41.9	42.1	42.4	43.5	45.5	50.0	70.7	102.9
950	39.6	39.6	39.7	39.7	39.7	39.8	40.0	40.3	41.4	43.5	48.3	73.5	115.5
900	37.6	37.6	37.6	37.6	37.6	37.7	37.9	38.3	39.4	41.5	46.8	80.2	138.7
850	35.5	35.6	35.6	35.6	35.6	35.7	35.9	36.3	37.4	39.7	45.6	96.7	178.8
800	33.6	33.6	33.6	33.6	33.6	33.7	33.9	34.3	35.5	37.9	44.9	129.6	223.2
750	31.6	31.6	31.6	31.6	31.7	31.8	32.0	32.3	33.6	36.3	45.2	202.5	258.3
700	29.7	29.7	29.7	29.7	29.8	29.9	30.1	30.4	31.8	35.0	47.5[a]	262.8	295.1
650	27.8	27.8	27.9	27.9	27.9	28.0	28.2	28.6	30.1	34.1	55.7	304.3	326.7
600	26.0	26.0	26.1	26.1	26.1	26.2	26.4	26.9	28.7	34.1	301.9	333.7	349.3
550	24.3	24.3	24.3	24.3	24.4	24.5	24.7	25.2	27.5	36.1	333.7	356.1	368.0
500	22.6	22.6	22.6	22.6	22.7	22.8	23.0	23.6	26.9	350.8	357.4	373.8	383.6
450	21.0	21.0	21.0	21.0	21.0	21.2	21.4	22.3	368.1	370.6	375.3	387.9	396.5
400	19.4	19.4	19.4	19.4	19.5	19.6	20.0	21.3	383.0	384.9	388.5	398.6	406.4
350	17.9	17.9	17.9	17.9	18.0	18.2	18.8	392.0	392.9	394.4	397.4	406.1	413.2
300	16.5	16.5	16.5	16.5	16.6	16.9	396.9	397.2	398.0	399.3	402.0	409.9	416.4
250	15.1	15.1	15.1	15.2	15.3	396.9	397.0	397.3	398.1	399.4	402.1	409.7	415.8
200	13.8	13.8	13.9	14.0	391.6	391.6	391.8	392.1	393.0	394.4	397.2	404.9	410.6
150	12.7	12.7	380.5	380.5	380.6	380.7	380.8	381.1	382.1	383.7	386.7	394.7	400.3
100	363.3	363.3	363.3	363.3	363.3	363.4	363.6	363.9	365.0	366.6	369.8	378.3	384.1
50	339.1	339.1	339.1	339.1	339.2	339.3	339.4	339.8	340.8	342.5	345.7	354.6	361.0
32	328.6	328.6	328.6	328.6	328.6	328.7	328.9	329.2	330.3	331.9	335.1	344.1	350.8

[a] Critical point.

Ammonia, NH_3

Temp. °F	Abs. press. sat. vap. lb./in.²	Abs. press. sat. vap. kg/cm²	Heat content abv. −40°F BTU/lb. Liq.	Heat content abv. −40°F BTU/lb. Vap.	Ht. of vaporiz. BTU/lb.	Heat content abv. −40°C g-cal./g Liq.	Heat content abv. −40°C g-cal./g Vap.	Ht. of vaporiz. g-cal./g	Temp. °C
−60	5.55	0.390	−21.2	589.6	510.8	−11.8	327.6	339.3	−51.11
−58	5.93	.417	−19.1	590.4	609.5	−10.6	328.0	338.6	−50.00
−56	6.33	.445	−17.0	591.2	608.2	− 9.44	328.4	337.9	−48.89
−54	6.75	.475	−14.8	592.1	606.9	− 8.22	328.9	337.2	−47.78
−52	7.20	.506	−12.7	592.9	605.6	− 7.06	329.4	336.4	−46.67
−50	7.67	0.539	−10.6	593.7	604.3	− 5.89	329.8	335.7	−45.56
−48	8.16	.574	− 8.5	594.4	602.9	− 4.7	330.2	334.9	−44.44
−46	8.68	.610	− 6.4	595.2	601.6	− 3.6	330.7	334.2	−43.33
−44	9.23	.649	− 4.3	596.0	600.3	− 2.4	331.1	333.5	−42.22
−42	9.81	.690	− 2.1	596.8	598.9	− 1.2	331.6	332.7	−41.11
−40	10.41	0.7319	0.0	597.6	597.6	0.0	332.0	332.0	−40.00
−38	11.04	.7762	+ 2.1	598.3	596.2	+ 1.2	332.4	331.2	−38.89
−36	11.71	.8233	4.3	599.1	594.8	2.4	332.8	330.4	−37.78
−34	12.41	.8725	6.4	599.9	593.5	3.6	333.3	329.7	−36.67
−32	13.14	.9238	8.5	600.6	592.1	4.7	333.7	328.9	−35.56
−30	13.90	0.9773	10.7	601.4	590.7	5.94	334.1	328.2	−34.44
−28	14.71	1.034	12.8	602.1	589.3	7.11	334.5	327.4	−33.33
−26	15.55	1.093	14.9	602.8	587.9	8.28	334.9	326.6	−32.22
−24	16.42	1.154	17.1	603.6	586.5	9.50	335.3	325.8	−31.11
−22	17.34	1.219	19.2	604.3	585.1	10.7	335.7	325.1	−30.00
−20	18.30	1.287	21.4	605.0	583.6	11.9	336.1	324.2	−28.89
−18	19.30	1.357	23.5	605.7	582.2	13.1	336.5	323.4	−27.78
−16	20.34	1.430	25.6	606.4	580.8	14.2	336.9	322.7	−26.67
−14	21.43	1.507	27.8	607.1	579.3	15.4	337.3	321.8	−25.56
−12	22.56	1.586	30.0	607.8	577.8	16.7	337.7	321.0	−24.44
−10	23.74	1.669	32.1	608.5	576.4	17.8	338.1	320.2	−23.33
− 8	24.97	1.756	34.3	609.2	574.9	19.1	338.4	319.4	−22.22
− 6	26.26	1.846	36.4	609.8	573.4	20.2	338.8	318.6	−21.11
− 4	27.59	1.940	38.6	610.5	571.9	21.4	339.2	317.7	−20.00
− 2	28.98	2.037	40.7	611.1	570.4	22.6	339.5	316.9	−18.89
0	30.42	2.139	42.9	611.8	568.9	23.8	339.9	316.1	−17.78
2	31.92	2.244	45.1	612.4	567.3	25.1	340.2	315.2	−16.67
4	33.47	2.353	47.2	613.0	565.8	26.2	340.6	314.3	−15.56
6	35.09	2.467	49.4	613.6	564.2	27.4	340.9	313.4	−14.44
8	36.77	2.585	51.6	614.3	562.7	28.7	341.3	312.6	−13.33
10	38.51	2.708	53.8	614.9	561.1	29.9	341.6	311.7	−12.22
12	40.31	2.834	56.0	615.5	559.5	31.1	341.9	310.8	−11.11
14	42.18	2.966	58.2	616.1	557.9	32.3	342.3	309.9	−10.00
16	44.12	3.102	60.3	616.6	556.3	33.5	342.6	309.1	− 8.89
18	46.13	3.243	62.5	617.2	554.7	34.7	342.9	308.2	− 7.78
20	48.21	3.390	64.7	617.8	553.1	35.9	343.2	307.3	− 6.67
22	50.36	3.541	66.9	618.3	551.4	37.2	343.5	306.3	− 5.56
24	52.59	3.697	69.1	618.9	549.8	38.4	343.8	305.4	− 4.44
26	54.90	3.860	71.3	619.4	548.1	39.6	344.1	304.5	− 3.33
28	57.28	4.027	73.5	619.9	546.4	40.8	344.4	303.6	− 2.22

THERMODYNAMIC PROPERTIES

Ammonia, NH_3

Temp. °F	Spec. vol. sat. vap. ft.³/lb.	Spec. vol. sat. vap. m³/kg	Density sat. vap. lb./ft.³	Density sat. vap. kg/m³	Dens. liq. lb./ft.³	Entropy from −40°F BTU/lb./°F Liq.	Entropy from −40°F BTU/lb./°F Vap.	Temp. °C
−60	44.73	2.792	0.02235	0.3580	43.91	−0.0517	1.4769	−51.11
−58	42.05	2.625	.02378	.3809		− .0464	1.4713	−50.00
−56	39.56	2.470	.02528	.4049		− .0412	1.4658	−48.89
−54	37.24	2.325	.02685	.4301		− .0360	1.4604	−47.78
−52	35.09	2.191	.02850	.4565		− .0307	1.4551	−46.67
−50	33.08	2.065	0.03023	0.4842	43.49	−0.0256	1.4497	−45.56
−48	31.20	1.948	.03205	.5134		− .0204	1.4445	−44.44
−46	29.45	1.839	.03395	.5438		− .0153	1.4393	−43.33
−44	27.82	1.737	.03595	.5758		− .0102	1.4342	−42.22
−42	26.29	1.641	.03804	.6093		− .0051	1.4292	−41.11
−40	24.86	1.552	0.04022	0.6442	43.08	0.0000	1.4242	−40.00
−38	23.53	1.469	.04251	.6809		.0051	1.4193	−38.89
−36	22.27	1.390	.04489	.7190		.0101	1.4144	−37.78
−34	21.10	1.317	.04739	.7591		.0151	1.4096	−36.67
−32	20.00	1.249	.04999	.8007		.0201	1.4048	−35.56
−30	18.97	1.184	0.05271	0.8443	42.65	0.0250	1.4001	−34.44
−28	18.00	1.124	.05555	.8898		.0300	1.3955	−33.33
−26	17.09	1.067	.05850	.9371		.0350	1.3909	−32.22
−24	16.24	1.014	.06158	.9864		.0399	1.3863	−31.11
−22	15.43	0.9633	.06479	1.038		.0448	1.3818	−30.00
−20	14.68	0.9164	0.06813	1.091	42.22	0.0497	1.3774	−28.89
−18	13.97	.8721	.07161	1.147		.0545	1.3729	−27.78
−16	13.29	.8297	.07522	1.205		.0594	1.3686	−26.67
−14	12.66	.7903	.07898	1.265		.0642	1.3643	−25.56
−12	12.06	.7529	.08289	1.328		.0690	1.3600	−24.44
−10	11.50	0.7179	0.08695	1.393	41.78	0.0738	1.3558	−23.33
− 8	10.97	.6848	.09117	1.460		.0786	1.3516	−22.22
− 6	10.47	.6536	.09555	1.531		.0833	1.3474	−21.11
− 4	9.991	.6237	.1001	1.603		.0880	1.3433	−20.00
− 2	9.541	.5956	.1048	1.679		.0928	1.3393	−18.89
0	9.116	0.5691	0.1097	1.757	41.34	0.0975	1.3352	−17.78
2	8.714	.5440	.1148	1.839		.1022	1.3312	−16.67
4	8.333	.5202	.1200	1.922		.1069	1.3273	−15.56
6	7.971	.4976	.1254	2.009		.1115	1.3234	−14.44
8	7.629	.4763	.1311	2.100		.1162	1.3195	−13.33
10	7.304	0.4560	0.1369	2.193	40.89	0.1208	1.3157	−12.22
12	6.996	.4367	.1429	2.289		.1254	1.3118	−11.11
14	6.703	.4185	.1492	2.390		.1300	1.3081	−10.00
16	6.425	.4011	.1556	2.492		.1346	1.3043	− 8.89
18	6.161	.3846	.1623	2.600		.1392	1.3006	− 7.78
20	5.910	0.3690	0.1692	2.710	40.43	0.1437	1.2969	− 6.67
22	5.671	.3540	.1763	2.824		.1483	1.2933	− 5.56
24	5.443	.3398	.1837	2.943		.1528	1.2897	− 4.44
26	5.227	.3263	.1913	3.064		.1573	1.2861	− 3.33
28	5.021	.3135	.1992	3.191		.1618	1.2825	− 2.22

THERMODYNAMIC PROPERTIES (Continued)

Ammonia, NH₃ (Continued)

Temp. °F	Abs. press. sat. vap. lb./in.²	Abs. press. sat. vap. kg/cm²	Heat content abv. −40°F BTU/lb. Liq.	Heat content abv. −40°F BTU/lb. Vap.	Ht. of vaporiz. BTU/lb.	Heat content abv. −40°C g-cal./g Liq.	Heat content abv. −40°C g-cal./g Vap.	Ht. of vaporiz. g-cal./g	Temp. °C
30	59.74	4.200	75.7	620.5	544.8	42.1	344.7	302.7	− 1.11
32	62.29	4.379	77.9	621.0	543.1	43.3	345.0	301.7	0.00
34	64.91	4.564	80.1	621.5	541.4	44.5	345.3	300.8	+ 1.11
36	67.63	4.755	82.3	622.0	539.7	45.7	345.6	299.8	2.22
38	70.43	4.952	84.6	622.5	537.9	47.0	345.8	298.8	3.33
40	73.32	5.155	86.8	623.0	536.2	48.2	346.1	297.9	4.44
42	76.31	5.365	89.0	623.4	534.4	49.4	346.3	296.9	5.56
44	79.38	5.581	91.2	623.9	532.7	50.7	346.6	295.9	6.67
46	82.55	5.804	93.5	624.4	530.9	51.9	346.9	294.9	7.78
48	85.82	6.034	95.7	624.8	529.1	53.2	347.1	293.9	8.89
50	89.19	6.271	97.9	625.2	527.3	54.4	347.3	292.9	10.00
52	92.66	6.515	100.2	625.7	525.5	55.67	347.6	291.9	11.11
54	96.23	6.766	102.4	626.1	523.7	56.89	347.8	290.9	12.22
56	99.91	7.024	104.7	626.5	521.8	58.17	348.1	289.9	13.33
58	103.7	7.291	106.9	626.9	520.0	59.39	348.3	288.9	14.44
60	107.6	7.565	109.2	327.3	518.1	60.67	348.5	287.8	15.56
62	111.6	7.846	111.5	627.7	516.2	61.94	348.7	286.8	16.67
64	115.7	8.135	113.7	628.0	514.3	63.17	348.9	285.7	17.78
66	120.0	8.437	116.0	628.4	512.4	64.44	349.1	284.7	18.89
68	124.3	8.739	118.3	628.8	510.5	65.72	349.3	283.6	20.00
70	128.8	9.056	120.5	629.1	508.6	66.94	349.5	282.6	21.11
72	133.4	9.379	122.8	629.4	506.6	68.22	349.7	281.4	22.22
74	138.1	9.709	125.1	629.8	504.7	69.50	349.9	280.4	23.33
76	143.0	10.05	127.4	630.1	502.7	70.78	350.1	279.3	24.44
78	147.9	10.40	129.7	630.4	500.7	72.06	350.2	278.2	25.56
80	153.0	10.76	132.0	630.7	498.7	73.33	350.4	277.1	26.67
82	158.3	11.13	134.3	631.0	496.7	74.61	350.6	275.9	27.78
84	163.7	11.51	136.6	631.3	494.7	75.89	350.7	274.8	28.89
86	169.2	11.90	138.9	631.5	492.6	77.17	350.8	273.7	30.00
88	174.8	12.29	141.2	631.8	490.6	78.44	351.0	272.6	31.11
90	180.6	12.70	143.5	632.0	488.5	79.72	351.1	271.4	32.22
92	186.6	13.12	145.8	632.2	486.4	81.00	351.2	270.2	33.33
94	192.7	13.55	148.2	632.5	484.3	82.33	351.4	269.1	34.44
96	198.9	13.98	150.5	632.6	482.1	83.61	351.4	267.8	35.56
98	205.3	14.43	152.9	632.9	480.0	84.94	351.6	266.7	36.67
100	211.9	14.90	155.2	633.0	477.8	86.22	351.7	265.4	37.78
102	218.6	15.37	157.6	633.2	475.6	87.56	351.8	264.2	38.89
104	225.4	15.85	159.9	633.4	473.5	88.83	351.9	263.1	40.00
106	232.5	16.35	162.3	633.5	471.2	90.17	351.9	261.8	41.11
108	239.7	16.85	164.6	633.6	469.0	91.44	352.0	260.6	42.22
110	247.0	17.37	167.0	633.7	466.7	92.78	352.1	259.3	43.33
112	254.5	17.89	169.4	633.8	464.4	94.11	352.1	258.0	44.44
114	262.2	18.43	171.8	633.9	462.1	95.44	352.2	256.7	45.56
116	270.1	18.99	174.2	634.0	459.8	96.78	352.2	255.4	46.67
118	278.2	19.56	176.6	634.0	457.4	98.11	352.2	254.1	47.78
120	286.4	20.14	179.0	634.0	455.0	99.45	352.2	252.8	48.89
122	294.8	20.73	181.4	634.0	452.6	100.8	352.2	251.4	50.00
124	303.4	21.33	183.9	634.0	450.1	102.2	352.2	250.1	51.11

Ammonia, NH₃ (Continued)

Temp. °F	Spec. vol. sat. vap. ft.³/lb.	Spec. vol. sat. vap. m³/kg	Density sat. vap. lb./ft.³	Density sat. vap. kg/m³	Dens. liq. lb./ft.³	Entropy from −40°F BTU/lb./°F Liq.	Entropy from −40°F BTU/lb./°F Vap.	Temp. °C
30	4.825	0.3012	0.2073	3.321	39.96	0.1663	1.2790	− 1.11
32	4.637	.2895	.2156	3.453		.1708	1.2755	0.00
34	4.459	.2784	.2243	3.593		.1753	1.2721	+ 1.11
36	4.289	.2678	.2332	3.735		.1797	1.2686	2.22
38	4.126	.2576	.2423	3.881		.1841	1.2652	3.33
40	3.971	0.2479	0.2518	4.033	39.49	0.1885	1.2618	4.44
42	3.823	.2387	.2616	4.190		.1930	1.2585	5.56
44	3.682	.2299	.2716	4.350		.1974	1.2552	6.67
46	3.547	.2214	.2819	4.515		.2018	1.2519	7.78
48	3.418	.2134	.2926	4.687		.2062	1.2486	8.89
50	3.294	0.2056	0.3036	4.863	39.00	0.2105	1.2453	10.00
52	3.176	.1983	.3149	5.044		.2149	1.2421	11.11
54	3.063	.1912	.3265	5.230		.2192	1.2389	12.22
56	2.954	.1844	.3385	5.422		.2236	1.2357	13.33
58	2.851	.1780	.3508	5.619		.2279	1.2325	14.44
60	2.751	0.1717	0.3635	5.823	38.50	0.2322	1.2294	15.56
62	2.656	.1658	.3765	6.031		.2365	1.2262	16.67
64	2.565	.1601	.3899	6.245		.2408	1.2231	17.78
66	2.477	.1546	.4037	6.466		.2451	1.2201	18.89
68	2.393	.1494	.4179	6.694		.2494	1.2170	20.00
70	2.312	0.1443	0.4325	6.928	38.00	0.2537	1.2140	21.11
72	2.235	.1395	.4474	7.166		.2579	1.2110	22.22
74	2.161	.1349	.4628	7.413		.2622	1.2080	23.33
76	2.089	.1304	.4786	7.666		.2664	1.2050	24.44
78	2.021	.1262	.4949	7.927		.2706	1.2020	25.56
80	1.955	0.1220	0.5115	8.193	37.48	0.2749	1.1991	26.67
82	1.892	.1181	.5287	8.469		.2791	1.1962	27.78
84	1.831	.1143	.5462	8.749		.2833	1.1933	28.89
86	1.772	.1106	.5643	9.039		.2875	1.1904	30.00
88	1.716	.1071	.5828	9.335		.2917	1.1875	31.11
90	1.661	0.1037	0.6019	9.641	36.95	0.2958	1.1846	32.22
92	1.609	.1004	.6214	9.954		.3000	1.1818	33.33
94	1.559	.09733	.6415	10.28		.3041	1.1789	34.44
96	1.510	.09427	.6620	10.60		.3083	1.1761	35.56
98	1.464	.09140	.6832	10.94		.3125	1.1733	36.67
100	1.419	0.08859	0.7048	11.29	36.40	0.3166	1.1705	37.78
102	1.375	.08584	.7270	11.65		.3207	1.1677	38.89
104	1.334	.08328	.7498	12.01		.3248	1.1649	40.00
106	1.293	.08072	.7732	12.39		.3289	1.1621	41.11
108	1.254	.07829	.7972	12.77		.3330	1.1593	42.22
110	1.217	0.07598	0.8219	13.17	35.84	0.3372	1.1566	43.33
112	1.180	.07367	.8471	13.57		.3413	1.1538	44.44
114	1.145	.07148	.8730	13.98		.3453	1.1510	45.56
116	1.112	.06942	.8996	14.41		.3495	1.1483	46.67
118	1.079	.06736	.9269	14.85		.3535	1.1455	47.78
120	1.047	0.06536	0.9549	15.30	35.26	0.3576	1.1427	48.89
122	1.017	.06349	.9837	15.76		.3618	1.1400	50.00
124	0.987	.0616	1.0132	16.229		.3659	1.1372	51.11

THERMODYNAMIC PROPERTIES (Continued)

Carbon Dioxide, CO_2

Temp. °F	Abs. press sat. vap lb./in.²	Abs. press sat. vap kg/cm²	Heat content abv. 32°F BTU/lb. Liq.	Heat content abv. 32°F BTU/lb. Vap.	Heat of vaporiz. BTU/lb.	Heat content abv 0°C g-cal./g Liq.	Heat content abv 0°C g-cal./g Vap.	Heat of vaporis. g-cal./g	Temp. °C
−20	220.6	15.51	−23.96	102.0	126.0	−13.31	56.67	70.00	−28.89
−18	228.4	16.06	−23.13	102.1	125.2	−12.85	56.72	69.56	−27.78
−16	236.4	16.62	−22.30	102.2	124.5	−12.39	56.78	69.17	−26.67
−14	244.6	17.20	−21.46	102.2	123.7	−11.92	56.78	68.72	−25.56
−12	253.0	17.79	−20.61	102.3	122.9	−11.45	56.83	68.28	−24.44
−10	261.7	18.40	−19.76	102.3	122.0	−10.98	56.83	67.78	−23.33
− 8	270.6	19.03	−18.90	102.3	121.2	−10.50	56.83	67.33	−22.22
− 6	279.7	19.66	−18.04	102.3	120.3	−10.02	56.83	66.83	−21.11
− 4	289.1	20.33	−17.17	102.3	119.5	− 9.539	56.83	66.39	−20.00
− 2	298.7	21.00	−16.29	102.3	118.6	− 9.050	56.83	65.89	−18.89
0	308.6	21.70	−15.41	102.2	117.7	− 8.561	56.78	65.39	−17.78
2	318.7	22.41	−14.51	102.2	116.7	− 8.061	56.78	64.83	−16.67
4	329.1	23.14	−13.61	102.1	115.8	− 7.561	56.72	64.33	−15.56
6	339.8	23.89	−12.71	102.1	114.8	− 7.061	56.72	63.78	−14.44
8	350.7	24.66	−11.79	102.0	113.8	− 6.550	56.67	63.22	−13.33
10	361.8	25.44	−10.87	101.9	112.8	− 6.039	56.61	62.67	−12.22
12	373.3	26.25	− 9.934	101.8	111.7	− 5.519	56.56	62.06	−11.11
14	385.0	27.07	− 8.992	101.7	110.7	− 4.996	56.50	61.50	−10.00
16	397.1	27.92	− 8.038	101.5	109.6	− 4.466	56.39	60.89	− 8.89
18	409.4	28.78	− 7.076	101.4	108.5	− 3.931	56.33	60.28	− 7.78
20	422.0	29.67	− 6.102	101.2	107.3	− 3.390	56.22	59.61	− 6.67
22	434.9	30.58	− 5.117	101.0	106.1	− 2.843	56.11	58.94	− 5.56
24	448.1	31.50	− 4.121	100.8	104.9	− 2.289	56.90	58.28	− 4.44
25	454.8	31.98	− 3.618	100.7	104.3	− 2.010	55.94	57.94	− 3.89
27	468.5	32.94	− 2.601	100.5	103.1	− 1.445	55.83	57.28	− 2.78
29	482.5	33.92	− 1.570	100.2	101.8	− 0.8722	55.67	56.56	− 1.67
31	496.8	34.93	− 0.525	99.98	100.5	− 0.292	55.54	55.83	− 0.56
33	511.4	35.95	+ 0.531	99.69	99.16	+ 0.295	55.38	55.09	+ 0.56
35	526.4	37.01	1.604	99.38	97.77	0.8911	55.21	54.32	1.67
37	541.7	38.09	2.697	99.05	96.35	1.498	55.03	53.53	2.78
39	557.4	39.19	3.806	98.69	94.88	2.114	54.83	52.71	3.89
41	573.4	40.31	4.932	98.31	93.37	2.740	54.62	51.87	5.00
43	589.8	41.47	6.080	97.90	91.82	3.378	54.39	51.01	6.11
45	606.5	42.64	7.251	97.46	90.21	4.028	54.14	50.12	7.22
47	623.6	43.84	8.443	96.99	88.55	4.691	53.88	49.19	8.33
49	641.1	45.07	9.664	96.50	86.83	5.369	53.61	48.24	9.44
51	659.0	46.33	10.91	95.97	85.06	6.061	53.32	47.26	10.56
53	677.3	47.62	12.19	95.40	83.21	6.772	53.00	46.23	11.67
55	695.9	48.93	13.49	94.78	81.29	7.494	52.66	45.16	12.78
57	714.9	50.26	14.84	94.13	79.30	8.244	52.29	44.06	13.89
59	734.3	51.63	16.22	93.44	77.22	9.011	51.91	42.90	15.00
61	754.2	53.03	17.65	92.69	75.04	9.806	51.49	41.69	16.11
63	774.5	54.45	19.13	91.88	72.75	10.63	51.04	40.42	17.22
65	795.1	55.90	20.66	91.01	70.35	11.48	50.56	39.08	18.33
67	816.2	57.38	22.25	90.07	67.81	12.36	50.04	37.67	19.44
69	837.8	58.90	23.92	89.04	65.12	13.29	49.47	36.18	20.56
71	859.8	60.45	25.67	87.92	62.25	14.26	48.84	34.58	21.67
73	882.2	62.02	27.52	86.69	59.17	15.29	48.16	32.87	22.78
75	905.1	63.63	29.50	85.33	55.83	16.39	47.41	31.02	23.89
77	928.4	65.27	31.62	83.80	52.17	17.57	46.56	28.98	25.00
79	952.2	66.95	33.95	82.06	48.11	18.86	45.59	26.73	26.11
81	976.5	68.65	36.54	80.03	43.49	20.30	44.46	24.16	27.22
83	1001.0	70.377	39.53	77.60	38.07	21.96	43.11	21.15	28.33
85	1027.0	72.205	43.18	74.47	31.29	23.99	41.37	17.38	29.44
86	1039.0	73.049	45.45	72.46	27.00	25.25	40.26	15.00	30.00
87	1052.0	73.963	48.32	69.84	21.52	26.84	38.80	11.96	30.56
88	1065.0	74.877	52.78	65.62	12.84	29.32	36.46	7.133	31.11

Carbon Dioxide, CO_2

Temp. °F	Spec. vol. sat. vap. ft.³/lb.	Spec. vol. sat. vap. m³/kg	Density sat. vap. lb./ft.³	Density sat. vap. kg/m³	Dens. liq. lb./ft.³	Entropy from 32°F BTU/lb./°F Liq.	Entropy from 32°F BTU/lb./°F Vap.	Temp. °C
−20	0.4166	0.02601	2.401	38.46	64.34	−0.0514	0.2353	−28.89
−18	.4018	.02508	2.489	39.87	64.15	− .0495	.2342	−27.78
−16	.3876	.02420	2.580	41.33	63.94	− .0476	.2331	−26.67
−14	.3739	.02334	2.674	42.83	63.73	− .0458	.2319	−25.56
−12	.3608	.02252	2.772	44.40	63.49	− .0439	.2307	−24.44
−10	0.3482	0.02174	2.872	46.00	63.25	−0.0420	0.2296	−23.33
− 8	.3360	.02098	2.976	47.67	63.01	− .0401	.2284	−22.22
− 6	.3243	.02025	3.083	49.38	62.76	− .0382	.2273	−21.11
− 4	.3131	.01955	3.194	51.16	62.50	− .0362	.2261	−20.00
− 2	.3022	.01887	3.309	53.00	62.23	− .0343	.2249	−18.89
0	0.2918	0.01822	3.427	54.89	61.95	−0.0324	0.2237	−17.78
2	.2817	.01759	3.550	56.86	61.65	− .0304	.2225	−16.67
4	.2720	.01698	3.676	58.88	61.36	− .0285	.2213	−15.56
6	.2627	.01640	3.807	60.98	61.07	− .0266	.2201	−14.44
8	.2537	.01584	3.942	63.14	60.77	− .0246	.2189	−13.33
10	0.2450	0.01529	4.082	65.39	60.48	−0.0226	0.2176	−12.22
12	.2366	.01477	4.227	67.71	60.18	− .0206	.2164	−11.11
14	.2285	.01426	4.377	70.11	59.88	− .0186	.2151	−10.00
16	.2207	.01378	4.532	72.59	59.58	− .0166	.2139	− 8.89
18	.2131	.01330	4.692	75.16	59.27	− .0146	.2126	− 7.78
20	0.2058	0.01285	4.859	77.83	58.95	−0.0126	0.2113	− 6.67
22	.1987	.01240	5.031	80.59	58.64	− .0105	.2100	− 5.56
24	.1919	.01198	5.211	83.47	58.31	− .0085	.2037	− 4.44
25	.1886	.01177	5.303	84.94	58.14	− .0074	.2080	− 3.89
27	.1821	.01137	5.492	87.97	57.81	− .0053	.2066	− 2.78
29	0.1758	0.01097	5.688	91.11	57.47	−0.0032	0.2053	− 1.67
31	.1697	.01059	5.892	94.38	57.12	− .0011	.2039	− 0.56
33	.1639	.01023	6.103	97.76	56.77	+ .0011	.2025	+ 0.56
35	.1581	.009870	6.323	101.3	56.41	.0033	.2010	1.67
37	.1526	.009527	6.553	105.0	56.03	.0055	.1996	2.78
39	0.1472	0.009189	6.792	108.8	55.65	0.0077	0.1981	3.89
41	.1420	.008865	7.040	112.8	55.25	.0099	.1965	5.00
43	.1370	.008553	7.300	116.9	54.84	.0122	.1950	6.11
45	.1321	.008247	7.571	121.3	54.41	.0146	.1934	7.22
47	.1273	.007947	7.854	125.8	53.97	.0169	.1918	8.33
49	0.1227	0.007660	8.151	130.6	53.51	0.0193	0.1901	9.44
51	.1182	.007379	8.461	135.5	53.04	.0218	.1884	10.56
53	.1138	.007104	8.787	140.8	52.55	.0243	.1867	11.67
55	.1095	.006836	9.132	146.3	52.05	.0268	.1849	12.78
57	.1053	.006574	9.497	152.1	51.53	.0294	.1830	13.89
59	0.1012	0.006318	9.880	158.3	50.99	0.0321	0.1811	15.00
61	.0972	.00607	10.29	164.8	50.42	.0348	.1790	16.11
63	.0933	.00582	10.72	171.7	49.80	.0377	.1770	17.22
65	.0894	.00558	11.18	179.1	49.14	.0406	.1748	18.33
67	.0856	.00534	11.67	186.9	48.44	.0436	.1725	19.44
69	0.0819	0.00511	12.21	195.6	47.69	0.0468	0.1701	20.56
71	.0782	.00488	12.82	205.4	46.87	.0501	.1675	21.67
73	.0745	.00465	13.43	215.1	45.99	.0536	.1647	22.78
75	.0708	.00442	14.13	226.3	45.05	.0573	.1618	23.89
77	.0671	.00419	14.90	238.7	44.06	.0613	.1585	25.00
79	0.0633	0.00395	15.81	253.2	43.04	0.0656	0.1550	26.11
81	.0592	.00370	16.90	270.7	41.95	.0704	.1509	27.22
83	.0548	.00342	18.25	292.3	40.62	.0759	.1461	28.33
85	.0500	.00312	20.00	320.4	38.76	.0826	.1401	29.44
86	.0474	.00296	21.09	337.8	37.41	.0868	.1363	30.00
87	0.0446	0.00278	22.42	359.1	35.34	0.0921	0.1314	30.56
88	.0401	.00250	24.95	399.6	32.79	.1002	.1237	31.11

THERMODYNAMIC PROPERTIES (Continued)

THERMODYNAMIC PROPERTIES (Continued)

Sulfur Dioxide, SO_2

Temp. °F	Abs. press. sat. vap. lb./in.²	Abs. press. sat. vap. kg/cm²	Heat content abv. −40°F BTU/lb. Liq.	Heat content abv. −40°F BTU/lb. Vap.	Ht. of vaporis BTU/lb.	Heat content abv. −40°C g-cal./g Liq.	Heat content abv. −40°C g-cal./g Vap.	Ht. of vaporis. g-cal./g	Temp. °C
−40	3.136	0.2205	0.00	178.61	178.61	0.00	99.228	99.228	−40 00
−30	4.331	.3045	2.93	179.90	176.97	1.63	99.945	98.317	−34.44
−20	5 883	.4136	5.98	181.07	175.09	3.32	100.59	97.272	−28.89
−10	7.863	.5528	9.16	182.13	172.97	5.09	101.18	96.095	−23.33
0	10.35	.7277	12.44	183.07	170.63	6.911	101.71	94.795	−17.78
2	10.91	0.7670	13.12	183.25	170.13	7.289	101.81	94.517	−16.67
4	11.50	.8085	13.78	183.41	169.63	7.656	101.89	94.239	−15.56
5	11.81	.8303	14.11	183.49	169.38	7.839	101.94	94.100	−15.00
6	12.12	.8521	14.45	183.57	169.12	8.028	101.98	93.956	−14.44
8	12.75	.8964	15.13	183.73	168.60	8.406	102.07	93.667	−13.33
10	13.42	0.9435	15.80	183.87	168.07	8.778	102.15	93.372	−12.22
12	14.12	.9927	16.48	184.01	167.53	9.156	102.23	93.072	−11.11
14	14.84	1.043	17.15	184.14	166.97	9.528	102.30	92.761	−10.00
16	15.59	1.096	17.84	184.28	166.44	9.911	102.38	92.467	− 8.89
18	16.37	1.1509	18.52	184.40	165.88	10.29	102.44	92.156	− 7.78
20	17.18	1.208	19.20	184.52	165.32	10.67	102.51	91.845	− 6.67
22	18.03	1.268	19.90	184.64	164.74	11.06	102.58	91.522	− 5.56
24	18.89	1.328	20.58	184.74	164.16	11.43	102.63	91.200	− 4.44
26	19.80	1.392	21.26	184.84	163.58	11.81	102.69	90.878	− 3.33
28	20.73	1.457	21.96	184.94	162.98	12.20	102.74	90.545	− 2.22
30	21.70	1.526	22.64	185.02	162.38	12.58	102.79	90.211	− 1.11
32	22.71	1.597	23.33	185.10	161.77	12.96	102.83	89.872	0.00
34	23.75	1.670	24.03	185.18	161.15	13.35	102.88	89.528	+ 1.11
36	24.82	1.745	24.72	185.25	160.53	13.73	102.92	89.183	2.22
38	25.95	1.824	25.42	185.31	159 89	14.12	102.95	88.828	3.33
40	27.10	1 905	26.12	185.37	159.25	14.51	102.98	88.472	4.44
42	28.29	1.989	26.81	185.42	158.61	14.89	103.01	88.117	5.56
44	29.52	2.075	27.51	185.46	157.95	15.28	103.03	87.750	6.67
46	30.79	2.165	28.21	185.50	157.29	15.67	103.06	87.383	7.78
48	32.10	2.257	28.92	185.54	156.62	16.07	103.08	87.011	8.89
50	33.45	2.352	29.61	185.56	155.95	16.45	103.09	86.639	10.00
52	34.86	2.451	30.31	185.58	155.27	16.84	103.10	86.261	11.11
54	36.31	2.553	31.00	185.59	154.59	17.22	103.11	85.883	12.22
56	37.80	2.658	31.72	185.61	153.89	17.62	103.12	85.495	13.33
58	39.33	2.765	32.42	185.61	153.19	18.01	103.12	85.106	14.44
60	40.93	2.878	33.10	185.59	152.49	18.39	103.11	84.717	15.56
62	42.58	2.994	33.79	185.57	151.78	18.77	103.09	84.322	16.67
64	44.27	3.112	34.49	185.55	151.06	19.16	103.08	83.922	17.78
66	46.00	3.234	35.19	185.53	150.34	19.55	103.07	83.522	18.89
68	47.78	3.359	35.88	185.50	149.62	19.93	103 06	83.122	20.00
70	49.62	3.489	36.58	185.46	148.88	20.32	103.03	82.711	21.11
72	51.54	3.624	37.28	185.42	148.14	20.71	103.01	82.300	22.22
74	53.48	3.760	37.97	185.37	147.40	21.09	102.98	81.889	23.33
76	55.48	3.901	38.67	185.31	146.64	21.48	102.95	81.467	24.44
78	57.56	4.047	39.36	185.24	145.88	21.87	102.91	81.045	25.56
80	59.68	4.196	40.05	185.17	145.12	22.25	102.87	80.622	26.67
82	61.88	4.351	40.73	185.09	144.36	22.63	102.83	80.200	27.78
84	64.14	4.509	41.43	185.01	143.58	23.02	102.78	79.767	28.89
86	66.45	4.672	42.12	184.92	142.80	23.40	102.73	79.333	30.00
88	68.84	4.840	42.80	184.82	142.02	23.78	102.68	78.900	31.11
90	71.25	5.009	43.50	184.72	141.22	24.17	102.62	78.456	32.22
92	73.70	5.182	44.19	184.61	140.42	24.55	102.56	78.011	33.33
94	76.30	5.364	44.86	184.49	139.62	24.92	102.49	77.567	34.44
96	79.03	5.556	45.54	184.37	138.83	25.30	102.43	77.128	35.56
98	81.77	5.749	46.22	184.25	138.03	25.68	102.36	76.683	36.67
100	84.52	5.942	46.90	184.10	137.20	26.06	102.28	76.222	37.78

Sulfur Dioxide, SO_2

Temp. °F	Spec. vol. sat. vap. ft.³/lb.	Spec. vol. sat. vap. m³/kg	Density sat. vap. lb./ft.³	Density sat. vap. kg/m³	Dens. liq. lb./ft.³	Entropy from −40°F BTU/lb./°F Liq.	Entropy from −40°F BTU/lb./°F Vap.	Temp. °C
−40	22.42	1.400	0.04460	0.7144	95.79	0.00000	0.42562	−40.00
−30	16.56	1.034	.06039	0.9673	94.94	.00674	.41864	−34.44
−20	12.42	0.7754	.08119	1.301	94.10	.01366	.41192	−28.89
−10	9.44	.5893	.1025	1.642	93.27	.02075	.40544	−23.33
0	7.280	.4545	.1374	2.201	92.42	.02795	.39917	−17.78
2	6.923	0.4322	0.1444	2.313	92.25	0.02941	0.39794	−16.67
4	6.584	.4110	.1501	2.404	92.08	.03084	.39670	−15.56
5	6.421	.4009	.1558	2.496	92.00	.03155	.39609	−15.00
6	6.266	.3912	.1596	2.556	91.91	.03228	.39547	−14.44
8	5.967	.3725	.1676	2.685	91.74	.03373	.39426	−13.33
10	5.682	0.3547	0.1760	2.819	91.58	0.03519	0.39306	−12.22
12	5.417	.3382	.1846	2.957	91.41	.03664	.39185	−11.11
14	5.164	.3224	.1936	3.101	91.24	.03808	.39065	−10.00
16	4.926	.3075	.2030	3.252	91.07	.03953	.38946	− 8.89
18	4.701	.2935	.2127	3.407	90.89	.04098	.38827	− 7.78
20	4.487	0.2801	0.2228	3.569	90.71	0.04241	0.38707	− 6.67
22	4.287	.2676	.2332	3.735	90.53	.04385	.38589	− 5.56
24	4.096	.2557	.2441	3.910	90.33	.04528	.38471	− 4.44
26	3.915	.2444	.2559	4.099	90.15	.04671	.38354	− 3.33
28	3.744	.2337	.2671	4.278	89.96	.04814	.38237	− 2.22
30	3.581	0.2236	0.2800	4.485	89.76	0.04956	0.38119	− 1.11
32	3.437	.2146	.2909	4.660	89.58	.05099	.38003	0.00
34	3.283	.2050	.3046	4.879	89.39	.05242	.37887	+ 1.11
36	3.144	.1963	.3181	5.095	89.18	.05384	.37772	2.22
38	3.013	.1881	.3319	5.316	89.00	.05527	.37657	3.33
40	2.887	0.1802	0.3464	5.549	88.81	0.05668	0.37541	4.44
42	2.769	.1729	.3611	5.784	88.62	.05809	.37425	5.56
44	2.656	.1658	.3765	6.031	88.43	.05949	.37311	6.67
46	2.548	.1591	.3925	6.287	88.24	.06090	.37197	7.78
48	2.446	.1527	.4088	6.548	88.05	.06231	.37083	8.89
50	2.348	0.1466	0.4259	6.822	87.87	0.06370	0.36969	10.00
52	2.256	.1408	.4433	7.101	87.67	.06509	.36857	11.11
54	2.167	.1353	.4615	7.392	87.51	.06646	.36743	12.22
56	2.083	.1300	.4801	7.690	87.31	.06785	.36629	13.33
58	2.003	.1250	.4992	7.996	87.13	.06923	.36517	14.44
60	1.926	0.1202	0.5194	8.320	86.95	0.07060	0.36405	15.56
62	1.853	.1157	.5396	8.643	86.77	.07196	.36293	16.67
64	1.783	.1113	.5609	8.984	86.59	.07333	.36181	17.78
66	1.716	.1071	.5827	9.334	86.41	.07469	.36070	18.89
68	1.652	.1031	.6054	9.697	86.22	.07602	.35958	20.00
70	1.590	0.09926	0.6290	10.08	86.02	0.07736	0.35846	21.11
72	1.532	.09564	.6527	10.45	85.82	.07871	.35736	22.22
74	1.476	.09214	.6777	10.86	85.62	.08003	.35624	23.33
76	1.422	.08877	.7030	11.26	85.42	.08135	.35512	24.44
78	1.371	.08559	.7295	11.69	85.23	.08268	.35401	25.56
80	1.321	0.08247	0.7570	12.13	85.03	0.08399	0.35291	26.67
82	1.274	.07953	.7850	12.57	84.84	.08525	.35177	27.78
84	1.229	.07672	.8140	13.04	84.64	.08653	.35065	28.89
86	1.185	.07398	.8440	13.52	84.44	.08783	.34954	30.00
88	1.144	.07142	.8740	14.00	84.25	.08910	.34843	31.11
90	1.104	0.06892	0.9058	14.51	84.05	0.09038	0.34731	32.22
92	1.065	.06649	.9390	15.04	83.86	.09165	.34620	33.33
94	1.028	.06418	.9730	15.59	83.67	.09389	.34508	34.44
96	0.9931	.06200	1.007	16.13	83.47	.09411	.34397	35.56
98	0.9591	.05987	1.043	16.71	83.27	.09532	.34285	36.67
100	0.9262	0.05782	1.080	17.30	83.07	0.09657	0.34173	37.78

Butane, $CH_3(CH_2)_2CH_3$

Temp. °F	Abs. press. sat. vap. lb./in.²	Abs. press. sat. vap. kg/cm²	Heat content abv. 32°F BTU/lb. Liq.	Heat content abv. 32°F BTU/lb. Vap.	Ht. of vaporiz. BTU/lb.	Heat content abv. 0°C g-cal./g Liq.	Heat content abv. 0°C g-cal./g Vap.	Ht. of vaporiz. g-cal./g	Temp. °C
0	7.3	0.51	−17.2	153.3	170.5	− 9.56	85.17	94.72	−17.78
10	9.2	0.65	−11.7	156.8	168.5	− 6.50	87.11	93.61	−12.22
20	11.6	0.816	− 6.7	160.3	167.0	− 3.7	89.06	92.78	− 6.67
30	14.4	1.01	− 1.2	164.3	165.5	− 0.67	91.28	91.94	− 1.11
40	17.7	1.24	+ 4.3	167.8	163.5	+ 2.4	93.22	90.83	+ 4.44
50	21.6	1.52	9.8	171.3	161.5	5.4	95.17	89.72	10.00
60	26.3	1.85	15.8	175.3	159.5	8.78	97.39	88.61	15.56
70	31.6	2.22	21.3	178.8	157.5	11.8	99.33	87.50	21.11
80	37.6	2.64	27.3	182.3	155.0	15.2	101.3	86.11	26.67
90	44.5	3.13	33.8	185.8	152.0	18.8	103.2	84.44	32.22
100	52.2	3.67	39.8	189.3	149.5	22.1	105.2	83.06	37.78
110	60.8	4.27	46.3	193.3	147.0	25.7	107.4	81.67	43.33
120	70.8	4.98	52.8	196.3	143.5	29.3	109.1	79.72	48.89
130	81.4	5.72	59.3	199.8	140.5	32.9	111.0	78.06	54.44
140	92.6	6.51	66.3	203.8	137.5	36.8	113.2	76.39	60.00

Isobutane, $(CH_3)_3CH$

Temp. °F	lb./in.²	kg/cm²	Liq.	Vap.	BTU/lb.	Liq.	Vap.	g-cal./g	Temp. °C
−20	7.50	0.527	−25.5	140.0	165.5	−14.2	77.78	91.94	−28.89
−10	9.28	0.652	−21.0	142.0	163.0	−11.7	78.89	90.56	−23.33
0	11.6	0.816	−16.5	144.0	160.5	− 9.17	80.00	89.17	−17.78
+10	14.6	1.03	−11.5	147.0	158.5	− 6.39	81.67	88.06	−12.22
20	18.2	1.28	− 6.5	149.5	156.0	− 3.6	83.06	86.67	− 6.67
30	22.3	1.57	− 1.0	152.5	153.5	− 0.56	84.72	85.28	− 1.11
40	26.9	1.89	+ 4.5	155.5	151.0	+ 2.5	86.39	83.89	+ 4.44
50	32.5	2.28	10.5	159.0	148.5	5.83	88.33	82.50	10.00
60	38.7	2.72	16.5	162.5	146.0	9.17	90.28	81.11	15.56
70	45.8	3.22	23.0	166.5	143.5	12.8	92.50	79.72	21.11
80	53.9	3.79	30.0	170.5	140.5	16.7	94.72	78.06	26.67
90	63.3	4.45	37.0	174.5	137.5	20.6	96.94	76.39	32.22
100	73.7	5.18	44.5	179.0	134.5	24.7	99.44	74.72	37.78
110	85.1	5.98	52.5	183.5	131.0	29.2	101.9	72.78	43.33
120	98.0	6.89	60.5	188.0	127.5	33.6	104.4	70.83	48.89
130	112.0	7.87	69.5	193.5	124.0	38.6	107.5	68.89	54.44
140	126.8	8.915	78.5	199.0	120.5	43.6	110.6	66.94	60.00

Propane, C_3H_8

Temp. °F	lb./in.²	kg/cm²	Liq.	Vap.	BTU/lb.	Liq.	Vap.	g-cal./g	Temp. °C
−70	7.37	0.518	−55.2	134.3	189.5	−30.7	74.61	105.3	−56.67
−60	9.72	0.683	−50.2	136.8	187.0	−27.9	76.00	103.9	−51.11
−50	12.6	0.886	−44.7	139.8	184.5	−24.8	77.67	102.5	−45.56
−40	16.2	1.14	−39.7	141.8	181.5	−22.1	78.78	100.8	−40.00
−30	20.3	1.43	−34.2	144.8	179.0	−19.0	80.44	99.44	−34.44
−20	25.4	1.79	−29.2	146.8	176.0	−16.2	81.56	97.78	−28.89
−10	31.4	2.21	−23.7	149.8	173.5	−13.2	83.22	96.39	−23.33
0	38.2	2.69	−18.2	152.3	170.5	−10.1	84.61	94.72	−17.78
+10	46.0	3.23	−12.7	155.3	168.0	− 7.06	86.28	93.33	−12.22
20	55.5	3.90	− 7.2	157.8	165.0	− 4.0	87.67	91.67	− 6.67
30	66.3	4.66	− 1.2	160.8	162.0	− 0.67	89.33	90.00	− 1.11
40	78.0	5.48	+ 4.8	163.8	159.0	+ 2.7	91.00	88.33	+ 4.44
50	91.8	6.45	10.8	166.8	156.0	6.00	92.67	86.67	10.00
60	107.1	7.530	16.8	169.8	153.0	9.33	94.33	85.00	15.56
70	124.0	8.718	22.8	172.3	149.5	12.7	95.72	83.06	21.11
80	142.8	10.04	29.3	175.3	146.0	16.3	97.39	81.11	26.67
90	164.0	11.53	35.8	178.3	142.5	19.9	99.06	79.17	32.22
100	187.0	13.15	42.3	180.8	138.5	23.5	100.4	76.94	37.78
110	213.0	14.98	48.8	182.8	134.0	27.1	101.6	74.44	43.33
120	240.0	16.87	55.3	184.3	129.0	30.7	102.4	71.67	48.89

THERMODYNAMIC PROPERTIES (Continued)

Butane, $CH_3(CH_2)_2CH_3$

Temp. °F	Spec. vol. sat. vap. ft.³/lb.	Spec. vol. sat. vap. m³/kg	Density of sat. vap. lb./ft.³	Density of sat. vap. kg/m³	Density of liq. lb./ft.³	Density of liq. kg/m³	Temp. °C
0	11.1	0.693	0.0901	1.44	38.59	618.1	−17.78
10	8.95	.559	.112	1.79	38.24	612.5	−12.22
20	7.23	.451	.138	2.21	37.89	606.9	− 6.67
30	5.90	.368	.169	2.71	37.54	601.3	− 1.11
40	4.88	.305	.205	3.28	37.19	595.7	+ 4.44
50	4.07	.254	.246	3.94	36.82	589.8	10.00
60	3.40	.212	.294	4.71	36.45	583.9	15.56
70	2.88	.180	.347	5.56	36.06	577.6	21.11
80	2.46	.154	.407	6.52	35.65	571.0	26.67
90	2.10	.131	.476	7.62	35.24	564.5	32.22
100	1.81	.113	.552	8.84	34.84	558.1	37.78
110	1.58	.0986	.633	10.1	34.41	551.2	43.33
120	1.38	.0862	.725	11.6	33.96	544.0	48.89
130	1.21	.0755	.826	13.2	33.49	536.4	54.44
140	1.07	.0668	.934	15.0	32.98	528.3	60.00

Isobutane, $(CH_3)_3CH$

Temp. °F	ft.³/lb.	m³/kg	lb./ft.³	kg/m³	lb./ft.³	kg/m³	Temp. °C
−20	10.5	0.655	0.0952	1.52	38.35	614.3	−28.89
−10	8.91	.556	.112	1.79	37.95	607.9	−23.33
0	7.17	.448	.139	2.23	37.60	602.3	−17.78
+10	5.75	.359	.174	2.79	37.20	595.9	−12.22
20	4.68	.292	.214	3.43	36.80	589.5	− 6.67
30	3.86	.241	.259	4.15	36.40	583.1	− 1.11
40	3.22	.201	.311	4.98	36.00	576.6	+ 4.44
50	2.71	.169	.369	5.91	35.60	570.2	10.00
60	2.28	.142	.439	7.03	35.20	563.8	15.56
70	1.94	.121	.515	8.25	34.80	557.4	21.11
80	1.66	.104	.602	9.64	34.35	550.2	26.67
90	1.42	.0886	.704	11.3	33.90	543.0	32.22
100	1.23	.0768	.813	13.0	33.45	535.8	37.78
110	1.07	.0668	.935	15.0	33.00	528.6	43.33
120	0.926	.0578	1.08	17.3	32.50	520.6	48.89
130	0.811	.0506	1.23	19.7	32.00	512.6	54.44
140	0.716	.0447	1.32	21.1	31.80	509.4	60.00

Propane, C_3H_8

Temp. °F	ft.³/lb.	m³/kg	lb./ft.³	kg/m³	lb./ft.³	kg/m³	Temp. °C
−70	12.9	0.805	0.0775	1.24	37.40	599.1	−56.67
−60	9.93	.620	.111	1.78	37.00	592.7	−51.11
−50	7.74	.483	.129	2.07	36.60	586.3	−45.56
−40	6.13	.383	.163	2.61	36.19	579.7	−40.00
−30	4.93	.308	.203	3.25	35.78	573.1	−34.44
−20	4.00	.250	.250	4.00	35.37	566.6	−28.89
−10	3.26	.204	.307	4.92	34.96	560.0	−23.33
0	2.71	.169	.369	5.91	34.54	553.3	−17.78
+10	2.27	.142	.441	7.06	34.12	546.5	−12.22
20	1.90	.119	.526	8.43	33.67	539.3	− 6.67
30	1.60	.0999	.625	10.0	33.20	531.8	− 1.11
40	1.37	.0855	.730	11.7	32.73	524.3	+ 4.44
50	1.18	.0737	.847	13.6	32.24	516.4	10.00
60	1.01	.0631	.990	15.9	31.75	508.6	15.56
70	0.883	.0551	1.13	18.1	31.24	500.4	21.11
80	0.770	.0481	1.30	20.8	30.70	491.8	26.67
90	0.673	.0420	1.49	23.9	30.15	482.9	32.22
100	0.591	.0369	1.69	27.1	29.58	473.8	37.78
110	0.519	.0324	1.96	31.4	28.85	462.1	43.33
120	0.459	.0287	2.18	34.9	28.30	453.3	48.89

Difluorodichloromethane, CCl_2F_2 ("F-12")

Temp. °F	Abs. press. sat. vap. lb./in.²	Abs. press. sat. vap. kg/cm²	Heat content abv. −40°F BTU/lb. Liq.	Heat content abv. −40°F BTU/lb. Vap.	Ht. of vaporiz. BTU/lb.	Heat content abv. −40°C g-cal./g Liq.	Heat content abv. −40°C g-cal./g Vap.	Ht. of vaporiz. g-cal./g	Temp. °C
−40	9.32	0.655	0	73.50	73.50	0	40.83	40.83	−40.00
−30	12.02	0.845	2.03	74.70	72.67	1.13	41.50	40.37	−34.44
−20	15.28	1.074	4.07	75.87	71.80	2.26	42.15	39.89	−28.89
−10	19.20	1.350	6.14	77.05	70.91	3.41	42.81	39.39	−23.33
0	23.87	1.678	8.25	78.21	69.96	4.58	43.45	38.87	−17.78
+10	29.35	2.064	10.39	79.36	68.97	5.772	44.09	38.32	−12.22
20	35.75	2.513	12.55	80.49	67.94	6.972	44.72	37.74	− 6.67
30	43.16	3.034	14.76	81.61	66.85	8.200	45.34	37.14	− 1.11
40	51.68	3.633	17.00	82.71	65.71	9.444	45.95	36.51	+ 4.44
50	61.39	4.316	19.27	83.78	64.51	10.71	46.54	35.84	10.00
60	72.41	5.091	21.57	84.82	63.25	11.98	47.12	35.14	15.56
70	84.82	5.963	23.90	85.82	61.92	13.28	47.68	34.40	21.11
80	98.76	6.944	26.28	86.80	60.52	14.60	48.22	33.62	26.67
90	114.3	8.036	28.70	87.74	59.04	15.94	48.74	32.80	32.22
100	131.6	9.252	31.16	88.62	57.46	17.31	49.23	31.92	37.78
110	150.7	10.60	33.65	89.43	55.78	18.69	49.68	30.99	43.33
120	171.8	12.08	36.16	90.15	53.99	20.09	50.08	29.99	48.89

Carbon Disulfide, CS_2

Temp. °F	Abs. press. sat. vap. lb./in.²	Abs. press. sat. vap. kg/cm²	Heat content abv. 32°F BTU/lb. Liq.	Heat content abv. 32°F BTU/lb. Vap.	Ht. of vaporiz. BTU/lb.	Heat content abv. 0°C g-cal./g Liq.	Heat content abv. 0°C g-cal./g Vap.	Ht. of vaporiz. g-cal./g	Temp. °C
0	1.10	0.0773	− 8.60	156.90	165.5	− 4.78	87.167	91.94	−17.78
10	1.46	.103	− 5.60	158.90	164.5	− 3.11	88.278	91.39	−12.22
20	1.89	.133	− 3.00	160.20	163.2	− 1.67	89.000	90.67	− 6.67
30	2.36	.166	− 0.50	161.70	162.2	− 0.28	89.833	90.11	− 1.11
40	3.03	.213	+ 2.05	163.25	161.2	+ 1.14	90.695	89.56	+ 4.44
50	3.90	274	4.24	164.24	160.0	2.36	91.245	88.89	10.00
60	4.95	.348	7 20	166.40	159.2	4.00	92.445	88.44	15.56
70	5.85	.411	9.80	167.90	158.1	5.44	93.278	87.83	21.11
80	7.30	.513	11.70	168.60	156.9	6.500	93.667	87.17	26.67
90	9.15	.643	13.80	169.40	155.6	7.667	94.111	86.44	32.22
100	11.08	.7790	16.15	170.55	154.4	8.972	94.750	85.78	37.78
110	13.50	.9491	18.30	171.50	153.2	10.17	95.278	85.11	43.33
120	16.10	1.132	20.01	172.01	152.0	11.12	95.561	84.44	48.89

Carbon Tetrachloride, CCl_4

Temp. °F	Abs. press. sat. vap. lb./in.²	Abs. press. sat. vap. kg/cm²	Heat content BTU/lb. Liq.	Heat content BTU/lb. Vap.	Ht. of vaporiz. BTU/lb.	Heat content g-cal./g Liq.	Heat content g-cal./g Vap.	Ht. of vaporiz. g-cal./g	Temp. °C
20	0.40	0.028	− 2.00	92.45	94.45	− 1.11	51.36	52.47	− 6.67
30	0.60	.042	− 0.25	93.45	93.70	− 0.14	51.92	52.06	− 1.11
40	0.84	.059	+ 1.60	94.80	93.20	+ 0.889	52.67	51.78	+ 4.44
60	1.42	.100	5.95	98.15	92.20	3.31	54.53	51.22	15.56
70	1.85	.130	8.20	99.53	91.40	4.56	55.29	50.78	21.11
80	2.40	.169	9.80	99.87	90.07	5.44	55.48	50.04	26.67
90	3.12	.219	11.60	101.62	90.02	6.444	56.46	50.01	32.22
100	4.00	.281	13.40	102.80	89.40	7.444	57.11	49.67	37.78
110	4.89	.344	15.80	104.50	88.70	8.778	58.06	49.28	43.33
120	5.95	.418	18.06	105.90	87.90	10.03	58.83	48.83	48.89

Ethyl Ether, $(C_2H_5)_2O$

Temp. °F	Abs. press. sat. vap. lb./in.²	Abs. press. sat. vap. kg/cm²	Heat content BTU/lb. Liq.	Heat content BTU/lb. Vap.	Ht. of vaporiz. BTU/lb.	Heat content g-cal./g Liq.	Heat content g-cal./g Vap.	Ht. of vaporiz. g-cal./g	Temp. °C
0	1.3	0.091	−18.00	153.00	171.0	−10.00	85.000	95.00	−17.78
10	1.8	.13	−12.0	158.43	170.4	− 6.67	88.017	94.67	−12.22
20	2.5	.18	− 6.50	163.50	170.0	− 3.61	90.833	94.44	− 6.67
30	3.4	.24	− 1.50	167.90	169.4	− 0.833	93.278	94.11	− 1.11
40	4.4	.31	+ 4.00	172.40	168.4	+ 2.22	95.778	93.56	+ 4.44
50	5.5	.39	9.57	177.17	167.6	5.32	98.428	93.11	10.00
70	8.8	.62	20.04	185.44	165.4	11.13	103.02	91.89	21.11
80	10.9	.766	26.40	190.60	164.2	14.67	105.89	91.22	26.67
90	13.4	.942	31.50	194.50	163.0	17.50	108.06	90.56	32.22
100	16.0	1.12	36.50	197.50	161.5	20.28	109.72	89.72	37.78

THERMODYNAMIC PROPERTIES (Continued)

Difluorodichloromethane, CCl_2F_2 ("F-12")

Temp. °F	Spec. vol sat. vap. ft.³/lb.	Spec. vol sat. vap. m³/kg	Density of vap. lb./ft.³	Density of vap. kg/m³	Dens. liq. lb./ft.³	Entropy from −40°F BTU/lb./°F Liq.	Entropy from −40°F BTU/lb./°F Vap.	Temp. °C
−40	3.911	0.2442	0.2557	4.096	94.58	0	0.17517	−40.00
−30	3.088	.1928	.3238	5.187	93.59	0.00471	.17387	−34.44
−20	2.474	.1544	.4042	6.474	92.58	.00940	.17275	−28.89
−10	2.003	.1250	.4993	7.998	91.57	.01403	.17175	−23.33
0	1.637	.1022	.6109	9.785	90.52	.01869	.17091	−17.78
+10	1.351	.08434	.7402	11.86	89.45	.02328	.17015	−12.22
20	1.121	.06998	.8921	14.29	88.37	.02783	.16949	− 6.67
30	0.939	.0586	1.065	17.06	87.24	.03233	.16887	− 1.11
40	.792	.0494	1.263	20.23	86.10	.03680	.16833	+ 4.44
50	.673	.0420	1.485	23.79	84.94	.04126	.16785	10.00
60	.575	.0359	1.740	27.87	83.78	.04568	.16741	15.56
70	.493	.0308	2.028	32.48	82.60	.05009	.16701	21.11
80	.425	.0265	2.353	37.69	81.39	.05446	.16662	26.67
90	.368	.0230	2.721	43.58	80.11	.05882	.16624	32.22
100	.319	.0199	3.135	50.22	78.80	.06316	.16584	37.78
110	.277	.0173	3.610	57.82	77.46	.06749	.16542	43.33
120	.240	.0150	4.167	66.75	76.02	.07180	.16495	48.89

Carbon Disulfide, CS_2

Temp. °F	Spec. vol. sat. vap. ft.³/lb.	Spec. vol. sat. vap. m³/kg	Density sat. vap. lb./ft.³	Density sat. vap. kg/m³	Temp. °C
0	53.76	3.356	0.0186	0.2979	−17.78
10	43.47	2.714	.0230	.3684	−12.22
20	34.84	2.175	.0287	.4597	− 6.67
30	29.49	1.841	.0339	.5430	− 1.11
40	23.52	1.468	.0425	.6808	+ 4.44
50	20.60	1.286	.0482	.7721	10.00
60	18.00	1.124	.0555	.8890	15.56
70	13.20	0.824	.0758	1.214	21.11
80	10.40	0.649	.0961	1.539	26.67
90	8.30	0.518	.1204	1.929	32.22
100	7.03	0.439	.1369	2.193	37.78
110	5.80	0.362	.1724	2.762	43.33
120	5.10	0.318	.1960	3.140	48.89

Carbon Tetrachloride. CCl_4

Temp. °F	Spec. vol. sat. vap. ft.³/lb.	Spec. vol. sat. vap. m³/kg	Density sat. vap. lb./ft.³	Density sat. vap. kg/m³	Temp. °C
20	69.5	4.34	0.01438	0.2303	− 6.67
30	53.0	3.31	.01886	.3021	− 1.11
40	40.0	2.50	.02500	.4005	+ 4.44
60	24.0	1.50	.04166	.6673	15.56
70	19.5	1.22	.05128	.8214	21.11
80	16.0	0.999	.06345	1.016	26.67
90	13.0	0.812	.07692	1.232	32.22
100	10.0	0.624	.1000	1.602	37.78
110	8.5	0.53	.1176	1.884	43.33
120	7.5	0.47	.1333	2.135	48.89

Ethyl Ether, $(C_2H_5)_2O$

Temp. °F	Spec. vol. sat. vap. ft.³/lb.	Spec. vol. sat. vap. m³/kg	Density sat. vap. lb./ft.³	Density sat. vap. kg/m³	Temp. °C
0	38.0	2.37	0.0263	0.4213	−17.78
10	32.5	2.03	.0332	.5318	−12.22
20	27.0	1.69	.0372	.5959	− 6.67
30	21.4	1.34	.0468	.7496	− 1.11
40	17.0	1.06	.0588	.9419	+ 4.44
50	13.2	0.824	.0757	1.213	10.00
70	7.8	0.49	.1280	2.050	21.11
80	6.2	0.39	.1620	2.595	26.67
90	5.1	0.32	.1960	3.140	32.22
100	4.5	0.28	.2220	3.556	37.78

THERMODYNAMIC PROPERTIES (Concluded)

Methyl Chloride, CH_3Cl

Temp. °F	Abs. press. sat. vap. lb./in.²	Abs. press. sat. vap. kg/cm²	Heat content abv. 32°F BTU/lb. Liq.	Heat content abv. 32°F BTU/lb. Vap.	Ht. of vaporiz. BTU/lb.	Heat content abv. 0°C g-cal./g Liq.	Heat content abv. 0°C g-cal./g Vap.	Ht. of vaporiz. g-cal./g	Temp. °C
−20	11.75	0.8261	−19.0	167.36	186.36	−10.6	92.978	103.53	−28.89
−10	15.0	1.055	−15.38	168.83	184.21	− 8.544	93.795	102.34	−23.33
− 5	16.79	1.180	−13.58	169.54	183.12	− 7.544	94.189	101.73	−20.56
0	18.8	1.32	−11.75	170.23	181.98	− 6.528	94.572	101.10	−17.78
+ 5	21.0	1.48	− 9.93	170.96	180.84	− 5.517	94.978	100.47	−15.00
10	23.3	1.64	− 8.06	171.58	179.65	− 4.478	95.322	99.806	−12.22
15	25.9	1.82	− 6.74	172.24	178.47	− 3.744	95.689	99.150	− 9.44
20	28.8	2.02	− 4.32	172.95	177.27	− 2.400	96.083	98.483	− 6.67
25	31.8	2.24	− 2.48	173.63	176.10	− 1.378	96.461	97.833	− 3.89
30	35.2	2.47	− 0.62	174.28	174.90	− 0.344	96.822	97.167	− 1.11
35	38.7	2.72	+ 1.75	174.92	173.77	+ 0.972	97.178	96.539	+ 1.67
40	42.6	3.00	3.15	175.57	172.42	1.75	97.539	95.789	4.44
45	46.9	3.30	5.04	176.20	171.16	2.80	97.889	95.089	7.22
50	51.5	3.62	6.88	176.78	169.90	3.82	98.211	94.389	10.00
55	56.4	3.97	8.80	177.45	168.65	4.89	98.583	93.695	12.78
60	61.6	4.33	10.70	178.05	167.35	5.944	98.917	92.972	15.56
65	67.3	4.73	12.62	178.64	166.02	7.011	99.245	92.233	18.33
70	73.3	5.15	14.52	179.17	164.65	8.067	99.539	91.472	21.11
75	79.2	5.57	16.46	179.78	163.28	9.144	99.878	90.711	23.89
80	85.3	6.00	18.36	180.24	161.88	10.20	100.13	89.933	26.67
85	94.1	6.62	20.12	180.74	160.48	11.18	100.41	89.156	29.44
90	102.1	7.178	22.13	181.22	159.09	12.29	100.68	88.383	32.22
95	110.3	7.755	24.07	181.76	157.70	13.37	100.98	87.611	35.00
100	118.8	8.352	26.06	182.36	156.30	14.48	101.31	86.833	37.78

Mercury, Hg

Temp. °F	Abs. press. sat. vap. lb./in.²	Abs. press. sat. vap. kg/cm²	Heat content abv. 32°F BTU/lb. Liq.	Heat content abv. 32°F BTU/lb. Vap.	Ht. of vaporiz. BTU/lb.	Heat content abv. 0°C g-cal./g Liq.	Heat content abv. 0°C g-cal./g Vap	Ht. of vaporiz. g-cal./g	Temp. °C
402	0.4	0.03	13.81	141.96	128.15	7.672	78.867	71.195	205.56
444	0.8	0.06	15.36	142.60	127.24	8.533	79.222	70.689	228.89
458	1.0	0.07	15.89	142.81	126.92	8.828	79.339	70.511	236.67
485	1.5	0.11	16.90	143.23	126.33	9.389	79.572	70.183	251.67
505	2.0	0.14	17.65	143.54	125.89	9.806	79.745	69.939	262.78
558	4.0	0.28	19.62	144.34	124.72	10.90	80.189	69.289	292.22
591	6.0	0.42	20.87	144.86	123.99	11.59	80.478	68.883	310.56
617	8.0	0.56	21.81	145.24	123.43	12.12	80.689	68.572	325.00
637	10.0	0.703	22.58	145.56	122.98	12.54	80.867	68.322	336.11
676	15.0	1.05	24.04	146.16	122.12	13.36	81.200	67.844	357.78
706	20.0	1.41	25.15	146.61	121.46	13.97	81.450	67.478	374.44
730	25.0	1.76	26.05	146.98	120.93	14.47	81.656	67.183	387.78
751	30.0	2.11	26.81	147.29	120.48	14.89	81.828	66.933	399.44
769	35.0	2.46	27.49	147.57	120.08	15.27	81.983	66.711	409.44
785	40.0	2.81	28.08	147.81	119.73	15.60	82.117	66.517	418.33
799	45.0	3.16	28.62	148.04	119.42	15.90	82.245	66.344	426.11
812	50	3.5	29.11	148.24	119.13	16.17	82.356	66.183	433.33
836	60	4.2	29.99	148.60	118.61	16.66	82.556	65.894	446.67
857	70	4.9	30.75	148.90	118.15	17.08	82.722	65.639	458.33
875	80	5.6	31.44	149.19	117.75	17.47	82.883	65.417	468.33
892	90	6.3	32.06	149.44	117.38	17.81	83.022	65.211	477.78
907	100	7.03	32.63	149.68	117.05	18.13	83.156	65.028	486.11
921	110	7.73	33.16	149.90	116.74	18.42	83.278	64.856	493.89
934	120	8.44	33.66	150.10	116.44	18.70	83.389	64.689	501.11
947	130	9.14	34.12	150.29	116.17	18.96	83.495	64.539	508.33
958	140	9.84	34.55	150.47	115.92	19.19	83.595	64.400	514.44
969	150	10.5	34.96	150.63	115.67	19.42	83.683	64.261	520.56
1000	180	12.7	36.09	151.10	115.01	20.05	83.945	63.894	537.78

Methyl Chloride, CH_3Cl

Temp. °F	Spec. vol. sat. vap. ft.³/lb.	Spec. vol. sat. vap. m³/kg	Spec. vol. liq ft.³/lb.	Spec. vol. liq m³/kg	Density sat. vap. lb./ft.³	Density sat. vap. kg/m³	Density of liq. lb./ft.³	Density of liq. kg/m³	Temp °C
−20	8.09	.505	.015827	.0009880	0.124	1.98	63.185	1012.1	−28.89
−10	6.46	.403	.015985	.0009979	.155	2.48	62.560	1002.1	−23.33
− 5	5.80	.362	.016013	.0009997	.172	2.76	62.450	1000.3	−20.56
0	5.18	.323	.016146	.001008	.193	3.09	61.936	992.09	−17.78
+ 5	4.68	.292	.016228	.001013	.214	3.42	61.623	987.08	−15.00
10	4.18	.261	.016310	.001018	.239	3.83	61.311	982.08	−12.22
15	3.88	.242	.016388	.001023	.258	4.13	61.022	977.45	− 9.44
20	3.41	.213	.016474	.001028	.293	4.70	60.702	972.32	− 6.67
25	3.09	.193	.016552	.001033	.324	5.18	60.415	967.73	− 3.89
30	2.81	.175	.016645	.001039	.356	5.70	60.077	962.31	− 1.11
35	2.50	.156	.016746	.001045	.400	6.41	59.715	956.51	+ 1.67
40	2.31	.144	.016809	.001049	.433	6.93	59.492	952.94	4.44
45	2.10	.131	.016929	.001057	.476	7.63	59.069	946.17	7.22
50	1.93	.120	.017023	.001063	.518	8.30	58.745	940.98	10.00
55	1.75	.109	.017118	.001069	.571	9.16	58.419	935.76	12.78
60	1.61	.101	.017219	.001075	.621	9.95	58.077	930.28	15.56
65	1.47	.0918	.017318	.001081	.680	10.9	57.742	924.91	18.33
70	1.34	.0837	.017421	.001088	.746	12.0	57.403	919.48	21.11
75	1.24	.0774	.017526	.001094	.806	12.9	57.058	913.96	23.89
80	1.14	.0712	.017632	.001101	.877	14.1	56.714	908.44	26.67
85	1.05	.0655	.017740	.001108	.952	15.3	56.369	902.92	29.44
90	0.98	.061	.017850	.001114	1.02	16.3	56.022	897.36	32.22
95	0.91	.057	.017961	.001121	1.10	17.6	55.675	891.80	35.00
100	0.85	.053	.018074	.001128	1.18	18.8	55.327	886.23	37.78

Mercury, Hg

Temp. °F	Spec. vol. sat. vap. ft.³/lb.	Spec. vol. sat. vap. m³/kg	Density of sat. vap. lb./ft.³	Density of sat. vap. kg/m³	Entropy above 32°F BTU/lb./°F Liq.	Entropy above 32°F BTU/lb./°F Vap.	Entropy above 32°F BTU/lb./°F Evap.	Temp °C
402	114.50	7.1480	0.008733	0.1399	.0209	.1696	.1487	205.56
444	59.72	3.728	.016745	0.26822	.0227	.1635	.1408	228.89
458	48.45	3.025	.02064	0.3306	.0233	.1616	.1383	236.67
485	33.14	2.069	.03017	0.4833	.0244	.1581	.1337	251.67
505	25.32	1.581	.03948	0.6324	.0251	.1556	.1305	262.78
558	13.26	0.8278	.07540	1.208	.0271	.1497	.1226	292.22
591	9.096	.5678	.10993	1.7609	.0283	.1462	.1179	310.56
617	6.9630	.43469	.14361	2.3003	.0292	.1439	.1147	325.00
637	5.6610	.35341	.17664	2.8294	.0299	.1420	.1121	336.11
676	3.8923	.24299	.25691	4.1152	.0312	.1387	.1075	357.78
706	2.983	.1862	.3352	5.369	.0322	.1364	.1042	374.44
730	2.429	.1516	.4117	6.595	.0330	.1346	.1016	387.78
751	2.053	.1282	.4871	7.802	.0336	.1331	.0995	399.44
769	1.7815	.11122	.5613	8.991	.0342	.1319	.0977	409.44
785	1.5762	.098399	.6344	10.16	.0346	.1308	.0962	418.33
799	1.4147	.088317	.7069	11.32	.0351	.1300	.0949	426.11
812	1.284	.08016	.7788	12.47	.0355	.1291	.0936	433.33
836	1.086	.06780	.9204	14.74	.0361	.1276	.0915	446.67
857	0.9436	.05891	1.0597	16.974	.0367	.1265	.0898	458.33
875	.8349	.05212	1.1977	19.185	.0372	.1254	.0882	468.33
892	.7497	.04680	1.3338	21.365	.0377	.1247	.0870	477.78
907	.6811	.04252	1.4682	23.518	.0381	.1237	.0856	486.11
921	.6242	.03897	1.6020	25.661	.0385	.1230	.0845	493.89
934	.5767	.03600	1.7340	27.775	.0389	.1224	.0835	501.11
947	.5360	.03346	1.8656	29.883	.0392	.1218	.0826	508.33
958	.5012	.03129	1.9952	31.959	.0395	.1213	.0818	514.44
969	.4706	.02938	2.125	34.04	.0398	.1207	.0809	520.56
1000	.3990	.02491	2.506	40.14	.0406	.1194	.0788	537.78

PHYSICAL PROPERTIES OF FLUOROCARBON REFRIGERANTS

Property No.	Refrigerant name		11	12	13	13B1	14	21
1	Formula		CCl_3F	CCl_2F_2	$CClF_3$	$CBrF_3$	CF_4	$CHCl_2F$
2	Molecular weight		137.37	120.91	104.46	148.92	88.01	102.92
3	Normal boiling point; °C		23.82	−29.79	−81.4	−57.75	−127.96	8.92
4	Normal freezing point; °C		−111	−158	−181	−168	−184	−135
5	Critical temperature; °C		198	112	28.9	67	−45.67	178.5
6	Critical pressure; atm		43.5	40.6	38.2	39.1	36.96	51
7	Critical volume; cm/mol		247	217	181	200	141	197
8	Critical density; g/cm		0.554	0.558	0.578	0.745	0.626	0.522
9	Density of liquid at 25°C; g/cm		1.467	1.311	$1.298^{-30°C}$	1.538	$1.317^{-80°C}$	1.366
10	Density of saturated vapor at B.P.; g/liter		5.86	6.33	7.01	8.71	7.62	4.57
11	Specific heat of liquid at 25°C; cal/g		0.208	0.232	$0.247^{-30°C}$	0.208	$0.294^{-80°C}$	0.256
12	Specific heat of vapor at 25°C and 1 atm; cal/g		$0.142^{38°C}$	0.145	0.158	0.112	0.169	0.140
13	Heat of vaporization at B.P.; cal/g		43.10	39.47	35.47	28.38	32.49	57.86
14	Thermal conductivity at 25°C; Btu/(hr)(ft)(°F)	liquid	0.050	0.041	0.020	0.025	$0.040^{-100°F}$	0.063
		vapor; 1 atm	0.00484	0.00557	—	—	—	0.0569
15	Viscosity at 25°C; centipoise	liquid	0.42	0.26	0.016	0.15	0.020	0.34
		vapor; 1 atm	0.011	0.013	—	0.016	—	0.011
16	Surface tension at 25°C; dyne cm		18	9	$14^{-73.3°C}$	4	$14^{-73.3°C}$	18
17	Refractive index of liquid at 25°C		1.374	1.287	$1.199^{-73.3°C}$	1.238	$1.151^{-73.3°C}$	1.354
18	Dielectric constant	liquid	$2.28^{29°C}$	$2.13^{29°C}$	—	—	—	$5.34^{28°C}$
		vapor at 0.5 atm	$1.10019^{29°C}$	$1.0016^{29°C}$	$1.0013^{29°C}$	—	$1.0006^{24.5°C}$	$1.0035^{30°C}$
19	Solubility in water at 25°C and 1 atm; wt %		0.011	0.028	0.009	0.03	0.0015	0.95
20	Solubility of water in compound at 25°C; wt %		0.011	0.009	—	$0.0095^{21.1°C}$	—	0.13
21	Toxicity; Group number (See separate table for definition of group number.)		5a	6	probably 6	6	probably 6	<4 >5

PHYSICAL PROPERTIES OF FLUOROCARBON REFRIGERANTS (Continued)

Property No.		22	23	112	113	114	114B2	115	116	500	502
1		$CHClF_2$	CHF_3	$C_2Cl_4F_2$	$C_2Cl_3F_3$	$C_2Cl_2F_4$	$C_2Br_2F_4$	C_2ClF_5	C_2F_6	*	**
2		80.47	70.01	203.83	187.38	170.91	259.83	154.47	138.01	105.5	120.7
3		−40.75	−82.03	92.8	47.57	3.77	47.26	−38.7	−78.2	−33.5	−45.6
4		−160	−155.2	26	−35	−94	−110.5	−106	−100.6	−158.9	—
5		96	25.9	278	214.1	145.7	214.5	80	19.7	105.4	179.89
6		49.12	47.7	34	33.7	32.2	34	30.8	29.4	43.7	590.3
7		165	133	370	325	293	329	259	225	—	290
8		0.525	0.525	0.55	0.576	0.582	0.790	0.596	0.612	0.497	0.56
9		1.194	0.670	$1.634^{30°C}$	1.565	1.456	2.163	1.291	$1.587^{-73.3°C}$	$1.138^{30°C}$	1.242
10		4.72	4.66	7.02	7.38	7.83	—	8.37	9.01	—	6.05
11		0.300	1.553	—	0.218	0.243	0.166	0.285	$0.232^{-73.3°C}$	$0.161^{30°C}$	—
12		0.157	0.176	—	$0.161^{60°C}$	0.170	—	0.164	$0.182^{0 mm}$	—	—
13		55.81	57.23	37 (est)	35.07	32.51	25 (est)	30.11	27.97	—	42.48
14		0.052	0.008	0.040	0.038	0.034	0.027	0.026	$0.045^{-100°F}$	—	0.038
		0.00678	—	—	$0.0045^{0.5 atm}$	0.00646	—	0.00803	0.0098 (est)	—	—
15		0.23	0.016 (est)	1.21	0.68	0.38	0.72	0.26	—	$0.292^{-15°C}$	0.25
		0.013	—	—	$0.01^{0.1 atm}$	0.011	—	0.013	—	—	–
16		8	$15^{-73.3°C}$	$23^{30°C}$	19	12	18	5	$16^{-73.3°C}$	—	8
17		1.256	$1.251^{-73.3°C}$	1.413	1.354	1.288	1.367	1.214	$1.206^{-73.3°C}$	—	1.235
18		$6.11^{24°C}$	—	$2.52^{25°C}$	$2.41^{25°C}$	$2.26^{25°C}$	$2.34^{25°C}$	—	—	—	—
		$1.0035^{25.4°C}$	—	—	—	$1.0021^{26.5°C}$	—	$1.0018^{27.4°C}$	—	—	—
19		0.30	0.10	0.12 (sat. pr)	0.017 (sat. pr)	0.013	—	0.006	—	—	—
20		0.13	—	—	0.011	0.009	—	—	—	—	0.560
21		5a	probably 6	<4 >5	<4 >5	6	5a	6	probably 6	5a	5a

* Azeotrope of CCl_2F_2 (73.8 wt %) and $C_2H_4F_2$ (26.2 wt %) ** Azeotrope of $CHClF_2$ (48.8 wt %) and C_2ClF_5 (51.2 wt %).

UNDERWRITERS' LABORATORIES' CLASSIFICATION OF COMPARATIVE LIFE HAZARD OF GASES AND VAPORS

(Group number definition)

Group	Definition	Examples
1	Gases or vapors which in concentrations of the order of ½ to 1 percent for durations of exposure of the order of 5 minutes are lethal or produce serious injury.	Sulfur dioxide
2	Gases or vapors which in concentrations of the order of ½ to 1 percent for durations of exposure of the order of ½ hour are lethal or produce serious injury.	Ammonia, Methyl bromide
3	Gases or vapors which in concentrations of the order of 2 to 2½ percent for durations of exposure of the order of 1 hour are lethal or produce serious injury.	Bromochloromethane, Carbon tetrachloride, Chloroform, Methyl formate
4	Gases or vapors which in concentrations of the order of 2 to 2½ percent for durations of exposure of the order of 2 hours are lethal or produce serious injury.	Dichloroethylene, Methyl chloride, Ethyl bromide
Between 4 and 5	Appear to classify as somewhat less toxic than group 4.	Methylene chloride, Ethyl chloride, Refrigerant 112*
	Much less toxic than group 4 but somewhat more toxic than group 5.	Refrigerant 113 Refrigerant 21
5a	Gases or vapors much less toxic than group 4 but more toxic than group 6.	Refrigerant 11 Refrigerant 22 Refrigerant 114B2 Refrigerant 502 Carbon dioxide
5b	Gases or vapors which available data indicate would classify as either group 5a or group 6.	Ethane, Propane, Butane
6	Gases or vapors which in concentrations up to at least about 20 percent by volume for durations of exposure of the order of 2 hours do not appear to produce injury.	Refrigerant 13B1 Refrigerant 12 Refrigerant 114 Refrigerant 115 Refrigerant 13* Refrigerant 14* Refrigerant 23* Refrigerant 116* Refrigerant C318*

* Not tested by U.L. but estimated to belong in group indicated.

THERMAL CONDUCTIVITY OF LIQUID FLUOROCARBONS

To convert from W/(m)(K) to Btu/(hr)(ft)(°F) divide by 1.7296.
To convert from W/(m)(K) to cal/(sec)(cm)(°C) divide by 418.4.

Fluorocarbon	Formula	Temperature, K	Conductivity, W/(m)(K)
12	CCl_2F_2	277.2	94.14
		298.1	97.49
		303.8	103.76
		317.9	110.04
		329.9	117.15
		346.8	126.36
22	$CHClF_2$	289.5	107.11
		302.8	114.64
		327.8	126.76
		346.8	140.16
114	$C_2Cl_2F_4$	303.6	113.39
		316.8	122.59
		328.7	130.96
		343.0	140.58
13B1	$CBrF_3$	277.3	91.21
		282.3	93.30
		303.5	103.34
		318.7	111.71
		331.9	118.41
		342.8	123.01
		346.9	128.03
C-318	C_4F_8	280.1	112.97
		287.3	117.57
		298.0	130.96
		310.7	141.42
		323.2	148.11
		332.2	156.48
		342.9	158.99
		348.4	165.69
		350.8	166.94

MISCELLANEOUS PROPERTIES OF COMMON REFRIGERANTS

Refrigerant	Formula	Flash Point °F.	Ignition Temp. °F.	Explosive Limits % by Volume: Lower	Explosive Limits % by Volume: Upper	Vapor Density (Air = 1)	Boiling Point °F	Threshold Limit Value* Parts per Million in Air	Water Soluble	Odor	Refrigerant
Ammonia	NH_3	....	1204	16	25	0.59	−28	100	yes	yes	Ammonia
Bromotrifluoromethane (Kulene-131)	CF_3Br		nonflammable			5.25	−73.6	....	no	yes	Bromotrifluoromethane (Kulene-131)
Butane	C_4H_{10}	−76	806	1.8	8.4	2.04	33	....	no	no	Butane
Carbon dioxide	CO_2		nonflammable			1.53	−108	5000	yes	no	Carbon dioxide
Carbon tetrachloride	CCl_4		nonflammable			5.32	170	25	no	yes	Carbon tetrachloride
Dichlorodifluoromethane (Freon-12)	CCl_2F_2		nonflammable			4.17	−21.6	....	no	no	Dichlorodifluoromethane (Freon-12)
Dichlorodifluoromethane, 73.8 % Ethylidene fluoride, 26.2 % (Carrene-7)	CCl_2F_2 CH_3CHF_2		nonflammable			3.24	−28.0	....	no	yes	Dichlorodifluoromethane, 73.8 % Ethylidene fluoride, 26.2 % (Carrene-7)
Dichloromonofluoromethane (Freon-21)	$CHCl_2F$	practically nonflammable				3.55	48	....	no	yes	Dichloromonofluoromethane (Freon-21)
Dichlorotetrafluoroethane (Freon-114)	$C_2Cl_2F_4$	practically nonflammable				5.89	38	....	no	no	Dichlorotetrafluoroethane (Freon-114)
Ethane	C_2H_6	<20	950	3.0	12.5	1.04	−128	....	no	no	Ethane
Ethylene	C_2H_4	<20	842	3.1	32	0.972	−155	....	yes	yes	Ethylene
Isobutane	$(CH_3)_3CH$	<20	1010	1.8	8.4	2.01	14	....	no	no	Isobutane
Methyl chloride	CH_3Cl	632	1170	10.7	11.4	1.78	−11	100	yes	yes	Methyl chloride
Monochlorodifluoromethane (Freon-22)	$CHClF_2$	practically nonflammable				2.9	−41	....	yes	no	Monochlorodifluoromethane (Freon-22)
Monochlorotrifluoromethane (Freon-13)	$CClF_3$		nonflammable			3.6	−112	....	.....	no	Monochlorotrifluoromethane (Freon-13)
Propane	C_3H_8	<20	871	2.2	9.5	1.56	−45	....	no	no	Propane
Propylene	C_3H_6	<20	927	2.4	10.3	1.49	−53	....	yes	yes	Propylene
Sulfur dioxide	SO_2		nonflammable			2.2	14	10	yes	yes	Sulfur dioxide
Tetrafluoromethane (Freon-14)	CF_4		nonflammable			3.0	−198	....	no	no	Tetrafluoromethane (Freon-14)
Trichloroethylene	C_2HCl_3	nonflammable at normal temperature				4.53	189	200	no	yes	Trichloroethylene
Trichloromonofluoromethane (Freon-11) (Carrene-2)	CCl_3F		nonflammable			4.7	75.3	....	no	no	Trichloromonofluoromethane (Freon-11) (Carrene-2)
Trichlorotrifluoroethane (Freon-113)	$C_2Cl_3F_3$	practically nonflammable				6.4	118	....	no	no	Trichlorotrifluoroethane (Freon-113)

* Maximum average atmospheric concentration of contaminants to which workers may be exposed for an eight-hour work day without injury to health. (American Conference of Governmental Industrial Hygienists: "Threshold Limit Values for 1954.) Where blanks appear in this column no published information was available on threshold limit values.

MISCELLANEOUS PROPERTIES OF COMMON REFRIGERANTS

HYGROMETRIC AND BAROMETRIC TABLES

CONVERSION TABLE FOR BAROMETRIC READINGS

U. S. inches to cm.

Inches.	.00	.01	.02	.03	.04	.05	.06	.07	.08	.09
27.0	68.580	.606	.631	.656	.682	.707	.733	.758	.783	.809
27.1	.834	.860	.885	.910	.936	.961	.987	*.012	*.037	*.063
27.2	69.088	.114	.139	.164	.190	.215	.241	.266	.291	.317
27.3	.342	.368	.393	.418	.444	.469	.495	.520	.545	.571
27.4	.596	.622	.647	.672	.698	.723	.749	.774	.799	.825
27.5	.850	.876	.901	.926	.952	.977	*.002	*.028	*.053	*.079
27.6	70.104	.130	.155	.180	.206	.231	.257	.282	.307	333
27.7	.358	.384	.409	.434	.460	.485	.511	.536	.561	.587
27.8	.612	.638	.663	.688	.714	.739	.765	.790	.815	.841
27.9	.866	.892	.917	.942	.968	.993	*.018	*.044	*.069	*.095
28.0	71.120	.146	.171	.196	.222	.247	.273	.298	.323	.349
28.1	.374	.400	.425	.450	.476	.501	.527	.552	.577	.603
28.2	.628	.654	.679	.704	.730	.755	.781	.806	.831	.857
28.3	.882	.908	.933	.958	.984	*.009	*.035	*.060	*.085	*.111
28.4	72.136	.162	.187	.212	.238	.263	.289	.314	.339	.365
28.5	390	.416	.441	.466	.492	.517	.543	.568	.593	.619
28.6	.644	.670	.695	.720	.746	.771	.797	.822	.847	.873
28.7	.898	.924	.949	.974	*.000	*.025	*.051	*.076	*.101	*.127
28.8	73.152	.178	.203	.228	.254	.279	.305	.330	.355	.381
28.9	.406	.432	.457	.482	.508	.533	.559	.584	.609	.635
29.0	.660	.686	.711	.736	.762	.787	.813	.838	.863	.889
29.1	.914	.940	.965	.990	*.016	*.041	*.067	*.092	* 117	*.143
29.2	74.168	.194	.219	.244	.270	.295	.321	.346	.371	.397
29.3	.422	.448	.473	.498	.524	.549	.575	.600	.625	.651
29.4	.676	.702	.727	.752	.778	.803	.829	.854	.879	.905
29.5	.930	.956	.981	*.006	*.032	*.057	*.083	*.108	*.133	*.159
29.6	75.184	.210	.235	.260	.286	.311	.337	.362	.387	.413
29.7	.438	.464	.489	.514	.540	.565	.591	.616	.641	.667
29.8	.692	.718	.743	.768	.794	.819	.845	.870	.895	.921
29.9	.946	.972	.997	*.022	*.048	*.073	*.099	*.124	*.149	*.175
30.0	76.200	.226	.251	.277	.302	.327	.353	.378	.404	.429
30.1	.454	.480	.505	.531	.556	.581	.607	.632	.658	.683
30.2	.708	.734	.759	.785	.810	.835	.861	.886	.912	.937
30.3	.962	.988	*.013	*.039	*.064	*.089	*.115	*.140	*.166	*.191
30.4	77.216	.242	.267	.293	.318	.343	.369	.394	.420	.445
30.5	.470	.496	.521	.547	.572	.597	.623	.648	.674	.699
30.6	.724	.750	.775	.801	.826	.851	.877	.902	.928	.953
30.7	.978	*.004	*.029	*.055	*.080	*.105	*.131	*.156	*.182	*.207
30.8	78.232	.258	.283	.309	.334	.359	.385	410	.436	.461
30.9	.486	.512	.537	.563	.588	.613	.639	.664	.690	.715

U. S. Inches to Millibars

Based on the relation 1 inch of mercury at 32°F represents a pressure of 33.8639 millibars.

Note:
Figures in last nine columns to be preceded by 7, 8, 9 or 10 as indicated in column 2.

Inches	.00	.01	.02	.03	.04	.05	.06	.07	.08	.09
23.0	7 78.87	79.21	79.55	79.89	80.22	80.56	80.90	81.24	81.58	81.92
23.1	7 82.26	82.59	82.93	83.27	83.61	83.95	84.29	84.63	84.97	85.30
23.2	7 85.64	85.98	86.32	86.66	87.00	87.34	87.67	88.01	88.35	88.69
23.3	7 89.03	89.37	89.71	90.04	90.38	90.72	91.06	91.40	91.74	92.08
23.4	7 92.42	92.75	93.09	93.43	93.77	94.11	94.45	94.79	95.12	95.46
23.5	7 95.80	96.14	96.48	96.82	97.16	97.49	97.83	98.17	98.51	98.85
23.6	7 99.19	99.53	99.87	*00.20	*00.54	*00.88	*01.22	*01.56	*01.90	*02.24
23.7	8 02.57	02.91	03.25	03.59	03.93	04.27	04.61	04.94	05.28	05.62
23.8	8 05.96	06.30	06.64	06.98	07.32	07.65	07.99	08.33	08.67	09.01
23.9	8 09.35	09.69	10.02	10.36	10.70	11.04	11.38	11.72	12.06	12.39
24.0	8 12.73	13.07	13.41	13.75	14.09	14.43	14.77	15.10	15.44	15.78
24.1	8 16.12	16.46	16.80	17.14	17.47	17.81	18.15	18.49	18.83	19.17
24.2	8 19.51	19.85	20.18	20.52	20.86	21.20	21.54	21.88	22.22	22.55
24.3	8 22.89	23.23	23.57	23.91	24.25	24.59	24.92	25.26	25.60	25.94
24.4	8 26.28	26.62	26.96	27.30	27.63	27.97	28.31	28.65	28.99	29.33
24.5	8 29.67	30.00	30.34	30.68	31.02	31.36	31.70	32.04	32.37	32.71
24.6	8 33.05	33.39	33.73	34.07	34.41	34.75	35.08	35.42	35.76	36.10
24.7	8 36.44	36.78	37.12	37.45	37.79	38.13	38.47	38.81	39.15	39.49
24.8	8 39.82	40.16	40.50	40.84	41.18	41.52	41.86	42.20	42.53	42.87
24.9	8 43.21	43.55	43.89	44.23	44.57	44.90	45.24	45.58	45.92	46.26
25.0	8 46.60	46.94	47.27	47.61	47.95	48.29	48.63	48.97	49.31	49.65
25.1	8 49.98	50.32	50.66	51.00	51.34	51.68	52.02	52.35	52.69	53.03
25.2	8 53.37	53.71	54.05	54.39	54.72	55.06	55.40	55.74	56.08	56.42
25.3	8 56.76	57.10	57.43	57.77	58.11	58.45	58.79	59.13	59.47	59.80
25.4	8 60.14	60.48	60.82	61.16	61.50	61.84	62.17	62.51	62.85	63.19

U. S. Inches to Millibars (Continued)

Inches	.00	.01	.02	.03	.04	.05	.06	.07	.08	.09
25.5	8 63.53	63.87	64.21	64.55	64.88	65.22	65.56	65.90	66.24	66.58
25.6	8 66.92	67.25	67.59	67.93	68.27	68.61	68.95	69.29	69.62	69.96
25.7	8 70.30	70.64	70.98	71.32	71.66	72.00	72.33	72.67	73.01	73.35
25.8	8 73.69	74.03	74.37	74.70	75.04	75.38	75.72	76.06	76.40	76.74
25.9	8 77.08	77.41	77.75	78.09	78.43	78.77	79.11	79.45	79.78	80.12
26.0	8 80.46	80.80	81.14	81.48	81.82	82.15	82.49	82.83	83.17	83.51
26.1	8 83.85	84.19	84.53	84.86	85.20	85.54	85.88	86.22	86.56	86.90
26.2	8 87.23	87.57	87.91	88.25	88.59	88.93	89.27	89.60	89.94	90.28
26.3	8 90.62	90.96	91.30	91.64	91.98	92.31	92.65	92.99	93.33	93.67
26.4	8 94.01	94.35	94.68	95.02	95.36	95.70	96.04	96.38	96.72	97.05
26.5	8 97.39	97.73	98.07	98.41	98.75	99.09	99.43	99.76	*00.10	*00.44
26.6	9 00.78	01.12	01.46	01.80	02.13	02.47	02.81	03.15	03.49	03.83
26.7	9 04.17	04.50	04.84	05.18	05.52	05.86	06.20	06.54	06.88	07.21
26.8	9 07.55	07.89	08.23	08.57	08.91	09.25	09.58	09.92	10.26	10.60
26.9	9 10.94	11.28	11.62	11.95	12.29	12.63	12.97	13.31	13.65	13.99
27.0	9 14.33	14.66	15.00	15.34	15.68	16.02	16.36	16.70	17.03	17.37
27.1	9 17.71	18.05	18.39	18.73	19.07	19.40	19.74	20.08	20.42	20.76
27.2	9 21.10	21.44	21.78	22.11	22.45	22.79	23.13	23.47	23.81	24.15
27.3	9 24.48	24.82	25.16	25 50	25.84	26.18	26.52	26.85	27.19	27.53
27.4	9 27.87	28 21	28.55	28.89	29.23	29.56	29.90	30.24	30.58	30.[illegible]
27.5	9 31.26	31.60	31.93	32.27	32.61	32.95	33.29	33.63	33.97	34.31
27.6	9 34.64	34.98	35.32	35.66	36.00	36.34	36.68	37.01	37.35	37.69
27.7	9 38.03	38.37	38.71	39.05	39.38	39.72	40.06	40.40	40.74	41.08
27.8	9 41.42	41.76	42.09	42.43	42.77	43.11	43.45	43.79	44.13	44.46
27.9	9 44.80	45.14	45.48	45.82	46.16	46.50	46.83	47.17	47.51	47.85
28.0	9 48.19	48.53	48.87	49.21	49.54	49.88	50.22	50.56	50.90	51.24
28.1	9 51.58	51.91	52.25	52.59	52.93	53.27	53.61	53.95	54.28	54.62
28.2	9 54.96	55.30	55.64	55.98	56.32	56.66	56.99	57.33	57.67	58.01
28.3	9 58.35	58.69	59.03	59.36	59.70	60.04	60.38	60.72	61.06	61.40
28.4	9 61.73	62.07	62.41	62.75	63.09	63.43	63.77	64.11	64.44	64.78
28.5	9 65.12	65.46	65.80	66.14	66.48	66.81	67.15	67.49	67.83	68.17
28.6	9 68.51	68.85	69.18	69.52	69.86	70.20	70.54	70.88	71.22	71.56
28.7	9 71.89	72.23	72.57	72.91	73.25	73.59	73.93	74.26	74.60	74.94
28.8	9 75.28	75.62	75.96	76.30	76.63	76.97	77.31	77.65	77.99	78.33
28.9	9 78.67	79.01	79.34	79.68	80.02	80.36	80.70	81.04	81.38	81.71
29.0	9 82.05	82.39	82.73	83.07	83.41	83.75	84.08	84.42	84.76	85.10
29.1	9 85.44	85.78	86.12	86.46	86.79	87.13	87.47	87.81	88.15	88.49
29.2	9 88.83	89.16	89.50	89.84	90.18	90.52	90.86	91.20	91.53	91.87
29.3	9 92.21	92.55	92.89	93.23	93.57	93.91	94.24	94.58	94.92	95.26
29.4	9 95.60	95.94	96.28	96.61	96.95	97.29	97.63	97.97	98.31	98.65
29.5	9 98.99	99.32	99.66	*00.00	*00.34	*00.68	*01.02	*01.36	*01.69	*02.03
29.6	10 02.37	02.71	03.05	03.39	03.73	04.06	04.40	04.74	05.08	05.42
29.7	10 05.76	06.10	06.44	06.77	07.11	07.45	07.79	08.13	08.47	08.81
29.8	10 09.14	09.48	09.82	10.16	10.50	10.84	11.18	11.51	11.85	12.19
29.9	10 12.53	12.87	13.21	13.55	13.89	14.22	14.56	14.90	15.24	15.58
30.0	10 15.92	16.26	16.59	16.93	17.27	17.61	17.95	18.29	18.63	18.96
30.1	10 19.30	19.64	19.98	20.32	20.66	21.00	21.34	21.67	22.01	22.35
30.2	10 22.69	23.03	23.37	23.71	24.04	24.38	24.72	25.06	25.40	25.74
30.3	10 26.08	26.41	26.75	27.09	27.43	27.77	28.11	28.45	28.79	29.12
30.4	10 29.46	29.80	30.14	30.48	30.82	31.16	31.49	31.83	32.17	32.51
30.5	10 32.85	33.19	33.53	33.86	34.20	34.54	34.88	35.22	35.56	35.90
30.6	10 36.24	36.57	36.91	37.25	37.59	37.93	38.27	38.61	38.94	39.28
30.7	10 39.62	39.96	40.30	40.64	40.98	41.31	41.65	41.99	42.33	42.67
30.8	10 43.01	43.35	43.69	44.02	44.36	44.70	45.04	45.38	45.72	46.06
30.9	10 46.39	46.73	47.07	47.41	47.75	48.09	48.43	48.76	49.10	49.44
31.0	10 49.78	50.12	50.46	50.80	51.14	51.47	51.81	52.15	52.49	52.83
31.1	10 53.17	53.51	53.84	54.18	54.52	54.86	55.20	55.54	55.88	56.22
31.2	10 56.55	56.89	57.23	57.57	57.91	58.25	58.59	58.92	59.26	59.60
31.3	10 59.94	60.28	60.62	60.96	61.29	61.63	61.97	62.31	62.65	62.99
31.4	10 63.33	63.67	64.00	64.34	64.68	65.02	65.36	65.70	66.04	66.37
31.5	10 66.71	67.05	67.39	67.73	68.07	68.41	68.74	69.08	69.42	69.76
31.6	10 70.10	70.44	70.78	71.12	71.45	71.79	72.13	72.47	72.81	73.15
31.7	10 73.49	73.82	74.16	74.50	74.84	75.18	75.52	75.86	76.19	76.53
31.8	10 76.87	77.21	77.55	77.89	78.23	78.57	78.90	79.24	79.58	79.92
31.9	10 80.26	80.60	80.94	81.27	81.61	81.95	82.29	82.63	82.97	83.31

Note: Figures in last nine columns to be preceded by 9 or 10 as indicated.

CONVERSION TABLE FOR BAROMETRIC READINGS (Continued)

Centimeters to Millibars

Based on the relation 1 centimeter of mercury at 0°C represents a pressure of 13.3322 millibars. Note: Figures in last nine columns to be preceded by 9.

Centi-meters	.00	.01	.02	.03	.04	.05	.06	.07	.08	.09
68.0	9 06.59	06.72	06.86	06.99	07.12	07.26	07.39	07.52	07.66	07.79
68.1	9 07.92	08.06	08.19	08.32	08.46	08.59	08.72	08.86	08.99	09.12
68.2	9 09.26	09.39	09.52	09.66	09.79	09.92	10.06	10.19	10.32	10.46
68.3	9 10.59	10.72	10.86	10.99	11.12	11.26	11.39	11.52	11.66	11.79
68.4	9 11.92	12.06	12.19	12.32	12.46	12.59	12.72	12.86	12.99	13.12
68.5	9 13.26	13.39	13.52	13.66	13.79	13.92	14.06	14.19	14.32	14.46
68.6	9 14.59	14.72	14.86	14.99	15.12	15.26	15.39	15.52	15.66	15.79
68.7	9 15.92	16.06	16.19	16.32	16.46	16.59	16.72	16.86	16.99	17.12
68.8	9 17.26	17.39	17.52	17.66	17.79	17.92	18.06	18.19	18.32	18.46
68.9	9 18.59	18.72	18.86	18.99	19.12	19.26	19.39	19.52	19.66	19.79
69.0	9 19.92	20.06	20.19	20.32	20.46	20.59	20.72	20.86	20.99	21.12
69.1	9 21.26	21.39	21.52	21.65	21.79	21.92	22.05	22.19	22.32	22.45
69.2	9 22.59	22.72	22.85	22.99	23.12	23.25	23.39	23.52	23.65	23.79
69.3	9 23.92	24.05	24.19	24.32	24.45	24.59	24.72	24.85	24.99	25.12
69.4	9 25.25	25.39	25.52	25.65	25.79	25.92	26.05	26.19	26.32	26.45
69.5	9 26.59	26.72	26.85	26.99	27.12	27.25	27.39	27.52	27.65	27.79
69.6	9 27.92	28.05	28.19	28.32	28.45	28.59	28.72	28.85	28.99	29.12
69.7	9 29.25	29.39	29.52	29.65	29.79	29.92	30.05	30.19	30.32	30.45
69.8	9 30.59	30.72	30.85	30.99	31.12	31.25	31.39	31.52	31.65	31.79
69.9	9 31.92	32.05	32.19	32.32	32.45	32.59	32.72	32.85	32.99	33.12
70.0	9 33.25	33.39	33.52	33.65	33.79	33.92	34.05	34.19	34.32	34.45
70.1	9 34.59	34.72	34.85	34.99	35.12	35.25	35.39	35.52	35.65	35.79
70.2	9 35.92	36.05	36.19	36.32	36.45	36.59	36.72	36.85	36.99	37.12
70.3	9 37.25	37.39	37.52	37.65	37.79	37.92	38.05	38.19	38.32	38.45
70.4	9 38.59	38.72	38.85	38.99	39.12	39.25	39.39	39.52	39.65	39.79
70.5	9 39.92	40.05	40.19	40.32	40.45	40.59	40.72	40.85	40.99	41.12
70.6	9 41.25	41.39	41.52	41.65	41.79	41.92	42.05	42.19	42.32	42.45
70.7	9 42.59	42.72	42.85	42.99	43.12	43.25	43.39	43.52	43.65	43.79
70.8	9 43.92	44.05	44.19	44.32	44.45	44.59	44.72	44.85	44.99	45.12
70.9	9 45.25	45.39	45.52	45.65	45.79	45.92	46.05	46.19	46.32	46.45
71.0	9 46.59	46.72	46.85	46.99	47.12	47.25	47.39	47.52	47.65	47.79
71.1	9 47.92	48.05	48.19	48.32	48.45	48.59	48.72	48.85	48.99	49.12
71.2	9 49.25	49.39	49.52	49.65	49.79	49.92	50.05	50.19	50.32	50.45
71.3	9 50.59	50.72	50.85	50.99	51.12	51.25	51.39	51.52	51.65	51.79
71.4	9 51.92	52.05	52.19	52.32	52.45	52.59	52.72	52.85	52.99	53.12
71.5	9 53.25	53.39	53.52	53.65	53.79	53.92	54.05	54.19	54.32	54.45
71.6	9 54.59	54.72	54.85	54.99	55.12	55.25	55.39	55.52	55.65	55.79
71.7	9 55.92	56.05	56.19	56.32	56.45	56.59	56.72	56.85	56.99	57.12
71.8	9 57.25	57.39	57.52	57.65	57.79	57.92	58.05	58.19	58.32	58.45
71.9	9 58.59	58.72	58.85	58.99	59.12	59.25	59.39	59.52	59.65	59.79
72.0	9 59.92	60.05	60.19	60.32	60.45	60.59	60.72	60.85	60.98	61.12
72.1	9 61.25	61.38	61.52	61.65	61.78	61.92	62.05	62.18	62.32	62.45
72.2	9 62.58	62.72	62.85	62.98	63.12	63.25	63.38	63.52	63.65	63.78
72.3	9 63.92	64.05	64.18	64.32	64.45	64.58	64.72	64.85	64.98	65.12
72.4	9 65.25	65.38	65.52	65.65	65.78	65.92	66.05	66.18	66.32	66.45
72.5	9 66.58	66.72	66.85	66.98	67.12	67.25	67.38	67.52	67.65	67.78
72.6	9 67.92	68.05	68.18	68.32	68.45	68.58	68.72	68.85	68.98	69.12
72.7	9 69.25	69.38	69.52	69.65	69.78	69.92	70.05	70.18	70.32	70.45
72.8	9 70.58	70.72	70.85	70.98	71.12	71.25	71.38	71.52	71.65	71.78
72.9	9 71.92	72.05	72.18	72.32	72.45	72.58	72.72	72.85	72.98	73.12

Centimeters to Millibars (Continued)

Note: Figures in last nine columns to be preceded by 9 or 10.

Centi-meters	.00	.01	.02	.03	.04	.05	.06	.07	.08	.09
73.0	9 73.25	73.38	73.52	73.65	73.78	73.92	74.05	74.18	74.32	74.45
73.1	9 74.58	74.72	74.85	74.98	75.12	75.25	75.38	75.52	75.65	75.78
73.2	9 75.92	76.05	76.18	76.32	76.45	76.58	76.72	76.85	76.98	77.12
73.3	9 77.25	77.38	77.52	77.65	77.78	77.92	78.05	78.18	78.32	78.45
73.4	9 78.58	78.72	78.85	78.98	79.12	79.25	79.38	79.52	79.65	79.78
73.5	9 79.92	80.05	80.18	80.32	80.45	80.58	80.72	80.85	80.98	81.12
73.6	9 81.25	81.38	81.52	81.65	81.78	81.92	82.05	82.18	82.32	82.45
73.7	9 82.58	82.72	82.85	82.98	83.12	83.25	83.38	83.52	83.65	83.78
73.8	9 83.92	84.05	84.18	84.32	84.45	84.58	84.72	84.85	84.98	85.12
73.9	9 85.25	85.38	85.52	85.65	85.78	85.92	86.05	86.18	86.32	86.45
74.0	9 86.58	86.72	86.85	86.98	87.12	87.25	87.38	87.52	87.65	87.78
74.1	9 87.92	88.05	88.18	88.32	88.45	88.58	88.72	88.85	88.98	89.12
74.2	9 89.25	89.38	89.52	89.65	89.78	89.92	90.05	90.18	90.32	90.45
74.3	9 90.58	90.72	90.85	90.98	91.12	91.25	91.38	91.52	91.65	91.78
74.4	9 91.92	92.05	92.18	92.32	92.45	92.58	92.72	92.85	92.98	93.12
74.5	9 93.25	93.38	93.52	93.65	93.78	93.92	94.05	94.18	94.32	94.45
74.6	9 94.58	94.72	94.85	94.98	95.12	95.25	95.38	95.52	95.65	95.78
74.7	9 95.92	96.05	96.18	96.32	96.45	96.58	96.72	96.85	96.98	97.12
74.8	9 97.25	97.38	97.52	97.65	97.78	97.92	98.05	98.18	98.32	98.45
74.9	9 98.58	98.72	98.85	98.98	99.12	99.25	99.38	99.52	99.65	99.78
75.0	9 99.92	*00.05	*00.18	*00.31	*00.45	*00.58	*00.71	*00.85	*00.98	*01.11
75.1	10 01.25	01.38	01.51	01.65	01.78	01.91	02.05	02.18	02.31	02.45
75.2	10 02.58	02.71	02.85	02.98	03.11	03.25	03.38	03.51	03.65	03.78
75.3	10 03.91	04.05	04.18	04.31	04.45	04.58	04.71	04.85	04.98	05.11
75.4	10 05.25	05.38	05.51	05.65	05.78	05.91	06.05	06.18	06.31	06.45
75.5	10 06.58	06.71	06.85	06.98	07.11	07.25	07.38	07.51	07.65	07.78
75.6	10 07.91	08.05	08.18	08.31	08.45	08.58	08.71	08.85	08.98	09.11
75.7	10 09.25	09.38	09.51	09.65	09.78	09.91	10.05	10.18	10.31	10.45
75.8	10 10.58	10.71	10.85	10.98	11.11	11.25	11.38	11.51	11.65	11.78
75.9	10 11.91	12.05	12.18	12.31	12.45	12.58	12.71	12.85	12.98	13.11
76.0	10 13.25	13.38	13.51	13.65	13.78	13.91	14.05	14.18	14.31	14.45
76.1	10 14.58	14.71	14.85	14.98	15.11	15.25	15.38	15.51	15.65	15.78
76.2	10 15.91	16.05	16.18	16.31	16.45	16.58	16.71	16.85	16.98	17.11
76.3	10 17.25	17.38	17.51	17.65	17.78	17.91	18.05	18.18	18.31	18.45
76.4	10 18.58	18.71	18.85	18.98	19.11	19.25	19.38	19.51	19.65	19.78
76.5	10 19.91	20.05	20.18	20.31	20.45	20.58	20.71	20.85	20.98	21.11
76.6	10 21.25	21.38	21.51	21.65	21.78	21.91	22.05	22.18	22.31	22.45
76.7	10 22.58	22.71	22.85	22.98	23.11	23.25	23.38	23.51	23.65	23.78
76.8	10 23.91	24.05	24.18	24.31	24.45	24.58	24.71	24.85	24.98	25.11
76.9	10 25.25	25.38	25.51	25.65	25.78	25.91	26.05	26.18	26.31	26.45
77.0	10 26.58	26.71	26.85	26.98	27.11	27.25	27.38	27.51	27.65	27.78
77.1	10 27.91	28.05	28.18	28.31	28.45	28.58	28.71	28.85	28.98	29.11
77.2	10 29.25	29.38	29.51	29.65	29.78	29.91	30.05	30.18	30.31	30.45
77.3	10 30.58	30.71	30.85	30.98	31.11	31.25	31.38	31.51	31.65	31.78
77.4	10 31.91	32.05	32.18	32.31	32.45	32.58	32.71	32.85	32.98	33.11
77.5	10 33.25	33.38	33.51	33.65	33.78	33.91	34.05	34.18	34.31	34.45
77.6	10 34.58	34.71	34.85	34.98	35.11	35.25	35.38	35.51	35.65	35.78
77.7	10 35.91	36.05	36.18	36.31	36.45	36.58	36.71	36.85	36.98	37.11
77.8	10 37.25	37.38	37.51	37.65	37.78	37.91	38.05	38.18	38.31	38.45
77.9	10 38.58	38.71	38.85	38.98	39.11	39.24	39.38	39.51	39.64	39.78

TEMPERATURE CORRECTION FOR BAROMETER READINGS

BRASS SCALE—METRIC UNITS

To reduce readings of a mercurial barometer with a brass scale to 0°C subtract the appropriate quantity as found in the table. These values are based on the coefficient of expansion of mercury $(181792 + 0.175t + 0.035116t^2) \times 10^{-9}$, and of brass 0.0000184 per °C. Corrections are in millimeters.

Temp. °C	Observed height in millimeters																	
	620	630	640	650	660	670	680	690	700	710	720	730	740	750	760	770	780	790
0	0.00	0.00	0.00	0.00	0.00	0.00	0.00	0.00	0.00	0.00	0.00	0.00	0.00	0.00	0.00	0.00	0.00	0.00
1	.10	.10	.10	.11	.11	.11	.11	.11	.11	.12	.12	.12	.12	.12	.12	.13	.13	.13
2	.20	.21	.21	.21	.22	.22	.22	.23	.23	.23	.24	.24	.24	.25	.25	.25	.25	.26
3	.30	.31	.31	.32	.32	.33	.33	.34	.34	.35	.35	.36	.36	.37	.37	.38	.38	.39
4	.40	.41	.42	.42	.43	.44	.44	.45	.46	.46	.47	.48	.48	.49	.50	.50	.51	.52
5	0.51	0.51	0.52	0.53	0.54	0.55	0.56	0.56	0.57	0.58	0.59	0.60	0.60	0.61	0.62	0.63	0.64	0.64
6	.61	.62	.63	.64	.65	.66	.67	.68	.69	.70	.71	.71	.72	.73	.74	.75	.76	.77
7	.71	.72	.73	.74	.75	.77	.78	.79	.80	.81	.82	.83	.85	.86	.87	.88	.89	.90
8	.81	.82	.84	.85	.86	.87	.89	.90	.91	.93	.94	.95	.97	.98	.99	1.01	1.02	1.03
9	.91	.92	.94	.95	.97	.98	1.00	1.01	1.03	1.04	1.06	1.07	1.09	1.10	1.12	1.13	1.15	1.16
10	1.01	1.03	1.04	1.06	1.08	1.09	1.11	1.13	1.14	1.16	1.17	1.19	1.21	1.22	1.24	1.26	1.27	1.29
11	1.11	1.13	1.15	1.17	1.18	1.20	1.22	1.24	1.26	1.27	1.29	1.31	1.33	1.35	1.36	1.38	1.40	1.42
12	1.21	1.23	1.25	1.27	1.29	1.31	1.33	1.35	1.37	1.39	1.41	1.43	1.45	1.47	1.49	1.51	1.53	1.55
13	1.31	1.34	1.36	1.38	1.40	1.42	1.44	1.46	1.48	1.50	1.53	1.55	1.57	1.59	1.61	1.63	1.65	1.67
14	1.41	1.44	1.46	1.48	1.51	1.53	1.55	1.57	1.60	1.62	1.64	1.67	1.69	1.71	1.73	1.76	1.78	1.80

CORRECTION FOR BAROMETER (Continued)

Brass Scale—Metric Units (Continued)

Temp. °C	620	630	640	650	660	670	680	690	700	710	720	730	740	750	760	770	780	790
	Observed height in millimeters																	
15	1.52	1.54	1.56	1.59	1.61	1.64	1.66	1.69	1.71	1.74	1.76	1.78	1.81	1.83	1.86	1.88	1.91	1.93
16	1.62	1.64	1.67	1.69	1.72	1.75	1.77	1.80	1.82	1.85	1.88	1.90	1.93	1.96	1.98	2.01	2.03	2.06
17	1.72	1.74	1.77	1.80	1.83	1.86	1.88	1.91	1.94	1.97	1.99	2.02	2.05	2.08	2.10	2.13	2.16	2.19
18	1.82	1.85	1.88	1.91	1.93	1.96	1.99	2.02	2.05	2.08	2.11	2.14	2.17	2.20	2.23	2.26	2.29	2.32
19	1.92	1.95	1.98	2.01	2.04	2.07	2.10	2.13	2.17	2.20	2.23	2.26	2.29	2.32	2.35	2.38	2.41	2.44
20	2.02	2.05	2.08	2.12	2.15	2.18	2.21	2.25	2.28	2.31	2.34	2.38	2.41	2.44	2.47	2.51	2.54	2.57
21	2.12	2.15	2.19	2.22	2.26	2.29	2.32	2.36	2.39	2.43	2.46	2.50	2.53	2.56	2.60	2.63	2.67	2.70
22	2.22	2.26	2.29	2.33	2.36	2.40	2.43	2.47	2.51	2.54	2.58	2.61	2.65	2.69	2.72	2.76	2.79	2.83
23	2.32	2.36	2.40	2.43	2.47	2.51	2.54	2.58	2.62	2.66	2.69	2.73	2.77	2.81	2.84	2.88	2.92	2.96
24	2.42	2.46	2.50	2.54	2.58	2.62	2.66	2.69	2.73	2.77	2.81	2.85	2.89	2.93	2.97	3.01	3.05	3.08
25	2.52	2.56	2.60	2.64	2.68	2.72	2.77	2.81	2.85	2.89	2.93	2.97	3.01	3.05	3.09	3.13	3.17	3.21
26	2.62	2.66	2.71	2.75	2.79	2.83	2.88	2.92	2.96	3.00	3.04	3.09	3.13	3.17	3.21	3.26	3.30	3.34
27	2.72	2.77	2.81	2.85	2.90	2.94	2.99	3.03	3.07	3.12	3.16	3.20	3.25	3.29	3.34	3.38	3.42	3.47
28	2.82	2.87	2.91	2.96	3.00	3.05	3.10	3.14	3.19	3.23	3.28	3.32	3.37	3.41	3.46	3.51	3.55	3.60
29	2.92	2.97	3.02	3.06	3.11	3.16	3.21	3.25	3.30	3.35	3.39	3.44	3.49	3.54	3.58	3.63	3.68	3.72
30	3.02	3.07	3.12	3.17	3.22	3.27	3.32	3.36	3.41	3.46	3.51	3.56	3.61	3.66	3.71	3.75	3.80	3.85
31	3.12	3.17	3.22	3.27	3.32	3.37	3.43	3.48	3.53	3.58	3.63	3.68	3.73	3.78	3.83	3.88	3.93	3.98
32	3.22	3.28	3.33	3.38	3.43	3.48	3.54	3.59	3.64	3.69	3.74	3.79	3.85	3.90	3.95	4.00	4.05	4.11
33	3.32	3.38	3.43	3.48	3.54	3.59	3.64	3.70	3.75	3.81	3.86	3.91	3.97	4.02	4.07	4.13	4.18	4.23
34	3.42	3.48	3.53	3.59	3.64	3.70	3.75	3.81	3.87	3.92	3.98	4.03	4.09	4.14	4.20	4.25	4.31	4.36
35	3.52	3.58	3.64	3.69	3.75	3.81	3.86	3.92	3.98	4.03	4.09	4.15	4.21	4.26	4.32	4.38	4.43	4.49

Brass Scale—English Units

Standard Temperature of scale 62° F; of mercury, 32° F. Zero correction at 28.5° F; subtract corrections above, add below. Owing to the difference in the standard temperature of English and metric scales, readings taken in inches to be reduced to centimeters should *first* be corrected for temperature.

Temp. °F	23.0 in.	23.5 in.	24.0 in.	24.5 in.	25.0 in.	25.5 in.	26.0 in.	26.5 in.	27.0 in.	27.5 in.	28.0 in.	28.5 in.	29.0 in.	29.5 in.	30.0 in.	30.5 in.	31.0 in.	31.5 in.
	Observed height in inches																	
0	+.060	+.061	+.063	+.064	+.065	+.067	+.068	+.069	+.070	.072	.073	.075	.076	.077	.078	.080	.081	.082
2	.056	.057	.058	.060	.061	.062	.063	.065	.065	.067	.068	.069	.070	.072	.073	.074	.075	.077
4	.052	.053	.054	.055	.056	.057	.058	.060	.061	.062	.063	.064	.065	.066	.067	.069	.070	.071
6	.047	.048	.049	.051	.052	.053	.054	.055	.056	.057	.058	.059	.060	.061	.062	.063	.064	.065
8	.043	.044	.045	.046	.047	.048	.049	.050	.051	.052	.053	.054	.054	.056	.056	.057	.058	.059
10	.039	.040	.041	.042	.042	.043	.044	.045	.046	.047	.047	.048	.049	.050	.051	.052	.053	.054
12	.035	.036	.036	.037	.038	.039	.039	.040	.041	.042	.042	.043	.044	.045	.045	.046	.047	.048
14	.031	.031	.032	.033	.033	.034	.035	.035	.036	.037	.037	.038	.039	.039	.040	.041	.041	.042
16	.026	.027	.028	.028	.029	.029	.030	.031	.031	.032	.032	.033	.033	.034	.034	.035	.036	.036
18	.022	.023	.023	.024	.024	.025	.025	.026	.026	.027	.027	.028	.028	.029	.029	.030	.030	.031
20	.018	.018	.019	.019	.020	.020	.020	.021	.021	.022	.022	.022	.023	.023	.024	.024	.024	.025
22	.014	.014	.014	.015	.015	.015	.016	.016	.016	.017	.017	.017	.017	.018	.018	.018	.019	.019
24	.010	.010	.010	.010	.011	.011	.011	.011	.011	.012	.012	.012	.012	.012	.013	.013	.013	.013
26	.005	.006	.006	.006	.006	.006	.006	.006	.006	.007	.007	.007	.007	.007	.007	.007	.007	.008
28	+.001	+.001	+.001	+.001	+.001	+.001	+.001	+.002	+.002	+.002	+.002	+.002	+.002	+.002	+.002	+.002	+.002	+.002
30	−.003	−.003	−.003	−.003	−.003	−.003	−.003	−.003	−.003	−.003	−.003	−.004	−.004	−.004	−.004	−.004	−.004	−.004
32	.007	.007	.007	.008	.008	.008	.008	.008	.008	.008	.009	.009	.009	.009	.009	.009	.009	.010
34	.011	.011	.012	.012	.012	.012	.013	.013	.013	.013	.014	.014	.014	.014	.015	.015	.015	.015
36	.015	.016	.016	.016	.017	.017	.017	.018	.018	.018	.019	.019	.019	.020	.020	.020	.021	.021
38	.020	.020	.020	.021	.021	.022	.022	.023	.023	.023	.024	.024	.025	.025	.026	.026	.026	.027
40	.024	.024	.025	.025	.026	.026	.027	.027	.028	.028	.029	.030	.030	.031	.031	.032	.032	.033
42	.028	.029	.029	.030	.030	.031	.032	.032	.033	.033	.034	.035	.035	.036	.036	.037	.038	.038
44	.032	.033	.033	.034	.035	.036	.036	.037	.038	.038	.039	.040	.040	.041	.042	.043	.043	.044
46	.036	.037	.038	.039	.039	.040	.041	.042	.043	.043	.044	.045	.046	.047	.047	.048	.049	.050
48	.040	.041	.042	.043	.044	.045	.046	.047	.047	.048	.049	.050	.051	.052	.053	.054	.054	.055
50	.045	.046	.046	.048	.048	.050	.050	.052	.052	.053	.054	.055	.056	.057	.058	.059	.060	.061
52	.049	.050	.051	.052	.053	.054	.055	.056	.057	.058	.059	.061	.061	.063	.064	.065	.066	.067
54	.053	.054	.055	.057	.057	.059	.060	.061	.062	.063	.064	.066	.067	.068	.069	.070	.071	.073
56	.057	.058	.060	.061	.062	.063	.064	.066	.067	.068	.069	.071	.072	.073	.074	.076	.077	.078
58	.061	.063	.064	.065	.066	.068	.069	.071	.072	.073	.074	.076	.077	.079	.080	.081	.082	.084
60	.065	.067	.068	.070	.071	.073	.074	.076	.077	.078	.080	.081	.082	.084	.085	.087	.088	.090
62	.069	.071	.073	.074	.076	.077	.079	.080	.082	.083	.085	.086	.088	.089	.091	.092	.094	.095
64	.074	.075	.077	.079	.080	.082	.083	.085	.086	.088	.090	.092	.093	.095	.096	.098	.099	.101
66	.078	.079	.081	.083	.085	.087	.088	.090	.091	.093	.095	.097	.098	.100	.101	.103	.105	.107
68	.082	.084	.085	.088	.089	.091	.093	.095	.096	.098	.100	.102	.103	.105	.107	.109	.110	.113
70	.086	.088	.090	.092	.094	.096	.097	.100	.101	.103	.105	.107	.109	.111	.112	.115	.116	.118
72	.090	.092	.094	.096	.098	.100	.102	.104	.106	.108	.110	.112	.114	.116	.118	.120	.122	.124
74	.094	.096	.098	.101	.103	.105	.107	.109	.111	.113	.115	.117	.119	.121	.123	.126	.127	.130
76	.098	.101	.103	.105	.107	.110	.111	.114	.116	.118	.120	.122	.124	.127	.128	.131	.133	.135
78	.103	.105	.107	.110	.112	.114	.116	.119	.120	.123	.125	.128	.129	.132	.134	.137	.138	.141
80	.107	.109	.111	.114	.116	.119	.121	.123	.125	.128	.130	.133	.135	.137	.139	.142	.144	.147
82	.111	.113	.116	.119	.121	.123	.125	.128	.130	.133	.135	.138	.140	.143	.145	.148	.149	.152
84	.115	.118	.120	.123	.125	.128	.130	.133	.185	.138	.140	.143	.145	.148	.150	.153	.155	.158
86	.119	.122	.124	.127	.130	.133	.135	.138	.140	.143	.145	.148	.150	.153	.155	.159	.161	.164
88	.123	.126	.129	.132	.134	.137	.139	.143	.145	.148	.150	.153	.155	.159	.161	.164	.166	.169
90	.127	.130	.133	.136	.138	.142	.144	.147	.150	.153	.155	.158	.161	.164	.166	.170	.172	.175
92	.132	.134	.137	.141	.143	.146	.149	.152	.154	.158	.160	.163	.166	.169	.172	.175	.177	.181
94	.136	.139	.142	.145	.147	.151	.153	.157	.159	.163	.165	.169	.171	.175	.177	.180	.183	.186
96	.140	.143	.146	.150	.152	.155	.158	.161	.134	.168	.170	.174	.176	.180	.182	.186	.188	.192
98	.144	.147	.150	.154	.156	.160	.163	.166	.169	.172	.175	.179	.181	.185	.188	.191	.194	.197
100	.148	.151	.154	.158	.161	.164	.167	.171	.174	.177	.180	.184	.187	.190	.193	.197	.200	.203

TEMPERATURE CORRECTION, GLASS SCALE

METRIC

To reduce readings of a mercurial barometer with a glass scale to 0° C. subtract the appropriate quantity as found in table.

Temp. °C.	Observed height in centimeters. 70 cm.	71 cm.	72 cm.	73 cm.	74 cm.	75 cm.	76 cm.	77 cm.	78 cm.	Temp. °C.	Observed height in centimeters. 70 cm.	71 cm.	72 cm.	73 cm.	74 cm.	75 cm.	76 cm.	77 cm.	78 cm.
0	0.000	0.000	0.000	0.000	0.000	0.000	0.000	0.000	0.000	15	0.181	0.184	0.186	0.189	0.191	0.193	0.196	0.198	0.201
1	.012	.012	.013	.013	.013	.013	.013	.013	.014	16	.194	.196	.199	.201	.204	.207	.209	.212	.214
2	.025	.025	.025	.026	.026	.026	.026	.027	.027	17	.205	.208	.210	.213	.216	.219	.221	.224	.227
3	.036	.036	.037	.037	.038	.038	.039	.039	.040	18	.217	.220	.223	.226	.229	.232	.235	.238	.241
4	.048	.049	.049	.050	.051	.051	.052	.053	.053	19	.230	.233	.236	.239	.242	.245	.248	.251	.254
5	0.060	0.061	0.062	0.063	0.064	0.064	0.065	0.066	0.067	20	0.242	0.245	0.248	0.252	0.255	0.258	0.261	0.264	0.268
6	.073	.074	.074	.076	.077	.077	.078	.079	.080	21	.254	.258	.261	.264	.268	.271	.275	.278	.281
7	.085	.086	.087	.088	.089	.091	.092	.093	.094	22	.266	.269	.273	.276	.280	.283	.287	.290	.294
8	.096	.098	.099	.100	.101	.103	.104	.105	.107	23	.278	.282	.285	.289	.293	.296	.300	.304	.308
9	.109	.110	.111	.113	.114	.116	.117	.119	.120	24	.290	.294	.298	.302	.306	.310	.313	.317	.321
10	0.121	0.122	0.124	0.126	0.127	0.129	0.130	0.132	0.134	25	0.303	0.307	0.311	0.315	0.319	0.323	0.327	0.331	0.335
11	.133	.135	.137	.138	.140	.142	.144	.146	.147	26	.315	.319	.323	.327	.332	.336	.340	.344	.348
12	.144	.146	.148	.150	.152	.154	.156	.158	.160	27	.326	.331	.335	.339	.344	.348	.352	.357	.361
13	.157	.159	.161	.163	.165	.167	.169	.171	.174	28	.339	.343	.348	.352	.357	.361	.366	.370	.375
14	.169	.171	.174	.176	.178	.180	.183	.185	.187	29	.351	.356	.360	.365	.370	.374	.379	.384	.388
										30	0.363	0.368	0.373	0.378	0.383	0.387	0.392	0.397	0.402

WEIGHT IN GRAMS OF A CUBIC METER OF SATURATED AQUEOUS VAPOR

(From Smithsonian Tables)

Mass in grams per cubic meter.

Temp. °C	0.0	1.0	2.0	3.0	4.0	5.0	6.0	7.0	8.0	9.0
−20	1.074	.988	.909	.836	.768	.705	.646	.592	.542	.496
−10	2.358	2.186	2.026	1.876	1.736	1.605	1.483	1.369	1.264	1.165
− 0	4.847	4.523	4.217	3.930	3.660	3.407	3.169	2.946	2.737	2.541
+ 0	4.847	5.192	5.559	5.947	6.360	6.797	7.260	7.750	8.270	8.819
+10	9.399	10.01	10.66	11.35	12.07	12.83	13.63	14.84	15.37	16.21
+20	17.30	18.34	19.43	20.58	21.78	23.05	24.38	25.78	27.24	28.78
+30	30.38	32.07	33.83	35.68	37.61	39.63	41.75	43.96	46.26	48.67

EFFICIENCY OF DRYING AGENTS

Compiled by John H. Yoe

A. Drying agents depending upon chemical action (absorption) for their efficiency:*

Substance	Residual water, mg per liter of dry air**	Reference
P_2O_5	<1 mg in 40,000 l,	Morley, Am. J. Sci., **34**, 199 (1887); J.A.C.S., **26**, 1171 (1904).
$Mg(ClO_4)_2$ anhyd.	"Unweighable" in 210 l,	Willard and Smith, J.A.C.S., **44**, 2255 (1922).
BaO	0.00065	Bower, Bur. Std. J. Res., **12**, 241 (1934).
KOH fused	0.002	Baxter and Starkweather, J.A.C.S., **38**, 2038 (1916).
CaO	0.003	Bower, loc. cit.
H_2SO_4	0.003	Baxter and Starkweather, loc. cit.
$CaSO_4$ anhyd.	0.005	Bower, loc. cit.
Al_2O_3	0.005	Ibid.
KOH sticks	0.014	Ibid.
NaOH fused	0.16	Baxter and Starkweather, loc. cit.
$CaBr_2$	0.18	Baxter and Warren, J.A.C.S., **33**, 340 (1911).
$CaCl_2$ fused	0.34	Baxter and Starkweather, loc. cit.
NaOH sticks	0.80	Bower, loc. cit.
$Ba(ClO_4)_2$	0.82	Ibid.
$ZnCl_2$	0.85	Baxter and Warren, loc. cit.
$ZnBr_2$	1.16	Ibid.
$CaCl_2$ granular	1.5	Bower, loc. cit.
$CuSO_4$ anhyd.	2.8	Ibid.

B. Drying agents depending upon physical action (adsorption) for their efficiency:* Alumina (low temperature fired), asbestos, charcoal, clay and porcelain (low temperature fired), glass wool, kieselguhr, silica gel, refrigeration.

*** It should be noted that the efficiency of some drying agents (e.g. Al_2O_3 and anhydrous $CaCl_2$, and probably also BaO, anhydrous $Mg(ClO_4)_2$, $Mg(ClO_4)_2 \cdot 3H_2O$, anhydrous $Ba(ClO_4)_2$, and $CaSO_4$) depends upon both adsorption and absorption.**

**** 30°C. for Bower's values; others 25°C. or room temp.**

REDUCTION OF BAROMETER TO SEA LEVEL

The correction to be added to reduce barometric readings to "sea level" values depends principally on three factors: The temperature of the air column (assumed) from the station to sea level, the altitude of the station, and the value of the reading itself. Two tables are provided. Table I is entered with the altitude and assumed temperature and a factor "2000 m" taken out. Table II is entered with the above factor and the approximate barometer reading and the final correction taken out.

The correction is to be added. If B_0 is the corrected or sea level value; B the barometer reading at the station; C the correction,—

$$C = B_0 - B = B\,(10^m - 1)$$

The actual barometer reading at the station should be corrected for temperature of the mercury column by the usual methods before entering the tables or applying the sea level correction.

A complete explanation of the theory of the corrections and a more extended set of tables will be found in the Smithsonian Meteorological Tables.

Latitude Factor

The influence of the latitude on the value of the correction is usually negligible, being overshadowed by uncertainties in the assumed temperature of the air column. For cases where this correction is desirable the table below is provided. The value of the temperature-altitude factor "2000 m" obtained in Table I is corrected for latitude by subtracting for latitudes 0-45° and adding for latitudes from 45-90° the values found. With this corrected value of "2000 m" Table II is entered for the value of the correction.

LATITUDE FACTOR

To be used in connection with Tables I and II, either English or metric units, to obtain latitude corrections to temperature-altitude factor. For latitudes 0-45° subtract the correction. For latitudes 45-90° add the correction.

Temp.—Alt. from Table I	Latitude			
	0°	15°	30°	45°
100	0.3	0.2	0.1	0.0
200	0.5	0.5	0.3	0.0
300	0.8	0.7	0.4	0.0
	90°	75°	60°	45°

METRIC UNITS—TABLE I

Values of the temperature-altitude factor (2000 m.) for entering table II.

Altitude in meters	Assumed temperature of air column °C									
	−16°	−8°	0°	+4°	+8°	+12°	+16°	+20°	+24°	+28°
10	1.2	1.1	1.1	1.1	1.0	1.0	1.0	1.0	1.0	1.0
50	5.8	5.6	5.4	5.3	5.2	5.2	5.1	5.0	4.9	4.9
100	11.5	11.2	10.8	10.7	10.5	10.3	10.2	10.0	9.9	9.7
150	17.3	16.7	16.2	16.0	15.7	15.5	15.3	15.0	14.8	14.6
200	23.0	22.3	21.6	21.3	21.0	20.7	20.3	20.0	19.7	19.5
250	28.8	27.9	27.0	26.6	26.2	25.8	25.4	25.0	24.7	24.3
300	34.5	33.5	32.5	32.0	31.5	31.0	30.5	30.1	29.6	29.2
350	40.3	39.0	37.9	37.3	36.7	36.2	35.6	35.1	34.6	34.0
400	46.0	44.6	43.3	42.6	42.0	41.3	40.7	40.1	39.5	38.9
450	51.8	50.2	48.7	47.9	47.2	46.5	45.8	45.1	44.4	43.8
500	57.5	55.8	54.1	53.3	52.4	51.6	50.9	50.1	49.4	48.6
550	63.3	61.4	59.5	58.6	57.7	56.8	55.9	55.1	54.3	53.5
600	69.0	66.9	64.9	63.9	62.9	62.0	61.0	60.1	59.2	58.3
650	74.8	72.5	70.3	69.2	68.2	67.1	66.1	65.1	64.2	63.2
700	80.6	78.1	75.7	74.6	73.4	72.3	71.2	70.1	69.1	68.1
750	86.3	83.7	81.1	79.9	78.7	77.5	76.3	75.1	74.0	72.9
800	92.1	89.2	86.5	85.2	83.9	82.6	81.4	80.1	79.0	77.8
850	97.8	94.8	92.0	90.5	89.2	87.8	86.4	85.2	83.9	82.7
900	103.6	100.4	97.4	95.9	94.4	93.0	91.5	90.2	88.8	87.5
950	109.3	106.0	102.8	101.2	99.6	98.1	96.6	95.2	93.8	92.4
1000	115.1	111.5	108.2	106.5	104.9	103.3	101.7	100.2	98.7	97.3
1050	120.8	117.1	113.6	111.8	110.1	108.4	106.8	105.2	103.6	102.1
1100	126.6	122.7	119.0	117.2	115.4	113.6	111.9	110.2	108.6	107.0
1150	132.3	128.3	124.4	122.5	120.6	118.8	117.0	115.2	113.5	111.8
1200	138.1	133.8	129.8	127.8	125.9	123.9	122.0	120.2	118.4	116.7
1250	143.8	139.4	135.2	133.1	131.1	129.1	127.1	125.2	123.4	121.6
1300	149.6	145.0	140.6	138.5	136.3	134.3	132.2	130.2	128.3	126.4
1350	155.3	150.6	146.0	143.8	141.6	139.4	137.3	135.2	133.2	131.3
1400	161.1	156.2	151.4	149.1	146.8	144.6	142.4	140.2	138.2	136.2
1450	166.8	161.7	156.8	154.5	152.1	149.7	147.5	145.3	143.1	141.0

METRIC UNITS—TABLE I (Continued)

Altitude in meters	Assumed temperature of air column °C									
	−16°	−8°	0°	+4°	+8°	+12°	+16°	+20°	+24°	+28°
1500	172.6	167.3	162.3	159.8	157.3	154.9	152.5	150.3	148.0	145.9
1550	178.3	172.9	167.7	165.1	162.6	160.1	157.6	155.3	153.0	150.7
1600	184.1	178.5	173.1	170.4	167.8	165.2	162.7	160.3	157.9	155.6
1650	189.8	184.0	178.5	175.7	173.0	170.4	167.8	165.3	162.8	160.5
1700	195.6	189.6	183.9	181.1	178.3	175.6	172.9	170.3	167.8	165.3
1750	201.4	195.2	189.3	186.4	183.5	180.7	178.0	175.3	172.7	170.2
1800	207.1	200.8	194.7	191.7	188.8	185.9	183.1	180.3	177.6	175.0
1850	212.9	206.3	200.1	197.0	194.0	191.0	188.1	185.3	182.6	179.9
1900	218.6	211.9	205.5	202.4	199.3	196.2	193.2	190.3	187.5	184.8
1950	224.4	217.5	210.9	207.7	204.5	201.4	198.3	195.3	192.4	189.6
2000	230.1	223.0	216.3	213.0	209.7	206.5	203.4	200.3	197.4	194.5
2050	235.9	228.6	221.7	218.3	215.0	211.7	208.5	205.3	202.3	199.3
2100	241.6	234.2	227.1	223.7	220.2	216.8	213.5	210.4	207.2	204.2
2150	247.4	239.8	232.5	229.0	225.5	222.0	218.6	215.4	212.2	209.1
2200	253.1	245.4	237.9	234.3	230.7	227.2	223.7	220.4	217.1	213.9
2250	258.9	250.9	243.4	239.6	225.9	232.3	228.8	225.4	222.0	218.8
2300	264.6	256.5	248.8	245.0	241.2	237.5	233.9	230.4	227.0	223.6
2350	270.4	262.1	254.2	250.3	246.4	242.7	239.0	235.4	231.9	228.5
2400	276.1	267.7	259.6	255.6	251.7	247.8	244.0	240.4	236.8	233.4
2450	281.9	273.2	265.0	260.9	256.9	253.0	249.1	245.4	241.8	238.2
2500	287.6	278.8	270.4	266.2	262.2	258.1	254.2	250.4	246.7	243.1
2550	293.4	284.4	275.8	271.6	267.4	263.3	259.3	255.4	251.6	247.9
2600	299.1	290.0	281.2	276.9	272.6	268.5	264.4	260.4	256.6	252.8
2650	304.9	295.5	286.6	282.2	277.9	273.6	269.5	265.4	261.5	257.7
2700	310.6	301.1	292.0	287.5	283.1	278.8	274.5	270.4	266.4	262.5
2750	316.4	306.7	297.4	292.9	288.4	283.9	279.6	275.4	271.4	267.4
2800	322.1	312.3	302.8	298.2	293.6	289.1	284.7	280.4	276.3	272.2
2850	327.9	317.8	308.2	303.5	298.8	294.3	289.8	285.4	281.2	277.1
2900	333.6	323.4	313.6	308.8	304.1	299.4	294.9	290.4	286.2	282.0
2950	339.4	329.0	319.0	314.2	309.3	304.6	299.9	295.5	291.1	286.8
3000	345.1	334.5	324.4	319.5	314.6	309.7	305.0	300.5	296.0	291.7

METRIC UNITS—TABLE II

Values of Correction to be Added

Temp.—alt.—factor	Barometer reading					
	780 mm	760 mm	740 mm	720 mm	700 mm	
1	0.9	0.9	0.9	0.8	0.8	
5	4.5	4.4	4.3	4.2	4.0	
10	9.0	8.8	8.6	8.3	8.1	
15	13.6	13.2	12.9	12.5	12.2	
20	18.2	17.7	17.2	16.8	16.3	
25	22.8	22.2	21.6	21.0	20.4	
30	27.4	26.7	26.0	25.3	24.6	
35	.....	31.2	30.4	29.6	28.8	
	760 mm	740 mm	720 mm	700 mm	680 mm	660 mm
40	35.8	34.9	33.9	33.0	32.0	31.1
45	40.4	39.3	38.3	37.2	36.2	35.1
50	45.0	43.8	42.7	41.5	40.3	39.1
55	49.7	48.4	47.1	45.8	44.5	43.1
60	.....	52.9	51.5	50.1	48.6	47.2
65	.....	57.5	55.9	54.4	52.8	51.3
70	.....	62.1	60.4	58.7	57.1	55.4
75	.....	66.7	64.9	63.1	61.3	59.5
	720 mm	700 mm	680 mm	660 mm	640 mm	
80	69.5	67.5	65.6	63.7	61.7	
85	74.0	72.0	69.9	67.9	65.8	
90	78.6	76.4	74.2	72.1	69.9	
95	83.2	80.9	78.6	76.3	74.0	
100	87.9	85.4	83.0	80.5	78.1	
105	.....	89.9	87.4	84.8	82.2	
110	.....	94.5	91.8	89.1	86.4	
115	.....	99.1	96.3	93.4	90.6	
120	.....	103.7	100.7	97.8	94.8	
125	.....	108.3	105.3	102.2	99.1	
	680 mm	660 mm	640 mm	620 mm	600 mm	
125	105.3	102.2	99.1	96.0	92.9	
130	109.8	106.6	103.3	100.1	96.9	
135	114.3	111.0	107.6	104.3	100.9	
140	118.9	115.4	111.9	108.4	104.9	
145	123.5	119.9	116.3	112.6	109.0	
150	128.2	124.4	120.6	116.9	113.1	
155	.....	128.9	125.0	121.1	117.2	
160	.....	133.5	129.4	125.4	121.4	
165	.....	138.1	133.9	129.7	125.5	
170	.....	142.7	138.4	134.0	129.7	

Temp.—alt.—factor	Barometer reading				
	640 mm	620 mm	600 mm	580 mm	560 mm
170	138.4	134.0	129.7	125.4	121.1
175	142.9	138.4	133.9	129.5	125.0
180	147.4	142.8	138.2	133.6	129.0
185	151.9	147.2	142.4	137.7	132.9
190	156.5	151.6	146.7	141.8	136.9
195	161.1	156.1	151.0	146.0	141.0
200	165.7	160.5	155.4	150.2	145.0
205	170.4	165.0	159.7	154.4	149.1
210	.....	169.6	164.1	158.6	153.2
215	.....	174.1	168.5	162.9	157.3
	620 mm	600 mm	580 mm	560 mm	540 mm
215	174.1	168.5	162.9	157.3	151.7
220	178.7	172.9	167.2	161.4	155.7
225	183.3	177.4	171.5	165.6	159.7
230	188.0	181.9	175.8	169.8	163.7
235	192.6	186.4	180.2	174.0	167.8
240	.....	191.0	184.6	178.2	171.9
245	.....	195.5	189.0	182.5	176.0
250	.....	200.1	193.4	186.8	180.1
255	.....	204.7	197.9	191.1	184.3
260	.....	209.4	202.4	195.4	188.4
	580 mm	560 mm	540 mm	520 mm	
260	202.4	195.4	188.4	181.5	
265	206.9	199.8	192.6	185.5	
270	211.5	204.2	196.9	189.6	
275	216.0	208.6	201.1	193.7	
280	220.6	213.0	205.4	197.8	
285	225.2	217.5	209.7	201.9	
290	229.9	222.0	214.0	206.1	
295	.....	226.5	218.4	210.3	
300	.....	231.0	222.8	214.5	
	560 mm	540 mm	520 mm	500 mm	480 mm
305	235.6	227.2	218.8	210.3	201.9
310	240.2	231.6	223.0	214.4	205.9
315	244.8	236.0	227.3	218.6	209.8
320	249.4	240.5	231.6	222.7	213.8
325	254.1	245.0	236.0	226.9	217.8
330	.....	249.6	240.3	231.1	221.8
335	.....	254.1	244.7	235.3	225.9
340	.....	258.7	249.1	239.6	230.0
345	.....	263.3	253.6	243.8	234.1

REDUCTION OF BAROMETER TO SEA LEVEL
(Continued)

ENGLISH UNITS—TABLE I

Values of the temperature-altitude factor (2000 m.) for entering table II.

Altitude feet	Assumed temperature of air column °F									
	−20	0	+10	+20	+30	+40	+50	+60	+70	+80
200	7.4	7.1	6.9	6.8	6.6	6.5	6.3	6.2	6.1	6.0
400	14.8	14.1	13.8	13.5	13.2	13.0	12.7	12.4	12.2	11.9
600	22.2	21.2	20.7	20.3	19.9	19.5	19.0	18.6	18.2	17.9
800	29.6	28.3	27.7	27.1	26.5	25.9	25.4	24.8	24.3	23.8
1000	37.0	35.3	34.6	33.8	33.1	32.4	31.7	31.1	30.4	29.8
1200	44.3	42.4	41.5	40.6	39.7	38.9	38.1	37.3	36.5	35.8
1400	51.7	49.5	48.4	47.4	46.4	45.4	44.4	43.5	42.6	41.7
1600	59.1	56.5	55.3	54.1	53.0	51.9	50.8	49.7	48.7	47.7
1800	66.5	63.6	62.2	60.9	59.6	58.4	57.1	55.9	54.7	53.6
2000	73.9	70.6	69.1	67.7	66.2	64.8	63.4	62.1	60.8	59.6
2200	81.3	77.7	76.0	74.4	72.9	71.3	69.8	68.3	66.9	65.5
2400	88.7	84.8	82.9	81.2	79.5	77.8	76.1	74.5	73.0	71.5
2600	96.1	91.8	89.9	87.9	86.1	84.3	82.5	80.7	79.1	77.5
2800	103.5	98.9	96.8	94.7	92.7	90.8	88.8	87.0	85.1	83.4
3000	110.9	106.0	103.7	101.5	99.3	97.2	95.2	93.2	91.2	89.4
3200	118.2	113.0	110.6	108.2	106.0	103.7	101.5	99.4	97.3	95.3
3400	125.6	120.1	117.5	115.0	112.6	110.2	107.9	105.6	103.4	101.3
3600	133.0	127.2	124.4	121.8	119.2	116.7	114.2	111.8	109.5	107.2
3800	140.4	134.2	131.3	128.5	125.8	123.2	120.5	118.0	115.5	113.2
4000	147.8	141.3	138.2	135.3	132.4	129.6	126.9	124.2	121.6	119.2
4200	155.2	148.3	145.1	142.1	139.1	136.1	133.2	130.4	127.7	125.1
4400	162.6	155.4	152.0	148.8	145.7	142.6	139.6	136.6	133.8	131.1
4600	170.0	162.5	159.0	155.6	152.3	149.1	145.9	142.8	139.9	137.0
4800	177.3	169.5	165.9	162.3	158.9	155.6	152.2	149.0	145.9	143.0
5000	184.7	176.6	172.8	169.1	165.6	162.0	158.6	155.2	152.0	148.9
5200	192.1	183.7	179.7	175.9	172.2	168.5	164.9	161.5	158.1	154.9
5400	199.5	190.7	186.6	182.6	178.8	175.0	171.3	167.7	164.2	160.8
5600	206.9	197.8	193.5	189.4	185.4	181.5	177.6	173.9	170.3	166.8
5800	214.3	204.8	200.4	196.2	192.0	188.0	184.0	180.1	176.3	172.8
6000	221.7	211.9	207.3	202.9	198.7	194.4	190.3	186.3	182.4	178.7
6200	229.1	219.0	214.2	209.7	205.3	200.9	196.6	192.5	188.5	184.7
6400	236.4	226.0	221.1	216.4	211.9	207.4	203.0	198.7	194.6	190.6
6600	243.8	233.1	228.0	223.2	218.5	213.9	209.3	204.9	200.7	196.6
6800	251.2	240.1	235.0	230.0	225.1	220.4	215.7	211.1	206.7	202.5
7000	258.6	247.2	241.9	236.7	231.8	226.8	222.0	217.3	212.8	208.5
7200	266.0	254.3	248.8	243.5	238.4	233.3	228.4	223.5	218.9	214.4
7400	273.4	261.3	255.7	250.2	245.0	239.8	234.7	229.7	225.0	220.4
7600	280.8	268.4	262.6	257.0	251.6	246.3	241.0	235.9	231.1	226.4
7800	288.1	275.4	269.5	263.8	258.2	252.8	247.4	242.2	237.1	232.3
8000	295.5	282.5	276.4	270.5	264.8	259.2	253.7	248.4	243.2	238.3
8200	302.9	289.6	283.3	277.3	271.5	265.7	260.1	254.6	249.3	244.2
8400	310.3	296.6	290.2	284.0	278.1	272.2	266.4	260.8	255.4	250.2
8600	317.7	303.7	297.1	290.8	284.7	278.7	272.7	267.0	261.4	256.1
8800	325.1	310.7	304.0	297.6	291.3	285.2	279.1	273.2	267.5	262.1
9000	332.5	317.8	310.9	304.3	297.9	291.6	285.4	279.4	273.6	268.0

ENGLISH UNITS—TABLE II

Value of Correction to be Added.

Temp alt. factor	Barometer reading					Temp. alt. factor	Barometer reading				
	31	30	29	28	27		26	25	24	23	22
	in.	in.	in.	in.	in.		in.	in.	in.	in.	in.
1	0.04	0.03	0.03	.....	.....	165	5.44	5.23	5.02		
5	0.18	0.17	0.17	.....	.....	170	5.62	5.40	5.19		
10	0.36	0.35	0.34	0.32	.....	175	.....	5.58	5.36		
15	0.54	0.52	0.51	0.49	.....	180	.....	5.76	5.53	5.30	
20	0.72	0.70	0.68	0.65	.....	185	.....	5.93	5.70	5.46	
25	.....	0.88	0.85	0.82	.....	190	.....	6.11	5.87	5.62	
30	.....	1.05	1.02	0.98	.....	195	.....	6.29	6.04	5.79	
35	.....	1.23	1.19	1.15	.....	200	.....	6.47	6.21	5.96	
40	.....	1.41	1.37	1.32	1.27	205	.....	.....	6.39	6.12	
45	.....	1.60	1.54	1.49	1.44	210	.....	.....	6.56	6.29	
50	.....	.....	1.72	1.66	1.60	215	.....	.....	6.74	6.46	
55	.....	.....	1.90	1.83	1.76	220	.....	.....	6.92	6.63	6.34
60	.....	.....	2.07	2.00	1.93	225	.....	.....	7.10	6.80	6.51
65	.....	.....	2.25	2.18	2.10	230	.....	.....	7.28	6.97	6.67
70	.....	.....	2.43	2.35	2.27	235	.....	.....	7.46	7.15	6.84
75	.....	.....	.....	2.53	2.43	240	.....	.....	.....	7.32	7.00
80	.....	.....	.....	2.70	2.60	245	.....	.....	.....	7.49	7 17

	28	27	26	25	24		23	22	21	20
	in.	in.	in.	in.	in.		in.	in.	in.	in.
75	2.53	2.43	2.34	.....	.....	250	7.67	7.34		
80	2.70	2.60	2.51	.....	.....	255	7.85	7.51		
85	2.88	2.78	2.67	.....	.....	260	8.03	7.68	7.33	
90	3.06	2.95	2.84	.....	.....	265	8.21	7.85	7.49	
95	3.24	3.12	3.01	.....	.....	270	8.39	8.02	7.66	
100	3.42	3.29	3.17	.....	....	275	8.57	8.19	7.82	
105	3.60	3.47	3.34	3.21	...	280	.....	8.37	7.99	

ENGLISH UNITS—TABLE II (Cont)

Value of Correction to be Added.

	28	27	26	25	24		23	22	21	20
110	.....	3.85	3.51	3.38	.....	285	.....	8.54	8.16	
115	.....	3.82	3.68	3.54	.....	290	.....	8.72	8.32	
120	.....	4.00	3.85	3.70	.....	295	.....	8.90	8.49	8.09
125	.....	4.18	4.02	3.87	.....	300	.....	9.08	8.66	8.25
130	.....	4.36	4.20	4.04	.....	305	.....	9.26	8.83	8.41
135	.....	4.54	4.37	4.20	.....	310	.....	9.44	9.01	8.58
140	.....	.....	4.55	4.37	4.20	315	.....	9.62	9.18	8.74
145			4.72	4.54	4.36	320	.....	9.80	9.35	8.91
150			4.90	4.71	4.52	325	.....	.....	9.53	9.08
155	.....	.....	5.08	4.88	4.69	330	.....	.....	9.71	9.24
160	.....	.....	5.26	5.06	4.85					

REDUCTION OF BAROMETER TO GRAVITY AT SEA LEVEL

METRIC UNITS

Correction to be subtracted given in millimeters

(From Smithsonian Physical Tables)

Height above sea level in meters	OBSERVED HEIGHT OF BAROMETER IN MILLIMETERS						
	500	550	600	650	700	750	800
100	.....	.....	.....	.....	.02	.02	.02
200	.....	.....	.....	.....	.04	.05	.05
300	.....	.....	.....	.....	.07	.07	.07
400	.....	.....	.....	.....	.09	.10	.10
500	.....	.....	.....	.....	.11	.12	.13
600	.....	.....	.....	.12	.13	.14	
700	.....	.....	.....	.14	.15	.16	
800	.....	.....	.....	.16	.18	.19	
900	.....	.....	.....	.18	.20	.22	
1000	.....	.18	.19	.20	.22	.24	
1100	.....	.19	.21	.22	.24		
1200	.....	.21	.23	.24	.26		
1300	.....	.22	.24	.26	.29		
1400	.....	.24	.26	.28	.31		
1500	.24	.26	.28	.30	.33		
1600	.25	.28	.30	.32			
1700	.27	.30	.32	.34			
1800	.28	.31	.34	.36			
1900	.30	.33	.36	.39			
2000	.31	.34	.38	.41			
2100	.33	.36	.40				
2200	.35	.38	.41				
2300	.36	.40	.43				
2400	.38	.42	.45				
2500	.39	.43	.47				

ENGLISH UNITS

Height above sea level in feet	OBSERVED HEIGHT IN INCHES						
	18	20	22	24	26	28	30
1000	.....	.....	.....	.....	.003	.003	.003
2000	.....	.....	.....	.004	.005	.005	.006
3000	.....	.....	.007	.007	.008	.008	
4000	.....	.....	.009	.009	.010		
4500	.....	.....	.010	.010	.011		
5000	.....	.010	.011	.011	.012		
5500	.....	.011	.012	.013			
6000	.....	.011	.013	.014			
6500	.011	.012	.014	.015			
7000	.012	.013	.015	.016			
7500	.013	.014	.016	.017			
8000	.014	.015	.017				
8500	.015	.016	.018				
9000	.016	.017	.019				
9500	.016	.018	.020				

REDUCTION OF BAROMETER TO LATITUDE 45°

METRIC SCALE

For latitudes below 45°, subtract the correction; for latitudes greater than 45° it is to be added. Corrections in cm.

(From Smithsonian Meteorological Tables.)

Latitude		Observed Height of Barometer in Centimeters					
		68	70	72	74	76	78
25°	65°	0.116	0.120	0.123	0.127	0.130	0.133
26	64	.111	.115	.118	.121	.125	.128
27	63	.106	.110	.113	.116	.119	.122
28	62	.101	.104	.107	.110	.113	.116
29	61	.096	.099	.102	.104	.107	.110
30	60	0.091	0.094	0.096	0.098	0.101	0.104
31	59	.085	.087	.090	.092	.095	.097
32	58	.079	.082	.084	.086	.089	.091
33	57	.074	.076	.078	.080	.082	.084
34	56	.068	.070	.072	.074	.076	.078
35	55	0.062	0.064	0.066	0.067	0.069	0.071
36	54	.056	.058	.059	.061	.063	.064
37	53	.050	.051	.053	.054	.056	.057
38	52	.044	.045	.046	.048	.049	.050
39	51	.038	.039	.040	.041	.042	.043
40	50	0.031	0.032	0.033	0.034	0.035	0.036
41	49	.025	.026	.027	.027	.028	.029
42	48	.019	.019	.020	.021	.021	.022
43	47	.013	.013	.013	.014	.014	.014
44	46	.006	.007	.007	.007	.007	.007

ENGLISH SCALE

Corrections in inches.

Latitude		Observed Height in Inches					
		25	26	27	28	29	30
25°	65°	0.043	0.044	0.046	0.048	0.050	0.051
26	64	.041	.043	.044	.046	.048	.049
27	63	.039	.041	.042	.044	.045	.047
28	62	.037	.039	.040	.042	.043	.045
29	61	.035	.037	.038	.039	.041	.042
30	60	0.033	0.035	0.036	0.037	0.039	0.040
31	59	.031	.032	.034	.035	.036	.037
32	58	.029	.030	.032	.033	.034	.035
33	57	.027	.028	.029	.030	.031	.032
34	56	.025	.026	.027	.028	.029	.030
35	55	0.023	0.024	0.025	0.025	0.026	0.027
36	54	.021	.021	.022	.023	.024	.025
37	53	.018	.019	.020	.021	.021	.022
38	52	.016	.017	.017	.018	.019	.019
39	51	.014	.014	.015	.015	.016	.017
40	50	0.012	0.012	0.012	0.013	0.013	0.014
41	49	.009	.010	.010	.010	.011	.011
42	48	.007	.007	.008	.008	.008	.008
43	47	.005	.005	.005	.005	.005	.006
44	46	.002	.002	.003	.003	.003	.003

RELATIVE HUMIDITY—DEW-POINT

The table gives the relative humidity of the air for temperature *t* and dewpoint *d*.

(From Smithsonian Meteorological Tables.)

Depression of dew-point $t-d°$ F	Dew-point (*d*). −10	0	+10	+20	+30
0.0	100%	100%	100%	100%	100%
0.2	98	99	99	99	99
0.4	97	97	97	98	98
0.6	95	96	96	96	97
0.8	94	94	95	95	96
1.0	92	93	94	94	94
1.2	91	92	92	93	93
1.4	90	90	91	92	92
1.6	88	89	90	91	91
1.8	87	88	89	90	90
2.0	86	87	88	88	89
2.2	84	85	86	87	88
2.4	83	84	85	86	87
2.6	82	83	84	85	86
2.8	80	82	83	84	85

RELATIVE HUMIDITY—DEW-POINT (Continued)

Depression of dew-point $t-d°$ F	Dew-point (*d*). −10	0	+10	+20	+30
3.0	79	81	82	83	84
3.2	78	80	81	82	83
3.4	77	79	80	81	82
3.6	76	77	79	80	82
3.8	75	76	78	79	81
4.0	73	75	77	78	80
4.2	72	74	76	77	79
4.4	71	73	75	77	78
4.6	70	72	74	76	77
4.8	69	71	73	75	76
5.0	68	70	72	74	75
5.2	67	69	71	73	75
5.4	66	68	70	72	74
5.6	65	67	69	71	73
5.8	64	66	69	70	72
6.0	63	66	68	70	71
6.2	62	65	67	69	71
6.4	61	64	66	68	70
6.6	60	63	65	67	69
6.8	60	62	64	66	68
7.0	59	61	63	66	68
7.2	58	60	63	65	67
7.4	57	60	62	64	66
7.6	56	59	61	63	65
7.8	55	58	60	63	65
8.0	54	57	60	62	64
8.2	54	56	59	61	63
8.4	53	56	58	60	63
8.6	52	55	57	60	62
8.8	51	54	57	59	61
9.0	51	53	56	58	61
9.2	50	53	55	58	60
9.4	49	52	55	57	59
9.6	48	51	54	56	59
9.8	48	51	53	56	58
10.0	47	50	53	55	57
10.5	45	48	51	54	
11.0	44	47	49	52	
11.5	42	45	48	51	
12.0	41	44	47	49	
12.5	39	42	45	48	
13.0	38	41	44	46	
13.5	37	40	43	45	
14.0	35	38	41	44	
14.5	34	37	40	43	
15.0	33	36	39	42	
15.5	32	35	38	40	
16.0	31	34	37	39	
16.5	30	33	36	38	
17.0	29	32	35	37	
17.5	28	31	34	36	
18.0	27	30	33	35	
18.5	26	29	32	34	
19.0	25	28	31	33	
19.5	24	27	30	33	
20.0	24	26	29	32	
21.0	22	25	27		
22.0	21	23	26		
23.0	19	22	24		
24.0	18	21	23		
25.0	17	19	22		
26.0	16	18	21		
27.0	15	17	20		
28.0	14	16	19		
29.0	13	15	18		
30.0	12	14	17		

RELATIVE HUMIDITY FROM WET AND DRY BULB THERMOMETER (CENT. SCALE)

This table gives the approximate relative humidity directly from the reading of the air temperature (dry bulb) (t°C) and the wet bulb (t'°C). It is computed for a barometric pressure of 74.27 cm Hg. Errors resulting from the use of this table for air temperatures above −10°C and between 77.5 and 71 cm Hg will usually be within the errors of observation.

Condensed from Bulletin of the U. S. Weather Bureau No. 1071

t \ $t - t'$	0.2	0.4	0.6	0.8	1.0	1.2	1.4	1.6	1.8	2.0	2.2	2.4	2.6	2.8	3.0	3.2	3.4	3 6	3.8	4.0	4.5	5.0	5.5	6.0	6.5	7.0	7.5	8.0	8.5	9.0	9.5	10.0	10.5	11 0
−10	93	87	80	74	67	61	54	48	41	35	28	22	16	9	..	..	..	..	..	..	..	..	..	..	..	..	..	..	..	..	..	..	..	..
−9	94	88	81	75	69	63	57	51	45	39	33	27	21	15	9	..	..	..	..	..	..	..	..	..	..	..	..	..	..	..	..	..	..	..
−8	94	88	83	77	71	65	60	54	48	43	37	32	26	20	15	10	..	..	..	..	..	..	..	..	..	..	..	..	..	..	..	..	..	..
−7	95	89	84	78	73	67	62	57	52	46	41	36	31	25	20	15	10	5	..	..	..	..	..	..	..	..	..	..	..	..	..	..	..	..
−6	95	90	85	79	74	69	64	59	54	49	45	40	35	30	25	20	15	11	6	..	..	..	..	..	..	..	..	..	..	..	..	..	..	..
−5	95	90	86	81	76	71	66	62	57	52	48	43	39	34	29	25	20	16	11	7	..	..	..	..	..	..	..	..	..	..	..	..	..	..
−4	95	91	86	82	77	73	68	64	59	55	51	46	42	38	33	29	25	21	17	12	..	..	..	..	..	..	..	..	..	..	..	..	..	..
−3	96	91	87	82	78	74	70	66	62	57	53	49	45	41	37	33	29	25	21	17	8	..	..	..	..	..	..	..	..	..	..	..	..	..
−2	96	92	88	84	79	75	71	68	64	60	56	52	48	44	40	37	33	29	25	22	12	..	..	..	..	..	..	..	..	..	..	..	..	..
−1	96	92	88	84	81	77	73	69	66	62	58	54	51	47	43	40	36	33	29	26	17	8	..	..	..	..	..	..	..	..	..	..	..	..
0	96	93	89	85	81	78	74	71	67	64	60	57	53	50	46	43	40	36	33	29	21	13	5	..	..	..	..	..	..	..	..	..	..	..
1	97	93	90	86	83	80	76	73	70	66	63	59	56	53	49	46	43	40	36	33	25	17	10	..	..	..	..	..	..	..	..	..	..	..
2	97	93	90	87	84	81	78	74	71	68	65	62	59	55	52	49	46	43	40	37	29	22	14	7	..	..	..	..	..	..	..	..	..	..
3	97	94	91	88	84	82	78	76	72	70	67	64	61	58	55	52	49	46	43	40	33	26	19	12	5	..	..	..	..	..	..	..	..	..
4	97	94	91	88	85	82	79	77	74	71	68	65	62	60	57	54	51	48	46	43	36	29	22	16	9	..	..	..	..	..	..	..	..	..
5	97	94	91	88	86	83	80	77	75	72	69	67	64	61	58	56	53	51	48	45	39	33	26	20	13	7	..	..	..	..	..	..	..	..
6	97	94	92	89	86	84	81	78	76	73	70	68	65	63	60	58	55	53	50	48	41	35	29	24	17	11	5	..	..	..	..	..	..	..
7	97	95	92	89	87	84	82	79	77	74	72	69	67	64	62	59	57	54	52	50	44	38	32	26	21	15	10	..	..	..	..	..	..	..
8	97	95	92	90	87	85	82	80	77	75	73	70	68	65	63	61	58	56	54	51	46	40	35	29	24	19	14	8	..	..	..	..	..	..
9	98	95	93	90	88	85	83	81	78	76	74	71	69	67	64	62	60	58	55	53	48	42	37	32	27	22	17	12	7	..	..	..	..	..
10	98	95	93	90	88	86	83	81	79	77	74	72	70	68	66	63	61	59	57	55	50	44	39	34	29	24	20	15	10	6	..	..	..	..
11	98	95	93	91	89	86	84	82	80	78	75	73	71	69	67	65	62	60	58	56	51	46	41	36	32	27	22	18	13	9	5	..	..	..
12	98	96	93	91	89	87	85	82	80	78	76	74	72	70	68	66	64	62	60	58	53	48	43	39	34	29	25	21	16	12	8	..	..	..
13	98	96	93	91	89	87	85	83	81	79	77	75	73	71	69	67	65	63	61	59	54	50	45	41	36	32	28	23	19	15	11	7	..	..
14	98	96	94	92	90	88	86	84	82	79	78	76	74	72	70	68	66	64	62	60	56	51	47	42	38	34	30	26	22	18	14	10	6	..
15	98	96	94	92	90	88	86	84	82	80	78	76	74	73	71	69	67	65	63	61	57	53	48	44	40	36	32	27	24	20	16	13	9	6

t \ $t - t'$	0.5	1.0	1.5	2.0	2.5	3.0	3.5	4.0	4.5	5.0	5.5	6.0	6.5	7 0	7.5	8.0	8.5	9.0	9.5	.0	10.5	11.0	11.5	12.0	12.5	13.0	13.5	14.0	14.5	15.0	16.0	17.0	18.0	19.0	20.0
16	95	90	85	81	76	71	67	63	58	54	50	46	42	38	34	30	26	23	19	15	12	8	5	..	..	..	..	..	..	..	..	..	..	..	..
17	95	90	86	81	76	72	68	64	60	55	51	47	43	40	36	32	28	25	21	18	14	11	8	..	..	..	..	..	..	..	..	..	..	..	..
18	95	91	86	82	77	73	69	65	61	57	53	49	45	41	38	34	30	27	23	20	17	14	10	7	..	..	..	..	..	..	..	..	..	..	..
19	95	91	87	82	78	74	70	65	62	58	54	50	46	43	39	36	32	29	26	22	19	16	13	10	7	..	..	..	..	..	..	..	..	..	..
20	96	91	87	83	78	74	70	66	63	59	55	51	48	44	41	37	34	31	28	24	21	18	15	12	9	6	..	..	..	..	..	..	..	..	..
21	96	91	87	83	79	75	71	67	64	60	56	53	49	46	42	39	36	32	29	26	23	20	17	14	12	9	6	..	..	..	..	..	..	..	..
22	96	92	87	83	80	76	72	68	64	61	57	54	50	47	44	40	37	34	31	28	25	22	19	17	14	11	8	6	..	..	..	..	..	..	..
23	96	92	88	84	80	76	72	69	65	62	58	55	52	48	45	42	39	36	33	30	27	24	21	19	16	13	11	8	6	..	..	..	..	..	..
24	96	92	88	84	80	77	73	69	66	62	59	56	53	49	46	43	40	37	34	31	29	26	23	20	18	15	13	10	8	5	..	..	..	..	..
25	96	92	88	84	81	77	74	70	67	63	60	57	54	50	47	44	41	39	36	33	30	28	25	22	20	17	15	12	10	8	..	..	..	..	..
26	96	92	88	85	81	78	74	71	67	64	61	58	54	51	49	46	43	40	37	34	32	29	26	24	21	19	17	14	12	10	5	..	..	..	..
27	96	92	89	85	82	78	75	71	68	65	62	58	56	52	50	47	44	41	38	36	33	31	28	26	23	21	18	16	14	12	7	..	..	..	..
28	96	93	89	85	82	78	75	72	69	65	62	59	56	53	51	48	45	42	40	37	34	32	29	27	25	22	20	18	16	13	9	5	..	..	..
29	96	93	89	86	82	79	76	72	69	66	63	60	57	54	52	49	46	43	41	38	36	33	31	28	26	24	22	19	17	15	11	7	..	..	..
30	96	93	89	86	83	79	76	73	70	67	64	61	58	55	52	50	47	44	42	39	37	35	32	30	28	25	23	21	19	17	13	9	5	..	..
31	96	93	90	86	83	80	77	73	70	67	64	61	59	56	53	51	48	45	43	40	38	36	33	31	29	27	25	22	20	18	14	11	7	..	..
32	96	93	90	86	83	80	77	74	71	68	65	62	60	57	54	51	49	46	44	41	39	37	35	32	30	28	26	24	22	20	16	12	9	5	..
33	97	93	90	87	83	80	77	74	71	68	66	63	60	57	55	52	50	47	45	42	40	38	36	33	31	29	27	25	23	21	17	14	10	7	..
34	97	93	90	87	84	81	78	75	72	69	66	63	61	58	56	53	51	48	46	43	41	39	37	35	32	30	28	26	24	23	19	15	12	8	5
35	97	94	90	87	84	81	78	75	72	69	67	64	61	59	56	54	51	49	47	44	42	40	38	36	34	32	30	28	26	24	20	17	13	10	7
36	97	94	90	87	84	81	78	75	73	70	67	64	62	59	57	54	52	50	48	45	43	41	39	37	35	33	31	29	27	25	21	18	15	11	8
37	97	94	91	87	84	82	79	76	73	70	68	65	63	60	58	55	53	51	48	46	44	42	40	38	36	34	32	30	28	26	23	19	16	13	10
38	97	94	91	88	84	82	79	76	74	71	68	66	63	61	58	56	54	51	49	47	45	43	41	39	37	35	33	31	29	27	24	20	17	14	11
39	97	94	91	88	85	82	79	77	74	71	69	66	64	61	59	57	54	52	50	48	46	43	42	39	38	36	34	32	30	28	25	22	18	15	12
40	97	94	91	88	85	82	80	77	74	72	69	67	64	62	59	57	54	53	51	48	46	44	42	40	38	36	35	33	31	29	26	23	20	16	14

REDUCTION OF PSYCHROMETRIC OBSERVATION

For the reduction of observations with the wet and dry bulb thermometer. Assuming the relative velocity of the air to the thermometer bulbs is at least three meters per second; if t is the temperature of the air as indicated by the dry bulb, t_w, the temperature of the wet bulb, B, the barometric pressure, and E_w, the vapor tension of water corresponding to t_w, then the actual vapor tension is

$$E = E_W - 0.00066B(t - t_W)\ [1 + 0.00115\ (t - t_W)]\ \text{millimeters}$$

The value of the term

$$0.00066B(t - t_W)\ [1 + 0.00115\ (t - t_W)]$$

is given in the following table.

(From Miller's Laboratory Physics, Ginn & Co., publishers, by permission.)

Barometric Pressure B in Millimeters

$t - t_w$	700	710	720	730	740	750	760	770
°C	mm	mm	mm	mm	mm	mm	mm	mm
1	0.463	0.469	0.476	0.482	0.489	0.496	0.502	0.509
2	0.926	0.939	0.953	0.966	0.979	0.992	1.006	1.019
3	1.391	1.411	1.431	1.450	1.470	1.490	1.510	1.530
4	1.857	1.883	1.910	1.936	1.963	1.989	2.016	2.042
5	2.323	2.356	2.390	2.423	2.456	2.489	2.522	2.556
6	2.791	2.831	2.871	2.911	2.951	2.990	3.030	3.070
7	3.260	3.307	3.353	3.400	3.446	3.493	3.539	3.586
8	3.730	3.783	3.837	3.890	3.943	3.996	4.050	4.103
9	4.201	4.261	4.321	4.381	4.441	4.501	4.561	4.621
10	4.673	4.740	4.807	4.873	4.940	5.007	5.074	5.140
11	5.146	5.220	5.293	5.367	5.440	5.514	5.587	5.661
12	5.621	5.701	5.781	5.861	5.942	6.022	6.102	6.183
13	6.096	6.183	6.270	6.357	6.444	6.531	6.618	6.705
14	6.572	6.666	6.760	6.854	6.948	7.042	7.135	7.229
15	7.050	7.150	7.251	7.352	7.452	7.553	7.654	7.754
16	7.528	7.636	7.743	7.851	7.958	8.066	8.173	8.281
17	8.008	8.122	8.236	8.351	8.465	8.580	8.694	8.808
18	8.488	8.609	8.731	8.852	8.973	9.094	9.216	9.337
19	8.970	9.098	9.226	9.354	9.482	9.610	9.739	9.867
20	9.453	9.588	9.723	9.858	9.993	10.128	10.263	10.398

CONSTANT HUMIDITY

The following table shows the % humidity and the aqueous tension at the given temperature within a closed space when an excess of the substance indicated is in contact with a saturated aqueous solution of the given solid phase.

Solid phase	t°C.	% humidity	Aq. tension mm Hg
$H_3PO_4.\frac{1}{2}H_2O$	24	9	1.99
$KC_2H_3O_2$	168	13	738
$LiCl.H_2O$	20	15	2.60
$KC_2H_3O_2$	20	20	3.47
KF	100	22.9	174
NaBr	100	22.9	174
NaCl, KNO_3 and $NaNO_3$	16.39	30.49	4.23
$CaCl_2.6H_2O$	24.5	31	7.08
$CaCl_2.6H_2O$	20	32.3	5.61
$CaCl_2.6H_2O$	18.5	35	5.54
CrO_3	20	35	6.08
$CaCl_2.6H_2O$	10	38	3.47
$CaCl_2.6H_2O$	5	39.8	2.59
$Zn(NO_3)_2.6H_2O$	20	42	7.29
$K_2CO_3.2H_2O$	24.5	43	9.82
$K_2CO_3.2H_2O$	18.5	44	6.96
KNO_2	20	45	7.81
KCNS	20	47	8.16
NaI	100	50.4	383
$Ca(NO_3)_2.4H_2O$	24.5	51	11.6
$NaHSO_4.H_2O$	20	52	9.03
$Na_2Cr_2O_7.2H_2O$	20	52	9.03
$Mg(NO_3)_2.6H_2O$	24.5	52	11.9
$NaClO_3$	100	54	410
$Ca(NO_3)_2.4H_2O$	18.5	56	8.86
$Mg(NO_3)_2.6H_2O$	18.5	56	8.86
KI	100	56.2	427
$NaBr.2H_2O$	20	58	10.1
$Mg(C_2H_3O_2)_2.4H_2O$	20	65	11.3
$NaNO_2$	20	66	11.5
NH_4Cl and KNO_3	30	68.6	21.6
KBr	100	69.2	526
NH_4Cl and KNO_3	25	71.2	16.7
NH_4Cl and KNO_3	20	72.6	12.6
$NaClO_3$	20	75	13.0
$(NH_4)_2SO_4$	108	75	754
$NaC_2H_3O_2.3H_2O$	20	76	13.2
$H_2C_2O_4.2H_2O$	20	76	13.2
$Na_2S_2O_3.5H_2O$	20	78	13.5
NH_4Cl	20	79.5	13.8
NH_4Cl	25	79.3	18.6
NH_4Cl	30	77.5	24.4
$(NH_4)_2SO_4$	20	81	14.1
$(NH_4)_2SO_4$	25	81.1	19.1
$(NH_4)_2SO_4$	30	81.1	25.6
KBr	20	84	14.6
Tl_2SO_4	104.7	84.8	768
$KHSO_4$	20	86	14.9
$Na_2CO_3.10H_2O$	24.5	87	20.9
$BaCl_2.2H_2O$	24.5	88	20.1
K_2CrO_4	20	88	15.3
$Pb(NO_3)_2$	103.5	88.4	760
$ZnSO_4.7H_2O$	20	90	15.6
$Na_2CO_3.10H_2O$	18.5	92	14.6
$NaBrO_3$	20	92	16.0
K_2HPO_4	20	92	16.0
$NH_4H_2PO_4$	30	92.9	29.3
$NH_4H_2PO_4$	25	93	21.9
$Na_2SO_4.10H_2O$	20	93	16.1
$NH_4H_2PO_4$	20	93.1	16.2
$ZnSO_4.7H_2O$	5	94.7	6.10
$Na_2SO_3.7H_2O$	20	95	16.5
$Na_2HPO_4.12H_2O$	20	95	16.5
NaF	100	96.6	734
$Pb(NO_3)_2$	20	98	17.0
$CuSO_4.5H_2O$	20	98	17.0
$TlNO_3$	100.3	98.7	759
TlCl	100.1	99.7	761

CONSTANT HUMIDITY WITH SULFURIC ACID SOLUTIONS

The relative humidity and pressure of aqueous vapor of air in equilibrium conditions above aqueous solutions of sulfuric acid are given below.

Density of acid solution	Relative humidity	Vapor pressure at 20°C	Density of acid solution	Relative humidity	Vapor pressure at 20°C
1.00	100.0	17.4	1.30	58.3	10.1
1.05	97.5	17.0	1.35	47.2	8.3
1.10	93.9	16.3	1.40	37.1	6.5
1.15	88.8	15.4	1.50	18.8	3.3
1.20	80.5	14.0	1.60	8.5	1.5
1.25	70.4	12.2	1.70	3.2	0.6

For concentration of sulfuric acid solution refer to tables relating density to percent composition.

VELOCITY OF SOUND

Compiled by Gordon E. Becker, Bell Telephone Laboratories

The data for the Velocity of Sound in Various Materials were compiled from a variety of sources. For more extensive tables one is referred to the following books:

AIP Handbook, Smithsonian Tables.
Mason: Physical Acoustics and the Properties of Solids (1958).
Chalmers and Quarrell: Physical Examination of Metals (1960).
Mason: Piezoelectric Crystals and their Application to Ultrasonics (1950).
Bergmann: Der Ultraschall (Hirzel, 1954).

Definition of Terms: V_l = Velocity of plane longitudinal wave in bulk material
V_s = Velocity of plane transverse (shear) wave
V_{ext} = Velocity of longitudinal wave (extensional wave) in thin rods.

Solids

Substance	Density gm/cc	V_l m/sec	V_s m/sec	V_{ext} m/sec
Metals:				
Aluminum, rolled	2.7	6420	3040	5000
Beryllium	1.87	12890	8880	12870
Brass (70 Cu, 30 Zn)	8.6	4700	2110	3480
Copper, annealed	8.93	4760	2325	3810
Copper, rolled	8.93	5010	2270	3750
Duralumin 17S	2.79	6320	3130	5150
Gold, hard-drawn	19.7	3240	1200	2030
Iron, electrolytic	7.9	5950	3240	5120
Iron, Armco	7.85	5960	3240	5200
Lead, annealed	11.4	2160	700	1190
Lead, rolled	11.4	1960	690	1210
Magnesium, drawn, annealed	1.74	5770	3050	4940
Molybdenum	10.1	6250	3350	5400
Monel metal	8.90	5350	2720	4400
Nickel (unmagnetized)	8.85	5480	2990	4800
Nickel	8.9	6040	3000	4900
Platinum	21.4	3260	1730	2800
Silver	10.4	3650	1610	2680
Steel, mild	7.85	5960	3235	5200
Steel, 347 Stainless	7.9	5790	3100	5000
Steel (1 % C)	7.84	5940	3220	5180
Steel (1 % C, hardened)	7.84	5854	3150	5070
Tin, rolled	7.3	3320	1670	2730
Titanium	4.5	6070	3125	5080
Tungsten, annealed	19.3	5220	2890	4620
Tungsten, drawn	19.3	5410	2640	4320
Tungsten Carbide	13.8	6655	3980	6220
Zinc, rolled	7.1	4210	2440	3850
Various:				
Fused silica	2.2	5968	3764	5760
Glass, pyrex	2.32	5640	3280	5170
Glass, heavy silicate flint	3.88	3980	2380	3720
Glass, light borate crown	2.24	5100	2840	4540
Lucite	1.18	2680	1100	1840
Nylon 6-6	1.11	2620	1070	1800
Polyethylene	0.90	1950	540	920
Polystyrene	1.06	2350	1120	2240
Rubber, butyl	1.07	1830		
Rubber, gum	0.95	1550		
Rubber, neoprene	1.33	1600		
Brick	1.8			3650
Clay rock	2.2			3480
Cork	0.25			500
Marble	2.6			3810
Paraffin	0.9			1300
Tallow				390
Woods:				
Ash, along the fiber				4670
Ash, across the rings				1390
Ash, along the rings				1260
Beech, along the fiber				3340
Elm, along the fiber				4120
Maple, along the fiber				4110
Oak, along the fiber				3850

Liquids

Substance	Formula	Density gm/cc	Velocity at 25°C m/sec	$-\Delta v/\Delta t$ m/sec °C
Acetone	C_3H_6O	0.79	1174	4.5
Benzene	C_6H_6	0.870	1295	4.65
Carbon disulphide	CS_2	1.26	1149	—
Carbon tetrachloride	CCl_4	1.595	926	2.7
Castor oil	$C_{11}H_{10}O_{10}$	0.969	1477	3.6
Chloroform	$CHCl_3$	1.49	987	3.4
Ethanol	C_2H_6O	0.79	1207	4.0
Ethanol amide	C_2H_7NO	1.018	1724	3.4
Ethyl ether	$C_4H_{10}O$	0.713	985	4.87
Ethylene glycol	$C_2H_6O_2$	1.113	1658	2.1
Glycerol	$C_3H_5O_3$	1.26	1904	2.2
Kerosene	—	0.81	1324	3.6
Mercury	Hg	13.5	1450	—
Methanol	CH_4O	0.791	1103	3.2
Nitrobenzene	$C_6H_5NO_2$	1.20	1463	3.6
Turpentine	—	0.88	1255	—
Water (distilled)	H_2O	0.998	1496.7±.2	−2.4
Water (sea)	—	1.025	1531	−2.4
Xylene hexafluoride	$C_8H_4F_6$	1.37	879	—

Gases and Vapors

Substance	Formula	Density gm/liter	Velocity m/sec	$\Delta v/\Delta t$ m/sec °C
Gases (0°C)				
Air, dry		1.293	331.45	0.59
Ammonia	NH_3	0.771	415	
Argon	Ar	1.783	319	0.56
Carbon monoxide	CO	1.25	338	0.6
Carbon dioxide	CO_2	1.977	259	0.4
Chlorine	Cl_2	3.214	206	
Deuterium	D_2		890	1.6
Ethane (10°C)	C_2H_6	1.356	308	
Ethylene	C_2H_4	1.260	317	
Helium	He	0.178	965	0.8
Hydrogen	H_2	0.0899	1284	2.2
Hydrogen bromide	HBr	3.50	200	
Hydrogen chloride	HCl	1.639	296	
Hydrogen iodide	HI	5.66	157	
Hydrogen sulfide	H_2S	1.539	289	
Illuminating (Coal gas)			453	
Methane	CH_4	0.7168	430	
Neon	Ne	0.900	435	0.8
Nitric oxide (10°C)	NO	1.34	324	
Nitrogen	N_2	1.251	334	0.6
Nitrous oxide	N_2O	1.977	263	0.5
Oxygen	O_2	1.429	316	0.56
Sulfur dioxide	SO_2	2.927	213	0.47
Vapors (97.1°C)				
Acetone	C_3H_6O		239	0.32
Benzene	C_6H_6		202	0.3
Carbon tetrachloride	CCl_4		145	
Chloroform	$CHCl_3$		171	0.24
Ethanol	C_2H_6O		269	0.4
Ethyl ether	$C_4H_{10}O$		206	0.3
Methanol	CH_4O		335	0.46
Water vapor (134°C)	H_2O		494	

SOUND VELOCITY IN WATER ABOVE 212°F

By permission from the Acoustical Society of America, Volume 31 (1959) and J. C. McDade, D. R. Pardue, A. L. Gedrich and F. Vrataric.

Temperature °F	Velocity m/sec.	Velocity ft/sec.
186.8	1552	5092
200	1548	5079
210	1544	5066
220	1538	5046
230	1532	5026
240	1524	5000
250	1516	4974
260	1507	4944
270	1497	4911
280	1487	4879
290	1476	4843
300	1465	4806
310	1453	4767
320	1440	4724
330	1426	4678
340	1412	4633
350	1398	4587
360	1383	4537
370	1368	4488
380	1353	4439
390	1337	4386
400	1320	4331
410	1302	4272
420	1283	4209
430	1264	4147
440	1244	4081
450	1220	4010
460	1200	3940
470	1180	3880

SOUND VELOCITY IN WATER ABOVE 212°F (Continued)

Temperature °F	Velocity m/sec.	Velocity ft/sec.
480	1160	3800
490	1140	3730
500	1110	3650
510	1090	3570
520	1070	3500
530	1040	3410
540	1010	3320
550	980	3230

MUSICAL SCALES

Equal Tempered Chromatic Scale
$A_4 = 440$

American Standard pitch. Adopted by the American Standards Association in 1936

Note	Frequency	Note	Frequency	Note	Frequency	Note	Frequency
C_0	16.35	C_2	65.41	C_4	261.63	C_6	1046.50
$C\#_0$	17.32	$C\#_2$	69.30	$C\#_4$	277.18	$C\#_6$	1108.73
D_0	18.35	D_2	73.42	D_4	293.66	D_6	1174.66
$D\#_0$	19.45	$D\#_2$	77.78	$D\#_4$	311.13	$D\#_6$	1244.51
E_0	20.60	E_2	82.41	E_4	329.63	E_6	1318.51
F_0	21.83	F_2	87.31	F_4	349.23	F_6	1396.91
$F\#_0$	23.12	$F\#_2$	92.50	$F\#_4$	369.99	$F\#_6$	1479.98
G_0	24.50	G_2	98.00	G_4	392.00	G_6	1567.98
$G\#_0$	25.96	$G\#_2$	103.83	$G\#_4$	415.30	$G\#_6$	1661.22
A_0	27.50	A_2	110.00	A_4	**440.00**	A_6	1760.00
$A\#_0$	29.14	$A\#_2$	116.54	$A\#_4$	466.16	$A\#_6$	1864.66
B_0	30.87	B_2	123.47	B_4	493.88	B_6	1975.53
C_1	32.70	C_3	130.81	C_5	523.25	C_7	2093.00
$C\#_1$	34.65	$C\#_3$	138.59	$C\#_5$	554.37	$C\#_7$	2217.46
D_1	36.71	D_3	146.83	D_5	587.33	D_7	2349.32
$D\#_1$	38.89	$D\#_3$	155.56	$D\#_5$	622.25	$D\#_7$	2489.02
E_1	41.20	E_3	164.81	E_5	659.26	E_7	2637.02
F_1	43.65	F_3	174.61	F_5	698.46	F_7	2793.83
$F\#_1$	46.25	$F\#_3$	185.00	$F\#_5$	739.99	$F\#_7$	2959.96
G_1	49.00	G_3	196.00	G_5	783.99	G_7	3135.96
$G\#_1$	51.91	$G\#_3$	207.65	$G\#_5$	830.61	$G\#_7$	3322.44
A_1	55.00	A_3	220.00	A_5	880.00	A_7	3520.00
$A\#_1$	58.27	$A\#_3$	233.08	$A\#_5$	932.33	$A\#_7$	3729.31
B_1	61.74	B_3	246.94	B_5	987.77	B_7	3951.07
						C_8	4186.01

Equal Tempered Chromatic Scale
$A_4 = 435$

International Pitch, adopted 1891

Note	Frequency	Note	Frequency	Note	Frequency	Note	Frequency
C_0	16.17	C_2	64.66	C_4	258.65	C_6	1034.61
$C\#_0$	17.13	$C\#_2$	68.51	$C\#_4$	274.03	$C\#_6$	1096.13
D_0	18.15	D_2	72.58	D_4	290.33	D_6	1161.31
$D\#_0$	19.22	$D\#_2$	76.90	$D\#_4$	307.59	$D\#_6$	1230.37
E_0	20.37	E_2	81.47	E_4	325.88	E_6	1303.53
F_0	21.58	F_2	86.31	F_4	345.26	F_6	1381.04
$F\#_0$	22.86	$F\#_2$	91.45	$F\#_4$	365.79	$F\#_6$	1463.16
G_0	24.22	G_2	96.89	G_4	387.54	G_6	1550.16
$G\#_0$	25.66	$G\#_2$	102.65	$G\#_4$	410.59	$G\#_6$	1642.34
A_0	27.19	A_2	108.75	A_4	**435.00**	A_6	1740.00
$A\#_0$	28.80	$A\#_2$	115.22	$A\#_4$	460.87	$A\#_6$	1843.47
B_0	30.52	B_2	122.07	B_4	488.27	B_6	1953.08
C_1	32.33	C_3	129.33	C_5	517.31	C_7	2069.22
$C\#_1$	34.25	$C\#_3$	137.02	$C\#_5$	548.07	$C\#_7$	2192.26
D_1	36.29	D_3	145.16	D_5	580.66	D_7	2322.62
$D\#_1$	38.45	$D\#_3$	153.80	$D\#_5$	615.18	$D\#_7$	2460.73
E_1	40.74	E_3	162.94	E_5	651.76	E_7	2607.05
F_1	43.16	F_3	172.63	F_5	690.52	F_7	2762.08
$F\#_1$	45.72	$F\#_3$	182.89	$F\#_5$	731.58	$F\#_7$	2926.32
G_1	48.44	G_3	193.77	G_5	775.08	G_7	3100.33
$G\#_1$	51.32	$G\#_3$	205.29	$G\#_5$	821.17	$G\#_7$	3284.68
A_1	54.38	A_3	217.50	A_5	870.00	A_7	3480.00
$A\#_1$	57.61	$A\#_3$	230.43	$A\#_5$	921.73	$A\#_7$	3686.93
B_1	61.03	B_3	244.14	B_5	976.54	B_7	3906.17
						C_8	4138.44

Scientific or Just Scale
$C_4 = 256$

Note	Frequency	Note	Frequency	Note	Frequency	Note	Frequency
C_0	16	C_2	64	C_4	**256**	C_6	1024
D_0	18	D_2	72	D_4	288	D_6	1152
E_0	20	E_2	80	E_4	320	E_6	1280
F_0	21.33	F_2	85.33	F_4	341.33	F_6	1365.33
G_0	24	G_2	96	G_4	384	G_6	1536
A_0	26.67	A_2	106.67	A_4	426.67	A_6	1706.67
B_0	30	B_2	120	B_4	480	B_6	1920
C_1	32	C_3	128	C_5	512	C_7	2048
D_1	36	D_3	144	D_5	576	D_7	2304
E_1	40	E_3	160	E_5	640	E_7	2560
F_1	42.67	F_3	170.67	F_5	682.67	F_7	2730.67
G_1	48	G_3	192	G_5	768	G_7	3072
A_1	53.33	A_3	213.33	A_5	853.33	A_7	3413.33
B_1	60	B_3	240	B_5	960	B_7	3840
						C_8	4096

ABSORPTION AND VELOCITY OF SOUND IN STILL AIR

The following data refer only to the temperature 20C (68F). They were abstracted from an extensive compilation prepared by L. B. Evans and H. E. Bass. The entire report, Tables of Absorption and Velocity of Sound in Still Air at 68F (20C), AD-738 576 is available from National Technical Information Service, U. S. Department of Commerce, 5285 Port Royal Road, Springfield, Va. 22151.

Frequency (Hz)	Absorption (dB/1000 ft)	Absorption (dB/Km)	Absorption (dB/sec)	Velocity (1000 ft/sec)
Relative Humidity = 0%				
20.	0.154	0.51	0.174	1.126892
40.	0.327	1.07	0.368	1.127013
50.	0.384	1.26	0.433	1.127050
63.	0.436	1.43	0.491	1.127085
100.	0.509	1.67	0.573	1.127131
200.	0.560	1.84	0.631	1.127161
400.	0.596	1.96	0.672	1.127169
630.	0.645	2.11	0.727	1.127171
800.	0.692	2.27	0.780	1.127172
1250.	0.861	2.82	0.970	1.127172
2000.	1.262	4.14	1.423	1.127173
4000.	2.696	8.84	3.039	1.127178
6300.	4.541	14.89	5.118	1.127182
10000.	8.013	26.28	9.032	1.127184
12500.	10.918	35.81	12.306	1.127186
16000.	15.901	52.15	17.923	1.127187
20000.	22.978	75.37	25.901	1.127187
40000.	81.405	267.01	91.759	1.127188
63000.	196.544	644.66	221.542	1.127188
80000.	314.677	1032.14	354.700	1.127188
Relative Humidity = 5%				
20.	0.031	0.10	0.034	1.126973
40.	0.074	0.24	0.083	1.126996
50.	0.092	0.30	0.104	1.127004
63.	0.114	0.37	0.129	1.127009
100.	0.179	0.59	0.202	1.127019
200.	0.449	1.47	0.506	1.127028
400.	1.451	4.76	1.635	1.127043
630.	3.211	10.53	3.619	1.127067
800.	4.774	15.66	5.380	1.127088
1250.	9.164	30.06	10.329	1.127147
2000.	15.175	49.77	17.106	1.127228
4000.	22.685	74.41	25.573	1.127321
6300.	26.245	86.08	29.587	1.127352
10000.	30.781	100.96	34.701	1.127365
12500.	34.306	112.52	38.676	1.127369
16000.	40.263	132.06	45.391	1.127372
20000.	48.653	159.58	54.850	1.127373
40000.	115.903	380.16	130.666	1.127378
63000.	242.070	793.99	272.905	1.127381
80000.	367.063	1203.97	413.821	1.127383
Relative Humidity = 10%				
20.	0.021	0.07	0.024	1.127167
40.	0.064	0.21	0.072	1.127183
50.	0.084	0.28	0.095	1.127191

ABSORPTION AND VELOCITY OF SOUND IN STILL AIR (*Continued*)

Frequency (Hz)	Absorption (dB/1000 ft)	Absorption (dB/Km)	Absorption (dB/sec)	Velocity (1000 ft/sec)
Relative Humidity = 10%				
63.	0.108	0.35	0.122	1.127199
100.	0.161	0.53	0.181	1.127213
200.	0.289	0.95	0.326	1.127225
400.	0.706	2.32	0.796	1.127230
630.	1.501	4.92	1.692	1.127234
800.	2.297	7.54	2.590	1.127238
1250.	5.155	16.91	5.811	1.127254
2000.	11.658	38.24	13.141	1.127287
4000.	31.023	101.76	34.975	1.127386
6300.	47.085	154.44	53.087	1.127462
10000.	61.578	201.98	69.431	1.127522
12500.	68.146	223.52	76.837	1.127540
16000.	76.231	250.04	85.955	1.127555
20000.	85.605	280.78	96.525	1.127563
40000.	151.938	498.36	171.321	1.127576
63000.	277.191	909.19	312.555	1.127581
80000.	403.662	1324.01	455.162	1.127583
Relative Humidity = 20%				
20.	0.013	0.04	0.014	1.127568
40.	0.045	0.15	0.051	1.127577
50.	0.066	0.22	0.074	1.127582
63.	0.093	0.30	0.105	1.127587
100.	0.164	0.54	0.185	1.127603
200.	0.285	0.93	0.321	1.127624
400.	0.476	1.56	0.537	1.127633
630.	0.789	2.59	0.890	1.127636
800.	1.103	3.62	1.244	1.127637
1250.	2.277	7.47	2.568	1.127640
2000.	5.310	17.42	5.987	1.127645
4000.	18.991	62.29	21.416	1.127670
6300.	41.151	134.98	46.406	1.127710
10000.	79.657	261.28	89.836	1.127778
12500.	103.004	337.85	116.170	1.127817
16000.	130.465	427.93	147.146	1.127859
20000.	155.809	511.05	175.736	1.127893
40000.	251.952	826.40	284.192	1.127960
63000.	382.062	1253.16	430.957	1.127977
80000.	508.369	1667.45	573.431	1.127982
Relative Humidity = 30%				
20.	0.009	0.03	0.010	1.127976
40.	0.034	0.11	0.038	1.127980
50.	0.051	0.17	0.057	1.127984
63.	0.075	0.25	0.085	1.127987
100.	0.151	0.50	0.170	1.127999
200.	0.309	1.01	0.349	1.128023
400.	0.484	1.59	0.546	1.128037
630.	0.682	2.24	0.770	1.128041
800.	0.868	2.85	0.979	1.128044
1250.	1.552	5.09	1.751	1.128045
2000.	3.333	10.93	3.760	1.128047
4000.	11.856	38.89	13.374	1.128056

ABSORPTION AND VELOCITY OF SOUND IN STILL AIR (*Continued*)

Frequency (Hz)	Absorption (dB/1000 ft)	Absorption (dB/Km)	Absorption (dB/sec)	Velocity (1000 ft/sec)
Relative Humidity = 30%				
6300.	27.626	90.61	31.164	1.128072
10000.	62.493	204.98	70.498	1.128105
12500.	89.659	294.08	101.148	1.128131
16000.	128.814	422.51	145.324	1.128166
20000.	171.847	563.66	193.879	1.128204
40000.	338.710	1110.97	382.172	1.128316
63000.	499.838	1639.47	563.998	1.128361
80000.	635.085	2083.08	716.614	1.128375
Relative Humidity = 40%				
20.	0.007	0.02	0.008	1.128386
40.	0.027	0.09	0.030	1.128388
50.	0.041	0.13	0.046	1.128390
63.	0.062	0.20	0.070	1.128392
100.	0.134	0.44	0.151	1.128402
200.	0.318	1.04	0.359	1.128424
400.	0.524	1.72	0.592	1.128443
630.	0.692	2.27	0.781	1.128449
800.	0.829	2.72	0.935	1.128451
1250.	1.309	4.29	1.477	1.128453
2000.	2.544	8.34	2.870	1.128456
4000.	8.523	27.96	9.618	1.128460
6300.	19.995	65.58	22.564	1.128467
10000.	47.390	155.44	53.478	1.128484
12500.	70.857	232.41	79.962	1.128499
16000.	108.308	355.25	122.228	1.128521
20000.	154.838	507.87	174.742	1.128549
40000.	380.371	1247.62	429.310	1.128663
63000.	596.091	1955.18	672.827	1.128733
80000.	753.514	2471.53	850.536	1.128758
Relative Humidity = 50%				
20.	0.006	0.02	0.006	1.128795
40.	0.022	0.07	0.025	1.128797
50.	0.034	0.11	0.038	1.128798
63.	0.052	0.17	0.058	1.128800
100.	0.117	0.38	0.132	1.128807
200.	0.313	1.03	0.353	1.128826
400.	0.563	1.85	0.636	1.128849
630.	0.734	2.41	0.828	1.128858
800.	0.851	2.79	0.961	1.128860
1250.	1.231	4.04	1.390	1.128862
2000.	2.176	7.14	2.457	1.128865
4000.	6.752	22.15	7.622	1.128867
6300.	15.643	51.31	17.659	1.128871
10000.	37.521	123.07	42.357	1.128881
12500.	57.009	186.99	64.357	1.128890
16000.	89.594	293.87	101.143	1.128903
20000.	132.719	435.32	149.830	1.128922
40000.	382.722	1255.33	432.101	1.129020
63000.	654.606	2147.11	739.115	1.129099
80000.	844.024	2768.40	953.016	1.129133

Frequency (Hz)	Absorption (dB/1000 ft)	Absorption (dB/Km)	Absorption (dB/sec)	Velocity (1000 ft/sec)
Relative Humidity = 60%				
20.	0.005	0.02	0.005	1.129207
40.	0.018	0.06	0.021	1.129208
50.	0.029	0.09	0.032	1.129209
63.	0.044	0.15	0.050	1.129210
100.	0.103	0.34	0.117	1.129215
200.	0.301	0.99	0.339	1.129232
400.	0.593	1.94	0.669	1.129254
630.	0.782	2.57	0.884	1.129266
800.	0.896	2.94	1.012	1.129269
1250.	1.223	4.01	1.381	1.129273
2000.	1.997	6.55	2.256	1.129275
4000.	5.711	18.73	6.449	1.129277
6300.	12.962	42.51	14.638	1.129279
10000.	31.050	101.84	35.064	1.129286
12500.	47.462	155.67	53.598	1.129292
16000.	75.544	247.78	85.312	1.129300
20000.	113.958	373.78	128.694	1.129313
40000.	364.443	1195.37	411.598	1.129389
63000.	677.023	2220.64	764.675	1.129467
80000.	899.912	2951.71	1016.458	1.129508
Relative Humidity = 70%				
20.	0.004	0.01	0.005	1.129618
40.	0.016	0.05	0.018	1.129619
50.	0.025	0.08	0.028	1.129619
63.	0.039	0.13	0.044	1.129620
100.	0.092	0.30	0.104	1.129623
200.	0.284	0.93	0.321	1.129638
400.	0.611	2.01	0.691	1.129662
630.	0.829	2.72	0.937	1.129673
800.	0.947	3.11	1.070	1.129678
1250.	1.250	4.10	1.412	1.129683
2000.	1.915	6.28	2.163	1.129685
4000.	5.056	16.58	5.712	1.129687
6300.	11.197	36.73	12.649	1.129689
10000.	26.624	87.33	30.077	1.129693
12500.	40.758	133.69	46.044	1.129697
16000.	65.246	214.01	73.708	1.129704
20000.	99.365	325.92	112.254	1.129712
40000.	339.153	1112.42	383.165	1.129770
63000.	673.667	2209.63	761.137	1.129842
80000.	924.481	3032.30	1044.556	1.129884
Relative Humidity = 80%				
20.	0.004	0.01	0.004	1.130030
40.	0.014	0.05	0.016	1.130030
50.	0.022	0.07	0.025	1.130031
63.	0.034	0.11	0.039	1.130032
100.	0.082	0.27	0.093	1.130034
200.	0.267	0.88	0.302	1.130047
400.	0.620	2.03	0.701	1.130069
630.	0.870	2.85	0.983	1.130082
800.	0.998	3.27	1.128	1.130088

ABSORPTION AND VELOCITY OF SOUND IN STILL AIR (*Continued*)

Frequency (Hz)	Absorption (dB/1000 ft)	Absorption (dB/Km)	Absorption (dB/sec)	Velocity (1000 ft/sec)
Relative Humidity = 80%				
1250.	1.293	4.24	1.461	1.130094
2000.	1.887	6.19	2.132	1.130096
4000.	4.626	15.17	5.228	1.130098
6300.	9.976	32.72	11.274	1.130100
10000.	23.466	76.97	26.519	1.130102
12500.	35.896	117.74	40.566	1.130106
16000.	57.588	188.89	65.081	1.130110
20000.	88.144	289.11	99.612	1.130116
40000.	313.525	1028.36	354.334	1.130161
63000.	655.282	2149.33	740.615	1.130223
80000.	925.566	3035.86	1046.134	1.130264
Relative Humidity = 90%				
20.	0.003	0.01	0.004	1.130443
40.	0.013	0.04	0.014	1.130443
50.	0.019	0.06	0.022	1.130444
63.	0.031	0.10	0.035	1.130444
100.	0.074	0.24	0.084	1.130446
200.	0.250	0.82	0.283	1.130457
400.	0.621	2.04	0.702	1.130478
630.	0.903	2.96	1.021	1.130493
800.	1.045	3.43	1.182	1.130498
1250.	1.344	4.41	1.520	1.130505
2000.	1.892	6.21	2.139	1.130508
4000.	4.337	14.22	4.903	1.130511
6300.	9.099	29.84	10.286	1.130512
10000.	21.133	69.32	23.891	1.130514
12500.	32.258	105.81	36.468	1.130516
16000.	51.762	169.78	58.518	1.130520
20000.	79.424	260.51	89.791	1.130526
40000.	290.117	951.58	327.995	1.130561
63000.	629.701	2065.42	711.948	1.130612
80000.	911.286	2989.02	1030.346	1.130651
Relative Humidity = 100%				
20.	0.003	0.01	0.003	1.130856
40.	0.011	0.04	0.013	1.130856
50.	0.018	0.06	0.020	1.130857
63.	0.028	0.09	0.031	1.130857
100.	0.068	0.22	0.076	1.130858
200.	0.234	0.77	0.265	1.130868
400.	0.616	2.02	0.696	1.130888
630.	0.928	3.05	1.050	1.130902
800.	1.087	3.57	1.230	1.130909
1250.	1.399	4.59	1.582	1.130917
2000.	1.917	6.29	2.168	1.130921
4000.	4.140	13.58	4.682	1.130924
6300.	8.451	27.72	9.557	1.130925
10000.	19.357	63.49	21.891	1.130926
12500.	29.461	96.63	33.318	1.130929
16000.	47.227	154.90	53.410	1.130931
20000.	72.538	237.93	82.036	1.130935
40000.	269.598	884.28	304.905	1.130963
63000.	601.714	1973.62	680.543	1.131007
80000.	888.113	2913.01	1004.492	1.131042

VELOCITY OF SOUND IN DRY AIR

Data in this table apply only to dry air. These data have been calculated with air being treated as a perfect gas.

Temp °C	0 m sec⁻¹	1 m sec⁻¹	2 m sec⁻¹	3 m sec⁻¹	4 m sec⁻¹	5 m sec⁻¹	6 m sec⁻¹	7 m sec⁻¹	8 m sec⁻¹	9 m sec⁻¹
60	366.05	366.60	367.14	367.69	368.24	368.78	369.33	369.87	370.42	370.96
50	360.51	361.07	361.62	362.18	362.74	363.29	363.84	364.39	364.95	365.50
40	354.89	355.46	356.02	356.58	357.15	357.71	358.27	358.83	359.39	359.95
30	349.18	349.75	350.33	350.90	351.47	352.04	352.62	353.19	353.75	354.32
20	343.37	343.95	344.54	345.12	345.70	346.29	346.87	347.44	348.02	348.60
10	337.46	338.06	338.65	339.25	339.84	340.43	341.02	341.61	342.20	342.78
0	331.45	332.06	332.66	333.27	333.87	334.47	335.07	335.67	336.27	336.87
−10	325.33	324.71	324.09	323.47	322.84	322.22	321.60	320.97	320.34	319.72
−20	319.09	318.45	317.82	317.19	316.55	315.92	315.28	314.64	314.00	313.36
−30	312.72	312.08	311.43	310.78	310.14	409.49	308.84	308.19	307.53	306.88
−40	306.22	305.56	304.91	304.25	303.58	302.92	302.26	301.59	300.92	300.25
−50	299.58	298.91	298.24	297.56	296.89	296.21	295.53	294.85	294.16	293.48
−60	292.79	292.11	291.42	290.73	290.03	289.34	288.64	287.95	287.25	286.55
−70	285.84	285.14	284.43	283.73	283.02	282.30	281.59	280.88	280.16	279.44
−80	278.72	278.00	277.27	276.55	275.82	275.09	274.36	273.62	272.89	272.15
−90	271.41	270.67	269.92	269.18	268.43	267.68	266.93	266.17	265.42	264.66

ELECTRICITY AND MAGNETISM

SPARK-GAP VOLTAGES

Based on results of the American Institute of Electric Engineers Air at 760 mm, 25° C.

Peak voltage, kilovolts	Diameter of spherical electrodes, cm				Needle points
	2.5	5	10	25	
	Length of spark gap cm				
5	0.13	0.15	0.15	0.16	0.42
10	0.27	0.29	0.30	0.32	0.85
15	0.42	0.44	0.46	0.48	1.30
20	0.58	0.60	0.62	0.64	1.75
25	0.76	0.77	0.78	0.81	2.20
30	0.95	0.94	0.95	0.98	2.69
35	1.17	1.12	1.12	1.15	3.20
40	1.41	1.30	1.29	1.32	3.81
45	1.68	1.50	1.47	1.49	4.49
50	2.00	1.71	1.65	1.66	5.20
60	2.82	2.17	2.02	2.01	6.81
70	4.05	2.68	2.42	2.37	8.81
80	......	3.26	2.84	2.74	11.1
90	......	3.94	3.28	3.11	13.3
100	......	4.77	3.75	3.49	15.5
110	......	5.79	4.25	3.88	17.7
120	......	7.07	4.78	4.28	19.8
130	......	......	5.35	4.69	22.0
140	......	......	5.97	5.10	24.1
150	......	......	6.64	5.52	26.1
160	......	......	7.37	5.95	28.1
170	......	......	8.16	6.39	30.1
180	......	......	9.03	6.84	32.0
190	......	......	10.0	7.30	33.9
200	......	......	11.1	7.76	35.7
210	......	......	12.3	8.24	37.6
220	......	......	13.7	8.73	39.5
230	......	......	15.3	9.24	41.4
240	......	......	......	9.76	43.3
250	......	......	......	10.3	45.2
300	......	......	......	13.3	54.7

CORRECTIONS FOR TEMPERATURE AND PRESSURE

Values found in the above table may be corrected for temperature and pressure by multiplying the values given by the appropriate correction factor found below:

Temp. °C.	Pressure mm 720	740	760	780
0	1.04	1.06	1.09	1.12
10	1.00	1.02	1.05	1.08
20	0.96	0.99	1.02	1.04
30	0.93	0.96	0.98	1.01

DIELECTRIC CONSTANTS

Dielectric Constants of Pure Liquids

The values listed in the following table were obtained from National Bureau of Standards Circular 514

The dielectric constants are intended to be the limiting values at low frequencies. the so-called static values. Temperature is the only variable considered explicitly. Usually pressure is atmospheric or insignificantly different with respect to its effect on the dielectric constant. Where data are listed above the normal boiling point, the pressure corresponds to the vapor pressure of the liquid unless otherwise noted in the footnote.

Symbols

ϵ = dielectric constant ($\epsilon_{vacuum=1}$)
t = temperature, Centigrade (°C)
T = temperature, Absolute (°K)
$a = -d\epsilon/dt$
$\alpha = -d \log_{10} \epsilon/dt$
t_1,t_2 = the limits of temperature between which a or α is considered applicable
mp = melting point
bp = boiling point
f = frequency of alternating current in cycles per second

Standard Liquids

		$\epsilon_{20°C}$	$\epsilon_{25°C}$	a (or α)*
C_6H_{12}	Cyclohexane	2.023	2.015	0.0016
CCl_4	Carbon tetrachloride	2.238	2.228	.0020
C_6H_6	Benzene	2.284	2.274	.0020
C_6H_5Cl	Chlorobenzene	5.708	5.621	.00133(α)
$C_2H_4Cl_2$	1,2-Dichloroethane	10.65	10.36	.00240(α)
CH_4O	Methanol	33.62	32.63	.00260(α)

DIELECTRIC CONSTANTS

Dielectric Constants of Pure Liquids (Continued)

Standard Liquids (Continued)

		$\epsilon_{20°C}$	$\epsilon_{25°C}$	a (or α)*
$C_6H_5NO_2$	Nitrobenzene	35.74	34.82	.00225(α)
H_2O	Water	80.37	78.54	.00200(α)
H_2	Hydrogen	1.228 at 20.4°K		.0034
O_2	Oxygen	1.507 at 80.0°K		.0024

* The values of a or α given in this table are derived from data in the vicinity of room temperature and are not necessarily identical with the values listed in the following sections of this table. They may be used to calculate values of dielectric constant between 15° and 30°C without introducing significant error.

Inorganic Liquids

	Substance	ϵ	t°C	a (or α) × 10²	Range t_1,t_2
A	Argon	1.53_8	−191	0.34	−191, −184
$AlBr_3$	Aluminum bromide	3.38	100	0.33	100,240
AsH_3	Arsine	2.50	−100	0.43	−116, −72
BBr_3	Boron bromide	2.58	0	0.28	−70,80
Br_2	Bromine	3.09	20	0.7	0,50
CO_2	Carbon dioxide	1.60[c]	20	....	....
Cl_2	Chlorine	2.10_1	−50	0.31	−65, −33
		1.91	14	0.32	−22,14
		1.7_3	77		
		1.5_4	142		
D_2	Deuterium	1.277	20°K	0.4	18.8,21.2°K
D_2O	Deuterium oxide	78.25	25	([d])	0.4,98
F_2	Fluorine	1.54	−202	0.19	−216, −190
$GeCl_4$	Germanium tetrachloride	2.43_0	25	0.240	0,55
HBr	Hydrogen bromide	7.00	−85	0.26(α)	−85, −70
HCl	Hydrogen chloride	6.35	−15	0.288(α)	−85, −15
		12.	−113	....	....
		4.6	28	....	....
HF	Hydrogen fluoride	17_5.	−73	....	....
		13_4.	−42		
		11_1.	−27		
		84.	0		
HI	Hydrogen iodide	3.39	−50	0.8	−51, −37
H_2	Hydrogen	1.228	20.4°K	0.34	14,21°K
H_2O	Water	78.54	25	([e])	0,100
		34.5_9	200	([h])	100,370
H_2O_2	Hydrogen peroxide	84.2	0	([i])	−30,20
H_2S	Hydrogen sulfide	9.26	−85.5	....	....
		9.05	−78.5	....	....
He	Helium	1.055_5	2.06°K	....	....
		1.055_9	2.30[f]		
		1.055_3	2.63		
		1.053_9	3.09		
		1.051_8	3.58		
		1.048	4.19		
I_2	Iodine	6.8	400	....	$\epsilon_r = 13.3 - 0.016 (\pm 0.002) T$
		$11._7$	140		
		$13._0$	168		
NH_3	Ammonia	25.	−77.7	....	....
		22.4	−33.4	....	....
		18.9	5	....	....
		17.8	15		
		16.9	25		
		16.3	35		
NOBr	Nitrosyl bromide	$13._4$	15	....	....
NOCl	Nitrosyl chloride	$18._2$	12	....	....
N_2	Nitrogen	1.454	−203	0.29	−210, −195
N_2H_4	Hydrazine	$52._9$	20	0.21(α)	0,25
N_2O	Dinitrogen oxide	1.97	−90	....	....
		1.61	0	0.6	−6,14
N_2O_4	Dinitrogen tetroxide	2.5_6[b]	15	....	....
O_2	Oxygen	1.507	−193	0.24	−218, −183
P	Phosphorus	4.10	34	....	....
		4.06	46		
		3.86	85		
PCl_3	Phosphorus trichloride	3.43	25	0.84	17,60
PCl_5	Phosphorus pentachloride	2.8_6	160	....	....
$POCl_3$	Phosphoryl chloride	$13._3$	22	....	....
$PSCl_3$	Thiophosphoryl chloride	5.8	22	....	....
$PbCl_4$	Lead tetrachloride	2.78	20	....	....
S	Sulfur	3.52	118	([g])	....
		3.48	231		
$SOBr_2$	Thionyl bromide	9.06	20	3.0	at 20

[c] At pressure of 50 atmospheres.

[d] $\epsilon = 78.25[1 - 4.617(10^{-3})(t - 25) + 1.22(10^{-5})(t - 25)^2 - 2.7(10^{-8})(t - 25)^3]$; av. dev. ±0.04%.

[b] $f = 3.6 \times 10^8$ cycles/sec.

[e] $\epsilon = 78.54[1 - 4.579(10^{-3})(t - 25) + 1.19(10^{-5})(t - 25)^2 - 2.8(10^{-8})(t - 25)^3]$; av. dev. ±0.03%.

[f] Liquid transition and discontinuity in variation of dielectric constant with temperature at 2.295°K.

[g] Graphical data in the range 118°–350°C show a minimum near 160° and a broad maximum near 200°.

[h] $\epsilon = 5321/T + 233.76 - 0.9297T + 0.001417T^2 - 0.0000008292T^3$.

[i] $\epsilon = 84.2 - 0.62t + 0.0032t^2$.

Dielectric Constants of Pure Liquids (Continued)

Inorganic Liquids (Continued)

	Substance	ε	t°C	a (or α) × 10²	Range t_1,t_2
$SOCl_2$	Thionyl chloride	9.25	20	3.9	at 20
SO_2	Sulfur dioxide	17.6	−20	0.287(α)	−65,−15
		15.0_8	0	....	....
		$14._1$	20	7.7	14,140
		2.1_0	154[h]		
SO_3	Sulfur trioxide	3.11	18	....	....
S_2Cl_2	Sulfur monochloride	4.79	15	0.146(α)	−41,15
SO_2Cl_2	Sulfuryl chloride	$10._0$	22	....	....
$SbCl_5$	Antimony pentachloride	3.22	20	0.46	2,47
Se	Selenium	5.40	250	0.25	237,301
$SiCl_4$	Silicon tetrachloride	2.4_0	16	....	....
$SnCl_4$	Tin tetrachloride	2.87	20	0.30	−30,20
$TiCl_4$	Titanium tetrachloride	2.80	20	0.20	−20,20

Organic Liquids

	Substance	ε	t°C	a (or α) × 10²	Range t_1,t_2
	C_1				
CCl_4	Carbon tetrachloride	2.238	20	0.200	−10,60
CN_4O_8	Tetranitromethane	2.52_1	25	....	....
CO_2	Carbon dioxide	1.60_4[c]	0	....	....
CS_2	Carbon disulfide	2.641	20	0.268	−90,130
		3.001	−110		
		2.19	180		
$CHBr_3$	Bromoform	4.39	20	0.105(α)	10,70
$CHCl_3$	Chloroform	4.806	20	0.160(α)	0,50
		6.76	−60	....	....
		6.12	−40		
		5.61	−20		
		3.7_1	100	....	....
		3.3_3	140		
		2.9_3	180		
CHN	Hydrocyanic acid	$158._1$	0	(i)	−13,18
		$114._9$	20	0.63(α)	18,26
CH_2Br_2	Dibromomethane	7.77	10	....	....
		6.68	40		
CH_2Cl_2	Dichloromethane	9.08	20	(j)	−80,25
CH_2I_2	Diiodomethane	5.32	25	....	....
CH_2O_2	Formic acid	$58._5$[a]	16	....	....
CH_3Br	Bromomethane	9.82	0	(k)	−80,0
CH_3Cl	Chloromethane	12.6	−20	(l)	−70,−20
CH_3I	Iodomethane	7.00	20	(m)	−70,40
CH_3NO	Formamide	109.	20	72.	18,25
CH_4	Methane	1.70	−173	0.2	−181,−159
CH_4O	Methanol	32.63	25	0.264(α)	5,55
		64.	−113	....	....
		54.	−80		
		40.	−20		
CH_5N	Methylamine	11.4	−10	0.26(α)	−30,−10
		9.4	25	....	....
	C_2				
C_2HCl_3	Trichloroethylene	3.4_2	ca 16	....	....
C_2HCl_3O	Chloral	4.9_4	20	0.17(α)	15,45
		7.6	−40		
		4.2	62		
$C_2HCl_3O_2$	Trichloroacetic acid	4.6	60	....	....
$C_2H_2Br_2$	*cis*-1,2-Dibromoethylene	7.7_2	0	....	....
		7.0_8	25		
	trans-1,2-Dibromoethylene	2.9_7	0	....	....
		2.8_8	25		
$C_2H_2Br_4$	1,1,2,2-Tetrabromoethane	8.6	3	....	....
		7.0	22		
$C_2H_2Cl_2$	1,1-Dichloroethylene	4.6_7	16	....	....
	cis-1,2-Dichloroethylene	9.20	25	....	....
	trans-1,2-Dichloroethylene	2.14	25	....	....
$C_2H_2Cl_2O_2$	Dichloroacetic acid	8.2	22	....	....
		7.8	61		
C_2H_3ClO	Acetyl chloride	$16._9$	2	....	....
		$15._8$	22		
$C_2H_3ClO_2$	Chloroacetic acid	12.3	60	2.	60,80
C_2H_4O	Ethylene oxide	$13._9$	−1	....	....
	Acetaldehyde	$21._8$[a]	10	....	....
$C_2H_4O_2$	Acetic acid	6.15	20	....	....
		6.29	40		
		6.62	70		
	Methyl formate	8.5	20	5.	0,20
C_2H_5ClO	2-Chloroethanol (Ethylene chlorohydrin)	$25._8$	25	....	....
		$13._2$	132		
C_2H_5NO	Acetamide	59.[a]	83	....	....
$C_2H_5NO_2$	Nitroethane	28.0_6	30	11.4	30,35
C_2H_6O	Ethanol	24.30	25	....	....
		24.3[x]	25	0.270(α)	−5,70
		41.0[x]	−60	0.297(α)	−110,−20
	Methyl ether	5.02	25	2.38	25,100
		2.97	110		
		2.64	120		
		2.37	125		
		2.26	126.1		
		1.90	127.6[p]		

Organic Liquids (Continued)

	Substance	ε	t°C	a (or α) × 10²	Range t_1,t_2
$(C_2H_6OSi)_n$					
n = 4	Octamethylcyclotetrasiloxane	2.39	20	....	....
n = 5	Decamethylcyclopentasiloxane	2.50	20	....	....
n = 6	Dodecamethylcyclohexasiloxane	2.59	20	....	....
n = 7	Tetradecamethylcycloheptasiloxane	2.68	20	....	....
n = 8	Hexadecamethylcyclooctasiloxane	2.74	20	....	....
$C_2H_6O_2$	Glycol	$37._7$	25	0.224(α)	20,100
C_2H_7N	Dimethylamine	6.32	0	....	....
		5.26	25		
	C_3				
C_3H_6	Propene	1.87_5	20	....	....
		1.79_5	45		
		1.69_0	65		
		1.53_0	85		
		1.44_1	90		
		1.33_1	91.9[h]		
C_3H_6O	2-Propen-1-ol (Allyl alcohol)	$21._6$	15	....	....
	Acetone	20.7_0	25	0.205(α)	−60,40
		17.7	56	....	....
	Propionaldehyde	$18._5$[a]	17	....	....
$C_3H_6O_2$	Propionic acid	3.30	10	....	....
		3.44	40		
	Ethyl formate	7.1_6	25	....	....
	Methyl acetate	6.68	25	2.2	25,40
$C_3H_6O_3$	*dl*-Lactic acid	22.	17	....	....
$C_3H_7NO_2$	Ethyl carbamate (Urethan)	14.2	50	5.2	50,70
C_3H_8	Propane	1.61	0	0.20	−90,15
C_3H_8O	1-Propanol	20.1	25	0.293(α)	20,90
		38.	−80	....	....
		29.	−34		
	2-Propanol	18.3	25	0.310(α)	20,70
$C_3H_8O_2$	1,2-Propanediol	$32._0$	20	0.27(α)	at 20
	1,3-Propanediol	$35._0$	20	0.23(α)	at 20
$C_3H_8O_3$	Glycerol	42.5	25	0.208(α)	0,100
C_3H_9N	Isopropylamine	5.5[b]	20	....	....
	Trimethylamine	2.44	25	0.52	0,25
	C_4				
$C_4H_2O_3$	Maleic anhydride	50.[a]	60	....	....
C_4H_4O	Furan	2.95	25	....	....
C_4H_4S	Thiophene	2.76	16	....	....
C_4H_5N	Pyrrole	7.48	18	....	....
C_4H_5NS	Allyl isothiocyanate	$17._2$[b]	18	....	....
C_4H_6O	Vinyl ether	3.94	20	....	....
$C_4H_6O_3$	Acetic anhydride	$22._4$	1	....	....
		$20._7$	19		
C_4H_8O	2-Butanone	18.5_1	20	0.207(α)	−60,60
	Butyraldehyde	13.4	26	....	....
		10.8	77		
$C_4H_8O_2$	Butyric acid	2.97	20	−0.23	10,70
	Isobutyric acid	2.71	10	....	....
		2.73	40		
	Propyl formate	7.7_2[a]	19	....	....
	Ethyl acetate	6.02	25	1.5	at 25
		5.3_0	77	....	....
	Methyl propionate	5.5[n]	19	....	....
	1,4-Dioxane	2.209	25	0.170	20,50
C_4H_9NO	Morpholine	7.33	25	....	....
$C_4H_{10}O$	1-Butanol	17.8	20	0.300(α)	−40,20
		17.1	25	0.335(α)	25,70
		8.2	118	....	....
	2-Methyl-1-propanol	17.7	25	0.377(α)	20,90
		34.	−80	....	....
		26.	−34		
	2-Butanol	15.8	25	....	....
	2-Methyl-2-propanol	10.9	30	....	....
		8.49	50		
		6.89	70		
	Ethyl ether	4.335	20	2.0	at 20
		4.34[x]	20	0.217(α)	−40,30
		10.4	−116	....	....
		3.97	40	0.170(α)	40,140
		2.1_2	180		
		1.8_9	190		
		1.5_3	193.3[h]		

[a] $f = 4 \times 10^8$ cycles/sec.
[b] $f = 3.6 \times 10^8$ cycles/sec.
[c] At pressure of 50 atmospheres.
[h] Critical temperature.
[i] $\log_{10} \epsilon = 2.199 - 0.0079t + 0.00005t^2$.
[j] $\epsilon = (3320/T) - 2.24$.
[k] $\epsilon = (3320/T) - 2.34$.
[l] $\epsilon = 12.6 - 0.061(t + 20) + 0.0005(t + 20)^2$.
[m] $\epsilon = (2160/T) - 0.39$.
[n] $f = 5 \times 10^8$ cycles/sec.
[p] Critical temperature = 126.9°C.
[t] Mixture of *cis-trans* isomers.
[x] Value chosen to conform with the remainder of the tabulated data for this substance.

Dielectric Constants of Pure Liquids (Continued)

ORGANIC LIQUIDS (Continued)

	Substance	ε	t°C	a (or α) × 10²	Range t_1,t_2
$C_4H_{10}Zn$	Diethyl zinc	2.5_5	20	....	....
	C_5				
$C_5H_4O_2$	Furfural	$46._9$	1	....	....
		$41._9$	20		
		$34._9$	50		
C_5H_5N	Pyridine	12.3	25	....	....
		9.4	116	....	....
C_5H_8	1,3-Pentadiene[t]	2.32	25	....	....
	2-Methyl-1,3-butadiene (Isoprene)	2.10	25	0.24	−75,25
$C_5H_8O_2$	2,4-Pentanedione (Acetylacetone)	$25._7$[a]	20	....	....
C_5H_{10}	1-Pentene	2.100	20	....	....
	2-Methyl-1-butene	2.197	20	....	....
	Cyclopentane	1.965	20	....	....
	Ethylcyclopropane	1.933	20	....	....
$C_5H_{10}O$	2-Pentanone	15.4_5	20	0.195(α)	−40,80
		22.0	−60		
	3-Pentanone	17.0_0	20	0.225(α)	0,80
		19.4	−20		
		19.8	−40		
$C_5H_{10}O_2$	Valeric acid	2.6_6	20	....	....
	Isovaleric acid	2.6_4	20	....	....
	Methyl butyrate	5.6[n]	20	....	....
$C_5H_{11}N$	Piperidine	5.8[b]	22	....	....
C_5H_{12}	n-Pentane	1.844	20	0.160	−50,30
		2.011	−90		
		1.984	−70		
	2-Methylbutane	1.843	20	....	....
$C_5H_{12}O$	1-Pentanol	13.9	25	0.23(α)	15,35
	3-Methyl-1-butanol	14.7	25	....	....
		5.8_2	132		
	2-Methyl-2-butanol	5.82	25	....	....
	C_6				
$C_6H_4Cl_2$	o-Dichlorobenzene	9.93	25	0.194(α)	0,50
	m-Dichlorobenzene	5.04	25	0.120(α)	0,50
	p-Dichlorobenzene	2.41	50	0.18	50,80
C_6H_5Br	Bromobenzene	5.40	25	0.115(α)	0,70
C_6H_5Cl	Chlorobenzene	5.708	20	....	....
		5.621	25		
		5.71	20	0.130(α)	0,80
		7.28	−50		
		6.30	−20		
		4.21	130		
C_6H_5ClO	o-Chlorophenol	6.31	25	2.7	25,58
	p-Chlorophenol	9.47	55	3.7	55,65
C_6H_5I	Iodobenzene	4.63	20	....	....
$C_6H_5NO_2$	Nitrobenzene	34.82	25	0.225(α)	10,80
		20.8	130	0.164(α)	130,211
		24.9	90		
		22.7	110		
$C_6H_5NO_3$	o-Nitrophenol	$17._3$	50	6.4	50,60
C_6H_6	Benzene	2.284	20	0.200	10,60
		2.073	129	....	....
		1.966	182		
C_6H_6BrN	m-Bromoaniline	$13._0$[n]	19	....	....
C_6H_6ClN	m-Chloroaniline	$13._4$[n]	19	....	....
$C_6H_6N_2O_2$	o-Nitroaniline	$34._5$	90	3.	90,110
	p-Nitroaniline	$56._3$	160	6.	160,180
C_6H_6O	Phenol	9.78	60	0.32(α)	40,70
C_6H_7N	Aniline	6.89	20	0.148(α)	0,50
		5.93	70	....	....
		4.54	184.6	....	....
	2-Methylpyridine (α-Picoline)	9.8[b]	20	....	....
$C_6H_8N_2$	Phenylhydrazine	7.2	23	....	....
C_6H_{10}	Cyclohexene	2.220	25	....	....
		2.6_0	−105	....	....
$C_6H_{10}O$	Cyclohexanone	18.3	20	....	....
		$19._9$	−40	....	....
$C_6H_{10}O_3$	Ethyl acetoacetate	$15._7$[a]	22	....	....
C_6H_{12}	Cyclohexane	2.023	20	0.160	10,60
	Methylcyclopentane	1.985	20	....	....
	Ethylcyclobutane	1.965	20	....	....
$C_6H_{12}O$	Cyclohexanol	15.0	25	0.437(α)	20,66
		7.2_4	100		
		4.8_8	150		
$C_6H_{12}O_2$	Butyl acetate	5.01	20	1.4	20,40
		6.8_5	−73		
	Ethyl butyrate	5.10	18	1.0	at 20
$C_6H_{12}O_3$	Paraldehyde	13.9	25	....	....
		6.29	128		
C_6H_{14}	n-Hexane	1.890	20	0.155	−10,50
		2.044	−90		
		1.990	−50		
$C_6H_{14}O$	1-Hexanol	13.3	25	0.35(α)	15,35
		8.5_5	75		
	Propyl ether	3.3_9	26	....	....
	Isopropyl ether	3.88	25	1.8	0,25
$C_6H_{15}Al$	Triethyl aluminum	2.9	20	....	....
$C_6H_{15}N$	Dipropylamine	2.9[b]	21	....	....
	Triethylamine	2.42	25	....	....
$C_6H_{15}N$	Dipropylamine	2.9[b]	21	....	....
	Triethylamine	2.42	25	....	....
$C_6H_{18}OSi_2$	$(CH_3)_3Si[OSi(CH_3)_2]_nCH_3$				
n = 1	Hexamethyldisiloxane	2.17	20	....	....
n = 2	Octamethyltrisiloxane	2.30	20	....	....
n = 3	Decamethyltetrasiloxane	2.39	20	....	....
n = 4	Dodecamethylpentasiloxane	2.46	20	....	....
n = 5	Tetradecamethylhexa-siloxane	2.50	20	....	....
n = 66[w]		2.72	20	....	....
	C_7				
C_7H_5ClO	Benzoyl chloride	29.	0	....	....
		23.	20		
C_7H_5NO	Phenyl isocyanate	8.8[b]	20	....	....
C_7H_5NS	Phenyl isothiocyanate	$10._4$[a]	20	....	....
C_7H_6O	Benzaldehyde	$19._7$	0	....	....
		$17._8$	20		
$C_7H_6O_2$	Salicylaldehyde	$17._1$	30	7.	30,40.
C_7H_7Br	o-Bromotoluene	4.28	58	....	....
	m-Bromotoluene	5.36	58	....	....
	p-Bromotoluene	5.49	58	....	....
C_7H_7Cl	o-Chlorotoluene	4.45	20	....	....
		4.16	58	....	....
	m-Chlorotoluene	5.55	20	....	....
		5.04	58	....	....
	p-Chlorotoluene	6.08	20	....	....
		5.55	58	....	....
	α-Chlorotoluene	7.0	13	....	....
$C_7H_7NO_2$	o-Nitrotoluene	27.4	20	15.	at 20
		$21._6$	58	....	....
		11.8	222	....	....
	m-Nitrotoluene	$23._8$	20	....	....
		$21._9$	58	....	....
	p-Nitrotoluene	$22._2$	58	....	....
C_7H_8	Toluene	2.438	0	0.0455(α)	−90,0
		2.379	25	0.243	0,90
		2.15_7	127		
		2.04_2	181		
C_7H_8O	Benzyl alcohol	13.1	20	....	....
		9.47	70	....	....
		6.6	132	....	....
	o-Cresol	11.5	25	11.	25,30
	m-Cresol	11.8	25	0.41(α)	15,50
	p-Cresol	9.9_1	58	....	....
C_7H_9N	o-Toluidine	6.34	18	....	....
		5.71	58	....	....
		4.00	200	....	....
	m-Toluidine	5.95	18	....	....
		5.45	58	....	....
	p-Toluidine	4.98	54	....	....
	N-Methylaniline	5.97	22	....	....
	1-Heptene	2.05	20	....	....
	C_8				
C_8H_8	Styrene (Phenylethylene)	2.43	25	....	....
		2.32	75		
	Acetophenone	17.39	25	4.	at 25
		8.64	202	....	....
$C_8H_8O_2$	Phenyl acetate	5.23	20	0.7	at 20
	Methyl benzoate	6.59	20	0.14(α)	20,50
$C_8H_8O_3$	Methyl salicylate	9.41	30	3.1	30,40
C_8H_{10}	Ethylbenzene	2.412	20	....	....
	o-Xylene	2.568	20	0.266	−20,130
	m-Xylene	2.374	20	0.195	−40,180
	p-Xylene	2.270	20	0.160	20,130
C_8H_{18}	n-Octane	1.948	20	0.130	−50,50
		1.879	70		
		1.817	110		
	2,2,3-Trimethylpentane	1.96	20	....	....
	2,2,4-Trimethylpentane	1.940	20	0.142	−100,100
$C_8H_{18}O$	1-Octanol	10.3_4	20	0.410(α)	20,60
$C_8H_{20}O_4Si$	Tetraethyl silicate	4.1[b]	ca 20	....	....
	C_9				
C_9H_7N	Quinoline	9.00	25	....	....
		5.05	238		
	Isoquinoline	10.7	25	....	....
C_9H_8O	Cinnamaldehyde	16.9	24	....	....
$C_9H_{10}O_2$	Benzyl acetate	5.1[n]	21	....	....
	Ethyl benzoate	6.02	20	2.1	20,40
$C_9H_{10}O_3$	Ethyl salicylate	7.99	30	2.	30,40
C_9H_{12}	Isopropylbenzene (Cumene)	2.38_0	20	....	....
	1,3,5-Trimethylbenzene (Mesitylene)	2.27_9	20	....	....

[a] $f = 4 \times 10^8$ cycles/sec.
[b] $f = 3.6 \times 10^8$ cycles/sec.
[h] Critical temperature.
[n] $f = 5 \times 10^8$ cycles/sec.
[t] Mixture of *cis-trans* isomers.
[s] Value chosen to conform with the remainder of the tabulated data for this substance.
[w] Silicone oil of average molecular weight corresponding to this formula.

ORGANIC LIQUIDS (Continued)

	Substance	ε	t°C	a (or α) × 10²	Range t_1,t_2
C_9H_{20}	n-Nonane	1.972	20	0.135	−10,90
		2.059	−50		
		1.847	110		
		1.787	150		
	C_{10}				
$C_{10}H_8$	Naphthalene	2.54	85	....	....
$C_{10}H_{10}O_4$	Dimethyl phthalate	8.5	24	....	....
$C_{10}H_{16}$	d-Camphene	2.33	ca 40	....	....
	d-Pinene	2.64	25	....	....
	l-Pinene	2.76	20	....	....
	Terpinene	2.7[b]	21	....	....
	d-Limonene	2.3_6	20	....	....
	dl-Limonene (Dipentene)	2.3_0	20	....	....
$C_{10}H_{22}$	n-Decane	1.991	20	0.130	10,110
		2.050	−30		
		1.844	130		
		1.783	170		
$C_{10}H_{22}O$	1-Decanol	8.1	20		....
$C_{11}H_{24}$	n-Undecane	2.005	20	0.125	10,130
		2.039	−10		
		1.838	150		
		1.781	190		
	C_{12}				
$C_{12}H_{10}$	Diphenyl	2.53	75	0.18	75,155
$C_{12}H_{10}O$	Phenyl ether	3.65	30	0.7	30,50
$C_{12}H_{11}N$	Diphenylamine	3.3	52	....	....
$C_{12}H_{26}$	n-Dodecane	2.014	20	0.120	10,150
		2.047	−10		
		1.776	210		
	C_{13}				
$C_{13}H_{10}O$	Benzophenone	11.4	50	....	....
	C_{14}				
$C_{14}H_{15}N$	Dibenzylamine	3.6[b]	20	....	....
	C_{16}				
$C_{16}H_{32}O_2$	Palmitic acid	2.30	71	....	....
	C_{18}				
$C_{18}H_{32}O_2$	Linoleic acid	2.61	0	....	....
		2.71	20		
		2.70	70		
		2.60	120		
$C_{18}H_{34}O_2$	Oleic acid	2.46	20	....	....
		2.45	60		
		2.41	100		
$C_{18}H_{36}O_2$	Stearic acid	2.29	70	....	....
		2.26	100	....	....
	Ethyl palmitate	3.20	20	0.4	20,40
		2.71	104	....	....
		2.46	182		
	C_{19}				
$C_{19}H_{16}$	Triphenylmethane	2.45	100	0.14	94,175
$C_{19}H_{38}O_4$	Monopalmitin	5.34	67	....	....
		5.09	80		
	C_{20}				
$C_{20}H_{38}O_2$	Ethyl Oleate	3.17	28	0.48	28,122
$C_{20}H_{40}O_2$	Ethyl Stearate	2.98	40	0.6	32,50
		2.69	100		
		2.48	167		
	C_{21}				
$C_{21}H_{21}O_4P$	Tricresyl phosphate	6.9	40	....	....
	C_{22}				
$C_{22}H_{42}O_2$	Butyl oleate	4.0	25	....	....
$C_{22}H_{44}O_2$	Butyl stearate	3.11_1	30	0.53	30,35

[b] $f = 3.6 \times 10^8$ cycles/sec.
[a] $f = 5 \times 10^8$ cycles/sec.

DIELECTRIC CONSTANTS

Dielectric Constants of Solids

Compiled by Earle C. Gregg, Jr.

SOLIDS* (17 to 22°C)

Material	Frequency	Dielectric constant	Material	Frequency	Dielectric constant
Acetamide	4×10^8	4.0	Mercuric chloride	10^6	3.2
Acetanilide	....	2.9	Mercurous chloride	10^6	9.4
Acetic acid (2°C)	4×10^8	4.1	Naphthalene	4×10^8	2.52
Aluminum oleate	4×10^8	2.40	Phenanthrene	4×10^8	2.80
Ammonium bromide	10^6	7.1	Phenol (10°C)	4×10^8	4.3
Ammonium chloride	10^6	7.0	Phosphorus, red	10^8	4.1
Antimony trichloride	10^8	5.34	Phosphorus, yellow	10^8	3.6
Apatite ⊥ optic axis	3×10^8	9.50	Potassium aluminum sulfate	10^6	3.8
" ‖ optic axis	3×10^8	7.41			
Asphalt	$<3 \times 10^6$	2.68	Potassium carbonate (15°C)	10^8	5.6
Barium chloride (anhyd.)	6×10^7	11.4			
" " ($2H_2O$)	6×10^7	9.4	Potassium chlorate	6×10^7	5.1
Barium nitrate	6×10^7	5.9	Potassium chloride	10^4	5.03

SOLIDS* (17 to 22°C) (Continued)

Material	Frequency	Dielectric constant	Material	Frequency	Dielectric constant
Barium sulfate (15°C)	10^8	11.4	Potassium chromate	6×10^7	7.3
Beryl ⊥ optic axis	10^4	7.02	Potassium iodide	6×10^7	5.6
" ‖ optic axis	10^4	6.08	Potassium nitrate	6×10^7	5.0
Calcite ⊥ optic axis	10^4	8.5	Potassium sulfate	6×10^7	5.9
" ‖ optic axis	10^4	8.0	Quartz ⊥ optic axis	3×10^7	4.34
Calcium carbonate	10^6	6.14	" ‖ optic axis	3×10^7	4.27
Calcium fluoride	10^4	7.36	Resorcinol	4×10^8	3.2
Calcium sulfate ($2H_2O$)	10^4	5.66	Ruby ⊥ optic axis	10^4	13.27
Cassiterite ⊥ optic axis	10^{12}	23.4	" ‖ optic axis	10^4	11.28
Cassiterite ‖ optic axis	10^{12}	24.	Rutile ⊥ optic axis	10^8	86
d-Cocaine	5×10^8	3.10	" ‖ optic axis	10^8	170
Cupric oleate	4×10^8	2.80	Selenium	10^8	6.6
Cupric oxide (15°C)	10^8	18.1	Silver bromide	10^6	12.2
Cupric sulfate (anhyd.)	6×10^7	10.3	Silver chloride	10^6	11.2
Cupric sulfate ($5H_2O$)	6×10^7	7.8	Silver cyanide	10^6	5.6
Diamond			Smithsonite ⊥ optic axis	10^{12}	9.3
"	10^8	5.5	" ‖ optic axis	10^{10}	9.4
Diphenylmethane	4×10^8	2.7	Sodium carbonate (anhyd.)	6×10^7	8.4
Dolomite ⊥ optic axis	10^8	8.0	Sodium carbonate ($10H_2O$)	6×10^7	5.3
" ‖	10^8	6.8	Sodium chloride	10^4	6.12
Ferrous oxide (15°C)	10^8	14.2	Sodium nitrate	....	5.2
Iodine	10^8	4	Sodium oleate	4×10^8	2.75
Lead acetate	10^6	2.6	Sodium perchlorate	6×10^7	5.4
Lead carbonate (15°C)	10^8	18.6	Sucrose (mean)	3×10^8	3.32
Lead chloride	10^6	4.2	Sulfur (mean)	....	4.0
Lead monoxide (15°C)	10^8	25.9	Thallium chloride	10^6	46.9
Lead nitrate	6×10^7	37.7	p-Toluidine	4×10^8	3.0
Lead oleate	4×10^8	3.27	Tourmaline ⊥ optic axis	10^4	7.10
Lead sulfate	10^6	14.3	" ‖ optic axis	10^4	6.3
Lead sulfide (15°)	16^6	17.9	Urea	4×10^8	3.5
Malachite (mean)	10^{12}	7.2	Zircon ⊥, ‖	10^8	12

* For plastics and other insulating materials, refer to table on Properties of Dielectrics.

DIELECTRIC CONSTANTS

Table of Dielectric Constants of Reference Gases at 20°C and 1 Atmosphere

The listed values $(\epsilon - 1)$ refer to the gas at a temperature of 20°C and pressure of 1 atmosphere. The values can be adjusted to slightly different conditions without introducing more than 0.1% error by use of the following equation:

$$\frac{(\epsilon - 1)_{t,p}}{(\epsilon - 1)_{20°,1\ atm}} = \frac{p}{760[1 + 0.003411(t - 20)]}$$

where p = pressure in mm Hg
t = degrees C

t should be between 10 and 30°C and p between 700 and 800 mm Hg. From National Bureau of Standards Circular 537

Substance	$(\epsilon - 1) \cdot 10^6$	Reference
Helium	Radio frequency	
	67.8	Watson
	63.7	Hector
	64.5	Jelatis
	Microwave	
	65.6	Birnbaum
	65.2	Essen
	Optical	
	64.6	Koch
	64.5	Cuthbertson
Hydrogen	Radio frequency	
	$254._0$	Watson
	Microwave	
	253.4	Essen
	Optical	
	254.1	Cuthbertson
	253.6	Koch
	253.7	Kirn
	254.3	Tausz
Oxygen	Radio frequency	
	$494._3$	Watson
	496.2	Jelatis
	Microwave[a]	
	494.9	Birnbaum
	495.0	Essen
	494.9	Essen
	Optical	
	494.5	Cuthbertson
	493.5	Lowery
	494.7	Tausz
	494.4	Ladenberg
Air (dry, CO_2 free)	Radio frequency	
	$537._0$	Watson
	Microwave[a]	
	536.5	Birnbaum
	536.6	Essen
	536.6	Essen
	Optical	
	536.9	Koch
	535.8	Meggers
	536.0	Traub
	536.7	Quarder
	536.4	Lowery
	536.5	Tausz
	536.1	Koster
	536.3	Perard
	535.8	Barrell
Nitrogen	Radio frequency	
	$547._2$	Watson
	Microwave	
	547.3	Birnbaum
	548.0	Essen
	548.0	Essen
	Optical	
	548.9	Cuthbertson
	548.7	Koch
	547.2	Tausz
Carbon dioxide	Radio frequency	
	$921._5$	Watson
	Microwave	
	922.4	Birnbaum
	$920._6$	Essen

[a] These values were derived from measurements of the refractive index after making allowance for the magnetic permeability of oxygen.

Table of Dielectric Constants of Reference Gases at 20°C and 1 Atmosphere

Substance	$(\epsilon - 1) \cdot 10^6$	Reference
Argon.......	Radio frequency	
	513.0	Watson
	516.4	Jelatis
	Microwave	
	517.7	Essen
	Optical	
	516.8	Cuthbertson
	517.8	Quarder
	517.0	Tausz
	516.7	Damköhler

[a] These values were derived from measurements of the refractive index after making allowance for the magnetic permeability of oxygen.

Dielectric Constants of Gases at 760 MM Pressure

Compiled by Earle C. Gregg, Jr.

GASES AT 760 MM PRESSURE

Material	Temperature °C	Frequency cycles/sec.	Dielectric Constant
Acetalde-	100	$<3 \times 10^6$	
hyde.....	0	2×10^6	1.0213
Acetone.....			1.0159
Acetyl chlo-	0	$<3 \times 10^6$	
ride......			1.0217
Acetylene...	0	$<3 \times 10^6$	1.00134
Air.........			1.000590
Ammonia...	0	$<3 \times 10^6$	1.0072
β-Amylene..			1.0028
Argon......	23	10^{10}	1.000545
Benzene....	400	3×10^8	1.0028
Bromine....	0	$<3 \times 10^6$	1.0128
Butylene....			1.00319
Carbon diox-	100	$<3 \times 10^6$	
ide.......			1.000985
Carbon di-	23	10^{10}	
sulfide....			1.0029
Carbon	23	10^{10}	
monoxide.			1.00070
Carbon tet-	100	$<3 \times 10^6$	
rachloride.			1.0030
Chloroform .	100	$<3 \times 10^6$	1.0042
Dichlorodi-			
fluorome-	100	$<3 \times 10^6$	
thane.....	0	$<3 \times 10^6$	1.00029
Dichloro-	0	$<3 \times 10^6$	
fluorome-			
thane.....	100	$<3 \times 10^6$	1.00049
Dimethyla-			
mine.....	0	2×10^6	1.00040
Ethane.....	0	$<3 \times 10^6$	1.00150
Ethyl alcohol	100	$<3 \times 10^6$	1.0061
Ethylamine.			1.00053
Ethyl bro-	20	$<3 \times 10^6$	
mide.....			1.0139
Ethyl chlo-	20	$<3 \times 10^6$	
ride......			1.0132
Ethyl ether.	23	10^{10}	1.0049
Ethyl for-	126	$<3 \times 10^6$	
mate.....			1.0083
Ethyl iodide.	20	$<3 \times 10^6$	1.0140
" " .			1.0089
Ethylene....	110	$<3 \times 10^6$	1.00144
Helium.....	140	$<3 \times 10^6$	1.0000684

Material	Temperature °C	Frequency cycles/sec.	Dielectric Constant
n-Heptane..	20	$<3 \times 10^6$	1.0035
Hydrogen..	100	$<3 \times 10^6$	1.000264
Hydrogen			
bromide..	20	$<3 \times 10^6$	1.00313
Hydrogen	0	$<3 \times 10^6$	
chloride...	0	$<3 \times 10^6$	1.0046
Hydrogen	0	$<10^6$	
iodide....	100	$<3 \times 10^6$	1.00234
Hydrogen	0	$<10^6$	
sulfide....	100	$<3 \times 10^6$	1.00030
Mercury....	180	$<3 \times 10^6$	1.00074
Methane....	0	$<3 \times 10^6$	1.000944
Methyl alco-			
hol.......	0	$<10^6$	1.0057
Methyl bro-			
mide.....	0	$<3 \times 10^6$	1.00095
Methyl chlo-			
ride......	0	$<3 \times 10^6$	1.00094
Methyl io-			
dide......	110	$<3 \times 10^6$	1.0063
Methyl-	120	$<3 \times 10^6$	
amine....			1.0038
Methylene			
chloride...	23	10^{10}	1.0065
Neon.......			1.000127
Nitrogen....			1.000580
Nitrome-	23	10^{10}	
thane.....			1.0247
Nitrous ox-	23	2.5×10^{10}	
ide (N_2O).	0	$<3 \times 10^6$	1.00113
Oxygen.....	100	$<3 \times 10^6$	1.000523
n-Pentane...	23	10^{10}	1.0025
n-Propyl			
chloride...	20	$<3 \times 10^6$	1.0143
iso-Propyl			
chloride...	20	$<3 \times 10^6$	1.0152
Sulfur diox-	100	$<3 \times 10^6$	
ide.......			1.00075
Toluene.....	100	$<3 \times 10^6$	1.0043
Vinyl bro-	20	$<3 \times 10^6$	
mide.....	100	$<3 \times 10^6$	1.0081
Water	0	$<3 \times 10^6$	
(steam)...	0	$<3 \times 10^6$	1.0126
" ...			1.00785

PROPERTIES OF DIELECTRICS

In most cases properties have been determined by A S T M (American Society for Testing Materials) test methods at room temperature under standard conditions. Values will in general change considerably with temperature.

DIELECTRIC CONSTANTS OF SOME PLASTICS AND RUBBERS

Name	°C	Frequency (Hertz)		
		1×10^3	1×10^6	1×10^8
Plastics				
Phenol-formaldehyde	25–27	5.15–8.61	4.45–5.05	4.1–4.5
	57	6.35	4.90	4.5
	88	8.5	5.2	4.7
Phenol-aniline-formaldehyde	25	4.50	4.31	4.11
	79	4.75	4.51	4.35
Melamine-formaldehyde	24–28	6.0–6.90	5.82–6.20	5.5–5.55
	57	6.95	5.40	4.90
	88	11.8	6.0	5.5
Urea-formaldehyde	24	6.7	6.0	5.2
	80	7.8	6.8	—
Polyamide resins:				
Nylon 66	25	3.75	3.33	3.16
Nylon 610	25	3.50	3.14	3.0
	84	11.2	4.4	3.4
Cellulose acetate	26	3.50–4.48	3.28–3.90	3.05–3.40
Cellulose nitrate	27	8.4	6.6	5.2
	78	7.5	6.2	5.2
Methyl cellulose	22	6.8	5.7	4.3
Ethyl cellulose	25	3.09	3.01	2.90
Silicone resins	25	3.79–3.91	3.79–3.82	3.82
Polyethylene	−12	2.37	2.35	2.33
	23	2.26	2.26	2.26
Polyisobutylene	25	2.23	2.23	2.23
Vinylite QYNA	20	3.10	2.88	2.85
	76	3.83	3.0	2.8
	110	8.6	—	—
Vinylite 5544	25	7.20	4.13	3.05
Vinylite 5901	25	5.5	3.4	3.0
Vinylite VU	24	5.65	3.30	2.80
	79	8.15	5.5	3.4
Vinylite VYHW	20	3.12	2.91	2.83
Vinylite VYNW	20	3.15	2.90	2.8
Polyvinyl chloride	25	4.55 (1×10^4)	3.3	—
Polyvinylidene and vinyl chloride	23	4.65	3.18	2.82
	84	4.94	4.40	3.2
Polychlorotrifluoroethylene	25	2.76	2.48	2.36
Polytetrafluoroethylene (Teflon)	22	2.1	2.1	2.1
	100	2.04	2.04	—
Polyvinylalcohol acetate	25	7.8	5.2	—
	85	100.	10.	—
Polyvinylacetals	26–27	3.02–3.12	2.86–2.92	2.67
	88	3.5	3.1	2.85
Polyacrylates:				
Lucite	−12	2.9	2.63	2.50
	23	2.84	2.63	2.58
	81	3.45	2.72	2.59
Plexiglas	27	3.12	2.76	—
Polystyrene	25	2.54–2.56	2.54–2.56	2.55
	80	2.54	2.54	2.54
Styrene copolymers	25	2.55–2.95	2.55–2.80	2.55–2.77
Polyesters	25	3.22–4.3	3.12–4.0	2.94–2.98
Alkyd resins:				
Alkyd isocyanate foam	25	1.223	1.218	1.20
Plaskon, clay filled	25	5.26	4.92	4.77
Plaskon, glass filled	25	5.04	4.73	4.50
Epoxy resins	25	3.63–3.67	3.52–3.62	3.32–3.35
Rubbers				
Hevea, vulcanized	27	2.94	2.74	2.42
Hevea compound	27	36.	9.	6.8
Gutta percha	25	2.60	2.53	2.47
Balata	25	2.50	2.50	2.42
Buna S	20	2.66	2.56	2.52
Butyl rubber compound	25	2.42	2.40	2.39
Neoprene	24	6.60	6.26	4.5
Silicon rubber	25	3.12–3.30	3.10–3.20	3.06–3.18

DIELECTRIC CONSTANTS OF CERAMICS

Material	Dielectric Constant 10^6 Cycles	Dielectric Strength volts/mil	Volume Resistivity Ohms-cm 23°C	Loss Factor*
Alumina	4.5–8.4	40–160	10^{11}–10^{14}	0.0002–0.01
Corderite	4.5–5.4	40–250	10^{12}–10^{14}	0.004–0.012
Forsterite	6.2	240	10^{14}	0.0004
Porcelain (Dry Process)	6.0–8.0	40–240	10^{12}–10^{14}	0.003–0.02
Porcelain (Wet Process)	6.0–7.0	90–400	10^{12}–10^{14}	0.006–0.01
Porcelain, Zircon	7.1–10.5	250–400	10^{13}–10^{15}	0.0002–0.008
Steatite	5.5–7.5	200–400	10^{13}–10^{15}	0.0002–0.004
Titanates (Ba, Sr, Ca, Mg, and Pb)	15–12,000	50–300	10^{8}–10^{15}	0.0001–0.02
Titanium Dioxide	14–110	100–210	10^{13}–10^{18}	0.0002–0.005

DIELECTRIC CONSTANTS OF WAXES

Acrawax C	2.4	...	...	0.005
Beeswax, white	2.75–3.0	...	5×10^{14}	0.025
Beeswax, yellow	2.9	...	8×10^{14}	0.029
Candelilla	2.25–2.50			
Carnauba	2.75–3.0			
Cerese, brown G	2.27	...	...	0.0025
Ceresine	2.25–2.50	...	$>5 \times 10^{18}$	0.0011
Halowax 1001	~4.10	...	2×10^{13}	0.014
Halowax 1013	~4.75	...	...	0.036
Halowax 1014	~4.40	...	...	0.035
Halowax 11-314	2.94	...	...	0.00094
Microcrystalline waxes	2.2–2.5	...	...	
Opalwax	3.1	...	...	0.34
Ozokerite wax	2.3	100–150	5×10^{14}	0.0018
Paraffin	2.0–2.5	250	10^{15}–10^{19}	0.003 (900 cps)
Parawax	2.25	...	10^{16}	0.00045
135 A.M.P. wax	2.25	...	...	0.00023

DIELECTRIC CONSTANTS OF GLASSES

Type	Dielectric Constant at 100 mc 20°C	Volume Resistivity 350°C megohm-cm	Loss Factor*
Corning 0010	6.32	10	0.015
Corning 0080	6.75	0.13	0.058
Corning 0120	6.65	100	0.012
Pyrex 1710	6.00	2,500	0.025
Pyrex 3320	4.71	...	0.019
Pyrex 7040	4.65	80	0.013
Pyrex 7050	4.77	16	0.017
Pyrex 7052	5.07	25	0.019
Pyrex 7060	4.70	13	0.018
Pyrex 7070	4.00	1,300	0.0048
Vycor 7230	3.83	...	0.0061
Pyrex 7720	4.50	16	0.014
Pyrex 7740	5.00	4	0.040
Pyrex 7750	4.28	50	0.011
Pyrex 7760	4.50	50	0.0081
Vycor 7900	3.9	130	0.0023
Vycor 7910	3.8	1,600	0.00091
Vycor 7911	3.8	4,000	0.00072
Corning 8870	9.5	5,000	0.0085
G. E. Clear (Silica Glass)	3.81	4,000–30,000	0.00038
Quartz (Fused)	3.75–4.1 (1 mc)	...	0.0002 (1 mc)

* Power factor × dielectric constant equals loss factor.

DIELECTRIC CONSTANT OF WATER

From NSRDS-NBS 24

W. J. Hamer

t°C	ε*	ε†	t°C	ε*	ε†
0	87.74	87.90	50	69.91	69.88
5	85.76	85.90	55	68.34	68.30
10	83.83	83.95	60	66.81	66.76
15	81.95	82.04	65	65.32	65.25
18	80.84	80.93	70	63.86	63.78
20	80.10	80.18	75	62.43	62.34
25	78.30	78.36	80	61.03	60.93
30	76.55	76.58	85	59.66	59.55
35	74.83	74.85	90	58.32	58.20
38	73.82	73.83	95	57.01	56.88
40	73.15	73.15	100	55.72	55.58
45	71.51	71.50			

* From data of Malmberg and Maryott (1956).
† From data of Owen, Miller, Milner and Cogan (1961).

DIELECTRIC CONSTANT OF DEUTERIUM OXIDE

t	ε	$-\frac{d\varepsilon}{dt}$	$-\frac{1}{\varepsilon}\frac{d\varepsilon}{dt}$
°C			
4	85.877	0.3974	4.627 × 10^{-3}
5	85.480	.3956	4.628
10	83.526	.3862	4.624
15	81.618	.3771	4.620
20	79.755	.3681	4.615
25	77.936	.3594	4.611
30	76.161	.3509	4.607
35	74.427	.3425	4.602
40	72.735	.3344	4.597
45	71.083	.3265	4.593
50	69.470	.3187	4.587
55	67.896	.3112	4.583
60	66.358	.3038	4.578
65	64.857	.2967	4.575
70	63.391	.2898	4.571
75	61.959	.2830	4.567
80	60.561	.2765	4.565
85	59.194	.2701	4.563
90	57.859	.2640	4.563
95	56.554	.2581	4.564
100	55.278	.2523	4.564

DIELECTRIC CONSTANTS (Continued)

Dielectric Constant of Liquid Parahydrogen vs. Temperature (°K) and Pressure (atm)

R. J. Corruccini

P atm \ T, °K	20	21	22	23	24	25	26	27	28	29	30	31	32
1	1.2297												
2	1.2302	1.2260	1.2216										
3	1.2306	1.2265	1.2221	1.2174	1.2122								
4	1.2311	1.2270	1.2227	1.2180	1.2129	1.2073	1.2010						
5	1.2315	1.2275	1.2233	1.2186	1.2136	1.2081	1.2020	1.1950					
6	1.2320	1.2280	1.2238	1.2192	1.2143	1.2089	1.2029	1.1962	1.1883				
7	1.2324	1.2285	1.2243	1.2198	1.2150	1.2097	1.2039	1.1973	1.1897	1.1805			
8	1.2329	1.2290	1.2249	1.2204	1.2157	1.2105	1.2048	1.1984	1.1911	1.1824			
9	1.2333	1.2295	1.2254	1.2210	1.2163	1.2112	1.2056	1.1994	1.1924	1.1842	1.1734		
10	1.2337	1.2300	1.2259	1.2216	1.2169	1.2119	1.2065	1.2004	1.1936	1.1857	1.1758	1.1621	
15	1.2358	1.2322	1.2284	1.2243	1.2200	1.2153	1.2103	1.2049	1.1990	1.1924	1.1847	1.1758	1.1645
20	1.2378	1.2343	1.2307	1.2268	1.2227	1.2184	1.2137	1.2088	1.2034	1.1976	1.1913	1.1839	1.1757
25	1.2396	1.2363	1.2328	1.2291	1.2253	1.2211	1.2168	1.2122	1.2073	1.2021	1.1964	1.1903	1.1832
30	1.2414	1.2382	1.2349	1.2313	1.2276	1.2237	1.2196	1.2153	1.2107	1.2059	1.2008	1.1952	1.1891
35	1.2431	1.2400	1.2368	1.2334	1.2298	1.2261	1.2222	1.2181	1.2138	1.2093	1.2046	1.1995	1.1942
40	1.2448	1.2418	1.2386	1.2354	1.2319	1.2284	1.2246	1.2208	1.2167	1.2124	1.2080	1.2033	1.1984
45	1.2464	1.2434	1.2404	1.2372	1.2339	1.2305	1.2269	1.2232	1.2193	1.2153	1.2111	1.2067	1.2021
50	1.2479	1.2450	1.2421	1.2390	1.2358	1.2325	1.2291	1.2255	1.2218	1.2179	1.2139	1.2098	1.2055
60	1.2508	1.2481	1.2453	1.2424	1.2394	1.2363	1.2331	1.2297	1.2263	1.2227	1.2191	1.2153	1.2114
70	1.2535	1.2510	1.2483	1.2455	1.2427	1.2397	1.2367	1.2336	1.2303	1.2270	1.2236	1.2201	1.2165
80	1.2561	1.2536	1.2511	1.2484	1.2457	1.2429	1.2400	1.2371	1.2340	1.2309	1.2277	1.2244	1.2211
90	1.2585	1.2561	1.2537	1.2512	1.2486	1.2459	1.2431	1.2403	1.2374	1.2345	1.2315	1.2284	1.2252
100	1.2608	1.2586	1.2562	1.2538	1.2513	1.2487	1.2461	1.2434	1.2406	1.2378	1.2349	1.2320	1.2290
120	1.2652	1.2631	1.2609	1.2586	1.2563	1.2539	1.2514	1.2489	1.2464	1.2438	1.2412	1.2385	1.2357
140	1.2693	1.2672	1.2651	1.2630	1.2608	1.2586	1.2563	1.2540	1.2516	1.2492	1.2467	1.2442	1.2417
160	1.2730	1.2711	1.2691	1.2671	1.2650	1.2629	1.2607	1.2585	1.2563	1.2540	1.2517	1.2494	1.2470
180	1.2766	1.2747	1.2728	1.2709	1.2689	1.2669	1.2649	1.2628	1.2606	1.2585	1.2563	1.2541	1.2518
200	1.2799	1.2781	1.2763	1.2745	1.2726	1.2707	1.2687	1.2667	1.2647	1.2626	1.2605	1.2584	1.2563
220	1.2831	1.2814	1.2796	1.2779	1.2760	1.2742	1.2723	1.2704	1.2685	1.2665	1.2645	1.2625	1.2605
240		1.2845	1.2828	1.2811	1.2793	1.2775	1.2757	1.2739	1.2720	1.2701	1.2682	1.2663	1.2643
260		1.2874	1.2858	1.2841	1.2824	1.2807	1.2790	1.2772	1.2754	1.2736	1.2717	1.2699	1.2680
280			1.2886	1.2870	1.2853	1.2837	1.2821	1.2803	1.2786	1.2768	1.2751	1.2733	1.2714
300			1.2914	1.2898	1.2882	1.2866	1.2850	1.2833	1.2817	1.2800	1.2782	1.2765	1.2747
320				1.2925	1.2910	1.2894	1.2878	1.2862	1.2846	1.2829	1.2813	1.2796	1.2779
340				1.2951	1.2936	1.2921	1.2905	1.2890	1.2874	1.2858	1.2842	1.2825	1.2809

Note: Values below the stepped line represent an extrapolation of p with density.

Selected Values of Electric Dipole Moments for Molecules in the Gas Phase

Ralph D. Nelson, Jr., David R. Lide, Jr., and Arthur A. Maryott

The following table was abstracted from the publication, "Selected Values of Electric Dipole Moments for Molecules in the Gas Phase" compiled by Nelson, Lide and Maryott and published as part of the National Reference Data Series—National Bureau of Standards (NSRDS—NBS 10). The publication is available from the Superintendent of Documents, U.S. Government Printing Office, Washington, D.C., 20402. Those desiring a complete listing of all compounds in the NSRDS—NBS 10, discussion of the bibliographic procedure and the principal methods of dipole moment measurement should obtain the publication.

Values of the dipole moment, μ, are expressed in the cgs system of units, since this is the system universally used by workers in the field. The numerical values are in debye units, D, ($1\ D = 10^{-18}$ electrostatic units of charge × centimeters). The conversion factor to the Système International is $1\ D = 3.33564 \times 10^{-30}$ coulomb-meter.

Code symbol	Estimated accuracy of value
A	$\pm$ 1% or, for $\mu < 1.0$ D, $\pm$ 0.01 D
B	$\pm$ 2% or, for $\mu < 1.0$ D, $\pm$ 0.02 D
C	$\pm$ 5% or, for $\mu < 1.0$ D, $\pm$ 0.05 D
S	$\mu \equiv 0$ on grounds of molecular symmetry

Code symbol	Estimated accuracy of value
i	The significance of these values may involve some ambiguity because of the possibility of different conformations or spatial isomers.

Compounds not containing carbon

Formula	Compound name	Selected moment (debyes)	
AgCl	Silver chloride	5.73	C
$AsCl_3$	Arsenic trichloride	1.59	C
AsF_3	Arsenic trifluoride	2.59	B
AsH_3	Arsine	0.20	C
BCl_3	Boron trichloride	0	S
BF_3	Boron trifluoride	0	S
B_2H_6	Diborane	0	S
$B_3H_6N_3$	Triborotriazine (Borazine)	0	S
B_5H_9	Pentaborane	2.13	B
BaO	Barium oxide	7.95	A
BrH	Hydrogen bromide	0.82	B
BrH_3Si	Bromosilane	1.33	B
BrK	Potassium bromide	10.41	B
BrLi	Lithium bromide	7.27	A
Br_2Hg	Mercury dibromide	0	S
Br_4Sn	Tin tetrabromide	0	S
ClCs	Cesium chloride	10.42	A
ClF	Chlorine fluoride	0.88	C
$ClFO_3$	Perchloryl fluoride	0.023	A
$ClGeH_3$	Chlorogermane	2.13	A
ClH	Hydrogen chloride	1.08	B
ClH_3Si	Chlorosilane	1.31	A
ClK	Potassium chloride	10.27	A
ClLi	Lithium chloride	7.13	A
ClNa	Sodium chloride	9.00	A
$ClNO_2$	Nitryl chloride	0.53	A
ClTl	Thallium chloride	4.44	B
Cl_2F_3P	Dichlorotrifluorophosphorus	0.68	C
Cl_2H_2Si	Dichlorosilane	1.17	B
Cl_2Hg	Mercury dichloride	0	S
Cl_2OS	Thionyl chloride	1.45	B
Cl_2O_2S	Sulfuryl chloride	1.81	B
Cl_3F_2P	Trichlorodifluorophosphorus	0	S
Cl_3HSi	Trichlorosilane	0.86	B
Cl_3P	Phosphorus trichloride	0.78	C
Cl_4FP	Tetrachlorofluorophosphorus	0.21	B
Cl_4Ge	Germanium tetrachloride	0	S
Cl_4Si	Silicon tetrachloride	0	S
Cl_4Sn	Tin tetrachloride	0	S
Cl_4Ti	Titanium tetrachloride	0	S
CsF	Cesium fluoride	7.88	A
FH	Hydrogen fluoride	1.82	A
FH_3Si	Fluorosilane	1.27	B
FH_5Si_2	Fluorodisilane	1.26	A
FK	Potassium fluoride	8.60	A
FLi	Lithium fluoride	6.33	A
FNO	Nitrosyl fluoride	1.81	B
FNa	Sodium fluoride	8.16	A
FRb	Rubidium fluoride	8.55	A
FTl	Thallium fluoride	4.23	A
F_2HN	Difluoramine	1.92	A
F_2H_2Si	Difluorosilane	1.55	A
F_2N_2	*cis*-Difluorodiazine	0.16	A
F_2O	Oxygen difluoride	0.297	A
F_2OS	Thionyl fluoride	1.63	A
F_2O_2	Dioxygen difluoride	1.44	C
F_2O_2S	Sulfuryl fluoride	1.12	B
F_2S_2	Sulfur monofluoride ($S = SF_2$ isomer)	1.03	C
F_2S_2	Sulfur monofluoride (FSSF isomer)	1.45	B
F_2Si	Silicon difluoride	1.23	B
F_3HSi	Trifluorosilane	1.27	B
F_3N	Nitrogen trifluoride	0.235	A
F_3NS	Nitridotrifluorosulfur	1.91	B
F_3OP	Phosphoryl fluoride	1.76	B
F_3P	Phosphorus trifluoride	1.03	A
F_3PS	Thiophosphoryl fluoride	0.64	B

Compounds not containing carbon—Continued

Formula	Compound name	Selected moment (debyes)	
F_4N_2	Tetrafluorohydrazine, *gauche* conformation	0.26	B
F_4S	Sulfur tetrafluoride	0.632	A
F_4Si	Silicon tetrafluoride	0	S
F_5P	Phosphorus pentafluoride	0	S
F_5I	Iodine pentafluoride	2.18	C
F_6S	Sulfur hexafluoride	0	S
F_6Se	Selenium hexafluoride	0	S
F_6Te	Tellurium hexafluoride	0	S
F_6U	Uranium hexafluoride	0	S
HI	Hydrogen iodide	0.44	B
HLi	Lithium hydride	5.88	A
HN	Imidyl radical		
HNO_3	Nitric acid	2.17	A
HO	Hydroxyl radical	1.66	A
H_2O	Water	1.85	A
H_2O_2	Hydrogen peroxide	2.2	D
H_2S	Hydrogen sulfide	0.97	A
H_3N	Ammonia	1.47	A
H_3P	Phosphine	0.58	A
H_3Sb	Stibine	0.12	C
H_4N_2	Hydrazine	1.75	C
H_4Si	Silane	0	S
H_6OSi_2	Disilyl ether (disiloxane)	0.24	B
H_6Si_2	Disilane	0	S
HgI_2	Mercury diiodide	0	S
ILi	Lithium iodide	7.43	A
I_4Sn	Tin tetraiodide	0	S
NO	Nitrogen monoxide (nitric oxide)	0.153	A
NO_2	Nitrogen dioxide	0.316	A
N_2O	Dinitrogen oxide (nitrous oxide)	0.167	A
OS	Sulfur monoxide	1.55	A
OS_2	Disulfur monoxide	1.47	B
OSr	Strontium oxide	8.90	A
O_2S	Sulfur dioxide	1.63	A
O_3	Ozone	0.53	B
O_3S	Sulfur trioxide	0	S
O_4Os	Osmium tetroxide	0	S

Compounds containing carbon

Formula	Compound name	Selected moment (debyes)	
$CBrF_3$	Bromotrifluoromethane	0.65	C
CBr_2F_2	Dibromodifluoromethane	0.66	C
$CClF_3$	Chlorotrifluoromethane	0.50	A
CClN	Cyanogen chloride	2.82	B
CCl_2F_2	Dichlorodifluoromethane	0.51	C
CCl_2O	Carbonyl chloride (phosgene)	1.17	A
CCl_2S	Thiocarbonyl chloride	0.29	C
CCl_3F	Trichlorofluoromethane	0.45	C
CCl_3NO_2	Trichloronitromethane	1.89	C
CCl_4	Carbon tetrachloride	0	S
CFN	Cyanogen fluoride	2.17	C
CF_2	Carbon difluoride	0.46	B
CF_2O	Carbonyl fluoride	0.95	A
CF_3I	Iodotrifluoromethane	0.92	C
CF_3NO_2	Trifluoronitromethane	1.44	C
CF_4	Carbon tetrafluoride	0	S
CN_4O_8	Tetranitromethane	0	S
CO	Carbon monoxide	0.112	A
COS	Carbonyl sulfide	0.712	A
COSe	Carbonyl selenide	0.73	B
CO_2	Carbon dioxide	0	S
CS	Carbon monosulfide	1.98	A
CSTe	Thiocarbonyl telluride	0.17	A
CS_2	Carbon disulfide	0	S
$CHBr_3$	Tribromomethane	0.99	B
$CHClF_2$	Chlorodifluoromethane	1.42	B

Selected Values of Electric Dipole Moments for Molecules in the Gas Phase (Continued)

Compounds containing carbon—Continued

Formula	Compound name	Selected moment (debyes)	
$CHCl_2F$	Dichlorofluoromethane	1.29	B
$CHCl_3$	Trichloromethane (chloroform)	1.01	B
$CHFO$	Formyl fluoride	2.02	A
CHF_3	Trifluoromethane	1.65	A
CHN	Hydrogen cyanide	2.98	A
CHP	Methylidyne phosphide (methinophosphide)	0.390	A
CH_2Br_2	Dibromomethane	1.43	B
CH_2ClF	Chlorofluoromethane	1.82	B
CH_2ClNO_2	Chloronitromethane	2.91	B
CH_2Cl_2	Dichloromethane	1.60	B
CH_2F_2	Difluoromethane	1.97	A
CH_2N_2	Cyanogen amide (cyanamide)	4.27	C
CH_2N_2	Diazomethane	1.50	A
CH_2N_2	Diazirine	1.59	C
CH_2O	Methanal (formaldehyde)	2.33	A
CH_2O_2	Methanoic acid (formic acid)	1.41	A
CH_3BF_2	Methyl difluoroborane	1.66	B
CH_3BO	Carbonyl borane	1.80	B
CH_3Br	Bromomethane	1.81	A
CH_3Cl	Chloromethane	1.87	A
CH_3F	Fluoromethane	1.85	A
CH_3I	Iodomethane	1.62	B
CH_3NO	Hydroxyliminomethane (formaldoxime)	0.44	A
CH_3NO	Formyl amide (formamide)	3.73	B
CH_3NOS	Methyl sulfinylamine	1.70	B
CH_3NO_2	Nitromethane	3.46	A
CH_3NO_3	Methyl nitrate	3.12	B
CH_3N_3	Methyl azide	2.17	B
CH_4	Methane	0	S
CH_4F_2Si	Methyl difluorosilane	2.11	A
CH_4O	Methanol	1.70	A
CH_4S	Methanethiol (methyl mercaptan)	1.52	C
CH_5FSi	Methyl monofluorosilane	1.71	A
CH_5N	Methyl amine	1.31	B
CH_5P	Methyl phosphine	1.10	A
CH_6Ge	Methyl germane	0.643	A
CH_6OSi	Methoxysilane	1.17	B
CH_6Si	Methyl silane	0.735	A
CH_6Sn	Methyl stannane	0.68	C
C_2ClF_5	Chloropentafluorethane	0.52	C
C_2F_6	Hexafluorethane	0	S
C_2N_2	Dicyanogen (cyanogen)	0	S
C_2N_2S	Dicyano sulfide	3.02	A
C_2HCl	Chloroacetylene	0.44	A
C_2HCl_5	Pentachloroethane	0.92	C
C_2HF	Fluoroacetylene	0.73	C
C_2HF_3	Trifluoroethylene	1.40	C
C_2HF_5	Pentafluoroethane	1.54	C
C_2H_2	Acetylene	0	S
$C_2H_2Cl_2$	1, 1-Dichloroethylene	1.34	A
$C_2H_2Cl_2$	*cis*-1, 2-Dichloroethylene	1.90	B
$C_2H_2Cl_2O$	Chloroacetyl chloride	2.23	Ci
$C_2H_2Cl_4$	1,1,2,2-Tetrachloroethane	1.32	Ci
C_2H_2FN	Fluorocyanomethane	3.43	C
$C_2H_2F_2$	1,1-Difluoroethylene	1.38	A
$C_2H_2F_2$	*cis*-1,2-Difluoroethylene	2.42	A
$C_2H_2N_2O$	1,2,5-Oxadiazole	3.38	A
$C_2H_2N_2O$	1,3,4-Oxadiazole	3.04	B
$C_2H_2N_2S$	1,2,5-Thiadiazole	1.56	A
$C_2H_2N_2S$	1.3.4-Thiadiazole	3.29	B
C_2H_2O	Methylene carbonyl (ketene)	1.42	B
C_2H_3Br	Bromoethylene	1.42	B
C_2H_3Cl	Chloroethylene	1.45	B
$C_2H_3ClF_2$	1-Chloro-1,1-difluoroethane	2.14	B
C_2H_3ClO	Acetyl chloride	2.72	C
$C_2H_3Cl_3$	1,1,1-Trichloroethane	1.78	B
C_2H_3F	Fluoroethylene	1.43	A
C_2H_3FO	Acetyl fluoride	2.96	A
$C_2H_3F_3$	1,1,1-Trifluoroethane	2.32	B
$C_2H_3F_3$	1,1,2-Trifluoroethane	1.58	B
C_2H_3N	Cyanomethane (acetonitrile)	3.92	A
C_2H_3N	Isocyanomethane	3.85	B
C_2H_4	Ethylene	0	S
C_2H_4ClF	1-Chloro-2-fluoroethane, *gauche* conformation	2.72	C
$C_2H_4ClNO_2$	1-Chloro-1-nitroethane	3.27	B
$C_2H_4Cl_2$	1,1-Dichloroethane	2.06	B
$C_2H_4F_2$	1,1-Difluoroethane	2.27	B
C_2H_4Ge	Germyl acetylene	0.136	A
C_2H_4O	Oxirane (ethylene oxide)	1.89	A
C_2H_4O	Ethanal (acetaldehyde)	2.69	B
$C_2H_4O_2$	Ethanoic acid (acetic acid)	1.74	C
$C_2H_4O_2$	Methyl methanoate (methly formate)	1.77	B
C_2H_4S	Thiirane (ethylene sulfide)	1.85	A
C_2H_4Si	Silyl acetylene	0.316	A
C_2H_5Br	Bromoethane	2.03	A
C_2H_5BrO	Bromomethoxymethane	2.05	Ci
C_2H_5Cl	Chloroethane	2.05	A
C_2H_5ClO	2-Chloroethanol	1.78	Ci
C_2H_5F	Fluoroethane	1.94	B
C_2H_5I	Iodoethane	1.91	B
C_2H_5N	Iminoethane (ethyleneimine)	1.90	A
C_2H_5N	Methyliminomethane ($CH_3N = CH_2$)	1.53	B

Compounds containing carbon—Continued

Formula	Compound name	Selected moment (debyes)	
C_2H_5NO	Acetyl amine (acetamide)	3.76	Bi
C_2H_5NO	Methylaminomethanal (*N*-methylformamide)	3.83	Bi
$C_2H_5NO_2$	Nitritoethane (ethyl nitrite)	2.40	Ci
$C_2H_5NO_2$	Nitroethane	3.65	B
C_2H_6	Ethane	0	S
C_2H_6BF	Dimethyl fluoroborane	1.32	C
C_2H_6O	Ethanol	1.69	Bi
C_2H_6O	Dimethyl ether	1.30	A
C_2H_6OS	Dimethylsulfoxide	3.96	A
$C_2H_6O_2$	1,2-Ethanediol (ethylene glycol)	2.28	Ci
$C_2H_6O_2S$	Dimethyl sulfoxylate (dimethyl sulfone)	4.49	B
C_2H_6S	Ethanethiol	1.58	Bi
C_2H_6S	Dimethyl sulfide	1.50	A
C_2H_6Si	Silyl ethylene	0.66	A
$C_2H_7B_5$	2,4-Dicarbaheptaborane	1.32	B
C_2H_7N	Aminoethane (ethyl amine)	1.22	Ci
C_2H_7N	Dimethyl amine	1.03	B
C_2H_7P	Ethyl phosphine	1.17	Bi
C_2H_7P	Dimethyl phosphine	1.23	A
$C_2H_8N_2$	1,2-Diaminoethane	1.99	Ci
C_2H_8Si	Dimethyl silane	0.75	A
C_2H_8Si	Ethyl silane	0.81	B
C_3O_2	Dicarbonyl carbon (carbon suboxide)	0	S
C_3HF_3	3,3,3-Trifluoropropyne	2.36	B
C_3HN	Cyanoacetylene	3.72	A
$C_3H_2N_2$	Dicyanomethane	3.73	A
C_3H_2O	Propynal	2.47	B
$C_3H_2O_3$	Vinylene carbonate	4.55	A
C_3H_3Br	3-Bromopropyne	1.54	C
C_3H_3Cl	3-Chloropropyne	1.68	C
$C_3H_3F_3$	3,3,3-Trifluoropropene	2.45	B
C_3H_3N	Cyanoethylene	3.87	B
C_3H_3NO	Acetyl cyanide	3.45	B
C_3H_3NS	Thiazole	1.62	B
C_3H_4	Cyclopropene	0.45	A
C_3H_4	Propyne	0.781	A
C_3H_4	Propadiene (allene)	0	S
$C_3H_4Cl_2$	1,1-Dichlorocyclopropane	1.58	B
C_3H_4O	Ethylidene carbonyl (methyl ketene)	1.79	B
C_3H_4O	Propenal, *trans* conformation (acrolein)	3.12	B
$C_3H_4O_2$	2-Oxoöxetane (β-propiolactone)	4.18	A
$C_3H_4O_2$	Vinyl formate	1.49	A
C_3H_5Cl	2-Chloropropene	1.66	B
C_3H_5Cl	*cis*-1-Chloropropene	1.67	C
C_3H_5Cl	*trans*-1-Chloropropene	1.97	C
C_3H_5Cl	3-Chloropropene	1.94	Ci
C_3H_5F	*cis*-1-Fluoropropene	1.46	B
C_3H_5F	2-Fluoropropene	1.61	B
C_3H_5F	3-Fluoropropene, *cis* conformation	1.76	A
C_3H_5F	3-Fluoropropene, *gauche* conformation.	1.94	A
C_3H_5N	Cyanoethane (propionitrile)	4.02	A
C_3H_6	Cyclopropane	0	S
C_3H_6	Propene	0.366	A
$C_3H_6ClNO_2$	1-Chloro-1-nitropropane	3.48	Bi
$C_3H_6Cl_2$	1,2-Dichloropropane		i
$C_3H_6Cl_2$	1,3-Dichloropropane	2.08	Bi
$C_3H_6Cl_2$	2,2-Dichloropropane	2.27	C
C_3H_6O	Oxetane (trimethylene oxide)	1.94	A
C_3H_6O	Methyl oxirane (propylene oxide)	2.01	A
C_3H_6O	Propanone (acetone)	2.88	A
C_3H_6O	2-Propen-1-ol (allyl alcohol)	1.60	C
C_3H_6O	Propanal, *cis* conformation (propionaldehyde)	2.52	B
$C_3H_6O_2$	Propanoic acid	1.75	Ci
$C_3H_6O_2$	Methyl acetate	1.72	Ci
$C_3H_6O_2$	Ethyl formate	1.93	Ci
$C_3H_6O_3$	1,3,5-Trioxane	2.08	A
C_3H_6S	Thietane (trimethylene sulfide)	1.85	C
C_3H_6S	Methyl thiirane (propylene sulfide)	1.95	A
C_3H_7Br	1-Bromopropane	2.18	Ci
C_3H_7Br	2-Bromopropane	2.21	C
C_3H_7Cl	1-Chloropropane	2.05	Bi
C_3H_7Cl	2-Chloropropane	2.17	C
C_3H_7F	1-Fluoropropane, *gauche* conformation	1.90	C
C_3H_7F	1-Fluoropropane, *trans* conformation	2.05	B
C_3H_7I	1-Iodopropane	2.04	Ci
C_3H_7NO	N,N-Dimethylformamide	3.82	Bi
C_3H_7NO	Acetyl methylamine (*N*-Methylacetamide)	3.73	Bi
$C_3H_7NO_2$	1-Nitropropane	3.66	Bi
$C_3H_7NO_2$	2-Nitropropane	3.73	B
C_3H_8	Propane	0.084	A
C_3H_8O	1-Propanol	1.68	Bi
C_3H_8O	2-Propanol	1.66	Bi
C_3H_8O	Methoxyethane (methyl ethyl ether)	1.23	Ci
C_3H_9As	Trimethyl arsine	0.86	B
C_3H_9N	Trimethyl amine	0.612	A
C_3H_9N	1-Aminopropane (*n*-propylamine)	1.17	Ci
C_3H_9P	Trimethyl phosphine	1.19	A

Selected Values of Electric Dipole Moments for Molecules in the Gas Phase (Continued)

Compounds containing carbon—Continued

Formula	Compound name	Selected moment (debyes)	
$C_3H_{10}Si$	Trimethyl silane	0.525	A
C_4F_8	Perfluorocyclobutane	0	S
$C_4H_2N_2$	*trans*-1,2-Dicyanoethylene	0	S
$C_4H_4Cl_2$	1,4-Dichloro-2-butyne	2.10	Bi
$C_4H_4F_2$	1,1-Difluoro-1,3-butadiene (*trans* conformation)	1.29	A
C_4H_4O	Furan	0.66	A
$C_4H_4O_2$	Diketene	3.53	B
C_4H_4S	Thiophene	0.55	C
C_4H_5Cl	4-Chloro-1,2-butadiene	2.02	Ci
C_4H_5Cl	1-Chloro-2-butyne	2.19	C
C_4H_5F	2-Fluoro-1,3-butadiene (*trans* conformation)	1.42	A
C_4H_5N	Pyrrole	1.84	C
C_4H_5N	*cis*-1-Cyanopropene	4.08	B
C_4H_5N	*trans*-1-Cyanopropene	4.50	B
C_4H_5N	2-Cyanopropene (methacrylonitrile)	3.69	C
C_4H_6	Cyclobutene	0.132	A
C_4H_6	1-Butyne	0.80	C
C_4H_6	1,2-Butadiene	0.403	A
C_4H_6	1,3-Butadiene	0	S
C_4H_6O	Cyclobutanone	2.99	B
C_4H_6O	*trans*-2-Butenal (crotonaldehyde)	3.67	Bi
C_4H_6O	2-Methylpropenal (methacrolein)	2.68	Ci
C_4H_6O	3-Butene-2-one	3.16	B
C_4H_7Cl	1-Chloro-2-methylpropene	1.95	Bi
$C_4H_7Cl_3$	1,1,2-Trichloro-2-methylpropane	1.86	Ci
C_4H_7F	Fluorocyclobutane	1.94	A
C_4H_7N	1-Cyanopropane	4.07	Bi
C_4H_8	1-Butene	0.34	Ci
C_4H_8	*trans*-2-Butene	0	S
C_4H_8	2-Methylpropene	0.50	A
$C_4H_8Cl_2$	1,4-Dichlorobutane	2.22	Ci
C_4H_8O	Tetrahydrofuran	1.63	C
C_4H_8O	*cis*-2,3-Dimethyloxirane	2.03	A
C_4H_8O	Butanal	2.72	Bi
$C_4H_8O_2$	1,4-Dioxane	0	S
$C_4H_8O_2$	Ethyl acetate	1.78	Ci
C_4H_9Br	1-Bromobutane	2.08	Ci
C_4H_9Br	2-Bromobutane	2.23	Ci
C_4H_9Cl	1-Chlorobutane	2.05	Bi
C_4H_9Cl	2-Chlorobutane	2.04	Ci
C_4H_9Cl	1-Chloro-2-methylpropane	2.00	Ci
C_4H_9Cl	2-Chloro-2-methylpropane	2.13	B
C_4H_9F	2-Fluoro-2-methylpropane	1.96	A
C_4H_9I	1-Iodobutane	2.12	Ci
C_4H_9NO	Propanoyl methylamine (N-methylpropionamide)	3.61	Bi
C_4H_9NO	Acetyl dimethylamine (N,N-dimethylacetamide)	3.81	Bi
$C_4H_9NO_2$	2-Nitrito-2-methylpropane (t-butyl nitrite)	2.74	Ci
$C_4H_9NO_2$	1-Nitrobutane	3.59	Bi
$C_4H_9NO_2$	2-Nitro-2-methylpropane	3.71	B
C_4H_{10}	Butane	$\leqslant$0.05	Ci
C_4H_{10}	2-Methylpropane	0.132	A
$C_4H_{10}O$	1-Butanol	1.66	Bi
$C_4H_{10}O$	2-Methylpropan-1-ol (isobutanol)	1.64	C
$C_4H_{10}O$	Diethyl ether	1.15	Bi
$C_4H_{10}S$	Diethyl sulfide	1.54	Ci
$C_4H_{11}N$	Diethyl amine	0.92	Ci
C_5H_5N	Pyridine	2.19	B
C_5H_5N	1-Cyano-1,3-butadiene	3.90	Ci
C_5H_6	1,3-Cyclopentadiene	0.419	A
C_5H_8	Cyclopentene	0.20	B
C_5H_8	1-Pentyne	0.81	Ci
C_5H_8	2-Methyl-1,3-butadiene (*trans* conformation)	0.25	A
$C_5H_8O_2$	Acetylacetone		Ci
C_5H_9N	1-Cyanobutane	4.12	Bi
C_5H_9N	2-Cyano-2-methylpropane	3.95	A
$C_5H_{10}O_3$	Diethyl carbonate	1.10	Ci
$C_5H_{11}Br$	1-Bromopentane	2.20	Ci
$C_5H_{11}Cl$	1-Chloropentane	2.16	Ci
C_5H_{12}	2-Methylbutane	0.13	C
C_5H_{12}	2,2-Dimethylpropane	0	S
$C_6H_2Cl_2O_2$	2,5-Dichloro-1,4-cyclohexadienedione	0	S
$C_6H_4ClNO_2$	*o*-Chloronitrobenzene	4.64	B
$C_6H_4ClNO_2$	*m*-Chloronitrobenzene	3.73	B
$C_6H_4ClNO_2$	*p*-Chloronitrobenzene	2.83	B
$C_6H_4Cl_2$	*o*-Dichlorobenzene	2.50	B
$C_6H_4Cl_2$	*m*-Dichlorobenzene	1.72	C
$C_6H_4Cl_2$	*p*-Dichlorobenzene	0	S
$C_6H_4FNO_2$	*p*-Fluoronitrobenzene	2.87	B
$C_6H_4F_2$	*m*-Difluorobenzene	1.58	B
$C_6H_4N_2O_4$	*p*-Dinitrobenzene	0	S
$C_6H_4O_2$	1,4-Cyclohexadienedione (*p*-benzoquinone)	0	S
C_6H_5Br	Bromobenzene	1.70	B
C_6H_5Cl	Chlorobenzene	1.69	B
C_6H_5ClO	*p*-Chlorophenol	2.11	C
C_6H_5F	Fluorobenzene	1.60	C
C_6H_5I	Iodobenzene	1.70	C
$C_6H_5NO_2$	Nitrobenzene	4.22	B
C_6H_6	Benzene	0	S

Compounds containing carbon—Continued

Formula	Compound name	Selected moment (debyes)	
C_6H_6O	Phenol	1.45	C
C_6H_7N	Aminobenzene (aniline)	1.53	C
C_6H_8	1,3-Cyclohexadiene	0.44	B
C_6H_{10}	1-Hexyne	0.83	Ci
C_6H_{10}	3, 3-Dimethyl-1-butyne	0.66	A
$C_6H_{10}Cl_2$	*cis*-1e,2a-Dichlorocyclohexane	3.11	C
$C_6H_{12}N_2$	Diisopropylidene hydrazine (dimethyl ketazine)	1.53	Bi
$C_6H_{12}O_2$	Pentyl formate (n-amyl formate)	1.90	Ci
$C_6H_{12}O_3$	2,4,6-Trimethyl-1,3,5-trioxane (paraldehyde)	1.43	C
$C_6H_{14}O$	Dipropyl ether	1.21	Ci
$C_6H_{14}O_2$	1,1-Diethoxyethane		i
$C_6H_{15}N$	Triethyl amine	0.66	Ci
$C_7H_4ClF_3$	*o*-Chloro(trifluoromethyl)benzene	3.46	B
$C_7H_4ClF_3$	*p*-Chloro(trifluoromethyl)benzene	1.58	C
$C_7H_5F_3$	(Trifluoromethyl)benzene	2.86	B
C_7H_5N	Cyanobenzene (benzonitrile)	4.18	B
C_7H_7Cl	*o*-Chlorotoluene	1.56	C
C_7H_7Cl	*p*-Chlorotoluene	2.21	B
C_7H_7F	*o*-Fluorotoluene	1.37	C
C_7H_7F	*m*-Fluorotoluene	1.86	C
C_7H_7F	*p*-Fluorotoluene	2.00	C
$C_7H_7NO_3$	*o*-Nitro(methoxy)benzene	4.83	Bi
$C_7H_7NO_3$	*m*-Nitro(methoxy)benzene	4.55	Bi
$C_7H_7NO_3$	*p*-Nitro(methoxy)benzene	5.26	B
C_7H_8	1,3,5-Cycloheptatriene	0.25	C
C_7H_8	Toluene	0.36	C
C_7H_8O	Phenylmethanol (benzyl alcohol)	1.71	C
C_7H_8O	Methoxybenzene (anisole)	1.38	C
C_7H_9NO	*o*-Amino(methoxy)benzene	1.61	Ci
$C_7H_{14}O_2$	Pentyl acetate (*n*-amyl acetate)	1.75	Ci
$C_7H_{15}Br$	1-Bromoheptane	2.16	Ci
$C_8H_4N_2$	*p*-Dicyanobenzene	0	S
C_8H_8O	Acetylbenzene (acetophenone)	3.02	B
$C_8H_8O_2$	2,5-Dimethyl-1,-4-cyclohexadienedione	0	S
C_8H_{10}	Ethylbenzene	0.59	C
C_8H_{10}	*o*-Xylene	0.62	C
C_8H_{10}	*p*-Xylene	0	S
$C_8H_{12}O_2$	Tetramethylcyclobutane-1,3-dione	0	S
$C_8H_{18}O$	Dibutyl ether	1.17	Ci
C_9H_7N	Quinoline	2.29	C
C_9H_7N	Isoquinoline	2.73	C
$C_9H_{10}O_2$	Ethyl phenylformate (ethyl benzoate)	2.00	Ci
$C_{10}H_8$	Azulene	0.80	B
$C_{10}H_{14}BeO_4$	Bis(2,4-pentanedionato) beryllium	0	S
$C_{12}H_9BrO$	*p*-Bromophenoxybenzene	1.98	C
$C_{12}H_9NO_3$	*p*-Nitrophenoxybenzene	4.54	B
$C_{12}H_{10}$	Phenylbenzene (diphenyl)	0	S
$C_{13}H_{11}BrO$	*p*-Bromophenoxy-*p*-toluene	2.45	C
$C_{14}H_{14}O$	Bis(*p*-tolyl) ether	1.54	C
$C_{15}H_{21}AlO_6$	Tris(2,4-pentanedionato) aluminum	0	S
$C_{15}H_{21}CrO_6$	Tris(2,4-pentanedionato) chromium (III)	0	S
$C_{15}H_{21}FeO_6$	Tris(2,4-pentanedionato) iron (III)	0	S
$C_{20}H_{28}O_8Th$	Tetrakis(2,4-pentanedionato) thorium	0	S

DIPOLE MOMENTS

The method of measurement of the dipole moments is indicated in the following **two tables** by the symbols:

- B benzene solution
- C carbon tetrachloride solution
- D 1,4-dioxane solution
- H n-heptane solution
- St measurement of Stark effect in microwave spectrum of gas

Dipole Moments for Some Inorganic Compounds

Compound	Dipole Moment × 10^{-18} e. s. u.	Method
$AlBr_3$	5.14	B
AlI_3	2.48	B
CsCl	10.42	St
CsF	7.875	St
HF	1.92 ± 0.02	..
HCl	1.084 ± 0.003–0.007	..
NBr	0.78	..
HDSe	0.62	St
HI	0.38	..
DCl	1.084 ± 0.003–0.007	..
HNO_3	2.16	St
$HgBr_2$	0.95	B
$HgCl_2$	1.23	B
H_2O	1.87	..
H_2O_2	2.13 ± 0.05	..
H_2S	1.10	..
SO_2	1.60	..
SO_3	0.00	..
SO_2F_2	1.110	St
NH_3	1.3	..
N_2H_4	1.84	..
NO	0.16	..
NO_2	0.29	..
N_2O_4	0.37	..
NOCl	1.83	..
NOBr	1.87	..
PCl_3	0.90–1.16	..
PCl_5	0.0	..
CO	0.10	..
CO_2	0.0	..
SiD_2F_2	1.53	St
SiH_2F_2	1.54	St
$SnCl_4$	0.95	B
SnI_4	0	B
$TiCl_4$	0	C

Dipole Moments for Some Organo-metallic Compounds

Compound	Dipole Moment × 10^{-18} e. s. u.	Method
Beryllium diethyl	1.0	H
Cadmium diethyl	0.3	H
Mercury diethyl	0.0	H
Magnesium diethyl	4.8	D
Zinc diethyl	0.0	H
Beryllium diphenyl	1.6	B
Cadmium diphenyl	0.6	B
Mercury diphenyl	0.2	B
Magnesium diphenyl	4.9	D
Zinc diphenyl	0.8	B
Chromium (0), diphenyl	0	B
Chromium, ditolyl	0	B
Cobalt, mononitrosyl tricarbonyl	0.72	B
Cyclopentadienyl, chromium dicarbonyl mono nitrosyl	3.23	B
Cyclopentadienyl, manganese tri-carbonyl	3.30	B
Cyclopentadienyl, cobalt ducarbonyl	2.87	B
Cyclopentadienyl, vanadium tetra-carbonyl	3.17	B
Penta cyclopentadienyl, dicobalt	0	B
Dicyclopentadienyl, iron (II)	0	B
Dicyclopentadienyl, lead (II)	1.63	B
Dicyclopentadienyl, tin (II)	1.02	B
Ethyl lithium	0.87	B
Iron, dinitrosyl dicarbonyl	0.95	B
Iron, tetracarbonyl-diiodide	3.68	B
Iron, tetracarbonyl mono-(methyl isonitrile)	5.07	B
Iron, pentacarbonyl	0.63	B
Iron, bis(p-chlorophenyl cyclo-pentadienyl	3.12	B
Ruthenium (II), di-indenyl	0	B

Dipole Moments

Dipole Moments of Amino Acid Esters

Substance	$\mu \cdot 10^{18}$ e. s. u.
Glycine ethyl ester	2.11
α-Alanine ethyl ester	2.09
α-Aminobutyric acid ethyl ester	2.13
α-Aminovaleric acid ethyl ester	2.13
Valine ethyl ester	2.11
α-Aminocaproic acid ethyl ester	2.13
β-Alanine ethyl ester	2.14
β-Aminobutyric acid ethyl ester	2.11

Accurate to ±0.01 . 10^{-18} e. s. u.
J. Wyman, Chem. Rev., 1936, **19**, 213.

Dipole Moments of Amides

Urea	4.56
Thiourea	4.89
Symm.-dimethylurea	4.8
Tetraethylurea	3.3
Propylurea	4.1
Acetamide	3.6
Sulfamide	3.9
Benzamide	3.6
Valeramide	3.7
Caproamide	3.9

For comprehensive list of dipole moments see Trans. Faraday Soc., 1934, **30**, General Discussion.

Dipole Moments of Some Hormones and Related Compounds in Dioxan

Cholestane-3(β) : 7(α)-diol	2.31
Cholestane-3(β) : 7(β)-diol	2.55
Cholestane	2.98
Δ^5-Cholestane-3(β)ol-7 one	3.79
Androsterone	3.70
β-Androsterone	2.95
Δ^5-Androstene-3(β) : 17(α)-diol	2.89
Δ^5-Androstene-3(β) : 17(β)-diol	2.69
Δ^5-Androstene-3(β)ol-17 one	2.46
Testosterone	4.32
cis-Testosterone	5.17
Δ^4-Androstene-3 : 17 dione	3.32
Isophorone	3.96

Ethylenic >C═C< in a six membered ring and conjugated with >C═0 increases the dipole moment approximately by 1 Debye. Non-conjugated >C═C< in sterols decreases the dipole moment by approximately 0.49. Biological activity is not correlated with dipole moment. W. D. Kumler and G. M. Fohlen, J. Am. Chem. Soc., 1945, **67**, 437.

ELECTRON AFFINITIES

F. M. Page

The tabulated values are given in kJ mol^{-1}, eV and kcal mol^{-1} for the convenience of users. All values are experimental observations and refer to the gas phase at 0°K unless otherwise stated. Extended tables of atomic electron affinities are given by L. M. Branscombe in "Atomic and Molecular Processes", Ed. D. R. Bates, Academic Press, New York (1962), and for radical and molecular electron affinities in "Negative Ions and the Magnetron", Wiley—Interscience—London (1969). See also H. O. Pritchard, Chem. Rev. **52**, 529 (1953); F. H. Field and J. L. Franklin, "Electron Impact Phenomena", Academic Press, New York (1957), N. S. Buchel'nikova, Usp, fiz. nauk **65**, 357 (1958), A.E.C. Tr3657, Oak Ridge (1959), V. I. Vedeneyev et al, "Bond Energies, Ionisation Potentials and Electron Affinities", Arnold, London (1966) and C. J. Schexnayder, N.A.S.A. Technical Note D1791, Washington (1963).

Atoms

Ion Formed	Electron Affinity		
	kJ mol^{-1}	eV	kcal mol^{-1}
Br^-	324	3.363	77.4
C^-	108	1.12	25.8
Cl^-	348	3.613	83.2
F^-	333	3.448	79.6
H^-	77	0.80	18.4
*H^- ($1s^2$)	72	0.747	17.2
I^-	296	3.063	70.8
O^-	142	1.466	33.8
S^-	200	2.07	47.7

* Quantum mechanical calculation.

Molecules

Ion Formed	Electron Affinity		
	kJ mol^{-1}	eV	kcal mol^{-1}
BF_3^-	255	2.65	61.0
(*p*-Benzoquinone)$^-$	129	1.34	30.9
(Chloranil)$^-$	231	2.40	55.3
(Fluoranil)$^-$	218	2.27	52.2
No_2^-	377	3.91	90.0
O_2^-	43	0.45	10.1
SF_6^-	138	1.43	33.0
(Tetracyanoethylene)$^-$	278	2.88	66.4
WF_6^-	264	2.74	63.0
UF_6^-	280	2.91	67.0

Radicals

Ion Formed	Electron Affinity		
	kJ mol^{-1}	eV	kcal mol^{-1}
Ch_3^-	104	1.08	24.8
$C_2H_5^-$	86	0.89	20.5
$C_6H_5^-$	210	2.20	49.6
CF_3^-	179	1.85	42.7
CCl_3^-	117	1.22	28.0
SiF_3^-	323	3.35	77.2
NH_2^-	108	1.12	25.7
$C_6H_5NH^-$	149	1.55	35.7
$(C_6H_5)_2N^-$	114	1.19	23.3
PH_2^-	154	1.60	36.8
OH^-	176	1.83	42.1
CH_3O^-	36	0.38	8.7
CF_3O^-	131	1.35	31.2
SH^-	211	2.19	50.4
CH_3S^-	127	1.32	30.4
CN^-	305	3.17	7.30
SCN^-	209	2.17	49.9
$SeCN^-$	255	2.64	60.9

IONIZATION POTENTIALS OF THE ELEMENTS

Different methods have been employed to measure ionization potentials. Abbreviations of the methods used for data listed in the following table are:

S; Vacuum ultraviolet spectroscopy
SI; Surface ionization, mass spectrometric
EI; Electron impact with mass analysis

El.	At. No.	Ionization potential in volts: I	II	III	IV	V	VI	VII	VIII	Meth.
Ar	18	15.755	27.62	40.9	59.79	75	91.3	124	143.46	S
Ac	89	6.9	12.1	20	......	......	......	......	......	S
Ag	47	7.574	21.48	34.82	......	......	......	......	......	S
Al	13	5.984	18.823	28.44	119.96	153.77	190.42	241.38	284.53	S
As	33	9.81	18.63	28.34	50.1	62.6	127.5	......	......	S
At	85	9.5	......	......	......	......	......	......	......	S
Au	79	9.22	20.5	......	......	......	......	......	......	S
B	5	8.296	25.149	37.92	259.298	340.127	......	......	......	S
Ba	56	5.21	10.001	35.5	......	......	......	......	......	S
Be	4	9.32	18.206	153.85	217.657	......	......	......	......	S
Bi	83	7.287	16.68	25.56	45.3	56	88.3	......	......	S
Br	35	11.84	21.6	35.9	47.3	59.7	88.6	103	193	S
C	6	11.256	24.376	47.871	64.476	391.986	489.84	......	......	S
Ca	20	6.111	11.868	51.21	67	84.39	109	128	143.3	S
Cb(Nb)	41	6.88	14.32	25.04	38.3	50	103	125	......	S
Cd	48	8.991	16.904	37.47	......	......	......	......	......	S
Ce	58	5.6	12.3	20	33.3	......	......	......	......	SI
Cl	17	13.01	23.8	39.9	53.5	67.8	96.7	114.27	348.3	S
Co	27	7.86	17.05	33.49	83.1	......	......	......	......	S
Cr	24	6.764	16.49	30.95	50	73	91	161	185	S
Cs	55	3.893	25.1	35	......	......	......	......	......	S
Cu	29	7.724	20.29	36.83	......	......	......	......	......	S
Dy	66	6.8	......	......	......	......	......	......	......	S
Er	68	6.08	......	......	......	......	......	......	......	SI
Eu	63	5.67	11.24	......	......	......	......	......	......	S
F	9	17.418	34.98	62.646	87.14	114.214	157.117	185.139	953.6	S
Fe	26	7.87	16.18	30.643	56.8	......	......	......	151	S
Fr	87	4	......	......	......	......	......	......	......	S
Ga	31	6	20.57	30.7	64.2	......	......	......	......	S
Gd	64	6.16	12	......	......	......	......	......	......	S
Ge	32	7.88	15.93	34.21	44.7	93.4	......	......	......	S
H	1	13.595	......	......	......	......	......	......	......	S
He	2	24.481	54.403	......	......	......	......	......	......	S
Hf	72	7	14.9	23.2	33.3	......	......	......	......	S
Hg	80	10.43	18.751	34.2	49.5**	......	67**	......	......	S
I	53	10.454	19.13	......	......	......	......	......	170	S
In	49	5.785	18.86	28.03	54.4	......	......	......	......	S
Ir	77	9	......	......	......	......	......	......	......	S
K	19	4.339	31.81	46	60.9	82.6	99.7	118	155	S
Kr	36	13.996	24.56	36.9	43.5**	63**	94**	......	......	S
La	57	5.61	11.43	19.17	......	......	......	......	......	S
Li	3	5.39	75.619	122.419	......	......	......	......	......	S
Lu	71	......	14.7	......	......	......	......	......	......	S
Mg	12	7.644	15.031	80.14	109.29	141.23	186.49	224.9	265.957	S
Mn	25	7.432	15.636	33.69	52	76	......	119	196	S
Mo	42	7.10	16.15	27.13	46.4	61.2	68	126	153	S
N	7	14.53	29.593	47.426	77.45	97.863	551.925	666.83	......	S
Na	11	5.138	47.29	71.715	98.88	138.37	172.09	208.444	264.155	S
Nb(Cb)	41	6.88	14.32	25.04	38.3	50	103	125	......	S
Nd	60	5.51	......	......	......	......	......	......	......	SI
Ne	10	21.559	41.07	63.5	97.02	126.3	157.91	......	......	S
Ni	28	7.633	18.15	35.16	......	......	......	......	......	S
O	8	13.614	35.108	54.886	77.394	113.873	138.08	739.114	871.12	S
Os	76	8.5	17	......	......	......	......	......	......	S
P	15	10.484	19.72	30.156	51.354	65.007	220.414	263.31	309.26	S
Pb	82	7.415	15.028	31.93	42.31	68.8	......	......	......	S
Pd	46	8.33	19.42	32.92	......	......	......	......	......	S
Po	84	8.43	......	......	......	......	......	......	......	S
Pr	59	5.46	......	......	......	......	......	......	......	SI
Pt	78	9.0	18.56	......	......	......	......	......	......	S
Pu	94	5.1	......	......	......	......	......	......	......	S
Ra	88	5.277	10.144	......	......	......	......	......	......	S
Rb	37	4.176	27.5	40	......	......	......	......	......	S
Re	75	7.87	16.6	......	......	......	......	......	......	S
Rh	45	7.46	18.07	31.05	......	......	......	......	......	S
Rn	86	10.746	......	......	......	......	......	......	......	S
Ru	44	7.364	16.76	28.46	......	......	......	......	......	S
S	16	10.357	23.4	35	47.29	72.5	88.029	280.99	328.8	S
Sb	51	8.639	16.5	25.3	44.1	56	108	......	......	S
Sc	21	6.54	12.8	24.75	73.9	92	111	139	159	S
Se	34	9.75	21.5	32	43	68	82	155	......	S
Si	14	8.149	16.34	33.488	45.13	166.73	205.11	246.41	303.07	S
Sm	62	5.6	11.2	......	......	......	......	......	......	S
Sn	50	7.342	14.628	30.49	40.72	72.3	......	......	......	S
Sr	38	5.692	11.027	......	57	......	......	......	......	S
Ta	73	7.88	16.2	......	......	......	......	......	......	S
Tb	65	5.98	......	......	......	......	......	......	......	SI
Tc	43	7.28	15.26	29.54	......	......	......	......	......	S
Te	52	9.01	18.6	31	38	60	72	12.7	......	S
Th	90	6.95	......	......	29.38	......	......	......	......	SI
Ti	22	6.82	13.57	27.47	43.24	99.8	120	141	172	S
Tl	81	6.106	20.42	29.8	50.7	......	......	......	......	S
Tm	69	5.81	......	......	......	......	......	......	......	S
U	92	6.08	......	......	......	......	......	......	......	SI
V	23	6.74	14.65	29.31	48	65	129	151	170	S
W	74	7.98	17.7	......	......	......	......	......	......	S
Xe	54	12.127*	21.2	31.3	42	53	58	135	......	EI
Y	39	6.38	12.23	20.5	......	77	......	......	......	S
Yb	70	6.2	12.10	......	......	......	......	......	......	S
Zn	30	9.391	17.96	39.7	......	......	......	......	......	S
Zr	40	6.84	13.13	22.98	34.33	......	99	......	......	S

*These steps by S method. **These steps by EI method.

NUCLEAR SPINS, MOMENTS, AND MAGNETIC RESONANCE FREQUENCIES

Kenneth Lee and Weston A. Anderson

1967

This table contains the published values for the nuclear spins, magnetic moments, and quadrupole moments, and the calculated values for the nuclear magnetic resonance (NMR) frequency and for the relative sensitivities. Only those isotopes with both published spin and magnetic moment values are tabulated. The magnetic and quadrupole moment values were selected from results published during the period from January, 1955 to June, 1967. Earlier references were obtained from H. E. Walchli, A Table of Nuclear Moment Data, U.S. Atomic Energy Commission Report ORNL—1469, Supplement I (1953) and Supplement II (1955), and D. Streminger, J. M. Hollander, and G. T. Seaborg, Table of Isotopes, Rev. Mod. Phys. **30**, 585 (1958). A table containing the known (1963) spin and electromagnetic moment values of nuclear ground and excited states has been compiled by I. Lindgren, Perturbed Angular Correlations; E. Karlsson, E. Mathias, and K. Siegbahn, editors; North-Holland Publishing Co. (1964). The magnetic moments given in this latter table are corrected for the diamagnetic effect. A more complete list of spin and moment results for nuclei in excited states are included in Lindgren's table.

In general, the results chosen for this table were selected with an inclination to NMR measurements and to the precision of the measurement. Only six significant figures are used in this table. Therefore, the number of figures may be less than those published. The experimental methods employed in determining the moments are indicated by the following symbols:

Ab = atomic beam magnetic resonance (hyperfine structure, double or triple resonance or other method)
E = electron spin resonance or electron-nuclear double resonance
M = microwave absorption in gases
Mb = molecular (or diamagnetic) beam magnetic resonance
Mc = miscellaneous
Mo = Mössbauer effect
N = nuclear magnetic resonance
No = nuclear orientation
O = optical spectroscopy (hyperfine structure, band structure, double resonance, or optical pumping)
Qr = quadrupole resonance

Other symbols used in the table are:

A = atomic weight (mass number)
El = element
I = nuclear spin in units of $h/2\pi$
μ = magnetic moment in units of the nuclear magneton $eh/4\pi Mc$
m = metastable excited state
Q = quadrupole moment in units of barns (10^{-24} cm^2)
Z = atomic number
* = radioactive isotope
• = magnetic moment observed by NMR
( ) = assumed or estimated values

Assuming a nuclear magneton value of 5.0505×10^{-24} erg/gauss, the NMR frequency was calculated for a total field of 10^4 gauss. The sensitivities, relative to the proton, are calculated from the following expressions:

$$\text{Sensitivity at constant field} = 7.652 \times 10^{-3}\mu^3 (I+1)/I^2$$
$$\text{Sensitivity at constant frequency} = 0.2387\ \mu(I+1).$$

These expressions assume an equal number of nuclei, a constant temperature, and $T_1 = T_2$ (the longitudinal relaxation time equals the transverse relaxation time). These sensitivities represent the ideal induced voltage in the receiver coil at saturation and with a constant noise source. The calculated values are therefore determined under complete optimum conditions and should be regarded as such.

Isotope			Spin I	NMR Frequency in MHz for a 10 kilogauss field	Natural Abundance %	Relative Sensitivity for Equal Number of Nuclei		Magnetic Moment μ			Electric Quadrupole Moment Q		
Z	El	A				At constant field	At constant frequency	In multiples of the nuclear magneton (eh/4πMc)	Reference	Method	In multiples of barns ($10^{-24}cm^2$)	Reference	Method
0	n	1*	1/2	29.167	—	0.322	0.685	−1.91315	1	Ab	—		
1	•H	1	1/2	42.5759	99.985	1.00	1.00	2.79268	2	N	—		
1	•H	2	1	6.53566	1.5x10^{-2}	9.65x10^{-3}	0.409	0.857387	3	N	2.73x10^{-3}	4	Ab
1	•H	3*	1/2	45.4129	—	1.21	1.07	2.97877	5	N	—		
2	•He	3	1/2	32.433	1.3x10^{-4}	0.442	0.762	−2.1274	6 7	N	—		
3	•Li	6	1	6.2653	7.42	8.50x10^{-3}	0.392	0.82192	8	N	6.9x10^{-4}	9	N
3	•Li	7	3/2	16.546	92.58	0.293	1.94	3.2560	8	N	−3x10^{-2}	10	O
3	Li	8*	2	6.300	—	2.59x10^{-2}	1.184	1.653	11	Ab			
4	•Be	9	3/2	5.9834	100	1.39x10^{-2}	0.703	−1.1774	12 13	N	5.2x10^{-2}	14	Ab
5	•B	10	3	4.5754	19.58	1.99x10^{-2}	1.72	1.8007	15	N	7.4x10^{-2}	16	Ab
5	•B	11	3/2	13.660	80.42	0.165	1.60	2.6880	17	N	3.55x10^{-2}	16	Ab
6	•C	13	1/2	10.7054	1.108	1.59x10^{-2}	0.251	0.702199	18	N	—		
7	N	13*	1/2	4.91	—	1.53x10^{-3}	0.115	(−)0.322	19	Ab	—		
7	•N	14	1	3.0756	99.63	1.01x10^{-3}	0.193	0.40347	20 21	N	1.6x10^{-2}		Mc
7	•N	15	1/2	4.3142	0.37	1.04x10^{-3}	0.101	−0.28298	20 21	N	—		
8	O	15*	1/2	11.0	—	1.70x10^{-2}	0.257	0.719	23	Ab	—		
8	•O	17	5/2	5.772	3.7x10^{-2}	2.91x10^{-2}	1.58	−1.8930	24	N	−2.6x10^{-2}	25	M
9	•F	17*	5/2	14.40	—	0.451	3.94	4.720	26	N			
9	•F	19	1/2	40.0541	100	0.833	0.941	2.62727	17 21	N	—		
9	•F	20*	2	7.977	—	5.26x10^{-2}	1.50	2.093	27	N			
10	Ne	19*	(1/2)	28.75	—	0.308	0.675	−1.886	28	Ab			
10	Ne	21	3/2	3.3611	0.257	2.50x10^{-3}	0.395	−0.66140	29	Ab			
11	Na	21*	3/2	12.126	—	0.116	1.42	2.3861	30	Ab			
11	Na	22*	3	4.436	—	1.81x10^{-2}	1.67	1.746	31	Ab			
11	•Na	23	3/2	11.262	100	9.25x10^{-2}	1.32	2.2161	17	N	0.14-0.15	32 33	O
11	Na	24*	4	3.221	—	1.15x10^{-2}	2.02	1.690	34	Ab			
12	•Mg	25	5/2	2.6054	10.13	2.67x10^{-3}	0.714	−0.85449	13	N			
13	•Al	27	5/2	11.094	100	0.206	3.04	3.6385	35	N	0.149	36	Ab
14	•Si	29	1/2	8.4578	4.70	7.84x10^{-3}	0.199	−0.55477	37	N	—		
15	•P	31	1/2	17.235	100	6.63x10^{-2}	0.405	1.1305	38	N	—		
15	P	32*	1	1.923	—	2.46x10^{-4}	0.120	−0.2523	39	E			
16	•S	33	3/2	3.2654	0.76	2.26x10^{-3}	0.383	0.64257	37	N	−6.4x10^{-2}	40	M
16	S	35*	3/2	5.08	—	8.50x10^{-3}	0.597	1.00	41	M	4.5x10^{-2}	40	M
17	•Cl	35	3/2	4.1717	75.53	4.70x10^{-3}	0.490	0.82091	8	N	−7.89x10^{-2}	42	Ab
17	•Cl	36*	2	4.8931	—	1.21x10^{-2}	0.919	1.2838	43	N	−1.72x10^{-2}	44	M
17	•Cl	37	3/2	3.472	24.47	2.71x10^{-3}	0.408	0.6833	45	N	−6.21x10^{-2}	42	Ab
18	Ar	37*	3/2	5.08	—	8.50x10^{-3}	0.597	1.0	240	O			
19	K	38*	3	3.491	—	8.82x10^{-3}	1.31	1.374	46	Ab			
19	•K	39	3/2	1.9868	93.10	5.08x10^{-4}	0.233	0.39097	47 48	N	0.11	49 50	O
19	K	40*	4	2.470	1.18x10^{-2}	5.21x10^{-3}	1.55	−1.296	51	Mb			
19	•K	41	3/2	1.0905	6.88	8.40x10^{-5}	0.128	0.21459	48	N			
19	K	42*	2	4.345	—	8.50x10^{-3}	0.816	−1.140	52	Ab			
19	K	43*	3/2	0.828	—	3.68x10^{-5}	9.73x10^{-2}	0.163	53	Ab			
20	•Ca	41*	7/2	3.4681	—	1.14x10^{-2}	1.71	−1.5924	54	N			
20	•Ca	43	7/2	2.8646	0.145	6.40x10^{-3}	1.41	−1.3153	55	N			
21	Sc	43*	7/2	10.04	—	0.275	4.95	4.61	56	Ab	−0.26	56	Ab
21	Sc	44*	2	9.76	—	9.63x10^{-2}	1.83	2.56	57	Ab	0.14	57	Ab
21	Sc	44^*_m	6	5.03	—	9.24x10^{-2}	6.62	3.96	57	Ab	0.37	57	Ab
21	•Sc	45	7/2	10.343	100	0.301	5.10	4.7492	58	N	−0.22	59	Ab
21	Sc	46*	4	5.77	—	6.65x10^{-2}	3.62	3.03	60	Ab	0.12	60	Ab
21	Sc	47*	7/2	11.6	—	0.426	5.73	5.33	56	Ab	−0.22	56	Ab
22	Ti	45*	7/2	0.207	—	2.40x10^{-6}	0.102	0.095	61	Ab	1.5x10^{-2}	61	Ab
22	•Ti	47	5/2	2.4000	7.28	2.09x10^{-3}	0.658	−0.78710	62	N			
22	•Ti	49	7/2	2.4005	5.51	3.76x10^{-3}	1.18	−1.1022	62	N			
23	V	49*	7/2	9.71	—	0.249	4.79	4.46	63	E			
23	•V	50	6	4.2450	0.24	5.55x10^{-2}	5.58	3.3413	8	N			
23	•V	51	7/2	11.19	99.76	0.382	5.52	5.139	64	N	−4x10^{-2}	65	O
24	•Cr	53	3/2	2.4065	9.55	9.03x10^{-4}	0.283	−0.47354	66	N			
25	Mn	52*	6	3.907	—	4.33x10^{-2}	5.14	3.075	67	Ab			
25	Mn	52^*_m	2	0.030	—	2.9x10^{-9}	5.73x10^{-3}	0.008	68	Ab			
25	Mn	53*	7/2	11.0	—	0.362	5.42	5.05	69	E			

Isotope Z	El	A	Spin I	NMR Frequency in MHz for a 10 kilogauss field	Natural Abundance %	Relative Sensitivity for Equal Number of Nuclei: At constant field	At constant frequency	Magnetic Moment μ: In multiples of the nuclear magneton (eh/4πMc)	Reference	Method	Electric Quadrupole Moment Q: In multiples of barns ($10^{-24}cm^2$)	Reference	Method
25	Mn	54*	(2) (3)	8.4 6.6	—	6.11x10^{-2} 5.98x10^{-2}	1.58 2.48	(2.2) (2.6)	70	No			
25	•Mn	55	5/2	10.501	100	0.175	2.88	3.444	71	E	0.55	72	M
25	Mn	56*	3	8.233	—	0.116	3.09	3.240	73	Ab			
26	•Fe	57	1/2	1.3758	2.19	3.37x10^{-5}	3.23x10^{-2}	0.09024	74	E	—		
27	Co	55*	7/2	10.0	—	0.274	4.94	4.6	75	No			
27	Co	56*	4	7.34	—	0.136	4.60	3.85	76	E			
27	Co	57*	7/2	10.1	—	0.283	4.99	4.65	76	E			
27	Co	58*	2	15.4	—	0.381	2.90	4.05	77	E			
27	•Co	59	7/2	10.054	100	0.277	4.96	4.6163	241 242	N	0.40	78	Ab
27	Co	60*	5	5.793	—	0.101	5.44	3.800	79	E			
28	•Ni	61	3/2	3.8047	1.19	3.57x10^{-3}	0.447	−0.74868	80	N			
29	Cu	61*	3/2	10.8	—	8.22x10^{-2}	1.27	2.13	81	Ab			
29	•Cu	63	3/2	11.285	69.09	9.31x10^{-2}	1.33	2.2206	8	N	−0.16	82	E
29	Cu	64*	1	3.1	—	9.79x10^{-4}	0.191	0.40	83	Ab			
29	•Cu	65	3/2	12.089	30.91	0.114	1.42	2.3789	8	N	−0.15	82	E
29	Cu	66*	1	1.65	—	1.54x10^{-4}	0.103	−0.216	81	Ab			
30	Zn	65*	5/2	2.345	—	1.95x10^{-3}	0.643	0.7692	83	O	−2.4x10^{-2}	83	O
30	•Zn	67	5/2	2.663	4.11	2.85x10^{-3}	0.730	0.8733	37	N	0.15	83	O
31	Ga	68*	1	0.0892	—	2.45x10^{-8}	5.59x10^{-3}	0.0117	84	Ab	3.1x10^{-2}	84	Ab
31	Ga	69	3/2	10.22	60.4	6.91x10^{-2}	1.20	2.011	8	N	0.178	85	Ab
31	Ga	71	3/2	12.984	39.6	0.142	1.52	2.5549	8	N	0.112	85	Ab
31	Ga	72*	3	0.33591	—	7.80x10^{-6}	0.126	−0.13220	86	Ab	0.72	86	Ab
32	Ge	71*	1/2	8.4	—	7.64x10^{-3}	0.197	0.55	87	Ab	—		
32	•Ge	73	9/2	1.4852	7.76	1.40x10^{-3}	1.15	−0.87679	62	N	−0.2	88	M
33	•As	75	3/2	7.2919	100	2.51x10^{-2}	0.856	1.4349	15	N	0.3	89	M
33	As	76*	2	3.45	—	4.27x10^{-3}	0.649	−0.906	90	No			
34	•Se	77	1/2	8.118	7.58	6.93x10^{-3}	0.191	0.5325	8	N	—		
34	Se	79	7/2	2.22	—	2.98x10^{-3}	1.10	−1.02	91	M	0.9	40	M
35	Br	76*	1	4.18	—	2.52x10^{-3}	0.262	(−)0.548	92	Ab	0.27	92	Ab
35	•Br	79	3/2	10.667	50.54	7.86x10^{-2}	1.25	2.0990	8	N	0.33	93	Ab
35	Br	80*	1	3.92	—	2.08x10^{-3}	0.245	0.514	94	Ab	0.20	94	Ab
35	Br	80^{*}_{m}	5	2.008	—	4.20x10^{-3}	1.89	1.317	94	Ab	0.76	94	Ab
35	•Br	81	3/2	11.498	49.46	9.85x10^{-2}	1.35	2.2626	8	N	0.28	93	Ab
35	Br	82*	5	2.479	—	7.90x10^{-3}	2.33	(+)1.626	95	Ab	(+)0.76	95	Ab
36	•Kr	83	9/2	1.638	11.55	1.88x10^{-3}	1.27	−0.9671	96	N	0.15	97	O
36	Kr	85*	9/2	1.6956	—	2.08x10^{-3}	1.31	−1.001	98	O	0.25	98	O
37	Rb	81*	3/2	10.4	—	7.32x10^{-2}	1.22	2.05	99	Ab			
37	Rb	82^{*}_{m}	5	2.29	—	6.20x10^{-3}	2.15	1.50	99	Ab			
37	Rb	83*	5/2	4.33	—	1.23x10^{-2}	1.19	1.42	99	Ab			
37	Rb	84*	2	5.03	—	1.32x10^{-2}	0.945	−1.32	99	Ab			
37	•Rb	85	5/2	4.1108	72.15	1.05x10^{-2}	1.13	1.3482	100	N	0.27	101	Ab
37	Rb	86*	2	6.44	—	2.77x10^{-2}	1.21	−1.69	102	Ab			
37	•Rb	87	3/2	13.931	27.85	0.175	1.64	2.7414	103	N	0.13	101	Ab
38	•Sr	87	9/2	1.8452	7.02	2.69x10^{-3}	1.43	−1.0893	66	N	0.2	104	E
39	•Y	89	1/2	2.0859	100	1.18x10^{-4}	4.90x10^{-2}	−0.13682	47	N	—		
39	Y	90*	2	6.17	—	2.44x10^{-2}	1.16	−1.62	105	Ab	−0.16	105	Ab
39	Y	91*	1/2	2.49	—	1.99x10^{-4}	5.84x10^{-2}	0.163	60	Ab	—		
40	•Zr	91	5/2	3.97249	11.23	9.48x10^{-3}	1.09	−1.30284	106	N			
41	•Nb	93	9/2	10.407	100	0.482	8.07	6.1435	107	N	−0.2	108	O
42	•Mo	95	5/2	2.774	15.72	3.23x10^{-3}	0.760	0.9097	45	N	0.12	109	N
42	•Mo	97	5/2	2.832	9.46	3.43x10^{-3}	0.776	−0.9289	45	N	1.1	109	N
43	•Tc	99*	9/2	9.5830	—	0.376	7.43	5.6572	110	N	0.3	111	O
44	Ru	99	3/2	1.44	12.72	1.95x10^{-4}	0.169	−0.284	112	Mo			
44	Ru	101	5/2	2.1	17.07	1.41x10^{-3}	0.576	−0.69	113	O			
45	•Rh	103	1/2	1.3401	100	3.11x10^{-5}	3.15x10^{-2}	−0.08790	43 114	N	—		
46	•Pd	105	5/2	1.95	22.23	1.12x10^{-3}	0.534	−0.639	115	N			
47	Ag	104*	5	6.1	—	0.118	5.73	4.0	116	Ab			
47	Ag	104^{*}_{m}	2	14.0	—	0.291	2.65	3.7	116	Ab			
47	Ag	105*	1/2	1.54	—	4.73x10^{-5}	3.62x10^{-2}	0.101	117	Ab	—		
47	•Ag	107	1/2	1.7229	51.82	6.62x10^{-5}	4.05x10^{-2}	−0.11301	47 118	N	—		
47	Ag	108*	1	32.0	—	1.13	2.01	4.2	119	Ab			
47	•Ag	109	1/2	1.9807	48.18	1.01x10^{-4}	4.65x10^{-2}	−0.12992	47 118	N	—		
47	Ag	110^{*}_{m}	6	4.557	—	6.87x10^{-2}	5.99	3.587	120	Ab			
47	Ag	111*	1/2	2.21	—	1.40x10^{-4}	5.19x10^{-2}	−0.145	121	Ab	—		
47	Ag	112*	2	0.2077	—	9.00x10^{-7}	3.90x10^{-2}	0.0545	122	Ab			
47	Ag	113*	1/2	2.41	—	1.81x10^{-4}	5.66x10^{-2}	0.158	122	Ab	—		
48	Cd	107*	5/2	1.879	—	1.00x10^{-3}	0.515	−0.6162	123	O	0.8	123	O
48	Cd	109*	5/2	2.529	—	2.44x10^{-3}	0.693	−0.8293	124	O	0.8	124	O
48	•Cd	111	1/2	9.028	12.75	9.54x10^{-3}	0.212	−0.5922	125 126	N	—		
48	•Cd	113	1/2	9.445	12.26	1.09x10^{-2}	0.222	−0.6195	125 126	N	—		
48	Cd	113^{*}_{m}	11/2	1.51	—	2.13x10^{-3}	1.69	−1.09	239	O	−0.79	239	O
48	Cd	115*	1/2	9.862	—	1.24x10^{-2}	0.232	−0.6469	127	O	—		
48	Cd	115^{*}_{m}	11/2	1.447	—	1.87x10^{-3}	1.62	−1.044	127	O	−0.61	127	O
49	•In	113	9/2	9.3099	4.28	0.345	7.22	5.4960	45	N	1.14	97	Ab
49	In	113^{*}_{m}	1/2	3.209	—	4.28x10^{-4}	7.54x10^{-2}	−0.2105	128	Ab	—		
49	In	114^{*}_{m}	5	7.2	—	0.191	6.73	4.7	129	Ab			
49	•In	115*	9/2	9.3301	95.72	0.347	7.23	5.5079	130	N	1.16	97	Ab
49	In	115^{*}_{m}	1/2	3.715	—	6.64x10^{-4}	8.73x10^{-2}	−0.2437	131	Ab	—		
49	In	116*	5	6.42	—	0.137	6.03	4.21	132	Ab			
49	In	116^{*}_{m}	5	6.7	—	0.156	6.30	4.4	129	Ab			
50	•Sn	115	1/2	13.922	0.35	3.50x10^{-2}	0.327	−0.91320	125	N	—		
50	•Sn	117	1/2	15.168	7.61	4.52x10^{-2}	0.356	−0.99490	125	N	—		
50	•Sn	119	1/2	15.869	8.58	5.18x10^{-2}	0.373	−1.0409	125	N	—		
51	•Sb	121	5/2	10.189	57.25	0.160	2.79	3.3415	133	N	−0.5	134	O
51	Sb	122*	2	7.24	—	3.94x10^{-2}	1.36	−1.90	135 136	No			
51	•Sb	123	7/2	5.5176	42.75	4.57x10^{-2}	2.72	2.5334	45	N	−0.7	134	O
52	Te	119*	1/2	4.12	—	9.04x10^{-4}	9.67x10^{-2}	0.27	137	Ab	—		
52	•Te	123	1/2	11.16	0.87	1.8x10^{-2}	0.262	−0.7319	37	N	—		
52	•Te	125	1/2	13.45	6.99	3.15x10^{-2}	0.316	−0.8824	37	N	—		
53	I	125*	5/2	9.0	—	0.116	2.51	3	138	M	−0.66	138	M
53	•I	127	5/2	8.5183	100	9.34x10^{-2}	2.33	2.7937	139	N	−0.69	134	O
53	•I	129*	7/2	5.6694	—	4.96x10^{-2}	2.80	2.6031	139	N	−0.48	140	Qr
53	I	131*	7/2	5.963	—	5.77x10^{-2}	2.94	2.738	141	Ab	−0.41	141	Ab
54	•Xe	129	1/2	11.777	26.44	2.12x10^{-2}	0.277	−0.77247	142	N	—		
54	•Xe	131	3/2	3.4911	21.18	2.76x10^{-3}	0.410	0.68697	142	N	−0.12	143	O
55	Cs	127*	1/2	21.8	—	0.134	0.512	1.43	144	Ab	—		
55	Cs	129*	1/2	22.4	—	0.146	0.526	1.47	144	Ab	—		
55	Cs	130*	1	10.7	—	4.20x10^{-2}	0.668	1.4	144	Ab			
55	Cs	131*	5/2	10.72	—	0.186	2.94	3.517	145	Ab			
55	Cs	132*	2	8.46	—	6.28x10^{-2}	1.59	2.22	144	Ab			
55	•Cs	133	7/2	5.58469	100	4.74x10^{-2}	2.75	2.56422	8	N	−3x10^{-3}	146 147	Ab
55	Cs	134*	4	5.666	—	6.28x10^{-2}	3.55	2.973	148	Ab	0.43	149	O
55	Cs	134^{*}_{m}	8	1.0447	—	1.42x10^{-3}	2.36	1.0964	150	Ab			
55	Cs	135*	7/2	5.9096	—	5.62x10^{-2}	2.91	2.7134	148	Ab			
55	Cs	137*	7/2	6.1459	—	6.32x10^{-2}	3.03	2.8219	148	Ab			
56	•Ba	135	3/2	4.2296	6.59	4.90x10^{-3}	0.497	0.83229	151 152	N	0.25	153	O
56	•Ba	137	3/2	4.7315	11.32	6.86x10^{-3}	0.556	0.93107	151 152	N	0.2	153	O
57	•La	138	5*	5.6171	0.089	9.19x10^{-2}	5.28	3.6844	154	N	2.7	154	N
57	•La	139	7/2	6.0144	99.911	5.92x10^{-2}	2.97	2.7615	107	N	0.21	155	Ab
58	Ce	137*	3/2	4.6	—	6.20x10^{-3}	0.537	0.9	156	No			
58	Ce	137^{*}_{m}	11/2	0.96	—	5.40x10^{-4}	1.07	0.69	157	No			
58	Ce	139*	3/2	5.1	—	8.50x10^{-3}	0.597	1.0	156	No			
58	Ce	141*	7/2	2.1	—	2.57x10^{-3}	1.04	0.97	158	E			

NUCLEAR SPINS, MOMENTS, AND MAGNETIC RESONANCE FREQUENCIES (Continued)

Isotope			Spin I	NMR Frequency in MHz for a 10 kilogauss field	Natural Abundance %	Relative Sensitivity for Equal Number of Nuclei		Magnetic Moment μ			Electric Quadrupole Moment Q		
Z	El	A				At constant field	At constant frequency	In multiples of the nuclear magneton (eh/4πMc)	Reference	Method	In multiples of barns ($10^{-24}cm^2$)	Reference	Method
58	Ce	143*	7/2	2.2	—	2.81x10^{-3}	1.07	1.0	156	No			
59	•Pr	141	5/2	12.5	100	0.293	3.42	4.09	159 160	O	−5.9x10^{-2}	158	Ab
59	Pr	142*	2	1.1	—	1.55x10^{-4}	0.215	0.30	161	Ab	4x10^{-2}	161	Ab
60	Nd	143	7/2	2.315	12.17	3.38x10^{-3}	1.14	−1.063	162	Ab	−0.48	162	Ab
60	Nd	145	7/2	1.42	8.30	7.86x10^{-4}	0.703	−0.654	162	Ab	−0.25	162	Ab
60	Nd	147*	5/2	1.77	—	8.32x10^{-4}	0.484	0.579	158	E			
61	Pm	143*	(5/2) (7/2)	11.6 8.5	—	0.235 0.167	3.17 4.19	(3.8) (3.9)	163	No			
61	Pm	144*	(5) (6)	2.6 2.3	—	9.02x10^{-3} 8.68x10^{-3}	2.43 3.01	(1.7) (1.8)	164	No			
61	Pm	147*	7/2	5.62	—	4.83x10^{-2}	2.77	2.58	165	O	0.7	165	O
61	Pm	148*	1	16	—	0.142	1.00	2.1	166	Ab	0.2	166	Ab
61	Pm	148$^*_m$	6	2.3	—	8.68x10^{-3}	3.01	1.8	163	No			
61	Pm	149*	7/2	7.2	—	0.101	3.54	3.3	163	No			
61	Pm	151*	5/2	5.5	—	2.50x10^{-2}	1.50	1.8	167	Ab	1.9	167	Ab
62	Sm	147	7/2	1.76	14.97	1.48x10^{-3}	0.867	−0.807	168	Ab	−0.208	158	Ab
62	Sm	149	7/2	1.40	13.83	7.47x10^{-4}	0.691	−0.643	158	Ab	6.0x10^{-2}	158	Ab
63	•Eu	151	5/2	10.559	47.82	0.178	2.89	3.4630	169	Ab	1.16	170	O
63	Eu	152*	3	4.858	—	2.38x10^{-2}	1.83	1.912	171	Ab			
63	•Eu	153	5/2	4.6627	52.18	1.53x10^{-2}	1.28	1.5292	169 172	Ab	2.9	170	O
63	Eu	154*	3	5.084	—	2.72x10^{-2}	1.91	2.001	158	E			
64	•Gd	155	3/2	1.6	14.73	2.79x10^{-4}	0.191	−0.32	173 174	O	1.6	173	O
64	•Gd	157	3/2	2.0	15.68	5.44x10^{-4}	0.239	−0.40	173 174	O	2	173	O
65	Tb	156*	3	3.8	—	1.15x10^{-2}	1.43	1.5	175	No	1.4	175	No
65	Tb	159	3/2	9.66	100	5.83x10^{-2}	1.13	1.90	158	E	1.3	176	O
65	Tb	160*	3	4.1	—	1.39x10^{-2}	1.53	1.6	177	No	1.9	177	No
66	Dy	155*	(3/2)	1.1	—	7.87x10^{-5}	0.125	0.21	178	No			
66	Dy	157*	(3/2)	1.6	—	2.79x10^{-4}	0.191	0.32	178	No			
66	Dy	161	5/2	1.4	18.88	4.17x10^{-4}	0.384	−0.46	158	E	1.4	158	E
66	Dy	163	5/2	2.0	24.97	1.12x10^{-3}	0.535	0.64	158	E	1.6	158	E
67	Ho	165	7/2	8.73	100	0.181	4.31	4.01	158	Ab	2.82	158	Ab
68	Er	165*	5/2	2.0	—	1.18x10^{-3}	0.543	0.65	166	Ab	2.2	166	Ab
68	Er	167	7/2	1.23	22.94	5.07x10^{-4}	0.607	−0.565	162	Ab	2.83	162	Ab
68	Er	169*	1/2	7.8	—	6.09x10^{-3}	0.183	0.51	179	Ab	—		
68	Er	171*	5/2	2.1	—	1.47x10^{-3}	0.585	0.70	180	Ab			
69	Tm	166*	2	0.19	—	7.17x10^{-7}	3.58x10^{-2}	0.05	181	Ab	4.6	181	Ab
69	•Tm	169	1/2	3.52	100	5.66x10^{-4}	8.27x10^{-2}	−0.231	182 160	Ab	—		
69	Tm	170*	1	2.0	—	2.69x10^{-4}	0.124	0.26	183	Ab	0.61	183	Ab
69	Tm	171*	1/2	3.46	—	5.37x10^{-4}	8.13x10^{-2}	0.227	180	Ab	—		
70	•Yb	171	1/2	7.4990	14.31	5.46x10^{-3}	0.176	0.49188	184 185	O	—		
70	Yb	173	5/2	2.0659	16.13	1.33x10^{-3}	0.566	−0.67755	184 185	O	2.8	186	O
70	Yb	175*	(7/2)	0.33	—	9.40x10^{-6}	0.161	−0.15	187	No			
71	•Lu	175	7/2	4.86	97.41	3.12x10^{-2}	2.40	2.23	188	N	5.68	189	Ab
71	Lu	176*	7	3.4	2.59	3.72x10^{-2}	5.92	3.1	190	Ab	8.0	190	Ab
71	Lu	177*	7/2	4.84	—	3.08x10^{-2}	2.38	2.22	191	Ab	5.51	191	Ab
72	Hf	177	7/2	1.3	18.50	6.38x10^{-4}	0.655	0.61	192 193	O	3	192 193	O
72	Hf	179	9/2	0.80	13.75	2.16x10^{-4}	0.617	−0.47	193	O	3	193	O
73	•Ta	181	7/2	5.096	99.988	3.60x10^{-2}	2.51	2.340	194	N	3	195	O
74	•W	183	1/2	1.7716	14.40	7.20x10^{-5}	4.16x10^{-2}	0.116205	196	N	—		
75	•Re	185	5/2	9.5855	37.07	0.133	2.63	3.1437	13	N	2.8	97 241	O
75	Re	186*	1	13.17	—	7.90x10^{-2}	0.825	1.728	197	Ab			
75	•Re	187*	5/2	9.6837	62.93	0.137	2.65	3.1759	13	N	2.6	97 241	O
75	Re	188*	1	13.55	—	8.59x10^{-2}	0.848	1.777	197	Ab			
76	•Os	187	1/2	0.98059	1.64	1.22x10^{-5}	2.30x10^{-2}	0.06432	198	N	—		
76	•Os	189	3/2	3.3034	16.1	2.34x10^{-3}	0.388	0.65004	199	N	0.8	195	O
77	•Ir	191	3/2	0.7318	37.3	2.53x10^{-5}	8.59x10^{-2}	0.1440	200 201	N	1.5	202	O
77	•Ir	193	3/2	0.7968	62.7	3.27x10^{-5}	9.36x10^{-2}	0.1568	200 201	N	1.5	202	O
78	•Pt	195	1/2	9.153	33.8	9.94x10^{-3}	0.215	0.6004	45	N	—		
79	Au	190*	1	0.496	—	4.20x10^{-6}	3.10x10^{-2}	0.065	203	Ab			
79	Au	194*	1	0.56	—	5.95x10^{-6}	3.49x10^{-2}	0.073	204	Ab			
79	Au	195*	3/2	0.742	—	2.65x10^{-5}	8.71x10^{-2}	0.146	204	Ab			
79	Au	196*	2	2.3	—	1.24x10^{-3}	0.430	0.6	204	Ab			
79	•Au	197	3/2	0.729188	100	2.51x10^{-5}	8.56x10^{-2}	0.143489	205 201	Ab	0.59	206	Ab
79	Au	198*	2	2.227	—	1.14x10^{-3}	0.418	0.5842	207	Ab			
79	Au	199*	3/2	1.358	—	1.62x10^{-4}	0.160	0.2673	207	Ab			
80	Hg	193*	3/2	3.1	—	1.93x10^{-3}	0.364	−0.61	208	O			
80	Hg	193$^*_m$	13/2	1.23	—	1.57x10^{-3}	1.88	−1.05	209	O	1.37	209	O
80	Hg	195*	1/2	8.1	—	6.84x10^{-3}	0.190	0.53	210 211	O	—		
80	Hg	195$^*_m$	13/2	1.22	—	1.53x10^{-3}	1.86	−1.04	209	O	1.41	209	O
80	Hg	197*	1/2	7.9	—	6.46x10^{-3}	0.186	0.52	212	O	—		
80	•Hg	199	1/2	7.59012	16.84	5.67x10^{-3}	0.178	0.497859	213 45	No			
80	•Hg	201	3/2	2.8099	13.22	1.44x10^{-3}	0.330	−0.55293	214 215	O	0.50	216	O
80	Hg	203*	5/2	2.5	—	2.45x10^{-3}	0.693	0.83	217	O	0.5	217	O
81	Tl	197*	1/2	23.6	—	0.171	0.555	1.55	218	O	—		
81	Tl	199*	1/2	23.9	—	0.178	0.562	1.57	219	O	—		
81	Tl	200*	2	0.57	—	1.94x10^{-5}	0.107	(0.15)	219	O			
81	Tl	201*	1/2	24.1	—	0.181	0.566	1.58	219	O	—		
81	Tl	202*	2	0.57	—	1.94x10^{-5}	0.107	(0.15)	219	O			
81	•Tl	203	1/2	24.332	29.50	0.187	0.571	1.5960	220	N	—		
81	Tl	204*	2	0.34	—	4.05x10^{-6}	6.37x10^{-2}	0.089	221	Ab			
81	•Tl	205	1/2	24.570	70.50	0.192	0.577	1.6116	8	N	—		
82	•Pb	207	1/2	8.90771	22.6	9.16x10^{-3}	0.209	0.584284	222	N	—		
83	Bi	203*	9/2	7.78	—	0.201	6.03	4.59	223	Ab	−0.64	223	Ab
83	Bi	204*	6	5.40	—	0.114	7.10	4.25	223	Ab	−0.41	223	Ab
83	Bi	205*	9/2	9.3	—	0.346	7.22	(5.5)	223	Ab			
83	Bi	206*	6	5.79	—	0.141	7.62	4.56	223	Ab	−0.19	223	Ab
83	•Bi	209*	9/2	6.84178	100	0.137	5.30	4.03896	224	N	−0.4	97	O
83	Bi	210*	1	0.337	—	1.32x10^{-6}	2.11x10^{-2}	0.0442	225	Ab	0.13	225	Ab
84	Po	205*	5/2	0.79	—	7.55x10^{-5}	0.217	0.26	226	Ab	0.17	226	Ab
84	Po	207*	5/2	0.82	—	8.43x10^{-5}	0.226	0.27	226	Ab	0.28	226	Ab
89	Ac	227*	3/2	5.6	—	1.13x10^{-2}	0.656	1.1	227	O	−1.7	227	O
90	Th	229*	5/2	1.2	—	2.74x10^{-4}	0.334	0.4	228	O	4.6	228	O
91	Pa	231*	3/2	9.96	—	6.40x10^{-2}	1.17	1.96	229	E			
91	Pa	233*	3/2	17	—	0.334	2.03	3.4	230	Ab	−3.0	230	Ab
92	U	233*	5/2	1.6	—	6.75x10^{-4}	0.451	0.54	231	E	3.5	231	E
92	U	235*	7/2	0.76	0.72	1.21x10^{-4}	0.376	0.35	231	E	4.1	231	E
93	Np	237*	5/2	18	—	0.926	5.01	(6)	232	E			
94	Pu	239*	1/2	3.05	—	3.67x10^{-4}	7.16x10^{-2}	0.200	233	Ab	—		
94	Pu	241*	5/2	2.09	—	1.38x10^{-3}	0.573	−0.686	234	O			
95	Am	241*	5/2	4.82	—	1.69x10^{-2}	1.32	1.58	235	Ab	4.9	236	O
95	Am	242*	1	2.90	—	8.46x10^{-4}	0.182	0.381	235	Ab	−2.8	237	Ab
95	Am	243*	5/2	4.79	—	1.66x10^{-2}	1.31	1.57	238	O	4.9	236	O
Free electron with g=2.00232			1/2	2.80246x 10^4	—	2.84x10^{-8}	657	−1836.09	—		—		

REFERENCES

(1) V. W. Cohen, et al., Phys. Rev. **104**, 283 (1956).
(2) H. Sommer, et al., Phys. Rev. **82**, 697 (1951).
(3) T. F. Wimett, Phys. Rev. **91**, 499A (1953).
(4) H. Kopfermann, et al., Z. Physik **144**, 9 (1956).
(5) F. Bloch, et al., Phys. Rev. **71**, 551 (1947).
(6) H. L. Anderson, Phys. Rev. **76**, 1460 (1949).
(7) M. P. Klein, et al., Phys. Rev. **106**, 837 (1957).
(8) H. E. Walchli, Thesis, M.S., U. of Tenn. (1954). AEC Report ORNL-1775.
(9) N. A. Schuster, et al., Phys. Rev. **81**, 157 (1951).
(10) K. C. Brog, et al., Phys. Rev. **153**, 91 (1967).
(11) D. Connor, Phys. Rev. Letters **3**, 429 (1959).
(12) L. C. Brown, et al., J. Chem. Phys. **24**, 751 (1956).
(13) F. Alder, et al., Phys. Rev. **82**, 105 (1951).
(14) A. G. Blachman, et al., Bull Am. Phys. Soc. **11**, 343 (1966).
(15) Y. Ting, et al., Phys. Rev. **89**, 595 (1953).
(16) G. Wessel, Phys. Rev. **92**, 1581 (1953).
(17) G. Lindström, Arkiv Fysik **4**, 1 (1951).
(18) V. Royden, Phys. Rev. **96**, 543 (1954).
(19) A. M. Bernstein, et al., Phys. Rev. **136**, B27 (1964).
(20) L. W. Anderson, et al., Phys. Rev. **116**, 87 (1959).
(21) M. R. Baker, et al., Phys. Rev. **133**, A1533 (1964).
(22) A. Bassompiere, Compt. Rend. **240**, 285 (1955).
(23) E. D. Commins, et al., Phys. Rev. **131**, 700 (1963).
(24) F. Alder, et al., Phys. Rev. **81**, 1067 (1951).
(25) M. J. Stevenson, et al., Phys. Rev. **107**, 635 (1957).
(26) K. Sugimoto, et al., J. Phys. Soc. Japan **21**, 213 (1966).
(27) T. Tsang, et al., Phys. Rev. **132**, 1141 (1963).
(28) E. D. Commins, et al., Phys. Rev. Letters **10**, 347 (1963).
(29) J. T. LaTourette, et al., Phys. Rev. **107**, 1202 (1957).
(30) O. Ames, et al., Phys. Rev. **137**, B1157 (1965).
(31) L. Davis, et al., Phys. Rev. **76**, 1068 (1949).
(32) H. Ackermann, Z. Physik. **194**, 253 (1966).
(33) M. Baumann, et al., Z. Physik **194**, 270 (1966).
(34) Y. W. Chan, et al., Phys. Rev. **150**, 933 (1966).
(35) L. C. Brown, et al., J. Chem. Phys. **24**, 751 (1956).
(36) H. Lew, et al., Phys. Rev. **90**, 1 (1953).
(37) H. E. Weaver, Phys. Rev. **89**, 923 (1953).
(38) T. Kanda, et al., Phys. Rev. **85**, 938 (1952).
(39) G. Feher, et al., Phys. Rev. **107**, 1462 (1957).
(40) G. R. Bird, et al., Phys. Rev. **94**, 1203 (1954).
(41) B. F. Burke, et al., Phys. Rev. **93**, 193 (1954).
(42) V. Jaccarino, et al., Phys. Rev. **83**, 471 (1951).
(43) P. B. Sogo, et al., Phys. Rev. **98**, 1316 (1955).
(44) C. H. Townes, et al., Phys. Rev. **76**, 691 (1949).
(45) W. G. Proctor, et al., Phys. Rev. **81**, 20 (1951).
(46) E. A. Phillips, et al., Phys. Rev. **138**, B773 (1965).
(47) E. Brun, et al., Phys. Rev. **93**, 172 (1954).
(48) O. Lutz, et al., Phys. Letters **24A**, 122 (1967).
(49) G. W. Series, Phys. Rev. **105**, 1128 (1957).
(50) G. J. Ritter, et al., Proc. Roy. Soc. (London) **238**, 473 (1957).
(51) J. T. Eisinger, et al., Phys. Rev. **86**, 73 (1952).
(52) J. M. Kahn, et al., Phys. Rev. **134**, A45 (1964).
(53) F. R. Petersen, et al., Phys. Rev. **116**, 734 (1959).
(54) E. Brun, et al., Phys. Rev. Letters **9**, 166 (1962).
(55) C. D. Jeffries, Phys. Rev. **90**, 1130 (1953).
(56) R. G. Cornwall, et al., Phys. Rev. **141**, 1106 (1966).
(57) D. L. Harris, et al., Phys. Rev. **132**, 310 (1963).
(58) D. M. Hunten, Can. J. Phys. **29**, 463 (1951).
(59) G. Fricke, et al., Z. Physik, **156**, 416 (1959).
(60) F. R. Petersen, et al., Phys. Rev. **128**, 1740 (1962).
(61) R. G. Cornwall, et al., Phys. Rev. **148**, 1157 (1966).
(62) C. D. Jeffries, Phys. Rev. **92**, 1262 (1953).
(63) M. M. Weiss, et al., Bull. Am. Phys. Soc. **2**, 31 (1957).
(64) H. E. Walchli, et al., Phys. Rev. **87**, 541 (1952).
(65) K. Murakawa, et al., J. Phys. Soc. Japan **21**, 1466 (1966).
(66) C. D. Jeffries, et al., Phys. Rev. **91**, 1286 (1953).
(67) K. E. Adelroth, et al., Arkiv Fysik **31**, 549 (1966).
(68) E. A. Phillips, et al., Phys. Rev. **140**, B555 (1965).
(69) W. Dobrowolski, et al., Phys. Rev. **104**, 1378 (1956).
(70) R W. Bauer, et al., Phys. Rev. **120**, 946 (1960).
(71) W. B. Mims, et al., Phys. Letters **24A**, 481 (1967).
(72) A. Javan, et al., Phys. Rev. **96**, 649 (1954).
(73) W. J. Childs, et al., Phys. Rev. **122**, 891 (1961).
(74) P. R. Locher, et al., Phys. Rev. **139**, A991 (1965).
(75) R. W. Bauer, et al., Nucl. Phys. **16**, 264 (1960).
(76) J. M. Baker, et al., Proc. Phys. Soc. (London), **A69**, 354 (1956).
(77) W. Dobrov, et al., Phys. Rev. **108**, 60 (1957).
(78) D. V. Ehrenstein, et al., Z. Physik **159**, 230 (1960).
(79) W. Dobrowolski, et al., Phys. Rev. **101**, 1001 (1956).
(80) L. E. Drain, Phys. Letters **11**, 114 (1964).
(81) B. M. Dodsworth, et al., Phys. Rev. **142**, 638 (1966).
(82) B. Bleaney, et al., Proc. Roy. Soc. (London) **228A**, 166 (1955).
(83) A Lemonick, et al., Phys. Rev. **95**, 1356 (1954).
(84) V. J. Ehlers, et al., Phys. Rev. **127**, 529 (1962).
(85) R. T. Daly, et al., Phys. Rev. **96**, 539 (1954).
(86) W. J. Childs, et al., Phys. Rev. **120**, 2138 (1960).
(87) W. J. Childs, et al., Phys. Rev. **141**, 15 (1966).
(88) J. M. Mays, et al., Phys. Rev. **81**, 940 (1951).
(89) B. P. Dailey, et al., Phys. Rev. **74**, 1245 (1948).
(90) F. M. Pipkin, et al., Phys. Rev. **106**, 1102 (1957).
(91) W. A. Hardy, et al., Phys. Rev. **92**, 1532 (1953).
(92) E. Lipworth, et al., Phys. Rev. **119**, 1053 (1960).
(93) J. G. King, et al., Phys. Rev. **94**, 1610 (1954).
(94) M. B. White, et al., Phys. Rev. **136**, B584 (1964).
(95) H. L. Garvin, et al., Phys. Rev. **116**, 393 (1959).
(96) E. Brun, et al., Helv. Phys. Acta **27**, 173A (1954).
(97) J. E. Mack, Rev. Mod. Phys. **22**, 64 (1950).
(98) E. Rasmussen, et al., Z. Physik **141**, 160 (1955).
(99) J. C. Hubbs, et al., Phys. Rev. **107**, 723 (1957).
(100) H. Walchli, et al., Phys. Rev. **85**, 922 (1952).
(101) B. Senitzky, et al., Phys. Rev. **103**, 315 (1956).
(102) E. H. Bellamy, et al., Phil. Mag. **44**, 33 (1953).
(103) E. Yasaitis, et al., Phys. Rev. **82**, 750 (1951).
(104) J. W. Culvahouse, et al., Phys. Rev. **140**, A1181 (1965).
(105) F. R. Petersen, et al., Phys. Rev. **125**, 284 (1962).
(106) E. Brun, et al., Phys. Rev. **105**, 1929 (1957).
(107) R. E. Sheriff, et al., Phys. Rev. **82**, 651 (1951).
(108) K. Murakawa, Phys. Rev. **98**, 1285 (1955).
(109) A. Narath, et al., Phys. Rev. **143**, 328 (1966).
(110) H. Walchli, et al., Phys. Rev. **85**, 479 (1952).
(111) K. G. Kessler, et al., Phys. Rev. **92**, 303 (1953).
(112) E. Matthias, et al., Phys. Rev. **139**, B532 (1965).
(113) K. Murakawa, J. Phys. Soc. Japan **10**, 919 (1955).
(114) J. A. Seitchik, et al., Phys. Rev. **138**, A148 (1965).
(115) J. A. Seitchik, et al., Phys. Rev. **136**, A1119 (1964).
(116) O. Ames, et al., Phys. Rev. **123**, 1793 (1960).
(117) W. B. Ewbank, et al., Phys. Rev. **129**, 1617 (1963).
(118) P. B. Sogo, et al., Phys. Rev. **93**, 174 (1954).
(119) G. K. Rochester, et al., Phys. Letters 8, **266** (1964).
(120) S. G. Schmelling, et al., Phys. Rev. **154**, 1142 (1967).
(121) G. K. Woodgate, et al., Proc. Phys. Soc. (London) **69**, 581 (1956).
(122) Y. W. Chan, et al., Phys. Rev. **133**, B1138 (1964).
(123) F. W. Byron, et al., Phys. Rev. **132**, 1181 (1963).
(124) P. Thaddeus, et al., Phys. Rev. **132**, 1186 (1963).
(125) W. G. Proctor, Phys. Rev. **79**, 35 (1950).
(126) M. Leduc, et al., Compt. Rend. **B262**, 736 (1966).
(127) M. N. McDermott, et al., Phys. Rev. **134**, B25 (1964).
(128) W. J. Childs, et al., Phys. Rev. **118**, 1578 (1960).
(129) L. S. Goodman, et al., Phys. Rev. **108**, 1524 (1957).
(130) M. Rice, et al., Phys. Rev. **106**, 953 (1957).
(131) J. A. Cameron, et al., Can. J. Phys. **40**, 931 (1962).
(132) P. B. Nutter, Phil. Mag. **1**, 587 (1956).
(133) V. W. Cohen, et al., Phys. Rev. **79**, 191 (1950).
(134) K. Murakawa, Phys. Rev. **100**, 1369 (1955).
(135) F. M. Pipkin, Bull. Am. Phys. Soc. **3**, 8 (1958).
(136) P. C. B. Fernando, et al., Phil. Mag. **5**, 1309 (1960).
(137) K. E. Adelroth, et al., Arkiv Fysik **30**, 111 (1965).
(138) P. C. Fletcher, et al., Phys. Rev. **110**, 536 (1958).
(139) H. Walchli, et al., Phys. Rev. **82**, 97 (1951).
(140) R. Livingston, et al., Phys. Rev. **90**, 609 (1953).
(141) E. Lipworth, et al., Phys. Rev. **119**, 2022 (1960).
(142) D. Brinkmann, Helv. Phys. Acta **36**, 413 (1963).
(143) A. Bohr, et al., Arkiv Fysik **4**, 455 (1952).
(144) W. A. Nierenberg, et al., Phys. Rev. **112**, 186 (1958).
(145) R. D. Worley, et al., Phys. Rev. **140**, B1483 (1965).
(146) P. Buck, et al., Phys. Rev. **104**, 553 (1956).
(147) K. H. Althoff, et al., Naturwiss. **41**, 368 (1954).
(148) H. H. Stroke, et al., Phys. Rev. **105**, 590 (1957).
(149) G. Heinzelmann, et al., Phys. Letters 21, 162 (1966).
(150) V. W. Cohen, et al., Phys. Rev. **127**, 517 (1962).
(151) H. E. Walchli, et al., Phys. Rev. **102**, 1334 (1956).
(152) L. Olschewski, et al., Z. Physik **196**, 77 (1966).
(153) N. I. Kaliteevskii, et al., Soviet Physics—JETP **12**, 661 (1961).
(154) P. B. Sogo, et al., Phys. Rev. **99**, 613 (1955).
(155) K. Murakawa, et al., J. Phys. Soc. Japan **16**, 2533 (1961).
(156) J. N. Haag, et al., Phys. Rev. **129**, 1601 (1963).
(157) J. Blok, et al., Phys. Rev. **143**, 78 (1966).
(158) B. Bleaney, Quantum Electronics Conf., Paris, 1963; P. Grivet and N. Bloembergen, editors; Columbia U. Press (1964).
(159) J. Reader, et al., Phys. Rev. **137**, B784 (1965).
(160) E. D. Jones, Phys. Rev. Letters **19**, 432 (1967).
(161) A. Y. Cabezas, et al., Phys. Rev. **126**, 1004 (1962).
(162) K. F. Smith, et al., Proc. Phys. Soc. (London) **86**, 1249 (1965).
(163) R. W. Grant, et al., Phys. Rev. **130**, 1100 (1963).
(164) D. A. Shirley, et al., Phys. Rev. **121**, 558 (1961).
(165) J. Reader, Phys. Rev. **141**, 1123 (1966).
(166) D. Ali, et al., Phys. Rev. **138**, B1356 (1965).
(167) B. Budick, et al., Phys. Rev. **132**, 723 (1963).
(168) G. K. Woodgate, Proc. Roy. Soc. (London) **A293**, 117 (1966).
(169) L. Evans, et al., Proc. Roy. Soc. (London) **A289**, 114 (1965).
(170) W. Müller, et al., Z. Physik **183**, 303 (1965).
(171) S. S. Alpert, Phys. Rev. **129**, 1344 (1963).
(172) E. L. Boyd, Phys. Rev. **145**, 174 (1966).
(173) N. I. Kaliteevskii, et al., Soviet Phys.—JETP **37**, 629 (1960).
(174) E. L. Boyd, et al., Phys. Rev. Letters **12**, 20 (1964).
(175) C. A. Lovejoy, et al., Nucl. Phys. **30**, 452 (1962).
(176) C. Arnoult, et al., J. Opt. Soc. Am. **56**, 177 (1966).
(177) C. E. Johnson, et al., Phys. Rev. **120**, 2108 (1960).
(178) Q. O. Navarro, et al., Phys. Rev. **123**, 186 (1961).
(179) W. M. Doyle, et al., Phys. Rev. **131**, 1586 (1963).
(180) B. Budick, et al., Phys. Rev. **135**, B1281 (1964).
(181) J. C. Walker, Phys. Rev. **127**, 1739 (1962).
(182) D. Giglberger, et al., Z. Physik **199**, 244 (1967).
(183) A. Y. Cabezas, et al., Phys. Rev. **120**, 920 (1960).
(184) L. Olschewski, et al., Z. Physik **200**, 224 (1967).
(185) A. C. Gossard, et al., Phys. Rev. **133**, A881 (1964).
(186) J. S. Ross, et al., Phys. Rev. **128**, 1159 (1962).
(187) M. A. Grace, et al., Phil. Mag. **2**, 1079 (1957).
(188) A. H. Reddoch, et al., Phys. Rev. **126**, 1493 (1962).
(189) G. J. Ritter, Phys. Rev. **126**, 240 (1962).
(190) I. J. Spalding, et al., Proc. Phys. Soc. (London) **79**, 787 (1962).
(191) F. R. Petersen, et al., Phys. Rev. **126**, 252 (1962).
(192) D. R. Speck, Bull. Am. Phys. Soc. **1**, 282 (1956).
(193) D. R. Speck, et al., Phys. Rev. **101**, 1831 (1956).
(194) L. H. Bennett, et al., Phys. Rev. **120**, 1812 (1960).
(195) K. Murakawa, et al., Phys. Rev. **105**, 671 (1957).
(196) M. P. Klein, et al., Bull. Am. Phys. Soc. **6**, 104 (1961).
(197) L. Armstrong, et al., Phys. Rev. **138**, B310 (1965).
(198) J. Kaufmann, et al., Phys. Letters **24A**, 115 (1967).
(199) H. R. Loeliger, et al., Phys. Rev. **95**, 291 (1954).

REFERENCES (Continued)

(200) A. Narath, (to be published in Phys. Rev.).
(201) A. Narath, et al., Bull. Am. Phys. Soc. **12**, 314 (1967).
(202) W. von Siemens, Ann. Physik **13**, 136 (1953).
(203) Y. W. Chan, et al., Phys. Rev. **144**, 1020 (1966).
(204) Y. W. Chan, et al., Phys. Rev. **137**, B1129 (1965).
(205) H. Dahmen, et al., Z. Physik **200**, 456 (1967).
(206) W. J. Childs, et al., Phys. Rev. **141**, 176 (1966).
(207) P. A. Vanden Bout, et al., Phys. Rev. **158**, 1078 (1967).
(208) H. Kleiman, et al., Phys. Letters **13**, 212 (1964).
(209) W. J. Tomlinson, et al., Nucl. Phys. **60**, 614 (1964).
(210) H. Kleiman, et al., J. Opt. Soc. Am. **53**, 822 (1963).
(211) W. J. Tomlinson, et al., J. Opt. Soc. Am. **53**, 828 (1963).
(212) F. Bitter, et al., Phys. Rev. **96**, 1531 (1954).
(213) B. Cagnac, et al., Compt. Rend. **249**, 77 (1959).
(214) B. Cagnac, Ann. Phys. **6**, 467 (1961).
(215) J. C. Lehmann, et al., Compt. Rend. **257**, 3152 (1963).
(216) J. Blaise, et al., J. Phys. Radium **18**, 193 (1957).
(217) O. Redi, et al., Phys. Letters **8**, 257 (1964).
(218) S. P. Davis, et al., J. Opt. Soc. Am. **56**, 1604 (1966).
(219) R. J. Hull, et al., Phys. Rev. **122**, 1574 (1961).
(220) H. L. Poss, Phys. Rev. **75**, 600 (1949).
(221) G. O. Brink, et al., Phys. Rev. **107**, 189 (1957).
(222) E. B. Baker, J. Chem. Phys. **26**, 960 (1957).
(223) I. Lindgren, et al., Arkiv Fysik **15**, 445 (1959).
(224) Y. Ting, et al., Phys. Rev. **89**, 595 (1953).
(225) S. S. Alpert, et al., Phys. Rev. **125**, 256 (1962).
(226) C. M. Olsmats, et al., Arkiv Fysik **19**, 469 (1961).
(227) M. Fred, et al., Phys. Rev. **98**, 1514 (1955).
(228) V. N. Egorov, Opt. Spectry. (USSR) **16**, 301 (1964).
(229) J. D. Axe, et al., Phys. Rev. **121**, 1630 (1961).
(230) R. Marrus, et al., Nucl. Phys. **23**, 90 (1961).
(231) P. B. Dorain, et al., Phys. Rev. **105**, 1307 (1957).
(232) B. Bleaney, et al., Phil. Mag. **45**, 992 (1954).
(233) J. Faust, et al., Phys. Letters **16**, 71 (1965).
(234) R. J. Champeau, et al., Compt. Rend. **257**, 1238 (1963).
(235) L. Armstrong, et al., Phys. Rev. **144**, 994 (1966).
(236) T. E. Manning, et al., Phys. Rev. **102**, 1108 (1956).
(237) R. Marrus, et al., Phys. Rev. **124**, 1904 (1961).
(238) M. Fred, et al., J. Opt. Soc. Am. **47**, 1076 (1957).
(239) B. Perry, et al., Bull. Am. Phys. Soc. **8**, 345 (1963).
(240) M. M. Robertson, et al., Phys. Rev. **140**, B820 (1965).
(241) R. E. Walstedt, et al., Phys. Rev. **162**, 301 (1967).
(242) R. Freeman, et al., Proc. Roy Soc. (London) **242A**, 455 (1957).

IONIZATION POTENTIALS OF MOLECULES

From data published up to July, 1966
Condensed by J. L. Franklin and Pat Haug from a compilation entitled
"Ionization Potentials, Appearance Potentials and Heats of Formation of Positive Ions"
by
J. L. Franklin, J. G. Dillard, H. M. Rosenstock, J. T. Herron, K. Draxl and F. H. Field
Published by National Standard Reference Data System

The following symbols are employed for the principal important methods:

CTS = Charge Transfer Spectra, EI = Electron Impact, PE = Photoelectron Spectroscopy, PI = Photoionization, S = Optical Spectroscopy.

Molecule	Ionization Potential ev	Method	ΔH_f of Ion Kcal/mole	Reference
H_2	15.427 ±0.002	S	356	*1*
D_2	15.46 ±0.01	PI	356	*2*
BH	9.77 ±0.05	S	333	*3*
BH_3	11.4 ±0.2	EI	279	*
B_2H_6	12.0	EI	286	*4, 5*
B_5H_9	10.5	EI	262	*
B_6H_{10}	9.3 ±0.1	EI	237	*6*
C_2	12.0 ±0.6	EI	475	*7*
C_3	12.6	EI	480	*7*
CH	11.13 ±0.22	S	399	*8*
CH_2	10.396 ±0.003	EI, S, PI	333	*9, 10, 11*
CH_3	9.83	S, PI	259	*12, 13*
CD_3	9.832 ±0.002	S	259	*12*
CH_4	12.6	PI	274	*
CD_4	12.888	PI	280	*11, 14*
C_2H_2	11.4	PI, PE	317	*
C_2D_2	11.416 ±0.006	PI	317	*15*
C_2H_3	9.4	EI	269	*
C_2H_4	10.5	S, PE	253	*16, 17*
C_2H_5	8.4	PI	219	*13*
C_2H_6	11.5	PI, PE	245	*14, 17*
$HC\equiv C-CH_2$	8.25	PI, EI		*18*
$CH_2=C=CH_2$	10.16 ±0.02	EI	280	*19*
$CH_3C\equiv CH$	10.36 ±0.01	PI	283	*20*
cyclo-C_3H_4	9.95	EI	296	*21*
C_3H_5 (allyl)	8.15	EI	216	*22, 8*
cyclo-C_3H_5	8.05	EI	239	*23*
C_3H_6	9.73	S, PI	229	*
cyclo-C_3H_6	10.09 ±0.02	PI	245	*24*
n-C_3H_7	8.1	PI	209	*13*
iso-C_3H_7	7.5	PI	190	*13*
C_3H_8	11.1	PI, PE	231	*
C_4H_2	10.2 ±0.1	EI		*25*
C_4H_4	9.87	EI	294	*25, 26*
$CH_3CH=C=CH_2$	9.57 ±0.02	EI	259	*19*
$CH_2=CHCH=CH_2$	9.07	PI, PE	236	*
$C_2H_5C\equiv CH$	10.18 ±0.01	PI	274	*20*
$CH_3C\equiv CCH_3$	9.9 ±0.1	EI	263	*25*
cyclo-C_4H_7	7.88 ±0.05	EI	213	*23*
$CH_3CH=CHCH_2$	7.71 ±0.05	EI	203	*27, 8*
$CH_2=C(CH_3)CH_2$	8.03 ±0.05	EI	203	*27, 8*
1-C_4H_8	9.6	PI	221	*
cis-2-C_4H_8	9.13	PI	209	*24, 28*
trans-2-C_4H_8	9.13	PI	208	*24, 28*
iso-C_4H_8	9.23 ±0.02	PI	209	*24, 28*
cyclo-C_4H_8	10.58	EI	250	*23*
n-C_4H_9	8.64 ±0.05	EI	218	*29*
sec-C_4H_9	7.93 ±0.05	EI	192	*29*
iso-C_4H_9	8.35 ±0.05	EI	205	*29*
tert-C_4H_9	7.42 ±0.07	EI	176	*29*
n-C_4H_{10}	10.63 ±0.03	PI	215	*30*
iso-C_4H_{10}	10.57	PI	212	*24*
cyclo-C_5H_6	8.97	EI	239	*31, 32*
$C_2H_5CH=C=CH_2$	9.42	EI	252	*21*
$CH_3CH=CHCH=CH_2$	8.68	EI	219	*21*
$CH_3CH=C=CHCH_3$	9.26	EI	247	*21*
$CH_2=CHCH_2CH=CH_2$	9.58	EI	246	*21*
$CH_2=CHC(CH_3)=CH_2$	8.845 ±0.005	PI	235	*24*
cyclo-C_5H_8	9.01 ±0.01	PI	216	*24*
cyclo-C_5H_9	7.79 ±0.02	EI	194	*23*
1-C_5H_{10}	9.50 ±0.02	PI	214	*24*
cis-2-C_5H_{10}	9.11	EI	203	*21*
trans-2-C_5H_{10}	9.06	EI	201	*21*
$(CH_3)_2CHCH=CH_2$	9.51 ±0.03	PI	212	*24*
$C_2H_5C(CH_3)=CH_2$	9.12 ±0.02	PI	202	*24*
$(CH_3)_2C=CHCH_3$	8.67 ±0.02	PI	189	*24*
cyclo-$C_3H_4(CH_3)_2$	9.77 ±0.02	EI	225	*33*
cyclo-C_5H_{10}	10.53 ±0.05	PI	224	*24*

Molecule	Ionization Potential ev	Method	ΔH_f of Ion Kcal/mole	Reference
tert-C_5H_{11}	7.12 ±0.1	EI	164	*34, 35*
neo-C_5H_{11}	8.33 ±0.1	EI	196	*34*
n-C_5H_{12}	10.35	PI	204	*24*
iso-C_5H_{12}	10.32	PI	201	*24*
neo-C_5H_{12}	10.35	PI	199	*24*
C_6H_4 (benzyne)	9.6	EI		*
cyclo-C_6H_5	9.2	PI, PE	284	*36, 37*
cyclo-C_6H_6	9.24	S, PI	233	*
$CH\equiv CCH=CHCH=CH_2$	9.50	EI	307	*38*
$C_2H_5C\equiv CC\equiv CH$	9.25	EI	307	*38*
$CH_3C\equiv CCH_2C\equiv CH$	9.75	EI	319	*38*
$CH_3C\equiv CC\equiv CCH_3$	9.20	EI	301	*38*
$CH\equiv CCH_2CH_2C\equiv CH$	10.35	EI	338	*38*
C_6H_8 (1-methylcyclopentadiene)	8.43 ±0.05	EI	218	*31*
C_6H_8 (2-methylcyclopentadiene)	8.46 ±0.05	EI	219	*31*
cyclo-C_6H_{10}	8.72	PE	199	*17*
cyclo-C_6H_{11}	7.7	EI	185	*23*
1-C_6H_{12}	9.45 ±0.02	PI	208	*35, 24*
$(CH_3)_2C=C(CH_3)_2$	8.30	PI	175	*28*
cyclo-C_6H_{12}	9.8	PI, PE	197	*17, 24*
n-C_6H_{14}	10.18	PI	195	*24*
iso-C_6H_{14}	10.12	PI	192	*24*
$(C_2H_5)_2CHCH_3$	10.08	PI	191	*24*
$C_2H_5C(CH_3)_3$	10.06	PI	188	*24*
$(CH_3)_2CHCH(CH_3)_2$	10.02	PI	189	*24*
cyclo-$C_6H_5CH_2$	7.76 ±0.08	EI	216	*39*
cyclo-C_7H_7	6.240 ±0.01	S	209	*40*
cyclo-$C_6H_5CH_3$	8.82 ±0.01	PI	215	*
cyclo-C_7H_8	8.5	EI	240	*32, 41, 42*
bicyclo-(2.2.1)C_7H_8	8.67	EI	267	*42*
bicyclo-(3.2.0)C_7H_8	9.37	EI	246	*41*
C_7H_{10} (1,2-dimethylcyclopentadiene)	8.1 ±0.1	EI	204	*31*
C_7H_{10} (5,5-dimethylcyclopentadiene)	8.22 ±0.05	EI	206	*31*
C_7H_{10} (1,3-cycloheptadiene)	8.55	EI	219	*41*
C_7H_{10} (norbornene)	8.95 ±0.15	EI	237	*43*
C_7H_{12} (4-methylcyclohexane)	8.91 ±0.01	PI	198	*24*
cyclo-$C_6H_{11}CH_3$	9.85 ±0.03	PI	190	*24*
n-C_7H_{16}	9.90 ±0.05	PI	183	*37*
cyclo-$C_6H_5C\equiv CH$	8.815 ±0.005	PI	279	*24*
C_8H_8 (styrene)	8.47 ±0.02	PI	232	*24*
C_8H_8 (cyclotatetraene)	8.0	PI, PE	255	*17, 24*
C_8H_8 (cubane)	8.74 ±0.15	EI	350	*44*
m-$C_6H_4CH_3CH_2$	7.65 ±0.03	EI	206	*39*
p-$C_6H_4CH_3CH_2$	7.46 ±0.03	EI	202	*39*
cyclo-$C_6H_5C_2H_5$	8.76 ±0.01	PI	209	*24*
o-$C_6H_4(CH_3)_2$	8.56	PI	202	*
m-$C_6H_4(CH_3)_2$	8.58	PI, PE	202	*
p-$C_6H_4(CH_3)_2$	8.44	PI	199	*
C_8H_{10} (7-methylcycloheptatriene)	8.39 ±0.1	EI	231	*45*
C_8H_{10} (1-methylspiroheptadiene)	8.02 ±0.1	EI	229	*45*
C_8H_{10} (2-methylspiroheptadiene)	8.07 ±0.1	EI	230	*45*
C_8H_{10} (6-methylspiroheptadiene)	8.40 ±0.1	EI	239	*45*
C_8H_{12} (1,2,3-trimethylcyclopentadiene)	7.96 ±0.05	EI	194	*31*
C_8H_{12} (1,5,5-trimethylcyclopentadiene)	8.00 ±0.1	EI	193	*31*

* Average of several values.

IONIZATION POTENTIALS OF MOLECULES (Continued)

Molecule	Ionization Potential ev	Method	ΔH_f of Ion Kcal/mole	Reference
C_8H_{12} (4-vinylcyclohexene)	8.93 ±0.02	PI	224	*24*
cis-1,2-cyclo-$C_6H_{10}(CH_3)_2$	10.08 ±0.02	EI	191	*46*
trans-1,2-cyclo-$C_6H_{10}(CH_3)_2$	10.08 ±0.03	EI	189	*46*
C_8H_{18} (2,2,4-trimethylpentane)	9.86	PI	174	*24*
C_8H_{18} (2,2,3,3-tetramethylbutane)	9.79	EI	184	*8*
C_9H_8 (indene)	8.81	EI	246	*47*
$C_6H_5C(CH_3){=}CH_2$	8.35 ±0.01	PI	220	*24*
cyclo-$C_6H_5(n\text{-}C_3H_7)$	8.72 ±0.01	PI	203	*24*
cyclo-$C_6H_5(iso\text{-}C_3H_7)$	8.69 ±0.01	PI	201	*24*
1,2,3 cyclo-$C_6H_3(CH_3)_3$	8.48	PI	193	*28*
1,2,4 cyclo-$C_6H_3(CH_3)_3$	8.27	PI	187	*28*
1,3,5 cyclo-$C_6H_3(CH_3)_3$	8.4	PI	190	*
$C_{10}H_8$ (naphthalene)	8.12	PI	220	*24, 48*
$C_{10}H_8$ (azulene)	7.42	S	243	*49, 50*
$C_6H_4C_3H_7CH_2$	7.42	EI	188	*39*
cyclo-C_6H_5 (n-C_4H_9)	8.69 ±0.01	PI	197	*24*
$C_{10}H_{14}$ (sec-butylbenzene)	8.68 ±0.01	PI	196	*24*
$C_{10}H_{14}$ (tert-butylbenzene)	8.68 ±0.01	PI	193	*24*
$C_{10}H_{14}$ (1,2,3,5-tetramethylbenzene)	8.47 ±0.05	EI	185	*24*
$C_{10}H_{14}$ (1,2,4,5-tetramethylbenzene)	8.03	PI, PE	174	*
$C_{10}H_{18}$ (cis-decaline)	9.61 ±0.02	EI	181	*51*
$C_{10}H_{18}$ (trans-decaline)	9.61 ±0.02	EI	178	*51*
$C_{11}H_9$ (1-naphthyl methyl)	7.35	EI	208	*52*
$C_{11}H_9$ (2-naphthyl methyl)	7.56 ±0.05	EI	217	*52*
$C_{11}H_{10}$ (methylnaphthalene)	7.96 ±0.01	PI	209	*24*
$C_{11}H_{10}$ (2-methylnaphthalene)	7.955 ±0.01	PI	209	*24*
$C_6H(CH_3)_5$	7.92 ±0.02	PI	155	*53*
$C_{11}H_{18}$ (hexamethylcyclopentadiene)	7.74 ±0.05	EI	165	*31*
$C_{12}H_{10}$ (biphenyl)	8.27 ±0.01	PI	230	*24*
cyclo-$C_6(CH_3)_6$	7.85 ±0.02	PI	152	*28*
$C_{13}H_{10}$ (fluorene)	8.63	EI	243	*47*
$C_{14}H_{10}$ (diphenylacetylene)	8.85 ±0.05	EI	303	*54*
$C_{14}H_{10}$ (anthracene)	7.55	EI	228	*55*
$C_{14}H_{10}$ (phenanthrene)	8.1	EI	233	*54, 55*
$C_{18}H_{12}$ (1,2-benzanthracene)	8.01	EI	251	*56*
$C_{18}H_{30}$ (1-phenyldodecane)	9.05 ±0.1	EI	165	*57*
$C_{18}H_{30}$ (3-phenyldodecane)	8.95 ±0.1	EI	163	*57*
$C_{19}H_{32}$ (7-phenyltridecane)	8.91 ±0.1	EI	157	*57*
$C_{26}H_{46}$ (1-phenylicosane)	9.34 ±0.1	EI	132	*57*
$C_{26}H_{46}$ (2-phenylicosane)	9.22 ±0.1	EI	129	*57*
$C_{26}H_{46}$ (3-phenylicosane)	8.95 ±0.1	EI	123	*57*
$C_{26}H_{46}$ (4-phenylicosane)	9.01 ±0.1	EI	125	*57*
$C_{26}H_{46}$ (5-phenylicosane)	9.04 ±0.1	EI	125	*57*
$C_{26}H_{46}$ (7-phenylicosane)	8.97 ±0.1	EI	124	*57*
$C_{26}H_{46}$ (9-phenylicosane)	9.06 ±0.1	EI	126	*57*
$(CH_3)_3B$	8.8 ±0.2	EI	173	*58*
$(C_2H_5)_3B$	9.0 ±0.2	EI	170	*58*
N_2	15.576	S	359	*59*
NH	13.10 ±0.05	EI	382	*60*
NH_2	11.4 ±0.1	EI	304	*61*
NH_3	10.2	S, PI, PE	223	*
N_2H_2	9.85 ±0.1	EI		*62*
N_2H_4	8.74 ±0.06	PI	224	*63*
CN	14.3	EI	430	*64, 65*
HCN	13.8	EI	351	*26, 66*

Molecule	Ionization Potential ev	Method	ΔH_f of Ion Kcal/mole	Reference
C_2N_2	13.6	EI	387	*8, 64*
CH_5N	8.97	PI	201	*67*
C_2H_2N	10.9	EI	298	*68*
CH_3CN	12.2	PI	302	*14, 24*
cyclo-C_2H_5N	9.94 ±0.15	EI	255	*69*
$C_2H_5NH_2$	8.86 ±0.02	PI	193	*24*
$(CH_3)_2NH$	8.24 ±0.02	PI	186	*24*
$CH_2{=}CHCN$	10.91 ±0.01	PI	296	*24*
C_2H_5CN	11.84 ±0.02	PI	289	*24*
C_3H_7N	9.1 ±0.15	EI	225	*70*
$(CH_2)_3NH$	9.1 ±0.15	EI	225	*70*
n-$C_3H_7NH_2$	8.78 ±0.02	PI	185	*24*
iso-$C_3H_7NH_2$	8.72 ±0.03	PI	183	*24*
$(CH_3)_3N$	7.82 ±0.02	PI	175	*24*
$CH_2{=}CHCH_2CN$	10.39 ±0.01	PI	281	*24*
C_4H_5N (pyrrole)	8.20 ±0.01	PI	215	*24*
$(CH_3)_2CCN$	9.15 ±0.1	EI	239	*68*
n-C_3H_7CN	11.67 ±0.05	PI	280	*24*
C_4H_9N (pyrrolidine)	8.41	PE	192	*17*
n-$C_4H_9NH_2$	8.71 ±0.03	PI	179	*24*
sec-$C_4H_9NH_2$	8.70	PI	177	*24*
iso-$C_4H_9NH_2$	8.70	PI	177	*24*
tert-$C_4H_9NH_2$	8.64	PI	173	*24*
$(C_2H_5)_2NH$	8.01 ±0.01	PI	163	*24*
C_5H_5N (pyridine)	9.3	S, PI	247	*24, 71*
C_6H_7N (aniline)	7.7	PI	202	*
C_6H_7N (2-picoline)	9.02 ±0.03	PI	232	*24*
C_6H_7N (3-picoline)	9.04 ±0.03	PI	236	*24*
C_6H_7N (4-picoline)	9.04 ±0.03	PI	233	*24*
$C_6H_{13}N$ (cyclohexylamine)	8.86	PE	181	*17*
$(n\text{-}C_3H_7)_2NH$	7.84 ±0.02	PI	153	*24*
$(iso\text{-}C_3H_7)_2NH$	7.73 ±0.03	PI	148	*24*
$(C_2H_5)_3N$	7.50 ±0.02	PI	147	*24*
cyclo-C_6H_5CN	9.705 ±0.01	PI	277	*24*
C_7H_9N (n-methylaniline)	7.32	PI	192	*72, 73*
C_7H_9N (m-toluidine)	7.50 ±0.02	PI	189	*48*
C_7H_9N (2,3-lutidine)	8.85 ±0.02	PI	218	*24*
C_7H_9N (2,4-lutidine)	8.85 ±0.03	PI	218	*24*
C_7H_9N (2,6-lutidine)	8.85 ±0.02	PI	218	*24*
$C_6H_5CH_2CN$	9.40 ±0.5	EI	259	*74*
m-$C_6H_4CH_3CN$	9.66 ±0.05	EI	271	*74*
p-$C_6H_4CH_3CN$	9.76	EI	273	*75*
cyclo-$C_6H_5NHC_2H_5$	7.56	CTS	193	*76*
cyclo-$C_6H_5N(CH_3)_2$	7.12	PI	185	*72, 73*
$(n\text{-}C_4H_9)_2NH$	7.69 ±0.03	PI	140	*24*
$C_9H_{13}N$ (N-n-propylaniline)	7.54	CTS	188	*76*
$C_9H_{13}N$ (N-ethyl-N-methylaniline)	7.37	CTS	185	*76*
$C_9H_{13}N$ (N,N-dimethyl-o-toluidine)	7.37	CTS	184	*76*
$C_9H_{13}N$ (N,N-dimethyl-m-toluidine)	7.35	CTS	181	*76*
$C_9H_{13}N$ (N,N-dimethyl-p-toluidine)	7.33	CTS	181	*76*
$(n\text{-}C_3H_7)_3N$	7.23	PI	207	*24*
$C_{10}H_{15}N$ (N-n-butylaniline)	7.53	CTS	183	*76*
$C_{10}H_{15}N$ (N,N-diethylaniline)	6.99	CTS	172	*76*
$C_{10}H_{15}N$ (N,N-dimethyl-p-ethylaniline)	7.38	CTS	177	*76*
$C_{10}H_{15}N$ (N,N-2,4-tetramethylaniline)	7.17	CTS	171	*76*
$C_{10}H_{15}N$ (N,N,2,6-tetramethylaniline)	7.22	CTS	173	*76*
$C_{10}H_{15}N$ (N,N,3,5-tetramethylaniline)	7.25	CTS	172	*76*
$C_{11}H_{17}N$ (N,N-diethyl-p-toluidene)	6.93	CTS	164	*76*
$C_{11}H_{17}N$ (N,N-dimethyl-p-isopropylaniline)	7.41	CTS	174	*76*
$(C_6H_5)_2NH$	7.25 ±0.03	PI	223	*77*
$C_{12}H_{19}N$ (N,N-di-n-propylaniline)	6.96	CTS	163	*76*
$C_{12}H_{19}N$ (N,N-dimethyl-p-tert-butylaniline)	7.43	CTS	165	*76*
$C_{14}H_{23}N$ (N,N-di-n-butylaniline)	6.95	CTS	153	*76*

* Average of several values.

Molecule	Ionization Potential ev	Method	ΔH_f of Ion Kcal/mole	Reference
$(C_6H_5)_3N$	6.86 ±0.03	PI	243	*77*
CH_2N_2 (diazirine)	10.18 ±0.05	EI	314	*78*
CH_2N_2 (diazomethane)	8.999 ±0.001	S	257	*79*
CH_6N_2 (methyl-hydrazine)	8.00 ±0.06	PI	207	*63*
$CH_3N{=}NCH_3$	8.65 ±0.1	EI	243	*80*
$C_2H_8N_2$ (1,1-dimethyl-hydrazine)	7.67 ±0.05	PI	197	*81*
$C_2H_8N_2$ (1,2-dimethyl-hydrazine)	7.75 ±0.1	EI	200	*82*
$(CH_3)_2N_2H$	7.93 ±0.1	EI	202	*82*
o-$C_4H_4N_2$ (o-diazine)	9.9	EI	275	*83*
m-$C_4H_4N_2$ (m-diazine)	9.9	EI	277	*83*
p-$C_4H_4N_2$ (p-diazine)	9.8	EI	274	*83*
1,1$(C_2H_5)_2N_2H_2$	7.59 ±0.05	PI	184	*63*
$(CH_3)_4N_2$	7.76 ±0.05	EI	196	*82*
$C_5H_4NNH_2$	8.97 ±0.05	EI	244	*84*
$C_5H_{14}N_2$ (1-methyl-1-n-butylhydrazine)	7.62 ±0.05	PI	180	*63*
$(CH_3)_2NC_6H_4N(CH_3)_2$ [p-bis(dimethylamino)-benzene]	6.9	CTS	180	*85*
CH_3N_3	9.5 ±0.1	EI	276	*86*
O_2	12.063 ±0.001	PI	278	*
O_3	12.3 ±0.1	PE	318	*87*
OH	13.17 ±0.1	EI	313	*88*
H_2O	12.6	PI	233	*
D_2O	12.6	PI	232	*37*
HO_2	11.53 ±0.02	EI	271	*89*
H_2O_2	11.0	EI	233	*90, 91*
Li_2O	6.8	EI	120	*92, 93*
CO	14.013 ±0.004	S	297	*94*
CO_2	13.769 ±0.03	S	223	*96*
NO	9.25	PI, S	235	*
N_2O	12.894	S	317	*95*
NO_2	9.79	PI	233	*97, 98*
CHO	9.8	EI	221	*
CH_2O	10.88	PI	223	*24, 48*
CH_3OH	10.84	PI, PE	202	*17, 24*
CH_3CO	10.3	PI	152	*
CH_3CHO	10.2	PI	196	*
C_2H_4O (ethylene oxide)	10.6	PI, S	231	*
C_2H_5OH	10.49	PI	185	*24, 30*
CH_3OCH_3	9.98	S, PI	186	*24, 99*
$CH_2{=}CHCHO$	10.10 ±0.01	PI	210	*24*
C_2H_5CHO	9.98	PI	181	*24*
CH_3COCH_3	9.69	PI	171	*
$CH_2{=}CHCH_2OH$	9.67 ±0.05	PI	191	*24*
$CH_2{=}CHOCH_3$	8.93 ±0.02	PI	178	*24*
C_3H_6O (propyleneoxide)	10.22 ±0.02	PI	214	*24*
C_3H_6O (trimethylene-oxide)	9.667 ±0.005	S	199	*100*
n-C_3H_7OH	10.1	PI	172	*24, 101*
iso-C_3H_7OH	10.15	PI	169	*24*
C_4H_4O (furan)	8.89	S, PI	197	*24, 102*
$CH_3CH{=}CHCHO$	9.73 ±0.01	PI	194	*24*
n-C_3H_7CHO	9.86 ±0.02	PI	174	*24*
iso-C_3H_7CHO	9.74 ±0.03	PI	169	*24*
$C_2H_5COCH_3$	9.5	PI	161	*
cyclo-C_4H_8O	9.42	S	174	*100*
n-C_4H_9OH	10.04	PI	165	*24*
$C_2H_5OC_2H_5$	9.6	PI	161	*24*
C_5H_8O (cyclo-pentanone)	9.26 ±0.01	PI	163	*24*
C_5H_8O (dihydropyran)	8.34 ±0.01	PI	164	*24*
n-C_4H_9CHO	9.82 ±0.05	PI	168	*24*
iso-C_4H_9CHO	9.71 ±0.05	PI	164	*24*
n-$C_3H_7COCH_3$	9.37 ±0.02	PI	154	*103*
iso-$C_3H_7COCH_3$	9.30 ±0.02	PI	151	*24, 103*
$(C_2H_5)_2CO$	9.32 ±0.01	PI	153	*24*
cyclo-$C_5H_{10}O$	9.25 ±0.01	S	161	*100*
C_6H_5O	8.84	EI	226	*36*
C_6H_6O (phenol)	8.51	PI	173	*24, 48*
$(CH_3)_2C{=}CHCOCH_3$	9.08 ±0.03	PI	168	*24*
$C_6H_{10}O$ (cyclo-hexanone)	9.14 ±0.01	PI	152	*24*
n-$C_4H_9COCH_3$	9.35	PI	149	*24, 103*
iso-$C_4H_9COCH_3$	9.30	PI	147	*24, 103*
tert-$C_4H_9COCH_3$	9.17 ±0.03	PI	140	*24*
$(n\text{-}C_3H_7)_2O$	9.27 ±0.05	PI	147	*24*
$(iso\text{-}C_3H_7)_2O$	9.20 ±0.05	PI	142	*24*
C_6H_5CHO	9.52	PI	209	*24*
C_7H_6O (tropone)	9.68 ±0.02	EI	240	*104*
cyclo-$C_6H_5CH_2OH$	9.14 ±0.05	EI	186	*74*
cyclo-$C_6H_5OCH_3$	8.21 ±0.02	PI	173	*24*
C_7H_8O (m-cresol)	8.52 ±0.05	EI	165	*74*
n-$C_5H_{11}COCH_3$	9.33 ±0.03	PI	143	*24*
C_8H_8O (acetophenone)	9.27 ±0.03	PI	191	*24*
C_8H_8O (p-methyl-benzaldehyde)	9.33 ±0.05	EI	194	*105*
$C_8H_{10}O$ (benzyl methyl ether)	8.85 ±0.03	PI	184	*30*
$C_8H_{10}O$ (phenyl ethyl ether)	8.13 ±0.02	PI	167	*24*
$C_8H_{10}O$ (m-methylanisole)	8.31 ±0.05	EI	169	*74*
$C_9H_{10}O$ (phenyl ethyl ketone)	9.27 ±0.05	EI	189	*74*
$C_9H_{10}O$ (m-methyl-acetophenone)	9.15 ±0.05	EI	182	*74*
$C_{12}H_{10}O$ (phenyl ether)	8.82 ±0.05	EI	220	*106*
$C_{13}H_{10}O$ (benzophenone)	9.4	EI	229	*106, 105*
$C_{14}H_{12}O$ (p-methyl-benzophenone)	9.13 ±0.05	EI	214	*105*
HCOOH	11.05 ±0.01	PI	164	*24*
CH_3COOH	10.36	PI	135	*24*
$HCOOCH_3$	10.815 ±0.005	PI	166	*24*
C_2H_5COOH	10.24 ±0.03	PI	127	*24*
$HCOOC_2H_5$	10.61 ±0.01	PI	156	*24*
CH_3COOCH_3	10.27 ±0.02	PI	138	*24*
$(CH_3O)_2CH_2$	10.00 ±0.05	PI	145	*24*
$CH_3COOCH{=}CH_2$	9.19 ±0.05	PI	137	*24*
$CH_3COCOCH_3$	9.24 ±0.03	PI	135	*24*
n-C_3H_7COOH	10.16 ±0.05	PI	121	*24*
iso-C_3H_7COOH	10.02 ±0.05	PI	115	*24*
$HCOOCH_2CH_2CH_3$	10.54 ±0.01	PI	149	*24*
$CH_3COOC_2H_5$	10.11 ±0.02	PI	126	*24*
$C_2H_5COOCH_3$	10.15 ±0.03	PI	127	*24*
$C_4H_8O_2$ (p-dioxane)	9.13 ±0.03	PI	126	*24*
$(CH_3O)_2CHCH_3$	9.65 ±0.03	PI	129	*24*
$C_4H_4O_2$ (2-furfur-aldehyde)	9.21 ±0.01	PI	187	*24*
$CH_3COCH_2COCH_3$	8.87 ±0.03	PI	122	*24*
$HCOO(CH_2)_3CH_3$	10.50 ±0.02	PI	144	*24*
$HCOOCH_2CH(CH_3)_2$	10.46 ±0.02	PI	139	*24*
$CH_3COOCH_2CH_2CH_3$	10.04 ±0.03	PI	121	*24*
$CH_3COOCH(CH_3)_2$	9.99 ±0.03	PI	116	*24*
$C_2H_5COOC_2H_5$	10.00 ±0.02	PI	119	*24*
n-$C_3H_7COOCH_3$	10.07 ±0.03	PI	125	*24*
iso-$C_3H_7COOCH_3$	9.98 ±0.02	PI	121	*24*
$(C_2H_5O)_2CH_2$	9.70 ±0.05	PI	134	*24*
p-$C_6H_4O_2$	9.67 ±0.02	PI	198	*48*
$CH_3COOC_4H_9$	9.56 ±0.03	PI	104	*48*
$CH_3COO(CH_2)_3CH_3$	10.01	PI	114	*24*
CH_3COOCH_2-$CH(CH_3)_2$	9.97	PI	111	*24*
CH_3COOCH-$(CH_3)C_2H_5$	9.91 ±0.03	PI	110	*24*
C_6H_5COOH	9.73 ±0.09	EI	152	*105*
p-HOC_6H_4CHO	9.32 ±0.02	EI	157	*105*
$C_8H_8O_2$ (α-hydroxy-acetophenone)	9.33 ±0.05	EI	159	*74*
$C_6H_5COOCH_3$	9.35 ±0.06	EI	144	*105*
$C_8H_8O_2$ (p-methoxy-benzaldehyde)	8.60 ±0.03	EI	150	*105*
$C_8H_8O_2$ (m-hydroxy-acetophenone)	8.67 ±0.05	EI	134	*74*
$C_8H_8O_2$ (p-hydroxy-acetophenone)	8.70 ±0.03	EI	135	*105*
$C_9H_{10}O_2$ (α-methoxy-acetophenone)	8.60 ±0.05	EI	142	*74*
$C_9H_{10}O_2$ (m-methoxy-acetophenone)	8.53 ±0.05	EI	140	*74*
$C_9H_{10}O_2$ (p-methoxy-acetophenone)	8.62 ±0.05	EI	142	*105*
$C_9H_{10}O_2$ (methyl p-toluate)	8.94 ±0.04	EI	130	*105*
$C_{13}H_{10}O_2$ (p-hydroxy-benzophenone)	8.59 ±0.05	EI	165	*105*
$C_{13}H_{10}O_2$ (phenyl benzoate)	8.98 ±0.05	EI	177	*106*
$C_{14}H_{10}O_2$ (benzil)	8.78 ±0.05	EI	181	*106*

* Average of several values.

IONIZATION POTENTIALS OF MOLECULES (Continued)

Molecule	Ionization Potential ev	Method	ΔH_f of Ion Kcal/mole	Reference
$CH_3OCH_2COOCH_3$	9.56 ±0.05	EI	88	*74*
$C_9H_{10}O_3$ (methyl p-methoxybenzoate)	8.43 ±0.04	EI	90	*105*
$(C_6H_5)_2CO_3$	9.01 ±0.05	EI	122	*106*
CH_3CONH_2	9.77 ±0.02	PI	171	*24*
$HCON(CH_3)_2$	9.12 ±0.02	PI	160	*24*
$CH_3CONHCH_3$	8.90 ±0.02	PI	150	*24*
$CH_3CON(CH_3)_2$	8.81 ±0.03	PI	145	*24*
C_5H_4NOH	9.70 ±0.05	EI	209	*84*
$HCON(C_2H_5)_2$	8.89 ±0.02	PI	145	*24*
C_6H_5NO (2-pyridinecarboxaldehyde)	9.75 ±0.05	EI	227	*84*
C_6H_5NO (4-pyridinecarboxaldehyde)	10.12 ±0.05	EI	235	*84*
$CH_3CON(C_2H_5)_2$	8.60 ±0.02	PI	130	*24*
C_6H_5NCO	8.77 ±0.02	PI	222	*24*
C_7H_7NO (benzamide)	9.4 ±0.2	EI	197	*107, 108*
C_7H_7NO (p-aminobenzaldehyde)	8.25 ±0.02	EI	182	*105*
C_7H_9NO (p-methoxyanaline)	7.82	EI	169	*75*
C_8H_9NO (acetanilide)	8.39 ±0.10	EI	171	*108*
C_8H_9NO (m-aminoacetophenone)	8.09 ±0.05	EI	171	*74*
C_8H_9NO (p-aminoacetophenone)	8.17 ±0.02	EI	172	*105*
C_9H_7NO (α-cyanoacetophenone)	9.56 ±0.05	EI	235	*74*
CH_3NO_2	11.1	PI	238	*14, 24*
$C_2H_5NO_2$	10.88 ±0.05	PI	226	*24*
n-$C_3H_7NO_2$	10.81 ±0.03	PI	221	*24*
iso-$C_3H_7NO_2$	10.71 ±0.05	PI	217	*24*
$C_6H_5NO_2$	9.92	PI	246	*24*
$C_7H_6NO_2$ (m-nitrobenzyl radical)	8.56 ±0.1	EI	227	*39*
$C_7H_7NO_2$ (m-nitrotoluene)	9.65 ±0.05	EI	233	*74*
$C_7H_7NO_2$ (p-nitrotoluene)	9.82	EI	237	*75*
$C_8H_9NO_2$	8.08 ±0.01	EI	122	*90*
$C_6H_6N_2O_2$ (o-nitroaniline)	8.66	EI	215	*75*
$C_6H_6N_2O_2$ (m-nitroaniline)	8.7	EI	216	*75*
$C_6H_6N_2O_2$ (p-nitroaniline)	8.85	EI	219	*75*
$C_2H_5NO_3$	11.22	PI	222	*24*
n-$C_3H_7ONO_2$	11.07 ±0.02	PI	213	*24*
$C_6H_5NO_3$ (p-nitrophenol)	9.52	EI	187	*75*
$C_7H_5NO_3$ (p-nitrobenzaldehyde)	10.27 ±0.01	EI	217	*105*
$C_8H_7NO_3$ (m-nitroacetophenone)	9.89 ±0.05	EI	201	*74*
$C_8H_7NO_3$ (p-nitroacetophenone)	10.07 ±0.02	EI	205	*105*
$C_8H_7NO_4$ (methyl-p-nitrobenzoate)	10.20 ±0.03	EI	160	*105*
F_2	15.7	S	362	*109*
HF	15.77 ±0.02	PI	299	*110*
BF	11.3	EI	233	*111, 112*
BF_3	15.5	EI	87	*
CF_2	11.8	EI	237	*113, 114*
C_2F_4	10.12	PI	78	*28*
C_6F_6	9.97	PI		*28*
NF_2	11.9	EI	284	*115, 116*
trans-N_2F_2	13.1 ±0.1	EI	322	*115*
NF_3	13.2 ±0.2	EI	275	*115, 123*
N_2F_4	12.04 ±0.10	EI	276	*124*
OF_2	13.6	EI	309	*117, 118*
XeF_2	11.5 ±0.2	S	239	*119*
CH_2F	9.35	EI	207	*29*
CH_3F	12.85 ±0.01	EI	229	*120*
C_2H_3F	10.37	PI	211	*28, 121*
cyclo-C_6H_5F	9.2	S, PI	186	*
C_7H_7F	8.9	PI	172	*24*
CHF_2	9.45	EI	143	*29, 122*
$C_2H_2F_2$	10.30	PI	159	*28, 121*
o-$C_6H_4F_2$	9.31	PI	147	*121*
p-$C_6H_4F_2$	9.15	PI	140	*121*
C_2HF_3	10.14	PI	122	*28*
$CH_2{=}CHCF_3$	10.9	PI	93	*28*
$C_7H_5F_3$	9.68	S, PI	84	*24, 125*
$C_7H_{11}F_3$ (trifluoromethylcyclohexane)	10.46 ±0.02	PI	37	*24*
C_6H_5OF (o-fluorophenol)	8.66 ±0.01	PI	132	*24*
Na_2	4.90 ±0.01	PI	147	*126*
AlF	9.8	EI	166	*127, 128*
$(CH_3)_4Si$	9.9	EI	171	*129, 130*
PH_3	9.98	PI	231	*131*
PF_3	9.71	PI	4	*131*
CH_3PH_2	9.72 ±0.15	EI	217	*132*
$C_2H_5PH_2$	9.47 ±0.5	EI	206	*132*
$(CH_3)_3P$	8.60 ±0.2	EI	175	*132*
$(C_6H_5)_3P$	7.36 ±0.05	PI	242	*77*
S_6	9.7	EI	248	*133*
S_7	9.2 ±0.3	EI		*133*
HS	10.50 ±0.1	EI	276	*134*
H_2S	10.4	S	235	*135*
CS_2	10.080	S	261	*95*
SO_2	12.34 ±0.02	PI	214	*30*
CH_3S	8.06 ±0.1	EI	218	*134*
CH_3SH	9.440 ±0.005	PI	212	*24*
C_2H_4S (ethylene sulfide)	8.87 ±0.15	EI	224	*69*
C_2H_5SH	9.285 ±0.005	PI	203	*24*
CH_3SCH_3	8.685 ±0.005	PI	191	*24*
C_3H_6S (propylene sulfide)	8.6 ±0.2	EI	218	*136*
$(CH_2)_3S$	8.9 ±0.15	EI	220	*70*
n-C_3H_7SH	9.195 ±0.005	PI	197	*24*
$C_2H_5SCH_3$	8.55 ±0.01	PI	183	*24*
C_4H_4S (thiophene)	8.860 ±0.005	PI	229	*24*
$CH_3SCH_2{-}CH{=}CH_2$	8.70 ±0.2	EI	211	*137*
$(CH_2)_4S$	8.57 ±0.75	EI	190	*70*
n-C_4H_9SH	9.14 ±0.02	PI	191	*24*
$C_2H_5SC_2H_5$	8.430 ±0.005	PI	175	*24*
n-$C_3H_7SCH_3$	8.8 ±0.15	EI	183	*138*
iso-$C_3H_7SCH_3$	8.7 ±0.2	EI	179	*137*
C_6H_5S	8.63 ±0.1	EI	250	*134*
C_6H_5SH	8.32 ±0.01	PI	217	*138*
C_6H_8S (2-ethylthiophene)	8.8 ±0.2	EI	215	*139*
$(n\text{-}C_3H_7)_2S$	8.5	PI, EI	170	*24, 143*
$C_6H_5S_2CH_3$	8.9	EI	229	*137*
$C_7H_{10}S$ (2-propylthiophene)	8.6 ±0.2	EI	205	*139*
$C_8H_{12}S$	8.8	EI	221	*137*
$C_8H_{12}S$ (2-butylthiophene)	8.5 ±0.2	EI	198	*139*
CH_3SSCH_3	8.46 ±0.03	PI	189	*24*
$C_2H_5SSC_2H_5$	8.27 ±0.03	PI	173	*24*
CH_3SSSCH_3	8.80 ±0.15	EI	203	*140*
COS	11.17 ±0.01	PI	224	*138*
SO_2F_2	13.3 ±0.1	EI	102	*141*
CH_3NCS	9.25 ±0.03	PI	245	*24*
CH_3SCN	10.065 ±0.01	PI	270	*24*
C_2H_5NCS	9.14 ±0.03	PI	237	*24*
C_2H_5SCN	9.89 ±0.01	PI	261	*24*
C_7H_5NS (phenylisothiocyanate)	8.520 ±0.005	PI	260	*24*
$C_6H_5CH_2SCN$	9.06 ±0.05	EI	274	*74*
NH_2CSNH_2	8.50 ±0.05	EI	194	*142*
$NH_2CSNHCH_3$	8.29 ±0.05	EI	188	*142*
$NH_2CSNHCH{=}CH_2$	8.29 ±0.05	EI	213	*142*
$NH_2CSN(CH_3)_2$	8.34 ±0.05	EI	186	*142*
$CH_3NHCSNHCH_3$	8.17 ±0.05	EI	184	*142*
$CH_3NHCSN(CH_3)_2$	7.93 ±0.05	EI	176	*142*
$C_5H_{12}N_2S$	7.98 ±0.05	EI	170	*142*
$(CH_3)_2NCSN(CH_3)_2$	7.95 ±0.05	EI	173	*142*
CH_3COSH	10.00 ±0.02	PI	179	*24*
Cl_2	11.48 ±0.01	PI	265	*24*
HCl	12.74	PI	272	*
LiCl	10.1	EI	186	*144*
CCl_3	8.78 ±0.05	EI	214	*
CCl_4	11.47 ±0.01	PI	240	*24*
C_2Cl_4	9.32 ±0.01	PI	212	*24, 28*
PCl_3	9.91	PI	160	*131*
CH_2Cl	9.32	EI	244	*29*
CH_3Cl	11.3	S, PI	239	*14, 145*
C_2H_3Cl	9.996	S, PI	239	*
C_2H_5Cl	10.97	PI	226	*24*
$CH_3C{\equiv}CCl$	9.9 ±0.1	EI	267	*25*
n-C_3H_7Cl	10.82 ±0.03	PI	219	*24*

* Average of several values.

IONIZATION POTENTIALS OF MOLECULES (Continued)

Molecule	Ionization Potential ev	Method	ΔH_f of Ion Kcal/mole	Reference
iso-C_3H_7Cl	10.78 ±0.02	PI	211	*24*
n-C_4H_9Cl	10.67 ±0.03	PI	210	*24*
sec-C_4H_9Cl	10.65 ±0.03	PI	210	*24*
iso-C_4H_9Cl	10.66 ±0.03	PI	208	*24*
tert-C_4H_9Cl	10.61 ±0.03	PI	202	*24*
C_6H_5Cl	9.07	PI	222	*24*
$C_6H_5CH_2Cl$	9.19 ±0.05	EI	219	*74*
o-$C_6H_4ClCH_3$	8.83 ±0.02	PI	208	*24*
m-$C_6H_4ClCH_3$	8.83 ±0.02	PI	207	*24*
p-$C_6H_4ClCH_3$	8.69 ±0.02	PI	204	*24*
C_7H_9Cl (endo-5-chloro-2-norbornene)	9.1 ±0.15	EI	233	*43*
C_7H_9Cl (exo-5-chloro-2-norbornene)	9.15 ±0.15	EI	234	*43*
C_7H_9Cl (3-chloro-nortricyclene)	9.51 ±0.15	EI	234	*43*
$CHCl_2$	9.30	EI	245	*29*
CH_2Cl_2	11.35 ±0.02	PI	240	*24*
cis-$C_2H_2Cl_2$	9.65	PI	223	*
trans-$C_2H_2Cl_2$	9.64	PI	224	*
CH_2ClCH_2Cl	11.12 ±0.05	PI	225	*24*
$CH_2{=}CClCH_2Cl$	9.82 ±0.03	PI	218	*24*
1,2-$C_3H_6Cl_2$	10.87 ±0.05	PI	215	*24*
1,3-$C_3H_6Cl_2$	10.85 ±0.05	PI	215	*24*
o-$C_6H_4Cl_2$	9.06	PI	217	*24, 28*
m-$C_6H_4Cl_2$	9.12 ±0.01	PI	217	*24*
p-$C_6H_4Cl_2$	8.95	PI	212	*28*
$CHCl_3$	11.42 ±0.03	PI	239	*24*
C_2HCl_3	9.45	PI	216	*
$CHCl_2CHCl_2$	11.10 ±0.05	EI	220	*146*
$CNCl$	12.49 ±0.04	EI	321	*147*
CF_3Cl	12.91 ±0.03	PI	132	*24*
C_2F_3Cl	10.4 ±0.2	EI	107	*148*
C_6F_5Cl	10.4 ±0.1	EI		*149*
CF_2Cl_2	12.31 ±0.05	PI	170	*24*
$CF_3CCl{=}CClCF_3$	10.36 ±0.1	PI		*24*
$CFCl_3$	11.77 ±0.02	PI	205	*24*
CF_3CCl_3	11.78 ±0.03	PI		*24*
$CFCl_2CF_2Cl$	11.99 ±0.02	PI	95	*24*
ClO_3F	13.6 ±0.2	EI	308	*150*
C_5H_4NCl (2-chloro-pyridene)	9.91 ±0.05	EI	255	*84*
C_5H_4NCl (4-chloro-pyridene)	10.15 ±0.05	EI	260	*84*
CH_3COCl	11.02 ±0.05	PI	196	*24*
CH_3COCH_2Cl	9.99	EI	173	*105, 151*
o-$C_6H_4(OH)Cl$	9.28	EI	181	*75*
p-$C_6H_4(OH)Cl$	9.07	EI	175	*75*
C_6H_5COCl	9.70 ±0.01	EI	195	*105*
p-C_6H_4ClCHO	9.61 ±0.01	EI	201	*105*
C_8H_7OCl (α-chloro-acetophenone)	9.5	EI	195	*74, 105*
C_8H_7OCl (p-chloro-acetophenone)	9.47 ±0.05	EI	190	*105*
C_8H_7OCl (p-methyl-benzylchloride)	9.37 ±0.01	EI	187	*105*
$C_6H_4ClCOC_6H_5$	9.68 ±0.01	EI	227	*105*
$CH_2ClCOOCH_3$	10.53 ±0.05	EI	138	*74*
p-$CH_3OC_6H_4COCl$	8.87 ±0.05	EI	149	*105*
p-ClC_6H_4COCl	9.58 ±0.03	EI	192	*105*
cis-C_2H_2FCl	9.86	PI	191	*24, 121*
trans-C_2H_2FCl	9.87	PI	191	*24, 121*
o-C_6H_4FCl	9.155 ±0.01	PI	180	*24*
m-C_6H_4FCl	9.21 ±0.01	PI	180	*24*
p-C_6H_4FCl	9.43 ±0.02	EI	185	*152*
CHF_2Cl	12.45 ±0.05	PI	174	*24*
cis-C_2HF_2Cl	9.86 ±0.02	PI	147	*121*
trans-C_2HF_2Cl	9.83 ±0.02	PI	147	*121*
CH_3CF_2Cl	11.98 ±0.01	PI		*24*
n-$C_3F_7CH_2Cl$	11.84 ±0.02	PI		*24*
$CHFCl_2$	12.39 ±0.20	EI	217	*153*
$(CH_3)_3SiCl$	10.58 ±0.04	EI	160	*130*
CH_3SiCl_3	11.36 ±0.03	EI	136	*153*
$CH_2{=}CHSiCl_3$	10.79 ±0.02	PI	148	*24*
$C_2H_5SiCl_3$	10.74 ±0.04	EI	117	*153*
iso-$C_3H_7SiCl_3$	10.28 ±0.1	EI	100	*153*
C_4H_3ClS	8.68 ±0.01	PI	217	*24*
$C_6H_4NO_2COCl$	10.66 ±0.01	EI	219	*105*
CaF	5.8	EI	75	*
$C_5H_5V(CO)_4$	8.2 ±0.3	EI	9	*154*
$Cr(CO)_6$	8.03 ±0.03	PI	−55	*155*
$C_5H_5Mn(CO)_3$	8.3 ±0.4	EI	83	*154*
$Fe(CO)_5$	7.95 ±0.03	PI	8	*155*
$C_5H_5Co(CO)_2$	8.3 ±0.2	EI	136	*154*
$Ni(CO)_4$	8.28 ±0.03	PI	47	*155*
$(CH_3)_4Ge$	9.2 ±0.2	EI	177	*156*
As_4	9.07 ±0.07	EI	244	*157*
AsH_3	10.03	PI	247	*131*
$AsCl_3$	11.7 ±0.1	EI	208	*158*
$(CH_3)_3As$	8.3 ±0.1	EI	202	*158*
$(C_6H_5)_3As$	7.34 ±0.07	PI	250	*77*
Br_2	10.54 ±0.03	PI	250	*24, 159*
HBr	11.62 ±0.03	PI	259	*24*
$MgBr_2$	10.65 ±0.15	EI	172	*160*
$BrCl$	11.1 ±0.2	EI	259	*161*
CH_3Br	10.53	S, PI	234	*
C_2H_3Br	9.80	PI	243	*
C_2H_5Br	10.29	S	222	*24, 162*
$CH_3C{\equiv}CBr$	10.1 ±0.1	EI	283	*25*
$CH_3CH{=}CHBr$	9.30 ±0.05	PI	224	*24*
n-C_3H_7Br	10.18 ±0.01	PI	216	*24*
iso-C_3H_7Br	10.075 ±0.01	PI	208	*24*
n-C_4H_9Br	10.125 ±0.01	PI	208	*24*
sec-C_4H_9Br	9.98 ±0.01	PI	206	*24*
iso-C_4H_9Br	10.09 ±0.02	PI	208	*24*
tert-C_4H_9Br	9.89 ±0.03	PI	201	*24*
n-$C_5H_{11}Br$	10.10 ±0.02	PI	205	*24*
C_6H_5Br	8.98 ±0.02	PI	231	*24*
o-$C_6H_4BrCH_3$	8.78 ±0.01	PI	218	*24*
m-$C_6H_4BrCH_3$	8.81 ±0.02	PI	218	*24*
p-$C_6H_4BrCH_3$	8.67 ±0.02	PI	215	*24*
CH_2Br_2	10.49 ±0.02	PI	241	*24*
cis-$C_2H_2Br_2$	9.45	PI	241	*
trans-$C_2H_2Br_2$	9.46	PI	240	*
CH_3CHBr_2	10.19 ±0.03	PI	240	*24*
1,3-$C_3H_6Br_2$	10.07 ±0.02	PI	221	*24*
$CHBr_3$	10.51 ±0.02	PI	246	*24*
C_2HBr_3	9.27	PI	240	*24, 28*
$CNBr$	11.95 ±0.08	EI	320	*147*
CF_3Br	11.89	EI	121	*
C_5H_4NBr (2-bromo-pyridine)	9.65 ±0.05	EI	261	*84*
C_5H_4NBr (4-bromo-pyridine)	9.94 ±0.05	EI	267	*84*
CH_3COBr	10.55 ±0.05	PI	197	*24*
C_6H_4BrOH	9.04	EI	187	*75*
$CH_2BrCOOCH_3$	10.37 ±0.05	EI	146	*74*
C_6H_4FBr	8.99 ±0.03	PI	187	*24*
CF_2BrCH_2Br	10.83 ±0.01	PI	160	*24*
C_4H_3BrS	8.63 ±0.01	PI	228	*24*
CH_2ClBr	10.77 ±0.01	PI	236	*24*
CH_2BrCH_2Cl	10.63 ±0.03	PI	227	*24*
$CHCl_2Br$	10.88 ±0.05	EI	237	*146*
$(CH_3)_3SiBr$	10.24 ±0.02	EI	171	*130*
$Mo(CO)_6$	8.12 ±0.03	PI	−31	*155*
RuO_4	12.33 ±0.23	EI	240	*163*
$(CH_3)_4Sn$	8.25 ±0.15	EI	186	*129*
SbH_3	9.58	PI	256	*131*
$(C_6H_5)_3Sb$	7.3 ±0.1	PI	255	*77*
I_2	9.28 ±0.02	PI	229	*24*
HI	10.39	PI	246	*24, 159*
ICl	10.31 ±0.02	EI	242	*164*
IBr	9.98 ±0.03	EI	240	*164*
CH_3I	9.54	S, PI	223	*24, 145*
C_2H_5I	9.33	S, PI	213	*24, 162*
n-C_3H_7I	9.26 ±0.01	PI	208	*24*
iso-C_3H_7I	9.17 ±0.02	PI	201	*24*
n-C_4H_9I	9.21 ±0.01	PI	202	*24*
sec-C_4H_9I	9.09 ±0.02	PI	198	*24*
iso-C_4H_9I	9.18 ±0.02	PI	200	*24*
tert-C_4H_9I	9.02 ±0.03	PI	193	*24*
n-$C_5H_{11}I$	9.19 ±0.01	PI	197	*24*
C_6H_5I	8.73 ±0.03	PI	238	*24*
o-$C_6H_4ICH_3$	8.62 ±0.01	PI	228	*24*
m-$C_6H_4ICH_3$	8.61 ±0.03	PI	226	*24*
p-$C_6H_4ICH_3$	8.50 ±0.01	PI	224	*24*
$W(CO)_6$	8.18 ±0.03	PI	−20	*155*
OsO_4	12.97 ±0.12	EI	219	*163*
$(CH_3)_2Hg$	9.0	EI	233	*143, 156*
$(C_2H_5)_2Hg$	8.5 ±0.1	EI	221	*143*
(iso-$C_3H_7)_2Hg$	7.6 ±0.1	EI	188	*143*
CH_3HgCl	11.5 ±0.2	EI	253	*143*
$(CH_3)_4Pb$	8.0 ±0.4	EI	217	*129*
$(C_6H_5)_3Bi$	7.3 ±0.1	PI	288	*77*

* Average of several values.

REFERENCES

1. Beutler, H. and Junger, H. O. *Zeit. f. Physik* **100**, 80 (1936).
2. Dibeler, V. H., Reese, R. M. and Krauss, M. *Adv. Mass Spectry.* **3**, 471 (1966).
3. Bauer, S. H., Herzberg, G. and Johns, J. W. C. *J. Mol. Spectr.* **13**, 256 (1964).
4. Margrave, J. L. *J. Phys. Chem.* **61**, 38 (1957).
5. Koski, W. S., Kaufman, J. J., Pachucki, C. F. and Shipko, F. J. *J. Am. Chem. Soc.* **80**, 3202 (1958).
6. Fehlner, T. P. and Koski, W. S. *J. Am. Chem. Soc.* **86**, 581 (1964).
7. Drowart, J., DeMaria, G. and Inghram, M. G. *J. Chem. Phys.* **29**, 1015 (1958).
8. Field, F. H. and Franklin, J. L. *Electron Impact Phenomena and the Properties of Gaseous Ions*, Academic Press, Inc., New York, N.Y. (1957).
9. Herzberg, G. *Can. J. Phys.* **39**, 1511 (1961).
10. Waldron, J. D., *Metropolitan Vickers Gazette* **27**, 66 (1956).
11. Dibeler, V. H., Krauss, M., Reese, R. M. and Harllee, F. N. *J. Chem. Phys.* **42**, 3791 (1965).
12. Herzberg, G. and Shoosmith, J., *Can J. Phys.* **34**, 523 (1956).
13. Elder, F. A., Giese, C., Steiner, B. and Inghram, M. *J. Chem. Phys.* **36**, 3292 (1962).
14. Nicholson, A. J. C., *J. Chem. Phys.* **43**, 1171 (1965).
15. Dibeler, V. H. and Reese, R. M. *J. Res. Natl. Bur. Std.* **A68**, 409 (1964).
16. Zelikoff, M. and Watanabe, K., *J. Opt. Soc. Am.* **43**, 756 (1953).
17. Al-Joboury, M. I. and Turner, D. W., *J. Chem. Soc. London* **4434** (1964).
18. Farmer, J. B. and Lossing, F. P., *Can. J. Chem.* **33**, 861 (1955).
19. Collin, J. and Lossing, F. P., *J. Am. Chem. Soc.* **79**, 5848 (1957).
20. Watanabe, K. and Namioka, T., *J. Chem. Phys.* **24**, 915 (1956).
21. Collin, J. and Lossing, F. P., *J. Am. Chem. Soc.* **81**, 2064 (1959).
22. Dorman, F. H. *J. Chem. Phys.* **43**, 3507 (1965).
23. Pottie, R. F., Harrison, A. G. and Lossing, F. P. *J. Am. Chem. Soc.* **83**, 3204 (1961).
24. Watanabe, K., Nakayama, T. and Mottl, J. *J. Quant. Spectrosc. Radiat. Transfer* **2**, 369 (1962).
25. Coats, F. H. and Anderson, R. C. *J. Am. Chem. Soc.* **79**, 1340 (1957).
26. Varsel, C. J., Morrell, F. A., Resnik, F. E. and Powell, W. A. *Anal. Chem.* **32**, 182 (1960).
27. McDowell, C. A., Lossing, F. P., Henderson, I. H. S. and Farmer, J. B. *Can. J. Chem.* **34**, 345 (1956).
28. Bralsford, R., Harris, P. V. and Price, W. C. *Proc. Roy. Soc.* **A258**, 459 (1960).
29. Lossing, F. P., Kebarle, P. and DeSousa, J. B. *Adv. Mass Spectrometry* **431** (1959).
30. Watanabe, K. *J. Chem. Phys.* **26**, 542 (1957).
31. Meyer, F. and Harrison, A. G. *Can. J. Chem.* **42**, 2256 (1964).
32. Harrison, A. G., Honnen, L. R., Dauben, H. J. and Lossing, F. P. *J. Am. Chem. Soc.* **82**, 5593 (1960).
33. Natalis, P. and Laune, *J. Bull. Soc. Chim. Belg.* **73**, 944 (1964).
34. Taubert, R. and Lossing, F. P. *J. Am. Chem. Soc.* **84**, 1523 (1962).
35. Steiner, B., Giese, C. F. and Inghram, M. G. *J. Chem. Phys.* **34**, 189 (1961).
36. Fisher, I. P., Palmer, T. F. and Lossing, F. P. *J. Am. Chem. Soc.* **86**, 2741 (1964).
37. Brehm, B. *Z. Naturforschg.* **21a**, 196 (1966).
38. Momigny, J., Brakier, L. and D Or, L. *Bull. Classe Sci. Acad. Roy. Belg.* **48**, 1002 (1962).
39. Harrison, A. G., Kebarle, P. and Lossing, F. P. *J. Am. Chem. Soc.* **83**, 777 (1961).
40. Thrush, B. A. and Zwolenik, J. J. *Disc. Faraday Soc.* **35**, 196 (1963).
41. Lifshitz, C. and Bauer, S. H. *J. Phys. Chem.* **67**, 1629 (1963).
42. Meyerson, S., McCollum, J. D. and Rylander, P. N. *J. Am. Chem. Soc.* **83**, 1401 (1961).
43. Steele, W. C., Jennings, B. H., Botyos, G. L. and Dudek, G. O. *J. Org. Chem.* **30**, 2886 (1965).
44. Kybett, B. D., Carroll, S., Natalis, P., Bonnell, D. W., Margrave, J. L. and Franklin, J. L. *J. Am. Chem. Soc.* **88**, 626 (1966).
45. Meyer, F., Haynes, P., McLean, S. and Harrison, A. G. *Can. J. Chem.* **43**, 211 (1965).
46. Natalis, P. *Bull. Soc. Chim. Belg.* **73**, 961 (1964).
47. Pottie, R. F. and Lossing, F. P. *J. Am. Chem. Soc.* **85**, 269 (1963).
48. Vilesov, F. I. and Terenin, A. N. *Dokl. Phys. Chem., Proc. Acad. Sci.* **USSR 115**, 539 (1957).
49. Clark, L. B. *J. Chem. Phys.* **43**, 2566 (1965).
50. Kitagawa, T., Harada, Y., Inokuchi, H. and Kodera, K. *J. Mol. Spectr.* **19**, 1 (1966).
51. Natalis, P. *Bull. Soc. Roy. Sci. Liege* **31**, 803 (1962).
52. Harrison, A. G. and Lossing, F. P. *J. Am. Chem. Soc.* **82**, 1052 (1960).
53. Vilesov, F. I. *J. Phys. Chem.* **USSR 35**, 986 (1961).
54. Natalis, P. and Franklin, J. L. *J. Phys. Chem.* **69**, 2935 (1965).
55. Wacks, M. E. and Dibeler, V. H. *J. Chem. Phys.* **31**, 1557 (1959).
56. Wacks, M. E. *J. Chem. Phys.* **41**, 1661 (1964).
57. King, A. B. *J. Chem. Phys.* **42**, 3526 (1965).
58. Law, R. W. and Margrave, J. L. *J. Chem. Phys.* **25**, 1086 (1956).
59. Lofthus, A. *The Molecular Spectrum of Nitrogen* Department of Physics, University of Oslo, Blindern, **Norway, Spectroscopic Report No. 2**, 1 (1960).
60. Reed, R. I. and Snedden, W. *J. Chem. Soc.* **4132** (1959).
61. Foner, S. N. and Hudson, R. L. *J. Chem. Phys.* **29**, 442 (1958).
62. Foner, S. N. and Hudson, R. L. *J. Chem. Phys.* **28**, 719 (1958).
63. Akopyan, M. E. and Vilesov, F. I. *Kinetics and Catalysis*, **4**, 32 (1963).
64. Dibeler, V. H., Reese, R. M. and Franklin, J. L. *J. Am. Chem. Soc.* **83**, 1813 (1961).
65. Berkowitz, J. *J. Chem. Phys.* **36**, 2533 (1962).
66. Morrison, J. D. and Nicholson, A. J. C. *J. Chem. Phys.* **20**, 1021 (1952).
67. Watanabe, K. and Mottl, J. R. *J. Chem. Phys.* **26**, 1773 (1957).
68. Pottie, R. F. and Lossing, F. P. *J. Am. Chem. Soc.* **83**, 4737 (1961)
69. Gallegos, E. and Kiser, R. W. *J. Phys. Chem.* **65**, 1177 (1961).
70. Gallegos, E. J. and Kiser, R. W. *J. Phys. Chem.* **66**, 136 (1962).
71. Amr El-Sayed, M. F., Kasha, M. and Tanaka, Y. *J. Chem. Phys.* **34**, 334 (1961).
72. Akopyan, M. E. and Vilesov, F. I. *Dokl. Phys. Chem., Proc. Acad. Sci.* **USSR 158**, 965 (1964).
73. Kurbatov, B. L., Vilesov, F. I. and Terenin, A. N. *Soviet Physics-Doklady* **6**, 883 (1962).
74. Pignataro, S., Foffani, A., Innorta, G. and Distefano, G. *Z. Phys. Chem. Neue Folge* **49**, 291 (1966).
75. Crable, G. F. and Kearns, G. L. *J. Phys. Chem.* **66**, 436 (1962).
76. Farrell, P. G. and Newton, J. *J. Phys. Chem.* **69**, 3506 (1965).
77. Vilesov, F. I. and Zaitsev, V. M. *Dokl. Phys. Chem., Proc. Acad. Sci.* **USSR 154**, 117 (1964).
78. Paulett, G. S. and Ettinger, R. *J. Chem. Phys.* **39**, 825 (1963).
79. Merer, A. J., *Can. J. Phys.* **42**, 1242 (1964).
80. Gowenlock, B. G., Majer, J. R. and Snelling, D. R. *Trans. Faraday Soc.* **58**, 670 (1962).
81. Akopyan, M. E., Vilesov, F. I. and Terenin, A. N. *Izv. Akad. Nauk USSR, Ser. Fiz.* **27**, 1083 (1963).
82. Dibeler, V. H., Franklin, J. L. and Reese, R. M. *J. Am. Chem. Soc.* **81**, 68 (1959).
83. Momigny, J., Urbain, J. and Wankenne, H. *Bull. Soc. Roy. Sci. Liege* **34**, 337 (1965).
84. Basila, M. R. and Clancy, D. J. *J. Phys. Chem.* **67**, 1551 (1963).
85. Finch, A. C. M. *J. Chem. Soc.* **2272** (1964).
86. Franklin, J. L., Dibeler, V. H., Reese, R. M. and Krauss, M. *J. Am. Chem. Soc.* **80**, 298 (1958).
87. Radwan, T. N. and Turner, D. W. *J. Chem. Soc.* **Sect. A85** (1966).
88. Foner, S. N. and Hudson, R. L. *Advances in Chem.* **Ser. 34** (1962).
89. Foner, S. N. and Hudson, R. L. *J. Chem. Phys.* **23**, 1364 (1955).
90. Lindeman, L. P. and Guffy, J. C. *J. Chem. Phys.* **29**, 247 (1958).
91. Foner, S. N. and Hudson, R. L. *J. Chem. Phys.* **36**, 2676 (1962).
92. Berkowitz, J., Chupka, W. A., Blue, G. D. and Margrave, J. L. *J. Phys. Chem.*, **63**, 644 (1959).
93. White, D., Seshadri, K. S., Dever, D. F., Mann, D. E. and Linevski, M. J. *J. Chem. Phys.* **39**, 2463 (1963).
94. Krupenie, P. H. *The Band Spectrum of Carbon Monoxide* Institute for Basic Standards, National Bureau of Standards, Washington, D.C., **NSRDS-NBS 5**, 1 (1966).
95. Tanaka, Y., Jursa, A. S. and LeBlanc, F. J. *J. Chem. Phys.* **32**, 1205 (1960).
96. Tanaka, Y., Jursa, A. S. and LeBlanc, F. J. *J, Chem. Phys.* **32**, 1199 (1960).
97. Nakayama, T., Kitamura, M. Y. and Watanabe, K. *J. Chem. Phys.* **30**, 1180 (1959).
98. Frost, D. C., Mak, D. and McDowell, C. A. *Can. J. Chem.* **40**, 1064 (1962).
99. Hernandez, G. J. *J. Chem. Phys.* **38**, 1644 (1963).
100. Hernandez, G. J. *J. Chem. Phys.* **38**, 2233 (1963).
101. Chupka, W. A. *J. Chem. Phys.* **30**, 191 (1959).
102. Watanabe, K. and Nakayama, T. *J. Chem. Phys.* **29**, 48 (1958).
103. Murad, E. and Inghram, M. G. *J. Chem. Phys*, **40**, 3263 (1964).
104. Higasi, K., Nozoe, T. and Omura, I. *Bull. Chem. Soc. Japan* **30**, 408 (1957).
105. Foffani, A., Pignataro, S., Cantone, B. and Grasso, F. *Z. Phys. Chem. Neue Folge* **42**, 221 (1964).
106. Natalis, P. and Franklin, J. L. *J. Phys. Chem.* **69**, 2943 (1965).
107. Cotter, J. L. *J. Chem. Soc.* **5742** (1965).
108. Cotter, J. L. *J. Chem. Soc.* **5477** (1964).
109. Iczkowski, R. P. and Margrave, J. L. *J. Chem. Phys.* **30**, 403 (1959).
110. Frost, D. C. and McDowell, C. A. *Can. J. Chem.* **36**, 39 (1958).
111. Hildenbrand, D. L. and Muran, E. *J. Chem. Phys.* **43**, 1400 (1965).
112. Marriott, J. and Craggs, J. D. *J. Electronics and Control* **3**, 194 (1957).
113. Fisher, I. P., Homer, J. B. and Lossing, F. P. *J. Am. Chem. Soc.* **87**, 957 (1965).
114. Pottie, R. F. *J. Chem. Phys.* **42**, 2607 (1965).
115. Herron, J. T. and Dibeler, V. H. *J. Res. Natl. Bur. Std.* **65**, 405 (1961).
116. Loughran, E. D. and Mader, C. *J. Chem. Phys.* **32**, 1578 (1960).
117. Frost, D. C. and McDowell, C. A. *The Determination of Ionization and Dissociation Potentials of Molecules by Radiation with Electrons* Department of Chemistry, University of British Columbia, Vancouver 8, B. C., Canada, **AFCRL-TR-60-423 1** (1960).
118. Dibeler, V. H., Reese, R. M. and Franklin, J. L. *J. Chem. Phys.* **27**, 1296 (1957).
119. Wilson, E. G. Jortner, J. and Rice, S. A. *J. Am. Chem. Soc.* **85**, 813 (1963).
120. Frost, D. C. and McDowell, C. A. *Proc. Roy. Soc.* **A241**, 194 (1957).
121. Momigny, J. *Nature* **199**, 1179 (1963).
122. Martin, R. H., Lampe, F. W. and Taft, R. W. *J. Am. Chem. Soc.* **88**, 1353 (1966).
123. Reese, R. M. and Dibeler, V. H. *J. Chem. Phys.* **24**, 1175 (1956).
124. Herron, J. T. and Dibeler, V. H. *J. Chem. Phys.* **33**, 1595 (1960).
125. Hammond, V. J., Price, W. C., Teegan, J. P. and Walsh, A. D. *Disc. Faraday Soc.* **9**, 53 (1950).
126. Hudson, R. D. *J. Chem. Phys.* **43**, 1790 (1965).
127. Margrave, J. L. *J. Chem. Phys.* **41**, 2250 (1964).

REFERENCES (Continued)

128. Ehlert, T. C. and Margrave, J. L. *J. Am. Chem. Soc.* **86**, 3901 (1964).
129. Hobrock, B. G. and Kiser, R. W. *J. Phys. Chem.* **65**, 2186 (1961).
130. Hess, G. G., Lampe, F. W. and Sommer, L. H. *J. Am. Chem. Soc.* **87**, 5327 (1965).
131. Price, W. C. and Passmore, T. R. *Disc. Faraday Soc.* **35**, 232 (1963).
132. Wada, Y. and Kiser, R. W. *J. Phys. Chem.* **68**, 2290 (1964).
133. Berkowitz, J. and Chupka, W. A. *J. Chem. Phys.* **40**, 287 (1964).
134. Palmer, T. F. and Lossing, F. P. *J. Am. Chem. Soc.* **84**, 4661 (1962).
135. Price, W. C. *Bull. Am. Phys. Soc.* **10**, 9 (1935).
136. Hobrock, B. G. and Kiser, R. W. *J. Phys. Chem.* **66**, 1551 (1962).
137. Hobrock, B. G. and Kiser, R. W. *J. Phys. Chem.* **67**, 648 (1963).
138. Hobrock, B. G. and Kiser, R. W. *J. Phys. Chem.* **66**, 1648 (1962).
139. Khvostenko, V. I. *Russ. J. Phys. Chem.* **36**, 197 (1962).
140. Hobrock, B. G. and Kiser, R. W. *J. Phys. Chem.* **67**, 1283 (1963).
141. Reese, R. M., Dibeler, V. H. and Franklin, J. L. *J. Chem. Phys.* **29**, 880 (1958).
142. Baldwin, M., Maccoll, A., Kirkien-Konasiewicz, A. and Saville, B. *Chem. Ind.* **286** (1966).
143. Gowenlock, B. G., Kay, J. and Majer, J. R. *Trans. Faraday Soc.* **59**, 2463 (1963).
144. Berkowitz, J., Tasman, H. A. and Chupka, W. A. *J. Chem. Phys.* **36**, 2170 (1962).
145. Price, W. C. *J. Chem. Phys.* **4**, 539 (1936).
146. Harrison, A. G. and Shannon, T. W. *Can. J. Chem.* **40**, 1730 (1962).
147. Herron, J. T. and Dibeler, V. H. *J. Am. Chem. Soc.* **82**, 1555 (1960).
148. Margrave, J. L. *J. Chem. Phys.* **31**, 1432 (1959).
149. Majer, J. R. and Patrick, C. R. *Trans. Faraday Soc.* **58**, 17 (1962).
150. Dibeler, V. H., Reese, R. M. and Mann, D. E. *J. Chem. Phys.* **27**, 176 (1957).
151. Foffani, A., Pignataro, S., Cantone, B. and Grasso, F. *Nuovo Cimento* **29**, 918 (1963).
152. Momigny, J. and Wirtz-Cordier, A. M. *Ann. Soc. Sci. Bruxelles* **76**, 164 (1962).
153. Hobrock, D. L. and Kiser, R. W. *J. Phys. Chem.* **68**, 575 (1964).
154. Winters, R. E. and Kiser, R. W. *J. Organometal. Chem.* **4**, 190 (1965).
155. Vilesov, F. I. and Kurbatov, B. L. *Dokl. Phys. Chem., Proc. Acad. Sci. USSR* **140**, 792 (1961).
156. Hobrock, B. G. and Kiser, R. W. *J. Phys. Chem.* **66**, 155 (1962).
157. Westmore, J. B., Mann, K. H. and Tickner, A. W. *J. Phys. Chem.* **68**, 606 (1964).
158. Cullen, W. R. and Frost, D. C. *Can. J. Chem.* **40**, 390 (1962).
159. Morrison, J. D., Hurzeler, H., Inghram, M. G. and Stanton, H. E. *J. Chem. Phys.* **33**, 821 (1960).
160. Berkowitz, J. and Marquart, J. R. *J. Chem. Phys.* **37**, 1853 (1962).
161. Irsa, A. P. and Friedman, L. *J. Inorg. Nucl. Chem.* **6**, 77 (1958).
162. Price, W. C. *J. Chem. Phys.* **4**, 547 (1936).
163. Dillard, J. G. and Kiser, R. W. *J. Phys. Chem.* **69**, 3893 (1965).
164. Frost, D. C. and McDowell, C. A. *Can. J. Chem.* **38**, 407 (1960).

ELECTRON WORK FUNCTIONS OF THE ELEMENTS

Compiled by Herbert B. Michaelson, 1977

The measured values cited for polycrystalline and single-crystal specimens are selected as being the best available data at this time. The selection is based on (1) The validity of the experimental technique (e.g., vacua of 10^{-9} or 10^{-10} Torr, clean surfaces, and identification of crystal-face distribution and other surface conditions), and (2) Best agreement with preferred values and theoretical values of the true work function (given variously by Fomenko,[1] Rivière,[2] Trasatti,[3] and Lang and Kohn[4]). Experimental data that are not well substantiated according to these criteria are listed in *italics*. Crystallographic directions for single-crystal data are indicated by parentheses.

Abbreviations apply to the experimental method: T, thermionic; P, photoelectric; CPD, contact potential difference; F, field emission. Important distinctions among such measurements are discussed in the Rivière[2] paper, pp. 180 to 198.

Element	Experimental value, ϕ (eV)	Experimental method	Ref.
Ag	*4.26*	P	5
	4.64 (100)	P	5
	4.52 (110)	P	5
	4.74 (111)	P	6
Al	4.28	P	7
	4.41 (100)	P	8
	4.06 (110)	P	7
	4.24 (111)	P	8
As	*3.75*	P	9
Au	5.1	P	10
	5.47 (100)	P	11
	5.37 (110)		11
	5.31 (111)		11
B	*4.45*	T	12
Ba	*2.7*	T	13
Be	4.98	P	14
Bi	*4.22*	P	15
C	*5.0*	CPD	16
Ca	2.87	P	17
Cd	*4.22*	CPD	18
Ce	2.9	P	10
Co	5.0	P	10
Cr	4.5	P	10
Cs	2.14	P	19
Cu	4.65	P	10
	4.59 (100)	P	20
	4.48 (110)	P	20
	4.94 (111)	P	20
	4.53 (112)	P	20
Eu	2.5	P	10
Fe	4.5	P	10
	4.67 (100)	P	21
	4.81α (111)	P	22
	4.70α	P	23
	4.62β	P	23
	4.68γ	P	23
Ga	4.2	CPD	24
Ge	5.0	CPD	25
	4.80 (111)	P	26
Gd	3.1	P	10
Hf	3.9	P	10
Hg	4.49	P	27
In	4.12	P	28
Ir	*5.27*	T	29
	5.42 (110)	F	30
	5.76 (111)	F	30
	5.67 (100)	F	31
	5.00 (210)	F	31
K	2.30	P	32
La	3.5	P	10
Li	*2.9*	F	33
Lu	*3.3*	CPD	34
Mg	*3.66*	P	35
Mn	4.1	P	10
Mo	4.6	P	10
	4.53 (100)	P	36
	4.95 (110)	P	36
	4.55 (111)	P	36
	4.36 (112)	P	36
	4.50 (114)	P	36
	4.55 (332)	P	36
Na	2.75	P	37

ELECTRON WORK FUNCTIONS OF THE ELEMENTS (continued)

Element	Experimental value, ϕ (eV)	Experimental method	Ref.
Nb	4.3	P	10
	4.02 (001)	T	38
	4.87 (110)	T	38
	4.36 (111)	T	38
	4.63 (112)	T	38
	4.29 (113)	T	38
	3.95 (116)	T	38
	4.18 (310)	T	38
Nd	3.2	P	10
Ni	5.15	P	10
	5.22 (100)	P	39
	5.04 (110)	P	39
	5.35 (111)	P	39
Os	*4.83*	T	29
Pb	4.25	P	40
Pd	5.12	P	31
	5.6 (111)	P	41
Pt	5.65	P	10
	5.7 (111)	P	41
Rb	*2.16*	P	27
Re	*4.96*	T	29
	5.75 (1011)	F	33
Rh	4.98	P	31
Ru	4.71	P	31
Sb	4.55 (amorph.)	–	42
	4.7 (100)	–	43
Sc	3.5	P	10
Se	5.9	P	44
Si	4.85n	CPD	40
	*4.91*p (100)	CPD	45
	4.60p (111)	P	46
Sm	2.7	P	10
Sn	*4.42*	CPD	47
Sr	*2.59*	T	48
Ta	4.25	T	29
	4.15 (100)	T	49
	4.80 (110)	T	49
	4.00 (111)	T	49
Tb	3.0	P	50
Te	4.95	P	44
Th	3.4	T	51
Ti	4.33	P	10
Tl	*3.84*	CPD	52
U	3.63	P & CPD	53
	3.73 (100)	P & CPD	54
	3.90 (110)	P & CPD	54
	3.67 (113)	P & CPD	54
V	4.3	P	10
W	4.55	CPD	55
	4.63 (100)	F	30
	5.25 (110)	F	30
	4.47 (111)	F	30
	4.18 (113)	CPD	56
	4.30 (116)	T	57
Y	3.1	P	10
Zn	*4.33*	P	15
	4.9 (0001)	CPD	58
Zr	4.05	P	10

REFERENCES

1. Fomenko, V. S., *Emission Properties of Materials,* 3rd ed., Naukova Dumka, Kiev, 1970 (in Russian).
2. Rivière, J. C., *Solid State Surface Science,* Green, M., Ed., Vol. 1, Marcel Dekker, 1969, chap. 4.
3. Trasatti, S., *Chim. Ind.* (Milan), 53(6), 559, 1971.
4. Lang, N. D. and Kohn, W., *Phys. Rev. B,* 3(4), 1215, 1971.
5. Dweydari, A. W. and Mee, C. H. B., *Phys. Status Solidi A,* 27, 223, 1975.
6. Dweydari, A. W. and Mee, C. H. B., *Phys. Status Solidi A,* 17, 247, 1973.
7. Eastment, R. M. and Mee, C. H. B., *J. Phys. F,* 3, 1738, 1973.
8. Grepstad, J. K., Gartland, P. O., and Slagsvold, B. J., *Surf. Sci.,* 57, 348, 1976.
9. Raisin, C. and Pinchaux, R., *Solid State Commun.,* 16, 941, 1975.
10. Eastman, D. E., *Phys. Rev. Sect. B,* 2, 1, 1970.
11. Potter, H. C. and Blakeley, J. M., *J. Vac. Sci. Technol.,* 12, 635, 1975 and Potter, H. C., Ph.D. thesis Cornell University, Materials Science Center Rep. No. 1353, 1970.
12. Adirovich, E. I. and Gol'dshtein, L. M., *Fiz. Tverdogo Tela* (Leningrad), 9, 1258, 1967.
13. Bondarenko, B. V. and Makhov, V. I., *Sov. Phys. Solid State,* 12(7), 1522, 1971.
14. Gustafsson, Broden, and Nilsson, *J. Phys. F,* 4, 2351, 1974.
15. Suhrmann, R. and Wedler, G., *Z. Angew. Phys.,* 14, 70, 1962.
16. Robrieux, B., Faure, R., and Dussaulcy, J. P., *C. R. Acad. Sci. Ser. B,* 278(14), 659, 1974.
17. Gaudart, L. and Riviora, R., *Appl. Opt.,* 10, 2336, 1971.
18. Anderson, P. A., *Phys. Rev.,* 98, 1739, 1955.
19. Boutry, G. A. and Dormont, H., *Philips Tech. Rev.,* 30, 225, 1969.
20. Gartland, P. O., *Phys. Norv.,* 6(3,4), 201, 1972.
21. Ueda, K. and Shimizu, R., *Jp. J. Appl. Phys.,* 11(6), 916, 1972.
22. Kobayashi, H. and Kato, S., *Surf. Sci.,* 18(2), 341, 1969.
23. Cardwell, A., *Phys. Rev.,* 92, 554, 1953.
24. Osipova, E. V., Shurmovskaya, N. A., and Burshtein, R. Kh., *Elektrokhimiya,* 5(10), 1139, 1969 (in Russian).
25. Boiko, B. A., Gorodetskii, D. A., and Yas'ko, A. A., *Sov. Phys. Solid State,* 15(11), 2101, 1974.
26. Gobeli, G. W. and Allen, F. G., *Surf. Sci.,* 2, 402, 1964.
27. Lazarev, V. B. and Malov, Y. I., *Fiz. Met. Metalloved.,* 24(3), 565, 1967.
28. Peisner, J., Roboz, P., and Barna, P. B., *Phys. Stat. A,* 4, K187, 1971.
29. Wilson, R. G., *J. Appl. Phys.,* 37, 3170, 1966.
30. Strayer, R. W., Mackie, W., and Swanson, L. W., *Surf. Sci.,* 34, 225, 1973.
31. Nieuwenhuys, Bouwman, and Sachtler, *Thin Solid Films,* 21, 51, 1974.
32. Van Oirschot, Th. G. J., van den Brink, M., and Sachtler, W. H. M., *Surf. Sci.,* 29, 189, 1972.
33. Ovchinnikov, A. P. and Tsarev, B. M., *Sov. Phys. Solid State,* 9(12), 2766, 1968.
34. Bondarenko, B. V. and Makhov, V. I., *Sov. Phys. Solid State,* 12, 2986, 1971.
35. Garron, R., *C. R. Acad. Sci.,* 258, 1458, 1964.
36. Berge, Gartland, and Slagsvold, *Surf. Sci.,* 43, 275, 1974.
37. Whitefield, R. J. and Brady, J. J., *Phys. Rev. Lett.,* 26(7), 380, 1971.
38. Leblanc, R. P., Vanbrugghe, B. C., and Girouard, F. E., *Can. J. Phys.,* 52, 1589, 1974.
39. Baker, B. G., Johnson, E. B., and Maire, G. I. C., *Surf. Sci.,* 24, 572, 1971.
40. Thanailakis, A., *Inst. Phys. Conf. Ser.,* p. 59, 1974.
41. Demuth, J. E., *Chem. Phys. Lett.,* 45, 12, 1977.
42. Gorodetskii, D. A. and Yas'ko, A. A., *Sov. Phys. Solid State,* 13(11), 2928, 1972.
43. Gorodetskii, D. A. and Yas'ko, A. A., *Sov. Phys. Solid State,* 13(5), 1085, 1971.
44. Williams, R. H. and Polanco, J. I., *J. Phys. C,* 7, 2745, 1974.
45. Allen, F. G., *J. Phys. Chem. Solids,* 8, 119, 1959.
46. Allen, F. G. and Gobeli, G. W., *J. Appl. Phys.,* 35, 597, 1964.
47. Simmons, J. G., *Phys. Rev. Lett.,* 10, 10, 1963.
48. Alleau, T., *Surface Phenomena in Thermionic Emitters, Round Table Conf.,* Inst. Tech. Phys. Julich Nucl. Res. Establ., Julich, Germany, 1969, p. 54 (in English).
49. Protopopov, Mikheeva, Shreinberg, and Shuppe, *Fiz. Tverdogo Tela,* 8(4), 1140, 1966.
50. Nemchenok, R. L., Strakovskaya, S. E., and Titenskii, A. I., *Fiz. Tverdogo Tela* 11(9), 2692, 1969.
51. Estrup, P. J., Anderson, J. R., and Danforth, W. E., *Surf. Sci.,* 4, 286, 1966.
52. Klein, O. and Lange, E., *Z. Elektrochem.,* 44, 542, 1938.
53. Hopkins, B. J. and Sargood, A. J., *Nuovo Cimento,* 5, 459, 1967.
54. Lea, C. and Mee, C. H. B., *J. Appl. Phys.,* 39, 5890, 1968.
55. Hopkins, B. J. and Rivière, J. C., *Proc. Phys. Soc.* (London), 81, 590, 1963.
56. Love, H. M. and Dyer, G. L., *Can. J. Phys.,* 40, 1837, 1962.
57. Sultanov, V. M., *Radio Eng. Electron.,* 9, 252, 1964 (English translation).
58. Baker, J. M. and Blakeley, J. M., *Surf. Sci.,* 32, 45, 1972.

PROPERTIES OF METALS AS CONDUCTORS

Metal.	Resistivity microhm-centimeters 20° C.	Temp. coefficient 20° C.	Specific gravity.	Tensile strength, lbs./in.	Melting point ° C.
*Advance. See *constantan*					
Aluminum	2.824	0.0039	2.70	30,000	659
Antimony	41.7	.0036	6.6	......	630
Arsenic	33.3	.0042	5.73	......	......
Bismuth	120	.004	9.8	......	271
Brass	7	.002	8.6	70,000	900
Cadmium	7.6	.0038	8.6	......	321
*Calido. See *nichrome*					
Climax	87	.0007	8.1	150,000	1250
Cobalt	9.8	.0033	8.71	......	1480
Constantan	49	.00001	8.9	120,000	1190
Copper: annealed	1.7241	.00393	8.89	30,000	1083
hard-drawn	1.771	.00382	8.89	60,000	
Eureka. See *constantan*					
Excello	92	.00016	8.9	95,000	1500
Gas Carbon	5000	−.0005	......	......	3500
German silver, 18% Ni	33	.0004	8.4	150,000	1100
Gold	2.44	.0034	19.3	20,000	1063
Ideal. See *constantan*					
Iron, 99.98% pure	10	.005	7.8	......	1530
Lead	22	.0039	11.4	3,000	327
Magnesium	4.6	.004	1.74	33,000	651
Manganin	44	.00001	8.4	150,000	910
Mercury	95.783	.00089	13.546	0	−38.9
Molybdenum, drawn	5.7	.004	9.0	......	2500
Monel metal	42	.0020	8.9	160,000	1300
*Nichrome	100	.0004	8.2	150,000	1500
Nickel	7.8	.006	8.9	120,000	1452
Palladium	11	.0033	12.2	39,000	1550
Phosphor bronze	7.8	.0018	8.9	25,000	750
Platinum	10	.003	21.4	50,000	1755
Silver	1.59	.0038	10.5	42,000	960
Steel, E. B. B.	10.4	.005	7.7	53,000	1510
Steel, B. B.	11.9	.004	7.7	58,000	1510
Steel, Siemens-Martin	18	.003	7.7	100,000	1510
Steel, manganese	70	.001	7.5	230,000	1260
Tantalum	15.5	.0031	16.6	......	2850
*Therlo	47	.00001	8.2		
Tin	11.5	.0042	7.3	4,000	232
Tungsten, drawn	5.6	.0045	19	500,000	3400
Zinc	5.8	.0037	7.1	10,000	419

* Trade mark.

Superconductivity*

B.W. Roberts
General Electric Research Laboratory, Schenectady, New York

The following tables on superconductivity include superconductive properties of chemical elements, thin films, a selected list of compounds and alloys, and high-magnetic-field superconductors.

The historically first observed and most distinctive property of a superconductive body is the near total loss of resistance at a critical temperature (T_c) that is characteristic of each material. Figure 1(a) below illustrates schematically two types of possible transitions. The sharp vertical discontinuity in resistance is indicative of that found for a single crystal of a very pure element or one of a few well annealed alloy compositions. The broad transition, illustrated by broken lines, suggests the transition shape seen for materials that are not homogeneous and contain unusual strain distributions. Careful testing of the resistivity limits for superconductors shows that it is less than 4×10^{-23} ohm-cm, while the lowest resistivity observed in metals is of the order of 10^{-13} ohm-cm. If one compares the resistivity of a superconductive body to that of copper at room temperature, the superconductive body is at least 10^{17} times less resistive.

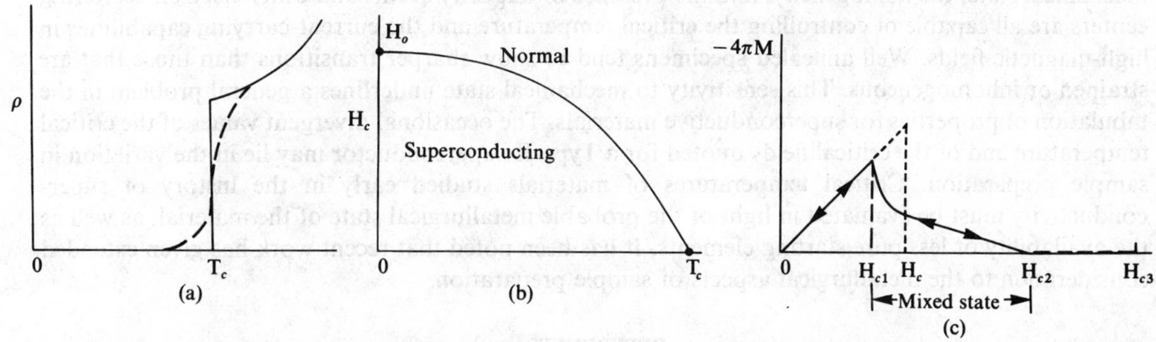

Figure 1. PHYSICAL PROPERTIES OF SUPERCONDUCTORS

(a) Resistivity versus temperature for a pure and perfect lattice (solid line). Impure and/or imperfect lattice (broken line).
(b) Magnetic-field temperature dependence for Type-I or "soft" superconductors.
(c) Schematic magnetization curve for "hard" or Type-II superconductors.

The temperature interval ΔT_c, over which the transition between the normal and superconductive states takes place, may be of the order of as little as 2×10^{-5} °K *or* several °K in width, depending on the material state. The narrow transition width was attained in 99.9999 percent pure gallium single crystals.

A Type-I superconductor below T_c, as exemplified by a pure metal, exhibits perfect diamagnetism and excludes a magnetic field up to some critical field H_c, whereupon it reverts to the normal state as shown in the H-T diagram of Figure 1(b).

The difference in entropy near absolute zero between the superconductive and normal states relates directly to the electronic specific heat, γ: $(S_s - S_n)_{T \to 0} = -\gamma T$.

The magnetization of a typical high-field superconductor is shown in Figure 1(c). The discovery of the large current-carrying capability of Nb_3Sn and other similar alloys has led to an extensive study of the physical properties of these alloys. In brief, a high-field superconductor, or Type-II superconductor, passes from the perfect diamagnetic state at low magnetic fields to a mixed state and finally to a sheathed state before attaining the normal resistive state of the metal. The magnetic field values separating the four stages are given as H_{c1}, H_{c2}, and H_{c3}. The superconductive state below H_{c1} is perfectly diamagnetic, identical to the state of most pure metals of the "soft" or Type-I

***Prepared for Office of Standard Reference Data, National Bureau of Standards, by Standard Reference Data Center on Superconductive Materials, Schenectady, N.Y.**

superconductor. Between H_{c1} and H_{c2} a "mixed superconductive state" is found in which fluxons (a minimal unit of magnetic flux) create lines of normal superconductor in a superconductive matrix. The volume of the normal state is proportional to $-4\pi M$ in the "mixed state" region. Thus at H_{c2} the fluxon density has become so great as to drive the interior volume of the superconductive body completely normal. Between H_{c2} and H_{c3} the superconductor has a sheath of current-carrying superconductive material at the body surface, and above H_{c3} the normal state exists. With several types of careful measurement, it is possible to determine H_{c1}, H_{c2}, and H_{c3}. Table 2-35 contains some of the available data on high-field superconductive materials.

High-field superconductive phenomena are also related to specimen dimension and configuration. For example, the Type-I superconductor, Hg, has entirely different magnetization behavior in high magnetic fields when contained in the very fine sets of filamentary tunnels found in an unprocessed Vycor glass. The great majority of superconductive materials are Type II. The elements in very pure form and a very few precisely stoichiometric and well annealed compounds are Type-I with the possible exceptions of vanadium and niobium.

Metallurgical Aspects. The sensitivity of superconductive properties to the material state is most pronounced and has been used in a reverse sense to study and specify the detailed state of alloys. The mechanical state, the homogeneity, and the presence of impurity atoms and other electron-scattering centers are all capable of controlling the critical temperature and the current-carrying capabilities in high-magnetic fields. Well annealed specimens tend to show sharper transitions than those that are strained or inhomogeneous. This sensitivity to mechanical state underlines a general problem in the tabulation of properties for superconductive materials. The occasional divergent values of the critical temperature and of the critical fields quoted for a Type-II superconductor may lie in the variation in sample preparation. Critical temperatures of materials studied early in the history of superconductivity must be evaluated in light of the probable metallurgical state of the material, as well as the availability of less pure starting elements. It has been noted that recent work has given extended consideration to the metallurgical aspects of sample preparation.

REFERENCES

References to the data presented in this section, to additional entries of superconductive materials, and to those materials specifically tested and found non-superconductive to some low temperature may be found in the following publications:

"Superconductive Materials and Some of Their Properties", *Progress in Cryogenics*, B.W. Roberts, Vol. IV, Heywood and Co., 1964, pp. 160–231.

"Superconductive Materials and Some of Their Properties", B.W. Roberts, National Bureau of Standards Technical Notes 408 and 482, U.S. Government Printing Office, 1966 and 1969.

SELECTED PROPERTIES OF THE SUPERCONDUCTIVE ELEMENTS

Conversion Factors

Oe × 79.57 = A/m; katm × 1.013 × 10^8 = N/m^2; kb × 1.0 × 10^8 = N/m^2

Element	T_c(K)	H_o(oersteds)	θ_D(K)	γ(mJmole^{-1} deg · K^2)
Al	1.175	104.93	420	1.35
Be	0.026			0.21
Cd	0.518, 0.52	29.6	209	0.688
Ga	1.0833	59.3	325	0.60
Ga (β)	5.90 , 6.2	560		
Ga (γ)	7.62	950		
Ga (δ)	7.85	815		
Hg (α)	4.154	411	87, 71.9	1.81
Hg (β)	3.949	339	93	1.37
In	3.405	281.53	109	1.672
Ir	0.14 , 0.11	19	425	3.27
La (α)	4.88	808, 798	142	10.0, 11.3
La (β)	6.00	1,096	139	11.3
Mo	0.916	90, 98	460	1.83
Nb	9.25	1,970	277, 238	7.80
Os	0.655	65	500	2.35
Pa	1.4			
Pb	7.23	803	96.3	3.0
Re	1.697	188, 211	415	2.35
Ru	0.493	66	580	3.0
Sb	2.6–2.7			
Sn	3.721	305	195	1.78
Ta	4.47	831	258	6.15
Tc	7.73 , 7.78	1,410	411	4.84, 6.28
Th	1.39	159.1	165	4.31
Ti	0.39	56, 100	429, 412	3.32
Tl	2.332, 2.39	181	78.5	1.47
V	5.43 , 5.31	1,100, 1,400	382	9.82
W	0.0154	1.15	550	0.90
Zn	0.875	55	319.7	0.633
Zr	0.53	47	290	2.78
Zr (ω)	0.65			

Thin Films Condensed at Various Temperatures

Element	T_c(K)
Al	1.18–~5.7
Be	~03, ~9.6; 6.5–10.6[a]; 10.2[b]
Bi	~2–~5, 6.11, 6.154, 6.173
Cd	0.53–0.91
Ga	6.4–6.8, 7.4–8.4, 8.56
In	3.43–4.5; 3.68–4.17[c]
La	5.0–6.74
Mo	3.3–3.8, 4–6.7
Nb	6.2–10.1
Pb	~2–7.7
Re	~7
Sn	3.6, 3.84–6.0
Ta	<1.7–4.25, 3.16–4.8
Ti	1.3
Tl	2.64
V	5.14–6.02
W	<1.0–4.1
Zn	0.77–1.48

[a]With KCl.
[b]With Zn etioporphyrin.
[c]In glass pores.

Data for Elements Studied Under Pressure

Element	T_c(K)	Pressure
As	0.31–0.5	220–140 kb
	0.2–0.25	~140–100 kb
Ba II	~1.3	55 kb
Ba III	3.05	85–88 kb
	~5.2	>140 kb
Bi II	3.916	25 katm
	3.90	25.2 katm
	3.86	26.8 katm
Bi III	6.55	~37 kb
	7.25	27–28.4 katm
Bi IV	7.0	43, 43–62 kb
Bi V	8.3, 8.55	81 kb
Bi VI	8.55	90, 92–101 kb
Ce	1.7	50 kb
Cs	~1.5	>~125 kb
Ga II	6.24, 6.38	⩾35 katm
Ga II′	7.5	⩾35 katm (P → 0)
Ge	4.85–5.4	~120 kb
	5.35	115 kb
La	~5.5–11.93	0–~140 kb
P	4.7	>100 kb
	5.8	170 kb
Pb II	3.55, 3.6	160 kb
Sb	3.55	85 kb
	3.52	93 kb
	3.53	100 kb
	3.40	~150 kb
Se II	6.75, 6.95	~130 kb
Si	6.7, 7.1	120 kb
Sn II	5.2	125 kb
	4.85	160 kb
Sn III	5.30	113 kb
Te II	2.05	43 kb
	3.4	50 kb
Te III	4.28	70 kb
Te IV	4.25	84 kb
Tl, cub.	1.45	35 kb
Tl, hex.	1.95	35 kb
U	2.3	10 kb
Y	~1.2, ~2.7	120–170 kb

From Roberts, B. W., Properties of Selected Superconductive Materials, 1974 Supplement, NBS Technical Note 825, U.S. Government Printing Office, Washington, D.C., 1974, 10.

SELECTED SUPERCONDUCTIVE COMPOUNDS AND ALLOYS

All compositions are denoted on an atomic basis, i.e., AB, AB_2, or AB_3 for compounds, unless noted. Solid solutions or odd compositions may be denoted as A_zB_{1-z} or A_zB. A series of three or more alloys is indicated as A_xB_{1-x} or by actual indication of the atomic fraction range, such as $A_{0-0.6}B_{1-0.4}$. The critical temperature of such a series of alloys is denoted by a range of values or possibly the maximum value.

The selection of the critical temperature from a transition in the effective permeability, or the change in resistance, or possibly the incremental changes in frequency observed by certain techniques is not often obvious from the literature. Most authors choose the mid-point of such curves as the probable critical temperature of the idealized material, while others will choose the highest temperature at which a deviation from the normal state property is observed. In view of the previous discussion concerning the variability of the superconductive properties as a function of purity and other metallurgical aspects, it is recommended that appropriate literature be checked to determine the most probable critical temperature or critical field of a given alloy.

A very limited amount of data on critical fields, H_o, is available for these compounds and alloys; these values are given at the end of the table.

SYMBOLS: n = number of normal carriers per cubic centimeter for semiconductor superconductors.

Substance	T_c, °K	*Crystal structure type*††
$Ag_xAl_yZn_{1-x-y}$	0.5–0.845	
$Ag_7BF_4O_8$	0.15	Cubic
$AgBi_2$	3.0–2.78	
$Ag_7F_{0.25}N_{0.75}O_{10.25}$	0.85–0.90	
Ag_7FO_8	0.3	Cubic
Ag_2F	0.066	
$Ag_{0.8-0.3}Ga_{0.2-0.7}$	6.5–8	
Ag_4Ge	0.85	Hex., c.p.
$Ag_{0.438}Hg_{0.562}$	0.64	$D8_2$
$AgIn_2$	~2.4	C16
$Ag_{0.1}In_{0.9}Te$ (n = 1.40×10^{22})	1.20–1.89	Bl
$Ag_{0.2}In_{0.8}Te$ (n = 1.07×10^{22})	0.77–1.00	Bl
AgLa (9.5 kbar)	1.2	B2
Ag_7NO_{11}	1.04	Cubic
Ag_xPb_{1-x}	7.2 max.	
Ag_xSn_{1-x} (film)	2.0–3.8	
Ag_xSn_{1-x}	1.5–3.7	
$AgTe_3$	2.6	Cubic
$AgTh_2$	2.26	Cl6
$Ag_{0.03}Tl_{0.97}$	2.67	
$Ag_{0.94}Tl_{0.06}$	2.32	
Ag_xZn_{1-x}	0.5–0.845	
Al (film)	1.3–2.31	
Al (1 to 21 katm)	1.170–0.687	Al
$AlAu_4$	0.4–0.7	Like Al3
Al_2CMo_3	10.0	Al3
Al_2CMo_3	9.8–10.2	Al3 + trace 2nd phase
Al_2CaSi	5.8	
$Al_{0.131}Cr_{0.088}V_{0.781}$	1.46	Cubic
$AlGe_2$	1.75	
$Al_{0.5}Ge_{0.5}Nb$	12.6	Al5
$Al_{\sim0.8}Ge_{\sim0.2}Nb_3$	20.7	Al5
$AlLa_3$	5.57	DO_{19}
Al_2La	3.23	Cl5
Al_3Mg_2	0.84	Cubic, f.c.
$AlMo_3$	0.58	Al5
$AlMo_6Pd$	2.1	
AlN	1.55	B4
Al_2NNb_3	1.3	Al3
$AlNb_3$	18.0	Al5
Al_xNb_{1-x}	<4.2–13.5	$D8_b$
Al_xNb_{1-x}	12–17.5	Al5
$Al_{0.27}Nb_{0.73-0.48}V_{0-0.25}$	14.5–17.5	Al5
$AlNb_xV_{1-x}$	<4.2–13.5	
AlOs	0.39	B2
Al_3Os	5.90	
AlPb (films)	1.2–7	
Al_2Pt	0.48–0.55	Cl
Al_5Re_{24}	3.35	Al2
Al_3Th	0.75	DO_{19}
$Al_xTi_yV_{1-x-y}$	2.05–3.62	Cubic
$Al_{0.108}V_{0.892}$	1.82	Cubic
Al_xZn_{1-x}	0.5–0.845	
$AlZr_3$	0.73	Ll_2
AsBiPb	9.0	
AsBiPbSb	9.0	
$As_{0.33}InTe_{0.67}$ (n = 1.24×10^{22})	0.85–1.15	Bl
$As_{0.5}InTe_{0.5}$ (n = 0.97×10^{22})	0.44–0.62	Bl
$As_{0.50}Ni_{0.06}Pd_{0.44}$	1.39	C2
AsPb	8.4	
$AsPd_2$ (low-temperature phase)	0.60	Hexagonal
$AsPd_2$ (high-temp. phase)	1.70	C22

††See key at end of table.

Substance	T_c, °K	*Crystal structure type*††
$AsPd_5$	0.46	Complex
AsRh	0.58	B31
$AsRh_{1.4-1.6}$	<0.03–0.56	Hexagonal
AsSn	4.10	
AsSn ($n = 2.14 \times 10^{22}$)	3.41–3.65	Bl
$As_{\sim 2}Sn_{\sim 3}$	3.5–3.6, 1.21–1.17	
As_3Sn_4 ($n = 0.56 \times 10^{22}$)	1.16–1.19	Rhombohedral
Au_5Ba	0.4–0.7	$D2_d$
AuBe	2.64	B20
Au_2Bi	1.80	Cl5
Au_5Ca	0.34–0.38	$Cl5_b$
AuGa	1.2	B31
$Au_{0.40-0.92}Ge_{0.60-0.08}$	<0.32–1.63	Complex
AuIn	0.4–0.6	Complex
AuLu	<0.35	B2
$AuNb_3$	11.5	Al5
$AuNb_3$	1.2	A2
$Au_{0-0.3}Nb_{1-0.7}$	1.1–11.0	
$Au_{0.02-0.98}Nb_3Rh_{0.98-0.02}$	2.53–10.9	Al5
$AuNb_{3(1-x)}V_{3x}$	1.5–11.0	Al5
$AuPb_2$	3.15	
$AuPb_2$ (film)	4.3	
$AuPb_3$	4.40	
$AuPb_3$ (film)	4.25	
Au_2Pb	1.18, 6–7	Cl5
$AuSb_2$	0.58	C2
AuSn	1.25	$B8_1$
Au_xSn_{1-x} (film)	2.0–3.8	
Au_5Sn	0.7–1.1	A3
Au_3Te_5	1.62	Cubic
$AuTh_2$	3.08	Cl6
AuTl	1.92	
AuV_3	0.74	Al5
Au_xZn_{1-x}	0.50–0.845	
$AuZn_3$	1.21	Cubic
Au_xZr_y	1.7–2.8	A3
$AuZr_3$	0.92	Al5
$BCMo_2$	5.4	Orthorhombic
$B_{0.03}C_{0.51}Mo_{0.47}$	12.5	
$BCMo_2$	5.3–7.0	Orthorhombic
BHf	3.1	Cubic
B_6La	5.7	
$B_{12}Lu$	0.48	
BMo	0.5 (extrapolated)	
BMo_2	4.74	Cl6
BNb	8.25	B_f
BRe_2	2.80, 4.6	
$B_{0.3}Ru_{0.7}$	2.58	$D10_2$
$B_{12}Sc$	0.39	
BTa	4.0	B_f
B_6Th	0.74	

Substance	T_c, °K	*Crystal structure type*††
BW_2	3.1	Cl6
B_6Y	6.5–7.1	
$B_{12}Y$	4.7	
BZr	3.4	Cubic
$B_{12}Zr$	5.82	
$BaBi_3$	5.69	Tetragonal
$Ba_xO_3Sr_{1-x}Ti$ ($n = 4.2–11 \times 10^{19}$)	<0.1–0.55	
$Ba_{0.13}O_3W$	1.9	Tetragonal
$Ba_{0.14}O_3W$	<1.25–2.2	Hexagonal
$BaRh_2$	6.0	Cl5
$Be_{22}Mo$	2.51	Cubic, like $Be_{22}Re$
$Be_8Nb_5Zr_2$	5.2	
$Be_{0.98-0.92}Re_{0.02-0.08}$ (quenched)	9.5–9.75	Cubic
$Be_{0.957}Re_{0.043}$	9.62	Cubic, like $Be_{22}Re$
BeTc	5.21	Cubic
$Be_{22}W$	4.12	Cubic, like $Be_{22}Re$
$Be_{13}W$	4.1	Tetragonal
Bi_3Ca	2.0	
$Bi_{0.5}Cd_{0.13}Pb_{0.25}Sn_{0.12}$ (weight fractions)	8.2	
BiCo	0.42–0.49	
Bi_2Cs	4.75	Cl5
Bi_xCu_{1-x} (electrodeposited)	2.2	
BiCu	1.33–1.40	
$Bi_{0.019}In_{0.981}$	3.86	
$Bi_{0.05}In_{0.95}$	4.65	α-phase
$Bi_{0.10}In_{0.90}$	5.05	α-phase
$Bi_{0.15-0.30}In_{0.85-0.70}$	5.3–5.4	α- and β-phases
$Bi_{0.34-0.48}In_{0.66-0.52}$	4.0–4.1	
Bi_3In_5	4.1	
$BiIn_2$	5.65	β-phase
Bi_2Ir	1.7–2.3	
Bi_2Ir (quenched)	3.0–3.96	
BiK	3.6	
Bi_2K	3.58	Cl5
BiLi	2.47	Ll_0, α-phase
$Bi_{4-9}Mg$	0.7–~1.0	
Bi_3Mo	3–3.7	
BiNa	2.25	Ll_0
$BiNb_3$ (high pressure and temperature)	3.05	Al5
BiNi	4.25	$B8_1$
Bi_3Ni	4.06	Orthorhombic
$Bi_{1-0}Pb_{0-1}$	7.26–9.14	
$Bi_{1-0}Pb_{0-1}$ (film)	7.25–8.67	
$Bi_{0.05-0.40}Pb_{0.95-0.60}$	7.35–8.4	Hexagonal, c.p., to ε-phase
BiPbSb	8.9	

††See key at end of table.

Substance	T_c, °K	*Crystal structure type*††
$Bi_{0.5}Pb_{0.31}Sn_{0.19}$ (weight fractions)	8.5	
$Bi_{0.5}Pb_{0.25}Sn_{0.25}$	8.5	
$BiPd_2$	4.0	
$Bi_{0.4}Pd_{0.6}$	3.7–4	Hexagonal, ordered
BiPd	3.7	Orthorhombic
Bi_2Pd	1.70	Monoclinic, α-phase
Bi_2Pd	4.25	Tetragonal, β-phase
BiPdSe	1.0	C2
BiPdTe	1.2	C2
BiPt	1.21	$B8_1$
BiPtSe	1.45	C2
BiPtTe	1.15	C2
Bi_2Pt	0.155	Hexagonal
Bi_2Rb	4.25	Cl5
$BiRe_2$	1.9–2.2	
BiRh	2.06	$B8_1$
Bi_3Rh	3.2	Orthorhombic, like NiB_3
Bi_4Rh	2.7	Hexagonal
Bi_3Sn	3.6–3.8	
BiSn	3.8	
Bi_xSn_y	3.85–4.18	
Bi_3Sr	5.62	Ll_2
Bi_3Te	0.75–1.0	
Bi_5Tl_3	6.4	
$Bi_{0.26}Tl_{0.74}$	4.4	Cubic, disordered
$Bi_{0.26}Tl_{0.74}$	4.15	Ll_2, ordered?
Bi_2Y_3	2.25	
Bi_3Zn	0.8–0.9	
$Bi_{0.3}Zr_{0.7}$	1.51	
$BiZr_3$	2.4–2.8	
CCs_x	0.020–0.135	Hexagonal
C_8K (gold)	0.55	
$CGaMo_2$	3.7–4.1	Hexagonal, H-phase
$CHf_{0.5}Mo_{0.5}$	3.4	Bl
$CHf_{0.3}Mo_{0.7}$	5.5	Bl
$CHf_{0.25}Mo_{0.75}$	6.6	Bl
$CHf_{0.7}Nb_{0.3}$	6.1	Bl
$CHf_{0.6}Nb_{0.4}$	4.5	Bl
$CHf_{0.5}Nb_{0.5}$	4.8	Bl
$CHf_{0.4}Nb_{0.6}$	5.6	Bl
$CHf_{0.25}Nb_{0.75}$	7.0	Bl
$CHf_{0.2}Nb_{0.8}$	7.8	Bl
$CHf_{0.9-0.1}Ta_{0.1-0.9}$	5.0–9.0	Bl
Ck (excess K)	0.55	Hexagonal
C_8K	0.39	Hexagonal
$C_{0.40-0.44}Mo_{0.60-0.56}$	9–13	
CMo	6.5, 9.26	
CMo_2	12.2	Orthorhombic

Substance	T_c, °K	*Crystal structure type*††
$C_{0.44}Mo_{0.56}$	1.3	Bl
$C_{0.5}Mo_xNb_{1-x}$	10.8–12.5	Bl
$C_{0.6}Mo_{4.8}Si_3$	7.6	$D8_8$
$CMo_{0.2}Ta_{0.8}$	7.5	Bl
$CMo_{0.5}Ta_{0.5}$	7.7	Bl
$CMo_{0.75}Ta_{0.25}$	8.5	Bl
$CMo_{0.8}Ta_{0.2}$	8.7	Bl
$CMo_{0.85}Ta_{0.15}$	8.9	Bl
CMo_xTi_{1-x}	10.2 max.	Bl
$CMo_{0.83}Ti_{0.17}$	10.2	Bl
CMo_xV_{1-x}	2.9–9.3	Bl
CMo_xZr_{1-x}	3.8–9.5	Bl
$C_{0.1-0.9}N_{0.9-0.1}Nb$	8.5–17.9	
$C_{0-0.38}N_{1-0.62}Ta$	10.0–11.3	
CNb (whiskers)	7.5–10.5	
$C_{0.984}Nb$	9.8	Bl
CNb (extrapolated)	~14	
$C_{0.7-1.0}Nb_{0.3-0}$	6–11	Bl
CNb_2	9.1	
CNb_xTa_{1-x}	8.2–13.9	
CNb_xTi_{1-x}	<4.2–8.8	Bl
$CNb_{0.6-0.9}W_{0.4-0.1}$	12.5–11.6	Bl
$CNb_{0.1-0.9}Zr_{0.9-0.1}$	4.2–8.4	Bl
CRb_x (gold)	0.023–0.151	Hexagonal
$CRe_{0.01-0.08}W$	1.3–5.0	
$CRe_{0.06}W$	5.0	
CTa	~11 (extrapolated)	
$C_{0.987}Ta$	9.7	
$C_{0.848-0.987}Ta$	2.04–9.7	
CTa (film)	5.09	Bl
CTa_2	3.26	L'_3
$CTa_{0.4}Ti_{0.6}$	4.8	Bl
$CTa_{1-0.4}W_{0-0.6}$	8.5–10.5	Bl
$CTa_{0.2-0.9}Zr_{0.8-0.1}$	4.6–8.3	Bl
CTc (excess C)	3.85	Cubic
$CTi_{0.5-0.7}W_{0.5-0.3}$	6.7–2.1	Bl
CW	1.0	
CW_2	2.74	L'_3
CW_2	5.2	Cubic, f.c.
$CaIr_2$	6.15	Cl5
$Ca_xO_3Sr_{1-x}Ti$ ($n = 3.7–11.0 \times 10^{19}$)	<0.1–0.55	
$Ca_{0.1}O_3W$	1.4–3.4	Hexagonal
CaPb	7.0	
$CaRh_2$	6.40	Cl5
$Cd_{0.3-0.5}Hg_{0.7-0.5}$	1.70–1.92	
CdHg	1.77, 2.15	Tetragonal
$Cd_{0.0075-0.05}In_{1-x}$	3.24–3.36	Tetragonal
$Cd_{0.97}Pb_{0.03}$	4.2	
CdSn	3.65	
$Cd_{0.17}Tl_{0.83}$	2.3	
$Cd_{0.18}Tl_{0.82}$	2.54	
$CeCo_2$	0.84	Cl5
$CeCo_{1.67}Ni_{0.33}$	0.46	Cl5

††See key at end of table.

Substance	T_c, °*K*	*Crystal structure type*††
$CeCo_{1.67}Rh_{0.33}$	0.47	C15
$Ce_xGd_{1-x}Ru_2$	3.2–5.2	C15
$CeIr_3$	3.34	
$CeIr_5$	1.82	
$Ce_{0.005}La_{0.995}$	4.6	
Ce_xLa_{1-x}	1.3–6.3	
$Ce_xPr_{1-x}Ru_2$	1.4–5.3	C15
Ce_xPt_{1-x}	0.7–1.55	
$CeRu_2$	6.0	C15
$Co_xFe_{1-x}Si_2$	1.4 max.	C1
$CoHf_2$	0.56	$E9_3$
$CoLa_3$	4.28	
$CoLu_3$	~0.35	
$Co_{0-0.01}Mo_{0.8}Re_{0.2}$	2–10	
$Co_{0.02-0.10}Nb_3Rh_{0.98-0.90}$	2.28–1.90	A15
$Co_xNi_{1-x}Si_2$	1.4 max.	C1
$Co_{0.5}Rh_{0.5}Si_2$	2.5	
$Co_xRh_{1-x}Si_2$	3.65 max.	
$Co_{\sim 0.3}Sc_{\sim 0.7}$	~0.35	
$CoSi_2$	1.40, 1.22	C1
Co_3Th_7	1.83	$D10_2$
Co_xTi_{1-x}	2.8 max.	Co in α-Ti
Co_xTi_{1-x}	3.8 max.	Co in β-Ti
$CoTi_2$	3.44	$E9_3$
CoTi	0.71	A2
CoU	1.7	B2, distorted
CoU_6	2.29	$D2_c$
$Co_{0.28}Y_{0.72}$	0.34	
CoY_3	<0.34	
$CoZr_2$	6.3	C16
$Co_{0.1}Zr_{0.9}$	3.9	A3
$Cr_{0.6}Ir_{0.4}$	0.4	Hexagonal, c.p.
$Cr_{0.65}Ir_{0.35}$	0.59	Hexagonal, c.p.
$Cr_{0.7}Ir_{0.3}$	0.76	Hexagonal, c.p.
$Cr_{0.72}Ir_{0.28}$	0.83	
Cr_3Ir	0.45	A15
$Cr_{0-0.1}Nb_{1-0.9}$	4.6–9.2	A2
$Cr_{0.80}Os_{0.20}$	2.5	Cubic
Cr_xRe_{1-x}	1.2–5.2	
$Cr_{0.40}Re_{0.60}$	2.15	$D8_b$
$Cr_{0.8-0.6}Rh_{0.2-0.4}$	0.5–1.10	A3
Cr_3Ru (annealed)	3.3	A15
Cr_2Ru	2.02	$D8_b$
$Cr_{0.1-0.5}Ru_{0.9-0.5}$	0.34–1.65	A3
Cr_xTi_{1-x}	3.6 max.	Cr in α-Ti
Cr_xTi_{1-x}	4.2 max.	Cr in β-Ti
$Cr_{0.1}Ti_{0.3}V_{0.6}$	5.6	
$Cr_{0.0175}U_{0.9825}$	0.75	β-phase
$Cs_{0.32}O_3W$	1.12	Hexagonal
$Cu_{0.15}In_{0.85}$ (film)	3.75	
$Cu_{0.04-0.08}In_{1-x}$	4.4	
CuLa	5.85	
Cu_xPb_{1-x}	5.7–7.7	
CuS	1.62	B18
CuS_2	1.48–1.53	C18
CuSSe	1.5–2.0	C18
$CuSe_2$	2.3–2.43	C18
CuSeTe	1.6–2.0	C18
Cu_xSn_{1-x}	3.2–3.7	
Cu_xSn_{1-x} (film) (made at 10°K)	3.6–7	
Cu_xSn_{1-x} (film) (made at 300°K)	2.8–3.7	
$CuTe_2$	<1.25–1.3	C18
$CuTh_2$	3.49	C16
$Cu_{0-0.027}V$	3.9–5.3	A2
Cu_xZn_{1-x}	0.5–0.845	
Er_xLa_{1-x}	1.4–6.3	
$Fe_{0-0.04}Mo_{0.8}Re_{0.2}$	1–10	
$Fe_{0.05}Ni_{0.05}Zr_{0.90}$	~3.9	
Fe_3Th_7	1.86	D10
Fe_xTi_{1-x}	3.2 max.	Fe in α-Ti
Fe_xTi_{1-x}	3.7 max.	Fe in β-Ti
$Fe_xTi_{0.6}V_{1-x}$	6.8 max.	
FeU_6	3.86	$D2_c$
$Fe_{0.1}Zr_{0.9}$	1.0	A3
$Ga_{0.5}Ge_{0.5}Nb_3$	7.3	A15
$GaLa_3$	5.84	
Ga_2Mo	9.5	
$GaMo_3$	0.76	A15
Ga_4Mo	9.8	
GaN (black)	5.85	B4
$GaNb_3$	14.5	A15
$Ga_xNb_3Sn_{1-x}$	14–18.37	A15
$Ga_{0.7}Pt_{0.3}$	2.9	C1
GaPt	1.74	B20
GaSb (120 kbar, 77°K, annealed)	4.24	A5
GaSb (unannealed)	~5.9	
$Ga_{0.1}Sn_{1-0}$ (quenched)	3.47–4.18	
$Ga_{0-1}Sn_{1-0}$ (annealed)	2.6–3.85	
Ga_5V_2	3.55	Tetragonal, Mn_2Hg_5 type
GaV_3	16.8	A15
$GaV_{2.1-3.5}$	6.3–14.45	A15
$GaV_{4.5}$	9.15	
Ga_3Zr	1.38	
Gd_xLa_{1-x}	<1.0–5.5	
$Gd_xOs_2Y_{1-x}$	1.4–4.7	
$Gd_xRu_2Th_{1-x}$	3.6 max.	C15
GeIr	4.7	B31
Ge_2La	1.49, 2.2	Orthorhombic, distorted $ThSi_2$-type
$GeMo_3$	1.43	A15
$GeNb_2$	1.9	
$GeNb_3$ (quenched)	6–17	A15
$Ge_{0.29}Nb_{0.71}$	6	A15
$Ge_xNb_3Sn_{1-x}$	17.6–18.0	A15
$Ge_{0.5}Nb_3Sn_{0.5}$	11.3	

††See key at end of table.

SELECTED SUPERCONDUCTIVE COMPOUNDS AND ALLOYS (Continued)

Substance	*T_c, °K*	*Crystal structure type††*
GePt	0.40	B31
Ge_3Rh_5	2.12	Orthorhombic, related to $InNi_2$
Ge_2Sc	1.3	
Ge_3Te_4 ($n = 1.06 \times 10^{22}$)	1.55–1.80	Rhombohedral
Ge_xTe_{1-x} ($n = 8.5-64 \times 10^{20}$)	0.07–0.41	Bl
GeV_3	6.01	A15
Ge_2Y	3.80	C_c
$Ge_{1.62}Y$	2.4	
$H_{0.33}Nb_{0.67}$	7.28	Cubic, b.c.
$H_{0.1}Nb_{0.9}$	7.38	Cubic, b.c.
$H_{0.05}Nb_{0.95}$	7.83	Cubic, b.c.
$H_{0.12}Ta_{0.88}$	2.81	Cubic, b.c.
$H_{0.08}Ta_{0.92}$	3.26	Cubic, b.c.
$H_{0.04}Ta_{0.96}$	3.62	Cubic, b.c.
$HfN_{0.989}$	6.6	Bl
$Hf_{0-0.5}Nb_{1-0.5}$	8.3–9.5	A2
$Hf_{0.75}Nb_{0.25}$	>4.2	
$HfOs_2$	2.69	C14
$HfRe_2$	4.80	C14
$Hf_{0.14}Re_{0.86}$	5.86	A12
$Hf_{0.99-0.96}Rh_{0.01-0.04}$	0.85–1.51	
$Hf_{0-0.55}Ta_{1-0.45}$	4.4–6.5	A2
HfV_2	8.9–9.6	C15
Hg_xIn_{1-x}	3.14–4.55	
HgIn	3.81	
Hg_2K	1.20	Orthorhombic
Hg_3K	3.18	
Hg_4K	3.27	
Hg_8K	3.42	
Hg_3Li	1.7	Hexagonal
Hg_2Na	1.62	Hexagonal
Hg_4Na	3.05	
Hg_xPb_{1-x}	4.14–7.26	
HgSn	4.2	
Hg_xTl_{1-x}	2.30–4.109	
Hg_5Tl_2	3.86	
Ho_xLa_{1-x}	1.3–6.3	
$InLa_3$	9.83, 10.4	$L1_2$
$InLa_3$ (0–35, kbar)	9.75–10.55	
$In_{1-0.86}Mg_{0-0.14}$	3.395–3.363	
$InNb_3$ (high pressure and temp.)	4–8, 9.2	A15
$In_{0-0.3}Nb_3Sn_{1-0.7}$	18.0–18.19	A15
$In_{0.5}Nb_3Zr_{0.5}$	6.4	
$In_{0.11}O_3W$	<1.25–2.8	Hexagonal
$In_{0.95-0.85}Pb_{0.05-0.15}$	3.6–5.05	
$In_{0.98-0.91}Pb_{0.02-0.09}$	3.45–4.2	
InPb	6.65	
InPd	0.7	B2
InSb (quenched from 170 kbar into liquid N_2)	4.8	Like A5

Substance	*T_c, °K*	*Crystal structure type††*
InSb	2.1	
$(InSb)_{0.95-0.10}Sn_{0.05-0.90}$ (various heat treatments)	3.8–5.1	
$(InSb)_{0-0.07}Sn_{1-0.93}$	3.67–3.74	
In_3Sn	~5.5	
In_xSn_{1-x}	3.4–7.3	
$In_{0.82-1}Te$ ($n = 0.83–1.71 \times 10^{22}$)	1.02–3.45	Bl
$In_{1.000}Te_{1.002}$	3.5–3.7	Bl
In_3Te_4 ($n = 0.47 \times 10^{22}$)	1.15–1.25	Rhombohedral
In_xTl_{1-x}	2.7–3.374	
$In_{0.8}Tl_{0.2}$	3.223	
$In_{0.62}Tl_{0.38}$	2.760	
$In_{0.78-0.69}Tl_{0.22-0.31}$	3.18–3.32	Tetragonal
$In_{0.69-0.62}Tl_{0.31-0.38}$	2.98–3.3	Cubic, f.c.
Ir_2La	0.48	C15
Ir_3La	2.32	$Dl0_2$
Ir_3La_7	2.24	$Dl0_2$
Ir_5La	2.13	
Ir_2Lu	2.47	C15
Ir_3Lu	2.89	C15
IrMo	<1.0	A3
$IrMo_3$	8.8	A15
$IrMo_3$	6.8	$D8_b$
$IrNb_3$	1.9	A15
$Ir_{0.4}Nb_{0.6}$	9.8	$D8_b$
$Ir_{0.37}Nb_{0.63}$	2.32	$D8_b$
IrNb	7.9	$D8_b$
$Ir_{0.02}Nb_3Rh_{0.98}$	2.43	A15
$Ir_{0.05}Nb_3Rh_{0.95}$	2.38	A15
$Ir_{0.287}O_{0.14}Ti_{0.573}$	5.5	$E9_3$
$Ir_{0.265}O_{0.035}Ti_{0.65}$	2.30	$E9_3$
Ir_xOs_{1-x}	0.3–0.98 (max.)–0.6	
IrOsY	2.6	C15
$Ir_{1.5}Os_{0.5}$	2.4	C14
Ir_2Sc	2.07	C15
$Ir_{2.5}Sc$	2.46	C15
$IrSn_2$	0.65–0.78	Cl
Ir_2Sr	5.70	C15
$Ir_{0.5}Te_{0.5}$	~3	
$IrTe_3$	1.18	C2
IrTh	<0.37	B_f
Ir_2Th	6.50	C15
Ir_3Th	4.71	
Ir_3Th_7	1.52	$Dl0_2$
Ir_5Th	3.93	$D2_d$
$IrTi_3$	5.40	A15
IrV_2	1.39	A15
IrW_3	3.82	
$Ir_{0.28}W_{0.72}$	4.49	
Ir_2Y	2.18, 1.38	C15
$Ir_{0.69}Y_{0.31}$	1.98, 1.44	C15
$Ir_{0.70}Y_{0.30}$	2.16	C15

††See key at end of table.

Substance	*T_c, °K*	*Crystal structure type*††
Ir_2Y	1.09	Cl5
Ir_2Y_3	1.61	
Ir_xY_{1-x}	0.3–3.7	
Ir_2Zr	4.10	Cl5
$Ir_{0.1}Zr_{0.9}$	5.5	A3
$K_{0.27-0.31}O_3W$	0.50	Hexagonal
$K_{0.40-0.57}O_3W$	1.5	Tetragonal
$La_{0.55}Lu_{0.45}$	2.2	Hexagonal, La type
$La_{0.8}Lu_{0.2}$	3.4	Hexagonal, La Type
$LaMg_2$	1.05	Cl5
LaN	1.35	
$LaOs_2$	6.5	Cl5
$LaPt_2$	0.46	Cl5
$La_{0.28}Pt_{0.72}$	0.54	Cl5
$LaRh_3$	2.60	
$LaRh_5$	1.62	
La_7Rh_3	2.58	$Dl0_2$
$LaRu_2$	1.63	Cl5
La_3S_4	6.5	$D7_3$
La_3Se_4	8.6	$D7_3$
$LaSi_2$	2.3	C_c
La_xY_{1-x}	1.7–5.4	
LaZn	1.04	B2
LiPb	7.2	
$LuOs_2$	3.49	Cl4
$Lu_{0.275}Rh_{0.725}$	1.27	Cl5
$LuRh_5$	0.49	
$LuRu_2$	0.86	Cl4
$Mg_{\sim 0.47}Tl_{\sim 0.53}$	2.75	B2
Mg_2Nb	5.6	
Mn_xTi_{1-x}	2.3 max.	Mn in α-Ti
Mn_xTi_{1-x}	1.1–3.0	Mn in β-Ti
MnU_6	2.32	$D2_c$
MoN	12	Hexagonal
Mo_2N	5.0	Cubic, f.c.
Mo_xNb_{1-x}	0.016–9.2	
Mo_3Os	7.2	Al5
$Mo_{0.62}Os_{0.38}$	5.65	$D8_b$
Mo_3P	5.31	DO_e
$Mo_{0.5}Pd_{0.5}$	3.52	A3
Mo_3Re	10.0	
Mo_xRe_{1-x}	1.2–12.2	
$MoRe_3$	9.25, 9.89	Al2
$Mo_{0.42}Re_{0.58}$	6.35	$D8_b$
$Mo_{0.52}Re_{0.48}$	11.1	
$Mo_{0.57}Re_{0.43}$	14.0	
$Mo_{\sim 0.60}Re_{0.395}$	10.6	
MoRh	1.97	A3
Mo_xRh_{1-x}	1.5–8.2	Cubic, b.c.
MoRu	9.5–10.5	A3
$Mo_{0.61}Ru_{0.39}$	7.18	$D8_b$
$Mo_{0.2}Ru_{0.8}$	1.66	A3
Mo_3Sb_4	2.1	
Mo_3Si	1.30	Al5
$MoSi_{0.7}$	1.34	
Mo_xSiV_{3-x}	4.54–16.0	Al5
Mo_xTc_{1-x}	10.8–15.8	
$Mo_{0.16}Ti_{0.84}$	4.18, 4.25	
$Mo_{0.913}Ti_{0.087}$	2.95	
$Mo_{0.04}Ti_{0.96}$	2.0	Cubic
$Mo_{0.025}Ti_{0.975}$	1.8	
Mo_xU_{1-x}	0.7–2.1	
Mo_xV_{1-x}	0–∼5.3	
Mo_2Zr	4.27–4.75	Cl5
NNb (whiskers)	10–14.5	
NNb (diffusion wires)	16.10	
NNb (film)	6–9	Bl
$N_{0.988}Nb$	14.9	Bl
$N_{0.824-0.988}Nb$	14.4–15.3	Bl
$N_{0.70-0.795}Nb$	11.3–12.9	Cubic and tetragonal
NNb_xO_y	13.5–17.0	Bl
NNb_xO_y	6.0–11	
$N_{100-42\ w/o}Nb_{0-58\ w/o}Ti$†	15–16.8	
$N_{100-75\ w/o}Nb_{0-25\ w/o}Zr$†	12.5–16.35	
NNb_xZr_{1-x}	9.8–13.8	Bl
$N_{0.93}Nb_{0.85}Zr_{0.15}$	13.8	Bl
$N_xO_yTi_z$	2.9–5.6	Cubic
$N_xO_yV_z$	5.8–8.2	Cubic
$N_{0.34}Re$	4–5	Cubic, f.c.
NTa	12–14 (extrapolated)	Bl
NTa (film)	4.84	Bl
$N_{0.6-0.987}Ti$	<1.17–5.8	Bl
$N_{0.82-0.99}V$	2.9–7.9	Bl
NZr	9.8	Bl
$N_{0.906-0.984}Zr$	3.0–9.5	Bl
$Na_{0.28-0.35}O_3W$	0.56	Tetragonal
$Na_{0.28}Pb_{0.72}$	7.2	
NbO	1.25	
$NbOs_2$	2.52	Al2
Nb_3Os	1.05	Al5
$Nb_{0.6}Os_{0.4}$	1.89, 1.78	$D8_b$
$Nb_3Os_{0.02-0.10}Rh_{0.98-0.90}$	2.42–2.30	Al5
$Nb_{0.6}Pd_{0.4}$	1.60	$D8_f$ plus cubic
$Nb_3Pd_{0.02-0.10}Rh_{0.98-0.90}$	2.49–2.55	Al5
$Nb_{0.62}Pt_{0.38}$	4.21	$D8_b$
Nb_3Pt	10.9	Al5
Nb_5Pt_3	3.73	$D8_b$
$Nb_3Pt_{0.02-0.98}Rh_{0.98-0.02}$	2.52–9.6	Al5
$Nb_{0.38-0.18}Re_{0.62-0.82}$	2.43–9.70	Al2
Nb_3Rh	2.64	Al5
$Nb_{0.60}Rh_{0.40}$	4.21	$D8_b$ plus other
$Nb_3Rh_{0.98-0.90}Ru_{0.02-0.10}$	2.42–2.44	Al5
Nb_xRu_{1-x}	1.2–4.8	
NbS_2	6.1–6.3	Hexagonal, $NbSe_2$ type

†*w/o* denotes weight percent. ††See key at end of table.

SELECTED SUPERCONDUCTIVE COMPOUNDS AND ALLOYS (Continued)

Substance	T_c, °K	Crystal structure type††	Substance	T_c, °K	Crystal structure type††
NbS_2	5.0–5.5	Hexagonal, three-layer type	Os_2Zr	3.0	C14
			Os_xZr_{1-x}	1.50–5.6	
			PPb	7.8	
$Nb_3Sb_{0-0.7}Sn_{1-0.3}$	6.8–18	A15	$PPd_{3.0-3.2}$	<0.35–0.7	DO_{11}
$NbSe_2$	5.15–5.62	Hexagonal, NbS_2 type	P_3Pd_7 (high temperature)	1.0	Rhombohedral
			P_3Pd_7 (low temp.)	0.70	Complex
$Nb_{1-1.05}Se_2$	2.2–7.0	Hexagonal, NbS_2 type	PRh	1.22	
			PRh_2	1.3	Cl
Nb_3Si	1.5	Ll_2	PW_3	2.26	DO_e
Nb_3SiSnV_3	4.0		Pb_2Pd	2.95	C16
Nb_3Sn	18.05	A15	Pb_4Pt	2.80	Related to C16
$Nb_{0.8}Sn_{0.2}$	18.18, 18.5	A15	Pb_2Rh	2.66	C16
Nb_xSn_{1-x} (film)	2.6–18.5		PbSb	6.6	
$NbSn_2$	2.60	Orthorhombic	PbTe (plus 0.1 *w/o* Pb)†	5.19	
Nb_3Sn_2	16.6	Tetragonal	PbTe (plus 0.1 *w/o* Tl)†	5.24–5.27	
$NbSnTa_2$	10.8	A15	$PbTl_{0.27}$	6.43	
Nb_2SnTa	16.4	A15	$PbTl_{0.17}$	6.73	
$Nb_{2.5}SnTa_{0.5}$	17.6	A15	$PbTl_{0.12}$	6.88	
$Nb_{2.75}SnTa_{0.25}$	17.8	A15	$PbTl_{0.075}$	6.98	
$Nb_{3x}SnTa_{3(1-x)}$	6.0–18.0		$PbTl_{0.04}$	7.06	
NbSnTaV	6.2	A15	$Pb_{1-0.26}Tl_{0-0.74}$	7.20–3.68	
$Nb_2SnTa_{0.5}V_{0.5}$	12.2	A15	$PbTl_2$	3.75–4.1	
$NbSnV_2$	5.5	A15	Pb_3Zr_5	4.60	$D8_8$
Nb_2SnV	9.8	A15	$PbZr_3$	0.76	A15
$Nb_{2.5}SnV_{0.5}$	14.2	A15	$Pd_{0.9}Pt_{0.1}Te_2$	1.65	C6
Nb_xTa_{1-x}	4.4–9.2	A2	$Pd_{0.05}Ru_{0.05}Zr_{0.9}$	~9	
$NbTc_3$	10.5	A12	$Pd_{2.2}S$ (quenched)	1.63	Cubic
Nb_xTi_{1-x}	0.6–9.8		$PdSb_2$	1.25	C2
$Nb_{0.6}Ti_{0.4}$	9.8		PdSb	1.50	$B8_1$
Nb_xU_{1-x}	1.95 max.		PdSbSe	1.0	C2
$Nb_{0.88}V_{0.12}$	5.7	A2	PdSbTe	1.2	C2
$Nb_{0.75}Zr_{0.25}$	10.8		Pd_4Se	0.42	Tetragonal
$Nb_{0.66}Zr_{0.33}$	10.8		$Pd_{6-7}Se$	0.66	Like Pd_4Te
$Ni_{0.3}Th_{0.7}$	1.98	$D10_2$	$Pd_{2.8}Se$	2.3	
$NiZr_2$	1.52		Pd_xSe_{1-x}	2.5 max.	
$Ni_{0.1}Zr_{0.9}$	1.5	A3	PdSi	0.93	B31
$O_3Rb_{0.27-0.29}W$	1.98	Hexagonal	PdSn	0.41	B31
O_3SrTi			$PdSn_2$	3.34	
($n = 1.7–12.0 \times 10^{19}$)	0.12–0.37		Pd_2Sn	0.41	C37
O_3SrTi			Pd_3Sn_2	0.47–0.64	$B8_2$
($n = 10^{18}–10^{21}$)	0.05–0.47		PdTe	2.3, 3.85	$B8_1$
O_3SrTi			$PdTe_{1.02-1.08}$	2.56–1.88	$B8_1$
($n = \sim 10^{20}$)	0.47		$PdTe_2$	1.69	C6
OTi	0.58		$PdTe_{2.1}$	1.89	C6
$O_3Sr_{0.08}W$	2–4	Hexagonal	$PdTe_{2.3}$	1.85	C6
$O_3Tl_{0.30}W$	2.0–2.14	Hexagonal	$Pd_{1.1}Te$	4.07	$B8_1$
OV_3Zr_3	7.5	$E9_3$	$PdTh_2$	0.85	C16
OW_3 (film)	3.35, 1.1	A15	$Pd_{0.1}Zr_{0.9}$	7.5	A3
OsReY	2.0	C14	PtSb	2.1	$B8_1$
Os_2Sc	4.6	C14	PtSi	0.88	B31
OsTa	1.95	A12	PtSn	0.37	$B8_1$
Os_3Th_7	1.51	$D10_2$	PtTe	0.59	Orthorhombic
Os_xW_{1-x}	0.9–4.1		PtTh	0.44	B_f
OsW_3	~3		Pt_3Th_7	0.98	$D10_2$
Os_2Y	4.7	C14	Pt_5Th	3.13	

†*w/o* denotes weight percent. ††See key at end of table.

Substance	T_c, °K	Crystal structure type††
$PtTi_3$	0.58	A15
$Pt_{0.02}U_{0.98}$	0.87	β-phase
$PtV_{2.5}$	1.36	A15
PtV_3	2.87–3.20	A15
$PtV_{3.5}$	1.26	A15
$Pt_{0.5}W_{0.5}$	1.45	A1
Pt_xW_{1-x}	0.4–2.7	
Pt_2Y_3	0.90	
Pt_2Y	1.57, 1.70	C15
Pt_3Y_7	0.82	$D10_2$
PtZr	3.0	A3
$Re_{0.64}Ta_{0.36}$	1.46	A12
$Re_{24}Ti_5$	6.60	A12
Re_xTi_{1-x}	6.6 max.	
$Re_{0.76}V_{0.24}$	4.52	$D8_b$
$Re_{0.92}V_{0.08}$	6.8	A3
$Re_{0.6}W_{0.4}$	6.0	
$Re_{0.5}W_{0.5}$	5.12	$D8_b$
Re_2Y	1.83	C14
Re_2Zr	5.9	C14
Re_6Zr	7.40	A12
$Rh_{17}S_{15}$	5.8	Cubic
$Rh_{\sim 0.24}Sc_{\sim 0.76}$	0.88, 0.92	
Rh_xSe_{1-x}	6.0 max.	
Rh_2Sr	6.2	C15
$Rh_{0.4}Ta_{0.6}$	2.35	$D8_b$
$RhTe_2$	1.51	C2
$Rh_{0.67}Te_{0.33}$	0.49	
Rh_xTe_{1-x}	1.51 max.	
RhTh	0.36	B_f
Rh_3Th_7	2.15	$D10_2$
Rh_5Th	1.07	
Rh_xTi_{1-x}	2.25–3.95	
$Rh_{0.02}U_{0.98}$	0.96	
RhV_3	0.38	A15
RhW	~3.4	A3
RhY_3	0.65	
Rh_2Y_3	1.48	
Rh_3Y	1.07	C15
Rh_5Y	0.56	
$RhZr_2$	10.8	C16
$Rh_{0.005}Zr$ (annealed)	5.8	
$Rh_{0-0.45}Zr_{1-0.55}$	2.1–10.8	
$Rh_{0.1}Zr_{0.9}$	9.0	Hexagonal, c.p.
Ru_2Sc	1.67	C14
Ru_2Th	3.56	C15
RuTi	1.07	B2
$Ru_{0.05}Ti_{0.95}$	2.5	
$Ru_{0.1}Ti_{0.9}$	3.5	
$Ru_xTi_{0.6}V_y$	6.6 max.	
$Ru_{0.45}V_{0.55}$	4.0	B2
RuW	7.5	A3
Ru_2Y	1.52	C14
Ru_2Zr	1.84	C14
$Ru_{0.1}Zr_{0.9}$	5.7	A3
SbSn	1.30–1.42, 1.42–2.37	B1 or distorted B1
$SbTi_3$	5.8	A15
Sb_2Tl_7	5.2	
$Sb_{0.01-0.03}V_{0.99-0.97}$	3.76–2.63	A2
SbV_3	0.80	A15
Si_2Th	3.2	C_c, α-phase
Si_2Th	2.4	C32, β-phase
SiV_3	17.1	A15
$Si_{0.9}V_3Al_{0.1}$	14.05	A15
$Si_{0.9}V_3B_{0.1}$	15.8	A15
$Si_{0.9}V_3C_{0.1}$	16.4	A15
$SiV_{2.7}Cr_{0.3}$	11.3	A15
$Si_{0.9}V_3Ge_{0.1}$	14.0	A15
$SiV_{2.7}Mo_{0.3}$	11.7	A15
$SiV_{2.7}Nb_{0.3}$	12.8	A15
$SiV_{2.7}Ru_{0.3}$	2.9	A15
$SiV_{2.7}Ti_{0.3}$	10.9	A15
$SiV_{2.7}Zr_{0.3}$	13.2	A15
Si_2W_3	2.8, 2.84	
$Sn_{0.174-0.104}Ta_{0.826-0.896}$	6.5–<4.2	A15
$SnTa_3$	8.35	A15, highly ordered
$SnTa_3$	6.2	A15, partially ordered
$SnTaV_2$	2.8	A15
$SnTa_2V$	3.7	A15
Sn_xTe_{1-x} ($n = 10.5–20 \times 10^{20}$)	0.07–0.22	B1
Sn_xTl_{1-x}	2.37–5.2	
SnV_3	3.8	A15
$Sn_{0.02-0.057}V_{0.98-0.943}$	2.87–~1.6	A2
$Ta_{0.025}Ti_{0.975}$	1.3	Hexagonal
$Ta_{0.05}Ti_{0.95}$	2.9	Hexagonal
$Ta_{0.05-0.75}V_{0.095-0.25}$	4.30–2.65	A2
$Ta_{0.8-1}W_{0.2-0}$	1.2–4.4	A2
$Tc_{0.1-0.4}W_{0.9-0.6}$	1.25–7.18	Cubic
$Tc_{0.50}W_{0.50}$	7.52	α plus σ
$Tc_{0.60}W_{0.40}$	7.88	σ plus α
Tc_6Zr	9.7	A12
$Th_{0-0.55}Y_{1-0.45}$	1.2–1.8	
$Ti_{0.70}V_{0.30}$	6.14	Cubic
Ti_xV_{1-x}	0.2–7.5	
$Ti_{0.5}Zr_{0.5}$ (annealed)	1.23	
$Ti_{0.5}Zr_{0.5}$ (quenched)	2.0	
V_2Zr	8.80	C15
$V_{0.26}Zr_{0.74}$	≈5.9	
W_2Zr	2.16	C15

††See key at end of table.

SELECTED SUPERCONDUCTIVE COMPOUNDS AND ALLOYS (Continued)

CRITICAL FIELD DATA

Substance	H_o, *oersteds*	*Substance*	H_o, *oersteds*
Ag_2F	2.5	InSb	1,100
Ag_7NO_{11}	57	In_xTl_{1-x}	252–284
Al_2CMo_3	1,700	$In_{0.8}Tl_{0.2}$	252
$BaBi_3$	740	$Mg_{\sim 0.47}Tl_{\sim 0.53}$	220
Bi_2Pt	10	$Mo_{0.16}Ti_{0.84}$	<985
Bi_3Sr	530	$NbSn_2$	620
Bi_5Tl_3	>400	$PbTl_{0.27}$	756
CdSn	>266	$PbTl_{0.17}$	796
$CoSi_2$	105	$PbTl_{0.12}$	849
$Cr_{0.1}Ti_{0.3}V_{0.6}$	1,360	$PbTl_{0.075}$	880
$In_{1-0.86}Mg_{0-0.14}$	272.4–259.2	$PbTl_{0.04}$	864

KEY TO CRYSTAL STRUCTURE TYPES

*"Struck-turbericht" type**	*Example*	*Class*	*"Struck-turbericht" type**	*Example*	*Class*
A1	Cu	Cubic, f.c.	$C15_b$	$AuBe_5$	Cubic
A2	W	Cubic, b.c.	C16	$CuAl_2$	Tetragonal, b.c.
A3	Mg	Hexagonal, close packed	C18	FeS_2	Orthorhombic
A4	Diamond	Cubic, f.c.	C22	Fe_2P	Trigonal
A5	White Sn	Tetragonal, b.c.	C23	$PbCl_2$	Orthorhombic
A6	In	Tetragonal, b.c. (f.c. cell usually used)	C32	AlB_2	Hexagonal
			C36	$MgNi_2$	Hexagonal
A7	As	Rhombohedral	C37	Co_2Si	Orthorhombic
A8	Se	Trigonal	C49	$ZrSi_2$	Orthorhombic
A10	Hg	Rhombohedral	C54	$TiSi_2$	Orthorhombic
A12	α-Mn	Cubic, b.c.	C_c	Si_2Th	Tetragonal, b.c.
A13	β-Mn	Cubic	DO_3	BiF_3	Cubic, f.c.
A15	"β-W" (WO_3)	Cubic	DO_{11}	Fe_3C	Orthorhombic
B1	NaCl	Cubic, f.c.	DO_{18}	Na_3As	Hexagonal
B2	CsCl	Cubic	DO_{19}	Ni_3Sn	Hexagonal
B3	ZnS	Cubic	DO_{20}	$NiAl_3$	Orthorhombic
B4	ZnS	Hexagonal	DO_{22}	$TiAl_3$	Tetragonal
$B8_1$	NiAs	Hexagonal	DO_e	Ni_3P	Tetragonal, b.c.
$B8_2$	Ni_2In	Hexagonal	$D1_3$	Al_4Ba	Tetragonal, b.c.
B10	PbO	Tetragonal	$D1_c$	$PtSn_4$	Orthorhombic
B11	γ-CuTi	Tetragonal	$D2_1$	CaB_6	Cubic
B17	PtS	Tetragonal	$D2_c$	MnU_6	Tetragonal, b.c.
B18	CuS	Hexagonal	$D2_d$	$CaZn_5$	Hexagonal
B20	FeSi	Cubic	$D5_2$	La_2O_3	Trigonal
B27	FeB	Orthorhombic	$D5_8$	Sb_2S_3	Orthorhombic
B31	MnP	Orthorhombic	$D7_3$	Th_3P_4	Cubic, b.c.
B32	NaTl	Cubic, f.c.	$D7_b$	Ta_3B_4	Orthorhombic
B34	PdS	Tetragonal	$D8_1$	Fe_3Zn_{10}	Cubic, b.c.
B_f	δ-CrB	Orthorhombic	$D8_2$	Cu_5Zn_8	Cubic, b.c.
B_g	MoB	Tetragonal, b.c.	$D8_3$	Cu_9Al_4	Cubic
B_h	WC	Hexagonal	$D8_8$	Mn_5Si_3	Hexagonal
B_i	γ-MoC	Hexagonal	$D8_b$	CrFe	Tetragonal
C1	CaF_2	Cubic, f.c.	$D8_i$	Mo_2B_5	Rhombohedral
$C1_b$	MgAgAs	Cubic, f.c.	$D10_2$	Fe_3Th_7	Hexagonal
C2	FeS_2	Cubic	$E2_1$	$CaTiO_3$	Cubic
C6	CdI_2	Trigonal	$E9_3$	Fe_3W_3C	Cubic, f.c.
C11b	$MoSi_2$	Tetragonal, b.c.	$L1_0$	CuAu	Tetragonal
C12	$CaSi_2$	Rhombohedral	$L1_2$	Cu_3Au	Cubic
C14	$MgZn_2$	Hexagonal	L'_{2b}	ThH_2	Tetragonal, b.c.
C15	Cu_2Mg	Cubic, f.c.	L'_3	Fe_2N	Hexagonal

*See "Handbook of Lattice Spacing and Structures of Metals", W.B. Pearson, Vol. I, Pergamon Press, 1958, p. 79, and Vol. II, Pergamon Press, 1967, p. 3.

HIGH CRITICAL MAGNETIC-FIELD SUPERCONDUCTIVE COMPOUNDS AND ALLOYS

With Critical Temperatures, H_{c1}, H_{c2}, H_{c3}, and the Temperature of Field Observations, T_{obs}

Substance	T_c, °K	H_{c1}, kg	H_{c2}, kg	H_{c3}, kg	T_{obs}, °K†
Al_2CMo_3	9.8–10.2	0.091	156		1.2
$AlNb_3$		0.375			
$Ba_xO_3Sr_{1-x}Ti$	<0.1–0.55	0.0039 max.			
$Bi_{0.5}Cd_{0.1}Pb_{0.27}Sn_{0.13}$			>24		3.06
Bi_xPb_{1-x}	7.35–8.4	0.122 max.	~30 max.		4.2
$Bi_{0.56}Pb_{0.44}$	8.8		15		4.2
$Bi_{7.5\ w/o}Pb_{92.5\ w/o}$‡			2.32		
$Bi_{0.099}Pb_{0.901}$		0.29	2.8		
$Bi_{0.02}Pb_{0.98}$		0.46	0.73		
$Bi_{0.53}Pb_{0.32}Sn_{0.16}$			>25		3.06
$Bi_{1-0.93}Sn_{0-0.07}$			0–0.032		3.7
Bi_5Tl_3	6.4		>5.56		3.35
C_8K (excess K)	0.55		0.160 (H⊥c)		0.32
			0.730 (H∥c)		0.32
C_8K	0.39		0.025 (H⊥c)		0.32
			0.250 (H∥c)		0.32
$C_{0.44}Mo_{0.56}$	12.5–13.5	0.087	98.5		1.2
CNb	8–10	0.12	16.9		4.2
$CNb_{0.4}Ta_{0.6}$	10–13.6	0.19	14.1		1.2
CTa	9–11.4	0.22	4.6		1.2
$Ca_xO_3Sr_{1-x}Ti$	<0.1–0.55	0.002–0.004			
$Cd_{0.1}Hg_{0.9}$ (by weight)		0.23	0.34		2.04
$Cd_{0.05}Hg_{0.95}$		0.28	0.31		2.16
$Cr_{0.10}Ti_{0.30}V_{0.60}$	5.6	0.071	84.4		0
GaN	5.85	0.725			4.2
Ga_xNb_{1-x}			>28		4.2
GaSb (annealed)	4.24		2.64		3.5
$GaV_{1.95}$	5.3		73***		
$GaV_{2.1-3.5}$	6.3–14.45		230–300**		0
GaV_3		0.4	350***		0
			500**		
$GaV_{4.5}$	9.15		121*		0
Hf_xNb_y			>52–>102		1.2
Hf_xTa_y			>28–>86		1.2
$Hg_{0.05}Pb_{0.95}$		0.235	2.3		
$Hg_{0.101}Pb_{0.899}$		0.23	4.3		4.2
$Hg_{0.15}Pb_{0.85}$	~6.75		>13		2.93
$In_{0.98}Pb_{0.02}$	3.45	0.1		0.12	2.76
$In_{0.96}Pb_{0.04}$	3.68	0.1	0.12	0.25	2.94
$In_{0.94}Pb_{0.06}$	3.90	0.095	0.18	0.35	3.12
$In_{0.913}Pb_{0.087}$	4.2	~0.17	0.55	2.65	
$In_{0.316}Pb_{0.684}$		0.155	3.7		4.2
$In_{0.17}Pb_{0.83}$			2.8	5.5	4.2
$In_{1.000}Te_{1.002}$	3.5–3.7		1.2*		0
$In_{0.95}Tl_{0.05}$		0.263	0.263		3.3
$In_{0.90}Tl_{0.10}$		0.257	0.257		3.25
$In_{0.83}Tl_{0.17}$		0.242	0.39		3.21
$In_{0.75}Tl_{0.25}$		0.216	0.50		3.16
LaN	1.35	0.45			0.76
La_3S_4	6.5	≈0.15	>25		1.3
La_3Se_4	8.6	≈0.2	>25		1.25
$Mo_{0.52}Re_{0.48}$	11.1		14–21	22–33	4.2
			18–28	37–43	1.3
$Mo_{0.6\pm0.05}Re_{0.395}$	10.6		14–20	20–37	4.2
			19–26	26–37	1.3
$Mo_{\sim0.5}Tc_{\sim0.5}$			~75*		0
$Mo_{0.16}Ti_{0.84}$	4.18	0.028	98.7*		0
			36–38		3.0
$Mo_{0.913}Ti_{0.087}$	2.95	0.060	~15		4.2
$Mo_{0.1-0.3}U_{0.9-0.7}$	1.85–2.06		>25		
$Mo_{0.17}Zr_{0.83}$			~30		
$N_{(12.8\ w/o)}Nb$	15.2		>9.5		13.2
NNb (wires)	16.1		153*		0
			132		4.2
			95		8
			53		12
NNb_xO_{1-x}	13.5–17.0		~38		
NNb_xZr_{1-x}	9.8–13.8		4–>130		4.2
$N_{0.93}Nb_{0.85}Zr_{0.15}$	13.8		>130		4.2
$Na_{0.086}Pb_{0.914}$		0.19	6.0		
$Na_{0.016}Pb_{0.984}$		0.28	2.05		

‡*w/o* denotes weight percent.

Substance	T_c, °K	H_{c1}, kg	H_{c2}, kg	H_{c3}, kg	T_{obs}, °K†
Nb	9.15		2.020		1.4
			1.710		4.2
Nb		0.4–1.1	3–5.5		4.2
Nb (unstrained)		1.1–1.8	3.40	6–9.1	4.2
Nb (strained)		1.25–1.92	3.44	6.0–8.7	4.2
Nb (cold-drawn wire)		2.48	4.10	≈10	4.2
Nb (film)			>25		4.2
NbSc			>30		
Nb_3Sn		0.170	221		4.2
			70		14.15
			54		15
			34		16
			17		17
$Nb_{0.1}Ta_{0.9}$		0.084	0.154		4.195
$Nb_{0.2}Ta_{0.8}$			10		4.2
$Nb_{0.65-0.73}Ta_{0.02-0.10}Zr_{0.25}$			>70–>90		4.2
Nb_xTi_{1-x}			148 max.		1.2
			120 max.		4.2
$Nb_{0.222}U_{0.778}$		1.98	23		1.2
Nb_xZr_{1-x}			127 max.		1.2
			94 max.		4.2
O_3SrTi	0.43	.0049*	.504*		0
O_3SrTi	0.33	.00195*	.420*		0
$PbSb_{1\ w/o}$ (quenched)			>1.5		4.2
$PbSb_{1\ w/o}$ (annealed)			>0.7		4.2
$PbSb_{2.8\ w/o}$ (quenched)			>2.3		4.2
$PbSb_{2.8\ w/o}$ (annealed)			>0.7		4.2
$Pb_{0.871}Sn_{0.129}$		0.45	1.1		
$Pb_{0.965}Sn_{0.035}$		0.53	0.56		
$Pb_{1-0.26}Tl_{0-0.74}$	7.20–3.68		2–6.9*		0
$PbTl_{0.17}$	6.73		4.5*		0
$Re_{0.26}W_{0.74}$			>30		
$Sb_{0.93}Sn_{0.07}$			0.12		3.7
SiV_3	17.0	0.55	156***		
Sn_xTe_{1-x}		0.00043–0.00236	0.005–0.0775		0.012–0.079
Ta (99.95%)		0.425	1.850		1.3
		0.325	1.425		2.27
		0.275	1.175		2.66
		0.090	0.375		3.72
$Ta_{0.5}Nb_{0.5}$			3.55		4.2
$Ta_{0.65-0}Ti_{0.35-1}$	4.4–7.8		>14–138		1.2
$Ta_{0.5}Ti_{0.5}$			138		1.2
Te	~3.3	0.25*			0
Tc_xW_{1-x}	5.75–7.88		8–44		4.2
Ti				2.7	4.2
$Ti_{0.75}V_{0.25}$	5.3	0.029*	199*		0
$Ti_{0.775}V_{0.225}$	4.7	0.024*	172*		0
$Ti_{0.615}V_{0.385}$	7.07	0.050	~34		4.2
$Ti_{0.516}V_{0.484}$	7.20	0.062	~28		4.2
$Ti_{0.415}V_{0.585}$	7.49	0.078	~25		4.2
$Ti_{0.12}V_{0.88}$			17.3	28.1	4.2
$Ti_{0.09}V_{0.91}$			14.3	16.4	4.2
$Ti_{0.06}V_{0.94}$			8.2	12.7	4.2
$Ti_{0.03}V_{0.97}$			3.8	6.8	4.2
Ti_xV_{1-x}			108 max.		1.2
V	5.31	~0.8	~3.4		1.79
		~0.75	~3.15		2
		~0.45	~2.2		3
		~0.30	~1.2		4
$V_{0.26}Zr_{0.74}$	≈5.9	0.238			1.05
		0.227			1.78
		0.185			3.04
		0.165			3.5
W (film)	1.7–4.1		>34		1

†Temperature of critical field measurement.
*Extrapolated.
**Linear extrapolation.
***Parabolic extrapolation.

TABLES OF PROPERTIES OF SEMICONDUCTORS

Compiled by Dr. Brian Randall Pamplin

The term "semiconductor" is applied to a material in which electric current is carried by electrons or holes and whose electrical conductivity when extremely pure rises exponentially with temperature and may be increased from this low "intrinsic" value by many orders of magnitude by "doping" with electrically active impurities.

Semiconductors are characterised by an energy gap in the allowed energies of electrons in the material which separates the normally filled energy levels of the *valence band* (where "missing" electrons behave like positively charged current carriers "holes") and the *conduction band* (where electrons behave rather like a gas of free negatively charged carriers with an effective mass dependent on the material and the direction of the electrons' motion). This energy gap depends on the nature of the material and varies with direction in anisotropic crystals. It is slightly dependent on temperature and pressure, and this dependence is usually almost linear at normal temperatures and pressures.

The data is presented in three tables. Table I "General Properties of Semiconductors" lists the main crystallographic and semiconducting properties of a large number of semiconducting materials in three main categories; "Tetrahedral Semiconductors" in which every atom is tetrahedrally co-ordinated to four nearest neighbour atoms (or atomic sites) as for example in the diamond structure; "Octahedral Semiconductors" in which every atom is octahedrally co-ordinated to six nearest neighbour atoms—as for example in the halite structure; and "Other Semiconductors".

Table II gives more detailed information about some better known semiconductors, while Table III gives some information about the electronic energy band structure parameters of the best known materials.

TABLE I

GENERAL PROPERTIES OF SEMICONDUCTORS

(listed by Crystal Structure)

Substance	Lattice Parameters (A° Room temperature)	Density (gm/cc)	Melting Point (°K)	Minimum Room Temperature Energy Gap (eV)	Thermal Conductivity	Heat of Formation k cal/mole	Mobility (Room Temperature) (cm²/V.s) Electrons	Holes	Remarks
PART A ADAMANTINE SEMICONDUCTORS									
§A1 Diamond Structure Elements (Strukturbericht symbol A4, Space Group Fd3m-O_h^7)									
C	3.5597	3.51	4300	5.4	2000	161	1800	1400	
Si	5.43072	2.3283	1685	1.107	1240	77.5	1900	500	
Ge	5.65754	5.3234	1231	0.67	640	69.5	3800	1820	
α-Sn	6.4912	5.765	503	0.08		64	2500	2400	
§A2 Sphalerite (Zinc Blende) Structure Compounds (Strukturbericht symbol B3 Space Group F $\overline{4}$ 3m-T_d^2)									
I VII Compounds									
CuF	4.255								
CuCl	5.4057	3.53	695						
CuBr	5.6905	4.72	770	2.94		115			
CuI	6.0427	5.63	878			105			
AgBr						102	4000		
AgI	6.473	5.67				93	30		
II VI Compounds									
BeS	4.865	2.36							
BeSe	5.139	4.315							
BeTe	5.626	5.090							
BePo	8.38	7.3							
Zn0	4.63								see § A3
ZnS	5.4093	4.079	2100	3.54	140	114	180	5(400°C)	see also § A3
ZnSe	5.6676	5.42	1790	2.58	140	101	540	28	
ZnTe	6.101	5.72	1568	2.26	140	90	340	100	
ZnPo									
CdS	5.5818		1750						see § A3
CdSe	6.05		1512						see § A3
CdTe	6.477	5.86	1365	1.44	55	81	1200	50	
CdPo									
HgS	5.8517	7.73	~2020						usually cinnabar
HgSe	6.084	8.25	1070	- 0.10	10	59	20000		
HgTe	6.460	8.17	943	-0.15	20	58	25000	350	
III V Compounds									
BN	3.615	3.49	3000	~4	200	195			
BP(L.T.)	4.538	2.9		~6			500	70	
BAs	4.777								
AlP	5.451	2.85	1770	2.5					
AlAs	5.6622	3.81	1870	2.16		150	1200	420	
AlSb	6.1355	4.218	1330	1.60	600	140	200–400	550	
GaP	5.4505	4.13	1750	2.24	1100	152	300	100	
GaAs	5.65315	5.316	1510	1.35	370	128	8800	400	
GaSb	6.0954	5.619	980	0.67	270	118	4000	1400	
InP	5.86875	4.787	1330	1.27	800	134	4600	150	
InAs	6.05838	5.66	1215	0.36	290	114	33000	460	
InSb	6.47877	5.775	798	0.165	160	107	78000	750	
Other Sphalerite Structure Compounds									
MnS	5.60								see also § A3
MnSe	5.82								see also § A3
β-SiC	4.348	3.21	3070	2.3			4000		
Ga_2Te_3	5.899	5.75	1063	~1.0	~14	65			
In_2Te_3(H.T.)	6.150	5.8	940	~1.0	~8	47.4	~10		
$MgGeP_2$	5.652								
$ZnSnP_2$	5.65			2.1					
$ZnSnAs_2$(H.T.)	5.851	5.53	1050	~0.7	70				
§ A3 Wurtzite (Zincite) Structure Compounds (Strukturbericht symbol B4, Space Group P 6_3mc-C_{6v}^4)									
I VII Compounds									
CuCl	3.91	6.42	T_c 680°K						
CuBr	4.06	6.66	T_c 658°K						
CuI	4.31	7.09							
AgI	4.580	7.494		2.63					

TABLE 1

GENERAL PROPERTIES OF SEMICONDUCTORS (Continued)

Substance	Lattice Parameters (A° Room temperature)		Density (gm/cc)	Melting Point (°K)	Minimum Room Temperature Energy Gap (eV)	Thermal Conductivity	Heat of Formation k cal/mole	Mobility (Room Temperature) (cm²/V.s) Electrons	Holes	Remarks
II VI Compounds										
BeO	2.698	4.380		2800						
MgTe	4.54	7.39	3.85	~2800						
ZnO	3.24950	5.2069	5.66	2250	3.2	6	154	180		
ZnS	3.8140	6.2576	4.1	2100	3.67		110			
ZnSe	3.996	6.626		1793						
ZnTe	4.27	6.99		1568						
CdS	4.1348	6.7490	4.82	1748	2.42		96	400		
CdSe	4.299	7.010	5.66	1512	1.74		90	650		
CdTe	4.57	7.47			1.50					
III V Compounds										
BP(H.T.)	3.562	5.900								
AlN	3.111	4.978	3.26	~2500	6.02		197			
GaN	3.190	5.189	6.10	1500	3.34		157			
InN	3.533	5.693	6.88	1200	2.0		133			
Other Wurtzite Structure Compounds										
MnS	3.985	6.45								
MnSe	4.12	6.72								
SiC	3.076	5.048								
MnTe	4.078	6.701			~1.0					
Al_2S_3	3.579	5.829	2.55		4.1		426			
Al_2Se_3	3.890	6.30	3.91		3.1		367			
§ $A\bar{4}$ Chalcopyrite Structure Compounds (Strukturbericht symbol $E1_1$, Space Group I $\bar{4}$ 2d $-D_{2d}^{12}$)										
I III VI_2 Compounds										
$CuAlS_2$	5.323	10.44	3.47		2.5					
$CuAlSe_2$	5.617	10.92	4.70	1270	1.1					
$CuAlTe_2$	5.976	11.80	5.50	1160	0.88					
$CuGaS_2$	5.360	10.49	4.35							
$CuGaSe_2$	5.618	11.01	5.56	1310	0.96, 1.63					
$CuGaTe_2$	6.013	11.93	5.99	1150	0.82, 1.0					
$CuInS_2$	5.528	11.08	4.75		1.2					
$CuInSe_2$	5.785	11.57	5.77	1250	0.86, 0.92	37				
$CuInTe_2$	6.179	12.365	6.10	970	0.95	49				
$CuTlS_2$	5.580	11.17	6.32							
$CuTlSe_2$(L.T.)	5.844	11.65	7.11	680	1.07					
$CuFeS_2$	5.25	10.32		1150	0.53					
$CuFeSe_2$				850	0.16					
$CuLaS_2$	5.65	10.86								
$AgAlS_2$	5.707	10.28	3.94							
$AgAlSe_2$	5.968	10.77	5.07	1220	0.7					
$AgAlTe_2$	6.309	11.85	6.18	1000	0.56					
$AgGaS_2$	5.755	10.28	4.72							
$AgGaSe_2$	5.985	10.90	5.84	1120	1.66					
$AgGaTe_2$	6.301	11.96	6.05	990	1.1	10				
$AgInS_2$(L.T.)	5.828	11.19	5.00		1.9					
$AgInSe_2$	6.102	11.69	5.81	1053	1.18	30				
$AgInTe_2$	6.42	12.59	6.12	965	0.96, 0.52					
$AgFeS_2$	5.66	10.30	4.53							
II IV V_2 Compounds										
$ZnSiP_2$	5.400	10.441	3.39	1640	2.3			1000		
$ZnGeP_2$	5.465	10.771	4.17	1295	2.2					
$CdSiP_2$	5.678	10.431	4.00	~1470	2.2			1000		
$CdGeP_2$	5.741	10.775	4.48	~1060	1.8					
$CdSnP_2$	5.900	11.518			1.5					
$ZnSiAs_2$	5.61	10.88	4.70	~1350	1.7				50	
$ZnGeAs_2$	5.672	11.153	5.32	~1150	0.85	110				
$ZnSnAs_2$	5.8515	11.704	5.53	~ 910	0.65	150			300	disorders at 910°K
$CdSiAs_2$	5.884	10.882								
$CdGeAs_2$	5.9427	11.2172	5.60	~ 903	0.53	40		70	25	disorders at 903°K
$CdSnAs_2$	6.0944	11.9182	5.72	880	0.26	70		22000	250	

TABLE I

GENERAL PROPERTIES OF SEMICONDUCTORS (Continued)

Substance	Lattice Parameters (A° Room temperature)	Density (gm/cc)	Melting Point (°K)	Minimum Room Temperature Energy Gap (eV)	Thermal Conductivity	Heat of Formation k cal/mole	Mobility (Room Temperature) (cm²/V.s) Electrons	Holes	Remarks
§ A5 "Defect Chalcopyrite" Structure Compounds (Strukturbericht symbol E3, Space Group $I\,\bar{4} - S_4^2$)									
$ZnAl_2Se_4$	5.503 10.90	4.37							
$ZnAl_2Te_4$(?)	5.904 12.05	4.95							
$ZnGa_2S_4$(?)	5.274 10.44	3.80							
$ZnGa_2Se_4$(?)	5.496 10.99	5.21							
$ZnGa_2Te_4$(?)	5.937 11.87	5.67		1.35					
$ZnIn_2Se_4$	5.711 11.42	5.44	1250	1.82					
$ZnIn_2Te_4$	6.122 12.24	5.83	1075	1.2					
$CdAl_2S_4$	5.564 10.32	3.06							
$CdAl_2Se_4$	5.747 10.68	4.54							
$CdAl_2Te_4$(?)	6.011 12.21	5.10							
$CdGa_2S_4$	5.577 10.08	4.03							
$CdGa_2Se_4$	5.743 10.73	5.32							
$CdGa_2Te_4$	6.093 11.81	5.77							
$CdIn_2Te_4$	6.205 12.41	5.9	1060	(1.26 or 0.9)					
$HgAl_2S_4$	5.488 10.26	4.11							
$HgAl_2Se_4$	5.708 10.74	5.05							
$HgAl_2Te_4$(?)	6.004 12.11	5.81							
$HgGa_2S_4$	5.507 10.23	5.00							
$HgGa_2Se_4$	5.715 10.78	6.18							
$HgIn_2Se_4$	5.764 11.80	6.3	1100	0.6					
$HgIn_2Te_4$(?)	6.186 12.37	6.3	980	0.86		200			
§ A6 Other Adamantine Compounds									
αSiC	3.0817 15.1183	3.21	3070	2.86			400		6H structure
$Hg_5Ga_2Te_8$	6.235								B3 with super lattice
$Hg_5In_2Te_8$	6.328			0.7			2000		B3 with super lattice
Cd In_2Se_4	a=c=5.823			1.55					
PART B OCTAHEDRAL SEMICONDUCTORS									
Halite Structure Semiconductors (Strukturbericht symbol B1, Space Group Fm3m − O_h^5)									
SnSe	6.020		1133						
SnTe	6.313		1080(max)	0.5	91				
PbS	5.9362	7.61	1390	0.37	23	104	600	600	
PbSe	6.1243	8.15	1340	0.26	17	94	1000	900	
PbTe	6.454	8.16	1180	0.25	23	94	1600	600	
Selected Other Binary Chalcides									
BiSe	5.99	7.98	880	0.4					
BiTe	6.47								
EuSe	6.191		2300	1.8	2.4				
GdSe	5.771		2400						
NiD	4.1684	6.6	2260	2.0 or 3.7			4	-	
CdO	4.6953		1700	2.5	7	127	100		
SrS	6.0199	3.643	3000	4.1					
Selected Multineny Compounds									
$AgSbSe_2$	5.786	6.60	910	0.58	10.5				
$AgSbTe_2$ (or $Ag_{19}Sb_{29}Te_{52}$)	6.078	7.12	830	0.7, 0.27	8.6, 0.3				
$AgBiS_2$(H.T.)	5.648								
$AgBiSe_2$(H.T.)	5.82								
$AgBiTe_2$(H.T.)	6.155								
Cu_2CdSnS_4	5.586	10.83		1.16			<2		

TABLE II

SEMICONDUCTING PROPERTIES OF SELECTED MATERIALS

Substance	Minimum Energy Gap (eV) R.T.	Minimum Energy Gap (eV) 0°K	$\frac{dE_g}{dT}$ x 10^4 eV/°C	$\frac{dE_g}{dP}$ x 10^6 eV.cm²/kg	Density of States Electron Effective Mass m_{d_n} (m_o)	Electron Mobility and Temperature Dependence μ_n cm²/V.s	Electron Mobility and Temperature Dependence −x	Density of States Hole Effective Mass m_{d_p} (m_o)	Hole Mobility and Temperature Dependence μ_p cm²/V.s	Hole Mobility and Temperature Dependence −x
Si	1.107	1.153	−2.3	−2.0	1.1	1,900	2.6	0.56	500	2.3
Ge	0.67	0.744	−3.7	+7.3	0.55	3,800	1.66	0.3	1,820	2.33
αSn	0.08	0.094	−0.5		0.02	2,500	1.65	0.3	2,400	2.0
Te	0.33				0.68	1,100		0.19	560	
III–V Compounds										
AlAs	2.2	2.3				1,200			420	
AlSb	1.6	1.7	−3.5	−1.6	0.09	200	1.5	0.4	500	1.8
GaP	2.24	2.40	−5.4	−1.7	0.35	300	1.5	0.5	150	1.5
GaAs	1.35	1.53	−5.0	+9.4	0.068	9,000	1.0	0.5	500	2.1
GaSb	0.67	0.78	−3.5	+12	0.050	5,000	2.0	0.23	1,400	0.9
InP	1.27	1.41	−4.6	+4.6	0.067	5,000	2.0		200	2.4
InAs	0.36	0.43	−2.8	+8	0.022	33,000	1.2	0.41	460	2.3
InSb	0.165	0.23	−2.8	+15	0.014	78,000	1.6	0.4	750	2.1
II–VI Compounds										
ZnO	3.2		−9.5	+0.6	0.38	180	1.5			
ZnS	3.54		−5.3	+5.7		180			5(400°C)	
ZnSe	2.58	2.80	−7.2	+6		540			28	
ZnTe	2.26			+6		340			100	
CdO	2.5 ± .1		−6		0.1	120				
CdS	2.42		−5	+3.3	0.165	400		0.8		
CdSe	1.74	1.85	−4.6		0.13	650	1.0	0.6		
CdTe	1.44	1.56	−4.1	+8	0.14	1,200		0.35	50	
HgSe	0.30				0.030	20,000	2.0			
HgTe	0.15		−1		0.017	25,000		0.5	350	
Halite Structure Compounds										
PbS	0.37	0.28	+4		0.16	800		0.1	1,000	2.2
PbSe	0.26	0.16	+4		0.3	1,500		0.34	1,500	2.2
PbTe	0.25	0.19	+4	−7	0.21	1,600		0.14	750	2.2
Others										
ZnSb	0.50	0.56			0.15	10				1.5
CdSb	0.45	0.57	−5.4		0.15	300			2,000	1.5
Bi_2S_3	1.3					200			1,100	
Bi_2Se_3	0.27					600			675	
Bi_2Te_3	0.13		−0.95		0.58	1,200	1.68	1.07	510	1.95
Mg_2Si		0.77	−6.4		0.46	400	2.5		70	
Mg_2Ge		0.74	−9			280	2		110	
Mg_2Sn	0.21	0.33	−3.5		0.37	320			260	
Mg_3Sb_2		0.32				20			82	
Zn_3As_2	0.93					10	1.1		10	
Cd_3As_2	0.55				0.046	100,000	0.88			
GaSe	2.05		3.8						20	
GaTe	1.66	1.80	−3.6			14	−5			
InSe	1.8					900				
TlSe	0.57		−3.9		0.3	30		0.6	20	1.5
$CdSnAs_2$	0.23				0.05	25,000	1.7			
Ga_2Te_3	1.1	1.55	−4.8							
α-In_2Te_3	1.1	1.2			0.7				50	1.1
β-In_2Te_3	1.0								5	
$Hg_5In_2Te_8$	0.5								11,000	
SnO_2									78	

TABLE III

PART A. DATA ON VALENCE BANDS OF SEMICONDUCTORS (Room Temperature data)

Substance	Band Curvature Effective Mass: Heavy Holes	Light Holes	"Split-off" Band Holes	Energy Separation of "Split-off" Band (eV)	Measured (Light) Hole Mobility $cm^2/V.s$
	(Expressed as fraction of free electron mass)				
Semiconductors with Valence Band Maximum at Centre of Brillouin Zone ("Γ")					
Si	0.52	0.16	0.25	0.044	500
Ge	0.34	0.043	0.08	0.3	1,820
Sn	0.3				2,400
AlAs					
AlSb	0.4			0.7	550
GaP				0.13	100
GaAs	0.8	0.12	0.20	0.34	400
GaSb	0.23	0.06		0.7	1,400
InP				0.21	150
InAs	0.41	0.025	0.083	0.43	460
InSb	0.4	0.015		0.85	750
CdTe	0.35				50
HgTe	0.5				350

Semiconductors with Multiple Valence Band Maxima

Substance	Number of Equivalent Valleys Direction	Curvature Effective Masses: Longitudinal m_L	Transverse m_T	Anisotropy $K = \frac{m_L}{m_T}$	Measured (Light) Hole Mobility $cm^2/V.s$
PbSe	4 "L" [111]	0.095	0.047	2.0	1,500
PbTe	4 "L" [111]	0.27	0.02	10	750
Bi_2Te_3	6	0.207	~0.045	4.5	515

PART B. DATA ON CONDUCTION BANDS OF SEMICONDUCTORS (Room Temperature Data)

Single Valley Semiconductors

Substance	Energy Gap (eV)	Effective Mass (m_o)	Mobility ($cm^2/V.s$)	Comments
GaAs	1.35	0.067	8,500	3(or 6?) equivalent [100] valleys 0.36 eV above this maximum with a mobility of ~50
InP	1.27	0.067	5,000	3(or 6?) equivalent [100] valleys 0.4 eV above this minimum.
InAs	0.36	0.022	33,000	equivalent valleys ~1.0 eV above this minimum.
InSb	0.165	0.014	78,000	
CdTe	1.44	0.11	1,000	4(or 8?) equivalent [111] valleys 0.51 eV above this minimum.

Multivalley Semiconductors

Substance	Energy Gap	Number of Equivalent Valleys and Direction	Band Curvature Effective Mass: Longitudinal m_L	Transverse m_T	Anisotropy $K = \frac{m_L}{m_T}$	Comments
Si	1.107	6 in [100] "Δ"	0.90	0.192	4.7	
Ge	0.67	4 in [111] at "L"	1.588	0.0815	19.5	
GaSb	0.67	as Ge (?)	~1.0	~0.2	~5	
PbSe	0.26	4 in [111] at "L"	0.085	0.05	1.7	
PbTe	0.25	4 in [111] at "L"	0.21	0.029	5.5	
Bi_2Te_3	0.13	6			~0.05	

STANDARD CALIBRATION TABLES FOR THERMOCOUPLES

The following tables which represent the Temperature-E. M. F. functions of various thermocouples should be used with appropriate correction curves if precise results are desired. These curves must be determined for each individual couple by plotting ΔE, the difference between the observed and the standard E. M. F., against the standard E. M. F. at three or more fixed temperature points. The value ΔE as shown by such a correction curve is then subtracted algebraically from the observed E. M. F. to give the true E. M. F. reading.

In the following tables the fixed or "cold junction" is at 0° C.; when the cold junction is not maintained at 0° C. the readings of the E. M. F. must be corrected as follows: $Et = E_{(t-tc)} + Etc$ where $E_{(t-tc)}$ is the observed reading, *Etc* is the E. M. F. for the temperature corresponding to the cold junction temperature as read from the standard table and *Et* is the E. M. F. produced by the hot junction corrected to the value which would be obtained with the cold junction at 0° C. The temperature corresponding to *Et* is then obtained by reference to the standard table.

Since the E. M. F.-temperature function is not linear the cold junction should be maintained at a temperature very close to that at which the thermocouple was calibrated. Otherwise considerable error will result despite the above correction.

PLATINUM VERSUS PLATINUM-10-PERCENT RHODIUM THERMOCOUPLES

(Electromotive Force in Absolute Millivolts. Temperatures in Degrees C (Int. 1948). Reference Junctions at 0° C.)

°C	0	10	20	30	40	50	60	70	80	90
					Millivolts					
0	0	0.06	0.11	0.17	0.24	0.30	0.36	0.43	0.50	0.57
100	0.64	0.72	0.79	0.87	0.95	1.03	1.11	1.19	1.27	1.35
200	1.44	1.52	1.61	1.69	1.78	1.87	1.96	2.05	2.14	2.23
300	2.32	2.41	2.50	2.59	2.69	2.78	2.87	2.97	3.06	3.16
400	3.25	3.35	3.44	3.54	3.64	3.73	3.83	3.93	4.02	4.12
500	4.22	4.32	4.42	4.52	4.62	4.72	4.82	4.92	5.02	5.12
600	5.22	5.33	5.43	5.53	5.64	5.74	5.84	5.95	6.05	6.16
700	6.26	6.37	6.47	6.58	6.68	6.79	6.90	7.01	7.11	7.22
800	7.33	7.44	7.55	7.66	7.77	7.88	7.99	8.10	8.21	8.32
900	8.43	8.55	8.66	8.77	8.88	9.00	9.11	9.23	9.34	9.46
1000	9.57	9.69	9.80	9.92	10.04	10.15	10.27	10.39	10.51	10.62
1100	10.74	10.86	10.98	11.10	11.22	11.34	11.46	11.58	11.70	11.82
1200	11.94	12.06	12.18	12.30	12.42	12.54	12.66	12.78	12.90	13.02
1300	13.14	13.26	13.38	13.50	13.62	13.74	13.86	13.98	14.10	14.22
1400	14.34	14.46	14.58	14.70	14.82	14.94	15.05	15.17	15.29	15.41
1500	15.53	15.65	15.77	15.89	16.01	16.12	16.24	16.36	16.48	16.60
1600	16.72	16.83	16.95	17.07	17.19	17.31	17.42	17.54	17.66	17.77
1700	17.89	18.01	18.12	18.24	18.36	18.47	18.59	...	...	...

CALIBRATION TABLES FOR THERMOCOUPLES (Continued)

PLATINUM VERSUS PLATINUM-10-PERCENT RHODIUM THERMOCOUPLES

(Electromotive Force in Absolute Millivolts. Temperatures in Degrees F.* Reference Junctions at 32° F.)

°F	0	10	20	30	40	50	60	70	80	90
					Millivolts					
0	...	...	...	...	0.02	0.06	0.09	0.12	0.15	0.19
100	0.22	0.26	0.29	0.33	0.36	0.40	0.44	0.48	0.52	0.56
200	0.60	0.64	0.68	0.72	0.76	0.80	0.84	0.89	0.93	0.97
300	1.02	1.06	1.11	1.15	1.20	1.24	1.29	1.33	1.38	1.43
400	1.47	1.52	1.57	1.62	1.66	1.71	1.76	1.81	1.86	1.91
500	1.96	2.01	2.06	2.11	2.16	2.21	2.26	2.31	2.36	2.41
600	2.46	2.51	2.56	2.61	2.66	2.72	2.77	2.82	2.87	2.92
700	2.98	3.03	3.08	3.14	3.19	3.24	3.29	3.35	3.40	3.45
800	3.51	3.56	3.61	3.67	3.72	3.78	3.83	3.88	3.94	3.99
900	4.05	4.10	4.16	4.21	4.26	4.32	4.37	4.43	4.49	4.54
1060	4.60	4.65	4.71	4.76	4.82	4.87	4.93	4.99	5.04	5.10
1100	5.16	5.21	5.27	5.33	5.38	5.44	5.50	5.56	5.61	5.67
1200	5.73	5.78	5.84	5.90	5.96	6.02	6.07	6.13	6.19	6.25
1300	6.31	6.37	6.42	6.48	6.54	6.60	6.66	6.72	6.78	6.84
1400	6.90	6.96	7.02	7.08	7.14	7.20	7.26	7.32	7.38	7.44
1500	7.50	7.56	7.62	7.68	7.74	7.80	7.86	7.93	7.99	8.05
1600	8.11	8.17	8.23	8.30	8.36	8.42	8.48	8.55	8.61	8.67
1700	8.73	8.80	8.86	8.92	8.98	9.05	9.11	9.17	9.24	9.30
1800	9.37	9.43	9.49	9.56	9.62	9.69	9.75	9.82	9.88	9.94
1900	10.01	10.07	10.14	10.20	10.27	10.33	10.40	10.47	10.53	10.60
2000	10.66	10.73	10.79	10.86	10.93	10.99	11.06	11.12	11.19	11.26
2100	11.32	11.39	11.46	11.52	11.59	11.66	11.72	11.79	11.86	11.92
2200	11.99	12.06	12.12	12.19	12.26	12.32	12.39	12.46	12.52	12.59
2300	12.66	12.72	12.79	12.86	12.92	12.99	13.06	13.12	13.19	13.26
2400	13.33	13.39	13.46	13.53	13.59	13.66	13.73	13.79	13.86	13.92
2500	13.99	14.06	14.12	14.19	14.26	14.32	14.39	14.46	14.52	14.59
2600	14.66	14.72	14.79	14.86	14.92	14.99	15.05	15.12	15.19	15.25
2700	15.32	15.39	15.45	15.52	15.58	15.65	15.72	15.78	15.85	15.91
2800	15.98	16.05	16.11	16.18	16.24	16.31	16.37	16.44	16.51	16.57
2900	16.64	16.70	16.77	16.83	16.90	16.97	17.03	17.10	17.16	17.23
3000	17.29	17.36	17.42	17.49	17.55	17.62	17.68	17.75	17.81	17.88
3100	17.94	18.01	18.07	18.14	18.20	18.27	18.33	18.40	18.46	18.53
3200	18.59	18.66	...	...	...	...	...	...	...	...

* Based on the International Temperature Scale of 1948.

CALIBRATION TABLES FOR THERMOCOUPLES (Continued)

PLATINUM VERSUS PLATINUM-13-PERCENT RHODIUM THERMOCOUPLES

(Electromotive Force in Absolute Millivolts. Temperatures in Degrees C (Int. 1948) Reference Junctions at 0° C.)

°C	0	10	20	30	40	50	60	70	80	90
	Millivolts									
0	0.00	0.06	0.11	0.17	0.23	0.30	0.36	0.43	0.50	0.57
100	0.65	0.72	0.80	0.88	0.96	1.04	1.12	1.21	1.29	1.38
200	1.47	1.55	1.64	1.73	1.83	1.92	2.01	2.11	2.20	2.30
300	2.40	2.49	2.59	2.69	2.79	2.89	2.99	3.09	3.19	3.30
400	3.40	3.50	3.61	3.71	3.82	3.92	4.03	4.13	4.24	4.35
500	4.46	4.56	4.67	4.78	4.89	5.00	5.12	5.23	5.34	5.45
600	5.56	5.68	5.79	5.91	6.02	6.14	6.25	6.37	6.49	6.60
700	6.72	6.84	6.96	7.08	7.20	7.32	7.44	7.56	7.68	7.80
800	7.92	8.05	8.17	8.29	8.42	8.54	8.67	8.80	8.92	9.05
900	9.18	9.30	9.43	9.56	9.69	9.82	9.95	10.08	10.21	10.34
1000	10.47	10.60	10.74	10.87	11.00	11.14	11.27	11.41	11.54	11.68
1100	11.82	11.95	12.09	12.23	12.37	12.50	12.64	12.78	12.92	13.06
1200	13.19	13.33	13.47	13.61	13.75	13.89	14.03	14.17	14.30	14.44
1300	14.58	14.72	14.86	15.00	15.14	15.28	15.42	15.55	15.69	15.83
1400	15.97	16.11	16.25	16.39	16.52	16.66	16.80	16.94	17.08	17.22
1500	17.36	17.49	17.63	17.77	17.91	18.04	18.18	18.32	18.45	18.59
1600	18.73	18.86	19.00	19.14	19.27	19.41	19.55	19.68	19.82	19.95
1700	20.09	...	...	...	...	...	...	...	...	...

(Electromotive Force in Absolute Millivolts. Temperatures in Degrees F.* Reference Junctions at 32° F.)

°F	0	10	20	30	40	50	60	70	80	90
	Millivolts									
0	...	...	...	...	0.02	0.06	0.09	0.12	0.15	0.19
100	0.22	0.26	0.29	0.33	0.36	0.40	0.44	0.48	0.52	0.56
200	0.60	0.64	0.68	0.72	0.76	0.81	0.85	0.89	0.94	0.98
300	1.03	1.08	1.12	1.17	1.21	1.26	1.31	1.36	1.41	1.46
400	1.50	1.55	1.60	1.65	1.70	1.75	1.81	1.86	1.91	1.96
500	2.01	2.07	2.12	2.17	2.22	2.28	2.33	2.38	2.44	2.49
600	2.55	2.60	2.66	2.71	2.77	2.82	2.88	2.94	2.99	3.05
700	3.10	3.16	3.22	3.27	3.33	3.39	3.45	3.50	3.56	3.62
800	3.68	3.74	3.79	3.85	3.91	3.97	4.03	4.09	4.15	4.21
900	4.26	4.32	4.38	4.44	4.50	4.56	4.62	4.69	4.75	4.81
1000	4.87	4.93	4.99	5.05	5.12	5.18	5.24	5.30	5.36	5.43
1100	5.49	5.55	5.61	5.68	5.74	5.81	5.87	5.93	6.00	6.06
1200	6.13	6.19	6.25	6.32	6.38	6.45	6.51	6.58	6.64	6.71
1300	6.77	6.84	6.90	6.97	7.04	7.10	7.17	7.24	7.30	7.37
1400	7.44	7.50	7.57	7.64	7.71	7.77	7.84	7.91	7.98	8.05
1500	8.12	8.18	8.25	8.32	8.39	8.46	8.53	8.60	8.67	8.74
1600	8.81	8.88	8.95	9.02	9.09	9.16	9.23	9.30	9.37	9.45
1700	9.52	9.59	9.66	9.73	9.80	9.87	9.95	10.02	10.09	10.16
1800	10.24	10.31	10.38	10.46	10.53	10.60	10.68	10.75	10.82	10.90
1900	10.97	11.05	11.12	11.20	11.27	11.35	11.42	11.50	11.58	11.65
2000	11.73	11.80	11.88	11.95	12.03	12.11	12.18	12.26	12.34	12.41
2100	12.49	12.56	12.64	12.72	12.80	12.87	12.95	13.03	13.10	13.18
2200	13.26	13.33	13.41	13.49	13.56	13.64	13.72	13.80	13.87	13.95
2300	14.03	14.10	14.18	14.26	14.34	14.41	14.49	14.57	14.64	14.72
2400	14.80	14.88	14.95	15.03	15.11	15.18	15.26	15.34	15.42	15.49
2500	15.57	15.65	15.72	15.80	15.88	15.95	16.03	16.11	16.19	16.26
2600	16.34	16.42	16.49	16.57	16.65	16.73	16.80	16.88	16.96	17.03
2700	17.11	17.19	17.26	17.34	17.42	17.49	17.57	17.65	17.72	17.80
2800	17.88	17.95	18.03	18.10	18.18	18.26	18.33	18.41	18.48	18.56
2900	18.64	18.71	18.79	18.86	18.94	19.02	19.09	19.17	19.24	19.32
3000	19.39	19.47	19.55	19.62	19.70	19.77	19.85	19.92	20.00	20.08
3100	20.15	...	...	...	...	...	...	...	...	...

CALIBRATION TABLES FOR THERMOCOUPLES (Continued)

CHROMEL-ALUMEL THERMOCOUPLES

(Electromotive Force in Absolute Millivolts. Temperatures in Degrees C (Int. 1948) Reference Junctions at 0° C.)

°C	0	1	2	3	4	5	6	7	8	9
−190	−5.60	−5.62	−5.63	−5.65	−5.67	−5.68	−5.70	−5.71	−5.73	−5.74
−180	−5.43	−5.45	−5.46	−5.48	−5.50	−5.52	−5.53	−5.55	−5.57	−5.58
−170	−5.24	−5.26	−5.28	−5.30	−5.32	−5.34	−5.35	−5.37	−5.39	−5.41
−160	−5.03	−5.05	−5.08	−5.10	−5.12	−5.14	−5.16	−5.18	−5.20	−5.22
−150	−4.81	−4.84	−4.86	−4.88	−4.90	−4.92	−4.95	−4.97	−4.99	−5.01
−140	−4.58	−4.60	−4.62	−4.65	−4.67	−4.70	−4.72	−4.74	−4.77	−4.79
−130	−4.32	−4.35	−4.37	−4.40	−4.42	−4.45	−4.48	−4.50	−4.52	−4.55
−120	−4.06	−4.08	−4.11	−4.14	−4.16	−4.19	−4.22	−4.24	−4.27	−4.30
−110	−3.78	−3.81	−3.84	−3.86	−3.89	−3.92	−3.95	−3.98	−4.00	−4.03
−100	−3.49	−3.52	−3.55	−3.58	−3.61	−3.64	−3.66	−3.69	−3.72	−3.75
−90	−3.19	−3.22	−3.25	−3.28	−3.31	−3.34	−3.37	−3.40	−3.43	−3.46
−80	−2.87	−2.90	−2.93	−2.96	−3.00	−3.03	−3.06	−3.09	−3.12	−3.16
−70	−2.54	−2.57	−2.61	−2.64	−2.67	−2.71	−2.74	−2.77	−2.80	−2.84
−60	−2.20	−2.24	−2.27	−2.30	−2.34	−2.37	−2.41	−2.44	−2.47	−2.51
−50	−1.86	−1.89	−1.93	−1.96	−2.00	−2.03	−2.07	−2.10	−2.13	−2.17
−40	−1.50	−1.54	−1.57	−1.61	−1.64	−1.68	−1.72	−1.75	−1.79	−1.82
−30	−1.14	−1.17	−1.21	−1.25	−1.28	−1.32	−1.36	−1.39	−1.43	−1.47
−20	−0.77	−0.80	−0.84	−0.88	−0.92	−0.95	−0.99	−1.03	−1.06	−1.10
−10	−0.39	−0.42	−0.46	−0.50	−0.54	−0.58	−0.62	−0.66	−0.69	−0.73
(−)0	−0.00	−0.04	−0.08	−0.12	−0.16	−0.19	−0.23	−0.27	−0.31	−0.35
(+)0	0.00	0.04	0.08	0.12	0.16	0.20	0.24	0.28	0.32	0.36
10	0.40	0.44	0.48	0.52	0.56	0.60	0.64	0.68	0.72	0.76
20	0.80	0.84	0.88	0.92	0.96	1.00	1.04	1.08	1.12	1.16
30	1.20	1.24	1.28	1.32	1.36	1.40	1.44	1.49	1.53	1.57
40	1.61	1.65	1.69	1.73	1.77	1.81	1.85	1.90	1.94	1.98
50	2.02	2.06	2.10	2.14	2.18	2.23	2.27	2.31	2.35	2.39
60	2.43	2.47	2.51	2.56	2.60	2.64	2.68	2.72	2.76	2.80
70	2.85	2.89	2.93	2.97	3.01	3.05	3.10	3.14	3.18	3.22
80	3.26	3.30	3.35	3.39	3.43	3.47	3.51	3.56	3.60	3.64
90	3.68	3.72	3.76	3.81	3.85	3.89	3.93	3.97	4.01	4.06
100	4.10	4.14	4.18	4.22	4.26	4.31	4.35	4.39	4.43	4.47
110	4.51	4.55	4.60	4.64	4.68	4.72	4.76	4.80	4.84	4.88
120	4.92	4.96	5.01	5.05	5.09	5.13	5.17	5.21	5.25	5.29
130	5.33	5.37	5.41	5.45	5.49	5.53	5.57	5.61	5.65	5.69
140	5.73	5.77	5.81	5.85	5.89	5.93	5.97	6.01	6.05	6.09
150	6.13	6.17	6.21	6.25	6.29	6.33	6.37	6.41	6.45	6.49
160	6.53	6.57	6.61	6.65	6.69	6.73	6.77	6.81	6.85	6.89
170	6.93	6.97	7.01	7.05	7.09	7.13	7.17	7.21	7.25	7.29
180	7.33	7.37	7.41	7.45	7.49	7.53	7.57	7.61	7.65	7.69
190	7.73	7.77	7.81	7.85	7.89	7.93	7.97	8.01	8.05	8.09
200	8.13	8.17	8.21	8.25	8.29	8.33	8.37	8.41	8.46	8.50
210	8.54	8.58	8.62	8.66	8.70	8.74	8.78	8.82	8.86	8.90
220	8.94	8.98	9.02	9.06	9.10	9.14	9.18	9.22	9.26	9.30
230	9.34	9.38	9.42	9.46	9.50	9.54	9.59	9.63	9.67	9.71
240	9.75	9.79	9.83	9.87	9.91	9.95	9.99	10.03	10.07	10.11
250	10.16	10.20	10.24	10.28	10.32	10.36	10.40	10.44	10.48	10.52
260	10.57	10.61	10.65	10.69	10.73	10.77	10.81	10.85	10.89	10.93
270	10.98	11.02	11.06	11.10	11.14	11.18	11.22	11.26	11.30	11.34
280	11.39	11.43	11.47	11.51	11.55	11.59	11.63	11.67	11.72	11.76
290	11.80	11.84	11.88	11.92	11.96	12.01	12.05	12.09	12.13	12.17
300	12.21	12.25	12.29	12.34	12.38	12.42	12.46	12.50	12.54	12.58
310	12.63	12.67	12.71	12.75	12.79	12.83	12.88	12.92	12.96	13.00
320	13.04	13.08	13.12	13.17	13.21	13.25	13.29	13.33	13.37	13.42
330	13.46	13.50	13.54	13.58	13.62	13.67	13.71	13.75	13.79	13.83
340	13.88	13.92	13.96	14.00	14.04	14.09	14.13	14.17	14.21	14.25
350	14.29	14.34	14.38	14.42	14.46	14.50	14.55	14.59	14.63	14.67
360	14.71	14.76	14.80	14.84	14.88	14.92	14.97	15.01	15.05	15.09
370	15.13	15.18	15.22	15.26	15.30	15.34	15.39	15.43	15.47	15.51
380	15.55	15.60	15.64	15.68	15.72	15.76	15.81	15.85	15.89	15.93
390	15.98	16.02	16.06	16.10	16.14	16.19	16.23	16.27	16.31	16.36
400	16.40	16.44	16.48	16.52	16.57	16.61	16.65	16.69	16.74	16.78
410	16.82	16.86	16.91	16.95	16.99	17.03	17.07	17.12	17.16	17.20
420	17.24	17.29	17.33	17.37	17.41	17.46	17.50	17.54	17.58	17.62
430	17.67	17.71	17.75	17.79	17.84	17.88	17.92	17.96	18.01	18.05
440	18.09	18.13	18.17	18.22	18.26	18.30	18.34	18.39	18.43	18.47
450	18.51	18.56	18.60	18.64	18.68	18.73	18.77	18.81	18.85	18.90
460	18.94	18.98	19.02	19.07	19.11	19.15	19.19	19.24	19.28	19.32
470	19.36	19.41	19.45	19.49	19.54	19.58	19.62	19.66	19.71	19.75
480	19.79	19.84	19.88	19.92	19.96	20.01	20.05	20.09	20.13	20.18
490	20.22	20.26	20.31	20.35	20.39	20.43	20.48	20.52	20.56	20.60
500	20.65	20.69	20.73	20.77	20.82	20.86	20.90	20.94	20.99	21.03
510	21.07	21.11	21.16	21.20	21.24	21.28	21.32	21.37	21.41	21.45
520	21.50	21.54	21.58	21.63	21.67	21.71	21.75	21.80	21.84	21.88
530	21.92	21.97	22.01	22.05	22.09	22.14	22.18	22.22	22.26	22.31
540	22.35	22.39	22.43	22.48	22.52	22.56	22.61	22.65	22.69	22.73
550	22.78	22.82	22.86	22.90	22.95	22.99	23.03	23.07	23.12	23.16

* Based on the International Temperature Scale of 1948.

CALIBRATION TABLES FOR THERMOCOUPLES (Continued)

CHROMEL-ALUMEL THERMOCOUPLES

Electromotive Force in Absolute Millivolts. Temperatures in Degrees C (Int. 1948). Reference Junctions at 0° C.)

°C	0	1	2	3	4	5	6	7	8	9
					Millivolts					
560	23.20	23.25	23.29	23.33	23.38	23.42	23.46	23.50	23.54	23.59
570	23.63	23.67	23.72	23.76	23.80	23.84	23.89	23.93	23.97	24.01
580	24.06	24.10	24.14	24.18	24.23	24.27	24.31	24.36	24.40	24.44
590	24.49	24.53	24.57	24.61	24.65	24.70	24.74	24.78	24.83	24.87
600	24.91	24.95	25.00	25.04	25.08	25.12	25.17	25.21	25.25	25.29
610	25.34	25.38	25.42	25.47	25.51	25.55	25.59	25.64	25.68	25.72
620	25.76	25.81	25.85	25.89	25.93	25.98	26.02	26.06	26.10	26.15
630	26.19	26.23	26.27	26.32	26.36	26.40	26.44	26.48	26.53	26.57
640	26.61	26.65	26.70	26.74	26.78	26.82	26.86	26.91	26.95	26.99
650	27.03	27.07	27.12	27.16	27.20	27.24	27.28	27.33	27.37	27.41
660	27.45	27.49	27.54	27.58	27.62	27.66	27.71	27.75	27.79	27.83
670	27.87	27.92	27.96	28.00	28.04	28.08	28.13	28.17	28.21	28.25
680	28.29	28.34	28.38	28.42	28.46	28.50	28.55	28.59	28.63	29.67
690	28.72	28.76	28.80	28.84	28.88	28.93	28.97	29.01	29.05	29.10
700	29.14	29.18	29.22	29.26	29.30	29.35	29.39	29.43	29.47	29.52
710	29.56	29.60	29.64	29.68	29.72	29.77	29.81	29.85	29.89	29.93
720	29.97	30.02	30.06	30.10	30.14	30.18	30.23	30.27	30.31	30.35
730	30.39	30.44	30.48	30.52	30.56	30.60	30.65	30.69	30.73	30.77
740	30.81	30.85	30.90	30.94	30.98	31.02	31.06	31.10	31.15	31.19
750	31.23	31.27	31.31	31.35	31.40	31.44	31.48	31.52	31.56	31.60
760	31.65	31.69	31.73	31.77	31.81	31.85	31.90	31.94	31.98	32.02
770	32.06	32.10	32.15	32.19	32.23	32.27	32.31	32.35	32.39	32.43
780	32.48	32.52	32.56	32.60	32.64	32.68	32.72	32.76	32.81	32.85
790	32.89	32 93	32.97	33.01	33.05	33.09	33.13	33.18	33.22	33.26
800	33.30	33.34	33.38	33.42	33.46	33.50	33.54	33.59	33.63	33.67
810	33.71	33.75	33.79	33.83	33.87	33.91	33.95	33.99	34.04	34.08
820	34.12	34.16	34.20	34.24	34.28	34.32	34.36	34.40	34.44	34.48
830	34.53	34.57	34.61	34.65	34.69	34.73	34.77	34.81	34.85	34.89
840	34.93	34.97	35.02	35.06	35.10	35.14	35.18	35.22	35.26	35.30
850	35.34	35.38	35.42	35.46	35.50	35.54	35.58	35.63	35.67	35.71
860	35.75	35.79	35.83	35.87	35.91	35.95	35.99	36.03	36.07	36.11
870	36.15	36.19	36.23	36.27	36.31	36.35	36.39	36.43	36.47	36.51
880	36.55	36.59	36.63	36.67	36.72	36.76	36.80	36.84	36.88	36.92
890	36.96	37.00	37.04	37.08	37.12	37.16	37.20	37.24	37.28	37.32
900	37.36	37.40	37.44	37.48	37.52	37.56	37.60	37.64	37.68	37.72
910	37.76	37.80	37.84	37.88	37.92	37.96	38.00	38.04	38.08	38.12
920	38.16	38.20	38.24	38.28	38.32	38.36	38.40	38.44	38.48	38.52
930	38.56	38.60	38.64	38.68	38.72	38.76	38.80	38.84	38.88	38.92
940	38.95	38.99	39.03	39.07	39.11	39.15	39.19	39.23	39.27	39.31
950	39.35	39.39	39.43	39.47	39.51	39.55	39.59	39.63	39.67	39.71
960	39.75	39.79	39.83	39.86	39.90	39.94	39.98	40.02	40.06	40.10
970	40.14	40.18	40.22	40.26	40.30	40.34	40.38	40.41	40.45	40.49
980	40.53	40.57	40.61	40.65	40.69	40.73	40.77	40.81	40.85	40.89
990	40.92	40.96	41.00	41.04	41.08	41.12	41.16	41 20	41.24	41.28
1000	41.31	41.35	41.39	41.43	41.47	41.51	41.55	41.59	41.63	41.67
1010	41.70	41.74	41.78	41.82	41.86	41.90	41.94	41.98	42.02	42.05
1020	42.09	42.13	42.17	42.21	42.25	42.29	42.33	42.36	42.40	42.44
1030	42.48	42.52	42.56	42.60	42.63	42.67	42.71	42.75	42.79	42.83
1040	42.87	42.90	42.94	42.98	43.02	43.06	43.10	43.14	43.17	43.21
1050	43.25	43.29	43.33	43.37	43.41	43.44	43.48	43.52	43.56	43.60
1060	43.63	43.67	43.71	43.75	43.79	43.83	43.87	43.90	43.94	43.98
1070	44.02	44.06	44.10	44.13	44.17	44.21	44.25	44.29	44.33	44.36
1080	44.40	44.44	44.48	44.52	44.55	44.59	44.63	44.67	44.71	44.74
1090	44.78	44.82	44.86	44.90	44.93	44.97	45.01	45.05	45.09	45.12
1100	45.16	45.20	45.24	45.27	45.31	45.35	45.39	45.43	45.46	45.50
1110	45.54	45.58	45.62	45.65	45.69	45.73	45.77	45.80	45.84	45.88
1120	45.92	45.96	45.99	46.03	46.07	46.11	46.14	46.18	46.22	46.26
1130	46.29	46.33	46.37	46.41	46.44	46.48	46.52	46.56	46.59	46.63
1140	46.67	46.70	46.74	46.78	46.82	46.85	46.89	46.93	46.97	47.00
1150	47.04	47.08	47.12	47.15	47.19	47.23	47.26	47.30	47.34	47.38
1160	47.41	47.45	47.49	47.52	47.56	47.60	47.63	47.67	47.71	47.75
1170	47.78	47.82	47.86	47.89	47.93	47.97	48.00	48.04	48.08	48.12
1180	48.15	48.19	48.23	48.26	48.30	48.34	48.37	48.41	48.45	48.48
1190	48.52	48.56	48.59	48.63	48.67	48.70	48.74	48.78	48.81	48.85
1200	48.89	48.92	48.96	49.00	49.03	49.07	49.11	49.14	49.18	49.22
1210	49.25	49.29	49.32	49.36	49.40	49.43	49.47	49.51	49.54	49.58
1220	49.62	49.65	49.69	49.72	49.76	49.80	49.83	49.87	49.90	49.94
1230	49.98	50.01	50.05	50.08	50.12	50.16	50.19	50.23	50.26	50.30
1240	50.34	50.37	50.41	50.44	50.48	50.52	50.55	50.59	50.62	50.66
1250	50.69	50.73	50.77	50.80	50.84	50.87	50.91	50.94	50.98	51.02
1260	51.05	51.09	51.12	51.16	51.19	51.23	51.27	51.30	51.34	51.37
1270	51.41	51.44	51.48	51.51	51.55	51.58	51.62	51.66	51.69	51.73
1280	51.76	51.80	51.83	51.87	51.90	51.94	51.97	52.01	52.04	52.08
1290	52.11	52.15	52.18	52.22	52.25	52.29	52.32	52.36	52.39	52.43
1300	52.46	52.50	52.53	52.57	52.60	52.64	52.67	52.71	52.74	52.78

* Based on the International Temperature Scale of 1948.

CALIBRATION TABLES FOR THERMOCOUPLES (Continued)

CHROMEL-ALUMEL THERMOCOUPLES

(Electromotive Force in Absolute Millivolts. Temperatures in Degrees C (Int. 1948). Reference Junctions at 0° C.)

°C	0	1	2	3	4	5	6	7	8	9
					Millivolts					
1300	52.46	52.50	52.53	52.57	52.60	52.64	52.67	52.71	52.74	52.78
1310	52.81	52.85	52.88	52.92	52.95	52.99	53.02	53.06	53.09	53.13
1320	53.16	53.20	53.23	53.27	53.30	53.34	53.37	53.41	53.44	53.47
1330	53.51	53.54	53.58	53.61	53.65	53.68	53.72	53.75	53.79	53.82
1340	53.85	53.89	53.92	53.96	53.99	54.03	54.06	54.10	54.13	54.16
1350	54.20	54.23	54.27	54.30	54.34	54.37	54.40	54.44	54.47	54.51
1360	54.54	54.57	54.61	54.64	54.68	54.71	54.74	54.78	54.81	54.85
1370	54.88	54.91	....	....	....	....	....	....	....	.....

CHROMEL-ALUMEL THERMOCOUPLES

(Electromotive Force in Absolute Millivolts. Temperatures in Degrees F.* Reference Junctions at 32° F.)

°F	0	1	2	3	4	5	6	7	8	9
					Millivolts					
−300	−5.51	−5.52	−5.53	−5.54	−5.54	−5.55	−5.56	−5.57	−5.58	−5.59
−290	−5.41	−5.42	−5.43	−5.44	−5.45	−5.46	−5.47	−5.48	−5.49	−5.50
−280	−5.30	−5.31	−5.32	−5.34	−5.35	−5.36	−5.37	−5.38	−5.39	−5.40
−270	−5.20	−5.21	−5.22	−5.23	−5.24	−5.25	−5.26	−5.27	−5.28	−5.29
−260	−5.08	−5.09	−5.10	−5.12	−5.13	−5.14	−5.15	−5.16	−5.17	−5.18
−250	−4.96	−4.97	−4.99	−5.00	−5.01	−5.02	−5.03	−5.04	−5.06	−5.07
−240	−4.84	−4.85	−4.86	−4.88	−4.89	−4.90	−4.91	−4.92	−4.94	−4.95
−230	−4.71	−4.72	−4.74	−4.75	−4.76	−4.77	−4.79	−4.80	−4.81	−4.82
−220	−4.58	−4.59	−4.60	−4.62	−4.63	−4.64	−4.66	−4.67	−4.68	−4.70
−210	−4.44	−4.45	−4.46	−4.48	−4.49	−4.51	−4.52	−4.53	−4.55	−4.56
−200	−4.29	−4.31	−4.32	−4.34	−4.35	−4.36	−4.38	−4.39	−4.41	−4.42
−190	−4.15	−4.16	−4.18	−4.19	−4.21	−4.22	−4.24	−4.25	−4.26	−4.28
−180	−4.00	−4.01	−4.03	−4.04	−4.06	−4.07	−4.09	−4.10	−4.12	−4.13
−170	−3.84	−3.86	−3.88	−3.89	−3.91	−3.92	−3.94	−3.95	−3.97	−3.98
−160	−3.69	−3.70	−3.72	−3.73	−3.75	−3.76	−3.78	−3.80	−3.81	−3.83
−150	−3.52	−3.54	−3.56	−3.57	−3.59	−3.60	−3.62	−3.64	−3.65	−3.67
−140	−3.36	−3.38	−3.39	−3.41	−3.42	−3.44	−3.46	−3.47	−3.49	−3.51
−130	−3.19	−3.20	−3.22	−3.24	−3.25	−3.27	−3.29	−3.31	−3.32	−3.34
−120	−3.01	−3.03	−3.05	−3.06	−3.08	−3.10	−3.12	−3.13	−3.15	−3.17
−110	−2.83	−2.85	−2.87	−2.89	−2.90	−2.92	−2.94	−2.96	−2.98	−2.99
−100	−2.65	−2.67	−2.69	−2.71	−2.72	−2.74	−2.76	−2.78	−2.80	−2.82
−90	−2.47	−2.49	−2.50	−2.52	−2.54	−2.56	−2.58	−2.60	−2.62	−2.63
−80	−2.28	−2.30	−2.32	−2.34	−2.36	−2.37	−2.39	−2.41	−2.43	−2.45
−70	−2.09	−2.11	−2.13	−2.15	−2.17	−2.18	−2.20	−2.22	−2.24	−2.26
−60	−1.90	−1.92	−1.94	−1.96	−1.97	−1.99	−2.01	−2.03	−2.05	−2.07
−50	−1.70	−1.72	−1.74	−1.76	−1.78	−1.80	−1.82	−1.84	−1.86	−1.88
−40	−1.50	−1.52	−1.54	−1.56	−1.58	−1.60	−1.62	−1.64	−1.66	−1.68
−30	−1.30	−1.32	−1.34	−1.36	−1.38	−1.40	−1.42	−1.44	−1.46	−1.48
−20	−1.10	−1.12	−1.14	−1.16	−1.18	−1.20	−1.22	−1.24	−1.26	−1.28
−10	−0.89	−0.91	−0.93	−0.95	−0.97	−0.99	−1.01	−1.03	−1.06	−1.08
(−)0	−0.68	−0.70	−0.72	−0.75	−0.77	−0.79	−0.81	−0.83	−0.85	−0.87
(+)0	−0.68	−0.66	−0.64	−0.62	−0.60	−0.58	−0.56	−0.54	−0.52	−0.49
10	−0.47	−0.45	−0.43	−0.41	−0.39	−0.37	−0.34	−0.32	−0.30	−0.28
20	−0.26	−0.24	−0.22	−0.19	−0.17	−0.15	−0.13	−0.11	−0.09	−0.07
30	−0.04	−0.02	0.00	0.02	0.04	0.07	0.09	0.11	0.13	0.15
40	0.18	0.20	0.22	0.24	0.26	0.29	0.31	0.33	0.35	0.37
50	0.40	0.42	0.44	0.46	0.48	0.51	0.53	0.55	0.57	0.60
60	0.62	0.64	0.66	0.68	0.71	0.73	0.75	0.77	0.80	0.82
70	0.84	0.86	0.88	0.91	0.93	0.95	0.97	1.00	1.02	1.04
80	1.06	1.09	1.11	1.13	1.15	1.18	1.20	1.22	1.24	1.27
90	1.29	1.31	1.33	1.36	1.33	1.40	1.43	1.45	1.47	1.49
100	1.52	1.54	1.56	1.58	1.61	1.63	1.65	1.68	1.70	1.72
110	1.74	1.77	1.79	1.81	1.84	1.86	1.88	1.90	1.93	1.95
120	1.97	2.00	2.02	2.04	2.06	2.09	2.11	2.13	2.16	2.18
130	2.20	2.23	2.25	2.27	2.29	2.32	2.34	2.36	2.39	2.41
140	2.43	2.46	2.48	2.50	2.52	2.55	2.57	2.59	2.62	2.64
150	2.66	2.69	2.71	2.73	2.75	2.78	2.80	2.82	2.85	2.87
160	2.89	2.92	2.94	2.96	2.98	3.01	3.03	3.05	3.08	3.10
170	3.12	3.15	3.17	3.19	3.22	3.24	3.26	3.29	3.31	3.33
180	3.36	3.38	3.40	3.43	3.45	3.47	3.49	3.52	3.54	3.56
190	3.59	3.61	3.63	3.66	3.68	3.70	3.73	3.75	3.77	3.80
200	3.82	3.84	3.87	3.89	3.91	3.94	3.96	3.98	4.01	4.03
210	4.05	4.08	4.10	4.12	4.15	4.17	4.19	4.21	4.24	4.26
220	4.28	4.31	4.33	4.35	4.38	4.40	4.42	4.44	4.47	4.49
230	4.51	4.54	4.56	4.58	4.61	4.63	4.65	4.67	4.70	4.72
240	4.74	4.77	4.79	4.81	4.83	4.86	4.88	4.90	4.92	4.95
250	4.97	4.99	5.02	5.04	5.06	5.08	5.11	5.13	5.15	5.17
260	5.20	5.22	5.24	5.26	5.29	5.31	5.33	5.35	5.38	5.40
270	5.42	5.44	5.47	5.49	5.51	5.53	5.56	5.58	5.60	5.62
280	5.65	5.67	5.69	5.71	5.73	5.76	5.78	5.80	5.82	5.85
290	5.87	5.89	5.91	5.93	5.96	5.98	6.00	6.02	6.05	6.07
300	6.09	6.11	6.13	6.16	6.18	6.20	6.22	6.25	6.27	6.29
310	6.31	6.33	6.36	6.38	6.40	6.42	6.45	6.47	6.49	6.51
320	6.53	6.56	6.58	6.60	6.62	6.65	6.67	6.69	6.71	6.73
330	6.76	6.78	6.80	6.82	6.84	6.87	6.89	6.91	6.93	6.96
340	6.98	7.00	7.02	7.04	7.07	7.09	7.11	7.13	7.15	7.18
350	7.20	7.22	7.24	7.26	7.29	7.31	7.33	7.35	7.38	7.40

* Based on the International Temperature Scale of 1948.

CALIBRATION TABLES FOR THERMOCOUPLES (Continued)
CHROMEL-ALUMEL THERMOCOUPLES

(Electromotive Force in Absolute Millivolts. Temperatures in Degrees F.* Reference Junctions at 32° F.)

°F	0	1	2	3	4	5	6	7	8	9
	Millivolts									
360	7.42	7.44	7.46	7.49	7.51	7.53	7.55	7.58	7.60	7.62
370	7.64	7.66	7.69	7.71	7.73	7.75	7.78	7.80	7.82	7.84
380	7.87	7.89	7.91	7.93	7.95	7.98	8.00	8.02	8.04	8.07
390	8.09	8.11	8.13	8.16	8.18	8.20	8.22	8.24	8.27	8.29
400	8.31	8.33	8.36	8.38	8.40	8.42	8.45	8.47	8.49	8.51
410	8.54	8.56	8.58	8.60	8.62	8.65	8.67	8.69	8.71	8.74
420	8.76	8.78	8.80	8.82	8.85	8.87	8.89	8.91	8.94	8.96
430	8.98	9.00	9.03	9.05	9.07	9.09	9.12	9.14	9.16	9.18
440	9.21	9.23	9.25	9.27	9.30	9.32	9.34	9.36	9.39	9.41
450	9.43	9.45	9.48	9.50	9.52	9.54	9.57	9.59	9.61	9.63
460	9.66	9.68	9.70	9.73	9.75	9.77	9.79	9.82	9.84	9.86
470	9.88	9.91	9.93	9.95	9.97	10.00	10.02	10.04	10.06	10.09
480	10.11	10.13	10.16	10.18	10.20	10.22	10.25	10.27	10.29	10.31
490	10.34	10.36	10.38	10.40	10.43	10.45	10.47	10.50	10.52	10.54
500	10.57	10.59	10.61	10.63	10.66	10.68	10.70	10.72	10.75	10.77
510	10.79	10.82	10.84	10.86	10.88	10.91	10.93	10.95	10.98	11.00
520	11.02	11.04	11.07	11.09	11.11	11.13	11.16	11.18	11.20	11.23
530	11.25	11.27	11.29	11.32	11.34	11.36	11.39	11.41	11.43	11.45
540	11.48	11.50	11.52	11.55	11.57	11.59	11.61	11.64	11.66	11.68
550	11.71	11.73	11.75	11.78	11.80	11.82	11.84	11.87	11.89	11.91
560	11.94	11.96	11.98	12.01	12.03	12.05	12.07	12.10	12.12	12.14
570	12.17	12.19	12.21	12.24	12.26	12.28	12.30	12.33	12.35	12.37
580	12.40	12.42	12.44	12.47	12.49	12.51	12.53	12.56	12.58	12.60
590	12.63	12.65	12.67	12.70	12.72	12.74	12.76	12.79	12.81	12.83
600	12.86	12.88	12.90	12.93	12.95	12.97	13.00	13.02	13.04	13.06
610	13.09	13.11	13.13	13.16	13.18	13.20	13.23	13.25	13.27	13.30
620	13.32	13.34	13.36	13.39	13.41	13.44	13.46	13.48	13.50	13.53
630	13.55	13.57	13.60	13.62	13.64	13.67	13.69	13.71	13.74	13.76
640	13.78	13.81	13.83	13.85	13.88	13.90	13.92	13.95	13.97	13.99
650	14.02	14.04	14.06	14.09	14.11	14.13	14.15	14.18	14.20	14.22
660	14.25	14.27	14.29	14.32	14.34	14.36	14.39	14.41	14.43	14.46
670	14.48	14.50	14.53	14.55	14.57	14.60	14.62	14.64	14.67	14.69
680	14.71	14.74	14.76	14.78	14.81	14.83	14.85	14.88	14.90	14.92
690	14.95	14.97	14.99	15.02	15.04	15.06	15.09	15.11	15.13	15.16
700	15.18	15.20	15.23	15.25	15.27	15.30	15.32	15.34	15.37	15.39
710	15.41	15.44	15.46	15.48	15.51	15.53	15.55	15.58	15.60	15.62
720	15.65	15.67	15.69	15.72	15.74	15.76	15.79	15.81	15.83	15.86
730	15.88	15.90	15.93	15.95	15.98	16.00	16.02	16.05	16.07	16.09
740	16.12	16.14	16.16	16.19	16.21	16.23	16.26	16.28	16.30	16.33
750	16.35	16.37	16.40	16.42	16.45	16.47	16.49	16.52	16.54	16.56
760	16.59	16.61	16.63	16.66	16.68	16.70	16.73	16.75	16.77	16.80
770	16.82	16.84	16.87	16.89	16.92	16.94	16.96	16.99	17.01	17.03
780	17.06	17.08	17.10	17.13	17.15	17.17	17.20	17.22	17.24	17.27
790	17.29	17.31	17.34	17.36	17.39	17.41	17.43	17.46	17.48	17.50
800	17.53	17.55	17.57	17.60	17.62	17.64	17.67	17.69	17.71	17.74
810	17.76	17.78	17.81	17.83	17.86	17.88	17.90	17.93	17.95	17.97
820	18.00	18.02	18.04	18.07	18.09	18.11	18.14	18.16	18.18	18.21
830	18.23	18.25	18.28	18.30	18.33	18.35	18.37	18.40	18.42	18.44
840	18.47	18.49	18.51	18.54	18.56	18.58	18.61	18.63	18.65	18.68
850	18.70	18.73	18.75	18.77	18.80	18.82	18.84	18.87	18.89	18.91
860	18.94	18.96	18.99	19.01	19.03	19.06	19.08	19.10	19.13	19.15
870	19.18	19.20	19.22	19.25	19.27	19.29	19.32	19.34	19.36	19.39
880	19.41	19.44	19.46	19.48	19.51	19.53	19.55	19.58	19.60	19.63
890	19.65	19.67	19.70	19.72	19.75	19.77	19.79	19.82	19.84	19.86
900	19.89	19.91	19.94	19.96	19.98	20.01	20.03	20.05	20.08	20.10
910	20.13	20.15	20.17	20.20	20.22	20.24	20.27	20.29	20.32	20.34
920	20.36	20.39	20.41	20.43	20.46	20.48	20.50	20.53	20.55	20.58
930	20.60	20.62	20.65	20.67	20.69	20.72	20.74	20.76	20.79	20.81
940	20.84	20.86	20.88	20.91	20.93	20.95	20.98	21.00	21.03	21.05
950	21.07	21.10	21.12	21.14	21.17	21.19	21.21	21.24	21.26	21.28
960	21.31	21.33	21.36	21.38	21.40	21.43	21.45	21.47	21.50	21.52
970	21.54	21.57	21.59	21.62	21.64	21.66	21.69	21.71	21.73	21.76
980	21.78	21.81	21.83	21.85	21.88	21.90	21.92	21.95	21.97	21.99
990	22.02	22.04	22.07	22.09	22.11	22.14	22.16	22.18	22.21	22.23
1000	22.26	22.28	22.30	22.33	22.35	22.37	22.40	22.42	22.44	22.47
1010	22.49	22.52	22.54	22.56	22.59	22.61	22.63	22.66	22.68	22.71
1020	22.73	22.75	22.78	22.80	22.82	22.85	22.87	22.90	22.92	22.94
1030	22.97	22.99	23.01	23.04	23.06	23.08	23.11	23.13	23.16	23.18
1040	23.20	23.23	23.25	23.27	23.30	23.32	23.35	23.37	23.39	23.42
1050	23.44	23.46	23.49	23.51	23.54	23.56	23.58	23.61	23.63	23.65
1060	23.68	23.70	23.72	23.75	23.77	23.80	23.82	23.84	23.87	23.89
1070	23.91	23.94	23.96	23.99	24.01	24.03	24.06	24.08	24.10	24.13
1080	24.15	24.18	24.20	24.22	24.25	24.27	24.29	24.32	24.34	24.36
1090	24.39	24.41	24.44	24.46	24.49	24.51	24.53	24.55	24.58	24.60
1100	24.63	24.65	24.67	24.70	24.72	24.74	24.77	24.79	24.82	24.84

* Based on the International Temperature Scale of 1948.

CALIBRATION TABLES FOR THERMOCOUPLES (Continued)
CHROMEL-ALUMEL THERMOCOUPLES

(Electromotive Force in Absolute Millivolts. Temperatures in Degrees F.* Reference Junctions at 32° F.)

°F	0	1	2	3	4	5	6	7	8	9
	Millivolts									
1100	24.63	24.65	24.67	24.70	24.72	24.74	24.77	24.79	24.82	24.84
1110	24.86	24.89	24.91	24.93	24.96	24.98	25.01	25.03	25.05	25.08
1120	25.10	25.12	25.15	25.17	25.20	25.22	25.24	25.27	25.29	25.31
1130	25.34	25.36	25.38	25.41	25.43	25.46	25.48	25.50	25.53	25.55
1140	25.57	25.60	25.62	25.65	25.67	25.69	25.72	25.74	25.76	25.79
1150	25.81	25.83	25.86	25.88	25.91	25.93	25.95	25.98	26.00	26.02
1160	26.05	26.07	26.09	26.12	26.14	26.16	26.19	26.21	26.24	26.26
1170	26.28	26.31	26.33	26.35	26.38	26.40	26.42	26.45	26.47	26.49
1180	26.52	26.54	26.56	26.59	26.61	26.63	26.66	26.68	26.70	26.73
1190	26.75	26.77	26.80	26.82	26.85	26.87	26.89	26.91	26.94	26.96
1200	26.98	27.01	27.03	27.06	27.08	27.10	27.12	27.15	27.17	27.20
1210	27.22	27.24	27.27	27.29	27.31	27.34	27.36	27.38	27.40	27.43
1220	27.45	27.48	27.50	27.52	27.55	27.57	27.59	27.62	27.64	27.66
1230	27.69	27.71	27.73	27.76	27.78	27.80	27.83	27.85	27.87	27.90
1240	27.92	27.94	27.97	27.99	28.01	28.04	28.06	28.08	28.11	28.13
1250	28.15	28.18	28.20	28.22	28.25	28.27	28.29	28.32	28.34	28.37
1260	28.39	28.41	28.44	28.46	28.48	28.50	28.53	28.55	28.58	28.60
1270	28.62	28.65	28.67	28.69	28.72	28.74	28.76	28.79	28.81	28.83
1280	28.86	28.88	28.90	28.93	28.95	28.97	29.00	29.02	29.04	29.07
1290	29.09	29.11	29.14	29.16	29.18	29.21	29.23	29.25	29.28	29.30
1300	29.32	29.35	29.37	29.39	29.42	29.44	29.46	29.49	29.51	29.53
1310	29.56	29.58	29.60	29.63	29.65	29.67	29.70	29.72	29.74	29.77
1320	29.79	29.81	29.84	29.86	29.88	29.91	29.93	29.95	29.97	30.00
1330	30.02	30.05	30.07	30.09	30.11	30.14	30.16	30.18	30.21	30.23
1340	30.25	30.28	30.30	30.32	30.35	30.37	30.39	30.42	30.44	30.46
1350	30.49	30.51	30.53	30.56	30.58	30.60	30.63	30.65	30.67	30.70
1360	30.72	30.74	30.77	30.79	30.81	30.83	30.86	30.88	30.90	30.93
1370	30.95	30.97	31.00	31.02	31.04	31.07	31.09	31.11	31.14	31.16
1380	31.18	31.21	31.23	31.25	31.28	31.30	31.32	31.34	31.37	31.39
1390	31.42	31.44	31.46	31.48	31.51	31.53	31.55	31.58	31.60	31.62
1400	31.65	31.67	31.69	31.72	31.74	31.76	31.78	31.81	31.83	31.85
1410	31.88	31.90	31.92	31.95	31.97	31.99	32.02	32.04	32.06	32.08
1420	32.11	32.13	32.15	32.18	32.20	32.22	32.25	32.27	32.29	32.31
1430	32.34	32.36	32.38	32.41	32.43	32.45	32.48	32.50	32.52	32.54
1440	32.57	32.59	32.61	32.64	32.66	32.68	32.70	32.73	32.75	32.77
1450	32.80	32.82	32.84	32.86	32.89	32.91	32.93	32.96	32.98	33.00
1460	33.02	33.05	33.07	33.09	33.12	33.14	33.16	33.18	33.21	33.23
1470	33.25	33.28	33.30	33.32	33.34	33.37	33.39	33.41	33.43	33.46
1480	33.48	33.50	33.53	33.55	33.57	33.59	33.62	33.64	33.66	33.69
1490	33.71	33.73	33.75	33.78	33.80	33.82	33.84	33.87	33.89	33.91
1500	33.93	33.96	33.98	34.00	34.03	34.05	34.07	34.09	34.12	34.14
1510	34.16	34.18	34.21	34.23	34.25	34.28	34.30	34.32	34.34	34.37
1520	34.39	34.41	34.43	34.46	34.48	34.50	34.53	34.55	34.57	34.59
1530	34.62	34.64	34.66	34.68	34.71	34.73	34.75	34.77	34.80	34.82
1540	34.84	34.87	34.89	34.91	34.93	34.96	34.98	35.00	35.02	35.05
1550	35.07	35.09	35.11	35.14	35.16	35.18	35.21	35.23	35.25	35.27
1560	35.29	35.32	35.34	35.36	35.39	35.41	35.43	35.45	35.48	35.50
1570	35.52	35.54	35.57	35.59	35.61	35.63	35.66	35.68	35.70	35.72
1580	35.75	35.77	35.79	35.81	35.84	35.86	35.88	35.90	35.93	35.95
1590	35.97	35.99	36.02	36.04	36.06	36.08	36.11	36.13	36.15	36.17
1600	36.19	36.19	36.22	36.24	36.29	36.31	36.33	36.35	36.37	36.40
1610	36.42	36.44	36.46	36.49	36.51	36.53	36.55	36.58	36.60	36.62
1620	36.64	36.67	36.69	36.71	36.73	36.76	36.78	36.80	36.82	36.84
1630	36.87	36.89	36.91	36.93	36.96	36.98	37.00	37.02	37.05	37.07
1640	37.09	37.11	37.14	37.16	37.18	37.20	37.23	37.25	37.27	37.29
1650	37.31	37.34	37.36	37.38	37.40	37.43	37.45	37.47	37.49	37.52
1660	37.54	37.56	37.58	37.60	37.63	37.65	37.67	37.69	37.72	37.74
1670	37.76	37.78	37.81	37.83	37.85	37.87	37.89	37.92	37.94	37.96
1680	37.98	38.01	38.03	38.05	38.07	38.09	38.12	38.14	38.16	38.18
1690	38.20	38.23	38.25	38.27	38.29	38.32	38.34	38.36	38.38	38.40
1700	38.43	38.45	38.47	38.49	38.51	38.54	38.56	38.58	38.60	38.62
1710	38.65	38.67	38.69	38.71	38.73	38.76	38.78	38.80	38.82	38.84
1720	38.87	38.89	38.91	38.93	38.95	38.98	39.00	39.02	39.04	39.06
1730	39.09	39.11	39.13	39.15	39.17	39.20	39.22	39.24	39.26	39.28
1740	39.31	39.33	39.35	39.37	39.39	39.42	39.44	39.46	39.48	39.50
1750	39.53	39.55	39.57	39.59	39.61	39.64	39.66	39.68	39.70	39.72
1760	39.75	39.77	39.79	39.81	39.83	39.86	39.88	39.90	39.92	39.94
1770	39.96	39.99	40.01	40.03	40.05	40.07	40.10	40.12	40.14	40.16
1780	40.18	40.20	40.23	40.25	40.27	40.29	40.31	40.34	40.36	40.38
1790	40.40	40.42	40.44	40.47	40.49	40.51	40.53	40.55	40.58	40.60
1800	40.62	40.64	40.66	40.68	40.71	40.73	40.75	40.77	40.79	40.82
1810	40.84	40.86	40.88	40.90	40.92	40.95	40.97	40.99	41.01	41.03
1820	41.05	41.08	41.10	41.12	41.14	41.16	41.18	41.21	41.23	41.25
1830	41.27	41.29	41.31	41.34	41.36	41.38	41.40	41.42	41.45	41.47
1840	41.49	41.51	41.53	41.55	41.57	41.60	41.62	41.64	41.66	41.68
1850	41.70	41.73	41.75	41.77	41.79	41.81	41.83	41.85	41.88	41.90

* Based on the International Temperature Scale of 1948.

CALIBRATION TABLES FOR THERMOCOUPLES (Continued)

CHROMEL-ALUMEL THERMOCOUPLES

(Electromotive Force in Absolute Millivolts. Temperatures in Degrees F.* Reference Junctions at 32° F.)

°F	0	1	2	3	4	5	6	7	8	9
	Millivolts									
1860	41.92	41.94	41.96	41.99	42.01	42.03	42.05	42.07	42.09	42.11
1870	42.14	42.16	42.18	42.20	42.22	42.24	42.26	42.29	42.31	42.33
1880	42.35	42.37	42.39	42.42	42.44	42.46	42.48	42.50	42.52	42.55
1890	42.57	42.59	42.61	42.63	42.65	42.67	42.69	42.72	42.74	42.76
1900	42.78	42.80	42.82	42.84	42.87	42.89	42.91	42.93	42.95	42.97
1910	42.99	43.01	43.04	43.06	43.08	43.10	43.12	43.14	43.17	43.19
1920	43.21	43.23	43.25	43.27	43.29	43.31	43.34	43.36	43.38	43.40
1930	43.42	43.44	43.47	43.49	43.51	43.53	43.55	43.57	43.59	43.61
1940	43.63	43.66	43.68	43.70	43.72	43.74	43.76	43.78	43.81	43.83
1950	43.85	43.87	43.89	43.91	43.93	43.95	43.98	44.00	44.02	44.04
1960	44.06	44.08	44.10	44.13	44.15	44.17	44.19	44.21	44.23	44.25
1970	44.27	44.30	44.32	44.34	44.36	44.38	44.40	44.42	44.44	44.47
1980	44.49	44.51	44.53	44.55	44.57	44.59	44.61	44.63	44.66	44.68
1990	44.70	44.72	44.74	44.76	44.78	44.80	44.82	44.85	44.87	44.89
2000	44.91	44.93	44.95	44.97	44.99	45.01	45.03	45.06	45.08	45.10
2010	45.12	45.14	45.16	45.18	45.20	45.22	45.24	45.27	45.29	45.31
2020	45.33	45.35	45.87	45.39	45.41	45.43	45.45	45.48	45.50	45.52
2030	45.54	45.56	45.58	45.60	45.62	45.64	45.66	45.69	45.71	45.73
2040	45.75	45.77	45.79	45.81	45.83	45.85	45.87	45.90	45.92	45.94
2050	45.96	45.98	46.00	46.02	46.04	46.06	46.08	46.11	46.13	46.15
2060	46.17	46.19	46.21	46.23	46.25	46.27	46.29	46.31	46.33	46.36
2070	46.38	46.40	46.42	46.44	46.46	46.48	46.50	46.52	46.54	46.56
2080	46.58	46.60	46.63	46.65	46.67	46.69	46.71	46.73	46.75	46.77
2090	46.79	46.81	46.83	46.85	46.87	46.90	46.92	46.94	46.96	46.98
2100	47.00	47.02	47.04	47.06	47.08	47.10	47.12	47.14	47.17	47.19
2110	47.21	47.23	47.25	47.27	47.29	47.31	47.33	47.35	47.37	47.39
2120	47.41	47.43	47.45	47.47	47.49	47.52	47.54	47.56	47.58	47.60
2130	47.62	47.64	47.66	47.68	47.70	47.72	47.74	47.76	47.78	47.80
2140	47.82	47.84	47.86	47.89	47.91	47.93	47.95	47.97	47.99	48.01
2150	48.03	48.05	48.07	48.09	48.11	48.13	48.15	48.17	48.19	48.21
2160	48.23	48.25	48.27	48.29	48.32	48.34	48.36	48.38	48.40	48.42
2170	48.44	48.46	48.48	48.50	48.52	48.54	48.56	48.58	48.60	48.62
2180	48.64	48.66	48.68	48.70	48.72	48.74	48.76	48.79	48.81	48.83
2190	48.85	48.87	48.89	48.91	48.93	48.95	48.97	48.99	49.01	49.03
2200	49.05	49.07	49.09	49.11	49.13	49.15	49.17	49.19	49.21	49.23
2210	49.25	49.27	49.29	49.31	49.33	49.35	49.37	49.39	49.41	49.43
2220	49.45	49.47	49.49	49.51	49.53	49.55	49.57	49.59	49.61	49.63
2230	49.65	49.67	49.69	49.71	49.73	49.76	49.78	49.80	49.82	49.84
2240	49.86	49.88	49.90	49.92	49.94	49.96	49.98	50.00	50.02	50.04
2250	50.06	50.08	50.10	50.12	50.14	50.16	50.18	50.20	50.22	50.24
2260	50.26	50.28	50.30	50.32	50.34	50.36	50.38	50.40	50.42	50.44
2270	50.46	50.48	50.50	50.52	50.54	50.56	50.57	50.59	50.61	50.63
2280	50.65	50.67	50.69	50.71	50.73	50.75	50.77	50.79	50.81	50.83
2290	50.85	50.87	50.89	50.91	50.93	50.95	50.97	50.99	51.01	51.03
2300	51.05	51.07	51.09	51.11	51.13	51.15	51.17	51.19	51.21	51.23
2310	51.25	51.27	51.29	51.31	51.33	51.35	51.37	51.39	51.41	51.43
2320	51.45	51.47	51.48	51.50	51.52	51.54	51.56	51.58	51.60	51.62
2330	51.64	51.66	51.68	51.70	51.72	51.74	51.76	51.78	51.80	51.82
2340	51.84	51.86	51.88	51.90	51.92	51.94	51.96	51.98	52.00	52.01
2350	52.03	52.05	52.07	52.09	52.11	52.13	52.15	52.17	52.19	52.21
2360	52.23	52.25	52.27	52.29	52.31	52.33	52.35	52.37	52.39	52.41
2370	52.42	52.44	52.46	52.48	52.50	52.52	52.54	52.56	52.58	52.60
2380	52.62	52.64	52.66	52.68	52.70	52.72	52.74	52.76	52.77	52.79
2390	52.81	52.83	52.85	52.87	52.89	52.91	52.93	52.95	52.97	52.99
2400	53.01	53.03	53.05	53.07	53.08	53.10	53.12	53.14	53.16	53.18
2410	53.20	53.22	53.24	53.26	53.28	53.30	53.32	53.34	53.35	53.37
2420	53.39	53.41	53.43	53.45	53.47	53.49	53.51	53.53	53.55	53.57
2430	53.59	53.60	53.62	53.64	53.66	53.68	53.70	53.72	53.74	53.76
2440	53.78	53.80	53.82	53.83	53.85	53.87	53.89	53.91	53.93	53.95
2450	53.97	53.99	54.01	54.03	54.04	54.06	54.08	54.10	54.12	54.14
2460	54.16	54.18	54.20	54.22	54.24	54.25	54.27	54.29	54.31	54.33
2470	54.35	54.37	54.39	54.41	54.43	54.44	54.46	54.48	54.50	54.52
2480	54.54	54.56	54.58	54.60	54.62	54.63	54.65	54.67	54.69	54.71
2490	54.73	54.75	54.77	54.79	54.81	54.82	54.84	54.86	54.88	54.90
2500	54.92	....	....	....	....	....	....	....	....	....

* Based on the International Temperature Scale of 1948.

CALIBRATION TABLES FOR THERMOCOUPLES (Continued)

IRON-CONSTANTAN THERMOCOUPLES (MODIFIED 1913)

(Electromotive Force in Absolute Millivolts. Temperatures in Degrees C (Int. 1948). Reference Junctions at 0° C.)

°C	0	1	2	3	4	5	6	7	8	9
	Millivolts									
−190	−7.66	−7.69	−7.71	−7.73	−7.76	−7.78				
−180	−7.40	−7.43	−7.46	−7.49	−7.51	−7.54	−7.56	−7.59	−7.61	−7.64
−170	−7.12	−7.15	−7.18	−7.21	−7.24	−7.27	−7.30	−7.32	−7.35	−7.38
−160	−6.82	−6.85	−6.88	−6.91	−6.94	−6.97	−7.00	−7.03	−7.06	−7.09
−150	−6.50	−6.53	−6.56	−6.60	−6.63	−6.66	−6.69	−6.72	−6.76	−6.79
−140	−6.16	−6.19	−6.22	−6.26	−6.29	−6.33	−6.36	−6.40	−6.43	−6.46
−130	−5.80	−5.84	−5.87	−5.91	−5.94	−5.98	−6.01	−6.05	−6.08	−6.12
−120	−5.42	−5.46	−5.50	−5.54	−5.58	−5.61	−5.65	−5.69	−5.72	−5.76
−110	−5.03	−5.07	−5.11	−5.15	−5.19	−5.23	−5.27	−5.31	−5.35	−5.38
−100	−4.63	−4.67	−4.71	−4.75	−4.79	−4.83	−4.87	−4.91	−4.95	−4.99
−90	−4.21	−4.25	−4.30	−4.34	−4.38	−4.42	−4.46	−4.50	−4.55	−4.59
−80	−3.78	−3.82	−3.87	−3.91	−3.96	−4.00	−4.04	−4.08	−4.13	−4.17
−70	−3.34	−3.38	−3.43	−3.47	−3.52	−3.56	−3.60	−3.65	−3.69	−3.74
−60	−2.89	−2.94	−2.98	−3.03	−3.07	−3.12	−3.16	−3.21	−3.25	−3.30
−50	−2.43	−2.48	−2.52	−2.57	−2.62	−2.66	−2.71	−2.75	−2.80	−2.84
−40	−1.96	−2.01	−2.06	−2.10	−2.15	−2.20	−2.24	−2.29	−2.34	−2.38
−30	−1.48	−1.53	−1.58	−1.63	−1.67	−1.72	−1.77	−1.82	−1.87	−1.91
−20	−1.00	−1.04	−1.09	−1.14	−1.19	−1.24	−1.29	−1.34	−1.39	−1.43
−10	−0.50	−0.55	−0.60	−0.65	−0.70	−0.75	−0.80	−0.85	−0.90	−0.95
(−)0	0.00	−0.05	−0.10	−0.15	−0.20	−0.25	−0.30	−0.35	−0.40	−0.45
(+)0	0.00	0.05	0.10	0.15	0.20	0.25	0.30	0.35	0.40	0.45
10	0.50	0.56	0.61	0.66	0.71	0.76	0.81	0.86	0.91	0.97
20	1.02	1.07	1.12	1.17	1.22	1.28	1.33	1.38	1.43	1.48
30	1.54	1.59	1.64	1.69	1.74	1.80	1.85	1.90	1.95	2.00
40	2.06	2.11	2.16	2.22	2.27	2.32	2.37	2.42	2.48	2.53
50	2.58	2.64	2.69	2.74	2.80	2.85	2.90	2.96	3.01	3.06
60	3.11	3.17	3.22	3.27	3.33	3.38	3.43	3.49	3.54	3.60
70	3.65	3.70	3.76	3.81	3.86	3.92	3.97	4.02	4.08	4.13
80	4.19	4.24	4.29	4.35	4.40	4.46	4.51	4.56	4.62	4.67
90	4.73	4.78	4.83	4.89	4.94	5.00	5.05	5.10	5.16	5.21
100	5.27	5.32	5.38	5.43	5.48	5.54	5.59	5.65	5.70	5.76
110	5.81	5.86	5.92	5.97	6.03	6.08	6.14	6.19	6.25	6.30
120	6.36	6.41	6.47	6.52	6.58	6.63	6.68	6.74	6.79	6.85
130	6.90	6.96	7.01	7.07	7.12	7.18	7.23	7.29	7.34	7.40
140	7.45	7.51	7.56	7.62	7.67	7.73	7.78	7.84	7.89	7.95
150	8.00	8.06	8.12	8.17	8.23	8.28	8.34	8.39	8.45	8.50
160	8.56	8.61	8.67	8.72	8.78	8.84	8.89	8.95	9.00	9.06
170	9.11	9.17	9.22	9.28	9.33	9.39	9.44	9.50	9.56	9.61
180	9.67	9.72	9.78	9.83	9.89	9.95	10.00	10.06	10.11	10.17
190	10.22	10.28	10.34	10.39	10.45	10.50	10.56	10.61	10.67	10.72
200	10.78	10.84	10.89	10.95	11.00	11.06	11.12	11.17	11.23	11.28
210	11.34	11.39	11.45	11.50	11.56	11.62	11.67	11.73	11.78	11.84
220	11.89	11.95	12.00	12.06	12.12	12.17	12.23	12.28	12.34	12.39
230	12.45	12.50	12.56	12.62	12.67	12.73	12.78	12.84	12.89	12.95
240	13.01	13.06	13.12	13.17	13.23	13.28	13.34	13.40	13.45	13.51
250	13.56	13.62	13.67	13.73	13.78	13.84	13.89	13.95	14.00	14.06
260	14.12	14.17	14.23	14.28	14.34	14.39	14.45	14.50	14.56	14.61
270	14.67	14.72	14.78	14.83	14.89	14.94	15.00	15.06	15.11	15.17
280	15.22	15.28	15.33	15.39	15.44	15.50	15.55	15.61	15.66	15.72
290	15.77	15.83	15.88	15.94	16.00	16.05	16.11	16.16	16.22	16.27
300	16.33	16.38	16.44	16.49	16.55	16.60	16.66	16.71	16.77	16.82
310	16.88	16.93	16.99	17.04	17.10	17.15	17.21	17.26	17.32	17.37
320	17.43	17.48	17.54	17.60	17.65	17.71	17.76	17.82	17.87	17.93
330	17.98	18.04	18.09	18.15	18.20	18.26	18.32	18.37	18.43	18.48
340	18.54	18.59	18.65	18.70	18.76	18.81	18.87	18.92	18.98	19.03
350	19.09	19.14	19.20	19.26	19.31	19.37	19.42	19.48	19.53	19.59
360	19.64	19.70	19.75	19.81	19.86	19.92	19.97	20.03	20.08	20.14
370	20.20	20.25	20.31	20.36	20.42	20.47	20.53	20.58	20.64	20.69
380	20.75	20.80	20.86	20.91	20.97	21.02	21.08	21.13	21.19	21.24
390	21.30	21.35	21.41	21.46	21.52	21.57	21.63	21.68	21.74	21.79
400	21.85	21.90	21.96	22.02	22.07	22.13	22.18	22.24	22.29	22.35
410	22.40	22.46	22.51	22.57	22.62	22.68	22.73	22.79	22.84	22.90
420	22.95	23.01	23.06	23.12	23.17	23.23	23.28	23.34	23.39	23.45
430	23.50	23.56	23.61	23.67	23.72	23.78	23.83	23.89	23.94	24.00
440	24.06	24.11	24.17	24.22	24.28	24.33	24.39	24.44	24.50	24.55
450	24.61	24.66	24.72	24.77	24.83	24.88	24.94	25.00	25.05	25.11
460	25.16	25.22	25.27	25.33	25.38	25.44	25.49	25.55	25.60	25.66
470	25.72	25.77	25.83	25.88	25.94	25.99	26.05	26.10	26.16	26.22
480	26.27	26.33	26.38	26.44	26.49	26.55	26.61	26.66	26.72	26.77
490	26.83	26.89	26.94	27.00	27.05	27.11	27.17	27.22	27.28	27.33
500	27.39	27.45	27.50	27.56	27.61	27.67	27.73	27.78	27.84	27.90
510	27.95	28.01	28.07	28.12	28.18	28.23	28.29	28.35	28.40	28.46
520	28.52	28.57	28.63	28.69	28.74	28.80	28.86	28.91	28.97	29.02
530	29.08	29.14	29.20	29.25	29.31	29.37	29.42	29.48	29.54	29.59
540	29.65	29.71	29.76	29.82	29.88	29.94	29.99	30.05	30.11	30.16
550	30.22	30.28	30.34	30.39	30.45	30.51	30.57	30.62	30.68	30.74

* Based on the International Temperature Scale of 1948.

CALIBRATION TABLES FOR THERMOCOUPLES (Continued)
IRON-CONSTANTAN THERMOCOUPLES (MODIFIED 1913)

(Electromotive Force in Absolute Millivolts. Temperatures in Degrees C (Int. 1948). Reference Junctions at 0° C.)

°C	0	1	2	3	4	5	6	7	8	9
	Millivolts									
560	30.80	30.85	30.91	30.97	31.02	31.08	31.14	31.20	31.26	31.31
570	31.37	31.43	31.49	31.54	31.60	31.66	31.72	31.78	31.83	31.89
580	31.95	32.01	32.06	32.12	32.18	32.24	32.30	32.36	32.41	32.47
590	32.53	32.59	32.65	32.71	32.76	32.82	32.88	32.94	33.00	33.06
600	33.11	33.17	33.23	33.29	33.35	33.41	33.46	33.52	33.58	33.64
610	33.70	33.76	33.82	33.88	33.94	33.99	34.05	34.11	34.17	34.23
620	34.29	34.35	34.41	34.47	34.53	34.58	34.64	34.70	34.76	34.82
630	34.88	34.94	35.00	35.06	35.12	35.18	35.24	35.30	35.36	35.42
640	35.48	35.54	35.60	36.66	35.72	35.78	35.84	35.90	35.96	36.02
650	36.08	36.14	36.20	36.26	36.32	36.38	36.44	36.50	36.56	36.62
660	36.69	36.75	36.81	36.87	36.93	36.99	37.05	37.11	37.18	37.24
670	37.30	37.36	37.42	37.48	37.54	37.60	37.66	37.73	37.79	37.85
680	37.91	37.97	38.04	38.10	38.16	38.22	38.28	38.34	38.41	38.47
690	38.53	38.59	38.66	38.72	38.78	38.84	38.90	38.97	39.03	39.09
700	39.15	39.22	39.28	39.34	39.40	39.47	39.53	39.59	39.65	39.72
710	39.78	39.84	39.91	39.97	40.03	40.10	40.16	40.22	40.28	40.35
720	40.41	40.48	40.54	40.60	40.66	40.73	40.79	40.86	40.92	40.98
730	41.05	41.11	41.17	41.24	41.30	41.36	41.43	41.49	41.56	41.62
740	41.68	41.75	41.81	41.87	41.94	42.00	42.07	42.13	42.19	42.26
750	42.32	42.38	42.45	42.51	42.58	42.64	42.70	42.77	42.83	42.90
760	42.96	....	....	....	....	....	....	....	....	....

(Electromotive Force in Absolute Millivolts. Temperatures in Degrees F.* Reference Junctions at 32° F.)

°F	0	1	2	3	4	5	6	7	8	9
	Millivolts									
−310	−7.66	−7.68	−7.69	−7.70	−7.71	−7.73	−7.74	−7.75	−7.76	−7.78
−300	−7.52	−7.54	−7.55	−7.57	−7.58	−7.59	−7.61	−7.62	−7.64	−7.65
−290	−7.38	−7.39	−7.40	−7.42	−7.44	−7.45	−7.46	−7.48	−7.49	−7.51
−280	−7.22	−7.24	−7.25	−7.27	−7.28	−7.30	−7.31	−7.33	−7.34	−7.36
−270	−7.06	−7.07	−7.09	−7.11	−7.12	−7.14	−7.15	−7.17	−7.19	−7.20
−260	−6.89	−6.90	−6.92	−6.94	−6.96	−6.97	−6.99	−7.01	−7.02	−7.04
−250	−6.71	−6.73	−6.75	−6.77	−6.78	−6.80	−6.82	−6.84	−6.85	−6.87
−240	−6.53	−6.55	−6.57	−6.59	−6.61	−6.62	−6.64	−6.66	−6.68	−6.70
−230	−6.35	−6.37	−6.38	−6.40	−6.42	−6.44	−6.46	−6.48	−6.50	−6.52
−220	−6.16	−6.18	−6.19	−6.21	−6.23	−6.25	−6.27	−6.29	−6.31	−6.33
−210	−5.96	−5.98	−6.00	−6.02	−6.04	−6.06	−6.08	−6.10	−6.12	−6.14
−200	−5.76	−5.78	−5.80	−5.82	−5.84	−5.86	−5.88	−5.90	−5.92	−5.94
−190	−5.55	−5.57	−5.59	−5.61	−5.63	−5.65	−5.67	−5.70	−5.72	−5.74
−180	−5.34	−5.36	−5.38	−5.40	−5.42	−5.44	−5.46	−5.49	−5.51	−5.53
−170	−5.12	−5.14	−5.16	−5.19	−5.21	−5.23	−5.25	−5.27	−5.30	−5.32
−160	−4.90	−4.92	−4.94	−4.97	−4.99	−5.01	−5.03	−5.06	−5.08	−5.10
−150	−4.68	−4.70	−4.72	−4.74	−4.76	−4.79	−4.81	−4.83	−4.86	−4.88
−140	−4.44	−4.47	−4.49	−4.51	−4.54	−4.56	−4.58	−4.61	−4.63	−4.65
−130	−4.21	−4.23	−4.26	−4.28	−4.30	−4.33	−4.35	−4.38	−4.40	−4.42
−120	−3.97	−4.00	−4.02	−4.04	−4.07	−4.09	−4.12	−4.14	−4.16	−4.19
−110	−3.73	−3.76	−3.78	−3.81	−3.83	−3.85	−3.88	−3.90	−3.93	−3.95
−100	−3.49	−3.51	−3.54	−3.56	−3.59	−3.61	−3.64	−3.66	−3.68	−3.71
−90	−3.24	−3.27	−3.29	−3.32	−3.34	−3.36	−3.39	−3.41	−3.44	−3.46
−80	−2.99	−3.02	−3.04	−3.07	−3.09	−3.12	−3.14	−3.17	−3.19	−3.22
−70	−2.74	−2.76	−2.79	−2.81	−2.84	−2.86	−2.89	−2.92	−2.94	−2.97
−60	−2.48	−2.51	−2.53	−2.56	−2.58	−2.61	−2.64	−2.66	−2.69	−2.71
−50	−2.22	−2.25	−2.27	−2.30	−2.33	−2.35	−2.38	−2.40	−2.43	−2.46
−40	−1.96	−1.99	−2.01	−2.04	−2.06	−2.09	−2.12	−2.14	−2.17	−2.20
−30	−1.70	−1.72	−1.75	−1.78	−1.80	−1.83	−1.86	−1.88	−1.91	−1.94
−20	−1.43	−1.46	−1.48	−1.51	−1.54	−1.56	−1.59	−1.62	−1.64	−1.67
−10	−1.16	−1.19	−1.21	−1.24	−1.27	−1.29	−1.32	−1.35	−1.38	−1.40
(−)0	−0.89	−0.91	−0.94	−0.97	−1.00	−1.02	−1.05	−1.08	−1.10	−1.13
+(0)	−0.89	−0.86	−0.83	−0.80	−0.78	−0.75	−0.72	−0.70	−0.67	−0.64
10	−0.61	−0.58	−0.56	−0.53	−0.50	−0.48	−0.45	−0.42	−0.39	−0.36
20	−0.34	−0.31	−0.28	−0.25	−0.22	−0.20	−0.17	−0.14	−0.11	−0.09
30	−0.06	−0.03	0.00	0.03	0.05	0.08	0.11	0.14	0.17	0.19
40	0.22	0.25	0.28	0.31	0.34	0.36	0.39	0.42	0.45	0.48
50	0.50	0.53	0.56	0.59	0.62	0.65	0.67	0.70	0.73	0.76
60	0.79	0.82	0.84	0.87	0.90	0.93	0.96	0.99	1.02	1.04
70	1.07	1.10	1.13	1.16	1.19	1.22	1.25	1.28	1.30	1.33
80	1.36	1.39	1.42	1.45	1.48	1.51	1.54	1.56	1.59	1.62
90	1.65	1.68	1.71	1.74	1.77	1.80	1.83	1.85	1.88	1.91
100	1.94	1.97	2.00	2.03	2.06	2.09	2.12	2.14	2.17	2.20
110	2.23	2.26	2.29	2.32	2.35	2.38	2.41	2.44	2.47	2.50
120	2.52	2.55	2.58	2.61	2.64	2.67	2.70	2.73	2.76	2.79
130	2.82	2.85	2.88	2.91	2.94	2.97	3.00	3.03	3.06	3.08
140	3.11	3.14	3.17	3.20	3.23	3.26	3.29	3.32	3.35	3.38
150	3.41	3.44	3.47	3.50	3.53	3.56	3.59	3.62	3.65	3.68
160	3.71	3.74	3.77	3.80	3.83	3.86	3.89	3.92	3.95	3.98
170	4.01	4.04	4.07	4.10	4.13	4.16	4.19	4.22	4.25	4.28
180	4.31	4.34	4.37	4.40	4.43	4.46	4.49	4.52	4.55	4.58
190	4.61	4.64	4.67	4.70	4.73	4.76	4.79	4.82	4.85	4.88
200	4.91	4.94	4.97	5.00	5.03	5.06	5.09	5.12	5.15	5.18

* Based on the International Temperature Scale of 1948.

CALIBRATION TABLES FOR THERMOCOUPLES (Continued)
IRON-CONSTANTAN THERMOCOUPLES (MODIFIED 1913)

(Electromotive Force in Absolute Millivolts. Temperatures in Degrees F.* Reference Junctions at 32° F.)

°F	0	1	2	3	4	5	6	7	8	9
	Millivolts									
200	4.91	4.94	4.97	5.00	5.03	5.06	5.09	5.12	5.15	5.18
210	5.21	5.24	5.27	5.30	5.33	5.36	5.39	5.42	5.45	5.48
220	5.51	5.54	5.57	5.60	5.63	5.66	5.69	5.72	5.75	5.78
230	5.81	5.84	5.87	5.90	5.93	5.96	5.99	6.02	6.05	6.08
240	6.11	6.14	6.17	6.20	6.24	6.27	6.30	6.33	6.36	6.39
250	6.42	6.45	6.48	6.51	6.54	6.57	6.60	6.63	6.66	6.69
260	6.72	6.75	6.78	6.81	6.84	6.87	6.90	6.93	6.96	7.00
270	7.03	7.06	7.09	7.12	7.15	7.18	7.21	7.24	7.27	7.30
280	7.33	7.36	7.39	7.42	7.45	7.48	7.51	7.54	7.58	7.61
290	7.64	7.67	7.70	7.73	7.76	7.79	7.82	7.85	7.88	7.91
300	7.94	7.97	8.00	8.04	8.07	8.10	8.13	8.16	8.19	8.22
310	8.25	8.28	8.31	8.34	8.37	8.40	8.44	8.47	8.50	8.53
320	8.56	8.59	8.62	8.65	8.68	8.71	8.74	8.77	8.80	8.84
330	8.87	8.90	8.93	8.96	8.99	9.02	9.05	9.08	9.11	9.14
340	9.17	9.20	9.24	9.27	9.30	9.33	9.36	9.39	9.42	9.45
350	9.48	9.51	9.54	9.58	9.61	9.64	9.67	9.70	9.73	9.76
360	9.79	9.82	9.85	9.88	9.92	9.95	9.98	10.01	10.04	10.07
370	10.10	10.13	10.16	10.19	10.22	10.25	10.28	10.32	10.35	10.38
380	10.41	10.44	10.47	10.50	10.53	10.56	10.60	10.63	10.66	10.69
390	10.72	10.75	10.78	10.81	10.84	10.87	10.90	10.94	10.97	11.00
400	11.03	11.06	11.09	11.12	11.15	11.18	11.21	11.24	11.28	11.31
410	11.34	11.37	11.40	11.43	11.46	11.49	11.52	11.55	11.58	11.62
420	11.65	11.68	11.71	11.74	11.77	11.80	11.83	11.86	11.89	11.92
430	11.96	11.99	12.02	12.05	12.08	12.11	12.14	12.17	12.20	12.23
440	12.26	12.30	12.33	12.36	12.39	12.42	12.45	12.48	12.51	12.54
450	12.57	12.60	12.64	12.67	12.70	12.73	12.76	12.79	12.82	12.85
460	12.88	12.91	12.94	12.98	13.01	13.04	13.07	13.10	13.13	13.16
470	13.19	13.22	13.25	13.28	13.31	13.34	13.38	13.41	13.44	13.47
480	13.50	13.53	13.56	13.59	13.62	13.65	13.68	13.72	13.75	13.78
490	13.81	13.84	13.87	13.90	13.93	13.96	13.99	14.02	14.05	14.08
500	14.12	14.15	14.18	14.21	14.24	14.27	14.30	14.33	14.36	14.39
510	14.42	14.45	14.48	14.52	14.55	14.58	14.61	14.64	14.67	14.70
520	14.73	14.76	14.79	14.82	14.85	14.88	14.91	14.94	14.98	15.01
530	15.04	15.07	15.10	15.13	15.16	15.19	15.22	15.25	15.28	15.31
540	15.34	15.37	15.40	15.44	15.47	15.50	15.53	15.56	15.59	15.62
550	15.65	15.68	15.71	15.74	15.77	15.80	15.84	15.87	15.90	15.93
560	15.96	15.99	16.02	16.05	16.08	16.11	16.14	16.17	16.20	16.23
570	16.26	16.30	16.33	16.36	16.39	16.42	16.45	16.48	16.51	16.54
580	16.57	16.60	16.63	16.66	16.69	16.72	16.75	16.78	16.82	16.85
590	16.88	16.91	16.94	16.97	17.00	17.03	17.06	17.09	17.12	17.15
600	17.18	17.21	17.24	17.28	17.31	17.34	17.37	17.40	17.43	17.46
610	17.49	17.52	17.55	17.58	17.61	17.64	17.68	17.71	17.74	17.77
620	17 80	17.83	17.86	17.89	17.92	17.95	17.98	18.01	18.04	18.08
630	18.11	18.14	18.17	18.20	18.23	18.26	18.29	18.32	18.35	18.38
640	18.41	18.44	18.47	18.50	18.54	18.57	18.60	18.63	18.66	18.69
650	18.72	18.75	18.78	18.81	18.84	18.87	18.90	18.94	18.97	19.00
660	19.03	19.06	19.09	19.12	19.15	19.18	19.21	19.24	19.27	19.30
670	19.34	19.37	19.40	19.43	19.46	19.49	19.52	19.55	19.58	19.61
680	19.64	19.67	19.70	19.74	19.77	19.80	19.83	19.86	19.89	19.92
690	19.95	19.98	20.01	20.04	20.07	20.10	20.13	20.16	20.20	20.23
700	20.26	20.29	20.32	20.35	20.38	20.41	20.44	20.47	20.50	20.53
710	20.56	20.59	20.62	20.66	20.69	20.72	20.75	20.78	20.81	20.84
720	20.87	20.90	20.93	20.96	20.99	21.02	21.05	21.08	21.11	21.14
730	21.18	21.21	21.24	21.27	21.30	21.33	21.36	21.39	21.42	21.45
740	21.48	21.51	21.54	21.57	21.60	21.64	21.67	21.70	21.73	21.76
750	21.79	21.82	21.85	21.88	21.91	21.94	21.97	22.00	22.03	22.06
760	22.10	22.13	22.16	22.19	22.22	22.25	22.28	22.31	22.34	22.37
770	22.40	22.43	22.46	22.49	22.52	22.55	22.58	22.62	22.65	22.68
780	22.71	22.74	22.77	22.80	22.83	22.86	22.89	22.92	22.95	22.98
790	23.01	23.04	23.08	23.11	23.14	23.17	23.20	23.23	23.26	23.29
800	23.32	23.35	23.38	23.41	23.44	23.47	23.50	23.53	23.56	23.60
810	23.63	23.66	23.69	23.72	23.75	23.78	23.81	23.84	23.87	23.90
820	23.93	23.96	23.99	24.02	24.06	24.09	24.12	24.15	24.18	24.21
830	24.24	24.27	24.30	24.33	24.36	24.39	24.42	24.45	24.48	24.52
840	24.55	24.58	24.61	24.64	24.67	24.70	24.73	24.76	24.79	24.82
850	24.85	24.88	24.91	24.94	24.98	25.01	25.04	25.07	25.10	25.13
860	25.16	25.19	25.22	25.25	25.28	25.32	25.35	25.38	25.41	25.44
870	25.47	25.50	25.53	25.56	25.59	25.62	25.65	25.68	25.72	25.75
880	25.78	25.81	25.84	25.87	25.90	25.93	25.96	25.99	26.02	26.06
890	26.09	26.12	26.15	26.18	26.21	26.24	26.27	26.30	26.33	26.36
900	26.40	26.43	26.46	26.49	26.52	26.55	26.58	26.61	26.64	26.67
910	26.70	26.74	26.77	26.80	26.83	26.86	26.89	26.92	26.95	26.98
920	27.02	27.05	27.08	27.11	27.14	27.17	27.20	27.23	27.26	27.30
930	27.33	27.36	27.39	27.42	27.45	27.48	27.51	27.54	27.58	27.61
940	27.64	27.67	27.70	27.73	27.76	27.80	27.83	27.86	27.89	27.92
950	27.95	27.98	28.02	28.05	28.08	28.11	28.14	28.17	28.20	28.23

* Based on the International Temperature Scale of 1948.

CALIBRATION TABLES FOR THERMOCOUPLES (Continued)
IRON-CONSTANTAN THERMOCOUPLES (MODIFIED 1913)

(Electromotive Force in Absolute Millivolts. Temperatures in Degrees F.* Reference Junctions at 32° F.)

°F	0	1	2	3	4	5	6	7	8	9
	Millivolts									
960	28.26	28.30	28.33	28.36	28.39	28.42	28.45	28.48	28.52	28.55
970	28.58	28.61	28.64	28.67	28.70	28.74	28.77	28.80	28.83	28.86
980	28.89	28.92	28.96	28.99	29.02	29.05	29.08	29.11	29.14	29.18
990	29.21	29.24	29.27	29.30	29.33	29.37	29.40	29.43	29.46	29.49
1000	29.52	29.56	29.59	29.62	29.65	29.68	29.71	29.75	29.78	29.81
1010	29.84	29.87	29.90	29.94	29.97	30.00	30.03	30.06	30.10	30.13
1020	30.16	30.19	30.22	30.25	30.28	30.32	30.35	30.38	30.41	30.44
1030	30.48	30.51	30.54	30.57	30.60	30.64	30.67	30.70	30.73	30.76
1040	30.80	30.83	30.86	30.89	30.92	30.96	30.99	31.02	31.05	31.08
1050	31.12	31.15	31.18	31.21	31.24	31.28	31.31	31.34	31.37	31.40
1060	31.44	31.47	31.50	31.53	31.56	31.60	31.63	31.66	31.69	31.72
1070	31.76	31.79	31.82	31.85	31.88	31.92	31.95	31.98	32.01	32.05
1080	32.08	32.11	32.14	32.18	32.21	32.24	32.27	32.30	32.34	32.37
1090	32.40	32.43	32.47	32.50	32.53	32.56	32.60	32.63	32.66	32.69
1100	32.72	32.76	32.79	32.82	32.86	32.89	32.92	32.95	32.98	33.02
1110	33.05	33.08	33.11	33.15	33.18	33.21	33.24	33.28	33.31	33.34
1120	33.37	33.41	33.44	33.47	33.50	33.54	33.57	33.60	33.64	33.67
1130	33.70	33.73	33.76	33.80	33.83	33.86	33.90	33.93	33.96	33.99
1140	34.03	34.06	34.09	34.12	34.16	34.19	34.22	34.26	34.29	34.32
1150	34.36	34.39	34.42	34.45	34.49	34.52	34.55	34.58	34.62	34.65
1160	34.68	34.72	34.75	34.78	34.82	34.85	34.88	34.92	34.95	34.98
1170	35.01	35.05	35.08	35.11	35.15	35.18	35.21	35.25	35.28	35.31
1180	35.35	35.38	35.41	35.45	35.48	35.51	35.54	35.58	35.61	35.64
1190	35.68	35.71	35.74	35.78	35.81	35.84	35.88	35.91	35.94	35.98
1200	36.01	36.05	36.08	36.11	36.15	36.18	36.21	36.25	36.28	36.31
1210	36.35	36.38	36.42	36.45	36.48	36.52	36.55	36.58	36.62	36.65
1220	36.69	36.72	36.75	36.79	36.82	36.86	36.89	36.92	36.96	36.99
1230	37.02	37.06	37.09	37.13	37.16	37.20	37.23	37.26	37.30	37.33
1240	37.36	37.40	37.43	37.47	37.50	37.54	37.57	37.60	37.64	37.67
1250	37.71	37.74	37.78	37.81	37.84	37.88	37.91	37.95	37.98	38.02
1260	38.05	38.08	38.12	38.15	38.19	38.22	38.26	38.29	38.32	38.36
1270	38.39	38.43	38.46	38.50	38.53	38.57	38.60	38.64	38.67	38.70
1280	38.74	38.77	38.81	38.84	38.88	38.91	38.95	38.98	39.02	39.05
1290	39.08	39.12	39.15	39.19	39.22	39.26	39.29	39.33	39.36	39.40
1300	39.43	39.47	39.50	39.54	39.57	39.61	39.64	39.68	39.71	39.75
1310	39.78	39.82	39.85	39.89	39.92	39.96	39.99	40.03	40.06	40.10
1320	40.13	40.17	40.20	40.24	40.27	40.31	40.34	40.38	40.41	40.45
1330	40.48	40.52	40.55	40.59	40.62	40.66	40.69	40.73	40.76	40.80
1340	40.83	40.87	40.90	40.94	40.98	41.01	41.05	41.08	41.12	41.15
1350	41.19	41.22	41.26	41.29	41.33	41.36	41.40	41.43	41.47	41.50
1360	41.54	41.58	41.61	41.65	41.68	41.72	41.75	41.79	41.82	41.86
1370	41.90	41.93	41.97	42.00	42.04	42.07	42.11	42.14	42.18	42.22
1380	42.25	42.29	42.32	42.36	42.39	42.43	42.46	42.50	42.53	42.57
1390	42.61	42.64	42.68	42.71	42.75	42.78	42.82	42.85	42.89	42.92
1400	42.96	....	....	....	....	....	....	....	....	....

* Based on the International Temperature Scale of 1948.

CALIBRATION TABLES FOR THERMOCOUPLES (Continued)
TEMPERATURE-E. M. F. VALUES FOR COPPER-CONSTANTAN

E. M. F. values are in millivolts; reference junctions at 0°C.; temperatures are in degrees Centigrade
Roeser and Wensel, National Bureau of Standards

°C	0°	10°	20°	30°	40°	50°	60°	70°	80°	90°
−200°	−5.54									
−100°	−3.35	−3.62	−3.89	−4.14	−4.38	−4.60	−4.82	−5.02	−5.20	−5.38
0°	0	−0.38	−0.75	−1.11	−1.47	−1.81	−2.14	−2.46	−2.77	−3.06
0°	0	0.39	0.79	1.19	1.61	2.03	2.47	2.91	3.36	3.81
100°	4.28	4.75	5.23	5.71	6.20	6.70	7.21	7.72	8.23	8.76
200°	9.29	9.82	10.36	10.91	11.46	12.01	12.57	13.14	13.71	14.28
300°	14.86	15.44	16.03	16.62	17.22	17.82	18.42	19.03	19.64	20.25
400°	20.87									

TEMPERATURE-E. M. F. VALUES FOR COPPER-CONSTANTAN

E. M. F. values are in millivolts; reference junctions at 32°F.; temperatures are in degrees Fahrenheit
Roeser and Wensel, National Bureau of Standards

°F	0°	10°	20°	30°	40°	50°	60°	70°	80°	90°
−300°	−5.28									
−200°	−4.11	−4.25	−4.38	−4.50	−4.63	−4.75	−4.86	−4.97	−5.08	−5.18
−100°	−2.56	−2.73	−2.90	−3.06	−3.22	−3.38	−3.53	−3.68	−3.83	−3.97
0°	−0.67	−0.87	−1.07	−1.27	−1.47	−1.66	−1.84	−2.03	−2.21	−2.39
0°	−0.67	−0.46	−0.26	−0.04	+0.17	0.39	0.61	0.83	1.06	1.29
100°	1.52	1.75	1.99	2.23	2.47	2.71	2.96	3.21	3.46	3.71
200°	3.97	4.22	4.48	4.75	5.01	5.28	5.55	5.82	6.09	6.37
300°	6.64	6.92	7.21	7.49	7.77	8.06	8.35	8.64	8.93	9.23
400°	9.52	9.82	10.12	10.42	10.72	11.03	11.33	11.64	11.95	12.26
500°	12.57	12.89	13.20	13.52	13.83	14.15	14.47	14.79	15.12	15.44
600°	15.77	16.10	16.42	16.75	17.08	17.42	17.75	18.08	18.42	18.75
700°	19.09	19.43	19.77	20.11	20.45	20.80				

REFERENCE TABLE FOR Pt TO Pt—10 PER CENT Rh THERMOCOUPLE

Emfs are expressed in microvolts and temperatures in °C. Cold junctions at 0°C.
ROESER AND WENSEL, NATIONAL BUREAU OF STANDARDS

E(μv)	0	1,000	2,000	3,000	4,000	5,000	6,000	7,000	8,000	E(μv)
0	0	146.9	265.0	373.7	477.7	578.1	675.3	769.5	861.0	0
	17.7	12.5	11.2	10.5	10.2	9.9	9.5	9.3	9.0	
100	17.7	159.4	276.2	384.2	487.9	588.0	684.8	778.8	870.0	100
	16.7	12.3	11.1	10 5	10.2	9.8	9.5	9.2	9.0	
200	34.4	171.7	287.3	394.7	498.1	597.8	694.3	788.0	879.0	200
	15.8	12.1	11.0	10.5	10 1	9 8	9.5	9.2	9.0	
300	50.2	183.8	298.3	405.2	508.2	607.6	703.8	797.2	888.0	300
	15.2	12.0	11.0	10.5	10.1	9.8	9.5	9.2	9.0	
400	65.4	195.8	309.3	415.7	518.3	617.4	713.3	806.4	897.0	400
	14.6	11.8	10.9	10 4	10.1	9.7	9.4	9 2	8.9	
500	80.0	207.6	320.2	426.1	528.4	627.1	722.7	815.6	905.9	500
	14.1	11.7	10.8	10.4	10.0	9.7	9.4	9 1	8.9	
600	94.1	219.3	331.0	436.5	538.4	636.8	732.1	824.7	914.8	600
	13.7	11 6	10 7	10.3	10.0	9.7	9.4	9.1	8.9	
700	107.8	230.9	341.7	446.8	548.4	646.5	741.5	833.8	923.7	700
	13.3	11.5	10 7	10.3	9.9	9.6	9.4	9.1	8.9	
800	121.1	242 4	352.4	457.1	558 3	656.1	750.9	842.9	932.6	800
	13.0	11 3	10.7	10.3	9.9	9.6	9.3	9.1	8.8	
900	134.1	253.7	363.1	467.4	568.2	665.7	760.2	852.0	941.4	900
	12.8	11.3	10.6	10.3	9 9	9.6	9.3	9.0	8.8	
1,000	146.9	265.0	373.7	477.7	578.1	675.3	769.5	861.0	950.2	1,000

REFERENCE TABLE FOR Pt TO Pt—10 PER CENT Rh THERMOCOUPLE (Continued)

Emfs are expressed in microvolts and temperatures in °C. Cold junctions at 0°C.

E(μv)	9,000	10,000	11,000	12,000	13,000	14,000	15,000	16,000	17,000	E(μv)
0	950.2	1,037.2	1,122.3	1,206.4	1,290.0	1,373.8	1,458.0	1,542.6	1,627.8	0
	8.8	8.6	8.5	8.3	8.3	8.4	8.4	8.5	8.6	
100	959.0	1,045.8	1,130.8	1,214.7	1,298.3	1,382.2	1,466.4	1,551.1	1,636.4	100
	8.8	8.6	8.4	8.4	8.4	8.4	8.4	8.5	8.5	
200	967.8	1,054.4	1,139.2	1,223.1	1,306.7	1,390.6	1,474.8	1,559.6	1,644.9	200
	8.7	8.5	8.4	8.3	8.4	8.4	8.5	8.5	8.6	
300	976.5	1,062.9	1,147.6	1,231.4	1,315.1	1,399.0	1,483.3	1,568.1	1,653.5	300
	8.8	8.6	8.4	8.4	8.4	8.4	8.5	8.5	8.6	
400	985.3	1,071.5	1,156.0	1,239.8	1,323.5	1,407.4	1,491.8	1,576.6	1,662.1	400
	8.7	8.5	8.4	8.4	8.3	8.4	8.4	8.5	8.6	
500	994.0	1,080.0	1,164.4	1,248.2	1,331.8	1,415.8	1,500.2	1,585.1	1,670.7	500
	8.7	8.5	8.4	8.3	8.4	8.4	8.5	8.6	8.6	
600	1,002.7	1,088.5	1,172.8	1,256.5	1,340.2	1,424.2	1,508.7	1,593.7	1,679.3	600
	8.6	8.5	8.4	8.4	8.4	8.5	8.5	8.5	8.6	
700	1,011.3	1,097.0	1,181.2	1,264.9	1,348.6	1,432.7	1,517.2	1,602.2	1,687.9	700
	8.7	8.4	8.4	8.3	8.4	8.4	8.4	8.5	8.6	
800	1,020.0	1,105.4	1,189.6	1,273.2	1,357.0	1,441.1	1,525.6	1,610.7	1,696.5	800
	8.6	8.5	8.4	8.4	8.4	8.4	8.5	8.6	8.6	
900	1,028.6	1,113.9	1,198.0	1,281.6	1,365.4	1,449.5	1,534.1	1,619.3	1,705.1	900
	8.6	8.4	8.4	8.4	8.4	8.5	8.5	8.5	8.6	
1,000	1,037.2	1,122.3	1,206.4	1,290.0	1,373.8	1,458.0	1,542.6	1,627.8	1,713.7	1,000

Emfs are expressed in microvolts and temperatures in °F. Cold junctions at 32°F.
ROESER AND WENSEL, NATIONAL BUREAU OF STANDARDS

E(μv)	0	1,000	2,000	3,000	4,000	5,000	6,000	7,000	8,000	E(μv)
0	32.0	296.4	509.0	704.7	891.9	1,072.6	1,247.5	1,417.1	1,581.8	0
	31.9	22.5	20.1	19.0	18.4	17.7	17.1	16.7	16.2	
100	63.9	318.9	529.1	723.7	910.3	1,090.3	1,264.6	1,433.8	1,598.0	100
	30.0	22.1	20.0	18.9	18.3	17.7	17.1	16.6	16.2	
200	93.9	341.0	549.1	742.6	928.6	1,108.0	1,281.7	1,450.4	1,614.2	200
	28.5	21.8	19.9	18.8	18.2	17.7	17.1	16.6	16.2	
300	122.4	362.8	569.0	761.4	946.8	1,125.7	1,298.8	1,467.0	1,630.4	300
	27.3	21.6	19.8	18.8	18.1	17.6	17.1	16.5	16.1	
400	149.7	384.4	588.8	780.2	964.9	1,143.3	1,315.9	1,483.5	1,646.5	400
	26.3	21.3	19.6	18.8	18.1	17.5	17.0	16.5	16.1	
500	176.0	405.7	608.4	799.0	983.0	1,160.8	1,332.9	1,500.0	1,662.6	500
	25.4	21.0	19.4	18.7	18.1	17.5	16.9	16.5	16.0	
600	201.4	426.7	627.8	817.7	1,001.1	1,178.3	1,349.8	1,516.5	1,678.6	600
	24.6	20.9	19.3	18.6	18.0	17.4	16.9	16.4	16.0	
700	226.0	447.6	647.1	836.3	1,019.1	1,195.7	1,366.7	1,532.9	1,694.6	700
	24.0	20.7	19.3	18.5	17.9	17.3	16.9	16.3	16.0	
800	250.0	468.3	666.4	854.8	1,037.0	1,213.0	1,383.6	1,549.2	1,710.6	800
	23.4	20.4	19.2	18.6	17.8	17.3	16.8	16.3	15.9	
900	273.4	488.7	685.6	873.4	1,054.8	1,230.3	1,400.4	1,565.5	1,726.5	900
	23.0	20.3	19.1	18.5	17.8	17.2	16.7	16.3	15.9	
1,000	296.4	509.0	704.7	891.9	1,072.6	1,247.5	1,417.1	1,581.8	1,742.4	1,000

Emfs are expressed in microvolts and temperatures in °F. Cold junctions at 32°F.

E(μv)	9,000	10,000	11,000	12,000	13,000	14,000	15,000	16,000	17,000	E(μv)
0	1,742.4	1,899.0	2,052.1	2,203.5	2,354.0	2,504.8	2,656.4	2,808.7	2,962.0	0
	15.8	15.5	15.2	15.0	15.0	15.2	15.1	15.3	15.4	
100	1,758.2	1,914.5	2,067.3	2,218.5	2,369.0	2,520.0	2,671.5	2,824.0	2,977.4	100
	15.8	15.4	15.2	15.0	15.1	15.1	15.2	15.3	15.4	
200	1,774.0	1,929.9	2,082.5	2,233.5	2,384.1	2,535.1	2,686.7	2,839.3	2,992.8	200
	15.7	15.4	15.2	15.0	15.1	15.1	15.2	15.3	15.5	
300	1,789.7	1,945.3	2,097.7	2,248.5	2,399.2	2,550.2	2,701.9	2,854.6	3,008.3	300
	15.8	15.4	15.1	15.1	15.1	15.1	15.3	15.3	15.5	
400	1,805.5	1,960.7	2,112.8	2,263.6	2,414.3	2,565.3	2,717.2	2,869.9	3,023.8	400
	15.7	15.3	15.1	15.1	15.0	15.1	15.2	15.3	15.5	
500	1,821.2	1,976.0	2,127.9	2,278.7	2,429.3	2,580.4	2,732.4	2,885.2	3,039.3	500
	15.6	15.3	15.1	15.0	15.1	15.2	15.3	15.4	15.4	
600	1,836.8	1,991.3	2,143.0	2,293.7	2,444.4	2,595.6	2,747.7	2,900.6	3,054.7	600
	15.6	15.2	15.2	15.1	15.1	15.2	15.3	15.4	15.5	
700	1,852.4	2,006.5	2,158.2	2,308.8	2,459.5	2,610.8	2,763.0	2,916.0	3,070.2	700
	15.6	15.2	15.1	15.0	15.1	15.2	15.2	15.3	15.5	
800	1,868.0	2,021.7	2,173.3	2,323.8	2,474.6	2,626.0	2,778.2	2,931.3	3,085.7	800
	15.5	15.2	15.1	15.1	15.1	15.2	15.2	15.4	15.5	
900	1,883.5	2,036.9	2,188.4	2,338.9	2,489.7	2,641.2	2,793.4	2,946.7	3,101.2	900
	15.5	15.2	15.1	15.1	15.1	15.2	15.3	15.3	15.5	
1,000	1,899.0	2,052.1	2,203.5	2,354.0	2,504.8	2,656.4	2,808.7	2,962.0	3,116.7	1,000

REFERENCE TABLE FOR Pt TO Pt—13 PER CENT Rh THERMOCOUPLE

Emfs are expressed in microvolts and temperatures in °C. Cold junctions at 0°C.
ROESER AND WENSEL, NATIONAL BUREAU OF STANDARDS

E(µv)	0	1,000	2,000	3,000	4,000	5,000	6,000	7,000	8,000	9,000	E(µv)
0	0	145.3	258.8	361.0	457.4	549.8	638.3	723.5	806.0	886.1	0
	17.9	12.2	10.6	9.9	9.5	9.0	8.7	8.4	8.1	7.9	
100	17.9	157.5	269.4	370.9	466.9	558.8	647.0	731.9	814.1	894.0	100
	16.7	12.0	10.5	9.8	9.4	9.0	8.6	8.4	8.1	7.8	
200	34.6	169.5	279.9	380.7	476.3	567.8	655.6	740.3	822.2	901.8	200
	15.8	11.7	10.4	9.8	9.4	9.0	8.6	8.3	8.1	7.9	
300	50.4	181.2	290.3	390.5	485.7	576.8	644.2	748.6	830.3	909.7	300
	15.1	11.5	10.3	9.7	9.3	8.9	8.6	8.3	8.0	7.8	
400	65.5	192.7	300.6	400.2	495.0	585.7	672.8	756.9	838.3	917.5	400
	14.5	11.4	10.2	9.7	9.3	8.9	8.5	8.2	8.0	7.8	
500	80.0	204.1	310.8	409.9	504.3	594.6	681.3	765.1	846.3	925.3	500
	13.9	11.2	10.2	9.6	9.2	8.8	8.5	8.2	8.0	7.8	
600	93.9	215.3	321.0	419.5	513.5	603.4	689.8	773.3	854.3	933.1	600
	13.4	11.1	10.1	9.5	9.1	8.8	8.5	8.2	8.0	7.8	
700	107.3	226.4	331.1	429.0	522.6	612.2	698.3	781.5	862.3	940.9	700
	13.0	10.9	10.0	9.5	9.1	8.7	8.4	8.2	8.0	7.8	
800	120.3	237.3	341.1	438.5	531.7	620.9	706.7	789.7	870.3	948.7	800
	12.6	10.8	10.0	9.5	9.1	8.7	8.4	8.1	7.9	7.7	
900	132.9	248.1	351.1	448.0	540.8	629.6	715.1	797.8	878.2	956.4	900
	12.4	10.7	9.9	9.4	9.0	8.7	8.4	8.2	7.9	7.7	
1,000	145.3	258.8	361.0	457.4	549.8	638.3	723.5	806.0	886.1	964.1	1,000

Emfs are expressed in microvolts and temperatures in °C. Cold junctions at 0°C.

E(µv)	10,000	11,000	12,000	13,000	14,000	15,000	16,000	17,000	18,000	19,000	E(µv)
0	964.1	1,040.0	1,113.9	1,186.9	1,259.3	1,331.8	1,404.3	1,476.9	1,550.0	1,623.6	0
	7.6	7.4	7.3	7.2	7.3	7.2	7.3	7.3	7.4	7.4	
100	971.7	1,047.4	1,121.2	1,194.1	1,266.6	1,339.0	1,411.6	1,484.2	1,557.4	1,631.0	100
	7.7	7.5	7.4	7.3	7.2	7.2	7.3	7.3	7.4	7.4	
200	979.4	1,054.9	1,128.6	1,201.4	1,273 8	1,346.2	1,418.9	1,491.5	1,564.8	1,638.4	200
	7.6	7.4	7.3	7.3	7.3	7.3	7.2	7.3	7.3	7.4	
300	987.0	1,062.3	1,135.9	1,208.7	1,281.1	1,353.5	1,426.1	1,498.8	1,572.1	1,645.8	300
	7.7	7.4	7.3	7.2	7.2	7.3	7.2	7.3	7.4	7.4	
400	994.7	1,069.7	1,143.2	1,215.9	1,288.3	1,360.8	1,433.3	1,506.1	1,579.5	1,653.2	400
	7.6	7.4	7.3	7.3	7 2	7 2	7.2	7.3	7.3	7.4	
500	1,002.3	1,077.1	1,150.5	1,223.2	1,295.5	1,368.0	1,440.5	1,513.4	1,586.8	1,660.6	500
	7.6	7.3	7.3	7.2	7.2	7.3	7.3	7.3	7.4	7.4	
600	1,009.9	1,084.4	1,157.8	1,230.4	1,302.7	1,375.3	1,447.8	1,520.7	1,594.2	1,668.0	600
	7.6	7.4	7.3	7.2	7.2	7.3	7.2	7.3	7.3	7.4	
700	1,017.5	1,091.8	1,165.1	1,237.6	1,309.9	1,382.6	1,455.0	1,528.0	1,601.5	1,675.4	700
	7.5	7.4	7.2	7.2	7.3	7.2	7.3	7.3	7.4	7.4	
800	1,025.0	1,099.2	1,172.3	1,244.8	1,317.2	1,389.8	1,462.3	1,535.3	1,608.9	1,682.8	800
	7.5	7.3	7.3	7.3	7.3	7.3	7.3	7.4	7.3	7.4	
900	1,032.5	1,106.5	1,179.6	1,252.1	1,324.5	1,397.1	1,469.6	1,542.7	1,616.2	1,690.2	900
	7.5	7.4	7.3	7.2	7.3	7.2	7.3	7.3	7.4	7.4	
1,000	1,040.0	1,113.9	1,186.9	1,259.3	1,331.8	1,404.3	1,476.9	1,550.0	1,623.6	1,697.6	1,000

Emfs are expressed in microvolts and temperatures in °F. Cold junctions at 32°F.
ROESER AND WENSEL, NATIONAL BUREAU OF STANDARDS

E(µv)	0	1,000	2,000	3,000	4,000	5,000	6,000	7,000	8,000	9,000	E(µv)
0	32.0	293.5	497.8	681.8	855.3	1,021.6	1,180.9	1,334.3	1,482.8	1,627.0	0
	32.2	22.0	19.1	17.8	17.0	16.2	15.6	15.1	14.6	14.2	
100	64.2	315.5	516.9	699.6	872.3	1,037.8	1,196.5	1,349.4	1,497.4	1,641.2	100
	30.1	21.6	18.9	17.7	17.0	16.2	15.6	15.1	14.6	14.1	
200	94.3	337.1	535.8	717.3	889.3	1,054.0	1,212.1	1,364.5	1,512.0	1,655.3	200
	28.4	21.1	18.7	17.6	16.9	16.2	15.5	15.0	14.5	14.1	
300	122.7	358.2	554.5	734.9	906.2	1,070.2	1,227.6	1,379.5	1,526.5	1,669.4	300
	27.2	20.7	18.6	17.5	16.8	16.1	15.4	14.9	14.4	14.1	
400	149.9	378.9	573.1	752.4	923.0	1,086.3	1,243.0	1,394.4	1,540.9	1,683.5	400
	26.1	20.4	18.4	17.4	16.7	16.0	15.3	14.8	14.4	14.0	
500	176.0	399.3	591.5	769.8	939.7	1,102.3	1,258.3	1,409.2	1,555.3	1,697.5	500
	25.0	20.2	18.3	17 3	16.6	15.9	15.3	14.8	14.4	14.1	
600	201.0	419.5	609.8	787.1	956.3	1,118.2	1,273.6	1,424.0	1,569.7	1,711.6	600
	24.1	20.0	18.2	17.2	16.4	15.8	15.3	14.8	14.4	14.0	
700	225.1	439.5	628.0	804.3	972.7	1,134.0	1,288.9	1,438.8	1,584.1	1,725.6	700
	23.4	19.7	18.0	17.1	16.4	15.7	15.2	14.7	14.4	14.0	
800	248.5	459.2	646.0	821.4	989.1	1,149.7	1,304.1	1,453.5	1,598.5	1,739.6	800
	22.7	19.4	18.0	17.0	16.3	15.6	15.1	14.6	14.3	13.9	
900	271.2	478.6	664.0	838.4	1,005.4	1,165.3	1,319.2	1,468.1	1,612.8	1,753.5	900
	22.3	19.2	17.8	16.9	16.2	15.6	15.1	14.7	14.2	13.9	
1,000	293.5	497.8	681.8	855.3	1,021.6	1,180.9	1,334.3	1,482.8	1,627.0	1,767.4	1,000

Emfs are expressed in microvolts and temperatures in °F. Cold junctions at 32°F.

E(µv)	10,000	11,000	12,000	13,000	14,000	15,000	16,000	17,000	18,000	19,000	E(µv)
0	1,767.4	1,904.0	2,037.0	2,168.4	2,298.7	2,429.2	2,559.7	2,690.4	2,822.0	2,954.5	0
	13.8	13.4	13.2	13.0	13.1	13.0	13.1	13.2	13.3	13.3	
100	1,781.2	1,917.4	2,050.2	2,181.4	2,311.8	2,442.2	2,572.8	2,703.6	2,835.3	2,967.8	100
	13.7	13.4	13.2	13.1	13.0	13.0	13.1	13.1	13.3	13.3	
200	1,794.9	1,930.8	2,063.4	2,194.5	2,324.8	2,455.2	2,585.9	2,716.7	2,848.6	2,981.1	200
	13.7	13.3	13.2	13.1	13.1	13.1	13.0	13.1	13.2	13.3	
300	1,808.6	1,944.1	2,076.6	2,207.6	2,337.9	2,468.3	2,598.9	2,729.8	2,861.8	2,994.4	300
	13.8	13.3	13.2	13.0	13.0	13.1	13.0	13.2	13.3	13.4	
400	1,822.4	1,957.4	2,089.8	2,220.6	2,350.9	2,481.4	2,611.9	2,743.0	2,875.1	3,007.8	400
	13.7	13.3	13.1	13.1	13.0	13.0	13.0	13.1	13.2	13.3	
500	1,836.1	1,970.7	2,102.9	2,233.7	2,363.9	2,494.4	2,624.9	2,756.1	2,888.3	3,021.1	500
	13.7	13 3	13.1	13.0	13.0	13.1	13.1	13.2	13.3	13.3	
600	1,849.8	1,984.0	2,116.0	2,246.7	2,376.9	2,507.5	2,638.0	2,769.3	2,901.6	3,034.4	600
	13.6	13.3	13.1	13.0	13.0	13.1	13.0	13.1	13.2	13.3	
700	1,863.4	1,997.3	2,129.1	2,259.7	2,389.9	2,520.6	2,651.0	2,782.4	2,914.8	3,047.7	700
	13.6	13.3	13.1	13.0	13.1	13.0	13.1	13.2	13.2	13.3	
800	1,877.0	2,010.6	2,142.2	2,272.7	2,403.0	2,533.6	2,664.1	2,795.6	2,928.0	3,061.0	800
	13.5	13.2	13.1	13.0	13.1	13.1	13.1	13.2	13.2	13.4	
900	1,890.5	2,023.8	2,155.3	2,285.7	2,416.1	2,546.7	2,677.2	2,808.8	2,941.2	3,074.4	900
	13.5	13.2	13.1	13.0	13.1	13.0	13.2	13.2	13.3	13.3	
1,000	1,904.0	2,037.0	2,168.4	2,298.7	2,429.2	2,559.7	2,690.4	2,822.0	2,954.5	3,087.7	1,000

VALUES FOR THE LANGEVIN FUNCTION $\mathcal{L}(u)$

Compiled by Allen L. King

Because of random thermal rotations a dipole or dipole-like element ordinarily has an average dipole moment of zero. If it is placed in an orienting field F however it tends to align itself with the field so that the average component of the dipole moment parallel to the field equals $\bar{p}$. Classically, if the system is in thermal equilibrium,

$$\bar{p} = p_0(\coth u - 1/u) = p_0\mathcal{L}(u)$$

Here p_0 is the moment of the dipole itself and $\mathcal{L}(u)$ is the Langevin function of the argument u which equals Fp_0/kT.

$$\text{For } u \ll 1 \quad \mathcal{L}(u) = \frac{u}{3}$$

The Langevin function applies to permanent electric and magnetic dipoles.

(u)	0	1	2	3	4	5	6	7	8	9	10	Diff.
0.0	0000	0033	0066	0100	0133	0166	0200	0233	0266	0300	0333	33
0.1	0333	0366	0400	0433	0466	0499	0532	0565	0598	0631	0665	33
0.2	0665	0698	0731	0764	0797	0830	0862	0896	0929	0961	0994	33
0.3	0994	1027	1059	1092	1124	1157	1190	1222	1254	1287	1319	32
0.4	1319	1352	1384	1416	1448	1480	1512	1544	1576	1608	1640	32
0.5	1640	1671	1703	1734	1765	1797	1829	1860	1892	1923	1953	31
0.6	1953	1985	2016	2046	2077	2108	2138	2169	2200	2230	2260	31
0.7	2260	2290	2321	2351	2381	2411	2441	2471	2500	2530	2559	30
0.8	2559	2589	2618	2647	2676	2705	2734	2763	2792	2821	2850	29
0.9	2850	2878	2906	2934	2963	2991	3019	3047	3075	3103	3130	28
1.0	3130	3158	3185	3212	3240	3267	3294	3321	3348	3375	3401	27
1.1	3401	3428	3454	3480	3507	3533	3559	3585	3611	3637	3662	26
1.2	3662	3688	3713	3738	3763	3788	3813	3838	3863	3888	3913	25
1.3	3913	3937	3961	3985	4009	4033	4057	4081	4105	4129	4152	24
1.4	4152	4176	4199	4222	4245	4268	4291	4313	4336	4359	4381	23
1.5	4381	4403	4426	4448	4469	4491	4514	4536	4557	4579	4600	22
1.6	4600	4621	4642	4663	4684	4705	4726	4747	4768	4788	4808	21
1.7	4808	4828	4849	4869	4889	4909	4928	4948	4967	4987	5006	20
1.8	5006	5025	5044	5064	5083	5102	5121	5139	5158	5176	5195	19
1.9	5195	5213	5231	5249	5267	5285	5303	5321	5338	5356	5373	18
2.0	5373	5391	5408	5425	5442	5459	5476	5493	5509	5526	5542	17
2.1	5542	5559	5575	5591	5608	5624	5640	5656	5672	5688	5704	16
2.2	5704	5719	5734	5750	5765	5780	5795	5810	5825	5840	5855	15
2.3	5855	5870	5885	5899	5913	5928	5943	5957	5971	5985	6000	14
2.4	6000	6014	6027	6041	6055	6068	6082	6095	6109	6122	6136	13.5
2.5	6136	6149	6162	6175	6188	6201	6214	6227	6239	6252	6265	13
2.6	6265	6278	6290	6302	6314	6326	6339	6351	6363	6375	6387	12
2.7	6387	6399	6411	6422	6434	6446	6457	6469	6480	6492	6503	11.5
2.8	6503	6514	6525	6536	6548	6559	6570	6581	6591	6602	6613	11
2.9	6613	6624	6634	6644	6655	6665	6676	6686	6696	6707	6717	10
3.0	6717	6727	6737	6747	6757	6767	6777	6787	6796	6806	6815	10
3.1	6815	6825	6834	6844	6853	6862	6871	6880	6890	6899	6908	9
3.2	6908	6917	6926	6935	6944	6953	6962	6971	6979	6988	6997	9
3.3	6997	7006	7014	7023	7031	7040	7048	7056	7064	7073	7081	8.5
3.4	7081	7089	7097	7105	7113	7121	7129	7137	7145	7153	7161	8
3.5	7161	7169	7176	7184	7192	7200	7207	7215	7223	7230	7237	8
3.6	7237	7245	7252	7259	7267	7274	7281	7288	7295	7302	7309	7
3.7	7309	7317	7324	7330	7337	7344	7351	7358	7364	7371	7378	7
3.8	7378	7385	7392	7398	7405	7412	7418	7425	7431	7437	7444	6.5
3.9	7444	7450	7457	7463	7470	7476	7482	7488	7494	7501	7507	6

Note: Higher values of the Langevin function are nearly equal to $(u - 1)/u$.

MAGNETIC PROPERTIES AND COMPOSITION OF SOME PERMANENT MAGNETIC ALLOYS

Name	Composition,* Weight percent					Remanence, B_r (Gauss)	Coercive force H_c (Oersteds)	Maximum energy product, $(BH)_{max}$ (Gauss-Oersteds $\times 10^{-6}$)
	Al	Ni	Co	Cu	Other			
U.S.A.								
Alnico I	12	20–22	5	...	......	7,100	440	1.4
Alnico II	10	17	12.5	6	......	7,200	540	1.6
Alnico III	12	24–26	...	3	......	6,900	470	1.35
Alnico IV	12	27–28	5	...	......	5,500	700	1.3
Alnico V†	8	14	24	3	......	12,500	600	5.0
Alnico V DG†	8	14	24	3	......	13,100	640	6.0
Alnico VI†	8	15	24	3	1.25 Ti	10,500	750	3.75
Alnico VII†	8.5	18	24	3	5 Ti	7,200	1,050	2.75
Alnico XII	6	18	35	...	8 Ti	5,800	950	1.6
Carbon steel	...	...	...	...	1 Mn 0.9 C	10,000	50	0.2
Chromium steel	...	...	...	...	3.5 Cr 0.9 C 0.3 Mn	9,700	65	0.3
Cobalt steel	...	...	17	...	2.5 Cr 8 W 0.75 C	9,500	150	0.65
Cunico	...	21	29	50	......	3,400	660	0.80
Cunife	...	20	...	60	......	5,400	550	1.5
Ferroxdur 1	$BaFe_{12}O_{19}$					2,200	1,800	1.0
Ferroxdur 2	$BaF_{12}O_{19}$ (oriented)					3,840	2,000	3.5
Platinum-Cobalt	...	...	23	...	77 Pt	6,000	4,300	7.5
Remalloy	...	...	12	...	17 Mo	10,500	250	1.1
Silmanol	4.4	...	...	...	86.6 Ag 8.8 Mn	550	6,000	0.075
Tungsten steel	...	...	...	...	5 W 0.3 Mn 0.7 C	10,300	70	0.32
Vicalloy I	...	...	52	...	10 V	8,800	300	1.0
Vicalloy II (wire)	...	...	52	...	14 V	10,000	510	3.5
Germany								
Alni 90	12	21	...	...	......	8,000	350	1.2
Alni 120	13	27	...	...	......	6,000	570	1.2
Alnico 130	12	23	5	...	......	6,300	620	1.4
Alnico 160	11	24	12	4	......	6,200	700	1.6
Alnico 190	12	21	15	4	......	7,000	700	1.8
Alnico 250	8	19	23	4	6 Ti	6,500	1,000	2.2
Alnico 400†	9	15	23	4	......	12,000	650	4.8
Alnico 580† (semicolumnar)	9	15	23	4	......	13,000	700	6.0
Oerstit 800	9	18	19	4	4 Ti	6,600	750	1.95
Great Britain								
Alcomax I	7.5	11	25	3	1.5 Ti	12,000	475	3.5
Alcomax II	8	11.5	24	4.5	......	12,400	575	4.7
Alcomax IISC (semicolumnar)	8	11	22	4.5	......	12,800	600	5.15
Alcomax III	8	13.5	24	3	0.8 Nb	12,500	670	5.10
Alcomax IIISC (semicolumnar)	8	13.5	24	3	0.8 Nb	13,000	700	5.80
Alcomax IV	8	13.5	24	3	2.5 Nb	11,200	750	4.30

* Remainder of unlisted composition is either iron or iron plus trace impurities.

† Cast anisotropic. Unmarked ones are cast isotropic.

MAGNETIC PROPERTIES AND COMPOSITION OF SOME PERMANENT MAGNETIC ALLOYS (Continued)

Name	Composition,* Weight percent					Remanence, B_r (Gauss)	Coercive force H_c (Oersteds)	Maximum energy product, $(BH)_{max}$ (Gauss-Oersteds × 10^{-6})
	Al	Ni	Co	Cu	Other			
Alcomax IVSC (semicolumnar)	8	13.5	24	3	2.5 Nb	11,700	780	5.10
Alni, high B_r	13	24	....	3.5	......	6,200	480	1.25
Alni, normal	.....	.....	.....	.....	......	5,600	580	1.25
Alni, high H_c	12	32	....	....	0–0.5 Ti	5,000	680	1.25
Alnico, high B_r	10	17	12	6	......	8,000	500	1.70
Alnico, normal	.....	.....	.....	.....	......	7,250	560	1.70
Alnico, high H_c	10	20	13.5	6	0.25 Ti	6,600	620	1.70
Columax (columnar)	similar to Alcomax III or IV					13,000–14,000	700–800	7.0–8.5
Hycomax	9	21	20	1.6	......	9,500	830	3.3

* Remainder of unlisted composition is either iron or iron plus trace impurities
† Cast anisotropic. Unmarked ones are cast isotropic.

HIGH PERMEABILITY MAGNETIC ALLOYS
(See separate table for magnetic properties)

Name	Composition,* Weight percent	Sp. gr., gm. per cc	Tensile strength		Remark	Use
			Kg/mm²**	Form		
Silicon iron AISI M 15	Si 4	7.68–7.64	44.3	....	Annealed 4 hrs 802–1093°C	Low core losses
Silicon iron AISI M 8	Si 3	7.68–7.64	44.2	Grain oriented	Annealed 4 hrs 802–1204°C	
45 Permalloy	Ni 45; Mn 0.3	8.17	....	....	....	Audio transformer, coils, relays
Monimax	Ni 47; Mo 3	8.27	....	....	....	High frequency coils
4-79 Permalloy	Ni 79; Mo 4; Mn 0.3	8.74	55.4	....	H_2 annealed 1121°C	Audio coils, transformers, magnetic shields
Sinimax	Ni 43; Si 3	7.70	....	....	....	High frequency coils
Nu-metal	Ni 75; Cr 2; Cu 5	8.58	44.8	....	H_2 annealed 1221°C	Audio coils, magnetic shields, transformers
Supermalloy	Ni 79; Mo 5; Mn 0.3	8.77	....	....	....	Pulse transformers, magnetic amplifiers, coils
2-V Permendur	Co 40; V 2	8.15	46.3	....	....	D-c electromagnets, pole tips

* Iron is additional alloying metal.
** kg/mm² × 1422.33 = lbs/in.²

CAST PERMANENT MAGNETIC ALLOYS

(See separate table for magnetic properties)

Alloy name, country of manufacture*	Composition,** weight percent	Sp. gr., gm per cc	Thermal expansion		Tensile strength		Remark†	Use
			$\frac{Cm \times 10^{-6}}{cm \times °C}$	Between °C	*** Kg/mm²	Form		
Alnico I (USA)*	Al 12; Ni 20–22; Co 5	6.9	12.6	20–300	2.9	Cast	i.	Permanent magnets
Alnico II (USA)*	Al 10; Ni 17; Cu 6; Co 12.5	7.1	12.4	20–300	2.1 45.7	Cast Sintered	i.	Temperature controls, magnetic toys and novelties
Alnico III (USA)*	Al 12; Ni 24–26; Cu 3	6.9	12	20–300	8.5	Cast	i.	Tractor magnetos
Alnico IV (USA)*	Al 12; Ni 27–28; Co 5	7.0	13.1	20–300	6.3 42.1	Cast Sintered	i.	Application requiring high coercive force
Alnico V (USA)*	Al 8; Ni 14; Co 24; Cu 3	7.3	11.3	......	3.8 35	Cast Sintered	a.	Application requiring high energy
Alnico V DG (USA)*	Al 8; Ni 14; Co 24; Cu 3	7.3	11.3		....		a., c.	
Alnico VI (USA)*	Al 8; Ni 15; Co 24; Cu 3; Ti 1.25	7.3	11.4	......	16.1	Cast	a.	Application requiring high energy
Alnico VII (USA)*	Al 8.5; Ni 18; Cu 3; Co 24; Ti 5	7.17	11.4	......	....	....	a.	
Alnico XII (USA)*	Al 6; Ni 18; Co 35; Ti 8	7.2	11	20–300	....	....	.....	Permanent magnets
Comol (USA)*	Co 12; Mo 17	8.16	9.3	20–300	88.6	....	.....	Permanent magnets
Cunife (USA)*	Cu 60; Ni 20	8.52	....		70.3	....	.....	Permanent magnets
Cunico (USA)*	Cu 50; Ni 21	8.31			70.3	....	.....	Permanent magnets
Barium ferrite Feroxdur (USA)*	Ba $Fe_{12}O_{19}$	4.7	10	......	70.3	....		Ceramics
Alcomax I (GB)*	Al 7.5; Ni 11; Co 25; Cu 3; Ti 1.5	....	....	......	....	....	a.	Permanent magnets
Alcomax II (GB)*	Al 8; Ni 11.5; Co 24; Cu 4.5	....	....	......	....	....	a.	Permanent magnets
Alcomax II SC(GB)*	Al 8; Ni 11; Co 22; Cu 4.5	7.3		......	....	....	a., sc.	
Alcomax III (GB)*	Al 8; Ni 13.5; Co 24; Nb 0.8	7.3	....	......	....	....	a.	Magnets for motors, loudspeakers
Alcomax IV (GB)*	Al 8; Ni 13.5; Cu 3; Co 24; Nb 2.5	....	....	......	....	....		Magnets for cycle-dynamos
Columax (GB)*	Similar to Alcomax III or IV	....	....	......	....	....	a., sc.	Permanent magnets, heat treatable
Hycomax (GB)*	Al 9; Ni 21; Co 20; Cu 1.6	....	....	......	....	....	a.	Permanent magnets
Alnico (high H_c) (GB)*	Al 10; Ni 20; Co 13.5; Cu 6; Ti 0.25	7.3	....	......	....	....	i.	

* USA—United States; GB—Great Britain; Ger—Germany.

** Iron is the additional alloying metal for each of the magnets listed.

*** kg/mm × 1422.33 = lbs/in.²

† i. = isotropic; a. = anisotropic; c. = columnar; sc. = semicolumnar.

CAST PERMANENT MAGNETIC ALLOYS (Continued)

Alloy name, country of manufacture	Composition,** weight percent	Sp. gr., gm per cc	Thermal expansion		Tensile strength		Remark†	Use
			Cm × 10⁻⁶ / cm × °C	Between °C	*** Kg/mm²	Form		
Alnico (high B_r) (GB)*	Al 10; Ni 17; Co 12; Cu 6	7.3	....	......	....	....	i.	
Alni (high H_c) (GB)*	Al 12; Ni 32; Ti 0–0.5	6.9	....	......	....	....	i.	
Alni (high B_r) (GB)*	Al 13; Ni 24; Cu 3.5	....	....	......	....	....	i.	
Alnico 580 (Ger)*	Al 9; Ni 15; Co 23; Cu 4	....	....	......	....	....	i.	
Alnico 400 (Ger)*	Al 9; Ni 15; Co 23; Cu 4	....	....	......	....	....	a.	
Oerstit 800 (Ger)*	Al 9; Ni 18; Co 19; Cu 4; Ti 4	....	....	......	....	....	i.	Permanent magnets
Alnico 250 (Ger)*	Al 8; Ni 19; Co 23; Cu 4; Ti 6	....	....	......	....	....	i.	
Alnico 190 (Ger)*	Al 12; Ni 21; Cu 4; Co 15	....	....	......	....	....	i.	
Alnico 160 (Austria)	Al 11; Ni 24; Co 12; Cu 4	....	....	......	....	....	i.	Permanent magnets, sintered
Alnico 130 (Ger)*	Al 12; Ni 23; Co 5	....	....	......	....	....	i.	
Alni 120 (Ger)*	Al 13; Ni 27	....	....	......	....	....	i.	
Alni 90 (Ger)*	Al 12; Ni 21	....	....	......	....	....	i.	

* USA—United States; GB—Great Britain; Ger—Germany.

** Iron is the additional alloying metal for each of the magnets listed.

*** kg/mm × 1422.33 = lbs/in.²

† i. = isotropic; a. = anisotropic; c. = columnar; sc. = semicolumnar.

Properties of Antiferromagnetic Compounds

Compound	Crystal Symmetry	θ_N(°K)	θ_P(°K)	$(P_A)_{eff}$ μ_B	P_A μ_B
$CoCl_2$	Rhombohedral	25	−38.1	5.18	3.1 ± 0.6
CoF_2	Tetragonal	38	50	5.15	3.0
CoO	Tetragonal	291	330	5.1	3.8
Cr	Cubic	475			
Cr_2O_3	Rhombohedral	307	485	3.73	3.0
CrSb	Hexagonal	723	550	4.92	2.7
$CuBr_2$	Monoclinic	189	246	1.9	
$CuCl_2·2H_2O$	Orthorhombic	4.3	4–5	1.9	
$CuCl_2$	Monoclinic	~70	109	2.08	
$FeCl_2$	Hexagonal	24	−48	5.38	4.4 ± 0.7
FeF_2	Tetragonal	79–90	117	5.56	4.64
FeO	Rhombohedral	198	507	7.06	3.32
α-Fe_2O_3	Rhombohedral	953	2940	6.4	5.0
α-Mn	Cubic	95			
$MnBr_2·4H_2O$	Monoclinic	2.1	{2.5, 1.3}	5.93	
$MnCl_2·4H_2O$	Monoclinic	1.66	1.8	5.94	
MnF_2	Tetragonal	72–75	113.2	5.71	5
MnO	Rhombohedral	122	610	5.95	5.0
β-MnS	Cubic	160	982	5.82	5.0
MnSe	Cubic	~173	361	5.67	
MnTe	Hexagonal	310–323	690	6.07	5.0
$NiCl_2$	Hexagonal	50	−68	3.32	
NiF_2	Tetragonal	78.5–83	115.6	3.5	2.0
NiO	Rhombohedral	533–650	~2000	4.6	2.0
$TiCl_3$		100			
V_2O_3		170			

1. θ_N = Néel temperature, determined from susceptibility maxima or from the disappearance of magnetic scattering.
2. θ_P = a constant in the Curie-Weiss law written in the form $\chi_A = C_A/(T + \theta_P)$, which is valid for antiferromagnetic material for $T > \theta_N$.
3. $(P_A)_{eff}$ = effective moment per atom, derived from the atomic Curie constant $C_A = (P_A)^2_{eff}(N^2/3R)$ and expressed in units of the Bohr magneton, $\mu_B = 0.9273 \times 10^{-20}$ erg gauss^{-1}.
4. P_A = magnetic moment per atom, obtained from neutron diffraction measurements in the ordered state.

SATURATION CONSTANTS AND CURIE POINTS OF FERROMAGNETIC ELEMENTS

Element	σ_s (20°C)	M_s (20°C)	σ_s (0°K)	n_B	Curie Point (°C)
Fe	218.0	1,714	221.9	2.219	770
Co	161	1,422	162.5	1.715	1,131
Ni	54.39	484.1	57.50	0.604	358
Gd	0	0	253.5	7.12	16

σ_s = saturation magnetic moment/gram; M_s = saturation magnetic moment/cm³, in cgs units. n_B = magnetic moment per atom in Bohr magnetons.

From American Institute of Physics Handbook, McGraw-Hill Company (1963) by permission.

MAGNETIC PROPERTIES OF TRANSFORMER STEELS

Ordinary Transformer Steel

B (Gauss)	H (Oersted)	Permeability = B/H
2,000	0.60	3,340
4,000	0.87	4,600
6,000	1.10	5,450
8,000	1.48	5,400
10,000	2.28	4,380
12,000	3.85	3,120
14,000	10.9	1,280
16,000	43.0	372
18,000	149	121

High Silicon Transformer Steel

B	H	Permeability
2,000	0.50	4,000
4,000	0.70	5,720
6,000	0.90	6,670
8,000	1.28	6,250
10,000	1.99	5,020
12,000	3.60	3,340
14,000	9.80	1,430
16,000	47.4	338
18.000	165	109

SATURATION CONSTANTS FOR MAGNETIC SUBSTANCES

Substance	Field intensity	Induced magnetization	Substance	Field intensity	Induced magnetization
	(For sat	uration)		(For sat	uration)
Cobalt..........	9000	1300	Nickel, hard....	8000	400
Iron, wrought....	2000	1700	annealed.....	7000	515
cast..........	4000	1200	Vicker's steel...	15000	1600
Manganese steel.	7000	200			

INITIAL PERMEABILITY OF HIGH PURITY IRON FOR VARIOUS TEMPERATURES

L. Alberts and B. J. Shepstone

Temperature °C	Permeability (gauss/oersted)
0	920
200	1040
400	1440
600	2550
700	3900
770	12580

MAGNETIC MATERIALS
High-permeability Materials

Material	Form	Approximate percent composition					Typical heat treatment °C	Permeability at $B = 20$ gausses	Maximum permeability	Saturation flux density B gausses	Hysteresis‡ loss, W_h ergs/cm³	Coercive‡ force H_c oersteds	Resistivity microhm-cm	Density, g/cm³
		Fe	Ni	Co	Mo	Other								
Cold rolled steel...	Sheet	98.5	—	—	—	—	950 Anneal	180	2,000	21,000	—	1.8	10	7.88
Iron............	Sheet	99.91	—	—	—	—	950 Anneal	200	5,000	21,500	5,000	1.0	10	7.88
Purified iron......	Sheet	99.95	—	—	—	—	1480 H_2 + 880	5,000	180,000	21,500	300	.05	10	7.88
4% Silicon-iron...	Sheet	96	—	—	—	4 Si	800 Anneal	500	7,000	19,700	3,500	.5	60	7.65
Grain oriented*.	Sheet	97	—	—	—	3 Si	800 Anneal	1,500	30,000	20,000	—	.15	47	7.67
45 Permalloy......	Sheet	54.7	45	—	—	.3 Mn	1050 Anneal	2,500	25,000	16,000	1,200	.3	45	8.17
45 Permalloy†.....	Sheet	54.7	45	—	—	.3 Mn	1200 H_2 Anneal	4,000	50,000	16,000	—	.07	45	8.17
Hipernik..........	Sheet	50	50	—	—	—	1200 H_2 Anneal	4,500	70,000	16,000	220	.05	50	8.25
Monimax.........	Sheet	—	—	—	—	—	1125 H_2 Anneal	2,000	35,000	15,000	—	.1	80	8.27
Sinimax..........	Sheet	—	—	—	—	—	1125 H_2 Anneal	3,000	35,000	11,000	—	—	90	—
78 Permalloy......	Sheet	21.2	78.5	—	—	.3 Mn	1050 + 600 Q§	8,000	100,000	10,700	200	.05	16	8.60
4-79 Permalloy...	Sheet	16.7	79	—	4	.3 Mn	1100 + Q	20,000	100,000	8,700	200	.05	55	8.72
Mu metal.........	Sheet	18	75	—	—	2 Cr, 5 Cu	1175 H_2	20,000	100,000	6,500	—	.05	62	8.58
Supermalloy......	Sheet	15.7	79	—	5	.3 Mn	1300 H_2 + Q	100,000	800,000	8,000	—	.002	60	8.77
Permendur........	Sheet	49.7	—	50	—	.3 Mn	800 Anneal	800	5,000	24,500	12,000	2.0	7	8.3
2V Permendur....	Sheet	49	—	49	—	2 V	800 Anneal	800	4,500	24,000	6,000	2.0	26	8.2
Hiperco...........	Sheet	64	—	34	—	Cr	850 Anneal	650	10,000	24,200	—	1.0	25	8.0
2-81 Permalloy...	Insulated powder	17	81	—	2	—	650 Anneal	125	130	8,000	—	<1.0	10^6	7.8
Carbonyl iron.....	Insulated powder	99.9	—	—	—	—	—	55	132	—	—	—	—	7.86
Ferroxcube III....	Sintered powder	$MnFe_2O_4 + ZnFe_2O_4$					—	1,000	1,500	2,500	—	.1	10^8	5.0

* Properties in direction of rolling.
† Similar properties for Nicaloi, 4750 alloy, Carpenter 49, Armco 48.
‡ At saturation.
§ Q, quench or controlled cooling.

Permanent Magnet Alloys

Material	Percent composition (remainder Fe)	Heat treatment* (temperature, °C)	Magnetizing force $H_{max.}$ oersteds	Coercive force H_c oersteds	Residual induction B_r gausses	Energy product $BH_{max.}$ $\times 10^{-6}$	Method of fabrication†	Mechanical properties‡	Weight lb/in.^{3}
Carbon steel	1 Mn, 0.9 C	Q 800	300	50	10,000	.20	HR, M, P	H, S	.280
Tungsten steel	5 W, 0.3 Mn, 0.7 C	Q 850	300	70	10,300	.32	HR, M, P	H, S	.292
Chromium steel	3.5 Cr, 0.9 C, 0.3 Mn	Q 830	300	65	9,700	.30	HR, M, P	H, S	.280
17% Cobalt steel	17 Co, 0.75 C, 2.5 Cr, 8 W	—	1,000	150	9,500	.65	HR, M, P	H, S	—
36% Cobalt steel	36 Co, 0.7 C, 4 Cr, 5 W	Q 950	1,000	240	9,500	.97	HR, M, P	H, S	.296
Remalloy or Comol	17 Mo, 12 Co	Q 1200, B 700	1,000	250	10,500	1.1	HR, M, P	H	.295
Alnico I	12 Al, 20 Ni, 5 Co	A 1200, B 700	2,000	440	7,200	1.4	C, G	H, B	.249
Alnico II	10 Al, 17 Ni, 2.5 Co, 6 Cu	A 1200, B 600	2,000	550	7,200	1.6	C, G	H, B	.256
Alnico II (sintered)	10 Al, 17 Ni, 2.5 Co, 6 Cu	A 1300	2,000	520	6,900	1.4	Sn, G	H	.249
Alnico IV	12 Al, 28 Ni, 5 Co	Q 1200, B 650	3,000	700	5,500	1.3	Sn, C, G	H	.253
Alnico V	8 Al, 14 Ni, 24 Co, 3 Cu	AF 1300, B 600	2,000	550	12,500	4.5	C, G	H, B	.264
Alnico VI	8 Al, 15 Ni, 24 Co, 3 Cu, 1 Ti	—	3,000	750	10,000	3.5	C, G	H, B	.268
Alnico XII	6 Al, 18 Ni, 35 Co, 8 Ti	—	3,000	950	5,800	1.5	C, G	H, B	.26
Vicalloy I	52 Co, 10 V	B 600	1,000	300	8,800	1.0	C, CR, M, P	D	.295
Vicalloy II (wire)	52 Co, 14 V	CW + B 600	2,000	510	10,000	3.5	C, CR, M, P	D	.292
Cunife (wire)	60 Cu, 20 Ni	CW + B 600	2,400	550	5,400	1.5	C, CR, M, P	D, M	.311
Cunico	50 Cu, 21 Ni, 29 Co	—	3,200	660	3,400	.80	C, CR, M, P	D, M	.300
Vectolite	30 Fe_2O_3, 44 Fe_3O_4, 26 C_2O_3	—	3,000	1,000	1,600	.60	Sn, G	W	.113
Silmanal	86.8 Ag, 8.8 Mn, 4.4 Al	—	20,000	6,000[a]	550	.075	C, CR, M, P	D, M	.325
Platinum-cobalt	77 Pt, 23 Co	Q 1200, B 650	15,000	3,600	5,900	6.5	C, CR, M	D	—
Hyflux	Fine powder	—	2,000	390	6,600	.97	—	—	.176

[a] Value given is intrinsic H_c.
* Q—Quenched in oil or water. A—Air cooled. B—Baked. F—Cooled in magnetic field. CW—Cold worked.
† HR—Hot rolled or forged. CR—Cold rolled or drawn. M—Machined. G—Must be ground. P—Punched. C—Cast. Sn—Sintered.
‡ H—Hard. B—Brittle. S—Strong. D—Ductile. M—Malleable. W—Weak.

MAGNETIC SUSCEPTIBILITY OF THE ELEMENTS AND INORGANIC COMPOUNDS

The following table lists the magnetic susceptibilities of one gram formula weight of a number of paramagnetic and diamagnetic inorganic compounds as well as the magnetic susceptibilities of the elements.

In each instance the magnetic moment is expressed in cgs units.

A more extensive listing of the magnetic susceptibilities of inorganic compounds as well as those for organic compounds may be found in Constantes Selectionnees Diamagnetisme et Paramagnetisme Relaxation Paramagnetique, Volume 7. This table is abridged from the above publication by permission of the publishers.

Substance	Formula	Temp. °K.	Susceptibility 10^{-6} cgs
Aluminum (s)	Al	ord.	+16.5
" (*l*)	Al	....	+12.0
Fluoride	AlF_3	302	−13.4
Oxide	Al_2O_3	ord.	−37.0
Sulfate	$Al_2(SO_4)_3$	ord.	−93.0
"	$Al_2(SO_4)_3 \cdot 18H_2O$	ord.	−323.0
Ammonia (g)	NH_3	ord.	−18.0
" (aq)	NH_3	ord.	−17.0
Ammonium			
Acetate	$NH_4C_2H_3O_2$	ord.	−41.1
Bromide	NH_4Br	ord.	−47.0
Carbonate	$(NH_4)_2CO_3$	ord.	−42.50
Chlorate	NH_4ClO_3	ord.	−42.1
Chloride	NH_4Cl	ord.	−36.7
Fluoride	NH_4F	ord.	−23.0
Hydroxide (aq)	NH_4OH	ord.	−31.5
Iodate	NH_4IO_3	ord.	−62.3
Iodide	NH_4I	ord.	−66.0
Nitrate	NH_4NO_3	ord.	−33.6
Sulfate	$(NH_4)_2SO_4$	ord.	−67.0
Thiocyanate	NH_4SCN	ord.	−48.1
Americium (s)	Am	300	+1000.0
Antimony (s)	Sb	293	−99.0
" (*l*)	Sb		−2.5
Bromide	$SbBr_3$	ord.	−115.0
Chloride, tri	$SbCl_3$	ord.	−86.7
Chloride, penta	$SbCl_5$	ord.	−120.0
Fluoride	SbF_3	ord.	−46.0
Iodide	SbI_3	ord.	−147.0
Oxide	Sb_2O_3	ord.	−69.4
Sulfide	Sb_2S_3	ord.	−86.0
Argon (g)	A	ord.	−19.6
Arsenic (α)	As	293	−5.5
" (β)	As	293	−23.7
" (γ)	As	293	−23.0
Bromide	$AsBr_3$	ord.	−106.0
Chloride	$AsCl_3$	ord.	−79.9
Iodide	AsI_3	ord.	−142.0
Sulfide	As_2S_3	ord.	−70.0
Arsenious Acid	H_3AsO_3	ord.	−51.2
Barium	Ba	ord.	+20.6
Acetate	$Ba(C_2H_3O_2)_2 \cdot H_2O$	ord.	−100.1
Bromate	$Ba(BrO_3)_2$	ord.	−105.8
Barium (continued)			
Bromide	$BaBr_2$	ord.	−92.0
"	$BaBr_2 \cdot 2H_2O$	ord.	−119.0
Carbonate	$Ba(CO_3)$	ord.	−58.9
Chlorate	$Ba(ClO_3)_2$	ord.	−87.5
Chloride	$BaCl_2$	ord.	−72.6
"	$BaCl_2 \cdot 2H_2O$	ord.	−100.0
Fluoride	BaF_2	ord.	−51.0
Hydroxide	$Ba(OH)_2$	ord.	−53.2
"	$Ba(OH)_2 \cdot 8H_2O$	ord.	−157.0
Iodate	$Ba(IO_3)_2$	ord.	−122.5
Iodide	BaI_2	ord.	−124.0
"	$BaI_2 \cdot 2H_2O$	ord.	−163.0
Nitrate	$Ba(NO_3)_2$	ord.	−66.5
Oxide	BaO	ord.	−29.1
"	BaO_2	ord.	−40.6
Sulfate	$BaSO_4$	ord.	−71.3
Beryllium (s)	Be	ord.	−9.0
Chloride	$BeCl_2$	ord.	−26.5
Hydroxide	$Be(OH)_2$	ord.	−23.1
Nitrate (aq)	$Be(NO_3)_2$	298	−41.0
Oxide	BeO	ord.	−11.9
Sulfate	$BeSO_4$	ord.	−37.0
Bismuth (s)	Bi	ord.	−280.1
" (*l*)	Bi	....	−10.5
Bromide	$BiBr_3$	ord.	−147.0
Chloride	$BiCl_3$	ord.	−26.5
Chromate	$Bi_2(CrO_4)_3$	ord.	+154.0
Fluoride	BiF_3	303	−61.0
Hydroxide	$Bi(OH)_3$	ord.	−65.8
Iodide	BiI_3	ord.	−200.5
Nitrate	$Bi(NO_3)_3$	ord.	−91.0
"	$Bi(NO_3)_3 \cdot 5H_2O$	ord.	−159.0
Oxide	BiO	ord.	−110.0
"	Bi_2O_3	ord.	−83.0
Phosphate	$BiPO_4$	ord.	−77.0
Sulfate	$Bi_2(SO_4)_3$	ord.	−199.0
Sulfide	Bi_2S_3	ord.	−123.0
Boric Acid	H_3BO_3	ord.	−34.1
Boron (s)	B	ord.	−6.7
Chloride	BCl_3	ord.	−59.9
Oxide	B_2O_3	ord.	−39.0
Bromine (*l*)	Br_2	....	−56.4
" (g)	Br_2	....	−73 5

Substance	Formula	Temp. °K.	Susceptibility 10^{-6} cgs
Bromine (*l*) (continued)			
Fluoride	BrF_3	ord.	−33.9
"	BrF_5	ord.	−45.1
Cadmium (s)	Cd	ord.	−19.8
" (*l*)	Cd	....	−18.0
Acetate	$Cd(C_2H_3O_2)_2$	ord.	−83.7
Bromide	$CdBr_2$	ord.	−87.3
"	$CdBr_2 \cdot 4H_2O$	ord.	−140.0
Carbonate	$CdCO_3$	ord.	−46.7
Chloride	$CdCl_2$	ord.	−68.7
"	$CdCl_2 \cdot 2H_2O$	ord.	−99.0
Chromate	$CdCrO_4$	ord.	−16.8
Cyanide	$Cd(CN)_2$	ord.	−54.0
Fluoride	CdF_2	ord.	−40.6
Hydroxide	$Cd(OH)_2$	ord.	−41.0
Iodate	$Cd(IO_3)_2$	ord.	−108.4
Iodide	CdI_2	ord.	−117.2
Nitrate	$Cd(NO_3)_2$	ord.	−55.1
"	$Cd(NO_3)_2 \cdot 4H_2O$	ord.	−140.0
Oxide	CdO	ord.	−30.0
Phosphate	$Cd_3(PO_4)_2$	ord.	−159.0
Sulfate	$CdSO_4$	ord.	−59.2
Sulfide	CdS	ord.	−50.0
Calcium (s)	Ca	....	+40.0
Acetate	$Ca(C_2H_3O_2)_2$	ord.	−70.5
Bromate	$Ca(BrO_3)_2$	ord.	−84.9
Bromide	$CaBr_2$	ord.	−73.8
"	$CaBr_2 \cdot 3H_2O$	ord.	−115.0
Carbonate	$CaCO_3$	ord.	−38.2
Chloride	$CaCl_2$	ord.	−54.7
Fluoride	CaF_2	ord.	−28.0
Hydroxide	$Ca(OH)_2$	ord.	−22.0
Iodate	$Ca(IO_3)_2$	ord.	−101.4
Iodide	CaI_2	ord.	−109.0
Nitrate (aq)	$Ca(NO_3)_2$	ord.	−45.9
Oxide	CaO	ord.	−15.0
"	CaO_2	ord.	−23.8
Sulfate	$CaSO_4$	ord.	−49.7
"	$CaSO_4 \cdot 2H_2O$	ord.	−74.0
Carbon (dia)	C	ord.	−5.9
" (graph)	C	ord.	−6.0
Dioxide	CO_2	ord.	−21.0
Monoxide	CO	ord.	−9.8
Cerium (α)	Ce	80.5	+5160.0
" (β)	Ce	293	+2450.0
" (β)	Ce	80.5	+6230.0
" (γ)	Ce	287.9	+2420.0
" (γ)	Ce	125.6	+4640.0
" (γ)	Ce	80.5	+5200.0
Chloride	$CeCl_3$	287	+2490.0
Fluoride	CeF_3	293	+2190.0
Nitrate	$Ce(NO_3)_3 \cdot 5H_2O$	292	+2310.0
Oxide	CeO_2	293	+26.0
Sulfate	$CeSO_4$	ord.	+37.0
"	$Ce_2(SO_4)_3 \cdot 5H_2O$	293	+4540.0
"	$Ce(SO_4)_2 \cdot 4H_2O$	293	−97.0
Sulfide	CeS	ord.	+2110.0
"	Ce_2S_3	292	+5080.0
Cesium (s)	Cs	ord.	+29.0
" (*l*)	Cs	....	+26.5
Bromate	$CsBrO_3$	ord.	−75.1
Bromide	CsBr	ord.	−67.2
Carbonate	Cs_2CO_3	ord.	−103.6
Chlorate	$CsClO_3$	ord.	−65.0
Chloride	CsCl	ord.	−86.7
Fluoride	CsF	ord.	−44.5
Iodate	$CsIO_3$	ord.	−83.1
Iodide	CsI	ord.	−82.6
Oxide	CsO_2	293	+1534.0
"	CsO_2	90	+4504.0
Sulfate	Cs_2SO_4	ord.	−116.0
Sulfide	Cs_2S	ord.	−104.0
Chlorine (*l*)	Cl_2	ord.	−40.5
Fluoride, tri	ClF_3	ord.	−26.5
Chromium	Cr	273	+180.0
"	Cr	1713	+224.0
Acetate	$Cr(C_2H_3O_2)_2$	293	+5104.0
Chloride	$CrCl_2$	293	+7230.0
"	$CrCl_3$	293	+6890.0
Fluoride	CrF_3	293	+4370.0
Oxide	Cr_2O_3	300	+1960.0
"	CrO_3	ord.	+40.0
Sulfate	$Cr_2(SO_4)_3$	293	+11,800.0
"	$Cr_2(SO_4)_3 \cdot 8H_2O$	290	+12,700.0
"	$Cr_2(SO_4)_3 \cdot 10H_2O$	290	+12,600.0
"	$Cr_2(SO_4)_3 \cdot 14H_2O$	290	+12,160.0
"	$CrSO_4 \cdot 6H_2O$	293	+9690.0
Sulfide	CrS	ord.	+2390.0
Cobalt	Co	....	ferro
Acetate	$Co(C_2H_3O_2)_2$	293	+11,000.0
Bromide	$CoBr_2$	293	+13,000.0
Chloride	$CoCl_2$	293	+12,660.0
"	$CoCl_2 \cdot 6H_2O$	293	+9710.0
Cyanide	$Co(CN)_2$	303	+3825.0
Cobalt (continued)			
Fluoride	CoF_2	293	+9490.0
"	CoF_3	293	+1900.0
Iodide	CoI_2	293	+10,760.0
Oxide	CoO	260	+4900.0
"	Co_2O_3	ord.	+4560.0
"	Co_3O_4	ord.	+7380.0
Phosphate	$Co_3(PO_4)_2$	291	+28,110.0
Sulfate	$CoSO_4$	293	+10,000.0
"	$Co_2(SO_4)_3$	297	+1000.0
Sulfide	CoS	688	+251.0
"	CoS	293	+225.0
Thiocyanate	$Co(SCN)_2$	303	+11,090.0
Copper (s)	Cu	296	−5.46
" (*l*)	Cu	....	−6.16
Bromide	CuBr	ord.	−49.0
"	$CuBr_2$	341.6	+653.3
"	$CuBr_2$	292.7	+685.5
"	$CuBr_2$	189	+736.9
"	$CuBr_2$	90	+658.7
Chloride	CuCl	ord.	−40.0
"	$CuCl_2$	373.3	+1030.0
"	$CuCl_2$	289	+1080.0
"	$CuCl_2$	170	+1815.0
"	$CuCl_2$	69.25	+2370.0
"	$CuCl_2 \cdot 2H_2O$	293	+1420.0
Cyanide	CuCN	ord.	−24.0
Fluoride	CuF_2	293	+1050.0
"	CuF_2	90	+1420.0
"	$CuF_2 \cdot 2H_2O$	293	+1600.0
Hydroxide	$Cu(OH)_2$	292	+1170.0
Iodide	CuI	ord.	−63.0
Nitrate	$Cu(NO_3)_2 \cdot 3H_2O$	293	+1570.0
"	$Cu(NO_3)_2 \cdot 6H_2O$	293	+1625.0
Oxide	Cu_2O	293	−20.0
"	CuO	780	+259.6
"	CuO	561	+267.3
"	CuO	397	+256.9
"	CuO	289.6	+238.9
"	CuO	120	+156.2
Phosphide	Cu_3P	ord.	−33.0
"	CuP_2	ord.	−35.0
Sulfate	$CuSO_4$	293	+1330.0
"	$CuSO_4 \cdot H_2O$	293	+1520.0
"	$CuSO_4 \cdot 3H_2O$	ord.	+1480.0
"	$CuSO_4 \cdot 5H_2O$	293	+1460.0
Sulfide	CuS	293	−2.0
Thiocyanate	CuSCN	ord.	−48.0
Dysprosium	Dy	293.2	103,500.0
Oxide	Dy_2O_3	287.2	+89,600.0
Sulfate	$Dy_2(SO_4)_3$	293	+91,400.0
"	$Dy_2(SO_4)_3 \cdot 8H_2O$	291.2	+92,760.0
Sulfide	Dy_2S_3	292	+95,200.0
Erbium	Er	291	+44,300.0
Oxide	Er_2O_3	286	+73,920.0
Sulfate	$Er_2(SO_4)_3 \cdot 8H_2O$	293	+74,600.0
Sulfide	Er_2S_3	292	+77,200.0
Europium	Eu	293	+34,000.0
Bromide	$EuBr_2$	292	+26,800.0
Chloride	$EuCl_2$	292	+26,500.0
Fluoride	EuF_2	292	+23,750.0
Iodide	EuI_2	292	+26,000.0
Oxide	Eu_2O_3	298	+10,100.0
Sulfate	$EuSO_4$	293	+25,730.0
"	$Eu_2(SO_4)_3$	293	+10,400.0
"	$Eu_2(SO_4)_3 \cdot 8H_2O$	293	+9,540.0
Sulfide	EuS	293	+23,800.0
"	EuS	195	+35,400.0
Gadolinium	Gd	300.6	+755,000.0
Chloride	$GdCl_3$	293	+27,930.0
Oxide	Gd_2O_3	293	+53,200.0
Sulfate	$Gd_2(SO_4)_3$	285.5	+54,200.0
"	$Gd_2(SO_4)_3 \cdot 8H_2O$	293	+53,280.0
Sulfide	Gd_2S_3	292	+55,500.0
Gallium (s)	Ga	80	−24.4
" (s)	Ga	290	−21.6
" (*l*)	Ga	313	+2.5
Chloride	$GaCl_3$	ord.	−63.0
Iodide	GaI_3	ord.	−149.0
Oxide	Ga_2O	ord.	−34.0
Sulfide	Ga_2S	ord.	−36.0
"	GaS	ord.	−23.0
"	Ga_2S_3	ord.	−80.0
Germanium	Ge	293	−76.84
Chloride	$GeCl_4$	ord.	−72.0
Fluoride	GeF_4	ord.	−50.0
Iodide	GeI_4	ord.	−174.0
Oxide	GeO	ord.	−28.8
"	GeO_2	ord.	−34.3
Sulfide	GeS	ord.	−40.9
"	GeS_2	ord.	−53.3
Gold (s)	Au	296	−28.0
" (*l*)	Au	....	−34.0

MAGNETIC SUSCEPTIBILITY OF THE ELEMENTS AND INORGANIC COMPOUNDS (Continued)

Substance	Formula	Temp. °K.	Susceptibility 10^{-6} cgs
Gold (continued)			
Bromide	$AuBr$	ord.	−61.0
Chloride	$AuCl$	ord.	−67.0
	$AuCl_3$	ord.	−112.0
Fluoride	AuF_3	ord.	+74.0
Iodide	AuI	ord.	−91.0
Phosphide	AuP_2	ord.	−107.0
Hafnium (s)	Hf	298	+75.0
" (s)	Hf	1673	+104.0
Oxide	HfO_2	ord.	−23.0
Helium (g)	He	ord.	−1.88
Holmium	Ho	....	....
Oxide	Ho_2O_3	293	+88,100.0
Sulfate	$Ho_2(SO_4)_3$	293	+91,700.0
"	$Ho_2(SO_4)_3 \cdot 8H_2O$	293	+91,600.0
Hydrogen (g)	H_2	....	−3.98
Bromide (*l*)	HBr	273	....
" (aq)	HBr	ord.	....
Chloride (*l*)	HCl	273	−22.6
" (aq)	HCl	300	−22.0
Fluoride (*l*)	HF	287	−8.6
" (aq)	HF	ord.	−9.3
Iodide (*l*)	HI	281	−47.7
" (*l*)	HI	233	−48.3
" (s)	HI	195	−47.3
" (aq)	HI	ord.	−50.2
Oxide—See Water			
Peroxide	H_2O_2	ord.	−17.7
Sulfide	H_2S	ord.	−25.5
Indium	In	ord.	−64.0
Bromide	$InBr_3$	ord.	−107.0
Chloride	$InCl$	ord.	−30.0
"	$InCl_2$	ord.	−56.0
"	$InCl_3$	ord.	−86.0
Fluoride	InF_3	ord.	−61.0
Oxide	In_2O	ord.	−47.0
"	In_2O_3	ord.	−56.0
Sulfide	In_2S	ord.	−50.0
"	InS	ord.	−28.0
"	In_2S_3	ord.	−98.0
Iodic Acid	HIO_3	ord.	−48.0
metaper	HIO_4	ord.	−56.5
orthoparaper	H_5IO_6	ord.	−71.4
Iodine (s)	I_2	ord.	−88.7
" (atomic)	I	1303	+869.0
" (atomic)	I	1400	+1120.0
Chloride	ICl	ord.	−54.6
"	ICl_3	ord.	−90.2
Fluoride	IF_5	ord.	−58.1
Oxide	I_2O_5	ord.	−79.4
Iridium	Ir	298	+25.6
"	Ir	698	+32.1
Chloride	$IrCl_3$	ord.	−14.4
Oxide	IrO_2	298	+224.0
Iron	Fe	....	ferro
Bromide	$FeBr_2$	ord.	+13,600.0
Carbonate	$FeCO_3$	293	+11,300.0
Chloride	$FeCl_2$	293	+14,750.0
"	$FeCl_2 \cdot 4H_2O$	293	+12,900.0
"	$FeCl_3$	293	+13,450.0
"	$FeCl_3$	398	+9,980.0
"	$FeCl_3 \cdot 6H_2O$	290	+15,250.0
Fluoride	FeF_2	293	+9,500.0
"	FeF_3	305	+13,760.0
"	$FeF_3 \cdot 3H_2O$	293	+7870.0
Iodide	FeI_2	ord.	+13,600.0
Nitrate	$Fe(NO_3)_3 \cdot 9H_2O$	293	+15,200.0
Oxide	FeO	293	+7200.0
"	Fe_2O_3	1033	+3586.0
Phosphate	$FePO_4$	ord.	+11,500.0
Sulfate	$FeSO_4$	293	+10,200.0
"	$FeSO_4 \cdot H_2O$	290	+10,500.0
"	$FeSO_4 \cdot 7H_2O$	293	+11,200.0
Sulfide	FeS	293	+1074.0
Krypton	Kr	....	−28.8
Lanthanum	La	ord.	+118.0
Oxide	La_2O_3	ord.	−78.0
Sulfate	$La_2(SO_4)_3 \cdot 9H_2O$	293	−262.0
Sulfide	La_2S_3	292	−37.0
"	La_2S_4	293	−100.0
Lead (s)	Pb	289	−23.0
" (*l*)	Pb	330	−15.5
Acetate	$Pb(C_2H_3O_2)_2$	ord.	−89.1
Bromide	$PbBr_2$	ord.	−90.6
Carbonate	$PbCO_3$	ord.	−61.2
Chloride	$PbCl_2$	ord.	−73.8
Chromate	$PbCrO_4$	ord.	−18.0
Fluoride	PbF_2	ord.	−58.1
Iodate	$Pb(IO_3)_2$	ord.	−131.0
Iodide	PbI_2	ord.	−126.5
Nitrate	$Pb(NO_3)_2$	ord.	−74.0
Oxide	PbO	ord.	−42.0
Phosphate	$Pb_3(PO_4)_2$	ord.	−182.0
Sulfate	$PbSO_4$	ord.	−69.7
Sulfide	PbS	ord.	−84.0
Lead (s) (continued)			
Thiocyanate	$Pb(CNS)_2$	ord.	−82.0
Lithium	Li	ord.	+14.2
Acetate	$LiC_2H_3O_2$	ord.	−34.0
Bromate	$LiBrO_3$	ord.	−39.0
Bromide	$LiBr$	ord.	−34.7
Carbonate	Li_2CO_3	ord.	−27.0
Chlorate (aq)	$LiClO_3$	ord.	−28.8
Chloride	$LiCl$	ord.	−24.3
Fluoride	LiF	ord.	−10.1
Hydride	LiH	ord.	−4.6
Hydroxide	$LiOH$	ord.	−12.3
Iodate	$LiIO_3$	ord.	−47.0
Iodide	LiI	ord.	−50.0
Nitrate	$LiNO_3 \cdot 3H_2O$	ord.	−62.0
Sulfate	Li_2SO_4	ord.	−40.0
Lutetium	Lu	ord.	>0.0
Magnesium	Mg	ord.	+13.1
Acetate	$Mg(C_2H_3O_2)_2 \cdot 4H_2O$	ord.	−116.0
Bromide	$MgBr_2$	ord.	−72.0
Carbonate	$MgCO_3$	ord.	−32.4
"	$MgCO_3 \cdot 3H_2O$	ord.	−72.7
Chloride	$MgCl_2$	ord.	−47.4
Fluoride	MgF_2	ord.	−22.7
Hydroxide	$Mg(OH)_2$	288	−22.1
Iodide	MgI_2	ord.	−111.0
Oxide	MgO	ord.	−10.2
Phosphate	$Mg_3(PO_4)_2 \cdot 4H_2O$	ord.	−167.0
Sulfate	$MgSO_4$	294	−50.0
"	$MgSO_4 \cdot H_2O$	ord.	−61.0
"	$MgSO_4 \cdot 5H_2O$	ord.	−109.0
"	$MgSO_4 \cdot 7H_2O$	ord.	−135.7
Manganese (α)	Mn	293	+529.0
" (β)	Mn	293	+483.0
Acetate	$Mn(C_2H_3O_2)_2$	293	+13,650.0
Bromide	$MnBr_2$	294	+13,900.0
Carbonate	$MnCO_3$	293	+11,400.0
Chloride	$MnCl_2$	293	+14,350.0
"	$MnCl_2 \cdot 4H_2O$	293	+14,600.0
Fluoride	MnF_2	290	+10,700.0
"	MnF_3	293	+10,500.0
Hydroxide	$Mn(OH)_2$	293	+13,500.0
Iodide	MnI_2	293	+14,400.0
Oxide	MnO	293	+4850.0
"	Mn_2O_3	293	+14,100.0
"	MnO_2	293	+2280.0
"	Mn_3O_4	298	+12,400.0
Sulfate	$MnSO_4$	293	+13,660.0
"	$MnSO_4 \cdot H_2O$	293	+14,200.0
"	$MnSO_4 \cdot 4H_2O$	293	+14,600.0
"	$MnSO_4 \cdot 5H_2O$	293	+14,700.0
Sulfide (α)	MnS	293	+5630.0
" (β)	MnS	293	+3850.0
Mercury (s)	Hg	....	−24.1
" (*l*)	Hg	293	−33.44
" (g)	Hg	....	−78.3
Acetate	$HgC_2H_3O_2$	ord.	−70.5
"	$Hg(C_2H_3O_2)_2$	ord.	−100.0
Bromate	$HgBrO_3$	ord.	−57.7
Bromide	$HgBr$	ord.	−57.2
"	$HgBr_2$	ord.	−94.2
Chloride	$HgCl$	ord.	−52.0
"	$HgCl_2$	ord.	−82.0
Chromate	Hg_2CrO_4	ord.	−63.0
"	$HgCrO_4$	ord.	−12.5
Cyanide	$Hg(CN)_2$	ord.	−67.0
Fluoride	HgF	ord.	−53.0
"	HgF_2	302	−62.0
Hydroxide	$Hg_2(OH)_2$	ord.	−100.0
Iodate	$HgIO_3$	ord.	−92.0
Iodide	HgI	ord.	−83.0
"	HgI_2	ord.	−128.6
Nitrate	$HgNO_3$	ord.	−55.9
"	$Hg(NO_3)_2$	ord.	−74.0
Oxide	Hg_2O	ord.	−76.3
"	HgO	ord.	−44.0
Sulfate	Hg_2SO_4	ord.	−123.0
"	$HgSO_4$	ord.	−78.1
Sulfide	HgS	ord.	−55.4
Thiocyanate	$Hg(SCN)_2$	ord.	−96.5
Molybdenum	Mo	298	+89.0
"	Mo	63.8	+108.0
"	Mo	20.4	+149.2
Bromide	$MoBr_3$	293	+525.0
"	$MoBr_4$	293	+520.0
"	Mo_2Br_6	290.5	−46.0
Chloride	$MoCl_3$	290	+43.0
"	$MoCl_4$	291	+1750.0
"	$MoCl_5$	289	+990.0
Fluoride	MoF_6	ord.	−26.0
Oxide	Mo_2O_3	ord.	−42.0
"	MoO_2	289	+41.0
"	MoO_3	292.5	+3.0
"	Mo_2O_5	ord.	+42.0
Sulfide	MoS_2	289	−63.0

MAGNETIC SUSCEPTIBILITY OF THE ELEMENTS AND INORGANIC COMPOUNDS (Continued)

Substance	Formula	Temp. °K.	Susceptibility 10^{-6} cgs
Neodymium	Nd	287.7	5628.0
Fluoride	NdF_3	293	+4980.0
Nitrate	$Nd(NO_3)_3$	293	+5020.0
Oxide	Nd_2O_3	292.0	+10,200.0
Sulfate	$Nd_2(SO_4)_3$	293	+9990.0
Sulfide	Nd_2S_3	292	+5550.0
Neon	Ne	ord.	−6.74
Nickel	Ni	....	ferro
Acetate	$Ni(C_2H_3O_2)_2$	293	+4690.0
Bromide	$NiBr_2$	293	+5600.0
Chloride	$NiCl_2$	293	+6145.0
"	$NiCl_2 \cdot 6H_2O$	293	+4240.0
Fluoride	NiF_2	293	+2410.0
Hydroxide	$Ni(OH)_2$	ord.	+4500.0
Iodide	NiI_2	293	+3875.0
Nitrate	$Ni(NO_3)_2 \cdot 6H_2O$	293.5	+4300.0
Oxide	NiO	293	+660.0
Sulfate	$NiSO_4$	293	+4005.0
Sulfide	NiS	293	+190.0
"	Ni_3S_2	ord.	+1030.0
Niobium	Nb	298	+195.0
Oxide	Nb_2O_5	ord.	−10.0
Nitric Acid	HNO_3	ord.	−19.9
Nitrogen	N_2	ord.	−12.0
Oxide	N_2O	285	−18.9
" (g)	NO	293	+1461.0
" (g)	NO	203.8	+1895.0
" (g)	NO	146.9	+2324.0
" (*l*)	NO	117.64	+114.2
" (s)	NO	90	+19.8
"	N_2O_3	291	−16.0
"	NO_2	408	+150.0
"	N_2O_4	303.6	−22.1
"	N_2O_4	295.1	−23.0
"	N_2O_4	257	−25.4
" (aq)	N_2O_5	289	−35.6
Osmium	Os	298	+9.9
Chloride	$OsCl_2$	ord.	+41.3
Oxygen (g)	O_2	293	+3449.0
" (*l*)	O_2	90.1	+7699.0
" (*l*)	O_2	70.8	+8685.0
" (s, γ)	O_2	54.3	+10,200.0
" (s, β)	O_2	....	....
" (s, α)	O_2	23.7	+1760.0
Ozone (*l*)	O_3	....	+6.7
Palladium	Pd	288	+567.4
Chloride	$PdCl_2$	291.3	−38.0
Fluoride	PdF_3	293	+1760.0
Hydride	PdH	ord.	+1077.0
"	Pd_4H	ord.	+2353.0
Phosphoric Acid (aq)	H_3PO_4	ord.	−43.8
Phosphorous Acid (aq)	H_3PO_3	ord.	−42.5
Phosphorous (red)	P	....	−20.8
" (black)	P	....	−26.6
Chloride	PCl_3	ord.	−63.4
Platinum	Pt	290.3	+201.9
Chloride	$PtCl_2$	298	−54.0
"	$PtCl_3$	ord.	−66.7
"	$PtCl_4$	ord.	−93.0
Fluoride	PtF_4	293	+455.0
Oxide	Pt_2O_3	ord.	−37.70
Plutonium	Pu	293	+610.0
Fluoride	PuF_4	301	+1760.0
"	PuF_6	295	+173.0
Oxide	PuO_2	300	+730.0
Potassium	K	ord.	+20.8
Acetate (aq)	$KC_2H_3O_2$	28	−45.0
Bromate	$KBrO_3$	ord.	−52.6
Bromide	KBr	ord.	−49.1
Carbonate	K_2CO_3	ord.	−59.0
Chlorate	$KClO_3$	ord.	−42.8
Chloride	KCl	ord.	−39.0
Chromate	K_2CrO_4	ord.	−3.9
"	$K_2Cr_2O_7$	293	+29.4
Cyanide	KCN	ord.	−37.0
Ferricyanide	$K_3Fe(CN)_6$	297	+2290.0
Ferrocyanide	$K_4Fe(CN)_6$	ord.	−130.0
"	$K_4Fe(CN)_6 \cdot 3H_2O$	ord.	−172.3
Fluoride	KF	ord.	−23.6
Hydroxide (aq)	KOH	ord.	−22.0
Iodate	KIO_3	ord.	−63.1
Iodide	KI	ord.	−63.8
Nitrate	KNO_3	ord.	−33.7
Nitrite	KNO_2	ord.	−23.3
Oxide	KO_2	293	+3230.0
"	KO_3	ord.	+1185.0
Permanganate	$KMnO_4$	ord.	+20.0
Sulfate	K_2SO_4	ord.	−67.0
"	$KHSO_4$	ord.	−49.8
Sulfide	K_2S	ord.	−60.0
"	K_2S_2	ord.	−71.0
"	K_2S_3	ord.	−80.0
"	K_2S_4	ord.	−89.0
"	K_2S_5	ord.	−98.0
Sulfite	K_2SO_3	ord.	−64.0
Potassium (continued)			
Thiocyanate	KSCN	ord.	−48.0
Praseodymium	Pr	293	+5010.0
Chloride	$PrCl_3$	307.1	+44.5
Oxide	PrO_2	293	+1930.0
"	Pr_2O_3	827	+4000.0
Oxide	Pr_2O_3	294.5	+8994.0
Sulfate	$Pr_2(SO_4)_3$	291	+9660.0
"	$Pr_2(SO_4)_3 \cdot 8H_2O$	289	+9880.0
Sulfide	Pr_2S_3	292	+10,770.0
Rhenium	Re	293	+67.6
Chloride	$ReCl_5$	293	+1225.0
Oxide	ReO_2	ord.	+44.0
"	$ReO_2 \cdot 2H_2O$	295	+74.0
"	ReO_3	ord.	+16.0
"	Re_2O_7	ord.	−16.0
Sulfide	ReS_2	ord.	+38.0
Rhodium	Rh	298	+111.0
"	Rh	723	+123.0
Chloride	$RhCl_3$	298	−7.5
Fluoride	RhF_4	293	+500.0
Oxide	Rh_2O_3	298	+104.0
Sulfate	$Rh_2(SO_4)_3 \cdot 6H_2O$	298	+104.0
"	$Rh_2(SO_4)_3 \cdot 14H_2O$	298	+149.0
Rubidium	Rb	303	+17.0
Bromide	RbBr	ord.	−56.4
Carbonate	Rb_2CO_3	ord.	−75.4
Chloride	RbCl	ord.	−46.0
Fluoride	RbF	ord.	−31.9
Iodide	RbI	ord.	−72.2
Nitrate	$RbNO_3$	ord.	−41.0
Oxide	RbO_2	293	+1527.0
Sulfate	Rb_2SO_4	ord.	−88.4
Sulfide	Rb_2S	ord.	−80.0
"	Rb_2S_3	ord.	−90.0
Ruthenium	Ru	298	+43.2
"	Ru	723	+50.2
Chloride	$RuCl_3$	290.9	+1998.0
Oxide	RuO_2	298	+162.0
Samarium	Sm	291	+1860.0
"	Sm	195	+2230.0
Bromide	$SmBr_2$	293	+5337.0
"	$SmBr_3$	293	+972.0
Oxide	Sm_2O_3	292	+1988.0
"	Sm_2O_3	170	+1960.0
"	Sm_2O_3	85	+2282.0
Sulfate	$Sm_2(SO_4)_3 \cdot 8H_2O$	293	+1710.0
Sulfide	Sm_2S_3	292	+3300.0
Scandium	Sc	292	+315.0
Selenic Acid	H_2SeO_4	ord.	−51.2
Selenious Acid	H_2SeO_3	ord.	−45.4
Selenium (s)	Se	ord.	−25.0
" (*l*)	Se	900	−24.0
Bromide	Se_2Br_2	ord.	−113.0
Chloride	Se_2Cl_2	ord.	−94.8
Fluoride	SeF_6	ord.	−51.0
Oxide	SeO_2	ord.	−27.2
Silicon	Si	ord.	−3.9
Bromide	$SiBr_4$	ord.	−128.6
Carbide	SiC	ord.	−12.8
Chloride	$SiCl_4$	ord.	−88.3
Hydroxide	$Si(OH)_4$	ord.	−42.6
Oxide	SiO_2	ord.	−29.6
Silver (s)	Ag	296	−19.5
" (*l*)	Ag	....	−24.0
Acetate	$AgC_2H_3O_2$	ord.	−60.4
Bromide	AgBr	283	−59.7
Carbonate	Ag_2CO_3	ord.	−80.90
Chloride	AgCl	ord.	−49.0
Chromate	Ag_2CrO_4	ord.	−40.0
Cyanide	AgCN	ord.	−43.2
Fluoride	AgF	ord.	−36.5
Iodide	AgI	ord.	−80.0
Nitrate	$AgNO_3$	ord.	−45.7
Nitrite	$AgNO_2$	ord.	−42.0
Oxide	Ag_2O	ord.	−134.0
"	AgO	287	−19.6
Permanganate	$AgMnO_4$	300	−63.0
Phosphate	Ag_3PO_4	ord.	−120.0
Sulfate	Ag_2SO_4	ord.	−92.90
Thiocyanate	AgSCN	ord.	−61.8
Sodium	Na	ord.	+16.0
Acetate	$NaC_2H_3O_2$	ord.	−37.6
Borate tetra	$Na_2B_4O_7$	ord.	−85.0
Bromate	$NaBrO_3$	ord.	−44.2
Bromide	NaBr	ord.	−41.0
Carbonate	Na_2CO_3	ord.	−41.0
Chlorate	$NaClO_3$	ord.	−34.7
Chloride	NaCl	ord.	−30.3
Chromate, di.	$Na_2Cr_2O_7$	ord.	+55.0
Fluoride	NaF	ord.	−16.4
Hydroxide (aq)	NaOH	300	−16.0
Iodate	$NaIO_3$	ord.	−53.0
Iodide	NaI	ord.	−57.0
Nitrate	$NaNO_3$	ord.	−25.6

MAGNETIC SUSCEPTIBILITY OF THE ELEMENTS AND INORGANIC COMPOUNDS (Continued)

Substance	Formula	Temp. °K.	Susceptibility 10^{-6} cgs
Sodium (continued)			
Nitrite	$NaNO_2$	ord.	−14.5
Oxide	Na_2O	ord.	−19.8
"	Na_2O_2	ord.	−28.10
Phosphate, meta	$NaPO_3$	ord.	−42.5
"	Na_2HPO_4	ord.	−56.6
Sulfate	Na_2SO_4	ord.	−52.0
"	$Na_2SO_4 \cdot 10H_2O$	ord.	−184.0
Sulfide	Na_2S	ord.	−39.0
"	Na_2S_2	ord.	−53.0
"	Na_2S_3	ord.	−68.0
"	Na_2S_4	ord.	−84.0
"	Na_2S_5	ord.	−99.0
Strontium	Sr	ord.	+92.0
Acetate	$Sr(C_2H_3O_2)_2$	ord.	−79.0
Bromate	$Sr(BrO_3)_2$	ord.	−93.5
Bromide	$SrBr_2$	ord.	−86.6
"	$SrBr_2 \cdot 6H_2O$	ord.	−160.0
Carbonate	$SrCO_3$	ord.	−47.0
Chlorate	$Sr(ClO_3)_2$	ord.	−73.0
Chloride	$SrCl_2$	ord.	−63.0
"	$SrCl_2 \cdot 6H_2O$	ord.	−145.0
Chromate	$SrCrO_4$	ord.	−5.1
Fluoride	SrF_2	ord.	−37.2
Hydroxide	$Sr(OH)_2$	ord.	−40.0
"	$Sr(OH)_2 \cdot 8H_2O$	ord.	−136.0
Iodate	$Sr(IO_3)_2$	ord.	−108.0
Iodide	SrI_2	ord.	−112.0
Nitrate	$Sr(NO_3)_2$	ord.	−57.2
"	$Sr(NO_3)_2 \cdot 4H_2O$	ord.	−106.0
Oxide	SrO	ord.	−35.0
"	SrO_2	ord.	−32.3
Sulfate	$SrSO_4$	ord.	−57.9
Sulfur (α)	S	ord.	−15.5
" (β)	S	ord.	−14.9
" (*l*)	S	....	−15.4
" (g)	S	828	+700.0
" (g)	S	1023	+464.0
Chloride	S_2Cl_2	ord.	−62.2
"	SCl_2	ord.	−49.4
"	SCl_3	ord.	−49.4
Fluoride	SF_6	ord.	−44.0
Iodide	SI	ord.	−52.7
Oxide (*l*)	SO_2	ord.	−18.2
Sulfuric Acid	H_2SO_4	ord.	−39.8
Tantalum	Ta	293	+154.0
"	Ta	2143	+124.0
Chloride	$TaCl_5$	304	+140.0
Fluoride	TaF_5	293	+795.0
Oxide	Ta_2O_5	ord.	−32.0
Technetium	Tc	402	250.0
"	Tc	298	270.0
"	Tc	78	290.0
Oxide	$TcO_2 \cdot 2H_2O$	300	+244.0
"	Tc_2O_7	298	−40.0
Tellurium (s)	Te	ord.	−39.5
" (*l*)	Te	....	−6.4
Bromide	$TeBr_2$	ord.	−106.0
Chloride	$TeCl_2$	ord.	−94.0
Fluoride	TeF_6	ord.	−66.0
Terbium	Tb	273	+146,000.0
Oxide	Tb_2O_3	288.1	+78,340.0
Sulfate	$Tb_2(SO_4)_3$	293	+78,200.0
"	$Tb_2(SO_4)_3 \cdot 8H_2O$	293	+76,500.0
Thallium (α)	Tl	ord.	−50.9
" (β)	Tl	>508	−32.3
" (*l*)	Tl	573	−26.8
Acetate	$TlC_2H_3O_2$	ord.	−69.0
Bromate	$TlBrO_3$	ord.	−75.9
Bromide	$TlBr$	ord.	−63.9
Carbonate	Tl_2CO_3	ord.	−101.6
Chlorate	$TlClO_3$	ord.	−65.5
Chloride	$TlCl$	ord.	−57.8
Chromate	Tl_2CrO_4	ord.	−39.3
Cyanide	$TlCN$	ord.	−49.0
Fluoride	TlF	ord.	−44.4
Iodate	$TlIO_3$	ord.	−86.8
Iodide	TlI	ord.	−82.2
Nitrate	$TlNO_3$	ord.	−56.5
Nitrite	$TlNO_2$	ord.	−50.8
Oxide	Tl_2O_3	ord.	+76.0
Phosphate	Tl_3PO_4	ord.	−145.2
Sulfate	Tl_2SO_4	ord.	−112.6
Sulfide	Tl_2S	ord.	−88.8
Thiocyanate	$TlCNS$	ord.	−66.7
Thorium	Th	293	+132.0
"	Th	90	+153.0
Chloride	$ThCl_4 \cdot 8H_2O$	305.2	−180.0
Nitrate	$Th(NO_3)_4$	ord.	−108.0
Oxide	ThO_2	ord.	−16.0
Thulium	Tm	291	+25,500.0
Oxide	Tm_2O_3	296.5	+51,444.0
Tin (white)	Sn	ord.	+3.1
" (gray)	Sn	280	−37.0
" (gray)	Sn	100	−31.7
Tin (continued)			
" (*l*)	Sn	....	−4.5
Bromide	$SnBr_4$	ord.	−149.0
Chloride	$SnCl_2$	ord.	−69.0
"	$SnCl_2 \cdot 2H_2O$	ord.	−91.4
" (*l*)	$SnCl_4$	ord.	−115.0
Hydroxide	$Sn(OH)_4$	ord.	−60.0
Oxide	SnO	ord.	−19.0
"	SnO_2	ord.	−41.0
Titanium	Ti	293	+153.0
"	Ti	90	+150.0
Bromide	$TiBr_2$	288	+640.0
"	$TiBr_3$	441	+520.0
"	$TiBr_3$	291	+660.0
"	$TiBr_3$	195	+680.0
"	$TiBr_3$	90	+220.0
Carbide	TiC	ord.	+8.0
Chloride	$TiCl_2$	288	+570.0
"	$TiCl_2$	685	+705.0
"	$TiCl_3$	373	+1030.0
"	$TiCl_3$	292	+1110.0
"	$TiCl_3$	212	+690.0
"	$TiCl_3$	90	+220.0
"	$TiCl_4$	ord.	−54.0
Fluoride	TiF_3	293	+1300.0
Iodide	TiI_2	288	+1790.0
"	TiI_3	434	+221.0
"	TiI_3	292	+160.0
"	TiI_3	195	+159.0
"	TiI_3	90	+167.0
Oxide	Ti_2O_3	382	+152.0
"	Ti_2O_3	298	+125.6
"	TiO_2	248	+132.4
"	TiO_2	ord.	+5.9
Sulfide	TiS	ord.	+432.0
Tungsten	W	298	+59.0
Bromide	WBr_5	293	+250.0
Carbide	WC	ord.	+10.0
Chloride	WCl_2	293	−25.0
"	WCl_5	293	+387.0
"	WCl_6	ord.	−71.0
Fluoride	WF_6	ord.	−40.0
Oxide	WO_2	ord.	+57.0
"	WO_3	ord.	−15.8
Sulfide	WS_2	303	+5850.0
Tungstic Acid, ortho	H_2WO_4	ord.	−28.0
Uranium (α)	U	78	+395.0
" (α)	U	298	+409.0
" (α)	U	623	+440.0
" (β)	U	....	....
" (γ)	U	1393	+514.0
Bromide	UBr_3	294	+4740.0
"	UBr_4	293	+3530.0
Chloride	UCl_3	300	+3460.0
"	UCl_4	294	+3680.0
Fluoride	UF_4	300	+3530.0
"	UF_6	ord.	+43.0
Hydrides	UH_3	462	+2821.0
"	UH_3	391	+3568.0
"	UH_3	295	+6244.0
"	UH_3	255	+9306.0
Iodide	UI_3	293	+4460.0
Oxide	UO	293	+1600.0
"	UO_2	293	+2360.0
"	UO_3	ord.	+128.0
Sulfate	$U(SO_4)_2$	ord.	+31.0
Sulfide (α)	US_2	290.5	+3137.0
" (β)	US_2	290.5	+3470.0
"	U_2S_3	ord.	+5206.0
"	U_3S_5	ord.	+11,220.0
Vanadium	V	298	+255.0
Bromide	VBr_2	293	+3230.0
"	VBr_2	195	+3760.0
"	VBr_2	90	+4470.0
"	VBr_3	293	+2890.0
"	VBr_3	195	+4110.0
"	VBr_3	90	+8540.0
Chloride	VCl_2	293	+2410.0
"	VCl_3	293	+3030.0
"	VCl_4	293	+1130.0
"	VCl_4	195	+1700.0
"	VCl_4	90	+4360.0
Fluoride	VF_3	293	+2730.0
Oxide	VO_2	290	+270.0
"	V_2O_3	293	+1976.0
"	V_2O_5	ord.	+128.0
Sulfide	VS	ord.	+600.0
"	V_2S_3	293	+1560.0
Water (g)	H_2O	>373	−13.1
" (*l*)	H_2O	373	−13.09
" (*l*)	H_2O	293	−12.97
" (*l*)	H_2O	273	−12.93
" (s)	H_2O	273	−12.65
" (s)	H_2O	223	−12.31

MAGNETIC SUSCEPTIBILITY OF THE ELEMENTS AND INORGANIC COMPOUNDS (Continued)

Substance	Formula	Temp. °K.	Susceptibility 10^{-6} cgs
Water (continued)			
" (*l*)	DHO	302	−12.97
" (*l*)	D_2O	293	−12.76
" (*l*)	D_2O	276.8	−12.66
" (*s*)	D_2O	276.8	−12.54
" (*s*)	D_2O	213	−12.41
Xenon	Xe	ord.	−43.9
Ytterbium	Yb	292	+249.0
"	Yb	90	+639.0
Sulfide	Yb_2S_3	292	+18,300.0
Yttrium	Y	292	+2.15
"	Y	90	+2.43
Oxide	Y_2O_3	293	+44.4
Sulfide	Y_2S_3	ord.	+100.0
Zinc (*s*)	Zn	ord.	−11.4
" (*l*)	Zn	....	−7.8
Acetate	$Zn(C_2H_3O_2)_2{\cdot}2H_2O$	ord.	−101.0
Carbonate	$ZnCO_3$	ord.	−34.0
Chloride	$ZnCl_2$	296	−65.0
Cyanide	$Zn(CN)_2$	ord.	−46.0
Fluoride	ZnF_2	299.6	−38.2
Hydroxide	$Zn(OH)_2$	ord.	−67.0
Iodide	ZnI_2	ord.	−98.0
Nitrate (aq)	$Zn(NO_3)_2$	ord.	−63.0
Oxide	ZnO	ord.	−46.0
Phosphate	$Zn_3(PO_4)_2$	ord.	−141.0
Sulfate	$ZnSO_4$	ord.	−45.0
"	$ZnSO_4{\cdot}H_2O$	ord.	−63.0
"	$ZnSO_4{\cdot}7H_2O$	ord.	−143.0
Sulfide	ZnS	ord.	−25.0
Zirconium	Zr	293	+122.0
"	Zr	90	+119.0
Carbide	ZrC	ord.	−26.0
Nitrate	$Zr(NO_3)_4{\cdot}5H_2O$	ord.	−77.0
Oxide	ZrO_2	ord.	−13.8

DIAMAGNETIC SUSCEPTIBILITIES OF ORGANIC COMPOUNDS

Compiled by George W. Smith

This table and its supplement contain values for the molar susceptibility of χ_M, specific susceptibility χ, and volumetric susceptibility K. The cgs Gaussian system of units is employed. In the Gaussian units the relation between magnetic induction B and magnetic field strength H is

$$B = H + 4\pi I \tag{1}$$

where I is the magnetization or magnetic moment per unit volume. Actually, the quantities involved in equation (1) are all vectors, but one may assume that all three are collinear, a reasonable assumption for organic diamagnetic substances in the liquid state. For crystals I may vary with crystal orientation. Equation (1) may be written

$$B = H + 4\pi KH = (I + 4\pi K)H \tag{2}$$

Here, K is the magnetic susceptibility, often called the volumetric susceptibility and is a unitless quantity.

Other susceptibilities of use to chemists and physicists are the specific or mass susceptibility which is defined

$$\chi = K/\rho \tag{3}$$

where ρ is the density of the sample in grams per cc., and the molar susceptibility which is defined as

$$\chi_M = M\chi = MK/\rho \tag{4}$$

where M is the molecular weight of the substance in grams.

Temperatures, when listed, are enclosed in parentheses and are listed in degrees C.

Literature references for values contained in this table may be found in General Motors Research Laboratories Bulletins GMR-317 and GMR-396.

Compound	$-\chi_M \times 10^6$	$-\chi \times 10^6$	$-K \times 10^6$
Acenaphthanthracene	184	.73	
Acenaphthene	109.3	(.709)	(.726) (99°)
Acetal	81.39	.688 (32°)	(.568) (32°)
Acetaldehyde	22.70	$.515_3$	(.403) (18°)
Acetamide	34.1	.577	(.618) (20°)
Acetic acid	31.54	.525 (32°)	(.551) (32°)
Acetic anhydride	(52.8)	.517	(.562) (15°)
Acetoaminofluorene	(141)	.63	
Acetone	33.7_8	$.581_4$	(.460) (20°)
Acetonitrile	28.0	(.682)	(.534) (20°)
Acetonylacetone	62.51	$.547_6$	(.531) (20°)
Acetophenone	72.05	$.599_8$	(.615) (20°)
Acetophenone oxime	79.9_0	$.592_0$	
Acetophenone oxime-O-methyl ether	92.3_1	$.618_8$	
Acetoxime	44.4_2	607_6	(.480) (20°)

DIAMAGNETIC SUSCEPTIBILITIES OF ORGANIC COMPOUNDS

Compound	$-\chi_M \times 10^6$	$-\chi \times 10^6$	$-K \times 10^6$
Acetoxime-O-benzyl ether	104.8_9	$.642_7$	
Acetoxime-O-methyl-ether	54.8_7	$.629_8$	
Acetylacetone	54.88	$.548_1$	(.535) (20°)
Acetyl chloride	38.9	.496	(.548) (20°)
Acetylene	12.5	(.480)	
Acetylphenylacetylene	86.9	.508	
Acetylthiophene	71.7	(.568)	
Acridine	(123.3)	.688	(.757) (20°)
Adonitol	91.30	.600	
Alanine	50.5	(.567)	
Allyl acetate	(56.7)	.566	(.525) (20°)
Allyl alcohol	36.70	.632	.540 (20°)
1-Allylpyrrole	73.80	.685 (20°)	
Aminoazobenzene	(118.3)	.600	
Aminoazotoluene	(142.2)	.631	
o-Aminoazotoluene	(138)	.61 ± .02	
α-Aminobutyric acid	62.1	(.602)	
Aminomethyldiethyldiazine	114.8	(.696)	
4-Aminostilbene	122.5	(.628)	
2-Aminothiazole	56.0	.564	
n-Amyl acetate	89.06	684_5	.5979 (20.7°)
iso-Amyl acetate	89.40	.687	.599 (20°)
n-Amyl alcohol	(67.5)	.766	(.624) (20°)
γ-Amyl alcohol	(71.0)	.8060	(.655) (25°)
Inactive Amyl alcohol	69.06	783_4 (25°)	
iso-Amyl alcohol	68.96	$.782_3$ (25°)	(.64) (15°)
sec-Amyl alcohol	69.1	.785	(.635) (20°)
tert-Amyl alcohol	(70.9)	.804	(,654) (15°)
n-Amylamine	69.4	(.796)	(.606) (20°)
iso-Amylamine	71.6	(.821)	(.616) (20°)
n-Amylbenzene	112.55	(.759)	(.652) (20°)
iso-Amyl bromide	(88.7)	.587	(.706) (20°)
iso-Amyl-n-butyrate	113.52	$.717_4$ (25°)	(.616) (25°)
iso-Amyl chloride	(79.0)	.741	(.662) (20°)
iso-Amyl cyanide	73.4	(.755)	(.609) (20°)
iso-Amylene	53.7	.766	
Amylene bromide	(114.5)	.498	
Amylene chloride	(95.2)	.675	
iso-Amyl ether	(129)	.813	(.635) (15°)
iso-Amyl formate	78.38	$.674_8$	(.591) (25°)
Amylidene chloride	(93.4)	.662	
Amyl iodide	(118.7)	.5996 (18°)	(.910) (20°)
n-Amyl methyl ketone	80.50	$.705_3$	(.580) (15°)
Amyl nitrate	(76.4)	.574	
iso-Amyl propionate	101.73	705_4 (25°)	(.609) (25°)
n-Amyl valerate	124.55	$.723_9$	(.638) (0°)
Anethole	(96.0)	.648	(.644) (15°)
Aniline	62.95	(.676)	(.691) (20°)
Anisidine	(80.5)	.654	(.72) (20°)
Anisole	72.79	(.673)	(.672) (15°)
Anthanthrene	204.2	.739	
Anthanthrone	178.1	.581	
Anthracene	(130)	.731	(.914) (27°)
Anthracenedinitrile	154.6	(.678)	
Anthracenonitrile	142.1	(.700)	
Anthraquinone	(119.6)	.575	(.825) (20°)
Anthrazine	245.7	.646	
Arabinose	85.70	.571	(.905) (20°)
Arbutoside	158.0	(.589)	
Asarone	131.4	(.631)	(.735) (18°)
Asparagine	69.5	(.526)	(.812) (15°)
Aspartic acid	64.2 ± .4	(.482)	(.800) (12°)
Aurin	(161.4)	.556	
p-Azoanisole	(147.7)	.610	
Azobenzene	(106.8)	.586	.611 (70.5°)
p-Azophenetole	(171.7)	.635	
m-Azotoluene	(127.8)	.608	.643 (58°)
Azulene	98.5	.768	
Barbituric acid (Anh.)	53.8	(.420)	
Barbituric acid (.$2H_2O$)	78.6	(4.79)	
Benzalazine	(123.7)	.594	
Benzaldehyde	60.78	(.573)	(.602) (15°)
Benzaldoxime	(69.8)	.576	(.639) (20°)
Benzamide	(72.3)	.597	(.801) (4°)
Benzanthrone	142.9	.620	
Benzene	54.84	.702 (32°)	.611
Benzidine	110.9	.603	(.754) (20°)
Benzil	(118.6)	.564	.616 (100°)
Benzoic acid	70.28	(.575)	(.728) (15°)
Benzoic anhydride	(124.9)	.552	(.662) (15°)
Benzonitrile	65.19	(.632)	(.638) (15°)
Benzophenone	109.60	$.601_3$	(.66) (50°)
3,4-Benzopyrene	135.7	.538	
Benzopyrene	194.0		
Benzoyl acetone	(95.0)	.586	(.639) (60°)
Benzoyl chloride	(75.8)	.539 (20°)	(.657) 15°)
Benzyl acetate	93.18	(.620)	(.655) (16°)
Benzyl alcohol	71.83	(.664)	(.697) (15°)
Benzylamine	75.26	(.702)	(.690) (19°)
Benzyl chloride	81.98	(.647)	(.713) (18°)
Benzyl formate	81.43	(.598)	(.646) (20°)
Benzylideneaniline	(100.4)	.554	
Benzylidene chloride	(97.9)	.608	(.763) (14°)
Benzylidenemethylamine	(73.1)	.613	
Benzyl methyl ketone	83.44	$.621_9$	(.624) (20°)
Bibenzyl	(126.8)	.696	.671 (54.5°)

DIAMAGNETIC SUSCEPTIBILITIES OF ORGANIC COMPOUNDS (Continued)

Compound	$-\chi_M \times 10^6$	$-\chi \times 10^6$	$-K \times 10^6$
3,4-Bis-(p-hydroxyphenyl)-2,4-hexadiene	(157)	.59	
m,m′-Bitolyl	(127.4)	.6993 (27.4°)	(.699) (16°)
m,m′-Bitolyl sulfide	(140.0)	.6530 (27.4°)	
Borneol	126.0	(.817)	(.826) (20°)
Bromobenzene	78.92	.5030 (20°)	(.753) (20°)
Bromobenzenediazocyanide	86.88	(.414)	
Bromochloromethane	55.0 ± .6	(.425)	(.846) (19°)
Bromodichloromethane	66.3 ± .3	(.405)	(.812) (15°)
Bromoform	82.60	.327	.948 (20°)
Bromonaphthalene	(123.8)	.598	
α-Bromonaphthalene	115.90	.560	.840 (20°)
m-Bromotoluene	(93.4)	.546	(.770) (20°)
Bromotrichloromethane	73.1 ± .7	(.369)	(.758) (0°)
Butane	57.4	(.988)	
iso-Butane	51.7	(.890)	
1,4-Butanediol	61.5	(.682)	(.696) (20°)
2-Butene (cis)	42.6	(.759)	
2-Butene (trans)	43.3	(.772)	
1-Butene-3,4-diacetate	95.5	(.555)	
2-Butene-1,4-diacetate (cis)	95.2	(.553)	
2-Butene-1,4-diacetate (trans)	95.1	(.552)	
2-Butene-1,4-diol (cis)	54.3	(.616)	
2-Butene-1,4-diol (trans)	53.5	(.607)	
n-Butyl acetate	77.47	$.666_9$ (25°)	(.583) (25°)
iso-Butyl acetate	78.52	$.676_9$	(.584) (25°)
n-Butyl alcohol	56.536 (20°)	(.7627)	(.6176) (20°)
iso-Butyl alcohol	57.704 (20°)	(.7785)	(.624) (20°)
sec-Butyl alcohol	57.683 (20°)	(.7782)	(.629) (20°)
tert-Butyl alcohol	57.42	$.774_7$ (25°)	(.611) (20°)
n-Butylamine	58.9	(.805)	(.596) (20°)
iso-Butylamine	59.8	(.818)	(.599) (20°)
9-Butyl anthracene	176.0	(.751)	
n-Butylbenzene	100.79	(.751)	(.646) (20°)
iso-Butylbenzene	101.81	(.759)	(.648) (20°)
tert-Butylbenzene	102.5	(.764)	(.662) (20°)
n-Butyl benzoate	116.69	$.654_8$	(.656) (25°)
Butyl bromide	77.14	.563 (20°)	(.730) (20°)
iso-Butyl bromide	79.88	.583 (20°)	(.737) (20°)
1-n-Butyl chloride	67.10	.725	.642 (20°)
2-n-Butyl chloride	67.40	.728	.635 (20°)
n-Butyl cyanide	(62.8)	.7558 (27.4°)	(.606) (20°)
tert-Butyl cyclohexane	115.09	.8205	.6670 (20°)
1,4-Butyl diacetate	103.4	(.594)	
n-Butyl ethyl ketone	80.73	$.707_2$	(.579) (20°)
n-Butyl formate	65.83	$.644_6$	(.571) (25°)
iso-Butyl formate	66.79	$.654_9$	(.574) (25°)
iso-Butylideneazine	(95.8)	.683	
Butyl iodide	(93.6)	.5086 (18°)	(.822) (20°)
iso-Butyl methyl ketone	70.05	$.699_5$	(.561) (20°)
tert-Butyl methyl ketone	69.86	$.697_9$	(.558) (16°)
n-Butyl perfluor-n-butyrate	126.7	(.469)	
p-tert-Butylphenol	108.0	(.719)	(.653) (114°)
Butyl sulfide	(113.7)	.7774 (27.4°)	(.652) (16°)
Butyl thiocyanate	(79.38)	.6891 (27.4°)	(.659) (25°)
2-Butyne-1,4-diacetate	95.9	(.564)	
2-Butyne-1,4-dibenzoate	169.0	(.561)	
2-Butyne-1,4-diol	50.3	(.584)	
n-Butyraldehyde	46.08	$.639_4$	(.522) (20°)
iso-Butyraldehyde	46.38	$.643_6$	(.511) (20°)
iso-Butyraldoxime	56.1_2	$.644_3$	(.576) (20°)
n-Butyric acid	55.10	.625	.598 (20°)
iso-Butyric acid	56.06	$.636_3$ (25°)	(.601) (25°)
Butyronitrile	49.4	(.715)	(.569) (15°)
Butyrylphenylacetylene	(106.4)	.618	
Cacodyl	(99.9)	.476	(.689) (15°)
Cacodylic acid	(79.9)	.579	
Camphor	(103)	.68	(.67) (25°)
Camphoric acid	129.0	(.644)	(.791) (20°)
Camphoric anhydride	(113)	.620	(.740) (20°)
n-Caproic acid	78.55	$.676_2$	(.624) (25°)
Caproylphenylacetylene	(130.4)	.651	
n-Caprylic acid	101.60	.7053	(.642) (20°)
Carbanilide	134.05	(.6316)	(.783) (20°)
Carbazole	117.4	(.702)	
Carbon disulfide	42.2	.554	(.699) (22°)
Carbon tetrabromide	93.73	.2826 (20°)	(.966)
Carbon tetrachloride	66.60	.433	.691 (20°)
Carbon tetraiodide	(136)	.261	(1.13) (20°)
Carvacrol	(109.1)	.726	(.709) (20°)
Carvone	(92.2)	.614	(.590) (20°)
Cetyl alcohol	(183.5)	.757 (17.5°)	(.619) (50°)
Cetyl mercaptan	390.4	(1.510)	
Chloral	(67.7)	.459	(.694) (20°)
Chloranil	(112.6)	.458	
Chloracetic acid	48.1	(.509)	(.804) (20°)
Chloroacetone	(50.9)	.550	(.633) (20°)
Chloroacetylchloride	53.7	(.475)	(.710) (0°)
p-Chloranisole	89.1	(.625)	
Chlorobenzene	69.97	(.6216)	(.688) (20°)
Chlorobenzene diazocyanide	65.02	(.393)	
Chlorodibromomethane	75.1 ± .4	(.361)	(.883) (15°)
Chlorodifluoromethane	38.6	.446	
1-Chloro-2,3-dihydroxypropane	(77.9)	.604	
Chlorodiphenylmethane	131.9	(.651)	
Chloroethylene	35.9	.574	(.528) (liq., 15°)
Chloroform	59.30	.497	.740 (20°)

Compound	$-\chi_M \times 10^6$	$-\chi \times 10^6$	$-K \times 10^6$
Chlorofumaric acid	67.02	(.445)	
p-Chloroiodo benzene	99.42	(.417)	(.786) (57°)
Chloromaleic acid	67.36	(.448)	
Chloromethylstilbene	144.8	(.633)	
α-Chloronaphthalene	107.60	.661	.789 (20°)
m-Chloronitrobenzene	(74.8)	.475	.638 (48°)
o-Chlorophenol	77.4	(.602)	(.747) (18°)
p-Chlorophenol	77.6	(.604)	(.789) (20°)
1-(o-Chlorophenylazo)-2-naphthol	161.0	(.570)	
1-(p-Chlorophenylazo)-2-naphthol	161.4	(.571)	
Chlorotrifluoroethylene	49.1	.422	
Chlorotrifluoromethane	45.3 ± 1.5	(.434)	
Cholesterol	(284.2)	.735	(.784) (20°)
Chrysene	166.67	.731	
Chrysoidine	(126.3)	.595	
Cinnamic acid	78.36	.529	(.660) (4°)
Cinnamic acid (α-trans)	78.2	(.528)	
Cinnamic acid (β-trans)	79.0	(.533)	
Cinnamic acid (cis-MP 68°)	77.6	(.524)	
Cinnamic acid (cis-MP 58°)	77.9	(.526)	
Cinnamic acid (cis-MP 42°)	83.2	(.562)	
Cinnamic aldehyde	(74.8)	.566	(.629) (15°)
Cinnamyl alcohol	(87.2)	.650	(.679) (20°)
Cinnamylideneaniline	123.2	(.595)	
Citral	(98.9)	.650	(.577) (20°)
Coronene	(243.3)	.810	
Coumarin	82.5	(.565)	(.528) (20°)
o-Cresol	72.90	.675	(.706) (20°)
m-Cresol	72.02	.667 (26°)	(.690) (20°)
p-Cresol	72.1	.667 (26°)	(.690) (20°)
o-Cresylmethyl ether	81.94	.671 (40°)	(.661) (15°)
m-Cresylmethyl ether	77.91	.638 (40°)	(.623) (15°)
p-Cresylmethyl ether	79.13	.648 (40°)	(.629) (19°)
Cumene	89.53	$.744_9$	(.642)
Cyamelide	(56.1)	.435	(.490) (15°)
Cyameluric acid	101.1 (10°)	(.457)	
9-Cyanoanthracene	142.1	(.699)	
Cyanogen	(21.6)	.415	(.359) (liq., 17°)
Cyanuric acid	61.5	.476	(.842) (0°)
Cyclobutanecarboxylic acid	58.16	.5816 (30°)	(.613) (30°)
1,3-Cyclohexadiene	48.6	(.607)	(.510) (20°)
1,4-Cyclohexadiene	48.7	(.608)	(.515) (20°)
Cyclohexane	68.13	.8100 (27.5°)	(.627) (20°)
Cyclohexanecarboxylic acid	83.24	.6499 (30°)	(.668) (30°)
Cyclohexanol	73.40	.732	.694 (20°)
Cyclohexanone	62.0_4	$.632_3$	(.599) (20°)
Cyclohexanone oxime	71.5_2	$.632_1$	
Cyclohexanoneoxime-O-methyl ether	82.9_6	$.652_3$	
Cyclohexene	57.5	(.700)	(.567) (20°)
Cyclohexenol	64.1	(.653)	
Cyclooctane	91.4	(.815)	(.684) (20°)
Cyclooctene	84.6	(.769)	(.654) (20°)
Cyclooctatetraene	(53.9)	.518	
Cyclopentane	59.18	.8439	.6290 (20°)
Cyclopentanecarboxylic acid	73.48	.6446 (30°)	(.677) (30°)
Cyclopentanone	51.63	.6141 (30°)	(.582) (30°)
Cyclopropane	39.9	(.948)	(.683) (−79°)
Cyclopropanecarboxylic acid	45.33	.5271 (30°)	(.569) (30°)
p-Cymene	102.8	.766 (20°)	(.656) (20°)
Decalin	106.70	.7718	.6814 (20°)
cis-Decalin	(107.0)	.774	(.686) (35°)
trans-Decalin	(107.7)	.779	(.670) (35°)
n-Decane	119.74	(.8416)	(.6143) (20°)
l-Deuterio pyrrole	48.75	.716 (20°)	
Deuteroindene	80.88	.690	(.692) (13°)
Diacetal	(153.8)	.668	
Di-iso-amylamine	(133.1)	.846	(.649) (21°)
Diazoacetic ester	57	(.50)	(.54) (24°)
Dibenzocoronene	289.4	.778	
1,2,5,6-Dibenzofluorene	184	.69 ± .03	
3,4,5,6-Dibenzophenanthrene	(203)	.73 ± .03	
Dibenzphenanthrone	200.5	.716	
Dibenzpyrene	213.6	.706	
Dibenzpyrenequinone	183.1	.551	
iso-Dibenzpyrenequinone	194.6	.586	
Dibenzyl ketone	131.70	$.626_2$	
p-Dibromobenzene	(101.4)	.430	.786 (100°)
2,3-Dibromo-2-butene-1,4-diol	94.2	(.383)	
Dibromodichloromethane	81.1 ± .4	(.334)	(.808) (25°)
1,2-Dibromodiiodoethylene	(140.1)	.320	
1,2-Dibromoethylene	(71.7)	.386	(.877) (17.5°)
1,2-Dibromo-2-fluoroethane	(78.0)	.379	(.855) (17°)
Dibromo-4-nitrophenol	(167.5)	.564	
1,2-Dibromotetrachloroethane	(126.0)	.387	(1.049)
Di-n-butylamine	103.7	(.802)	(.767) (20°)
Di-iso-butylamine	105.7	(.817)	(.609) (20°)
Di-sec-butylamine	105.9	(.819)	(.641) (0°)
Di-iso-butyl ketone	104.30	$.733_5$	(.591) (20°)
Di-tert-butyl ketone	104.06	$.732_1$	
2,6-Di-tert-butyl-4-methyl phenol	165.3	(.750)	
2,4-Di-tert-butyl phenol	155.6	(.754)	
Dibutyl phthalate	175.1	(.629)	(.657) (21°)
Di-iso-butyralacetylene	(125.6)	.738	
Dicetyl sulfide	401.7	(.832)	
Dichloroacetic acid	58.2	(.451)	(.705) (20°)
Dichloroacetyl chloride	69.0	(.468)	

Compound	$-\chi_M \times 10^6$	$-\chi \times 10^6$	$-K \times 10^6$
o-Dichlorobenzene	84.26	.5734	(.748) (20°)
m-Dichlorobenzene	83.19	.5661	(.729) (20°)
p-Dichlorobenzene	82.93	.5644	(.823) (20.5°)
1,4-Dichloro-2-butyne	74.2	(.603)	
1,2-Dichloro-1,2-dibromoethane	(108.6)	.423	
1,1-Dichloro-difluoroethylene	60.0	.451	
Dichlorodifluoromethane	52.2	.432	(.642) (−30°)
1,1-Dichloroethylene	49.2	.508	(.635) (15°)
cis-1,2-Dichloroethylene	51.0	.526	(.679) (15°)
trans-1,2-Dichloroethylene	48.9	.504	(.638) (15°)
1,3-Dichloro-2-hydroxypropane	(80.1)	.621	
Dicyandiamide	44.55	(.530)	(.742) (14°)
Dicyclohexanol acetylene	(151.6)	.682	
Dicyclohexyl	129.31	.7776	.6889 (20°)
1,1-Dicyclohexylnonane	231.98	.7930	.7001 (20°)
Diethanolacetylene	(75.3)	.660	
Diethyl acetaldehyde	70.71	.705$_9$	(.576) (20°)
Diethylallylacetophenone	146.2	(.676)	(.663) (16°)
Diethyl allylmalonate	118.8	(.593)	(.602) (14°)
Diethylamine	56.8	(.777)	(.552) (18°)
Diethylcyclohexylamine	(124.5)	.802	(.699) (0°)
Diethyl ethylmalonate	115.2	(.612)	(.614) (20°)
Diethyl ketone	58.1$_4$	.675$_1$	(.551) (19°)
Diethyl ketoxime	68.3$_1$	.6754	
Diethyl malonate	(92.6)	.5782	(.611) (20°)
Diethyl-3-(1-methyl butane) ethylmalonate	175	(.677)	
Diethyl oxalate	81.71	.5595	(.603) (15°)
Diethyl phthalate	127.5	(.574)	(.645) (25°)
Diethyl sebacate	(177.0)	.685	(.661) (20°)
Diethylstilbestrol	172.0	(.547)	
Diethylstilbestrol dipropionate	265.2	.720	
Diethyl succinate	105.07	.6035	(.628) (20°)
Diethyl sulfate	(86.8)	.563	(.667) (15°)
Diethyl sulfide	(67.9)	.753	(.630) (20°)
Diethyl tartrate	(113.4)	.550	(.662) (20°)
Difluoroacetamide	(41.2)	.433	
1-Difluoro-2-dibromoethane	(85.5)	.382	(.883) (20°)
1,1-Difluoro-2,2-dichloroethyl amyl ether	129.84	(.587)	(.694) (20°)
1,1-Difluoro-2,2-dichloroethyl butyl ether	119.48	(.577)	(.703) (20°)
1,1-Difluoro-2,2-dichloroethyl ethyl ether	96.13	(.537)	(.723) (20°)
1,1-Difluoro-2,2-dichloroethyl methyl ether	80.68	(.489)	(.696) (20°)
1,1-Difluoro-2,2-dichloroethyl propyl ether	107.19	(.555)	(.701) (20°)
Difluoroethanol	(41.3)	.503	
Di-n-heptylamine	171.5	(.805)	
Di-n-hexylamine	148.9	(.803)	
Dihydronaphthalene	(85.1)	.654	(.652) (12°)
o-Dimethoxybenzene	87.39	.6329	(.686) (25°)
m-Dimethoxybenzene	87.21	.6316	(.682) (0°)
p-Dimethoxybenzene	86.65	.6275	(.661) (55°)
o-(2,5-Dimethoxybenzoyl)-benzoic acid	161.0	(.562)	
Dimethoxymethane	(47.3)	.621	(.532)
Dimethylacetophenone	96.8	(.653)	(.645) (16°)
Dimethylallylacetophenone	122.4	(.650)	(.635) (16°)
Dimethylaniline	89.66	(.740)	
2,2-Dimethylbutane	76.24	.8848	.5744 (20°)
2,3-Dimethylbutane	76.22	.8845	.5853 (20°)
2,3-Dimethyl-2-butene	65.9	(.783)	(.557)
Dimethylcyclohexanone	(84.8)	.672	
1,2 and 1,3 Dimethylcyclopentanes	81.31	.8281	.6224 (20°)
2,5-Dimethyl-2,5-dibromo-3-hexine	(135.6)	.506	
Dimethyl diethylketo tetrahydrofurfurane	(116.2)	.753	
2,5-Dimethyl-1-ethylpyrrole	94.61	.768 (20°)	
2,5-Dimethyl-3-ethylpyrrole	93.87	.762 (20°)	
2,5-Dimethylfuran	66.37	.687 (20°)	(.620) (18°)
Dimethyl furazan	57.27	.584	
2,5-Dimethyl-4-heptene	100.6	(.797)	
2,4-Dimethyl-2,4-hexadiene	(78.7)	.714	
2,3-Dimethylhexane	98.77	.8648	.6164 (20°)
2,5-Dimethylhexane	98.15	.8593	.5969 (20°)
3,4-Dimethylhexane	99.06	.8673	.6240 (20°)
2,6-Dimethyl-4 hexanol	116.9	.812	
2,5-Dimethyl-3-hexine-2,5-diol	(103.0)	.724	
Dimethyl isoxazole	59.7	(.615)	
Dimethylketo tetrahydrofurfurane	(68.5)	.600	
Dimethyl malonate	69.69	.5277	(.609) (20°)
1,6-Dimethylnaphthalene	113.3	(.725)	
2,4-Dimethylnonane	134.68	(.862)	(.636) (20°)
3,4-Dimethylnonane	134.70	(.862)	(.647) (20°)
4,5-Dimethylnonane	134.52	(.861)	(.647) (20°)
2,6-Dimethyl-2,6,8-nonatriene	(108.8)	.724	
Dimethyl-2,4-nonatriene	(148.8)	.990	
2,6-Dimethyloctane	122.54	(.861)	(.627) (20°)
3,4-Dimethyloxadiazole	57.17	.583	
Dimethyl oxalate	(55.7)	.472	(.542) (54°)
Dimethyl oxamide	(63.2)	.544	
2,2-Dimethylpentane	86.97	.8680	.5849 (20°)
2,3-Dimethylpentane	87.51	.8733	.6070 (20°)
2,4-Dimethylpentane	87.48	.8732	.5876 (20°)
2,2-Dimethylpropane	63.1	(.875)	(.536) (0°)
2,5-Dimethyl-3-propyl-pyrrole	106.07	.773 (20°)	
2,4-Dimethylpyrrole	69.64	.732 (20°)	(.679) (14°)

Compound	$-\chi_M \times 10^6$	$-\chi \times 10^6$	$-K \times 10^6$
2,5-Dimethylpyrrole	71.92	.756 (20°)	(.707) (20°)
α,ω-Dimethyl styrene	(90.7)	.686	
Dimethyl succinate	81.50	.5581	(.625) (18°)
Dimethyl sulfate	(62.2)	.493	(.657) (20°)
Dimethyl sulfide	(44.9)	.723	(.612) (21°)
Dimethyltrichloromethylcarbinol	(105)	.59	
N,N-Dimethyl urea	55.1	(.625)	(.784)
N,N′-Dimethyl urea	56.3	(.639)	(.730)
o-Dinitrobenzene	65.98	.3921	(.614) (17°)
m-Dinitrobenzene	70.53	.4197	(.659) (0°)
p-Dinitrobenzene	68.30	.4064	(.660) (30°)
2,4-Dinitrophenol	(73.1)	.397	(.668) (24°)
Dinitroresorcinol	(62.4)	.312	
1,4-Dioxane	52.16	.592 (32°)	(.606) (32°)
Diphenyl	103.25	.6695	(.664) (73°)
1,1-Diphenylallyl-3-chloride	146.1	(.639)	
1,3-Diphenylallyl-3-chloride	140.7	(.615)	
Diphenylamine	(109.7)	.648	.686 (55.5°)
Diphenyl-bis-diazo cyanide	85.03	(.327)	
Diphenylbutadiene	129.6	(.629)	
Diphenylchloroarsine	(145.5)	.550	(.871) (40°)
Diphenyldecapentaene	180.5	(.635)	
Diphenyldiacetylene	(134.6)	.640	
Diphenyldiazomethane	115	(.592)	
Diphenyldihydrotetrazine	129.9	(.545)	
1,1-Diphenylethylene	(118.0)	.655	(.680) (14°)
1,6-Diphenylhexane	171.81	.7208	.6877 (20°)
Diphenylhexatriene	146.9	(.632)	
Diphenylmethane	(115.7)	.688	.684 (35.5°)
Diphenylmethanol	119.1	.647	
1,1-Diphenylnonane	206.32	.7357	.6935 (20°)
Diphenyloctatetraene	164.3	(.636)	
Diphenylphenoxyarsine	(225.2)	.567	
N,N-Diphenyl urea	126.3	(.595)	(.759)
N,N′-Diphenyl urea	127.5	(.600)	(.743) (20°)
Di-n-propyl ketone	80.45	.705$_0$	(.576) (20°)
Di-iso-propyl ketone	81.14	.711$_0$	(.573) (20°)
Dipropyl oxalate	105.27	.6046	(.628) (0°)
Di-iso-propyl oxalate	106.02	.6089	
Dodecyl alcohol	147.70	.7849 (20.7°)	(.652) (24°)
Dulcitol	112.40	.617	(.905) (15°)
Elaidic acid	204.8	(.725)	(.619) (79°)
Erythritol	73.80	.604	(.876) (20°)
Ethane	27.3$_7$	(.910)	(.511) (−100°)
4-Ethoxy-3-methoxybenzyl acetate	138.5	.619	
4-Ethoxy-3-methoxybenzyl benzoate	177.3	.620	
1-Ethoxynaphthalene	119.9	(.696)	(.738) (20°)
2-Ethoxynaphthalene	119.2	(.692)	(.734) (25°)
Ethyl acetate	54.10	.614	.554 (20°)
Ethyl acetoacetate	71.67	.550$_8$	(.565) (20°)
Ethylacetophenone	95.5	(.644)	(.639) (16°)
Ethyl alcohol	33.60	.728	.575 (20°)
Ethylally acetophenone	122.5	(.651)	(.634) (16°)
Ethyl amylpropiolate	(112.7)	.670	
Ethylaniline	89.30	(.737)	(.709)
9-Ethyl anthracene	153.0	(.741)	(.771) (99°)
Ethylbenzene	77.20	.7272	.6341 (20°)
Ethyl benzoate	93.32	.621$_1$	(.648) (25°)
Ethyl benzoylacetate	(115.3)	.600	(.673) (20°)
Ethyl benzylidenecyanoacetate	(116.3)	.578	
Ethyl benzylmalonate	(154.5)	.6172	(.663) (20°)
Ethyl bromide	54.70	.502	.719 (20°)
Ethyl bromoacetate	(82.8)	.496	(.747) (20°)
Ethyl-1-isobutylacetoacetate	(121.4)	.652	
Ethyl butylmalonate	139.3	.644$_2$	(.629) (20°)
Ethyl-n-butyrate	(77.7)	.6693	(.585) (25°)
Ethyl-iso-butyrate	78.32	.674$_3$	(.583) (25°)
Ethyl chloroacetate	(72.3)	.590	(.684) (20°)
Ethyl cinnamate	(107.5)	.610	(.640) (20°)
Ethyl-iso-cyanate	(45.6)	.642	(.582) (16°)
Ethyl cyanoacetate	(67.3)	.595	(.632) (20°)
Ethylcyclohexane	91.09	.8118	.6324 (20°)
Ethyldiallylacetophenone	147.4	(.646)	(.636) (16°)
Ethyl dibromocinnamate	174.5	.519	
Ethyl dichloroacetate	85.2	(.543)	(.696) (20°)
Ethyl diethylacetoacetate	(117.9)	.6328	(.615) (20°)
Ethyl diethylmalonate	(140.4)	.6492	(.641)(20°)
Ethyl dithiolacetate	(71.0)	.5904 (27.4°)	
Ethylene	12.0	(.428)	(.242) (−102°)
Ethylene	15.30	.546 (32°)	(.309) (−102°)
Ethylene bromide	78.80	.419	.915 (20°)
Ethylene chloride	59.62	.602 (32°)	(.757) (20°)
Ethylenediamine	46.26	.771 (32°)	(.686) (20°)
Ethylene iodide	104.7	.371 (32°)	(.791) (10°)
Ethylene oxide	30.7	(.697)	(.618) (7°)
Ethyl ether	55.10	.743	.534 (20°)
Ethyl ethylacetoacetate	93.9	.5937	(.582) (20°)
Ethyl ethylbutylmalonate	(163.3)	.6683	(.650) (20°)
Ethyl ethylpropylmalonate	(152.4)	.6619	(.648) (20°)
Ethyl formate	43.00	.580	.531 (20°)
Ethyl hexylpropiolate	129.9	.713	
Ethyl hydroxylamine	(43.0)	.704	
Ethylidene chloride	(57.4)	.580	(.681) (20°)
Ethyl iodide	(69.7)	.4470 (17.5°)	(.864) (20°)
Ethyl iodoacetate	(97.6)	.456	(.829) (13°)
Ethyl lactate	(72.6)	.615	(.633) (25°)

Compound	$-\chi_M \times 10^6$	$-\chi \times 10^6$	$-K \times 10^6$
Ethyl methylacetoacetate	(81.9)	.5684	(.569) (20°)
Ethyl methyl ketoxime	57.3_2	$.658_0$	(.530) (20°)
Ethyl-1-methyl-2-oxocyclohexane-carboxylate	112.1	(.608)	
Ethyl methylphenylmalonate	(153.2)	.6121	(.658) (20°)
Ethyl nitrophenylpropiolate	114.6	.523	
Ethyl oxamate	62.0	(.529)	(.427) (19°)
Ethyl perfluor-n-butyrate	103.5	(.427)	
Ethyl phenylacetate	104.27	(.635)	(.656) (20°)
Ethyl phenylmalonate	(142.2)	.6017	(.659) (20°)
Ethyl phenylpropiolate	(104.2)	.598	(.636) (13°)
Ethyl phosphate	(98.2)	.539	(.576) (25°)
Ethyl propionate	(66.5)	.6514	(.584) (15°)
Ethyl propylacetoacetate	(105.7)	.6135	(.593) (20°)
Ethyl-n-propyl ketone	69.03	$.689_1$	(.560) (22°)
Ethyl succinimide	(72.0)	.566	
Ethyl sulfine	(67.0)	.631	
Ethyl sulfite	(75.4)	.546	(.604) (0°)
Ethylsulfone ethyl ether	(81.8)	.592	
Ethyl thiocyanate	(55.7)	.6392 (27.4°)	(.637) (25°)
Ethyl isothiocyanate	(59.0)	.6772 (27.4°)	(.680) (15°)
Ethyl thiolacetate	(62.7)	.6019 (27.4°)	(.586) (25°)
Ethyl thionacetate	(63.5)	.6098 (27.4°)	
Ethyl tribromoacetate	(119.5)	.368	(.821) (20°)
Ethyl trichloroacetate	(99.6)	.520	(.719) (20°)
N-Ethyl urea	55.5	(.630)	(.764) (18°)
Ethyl iso-valerate	(91.1)	.700	(.607) (20°)
Eucalyptol	(116.3)	.754	(.699) (20°)
Eugenol and iso-eugenol	(102.1)	.622	(.663) (20°)
Flavanthrone	241.0	.590	
Fluorene	110.5	.655	
Fluorenone	99.4	.552	(.623) (100°)
Fluorobenzene	(58.4)	.608	(.623) (20°)
Fluorobromoacetic acid	(59.5)	.379	
Fluorodichloromethane	48.8	.474	(.676) (0°)
p-Fluorophenetole	(88.0)	.628	
Fluoro trichloroethylene	72.5	.485	(.742) (25°)
Fluorotrichloromethane	58.7	.427	(.638) (17°)
Formaldehyde	(18.6)	.62	(.51) (−20°)
Formamide	(21.9)	.486	(.551) (20°)
Formic acid	19.90	.432	.527 (20°)
β-Formylpropionic acid	55.3	(.542)	
Fructose	102.60	.570	
Fulvene (Benzene χ_M measured to be 49)	42.9	(.549)	(.452) (20°)
Fulvene $\left(\chi \frac{54.8}{49}\right)$	48.0	(.614)	(.505) (20°)
Fumaric acid	49.11	(.423)	(.692) (20°)
Furan	43.09	.633 (20°)	(.598) (15°)
Furfural	47.1	(.490)	(.568) (20°)
Galactose	103.00	.572	
Gallic acid	90.0	(.529)	(.896) (4°)
Geraniol formate	(119.9)	.658	(.610) (20°)
Glucose	102.60	.570	
D-Glucose	101.5	(.563)	(.869) (25°)
Glutamic acid	78.5	(.533)	
Glycerol	57.06	.619	.779 (20°)
Glycine	40.3	(.537)	(.846) (50°)
Glycol	38.80	.624	.698 (20°)
Guaiacol	(79.2)	.638	(.720) (21°)
Helianthrone	189.9	.497	
1,2-Heptadiene	73.5	(.764)	
2,3-Heptadiene	72.1	(.749)	
Heptaldehyde	81.02	$.709_6$	(.603) (20°)
n-Heptane	85.24	.8507	.5817 (20°)
4-Heptanol	91.5	.789	(.647) (20°)
n-Heptanoic acid	88.60	.680	.626 (20°)
n-Heptyl amine	93.1	(.808)	(.628) (20°)
n-Heptyl benzene	134.41	.7625	.6528 (20°)
Heptyl cyclohexane	147.40	.8084	.6559 (20°)
n-Heptylic acid	89.74	.6900	(.630) (25°)
1-Heptyne	77.0	(.801)	(.584) (25°)
2-Heptyne	79.5	(.826)	(.615) (25°)
Hexabromoethane	(148.0)	.294	(1.124) (20°)
Hexachlorobenzene	(147.5)	.518	(1.059) (24°)
Hexachloroethane	(112.7)	.476	(.995) (20°)
Hexachlorohexatrione	(145.0)	.433	
n-Hexadecane	187.63	.8286	.6421 (20°)
1,5-Hexadiene	(55.1)	.671	(.462) (20°)
2,3-Hexadiene	60.9	(.741)	
n-Hexaldehyde	69.40	$.693_0$	
2,2,4,7,9,9-Hexamethyldecane	191.52	.8458	.6596 (20°)
Hexamethyl disiloxane	118.9	.7324	
Hexamethylene glycol	84.30	.713	
n-Hexane	(74.6)	.8654 (27.4°)	(.565)
Hexene	65.7	(.781)	
Hexestrol	(165)	.61 ± .02	
n-Hexyl alcohol	79.20	.774	.637 (20°)
n-Hexyl benzene	124.23 (20°)	(.767)	(.658) (20°)
n-Hexyl methyl ketone	91.4_2	$.713_1$	(.583)
n-Hexyl methyl ketoxime	102.5_8	$.716_2$	(.634) (20°)
Hexylpropiolamide	(103.7)	.677	
Hydrindene	(78.5)	.664	(.639) (16°)
Hydroquinone	64.63	.587	(.797) (20°)
Hydroxyazobenzene	(99.7)	.503	
p-Hydroxybenzaldehyde	66.8	(.547)	(.618) (130°)
4-Hydroxy-2-butanone	48.5	.55	(.573) (14°)
Indene (natural)	84.79	(.730)	(.723) 25°)
Indene (synthetic)	80.89	(.696)	(.690) (25°)
Indole	85.0	(.726)	
Iodobenzene	92.00	.451	.826 (20°)
Iodoform (in sol'n)	117.1	.2974 (20°)	(1.192) (17°)
1-Iodo-2-phenylacetylene	(110.1)	.483	
o-Iodotoluene	(112.2)	.5145 (30°)	(.874) (20°)
m-Iodotoluene	(112.3)	.5152 (30°)	(.875) (20°)
p-Iodotoluene	101.31	(.465)	(.780) (40°)
Leucine	84.9	(.647)	
iso-Leucine	84.9	(.647)	
Maleic acid	49.71	(.428)	(.681) (20°)
Maleic anhydride	(35.8)	.365	(.341) (20°)
Malonic acid	(46.3)	.4453	(.726) (15°)
Mannitol	111.20	.610	(.908) (20°)
Mannose	102.90	.571	(.879)
Mesitylene	92.32	.7682 (20°)	(.665) (20°)
Methane	12.2_7	(.765)	
Methione	91.0	(.610)	
p-Methoxyazobenzene	(118.9)	.560	
o-Methoxybenzaldehyde	76.0	(.558)	(.632) (20°)
p-Methoxybenzaldehyde	78.0	(.572)	(.642) (20°)
o-Methoxybenzyl alcohol	87.9	.637	(.664) (25°)
1-Methoxynaphthalene	107.0	(.676)	(.741) (14°)
2-Methoxynaphthalene	107.6	(.680)	
1-(o-Methoxyphenylazo)-2-naphthol	163.6	(.588)	
Methoxysaligenin acetate	110.3	.613	
Methyl acetate	42.60	.575	.537 (20°)
Methyl acetoacetate	59.60	$.513_2$	(.553) (20°)
Methylacetylacetone	(65.0)	.569	
Methyl alcohol	21.40	.668	.530 (20°)
Methylallylaketone	111.9	(1.330)	
Methylamine	(27.0)	.870	(.608) (−11°)
N-Methylaniline	82.74	(.773)	(.762) (20°)
9-Methylanthracene	146.5	.762)	(.812) (99°)
Methyl benzoate	81.59	$.599_3$	(.651) (25°)
Methyl-o-benzoylbenzoate	139.4	(.580)	(.690) (19°)
Methylbenzylaniline	(132.2)	.670	
Methyl bromide	42.8	.451	(1.044) (0°)
2-Methylbutane	64.40	.8925	.5531 (20°)
2-Methyl-2-butene	54.14	(.772)	(.516) (13°)
Methyl butyl ketone	(69.1)	.690	(.563) (15°)
Methyl iso-butyl ketone	(69.3)	.692	(.554) (20°)
Methyl tert-butyl ketone	(70.4)	.703	(.562) (16°)
p-Methyl-o-tert-butylphenol	120.3	(.732)	
Methyl butyrate	(66.4)	.6498	(.588) (16°)
o-Methylcarbanilide	154.0	(.681)	
Methyl chloride	(32.0)	.633	
Methyl chloroacetate	58.1	(.535)	(.661) (20°)
Methylcholanthrene	(182)	.68 ± .04	
3-Methylcholanthrene	194.0	(.723)	
Methylcyclohexane	78.91	.8038	.6181 (20°)
2-Methylcyclohexanone	(74.0)	.660	(.610) (18°)
3-Methylcyclohexanone	(74.8)	.667	(.610) (20°)
4-Methylcyclohexanone	(63.5)	.566	(.516) (24°)
Methylcyclopentane	70.17	.8338	.6245 (20°)
4-Methyl-2,6-di-tert-butylphenol	167.6	(.761)	
Methyl dichloroacetate	73.1	(.511)	
Methyldiphenoxyphosphine oxide	(152.9)	.616	
Methyldiphenyltriazine	(155.1)	.627	
Methylene bromide	65.10	.375	.935 (20°)
Methylene chloride	(46.6)	.549	(.733) (20°)
Methylene iodide	93.10	.348	1.156 (20°)
Methylene succinic acid	57.57	(.443)	(.723)
Methyl ether	26.3	.571	
Methylethylallyacetophenone	133.3	(.659)	(.643) (16°)
Methyl ethyl ketone	45.5_8	$.632_2$	(.509) (20°)
Methyl formate	(32.0)	.5327	(.519) (20°)
Methylfumaric acid	56.98	(.438)	(.642)
3-Methylheptane	97.99	.8580	.6056 (20°)
2-Methyl-4-heptene	88.0	(.784)	
5-Methyl-1,2-hexadiene	73.6	(.765)	(.553) (19°)
2-Methylhexane	86.24	.8607	.5841 (20°)
Methyl hexyl ketone	(93.3)	.728	(.596) (20°)
Methyl-m-hydroxybenzoate	88.4	(.581)	
Methyl-p-hydroxybenzoate	88.7	(.583)	
Methyl iodide	(57.2)	.403	(.918) (20°)
Methylmaleic acid	57.84	(.446)	(.721)
9-Methyl-10-methoxyanthracene	158.1	(.711)	
Methyl-o-methoxybenzoate	95.6	(.575)	(.665) (19°)
Methyl-p-methoxybenzoate	98.6	(.593)	
Methyl-α-methoxy-isobutyrate	(81.9)	.620	
1-Methylnaphthalene	102.8	(.723)	(.741) (14°)
2-Methylnaphthalene	102.6	(.722)	(.743) (20°)
4-Methylnonane	121.39	(.853)	(.625) (20°)
5-Methyl-5-nonene	111.6	(.796)	
4-Methyloctane	109.63	(.855)	(.618) (20°)
Methylol urea	48.3	(.493)	
2-Methylpentane	75.26	.8734	.5705 (20°)
3-Methylpentane	75.52	.8764	.5823 (20°)
4-Methyl-2-pentanol	80.4	.788	(.641) (20°)
Methyl perfluor-n-butyrate	92.5	(.406)	
Methyl phenylacetate	92.73	(.618)	(.645) (16°)
Methyl phenylpropiolate	95.6	.597	
2-Methylpropene	44.4	(.791)	
Methyl propionate	(55.0)	.6240	(.571) (20°)

Compound	$-\chi_M \times 10^6$	$-\chi \times 10^6$	$-K \times 10^6$
Methyl-n-propyl ketone	57.41	.666$_4$	(.541) (15°)
Methyl-iso-propyl ketone	58.45	.679$_0$	(.545) (20°)
1-Methylpyrrole	58.56	.722 (20°)	(.664) (10°)
2-Methylpyrrole	60.10	.741 (20°)	(.700)
Methyl salicylate	86.30	.567	.668 (20°)
Methyl silicone	(172.7)	.730	
α-Methyl styrene	(80.1)	.678	(.620) (20°)
2-Methylthiazole	59.56	.601 (20°)	
2-Methylthiophene	66.35	.676 (20°)	(.689) (20°)
Methyl trichloroacetate	84.2	(.475)	(.707) (19°)
N-Methyl urea	44.6	(.602)	(.725)
Morpholine	55.0	(.631)	(.631)
Myleran	169.7	.69	(.834) (4°)
Myristic acid	176.0	(.771)	(.661) (60°)
Naphthalaldehydic acid	117.6	(.588)	
Naphthalene	(91.9)	.717	(.821) (20°)
Naphthalene picrate	185.9	(.523)	
2-Naphthalenesulfonylamine	127.6	(.616)	
2-Naphthalenesulfonyl chloride	121.91	(.538)	
meso-Naphthodianthrene	214.6	.612	
meso-Naphthodianthrone	221.8	.583	
1-Naphthol	98.2	.681	(.834) (4°)
	97.0	.673	(.819) (4°)
2-Naphthol	98.25	(.682)	(.829) (4°)
α-Naphthonitrile	103.3	(.674)	(.753) (5°)
β-Naphthonitrile	101.0	(.659)	(.721) (60°)
α-Naphthoquinone	73.5	(.465)	(.661)
β-Naphthoquinone	67.9	(.429)	
N-1-Naphthylacetamide	117.8	(.636)	
N-2-Naphthylacetamide	117.8	(.636)	
1-Naphthylamine	98.8	.690	.757 (54°)
2-Naphthylamine	98.00	(.684)	(.726) (98°)
1-Naphthylamine hydrochloride	(127.6)	.710	
Nicotine	113.328	(.699)	(.705) (20°)
o-Nitroaniline	66.47	(.481)	(.694) (15°)
m-Nitroaniline	70.09	(.507)	(.725) (20°)
p-Nitroaniline	66.43	(.481)	(.691) (14°)
o-Nitrobenzaldehyde	68.23	.4517	
m-Nitrobenzaldehyde	68.55	.4538	
p-Nitrobenzaldehyde	66.57	.4407	(.507) (0°)
Nitrobenzene	61.80	.502	.604 (20°)
Nitrobenzene diazo cyanide	59.22	(.336)	
o-Nitrobenzoic acid	76.11	.4556	(.718) (4°)
m-Nitrobenzoic acid	80.22	.4802	(.717) (4°)
p-Nitrobenzoic acid	78.81	.4718	(.731) (32°)
o-Nitrobromobenzene	87.3	(.432)	(.700) (80°)
m-Nitrobromobenzene	89.5	(.443)	(.755) (20°)
p-Nitrobromobenzene	89.6	(.444)	(.859) (22°)
m-Nitro carbanilide	148.1	(.576)	
Nitroethane	(35.4)	.472	(.497) (20°)
Nitromethane	21.1	.3457	(.391) (25°)
1-Nitronaphthalene	98.47	(.569)	(.696) (62°)
o-Nitrophenol	73.3	.527 (24°)	(.873) (20°)
m-Nitrophenol	70.8	.509 (25°)	(.756)
p-Nitrophenol	69.5	.500 (22°)	(.740)
1-(m-Nitrophenylazo)-2-naphthol	142.0	(.484)	
1-(p-Nitrophenylazo)-2-naphthol	141.7	(.483)	
Nitrophenylfluoroform	(84.1)	.440	
2-Nitropropane	45.73	.5135	(.509) (20°)
Nitrosobenzene	59.1	(.552)	
N-Nitrosodiethylamine	59.3	(.580)	(.546) (20°)
p-Nitrosodiethylaniline	92.6	(.520)	(.644) (15°)
p-Nitrosodimethylaniline	73.3	(.488)	
N-Nitrosodiphenylamine	110.7	(.558)	
1-Nitroso-2-naphthol	83.9	(.485)	
2-Nitroso-1-naphthol	82.7	(.478)	
4-Nitroso-1-naphthol	91.8	(.530)	
m-Nitrosonitrobenzene	66.0	(.433)	
p-Nitrosonitrobenzene	65.8	(.433)	
p-Nitrosophenol	50.7	(.412)	
Nitrosopiperidine	(63.4)	.555	(.590) (20°)
p-Nitrosotoluene	70.4	(.581)	
o-Nitrotoluene	72.28	.5272	(.613) (20°)
m-Nitrotoluene	72.71	.5304	(.614) (20°)
p-Nitrotoluene (in sol'n)	72.06	.5257	(.676) (20°)
n-Nonane	108.13	.8431	.6057 (20°)
1,2-Octadiene	83.6	(.759)	
n-Octane	96.63	.8460	.5949 (20°)
Octanonoxime	(102.7)	.717	
Octyl alcohol	102.65	.7766 (20°)	(.640) (20°)
Octyl chloride	(114.9)	.773	(.676) (20°)
Octylcyclohexane	158.09	.8051	.6578 (20°)
Octylene	(89.5)	.798	(.576) (17°)
Octylene bromide	(150.4)	.553	
n-Octyl mercaptan	(115.1)	.7866 (27.4°)	
Oenanthylidene chloride	(116.5)	.689	
Oleic acid	208.5	(.738)	(.661) (18°)
Opianic acid	111.5	(.530)	
Ovalene	353.8	.888	
Oxalic acid (anh.)	33.8	(.375)	
Oxalic acid	60.05	(.4763)	(.787)
Oxamide	(39.0)	.443	(.738)
Palmitic acid	198.6	(.775)	(.661) (62°)
Paraldehyde	(86.2)	.652	(.648) (20°)
Pentabromophenol	(194.0)	.397	
Pentacene	(205.4)	.738	
Pentachloroethane	(99.1)	.490	(.819) (25°)
Pentachlorohexadione	(129.5)	.452	
2,3-Pentadiene	49.1	(.721)	(.501) (20°)
n-Pentane	63.05	.8739	.5472 (20°)
2,4-Pentanediol	70.4	.677	
Perfluoroacetic acid	43.3	(.380)	
Perfluoro-n-butyric acid	81.0	(.378)	
Perfluorobutyric anhydride	149.4	(.387)	
Perfluorocyclooctane oxide	157.6	(.379)	
Perfluoropropionic acid	61.0	(.372)	
Perhydroanthracene	146.01	.7592	.7178 (20°)
Perylene	166.8	.662	
Phenanthrene	(127.9)	.718	(.763) (100°)
Phenanthrenequinone	104.5	.502	(.698)
Phenanthrenonitrile	139.0	(.685)	
o-Phenetidine	(101.7)	.741 (25°)	
p-Phenetidine	(96.8)	.706 (25°)	(.749) (15°)
Phenetole	(84.5)	.692	(.689) (20°)
Phenol	60.21	(.640)	(.675) (45°)
Phenothiazine	114.8	(.576)	
Phenylacetaldehyde	72.01	.599$_4$	(.614) (20°)
Phenyl acetate	82.04	(.603)	(.647) (25°)
Phenylacetic acid	82.72	(.608)	(.657) (80°)
Phenylacetylene	72.01	(.705)	(.655) (20°)
1-Phenylazo-2-naphthol	137.6	(.554)	
2-Phenylbensofuran	130.5	(.672)	
1-Phenyl-4-benzoyl-1,3-butadiene	(140.3)	.599	
Phenylbutadiene	(85.7)	.658	
4-Phenyl-1-butene	93.49	(.7077)	(.6239) (20°)
Phenylbutyl acetate	134.5	.653	
Phenyl n-butyrate	105.46	(.643)	
Phenyl iso-cyanate	(72.7)	.610	(.699) (20°)
o-Phenylenediamine	71.98	.6662	
m-Phenylenediamine	70.53	.6529	(.723) (58°)
p-Phenylenediamine	70.28	.6503	
Phenyl ether	(108.1)	.635	(.681) (20°)
Phenylethyl sulfide	(94.4)	.6826 (27.4°)	
Phenylfluoroform	(77.3)	.529	
Phenylhydrazine	67.82	(.627)	(.688) (23°)
Phenylhydroxylamine	(68.2)	.625	
Phenyl mercaptan	(70.8)	.6425 (27.4°)	(.693) (20°)
1-Phenyl-2-Methylbutane	113.53	(.766)	(.660) (20°)
Phenylmethyl sulfide	(83.2)	.6695 (27.4°)	
Phenylpropiolamide	(83.3)	.574	
Phenyl propionate	93.79	(.625)	(.654) (25°)
Phenylsulfone	(129.0)	.591	(.740) (20°)
Phenyl thiocyanate	(81.5)	.6027 (27.4°)	(.677) (24°)
Phenyl isothiocyanate	(86.0)	.6365 (27.4°)	(.719) (24°)
1-Phenyl-4,6,6-trimethylheptane	173.90	(.796)	(.682) (20°)
N-Phenyl urea	82.1	(.603)	(.785)
Phloroglucinol	(73.4)	.582	
Phthalamide	(91.3)	.556	
Phthalic acid	83.61	.5035	(.802) (20°)
iso-Phthalic acid	84.64	.5097	
tere-Phthalic acid	83.51	.5029	(.759)
Phthalic anhydride	67.31	(.454)	(.694) (4°)
Phthalimide	(78.4)	.533	
Picric acid	84.38	(.368)	(.649)
Piperazine	56.8	(.659)	
Piperidine	64.2	(.754)	(.650) (20°)
Propane	40.5	(.919)	(.538) (−45°)
Propene	31.5	(.749)	(.456) (−47°)
Propionaldehyde	34.32	.591$_0$	(.477) (20°)
Propionic acid	43.50	.586	.582 (20°)
Propionitrile	38.5	(.699)	(.547) (21°)
Propionylphenylacetylene	(95.1)	.601	
Propiophenone	83.73	.624$_0$	(.631) (20°)
n-Propyl acetate	65.91	.645$_3$ (25°)	(.569) (25°)
iso-Propyl acetate	67.04	.656$_4$	(.566) (25°)
n-Propyl alcohol	45.176 (20°)	(.7518)	(.6047) (20°)
iso-Propyl alcohol	45.794 (20°)	(.7621)	(.5985) (20°)
9-Propylanthracene	164.0	(.744)	
n-Propylbenzene	89.24	(.742)	(.640) (20°)
n-Propyl benzoate	105.00	(.640)	(.646) (25°)
n-Propyl bromide	(65.6)	.533	(.721) (20°)
iso-Propyl bromide	(65.1)	.529	(.693) (20°)
Propyl butyrate	(89.4)	.6867	(.604) (15°)
prim-Propyl chloride	56.10	.715	.633 (20°)
iso-Propylcyclohexane	102.65	.8131	.6528 (20°)
Propylenediamine	(58.1)	.784	(.688) (15°)
Propylene oxide	42.5	(.732)	(.629) (0°)
Propyl formate	(55.0)	.6248	(.563) (20°)
Propyl hexylpropiolate	(136.8)	.697	
Propyl iodide	(84.3)	.4958 (30°)	(.864) (20°)
Propyl propionate (extrap.)	(77.95)	.6711	(.593) (20°)
Propyl sulfide	(92.1)	.7787 (27.4°)	(.634) (17°)
N-Propyl urea	67.4	(.660)	
Pseudocumene	(101.6)	.845 (20°)	(.740) (20°)
Pyramidone	149.0	(.645)	
Pyranthrene	266.9	.709	
Pyranthrone	250.3	.616	
Pyrazine	37.6	(.469)	(.484) (61°)
Pyrene	147.9	.731	(.933) (0°)
Pyridine	49.21	(.622)	(.611) (20°)
Pyrocatechol	68.76	.6248	(.857) (15°)
Pyrrole	47.6	(.709)	(.688) (20°)

Compound	$-\chi_M \times 10^6$	$-\chi \times 10^6$	$-K \times 10^6$
Pyrrolidine	54.8	(.771)	(.657) (23°)
Quinoline	86.0	(.666)	(.729) (20°)
Quinone	38.4	(.355)	(.468) (20°)
Quinonoxime	(50.4)	.409	
Resorcinol	67.26	.6112	(.785) (15°)
Rhamnose	99.20	.605	(.890) (20°)
Safrol and iso-Safro	(97.5)	.601	(.66) (20°)
Salicylaldehyde	64.4	(.527)	(.615) (20°)
Salicylic acid	72.23	.523	(.755) (20°)
Saligenin	76.9	.620	(.720) (25°)
Salol	(123.2)	.575	.678 (45°)
Salvarsan dihydrochloride	(246.1)	.518	
Selenophene	66.82	.510	
iso-Selenophene	110–111	.84–.85	
trans-Selenophene	70–77	.53–.59	
Sorbitol	107.80	.592	
Stearic acid	220.8	(.776)	(.657) (69°)
Stilbene	(120.0)	.666	(.646) (125°)
Stilbestrol	(130)	.62, .63	
Styrene	(68.2)	.655	(.594) (20°)
Succinic acid	(57.9)	.4902	(.767) (15°)
Succinic anhydride	(47.5)	.475	(.524)
Succinimide	(47.3)	.477	(.674) (16°)
Sulfamide	44.4	(.462)	(.832)
p-Sulfanilamide	80.15	(.465)	
Terpineol	111.9	(.725)	(.678) (room temp.)
Tetrabenzylmonosilane	266.2	(.678)	
1,1,2,2-Tetrabromoethane	(123.4)	.357	(1.058) (20°)
Tetrabromoethylene	(114.8)	.334	
Tetracene	(168.0)	.736	
1,1,2,2-Tetrachloroethane	(89.8)	.535	(.856) (20°)
Tetrachloroethylene	81.6	.492	(.802) (15°)
Tetrahydroquinoline	(89.0)	.668	(.715) (4°)
Tetraiodoethylene	(164.3)	.309	(.922) (20°)
Tetraiodopyrrole	(188.9)	.331	
Tetramethylketotetrahydrofurfurane	(104.7)	.736	
Tetranitromethane	43.02	.2195	(.360) (20°)
Tetraphenylbutadiene	228.0	(.636)	
Tetraphenyldecapentaene	280.8	(.643)	
Tetraphenylhexatriene	246.4	(.641)	
Tetraphenyloctatetraene	264.1	(.643)	
Tetraphenylrubene	344.0	(.646)	
Tetra-p-tolylmonosilane	276.4	(.704)	
Tetrolic acetal	(97.8)	.688	
Tetronic acid	(52.5)	.525	
Thiacoumerin	93.6	.577	
Thiazole	50.55	.595 (20°)	(714) (17°)
Thiobarbituric acid	72.9	(.506)	
Thiophene	57.38	.682 (20°)	(.726) (20°)
Tolane	(118.9)	.667	(.644) (100°)
Toluene	66.11	.7176	.6179 (20°)
o-Toluidine	76.0	.710 (24°)	(.709) (20°)
m-Toluidine	74.6	.697 (25°)	(.689) (20°)
p-Toluidine	72.1	.673 (25°)	(.704) (20°)
α-Tolunitrile	76.87	(.656)	(.666) (18°)
1-(o-Tolylazo)-2-naphthol	148.7	(.567)	
1-(p-Tolylazo)-2-naphthol	157.6	(.601)	
Triallylacetophenone	152.5	(.634)	
Tri-iso-amylamine	192	.845	(.647) (25°)

Compound	$-\chi_M \times 10^6$	$-\chi \times 10^6$	$-K \times 10^6$
Trianilinophosphine oxide	(201.7)	.624	
1,2,3-Tribromopropane	(117.9)	.420	(1.023) (23°)
Tri-iso-butylamine	(156.8)	.846	(.646) (25°)
Trichloroacetic acid (in sol'n)	73.0	(.44)	(.723) (46°)
Trichlorobenzene	(106.5)	.587	
Trichloro-tert-butyl alcohol (in sol'n)	98.01	.552	
Trichloroethylene	65.8	.501	(.734) (20°)
Trichloronitromethane	(75.3)	.458	(.756) (20°)
Triethylamine	81.4	(.804)	(.586) (20°)
Triethyl citrate	(161.9)	.586	(.666) (20°)
Triethyl phosphate	(125.3)	.688	(.735) (20°)
Triethylphosphine	(90.0)	.762	(.610) (15°)
Triethylphosphine oxide	(91.6)	.683	
Triethyl phosphite	(104.8)	.631	(.611) (20°)
Triethyl triazinetricarbonate	(164.1)	.552	
Trifluorocresol	(83.8)	.517	
Tri-n-heptylamine	251.3	(.806)	
Tri-n-hexylamine	221.7	(.823)	
Trimethylacetophenone	108.2	(.667)	(.648) (16°)
2,2,3-Trimethylbutane	88.36	.8818	.6086 (20°)
2,2,3-Trimethylpentane	99.86	.8743	.6261 (20°)
2,2,4-Trimethylpentane	98.34	.8610	.5958 (20°)
2,3,5-Trimethylpyrrole	82.31	.754 (20°)	
1,3,5-Trinitrobenzene	74.55	(.350)	(.591) (20°)
Triperfluorobutylamine	253.0	(.377)	
Triphenoxyarsine	(195.2)	.551	
Triphenylarsine	(177.0)	.578	
Triphenylarsine dihydroxide	(270.5)	.795	
Triphenylarsine oxide	(199.1)	.618	
Triphenylbismuthine	(196.8)	.447	(.708) (20°)
Triphenylbismuthine dinitrate	(254.5)	.451	
Triphenylcarbinol	(175.7)	.675	(.802) (20°)
Triphenylmethane	(165.6)	.678	.686 (100°)
Triphenylphosphine	(166.8)	.636	(.759)
Triphenyl phosphite	(183.7)	.592	(.701) (18°)
Triphenylstibine	(182.2)	.516	
Triphenylstibine dihydroxide	(238.5)	.616	
N,N′,N-Triphenyl urea	176.5	(.613)	
Triquinoyl	(133.0)	.426	
Tropolone	61	.50	
Tryptophan	132.0	(.646)	
Tyrosine	105.3	(.581)	
Undecane	131.84	(.8435)	(.6247) (20°)
Urea	33.4	(.556)	(.742) (20°)
Urethan	(57)	.64	(.63) (21°)
iso-Valeraldehyde	(57.5)	.668	(.536) (17°)
n-Valeric acid	66.85	.6548	(.617) (20°)
iso-Valeric acid	(67.7)	.663	(.621) (15°)
Valerylphenylacetylene	(119.0)	.639	
Valine	74.3	(.634)	
Violanthrene	273.5	.641	
Violanthrone	204.8	.449	
iso-Violanthrone	215.9	.473	
Water	(13.00)	.7218 (20°)	(.7205) (20°)
Water (value usually used as standard)	(12.97)	.720 (20°)	(.719) (20°)
Xanthone	(108.1)	.551	
o-Xylene	77.78	.7327	.6440 (20°)
m-Xylene	76.56	.7212	.6235 (20°)
p-Xylene	76.78	.7232	.6226 (20°)
Xylose	84.80	.565	(.862) (20°)

DIAMAGNETIC SUSCEPTIBILITIES OF ORGANIC COMPOUNDS (Continued)

Supplementary Table

Formula	Compound	$-\chi_M \times 10^6$	$-\chi \times 10^6$	$-K \times 10^6$
$H_6N_3B_3$	Borazine	49.6	.616	
CO	Carbon monoxide	9.8	.350	
COS	Carbon oxysulfide	32.4	.539	
$COCl_2$	Phosgene	48	.485	(.675) (19°)
CO_2	Carbon dioxide	20	.45	
$CNCl$	Cyanogen chloride	32.4	.527	(.642) (4°)
CCl_2S	Thiophosgene	50.6	.440	(.664) (15°)
CH_2N_2	Cyanamide	24.8	.590	(.639)
CH_3SiCl_3	Methyl trichlorsilane	87.45	$.584_8$	
CH_4N_2S	Thiourea	42.4	.557	(.782) (20°)
C_2HF_3	Trifluoroethylene	32.2	.393	
$C_2HF_3Cl_2$	1,2-Dichloro-1,1,2-trifluoroethane	66.2	.433	
$C_2HF_3Br_2$	1,2-Dibromo-1,1,2-trifluoroethane	90.9	.376	
$C_2H_3O_2N_3$	Urazole	46.2	.456	
C_2H_4OS	Thioacetic acid	38.4	.505	(.542) (10°)
$C_2H_4O_2S$	Mercaptoacetic acid	50.0	.543	(.720) (20°)
C_2H_5NS	Thioacetamide	42.45	.565	
$C_2H_5SiCl_3$	Ethyl trichlorosilane	98.84	$.604_6$	
$C_2H_6O_2S_2$	Methyl thiosulfite	62.3	.494	
$C_2H_6O_3S$	Methyl sulfite	54.2	.492	
$C_2H_6N_2S$	N-Methyl thiourea	53.6	.595	
C_2H_6S	Ethyl mercaptan	47.0	.756	(.635)
$C_2H_6SiCl_2$	Dimethyl dichlorosilane	82.45	$.639_2$	
$C_2H_8N_2$. HCl	1,2-Diamino ethane hydrochloride	76.2	.789	
C_3H_3ON	Pyruvic nitrile	33.2	.481	
$C_3H_5OF_3$	1-Methoxy-1,1,2-trifluoroethane	55.9	.490	
C_3H_5OCl	Propionyl chloride	51	.55	(.59)

Formula	Compound	$-\chi_M \times 10^6$	$-\chi \times 10^6$	$-K \times 10^6$
$C_3H_5O_2Br$	Methylbromoacetate	71.1	.465	
C_3H_5Cl	1-Chloro-2-propene	47.8	.625	
C_3H_5Br	1-Bromo-2-propene	58.6	.484	
C_3H_5I	1-Iodo-2-propene	72.8	.433	
$C_3H_6N_6$	Melamine	61.8	.490	(.771) (250°)
C_3H_7N	1-Amino-2-propene	40.1	.702	
$C_3H_7SiCl_3$	n-Propyl trichlorosilane	110.2	$.620_0$	
$C_3H_8O_2N_2$	N-Hydroxymethyl-N-methylurea	68.0	.653	
$C_3H_8N_2S$	N,N-Dimethyl thiourea	64.2	.616	
$C_3H_8N_2S$	N,N'-Dimethyl thiourea	64.7	.621	
$C_3H_8N_2S$	N-Ethyl thiourea	62.8	.603	
C_3H_8S	Propyl mercaptan	58.5	.768	(.642) (25°)
C_3H_9SiCl	Trimethyl chlorosilane	77.36	$.712_2$	
$C_3H_{12}N_3B_3$	N-Trimethyl borazine	78.6	.641	
$C_4H_5ON_3$	Cytosine	55.8	.502	
$C_4H_6O_2$	Vinylacetate	46.4	.539	(.502)
$C_4H_6O_2NCl_3$	Nitro tri(chloromethyl) methane	107.8	.522	
$C_4H_6O_6$	Tartaric acid	67.5	.402	
C_4H_7OCl	Butyryl chloride	62.1	.582	(.598) (20°)
C_4H_7OCl	iso-Butyryl chloride	63.9	.599	
$C_4H_8N_2S$	N-Allyl thiourea	69.0	.595	(.725) (20°)
$C_4H_9ON_3$	Acetone semicarbazone	66.29	$.575_8$	(.716) (20°)
$C_4H_{10}O_2S_2$	Ethyl thiosulfite	86.2	.559	
$C_4H_{10}S_2$	Ethyl disulfide	83.6	.684	(.679) (20°)
$C_4H_{10}SiCl_2$	Diethyl dichlorosilane	105.80	$.673_5$	
$C_5H_4O_3N_4$	Uric acid	66.2	.394	(.746)
C_5H_6	Cyclopentadiene	44.5	.673	
$C_5H_6O_2N_2$	Thymine	57.1	.453	
$C_5H_8O_2$	Methyl methacrylate	57.3	.572	(.535) (20°)
$C_5H_8N_2$	3,5-Dimethyl pyrazole	56.2	.585	
$C_5H_8Cl_4$	Tetra (chloromethyl) methane	129	.652	
$C_5H_9S_3Na$	Sodium butyl thiocarbonate	104.5	.555	
$C_5H_9Cl_3$	1,1,1-Tri(chloromethyl) ethane	114	.650	
$C_5H_{10}O$	Cyclopentanol	64.0	.743	(.705) (20°)
$C_5H_{10}O_2$	Ethyl carbonate	75.4	.738	
$C_5H_{10}O_5$	D-Ribose	84.6	.564	
$C_5H_{10}NS_2Na$	Sodium diethyldithiocarbomate	99.2	.579	
$C_5H_{10}Cl_2$	2,2-Di-(Chloromethyl) propane	96.9	.687	
$C_5H_{11}ON$	Methyl n-propyl ketoxime	68.82	$.679_6$	(.618) (20°)
$C_5H_{11}ON_3$	Ethyl methyl ketone semicarbazone	77.93	$.603_4$	(.710) (20°)
$C_5H_{12}ON_2$	N,N'-Diethyl urea	74.1	.638	
$C_5H_{12}ON_2$	N,N,N',N'-Tetramethyl urea	75.7	.652	(.634) (15°)
$C_5H_{14}O_2Si$	Methyl diethoxysilane	92.99	$.692_4$	
$C_6H_3O_4N_2Cl$	1-Chloro-2,4-dinitrobenzene	84.4	.417	(.708) (22°)
$C_6H_3Cl_2I$	3,4-Dichloro-1-iodobenzene	118	.432	
$C_6H_5SiCl_3$	Phenyl trichlorosilane	120.4	$.569_4$	
C_6H_6NBr	o-Bromoaniline	87.32	.508	
C_6H_6NBr	m-Bromoaniline	84.89	.494	(.780) (20.4°)
C_6H_6NBr	p-Bromoaniline	84.06	.489	(.880)
$C_6H_7O_3NS . H_2O$	Sulfanilic acid	90.1	.471	
C_6H_7N	2-Methyl pyridine	60.3	.648	(.616) (15°)
C_6H_7N	3-Methyl pyridine	59.8	.642	(.617) (15°)
C_6H_7N	4-Methyl pyridine	59.8	.642	(.614) (15°)
$C_6H_7N . HI$	Aniline hydroiodide	113.6	.514	
C_6H_{10}	2,3-Dimethyl-1,3-butadiene	57.2	.696	(.539) (0°)
$C_6H_{11}OCl$	2-Ethyl butyryl chloride	87.2	.648	
$C_6H_{11}SiCl_3$	Cyclohexyltrichlorosilane	138.1	$.634_3$	
$C_6H_{12}N_2S_3$	Tetramethyl thiuram monosulfide	118.4	.568	(.795)
$C_6H_{12}N_2S_4$	Tetramethl thiuram disulfide	140.6	.585	(.755) (20°)
$C_6H_{13}ON$	Methyl-n-butyl ketoxime	79.97	$.694_2$	(.623) (20°)
$C_6H_{13}ON_3$	Methyl-n-propyl ketone semicarbazone	89.47	$.624_9$	(.669) (20°)
$C_6H_{13}ON_3$	Diethyl ketone semicarbazone	90.68	$.633_3$	(.730) (20°)
C_6H_{14}	3-Ethyl pentane	86.21	$.861_0$	(.601) (20°)
$C_6H_{14}O$	iso-Propyl ether	79.4	.777	(.564) (20°)
$C_6H_{14}S_2$	Propyl disulfide	106.2	.707	
$C_6H_{16}O_2Si$	Dimethyl diethoxysilane	104.6	$.705_2$	
$C_6H_{18}O_3Si_3$	Hexamethylcyclotrisiloxane	140.7	$.632_4$	
$C_6H_{18}N_3B_3$	Hexamethylborazine	119	.723	
$C_7H_5O_2Cl$	o-Chlorobenzoic acid	83.56	.534	(.824) (20°)
$C_7H_5O_4N$	Quinoleic acid	72.3	.433	
$C_7H_5NS_2$	2-Mercaptobenzothiazole	99.4	.594	(.843) (20°)
$C_7H_5F_2Cl$	Chlorodifluoromethylbenzene	87.2	.536	
$C_7H_7O_2N$	o-Aminobenzoic acid	77.18	.563	
C_7H_7Cl	o-Chlorotoluene	81.98	.648	(.701) (20°)
C_7H_7Cl	m-Chlorotoluene	80.07	.633	(.679) (20°)
C_7H_7Cl	p-Chlorotoluene	80.07	.633	(.677) (20°)
$C_7H_8N_2S$	N-phenylthiourea	87.4	.574	(.75)
C_7H_9ON	o-Anisidine	80.44	.654	(.714) (20°)
C_7H_9ON	m-Anisidine	79.95	.650	(.712) (20°)
C_7H_9ON	p-Anisidine	80.56	.655	(.702) (55°)
C_7H_9N	2,4-Dimethyl pyridine	71.50	.667	(.633) (0°)
C_7H_9N	2,6-Dimethyl pyridine	71.72	.669	(.630) (0°)
$C_7H_{15}ON$	Methyl amyl ketoxime	91.24	$.706_2$	(.630) (20°)
$C_7H_{15}ON_3$	Methyl-n-butyl ketone semicarbazone	100.40	$.638_6$	(.643) (20°)
$C_7H_{16}O$	1-Heptanol	91.7	.790	(.649) (20°)
$C_7H_{18}O_3Si$	Methyl triethoxysilane	120.6	$.676_2$	
C_8H_7OCl	Phenylacetyl chloride	88.5	.572	(.668) (20°)
$C_8H_8O_2$	o-Toluic acid	80.83	.594	(.626) (112°)
$C_8H_{10}ON_2$	N-Methyl, N'Phenyl urea	93.35	.622	
$C_8H_{10}O_4$	Butyne diacetate	95.9	.563	
$C_8H_{11}ON$	m-Phenetidine	90.28	.659	
$C_8H_{11}N$	2,4,6-Trimethyl pyridine	83.22	.687	(.630) (20°)
$C_8H_{16}O_2$	Hexylacetate	100.9	.700	(.623) (0°)
$C_8H_{17}ON_3$	Methyl amyl ketone semicarbazone	112.25	$.655_5$	(.687) (20°)
C_8H_{18}	3-Methyl-3-ethyl pentane	99.9	.875	(.623)
C_8H_{18}	4-Methyl heptane	97.30	$.851_7$	(.614)
C_8H_{18}	3-Ethyl hexane	97.76	$.855_8$	(.614) (20°)
C_8H_{18}	2,3,4-Trimethyl pentane	99.75	$.872_3$	

DIAMAGNETIC SUSCEPTIBILITIES OF ORGANIC COMPOUNDS (Continued)

Formula	Compound	$-\chi_M \times 10^6$	$-\chi \times 10^6$	$-K \times 10^6$
$C_8H_{18}S_2$	Butyl disulfide	129.6	.727	
$C_8H_{18}S_3$	Butyl trisulfide	144.5	.687	
$C_8H_{18}S_4$	Butyl tetrasulfide	158.5	.653	
$C_8H_{20}O_4Si$	Tetraethoxysilane	137.1	$.657_9$	
$C_8H_{24}O_4Si_4$	Octamethyl cyclotetrasiloxane	187.4	$.632_2$	
C_9H_9N	Phenyl propionitrile	78.2	.615	
$C_9H_6O_2$	Phenyl propiolic acid	81.0	.554	
$C_9H_6O_3$	7-Hydroxycoumarin	88.22	$.544_9$	
C_9H_7N	iso-Quinoline	83.9	.650	(.714) (20°)
$C_9H_{10}O_4N_4$	Acetone-2,4-dinitrophenylhydrazone	110.62	$.464_4$	(.653) (20°)
$C_9H_{11}ON$	Benzylideneamino-1-hydroxyethane	91.0	.610	
$C_9H_{11}N$	Benzylideneaminoethane	85.5	.642	
$C_9H_{12}O_3S$	Ethyl-p-toluene sulfonate	115	.574	
$C_9H_{12}O_6N_2$	Unidine	106.9	.438	
$C_9H_{13}O_5N_3$	Cytidine	123.7	.509	
$C_9H_{14}O_3$	1-Acetyl-1 ethoxycarbonyl cyclobutane	103.4	.608	
$C_9H_{14}O_4$	1,1-Di(ethoxycarbonyl) cyclopropane	110.4	.593	
$C_9H_{14}Si$	Trimethylphenylsilane	109.1	.726	
$C_9H_{19}ON_3$	Methyl-n-hexyl ketone semicarbazone	123.60	$.666_2$	(.716) (20°)
$C_9H_{20}O$	Di-isobutyl carbinol	116.9	.810	(.667) (0°)
$C_9H_{20}ON_2$	N,N,N,'N'-Tetraethylurea	122.4	.710	
$C_9H_{20}N_2S$	N,N,N',N'-Tetraethylthiourea	132.6	.704	
$C_9H_{24}N_3B_3$	B-Triethyl-N-Trimethyl borazine	146	.706	
$C_{10}H_6O_7N_2$	5,7-Dinitro-7-hydroxy-4-methylcoumarin	111.9	$.420_6$	
$C_{10}H_7O_3Br$	7-Hydroxy-4-methyl-3-bromocoumarin	127.3	$.498_7$	
$C_{10}H_7O_5N$	5-Nitro-6-hydroxy-4-methylcoumarin	105.6	$.477_9$	
$C_{10}H_7O_5N$	6-Nitro-7-hydroxy-4-methylcoumarin	105.5	$.477_5$	
$C_{10}H_7O_5N$	8-Nitro-7-hydroxy-4-methylcoumarin	106.0	$.479_8$	
$C_{10}H_8O_3$	6-Hydroxy-4-methylcoumarin	98.69	$.560_7$	
$C_{10}H_8O_3$	7-Hydroxy-4-methylcoumarin	99.96	$.563_7$	
$C_{10}H_8O_4$	Benzylidene malonic acid	97.5	.507	
$C_{10}H_8O_4$	5,7-Dihydroxy-4-methylcoumarin	106.9	$.556_7$	
$C_{10}H_8O_4$	7,8-Dihydroxy-4-methylcoumarin	105.1	$.550_4$	
$C_{10}H_8S$	1-Mercaptonaphthalene	109.5	.684	
$C_{10}H_8S_2$	1,8-Dimercaptonaphthalene	118.0	.614	
$C_{10}H_{10}O_4$	Methyl terephthalate	101.6	.523	
$C_{10}H_{10}Os$	Bis cyclopentadienyl osmium	193(20°)	.602	
$C_{10}H_{12}$	1,2,3,4-Tetrahydronaphthalene	93.3	.706	(.685)
$C_{10}H_{12}O_4N_4$	Ethyl methyl ketone-2,4-dinitrophenylhydrazone	124.11	$.492_1$	(.631) (20°)
$C_{10}H_{13}O_4N_5$	Adenosine	137.5	.514	
$C_{10}H_{13}O_5N_5$	Guanosine	149.1	.526	
$C_{10}H_{14}$	Durene	101.2	.754	(.632) (81°)
$C_{10}H_{14}$	sec-Butyl benzene	101.31	$.754_9$	(.651) (20°)
$C_{10}H_{16}O_4$	1,1-Di(ethoxycarbonyl) cyclobutane	118	.589	
$C_{10}H_{19}N$	Camphylamine	122.0	.796	
$C_{11}H_9BrS$	1-Bromo-4-Methylthionaphthalene	147.0	.581	
$C_{11}H_{10}O_2$	4,6-Dimethylcoumarin	106.2	$.610_3$	
$C_{11}H_{10}O_2$	4,7-Dimethylcoumarin	107.6	$.619_0$	
$C_{11}H_{10}O_3$	5-Hydroxy-4,7-dimethylcoumarin	113.5	$.597_3$	
$C_{11}H_{10}O_3$	6-Methoxy-4-methylcoumarin	109.5	$.576_3$	
$C_{11}H_{10}O_3$	7-Methoxy-4-methylcoumarin	110.5	$.581_4$	
$C_{11}H_{10}S$	Methyl-α-naphthyl sulfide	120.2	.690	
$C_{11}H_{14}O_2$	Benzyl butyrate	116.3	.653	(.663) (17.5°)
$C_{11}H_{14}O_4N_4$	Diethyl ketone-2,4-dinitrophenylhydrazone	135.16	$.507_6$	(.670) (20°)
$C_{11}H_{14}O_4N_4$	Methyl-n-propyl ketone-2,4-dinitrophenylhydrazone	135.70	$.509_7$	(.623) (20°)
$C_{11}H_{16}ON_2$	N,N-Diethyl-N'-phenyl urea	126.4	$.657_3$	
$C_{12}H_6O_4N_2Cl_2$	4,4'-Dichloro-2,2'-dinitro-1,1'-biphenyl	150	.479	
$C_{12}H_8$	Acenaphthylene	111.6	.733	(.659) (16°)
$C_{12}H_8O_4N_2$	2,4'-Dinitro-1,1'-biphenyl	116.5	.477	(.703) (20°)
$C_{12}H_8Cl_2$	4,4'-Dichloro-1,1'-biphenyl	133.1	.597	(.859) (20°)
$C_{12}H_8Br_2$	4,4'-Dibromo-1,1'-biphenyl	151.3	.485	(.920) (20°)
$C_{12}H_9O_2N$	2-Nitro-1,1'-biphenyl	109	.547	(.788) (20°)
$C_{12}H_9O_2N$	5-Nitroacenaphthene	116.0	.582	
$C_{12}H_{10}O_4$	Cinnamylidene malonic acid	105.4	.483	
$C_{12}H_{10}O_4$	Quinhydrone	105	.481	(.674) (20°)
$C_{12}H_{10}S$	Phenyl sulfide	119.2	.640	(.716) (20°)
$C_{12}H_{10}S_2$	Phenyl disulfide	122.5	.561	
$C_{12}H_{10}Cl_2Si$	Diphenyl dichlorosilane	153.5	$.606_3$	
$C_{12}H_{12}O_2Si$	Diphenyl silanediol	131.6	$.608_2$	
$C_{12}H_{16}O_2$	Amyl benzoate	128.5	.668	
$C_{12}H_{16}O_4N_4$	Methyl-n-butyl ketone-2,4-dinitrophenylhydrazone	147.65	$.526_8$	(.643) (20°)
$C_{12}H_{18}$	Hexamethyl benzene	122.5	.755	
$C_{12}H_{18}$	Diisopropyl benzene	124.77	$.767_1$	
$C_{12}H_{22}O_{11}$	Sucrose (saccharose)	189.1	.552	(.877) (15°)
$C_{12}H_{25}Cl_3Si$	Dodecyl trichlorosilane	209.2	$.688_9$	
$C_{12}H_{30}N_3B_3$	Hexaethylborazine	189	.759	
$C_{13}H_8OS$	Thioxanthone	130	.612	
$C_{13}H_{10}O_2$	Phenyl benzoate	117.3	.592	
$C_{13}H_{10}S$	Thiobenzophenone	118.1	.596	
$C_{13}H_{11}NS$	N-Thiobenzoyl aniline	123.0	.577	
$C_{13}H_{12}N_2S$	N,N'-Diphenyl thiourea	136.5	.598	
$C_{13}H_{12}N_2S$	N,N-Diphenyl thiourea	134.9	.591	
$C_{13}H_{18}O_4N_4$	Methyl amyl ketone-2,4-dinitrophenylhydrazone	156.84	$.532_9$	(.651) (20°)
$C_{14}H_8N_2S_4$	Di-2-benzothiazolyl disulfide	189.0	.568	
$C_{14}H_{10}O$	Anthrone	118	.608	
$C_{14}H_{12}O_2$	Benzyl benzoate	132.2	.622	(.693) (18°)
$C_{14}H_{12}O_2$	Diphenyl acetic acid	124.5	.587	
$C_{14}H_{14}O_2S_2$	p-Tolyl thiosulfonate	157	.564	
$C_{14}H_{14}N_2$	p-Azotoluene	135.1	.643	
$C_{14}H_{16}Si$	Dimethyl diphenyl silane	146.6	.690	
$C_{14}H_{20}O_4N_4$	Methyl n-hexyl ketone-2,4-dinitrophenylhydrazone	171.73	$.557_0$	(.655) (20°)
$C_{15}H_{10}O_2$	Flavone	120	.540	
$C_{15}H_{12}$	9-Ethylidene fluorene	124.8	.649	
$C_{15}H_{12}O$	Chalcone	125.7	.604	(.646) (62°)
$C_{15}H_{14}O_2$	Benzyl phenylacetate	143.7	.635	
$C_{15}H_{16}ON_2$	N,N'-Dimethyl-N,N'-Diphenyl urea	148.9	.620	
$C_{16}H_{10}$	Fluoranthene	138.0	.682	
$C_{16}H_{12}$	1-Benzylidene indene	130.5	.716	

DIAMAGNETIC SUSCEPTIBILITIES OF ORGANIC COMPOUNDS (Continued)

Formula	Compound	$-\chi_M \times 10^6$	$-\chi \times 10^6$	$-k \times 10^6$
$C_{16}H_{16}$	1,2-Dibenzylamine ethane	164	.787	
$C_{16}H_{33}SiCl_3$	Hexadecyl trichlorosilane	252.4	.701$_3$	
$C_{16}H_{48}O_6Si_7$	Hexadecamethyl heptasiloxane	351.1	.658$_8$	
$C_{17}H_{16}$	9-Butylidene fluorene	147.5	.670	
$C_{18}H_8S_4$	Tetrathiotetracene	202	.573	
$C_{18}H_{10}Cl_2$	5,11-Dichloro tetracene	202	.680	
$C_{18}H_{10}Br_2$	5,11-Dibromo tetracene	216	.559	
$C_{18}H_{12}$	Triphenylene	156.6	.686	
$C_{18}H_{14}$	o-Diphenyl benzene	150.4	.653	
$C_{18}H_{14}$	p-Diphenyl benzene (Terphenyl)	152	.660	
$C_{18}H_{15}SiCl$	Triphenyl chlorosilane	186.7	.633$_3$	(.814) (0°)
$C_{18}H_{16}OSi$	Triphenyl silanol	176.6	.634$_5$	
$C_{18}H_{16}Si$	Triphenylsilane	174	.668	
$C_{18}H_{20}$	5-Cyclohexyl acenaphthene	165.2	.699	
$C_{18}H_{22}$	Dimesitylene	171.8	.721	
$C_{18}H_{24}O_2$	Estradiol	186.6	.717	
$C_{18}H_{37}SiCl_3$	Octadecyl trichlorosilane	273.7	.705$_7$	
$C_{19}H_{14}O$	4-Benzoyl-1,1′-Biphenyl	158	.612	
$C_{19}H_{18}Si$	Methyl triphenyl silane	186	.678	
$C_{20}H_{14}O_2$	1,4-Dibenzoyl benzene	153	.534	
$C_{21}H_{24}N_3B_3$	B-Trimethyl-N-triphenylborazine	234	.667	
$C_{24}H_{18}$	4,4′-Diphenyl-1,1′-Biphenyl	201.3	.657	
$C_{24}H_{20}Si$	Tetraphenyl silane	224	.666	
$C_{24}H_{30}N_3B_3$	B-Triethyl-N-triphenylborazine	264	.693	
$C_{24}H_{40}O_4$	Desoxycholic acid	272.0	.715	
$C_{24}H_{40}O_5$	Cholic acid	282.3	.691	
$C_{25}H_{20}N_2S$	N,N,N′,N′-Tetraphenylthiourea	229.7	.604	
$C_{26}H_{20}$	Tetraphenyl ethylene	217.4	.654	
$C_{28}H_{16}O_2$	Bianthrone (Dianthraquinone)	220	.572	
$C_{28}H_{18}O_2$	Dianthrone (Dianthronol)	229	.593	
$C_{28}H_{20}$	1,1,4,4-Tetraphenyl-1,2,3-butatriene	227	.659	
$C_{28}H_{44}O$	Calciferol	273.3	.689	
$C_{28}H_{44}O$	Ergosterol	279.6	.704	
$C_{36}H_{30}O_3Si_3$	Hexaphenyl cyclotrisiloxane	364.1	.612$_9$	
$C_{48}H_{40}O_4Si_4$	Octaphenyl cyclotetrasiloxane	485.3	.613$_2$	
$C_{60}H_{36}$	Hexabenzocoronene	346.0	.457	

DIAMAGNETIC SUSCEPTIBILITY DATA OF ORGANOSILICON COMPOUNDS

R. R. Gupta

The molecular susceptibility is represented by χ_M and expressed in c.g.s. units. For some of the compounds volume or specific susceptibility is available in the literature and in such cases χ_M has been calculated by the relation[1]

$$\chi_M = M \times \chi = M\,\kappa/\zeta$$

where M is the molecular weight of the substance in grams, χ is the specific or mass susceptibility, κ is the volume susceptibility and ζ is the density of the substance in grams per cubic centimeter.

The compounds are arranged in the increasing order of C, H, Si, O, Cl, Br, I, S, and N.

DIAMAGNETIC SUSCEPTIBILITY DATA OF ORGANO-SILICON COMPOUNDS

Compound	Molecular formula	Structural formula	$-\chi_M \times 10^{-6}$	Ref.
Methyl trichlorosilane	CH_3SiCl_3	$CH_3{-}Si{-}Cl_3$	85.40, 87.45	2, 8
Methyl tribromosilane	CH_3SiBr_3	$CH_3{-}Si{-}Br_3$	126.00, 128.00, 155.50	10, 11, 4
Methyl dichlorosilane	CH_4SiCl_2	$CH_3{-}SiH{-}Cl_2$	66.90	4
Ethyl trichlorosilane	$C_2H_5SiCl_3$	$C_2H_5{-}Si{-}Cl_3$	98.00, 98.84, 96.40	2, 9, 7
Ethyl tribromosilane	$C_2H_5SiBr_3$	$C_2H_5{-}Si{-}Br_3$	147.50	4
Ethyl triiodosilane	$C_2H_5SiI_3$	$C_2H_5{-}Si{-}I_3$	171.50	4
Ethyl dichlorosilane	$C_2H_6SiCl_2$	$C_2H_5{-}SiH{-}Cl_2$	78.90	4
Dimethyl dichlorosilane	$C_2H_6SiCl_2$	$(CH_3)_2{-}Si{-}Cl_2$	81.30, 82.45	4, 8
Dimethyl diiodosilane	$C_2H_6SiI_2$	$(CH_3)_2{-}Si{-}I_2$	131.80	4
Dimethyl silanediol	$C_2H_8SiO_2$	$(CH_3)_2{-}Si{-}(OH)_2$	58.40	17, 18
n-Propyl trichlorosilane	$C_3H_7SiCl_3$	$CH_3{-}(CH_2)_2{-}Si{-}Cl_3$	108.00, 110.20	4, 8
Trimethyl chlorosilane	C_3H_9SiCl	$(CH_3)_3{-}SiCl$	79.00, 73.76	4, 8
Trimethyl bromosilane	C_3H_9SiBr	$(CH_3)_3{-}Si{-}Br$	91.30	4
Trimethyl iodosilane	C_3H_9SiI	$(CH_3)_3{-}Si{-}I$	104.10	4
1,4,6,9-Tetraoxa-5-silaspiro[4,4]nonane	$C_4H_8SiO_4$	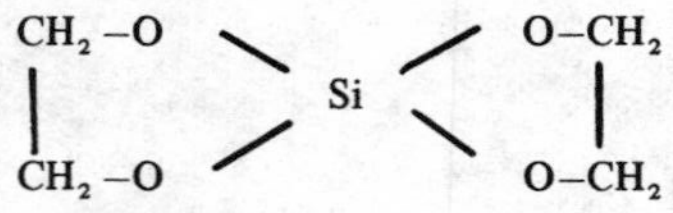	80.94	5
Diethyl dichlorosilane	$C_4H_{10}SiCl_2$	$(C_2H_5)_2{-}SiCl_2$	105.80	8

DIAMAGNETIC SUSCEPTIBILITY DATA OF ORGANO-SILICON COMPOUNDS (Continued)

Compound	Molecular formula	Structural formula	$^{x}Mx-10^{-6}$	Ref.
Tetramethylsilane	$C_4H_{12}Si$	$Si-(CH_3)_4$	74.80	1
Trimethyl methoxysilane	$C_4H_{12}SiO$	$(CH_3)_3-Si-OCH_3$	78.80	1
Dimethyl dimethoxysilane	$C_4H_{12}SiO_2$	$(CH_3)_2-Si-(OCH_3)_2$	81.70, 81.60, 81.95	1, 17, 5
Diethyl silanediol	$C_4H_{12}SiO_2$	$(C_2H_5)_2-Si-(OH)_2$	81.00	17, 18
Methyl trimethoxysilane	$C_4H_{12}SiO_3$	$CH_3-Si-(OCH_3)_3$	85.6	17
Tetramethoxysilane	$C_4H_{12}SiO_4$	$Si-(OCH_3)_4$	83.68	5
Trimethylacetoxysilane	$C_5H_{12}SiO_2$	$(CH_3)_3-Si-OCOCH_3$	85.00, 86.09	16, 5
Trimethylsilylchloromethylamidoxime	$C_5H_{13}SiOClN_2$	$NH_2-C-(CH_2Cl)=N-O-Si-(CH_3)_3$	109.60	15
Trimethylethylsilane	$C_5H_{14}Si$	$C_2H_5-Si-(CH_3)_3$	86.00	2
Trimethylethoxysilane	$C_5H_{14}SiO$	$(CH_3)_3-Si-OC_2H_5$	89.50	1, 17
Phenyl trichlorosilane	$C_6H_5SiCl_3$	$C_6H_5-Si-Cl_3$	120.4	8
Dimethyl diacetoxysilane	$C_6H_{12}SiO_4$	$(CH_3)_2-Si-(OCOCH_3)_2$	98.3, 101.18	16, 5
1,4,6,9-Tetraoxa-2,7-dimethyl-5-silaspiro[4,4]nonane	$C_6H_{12}SiO_4$	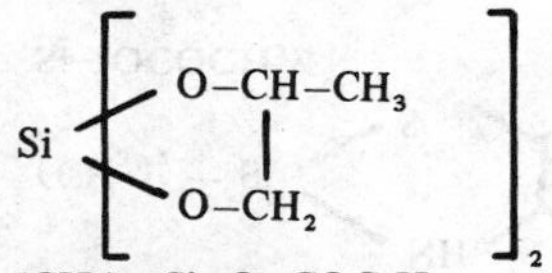	104.34	5
Trimethylpropionoxysilane	$C_6H_{14}Si-O_2$	$(CH_3)_3-Si-O-COC_2H_5$	99.09	5
Dimethyldiethylsilane	$C_6H_{16}Si$	$(C_2H_5)_2-Si-(CH_3)_2$	97.90	1
Triethylsilane	$C_6H_{16}Si$	$(C_2H_5)_3-Si-H$	95.00	3
Trimethylepropoxysilane	$C_6H_{16}SiO$	$(CH_3)_3-Si-O-(CH_2)_2-CH_3$	101.1	1
Trimethylsilylethylamidoxime	$C_6H_{16}SiON_2$	$(NH_2)-C(C_2H_5)=N-O-Si(CH_3)_3$	101.00	15
Dimethyldiethoxysilane	$C_6H_{16}SiO_2$	$(CH_3)_2-Si-(OC_2H_5)_2$	103.60, 104.70, 104.1, 104.6	1, 18, 5, 8
Trimethylthio-n-propylsilane	$C_6H_{16}SiS$	$(CH_3)_3-Si-S-(CH_2)_2-CH_3$	114.09	12
Trimethylthio-isopropylsilane	$C_6H_{16}SiS$	$(CH_3)_3-Si-S-(CH_3)_2$	115.11	12
Hexamethyldisiloxane	$C_6H_{18}Si_2O$	$(CH_3)_3-Si-O-Si-(CH_3)_2$	118.9	8
Hexamethylcyclotrisiloxane	$C_6H_{18}Si_3O_3$	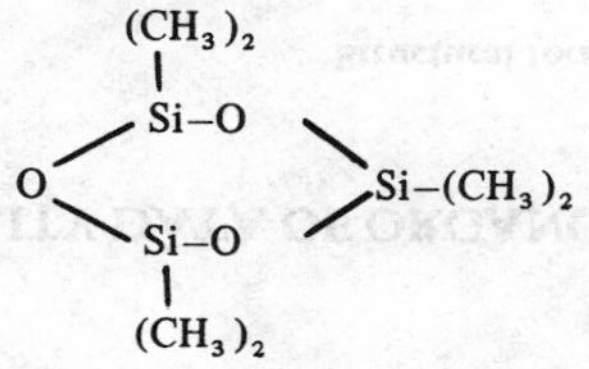	140.5	8

DIAMAGNETIC SUSCEPTIBILITY DATA OF ORGANO-SILCON COMPOUNDS (Continued)

Compound	Molecular formula	Structural formula	$^{x}M_x - 10^{-6}$	Ref.
Methyltriacetoxysilane	$C_7H_{12}SiO_6$	$(CH_3)_3-Si-(OCOCH_3)_3$	112.50	5
Triethylmethylsilane	$C_7H_{18}Si$	$(C_2H_5)_3-Si-CH_3$	109.30	2
Trimethylbutylsilane	$C_7H_{18}Si$	$(CH_3)_3-Si-(CH_2)_3-CH_3$	109.00	2
Trimethylbutoxysilane	$C_7H_{18}SiO$	$(CH_3)_3-Si-O-(CH_2)_3-CH_3$	114.2	5
Trimethylsilyl-*n*-propylamidoxime	$C_7H_{18}SiON_2$	$NH_2-C-(CH_2-CH_2-CH_3)=N-O-Si-(CH_3)_3$	123.8	15
Methyltriethoxysilane	$C_7H_{18}SiO_3$	$CH_3-Si-(OC_2H_5)_3$	120.3	17
Trimethylthio-*n*-butylsilane	$C_7H_{18}SiS$	$(CH_3)_3-Si-S-(CH_2)_3CH_3$	125.73	12
Trimethylthio-*tert*-butylsilane	$C_7H_{18}SiS$	$(CH_3)_3-Si-S-C-(CH_3)_3$	126.55	12
Trimethyl*N*-dimethylsilane	$C_7H_{19}SiN$	$(CH_3)_3-Si-N-(CH_3)_2$	91.3	1
2,2-Dimethyl-2-sila-2,3-dihydrobenzothiazole	$C_8H_{11}SiN$	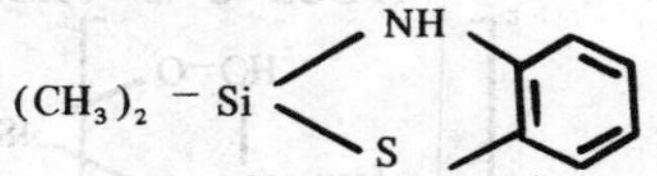	144.44	12
Tetraacetoxysilane	$C_8H_{12}SiO$	$Si-(OCOCH_3)_4$	129.26	5
Dimethyldiproprionoxysilane	$C_8H_{16}SiO$	$(CH_3)_2-Si-(OCOC_2H_5)_2$	122.0, 123.97	16, 5
1,4,6,9-Tetraoxo-2,3,7,8-tetramethyl-5-silaspiro[4,4]nonane	$C_8H_{16}SiO_4$	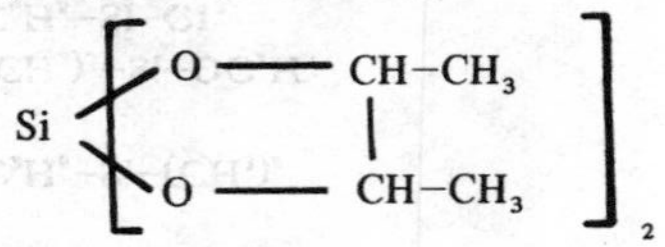	127.12	5
1,5,7,11-Tetraoxa-2,8-dimethyl-6-silaspiro[5,5]undecane	$C_8H_{16}SiO_4$	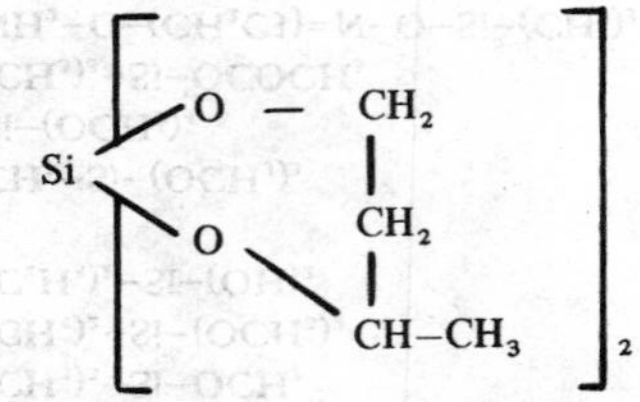	126.34	5
Dimethyldi-*n*-propylsilane	C_8H_2OSi	$(CH_3)_2-Si-[(CH_2)_2-CH_3]_2$	121.70	1
Tetraethylsilane	C_8H_20Si	$Si-(C_2H_5)_4$	120.30, 120.40, 117.00	1, 4, 5
Dimethyl-di-*n*-propoxysilane	$C_8H_2OSiO_2$	$(CH_3)_2-Si-[O(CH_2)_2-CH_3]_2$	125.70, 126.90	1, 5
Tetraethoxysilane	$C_8H_2OSiO_4$	$Si-(OC_2H_5)_4$	83.68, 134.50	5, 17

DIAMAGNETIC SUSCEPTIBILITY DATA OF ORGANO-SILICON COMPOUNDS (Continued)

Compound	Molecular formula	Structural formula	$-\chi_M \times 10^{-6}$	Ref.
Bis-(trimethylsilyl)-methylamidoxime	$C_8H_{22}SiO_2N_2$	$(CH_3)_3-Si-NH-C(CH_3)=N-O-Si-(CH_3)_3$	156.60	15
Octamethyltrisiloxane	$C_8H_{24}Si_3O_2$	$(CH_3)_3-Si-O-Si-O-Si(CH_3)_3(CH_3)_2$	165.00	16
Octamethyltetracyclosiloxane	$C_8H_{24}Si_4O_4$	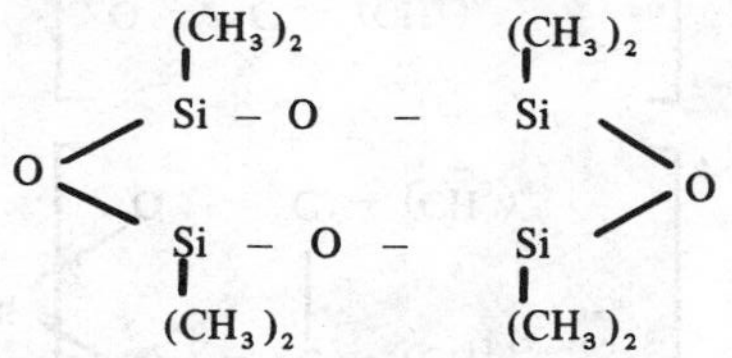	188.00	5
Tri-*n*-propylsilane	$C_9H_{22}Si$	$CH_3-(CH_2)_{23}-Si-H$	130.00	3
Trimethyl *N*-(*n*-propyl)silane	$C_9H_{23}SiN$	$(CH_3)_3-Si-N-[(CH_2)_2-CH_3]_2$	113.4	1
Bis(trimethyl)-ethylamidoxime	$C_9H_{24}Si_2ON_2$	$(CH_3)_3-Si-NH-C(C_2H_5)=N-O-Si(CH_3)_3$	168.00	15
Bis(trimethyl)-phenylamidoxime	$C_{10}H_{16}SiON_2$	$NH_2-C-(C_6H_5)=N-O-Si(CH_3)_3$	141.7	15
2,2-Diacetoxy-4,4,6,6,tetramethyl-1,3-dioxa-2-sila-cyclohexane	$C_{10}H_{20}SiO_4$	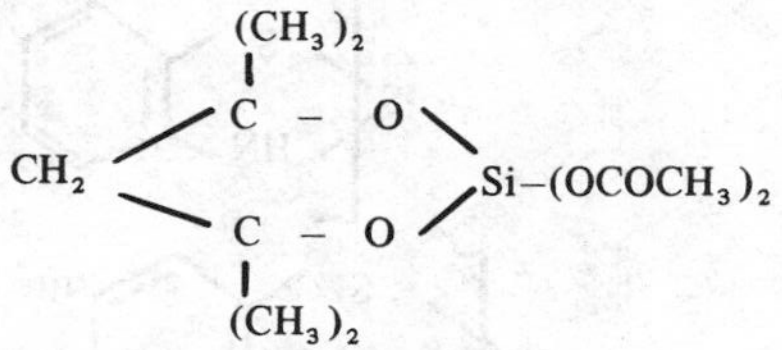	150.65	5
Dimethyl di-*n*-butylsilane	$C_{10}H_{24}Si$	$(CH_3)_2-Si-[(CH_2)_3-CH_3]_2$	145.80	1
Methyl tri-*n*-butylsilane	$C_{10}H_{24}Si$	$CH_3-Si-[(CH_2)_2-CH_3]_3$	143.90	1
Dimethyl di-*n*-butoxysilane	$C_{10}H_{24}SiO_2$	$(CH_3)_2-Si-[O(CH_2)_3CH_3]_2$	147.40, 149.40	1, 5
Dimethyl di-iso-butoxysilane	$C_{10}H_{24}SiO_2$	$(CH_3)_2-Si-[O-CH_2-CH-(CH_3)_2]_2$	150.73	5
Dimethyl di-(1-methylpropoxy)silane	$C_{10}H_{24}SiO_2$	$(CH_3)_2-Si-[O-CH(CH_3)-C_2H_5]_2$	150.83	5
Di-*n*-propyldiethoxysilane	$C_{10}H_{24}SiO_2$	$[CH_3-(CH_2)_2]_2-Si-(OC_2H_5)_2$	150.80	17, 18
Bis(trimethylsilyl) *n*-propyl amidoxime	$C_{10}H_{26}SiON_2$	$(CH_3)_3-Si-NH-C(CH_2)_2-CH_3=N-O-Si-(CH_3)_3$	179.00	15
Trimethyl-*n*-octylsilane	$C_{11}H_{26}Si$	$(CH_3)_3-Si-(CH_2)_7-CH_3$	156.00	2
Trimethyl-*n*-octoxysilane	$C_{11}H_{26}SiO$	$(CH_3)_3-Si-O(CH_2)_7-CH_3$	149.00	1
Trimethyl *N*-(di-*N*-butyl)silane	$C_{11}H_{27}SiN$	$(CH_3)_3-Si-N-[(CH_2)_3-CH_3]_2$	158.5	1

DIAMAGNETIC SUSCEPTIBILITY DATA OF ORGANO-SILICON COMPOUNDS (Continued)

Compound	Molecular formula	Structural formula	$\chi_M \times 10^{-6}$	Ref.
2,2′Spirobis(6-chloro-2,3-dihydrobenzothiazole)silane	$C_{12}H_8SiCl_2S_2N_2$	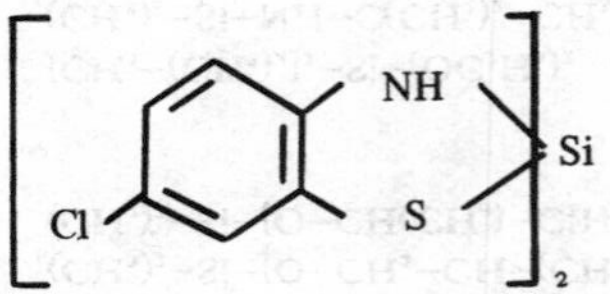	183.29	12
2,2′Spirobis(6-bromo-2,3-dihydrobenzothiazole)silane	$C_{12}H_8SiBr_2S_2N_2$	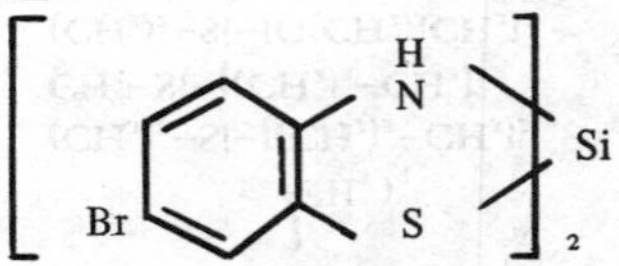	199.37	12
2,2′Spirobis(2,3-dihydrobenzothiazole)silane	$C_{12}H_{10}SiS_2N_2$	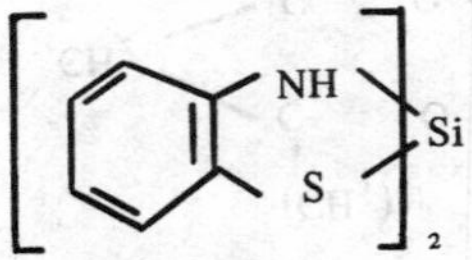	153.71	12
Diphenylsilanediol	$C_{12}H_{12}SiO_2$	$(C_6H_5)_2-Si-(OH)_2$	131.60	8
Tetrapropionoxysilane	$C_{12}H_{20}SiO_8$	$Si-(OCOC_2H_5)_4$	129.26	5
1,4,6,9-Tetraoxo-2,3,7,8-octamethyl-1-5-silaspiro[4,4]nonane	$C_{12}H_{24}SiO_4$	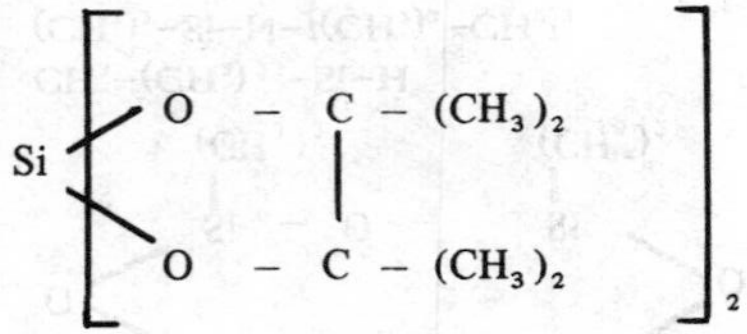	127.45	5
1,5,7,11-Tetraoxa-2,4,8,10-hexamethyl-1-6-silaspiro[5,5]undecane	$C_{12}H_{24}SiO_4$	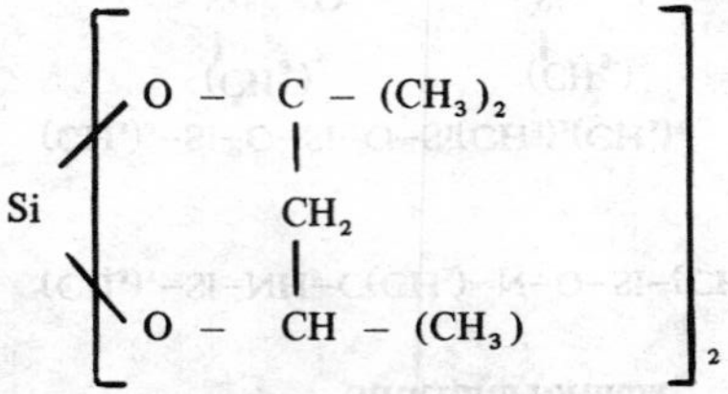	172.48	5
Dodeconyltrichlorosilane	$C_{12}H_{25}SiCl_3$	$CH_3-(CH_2)_{11}-Si-Cl_3$	210.40, 209.20	2, 8
Dimethyl di-*n*-pentylsilane	$C_{12}H_{28}Si$	$(CH_3)_2-Si-[(CH_2)_4CH_3]_2$	167.60	1
Tetra-*n*-propylsilane	$C_{12}H_{28}Si$	$Si-[(CH_2)_2-CH_3]_4$	165.80	1
Tri-*n*-butylsilane	$C_{12}H_{28}Si$	$H-Si-[(CH_2)_3-CH_3]_3$	172.00	3

DIAMAGNETIC SUSCEPTIBILITY DATA OF ORGANO-SILCON COMPOUNDS (Continued)

Compound	Molecular formula	Structural formula	$^{x}Mx-10^{-6}$	Ref.
Dimethyl di-*n*-pentoxysilane	$C_{12}H_{28}SiO_2$	$(CH_3)_2-Si[O-(CH_2)_4-CH_3]_2$	172.37	5
Hexaethoxydisiloxane	$C_{12}H_{30}Si_2O_7$	$(H_5C_20)_3-Si-O-Si-(OC_2H_5)_3$	217.00	13
Bis-(trimethylsilyl)-phenylamidoxime	$C_{13}H_{24}Si_2ON_2$	$(CH_3)_3-Si-NH-C(C_6H_5)=N-O-Si-(CH_3)_3$	199.50	15
Methyltri-*n*-butylsilane	$C_{13}H_{30}Si$	$CH_3-Si-[(CH_2)_3-CH_3]_3$	178.50	1
2,2′-Spirobis(6-methyl-2,3-dihydrobenzothiazole)silane	$C_{14}H_{14}S_2N_2$	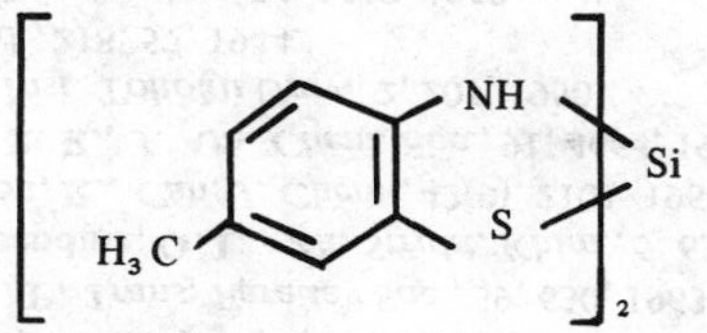	170.06	12
Diphenyldimethoxysilane	$C_{14}H_{16}SiO_2$	$(C_6H_5)_2-Si-(OCH_3)_2$	156.80	19
Trimethyldodeconylsilane	$C_{15}H_{34}Si$	$(CH_3)_3-Si-(CH_2)_{11}-CH_3$	192.00	2
Trimethyl-*n*-dodeconoxysilane	$C_{15}H_{34}SiO$	$(CH_3)_3-Si-O-(CH_2)_{11}-CH_3$	205.6	1
Diphenyl diethoxysilane	$C_{16}H_{20}SiO_2$	$(C_6H_5)_2-Si-(OC_2H_5)_2$	179.25	19
Hexadodeconyltrichlorosilane	$C_{16}H_{33}SiCl_3$	$C_{16}H_{33}-Si-Cl_3$	259.00, 252.40	2, 8
Tetra-*n*-butylsilane	$C_{16}H_{36}Si$	$Si-[(CH_2)_3-CH_3]_4$	202.50	3
Hexa-dodecamethylheptasiloxane	$C_{16}H_{48}O_6$	$(CH_3)_3-Si-(O-Si)_5-O-Si-(CH_3)_3$, with $(CH_3)_2$ on the $(O-Si)_5$ Si	351.3	8
Triphenylhydroxysilane	$C_{18}H_{16}SiO$	$(C_6H_5)_3-Si-OH$	176.0	8
Dephenyl di-*n*-propoxysilane	$C_{18}H_{24}SiO_2$	$(C_6H_5)_2-Si-[O(CH_2)_2-CH_3]_2$	201.58	19
Diphenyl di-isopropoxysilane	$C_{18}H_{24}SiO_2$	$(C_6H_5)_2-Si-[OCH-(CH_3)]_2$	202.46	19
n-Octadeconyltrichlorosilane	$C_{18}H_7SiCl_3$	$CH_3-(CH_2)_{17}-Si-Cl_3$	278.00, 273.7	2, 8
Diphenyl di-*n*-butoxysilane	$C_{20}H_{28}SiO_2$	$(C_6H_5)_2-Si-[O-(CH_2)_3-CH_3]_2$	223.62	19
Diphenyl di-iso-butoxy silane	$C_{20}H_{28}SiO_2$	$(C_6H_5)_2-Si-[O-CH_2-CH(CH_3)]_2$	224.46	19
Trimethyl*n*-octadeconyl silane	$C_{21}H_{46}Si$	$CH_3-(CH_2)_{17}-Si-(CH_3)_3$	278.00	2
Tetraphenylsilane	$C_{24}H_{20}Si$	$Si-(C_6H_5)_4$	212.20	7
Tetrabenzylsilane	$C_{28}H_{28}Si$	$Si-(CH_2-C_6H_5)_4$	266.20	6
Tetra-*p*-tolylsilane	$C_{28}H_{28}Si$	$Si-(C_6H_4-CH_3)_4$	276.40	6

DIAMAGNETIC SUSCEPTIBILITY DATA OF ORGANO-SILICON COMPOUNDS (Continued)

Compound	Molecular formula	Structural formula	$^{x}Mx-10^{-6}$	Ref.
Other silicon compounds				
Carborundum (silicon carbide)	CSi	C–Si	12.80	20
Silane	SiH_4	$Si-H_4$	20.4	13
Silicon dioxide	SiO_2	O=Si=O	29.6	11
Tetrachlorosilane	$SiCl_4$	$Si-Cl_4$	87.40, 84.60, 88.30	4, 11, 10
Tetrabromosilane	$SiBr_4$	$Si-Br_4$	123.30	4
Tetraiodosilane	SiI_4	$Si-I_4$	186.20	4
Trichlorosilane	$HSiCl_3$	$H-Si-Cl_3$	70.50, 71.30, 70.60	2, 7, 4
Silicic acid	H_2SiO_3	$O=S-(OH)_2$	33.30	11
Ortho silicic acid	H_4SiO_4	$Si-(OH)_4$	42.60	21
Disilane	Si_2H_6	$H_3-Si-Si-H_3$	37.3	4
Hexachlorodisilane	Si_2Cl_6	$Cl_3-Si-Si-Cl_3$	138.00	7

REFERENCES

1. **Abel, E. W., Bush, R. P., Jenkins, C. R., and Zobel, T.** *Trans. Faraday Soc.*, 60, 1214, 1964.
2. **Abel,E. W. and Bush, R. P.** *Trans. Faraday Soc.*, 59, 630, 1963.
3. **Dorfman, Y. G. and Lependina, O. L.**, *Zh. Strukt. Khim.*, 5, 632, 1964.
4. **Lister, M. W. and Marson, R.**, *Can. J. Chem.*, 42(9), 2101, 1964.
5. **Mital, R. L. and Gupta, R. R.**, *J. Am. Chem. Soc.*, 91, 4664, 1969.
6. **Asai, K.**, *Sci. Rep. Res. Inst. Tohoku Univ.*, 2, 205, 1950.
7. **Pascal, P.**, *Compt. Rend.*, 218, 57, 1944.
8. **Mathur, R. M.**, *Trans. Faraday Soc.*, 54, 1577, 1958.
9. **Mathur, R. M.**, *J. Vikram Univ.*, 2, 55, 1958.
10. **Nevgi, M. B.**, *J. Univ. Bombay*, 7(3), 82, 1958.
11. **Fox, G., Gorter, C. J., and Smits, L. J.**, *Tables Constantes Selectioness Diamagnetisme et Paramagnetisme*, Masson et cie, Paris, 1957, 122.
12. **Mital, R. L., Goyal, R. D., and Gupta, R. R.**, *Inorg. Chem.*, 11, 1924, 1972.
13. **Barter, C., Meisenheimer, R. G., and Stevenson, D. P.**, *J. Phys. Chem.*, 64, 1312, 1960.
14. **Foex, G., Gorter, C. J., and Smits, L. J.**, *Compt. Rend.*, 175, 125, 1922.
15. **Goel, A. B., and Gupta, V. D.**, *Indian J. Chem.*, 13(2), 181, 1975.
16. **Pacault, A.**, *Compt. Rend.*, 232, 1352, 1951.
17. **Mital, R. L.**, *Bull. Chem. Soc. Jpn.*, 37(10), 1440, 1964.
18. **Mital, R. L.**, *J. Phy. Chem.*, 68, 1613, 1964.
19. **Goyal, R. D., Gupta, R. R., and Mital, R. L.**, *J. Phys. Chem.*, 76, 1579, 1972.
20. **Sigamony, A.**, *Proc. Indian Acad. Sci.*, 19A, 377, 1944.
21. **Pascal, P.**, *Compt. Rend.*, 175, 814, 1922.

MASS ATTENUATION COEFFICIENTS

Wavelength	Formvar $(C_5H_7O_2)_x$	Collodion $(C_{12}H_{11}O_{22}N_6)_x$	Polypropylene $(CH_2)_x$	Cellulose acetate $(C_{10}H_{21}O_{15})_x$	Mylar $(C_{10}H_8O_4)_x$	Teflon $(CF_2)_x$	Energy (eV)
2.0	14	20	8	19	14	28	6199.0
4.0	113	156	69	150	116	220	3099.5
6.0	372	510	234	489	384	700	2066.3
8.0	850	1140	550	1100	870	1540	1549.8
10.0	1580	2110	1040	2020	1630	2800	1239.8
12.0	2600	3450	1740	3310	2680	4520	1033.2
14.0	3920	5200	2660	4950	4040	6700	885.6
16.0	5600	7300	3830	7000	5800	9400	774.9
18.0	7500	9800	5200	9400	7800	12600	688.8
20.0	9900	12800	6900	12300	10200	2780	619.9
22.0	12500	16200	8800	15500	12900	3540	563.5
24.0	8200	6400	11000	4850	8500	4430	516.6
26.0	10100	7900	13500	5900	10400	5400	476.8
28.0	12000	9400	16200	7100	12400	6500	442.8
30.0	14300	11100	19200	8400	14700	7800	413.3
32.0	16700	8200	22400	9900	17200	9100	387.4
34.0	19300	9500	25900	11400	19900	10600	364.6
36.0	22100	10800	29600	13100	22800	12100	344.4
38.0	25000	12300	33600	14900	25800	13800	326.3
40.0	28200	13900	37800	16800	29100	15600	309.9
42.0	31500	15500	42200	18700	32500	17500	295.2
44.0	3250	4540	1940	4370	3350	6900	281.8
46.0	3640	5100	2180	4900	3760	7800	269.5
48.0	4050	5600	2430	5400	4170	8600	258.3
50.0	4450	6200	2690	6000	4590	9500	248.0
52.0	4910	6800	2960	6600	5100	10500	238.4
54.0	5400	7500	3240	7200	5600	11500	229.6
56.0	5900	8200	3540	7900	6100	12500	221.4
58.0	6400	8900	3860	8600	6600	13600	213.8
60.0	7000	9700	4190	9300	7200	14800	206.6
62.0	7500	10500	4540	10100	7800	16000	200.0
64.0	8100	11300	4880	10900	8400	17300	193.7
66.0	8700	12100	5200	11700	9000	18600	187.8
68.0	9400	13100	5700	12600	9700	19900	182.3
70.0	10000	14000	6000	13500	10300	21300	177.1
72.0	10700	14900	6400	14400	11100	22700	172.2
74.0	11400	15900	6800	15400	11800	24200	167.5
76.0	12200	17000	7300	16300	12500	25700	163.1
78.0	12900	18000	7700	17300	13300	27200	158.9
80.0	13600	19100	8100	18400	14100	28800	155.0
82.0	14500	20200	8600	19500	14900	30500	151.2
84.0	15300	21400	9100	20600	15800	32100	147.6
86.0	16100	22600	9600	21700	16600	33800	144.2
88.0	17000	23700	10100	22900	17500	35500	140.9
90.0	17800	25000	10600	24100	18400	37300	137.8
92.0	18800	26300	11100	25300	19400	39100	134.8
94.0	19700	27600	11700	26500	20300	40900	131.9
96.0	20600	28900	12200	27800	21300	43000	129.1
98.0	21600	30300	12800	29200	22300	44600	126.5
100.0	22600	31700	13400	30500	23300	46300	124.0
105.0	25200	35300	15000	34000	26000	51000	118.1
110.0	27900	39100	16600	37600	28800	56000	112.7
115.0	30700	43000	18200	41300	31600	61000	107.8
120.0	33600	47000	20000	45100	34600	67000	103.3
125.0	36800	52000	21900	49600	38000	72000	99.2
130.0	39800	56000	23900	53000	41100	78000	95.4
135.0	43200	60000	25900	58000	44600	83000	91.8
140.0	46700	65000	28100	63000	48200	89000	88.6
145.0	50000	70000	30300	67000	52000	95000	85.5
150.0	54000	75000	32600	72000	55000	101000	82.7
155.0	57000	80000	35000	76000	59000	106000	80.0
160.0	61000	85000	37600	82000	63000	112000	77.5
165.0	65000	90000	40200	87000	67000	118000	75.1
170.0	69000	96000	42800	92000	72000	124000	72.9
175.0	73000	101000	45400	97000	75000	130000	70.8
180.0	77000	107000	47900	102000	80000	136000	68.9
185.0	82000	113000	51000	108000	84000	141000	67.0
190.0	86000	118000	54000	112000	88000	147000	65.3
195.0	90000	124000	57000	118000	93000	153000	63.6
200.0	95000	130000	61000	124000	98000	159000	62.0

MASS ATTENUATION COEFFICIENTS (Continued)

Wavelength	Polystyrene $(CH)_x$	Nylon $(C_{12}H_{22}O_3N_2)_x$	Vyns $(C_{22}H_{33}O_2Cl_9)_x$	Saran $(C_2H_2Cl_2)_x$	Aluminum oxide Al_2O_3	Quartz $(SiO_2)_x$	Energy (eV)
2.0	9	12	114	164	68	77	6199.0
4.0	74	96	750	1070	480	530	3099.5
6.0	252	318	350	375	1440	1600	2066.3
8.0	590	730	780	820	860	980	1549.8
10.0	1120	1370	1440	1520	1580	1810	1239.8
12.0	1870	2270	2370	2490	2570	2950	1033.2
14.0	2870	3440	3590	3760	3840	4400	885.6
16.0	4120	4910	5100	5400	5500	6200	774.9
18.0	5600	6700	6900	7200	7300	8400	688.8
20.0	7500	8800	9000	9300	9600	11000	619.9
22.0	9500	11100	11400	11800	12100	13800	563.5
24.0	11900	10600	12800	14400	4290	4920	516.6
26.0	14600	13000	15500	17300	5200	6000	476.8
28.0	17400	15500	18400	20400	6300	7200	442.8
30.0	20700	18400	21400	23700	7500	8500	413.3
32.0	24200	17300	24700	27100	8700	9900	387.4
34.0	28000	20000	28100	30800	10100	11300	364.6
36.0	31900	22800	31600	34400	11500	12900	344.4
38.0	36200	25800	35400	38200	13100	14500	326.3
40.0	40700	29100	39100	42100	14700	16200	309.9
42.0	45500	32500	43100	46000	16300	18000	295.2
44.0	2090	2730	25700	37000	18100	19800	281.8
46.0	2350	3060	23600	33600	19900	21700	269.5
48.0	2620	3400	25300	36100	21800	23600	258.3
50.0	2900	3750	27300	38800	23700	25500	248.0
52.0	3190	4130	28900	41100	25700	27500	238.4
54.0	3500	4540	30600	43400	27700	29400	229.6
56.0	3820	4950	32200	45700	29800	31800	221.4
58.0	4160	5400	33900	47900	31800	33700	213.8
60.0	4510	5800	35100	49500	33900	35600	206.6
62.0	4890	6300			35900	37500	200.0
64.0	5300	6800			38000	39900	193.7
66.0	5600	7300			40600	41900	187.8
68.0	6100	7900			42800	44000	182.3
70.0	6500	8400			44900	46000	177.1
72.0	6900	9000			47100	48000	172.2
74.0	7400	9600			49300	50000	167.5
76.0	7800	10200			51000	52000	163.1
78.0	8300	10800			54000	54000	158.9
80.0	8800	11400			56000	56000	155.0
82.0	9300	12100			58000	52000	151.2
84.0	9800	12800			61000	53000	147.6
86.0	10300	13500			62000	56000	144.2
88.0	10900	14200			65000	57000	140.9
90.0	11400	15000			67000	59000	137.8
92.0	12000	15700			69000	61000	134.8
94.0	12600	16500			71000	63000	131.9
96.0	13200	17300			73000	65000	129.1
98.0	13800	18100			75000	67000	126.5
100.0	14500	19000			77000	69000	124.0
105.0	16100	21100			83000	73000	118.1
110.0	17900	23400			79000	78000	112.7
115.0	19600	25800			84000	82000	107.8
120.0	21600	28300			88000	86000	103.3
125.0	23600	31000			93000		99.2
130.0	25700	33600			97000		95.4
135.0	28000	36500			101000		91.8
140.0	30300	39400			106000		88.6
145.0	32700	42400			110000		85.5
150.0	35100	45600			114000		82.7
155.0	37700	48900			118000		80.0
160.0	40500	52000			122000		77.5
165.0	43300	56000			125000		75.1
170.0	46100	59000					72.9
175.0	48900	63000					70.8
180.0	52000	66000					68.9
185.0	55000	70000					67.0
190.0	58000	74000					65.3
195.0	62000	78000					63.6
200.0	65000	82000					62.0

MASS ATTENUATION COEFFICIENTS (Continued)

Wavelength	Stearate $CH_3(CH_2)_{16}COO^-$	Animal proteins C = 52.5 %, H = 7 %, S = 1.5 %, O = 22.5 %, N = 16.5 %	Air O = 21 %, N = 78 %, Ar = 1 %	P 10 CH_4 = 10 %, Ar = 90 %	Methane CH_4	Q Gas C_4H_{10} = 1.3 %, He = 98.7 %	Energy (eV)
2.0	10	16	21	230	7	1	6199.0
4.0	84	126	148	162	60	12	3099.5
6.0	281	361	481	467	205	41	2066.3
8.0	650	820	1090	1020	479	97	1549.8
10.0	1220	1530	2020	1850	910	185	1239.8
12.0	2030	2530	3310	3010	1520	313	1033.2
14.0	3080	3830	4980	4500	2330	484	885.6
16.0	4410	5500	7100	6400	3350	700	774.9
18.0	6000	7400	9500	8400	4570	970	688.8
20.0	7900	9700	12400	10900	6100	1300	619.9
22.0	10100	12300	15700	13500	7700	1670	563.5
24.0	10000	10300	14100	16400	9700	2110	516.6
26.0	12200	12600	17100	19600	11800	2620	476.8
28.0	14600	15100	20400	22800	14100	3160	442.8
30.0	17300	17800	24000	26300	16800	3780	413.3
32.0	20300	15500	2290	29700	19600	4460	387.4
34.0	23400	17900	2650	33300	22700	5200	364.6
36.0	26800	20400	3040	36900	25900	6000	344.4
38.0	30300	23100	3460	40500	29300	6900	326.3
40.0	34100	26000	3810	37600	33000	7800	309.9
42.0	38200	29100	4270	40900	36900	8800	295.2
44.0	2390	3750	4780	42600	1700	2960	281.8
46.0	2680	4180	5300	45600	1910	3380	269.5
48.0	2980	4610	5900	48900	2130	3840	258.3
50.0	3290	5000	6400	52000	2350	4340	248.0
52.0	3620	5500	6300		2590	4910	238.4
54.0	3970	5900	7000		2840	5500	229.6
56.0	4340	6400	7600		3100	6100	221.4
58.0	4730	6900	8200		3380	6700	213.8
60.0	5100	7500	8900		3660	7400	206.6
62.0	5600	8100	9700		3970	8200	200.0
64.0	6000	8700	10400		4270	8900	193.7
66.0	6400	9300	11200		4570	9700	187.8
68.0	6900	10000	12100		4940	10600	182.3
70.0	7400	10600	12900		5200	11500	177.1
72.0	7900	11300	13800		5600	12500	172.2
74.0	8400	12000	14700		6000	13500	167.5
76.0	8900	11600	15600		6400	14600	163.1
78,0	9500	12300	16600		6700	15600	158.9
80.0	10000	13000	17600		7100	16800	155.0
82.0	10600	13800	18600		7600	18000	151.2
84.0	11200	14600	19700		7900	19300	147.6
86.0	11800	15400	20800		8400	20500	144.2
88.0	12400	16200	21900		8800	21900	140.9
90.0	13100	17000	23100		9300	23300	137.8
92.0	13700	17900	24200		9700	24700	134.8
94.0	14400	18800	25400		10300	26200	131.9
96.0	15100	19700	26700		10700	27800	129.1
98.0	15800	20700	28000		11200	29400	126.5
100.0	16600	21600	29200		11800	31000	124.0
105.0	18400	24100	32700		13100	35400	118.1
110.0	20400	26700	36200		14500	40100	112.7
115.0	22500	29300	39900		15900	44800	107.8
120.0	24600	32200	43800		17500	50000	103.3
125.0	27000	35300	48000		19200	56000	99.2
130.0	29300	38200	52000		20900	62000	95.4
135.0	31800	41500	57000		22700	68000	91.8
140.0	34400	44800	61000		24600	75000	88.6
145.0	37000	48200	65000		26500	82000	85.5
150.0	39800	52000	71000		28500	89000	82.7
155.0	42600	55000	76000		30600	96000	80.0
160.0	45700	59000	80000		32900	104000	77.5
165.0	48600	63000	86000		35100	112000	75.1
170.0	52000	67000	91000		37400	121000	72.9
175.0	55000	71000	96000		39700	130000	70.8
180.0	58000	75000	102000		41900	138000	68.9
185.0	62000	79000	108000		44900	147000	67.0
190.0	65000	83000	114000		47200	157000	65.3
195.0	68000	88000	119000		50000	167000	63.6
200.0	72000	92000	125000		53000	177000	62.0

INTRODUCTION TO X-RAY CROSS SECTIONS

Alex F. Burr

These tables are part of an extensive report published by W. H. McMaster, et al. as UCRL 50174 and available from the National Technical Information Service, Springfield, Va. 22151. Section I of UCRL 50174 describes the data base and the treatment given it, Section II contains the total cross sections between 1 and 1000 keV for all the elements, Section III contains results used in producing Section II, and Section IV contains total cross sections for selected energies and is reproduced in part here. To obtain these values existing experimental x-ray total cross section data and theoretical cross section calculations were surveyed. The coherent (Rayleigh) scattering cross sections and the incoherent (Compton) scattering cross sections were computed. The photo-electric cross sections were obtained by least squares fitting of experimental data, theory, and interpolation of experiment and theory. The following table contains cross sections interpolated from the basic compilation at those wavelengths of most use to x-ray crystallographers. The wavelengths chosen were selected to correspond to those given in the International Tables for X-Ray Crystallography. The energy-to-wavelength conversion is given below.

Table I. Energy-to-wavelength conversion

Target radiation	Å	keV	Target radiation	Å	keV
Ag $K\bar{\alpha}$	0.5608	22.105	Ni $K\bar{\alpha}$	1.6591	7.472
$K\beta_1$	0.4970	24.942	$K\beta_1$	1.5001	8.265
Pd $K\bar{\alpha}$	0.5869	21.125	Co $K\bar{\alpha}$	1.7902	6.925
$K\beta_1$	0.5205	23.819	$K\beta_1$	1.6208	7.649
Rh $K\bar{\alpha}$	0.6147	20.169	Fe $K\bar{\alpha}$	1.9373	6.400
$K\beta_1$	0.5456	22.724	$K\beta_1$	1.7565	7.058
Mo $K\bar{\alpha}$	0.7107	17.444	Mn $K\bar{\alpha}$	2.1031	5.895
$K\beta_1$	0.6323	19.608	$K\beta_1$	1.9102	6.490
Zn $K\bar{\alpha}$	1.4364	8.631	Cr $K\bar{\alpha}$	2.2909	5.412
$K\beta_1$	1.2952	9.572	$K\beta_1$	2.0848	5.947
Cu $K\bar{\alpha}$	1.5418	8.041	Ti $K\bar{\alpha}$	2.7496	4.509
$K\beta_1$	1.3922	8.905	$K\beta_1$	2.5138	4.932

Table II. Total Cross Section in cm^2/g

Z KEV	1 H	2 He	3 Li	4 Be	5 B	6 C	7 N	8 O	9 F	10 Ne
4.51	.432	.661	2.10	5.63	12.9	25.6	43.0	64.8	91.8	125
4.93	.421	.550	1.62	4.25	9.69	19.4	32.6	49.4	70.3	95.9
5.41	.412	.465	1.24	3.18	7.23	14.5	24.4	37.2	53.1	72.7
5.90	.405	.405	.986	2.45	5.53	11.1	18.7	28.6	41.1	56.4
5.95	.405	.400	.964	2.39	5.39	10.8	18.2	27.9	40.0	54.9
6.40	.400	.362	.798	1.92	4.28	8.55	14.5	22.2	32.0	44.0
6.49	.400	.355	.770	1.84	4.10	8.18	13.9	21.3	30.6	42.2
6.93	.397	.329	.659	1.52	3.36	6.68	11.3	17.4	25.1	34.7
7.06	.396	.322	.631	1.44	3.17	6.30	10.7	16.5	23.7	32.8
7.47	.394	.303	.555	1.23	2.67	5.28	8.96	13.8	19.9	27.6
7.65	.393	.297	.528	1.15	2.49	4.92	8.33	12.9	18.6	25.7
8.04	.391	.284	.477	1.01	2.14	4.22	7.14	11.0	16.0	22.1
8.27	.390	.277	.452	.936	1.98	3.88	6.56	10.1	14.7	20.4
8.63	.389	.268	.417	.837	1.74	3.40	5.74	8.87	12.8	17.9
8.91	.388	.262	.394	.774	1.59	3.09	5.22	8.06	11.7	16.2
9.57	.386	.250	.349	.651	1.30	2.49	4.19	6.47	9.36	13.1
17.44	.373	.202	.197	.245	.345	.535	.790	1.15	1.58	2.21
19.61	.370	.197	.187	.222	.293	.429	.605	.855	1.15	1.60
20.17	.369	.196	.185	.217	.283	.408	.570	.799	1.07	1.48
21.13	.368	.195	.182	.210	.268	.379	.519	.717	.952	1.31
22.11	.366	.193	.179	.205	.256	.354	.476	.648	.851	1.16
22.72	.366	.192	.177	.201	.249	.340	.452	.610	.795	1.08
23.82	.364	.191	.175	.196	.239	.319	.416	.553	.711	.963
24.94	.363	.189	.173	.192	.229	.301	.385	.504	.640	.861

Table II. Total Cross Section in cm²/g (Continued)

Z KEV	11 Na	12 Mg	13 Al	14 Si	15 P	16 S	17 Cl	18 Ar	19 K	20 Ca
4.51	168	220	264	336	386	464	518	577	679	805
4.93	129	171	206	263	304	364	411	456	542	638
5.41	98.5	131	158	203	236	282	322	356	427	500
5.90	76.6	102	124	160	186	223	256	282	342	398
5.95	74.7	99.6	121	156	182	217	250	276	334	389
6.40	59.9	80.2	97.5	126	148	177	205	225	275	319
6.49	57.5	77	93.7	121	142	170	197	217	264	307
6.93	47.3	63.5	77.5	100	118	141	165	181	222	257
7.06	44.7	60.1	73.4	95.1	112	134	156	172	211	244
7.47	37.7	50.8	62.2	80.7	95.2	114	134	147	181	209
7.65	35.2	47.4	58.1	75.4	89.1	107	125	137	170	196
8.04	30.3	40.9	50.2	65.3	77.3	92.5	109	120	148	171
8.27	27.9	37.6	46.3	60.2	71.3	85.5	101	111	138	159
8.63	24.4	33.0	40.7	53.0	62.9	75.4	89.4	97.7	122	141
8.91	22.2	30.1	37.1	48.4	57.4	68.9	81.8	89.3	112	129
9.57	17.9	24.2	30.0	39.1	46.6	55.9	66.7	72.7	91.4	106
17.44	2.94	3.98	5.04	6.53	7.87	9.63	11.6	12.6	16.2	19
19.61	2.10	2.83	3.59	4.62	5.57	6.84	8.26	8.95	11.5	13.6
20.17	1.94	2.60	3.30	4.26	5.13	6.30	7.61	8.24	10.6	12.5
21.13	1.70	2.28	2.89	3.71	4.47	5.50	6.63	7.18	9.24	10.9
22.11	1.50	2.00	2.54	3.25	3.91	4.82	5.80	6.28	8.08	9.57
22.72	1.39	1.86	2.35	3.00	3.61	4.45	5.35	5.79	7.45	8.84
23.82	1.23	1.63	2.06	2.62	3.15	3.88	4.66	5.04	6.49	7.71
24.94	1.09	1.44	1.81	2.30	2.76	3.40	4.08	4.41	5.67	6.75

Z KEV	21 Sc	22 Ti	23 V	24 Cr	25 Mn	26 Fe	27 Co	28 Ni	29 Cu	30 Zn
4.51	819	111	125	143	160	188	206	240	257	280
4.93	658	86.8	97.3	111	125	147	161	188	201	220
5.41	521	571	75.1	85.7	96.1	113	125	146	155	172
5.90	420	459	513	67.4	75.6	88.9	98.4	115	123	137
5.95	411	449	501	65.8	73.8	86.8	96.1	113	120	134
6.40	339	370	411	462	59.9	70.4	78.3	91.8	97.4	110
6.49	327	357	396	445	57.6	67.7	75.3	88.3	93.6	106
6.93	276	301	333	375	405	56.3	62.9	73.8	78.1	88.7
7.06	262	286	316	357	385	53.3	59.6	70.0	74.1	84.3
7.47	226	246	271	307	331	367	50.9	59.8	63.2	72.4
7.65	212	231	255	288	311	346	47.7	56.1	59.2	68.0
8.04	186	202	223	252	273	304	339	48.8	51.5	59.5
8.27	173	188	206	234	253	283	315	45.2	47.7	55.3
8.63	153	167	183	208	225	253	281	306	42.3	49.2
8.91	141	153	168	191	207	234	259	283	38.7	45.3
9.57	116	126	138	157	170	194	214	236	245	37.4
17.44	21.0	23.3	25.2	29.3	31.9	37.7	41.0	47.2	49.3	55.5
19.61	15.0	16.7	18.1	21.0	22.9	27.2	29.5	34.2	35.8	40.3
20.17	13.8	15.4	16.7	19.4	21.1	25.1	27.2	31.6	33.1	37.2
21.13	12.1	13.4	14.6	17.0	18.5	22.0	23.8	27.7	29.0	32.7
22.11	10.5	11.8	12.8	14.9	16.2	19.3	20.9	24.3	25.5	28.7
22.72	9.72	10.9	11.8	13.7	15.0	17.8	19.3	22.5	23.6	26.6
23.82	8.47	9.48	10.3	12.0	13.1	15.6	16.9	19.7	20.7	23.3
24.94	7.40	8.30	9.02	10.5	11.5	13.7	14.8	17.3	18.2	20.4

Table II. Total Cross Section in cm²/g (Continued)

Z KEV	31 Ga	32 Ge	33 As	34 Se	35 Br	36 Kr	37 Rb	38 Sr	39 Y	40 Zr
4.51	309	329	368	403	435	464	508	552	599	648
4.93	242	258	288	317	342	366	400	436	473	511
5.41	187	200	224	246	266	285	312	339	369	399
5.90	148	158	178	195	211	226	248	270	294	317
5.95	144	155	173	190	206	221	242	263	287	310
6.40	117	126	142	156	169	181	198	215	235	254
6.49	113	122	136	150	162	174	191	207	227	244
6.93	94.2	102	114	125	136	146	160	174	190	205
7.06	89.3	96.8	108	119	129	138	152	165	181	195
7.47	76.2	82.9	92.5	101	110	119	130	141	155	167
7.65	71.4	77.8	86.8	95.1	104	111	122	132	145	157
8.04	62.1	67.9	75.7	82.9	90.3	97	106	115	127	137
8.27	57.5	63.0	70.1	77.8	83.7	89.9	98.5	107	118	127
8.63	50.9	56.0	62.2	68.0	74.2	79.8	87.4	94.7	105	113
8.91	46.7	51.4	57.0	62.3	68.1	73.2	80.2	86.8	96.2	103
9.57	38.1	42.3	46.7	51.0	55.8	60.0	65.7	71.0	79.0	84.8
17.44	56.9	60.5	66.0	68.8	74.7	79.1	83.0	88.0	97.6	16.1
19.61	41.7	44.3	48.6	51.2	55.6	58.6	62.1	65.6	72.6	75.2
20.17	38.6	41.0	45.1	47.6	51.7	54.5	57.8	61.0	67.5	70.1
21.13	34.0	36.1	39.7	42.1	45.6	48.1	51.1	54.0	59.7	62.1
22.11	29.9	31.8	35.0	37.2	40.4	42.5	45.3	47.9	53.0	55.2
22.72	27.7	29.4	32.4	34.6	37.5	39.5	42.1	44.5	49.2	51.4
23.82	24.3	25.8	28.5	30.5	33.1	34.8	37.2	39.3	43.5	45.5
24.94	21.4	22.7	25.1	26.9	29.2	30.7	32.9	34.8	38.5	40.4

Z KEV	41 Nb	42 Mo	43 Te	44 Ru	45 Rh	46 Pd	47 Ag	48 Cd	49 In	50 Sn
4.51	697	738	786	832	892	928	987	1064	1151	1128
4.93	552	585	621	660	708	739	785	842	906	899
5.41	432	457	486	518	555	581	617	659	706	709
5.90	344	365	387	414	444	466	495	526	561	569
5.95	336	357	378	404	434	455	484	514	548	557
6.40	276	293	310	333	357	375	398	422	449	460
6.49	266	282	299	320	344	361	384	407	433	443
6.93	223	237	251	269	289	304	324	342	363	374
7.06	212	225	238	256	275	289	308	325	345	356
7.47	182	193	204	220	236	249	265	279	295	307
7.65	170	181	192	207	222	234	249	262	277	289
8.04	149	158	168	181	194	205	218	229	242	253
8.27	138	147	156	168	180	190	203	213	225	236
8.63	122	130	138	149	160	169	180	189	200	210
8.91	112	120	127	137	147	156	166	174	183	193
9.57	92.0	98.2	104	113	121	128	137	143	151	159
17.44	17.0	18.4	19.8	21.3	23.1	24.4	26.4	27.7	29.1	31.2
19.61	81.2	13.3	14.3	15.4	16.7	17.6	19.1	20.1	21.2	22.6
20.17	75.6	79.3	13.2	14.2	15.4	16.3	17.7	18.6	19.6	20.9
21.13	67.1	70.3	73.0	12.5	13.5	14.3	15.5	16.4	17.3	18.4
22.11	59.6	62.5	65.0	11.0	11.9	12.6	13.7	14.5	15.3	16.2
22.72	55.5	58.2	60.6	63.8	11.0	11.6	12.7	13.4	14.1	15.0
23.82	49.1	51.5	53.7	56.6	58.3	10.2	11.1	11.8	12.4	13.2
24.94	43.6	45.7	47.7	50.3	52.0	57.0	97.8	10.4	11.0	11.6

Table II. Total Cross Section in cm²/g (Continued)

Z KEV	51 Sb	52 Te	53 I	54 Xe	55 Cs	56 Ba	57 La	58 Ce	59 Pr	60 Nd
4.51	997	753	293	300	324	334	355	383	414	433
4.93	926	843	921	683	259	266	282	304	330	344
5.41	733	769	835	755	803	587	223	240	261	271
5.90	592	617	666	701	742	660	677	521	210	218
5.95	579	603	651	685	725	645	662	509	205	213
6.40	479	497	535	565	597	615	636	592	450	464
6.49	462	479	516	545	575	593	613	571	610	447
6.93	391	404	434	459	484	499	519	559	596	532
7.06	373	385	413	437	460	475	494	532	567	506
7.47	322	331	355	376	395	408	427	459	488	506
7.65	303	312	333	353	372	384	402	432	459	476
8.04	267	273	292	310	325	336	354	379	402	418
8.27	248	254	271	288	302	312	329	352	374	389
8.63	221	226	241	256	269	278	294	314	333	347
8.91	204	208	222	236	248	256	271	290	307	320
9.57	168	172	183	195	204	211	224	240	253	265
17.44	33.0	33.9	36.3	38.3	40.4	42.4	45.3	48.6	50.8	53.3
19.61	23.9	24.7	26.5	27.9	29.5	31.0	33.1	35.5	37.1	38.9
20.17	22.1	22.8	24.6	25.8	27.3	28.7	30.7	33.0	34.4	36.0
21.13	19.4	20.1	21.7	22.7	24.1	25.4	27.0	29.1	30.3	31.7
22.11	17.1	17.7	19.2	20.0	21.3	22.5	23.9	25.7	26.8	28.0
22.72	15.8	16.4	17.8	18.6	19.8	20.9	22.1	23.9	24.9	26.0
23.82	13.8	14.4	15.7	16.3	17.4	18.4	19.5	21.0	21.9	22.9
24.94	12.1	12.7	13.9	14.4	15.4	16.3	17.2	18.6	19.3	20.2

Z KEV	61 Pm	62 Sm	63 Eu	64 Gd	65 Tb	66 Dy	67 Ho	68 Er	69 Tm	70 Yb
4.51	455	473	503	510	546	568	589	615	644	664
4.93	361	375	398	405	432	449	465	485	508	524
5.41	285	295	313	319	339	352	363	380	397	410
5.90	229	237	251	256	271	281	290	303	317	327
5.95	224	232	245	250	265	275	283	296	310	319
6.40	186	192	203	207	219	227	234	244	255	263
6.49	476	185	195	200	211	219	225	235	246	254
6.93	401	412	165	170	179	185	190	198	207	214
7.06	536	392	420	161	170	176	181	189	197	204
7.47	535	475	361	369	147	152	156	163	170	175
7.65	503	446	477	347	367	143	146	153	160	165
8.04	441	454	418	427	322	337	128	134	140	145
8.27	410	422	450	397	420	313	333	125	131	135
8.63	366	376	401	410	375	392	296	318	117	120
8.91	336	347	369	377	400	360	272	292	289	111
9.57	278	287	305	312	330	344	364	336	239	232
17.44	55.5	58.0	61.2	62.8	66.8	68.9	72.1	75.6	79.0	80.2
19.61	40.5	42.4	44.7	46.0	48.9	50.4	52.8	55.1	57.9	59.2
20.17	37.6	39.3	41.5	42.6	45.3	46.7	48.9	51.0	53.8	55.0
21.13	33.1	34.7	36.6	37.6	40.0	41.2	43.2	45.0	47.5	48.7
22.11	29.3	30.7	32.4	33.3	35.4	36.5	38.3	39.8	42.1	43.2
22.72	27.1	28.5	30.1	30.9	32.9	33.9	35.6	37.0	39.1	40.2
23.82	23.9	25.1	26.5	27.3	29.0	29.9	31.4	32.6	34.5	35.5
24.94	21.1	22.2	23.4	24.1	25.6	26.4	27.8	28.8	30.6	31.5

Table II. Total Cross Section in cm^2/g (Continued)

Z KEV	71 Lu	72 Hf	73 Ta	74 W	75 Re	76 Os	77 Ir	78 Pt	79 Au	80 Hg
4.51	696	720	736	753	796	824	871	934	906	958
4.93	549	568	581	598	631	653	688	734	720	760
5.41	430	445	455	470	496	512	540	572	568	598
5.90	343	355	363	378	397	410	431	455	457	480
5.95	335	347	355	369	388	401	422	444	447	469
6.40	276	286	293	306	321	332	348	365	371	388
6.49	266	276	282	295	310	320	336	351	358	375
6.93	225	233	238	250	262	270	283	295	303	317
7.06	214	222	227	238	249	257	270	281	289	302
7.47	184	191	195	206	215	222	233	241	250	261
7.65	173	180	184	194	203	209	219	226	236	246
8.04	152	158	162	171	178	184	192	198	208	216
8.27	142	147	150	159	166	171	179	184	194	202
8.63	126	131	134	142	149	153	160	164	174	180
8.91	117	121	124	132	137	141	148	151	161	167
9.57	247	236	103	110	114	118	123	125	134	139
17.44	84.2	86.3	89.5	95.8	98.7	100	103	109	111	115
19.61	62.0	64.2	66.1	70.6	72.5	74.1	77.2	80.2	82.3	85.3
20.17	57.6	59.7	61.4	65.5	67.3	68.9	71.9	74.6	76.5	79.4
21.13	51.0	52.9	54.4	58.0	59.5	61.1	63.8	66.0	67.8	70.4
22.11	45.3	46.9	48.3	51.4	52.8	54.2	56.8	58.6	60.2	62.6
22.72	42.1	43.7	44.9	47.8	49.1	50.5	52.9	54.5	56.0	58.3
23.82	37.2	38.6	39.7	42.2	43.3	44.6	46.8	48.2	49.5	51.7
24.94	33.0	34.2	35.2	37.4	38.4	39.6	41.6	42.7	43.9	45.9

Z KEV	81 Tl	82 Pb	83 Bi	86 Rn	90 Th	92 U	95 Pu
4.51	991	1035	1066	1174	1098	1084	960
4.93	785	820	847	930	993	862	1023
5.41	617	645	667	731	844	774	803
5.90	494	517	536	586	678	672	731
5.95	483	505	524	573	663	657	772
6.40	400	418	435	474	549	545	638
6.49	386	403	419	458	530	526	615
6.93	326	341	355	387	449	446	520
7.06	311	325	338	369	428	425	495
7.47	268	280	293	318	370	368	427
7.65	253	264	276	300	348	347	402
8.04	222	232	243	264	307	306	353
8.27	207	216	227	246	286	285	329
8.63	185	194	203	220	256	256	294
8.91	171	179	188	203	237	236	271
9.57	142	149	156	169	197	197	225
17.44	119	123	126	117	99.5	96.7	48.8
19.61	88.3	90.6	93.5	101	73.3	72.6	79.0
20.17	82.0	84.1	87.0	93.8	95.0	67.8	73.7
21.13	72.7	74.5	77.2	83.3	97.2	84.2	65.7
22.11	64.6	66.1	68.7	74.1	86.3	86.9	58.8
22.72	60.1	61.5	64.0	69.1	80.3	81.1	76.4
23.82	53.2	54.4	56.7	61.2	71.0	72.1	78.3
24.94	47.2	48.2	50.4	54.4	62.9	64.2	69.8

X-RAY WAVELENGTHS

J. A. Bearden

These tables were originally published as the final report to the U.S. Atomic Energy Commission as Report NYO-10586 in partial fulfillment of Contract AT(30-1)-2543. The tables were later reproduced in *Review of Modern Physics*. The data may also be obtained from the Superintendent of Documents, U.S. Government Printing Office, Washington, D. C. 20402 in the publication NSRDS-NBS 14. Persons seeking discussion of the experimental work, conventions, secondary standards, etc. will find these in *Review of Modern Physics*, Vol. 39, No. 1, 78-124, January 1967.

THE W $K\alpha_1$ WAVELENGTH STANDARD

A wavelength standard should possess characteristics which permit its ready redetermination in other laboratories by different techniques. Considering all of the factors involved in the selection of a wavelength standard, the W $K\alpha_1$ line is superior to any other x-ray or γ-ray wavelength. Its advantages as the x-ray wavelength standard are discussed in *Review of Modern Physics* Vol. 39, page 82 (1967).

$$\lambda \mathrm{W}\ K\alpha_1 = 0.2090100\ \text{Å} \pm 5\ \text{ppm}.$$

This numerical value of the wavelength of the W $K\alpha_1$ line is used to define the *x-ray wavelength standard* by the relation

$$\lambda(\mathrm{W}\ K\alpha_1) = 0.2090100\ \text{Å}^*.$$

This is a new unit of length which may differ from the angstrom by ±5 ppm (probable error), *but as a wavelength standard it has no error*. In order to clearly indicate that this unit is not exactly an angstrom, it has been designated Å*.

Wavelengths tabulated normally refer to the pure element in its solid form. However, there are many instances in which such data are not available. For example, rare gases are of necessity almost always used in the gaseous form, while the rare-earth elements were customarily used in the form of salts.

In high precision work there is some ambiguity as to exactly what feature of a line profile should be taken to be the "true wavelength." In double-crystal work the line peak is usually employed. In crystallography the centroid is widely used; in photographic work with visual observation of the plates, there is involved some subjective criterion of the observer which it is difficult to define precisely. In this survey the peak of the line profile has been adopted as the standard criterion.

In the study of the X-ray literature, the wavelengths of a number of lines were noted which appeared inconsistent with the remaining data. A Moseley-type diagram was constructed, and if the value was clearly outside estimated probable error, it was assumed that an experimental or typographical error had occurred, and the interpolated value was listed in the table. Such cases are marked with a dagger (†) as a superscript to the wavelength. For elements of atomic number 85 through 89 and 91, there are no measured lines of the *K* series and very few of other series except for 88 radium and 91 protactinium. Likewise there are very few measurements for 43 technetium and 54 xenon. In these cases, interpolated values are listed for the more prominent lines and marked with a dagger (†).

X-RAY WAVELENGTHS

X-ray wavelengths in Å* units and in keV. The probable error (p.e.) is the error in the last digit of wavelength. Designation indicates both conventional Siegbahn notation (if applicable) and transition, e.g., β_1 $L_{II}M_{IV}$ denotes a transition between the L_{II} and M_{IV} levels, which is the $L\beta_1$ line in Siegbahn notation.

Designation	Å*	p.e.	keV	Å*	p.e.	keV
	3 Lithium			**4 Beryllium**		
$\alpha\ KL$	228.	1	0.0543	114.	1	0.1085
	5 Boron			**6 Carbon**		
$\alpha\ KL$	67.6	3	0.1833	44.7	3	0.277
	7 Nitrogen			**8 Oxygen**		
$\alpha\ KL$	31.6	4	0.3924	23.62	3	0.5249
	9 Fluorine			**10 Neon**		
$\alpha_{1,2}\ KL_{II,III}$	18.32	2	0.6768	14.610	3	0.8486
$\beta\ KM$				14.452	5	0.8579
	11 Sodium			**12 Magnesium**		
$\alpha_{1,2}\ KL_{II,III}$	11.9101	9	1.0410	9.8900	2	1.25360
$\beta\ KM$	11.575	2	1.0711	9.521	2	1.3022
$L_{II,III}M$	407.1	5	0.03045	251.5	5	0.0493
$L_IL_{II,III}$	376	1	0.0330	317	1	0.0392
	13 Aluminum			**14 Silicon**		
$\alpha_2\ KL_{II}$	8.34173	9	1.48627	7.12791	9	1.73938
$\alpha_1\ KL_{III}$	8.33934	9	1.48670	7.12542	9	1.73998
$\beta\ KM$	7.960	2	1.5574	6.753	1	1.8359
$L_{II,III}$	171.4	5	0.0724	135.5	4	0.0915
$L_IL_{II,III}$	290.	1	0.0428			
	15 Phosphorus			**16 Sulfur**		
$\alpha_2\ KL_{II}$	6.160†	1	2.0127	5.37496	8	2.30664
$\alpha_1\ KL_{III}$	6.157†	1	2.0137	5.37216	7	2.30784
$\beta\ KM$	5.796	2	2.1390			
$\beta_1\ KM$				5.0316	2	2.4640
$\beta_x\ KM$				5.0233	3	2.4681
$L_{II,III}M$	103.8	4	0.1194			
$l, \eta\ L_{II,III}M_I$				83.4	3	0.1487
	17 Chlorine			**18 Argon**		
$\alpha_2\ KL_{II}$	4.7307	1	2.62078	4.19474	5	2.95563
$\alpha_1\ KL_{III}$	4.7278	1	2.62239	4.19180	5	2.95770
$\beta\ KM$	4.4034	3	2.8156			
$\beta_{1,3}\ KM_{II,III}$				3.8860	2	3.1905
$\eta\ L_{II}M_I$	67.33	9	0.1841	55.9†	1	0.2217
$l\ L_{III}M_I$	67.90	9	0.1826	56.3†	1	0.2201
	19 Potassium			**20 Calcium**		
$\alpha_2\ KL_{II}$	3.7445	2	3.3111	3.36166	3	3.68809
$\alpha_1\ KL_{III}$	3.7414	2	3.3138	3.35839	3	3.69168
$\beta_{1,3}\ KM_{II,III}$	3.4539	2	3.5896	3.0897	2	4.0127
$\beta_5\ KM_{IV,V}$	3.4413	4	3.6027	3.0746	3	4.0325

Designation	Å*	p.e.	keV	Å*	p.e.	keV
	19 Potassium (*Cont.*)			**20 Calcium** (*Cont.*)		
$\eta\ L_{II}M_I$	47.24	2	0.2625	40.46	2	0.3064
β_1				35.94	2	0.3449
$l\ L_{III}M_I$	47.74	1	0.25971	40.96	2	0.3027
$\alpha_{1,2}\ L_{III}M_{IV,V}$				36.33	2	0.3413
$M_{II,III}N_I$	692	9	0.0179	525.	9	0.0236
	21 Scandium			**22 Titanium**		
$\alpha_2\ KL_{II}$	3.0342	1	4.0861	2.75216	2	4.50486
$\alpha_1\ KL_{III}$	3.0309†	1	4.0906	2.74851	2	4.51084
$\beta_{1,3}\ KM_{II,III}$	2.7796	2	4.4605	2.51391	2	4.93181
$\beta_5\ KM_{IV,V}$	2.7634	3	4.4865	2.4985	2	4.9623
$\eta\ L_{II}M_I$	35.13	2	0.3529	30.89	3	0.4013
$\beta_1\ L_{II}M_{IV}$	31.02	2	0.3996	27.05	2	0.4584
$l\ L_{III}M_I$	35.59	3	0.3483	31.36	2	0.3953
$\alpha_{1,2}\ L_{III}M_{IV,V}$	31.35	3	0.3954	27.42	2	0.4522
	23 Vanadium			**24 Chromium**		
$\alpha_2\ KL_{II}$	2.50738	2	4.94464	2.293606	3	5.40551
$\alpha_1\ KL_{III}$	2.50356	2	4.95220	2.28970	2	5.41472
$\beta_{1,3}\ KM_{II,III}$	2.28440	2	5.42729	2.08487	2	5.94671
$\beta_5\ KM_{IV,V}$	2.26951	6	5.4629	2.07087	6	5.9869
$\beta_{3,4}\ L_IM_{II,III}$	21.19†	9	0.585	18.96	2	0.654
$\eta\ L_{II}M_I$	27.34	3	0.4535	24.30	3	0.5102
$\beta_1\ L_{II}M_{IV}$	23.88	4	0.5192	21.27	1	0.5828
$l\ L_{III}M_I$	27.77	1	0.4465	24.78	1	0.5003
$\alpha_{1,2}\ L_{III}M_{IV,V}$	24.25	3	0.5113	21.64	3	0.5728
$M_{II,III}M_{IV,V}$	337.	9	0.037	309.	9	0.040
	25 Manganese			**26 Iron**		
$\alpha_2\ KL_{II}$	2.10578	2	5.88765	1.939980	9	6.39084
$\alpha_1\ KL_{III}$	2.101820	9	5.89875	1.936042	9	6.40384
$\beta_{1,3}\ KM_{II,III}$	1.91021	2	6.49045	1.75661	2	7.05798
$\beta_5\ KM_{IV,V}$	1.8971	1	6.5352	1.7442	1	7.1081
$\beta_{3,4}\ L_IM_{II,III}$	17.19	2	0.721	15.65	2	0.792
$\eta\ L_{II}M_I$	21.85	2	0.5675	19.75	4	0.628
$\beta_1\ L_{II}M_{IV}$	19.11	2	0.6488	17.26	1	0.7185
$l\ L_{III}M_I$	22.29	1	0.5563	20.15	1	0.6152
$\alpha_{1,2}\ L_{III}M_{IV,V}$	19.45	1	0.6374	17.59	2	0.7050
$M_{II,III}M_{IV,V}$	273.	6	0.045	243.	5	0.051
	27 Cobalt			**28 Nickel**		
$\alpha_2\ KL_{II}$	1.792850	9	6.91530	1.661747	8	7.46089
$\alpha_1\ KL_{III}$	1.788965	9	6.93032	1.657910	8	7.47815
$\beta_{1,3}\ KM_{II,III}$	1.62079	2	7.64943	1.500135	8	8.26466
$\beta_5\ KM_{IV,V}$	1.60891	3	7.7059	1.48862	4	8.3286
$\beta_{3,4}\ L_IM_{II,III}$	14.31	3	0.870	13.18	1	0.941
$\eta\ L_{II}M_I$	17.87	3	0.694	16.27	3	0.762
$\beta_1\ L_{II}M_{IV}$	15.666	8	0.7914	14.271	6	0.8688
$l\ L_{III}M_I$	18.292	8	0.6778	16.693	9	0.7427
$\alpha_{1,2}\ L_{III}M_{IV,V}$	15.972	6	0.7762	14.561	3	0.8515
$M_{II,III}M_{IV,V}$	214.	6	0.058	190.	2	0.0651

Designation	Å*	p.e.	keV	Å*	p.e.	keV
	29 Copper			**30 Zinc**		
$\alpha_2\ KL_{II}$	1.544390	2	8.02783	1.439000	8	8.61578
$\alpha_1\ KL_{III}$	1.540562	2	8.04778	1.435155	7	8.63886
$\beta_3\ KM_{II}$	1.3926	1	8.9029			
$\beta_{1,3}\ KM_{II,III}$	1.392218	9	8.90529	1.29525	2	9.5720
$\beta_2\ KN_{II,III}$				1.28372	2	9.6580
$\beta_5\ KM_{IV,V}$	1.38109	3	8.9770	1.2848	1	9.6501
$\beta_{3,4}\ L_IM_{II,III}$	12.122	8	1.0228	11.200	7	1.1070
$\eta\ L_{II}M_I$	14.90	2	0.832	13.68	2	0.906
$\beta_1\ L_{II}M_{IV}$	13.053	3	0.9498	11.983	3	1.0347
$l\ L_{III}M_I$	15.286	9	0.8111	14.02	2	0.884
$\alpha_{1,2}\ L_{III}M_{IV,V}$	13.336	3	0.9297	12.254	3	1.0117
$M_{II,III}M_{V,V}$	173.	3	0.072	157.	3	0.079
	31 Gallium			**32 Germanium**		
$\alpha_2\ KL_{II}$	1.34399	1	9.22482	1.258011	9	9.85532
$\alpha_1\ KL_{III}$	1.340083	9	9.25174	1.254054	9	9.88642
$\beta_3\ KM_{II}$	1.20835	5	10.2603	1.12936	9	10.9780
$\beta_1\ KM_{III}$	1.20789	2	10.2642	1.12894	2	10.9821
$\beta_2\ KN_{II,III}$	1.19600	2	10.3663	1.11686	2	11.1008
$\beta_5\ KM_{IV,V}$	1.1981	2	10.348	1.1195	1	11.0745
$\beta_4\ L_IM_{II}$				9.640	2	1.2861
$\beta_3\ L_IM_{III}$				9.581	2	1.2941
$\beta_{3,4}\ L_IM_{II,III}$	10.359†	8	1.197			
$\eta\ L_{II}M_I$	12.597	2	0.9842	11.609	2	1.0680
$\beta_1\ L_{II}M_{IV}$	11.023	2	1.1248	10.175	1	1.2185
$l\ L_{III}M_I$	12.953	2	0.9572	11.965	4	1.0362
$\alpha_{1,2}\ L_{III}M_{IV,V}$	11.292	1	1.09792	10.4361	8	1.18800
	33 Arsenic			**34 Selenium**		
$\alpha_2\ KL_{II}$	1.17987	1	10.50799	1.10882	2	11.1814
$\alpha_1\ KL_{III}$	1.17588	1	10.54372	1.10477	2	11.2224
$\beta_3\ KM_{II}$	1.05783	5	11.7203	0.99268	5	12.4896
$\beta_1\ KM_{III}$	1.05730	2	11.7262	0.99218	3	12.4959
$\beta_2\ KN_{II,III}$	1.04500	3	11.8642	0.97992	5	12.6522
$\beta_5\ KM_{IV,V}$	1.0488	1	11.822	0.9843	1	12.595
$\beta_{3,4}\ L_IM_{II,III}$	8.929	1	1.3884	8.321†	9	1.490
$\eta\ L_{II}M_I$	10.734	1	1.1550	9.962	1	1.2446
$\beta_1\ L_{II}M_{IV}$	9.4141	8	1.3170	8.7358	5	1.41923
$l\ L_{III}M_I$	11.072	1	1.1198	10.294	1	1.2044
$\alpha_{1,2}\ L_{III}M_{IV,V}$	9.6709	8	1.2820	8.9900	5	1.37910
M_VN_{III}				230.	2	0.0538
	35 Bromine			**36 Krypton**		
$\alpha_2\ KL_{II}$	1.04382	2	11.8776	0.9841	1	12.598
$\alpha_1\ KL_{III}$	1.03974	2	11.9242	0.9801	1	12.649
$\beta_3\ KM_{II}$	0.93327	5	13.2845	0.8790	1	14.104
$\beta_1\ KM_{III}$	0.93279	2	13.2914	0.8785	1	14.112
$\beta_2\ KN_{II,III}$	0.92046	2	13.4695	0.8661	1	14.315
$\beta_5\ KM_{IV,V}$	0.9255	1	13.396	0.8708	2	14.238
$\beta_4\ KN_{IV,V}$				0.8653	2	14.328
$\beta_4\ L_IM_{II}$				7.304	5	1.697
$\beta_3\ L_IM_{III}$				7.264	5	1.707

Designation	Å*	p.e.	keV	Å*	p.e.	keV
	35 Bromine (*Cont.*)			**36 Krypton** (*Cont.*)		
$\beta_{3,4}\ L_IM_{II,III}$	7.767†	9	1.596			
$\eta\ L_{II}M_I$	9.255	1	1.3396			
$\beta_1\ L_{II}M_{IV}$	8.1251	5	1.52590	7.576†	3	1.6366
γ_5				7.279	5	1.703
$l\ L_{III}M_I$	9.585	1	1.2935			
$\alpha_{1,2}\ L_{III}M_{IV,V}$	8.3746	5	1.48043	7.817†	3	1.5860
β_6				7.510	4	1.6510
$L_{II}N_{III}$				7.250	5	1.710
M_IM_{II}	184.6	3	0.0672			
M_IM_{III}	164.7	3	0.0753			
$M_{II}M_{IV}$	109.4	3	0.1133			
$M_{II}N_I$	76.9	2	0.1613			
$M_{III}M_{IV,V}$	113.8	3	0.1089			
	79.8	3	0.1554			
$\zeta_2\ M_{IV}N_{II}$	191.1	2	0.06488			
$M_{IV}N_{III}$	189.5	3	0.0654			
$\zeta_1\ M_VN_{III}$	192.6	2	0.06437			
	37 Rubidium			**38 Strontium**		
$\alpha_2\ KL_{II}$	0.92969	1	13.3358	0.87943	1	14.0979
$\alpha_1\ KL_{III}$	0.925553	9	13.3953	0.87526	1	14.1650
$\beta_3\ KM_{II}$	0.82921	3	14.9517	0.78345	3	15.8249
$\beta_1\ KM_{III}$	0.82868	2	14.9613	0.78292	2	15.8357
$\beta_2\ KN_{II,III}$	0.81645	3	15.1854	0.77081	3	16.0846
$\beta_5\ KM_{IV,V}$	0.8219	1	15.085	0.7764	1	15.969
$\beta_4\ KN_{IV,V}$	0.8154	2	15.205	0.76989	5	16.104
$\beta_4\ L_IM_{II}$	6.8207	3	1.81771	6.4026	3	1.93643
$\beta_3\ L_IM_{III}$	6.7876	3	1.82659	6.3672	3	1.94719
$\gamma_{2,3}\ L_IN_{II,III}$	6.0458	3	2.0507	5.6445	3	2.1965
$\eta\ L_{II}M_I$	8.0415	4	1.54177	7.5171	3	1.64933
$\beta_1\ L_{II}M_{IV}$	7.0759	3	1.75217	6.6239	3	1.87172
$\gamma_5\ L_{II}N_{IV}$	6.7553	3	1.83532	6.2961	3	1.96916
$l\ L_{III}M_I$	8.3636	4	1.48238	7.8362	3	1.58215
$\alpha_2\ L_{III}M_{IV}$	7.3251	3	1.69256	6.8697	3	1.80474
$\alpha_1\ L_{III}M_V$	7.3183	2	1.69413	6.8628	2	1.80656
$\beta_6\ L_{III}N_I$	6.9842	3	1.77517	6.5191	3	1.90181
M_IM_{III}	144.4	3	0.0859			
$M_{II}M_{IV}$	91.5	2	0.1355	85.7	2	0.1447
$M_{II}N_I$	57.0	2	0.2174	51.3	1	0.2416
$M_{III}M_{IV,V}$	96.7	2	0.1282	91.4	2	0.1357
$M_{III}N_I$	59.5	2	0.2083	53.6	1	0.2313
$\zeta_2\ M_{IV}N_{II}$	127.8	2	0.0970			
$M_{IV}N_{III}$	126.8	2	0.0978			
$\zeta_2\ M_{IV}N_{II,III}$				108.0	2	0.1148
$\zeta_1\ M_VN_{III}$	128.7	2	0.0964	108.7	1	0.1140
	39 Yttrium			**40 Zirconium**		
$\alpha_2\ KL_{II}$	0.83305	1	14.8829	0.79015	1	15.6909
$\alpha_1\ KL_{III}$	0.82884	1	14.9584	0.78593	1	15.7751
$\beta_3\ KM_{II}$	0.74126	3	16.7258	0.70228	4	17.654
$\beta_1\ KM_{III}$	0.74072	2	16.7378	0.70173	3	17.6678
$\beta_2\ KN_{II,III}$	0.72864	4	17.0154	0.68993	4	17.970
$\beta_5\ KM_{IV,V}$	0.7345	1	16.879	0.6959	1	17.815

Designation	Å*	p.e.	keV	Å*	p.e.	keV
	39 Yttrium (*Cont.*)			**40 Zirconium** (*Cont.*)		
β_4 $KN_{IV,V}$	0.72776	5	17.036	0.68901	5	17.994
β_4 L_IM_{II}	6.0186	3	2.0600	5.6681	3	2.1873
β_3 L_IM_{III}	5.9832	3	2.0722	5.6330	3	2.2010
$\gamma_{2,3}$ $L_IN_{II,III}$	5.2830	3	2.3468	4.9536	3	2.5029
η L_IM_I	7.0406	3	1.76095	6.6069	3	1.87654
β_1 $L_{III}M_{IV}$	6.2120	3	1.99584	5.8360	3	2.1244
γ_5 $L_{II}N_I$	5.8754	3	2.1102	5.4977	3	2.2551
γ_1 $L_{II}N_{IV}$				5.3843	3	2.3027
l $L_{III}M_I$	7.3563	3	1.68536	6.9185	3	1.79201
α_2 $L_{III}M_{IV}$	6.4558	3	1.92047	6.0778	3	2.0399
α_1 $L_{III}M_V$	6.4488	2	1.92256	6.0705	2	2.04236
β_6 $L_{III}N_I$	6.0942	3	2.0344	5.7101	3	2.1712
$\beta_{2,15}$				5.5863	3	2.2194
$M_{II}M_{IV}$	81.5	2	0.1522	76.7	2	0.1617
$M_{II}N_I$	46.48	9	0.267			
$M_{III}M_V$				80.9	3	0.1533
$M_{III}N_I$	48.5	2	0.256			
$M_{III}M_{IV,V}$	86.5	2	0.1434			
ζ $M_{IV,V}N_{II,III}$	93.4	2	0.1328	82.1	2	0.1511
$M_{IV,V}O_{II,III}$				70.0	4	0.177
	41 Niobium			**42 Molybdenum**		
α_2 KL_{II}	0.75044	1	16.5210	0.713590	6	17.3743
α_1 KL_{III}	0.74620	1	16.6151	0.709300	1	17.47934
β_3 KM_{II}	0.66634	3	18.6063	0.632872	9	19.5903
β_1 KM_{III}	0.66576	2	18.6225	0.632288	9	19.6083
β_2^{II}				0.62107	5	19.963
β_2 $KN_{II,III}$	0.65416	4	18.953	0.62099	2	19.9652
β_4 $KN_{IV,V}$	0.65318	5	18.981			
β_5^{II} KM_{IV}				0.62708	5	19.771
β_5^{I} KM_V				0.62692	5	19.776
β_4 $KN_{IV,V}$				0.62001	9	19.996
β_4 L_IM_{II}	5.3455	3	2.3194	5.0488	3	2.4557
β_3 L_IM_{III}	5.3102	3	2.3348	5.0133	3	2.4730
$\gamma_{2,3}$ $L_IN_{II,III}$	4.6542	2	2.6638	4.3800	2	2.8306
η $L_{II}M_I$	6.2109	3	1.99620	5.8475	3	2.1202
β_1 $L_{II}M_{IV}$	5.4923	3	2.2574	5.17708	8	2.39481
γ_5 $L_{II}N_I$	5.1517	3	2.4066	4.8369	2	2.5632
γ_1 $L_{II}N_{IV}$	5.0361	3	2.4618	4.7258	2	2.6235
l $L_{III}M_I$	6.5176	3	1.90225	6.1508	3	2.01568
α_2 $L_{III}M_{IV}$	5.7319	3	2.1630	5.41437	8	2.28985
α_1 $L_{III}M_V$	5.7243	2	2.16589	5.40655	8	2.29316
β_6 $L_{III}N_I$	5.3613	3	2.3125	5.0488	5	2.4557
$\beta_{2,15}$ $L_{III}N_{IV,V}$	5.2379	3	2.3670	4.9232	2	2.5183
$M_{II}M_{IV}$	72.1	3	0.1718	68.9	2	0.1798
$M_{II}N_I$	38.4	3	0.323	35.3	3	0.351
$M_{II}N_{IV}$	33.1	2	0.375			
$M_{III}M_V$	78.4	2	0.1582	74.9	1	0.1656
$M_{III}N_I$	40.7	2	0.305	37.5	2	0.331
γ $M_{III}N_{IV,V}$	34.9	2	0.356			
ζ $M_{IV,V}N_{II,III}$	72.19	9	0.1717	64.38	7	0.1926
$M_{IV,V}O_{II,III}$	61.9	2	0.2002	54.8	2	0.2262

Designation	Å*	p.e.	keV	Å*	p.e.	keV
	43 Technetium			**44 Ruthenium**		
α_2 KL_{II}	0.67932†	3	18.2508	0.647408	5	19.1504
α_1 KL_{III}	0.67502†	3	18.3671	0.643083	4	19.2792
β_3 KM_{II}	0.60188†	4	20.599	0.573067	4	21.6346
β_1 KM_{III}	0.60130†	4	20.619	0.572482	4	21.6568
β_2 $KN_{II,III}$	0.59024†	5	21.005	0.56166	3	22.074
β_5^{II} KM_{IV}				0.5680	2	21.829
β_5^{I} KM_V				0.56785	9	21.834
β_4				0.56089	9	22.104
β_4 $L_{II}M_{IV}$				4.5230	2	2.7411
β_3 L_IM_{III}				4.4866	3	2.7634
$\gamma_{2,3}$ $L_IN_{II,III}$				3.8977	2	3.1809
η $L_{II}M_I$				5.2050	2	2.38197
β_1 $L_{II}M_{IV}$	4.8873†	8	2.5368	4.62058	3	2.68323
γ_5 $L_{II}N_I$				4.2873	2	2.8918
γ_1 $L_{II}N_{IV}$				4.1822	2	2.9645
l $L_{III}M_I$				5.5035	3	2.2528
α_2 $L_{III}M_{IV}$				4.85381	7	2.55431
α_1 $L_{III}M_V$	5.1148†	3	2.4240	4.84575	5	2.55855
β_6 $L_{III}N_I$				4.4866	3	2.7634
$\beta_{2,15}$ $L_{III}N_{IV,V}$				4.3718	2	2.8360
$M_{II}M_{IV}$				62.2	1	0.1992
$M_{II}N_I$				32.3	2	0.384
$M_{II}N_{IV}$				25.50	9	0.486
$M_{III}M_V$				68.3	1	0.1814
γ $M_{III}N_{IV,V}$				26.9	1	0.462
ζ $M_{IV,V}N_{II,III}$				52.34	7	0.2369
$M_{IV,V}O_{II,III}$				44.8	1	0.2768
	45 Rhodium			**46 Palladium**		
α_2 KL_{II}	0.617630	4	20.0737	0.589821	3	21.0201
α_1 KL_{III}	0.613279	4	20.2161	0.585448	3	21.1771
β_3 KM_{II}	0.546200	4	22.6989	0.521123	4	23.7911
β_1 KM_{III}	0.545605	4	22.7236	0.520520	4	23.8187
β_2^{II} KN_{II}	0.53513	5	23.168			
β_2 $KN_{II,III}$	0.53503	2	23.1728	0.510228	4	24.2991
β_5^{II} KM_{IV}	0.54118	9	22.909			
β_5^{I} KM_V	0.54101	9	22.917			
β_4 $KN_{IV,V}$	0.53401	9	23.217	0.5093	2	24.346
β_5 $KM_{IV,V}$				0.51670	9	23.995
β_4 L_IM_{II}	4.2888	2	2.8908	4.0711	2	3.0454
β_3 L_IM_{III}	4.2522	2	2.9157	4.0346	2	3.0730
$\gamma_{2,3}$ $L_IN_{II,III}$	3.6855	2	3.3640	3.4892	2	3.5533
η $L_{II}M_I$	4.9217	2	2.5191	4.6605	2	2.6603
β_1 $L_{II}M_{IV}$	4.37414	4	2.83441	4.14622	5	2.99022
γ_5 $L_{II}N_I$	4.0451	2	3.0650	3.8222	2	3.2437
γ_1 $L_{II}N_{IV}$	3.9437	2	3.1438	3.7246	2	3.3287
l $L_{III}M_I$	5.2169	3	2.3765	4.9525	3	2.5034
α_2 $L_{III}M_{IV}$	4.60545	9	2.69205	4.37588	7	2.83329
α_1 $L_{III}M_V$	4.59743	9	2.69674	4.36767	5	2.83861
β_6 $L_{III}N_I$	4.2417	2	2.9229	4.0162	2	3.0870
$\beta_{2,15}$ $L_{III}N_{IV,V}$	4.1310	2	3.0013	3.90887	4	3.17179
β_{10} L_IM_{IV}				3.7988	2	3.2637

Designation	Å*	p.e.	keV	Å*	p.e.	keV
	45 Rhodium (*Cont.*)			**46 Palladium** (*Cont.*)		
$\beta_9\ L_IM_V$				3.7920	2	3.2696
$M_IN_{II,III}$				20.1	2	0.616
$M_{II}M_{IV}$	59.3	1	0.2090	56.5	1	0.2194
$M_{II}N_I$	28.1	2	0.442	26.2	2	0.474
$M_{II}N_{IV}$				22.1	1	0.560
$M_{III}M_V$	65.5	1	0.1892	62.9	1	0.1970
$M_{III}N_I$	29.8	1	0.417	27.9	1	0.445
$\gamma\ M_{III}N_{IV,V}$	25.01	9	0.496	23.3†	1	0.531
$\zeta\ M_{IV,V}N_{II,III}$	47.67	9	0.2601	43.6	1	0.2844
$M_{IV,V}O_{II,III}$	40.9	2	0.303	37.4	2	0.332
	47 Silver			**48 Cadmium**		
$\alpha_2\ KL_{II}$	0.563798	4	21.9903	0.539422	3	22.9841
$\alpha_1\ KL_{III}$	0.5594075	6	22.16292	0.535010	3	23.1736
$\beta_3\ KM_{II}$	0.497685	4	24.9115	0.475730	5	26.0612
$\beta_1\ KM_{III}$	0.497069	4	24.9424	0.475105	6	26.0955
$\beta_2\ KN_{II,III}$	0.487032	4	25.4564	0.465328	7	26.6438
$\beta_5\ KM_{IV,V}$	0.49306	2	25.145			
$\beta_4\ KN_{IV,V}$	0.48598	3	25.512			
$\beta_4\ L_IM_{II}$	3.87023	5	3.20346	3.68203	9	3.36719
$\beta_3\ L_IM_{III}$	3.83313	9	3.23446	3.64495	9	3.40145
$\gamma_2\ L_IN_{II}$	3.31216	9	3.7432	3.1377	2	3.9513
$\gamma_3\ L_IN_{III}$	3.30635	9	3.7498			
$\eta\ L_{II}M_I$	4.4183	2	2.8061	4.19315	9	2.95675
$\beta_1\ L_{II}M_{IV}$	3.93473	3	3.15094	3.73823	4	3.31657
$\gamma_5\ L_{II}N_I$	3.61638	9	3.42832	3.42551	9	3.61935
$\gamma_1\ L_{II}N_{IV}$	3.52260	4	3.51959	3.33564	6	3.71686
$l\ L_{III}M_I$	4.7076	2	2.6337	4.48014	9	2.76735
$\alpha_2\ L_{III}M_{IV}$	4.16294	5	2.97821	3.96496	6	3.12691
$\alpha_1\ L_{III}M_V$	4.15443	3	2.98431	3.95635	4	3.13373
$\beta_6\ L_{III}N_I$	3.80774	9	3.25603	3.61467	9	3.42994
$\beta_{2,15}\ L_{III}N_{IV,V}$	3.70335	3	3.34781	3.51408	4	3.52812
$\beta_{10}\ L_IM_{IV}$	3.61158	9	3.43287	3.4367	2	3.6075
$\beta_9\ L_IM_V$	3.60497	9	3.43917	3.43015	9	3.61445
$M_IN_{II,III}$	18.8	2	0.658			
$M_{II}M_{IV}$	54.0	1	0.2295	52.0	2	0.2384
$M_{II}N_I$				22.9	2	0.540
$M_{II}N_{IV}$	20.66	7	0.600	19.40	7	0.639
$M_{III}M_V$	60.5	1	0.2048	58.7	2	0.2111
$M_{III}N_I$	26.0	1	0.478	24.5	1	0.507
$\gamma\ M_{III}N_{IV,V}$	21.82	7	0.568	20.47	7	0.606
$M_{IV}O_{II,III}$				30.4	1	0.408
$\zeta\ M_{IV,V}N_{II,III}$	39.77	7	0.3117	36.8	1	0.3371
M_VN_I	24.4	2	0.509			
M_VO_{III}				30.8	1	0.403
$M_{IV,V}O_{II,III}$	33.5	3	0.370			
	49 Indium			**50 Tin**		
$\alpha_2\ KL_{II}$	0.516544	3	24.0020	0.495053	3	25.0440
$\alpha_1\ KL_{III}$	0.512113	3	24.2097	0.490599	3	25.2713
$\beta_3\ KM_{II}$	0.455181	4	27.2377	0.435877	5	28.4440

Designation	Å*	p.e.	keV	Å*	p.e.	keV
	49 Indium (*Cont.*)			**50 Tin** (*Cont.*)		
$\beta_1\ KM_{III}$	0.454545	4	27.2759	0.435236	5	28.4860
$\beta_2\ KN_{II,III}$	0.44500	1	27.8608	0.425915	8	29.1093
$KO_{II,III}$	0.44374	3	27.940	0.42467	3	29.195
$\beta_5^{II}\ KM_{IV}$	0.45098	2	27.491	0.43184	3	28.710
$\beta_5^{I}\ KM_V$	0.45086	2	27.499	0.43175	3	28.716
$\beta_4\ KN_{IV,V}$	0.44393	4	27.928	0.42495	3	29.175
$\beta_4\ L_IM_{II}$	3.50697	9	3.5353	3.34335	9	3.7083
$\beta_3\ L_IM_{III}$	3.46984	9	3.5731	3.30585	3	3.7500
$\gamma_{2,3}\ L_IN_{II,III}$	2.9800	2	4.1605	2.8327	2	4.3768
$\gamma_4\ L_IO_{II,III}$	2.9264	2	4.2367	2.7775	2	4.4638
$\eta\ L_{II}M_I$	3.98327	9	3.11254	3.78876	9	3.27234
$\beta_1\ L_{II}M_{IV}$	3.55531	4	3.48721	3.38487	3	3.66280
$\gamma_5\ L_{II}N_I$	3.24907	9	3.8159	3.08475	9	4.0192
$\gamma_1\ L_{II}N_{IV}$	3.16213	4	3.92081	3.00115	3	4.13112
$l\ L_{III}M_I$	4.26873	9	2.90440	4.07165	9	3.04499
$\alpha_2\ L_{III}M_{IV}$	3.78073	6	3.27929	3.60891	4	3.43542
$\alpha_1\ L_{III}M_V$	3.77192	4	3.28694	3.59994	3	3.44398
$\beta_6\ L_{III}N_I$	3.43606	9	3.60823	3.26901	9	3.7926
$\beta_{2,15}\ L_{III}N_{IV,V}$	3.33838	3	3.71381	3.17505	3	3.90486
$\beta_7\ L_{III}O_I$	3.324	4	3.730	3.1564	3	3.9279
$\beta_{10}\ L_IM_{IV}$	3.27404	9	3.7868	3.12170	9	3.9716
$\beta_9\ L_IM_V$	3.26763	9	3.7942	3.11513	9	3.9800
$M_{II}M_{IV}$				47.3	1	0.2621
$M_{II}N_I$				20.0	1	0.619
$M_{II}N_{IV}$				16.93	5	0.733
$M_{III}M_V$				54.2	1	0.2287
$M_{III}N_I$				21.5	1	0.575
$\gamma\ M_{III}N_{IV,V}$				17.94	5	0.691
$M_{IV}O_{II,III}$				25.3	1	0.491
$\zeta\ M_{IV,V}N_{II,III}$				31.24	9	0.397
M_VO_{III}				25.7	1	0.483
	51 Antimony			**52 Tellurium**		
$\alpha_2\ KL_{II}$	0.474827	3	26.1108	0.455784	3	27.2017
$\alpha_1\ KL_{III}$	0.470354	3	26.3591	0.451295	3	27.4723
$\beta_3\ KM_{II}$	0.417737	4	29.6792	0.400659	4	30.9443
$\beta_1\ KM_{III}$	0.417085	3	29.7256	0.399995	5	30.9957
$\beta_2\ KN_{II,III}$	0.407973	5	30.3895	0.391102	6	31.7004
$KO_{II,III}$	0.40666	1	30.4875	0.38974	1	31.8114
$\beta_5^{II}\ KM_{IV}$	0.41388	1	29.9560			
$\beta_5^{I}\ KM_V$	0.41378	1	29.9632			
$\beta_4\ KN_{IV,V}$	0.40702	1	30.4604			
$\beta_4\ L_IM_{II}$	3.19014	9	3.8864	3.04661	9	4.0695
$\beta_3\ L_IM_{III}$	3.15258	9	3.9327	3.00893	9	4.1204
$\gamma_{2,3}\ L_IN_{II,III}$	2.6953	2	4.5999	2.5674	2	4.8290
$\gamma_4\ L_IO_{II,III}$	2.6398	2	4.6967	2.5113	2	4.9369
$\eta\ L_{II}M_I$	3.60765	9	3.43661	3.43832	9	3.60586
$\beta_1\ L_{II}M_{IV}$	3.22567	4	3.84357	3.07677	6	4.02958
$\gamma_5\ L_{II}N_I$	2.93187	9	4.2287	2.79007	9	4.4437
$\gamma_1\ L_{II}N_{IV}$	2.85159	3	4.34779	2.71241	6	4.5709
$l\ L_{III}M_I$	3.88826	9	3.18860	3.71696	9	3.33555
$\alpha_2\ L_{III}M_{IV}$	3.44840	6	3.59532	3.29846	9	3.7588

X-Ray Wavelengths
TABLE A (*Continued*)

Designation	Å*	p.e.	keV	Å*	p.e.	keV
	51 Antimony (*Cont.*)			**52 Tellurium** (*Cont.*)		
$\alpha_1\ L_{III}M_V$	3.43941	4	3.60472	3.28920	6	3.76933
$\beta_6\ L_{III}N_I$	3.11513	9	3.9800	2.97088	9	4.1732
$\beta_{2,15}\ L_{III}N_{IV,V}$	3.02335	3	4.10078	2.88217	8	4.3017
$\beta_7\ L_{III}O_I$	3.0052	3	4.1255	2.8634	3	4.3298
$\beta_{10}\ L_IM_{IV}$	2.97917	9	4.1616	2.84679	9	4.3551
$\beta_9\ L_IM_V$	2.97261	9	4.1708	2.83897	9	4.3671
$M_{II}M_{IV}$	45.2	1	0.2743			
$M_{II}N_I$	18.8	1	0.658	17.6	1	0.703
$M_{II}N_{IV}$	15.98	5	0.776			
$M_{III}M_V$	52.2	1	0.2375	50.3	1	0.2465
$M_{III}N_I$	20.2	1	0.612	19.1	1	0.648
$\gamma\ M_{III}N_{IV,V}$	16.92	4	0.733	15.93	4	0.778
$M_{IV}O_{II,III}$				21.34	5	0.581
$\zeta\ M_{IV,V}N_{II,III}$	28.88	8	0.429	26.72	9	0.464
M_VO_{III}				21.78	5	0.569
	53 Iodine			**54 Xenon**		
$\alpha_2\ KL_{II}$	0.437829	7	28.3172	0.42087†	2	29.458
$\alpha_1\ KL_{III}$	0.433318	5	28.6120	0.41634†	2	29.779
$\beta_3\ KM_{II}$	0.384564	4	32.2394	0.36941†	2	33.562
$\beta_1\ KM_{III}$	0.383905	4	32.2947	0.36872†	2	33.624
$\beta_2\ KN_{II,III}$	0.37523†	2	33.042	0.36026†	3	34.415
$\beta_4\ L_IM_{II}$	2.91207	9	4.2575			
$\beta_3\ L_IM_{III}$	2.87429	9	4.3134			
$\gamma_{2,3}\ L_IN_{II,III}$	2.4475	2	5.0657			
$\gamma_4\ L_IO_{II,III}$	2.3913	2	5.1848			
$\eta\ L_{II}M_I$	3.27979	9	3.7801			
$\beta_1\ L_{II}M_{IV}$	2.93744	6	4.22072			
$\gamma_5\ L_{II}N_I$	2.65710	9	4.6660			
$\gamma_1\ L_{II}N_{IV}$	2.58244	8	4.8009			
$l\ L_{III}M_I$	3.55754	9	3.48502			
$\alpha_2\ L_{III}M_{IV}$	3.15791	6	3.92604			
$\alpha_1\ L_{III}M_V$	3.14860	6	3.93765	3.0166†	2	4.1099
$\beta_6\ L_{III}N_I$	2.83672	9	4.3706			
$\beta_{2,15}\ L_{III}N_{IV,V}$	2.75053	8	4.5075			
$\beta_7\ L_{III}O_I$	2.7288	3	4.5435			
$\beta_{10}\ L_IM_{IV}$	2.72104	9	4.5564			
$\beta_9\ L_IM_V$	2.71352	9	4.5690			
	55 Cesium			**56 Barium**		
$\alpha_2\ KL_{II}$	0.404835	4	30.6251	0.389668	5	31.8171
$\alpha_1\ KL_{III}$	0.400290	4	30.9728	0.385111	4	32.1936
$\beta_3\ KM_{II}$	0.355050	4	34.9194	0.341507	4	36.3040
$\beta_1\ KM_{III}$	0.354364	7	34.9869	0.340811	3	36.3782
$\beta_2\ KN_{II,III}$	0.34611	2	35.822	0.33277	1	37.257
$KO_{II,III}$				0.33127	2	37.426
$\beta_5^{II}\ KM_{IV}$				0.33835	2	36.643
$\beta_5^{I}\ KM_V$				0.33814	2	36.666
$\beta_4\ KN_{IV,V}$				0.33229	2	37.311
$\beta_4\ L_IM_{II}$	2.6666	2	4.6494	2.5553	2	4.8519
$\beta_3\ L_IM_{III}$	2.6285	2	4.7167	2.5164	2	4.9269
$\gamma_2\ L_IN_{II}$	2.2371	2	5.5420	2.1387	2	5.7969
$\gamma_3\ L_IN_{III}$	2.2328	2	5.5527	2.1342	2	5.8092

Designation	Å*	p.e.	keV	Å*	p.e.	keV
	55 Cesium (*Cont.*)			**56 Barium** (*Cont.*)		
$\gamma_4\ L_IO_{II,III}$	2.1741	2	5.7026	2.0756	3	5.9733
$\eta\ L_{II}M_I$	2.9932	2	4.1421	2.8627	3	4.3309
$\beta_1\ L_{II}M_{IV}$	2.6837	2	4.6198	2.56821	5	4.82753
$\gamma_5\ L_{II}N_I$	2.4174	2	5.1287	2.3085	3	5.3707
$\gamma_1\ L_{II}N_{IV}$	2.3480	2	5.2804	2.2415	2	5.5311
$l\ L_{III}M_I$	3.2670	2	3.7950	3.1355	2	3.9541
$\alpha_2\ L_{III}M_{IV}$	2.9020	2	4.2722	2.78553	5	4.45090
$\alpha_1\ L_{III}M_V$	2.8924	2	4.2865	2.77595	5	4.46626
$\beta_6\ L_{III}N_I$	2.5932	2	4.7811	2.4826	2	4.9939
$\beta_{2,15}\ L_{III}N_{IV,V}$	2.5118	2	4.9359	2.40435	6	5.1565
$\beta_7\ L_{III}O_I$	2.4849	2	4.9893	2.3806	2	5.2079
$\beta_{10}\ L_IM_{IV}$	2.4920	2	4.9752	2.3869	2	5.1941
$\beta_9\ L_IM_V$	2.4783	2	5.0026	2.3764	2	5.2171
$\gamma\ M_{III}N_{IV,V}$				12.75	3	0.973
$M_{IV}O_{II}$				15.91	5	0.779
$M_{IV}O_{III}$				15.72	9	0.789
$\zeta\ M_VN_{III}$				20.64	4	0.601
M_VO_{III}				16.20	5	0.765
$N_{IV}O_{II}$	188.6	1	0.06574	163.3	2	0.07590
$N_{IV}O_{III}$	183.8	1	0.06746	159.0	2	0.07796
N_VO_{III}	190.3	1	0.06515	164.6	2	0.07530
	57 Lanthanum			**58 Cerium**		
$\alpha_2\ KL_{II}$	0.375313	2	33.0341	0.361683	2	34.2789
$\alpha_1\ KL_{III}$	0.370737	2	33.4418	0.357092	2	34.7197
$\beta_3\ KM_{II}$	0.328686	4	37.7202	0.316520	4	39.1701
$\beta_1\ KM_{III}$	0.327983	3	37.8010	0.315816	2	39.2573
$\beta_2\ KN_{II,III}$	0.320117	7	38.7299	0.30816	1	40.233
$KO_{II,III}$	0.31864	2	38.909	0.30668	2	40.427
$\beta_5^{II}\ KM_{IV}$	0.32563	2	38.074	0.31357	2	39.539
$\beta_5^{I}\ KM_V$	0.32546	2	38.094	0.31342	2	39.558
$\beta_4\ KN_{IV,V}$	0.31931	2	38.828	0.30737	2	40.337
$\beta_4\ L_IM_{II}$	2.4493	3	5.0620	2.3497	4	5.2765
$\beta_3\ L_IM_{III}$	2.4105	3	5.1434	2.3109	3	5.3651
$\gamma_2\ L_IN_{II}$	2.0460	4	6.060	1.9602	3	6.3250
$\gamma_3\ L_IN_{III}$	2.0410	4	6.074	1.9553	3	6.3409
$\gamma_4\ L_IO_{II,III}$	1.9830	4	6.252	1.8991	4	6.528
$\eta\ L_{II}M_I$	2.740	3	4.525	2.6203	4	4.7315
$\beta_1\ L_{II}M_{IV}$	2.45891	5	5.0421	2.3561	3	5.2622
$\gamma_5\ L_{II}N_I$	2.2056	4	5.621	2.1103	3	5.8751
$\gamma_1\ L_{II}N_{IV}$	2.1418	3	5.7885	2.0487	4	6.052
$\gamma_8\ L_{II}O_I$				2.0237	4	6.126
$l\ L_{III}M_I$	3.006	3	4.124	2.8917	4	4.2875
$\alpha_2\ L_{III}M_{IV}$	2.67533	5	4.63423	2.5706	3	4.8230
$\alpha_1\ L_{III}M_V$	2.66570	5	4.65097	2.5615	2	4.8402
$\beta_6\ L_{III}N_I$	2.3790	4	5.2114	2.2818	3	5.4334
$\beta_{2,15}\ L_{III}N_{IV,V}$	2.3030	3	5.3835	2.2087	2	5.6134
$\beta_7\ L_{III}O_I$	2.275	3	5.450	2.1701	2	5.7132
$\beta_{10}\ L_IM_{IV}$	2.290	3	5.415	2.1958	5	5.646
$\beta_9\ L_IM_V$	2.282	3	5.434	2.1885	3	5.6650
$\gamma\ M_{III}N_{IV,V}$	12.08	4	1.027	11.53	1	1.0749
$\beta\ M_{IV}N_{VI}$	14.51	5	0.854	13.75	4	0.902
$\zeta\ M_VN_{III}$	19.44	5	0.638	18.35	4	0.676
$\alpha\ M_VN_{VI,VII}$	14.88	5	0.833	14.04	2	0.883

Designation	Å*	p.e.	keV	Å*	p.e.	keV
	57 Lanthanum (*Cont.*)			**58 Cerium** (*Cont.*)		
$M_VO_{II,III}$				14.39	5	0.862
$N_{IV,V}O_{II,III}$	152.6	6	0.0812	144.4	6	0.0859
	59 Praseodymium			**60 Neodymium**		
$\alpha_2\ KL_{II}$	0.348749	2	35.5502	0.336472	2	36.8474
$\alpha_1\ KL_{III}$	0.344140	2	36.0263	0.331846	2	37.3610
$\beta_3\ KM_{II}$	0.304975	5	40.6529	0.294027	3	42.1665
$\beta_1\ KM_{III}$	0.304261	4	40.7482	0.293299	2	42.2713
$\beta_2\ KN_{II,III}$	0.29679	2	41.773	0.2861†	1	43.33
$\beta_4\ L_IM_{II}$	2.2550	4	5.4981	2.1669	3	5.7216
$\beta_3\ L_IM_{III}$	2.2172	3	5.5918	2.1268	2	5.8294
$\gamma_2\ L_IN_{II}$	1.8791	4	6.598	1.8013	4	6.883
$\gamma_3\ L_IN_{III}$	1.8740	4	6.616	1.7964	4	6.902
$\gamma_4\ L_IO_{II,III}$	1.8193	4	6.815	1.7445	4	7.107
$\eta\ L_{II}M_I$	2.512	3	4.935	2.4094	4	5.1457
$\beta_1\ L_{II}M_{IV}$	2.2588	3	5.4889	2.1669	2	5.7216
$\gamma_5\ L_{II}N_I$	2.0205	4	6.136	1.9355	4	6.406
$\gamma_1\ L_{II}N_{IV}$	1.9611	3	6.3221	1.8779	2	6.6021
$\gamma_8\ L_{II}O_I$	1.9362	4	6.403	1.8552	5	6.683
$l\ L_{III}M_I$	2.7841	4	4.4532	2.6760	4	4.6330
$\alpha_2\ L_{III}M_{IV}$	2.4729	3	5.0135	2.3807	3	5.2077
$\alpha_1\ L_{III}M_V$	2.4630	2	5.0337	2.3704	2	5.2304
$\beta_6\ L_{III}N_I$	2.1906	4	5.660	2.1039	3	5.8930
$\beta_{2,15}\ L_{III}N_{IV,V}$	2.1194	4	5.850	2.0360	3	6.0894
$\beta_7\ L_{III}O_I$	2.0919	4	5.927	2.0092	3	6.1708
$\beta_{10}\ L_IM_{IV}$	2.1071	4	5.884	2.0237	3	6.1265
$\beta_9\ L_IM_V$	2.1004	4	5.903	2.0165	3	6.1484
$\gamma\ M_{III}N_{IV,V}$	10.998	9	1.1273	10.505	9	1.180
$\beta\ M_{IV}N_{VI}$	13.06	2	0.950	12.44	2	0.997
$\zeta\ M_VN_{III}$	17.38	4	0.714	16.46	4	0.753
$\alpha\ M_VN_{VI,VII}$	13.343	5	0.9292	12.68	2	0.978
$N_{IV,V}N_{VI,VII}$	113.	1	0.1095	107.	1	0.116
$N_{IV,V}O_{II,III}$	136.5	4	0.0908	128.9	7	0.0962
	61 Promethium			**62 Samarium**		
$\alpha_2\ KL_{II}$	0.324803	4	38.1712	0.313698	2	39.5224
$\alpha_1\ KL_{III}$	0.320160	4	38.7247	0.309040	2	40.1181
$\beta_3\ KM_{II}$	0.28363†	4	43.713	0.27376	2	45.289
$\beta_1\ KM_{III}$	0.28290†	3	43.826	0.27301	2	45.413
$\beta_2\ KN_{II,III}$	0.2759†	1	44.94	0.2662	1	46.58
$KO_{II,III}$				0.26491	3	46.801
$\beta_5\ KM_{IV,V}$				0.27111	3	45.731
$\beta_4\ L_IM_{II}$				2.00095	6	6.1963
$\beta_3\ L_IM_{III}$	2.0421	4	6.071	1.96241	3	6.3180
$\gamma_2\ L_IN_{II}$				1.66044	6	7.4668
$\gamma_3\ L_IN_{III}$				1.65601	3	7.4867
$\gamma_4\ L_IO_{II,III}$				1.60728	3	7.7137
$\eta\ L_{II}M_I$				2.21824	3	5.5892
$\beta_1\ L_{II}M_{IV}$	2.0797	4	5.961	1.99806	3	6.2051
$\gamma_5\ L_{II}N_I$				1.77934	3	6.9678
$\gamma_1\ L_{II}N_{IV}$	1.7989	9	6.892	1.72724	3	7.1780
$\gamma_6\ L_{II}O_{IV}$				1.6966	9	7.3076
$l\ L_{III}M_I$				2.4823	4	4.9945

Designation	Å*	p.e.	keV	Å*	p.e.	keV
	61 Promethium (*Cont.*)			**62 Samarium** (*Cont.*)		
$\alpha_2\ L_{III}M_{IV}$	2.2926	4	5.4078	2.21062	3	5.6084
$\alpha_1\ L_{III}M_V$	2.2822	3	5.4325	2.1998	2	5.6361
$\beta_6\ L_{III}N_I$				1.94643	3	6.3697
$\beta_{2,15}\ L_{III}N_{IV,V}$	1.9559	6	6.339	1.88221	3	6.5870
$\beta_7\ L_{III}O_I$				1.85626	3	6.6791
$\beta_5\ L_{III}O_{IV,V}$				1.84700	9	6.7126
$\beta_{10}\ L_IM_{IV}$				1.86990	3	6.6304
$\beta_9\ L_IM_V$				1.86166	3	6.6597
$\gamma\ M_{III}N_{IV,V}$				9.600	9	1.291
$\beta\ M_{IV}N_{VI}$				11.27	1	1.0998
$\zeta\ M_VN_{III}$				14.91	4	0.831
$\alpha\ M_VN_{VI,VII}$				11.47	3	1.081
$N_{IV,V}N_{VI,VII}$				98.	1	0.126
$N_{IV,V}O_{II,III}$				117.4	4	0.1056
	63 Europium			**64 Gadolinium**		
$\alpha_2\ KL_{II}$	0.303118	2	40.9019	0.293038	2	42.3089
$\alpha_1\ KL_{III}$	0.298446	2	41.5422	0.288353	2	42.9962
$\beta_3\ KM_{II}$	0.264332	5	46.9036	0.25534	2	48.555
$\beta_1\ KM_{III}$	0.263577	5	47.0379	0.25460	2	48.697
$\beta_2\ KN_{II,III}$	0.256923	8	48.256	0.24816	3	49.959
$KO_{II,III}$	0.255645	7	48.497	0.24687	3	50.221
$\beta_5\ KM_{IV,V}$				0.25275	3	49.052
$\beta_4\ L_IM_{II}$	1.9255	2	6.4389	1.8540	2	6.6871
$\beta_3\ L_IM_{III}$	1.8867	2	6.5713	1.8150	2	6.8311
$\gamma_2\ L_IN_{II}$	1.5961	2	7.7677	1.5331	2	8.087
$\gamma_3\ L_IN_{III}$	1.5903	2	7.7961	1.5297	2	8.105
$\gamma_4\ L_IO_{II,III}$	1.5439	1	8.0304	1.4839	2	8.355
$\eta\ L_{II}M_I$	2.1315	2	5.8166	2.0494	1	6.0495
$\beta_1\ L_{II}M_{IV}$	1.9203	2	6.4564	1.8468	2	6.7132
$\gamma_5\ L_{II}N_I$	1.7085	2	7.2566	1.6412	2	7.5543
$\gamma_1\ L_{II}N_{IV}$	1.6574	2	7.4803	1.5924	2	7.7858
$\gamma_8\ L_{II}O_I$	1.6346	2	7.5849	1.5707	2	7.894
$\gamma_6\ L_{II}O_{IV}$	1.6282	2	7.6147	1.5644	2	7.925
$l\ L_{III}M_I$	2.3948	2	5.1772	2.3122	2	5.3621
$\alpha_2\ L_{III}M_{IV}$	2.1315	2	5.8166	2.0578	2	6.0250
$\alpha_1\ L_{III}M_V$	2.1209	2	5.8457	2.0468	2	6.0572
$\beta_6\ L_{III}N_I$	1.8737	2	6.6170	1.8054	2	6.8671
$\beta_{2,15}\ L_{III}N_{IV,V}$	1.8118	2	6.8432	1.7455	2	7.1028
$\beta_7\ L_{III}O_I$	1.7851	2	6.9453	1.7203	2	7.2071
$\beta_5\ L_{III}O_{IV,V}$	1.7772	2	6.9763	1.7130	2	7.2374
$\beta_{10}\ L_IM_{IV}$	1.7993	3	6.890	1.7315	3	7.160
$\beta_9\ L_IM_V$	1.7916	3	6.920	1.7240	3	7.192
$L_IO_{IV,V}$				1.4807	3	8.373
$\gamma\ M_{III}N_{IV,V}$	9.211	9	1.346	8.844	9	1.402
$\beta\ M_{IV}N_{VI}$	10.750	7	1.1533	10.254	6	1.2091
$\zeta\ M_VN_{III}$	14.22	2	0.872	13.57	2	0.914
$\alpha\ M_VN_{VI,VII}$	10.96	3	1.131	10.46	3	1.185
$N_{IV,V}O_{II,III}$	112.0	6	0.1107			
	65 Terbium			**66 Dysprosium**		
$\alpha_2\ KL_{II}$	0.283423	2	43.7441	0.274247	2	45.2078
$\alpha_1\ KL_{III}$	0.278724	2	44.4816	0.269533	2	45.9984
$\beta_3\ KM_{II}$	0.24683	2	50.229	0.23862	2	51.957

Designation	Å*	p.e.	keV	Å*	p.e.	keV
	65 Terbium (*Cont.*)			**66 Dysprosium** (*Cont.*)		
$\beta_1\ KM_{III}$	0.24608	2	50.382	0.23788	2	52.119
$\beta_2\ KN_{II,III}$	0.2397†	2	51.72	0.2317†	2	53.51
$KO_{II,III}$	0.23858	3	51.965	0.23056	3	53.774
$\beta_5\ KM_{IV,V}$				0.23618	3	52.494
$\beta_4\ L_IM_{II}$	1.7864	2	6.9403	1.72103	7	7.2039
$\beta_3\ L_IM_{III}$	1.7472	2	7.0959	1.6822	2	7.3702
$\gamma_2\ L_IN_{II}$	1.4764	2	8.398	1.42278	7	8.7140
$\gamma_3\ L_IN_{III}$	1.4718	2	8.423	1.41640	7	8.7532
$\gamma_4\ L_IO_{II,III}$	1.4276	2	8.685	1.37459	7	9.0195
$\eta\ L_{II}M_I$	1.9730	2	6.2839	1.89743	7	6.5342
$\beta_1\ L_{II}M_{IV}$	1.7768	3	6.978	1.71062	7	7.2477
$\gamma_5\ L_{II}N_I$	1.5787	2	7.8535	1.51824	7	8.1661
$\gamma_1\ L_{II}N_{IV}$	1.5303	2	8.102	1.47266	7	8.4188
$\gamma_8\ L_{II}O_I$	1.5097	2	8.212			
$\gamma_6\ L_{II}O_{IV}$	1.5035	2	8.246	1.44579	7	8.5753
$l\ L_{III}M_I$	2.2352	2	5.5467	2.15877	7	5.7431
$\alpha_2\ L_{III}M_{IV}$	1.9875	2	6.2380	1.91991	3	6.4577
$\alpha_1\ L_{III}M_V$	1.9765	2	6.2728	1.90881	3	6.4952
$\beta_6\ L_{III}N_I$	1.7422	2	7.1163	1.68213	7	7.3705
$\beta_{2,15}\ L_{III}N_{IV,V}$	1.6830	2	7.3667	1.62369	7	7.6357
$\beta_7\ L_{III}O_I$	1.6585	2	7.4753	1.60447	7	7.7272
$\beta_5\ L_{III}O_{IV,V}$	1.6510	2	7.5094	1.58837	7	7.8055
$\beta_{10}\ L_IM_{IV}$	1.6673	3	7.436	1.60743	9	7.7130
$\beta_9\ L_IM_V$				1.59973	9	7.7501
$L_IO_{IV,V}$	1.4228	3	8.714			
$\gamma\ M_{III}N_{IV,V}$	8.486	9	1.461	8.144	9	1.522
$\beta\ M_{IV}N_{VI}$	9.792	6	1.2661	9.357	6	1.3250
$\zeta\ M_VN_{III}$	12.98	2	0.955	12.43	2	0.998
$\alpha\ M_VN_{VI,VII}$	10.00	2	1.240	9.59	2	1.293
$N_{IV,V}N_{VI,VII}$	86.	1	0.144	83.	1	0.149
$N_{IV,V}O_{II,III}$	102.2	4	0.1213	97.2	8	0.128
	67 Holmium			**68 Erbium**		
$\alpha_2\ KL_{II}$	0.265486	2	46.6997	0.257110	2	48.2211
$\alpha_1\ KL_{III}$	0.260756	2	47.5467	0.252365	2	49.1277
$\beta_3\ KM_{II}$	0.23083	2	53.711	0.22341	2	55.494
$\beta_1\ KM_{III}$	0.23012	2	53.877	0.22266	2	55.681
$\beta_2\ KN_{II,III}$	0.2241†	2	55.32	0.2167†	2	57.21
$KO_{II,III}$	0.22305	3	55.584	0.21581	3	57.450
$\beta_5\ KM_{IV,V}$	0.22855	3	54.246	0.22124	3	56.040
$\beta_4\ L_IM_{II}$	1.6595	2	7.4708	1.6007	1	7.7453
$\beta_3\ L_IM_{III}$	1.6203	2	7.6519	1.5616	1	7.9392
$\gamma_2\ L_IN_{II}$	1.3698	2	9.051	1.3210	2	9.385
$\gamma_3\ L_IN_{III}$	1.3643	2	9.087	1.3146	1	9.4309
$\gamma_4\ L_IO_{II,III}$	1.3225	2	9.374	1.2752	2	9.722
$\eta\ L_{II}M_I$	1.8264	2	6.7883	1.7566	1	7.0579
$\beta_1\ L_{II}M_{IV}$	1.6475	2	7.5253	1.5873	1	7.8109
$\gamma_5\ L_{II}N_I$	1.4618	2	8.481	1.4067	3	8.814
$\gamma_1\ L_{II}N_{IV}$	1.4174	2	8.747	1.3641	2	9.089
$\gamma_8\ L_{II}O_I$	1.3983	2	8.867			
$\gamma_6\ L_{II}O_{IV}$	1.3923	2	8.905	1.3397	3	9.255
$l\ L_{III}M_I$	2.0860	2	5.9434	2.015	1	6.152
$\alpha_2\ L_{III}M_{IV}$	1.8561	2	6.6795	1.7955	2	6.9050
$\alpha_1\ L_{III}M_V$	1.8450	2	6.7198	1.78425	9	6.9487
$\beta_6\ L_{III}N_I$	1.6237	2	7.6359	1.5675	2	7.909
$\beta_{2,15}\ L_{III}N_{IV,V}$	1.5671	2	7.911	1.51399	9	8.1890
$\beta_7\ L_{III}O_I$				1.4941	3	8.298
$\beta_5\ L_{III}O_{IV,V}$	1.5378	2	8.062	1.4848	3	8.350

Designation	Å*	p.e.	keV	Å*	p.e.	keV
	67 Holmium (*Cont.*)			**68 Erbium** (*Cont.*)		
$\beta_{10}\ L_IM_{IV}$	1.5486	3	8.006	1.4941	3	8.298
$L_IO_{IV,V}$	1.3208	3	9.387			
$\beta_9\ L_IM_V$				1.4855	5	8.346
$M_{II}N_{IV}$				7.60	1	1.632
$\gamma\ M_{III}N_{IV,V}$	7.865	9	1.576			
$\gamma\ M_{III\ V}$				7.546	8	1.643
$\beta\ M_{IV}N_{VI}$	8.965	4	1.3830	8.592	3	1.4430
$\zeta\ M_VN_{III}$	11.86	1	1.0450	11.37	1	1.0901
$\alpha\ M_VN_{VI,VII}$	9.20	2	1.348	8.82	1	1.406
$N_{IV}N_{VI}$				72.7	9	0.171
$N_VN_{VI,VII}$				76.3	7	0.163
	69 Thulium			**70 Ytterbium**		
$\alpha_2\ KL_{II}$	0.249095	2	49.7726	0.241424	2	51.3540
$\alpha_1\ KL_{III}$	0.244338	2	50.7416	0.236655	2	52.3889
$\beta_3\ KM_{II}$	0.21636	2	57.304	0.2096†	1	59.14
$\beta_1\ KM_{III}$	0.21556	2	57.517	0.20884	8	59.37
$\beta_2\ KN_{II,III}$	0.2098†	2	59.09	0.2033†	2	60.98
$KO_{II,III}$	0.20891	2	59.346	0.20226	2	61.298
$\beta_5\ KM_{IV,V}$	0.21404	2	57.923	0.20739	2	59.782
$\beta_4\ L_IM_{II}$	1.5448	2	8.026	1.49138	3	8.3132
$\beta_3\ L_IM_{III}$	1.5063	2	8.231	1.45233	5	8.5367
$\gamma_2\ L_IN_{II}$	1.2742	2	9.730	1.22879	7	10.0897
$\gamma_3\ L_IN_{III}$	1.2678	2	9.779	1.22232	5	10.1431
$\gamma_4\ L_IO_{II,III}$	1.2294	2	10.084	1.1853	1	10.4603
$\eta\ L_{II}M_I$	1.6963	2	7.3088	1.63560	5	7.5802
$\beta_1\ L_{II}M_{IV}$	1.5304	2	8.101	1.47565	5	8.4018
$\gamma_5\ L_{II}N_I$	1.3558	2	9.144	1.3063	1	9.4910
$\gamma_1\ L_{II}N_{IV}$	1.3153	2	9.426	1.26769	5	9.8701
$\gamma_8\ L_{II}O_I$				1.24923	5	9.9246
$\gamma_6\ L_{II}O_{IV}$	1.2905	2	9.607	1.24271	3	9.9766
$l\ L_{III}M_I$	1.9550	2	6.3419	1.89415	5	6.5455
$\alpha_2\ L_{III}M_{IV}$	1.7381	2	7.1331	1.68285	5	7.3673
$\alpha_1\ L_{III}M_V$	1.7268†	2	7.1799	1.67189	4	7.4156
$\beta_6\ L_{III}N_I$	1.5162	2	8.177	1.4661	1	8.4563
$\beta_{2,15}\ L_{III}N_{IV,V}$	1.4640	2	8.468	1.41550	5	8.7588
$\beta_7\ L_{III}O_I$				1.3948	1	8.8889
$\beta_5\ L_{III}O_{IV,V}$	1.4349	2	8.641	1.38696	7	8.9390
$\beta_{10}\ L_IM_{IV}$	1.4410	3	8.604	1.3915	1	8.9100
$\beta_9\ L_IM_V$	1.4336	3	8.648	1.3838	1	8.9597
L_IO_I				1.1886	1	10.4312
$L_IO_{IV,V}$	1.2263	3	10.110	1.1827	1	10.4833
$L_{II}M_{II}$				1.58844	9	7.8052
$L_{II}O_{II,III}$				1.2453	1	9.9561
$t\ L_{III}M_{II}$				1.83091	9	6.7715
$L_{III}O_{II,III}$				1.3898	1	8.9209
$M_{III}N_I$				8.470	9	1.464
$\gamma\ M_{III}N_V$				7.024	8	1.765
$\beta\ M_{IV}N_{VI}$	8.249	7	1.503	7.909	2	1.5675
$\zeta\ M_VN_{III}$				10.48	1	1.183
$\alpha\ M_VN_{VI,VII}$	8.48	1	1.462	8.149	5	1.5214
$N_{IV}N_{VI}$				65.1	7	0.190
$N_VN_{VI,VII}$				69.3	5	0.179
	71 Lutetium			**72 Hafnium**		
$\alpha_2\ KL_{II}$	0.234081	2	52.9650	0.227024	3	54.6114
$\alpha_1\ KL_{III}$	0.229298	2	54.0698	0.222227	3	55.7902
$\beta_3\ KM_{II}$	0.20309†	4	61.05	0.19686†	4	62.98

Designation	Å*	p.e.	keV	Å*	p.e.	keV
	71 Lutetium (*Cont.*)			**72 Hafnium** (*Cont.*)		
β_1 KM_{III}	0.20231†	3	61.283	0.19607†	3	63.234
β_2 $KN_{II,III}$	0.1969†	2	62.97	0.1908†	2	64.98
$KO_{II,III}$	0.19589	2	63.293			
β_5 $KM_{IV,V}$	0.20084	2	61.732			
β_4 L_IM_{II}	1.44056	5	8.6064	1.39220	5	8.9054
β_3 L_IM_{III}	1.40140	5	8.8469	1.35300	5	9.1634
γ_2 L_IN_{II}	1.1853	2	10.460	1.14442	5	10.8335
γ_3 L_IN_{III}	1.17953	4	10.5110	1.13841	5	10.8907
γ'_4 L_IO_{II}				1.10376	5	11.2326
γ_4 $L_IO_{II,III}$	1.1435	1	10.8425	1.10303	5	11.2401
η $L_{II}M_I$	1.5779	1	7.8575	1.52325	5	8.1393
β_1 $L_{II}M_{IV}$	1.42359	3	8.7090	1.37410	5	9.0227
γ_5 $L_{II}N_I$	1.2596	1	9.8428	1.21537	5	10.2011
γ_1 $L_{II}N_{IV}$	1.22228	4	10.1434	1.17900	5	10.5158
γ_8 $L_{II}O_I$	1.2047	1	10.2915	1.16138	5	10.6754
γ_6 $L_{II}O_{IV}$	1.1987	1	10.3431	1.15519	5	10.7325
l $L_{III}M_I$	1.8360	1	6.7528	1.78145	5	6.9596
α_2 $L_{III}M_{IV}$	1.63029	5	7.6049	1.58046	5	7.8446
α_1 $L_{III}M_{IV}$	1.61951	3	7.6555	1.56958	5	7.8990
β_6 $L_{III}N_I$	1.4189	1	8.7376	1.37410	5	9.0227
β_{15} $L_{III}N_{IV}$	1.3715	1	9.0395	1.32783	5	9.3371
β_2 $L_{III}N_V$	1.37012	3	9.0489	1.32639	5	9.3473
β_7 $L_{III}O_I$	1.34949	5	9.1873	1.30564	5	9.4958
β_5 $L_{III}O_{IV,V}$	1.34183	7	9.2397	1.29761	5	9.5546
L_IM_I				1.43025	9	8.6685
β_{10} L_IM_{IV}	1.3430	2	9.232	1.29819	9	9.5503
β_9 L_IM_V	1.3358	1	9.2816	1.29025	9	9.6090
L_IN_{IV}	1.16227	9	10.6672	1.12250	9	11.0451
γ_{11} L_IN_V	1.16107	9	10.6782	1.12146	9	11.0553
L_IO_I				1.10664	9	11.2034
L_IO_{IV}				1.10086	9	11.2622
$L_{II}M_{II}$	1.53333	9	8.0858	1.48064	9	8.3735
β_{17} $L_{II}M_{III}$				1.43643	9	8.6312
$L_{II}N_V$				1.17788	9	10.5258
v $L_{II}N_{VI}$				1.15830	9	10.7037
$L_{II}O_{II,III}$	1.2014	1	10.3198			
t $L_{III}M_{II}$	1.7760	1	6.9810	1.72305	9	7.1954
s $L_{III}M_{III}$				1.66346	9	7.4532
$L_{III}N_{II}$				1.35887	9	9.1239
$L_{III}N_{III}$				1.35053	9	9.1802
u $L_{III}N_{VI,VII}$				1.30165	9	9.5249
$L_{III}O_{II,III}$	1.34524	9	9.2163			
$M_{III}N_I$				7.887	9	1.572
γ $M_{III}N_V$	6.768	6	1.832	6.544	4	1.895
ζ_2				9.686	7	1.2800
β $M_{IV}N_{VI}$	7.601	2	1.6312	7.303	1	1.6976
ζ_1				9.686	7	1.2800
α $M_VN_{VI,VII}$	7.840	2	1.5813	7.539	1	1.6446
$N_{IV}N_{VI}$	63.0	5	0.197			
$N_VN_{VI,VII}$	65.7	2	0.1886			
	73 Tantalum			**74 Tungsten**		
α_2 KL_{II}	0.220305	8	56.277	0.213828	2	57.9817
α_1 KL_{III}	0.215497	4	57.532	0.2090100	Std	59.31824
β_3 KM_{II}	0.190890	2	64.9488	0.185181	2	66.9514
β_1 KM_{III}	0.190089	4	65.223	0.184374	2	67.2443
β_2^{II} KN_{II}	0.185188	9	66.949	0.17960	1	69.031
β_2^{I} KN_{III}	0.185011	8	67.013	0.179421	7	69.101

Designation	Å*	p.e.	keV	Å*	p.e.	keV
	73 Tantalum (*Cont.*)			**74 Tungsten** (*Cont.*)		
$KO_{II,III}$	0.184031	7	67.370	0.178444	5	69.479
KL_I				0.21592	4	57.42
β_5^{II} KM_{IV}	0.188920	6	65.626	0.183264	5	67.652
β_5^{I} KM_V	0.188757	6	65.683	0.183092	7	67.715
β_4 $KN_{IV,V}$	0.18451	1	67.194	0.17892	2	69.294
β_4 L_IM_{II}	1.34581	3	9.2124	1.30162	5	9.5252
β_3 L_IM_{III}	1.30678	3	9.4875	1.26269	5	9.8188
γ_2 L_IN_{II}	1.1053	1	11.217	1.06806	3	11.6080
γ_3 L_IN_{III}	1.09936	4	11.2776	1.06200	6	11.6743
γ'_4 L_IO_{II}	1.06544	3	11.6366	1.02863	3	12.0530
γ_4 L_IO_{III}	1.06467	3	11.6451	1.02775	3	12.0634
η $L_{II}M_I$	1.47106	5	8.4280	1.42110	3	8.7243
β_1 $L_{II}M_{IV}$	1.32698	3	9.3431	1.281809	9	9.67235
γ_5 $L_{II}N_I$	1.1729	1	10.5702	1.13235	3	10.9490
γ_1 $L_{II}N_{IV}$	1.13794	3	10.8952	1.09855	3	11.2859
γ_8 $L_{II}O_I$	1.1205	1	11.0646	1.08113	4	11.4677
γ_6 $L_{II}O_{IV}$	1.11388	3	11.1306	1.07448	5	11.5387
l $L_{III}M_I$	1.72841	5	7.1731	1.6782	1	7.3878
α_2 $L_{III}M_{IV}$	1.53293	2	8.0879	1.48743	2	8.3352
α_1 $L_{III}M_V$	1.52197	2	8.1461	1.47639	2	8.3976
β_6 $L_{III}N_I$	1.33094	8	9.3153	1.28989	7	9.6117
β_{15} $L_{III}N_{IV}$	1.28619	5	9.6394	1.24631	3	9.9478
β_2 $L_{III}N_V$	1.28454	2	9.6518	1.24460	3	9.9615
β_7 $L_{III}O_I$	1.26385	5	9.8098	1.22400	4	10.1292
β_5 $L_{III}O_{IV,V}$	1.2555	1	9.8750	1.21545	3	10.2004
L_IM_I				1.3365	3	9.277
β_{10} L_IM_{IV}	1.2537	2	9.889	1.21218	3	10.2279
β_9 L_IM_V	1.2466	2	9.946	1.20479	7	10.2907
L_IN_I	1.11521	9	11.1173			
L_IN_{IV}	1.08377	7	11.4398	1.0468	2	11.844
γ_{11} L_IN_V	1.08205	7	11.4580	1.0458	1	11.856
$L_IN_{VI,VII}$	1.06357	9	11.6570			
L_IO_I	1.06771	9	11.6118	1.0317	3	12.017
$L_IO_{IV,V}$	1.06192	9	11.6752	1.0250	2	12.095
$L_{II}M_{II}$	1.43048	9	8.6671			
β_{17} $L_{II}M_{III}$	1.3864	1	8.9428	1.3387	2	9.261
$L_{II}M_V$	1.31897	9	9.3998	1.2728	2	9.741
$L_{II}N_{II}$	1.1600	2	10.688	1.1218	3	11.052
$L_{II}N_{III}$	1.1553	1	10.7316	1.1149	2	11.120
$L_{II}N_V$	1.13687	9	10.9055			
v $L_{II}N_{VI}$	1.1158	1	11.1113	1.0771	1	11.510
$L_{II}O_{II}$	1.11789	9	11.0907			
$L_{II}O_{III}$	1.11693	9	11.1001	1.0792	2	11.488
t $L_{III}M_{II}$	1.67265	9	7.4123	1.6244	3	7.632
s $L_{III}M_{III}$	1.61264	9	7.6881	1.5642	3	7.926
$L_{III}N_{II}$	1.3167	1	9.4158	1.2765	2	9.712
$L_{III}N_{III}$	1.3086	1	9.4742	1.2672	2	9.784
u $L_{III}N_{VI,VII}$	1.25778	4	9.8572	1.21868	5	10.1733
$L_{III}O_{II,III}$	1.2601	3	9.839	1.2211	2	10.153
M_IN_{III}	5.40	2	2.295	5.172	9	2.397
$M_IO_{II,III}$				4.44	2	2.79
$M_{II}N_I$				6.28	2	1.973
$M_{II}N_{IV}$	5.570	4	2.226	5.357	4	2.314
$M_{III}N_I$	7.612	9	1.629	7.360	8	1.684
$M_{III}N_{IV}$	6.353	5	1.951	6.134	4	2.021
γ $M_{III}N_V$	6.312	4	1.964	6.092	3	2.035
$M_{III}O_I$	5.83	2	2.126	5.628	8	2.203
$M_{III}O_{IV,V}$	5.67	3	2.19			

Designation	Å*	p.e.	keV	Å*	p.e.	keV
	73 Tantalum (*Cont.*)			**74 Tungsten** (*Cont.*)		
$\zeta_2\ M_{IV}N_{II}$	9.330	5	1.3288	8.993	5	1.3787
$M_{IV}N_{III}$	8.90	2	1.393	8.573	8	1.446
$\beta\ M_{IV}N_{VI}$	7.023	1	1.7655	6.757	1	1.8349
$M_{IV}O_{II}$	7.09	2	1.748	6.806	9	1.822
$\zeta_1\ M_VN_{III}$	9.316	4	1.3308	8.962	4	1.3835
$\alpha\ M_VN_{VI,VII}$	7.252	1	1.7096			
$\alpha_2\ M_VN_{VI}$				6.992	2	1.7731
$\alpha_1\ M_VN_{VII}$				6.983	1	1.7754
M_VO_{III}	7.30	2	1.700	7.005	9	1.770
$N_{II}N_{IV}$				54.0	2	0.2295
$N_{IV}N_{VI}$	58.2	1	0.2130	55.8	1	0.2221
$N_VN_{VI,VII}$	61.1	2	0.2028			
N_VN_{VI}				59.5	3	0.208
N_VN_{VII}				58.4	1	0.2122

Designation	Å*	p.e.	keV	Å*	p.e.	keV
	75 Rhenium			**76 Osmium**		
$\alpha_2\ KL_{II}$	0.207611	1	59.7179	0.201639	2	61.4867
$\alpha_1\ KL_{III}$	0.202781	2	61.1403	0.196794	2	63.0005
$\beta_3\ KM_{II}$	0.179697	3	68.994	0.174431	3	71.077
$\beta_1\ KM_{III}$	0.178880	3	69.310	0.173611	3	71.413
$\beta_2^{II}\ KN_{II}$	0.17425	1	71.151	0.16910	1	73.318
$\beta_2^{I}\ KN_{III}$	0.174054	6	71.232	0.168906	6	73.402
$KO_{II,III}$	0.17308	1	71.633	0.16798	1	73.808
$\beta_5^{II}\ KM_{IV}$	0.17783	1	69.719	0.17262	1	71.824
$\beta_5^{I}\ KM_V$	0.17766	1	69.786	0.17245	1	71.895
$\beta_4\ KN_{IV,V}$	0.17362	2	71.410	0.16842	2	73.615
$\beta_4\ L_IM_{II}$	1.25917	5	9.8463	1.21844	5	10.1754
$\beta_3\ L_IM_{III}$	1.22031	5	10.1598	1.17955	7	10.5108
$\gamma_2\ L_IN_{II}$	1.03233	5	12.0098	0.99805	5	12.4224
$\gamma_3\ L_IN_{III}$	1.02613	7	12.0824	0.99186	5	12.4998
$\gamma'_4\ L_IO_{II}$	0.99334	5	12.4813	0.96033	8	12.910
$\gamma_4\ L_IO_{III}$	0.99249	5	12.4920	0.95938	8	12.923
$\eta\ L_{II}M_I$	1.37342	5	9.0272	1.32785	7	9.3370
$\beta_1\ L_{II}M_{IV}$	1.23858	2	10.0100	1.19727	7	10.3553
$\gamma_5\ L_{II}N_I$	1.09388	5	11.3341	1.05693	5	11.7303
$\gamma_1\ L_{II}N_{IV}$	1.06099	5	11.6854	1.02503	5	12.0953
$\gamma_8\ L_{II}O_I$	1.04398	5	11.8758	1.00788	5	12.3012
$\gamma_6\ L_{II}O_{IV}$	1.03699	9	11.956	1.00107	5	12.3848
$l\ L_{III}M_I$	1.63056	5	7.6036	1.58498	7	7.8222
$\alpha_2\ L_{III}M_{IV}$	1.44396	5	8.5862	1.40234	5	8.8410
$\alpha_1\ L_{III}M_V$	1.43290	4	8.6525	1.39121	5	8.9117
$\beta_6\ L_{III}N_I$	1.25100	5	9.9105	1.21349	5	10.2169
$\beta_{15}\ L_{III}N_{IV}$	1.20819	5	10.2617	1.17167	5	10.5816
$\beta_2\ L_{III}N_V$	1.20660	4	10.2752	1.16979	8	10.5985
$\beta_7\ L_{III}O_I$	1.18610	5	10.4529	1.14933	8	10.7872
$\beta_5\ L_{III}O_{IV,V}$	1.17721	5	10.5318	1.1405	1	10.8711
$\beta_{10}\ L_IM_{IV}$	1.17218	5	10.5770	1.13353	5	10.9376
$\beta_9\ L_IM_V$	1.16487	4	10.6433	1.12637	6	11.0071
L_IN_I	1.0420	1	11.899			
L_IN_{IV}	1.0119	1	12.252	0.9772	3	12.687
$\gamma_{11}\ L_IN_V$	1.0108	1	12.266	0.9765	3	12.696
L_IO_I	0.9965	1	12.442	0.96318	7	12.8721
$L_IO_{IV,V}$	0.9900	1	12.524	0.95603	5	12.9683
$L_{II}M_{II}$	1.3366	1	9.2761	1.2934	2	9.586
$\beta_{17}\ L_{II}M_{III}$	1.2927	1	9.5910	1.2480	2	9.934

Designation	Å*	p.e.	keV	Å*	p.e.	keV
	75 Rhenium (*Cont.*)			**76 Osmium** (*Cont.*)		
$L_{II}M_V$	1.2305	1	10.0753	1.18977	7	10.4205
$L_{II}N_{II}$	1.0839	1	11.438			
$L_{II}N_{III}$	1.0767	1	11.515	1.03973	5	11.9243
$v\ L_{II}N_{VI}$	1.0404	1	11.917	1.0050	2	12.337
$L_{II}O_{III}$	1.0397	1	11.925	1.0047	2	12.340
$t\ L_{III}M_{II}$	1.5789	1	7.8525	1.5347	2	8.079
$s\ L_{III}M_{III}$	1.5178	1	8.1682	1.4735	2	8.414
$L_{III}N_I$				1.20086	7	10.3244
$L_{III}N_{III}$	1.2283	1	10.0933			
$u\ L_{III}N_{VI,VII}$	1.1815	1	10.4931	1.14537	7	10.8245
M_IN_{III}				4.79	2	2.59
$M_{II}N_I$				5.81	2	2.133
$M_{II}N_{IV}$				4.955	4	2.502
$M_{III}N_I$				6.89	2	1.798
$M_{III}N_{IV}$	5.931	5	2.090	5.724	5	2.166
$\gamma\ M_{III}N_V$	5.885	2	2.1067	5.682	4	2.182
$\zeta_2\ M_{IV}N_{II}$	8.664	5	1.4310	8.359	5	1.4831
$M_{IV}N_{III}$	8.239	8	1.505			
$\beta\ M_{IV}N_{VI}$	6.504	1	1.9061	6.267	1	1.9783
$\zeta_1\ M_VN_{III}$	8.629	4	1.4368	8.310	4	1.4919
$\alpha\ M_VN_{VI,VII}$	6.729	1	1.8425	6.490	1	1.9102
$N_{IV}N_{VI}$				51.9	1	0.2388
$N_VN_{VI,VII}$				54.7	2	0.2266

Designation	Å*	p.e.	keV	Å*	p.e.	keV
	77 Iridium			**78 Platinum**		
$\alpha_2\ KL_{II}$	0.195904	2	63.2867	0.190381	4	65.122
$\alpha_1\ KL_{III}$	0.191047	2	64.8956	0.185511	4	66.832
$\beta_3\ KM_{II}$	0.169367	2	73.2027	0.164501	3	75.368
$\beta_1\ KM_{III}$	0.168542	2	73.5608	0.163675	3	75.748
$\beta_2^{II}\ KN_{II}$	0.16415	1	75.529	0.15939	1	77.785
$\beta_2^{I}\ KN_{III}$	0.163956	7	75.619	0.15920	1	77.878
$KO_{II,III}$	0.163019	5	76.053	0.15826	1	78.341
$\beta_5^{II}\ KM_{IV}$	0.16759	2	73.980	0.16271	2	76.199
$\beta_5^{I}\ KM_V$	0.167373	9	74.075	0.16255	3	76.27
$\beta_4\ KN_{IV,V}$	0.16352	2	75.821	0.15881	2	78.069
$\beta_4\ L_IM_{II}$	1.17958	3	10.5106	1.14223	5	10.8543
$\beta_3\ L_IM_{III}$	1.14085	3	10.8674	1.10394	5	11.2308
$\gamma_2\ L_IN_{II}$	0.96545	3	12.8418	0.93427	5	13.2704
$\gamma_3\ L_IN_{III}$	0.95931	5	12.9240	0.92791	5	13.3613
$\gamma'_4\ L_IO_{II}$	0.92831	3	13.3555	0.89747	4	13.8145
$\gamma_4\ L_IO_{III}$	0.92744	3	13.3681	0.89659	4	13.8281
$\eta\ L_{II}M_I$	1.28448	3	9.6522	1.2429	2	9.975
$\beta_1\ L_{II}M_{IV}$	1.15781	3	10.7083	1.11990	2	11.0707
$\gamma_5\ L_{II}N_I$	1.02175	5	12.1342	0.9877	2	12.552
$\gamma_1\ L_{II}N_{IV}$	0.99085	3	12.5126	0.95797	3	12.9420
$\gamma_8\ L_{II}O_I$	0.97409	3	12.7279	0.9411	1	13.173
$\gamma_6\ L_{II}O_{IV}$	0.96708	4	12.8201	0.9342	2	13.271
$l\ L_{III}M_I$	1.54094	3	8.0458	1.4995	2	8.268
$\alpha_2\ L_{III}M_{IV}$	1.36250	5	9.0995	1.32432	2	9.3618
$\alpha_1\ L_{III}M_V$	1.35128	3	9.1751	1.31304	3	9.4423
$\beta_6\ L_{III}N_I$	1.17796	3	10.5251	1.14355	5	10.8418
$\beta_{15}\ L_{III}N_{IV}$	1.13707	3	10.9036			
$\beta_2\ L_{III}N_V$	1.13532	3	10.9203	1.10200	3	11.2505
$\beta_7\ L_{III}O_I$	1.11489	3	11.1205	1.08168	3	11.4619

Designation	Å*	p.e.	keV	Å*	p.e.	keV
	77 Iridium (*Cont.*)			**78 Platinum** (*Cont.*)		
$\beta_5\ L_{III}O_{IV,V}$	1.10585	3	11.2114	1.0724	2	11.561
L_IM_I	1.2102	2	10.245	1.16962	9	10.6001
$\beta_{10}\ L_IM_{IV}$	1.09702	4	11.3016	1.06183	7	11.6762
$\beta_9\ L_IM_V$	1.08975	5	11.3770	1.05446	5	11.7577
L_IN_I	0.9766	2	12.695	0.9455	2	13.113
L_IN_{IV}	0.9459	2	13.108			
$\gamma_{11}\ L_IN_V$	0.9446	2	13.126	0.9143	2	13.560
$L_IO_{IV,V}$	0.9243	3	13.413			
L_IO_I				0.8995	2	13.784
L_IO_{IV}				0.8943	1	13.864
L_IO_V				0.8934	1	13.878
$L_{II}M_{II}$	1.2502	3	9.917	1.213	1	10.225
$\beta_{17}\ L_{II}M_{III}$	1.2069	2	10.273	1.1667	1	10.6265
$L_{II}M_V$	1.1489	2	10.791	1.1129	2	11.140
$L_{II}N_{II}$	1.0120	2	12.251	0.9792	2	12.661
$L_{II}N_{III}$	1.0054	3	12.332	0.97173	4	12.7588
$v\ L_{II}N_{VI}$	0.97161	6	12.7603	0.93931	5	13.1992
$L_{II}O_{III}$	0.96979	5	12.7843			
$t\ L_{III}M_{II}$	1.4930	3	8.304	1.4530	2	8.533
$s\ L_{III}M_{III}$	1.4318	2	8.659	1.3895	2	8.923
$L_{III}N_{II}$	1.16545	5	10.6380	1.1310	2	10.962
$L_{III}N_{III}$	1.1560	3	10.725	1.1226	2	11.044
$u\ L_{III}N_{VI,VII}$	1.11145	4	11.1549	1.07896	5	11.4908
$L_{III}O_{II,III}$	1.10923	6	11.1772	1.0761	3	11.521
M_IN_{III}	4.631†	9	2.677	4.460	9	2.780
$M_{II}N_{IV}$	4.780	4	2.594	4.601	4	2.695
$M_{III}N_I$	6.669	9	1.859	6.455	9	1.921
$M_{III}N_{IV}$	5.540	5	2.238	5.357	5	2.314
$\gamma\ M_{III}N_V$	5.500	4	2.254	5.319	4	2.331
$M_{III}O_I$				4.876	9	2.543
$M_{III}O_{IV,V}$	4.869	9	2.546	4.694	8	2.641
$\zeta_2\ M_{IV}N_{II}$	8.065	5	1.5373	7.790	5	1.592
$M_{IV}N_{III}$	7.645	8	1.622	7.371	8	1.682
$\beta\ M_{IV}N_{VI}$	6.038	1	2.0535	5.828	1	2.1273
$\zeta_1\ M_VN_{III}$	8.021	4	1.5458	7.738	4	1.6022
$\alpha_2\ M_VN_{VI}$	6.275	3	1.9758	6.058	3	2.047
$\alpha_1\ M_VN_{VII}$	6.262	1	1.9799	6.047	1	2.0505
M_VO_{III}				5.987	9	2.071
$N_{IV}N_{VI}$	50.2	1	0.2470	48.1	2	0.258
$N_VN_{VI,VII}$	52.8	1	0.2348	50.9	1	0.2436
	79 Gold			**80 Mercury**		
$\alpha_2\ KL_{II}$	0.185075	2	66.9895	0.179958	3	68.895
$\alpha_1\ KL_{III}$	0.180195	2	68.8037	0.175068	3	70.819
$\beta_3\ KM_{II}$	0.159810	2	77.580	0.155321	3	79.822
$\beta_1\ KM_{III}$	0.158982	3	77.984	0.154487	3	80.253
$\beta_2^{II}\ KN_{II}$	0.15483	2	80.08	0.15040	2	82.43
$\beta_2^{I}\ KN_{III}$	0.154618	9	80.185	0.15020	2	82.54
$KO_{II,III}$	0.153694	7	80.667	0.14931	2	83.04
KL_I	0.18672	4	66.40			
$\beta_5^{II}\ KM_{IV}$	0.158062	7	78.438			
$\beta_5^{I}\ KM_V$	0.157880	5	78.529			
$\beta_5\ KM_{IV,V}$				0.15353	2	80.75
$\beta_4\ KN_{IV,V}$	0.154224	5	80.391	0.14978	2	82.78
$\beta_4\ L_IM_{II}$	1.10651	3	11.2047	1.07222	7	11.5630
$\beta_3\ L_IM_{III}$	1.06785	9	11.6103	1.03358	7	11.9953

Designation	Å*	p.e.	keV	Å*	p.e.	keV
	79 Gold (*Cont.*)			**80 Mercury** (*Cont.*)		
$\gamma_2\ L_IN_{II}$	0.90434	3	13.7095	0.87544	7	14.162
$\gamma_3\ L_IN_{III}$	0.89783	5	13.8090	0.86915	7	14.265
$\gamma'_4\ L_IO_{II}$	0.86816	4	14.2809	0.84013	7	14.757
$\gamma_4\ L_IO_{III}$	0.86703	4	14.2996	0.83894	7	14.778
$\eta\ L_{II}M_I$	1.20273	3	10.3083	1.1640	1	10.6512
$\beta_1\ L_{II}M_{IV}$	1.08353	3	11.4423	1.04868	5	11.8226
$\gamma_5\ L_{II}N_I$	0.95559	3	12.9743	0.92453	7	13.410
$\gamma_1\ L_{II}N_{IV}$	0.92650	3	13.3817	0.89646	5	13.8301
$\gamma_8\ L_{II}O_I$	0.90989	5	13.6260	0.87995	7	14.090
$\gamma_6\ L_{II}O_{IV}$	0.90297	3	13.7304	0.87319	7	14.199
$l\ L_{III}M_I$	1.45964	9	8.4939	1.4216	1	8.7210
$\alpha_2\ L_{III}M_{IV}$	1.28772	3	9.6280	1.25264	7	9.8976
$\alpha_1\ L_{III}M_V$	1.27640	3	9.7133	1.24120	5	9.9888
$\beta_6\ L_{III}N_I$	1.11092	3	11.1602	1.07975	7	11.4824
$\beta_{15}\ L_{III}N_{IV}$	1.07188	5	11.5667	1.04151	7	11.9040
$\beta_2\ L_{III}N_V$	1.07022	3	11.5847	1.03975	7	11.9241
$\beta_7\ L_{III}O_I$	1.04974	8	11.8106	1.01937	7	12.1625
$\beta_5\ L_{III}O_{IV,V}$	1.04044	3	11.9163	1.00987	7	12.2769
L_IM_I	1.13525	5	10.9210	1.0999	2	11.272
$\beta_{10}\ L_IM_{IV}$	1.02789	7	12.0617	0.9962	2	12.446
$\beta_9\ L_IM_V$	1.02063	7	12.1474	0.9871	2	12.560
L_IN_I	0.9131	1	13.578	0.8827	2	14.045
L_IN_{IV}	0.88563	7	13.999			
$\gamma_{11}\ L_IN_V$	0.88433	7	14.020	0.85657	7	14.474
L_IO_I	0.87074	5	14.2385	0.8452	2	14.670
$L_IO_{IV,V}$	0.86400	5	14.3497	0.8350	2	14.847
$L_{II}M_{II}$	1.1708	1	10.5892	1.1387	5	10.888
$\beta_{17}\ L_{II}M_{III}$	1.12798	5	10.9915	1.0916	5	11.358
$L_{II}M_V$	1.0756	2	11.526			
$L_{II}N_{III}$	0.9402	2	13.186	0.90894	7	13.640
$v\ L_{II}N_{VI}$	0.90837	5	13.6487	0.87885	7	14.107
$L_{II}O_{II}$	0.90746	7	13.662	0.8784	1	14.114
$L_{II}O_{III}$	0.90638	7	13.679	0.8758	1	14.156
$t\ L_{III}M_{II}$	1.41366	7	8.7702	1.3746	2	9.019
$s\ L_{III}M_{III}$	1.35131	7	9.1749	1.3112	2	9.455
$L_{III}N_{II}$	1.09968	7	11.2743	1.0649	2	11.642
$L_{III}N_{III}$	1.09026	7	11.3717	1.0585	1	11.713
$u\ L_{III}N_{VI,VII}$	1.04752	5	11.8357			
$u'\ L_{III}N_{VI}$				1.01769	7	12.1826
$u\ L_{III}N_{VII}$				1.01674	7	12.1940
$L_{III}O_{II,III}$	1.0450	2	11.865			
$L_{III}O_{II}$				1.01558	7	12.2079
$L_{III}O_{III}$				1.01404	7	12.2264
$L_{III}P_{II,III}$	1.03876	7	11.9355			
M_IN_{III}	4.300	9	2.883			
$M_{II}N_{IV}$	4.432	4	2.797			
$M_{III}N_I$	6.259	9	1.981	6.09	2	2.036
$M_{III}N_{IV}$	5.186	5	2.391			
$\gamma\ M_{III}N_V$	5.145	4	2.410	4.984†	2	2.4875
$M_{III}O_I$	4.703	9	2.636			
$M_{III}O_{IV,V}$	4.522	6	2.742			
$\zeta_2\ M_{IV}N_{II}$	7.523	5	1.648			
$M_{IV}N_{III}$	7.101	8	1.746	6.87	2	1.805
$\beta\ M_{IV}N_{VI}$	5.624	1	2.2046	5.4318†	9	2.2825
$\zeta_1\ M_VN_{III}$	7.466	4	1.6605			
$\alpha_2\ M_VN_{VI}$	5.854	3	2.118			

Designation	Å*	p.e.	keV	Å*	p.e.	keV
	79 Gold (*Cont.*)			**80 Mercury** (*Cont.*)		
$\alpha_1\ M_{V}N_{VII}$	5.840	1	2.1229	5.6476†	9	2.1953
$M_{V}O_{III}$	5.767	9	2.150			
$N_{IV}N_{VI}$	46.8	2	0.265	45.2†	3	0.274
$N_{V}N_{VI,VII}$	49.4	1	0.2510	47.9†	3	0.259

Designation	Å*	p.e.	keV	Å*	p.e.	keV
	81 Thallium			**82 Lead**		
$\alpha_2\ KL_{II}$	0.175036	2	70.8319	0.170294	2	72.8042
$\alpha_1\ KL_{III}$	0.170136	2	72.8715	0.165376	2	74.9694
$\beta_3\ KM_{II}$	0.150980	6	82.118	0.146810	4	84.450
$\beta_1\ KM_{III}$	0.150142	5	82.576	0.145970	6	84.936
$\beta_2^{II}\ KN_{II}$	0.14614	1	84.836	0.14212	2	87.23
$\beta_2^{I}\ KN_{III}$	0.14595	1	84.946	0.14191	1	87.364
$KO_{II,III}$	0.14509	1	85.451	0.141012	8	87.922
KP				0.1408	1	88.06
$\beta_5\ KM_{IV,V}$	0.14917	1	83.114			
$\beta_5^{II}\ KM_{IV}$				0.14512	2	85.43
$\beta_5^{I}\ KM_{V}$				0.14495	3	85.53
$\beta_4\ KN_{IV,V}$	0.14553	2	85.19	0.14155	3	87.59
$\beta_4\ L_{I}M_{II}$	1.03918	3	11.9306	1.0075	1	12.306
$\beta_3\ L_{I}M_{III}$	1.00062	3	12.3904	0.96911	7	12.7933
$\gamma_2\ L_{I}N_{II}$	0.84773	5	14.6251	0.8210	2	15.101
$\gamma_3\ L_{I}N_{III}$	0.84130	4	14.7368	0.8147	1	15.218
$\gamma'_4\ L_{I}O_{II}$	0.81308	5	15.2482	0.78706	7	15.752
$\gamma_4\ L_{I}O_{III}$	0.81184	5	15.2716	0.7858	1	15.777
$\eta\ L_{II}M_{I}$	1.12769	3	10.9943	1.09241	7	11.3493
$\beta_1\ L_{II}M_{IV}$	1.01513	4	12.2133	0.98291	3	12.6137
$\gamma_5\ L_{II}N_{I}$	0.89500	4	13.8526	0.86655	5	14.3075
$\gamma_1\ L_{II}N_{IV}$	0.86752	3	14.2915	0.83973	3	14.7644
$\gamma_8\ L_{II}O_{I}$	0.8513	2	14.564	0.82365	5	15.0527
$\gamma_6\ L_{II}O_{IV}$	0.8442	2	14.685	0.81683	5	15.1783
$L_{II}P_{I}$				0.81583	5	15.1969
$l\ L_{III}M_{I}$	1.38477	3	8.9532	1.34990	7	9.1845
$\alpha_2\ L_{III}M_{IV}$	1.21875	3	10.1728	1.18648	5	10.4495
$\alpha_1\ L_{III}M_{V}$	1.20739	4	10.2685	1.17501	2	10.5515
$\beta_6\ L_{III}N_{I}$	1.04963	5	11.8118	1.0210	1	12.143
$\beta_{15}\ L_{III}N_{IV}$	1.01201	3	12.2510	0.98389	7	12.6011
$\beta_2\ L_{III}N_{V}$	1.01031	3	12.2715	0.98221	7	12.6226
$\beta_7\ L_{III}O_{I}$	0.99017	5	12.5212	0.9620	1	12.888
$\beta_5\ L_{III}O_{IV,V}$	0.98058	3	12.6436	0.9526	1	13.015
$L_{I}M_{I}$	1.0644	2	11.648	1.0323	2	12.010
$\beta_{10}\ L_{I}M_{IV}$	0.96389	7	12.8626	0.9339	2	13.275
$\beta_9\ L_{I}M_{V}$	0.95675	7	12.9585	0.9268	1	13.377
$L_{I}N_{I}$	0.8549	1	14.503	0.82859	7	14.963
$L_{I}N_{IV}$	0.83001	7	14.937	0.80364	7	15.427
$\gamma_{11}\ L_{I}N_{V}$	0.82879	5	14.9593	0.80233	9	15.453
$L_{I}N_{VI,VII}$				0.7884	1	15.725
$L_{I}O_{I}$	0.8158	1	15.198	0.7897	1	15.699
$L_{I}O_{IV,V}$	0.80861	5	15.3327	0.78257	7	15.843
$L_{II}M_{II}$	1.0997	1	11.274	1.0644	2	11.648
$\beta_{17}\ L_{II}M_{III}$	1.05609	7	11.7397	1.0223	1	12.127
$L_{II}M_{V}$	1.00722	5	12.3093	0.9747	1	12.720
$L_{II}N_{II}$	0.882	2	14.057	0.8585	3	14.442

Designation	Å*	p.e.	keV	Å*	p.e.	keV
	81 Thallium (*Cont.*)			**82 Lead** (*Cont.*)		
$L_{II}N_{III}$	0.87996	5	14.0893	0.85192	7	14.553
$L_{II}N_{V}$				0.8382	2	14.791
$v\ L_{II}N_{VI}$	0.85048	5	14.5777	0.82327	7	15.060
$L_{II}O_{II}$	0.8490	1	14.604			
$L_{II}O_{III}$				0.8200	1	15.120
$t\ L_{III}M_{II}$	1.34154	5	9.2417	1.30767	7	9.4811
$s\ L_{III}M_{III}$	1.27807	5	9.7007	1.24385	7	9.9675
$L_{III}N_{II}$				1.01040	7	12.2705
$L_{III}N_{III}$	1.0286	1	12.053	1.0005	1	12.392
$u\ L_{III}N_{VI,VII}$	0.9888	1	12.538	0.96133	7	12.8968
$L_{III}O_{II}$	0.98738	5	12.5566	0.9586	1	12.934
$L_{III}O_{III}$	0.98538	5	12.5820	0.9578	1	12.945
$L_{III}P_{II,III}$	0.97926	5	12.6607	0.95118	7	13.0344
$M_{I}N_{III}$	4.013	9	3.089	3.872	9	3.202
$M_{II}N_{I}$				4.655	8	2.664
$M_{II}N_{IV}$	4.116	4	3.013	3.968	5	3.124
$M_{III}N_{I}$	5.884	8	2.107	5.704	8	2.174
$M_{III}N_{IV}$	4.865	5	2.548	4.715	3	2.630
$\gamma\ M_{III}N_{V}$	4.823	4	2.571	4.674	1	2.6527
$M_{III}O_{I}$				4.244	9	2.921
$M_{II}O_{IV,V}$	4.216	6	2.941	4.069	6	3.047
$\zeta_2\ M_{IV}N_{II}$	7.032	5	1.763	6.802	5	1.823
$M_{IV}N_{III}$				6.384	7	1.942
$\beta\ M_{IV}N_{VI}$	5.249	1	2.3621	5.076	1	2.4427
$M_{IV}O_{II}$	5.196	9	2.386	5.004	9	2.477
$\zeta_1\ M_{V}N_{III}$	6.974	4	1.778	6.740	3	1.8395
$\alpha_2\ M_{V}N_{VI}$	5.472	2	2.2656	5.299	2	2.3397
$\alpha_1\ M_{V}N_{VII}$	5.460	1	2.2706	5.286	1	2.3455
$M_{V}O_{III}$				5.168	9	2.399
$N_{IV}N_{VI}$				42.3	2	0.293
$N_{V}N_{VI,VII}$	46.5	2	0.267	45.0	1	0.2756
$N_{VI}O_{IV}$	115.3	2	0.1075	102.4	1	0.1211
$N_{VI}O_{V}$	113.0	1	0.10968	100.2	2	0.1237
$N_{VII}O_{V}$	117.7	1	0.10530	104.3	1	0.1189
	83 Bismuth			**84 Polonium**		
$\alpha_2\ KL_{II}$	0.165717	2	74.8148	0.16130†	1	76.862
$\alpha_1\ KL_{III}$	0.160789	2	77.1079	0.15636†	1	79.290
$\beta_3\ KM_{II}$	0.142779	7	86.834	0.13892†	2	89.25
$\beta_1\ KM_{III}$	0.141948	3	87.343	0.13807†	2	89.80
$\beta_2^{II}\ KN_{II}$	0.13817	1	89.733	0.13438†	2	92.26
$\beta_2^{I}\ KN_{III}$	0.13797	1	89.864	0.13418†	2	92.40
$KO_{II,III}$	0.13709	1	90.435			
$\beta_5\ KM_{IV,V}$	0.14111	1	87.860			
$\beta_4\ KN_{IV,V}$	0.13759	2	90.11			
$\beta_4\ L_{I}M_{II}$	0.97690	4	12.6912	0.9475	3	13.086
$\beta_3\ L_{I}M_{III}$	0.93855	3	13.2098	0.9091	3	13.638
$\gamma_2\ L_{I}N_{II}$	0.79565	3	15.5824	0.772	1	16.07
$\gamma_3\ L_{I}N_{III}$	0.78917	5	15.7102			
$\gamma'_4\ L_{I}O_{II}$	0.76198	3	16.2709			
$\gamma_4\ L_{I}O_{III}$	0.76087	3	16.2947			
$\gamma_{13}\ L_{I}P_{II,III}$	0.75690	3	16.3802			
$\eta\ L_{II}M_{I}$	1.05856	3	11.7122			
$\beta_1\ L_{II}M_{IV}$	0.951978	9	13.0235	0.9220	2	13.447
$\gamma_5\ L_{II}N_{I}$	0.83923	5	14.7732			

Designation	Å*	p.e.	keV	Å*	p.e.	keV
	83 Bismuth (*Cont.*)			84 Polonium (*Cont.*)		
$\gamma_1\ L_{II}N_{IV}$	0.81311	2	15.2477	0.78748	9	15.744
$\gamma_8\ L_{II}O_{I}$	0.7973	1	15.551			
$\gamma_6\ L_{II}O_{IV}$	0.79043	3	15.6853	0.7645	2	16.218
$l\ L_{III}M_{I}$	1.31610	7	9.4204	1.2829	5	9.664
$\alpha_2\ L_{III}M_{IV}$	1.15536	1	10.73091	1.12548†	5	11.0158
$\alpha_1\ L_{III}M_{V}$	1.14386	2	10.8388	1.11386	4	11.1308
$\beta_6\ L_{III}N_{I}$	0.99331	3	12.4816	0.9672	2	12.819
$\beta_{15}\ L_{III}N_{IV}$	0.95702	5	12.9549	0.9312	2	13.314
$\beta_2\ L_{III}N_{V}$	0.95518	4	12.9799	0.92937	5	13.3404
$\beta_7\ L_{III}O_{I}$	0.93505	5	13.2593			
$\beta_5\ L_{III}O_{IV,V}$	0.92556	3	13.3953	0.8996	2	13.782
$L_{I}M_{I}$	1.0005	9	12.39			
$\beta_{10}\ L_{I}M_{IV}$	0.90495	4	13.7002			
$\beta_9\ L_{I}M_{V}$	0.89791	3	13.8077			
$L_{I}N_{I}$	0.8022	1	15.456			
$L_{I}N_{IV}$	0.7795	5	15.904			
$\gamma_{11}\ L_{I}N_{V}$	0.77728	5	15.951			
$L_{I}N_{VI,VII}$	0.7641	5	16.23			
$L_{I}O_{IV,V}$	0.75791	5	16.358			
$L_{II}M_{II}$	1.0346	9	11.98			
$\beta_{17}\ L_{II}M_{III}$	0.98913	5	12.5344			
$L_{II}M_{V}$	0.94419	5	13.1310			
$L_{II}N_{II}$	0.8344	9	14.86			
$L_{II}N_{III}$	0.8248	1	15.031			
$v\ L_{II}N_{VI}$	0.79721	9	15.552			
$L_{II}O_{III}$	0.79384	5	15.6178			
$t\ L_{III}M_{II}$	1.2748	1	9.7252			
$s\ L_{III}M_{III}$	1.2105	1	10.2421			
$L_{III}N_{II}$	0.98280	5	12.6151			
$L_{III}N_{III}$	0.97321	5	12.7394			
$u\ L_{III}N_{VI,VII}$	0.93505	5	13.2593			
$L_{III}O_{II}$	0.9323	2	13.298			
$L_{III}O_{III}$	0.9302	2	13.328			
$L_{III}P_{II,III}$	0.92413	4	13.4159			
$M_{I}N_{II}$	3.892	9	3.185			
$M_{I}N_{III}$	3.740	9	3.315			
$M_{II}N_{IV}$	3.834	4	3.234			
$M_{III}N_{I}$	5.537	8	2.239			
$M_{III}N_{IV}$	4.571	5	2.712			
$\gamma\ M_{III}N_{V}$	4.532	2	2.735			
$M_{III}O_{I}$	4.105	9	3.021			
$M_{III}O_{IV,V}$	3.932	6	3.153			
$\zeta_2\ M_{IV}N_{II}$	6.585	5	1.883			
$M_{IV}N_{III}$	6.162	8	2.012			
$\beta\ M_{IV}N_{VI}$	4.909	1	2.5255			
$M_{IV}O_{II}$	4.823	3	2.571			
$M_{IV}P_{II,III}$	4.59	2	2.70			
$\zeta_1\ M_{V}N_{III}$	6.521	4	1.901			
$\alpha_2\ M_{V}N_{VI}$	5.130	2	2.4170			
$\alpha_1\ M_{V}N_{VII}$	5.118	1	2.4226			
$N_{I}P_{II,III}$	13.30	6	0.932			
$N_{VI}O_{IV}$	91.6	1	0.1354			
$N_{VII}O_{V}$	93.2	1	0.1330			

Designation	Å*	p.e.	keV	Å*	p.e.	keV
	85 Astatine			**86 Radon**		
$\alpha_2\ KL_{II}$	0.15705†	2	78.95	0.15294†	3	81.07
$\alpha_1\ KL_{III}$	0.15210†	2	81.52	0.14798†	3	83.78
$\beta_3\ KM_{II}$	0.13517†	4	91.72	0.13155†	5	94.24
$\beta_1\ KM_{III}$	0.13432†	4	92.30	0.13069†	5	94.87
$\beta_2^{II}\ KN_{II}$	0.13072†	4	94.84	0.12719†	5	97.47
$\beta_2^{I}\ KN_{III}$	0.13052†	4	94.99	0.12698†	5	97.64
$\beta_3\ L_{I}M_{III}$	0.88135†	9	14.067	0.85436†	9	14.512
$\beta_1\ L_{II}M_{IV}$	0.89349†	9	13.876	0.86605†	9	14.316
$\gamma_1\ L_{II}N_{IV}$	0.76289†	9	16.251	0.73928†	9	16.770
$\alpha_2\ L_{III}M_{IV}$	1.09671†	5	11.3048	1.06899†	5	11.5979
$\alpha_1\ L_{III}M_{V}$	1.08500†	5	11.4268	1.05723†	5	11.7270
	87 Francium			**88 Radium**		
$\alpha_2\ KL_{II}$	0.14896†	3	83.23	0.14512†	2	85.43
$\alpha_1\ KL_{III}$	0.14399†	3	86.10	0.14014†	2	88.47
$\beta_3\ KM_{II}$	0.12807†	5	96.81	0.12469†	3	99.43
$\beta_1\ KM_{III}$	0.12719†	5	97.47	0.12382†	3	100.13
$\beta_2^{II}\ KN_{II}$	0.12379†	5	100.16	0.12050†	3	102.89
$\beta_2^{I}\ KN_{III}$	0.12358†	5	100.33	0.12029†	3	103.07
$\beta_4\ L_{I}M_{II}$				0.84071	5	14.7472
$\beta_3\ L_{I}M_{III}$	0.82789†	9	14.976	0.80273	5	15.4449
$\gamma_2\ L_{I}N_{II}$				0.68199	5	18.179
$\gamma_3\ L_{I}N_{III}$				0.67538	5	18.357
$\gamma'_4\ L_{I}O_{II}$				0.65131	5	19.036
$\gamma_4\ L_{I}O_{III}$				0.64965	5	19.084
$\gamma_{13}\ L_{I}P_{II,III}$				0.64513	5	19.218
$\eta\ L_{II}M_{I}$				0.90742	5	13.6630
$\beta_1\ L_{II}M_{IV}$	0.83940†	9	14.770	0.81375	5	15.2358
$\gamma_5\ L_{II}N_{I}$				0.71774	5	17.274
$\gamma_1\ L_{II}N_{IV}$	0.71652†	9	17.303	0.69463	5	17.849
γ_8				0.6801	1	18.230
$\gamma_6\ L_{II}O_{IV}$				0.67328	5	18.414
$L_{II}P_{I}$				0.6724	1	18.439
$l\ L_{III}M_{I}$				1.16719	5	10.6222
$\alpha_2\ L_{III}M_{IV}$	1.04230	5	11.8950	1.01656	5	12.1962
$\alpha_1\ L_{III}M_{V}$	1.03049	5	12.0313	1.00473	5	12.3397
$\beta_6\ L_{III}N_{I}$				0.87088	5	14.2362
$\beta_{15}\ L_{III}N_{IV}$				0.83722	5	14.8086
$\beta_2\ L_{III}N_{V}$	0.858	2	14.45	0.83537	5	14.8414
$\beta_7\ L_{III}O_{I}$				0.8162	1	15.190
$\beta_5\ L_{III}O_{IV,V}$				0.80627	5	15.3771
$L_{III}P_{I}$				0.8050	1	15.402
$\beta_{10}\ L_{I}M_{IV}$				0.77546	5	15.988
$\beta_9\ L_{I}M_{V}$				0.76857	5	16.131
$L_{I}N_{I}$				0.6874	1	18.036
$L_{I}N_{IV}$				0.6666	1	18.600
$\gamma_{11}\ L_{I}N_{V}$				0.6654	1	18.633
$L_{I}O_{IV,V}$				0.6468	1	19.167
$\beta_{17}\ L_{II}M_{III}$				0.8438	1	14.692
$L_{II}N_{III}$				0.7043	1	17.604
$L_{II}N_{V}$				0.6932	1	17.884
$L_{II}O_{II}$				0.6780	1	18.286

Designation	Å*	p.e.	keV	Å*	p.e.	keV
	87 Francium (*Cont.*)			**88 Radium** (*Cont.*)		
$L_{II}O_{III}$				0.6764	1	18.330
$L_{II}P_{II,III}$				0.6714	1	18.466
$L_{III}N_{II}$				0.8618	1	14.387
$L_{III}N_{III}$				0.8512	1	14.566
$u\ L_{III}N_{VI,VII}$				0.8186	1	15.146
$L_{III}P_{II,III}$				0.8038	1	15.425
	89 Actinium			**90 Thorium**		
$\alpha_2\ KL_{II}$	0.14141†	2	87.67	0.137829	2	89.953
$\alpha_1\ KL_{III}$	0.136417†	8	90.884	0.132813	2	93.350
$\beta_3\ KM_{II}$	0.12143†	2	102.10	0.118268	3	104.831
$\beta_1\ KM_{III}$	0.12055†	2	102.85	0.117396	9	105.609
$\beta_2^{II}\ KN_{II}$	0.11732†	2	105.67	0.11426	1	108.511
$\beta_2^{I}\ KN_{III}$	0.11711†	2	105.86	0.114040	9	108.717
$KO_{II,III}$				0.11322	1	109.500
$\beta_5\ KM_{IV,V}$				0.116667	9	106.269
$\beta_4\ KN_{IV,V}$				0.11366	2	109.08
$\beta_4\ L_IM_{II}$				0.79257	4	15.6429
$\beta_3\ L_IM_{III}$	0.77822†	9	15.931	0.75479	3	16.4258
$\gamma_2\ L_IN_{II}$				0.64221	4	19.305
$\gamma_3\ L_IN_{III}$				0.63559	4	19.507
$\gamma'_4\ L_IO_{II}$				0.61251	4	20.242
$\gamma_4\ L_IO_{III}$				0.61098	4	20.292
$\gamma_{13}\ L_IP_{II,III}$				0.60705	8	20.424
$\eta\ L_{II}M_I$				0.85446	4	14.5099
$\beta_1\ L_{II}M_{IV}$	0.78903†	9	15.713	0.765210	9	16.2022
$\gamma_5\ L_{II}N_I$				0.67491	4	18.370
$\gamma_1\ L_{II}N_{IV}$	0.67351†	9	18.408	0.65313	3	18.9825
$\gamma_8\ L_{II}O_I$				0.63898	5	19.403
$\gamma_6\ L_{II}O_{IV}$				0.63258	4	19.599
$L_{II}P_I$				0.6316	1	19.629
$L_{II}P_{IV}$				0.62991	9	19.682
$l\ L_{III}M_I$				1.11508	4	11.1186
$\alpha_2\ L_{III}M_{IV}$	0.99178†	5	12.5008	0.96788	2	12.8096
$\alpha_1\ L_{III}M_V$	0.97993†	5	12.6520	0.95600	3	12.9687
$\beta_6\ L_{III}N_I$				0.82790	8	14.975
$\beta_{15}\ L_{III}N_{IV}$				0.79539	5	15.5875
$\beta_2\ L_{III}N_V$				0.79354	3	15.6237
$\beta_7\ L_{III}O_I$				0.77437	4	16.0105
$\beta_5\ L_{III}O_{IV,V}$				0.76468	5	16.213
$L_{III}P_I$				0.76338	5	16.241
$L_{III}P_{IV,V}$				0.76087	9	16.295
$\beta_{10}\ L_IM_{IV}$				0.7301	1	16.981
$\beta_9\ L_IM_V$				0.7234	1	17.139
L_IN_I				0.64755	5	19.146
L_IN_{IV}				0.6276	1	19.755
$\gamma_{11}\ L_IN_V$				0.62636	9	19.794
$L_IN_{VI,VII}$				0.6160	1	20.128
L_IO_I				0.6146	1	20.174
$L_IO_{IV,V}$				0.6083	1	20.383
$L_{II}M_{II}$				0.8338	1	14.869
$\beta_{17}\ L_{II}M_{III}$				0.79257	4	15.6429
$L_{II}M_V$				0.7579	1	16.359
$L_{II}N_{III}$				0.6620	1	18.729
$L_{II}N_V$				0.6521	1	19.014

Designation	Å*	p.e.	keV	Å*	p.e.	keV
	89 Actinium (*Cont.*)			**90 Thorium** (*Cont.*)		
$v\ L_{II}N_{VI}$				0.64064	9	19.353
$L_{II}O_{II}$				0.6369	1	19.466
$L_{II}O_{III}$				0.6356	1	19.506
$L_{II}P_{II,III}$				0.6312	1	19.642
$t\ L_{III}M_I$				1.08009	9	11.4788
$s\ L_{III}M_{II}$				1.0112	1	12.261
$L_{III}N_{II}$				0.8190	2	15.138
$L_{III}N_{III}$				0.8082	1	15.341
$u\ L_{III}N_{VI,VII}$				0.77661	5	15.964
$L_{III}O_{II}$				0.7713	1	16.074
$L_{III}O_{III}$				0.7690	1	16.123
$L_{III}P_{II,III}$				0.7625	2	16.260
M_IN_{III}				2.934	8	4.23
M_IO_{III}				2.442	9	5.08
$M_{II}N_I$				3.537	9	3.505
$M_{II}N_{IV}$				3.011	2	4.117
$M_{II}O_{IV}$				2.618	5	4.735
$M_{III}N_I$				4.568	5	2.714
$M_{III}N_{IV}$				3.718	3	3.335
$\gamma\ M_{II}N_V$				3.679	2	3.370
$M_{III}O_I$				3.283	9	3.78
$M_{III}O_{IV,V}$				3.131	3	3.959
$\zeta_2\ M_VN_{II}$				5.340	5	2.322
$M_{IV}N_{III}$				4.911	5	2.524
$\beta\ M_{IV}N_{VI}$				3.941	1	3.1458
$M_{IV}O_{II}$				3.808	4	3.256
$\zeta_1\ M_VN_{III}$				5.245	5	2.364
$\alpha_2\ M_VN_{VI}$				4.151	2	2.987
$\alpha_1\ M_VN_{VII}$				4.1381	9	2.9961
M_VP_{III}				3.760	9	3.298
$N\ P_{II}$				9.44	7	1.313
$N\ P_{III}$				9.40	7	1.1319
$N_{II}O_{IV}$				11.56	5	1.072
N_IP_I				11.07	7	1.120
$N_{III}O_V$				13.8	1	0.897
$N_{IV}N_{VI}$				33.57	9	0.3693
$N_VN_{VI,VII}$				36.32	9	0.3414
$N_{VI}O_{IV}$				49.5	1	0.2505
$N_{VI}O_V$				48.2	1	0.2572
$N_{VII}O_V$				50.0	1	0.2479
$O_{III}P_{IV,V}$				68.2	3	0.1817
$O_{IV,V}Q_{II,III}$				181.	5	0.068
	91 Protactinium			**92 Uranium**		
$\alpha_2\ KL_{II}$	0.134343†	9	92.287	0.130968	4	94.665
$\alpha_1\ KL_{III}$	0.129325†	3	95.868	0.125947	3	98.439
$\beta_3\ KM_{II}$	0.11523†	2	107.60	0.112296	4	110.406
$\beta_1\ KM_{III}$	0.114345†	8	108.427	0.111394	5	111.300
$\beta_2^{II}\ KN_{II}$	0.11129†	2	111.40	0.10837	1	114.40
$\beta_2^{I}\ KN_{III}$	0.11107†	2	111.62	0.10818	1	114.60
$KO_{II,III}$				0.10744	1	115.39
$\beta_5\ KM_{IV,V}$				0.11069	1	112.01
$\beta_4\ KN_{IV,V}$				0.10780	2	115.01
$\beta_4\ L_IM_I$	0.7699	1	16.104	0.747985	9	16.5753
$\beta_3\ L_IM_{III}$	0.73230	5	16.930	0.71029	2	17.4550

Designation	Å*	p.e.	keV	Å*	p.e.	keV
	91 Protactinium (*Cont.*)			**92 Uranium** (*Cont.*)		
$\gamma_2\ L_IN_{II}$	0.6239	1	19.872	0.605237	9	20.4847
$\gamma_3\ L_IN_{III}$	0.6169	1	20.098	0.598574	9	20.7127
$\gamma'_4\ L_IO_{II}$				0.576700	9	21.4984
$\gamma_4\ L_IO_{II,III}$	0.5937	1	20.882	0.57499	9	21.562
γ_{13}				0.5706	1	21.729
$\eta\ L_{II}M_I$	0.8295	1	14.946	0.80509	2	15.3997
$\beta_1\ L_{II}M_{IV}$	0.74232	5	16.702	0.719984	8	17.2200
$\gamma_5\ L_{II}N_I$	0.6550	1	18.930	0.63557	2	19.5072
$\gamma_1\ L_{II}N_{IV}$	0.63358†	9	19.568	0.614770	9	20.1671
$\gamma_8\ L_{II}O_I$				0.60125	5	20.621
$\gamma_6\ L_{II}O_{IV}$	0.6133	1	20.216	0.594845	9	20.8426
$L_{II}P_{IV}$				0.59203	5	20.942
$l\ L_{III}M_I$	1.0908	1	11.366	1.06712	2	11.6183
$\alpha_2\ L_{III}M_{IV}$	0.94482†	5	13.1222	0.922558	9	13.4388
$\alpha_1\ L_{III}M_V$	0.93284	5	13.2907	0.910639	9	13.6147
$\beta_6\ L_{III}N_I$	0.8079	1	15.347	0.78838	2	15.7260
$\beta_{15}\ L_{III}N_{IV}$				0.756642	9	16.3857
$\beta_2\ L_{III}N_V$	0.7737	1	16.024	0.754681	9	16.4283
$\beta_7\ L_{III}O_I$	0.7546	2	16.431	0.73602	6	16.845
$\beta_5\ L_{III}O_{IV,V}$	0.7452	2	16.636	0.726305	9	17.0701
$L_{III}P_I$				0.72521	5	17.096
$L_{III}P_{IV,V}$				0.72240	5	17.162
$\beta_{10}\ L_IM_{IV}$	0.7088	2	17.492	0.68760	5	18.031
$\beta_9\ L_IM_V$	0.7018	1	17.667	0.681014	8	18.2054
L_IN_{IV}				0.59096	5	20.979
$\gamma_{11}\ L_IN_V$				0.58986	5	21.019
$L_IO_{IV,V}$				0.5725	1	21.657
$\beta_{17}\ L_{II}M_{III}$				0.74503	5	16.641
$L_{II}N_{III}$				0.6228	1	19.907
$v\ L_{II}N_{VI}$				0.6031	1	20.556
$L_{III}O_{III}$				0.59728	5	20.758
$L_{II}P_{II,III}$				0.5930	2	20.906
$t\ L_{III}M_{II}$				1.0347	1	11.982
$s\ L_{III}M_{III}$				0.9636	1	12.866
$L_{III}N_{II}$				0.78017	9	15.892
$L_{III}N_{III}$				0.7691	1	16.120
$u\ L_{III}N_{VI,VII}$				0.738603	9	16.7859
$L_{III}O_{II}$				0.7333	1	16.907
$L_{III}O_{III}$				0.7309	1	16.962
$L_{III}P_{II,III}$				0.72426	5	17.118
M_IN_{II}				2.92	2	4.25
M_IN_{III}				2.753	8	4.50
M_IO_{III}				2.304	7	5.38
M_IP_{III}				2.253	6	5.50
$M_{II}N_I$	3.441	5	3.603	3.329	4	3.724
$M_{II}N_{IV}$	2.910	2	4.260	2.817	2	4.401
$M_{II}O_{IV}$	2.527	4	4.906	2.443	4	5.075
$M_{III}N_I$	4.450	4	2.786	4.330	2	2.863
$M_{III}N_{IV}$	3.614	2	3.430	3.521	2	3.521
$\gamma\ M_{III}N_V$	3.577	1	3.4657	3.479	1	3.563
$M_{III}O_I$	3.245	9	3.82	3.115	7	3.980
$M_{III}O_{IV,V}$	3.038	2	4.081	2.948	2	4.205
$\zeta_2\ M_{IV}N_{II}$	5.193	2	2.3876	5.050	2	2.4548
$M_{IV}N_{III}$				4.625	5	2.681
$\beta\ M_{IV}N_{VI}$	3.827	1	3.2397	3.716	1	3.3367

Designation	Å*	p.e.	keV	Å*	p.e.	keV
	91 Protactinium (*Cont.*)			**92 Uranium** (*Cont.*)		
$M_{IV}O_{II}$	3.691	2	3.359	3.576	1	3.4666
$\zeta_1\ M_VN_{III}$	5.092	2	2.4350	4.946	2	2.507
$\alpha_2\ M_VN_{VI}$	4.035	3	3.072	3.924	1	3.1595
$\alpha_1\ M_VN_{VII}$	4.022	1	3.0823	3.910	1	3.1708
N_IO_{III}				10.09	7	1.229
N_IP_{II}				8.81	7	1.41
N_IP_{III}				8.76	7	1.42
$N_{II}P_I$				10.40	7	1.192
$N_{III}O_V$				12.90	9	0.961
$N_{IV}N_{VI}$				31.8	1	0.390
$N_VN_{VI,VII}$				34.8	1	0.357
$N_{IV}O_{IV}$				43.3	2	0.286
$N_{VI}O_V$				42.1	2	0.295
$N_IP_{IV,V}$				8.60	7	1.44
	93 Neptunium			**94 Plutonium**		
$\beta_4\ L_IM_{II}$	0.72671	2	17.0607	0.70620	2	17.5560
$\beta_3\ L_IM_{III}$	0.68920†	9	17.989	0.66871	2	18.5405
$\gamma_2\ L_IN_{II}$	0.5873	5	21.11	0.57068	2	21.7251
$\gamma_3\ L_IN_{III}$	0.5810	5	21.34	0.564001	9	21.9824
$\gamma'_4\ L_IO_{II}$				0.5432	1	22.823
$\gamma_4\ L_IO_{II,III}$	0.5585	5	22.20	0.5416	1	22.891
$\eta\ L_{II}M_I$	0.7809	2	15.876	0.7591	1	16.333
$\beta_1\ L_{II}M_{IV}$	0.698478	9	17.7502	0.67772	2	18.2937
$\gamma_5\ L_{II}N_I$	0.616	1	20.12	0.5988	1	20.704
$\gamma_1\ L_{II}N_{IV}$	0.596498	9	20.7848	0.578882	9	21.4173
γ_8				0.5658	1	21.914
$\gamma_6\ L_{II}O_{IV}$	0.57699	5	21.488	0.55973	2	22.1502
$l\ L_{III}M_I$	1.0428	6	11.890	1.0226	1	12.124
$\alpha_2\ L_{III}M_{IV}$	0.901045	9	13.7597	0.88028	2	14.0842
$\alpha_1\ L_{III}M_V$	0.889128	9	13.9441	0.86830	2	14.2786
$\beta_6\ L_{III}N_I$	0.769	1	16.13	0.75148	2	16.4983
$\beta_{15}\ L_{III}N_{IV}$				0.7205	1	17.208
$\beta_2\ L_{III}N_V$	0.736230	9	16.8400	0.71851	2	17.2553
$\beta_7\ L_{III}O_I$				0.7003	1	17.705
$\beta_5\ L_{III}O_{IV,V}$	0.70814	2	17.5081	0.69068	2	17.9506
$\beta_{10}\ L_IM_{IV}$				0.6482	1	19.126
$\beta_9\ L_IM_V$				0.6416	1	19.323
$u\ L_{III}N_{VI,VII}$				0.7031	1	17.635
	95 Americium					
$\beta_4\ L_IM_{II}$	0.68639	2	18.0627			
$\beta_3\ L_IM_{III}$	0.64891	2	19.1059			
$\gamma_2\ L_IN_{II}$	0.5544	2	22.361			
$\beta_1\ L_{II}M_{IV}$	0.657655	9	18.8520			
$\gamma_1\ L_{II}N_{IV}$	0.561886	9	22.0652			
$\gamma_6\ L_{II}O_{IV}$	0.54311	2	22.8282			
$l\ L_{III}M_I$	1.0012	6	12.384			
$\alpha_2\ L_{III}M_{IV}$	0.860266	9	14.4119			
$\alpha_1\ L_{III}M_V$	0.848187	9	14.6172			
$\beta_6\ L_{III}N_I$	0.73418	2	16.8870			
$\beta_{15}\ L_{III}N_{IV}$	0.70341	2	17.6258			
$\beta_2\ L_{III}N_V$	0.701390	9	17.6765			
$\beta_5\ L_{III}O_{IV,V}$	0.67383	2	18.3996			

TABLE B. Wavelengths in numerical order of the emission lines and absorption edges.

Wavelength Å*	p.e.	Element	Designation		keV
0.10723	1	92 U	K	Abs. Edge	115.62
0.10744	1	92 U		$KO_{II,III}$	115.39
0.10780	2	92 U	$K\beta_4$	$KN_{IV,V}$	115.01
0.10818	1	92 U	$K\beta_2^{I}$	KN_{III}	114.60
0.10837	1	92 U	$K\beta_2^{II}$	KN_{II}	114.40
0.11069	1	92 U	$K\beta_5$	$KM_{IV,V}$	112.01
0.11107	2	91 Pa	$K\beta_2^{I}$	KN_{III}	111.62
0.11129	2	91 Pa	$K\beta_2^{II}$	KN_{II}	111.40
0.111394	5	92 U	$K\beta_1$	KM_{III}	111.300
0.112296	4	92 U	$K\beta_3$	KM_{II}	110.406
0.11307	1	90 Th	K	Abs. Edge	109.646
0.11322	1	90 Th		$KO_{II,III}$	109.500
0.11366	2	90 Th	$K\beta_4$	$KN_{IV,V}$	109.08
0.114040	9	90 Th	$K\beta_2^{I}$	KN_{III}	108.717
0.11426	1	90 Th	$K\beta_2^{II}$	KN_{II}	108.511
0.114345	8	91 Pa	$K\beta_1$	KM_{III}	108.427
0.11523	2	91 Pa	$K\beta_3$	KM_{II}	107.60
0.116667	9	90 Th	$K\beta_5$	$KM_{IV,V}$	106.269
0.11711	2	89 Ac	$K\beta_2^{I}$	KN_{III}	105.86
0.11732	2	89 Ac	$K\beta_2^{II}$	KN_{II}	105.67
0.117396	9	90 Th	$K\beta_1$	KM_{III}	105.609
0.118268	3	90 Th	$K\beta_3$	KM_{II}	104.831
0.12029	3	88 Ra	$K\beta_2^{I}$	KN_{III}	103.07
0.12050	3	88 Ra	$K\beta_2^{II}$	KN_{II}	102.89
0.12055	2	89 Ac	$K\beta_1$	KM_{III}	102.85
0.12143	2	89 Ac	$K\beta_3$	KM_{II}	102.10
0.12358	5	87 Fr	$K\beta_2^{I}$	KN_{III}	100.33
0.12379	5	87 Fr	$K\beta_2^{II}$	KN_{II}	100.16
0.12382	3	88 Ra	$K\beta_1$	KM_{III}	100.13
0.12469	3	88 Ra	$K\beta_3$	KM_{II}	99.43
0.125947	3	92 U	$K\alpha_1$	KL_{III}	98.439
0.12698	5	86 Rn	$K\beta_2^{I}$	KN_{III}	97.64
0.12719	5	86 Rn	$K\beta_2^{II}$	KN_{II}	97.47
0.12719	5	87 Fr	$K\beta_1$	KM_{III}	97.47
0.12807	5	87 Fr	$K\beta_3$	KM_{II}	96.81
0.129325	3	91 Pa	$K\alpha_1$	KL_{III}	95.868
0.13052	4	85 At	$K\beta_2^{I}$	KN_{III}	94.99
0.13069	5	86 Rn	$K\beta_1$	KM_{III}	94.87
0.13072	4	85 At	$K\beta_2^{II}$	KN_{II}	94.84
0.130968	4	92 U	$K\alpha_2$	KL_{II}	94.665
0.13155	5	86 Rn	$K\beta_3$	KM_{II}	94.24
0.132813	2	90 Th	$K\alpha_1$	KL_{III}	93.350
0.13418	2	84 Po	$K\beta_2^{I}$	KN_{III}	92.40
0.13432	4	85 At	$K\beta_1$	KM_{III}	92.30
0.134343	9	91 Pa	$K\alpha_2$	KL_{II}	92.287
0.13438	2	84 Po	$K\beta_2^{II}$	KN_{II}	92.26
0.13517	4	85 At	$K\beta_3$	KM_{II}	91.72
0.136417	8	89 Ac	$K\alpha_1$	KL_{III}	90.884
0.13694	1	83 Bi	K	Abs. Edge	90.534
0.13709	1	83 Bi		$KO_{II,III}$	90.435
0.13759	2	83 Bi	$K\beta_4$	$KN_{IV,V}$	90.11
0.137829	2	90 Th	$K\alpha_2$	KL_{II}	89.953
0.13797	1	83 Bi	$K\beta_2^{I}$	KN_{III}	89.864
0.13807	2	84 Po	$K\beta_1$	KM_{III}	89.80
0.13817	1	83 Bi	$K\beta_2^{II}$	KN_{II}	89.733
0.13892	2	84 Po	$K\beta_3$	KM_{II}	89.25
0.14014	2	88 Ra	$K\alpha_1$	KL_{III}	88.47
0.1408	1	82 Pb		KP	88.06
0.140880	5	82 Pb	K	Abs. Edge	88.005
0.141012	8	82 Pb		$KO_{II,III}$	87.922
0.14111	1	83 Bi	$K\beta_5$	$KM_{IV,V}$	87.860
0.14141	2	89 Ac	$K\alpha_2$	KL_{II}	87.67
0.14155	3	82 Pb	$K\beta_4$	$KN_{IV,V}$	87.59
0.14191	1	82 Pb	$K\beta_2^{I}$	KN_{III}	87.364
0.141948	3	83 Bi	$K\beta_1$	KM_{III}	87.343
0.14212	2	82 Pb	$K\beta_2^{II}$	KN_{II}	87.23
0.142779	7	83 Bi	$K\beta_3$	KM_{II}	86.834
0.14399	3	87 Fr	$K\alpha_1$	KL_{III}	86.10
0.14495	1	81 Tl	K	Abs. Edge	85.533
0.14495	3	82 Pb	$K\beta_5^{I}$	KM_{V}	85.53
0.14509	1	81 Tl		$KO_{II,III}$	85.451
0.14512	2	82 Pb	$K\beta_5^{II}$	KM_{IV}	85.43
0.14512	2	88 Ra	$K\alpha_2$	KL_{II}	85.43
0.14553	2	81 Tl	$K\beta_4$	$KN_{IV,V}$	85.19
0.14595	1	81 Tl	$K\beta_2^{I}$	KN_{III}	84.946
0.145970	6	82 Pb	$K\beta_1$	KM_{III}	84.936
0.14614	1	81 Tl	$K\beta_2^{II}$	KN_{II}	84.836
0.146810	4	82 Pb	$K\beta_3$	KM_{II}	84.450
0.14798	3	86 Rn	$K\alpha_1$	KL_{III}	83.78
0.14896	3	87 Fr	$K\alpha_2$	KL_{II}	83.23
0.14917	1	81 Tl	$K\beta_5$	$KM_{IV,V}$	83.114
0.14918	1	80 Hg	K	Abs. Edge	83.109
0.14931	2	80 Hg		$KO_{II,III}$	83.04
0.14978	2	80 Hg	$K\beta_4$	$KN_{IV,V}$	82.78
0.150142	5	81 Tl	$K\beta_1$	KM_{III}	82.576
0.15020	2	80 Hg	$K\beta_2^{I}$	KN_{III}	82.54
0.15040	2	80 Hg	$K\beta_2^{II}$	KN_{II}	82.43
0.150980	6	81 Tl	$K\beta_3$	KM_{II}	82.118
0.15210	2	85 At	$K\alpha_1$	KL_{III}	81.52
0.15294	3	86 Rn	$K\alpha_2$	KL_{II}	81.07
0.15353	2	80 Hg	$K\beta_5$	$KM_{IV,V}$	80.75
0.153593	5	79 Au	K	Abs. Edge	80.720
0.153694	7	79 Au		$KO_{II,III}$	80.667
0.154224	5	79 Au	$K\beta_4$	$KN_{IV,V}$	80.391
0.154487	3	80 Hg	$K\beta_1$	KM_{III}	80.253
0.154618	9	79 Au	$K\beta_2^{I}$	KN_{III}	80.185
0.15483	2	79 Au	$K\beta_2^{II}$	KN_{II}	80.08
0.155321	3	80 Hg	$K\beta_3$	KM_{II}	79.822
0.15636	1	84 Po	$K\alpha_1$	KL_{III}	79.290
0.15705	2	85 At	$K\alpha_2$	KL_{II}	78.95
0.157880	5	79 Au	$K\beta_5^{I}$	KM_{V}	78.529
0.158062	7	79 Au	$K\beta_5^{II}$	KM_{IV}	78.438
0.15818	1	78 Pt	K	Abs. Edge	78.381
0.15826	1	78 Pt		$KO_{II,III}$	78.341
0.15881	2	78 Pt	$K\beta_4$	$KN_{IV,V}$	78.069
0.158982	3	79 Au	$K\beta_1$	KM_{III}	77.984
0.15920	1	78 Pt	$K\beta_2^{I}$	KN_{III}	77.878
0.15939	1	78 Pt	$K\beta_2^{II}$	KN_{II}	77.785
0.159810	2	79 Au	$K\beta_3$	KM_{II}	77.580
0.160789	2	83 Bi	$K\alpha_1$	KL_{III}	77.1079
0.16130	1	84 Po	$K\alpha_2$	KL_{II}	76.862
0.16255	3	78 Pt	$K\beta_5^{I}$	KM_{V}	76.27
0.16271	2	78 Pt	$K\beta_5^{II}$	KM_{IV}	76.199
0.16292	1	77 Ir	K	Abs. Edge	76.101

TABLE B (*Continued*)

Wavelength Å*	p.e.	Element	Designation		keV
0.163019	5	77 Ir		$KO_{II,III}$	76.053
0.16352	2	77 Ir	$K\beta_4$	$KN_{IV,V}$	75.821
0.163675	3	78 Pt	$K\beta_1$	KM_{III}	75.748
0.163956	7	77 Ir	$K\beta_2^{I}$	KN_{III}	75.619
0.16415	1	77 Ir	$K\beta_2^{II}$	KN_{II}	75.529
0.164501	3	78 Pt	$K\beta_3$	KM_{II}	75.368
0.165376	2	82 Pb	$K\alpha_1$	KL_{III}	74.9694
0.165717	2	83 Bi	$K\alpha_2$	KL_{II}	74.8148
0.167373	9	77 Ir	$K\beta_5^{I}$	KM_{V}	74.075
0.16759	2	77 Ir	$K\beta_5^{II}$	KM_{IV}	73.980
0.16787	1	76 Os	K	Abs. Edge	73.856
0.16798	1	76 Os		$KO_{II,III}$	73.808
0.16842	2	76 Os	$K\beta_4$	$KN_{IV,V}$	73.615
0.168542	2	77 Ir	$K\beta_1$	KM_{III}	73.5608
0.168906	6	76 Os	$K\beta_2^{I}$	KN_{III}	73.402
0.16910	1	76 Os	$K\beta_2^{II}$	KN_{II}	73.318
0.169367	2	77 Ir	$K\beta_3$	KM_{II}	73.2027
0.170136	2	81 Tl	$K\alpha_1$	KL_{III}	72.8715
0.170294	2	82 Pb	$K\alpha_2$	KL_{II}	72.8042
0.17245	1	76 Os	$K\beta_5^{I}$	KM_{V}	71.895
0.17262	1	76 Os	$K\beta_5^{II}$	KM_{IV}	71.824
0.17302	1	75 Re	K	Abs. Edge	71.658
0.17308	1	75 Re		$KO_{II,III}$	71.633
0.173611	3	76 Os	$K\beta_1$	KM_{III}	71.413
0.17362	2	75 Re	$K\beta_4$	$KN_{IV,V}$	71.410
0.174054	6	75 Re	$K\beta_2^{I}$	KN_{III}	71.232
0.17425	1	75 Re	$K\beta_2^{II}$	KN_{II}	71.151
0.174431	3	76 Os	$K\beta_3$	KM_{II}	71.077
0.175036	2	81 Tl	$K\alpha_2$	KL_{II}	70.8319
0.175068	3	80 Hg	$K\alpha_1$	KL_{III}	70.819
0.17766	1	75 Re	$K\beta_5^{I}$	KM_{V}	69.786
0.17783	1	75 Re	$K\beta_5^{II}$	KM_{IV}	69.719
0.17837	1	74 W	K	Abs. Edge	69.508
0.178444	5	74 W		$KO_{II,III}$	69.479
0.178880	3	75 Re	$K\beta_1$	KM_{III}	69.310
0.17892	2	74 W	$K\beta_4$	$KN_{IV,V}$	69.294
0.179421	7	74 W	$K\beta_2^{I}$	KN_{III}	69.101
0.17960	1	74 W	$K\beta_2^{II}$	KN_{II}	69.031
0.179697	3	75 Re	$K\beta_3$	KM_{II}	68.994
0.179958	3	80 Hg	$K\alpha_2$	KL_{II}	68.895
0.180195	2	79 Au	$K\alpha_1$	KL_{III}	68.8037
0.183092	7	74 W	$K\beta_5^{I}$	KM_{V}	67.715
0.183264	5	74 W	$K\beta_5^{II}$	KM_{IV}	67.652
0.18394	1	73 Ta	K	Abs. Edge	67.403
0.184031	7	73 Ta		$KO_{II,III}$	67.370
0.184374	2	74 W	$K\beta_1$	KM_{III}	67.2443
0.18451	1	73 Ta	$K\beta_4$	$KN_{IV,V}$	67.194
0.185011	8	73 Ta	$K\beta_2^{I}$	KN_{III}	67.013
0.185075	2	79 Au	$K\alpha_2$	KL_{II}	66.9895
0.185181	2	74 W	$K\beta_3$	KM_{II}	66.9514
0.185188	9	73 Ta	$K\beta_2^{II}$	KN_{II}	66.949
0.185511	4	78 Pt	$K\alpha_1$	KL_{III}	66.832
0.18672	4	79 Au		KL_{I}	66.40
0.188757	6	73 Ta	$K\beta_5^{I}$	KM_{V}	65.683
0.188920	6	73 Ta	$K\beta_5^{II}$	KM_{IV}	65.626
0.18982	5	72 Hf	K	Abs. Edge	65.31
0.190089	4	73 Ta	$K\beta_1$	KM_{III}	65.223
0.190381	4	78 Pt	$K\alpha_2$	KL_{II}	65.122
0.1908	2	72 Hf	$K\beta_2$	$KN_{II,III}$	64.98
0.190890	2	73 Ta	$K\beta_3$	KM_{II}	64.9488
0.191047	2	77 Ir	$K\alpha_1$	KL_{III}	64.8956
0.19585	5	71 Lu	K	Abs. Edge	63.31
0.19589	2	71 Lu		$KO_{II,III}$	63.293
0.195904	2	77 Ir	$K\alpha_2$	KL_{II}	63.2867
0.19607	3	72 Hf	$K\beta_1$	KM_{III}	63.234
0.196794	2	76 Os	$K\alpha_1$	KL_{III}	63.0005
0.19686	4	72 Hf	$K\beta_3$	KM_{II}	62.98
0.1969	2	71 Lu	$K\beta_2$	$KN_{II,III}$	62.97
0.20084	2	71 Lu	$K\beta_5$	$KM_{IV,V}$	61.732
0.201639	2	76 Os	$K\alpha_2$	KL_{II}	61.4867
0.20224	5	70 Yb	K	Abs. Edge	61.30
0.20226	2	70 Yb		$KO_{II,III}$	61.298
0.20231	3	71 Lu	$K\beta_1$	KM_{III}	61.283
0.202781	2	75 Re	$K\alpha_1$	KL_{III}	61.1403
0.20309	4	71 Lu	$K\beta_3$	KM_{II}	61.05
0.2033	2	70 Yb	$K\beta_2$	$KN_{II,III}$	60.89
0.20739	2	70 Yb	$K\beta_5$	$KM_{IV,V}$	59.782
0.207611	1	75 Re	$K\alpha_2$	KL_{II}	59.7179
0.20880	5	69 Tm	K	Abs. Edge	59.38
0.20884	8	70 Yb	$K\beta_1$	KM_{III}	59.37
0.20891	2	69 Tm		$KO_{II,III}$	59.346
0.2090100	Std.	74 W	$K\alpha_1$	KL_{III}	59.31824
0.2096	1	70 Yb	$K\beta_3$	KM_{II}	59.14
0.2098	2	69 Tm	$K\beta_2$	$KN_{II,III}$	59.09
0.213828	2	74 W	$K\alpha_2$	KL_{II}	57.9817
0.21404	2	69 Tm	$K\beta_5$	$KM_{IV,V}$	57.923
0.215497	4	73 Ta	$K\alpha_1$	KL_{III}	57.532
0.21556	2	69 Tm	$K\beta_1$	KM_{III}	57.517
0.21567	1	68 Er	K	Abs. Edge	57.487
0.21581	3	68 Er		$KO_{II,III}$	57.450
0.21592	4	74 W		KL_{I}	57.42
0.21636	2	69 Tm	$K\beta_3$	KM_{II}	57.304
0.2167	2	68 Er	$K\beta_2$	$KN_{II,III}$	57.21
0.220305	8	73 Ta	$K\alpha_2$	KL_{II}	56.277
0.22124	3	68 Er	$K\beta_5$	$KM_{IV,V}$	56.040
0.222227	3	72 Hf	$K\alpha_1$	KL_{III}	55.7902
0.22266	2	68 Er	$K\beta_1$	KM_{III}	55.681
0.22291	1	67 Ho	K	Abs. Edge	55.619
0.22305	3	67 Ho		$KO_{II,III}$	55.584
0.22341	2	68 Er	$K\beta_3$	KM_{II}	55.494
0.2241	2	67 Ho	$K\beta_2$	$KN_{II,III}$	55.32
0.227024	3	72 Hf	$K\alpha_2$	KL_{II}	54.6114
0.22855	3	67 Ho	$K\beta_5$	$KM_{IV,V}$	54.246
0.229298	2	71 Lu	$K\alpha_1$	KL_{III}	54.0698
0.23012	2	67 Ho	$K\beta_1$	KM_{III}	53.877
0.23048	1	66 Dy	K	Abs. Edge	53.793
0.23056	3	66 Dy		$KO_{II,III}$	53.774
0.23083	2	67 Ho	$K\beta_3$	KM_{II}	53.711
0.2317	2	66 Dy	$K\beta_2$	$KN_{II,III}$	53.47
0.234081	2	71 Lu	$K\alpha_2$	KL_{II}	52.9650
0.23618	3	66 Dy	$K\beta_5$	$KM_{IV,V}$	52.494
0.236655	2	70 Yb	$K\alpha_1$	KL_{III}	52.3889
0.23788	2	66 Dy	$K\beta_1$	KM_{III}	52.119
0.23841	1	65 Tb	K	Abs. Edge	52.002

Wavelength Å*	p.e.	Element		Designation	keV
0.23858	3	65 Tb		$KO_{II,III}$	51.965
0.23862	2	66 Dy	$K\beta_3$	KM_{II}	51.957
0.2397	2	65 Tb	$K\beta_2$	$KN_{II,III}$	51.68
0.241424	2	70 Yb	$K\alpha_2$	KL_{II}	51.3540
0.244338	2	69 Tm	$K\alpha_1$	KL_{III}	50.7416
0.24608	2	65 Tb	$K\beta_1$	KM_{III}	50.382
0.24681	1	64 Gd	K	Abs. Edge	50.233
0.24683	2	65 Tb	$K\beta_3$	KM_{II}	50.229
0.24687	3	64 Gd		$KO_{II,III}$	50.221
0.24816	3	64 Gd	$K\beta_2$	$KN_{II,III}$	49.959
0.249095	2	69 Tm	$K\alpha_2$	KL_{II}	49.7726
0.252365	2	68 Er	$K\alpha_1$	KL_{III}	49.1277
0.25275	3	64 Gd	$K\beta_5$	$KM_{IV,V}$	49.052
0.25460	2	64 Gd	$K\beta_1$	KM_{III}	48.697
0.25534	2	64 Gd	$K\beta_3$	KM_{II}	48.555
0.25553	1	63 Eu	K	Abs. Edge	48.519
0.255645	7	63 Eu		$KO_{II,III}$	48.497
0.256923	8	63 Eu	$K\beta_2^{I}$	$KN_{II,III}$	48.256
0.257110	2	68 Er	$K\alpha_2$	KL_{II}	48.2211
0.260756	2	67 Ho	$K\alpha_1$	KL_{III}	47.5467
0.263577	5	63 Eu	$K\beta_1$	KM_{III}	47.0379
0.264332	5	63 Eu	$K\beta_3$	KM_{II}	46.9036
0.26464	5	62 Sm	K	Abs. Edge	46.849
0.26491	3	62 Sm		$KO_{II,III}$	46.801
0.265486	2	67 Ho	$K\alpha_2$	KL_{II}	46.6997
0.2662	1	62 Sm	$K\beta_2$	$KN_{II,III}$	46.57
0.269533	2	66 Dy	$K\alpha_1$	KL_{III}	45.9984
0.27111	3	62 Sm	$K\beta_5$	$KM_{IV,V}$	45.731
0.27301	2	62 Sm	$K\beta_1$	KM_{III}	45.413
0.27376	2	62 Sm	$K\beta_3$	KM_{II}	45.289
0.274247	2	66 Dy	$K\alpha_2$	KL_{II}	45.2078
0.27431	5	61 Pm	K	Abs. Edge	45.198
0.2759	1	61 Pm	$K\beta_2$	$KN_{II,III}$	44.93
0.278724	2	65 Tb	$K\alpha_1$	KL_{III}	44.4816
0.28290	3	61 Pm	$K\beta_1$	KM_{III}	43.826
0.283423	2	65 Tb	$K\alpha_2$	KL_{II}	43.7441
0.28363	4	61 Pm	$K\beta_3$	KM_{II}	43.713
0.28453	5	60 Nd	K	Abs. Edge	43.574
0.2861	1	60 Nd	$K\beta_2$	$KN_{II,III}$	43.32
0.288353	2	64 Gd	$K\alpha_1$	KL_{III}	42.9962
0.293038	2	64 Gd	$K\alpha_2$	KL_{II}	42.3089
0.293299	2	60 Nd	$K\beta_1$	KM_{III}	42.2713
0.294027	3	60 Nd	$K\beta_3$	KM_{II}	42.1665
0.29518	5	59 Pr	K	Abs. Edge	42.002
0.29679	2	59 Pr	$K\beta_2$	$KN_{II,III}$	41.773
0.298446	2	63 Eu	$K\alpha_1$	KL_{III}	41.5422
0.303118	2	63 Eu	$K\alpha_2$	KL_{II}	40.9019
0.304261	4	59 Pr	$K\beta_1$	KM_{III}	40.7482
0.304975	5	59 Pr	$K\beta_3$	KM_{II}	40.6529
0.30648	5	58 Ce	K	Abs. Edge	40.453
0.30668	2	58 Ce		$KO_{II,III}$	40.427
0.30737	2	58 Ce	$K\beta_4^{I}$	$KN_{IV,V}$	40.337
0.30816	1	58 Ce	$K\beta_2$	$KN_{II,III}$	40.233
0.309040	2	62 Sm	$K\alpha_1$	KL_{III}	40.1181
0.31342	2	58 Ce	$K\beta_5^{I}$	KM_{V}	39.558
0.31357	2	58 Ce	$K\beta_5^{II}$	KM_{IV}	39.539
0.313698	2	62 Sm	$K\alpha_2$	KL_{II}	39.5224
0.315816	2	58 Ce	$K\beta_1$	KM_{III}	39.2573
0.316520	4	58 Ce	$K\beta_3$	KM_{II}	39.1701
0.31844	5	57 La	K	Abs. Edge	38.934
0.31864	2	57 La		$KO_{II,III}$	38.909
0.31931	2	57 La	$K\beta_4^{I}$	$KN_{IV,V}$	38.828
0.320117	7	57 La	$K\beta_2$	$KN_{II,III}$	38.7299
0.320160	4	61 Pm	$K\alpha_1$	KL_{III}	38.7247
0.324803	4	61 Pm	$K\alpha_2$	KL_{II}	38.1712
0.32546	2	57 La	$K\beta_5^{I}$	KM_{V}	38.094
0.32563	2	57 La	$K\beta_5^{II}$	KM_{IV}	38.074
0.327983	3	57 La	$K\beta_1$	KM_{III}	37.8010
0.328686	4	57 La	$K\beta_3$	KM_{II}	37.7202
0.33104	1	56 Ba	K	Abs. Edge	37.452
0.33127	2	56 Ba		$KO_{II,III}$	37.426
0.331846	2	60 Nd	$K\alpha_1$	KL_{III}	37.3610
0.33229	2	56 Ba	$K\beta_4^{II}$	KN_{IV}	37.311
0.33277	1	56 Ba	$K\beta_2$	$KN_{II,III}$	37.257
0.336472	2	60 Nd	$K\alpha_2$	KL_{II}	36.8474
0.33814	2	56 Ba	$K\beta_5^{I}$	KM_{V}	36.666
0.33835	2	56 Ba	$K\beta_5^{II}$	KM_{IV}	36.643
0.340811	3	56 Ba	$K\beta_1$	KM_{III}	36.3782
0.341507	4	56 Ba	$K\beta_3$	KM_{II}	36.3040
0.344140	2	59 Pr	$K\alpha_1$	KL_{III}	36.0263
0.34451	1	55 Cs	K	Abs. Edge	35.987
0.34611	2	55 Cs	$K\beta_2$	$KN_{II,III}$	35.822
0.348749	2	59 Pr	$K\alpha_2$	KL_{II}	35.5502
0.354364	7	55 Cs	$K\beta_1$	KM_{III}	34.9869
0.355050	4	55 Cs	$K\beta_3$	KM_{II}	34.9194
0.357092	2	58 Ce	$K\alpha_1$	KL_{III}	34.7197
0.3584	5	54 Xe	K	Abs. Edge	34.59
0.36026	3	54 Xe	$K\beta_2$	$KN_{II,III}$	34.415
0.361683	2	58 Ce	$K\alpha_2$	KL_{II}	34.2789
0.36872	2	54 Xe	$K\beta_1$	KM_{III}	33.624
0.36941	2	54 Xe	$K\beta_3$	KM_{II}	33.562
0.370737	2	57 La	$K\alpha_1$	KL_{III}	33.4418
0.37381	1	53 I	K	Abs. Edge	33.1665
0.37523	2	53 I	$K\beta_2$	$KN_{II,III}$	33.042
0.375313	2	57 La	$K\alpha_2$	KL_{II}	33.0341
0.383905	4	53 I	$K\beta_1$	KM_{III}	32.2947
0.384564	4	53 I	$K\beta_3$	KM_{II}	32.2394
0.385111	4	56 Ba	$K\alpha_1$	KL_{III}	32.1936
0.389668	5	56 Ba	$K\alpha_2$	KL_{II}	31.8171
0.38974	1	52 Te		$KO_{II,III}$	31.8114
0.38974	1	52 Te	K	Abs. Edge	31.8114
0.391102	6	52 Te	$K\beta_2$	$KN_{II,III}$	31.7004
0.399995	5	52 Te	$K\beta_1$	KM_{III}	30.9957
0.400290	4	55 Cs	$K\alpha_1$	KL_{III}	30.9728
0.400659	4	52 Te	$K\beta_3$	KM_{II}	30.9443
0.404835	4	55 Cs	$K\alpha_2$	KL_{II}	30.6251
0.40666	1	51 Sb		$KO_{II,III}$	30.4875
0.40668	1	51 Sb	K	Abs. Edge	30.4860
0.40702	1	51 Sb	$K\beta_4^{I}$	$KN_{IV,V}$	30.4604
0.407973	5	51 Sb	$K\beta_2$	$KN_{II,III}$	30.3895
0.41378	1	51 Sb	$K\beta_5^{I}$	KM_{V}	29.9632
0.41388	1	51 Sb	$K\beta_5^{II}$	KM_{IV}	29.9560
0.41634	2	54 Xe	$K\alpha_1$	KL_{III}	29.779
0.417085	3	51 Sb	$K\beta_1$	KM_{III}	29.7256

Wavelength Å*	p.e.	Element	Designation		keV
0.417737	4	51 Sb	$K\beta_3$	KM_{II}	29.6792
0.42087	2	54 Xe	$K\alpha_2$	KL_{II}	29.458
0.42467	3	50 Sn		$KO_{II,III}$	29.195
0.42467	1	50 Sn	K	Abs. Edge	29.1947
0.42495	3	50 Sn	$K\beta_4^I$	$KN_{IV,V}$	29.175
0.425915	8	50 Sn	$K\beta_2$	$KN_{II,III}$	29.1093
0.43175	3	50 Sn	$K\beta_5^I$	KM_V	28.716
0.43184	3	50 Sn	$K\beta_5^{II}$	KM_{IV}	28.710
0.433318	5	53 I	$K\alpha_1$	KL_{III}	28.6120
0.435236	5	50 Sn	$K\beta_1$	KM_{III}	28.4860
0.435877	5	50 Sn	$K\beta_3$	KM_{II}	28.4440
0.437829	7	53 I	$K\alpha_2$	KL_{II}	28.3172
0.44371	1	49 In	K	Abs. Edge	27.9420
0.44374	3	49 In		$KO_{II,III}$	27.940
0.44393	4	49 In	$K\beta_4^I$	$KN_{IV,V}$	27.928
0.44500	1	49 In	$K\beta_2$	$KN_{II,III}$	27.8608
0.45086	2	49 In	$K\beta_5^I$	KM_V	27.499
0.45098	2	49 In	$K\beta_5^{II}$	KM_{IV}	27.491
0.451295	3	52 Te	$K\alpha_1$	KL_{III}	27.4723
0.454545	4	49 In	$K\beta_1$	KM_{III}	27.2759
0.455181	4	49 In	$K\beta_3$	KM_{II}	27.2377
0.455784	3	52 Te	$K\alpha_2$	KL_{II}	27.2017
0.46407	1	48 Cd	K	Abs. Edge	26.7159
0.465328	7	48 Cd	$K\beta_2$	$KN_{II,III}$	26.6438
0.470354	3	51 Sb	$K\alpha_1$	KL_{III}	26.3591
0.474827	3	51 Sb	$K\alpha_2$	KL_{II}	26.1108
0.475105	6	48 Cd	$K\beta_1$	KM_{III}	26.0955
0.475730	5	48 Cd	$K\beta_3$	KM_{II}	26.0612
0.48589	1	47 Ag	K	Abs. Edge	25.5165
0.4859	9	47 Ag	$K\beta_4$	$KN_{IV,V}$	25.512
0.487032	4	47 Ag	$K\beta_2$	$KN_{II,III}$	25.4564
0.490599	3	50 Sn	$K\alpha_1$	KL_{III}	25.2713
0.49306	2	47 Ag	$K\beta_5$	$KM_{IV,V}$	25.145
0.495053	3	50 Sn	$K\alpha_2$	KL_{II}	25.0440
0.497069	4	47 Ag	$K\beta_1$	KM_{III}	24.9424
0.497685	4	47 Ag	$K\beta_3$	KM_{II}	24.9115
0.5092	1	46 Pd	K	Abs. Edge	24.348
0.5093	2	46 Pd	$K\beta_4$	$KN_{IV,V}$	24.346
0.510228	4	46 Pd	$K\beta_2$	$KN_{II,III}$	24.2991
0.512113	3	49 In	$K\alpha_1$	KL_{III}	24.2097
0.516544	3	49 In	$K\alpha_2$	KL_{II}	24.0020
0.51670	9	46 Pd	$K\beta_5$	$KM_{IV,V}$	23.995
0.520520	4	46 Pd	$K\beta_1$	KM_{III}	23.8187
0.521123	4	46 Pd	$K\beta_3$	KM_{II}	23.7911
0.53395	1	45 Rh	K	Abs. Edge	23.2198
0.53401	9	45 Rh	$K\beta_4^I$	$KN_{IV,V}$	23.217
0.535010	3	48 Cd	$K\alpha_1$	KL_{III}	23.1736
0.53503	2	45 Rh	$K\beta_2$	$KN_{II,III}$	23.1728
0.53513	5	45 Rh	$K\beta_2^{II}$	KN_{II}	23.168
0.5365	1	94 Pu	L_I	Abs. Edge	23.109
0.539422	3	48 Cd	$K\alpha_2$	KL_{II}	22.9841
0.54101	9	45 Rh	$K\beta_5^I$	KM_V	22.917
0.54118	9	45 Rh	$K\beta_5^{II}$	KM_{IV}	22.909
0.5416	1	94 Pu	$L\gamma_4$	L_IO_{III}	22.891
0.54311	2	95 Am	$L\gamma_6$	$L_{II}O_{IV}$	22.8282
0.5432	1	94 Pu	$L\gamma_4'$	L_IO_{II}	22.823
0.545605	4	45 Rh	$K\beta_1$	KM_{III}	22.7236
0.546200	4	45 Rh	$K\beta_3$	KM_{II}	22.6989
0.5544	2	95 Am	$L\gamma_2$	L_IN_{II}	22.361
0.5572	1	94 Pu	L_{II}	Abs. Edge	22.253
0.5585	5	93 Np	$L\gamma_4$	$L_IO_{II,III}$	22.20
0.5594075	6	47 Ag	$K\alpha_1$	KL_{III}	22.16292
0.55973	2	94 Pu	$L\gamma_6$	$L_{II}O_{IV}$	22.1502
0.56051	1	44 Ru	K	Abs. Edge	22.1193
0.56089	9	44 Ru	$K\beta_4$	$KN_{IV,V}$	22.104
0.56166	3	44 Ru	$K\beta_2$	$KN_{II,III}$	22.074
0.561886	9	95 Am	$L\gamma_1$	$L_{II}N_{IV}$	22.0652
0.563798	4	47 Ag	$K\alpha_2$	KL_{II}	21.9903
0.564001	9	94 Pu	$L\gamma_3$	L_IN_{III}	21.9824
0.5658	1	94 Pu	$L\gamma_8$	$L_{II}O_I$	21.914
0.56785	9	44 Ru	$K\beta_5^I$	KM_V	21.834
0.5680	2	44 Ru	$K\beta_5^{II}$	KM_{IV}	21.829
0.5695	1	92 U	L_I	Abs. Edge	21.771
0.5706	1	92 U	$L\gamma_{13}$	$L_IP_{II,III}$	21.729
0.57068	2	94 Pu	$L\gamma_2$	L_IN_{II}	21.1251
0.572482	4	44 Ru	$K\beta_1$	KM_{III}	21.6568
0.5725	1	92 U		$L_IO_{IV,V}$	21.657
0.573067	4	44 Ru	$K\beta_3$	KM_{II}	21.6346
0.57499	9	92 U	$L\gamma_4$	L_IO_{III}	21.562
0.576700	9	92 U	$L\gamma_4'$	L_IO_{II}	21.4984
0.57699	5	93 Np	$L\gamma_6$	$L_{II}O_{IV}$	21.488
0.578882	9	94 Pu	$L\gamma_1$	$L_{II}N_{IV}$	21.4173
0.5810	5	93 Np	$L\gamma_3$	L_IN_{III}	21.34
0.585448	3	46 Pd	$K\alpha_1$	KL_{III}	21.1771
0.5873	5	93 Np	$L\gamma_2$	L_IN_{II}	21.11
0.58906	1	43 Tc	K	Abs. Edge	21.0473
0.589821	3	46 Pd	$K\alpha_2$	KL_{II}	21.0201
0.58986	5	92 U	$L\gamma_{11}$	L_IN_V	21.019
0.59024	5	43 Tc	$K\beta_2$	$KN_{II,III}$	21.005
0.59096	5	92 U		L_IN_{IV}	20.979
0.5919	1	92 U	L_{II}	Abs. Edge	20.945
0.59203	5	92 U		$L_{II}P_{IV}$	20.942
0.5930	2	92 U		$L_{II}P_{II,III}$	20.906
0.5937	1	91 Pa	$L\gamma_4$	$L_IO_{II,III}$	20.882
0.594845	9	92 U	$L\gamma_6$	$L_{II}O_{IV}$	20.8426
0.596498	9	93 Np	$L\gamma_1$	$L_{II}N_{IV}$	20.7848
0.59728	5	92 U		$L_{II}O_{III}$	20.758
0.598574	9	92 U	$L\gamma_3$	L_IN_{III}	20.7127
0.5988	1	94 Pu	$L\gamma_5$	$L_{II}N_I$	20.704
0.60125	5	92 U	$L\gamma_8$	$L_{II}O_I$	20.621
0.60130	4	43 Tc	$K\beta_1$	KM_{III}	20.619
0.60188	4	43 Tc	$K\beta_3$	KM_{II}	20.599
0.6031	1	92 U	Lv	$L_{II}N_{VI}$	20.556
0.605237	9	92 U	$L\gamma_2$	L_IN_{II}	20.4847
0.6059	1	90 Th	L_I	Abs. Edge	20.464
0.60705	8	90 Th	$L\gamma_{13}$	$L_IP_{II,III}$	20.424
0.6083	1	90 Th		$L_IO_{IV,V}$	20.383
0.61098	4	90 Th	$L\gamma_4$	L_IO_{III}	20.292
0.61251	4	90 Th	$L\gamma_4'$	L_IO_{II}	20.242
0.6133	1	91 Pa	$L\gamma_6$	$L_{II}O_{IV}$	20.216
0.613279	4	45 Rh	$K\alpha_1$	KL_{III}	20.2161
0.6146	1	90 Th		L_IO_I	20.174
0.614770	9	92 U	$L\gamma_1$	$L_{II}N_{IV}$	20.1671
0.6160	1	90 Th		$L_IN_{VI,VII}$	20.128

Wavelength Å*	p.e.	Element		Designation	keV
0.616	1	93 Np	$L\gamma_5$	$L_{II}N_{I}$	20.12
0.6169	1	91 Pa	$L\gamma_3$	$L_{I}N_{III}$	20.098
0.617630	4	45 Rh	$K\alpha_2$	KL_{II}	20.0737
0.61978	1	42 Mo	K	Abs. Edge	20.0039
0.62001	9	42 Mo	$K\beta_4^{I}$	$KN_{IV,V}$	19.996
0.62099	2	42 Mo	$K\beta_2$	$KN_{II,III}$	19.9652
0.62107	5	42 Mo	$K\beta_2^{II}$	KN_{II}	19.963
0.6228	1	92 U		$L_{II}N_{III}$	19.907
0.6239	1	91 Pa	$L\gamma_2$	$L_{I}N_{II}$	19.872
0.62636	9	90 Th	$L\gamma_{11}$	$L_{I}N_{V}$	19.794
0.62692	5	42 No	$K\beta_5^{I}$	KM_{V}	19.776
0.62708	5	42 Mo	$K\beta_5^{II}$	KM_{IV}	19.771
0.6276	1	90 Th		$L_{I}N_{IV}$	19.755
0.6299	1	90 Th	L_{II}	Abs. Edge	19.683
0.62991	9	90 Th		$L_{II}P_{IV}$	19.682
0.6312	1	90 Th		$L_{II}P_{II,III}$	19.642
0.6316	1	90 Th		$L_{II}P_{I}$	19.629
0.632288	9	42 Mo	$K\beta_1$	KM_{III}	19.6083
0.63258	4	90 Th	$L\gamma_6$	$L_{II}O_{IV}$	19.599
0.632872	2	42 Mo	$K\beta_3$	KM_{II}	19.5903
0.63358	9	91 Pa	$L\gamma_1$	$L_{II}N_{IV}$	19.568
0.63557	2	92 U	$L\gamma_5$	$L_{II}N_{I}$	19.5072
0.63559	4	90 Th	$L\gamma_3$	$L_{I}N_{III}$	19.507
0.6356	1	90 Th		$L_{II}O_{III}$	19.506
0.6369	1	90 Th		$L_{II}O_{II}$	19.466
0.63898	5	90 Th	$L\gamma_8$	$L_{II}O_{I}$	19.403
0.64064	9	90 Th	Lv	$L_{II}N_{VI}$	19.353
0.6416	1	94 Pu	$L\beta_9$	$L_{I}M_{V}$	19.323
0.64221	4	90 Th	$L\gamma_2$	$L_{I}N_{II}$	19.305
0.643083	4	44 Ru	$K\alpha_1$	KL_{III}	19.2792
0.6445	1	88 Ra	L_{I}	Abs. Edge	19.236
0.64513	5	88 Ra	$L\gamma_{13}$	$L_{I}P_{II,III}$	19.218
0.6468	1	88 Ra		$L_{I}O_{IV,V}$	19.167
0.647408	5	44 Ru	$K\alpha_2$	KL_{II}	19.1504
0.64755	5	90 Th		$L_{I}N_{I}$	19.146
0.6482	1	94 Pu	$L\beta_{10}$	$L_{I}M_{IV}$	19.126
0.64891	2	95 Am	$L\beta_3$	$L_{I}M_{III}$	19.1059
0.64965	5	88 Ra	$L\gamma_4$	$L_{I}O_{III}$	19.084
0.65131	5	88 Ra	$L\gamma_4'$	$L_{I}O_{II}$	19.036
0.6521	1	90 Th		$L_{II}N_{V}$	19.014
0.65298	1	41 Nb	K	Abs. Edge	18.9869
0.65313	3	90 Th	$L\gamma_1$	$L_{II}N_{IV}$	18.9825
0.65318	5	41 Nb	$K\beta_4$	$KN_{IV,V}$	18.981
0.65416	4	41 Nb	$K\beta_2$	$KN_{II,III}$	18.953
0.6550	1	91 Pa	$L\gamma_5$	$L_{II}N_{I}$	18.930
0.657655	9	95 Am	$L\beta_1$	$L_{II}M_{IV}$	18.8520
0.6620	1	90 Th		$L_{II}N_{III}$	18.729
0.6654	1	88 Ra	$L\gamma_{11}$	$L_{I}N_{V}$	18.633
0.66576	2	41 Nb	$K\beta_1$	KM_{III}	18.6225
0.66634	3	41 Nb	$K\beta_3$	KM_{II}	18.6063
0.6666	1	88 Ra		$L_{I}N_{IV}$	18.600
0.66871	2	94 Pu	$L\beta_3$	$L_{I}M_{III}$	18.5405
0.6707	1	88 Ra	L_{II}	Abs. Edge	18.486
0.6714	1	88 Ra		$L_{II}P_{II,III}$	18.466
0.6724	1	88 Ra		$L_{II}P_{I}$	18.439
0.67328	5	88 Ra	$L\gamma_6$	$L_{II}O_{IV}$	18.414
0.67351	9	89 Ac	$L\gamma_1$	$L_{II}N_{IV}$	18.408
0.67383	2	95 Am	$L\beta_5$	$L_{III}O_{IV,V}$	18.3996
0.67491	4	90 Th	$L\gamma_5$	$L_{II}N_{I}$	18.370
0.67502	3	43 Tc	$K\alpha_1$	KL_{III}	18.3671
0.67538	5	88 Ra	$L\gamma_3$	$L_{I}N_{III}$	18.357
0.6764	1	88 Ra		$L_{II}O_{III}$	18.330
0.67772	2	94 Pu	$L\beta_1$	$L_{II}M_{IV}$	18.2937
0.6780	1	88 Ra		$L_{II}O_{II}$	18.286
0.67932	3	43 Tc	$K\alpha_2$	KL_{II}	18.2508
0.6801	1	88 Ra	$L\gamma_8$	$L_{II}O_{I}$	18.230
0.681014	8	92 U	$L\beta_9$	$L_{I}M_{V}$	18.2054
0.68199	5	88 Ra	$L\gamma_2$	$L_{I}N_{II}$	18.179
0.68639	2	95 Am	$L\beta_4$	$L_{I}M_{II}$	18.0627
0.6867	1	94 Pu	L_{III}	Abs. Edge	18.054
0.6874	1	88 Ra		$L_{I}N_{I}$	18.036
0.68760	5	92 U	$L\beta_{10}$	$L_{I}M_{IV}$	18.031
0.68883	1	40 Zr	K	Abs. Edge	17.9989
0.68901	5	40 Zr	$K\beta_4$	$KN_{IV,V}$	17.994
0.68920	9	93 Np	$L\beta_3$	$L_{I}M_{III}$	17.989
0.68993	4	40 Zr	$K\beta_2$	$KN_{II,III}$	17.970
0.69068	2	94 Pu	$L\beta_5$	$L_{III}O_{IV,V}$	17.9506
0.6932	1	88 Ra		$L_{II}N_{V}$	17.884
0.69463	5	88 Ra	$L\gamma_1$	$L_{II}N_{IV}$	17.849
0.6959	1	40 Zr	$K\beta_5$	$KM_{IV,V}$	17.815
0.698478	9	93 Np	$L\beta_1$	$L_{II}M_{IV}$	17.7502
0.7003	1	94 Pu	$L\beta_7$	$L_{III}O_{I}$	17.705
0.701390	9	95 Am	$L\beta_2$	$L_{III}N_{V}$	17.6765
0.70173	3	40 Zr	$K\beta_1$	KM_{III}	17.6678
0.7018	1	91 Pa	$L\beta_9$	$L_{I}M_{V}$	17.667
0.70228	4	40 Zr	$K\beta_3$	KM_{II}	17.654
0.7031	1	94 Pu	Lu	$L_{III}N_{VI,VII}$	17.635
0.70341	2	95 Am	$L\beta_{15}$	$L_{III}N_{IV}$	17.6258
0.7043	1	88 Ra		$L_{II}N_{III}$	17.604
0.70620	2	94 Pu	$L\beta_4$	$L_{I}M_{II}$	17.5560
0.70814	2	93 Np	$L\beta_5$	$L_{III}O_{IV,V}$	17.5081
0.7088	2	91 Pa	$L\beta_{10}$	$L_{I}M_{IV}$	17.492
0.709300	1	42 Mo	$K\alpha_1$	KL_{III}	17.47934
0.71029	2	92 U	$L\beta_3$	$L_{I}M_{III}$	17.4550
0.713590	6	42 Mo	$K\alpha_2$	KL_{II}	17.3743
0.71652	9	87 Fr	$L\gamma_1$	$L_{II}N_{IV}$	17.303
0.71774	5	88 Ra	$L\gamma_5$	$L_{II}N_{I}$	17.274
0.71851	2	94 Pu	$L\beta_2$	$L_{III}N_{V}$	17.2553
0.719984	8	92 U	$L\beta_1$	$L_{II}M_{IV}$	17.2200
0.7205	1	94 Pu	$L\beta_{15}$	$L_{III}N_{IV}$	17.208
0.7223	1	92 U	L_{III}	Abs. Edge	17.165
0.72240	5	92 U		$L_{III}P_{IV,V}$	17.162
0.7234	1	90 Th	$L\beta_9$	$L_{I}M_{V}$	17.139
0.72426	5	92 U		$L_{III}P_{II,III}$	17.118
0.72521	5	92 U		$L_{III}P_{I}$	17.096
0.726305	9	92 U	$L\beta_5$	$L_{III}O_{IV,V}$	17.0701
0.72671	2	93 Np	$L\beta_4$	$L_{I}M_{II}$	17.0607
0.72766	5	39 Y	K	Abs. Edge	17.038
0.72776	5	39 Y	$K\beta_4$	$KN_{IV,V}$	17.036
0.72864	4	39 Y	$K\beta_2$	$KN_{II,III}$	17.0154
0.7301	1	90 Th	$L\beta_{10}$	$L_{I}M_{IV}$	16.981
0.7309	1	92 U		$L_{III}O_{III}$	16.962
0.73230	5	91 Pa	$L\beta_3$	$L_{I}M_{III}$	16.930
0.7333	1	92 U		$L_{III}O_{II}$	16.907

Wavelength Å*	p.e.	Element		Designation	keV
0.73418	2	95 Am	$L\beta_6$	$L_{III}N_I$	16.8870
0.7345	1	39 Y	$K\beta_5$	$KM_{IV,V}$	16.879
0.73602	6	92 U	$L\beta_7$	$L_{III}O_I$	16.845
0.736230	9	93 Np	$L\beta_2$	$L_{III}N_V$	16.8400
0.738603	9	92 U	Lu	$L_{III}N_{VI,VII}$	16.7859
0.73928	9	86 Rn	$L\gamma_1$	$L_{II}N_{IV}$	16.770
0.74072	2	39 Y	$K\beta_1$	KM_{III}	16.7378
0.74126	3	39 Y	$K\beta_3$	KM_{II}	16.7258
0.74232	5	91 Pa	$L\beta_1$	$L_{II}M_{IV}$	16.702
0.74503	5	92 U	$L\beta_{17}$	$L_{II}M_{III}$	16.641
0.7452	2	91 Pa	$L\beta_5$	$L_{III}O_{IV,V}$	16.636
0.74620	1	41 Nb	$K\alpha_1$	KL_{III}	16.6151
0.747985	9	92 U	$L\beta_4$	L_IM_{II}	16.5753
0.75044	1	41 Nb	$K\alpha_2$	KL_{II}	16.5210
0.75148	2	94 Pu	$L\beta_6$	$L_{III}N_I$	16.4983
0.7546	2	91 Pa	$L\beta_7$	$L_{III}O_I$	16.431
0.754681	9	92 U	$L\beta_2$	$L_{III}N_V$	16.4283
0.75479	3	90 Th	$L\beta_3$	L_IM_{III}	16.4258
0.756642	9	92 U	$L\beta_{15}$	$L_{III}N_{IV}$	16.3857
0.75690	3	83 Bi	$L\gamma_{13}$	$L_IP_{II,III}$	16.3802
0.7571	1	83 Bi	L_I	Abs. Edge	16.376
0.7579	1	90 Th		$L_{II}M_V$	16.359
0.75791	5	83 Bi		$L_IO_{IV,V}$	16.358
0.7591	1	94 Pu	$L\eta$	$L_{II}M_I$	16.333
0.7607	1	90 Th	L_{III}	Abs. Edge	16.299
0.76087	9	90 Th		$L_{III}P_{IV,V}$	16.295
0.76087	3	83 Bi	$L\gamma_4$	L_IO_{III}	16.2947
0.76198	3	83 Bi	$L\gamma_4'$	L_IO_{II}	16.2709
0.7625	2	90 Th		$L_{III}P_{II,III}$	16.260
0.76289	9	85 At	$L\gamma_1$	$L_{II}N_{IV}$	16.251
0.76338	5	90 Th		$L_{III}P_I$	16.241
0.7641	5	83 Bi		$L_IN_{VI,VII}$	16.23
0.7645	2	84 Po	$L\gamma_6$	$L_{II}O_{IV}$	16.218
0.76468	5	90 Th	$L\beta_5$	$L_{III}O_{IV,V}$	16.213
0.765210	9	90 Th	$L\beta_1$	$L_{II}M_{IV}$	16.2022
0.76857	5	88 Ra	$L\beta_9$	L_IM_V	16.131
0.769	1	93 Np	$L\beta_6$	$L_{III}N_I$	16.13
0.7690	1	90 Th		$L_{III}O_{III}$	16.123
0.7691	1	92 U		$L_{III}N_{III}$	16.120
0.76973	5	38 Sr	K	Abs. Edge	16.107
0.7699	1	91 Pa	$L\beta_4$	L_IM_{II}	16.104
0.76989	5	38 Sr	$K\beta_4$	$KN_{IV,V}$	16.104
0.77081	3	38 Sr	$K\beta_2$	$KN_{II,III}$	16.0846
0.7713	1	90 Th		$L_{III}O_{II}$	16.074
0.772	1	84 Po	$L\gamma_2$	L_IN_{II}	16.07
0.7737	1	91 Pa	$L\beta_2$	$L_{III}N_V$	16.024
0.77437	4	90 Th	$L\beta_7$	$L_{III}O_I$	16.0105
0.77546	5	88 Ra	$L\beta_{10}$	L_IM_{IV}	15.988
0.7764	1	38 Sr	$K\beta_5$	$KM_{IV,V}$	15.969
0.77661	5	90 Th	Lu	$L_{III}N_{VI,VII}$	15.964
0.77728	5	83 Bi	$L\gamma_{11}$	L_IN_V	15.951
0.77822	9	89 Ac	$L\beta_3$	L_IM_{III}	15.931
0.77954	5	83 Bi		L_IN_{IV}	15.904
0.78017	9	92 U		$L_{III}N_{II}$	15.892
0.7809	2	93 Np	$L\eta$	$L_{II}M_I$	15.876
0.78196	5	82 Pb	L_I	Abs. Edge	15.855
0.78257	7	82 Pb		$L_IO_{IV,V}$	15.843
0.78292	2	38 Sr	$K\beta_1$	KM_{III}	15.8357
0.78345	3	38 Sr	$K\beta_3$	KM_{II}	15.8249
0.7858	1	82 Pb	$L\gamma_4$	L_IO_{III}	15.777
0.78593	1	40 Zr	$K\alpha_1$	KL_{III}	15.7751
0.78706	7	82 Pb	$L\gamma_4'$	L_IO_{II}	15.752
0.78748	9	84 Po	$L\gamma_1$	$L_{II}N_{IV}$	15.744
0.78838	2	92 U	$L\beta_6$	$L_{III}N_I$	15.7260
0.7884	1	82 Pb		$L_IN_{VI,VII}$	15.725
0.7887	1	83 Bi	L_{II}	Abs. Edge	15.719
0.78903	9	89 Ac	$L\beta_1$	$L_{II}M_{IV}$	15.713
0.78917	5	83 Bi	$L\gamma_3$	L_IN_{III}	15.7102
0.7897	1	82 Pb		L_IO_I	15.699
0.79015	1	40 Zr	$K\alpha_2$	KL_{II}	15.6909
0.79043	3	83 Bi	$L\gamma_6$	$L_{II}O_{IV}$	15.6853
0.79257	4	90 Th	$L\beta_4$	L_IM_{II}	15.6429
0.79257	4	90 Th	$L\beta_{17}$	$L_{II}M_{III}$	15.6429
0.79354	3	90 Th	$L\beta_2$	$L_{III}M_V$	15.6237
0.79384	5	83 Bi		$L_{II}O_{III}$	15.6178
0.79539	5	90 Th	$L\beta_{15}$	$L_{III}N_{IV}$	15.5875
0.79565	3	83 Bi	$L\gamma_2$	L_IN_{II}	15.5824
0.79721	9	83 Bi	Lv	$L_{II}N_{VI}$	15.552
0.7973	1	83 Bi	$L\gamma_8$	$L_{II}O_I$	15.551
0.8022	1	83 Bi		L_IN_I	15.456
0.80233	9	82 Pb	$L\gamma_{11}$	L_IN_V	15.453
0.80273	5	88 Ra	$L\beta_3$	L_IM_{III}	15.4449
0.8028	1	88 Ra	L_{III}	Abs. Edge	15.444
0.80364	7	82 Pb		L_IN_{IV}	15.427
0.8038	1	88 Ra		$L_{III}P_{II,III}$	15.425
0.8050	1	88 Ra		$L_{III}P_I$	15.402
0.80509	2	92 U	$L\eta$	$L_{II}M_I$	15.3997
0.80627	5	88 Ra	$L\beta_5$	$L_{III}O_{IV,V}$	15.3771
0.8079	1	91 Pa	$L\beta_6$	$L_{III}N_I$	15.347
0.8081	1	81 Tl	L_I	Abs. Edge	15.343
0.8082	1	90 Th		$L_{III}N_{III}$	15.341
0.80861	5	81 Tl		$L_IO_{IV,V}$	15.3327
0.81163	9	90 Th		L_IM_I	15.276
0.81184	5	81 Tl	$L\gamma_4$	L_IO_{III}	15.2716
0.81308	5	81 Tl	$L\gamma_4'$	L_IO_{II}	15.2482
0.81311	2	83 Bi	$L\gamma_1$	$L_{II}N_{IV}$	15.2477
0.81375	5	88 Ra	$L\beta_1$	$L_{II}M_{IV}$	15.2358
0.8147	1	82 Pb	$L\gamma_3$	L_IN_{III}	15.218
0.81538	5	82 Pb	L_{II}	Abs. Edge	15.2053
0.8154	2	37 Rb	$K\beta_4$	$KN_{IV,V}$	15.205
0.81554	5	37 Rb	K	Abs. Edge	15.2023
0.8158	1	81 Tl		L_IO_I	15.198
0.81583	5	82 Pb		$L_{II}P_I$	15.1969
0.8162	1	88 Ra	$L\beta_7$	$L_{III}O_I$	15.190
0.81645	3	37 Rb	$K\beta_2$	$KN_{II,III}$	15.1854
0.81683	5	82 Pb	$L\gamma_6$	$L_{II}O_{IV}$	15.1783
0.8186	1	88 Ra	Lu	$L_{III}N_{VI,VII}$	15.146
0.8190	2	90 Th		$L_{III}N_{II}$	15.138
0.8200	1	82 Pb		$L_{II}O_{III}$	15.120
0.8210	2	82 Pb	$L\gamma_2$	L_IN_{II}	15.101
0.8219	1	37 Rb	$K\beta_5$	$KM_{IV,V}$	15.085
0.82327	7	82 Pb	Lv	$L_{II}N_{VI}$	15.060
0.82365	5	82 Pb	$L\gamma_8$	$L_{II}O_I$	15.0527
0.8248	1	83 Bi		$L_{II}N_{III}$	15.031

Wavelength Å*	p.e.	Element	Designation		keV
0.82789	9	87 Fr	$L\beta_{3}$	$L_{I}M_{III}$	14.976
0.82790	8	90 Th	$L\beta_{6}$	$L_{III}N_{I}$	14.975
0.82859	7	82 Pb		$L_{I}N_{I}$	14.963
0.82868	2	37 Rb	$K\beta_{1}$	KM_{III}	14.9613
0.82879	5	81 Tl	$L\gamma_{11}$	$L_{I}N_{V}$	14.9593
0.82884	1	39 Y	$K\alpha_{1}$	KL_{III}	14.9584
0.82921	3	37 Rb	$K\beta_{3}$	KM_{II}	14.9517
0.8295	1	91 Pa	$L\eta$	$L_{II}M_{I}$	14.946
0.83001	7	81 Tl		$L_{I}N_{IV}$	14.937
0.83305	1	39 Y	$K\alpha_{2}$	KL_{II}	14.8829
0.8338	1	90 Th		$L_{II}M_{II}$	14.869
0.8344	9	83 Bi		$L_{II}N_{II}$	14.86
0.8350	2	80 Hg		$L_{I}O_{IV,V}$	14.847
0.8353	1	80 Hg	L_{I}	Abs. Edge	14.842
0.83537	5	88 Ra	$L\beta_{2}$	$L_{III}N_{V}$	14.8414
0.83722	5	88 Ra	$L\beta_{15}$	$L_{III}N_{IV}$	14.8086
0.8382	2	82 Pb		$L_{II}N_{V}$	14.791
0.83894	7	80 Hg	$L\gamma_{4}$	$L_{I}O_{III}$	14.778
0.83923	5	83 Bi	$L\gamma_{5}$	$L_{II}N_{I}$	14.7732
0.83940	9	87 Fr	$L\beta_{1}$	$L_{II}M_{IV}$	14.770
0.83973	3	82 Pb	$L\gamma_{1}$	$L_{II}N_{IV}$	14.7644
0.84013	7	80 Hg	$L\gamma_{4}'$	$L_{I}O_{II}$	14.757
0.84071	5	88 Ra	$L\beta_{4}$	$L_{I}M_{II}$	14.7472
0.84130	4	81 Tl	$L\gamma_{3}$	$L_{I}N_{III}$	14.7368
0.8434	1	81 Tl	L_{II}	Abs. Edge	14.699
0.8438	1	88 Ra	$L\beta_{17}$	$L_{II}M_{III}$	14.692
0.8442	2	81 Tl	$L\gamma_{6}$	$L_{II}O_{IV}$	14.685
0.8452	2	80 Hg		$L_{I}O_{I}$	14.670
0.84773	5	81 Tl	$L\gamma_{2}$	$L_{I}N_{II}$	14.6251
0.848187	9	95 Am	$L\alpha_{1}$	$L_{III}M_{V}$	14.6172
0.8490	1	81 Tl		$L_{II}O_{II}$	14.604
0.85048	5	81 Tl	Lv	$L_{II}N_{VI}$	14.5777
0.8512	1	88 Ra		$L_{III}N_{III}$	14.566
0.8513	2	81 Tl	$L\gamma_{8}$	$L_{II}O_{I}$	14.564
0.85192	7	82 Pb		$L_{II}N_{III}$	14.553
0.85436	9	86 Rn	$L\beta_{3}$	$L_{I}M_{III}$	14.512
0.85446	4	90 Th	$L\eta$	$L_{II}M_{I}$	14.5099
0.8549	1	81 Tl		$L_{I}N_{I}$	14.503
0.85657	7	80 Hg	$L\gamma_{11}$	$L_{I}N_{V}$	14.474
0.858	2	87 Fr	$L\beta_{2}$	$L_{III}N_{V}$	14.45
0.8585	3	82 Pb		$L_{II}N_{II}$	14.442
0.860266	9	95 Am	$L\alpha_{2}$	$L_{III}M_{IV}$	14.4119
0.8618	1	88 Ra		$L_{III}N_{II}$	14.387
0.86376	5	79 Au	L_{I}	Abs. Edge	14.3537
0.86400	5	79 Au		$L_{I}O_{IV,V}$	14.3497
0.8653	2	36 Kr	$K\beta_{4}$	$KN_{IV,V}$	14.328
0.86552	1	36 Kr	K	Abs. Edge	14.3244
0.86605	9	86 Rn	$L\beta_{1}$	$L_{II}M_{IV}$	14.316
0.8661	1	36 Kr	$K\beta_{2}$	$KN_{II,III}$	14.315
0.86655	5	82 Pb	$L\gamma_{5}$	$L_{II}N_{I}$	14.3075
0.86703	4	79 Au	$L\gamma_{4}$	$L_{I}O_{III}$	14.2996
0.86752	3	81 Tl	$L\gamma_{1}$	$L_{II}N_{IV}$	14.2915
0.86816	4	79 Au	$L\gamma_{4}'$	$L_{I}O_{II}$	14.2809
0.86830	2	94 Pu	$L\alpha_{1}$	$L_{III}M_{V}$	14.2786
0.86915	7	80 Hg	$L\gamma_{3}$	$L_{I}N_{III}$	14.265
0.87074	5	79 Au		$L_{I}O_{I}$	14.2385
0.8708	2	36 Kr	$K\beta_{5}$	$KM_{IV,V}$	14.238
0.87088	5	88 Ra	$L\beta_{6}$	$L_{III}N_{I}$	14.2362
0.8722	1	80 Hg	L_{II}	Abs. Edge	14.215
0.87319	7	80 Hg	$L\gamma_{6}$	$L_{II}O_{IV}$	14.199
0.87526	1	38 Sr	$K\alpha_{1}$	KL_{III}	14.1650
0.87544	7	80 Hg	$L\gamma_{2}$	$L_{I}N_{II}$	14.162
0.8758	1	80 Hg		$L_{II}O_{III}$	14.156
0.8784	1	80 Hg		$L_{II}O_{II}$	14.114
0.8785	1	36 Kr	$K\beta_{1}$	KM_{III}	14.112
0.87885	7	80 Hg	Lv	$L_{II}N_{VI}$	14.107
0.8790	1	36 Kr	$K\beta_{3}$	KM_{II}	14.104
0.87943	1	38 Sr	$K\alpha_{2}$	KL_{II}	14.0979
0.87995	7	80 Hg	$L\gamma_{8}$	$L_{II}O_{I}$	14.090
0.87996	5	81 Tl		$L_{II}N_{III}$	14.0893
0.88028	2	94 Pu	$L\alpha_{2}$	$L_{III}M_{IV}$	14.0842
0.88135	9	85 At	$L\beta_{3}$	$L_{I}M_{III}$	14.067
0.8827	2	80 Hg		$L_{I}N_{I}$	14.045
0.88433	7	79 Au	$L\gamma_{11}$	$L_{I}N_{V}$	14.020
0.88563	7	79 Au		$L_{I}N_{IV}$	13.999
0.8882	2	81 Tl		$L_{II}M_{II}$	13.959
0.889128	9	93 Np	$L\alpha_{1}$	$L_{III}M_{V}$	13.9441
0.8931	1	78 Pt	L_{I}	Abs. Edge	13.883
0.8934	1	78 Pt		$L_{I}O_{V}$	13.878
0.89349	9	85 At	$L\beta_{1}$	$L_{II}M_{IV}$	13.876
0.8943	1	78 Pt		$L_{I}O_{IV}$	13.864
0.89500	4	81 Tl	$L\gamma_{5}$	$L_{II}N_{I}$	13.8526
0.89646	5	80 Hg	$L\gamma_{1}$	$L_{II}N_{IV}$	13.8301
0.89659	4	78 Pt	$L\gamma_{4}$	$L_{I}O_{III}$	13.8281
0.89747	4	78 Pt	$L\gamma_{4}'$	$L_{I}O_{II}$	13.8145
0.89783	5	79 Au	$L\gamma_{3}$	$L_{I}N_{III}$	13.8090
0.89791	3	83 Bi	$L\beta_{9}$	$L_{I}M_{V}$	13.8077
0.8995	2	78 Pt		$L_{I}O_{I}$	13.784
0.8996	2	84 Po	$L\beta_{5}$	$L_{III}O_{IV,V}$	13.782
0.901045	9	93 Np	$L\alpha_{2}$	$L_{III}M_{IV}$	13.7597
0.90259	5	79 Au	L_{II}	Abs. Edge	13.7361
0.90297	3	79 Au	$L\gamma_{6}$	$L_{II}O_{IV}$	13.7304
0.90434	3	79 Au	$L\gamma_{2}$	$L_{I}N_{II}$	13.7095
0.90495	4	83 Bi	$L\beta_{10}$	$L_{I}M_{IV}$	13.7002
0.90638	7	79 Au		$L_{II}O_{III}$	13.679
0.90742	5	88 Ra	$L\eta$	$L_{II}M_{I}$	13.6630
0.90746	7	79 Au		$L_{II}O_{II}$	13.662
0.90837	5	79 Au	Lv	$L_{II}N_{VI}$	13.6487
0.90894	7	80 Hg		$L_{II}N_{III}$	13.640
0.9091	3	84 Po	$L\beta_{3}$	$L_{I}M_{III}$	13.638
0.90989	5	79 Au	$L\gamma_{8}$	$L_{II}O_{I}$	13.6260
0.910639	9	92 U	$L\alpha_{1}$	$L_{III}M_{V}$	13.6147
0.9131	1	79 Au		$L_{I}N_{I}$	13.578
0.9143	2	78 Pt	$L\gamma_{11}$	$L_{I}N_{V}$	13.560
0.9204	1	35 Br	K	Abs. Edge	13.470
0.92046	2	35 Br	$K\beta_{2}$	$KN_{II,III}$	13.4695
0.9220	2	84 Po	$L\beta_{1}$	$L_{II}M_{IV}$	13.447
0.922558	9	92 U	$L\alpha_{2}$	$L_{III}M_{IV}$	13.4388
0.9234	1	83 Bi	L_{III}	Abs. Edge	13.426
0.9236	1	77 Ir	L_{I}	Abs. Edge	13.423
0.92413	4	83 Bi		$L_{III}P_{II,III}$	13.4159
0.9243	3	77 Ir		$L_{I}O_{IV,V}$	13.413
0.92453	7	80 Hg	$L\gamma_{5}$	$L_{II}N_{I}$	13.410
0.9255	1	35 Br	$K\beta_{5}$	$KM_{IV,V}$	13.396

Wavelength Å*	p.e.	Element	Designation		keV
0.925553	9	37 Rb	$K\alpha_1$	KL_{III}	13.3953
0.92556	3	83 Bi	$L\beta_5$	$L_{III}O_{IV,V}$	13.3953
0.92650	3	79 Au	$L\gamma_1$	$L_{II}N_{IV}$	13.3817
0.9268	1	82 Pb	$L\beta_9$	L_IM_V	13.377
0.92744	3	77 Ir	$L\gamma_4$	L_IO_{III}	13.3681
0.92791	5	78 Pt	$L\gamma_3$	L_IN_{III}	13.3613
0.92831	3	77 Ir	$L\gamma_4'$	L_IO_{II}	13.3555
0.92937	5	84 Po	$L\beta_2$	$L_{III}N_V$	13.3404
0.92969	1	37 Rb	$K\alpha_2$	KL_{II}	13.3358
0.9302	2	83 Bi		$L_{III}O_{III}$	13.328
0.9312	2	84 Po	$L\beta_{15}$	$L_{III}N_{IV}$	13.314
0.9323	2	83 Bi		$L_{III}O_{II}$	13.298
0.93279	2	35 Br	$K\beta_1$	KM_{III}	13.2914
0.93284	5	91 Pa	$L\alpha_1$	$L_{III}M_V$	13.2907
0.93327	5	35 Br	$K\beta_3$	KM_{II}	13.2845
0.9339	2	82 Pb	$L\beta_{10}$	L_IM_{IV}	13.275
0.93414	5	78 Pt	L_{II}	Abs. Edge	13.2723
0.9342	2	78 Pt	$L\gamma_6$	$L_{II}O_{IV}$	13.271
0.93427	5	78 Pt	$L\gamma_2$	L_IN_{II}	13.2704
0.93505	5	83 Bi	$L\beta_7$	$L_{III}O_I$	13.2593
0.93505	5	83 Bi	Lu	$L_{III}N_{VI,VII}$	13.2593
0.93855	3	83 Bi	$L\beta_3$	L_IM_{III}	13.2098
0.93931	5	78 Pt	Lv	$L_{II}N_{VI}$	13.1992
0.9402	2	79 Au		$L_{II}N_{III}$	13.186
0.9411	1	78 Pt	$L\gamma_8$	$L_{II}O_I$	13.173
0.94419	5	83 Bi		$L_{II}M_V$	13.1310
0.9446	2	77 Ir	$L\gamma_{11}$	L_IN_V	13.126
0.94482	5	91 Pa	$L\alpha_2$	$L_{III}M_{IV}$	13.1222
0.9455	2	78 Pt		L_IN_I	13.113
0.9459	2	77 Ir		L_IN_{IV}	13.108
0.9475	3	84 Po	$L\beta_4$	L_IM_{II}	13.086
0.95073	5	82 Pb	L_{III}	Abs. Edge	13.0406
0.95118	7	82 Pb		$L_{III}P_{II,III}$	13.0344
0.951978	9	83 Bi	$L\beta_1$	$L_{II}M_{IV}$	13.0235
0.9526	1	82 Pb	$L\beta_5$	$L_{III}O_{IV,V}$	13.015
0.95518	4	83 Bi	$L\beta_2$	$L_{III}N_V$	12.9799
0.95559	3	79 Au	$L\gamma_5$	$L_{II}N_I$	12.9743
0.9558	1	76 Os	L_I	Abs. Edge	12.972
0.95600	3	90 Th	$L\alpha_1$	$L_{III}M_V$	12.9687
0.95603	5	76 Os		$L_IO_{IV,V}$	12.9683
0.95675	7	81 Tl	$L\beta_9$	L_IM_V	12.9585
0.95702	5	83 Bi	$L\beta_{15}$	$L_{III}N_{IV}$	12.9549
0.9578	1	82 Pb		$L_{III}O_{III}$	12.945
0.95797	3	78 Pt	$L\gamma_1$	$L_{II}N_{IV}$	12.9420
0.9586	1	82 Pb		$L_{III}O_{II}$	12.934
0.95931	5	77 Ir	$L\gamma_3$	L_IN_{III}	12.9240
0.95938	8	76 Os	$L\gamma_4$	L_IO_{III}	12.923
0.96033	8	76 Os	$L\gamma_4'$	L_IO_{II}	12.910
0.96133	7	82 Pb	Lu	$L_{III}N_{VI,VII}$	12.8968
0.9620	1	82 Pb	$L\beta_7$	$L_{III}O_I$	12.888
0.96318	7	76 Os		L_IO_I	12.8721
0.9636	1	92 U	Ls	$L_{III}M_{III}$	12.866
0.96389	7	81 Tl	$L\beta_{10}$	L_IM_{IV}	12.8626
0.96545	3	77 Ir	$L\gamma_2$	L_IN_{II}	12.8418
0.96708	4	77 Ir	$L\gamma_6$	$L_{II}O_{IV}$	12.8201
0.9671	1	77 Ir	L_{II}	Abs. Edge	12.820
0.9672	2	84 Po	$L\beta_6$	$L_{III}N_I$	12.819
0.96788	2	90 Th	$L\alpha_2$	$L_{III}M_{IV}$	12.8096
0.96911	7	82 Pb	$L\beta_3$	L_IM_{III}	12.7933
0.96979	5	77 Ir		$L_{II}O_{III}$	12.7843
0.97161	6	77 Ir	Lv	$L_{II}N_{VI}$	12.7603
0.97173	4	78 Pt		$L_{II}N_{III}$	12.7588
0.97321	5	83 Bi		$L_{III}N_{III}$	12.7394
0.97409	3	77 Ir	$L\gamma_8$	$L_{II}O_I$	12.7279
0.9747	1	82 Pb		$L_{II}M_V$	12.720
0.9765	3	76 Os	$L\gamma_{11}$	L_IN_V	12.696
0.9766	2	77 Ir		L_IN_I	12.695
0.97690	4	83 Bi	$L\beta_4$	L_IM_{II}	12.6912
0.9772	3	76 Os		L_IN_{IV}	12.687
0.9792	2	78 Pt		$L_{II}N_{II}$	12.661
0.97926	5	81 Tl		$L_{III}P_{II,III}$	12.6607
0.9793	1	81 Tl	L_{III}	Abs. Edge	12.660
0.97974	1	34 Se	K	Abs. Edge	12.6545
0.97992	5	34 Se	$K\beta_2$	$KN_{II,III}$	12.6522
0.97993	5	89 Ac	$L\alpha_1$	$L_{III}M_V$	12.6520
0.9801	1	36 Kr	$K\alpha_1$	KL_{III}	12.649
0.98058	3	81 Tl	$L\beta_5$	$L_{III}O_{IV,V}$	12.6436
0.98221	7	82 Pb	$L\beta_2$	$L_{III}N_V$	12.6226
0.98280	5	83 Bi		$L_{III}N_{II}$	12.6151
0.98291	3	82 Pb	$L\beta_1$	$L_{II}M_{IV}$	12.6137
0.98389	7	82 Pb	$L\beta_{15}$	$L_{III}N_{IV}$	12.6011
0.9841	1	36 Kr	$K\alpha_2$	KL_{II}	12.598
0.9843	1	34 Se	$K\beta_5$	$KM_{IV,V}$	12.595
0.98538	5	81 Tl		$L_{III}O_{III}$	12.5820
0.9871	2	80 Hg	$L\beta_9$	L_IM_V	12.560
0.98738	5	81 Tl		$L_{III}O_{II}$	12.5566
0.9877	2	78 Pt	$L\gamma_5$	$L_{II}N_I$	12.552
0.9888	1	81 Tl	Lu	$L_{III}N_{VI,VII}$	12.538
0.98913	5	83 Bi	$L\beta_{17}$	$L_{II}M_{III}$	12.5344
0.9894	1	75 Re	L_I	Abs. Edge	12.530
0.9900	1	75 Re		$L_IO_{IV,V}$	12.524
0.99017	5	81 Tl	$L\beta_7$	$L_{III}O_I$	12.5212
0.99085	3	77 Ir	$L\gamma_1$	$L_{II}N_{IV}$	12.5126
0.99178	5	89 Ac	$L\alpha_2$	$L_{III}M_{IV}$	12.5008
0.99186	5	76 Os	$L\gamma_3$	L_IN_{III}	12.4998
0.99218	3	34 Se	$K\beta_1$	KM_{III}	12.4959
0.99249	5	75 Re	$L\gamma_4$	L_IO_{III}	12.4920
0.99268	5	34 Se	$K\beta_3$	KM_{II}	12.4896
0.99331	3	83 Bi	$L\beta_6$	$L_{III}N_I$	12.4816
0.99334	5	75 Re	$L\gamma_4'$	L_IO_{II}	12.4813
0.9962	2	80 Hg	$L\beta_{10}$	L_IM_{IV}	12.446
0.9965	1	75 Re		L_IO_I	12.442
0.99805	5	76 Os	$L\gamma_2$	L_IN_{II}	12.4224
1.0005	1	82 Pb		$L_{III}N_{III}$	12.392
1.0005	9	83 Bi		L_IM_I	12.39
1.00062	3	81 Tl	$L\beta_3$	L_IM_{III}	12.3904
1.00107	5	76 Os	$L\gamma_6$	$L_{II}O_{IV}$	12.3848
1.0012	6	95 Am	Ll	$L_{III}M_I$	12.384
1.0014	1	76 Os	L_{II}	Abs. Edge	12.381
1.0047	2	76 Os		$L_{II}O_{III}$	12.340
1.00473	5	88 Ra	$L\alpha_1$	$L_{III}M_V$	12.3397
1.0050	2	76 Os	Lv	$L_{II}N_{VI}$	12.337
1.0054	3	77 Ir		$L_{II}N_{III}$	12.332
1.00722	5	81 Tl		$L_{II}M_V$	12.3093

Wavelength Å*	p.e.	Element	Designation		keV
1.0075	1	82 Pb	$L\beta_4$	L_IM_{II}	12.306
1.00788	5	76 Os	$L\gamma_8$	$L_{II}O_I$	12.3012
1.0091	1	80 Hg	L_{III}	Abs. Edge	12.286
1.00987	7	80 Hg	$L\beta_5$	$L_{III}O_{IV,V}$	12.2769
1.01031	3	81 Tl	$L\beta_2$	$L_{III}N_V$	12.2715
1.01040	7	82 Pb		$L_{III}N_{II}$	12.2705
1.0108	1	75 Re	$L\gamma_{11}$	L_IN_V	12.266
1.0112	1	90 Th	Ls	$L_{III}M_{III}$	12.261
1.0119	1	75 Re		L_IN_{IV}	12.252
1.0120	2	77 Ir		$L_{II}N_{II}$	12.251
1.01201	3	81 Tl	$L\beta_{15}$	$L_{III}N_{IV}$	12.2510
1.01404	7	80 Hg		$L_{III}O_{III}$	12.2264
1.01513	4	81 Tl	$L\beta_1$	$L_{II}M_{IV}$	12.2133
1.01558	7	80 Hg		$L_{III}O_{II}$	12.2079
1.01656	5	88 Ra	$L\alpha_2$	$L_{III}M_{IV}$	12.1962
1.01674	7	80 Hg	Lu	$L_{III}N_{VII}$	12.1940
1.01769	7	80 Hg	Lu'	$L_{III}N_{VI}$	12.1826
1.01937	7	80 Hg	$L\beta_7$	$L_{III}O_I$	12.1625
1.02063	7	79 Au	$L\beta_9$	L_IM_V	12.1474
1.0210	1	82 Pb	$L\beta_6$	$L_{III}N_I$	12.143
1.02175	5	77 Ir	$L\gamma_5$	$L_{II}N_I$	12.1342
1.0223	1	82 Pb	$L\beta_{17}$	$L_{II}M_{III}$	12.127
1.0226	1	94 Pu	Ll	$L_{III}M_I$	12.124
1.02467	5	74 W	L_I	Abs. Edge	12.0996
1.0250	2	74 W		$L_IO_{IV,V}$	12.095
1.02503	5	76 Os	$L\gamma_1$	$L_{II}N_{IV}$	12.0953
1.02613	7	75 Re	$L\gamma_3$	L_IN_{III}	12.0824
1.02775	3	74 W	$L\gamma_4$	L_IO_{III}	12.0634
1.02789	7	79 Au	$L\beta_{10}$	L_IM_{IV}	12.0617
1.0286	1	81 Tl		$L_{III}N_{III}$	12.053
1.02863	3	74 W	$L\gamma_4'$	L_IO_{II}	12.0530
1.03049	5	87 Fr	$L\alpha_1$	$L_{III}M_V$	12.0313
1.0317	3	74 W		L_IO_I	12.017
1.03233	5	75 Re	$L\gamma_2$	L_IN_{II}	12.0098
1.0323	2	82 Pb		L_IM_I	12.010
1.03358	7	80 Hg	$L\beta_3$	L_IM_{III}	11.9953
1.0346	9	83 Bi		L_IM_{II}	11.98
1.0347	1	92 U	Lt	$L_{III}M_{II}$	11.982
1.03699	9	75 Re	$L\gamma_6$	$L_{II}O_{IV}$	11.956
1.0371	1	75 Re	L_{II}	Abs. Edge	11.954
1.03876	7	79 Au		$L_{III}P_{II,III}$	11.9355
1.03918	3	81 Tl	$L\beta_4$	L_IM_{II}	11.9306
1.0397	1	75 Re		$L_{II}O_{III}$	11.925
1.03973	5	76 Os		$L_{II}N_{III}$	11.9243
1.03974	2	35 Br	$K\alpha_1$	KL_{III}	11.9242
1.03975	7	80 Hg	$L\beta_2$	$L_{III}N_V$	11.9241
1.04000	5	79 Au	L_{III}	Abs. Edge	11.9212
1.0404	1	75 Re	Lv	$L_{II}N_{VI}$	11.917
1.04044	3	79 Au	$L\beta_5$	$L_{III}O_{IV,V}$	11.9163
1.04151	7	80 Hg	$L\beta_{15}$	$L_{III}N_{IV}$	11.9040
1.0420	1	75 Re		L_IN_I	11.899
1.04230	5	87 Fr	$L\alpha_2$	$L_{III}M_{IV}$	11.8950
1.0428	6	93 Np	Ll	$L_{III}M_I$	11.890
1.04382	2	35 Br	$K\alpha_2$	KL_{II}	11.8776
1.04398	5	75 Re	$L\gamma_8$	$L_{II}O_I$	11.8758
1.0450	2	79 Au		$L_{III}O_{II,III}$	11.865
1.0450	1	33 As	K	Abs. Edge	11.865
1.04500	3	33 As	$K\beta_2$	$KN_{II,III}$	11.8642
1.0458	1	74 W	$L\gamma_{11}$	L_IN_V	11.856
1.0468	2	74 W		L_IN_{IV}	11.844
1.04752	5	79 Au	Lu	$L_{III}N_{VI,VII}$	11.8357
1.04868	5	80 Hg	$L\beta_1$	$L_{II}M_{IV}$	11.8226
1.0488	1	33 As	$K\beta_5$	$KM_{IV,V}$	11.822
1.04963	5	81 Tl	$L\beta_6$	$L_{III}N_I$	11.8118
1.04974	8	79 Au	$L\beta_7$	$L_{III}O_I$	11.8106
1.05446	5	78 Pt	$L\beta_9$	L_IM_V	11.7577
1.05609	7	81 Tl	$L\beta_{17}$	$L_{II}M_{III}$	11.7397
1.05693	5	76 Os	$L\gamma_5$	$L_{II}N_I$	11.7303
1.05723	5	86 Rn	$L\alpha_1$	$L_{III}M_V$	11.7270
1.05730	2	33 As	$K\beta_1$	KM_{III}	11.7262
1.05783	5	33 As	$K\beta_3$	KM_{II}	11.7203
1.0585	1	80 Hg		$L_{III}N_{III}$	11.713
1.05856	3	83 Bi	$L\eta$	$L_{II}M_I$	11.7122
1.06099	5	75 Re	$L\gamma_1$	$L_{II}N_{IV}$	11.6854
1.0613	1	73 Ta	L_I	Abs. Edge	11.682
1.06183	7	78 Pt	$L\beta_{10}$	L_IM_{IV}	11.6762
1.06192	9	73 Ta		$L_IO_{IV,V}$	11.6752
1.06200	6	74 W	$L\gamma_3$	L_IN_{III}	11.6743
1.06357	9	73 Ta		$L_IN_{VI,VII}$	11.6570
1.0644	2	82 Pb		$L_{II}M_{II}$	11.648
1.0644	2	81 Tl		L_IM_I	11.648
1.06467	3	73 Ta	$L\gamma_4$	L_IO_{III}	11.6451
1.0649	2	80 Hg		$L_{III}N_{II}$	11.642
1.06544	3	73 Ta	$L\gamma_4'$	L_IO_{II}	11.6366
1.06712	2	92 U	Ll	$L_{III}M_I$	11.6183
1.06771	9	73 Ta		L_IO_I	11.6118
1.06785	9	79 Au	$L\beta_3$	L_IM_{III}	11.6103
1.06806	3	74 W	$L\gamma_2$	L_IN_{II}	11.6080
1.06899	5	86 Rn	$L\alpha_2$	$L_{III}M_{IV}$	11.5979
1.07022	3	79 Au	$L\beta_2$	$L_{III}N_V$	11.5847
1.07188	5	79 Au	$L\beta_{15}$	$L_{III}N_{IV}$	11.5667
1.07222	7	80 Hg	$L\beta_4$	L_IM_{II}	11.5630
1.0723	1	78 Pt	L_{III}	Abs. Edge	11.562
1.0724	2	78 Pt	$L\beta_5$	$L_{III}O_{IV,V}$	11.561
1.07448	5	74 W	$L\gamma_6$	$L_{II}O_{IV}$	11.5387
1.0745	1	74 W	L_{II}	Abs. Edge	11.538
1.0756	2	79 Au		$L_{II}M_V$	11.526
1.0761	3	78 Pt		$L_{III}C_{II,III}$	11.521
1.0767	1	75 Re		$L_{II}N_{III}$	11.515
1.0771	1	74 W	Lv	$L_{II}N_{VI}$	11.510
1.07896	5	78 Pt	Lu	$L_{III}N_{VI,VII}$	11.4908
1.0792	2	74 W		$L_{II}O_{III}$	11.488
1.07975	7	80 Hg	$L\beta_6$	$L_{III}N_I$	11.4824
1.08009	9	90 Th	Lt	$L_{III}M_{II}$	11.4788
1.08113	4	74 W	$L\gamma_8$	$L_{II}O_I$	11.4677
1.08168	3	78 Pt	$L\beta_7$	$L_{III}O_I$	11.4619
1.08205	7	73 Ta	$L\gamma_{11}$	L_IN_V	11.4580
1.08353	3	79 Au	$L\beta_1$	$L_{II}M_{IV}$	11.4423
1.08377	7	73 Ta		L_IN_{IV}	11.4398
1.0839	1	75 Re		$L_{II}N_{II}$	11.438
1.08500	5	85 At	$L\alpha_1$	$L_{III}M_V$	11.4268
1.08975	5	77 Ir	$L\beta_9$	L_IM_V	11.3770
1.09026	7	79 Au		$L_{III}N_{III}$	11.3717
1.0908	1	91 Pa	Ll	$L_{III}M_I$	11.366

Wavelength Å*	p.e.	Element	Designation		keV
1.0916	5	80 Hg	$L\beta_{17}$	$L_{II}M_{III}$	11.358
1.09241	7	82 Pb	$L\eta$	$L_{II}M_{I}$	11.3493
1.09388	5	75 Re	$L\gamma_{5}$	$L_{II}N_{I}$	11.3341
1.09671	5	85 At	$L\alpha_{2}$	$L_{III}M_{IV}$	11.3048
1.09702	4	77 Ir	$L\beta_{10}$	$L_{I}M_{IV}$	11.3016
1.09855	3	74 W	$L\gamma_{1}$	$L_{II}N_{IV}$	11.2859
1.09936	4	73 Ta	$L\gamma_{3}$	$L_{I}N_{III}$	11.2776
1.0997	1	81 Tl		$L_{II}M_{II}$	11.274
1.0997	1	72 Hf	L_{I}	Abs. Edge	11.274
1.09968	7	79 Au		$L_{III}N_{II}$	11.2743
1.0999	2	80 Hg		$L_{I}M_{I}$	11.272
1.10086	9	72 Hf		$L_{I}O_{IV}$	11.2622
1.10200	3	78 Pt	$L\beta_{2}$	$L_{III}N_{V}$	11.2505
1.10303	5	72 Hf	$L\gamma_{4}$	$L_{I}O_{III}$	11.2401
1.10376	5	72 Hf	$L\gamma_{4}'$	$L_{I}O_{II}$	11.2326
1.10394	5	78 Pt	$L\beta_{3}$	$L_{I}M_{III}$	11.2308
1.10477	2	34 Se	$K\alpha_{1}$	KL_{III}	11.2224
1.1053	1	73 Ta	$L\gamma_{2}$	$L_{I}N_{II}$	11.217
1.1058	1	77 Ir	L_{III}	Abs. Edge	11.212
1.10585	3	77 Ir	$L\beta_{5}$	$L_{III}O_{IV,V}$	11.2114
1.10651	3	79 Au	$L\beta_{4}$	$L_{I}M_{II}$	11.2047
1.10664	9	72 Hf		$L_{I}O_{I}$	11.2034
1.10882	2	34 Se	$K\alpha_{2}$	KL_{II}	11.1814
1.10923	6	77 Ir		$L_{III}O_{II,III}$	11.1772
1.11092	3	79 Au	$L\beta_{6}$	$L_{III}N_{I}$	11.1602
1.11145	4	77 Ir	Lu	$L_{III}N_{VI,VII}$	11.1549
1.1129	2	78 Pt		$L_{II}M_{V}$	11.140
1.1137	1	73 Ta	L_{II}	Abs. Edge	11.132
1.11386	4	84 Po	$L\alpha_{1}$	$L_{III}M_{V}$	11.1308
1.11388	3	73 Ta	$L\gamma_{6}$	$L_{II}O_{IV}$	11.1306
1.11489	3	77 Ir	$L\beta_{7}$	$L_{III}O_{I}$	11.1205
1.1149	2	74 W		$L_{II}N_{III}$	11.120
1.11508	4	90 Th	Ll	$L_{III}M_{I}$	11.1186
1.11521	9	73 Ta		$L_{I}N_{I}$	11.1173
1.1158	1	73 Ta	Lv	$L_{II}N_{VI}$	11.1113
1.11658	5	32 Ge	K	Abs. Edge	11.1036
1.11686	2	32 Ge	$K\beta_{2}$	$KN_{II,III}$	11.1008
1.11693	9	73 Ta		$L_{II}O_{III}$	11.1001
1.11789	9	73 Ta		$L_{II}O_{II}$	11.0907
1.1195	1	32 Ge	$K\beta_{5}$	$KM_{IV,V}$	11.0745
1.11990	2	78 Pt	$L\beta_{1}$	$L_{II}M_{IV}$	11.0707
1.1205	1	73 Ta	$L\gamma_{8}$	$L_{II}O_{I}$	11.0646
1.12146	9	72 Hf	$L\gamma_{11}$	$L_{I}N_{V}$	11.0553
1.1218	3	74 W		$L_{II}N_{II}$	11.052
1.12250	9	72 Hf		$L_{I}N_{IV}$	11.0451
1.1226	2	78 Pt		$L_{III}N_{III}$	11.044
1.12548	5	84 Po	$L\alpha_{2}$	$L_{III}M_{IV}$	11.0158
1.12637	6	76 Os	$L\beta_{9}$	$L_{I}M_{V}$	11.0071
1.12769	3	81 Tl	$L\eta$	$L_{II}M_{I}$	10.9943
1.12798	5	79 Au	$L\beta_{17}$	$L_{II}M_{III}$	10.9915
1.12894	2	32 Ge	$K\beta_{1}$	KM_{III}	10.9821
1.12936	9	32 Ge	$K\beta_{3}$	KM_{II}	10.9780
1.1310	2	78 Pt		$L_{III}N_{II}$	10.962
1.13235	3	74 W	$L\gamma_{5}$	$L_{II}N_{I}$	10.9490
1.13353	5	76 Os	$L\beta_{10}$	$L_{I}M_{IV}$	10.9376
1.13525	5	79 Au		$L_{I}M_{I}$	10.9210
1.13532	3	77 Ir	$L\beta_{2}$	$L_{III}N_{V}$	10.9203
1.13687	9	73 Ta		$L_{II}N_{V}$	10.9055
1.13707	3	77 Ir	$L\beta_{15}$	$L_{III}N_{IV}$	10.9036
1.13794	3	73 Ta	$L\gamma_{1}$	$L_{II}N_{IV}$	10.8952
1.13841	5	72 Hf	$L\gamma_{3}$	$L_{I}N_{III}$	10.8907
1.1387	5	80 Hg		$L_{II}M_{II}$	10.888
1.1402	1	71 Lu	L_{I}	Abs. Edge	10.8740
1.1405	1	76 Os	$L\beta_{5}$	$L_{III}O_{IV,V}$	10.8711
1.1408	1	76 Os	L_{III}	Abs. Edge	10.8683
1.14085	3	77 Ir	$L\beta_{3}$	$L_{I}M_{III}$	10.8674
1.14223	5	78 Pt	$L\beta_{4}$	$L_{I}M_{II}$	10.8543
1.1435	1	71 Lu	$L\gamma_{4}$	$L_{I}O_{II,III}$	10.8425
1.14355	5	78 Pt	$L\beta_{6}$	$L_{III}N_{I}$	10.8418
1.14386	2	83 Bi	$L\alpha_{1}$	$L_{III}M_{V}$	10.8388
1.14442	5	72 Hf	$L\gamma_{2}$	$L_{I}N_{II}$	10.8335
1.14537	7	76 Os	Lu	$L_{III}N_{VI,VII}$	10.8245
1.1489	2	77 Ir		$L_{II}M_{V}$	10.791
1.14933	8	76 Os	$L\beta_{7}$	$L_{III}O_{I}$	10.7872
1.1548	1	72 Hf	L_{II}	Abs. Edge	10.7362
1.15519	5	72 Hf	$L\gamma_{6}$	$L_{II}O_{IV}$	10.7325
1.1553	1	73 Ta		$L_{II}N_{III}$	10.7316
1.15536	1	83 Bi	$L\alpha_{2}$	$L_{III}M_{IV}$	10.73091
1.1560	3	77 Ir		$L_{III}N_{III}$	10.725
1.15781	3	77 Ir	$L\beta_{1}$	$L_{II}M_{IV}$	10.7083
1.15830	9	72 Hf	Lv	$L_{II}N_{VI}$	10.7037
1.1600	2	73 Ta		$L_{II}N_{II}$	10.688
1.16107	9	71 Lu	$L\gamma_{11}$	$L_{I}N_{V}$	10.6782
1.16138	5	72 Hf	$L\gamma_{8}$	$L_{II}O_{I}$	10.6754
1.16227	9	71 Lu		$L_{I}N_{IV}$	10.6672
1.1640	1	80 Hg	$L\eta$	$L_{II}M_{I}$	10.6512
1.16487	4	75 Re	$L\beta_{9}$	$L_{I}M_{V}$	10.6433
1.16545	5	77 Ir		$L_{III}N_{II}$	10.6380
1.1667	1	78 Pt	$L\beta_{17}$	$L_{II}M_{III}$	10.6265
1.16719	5	88 Ra	Ll	$L_{III}M_{I}$	10.6222
1.16962	9	78 Pt		$L_{I}M_{I}$	10.6001
1.16979	8	76 Os	$L\beta_{2}$	$L_{III}N_{V}$	10.5985
1.1708	1	79 Au		$L_{II}M_{II}$	10.5892
1.17167	5	76 Os	$L\beta_{15}$	$L_{III}N_{IV}$	10.5816
1.17218	5	75 Re	$L\beta_{10}$	$L_{I}M_{IV}$	10.5770
1.1729	1	73 Ta	$L\gamma_{5}$	$L_{II}N_{I}$	10.5702
1.17501	2	82 Pb	$L\alpha_{1}$	$L_{III}M_{V}$	10.5515
1.17588	1	33 As	$K\alpha_{1}$	KL_{III}	10.54372
1.17721	5	75 Re	$L\beta_{5}$	$L_{III}O_{IV,V}$	10.5318
1.1773	1	75 Re	L_{III}	Abs. Edge	10.5306
1.17788	9	72 Hf		$L_{II}N_{V}$	10.5258
1.17796	3	77 Ir	$L\beta_{6}$	$L_{III}N_{I}$	10.5251
1.17900	5	72 Hf	$L\gamma_{1}$	$L_{II}N_{IV}$	10.5158
1.17953	4	71 Lu	$L\gamma_{3}$	$L_{I}N_{III}$	10.5110
1.17955	7	76 Os	$L\beta_{3}$	$L_{I}M_{III}$	10.5108
1.17958	3	77 Ir	$L\beta_{4}$	$L_{I}M_{II}$	10.5106
1.17987	1	33 As	$K\alpha_{2}$	KL_{II}	10.50799
1.1815	1	75 Re	Lu	$L_{III}N_{VI,VII}$	10.4931
1.1818	1	70 Yb	L_{I}	Abs. Edge	10.4904
1.1827	1	70 Yb		$L_{I}O_{IV,V}$	10.4833
1.1853	1	70 Yb	$L\gamma_{4}$	$L_{I}O_{II,III}$	10.4603
1.1853	2	71 Lu	$L\gamma_{2}$	$L_{I}N_{II}$	10.460
1.18610	5	75 Re	$L\beta_{7}$	$L_{III}O_{I}$	10.4529
1.18648	5	82 Pb	$L\alpha_{2}$	$L_{III}M_{IV}$	10.4495

Wavelength Å*	p.e.	Element		Designation	keV
1.1886	1	70 Yb		$L_{I}O_{I}$	10.4312
1.18977	7	76 Os		$L_{II}M_{V}$	10.4205
1.1958	1	31 Ga	K	Abs. Edge	10.3682
1.19600	2	31 Ga	$K\beta_{2}$	$KN_{II,III}$	10.3663
1.19727	7	76 Os	$L\beta_{1}$	$L_{II}M_{IV}$	10.3553
1.1981	2	31 Ga	$K\beta_{5}$	$KM_{IV,V}$	10.348
1.1985	1	71 Lu	L_{II}	Abs. Edge	10.3448
1.1987	1	71 Lu	$L\gamma_{6}$	$L_{II}O_{IV}$	10.3431
1.20086	7	76 Os		$L_{III}N_{II}$	10.3244
1.2014	1	71 Lu		$L_{II}O_{II,III}$	10.3198
1.20273	3	79 Au	$L\eta$	$L_{II}M_{I}$	10.3083
1.2047	1	71 Lu	$L\gamma_{8}$	$L_{II}O_{I}$	10.2915
1.20479	7	74 W	$L\beta_{9}$	$L_{I}M_{V}$	10.2907
1.20660	4	75 Re	$L\beta_{2}$	$L_{III}N_{V}$	10.2752
1.2069	2	77 Ir	$L\beta_{17}$	$L_{II}M_{III}$	10.273
1.20739	4	81 Tl	$L\alpha_{1}$	$L_{III}M_{V}$	10.2685
1.20789	2	31 Ga	$K\beta_{1}$	KM_{III}	10.2642
1.20819	5	75 Re	$L\beta_{15}$	$L_{III}N_{IV}$	10.2617
1.20835	5	31 Ga	$K\beta_{3}$	KM_{II}	10.2603
1.2102	2	77 Ir		$L_{I}M_{I}$	10.245
1.2105	1	83 Bi	Ls	$L_{III}M_{III}$	10.2421
1.21218	3	74 W	$L\beta_{10}$	$L_{I}M_{IV}$	10.2279
1.213	1	78 Pt		$L_{II}M_{II}$	10.225
1.21349	5	76 Os	$L\beta_{6}$	$L_{III}N_{I}$	10.2169
1.21537	5	72 Hf	$L\gamma_{5}$	$L_{II}N_{I}$	10.2011
1.21545	3	74 W	$L\beta_{5}$	$L_{III}O_{IV,V}$	10.2004
1.2155	1	74 W	L_{III}	Abs. Edge	10.1999
1.21844	5	76 Os	$L\beta_{4}$	$L_{I}M_{II}$	10.1754
1.21868	5	74 W	Lu	$L_{III}N_{VI,VII}$	10.1733
1.21875	3	81 Tl	$L\alpha_{2}$	$L_{III}M_{IV}$	10.1728
1.22031	5	75 Re	$L\beta_{3}$	$L_{I}M_{III}$	10.1598
1.2211	2	74 W		$L_{III}O_{II,III}$	10.153
1.22228	4	71 Lu	$L\gamma_{1}$	$L_{II}N_{IV}$	10.1434
1.22232	5	70 Yb	$L\gamma_{3}$	$L_{I}N_{III}$	10.1431
1.22400	4	74 W	$L\beta_{7}$	$L_{III}O_{I}$	10.1292
1.2250	1	69 Tm	L_{I}	Abs. Edge	10.1206
1.2263	3	69 Tm		$L_{I}O_{IV,V}$	10.110
1.2283	1	75 Re		$L_{III}N_{III}$	10.0933
1.22879	7	70 Yb	$L\gamma_{2}$	$L_{I}N_{II}$	10.0897
1.2294	2	69 Tm	$L\gamma_{4}$	$L_{I}O_{II,III}$	10.084
1.2305	1	75 Re		$L_{II}M_{V}$	10.0753
1.23858	2	75 Re	$L\beta_{1}$	$L_{II}M_{IV}$	10.0100
1.24120	5	80 Hg	$L\alpha_{1}$	$L_{III}M_{V}$	9.9888
1.24271	3	70 Yb	$L\gamma_{6}$	$L_{II}O_{IV}$	9.9766
1.2428	1	70 Yb	L_{II}	Abs. Edge	9.9761
1.2429	2	78 Pt	$L\eta$	$L_{II}M_{I}$	9.975
1.24385	7	82 Pb	Ls	$L_{III}M_{III}$	9.9675
1.24460	3	74 W	$L\beta_{2}$	$L_{III}N_{V}$	9.9615
1.2453	1	70 Yb		$L_{II}O_{II,III}$	9.9561
1.24631	3	74 W	$L\beta_{15}$	$L_{III}N_{IV}$	9.9478
1.2466	2	73 Ta	$L\beta_{9}$	$L_{I}M_{V}$	9.946
1.2480	2	76 Os	$L\beta_{17}$	$L_{II}M_{III}$	9.934
1.24923	5	70 Yb	$L\gamma_{8}$	$L_{II}O_{I}$	9.9246
1.2502	3	77 Ir		$L_{II}M_{II}$	9.917
1.25100	5	75 Re	$L\beta_{6}$	$L_{III}N_{I}$	9.9105
1.25264	7	80 Hg	$L\alpha_{2}$	$L_{III}M_{IV}$	9.8976
1.2537	2	73 Ta	$L\beta_{10}$	$L_{I}M_{IV}$	9.889
1.254054	9	32 Ge	$K\alpha_{1}$	KL_{III}	9.88642
1.2553	1	73 Ta	L_{III}	Abs. Edge	9.8766
1.2555	1	73 Ta	$L\beta_{5}$	$L_{III}O_{IV,V}$	9.8750
1.25778	4	73 Ta	Lu	$L_{III}N_{VI,VII}$	9.8572
1.258011	9	32 Ge	$K\alpha_{2}$	KL_{II}	9.85532
1.25917	5	75 Re	$L\beta_{4}$	$L_{I}M_{II}$	9.8463
1.2596	1	71 Lu	$L\gamma_{5}$	$L_{II}N_{I}$	9.8428
1.2601	3	73 Ta		$L_{III}O_{II,III}$	9.839
1.26269	5	74 W	$L\beta_{3}$	$L_{I}M_{III}$	9.8188
1.26385	5	73 Ta	$L\beta_{7}$	$L_{III}O_{I}$	9.8098
1.2672	2	74 W		$L_{III}N_{III}$	9.784
1.26769	5	70 Yb	$L\gamma_{1}$	$L_{II}N_{IV}$	9.7801
1.2678	2	69 Tm	$L\gamma_{3}$	$L_{I}N_{III}$	9.779
1.2706	1	68 Er	L_{I}	Abs. Edge	9.7574
1.2728	2	74 W		$L_{II}M_{V}$	9.741
1.2742	2	69 Tm	$L\gamma_{2}$	$L_{I}N_{II}$	9.730
1.2748	1	83 Bi	Lt	$L_{III}M_{II}$	9.7252
1.2752	2	68 Er	$L\gamma_{4}$	$L_{I}O_{II,III}$	9.722
1.27640	3	79 Au	$L\alpha_{1}$	$L_{III}M_{V}$	9.7133
1.2765	2	74 W		$L_{III}N_{II}$	9.712
1.27807	5	81 Tl	Ls	$L_{III}M_{III}$	9.7007
1.281809	9	74 W	$L\beta_{1}$	$L_{II}M_{IV}$	9.67235
1.2829	5	84 Po	Ll	$L_{III}M_{I}$	9.664
1.2834	1	30 Zn	K	Abs. Edge	9.6607
1.28372	2	30 Zn	$K\beta_{2}$	$KN_{II,III}$	9.6580
1.28448	3	77 Ir	$L\eta$	$L_{II}M_{I}$	9.6522
1.28454	2	73 Ta	$L\beta_{2}$	$L_{III}N_{V}$	9.6518
1.2848	1	30 Zn	$K\beta_{5}$	$KM_{IV,V}$	9.6501
1.28619	5	73 Ta	$L\beta_{15}$	$L_{III}N_{IV}$	9.6394
1.28772	3	79 Au	$L\alpha_{2}$	$L_{III}M_{IV}$	9.6280
1.2892	1	69 Tm	L_{II}	Abs. Edge	9.6171
1.28989	7	74 W	$L\beta_{6}$	$L_{III}N_{I}$	9.6117
1.29025	9	72 Hf	$L\beta_{9}$	$L_{I}M_{V}$	9.6090
1.2905	2	69 Tm	$L\gamma_{6}$	$L_{II}O_{IV}$	9.607
1.2927	1	75 Re	$L\beta_{17}$	$L_{II}M_{III}$	9.5910
1.2934	2	76 Os		$L_{II}M_{II}$	9.586
1.29525	2	30 Zn	$K\beta_{1,3}$	$KM_{II,III}$	9.5720
1.2972	1	72 Hf	L_{III}	Abs. Edge	9.5577
1.29761	5	72 Hf	$L\beta_{5}$	$L_{III}O_{IV,V}$	9.5546
1.29819	9	72 Hf	$L\beta_{10}$	$L_{I}M_{IV}$	9.5503
1.30162	5	74 W	$L\beta_{4}$	$L_{I}M_{II}$	9.5252
1.30165	9	72 Hf	Lu	$L_{III}N_{VI,VII}$	9.5249
1.30564	5	72 Hf	$L\beta_{7}$	$L_{III}O_{I}$	9.4958
1.3063	1	70 Yb	$L\gamma_{5}$	$L_{II}N_{I}$	9.4910
1.30678	3	73 Ta	$L\beta_{3}$	$L_{I}M_{III}$	9.4875
1.30767	7	82 Pb	Lt	$L_{III}M_{II}$	9.4811
1.3086	1	73 Ta		$L_{III}N_{III}$	9.4742
1.3112	2	80 Hg	Ls	$L_{III}M_{III}$	9.455
1.31304	3	78 Pt	$L\alpha_{1}$	$L_{III}M_{V}$	9.4423
1.3146	1	68 Er	$L\gamma_{3}$	$L_{I}N_{III}$	9.4309
1.3153	2	69 Tm	$L\gamma_{1}$	$L_{II}N_{IV}$	9.426
1.31610	7	83 Bi	Ll	$L_{III}M_{I}$	9.4204
1.3167	1	73 Ta		$L_{III}N_{II}$	9.4158
1.31897	9	73 Ta		$L_{II}M_{V}$	9.3998
1.3190	1	67 Ho	L_{I}	Abs. Edge	9.3994
1.3208	3	67 Ho		$L_{I}O_{IV,V}$	9.387
1.3210	2	68 Er	$L\gamma_{2}$	$L_{I}N_{II}$	9.385

Wavelength Å*	p.e.	Element	Designation		keV
1.3225	2	67 Ho	$L\gamma_4$	$L_IO_{II,III}$	9.374
1.32432	2	78 Pt	$L\alpha_2$	$L_{III}M_{IV}$	9.3618
1.32639	5	72 Hf	$L\beta_2$	$L_{III}N_V$	9.3473
1.32698	3	73 Ta	$L\beta_1$	$L_{II}M_{IV}$	9.3431
1.32783	5	72 Hf	$L\beta_{15}$	$L_{III}N_{IV}$	9.3371
1.32785	7	76 Os	$L\eta$	$L_{II}M_I$	9.3370
1.33094	8	73 Ta	$L\beta_6$	$L_{III}N_I$	9.3153
1.3358	1	71 Lu	$L\beta_9$	L_IM_V	9.2816
1.3365	3	74 W		L_IM_I	9.277
1.3366	1	75 Re		$L_{II}M_{II}$	9.2761
1.3386	1	68 Er	L_{II}	Abs. Edge	9.2622
1.3387	2	74 W	$L\beta_{17}$	$L_{II}M_{III}$	9.261
1.3397	3	68 Er	$L\gamma_6$	$L_{II}O_{IV}$	9.255
1.340083	9	31 Ga	$K\alpha_1$	KL_{III}	9.25174
1.3405	1	71 Lu	L_{III}	Abs. Edge	9.2490
1.34154	5	81 Tl	Lt	$L_{III}M_{II}$	9.2417
1.34183	7	71 Lu	$L\beta_5$	$L_{III}O_{IV,V}$	9.2397
1.3430	2	71 Lu	$L\beta_{10}$	L_IM_{IV}	9.232
1.34399	1	31 Ga	$K\alpha_2$	KL_{II}	9.22482
1.34524	9	71 Lu		$L_{III}O_{II,III}$	9.2163
1.34581	3	73 Ta	$L\beta_4$	L_IM_{II}	9.2124
1.34949	5	71 Lu	$L\beta_7$	$L_{III}O_I$	9.1873
1.34990	7	82 Pb	Ll	$L_{III}M_I$	9.1845
1.35053	9	72 Hf		$L_{III}N_{III}$	9.1802
1.35128	3	77 Ir	$L\alpha_1$	$L_{III}M_V$	9.1751
1.35131	7	79 Au	Ls	$L_{III}M_{III}$	9.1749
1.35300	5	72 Hf	$L\beta_3$	L_IM_{III}	9.1634
1.3558	2	69 Tm	$L\gamma_5$	$L_{II}N_I$	9.144
1.35887	9	72 Hf		$L_{III}N_{II}$	9.1239
1.36250	5	77 Ir	$L\alpha_2$	$L_{III}M_{IV}$	9.0995
1.3641	2	68 Er	$L\gamma_1$	$L_{II}N_{IV}$	9.089
1.3643	2	67 Ho	$L\gamma_3$	L_IN_{III}	9.087
1.3692	1	66 Dy	L_I	Abs. Edge	9.0548
1.3698	2	67 Ho	$L\gamma_2$	L_IN_{II}	9.051
1.37012	3	71 Lu	$L\beta_2$	$L_{III}N_V$	9.0489
1.3715	1	71 Lu	$L\beta_{15}$	$L_{III}N_{IV}$	9.0395
1.37342	5	75 Re	$L\eta$	$L_{II}M_I$	9.0272
1.37410	5	72 Hf	$L\beta_1$	$L_{II}M_{IV}$	9.0227
1.37410	5	72 Hf	$L\beta_6$	$L_{III}N_I$	9.0227
1.37459	7	66 Dy	$L\gamma_4$	$L_IO_{II,III}$	9.0195
1.3746	2	80 Hg	Lt	$L_{III}M_{II}$	9.019
1.38059	5	29 Cu	K	Abs. Edge	8.9803
1.38109	3	29 Cu	$K\beta_2$	$KM_{IV,V}$	8.9770
1.3838	1	70 Yb	$L\beta_9$	L_IM_V	8.9597
1.38477	3	81 Tl	Ll	$L_{III}M_I$	8.9532
1.3862	1	70 Yb	L_{III}	Abs. Edge	8.9441
1.3864	1	73 Ta	$L\beta_{17}$	$L_{II}M_{III}$	8.9428
1.38696	7	70 Yb	$L\beta_5$	$L_{III}O_{IV,V}$	8.9390
1.3895	2	78 Pt	Ls	$L_{III}M_{III}$	8.923
1.3898	1	70 Yb		$L_{III}O_{II,III}$	8.9209
1.3905	1	67 Ho	L_{II}	Abs. Edge	8.9164
1.39121	5	76 Os	$L\alpha_1$	$L_{III}M_V$	8.9117
1.3915	1	70 Yb	$L\beta_{10}$	L_IM_{IV}	8.9100
1.39220	5	72 Hf	$L\beta_4$	L_IM_{II}	8.9054
1.392218	9	29 Cu	$K\beta_{1,3}$	$KM_{II,III}$	8.90529
1.3923	2	67 Ho	$L\gamma_6$	$L_{II}O_{IV}$	8.905
1.3926	1	29 Cu	$K\beta_3$	KM_{II}	8.9029
1.3948	1	70 Yb	$L\beta_7$	$L_{III}O_I$	8.8889
1.3983	2	67 Ho	$L\gamma_8$	$L_{II}O_I$	8.867
1.40140	5	71 Lu	$L\beta_3$	L_IM_{III}	8.8469
1.40234	5	76 Os	$L\alpha_2$	$L_{III}M_{IV}$	8.8410
1.4067	3	68 Er	$L\gamma_5$	$L_{II}N_I$	8.814
1.41366	7	79 Au	Lt	$L_{III}M_{II}$	8.7702
1.41550	5	70 Yb	$L\beta_{2,15}$	$L_{III}N_{IV,V}$	8.7588
1.41640	7	66 Dy	$L\gamma_3$	L_IN_{III}	8.7532
1.4174	2	67 Ho	$L\gamma_1$	$L_{II}N_{IV}$	8.747
1.4189	1	71 Lu	$L\beta_6$	$L_{III}N_I$	8.7376
1.42110	3	74 W	$L\eta$	$L_{II}M_I$	8.7243
1.4216	1	80 Hg	Ll	$L_{III}M_I$	8.7210
1.4223	1	65 Tb	L_I	Abs. Edge	8.7167
1.42278	7	66 Dy	$L\gamma_2$	L_IN_{II}	8.7140
1.4228	3	65 Tb		$L_IO_{IV,V}$	8.714
1.42359	3	71 Lu	$L\beta_1$	$L_{II}M_{IV}$	8.7090
1.4276	2	65 Tb	$L\gamma_4$	$L_IO_{II,III}$	8.685
1.43025	9	72 Hf		L_IM_I	8.6685
1.43048	9	73 Ta		$L_{II}M_{II}$	8.6671
1.4318	2	77 Ir	Ls	$L_{III}M_{III}$	8.659
1.43290	4	75 Re	$L\alpha_1$	$L_{III}M_V$	8.6525
1.4334	1	69 Tm	L_{III}	Abs. Edge	8.6496
1.4336	3	69 Tm	$L\beta_9$	L_IM_V	8.648
1.4349	2	69 Tm	$L\beta_5$	$L_{III}O_{IV,V}$	8.641
1.435155	7	30 Zn	$K\alpha_1$	KL_{III}	8.63886
1.43643	9	72 Hf	$L\beta_{17}$	$L_{II}M_{III}$	8.6312
1.439000	8	30 Zn	$K\alpha_2$	KL_{II}	8.61578
1.44056	5	71 Lu	$L\beta_4$	L_IM_{II}	8.6064
1.4410	3	69 Tm	$L\beta_{10}$	L_IM_{IV}	8.604
1.44396	5	75 Re	$L\alpha_2$	$L_{III}M_{IV}$	8.5862
1.4445	1	66 Dy	L_{II}	Abs. Edge	8.5830
1.44579	7	66 Dy	$L\gamma_6$	$L_{II}O_{IV}$	8.5753
1.45233	5	70 Yb	$L\beta_3$	L_IM_{III}	8.5367
1.4530	2	78 Pt	Lt	$L_{III}M_{II}$	8.533
1.45964	9	79 Au	Ll	$L_{III}M_I$	8.4939
1.4618	2	67 Ho	$L\gamma_5$	$L_{II}N_I$	8.481
1.4640	2	69 Tm	$L\beta_{2,15}$	$L_{III}N_{IV,V}$	8.468
1.4661	1	70 Yb	$L\beta_6$	$L_{III}N_I$	8.4563
1.47106	5	73 Ta	$L\eta$	$L_{II}M_I$	8.4280
1.4718	2	65 Tb	$L\gamma_3$	L_IN_{III}	8.423
1.47266	7	66 Dy	$L\gamma_1$	$L_{II}N_{IV}$	8.4188
1.4735	2	76 Os	Ls	$L_{III}M_{III}$	8.414
1.47565	5	70 Yb	$L\beta_1$	$L_{II}M_{IV}$	8.4018
1.4764	2	65 Tb	$L\gamma_2$	L_IN_{II}	8.398
1.47639	2	74 W	$L\alpha_1$	$L_{III}M_V$	8.3976
1.4784	1	64 Gd	L_I	Abs. Edge	8.3864
1.48064	9	72 Hf		$L_{II}M_{II}$	8.3735
1.4807	3	64 Gd		$L_IO_{IV,V}$	8.373
1.4835	1	68 Er	L_{III}	Abs. Edge	8.3575
1.4839	2	64 Gd	$L\gamma_4$	$L_IO_{II,III}$	8.355
1.4848	3	68 Er	$L\beta_5$	$L_{III}O_{IV,V}$	8.350
1.4855	5	68 Er	$L\beta_9$	L_IM_V	8.346
1.48743	2	74 W	$L\alpha_2$	$L_{III}M_{IV}$	8.3352
1.48807	1	28 Ni	K	Abs. Edge	8.33165
1.48862	4	28 Ni	$K\beta_5$	$KM_{IV,V}$	8.3286
1.49138	3	70 Yb	$L\beta_4$	L_IM_{II}	8.3132
1.4930	3	77 Ir	Lt	$L_{III}M_{II}$	8.304

Wavelength Å*	p.e.	Element	Designation		keV
1.4941	3	68 Er	$L\beta_7$	$L_{III}O_I$	8.298
1.4941	3	68 Er	$L\beta_{10}$	L_IM_{IV}	8.298
1.4995	2	78 Pt	Ll	$L_{III}M_I$	8.268
1.500135	8	28 Ni	$K\beta_{1,3}$	$KM_{II,III}$	8.26466
1.5023	1	65 Tb	L_{II}	Abs. Edge	8.2527
1.5035	2	65 Tb	$L\gamma_6$	$L_{II}O_{IV}$	8.246
1.5063	2	69 Tm	$L\beta_3$	L_IM_{III}	8.231
1.5097	2	65 Tb	$L\gamma_8$	$L_{II}O_I$	8.212
1.51399	9	68 Er	$L\beta_{2,15}$	$L_{III}N_{IV,V}$	8.1890
1.5162	2	69 Tm	$L\beta_6$	$L_{III}N_I$	8.177
1.5178	1	75 Re	Ls	$L_{III}M_{III}$	8.1682
1.51824	7	66 Dy	$L\gamma_5$	$L_{II}N_I$	8.1661
1.52197	2	73 Ta	$L\alpha_1$	$L_{III}M_V$	8.1461
1.52325	5	72 Hf	$L\eta$	$L_{II}M_I$	8.1393
1.5297	2	64 Gd	$L\gamma_3$	L_IN_{III}	8.105
1.5303	2	65 Tb	$L\gamma_1$	$L_{II}N_{IV}$	8.102
1.5304	2	69 Tm	$L\beta_1$	$L_{II}M_{IV}$	8.101
1.53293	2	73 Ta	$L\alpha_2$	$L_{III}M_{IV}$	8.0879
1.5331	2	64 Gd	$L\gamma_2$	L_IN_{II}	8.087
1.53333	9	71 Lu		$L_{II}M_{II}$	8.0858
1.5347	2	76 Os	Lt	$L_{III}M_{II}$	8.079
1.5368	1	67 Ho	L_{III}	Abs. Edge	8.0676
1.5378	2	67 Ho	$L\beta_5$	$L_{III}O_{IV,V}$	8.062
1.5381	1	63 Eu	L_I	Abs. Edge	8.0607
1.540562	2	29 Cu	$K\alpha_1$	KL_{III}	8.04778
1.54094	3	77 Ir	Ll	$L_{III}M_I$	8.0458
1.5439	1	63 Eu	$L\gamma_4$	$L_IO_{II,III}$	8.0304
1.544390	9	29 Cu	$K\alpha_2$	KL_{II}	8.02783
1.5448	2	69 Tm	$L\beta_4$	L_IM_{II}	8.026
1.5486	3	67 Ho	$L\beta_{10}$	L_IM_{IV}	8.006
1.5616	1	68 Er	$L\beta_3$	L_IM_{III}	7.9392
1.5632	1	64 Gd	L_{II}	Abs. Edge	7.9310
1.5642	3	74 W	Ls	$L_{III}M_{III}$	7.926
1.5644	2	64 Gd	$L\gamma_6$	$L_{II}O_{IV}$	7.925
1.5671	2	67 Ho	$L\beta_{2,15}$	$L_{III}N_{IV,V}$	7.911
1.5675	2	68 Er	$L\beta_6$	$L_{III}N_I$	7.909
1.56958	5	72 Hf	$L\alpha_1$	$L_{III}M_V$	7.8990
1.5707	2	64 Gd	$L\gamma_8$	$L_{II}O_I$	7.894
1.5779	1	71 Lu	$L\eta$	$L_{II}M_I$	7.8575
1.5787	2	65 Tb	$L\gamma_5$	$L_{II}N_I$	7.8535
1.5789	1	75 Re	Lt	$L_{III}M_{II}$	7.8525
1.58046	5	72 Hf	$L\alpha_2$	$L_{III}M_{IV}$	7.8446
1.58498	7	76 Os	Ll	$L_{III}M_I$	7.8222
1.5873	1	68 Er	$L\beta_1$	$L_{II}M_{IV}$	7.8109
1.58837	7	66 Dy	$L\beta_5$	$L_{III}O_{IV,V}$	7.8055
1.58844	9	70 Yb		$L_{II}M_{II}$	7.8052
1.5903	2	63 Eu	$L\gamma_3$	L_IN_{III}	7.7961
1.5916	1	66 Dy	L_{III}	Abs. Edge	7.7897
1.5924	2	64 Gd	$L\gamma_1$	$L_{II}N_{IV}$	7.7858
1.5961	2	63 Eu	$L\gamma_2$	L_IN_{II}	7.7677
1.59973	9	66 Dy	$L\beta_9$	L_IM_V	7.7501
1.6002	1	62 Sm	L_I	Abs. Edge	7.7478
1.6007	1	68 Er	$L\beta_4$	L_IM_{II}	7.7453
1.60447	7	66 Dy	$L\beta_7$	$L_{III}O_I$	7.7272
1.60728	3	62 Sm	$L\gamma_4$	$L_IO_{II,III}$	7.714
1.60743	9	66 Dy	$L\beta_{10}$	L_IM_{IV}	7.7130
1.60815	1	27 Co	K	Abs. Edge	7.70954
1.60891	3	27 Co	$K\beta_5$	$KM_{IV,V}$	7.7059
1.61264	9	73 Ta	Ls	$L_{III}M_{III}$	7.6881
1.61951	3	71 Lu	$L\alpha_1$	$L_{III}M_V$	7.6555
1.6203	2	67 Ho	$L\beta_3$	L_IM_{III}	7.6519
1.62079	2	27 Co	$K\beta_{1,3}$	$KM_{II,III}$	7.64943
1.6237	2	67 Ho	$L\beta_6$	$L_{III}N_I$	7.6359
1.62369	7	66 Dy	$L\beta_{2,15}$	$L_{III}N_{IV,V}$	7.6357
1.6244	3	74 W	Lt	$L_{III}M_{II}$	7.6324
1.6271	1	63 Eu	L_{II}	Abs. Edge	7.6199
1.6282	2	63 Eu	$L\gamma_6$	$L_{II}O_{IV}$	7.6147
1.63029	5	71 Lu	$L\alpha_2$	$L_{III}M_{IV}$	7.6049
1.63056	5	75 Re	Ll	$L_{III}M_I$	7.6036
1.6346	2	63 Eu	$L\gamma_8$	$L_{II}O_I$	7.5849
1.63560	5	70 Yb	$L\eta$	$L_{II}M_I$	7.5802
1.6412	2	64 Gd	$L\gamma_5$	$L_{II}N_I$	7.5543
1.6475	2	67 Ho	$L\beta_1$	$L_{II}M_{IV}$	7.5253
1.6497	1	65 Tb	L_{III}	Abs. Edge	7.5153
1.6510	2	65 Tb	$L\beta_5$	$L_{III}O_{IV,V}$	7.5094
1.65601	3	62 Sm	$L\gamma_3$	L_IN_{III}	7.487
1.6574	2	63 Eu	$L\gamma_1$	$L_{II}N_{IV}$	7.4803
1.657910	8	28 Ni	$K\alpha_1$	KL_{III}	7.47815
1.6585	2	65 Tb	$L\beta_7$	$L_{III}O_I$	7.4753
1.6595	2	67 Ho	$L\beta_4$	L_IM_{II}	7.4708
1.66044	6	62 Sm	$L\gamma_2$	L_IN_{II}	7.467
1.661747	8	28 Ni	$K\alpha_2$	KL_{II}	7.46089
1.66346	9	72 Hf	Ls	$L_{III}M_{III}$	7.4532
1.6673	3	65 Tb	$L\beta_{10}$	L_IM_{IV}	7.436
1.6674	5	61 Pm	L_I	Abs. Edge	7.436
1.67189	4	70 Yb	$L\alpha_1$	$L_{III}M_V$	7.4156
1.67265	9	73 Ta	Lt	$L_{III}M_{II}$	7.4123
1.6782	1	74 W	Ll	$L_{III}M_I$	7.3878
1.68213	7	66 Dy	$L\beta_6$	$L_{III}N_I$	7.3705
1.6822	2	66 Dy	$L\beta_3$	L_IM_{III}	7.3702
1.68285	5	70 Yb	$L\alpha_2$	$L_{III}M_{IV}$	7.3673
1.6830	2	65 Tb	$L\beta_{2,15}$	$L_{III}N_{IV,V}$	7.3667
1.6953	1	62 Sm	L_{II}	Abs. Edge	7.3132
1.6963	2	69 Tm	$L\eta$	$L_{II}M_I$	7.3088
1.6966	9	62 Sm	$L\gamma_6$	$L_{II}O_{IV}$	7.308
1.7085	2	63 Eu	$L\gamma_5$	$L_{II}N_I$	7.2566
1.71062	7	66 Dy	$L\beta_1$	$L_{II}M_{IV}$	7.2477
1.7117	1	64 Gd	L_{III}	Abs. Edge	7.2430
1.7130	2	64 Gd	$L\beta_5$	$L_{III}O_{IV,V}$	7.2374
1.7203	2	64 Gd	$L\beta_7$	$L_{III}O_I$	7.2071
1.72103	7	66 Dy	$L\beta_4$	L_IM_{II}	7.2039
1.72305	9	72 Hf	Lt	$L_{III}M_{II}$	7.1954
1.7240	3	64 Gd	$L\beta_9$	L_IM_V	7.192
1.72724	3	62 Sm	$L\gamma_1$	$L_{II}N_{IV}$	7.178
1.7268	2	69 Tm	$L\alpha_1$	$L_{III}M_V$	7.1799
1.72841	5	73 Ta	Ll	$L_{III}M_I$	7.1731
1.7315	3	64 Gd	$L\beta_{10}$	L_IM_{IV}	7.160
1.7381	2	69 Tm	$L\alpha_2$	$L_{III}M_{IV}$	7.1331
1.7390	1	60 Nd	L_I	Abs. Edge	7.1294
1.7422	2	65 Tb	$L\beta_6$	$L_{III}N_I$	7.1163
1.74346	1	26 Fe	K	Abs. Edge	7.11120
1.7442	1	26 Fe	$K\beta_5$	$KM_{IV,V}$	7.1081
1.7445	4	60 Nd	$L\gamma_4$	$L_IO_{II,III}$	7.107
1.7455	2	64 Gd	$L\beta_{2,15}$	$L_{III}N_{IV,V}$	7.1028

Wavelength Å*	p.e.	Element	Designation		keV
1.7472	2	65 Tb	$L\beta_3$	$L_{I}M_{III}$	7.0959
1.75661	2	26 Fe	$K\beta_{1,3}$	$KM_{II,III}$	7.05798
1.7566	1	68 Er	$L\eta$	$L_{II}M_{I}$	7.0579
1.7676	5	61 Pm	L_{II}	Abs. Edge	7.014
1.7760	1	71 Lu	Lt	$L_{III}M_{II}$	6.9810
1.7761	1	63 Eu	L_{III}	Abs. Edge	6.9806
1.7768	3	65 Tb	$L\beta_1$	$L_{II}M_{IV}$	6.978
1.7772	2	63 Eu	$L\beta_5$	$L_{III}O_{IV,V}$	6.9763
1.77934	3	62 Sm	$L\gamma_5$	$L_{II}N_{I}$	6.968
1.78145	5	72 Hf	Ll	$L_{III}M_{I}$	6.9596
1.78425	9	68 Er	$L\alpha_1$	$L_{III}M_{V}$	6.9487
1.7851	2	63 Eu	$L\beta_7$	$L_{III}O_{I}$	6.9453
1.7864	2	65 Tb	$L\beta_4$	$L_{I}M_{II}$	6.9403
1.788965	9	27 Co	$K\alpha_1$	KL_{III}	6.93032
1.7916	3	63 Eu	$L\beta_9$	$L_{I}M_{V}$	6.920
1.792850	9	27 Co	$K\alpha_2$	KL_{II}	6.91530
1.7955	2	68 Er	$L\alpha_2$	$L_{III}M_{IV}$	6.9050
1.7964	4	60 Nd	$L\gamma_3$	$L_{I}N_{III}$	6.902
1.7989	9	61 Pm	$L\gamma_1$	$L_{II}N_{IV}$	6.892
1.7993	3	63 Eu	$L\beta_{10}$	$L_{I}M_{IV}$	6.890
1.8013	4	60 Nd	$L\gamma_2$	$L_{I}N_{II}$	6.883
1.8054	2	64 Gd	$L\beta_6$	$L_{III}N_{I}$	6.8671
1.8118	2	63 Eu	$L\beta_{2,15}$	$L_{III}N_{IV,V}$	6.8432
1.8141	5	59 Pr	L_{I}	Abs. Edge	6.834
1.8150	2	64 Gd	$L\beta_3$	$L_{I}M_{III}$	6.8311
1.8193	4	59 Pr	$L\gamma_4$	$L_{I}O_{II,III}$	6.815
1.8264	2	67 Ho	$L\eta$	$L_{II}M_{I}$	6.7883
1.83091	9	70 Yb	Lt	$L_{III}M_{II}$	6.7715
1.8360	1	71 Lu	Ll	$L_{III}M_{I}$	6.7528
1.8440	1	60 Nd	L_{II}	Abs. Edge	6.7234
1.8450	2	67 Ho	$L\alpha_1$	$L_{III}M_{V}$	6.7198
1.8457	1	62 Sm	L_{III}	Abs. Edge	6.7172
1.8468	2	64 Gd	$L\beta_1$	$L_{II}M_{IV}$	6.7132
1.84700	9	62 Sm	$L\beta_5$	$L_{III}O_{IV,V}$	6.7126
1.8540	2	64 Gd	$L\beta_4$	$L_{I}M_{II}$	6.6871
1.8552	5	60 Nd	$L\gamma_8$	$L_{II}O_{I}$	6.683
1.8561	2	67 Ho	$L\alpha_2$	$L_{III}M_{IV}$	6.6795
1.85626	3	62 Sm	$L\beta_7$	$L_{III}O_{I}$	6.679
1.86166	3	62 Sm	$L\beta_9$	$L_{I}M_{V}$	6.660
1.86990	3	62 Sm	$L\beta_{10}$	$L_{I}M_{IV}$	6.634
1.8737	2	63 Eu	$L\beta_6$	$L_{III}N_{I}$	6.6170
1.8740	4	59 Pr	$L\gamma_3$	$L_{I}N_{III}$	6.616
1.8779	2	60 Nd	$L\gamma_1$	$L_{II}N_{IV}$	6.6021
1.8791	4	59 Pr	$L\gamma_2$	$L_{I}N_{II}$	6.598
1.8821	3	62 Sm	$L\beta_{2,15}$	$L_{III}N_{IV,V}$	6.586
1.8867	2	63 Eu	$L\beta_3$	$L_{I}M_{III}$	6.5713
1.8934	5	58 Ce	L_{I}	Abs. Edge	6.548
1.89415	5	70 Yb	Ll	$L_{III}M_{I}$	6.5455
1.89643	5	25 Mn	K	Abs. Edge	6.5376
1.8971	1	25 Mn	$K\beta_5$	$KM_{IV,V}$	6.5352
1.89743	7	66 Dy	$L\eta$	$L_{II}M_{I}$	6.5342
1.8991	4	58 Ce	$L\gamma_4$	$L_{I}O_{II,III}$	6.528
1.90881	3	66 Dy	$L\alpha_1$	$L_{III}M_{V}$	6.4952
1.91021	2	25 Mn	$K\beta_{1,3}$	$KM_{II,III}$	6.49045
1.9191	1	61 Pm	L_{III}	Abs. Edge	6.4605
1.91991	3	66 Dy	$L\alpha_2$	$L_{III}M_{IV}$	6.4577
1.9203	2	63 Eu	$L\beta_1$	$L_{II}M_{IV}$	6.4564

Wavelength Å*	p.e.	Element	Designation		keV
1.9255	2	63 Eu	$L\beta_4$	$L_{I}M_{II}$	6.4389
1.9255	5	59 Pr	L_{II}	Abs. Edge	6.439
1.9355	4	60 Nd	$L\gamma_5$	$L_{II}N_{I}$	6.406
1.936042	9	26 Fe	$K\alpha_1$	KL_{III}	6.40384
1.9362	4	59 Pr	$L\gamma_8$	$L_{II}O_{I}$	6.403
1.939980	9	26 Fe	$K\alpha_2$	KL_{II}	6.39084
1.94643	3	62 Sm	$L\beta_6$	$L_{III}N_{I}$	6.3693
1.9550	2	69 Tm	Ll	$L_{III}M_{I}$	6.3419
1.9553	3	58 Ce	$L\gamma_3$	$L_{I}N_{III}$	6.3409
1.9559	6	61 Pm	$L\beta_{2,15}$	$L_{III}N_{IV,V}$	6.339
1.9602	3	58 Ce	$L\gamma_2$	$L_{I}N_{II}$	6.3250
1.9611	3	59 Pr	$L\gamma_1$	$L_{II}N_{IV}$	6.3221
1.96241	3	62 Sm	$L\beta_3$	$L_{I}M_{III}$	6.318
1.9730	2	65 Tb	$L\eta$	$L_{II}M_{I}$	6.2839
1.9765	2	65 Tb	$L\alpha_1$	$L_{III}M_{V}$	6.2728
1.9780	5	57 La	L_{I}	Abs. Edge	6.268
1.9830	4	57 La	$L\gamma_4$	$L_{I}O_{II,III}$	6.252
1.9875	2	65 Tb	$L\alpha_2$	$L_{III}M_{IV}$	6.2380
1.9967	1	60 Nd	L_{III}	Abs. Edge	6.2092
1.99806	3	62 Sm	$L\beta_1$	$L_{II}M_{IV}$	6.2051
2.00095	6	62 Sm	$L\beta_4$	$L_{I}M_{II}$	6.196
2.0092	3	60 Nd	$L\beta_7$	$L_{III}O_{I}$	6.1708
2.0124	5	58 Ce	L_{II}	Abs. Edge	6.161
2.015	1	68 Er	Ll	$L_{III}M_{I}$	6.152
2.0165	3	60 Nd	$L\beta_9$	$L_{I}M_{V}$	6.1484
2.0205	4	59 Pr	$L\gamma_5$	$L_{II}N_{I}$	6.136
2.0237	4	58 Ce	$L\gamma_8$	$L_{II}O_{I}$	6.126
2.0237	3	60 Nd	$L\beta_{10}$	$L_{I}M_{IV}$	6.1265
2.0360	3	60 Nd	$L\beta_{2,15}$	$L_{III}N_{IV,V}$	6.0894
2.0410	4	57 La	$L\gamma_3$	$L_{I}N_{III}$	6.074
2.0421	4	61 Pm	$L\beta_3$	$L_{I}M_{III}$	6.071
2.0460	4	57 La	$L\gamma_2$	$L_{I}N_{II}$	6.060
2.0468	2	64 Gd	$L\alpha_1$	$L_{III}M_{V}$	6.0572
2.0487	4	58 Ce	$L\gamma_1$	$L_{II}N_{IV}$	6.052
2.0494	1	64 Gd	$L\eta$	$L_{II}M_{I}$	6.0495
2.0578	2	64 Gd	$L\alpha_2$	$L_{III}M_{IV}$	6.0250
2.0678	5	56 Ba	L_{I}	Abs. Edge	5.996
2.07020	5	24 Cr	K	Abs. Edge	5.9888
2.07087	6	24 Cr	$K\beta_5$	$KM_{IV,V}$	5.9869
2.0756	3	56 Ba	$L\gamma_4$	$L_{I}O_{II,III}$	5.9733
2.0791	5	59 Pr	L_{III}	Abs. Edge	5.963
2.0797	4	61 Pm	$L\beta_1$	$L_{II}M_{IV}$	5.961
2.08487	2	24 Cr	$K\beta_{1,3}$	$KM_{II,III}$	5.94671
2.0860	2	67 Ho	Ll	$L_{III}M_{I}$	5.9434
2.0919	4	59 Pr	$L\beta_7$	$L_{III}O_{I}$	5.927
2.1004	4	59 Pr	$L\beta_9$	$L_{I}M_{V}$	5.903
2.101820	9	25 Mn	$K\alpha_1$	KL_{III}	5.89875
2.1039	3	60 Nd	$L\beta_6$	$L_{III}N_{I}$	5.8930
2.1053	5	57 La	L_{II}	Abs. Edge	5.889
2.10578	2	25 Mn	$K\alpha_2$	KL_{II}	5.88765
2.1071	4	59 Pr	$L\beta_{10}$	$L_{I}M_{IV}$	5.884
2.1103	3	58 Ce	$L\gamma_5$	$L_{II}N_{I}$	5.8751
2.1194	4	59 Pr	$L\beta_{2,15}$	$L_{III}N_{IV,V}$	5.850
2.1209	2	63 Eu	$L\alpha_1$	$L_{III}M_{V}$	5.8457
2.1268	2	60 Nd	$L\beta_3$	$L_{I}M_{III}$	5.8294
2.1315	2	63 Eu	$L\eta$	$L_{II}M_{I}$	5.8166
2.1315	2	63 Eu	$L\alpha_2$	$L_{III}M_{IV}$	5.8166

Wavelength Å*	p.e.	Element	Designation		keV	Wavelength Å*	p.e.	Element	Designation		keV
2.1342	2	56 Ba	$L\gamma_3$	$L_I N_{III}$	5.8092	2.3913	2	53 I	$L\gamma_4$	$L_I O_{II,III}$	5.1848
2.1387	2	56 Ba	$L\gamma_2$	$L_I N_{II}$	5.7969	2.3948	2	63 Eu	Ll	$L_{III} M_I$	5.1772
2.1418	3	57 La	$L\gamma_1$	$L_{II} N_{IV}$	5.7885	2.40435	6	56 Ba	$L\beta_{2,15}$	$L_{III} N_{IV,V}$	5.1565
2.15877	7	66 Dy	Ll	$L_{III} M_I$	5.7431	2.4094	4	60 Nd	$L\eta$	$L_{II} M_I$	5.1457
2.166	1	58 Ce	L_{III}	Abs. Edge	5.723	2.4105	3	57 La	$L\beta_3$	$L_I M_{III}$	5.1434
2.1669	3	60 Nd	$L\beta_4$	$L_I M_{II}$	5.7216	2.4174	2	55 Cs	$L\gamma_5$	$L_{II} N_I$	5.1287
2.1669	2	60 Nd	$L\beta_1$	$L_{II} M_{IV}$	5.7216	2.4292	1	54 Xe	L_{II}	Abs. Edge	5.1037
2.1673	5	55 Cs	L_I	Abs. Edge	5.721	2.442	9	90 Th		$M_I O_{III}$	5.08
2.1701	2	58 Ce	$L\beta_7$	$L_{III} O_I$	5.7132	2.443	4	92 U		$M_{II} O_{IV}$	5.075
2.1741	2	55 Cs	$L\gamma_4$	$L_I O_{II,III}$	5.7026	2.4475	2	53 I	$L\gamma_{2,3}$	$L_I N_{II,III}$	5.0657
2.1885	3	58 Ce	$L\beta_9$	$L_I M_V$	5.6650	2.4493	3	57 La	$L\beta_4$	$L_I M_{II}$	5.0620
2.1906	4	59 Pr	$L\beta_6$	$L_{III} N_I$	5.660	2.45891	5	57 La	$L\beta_1$	$L_{II} M_{IV}$	5.0421
2.1958	5	58 Ce	$L\beta_{10}$	$L_I M_{IV}$	5.646	2.4630	2	59 Pr	$L\alpha_1$	$L_{III} M_V$	5.0337
2.1998	2	62 Sm	$L\alpha_1$	$L_{III} M_V$	5.6361	2.4729	3	59 Pr	$L\alpha_2$	$L_{III} M_{IV}$	5.0135
2.2048	1	56 Ba	L_{II}	Abs. Edge	5.6233	2.4740	1	55 Cs	L_{III}	Abs. Edge	5.0113
2.2056	4	57 La	$L\gamma_5$	$L_{II} N_I$	5.621	2.4783	2	55 Cs	$L\beta_9$	$L_I M_V$	5.0026
2.2087	2	58 Ce	$L\beta_{2,15}$	$L_{III} N_{IV,V}$	5.6134	2.4823	4	62 Sm	Ll	$L_{III} M_I$	4.9945
2.21062	3	62 Sm	$L\alpha_2$	$L_{III} M_{IV}$	5.6090	2.4826	2	56 Ba	$L\beta_6$	$L_{III} N_I$	4.9939
2.2172	3	59 Pr	$L\beta_3$	$L_I M_{III}$	5.5918	2.4849	2	55 Cs	$L\beta_7$	$L_{III} O_I$	4.9893
2.21824	3	62 Sm	$L\eta$	$L_{II} M_I$	5.589	2.4920	2	55 Cs	$L\beta_{10}$	$L_I M_{IV}$	4.9752
2.2328	2	55 Cs	$L\gamma_3$	$L_I N_{III}$	5.5527	2.49734	5	22 Ti	K	Abs. Edge	4.96452
2.2352	2	65 Tb	Ll	$L_{III} M_I$	5.5467	2.4985	2	22 Ti	$K\beta_5$	$K M_{IV,V}$	4.9623
2.2371	2	55 Cs	$L\gamma_2$	$L_I N_{II}$	5.5420	2.50356	2	23 V	$K\alpha_1$	$K L_{III}$	4.95220
2.2415	2	56 Ba	$L\gamma_1$	$L_{II} N_{IV}$	5.5311	2.50738	2	23 V	$K\alpha_2$	$K L_{II}$	4.94464
2.253	6	92 U		$M_I P_{III}$	5.50	2.5099	1	52 Te	L_I	Abs. Edge	4.9397
2.2550	4	59 Pr	$L\beta_4$	$L_I M_{II}$	5.4981	2.5113	2	52 Te	$L\gamma_4$	$L_I O_{II,III}$	4.9369
2.2588	3	59 Pr	$L\beta_1$	$L_{II} M_{IV}$	5.4889	2.5118	2	55 Cs	$L\beta_{2,15}$	$L_{III} N_{IV,V}$	4.9359
2.261	1	57 La	L_{III}	Abs. Edge	5.484	2.512	3	59 Pr	$L\eta$	$L_{II} M_I$	4.935
2.2691	1	23 V	K	Abs. Edge	5.4639	2.51391	2	22 Ti	$K\beta_{1,3}$	$K M_{II,III}$	4.93181
2.26951	6	23 V	$K\beta_5$	$K M_{IV,V}$	5.4629	2.5164	2	56 Ba	$L\beta_3$	$L_I M_{III}$	4.9269
2.2737	1	54 Xe	L_I	Abs. Edge	5.4528	2.527	4	91 Pa		$M_{II} O_{IV}$	4.906
2.275	3	57 La	$L\beta_7$	$L_{III} O_I$	5.450	2.5542	5	53 I	L_{II}	Abs. Edge	4.8540
2.282	3	57 La	$L\beta_9$	$L_I M_V$	5.434	2.5553	2	56 Ba	$L\beta_4$	$L_I M_{II}$	4.8519
2.2818	3	58 Ce	$L\beta_6$	$L_{III} N_I$	5.4334	2.5615	2	58 Ce	$L\alpha_1$	$L_{III} M_V$	4.8402
2.2822	3	61 Pm	$L\alpha_1$	$L_{III} M_V$	5.4325	2.5674	2	52 Te	$L\gamma_{2,3}$	$L_I N_{II,III}$	4.8290
2.28440	2	23 V	$K\beta_{1,3}$	$K M_{II,III}$	5.42729	2.56821	5	56 Ba	$L\beta_1$	$L_{II} M_{IV}$	4.82753
2.28970	2	24 Cr	$K\alpha_1$	$K L_{III}$	5.41472	2.5706	3	58 Ce	$L\alpha_2$	$L_{III} M_{IV}$	4.8230
2.290	3	57 La	$L\beta_{10}$	$L_I M_{IV}$	5.415	2.58244	8	53 I	$L\gamma_1$	$L_{II} N_{IV}$	4.8009
2.2926	4	61 Pm	$L\alpha_2$	$L_{III} M_{IV}$	5.4078	2.5926	1	54 Xe	L_{III}	Abs. Edge	4.7822
2.293606	3	24 Cr	$K\alpha_2$	$K L_{II}$	5.405509	2.5932	2	55 Cs	$L\beta_6$	$L_{III} N_I$	4.7811
2.3030	3	57 La	$L\beta_{2,15}$	$L_{III} N_{IV,V}$	5.3835	2.618	5	90 Th		$M_{II} O_{IV}$	4.735
2.304	7	92 U		$M_I O_{III}$	5.38	2.6203	4	58 Ce	$L\eta$	$L_{II} M_I$	4.7315
2.3085	3	56 Ba	$L\gamma_5$	$L_{II} N_I$	5.3707	2.6285	2	55 Cs	$L\beta_3$	$L_I M_{III}$	4.7167
2.3109	3	58 Ce	$L\beta_3$	$L_I M_{III}$	5.3651	2.6388	1	51 Sb	L_I	Abs. Edge	4.6984
2.3122	2	64 Gd	Ll	$L_{III} M_I$	5.3621	2.6398	2	51 Sb	$L\gamma_4$	$L_I O_{II,III}$	4.6967
2.3139	1	55 Cs	L_{II}	Abs. Edge	5.3581	2.65710	9	53 I	$L\gamma_5$	$L_{II} N_I$	4.6660
2.3480	2	55 Cs	$L\gamma_1$	$L_{II} N_{IV}$	5.2804	2.66570	5	57 La	$L\alpha_1$	$L_{III} M_V$	4.65097
2.3497	4	58 Ce	$L\beta_4$	$L_I M_{II}$	5.2765	2.6666	2	55 Cs	$L\beta_4$	$L_I M_{II}$	4.6494
2.3561	3	58 Ce	$L\beta_1$	$L_{II} M_{IV}$	5.2622	2.67533	5	57 La	$L\alpha_2$	$L_{III} M_{IV}$	4.63423
2.3629	1	56 Ba	L_{III}	Abs. Edge	5.2470	2.6760	4	60 Nd	Ll	$L_{III} M_I$	4.6330
2.3704	2	60 Nd	$L\alpha_1$	$L_{III} M_V$	5.2304	2.6837	2	55 Cs	$L\beta_1$	$L_{II} M_{IV}$	4.6198
2.3764	2	56 Ba	$L\beta_9$	$L_I M_V$	5.2171	2.6879	1	52 Te	L_{II}	Abs. Edge	4.6126
2.3790	4	57 La	$L\beta_6$	$L_{III} N_I$	5.2114	2.6953	2	51 Sb	$L\gamma_{2,3}$	$L_I N_{II,III}$	4.5999
2.3806	2	56 Ba	$L\beta_7$	$L_{III} O_I$	5.2079	2.71241	6	52 Te	$L\gamma_1$	$L_{II} N_{IV}$	4.5709
2.3807	3	60 Nd	$L\alpha_2$	$L_{III} M_{IV}$	5.2077	2.71352	9	53 I	$L\beta_9$	$L_I M_V$	4.5690
2.3869	2	56 Ba	$L\beta_{10}$	$L_I M_{IV}$	5.1941	2.7196	5	53 I	L_{III}	Abs. Edge	4.5587
2.3880	5	53 I	L_I	Abs. Edge	5.192	2.72104	9	53 I	$L\beta_{10}$	$L_I M_{IV}$	4.5564

Wavelength Å*	p.e.	Element	Designation		keV
2.7288	3	53 I	$L\beta_7$	$L_{III}O_{I}$	4.5435
2.740	3	57 La	$L\eta$	$L_{II}M_{I}$	4.525
2.74851	2	22 Ti	$K\alpha_1$	KL_{III}	4.51084
2.75053	8	53 I	$L\beta_{2,15}$	$L_{III}N_{IV,V}$	4.5075
2.75216	2	22 Ti	$K\alpha_2$	KL_{II}	4.50486
2.753	8	92 U		$M_{I}N_{III}$	4.50
2.762	1	21 Sc	K	Abs. Edge	4.489
2.7634	3	21 Sc	$K\beta_5$	$KM_{IV,V}$	4.4865
2.77595	5	56 Ba	$L\alpha_1$	$L_{III}M_{V}$	4.46626
2.7769	1	50 Sn	L_{I}	Abs. Edge	4.4648
2.7775	2	50 Sn	$L\gamma_4$	$L_{I}O_{II,III}$	4.4638
2.7796	2	21 Sc	$K\beta_{1,3}$	$KM_{II,III}$	4.4605
2.7841	4	59 Pr	Ll	$L_{III}M_{I}$	4.4532
2.78553	5	56 Ba	$L\alpha_2$	$L_{III}M_{IV}$	4.45090
2.79007	9	52 Te	$L\gamma_5$	$L_{II}N_{I}$	4.4437
2.817	2	92 U		$M_{II}N_{IV}$	4.401
2.8294	5	51 Sb	L_{II}	Abs. Edge	4.3819
2.8327	2	50 Sn	$L\gamma_{2,3}$	$L_{I}N_{II,III}$	4.3768
2.83672	9	53 I	$L\beta_6$	$L_{III}N_{I}$	4.3706
2.83897	9	52 Te	$L\beta_9$	$L_{I}M_{V}$	4.3671
2.84679	9	52 Te	$L\beta_{10}$	$L_{I}M_{IV}$	4.3551
2.85159	3	51 Sb	$L\gamma_1$	$L_{II}N_{IV}$	4.34779
2.8555	1	52 Te	L_{III}	Abs. Edge	4.3418
2.8627	3	56 Ba	$L\eta$	$L_{II}M_{I}$	4.3309
2.8634	3	52 Te	$L\beta_7$	$L_{III}O_{I}$	4.3298
2.87429	9	53 I	$L\beta_3$	$L_{I}M_{III}$	4.3134
2.88217	8	52 Te	$L\beta_{2,15}$	$L_{III}N_{IV,V}$	4.3017
2.884	5	92 U	M_{III}	Abs. Edge	4.299
2.8917	4	58 Ce	Ll	$L_{III}M_{I}$	4.2875
2.8924	2	55 Cs	$L\alpha_1$	$L_{III}M_{V}$	4.2865
2.9020	2	55 Cs	$L\alpha_2$	$L_{III}M_{IV}$	4.2722
2.910	2	91 Pa		$M_{II}N_{IV}$	4.260
2.91207	9	53 I	$L\beta_4$	$L_{I}M_{II}$	4.2575
2.92	2	92 U		$M_{I}N_{II}$	4.25
2.9260	1	49 In	L_{I}	Abs. Edge	4.2373
2.9264	2	49 In	$L\gamma_4$	$L_{I}O_{II,III}$	4.2367
2.93187	9	51 Sb	$L\gamma_5$	$L_{II}N_{I}$	4.2287
2.934	8	90 Th		$M_{I}N_{III}$	4.23
2.93744	6	53 I	$L\beta_1$	$L_{II}M_{IV}$	4.22072
2.948	2	92 U		$M_{III}O_{IV,V}$	4.205
2.97088	9	52 Te	$L\beta_6$	$L_{III}N_{I}$	4.1732
2.97261	9	51 Sb	$L\beta_9$	$L_{I}M_{V}$	4.1708
2.97917	9	51 Sb	$L\beta_{10}$	$L_{I}M_{IV}$	4.1616
2.9800	2	49 In	$L\gamma_{2,3}$	$L_{I}N_{II,III}$	4.1605
2.9823	1	50 Sn	L_{II}	Abs. Edge	4.1573
2.9932	2	55 Cs	$L\eta$	$L_{II}M_{I}$	4.1421
3.0003	1	51 Sb	L_{III}	Abs. Edge	4.1323
3.00115	3	50 Sn	$L\gamma_1$	$L_{II}N_{IV}$	4.13112
3.0052	3	51 Sb	$L\beta_7$	$L_{III}O_{I}$	4.1255
3.006	3	57 La	Ll	$L_{III}M_{I}$	4.124
3.00893	9	52 Te	$L\beta_3$	$L_{I}M_{III}$	4.1204
3.011	2	90 Th		$M_{II}N_{IV}$	4.117
3.0166	2	54 Xe	$L\alpha_1$	$L_{III}M_{V}$	4.1099
3.02335	3	51 Sb	$L\beta_{2,15}$	$L_{III}N_{IV,V}$	4.10078
3.0309	1	21 Sc	$K\alpha_1$	KL_{III}	4.0906
3.0342	1	21 Sc	$K\alpha_2$	KL_{II}	4.0861
3.038	2	91 Pa		$M_{III}O_{IV,V}$	4.081
3.04661	9	52 Te	$L\beta_4$	$L_{I}M_{II}$	4.0695
3.068	5	90 Th	M_{III}	Abs. Edge	4.041
3.0703	1	20 Ca	K	Abs. Edge	4.0381
3.0746	3	20 Ca	$K\beta_5$	$KM_{IV,V}$	4.0325
3.07677	6	52 Te	$L\beta_1$	$L_{II}M_{IV}$	4.02958
3.08475	9	50 Sn	$L\gamma_5$	$L_{II}N_{I}$	4.0192
3.0849	1	48 Cd	L_{I}	Abs. Edge	4.0190
3.0897	2	20 Ca	$K\beta_{1,3}$	$KM_{II,III}$	4.0127
3.094	5	83 Bi	M_{I}	Abs. Edge	4.007
3.11513	9	50 Sn	$L\beta_9$	$L_{I}M_{V}$	3.9800
3.11513	9	51 Sb	$L\beta_6$	$L_{III}N_{I}$	3.9800
3.115	7	92 U		$M_{III}O_{I}$	3.980
3.12170	9	50 Sn	$L\beta_{10}$	$L_{I}M_{IV}$	3.9716
3.131	3	90 Th		$M_{III}O_{IV,V}$	3.959
3.1355	2	56 Ba	Ll	$L_{III}M_{I}$	3.9541
3.1377	2	48 Cd	$L\gamma_2$	$L_{I}N_{II}$	3.9513
3.1473	1	49 In	L_{II}	Abs. Edge	3.9393
3.14860	6	53 I	$L\alpha_1$	$L_{III}M_{V}$	3.93765
3.15258	9	51 Sb	$L\beta_3$	$L_{I}M_{III}$	3.9327
3.1557	1	50 Sn	L_{III}	Abs. Edge	3.9288
3.1564	3	50 Sn	$L\beta_7$	$L_{III}O_{I}$	3.9279
3.15791	6	53 I	$L\alpha_2$	$L_{III}M_{IV}$	3.92604
3.16213	4	49 In	$L\gamma_1$	$L_{II}N_{IV}$	3.92081
3.17505	3	50 Sn	$L\beta_{2,15}$	$L_{III}N_{IV,V}$	3.90486
3.19014	9	51 Sb	$L\beta_4$	$L_{I}M_{II}$	3.8364
3.217	5	82 Pb	M_{I}	Abs. Edge	3.854
3.22567	4	51 Sb	$L\beta_1$	$L_{II}M_{IV}$	3.84357
3.245	9	91 Pa		$M_{III}O_{I}$	3.82
3.24907	9	49 In	$L\gamma_5$	$L_{II}N_{I}$	3.8159
3.2564	1	47 Ag	L_{I}	Abs. Edge	3.8072
3.2670	2	55 Cs	Ll	$L_{III}M_{I}$	3.7950
3.26763	9	49 In	$L\beta_9$	$L_{I}M_{V}$	3.7942
3.26901	9	50 Sn	$L\beta_6$	$L_{III}N_{I}$	3.7926
3.27404	9	49 In	$L\beta_{10}$	$L_{I}M_{IV}$	3.7868
3.27979	9	53 I	$L\eta$	$L_{II}M_{I}$	3.7801
3.283	9	90 Th		$M_{III}O_{I}$	3.78
3.28920	6	52 Te	$L\alpha_1$	$L_{III}M_{V}$	3.76933
3.29846	9	52 Te	$L\alpha_2$	$L_{III}M_{IV}$	3.7588
3.30585	3	50 Sn	$L\beta_3$	$L_{I}M_{III}$	3.7500
3.30635	9	47 Ag	$L\gamma_3$	$L_{I}N_{III}$	3.7498
3.31216	9	47 Ag	$L\gamma_2$	$L_{I}N_{II}$	3.7432
3.3237	1	49 In	L_{III}	Abs. Edge	3.7302
3.324	4	49 In	$L\beta_7$	$L_{III}O_{I}$	3.730
3.3257	1	48 Cd	L_{II}	Abs. Edge	3.7280
3.329	4	92 U		$M_{II}N_{I}$	3.724
3.333	5	92 U	M_{IV}	Abs. Edge	3.720
3.33564	6	48 Cd	$L\gamma_1$	$L_{II}N_{IV}$	3.71686
3.33838	3	49 In	$L\beta_{2,15}$	$L_{III}N_{IV,V}$	3.71381
3.34335	9	50 Sn	$L\beta_4$	$L_{I}M_{II}$	3.7083
3.346	5	81 Tl	M_{I}	Abs. Edge	3.705
3.35839	3	20 Ca	$K\alpha_1$	KL_{III}	3.69168
3.359	5	83 Bi	M_{II}	Abs. Edge	3.691
3.36166	3	20 Ca	$K\alpha_2$	KL_{II}	3.68809
3.38487	3	50 Sn	$L\beta_1$	$L_{II}M_{IV}$	3.66280
3.42551	9	48 Cd	$L\gamma_5$	$L_{II}N_{I}$	3.61935
3.43015	9	48 Cd	$L\beta_9$	$L_{I}M_{V}$	3.61445
3.43606	9	49 In	$L\beta_6$	$L_{III}N_{I}$	3.60823

Wavelength Å*	p.e.	Element	Designation		keV
3.4365	1	19 K	K	Abs. Edge	3.6078
3.4367	2	48 Cd	$L\beta_{10}$	L_IM_{IV}	3.6075
3.437	1	46 Pd	L_I	Abs. Edge	3.607
3.43832	9	52 Te	$L\eta$	$L_{II}M_I$	3.60586
3.43941	4	51 Sb	$L\alpha_1$	$L_{III}M_V$	3.60472
3.441	5	91 Pa		$M_{II}N_I$	3.603
3.4413	4	19 K	$K\beta_5$	$KM_{IV,V}$	3.6027
3.44840	6	51 Sb	$L\alpha_2$	$L_{III}M_{IV}$	3.59532
3.4539	2	19 K	$K\beta_{1,3}$	$KM_{II,III}$	3.5896
3.46984	9	49 In	$L\beta_3$	L_IM_{III}	3.57311
3.478	5	80 Hg	M_I	Abs. Edge	3.565
3.479	1	92 U	$M\gamma$	$M_{III}N_V$	3.563
3.4892	2	46 Pd	$L\gamma_{2,3}$	$L_IN_{II,III}$	3.5533
3.492	5	82 Pb	M_{II}	Abs. Edge	3.550
3.497	5	92 U	M_V	Abs. Edge	3.545
3.5047	1	48 Cd	L_{III}	Abs. Edge	3.5376
3.50697	9	49 In	$L\beta_4$	L_IM_{II}	3.53528
3.51408	4	48 Cd	$L\beta_{2,15}$	$L_{III}N_{IV,V}$	3.52812
3.5164	1	47 Ag	L_{II}	Abs. Edge	3.5258
3.521	2	92 U		$M_{III}N_{IV}$	3.521
3.52260	4	47 Ag	$L\gamma_1$	$L_{II}N_{IV}$	3.51959
3.537	9	90 Th		$M_{II}N_I$	3.505
3.55531	4	49 In	$L\beta_1$	$L_{II}M_{IV}$	3.48721
3.557	5	90 Th	M_{IV}	Abs. Edge	3.485
3.55754	9	53 I	Ll	$L_{III}M_I$	3.48502
3.576	1	92 U		$M_{IV}O_{II}$	3.4666
3.577	1	91 Pa	$M\gamma$	$M_{III}N_V$	3.4657
3.59994	3	50 Sn	$L\alpha_1$	$L_{III}M_V$	3.44398
3.60497	9	47 Ag	$L\beta_9$	L_IM_V	3.43917
3.60765	9	51 Sb	$L\eta$	$L_{II}M_I$	3.43661
3.60891	4	50 Sn	$L\alpha_2$	$L_{III}M_{IV}$	3.43542
3.61158	9	47 Ag	$L\beta_{10}$	L_IM_{IV}	3.43287
3.614	2	91 Pa		$M_{III}N_{IV}$	3.430
3.61467	9	48 Cd	$L\beta_6$	$L_{III}N_I$	3.42994
3.61638	9	47 Ag	$L\gamma_5$	$L_{II}N_I$	3.42832
3.616	5	79 Au	M_I	Abs. Edge	3.428
3.629	5	45 Rh	L_I	Abs. Edge	3.417
3.634	5	81 Tl	M_{II}	Abs. Edge	3.412
3.64495	9	48 Cd	$L\beta_3$	L_IM_{III}	3.40145
3.679	2	90 Th	$M\gamma$	$M_{III}N_V$	3.370
3.68203	9	48 Cd	$L\beta_4$	L_IM_{II}	3.36719
3.6855	2	45 Rh	$L\gamma_{2,3}$	$L_IN_{II,III}$	3.3640
3.691	2	91 Pa		$M_{IV}O_{II}$	3.359
3.6999	1	47 Ag	L_{III}	Abs. Edge	3.35096
3.70335	3	47 Ag	$L\beta_{2,15}$	$L_{III}N_{IV,V}$	3.34781
3.716	1	92 U	$M\beta$	$M_{IV}N_{VI}$	3.3367
3.71696	9	52 Te	Ll	$L_{III}M_I$	3.33555
3.718	3	90 Th		$M_{III}N_{IV}$	3.335
3.7228	1	46 Pd	L_{II}	Abs. Edge	3.33031
3.7246	2	46 Pd	$L\gamma_1$	$L_{II}N_{IV}$	3.3287
3.729	5	90 Th	M_V	Abs. Edge	3.325
3.73823	4	48 Cd	$L\beta_1$	$L_{II}M_{IV}$	3.31657
3.740	9	83 Bi		M_IN_{III}	3.315
3.7414	2	19 K	$K\alpha_1$	KL_{III}	3.3138
3.7445	2	19 K	$K\alpha_2$	KL_{II}	3.3111
3.760	9	90 Th		M_VP_{III}	3.298
3.762	5	78 Pt	M_I	Abs. Edge	3.296
3.77192	4	49 In	$L\alpha_1$	$L_{III}M_V$	3.28694
3.78073	6	49 In	$L\alpha_2$	$L_{III}M_{IV}$	3.27929
3.783	5	80 Hg	M_{II}	Abs. Edge	3.277
3.78876	9	50 Sn	$L\eta$	$L_{II}M_I$	3.27234
3.7920	2	46 Pd	$L\beta_9$	L_IM_V	3.2696
3.7988	2	46 Pd	$L\beta_{10}$	L_IM_{IV}	3.2637
3.80774	9	47 Ag	$L\beta_6$	$L_{III}N_I$	3.25603
3.808	4	90 Th		$M_{IV}O_{II}$	3.256
3.8222	2	46 Pd	$L\gamma_5$	$L_{II}N_I$	3.2437
3.827	1	91 Pa	$M\beta$	$M_{IV}N_{VI}$	3.2397
3.83313	9	47 Ag	$L\beta_3$	L_IM_{III}	3.23446
3.834	4	83 Bi		$M_{II}N_{IV}$	3.234
3.835	5	44 Ru	L_I	Abs. Edge	3.233
3.87023	5	47 Ag	$L\beta_4$	L_IM_{II}	3.20346
3.87090	5	18 A	K	Abs. Edge	3.20290
3.872	9	82 Pb		M_IN_{III}	3.202
3.8860	2	18 A	$K\beta_{1,3}$	$KM_{II,III}$	3.1905
3.88826	9	51 Sb	Ll	$L_{III}M_I$	3.18860
3.892	9	83 Bi		M_IN_{II}	3.185
3.8977	2	44 Ru	$L\gamma_{2,3}$	$L_IN_{II,III}$	3.1809
3.904	5	83 Bi	M_{III}	Abs. Edge	3.176
3.9074	1	46 Pd	L_{III}	Abs. Edge	3.17298
3.90887	4	46 Pd	$L\beta_{2,15}$	$L_{III}N_{IV,V}$	3.17179
3.910	1	92 U	$M\alpha_1$	M_VN_{VII}	3.1708
3.915	5	77 Ir	M_I	Abs. Edge	3.167
3.924	1	92 U	$M\alpha_2$	M_VN_{VI}	3.1595
3.932	6	83 Bi		$M_{III}O_{IV,V}$	3.153
3.93473	3	47 Ag	$L\beta_1$	$L_{II}M_{IV}$	3.15094
3.936	5	79 Au	M_{II}	Abs. Edge	3.150
3.941	1	90 Th	$M\beta$	$M_{IV}N_{VI}$	3.1458
3.9425	5	45 Rh	L_{II}	Abs. Edge	3.1448
3.9437	2	45 Rh	$L\gamma_1$	$L_{II}N_{IV}$	3.1438
3.95635	4	48 Cd	$L\alpha_1$	$L_{III}M_V$	3.13373
3.96496	6	48 Cd	$L\alpha_2$	$L_{III}M_{IV}$	3.12691
3.968	5	82 Pb		$M_{II}N_{IV}$	3.124
3.98327	9	49 In	$L\eta$	$L_{II}M_I$	3.11254
4.013	9	81 Tl		M_IN_{III}	3.089
4.0162	2	46 Pd	$L\beta_6$	$L_{III}N_I$	3.0870
4.022	1	91 Pa	$M\alpha_1$	M_VN_{VII}	3.0823
4.0346	2	46 Pd	$L\beta_3$	L_IM_{III}	3.0730
4.035	3	91 Pa	$M\alpha_2$	M_VN_{VI}	3.072
4.0451	2	45 Rh	$L\gamma_5$	$L_{II}N_I$	3.0650
4.047	1	82 Pb	M_{III}	Abs. Edge	3.0632
4.058	5	43 Tc	L_I	Abs. Edge	3.055
4.069	6	82 Pb		$M_{III}O_{IV,V}$	3.047
4.0711	2	46 Pd	$L\beta_4$	L_IM_{II}	3.0454
4.071	5	76 Os	M_I	Abs. Edge	3.045
4.07165	9	50 Sn	Ll	$L_{III}M_I$	3.04499
4.093	5	78 Pt	M_{II}	Abs. Edge	3.029
4.105	9	83 Bi		$M_{III}O_I$	3.021
4.116	4	81 Tl		$M_{II}N_{IV}$	3.013
4.1299	5	45 Rh	L_{III}	Abs. Edge	3.0021
4.1310	2	45 Rh	$L\beta_{2,15}$	$L_{III}N_{IV,V}$	3.0013
4.1381	9	90 Th	$M\alpha_1$	M_VN_{VII}	2.9961
4.14622	5	46 Pd	$L\beta_1$	$L_{II}M_{IV}$	2.99022
4.151	2	90 Th	$M\alpha_2$	M_VN_{VI}	2.987
4.15443	3	47 Ag	$L\alpha_1$	$L_{III}M_V$	2.98431

Wavelength Å*	p.e.	Element		Designation	keV
4.16294	5	47 Ag	$L\alpha_2$	$L_{III}M_{IV}$	2.97821
4.180	1	44 Ru	L_{II}	Abs. Edge	2.9663
4.1822	2	44 Ru	$L\gamma_1$	$L_{II}N_{IV}$	2.9645
4.19180	5	18 A	$K\alpha_1$	KL_{III}	2.95770
4.19315	9	48 Cd	$L\eta$	$L_{II}M_{I}$	2.95675
4.19474	5	18 A	$K\alpha_2$	KL_{II}	2.95563
4.198	1	81 Tl	M_{III}	Abs. Edge	2.9535
4.216	6	81 Tl		$M_{III}O_{IV,V}$	2.941
4.236	5	75 Re	M_{I}	Abs. Edge	2.927
4.2417	2	45 Rh	$L\beta_6$	$L_{III}N_{I}$	2.9229
4.244	9	82 Pb		$M_{III}O_{I}$	2.921
4.2522	2	45 Rh	$L\beta_3$	$L_{I}M_{III}$	2.9157
4.260	5	77 Ir	M_{II}	Abs. Edge	2.910
4.26873	9	49 In	Ll	$L_{III}M_{I}$	2.90440
4.2873	2	44 Ru	$L\gamma_5$	$L_{II}N_{I}$	2.8918
4.2888	2	45 Rh	$L\beta_4$	$L_{I}M_{II}$	2.8908
4.300	9	79 Au		$M_{I}N_{III}$	2.883
4.304	5	42 Mo	L_{I}	Abs. Edge	2.881
4.330	2	92 U		$M_{III}N_{I}$	2.863
4.355	1	80 Hg	M_{III}	Abs. Edge	2.8469
4.36767	5	46 Pd	$L\alpha_1$	$L_{III}M_{V}$	2.83861
4.369	1	44 Ru	L_{III}	Abs. Edge	2.8377
4.3718	2	44 Ru	$L\beta_{2,15}$	$L_{III}N_{IV,V}$	2.8360
4.37414	4	45 Rh	$L\beta_1$	$L_{II}M_{IV}$	2.83441
4.37588	7	46 Pd	$L\alpha_2$	$L_{III}M_{IV}$	2.83329
4.3800	2	42 Mo	$L\gamma_{2,3}$	$L_{I}N_{II,III}$	2.8306
4.3971	1	17 Cl	K	Abs. Edge	2.81960
4.4034	3	17 Cl	$K\beta$	KM	2.8156
4.407	5	74 W	M_{I}	Abs. Edge	2.813
4.4183	2	47 Ag	$L\eta$	$L_{II}M_{I}$	2.8061
4.432	4	79 Au		$M_{II}N_{IV}$	2.797
4.433	5	76 Os	M_{II}	Abs. Edge	2.797
4.436	1	43 Te	L_{II}	Abs. Edge	2.7948
4.44	2	74 W		$M_{I}O_{II,III}$	2.79
4.450	4	91 Pa		$M_{III}N_{I}$	2.786
4.460	9	78 Pt		$M_{I}N_{III}$	2.780
4.48014	9	48 Cd	Ll	$L_{III}M_{I}$	2.76735
4.4866	3	44 Ru	$L\beta_3$	$L_{I}M_{III}$	2.7634
4.4866	3	44 Ru	$L\beta_6$	$L_{III}N_{I}$	2.7634
4.518	1	79 Au	M_{III}	Abs. Edge	2.7439
4.522	6	79 Au		$M_{III}O_{IV,V}$	2.742
4.5230	2	44 Ru	$L\beta_4$	$L_{I}M_{II}$	2.7411
4.532	2	83 Bi	$M\gamma$	$M_{III}N_{V}$	2.735
4.568	5	90 Th		$M_{III}N_{I}$	2.714
4.571	5	83 Bi		$M_{III}N_{IV}$	2.712
4.572	5	83 Bi	M_{IV}	Abs. Edge	2.711
4.575	5	41 Nb	L_{I}	Abs. Edge	2.710
4.585	5	73 Ta	M_{I}	Abs. Edge	2.704
4.59	2	83 Bi		$M_{IV}P_{II,III}$	2.70
4.59743	9	45 Rh	$L\alpha_1$	$L_{III}M_{V}$	2.69674
4.601	4	78 Pt		$M_{II}N_{IV}$	2.695
4.60545	9	45 Rh	$L\alpha_2$	$L_{III}M_{IV}$	2.69205
4.620	5	75 Re	M_{II}	Abs. Edge	2.684
4.62058	3	44 Ru	$L\beta_1$	$L_{II}M_{IV}$	2.68323
4.625	5	92 U		$M_{IV}N_{III}$	2.681
4.630	1	43 Tc	L_{III}	Abs. Edge	2.6780
4.631	9	77 Ir		$M_{I}N_{III}$	2.677
4.6542	2	41 Nb	$L\gamma_{2,3}$	$L_{I}N_{II,III}$	2.6638
4.655	8	82 Pb		$M_{II}N_{I}$	2.664
4.6605	2	46 Pd	$L\eta$	$L_{II}M_{I}$	2.6603
4.674	1	82 Pb	$M\gamma$	$M_{III}N_{V}$	2.6527
4.686	1	78 Pt	M_{III}	Abs. Edge	2.6459
4.694	8	78 Pt		$M_{III}O_{IV,V}$	2.641
4.703	9	79 Au		$M_{III}O_{I}$	2.636
4.7076	2	47 Ag	Ll	$L_{III}M_{I}$	2.6337
4.715	3	82 Pb		$M_{III}N_{IV}$	2.630
4.719	1	42 Mo	L_{II}	Abs. Edge	2.6274
4.7258	2	42 Mo	$L\gamma_1$	$L_{II}N_{IV}$	2.6235
4.7278	1	17 Cl	$K\alpha_1$	KL_{III}	2.62239
4.7307	1	17 Cl	$K\alpha_2$	KL_{II}	2.62078
4.757	5	82 Pb	M_{IV}	Abs. Edge	2.606
4.764	5	83 Bi	M_{V}	Abs. Edge	2.603
4.780	4	77 Ir		$M_{II}N_{IV}$	2.594
4.79	2	76 Os		$M_{I}N_{III}$	2.59
4.815	5	74 W	M_{II}	Abs. Edge	2.575
4.823	3	83 Bi		$M_{IV}O_{II}$	2.571
4.823	4	81 Tl	$M\gamma$	$M_{III}N_{V}$	2.571
4.8369	2	42 Mo	$L\gamma_5$	$L_{II}N_{I}$	2.5632
4.84575	5	44 Ru	$L\alpha_1$	$L_{III}M_{V}$	2.55855
4.85381	7	44 Ru	$L\alpha_2$	$L_{III}M_{IV}$	2.55431
4.861	1	77 Ir	M_{III}	Abs. Edge	2.5505
4.865	5	81 Tl		$M_{III}N_{IV}$	2.548
4.869	9	77 Ir		$M_{III}O_{IV,V}$	2.546
4.876	9	78 Pt		$M_{III}O_{I}$	2.543
4.879	5	40 Zr	L_{I}	Abs. Edge	2.541
4.8873	8	43 Tc	$L\beta_1$	$L_{II}M_{IV}$	2.5368
4.909	1	83 Bi	$M\beta$	$M_{IV}N_{VI}$	2.5255
4.911	5	90 Th		$M_{IV}N_{III}$	2.524
4.913	1	42 Mo	L_{III}	Abs. Edge	2.5234
4.9217	2	45 Rh	$L\eta$	$L_{II}M_{I}$	2.5191
4.9232	2	42 Mo	$L\beta_{2,15}$	$L_{III}N_{IV,V}$	2.5183
4.946	2	92 U	$M\zeta_1$	$M_{V}N_{III}$	2.507
4.952	5	81 Tl	M_{IV}	Abs. Edge	2.504
4.9525	3	46 Pd	Ll	$L_{III}M_{I}$	2.5034
4.9536	3	40 Zr	$L\gamma_{2,3}$	$L_{I}N_{II,III}$	2.5029
4.955	4	76 Os		$M_{II}N_{IV}$	2.502
4.955	5	82 Pb	M_{V}	Abs. Edge	2.502
4.984	2	80 Hg	$M\gamma$	$M_{III}N_{V}$	2.4875
5.004	9	82 Pb		$M_{IV}O_{II}$	2.477
5.0133	3	42 Mo	$L\beta_3$	$L_{I}M_{III}$	2.4730
5.0185	1	16 S	K	Abs. Edge	2.47048
5.020	5	73 Ta	M_{II}	Abs. Edge	2.470
5.0233	3	16 S	$K\beta_x$	KM	2.4681
5.031	1	41 Nb	L_{II}	Abs. Edge	2.4641
5.0316	2	16 S	$K\beta_1$	KM	2.46404
5.0361	3	41 Nb	$L\gamma_1$	$L_{II}N_{IV}$	2.4618
5.043	5	76 Os	M_{III}	Abs. Edge	2.458
5.0488	3	42 Mo	$L\beta_4$	$L_{I}M_{II}$	2.4557
5.0488	5	42 Mo	$L\beta_6$	$L_{III}N_{I}$	2.4557
5.050	2	92 U	$M\zeta_2$	$M_{IV}N_{II}$	2.4548
5.076	1	82 Pb	$M\beta$	$M_{IV}N_{VI}$	2.4427
5.092	2	91 Pa	$M\zeta_1$	$M_{V}N_{III}$	2.4350
5.1148	3	43 Tc	$L\alpha_1$	$L_{III}M_{V}$	2.4240
5.118	1	83 Bi	$M\alpha_1$	$M_{V}N_{VII}$	2.4226

Wavelength Å*	p.e.	Element	Designation		keV
5.130	2	83 Bi	$M\alpha_2$	M_VN_{VI}	2.4170
5.145	4	79 Au	$M\gamma$	$M_{III}N_V$	2.410
5.1517	3	41 Nb	$L\gamma_5$	$L_{II}N_I$	2.4066
5.153	5	81 Tl	M_V	Abs. Edge	2.406
5.157	5	80 Hg	M_{IV}	Abs. Edge	2.404
5.168	9	82 Pb		M_VO_{III}	2.399
5.172	9	74 W		M_IN_{III}	2.397
5.17708	8	42 Mo	$L\beta_1$	$L_{II}M_{IV}$	2.39481
5.186	5	79 Au		$M_{III}N_{IV}$	2.391
5.193	2	91 Pa	$M\zeta_2$	$M_{IV}N_{II}$	2.3876
5.196	9	81 Tl		$M_{IV}O_{II}$	2.386
5.2050	2	44 Ru	$L\eta$	$L_{II}M_I$	2.38197
5.217	5	39 Y	L_I	Abs. Edge	2.377
5.2169	3	45 Rh	Ll	$L_{III}M_I$	2.3765
5.230	1	41 Nb	L_{III}	Abs. Edge	2.3706
5.234	5	75 Re	M_{III}	Abs. Edge	2.369
5.2379	3	41 Nb	$L\beta_{2,15}$	$L_{III}N_{IV,V}$	2.3670
5.245	5	90 Th	$M\zeta_1$	M_VN_{III}	2.364
5.249	1	81 Tl	$M\beta$	$M_{IV}N_{VI}$	2.3621
5.2830	3	39 Y	$L\gamma_{2,3}$	$L_IN_{II,III}$	2.3468
5.286	1	82 Pb	$M\alpha_1$	M_VN_{VII}	2.3455
5.299	2	82 Pb	$M\alpha_2$	M_VN_{VI}	2.3397
5.3102	3	41 Nb	$L\beta_3$	L_IM_{III}	2.3348
5.319	4	78 Pt	$M\gamma$	$M_{III}N_V$	2.331
5.340	5	90 Th	$M\zeta_2$	$M_{IV}N_{II}$	2.322
5.3455	3	41 Nb	$L\beta_4$	L_IM_{II}	2.3194
5.357	4	74 W		$M_{II}N_{IV}$	2.314
5.357	5	78 Pt		$M_{III}N_{IV}$	2.314
5.36	1	80 Hg	M_V	Abs. Edge	2.313
5.3613	3	41 Nb	$L\beta_6$	$L_{III}N_I$	2.3125
5.37216	7	16 S	$K\alpha_1$	KL_{III}	2.30784
5.374	5	79 Au	M_{IV}	Abs. Edge	2.307
5.37496	8	16 S	$K\alpha_2$	KL_{II}	2.30664
5.378	1	40 Zr	L_{II}	Abs. Edge	2.3053
5.3843	3	40 Zr	$L\gamma_1$	$L_{II}N_{IV}$	2.3027
5.40	2	73 Ta		M_IN_{III}	2.295
5.40655	8	42 Mo	$L\alpha_1$	$L_{III}M_V$	2.29316
5.41437	8	42 Mo	$L\alpha_2$	$L_{III}M_{IV}$	2.28985
5.4318	9	80 Hg	$M\beta$	$M_{IV}N_{VI}$	2.2825
5.435	1	74 W	M_{III}	Abs. Edge	2.2811
5.460	1	81 Tl	$M\alpha_1$	M_VN_{VII}	2.2706
5.472	2	81 Tl	$M\alpha_2$	M_VN_{VI}	2.2656
5.4923	3	41 Nb	$L\beta_1$	$L_{II}M_{IV}$	2.2574
5.4977	3	40 Zr	$L\gamma_5$	$L_{II}N_I$	2.2551
5.500	4	77 Ir	$M\gamma$	$M_{III}N_V$	2.254
5.5035	3	44 Ru	Ll	$L_{III}M_I$	2.2528
5.537	8	83 Bi		$M_{III}N_I$	2.239
5.540	5	77 Ir		$M_{III}N_{IV}$	2.238
5.570	4	73 Ta		$M_{II}N_{IV}$	2.226
5.579	1	40 Zr	L_{III}	Abs. Edge	2.2225
5.584	5	79 Au	M_V	Abs. Edge	2.220
5.5863	3	40 Zr	$L\beta_{2,15}$	$L_{III}N_{IV,V}$	2.2194
5.59	1	78 Pt	M_{IV}	Abs. Edge	2.217
5.592	5	38 Sr	L_I	Abs. Edge	2.217
5.624	1	79 Au	$M\beta$	$M_{IV}N_{VI}$	2.2046
5.628	8	74 W		$M_{III}O_I$	2.203
5.6330	3	40 Zr	$L\beta_3$	L_IM_{III}	2.2010
5.6445	3	38 Sr	$L\gamma_{2,3}$	$L_IN_{II,III}$	2.1965
5.6476	9	80 Hg	$M\alpha_1$	M_VN_{VII}	2.1953
5.650	5	73 Ta	M_{III}	Abs. Edge	2.194
5.6681	3	40 Zr	$L\beta_4$	L_IM_{II}	2.1873
5.67	3	73 Ta		$M_{III}O_{IV,V}$	2.19
5.682	4	76 Os	$M\gamma$	$M_{III}N_V$	2.182
5.704	8	82 Pb		$M_{III}N_I$	2.174
5.7101	3	40 Zr	$L\beta_6$	$L_{III}N_I$	2.1712
5.724	5	76 Os		$M_{III}N_{IV}$	2.166
5.7243	2	41 Nb	$L\alpha_1$	$L_{III}M_V$	2.16589
5.7319	3	41 Nb	$L\alpha_2$	$L_{III}M_{IV}$	2.1630
5.756	1	39 Y	L_{II}	Abs. Edge	2.1540
5.767	9	79 Au		M_VO_{III}	2.150
5.784	1	15 P	K	Abs. Edge	2.1435
5.796	2	15 P	$K\beta$	KM	2.1391
5.81	2	76 Os		$M_{II}N_I$	2.133
5.81	1	78 Pt	M_V	Abs. Edge	2.133
5.828	1	78 Pt	$M\beta$	$M_{IV}N_{VI}$	2.1273
5.83	2	73 Ta		$M_{III}O_I$	2.126
5.83	1	77 Ir	M_{IV}	Abs. Edge	2.126
5.8360	3	40 Zr	$L\beta_1$	$L_{II}M_{IV}$	2.1244
5.840	1	79 Au	$M\alpha_1$	M_VN_{VII}	2.1229
5.8475	3	42 Mo	$L\eta$	$L_{II}M_I$	2.1202
5.854	3	79 Au	$M\alpha_2$	M_VN_{VI}	2.118
5.8754	3	39 Y	$L\gamma_5$	$L_{II}N_I$	2.1102
5.884	8	81 Tl		$M_{III}N_I$	2.107
5.885	2	75 Re	$M\gamma$	$M_{III}N_V$	2.1067
5.931	5	75 Re		$M_{III}N_{IV}$	2.090
5.962	1	39 Y	L_{III}	Abs. Edge	2.0794
5.9832	3	39 Y	$L\beta_3$	L_IM_{III}	2.0722
5.987	9	78 Pt		M_VO_{III}	2.071
6.008	5	37 Rb	L_I	Abs. Edge	2.063
6.0186	3	39 Y	$L\beta_4$	L_IM_{II}	2.0600
6.038	1	77 Ir	$M\beta$	$M_{IV}N_{VI}$	2.0535
6.0458	3	37 Rb	$L\gamma_{2,3}$	$L_IN_{II,III}$	2.0507
6.047	1	78 Pt	$M\alpha_1$	M_VN_{VII}	2.0505
6.05	1	77 Ir	M_V	Abs. Edge	2.048
6.058	3	78 Pt	$M\alpha_2$	M_VN_{VI}	2.047
6.0705	2	40 Zr	$L\alpha_1$	$L_{III}M_V$	2.04236
6.073	5	76 Os	M_{IV}	Abs. Edge	2.042
6.0778	3	40 Zr	$L\alpha_2$	$L_{III}M_{IV}$	2.0399
6.09	2	80 Hg		$M_{III}N_I$	2.036
6.092	3	74 W	$M\gamma$	$M_{III}N_V$	2.035
6.0942	3	39 Y	$L\beta_6$	$L_{III}N_I$	2.0344
6.134	4	74 W		$M_{III}N_{IV}$	2.021
6.1508	3	42 Mo	Ll	$L_{III}M_I$	2.01568
6.157	1	15 P	$K\alpha_1$	KL_{III}	2.0137
6.160	1	15 P	$K\alpha_2$	KL_{II}	2.0127
6.162	8	83 Bi		$M_{IV}N_{III}$	2.012
6.173	1	38 Sr	L_{II}	Abs. Edge	2.0085
6.2109	3	41 Nb	$L\eta$	$L_{II}M_I$	1.99620
6.2120	3	39 Y	$L\beta_1$	$L_{II}M_{IV}$	1.99584
6.259	9	79 Au		$M_{III}N_I$	1.981
6.262	1	77 Ir	$M\alpha_1$	M_VN_{VII}	1.9799
6.267	1	76 Os	$M\beta$	$M_{IV}N_{VI}$	1.9783
6.275	3	77 Ir	$M\alpha_2$	M_VN_{VI}	1.9758
6.28	2	74 W		$M_{II}N_I$	1.973

Wavelength Å*	p.e.	Element	Designation		keV
6.2961	3	38 Sr	$L\gamma_5$	$L_{II}N_I$	1.96916
6.30	1	76 Os	M_V	Abs. Edge	1.967
6.312	4	73 Ta	$M\gamma$	$M_{III}N_V$	1.964
6.33	1	75 Re	M_{IV}	Abs. Edge	1.958
6.353	5	73 Ta		$M_{III}N_{IV}$	1.951
6.3672	3	38 Sr	$L\beta_3$	L_IM_{III}	1.94719
6.384	7	82 Pb		$M_{IV}N_{III}$	1.942
6.387	1	38 Sr	L_{III}	Abs. Edge	1.9411
6.4026	3	38 Sr	$L\beta_4$	L_IM_{II}	1.93643
6.4488	2	39 Y	$L\alpha_1$	$L_{III}M_V$	1.92256
6.455	9	78 Pt		$M_{III}N_I$	1.921
6.4558	3	39 Y	$L\alpha_2$	$L_{III}M_{IV}$	1.92047
6.47	1	36 Kr	L_I	Abs. Edge	1.915
6.490	1	76 Os	$M\alpha$	$M_VN_{VI,VII}$	1.9102
6.504	1	75 Re	$M\beta$	$M_{IV}N_{VI}$	1.9061
6.5176	3	41 Nb	Ll	$L_{III}M_I$	1.90225
6.5191	3	38 Sr	$L\beta_6$	$L_{III}N_I$	1.90181
6.521	4	83 Bi	$M\zeta_1$	M_VN_{III}	1.901
6.544	4	72 Hf	$M\gamma$	$M_{III}N_V$	1.895
6.560	5	75 Re	M_V	Abs. Edge	1.890
6.585	5	83 Bi	$M\zeta_2$	$M_{IV}N_{II}$	1.883
6.59	1	74 W	M_{IV}	Abs. Edge	1.880
6.6069	3	40 Zr	$L\eta$	$L_{II}M_I$	1.87654
6.6239	3	38 Sr	$L\beta_1$	$L_{II}M_{IV}$	1.87172
6.644	1	37 Rb	L_{II}	Abs. Edge	1.8661
6.669	9	77 Ir		$M_{III}N_I$	1.859
6.729	1	75 Re	$M\alpha$	$M_VN_{VI,VII}$	1.8425
6.738	1	14 Si	K	Abs. Edge	1.8400
6.740	3	82 Pb	$M\zeta_1$	M_VN_{III}	1.8395
6.7530	1	14 Si	$K\beta$	KM	1.83594
6.755	3	37 Rb	$L\gamma_5$	$L_{II}N_{IV}$	1.83532
6.757	1	74 W	$M\beta$	$M_{IV}N_{VI}$	1.8349
6.768	6	71 Lu	$M\gamma$	$M_{III}N_V$	1.832
6.7876	3	37 Rb	$L\beta_3$	L_IM_{III}	1.82659
6.802	5	82 Pb	$M\zeta_2$	$M_{IV}N_{II}$	1.823
6.806	9	74 W		$M_{IV}O_{II}$	1.822
6.8207	3	37 Rb	$L\beta_4$	L_IM_{II}	1.81771
6.83	1	74 W	M_V	Abs. Edge	1.814
6.862	1	37 Rb	L_{III}	Abs. Edge	1.8067
6.8628	2	38 Sr	$L\alpha_1$	$L_{III}M_V$	1.80656
6.8697	3	38 Sr	$L\alpha_2$	$L_{III}M_{IV}$	1.80474
6.87	1	73 Ta	M_{IV}	Abs. Edge	1.804
6.87	2	80 Hg	δ	$M_{IV}N_{III}$	1.805
6.89	2	76 Os		$M_{III}N_I$	1.798
6.9185	3	40 Zr	Ll	$L_{III}M_I$	1.79201
6.959	5	35 Br	L_I	Abs. Edge	1.781
6.974	4	81 Tl	$M\zeta_1$	M_VN_{III}	1.778
6.983	1	74 W	$M\alpha_1$	M_VN_{VII}	1.7754
6.9842	3	37 Rb	$L\beta_6$	$L_{III}N_I$	1.77517
6.992	2	74 W	$M\alpha_2$	M_VN_{VI}	1.7731
7.005	9	74 W		M_VO_{III}	1.770
7.023	1	73 Ta	$M\beta$	$M_{IV}N_{VI}$	1.7655
7.024	8	70 Yb	$M\gamma$	$M_{III}N_V$	1.765
7.032	5	81 Tl	$M\zeta_2$	$M_{IV}N_{II}$	1.763
7.0406	3	39 Y	$L\eta$	$L_{II}M_I$	1.76095
7.0759	3	37 Rb	$L\beta_1$	$L_{II}M_{IV}$	1.75217
7.09	2	73 Ta		$M_{IV}O_{II,III}$	1.748
7.101	8	79 Au		$M_{IV}N_{III}$	1.746
7.11	1	73 Ta	M_V	Abs. Edge	1.743
7.12542	9	14 Si	$K\alpha_1$	KL_{III}	1.73998
7.12791	9	14 Si	$K\alpha_2$	KL_{II}	1.73938
7.168	1	36 Kr	L_{II}	Abs. Edge	1.7297
7.250	5	36 Kr		$L_{II}N_{III}$	1.710
7.252	1	73 Ta	$M\alpha$	$M_VN_{VI,VII}$	1.7096
7.264	5	36 Kr	$L\beta_3$	L_IM_{III}	1.707
7.279	5	36 Kr	$L\gamma_5$	$L_{II}N_I$	1.703
7.30	2	73 Ta		M_VO_{III}	1.700
7.303	1	72 Hf	$M\beta$	$M_{IV}N_{VI}$	1.6976
7.304	5	36 Kr	$L\beta_4$	L_IM_{II}	1.697
7.3183	2	37 Rb	$L\alpha_1$	$L_{III}M_V$	1.69413
7.3251	3	37 Rb	$L\alpha_2$	$L_{III}M_{IV}$	1.69256
7.3563	3	39 Y	Ll	$L_{III}M_I$	1.68536
7.360	8	74 W		$M_{III}N_I$	1.684
7.371	8	78 Pt		$M_{IV}N_{III}$	1.682
7.392	1	36 Kr	L_{III}	Abs. Edge	1.6772
7.466	4	79 Au	$M\zeta_1$	M_VN_{III}	1.6605
7.503	1	34 Se	L_I	Abs. Edge	1.6525
7.510	4	36 Kr	$L\beta_6$	$L_{III}N_I$	1.6510
7.5171	3	38 Sr	$L\eta$	$L_{II}M_I$	1.64933
7.523	5	79 Au	$M\zeta_2$	$M_{IV}N_{II}$	1.648
7.539	1	72 Hf	$M\alpha$	$M_VN_{VI,VII}$	1.6446
7.546	8	68 Er	$M\gamma$	$M_{III}N_V$	1.643
7.576	3	36 Kr	$L\beta_1$	$L_{II}M_{IV}$	1.6366
7.60	1	68 Er		$M_{III}N_{IV}$	1.632
7.601	2	71 Lu	$M\beta$	$M_{IV}N_{VI}$	1.6312
7.612	9	73 Ta		$M_{III}N_I$	1.629
7.645	8	77 Ir		$M_{IV}N_{III}$	1.622
7.738	4	78 Pt	$M\zeta_1$	M_VN_{III}	1.6022
7.753	5	35 Br	L_{II}	Abs. Edge	1.599
7.767	9	35 Br	$L\beta_{3,4}$	$L_IM_{II,III}$	1.596
7.790	5	78 Pt	$M\zeta_2$	$M_{IV}N_{II}$	1.592
7.817	3	36 Kr	$L\alpha_{1,2}$	$L_{III}M_{IV,V}$	1.5860
7.8362	3	38 Sr	Ll	$L_{III}M_I$	1.58215
7.840	2	71 Lu	$M\alpha$	$M_VN_{VI,VII}$	1.5813
7.865	9	67 Ho	$M\gamma$	$M_{III}N_{IV,V}$	1.576
7.887	9	72 Hf		$M_{III}N_I$	1.572
7.909	2	70 Yb	$M\beta$	$M_{IV}N_{VI}$	1.5675
7.94813	5	13 Al	K	Abs. Edge	1.55988
7.960	2	13 Al	$K\beta$	KM	1.55745
7.984	5	35 Br	L_{III}	Abs. Edge	1.5530
8.021	4	77 Ir	$M\zeta_1$	M_VN_{III}	1.5458
8.0415	4	37 Rb	$L\eta$	$L_{II}M_I$	1.54177
8.065	5	77 Ir	$M\zeta_2$	$M_{IV}N_{II}$	1.5373
8.107	1	33 As	L_I	Abs. Edge	1.5293
8.1251	5	35 Br	$L\beta_1$	$L_{II}M_{IV}$	1.52590
8.144	9	66 Dy	$M\gamma$	$M_{III}N_{IV,V}$	1.522
8.149	5	70 Yb	$M\alpha$	$M_VN_{VI,VII}$	1.5214
8.239	8	75 Re		$M_{IV}N_{III}$	1.505
8.249	7	69 Tm	$M\beta$	$M_{IV}N_{VI}$	1.503
8.310	4	76 Os	$M\zeta_1$	M_VN_{III}	1.4919
8.321	9	34 Se	$L\beta_{3,4}$	$L_IM_{II,III}$	1.490
8.33934	9	13 Al	$K\alpha_1$	KL_{III}	1.48670
8.34173	9	13 Al	$K\alpha_2$	KL_{II}	1.48627
8.359	5	76 Os	$M\zeta_2$	$M_{IV}N_{II}$	1.4831

Wavelength Å*	p.e.	Element	Designation		keV
8.3636	4	37 Rb	Ll	$L_{III}M_{I}$	1.48238
8.3746	5	35 Br	$L\alpha_{1,2}$	$L_{III}M_{IV,V}$	1.48043
8.407	1	34 Se	L_{II}	Abs. Edge	1.4747
8.470	9	70 Yb		$M_{III}N_{I}$	1.464
8.48	1	69 Tm	$M\alpha$	$M_{V}N_{VI,VII}$	1.462
8.486	9	65 Tb	$M\gamma$	$M_{III}N_{IV,V}$	1.461
8.487	5	69 Tm	M_{V}	Abs. Edge	1.4609
8.573	8	74 W		$M_{IV}N_{III}$	1.446
8.592	3	68 Er	$M\beta$	$M_{IV}N_{VI}$	1.4430
8.60	7	92 U		$N_{I}P_{IV,V}$	1.44
8.601	5	68 Er	M_{IV}	Abs. Edge	1.4415
8.629	4	75 Re	$M\zeta_1$	$M_{V}N_{III}$	1.4368
8.646	1	34 Se	L_{III}	Abs. Edge	1.4340
8.664	5	75 Re	$M\zeta_2$	$M_{IV}N_{II}$	1.4310
8.7358	5	34 Se	$L\beta_1$	$L_{II}M_{IV}$	1.41923
8.76	7	92 U		$N_{I}P_{III}$	1.42
8.773	1	32 Ge	L_{I}	Abs. Edge	1.4132
8.81	7	92 U		$N_{I}P_{II}$	1.41
8.82	1	68 Er	$M\alpha$	$M_{V}N_{VI,VII}$	1.406
8.844	9	64 Gd	$M\gamma$	$M_{III}N_{IV,V}$	1.402
8.847	5	68 Er	M_{V}	Abs. Edge	1.4013
8.90	2	73 Ta		$M_{IV}N_{III}$	1.393
8.929	1	33 As	$L\beta_{3,4}$	$L_{I}M_{II,III}$	1.3884
8.962	4	74 W	$M\zeta_1$	$M_{V}N_{III}$	1.3835
8.965	4	67 Ho	$M\beta$	$M_{IV}N_{VI}$	1.3830
8.9900	5	34 Se	$L\alpha_{1,2}$	$L_{III}M_{IV,V}$	1.37910
8.993	5	74 W	$M\zeta_2$	$M_{IV}N_{II}$	1.3787
9.125	1	33 As	L_{II}	Abs. Edge	1.3587
9.20	2	67 Ho	$M\alpha$	$M_{V}N_{VI,VII}$	1.348
9.211	9	63 Eu	$M\gamma$	$M_{III}N_{IV,V}$	1.346
9.255	1	35 Br	$L\eta$	$L_{II}M_{I}$	1.3396
9.316	4	73 Ta	$M\zeta_1$	$M_{V}N_{III}$	1.3308
9.330	5	73 Ta	$M\zeta_2$	$M_{IV}N_{II}$	1.3288
9.357	6	66 Dy	$M\beta$	$M_{IV}N_{VI}$	1.3250
9.367	1	33 As	L_{III}	Abs. Edge	1.3235
9.40	7	90 Th		$N_{I}P_{III}$	1.319
9.4141	8	33 As	$L\beta_1$	$L_{II}M_{IV}$	1.3170
9.44	7	90 Th		$N_{I}P_{II}$	1.313
9.5122	1	12 Mg	K	Abs. Edge	1.30339
9.517	5	31 Ga	L_{I}	Abs. Edge	1.3028
9.521	2	12 Mg	$K\beta$	KM	1.3022
9.581	2	32 Ge	$L\beta_3$	$L_{I}M_{III}$	1.2941
9.585	1	35 Br	Ll	$L_{III}M_{I}$	1.2935
9.59	2	66 Dy	$M\alpha$	$M_{V}N_{VI,VII}$	1.293
9.600	9	62 Sm	$M\gamma$	$M_{III}N_{IV,V}$	1.291
9.640	2	32 Ge	$L\beta_4$	$L_{I}M_{II}$	1.2861
9.6709	8	33 As	$L\alpha_{1,2}$	$L_{III}M_{IV,V}$	1.2820
9.686	7	72 Hf	$M\zeta_2$	$M_{IV}N_{II}$	1.2800
9.686	7	72 Hf	$M\zeta_1$	$M_{V}N_{III}$	1.2800
9.792	6	65 Tb	$M\beta$	$M_{IV}N_{VI}$	1.2661
9.8900	2	12 Mg	$K\alpha_{1,2}$	$KL_{II,III}$	1.25360
9.924	1	32 Ge	L_{II}	Abs. Edge	1.2494
9.962	1	34 Se	$L\eta$	$L_{II}M_{I}$	1.2446
10.00	2	65 Tb	$M\alpha$	$M_{V}N_{VI,VII}$	1.240
10.09	7	92 U		$N_{I}O_{III}$	1.229
10.175	1	32 Ge	$L\beta_1$	$L_{II}M_{IV}$	1.2185
10.187	1	32 Ge	L_{III}	Abs. Edge	1.2170
10.254	6	64 Gd	$M\beta$	$M_{IV}N_{VI}$	1.2091
10.294	1	34 Se	Ll	$L_{III}M_{I}$	1.2044
10.359	9	31 Ga	$L\beta_{3,4}$	$L_{I}M_{II,III}$	1.197
10.40	7	92 U		$N_{II}P_{I}$	1.192
10.4361	8	32 Ge	$L\alpha_{1,2}$	$L_{III}M_{IV,V}$	1.18800
10.46	3	64 Gd	$M\alpha$	$M_{V}N_{VI,VII}$	1.185
10.48	1	70 Yb	$M\zeta$	$M_{V}N_{III}$	1.183
10.505	9	60 Nd	$M\gamma$	$M_{III}N_{IV,V}$	1.180
10.711	5	63 Eu	M_{IV}	Abs. Edge	1.1575
10.734	1	33 As	$L\eta$	$L_{II}M_{I}$	1.1550
10.750	7	63 Eu	$M\beta$	$M_{IV}N_{VI}$	1.1533
10.828	5	31 Ga	L_{II}	Abs. Edge	1.1450
10.96	3	63 Eu	$M\alpha$	$M_{V}N_{VI,VII}$	1.131
10.998	9	59 Pr	$M\gamma$	$M_{III}N_{IV,V}$	1.1273
11.013	5	63 Eu	M_{V}	Abs. Edge	1.1258
11.023	2	31 Ga	$L\beta_1$	$L_{II}M_{IV}$	1.1248
11.072	1	33 As	Ll	$L_{III}M_{I}$	1.1198
11.07	7	90 Th		$N_{II}P_{I}$	1.120
11.100	1	31 Ga	L_{III}	Abs. Edge	1.1169
11.200	7	30 Zn	$L\beta_{3,4}$	$L_{I}M_{II,III}$	1.1070
11.27	1	62 Sm	$M\beta$	$M_{IV}N_{VI}$	1.0998
11.288	5	62 Sm	M_{IV}	Abs. Edge	1.0983
11.292	1	31 Ga	$L\alpha_{1,2}$	$L_{III}M_{IV,V}$	1.09792
11.37	1	68 Er	$M\zeta$	$M_{V}N_{III}$	1.0901
11.47	3	62 Sm	$M\alpha$	$M_{V}N_{VI,VII}$	1.081
11.53	1	58 Ce	$M\gamma$	$M_{III}N_{IV,V}$	1.0749
11.552	5	62 Sm	M_{V}	Abs. Edge	1.0732
11.56	5	90 Th		$N_{II}O_{IV}$	1.072
11.569	1	11 Na	K	Abs. Edge	1.07167
11.575	2	11 Na	$K\beta$	KM	1.0711
11.609	2	32 Ge	$L\eta$	$L_{II}M_{I}$	1.0680
11.862	1	30 Zn	L_{II}	Abs. Edge	1.04523
11.86	1	67 Ho	$M\zeta$	$M_{V}N_{III}$	1.0450
11.9101	9	11 Na	$K\alpha_{1,2}$	$KL_{II,III}$	1.04098
11.965	2	32 Ge	Ll	$L_{III}M_{I}$	1.0362
11.983	3	30 Zn	$L\beta_1$	$L_{II}M_{IV}$	1.0347
12.08	4	57 La	$M\gamma$	$M_{III}N_{IV,V}$	1.027
12.122	3	29 Cu	$L\beta_{3,4}$	$L_{I}M_{II,III}$	1.0228
12.131	1	30 Zn	L_{III}	Abs. Edge	1.02201
12.254	3	30 Zn	$L\alpha_{1,2}$	$L_{III}M_{IV,V}$	1.0117
12.43	2	66 Dy	$M\zeta$	$M_{V}N_{III}$	0.998
12.44	2	60 Nd	$M\beta$	$M_{IV}N_{VI}$	0.997
12.459	5	60 Nd	M_{IV}	Abs. Edge	0.9951
12.597	2	31 Ga	$L\eta$	$L_{II}M_{I}$	0.9842
12.68	2	60 Nd	$M\alpha$	$M_{V}N_{VI,VII}$	0.978
12.737	5	60 Nd	M_{V}	Abs. Edge	0.9734
12.75	3	56 Ba	$M\gamma$	$M_{III}N_{IV,V}$	0.973
12.90	9	92 U		$N_{III}O_{V}$	0.961
12.953	2	31 Ga	Ll	$L_{III}M_{I}$	0.9572
12.98	2	65 Tb	$M\zeta$	$M_{V}N_{III}$	0.955
13.014	1	29 Cu	L_{II}	Abs. Edge	0.95268
13.053	3	29 Cu	$L\beta_1$	$L_{II}M_{IV}$	0.9498
13.06	2	59 Pr	$M\beta$	$M_{IV}N_{VI}$	0.950
13.06	1	30 Zn	L_{I}	Abs. Edge	0.9495
13.122	5	59 Pr	M_{IV}	Abs. Edge	0.9448
13.18	2	28 Ni	$L\beta_{3,4}$	$L_{I}M_{II,III}$	0.941
13.288	1	29 Cu	L_{III}	Abs. Edge	0.93306

Wavelength Å*	p.e.	Element	Designation		keV
13.30	6	83 Bi		$N_IP_{II,III}$	0.932
13.336	3	29 Cu	$L\alpha_{1,2}$	$L_{III}M_{IV,V}$	0.9297
13.343	5	59 Pr	$M\alpha$	$M_VN_{VI,VII}$	0.9292
13.394	5	59 Pr	M_V	Abs. Edge	0.9257
13.57	2	64 Gd	$M\zeta$	M_VN_{III}	0.914
13.68	2	30 Zn	$L\eta$	$L_{II}M_I$	0.906
13.75	4	58 Ce	$M\beta$	$M_{IV}N_{VI}$	0.902
13.8	1	90 Th		$N_{III}O_V$	0.897
14.02	2	30 Zn	Ll	$L_{III}M_I$	0.884
14.04	2	58 Ce	$M\alpha$	$M_VN_{VI,VII}$	0.883
14.22	2	63 Eu	$M\zeta$	M_VN_{III}	0.872
14.242	5	28 Ni	L_{II}	Abs. Edge	0.8706
14.271	6	28 Ni	$L\beta_1$	$L_{II}M_{IV}$	0.8688
14.3018	1	10 Ne	K	Abs. Edge	0.866889
14.31	3	27 Co	$L\beta_{3,4}$	$L_IM_{II,III}$	0.870
14.39	5	58 Ce		$M_VO_{II,III}$	0.862
14.452	5	10 Ne	$K\beta$	KM	0.8579
14.51	5	57 La	$M\beta$	$M_{IV}N_{VI}$	0.854
14.525	5	28 Ni	L_{III}	Abs. Edge	0.8536
14.561	3	28 Ni	$L\alpha_{1,2}$	$L_{III}M_{IV,V}$	0.8515
14.610	3	10 Ne	$K\alpha_{1,2}$	$KL_{II,III}$	0.8486
14.88	5	57 La	$M\alpha$	$M_VN_{VI,VII}$	0.833
14.90	2	29 Cu	$L\eta$	$L_{II}M_I$	0.832
14.91	4	62 Sm	$M\zeta$	M_VN_{III}	0.831
15.286	9	29 Cu	Ll	$L_{III}M_I$	0.8111
15.56	1	56 Ba	M_{IV}	Abs. Edge	0.7967
15.618	5	27 Co	L_{II}	Abs. Edge	0.7938
15.65	4	26 Fe	$L\beta_{3,4}$	$L_IM_{II,III}$	0.792
15.666	8	27 Co	$L\beta_1$	$L_{II}M_{IV}$	0.7914
15.72	9	56 Ba		$M_{IV}O_{III}$	0.789
15.89	1	56 Ba	M_V	Abs. Edge	0.7801
15.91	5	56 Ba		$M_{IV}O_{II}$	0.779
15.915	5	27 Co	L_{III}	Abs. Edge	0.7790
15.93	4	52 Te	$M\gamma$	$M_{III}N_{IV,V}$	0.778
15.972	6	27 Co	$L\alpha_{1,2}$	$L_{III}M_{IV,V}$	0.7762
15.98	5	51 Sb		$M_{II}N_{IV}$	0.776
16.20	5	56 Ba		M_VO_{III}	0.765
16.27	3	28 Ni	$L\eta$	$L_{II}M_I$	0.762
16.46	4	60 Nd	$M\zeta$	M_VN_{III}	0.753
16.693	9	28 Ni	Ll	$L_{III}M_I$	0.7427
16.7	1	24 Cr	L_I	Abs. Edge	0.741
16.92	4	51 Sb	$M\gamma$	$M_{III}N_{IV,V}$	0.733
16.93	5	50 Sn		$M_{II}N_{IV}$	0.733
17.19	4	25 Mn	$L\beta_{3,4}$	$L_IM_{II,III}$	0.721
17.202	5	26 Fe	L_{II}	Abs. Edge	0.7208
17.26	1	26 Fe	$L\beta_1$	$L_{II}M_{IV}$	0.7185
17.38	4	59 Pr	$M\zeta$	M_VN_{III}	0.714
17.525	5	26 Fe	L_{III}	Abs. Edge	0.7074
17.59	2	26 Fe	$L\alpha_{1,2}$	$L_{III}M_{IV,V}$	0.7050
17.6	1	52 Te		$M_{II}N_I$	0.703
17.87	3	27 Co	$L\eta$	$L_{II}M_I$	0.694
17.94	5	50 Sn	$M\gamma$	$M_{III}N_{IV,V}$	0.691
17.9	1	24 Cr	L_{II}	Abs. Edge	0.691
18.292	8	27 Co	Ll	$L_{III}M_I$	0.6778
18.32	2	9 F	$K\alpha$	KL	0.6768
18.35	4	58 Ce	$M\zeta$	M_VN_{III}	0.676
18.8	1	51 Sb		$M_{II}N_I$	0.658
18.8	2	47 Ag		$M_IN_{II,III}$	0.658
18.96	4	24 Cr	$L\beta_{3,4}$	$L_IM_{II,III}$	0.654
19.11	2	25 Mn	$L\beta_1$	$L_{II}M_{IV}$	0.6488
19.1	1	52 Te		$M_{III}N_I$	0.648
19.40	7	48 Cd		$M_{II}N_{IV}$	0.639
19.44	5	57 La	$M\zeta$	M_VN_{III}	0.638
19.45	1	25 Mn	$L\alpha_{1,2}$	$L_{III}M_{IV,V}$	0.6374
19.66	5	53 I	$M_{IV,V}$	Abs. Edge	0.631
19.75	4	26 Fe	$L\eta$	$L_{II}M_I$	0.628
20.0	1	50 Sn		$M_{II}N_I$	0.619
20.1	2	46 Pd		$M_IN_{II,III}$	0.616
20.15	1	26 Fe	Ll	$L_{III}M_I$	0.6152
20.2	1	51 Sb		$M_{III}N_I$	0.612
20.47	7	48 Cd	$M\gamma$	$M_{III}N_{IV,V}$	0.606
20.64	4	56 Ba	$M\zeta$	M_VN_{III}	0.601
20.66	7	47 Ag		$M_{II}N_{IV}$	0.600
20.7	1	24 Cr	L_{III}	Abs. Edge	0.598
21.19	5	23 Va	$L\beta_{3,4}$	$L_IM_{II,III}$	0.585
21.27	1	24 Cr	$L\beta_1$	$L_{II}M_{IV}$	0.5828
21.34	5	52 Te		$M_{IV}O_{II,III}$	0.581
21.5	1	50 Sn		$M_{III}N_I$	0.575
21.64	3	24 Cr	$L\alpha_{1,2}$	$L_{III}M_{IV,V}$	0.5728
21.78	5	52 Te		M_VO_{III}	0.569
21.82	7	47 Ag	$M\gamma$	$M_{III}N_{IV,V}$	0.568
21.85	2	25 Mn	$L\eta$	$L_{II}M_I$	0.5675
22.1	1	46 Pd		$M_{II}N_{IV}$	0.560
22.29	1	25 Mn	Ll	$L_{III}M_I$	0.5563
22.9	2	48 Cd		$M_{II}N_I$	0.540
23.32	1	8 O	K	Abs. Edge	0.5317
23.3	1	46 Pd	$M\gamma$	$M_{III}N_{IV,V}$	0.531
23.62	3	8 O	$K\alpha$	KL	0.5249
23.88	4	23 Va	$L\beta_1$	$L_{II}M_{IV}$	0.5192
24.25	3	23 Va	$L\alpha_{1,2}$	$L_{III}M_{IV,V}$	0.5113
24.28	5	50 Sn	$M_{IV,V}$	Abs. Edge	0.511
24.30	3	24 Cr	$L\eta$	$L_{II}M_I$	0.5102
24.4	2	47 Ag		M_VN_I	0.509
24.5	1	48 Cd		$M_{III}N_I$	0.507
24.78	1	24 Cr	Ll	$L_{III}M_I$	0.5003
25.01	9	45 Rh	$M\gamma$	$M_{III}N_{IV,V}$	0.496
25.3	1	50 Sn		$M_{IV}O_{II,III}$	0.491
25.50	9	44 Ru		$M_{II}N_{IV}$	0.486
25.7	1	50 Sn		M_VO_{III}	0.483
26.0	1	47 Ag		$M_{III}N_I$	0.478
26.2	2	46 Pd		$M_{II}N_I$	0.474
26.72	9	52 Te	$M\zeta$	$M_{IV,V}N_{II,III}$	0.464
26.9	1	44 Ru	$M\gamma$	$M_{III}N_{IV,V}$	0.462
27.05	2	22 Ti	$L\beta_1$	$L_{II}M_{IV}$	0.4584
27.29	1	22 Ti	$L_{II,III}$	Abs. Edge	0.4544
27.34	3	23 Va	$L\eta$	$L_{II}M_I$	0.4535
27.42	2	22 Ti	$L\alpha_{1,2}$	$L_{III}M_{IV,V}$	0.4522
27.77	1	23 Va	Ll	$L_{III}M_I$	0.4465
27.9	1	46 Pd		$M_{III}N_I$	0.445
28.1	2	45 Rh		$M_{II}N_I$	0.442
28.13	5	48 Cd	$M_{IV,V}$	Abs. Edge	0.4408
28.88	8	51 Sb	$M\zeta$	$M_{IV,V}N_{II,III}$	0.429
29.8	1	45 Rh		$M_{III}N_I$	0.417
30.4	1	48 Cd		$M_{IV}O_{II,III}$	0.408

TABLE B (*Continued*)

Wavelength Å*	p.e.	Element		Designation	keV
30.8	1	48 Cd		M_VO_{III}	0.403
30.82	5	47 Ag	M_{IV}	Abs. Edge	0.4022
30.89	3	22 Ti	$L\eta$	$L_{II}M_I$	0.4013
30.99	1	7 N	K	Abs. Edge	0.4000
31.02	2	21 Sc	$L\beta_1$	$L_{II}M_{IV}$	0.3996
31.14	5	47 Ag	M_V	Abs. Edge	0.3981
31.24	9	50 Sn	$M\zeta$	$M_{IV,V}N_{II,III}$	0.397
31.35	3	21 Sc	$L\alpha_{1,2}$	$L_{III}M_{IV,V}$	0.3954
31.36	2	22 Ti	Ll	$L_{III}M_I$	0.3953
31.60	4	7 N	$K\alpha$	KL	0.3924
31.8	1	92 U		$N_{IV}N_{VI}$	0.390
32.3	2	44 Ru		$M_{II}N_I$	0.384
33.1	2	41 Nb		$M_{II}N_{IV}$	0.375
33.5	3	47 Ag		$M_{IV,V}O_{II,III}$	0.370
33.57	9	90 Th		$N_{IV}N_{VI}$	0.3693
34.8	1	92 U		$N_VN_{VI,VII}$	0.357
34.9	2	41 Nb	$M\gamma$	$M_{III}N_{IV,V}$	0.356
35.13	2	21 Sc	$L\eta$	$L_{II}M_I$	0.3529
35.13	1	20 Ca	L_{II}	Abs. Edge	0.3529
35.3	3	42 Mo		$M_{II}N_I$	0.351
35.49	1	20 Ca	L_{III}	Abs. Edge	0.34931
35.59	3	21 Sc	Ll	$L_{III}M_I$	0.3483
35.63	1	20 Ca	$L_{II,III}$	Abs. Edge	0.34793
35.94	2	20 Ca	$L\beta_1$	$L_{II}M_{IV}$	0.3449
36.32	9	90 Th		$N_VN_{VI,VII}$	0.3414
36.33	2	20 Ca	$L\alpha_{1,2}$	$L_{III}M_{IV,V}$	0.3413
36.8	1	48 Cd	$M\zeta$	$M_{IV,V}N_{II,III}$	0.3371
37.4	2	46 Pd		$M_{IV,V}O_{II,III}$	0.332
37.5	2	42 Mo		$M_{III}N_I$	0.331
38.4	3	41 Nb		$M_{II}N_I$	0.323
39.77	7	47 Ag	$M\zeta$	$M_{IV,V}N_{II,III}$	0.3117
40.46	2	20 Ca	$L\eta$	$L_{II}M_I$	0.3064
40.7	2	41 Nb		$M_{III}N_I$	0.305
40.9	2	45 Rh		$M_{IV,V}O_{II,III}$	0.303
40.96	2	20 Ca	Ll	$L_{III}M_I$	0.3027
42.1	2	92 U		$N_{VI}O_V$	0.295
42.1	1	19 K	$L_{II,III}$	Abs. Edge	0.2946
42.3	2	82 Pb		$N_{IV}N_{VI}$	0.293
43.3	2	92 U		$N_{VI}O_{IV}$	0.286
43.6	1	46 Pd	$M\zeta$	$M_{IV,V}N_{II,III}$	0.2844
43.68	1	6 C	K	Abs. Edge	0.28384
44.7	3	6 C	$K\alpha$	KL	0.277
44.8	1	44 Ru		$M_{IV,V}O_{II,III}$	0.2768
45.0	1	82 Pb		$N_VN_{VI,VII}$	0.2756
45.2	3	80 Hg		$N_{IV}N_{VI}$	0.274
45.2	1	51 Sb		$M_{II}M_{IV}$	0.2743
46.48	9	39 Y		$M_{II}N_I$	0.267
46.5	2	81 Tl		$N_VN_{VI,VII}$	0.267
46.8	2	79 Au		$N_{IV}N_{VI}$	0.265
47.24	2	19 K	Ll	$L_{II}M_I$	0.2625
47.3	1	50 Sn		$M_{II}M_{IV}$	0.2621
47.67	9	45 Rh	$M\zeta$	$M_{IV,V}N_{II,III}$	0.2601
47.74	1	19 K	Ll	$L_{III}M_I$	0.25971
47.9	3	80 Hg		$N_VN_{VI,VII}$	0.259
48.1	2	78 Pt		$N_{IV}N_{VI}$	0.258
48.2	1	90 Th		$N_{VI}O_V$	0.2572
48.5	2	39 Y		$M_{III}N_I$	0.256
49.4	1	79 Au		$N_VN_{VI,VII}$	0.2510
49.5	1	90 Th		$N_{VI}O_{IV}$	0.2505
50.0	1	90 Th		$N_{VII}O_V$	0.2479
50.2	1	77 Ir		$N_{IV}N_{VI}$	0.2470
50.3	1	52 Te		$M_{III}M_V$	0.2465
50.9	1	78 Pt		$N_VN_{VI,VII}$	0.2436
51.3	1	38 Sr		$M_{II}N_I$	0.2416
51.9	1	76 Os		$N_{IV}N_{VI}$	0.2388
52.0	2	48 Cd		$M_{II}M_{IV}$	0.2384
52.2	1	51 Sb		$M_{III}M_V$	0.2375
52.34	7	44 Ru	$M\zeta$	$M_{IV,V}N_{II,III}$	0.2369
52.8	1	77 Ir		$N_VN_{VI,VII}$	0.2348
53.6	1	38 Sr		$M_{III}N_I$	0.2313
54.0	2	74 W		$N_{II}N_{IV}$	0.2295
54.0	1	47 Ag		$M_{II}M_{IV}$	0.2295
54.2	1	50 Sn		$M_{III}M_V$	0.2287
54.7	2	76 Os		$N_VN_{VI,VII}$	0.2266
54.8	2	42 Mo		$M_{IV,V}O_{II,III}$	0.2262
55.8	1	74 W		$N_{IV}N_{VI}$	0.2221
55.9	1	18 A	$L\eta$	$L_{II}M_I$	0.2217
56.3	1	18 A	Ll	$L_{III}M_I$	0.2201
56.5	1	46 Pd		$M_{II}M_{IV}$	0.2194
57.0	2	37 Rb		$M_{II}N_I$	0.2174
58.2	1	73 Ta		$N_{IV}N_{VI}$	0.2130
58.4	1	74 W		N_VN_{VII}	0.2122
58.7	2	48 Cd		$M_{III}M_V$	0.2111
59.3	1	45 Rh		$M_{II}M_{IV}$	0.2090
59.5	3	74 W		N_VN_{VI}	0.208
59.5	2	37 Rb		$M_{III}N_I$	0.2083
60.5	1	47 Ag		$M_{III}M_V$	0.2048
61.1	2	73 Ta		$N_VN_{VI,VII}$	0.2028
61.9	2	41 Nb		$M_{IV,V}O_{II,III}$	0.2002
62.2	1	44 Ru		$M_{II}M_{IV}$	0.1992
62.9	1	46 Pd		$M_{III}M_V$	0.1970
63.0	5	71 Lu		$N_{IV}N_{VI}$	0.197
64.38	7	42 Mo	$M\zeta$	$M_{IV,V}N_{II,III}$	0.1926
65.1	7	70 Yb		$N_{IV}N_{VI}$	0.190
65.5	1	45 Rh		$M_{III}M_V$	0.1892
65.7	2	71 Lu		$N_VN_{VI,VII}$	0.1886
67.33	9	17 Cl	$L\eta$	$L_{II}M_I$	0.1841
67.6	3	5 B	$K\alpha$	KL	0.1833
67.90	9	17 Cl	Ll	$L_{III}M_I$	0.1826
68.2	3	90 Th		$O_{III}P_{IV,V}$	0.1817
68.3	1	44 Ru		$M_{III}M_V$	0.1814
68.9	2	42 Mo		$M_{II}M_{IV}$	0.1798
69.3	5	70 Yb		$N_VN_{VI,VII}$	0.179
70.0	4	40 Zr		$M_{IV,V}O_{II,III}$	0.177
72.1	3	41 Nb		$M_{II}M_{IV}$	0.1718
72.19	9	41 Nb	$M\zeta$	$M_{IV,V}N_{II,III}$	0.1717
72.7	9	68 Er		$N_{IV}N_{VI}$	0.171
74.9	1	42 Mo		$M_{III}M_V$	0.1656
76.3	7	68 Er		$N_VN_{VI,VII}$	0.163
76.7	2	40 Zr		$M_{II}M_{IV}$	0.1617
76.9	2	35 Br		$M_{II}N_I$	0.1613
78.4	2	41 Nb		$M_{III}M_V$	0.1582
79.8	3	35 Br		$M_{III}N_I$	0.1554
80.9	3	40 Zr		$M_{III}M_V$	0.1533

TABLE B (*Continued*)

Wavelength Å*	p.e.	Element	Designation		keV
81.5	2	39 Y		$M_{II}M_{IV}$	0.1522
82.1	2	40 Zr	$M\zeta$	$M_{IV,V}N_{II,III}$	0.1511
83.	1	66 Dy		$N_{IV,V}N_{VI,VII}$	0.149
83.4	3	16 S	Ll, η	$L_{II,III}M_I$	0.1487
85.7	2	38 Sr		$M_{II}M_{IV}$	0.1447
86.	1	65 Tb		$N_{IV,V}N_{VI,VII}$	0.144
86.5	2	39 Y		$M_{III}M_{IV,V}$	0.1434
91.4	2	38 Sr		$M_{III}M_{IV,V}$	0.1357
91.5	2	37 Rb		$M_{II}M_{IV}$	0.1355
91.6	1	83 Bi		$N_{VI}O_{IV}$	0.1354
93.2	1	83 Bi		$N_{VII}O_V$	0.1330
93.4	2	39 Y	$M\zeta$	$M_{IV,V}N_{II,III}$	0.1328
94.	1	15 P	$L_{II,III}$	Abs. Edge	0.132
96.7	2	37 Rb		$M_{III}M_{IV,V}$	0.1282
97.2	8	66 Dy		$N_{IV,V}O_{II,III}$	0.128
98.	1	62 Sm		$N_{IV,V}N_{VI,VII}$	0.126
100.2	2	82 Pb		$N_{VI}O_V$	0.1237
102.2	4	65 Tb		$N_{IV,V}O_{II,III}$	0.1213
102.4	1	82 Pb		$N_{VI}O_{IV}$	0.1211
103.8	4	15 P		$L_{II,III}M$	0.1194
104.3	1	82 Pb		$N_{VII}O_V$	0.1189
107.	1	60 Nd		$N_{IV,V}N_{VI,VII}$	0.116
108.0	2	38 Sr	$M\zeta_2$	$M_{IV}N_{II,III}$	0.1148
108.7	1	38 Sr	$M\zeta_1$	M_VN_{III}	0.1140
109.4	3	35 Br		$M_{II}M_{IV}$	0.1133
110.6	5	29 Cu	M_I	Abs. Edge	0.1121
111.	1	4 Be	K	Abs. Edge	0.111
112.0	6	63 Eu		$N_{IV,V}O_{II,III}$	0.1107
113.0	1	81 Tl		$N_{VI}O_V$	0.10968
113.	1	59 Pr		$N_{IV,V}N_{VI,VII}$	0.1095
113.8	3	35 Br		$M_{III}M_{IV,V}$	0.1089
114.	1	4 Be	$K\alpha$	KL	0.1085
115.3	2	81 Tl		$N_{VI}O_{IV}$	0.1075
117.4	4	62 Sm		$N_{IV,V}O_{II,III}$	0.1056
117.7	1	81 Tl		$N_{VII}O_V$	0.10530
123.	1	14 Si	$L_{II,III}$	Abs. Edge	0.1006
126.8	2	37 Rb		$M_{IV}N_{III}$	0.0978
127.8	2	37 Rb	$M\zeta_2$	$M_{IV}N_{II}$	0.0970
128.7	2	37 Rb	$M\zeta_1$	M_VN_{III}	0.0964
128.9	7	60 Nd		$N_{IV,V}O_{II,III}$	0.0962
135.5	4	14 Si		$L_{II,III}M$	0.0915
136.5	4	59 Pr		$N_{IV,V}O_{II,III}$	0.0908
137.0	5	30 Zn	M_{II}	Abs. Edge	0.0905
142.5	1	13 Al	L_I	Abs. Edge	0.08701
143.9	5	30 Zn	M_{III}	Abs. Edge	0.0862
144.4	6	58 Ce		$N_{IV,V}O_{II,III}$	0.0859
144.4	3	37 Rb		M_IM_{III}	0.0859
152.6	6	57 La		$N_{IV,V}O_{II,III}$	0.0812
157.	3	30 Zn		$M_{II,III}M_{IV,V}$	0.079
159.0	2	56 Ba		$N_{IV}O_{III}$	0.07796
159.5	5	29 Cu	M_{II}	Abs. Edge	0.0777
163.3	2	56 Ba		$N_{IV}O_{II}$	0.07590
164.6	2	56 Ba		N_VO_{III}	0.07530
164.7	3	35 Br		M_IM_{III}	0.0753
166.0	5	29 Cu	M_{III}	Abs. Edge	0.0747
170.4	1	13 Al	$L_{II,III}$	Abs. Edge	0.07278
171.4	5	13 Al		$L_{II,III}M$	0.0724
173.	3	29 Cu		$M_{II,III}M_{IV,V}$	0.072
181.	5	90 Th		$O_{IV,V}Q_{II,III}$	0.068
183.8	1	55 Cs		$N_{IV}O_{III}$	0.06746
184.6	3	35 Br		M_IM_{II}	0.0672
188.4	1	28 Ni	M_{III}	Abs. Edge	0.06581
188.6	1	55 Cs		$N_{IV}O_{II}$	0.06574
189.5	3	35 Br		$M_{IV}N_{III}$	0.0654
190.3	1	55 Cs		N_VO_{III}	0.06515
190.	2	28 Ni		$M_{II,III}M_{IV,V}$	0.0651
191.1	2	35 Br	$M\zeta_2$	$M_{IV}N_{II}$	0.06488
192.6	2	35 Br	$M\zeta_1$	M_VN_{III}	0.06437
197.3	1	12 Mg	L_I	Abs. Edge	0.06284
202.	5	27 Co	$M_{II,III}$	Abs. Edge	0.061
203.	1	16 S		$L_IL_{II,III}$	0.061
214.	6	27 Co		$M_{II,III}M_{IV,V}$	0.058
224.	1	53 I	$N_{IV,V}$	Abs. Edge	0.0552
226.5	1	3 Li	K	Abs. Edge	0.05475
227.8	1	34 Se	M_V	Abs. Edge	0.05443
228.	1	3 Li	$K\alpha$	KL	0.0543
230.	2	34 Se		M_VN_{III}	0.0538
230.	1	26 Fe	$M_{II,III}$	Abs. Edge	0.0538
243.	5	26 Fe		$M_{II,III}M_{IV,V}$	0.051
249.3	1	12 Mg	L_{II}	Abs. Edge	0.04973
250.7	1	12 Mg	L_{III}	Abs. Edge	0.04945
251.5	5	12 Mg		$L_{II,III}M$	0.04929
273.	6	25 Mn		$M_{II,III}M_{IV,V}$	0.045
290.	1	13 Al		$L_IL_{II,III}$	0.0428
309.	9	24 Cr		$M_{II,III}M_{IV,V}$	0.040
317.	1	12 Mg		$L_IL_{II,III}$	0.0392
337.	9	23 V		$M_{II,III}M_{IV,V}$	0.0368
376.	1	11 Na		$L_IL_{II,III}$	0.03299
399.	5	35 Br	N_I	Abs. Edge	0.0311
405.	5	11 Na	$L_{II,III}$	Abs. Edge	0.0306
407.1	5	11 Na		$L_{II,III}M$	0.03045
417.	5	17 Cl	M_I	Abs. Edge	0.0297
444.	5	53 I	O_I	Abs. Edge	0.0279
525.	9	20 Ca		$M_{II,III}N_I$	0.0236
692.	9	19 K		$M_{II,III}N_I$	0.0179

X-RAY ATOMIC ENERGY LEVELS

J. A. Bearden and A. F. Burr

These tables were originally published as the final report to the U.S. Atomic Energy Commission as Report NYO-2543-1 in partial fulfillment of Contract AT(30-1)-2543. The tables were later reproduced in *Review of Modern Physics*. The data may also be obtained from the Superintendent of Documents, U.S. Government Printing Office, Washington, D. C. 20402 in the publication NSRDS-NBS 14. Persons seeking discussion of the details of calculations, sources of energy level information and the problem of properly interpreting the experimental measurements should refer to the original publication or to *Review of Modern Physics*, Vol. 39, 125–142, January 1967.

All of the x-ray emission wavelengths have recently been reevaluated and placed on a consistent Å* scale. For most elements these data give a highly overdetermined set of equations for energy level differences, which have been solved by least-squares adjustment for each case. This procedure makes "best" use of all x-ray wavelength data, and also permits calculation of the probable error for each energy difference. Photoelectron measurements of absolute energy levels are more precise than x-ray absorption edge data. These have been used to establish the absolute scale for eighty-one elements and, in many cases, to provide additional energy level difference data. The x-ray absorption wavelengths were used for eight elements and ionization measurements for two; the remaining five were interpolated by a Moseley diagram involving the output values of energy levels from adjacent elements. Probable errors are listed on an absolute energy basis. In the original source of the present data, a table of energy levels in Rydberg units is given. Difference tables in volts, Rydbergs, and milli-Å* wavelength units, with the respective probable errors, are also included there.

X-Ray Atomic Energy Levels

Recommended values of the atomic energy levels, and probable errors in eV. Where available, photoelectron direct measurements are listed in brackets [ ] immediately under the recommended values. The measured values of the x-ray absorption energies are shown in parentheses ( ). Interpolated values are enclosed in angle brackets ⟨ ⟩.

Level	1 H	2 He	3 Li	4 Be	5 B	6 C	7 N	8 O
K	13.59811[a]	24.58678[b]	54.75±0.02 (54.75)	111.0±1.0 (111.0)	188.0±0.4 [188.0][c]	283.8±0.4 [283.8][c] (283.8)	401.6±0.4 [401.6][c]	532.0±0.4 [532.0][c]
L_I								23.7±0.4 [23.7][d]
$L_{II,III}$					4.7±0.9	6.4±1.9	9.2±0.6	7.1±0.8
	9 F	**10 Ne**	**11 Na**	**12 Mg**	**13 Al**	**14 Si**	**15 P**	**16 S**
K	685.4±0.4 [685.4][c]	866.9±0.3 (866.9)	1072.1±0.4 [1072.1][c] (1072.)	1305.0±0.4 [1305.0][c] (1303.)	1559.6±0.4 [1559.6][c] (1559.8)	1838.9±0.4 [1838.9][c]	2145.5±0.4 [2145.5][d]	2472.0±0.4 [2472.0][c] (2470.)
L_I	⟨31.⟩	⟨45.⟩	63.3±0.4 [63.3][d]	89.4±0.4 [89.4][d] (63.)	117.7±0.4 [117.7][d] (87.)	148.7±0.4 [148.7][d]	189.3±0.4 [189.3][d]	229.2±0.4 [229.2][d]
$L_{II,III}$	8.6±0.8	18.3±0.4	31.1±0.4 (31.)	51.4±0.5 (50.)	73.1±0.5 (72.8)	99.2±0.5 (100.6)	132.2±0.5 (132.)	164.8±0.7
	17 Cl	**18 Ar**	**19 K**	**20 Ca**	**21 Sc**	**22 Ti**	**23 V**	**24 Cr**
K	2822.4±0.3 [2822.4][c] (2020.)	3202.9±0.3 (3202.9)	3607.4±0.4 [3607.4][c] (3607.8)	4038.1±0.4 [4038.1][c] (4038.1)	4492.8±0.4 [4492.8][c]	4966.4±0.4 [4966.4][d] (4964.5)	5465.1±0.3 [5465.1][c] (5464.)	5989.2±0.3 [5989.2][c] (5989.)
L_I	270.2±0.4 [270.2][d]	320. ⟨320.⟩[d]	377.1±0.4 [377.1][d]	437.8±0.4 [437.8][d]	500.4±0.4 [500.4][d]	563.7±0.4 [563.7][d]	628.2±0.4 [628.2][d]	694.6±0.4 [694.6][d]
L_{II}	201.6±0.3	247.3±0.3	296.3±0.4	350.0±0.4	406.7±0.4	461.5±0.4	520.5±0.3	583.7±0.3
L_{III}	200.0±0.3	245.2±0.3	293.6±0.4	346.4±0.4	402.2±0.4	455.5±0.4	512.9±0.3	574.5±0.3
M_I	17.5±0.4	25.3±0.4	33.9±0.4	43.7±0.4	53.8±0.4	60.3±0.4	66.5±0.4	74.1±0.4
$M_{II,III}$	6.8±0.4	12.4±0.3	17.8±0.4	25.4±0.4	32.3±0.5	34.6±0.4	37.8±0.3	42.5±0.3
$M_{IV,V}$					6.6±0.5	3.7	2.2±0.3	2.3±0.4

X-RAY ATOMIC ENERGY LEVELS (*Continued*)

	25 Mn	26 Fe	27 Co	28 Ni	29 Cu	30 Zn	31 Ga	32 Ge
K	6539.0±0.4 [6539.0][e] (6538.)	7112.0±0.9 [7111.3][e,f] (7111.2)	7708.9±0.3 [7708.9][e] (7709.5)	8332.8±0.4 [8332.8][e] (8331.6)	8978.9±0.4 [8978.9][e,g] (8980.3)	9658.6±0.6 [9658.6][g] (9660.7)	10367.1±0.5 [10367.1][g] (10368.2)	11103.1±0.7 [11103.8][g] (11103.6)
L_I	769.0±0.4 [769.0][d]	846.1±0.4 [846.1][d]	925.6±0.4 [925.6][d]	1008.1±0.4 [1008.1][d]	1096.1±0.4 [1096.0][d]	1193.6±0.9	1297.7±1.1	1414.3±0.7 [1413.6][g]
L_{II}	651.4±0.4	721.1±0.9 (720.8)	793.6±0.3 (793.8)	871.9±0.4 (870.6)	951.0±0.4 [950.0][h] (953.)	1042.8±0.6 (1045.)	1142.3±0.5	1247.8±0.7 (1249.)
L_{III}	640.3±0.4	708.1±0.9 (707.4)	778.6±0.3 (779.0)	854.7±0.4 (853.6)	931.1±0.4 [931.4][h] (933.)	1019.7±0.6 (1022.)	1115.4±0.5 (1117.)	1216.7±0.7 (1217.0)
M_I	83.9±0.5	92.9±0.9	100.7±0.4	111.8±0.6	119.8±0.6	135.9±1.1	158.1±0.5	180.0±0.8
M_{II} } M_{III}	48.6±0.4	54.0±0.9 (54.)	59.5±0.3 (61.)	68.1±0.4 (66.)	73.6±0.4 (75.)	86.6±0.6 (86.)	106.8±0.7 102.9±0.5	127.9±0.9 120.8±0.7
$M_{IV,V}$	3.3±0.5	3.6±0.9	2.9±0.3	3.6±0.4	1.6±0.4	8.1±0.6	17.4±0.5	28.7±0.7

	33 As	34 Se	35 Br	36 Kr	37 Rb	38 Sr	39 Y	40 Zr
K	11866.7±0.7 [11866.7][i] (11865.)	12657.8±0.7 [12657.8][g] (12654.5)	13473.7±0.4 (13470.)	14325.6±0.8 (14324.4)	15199.7±0.3 (15202.)	16104.6±0.3 (16107.)	17038.4±0.3 (17038.)	17997.6±0.4 (17999.)
L_I	1526.5±0.8 (1529.)	1653.9±3.5 (1652.5)	1782.0±0.4 [1782.0][j]	1921.0±0.6 [1921.2][k]	2065.1±0.3 [2065.4][j]	2216.3±0.3 [2216.2][l]	2372.5±0.3 [2372.7][l]	2531.6±0.3 [2531.6][l]
L_{II}	1358.6±0.7 (1358.7)	1476.2±0.7 (1474.7)	1596.0±0.4 [1596.2][j]	1727.2±0.5 [1727.2][k] (1730.)	1863.9±0.3 [1863.4][j]	2006.8±0.3 [2006.6][l] (2008.5)	2155.5±0.3 [2155.0][l] (2154.0)	2306.7±0.3 [2306.5][l] (2305.3)
L_{III}	1323.1±0.7 (1323.5)	1435.8±0.7 (1434.0)	1549.9±0.4 [1549.7][j]	1674.9±0.5 [1674.8][k] (1677.)	1804.4±0.3 [1804.6][j]	1939.6±0.3 [1939.9][l] (1941.)	2080.0±0.3 [2080.2][l] (2079.4)	2222.3±0.3 [2222.5][l] (2222.5)
M_I	203.5±0.7	231.5±0.7	256.5±0.4		322.1±0.3	357.5±0.3	393.6±0.3	430.3±0.3
M_{II}	146.4±1.2	168.2±1.3	189.3±0.4	222.7±1.1	247.4±0.3	279.8±0.3	312.4±0.4	344.2±0.4

X-RAY ATOMIC ENERGY LEVELS (*Continued*)

	33 As	34 Se	35 Br	36 Kr	37 Rb	38 Sr	39 Y	40 Zr
M_{III}	140.5±0.8	161.9±1.0	181.5±0.4	213.8±1.1	238.5±0.3	269.1±0.3	300.3±0.4	330.5±0.4
M_{IV}	41.2±0.7	56.7±0.8	70.1±0.4	88.9±0.8	111.8±0.3	135.0±0.3	159.6±0.3	182.4±0.3
M_V			69.0±0.4		110.3±0.3	133.1±0.3	157.4±0.3	180.0±0.3
N_I			27.3±0.5	24.0±0.8	29.3±0.3	37.7±0.3	45.4±0.3	51.3±0.3
N_{II}	2.5±1.0	5.6±1.3	5.2±0.4	10.6±1.9	14.8±0.4	19.9±0.3	25.6±0.4	28.7±0.4
N_{III}			4.6±0.4		14.0±0.3			

	41 Nb	42 Mo	43 Tc	44 Ru	45 Rh	46 Pd	47 Ag	48 Cd
K	18985.6±0.4 (18987.)	19999.5±0.3 (20004.)	21044.0±0.7	22117.2±0.3 (22119.)	23219.9±0.3 (23219.8)	24350.3±0.3 (24348.)	25514.0±0.3 (25516.)	26711.2±0.3 (26716.)
L_I	2697.7±0.3 [2697.7][l]	2865.5±0.3 [2866.0][l]	3042.5±0.4 [3042.5][l]	3224.0±0.3 [3224.3][l]	3411.9±0.3 [3412.0][l] (3417.)	3604.3±0.3 [3604.6][l] (3607.)	3805.8±0.3 [3806.2][m] (3807.)	4018.0±0.3 [4018.1][m] (4019.)
L_{II}	2464.7±0.3 [2464.7][l]	2625.1±0.3 [2624.5][l] (2627.)	2793.2±0.4 [2973.2][l]	2966.9±0.3 [2966.8][l] (2966.3)	3146.1±0.3 [3146.3][l] (3145.)	3330.3±0.3 [3330.3][l] (3330.3)	3523.7±0.3 [3523.6][g,m] (3526.)	3727.0±0.3 [3727.1][m] (3728.)
L_{III}	2370.5±0.3 [2370.6][l]	2520.2±0.3 [2520.2][l] (2523.2)	2676.9±0.4 [2676.9][l]	2837.9±0.3 [2837.7][l] (2837.7)	3003.8±0.3 [3003.5][g,l] (3002.)	3173.3±0.3 [3173.0][g,l] (3173.0)	3351.1±0.3 [3350.8][g] (3351.0)	3537.5±0.3 [3537.3][g] (3537.6)
M_I	468.4±0.3	504.6±0.3		585.0±0.3	627.1±0.3	669.9±0.3	717.5±0.3	770.2±0.3
M_{II}	378.4±0.4	409.7±0.4	444.9±1.5	482.8±0.3	521.0±0.3	559.1±0.3	602.4±0.3	650.7±0.3
M_{III}	363.0±0.4	392.3±0.3	425.0±1.5	460.6±0.3	496.2±0.3	531.5±0.3	571.4±0.3	616.5±0.3
M_{IV}	207.4±0.3	230.3±0.3	256.4±0.5	283.6±0.3	311.7±0.3	340.0±0.3	372.8±0.3	410.5±0.3
M_V	204.6±0.3	227.0±0.3	252.9±0.4	279.4±0.3	307.0±0.3	334.7±0.3	366.7±0.3	403.7±0.3
N_I	58.1±0.3	61.8±0.3		74.9±0.3	81.0±0.3	86.4±0.3	95.2±0.3	107.6±0.3
N_{II}	33.9±0.4	34.8±0.4	38.9±1.9	43.1±0.4	47.9±0.4	51.1±0.4	62.6±0.3	66.9±0.4
N_{III}							55.9±0.3	
$N_{IV,V}$	3.2±0.3	1.8±0.3		2.0±0.3	2.5±0.4	1.5±0.3	3.3±0.3	9.3±0.3

X-RAY ATOMIC ENERGY LEVELS (*Continued*)

	49 In	50 Sn	51 Sb	52 Te	53 I	54 Xe	55 Cs	56 Ba
K	27939.9±0.3	29200.1±0.4 (29195.)	30491.2±0.3 (30486.)	31813.8±0.3 (31811.)	33169.4±0.4 (33167.)	34561.4±1.1 (34590.)	35984.6±0.4 (35987.)	37440.6±0.4 (37452.)
L_I	4237.5±0.3 [4237.7][m] (4237.3)	4464.7±0.3 [4464.5][g] (4464.8)	4698.3±0.3 [4698.3][m] (4698.4)	4939.2±0.3 [4939.3][m] (4939.7)	5188.1±0.3 [5188.1][j]	5452.8±0.4 (5452.8)	5714.3±0.4 [5712.7][j] (5721.)	5988.8±0.4 [5986.8][j] (5996.)
L_{II}	3938.0±0.3 [3937.8][m] (3939.3)	4156.1±0.3 [4156.2][g] (4157.)	4380.4±0.3 [4380.6][m] (4382.)	4612.0±0.3 [4612.0][m] (4612.6)	4852.1±0.3 [4852.0][j]	5103.7±0.4 (5103.7)	5359.4±0.3 [5359.5][j] (5358.)	5623.6±0.3 [5623.6][j] (5623.3)
L_{III}	3730.1±0.3 [3730.0][g] (3730.2)	3928.8±0.3 [3928.8][g] (3928.8)	4132.2±0.3 [4132.2][g] (4132.3)	4341.4±0.3 [4341.2][g] (4341.8)	4557.1±0.3 [4557.1][j]	4782.2±0.4 (4782.2)	5011.9±0.3 [5012.0][j] (5011.3)	5247.0±0.3 [5247.3][j] (5247.0)
M_I	825.6±0.3	883.8±0.3	943.7±0.3	1006.0±0.3	1072.1±0.3		1217.1±0.4	1292.8±0.4
M_{II}	702.2±0.3	756.4±0.4	811.9±0.3	869.7±0.3	930.5±0.3	999.0±2.1	1065.0±0.5	1136.7±0.5
M_{III}	664.3±0.3	714.4±0.3	765.6±0.3	818.7±0.3	874.6±0.3	937.0±2.1	997.6±0.5	1062.2±0.5
M_{IV}	450.8±0.3	493.3±0.3	536.9±0.3	582.5±0.3	631.3±0.3		739.5±0.4	796.1±0.3
M_V	443.1±0.3	484.8±0.3	527.5±0.3	572.1±0.3	619.4±0.3	672.3±0.5	725.5±0.5	780.7±0.3
N_I	121.9±0.3	136.5±0.4	152.0±0.3	168.3±0.3	186.4±0.3		230.8±0.4	253.0±0.5
N_{II}, N_{III}	77.4±0.4	88.6±0.4	98.4±0.5	110.2±0.5	122.7±0.5	146.7±3.1		
N_{II}							172.3±0.6	191.8±0.7
N_{III}							161.6±0.6	179.7±0.6
N_{IV}, N_V	16.2±0.3	23.9±0.3	31.4±0.3	39.8±0.3	49.6±0.3			
N_{IV}							78.8±0.5	92.5±0.5
N_V							76.5±0.5	89.9±0.5
O_I	0.1±4.5	0.9±0.5	6.7±0.5	11.6±0.6	13.6±0.6		22.7±0.5	39.1±0.6
O_{II}, O_{III}	0.8±0.4	1.1±0.5	2.1±0.4	2.3±0.5	3.3±0.5			
O_{II}							13.1±0.5	16.6±0.5
O_{III}							11.4±0.5	14.6±0.5

	57 La	58 Ce	59 Pr	60 Nd	61 Pm	62 Sm	63 Eu	64 Gd
K	38924.6±0.4 (38934.)	40443.0±0.4 (40453.)	41990.6±0.5 (42002.)	43568.9±0.4 (43574.)	45184.0±0.7 (45198.)	46834.2±0.5 (46849.)	48519.0±0.4 (48519.)	50239.1±0.5 (50233.)

X-Ray Atomic Energy Levels (*Continued*)

	57 La	58 Ce	59 Pr	60 Nd	61 Pm	62 Sm	63 Eu	64 Gd
L_I	6266.3±0.5 [6266.3][n]	6548.8±0.5 [6548.5][n]	6834.8±0.5 [6834.9][n]	7126.0±0.4 [7125.8][n] (7129.)	7427.9±0.8 [7427.9][o]	7736.8±0.5 [7736.2][n] (7748.)	8052.0±0.4 [8051.7][n] (8061.)	8375.6±0.5 [8375.4][n] (8386.)
L_{II}	5890.6±0.4 [5890.7][n]	6164.2±0.4 [6164.3][n]	6440.4±0.5 [6440.2][n]	6721.5±0.4 [6721.8][n] (6723.)	7012.8±0.6 [7012.8][o]	7311.8±0.4 [7312.0][n] (7313.)	7617.1±0.4 [7617.6][n] (7620.)	7930.3±0.4 [7930.5][n] (7931.)
L_{III}	5482.7±0.4 [5482.6][n]	5723.4±0.4 [5723.6][n]	5964.3±0.4 [5964.3][n]	6207.9±0.4 [6208.0][n] (6209.)	6459.3±0.6 [6459.4][o]	6716.2±0.5 [6716.8][n] (6717.)	6976.9±0.4 [6976.7][n] (6981.)	7242.8±0.4 [7242.8][n] (7243.)
M_I	1361.3±0.3	1434.6±0.6	1511.0±0.8	1575.3±0.7		1722.8±0.8	1800.0±0.5	1880.8±0.5
M_{II}	1204.4±0.6	1272.8±0.6	1337.4±0.7	1402.8±0.6	1471.4±6.2	1540.7±1.2	1613.9±0.7	1688.3±0.7
M_{III}	1123.4±0.5	1185.4±0.5	1242.2±0.6	1297.4±0.5	1356.9±1.4	1419.8±1.1	1480.6±0.6	1544.0±0.8
M_{IV}	848.5±0.4	901.3±0.6	951.1±0.6	999.9±0.6	1051.5±0.9	1106.0±0.8	1160.6±0.6	1217.2±0.6
M_V	831.7±0.4	883.3±0.5	931.0±0.6	977.7±0.6	1026.9±1.0	1080.2±0.6	1130.9±0.6	1185.2±0.6
N_I	270.4±0.8	289.6±0.7	304.5±0.9	315.2±0.8		345.7±0.9	360.2±0.7	375.8±0.7
N_{II}	205.8±1.2	223.3±1.1	236.3±1.5	243.3±1.6	242.±16.	265.6±1.9	283.9±1.0	288.5±1.2
N_{III}	191.4±0.9	207.2±0.9	217.6±1.1	224.6±1.3		247.4±1.5	256.6±0.8	270.9±0.9
$N_{IV,V}$	98.9±0.8	110.0±0.6	113.2±0.7	117.5±0.7	120.4±2.0	129.0±1.2	133.2±0.6	140.5±0.8
$N_{VI,VII}$		0.1±1.2	2.0±0.6	1.5±0.9		5.5±1.1	0.0±3.2	0.1±3.5
O_I	32.3±7.2	37.8±1.3	37.4±1.0	37.5±0.9		37.4±1.5	31.8±0.7	36.1±0.8
$O_{II,III}$	14.4±1.2	19.8±1.2	22.3±0.7	21.1±0.8		21.3±1.5	22.0±0.6	20.3±1.2

	65 Tb	66 Dy	67 Ho	68 Er	69 Tm	70 Yb	71 Lu	72 Hf
K	51995.7±0.5 (52002.)	53788.5±0.5 (53793.)	55617.7±0.5 (55619.)	57485.5±0.5 (57487.)	59389.6±0.5	61332.3±0.5 (61300.)	63313.8±0.5 (63310.)	65350.8±0.6 (65310.)
L_I	8708.0±0.5 [8707.6][n] (8717.)	9045.8±0.5 [9046.5][n]	9394.2±0.4 [9394.3][n] (9399.)	9751.3±0.4 [9751.5][n] (9757.)	10115.7±0.4 [10115.6][n] (10121.)	10486.4±0.4 [10487.3][n] (10490.)	10870.4±0.4 [10870.1][n] (10874.)	11270.7±0.4 [11271.6][o] (11274.)

X-RAY ATOMIC ENERGY LEVELS (*Continued*)

	65 Tb	66 Dy	67 Ho	68 Er	69 Tm	70 Yb	71 Lu	72 Hf
L_{II}	8251.6±0.4 [8251.8][n] (8253.)	8580.6±0.4 [8580.4][n] (8583.)	8917.8±0.4 [8918.2][n] (8916.)	9264.3±0.4 [9264.3][n] (9262.)	9616.9±0.4 [9617.1][n] (9617.1)	9978.2±0.4 [9977.9][n] (9976.)	10348.6±0.4 [10349.0][n] (10345.)	10739.4±0.4 [10738.9][o] (10736.)
L_{III}	7514.0±0.4 [7514.2][n] (7515.)	7790.1±0.4 [7789.6][n] (7789.7)	8071.1±0.4 [8070.6][n] (8068.)	8357.9±0.4 [8357.6][n] (8357.5)	8648.0±0.4 [8647.8][n] (8649.6)	8943.6±0.4 [8942.6][n] (8944.1)	9244.1±0.4 [9243.8][n]	9560.7±0.4 [9560.4][o] (9558.)
M_I	1967.5±0.6	2046.8±0.4	2128.3±0.6	2206.5±0.6	2306.8±0.7	2398.1±0.4	2491.2±0.5	2600.9±0.4
M_{II}	1767.7±0.9	1841.8±0.5	1922.8±1.0	2005.8±0.6	2089.8±1.1	2173.0±0.4	2263.5±0.4	2365.4±0.4
M_{III}	1611.3±0.8	1675.6±0.9	1741.2±0.9	1811.8±0.6	1884.5±1.1	1949.8±0.5	2023.6±0.5	2107.6±0.4
M_{IV}	1275.0±0.6	1332.5±0.4	1391.5±0.7	1453.3±0.5	1514.6±0.7	1576.3±0.4	1639.4±0.4	1716.4±0.4
M_V	1241.2±0.7	1294.9±0.4	1351.4±0.8	1409.3±0.5	1467.7±0.9	1527.8±0.4	1588.5±0.4	1661.7±0.4
N_I	397.9±0.8	416.3±0.5	435.7±0.8	449.1±1.0	471.7±0.9	487.2±0.6	506.2±0.6	538.1±0.4
N_{II}	310.2±1.2	331.8±0.6	343.5±1.4	366.2±1.5	385.9±1.6	396.7±0.7	410.1±1.8	437.0±0.5
N_{III}	385.0±1.0	292.9±0.6	306.6±0.9	320.0±0.7	336.6±1.6	343.5±0.5	359.3±0.5	380.4±0.5
N_{IV} }	147.0±0.8	154.2±0.5	161.0±1.0	176.7±1.2 }	179.6±1.2	{ 198.1±0.5	204.8±0.5	223.8±0.4
N_V }				167.6±1.5 }		{ 184.9±1.3	195.0±0.4	213.7±0.5
$N_{VI,VII}$	2.6±1.5	4.2±1.6	3.7±3.0	4.3±1.4	5.3±1.9	6.3±1.0	6.9±0.5	17.1±0.5
O_I	39.0±0.8	62.9±0.5	51.2±1.3	59.8±1.7	53.2±3.0	54.1±0.5	56.8±0.5	64.9±0.4
O_{II} }	25.4±0.8	26.3±0.6	20.3±1.5	29.4±1.6	32.3±1.6	23.4±0.6	28.0±0.6	{ 38.1±0.6
O_{III} }								{ 30.6±0.6

	73 Ta	74 W	75 Re	76 Os	77 Ir	78 Pt	79 Au	80 Hg
K	67416.4±0.6 (67403.)	69525.0±0.3 (69508.)	71676.4±0.4 (71658.)	73870.8±0.5	76111.0±0.5	78394.8±0.7 (78381.)	80724.9±0.5 (80720.)	83102.3±0.8
L_I	11681.5±0.3 [11680.2][p] (11682.)	12099.8±0.3 [12098.2][p] (12099.6)	12526.7±0.4 (12530.)	12968.0±0.4 (12972.)	13418.5±0.3 (13423.)	13879.9±0.4 (13883.)	14352.8±0.4 (14353.7)	14839.3±1.0 (14842.)

X-RAY ATOMIC ENERGY LEVELS (*Continued*)

	73 Ta	74 W	75 Re	76 Os	77 Ir	78 Pt	79 Au	80 Hg
L_{II}	11136.1±0.3 [11136.1]^{p} (11132.)	11544.0±0.3 [11541.4]^{p} (11538.)	11958.7±0.3 [11956.9]^{p} (11954.)	12385.0±0.4 (12381.)	12824.1±0.3 [12824.0]^{e,p} (12820.)	13272.6±0.3 [13272.6]^{e,p} (13272.3)	13733.6±0.3 [13733.5]^{e,p} (13736.)	14208.7±0.7 (14215.)
L_{III}	9881.1±0.3 [9880.3]^{p} (9877.7)	10206.8±0.3 [10204.2]^{p} (10200.)	10535.3±0.3 [10534.2]^{p} (10531.)	10870.9±0.3 [10870.7]^{p} (10868.)	11215.2±0.3 [11215.1]^{e,p} (11212.)	11563.7±0.3 [11563.7]^{e,p} (11562.)	11918.7±0.3 [11918.2]^{e,p} (11921.)	12283.9±0.4 [12284.0]^{e,p} (12286.)
M_{I}	2708.0±0.4	2819.6±0.4	2931.7±0.4	3048.5±0.4	3173.7±1.7	3296.0±0.9	3424.9±0.3 [3424.8]^{p}	3561.6±1.1
M_{II}	2468.7±0.3 [2468.6]^{p}	2574.9±0.3 [2575.0]^{p}	2681.6±0.4	2792.2±0.3 [2791.9]^{p}	2908.7±0.3 [2909.1]^{p}	3026.5±0.4 [3026.5]^{p} (3029.)	3147.8±0.4 [3149.5]^{p}	3278.5±1.3
M_{III}	2194.0±0.3 [2194.1]^{p}	2281.0±0.3 [2281.0]^{p}	2367.3±0.3 [2367.3]^{p}	2457.2±0.4 [2457.4]^{p}	2550.7±0.3 [2550.5]^{p} (2550.5)	2645.4±0.4 [2645.5]^{p} (2645.9)	2743.0±0.3 [2743.1]^{p} (2744.0)	2847.1±0.4 [2847.1]^{p}
M_{IV}	1793.2±0.3 [1793.1]^{p}	1871.6±0.3 [1871.4]^{p}	1948.9±0.3 [1948.9]^{p}	2030.8±0.3 [2031.0]^{p}	2116.1±0.3 [2116.1]^{p}	2201.9±0.3 [2201.9]^{p}	2291.1±0.3 [2291.2]^{p} (2307.)	2384.9±0.3 [2384.9]^{p}
M_{V}	1735.1±0.3 [1735.2]^{p}	1809.2±0.3 [1809.3]^{p}	1882.9±0.3 [1882.9]^{p}	1960.1±0.3 [1960.2]^{p}	2040.4±0.3 [2040.5]^{p}	2121.6±0.3 [2121.6]^{p}	2205.7±0.3 [2206.1]^{p} (2220.)	2294.9±0.3 [2294.9]^{p}
N_{I}	565.5±0.5	595.0±0.4	625.0±0.4	654.3±0.5	690.1±0.4	722.0±0.6	758.8±0.4	800.3±1.0
N_{II}	464.8±0.5	491.6±0.4	517.9±0.5	546.5±0.5	577.1±0.4	609.2±0.6	643.7±0.5	676.9±2.4
N_{III}	404.5±0.4	425.3±0.5	444.4±0.5	468.2±0.6	494.3±0.6	519.0±0.6	545.4±0.5	571.0±1.4
N_{IV}	241.3±0.4	258.8±0.4	273.7±0.5	289.4±0.5	311.4±0.4	330.8±0.5	352.0±0.4	378.3±1.0
N_{V}	229.3±0.3	245.4±0.4	260.2±0.4	272.8±0.6	294.9±0.4	313.3±0.4	333.9±0.4	359.8±1.2
N_{VI}	25.0±0.4	36.5±0.4	40.6±0.4	46.3±0.6	63.4±0.4	74.3±0.4	86.4±0.4	102.2±0.5
N_{VII}		33.6±0.4			60.5±0.4	71.1±0.5	82.8±0.5	98.5±0.5
O_{I}	71.1±0.5	77.1±0.4	82.8±0.5	83.7±0.6	95.2±0.4	101.7±0.4	107.8±0.7	120.3±1.3
O_{II}	44.9±0.4	46.8±0.5	45.6±0.7	58.0±1.1	63.0±0.6	65.3±0.7	71.7±0.7	80.5±1.3
O_{III}	36.4±0.4	35.6±0.5	34.6±0.6	45.4±1.0	50.5±0.6	51.7±0.7	53.7±0.7	57.6±1.3
$O_{IV,V}$	5.7±0.4	6.1±0.4	3.5±0.5		3.8±0.4	2.2±1.3	2.5±0.5	6.4±1.4

X-RAY ATOMIC ENERGY LEVELS (*Continued*)

	81 Tl	82 Pb	83 Bi	84 Po	85 At	86 Rn	87 Fr	88 Ra
K	85530.4±0.6	88004.5±0.7 (88005.)	90525.9±0.7 (90534.)	93105.0±3.8	95729.9±7.7	98404.±12.	101137.±13.	103921.9±7.2
L_{I}	15346.7±0.4 (15343.)	15860.8±0.5 (15855.)	16387.5±0.4 (16376.)	16939.3±9.8	17493.±29.	18049.±38.	18639.±40.	19236.7±1.5 (19236.0)
L_{II}	14697.9±0.3 [14697.3][p] (14699.)	15200.0±0.4 (15205.)	15711.1±0.3 [15708.4][p] (15719.)	16244.3±2.4	16784.7±2.5	17337.1±3.4	17906.5±3.5	18484.3±1.5 (18486.0)
L_{III}	12657.5±0.3 [12656.3][e,p] (12660.)	13035.2±0.3 [13034.9][e,p] (13041.)	13418.6±0.3 [13418.3][e,p] (13426.)	13813.8±1.0 ⟨13813.8⟩	14213.5±2.0 ⟨14213.5⟩	14619.4±3.0 ⟨14619.4⟩	15031.2±3.0 ⟨15031.2⟩	15444.4±1.5 (15444.0)
M_{I}	3704.1±0.4	3850.7±0.5	3999.1±0.3 [3999.1][p]	4149.4±3.9	⟨4317.⟩	⟨4482.⟩	⟨4652.⟩	4822.0±1.5
M_{II}	3415.7±0.3 [3415.7][p]	3554.2±0.3 [3554.2][p]	3696.3±0.3 [3696.4][p]	3854.1±9.8	4008.±28.	4159.±38.	4327.±40.	4489.5±1.8
M_{III}	2956.6±0.3 [2956.5][p]	3066.4±0.4 [3066.3][p]	3176.9±0.3 [3176.8][p]	3301.9±9.9	3426.±29.	3538.±38.	3663.±40.	3791.8±1.7
M_{IV}	2485.1±0.3 [2485.2][p]	2585.6±0.3 [2585.5][p] (2606.)	2687.6±0.3 [2687.4][p]	2798.0±1.2	2908.7±2.1	3021.5±3.1	3136.2±3.1	3248.4±1.6
M_{V}	2389.3±0.3 [2389.4][p]	2484.0±0.3 [2484.2][p] (2502.)	2579.6±0.3 [2579.5][p]	2683.0±1.1	2786.7±2.1	2892.4±3.1	2999.9±3.1	3104.9±1.6
N_{I}	845.5±0.5	893.6±0.7	938.2±0.3 [938.7][p]	995.3±2.9	⟨1042.⟩	⟨1097.⟩	⟨1153.⟩	1208.4±1.6
N_{II}	721.3±0.8	763.9±0.8	805.3±0.3 [805.3][p]	851.±12.	886.±30.	929.±40.	980±42.	1057.6±1.8
N_{III}	609.0±0.5	644.5±0.6	678.9±0.3 [678.9][p]	705.±14.	740.±30.	768.±40.	810±43.	879.1±1.8
N_{IV}	406.6±0.4	435.2±0.5	463.6±0.3 [463.6][p]	500.2±2.4	533.2±3.2	566.6±4.0	603.3±4.1	635.9±1.6
N_{V}	386.2±0.5	412.9±0.6	440.0±0.3 [440.1][p]	473.4±1.3			577.±34.	602.7±1.7

X-RAY ATOMIC ENERGY LEVELS (*Continued*)

	81 Tl	82 Pb	83 Bi	84 Po	85 At	86 Rn	87 Fr	88 Ra
N_{VI}	122.8±0.4	142.9±0.4	161.9±0.5					298.9±2.4
N_{VII}	118.5±0.4	138.1±0.4	157.4±0.6					
O_{I}	136.3±0.7	147.3±0.8	159.3±0.7					254.4±2.1
O_{II}	99.6±0.6	104.8±1.0	116.8±0.7					200.4±2.0
O_{III}	75.4±0.6	86.0±1.0	92.8±0.6					152.8±2.0
O_{IV}	15.3±0.4	21.8±0.4	26.5±0.5	31.4±3.2				67.2±1.7
O_{V}	13.1±0.4	19.2±0.4	24.4±0.6					
P_{I}		3.1±1.0						43.5±2.2
$P_{II,III}$		0.7±1.0	2.7±0.7					18.8±1.8

	89 Ac	90 Th	91 Pa	92 U	93 Np	94 Pu	95 Am	96 Cm
K	106755.3±5.3	109650.9±0.9	112601.4±2.4	115606.1±1.6	118678.±33.	121818.±44.	125027.±55.	128220
L_{I}	19840.±18.	20472.1±0.5 (20464.)	21104.6±1.8 (21128.)	21757.4±0.3 (21771.)	22426.8±0.9	23097.2±1.6 (23109.)	23772.9±2.0 ⟨23772.9⟩	24460
L_{II}	19083.2±2.8	19693.2±0.4 (19683.)	20313.7±1.5 (20319.)	20947.6±0.3 (20945.)	21600.5±0.4	22266.2±0.7 (22253.)	22944.0±1.0	23779
L_{III}	15871.0±2.0 ⟨15871.0⟩	16300.3±0.3 [16299.6][q] (16299.)	16733.1±1.4 (16733.)	17166.3±0.3 [17168.5][r] (17165.)	17610.0±0.4 (17606.2)	18056.8±0.6 (18053.1)	18504.1±0.9 (18504.1)	18930
M_{I}	⟨5002.⟩	5182.3±0.3 [5182.3][q]	5366.9±1.6	5548.0±0.4	5723.2±3.6	5932.9±1.4	6120.5±7.5	6288
M_{II}	4656.±18.	4830.4±0.4 [4830.6][q]	5000.9±2.3	5182.2±0.4 [5180.9][r]	5366.2±0.7 [5366.4][s]	5541.2±1.7	5710.2±2.1	5895
M_{III}	3909.±18.	4046.1±0.4 [4046.1][q] (4041.)	4173.8±1.8	4303.4±0.3 [4303.6][r] (4299.)	4434.7±0.5 [4434.6][s]	4556.6±1.5	4667.0±2.1	4797
M_{IV}	3370.2±2.1	3490.8±0.3 [3490.7][q] (3485.)	3611.2±1.4 (3608.)	3727.6±0.3 [3728.1][r] (3720.)	3850.3±0.4 [3849.8][s]	3972.6±0.6 [3972.7][t]	4092.1±1.0	4227

X-RAY ATOMIC ENERGY LEVELS (*Continued*)

	89 Ac	90 Th	91 Pa	92 U	93 Np	94 Pu	95 Am	96 Cm
M_V	3219.0±2.1	3332.0±0.3 [3332.1][q] (3325.)	3441.8±1.4 (3436.)	3551.7±0.3 [3551.7][r] (3545.)	3665.8±0.4 [3664.2][s]	3778.1±0.6 [3778.0][t]	3886.9±1.0	3971
N_I	⟨1269.⟩	1329.5±0.4 [1329.8][q]	1387.1±1.9	1440.8±0.4 [1441.3][r]	1500.7±0.8 [1500.7][s]	1558.6±0.8	1617.1±1.1	1643
N_{II}	1080.±19.	1168.2±0.4 [1168.3][q]	1224.3±1.6	1272.6±0.3 [1272.5][r]	1327.7±0.8 [1327.7][s]	1372.1±1.8	1411.8±8.3	1440
N_{III}	890.±19.	967.3±0.4 [967.6][q]	1006.7±1.7	1044.9±0.3 [1044.9][r]	1086.8±0.7 [1086.8][s]	1114.8±1.6	⟨1135.7⟩	1154
N_{IV}	674.9±3.7	714.1±0.4 [714.4][q]	743.4±2.1	780.4±0.3 [779.7][r]	815.9±0.5 [817.1][s]	848.9±0.6 [848.9][t]	878.7±1.0	
N_V		676.4±0.4 [676.4][q]	708.2±1.8	737.7±0.3 [737.6][r]	770.3±0.4 [773.2][s]	801.4±0.6 [801.4][t]	827.6±1.0	
N_{VI}		344.4±0.3 [344.2][q]	371.2±1.6	391.3±0.6	415.0±0.8 [415.0][s]	445.8±1.7		
N_{VII}		335.2±0.4 [335.0][q]	359.5±1.6	380.9±0.9	404.4±0.5 [404.4][s]	432.4±2.1		
O_I		290.2±0.8	309.6±4.3	323.7±1.1		351.9±2.4		385
O_{II}		229.4±1.1	222.9±3.9	259.3±0.5	283.4±0.8 [283.4][s]	274.1±4.7		
O_{III}		181.8±0.4 [181.8][q]		195.1±1.3	206.1±0.7 [206.1][s]	206.5±4.7		
O_{IV}		94.3±0.4 [94.4][q]	94.1±2.8	105.0±0.5	109.3±0.7 [108.8][s]	116.0±1.2	115.8±1.3	
O_V		87.9±0.3 [88.1][q]		96.3±1.4	101.3±0.5 [101.4][s]	105.4±1.0	103.3±1.1	
P_I		59.5±1.1		70.7±1.2				
P_{II}		49.0±2.5		42.3±9.0				
P_{III}		43.0±2.5		32.3±9.0				

X-RAY ATOMIC ENERGY LEVELS (*Continued*)

	97 Bk	98 Cf	99 Es	100 Fm	101 Md	102 No	103 Lw
K	[131590±40][u]	135960	139490	143090	146780	150540	154380
L_I	[25275±17][u]	26110	26900	27700	28530	29380	30240
L_{II}	[24385±17][u]	25250	26020	26810	27610	28440	29280
L_{III}	[19452±20][u]	19930	20410	20900	21390	21880	22360
M_I	[6556±21][u]	6754	6977	7205	7441	7675	7900
M_{II}	[6147±31][u]	6359	6574	6793	7019	7245	7460
M_{III}	[4977±31][u]	5109	5252	5397	5546	5688	5710
M_{IV}	4366	4497	4630	4766	4903	5037	5150
M_V	4132	4253	4374	4498	4622	4741	4860
N_I	[1755±22][u]	1799	1868	1937	2010	2078	2140
N_{II}	1554	1616	1680	1747	1814	1876	1930
N_{III}	1235	1279	1321	1366	1410	1448	1480
O_I	[398±22][u]	419	435	454	472	484	490

[a] J. E. Mack, 1949, as given in C. E. Moore, *Atomic Energy Levels* (U. S. National Bureau of Standards, Washington, D. C., 1949), Vol. 1, p. 1.
[b] G. Herzberg, 1957, as given in C. E. Moore, *Atomic Energy Levels* (U. S. National Bureau of Standards, Washington, D. C., 1958), Vol. 3, p. 238.
[c] See Ref. 18.
[d] A. Fahlman, D. Hamrin, R. Nordberg, C. Nordling, and K. Siegbahn, Phys. Rev. Letters **14**, 127 (1965). See also Ref. 26.
[e] See Ref. 15.
[f] See Ref. 11.
[g] C. Nordling, Arkiv Fysik **15**, 397 (1959).
[h] E. Sokolowski, C. Nordling, and K. Siegbahn, Arkiv Fysik **12**, 301 (1957).
[i] C. Nordling and S. Hagström, Arkiv Fysik **16**, 515 (1960).
[j] I. Andersson and S. Hagström, Arkiv Fysik **27**, 161 (1964).
[k] M. O. Krause, Phys. Rev. **140**, A1845 (1965).
[l] A. Fahlman, O. Hörnfeldt, and C. Nordling, Arkiv Fysik **23**, 75 (1962).
[m] P. Bergvall, O. Hörnfeldt, and C. Nordling, Arkiv Fysik **17**, 113 (1960).
[n] P. Bergvall and S. Hagström, Arkiv Fysik **17**, 61 (1960).
[o] S. Hagström, Z. Physik **178**, 82 (1964).
[p] A. Fahlman and S. Hagström, Arkiv Fysik **27**, 69 (1964).
[q] C. Nordling and S. Hagström, Z. Physik **178**, 418 (1964).
[r] C. Nordling and S. Hagström, Arkiv Fysik **15**, 431 (1959).
[s] S. Hagström, Bull. Am. Phys. Soc. **11**, 389 (1966).
[t] A. Fahlman, K. Hamrin, R. Nordberg, C. Nordling, K. Siegbahn, and L. W. Holm, Phys. Letters **19**, 643 (1966).
[u] J. M. Hollander, M. D. Holtz, T. Novakov, and R. L. Graham ,Arkiv Fysik **28**, 375 (1965).

LATTICE SPACING OF COMMON ANALYSING CRYSTALS

Crystal	Reflection plane	*d* Spacing
ADP[a]	101	5.31
ADP[a]	110	5.325
ADP[a]	200	3.75
Beryl	$10\bar{1}0$	7.98
Calcite	100	3.036
EDDT[b]	020	4.404
Germanium	111	3.265
Graphite	001	6.69
Gypsum	010	7.600
KAP[c]	001	13.32
KBr	200	3.29
KCl	200	3.14
Lead stearate		51
LiF	200	2.014
LiF	220	1.424
Mica	002	9.96
NaCl	200	2.820
Oxalic acid	001	5.85
PET[d]	002	4.371
Quartz	$10\bar{1}0$	4.255
Quartz	$10\bar{1}1$	3.343
Quartz	$11\bar{2}0$	2.456
Silicon	111	3.13
Topaz	303	1.356

While several of the above spacings have been measured to more than four significant figures, no more than four figures are given here because complications introduced by the index of refraction, anomalous dispersion, temperature coefficient of expansion, and crystal impurities must be considered before the additional figures are useful.

[a] Ammonium dihydrogen phosphate.
[b] Ethylenediamine d-tartrate.
[c] Potassium acid phthalate.
[d] Pentaerythritol.

RADIATIVE TRANSITION PROBABILITIES FOR *K* X-RAY LINES

$K\beta_1' = KM_2 + KM_3 + KM_{4,5}$ $\quad K\beta_2' = KN_{2,3} + KO_{2,3}$

$K\alpha = K\alpha_1 + K\alpha_2$ $\quad K\beta = K\beta_1' + K\beta_2'$

Element	$K\alpha_2/K\alpha_1$	$K\alpha_3/K\alpha_1$	$K\beta_1/K\alpha_1$	$K\beta'_2/K\alpha_1$	$K\beta_4/K\alpha_1$	$K\beta_5/K\alpha_1$	$K\beta_3/K\beta_1$	$K\beta/K\alpha$
$_{12}$Mg								0.013
$_{14}$Si								0.027
$_{16}$S								0.059
$_{18}$Ar								0.105
$_{20}$Ca	0.502							0.128
$_{22}$Ti	0.503							0.134
$_{24}$Cr	0.504							0.135
$_{26}$Fe	0.506							0.135
$_{28}$Ni	0.508							0.135
$_{30}$Zn	0.510							0.138
$_{32}$Ge	0.513							0.147
$_{34}$Se	0.515							0.157
$_{36}$Kr	0.517			0.019				0.172
$_{38}$Sr	0.520			0.030				0.180
$_{40}$Br	0.523			0.037				0.190
$_{42}$Mo	0.525			0.041				0.197
$_{44}$Ru	0.527			0.045				0.204
$_{46}$Pd	0.529			0.048				0.210
$_{48}$Cd	0.532			0.053			0.519	0.213
$_{50}$Sn	0.534			0.055			0.519	0.220
$_{52}$Te	0.537			0.058			0.519	0.225
$_{54}$Xe	0.539			0.064			0.518	0.232
$_{56}$Ba	0.543			0.070			0.518	0.237
$_{58}$Ce	0.546			0.076			0.518	0.242
$_{60}$Nd	0.549	0.11×10^{-3}		0.083			0.518	0.247
$_{62}$Sm	0.552	0.14×10^{-3}		0.086			0.517	0.250
$_{64}$Gd	0.556	0.17×10^{-3}	0.192	0.089	0.85×10^{-3}	3.02×10^{-3}	0.517	0.255
$_{66}$Dy	0.560	0.21×10^{-3}	0.198	0.089	0.92×10^{-3}	3.43×10^{-3}	0.517	0.257
$_{68}$Er	0.564	0.26×10^{-3}	0.202	0.088	0.96×10^{-3}	3.85×10^{-3}	0.518	0.260
$_{70}$Yb	0.567	0.30×10^{-3}	0.207	0.087	1.04×10^{-3}	4.23×10^{-3}	0.518	0.264
$_{72}$Hf	0.572	0.36×10^{-3}	0.212	0.085	1.16×10^{-3}	4.62×10^{-3}	0.518	0.267
$_{74}$W	0.576	0.43×10^{-3}	0.216	0.086	1.28×10^{-3}	5.04×10^{-3}	0.518	0.269
$_{76}$Os	0.580	0.51×10^{-3}	0.222	0.087	1.43×10^{-3}	5.44×10^{-3}	0.519	0.273
$_{78}$Pt	0.583	0.63×10^{-3}	0.226	0.091	1.61×10^{-3}	5.84×10^{-3}	0.520	0.275
$_{80}$Hg	0.588	0.76×10^{-3}	0.228	0.096	1.80×10^{-3}	6.24×10^{-3}	0.520	0.278
$_{82}$Pb	0.593	0.91×10^{-3}	0.228	0.102	2.02×10^{-3}	6.64×10^{-3}	0.521	0.280
$_{84}$Po	0.597	1.12×10^{-3}	0.228	0.108	2.26×10^{-3}	7.05×10^{-3}	0.522	0.283
$_{86}$Rn	0.602	1.32×10^{-3}	0.228	0.113	2.52×10^{-3}	7.48×10^{-3}	0.523	0.286
$_{88}$Ra	0.608	1.58×10^{-3}	0.230	0.117	2.80×10^{-3}	7.80×10^{-3}	0.524	0.287
$_{90}$Th	0.613	1.85×10^{-3}	0.232	0.120	3.13×10^{-3}	8.25×10^{-3}	0.525	0.288
$_{92}$U	0.619	2.15×10^{-3}	0.234	0.123	3.47×10^{-3}	8.65×10^{-3}	0.527	0.289
$_{94}$Pu	0.625		0.234	0.125			0.528	0.291
$_{96}$Cm	0.632		0.234	0.128			0.529	0.293
$_{98}$Cf	0.642		0.238	0.132			0.531	0.295
$_{100}$Fm	0.648		0.240	0.135			0.533	0.297

From Salem, S. I., Panossian, S. L., and Krause, R. A., *At. Data Nucl. Data Tables*, 14, 91, 1974. Reproduced by permission of the copyright owner, Academic Press.

RADIATIVE TRANSITION PROBABILITIES FOR *L* X-RAY LINES

The following three tables present data for the radiative transition probabilities for the L_1, L_2, and L_3 X-ray lines. The data are normalized respectively to $L\beta_3$ = 100, $L\beta_1$ = 100 and $L\alpha_1$ = 100.

L_1 X-RAY LINES NORMALIZED TO $L\beta_3$ = 100

Element	$L\beta_3$	$L\beta_4$	$L\gamma_2$	$L\gamma_3$	Element	$L\beta_3$	$L\beta_4$	$L\gamma_2$	$L\gamma_3$
$_{32}$Ge	100			17.3	$_{66}$Dy	100	61.8	19.5	28.0
$_{34}$Se	100			18.0	$_{68}$Er	100	63.5	19.8	29.0
$_{36}$Kr	100			18.2	$_{70}$Yb	100	65.5	20.7	29.8
$_{38}$Sr	100			18.8	$_{72}$Hf	100	67.8	21.2	30.7
$_{40}$Zr	100			19.0	$_{74}$W	100	70.5	21.8	31.8
$_{42}$Mo	100	70.6		19.6	$_{76}$Os	100	73.2	23.0	32.8
$_{44}$Ru	100	67.8		20.2	$_{78}$Pt	100	76.5	24.5	33.8
$_{46}$Pd	100	65.5		20.6	$_{80}$Hg	100	80.3	26.3	35.0
$_{48}$Cd	100	63.5		21.3	$_{82}$Pb	100	84.2	28.6	36.0
$_{50}$Sn	100	62.1		22.0	$_{84}$Po	100	88.5	31.3	37.2
$_{52}$Te	100	60.7		22.6	$_{86}$Rn	100	93.4	34.2	38.2
$_{54}$Xe	100	59.8		23.3	$_{88}$Ra	100	98.9	37.5	39.6
$_{56}$Ba	100	59.5		24.0	$_{90}$Th	100	104.5	41.2	41.0
$_{58}$Ce	100	59.2		24.6	$_{92}$U	100	110.2	45.0	42.6
$_{60}$Nd	100	59.4		25.4	$_{94}$Pu	100	116.2	49.5	44.0
$_{62}$Sm	100	60.0		26.3	$_{96}$Cm	100	123.0	55.7	45.7
$_{64}$Gd	100	60.8	19.2	27.0					

L_2 X-RAY LINES NORMALIZED TO $L\beta_1$ = 100

Element	$L_{\beta 1}$	L_{η}	$L_{\gamma 1}$	$L_{\gamma 6}$	Element	$L_{\beta 1}$	L_{η}	$L_{\gamma 1}$	$L_{\gamma 6}$
$_{28}$Ni	100	7.60			$_{64}$Gd	100	2.35	17.00	
$_{30}$Zn	100	6.80			$_{66}$Dy	100	2.25	17.40	
$_{32}$Ge	100	6.28			$_{68}$Er	100	2.16	17.80	
$_{34}$Se	100	5.80			$_{70}$Yb	100	2.10	18.17	
$_{36}$Kr	100	5.35			$_{72}$Hf	100	2.08	18.43	
$_{38}$Sr	100	4.93			$_{74}$W	100	2.10	18.80	0.72
$_{40}$Zr	100	4.60	3.30		$_{76}$Os	100	2.12	19.34	1.65
$_{42}$Mo	100	4.30	5.50		$_{78}$Pt	100	2.18	19.73	2.40
$_{44}$Ru	100	4.00	7.33		$_{80}$Hg	100	2.25	20.35	3.10
$_{46}$Pd	100	3.75	10.67		$_{82}$Pb	100	2.30	20.93	3.65
$_{48}$Cd	100	3.55	10.60		$_{84}$Po	100	2.40	21.54	4.15
$_{50}$Sn	100	3.35	11.80		$_{86}$Rn	100	2.46	22.20	4.55
$_{52}$Te	100	3.20	12.70		$_{88}$Ra	100	2.50	22.87	4.87
$_{54}$Xe	100	3.00	14.00		$_{90}$Th	100	2.60	23.43	5.02
$_{56}$Ba	100	2.85	14.50		$_{92}$U	100	2.65	24.10	5.12
$_{58}$Ce	100	2.70	15.30		$_{94}$Pu	100	2.70	24.40	5.16
$_{60}$Nd	100	2.60	16.00		$_{96}$Cm	100	2.75	25.07	5.20
$_{62}$Sn	100	2.45	16.50						

RADIATIVE TRANSITION PROBABILITIES FOR *L* X-RAY LINES (*Continued*)

L_3 X-RAY LINES NORMALIZED TO $L\alpha_1$ = 100

Element	$L_{\alpha 1}$	$L_{\beta 2,15}$	$L_{\alpha 2}$	$L_{\beta 5}$	$L_{\beta 6}$	L_{ℓ}
$_{26}$Fe	100					12.22
$_{28}$Ni	100					8.95
$_{30}$Zn	100					7.34
$_{32}$Ge	100					6.45
$_{34}$Se	100					7.76
$_{36}$Kr	100					5.28
$_{38}$Sr	100					4.92
$_{40}$Zr	100	0.70	11.10			4.67
$_{42}$Mo	100	5.17	11.10			4.45
$_{44}$Ru	100	9.30	11.12			4.28
$_{46}$Pd	100	11.80	11.12			4.11
$_{48}$Cd	100	14.33	11.12			4.07
$_{50}$Sn	100	16.00	11.13			4.00
$_{52}$Te	100	18.00	11.13			4.00
$_{54}$Xe	100	19.40	11.13			4.00
$_{56}$Ba	100	20.67	11.13			4.02
$_{58}$Ce	100	21.00	11.14			4.09
$_{60}$Nd	100	21.33	11.14		0.875	4.13
$_{62}$Sm	100	21.07	11.14		0.925	4.16
$_{64}$Gd	100	20.83	11.14		0.99	4.20
$_{66}$Dy	100	20.50	11.14		1.05	4.26
$_{68}$Er	100	20.04	11.15		1.12	4.33
$_{70}$Yb	100	19.40	11.15		1.17	4.47
$_{72}$Hf	100	21.33	11.15	0.30	1.21	4.59
$_{74}$W	100	22.74	11.16	0.50	1.25	4.76
$_{76}$Os	100	23.40	11.16	1.32	1.37	4.95
$_{78}$Pt	100	24.00	11.16	1.98	1.43	5.14
$_{80}$Hg	100	24.50	11.17	2.62	1.50	5.37
$_{82}$Pb	100	24.83	11.17	3.21	1.56	5.58
$_{84}$Po	100	25.13	11.17	3.73	1.62	5.80
$_{86}$Rn	100	25.60	11.18	4.25	1.68	6.00
$_{88}$Ra	100	25.92	11.18	4.73	1.76	6.26
$_{90}$Th	100	26.17	11.18	5.18	1.82	6.54
$_{92}$U	100	26.40	11.18	5.58	1.89	6.79
$_{94}$Pu	100	26.67	11.18	5.92	1.95	7.02
$_{96}$Cm	100	26.93	11.18	6.26	2.01	7.34

From Salem, S. I., Panossian, S. L., and Krause, R. A., *At. Data Nucl. Data Tables,* 14, 91, 1974. Reproduced by permission of the copyright owner. Academic Press.

K AND L X-RAY ABSORPTION EDGES

The wavelengths in Angstroms for the K absorption edge of elements having atomic numbers 3 to 11 and 83 to 100 were calculated from critical X-ray absorption energies listed in the Review of Scientific Instruments, **23,** 523–528. (1952.)

All wavelengths for the L absorption edges were calculated from absorption energies given in the above reference.

Values for the M, N, and O absorption edges are also listed in the above reference. The wavelengths in Angstroms for the K absorption edge of the elements having atomic numbers 12 to 82 are reproduced by permission from Structure of Metals, Barrett, 2nd edition, McGraw-Hill Book Company, Inc. (1952.)

ABSORPTION EDGES—ANGSTROMS

Atomic Number	Element	K	L_I	L_{II}	L_{III}
3	Li	226.62			
4	Be	110.68			
5	B	66.289			
6	C	43.648			
7	N	30.990			
8	O	23.301			
9	F	17.913			
10	Ne	14.183			
11	Na	11.478			
12	Mg	9.5117	197.39	247.92	
13	Al	7.9511	142.48	172.16	
14	Si	6.7446	105.05	126.48	
15	P	5.7866	81.02	96.843	
16	S	5.0182	64.228	76.049	76.519
17	Cl	4.3969	52.084	61.366	61.672
18	A	3.87068	43.192	50.390	50.803
19	K	3.43645	36.352	42.020	42.452
20	Ca	3.07016	31.068	35.417	35.827
21	Sc	2.7572	26.831	30.161	30.457
22	Ti	2.497$_{30}$	23.389	26.831	27.184
23	Va	2.269$_{02}$	20.523	23.702	24.0699
24	Cr	2.070$_{12}$	18.256	21.226	21.5958
25	Mn	1.896$_{36}$	16.268	18.896	19.2484
26	Fe	1.743$_{34}$	14.601	17.169	17.4838
27	Co	1.608$_{11}$	13.343	15.534	15.8314
28	Ni	1.488$_{02}$	12.267	14.135	14.4476
29	Cu	1.380$_{43}$	11.269	12.994	13.2578
30	Zn	1.283$_{3}$	10.330	11.8395	12.1055
31	Ga	1.195$_{67}$	9.535	10.6130	10.8546
32	Ge	1.116$_{52}$	8.729	9.9646	10.2277
33	As	1.044$_{97}$	8.107	9.1281	9.3767
34	Se	0.979$_{78}$	7.467	8.4212	8.6624
35	Br	0.91994	6.925	7.7523	7.9871
36	Kr	0.86546	6.456	7.1653	7.4227
37	Rb	0.81549	6.006	6.6538	6.8752
38	Sr	0.76969	5.604	6.1856	6.3996
39	Y	0.72762	5.19312	5.70981	5.91412
40	Zr	0.68877	4.89380	5.37088	5.57374
41	Nb	0.65291	4.59111	5.02472	5.22596
42	Mo	0.61977	4.32066	4.71330	4.90930
43	Tc	0.58888	4.06426	4.42714	4.62537
44	Ru	0.560$_{47}$	3.84133	4.17654	4.36632
45	Rh	0.533$_{78}$	3.64159	3.94902	4.13889
46	Pd	0.509$_{15}$	3.42999	3.71360	3.89688
47	Ag	0.4858$_{2}$	3.23824	3.49478	3.67288
48	Cd	0.46408	3.08434	3.32243	3.50070
49	In	0.44397	2.93327	3.15500	3.32154
50	Sn	0.42468	2.78875	2.99492	3.16952

Atomic Number	Element	K	L_I	L_{II}	L_{III}
51	Sb	0.40663	2.63296	2.82304	2.99637
52	Te	0.38972	2.50272	2.68253	2.85162
53	I	0.373$_{79}$	2.38982	2.55324	2.71901
54	Xe	0.35849	2.27449	2.43058	2.59330
55	Cs	0.34473	2.17245	2.31268	2.47227
56	Ba	0.33137	2.08336	2.20216	2.36114
57	La	0.31842	1.97892	2.10030	2.25792
58	Ce	0.30647	1.89078	2.00940	2.16410
59	Pr	0.29517	1.81307	1.92305	2.07707
60	Nd	0.28451	1.73759	1.84244	1.99452
61	Pm	0.27425	1.66837	1.76581	1.91888
62	Sm	0.26462	1.60113	1.69436	1.84464
63	Eu	0.25552	1.53815	1.62591	1.77491
64	Gd	0.24680	1.47870	1.56081	1.70955
65	Tb	0.23840	1.42270	1.50108	1.64840
66	Dy	0.23046	1.36926	1.44357	1.59025
67	Ho	0.22290	1.31942	1.38999	1.53529
68	Er	0.21565	1.27086	1.33721	1.48242
69	Tm	0.2089	1.22708	1.28856	1.43140
70	Yb	0.20223	1.18136	1.24146	1.38518
71	Lu	0.19584	1.14007	1.19745	1.34039
72	Hf	0.18981	1.09864	1.15311	1.29570
73	Ta	0.18393	1.06084	1.11264	1.25427
74	W	0.17837	1.02497	1.07436	1.21529
75	Re	0.17311	0.99009	1.03645	1.17720
76	Os	0.16780	0.95574	1.00129	1.14143
77	Ir	0.16286	0.92425	0.96700	1.10599
78	Pt	0.15817	0.89405	0.93484	1.07306
79	Au	0.15344	0.86378	0.90277	1.04028
80	Hg	0.14923	0.83531	0.87790	1.00944
81	Tl	0.14470	0.80787	0.84355	0.97968
82	Pb	0.14077	0.78153	0.81552	0.95112
83	Bi	0.13691	0.75649	0.78910	0.92459
84	Po	0.13306	0.73219	0.76377	0.89761
85	At	0.12949	0.70915	0.73873	0.87234
86	Rn	0.12591	0.68675	0.71529	0.84845
87	Fr	0.12261	0.66537	0.69290	0.82529
88	Ra	0.11931	0.64461	0.67114	0.80284
89	Ac	0.11618	0.62479	0.65002	0.78158
90	Th	0.11290	0.60610	0.63010	0.76151
91	Pa	0.11028	0.58748	0.61064	0.74138
92	U	0.10775	0.56974	0.59186	0.72225
93	Np	0.10487	0.55314	0.57415	0.70391
94	Pu	0.10228	0.53662	0.55712	0.68637
95	Am	0.09972	0.52084	0.54036	0.66933
96	Cm	0.09674	0.50595	0.52458	0.65276
97	Bk	0.09506	0.49131	0.50928	0.63667
98	Cf	0.09278	0.47713	0.49445	0.62135
99		0.09068	0.46357	0.48009	0.60645
100		0.08861	0.45059	0.46654	0.59225

NATURAL WIDTH, IN eV OF THE INDICATED K X-RAY LINES

Element	$K\alpha_1$	$K\alpha_2$	$K\beta_1$	$K\beta_3$	Element	$K\alpha_1$	$K\alpha_2$	$K\beta_1$	$K\beta_3$
20Ca	1.00	0.98			60Nd	21.50	21.50	23.25	21.33
22Ti	1.45	2.13			62Sm	26.00	24.70	25.65	24.65
24Cr	2.05	2.64			64Gd	29.50	28.00	29.37	28.00
26Fe	2.45	3.20			66Dy	33.90	32.20	32.73	32.00
28Ni	3.00	3.70			68Er	37.40	35.50	36.20	35.70
30Zn	3.40	3.96			70Yb	42.00	40.60	41.43	41.15
32Ge	3.75	4.18			72Hf	45.30	44.30	46.00	46.10
34Se	4.10	4.43			74W	47.75	48.00	51.83	51.50
36Kr	4.23	4.62			76Os	53.00	49.40	55.90	55.95
38Sr	5.17	4.97			78Pt	60.30	54.30	59.98	62.13
40Zn	5.70	5.25			80Hg	64.75	68.20	65.75	68.95
42Mo	6.82	6.80			82Pb	68.30	79.00	72.20	74.90
44Ru	7.41	7.96			84Po	73.20	86.30	78.60	82.85
46Pd	8.80	9.20			86Rn	80.00	89.50	85.50	91.20
48Cd	9.80	10.40			88Ra	87.00	91.20	94.20	98.95
50Sn	11.20	12.40	11.80	11.00	90Th	94.70	97.00	99.70	105.00
52Te	12.80	14.20	13.30	12.30	92U	103.00	106.00	115.00	120.00
54Xe	14.20	15.10	15.30	13.43					
56Ba	16.10	16.80	18.15	16.00					
58Ce	18.60	19.50	20.60	17.95					

NATURAL WIDTH, IN eV OF THE INDICATED L X-RAY LINES

Element	$L\alpha_1$	$L\alpha_2$	$L\beta_1$	$L\beta_2$	$L\beta_3$	$L\beta_4$	$L\gamma_1$
40Zn	1.68	1.52	1.87	5.13	5.50	5.60	3.34
42Mo	1.86	1.80	2.03	5.30	5.90	5.78	3.76
44Ru	2.03	1.98	2.18	5.45	6.35	5.96	4.15
46Pd	2.21	2.16	2.36	5.63	6.80	6.18	4.50
48Cd	2.43	2.40	2.54	5.82	7.23	6.28	4.83
50Sn	2.62	2.62	2.75	6.10	7.70	6.60	5.23
52Tc	2.88	2.88	2.96	6.25	8.22	6.82	5.60
54Xe	3.15	3.15	3.20	6.43	8.70	7.15	5.95
56Ba	3.39	3.45	3.45	6.70	9.20	7.42	6.35
58Ce	3.70	3.78	3.73	6.86	9.70	7.82	6.75
60Nd	3.93	4.08	4.00	7.18	10.30	8.15	7.16
62Sm	4.13	4.50	4.33	7.42	10.80	8.60	7.50
64Gd	4.46	4.90	4.63	7.70	11.20	9.08	7.83
66Dy	4.81	5.35	5.03	7.90	11.50	9.60	8.30
68Er	5.17	5.73	5.45	8.28	11.85	10.03	8.75
70Yb	5.40	6.22	5.90	8.58	12.20	11.00	9.20
72Hf	5.83	6.70	6.36	8.92	12.40	12.80	9.63
74W	6.50	7.20	6.90	9.06	13.10	14.60	10.20
76Os	7.04	7.70	7.42	9.60	14.60	16.50	10.65
78Pt	7.60	8.28	8.00	9.95	16.10	18.00	11.20
80Hg	8.10	8.80	8.70	10.40	17.40	19.70	11.80
82Pb	8.82	9.35	9.35	10.75	18.65	21.30	12.30
84Po	9.50	9.95	10.10	11.25	19.90	22.70	13.05
86Rn	10.03	10.50	10.65	11.65	21.00	24.00	13.55
88Ra	11.00	11.20	11.60	12.20	22.00	25.20	14.30
90Th	11.90	11.80	12.40	12.80	22.85	26.35	15.00
92U	12.40	12.40	13.50	13.30	23.70	27.50	15.70
94Pu	13.20	13.00	14.10	13.90	24.10	28.30	16.40
96Cm	14.80	13.60	15.70	14.60	25.00	29.40	17.10

DIFFRACTION DATA FOR CUBIC ISOMORPHS

From Volume 14, pages 689, 690, and 691 of the Analytical Edition of Industrial and Engineering Chemistry. (With permission.)

X Units	Substance
A 4	
3.56	C (diamond)
5.42	Si
5.62	Ge
6.46	α-Sn
A 1	
3.517	Ni
3.554	α-Co
3.60	Taenite (57.7 % Fe, 40.8 % Ni, 0.5 % P)
3.608	Cu
3.63	γ-Fe (1370°K.)
3.797	Rh
3.831	Ir
3.880	Pd
3.88–4.04	Pd-H
3.912	Pt
4.041	Al
4.070	Au
4.077	Ag
4.30	Co-N
4.40–4.46	Ti-H
4.52	Ne (4°K.)
4.66	Zr-H
4.84	β-Tl
4.939	Pb
5.08	Th
5.14	α-Ce
5.296	β-La
5.43	A (4°K.)
5.56	Ca
5.59	Kr (20°K.)
5.70	Kr (92°K.)
6.05	Sr
6.20	X (88°K.)
A 2	
2.861	α-Fe
2.875	α-Cr
2.90	β-Fe (1070°K.)
2.93	δ-Fe (1700°K.)
3.03	V
3.03–3.41	V-C
3.140	Mo
3.157	W
3.295	Cb
3.30	Ta
3.32	β-Ti (1200°K.)
3.46	Li (~80°K.)
3.50	Li
3.61	β-Zr (1120°K.)
4.24	Na (~80°K.)
4.29	Na
5.02	Ba
5.20	K (120°K.)
5.33	K
5.62	Rb (~80°K.)
6.05	Cs (~80°K.)
B 1	
4.018	LiF
4.065	LiD
4.08	VO
4.09	LiH
4.12	(Li_2TiO_3)
4.12–4.20	(Li_2TiO_3-MgO)
4.13	VN
4.14	CrN
4.14	VC
4.142	($63Li_2Fe_2O_4 \cdot 37Li_2TiO_3$)
4.173	NiO
4.207	MgO
4.282	MgO (1570°K.)
4.225	TiN
4.235	TiO
4.24	80 TiN-20 TiC
4.27	CoO
4.28	V-N
4.283	FeO (160°K.)
4.290	FeO (299°K.)
4.30	VC (ε-phase)
4.315	TiC
4.40	CbC
4.41	CbN
4.426	MnO (117°K.)
4.436	MnO (299°K.)
4.44	ScN
4.446	TaC
4.458	HfC
4.615	NaF
4.62	ZrN
4.69	CdO
4.69	ZrC
4.80	CaO
4.82 **B 1 (Continued)**	(Na_2CeO_3)
4.84	(Na_2PrO_3)
4.88	NaH
4.92	AgF
5.006	CaNH
5.13	SrO
5.14	LiCl
5.14	NdN
5.19	MgS
5.192	MnS (130°K.)
5.210	MnS (299°K.)
5.33	KF
5.45	MgSe
5.45	MnSe
5.45	SrNH
5.49	LiBr
5.52	BaO
5.545	AgCl
5.55–5.76	AgCl-AgBr
5.627	NaCl
5.63	RbF
5.68	CaS
5.69	SnAs
5.70	KH
5.755	AgBr
5.76–5.92	AgBr-AgI
5.83	NdP
5.83	NaCN
5.84	BaNH
5.87	SrS
5.91	CaSe
5.94	PbS
5.95	NaBr
5.957	EuS
5.96	NdAs
6.00	PrAs
6.00	LiI
6.01	CsF
6.04	RbH
6.05	β-NaSH (>360°K.)
6.06	CeAs
6.13	LaAs
6.14	PbSe
6.23	SrSe
6.278	KCl
6.285	SnTe
6.31	NdSb
6.345	CaTe
6.35	PrSb
6.36	BaS
6.38	CsH
6.40	CeSb
6.44	PbTe
6.462	NaI
6.45	PrBi
6.48	LaSb
6.49	CeBi
6.53	KCN
6.53	NH_4Cl (>457°K.)
6.56	RbCl
6.57	LaBi
6.58	KBr
6.59	BaSe
6.60	β-KSH (>440°K.)
6.65	SrTe
6.82	RbCN
6.86	RbBr
6.90	NH_4Br (>411°K.)
6.93	β-RbSH (470°K.)
6.99	BaTe
7.052	KI
7.10	β-CsCl (>730°K.)
7.24	NH_4I (>255°K.)
7.325	RbI
H 0_5	
6.96 ± 0.04	$AgClO_4$ (453 ± 20°K.)
7.16 ± 0.10	$NaClO_4$ (618 ± 35°K.)
7.49 ± 0.02	$KClO_4$ (598 ± 15°K.)
7.65 ± 0.05	$TlClO_4$ (553°K.)
7.65 ± 0.02	NH_4ClO_4 (528 ± 15°K.)
7.68 ± 0.03	$RbClO_4$ (583 ± 10°K.)
7.97 ± 0.01	$CsClO_4$ (513 ± 10°K.)
B 3	
4.255	CuF
4.36	CSi IV
4.855	BeS
5.10	BeSe
5.304	(Cu, Fe, Mo, Sn)$_4$(S, As, Te)$_{3-4}$, cousite
5.41	CuCl
5.425	β-ZnS
5.43	AlP
5.44	GaP
5.58	BeTe
5.60	MnS (red)

DIFFRACTION DATA FOR CUBIC ISOMORPHS
(Continued)

X Units		Substance
	B 3 (Continued)	
5.63		AlAs
5.635		GaAs
5.655		ZnSe
5.68		CuBr
5.82		β-CdS
5.84		HgS
5.86		InP
6.04		CdSe
6.04		InAs
6.05		CuI
6.07		HgSe
6.08		ZnTe
6.103		α-Cu_2HgI_4
6.12		AlSb
6.12		GaSb
6.13		SnSb
6.383		α-Ag_2HgI_4
6.40		HgTe
6.43		CdTe
6.45		InSb
6.48		AgI
	B 32	
6.195		LiGa
6.209		LiZn
6.36		LiAl
6.687		LiCd
6.786		LiIn
7.297		NaIn
7.373		($CeMg_3$)
7.373		($PrMg_3$)
7.473		NaTl
	B 20	
4.437		NiSi
4.438		FeSi
4.438		CoSi
4.548		MnSi
4.620		CrSi
	B 2	
2.603		NiBe
2.606		CoBe
2.69		CuBe
2.813		PdBe
2.82		AlNi
2.945		CuZn
2.989		CuPd
3.146		AuZn
3.156		AgZn
3.168		AgLi
3.259		AuMg
3.275		AgMg
3.287		HgLi
3.325		AgCd
3.34		AuCd (670°K.)
3.424		LiTl
3.442		HgMg
3.628		MgTl
3.67		PrZn
3.70		CeZn
3.73		AlNd
3.74		α-RbCl (83°K.)
3.75		LaZn
3.82		TlCN
3.82		PrCd
3.84		TlSb
3.835		TlCl
3.847		CaTl
3.86		NH_4Cl (<457°K.)
3.86		CeCd
3.88		MgPr
3.90		LaCd
3.97		TlBr
3.98		TlBi
4.024		SrTl
4.05		NH_4Br (<411°K.)
4.112		CsCl
4.20		TlI
4.20		CsCl (<720°K.)
4.25		CsCN
4.287		CsBr
4.29		CsSH
4.37		NH_4I (290°K.)
4.56		CsI
	D 2_1	
4.07		YB_6
4.07		ErB_6
4.10		NdB_6
4.12		GdB_6
4.12		PrB_6
4.13		CeB_6
4.13		YbB_6
4.14		CaB_6
4.15		LaB_6
4.15		ThB_6
4.19		SrB_6
4.33		BaB_6

X Units		Substance
	C 1	
4.33		Be_2C
4.619		Li_2O
5.06		($3ZrO_2 \cdot MgO$)
5.07		ZrO_2
5.08		($95ZrO_2 \cdot 5CeO_2$)
5.13		($95HfO_2 \cdot 5CeO_2$)
5.38		PrO_2
5.40		CeO_2
5.40		CdF_2
5.406		CuF_2
5.45		CaF_2
5.47		UO_2
5.526		($66CaF_2 \cdot 33YF_3$)
5.53		($91CaF_2 \cdot 9ThF_4$)
5.54		HgF_2
5.55		Na_2O
5.58		ThO_2
5.59		Cu_2S
5.704		Li_2S
5.749		Cu_2Se
5.782		SrF_2
5.796		EuF_2
5.838		($66SrF_2 \cdot 33LaF_3$)
5.91		$PtAl_2$
5.91		$PtGa_2$
5.935		β-PbF_2 (520°K.)
5.99		Al_2Au
6.005		Li_2Se
6.06		$AuGa_2$
6.19		BaF_2
6.34		Mg_2Si
6.35		$PtIn_2$
6.368		RaF_2
6.379		Mg_2Ge
6.436		K_2O
6.50		Li_2Te
6.50		$AuIn_2$
6.526		Na_2S
6.763		Mg_2Sn
6.809		Na_2Se
6.81		Mg_2Pb
6.98		$SrCl_2$
7.314		Na_2Te
7.38		K_2S
7.65		RbS_2
7.676		K_2Se
8.152		K_2Te
	C 15	
5.94		Be_2Cu
6.287		Be_2Ag
6.435		Be_2Ti
6.96		MgNiZn
7.03		Cu_2Mg
7.61		W_2Zr
7.79		Au_2Na
7.91		Au_2Pb
7.94		Au_2Bi
8.02		Al_2Ca
8.04		Al_2Ce
8.16		Al_2La
9.50		Bi_2K
	C 2	
5.41		FeS_2
5.42		(Fe, Ni)S_2 (6.5 % Ni)
5.57		RbS_2
5.57		RuS_2
5.57		Bravoite (53.8 % NiS_2, 39.1 % FeS_2, 7.1 % CoS_2)
5.62		OsS_2
5.64		CoS_2
5.65		(Cu, Ni, Co, Fe)(S, Se)$_2$
5.68		PtP_2
5.74		NiS_2
5.85		$CoSe_2$
5.92		$RuSe_2$
5.93		$OsSe_2$
5.94		$PtAs_2$
5.97		$PdAs_2$
6.02		$NiSe_2$
6.096		MnS_2
6.36		$RuTe_2$
6.37		$OsTe_2$
6.43		$PtSb_2$
6.44		$PdSb_2$
6.64		$AuSb_2$
6.94		$MnTe_2$
	C 3	
4.25		Cu_2O
4.73		Ag_2O
	F 1	
5.55		CoAsS
5.68		NiAsS
5.90		NiSbS
		(Ni, Fe)AsS, plessite
		Ni(As, Sb)S, corynite
		Ni(Sb, Bi)S, kallilite
		(Co, Ni)SbS, willyamite

DIFFRACTION DATA FOR CUBIC ISOMORPHS (Continued)

X Units	Substance
D 5_3	
8.13	Be_3N_2
9.37	$(Mn, Fe)_2O_3$
9.42	Mn_2O_3
9.74	Zn_3N_2
9.79	Sc_2O_3
9.94	Mg_3N_2
10.12	In_2O_3
10.15	Be_3P_2
10.37	Lu_2O_3
10.39	Yb_2O_3
10.52	Tm_2O_3
10.54	Er_2O_3
10.57	Tl_2O_3
10.58	Ho_2O_3
10.60	Y_2O_3
10.63	Dy_2O_3
10.70	Tb_2O_3
10.79	Gd_2O_3
10.79	Cd_3N_2
10.84	Eu_2O_3
10.85	Sm_2O_3
11.05	Nd_2O_3
11.40	α-Ca_3N_2
12.02	Mg_3P_2
12.33	Mg_3As_2
D 6_1	
11.05	As_4O_6
11.14	Sb_4O_6
D 1_1	
10.32	$ZrCl_4$
11.25	$TiBr_4$
(11.34)	(CBr_4) (>320°K.)
(11.62)	(CI_4)
11.89	GeI_4
11.99	SiI_4
12.00	TiI_4
12.23	SnI_4
E 2_1	
3.67	$YAlO_3$
3.75	$CdTiO_3$
3.78	$LaAlO_3$
3.80	$CaTiO_3$
3.83	$NaWO_3$
3.85	$(Na, Ce, Ca)(Ti, Cb)O_3$ Loparite
3.88	$NaTaO_3$
3.89	$LaGaO_3$
3.89	$NaCbO_3$
3.91	$SrTiO_3$
3.92	$CaSnO_3$
3.97	$BaTiO_3$
3.98	$KTaO_3$
3.99	$CaZrO_3$
4.00	$KMgF_3$
4.005	$KNiF_3$
4.01	$KCbO_3$
4.03	$SrSnO_3$
4.05	$KZnF_3$
4.07	$KCoF_3$
4.07	$SrHfO_3$
4.09	$SrZrO_3$
4.18	$BaZrO_3$
4.35	$BaPrO_3$
4.38	$BaCeO_3$
4.46	KIO_3
4.48	$BaThO_3$
4.5	NH_4IO_3
4.52	$RbIO_3$
4.66	$CsIO_3$
5.12	$MgZrO_3$
5.20	$CsCdCl_3$
5.33	$CsCdBr_3$
5.44	$CsHgCl_3$
5.77	$CsHgBr_3$
G 0_3	
6.57	$NaClO_3$
6.71	$NaBrO_3$
G 2_1	
7.60	$Ca(NO_3)_2$
7.81	$Sr(NO_3)_2$
7.84	$Pb(NO_3)_2$
8.11	$Ba(NO_3)_2$
H 1_1	
8.045	$NiAl_2O_4$
8.07	$CuAl_2O_4$
8.07	$CoCo_2O_4$
8.07	$MgAl_2O_4$
8.08	$CoAl_2O_4$
8.08	$ZnAl_2O_4$
8.10	$FeAl_2O_4$
8.11	$(Ni, Co)(Co, Ni)_2O_4$
8.11	$(Zn, Co)Co_2O_4$
8.11	$MgCo_2O_4$
8.27	$MnAl_2O_4$
8.27	$(Mn, Co)(Co, Mn)_2O_4$

X Units	Substance
H 1_1	
8.28	$MgGa_2O_4$
8.30	$NiCr_2O_4$
8.30	$MgCr_2O_4$
8.31	$ZnCr_2O_4$
8.32	$CoCr_2O_4$
8.32	$ZnGa_2O_4$
8.35	$NiFe_2O_4$
8.35	$Cu_2Cr_2O_4$
8.35	$FeCr_2O_4$
8.36	$MgFe_2O_4$
8.38	$CoFe_2O_4$
8.38	$NiMn_2O_4$
8.40	$ZnFe_2O_4$
8.40	$FeFe_2O_4$
8.42	$(Mn, Mg)Fe_2O_4$
8.42	$TiCo_2O_4$
8.43	$MnCr_2O_4$
8.43	$TiMg_2O_4$
8.43	$TiZn_2O_4$
8.44	$CuFe_2O_4$
8.47	FeV_2O_4
8.49	$MnCr_2O_4$
8.50	$TiFe_2O_4$
8.54	$MnFe_2O_4$
8.58	$CdCr_2O_4$
8.58	$SnMg_2O_4$
8.61	$SnCo_2O_4$
8.63	$SnZn_2O_4$
8.67	$CdFe_2O_4$
8.67	$TiMn_2O_4$
8.81	$MgIn_2O_4$
9.26	Ag_2MoO_4
9.4	$CoCo_2S_4$
9.45	$(Co, Ni)_3S_4$
9.46	$CuCo_2O_4$
9.5	NiN_2S_4
9.92	$ZnCr_2S_4$
10.05	$MnCr_2S_4$
10.19	$CdCr_2S_4$
12.54	$K_2Zn(Cn)_4$
12.76	$K_2Hg(CN)_4$
12.84	$K_2Cd(CN)_4$
H 5_8	
10.08	$2Na_2SO_4 \cdot NaCl \cdot NaF$
H 4_{13}	
12.11	$KCr(SO_4)_2 \cdot 12H_2O$
12.12	$KAl(SO_4)_2 \cdot 12H_2O$
12.15	$NH_4Al(SO_4)_2 \cdot 12H_2O$
12.15	$NH_4Fe(SO_4)_2 \cdot 12H_2O$
12.20	$RbAl(SO_4)_2 \cdot 12H_2O$
12.21	$TlAl(SO_4)_2 \cdot 12H_2O$
12.31	$CsAl(SO_4)_2 \cdot 12H_2O$
12.44	$NH_3 \cdot CH_3Al(SO_4)_2 \cdot 12H_2O$ (β-alum)
Langbeinite	
9.93	$K_2Mg(SO_4)_3$
10.2	$K_2(Ca, Mg)(SO_3)_3$
H 2_1	
6.00	Ag_3PO_4
6.120	Ag_3AsO_4 (90°K.)
6.130	Ag_3AsO_4 (380°K.)
H 2_4	
5.37	Cu_3VS_4
J 1_1	
8.17	K_2SiF_6
8.35	$(NH_4)_2SiF_6$
8.38	$Rb_2CrF_5H_2O$
8.41	$Tl_2CrF_5H_2O$
8.42	$(NH_4)_2VF_5H_2O$
8.42	$Rb_2VF_5H_2O$
8.45	$Tl_2VF_5H_2O$
8.45	Rb_2SiF_6
8.58	Tl_2SiF_6
8.87	Cs_2SiF_6
8.99	Cs_2GeF_6
9.73	K_2PtCl_6
9.73	K_2OsCl_6
9.76	Tl_2PtCl_6
9.84	$(NH_4)_2PtCl_6$
9.86	K_2ReCl_6
9.88	Rb_2PtCl_6
9.92	Rb_2TiCl_6
9.94	$(NH_4)_2SeCl_6$
9.97	K_2SnCl_6
9.97	Tl_2SnCl_6
9.98	Rb_2SeCl_6
10.02	Rb_2PdBr_6
10.04	$(NH_4)_2SnCl_6$
10.08	$Ni(NH_3)_6Cl_2$
10.10	Rb_2SnCl_6
10.10	$Co(NH_3)_6Cl_2$
10.11	Tl_2TeCl_6
10.14	$(NH_4)_2PbCl_6$
10.14	K_2TeCl_6
10.15	$Fe(NH_3)_6Cl_2$
10.16	$Mg(NH_3)_6Cl_2$
10.17	Cs_2PtCl_6

X Units	Substance
J 1_1	
10.18	Rb_2ZrCl_6
10.18	$(NH_4)_2TeCl_6$
10.20	$Mn(NH_3)_6Cl_2$
10.20	Rb_2PbCl_6
10.22	Cs_2TiCl_6
10.23	Rb_2TeCl_6
10.25	$Zn(NH_3)_6(ClO_4)_2$
10.26	Cs_2SeCl_6
10.30	K_2OSBr_6
10.35	Cs_2SnCl_6
10.36	K_2SeBr_6
10.36	K_2PtBr_6
10.39	$Co(NH_3)_6Br_2$
10.4	$Ni(NH_3)_6Br_2$
10.41	Cs_2ZrCl_6
10.42	Cs_2PbCl_6
10.45	Cs_2TeCl_6
10.45	$Co(NH_3)_5H_2OSO_4Br$
10.46	$(NH_4)_2SeBr_6$
10.46	$Zn(NH_3)_6Br_2$
10.47	$Fe(NH_3)_6Br_2$
10.47	$Mg(NH_3)_6Br_2$
10.48	K_2SnBr_6
10.51	$Co(NH_3)_6SO_4Br$
10.52	$Mn(NH_3)_6Br_2$
10.54	$Sr_2Ni(NO_2)_6$
10.55	$Pb_2Ni(NO_2)_6$
10.57	$(NH_4)_2SnBr_6$
10.58	Rb_2SnBr_6
10.62	$Co(NH_3)_5H_2OSO_4I$
10.63	$Co(NH_3)_6SeO_4Br$
10.67	$Ba_2Ni(NO_2)_6$
10.71	$Ca(NH_3)_6Br_2$
10.71	$Co(NH_3)_6SO_4I$
10.77	Cs_2SnBr_6
10.79	$Co(NH_3)_6SeO_4I$
10.9	$Ni(NH_3)_6I_2$
10.91	$Co(NH_3)_6I_2$
10.96	$Zn(NH_3)_6I_2$
10.97	$Fe(NH_3)_6I_2$
10.98	$Mg(NH_3)_6I_2$
11.04	$Mn(NH_3)_6I_2$
11.04	$Cd(NH_3)_6I_2$
11.24	$Ca(NH_3)_6I_2$
11.27	$Ni(NH_3)_6(BF_4)_2$
11.3	$Co(NH_3)_6(BF_4)_2$
(11.3)	$Zn(NH_3)_6(ClO_4)_2$
11.34	$Mg(NH_3)_6(BF_4)_2$
11.34	$Fe(NH_3)_6(BF_4)_2$
11.37	$Mn(NH_3)_6(BF_4)_2$
11.38	$Cd(NH_3)_6(BF_4)_2$
11.41	$Ni(NH_3)_6(ClO_4)_2$
11.43	$Co(NH_3)_6(ClO_4)_2$
11.46	$Ni(NH_3)_6(SO_3F)_2$
11.49	$Co(NH_3)_6(SO_3F)_2$
11.52	$Fe(NH_3)_6(ClO_4)_2$
11.53	$Mg(NH_3)_6(ClO_4)_2$
11.54	$Cd(NH_3)_6Br_2$
11.54	$Fe(NH_3)_6(SO_3F)_2$
11.58	$Mn(NH_3)_6(ClO_4)_2$
11.59	$Cd(NH_3)_6(ClO_4)_2$
11.59	$Mn(NH_3)_6(SO_3F)_2$
11.62	$Cd(NH_3)_6(SO_3F)_2$
11.91	$Ni(NH_3)_6(PF_6)_2$
11.94	$Co(NH_3)_6(PF_6)_2$
12.03	$Ni(NH_2{\cdot}CH_3)_6I_2$
12.05	$Co(NH_2{\cdot}CH_3)_6I_2$
12.19	$[NH(CH_3)_3]_2SnCl_6$
12.41	$[S(CH_3)_3]_2SnCl_6$
12.65	$[N(CH_3)_4]_2PtCl_6$
12.80	$[S(CH_3)_2C_2H_5]_2SnCl_6$
12.87	$[N(CH_3)_4]_2SnCl_6$
13.17	$[N(CH_3)_3C_2H_5]_2SnCl_6$
13.51	$[N(CH_3)(C_2H_5)_3]_2SnCl_6$
13.93	$[P(CH_3)(C_2H_5)_3]_2SnCl_6$
J 2_1 and Related Structures	
8.88	Li_3FeF_6
8.90	$(NH_4)_3AlF_6$
9.01	$(NH_4)_3CrF_6$
9.04	$(NH_4)_3VF_6$
9.10	$(NH_4)_3FeF_6$
9.10	$(NH_4)MoO_3F_3$
9.26	Na_3FeF_6
9.93	K_3FeF_6
9.96	$CuLi_2Fe(CN)_6$
10.0	$CuR_2Fe(CN)_6$
	R = Na, K, Rb, NH_4, Tl
10.15	$K_2CdFe(NO_2)_6$
10.17	$K_2CaCo(NO_2)_6$
10.19	$K_2CaFe(NO_2)_6$
10.2	$Fe^{III}RFe^{II}(CN)_6$
	R = Na, K, Rb, NH_4
10.22	$K_2HgFe(NO_2)_6$
10.23	$K_2SrCo(NO_2)_6$
10.25	$(NH_4)_2CaFe(NO_2)_6$
10.25	$NaTl_2Co(NO_2)_6$
10.28	$K_2CdNi(NO_2)_6$
10.28	$(NH_4)_2CdFe(NO_2)_6$
10.29	$K_2HgNi(NO_2)_6$
10.30	$K_2SrFe(NO_2)_6$
10.30	$Tl_2CaFe(NO_2)_6$
10.31	$K_2PbFe(NO_2)_6$
10.32	$K_2CaNi(NO_2)_6$
10.34	$(NH_4)_2SrFe(NO_2)_6$
10.37	$(NH_4)_2PbFe(NO_2)_6$
10.37	$Tl_2CdNi(NO_2)_6$
10.39	$Tl_2PbFe(NO_2)_6$
10.39	$NaRb_2Co(NO_2)_6$
10.40	$Tl_2SrFe(NO_2)_6$
10.4	$K_2PbCo(NO_2)_6$
10.41	$(NH_4)_2CdNi(NO_2)_6$
10.42	$Tl_2HgNi(NO_2)_6$
10.43	$K_2BaFe(NO_2)_6$
10.45	$K_2BaCo(NO_2)_6$
10.45	$K_3Co(NO_2)_6$
10.46	$(NH_4)_2HgNi(NO_2)_6$
10.47	$Rb_2HgNi(NO_2)_6$
10.49	$K_2SrNi(NO_2)_6$
10.49	$K_4Ni(NO_2)_6$
10.50	$(NH_4)_2BaFe(NO_2)_6$
10.54	$K_2LiBi(NO_2)_6$
10.55	$K_2PbNi(NO_2)_6$
10.55	$Tl_2BaFe(NO_2)_6$
10.58	$Rb_2CdNi, Cd(NO_2)_6$
10.58	$K_3Ir(NO_2)_6$
10.59	$Rb_2LiBi(NO_2)_6$
10.6	$K_2PbCu(NO_2)_6$
10.63	$K_3Rh(NO_2)_6$
10.63	$(NH_4)_2LiBi(NO_2)_6$
10.64	$Tl_2LiBi(NO_2)_6$
10.67	$K_2BaNi(NO_2)_6$
10.70	$NaCs_2Co(NO_2)_6$
10.70	$Ba_3[Rh(NO_2)_6]_2$
10.72	$Tl_3Co(NO_2)_6$
10.73	$Rb_3Co(NO_2)_6$
10.73	$(NH_4)_3Ir(NO_2)_6$
10.73	$Tl_3Ir(NO_2)_6$
10.77	$Rb_3Ir(NO_2)_6$
10.8	$(NH_4)_2Co(NO_2)_6$
10.81	$Cs_2Cd[Ni, Cd(NO_2)_6]$
10.82	$Co(NH_3)_5H_2OI_3$
10.83	$Rb_3Rh(NO_2)_6$
10.88	$K_2NaBi(NO_2)_6$
10.89	$Co(NH_3)_6I_3$
10.91	$Tl_3Rh(NO_2)_6$
10.91	$(NH_4)_3Rh(NO_2)_6$
10.94	$Cs_2LiBi(NO_2)_6$
10.95	$K_2AgBi(NO_2)_6$
10.98	$Rb_2NaBi(NO_2)_6$
10.99	$(NH_4)_2NaBi(NO_2)_6$
11.01	$Tl_2NaBi(NO_2)_6$
11.05	$Rb_2AgBi(NO_2)_6$
11.06	$Tl_2AgBi(NO_2)_6$
11.10	$(NH_4)_2AgBi(NO_2)_6$
11.15	$Cs_2NaBi(NO_2)_6$
11.15	$Cs_3Co(NO_2)_6$
11.17	$Cs_3Ir(NO_2)_6$
11.19	$Cs_3Bi(NO_2)_6$
11.19	$Cs_2AgBi(NO_2)_6$
11.21	$Co(NH_3)_6(BF_4)_3$
11.30	$Cs_3Rh(NO_2)_6$
11.32	$[Co(NH_3)_5{\cdot}H_2O](ClO_4)_3$
11.39	$Co(NH_3)_6(ClO_4)_3$
11.67	$Co(NH_3)_6(PF_6)$
K 6_1	
7.46	SiP_2O_7
7.80	TiP_2O_7
7.98	SnP_2O_7
8.18	HfP_2O_7
8.20	ZrP_2O_7
8.61	UP_2O_7
S 1_4	
11.51	$Al_2(Mg, Fe)_3(SiO_4)_3$, pyrope
11.51	$Al_2Fe_3(SiO_4)_3$, almandite
11.60	$Al_2Mn_3(SiO_4)_3$, spessartite
11.87	$Al_2Ca_3(SiO_4)_3$, grossularite
11.89	$(Al, Fe)_2Ca_3(SiO_4)_3$, hessonite
11.95	$Cr_2Ca_3(SiO_4)_3$, uvarovite
12.03	$Fe_2Ca_3(SiO_4)_3$, andradite
12.10	$(Na, Li)_3AlF_6$, cryolithionite
12.35–12.46	$(Mg, Mn)_2(Ca, Na)_3AsO_4)_3$, berzelite
S 6_1	
13.68	$NaAlSi_2O_6H_2O$
S 0_8	
13.82	$Al_{13}Si_5O_{20}(OH, F)_{18}Cl$, zunyite
S 6_2	
8.87	$Na_4(AlSiO_4)_3Cl$, sodalite
Tetrahedrite	
10.19	$(Cu, Fe)_{12}As_4S_{13}$, binnite
10.2–10.6	$(Cu, Ag)_{10}(Zn, Fe)_2(Sb, As)_4S_{13}$

PHOTOMETRIC QUANTITIES, UNITS AND STANDARDS

Photometric quantities and units are also given in the section Quantities and Units under the sub-division Light.

Candela-cd (formerly candle): The International System (SI) unit of luminous intensity. One candela is defined as the luminous intensity of one sixtieth of one square centimeter of projected area of a black body radiator operating at the temperature of freezing platinum (2042K).

Footcandle-fc: The unit of illumination when a foot is the unit of length. Footcandle is the illumination on a surface one square foot in area on which there is a uniformly distributed flux of one lumen, or the illumination on a surface all points of which are at a distance of one foot from a directionally uniform point source of one candela. (1 footcandle = 10.76 lux.)

Lumen-lm: The unit of luminous flux. It is equal to the flux through a unit solid angle (steradian), from a uniform point source of one candela, or to the flux on a unit surface all points of which are at unit distance from a uniform source of one candela.

Lux-lx: The International System (SI) unit of illumination in which the meter is the unit of length. (See Footcandle above.) 1 Lux = 0.0929 footcandles.

Luminous Efficacy of a Source of Light: The quotient of the total luminous flux emitted by the total lamp power input. It is expressed in lumens per watt.

Luminance-L: The luminous intensity (photometric brightness) of any surface in a given direction per unit of projected area of the surface as viewed from that direction. In International System (SI) units the luminance of light sources is commonly expressed in candelas per square centimeter.

EFFICACIES OF ILLUMINANTS

The theoretical maximum efficacy of 680 lm/W is that which would be obtained if all the power input were emitted as green light at a wavelength of 555 nm (at which the eye is most sensitive). If all power could be emitted uniformly over the visible spectrum as white light, the efficacy would be of the order of 220 lm/W.

Fluorescent life-ratings are based upon 3 hr of operation per start. Life increases as lamps are burned for longer periods.

Lamp	Lamp watts	Rated average life, hours	Approximate initial lumens	Approximate initial efficacy, lm/lamp W
A. Tungsten filament (120 V)				
Vacuum	25	2,500	235	9.4
Gas filled	40	1,500	455	11.4
Gas filled	60	1,000	870	14.5
Gas filled	75	750	1,190	15.9
Gas filled	100	750	1,750	17.5
Gas filled	200	750	4,010	20.1
Gas filled	500	1,000	10,850	21.7
Gas filled	1,000	1,000	23,740	23.7
Gas filled (3,200 K)	5,000	150	145,000	29.0
Gas filled (3,200 K)	10,000	75	335,000	33.5
B. Fluorescent (*cool* white)				
Preheat (T12)	20	9,000	1,300	65.0
Rapid start (T12)	40	20,000	3,150	78.8
Rapid start (T12, shielded cathode)	40	15,000	3,250	81.3
Rapid start (T12, U shaped, 3 5/8-in. leg)	40	12,000	2,900	72.5
Slimline (96T12, 425 mA)	75	12,000	6,300	84.0
Preheat (T17)	90	9,000	6,400	71.1
High output (96T12, 800 mA)	110	12,000	9,200	83.6
Grooved lamp (96PG17, 1500 mA)	215	9,000	16,000	74.4
C. Fluorescent (other)				
Rapid start (T12, deluxe cool white)	40	20,000+	2,200	55.0
Rapid start (T12, white)	40	20,000+	3,200	80.0
Rapid start (T12, warm white)	40	20,000+	3,150	78.7
Rapid start (T12, deluxe warm-white)	40	20,000+	2,150	53.7
Rapid start (T12, daylight)	40	20,000+	2,600	65.0
Rapid start (T12, natural)	40	20,000+	2,100	52.5
Rapid start (T12, 5,000 K)	40	20,000+	2,200	55.0
Rapid start (T12, 7,500 K)	40	20,000+	2,000	50.0
Rapid start (T12, blue)	40	20,000+	1,160	29.0
Rapid start (T12, gold)	40	20,000+	2,400	60.0
Rapid start (T12, green)	40	20,000+	4,500	112.5
Rapid start (T12, pink)	40	20,000+	1,160	29.0
Rapid start (T12, red)	40	20,000+	200	5.0
D. High-intensity discharge (HID)				
1. Mercury vapor (E-37)				
Clear	400	24,000	21,000	52.5
Color improved	400	24,000	20,500	51.3
Deluxe white	400	24,000	22,500	56.3
Warm deluxe white	400	24,000	20,000	50.0
2. Mercury vapor (BT-56)				
Clear	1,000	24,000	57,000	57.4
Color improved	1,000	24,000	55,000	55.0
Deluxe white	1,000	24,000	63,000	63.0
Warm deluxe white	1,000	24,000	58,000	58.0
3. Metal halide				
E-37	400	10,000	34,000	85.0
BT-56	1,000	10,000	100,000	100.0
BT-56	1,500	1,500	155,000	103.3
4. High-pressure sodium (Lucalox®)				
	100	12,000	9,500	95.0
	150	12,000	16,000	106.7
	250	15,000	30,000	120.0
	400	20,000	50,000	125.0
	1,000	15,000	140,000	140.0
E. Electric arc				
High intensity	11,700	–	36,800	31.4

Table compiled by C. J. Allen and G. D. Rowe.

PHOTOMETRIC QUANTITIES, UNITS AND STANDARDS (*Continued*)

APPROXIMATE LUMINANCE OF VARIOUS LIGHT SOURCES

Compiled by C.J. Allen and G.D. Rowe

Luminance of source is given in candelas per square centimeter

Source		Approx. avg. luminance* ($Cd\ cm^{-2}$)
Natural sources		
Clear sky	Average luminance	0.8
Sun (as observed from earth's surface)	At meridian	160000
Sun (as observed from earth's surface)	Near horizon	600
Moon (as observed from earth's surface)	Bright spot	0.25
Combustion sources		
Candle flame (sperm)	Bright spot	1.0
Welsbach mantle	Bright spot	6.2
Acetylene flame	Mees burner	10.5
Photoflash lamps		16000-40000 peak
Incandescent electric lamps		
Carbon filament	3 Lumens per watt	52
Tungsten filament	Vacuum lamp– 10 lumens per watt	200
Tungsten filament	Gas-filled lamp– 20 lumens per watt	1200
Tungsten filament	750-watt projection lamp–26 lumens per watt	7500
Fluorescent lamps (cool white)		
Rapid start (40wT12)	430 mA	0.82
High-output (96T12)	800 mA	1.13
Grooved bulb (96T17)	1500 mA	1.50
Fluorescent lamps (other than cool white)		
Daylight (40wT12)	430 mA	0.62
Blue (40wT12)	430 mA	0.30
Green (40wT12)	430 mA	1.17
Red (40wT12)	430 mA	0.05
High-intensity discharge (HID)		
Mercury-vapor (E-37)		
Clear	400 watt	970
Color-improved	400 watt	11.0
Deluxe-white	400 watt	12.1
Mercury-vapor (BT-56)		
Clear	1000 watt	980
Color-improved	1000 watt	15.0
Deluxe-white	1000 watt	17.2
Metal halide (E-37)		
Clear	400 watt	810
Color-improved	1000 watt	930
Deluxe white	1500 watt	1620
High pressure sodium (Lucalox®)	250 watt	520
	400 watt	780
	1000 watt	810

®Registered trademark of the General Electric Company

*Luminance values perpendicular to lamp axis

FLAME STANDARDS

Value of Various Former Standards in International Candles

Standard Pentane lamp, burning pentane........	10.0 candles
Standard Hefner lamp, burning amyl acetate.....	0.9 candles
Standard Carcel lamp, burning colza oil.........	9.6 candles

The *Carcel unit* is the horizontal intensity of the carcel lamp, burning 42 grams of colza oil per hour. For a consumption between 38 and 46 grams per hour the intensity may be considered proportional to the consumption.

The *Hefner unit* is the horizontal intensity of the Hefner lamp burning amyl acetate, with a flame 4 cm. high. If the flame is l mm. high, the intensity $I = 1 + 0.027(l - 40)$.

WAVE LENGTHS OF VARIOUS RADIATIONS

	Ångstroms
Cosmic rays....................................	0.0005
Gamma rays....................................	0.005–1.40
X-rays...	0.1–100
Ultra violet, below.............................	4000
Limit of sun's U.V. at earth's surface............	2920
Visible spectrum...............................	4000–7000
Violet, representative, 4100, limits..............	4000–4240
Blue, representative, 4700, limits...............	4240–4912
Green, representative, 5200, limits..............	4912–5750
Maximum visibility............................	5560
Yellow, representative, 5800, limits.............	5750–5850
Orange, representative, 6000, limits.............	5850–6470
Red, representative, 6500, limits................	6470–7000
Infra red, greater than.........................	7000
Hertzian waves, beyond.........................	2.20×10^6

BRIGHTNESS OF TUNGSTEN

Characteristics of Straight Tungsten Wire in a Vacuum
(Forsythe and Worthing, 1924).

Temperature °K			Brightness candles/cm²	$\dfrac{B}{dB}\dfrac{dt}{T}$
Absolute	Brightness	Color		
1000	966	1006	0.00012	22.0
1200	1149	1210	0.006	20.0
1400	1330	1414	0.11	17.2
1600	1509	1619	0.92	15.2
1800	1684	1825	5.05	13.7
2000	1857	2033	20.0	12.3
2200	2026	2242	61.3	11.2
2400	2192	2452	157.0	10.3
2600	2356	2663	347.0	9.6
2800	2516	2878	694.0	8.9
3000	2673	3094	1257.0	8.3
3200	2827	3311	2110.0	7.8
3400	2978	3533	3370.0	7.6
3655*	3165	3817	5740.0	7.3

* Melting-point of tungsten.

WAVE LENGTHS OF THE FRAUNHOFER LINES

Sun's Spectrum

At 15°C and 76 cm pressure. Wave length in Ångström units
(Fabry and Buisson system).

Line	Due to	Wave length	Line	Due to	Wave length
U	Fe	2947.9	h	H	4101.750
t	Fe	2994.4	g	Ca	4226.742
T	Fe	3021.067	G	{ Fe	4307.914
s	Fe	3047.623		{ Ca	4307.749
S_1 }	{ Fe	3100.683	G'	H	4340.477
S_2 }	{ Fe	3100.326	F	H	4861.344
	{ Fe	3099.943	b_4	{ Fe	5167.510
R	{ Ca	3181.277		{ Mg	5167.330
	{ Ca	3179.343	b_2	Mg	5172.700
Q	Fe	3286.773	b_1	Mg	5183.621
P	Ti	3361.194	E_2	Fe	5269.557
O	Fe	3441.020	D_2	Na	5889.977
N	Fe	3581.210	D_1	Na	5895.944
M	Fe	3727.636	C	H	6562.816
L	Fe	3820.438	B	O	6869.955
K	Ca	3933.684	A	{ O	7621
H	Ca	3968.494		{ O	7594
			Z	....	8228.5
			Y	....	8990.0

WAVE LENGTHS FOR SPECTROSCOPE CALIBRATION

Source	Wave length	Source	Wave length
Potassium flame........	0.7699μ	E, solar..............	0.5270μ
Potassium flame........	0.7665	b_1, solar or magnesium	
Mercury I arc..........	0.6907	flame...............	0.5184
B, solar..............	0.6869	b_2, solar or magnesium	
Lithium flame..........	0.6708	flame...............	0.5173
C, solar or hydrogen tube .	0.6563	Mercury I arc.........	0.4960
Mercury I arc..........	0.6234	Mercury I arc.........	0.4916
D_1, solar or sodium flame .	0.5896	F, solar or hydrogen tube	0.4861
D_2, solar or sodium flame .	0.5890	Strontium flame.......	0.4608
Mercury I arc..........	0.5791	Mercury I arc.........	0.4358
Mercury I arc..........	0.5770	G', solar or hydrogen tube	0.4340
Mercury I arc..........	0.5461	Mercury I arc.........	0.4047
Thallium flame.........	0.5351	H_1, solar.............	0.3969
		K, solar.............	0.3934

STANDARD UNITS, SYMBOLS, AND DEFINING EQUATIONS FOR FUNDAMENTAL PHOTOMETRIC AND RADIOMETRIC QUANTITIES

Submitted by Abraham Abramowitz
from Z-7.1-1967

Radiometric Quantities
(See Note at bottom of page)

Quantity*	Symbol*	Defining Equation**	Commonly Used Units	Symbol
Radiant energy	$Q, (Q_e)$		erg †joule kilowatt-hour	 J kWh
Radiant density	$w, (w_e)$	$w = dQ/dV$	†joule per cubic meter erg per cubic centimeter	J/m^3 erg/cm^3
Radiant flux	$\Phi, (\Phi_e)$	$\Phi = dQ/dt$	erg per second †watt	erg/s W
Radiant flux density at a surface				
Radiant exitance (Radiant emittance)†	$M, (M_e)$	$M = d\Phi/dA$	watt per square centimeter	W/cm^2
Irradiance	$E, (E_e)$	$E = d\Phi/dA$	‡watt per square meter, etc.	W/m^2
Radiant intensity	$I, (I_e)$	$I = d\Phi/d\omega$ (ω = solid angle through which flux from point source is radiated)	‡watt per steradian	W/sr
Radiance	$L, (L_e)$	$L = d^2\Phi/d\omega\,(dA\cos\theta)$ $= dI/(dA\cos\theta)$ (θ = angle between line of sight and normal to surface considered)	watt per steradian and square centimeter †watt per steradian and square meter	$W \cdot sr^{-1}\,cm^{-2}$ $W \cdot sr^{-1}\,m^{-2}$
Emissivity	ε	$\varepsilon = M/M_{\text{blackbody}}$ (M and $M_{\text{blackbody}}$ are respectively the radiant exitance of the measured specimen and that of a blackbody at the same temperature as the specimen)	one (numeric)	—

Note: The symbols for photometric quantities (see following table) are the same as those for the corresponding radiometric quantities (see above). When it is necessary to differentiate them the subscripts v and e respectively should be used, e.g., Q_v and Q_e.

*Quantities may be restricted to a narrow wavelength band by adding the word spectral and indicating the wavelength. The corresponding symbols are changed by adding a subscript λ, e.g., Q_λ for a spectral concentration or a λ in parentheses, e.g., $K(\lambda)$, for a function of wavelength.

**The equations in this column are given merely for identification.

***Φ_i = incident flux
Φ_a = absorbed flux
Φ_r = reflected flux
Φ_t = transmitted flux

†to be deprecated.

‡International System (SI) unit.

PHOTOMETRIC QUANTITIES

Quantity*	Symbol*	Defining Equation**	Commonly Used Units	Symbol
Absorptance	$\alpha, (\alpha_v, \alpha_e)$	$\alpha = \Phi_a/\Phi_i$***	one (numeric)	—
Reflectance	$\rho, (\rho_v, \rho_e)$	$\rho = \Phi_r/\Phi_i$***	one (numeric)	—
Transmittance	$\tau, (\tau_v, \tau_e)$	$\tau = \Phi_t/\Phi_i$***	one (numeric)	—
Luminous energy (quantity of light)	$Q, (Q_v)$	$Q_v = \int_{380}^{760} K(\lambda)\, Q_e \lambda\, d\lambda$	lumen-hour ‡lumen-second (talbot)	lm · h lm · s
Luminous density	$w, (w_v)$	$w = dQ/dV$	‡lumen-second per cubic meter	$\text{lm} \cdot \text{s} \cdot \text{m}^{-3}$
Luminous flux	$\Phi, (\Phi_v)$	$\Phi = dQ/dt$	‡lumen	lm
Luminous flux density at a surface				
Luminous exitance (Luminous emittance)†	$M, (M_v)$	$M = d\Phi/dA$	lumen per square foot	lm/ft^2
Illumination (illuminance)	$E, (E_v)$	$E = d\Phi/dA$	footcandle (lumen per square foot) ‡lux (lm/m^2) phot (lm/cm^2)	fc lx ph
Luminous intensity (candlepower)	$I, (I_v)$	$I = d\Phi/d\omega$ (ω = solid angle through which flux from point source is radiated)	‡candela (lumen per steradian)	cd
Luminance (photometric brightness)	$L, (L_v)$	$L = d^2\Phi/d\omega\,(dA \cos\theta)$ $= dI/(dA \cos\theta)$ (θ = angle between line of sight and normal to surface considered)	candela per unit area stilb (cd/cm^2) nit (†cd/m^2) footlambert ($\text{cd}/\pi\text{ft}^2$) lambert ($\text{cd}/\pi\text{cm}^2$) apostilb ($\text{cd}/\pi\text{m}^2$)	cd/in^2, etc. sb nt fL L asb
Luminous efficacy	K	$K = \Phi_v/\Phi_e$	‡lumen per watt	lm/W
Luminous efficiency	V	$V = K/K_{maximum}$ ($K_{maximum}$ = maximum value of $K(\lambda)$ function)	one (numeric)	—

ILLUMINATION CONVERSION FACTORS

1 lumen = 1/680 lightwatt (sec 3.7.2)	1 watt-second = 1 joule = 10^7 ergs
1 lumen-hour = 60 lumen-minutes	1 phot = 1 lumen/cm^2
1 footcandle = 1 lumen/ft^2	1 lux = 1 lumen/m^2

Number of → Multiplied by↘ Equals Number of ↓	Footcandles	*Lux	Phots	Milliphots
Footcandles	1	0.0929	929	0.929
*Lux	10.76	1	10,000	10
Phot	0.00108	0.0001	1	0.001
Milliphot	1.076	0.1	1,000	1

*The International Standard (SI) unit.

LINE SPECTRA OF THE ELEMENTS

Edited by Joseph Reader and Charles H. Corliss

These tables were prepared under the auspices of the Committee on Line Spectra of the Elements of the National Academy of Sciences — National Research Council. They contain the outstanding spectral lines of neutral (I), singly ionized (II), doubly ionized (III), triply ionized (IV), and quadruply ionized (V) atoms. Listed are lines that appear in emission from the vacuum ultraviolet to the far infrared. For most atoms these lines were selected from much larger lists in such a way as to include the stronger observed lines in each spectral region. In a few cases prominent monoxide band heads are also given. The present tables constitute the first phase of a projected set of tables that will include the I to V spectra of all atoms for which data are available.

The data were compiled by the following contributors, whose initials are given in the headings of the tables that they prepared:

- K. L. Andrew — Purdue University
- J. G. Conway — Lawrence Berkeley Laboratory
- C. H. Corliss — National Bureau of Standards
- R. D. Cowan — Los Alamos Scientific Laboratory
- C. R. Cowley — University of Michigan
- Henry M. and Hannah Crosswhite — Argonne National Laboratory
- S. P. Davis — University of California, Berkeley
- V. Kaufman — National Bureau of Standards
- R. L. Kelly — Naval Postgraduate School
- J. F. Kielkopf — University of Louisville
- W. C. Martin — National Bureau of Standards
- T. K. McCubbin — Pennsylvania State University
- L. J. Radziemski — Los Alamos Scientific Laboratory
- J. Reader — National Bureau of Standards
- G. V. Shalimoff — Lawrence Berkeley Laboratory
- R. W. Stanley — Purdue University
- J. O. Stoner, Jr. — University of Arizona
- H. H. Stroke — New York University
- D. R. Wood — Wright State University
- E. F. Worden — Lawrence Livermore Laboratory
- J. J. Wynne — International Business Machines Corporation

The literature references are collected at the end of the entire set of tables.

All wavelengths are given in Angstroms. Below 2000 A the wavelengths are in vacuum; above 2000 A the wavelengths are in air. Wavelengths given to three decimal places have an uncertainty of less than 0.001 A and are therefore suitable for the calibration of most spectrographs. In the air region, the elements used most commonly for calibration purposes are Ne, Ar, Kr, Fe, Th, and Hg; in the vacuum region, the most common are C, N, O, Si, and Cu.

A large number of the lines for the neutral and singly ionized atoms were extracted from the National Bureau of Standards (NBS) Tables of Spectral-Line Intensities.[1] The intensities of these lines represent quantitative estimates of relative line strengths that take account of varying detection sensitivity at different wave-

lengths. They are on a linear scale. For nearly all of the other lines the intensities represent qualitative estimates of the relative strengths of lines not greatly separated in wavelength. Because different observers frequently use different scales for their intensity estimates, these intensities are useful only as a rough indication of the appearance of a spectrum. In some cases the intensity scale is not intended to be linear. In the tables of first and second spectra the intensities of the lines of the singly ionized atom relative to those of the neutral atom should be used with caution, inasmuch as the concentration of ions in a light source depends greatly on the excitation conditions.

Descriptive symbols used in the tables have the following meanings:

- c — complex
- d — line consists of two unresolved lines
- h — hazy
- l — shaded to longer wavelengths
- s — shaded to shorter wavelengths
- p — perturbed by a close line
- b — band head
- r — easily reversed
- w — wide

ACTINIUM (AC)

Z = 89

Ac I and II

Ref. 193 — J.G.C.

Intensity		Wavelength	
		Air	
8	h	2,100.00	II
20		2,712.50	II
10		2,726.23	II
10	h	2,760.18	II
10	h	2,781.56	II
20		2,797.59	II
20		2,806.76	II
8		2,833.47	II
150	h	2,847.16	II
8		2,895.20	II
30		2,896.82	II
30		2,923.02	II
200		2,994.17	II
500		3,043.30	II
200		3,069.36	II
100		3,078.07	II
100		3,086.04	II
100		3,087.37	II
200		3,112.83	II
100		3,120.16	II
500	s	3,153.09	II
600	s	3,154.41	II
200	s	3,164.81	II
300	s	3,230.59	II
150	s	3,237.70	II
500		3,260.91	II
100	s	3,318.01	II
200	s	3,383.53	II
200		3,413.84	II
500		3,417.77	II
500	s	3,481.16	II
200		3,489.53	II
100		3,529.24	II
100		3,534.63	II
200	s	3,554.99	II
1,000	s	3,565.59	II
100		3,694.88	II
300	s	3,756.67	II
200		3,799.82	II
2,000	s	3,863.12	II
100		3,914.47	II
400	s	4,061.60	II
3,000	s	4,088.44	II
3,000	s	4,168.40	II
100		4,179.98	I
20		4,183.12	I
20		4,194.40	I
300	s	4,209.69	II
300		4,359.13	II
20	l	4,384.53	I
1,000	l	4,386.41	II
20		4,396.71	I
20		4,462.73	I
1,000	l	4,507.20	II
500		4,605.45	II
10		4,716.58	I
400	s	4,720.16	II
300		4,812.22	II
100		4,945.18	II
100		4,958.23	II
100		4,960.87	II
150		5,446.38	II
300	l	5,732.05	II
400		5,758.97	II
1,000		5,910.85	II
600	l	6,164.75	II
200	l	6,167.83	II
400		6,242.83	II
20		6,359.86	I
20	l	6,691.27	I
6		7,290.40	I
6	l	7,866.10	I

Ac III

Ref. 193 — J.G.C.

Intensity		Wavelength	
		Air	
1,000	h	2,626.44	III
50	h	2,682.90	III
2,000	h	2,952.55	III
2,000	h	3,392.78	III
3,000		3,487.59	III
2,000	h	4,413.09	III
3,000	h	4,569.87	III
8	h	5,193.21	III

Ac IV

Ref. 193 — J.G.C.

Intensity		Wavelength	
		Air	
20	h	2,062.00	IV
30	h	2,502.12	IV
100	h	2,558.08	IV
5	h	2,790.83	IV
50	h	2,793.90	IV
20	l	3,224.7	IV

ALUMINUM (AL)

Z = 13

Al I and II

Ref. 81, 89, 144, 227, ,228, 282 — E.F.W.

Intensity		Wavelength	
		Vacuum	
40		1,177.43	II
50		1,191.812	II
150		1,350.18	II
800		1,539.830	II
100		1,569.385	II
125		1,596.059	II
150		1,625.627	II
100		1,644.235	II
100		1,644.809	II
1,000		1,670.787	II
100		1,686.250	II
800		1,719.440	II
500		1,721.244	II
900		1,721.271	II
500		1,724.952	II
900		1,724.984	II
350		1,760.104	II
300		1,761.975	II
290		1,763.00	I
500		1,763.869	II
700		1,763.952	II
450		1,765.64	I
300		1,765.815	II
450		1,766.38	I
400		1,767.731	II
450		1,769.14	I
600		1,828.588	II
400		1,832.837	II
250		1,834.808	II
300		1,855.929	II
700		1,858.026	II
120		1,859.980	II
1,000		1,862.311	II
200		1,929.978	II
150		1,931.048	II
200		1,932.377	II
400		1,934.503	II
150		1,934.713	II
150		1,936.907	II
220		1,939.261	II
700		1,990.531	II
		Air	
150		2,016.052	II
150		2,016.234	II
100		2,016.368	II
200		2,074.008	II
700		2,094.264	II
150		2,094.744	II
300		2,094.791	II
100		2,095.104	II
200		2,095.141	II
60		2,150.70	I
60		2,181.00	I
400		2,269.10	I
120		2,269.22	I
60		2,312.49	I
70		2,313.53	I
90		2,317.48	I
60		2,319.06	I
140		2,321.56	I
460		2,367.05	I
110		2,367.61	I
110		2,368.11	I
180		2,369.30	I
140		2,370.22	I
70		2,370.73	I
160		2,372.07	I
850		2,373.12	I
170		2,373.35	I
110		2,373.57	I
60		2,378.40	I
60		2,513.30	I
240		2,567.98	I
480		2,575.10	I
60		2,575.40	I
80		2,631.55	II
110		2,637.70	II
150		2,652.48	I
200		2,660.39	I
160		2,669.17	II
650		2,816.19	II
90		2,837.96	I
90		2,840.10	I
150		3,041.28	II
360		3,050.07	I
60		3,054.68	I
450		3,057.14	I
90		3,064.29	I
60		3,066.14	I
150		3,074.64	II
4,500	r	3,082.153	I
7,200	r	3,092.710	I
1,800	r	3,092.839	I
150		3,428.92	II
70		3,439.35	I

Intensity		Wavelength	
150		3,443.64	I
70		3,444.86	I
70		3,458.22	I
60		3,479.81	I
60		3,482.63	I
450		3,586.56	II
360		3,587.07	II
290		3,587.45	II
220		3,651.06	II
110		3,651.10	II
150		3,654.98	II
290		3,655.00	II
450		3,900.68	II
60		3,932.00	I
4,500	r	3,944.006	I
9,000	r	3,961.520	I
110		3,995.86	II
290		4,226.81	II
150		4,585.82	II
110		4,588.19	II
550		4,666.80	II
110		4,898.76	II
110		4,902.77	II
150		5,280.21	II
70		5,107.52	I
290		5,283.77	II
150		5,285.85	II
110		5,312.32	II
220		5,316.07	II
150		5,371.84	II
180		5,557.06	I
110		5,557.95	I
450		5,593.23	II
110		5,853.62	II
220		5,971.94	II
290		6,001.76	II
220		6,001.88	II
450		6,006.42	II
150		6,061.11	II
290		6,068.43	II
110		6,068.53	II
450		6,073.23	II
110		6,181.57	II
150		6,181.68	II
290		6,182.28	II
220		6,182.45	II
450	h	6,183.42	II
450		6,201.52	II
360		6,201.70	II
290		6,226.18	II
360		6,231.78	II
450		6,243.36	II
450		6,335.74	II
360		6,696.02	I
230		6,698.67	I
60		7,083.97	I
70		7,084.64	I
110		7,361.57	I
140		7,362.30	I
60		7,606.16	I
90		7,614.82	I
230		7,835.31	I
290		7,836.13	I
60		7,993.05	I
90		8,003.19	I
70		8,065.97	I
110		8,075.35	I
290		8,640.70	II
360		8,772.87	I
450		8,773.90	I
110		8,828.91	I
180		8,841.28	I
90		8,912.90	I
140		8,923.56	I
60		9,089.91	I
70		9,139.95	I
150		9,290.65	II
110		9,290.75	II
150		10,076.29	II
110		10,768.36	I
140		10,782.04	I
110		10,872.98	I
230		10,891.73	I
450		11,253.19	I
570		11,254.88	I
570		13,123.41	I
450		13,150.76	I
230		16,718.96	I
300		16,750.56	I
140		16,763.36	I
300		21,093.04	I
360		21,163.75	I

Al III

Ref. 127 — E.F.W.

Intensity		Wavelength	
		Vacuum	
70		486.884	III
30		486.912	III
250		511.138	III
150		511.191	III
500		560.317	III
200		560.433	III
100		670.068	III
200		671.118	III
500		695.829	III
400		696.217	III
200		725.683	III
300		726.915	III
400		855.034	III
500		856.746	III
400		892.024	III
50		893.887	III
450		893.897	III
10		1,162.59	III
5		1,162.62	III
100		1,352.81	III
5		1,352.82	III
70		1,352.86	III
600		1,379.67	III
800		1,384.13	III
700		1,605.766	III
100		1,611.814	III
800		1,611.874	III
1,000		1,854.716	III
600		1,862.790	III
300		1,935.840	III
15		1,935.86	III
200		1,935.949	III
		Air	
110		2,399.00	III
285		2,762.77	III
220		2,762.87	III
450		2,906.93	III
360		3,348.52	III
290		3,350.88	III
870		3,601.63	III
550		3,601.93	III
750		3,612.36	III
450		3,702.11	III
550		3,713.12	III
110	h	3,980.14	III
110		4,082.45	III
150		4,088.61	III
110	h	4,142.37	III
650		4,149.92	III
650		4,150.17	III
110	h	4,364.64	III
650		4,479.89	III
650		4,479.97	III
760		4,512.56	III
550		4,528.94	III
870		4,529.19	III
110		4,701.15	III
150		4,701.41	III
110	h	4,904.10	III
110	h	5,151.01	III
110	h	5,163.89	III
1,200		5,696.60	III
1,000		5,722.73	III
110		6,055.21	III
220	h	7,635.37	III
150		7,660.26	III
220		7,681.97	III
360		7,881.79	III
150		7,882.52	III
290		7,905.51	III
290	h	8,243.59	III
360	h	8,275.11	III
290		9,571.52	III
360		9,605.99	III

Al IV

Ref. 8, 146 — E.F.W.

Intensity		Wavelength	
		Vacuum	
400		124.03	IV
700		129.73	IV
800		160.07	IV
700		161.69	IV
500		1,027.34	IV
800		1,042.17	IV
700		1,048.52	IV
500		1,058.90	IV
500		1,061.43	IV
600		1,064.89	IV
500		1,066.57	IV
600		1,069.44	IV
400		1,105.74	IV
600		1,118.82	IV
500		1,125.61	IV
500		1,136.82	IV
400		1,198.50	IV
400		1,220.55	IV
900		1,237.19	IV
600		1,240.21	IV
700		1,240.86	IV
700		1,248.79	IV
900		1,257.62	IV
800		1,264.18	IV
1,000		1,272.76	IV
400		1,337.90	IV
500		1,376.62	IV
500		1,388.79	IV
600		1,431.94	IV
700		1,441.82	IV
800		1,447.51	IV
600		1,457.96	IV
700		1,486.89	IV
800		1,494.79	IV
400		1,519.07	IV
800		1,537.54	IV
500		1,550.19	IV
1,000		1,557.25	IV
500		1,559.03	IV
700		1,564.16	IV
900		1,582.04	IV
800		1,584.46	IV
400		1,589.28	IV
400		1,606.65	IV
400		1,617.81	IV
600		1,627.54	IV
500		1,636.82	IV
800		1,639.06	IV
1,000		1,818.56	IV
700		1,881.16	IV
		Air	
400		2,515.87	IV
500		3,208.20	IV
500		3,267.21	IV
600		3,285.13	IV
400		3,344.46	IV
500		3,473.54	IV
900		3,492.23	IV
800		3,508.46	IV
500		3,511.28	IV
700		3,517.56	IV
400		3,527.03	IV
500		3,541.08	IV

Al V

Ref. 6 — E.F.W.

Intensity		Wavelength	
		Vacuum	
300		103.80	V
400		103.88	V
250		104.07	V
250		104.18	V
600		107.95	V
300		108.06	V
300		108.11	V
250		118.50	V
900		125.53	V
800		126.07	V
800		130.41	V
1,000		130.85	V
900		131.00	V
900		131.44	V
500		132.63	V
1,000		278.69	V
900		281.39	V
250		1,068.26	V
500		1,088.67	V
300		1,090.14	V
300		1,150.30	V
350		1,165.42	V
250		1,168.48	V
500		1,287.70	V
400		1,330.06	V
400		1,350.52	V
400		1,363.35	V
600		1,369.20	V
300		1,373.70	V
400		1,445.87	V
300		1,455.26	V
600		1,475.64	V
300		1,486.05	V
700		1,508.37	V
1,000		1,526.14	V
500		1,539.12	V
300		1,577.90	V
350		1,589.87	V

AMERICIUM (AM)

Z = 95

Am I and II

Ref. 92 — J.G.C.

Intensity		Wavelength	
		Air	
100	s	2,706.35	II
100	s	2,728.69	II
200	s	2,756.55	II
100	l	2,812.10	II
200	s	2,812.92	II
1,000	l	2,815.28	II
100	s	2,815.98	II
100	l	2,831.24	II
5,000	s	2,832.26	II
100	l	2,833.95	II
100	l	2,861.92	II
100	s	2,866.20	II
1,000	l	2,888.51	II
100	l	2,893.29	II
200	l	2,899.56	II
100	l	2,909.86	II
200	l	2,911.13	II
1,000	s	2,920.59	II
200	l	2,927.53	II
200	l	2,936.99	II
100	l	2,939.08	II
500	l	2,950.39	II
100	s	2,957.05	II
100	l	2,958.39	II
100	l	2,963.02	II
1,000	l	2,966.71	II
1,000	l	2,969.29	II
1,000	s	2,987.24	II
500	l	2,993.51	II
1,000	s	3,004.25	II
500	s	3,027.99	II
100	l	3,028.86	II
500	l	3,038.36	II
200	s	3,053.69	II
2,000	s	3,120.49	II
200	s	3,161.83	II
100	s	3,167.86	II
100	l	3,203.26	II
500	s	3,282.32	II
200	l	3,286.67	II
100	l	3,343.87	I
500	s	3,362.55	II
200	l	3,395.01	I
200	s	3,419.66	II
200		3,446.19	I
1,000	l	3,452.10	II
5,000	l	3,483.31	II
5,000		3,510.13	I
1,000	l	3,530.95	I
200	s	3,562.68	II
5,000		3,569.16	I
100	s	3,596.07	II
500		3,603.41	I
5,000		3,673.12	I
100	l	3,684.57	II

1,000	s	3,696.42	II
100	l	3,707.86	II
5,000	l	3,777.50	II
5,000	l	3,926.25	II
1,000	l	3,952.58	II
100	s	4,020.25	I
100	s	4,035.81	II
500	s	4,036.37	II
5,000	s	4,089.29	II
100	l	4,089.32	II
100	s	4,140.96	I
1,000	s	4,188.12	II
1,000		4,265.55	II
5,000		4,289.26	I
200	s	4,309.65	II
2,000	s	4,324.57	II
2,000	s	4,441.36	II
5,000	l	4,509.45	II
5,000	l	4,575.59	II
1,000	s	4,593.31	II
100	l	4,649.12	I
100	l	4,653.45	I
5,000	l	4,662.79	I
2,000	l	4,681.65	I
2,000	l	4,699.70	II
1,000	l	4,706.80	I
2,000	l	4,872.22	II
200	l	4,990.79	I
100	s	5,000.21	I
1,000	l	5,020.96	II
200	l	5,215.99	II
1,000	s	5,402.62	I
1,000	s	5,424.70	I
1,000	l	5,584.21	II
1,000	s	5,598.13	I
10,000	l	6,054.64	I
1,000	l	6,405.11	I
500	l	6,544.16	I
500	s	6,955.58	I

ANTIMONY (SB)
Z = 51

Sb I and II

Ref. 167, 194 — L.J.R. and J.R.

Intensity		Wavelength	
		Vacuum	
1		691.20	II
1		764.43	II
1		814.85	II
1		849.39	II
2		855.08	II
4		876.84	II
4		921.07	II
6		983.57	II
6		1,001.13	II
6		1,009.43	II
6		1,052.21	II
8		1,056.27	II
8		1,057.32	II
6		1,073.81	II
6		1,230.30	II
8		1,274.98	II
8		1,327.40	II
6		1,358.04	II
8		1,384.70	II
6		1,407.83	II
10		1,430.76	I
8		1,436.49	II
10	h	1,464.19	I
20	r	1,486.57	I
40	h	1,491.36	I
50	r	1,512.57	I
120	r	1,532.74	I
80	r	1,535.06	I
6		1,565.51	II
8		1,576.11	II
7		1,581.36	II
80	r	1,599.96	I
10		1,606.98	II
200	w	1,612.8	I
100	w	1,623.3	I
50	h	1,651.20	I
20		1,657.04	II
100	w	1,662.6	I
50		1,698.85	I
80	r	1,716.93	I
150	r	1,717.45	I
150	r	1,723.43	I
100	r	1,736.19	I
8		1,736.43	II
50		1,757.79	I
100	h	1,765.76	I
100	r	1,780.87	I
100	r	1,788.24	I
150		1,800.18	I
50	r	1,810.50	I
80	r	1,814.20	I
100		1,829.50	I
60		1,858.89	I
50	r	1,868.17	I
300	r	1,871.15	I
150	r	1,882.56	I
70		1,891.28	I
70		1,899.39	I
100		1,927.08	I
200	r	1,950.39	I
80	h	1,964.3	I
60		1,986.05	I
6		1,990.60	II
		Air	
50		2,024.00	I
60	r	2,029.49	I
70	r	2,039.77	I
150	r	2,049.57	I
50		2,063.43	I
1,000	r	2,068.33	I
100		2,079.56	I
50	r	2,098.41	I
80	r	2,118.48	I
100	r	2,127.39	I
50	r	2,137.05	I
100	r	2,139.69	I
10		2,141.80	II
50	r	2,141.83	I
100	r	2,144.86	I
50		2,158.91	I
1,500	r	2,175.81	I
250	r	2,179.19	I
6		2,179.25	II
200	r	2,201.32	I
300	r	2,208.45	I
6		2,208.50	II
150	r	2,220.73	I
100		2,221.98	I
120	r	2,224.93	I
6		2,225.15	II
300	r	2,262.51	I
120		2,288.98	I
150	r	2,293.44	I
300	r	2,306.46	I
2,500	r	2,311.47	I
150		2,315.89	I
400	h	2,373.67	I
300	h	2,383.64	I
100		2,395.22	I
150		2,422.13	I
250		2,426.35	I
400	r	2,445.51	I
400		2,478.32	I
150		2,480.44	I
8		2,480.46	II
100		2,510.54	I
2,000	r	2,528.52	I
15		2,528.54	II
10		2,567.75	II
150		2,574.06	I
1,500	r	2,598.05	I
500	r	2,598.09	I
300	r	2,612.31	I
200	r	2,652.60	I
12		2,656.55	II
300	r	2,670.64	I
200	r	2,682.76	I
120		2,692.25	I
150	r	2,718.90	I
400	r	2,769.95	I
12		2,851.09	II
100		2,851.11	I
1,000	r	2,877.92	I
12		2,966.10	II
15		2,980.96	II
500	r	3,029.83	I
12		3,034.01	II
12		3,040.67	II
600	r	3,232.52	I
20		3,241.28	II
700	r	3,267.51	I
12		3,383.09	II
100		3,383.15	I
15		3,498.46	II
12		3,520.47	II
25		3,637.80	II
250		3,637.83	I
20		3,722.78	II
200	r	3,722.79	I
20		3,850.22	II
200		4,033.55	I
20		4,033.56	II
20		4,133.63	II
15		4,140.54	II
15		4,195.17	II
20		4,219.07	II
20		4,314.32	II
12		4,344.83	II
12		4,411.42	II
12		4,446.48	II
12		4,506.92	II
15		4,514.50	II
30		4,596.90	II
20		4,599.09	II
15		4,604.77	II
30		4,647.32	II
20		4,675.74	II
40		4,711.26	II
12		4,735.44	II
20		4,757.81	II
20		4,765.36	II
12		4,766.91	II
30		4,784.03	II
20		4,802.01	II
20		4,832.82	II
20		4,877.24	II
15		4,947.40	II
15		5,044.56	II
12		5,166.32	II
15		5,176.55	II
20		5,238.94	II
20		5,354.24	II
15		5,464.08	II
30	h	5,490.32	I
40	h	5,556.10	I
15		5,568.13	II
30	l	5,602.19	I
100	l	5,632.02	I
30		5,639.75	II
60	h	5,830.34	I
15		5,895.09	II
100		6,005.21	II
20		6,053.41	II
30		6,079.80	II
50		6,130.04	II
20		6,154.94	II
12		6,302.76	II
20		6,611.49	I
30		6,647.44	II
15		6,688.01	II
6		6,806.67	II
30	h	7,648.28	I
80		7,844.44	I
200		7,924.65	I
40	h	7,969.55	I
60		8,411.69	I
150		8,572.64	I
100		8,619.55	I
30	h	8,682.7	I
30		9,132.21	I
400		9,518.68	I
30		9,866.78	I
400		9,949.14	I
200		10,078.49	I
300		10,261.01	I
50		10,364.33	I
50	h	10,488.3	I
200		10,585.60	I
1,000		10,677.41	I
800		10,741.94	I
80		10,794.11	I
600		10,839.73	I
200		10,868.58	I
400		10,879.55	I
300		11,012.79	I
40	h	11,079.95	I
30	h	11,084.98	I
50	h	11,104.84	I
50	h	11,108.52	I
30		11,189.61	I
150		11,266.23	I
30		11,863.37	I
1		11,957.7	I
5		12,116.06	I
2		12,276.6	I
2		12,466.75	I

Sb III

Ref. 164 — L.J.R. and J.R.

Intensity	Wavelength	
	Vacuum	
10	691.18	III
10	698.69	III
15	722.86	III
8	724.81	III
15	732.33	III
15	999.62	III
40	1,011.94	III
10	1,056.58	III
40	1,065.90	III
20	1,069.93	III
20	1,070.43	III
5	1,073.76	III
30	1,075.82	III
5	1,078.10	III
20	1,084.06	III
10	1,098.34	III
10	1,135.43	III
30	1,151.49	III
40	1,157.74	III
12	1,166.96	III
50	1,205.20	III
50	1,210.64	III
20	1,306.69	III
8	1,379.58	III
20	1,404.18	III
10	1,429.57	III
15	1,673.89	III
3	1,710.23	III
15	1,711.84	III
15	1,725.33	III
12	1,762.30	III
12	1,839.32	III
10	1,946.13	III
	Air	
3	2,054.10	III
2	2,091.85	III
5	2,127.00	III
5	2,507.71	III
15	2,590.13	III
1	2,614.20	III
12	2,617.17	III
1	2,617.63	III
20	2,669.39	III
5	2,785.87	III
20	2,790.27	III
20	3,336.61	III
50	3,504.07	III
15	3,519.06	III
15	3,533.45	III
40	3,559.18	III
40	3,566.25	III
30	3,738.90	III
40	4,265.09	III
50	4,352.16	III
30	4,591.89	III
30	4,692.91	III
1	5,247.71	III
1	5,690.8	III
1	5,717.3	III
3	5,845.5	III
5	6,246.7	III
3	6,287.6	III

ARGON (AR)
Z = 18

Ar I and II

Ref. 190, 203, 204, 219

E.F.W.

Intensity	Wavelength

		Vacuum	
30		487.227	II
50		490.650	II
30		490.701	II
30		519.327	II
30		542.912	II
200		543.203	II
70		547.461	II
70		556.817	II
70		573.362	II
30		576.736	II
70		580.263	II
30		583.437	II
70		597.700	II
30		602.858	II
30		612.372	II
500		661.867	II
30		664.562	II
200		666.011	II
1,000		670.946	II
3,000		671.851	II
70		676.242	II
30		677.952	II
30		679.218	II
200		679.401	II
200		718.090	II
3,000		723.361	II
500		725.548	II
70		730.930	II
200		740.269	II
200		744.925	II
70		745.322	II
20		802.859	I
100		806.471	I
60		806.869	I
30		807.218	I
40		807.653	I
50		809.927	I
120		816.232	I
70		816.464	I
80		820.124	I
120		825.346	I
120		826.365	I
150		834.392	I
100		835.002	I
100		842.805	I
180		866.800	I
150		869.754	I
180	r	876.058	I
180	r	879.947	I
150		894.310	I
1,000		919.781	II
1,000		932.054	II
1,000	r	1,048.220	I
500	r	1,066.660	I
		Air	
5		2,420.456	II
10		2,516.789	II
10		2,534.709	II
15		2,562.087	II
25		2,891.612	II
200		2,942.893	II
100		2,979.050	II
50		3,033.508	II
50		3,093.402	II
8		3,200.37	I
20		3,243.689	II
25		3,293.640	II
20		3,307.228	II
7		3,319.34	I
25		3,350.924	II
7		3,373.47	I
25		3,376.436	II
25		3,388.531	II
7		3,393.73	I
7		3,461.07	I
70		3,476.747	II
20		3,478.232	II
50		3,491.244	II
100		3,491.536	II
70		3,509.778	II
70		3,514.388	II
70		3,545.596	II
70		3,545.845	II
7		3,554.306	I
100		3,559.508	II
100		3,561.030	II

70	3,576.616	II
25	3,581.608	II
50	3,582.355	II
70	3,588.441	II
7	3,606.522	I
25	3,622.138	II
20	3,639.833	II
35	3,718.206	II
70	3,729.309	II
50	3,737.889	II
150	3,765.270	II
50	3,766.119	II
20	3,770.369	I
20	3,770.520	II
25	3,780.840	II
25	3,803.172	II
50	3,809.456	II
7	3,834.679	I
70	3,850.581	II
35	3,868.528	II
35	3,925.719	II
50	3,928.623	II
25	3,932.547	II
70	3,946.097	II
7	3,947.505	I
35	3,948.979	I
20	3,979.356	II
35	3,994.792	II
50	4,013.857	II
50	4,033.809	II
20	4,035.460	II
150	4,042.894	II
50	4,044.418	I
100	4,052.921	II
200	4,072.005	II
70	4,072.385	II
25	4,076.628	II
35	4,079.574	II
25	4,082.387	II
150	4,103.912	II
300	4,131.724	II
35	4,156.086	II
400	4,158.590	I
50	4,164.180	I
35	4,179.297	II
50	4,181.884	I
100	4,190.713	I
50	4,191.029	I
200	4,198.317	I
400	4,200.674	I
25	4,218.665	II
25	4,222.637	II
25	4,226.988	II
100	4,228.158	II
100	4,237.220	II
25	4,251.185	I
200	4,259.362	I
100	4,266.286	I
70	4,266.527	II
150	4,272.169	I
550	4,277.528	II
20	4,282.898	II
100	4,300.101	I
25	4,300.650	II
70	4,309.239	II
200	4,331.200	II
150	4,332.030	II
100	4,333.561	I
50	4,335.338	I
25	4,345.168	I
800	4,348.064	II
150	4,352.205	II
25	4,362.066	II
50	4,367.832	II
200	4,370.753	II
70	4,371.329	II
50	4,375.954	II
150	4,379.667	II
50	4,385.057	II
70	4,400.097	II
200	4,400.986	II
400	4,426.001	II
150	4,430.189	II
50	4,430.996	II
50	4,433.838	II
20	4,439.461	II
35	4,448.879	II
100	4,474.759	II
200	4,481.811	II
100	4,510.733	I

20	4,522.323	I
20	4,530.552	II
400	4,545.052	II
20	4,564.405	II
400	4,579.350	II
400	4,589.898	II
15	4,596.097	I
550	4,609.567	II
7	4,628.441	I
35	4,637.233	II
400	4,657.901	II
15	4,702.316	I
20	4,721.591	II
550	4,726.868	II
50	4,732.053	II
300	4,735.906	II
800	4,764.865	II
550	4,806.020	II
150	4,847.810	II
50	4,865.910	II
800	4,879.864	II
70	4,889.042	II
20	4,904.752	II
35	4,933.209	II
200	4,965.080	II
50	5,009.334	II
70	5,017.163	II
70	5,062.037	II
20	5,090.495	II
100	5,141.783	II
70	5,145.308	II
5	5,151.391	I
15	5,162.285	I
25	5,165.773	II
20	5,187.746	I
20	5,216.814	II
7	5,221.271	I
5	5,421.352	I
10	5,451.652	I
25	5,495.874	I
5	5,506.113	I
25	5,558.702	I
10	5,572.541	I
35	5,606.733	I
20	5,650.704	I
10	5,739.520	I
5	5,834.263	I
10	5,860.310	I
15	5,882.624	I
25	5,888.584	I
50	5,912.085	I
15	5,928.813	I
5	5,942.669	I
7	5,987.302	I
5	5,998.999	I
5	6,025.150	I
70	6,032.127	I
35	6,043.223	I
10	6,052.723	I
20	6,059.372	I
7	6,098.803	I
10	6,105.635	I
100	6,114.923	II
10	6,145.441	I
7	6,170.174	I
150	6,172.278	II
10	6,173.096	I
10	6,212.503	I
5	6,215.938	I
25	6,243.120	II
7	6,296.872	I
15	6,307.657	I
7	6,369.575	I
20	6,384.717	I
70	6,416.307	I
25	6,483.082	II
15	6,538.112	I
15	6,604.853	I
25	6,638.221	II
20	6,639.740	II
50	6,643.698	II
5	6,660.676	I
5	6,664.051	I
25	6,666.359	II
100	6,677.282	I
35	6,684.293	II
150	6,752.834	I
5	6,756.163	I
15	6,766.612	I
20	6,861.269	II

150	6,871.289	I
5	6,879.582	I
10	6,888.174	I
50	6,937.664	I
7	6,951.478	I
7	6,960.250	I
10,000	6,965.431	I
150	7,030.251	I
10,000	7,067.218	I
100	7,068.736	I
25	7,107.478	I
25	7,125.820	I
1,000	7,147.042	I
15	7,158.839	I
70	7,206.980	I
15	7,265.172	I
7	7,270.664	I
2,000	7,272.936	I
35	7,311.716	I
25	7,316.005	I
5	7,350.814	I
70	7,353.293	I
200	7,372.118	I
20	7,380.426	II
10,000	7,383.980	I
20	7,392.980	I
15	7,412.337	I
10	7,425.294	I
25	7,435.368	I
10	7,436.297	I
20,000	7,503.869	I
15,000	7,514.652	I
25,000	7,635.106	I
15,000	7,723.761	I
10,000	7,724.207	I
10	7,891.075	I
20,000	7,948.176	I
20,000	8,006.157	I
25,000	8,014.786	I
7	8,053.308	I
20,000	8,103.693	I
35,000	8,115.311	I
10,000	8,264.522	I
20	8,392.27	I
15,000	8,408.210	I
20,000	8,424.648	I
15,000	8,521.442	I
7	8,605.776	I
4,500	8,667.944	I
20	8,771.860	II
180	8,849.91	I
20	9,075.394	I
35,000	9,122.967	I
550	9,194.638	I
15,000	9,224.499	I
400	9,291.531	I
1,600	9,354.220	I
25,000	9,657.786	I
4,500	9,784.503	I
180	10,052.06	I
30	10,332.72	I
100	10,467.177	II
1,600	10,470.054	I
13	10,478.034	I
180	10,506.50	I
200	10,673.565	I
11	10,681.773	I
7	10,683.034	II
30	10,733.87	I
30	10,759.16	I
7	10,812.896	II
11	11,078.869	I
30	11,106.46	I
12	11,441.832	I
400	11,488.109	I
200	11,668.710	I
12	11,719.488	I
200	12,112.326	I
50	12,139.738	I
50	12,343.393	I
200	12,402.827	I
200	12,439.321	I
100	12,456.12	I
200	12,487.663	I
150	12,702.281	I
30	12,733.418	I
12	12,746.232	I
200	12,802.739	I
50	12,933.195	I
500	12,956.659	I

Intensity		Wavelength	
200		13,008.264	I
200		13,213.99	I
200		13,228.107	I
100		13,230.90	I
500		13,272.64	I
1,000		13,313.210	I
1,000		13,367.111	I
30		13,499.41	I
1,000		13,504.191	I
11		13,573.617	I
30		13,599.333	I
400		13,622.659	I
200		13,678.550	I
1,000		13,718.577	I
10		13,825.715	I
10		13,907.478	I
200		14,093.640	I
100		15,046.50	I
25		15,172.69	I
10		15,329.34	I
30		15,989.49	I
30		16,519.96	I
500		16,940.58	I
12		18,427.76	I
50		20,616.23	I
30		20,986.11	I
20		23,133.20	I
20		23,966.52	I

ARSENIC (AS)

Z = 33

As I and II

Ref. 168, 197 — R.L.K.

Intensity		Wavelength	
		Vacuum	
165		761.24	II
165		802.83	II
340		1,021.96	II
340		1,082.35	II
500		1,139.40	II
615		1,149.31	II
555		1,181.51	II
555		1,189.87	II
615		1,196.38	II
615		1,196.56	II
340		1,207.44	II
800		1,211.17	II
800		1,218.10	II
340		1,223.15	II
760		1,241.31	II
965		1,243.08	II
870		1,245.67	II
800		1,258.58	II
965		1,263.77	II
800		1,266.34	II
800		1,267.59	II
715		1,280.99	II
715		1,287.54	II
715		1,305.70	II
340		1,307.74	II
760		1,333.15	II
965		1,341.55	II
760		1,355.93	II
965		1,369.77	II
800		1,373.65	II
1,000		1,375.07	II
760		1,375.78	II
800		1,394.64	II
800		1,400.31	II
500		1,448.59	II
500		1,558.88	II
500		1,570.99	II
100	r	1,593.60	I
500		1,660.55	II
100		1,758.60	I
170		1,806.15	I
340		1,860.34	II
1,000	r	1,890.42	I
500		1,912.94	II
800	r	1,937.59	I
585	r	1,972.62	I
170	r	1,990.35	I
100	r	1,991.13	I
100	r	1,995.43	I
		Air	
230	r	2,003.34	I
100	r	2,009.19	I
100		2,013.32	I
100		2,112.99	I
100		2,144.08	I
135		2,165.52	I
350	r	2,288.12	I
350	r	2,349.84	I
100	r	2,370.77	I
135	r	2,381.18	I
170	r	2,456.53	I
340		2,602.00	II
170	r	2,780.22	I
300		2,830.359	II
300		2,831.164	II
100	r	2,860.44	I
300		2,884.406	II
615		2,959.572	II
300		3,003.819	II
300		3,116.516	II
340		3,842.60	II
715		4,190.082	II
615		4,197.40	II
615		4,242.982	II
500		4,315.657	II
500		4,323.867	II
500		4,336.64	II
500		4,352.145	II
425		4,352.864	II
375		4,371.17	II
615		4,427.106	II
615		4,431.562	II
715		4,458.469	II
340		4,461.075	II
715		4,466.348	II
500		4,474.46	II
800		4,494.230	II
850		4,507.659	II
615		4,539.74	II
715		4,543.483	II
615		4,602.427	II
340		4,629.787	II
340		4,707.586	II
340		4,730.67	II
340		4,888.557	II
100		5,068.98	I
340		5,105.58	II
500		5,107.55	II
100		5,121.34	I
100		5,141.63	I
425		5,231.38	II
500		5,331.23	II
100		5,408.13	I
135		5,451.32	I
340		5,497.727	II
425		5,558.09	II
425		5,651.32	II
425		6,110.07	II
500		6,170.27	II
300		6,511.74	II
300		7,092.27	II
300		7,102.72	II
340		7,990.53	II
300		8,174.51	II
100		8,428.91	I
100		8,564.71	I
100		8,654.14	I
135		8,821.73	I
100		8,869.66	I
135		9,267.28	I
200		9,300.61	I
230		9,597.95	I
290		9,626.70	I
230		9,833.76	I
100		9,886.05	I
140		9,900.55	I
170		9,915.71	I
290		9,923.05	I
100		10,010.63	I
290		10,024.04	I
100		10,453.09	I
100		10,575.02	I
170		10,614.07	I

As III

Ref. 163 — R.L.K.

Intensity	Wavelength	
	Vacuum	
185	849.9	III
185	866.3	III
510	871.7	III
325	889.0	III
325	927.5	III
325	953.6	III
325	937.2	III
325	963.8	III
120	1,172.2	III
185	1,209.3	III
	Air	
80	2,926.3	III
185	2,982.0	III
325	3,922.6	III
185	4,037.2	III

As IV

Ref. 244 — R.L.K.

Intensity	Wavelength	
	Air	
150	2,253.1	IV
200	2,263.2	IV
200	2,301.0	IV
250	2,417.5	IV
150	2,446.1	IV
250	2,454.0	IV
200	2,461.4	IV
150	3,108.8	IV

As V

Ref. 280 — R.L.K.

Intensity	Wavelength	
	Vacuum	
25	600.7	V
40	616.0	V
120	715.5	V
150	734.8	V
60	737.2	V
250	987.7	V
250	1,029.5	V
40	1,051.6	V
60	1,056.6	V

ASTATINE (AT)

Z = 85

At I

Ref. 188 — E.F.W.

Intensity	Wavelength	
	Air	
8	2,162.25	I
10	2,244.01	I

BARIUM (BA)

Z = 56

Ba I and II

Ref. 1, 252, 277, 279 — J.J.W.

Intensity		Wavelength	
		Vacuum	
200		1,487.00	II
400		1,503.90	II
300		1,544.50	II
		1,554.38	II
		1,572.73	II
200		1,572.90	II
		1,573.92	II
		1,607.16	II
		1,630.40	II
100		1,674.39	II
		1,674.51	II
400		1,694.31	II
		1,694.37	II
		1,761.75	II
		1,771.03	II
		1,786.93	II
100		1,904.16	II
500		1,924.77	II
		1,985.60	II
		1,487.00	II
		1,998.87	II
300		1,999.54	II
		Air	
		2,009.20	II
400		2,014.18	II
		2,052.68	II
		2,054.57	II
500		2,214.70	II
800		2,245.61	II
1,000		2,254.73	II
		2,304.22	II
1,400		2,304.24	II
		2,335.25	II
2,000		2,335.27	II
190		2,347.58	II
		2,347.58	II
60		2,528.51	II
8	h	2,596.64	I
100		2,634.78	II
8		2,702.63	I
18		2,771.36	II
15		2,785.28	I
100	r	3,071.58	I
10	h	3,108.21	I
8		3,132.60	I
8	h	3,135.72	I
10		3,137.70	I
10		3,155.34	I
10		3,155.67	I
12		3,158.05	I
12	h	3,158.54	I
25		3,165.60	I
15	h	3,173.69	I
30		3,183.16	I
15		3,183.96	I
10		3,193.91	I
25	h	3,203.70	I
30		3,221.63	I
40		3,222.19	I
50		3,261.96	I
60	r	3,262.34	I
40		3,281.50	I
15		3,281.77	I
50		3,322.80	I
80	h	3,356.80	I
60	r	3,377.08	I
20		3,377.39	I
70	r	3,420.32	I
25		3,421.01	I
30	h	3,421.48	I
40		3,463.74	I
200	r	3,501.11	I
80	h	3,524.97	I
30	h	3,531.35	I
80	h	3,544.66	I
20	h	3,547.68	I
100		3,552.45	II
200		3,567.73	II
100		3,576.28	II
30		3,577.62	I
80	h	3,579.67	I
200		3,596.57	II
40		3,630.64	I
40	h	3,636.83	I
20	h	3,688.47	I
400		3,735.75	II
200		3,816.69	II
200		3,842.80	II
100		3,854.76	II
20		3,889.33	I
1,400		3,891.78	II
20		3,892.65	I
40		3,909.91	I
500		3,914.73	II
50		3,935.72	I
20		3,937.87	I
200		3,939.67	II
500		3,949.51	II
80		3,993.40	I
30		3,995.66	I
300		4,036.26	II
200		4,083.77	II
30	h	4,084.86	I
1,500	h	4,130.66	II
20		4,132.43	I
200		4,166.00	II
500		4,216.04	II
800		4,267.95	II

100		4,283.10	I
300		4,287.80	II
200		4,297.60	II
800		4,309.32	II
20	h	4,323.00	I
600		4,325.73	II
200		4,326.74	II
300		4,329.62	II
80		4,350.33	I
60		4,402.54	I
400		4,405.23	II
40		4,431.89	I
60	h	4,488.98	I
50	h	4,493.64	I
40		4,505.92	I
200		4,509.63	II
60	h	4,523.17	I
130		4,524.93	II
65,000		4,554.03	II
40		4,573.85	I
80		4,579.64	I
30		4,599.75	I
20	h	4,619.92	I
25	h	4,628.33	I
300		4,644.10	II
30		4,673.62	I
35		4,691.62	I
20		4,700.43	I
800		4,708.94	II
40		4,726.44	I
800		4,843.46	II
300		4,847.14	II
200		4,850.84	II
30	h	4,877.65	I
400		4,899.97	II
15		4,902.90	I
20,000		4,934.09	II
8		4,947.35	I
1,000		4,957.15	II
300		4,997.81	II
1,000		5,013.00	II
20		5,159.94	I
20		5,267.03	I
800		5,361.35	II
1,000		5,391.60	II
200		5,421.05	II
100		5,424.55	I
200		5,428.79	II
300		5,480.30	II
200		5,519.05	I
1,000	r	5,535.48	I
20	h	5,620.40	I
10		5,680.18	I
400		5,777.62	I
800		5,784.18	II
100		5,800.23	I
20		5,805.69	I
150		5,826.28	I
2,800		5,853.68	II
15		5,907.64	I
100		5,971.70	I
800		5,981.25	II
100		5,997.09	I
300		5,999.85	II
100		6,019.47	I
200		6,063.12	I
300		6,110.78	I
400		6,135.83	II
20,000		6,141.72	II
150		6,341.68	I
500		6,378.91	II
90		6,450.85	I
150		6,482.91	I
12,000		6,496.90	II
300		6,498.76	I
150		6,527.31	I
3,000		6,595.33	I
150		6,654.10	I
1,500		6,675.27	I
1,800		6,693.84	I
1,000		6,769.62	II
600		6,865.69	I
300	h	6,867.85	I
1,000		6,874.09	II
6,000		7,059.94	I
2,400	h	7,120.33	I
600		7,195.24	I
600	h	7,228.84	I
3,000		7,280.30	I
1,200		7,392.41	I
300		7,417.53	I
900	h	7,459.78	I
600		7,488.08	I
450	h	7,636.90	I
600	h	7,642.91	I
1,800		7,672.09	I
1,200		7,780.48	I
180	h	7,839.57	I
1,500		7,905.75	I
600		7,911.34	I
900	h	8,210.24	I
1,800	h	8,559.97	I
100		8,710.74	II
100		8,737.71	II
300	h	8,799.76	I
300		8,860.98	I
450		8,914.99	I
300		9,219.69	I
300		9,308.08	I
300	h	9,324.58	I
1,500		9,370.06	I
300		9,455.92	I
450		9,589.37	I
900		9,608.88	I
300	h	9,645.72	I
1,500	h	9,830.37	I
900		10,001.08	I
600		10,032.10	I
1,200	h	10,233.23	I
300		10,471.26	I
120	h	10,791.25	I
180	h	11,012.69	I
150	h	11,114.42	I
240		11,303.04	I
120	h	11,697.45	I
120		13,207.30	I
120		13,810.50	I
120		14,077.90	I
120		15,000.40	I
120		20,712.00	I
150		25,515.70	I
150		29,223.90	I

Ba III

Ref. 111 — J.J.W.

Intensity		Wavelength	
		Vacuum	
5		403.82	III
2		407.12	III
7		420.12	III
4		423.84	III
9		448.95	III
8		456.96	III
14		555.48	III
14		587.57	III
18		647.27	III
9		653.36	III
15		743.12	III
12		1,097.41	III
15		1,113.67	III
11		1,116.01	III
14		1,133.05	III
12		1,151.76	III
12		1,170.62	III
13		1,207.29	III
11		1,218.92	III
12		1,224.55	III
12		1,288.53	III
11		1,299.18	III
11		1,307.40	III
12		1,308.87	III
12		1,315.72	III
12		1,334.01	III
11		1,354.71	III
11		1,369.53	III
11		1,416.61	III
12		1,478.85	III
12		1,510.68	III
12		1,514.22	III
12		1,565.61	III
12		1,566.12	III
12		1,574.55	III
12		1,596.80	III
12		1,610.95	III
12		1,615.78	III
12		1,711.53	III
12		1,861.74	III
12		1,883.92	III
11		1,974.76	III
		Air	
10		2,001.30	III
15		2,008.40	III
13		2,022.45	III
10		2,038.84	III
12		2,070.43	III
12		2,071.68	III
10		2,076.00	III
12		2,081.35	III
10		2,134.87	III
16		2,156.37	III
10		2,160.76	III
20		2,230.33	III
30		2,280.68	III
35		2,323.51	III
60		2,331.10	III
25		2,476.73	III
25		2,505.07	III
40		2,512.28	III
40		2,523.83	III
25		2,530.92	III
50		2,559.54	III
25		2,570.48	III
40		2,681.89	III
30		2,745.78	III
25		2,938.95	III
25		2,960.05	III
30		2,962.48	III
20		3,014.22	III
30		3,043.42	III
40		3,079.14	III
30		3,103.92	III
30		3,119.22	III
30		3,152.70	III
25		3,195.17	III
25		3,235.04	III
25		3,281.65	III
20		3,286.79	III
50		3,368.18	III
30		3,369.68	III
25		3,649.18	III
25		3,926.85	III
25		3,993.06	III
18	p	4,053.71	III
15		4,697.44	III
10		5,049.55	III
10		5,097.54	III
12	p	5,102.25	III
10		5,134.54	III
10		5,998.00	III
13		6,101.99	III
10		6,377.11	III
10		6,383.76	III
8		6,526.17	III
8		7,095.49	III
8		8,308.69	III
8		9,521.76	III

Ba IV

Ref. 78 — J.J.W.

Intensity	Wavelength	
	Vacuum	
40,000	794.89	IV
50,000	923.74	IV

BERKELIUM (BK)
Z = 97

Bk I and II

Ref. 53, 331 — J.G.C.

Intensity	Wavelength	
	Air	
s	2,748.02	II
s	2,827.57	II
l	2,872.11	II
s	2,878.57	II
l	2,884.77	II
s	2,889.80	II
s	2,893.66	II
l	2,910.65	II
s	2,926.49	
s	2,927.91	
l	2,941.71	II
l	2,951.76	II
l	2,969.13	II
l	2,987.76	II
l	3,178.47	II
	3,239.72	I
s	3,247.26	II
	3,252.19	I
s	3,263.47	II
l	3,288.75	I
	3,289.35	I
s	3,302.35	II
	3,335.26	I
s	3,387.45	II
	3,408.28	I
l	3,412.13	II
	3,426.95	I
l	3,432.62	I
l	3,437.47	I
	3,442.66	I
s	3,451.24	II
l	3,453.90	I
s	3,464.13	II
s	3,472.02	II
s	3,477.62	II
	3,528.72	I
l	3,531.40	I
l	3,535.73	I
s	3,542.19	II
	3,553.60	I
s	3,555.88	I
	3,556.52	I
	3,565.41	
l	3,567.25	II
	3,590.32	I
	3,595.88	I
	3,601.12	I
s	3,603.20	II
	3,604.78	
l	3,608.49	
	3,609.61	
	3,611.03	
l	3,611.93	
	3,613.91	
	3,616.62	
l	3,617.41	
	3,619.37	
s	3,621.81	
	3,627.61	I
	3,633.28	
l	3,637.05	I
	3,640.26	I
s	3,640.93	II
l	3,675.59	I
s	3,681.22	II
l	3,684.43	I
s	3,685.21	I
l	3,686.74	I
	3,692.73	I
	3,695.37	I
	3,703.28	I
s	3,704.02	I
	3,705.26	I
s	3,711.14	II
	3,712.93	I
	3,725.39	I
	3,739.92	I
	3,743.05	I
	3,744.37	I
	3,750.08	I
	3,751.91	I
	3,757.35	I
	3,757.85	I
s	3,771.06	II
	3,780.72	I
	3,781.17	I
	3,785.38	I
	3,788.21	I
	3,791.42	I
	3,796.21	I
	3,797.12	
	3,798.63	I
	3,801.08	II
s	3,802.35	I
	3,802.47	I
	3,815.29	I
s	3,823.10	II
s	3,824.08	II
	3,825.19	I
s	3,825.84	II
	3,827.41	I

	3,830.55	I
l	3,831.57	II
s	3,833.48	
s	3,835.97	II
	3,842.19	I
l	3,846.62	I
	3,847.63	I
	3,855.03	I
l	3,859.89	II
l	3,877.94	II
	3,880.11	I
l	3,882.60	I
l	3,894.55	II
s	3,906.09	II
l	3,912.16	II
l	3,916.37	II
	3,921.42	I
l	3,928.05	II
l	4,147.13	II
s	4,189.69	II
s	4,197.44	II
l	4,329.58	I
	4,351.50	I
	4,363.64	I
	4,423.01	I
	4,466.46	I
	4,685.70	
l	4,765.40	
s	5,056.73	I
l	5,118.24	I
s	5,135.53	II
	5,170.61	I
	5,197.55	I
	5,212.53	I
	5,271.95	I
	5,392.03	I
	5,394.24	I
l	5,404.62	I
	5,449.63	I
l	5,467.47	I
s	5,484.58	I
	5,512.22	II
l	5,537.93	I
l	5,556.80	I
s	5,557.09	I
	5,581.21	I
l	5,656.54	I
	5,659.03	I
	5,702.24	I
	5,910.71	I
l	7,040.85	I
	7,107.85	I
l	7,176.22	I
	7,249.26	
	7,252.50	I
l	7,257.21	I
	7,306.94	I
	7,394.18	
l	7,511.26	I
	7,551.12	I
	7,579.77	I
l	7,729.93	I
s	7,903.90	I
s	9,319.30	I
s	9,429.13	I
l	9,801.18	I
l	9,862.39	II
l	9,879.29	I
l	9,892.38	I
	10,126.20	I
l	10,186.58	I
	10,292.44	I
l	10,527.71	I
l	10,570.53	I
l	11,293.14	I
l	11,500.30	I
s	11,575.34	I
s	11,793.09	I
s	12,159.05	I
	13,061.13	I
	13,498.36	I
s	14,196.93	I
	15,136.24	I
	18,352.31	
l	19,273.87	I
s	19,653.22	I
s	23,902.85	I
s	24,192.62	I

BERYLLIUM (BE)

Z = 4

Be I and II

Ref. 15, 44, 115, 134, 135, 198, 335 — J.O.S.

Intensity		Wavelength	
		Vacuum	
		82.58	II
		83.66	II
		89.16	I
		89.80	II
		90.04	II
		90.21	I
		90.67	I
		91.06	II
		91.36	II
		91.74	II
		92.19	I
		92.61	II
		93.14	II
		93.42	II
		93.93	II
		94.78	II
		95.76	II
		96.29	I
		97.24	I
		97.44	I
		97.86	I
		97.97	I
		98.12	I
		98.37	I
		98.66	I
		98.94	I
		99.19	I
		100.86	I
		101.20	I
		102.13	I
		102.49	II
		104.40	II
		104.67	I
		105.80	I
		107.26	I
		107.38	I
		714.0	II
5		725.71	II
5		743.58	II
8		775.37	II
20		842.06	II
		865.3	II
2		925.25	II
10		943.56	II
10		973.27	II
		981.4	II
		1,020.1	II
8		1,026.93	II
5		1,036.32	II
15		1,048.23	II
20		1,143.03	II
		1,155.9	II
60		1,197.19	II
		1,426.12	I
		1,491.76	I
20		1,512.30	II
60		1,512.43	II
100		1,661.49	I
15		1,776.12	II
20		1,776.34	II
		1,907.	I
		1,909.0	II
		1,912.	I
		1,919.	I
5		1,929.67	I
10		1,943.68	I
		1,956.	I
50		1,964.59	I
5		1,985.13	I
		1,997.95	I
		1,997.98	I
60		1,998.01	I
		Air	
		2,033.25	I
		2,033.28	I
		2,033.38	I
50		2,055.90	I
100		2,056.01	I
10		2,125.57	I
20		2,125.68	I
25		2,145.	I
55		2,174.99	I
55		2,175.10	I
		2,273.5	II
		2,324.6	II
		2,337.0	I
950		2,348.61	I
20		2,350.66	I
60		2,350.71	I
200		2,350.83	I
2		2,413.34	II
16		2,413.46	II
20		2,453.84	II
		2,480.6	I
35		2,494.54	I
35		2,494.58	I
100		2,494.73	I
16		2,507.43	II
5		2,617.99	II
20		2,618.13	II
100		2,650.45	I
60		2,650.55	I
200		2,650.62	I
60		2,650.69	I
100		2,650.76	I
5		2,697.46	II
20		2,697.58	II
20		2,728.88	II
30		2,738.05	I
		2,764.2	II
20		2,898.13	I
10		2,898.19	I
20		2,898.25	I
30		2,986.06	I
10		2,986.42	I
60		3,019.33	I
30		3,019.49	I
30		3,019.53	I
20		3,019.60	I
10		3,046.52	II
30		3,046.69	II
		3,090.3	I
10		3,110.81	I
10		3,110.92	I
20		3,110.99	I
		3,120.	I
480		3,130.42	II
320		3,131.07	II
		3,136.	I
		3,150.	I
		3,160.6	I
		3,163.	I
		3,168.	I
		3,180.7	II
		3,187.	I
20		3,193.81	I
20		3,197.10	II
30		3,197.15	II
20		3,208.60	I
		3,220.	I
60		3,229.63	I
2		3,233.52	II
10		3,241.62	II
30		3,241.83	II
15		3,269.02	I
100		3,274.58	II
30		3,274.67	II
30		3,282.91	I
30		3,321.01	I
30		3,321.09	I
220		3,321.34	I
20		3,345.43	I
60		3,367.63	I
		3,405.6	II
5		3,451.37	I
300		3,455.18	I
20		3,476.56	I
300		3,515.54	I
10		3,555.	I
100		3,736.30	I
700		3,813.45	I
40		3,865.13	I
80		3,865.42	I
1		3,865.51	I
6		3,865.72	I
100		3,866.03	I
100		4,253.05	I
60		4,253.76	I
300		4,360.66	II
500		4,360.99	II
400		4,407.94	I
		4,526.6	I
		4,548.	I
12		4,572.66	I
700		4,673.33	II
1,000		4,673.42	II
6		4,709.37	I
200		4,828.16	II
40		4,849.16	I
2	h	4,858.22	II
80		5,087.75	I
8		5,218.12	II
20		5,218.33	II
3		5,255.86	II
64		5,270.28	II
500		5,270.81	II
20		5,403.04	II
20		5,410.21	II
		5,558.	I
10		6,229.11	I
16		6,279.43	II
30		6,279.73	II
30		6,473.54	I
60		6,547.89	II
60		6,558.36	II
30		6,564.52	I
2	h	6,636.44	II
1		6,756.72	II
2		6,757.13	II
30		6,786.56	I
1	h	6,884.22	I
6	h	6,884.44	I
100		6,982.75	I
6	h	7,154.40	I
40	h	7,154.65	I
100		7,209.13	I
3		7,401.20	II
2		7,401.43	II
10		7,551.90	I
10	h	7,618.68	I
20	h	7,618.88	I
60		8,090.06	I
5	h	8,158.99	I
10	h	8,159.24	I
4		8,254.07	I
10	h	8,287.07	I
30		8,547.36	I
60		8,547.67	I
300		8,801.37	I
6		8,882.18	I
40		9,190.45	I
20	h	9,243.92	I
1	h	9,343.89	II
40		9,392.74	I
2		9,476.43	II
16		9,477.03	II
20		9,847.32	I
10	h	9,895.63	I
20	h	9,895.96	I
80		9,939.78	I
16		10,095.52	II
20		10,095.73	II
60		10,119.92	II
80		10,331.03	I
20		11,066.06	I
30		11,066.46	I
		11,173.	II
1		11,173.73	II
120		11,496.39	I
2	h	11,625.16	II
		11,659.	II
2		11,660.25	II
100		12,095.36	II
30		12,098.18	II
100		14,643.92	I
60		14,644.75	I
200		16,157.72	I
80		17,855.38	I
120		17,856.63	I
100		18,143.54	I

160		31,755.05	I
200		31,778.70	I

Be III
Ref. 73, 102, 175 — J.O.S.

Intensity		Wavelength	
		Vacuum	
1	h	76.10	III
2		76.48	III
3		78.53	III
4		78.66	III
1	h	78.92	III
5		81.89	III
10		82.38	III
20		83.20	III
30		84.76	III
50		88.31	III
100		100.25	III
3		509.99	III
2		549.31	III
6		582.08	III
4		661.32	III
8		675.59	III
4		725.59	III
7		746.23	III
2		767.75	III
1		1,114.69	III
2		1,213.12	III
1		1,214.32	III
2		1,362.25	III
1		1,401.52	III
10		1,421.26	III
5		1,422.86	III
1		1,435.17	III
2		1,440.77	III
2	h	1,754.69	III
3		1,917.03	III
60	h	1,954.97	III
		Air	
75	h	2,076.94	III
60	h	2,080.38	III
25		2,118.56	III
15	h	2,122.27	III
15	h	2,127.20	III
5		2,137.25	III
5		2,191.57	III
100		3,720.36	III
		3,720.92	III
		3,722.98	III
90	h	4,249.14	III
2		4,485.52	III
100	h	4,487.30	III
1		4,495.09	III
140	h	4,497.8	III
140	h	6,142.01	III

Be IV
Ref. 272 — J.O.S.

Intensity	Wavelength	
	Vacuum	
	58.13	IV
	58.57	IV
	59.32	IV
	60.74	IV
	64.06	IV
	75.93	IV

BISMUTH (BI)
Z = 83

Bi I and II
Ref. 1, 357-359 — C.H.C.

Intensity		Wavelength	
		Vacuum	
15		1,058.88	II
20		1,085.47	II
10		1,099.20	II
8		1,163.19	II
8		1,167.06	II
10		1,225.43	II
15		1,232.78	II
10		1,241.05	II
10		1,265.35	II
15		1,283.73	II
10		1,306.18	II
20		1,325.46	II
20		1,329.47	II
20		1,350.07	II
25		1,372.61	II
15		1,376.02	II
20		1,393.92	II
45		1,436.83	II
25		1,447.94	II
50		1,455.11	II
25		1,462.14	II
35		1,486.93	II
20		1,502.50	II
40		1,520.57	II
40		1,533.17	II
30		1,536.77	II
35		1,538.06	II
20		1,563.67	II
40		1,573.70	II
60		1,591.79	II
25		1,601.58	II
40		1,609.70	II
40		1,611.38	II
20		1,652.81	II
20		1,749.29	II
80		1,777.11	II
60		1,787.47	II
70		1,791.93	II
70		1,823.80	II
100		1,902.41	II
9,000		1,954.53	I
7,000		1,960.13	I
25		1,989.35	II
		Air	
7,000		2,021.21	I
9,000		2,061.70	I
45	h	2,068.9	II
4,600		2,110.26	I
2,500		2,133.63	I
15		2,143.40	II
15		2,143.46	II
60		2,186.9	II
40	h	2,214.0	II
360		2,228.25	I
1,700		2,230.61	I
340		2,276.58	I
16		2,368.12	II
12		2,368.25	II
190		2,400.88	I
10		2,501.0	II
25		2,515.69	I
70		2,524.49	I
20	h	2,544.5	II
700		2,627.91	I
12		2,693.0	II
280	c	2,696.76	I
20		2,713.3	II
140	d	2,730.50	I
360		2,780.52	I
15		2,803.42	II
11		2,803.70	II
12		2,805.3	II
140	c	2,809.62	I
4,000		2,897.98	I
15		2,936.7	II
3,200		2,938.30	I
20		2,950.4	II
12		2,963.4	II
2,800		2,989.03	I
700		2,993.34	I
2,400		3,024.64	I
60		3,034.87	I
9,000	c	3,067.72	I
140		3,076.66	I
550	c	3,397.21	I
10		3,430.83	II
12		3,431.23	II
500	c	3,510.85	I
380	c	3,596.11	I
12		3,654.2	II
70	h	3,792.5	II
12		3,811.1	II
20		3,815.8	II
10		3,845.8	II
30		3,863.9	II
40	h	4,079.1	II
10		4,097.2	II
140		4,121.53	I
140		4,121.86	I
75	h	4,259.4	II
25		4,272.0	II
70	h	4,301.7	II
12	h	4,339.8	II
25	h	4,340.5	II
12	h	4,379.4	II
25	h	4,476.8	II
60	h	4,705.3	II
600	c	4,722.52	I
30		4,730.3	II
20		4,749.7	II
12		4,908.2	II
10		4,916.6	II
12		4,969.7	II
20		4,993.6	II
10		5,091.6	II
50	h	5,124.3	II
60	h	5,144.3	II
20		5,201.5	II
75	h	5,209.2	II
40	h	5,270.3	II
10		5,397.8	II
10	c	5,552.35	I
3		5,599.41	I
20		5,655.2	II
40	h	5,719.2	II
6		5,742.55	I
12		5,818.3	II
20		5,860.2	II
20		5,973.0	II
15		6,059.1	II
15		6,128.0	II
6		6,134.82	I
3		6,475.73	I
3		6,476.24	I
15		6,497.7	II
10		6,577.2	II
40	h	6,600.2	II
50	h	6,808.6	II
4	h	6,991.12	I
12		7,033.	II
2		7,036.15	I
10	h	7,381.	II
2		7,502.33	I
10	h	7,637.	II
10		7,750.	II
3		7,838.70	I
2		7,840.33	I
20		7,965.	II
12	h	8,050.	II
15		8,328.	II
15		8,388.	II
30		8,532.	II
2		8,544.54	I
1		8,579.74	I
25		8,653.	II
2		8,754.88	I
3		8,761.54	I
25		8,863.	II
2		8,907.81	I
2,000	d	9,657.04	I
40		9,827.78	I
20		10,104.5	I
15		10,138.8	I
20		10,300.6	I
20		10,536.19	I
50		11,072.44	I
15		11,551.6	I
1,500	d	11,710.37	I
40		11,999.49	I
200		12,165.08	I
10		12,374.64	I
200		12,690.04	I
100		12,817.8	I
200		14,330.5	I
50		16,001.5	I
60		22,551.6	I

Bi III
Ref. 359 — C.H.C.

Intensity		Wavelength	
		Vacuum	
1		590.73	III
5		670.76	III
4		775.16	III
6		803.65	III
7		920.93	III
6		925.48	III
25		1,039.99	III
50	h	1,045.76	III
30		1,051.81	III
20		1,139.01	III
15		1,145.91	III
50		1,224.64	III
40		1,326.84	III
60		1,346.12	III
35		1,423.33	III
35		1,423.52	III
60	h	1,461.00	III
60	h	1,606.40	III
20	h	1,691.5	III
20		1,834.32	III
10		1,863.9	III
10		1,912.12	III
10		1,988.26	III
		Air	
20		2,020.75	III
20		2,021.15	III
10		2,073.22	III
14		2,073.37	III
15		2,103.42	III
30		2,213.55	III
75	h	2,414.6	III
10		2,437.6	III
30	h	2,847.4	III
80	h	2,855.6	III
35		3,115.0	III
40	h	3,451.0	III
40		3,473.8	III
35		3,485.5	III
15		3,540.8	III
45		3,613.4	III
50		3,695.32	III
50		3,695.68	III
12		4,224.6	III
25		4,327.8	III
30		4,560.84	III
30		4,561.54	III
40	h	4,797.4	III
45	h	5,079.3	III
12		6,623.4	III
10	h	7,381.	III
12		7,551.	III
25		7,598.	III
10	h	7,637.	III
40		8,008.	III
50		8,070.	III
20		8,100.	III
15		8,671.	III
20		8,934.	III

Bi IV
Ref. 360 — C.H.C.

Intensity	Wavelength	
	Vacuum	
6	420.7	IV
6	431.2	IV
6	790.5	IV
6	790.6	IV
8	792.5	IV
10	820.3	IV
9	822.9	IV
12	824.9	IV
15	872.6	IV
8	876.8	IV
9	916.7	IV
12	923.9	IV
15	943.3	IV
9	967.6	IV
8	968.8	IV
8	989.8	IV
24	1,103.4	IV
7	1,128.8	IV
6	1,138.6	IV
6	1,139.8	IV
7	1,149.7	IV
60	1,317.0	IV
30	1,910.0	IV
	Air	
30	2,093.	IV
100	2,311.	IV

100	2,326.	IV
100	2,376.	IV
100	2,629.	IV
100	2,677.	IV
100	2,767.	IV
100	2,772.	IV
100	2,786.	IV
100	2,842.	IV
100	2,924.	IV
100	2,933.	IV
100	2,936.	IV
100	3,012.	IV
100	3,042.	IV
100	3,239.	IV
100	3,643.	IV
100	3,682.	IV
100	3,734.	IV
100	3,868.	IV
30	4,342.	IV
30	5,347.	IV

Bi V

Ref. 361 — C.H.C.

Intensity		Wavelength	
		Vacuum	
1		355.77	V
1		369.52	V
1		429.78	V
1		435.63	V
2		488.39	V
1		492.72	V
3		563.62	V
2		678.87	V
6		686.88	V
1		706.54	V
5		730.71	V
10		738.17	V
6		849.86	V
5		855.68	V
15	d	864.45	V
6		880.17	V
6		929.81	V
15	d	1,139.46	V

BORON (B)

Z = 5

B I and II

Ref. 66, 104, 171, 222 — R.L.K.

Intensity	Wavelength	
	Vacuum	
70	693.95	II
40	731.36	II
40	731.44	II
110	882.54	II
110	882.68	II
40	984.67	II
110	1,081.88	II
110	1,082.07	II
110	1,230.16	II
220	1,362.46	II
70	1,600.46	I
120	1,600.73	I
160	1,623.58	II
110	1,623.77	II
220	1,624.02	II
70	1,624.16	II
160	1,624.34	II
100	1,663.04	I
150	1,666.87	I
200	1,667.29	I
150	1,817.86	I
200	1,818.37	I
300	1,825.91	I
300	1,826.41	I
110	1,842.81	II
	Air	
250	2,066.38	I
250	2,066.65	I
100	2,066.93	I
300	2,067.19	I
500	2,088.91	I
500	2,089.57	I
70	2,220.30	II
40	2,323.03	II
40	2,328.67	II
40	2,393.20	II
220	2,395.05	II
40	2,459.69	II
40	2,459.90	II
1,000	2,496.77	I
1,000	2,497.73	I
160	2,918.08	II
110	3,032.26	II
70	3,179.33	II
110	3,323.18	II
110	3,323.60	II
450	3,451.29	II
285	4,121.93	II
110	4,194.79	II
110	4,472.10	II
110	4,472.85	II
70	4,784.21	II
110	4,940.38	II
110	6,080.44	II
70	6,285.47	II
70	7,030.20	II
40	7,031.90	II
70	8,668.57	I
20	8,667.22	I
800	11,660.04	I
570	11,662.47	I
125	15,629.08	I
200	16,240.38	I
250	16,244.67	I
235	18.994.33	I

B III

Ref. 69, 221 — R.L.K.

Intensity	Wavelength	
	Vacuum	
150	518.24	III
75	518.27	III
40	411.80	III
20	510.77	III
40	510.85	III
110	677.00	III
160	677.14	III
40	758.48	III
70	758.67	III
20	1,953.83	III
	Air	
550	2,065.78	III
450	2,067.23	III
160	2,077.09	III
40	2,234.09	III
70	2,234.59	III
40	4,242.98	III
70	4,243.61	III
220	4,487.05	III
360	4,497.73	III
110	7,835.25	III
70	7,841.41	III

B IV

Ref. 74 — R.L.K.

Intensity	Wavelength	
	Vacuum	
10	52.68	IV
30	60.31	IV
160	344.0	IV
450	385.0	IV
285	418.7	IV
70	1,112.2	IV
450	1,168.9	IV
70	1,170.9	IV
	Air	
70	2,524.7	IV
160	2,530.3	IV
450	2,821.68	IV
70	2,824.57	IV
285	2,825.85	IV

B V

Ref. 94 — R.L.K.

Intensity	Wavelength	
	Vacuum	
	41.00	V
30	48.59	V
	194.37	V
	262.37	V
	512.52	V
	749.74	V

BROMINE (BR)

Z = 35

Br I and II

Ref. 122, 124, 240, 248, 316

G.V.S.

Intensity		Wavelength	
		Vacuum	
300		711.68	II
250		815.48	II
350		856.19	II
1,000		889.23	II
500		896.64	II
500		905.99	II
300		922.56	II
1,000		948.97	II
500		984.93	II
500		1,012.10	II
1,000		1,015.54	II
500		1,037.02	II
1,000		1,049.00	II
450		1,064.76	II
500		1,071.87	II
250		1,101.50	I
300		1,134.59	I
250		1,136.29	I
250		1,177.23	I
400		1,178.90	I
1,000		1,189.28	I
250		1,189.38	I
1,000		1,189.50	I
500		1,198.37	I
800		1,209.76	I
1,000		1,210.73	I
750		1,216.01	I
1,000		1,221.13	I
900		1,221.87	I
1,000		1,223.24	I
1,200		1,224.41	I
1,200		1,226.90	I
750		1,228.05	I
7,500		1,232.43	I
1,200		1,243.90	I
800		1,249.59	I
1,500		1,251.66	I
1,000		1,255.80	I
1,500		1,259.20	I
1,200		1,261.66	I
1,200		1,266.20	I
1,000		1,279.48	I
1,000		1,286.26	I
3,000		1,309.91	I
3,000		1,316.74	I
1,000		1,317.37	I
2,000		1,317.70	I
12,000		1,384.60	I
3,000		1,449.90	I
50,000		1,488.45	I
30,000		1,531.74	I
25,000		1,540.65	I
30,000		1,574.84	I
20,000		1,576.39	I
25,000		1,582.31	I
75,000		1,633.40	I
		Air	
350		2,285.17	II
350		2,287.60	II
500	h	2,317.30	II
400		2,336.93	II
350		2,386.45	II
500		2,386.70	II
300		2,388.69	II
450		2,388.96	II
500		2,389.69	II
350		2,392.21	II
400		2,392.42	II
300		2,488.50	II
300	h	2,495.22	II
450		2,521.70	II
400		2,541.48	II
400		2,556.92	II
350		2,690.17	II
400		2,713.77	II
350	h	2,746.52	II
300	h	2,807.55	II
400	h	2,893.40	II
400	h	2,917.18	II
400	h	2,967.21	II
500	h	2,972.26	II
300		2,981.86	II
300	h	2,985.87	II
300	h	2,986.53	II
300	h	3,016.48	II
350		3,423.82	II
300	h	3,606.80	II
350		3,714.30	II
1,200		3,815.65	I
350		3,834.69	II
300		3,871.21	II
400		3,891.63	II
300		3,901.24	II
300		3,914.20	II
500		3,914.38	II
350		3,919.51	II
400		3,924.09	II
300		3,929.55	II
350		3,939.69	II
350		3,950.61	II
500		3,980.38	II
1,500		3,992.36	II
300		4,024.04	II
300		4,135.66	II
300		4,140.20	II
400		4,179.63	II
300		4,193.45	II
1,000		4,223.89	II
300		4,236.89	II
300		4,291.39	II
2,000		4,365.14	I
1,000		4,365.60	II
1,500		4,425.14	I
10,000		4,441.74	I
10,000		4,472.61	I
20,000		4,477.72	I
1,000		4,490.42	II
3,000		4,513.44	I
15,000		4,525.59	I
300		4,529.60	II
500		4,542.92	II
3,000		4,575.74	I
300		4,601.36	II
2,500		4,614.58	I
350		4,622.70	II
300		4,642.02	II
300		4,651.98	II
500		4,678.70	II
400		4,693.17	II
500		4,704.85	II
350		4,719.76	II
400		4,720.36	II
300		4,728.20	II
300		4,735.41	II
400		4,742.64	II
2,500		4,752.28	II
350		4,766.00	II
400		4,779.40	II
4,000		4,780.31	I
1,600		4,785.19	I
500		4,785.50	II
300		4,802.33	II
500		4,816.70	II
300		4,818.46	II
350		4,844.81	II
350		4,848.75	II
400		4,921.12	II
400		4,928.79	II
450		4,930.66	II
300		4,945.51	II
4,000		4,979.76	I
300		5,038.74	II
300		5,054.64	II
400		5,164.38	II
300		5,180.01	II
500		5,182.35	II

300		5,193.90	II
500		5,238.23	II
300		5,272.68	II
350		5,304.10	II
400		5,330.57	II
500		5,332.05	II
1,200		5,395.48	I
400		5,422.78	II
350		5,424.99	II
300		5,435.07	II
1,200		5,466.22	I
350		5,478.47	II
300		5,488.79	II
300		5,495.06	II
500		5,506.69	II
350		5,589.94	II
300		5,718.71	II
300		5,830.78	II
1,800		5,852.08	I
1,600		5,940.48	I
2,400		6,122.14	I
40,000		6,148.60	I
300		6,161.74	II
2,000		6,177.39	I
1,500		6,335.48	I
60,000		6,350.73	I
400		6,352.94	II
2,500		6,410.32	I
1,800		6,483.56	I
1,000		6,514.62	I
20,000		6,544.57	I
1,500		6,548.09	I
50,000	c	6,559.80	I
1,000		6,571.31	I
1,800		6,579.14	I
20,000		6,582.17	I
1,500		6,620.47	I
50,000	c	6,631.62	I
20,000		6,682.28	I
10,000		6,692.13	I
8,000		6,728.28	I
2,000		6,760.06	I
2,000		6,779.48	I
2,200		6,786.74	I
6,500		6,790.04	I
1,600	c	6,791.48	I
1,800		6,861.15	I
10,000		7,005.19	I
2,000		7,260.45	I
10,000		7,348.51	I
40,000		7,512.96	I
1,600		7,591.61	I
1,800		7,595.07	I
2,000		7,616.41	I
30,000		7,803.02	I
1,200		7,827.23	I
2,500	s	7,881.45	I
2,500		7,881.57	I
2,500		7,925.81	I
30,000	c	7,938.68	I
3,000		7,947.94	I
3,000		7,950.18	I
8,000		7,978.44	I
10,000		7,978.57	I
30,000		7,989.94	I
2,000		8,026.35	I
2,500		8,026.54	I
30,000		8,131.52	I
1,000	c	8,152.65	I
10,000		8,153.75	I
25,000		8,154.00	I
5,000		8,246.86	I
15,000		8,264.96	I
75,000	c	8,272.44	I
20,000		8,334.70	I
10,000		8,343.70	I
1,200		8,384.04	I
40,000		8,446.55	I
4,000		8,477.45	I
1,500		8,513.38	I
1,000		8,557.73	I
1,000		8,566.28	I
20,000		8,638.66	I
4,000		8,698.53	I
10,000	c	8,793.47	I
15,000		8,819.96	I
25,000		8,825.22	I
4,000		8,888.98	I
30,000		8,897.62	I
6,000		8,932.40	I
1,800		8,949.39	I
9,000		8,964.00	I
350		9,024.42	II
30,000		9,166.06	I
15,000		9,173.63	I
20,000		9,178.16	I
40,000		9,265.42	I
15,000		9,320.86	I
300		9,434.04	II
6,000		9,793.48	I
10,000		9,896.40	I
3,000		10,140.08	I
6,000		10,237.74	I
1,000		10,299.62	I
1,500		10,377.65	I
30,000		10,457.96	I
1,000		10,742.14	I
3,000		10,755.92	I
1,700		13,217.17	I
1,800		14,354.57	I
1,250		14,888.70	I
1,800		16,731.19	I
1,200		18,568.31	I
3,500		19,733.62	I
1,000		20,281.73	I
1,000		20,624.67	I
1,200		21,787.24	I
4,000		22,865.65	I
1,000		23,513.15	I
500		28,346.50	I
500		30,380.85	I
600		31,630.13	I
120		32,693.90	I
150		34,181.87	I
150		38,345.75	I
120		39,964.36	I

Br III

Ref. 246, 250 — G.V.S.

Intensity		Wavelength	
		Vacuum	
450		611.1	III
300		620.4	III
500		665.54	III
500		677.19	III
300		677.8	III
450		687.68	III
400		690.2	III
350		696.99	III
300		727.0	III
300		736.4	III
250		769.63	III
250		817.79	III
350		949.0	III
400		960.4	III
450		984.9	III
250		1,313.5	III
250		1,402.9	III
		Air	
400	h	2,293.44	III
300		2,313.29	III
300		2,462.39	III
300		2,482.60	III
350		2,499.25	III
350		2,529.49	III
350	h	2,551.09	III
350	h	2,570.83	III
300	h	2,573.17	III
400	h	2,584.99	III
500		2,589.14	III
300	h	2,594.48	III
400	h	2,595.98	III
450	h	2,606.20	III
350	h	2,608.15	III
500	h	2,613.13	III
350	h	2,616.26	III
500	h	2,626.52	III
350	h	2,629.23	III
350	h	2,639.60	III
350	h	2,671.53	III
350	h	2,735.83	III
300		2,770.50	III
300	h	2,785.28	III
300		2,804.16	III
400		2,926.96	III
300		2,936.22	III
350		2,969.00	III
400		2,994.04	III
500		3,020.76	III
300		3,033.63	III
350		3,036.45	III
500		3,074.42	III
350		3,091.94	III
350		3,117.29	III
300		3,147.81	III
400		3,174.08	III
300		3,321.08	III
450		3,333.07	III
500		3,349.64	III
300		3,385.25	III
450		3,447.36	III
400		3,487.58	III
300		3,506.47	III
450		3,517.36	III
500		3,540.16	III
300		3,551.08	III
500		3,562.43	III
450		3,600.71	III
250		3,693.53	III
450		3,820.26	III
200		3,903.95	III
350		4,506.55	III
200		4,519.74	III
150		5,175.87	III
100		5,446.80	III
100		7,192.8	III
100		7,673.1	III

Br IV

Ref. 139, 142, 243, 249
G.V.S.

Intensity		Wavelength	
		Vacuum	
700		379.73	III
700		400.37	III
1,000		545.43	III
1,000		559.76	III
1,000		569.19	III
1,000		576.59	III
1,000		585.10	III
1,000		586.71	III
1,000		597.51	III
1,000		600.09	III
1,000		601.27	III
1,000		607.03	III
1,000		617.85	III
1,000		619.87	III
1,000		630.14	III
1,000		642.23	III
1,000		661.53	III
1,000		683.51	III
1,000		697.72	III
1,000		715.39	III
1,000		731.00	III
1,000		800.12	III
1,000		813.66	III
900		1,274.82	III
1,000		1,703.51	III
		Air	
1,000		2,091.82	III
1,000		2,133.79	III
1,000		2,145.02	III
1,000		2,257.21	III
1,000		2,272.73	III
1,000		2,307.40	III
1,000		2,408.16	III
1,000		2,411.58	III
700		2,491.14	III
1,000		2,581.19	III
600		2,661.40	III
700		2,820.87	III
1,000		2,842.88	III
1,100	h	2,907.71	III
500		3,041.18	III
500		3,380.56	III

Br V

Ref. 49 — G.V.S.

Intensity		Wavelength	
		Vacuum	
600		468.37	V
800		482.11	V
900		531.97	V
1,000		547.90	V
700		549.77	V
800		621.03	V
800		632.22	V
700		645.44	V
400		652.64	V
800		657.54	V
800		679.62	V
700		812.95	V
1,000		850.81	V
150		855.27	V
600		1,041.60	V
1,000		1,069.15	V
500		1,080.54	V
900		1,112.13	V
1,000		1,143.56	V
150		1,429.75	V
400		1,442.60	V
150		1,470.35	V

CADMIUM (CD)

Z = 48

Cd I and II

Ref. 44, 285, 296 — R.D.C.

Intensity		Wavelength	
		Vacuum	
100		1,256.00	II
150		1,296.43	II
100		1,326.50	II
150		1,370.91	II
200		1,514.26	II
200		1,571.58	II
100		1,668.60	II
50		1,702.47	II
50		1,724.41	II
100		1,785.84	II
100		1,827.70	II
300		1,922.23	II
100		1,943.54	II
40		1,965.54	II
30		1,986.89	II
200		1,995.43	II
		Air	
100		2,007.49	II
50		2,032.45	II
75		2,036.23	II
150		2,096.00	II
1,000	r	2,144.41	II
50		2,155.06	II
100		2,187.79	II
1,000		2,194.56	II
1,000		2,265.02	II
1,500	r	2,288.022	I
1,000		2,312.77	II
200		2,321.07	II
40		2,376.82	II
50		2,418.69	II
50		2,469.73	II
40		2,487.93	II
3		2,491.00	I
40		2,495.58	II
10		2,508.91	I
50		2,509.11	II
30		2,516.22	II
15	h	2,518.59	I
25	h	2,525.196	I
50		2,544.613	I
50		2,551.98	II
25		2,553.465	I
3		2,565.789	I
500		2,572.93	II
50		2,580.106	I
3		2,584.87	I
30		2,592.026	I
25	h	2,602.048	I
50		2,628.979	I
40		2,632.190	I
75		2,639.420	I
40		2,659.23	II
50	h	2,660.325	I
25		2,668.20	II

Intensity		Wavelength	
50		2,672.62	II
100		2,677.540	I
25		2,677.748	I
50		2,707.00	II
75		2,712.505	I
50		2,733.820	I
1,000		2,748.54	II
100	h	2,763.894	I
50	h	2,764.230	I
50		2,774.958	I
30		2,823.19	II
200		2,836.900	I
25		2,856.46	II
100		2,868.180	I
200	r	2,880.767	I
50	r	2,881.224	I
200		2,914.67	II
50		2,927.87	II
200		2,929.27	II
1,000	r	2,980.620	I
200	r	2,981.362	I
50		2,981.845	I
50		3,030.60	II
150		3,080.822	I
25		3,081.48	II
30		3,082.593	I
100		3,092.34	II
200		3,133.167	I
50		3,146.79	II
150		3,250.33	II
300		3,252.524	I
300		3,261.055	I
50		3,343.21	II
50		3,385.49	II
30		3,388.88	II
800		3,403.652	I
50		3,417.49	II
50		3,442.42	II
100		3,464.43	II
1,000		3,466.200	I
800		3,467.655	I
25		3,483.08	II
150		3,495.44	II
25		3,499.952	I
100		3,524.11	II
100		3,535.69	II
1,000		3,610.508	I
800		3,612.873	I
60		3,614.453	I
20		3,649.558	I
10		3,981.926	I
100		4,029.12	II
200		4,134.77	II
50		4,141.49	II
100		4,285.08	II
8		4,306.672	I
100		4,412.41	II
3		4,412.989	I
1,000		4,415.63	II
30		4,440.45	II
8		4,662.352	I
200		4,678.149	I
30		4,744.69	II
300		4,799.912	I
50		4,881.72	II
50		5,025.50	II
1,000	h	5,085.822	I
6		5,154.660	I
100		5,268.01	II
100		5,271.60	II
1,000		5,337.48	II
1,000		5,378.13	II
200		5,381.89	II
40		5,843.30	II
50		5,880.22	II
300		6,099.142	I
100		6,111.49	I
100		6,325.166	I
30		6,330.013	I
400		6,354.72	II
500		6,359.98	II
2,000		6,438.470	I
400		6,464.94	II
25		6,567.65	II
500		6,725.78	II
100		6,759.19	II
30		6,778.116	I
50		7,237.01	II
100		7,284.38	II
1,000		7,345.670	I
50		8,066.99	II
5		8,200.309	I
20		9,292	I
15		11,655.	I
35		14,491.	I
80		15,712.	I
55	d	19,125.	I
25		24,378.	I
35		25,455.	I

CALCIUM (CA)

Z = 20

Ca I and II

Ref. 70, 150, 270 — J.J.W. and H.H.S.

Intensity	Wavelength	
	Vacuum	
24	1,341.89	II
12	1,342.54	II
12	1,432.50	II
20	1,433.75	II
20	1,553.18	II
32	1,554.64	II
4	1,642.80	II
20	1,643.77	II
36	1,644.44	II
60	1,649.86	II
32	1,651.99	II
12	1,673.86	II
20	1,680.05	II
2	1,680.13	II
8	1,691.78	II
16	1,698.18	II
20	1,807.34	II
40	1,814.50	II
4	1,814.65	II
60	1,840.06	II
40	1,838.01	II
20	1,843.09	II
40	1,850.69	II
	Air	
	2,103.24	II
	2,112.76	II
	2,113.15	II
	2,128.75	II
	2,131.51	II
	2,132.30	II
2	2,150.80	I
	2,197.79	II
5	2,200.73	I
	2,208.61	II
6	2,275.46	I
8	2,398.56	I
7	2,721.65	I
9	2,994.96	I
8	2,997.31	I
8	2,999.64	I
9	3,000.86	I
10	3,006.86	I
9	3,009.21	I
2	3,024.94	I
2	3,034.54	I
2	3,045.74	I
3	3,055.32	I
2	3,071.57	I
2	3,076.95	I
2	3,080.79	I
2	3,099.30	I
10	3,125.18	II
5	3,136.02	I
6	3,140.79	I
7	3,150.75	I
170	3,158.87	II
180	3,179.33	II
5	3,180.52	I
150	3,181.28	II
7	3,209.96	I
8	3,215.17	I
6	3,215.34	I
9	3,225.90	I
6	3,226.15	I
5	3,274.67	I
6	3,286.07	I
10	3,308.02	II
20	3,316.51	II
10	3,344.51	I
10	3,347.04	II
11	3,350.21	I
9	3,350.36	I
12	3,361.92	I
9	3,362.14	I
10	3,452.66	II
20	3,461.87	II
9	3,468.48	I
11	3,474.76	I
10	3,485.61	II
13	3,487.60	I
10	3,495.16	II
15	3,624.11	I
17	3,630.75	I
14	3,630.97	I
20	3,644.41	I
14	3,644.77	I
8	3,644.99	I
5	3,675.29	I
6	3,678.21	I
30	3,683.70	II
40	3,694.11	II
10	3,694.36	II
170	3,706.03	II
180	3,736.90	II
10	3,739.38	II
6	3,748.35	I
8	3,750.29	I
9	3,753.34	I
20	3,755.67	II
30	3,758.39	II
9	3,870.48	I
11	3,872.54	I
11	3,872.56	I
12	3,875.78	I
12	3,875.80	I
6	3,889.10	I
6	3,923.48	I
230	3,933.66	II
9	3,935.29	I
6	3,946.04	I
15	3,948.90	I
17	3,957.05	I
220	3,968.47	II
8	3,972.57	I
18	3,973.71	I
50	4,097.10	II
15	4,098.53	I
15	4,098.57	I
60	4,109.82	II
30	4,110.28	II
40	4,206.18	II
50	4,220.07	II
50	4,226.73	I
15	4,240.46	I
24	4,283.01	I
22	4,289.36	I
22	4,298.99	I
25	4,302.53	I
23	4,307.74	I
22	4,318.65	I
20	4,355.08	I
25	4,425.44	I
26	4,434.96	I
25	4,435.69	I
30	4,454.78	I
28	4,455.89	I
20	4,456.61	I
20	4,472.04	II
10	4,479.23	II
20	4,489.18	II
23	4,526.94	I
22	4,578.55	I
23	4,581.40	I
23	4,581.47	I
24	4,585.87	I
24	4,585.96	I
20	4,685.27	I
30	4,716.74	II
40	4,721.03	II
40	4,799.97	II
25	4,878.13	I
70	5,001.48	II
80	5,019.97	II
40	5,021.14	II
23	5,041.62	I
25	5,188.85	I
22	5,261.71	I
23	5,262.24	I
22	5,264.24	I
24	5,265.56	I
25	5,270.27	I
60	5,285.27	II
70	5,307.22	II
50	5,339.19	II
27	5,349.47	I
23	5,512.98	I
25	5,581.97	I
27	5,588.76	I
24	5,590.12	I
26	5,594.47	I
25	5,598.49	I
24	5,601.29	I
24	5,602.85	I
30	5,857.45	I
10	5,922.72	II
10	5,923.69	II
27	6,102.72	I
29	6,122.22	I
22	6,161.29	I
30	6,162.17	I
22	6,163.76	I
24	6,166.44	I
26	6,169.06	I
28	6,169.56	I
35	6,439.07	I
30	6,449.81	I
22	6,455.60	I
80	6,456.87	II
34	6,462.57	I
29	6,471.66	I
32	6,493.78	I
28	6,499.65	I
23	6,572.78	I
30	6,717.69	I
33	7,148.15	I
31	7,202.19	I
	7,291.47	II
	7,323.89	II
33	7,326.15	I
30	7,575.81	II
60	7,581.11	II
80	7,601.30	II
20	7,602.32	II
40	7,820.78	II
60	7,843.38	II
20	8,017.50	II
20	8,020.50	II
70	8,133.05	II
100	8,201.72	II
110	8,248.80	II
70	8,254.73	II
14	8,256.67	I
10	8,338.04	I
12	8,339.12	I
10	8,352.39	I
11	8,357.17	I
130	8,498.02	II
170	8,542.09	II
10	8,633.95	I
160	8,662.14	II
12	8,842.61	I
15	8,909.18	I
100	8,912.07	II
110	8,927.36	II
12	8,967.47	I
16	9,099.10	I
13	9,105.62	I
12	9,108.82	I
10	9,171.14	I
110	9,213.90	II
90	9,312.00	II
100	9,319.56	II
110	9,320.65	II
25	9,416.97	I
10	9,456.80	I
10	9,534.88	I
11	9,548.38	I
100	9,567.97	II
110	9,599.24	II
80	9,601.82	II
10	9,604.28	I
12	9,663.65	I
10	9,664.41	I
14	9,676.30	I

14	9,688.67	I
13	9,701.94	I
80	9,854.74	II
110	9,890.63	II
90	9,931.39	II
100	10,2223.04	II
20	10,343.81	I
13	10,838.97	I
13	10,861.58	I
13	10,863.87	I
14	10,869.50	I
14	10,879.87	I
20	11,838.99	I
10	11,949.72	I
25	12,816.04	I
24	12,823.86	I
25	12,909.10	I
30	13,033.57	I
21	13,086.44	I
24	13,134.95	I
20	16,150.77	I
22	16,157.36	I
21	16,197.04	I
20	18,925.47	I
24	18,970.14	I
30	19,046.14	I
48	19,309.20	I
49	19,452.99	I
47	19,505.72	I
50	19,776.79	I
35	19,853.10	I
34	19,862.22	I
23	19,917.19	I
24	19,933.70	I
	21,389.00	II
	21,428.90	II
20	22,607.93	I
25	22,624.93	I
30	22,651.23	I

Ca III

Ref. 25, 26 — J.J.W. and H.H.S.

Intensity	Wavelength	
	Vacuum	
6	296.96	III
9	403.72	III
7	409.95	III
5	439.69	III
5	633.59	III
5	685.41	III
5	697.55	III
5	699.09	III
5	699.89	III
6	701.39	III
5	727.66	III
8	740.55	III
6	746.25	III
5	747.98	III
5	779.61	III
5	800.30	III
5	809.93	III
5	817.06	III
6	821.57	III
6	840.56	III
6	1,020.07	III
5	1,034.65	III
5	1,187.30	III
8	1,188.61	III
8	1,188.61	III
5	1,190.86	III
10	1,262.65	III
11	1,278.39	III
10	1,281.55	III
12	1,286.52	III
12	1,298.04	III
11	1,317.70	III
10	1,328.95	III
11	1,335.13	III
10	1,360.01	III
11	1,385.43	III
11	1,397.69	III
13	1,453.16	III
12	1,459.79	III
11	1,461.88	III
15	1,463.34	III
16	1,484.87	III
12	1,496.88	III
11	1,506.88	III
20	1,545.29	III
15	1,555.53	III
18	1,562.47	III
13	1,571.27	III
13	1,586.13	III
10	1,762.26	III
10	1,783.93	III
10	1,794.22	III
12	1,800.21	III
13	1,807.89	III
14	1,812.15	III
11	1,813.59	III
12	1,830.06	III
10	1,860.43	III
14	1,870.26	III
14	1,872.37	III
10	1,894.12	III
11	1,910.10	III
12	1,935.72	III
10	1,939.68	III
11	1,943.01	III
12	1,948.26	III
10	1,953.55	III
10	1,958.97	III
13	1,964.61	III
13	1,967.94	III
12	1,972.82	III
10	1,977.01	III
10	1,978.55	III
11	1,981.19	III
	Air	
12	2,033.36	III
12	2,041.53	III
13	2,078.92	III
13	2,098.49	III
15	2,114.41	III
17	2,123.03	III
14	2,129.19	III
14	2,133.96	III
13	2,140.36	III
16	2,152.43	III
12	2,171.57	III
12	2,276.52	III
15	2,312.08	III
14	2,497.74	III
15	2,541.50	III
13	2,587.15	III
12	2,590.41	III
15	2,620.82	III
15	2,634.14	III
12	2,686.72	III
16	2,687.76	III
15	2,704.86	III
14	2,771.28	III
15	2,791.59	III
16	2,813.88	III
17	2,866.54	III
18	2,869.95	III
19	2,881.78	III
21	2,899.79	III
19	2,924.33	III
20	2,988.63	III
18	2,989.27	III
15	3,028.59	III
19	3,119.67	III
15	3,367.79	III
19	3,372.67	III
18	3,537.77	III
15	4,081.77	III
15	4,153.57	III
15	4,164.31	III
15	4,184.20	III
18	4,207.24	III
17	4,233.74	III
16	4,240.74	III
15	4,284.39	III
20	4,302.81	III
15	4,329.19	III
16	4,333.57	III
15	4,358.38	III
19	4,399.59	III
17	4,406.29	III
17	4,431.30	III
19	4,499.88	III
18	4,516.59	III
18	4,572.12	III
11	4,708.83	III
11	4,716.27	III
10	4,859.17	III
10	5,008.95	III
10	5,050.07	III
10	5,231.82	III
11	5,247.37	III
13	5,271.98	III
10	5,301.32	III
11	5,321.29	III
10	5,328.06	III
11	5,570.58	III
10	5,579.06	III
13	6,069.98	III
10	6,173.22	III
12	6,213.98	III
11	6,294.89	III
11	6,370.11	III
10	6,387.55	III
12	6,424.51	III
12	6,485.35	III
10	6,538.78	III
10	6,542.24	III
10	7,308.69	III
10	7,843.06	III
12	7,898.46	III
10	8,217.20	III

Ca IV

Ref. 150 — J.J.W. and H.H.S.

Intensity	Wavelength	
	Vacuum	
150	249.41	IV
150	250.15	IV
150	251.35	IV
250	296.55	IV
200	299.32	IV
200	318.09	IV
50	318.39	IV
120	321.59	IV
250	329.12	IV
150	329.39	IV
200	331.44	IV
250	331.99	IV
235	332.53	IV
150	332.81	IV
200	338.83	IV
150	339.79	IV
200	340.29	IV
200	341.29	IV
200	341.46	IV
250	342.45	IV
100	343.19	IV
200	343.44	IV
250	343.93	IV
200	344.96	IV
215	345.13	IV
250	374.74	IV
600	434.57	IV
100	437.27	IV
250	437.77	IV
200	438.93	IV
750	443.82	IV
50	445.02	IV
500	450.57	IV
50	454.55	IV
250	456.98	IV
250	461.09	IV
150	565.46	IV
750	656.00	IV
500	669.70	IV

Ca V

Ref. 150 — J.J.W. and H.H.S.

Intensity	Wavelength	
	Vacuum	
200	190.36	V
250	190.46	V
250	196.97	V
300	199.55	V
250	200.51	V
265	257.98	V
165	260.45	V
400	267.77	V
300	270.31	V
200	271.14	V
250	272.27	V
200	272.98	V
400	280.99	V
300	284.98	V
450	286.96	V
500	322.17	V
250	322.76	V
300	323.22	V
250	324.48	V
250	325.28	V
300	330.94	V
200	333.44	V
200	333.57	V
300	334.55	V
250 c	335.34	V
200	336.55	V
200	337.54	V
250	338.06	V
200	343.64	V
450	352.92	V
250	356.25	V
250	377.18	V
200	387.08	V
750	425.00	V
500	558.60	V
400	637.93	V
300	643.12	V
400	646.57	V
250	647.88	V
250	651.55	V
300	656.76	V

CALIFORNIUM (CF)

Z = 98

Cf I and II

Ref. 52, 331 — J.G.C.

Intensity		Wavelength	
		Air	
		2,739.31	
	s	2,759.10	
		2,774.52	
	l	2,852.03	
	s	2,855.23	
		3,298.14	
		3,352.71	
	l	3,367.79	
		3,392.22	I
		3,481.07	
	l	3,513.47	
		3,531.49	I
		3,540.98	I
		3,598.77	I
		3,605.32	I
		3,612.11	II
		3,617.49	I
	s	3,626.76	II
		3,659.46	
		3,662.70	I
	s	3,699.49	
	l	3,722.11	II
		3,739.35	I
		3,785.61	I
	l	3,789.04	II
	s	3,893.23	I
	l	3,993.57	II
		4,035.45	
		4,099.12	I
		4,242.38	I
		4,329.03	I
		4,335.22	I
		5,173.96	I
		5,179.08	I
		5,219.24	I
	s	5,279.01	
	s	5,320.09	
	s	5,339.13	
		5,408.88	I
		5,726.05	I
		6,622.83	I
		6,631.26	I
		6,677.90	I
		6,894.59	I
		6,927.10	II
	l	7,074.52	I

	s	7,307.90	I
		8,141.29	I
		8,241.77	I
		8,333.85	II
		8,423.49	II
		8,568.83	II
	l	9,228.52	
		9,337.70	
		9,649.51	
	s	10,308.41	
	l	10,568.83	
	s	10,614.84	
		11,300.19	
		11,681.85	
		11,941.33	
	l	12,183.05	
	s	12,352.72	
	s	12,437.48	
	l	12,789.05	
	l	13,329.98	
	s	13,362.98	
	l	13,376.89	
	l	13,474.44	
	l	14,772.49	
	s	15,281.32	
		15,587.12	
		15,675.92	
		16,759.06	
	s	17,626.25	
	l	18,718.69	
	h	19,068.72	
	l	19,336.96	
	l	19,576.84	
	l	20,393.38	
	s	20,869.98	

CARBON (C)

Z = 6

C I and II

Ref. 211 — R.L.K.

Intensity		Wavelength	
		Vacuum	
9		595.022	II
30		687.053	II
50		687.345	II
10		858.092	II
20		858.559	II
30		903.624	II
60		903.962	II
150		904.142	II
30		904.480	II
9		1,009.86	II
10		1,010.08	II
10		1,010.37	II
80		1,036.337	II
150		1,037.018	II
150		1,157.910	I
150		1,158.019	I
150		1,158.035	I
150		1,188.992	I
150		1,189.447	I
200		1,189.631	I
300		1,193.009	I
300		1,193.031	I
300		1,193.240	I
300		1,193.264	I
100		1,193.393	I
150		1,193.649	I
150		1,193.679	I
100		1,194.064	I
100		1,194.488	I
100		1,261.552	I
250		1,277.245	I
250		1,277.282	I
300		1,277.513	I
300		1,277.550	I
200		1,280.333	I
100		1,311.363	I
9		1,323.951	II
120		1,329.578	I
120		1,329.600	I
150		1,334.532	II
300		1,335.708	II
100		1,354.288	I
150		1,355.84	I
120		1,364.164	I
100		1,459.032	I
200		1,463.336	I
120		1,467.402	I
150		1,481.764	I
150		1,560.310	I
400		1,560.683	I
400		1,560.708	I
100		1,561.341	I
400		1,561.438	I
150		1,656.266	I
120		1,656.928	I
300		1,657.008	I
120		1,657.380	I
120		1,657.907	I
150		1,658.122	I
500		1,751.823	I
1,000		1,930.905	I
		Air	
800		2,478.56	I
250		2,509.12	II
350		2,512.06	II
250	h	2,574.83	II
350	l	2,741.28	II
250		2,746.49	II
1,000		2,836.71	II
800		2,837.60	II
800	h	2,992.62	II
350		3,876.19	II
350		3,876.41	II
350		3,876.66	II
570		3,918.98	II
800		3,920.69	II
250		4,074.52	II
350	l	4,075.85	II
800		4,267.00	II
1,000		4,267.26	II
200		4,771.75	I
200		4,932.05	I
200		5,052.17	I
350		5,132.94	II
350		5,133.28	II
350		5,143.49	II
570		5,145.16	II
400		5,151.09	II
300		5,380.34	I
250		5,648.07	II
350		5,662.47	II
570		5,889.77	II
350		5,891.59	II
200		6,001.13	I
250		6,006.03	I
110		6,007.18	I
150		6,010.68	I
300		6,013.22	I
250		6,014.84	I
800		6,578.05	II
570		6,582.88	II
200		6,587.61	I
250		6,783.90	II
250		7,113.18	I
250		7,115.19	I
250		7,115.63	II
200		7,116.99	I
350		7,119.90	II
800		7,231.32	II
1,000		7,236.42	II
200		7,860.89	I
200		8,058.62	I
520		8,335.15	I
250		9,061.43	I
200		9,062.47	I
200		9,078.28	I
250		9,088.51	I
450		9,094.83	I
300		9,111.80	I
800		9,405.73	I
150		9,603.03	I
250		9,620.80	I
300		9,658.44	I
200		10,683.08	I
300		10,691.25	I
12		11,619.29	I
23		11,628.83	I
13		11,658.85	I
47		11,659.68	I
24		11,669.63	I
85		11,748.22	I
142		11,753.32	I
114		11,754.76	I
11		11,777.54	I
17		11,892.91	I
30		11,895.75	I
26		12,614.10	I
20		13,502.27	I
38		14,399.65	I
16		14,403.25	I
61		14,420.12	I
12		14,429.03	I
13		14,442.24	I
12		16,559.66	I
50		16,890.38	I
10		17,338.56	I
11		17,448.60	I
13		18,139.80	I
23		19,721.99	I

C III

Ref. 22, 211 — R.L.K.

Intensity		Wavelength	
		Vacuum	
250		371.69	III
250		371.75	III
150		371.78	III
500		386.203	III
200		450.734	III
400		459.46	III
500		459.52	III
570		459.63	III
250		511.522	III
250		535.288	III
300		538.080	III
350		538.149	III
400		538.312	III
350		574.281	III
800		977.03	III
370		1,174.93	III
350		1,175.26	III
330		1,175.59	III
500		1,175.71	III
350		1,175.99	III
370		1,176.37	III
		Air	
250		2,162.94	III
800		2,296.87	III
150		2,697.75	III
110	l	2,724.85	III
150	l	2,725.30	III
150	l	2,725.90	III
200		2,982.11	III
150		4,056.06	III
200		4,067.94	III
250		4,068.91	III
250		4,070.26	III
150		4,162.86	III
250	h	4,186.90	III
200		4,325.56	III
600		4,647.42	III
520		4,650.25	III
375		4,651.47	III
200		4,665.86	III
450		5,695.92	III
150		5,826.42	III
150		6,744.38	III
150	h	7,037.25	III
150		7,612.65	III
300	h	8,196.48	III
150		8,332.99	III
300		8,500.32	III

C IV

Ref. 66, 211 — R.L.K.

Intensity		Wavelength	
		Vacuum	
250		244.91	IV
200		289.14	IV
250		289.23	IV
570		312.42	IV
500		312.46	IV
650		384.03	IV
700		384.18	IV
400		419.52	IV
500		419.71	IV
1,000		1,548.202	IV
900		1,550.774	IV
		Air	
200	l	2,524.41	IV
300	s	2,529.98	IV
200	w	4,658.30	IV
250		5,801.33	IV
200		5,811.98	IV
90	w	7,726.2	IV

C V

Ref. 211 — R.L.K.

Intensity		Wavelength	
		Vacuum	
110		34.973	V
450		40.268	V
110		227.19	V
160		248.66	V
160		248.74	V
		Air	
40		2,270.91	V
5		2,277.25	V
20		2,277.92	V
5		4,943.88	V
5		4,944.56	V

CERIUM (CE)

Z = 58

Ce I and II

Ref. 1 — C.H.C.

Intensity		Wavelength	
		Air	
130		2,462.97	II
110		2,518.51	II
200		2,548.68	II
340		2,651.01	II
120		2,696.07	II
120		2,706.88	II
120		2,723.38	II
110		2,741.96	II
100		2,750.89	II
150		2,761.42	II
120		2,784.27	II
100		2,785.35	II
100		2,790.53	II
140		2,791.42	II
270		2,830.90	II
100		2,833.31	II
250		2,874.14	II
110		2,908.42	II
100		2,918.67	II
120		2,955.94	II
110		2,964.80	II
100		2,972.58	II
400		2,976.91	II
150		2,977.46	II
120		2,980.41	II
250		2,990.87	II
110		2,994.42	II
320		2,995.64	II
400		3,008.79	II
370		3,017.20	II
210		3,037.73	II
200		3,051.98	II
350		3,055.24	II
320		3,056.78	II
680		3,063.01	II
320		3,083.67	II
250		3,084.44	II
200		3,090.37	II
370		3,103.38	II
200		3,107.47	II
320		3,110.28	II
300		3,111.17	II
220		3,127.53	II
200		3,130.33	II

240	3,130.87	II	440	3,659.23	II	310	3,931.37	II	450	4,120.83	II
200	3,144.60	II	350	3,659.97	II	230	3,931.83	II	510	4,123.24	II
290	3,145.28	II	880	3,660.64	II	310	3,933.73	II	510	4,123.49	II
290	3,146.41	II	880	3,667.98	II	560	3,938.09	II	980	4,123.87	II
290	3,164.15	II	220	3,672.18	I	770	3,940.34	II	510	4,124.79	II
290	3,169.18	II	350	3,672.79	II	310	3,940.97	II	980	4,127.37	II
290	3,171.61	II	220	3,679.42	II	2,000	3,942.15	II	250	4,127.74	II
480	3,183.52	II	300	3,694.91	II	2,700	3,942.75	II	200	4,128.07	II
240	3,186.13	II	220	3,704.98	II	770	3,943.89	II	530	4,130.71	II
200	3,190.34	II	1,000	3,709.29	II	310	3,947.97	II	480	4,131.10	II
710	3,194.83	II	1,000	3,709.93	II	3,100	3,952.54	II	2,700	4,133.80	II
200	3,199.28	II	1,400	3,716.37	II	340	3,953.66	II	270	4,135.44	II
990	3,201.71	II	420	3,718.19	II	310	3,955.36	II	270	4,137.47	II
200	3,218.38	II	420	3,718.38	II	230	3,956.06	II	2,000	4,137.65	II
710	3,218.94	II	210	3,719.80	II	980	3,956.28	II	270	4,138.10	II
880	3,221.17	II	420	3,725.68	II	230	3,958.27	II	210	4,138.35	II
330	3,225.67	II	490	3,728.02	II	230	3,958.87	II	770	4,142.40	II
710	3,227.11	II	800	3,728.42	II	770	3,960.91	II	390	4,144.49	II
240	3,229.36	II	320	3,748.06	II	390	3,964.50	II	670	4,145.00	II
480	3,231.24	II	250	3,751.45	II	770	3,967.05	II	480	4,146.23	II
710	3,234.16	II	200	3,755.43	II	450	3,971.68	II	280	4,148.90	II
330	3,234.89	II	300	3,762.98	II	270	3,972.07	II	420	4,149.79	II
390	3,236.74	II	680	3,764.12	II	270	3,975.07	II	980	4,149.94	II
390	3,243.37	II	200	3,765.04	II	770	3,978.65	II	420	4,150.91	II
200	3,246.67	II	300	3,768.76	II	560	3,980.88	II	1,400	4,151.97	II
200	3,260.98	II	210	3,770.76	II	560	3,982.89	II	230	4,153.13	II
200	3,263.88	II	300	3,771.60	II	310	3,983.29	II	450	4,159.03	II
990	3,272.25	II	250	3,776.61	II	770	3,984.68	II	310	4,163.52	II
330	3,274.86	II	620	3,781.62	II	370	3,989.44	II	1,300	4,165.61	II
200	3,279.84	II	440	3,782.52	II	700	3,992.39	II	620	4,166.88	II
330	3,285.22	II	200	3,783.58	II	370	3,992.91	II	250	4,167.80	II
240	3,295.28	II	860	3,786.63	II	910	3,993.82	II	320	4,169.77	II
200	3,296.88	II	520	3,788.75	II	2,800	3,999.24	II	320	4,169.88	II
220	3,300.15	II	300	3,792.32	II	230	4,001.56	II	340	4,176.70	II
240	3,304.84	II	2,500	3,801.52	II	910	4,003.77	II	340	4,181.08	II
200	3,312.22	II	800	3,803.09	II	370	4,005.64	II	340	4,185.33	II
240	3,314.72	II	1,000	3,808.11	II	210	4,007.59	II	3,500	4,186.60	II
200	3,317.80	II	490	3,809.21	II	2,700	4,012.39	II	530	4,187.32	II
200	3,334.46	II	250	3,812.20	II	910	4,014.90	II	560	4,193.09	II
240	3,341.87	II	490	3,815.85	II	250	4,015.88	II	370	4,193.28	II
330	3,343.86	II	470	3,817.46	II	200	4,019.04	II	370	4,193.87	II
440	3,344.76	II	300	3,819.02	II	240	4,022.27	II	630	4,196.34	II
200	3,355.02	II	470	3,823.90	II	840	4,024.49	II	280	4,198.00	II
240	3,357.22	II	470	3,830.55	II	240	4,025.15	II	280	4,198.67	II
200	3,360.54	II	490	3,831.08	II	840	4,028.41	II	840	4,198.72	II
240	3,366.55	II	490	3,834.55	II	250	4,030.34	II	240	4,201.24	II
200	3,371.18	II	270	3,836.10	II	840	4,031.34	II	910	4,202.94	II
200	3,373.46	II	1,100	3,838.54	II	340	4,037.67	II	270	4,209.41	II
200	3,373.73	II	200	3,843.76	II	2,100	4,040.76	II	370	4,214.04	II
480	3,377.13	II	220	3,846.52	II	910	4,042.58	II	310	4,217.59	II
200	3,383.68	II	250	3,848.10	II	230	4,045.21	II	1,500	4,222.60	II
200	3,404.91	II	860	3,848.59	II	620	4,046.34	II	770	4,227.75	II
240	3,405.98	II	860	3,853.15	II	210	4,051.43	II	390	4,231.74	II
290	3,417.45	II	1,200	3,854.18	II	210	4,051.99	II	240	4,234.21	II
600	3,422.71	II	1,200	3,854.31	II	700	4,053.51	II	200	4,236.02	II
390	3,426.21	II	620	3,855.29	II	450	4,054.99	II	980	4,239.92	II
290	3,441.21	II	390	3,857.02	II	280	4,062.22	II	390	4,242.72	II
480	3,476.84	II	370	3,857.64	II	230	4,062.94	II	310	4,245.89	II
240	3,482.35	II	200	3,862.46	II	280	4,067.28	II	310	4,245.98	II
710	3,485.05	II	200	3,868.13	II	420	4,068.84	II	390	4,246.72	II
210	3,507.94	II	270	3,874.68	II	1,100	4,071.81	II	1,100	4,248.68	II
600	3,517.38	II	620	3,876.97	II	270	4,072.92	II	390	4,253.37	II
210	3,520.52	II	1,100	3,878.36	II	1,800	4,073.48	II	620	4,255.79	II
330	3,521.88	II	1,500	3,882.45	II	210	4,073.74	II	200	4,263.43	II
210	3,526.68	II	1,000	3,889.98	II	1,500	4,075.71	II	620	4,270.19	II
600	3,534.05	II	210	3,890.75	II	1,500	4,075.85	II	390	4,270.72	II
770	3,539.08	II	210	3,890.98	II	210	4,076.24	II	200	4,278.86	II
210	3,545.60	II	620	3,895.11	II	420	4,077.47	II	280	4,285.37	II
290	3,546.19	II	590	3,896.80	II	530	4,078.32	II	200	4,288.66	II
240	3,552.73	II	490	3,898.27	II	270	4,078.52	II	200	4,289.44	II
420	3,555.00	II	270	3,898.94	II	270	4,080.44	II	2,000	4,289.94	II
1,200	3,560.80	II	200	3,903.34	II	670	4,081.22	II	200	4,296.07	II
210	3,576.23	II	250	3,904.34	II	910	4,083.23	II	1,500	4,296.67	II
1,000	3,577 45	II	200	3,906.92	II	450	4,085.23	II	420	4,296.78	II
330	3,590.60	II	770	3,907.29	II	250	4,087.36	II	590	4,299.36	II
390	3,607.63	II	560	3,908.41	II	230	4,088.85	II	770	4,300.33	II
550	3,609.69	II	390	3,908.54	II	450	4,101.77	II	420	4,305.14	II
420	3,613.70	II	270	3,909.31	II	250	4,105.00	II	770	4,306.72	II
440	3,622.15	II	230	3,912.19	II	510	4,107.42	II	390	4,309.74	II
380	3,623.74	II	980	3,912.44	II	200	4,110.38	II	560	4,320.72	II
440	3,623.84	II	390	3,915.52	II	250	4,111.39	II	310	4,330.45	II
200	3,631.19	II	390	3,916.14	II	420	4,115.37	II	310	4,332.71	II
350	3,646.97	II	230	3,917.64	II	250	4,117.01	II	240	4,336.23	II
260	3,647.75	II	770	3,918.28	II	200	4,117.29	II	980	4,337.77	II
260	3,647.95	II	480	3,919.81	II	200	4,117.59	II	340	4,339.31	II
420	3,653.11	II	590	3,921.73	II	770	4,118.14	II	700	4,349.79	II
660	3,653.67	II	560	3,923.11	II	250	4,119.02	II	560	4,352.71	II
310	3,654.97	II	450	3,924.64	II	310	4,119.79	II	910	4,364.66	II
1,800	3,655.85	II	770	3,931.09	II	310	4,119.88	II	350	4,373.82	II

530		4,375.92	II
910		4,382.17	II
700		4,386.84	II
310		4,388.01	II
1,700		4,391.66	II
200		4,398.79	II
510		4,399.20	II
350		4,410.64	II
350		4,410.76	II
310		4,416.90	II
980		4,418.78	II
200		4,423.68	II
310		4,427.07	II
480		4,427.92	II
310		4,428.44	II
650		4,429.27	II
480		4,444.39	II
450		4,444.70	II
770		4,449.34	II
620		4,450.73	II
2,400		4,460.21	II
450		4,461.14	II
420		4,463.41	II
280		4,467.54	II
1,400		4,471.24	II
450		4,472.72	II
700		4,479.36	II
700		4,483.90	II
840		4,486.91	II
250		4,497.85	II
100		4,506.41	I
110		4,515.86	II
100		4,519.59	II
770		4,523.08	II
840		4,527.35	II
840		4,528.47	II
110		4,532.49	II
110		4,539.07	II
840		4,539.75	II
210		4,544.96	II
250		4,551.30	II
650		4,560.28	II
310		4,560.96	II
2,100		4,562.36	II
420		4,565.84	II
1,100		4,572.28	II
420		4,582.50	II
130		4,591.12	II
840		4,593.93	II
420		4,606.40	II
420		4,624.90	II
1,700		4,628.16	II
170		4,632.32	II
110		4,650.51	I
130		4,654.29	II
110		4,669.50	II
150		4,680.13	II
270		4,684.61	II
200		4,714.00	II
100		4,714.81	II
110		4,725.09	II
100		4,733.52	II
310		4,737.28	II
100		4,739.53	II
160		4,747.17	II
110		4,757.84	II
100		4,768.77	II
230		4,773.94	II
110		4,822.55	I
140		4,847.77	I
180		4,882.46	II
110		4,943.44	I
130		4,971.50	II
130		4,994.63	I
210		5,009.10	I
100		5,011.77	II
120		5,022.87	II
120		5,037.78	II
120		5,040.85	I
180		5,044.02	II
120		5,071.78	I
240		5,075.35	II
470		5,079.68	II
130		5,112.70	I
160		5,117.17	II
170		5,129.57	I
110		5,147.57	II
100		5,149.99	I
280		5,159.69	I
280		5,161.48	I
190		5,174.55	I
370		5,187.46	II
210		5,191.66	II
190		5,211.92	I
260		5,223.46	I
180		5,229.75	I
140		5,232.92	II
260		5,245.92	I
130		5,265.71	II
340		5,274.23	II
130		5,296.56	I
130		5,328.08	I
190		5,330.54	II
450		5,353.53	II
300		5,393.40	II
150		5,397.64	I
280		5,409.23	II
110		5,420.38	I
140		5,449.24	I
140		5,468.37	II
140		5,472.29	II
260		5,512.08	II
110		5,556.25	I
170		5,564.97	I
130		5,565.97	I
100		5,595.88	I
240		5,601.28	I
190		5,655.14	I
240		5,669.96	I
120		5,677.75	I
120		5,692.94	I
300		5,696.99	I
370		5,699.23	I
240		5,719.03	I
140		5,773.12	I
120		5,788.15	I
120		5,812.92	I
230		5,940.86	I
11		6,000.18	I
55		6,001.90	I
55		6,005.86	I
15		6,006.20	I
55		6,006.82	I
19		6,007.37	I
75		6,013.42	I
23		6,016.59	I
110		6,024.20	I
15		6,027.16	I
11		6,031.26	I
23		6,033.58	II
35		6,034.20	II
23		6,034.41	I
35		6,035.49	II
110		6,043.39	II
28		6,045.42	I
55		6,047.40	I
19		6,051.80	II
23		6,057.50	I
35		6,058.00	I
23		6,066.75	I
19		6,069.46	I
35		6,069.48	I
35		6,072.00	I
35		6,076.61	I
17		6,077.16	I
17		6,080.37	I
17		6,081.28	I
19		6,088.86	I
19		6,088.96	I
35		6,093.19	I
45		6,098.34	II
11		6,099.80	I
28		6,108.74	II
15		6,118.56	I
17		6,118.90	I
45		6,123.67	I
19		6,132.00	II
19		6,132.18	I
11		6,135.45	I
23		6,139.03	I
15		6,142.92	I
35		6,143.36	II
23		6,146.43	I
19		6,147.84	I
23		6,151.72	I
19		6,159.82	I
19		6,162.14	I
19		6,165.45	I
19		6,175.28	I
35		6,186.17	I
15		6,187.97	I
15		6,195.23	I
19		6,195.53	I
19		6,198.05	I
35		6,208.98	I
11		6,216.82	I
35		6,228.94	I
19		6,229.13	I
23		6,232.45	II
28		6,237.45	I
13		6,238.71	I
11		6,241.87	I
13		6,242.91	I
15		6,253.65	I
13		6,257.99	I
15		6,264.27	I
45		6,272.05	II
15		6,276.47	I
35		6,295.58	I
28		6,299.51	II
23		6,300.21	I
13		6,306.64	I
35		6,310.01	I
15		6,335.40	I
11		6,337.21	I
13		6,340.70	I
35		6,343.95	II
35		6,371.11	II
28		6,386.84	I
23		6,393.02	II
11		6,395.16	I
11		6,425.29	I
35		6,430.07	I
19		6,434.39	I
23		6,436.40	I
19		6,446.12	I
35		6,458.03	I
19		6,466.88	II
28		6,467.39	I
35		6,473.72	I
17		6,490.97	I
11		6,503.27	I
11		6,507.16	II
23		6,513.59	I
19		6,517.31	I
19		6,551.70	I
45		6,555.65	I
23		6,579.10	I
15		6,606.35	I
15		6,606.86	II
22		6,612.06	I
10		6,623.00	I
30		6,628.93	I
13		6,650.89	I
22		6,652.72	II
10		6,661.41	I
13		6,665.59	I
10		6,675.54	II
15		6,686.60	I
26		6,700.66	I
35		6,704.27	I
13		6,704.52	II
10		6,706.04	II
15		6,728.71	I
15		6,729.57	I
15		6,744.70	I
10		6,746.90	I
30		6,774.28	II
35		6,775.59	I
10		6,778.28	I
18		6,807.81	I
10		6,808.82	I
15		6,818.23	I
10		6,829.73	II
13		6,847.25	I
12		6,856.55	I
10		6,893.66	I
10		6,898.45	II
30		6,924.81	I
10		6,939.45	I
19		6,973.50	II
10		6,983.82	II
30		6,986.02	II
12		7,054.51	I
11		7,058.68	II
11		7,060.00	I
35		7,061.75	II
11		7,064.49	I
35		7,086.35	II
11		7,105.04	II
11		7,115.08	II
10		7,124.73	I
16		7,141.42	I
19		7,150.23	II
10		7,151.67	I
16		7,155.25	I
16		7,156.99	II
16		7,189.40	II
10		7,191.72	I
11		7,201.56	II
16		7,201.89	I
10		7,203.55	I
12		7,210.67	I
19		7,217.36	I
16		7,235.71	II
22		7,238.36	II
12		7,241.73	I
25		7,252.75	I
12		7,262.64	I
11	h	7,277.90	I
11		7,296.17	I
19		7,301.42	II
19		7,313.45	II
25		7,329.91	I
16		7,334.68	II
12		7,343.44	I
25		7,397.77	I
11		7,401.27	I
12		7,417.94	II
11		7,424.70	C
12		7,433.08	I
11		7,438.56	I
12		7,444.44	I
10		7,472.41	I
16		7,486.57	II
11		7,527.46	I
11		7,527.68	I
10		7,533.73	I
10		7,551.25	I
12		7,562.44	I
10		7,562.86	I
10	h	7,563.60	I
10		7,603.10	I
25		7,616.11	II
12		7,646.08	I
10		7,647.88	I
12		7,682.47	I
25		7,689.17	II
10		7,732.33	I
16		7,748.35	I
10		7,797.70	I
12		7,842.59	I
22		7,844.94	II
16		7,850.02	II
16		7,851.18	II
22		7,857.54	II
12		7,864.49	I
10		7,866.04	I
16		7,898.96	II
11		7,913.52	I
10		7,927.30	C
10		7,927.72	I
10		7,934.50	II
30		8,025.56	II
16		8,070.71	I
10		8,094.43	I
16		8,120.36	I
10		8,241.55	II
12		8,261.09	I
16		8,418.23	II
11		8,495.82	I
12		8,539.08	II
10		8,612.64	I
10	h	8,647.66	I
11	h	8,702.38	II
25		8,772.14	II
12		8,810.84	I
30		8,891.20	II

Ce III

Ref. 136, 305 — J.R.

Intensity	Wavelength	
	Vacuum	
100	840.24	III

Intensity	Wavelength	
20	844.11	III
40	845.02	III
20	847.88	III
200	851.18	III
200	852.63	III
200	853.47	III
60	853.78	III
200	855.16	III
200	858.30	III
400	860.15	III
200	862.25	III
40	868.74	III
20	869.51	III
40	869.84	III
20	871.15	III
40	871.27	III
20	880.68	III
60	881.75	III
60	884.04	III
20	885.22	III
30	888.39	III
80	892.75	III
20	899.32	III
100	912.77	III
40	937.04	III
40	999.26	III
20	1,025.25	III
20	1,025.29	III
20	1,026.28	III
40	1,029.37	III
20	1,034.55	III
20	1,041.14	III
100	1,042.74	III
50	1,051.61	III
70	1,057.40	III
100	1,057.66	III
100	1,058.46	III
50	1,062.99	III
30	1,063.26	III
50	1,063.51	III
100	1,067.76	III
100	1,068.69	III
20	1,070.54	III
200	1,072.79	III
40	1,073.69	III
30	1,079.35	III
20	1,080.82	III
30	1,088.70	III
30	1,090.03	III
20	1,092.48	III
20	1,099.25	III
40	1,100.71	III
20	1,107.09	III
20	1,111.19	III
20	1,116.30	III
20	1,125.58	III
20	1,129.73	III
20	1,132.74	III
100	1,142.55	III
20	1,192.41	III
50	1,201.87	III
20	1,204.05	III
20	1,719.43	III
20	1,796.89	III
20	1,836.66	III
20	1,836.99	III
30	1,862.32	III
30	1,950.36	III
20	1,990.54	III
	Air	
100	2,033.34	III
100	2,057.65	III
100	2,077.87	III
200	2,083.32	III
100	2,089.96	III
400	2,109.07	III
100	2,122.55	III
500	2,136.95	III
1,000	2,151.44	III
3,000	2,166.88	III
2,000	2,169.48	III
5,000	2,180.64	III
1,000	2,183.71	III
2,000	2,203.15	III
3,000	2,218.11	III
5,000	2,222.01	III

Intensity	Wavelength	
5,000	2,225.08	III
3,000	2,227.84	III
3,000	2,228.05	III
5,000	2,242.29	III
2,000	2,249.25	III
2,000	2,264.85	III
2,000	2,268.20	III
2,000	2,287.82	III
2,000	2,298.70	III
3,000	2,300.65	III
4,000	2,302.09	III
5,000	2,317.34	III
10,000	2,318.64	III
5,000	2,324.31	III
2,000	2,337.66	III
5,000	2,350.10	III
2,000	2,362.54	III
2,000	2,367.77	III
10,000	2,372.34	III
5,000	2,377.07	III
5,000	2,377.48	III
10,000	2,380.12	III
3,000	2,382.28	III
3,000	2,385.06	III
5,000	2,395.04	III
3,000	2,406.15	III
4,000	2,408.08	III
2,000	2,410.26	III
5,000	2,415.60	III
2,000	2,417.01	III
2,000	2,423.02	III
3,000	2,428.64	III
5,000	2,430.24	III
10,000	2,431.45	III
15,000	2,439.80	III
3,000	2,441.55	III
2,000	2,444.78	III
10,000	2,454.32	III
10,000	2,469.95	III
3,000	2,471.66	III
5,000	2,477.25	III
8,000	2,479.44	III
3,000	2,479.51	III
10,000	2,483.82	III
10,000	2,497.50	III
3,000	2,503.56	III
2,000	2,504.43	III
20,000	2,531.99	III
3,000	2,539.27	III
3,000	2,557.49	III
4,000	2,577.67	III
2,000	2,578.30	III
2,000	2,584.71	III
10,000	2,603.59	III
2,000	2,607.96	III
2,000	2,615.79	III
2,000	2,649.38	III
2,000	2,662.81	III
3,000	2,719.30	III
2,000	2,730.04	III
3,000	2,743.71	III
4,000	2,748.90	III
4,000	2,754.87	III
4,000	2,768.28	III
3,000	2,849.40	III
2,000	2,861.39	III
4,000	2,907.05	III
10,000	2,923.81	III
5,000	2,925.26	III
10,000	2,931.54	III
2,000	2,948.53	III
5,000	2,973.72	III
10,000	3,022.75	III
50,000	3,031.58	III
95,000	3,055.59	III
20,000	3,056.56	III
40,000	3,057.23	III
20,000	3,057.58	III
40,000	3,085.10	III
20,000	3,106.98	III
30,000	3,110.53	III
30,000	3,121.56	III
20,000	3,141.29	III
20,000	3,143.96	III
20,000	3,147.06	III
20,000	3,228.57	III
3,000	3,234.20	III
4,000	3,267.76	III

Intensity	Wavelength	
3,000	3,267.94	III
20,000	3,353.29	III
10,000	3,395.77	III
4,000	3,398.91	III
30,000	3,427.36	III
40,000	3,443.63	III
30,000	3,454.39	III
40,000	3,459.39	III
60,000	3,470.92	III
50,000	3,497.81	III
60,000	3,504.64	III
500	3,514.41	III
50,000	3,544.07	III
3,000	3,784.29	III
800	3,936.80	III
300	3,957.10	III
500	4,169.42	III
300	4,191.70	III
500	4,194.83	III
300	4,213.26	III
300	4,217.13	III
400	4,284.77	III
300	4,304.71	III
600	4,346.35	III
400	4,389.97	III
600	4,448.32	III
500	4,485.27	III
1,000	4,521.92	III
1,000	4,535.73	III
300	4,576.90	III
500	4,627.60	III
300	4,766.07	III
500	4,976.45	III
500	5,650.97	III
1,000	5,664.20	III
500	5,691.08	III
300	5,710.59	III
500	5,749.47	III
500	5,949.83	III
2,000	5,962.22	III
500	5,962.71	III
400	5,979.56	III
1,000	5,983.40	III
3,000	6,002.63	III
10,000	6,032.54	III
10,000	6,060.91	III
500	6,061.79	III
500	6,097.35	III
500	6,098.87	III
500	6,135.10	III
300	6,287.79	III
500	6,308.16	III
300	6,341.75	III
1,000	6,944.94	III
700	7,739.04	III
300	7,758.27	III
500	7,826.80	III
300	7,948.64	III
500	7,960.31	III
500	7,991.01	III
400	8,030.80	III
300	8,084.12	III
400	8,177.33	III
300	8,186.03	III
300	8,222.16	III
300	9,056.53	III
300	9,079.58	III
400	9,328.20	III
300	9,367.03	III
300	9,567.37	III
400	10,458.37	III
400	10,494.42	III
300	10,534.36	III
400	10,684.46	III
15	12,756.96	III
12	12,821.62	III
80	15,847.58	III
80	15,956.79	III
12	15,960.59	III
87	16,128.75	III
42	18,579.82	III
38	19,141.29	III
27	19,377.15	III
26	19,466.14	III
20	19,498.14	III
55	19,524.18	III
30	20,685.63	III
12	21,380.23	III

Ce IV

Ref. 166 — J.R.

Intensity	Wavelength	
	Vacuum	
2	447.58	IV
1	443.11	IV
8	558.92	IV
8	571.59	IV
40	741.79	IV
30	754.60	IV
12	755.75	IV
6	975.20	IV
5	1,009.31	IV
2	1,022.12	IV
9	1,057.67	IV
1	1,059.64	IV
50	1,289.41	IV
75	1,332.16	IV
75	1,372.72	IV
2	1,577.60	IV
1	1,572.62	IV
15	1,641.58	IV
20	1,775.30	IV
20	1,779.03	IV
35	1,914.75	IV
10	1,937.21	IV
	Air	
100	2,000.42	IV
35	2,003.11	IV
100	2,009.94	IV
3	2,433.50	IV
5	2,445.50	IV

Ce V

Ref. 261 — J.R.

Intensity	Wavelength	
	Vacuum	
100	365.66	V
300	399.36	V
150	404.21	V
200	482.96	V
100	552.13	V

CESIUM (CS)

Z = 55

Cs I and II

Ref. 82, 130, 154, 155, 200, 201, 223, 263, 266, 267, 300, 325, 326 K.C.A.

Intensity	Wavelength	
	Vacuum	
25	591.044	II
5	607.291	II
4	612.786	II
200	639.356	II
10	657.112	II
50	668.386	II
1,500	718.138	II
1,500	808.761	II
1,500	813.837	II
3,500	901.270	II
4,000	926.657	II
130	1,178.65	II
80	1,191.55	II
	Air	
25	2,025.05	II
60	2,035.15	II
80	2,077.43	II
80	2,080.05	II
80	2,088.71	II
80	2,091.97	II
15	2,099.50	II
25	2,112.65	II
100	2,146.75	II
130	2,179.60	II
25	2,182.14	II
130	2,189.47	II

25		2,213.15	II
100		2,220.51	II
130		2,228.88	II
250		2,254.58	II
200		2,257.82	II
25		2,258.35	II
350		2,267.61	II
350		2,273.83	II
25		2,286.68	II
25		2,307.71	II
40		2,315.68	II
130		2,321.07	II
80		2,343.13	II
130		2,354.44	II
25		2,357.85	II
130		2,364.81	II
250		2,392.86	II
80		2,414.89	II
25		2,432.71	II
25		2,443.24	II
130		2,476.07	II
40		2,480.41	II
130		2,515.72	II
15		2,523.66	II
130		2,539.08	II
25		2,539.17	II
60		2,550.65	II
130		2,551.17	II
130		2,568.17	II
250		2,568.69	II
1,000		2,573.03	II
130		2,574.54	II
130		2,576.74	II
130		2,590.09	II
250		2,609.44	II
130		2,616.27	II
25		2,627.95	II
80		2,637.14	II
25		2,644.69	II
130		2,648.07	II
200		2,651.71	II
25		2,660.24	II
130		2,669.79	II
15		2,671.17	II
40		2,673.24	II
130		2,686.60	II
25		2,689.41	II
15		2,701.19	II
130		2,724.21	II
25		2,730.07	II
25		2,733.88	II
250		2,748.23	II
80		2,749.84	II
60		2,757.81	II
80		2,761.97	II
25		2,766.10	II
250		2,776.99	II
130		2,788.24	II
130		2,789.80	II
25		2,793.32	II
130		2,794.50	II
130		2,799.41	II
500		2,816.94	II
25		2,820.27	II
25		2,829.04	II
25		2,829.42	II
130		2,846.19	II
80		2,852.42	II
80		2,866.37	II
250		2,881.19	II
25		2,883.74	II
80		2,899.75	II
80		2,914.65	II
500		2,931.09	II
500		2,940.95	II
80		2,942.25	II
30		2,949.80	II
20		2,968.38	II
25		2,970.85	II
100		3,001.27	II
80		3,012.04	II
15		3,020.37	II
15		3,060.98	II
100		3,066.60	II
25		3,080.87	II
100		3,092.31	II
80		3,180.94	II
800		3,265.92	II
800		3,267.13	II
500		3,271.63	II
100		3,329.43	II
800		3,368.56	II
100		3,559.80	II
100		3,565.11	II
15		3,651.07	II
15		3,680.10	II
15		3,687.64	II
100		3,699.48	II
15		3,732.54	II
100		3,734.34	II
15		3,751.40	II
350		3,785.42	II
500		3,805.10	II
15		3,870.16	II
150	c	3,876.143	I
80	c	3,888.608	I
1,000		3,896.98	II
350		3,906.93	II
500		3,925.58	II
350		3,959.50	II
500		3,965.19	II
150		3,978.00	II
350		4,047.18	II
200		4,053.96	II
500		4,067.96	II
500		4,068.77	II
70		4,073.36	II
60		4,119.29	II
500		4,121.21	II
100		4,132.00	II
350		4,151.27	II
350		4,158.61	II
60		4,193.20	II
500		4,213.13	II
200		4,221.12	II
500		4,232.19	II
350		4,234.41	II
100		4,241.97	II
100		4,271.74	II
2,000		4,277.10	II
1,000		4,288.35	II
200		4,292.00	II
500		4,300.64	II
50		4,307.94	II
50		4,327.58	II
350		4,330.24	II
2,000		4,363.28	II
500		4,373.02	II
400		4,384.43	II
100		4,388.76	II
200		4,396.91	II
350		4,399.50	II
350		4,403.85	II
1,000		4,405.25	II
350		4,410.21	II
100		4,424.05	II
350		4,435.71	II
100		4,444.00	II
200		4,453.44	II
200		4,457.68	II
200		4,459.18	II
100		4,493.66	II
1,000		4,501.52	II
200		4,506.71	II
100		4,506.83	II
100		4,515.50	II
150		4,522.85	II
1,000		4,526.72	II
800		4,538.94	II
400	c	4,555.276	I
150		4,566.98	II
150		4,571.79	II
200	c	4,593.169	I
2,500		4,603.76	II
100		4,609.99	II
150		4,616.13	II
350		4,623.09	II
500		4,646.51	II
150		4,656.54	II
350		4,670.28	II
100		4,695.61	II
500		4,701.79	II
350		4,732.98	II
350		4,739.66	II
100		4,749.13	II
500		4,763.62	II
200		4,786.36	II
800		4,830.16	II
800		4,870.02	II
800		4,952.84	II
500		4,972.59	II
800		5,043.80	II
500		5,052.70	II
500		5,059.87	II
100		5,081.77	II
400		5,096.60	II
1,500		5,227.00	I
800		5,249.37	II
400		5,274.04	II
350		5,306.61	II
350		5,349.16	II
800		5,370.97	II
500		5,419.69	II
5	c	5,465.944	I
5		5,502.884	I
1,000		5,563.02	II
11		5,635.212	I
30	c	5,664.018	I
8		5,745.724	I
350		5,814.18	II
500		5,831.16	II
5	c	5,838.835	I
30		5,845.141	I
500		5,925.65	II
80	c	6,010.490	I
30		6,034.089	I
400		6,128.62	II
120		6,213.100	I
15		6,217.599	I
50	c	6,354.555	I
100		6,419.54	II
15		6,431.969	I
15		6,472.623	I
200		6,495.53	II
200		6,536.44	II
30		6,586.02	I
200		6,586.510	I
35		6,628.660	I
200	c	6,723.284	I
100		6,824.652	I
	c	6,848.91	I
100		6,870.455	I
	c	6,895.01	I
400		6,955.52	II
200		6,973.297	I
35		6,983.491	I
200		7,228.536	I
60		7,279.90	I
200		7,279.957	I
150	c	7,608.903	I
300		7,943.882	I
60	s	7,990.68	I
100		8,015.724	I
60	s	8,053.35	I
60		8,078.92	I
350		8,079.033	I
1,000	c	8,521.122	I
200	c	8,761.415	I
600	c	8,943.46	I
350		9,172.322	I
100		9,208.538	I
350		10,024.359	I
100		10,123.414	I
400		10,123.602	I
140		13,588.31	I
15		13,602.57	I
15		13,758.83	I
350		14,694.93	I
12		17,923.62	I
9		17,924.21	I
25		19,162.53	I
20		19,163.20	I
50		21,311.46	I
45		21,312.29	I
100		25,763.49	I
90		25,764.70	I
200		39,421.22	I
180		39,424.08	I

Cs III

Ref. 78 — J.R.

Intensity		Wavelength	
		Vacuum	
75		556.91	III
250		584.15	III
50		584.40	III
1,200		603.01	III
600		607.85	III
800		607.94	III
10,000		614.01	III
1,500		621.15	III
600		635.86	III
300		637.67	III
2,000		638.17	III
25		657.94	III
450		663.82	III
450		664.60	III
2,500		666.25	III
1,800		673.06	III
400		679.60	III
800		687.55	III
5,000		691.60	III
800		699.43	III
3,500		703.89	III
1,000		710.25	III
20,000		721.79	III
20,000		722.20	III
5,000		731.56	III
300		731.95	III
1,000	p	736.66	III
12,000		740.29	III
150		742.23	III
1,000		749.94	III
1,500		750.38	III
1,500		755.18	III
1,200		758.82	III
6		759.90	III
300		787.73	III
75		801.95	III
200		814.02	III
15		817.35	III
800		820.34	III
200		825.99	III
7,500		830.39	III
100		837.39	III
4		843.37	III
250		847.92	III
15,000		920.35	III
500		921.66	III
25		932.67	III
25,000	c	1,054.79	III

Cs IV

Ref. 259 — J.R.

Intensity		Wavelength	
		Vacuum	
150		703.40	IV
7		773.61	IV
4		823.64	IV
150		824.80	IV
15		861.83	IV
200		874.84	IV
150		896.92	IV
100		923.02	IV
200		986.14	IV
50	p	995.14	IV
150		1,068.91	IV

CHLORINE (CL)
Z = 17

Cl I and II

Ref. 238, 239 — L.J.R.

Intensity	Wavelength	
	Vacuum	
350	559.305	II
400	571.904	II
800	574.406	II
500	586.24	II
700	618.057	II
600	619.982	II
800	620.298	II
700	626.735	II
800	635.881	II
1,000	636.626	II
1,000	650.894	II
1,000	659.811	II
1,300	661.841	II
2,000	663.074	II

1,500		682.053	II
1,500		687.656	II
1,500		693.594	II
2,000		725.271	II
2,500		728.951	II
2,000		777.562	II
5,000		787.580	II
5,000		788.740	II
5,000		793.342	II
6,000		839.297	II
8,000		839.599	II
7,000	p	841.41	II
5,000		851.691	II
2,000		888.026	II
2,000		893.549	II
2,000		961.499	II
30		969.92	I
40		978.284	I
25		998.372	I
25		998.432	I
75		1,002.346	I
150		1,013.664	I
90		1,025.553	I
6,000		1,063.831	II
3,000		1,067.945	II
9,000		1,071.036	II
6,000		1,071.767	II
5,000		1,075.230	II
5,000		1,079.080	II
200		1,084.667	I
200		1,085.171	I
250		1,085.304	I
400		1,088.06	I
350		1,090.271	I
250		1,090.982	I
250		1,092.437	I
400		1,094.769	I
350		1,095.148	I
350		1,095.662	I
400		1,095.797	I
250		1,096.810	I
300		1,097.369	I
200		1,098.068	I
200		1,099.523	I
500		1,107.528	I
800		1,139.214	II
800		1,167.148	I
3,000		1,179.293	I
1,200		1,188.774	I
900		1,201.353	I
3,000		1,335.726	I
10,000		1,347.240	I
5,000		1,351.657	I
12,000		1,363.447	I
2,500		1,373.116	I
20,000		1,379.528	I
25,000		1,389.693	I
20,000		1,389.957	I
12,000		1,396.527	I
500		1,441.470	II
500		1,528.569	II
500		1,542.942	II
500		1,558.144	II
500		1,565.050	II
500		1,857.488	II
450	h	1,997.370	II

Air

450		2,032.116	II
350	h	2,088.583	II
350	h	2,091.458	II
170		2,427.79	II
360		2,434.07	II
340		2,498.53	II
470		2,502.74	II
260		2,546.96	II
500		2,549.88	II
460		2,564.84	II
320		2,603.31	II
950		2,658.72	II
750		2,676.95	II
1,200		2,688.04	II
410		2,912.05	II
950		2,996.65	II
500		3,006.06	II
950		3,057.96	II
1,300		3,071.32	II
1,400		3,092.19	II
1,200		3,123.72	II
1,900		3,315.43	II
1,200		3,329.10	II
2,500		3,353.35	II
20		3,726.54	I
1,200		3,749.96	II
1,000		3,781.17	II
1,500		3,798.76	II
1,900		3,805.18	II
1,300		3,809.46	II
1,700		3,820.20	II
2,800		3,827.59	II
4,500		3,833.35	II
2,500		3,843.20	II
3,100		3,845.37	II
3,900		3,845.65	II
1,500		3,845.80	II
10,000		3,850.99	II
7,900		3,851.37	II
1,200		3,851.65	II
25,000		3,860.83	II
4,400		3,860.99	II
1,000		3,861.37	II
1,500		3,913.87	II
1,100		3,916.63	II
20		3,944.82	I
20		4,104.79	I
10,000	h	4,132.50	II
65		4,209.67	I
50		4,226.42	I
60		4,264.58	I
100		4,363.27	I
100		4,369.50	I
5,000		4,372.93	II
100		4,379.90	I
100		4,389.75	I
90		4,390.40	I
90		4,403.03	I
100		4,438.49	I
90		4,475.30	I
1,500		4,489.91	II
100		4,526.19	I
80		4,600.98	I
40		4,623.938	I
50		4,654.040	I
80		4,661.208	I
45		4,691.523	I
40		4,721.255	I
45		4,740.729	I
4,300		4,768.65	II
13,000		4,781.32	II
99,000		4,794.55	II
29,000		4,810.06	II
16,000		4,819.47	II
81,000		4,896.77	II
47,000		4,904.78	II
26,000		4,917.73	II
10,000		4,995.48	II
26,000		5,078.26	II
30		5,099.789	I
56,000		5,217.94	II
23,000		5,221.36	II
15,000		5,392.12	II
99,000		5,423.23	II
10,000		5,423.51	II
19,000		5,443.37	II
10,000		5,444.21	II
5,600		5,457.02	II
40		5,532.162	I
50	d	5,796.305	I
45		5,799.914	I
30		5,856.742	I
100	d	5,948.58	I
50		6,019.812	I
35		6,082.61	I
1,900		6,094.69	II
160		6,114.43	I
200		6,140.245	I
160		6,194.757	I
160		6,398.66	I
150		6,434.833	I
150		6,531.43	I
1,400		6,661.67	II
150		6,678.43	I
1,300		6,686.02	II
1,200		6,713.41	II
150		6,840.29	I
300		6,932.903	I
300		6,981.886	I
600		7,086.814	I
7,500		7,256.62	I
5,000		7,414.11	I
550		7,462.370	I
550		7,489.47	I
700		7,492.118	I
11,000		7,547.072	I
2,300		7,672.42	I
450		7,702.828	I
7,000		7,717.581	I
10,000		7,744.97	I
2,200		7,769.16	I
650		7,771.09	I
2,200		7,821.36	I
1,700		7,830.75	I
3,000		7,878.22	I
220		7,893.34	I
2,300		7,899.31	I
1,800		7,915.08	I
3,000		7,924.645	I
2,100		7,933.89	I
1,700		7,935.012	I
650		7,952.52	I
1,500		7,974.72	I
1,300		7,976.97	I
600		7,980.60	I
2,900		7,997.85	I
2,200		8,015.61	I
1,100		8,023.33	I
400		8,051.07	I
1,700		8,084.51	I
2,200		8,085.56	I
3,000		8,086.67	I
1,300		8,087.73	I
250		8,094.67	I
2,500		8,194.42	I
2,200		8,199.13	I
2,200		8,200.21	I
800		8,203.78	I
18,000		8,212.04	I
3,000		8,220.45	I
20,000		8,221.74	I
18,000		8,333.31	I
1,000		8,360.71	II
560		8,361.84	II
99,900		8,375.94	I
180		8,382.67	II
100		8,392.02	II
400		8,406.199	I
15,000		8,428.25	I
2,200		8,467.34	I
2,200		8,550.44	I
20,000		8,575.24	I
750		8,578.02	I
75,000		8,585.97	I
450		8,628.54	I
300		8,641.71	I
3,500		8,686.26	I
2,200		8,912.92	I
3,000		8,948.06	I
2,000		9,038.982	I
2,500		9,045.43	I
1,000		9,069.656	I
2,000		9,073.17	I
7,500		9,121.15	I
3,000		9,191.731	I
500		9,197.596	I
4,000		9,288.86	I
1,500		9,393.862	I
3,500		9,452.10	I
500		9,486.964	I
1,000		9,584.801	I
3,500		9,592.22	I
250		9,632.509	I
1,000		9,702.439	I
250		9,744.426	I
200		9,807.057	I
400		9,875.970	I
331		10,392.549	I
38		10,432.83	II
10		10,506.62	II
14		10,509.12	II
19		10,512.46	II
25		10,514.17	II
9		10,801.47	II
5		10,885.42	II
1		10,955.71	II
300		11,123.05	I
231		11,392.62	I
269		11,409.69	I
1,000		11,436.33	I
180		11,720.56	I
195		11,866.76	I
172		12,021.7	I
350		13,243.8	I
310		13,296.0	I
550		13,346.8	I
525		13,821.7	I
148		14,369.7	I
294		14,931.7	I
269		15,108.0	I
381		15,465.1	I
169		15,467.6	I
1,094		15,520.3	I
1,487		15,730.1	I
193		15,818.4	I
2,780		15,869.7	I
277		15,883.3	I
342		15,928.9	I
735		15,960.0	I
283		15,970.5	I
129		16,077.6	I
259		16,198.5	I
227		19,370.3	I
717		19,755.3	I
185		19,766.8	I
227		20,199.4	I
85		20,370.1	I
100		24,470.0	I
		39,603.7	I
		39,615.3	I
		39,716.0	I
		39,744.0	I
		39,750.0	I
		39,875.3	I
		39.881.0	I
		39,985.7	I
		40,085.5	I
		40,089.5	I
		40,171.0	I
		40,310.3	I
		40,335.4	I
		40,532.2	I

Cl III
Ref. 28, 30 — L.J.R.

Intensity	Wavelength	
	Vacuum	
100	406.27	III
400	411.37	III
400	411.81	III
600	556.23	III
700	556.61	III
700	557.12	III
700	561.53	III
700	561.68	III
700	561.74	III
500	606.35	III
400	621.28	III
300	670.38	III
300	673.13	III
100	936.28	III
500	1,005.28	III
600	1,008.78	III
700	1,015.02	III
600	1,822.50	III
500	1,828.40	III
500	1,901.61	III
500	1,983.61	III
	Air	
400	2,006.84	III
700	2,253.07	III
500	2,268.95	III
500	2,278.34	III
700	2,283.93	III
600	2,323.50	III
500	2,336.45	III
600	2,340.64	III
600	2,359.67	III
600	2,370.37	III
700	2,416.42	III
600	2,447.14	III
600	2,448.58	III
500	2,486.91	III

500	2,532.48	III
600	2,580.67	III
500	2,603.59	III
500	2,632.67	III
500	2,633.18	III
600	2,665.54	III
700	2,710.37	III
600	2,965.56	III
600	3,104.46	III
800	3,139.34	III
900	3,191.45	III
700	3,289.80	III
700	3,320.57	III
800	3,329.06	III
900	3,340.42	III
800	3,392.89	III
800	3,393.45	III
900	3,530.03	III
800	3,560.68	III
900	3,602.10	III
800	3,612.85	III
700	3,622.69	III
700	3,656.95	III
700	3,670.28	III
700	3,682.05	III
600	3,705.45	III
600	3,707.34	III
800	3,720.45	III
800	3,748.81	III
500	3,779.35	III
500	3,925.87	III
700	3,991.50	III
600	4,018.50	III
600	4,059.07	III
500	4,104.23	III
500	4,106.83	III
400	4,370.91	III
500	4,608.21	III
300	4,703.14	III
100	4,863.75	III
10	4,971.64	III

Cl IV

Ref. 11, 28, 30, 31 — L.J.R.

Intensity	Wavelength	
	Vacuum	
300	319.62	IV
200	331.84	IV
400	437.83	IV
400	464.86	IV
800	486.17	IV
800	534.73	IV
700	535.67	IV
600	536.15	IV
900	537.61	IV
600	538.12	IV
500	549.22	IV
400	550.02	IV
700	552.02	IV
600	553.30	IV
700	554.62	IV
500	601.50	IV
500	604.59	IV
400	608.90	IV
400	612.07	IV
400	653.70	IV
400	745.21	IV
400	831.43	IV
500	834.84	IV
500	834.97	IV
400	840.81	IV
600	840.93	IV
	865.3	IV
500	973.21	IV
600	977.56	IV
400	977.90	IV
700	984.95	IV
400	985.75	IV
300	1,537.21	IV
200	1,539.30	IV
200	1,545.19	IV
200	1,549.15	IV
200	1,622.86	IV
	Air	
400	2,701.36	IV
500	2,724.03	IV
500	2,751.23	IV
400	2,770.64	IV
700	2,782.47	IV
400	2,835.4	IV
500	3,063.13	IV
600	3,076.68	IV
200	3,167.87	IV

Cl V

Ref. 11, 28, 30, 85, 233

L.J.R.

Intensity	Wavelength	
	Vacuum	
300	287.33	V
300	373.78	V
400	390.15	V
500	392.43	V
300	536.53	V
400	537.01	V
300	537.46	V
500	538.03	V
400	538.68	V
800	542.23	V
600	542.30	V
400	542.87	V
1,000	545.11	V
600	546.33	V
1,000	547.63	V
400	633.19	V
400	635.32	V
400	681.92	V
400	683.17	V
400	688.93	V
	715.55	V
	716.19	V
400	883.13	V
400	894.34	V
100	894.91	V

CHROMIUM (CR)

Z = 24

Cr I and II

Ref. 1 — C.H.C.

Intensity		Wavelength	
		Vacuum	
19,000		2,055.52	II
14,000		2,061.49	II
8,900		2,065.42	II
80	h	2,364.71	I
130		2,383.33	I
140		2,408.62	I
170		2,496.31	I
110		2,502.53	I
190		2,504.31	I
50		2,508.11	I
60		2,508.98	I
40		2,513.62	I
110		2,516.92	I
80		2,518.71	I
390		2,519.52	I
190		2,527.12	I
40		2,530.45	I
70		2,534.34	II
50		2,545.64	I
160		2,549.54	I
40		2,553.06	I
80		2,557.15	I
130		2,560.69	I
150		2,571.74	I
100		2,577.65	I
50		2,588.20	I
380		2,591.85	I
35		2,603.57	I
35		2,622.86	I
22		2,625.32	I
18		2,626.60	I
18		2,629.82	I
35		2,642.12	I
250		2,653.59	II
250		2,658.59	II
70		2,661.73	II
320		2,663.42	II
70		2,663.68	II
440		2,666.02	II
280		2,668.71	II
350		2,671.81	II
280		2,672.83	II
1,800		2,677.16	II
35		2,678.16	I
320		2,678.79	II
18		2,680.34	II
230		2,687.09	II
60		2,688.04	I
55		2,688.29	II
26		2,690.26	I
280		2,691.04	II
35		2,693.52	II
35		2,697.91	II
180		2,698.41	II
180		2,698.69	II
18		2,700.60	I
110		2,701.99	I
18		2,702.53	I
70		2,703.48	I
		2,703.55	II
35		2,703.86	II
18		2,705.43	I
60		2,708.79	II
35		2,709.31	II
140		2,712.31	II
45		2,716.18	I
55		2,717.51	II
45		2,718.43	II
170		2,722.75	II
18		2,724.04	II
420	h	2,726.51	I
45		2,727.26	II
280	h	2,731.91	I
170	h	2,736.47	I
70		2,739.38	I
70		2,740.10	II
95		2,741.07	I
95		2,742.03	II
95		2,742.17	I
250		2,743.64	II
35		2,746.21	II
110	h	2,748.29	I
330		2,748.98	II
390		2,750.73	II
45		2,751.60	I
280		2,751.87	II
110	h	2,752.88	I
35		2,754.28	II
22		2,754.90	I
22		2,755.27	I
22		2,756.75	I
150		2,757.10	I
350		2,757.72	II
60		2,758.98	II
80		2,759.39	II
45		2,759.73	II
90	h	2,761.76	I
750		2,762.59	II
22		2,763.06	I
80	h	2,764.35	I
750		2,766.54	II
22		2,767.54	I
250	h	2,769.92	I
18		2,771.45	I
45		2,778.06	II
22		2,779.14	I
80		2,780.30	II
610		2,780.70	I
70		2,785.70	II
35		2,787.63	II
35		2,787.84	I
90		2,792.16	II
55		2,798.67	II
70		2,800.77	II
80		2,812.01	II
60		2,818.36	II
45		2,822.01	II
180		2,822.37	II
22		2,826.75	I
180		2,830.47	II
70		2,834.26	II
2,500		2,835.63	II
45		2,836.48	II
55		2,838.79	II
110		2,840.02	II
1,700		2,843.25	II
22		2,846.02	I
45		2,849.29	I
1,200		2,849.84	II
120		2,851.36	II
55		2,853.22	II
55		2,855.07	II
880		2,855.68	II
90		2,856.77	II
70		2,857.40	II
610		2,858.91	II
440		2,860.93	II
790		2,862.57	II
750		2,865.11	II
55		2,865.33	II
610		2,866.74	II
90		2,867.10	II
480		2,867.65	II
210		2,870.44	II
110		2,871.63	I
160		2,873.48	II
90		2,873.82	II
320		2,875.99	II
230		2,876.24	II
180		2,877.98	II
70		2,878.45	II
120		2,879.27	I
95		2,880.87	II
30		2,881.14	I
170		2,887.00	I
55		2,888.74	II
700		2,889.29	I
55		2,889.82	II
55		2,891.42	I
370		2,893.25	I
190		2,894.17	I
55		2,896.46	II
210		2,896.75	I
55	d	2,897.67	II
		2,897.73	II
90		2,898.54	II
80		2,899.21	I
55		2,899.48	II
26		2,903.97	II
55		2,904.68	I
180		2,905.49	I
260		2,909.05	I
260		2,910.90	I
250		2,911.14	I
45		2,911.68	II
60		2,913.73	I
22		2,915.23	II
22		2,915.46	II
90		2,921.24	II
60		2,921.82	II
60		2,927.08	II
80		2,928.15	II
95		2,928.30	II
26		2,929.44	II
35		2,930.85	II
26		2,932.70	II
55		2,933.97	II
90		2,935.14	II
45		2,940.22	II
60		2,946.84	II
55		2,953.36	II
45		2,953.71	II
55		2,961.73	II
45		2,966.05	II
480		2,967.64	I
480		2,971.11	I
210		2,971.91	II
480		2,975.48	I
30		2,976.72	II
190		2,979.74	II
350		2,980.79	I
110		2,985.32	II
480		2,985.85	I
1,500		2,986.00	I
2,100		2,986.47	I
660		2,988.65	I
160		2,989.19	II
480		2,991.89	I
230		2,994.07	I
300		2,995.10	I
700		2,996.58	I
210		2,998.79	I
1,100		3,000.89	I
750		3,005.06	I
140		3,013.03	I

710		3,013.71	I
710		3,014.76	I
1,400		3,014.92	I
710		3,015.19	I
2,800		3,017.57	I
430		3,018.50	I
240		3,018.82	I
430		3,020.67	I
2,800		3,021.56	I
1,100		3,024.35	I
85		3,026.65	II
170		3,029.16	I
710		3,030.24	I
140		3,031.35	I
28		3,032.93	II
390		3,034.19	I
550		3,037.04	I
80		3,039.78	I
550		3,040.85	I
		3,040.91	II
55		3,041.74	II
110		3,050.14	II
710		3,053.88	I
24		3,059.52	II
85		3,065.07	I
28		3,067.16	II
85		3,073.68	I
55		3,077.83	I
28		3,095.86	I
28		3,109.34	I
28		3,110.86	I
240		3,118.65	II
45		3,119.25	I
40		3,119.71	I
430		3,120.37	II
28		3,122.60	II
470		3,124.94	II
		3,125.02	II
120		3,128.70	II
590		3,132.06	II
140		3,136.68	II
140		3,147.23	II
85		3,148.44	I
100		3,155.15	I
100		3,163.76	I
240		3,180.70	II
30		3,181.43	II
65	h	3,188.01	I
220		3,197.08	II
24		3,198.11	I
30		3,208.59	II
170		3,209.18	II
140		3,217.40	II
30		3,229.20	I
28		3,234.06	II
65		3,237.73	I
120		3,245.54	I
130		3,251.84	I
130		3,257.82	I
95		3,259.98	I
30		3,295.43	II
24		3,307.02	II
55		3,324.06	II
28		3,326.59	I
30		3,328.35	II
30		3,329.05	I
95		3,336.33	II
130		3,339.80	II
110		3,342.59	II
30		3,343.34	II
95		3,346.02	I
95		3,346.74	I
95		3,347.84	II
65		3,349.07	I
55		3,349.32	I
30		3,351.60	I
55		3,351.97	I
55	h	3,353.03	I
		3,353.13	II
170		3,358.50	II
160		3,360.30	II
65		3,361.77	II
55		3,362.21	I
430		3,368.05	II
30		3,376.40	I
55		3,378.34	II
30		3,379.17	I
30		3,379.37	II
95		3,379.83	II
140		3,382.68	II
95		3,391.43	II
55		3,392.99	II
70		3,393.84	II
55		3,394.30	II
30		3,402.40	II
170		3,403.32	II
360		3,408.76	II
210		3,421.21	II
270		3,422.74	II
140		3,433.31	II
270		3,433.60	I
55		3,434.11	I
160		3,436.19	I
70		3,441.12	I
140		3,441.44	I
30		3,443.79	I
170		3,445.62	I
30		3,447.02	I
170		3,447.43	I
70		3,447.76	I
190		3,453.33	I
40		3,453.74	I
130		3,455.60	I
100		3,460.43	I
65		3,465.25	I
40		3,467.02	I
70		3,467.72	I
45		3,469.59	I
16		3,472.76	I
24		3,472.91	I
40		3,473.61	I
70		3,481.30	I
55		3,481.54	I
55		3,494.97	I
40		3,495.38	II
80		3,510.54	I
40		3,511.84	II
120		3,550.64	I
80		3,558.52	I
130		3,566.16	I
130		3,573.64	I
80		3,574.04	I
330	h	3,574.80	I
		3,574.94	I
19,000		3,578.69	I
160	h	3,584.33	I
130		3,585.30	II
17,000		3,593.49	I
350		3,601.67	I
40		3,602.57	I
85		3,603.74	I
		3,603.78	I
13,000		3,605.33	I
40		3,608.40	I
40		3,609.48	I
40		3,610.05	I
70		3,612.61	I
85		3,615.64	I
130		3,632.84	I
350		3,636.59	I
630		3,639.80	I
85		3,640.39	I
70		3,641.47	I
220		3,641.83	I
45		3,646.16	I
85		3,648.53	I
220		3,649.00	I
170		3,653.91	I
220		3,656.26	I
45		3,662.84	I
130		3,663.21	I
45		3,665.98	I
95		3,666.64	I
55		3,668.03	I
65		3,676.32	I
40		3,677.68	II
55		3,677.89	II
40		3,679.82	I
19		3,681.69	I
120		3,685.55	I
130		3,686.80	I
130		3,687.25	I
75		3,687.54	I
19		3,688.46	I
75		3,712.95	II
40		3,716.53	I
130		3,730.81	I
150		3,732.03	I
95		3,742.97	I
480		3,743.58	I
570		3,743.88	I
85		3,744.49	I
55		3,748.61	I
340		3,749.00	I
50		3,757.17	I
230		3,757.66	I
60		3,758.04	I
24		3,767.43	I
260		3,768.24	I
95		3,768.73	I
95		3,788.86	I
95		3,790.45	I
130		3,791.38	I
130		3,792.14	I
120		3,793.29	I
130		3,793.88	I
85		3,794.61	I
140		3,797.13	I
200		3,797.72	I
530		3,804.80	I
110		3,806.83	I
110		3,807.93	I
180		3,815.43	I
70		3,818.48	I
180		3,819.56	I
70		3,823.52	I
130		3,826.42	I
130		3,830.03	I
380		3,841.28	I
190		3,848.98	I
140		3,849.36	I
290		3,850.04	I
140		3,852.22	I
190		3,854.22	I
110		3,855.29	I
140		3,855.57	I
260		3,857.63	I
70		3,874.53	I
660		3,883.29	I
50		3,883.66	I
570		3,885.22	I
380		3,886.79	I
60		3,891.93	I
260		3,894.04	I
40		3,897.65	I
35		3,902.11	I
360		3,902.92	I
60		3,903.16	I
960		3,908.76	I
120	d	3,911.82	I
		3,912.00	I
120		3,915.84	I
190		3,916.24	I
35		3,917.60	I
1,900		3,919.16	I
600		3,921.02	I
30		3,926.65	I
600		3,928.64	I
410		3,941.49	I
30		3,951.10	I
40		3,952.40	I
35		3,953.16	I
1,900		3,963.69	I
120		3,969.06	I
1,600		3,969.75	I
85		3,971.26	I
1,600		3,976.66	I
85		3,978.68	I
40		3,979.80	I
85		3,981.23	I
960		3,983.91	I
190		3,984.34	I
160		3,989.99	I
960		3,991.12	I
160		3,991.67	I
190		3,992.84	I
40		3,993.97	I
160		4,001.44	I
120		4,012.47	II
30		4,014.67	I
85		4,022.26	I
70		4,025.01	I
120		4,026.17	I
85		4,027.10	I
85		4,030.68	I
190		4,039.10	I
160		4,048.78	I
120		4,058.77	I
40		4,065.72	I
85		4,066.94	I
35		4,074.86	I
40		4,076.06	I
40		4,077.09	I
40		4,077.68	I
40		4,104.87	I
40		4,109.58	I
40		4,120.61	I
40		4,121.82	I
35		4,122.16	I
40		4,123.39	I
140		4,126.52	I
35		4,127.30	I
40		4,127.64	I
40		4,131.36	I
30		4,152.78	I
120		4,153.82	I
85		4,161.42	I
140		4,163.62	I
70		4,165.52	I
40		4,169.84	I
35		4,170.20	I
40		4,172.77	I
170		4,174.80	I
30		4,175.94	I
170		4,179.26	I
35		4,184.90	I
30		4,186.36	I
35		4,190.13	I
85		4,191.27	I
35		4,192.10	I
85		4,193.66	I
70		4,194.95	I
40		4,197.23	I
85		4,198.52	I
60		4,203.59	I
40		4,204.47	I
35		4,208.36	I
110		4,209.37	I
40		4,209.76	I
40		4,211.35	I
40		4,216.36	I
85		4,217.63	I
40		4,221.57	I
40		4,222.73	I
40		4,238.96	I
60		4,240.70	I
20,000		4,254.35	I
70		4,255.50	I
60		4,261.35	I
110		4,263.14	I
30		4,271.06	I
40		4,272.91	I
16,000		4,274.80	I
85		4,280.40	I
10,000		4,289.72	I
40		4,291.96	I
85		4,295.76	I
70		4,297.74	I
35		4,300.51	I
50		4,301.18	I
30		4,305.45	I
35		4,319.64	I
60		4,325.08	I
780		4,337.57	I
1,100		4,339.45	I
380		4,339.72	I
60		4,340.13	I
1,900		4,344.51	I
70		4,346.83	I
380		4,351.05	I
2,300		4,351.77	I
570		4,359.63	I
70		4,363.13	I
530		4,371.28	I
70		4,373.25	I
110		4,374.16	I
70		4,375.33	I
50		4,381.11	I
530		4,384.98	I
60		4,387.50	I
70		4,391.75	I
60		4,403.50	I
24		4,410.30	I
60		4,411.09	I
35		4,412.25	I
50		4,413.87	I

60		4,424.28	I
24		4,428.50	I
50		4,430.49	I
50		4,432.18	I
110		4,458.54	I
30		4,459.74	I
30		4,465.36	I
30		4,482.88	I
40		4,488.05	I
50		4,489.47	I
60		4,492.31	I
660		4,496.86	I
50		4,498.73	I
70		4,500.30	I
50		4,501.11	I
22		4,501.79	I
24		4,506.85	I
95		4,511.90	I
12		4,514.37	I
35		4,514.53	I
24		4,521.14	I
24		4,526.11	I
380		4,526.47	I
70	d	4,527.34	I
		4,527.47	I
24		4,529.85	I
380		4,530.74	I
50		4,535.15	I
240		4,535.72	I
40		4,539.79	I
240		4,540.50	I
240		4,540.72	I
35		4,541.07	I
19		4,541.51	I
24		4,542.62	I
140		4,544.62	I
24		4,545.34	I
600		4,545.96	I
50		4,556.17	I
22		4,558.66	II
19		4,564.17	I
120		4,565.51	I
95		4,569.64	I
120		4,571.68	I
22		4,575.12	I
360		4,580.06	I
24		4,586.14	I
360		4,591.39	I
70		4,595.59	I
50		4,600.10	I
480		4,600.75	I
50		4,601.02	I
240		4,613.37	I
600		4,616.14	I
70		4,619.55	I
85		4,621.96	I
70		4,622.49	I
24		4,622.76	I
550		4,626.19	I
24		4,632.18	I
40		4,637.18	I
50		4,637.77	I
50	d	4,639.52	I
		4,639.70	I
1,600		4,646.17	I
24		4,646.81	I
24		4,648.13	I
24		4,648.87	I
35		4,649.46	I
570		4,651.28	I
840		4,652.16	I
35		4,654.74	I
19		4,656.19	I
40		4,663.33	I
70		4,663.83	I
95		4,664.80	I
35		4,665.90	I
22		4,666.22	I
70		4,666.51	I
50		4,669.34	I
40		4,680.54	I
19		4,680.87	I
70		4,689.37	I
60		4,693.95	I
24		4,695.15	I
60		4,697.06	I
240	d	4,698.46	I
		4,698.62	I
35		4,700.61	I
190		4,708.04	I
240		4,718.43	I
50		4,723.10	I
50		4,724.42	I
50		4,727.15	I
24		4,729.72	I
120		4,730.71	I
140		4,737.35	I
19		4,745.31	I
70		4,752.08	I
340		4,756.11	I
50		4,764.29	I
22		4,766.63	I
30		4,767.86	I
190		4,789.32	I
95		4,792.51	I
120		4,801.03	I
110		4,829.38	I
14		4,836.86	I
17		4,861.20	I
70		4,861.84	I
140		4,870.80	I
35		4,885.78	I
19		4,885.96	I
130		4,887.01	I
19		4,888.53	I
35		4,903.24	I
260		4,922.27	I
110		4,936.33	I
70		4,942.50	I
110		4,954.81	I
35		4,964.93	I
60		5,013.32	I
17		5,051.90	I
17		5,065.91	I
40		5,067.71	I
40		5,072.92	I
30		5,110.75	I
17		5,113.13	I
17		5,123.46	I
50		5,139.65	I
14		5,144.67	I
70		5,166.23	I
35		5,177.43	I
70		5,184.59	I
70		5,192.00	I
12		5,193.49	I
85		5,196.44	I
35		5,200.19	I
5,300		5,204.52	I
8,400		5,206.04	I
11,000		5,208.44	I
19		5,214.13	I
30		5,221.75	I
85		5,224.94	I
12		5,226.89	I
19		5,238.97	I
30		5,243.40	I
290		5,247.56	I
60		5,254.92	I
60		5,255.13	I
19		5,261.75	I
530		5,264.15	I
30		5,265.16	I
180		5,265.72	I
35		5,272.01	I
30		5,273.44	I
95	h	5,275.17	I
35	h	5,275.69	I
70	h	5,276.03	I
19		5,280.29	I
10		5,287.19	I
340		5,296.69	I
70	h	5,297.36	I
660		5,298.27	I
85		5,300.75	I
17		5,304.21	I
24		5,312.88	I
24		5,318.78	I
340	h	5,328.34	I
70	h	5,329.17	I
17	h	5,329.72	I
14		5,340.44	I
10		5,344.76	I
780		5,345.81	I
380		5,348.32	I
30		5,386.98	I
22		5,387.57	I
10		5,390.39	I
40		5,400.61	I
22		5,405.00	I
1,400		5,409.79	I
12		5,442.41	I
19		5,463.97	I
19		5,480.50	I
24		5,628.64	I
7		5,642.36	I
12	h	5,649.37	I
24		5,664.04	I
7	h	5,681.20	I
7	h	5,682.48	I
24		5,694.73	I
40		5,698.33	I
24		5,702.31	I
12		5,712.64	I
24		5,712.78	I
7		5,719.82	I
7		5,746.43	I
7		5,753.69	I
12	h	5,781.20	I
6	h	5,781.81	I
24	h	5,783.11	I
30	h	5,783.93	I
24	h	5,785.00	I
19	h	5,785.82	I
60	h	5,787.99	I
180	h	5,791.00	I
35		6,330.10	I
22		6,362.87	I
19		6,661.08	I
11		6,669.26	I
5	h	6,881.62	I
10	h	6,882.38	I
21	h	6,883.03	I
27	h	6,924.13	I
17	h	6,925.20	I
30	h	6,978.48	I
11	h	6,979.82	I
7		7,185.52	I
6	h	7,236.20	I
85		7,355.90	I
130		7,400.21	I
150		7,462.31	I
11	h	7,942.04	I
5	h	8,163.18	I
9		8,348.28	I
6		8,450.26	I
3		8,455.24	I
6		8,548.86	I
40		8,947.15	I
19		8,976.83	I

COBALT (CO)

Z = 27

Co I and II

Ref. 1, 125, 276 — C.R.C.

Intensity		Wavelength	
		Vacuum	
20		1,265.93	II
40		1,271.94	II
20		1,276.90	II
30		1,293.97	II
25		1,295.53	II
30		1,295.86	II
20		1,297.10	II
80		1,299.58	II
30		1,302.39	II
20		1,306.76	II
80		1,306.95	II
40		1,311.12	II
40		1,311.86	II
30		1,315.42	II
30		1,316.09	II
20		1,318.86	II
30		1,318.60	II
20		1,319.84	II
20		1,409.33	II
20		1,466.21	II
20		1,471.87	II
30		1,472.90	II
30		1,475.81	II
20		1,484.26	II
30		1,486.50	II
20		1,509.23	II
20		1,590.54	II
20		1,595.77	II
20		1,599.30	II
20		1,693.34	II
8		1,706.05	II
10		1,723.01	II
15	h	1,740.55	II
15		1,743.39	II
8		1,754.21	II
20	d	1,808.01	II
10		1,837.56	II
15		1,839.37	II
1,500		1,842.34	I
1,800		1,847.89	I
1,800		1,852.71	I
2,400		1,855.05	I
1,500		1,878.28	I
10		1,917.62	II
1,800		1,936.58	I
30		1,940.63	II
1,500		1,946.79	I
30		1,950.09	II
1,500		1,951.90	I
1,800		1,954.22	I
1,800		1,955.17	I
30		1,957.42	II
1,500		1,958.55	I
1,500		1,961.59	I
1,500	h	1,968.69	I
1,500	h	1,968.93	I
3,000		1,970.71	I
1,800	h	1,971.16	I
1,800	h	1,972.52	I
1,500		1,973.85	I
1,800		1,976.97	I
2,400	h	1,980.89	I
1,500		1,989.80	I
1,800		1,990.34	I
1,500	l	1,998.49	I
		Air	
20		2,000.79	II
1,500		2,002.32	I
900		2,008.04	I
50		2,011.51	II
1,200	h	2,014.58	I
900		2,016.17	I
50		2,022.35	II
40		2,025.76	II
50		2,027.04	II
900		2,031.96	I
30		2,036.58	II
1,500		2,039.95	I
1,200		2,041.11	I
20		2,049.17	II
40		2,058.82	II
40		2,063.78	II
50		2,065.54	II
1,500	h	2,077.76	I
900		2,085.67	I
900		2,087.55	I
900		2,089.35	I
900		2,093.40	I
900		2,094.86	I
900		2,095.77	I
1,200		2,097.51	I
1,500		2,104.73	I
1,500		2,106.80	I
900		2,108.98	I
30		2,111.44	II
900	s	2,117.68	I
20		2,128.79	II
900		2,137.78	I
900		2,138.97	I
900		2,163.03	I
20		2,164.44	II
30		2,173.33	II
1,100		2,174.60	I
20		2,181.99	II
30		2,187.01	II
20		2,190.68	II
20		2,192.50	II
200		2,193.60	II
20		2,200.40	II
40	p	2,202.95	II
200		2,256.73	II
150		2,260.00	II
200		2,283.52	II

1,000		2,286.15	II
200		2,291.98	II
300	d	2,293.38	II
300		2,301.40	II
800	d	2,307.85	II
2,600		2,309.02	I
500		2,311.60	II
500		2,314.05	II
300		2,314.96	II
200		2,317.06	II
2,400		2,323.14	I
300	p	2,324.31	II
200	d	2,326.11	II
500		2,326.47	II
1,400		2,335.99	I
1,600		2,338.67	I
200		2,347.39	II
1,600		2,352.85	I
200	d	2,353.41	II
2,000		2,353.42	I
500		2,363.80	II
400		2,378.62	II
1,400		2,380.48	I
200		2,381.76	II
300	p	2,383.45	II
1,400		2,384.86	I
200		2,386.36	II
500		2,388.92	II
200		2,397.38	II
1,100	d	2,402.06	I
200	p	2,404.16	II
5,300		2,407.25	I
5,300		2,411.62	I
1,600		2,412.76	I
4,800		2,414.46	I
4,800		2,415.30	I
300		2,417.65	II
4,100		2,424.93	I
3,300		2,432.21	I
2,900		2,436.66	I
2,400		2,439.05	I
200		2,442.63	II
200	d	2,446.03	II
200	p	2,447.69	II
200		2,450.00	II
200		2,464.20	II
200		2,486.44	II
200		2,498.82	II
570		2,504.52	I
500		2,506.46	II
360		2,506.88	I
200		2,511.16	II
860		2,517.87	I
500		2,519.82	II
4,300		2,521.36	I
200	h	2,524.65	II
300		2,524.97	II
500		2,528.62	II
2,900		2,528.97	I
200	p	2,530.09	II
720		2,530.13	I
860		2,532.18	I
200	d	2,533.82	II
2,900		2,535.96	I
860		2,536.49	I
300		2,541.94	II
1,700		2,544.25	I
200		2,546.74	II
340		2,548.34	I
310		2,553.37	I
310		2,555.07	I
300		2,559.41	II
200		2,560.03	II
960		2,562.15	I
500		2,564.04	II
1,100		2,567.35	I
960		2,574.35	I
800		2,580.32	II
300	d	2,582.22	II
500		2,587.22	II
500		2,587.52	II
200		2,588.91	II
100	p	2,605.71	II
100		2,612.50	II
100		2,614.36	II
100	p	2,628.77	II
100		2,632.26	II
100		2,636.07	II
310		2,646.42	I

770		2,648.64	I
100		2,653.72	II
100		2,663.53	II
200		2,666.73	II
100		2,675.85	II
100		2,684.42	II
100		2,702.02	II
200		2,706.62	II
200		2,707.35	II
190		2,715.99	I
100		2,727.78	II
80		2,734.54	II
190		2,745.10	I
100		2,748.17	II
100		2,753.22	II
190		2,764.19	I
100		2,766.70	II
100		2,774.97	II
100		2,791.00	II
100		2,793.73	II
150		2,815.56	I
80		2,835.63	II
80		2,847.35	II
80		2,871.22	II
190		2,886.44	I
100		2,918.38	II
100		2,930.24	II
100		2,954.73	II
690		2,987.16	I
690		2,989.59	I
20		3,008.86	II
60		3,022.59	II
30		3,035.13	II
3,100		3,044.00	I
1,700		3,061.82	I
20		3,352.79	II
80		3,387.70	II
1,100		3,388.17	I
2,200		3,395.38	I
11,000		3,405.12	I
4,500		3,409.18	I
6,700		3,412.34	I
2,200		3,412.63	I
30		3,415.77	II
2,700		3,417.16	I
50		3,423.84	II
2,500		3,431.58	I
4,500		3,433.04	I
1,600		3,442.93	I
8,800		3,443.64	I
50		3,446.39	II
4,100		3,449.17	I
2,100		3,449.44	I
21,000		3,453.50	I
1,000		3,455.23	I
5,100		3,462.80	I
5,100		3,465.80	I
8,000		3,474.02	I
1,900		3,483.41	I
4,800		3,489.40	I
2,400		3,495.69	I
40		3,497.33	II
50		3,501.72	II
9,600		3,502.28	I
7,000		3,506.32	I
50		3,507.77	II
2,900		3,509.84	I
1,400		3,510.43	I
4,800		3,512.64	I
3,800		3,513.48	I
10		3,514.23	II
30		3,517.50	II
4,800		3,518.35	I
1,300		3,520.08	I
2,700		3,521.57	I
3,800		3,523.43	I
60		3,523.51	II
6,400		3,526.85	I
2,700		3,529.03	I
7,300		3,529.81	I
1,900		3,533.36	I
40		3,535.92	II
50		3,545.03	II
20		3,555.93	II
1,100		3,560.89	I
80		3,561.07	II
20		3,566.98	II
8,800		3,569.38	I
50		3,574.95	II

1,600		3,574.96	I
60		3,575.32	II
2,500		3,575.36	I
60		3,577.96	II
1,000		3,585.16	I
6,700		3,587.19	I
1,900		3,594.87	I
1,600		3,602.08	I
100		3,621.21	II
1,000		3,627.81	I
80		3,643.61	II
10		3,656.75	II
60		3,681.35	II
5		3,695.32	II
5		3,714.73	II
1,100		3,745.50	I
20		3,754.69	II
1,400		3,842.05	I
6,900		3,845.47	I
5,500		3,873.12	I
2,800		3,873.96	I
7,900		3,894.08	I
20	h	3,911.40	II
1,500		3,935.97	I
80	h	3,963.10	II
40	h	3,976.74	II
10		3,983.02	II
6,000		3,995.31	I
970		3,997.91	I
350		4,020.90	I
10	h	4,036.14	II
20	h	4,037.37	II
4		4,040.02	II
370		4,045.39	I
5	h	4,050.23	II
10	h	4,052.40	II
20	h	4,062.73	II
2	h	4,064.50	II
350		4,066.37	I
5	h	4,074.34	II
830		4,092.39	I
1		4,096.57	II
550		4,110.54	I
2,800		4,118.77	I
4,400		4,121.32	I
3		4,130.88	II
3		4,145.13	II
3	d	4,160.67	II
1		4,181.13	II
90		4,190.71	I
1		4,208.61	II
30	s	4,244.25	II
8		4,272.33	II
20	h	4,288.25	II
3		4,328.86	II
2		4,384.26	II
2	h	4,396.94	II
3		4,413.91	II
90		4,469.56	I
10		4,482.50	II
2	h	4,489.12	II
4	h	4,497.44	II
10	d	4,500.54	II
0		4,516.65	II
690		4,530.96	I
2	h	4,533.22	II
0	h	4,537.95	II
90		4,549.66	I
1	h	4,559.29	II
140		4,565.59	I
1		4,569.26	II
190		4,581.60	I
5		4,616.30	II
120		4,629.38	I
25	h	4,660.66	II
85		4,663.41	I
110		4,792.86	I
10	d	4,831.16	II
100		4,840.27	I
150		4,867.88	I
80	h	4,964.18	II
10	h	4,970.05	II
10	h	4,990.47	II
20	h	4,995.98	II
35		5,146.74	I
50		5,212.71	I
50		5,230.22	I
45		5,235.21	I
50		5,247.93	I

26		5,266.30	I
45		5,266.49	I
26		5,268.52	I
45	h	5,280.65	I
26		5,301.06	I
50		5,342.71	I
26		5,343.39	I
50		5,352.05	I
26		5,353.48	I
35		5,369.58	I
45		5,483.34	I
17		5,530.77	I
17		5,647.22	I
17		5,991.88	I
17		6,082.44	I
17		6,282.63	I
45		6,450.24	I
21		6,455.00	I
15		6,563.42	I
15		6,632.44	I
14		6,814.94	I
14		6,872.40	I
21		7,052.89	I
45		7,084.99	I
8		7,417.38	I
8		7,712.68	I
7		7,908.71	I
9		7,987.38	I
13		8,007.27	I
9		8,093.96	I
9		8,372.84	I
4		8,575.35	I
3		8,819.15	I

Co III
Ref. 291 — C.R.C.

Intensity		Wavelength	
		Vacuum	
1,000		1,696.01	III
800		1,697.99	III
500		1,702.79	III
1,000		1,707.35	III
500		1,707.95	III
500		1,723.97	III
400		1,745.67	III
500		1,755.98	III
5,000		1,760.35	III
500		1,769.96	III
500		1,773.22	III
5,000		1,773.57	III
500		1,774.42	III
1,000		1,777.14	III
2,000		1,780.05	III
3,000		1,782.97	III
500		1,784.06	III
1,000		1,787.08	III
1,000		1,789.07	III
500		1,790.26	III
500		1,791.28	III
500		1,798.06	III
500		1,805.54	III
400		1,811.47	III
400		1,821.26	III
400		1,821.69	III
400		1,821.77	III
1,000		1,823.08	III
400		1,825.36	III
750		1,825.95	III
2,000		1,830.09	III
2,000		1,831.44	III
750		1,831.92	III
400		1,832.20	III
5,000		1,835.00	III
1,000		1,837.63	III
500		1,846.16	III
500		1,852.92	III
400		1,854.39	III
400		1,854.76	III
1,000		1,861.78	III
2,000		1,863.83	III
400		1,864.19	III
500		1,871.87	III
1,000		1,881.70	III
500		1,895.37	III
500		1,919.12	III
500		1,928.57	III
500		1,940.15	III

500	1,953.94	III
500	1,959.41	III
400	1,989.60	III

Air

100	2,001.09	III
200	2,011.62	III
200	2,013.88	III
100	2,031.81	III
200	2,053.11	III
100	2,056.21	III
10	2,062.17	III
10	2,079.74	III
15	2,088.58	III
10	2,090.50	III
10	2,097.64	III
10	2,134.15	III
10	2,452.16	III
20	2,811.75	III
10	2,888.31	III
10	2,933.27	III
10	2,978.01	III
20	2,991.89	III
25	3,010.92	III
10	3,116.68	III
2	3,151.40	III
2	3,180.64	III
20	3,232.11	III
2	3,249.24	III
20	3,259.68	III
2	3,269.23	III
10	3,287.68	III
15	3,305.38	III
10	3,451.25	III
2	3,526.24	III
1	3,634.21	III
1	3,636.31	III
2	3,667.52	III
2	3,677.23	III
3	3,680.74	III
1	3,762.50	III
15	3,782.27	III

Co IV

Ref. 236 — C.R.C.

Intensity	Wavelength	
	Vacuum	
81	606.79	IV
74	607.59	IV
55	608.24	IV
66	609.16	IV
70	609.21	IV
64	609.28	IV
43	610.04	IV
37	610.25	IV
24	610.79	IV

Co V

Ref. 100, 159 — C.R.C.

Intensity	Wavelength	
	Vacuum	
20	355.52	V
18	355.88	V
12	356.06	V
4	1,006.86	V
10	1,007.51	V
15	1,009.02	V
10	1,010.94	V
10	1,013.80	V
1	1,017.43	V
10	1,018.36	V
10	1,021.14	V
1	1,028.08	V
3	1,226.31	V
8	1,228.19	V
15	1,231.73	V
2	1,234.55	V
20	1,236.95	V
2	1,239.85	V
8	1,246.91	V
6	1,258.61	V
5	1,263.28	V
20	1,270.70	V
20	1,272.23	V
2	1,275.52	V
50	1,277.01	V
30	1,281.63	V
15	1,284.00	V
28	1,286.95	V
25	1,295.55	V
40	1,295.87	V
35	1,301.12	V
50	1,345.67	V
6	1,351.22	V
15	1,353.42	V
40	1,355.20	V
30	1,357.67	V
20	1,361.32	V
30	1,362.46	V
30	1,364.17	V
30	1,368.24	V
4	1,369.30	V
10	1,371.01	V
30	1,373.09	V
30	1,375.20	V
25	1,378.12	V
10	1,379.05	V
10	1,380.21	V
32	1,389.11	V
15	1,459.77	V
35	1,468.98	V
30	1,476.65	V
25	1,482.62	V
20	1,482.91	V
25	1,486.02	V
20	1,488.73	V

COPPER (CU)

Z = 29

Cu I and II

Ref. 273, 290 — V.K.

Intensity	Wavelength	
	Vacuum	
80	685.141	II
100	709.313	II
100	718.179	II
150	724.489	II
200	735.520	II
250	736.032	II
80	779.295	II
100	797.455	II
150	810.998	II
200	813.883	II
300	826.996	II
150	848.808	II
250	851.303	II
250	858.487	II
400	861.994	II
400	865.390	II
250	869.336	II
150	873.263	II
200	876.723	II
250	877.012	II
200	877.555	II
500	878.699	II
100	884.133	II
250	885.847	II
600	886.943	II
600	890.567	II
500	892.414	II
800	893.678	II
400	894.227	II
600	896.759	II
400	896.976	II
600	901.073	II
400	906.113	II
800	914.213	II
600	922.019	II
500	924.239	II
400	935.232	II
600	935.898	II
600	943.335	II
600	945.525	II
500	945.965	II
200	954.383	II
250	956.290	II
400	958.154	II
200	960.414	II
250	968.042	II
200	974.759	II
250	977.567	II
100	987.657	II
250	992.953	II
300	1,004.055	II
300	1,008.569	II
300	1,008.728	II
300	1,010.269	II
250	1,012.597	II
500	1,018.707	II
500	1,027.831	II
250	1,028.328	II
200	1,030.263	II
600	1,036.470	II
600	1,039.348	II
600	1,039.582	II
800	1,044.519	II
800	1,044.744	II
500	1,049.755	II
600	1,054.690	II
400	1,055.797	II
600	1,056.955	II
400	1,058.799	II
600	1,059.096	II
600	1,060.634	II
600	1,063.005	II
200	1,065.782	II
200	1,066.134	II
500	1,069.195	II
300	1,073.745	II
200	1,088.395	II
300	1,094.402	II
250	1,097.053	II
150	1,119.947	II
200	1,142.640	II
300	1,144.856	II
100	1,250.048	II
150	1,265.506	II
300	1,275.572	II
150	1,282.455	II
150	1,287.468	II
150	1,298.395	II
300	1,308.297	II
300	1,314.337	II
100	1,320.686	II
100	1,326.395	II
150	1,350.594	II
250	1,351.837	II
150	1,355.305	II
300	1,358.773	II
200	1,359.009	II
200	1,362.560	II
250	1,367.951	II
200	1,371.840	II
100	1,393.128	II
100	1,398.642	II
150	1,402.777	II
150	1,407.169	II
100	1,414.898	II
250	1,418.426	II
250	1,421.759	II
200	1,427.829	II
400	1,430.243	II
250	1,434.904	II
150	1,436.236	II
150	1,442.139	II
200	1,445.984	II
200	1,449.058	II
250	1,450.304	II
200	1,452.294	II
300	1,458.002	II
250	1,459.412	II
200	1,463.752	II
200	1,463.838	II
400	1,466.070	II
400	1,470.697	II
200	1,472.395	II
250	1,473.978	II
200	1,474.935	II
150	1,476.059	II
200	1,481.544	II
200	1,485.328	II
750	1,488.831	II
300	1,492.834	II
250	1,493.366	II
250	1,495.430	II
350	1,496.687	II
150	1,503.368	II
250	1,504.757	II
200	1,505.388	II
300	1,508.632	II
350	1,510.506	II
200	1,512.465	II
200	1,513.366	II
500	1,514.492	II
200	1,517.631	II
500	1,519.492	II
600	1,519.837	II
200	1,520.540	II
200	1,524.860	II
150	1,525.764	II
500	1,531.856	II
300	1,532.131	II
250	1,533.986	II
250	1,535.002	II
500	1,537.559	II
200	1,540.239	II
300	1,540.389	II
300	1,540.588	II
750	1,541.703	II
400	1,544.677	II
100	1,547.958	II
300	1,550.653	II
300	1,551.389	II
500	1,552.646	II
250	1,553.896	II
400	1,555.134	II
500	1,555.703	II
300	1,558.345	II
400	1,565.924	II
400	1,566.415	II
100	1,569.416	II
300	1,579.492	II
300	1,580.626	II
400	1,581.995	II
500	1,583.682	II
400	1,590.165	II
600	1,593.556	II
400	1,598.402	II
400	1,602.388	II
200	1,604.848	II
300	1,605.281	II
400	1,606.834	II
250	1,608.639	II
150	1,610.296	II
200	1,617.915	II
600	1,621.426	II
400	1,622.428	II
250	1,630.268	II
100	1,636.605	II
250	1,649.458	II
30	r 1,655.32	I
200	1,656.322	II
200	1,660.001	II
300	1,663.002	II
100	1,672.776	II
30	1,688.09	I
30	1,691.08	I
30	r 1,703.84	I
50	r 1,713.36	I
150	1,717.721	II
50	r 1,725.66	I
100	1,736.551	II
50	r 1,741.57	I
150	1,753.281	II
200	r 1,774.82	I
100	r 1,825.35	I
250	1,929.751	II
250	1,944.597	II
100	1,946.493	II
200	1,957.518	II
150	1,970.495	II
150	1,977.027	II
500	1,979.956	II
300	1,989.855	II
250	1,999.698	II

Air

270	2,035.854	II
250	2,037.127	II
350	2,043.802	II
300	2,054.980	II
100	2,078.663	II
110	2,098.398	II
320	2,104.797	II
300	2,112.100	II
320	2,117.310	II

350		2,122.980	II
350		2,126.044	II
420		2,134.341	II
900		2,135.981	II
400		2,148.984	II
150		2,161.320	II
1,300	r	2,165.09	I
250		2,174.982	II
1,600	r	2,178.94	I
700		2,179.410	II
1,700	r	2,181.72	I
700		2,189.630	II
900		2,192.268	II
400		2,195.683	II
1,700	r	2,199.58	I
1,300	r	2,199.75	I
100		2,200.509	II
200		2,209.806	II
750		2,210.268	II
1,600	r	2,214.58	I
250		2,215.106	II
1,000	r	2,215.65	I
750		2,218.108	II
2,100	r	2,225.70	I
150		2,226.780	II
1,600	r	2,227.78	I
350		2,228.868	II
2,500	r	2,230.08	I
1,100	r	2,238.45	I
900		2,242.618	II
2,300	r	2,244.26	I
1,000		2,247.002	II
1,300	r	2,260.53	I
2,200	r	2,263.08	I
150		2,263.786	II
200		2,276.258	II
100		2,286.645	II
2,500	r	2,293.84	I
170		2,294.368	II
1,000		2,303.12	I
150		2,369.890	II
2,500	r	2,392.63	I
120		2,403.337	II
1,500		2,406.66	I
1,000		2,441.64	I
100		2,485.792	II
2,000	r	2,492.15	I
150		2,506.273	II
120		2,526.593	II
300		2,544.805	II
100		2,571.756	II
150		2,590.529	II
200		2,600.270	II
2,500	r	2,618.37	I
200		2,666.291	II
750		2,689.210	II
700		2,700.962	II
650		2,703.184	II
700		2,713.508	II
650		2,718.778	II
300		2,721.677	II
120		2,737.342	II
270		2,745.271	II
2,500	r	2,766.37	I
800		2,769.669	II
200		2,791.795	II
170		2,799.528	II
100		2,810.804	II
1,250	r	2,824.37	I
350		2,837.368	II
100		2,857.748	II
600		2,877.100	II
270		2,884.196	II
2,500	r	2,961.16	I
100		2,986.335	II
2,000		2,997.36	I
2,000		3,010.84	I
2,500		3,036.10	I
2,500		3,063.41	I
1,400		3,073.80	I
1,500		3,093.99	I
1,250		3,099.93	I
2,000		3,108.60	I
1,400	h	3,126.11	I
1,500		3,194.10	I
1,400		3,208.23	I
1,500	h	3,243.16	I
10,000	r	3,247.54	I
10,000	r	3,273.96	I
1,400	h	3,282.72	I
400		3,290.418	II
1,500	h	3,290.54	I
110		3,300.881	II
250		3,301.229	II
2,500	h	3,307.95	I
200		3,316.276	II
1,500		3,337.84	I
150		3,338.648	II
200		3,365.648	II
450		3,370.454	II
300		3,374.952	II
200		3,380.712	II
100		3,384.945	II
1,250	h	3,483.76	I
1,250		3,524.23	I
2,000		3,530.38	I
1,400		3,599.13	I
1,400		3,602.03	I
1,000		3,686.555	II
150		3,786.270	II
170		3,797.849	II
100		3,818.879	II
140		3,826.921	II
160		3,864.137	II
280		3,884.131	II
150		3,892.924	II
170		3,903.177	II
140		3,920.654	II
120		3,933.268	II
120		3,987.024	II
150		3,993.302	II
140		4,003.476	II
1,250		4,022.63	I
100		4,032.647	II
600		4,043.484	II
500		4,043.751	II
2,000		4,062.64	I
120		4,068.106	II
500		4,131.363	II
200		4,143.017	II
300		4,153.623	II
500		4,161.140	II
370		4,164.284	II
400		4,171.851	II
500		4,179.512	II
500		4,211.866	II
320		4,230.449	II
200		4,255.635	II
950		4,275.11	I
300		4,279.962	II
500		4,292.470	II
400		4,365.373	II
100		4,444.831	II
400		4,506.002	II
150		4,516.049	II
150		4,541.032	II
500		4,555.920	II
100		4,596.906	II
120		4,649.271	II
2,000		4,651.12	I
120		4,661.363	II
320		4,671.702	II
300		4,673.577	II
450		4,681.994	II
100		4,758.433	II
400		4,812.948	II
120		4,851.262	II
300		4,854.988	II
100		4,873.304	II
150		4,901.428	II
1,000		4,909.734	II
500		4,918.376	II
200		4,926.424	II
900		4,931.698	II
120		4,943.026	II
700		4,953.724	II
500		4,985.506	II
400		5,006.801	II
350		5,009.851	II
400		5,012.620	II
350		5,021.279	II
200		5,039.016	II
300		5,047.348	II
900		5,051.793	II
400		5,058.910	II
500		5,065.459	II
450		5,067.094	II
350		5,072.302	II
450		5,088.277	II
420		5,093.816	II
350		5,100.067	II
1,500		5,105.54	I
250		5,124.476	II
2,000		5,153.24	I
100		5,158.093	II
100		5,183.367	II
2,500		5,218.20	I
100		5,269.991	II
100		5,276.525	II
1,650		5,292.52	I
100		5,368.383	II
1,500		5,700.24	I
1,500		5,782.13	I
150		5,805.989	II
100		5,833.515	II
200		5,897.971	II
120		5,937.577	II
400		5,941.196	II
100		5,993.260	II
650		6,000.120	II
100		6,023.264	II
250		6,072.218	II
150		6,080.343	II
150		6,099.990	II
160		6,107.412	II
300		6,114.493	II
600		6,150.384	II
750		6,154.222	II
500		6,172.037	II
550		6,186.884	II
400		6,188.676	II
300		6,198.092	II
470		6,204.261	II
450		6,208.457	II
750		6,216.939	II
700		6,219.844	II
500		6,261.848	II
1,000		6,273.349	II
350		6,288.696	II
900		6,301.009	II
550		6,305.972	II
400		6,312.492	II
120		6,326.466	II
400		6,373.268	II
750		6,377.840	II
400		6,403.384	II
850		6,423.884	II
200		6,442.965	II
750		6,448.559	II
170		6,466.246	II
950		6,470.168	II
750		6,481.437	II
400		6,484.421	II
220		6,517.317	II
400		6,530.083	II
120		6,551.286	II
200		6,577.080	II
750		6,624.292	II
800		6,641.396	II
450		6,660.962	II
100		6,770.362	II
300		6,806.216	II
400		6,809.647	II
320		6,823.202	II
250		6,844.157	II
320		6,868.791	II
270		6,872.231	II
270		6,879.404	II
220		6,937.553	II
150		6,952.871	II
150		6,977.572	II
200		7,022.860	II
300		7,194.896	II
400		7,326.008	II
300		7,331.694	II
250		7,382.277	II
1,000		7,404.354	II
270		7,434.156	II
500		7,562.015	II
700		7,652.333	II
1,000		7,664.648	II
150		7,681.788	II
450		7,744.097	II
800		7,778.738	II
750		7,805.184	II
1,500		7,807.659	II
1,000		7,825.654	II
350		7,860.577	II
300		7,890.567	II
700		7,902.553	II
1,500		7,933.13	I
400		7,944.438	II
400		7,972.033	II
1,200		7,988.163	II
2,000		8,092.63	I
500		8,277.560	II
800		8,283.160	II
250		8,503.396	II
750		8,511.061	II
200		8,609.134	II
500		9,813.213	II
250		9,827.978	II
200		9,830.798	II
600		9,861.280	II
600		9,864.137	II
200		9,883.969	II
550		9,916.419	II
500		9,917.954	II
550		9,925.594	II
450		9,938.998	II
500		9,960.354	II
450		10,006.588	II
550		10,022.969	II
550		10,038.093	II
650		10,054.938	II
450		10,080.354	II

Cu III

Ref. 295 — V.K.

Intensity		Wavelength	
		Vacuum	
75		542.90	I
200		615.67	I
150		616.03	I
150		687.98	I
150		715.53	I
125		730.38	I
250		788.07	I
250		788.46	I
250		791.36	I
150		801.14	I
100		829.34	I
40		1,048.88	I
50		1,186.80	I
50		1,200.96	I
300		1,219.30	I
200		1,244.38	I
100		1,279.14	I
200		1,312.39	I
300		1,332.97	I
200		1,339.48	I
150		1,363.08	I
300	r	1,376.79	I
200	r	1,337.49	I
150		1,423.48	I
300	r	1,481.23	I
200		1,543.46	I
500	r	1,593.75	I
1,000	r	1,642.21	I
300		1,679.14	I
400		1,702.10	I
500		1,722.37	I
600		1,741.37	I
200		1,768.86	I
200		1,840.91	I
100		1,971.95	I
		Air	
200		2,013.22	I
150		2,157.28	I
100		2,299.47	I
500		2,368.17	I
400		2,391.74	I
800		2,405.50	I
700		2,412.34	I
2,000		2,444.44	I
500		2,468.41	I
1,000		2,482.36	I
700		2,486.46	I
500		2,508.49	I
500		2,522.38	I
500		2,538.66	I
400		2,566.37	I

400		2,573.33	I
500		2,609.32	I
200		2,643.92	I
200		2,696.38	I
20		2,751.33	I
100		2,812.94	I
100		2,978.87	I
75		3,548.87	I
100		3,639.42	I
500		3,702.92	I
800		3,744.70	I
400		3,748.27	I
600		3,752.06	I
1,000		3,776.97	I
800		3,790.80	I
600		3,804.13	I
600		3,809.18	I
300		3,881.68	I
150		3,953.81	I
100		4,090.49	I
200		4,283.40	I
500		4,351.97	I
1,000		4,352.80	I
500		4,355.24	I
500		4,370.84	I
500		4,371.40	I
500		4,373.43	I
1,000		4,377.11	I
200		4,386.42	I
150		4,927.41	I
400		5,094.28	I
200		5,168.97	I
400		5,208.34	I
600		5,219.21	I
200		5,268.59	I
400		5,317.78	I
300		5,369.79	I
350		5,418.48	I
250	d	5,494.94	I
50		5,573.94	I
100		5,609.00	I
75		5,702.12	I
100		5,768.56	I
100		5,850.72	I
200		5,965.25	I
30		6,100.87	I
50		6,369.27	I
20		6,512.54	I
20		6,644.13	I
50		6,793.20	I

Cu IV

Ref. 199 — V.K.

Intensity		Wavelength	
		Vacuum	
30		360.86	I
20		374.40	I
30		405.24	I
80		406.45	I
40		413.45	I
70		443.68	I
80		451.16	I
80		463.72	I
80		484.53	I
90		467.00	I
90		504.60	I
70		509.38	I
40		519.51	I
30		540.65	I
60		550.92	I
20		584.85	I
60		1,056.13	I
30		1,074.72	I
50		1,091.65	I
30		1,105.50	I
25		1,119.43	I
40	p	1,152.18	I
60		1,227.44	I
70		1,228.87	I
70		1,258.69	I
90		1,274.84	I
90		1,293.46	I
90		1,309.41	I
70		1,321.17	I
100	d	1,340.08	I
100		1,350.42	I
100		1,362.05	I
100		1,372.14	I
100		1,377.82	I
100		1,388.80	I
90		1,405.49	I
90		1,415.27	I
90		1,434.34	I
90		1,449.69	I
80		1,466.18	I
70		1,482.77	I
80		1,499.81	I
90		1,515.28	I
90		1,535.12	I
80		1,551.12	I
90		1,567.35	I
80		1,583.47	I
70		1,595.12	I
80	p	1,608.14	I
90		1,639.75	I
20		1,650.16	I
70		1,704.37	I
30		1,797.99	I
70		1,817.56	I
70		1,819.23	I
60		1,837.04	I
80		1,849.62	I
30		1,867.24	I
30		1,918.71	I
40		1,966.31	I

Cu V

Ref. 324 — V.K.

Intensity		Wavelength	
		Vacuum	
9	h	258.95	V
49		271.33	V
49		283.97	V
22		293.41	V
56		299.64	V
65		305.83	V
51		312.51	V
66		321.05	V
74		326.57	V
82		333.56	V
81		339.88	V
81		346.00	V
86		355.41	V
77		363.96	V
65		370.63	V
74		377.76	V
70		387.40	V
51		396.06	V
25		406.94	V
13		1,097.10	V
42		1,106.24	V
77		1,113.22	V
67		1,121.20	V
76		1,128.80	V
59		1,133.86	V
63		1,142.38	V
54		1,149.06	V
72		1,157.54	V
64		1,167.35	V
77		1,176.53	V
84		1,183.63	V
76		1,192.54	V
83		1,201.22	V
77		1,204.90	V
71		1,214.36	V
70		1,221.34	V
76		1,230.11	V
80		1,239.73	V
79		1,246.99	V
65		1,253.07	V
70		1,260.24	V
77		1,269.35	V
78		1,278.20	V
68		1,286.13	V
67		1,292.08	V
73		1,299.22	V
65		1,309.72	V
55		1,318.89	V
65		1,323.28	V
64		1,329.22	V

CURIUM (CM)

Z = 96

Cm I and II

Ref. 51, 332 — J.G.C.

Intensity		Wavelength	
		Vacuum	
		2,462.76	II
		2,617.17	II
		2,636.28	II
		2,651.17	II
		2,653.80	II
		2,725.68	II
		2,736.89	II
		2,748.04	II
		2,784.83	II
		2,811.62	II
		2,824.20	II
		2,833.58	II
		2,899.90	II
		2,912.97	II
		2,928.92	II
		2,996.18	II
		2,999.39	I
	b	3,014.87	II
		3,044.85	II
		3,109.69	I
		3,116.41	I
		3,137.16	I
		3,147.33	I
		3,155.10	I
		3,158.60	I
		3,169.98	II
		3,177.55	I
		3,179.10	I
		3,186.41	I
		3,188.11	I
		3,207.12	II
		3,207.71	I
		3,209.89	II
		3,209.94	I
		3,210.05	II
		3,220.76	II
		3,224.23	I
		3,225.11	I
		3,226.41	II
		3,230.28	I
		3,230.35	II
		3,236.74	I
		3,238.55	II
		3,242.66	II
		3,246.25	I
		3,252.68	I
		3,265.81	I
		3,280.45	I
		3,296.71	II
		3,304.85	I
		3,317.14	I
		3,374.70	I
		3,452.92	I
		3,458.34	I
		3,510.28	I
		3,522.36	I
		3,524.94	I
		3,542.06	I
		3,547.02	I
		3,547.92	I
		3,561.44	I
		3,572.95	II
		3,600.62	I
		3,639.94	I
		3,664.34	I
		3,709.43	I
		3,729.00	I
		3,732.35	I
		3,747.86	I
		3,763.05	I
		3,775.75	I
		3,816.30	I
		3,825.14	I
		3,833.32	I
		3,837.59	I
		3,842.00	I
		3,849.92	I
		3,854.11	I
		3,900.25	I
		3,904.06	II
		3,908.24	II
		3,936.67	I
		3,942.03	I
		3,944.15	I
		3,948.68	I
		3,953.36	I
		3,964.83	I
		3,995.10	I
		4,016.17	I
		4,031.76	I
		4,048.29	I
		4,049.65	I
		4,113.29	I
		4,129.71	I
		4,207.66	II
		4,211.62	I
		4,266.45	I
		4,293.00	I
		4,330.82	I
		4,345.69	I
		4,447.77	I
		4,459.16	I
		4,608.40	I
		5,846.07	I
		5,952.41	I
		6,058.90	I
		6,243.35	I
		6,376.71	I
		6,510.16	I
		6,554.41	I
		6,640.17	I
		6,663.25	I
		6,686.87	I
		6,706.85	I
		6,726.68	I
		6,793.15	I
		7,162.69	I
		7,577.80	I
		7,673.79	I
		7,720.47	I
		8,392.37	I
		9,293.25	I
		9,567.08	I
		9,657.12	I
		10,310.83	I
		10,351.73	I
		10,424.49	I
		10,508.11	I
		10,542.98	I
		10,792.25	I
		10,897.45	I
		11,507.45	I
		11,707.73	I
		11,780.95	I
		11,834.28	I
		12,017.85	I
		12,394.16	I
		12,454.98	I
		12,464.99	I
		13,004.56	I
		13,258.18	I
		13,289.84	I
		13,344.62	I
		13,480.54	I
		13,590.01	I
		13,644.77	I
		13,789.52	I
		13,840.18	I
		13,908.46	I
		13,964.14	I
		14,235.27	I
		14,334.52	I
		14,563.41	I
		14,580.23	I
		15,018.13	I
		15,222.27	I
		15,642.59	I
		15,757.23	I
		15,793.31	I
		16,008.41	I
		17,148.22	I
		17,453.18	I
		17,619.28	I
		18,069.02	I
		19,572.62	I
		19,975.98	I
		20,526.32	I
		20,853.49	I
		20,911.52	I
		20,968.11	I
		21,241.06	I
		21,393.23	I

DYSPROSIUM (DY)

Z = 66

Dy I and II

Ref. 1 — C.H.C.

Intensity	Wavelength	
	Vacuum	
260	2,356.91	II
65	2,381.95	
130	2,387.36	II
150	2,392.15	
180	2,402.29	II
240	2,410.01	II
150	2,422.75	II
260	2,439.84	II
90	2,455.15	II
110	2,459.99	II
90	2,471.40	II

110		2,480.93	II
170		2,490.61	II
90		2,510.31	II
170		2,513.55	II
170		2,517.61	II
130		2,543.81	II
90		2,545.12	II
150		2,552.29	II
180		2,557.94	II
90		2,560.21	II
90		2,566.25	II
220		2,585.30	I
90		2,591.56	II
75		2,592.54	II
130		2,600.16	II
130		2,600.76	II
75		2,608.69	II
370		2,623.69	I
440		2,634.80	II
110		2,642.15	I
110		2,645.35	II
110		2,667.94	I
55		2,676.84	II
50		2,677.34	II
85		2,689.31	II
85		2,692.83	II
55		2,709.01	II
55		2,727.17	II
85		2,729.50	II
55		2,735.79	I
40		2,739.30	II
85		2,740.70	II
220		2,755.75	II
55		2,757.08	II
70		2,766.50	II
70		2,772.42	II
110		2,772.61	II
40		2,779.58	II
55		2,791.44	II
120		2,800.33	II
110		2,800.53	II
110		2,801.41	II
300		2,816.39	II
140		2,825.42	II
140		2,862.70	I
190		2,877.88	II
110		2,884.28	II
120		2,885.53	I
120		2,890.74	II
120		2,900.82	II
110		2,904.62	II
190		2,906.39	II
390		2,913.95	II
110		2,934.31	II
250		2,934.52	II
110		2,941.05	II
140		2,944.56	II
150		2,947.06	II
150		2,947.21	II
250		2,948.31	II
170		2,950.33	II
110		2,952.12	II
140		2,953.70	II
220		2,964.60	I
110		2,977.42	II
110		2,985.97	II
220		3,015.68	II
390		3,026.16	II
210		3,029.81	II
610		3,038.28	II
280		3,043.13	II
210		3,047.56	II
280		3,060.64	II
390		3,062.62	II
220		3,066.99	II
330		3,071.91	II
280		3,073.54	II
220		3,078.68	II
280		3,101.93	II
220		3,103.24	II
410		3,109.76	II
330		3,128.41	II
830		3,135.38	II
360		3,140.64	II
500		3,141.14	II
220		3,143.83	II
250		3,146.16	II
1,200		3,156.52	II
670		3,162.83	II
1,000		3,169.99	II
400		3,177.89	II
220		3,178.37	II
200		3,184.79	II
330		3,186.38	II
240		3,187.68	II
330		3,193.30	II
240		3,206.40	II
220		3,207.12	II
290		3,208.85	II
470		3,215.19	II
830		3,216.63	II
240		3,221.49	II
290		3,223.28	II
240		3,225.08	II
330		3,225.95	II
490		3,235.89	II
290		3,236.69	II
490		3,245.12	II
200		3,248.36	II
1,200		3,251.27	II
200		3,252.19	II
290		3,256.26	II
200		3,266.21	II
240		3,269.11	II
200		3,272.73	II
890		3,280.09	II
490		3,282.77	II
200		3,287.94	II
200		3,293.88	II
200		3,296.30	II
200		3,305.40	II
200		3,305.51	II
240		3,306.19	II
440		3,308.79	II
1,100		3,308.88	II
510		3,312.72	II
780		3,316.32	II
240		3,317.12	II
1,000		3,319.88	II
270		3,326.19	II
780		3,341.00	II
270		3,341.88	II
200		3,347.83	II
270		3,352.69	II
510		3,353.58	II
240		3,359.46	II
510		3,368.11	II
5,300		3,385.02	II
210		3,386.57	II
610		3,388.85	II
210		3,391.96	II
3,800		3,393.57	II
1,300		3,396.16	II
380		3,407.16	II
5,300		3,407.80	II
420		3,408.14	II
1,300		3,413.78	II
530		3,414.82	II
780		3,419.63	II
530		3,425.06	II
420		3,429.44	II
1,900		3,434.37	II
330		3,438.94	II
560		3,440.93	II
1,300		3,441.45	II
3,800		3,445.57	II
830		3,446.99	II
440		3,449.89	II
2,700		3,454.32	II
440		3,454.51	II
1,300		3,456.56	II
4,400		3,460.97	II
720		3,468.43	II
560		3,471.14	II
560	d	3,471.53	II
380		3,473.70	II
1,300		3,477.07	II
4,400		3,494.49	II
560		3,496.34	II
400		3,497.81	II
830		3,498.71	II
400		3,501.50	II
830		3,504.53	II
830		3,505.45	II
1,300		3,506.81	II
560		3,517.26	II
4,400		3,523.98	II
22,000		3,531.70	II
4,400		3,534.96	II
5,500		3,536.02	II
4,400		3,538.52	II
400		3,539.37	II
1,700		3,542.33	II
400		3,544.20	II
400		3,544.35	II
1,400		3,546.83	II
330		3,548.19	II
4,400		3,550.22	II
2,200		3,551.62	II
440	h	3,558.23	II
440		3,559.30	II
2,200		3,563.15	II
560		3,563.69	II
780		3,573.83	II
1,400		3,574.15	II
4,400		3,576.24	II
1,700		3,576.87	II
830		3,577.98	II
440		3,580.04	II
400		3,584.42	II
3,300		3,585.06	II
1,400		3,585.78	II
560		3,586.11	II
360		3,590.07	II
1,100		3,591.41	II
560		3,591.81	II
560		3,592.11	II
1,800		3,595.04	II
400		3,596.06	II
560		3,600.38	II
360		3,602.82	II
1,800		3,606.12	II
440		3,618.51	II
560		3,620.16	II
470		3,624.27	II
1,100		3,629.42	II
4,000		3,630.24	II
440		3,632.78	II
400		3,635.27	II
360		3,637.28	II
1,100		3,640.25	II
400		3,643.92	II
11,000		3,645.40	II
360		3,645.86	II
1,000		3,648.78	II
700		3,664.62	II
400		3,666.84	I
990		3,672.30	II
420		3,672.70	II
400		3,673.14	II
1,400		3,674.08	II
2,200		3,676.59	II
640		3,678.51	I
820		3,684.85	I
1,300		3,685.78	I
4,700		3,694.81	II
370		3,697.31	II
990		3,698.21	II
540		3,701.63	II
330		3,707.40	II
440		3,707.57	II
440		3,708.22	II
420		3,710.07	II
330		3,711.66	II
1,600		3,724.45	II
300		3,728.00	I
930		3,739.34	I
1,200		3,747.82	II
1,400		3,753.51	II
1,400		3,753.75	II
1,200		3,757.05	I
4,700		3,757.37	II
640		3,767.63	I
330		3,771.11	I
640		3,773.05	I
370		3,774.71	I
420		3,781.47	I
330		3,785.41	II
3,300		3,786.18	II
1,600		3,788.44	II
700		3,791.87	II
510		3,804.14	II
580		3,806.27	II
470		3,812.27	I
470		3,813.67	II
1,400		3,816.76	II
700		3,825.68	II
2,300		3,836.50	II
370		3,840.89	I
1,400		3,841.31	II
330		3,842.00	II
330		3,844.36	I
420		3,846.34	II
420		3,847.02	I
330		3,849.39	II
1,200		3,853.03	II
420		3,858.40	I
370		3,866.58	II
560		3,868.45	II
1,600		3,868.81	I
300		3,869.42	II
820		3,869.86	II
7,000		3,872.11	II
1,200		3,873.99	II
470		3,879.11	II
300		3,881.99	II
5,800		3,898.53	II
540		3,914.87	II
540		3,915.59	II
540	d	3,917.29	I
320		3,923.38	II
420		3,927.86	I
540		3,930.14	I
2,100		3,931.52	II
320		3,932.22	II
370		3,933.00	II
320		3,934.21	II
420		3,936.70	I
540		3,942.53	II
10,000		3,944.68	II
420		3,946.93	II
540		3,950.39	II
420		3,954.55	II
800		3,957.79	II
370		3,962.59	I
320		3,967.51	I
14,000		3,968.39	II
2,700		3,978.57	II
1,400		3,981.92	II
1,600		3,983.65	II
800		3,984.21	II
540		3,991.32	II
1,600		3,996.69	II
8,000		4,000.45	II
420		4,005.84	I
320		4,006.07	I
540		4,011.29	II
540		4,013.82	I
540		4,014.70	II
370		4,023.71	I
420		4,027.78	II
520	d	4,028.32	II
520		4,032.47	II
420		4,033.65	II
420		4,036.32	II
320		4,041.98	II
12,000		4,045.97	I
1,600		4,050.56	II
520		4,055.14	II
2,500		4,073.12	II
7,400		4,077.96	II
370		4,085.34	I
390		4,096.10	I
3,900		4,103.30	II
860		4,103.87	I
1,500		4,111.34	II
490		4,124.63	II
390		4,128.24	II
350		4,129.12	I
990		4,129.42	II
350		4,130.35	I
390		4,133.85	I
470		4,141.50	II
1,200		4,143.10	II
990		4,146.06	I
5,700		4,167.97	I
370		4,171.93	I
930		4,183.72	I
12,000		4,186.82	I
320		4,190.94	I
2,200		4,191.64	I
6,800		4,194.84	I
320		4,195.19	II
800		4,198.02	I
680		4,201.30	I

680		4,202.24	I
230		4,205.06	I
370		4,206.54	II
440		4,211.24	I
16,000		4,211.72	I
1,800		4,213.18	I
3,700		4,215.16	I
4,400		4,218.09	I
4,400		4,221.11	I
540		4,222.21	I
2,700		4,225.16	I
680		4,232.02	I
680		4,239.85	I
440		4,245.91	I
440		4,256.33	II
250		4,276.69	I
370	d	4,294.93	II
		4,295.04	I
1,000		4,308.63	II
320		4,325.86	I
200		4,358.44	II
320		4,374.24	II
320		4,374.76	II
540		4,409.38	II
150		4,444.58	I
740		4,449.70	II
110		4,455.60	II
250		4,468.14	II
100	d	4,527.58	I
		4,527.76	II
100		4,541.66	II
140		4,565.09	I
420		4,577.78	I
2,100		4,589.36	I
990		4,612.26	I
50		4,613.83	I
50		4,614.82	I
60		4,617.26	II
140		4,620.03	II
50		4,662.72	I
110		4,664.66	II
85		4,673.60	II
50		4,682.03	II
50		4,689.75	II
95		4,698.68	II
85		4,721.22	I
70		4,727.13	II
170		4,731.84	II
40		4,745.73	II
60		4,754.99	II
50		4,760.04	II
60		4,771.94	I
50		4,774.80	I
120	h	4,775.79	I
75		4,786.92	II
95		4,791.29	I
29		4,800.64	I
50		4,807.94	I
40		4,810.28	I
50		4,812.80	I
75		4,819.04	I
85		4,824.96	I
75		4,828.88	I
50		4,829.68	II
70		4,832.38	I
35		4,833.75	II
75		4,841.75	I
40		4,856.24	II
40		4,868.05	II
40		4,875.93	I
85		4,880.16	I
40		4,884.55	I
95		4,888.08	I
40		4,889.33	II
75		4,890.10	II
50		4,893.68	I
24		4,899.24	I
55		4,916.41	I
50		4,922.22	II
65		4,923.16	II
480		4,957.34	II
24		4,959.59	I
28		4,973.57	I
40		4,985.52	I
50		5,003.87	I
55		5,004.28	II
24		5,010.60	I
24		5,017.98	II
70		5,022.12	I
30		5,024.03	I
24		5,024.54	I
40		5,027.87	I
50		5,033.00	I
160		5,042.63	I
24		5,047.25	I
50	h	5,050.21	I
30		5,053.35	I
24		5,055.46	I
95		5,070.68	I
120		5,077.67	I
80		5,090.38	II
80		5,110.32	I
130	h	5,120.04	I
30		5,135.02	I
190		5,139.60	II
40		5,161.03	II
40		5,164.12	II
50		5,165.34	II
110		5,169.69	II
20		5,172.90	II
80		5,185.30	II
40		5,188.45	II
290		5,192.86	II
95		5,197.66	II
50		5,246.94	II
70		5,259.88	II
130		5,260.56	I
55	b	5,263.3	DYO
65		5,267.11	I
50		5,272.25	II
50		5,275.29	II
50		5,279.70	II
55		5,282.07	I
28		5,284.99	II
40		5,297.82	II
160		5,301.58	I
40		5,309.02	II
50		5,324.69	II
24		5,337.43	II
65		5,340.30	I
30		5,352.11	I
30		5,368.20	II
20		5,385.63	II
85		5,389.58	II
40		5,395.57	I
20	h	5,398.26	DYO
24		5,399.93	II
50		5,404.19	II
80		5,419.13	I
70		5,423.32	I
30		5,424.27	I
40		5,426.70	II
30		5,443.34	II
95		5,451.11	I
30		5,455.47	II
24		5,469.10	II
28		5,496.83	II
24		5,502.79	I
28		5,506.52	I
24		5,515.41	II
30		5,528.01	I
65		5,547.27	I
40	d	5,600.65	II
24		5,605.53	I
30		5,613.23	I
20		5,627.49	I
100		5,639.50	I
55	h	5,645.99	I
80		5,652.01	I
24		5,685.58	I
28	h	5,693.67	DYO
24	h	5,694.10	DYO
28	b	5,694.54	DYO
28		5,698.72	II
24		5,702.91	I
70	h	5,718.46	I
28	h	5,725.84	DYO
55	h	5,728.64	DYO
24	h	5,738.73	DYO
50		5,740.20	I
55		5,745.53	I
24		5,750.48	I
24		5,758.79	I
80	h	5,832.01	DYO
55	h	5,833.85	DYO
40	h	5,834.86	DYO
28	h	5,844.41	DYO
24		5,845.65	DYO
40	h	5,848.05	DYO
40		5,855.56	DYO
55	h	5,868.11	II
40		5,915.16	II
20		5,924.56	II
70		5,945.80	I
50	l	5,964.46	I
120		5,974.49	I
24		5,984.86	I
140		5,988.56	I
24	h	6,005.75	DYO
24	h	6,006.54	DYO
24	h	6,006.97	DYO
30		6,008.94	I
65		6,010.82	I
24		6,017.26	I
24		6,030.98	I
24	c	6,042.49	DYO
24		6,058.18	I
30		6,085.06	I
140		6,088.26	I
24		6,127.15	I
24		6,133.64	I
24		6,158.28	I
100		6,168.43	I
20		6,196.23	II
270		6,259.09	I
30		6,260.36	I
14		6,343.32	I
40		6,386.80	I
24		6,396.60	II
50		6,421.92	I
13	h	6,436.55	I
8		6,460.83	I
10		6,468.58	II
11		6,474.91	I
20		6,483.59	II
28		6,486.59	I
20		6,558.02	I
160		6,579.37	I
14		6,594.14	II
15		6,643.37	I
22		6,658.36	I
29		6,661.64	I
75		6,667.86	I
10		6,700.64	II
29		6,747.93	I
10		6,757.62	I
45		6,765.89	I
12		6,818.20	I
180		6,835.42	I
80		6,852.96	I
22		6,856.46	I
22		6,888.83	I
15		6,897.97	II
65		6,899.32	II
22		6,906.53	II
15		6,929.55	I
29		6,950.28	II
11		6,951.42	I
40		6,958.08	I
13	h	6,982.44	I
13		6,991.30	I
45		6,998.10	I
20		7,017.42	I
35		7,055.95	II
24		7,075.14	II
17		7,109.26	II
11		7,120.81	II
13		7,175.11	II
11		7,213.27	I
17	h	7,230.04	I
13		7,250.01	I
17		7,345.13	II
11		7,370.23	II
20		7,376.04	II
11		7,407.59	I
24		7,412.37	I
55		7,426.86	II
20		7,457.05	II
17		7,516.61	II
55		7,543.73	I
17	h	7,553.00	I
27		7,559.78	I
40		7,562.96	II
20	h	7,577.46	II
27	h	7,591.30	I
13	h	7,611.55	I
11	h	7,617.70	I
35	h	7,641.09	I
17		7,645.86	I
13		7,646.64	I
80		7,662.36	I
11		7,666.78	II
35		7,715.33	I
45		7,729.76	II
20		7,751.62	II
35		7,812.06	II
27		7,909.38	I
11		7,968.63	II
12		7,982.85	II
13		8,147.29	I
27		8,198.77	II
100		8,201.57	II
11		8,218.62	II
20		8,265.53	II
35		8,326.10	I
35		8,392.01	II
12		8,405.85	II
20		8,416.64	II
24		8,438.58	II
11		8,630.12	I
27		8,655.94	II
17		8,657.68	II
17		8,678.49	II
11		8,696.83	II
11	h	8,715.95	II
20		8,750.40	II
12		8,780.83	I
45		8,791.39	II
13		8,833.08	II
24		8,850.37	II

EINSTEINIUM (ES)

Z = 99

Es I and II
Ref. 333 — J.G.C.

Intensity		Wavelength	
		Air	
300	l	2,694.32	II
100	l	2,703.84	
10,000	s	2,708.66	II
100	s	2,716.02	II
1,000	s	2,724.57	II
1,000	l	2,765.76	
3,000	l	2,787.10	II
1,000	l	2,796.11	II
3,000	l	2,815.15	
100	s	2,885.84	II
100	s	2,886.44	
3,000		2,907.03	
100		3,003.28	
1,000	l	3,065.40	
3,000		3,135.25	I
1,000	l	3,154.27	II
10,000		3,413.17	
300	l	3,423.12	I
100	l	3,424.28	
10,000	s	3,428.48	I
300		3,437.31	
100		3,437.34	
3,000	s	3,445.25	
300	l	3,446.93	
3,000		3,452.36	
100		3,453.16	
300	s	3,470.77	
3,000	s	3,484.59	I
300		3,494.30	
10,000	s	3,498.11	I
10,000	l	3,514.33	I
10,000	s	3,521.38	
10,000	s	3,523.49	I
300	l	3,528.58	
10,000	s	3,536.01	
10,000	s	3,547.75	II
300		3,549.97	
10,000	l	3,555.34	I
100	s	3,555.53	
3,000	s	3,556.65	
3,000	s	3,560.92	
1,000	s	3,575.68	
100	s	3,578.56	
300	s	3,579.38	

1,000	s	3,582.95	
3,000		3,590.28	
1,000		3,595.47	
10,000	s	3,602.43	II
300		3,605.58	
1,000	s	3,606.75	
100	l	3,624.52	
3,000	l	3,631.09	
10,000	s	3,632.87	
100	l	3,634.41	
1,000	l	3,651.94	
10,000	l	3,670.01	II
100	l	3,672.32	
100	s	3,713.56	
1,000	l	3,720.56	
300		3,722.32	
10,000	l	3,728.55	II
100	h	3,737.47	
300	s	3,776.27	
10,000	l	3,792.99	
10,000	l	3,801.49	
300	s	3,929.10	
3,000		3,930.77	
100		3,957.19	
100		3,995.35	
300		4,077.71	
10,000	s	4,082.24	
3,000	l	4,107.59	
1,000	l	4,176.94	
100	s	4,496.25	
100	l	4,631.66	
300	l	4,650.86	
1,000		4,789.93	
300	h	4,802.17	
1,000	h	4,802.21	
3,000	h	4,958.29	
10,000	s	5,052.08	
1,000	s	5,102.93	
100	s	5,155.82	
10,000	s	5,161.74	I
10,000	l	5,204.40	I
3000	l	5,615.51	I
100	s	6,539.71	

ERBIUM (ER)

Z = 68

Er I and II

Ref. 1 — C.H.C.

Wavelength			Intensity
		Air	
110		2,358.51	II
100		2,386.58	II
120		2,387.17	II
110		2,396.38	III
140		2,446.39	II
100		2,537.02	II
110		2,547.28	II
290		2,586.73	II
110		2,587.04	II
130		2,592.57	II
120		2,595.03	II
140		2,624.18	II
490		2,670.26	II
330		2,672.25	II
100		2,675.35	II
270		2,739.31	II
310		2,750.19	II
230		2,755.01	II
610		2,755.63	II
510		2,770.02	II
230		2,778.97	II
230		2,802.53	II
310		2,804.35	II
410		2,820.19	II
270	d	2,833.91	II
390		2,838.71	II
270		2,848.37	II
250		2,855.41	II
310		2,859.84	II
310		2,896.96	II
390		2,897.52	II
1,000		2,904.47	II
210		2,909.58	II
1,500		2,910.36	II
270		2,915.62	II
350		2,929.27	II
270		2,945.28	II
230		2,946.62	II
1,500		2,964.52	II
410		2,968.76	II
210		2,974.47	II
230		2,975.68	II
270		2,983.80	II
1,200		3,002.41	II
310		3,002.65	II
230		3,012.47	II
230		3,016.84	II
290		3,025.95	II
270		3,028.27	II
370		3,031.31	II
310		3,036.22	II
210		3,054.42	II
230		3,066.22	II
450		3,070.74	II
560		3,072.53	II
610		3,073.34	II
210		3,078.87	II
720		3,082.08	II
610		3,084.02	II
370		3,099.19	II
230		3,106.78	II
310	d	3,113.43	II
		3,113.54	II
770		3,122.72	II
290		3,132.52	II
470		3,132.77	II
410		3,141.10	II
		3,141.15	II
250		3,144.33	II
410		3,154.29	II
870		3,181.92	II
410		3,183.42	II
250		3,185.25	II
310		3,200.58	II
230		3,205.15	II
270		3,214.44	II
870		3,220.73	II
610		3,223.31	II
210		3,227.16	II
2,300		3,230.58	II
250		3,232.03	II
330		3,237.98	II
330		3,249.34	II
560		3,259.05	II
		3,259.11	II
2,700		3,264.78	II
430		3,267.10	II
		3,267.18	II
330		3,269.41	II
250		3,278.22	II
720		3,279.33	II
720		3,280.22	II
470		3,286.77	II
330	d	3,303.88	II
		3,303.95	
370		3,305.56	II
2,300		3,312.42	II
560		3,316.39	II
770		3,323.19	II
290		3,329.66	II
770		3,332.70	II
370		3,337.25	II
290		3,337.79	II
250		3,340.03	II
290		3,341.84	II
1,300		3,346.04	II
470		3,350.06	II
350		3,350.26	II
1,400		3,364.08	II
1,400	d	3,368.02	II
		3,368.13	I
450		3,370.55	II
7,700		3,372.71	II
970		3,374.17	II
290		3,381.32	II
230		3,382.06	I
1,700		3,385.08	II
450		3,389.74	II
2,300		3,392.00	II
350		3,396.07	II
290		3,396.84	II
390		3,401.83	II
350		3,417.63	II
490		3,428.39	II
770		3,441.13	II
390		3,442.68	I
490		3,469.51	I
970		3,471.71	II
610		3,479.41	II
970		3,485.85	II
350		3,486.82	II
350		3,496.86	II
6,700		3,499.10	II
610		3,502.78	I
390		3,508.38	II
490		3,514.89	II
390		3,518.18	II
610		3,524.91	II
410		3,539.59	I
310		3,548.26	II
820		3,549.84	II
310		3,553.20	II
1,500		3,558.02	I
510		3,558.71	I
1,000		3,559.90	II
310		3,565.17	I
920		3,570.75	II
310		3,578.24	I
1,000		3,580.52	II
370		3,586.60	I
610		3,590.76	I
410		3,595.84	I
610		3,599.50	II
1,000		3,599.83	II
510		3,604.90	II
410		3,607.42	I
3,100		3,616.56	II
510		3,617.85	II
510		3,618.92	II
720		3,628.04	I
310		3,629.37	I
1,000		3,633.54	II
510		3,634.67	I
1,600		3,638.68	I
900		3,645.94	II
520		3,650.41	II
360		3,652.58	II
500		3,652.87	II
360		3,664.45	I
470		3,669.02	II
500		3,682.70	II
320		3,684.01	I
380		3,684.28	II
7,900		3,692.65	II
450		3,696.25	II
380		3,697.68	I
540		3,700.72	II
520		3,707.64	II
520		3,712.39	II
320		3,719.35	I
1,300		3,729.52	II
450		3,731.26	II
540		3,738.16	II
340		3,741.10	II
900		3,742.64	II
900		3,747.43	I
540		3,756.05	I
410		3,781.01	II
1,800		3,786.84	II
560		3,787.86	II
560		3,791.83	II
500		3,792.79	I
560		3,797.06	II
1,600		3,810.33	I
3,600		3,830.48	II
540		3,849.91	I
320		3,851.60	II
680		3,855.90	I
540		3,858.39	II
7,500		3,862.85	I
1,500		3,880.61	II
1,200		3,882.89	II
400		3,890.61	II
4,200		3,892.68	I
5,200		3,896.23	II
810		3,902.76	II
250		3,903.98	I
250		3,904.56	II
1,200		3,905.40	I
11,000		3,906.31	II
280		3,918.05	I
210		3,918.35	II
280		3,921.88	II
810		3,932.25	II
3,200		3,937.01	I
2,100		3,938.63	II
3,200		3,944.42	I
550		3,948.06	I
250		3,951.48	I
320		3,956.42	I
280		3,966.35	I
2,700		3,973.04	I
3,200		3,973.58	I
1,400		3,974.72	II
280		3,976.73	I
810		3,977.02	I
1,100		3,982.33	I
280		3,987.53	I
810		3,987.66	I
230		3,991.15	I
230		4,004.05	I
14,000		4,007.96	I
230		4,008.18	II
280		4,009.16	II
1,100		4,012.58	I
350		4,015.57	II
3,000		4,020.51	I
450		4,021.55	I
230		4,043.01	II
1,000		4,046.96	I
280		4,048.34	II
200		4,049.49	II
940		4,055.47	II
550		4,059.51	I
690		4,059.78	II
420		4,077.88	I
550		4,081.24	II
3,500		4,087.63	I
210		4,092.90	I
1,100		4,098.10	I
350		4,100.56	II
320		4,116.36	I
320		4,118.55	I
600		4,131.50	I
550		4,142.91	II
6,900		4,151.11	I
280		4,189.98	II
1,000		4,190.70	I
130		4,205.32	I
1,400		4,218.43	I
200		4,220.99	I
320		4,230.20	II
140		4,234.78	II
200		4,251.94	II
140		4,276.48	II
690		4,286.56	I
320		4,298.91	I
320		4,301.60	II
140		4,303.81	II
110		4,319.94	II
130		4,328.81	I
110		4,331.36	I
140		4,340.92	I
190		4,348.34	I
110		4,369.39	II
160		4,382.17	I
300		4,384.70	II
300		4,386.40	I
100		4,403.17	II
810		4,409.34	I
180		4,418.70	I
570		4,419.61	II
110		4,422.51	II
320		4,424.57	I
370		4,426.77	I
110		4,437.66	I
100		4,459.24	II
100		4,473.50	II
130		4,496.39	I
200		4,500.75	II
130		4,522.74	I
160		4,563.26	II
1,000		4,606.61	I
160		4,630.88	II
110		4,640.60	II
110		4,665.44	II
310		4,673.16	I
570		4,675.62	II
150		4,679.06	II
230		4,722.69	I
150		4,729.05	I
130		4,751.52	II
170		4,759.65	II

190		4,820.35	II
140		4,857.44	I
150		4,872.09	II
210		4,900.08	II
210		4,934.11	II
130		4,944.36	I
180		4,951.74	II
130		4,976.42	I
250		5,007.25	I
140		5,028.33	I
120		5,028.91	II
200		5,035.94	I
210		5,042.05	II
130		5,043.86	I
130		5,044.89	I
130		5,077.59	II
120		5,124.56	I
130		5,127.41	II
120		5,131.53	I
130		5,133.83	II
170		5,164.77	II
130		5,172.78	I
160		5,188.90	II
150		5,206.52	I
60		5,212.91	II
30		5,215.13	II
30		5,218.26	II
45		5,229.34	II
140		5,255.93	II
22		5,256.47	II
27		5,257.02	II
35		5,264.77	II
80		5,272.91	I
55		5,277.71	I
27		5,279.34	II
45		5,302.30	II
55		5,333.06	I
27		5,333.33	II
27		5,334.23	II
22		5,343.94	II
30		5,344.50	II
90		5,348.06	I
45		5,350.47	I
35		5,368.85	I
35		5,395.87	II
60		5,414.63	II
18		5,422.81	II
18	h	5,451.30	I
35		5,454.27	II
180		5,456.62	I
35		5,462.43	II
90		5,468.32	I
18		5,477.47	II
80		5,485.97	II
27		5,497.44	II
27		5,516.02	I
80		5,593.46	I
45	d	5,601.14	I
		5,601.32	I
45	h	5,609.94	I
60		5,611.82	I
70		5,622.01	I
80		5,626.53	II
30		5,636.20	II
90		5,640.36	I
22		5,641.42	I
22	h	5,658.63	I
70		5,664.95	I
45		5,665.44	II
55		5,675.48	I
14		5,695.53	II
27		5,710.87	II
55		5,717.48	I
70		5,719.55	I
55		5,726.97	I
22		5,733.43	II
22		5,736.56	I
22		5,736.94	I
100		5,739.19	I
35		5,740.61	I
60		5,748.65	I
55		5,752.53	I
70		5,757.63	II
290		5,762.80	I
70		5,769.92	I
45		5,782.82	I
70		5,784.66	I
22		5,791.15	II
70		5,800.79	I
22		5,806.10	I
430		5,826.79	I
45		5,835.84	I
100		5,850.07	I
120		5,855.31	I
140		5,872.35	I
120		5,881.14	I
27		5,886.30	II
27		5,902.08	II
55		5,906.06	I
45		5,909.24	I
35		5,933.50	I
22		5,946.37	I
55		5,968.68	I
27		5,975.49	I
35		6,006.79	II
22		6,008.75	II
55		6,014.83	I
35		6,015.74	II
70		6,022.56	I
22		6,032.12	II
22		6,045.63	II
22		6,048.14	II
45		6,054.85	I
70		6,061.25	I
60		6,076.45	II
35		6,116.01	I
35		6,125.32	I
30		6,170.06	II
27		6,183.21	II
360		6,221.02	I
35		6,230.90	I
55		6,262.56	I
45		6,267.93	I
60		6,268.87	I
35		6,274.94	I
30		6,286.86	I
45		6,299.42	I
130		6,308.77	I
55		6,326.13	I
22		6,347.16	II
45		6,388.19	I
22		6,432.53	I
27		6,485.87	I
55		6,492.35	I
22		6,541.57	I
60		6,583.48	I
70		6,601.11	I
27		6,721.91	I
70		6,759.87	I
22		6,762.92	I
22		6,773.37	I
35		6,790.92	I
22		6,825.44	I
22		6,825.98	I
70		6,848.10	I
55		6,865.13	I
27		6,879.98	I
22		7,001.40	I
12		7,058.55	I
12	h	7,065.04	I
11		7,070.99	II
18		7,101.27	I
8		7,109.67	I
11		7,155.40	II
5		7,161.91	I
14	h	7,197.00	I
7	h	7,264.82	II
7	h	7,283.95	I
14		7,329.73	II
18		7,355.37	I
11		7,356.34	I
18		7,428.67	I
55		7,459.55	I
9		7,460.42	I
120		7,469.51	I
22		7,532.34	I
6	h	7,539.18	I
27		7,556.26	I
6	h	7,574.21	I
5	h	7,590.51	I
11		7,597.33	I
6		7,607.23	I
11		7,613.52	I
6		7,623.48	I
16	h	7,645.67	I
8		7,650.63	I
22		7,654.45	II
12	h	7,658.05	I
22		7,659.25	I
4		7,665.64	I
35		7,680.01	I
9		7,722.14	I
8		7,726.19	II
11	h	7,747.44	I
22		7,754.63	I
4		7,762.16	I
9		7,796.69	I
35		7,797.47	I
9		7,838.80	I
11		7,844.00	I
16		7,847.55	I
5		7,875.36	I
5		7,879.36	I
18		7,899.55	I
8		7,913.08	I
35		7,921.85	I
30		7,937.84	I
8		7,952.93	I
12		7,964.51	I
8		7,979.03	II
8		7,980.87	I
5		8,023.03	I
12		8,035.91	I
12		8,181.85	I
35		8,312.82	I
18		8,328.57	II
5		8,367.58	II
55		8,409.90	I
11		8,466.18	II
35		8,472.42	I
14		8,517.71	II
18		8,521.37	II
22		8,768.64	I
11	h	8,776.63	II
9		8,866.84	II

Er III

Ref. 301 — J.R.

Intensity	Wavelength	
	Vacuum	
2	2,165.26	III
3	2,190.77	III
10	2,198.15	III
1	2,223.98	III
4	2,232.35	III
60	2,235.28	III
8	2,245.60	III
2	2,255.95	III
80	2,269.36	III
600	2,277.65	III
100	2,309.19	III
40	2,358.69	III
10	2,358.79	III
50	2,359.33	III
200	2,367.64	III
20	2,375.50	III
10	2,377.07	III
80	2,381.25	III
20	2,381.40	III
40	2,381.75	III
6	2,391.96	III
60	2,393.08	III
5	2,393.60	III
250	2,396.40	III
10	2,398.91	III
2	2,402.75	III
80	2,404.58	III
100	2,410.47	III
200	2,419.81	III
200	2,422.47	III
40	2,431.51	III
60	2,464.60	III
2	2,492.04	III
100	2,508.59	III
2	2,531.03	III
100	2,532.36	III
8	2,536.76	III
80	2,540.91	III
3	2,543.31	III
10	2,545.95	III
50	2,557.22	III
80	2,570.74	III
40	2,580.02	III
2	2,589.55	III
80	2,590.72	III
20	2,591.56	III
200	2,591.83	III
20	2,598.39	III
3	2,599.18	III
100	2,603.62	III
40	2,604.91	III
25	2,614.53	III
30	2,617.64	III
2	2,618.40	III
2	2,618.94	III
8	2,625.19	III
20	2,626.37	III
4	2,637.52	III
200	2,637.77	III
5	2,651.49	III
25	2,683.10	III
400	2,723.29	III
100	2,738.53	III
500	2,739.27	III
8	2,741.41	III
80	2,746.03	III
6	2,752.20	III
80	2,756.20	III
400	2,759.23	III
150	2,761.92	III
60	2,762.66	III
15	2,767.11	III
100	2,768.72	III
60	2,772.07	III
60	2,774.80	III
20	2,775.55	III
2	2,780.60	III
80	2,783.11	III
500	2,792.54	III
10	2,804.10	III
100	2,805.87	III
6	2,808.44	III
50	2,824.75	III
150	2,830.34	III
1	2,831.95	III
8	2,833.03	III
8	2,845.29	III
60	2,846.08	III
6	2,849.63	III
1	2,869.52	III
8	2,878.24	III
1	2,955.93	III
1	2,958.63	III
1,000	3,055.10	III
1,000	3,070.40	III
500	3,100.40	III
1,500	3,166.25	III
3	3,172.47	III
1	3,173.45	III
50	3,175.74	III
400	3,214.95	III
2,000	3,301.23	III
8	3,341.00	III
200	3,480.54	III
8	3,592.96	III
600	3,715.67	III
200	3,739.43	III
4,000	3,816.78	III
600	3,962.87	III
40	4,009.70	III
2	4,088.58	III
1,000	4,288.18	III
40,000	4,290.06	III
300	4,338.24	III
20,000	4,386.86	III
30	4,612.93	III
15,000	4,735.56	III
2,000	4,783.12	III
8	4,876.07	III
8,000	5,903.30	III

EUROPIUM (EU)

Z = 63

Eu I and II

Ref. 1 — C.H.C.

Intensity	Wavelength	
	Vacuum	
21	2,499.39	II
26	2,554.78	II
26	2,559.18	II

160		2,564.17	II	45		3,246.03	I	150		3,884.75	I	75	h	4,464.97	II
110		2,568.17	II	45		3,247.32	II	23		3,896.78	I	24		4,485.15	II
26		2,574.76	II	100		3,247.55	I	23		3,900.18	I	3,000		4,522.57	II
230		2,577.14	II	100		3,266.39	II	70		3,900.51	I	45	h	4,535.59	I
26		2,581.86	II	150		3,272.77	II	28,000	c	3,907.10	II	11,000		4,594.03	I
26		2,604.61	I	210		3,277.78	II	45		3,915.24	II	21		4,602.63	I
30		2,635.50	II	150		3,301.95	II	45		3,916.00	I	9,800		4,627.22	I
1,000		2,638.77	II	45		3,304.50	II	230		3,917.29	I	8,300		4,661.88	I
380		2,641.27	II	140		3,308.02	II	23		3,917.70	II	30		4,713.59	I
40		2,653.61	II	140		3,313.33	II	40		3,918.52	I	27		4,740.50	I
640		2,668.34	II	65		3,319.89	II	100		3,919.09	II	45		4,792.59	I
110		2,673.42	II	95		3,321.86	II	40		3,928.87	II	40	h	4,829.30	I
250		2,678.29	II	85		3,322.26	I	32,000	c	3,930.48	II	60		4,830.33	I
250		2,685.66	II	950		3,334.33	I	55		3,941.56	II	40	h	4,840.47	I
550		2,692.03	II	45		3,338.75	II	30	h	3,942.21	II	60	h	4,849.64	I
700		2,701.14	II	110		3,350.40	I	60		3,942.94	II	110		4,867.62	I
800		2,701.90	II	40		3,351.56	II	120		3,943.08	II	40	h	4,884.05	I
240		2,705.28	II	40		3,354.38	II	30		3,944.59	II	90		4,894.68	I
180		2,709.99	I	45		3,367.64	II	30		3,945.67	II	60		4,900.86	I
700		2,716.98	II	140		3,369.06	II	30		3,949.13	II	150		4,907.18	I
70		2,723.96	I	65		3,380.25	II	60		3,949.60	I	180		4,911.40	I
4,200		2,727.78	II	75		3,390.78	II	45		3,950.76	II	55		4,953.52	I
190		2,729.33	II	190		3,391.99	II	55		3,951.33	II	55		4,960.21	I
380		2,729.44	II	280		3,396.58	II	60		3,955.75	I	55		4,962.55	I
50		2,731.37	I	45		3,419.84	II	40		3,957.92	II	45		4,975.76	I
40		2,732.61	I	65		3,423.09	II	30		3,963.61	I	180		5,013.17	I
80		2,735.25	I	150		3,425.02	II	120		3,964.90	II	170		5,022.91	I
160		2,740.62	II	45		3,426.44	II	150		3,966.59	II	110		5,029.54	I
70		2,743.28	I	45		3,435.05	II	45		3,967.18	I	90		5,033.55	I
120		2,744.26	II	65		3,435.20	II	30,000	c	3,971.96	II	75		5,067.95	I
40		2,745.61	I	40		3,435.72	II	60		3,978.42	I	75	h	5,092.69	I
70		2,747.29	II	45		3,440.82	II	30		3,979.63	II	90	h	5,096.44	I
80		2,747.83	I	150		3,441.00	II	55	h	3,986.60	I	170		5,114.37	I
90		2,752.17	II	45		3,445.18	II	40		3,988.24	II	90		5,124.77	I
480		2,781.89	II	85		3,457.05	I	30		3,993.93	II	170		5,129.10	I
1,900		2,802.84	II	45		3,457.56	II	55		3,995.98	II	90		5,130.08	I
220		2,811.75	II	130		3,461.38	II	60		4,003.71	II	210		5,133.52	I
30		2,813.08	II	85		3,467.88	I	180		4,011.69	II	270		5,160.07	I
3,400		2,813.94	II	75		3,477.07	I	150		4,017.58	II	210		5,166.70	I
550		2,816.18	II	75	h	3,505.30	II	120		4,039.19	I	60		5,193.74	I
2,000		2,820.78	II	470	c	3,521.09	II	45	h	4,078.24	I	200		5,199.85	I
400	c	2,828.72	II	75		3,531.15	II	120		4,085.38	II	110		5,200.96	I
120		2,829.30	II	45		3,532.23	II	75		4,096.80	II	120		5,206.44	I
140		2,833.26	II	65		3,538.08	II	60		4,106.88	I	750		5,215.10	I
80		2,843.96	II	150		3,542.15	II	90	h	4,112.04	II	300		5,223.49	I
60		2,852.05	II	85		3,543.85	II	45		4,119.30	II	120		5,239.24	I
260		2,859.67	II	45		3,549.71	II	75		4,127.28	I	200		5,266.40	I
280		2,862.57	II	180		3,552.52	II	33,000	c	4,129.70	II	390		5,271.96	I
25		2,864.42	II	75		3,589.27	I	30		4,136.59	II	110		5,272.48	I
60		2,876.06	II	45		3,591.31	II	40		4,137.07	I	150		5,282.82	I
100		2,878.87	I	150		3,603.20	II	30		4,141.02	II	55		5,287.25	I
80		2,887.85	II	75		3,611.57	II	60		4,141.72	II	60		5,289.25	I
200		2,892.54	I	45		3,616.15	II	30		4,151.52	II	120		5,291.26	I
140		2,893.03	I	95		3,622.54	II	45		4,151.64	II	60		5,293.68	I
360		2,893.83	I	95		3,632.18	II	30		4,157.72	I	120		5,294.64	I
			II	45		3,673.19	II	110		4,172.80	II	90		5,303.85	I
3,200		2,906.68	II	45		3,674.63	II	30		4,175.16	II	30	h	5,350.41	I
160		2,908.99	I	45		3,678.26	II	110		4,182.22	I	75	h	5,351.69	I
30		2,917.44	II	6,400		3,688.42	II	40		4,195.36	II	40		5,352.84	I
850		2,925.04	II	60		3,710.87	II	40		4,196.18	II	90		5,355.10	I
60		2,947.29	II	95		3,713.45	II	60,000	c	4,205.05	II	540		5,357.61	I
200	c	2,952.68	II	95		3,714.90	II	45		4,221.08	II	60		5,360.83	I
30		2,958.91	I	35		3,716.94	II	40		4,223.88	II	120		5,361.61	I
35		2,959.47	II	35		3,717.69	II	90	h	4,227.40	II	110		5,376.94	I
260		2,960.21	II	40		3,719.16	I	75		4,229.33	II	120		5,392.94	I
300		2,991.33	II	20,000	c	3,724.94	II	75		4,232.45	II	450		5,402.77	I
35		2,995.22	II	45		3,729.68	II	90		4,237.51	II	45		5,405.33	I
40		3,006.26	II	45		3,729.74	II	45		4,238.69	II	45		5,411.86	I
35		3,022.15	I	21		3,732.20	I	45		4,244.74	I	55		5,421.07	I
30		3,040.77	II	45		3,738.08	II	45		4,247.06	II	90		5,426.94	I
320	c	3,054.94	II	350		3,741.31	II	45		4,253.80	II	40		5,443.56	I
120		3,058.98	I	100		3,743.56	II	30		4,270.24	II	380		5,451.51	I
35		3,069.11	II	260		3,761.12	II	150		4,298.73	I	260		5,452.94	I
35		3,076.07	II	95		3,765.93	II	90		4,329.36	I	40		5,457.62	I
220		3,077.36	II	40		3,774.10	I	75		4,329.97	I	90		5,472.32	I
35		3,089.35	II	60		3,781.40	II	60		4,330.61	II	120		5,488.65	I
120		3,097.45	II	40		3,788.76	II	40		4,331.18	I	45		5,495.20	I
320		3,106.18	I	45		3,791.50	II	90		4,337.68	I	15		5,500.83	I
950		3,111.43	I	130		3,799.01	II	240		4,355.09	II	120		5,510.52	I
120		3,130.73	II	70		3,801.36		27		4,361.57	II	30		5,526.63	I
40		3,132.16	I	95		3,807.54	II	55		4,369.47	II	30		5,533.25	I
45		3,149.88	II	120		3,811.33	I	45	h	4,372.20	II	30		5,542.54	I
85		3,173.61	II	120		3,815.50	II	75		4,383.17	II	200		5,547.44	I
40		3,185.54	I	39,000	c	3,819.67	II	90		4,387.88	I	150		5,570.33	I
420		3,210.57	I	120		3,826.68	II	21	h	4,405.27	II	200		5,577.14	I
1,000		3,212.81	I	120		3,844.23	II	55		4,407.07	II	75		5,579.63	I
420		3,213.75	I	190		3,865.57	I	18		4,419.66	II	120		5,580.03	I
45		3,235.13	I	45		3,872.72	I	120		4,434.81	II	90		5,586.24	I
95		3,241.40	I	70		3,877.27	II	14,000	c	4,435.56	II	75		5,586.83	I

18		5,592.25	I
18		5,599.80	I
18		5,605.86	I
40		5,618.81	I
60		5,622.44	I
75		5,632.54	I
210		5,645.80	I
15		5,651.11	I
60		5,673.85	I
27		5,681.10	I
27		5,684.24	I
60		5,730.87	I
60		5,739.00	I
330		5,765.20	I
180		5,783.69	I
15		5,792.72	I
60		5,800.27	I
170		5,818.74	II
600	c	5,830.98	I
27		5,845.77	I
27		5,860.97	I
15		5,864.77	I
90		5,872.98	II
15		5,895.31	I
27		5,902.97	I
12		5,909.94	I
75		5,915.74	I
12		5,925.30	I
27		5,926.52	I
45		5,942.72	I
27		5,953.49	I
27		5,953.84	II
30		5,954.28	I
90		5,963.76	I
330		5,966.07	II
480	c	5,967.10	I
15	h	5,968.43	I
30		5,971.69	I
170		5,972.75	I
15		5,980.47	I
27		5,983.14	I
27		5,983.78	I
240		5,992.83	I
60		6,004.36	I
15	h	6,005.61	I
60	h	6,012.20	I
110		6,012.56	I
60		6,015.58	I
420		6,018.15	I
60		6,023.15	I
170		6,029.00	I
60		6,044.66	I
420		6,049.51	II
140		6,057.36	I
90		6,075.58	I
30		6,077.38	I
240		6,083.84	I
240		6,099.35	I
60		6,108.15	I
120		6,118.78	I
60		6,124.67	I
330		6,173.05	II
110		6,178.76	I
260	c	6,188.13	I
140		6,195.07	I
15	h	6,207.60	I
15		6,230.51	I
90	h	6,233.73	I
55		6,250.47	I
240		6,262.25	I
55		6,266.95	I
15	h	6,285.95	I
60		6,291.34	I
170		6,299.77	I
230		6,303.41	II
24	h	6,313.78	I
15		6,318.58	I
75		6,335.82	I
120	c	6,350.04	I
60		6,355.89	I
60		6,369.25	I
55		6,382.73	I
75		6,383.86	I
120	c	6,400.93	I
40		6,406.11	I
180		6,410.04	I
140		6,411.32	I
55		6,428.29	I
830		6,437.64	II
18		6,439.93	I
120		6,457.96	I
12		6,470.70	I
18		6,483.02	I
45		6,501.55	I
60		6,519.59	I
15		6,522.72	I
8	h	6,549.12	I
75		6,567.87	I
45		6,593.79	I
18	h	6,603.55	I
1,400		6,645.11	II
26		6,685.21	I
95		6,693.96	I
7	h	6,701.06	I
12	h	6,710.45	I
30		6,744.88	I
30	h	6,782.54	I
14	h	6,787.48	I
140		6,802.72	I
35		6,816.06	I
11	h	6,834.30	I
17		6,840.93	I
17	h	6,844.83	I
14	h	6,847.04	I
360		6,864.54	I
21		6,898.21	I
60	h	6,903.67	I
14	h	6,910.17	I
30	h	6,914.82	I
120		7,040.20	I
12		7,074.54	I
330		7,077.10	II
100		7,106.48	I
6		7,164.66	I
30		7,175.55	I
570		7,194.81	II
570		7,217.55	II
11	h	7,224.68	I
15		7,258.72	I
30		7,262.77	I
11	h	7,281.53	I
6	h	7,297.56	I
540		7,301.17	II
11		7,310.46	I
12		7,313.63	I
55	c	7,336.18	I
4		7,346.25	I
4		7,356.65	I
11		7,362.25	I
55	c	7,369.60	I
720		7,370.22	II
4		7,387.36	I
12		7,389.16	I
11	h	7,404.41	I
300		7,426.57	II
21	h	7,436.59	I
8		7,470.53	I
5	h	7,491.00	I
50	c	7,528.70	I
5	h	7,533.02	I
6		7,547.32	I
160		7,583.91	I
60	c	7,742.57	I
70		7,746.19	I
8	h	7,803.32	I
8		7,818.21	I
35		7,887.99	I
7		8,015.47	I
24	c	8,209.80	I
15	c	8,226.81	I
6	h	8,464.71	I
21	c	8,642.67	I
7		8,727.77	I
6		8,782.46	I
12	c	8,790.88	I
18		8,870.30	I

Eu III

Ref. 312 — J.R.

Intensity		Wavelength	
		Vacuum	
10		2,073.40	III
10		2,093.50	III
30		2,124.69	III
10		2,167.12	III
10		2,173.59	III
10		2,184.68	III
10		2,190.59	III
10		2,194.81	III
20		2,211.85	III
20		2,212.63	III
20		2,214.66	III
10		2,215.34	III
30		2,217.23	III
30		2,219.33	III
20		2,219.42	III
10		2,223.13	III
10		2,235.17	III
10		2,240.14	III
10		2,261.88	III
20		2,265.74	III
10		2,269.39	III
20		2,276.85	III
40		2,291.62	III
20		2,304.37	III
10		2,311.92	III
10		2,327.69	III
10		2,334.56	III
10		2,336.96	III
10		2,339.84	III
10		2,343.10	III
10		2,346.83	III
10		2,347.64	III
10		2,350.38	III
200		2,350.51	III
10		2,352.28	III
10		2,357.87	III
10		2,359.08	III
10		2,360.65	III
20		2,363.76	III
20		2,368.04	III
20		2,374.08	III
10		2,375.20	III
4,000		2,375.46	III
10		2,376.42	III
20		2,377.23	III
10		2,381.81	III
10		2,383.62	III
20		2,387.29	III
20		2,389.11	III
10		2,389.98	III
20		2,391.11	III
20		2,391.90	III
10		2,392.59	III
10		2,394.66	III
20		2,395.62	III
20		2,398.79	III
20		2,401.00	III
20		2,402.34	III
20		2,404.08	III
10		2,406.14	III
20		2,407.30	III
20		2,408.32	III
10		2,409.63	III
20		2,410.08	III
40		2,412.02	III
20		2,412.96	III
20		2,413.26	III
10		2,413.41	III
10		2,419.11	III
10		2,419.25	III
10		2,419.58	III
30		2,422.00	III
10		2,422.90	III
10		2,425.33	III
50		2,425.68	III
40		2,427.67	III
40		2,429.32	III
10	d	2,429.66	III
10		2,430.04	III
10		2,431.49	III
10		2,431.76	III
10		2,432.55	III
10		2,433.65	III
10		2,434.19	III
100	d	2,435.14	III
20		2,436.39	III
10		2,436.77	III
10		2,438.83	III
10		2,440.26	III
50		2,440.67	III
1,000		2,444.38	III
4,000		2,445.99	III
30	h	2,446.43	III
20		2,448.57	III
20		2,451.24	III
10		2,451.73	III
30		2,455.22	III
10		2,461.79	III
30		2,463.30	III
30		2,464.47	III
40		2,470.51	III
10		2,474.94	III
10		2,476.24	III
20		2,476.45	III
10		2,477.78	III
20		2,480.02	III
20		2,483.29	III
10		2,486.92	III
10		2,488.91	III
10		2,490.50	III
10		2,491.08	III
10		2,492.48	III
10		2,496.92	III
10		2,499.17	III
2,000		2,513.76	III
20		2,517.94	III
200		2,522.14	III
10		2,539.14	III
10		2,548.30	III
20		2,548.59	III
10		2,554.50	III
10		2,558.07	III
10		2,560.36	III
10		2,594.71	III
20		2,594.76	III
10		2,596.34	III
10		2,604.44	III
30		2,608.34	III
30		2,610.09	III
50	c	2,616.11	III
20		2,616.26	III
10		2,616.33	III
10		2,616.35	III
10		2,620.79	III
20		2,623.33	III
10		2,626.98	III
20	c	2,628.46	III
10		2,628.82	III
10		2,631.98	III
30	c	2,642.27	III
20		2,645.22	III
20	c	2,650.93	III
10		2,653.19	III
10		2,655.09	III
10		2,662.24	III
20	c	2,666.86	III
20	c	2,668.21	III
40	c	2,676.09	III
20	c	2,683.21	III
20		2,686.13	III
20	c	2,687.74	III
40	c	2,693.51	III
10		2,694.80	III
20		2,699.87	III
20	c	2,700.78	III
10	c	2,708.25	III
20	c	2,708.84	III
10		2,712.08	III
50	c	2,720.67	III
20	c	2,725.54	III
10		2,743.94	III
10		2,752.68	III
20		2,755.12	III
10		2,757.75	III
20	c	2,760.21	III
10		2,761.72	III
10	c	2,766.26	III
20	c	2,768.38	III
10	c	2,768.54	III
10		2,769.71	III
20	c	2,780.48	III
20		2,792.51	III
10		2,808.09	III
10		2,817.58	III
20	c	2,839.56	III
20		2,844.99	III
10	c	2,848.44	III
20		2,850.39	III
10	c	2,892.60	III
40		2,912.23	III
40	c	2,912.64	III
10		2,913.04	III
10		2,928.91	III

20	c	2,931.00	III
10	c	2,950.20	III
20		2,956.74	III
10		2,956.90	III
10	c	2,972.30	III
30	c	2,982.29	III
20	c	3,000.11	III
20		3,006.37	III
20	c	3,013.28	III
10	c	3,018.43	III
20	c	3,022.08	III
50	c	3,022.69	III
20	c	3,023.40	III
100	c	3,023.93	III
10		3,025.32	III
10	c	3,026.09	III
200	c	3,026.79	III
50	c	3,029.92	III
20	c	3,031.24	III
40	c	3,032.84	III
20	c	3,036.98	III
10		3,038.64	III
10		3,039.05	III
20	c	3,039.98	III
10		3,054.07	III
10	c	3,054.97	III
20		3,076.43	III
10	c	3,089.09	III
10		3,105.25	III
10		3,109.67	III
10		3,129.31	III
10		3,142.54	III
50	c	3,171.00	III
10		3,178.08	III
20	h	3,178.87	III
50	c	3,183.78	III
10	h	3,191.46	III
20	c	3,194.34	III
10		3,206.30	III
10		3,208.95	III
10	h	3,213.84	III
10		4,837.98	III
50		6,666.35	III
30		7,221.84	III
20		7,690.44	III
10		8,079.07	III

FLUORINE (F)

Z = 9

F I and II

Ref. 169, 224 — G.V.S.

Wavelength		Intensity	
		Vacuum	
30		375.30	II
30		380.90	II
40		407.04	II
50		430.91	II
40		431.55	II
40		435.64	II
70		457.18	II
40		471.95	II
60		472.00	II
50		472.71	II
40		473.02	II
90		484.60	II
50		513.64	II
70		514.94	II
70		546.85	II
60		547.87	II
50		548.32	II
40		548.52	II
90		605.67	II
80		606.29	II
100		606.80	II
70		606.92	II
80		607.47	II
90		608.06	II
15		780.39	I
10		780.52	I
10		782.38	I
12		791.88	I
10		792.54	I
10		794.42	I
150		806.96	I
125		809.60	I
500		951.87	I
1,000		954.83	I
750		955.55	I
500		958.52	I
20		972.40	I
350		973.90	I
100		976.22	I
40		976.51	I
100		977.75	I
40		1,129.76	II
40		1,327.06	II
50		1,328.11	II
40		1,333.59	II
50		1,343.60	II
40		1,344.04	II
50		1,400.61	II
40		1,407.14	II
60		1,493.09	II
50		1,493.24	II
40		1,493.31	II
40		1,702.13	II
40		1,744.75	II
50		1,745.55	II
60		1,747.39	II
		Air	
100		2,556.11	II
100		2,871.40	II
120		3,059.99	II
140		3,153.49	II
170		3,202.76	II
140		3,264.08	II
140		3,414.65	II
150		3,416.45	II
140		3,416.80	II
160		3,417.00	II
160		3,472.96	II
150		3,473.31	II
170		3,474.78	II
190		3,501.39	II
200		3,501.45	II
200		3,501.57	II
180		3,502.84	II
200		3,502.96	II
210		3,503.11	II
170		3,505.37	II
200		3,505.52	II
220		3,505.63	II
160		3,522.89	II
150		3,536.87	II
160		3,541.77	II
160		3,590.52	II
6		3,594.10	I
170		3,598.69	II
180		3,601.39	II
190		3,602.84	II
12		3,668.17	I
180		3,704.53	II
160		3,710.35	II
160		3,739.57	II
140		3,805.83	II
270		3,847.09	II
260		3,849.99	II
250		3,851.67	II
5		3,898.48	I
190		3,898.83	II
180		3,901.93	II
170		3,903.82	II
8		3,930.69	I
5		3,934.26	I
5		3,948.56	I
150		3,972.04	II
160		3,972.67	II
170		3,974.78	II
240		4,024.73	II
220		4,025.01	II
230		4,025.49	II
160		4,083.91	II
190		4,103.07	II
170		4,103.22	II
200		4,103.51	II
180		4,103.71	II
170		4,103.87	II
170		4,109.16	II
160		4,116.54	II
150		4,119.21	II
140		4,207.15	II
170	h	4,225.16	II
150	h	4,244.12	II
200		4,246.23	II
190		4,246.39	II
180		4,246.59	II
170		4,246.77	II
160		4,246.84	II
170	h	4,275.36	II
160	h	4,277.53	II
160	h	4,278.93	II
200		4,299.17	II
160		4,446.53	II
170		4,446.72	II
180		4,447.19	II
140		4,734.38	II
170		4,859.39	II
160		4,933.26	II
6		4,960.65	I
140		5,002.00	II
150		5,173.25	II
15		5,230.41	I
12		5,279.01	I
18		5,540.52	I
12		5,552.43	I
10		5,577.33	I
160		5,589.27	II
20		5,624.06	I
12		5,626.93	I
15		5,659.15	I
40		5,667.53	I
90		5,671.67	I
18		5,689.14	I
25		5,700.82	I
25		5,707.31	I
12		5,950.15	I
25		5,959.19	I
70		5,965.28	I
50		5,994.43	I
150		6,015.83	I
80		6,038.04	I
900		6,047.54	I
100		6,080.11	I
800		6,149.76	I
400		6,210.87	I
13,000		6,239.65	I
140		6,247.90	II
10,000		6,348.51	I
8,000		6,413.65	I
450		6,569.69	I
300		6,580.39	I
400		6,650.41	I
1,800		6,690.48	I
400		6,708.28	I
7,000		6,773.98	I
1,500		6,795.53	I
9,000		6,834.26	I
50,000		6,856.03	I
8,000		6,870.22	I
15,000		6,902.48	I
6,000		6,909.82	I
4,000		6,966.35	I
45,000		7,037.47	I
30,000		7,127.89	I
130		7,179.90	II
15,000		7,202.36	I
130	h	7,211.79	II
1,000		7,309.03	I
15,000		7,311.02	I
700		7,314.30	I
5,000		7,331.96	I
10,000		7,398.69	I
4,000		7,425.65	I
2,200		7,482.72	I
2,500		7,489.16	I
900		7,514.92	I
5,000		7,552.24	I
5,000		7,573.38	I
7,000		7,607.17	I
18,000		7,754.70	I
15,000		7,800.21	I
300		7,879.18	I
500		7,898.59	I
350		7,936.31	I
300		7,956.32	I
80		8,016.01	II
1,000		8,040.93	I
900		8,075.52	I
350		8,077.52	I
350		8,126.56	I
600		8,129.26	I
300		8,159.51	I
600		8,179.34	I
300		8,191.24	I
350		8,208.63	I
2,500		8,214.73	I
3,000		8,230.77	I
500		8,232.19	I
1,500		8,274.62	I
2,000		8,298.58	I
600		8,302.40	I
900		8,807.58	I
1,000		8,900.92	I
300		8,912.78	I
350		9,025.49	I
400		9,042.10	I
350		9,178.68	I
200		9,433.67	I
25		9,505.30	I
12		9,662.04	I
25		9,734.34	I
15		9,822.11	I
12		9,902.65	I
80	h	10,047.98	II
15		10,285.45	I
20		10,862.31	

F III

Ref. 225 — G.V.S.

Intensity		Wavelength	
		Vacuum	
50	h	230.12	III
50		255.72	III
60		255.77	III
70		255.86	III
70		261.71	III
60		261.75	III
80		263.81	III
70		279.69	III
80		315.22	III
70		315.54	III
60		315.75	III
100		429.51	III
110		430.15	III
80		430.22	III
90		464.29	III
100		465.11	III
120		508.39	III
120		567.69	III
110		567.75	III
80		630.14	III
90		630.20	III
120		656.12	III
130		656.87	III
140		658.33	III
80		1,219.03	III
80		1,266.87	III
90		1,267.71	III
70		1,297.54	III
70		1,359.92	III
110		1,498.93	III
120		1,502.01	III
110		1,504.18	III
140		1,504.79	III
130		1,506.30	III
110		1,506.77	III
100		1,553.02	III
110		1,557.59	III
100		1,563.73	III
100		1,565.54	III
100		1,623.40	III
100		1,650.76	III
130		1,670.39	III
140		1,677.40	III
100		1,716.99	III
120		1,770.09	III
150		1,770.67	III
110		1,772.93	III
140		1,773.36	III
160		1,791.65	III
110		1,803.03	III
100		1,804.70	III
170		1,805.90	III
110		1,839.30	III
120		1,839.97	III
110		1,840.14	III
80		1,900.76	III
		Air	
100		2,027.44	III
120		2,030.32	III
120		2,217.17	III
120		2,452.07	III

130		2,464.85	III
130		2,470.29	III
120		2,478.73	III
150		2,484.37	III
120		2,542.77	III
120		2,580.04	III
130		2,583.81	III
120		2,593.23	III
130		2,595.53	III
140		2,599.28	III
130		2,625.01	III
140		2,629.70	III
120		2,656.44	III
130		2,755.55	III
160		2,759.63	III
120		2,788.15	III
160		2,811.45	III
140		2,833.99	III
150		2,835.63	III
150		2,860.33	III
120		2,862.86	III
140		2,887.58	III
150		2,889.45	III
120		2,905.30	III
140		2,913.29	III
160		2,916.34	III
140		2,932.49	III
140		2,994.28	III
120	h	2,997.21	III
130		2,997.53	III
120		2,999.47	III
130		3,039.25	III
120		3,039.75	III
160		3,042.80	III
150		3,049.14	III
140		3,113.62	III
160		3,115.70	III
180		3,121.54	III
140		3,124.79	III
140		3,134.23	III
140		3,146.99	III
180		3,174.17	III
170		3,174.76	III
120		3,214.00	III
140	h	4,420.30	III
120	h	4,427.35	III
120	h	4,432.32	III
140	h	4,479.99	III
150		5,012.54	III
160		5,110.99	III
140		5,753.17	III
120		5,761.20	III
150		6,091.82	III
140		6,125.50	III
130		6,233.57	III
140		6,363.05	III
120		7,336.77	III
130		7,354.94	III

F IV

Ref. 68, 226 — G.V.S.

Intensity	Wavelength	
	Vacuum	
30	169.79	IV
30	169.84	IV
30	171.07	IV
40	176.37	IV
40	181.52	IV
40	181.57	IV
30	187.24	IV
50	196.39	IV
60	196.45	IV
50	199.76	IV
50	199.80	IV
50	199.85	IV
50	199.93	IV
50	200.00	IV
70	200.09	IV
60	201.01	IV
70	201.06	IV
60	201.10	IV
80	201.16	IV
60	201.22	IV
90	208.25	IV
70	213.85	IV
70	214.06	IV
70	220.77	IV
60	226.94	IV
50	227.10	IV
60	233.22	IV
50	233.39	IV
70	239.86	IV
70	240.02	IV
90	240.08	IV
70	240.15	IV
70	240.28	IV
70	240.37	IV
100	251.03	IV
60	270.23	IV
140	419.65	IV
150	420.05	IV
160	420.73	IV
150	430.76	IV
130	490.57	IV
160	491.00	IV
50	497.38	IV
60	497.83	IV
70	498.80	IV
140	570.64	IV
140	571.30	IV
150	571.39	IV
160	572.66	IV
140	676.12	IV
130	677.15	IV
150	677.22	IV
130	678.99	IV
160	679.21	IV
	Air	
40	2,171.44	IV
50	2,298.29	IV
40	2,451.58	IV
50	2,456.92	IV
40	2,820.74	IV
50	2,826.13	IV

F V

Ref. 68, 226 — G.V.S.

Intensity	Wavelength	
	Vacuum	
40	134.54	V
40	147.95	V
50	148.00	V
40	152.51	V
40	158.54	V
40	162.27	V
40	163.50	V
50	163.56	V
90	165.98	V
100	166.18	V
40	174.70	V
50	178.43	V
40	178.59	V
40	182.98	V
40	186.72	V
40	186.79	V
50	186.84	V
40	186.97	V
40	187.01	V
60	190.57	V
70	190.84	V
40	191.97	V
40	205.55	V
100	464.37	V
110	465.37	V
120	465.98	V
100	466.99	V
90	506.16	V
100	508.08	V
70	513.97	V
60	514.08	V
80	524.59	V
90	525.29	V
100	526.30	V
70	647.67	V
100	647.77	V
110	647.87	V
70	647.97	V
130	654.03	V
110	657.23	V
140	657.33	V
60	757.04	V
60	1,082.31	V
70	1,088.39	V
	Air	
10	2,229.18	V
20	2,252.72	V
20	2,450.63	V
10	2,461.33	V
10	2,693.98	V
10	2,702.30	V
10	2,703.96	V
20	2,707.17	V

GADOLINIUM (GD)
Z = 64

Gd I and II

Ref. 1 — C.H.C.

Intensity	Wavelength	
	Air	
100	2,468.22	II
55	2,471.58	II
35	2,485.67	II
70	2,487.46	II
110	2,488.72	II
55	2,493.29	II
35	2,496.35	II
45	2,499.04	II
28	2,543.68	II
28	2,586.13	II
28	2,661.50	II
70	2,720.50	II
430	2,750.22	II
460	2,764.08	II
40	2,768.51	II
320	2,769.81	II
230	2,770.17	II
21	2,770.98	II
45	2,778.76	I
45	2,779.14	II
440	2,781.40	II
70	2,787.68	I
390	2,791.96	II
100	2,794.66	II
930	2,796.93	II
60	2,808.38	II
750	2,809.72	II
160	2,810.93	II
45	2,814.01	II
300	2,833.75	II
35	2,836.69	II
70	2,837.00	II
560	2,840.23	II
140	2,841.33	II
40	2,853.91	II
60	2,856.52	II
19	2,859.78	II
120	2,862.48	II
60	2,865.06	II
40	2,866.33	II
40	2,871.75	II
460	2,881.33	II
40	2,882.13	II
130	2,885.60	II
35	2,907.44	II
170	2,910.53	II
60	2,913.08	II
45	2,918.52	II
95	2,923.32	II
35	2,924.25	II
35	2,928.34	II
35	2,947.80	II
70	2,948.01	II
35	2,952.43	I
35	2,955.60	II
70	2,960.93	II
130	2,963.60	II
80	2,965.43	II
29	2,972.74	II
560	2,980.15	II
35	2,983.74	II
40	2,991.52	II
95	2,993.04	II
1,200	2,999.04	II
370	3,002.86	II
100	3,005.09	II
2,100	3,010.13	II
130	3,012.19	II
1,900	3,027.60	II
120	3,028.98	II
2,100	3,032.84	II
1,600	3,034.05	II
130	3,043.01	I
160	3,046.48	I
280	3,053.57	II
100	3,059.92	I
1,000	3,068.64	II
560	3,072.56	II
640	3,076.92	II
150	3,077.08	II
2,100	3,081.99	II
140	3,084.01	II
280	3,089.95	II
140	3,092.06	II
460	3,098.64	II
190	3,098.90	II
3,500	3,100.50	II
120	3,101.18	II
230	3,101.91	II
580	3,102.55	II
130	3,108.36	II
170	3,111.19	I
160	3,113.17	II
120	3,118.60	II
120	3,119.01	I
510	3,119.94	II
100	3,120.18	II
370	3,123.99	II
120	3,124.25	II
130	3,128.56	II
130	3,130.81	II
100	3,133.09	II
460	3,133.85	II
210	3,135.03	II
190	3,136.93	I
190	3,137.30	I
120	3,138.71	I
230	3,143.13	II
930	3,145.00	II
370	3,145.52	II
230	3,146.88	II
980	3,156.53	II
200	3,158.63	I
140	3,160.69	II
980	3,161.37	II
220	3,190.28	I
220	3,199.30	I
160	3,199.58	I
110	3,203.41	I
690	3,223.74	II
	3,223.78	I
110	3,225.46	II
160	3,226.32	II
220	3,232.78	I
100	3,250.19	II
110	3,259.25	II
540	3,266.73	I
250	3,267.64	I
140	3,268.34	II
110	3,274.18	II
110	3,279.53	II
100	3,281.61	II
250	3,282.25	I
	3,282.30	II
430	3,291.48	I
370	3,292.21	II
430	3,294.08	I
330	3,313.73	II
200	3,315.59	II
430	3,330.34	II
1,400	3,331.38	II
830	3,332.13	II
1,100	3,336.18	II
590	3,345.98	II
200	3,350.10	II
5,400	3,350.47	II
220	3,357.61	I
270	3,358.43	II
4,300	3,358.62	II
780	3,360.71	II
5,400	3,362.23	II
270	3,364.24	II
200	3,365.59	II
220	3,374.69	II
220	3,379.76	II

220		3,380.52	II	450		3,650.95	II	2,200		3,916.51	II	2,200		4,130.37	II
1,100		3,392.53	II	620		3,652.54	II	450		3,923.25	II	270		4,131.48	II
540		3,395.12	II	3,900		3,654.62	II	1,200		3,934.79	I	1,100		4,132.28	II
220	d	3,397.22	I	3,100		3,656.15	II			3,934.82	II	750		4,134.16	I
		3,397.32	I	210		3,658.19	I	220		3,935.38	I	410		4,137.10	II
200		3,399.41	II	1,400		3,662.26	II	450		3,941.80	I	280		4,148.86	I
540		3,399.99	II	2,700		3,664.60	II	590		3,942.63	I	540		4,162.73	II
540		3,402.07	II	2,000		3,671.20	II	270		3,943.24	I	280		4,163.09	II
200		3,406.92	I	1,000		3,674.05	I	220		3,943.62	I	280		4,167.16	II
1,100	d	3,407.56	II	350		3,679.21	I	1,400		3,945.54	I			4,167.27	II
		3,407.61	II	2,000		3,684.13	I	300		3,952.00	II	2,400		4,175.54	I
250		3,409.30	II	720		3,686.33	II	590		3,953.37	I	2,400		4,184.25	II
220		3,411.02	I	3,100		3,687.74	II	1,200		3,957.67	II	2,200		4,190.78	I
220		3,413.27	II	210		3,694.03	II	750		3,959.44	II	750		4,191.07	II
1,400		3,416.95	II	2,000		3,697.73	II			3,959.52	II	750		4,191.63	I
1,400		3,418.73	II	1,300		3,699.73	II	220		3,963.66	II	450		4,197.68	II
6,900		3,422.47	II	2,700		3,712.70	II	590		3,966.28	I	590		4,204.86	II
390		3,422.75	II	2,000		3,713.57	I	590		3,968.26	II	1,300		4,212.00	II
1,100		3,423.90	I	1,400		3,716.36	II	750		3,969.00	I	970		4,215.02	II
		3,423.92	II	2,000		3,717.48	I	270		3,969.29	II	650		4,217.20	II
830		3,424.59	II	1,800	d	3,719.45	II	450		3,971.75	II	320		4,225.03	I
390		3,425.93	II			3,719.53	II	390		3,972.71	I	4,800		4,225.85	I
220		3,428.47	II	250		3,722.07	II	590		3,973.98	II	220		4,227.14	II
690		3,432.99	II	430		3,725.47	II	300		3,974.81	I	220		4,229.80	II
1,700		3,439.21	II	1,500		3,730.84	II	750		3,979.33	I	650		4,238.78	II
830		3,439.78	II	270		3,732.32	I	450		3,987.21	II	200		4,246.57	II
2,700		3,439.99	II	230		3,732.45	II	470		3,987.84	I	1,700		4,251.73	II
390		3,449.62	II	230		3,732.67	I	320		3,992.69	I	860		4,253.37	II
1,400		3,450.38	II	510		3,733.08	II	220		3,993.21	II	650		4,253.61	II
1,100		3,451.23	II	490		3,739.76	I	650		3,994.16	II	810		4,260.12	I
540		3,454.14	II	330		3,740.02	II	700		3,996.32	II	1,600		4,262.09	I
880		3,454.90	II	4,500		3,743.47	II	320		3,997.76	II	650		4,266.60	I
200		3,455.27	I	620		3,744.83	I	470		4,001.26	II	470		4,267.00	I
200		3,457.05	II	230		3,757.74	II	260		4,004.94	II	300		4,274.17	I
220		3,461.95	II	1,000		3,757.94	I	320		4,008.33	I	910		4,280.49	II
220		3,463.00	II	1,400		3,758.31	II	300		4,008.91	II	430		4,285.82	I
2,700		3,463.98	II	820		3,759.00	II	300		4,013.80	II	300		4,286.12	I
330		3,466.95	II	620		3,760.71	II	200		4,015.58	I	540		4,296.08	II
1,700		3,467.27	II	290		3,760.92	II	300		4,017.25	I	220		4,297.17	II
1,700		3,468.99	II	870		3,762.20	II	430		4,017.71	I	430		4,299.29	I
1,400		3,473.22	II	210		3,763.33	II	300		4,019.73	I	1,100		4,306.34	I
2,200		3,481.28	II	370		3,764.20	II	300		4,022.33	II	260		4,309.29	I
1,700		3,481.80	II	870		3,767.04	II	1,100		4,023.14	I	1,800		4,313.84	I
490		3,482.60	II	8,700		3,768.39	II	810		4,023.35	I	520		4,314.40	I
220		3,486.20	I	620		3,769.45	II	220		4,027.61	I	520		4,316.05	II
980		3,491.95	II	1,400		3,770.69	II	1,100		4,028.15	I	370		4,320.52	I
1,700		3,494.40	II	250		3,771.26	I	860		4,030.88	I	750		4,321.11	II
1,400		3,505.51	II	210		3,773.45	I	700		4,033.49	I			4,321.20	I
780		3,512.22	II	210		3,776.83	I	340		4,035.40	I	2,600	d	4,325.57	II
1,100		3,512.50	II	1,000		3,782.34	II	260		4,036.84	I			4,325.69	I
830		3,513.65	I	2,900		3,783.05	II	1,400		4,037.33	II	1,900		4,327.12	I
980		3,524.20	II	1,100		3,787.56	II	700		4,037.90	II	370		4,329.58	I
430		3,528.54	II	200		3,790.63	I	410		4,043.71	I	340		4,330.61	II
540		3,542.77	II	770		3,791.17	II	1,600		4,045.01	I	240		4,331.38	I
4,300		3,545.80	II	490		3,792.39	II	270		4,046.84	II	450		4,341.28	II
3,900		3,549.36	II	5,100		3,796.37	II	270		4,047.09	I	910		4,342.18	II
1,400		3,557.05	II	720		3,801.29	II	270		4,049.20	I	1,000		4,344.30	II
540		3,558.19	II	210		3,804.39	I	1,300		4,049.43	II	2,200		4,346.46	I
430		3,558.47	II	210		3,805.09	II	2,200		4,049.86	II	910		4,346.62	I
200		3,564.05	II	560		3,805.52	II	270		4,050.37	I	220		4,347.31	II
690		3,571.93	II	3,700		3,813.97	II	810		4,053.29	II	300		4,369.77	II
330		3,574.74	II	430		3,814.74	II	2,600		4,053.64	I	970		4,373.83	I
390		3,578.36	II	770		3,816.64	II	810		4,054.72	I	280		4,392.06	I
980		3,581.91	II	430		3,818.75	II	2,600		4,058.22	I	1,400		4,401.86	I
5,400		3,584.96	II	350		3,826.05	II	650		4,059.88	I	520		4,403.14	I
540		3,590.47	II	230		3,827.33	II	270		4,061.30	II	260		4,406.67	II
1,100		3,592.71	II	230		3,829.46	II	650		4,062.59	II	260		4,408.25	II
200		3,593.44	II	370		3,831.80	II	1,900		4,063.39	II	220		4,409.25	I
540		3,600.96	II	210		3,832.97	I	540		4,063.59	II	520		4,411.16	I
1,100		3,604.87	I	330		3,834.99	II	260		4,066.04	I	860		4,414.16	I
270		3,605.26	II	970		3,836.91	II	520		4,068.35	I	700		4,414.73	I
250		3,605.66	II	1,000		3,839.64	II	260		4,068.74	I	340		4,419.03	II
830		3,608.75	II	1,200		3,842.20	II	750		4,070.29	II	1,400		4,422.41	I
830		3,610.76	II	1,400		3,843.28	II			4,070.39	II	1,100		4,430.63	I
220		3,610.91	II	1,400		3,844.58	II	650		4,073.20	II	240	d	4,436.10	I
540		3,613.39	II	3,300		3,850.69	II	300		4,073.76	II			4,436.22	II
270	d	3,614.21	II	5,100		3,850.97	II	1,300		4,078.44	II	300		4,464.74	I
		3,614.42	I	4,300		3,852.45	II	2,800		4,078.70	I	300		4,466.55	II
430		3,617.16	II	470		3,855.56	II	520		4,083.70	I			4,466.60	I
390		3,620.46	II	250		3,863.05	II	1,500		4,085.56	II	520		4,467.08	I
270		3,624.89	II	1,600		3,866.99	I	260		4,087.69	II	700		4,474.13	I
250		3,629.51	II	250		3,873.57	I	650		4,090.41	I	860		4,476.12	I
330		3,634.76	II	220		3,875.46	II	1,100		4,092.71	I	220		4,478.80	II
220		3,639.05	II	1,500		3,894.70	II	260		4,093.72	I	280		4,481.06	II
250		3,640.18	II	450		3,895.79	II	260		4,094.48	II	220		4,483.33	II
330		3,641.39	II	750		3,902.40	II	2,600		4,098.61	II	220		4,484.70	I
870		3,645.62	II	300		3,902.71	I	520		4,098.90	II	280		4,486.90	I
6,100		3,646.19	II	240		3,904.29	I	650		4,100.26	I	500		4,497.13	I
310		3,649.44	II	450		3,905.65	I	390		4,111.44	II	220		4,497.32	I

430		4,506.21	I
140		4,506.33	II
140		4,514.50	II
1,100		4,519.66	I
300		4,522.82	II
150		4,524.12	I
910		4,537.81	I
220		4,540.02	II
300		4,542.03	I
240		4,548.00	I
120		4,558.08	II
130		4,573.81	I
260		4,575.91	I
280		4,579.59	I
410		4,581.29	I
130		4,582.53	II
410		4,583.07	I
160		4,586.99	I
220		4,596.98	II
320		4,597.91	II
410		4,598.90	I
340		4,601.05	II
240		4,602.93	I
520		4,614.50	I
140		4,624.42	I
430		4,636.64	I
110		4,639.00	II
170		4,640.04	I
170		4,646.00	I
170		4,647.64	I
170	d	4,648.59	I
		4,648.70	I
430		4,653.54	I
140	h	4,670.87	I
170		4,679.18	I
260		4,680.04	I
430		4,683.33	I
140		4,688.12	I
700		4,694.33	I
170		4,695.49	I
430		4,697.42	I
170		4,703.13	I
200		4,709.78	I
110		4,721.46	I
150		4,728.47	II
220		4,732.60	II
260		4,735.75	I
410		4,743.65	I
110		4,745.82	I
320		4,758.70	I
110		4,760.74	I
130		4,763.82	I
470		4,767.24	I
180		4,781.92	I
300		4,784.62	I
110		4,786.75	I
140		4,801.05	II
220		4,807.45	I
320		4,821.69	I
130		4,835.26	I
110		4,848.10	I
110		4,862.59	I
170		4,865.02	II
120		4,871.50	I
280		4,934.12	I
220		4,938.61	I
110		4,952.47	I
130		4,958.79	I
65		5,010.82	II
55		5,011.74	I
750		5,015.04	I
55		5,023.13	II
65		5,031.29	II
75		5,039.09	I
65		5,050.88	II
55		5,073.74	I
55		5,082.80	I
95		5,092.25	II
65		5,096.06	II
130		5,098.38	II
55		5,100.94	II
910		5,103.45	I
180		5,108.91	II
120		5,125.56	II
65		5,130.28	II
65		5,135.59	I
75		5,136.04	I
85		5,140.84	II
75		5,141.50	I
75		5,142.68	I
860		5,155.84	I
55		5,156.76	II
75		5,158.48	I
75		5,163.70	I
55		5,164.54	II
190		5,176.28	II
55		5,187.24	II
55		5,187.88	I
55		5,191.08	II
410		5,197.77	I
55		5,210.49	II
85		5,217.48	I
280		5,219.40	I
75		5,220.30	II
130		5,233.93	I
65		5,246.87	I
320		5,251.18	I
120		5,252.14	II
85		5,254.75	I
140		5,255.80	I
65		5,268.78	I
55		5,272.91	I
55		5,282.48	I
280		5,283.08	I
280		5,301.67	I
220		5,302.76	I
55		5,306.70	I
280		5,307.30	I
130		5,321.50	I
280		5,321.78	I
110		5,327.32	I
65		5,328.30	I
170		5,333.30	I
55		5,337.53	I
300		5,343.00	I
85		5,345.13	I
75		5,345.68	I
200		5,348.67	I
300		5,350.38	I
240		5,353.26	I
55		5,361.66	I
95		5,365.38	I
95		5,369.92	I
150		5,370.63	I
85		5,389.50	I
85		5,413.20	I
85		5,415.69	I
65		5,453.46	I
55		5,583.68	II
55	d	5,591.85	I
190		5,617.91	I
65		5,629.55	I
110		5,632.25	I
260		5,643.24	I
55	b	5,680.89	GDO
390		5,696.22	I
95		5,701.35	I
65		5,709.42	I
120		5,733.86	II
85		5,746.36	I
85	d	5,754.17	I
75		5,776.02	I
240		5,791.38	I
65	h	5,796.80	I
55		5,802.92	I
55	h	5,807.72	I
55	h	5,809.22	I
55		5,815.85	II
65	h	5,819.51	GDO
55		5,840.47	II
220		5,851.63	I
55		5,855.24	II
280		5,856.22	I
55		5,860.73	II
65		5,877.26	II
55		5,886.46	I
55		5,904.07	II
110		5,904.56	I
170		5,911.45	II
65		5,913.55	II
55		5,916.77	I
85		5,930.29	I
85		5,936.84	I
65		5,937.71	I
55	h	5,940.95	GDO
55	h	5,942.78	GDO
55		5,951.60	II
55		5,956.48	II
85		5,977.25	I
110	h	5,988.02	I
85		5,999.08	I
65		6,000.96	GDO
75	h	6,001.87	GDO
55		6,004.57	II
55		6,008.71	I
55		6,021.13	I
55		6,080.65	II
430		6,114.07	I
55		6,180.42	II
110	b	6,182.68	GDO
110	b	6,200.86	GDO
110	b	6,211.71	GDO
110	b	6,220.93	GDO
55	b	6,231.62	GDO
75	b	6,241.66	GDO
55	b	6,252.12	GDO
55	b	6,262.64	GDO
45	b	6,273.00	GDO
85		6,289.73	II
30		6,292.87	I
75		6,305.15	II
30		6,309.11	II
27		6,317.19	I
40		6,331.35	I
17		6,333.75	I
17		6,336.34	I
27		6,346.65	II
27	h	6,351.72	I
17		6,363.23	I
40		6,380.95	II
17		6,382.19	II
22		6,408.55	I
22		6,422.42	II
17		6,424.52	I
19	h	6,470.29	I
15		6,480.11	II
40	h	6,538.15	I
22		6,549.25	I
55		6,564.78	I
10		6,568.00	II
10		6,573.80	I
30		6,591.60	I
15		6,593.42	I
10		6,610.04	II
50		6,634.36	II
35		6,640.08	I
10		6,642.76	I
30		6,643.98	I
10		6,646.85	I
10		6,653.55	I
10		6,679.56	II
35		6,681.23	II
10		6,692.86	I
10		6,704.18	II
14		6,718.14	II
17		6,727.83	II
85		6,730.73	I
50		6,752.67	II
14		6,753.91	II
14		6,783.39	I
26		6,786.33	II
10		6,787.18	I
12		6,814.56	I
26		6,816.49	I
17		6,820.90	I
100		6,828.25	I
35		6,846.60	II
30		6,857.13	II
15		6,864.25	I
21		6,887.63	II
14		6,900.73	II
100		6,916.57	I
21		6,920.62	II
15		6,924.99	II
21		6,926.49	I
17		6,945.98	II
15		6,957.74	II
15		6,959.24	II
14		6,964.33	I
15		6,971.66	II
12		6,976.35	II
10		6,978.27	II
26	h	6,980.86	I
50		6,985.89	II
10		6,988.75	II
75		6,991.92	I
21		6,993.18	I
60		6,996.76	II
17		7,000.75	II
45		7,006.16	II
10		7,016.60	I
21		7,037.26	II
14		7,051.00	II
13		7,054.62	II
10		7,058.02	II
10		7,068.09	II
18		7,071.00	I
18		7,073.63	I
14		7,098.11	I
14		7,098.73	I
10	h	7,116.77	II
21		7,118.86	II
35		7,122.57	I
13		7,135.73	II
18		7,147.31	II
13		7,158.28	I
170		7,168.37	I
21		7,172.26	II
28		7,189.57	II
13		7,197.08	II
13		7,201.41	II
10		7,228.02	I
25		7,233.45	I
14		7,252.70	II
28		7,262.66	I
14		7,291.35	I
21		7,313.28	I
18		7,324.89	II
14		7,373.81	I
14		7,376.41	I
13		7,377.27	II
13		7,380.28	I
13		7,394.90	II
13		7,430.19	I
35		7,441.85	I
40		7,464.36	I
55		7,562.97	I
10		7,563.19	II
10		7,588.20	I
10		7,611.78	I
21		7,621.96	I
21		7,650.32	I
25		7,672.56	I
10		7,676.06	I
13		7,694.45	I
80		7,733.50	I
35		7,749.30	I
10		7,755.97	I
10		7,766.48	II
11	h	7,844.87	I
10		7,845.80	I
35		7,846.35	II
35		7,856.93	I
14		7,869.72	I
25		7,930.25	II
13		8,077.59	I
18		8,146.15	I
11		8,218.08	I
10		8,275.42	I
10		8,349.73	I
11		8,398.30	I
10		8,445.47	I
13	h	8,527.88	I
21		8,668.63	I
11		8,770.36	I
13		8,784.85	I
10		8,795.76	I
21	h	8,832.06	II
14	h	8,849.14	I
18	h	8,867.31	I

Gd III

Ref. 46, 137, 151 — J.F.K.

Intensity	Wavelength	
	Vacuum	
600	1,813.47	III
900	1,946.26	III
1,100	1,974.34	III
2,200	1,975.24	III

Air

900	2,008.79	III
3,400	2,018.07	III
1,800	2,027.82	III
800	2,046.02	III
500	2,057.79	III
1,500	2,080.08	III
1,800	2,098.20	III
1,300	2,125.68	III
1,400	2,148.03	III
1,700	2,176.84	III
1,700	2,223.95	III
1,700	2,236.73	III
1,200	2,239.84	III
1,500	2,243.75	III
1,700	2,250.18	III
1,300	2,257.05	III
1,300	2,292.51	III
1,000	2,300.38	III
1,200	2,303.72	III
1,500	2,307.03	III
1,100	2,313.50	III
1,700	2,313.56	III
1,000	2,315.09	III
1,400	2,323.12	III
1,700	2,323.18	III
2,200	2,323.78	III
1,900	2,329.35	III
2,100	2,335.01	III
1,600	2,336.02	III
1,900	2,338.97	III
1,600	2,339.88	III
2,100	2,342.74	III
2,500	2,346.52	III
2,800	2,359.31	III
1,900	2,360.87	III
2,300	2,361.91	III
1,400	2,362.38	III
2,100	2,363.26	III
1,600	2,365.22	III
2,000	2,373.38	III
1,300	2,374.29	III
1,300	2,381.38	III
1,600	2,387.82	III
2,000	2,388.77	III
1,200	2,393.86	III
1,200	2,397.34	III
1,200	2,405.03	III
1,300	2,408.41	III
1,200	2,409.35	III
1,500	2,466.84	III
1,100	2,469.14	III
1,300	2,499.53	III
1,600	2,520.38	III
1,600	2,534.11	III
1,300	2,536.10	III
1,600	2,551.56	III
2,200	2,553.90	III
2,100	2,554.04	III
2,500	2,563.33	III
2,100	2,564.46	III
1,000	2,565.04	III
2,400	2,565.95	III
1,800	2,569.27	III
1,800	2,573.57	III
2,000	2,576.06	III
1,300	2,576.15	III
1,400	2,578.13	III
1,600	2,578.76	III
1,700	2,583.62	III
2,000	2,588.21	III
2,000	2,588.46	III
1,300	2,595.81	III
1,800	2,609.77	III
1,200	2,619.40	III
1,200	2,621.52	III
1,400	2,623.52	III
1,400	2,625.48	III
2,000	2,628.10	III
1,300	2,628.99	III
2,400	2,629.83	III
2,100	2,632.30	III
1,800	2,633.32	III
1,400	2,635.71	III
1,600	2,636.44	III
1,700	2,637.15	III
2,100	2,637.97	III
2,100	2,638.06	III
2,200	2,640.53	III
1,600	2,641.65	III
2,100	2,643.71	III
1,600	2,644.52	III
1,800	2,646.04	III
1,800	2,646.84	III
1,600	2,651.48	III
2,000	2,655.59	III
1,900	2,656.55	III
2,200	2,660.83	III
1,800	2,675.75	III
1,800	2,679.44	III
1,700	2,680.63	III
1,800	2,682.52	III
1,500	2,683.91	III
1,600	2,692.78	III
1,900	2,692.86	III
1,500	2,694.43	III
2,800	2,697.39	III
1,500	2,702.91	III
2,800	2,703.28	III
1,600	2,704.53	III
1,800	2,717.35	III
2,700	2,727.89	III
450	2,751.24	III
1,800	2,833.83	III
9,000	2,904.73	III
1,800	2,918.40	III
9,500	2,955.53	III
1,000	2,975.42	III
1,000	2,984.10	III
1,000	3,116.59	III
2,500	3,118.04	III
4,000	3,176.66	III
400	3,253.53	III
400	3,330.34	III
400	3,371.05	III
400	3,402.97	III
450	3,624.90	III
250	3,700.47	III
300	3,831.73	III
300	3,910.24	III
300	4,016.91	III
600	4,177.26	III
400	4,279.96	III
300	4,314.28	III
300	4,445.91	III
600	4,684.25	III
600	4,715.06	III
600	4,782.79	III
250	4,976.72	III
5,000	5,091.70	III
300	5,124.06	III
1,800	5,347.95	III
3,000	5,365.96	III
1,100	5,412.62	III
4,000	5,553.30	III
3,000	5,587.88	III
3,000	5,658.98	III
1,800	5,786.96	III
1,500	5,862.09	III
1,500	5,987.85	III
5,000	14,332.88	III
2,000	17,474.78	III
800	19,996.34	III
800	21,259.44	III
600	22,493.33	III

Gd IV
Ref. 152 — J.F.K.

Intensity	Wavelength	
	Vacuum	
1,000	967.92	IV
1,000	983.42	IV
1,000	987.10	IV
1,000	987.91	IV
1,000	995.04	IV
1,000	995.80	IV
1,000	996.49	IV
1,000	999.24	IV
1,000	1,000.36	IV
1,000	1,002.73	IV
1,000	1,004.46	IV
1,000	1,005.66	IV
1,000	1,006.55	IV
1,200	1,007.24	IV
1,200	1,063.84	IV
500	1,228.37	IV
500	1,307.23	IV
500	1,313.29	IV
500	1,316.71	IV
600	1,321.42	IV
500	1,330.79	IV
1,100	1,393.24	IV
1,600	1,476.98	IV
1,500	1,705.03	IV
1,600	1,706.01	IV
2,000	1,736.24	IV
1,500	1,815.32	IV
400	1,997.89	IV
	Air	
800	2,049.28	IV
800	2,061.30	IV
800	2,070.40	IV
800	2,076.66	IV
1,000	2,094.29	IV
1,000	2,296.89	IV
1,000	2,352.66	IV
800	2,379.17	IV
900	2,385.65	IV
500	2,390.07	IV
700	2,392.30	IV
700	2,393.29	IV
700	2,395.76	IV
500	2,396.22	IV
600	2,396.27	IV
1,400	2,397.87	IV
700	2,402.70	IV
500	2,412.21	IV
900	2,419.26	IV
500	2,439.84	IV
600	2,440.38	IV
600	2,468.60	IV

GALLIUM (GA)

Z = 31

Ga I and II
Ref. 19, 132, 195, 281 — L.J.R.

Intensity		Wavelength	
		Vacuum	
2		829.60	II
2		958.67	II
1		960.57	II
2		969.19	II
2		998.52	II
3		1,002.95	II
5		1,012.38	II
3		1,019.10	II
5		1,023.80	II
8		1,033.69	II
1		1,113.87	II
3		1,119.25	II
5		1,130.81	II
1		1,167.62	II
2		1,173.78	II
3		1,186.81	II
1		1,227.13	II
5		1,286.38	II
3		1,327.81	II
20		1,414.44	II
5		1,449.49	II
2		1,463.65	II
3		1,473.73	II
3		1,483.52	II
3		1,485.95	II
3		1,495.21	II
3		1,504.41	II
3		1,505.01	II
5		1,514.57	II
3		1,515.19	II
8		1,535.40	II
5		1,536.37	II
1		1,536.91	II
3		1,669.83	II
5		1,695.85	II
5		1,799.42	II
10		1,813.98	II
15		1,845.30	II
		Air	
20		2,091.34	II
1		2,218.04	I
1		2,255.03	I
1		2,259.23	I
2		2,294.19	I
1		2,297.87	I
3		2,338.24	I
1		2,338.60	I
3		2,371.29	I
2		2,377.53	II
4		2,418.69	I
5		2,438.88	II
6		2,450.08	I
7		2,500.19	I
3		2,500.71	I
5		2,513.55	II
3		2,514.15	II
2		2,551.26	II
3		2,552.87	II
4		2,555.28	II
5		2,607.47	I
8		2,624.82	I
10		2,632.66	I
3		2,659.87	I
10		2,665.05	I
8		2,691.29	I
20		2,700.47	II
3		2,719.66	I
15		2,780.15	II
6		2,874.24	I
1		2,886.45	II
		2,893.65	II
2		2,910.77	II
6		2,943.64	I
6		2,944.17	I
3		2,969.41	II
1		2,971.01	II
3		2,971.60	II
5		2,974.77	II
1		2,992.84	II
2		3,011.90	II
1		3,158.18	II
4		3,374.94	II
1		3,375.95	II
2		3,436.66	II
3		3,446.46	II
2		3,447.26	II
5		3,470.34	II
1		3,471.46	II
1		3,472.52	II
2		3,583.60	II
1		3,693.93	II
2		3,705.85	II
4		3,734.85	II
9		3,924.39	II
10		4,032.99	I
10		4,172.04	I
4		4,251.11	II
15		4,251.16	II
10	h	4,254.04	II
4	h	4,255.64	II
5		4,255.70	II
10		4,255.77	II
40		4,262.00	II
3		5,218.21	II
1		5,338.3	II
2		5,353.49	I
2		5,360.6	II
1		5,363.5	II
3		5,416.8	II
1		5,421.6	II
1		5,425.6	II
10		6,334.2	II
2,000		6,396.56	I
1,000		6,413.44	I
5		6,419.4	II
3		6,456.3	II
1		7,000.0	II
3	h	7,051.24	I
5	h	7,106.82	I
1	h	7,116.3	I
2	h	7,172.9	I
5	h	7,193.6	I
7		7,198.7	II
10	h	7,251.4	I
3	h	7,289.6	I

5	h	7,349.3	I
20	h	7,403.0	I
30	h	7,464.0	I
6	h	7,556.6	I
10	h	7,620.5	I
50	h	7,734.77	I
2		7,793.0	II
100	h	7,800.01	I
4	h	7,801.6	I
15	h	8,002.55	I
20	h	8,074.25	I
3	h	8,167.5	I
5	h	8,171.6	I
100	h	8,311.86	I
200	h	8,386.49	I
10	h	8,389.30	I
7	h	8,415.51	I
10	h	8,419.91	I
20	h	8,808.75	I
30	h	8,813.56	I
20	h	8,856.37	I
30	h	8,944.33	I
200	h	9,492.92	I
200	h	9,493.12	I
300	h	9,589.36	I
20	h	9,594.25	I
60	h	10,898.10	I
100	h	10,905.95	I
10		10,968.27	I
20		11,103.51	I
400		11,949.12	I
200		12,109.78	I
40		12,885.05	I
50		13,057.50	I
50		14,982.75	I
60		14,996.64	I
20		17,756.91	I
10		17,868.96	I
60		22,016.81	I
70		22,568.71	I

Ga III

Ref. 141 — L.J.R.

Intensity	Wavelength	
	Vacuum	
50	620.00	III
40	622.01	III
90	806.51	III
90	817.30	III
50	828.70	III
80	1,085.00	III
60	1,105.61	III
90	1,150.27	III
90	1,267.16	III
80	1,293.46	III
60	1,295.36	III
60	1,323.15	III
70	1,353.92	III
90	1,495.07	III
50	1,534.46	III
	Air	
90	2,417.70	III
90	2,423.98	III
15	2,424.36	III
50	3,521.77	III
80	3,581.19	III
100	3,589.34	III
10	3,731.10	III
10	3,806.60	III
100	4,380.69	III
150	4,381.76	III
100	4,863.00	III
150	4,993.78	III
10	5,808.28	III
20	5,848.25	III
15	5,993.51	III

Ga IV

Ref. 141, 143 — L.J.R.

Intensity	Wavelength	
	Vacuum	
14	294.53	IV
61	295.67	IV
41	304.99	IV
4	422.12	IV
25	423.18	IV
16	439.92	IV
67	1,137.06	IV
70	1,156.10	IV
70	1,163.60	IV
75	1,170.58	IV
48	1,171.71	IV
68	1,185.23	IV
40	1,186.06	IV
73	1,190.89	IV
73	1,193.02	IV
75	1,195.02	IV
69	1,201.54	IV
72	1,206.89	IV
63	1,216.15	IV
50	1,228.03	IV
60	1,236.38	IV
60	1,238.59	IV
45	1,241.81	IV
75	1,245.53	IV
83	1,258.77	IV
81	1,264.66	IV
82	1,267.15	IV
81	1,279.24	IV
80	1,285.33	IV
82	1,295.86	IV
83	1,299.46	IV
82	1,303.53	IV
80	1,309.68	IV
80	1,314.82	IV
85	1,338.09	IV
77	1,347.03	IV
76	1,351.06	IV
74	1,364.63	IV
60	1,395.54	IV
77	1,402.55	IV
70	1,405.32	IV
73	1,465.87	IV

Ga V

Ref. 2, 62, 140 — L.J.R.

Intensity	Wavelength	
	Vacuum	
5	290.53	V
1	296.13	V
5	296.82	V
30	298.44	V
20	299.47	V
30	300.01	V
25	300.57	V
10	300.78	V
20	301.19	V
30	302.86	V
20	303.84	V
30	307.03	V
30	308.26	V
15	309.64	V
30	311.79	V
25	312.41	V
30	313.68	V
15	315.95	V
20	316.48	V
40	319.41	V
12	320.53	V
40	322.31	V
50	322.99	V
30	323.10	V
40	324.25	V
40	324.95	V
40	326.14	V
30	326.77	V
30	328.65	V
5	336.61	V
20	878.17	V
40	973.21	V
10	977.89	V
15	979.60	V
20	984.95	V
40	989.75	V
90	1,014.47	V
90	1,019.71	V
20	1,033.55	V
30	1,038.76	V
30	1,047.50	V
120	1,050.48	V
80	1,054.56	V
90	1,058.12	V
80	1,066.69	V
35	1,068.59	V
30	1,069.45	V
60	1,069.60	V
55	1,071.19	V
45	1,071.41	V
80	1,073.77	V
90	1,078.83	V
110	1,079.60	V
60	1,080.99	V
250	1,085.01	V
80	1,087.37	V
40	1,090.53	V
90	1,091.71	V
100	1,094.36	V
80	1,095.10	V
70	1,101.62	V
160	1,102.83	V
140	1,103.03	V
60	1,104.93	V
75	1,105.62	V
70	1,106.17	V
40	1,115.55	V
80	1,118.34	V
55	1,123.18	V
80	1,123.66	V
120	1,126.40	V
80	1,127.75	V
130	1,128.10	V
120	1,128.53	V
100	1,129.94	V
80	1,131.43	V
40	1,133.91	V
130	1,136.07	V
65	1,138.20	V
60	1,144.30	V
50	1,145.70	V
30	1,148.42	V
45	1,150.09	V
130	1,150.23	V
120	1,156.51	V
35	1,157.74	V
25	1,169.40	V
40	1,178.95	V
80	1,213.17	V
30	1,265.45	V
30	1,276.85	V
15	1,283.64	V
10	1,311.35	V

GERMANIUM (GE)

Z = 32

Ge I and II

Ref. 5, 119, 293, 340 — C.H.C.

Intensity		Wavelength	
		Vacuum	
1		822.97	II
3		835.08	II
10		850.50	II
10		862.234	II
15		875.493	II
15		905.977	II
20		920.554	II
50		999.101	II
100		1,016.638	II
100		1,075.072	II
300		1,085.51	II
200		1,098.71	II
500		1,106.74	II
500		1,120.46	II
200		1,164.27	II
500		1,181.19	II
500		1,181.65	II
200		1,188.73	II
100		1,189.62	II
300		1,191.26	II
50		1,191.72	II
500		1,237.059	II
500		1,261.905	II
100		1,264.710	II
100		1,380.42	II
50		1,392.26	II
200		1,401.24	II
200		1,538.091	II
500		1,576.855	II
75		1,581.070	II
100		1,602.486	II
3	r	1,615.57	I
2	r	1,624.130	I
2	r	1,630.173	I
3	r	1,636.31	I
2		1,638.96	I
4	r	1,639.730	I
2		1,647.531	I
200		1,649.194	II
2		1,651.528	I
4	r	1,651.955	I
3		1,661.345	I
4	r	1,663.539	I
10	h	1,665.275	I
4		1,667.802	I
3	r	1,670.608	I
100	r	1,691.090	I
200	r	1,716.784	I
100	h	1,739.102	I
100		1,742.195	I
50		1,746.065	I
200		1,750.043	I
100		1,758.279	I
100	h	1,764.185	I
100	h	1,765.284	I
50	h	1,766.433	I
200		1,774.176	I
200		1,785.046	I
100	h	1,793.071	I
75	h	1,801.432	I
200	h	1,841.328	I
200	h	1,842.410	I
100	h	1,844.410	I
100	h	1,845.872	I
100	h	1,846.958	I
200		1,853.134	I
500	r	1,860.086	I
100		1,865.052	I
300	r	1,874.256	I
100		1,895.197	I
500	r	1,904.702	I
50	h	1,908.434	I
30		1,912.409	I
300	r	1,917.592	I
100	h	1,923.467	I
500	r	1,929.826	I
10	h	1,934.048	I
100	r	1,937.483	I
500		1,938.008	II
100	r	1,938.300	I
500		1,938.891	II
30	s	1,944.116	I
200		1,944.731	I
200		1,955.115	I
500		1,962.013	I
30	h	1,963.373	I
30		1,965.383	I
200		1,970.880	I
200		1,979.274	II
300	h	1,987.849	I
300		1,988.267	I
500	r	1,998.887	I
		Air	
50		2,007.04	II
200		2,011.29	I
1,700		2,019.068	I
2,400	r	2,041.712	I
1,600	r	2,043.770	I
420		2,054.461	I
220	h	2,057.238	I
750	r	2,065.215	I
2,600	r	2,068.656	I
420		2,086.021	I
2,000	r	2,094.258	I
240		2,105.824	I
95	h	2,124.744	I
50	h	2,186.451	I
100		2,197.62	II
340	r	2,198.714	I
100		2,205.85	II
15		2,220.375	I
18		2,256.001	I
18		2,314.201	I
24		2,327.918	I

Intensity		Wavelength	
15		2,359.233	I
20		2,379.144	I
10		2,389.472	I
15		2,397.885	I
130		2,417.367	I
30		2,436.412	I
100		2,478.66	II
90		2,497.962	I
500		2,500.54	II
70		2,533.230	I
3		2,556.298	I
28		2,589.188	I
500		2,592.534	I
8		2,644.184	I
1,200		2,651.172	I
500		2,651.568	I
500		2,691.341	I
200		2,704.03	II
850		2,709.624	I
400		2,729.78	II
40		2,740.426	I
650		2,754.588	I
50		2,770.59	II
75		2,772.35	II
70		2,793.925	I
80		2,829.008	I
1,000		2,831.843	II
50		2,834.28	II
75		2,839.68	II
1,000		2,845.527	II
75		2,853.97	II
750		3,039.067	I
600		3,067.021	I
20		3,124.816	I
50		3,186.72	II
100		3,221.64	II
110		3,269.489	I
50		3,312.56	II
75		3,323.64	II
100		3,455.72	II
300		3,499.21	II
30		3,845.11	II
70		4,226.562	I
10		4,685.829	I
75	h	4,689.87	II
50	h	4,690.02	II
1,000		4,741.806	II
1,000		4,814.608	II
50		4,824.097	II
100		5,131.752	II
200		5,178.648	II
3		5,194.583	I
6		5,265.892	I
6		5,513.263	I
8		5,564.741	I
8		5,607.010	I
6		5,616.135	I
7		5,621.426	I
8		5,655.96	I
6		5,664.226	I
5		5,664.842	I
9		5,691.954	I
6		5,701.776	I
5		5,717.877	I
6		5,801.029	I
9		5,802.093	I
1,000		5,893.389	II
500		6,021.041	II
150		6,078.39	II
50		6,267.14	II
150		6,268.07	II
100		6,268.34	II
75		6,283.452	II
100		6,336.377	II
100		6,484.181	II
6		6,557.488	I
50		6,780.51	II
50		7,049.369	II
6		7,130.12	I
30		7,145.390	II
7		7,330.38	I
5		7,353.334	I
7		7,384.208	I
6		7,402.64	I
7		7,511.57	I
5		7,776.20	I
10		7,833.575	I
7		7,837.63	I
6		7,853.77	I
7		7,878.12	I
5		7,962.26	I
5		7,983.33	I
10		8,031.039	I
6		8,044.165	I
5		8,095.29	I
5		8,225.22	I
7		8,226.09	I
10		8,256.013	I
5		8,264.15	I
5		8,280.09	I
6		8,281.04	I
8		8,367.81	I
7		8,391.70	I
5		8,396.36	I
5		8,429.42	I
10		8,482.21	I
8		8,506.70	I
5		8,507.66	I
8		8,564.89	I
6		8,599.27	I
6		8,652.42	I
5		8,669.60	I
9		8,700.60	I
5		8,712.90	I
6		8,734.78	I
6		8,789.88	I
5		9,068.785	I
5		9,095.957	I
6		9,398.868	I
20		9,474.993	II
20		9,475.645	II
4		9,492.559	I
7		9,625.664	I
5		10,039.436	I
4		10,200.952	I
10		10,382.427	I
10		10,404.913	I
8		10,734.068	I
8		10,947.416	I
10		11,125.130	I
230		11,252.83	I
24		11,293.40	I
33		11,318.13	I
55		11,459.05	I
150		11,483.77	I
175		11,614.81	I
600		11,714.76	I
10		11,839.77	I
55		11,917.01	I
10		12,025.64	I
10		12,055.49	I
30		12,061.41	I
45		12,065.76	I
1,300		12,069.20	I
30		12,198.88	I
20		12,207.73	I
60		12,286.75	I
55		12,338.76	I
1,050		12,391.58	I
48		12,540.41	I
15		12,636.80	I
150		12,676.58	I
40		12,681.28	I
115		12,800.66	I
175		12,836.38	I
12		12,847.92	I
120		12,955.73	I
15		13,028.64	I
235		13,107.61	I
20		13,492.28	I
42		13,534.85	I
28		13,724.48	I
42		14,116.70	I
42		14,297.15	I
40		14,569.84	I
12		14,667.52	I
470		14,822.38	I
16		14,921.97	I
15		15,001.75	I
13		15,041.21	I
20		15,504.34	I
14		16,424.77	I
12		16,626.64	I
70		16,699.29	I
150		16,759.79	I
135		17,214.34	I
16		18,428.30	I
35		18,495.54	I
10		18,764.11	I
70		18,811.86	I
62		19,279.24	I
28		20,673.64	I
4		21,518.30	I
9		22,091.84	I
5		23,921.92	I

Ge III

Ref. 341 — C.H.C.

Intensity	Wavelength	
	Vacuum	
2	542.90	III
2	663.77	III
3	670.88	III
2	680.28	III
2	952.76	III
12	988.96	III
15	995.72	III
10	996.50	III
15	1,011.21	III
10	1,012.31	III
8	1,032.62	III
12	1,040.99	III
12	1,058.91	III
40	1,088.45	III
10	1,137.92	III
12	1,150.55	III
8	1,159.15	III
8	1,159.62	III
8	1,160.79	III
10	1,173.78	III
8	1,212.47	III
4	1,323.24	III
10	1,525.32	III
2	1,527.15	III
9	1,600.09	III
6	1,883.26	III
2	1,978.22	III
	Air	
2	2,019.22	III
4	2,022.25	III
3	2,062.14	III
15	2,100.05	III
15	2,102.42	III
25	2,104.45	III
3	2,922.86	III
25	3,197.56	III
35	3,211.86	III
25	3,214.95	III
40	3,255.05	III
20	3,259.90	III
5	3,369.57	III
20	3,414.27	III
40	3,434.03	III
8	3,464.59	III
40	3,489.08	III
2	3,724.51	III
15	3,884.78	III
200	4,178.96	III
12	4,245.41	III
200	4,260.85	III
150	4,291.71	III
10	4,674.36	III
10	5,016.88	III
18	5,134.75	III
5	5,229.37	III
3	5,256.61	III

Ge IV

Ref. 341 — C.H.C.

Intensity	Wavelength	
	Vacuum	
1	440.11	IV
1	441.95	IV
3	847.80	IV
3	868.30	IV
8	915.00	IV
8	936.70	IV
4	938.90	IV
1	1,073.44	IV
20	1,188.99	IV
20	1,229.81	IV
2	1,494.89	IV
6	1,500.61	IV
3	1,648.14	IV
	Air	
2	2,293.0	IV
2	2,343.37	IV
15	2,445.38	IV
15	2,445.71	IV
30	2,488.25	IV
20	2,542.44	IV
5	2,631.78	IV
3	2,698.08	IV
15	2,717.44	IV
30	2,736.09	IV
30	2,788.61	IV
5	3,071.84	IV
60	3,554.19	IV
50	3,676.65	IV

Ge V

Ref. 342 — C.H.C.

Intensity	Wavelength	
	Vacuum	
700	294.51	V
1,000	295.64	V
200	304.98	V
20	621.52	V
35	716.26	V
50	724.21	V
35	733.54	V
35	735.35	V
35	741.52	V
60	746.88	V
40	750.26	V
35	755.84	V
60	760.05	V
60	958.51	V
300	971.35	V
150	984.92	V
200	988.13	V
300	990.66	V
300	1,004.38	V
300	1,016.66	V
250	1,038.40	V
900	1,045.71	V
400	1,050.05	V
300	1,054.59	V
300	1,068.43	V
400	1,069.13	V
700	1,072.66	V
600	1,086.65	V
500	1,087.85	V
800	1,089.49	V
300	1,092.09	V
1,000	1,116.94	V
300	1,122.01	V
700	1,163.39	V
300	1,165.26	V
200	1,176.69	V
700	1,222.30	V

GOLD (AU)

Z = 79

Au I and II

Ref. 38, 72, 234 — C.H.C.

Intensity	Wavelength	
	Vacuum	
	925.72	II
	946.03	II
	950.39	II
20	957.78	II
3	967.94	II
3	974.47	II
2	982.24	II
	1,062.67	II
8	1,066.96	II
	1,085.00	II
8	1,090.78	II
5	1,094.92	II
20	1,103.31	II
3	1,166.76	II

5		1,210.86	II
20		1,224.57	II
40	h	1,305.34	I
25		1,310.47	I
100	h	1,328.37	I
3		1,336.26	II
10	h	1,338.37	I
10	h	1,342.80	I
20		1,350.09	I
20		1,350.84	I
22		1,351.74	I
25		1,352.82	I
25		1,354.14	I
30		1,355.79	I
35		1,357.86	I
40		1,360.51	I
6		1,362.33	II
20		1,362.47	I
8		1,363.15	II
50		1,363.98	I
25		1,364.15	I
10		1,364.74	II
60		1,368.62	I
35		1,368.98	I
70		1,374.82	I
50		1,375.76	I
30		1,378.87	I
8		1,380.53	II
80		1,382.75	I
50		1,385.33	I
20		1,389.14	I
60		1,392.27	I
6		1,393.80	II
50		1,402.12	I
		1,405.12	II
70		1,407.38	I
100		1,408.45	I
		1,410.69	II
		1,415.22	II
80		1,429.19	I
50		1,435.79	I
20		1,436.61	II
10		1,468.85	II
10		1,469.17	II
10		1,469.28	II
100		1,481.76	I
25		1,486.55	II
20		1,532.82	I
20		1,532.86	I
10		1,562.04	II
200		1,587.16	I
12		1,593.41	II
70		1,598.24	I
12		1,611.11	II
		1,616.65	II
2		1,622.83	II
100		1,624.34	I
2		1,632.53	II
50		1,639.90	I
150		1,646.67	I
10		1,656.99	II
100		1,665.76	I
25		1,673.59	II
7		1,694.38	II
2		1,698.65	II
200		1,699.34	I
30		1,700.69	II
10		1,720.04	II
25		1,725.75	II
45		1,740.52	II
10		1,749.80	II
35		1,756.15	II
60		1,783.22	II
35		1,793.31	II
35		1,800.58	II
25		1,823.24	II
100		1,879.83	I
20		1,919.64	I
20		1,921.64	II
45		1,942.31	I
25		1,951.93	I
30		1,978.19	I
		Air	
25		2,000.81	II
11,000		2,012.00	I
2,600		2,021.38	I
50		2,044.54	II
150		2,082.09	II
35		2,095.13	II
20		2,098.14	II
60		2,110.68	II
30		2,125.29	II
15		2,126.63	I
20		2,170.75	I
35		2,188.81	II
25		2,201.32	II
35		2,215.63	II
45		2,228.88	II
30		2,231.18	II
25		2,240.16	II
70		2,248.56	II
80		2,263.62	II
18		2,263.88	II
25		2,277.52	II
25		2,283.30	II
25		2,291.40	II
45		2,304.69	II
25		2,314.55	II
20		2,315.75	II
25		2,340.06	II
180		2,352.65	I
20		2,376.28	I
120		2,387.75	I
2,600		2,427.95	I
60		2,533.52	II
16		2,544.19	I
45		2,552.67	II
20		2,589.25	I
30		2,590.04	I
50		2,616.40	II
20		2,627.02	II
250		2,641.48	I
3,400		2,675.95	I
20		2,687.63	II
20		2,688.16	II
30		2,688.71	I
80		2,700.89	I
1,100		2,748.25	I
20		2,748.71	II
100		2,780.82	I
30		2,800.93	I
1,000		2,802.04	II
300		2,819.79	II
100		2,822.55	II
30		2,823.13	II
100	h	2,825.44	II
30		2,833.03	II
300		2,837.85	II
100		2,846.92	II
100		2,856.74	II
3		2,872.36	I
300		2,883.45	I
3		2,886.96	I
10		2,888.40	I
300		2,891.96	I
100		2,893.25	II
3		2,905.74	I
30		2,905.90	I
100		2,907.04	II
300		2,913.52	II
3		2,914.82	I
300		2,918.24	II
16		2,932.19	
30		2,940.67	I
100		2,954.22	II
10		2,973.33	I
100	h	2,990.27	II
300		2,994.80	II
10		3,002.65	I
3		3,005.85	I
10		3,024.67	I
320		3,029.20	I
30		3,033.25	I
300		3,065.42	I
10		3,102.63	I
10	h	3,117.01	I
100		3,122.50	II
1,600		3,122.78	I
30		3,126.86	II
30		3,127.03	I
10		3,164.88	I
10		3,172.35	II
30		3,191.76	I
100		3,194.72	I
30		3,200.37	I
30		3,204.74	I
1		3,221.86	I
30		3,225.25	I
300		3,230.63	I
10		3,253.94	I
10	h	3,265.10	I
30	h	3,267.07	I
10		3,271.63	I
10		3,273.47	I
300		3,308.30	I
300		3,309.64	I
100		3,320.12	I
100		3,355.15	I
30	h	3,368.44	I
10		3,381.90	I
100	h	3,391.31	I
100		3,395.40	I
30	h	3,440.36	I
100		3,467.21	I
30		3,471.61	I
30	h	3,509.04	I
10		3,510.82	I
3		3,523.34	II
3		3,545.61	I
30		3,553.57	I
300		3,557.36	I
10		3,558.22	I
30		3,565.97	I
30	h	3,584.37	I
300		3,586.73	I
30	h	3,588.79	I
30		3,598.06	I
100		3,611.57	I
30	h	3,614.00	I
30		3,622.74	I
100	h	3,631.31	I
10		3,633.22	II
30		3,634.53	I
10		3,635.12	I
300		3,637.90	I
3		3,639.87	I
100	h	3,645.02	I
10		3,649.09	I
100		3,650.74	I
10		3,653.53	I
3		3,654.69	I
10		3,655.30	I
10		3,656.90	I
30		3,706.55	II
100		3,709.62	I
10		3,766.61	I
10		3,770.76	I
100		3,796.01	I
30		3,801.92	I
30		3,804.01	II
10		3,821.85	I
10		3,825.70	I
100		3,874.73	I
30		3,880.25	I
30		3,889.48	I
100	h	3,892.26	I
400		3,897.86	I
30		3,901.09	I
300		3,909.38	I
100		3,927.69	I
10	h	3,959.10	I
30		3,966.23	I
30		3,976.65	I
30		3,979.68	I
30		3,991.37	I
3		4,012.57	I
10		4,016.07	II
400		4,040.93	I
30		4,052.79	II
700		4,065.07	I
10		4,076.35	II
3		4,083.28	II
100		4,084.10	I
30		4,101.70	I
30		4,128.59	I
30		4,201.13	I
30	h	4,227.88	I
100		4,241.80	I
200		4,315.11	I
30		4,361.04	II
10		4,420.61	II
120	h	4,437.27	I
250		4,488.25	I
900	h	4,607.51	I
100	h	4,620.56	I
1		4,663.92	I
3		4,663.97	I
10		4,694.69	I
3		4,760.17	II
500		4,792.58	I
100		4,811.60	I
10		4,822.96	I
30	h	4,950.82	I
30		5,064.59	I
30	h	5,108.84	I
100		5,147.44	I
300		5,230.26	I
100	h	5,261.76	I
100		5,655.77	I
100	h	5,721.36	I
300		5,837.37	I
100	h	5,862.93	I
300	h	5,956.96	I
30	h	5,962.68	I
600		6,278.17	I
100		6,562.68	I
30		6,652.89	I
600		7,510.73	I
10		8,145.06	I
10		9,254.28	I

HAFNIUM (HF)

Z = 72

Hf I and II
Ref. 1 — C.H.C.

Intensity		Wavelength	
		Air	
6,200		2,012.78	II
8,500		2,028.18	II
1,200		2,096.18	II
540		2,210.82	II
320		2,254.01	II
160		2,255.15	II
250		2,266.83	II
620		2,277.16	II
230		2,321.14	II
580		2,322.47	II
300		2,323.25	II
120		2,324.50	II
300		2,324.89	II
200		2,332.97	II
200		2,337.33	II
230		2,343.32	II
320		2,347.44	II
540		2,351.22	II
110		2,353.02	I
90		2,365.98	II
250		2,380.30	II
100		2,381.00	II
170		2,393.18	II
450		2,393.36	II
670		2,393.83	II
130		2,400.78	II
70		2,404.56	II
540		2,405.42	II
130		2,406.44	II
370		2,410.14	II
90		2,413.33	II
55		2,415.96	II
320		2,417.69	II
120		2,425.98	II
45		2,428.75	I
120		2,428.99	II
130		2,433.57	II
45		2,434.74	II
35		2,444.99	I
390		2,447.25	II
140		2,449.44	II
35		2,452.30	II
110		2,453.34	II
450		2,460.49	II
70		2,463.97	II
430		2,464.19	II
90		2,465.06	II
35		2,465.67	I
140		2,467.97	II
210		2,469.18	II
100		2,473.92	II
55		2,481.44	II
55		2,482.65	I
55		2,487.16	I

290		2,496.99	II
580		2,512.69	II
580		2,513.03	II
130		2,515.48	II
890		2,516.88	II
340		2,531.19	II
200		2,537.33	II
110		2,548.20	II
320		2,551.40	II
130		2,559.19	II
250		2,563.61	II
890		2,571.67	II
320		2,573.90	II
320		2,576.82	II
300		2,578.14	II
320		2,582.54	II
130		2,591.33	II
390		2,606.37	II
450		2,607.03	II
120		2,608.45	I
230		2,613.60	II
450		2,622.74	II
160		2,637.00	I
1,100		2,638.71	II
1,100		2,641.41	II
160		2,642.75	I
670		2,647.29	II
100		2,651.16	II
160		2,657.84	II
210		2,661.88	II
290		2,683.35	II
670		2,705.61	I
110		2,706.73	II
210		2,712.42	II
140		2,713.84	I
250		2,718.59	I
120		2,730.85	I
710		2,738.76	II
200		2,743.64	I
360		2,751.81	II
450		2,761.63	I
160		2,766.96	I
170		2,773.02	I
980		2,773.36	II
180		2,774.02	II
390		2,779.37	I
100		2,789.50	II
140		2,789.73	II
230		2,808.00	II
230		2,813.86	II
170		2,814.48	II
230		2,817.68	I
140		2,818.94	I
200		2,819.74	I
1,200		2,820.22	II
490		2,822.68	II
180		2,833.28	I
110		2,834.13	I
410		2,845.83	I
270		2,849.21	II
270		2,850.96	I
180		2,851.21	II
180		2,860.56	I
760		2,861.01	II
760		2,861.70	II
2,100		2,866.37	I
130		2,869.82	II
150		2,876.33	II
210		2,887.14	I
100		2,887.54	I
800		2,889.62	I
1,800		2,898.26	I
130		2,898.71	II
1,200		2,904.41	I
890		2,904.75	I
140		2,909.91	II
2,000		2,916.48	I
580		2,918.58	I
320		2,919.59	II
180		2,924.62	I
490		2,929.63	II
450		2,929.90	I
710		2,937.80	II
2,000		2,940.77	I
160		2,944.71	I
1,200		2,950.68	I
1,100		2,954.20	I
540		2,958.02	I
120		2,961.80	II
1,400		2,964.88	I
620		2,966.93	I
140		2,967.23	II
710		2,968.81	II
110		2,973.37	I
890		2,975.88	II
150		2,979.28	I
1,100		2,980.81	I
210		2,982.72	I
170		3,000.10	II
800		3,005.56	I
1,100		3,012.90	II
540		3,016.78	I
1,100		3,016.94	II
980		3,018.31	I
1,200		3,020.53	I
140		3,025.29	II
410		3,031.16	II
110		3,046.08	II
710		3,050.76	I
1,100		3,057.02	I
130		3,063.78	I
130		3,064.68	II
850		3,067.41	I
2,100		3,072.88	I
170		3,074.10	I
250		3,074.79	I
150		3,080.66	II
430		3,080.84	I
200		3,096.76	I
340		3,101.40	II
710		3,109.12	II
130		3,110.87	II
130		3,119.98	I
710		3,131.81	I
850		3,134.72	II
130		3,137.51	I
170		3,139.65	II
120		3,140.76	I
220		3,145.32	II
220		3,148.41	I
120		3,151.63	I
450		3,156.63	I
270		3,159.82	I
710		3,162.61	II
450		3,168.39	I
890		3,172.94	I
450		3,176.86	II
220		3,181.01	I
120		3,181.15	I
130		3,189.62	I
360		3,193.53	II
670		3,194.19	II
200		3,196.93	I
130		3,199.99	II
310		3,206.11	I
180		3,210.98	I
180		3,217.30	II
180		3,220.61	II
130		3,230.06	I
130		3,239.44	I
130		3,243.35	I
360		3,247.66	I
220		3,249.53	I
890		3,253.70	II
270		3,255.28	II
120		3,262.47	I
180		3,273.66	II
270		3,279.98	II
160		3,291.05	I
210		3,306.12	I
120		3,309.19	I
340		3,310.27	I
670		3,312.86	I
180		3,317.99	II
130		3,328.21	II
890		3,332.73	I
370		3,352.06	II
130		3,356.78	I
230		3,358.91	I
180		3,360.06	I
140		3,366.68	I
180		3,378.93	I
140		3,384.14	II
230		3,384.70	II
170		3,386.21	I
800		3,389.83	II
230		3,392.81	I
230		3,394.59	II
140		3,394.98	II
230		3,397.26	I
230		3,397.60	I
2,300		3,399.80	II
170		3,400.21	I
180		3,402.51	I
140		3,407.76	II
230		3,410.17	II
230		3,417.34	I
410		3,419.18	I
140		3,427.44	I
200		3,428.37	II
250		3,438.24	II
140		3,438.43	I
100		3,441.84	I
100		3,452.31	I
140		3,462.64	II
140		3,467.60	I
710		3,472.40	I
200		3,478.99	II
480		3,479.28	II
250		3,495.75	II
250		3,497.16	I
980		3,497.49	I
100		3,498.98	I
1,200		3,505.23	II
150		3,513.28	I
130		3,518.75	II
980		3,523.02	I
100		3,530.87	I
100		3,531.23	I
980		3,535.54	II
760		3,536.62	I
180		3,548.81	I
540		3,552.70	II
150		3,554.00	I
1,300		3,561.66	II
150		3,564.31	I
270		3,567.36	I
1,100		3,569.04	II
150		3,579.90	I
110		3,583.28	I
210		3,597.42	II
540		3,599.87	I
110		3,615.04	I
800		3,616.89	I
110		3,617.68	I
110		3,624.00	II
320		3,630.87	II
100		3,635.43	I
800		3,644.36	II
320		3,649.10	I
200		3,651.84	I
140		3,661.05	II
220		3,665.35	II
100		3,668.21	I
200		3,672.27	I
480		3,675.74	I
2,200		3,682.24	I
280		3,696.51	I
100		3,698.40	II
240		3,699.72	II
340		3,701.15	II
100		3,704.92	I
120		3,705.40	II
1,000		3,717.80	I
650		3,719.28	II
140		3,726.49	I
160		3,729.10	I
460		3,733.79	I
160		3,737.88	II
120		3,739.04	I
100		3,744.98	II
400		3,746.80	I
140		3,753.22	I
100		3,765.05	I
100		3,765.56	I
170		3,766.92	II
200		3,768.25	I
1,400		3,777.64	I
1,400		3,785.46	I
650		3,793.37	II
100		3,798.66	I
850	d	3,800.38	I
140		3,806.07	II
320		3,811.78	I
100		3,817.20	II
100		3,819.38	I
1,300		3,820.73	I
140		3,829.67	I
280		3,830.02	I
800		3,849.18	I
140		3,849.52	II
600		3,858.31	I
230		3,860.91	I
200		3,872.55	II
160		3,877.10	II
380		3,880.82	II
200		3,882.52	I
150		3,883.77	II
200		3,889.23	I
200		3,889.33	I
620		3,899.94	I
620		3,918.09	II
200		3,923.90	II
120		3,926.42	I
150		3,927.57	I
110		3,929.54	II
320		3,931.38	I
120		3,935.65	II
120		3,939.04	I
410		3,951.83	I
160		3,968.01	I
150	b	3,970.05	H
200		3,973.48	I
180		4,032.27	I
100		4,047.96	II
230		4,062.84	I
140		4,066.21	I
180		4,083.35	I
540		4,093.16	II
110		4,104.23	I
140		4,106.58	I
110		4,113.53	II
110		4,118.60	I
150		4,127.80	II
140		4,145.76	I
150		4,162.36	II
110		4,162.69	I
1,100		4,174.34	I
120		4,190.95	I
160		4,206.58	II
190		4,209.70	I
170		4,228.08	I
170		4,232.44	II
120	b	4,252.08	H
170		4,260.98	I
200		4,263.39	I
170		4,272.85	II
320		4,294.79	I
120		4,318.14	I
160		4,330.27	I
180		4,336.66	II
150		4,350.51	II
250		4,356.33	I
110		4,367.90	II
180		4,370.97	II
120		4,417.35	II
160		4,417.91	I
200		4,438.04	I
140		4,457.34	I
140		4,461.18	I
140		4,540.93	I
250		4,565.94	I
500	d	4,598.80	I
230		4,620.86	I
210		4,655.19	I
120		4,699.01	I
160		4,782.74	I
310		4,800.50	I
130		4,859.24	I
120		4,975.25	I
95		5,018.20	I
15		5,021.75	I
55		5,040.82	II
95		5,047.45	I
55	b	5,074.74	HOO
30		5,079.65	II
55	b	5,093.88	HOO
15		5,112.13	I
19		5,128.53	I
30		5,157.96	I
55		5,167.42	I
75		5,170.18	I
230		5,181.86	I
30		5,186.84	I
30		5,187.75	II
110		5,243.99	I

Intensity		Wavelength	
55		5,247.10	II
25		5,260.44	II
30		5,264.95	II
55		5,275.04	I
22		5,286.09	I
120		5,294.87	I
45		5,298.06	II
30		5,307.82	I
45		5,309.68	I
55		5,311.60	II
12		5,324.26	II
9		5,346.30	II
110		5,354.73	I
110		5,373.86	I
40		5,389.34	I
19		5,391.36	II
19		5,404.47	I
28		5,424.02	I
12		5,435.78	I
40		5,438.74	I
14		5,444.07	II
75		5,452.92	I
30		5,463.38	II
15		5,497.30	I
15		5,510.12	I
15		5,510.45	I
19		5,524.35	II
45		5,538.02	I
28		5,538.26	I
230		5,550.60	I
230		5,552.12	I
55		5,575.86	I
14		5,600.77	I
95		5,613.27	I
25		5,614.01	I
8		5,628.27	I
19		5,650.83	I
40	b	5,698.03	H
25		5,713.28	I
160		5,719.18	I
25	b	5,720.16	H
12		5,748.72	I
14		5,767.18	II
12		5,809.50	II
19		5,817.47	I
25		5,842.23	II
25		5,845.87	I
19		5,847.77	I
22		5,883.66	I
15		5,926.47	I
60		5,933.69	I
75		5,974.28	I
25		5,974.72	I
60		5,978.66	I
25		5,992.96	I
45	c	6,016.79	I
28	b	6,021.12	H
25	b	6,043.19	H
25		6,054.17	I
95		6,098.67	I
95		6,185.13	I
55		6,210.70	I
28		6,216.82	I
45		6,238.58	I
60		6,248.95	II
22	h	6,299.54	I
25	h	6,311.85	I
19		6,318.33	I
30		6,338.10	I
19	h	6,380.19	I
60		6,386.23	I
19	h	6,409.52	I
15	h	6,556.50	I
28		6,587.23	I
45		6,644.60	II
19		6,647.06	II
11		6,659.40	I
30		6,713.48	I
17		6,754.61	II
11		6,769.95	I
85		6,789.27	I
160		6,818.94	I
15		6,826.56	I
13		6,850.07	I
35		6,858.70	I
45		6,911.40	I
10		6,926.19	I
19		6,979.59	I
21		6,980.91	II
7		7,019.25	I
7		7,030.33	II
7		7,035.13	I
11		7,061.90	I
15		7,062.87	I
160		7,063.83	I
11	h	7,094.40	I
15		7,100.54	I
55		7,119.52	I
570		7,131.81	I
650		7,237.10	I
410		7,240.87	I
6		7,262.62	I
75		7,320.05	I
16		7,321.76	I
6		7,356.10	I
6		7,365.28	I
20		7,390.70	I
6		7,423.69	I
25		7,437.56	I
13		7,463.86	I
7		7,484.56	I
15		7,556.37	I
75		7,562.93	I
15		7,564.22	I
11		7,576.95	I
11		7,592.96	I
13		7,608.59	I
360		7,624.40	I
20		7,645.64	I
110		7,740.17	I
8		7,743.57	I
5		7,757.89	II
40		7,790.90	I
7		7,796.81	I
35		7,814.55	I
310		7,845.35	I
7		7,846.56	I
130		7,920.71	I
29		7,938.06	I
250		7,994.73	I
7		8,010.58	I
25		8,056.52	I
25		8,080.32	I
16		8,173.89	I
130		8,204.58	I
7		8,248.81	I
55		8,276.95	I
13		8,305.91	II
25		8,344.25	I
5		8,380.06	I
5		8,382.98	I
35		8,460.01	I
150		8,546.48	I
160		8,640.06	I
40		8,711.24	I
65		9,004.73	I

HELIUM (HE)
Z = 2

He I and II

Ref. 16, 94, 173, 183, 317

W.C.M.

Intensity		Wavelength	
		Vacuum	
15		231.454	II
20		232.584	II
30		234.347	II
50		237.331	II
100		243.027	II
300		256.317	II
1,000		303.780	II
500		303.786	II
10		320.293	I
2		505.500	I
3		505.684	I
4		505.912	I
5		506.200	I
7		506.570	I
10		507.058	I
15		507.718	I
20		508.643	I
25		509.998	I
35		512.098	I
50		515.616	I
100		522.213	I
400		537.030	I
1,000		584.334	I
50		591.412	I
5		958.70	II
6		972.11	II
8		992.36	II
15		1,025.27	II
30		1,084.94	II
35		1,215.09	II
50		1,215.17	II
120		1,640.34	II
180		1,640.47	II
		Air	
7		2,385.40	II
9		2,511.20	II
50		2,577.6	I
1		2,723.19	I
12		2,733.30	II
2		2,763.80	I
10		2,818.2	I
4		2,829.08	I
10		2,945.11	I
40		3,013.7	I
20		3,187.74	I
3		3,202.96	II
15		3,203.10	II
1		3,354.55	I
2		3,447.59	I
1		3,587.27	I
3		3,613.64	I
2		3,634.23	I
3		3,705.00	I
1		3,732.86	I
10		3,819.607	I
1		3,819.76	I
500		3,888.65	I
20		3,964.729	I
1		4,009.27	I
50		4,026.191	I
5		4,026.36	I
12		4,120.82	I
2		4,120.99	I
3		4,143.76	I
10		4,387.929	I
3		4,437.55	I
200		4,471.479	I
25		4,471.68	I
6		4,685.4	II
30		4,685.7	II
30		4,713.146	I
4		4,713.38	I
20		4,921.931	I
100		5,015.678	I
10		5,047.74	I
5		5,411.52	II
500		5,875.62	I
100		5,875.97	I
8		6,560.10	II
100		6,678.15	I
3		6,867.48	I
200		7,065.19	I
30		7,065.71	I
50		7,281.35	I
1		7,816.15	I
2		8,361.69	I
2		9,063.27	I
2		9,210.34	I
10		9,463.61	I
4		9,516.60	I
3		9,526.17	I
1		9,529.27	I
1		9,603.42	I
3		9,702.60	I
6		10,027.73	I
2		10,031.16	I
15		10,123.6	I
1		10,138.50	I
10		10,311.23	I
2		10,311.54	I
3		10,667.65	I
300		10,829.09	I
1,000		10,830.25	I
2,000		10,830.34	I
9		10,913.05	I
3		10,917.10	I
4		11,626.4	II
30		11,969.12	I
20		12,527.52	I
50		12,784.99	I
20		12,790.57	I
7		12,845.96	I
10		12,968.45	I
2		12,984.89	I
12		15,083.64	I
200		17,002.47	I
1		18,555.55	I
6		18,686.8	II
500		18,685.34	I
200		18,697.23	I
100		19,089.38	I
20		19,543.08	I
1,000		20,581.30	I
80		21,120.07	I
10		21,121.43	I
20		21,132.03	I
3		30,908.5	II
4		40,478.90	I

HOLMIUM (HO)
Z = 67

Ho I and II

Ref. 1 — C.H.C.

Intensity		Wavelength	
		Vacuum	
170		2,502.91	II
80		2,508.53	II
110		2,513.55	II
95		2,518.73	II
170		2,533.80	I
130		2,536.86	II
80		2,556.84	I
80		2,567.73	II
80		2,586.52	I
60		2,591.05	II
95		2,592.99	I
190		2,605.86	II
110		2,610.51	II
95		2,613.99	II
60		2,625.20	II
80		2,640.09	I
80		2,640.30	II
60		2,649.68	II
80		2,666.24	II
70		2,689.03	II
210		2,713.65	II
230		2,733.95	II
270		2,750.35	II
110	c	2,759.35	II
110		2,766.85	II
270		2,769.89	II
110		2,772.83	II
140		2,777.10	II
140		2,794.41	II
100		2,799.99	II
100		2,806.72	II
160	c	2,809.99	II
220		2,811.36	II
180		2,812.00	II
190		2,814.74	II
300		2,824.20	II
140		2,826.64	II
270	c	2,831.69	II
210		2,834.99	II
110		2,835.85	II
110		2,844.18	II
100		2,844.68	II
270		2,849.10	II
100		2,861.23	II
250		2,861.49	II
150		2,862.72	II
210		2,871.99	II
230		2,874.06	II
160		2,874.43	II
360		2,880.26	II
460		2,880.98	II
340		2,894.99	II
160		2,895.62	II
170		2,900.84	II
570	c	2,909.41	II
170		2,915.82	II
300		2,919.62	II
110		2,925.35	II

160		2,926.09	II
300	c	2,928.30	II
220		2,942.05	II
300		2,944.49	II
250	c	2,953.11	II
390		2,973.00	II
410	c	2,979.63	II
180		2,981.46	II
140		2,985.48	II
410		2,987.64	II
250		2,990.27	II
110		2,995.86	II
320	c	3,008.10	II
220		3,014.60	II
270		3,038.69	II
480	c	3,049.38	II
410	c	3,054.00	II
500	c	3,057.45	II
230		3,074.30	II
500	c	3,082.34	II
910		3,084.36	II
430	c	3,086.54	II
200		3,108.31	II
200	c	3,109.91	II
760		3,118.50	II
300	c	3,130.99	II
200	c	3,134.39	II
300	c	3,144.36	II
200		3,156.18	II
270		3,156.97	II
200	c	3,159.67	II
580	c	3,166.62	II
390	d	3,171.72	II
810		3,173.78	II
390		3,174.84	II
270	c	3,176.97	II
810	c	3,181.50	II
390		3,183.84	II
270	c	3,184.48	II
200		3,186.37	I
390	c	3,197.83	II
390		3,201.76	II
200		3,206.86	II
270	c	3,210.41	II
200	c	3,221.42	II
320		3,233.34	II
200		3,236.90	II
200		3,237.40	II
200	c	3,257.45	II
390	c	3,278.15	II
270		3,279.25	II
980	c	3,281.97	II
390		3,288.46	II
270	c	3,290.96	II
200	c	3,305.16	II
200		3,319.87	II
230		3,320.25	II
200		3,331.93	II
630	c	3,337.23	II
390	c	3,338.86	II
980	c	3,343.58	II
200		3,344.47	II
360		3,350.49	II
320		3,352.10	II
320	c	3,353.55	II
320		3,354.58	II
320		3,357.91	II
320		3,364.27	II
290		3,370.87	II
230		3,374.16	II
290	c	3,390.75	II
320	c	3,394.60	II
8,100	c	3,398.98	II
810	c	3,410.26	II
1,200		3,421.63	II
3,200		3,453.14	II
390	c	3,410.65	II
1,400	c	3,414.90	II
5,400		3,416.46	II
2,000	c	3,425.34	II
2,000	c	3,428.13	II
630	c	3,429.18	II
320	c	3,432.10	II
390		3,449.35	I
810	c	3,455.70	II
16,000	c	3,456.00	II
1,600		3,461.97	II
360	c	3,467.07	II
810	c	3,473.91	II

5,400	c	3,474.26	II
6,300		3,484.84	II
490		3,489.58	II
580	c	3,493.09	II
2,500	c	3,494.76	II
810	c	3,498.88	II
410	c	3,506.95	II
320		3,509.37	II
810		3,510.73	I
4,100	c	3,515.59	II
410	c	3,519.94	II
630		3,540.76	II
1,600		3,546.05	II
1,100	c	3,556.78	II
410		3,560.15	II
410	c	3,573.24	II
630	c	3,574.80	II
810		3,579.12	I
410		3,580.75	II
410		3,581.83	II
630	c	3,592.23	II
1,100	c	3,598.77	II
340		3,599.48	I
540	c	3,600.95	II
340		3,613.31	II
410		3,618.43	I
430	c	3,626.69	II
490		3,627.25	II
430	c	3,631.76	II
430	c	3,638.30	II
1,600	c	3,662.29	I
430		3,662.99	I
720		3,666.65	I
1,400		3,667.97	I
320		3,669.05	II
450		3,669.52	I
450	c	3,674.77	II
720		3,679.19	I
670		3,679.70	I
720		3,682.65	I
430		3,685.16	II
580		3,690.65	I
340		3,691.95	I
410		3,700.04	I
490	c	3,702.35	II
320		3,709.76	I
430		3,712.88	I
450		3,720.72	I
1,100		3,731.40	I
360		3,732.09	I
810		3,736.35	I
3,200	c	3,748.17	II
320	c	3,753.73	II
340		3,769.09	I
320		3,788.08	II
8,900	c	3,796.75	II
8,900	c	3,810.73	II
490		3,811.86	I
900	c	3,813.25	II
300		3,821.73	II
390		3,829.27	I
320	c	3,831.9	II
410	c	3,835.35	II
1,300	c	3,837.51	II
410	c	3,842.05	II
1,100		3,843.86	II
490	c	3,846.73	II
300		3,849.88	I
320		3,852.40	II
1,800	c	3,854.07	II
390	c	3,856.94	II
720		3,857.72	II
2,700	c	3,861.68	II
540		3,862.62	I
360		3,872.05	II
320	c	3,874.09	II
630		3,874.68	II
540		3,881.61	II
3,000	c	3,888.96	II
490		3,890.42	I
13,000	c	3,891.02	II
540		3,896.76	II
290		3,902.23	II
320		3,904.44	I
1,300	c	3,905.68	II
320		3,911.80	I
320		3,919.45	I
320	c	3,936.44	II
220		3,938.85	I

320	c	3,940.53	II
220		3,950.56	I
580		3,955.73	I
230	c	3,959.51	II
490		3,959.68	I
220		3,975.88	I
390	c	3,976.93	I
220	c	3,985.71	II
220		3,993.73	II
380		3,999.58	I
160	c	4,002.59	II
220		4,003.39	I
110		4,013.50	I
320		4,014.20	II
160	c	4,018.09	II
160	c	4,022.76	II
160	c	4,023.94	II
110		4,025.39	I
320		4,027.21	I
270		4,028.86	I
180	c	4,031.80	I
220		4,037.62	I
220	c	4,038.87	II
2,700		4,040.81	I
5,400	c	4,045.44	II
220	c	4,047.52	I
8,100		4,053.93	I
540		4,054.48	II
270		4,057.55	I
220		4,060.31	I
1,700		4,065.09	II
170		4,067.57	I
720		4,068.05	I
270		4,071.83	I
270		4,073.13	I
290		4,073.51	I
120	c	4,080.23	II
230		4,083.67	I
140		4,085.09	I
170		4,087.35	I
200		4,087.59	I
140		4,091.64	I
120		4,094.78	I
230		4,100.22	I
8,900		4,103.84	I
120		4,105.04	I
270		4,106.50	I
100		4,107.36	I
2,900		4,108.62	I
300		4,112.00	I
100		4,112.72	I
270		4,116.73	I
1,500		4,120.20	I
1,300		4,125.65	I
4,300		4,127.16	I
300		4,134.54	I
1,500		4,136.22	I
130		4,139.34	I
230		4,142.19	I
290		4,148.97	I
980	c	4,152.61	II
8,100		4,163.03	I
160		4,172.23	I
2,500		4,173.23	I
540		4,194.35	I
100		4,198.08	I
130		4,203.21	I
100		4,211.30	II
290		4,222.29	I
290		4,223.47	I
2,000		4,227.04	I
390		4,229.52	II
130	h	4,231.24	I
290		4,243.78	I
1,300	c	4,254.43	I
130	c	4,258.61	II
490		4,264.05	I
300		4,266.04	I
100		4,273.63	II
200		4,311.04	I
250		4,330.64	II
300		4,337.13	II
100	c	4,346.84	II
1,300		4,350.73	I
290		4,356.73	II
140		4,363.93	II
170		4,379.14	II
180	c	4,384.83	II
150		4,400.55	II

120		4,401.24	II
180		4,403.27	I
200		4,420,56	II
130		4,444.63	I
100		4,473.59	II
300		4,477.64	II
120		4,484.57	II
140		4,510.82	I
100	c	4,526.14	II
170		4,530.08	II
170	c	4,531.28	I
130	c	4,531.65	II
170		4,534.58	I
200		4,562.52	I
120	c	4,609.32	II
130		4,613.37	I
100		4,618.84	I
100	c	4,628.22	I
290		4,629.10	II
200	c	4,649.77	II
130	c	4,661.33	II
140	c	4,674.62	II
70		4,701.17	II
80		4,701.69	II
130		4,709.84	II
65		4,711.39	I
130	c	4,717.52	I
35	c	4,728.72	II
35		4,738.00	II
290		4,742.04	II
35	c	4,749.09	II
35		4,751.40	I
100	c	4,757.01	I
35		4,762.39	II
35		4,763.57	II
55		4,777.48	II
30		4,779.42	I
70	c	4,781.19	I
65		4,782.92	I
55	c	4,786.29	I
35		4,791.48	II
35		4,795.92	II
45	h	4,798.87	I
27		4,812.92	II
55		4,832.31	II
30		4,833.32	I
30		4,855.54	II
45		4,860.39	I
27	c	4,889.67	II
30		4,892.35	I
35		4,896.44	II
55		4,906.99	II
45		4,922.73	I
55	c	4,934.89	I
290		4,939.01	I
27	c	4,946.80	I
45		4,948.18	II
65	c	4,959.42	II
35		4,961.03	II
55	c	4,966.73	II
250	c	4,967.21	II
220		4,979.97	I
35	c	4,988.96	I
90		4,995.05	I
35	c	5,012.42	I
55		5,013.28	II
65		5,026.53	I
30		5,028.17	I
55		5,032.95	II
65	c	5,037.60	I
130		5,042.37	I
35		5,044.73	I
30		5,051.44	II
30		5,054.92	II
35	c	5,060.75	I
65		5,074.34	I
80		5,093.07	I
140		5,127.81	I
55		5,129.27	II
130		5,142.59	II
110		5,143.22	II
160		5,149.59	II
90	c	5,167.88	I
130	c	5,182.11	I
55		5,187.85	I
90		5,190.11	II
18		5,195.23	I
45		5,221.54	I
35	c	5,244.47	I

65		5,251.82	I
55		5,275.48	I
90		5,301.25	I
35		5,319.24	I
35		5,319.65	I
80		5,330.11	I
90		5,359.99	I
55		5,381.40	I
30		5,384.56	I
30		5,384.97	I
18	h	5,393.85	I
70		5,403.17	I
100		5,407.08	I
14		5,413.62	II
16		5,434.39	II
18		5,435.87	I
30		5,445.39	I
18		5,449.8	II
30	h	5,451.90	I
14		5,454.0	II
30	c	5,498.57	I
30		5,504.51	I
27		5,515.56	II
18		5,516.45	II
30		5,534.33	I
27		5,553.14	I
35	c	5,560.94	I
35	b	5,563.6	HOO
70		5,566.52	I
18		5,573.96	II
35	b	5,584.7	HOO
55	b	5,591.1	HOO
55	b	5,592.3	HOO
30	b	5,607.1	HOO
27		5,613.64	I
45	b	5,626.4	HOO
65		5,627.60	I
30		5,628.24	II
55		5,640.62	I
70	b	5,655.9	HOO
65	b	5,658.9	HOO
140		5,659.58	I
70	c	5,671.84	
65		5,674.70	I
140	c	5,691.47	I
70	b	5,696.3	HOO
140	c	5,696.57	I
27		5,734.02	I
45		5,736.4	HOO
55		5,739.24	I
22		5,749.58	I
30		5,766.64	
27	b	5,803.8	HOO
45	b	5,819.2	HOO
27	h	5,821.90	I
22		5,839.47	I
45	b	5,849.4	HOO
140	c	5,860.28	I
27	h	5,864.42	I
45		5,870.85	I
27	b	5,879.6	HOO
70	c	5,882.99	I
35	c	5,892.56	I
22		5,904.29	I
70		5,921.76	I
30	c	5,933.71	I
70	c	5,948.03	I
45		5,955.98	I
70		5,972.76	I
90		5,973.52	I
22		5,981.43	I
230	c	5,982.90	I
55		6,002.04	I
27		6,005.33	I
35		6,021.43	I
16		6,038.97	I
27		6,050.71	I
45		6,060.31	I
120		6,081.79	I
70	c	6,133.60	I
35		6,156.38	I
27		6,156.58	I
55		6,191.68	I
70		6,208.65	I
18		6,234.17	I
45	c	6,255.75	I
70	c	6,305.36	I
22		6,306.68	I
30		6,321.94	
30	c	6,354.35	I
30	c	6,372.59	I
14	h	6,373.86	I
22	h	6,413.41	I
27	c	6,471.77	I
13		6,479.17	I
11		6,515.30	I
11	h	6,538.99	I
70		6,550.97	I
15		6,560.08	I
35	d	6,600.58	I
260		6,604.94	I
55		6,607.47	I
13		6,628.35	I
120		6,628.99	I
15		6,632.24	I
9	h	6,652.98	I
15		6,662.52	I
19	c	6,680.46	I
24	c	6,681.62	I
15	h	6,682.02	
55	c	6,694.32	I
15	c	6,722.34	I
40		6,745.05	I
13		6,766.74	I
28	c	6,774.68	I
55	c	6,785.43	I
13		6,793.7	I
13	c	6,811.04	I
15	c	6,820.38	I
24		6,821.64	I
17	c	6,825.72	I
8	h	6,826.62	
8	h	6,852.97	
17	c	6,865.85	I
9		6,883.36	I
13		6,888.50	I
15	c	6,892.96	I
17		6,897.95	I
15	h	6,903.80	I
15	c	6,913.47	I
9		6,916.70	I
40	c	6,939.49	I
45	c	6,950.39	I
13	h	6,955.3	I
19		6,976.7	II
10		6,985.11	
9		6,994.38	
14	h	7,000.71	
10		7,079.07	
12		7,098.58	
9		7,242.08	I
9		7,250.60	I
14		7,308.55	
25		7,341.43	
18		7,389.40	
5	h	7,496.20	I
10	h	7,510.74	I
140		7,555.09	I
18		7,589.20	I
25		7,591.87	I
9	h	7,593.64	I
7	h	7,594.35	I
12		7,602.31	II
16		7,605.35	I
12		7,617.05	I
14		7,627.98	I
40	c	7,628.42	I
9	c	7,641.14	I
4		7,648.16	I
14	c	7,653.80	I
12	c	7,667.30	I
20		7,690.43	I
50	c	7,693.15	I
40	c	7,715.06	I
16		7,719.05	I
16	h	7,738.98	I
8	c	7,752.01	I
60	c	7,815.48	I
40	c	7,823.63	I
8	h	7,879.22	I
60		7,894.64	I
10	h	8,464.66	I
10	h	8,482.67	I
50		8,512.94	I
20		8,545.61	II
18		8,601.84	II
40		8,670.19	I
8	h	8,697.32	
16	h	8,805.48	II
20	c	8,834.49	I
90		8,915.98	II

HYDROGEN (H)

Z = 1

H I

Ref. 214 — W.C.M.

Intensity		Wavelength	
		Vacuum	
15		926.226	I
20		930.748	I
30		937.803	I
50		949.743	I
100		972.537	I
300		1,025.722	I
1,000		1,215.668	I
500		1,215.674	I
		Air	
5		3,835.384	I
6		3,889.049	I
8		3,970.072	I
15		4,101.74	I
30		4,340.47	I
80		4,861.33	I
120		6,562.72	I
180		6,562.852	I
5		9,545.97	I
7		10,049.4	I
12		10,938.1	I
20		12,818.1	I
40		18,751.0	I
5		21,655.3	I
8		26,251.5	I
15		40,511.6	I
4		46,525.1	I
6		74,578	I
3		123,685	I

INDIUM (IN)

Z = 49

In I and II

Ref. 1, 132, 348—350 — C.H.C.

Intensity		Wavelength	
2		1,648.00	I
1	h	1,676.16	I
5	h	1,711.54	I
2	h	1,741.23	I
1	h	1,758.49	I
		Air	
10		2,103.89	II
10		2,166.88	II
2		2,179.90	I
2		2,182.40	I
2		2,187.40	I
2		2,190.84	I
15		2,195.67	II
2		2,197.41	I
2		2,202.24	I
50		2,205.28	II
3		2,211.14	I
5		2,230.70	I
3		2,241.66	I
30		2,255.79	II
10		2,259.99	I
5		2,278.20	I
40		2,281.64	II
2		2,283.75	I
2		2,298.33	I
2		2,298.70	I
2		2,302.49	I
100	c	2,306.05	II
25		2,306.86	I
3		2,309.32	I
2		2,309.75	I
90	d	2,313.21	II
2		2,315.09	I
30		2,323.40	II
5		2,324.41	I
3		2,324.92	I
70	d	2,327.95	II
3		2,332.76	I
80	h	2,334.57	II
10		2,340.19	I
8		2,345.90	I
5		2,346.56	I
50	d	2,350.75	II
5		2,358.70	I
15		2,378.14	I
10		2,379.00	I
110	d	2,382.63	II
40		2,389.54	I
40		2,393.18	II
10		2,399.18	I
50	h	2,406.47	II
50		2,408.76	II
50		2,419.06	II
50		2,419.20	II
70	h	2,427.20	II
20		2,429.86	I
10		2,430.99	I
50		2,432.73	II
60		2,442.63	II
100		2,447.90	II
60		2,453.23	II
60		2,460.08	I
30	h	2,468.02	I
70		2,486.15	II
110	d	2,488.62	II
90		2,488.95	II
80		2,498.59	II
100		2,499.60	II
90	d	2,500.99	II
60		2,508.16	II
110	d	2,512.31	II
100		2,521.37	I
10		2,522.98	I
70		2,553.56	II
160	d	2,554.44	II
1,100		2,560.15	I
70		2,565.13	II
70	d	2,598.75	II
200		2,601.76	I
50		2,604.04	II
90	d	2,654.70	II
100	d	2,662.63	II
140	d	2,668.65	II
140	d	2,674.56	II
80		2,683.12	II
1,600		2,710.26	I
300		2,713.94	I
130	d	2,749.75	II
700		2,753.88	I
40		2,775.37	I
60		2,798.76	II
90	d	2,818.97	II
180	c	2,836.92	I
30	c	2,858.14	I
80		2,865.68	II
120	d	2,890.18	II
1,100		2,932.63	I
100		2,941.05	II
20	c	2,957.01	I
60	d	2,966.17	II
110	c	2,999.40	II
8,000		3,039.36	I
8	d	3,051.15	I
110	d	3,099.80	II
180	c	3,101.8	II
130	c	3,138.60	II
80	c	3,142.75	II
130	d	3,146.70	II
150		3,155.77	II
100	c	3,158.40	II
90	c	3,176.30	II
90	d	3,198.11	II
13,000		3,256.09	I
3,000		3,258.56	I
90	c	3,338.50	II
75	c	3,376.59	II
100	c	3,404.28	II
110	d	3,438.40	II
180	c	3,693.91	II
95	c	3,708.13	II
380	w	3,716.14	II

120	c	3,718.30	II
160	c	3,718.72	II
160	c	3,723.40	II
170	w	3,795.21	II
230	c	3,799.21	II
250	c	3,834.65	II
200	c	3,842.18	II
100		3,889.78	II
100	c	3,902.07	II
60	d	3,922.12	II
65	c	3,934.40	II
250	w	3,962.35	II
120	c	4,004.66	II
140	d	4,013.92	II
410	w	4,056.94	II
17,000		4,101.76	I
140	c	4,205.14	II
100	d	4,213.04	II
110	d	4,219.66	II
150	d	4,372.87	II
150	c	4,500.78	II
18,000		4,511.31	I
110	c	4,549.01	II
140	c	4,570.85	II
180	w	4,578.02	II
180	w	4,578.40	II
140	c	4,616.08	II
170	c	4,617.17	II
250	c	4,620.14	II
150	w	4,620.70	II
170	c	4,627.30	II
140	c	4,637.04	II
380	c	4,638.16	II
220	c	4,644.58	II
360	c	4,655.62	II
320	w	4,656.74	II
190	c	4,681.11	II
450	w	4,684.8	II
3		4,878.37	I
90	d	4,907.06	II
70	h	4,924.93	II
150	c	4,973.77	II
80	h	5,109.36	II
100	w	5,115.14	II
140	c	5,117.40	II
270	c	5,120.80	II
200	w	5,121.75	II
80	d	5,129.85	II
240	c	5,175.42	II
140	c	5,184.44	II
30		5,254.32	I
12		5,262.74	I
150	c	5,309.45	II
80		5,411.41	II
140	c	5,418.45	II
220	w	5,436.70	II
130	c	5,497.50	II
140	c	5,507.08	II
320	c	5,513.00	II
250	w	5,523.28	II
130	c	5,536.50	II
190	w	5,555.45	II
240	c	5,576.90	II
200	w	5,636.70	II
160	c	5,708.50	II
50		5,709.91	I
100	c	5,721.80	II
50		5,727.68	I
210	c	5,853.15	II
490	w	5,903.4	II
260	w	5,915.4	II
120	c	5,918.78	II
130	c	6,062.9	II
250	c	6,095.95	II
210	c	6,108.66	II
180	w	6,115.9	II
230	w	6,128.7	II
240	w	6,129.4	II
320	w	6,132.1	II
150	c	6,140.0	II
90		6,143.23	II
140	c	6,148.10	II
190	w	6,149.5	II
80		6,161.15	II
180	w	6,162.45	II
100	c	6,224.28	II
280	w	6,228.3	II
140	w	6,231.1	II
270	w	6,304.8	II
290	w	6,362.3	II
300	w	6,469.0	II
210	c	6,541.20	II
190	c	6,751.88	II
180	c	6,765.9	II
100	c	6,783.72	II
8	h	6,847.44	I
320	w	6,891.5	II
4	h	6,900.13	I
380	w	7,182.9	II
180	c	7,255.0	II
210	c	7,276.5	II
180	c	7,303.4	II
320	c	7,350.6	II
100	c	7,632.7	II
100	c	7,682.9	II
210	c	7,740.7	II
100	c	7,776.96	II
180	c	7,789.0	II
70	c	7,806.8	II
70	c	7,814.5	II
90	c	7,840.9	II
20	h	8,050.78	I
240	c	8,227.0	II
30	h	8,238.66	I
15	h	8,314.92	I
50	c	8,434.55	II
30		8,678.95	I
20		8,682.63	I
50		8,700.25	I
100	w	8,813.5	II
80	c	8,832.6	II
40		8,894.47	I
10		9,170.08	I
120	c	9,197.7	II
120	c	9,202.0	II
220	w	9,213.0	II
160	d	9,241.1	II
40	h	9,349.83	I
60	h	9,370.27	I
20		9,427.99	I
100		9,977.86	I
200		10,257.03	I
60	h	10,717.42	I
100	h	10,744.31	I
20		11,334.72	I
20		11,731.48	I
10		12,912.59	I
9		13,429.96	I
5		13,824.48	I
6		14,316.25	I
3		14,419.20	I
6		14,668.66	I
7		14,719.08	I
2		16,504.31	I
6		22,291.06	I
7		23,879.13	I

In III

Ref. 351 — C.H.C.

Intensity	Wavelength	
	Vacuum	
7	685.31	III
5	691.62	III
1	782.17	III
10	882.24	III
10	890.84	III
10	915.87	III
2	917.45	III
5	926.83	III
30	1,403.08	III
30	1,434.85	III
20	1,487.70	III
20	1,494.14	III
10	1,524.78	III
20	1,530.21	III
30	1,532.95	III
100	1,625.42	III
20	1,642.28	III
20	1,702.53	III
100	1,748.83	III
2	1,767.88	III
1	1,810.71	III
30	1,842.41	III
40	1,850.30	III
15	1,862.98	III
	Air	
30	2,154.08	III
2	2,154.42	III
10	2,199.52	III
5	2,232.18	III
20	2,261.26	III
5	2,266.26	III
5	2,272.41	III
5	2,272.84	III
10	2,300.90	III
100	2,527.41	III
50	2,725.52	III
80	2,726.15	III
100	2,982.80	III
100	3,008.08	III
30	3,008.82	III
30	3,293.55	III
8	3,350.91	III
5	3,562.32	III
100	3,852.82	III
100	4,023.77	III
150	4,032.32	III
50	4,062.30	III
100	4,071.57	III
100	4,072.93	III
100	4,252.68	III
40	4,509.58	III
200	5,248.77	III
100	5,645.15	III
40	5,723.17	III
100	5,819.50	III
200	6,197.72	III

In IV

Ref. 352 — C.H.C.

Intensity	Wavelength	
	Vacuum	
8	677.33	IV
15	684.53	IV
8	720.64	IV
4	1,470.34	IV
5	1,512.62	IV
5	1,523.50	IV
7	1,532.03	IV
8	1,568.92	IV
3	1,569.33	IV
12	1,601.50	IV
7	1,606.59	IV
10	1,628.48	IV
8	1,651.80	IV
10	1,655.62	IV
1	1,667.38	IV
5	1,670.64	IV
12	1,678.07	IV
4	1,702.02	IV
15	1,707.11	IV
7	1,721.89	IV
12	1,747.65	IV
10	1,748.10	IV
10	1,768.77	IV
12	1,773.01	IV
8	1,789.15	IV
12	1,793.38	IV
12	1,844.67	IV
9	1,851.10	IV
9	1,855.84	IV
15	1,856.64	IV
15	1,874.08	IV
5	1,903.52	IV
5	1,939.58	IV
3	1,943.31	IV
	Air	
6	2,004.73	IV
1	2,101.17	IV
8	2,112.31	IV
4	2,225.13	IV
1	1,605.08	IV

In V

Ref. 353 — C.H.C.

Intensity	Wavelength	
	Vacuum	
6	368.67	V
6	370.10	V
10	372.82	V
10	372.94	V
2	374.95	V
6	375.84	V
6	376.07	V
10	376.79	V
17	378.61	V
3	379.24	V
9	380.27	V
11	381.56	V
9	382.14	V
11	382.76	V
10	383.05	V
17	386.21	V
10	386.70	V
3	388.66	V
14	388.91	V
11	390.03	V
11	390.92	V
9	392.29	V
9	392.46	V
1	393.60	V
25	393.89	V
11	395.74	V
3	397.73	V
10	399.79	V
9	400.05	V
25	400.57	V
25	402.39	V
3	405.33	V
9	407.28	V
3	407.36	V
9	407.95	V
9	417.43	V
2	418.45	V
2	423.16	V

IODINE (I)

Z = 53

I I and II

Ref. 124, 153, 176, 184

L.J.R.

Intensity	Wavelength	
	Vacuum	
2	655.80	II
6	659.00	II
8	663.98	II
8	664.52	II
8	665.06	II
150	665.70	II
1,000	719.55	II
1,000	722.98	II
1,000	798.16	II
1,200	834.10	II
600	847.80	II
1,500	873.49	II
1,000	875.94	II
2,000	879.84	II
1,500	881.88	II
1,000	891.00	II
1,000	893.17	II
1,200	1,000.57	II
1,000	1,003.35	II
4,000	1,018.58	II
10,000	1,034.66	II
1,500	1,054.74	II
2,000	1,066.34	II
3,000	1,075.21	II
5,000	1,105.00	II
2,500	1,111.16	II
1,500	1,117.22	II
3,500	1,125.25	II
2,000	1,131.50	II
1,200	1,139.75	II
10,000	1,139.80	II
1,500	1,154.67	II
1,000	1,159.87	II
10,000	1,160.56	II
20,000	1,166.48	II
1,500	1,167.05	II
5,000	1,175.84	II
10,000	1,178.65	II
15,000	1,187.34	II
10,000	1,190.85	II
15	1,195.29	I
5,000	1,198.88	II
7,000	1,200.22	II
200	1,218.41	I
20,000	1,220.89	II
600	1,224.05	I

600		1,224.08	I
500		1,228.89	I
20,000		1,234.06	II
600		1,251.34	I
2,500		1,259.15	I
3,000		1,259.51	I
800		1,261.27	I
600		1,267.57	I
600		1,267.60	I
1,500		1,275.26	I
3,000		1,289.40	I
10,000		1,300.34	I
3,000		1,302.98	I
3,000		1,313.95	I
3,000		1,317.54	I
2,000		1,330.19	I
20,000		1,336.52	II
5,000		1,355.10	I
3,000		1,357.97	I
5,000		1,360.97	I
3,000		1,361.11	I
2,500		1,367.71	I
2,500		1,368.22	I
4,000		1,383.23	I
3,000		1,390.75	I
2,000		1,392.90	I
2,000		1,400.01	I
8,000		1,425.49	I
5,000		1,446.26	I
5,000		1,453.18	I
5,000		1,457.39	I
5,000		1,457.47	I
10,000		1,457.98	I
2,500		1,458.79	I
4,000		1,459.15	I
2,500		1,465.83	I
1,000		1,485.92	I
5,000		1,492.89	I
5,000		1,507.04	I
5,000		1,514.68	I
15,000		1,518.05	I
2,500		1,526.45	I
5,000		1,593.58	I
5,000		1,617.60	I
2,500		1,640.78	I
15,000		1,702.07	I
12,000		1,782.76	I
5,000		1,799.09	I
75,000		1,830.38	I
15,000		1,844.45	I

Air

2,000		2,061.63	I
100		2,408.01	II
100		2,419.18	II
100		2,494.74	II
100		2,533.60	II
200		2,534.27	II
1,000		2,566.24	II
2,000		2,582.79	II
300		2,593.46	II
200	c	2,688.98	II
500		2,730.12	II
20		2,765.15	II
200		2,808.59	II
1,500		2,878.63	II
1,000		2,993.87	II
5,000		3,078.75	II
200		3,161.03	II
1,000		3,175.07	II
300		3,355.53	II
250	c	3,424.99	II
300	c	3,497.41	II
500		3,526.90	II
200		3,742.14	II
200		4,102.23	I
200		4,129.21	I
100	d	4,134.15	I
500	d	4,321.84	I
300		4,452.86	II
200		4,599.77	II
300	c	4,632.45	II
500	d	4,666.48	II
1,000		4,675.53	II
250		4,763.31	I
1,000		4,862.32	I
200		4,916.94	I
1,000		4,986.92	II
400	c	5,065.37	II
10,000		5,119.29	I
200		5,149.73	II
3,000	c	5,161.20	II
300		5,176.19	II
600		5,216.27	II
500	d	5,228.97	II
1,000		5,234.57	I
3,000	c	5,245.71	II
500		5,269.36	II
400		5,299.78	II
400		5,322.80	II
10,000		5,338.22	II
5,000	c	5,345.15	II
1,000	c	5,369.86	II
800	c	5,405.42	II
800	c	5,407.36	II
600	c	5,427.06	I
3,000		5,435.83	II
1,000		5,438.00	II
2,000	c	5,464.62	II
800		5,491.50	II
1,000	c	5,496.94	II
1,000		5,504.72	II
600	c	5,522.06	II
600	c	5,598.52	II
1,000		5,600.32	II
1,500		5,612.89	II
10,000		5,625.69	II
1,000		5,678.08	II
2,000	c	5,690.91	II
500		5,702.05	II
4,000	c	5,710.53	II
1,000		5,738.27	II
1,000		5,760.72	II
1,000	d	5,764.33	I
500	c	5,774.83	II
500		5,787.02	II
2,000		5,894.03	I
5,000		5,950.25	II
300		5,984.86	I
2,000	d	6,024.08	I
500		6,068.93	II
2,000	c	6,074.98	II
1,000		6,082.43	I
2,000	c	6,127.49	II
800		6,191.88	I
1,000		6,204.86	II
500		6,213.10	I
800		6,244.48	I
900	c	6,257.49	II
1,000		6,293.98	I
500		6,313.13	I
800		6,330.37	I
400		6,333.50	I
2,000		6,337.85	I
1,000		6,339.44	I
500		6,359.16	I
1,000		6,566.49	I
2,000		6,583.75	I
1,000		6,585.27	I
5,000		6,619.66	I
500		6,661.11	I
600		6,665.96	II
500	c	6,697.29	I
300		6,718.83	II
400		6,732.03	I
4,000		6,812.57	II
1,000		6,958.78	II
500		6,989.78	I
200	c	7,085.21	II
500		7,120.05	I
1,200		7,122.05	I
2,000		7,142.06	I
1,000		7,164.79	II
400	d	7,191.66	I
700		7,227.30	I
1,000		7,236.78	I
500		7,237.84	I
500		7,351.35	II
5,000		7,402.06	I
1,000		7,410.50	I
500		7,416.48	I
5,000		7,468.99	I
500	c	7,490.52	I
2,000		7,554.18	I
500	d	7,556.65	I
2,000	c	7,700.20	I
500		7,798.98	II
600		7,897.98	I
500		7,969.48	I
1,000		8,003.63	I
99,000		8,043.74	I
300	d	8,065.70	I
1,000		8,090.76	I
800	c	8,169.38	I
500	d	8,222.57	I
4,000		8,240.05	I
10,000	c	8,393.30	I
150		8,414.60	II
1,000		8,486.11	I
1,500	c	8,664.95	I
500	c	8,700.80	I
250	d	8,748.22	I
1,000		8,853.24	I
2,000		8,853.80	I
3,000		8,857.50	I
1,000	d	8,898.50	I
400		8,964.69	I
400		8,993.13	I
5,000		9,022.40	I
15,000		9,058.33	I
1,000		9,098.86	I
12,000		9,113.91	I
600		9,128.03	I
30		9,195.30	II
600		9,227.74	I
1,000		9,335.05	I
4,000		9,426.71	I
3,000		9,427.15	I
10	c	9,480.33	II
2,000		9,598.22	I
2,000		9,649.61	I
3,000	d	9,653.06	I
5,000		9.731.73	I
500		10,003.05	I
750		10,131.16	I
1,000		10,238.82	I
400		10,375.20	I
400		10,391.74	I
6		10,405.49	II
5,000		10,466.54	I
1		11,084.68	II
400		11,236.56	I
350		11,558.46	I
320		11,778.34	I
450		11,996.86	I
300		12,033.69	I
150		12,304.58	I
60		13,149.16	I
140		13,958.27	I
200		14,287.02	I
100		14,460.00	I
225		15,032.57	I
105		15,528.65	I
150		16,037.33	I
15		18,275.71	I
20		18,348.52	I
15		18,982.41	I
35		19,070.17	I
110		19,105.12	I
50		19,370.02	I
10		20,648.69	I
220		22,183.03	I
150		22,226.53	I
30		22,309.21	I
32		24,420.82	I
12		27,365.42	I
9		27,573.05	I
10		30,361.93	I
8		30,383.88	I
10		34,295.73	I
9		34,513.11	I
3		40,228.54	I
2		41,633.80	I

I III

Ref. 20, 21, 161 — L.J.R.

Intensity	Wavelength	
	Vacuum	
6	666.81	III
8	705.11	III
7	784.64	III
7	784.80	III
8	795.52	III
5	865.97	III
5	920.38	III
6	961.17	III
6	1,078.58	III
8	1,094.20	III
4	1,244.66	III
8	1,252.35	III
5	1,306.93	III
	Air	
1	2,224.43	III
1	2,238.12	III
3	2,249.31	III
2	2,309.38	III
3	2,340.85	III
3	2,350.43	III
2	2,353.46	III
4	2,367.74	III
2	2,371.45	III
3	2,372.45	III
4	2,376.47	III
4	2,387.12	III
3	2,392.01	III
2	2,403.06	III
2	2,403.63	III
2	2,414.85	III
2	2,418.49	III
2	2,418.85	III
2	2,423.91	III
5	2,426.12	III
3	2,434.88	III
2	2,462.50	III
3	2,466.69	III
3	2,466.99	III
6	2,475.36	III
4	2,489.27	III
2	2,493.21	III
2	2,494.27	III
3	2,495.16	III
2	2,496.07	III
3	2,501.41	III
2	2,516.82	III
6	2,519.75	III
4	2,521.72	III
3	2,531.99	III
2	2,537.56	III
7	2,545.71	III
4	2,640.77	III
4	2,642.11	III
6	2,652.25	III
2	2,818.48	III
2	2,839.44	III
4	2,864.67	III
4	2,885.15	III
3	2,910.98	III
3	2,917.35	III
2	2,931.11	III
2	3,005.68	III
3	3,069.23	III
3	3,153.88	III
3	3,170.14	III
3	3,181.66	III
3	3,210.14	III
4	3,213.49	III
4	3,224.93	III
2	3,300.47	III
2	3,479.53	III
3	3,546.92	III
3	3,613.81	III
2	3,754.40	III
2	3,754.55	III
3	3,963.16	III
3	4,077.14	III

I IV

Ref. 21, 58 — L.J.R.

Intensity	Wavelength	
	Vacuum	
5	601.86	IV
6	612.46	IV
4	615.17	IV
4	654.22	IV
4	654.56	IV
7	919.28	IV
	Air	
5	2,249.30	IV
4	2,340.84	IV
7	2,361.13	IV
5	2,367.75	IV

6	2,372.45	IV
7	2,376.46	IV
4	2,385.28	IV
8	2,387.11	IV
6	2,392.00	IV
4	2,403.05	IV
2	2,418.45	IV
3	2,423.89	IV
9	2,426.10	IV
6	2,434.85	IV
3	2,466.68	IV
3	2,466.96	IV
8	2,475.35	IV
4	2,485.51	IV
5	2,489.24	IV
4	2,493.20	IV
2	2,501.38	IV
3	2,513.74	IV
8	2,519.74	IV
6	2,521.72	IV
4	2,531.98	IV
5	2,537.54	IV
8	2,545.67	IV
4	2,640.77	IV
5	2,642.11	IV
8	2,652.23	IV
3	2,818.45	IV
6	2,864.68	IV
4	2,910.97	IV
5	2,917.33	IV
4	3,069.17	IV
4	3,170.11	IV
4	3,181.64	IV
4	3,210.12	IV
6	3,213.48	IV
6	3,224.90	IV
4	3,546.90	IV

I V

Ref. 84 — L.J.R.

Intensity	Wavelength	
	Vacuum	
30	363.78	V
36	380.74	V
45	565.53	V
50	607.57	V

IRIDIUM (IR)

$Z = 77$

Ir I and II

Ref. 1 — C.H.C.

Intensity	Wavelength	
	Air	
9,900	2,010.65	I
8,700	2,022.35	I
15,000	2,033.57	I
6,200	2,052.22	I
5,000	2,060.64	I
3,700	2,083.22	I
3,100	2,085.74	I
17,000	2,088.82	I
14,000	2,092.63	I
2,700	2,112.68	I
1,800	2,119.54	I
2,000	2,125.44	I
4,500	2,126.81	II
2,000	2,127.52	I
4,500	2,127.94	I
3,700	2,148.22	I
2,500	2,150.54	I
3,500	2,152.68	II
2,900	2,155.81	I
7,900	2,158.05	I
2,100	2,162.88	I
5,800	2,169.42	II
4,500	2,175.24	I
2,700	2,178.17	I
1,600	2,187.43	II
1,100	2,190.38	II
740	2,191.64	I
910	2,208.09	II
1,300	2,220.37	I
790	2,221.07	II
2,500	2,242.68	II
620	2,245.76	II
2,100	2,253.38	I
	2,253.49	I
2,100	2,255.10	I
1,400	2,255.81	I
350	2,258.51	I
1,400	2,258.86	I
830	2,264.61	I
1,100	2,266.33	I
1,000	2,268.90	I
660	2,280.00	I
950	2,281.02	II
660	2,281.91	I
330	2,284.60	I
330	2,295.08	I
790	2,298.05	I
	2,298.16	I
460	2,299.53	I
910	2,300.50	I
2,700	2,304.22	I
410	2,305.47	I
210	2,307.27	I
910	2,308.93	I
460	2,315.38	I
410	2,321.45	I
410	2,321.58	I
210	2,327.98	I
540	2,333.30	I
740	2,333.84	I
580	2,334.50	I
1,600	2,343.18	I
740	2,343.61	I
100	2,352.62	I
580	2,355.00	I
230	2,357.53	II
410	2,358.16	I
500	2,360.73	I
2,500	2,363.04	I
370	2,368.04	II
3,500	2,372.77	I
290	2,375.09	II
250	2,377.28	I
250	2,377.98	I
500	2,379.38	I
540	2,381.62	I
210	2,383.17	I
120	2,386.58	II
1,300	2,386.89	I
2,500	2,390.62	I
2,700	2,391.18	I
230	2,407.59	I
290	2,409.37	I
290	2,410.17	I
290	2,410.73	I
540	2,413.31	I
370	2,415.86	I
620	2,418.11	I
120	2,424.32	I
120	2,424.66	I
210	2,424.89	I
370	2,424.99	I
290	2,425.66	I
170	2,426.53	II
540	2,427.61	I
540	2,431.24	I
1,300	2,431.94	I
170	2,432.36	I
100	2,432.58	I
270	2,435.14	I
250	2,445.34	I
250	2,447.76	I
190	2,448.23	I
910	2,452.81	I
1,300	2,455.61	I
230	2,455.87	I
210	2,457.03	I
210	2,457.23	I
120	2,465.09	I
870	2,467.30	I
3,300	2,475.12	I
210	2,478.11	I
2,100	2,481.18	I
100	2,485.38	I
620	2,493.08	I
210	2,496.27	I
250	2,502.63	I
4,100	2,502.98	I
170	2,504.37	I
120	2,505.74	I
120	2,507.63	I
170	2,509.71	I
170	2,511.94	I
170	2,512.58	II
210	2,513.71	I
120	2,515.36	I
40	2,524.88	II
170	2,525.05	I
120	2,532.52	I
990	2,533.13	I
1,100	2,534.46	I
580	2,537.22	I
170	2,537.68	I
100	2,541.48	I
580	2,542.02	I
40	2,542.80	II
7,900	2,543.97	I
150	2,545.54	I
790	2,546.03	I
120	2,547.20	I
120	2,547.69	I
210	2,551.40	I
190	2,554.40	I
210	2,555.35	I
170	2,555.88	I
150	2,563.28	I
910	2,564.18	I
210	2,569.88	I
100	2,570.62	I
230	2,572.70	I
740	2,577.26	I
100	2,578.71	I
35	2,579.49	II
740	2,592.06	I
740	2,599.04	I
150	2,602.04	I
190	2,604.55	I
190	2,607.52	I
700	2,608.25	I
1,800	2,611.30	I
210	2,614.98	I
330	2,617.78	I
210	2,619.88	I
70	2,623.64	II
250	2,625.32	I
100	2,626.76	I
700	2,634.17	I
170	2,635.27	I
250	2,639.42	I
3,500	2,639.71	I
210	2,644.19	I
170	2,653.76	I
100	2,656.81	I
1,800	2,661.98	I
350	2,662.63	I
2,700	2,664.79	I
140	2,668.99	I
520	2,669.91	I
520	2,671.84	I
330	2,673.61	I
120	2,676.83	I
110	2,684.04	I
270	2,692.34	I
3,000	2,694.23	I
110	2,704.03	I
160	2,712.74	I
140	2,744.00	I
330	2,772.46	I
250	2,775.55	I
520	2,781.29	I
330	2,785.22	I
540	2,797.35	I
1,600	2,797.70	I
380	2,798.18	I
410	2,800.82	I
680	2,823.18	I
1,200	2,824.45	I
110	2,833.24	II
110	2,835.66	I
820	2,836.40	I
160	2,837.33	I
1,100	2,839.16	I
820	2,840.22	I
160	2,842.28	I
3,800	2,849.72	I
110	2,863.84	I
380	2,875.60	I
380	2,875.98	I
270	2,877.68	I
140	2,879.41	I
820	2,882.64	I
650	2,897.15	I
260	2,901.95	I
260	2,904.80	I
200	2,907.24	I
440	2,916.36	I
230	2,918.57	I
4,400	2,924.79	I
1,200	2,934.64	I
880	2,936.68	I
250	2,938.47	I
190	2,939.27	I
140	2,940.54	I
2,700	2,943.15	I
230	2,946.97	I
200	2,949.76	I
1,200	2,951.22	I
150	2,962.99	I
200	2,974.95	I
440	2,980.65	I
150	2,985.80	I
190	2,990.62	I
300	2,996.08	I
180	2,997.41	I
220	3,002.25	I
600	3,003.63	I
160	3,011.69	I
120	3,016.43	I
270	3,017.31	I
140	3,019.23	I
110	3,025.82	I
380	3,029.36	I
330	3,039.26	I
35	3,042.65	II
300	3,047.16	I
300	3,049.44	I
300	3,057.28	I
1,600	3,068.89	I
190	3,069.09	I
190	3,069.71	I
170	3,076.69	I
320	3,083.22	I
240	3,086.44	I
390	3,088.04	I
510	3,100.29	I
510	3,100.45	I
340	3,120.76	I
200	3,121.78	I
3,400	3,133.32	I
190	3,150.61	I
190	3,154.74	I
190	3,159.15	I
140	3,168.18	I
490	3,168.88	I
370	3,177.58	I
170	3,180.35	I
370	3,198.92	I
610	3,212.12	I
370	3,219.51	I
5,100	3,220.78	I
100	3,221.28	I
300	3,229.28	I
100	3,230.76	I
470	3,241.52	I
200	3,262.01	I
390	3,266.44	I
160	3,277.28	I
100	3,287.59	I
160	3,310.52	I
200	3,322.60	I
130	3,334.16	I
560	3,368.48	I
660	3,437.02	I
100	3,437.50	I
410	3,448.97	I
3,200	3,513.64	I
220	3,515.95	I
410	3,522.03	I
160	3,557.17	I
320	3,558.99	I
1,200	3,573.72	I
320	3,594.39	I
220	3,609.77	I
190	3,617.21	I
160	3,626.29	I
660	3,628.67	I
220	3,636.20	I

300		3,661.71	I
300		3,664.62	I
320		3,674.98	I
200		3,687.08	I
140		3,725.38	I
200		3,731.36	II
130		3,738.53	I
530		3,747.20	I
120		3,793.79	I
3,100		3,800.12	I
230		3,817.24	I
170		3,865.64	I
480		3,902.51	I
480		3,915.38	I
400		3,934.84	I
120		3,946.27	I
590		3,976.31	I
460		3,992.12	I
180		4,020.03	I
350		4,033.76	I
130		4,040.08	I
370		4,069.92	I
150		4,070.68	I
100		4,092.61	I
140		4,115.78	I
23		4,127.92	I
27		4,155.70	I
15		4,166.04	I
90		4,172.56	I
35		4,182.47	I
15	h	4,183.21	I
18		4,185.66	I
23		4,197.54	I
27		4,217.76	I
13		4,220.80	I
75		4,259.11	I
27		4,265.30	I
260		4,268.10	I
23		4,286.62	I
75		4,301.60	I
55		4,310.59	I
220		4,311.50	I
18		4,351.30	I
18		4,352.56	I
18		4,392.59	I
160		4,399.47	I
65		4,403.78	I
110		4,426.27	I
15		4,450.18	I
55		4,478.48	I
16		4,495.35	I
11	h	4,496.03	I
55		4,545.68	I
30		4,548.48	I
13		4,550.78	I
35		4,568.09	I
18		4,570.02	I
18		4,604.48	I
75		4,616.39	I
26		4,656.18	I
17	h	4,668.99	I
21		4,708.88	I
50		4,728.86	I
21		4,731.86	I
26		4,756.46	I
13		4,757.96	I
65		4,778.16	I
30		4,795.67	I
10		4,807.14	I
21		4,809.47	I
10		4,840.77	I
17		4,845.38	I
50		4,938.09	I
26		4,970.48	I
25		4,999.74	I
25		5,002.74	I
17		5,009.17	I
30		5,014.98	I
17		5,046.06	I
30		5,123.66	I
20		5,177.95	I
22		5,238.92	I
12		5,340.74	I
35		5,364.32	I
75		5,449.50	I
30		5,454.50	I
7		5,469.40	I
10		5,620.04	I
45		5,625.55	I
10		5,828.55	I
10		5,882.30	I
7		5,887.36	I
35		5,894.06	I
7		6,026.10	I
12		6,067.83	I
20		6,110.67	I
12		6,288.28	I
7		6,334.44	I
5		6,624.73	I
10		6,686.08	I
5		6,830.01	I
5		6,929.88	I
4		7,183.71	I
6		7,834.32	I

IRON (FE)

Z = 26

Fe I and II

Ref. 56, 63, 105, 138, 174, 278

— H.M.C. and H.C.

Intensity	Wavelength	
	Vacuum	
12	1,055.27	II
15	1,068.36	II
15	1,071.60	II
15	1,096.89	II
12	1,099.12	II
18	1,112.09	II
12	1,121.99	II
12	1,122.86	II
12	1,128.07	II
12	1,130.43	II
15	1,133.41	II
12	1,133.68	II
12	1,138.64	II
12	1,142.33	II
12	1,143.23	II
18	1,144.95	II
12	1,147.41	II
15	1,148.29	II
12	1,151.16	II
12	1,267.44	II
12	1,272.00	II
12	1,371.02	II
12	1,563.79	II
12	1,580.62	II
18	1,608.46	II
18	1,618.47	II
15	1,621.68	II
15	1,629.15	II
15	1,631.12	II
18	1,635.40	II
15	1,636.32	II
15	1,639.40	II
12	1,641.76	II
12	1,647.16	II
12	1,670.74	II
12	1,702.04	II
12	1,761.38	II
20	1,785.26	II
20	1,786.74	II
18	1,788.07	II
30	1,934.538	I
25	1,937.269	I
50	1,946.988	I
25	1,951.571	I
30	1,952.59	I
30	1,953.005	I
60	1,957.823	I
60	1,960.144	I
30	1,961.25	I
50	1,962.111	I
12	1,963.11	II
	Air	
100	2,084.122	I
50	2,157.794	I
15	2,162.02	II
40	2,166.773	I
300	2,178.118	I
250	2,186.486	I
60	2,186.892	I
120	2,187.195	I
250	2,191.839	I
150	2,196.043	I
80	2,200.390	I
80	2,200.724	I
15	2,208.41	II
20	2,213.65	II
12	2,218.26	II
20	2,220.38	II
25	2,245.58	II
50	2,250.790	I
60	2,251.874	I
25	2,255.77	II
300	2,259.511	I
60	2,264.389	I
80	2,267.085	I
80	2,267.469	I
50	2,270.862	I
150	2,272.070	I
150	2,276.026	I
80	2,279.937	I
150	2,284.086	I
150	2,287.250	I
300	2,292.524	I
80	2,294.41	I
200	2,297.787	I
600	2,298.169	I
80	2,299.220	I
300	2,300.142	I
50	2,301.684	I
100	2,303.424	I
150	2,303.581	I
120	2,308.999	I
150	2,313.104	I
200	2,320.358	I
100	2,327.40	II
15	2,327.88	II
100	2,331.31	II
15	2,331.97	II
300	2,332.80	II
200	2,338.01	II
600	2,343.49	II
80	2,343.96	II
150	2,344.28	II
25	2,344.98	II
50	2,345.34	II
200	2,348.11	II
250	2,348.30	II
50	2,351.20	II
15	2,351.67	II
25	2,352.31	II
30	2,353.47	II
15	2,353.68	II
50	2,354.48	II
40	2,354.89	II
200	2,359.12	II
15	2,359.59	II
150	2,360.00	II
120	2,360.29	II
30	2,360.51	II
40	2,362.02	II
60	2,363.86	II
200	2,364.83	II
80	2,365.76	II
25	2,366.59	II
80	2,368.59	II
80	2,369.456	I
80	2,369.95	II
25	2,370.50	II
120	2,371.430	I
300	2,373.624	I
150	2,373.74	II
120	2,374.518	I
60	2,375.19	II
120	2,376.43	II
20	2,378.13	II
80	2,379.27	II
20	2,379.41	II
40	2,380.20	II
120	2,380.76	II
150	2,381.835	I
1,000	2,382.04	II
20	2,382.90	II
20	2,383.06	II
60	2,383.25	II
50	2,384.39	II
40	2,388.37	II
300	2,388.63	II
200	2,389.973	I
30	2,390.10	II
20	2,390.77	II
15	2,391.48	II
20	2,392.58	II
40	2,395.42	II
1,000	2,395.62	II
15	2,396.72	II
300	2,399.24	II
20	2,400.05	II
15	2,401.29	II
50	2,404.43	II
800	2,404.88	II
250	2,406.66	II
80	2,406.97	II
300	2,410.52	II
200	2,411.07	II
50	2,411.81	II
150	2,413.31	II
20	2,416.45	II
80	2,417.87	II
15	2,418.44	II
60	2,420.396	I
60	2,422.69	II
60	2,423.089	I
40	2,423.21	II
150	2,424.14	II
15	2,424.39	II
30	2,424.59	II
30	2,428.29	II
120	2,428.36	II
25	2,428.80	II
25	2,429.03	II
20	2,429.39	II
30	2,429.86	II
120	2,430.08	II
25	2,431.02	II
80	2,432.26	II
60	2,432.87	II
25	2,434.06	II
20	2,434.24	II
20	2,434.65	II
50	2,434.73	II
50	2,434.95	II
25	2,436.62	II
60	2,438.182	I
150	2,439.30	II
150	2,439.74	II
80	2,440.11	I
40	2,440.42	II
30	2,442.37	II
100	2,442.57	I
60	2,443.71	II
250	2,443.872	I
100	2,444.51	II
50	2,445.11	II
50	2,445.212	I
100	2,445.57	II
40	2,445.80	II
50	2,446.11	II
30	2,446.47	II
40	2,447.20	II
25	2,447.33	II
60	2,447.709	I
30	2,447.75	II
25	2,449.96	II
25	2,450.20	II
100	2,453.476	I
20	2,453.98	II
30	2,454.58	II
15	2,455.71	II
15	2,455.90	II
15	2,457.09	II
1,500	2,457.598	I
150	2,458.78	II
40	2,458.97	II
60	2,460.44	II
80	2,461.28	II
100	2,461.86	II
100	2,462.181	I
1,500	2,462.647	I
50	2,463.29	II
50	2,463.730	I
40	2,464.01	II
40	2,464.90	II
800	2,465.149	I
50	2,465.91	II
15	2,466.50	II
60	2,466.67	II
60	2,466.82	II

60	2,467.732	I	200	2,523.66	I	40	2,621.67	II	25	2,769.35	II
15	2,468.29	II	500	2,524.293	I	400	2,623.53	I	300	2,772.07	I
600	2,468.879	I	100	2,525.02	I	50	2,625.49	II	50	2,773.23	I
60	2,469.51	II	200	2,525.39	II	200	2,625.67	II	20	2,774.69	II
25	2,470.41	II	25	2,526.07	II	150	2,628.29	II	15	2,776.91	II
80	2,470.67	II	300	2,526.29	II	20	2,630.07	II	60	2,778.07	I
80	2,470.965	I	20	2,527.10	II	250	2,631.05	II	600	2,778.220	I
800	2,472.336	I	2,000	2,527.435	I	250	2,631.32	II	40	2,779.30	II
40	2,472.43	II	30	2,527.70	II	50	2,631.61	II	50	2,783.69	II
40	2,472.60	II	20	2,528.88	II	100	2,632.237	I	30	2,785.19	II
1,000	2,472.895	I	20	2,529.08	II	300	2,635.809	I	3,000	2,788.10	I
200	2,473.16	I	800	2,529.135	I	50	2,641.646	I	20	2,793.89	II
50	2,473.32	II	25	2,529.23	II	200	2,643.998	I	200	2,797.78	I
30	2,474.05	II	80	2,529.31	I	60	2,664.66	II	30	2,799.29	II
600	2,474.814	I	250	2,529.55	II	30	2,666.64	II	400	2,804.521	I
50	2,475.12	II	150	2,529.836	I	300	2,666.812	I	1,500	2,806.98	I
40	2,475.54	II	40	2,530.11	II	60	2,666.965	I	2,500	2,813.287	I
15	2,476.26	II	200	2,530.687	I	600	2,679.062	I	300	2,823.276	I
60	2,476.657	I	20	2,531.87	II	500	2,684.75	II	600	2,825.56	I
25	2,477.34	II	120	2,533.63	II	400	2,689.212	I	50	2,825.687	I
60	2,478.57	II	60	2,533.80	I	60	2,692.60	II	120	2,828.808	I
120	2,479.480	I	100	2,534.42	II	50	2,696.28	I	25	2,831.56	II
1,200	2,479.776	I	120	2,535.49	II	200	2,699.106	I	1,500	2,832.436	I
100	2,480.16	II	400	2,535.607	I	60	2,703.99	II	120	2,835.950	I
15	2,481.05	II	60	2,536.67	II	80	2,706.012	I	200	2,838.119	I
80	2,482.12	II	200	2,536.792	I	400	2,706.582	I	30	2,839.51	II
25	2,482.32	II	200	2,536.80	II	60	2,708.571	I	20	2,839.80	II
100	2,482.66	II	50	2,536.84	II	20	2,709.05	II	15	2,840.65	II
15	2,482.87	II	50	2,537.14	II	200	2,711.655	I	200	2,843.631	I
10,000	2,483.271	I	50	2,538.20	II	80	2,714.41	II	1,000	2,843.977	I
300	2,483.533	I	40	2,538.50	II	50	2,716.22	II	100	2,845.594	I
15	2,483.72	II	20	2,538.68	II	50	2,716.257	I	15	2,848.11	II
1,000	2,484.185	I	100	2,538.80	II	50	2,717.786	I	15	2,848.32	II
60	2,484.24	II	100	2,538.91	II	50	2,717.87	II	800	2,851.797	I
30	2,484.44	II	150	2,538.99	II	250	2,718.436	I	30	2,856.91	II
50	2,485.990	I	50	2,539.357	I	4,000	2,719.027	I	25	2,858.34	II
800	2,486.373	I	20	2,540.52	II	100	2,719.420	I	50	2,869.307	I
100	2,486.691	I	200	2,540.66	II	50	2,720.197	I	50	2,872.334	I
100	2,487.066	I	600	2,540.972	I	1,500	2,720.903	I	80	2,874.172	I
120	2,487.370	I	80	2,541.10	II	400	2,723.578	I	50	2,894.504	I
4,000	2,488.143	I	60	2,541.84	II	30	2,724.88	II	120	2,912.157	I
100	2,488.945	I	300	2,542.10	I	150	2,724.953	I	120	2,929.007	I
80	2,489.48	II	25	2,542.78	II	80	2,726.05	I	1,200	2,936.903	I
1,000	2,489.750	I	60	2,543.38	II	50	2,726.235	I	60	2,941.343	I
50	2,489.83	II	250	2,543.92	I	25	2,727.38	II	12	2,944.40	II
50	2,489.913	I	150	2,544.70	I	80	2,727.54	II	1,000	2,947.876	I
3,000	2,490.644	I	40	2,544.97	II	200	2,728.020	I	60	2,950.24	I
100	2,490.71	II	40	2,545.22	II	50	2,728.820	I	600	2,953.940	I
60	2,490.86	II	20	2,545.44	II	80	2,728.90	II	250	2,957.364	I
2,000	2,491.155	I	800	2,545.978	I	40	2,730.73	II	80	2,959.99	I
100	2,491.40	II	40	2,546.44	II	1,000	2,733.581	I	150	2,965.254	I
25	2,492.34	II	80	2,546.67	II	60	2,734.005	I	1,500	2,966.898	I
100	2,493.18	II	80	2,546.87	I	50	2,734.268	I	120	2,969.36	I
500	2,493.26	II	20	2,548.59	II	500	2,735.475	I	800	2,970.099	I
20	2,493.88	II	100	2,548.74	II	50	2,735.612	I	15	2,970.52	II
60	2,494.000	I	80	2,549.08	II	500	2,737.310	I	1,200	2,973.132	I
50	2,494.251	I	80	2,549.39	II	120	2,737.83	I	500	2,973.235	I
100	2,495.87	I	60	2,549.46	II	400	2,739.55	II	600	2,981.445	I
600	2,496.533	I	600	2,549.613	I	250	2,742.254	I	1,000	2,983.570	I
50	2,497.82	II	40	2,549.77	II	800	2,742.405	I	60	2,984.77	I
150	2,498.90	I	60	2,550.03	II	200	2,743.20	II	50	2,984.82	II
40	2,500.92	II	25	2,550.15	II	150	2,743.565	I	13	2,985.54	II
1,000	2,501.132	I	50	2,550.68	II	200	2,744.068	I	1,000	2,994.427	I
40	2,501.31	II	40	2,560.28	II	80	2,744.527	I	250	2,994.502	I
50	2,501.693	I	25	2,562.09	II	300	2,746.48	II	500	2,999.512	I
60	2,502.39	II	400	2,562.53	II	100	2,749.32	II	120	3,000.451	I
40	2,503.33	II	200	2,563.48	II	500	2,749.48	II	800	3,000.948	I
20	2,503.57	II	20	2,566.22	II	1,200	2,750.140	I	60	3,001.655	I
60	2,503.87	II	60	2,566.91	II	20	2,751.13	II	15	3,002.64	II
80	2,506.09	II	25	2,570.52	II	20	2,752.15	II	200	3,007.282	I
40	2,506.80	II	30	2,570.85	II	80	2,753.29	II	500	3,008.14	I
500	2,507.900	I	150	2,574.36	II	50	2,753.69	I	120	3,009.569	I
30	2,508.34	II	50	2,575.74	I	150	2,754.032	I	60	3,017.627	I
50	2,508.753	I	300	2,576.691	I	100	2,754.426	I	60	3,018.983	I
1,000	2,510.835	I	25	2,576.86	II	30	2,754.89	II	60	3,020.01	II
120	2,511.76	II	60	2,577.92	II	800	2,755.73	II	500	3,020.491	I
80	2,512.275	I	100	2,582.58	II	250	2,756.328	I	1,500	3,020.639	I
400	2,512.365	I	1,500	2,584.54	I	100	2,757.316	I	600	3,021.073	I
50	2,514.38	II	60	2,593.51	I	50	2,759.81	I	500	3,024.032	I
80	2,516.570	I	20	2,605.34	II	120	2,761.780	I	150	3,025.638	I
50	2,517.13	II	20	2,605.42	II	150	2,761.81	II	500	3,025.842	I
300	2,517.661	I	60	2,605.657	I	150	2,762.026	I	80	3,030.148	I
800	2,518.102	I	300	2,606.51	II	120	2,762.772	I	60	3,031.214	I
60	2,519.05	II	800	2,606.827	I	120	2,763.109	I	60	3,034.484	I
150	2,519.629	I	20	2,611.07	II	20	2,763.66	II	40	3,036.96	II
40	2,521.09	II	600	2,611.87	II	25	2,765.13	II	800	3,037.389	I
30	2,521.82	II	250	2,618.018	I	80	2,766.910	I	80	3,041.637	I
50	2,522.480	I	20	2,619.07	II	250	2,767.522	I	800	3,047.604	I
4,000	2,522.849	I	20	2,620.69	II	50	2,769.30	I	600	3,057.446	I

1,000	3,059.086	I	200	3,556.878	I	150	3,794.34	I	120	4,187.795	I
250	3,067.244	I	400	3,558.515	I	400	3,795.002	I	80	4,191.430	I
120	3,075.719	I	1,000	3,565.379	I	120	3,797.518	I	40	4,195.329	I
120	3,091.577	I	1,200	3,570.097	I	250	3,798.511	I	150	4,198.304	I
80	3,098.189	I	800	3,570.25	I	400	3,799.547	I	40	4,199.095	I
100	3,099.895	I	120	3,571.996	I	200	3,805.345	I	300	4,202.029	I
100	3,099.968	I	100	3,573.393	I	80	3,806.696	I	40	4,203.984	I
60	3,100.303	I	60	3,573.829	I	600	3,812.964	I	80	4,206.696	I
100	3,100.665	I	60	3,573.888	I	60	3,813.059	I	80	4,210.343	I
12	3,154.20	II	4,000	3,581.19	I	1,500	3,815.840	I	400	4,216.183	I
80	3,175.445	I	150	3,582.199	I	2,500	3,820.425	I	100	4,219.360	I
150	3,184.895	I	150	3,584.660	I	150	3,821.179	I	50	4,222.212	I
250	3,191.659	I	120	3,584.929	I	80	3,824.306	I	50	4,225.956	I
500	3,193.226	I	300	3,585.319	I	2,500	3,824.444	I	200	4,227.423	I
800	3,193.299	I	150	3,585.705	I	1,500	3,825.880	I	11	4,233.17	II
12	3,196.08	II	200	3,586.103	I	1,200	3,827.823	I	100	4,233.602	I
200	3,196.928	I	400	3,586.984	I	1,000	3,834.222	I	250	4,235.936	I
80	3,199.500	I	100	3,594.633	I	120	3,839.257	I	50	4,238.809	I
60	3,200.47	I	150	3,603.204	I	500	3,840.437	I	50	4,247.425	I
50	3,205.398	I	200	3,605.454	I	800	3,841.047	I	200	4,250.118	I
50	3,211.67	I	500	3,606.680	I	120	3,843.256	I	300	4,250.787	I
100	3,211.88	I	1,500	3,608.859	I	80	3,846.800	I	40	4,258.315	I
13	3,213.31	II	250	3,610.16	I	200	3,849.96	I	800	4,260.473	I
200	3,214.011	I	60	3,612.068	I	120	3,850.817	I	250	4,271.153	I
200	3,214.396	I	150	3,617.788	I	2,500	3,856.372	I	1,200	4,271.759	I
60	3,215.938	I	1,500	3,618.768	I	150	3,859.212	I	1,200	4,282.402	I
50	3,217.377	I	200	3,621.462	I	10,000	3,859.911	I	80	4,291.462	I
80	3,219.583	I	150	3,622.004	I	150	3,865.523	I	250	4,299.234	I
60	3,219.766	I	150	3,623.19	I	60	3,867.215	I	1,200	4,307.901	I
300	3,222.045	I	100	3,631.096	I	250	3,872.501	I	150	4,315.084	I
600	3,225.78	I	1,200	3,631.463	I	150	3,873.761	I	1,500	4,325.761	I
13	3,227.73	II	60	3,632.041	I	250	3,878.018	I	80	4,352.734	I
80	3,227.796	I	100	3,638.298	I	2,000	3,878.573	I	80	4,369.771	I
20	3,230.42	II	200	3,640.389	I	4,000	3,886.282	I	800	4,375.929	I
80	3,233.05	I	80	3,643.717	I	200	3,887.048	I	3,000	4,383.544	I
50	3,233.967	I	1,500	3,647.842	I	300	3,888.513	I	1,200	4,404.750	I
120	3,234.613	I	250	3,649.506	I	800	3,895.656	I	300	4,415.122	I
300	3,236.222	I	80	3,650.279	I	1,200	3,899.707	I	600	4,427.299	I
100	3,239.433	I	200	3,651.467	I	400	3,902.945	I	400	4,461.652	I
80	3,244.187	I	120	3,670.024	I	250	3,906.479	I	120	4,466.551	I
80	3,246.005	I	150	3,670.089	I	80	3,916.731	I	80	4,476.017	I
60	3,254.36	I	100	3,676.311	I	600	3,920.258	I	80	4,482.169	I
80	3,265.046	I	150	3,677.629	I	1,200	3,922.911	I	200	4,482.252	I
50	3,265.617	I	1,500	3,679.913	I	1,200	3,927.920	I	50	4,489.739	I
50	3,271.000	I	200	3,682.242	I	2,000	3,930.296	I	50	4,528.613	I
50	3,280.26	I	120	3,683.054	I	60	3,948.774	I	11	4,583.83	II
150	3,286.75	I	150	3,684.107	I	60	3,949.953	I	30	4,647.433	I
120	3,305.97	I	120	3,685.998	I	50	3,951.164	I	30	4,736.771	I
200	3,306.343	I	500	3,687.456	I	50	3,952.601	I	50	4,859.741	I
400	3,355.227	I	120	3,689.477	I	60	3,956.454	I	120	4,871.317	I
80	3,355.517	I	150	3,694.008	I	250	3,956.68	I	60	4,872.136	I
60	3,369.546	I	120	3,695.051	I	60	3,966.614	I	30	4,878.208	I
120	3,370.783	I	150	3,701.086	I	100	3,969.257	I	100	4,890.754	I
50	3,378.678	I	80	3,704.462	I	80	3,977.741	I	250	4,891.492	I
50	3,380.110	I	1,200	3,705.566	I	40	3,981.771	I	30	4,903.309	I
60	3,383.978	I	60	3,707.041	I	50	3,983.956	I	150	4,918.992	I
12	3,388.13	II	150	3,707.821	I	60	3,994.114	I	500	4,920.502	I
50	3,392.304	I	300	3,707.919	I	200	3,997.392	I	12	4,923.92	II
150	3,392.651	I	600	3,709.246	I	40	3,998.053	I	1,500	4,957.597	I
150	3,399.333	I	120	3,716.442	I	400	4,005.241	I	11	4,990.50	II
80	3,404.353	I	8,000	3,719.935	I	60	4,009.713	I	80	5,001.862	I
500	3,407.458	I	1,500	3,722.563	I	80	4,014.53	I	18	5,001.91	II
250	3,413.131	I	120	3,724.377	I	100	4,021.867	I	11	5,004.20	II
60	3,424.284	I	60	3,725.491	I	50	4,040.638	I	30	5,005.711	I
500	3,427.119	I	60	3,727.093	I	4,000	4,045.813	I	100	5,006.117	I
60	3,428.748	I	500	3,727.619	I	1,500	4,063.594	I	60	5,012.067	I
6,000	3,440.606	I	150	3,732.396	I	50	4,066.975	I	30	5,014.941	I
2,500	3,440.989	I	1,200	3,733.317	I	50	4,067.977	I	12	5,018.43	II
1,000	3,443.876	I	5,000	3,734.864	I•	1,200	4,071.737	I	11	5,030.64	II
200	3,445.149	I	120	3,735.324	I	40	4,076.629	I	25	5,030.77	I
15	3,453.61	II	6,000	3,737.131	I	40	4,100.737	I	12	5,035.71	II
1,200	3,465.860	I	100	3,738.306	I	40	4,107.489	I	150	5,041.755	I
2,000	3,475.450	I	400	3,743.362	I	150	4,118.544	I	30	5,049.819	I
500	3,476.702	I	80	3,743.47	I	40	4,127.608	I	30	5,051.634	I
2,500	3,490.574	I	6,000	3,745.561	I	400	4,132.058	I	25	5,074.748	I
500	3,497.840	I	1,200	3,745.899	I	80	4,134.676	I	18	5,100.73	II
250	3,513.817	I	3,000	3,748.262	I	40	4,136.997	I	15	5,100.95	II
300	3,521.261	I	80	3,748.964	I	200	4,143.415	I	150	5,110.357	I
400	3,526.040	I	3,000	3,749.485	I	800	4,143.869	I	40	5,133.69	I
100	3,526.166	I	1,500	3,758.232	I	40	4,153.898	I	40	5,139.251	I
60	3,526.237	I	400	3,760.05	I	50	4,154.500	I	100	5,139.462	I
60	3,526.381	I	1,500	3,763.788	I	60	4,156.799	I	11	5,144.36	II
60	3,526.467	I	400	3,765.54	I	50	4,172.744	I	12	5,149.46	II
100	3,533.199	I	600	3,767.191	I	60	4,174.912	I	25	5,151.910	I
200	3,536.556	I	60	3,776.452	I	50	4,175.635	I	30	5,162.27	I
300	3,541.083	I	250	3,785.95	I	50	4,177.593	I	80	5,166.281	I
250	3,542.075	I	100	3,786.68	I	120	4,181.754	I	2,500	5,167.487	I
80	3,553.739	I	250	3,787.880	I	50	4,184.891	I	80	5,168.897	I
400	3,554.925	I	250	3,790.092	I	120	4,187.038	I	12	5,169.03	II

500	5,171.595	I
50	5,191.454	I
80	5,192.343	I
200	5,194.941	I
30	5,204.582	I
25	5,215.179	I
150	5,216.274	I
18	5,216.85	II
60	5,226.862	I
1,000	5,227.150	I
13	5,227.49	II
250	5,232.939	I
13	5,247.95	II
13	5,251.23	II
18	5,260.26	II
11	5,264.18	II
100	5,266.555	I
1,200	5,269.537	I
800	5,270.357	I
30	5,281.789	I
60	5,283.621	I
25	5,302.299	I
11	5,306.18	II
13	5,316.23	II
150	5,324.178	I
800	5,328.038	I
300	5,328.531	I
100	5,332.899	I
14	5,339.59	II
80	5,339.928	I
500	5,341.023	I
25	5,364.87	I
40	5,367.47	I
50	5,369.96	I
400	5,371.489	I
60	5,383.37	I
14	5,387.06	II
40	5,393.167	I
12	5,395.86	II
300	5,397.127	I
15	5,402.06	II
60	5,404.12	I
250	5,405.774	I
30	5,410.91	I
60	5,415.20	I
60	5,424.07	I
30	5,427.83	II
250	5,429.695	I
13	5,429.99	II
100	5,434.523	I
200	5,446.871	I
25	5,455.45	I
120	5,455.609	I
16	5,465.93	II
20	5,466.94	II
16	5,482.31	II
14	5,493.83	II
25	5,497.516	I
20	5,501.464	I
18	5,506.20	II
30	5,506.778	I
12	5,510.78	II
12	5,529.06	II
13	5,544.76	II
30	5,569.618	I
60	5,572.841	I
120	5,586.755	I
200	5,615.644	I
20	5,624.541	I
12	5,645.40	II
50	5,662.515	I
20	5,762.990	I
11	5,783.63	II
30	5,862.353	I
13	5,885.02	II
16	5,902.82	II
30	5,914.114	I
14	5,955.70	II
30	5,986.956	I
18	5,961.71	II
30	5,962.4	II
13	5,965.63	II
40	6,065.482	I
30	6,102.159	I
40	6,136.614	I
40	6,137.694	I
30	6,147.73	II
20	6,149.24	II
15	6,175.16	II
40	6,191.558	I
30	6,213.429	I
30	6,219.279	I
40	6,230.726	I
20	6,238.37	II
20	6,246.317	I
80	6,247.56	II
30	6,252.554	I
15	6,305.32	II
12	6,331.97	II
15	6,383.75	II
20	6,393.602	I
30	6,399.999	I
20	6,411.647	I
20	6,416.90	II
20	6,421.349	I
30	6,430.844	I
20	6,446.43	II
200	6,456.38	II
60	6,494.981	I
20	6,516.05	II
20	6,546.239	I
20	6,592.913	I
40	6,677.989	I
15	6,855.18	I
15	6,945.21	I
20	7,067.44	II
15	7,130.94	I
25	7,164.443	I
80	7,187.313	I
30	7,207.381	I
12	7,224.51	II
50	7,307.97	II
40	7,320.70	II
20	7,376.46	II
30	7,445.746	I
20	7,462.38	II
40	7,495.059	I
60	7,511.045	I
15	7,586.04	I
15	7,711.71	II
30	7,780.59	I
40	7,832.22	I
80	7,937.131	I
60	7,945.984	I
80	7,998.939	I
60	8,046.047	I
50	8,085.176	I
150	8,220.41	I
120	8,327.053	I
20	8,331.908	I
120	8,387.770	I
30	8,468.404	I
15	8,514.069	I
60	8,661.898	I
150	8,688.621	I
12	8,793.38	I
12	8,824.23	I
20	8,866.96	I
15	8,999.56	I
15	10,216.32	I
13	10,469.65	I
21	11,119.80	I
14	11,374.08	I
52	11,422.32	I
87	11,439.12	I
91	11,593.59	I
255	11,607.57	I
160	11,638.26	I
230	11,689.98	I
160	11,783.26	I
580	11,882.84	I
225	11,884.08	I
1,030	11,973.05	I
15	12,638.71	I
14	12,879.76	I
17	13,565.04	I
30	14,236.25	I
24	14,285.11	I
14	14,292.38	I
16	14,308.69	I
96	14,400.56	I
20	14,442.28	I
72	14,512.23	I
50	14,555.06	I
14	14,565.95	I
40	14,826.43	I
37	15,051.77	I
28	15,207.55	I
94	15,294.58	I
16	15,335.40	I
30	15,621.67	I
25	15,631.97	I
14	15,723.59	I
41	15,769.42	I
28	15,813.13	I
13	16,444.82	I
20	16,486.69	I
105	18,856.65	I
47	18,987.01	I
25	19,113.68	I
22	19,791.88	I
14	22,380.82	I
21	22,619.85	I
38	26,222.04	I
17	26,659.22	I

Fe III

Ref. 71, 101 — J.R.

Intensity		Wavelength	
		Vacuum	
6		728.81	III
5		730.00	III
5		737.71	III
5		739.26	III
9		807.55	III
8		807.86	III
8		808.84	III
8	p	811.28	III
10		813.38	III
8		838.05	III
10		844.28	III
9		845.41	III
8	w	847.42	III
8		859.72	III
8	p	861.76	III
10	p	861.83	III
8		873.46	III
9		890.76	III
10		891.17	III
8		891.44	III
8		899.42	III
10		950.33	III
10		981.37	III
10	w	983.88	III
8		985.82	III
9		991.23	III
9		1,017.25	III
8		1,017.74	III
8		1,018.29	III
8		1,032.12	III
8		1,063.87	III
9		1,122.53	III
9		1,124.88	III
8		1,128.02	III
10	h	1,505.17	III
10	h	1,538.63	III
12	h	1,550.20	III
10	h	1,601.21	III
10		1,869.83	III
12		1,877.99	III
10		1,882.05	III
12		1,886.76	III
13		1,890.67	III
11		1,893.98	III
20		1,895.46	III
10	s	1,907.58	III
19		1,914.06	III
15		1,915.08	III
15		1,922.79	III
10	p	1,926.01	III
18		1,926.30	III
15		1,930.39	III
14		1,931.51	III
14		1,937.34	III
10	l	1,938.90	III
14	s	1,943.48	III
12		1,945.34	III
10		1,950.33	III
12		1,951.01	III
11		1,952.65	III
13		1,953.32	III
10		1,953.49	III
10	w	1,954.22	III
11		1,958.58	III
13		1,960.32	III
15		1,987.50	III
14		1,991.61	III
13		1,994.07	III
12		1,995.56	III
12		1,996.42	III
		Air	
10		2,061.55	III
12		2,068.24	III
14		2,078.99	III
10		2,084.35	III
12		2,090.14	III
15		2,097.48	III
12		2,097.69	III
12		2,103.80	III
10		2,107.32	III
15		2,151.78	III
12		2,157.71	III
12		2,158.47	III
10		2,161.27	III
12		2,166.95	III
12		2,171.04	III
15		2,174.66	III
12		2,180.41	III
10	p	2,208.85	III
10		2,221.83	III
10		2,229.27	III
10		2,232.43	III
10		2,232.69	III
10		2,235.91	III
10		2,238.16	III
12	p	2,241.54	III
12		2,261.59	III
10		2,267.42	III
10		2,293.06	III
15		2,295.86	III
10	p	2,317.70	III
10		2,319.22	III
10	p	2,321.71	III
10		2,326.95	III
10	p	2,336.77	III
10		2,338.96	III
8		2,389.53	III
8		2,438.17	III
8	p	2,582.37	III
8		2,595.62	III
8		2,617.15	III
9	p	2,645.39	III
10	h	2,695.13	III
9	h	2,695.34	III
8	h	2,700.02	III
8	h	2,701.13	III
8		2,773.31	III
10	p	2,813.24	III
8	p	2,895.08	III
9	p	2,902.47	III
12		2,904.43	III
8	p	2,905.80	III
10		2,907.50	III
12		2,907.70	III
8		2,923.90	III
8		2,948.39	III
8		2,963.23	III
12		3,001.62	III
12	h	3,007.28	III
15		3,013.17	III
10	p	3,136.43	III
10		3,174.09	III
10		3,175.99	III
10		3,178.01	III
13		3,266.88	III
11		3,276.08	III
10		3,288.81	III
9		3,305.22	III
9		3,339.39	III
9		3,499.59	III
9		3,500.28	III
10		3,501.76	III
10		3,586.04	III
11		3,600.94	III
11		3,603.88	III
16		3,954.33	III
11		3,968.72	III
9		3,969.49	III
10	w	3,979.42	III
10		4,035.42	III
11		4,053.11	III
12		4,081.00	III

Intensity		Wavelength	Spectrum
10		4,120.90	III
11		4,122.02	III
11		4,122.78	III
15		4,137.76	III
13		4,139.35	III
9		4,140.48	III
9		4,154.96	III
18		4,164.73	III
9		4,164.92	III
13		4,166.84	III
13		4,174.26	III
9		4,210.67	III
11		4,222.27	III
13		4,235.56	III
9		4,238.62	III
12		4,243.75	III
12	h	4,273.40	III
12		4,279.72	III
14	h	4,286.16	III
16	h	4,296.85	III
18	h	4,304.78	III
20	h	4,310.36	III
9		4,323.68	III
9	h	4,372.04	III
9	h	4,372.14	III
11	h	4,372.31	III
14	h	4,372.53	III
18	h	4,372.81	III
9		4,395.76	III
12		4,419.60	III
9		4,431.02	III
9		5,111.07	III
9		5,127.35	III
12		5,156.12	III
10		5,199.08	III
10		5,235.66	III
18		5,243.31	III
13	l	5,260.34	III
9		5,272.37	III
14		5,272.98	III
15		5,276.48	III
16		5,282.30	III
12		5,284.83	III
11		5,298.12	III
12		5,299.93	III
14	w	5,302.60	III
10		5,306.76	III
9		5,310.88	III
10		5,322.74	III
11		5,346.88	III
12		5,353.77	III
12		5,363.76	III
10		5,368.06	III
11	l	5,375.47	III
11		5,719.88	III
9		5,744.19	III
10		5,756.38	III
18		5,833.93	III
9		5,848.76	III
10		5,854.62	III
9		5,876.26	III
15		5,891.91	III
9		5,898.68	III
9		5,918.96	III
10	p	5,920.13	III
18	p	5,929.69	III
10		5,952.31	III
14		5,953.62	III
9		5,968.48	III
12		5,979.32	III
9	h	5,981.01	III
12	h	5,989.08	III
18		5,999.54	III
9		6,031.02	III
16		6,032.59	III
13		6,036.56	III
11		6,048.72	III
11		6,054.18	III
9		6,056.36	III
9		6,149.99	III
9		6,169.74	III
9		6,185.26	III
7		6,186.56	III
7		6,194.79	III
6		6,195.43	III
6		6,201.37	III
5	s	6,203.04	III
5		6,259.81	III
6	p	6,294.50	III
5		6,357.81	III
5	h	7,317.63	III
6	h	7,320.14	III
5	w	7,921.17	III
5	w	8,230.88	III
5	w	8,231.79	III
9	w	8,235.45	III
8	w	8,236.75	III
6	w	8,238.98	III
5		8,563.49	III

KRYPTON (KR)

Z = 36

Kr I and II

Ref. 61, 121, 123, 147, 208, 232

— E.F.W.

Note: Wavelengths given to three decimal places below 11,000 Å are for the 86 isotope. Because of the small isotope shifts in Kr, the wavelengths applicable to natural Kr should not differ from those for Kr^{86} by more than about two units in the last decimal place given.

Intensity		Wavelength	
		Vacuum	
60		729.40	II
200		761.18	II
100		763.98	II
60		766.20	II
200		771.03	II
60	p	773.69	II
200		782.10	II
100		783.72	II
60		818.15	II
60		830.38	II
100		844.06	II
60		864.82	II
60		868.87	II
200		884.14	II
1,000		886.30	II
400		891.01	II
200		911.39	II
2,000		917.43	II
50		945.44	I
50		946.54	I
20		951.06	I
50		953.40	I
50		963.37	I
2,000		964.97	II
100		1,001.06	I
100		1,003.55	I
100		1,030.02	I
200		1,164.87	I
650		1,235.84	I
		Air	
100	h	2,464.77	II
60		2,492.48	II
80	h	2,712.40	II
100		2,833.00	II
100	h	3,607.88	II
200		3,631.889	II
250		3,653.928	II
80		3,665.324	I
150		3,669.01	II
100		3,679.559	I
80		3,686.182	II
300	h	3,718.02	II
200		3,718.595	II
150		3,721.350	II
200		3,741.638	II
150		3,744.80	II
80		3,754.245	II
500		3,778.089	II
500		3,783.095	II
150	h	3,875.44	II
150		3,906.177	II
200		3,920.081	II
100		3,994.840	II
100	h	3,997.793	II
300		4,057.037	II
300		4,065.128	II
500		4,088.337	II
250		4,098.729	II
100		4,109.248	II
250		4,145.122	II
150		4,250.580	II
1,000		4,273.969	I
100		4,282.967	I
600		4,292.923	II
200		4,300.49	II
500	h	4,317.81	II
400		4,318.551	I
1,000		4,319.579	I
150	h	4,322.98	II
100		4,351.359	I
3,000		4,355.477	II
500		4,362.641	II
200		4,369.69	II
800		4,376.121	I
300	h	4,386.54	II
200		4,399.965	I
100		4,425.189	I
500		4,431.685	II
600		4,436.812	II
600		4,453.917	I
800		4,463.689	I
800		4,475.014	II
400	h	4,489.88	II
600		4,502.353	I
400	h	4,523.14	II
200	h	4,556.61	II
800		4,577.209	II
300		4,582.978	II
150	h	4,592.80	II
500		4,615.292	II
1,000		4,619.166	II
800		4,633.885	II
2,000		4,658.876	II
500		4,680.406	II
100		4,691.301	II
200		4,694.360	II
3,000		4,739.002	II
300		4,762.435	II
1,000		4,765.744	II
300		4,811.76	II
300		4,825.18	II
800		4,832.077	II
700		4,846.612	II
150		4,857.20	II
300		4,945.59	II
200		5,022.40	II
250		5,086.52	II
400	h	5,125.73	II
500		5,208.32	II
200		5,308.66	II
500		5,333.41	II
200		5,468.17	II
500		5,562.224	I
2,000		5,570.288	I
80		5,580.386	I
100		5,649.561	I
400		5,681.89	II
200	h	5,690.35	II
100		5,832.855	I
3,000		5,870.914	I
200		5,992.22	II
60		5,993.849	I
60		6,056.125	I
300		6,420.18	II
100		6,421.026	I
200		6,456.288	I
150		6,570.07	II
60		6,699.228	I
100		6,904.678	I
250		7,213.13	II
100		7,224.104	I
80		7,287.258	I
400		7,289.78	II
400		7,407.02	II
60		7,425.541	I
200		7,435.78	II
100		7,486.862	I
300		7,524.46	II
1,000		7,587.411	I
2,000		7,601.544	I
150		7,641.16	II
1,000		7,685.244	I
1,200		7,694.538	I
250		7,735.69	II
150		7,746.827	I
800		7,854.821	I
200		7,913.423	I
180		7,928.597	I
200		7,933.22	II
120		7,973.62	II
100		7,982.401	I
1,500		8,059.503	I
4,000		8,104.364	I
6,000		8,112.899	I
60		8,132.967	I
3,000		8,190.054	I
200		8,202.72	II
80		8,218.365	I
3,000		8,263.240	I
100		8,272.353	I
5,000		8,298.107	I
1,500		8,281.050	I
100		8,412.430	I
3,000		8,508.870	I
150		8,764.110	I
6,000		8,776.748	I
2,000		8,928.692	I
500		9,238.48	II
500	h	9,293.82	II
200	h	9,320.99	II
300		9,361.95	II
100		9,362.082	I
200	h	9,402.82	II
200	h	9,470.93	II
500		9,577.52	II
500	h	9,605.80	II
400	h	9,619.61	II
200		9,663.34	II
200	h	9,711.60	II
2,000		9,751.758	I
500		9,803.14	II
500		9,856.314	I
1,000		10,221.46	II
100		11,187.108	I
200		11,257.711	I
150		11,259.126	I
500		11,457.481	I
150		11,792.425	I
1,500		11,819.377	I
600		11,997.105	I
160		12,077.224	I
100		12,861.892	I
1,100		13,177.412	I
1,000		13,622.415	I
2,400		13,634.220	I
800		13,658.394	I
200		13,711.036	I
600		13,738.851	I
150		13,974.027	I
550		14,045.657	I
140		14,104.298	I
180		14,402.22	I
2,000		14,426.793	I
100		14,517.84	I
1,600		14,734.436	I
550		14,762.672	I
450		14,765.472	I
400		14,961.894	I
120		15,005.307	I
140		15,209.526	I
1,700		15,239.615	I
130		15,326.480	I
1,500		15,334.958	I
700		15,372.037	I
200		15,474.026	I
180		15,681.02	I
120		15,820.09	I
200		16,726.513	I
2,000		16,785.128	I
1,000		16,853.488	I
2,400		16,890.441	I
1,600		16,896.753	I
1,800		16,935.806	I
600		17,098.771	I

700	17,367.606	I
120	17,404.443	I
150	17,616.854	I
650	17,842.737	I
700	18,002.229	I
2,600	18,167.315	I
100	18,399.786	I
150	18,580.896	I
300	18,696.294	I
170	18,785.460	I
200	18,797.703	I
140	20,209.878	I
300	20,423.964	I
140	20,446.971	I
600	21,165.471	I
1,800	21,902.513	I
120	22,485.775	I
180	23,340.416	I
120	24,260.506	I
180	24,292.221	I
600	25,233.820	I
180	28,610.55	I
1,000	28,655.72	I
150	28,769.71	I
140	28,822.49	I
300	29,236.69	I
300	30,663.54	I
300	30,979.16	I
500	39,300.6	I
1,100	39,486.52	I
220	39,557.25	I
100	39,572.60	I
1,400	39,588.4	I
1,100	39,589.6	I
500	39,954.8	I
300	39,966.6	I
1,300	40,306.1	I
250	40,685.16	I

LANTHANUM (LA)

Z = 57

La I and II

Ref. 1 — C.H.C.

Intensity		Wavelength	
		Air	
240		2,187.87	II
770		2,256.76	II
200		2,319.44	II
400		2,610.34	II
420		2,808.39	II
130		2,885.14	II
160		2,893.07	II
110		2,950.50	II
180		3,104.59	II
130		3,142.76	II
510		3,245.13	II
260		3,249.35	II
550		3,265.67	II
800		3,303.11	II
1,500		3,337.49	II
870		3,344.56	II
200		3,376.33	II
1,500		3,380.91	II
130		3,452.18	II
180		3.453.17	II
200		3,574.43	I
320		3,628.83	II
120		3,637.15	II
170	d	3,641.53	I
		3,641.66	II
1,000		3,645.42	II
390		3,650.18	II
170		3,662.08	II
120		3,704.54	I
320		3,705.82	II
550		3,713.54	II
140		3,714.87	II
270		3,715.53	II
2,400		3,759.08	II
120		3,780.67	II
3,700		3,790.83	II
3,900		3,794.78	II
190		3,835.08	II
600		3,840.72	II
120		3,846.00	II
1,600		3,849.02	II
130		3,854.91	II
3,400		3,871.64	II
1,700		3,886.37	II
1,300		3,916.05	II
1,100		3,921.54	II
160		3,927.56	I
2,200		3,929.22	II
180		3,936.22	II
9,000		3,949.10	II
4,400		3,988.52	II
3,600		3,995.75	II
180		4,015.39	I
250		4,025.88	II
2,800		4,031.69	II
140		4,037.21	I
3,000		4,042.91	II
320		4,050.08	II
220		4,060.33	I
160		4,064.79	I
850		4,067.39	II
110		4,076.71	II
2,800		4,077.35	II
120		4,079.18	I
5,500		4,086.72	II
180		4,089.61	I
280		4,099.54	II
110		4,104.87	I
4,400		4,123.23	II
110		4,137.04	I
550		4,141.74	II
1,100		4,151.97	II
220		4,152.78	II
100		4,160.26	I
280		4,187.32	I
280		4,192.36	II
1,500		4,196.55	II
240		4,204.04	II
300		4,217.56	II
200		4,230.95	II
1,600		4,238.38	II
140		4,249.99	II
320		4,263.59	II
480		4,269.50	II
240		4,275.64	II
300		4,280.27	I
600		4,286.97	II
600		4,296.05	II
120		4,300.44	II
440		4,322.51	II
4,600		4,333.74	II
550		4,354.40	II
110		4,364.67	II
110	b	4,371.97	LAO
110	b	4,375.84	LAO
110		4,378.10	II
280		4,383.44	LAO
100		4,385.20	LAO
220	b	4,418.24	LAO
160	b	4,423.17	LAO
160		4,423.90	I
260		4,427.55	II
100	b	4,428.10	LAO
2,000		4,429.90	II
160	b	4,432.98	LAO
100	b	4,438.01	LAO
100		4,452.15	I
100		4,455.80	II
850		4,522.37	II
170		4,525.31	II
420		4,526.12	II
400		4,558.46	II
110		4,559.29	II
160		4,567.91	I
200		4,570.02	I
400		4,574.88	II
200		4,580.06	II
160		4,605.78	II
410		4,613.39	II
410		4,619.88	II
110		4,645.28	II
540		4,655.50	II
360		4,662.51	II
230		4,663.76	II
200		4,668.91	II
160		4,671.83	II
230		4,692.50	II
140		4,703.28	II
170		4,716.44	II
140		4,719.94	II
230		4,728.42	II
500		4,740.28	II
390		4,743.09	II
320		4,748.73	II
160		4,766.89	I
160		4,804.04	II
160		4,809.01	II
200		4,824.06	II
320		4,860.91	II
850		4,899.92	II
1,000		4,920.98	II
1,000		4,921.79	II
140		4,934.83	II
110		4,946.47	II
370		4,949.77	I
340		4,970.39	II
370		4,986.83	II
140		4,991.28	II
720		4,999.47	II
140		5,046.88	I
210		5,050.57	I
170		5,056.46	I
200		5,106.23	I
470		5,114.56	II
470		5,122.99	II
450		5,145.42	I
180		5,156.74	II
180		5,157.43	II
290		5,158.69	I
120		5,163.62	II
580		5,177.31	I
850		5,183.42	II
260		5,188.22	II
170		5,204.15	II
720		5,211.86	I
520		5,234.27	I
340		5,253.46	I
110		5,259.39	II
370		5,271.19	I
140		5,290.84	II
370		5,301.98	II
140		5,302.62	II
180		5,303.55	II
110		5,340.67	II
110		5,357.86	I
130		5,377.09	II
140		5,380.99	II
500		5,455.15	I
470		5,501.34	I
110	b	5,602.50	LAO
160		5,631.22	I
240		5,648.25	I
130		5,657.72	I
180		5,740.66	I
160		5,744.41	I
160		5,761.84	I
160		5,769.07	II
370		5,769.34	I
320		5,789.24	I
450		5,791.34	I
220		5,797.58	II
160		5,805.78	II
140		5,821.99	I
320		5,930.62	I
720		6,249.93	I
260	d	6,262.30	II
180		6,296.09	II
160		6,320.39	II
110		6,325.91	I
170		6,390.48	II
450		6,394.23	I
210		6,410.99	I
250		6,455.99	I
110		6,526.99	II
130		6,543.16	I
140		6,578.51	I
180		6,709.50	I
120		6,774.26	II
13	b	7,011.22	LAO
75		7,023.67	I
26		7,032.05	I
26	b	7,040.84	LAO
110		7,045.96	I
13	b	7,054.80	LAO
160		7,066.23	II
65		7,068.37	I
21	b	7,070.79	LAO
13		7,076.38	I
21	b	7,085.40	LAO
26	l	7,101.02	LAO
10	h	7,116.8	II
19	l	7,131.58	LAO
10		7,149.77	I
40	h	7,158.08	I
50		7,161.25	I
10		7,162.60	LAO
21		7,219.91	I
10	b	7,257.16	LAO
26		7,270.09	I
10		7,270.30	I
110	c	7,282.34	II
10		7,320.91	I
110	c	7,334.18	I
65		7,345.34	I
50	l	7,379.71	LAO
85	l	7,380.08	LAO
35		7,382.73	I
110	l	7,403.52	LAO
210	l	7,403.75	LAO
50	b	7,411.34	LAO
65	l	7,434.28	LAO
110	l	7,434.36	LAO
30	b	7,442.92	LAO
50	h	7,463.08	I
50	l	7,465.25	LAO
95	l	7,465.48	LAO
75	c	7,483.50	II
40	l	7,496.50	LAO
95	l	7,496.78	LAO
50		7,498.83	I
30	b	7,506.79	LAO
19	l	7,528.21	LAO
50	l	7,528.39	LAO
30		7,533.59	I
85		7,539.23	I
35	l	7,560.09	LAO
35	l	7,592.26	LAO
19	h	7,612.94	II
19	b	7,624.99	LAO
21		7,664.34	I
15	h	7,841.80	I
21	b	7,876.87	LAO
75	l	7,877.22	LAO
75	l	7,910.19	LAO
150	l	7,910.54	LAO
50	b	7,944.61	LAO
110	l	7,944.95	LAO
40		7,964.83	I
35	b	7,979.34	LAO
75	l	7,979.70	LAO
35	h	8,001.89	I
21	b	8,014.43	LAO
65	l	8,014.79	LAO
30	b	8,019.48	LAO
35	c	8,051.39	I
75		8,086.05	I
15	b	8,122.20	LAO
15	b	8,159.02	LAO
7	h	8,203.38	I
50		8,247.44	I
13	h	8,316.04	I
85		8,324.69	I
95		8,346.53	I
8	h	8,379.80	I
8	b	8,453.55	LAO
8	h	8,467.62	I
26		8,476.48	I
13	h	8,507.37	I
13	h	8,513.57	I
8	h	8,514.65	II
17	b	8,526.59	LAO
17	c	8,543.46	I
65		8,545.44	I
15	b	8,563.54	LAO
9	h	8,590.94	I
9	b	8,600.81	LAO
7	h	8,624.22	I
15		8,638.47	I
19	c	8,672.11	I
40		8,674.43	I
13	h	8,720.41	I
35		8,748.38	I
19		8,818.93	I
35		8,825.82	I

21		8,839.63	I

La III

Ref. 220, 309 — J.R.

Intensity		Wavelength	
		Vacuum	
3		744.19	III
10		753.03	III
1		786.64	III
200		787.14	III
1		796.03	III
400		796.99	III
1		797.20	III
10		835.03	III
30		845.62	III
1		850.73	III
1		860.39	III
2		860.88	III
5		865.04	III
2,000		870.40	III
30		872.43	III
1,000		882.34	III
20		882.72	III
200		929.71	III
400		942.86	III
30		967.69	III
10		974.33	III
50		979.99	III
10		980.29	III
200		1,058.63	III
1,000		1,072.59	III
5,000		1,076.91	III
50,000		1,081.61	III
95,000		1,099.73	III
5,000		1,100.70	III
30		1,208.80	III
30		1,212.29	III
200		1,236.54	III
100		1,253.99	III
2,000		1,255.63	III
100		1,259.55	III
100		1,322.42	III
5,000		1,330.04	III
10,000		1,349.18	III
5,000		1,459.49	III
2,000		1,466.44	III
10,000		1,523.79	III
500		1,528.55	III
5,000		1,536.17	III
200		1,923.34	III
500		1,938.57	III
		Air	
60		2,216.07	III
20		2,238.36	III
25		2,258.61	III
5		2,260.30	III
250		2,297.74	III
400		2,379.37	III
10		2,387.99	III
20		2,392.49	III
100		2,476.60	III
50		2,478.65	III
2		2,513.43	III
4		2,588.87	III
2		2,604.83	III
400		2,651.50	III
100		2,682.34	III
150		2,684.76	III
110		2,897.88	III
160		2,904.58	III
7		2,950.84	III
10		2,953.77	III
40		2,992.10	III
4		3,006.19	III
15		3,009.22	III
100		3,075.17	III
4		3,085.38	III
25		3,093.03	III
15		3,096.26	III
200		3,111.97	III
50		3,116.74	III
1,000		3,171.63	III
1,500		3,171.74	III
50		3,172.69	III
20		2,196.84	III
70		3,289.11	III
15		3,301.48	III
35		3,327.66	III
500		3,517.09	III
600		3,517.22	III
3		4,129.24	III
5		4,137.43	III
200		4,482.97	III
300		4,499.05	III
5		5,145.73	III
8		5,158.41	III
6		5,467.81	III
55		5,491.90	III
2		5,511.72	III
1		5,518.19	III
45		5,529.54	III
1		5,744.09	III
200		5,778.14	III
2		5,813.45	III
3		5,875.63	III
55		5,888.62	III
3		5,932.71	III
2		6,017.11	III
20		6,055.84	III
35		6,119.25	III
120		6,141.99	III
55		6,220.00	III
60		6,348.21	III
3		8,114.42	III
2		8,135.96	III
250		8,252.60	III
100		8,275.39	III
200		8,287.75	III
250		8,321.11	III
300		8,583.45	III
120		9,184.38	III
100		9,212.63	III
80		9,923.99	III
140		10,284.79	III
20		10,370.34	III
12		10,937.90	III

La IV

Ref. 79 — J.R.

Intensity		Wavelength	
		Vacuum	
100		344.12	IV
7,000		453.50	IV
10,000		463.14	IV
15,000		499.54	IV
40,000		552.02	IV
30,000		631.26	IV
25		724.92	IV
15		733.29	IV
10		797.03	IV
10	c	980.03	IV
50		1,039.30	IV
60	p	1,062.09	IV
75		1,158.35	IV
50		1,164.29	IV
400		1,230.90	IV
75	p	1,260.79	IV
300		1,261.12	IV
150		1,283.19	IV
2,000		1,302.31	IV
1,200		1,333.53	IV
3,500		1,334.96	IV
1,000		1,352.76	IV
25,000		1,368.04	IV
8,000		1,377.49	IV
3,000		1,394.32	IV
5,000		1,414.58	IV
7,000		1,432.55	IV
7,000		1,441.63	IV
7,500		1,462.15	IV
20,000		1,463.47	IV
7,500		1,467.54	IV
15,000		1,507.87	IV
5,000		1,527.19	IV
2,500		1,575.92	IV
1,500		1,583.61	IV
750		1,585.11	IV
750		1,637.42	IV
750	d	1,645.21	IV
1,000		1,664.84	IV
750		1,684.17	IV
2,000	p	1,767.65	IV
4,000		1,808.66	IV
1,000		1,851.81	IV
1,500		1,852.77	IV
750		1,879.79	IV
1,000		1,881.57	IV
800		1,889.22	IV
1,000		1,891.47	IV
5,000		1,902.97	IV
800		1,907.44	IV
1,500	c	1,950.80	IV
1,200	c	1,957.57	IV
		Air	
3,000		2,012.42	IV
750		2,037.43	IV
2,000	c	2,066.50	IV
3,000	c	2,073.18	IV
1,500		2,143.23	IV
4,000	c	2,197.45	IV
1,000	w	2,221.12	IV
900		2,227.34	IV
3,000		2,244.95	IV
7,500	w	2,265.91	IV
2,000		2,315.89	IV
750		2,348.36	IV
750		2,355.31	IV
2,000	c	2,407.10	IV
25,000	w	2,417.58	IV
1,200	p	2,443.92	IV
18,000	c	2,502.81	IV
15,000		2,515.02	IV
50,000		2,532.75	IV
900	d	2,535.76	IV
45,000		2,582.05	IV
18,000	c	2,591.30	IV
95,000	w	2,597.50	IV
5,000	c	2,608.01	IV
70,000	w	2,662.75	IV
50,000	w	2,848.30	IV
12,000	c	2,863.30	IV
30,000	c	2,962.58	IV
70,000	w	3,009.51	IV
90,000	c	3,056.68	IV
3,500		3,522.28	IV
2,000		3,650.40	IV
2,000	p	4,270.76	IV
1,500	w	4,549.80	IV
500		4,836.89	IV

La V

Ref. 78 — J.R.

Intensity		Wavelength	
		Vacuum	
2		389.03	V
400		390.72	V
1		398.53	V
30		399.34	V
350		405.10	V
50		416.13	V
3		421.55	V
50		423.07	V
400		424.78	V
1,000		432.11	V
2,500		435.28	V
700		436.14	V
700		436.84	V
300		437.11	V
700		437.55	V
20		444.01	V
10		444.07	V
1,250		450.40	V
600		457.30	V
1,000		463.85	V
150		476.67	V
5,000		482.16	V
200		482.43	V
2,000		483.30	V
7,000		498.08	V
4,000		499.03	V
10,000		503.58	V
40		508.15	V
1,500		525.71	V
12,000		526.76	V
10,000		531.07	V
15,000		533.23	V
4,000		540.20	V
6,000		544.80	V
8,000		547.44	V
3,000		570.90	V
2,500		593.18	V
750		597.70	V
2,000		600.01	V
5,000		600.24	V
700		611.70	V
500		617.60	V

LEAD (PB)

Z = 82

Pb I and II

Ref. 64, 274, 283, 329, 330 — D.R.W.

Intensity		Wavelength	
		Vacuum	
2		846.04	II
2	h	849.88	II
3		855.57	II
3		863.00	II
6		873.71	II
2		877.96	II
8		889.68	II
3		896.30	II
5		926.44	II
2		958.76	II
2		960.21	II
3		965.36	II
10		967.23	II
9		972.56	II
8		982.17	II
10		986.71	II
10		995.89	II
6		1,001.81	II
10		1,016.61	II
10		1,049.82	II
10		1,050.77	II
10		1,060.66	II
9		1,065.58	II
10		1,103.94	II
10		1,108.43	II
10		1,109.84	II
10		1,119.57	II
10		1,121.36	II
10		1,133.14	II
4		1,145.91	II
10		1,203.63	II
10		1,231.20	II
10		1,331.65	II
10		1,335.20	II
10		1,348.37	II
10		1,433.96	II
3		1,449.35	II
10		1,512.42	II
10		1,671.53	II
10		1,682.15	II
20		1,726.75	II
2		1,740.00	I
2		1,766.64	I
2		1,794.67	I
10		1,796.670	II
5		1,812.97	I
10		1,822.050	II
4		1,868.76	I
10		1,904.77	I
7		1,921.471	II
4		1,972.44	I
2		1,977.88	I
2		1,991.60	I
2		1,992.31	I
		Air	
5	r	2,022.02	I
5		2,050.88	I
8	r	2,053.28	I
6		2,111.758	I
10		2,115.066	I
500	r	2,170.00	I
7		2,175.580	I
7		2,187.888	I
8		2,189.603	I
10		2,203.534	II

Intensity		Wavelength	
20		2,237.425	I
20		2,246.86	I
25		2,246.89	I
150		2,332.418	I
180		2,388.797	I
550	r	2,393.792	I
140		2,399.597	I
320	r	2,401.940	I
320	r	2,411.734	I
150	r	2,443.829	I
160	r	2,446.181	I
130	r	2,476.378	I
8	c	2,526.69	II
8	c	2,576.60	II
80	r	2,577.260	I
2	c	2,608.38	II
500	r	2,613.655	I
900	r	2,614.175	I
160		2,628.262	I
4		2,634.256	II
10		2,657.094	I
700		2,663.154	I
10		2,697.541	I
25,000	r	2,801.995	I
100		2,822.58	I
14,000	r	2,823.189	I
35,000	r	2,833.053	I
6		2,840.557	II
14,000	r	2,873.311	I
3	c	2,887.30	II
3		2,914.442	II
2	c	2,947.43	II
3	c	2,948.53	II
15		2,966.460	I
15		2,972.991	I
15		2,980.157	I
4		2,986.876	II
10	c	3,016.39	II
150		3,118.894	I
600		3,220.528	I
100		3,229.613	I
400		3,240.186	I
200		3,262.355	I
35,000		3,572.729	I
50,000	r	3,639.568	I
20,000		3,671.491	I
70,000	r	3,683.462	I
10		3,713.982	II
25,000		3,739.935	I
15,000		4,019.632	I
95,000		4,057.807	I
45,000		4,062.136	I
5		4,110.76	II
4		4,113.35	II
10		4,152.82	II
10		4,157.814	I
10,000		4,168.033	I
9	c	4,242.14	II
20	c	4,244.92	II
7		4,293.82	II
6		4,296.65	II
200		4,340.413	I
10		4,352.74	II
20	c	4,386.46	II
10		4,579.051	II
10		4,582.27	II
1,000		5,005.416	I
100		5,006.572	I
50		5,042.58	II
10		5,070.58	II
10		5,074.53	II
10		5,076.35	I
50		5,089.484	I
20		5,090.01	I
10		5,107.242	I
10		5,111.64	II
2,000		5,201.437	I
10		5,367.64	II
10		5,372.099	II
10	c	5,544.25	II
20	c	5,608.85	II
40		5,692.346	I
200		5,895.624	I
2,000		6,001.862	I
9	c	6,009.58	II
500		6,011.667	I
8	c	6,041.17	II
500		6,059.356	I
40		6,075.74	II
40		6,081.409	II
50		6,110.520	I
10		6,159.89	II
100		6,235.266	I
50	c	6,660.20	II
10		6,892.11	I
5		7,128.94	I
20		7,193.60	II
20,000		7,228.965	I
5		7,304.68	I
8		7,330.15	I
10		7,346.676	I
10		7,558.97	II
10		7,632.56	II
4		7,732.96	II
20		7,809.259	I
5		7,817.97	I
6		7,829.01	I
5		7,896.737	I
2	d	8,156.91	II
10		8,168.001	I
6		8,191.886	I
5		8,217.711	I
8		8,255.61	I
40		8,272.690	I
6		8,335.54	II
10		8,395.68	II
20		8,409.384	I
10		8,478.492	I
8		8,532.17	I
7	c	8,544.95	II
7		8,709.90	II
5		8,719.39	II
5		8,722.810	I
10		8,857.457	I
10		9,050.82	II
10		9,063.43	II
2	d	9,245.28	II
8		9,293.476	I
5		9,384.35	I
5		9,385.89	I
15		9,438.05	I
15		9,604.297	I
6		9,608.73	I
15		9,674.351	I
200		10,290.458	I
5		10,434.32	I
100		10,498.965	I
50		10,649.249	I
5		10,759.41	I
7		10,759.74	I
15		10,886.688	I
40		10,969.53	I
6		11,059.22	I
3		11,333.08	I
2	d	11,479.49	II
2		11,488.76	I
5		11,627.91	II
1		12,561.37	I
		13,495.3	I
		13,498.2	I
		13,512.6	I
		14,722.8	I
		14,742.1	I
		14,743.0	I
		15,314.8	I
		15,327.6	I
		15,331.0	I
		15,349.6	I
		38,831.1	I
		38,950.1	I
		38,958.6	I
		39,039.4	I

Pb III

Ref. 54, 256, 297 — D.R.W.

Intensity	Wavelength	
	Vacuum	
1	961.01	III
3	1,030.5	III
12	1,048.9	III
4	1,069.2	III
3	1,074.7	III
4	1,118.67	III
4	1,167.0	III
4	1,250.6	III
1	1,266.9	III
20	1,553.1	III
1	1,610.1	III
4	1,711.23	III
	Air	
10	3,043.85	III
4	3,089.08	III
4	3,102.74	III
10	3,137.81	III
10	3,176.50	III
5	3,242.84	III
1	3,530.17	III
7	3,589.87	III
7	3,689.31	III
3	3,706.02	III
5	3,728.69	III
12	3,854.08	III
8	3,951.92	III
3	4,031.16	III
3	4,094.54	III
2	4,128.11	III
8	4,272.66	III
6	4,499.34	III
7	4,571.21	III
1	4,596.45	III
6	4,761.12	III
4	4,798.59	III
1	4,826.86	III
2	4,855.06	III
3	5,065.12	III
4	5,191.56	III
5	5,523.97	III
3	5,779.41	III
6	5,857.96	III

Pb IV

Ref. 106 — D.R.W.

Intensity	Wavelength	
	Vacuum	
8	475.36	IV
7	478.35	IV
10	496.38	IV
12	499.94	IV
9	515.07	IV
14	529.78	IV
20	570.16	IV
8	573.90	IV
8	584.52	IV
10	648.50	IV
9	656.10	IV
10	761.09	IV
18	802.07	IV
12	802.82	IV
10	812.59	IV
8	822.07	IV
10	827.41	IV
12	832.60	IV
8	840.99	IV
8	842.88	IV
12	845.94	IV
18	857.64	IV
8	859.02	IV
16	862.33	IV
14	870.44	IV
12	879.96	IV
7	880.35	IV
14	884.96	IV
14	884.99	IV
16	890.72	IV
12	908.51	IV
12	917.90	IV
10	922.12	IV
12	922.49	IV
7	924.52	IV
10	927.64	IV
14	932.20	IV
8	937.00	IV
7	952.85	IV
8	1,012.44	IV
14	1,028.61	IV
20	1,032.05	IV
16	1,041.24	IV
18	1,044.14	IV
15	1,056.53	IV
12	1,072.09	IV
7	1,079.88	IV
18	1,080.81	IV
20	1,084.17	IV
6	1,089.94	IV
7	1,099.47	IV
6	1,115.30	IV
20	1,116.08	IV
18	1,137.84	IV
8	1,142.77	IV
14	1,144.93	IV
20	1,189.95	IV
8	1,267.55	IV
8	1,290.82	IV
10	1,291.10	IV
20	1,313.05	IV
8	1,323.92	IV
12	1,343.06	IV
16	1,388.94	IV
6	1,397.02	IV
18	1,400.26	IV
10	1,404.34	IV
7	1,510.76	IV
14	1,535.71	IV
8	1,798.39	IV
8	1,893.19	IV
12	1,959.34	IV
16	1,973.16	IV
	Air	
10	2,042.58	IV
12	2,049.34	IV
12	2,079.22	IV
8	2,151.96	IV
15	2,154.01	IV
12	2,177.46	IV
16	2,359.53	IV
4	2,864.24	IV
4	2,864.50	IV
16	2,417.61	IV
4	2,978.14	IV
4	3,052.56	IV
4	3,221.17	IV
4	3,962.48	IV
4	4,049.80	IV
10	4,496.15	IV
16	4,534.60	IV
8	4,605.40	IV
2	5,914.54	IV

Pb V

Ref. 106 — D.R.W.

Intensity	Wavelength	
	Vacuum	
2	367.40	V
2	372.53	V
2	387.87	V
2	394.38	V
5	424.64	V
3	431.03	V
3	436.60	V
2	438.47	V
6	438.91	V
4	453.45	V
3	461.70	V
3	496.20	V
4	694.42	V
8	696.20	V
20	703.73	V
4	706.29	V
6	707.66	V
5	730.85	V
12	749.46	V
10	752.52	V
10	755.80	V
6	762.76	V
10	765.87	V
18	767.45	V
18	769.49	V
14	771.42	V
14	782.79	V
10	787.05	V
15	797.02	V
5	799.80	V
18	809.63	V
5	812.32	V
8	814.10	V
8	820.09	V

Intensity		Wavelength	
5		825.52	V
8		829.32	V
5		851.98	V
20		863.97	V
10		867.10	V
6		880.50	V
18		883.90	V
14		888.37	V
14		894.40	V
12		896.08	V
8		915.09	V
14		915.71	V
12		918.09	V
12		920.28	V
12		920.66	V
6		940.74	V
8		946.20	V
6		950.93	V
12		954.35	V
4		954.95	V
10		955.28	V
4		964.38	V
6		989.14	V
8		1,005.42	V
10		1,051.26	V
4		1,059.26	V
10		1,088.86	V
9		1,096.52	V
6		1,104.79	V
4		1,121.33	V
10		1,137.50	V
4		1,152.36	V
12		1,157.88	V
14		1,185.43	V
11		1,233.50	V
10		1,248.47	V
8		1,635.75	V
2		1,802.87	V
2		1,843.00	V
2		1,888.67	V
2		1,897.02	V
2		1,914.33	V
4		1,919.74	V
5		1,957.96	V
2		1,998.58	V
10		1,998.83	V
		Air	
8		2,078.45	V
10		2,142.55	V
10		2,167.97	V
20		2,259.01	V
10		2,276.66	V
8		2,301.49	V
15		2,424.81	V
4		4,809.36	V
5		6,650.99	V
4		6,753.20	V

LITHIUM (LI)

Z = 3

Li I and II

Ref. 3, 15, 17, 18, 37, 4

Intensity		Wavelength	
		Vacuum	
		125.5	II
		136.5	II
		140.5	II
		167.21	II
		168.74	II
		171.58	II
		178.02	II
		199.28	II
		207.5	II
		456.	II
		483.	II
		540.	II
		729.	II
		800.	II
		820.	II
		861.	II
		905.5	II
		917.5	II
		936.	II
		945.	II
		965.	II
		972.	II
		988.	II
		1,018.	II
		1,032.	II
		1,036.	II
		1,093.	II
		1,103.	II
		1,109.	II
		1,116.	II
		1,132.1	II
		1,141.	II
		1,166.4	II
		1,198.09	II
		1,215.	II
		1,238.	II
		1,253.8	II
		1,420.89	II
		1,424.	II
3		1,492.93	II
5		1,492.97	II
1		1,493.04	II
		1,555.	II
3		1,653.08	II
5		1,653.13	II
1		1,653.21	II
		1,681.66	II
		Air	
		2,009.	II
		2,039.	I
		2,068.	II
		2,131.	II
		2,164.	II
		2,173.4	I
		2,183.	II
		2,214.	II
		2,222.	II
		2,237.	II
		2,249.21	II
		2,286.82	II
		2,302.57	II
		2,303.33	II
		2,304.59	I
		2,304.92	I
		2,305.36	I
		2,305.83	I
		2,306.29	I
		2,306.82	I
		2,307.44	I
		2,308.97	I
		2,309.88	I
		2,310.94	I
		2,312.11	I
		2,313.49	I
		2,315.08	I
		2,316.95	I
		2,319.18	I
		2,321.88	I
		2,325.11	I
		2,329.02	I
		2,329.84	II
		2,333.94	I
3		2,336.88	II
5		2,336.91	II
2		2,337.00	II
		2,340.15	I
		2,348.22	I
		2,358.93	I
		2,373.54	I
		2,381.54	II
		2,383.20	II
1		2,394.39	I
		2,402.33	II
		2,410.84	II
3		2,425.43	I
		2,429.81	II
		2,460.2	I
10		2,475.06	I
		2,506.94	II
		2,508.78	II
		2,518.	I
		2,539.49	II
24		2,551.7	II
		2,559.	II
15		2,562.31	I
		2,605.08	II
		2,640.	
2		2,657.29	II
3		2,657.30	II
		2,674.46	II
0		2,728.24	II
5		2,728.29	II
2		2,728.32	II
3		2,730.47	II
1		2,730.55	II
5		2,741.20	I
		2,766.99	II
		2,790.31	II
		2,801.	I
		2,846.	I
		2,868.	I
		2,895.	I
2		2,934.02	II
2		2,934.07	II
5		2,934.12	II
1		2,934.25	II
		2,968.	I
3		3,029.12	II
3		3,029.14	II
		3,144.	I
3		3,155.31	II
4		3,155.33	II
1		3,196.26	II
9		3,196.33	II
4		3,196.36	II
5		3,199.33	II
2		3,199.43	II
17		3,232.66	I
		3,249.87	II
		3,306.28	II
		3,393.	
		3,488.	I
		3,579.8	I
		3,618.	I
		3,662.	I
		3,684.32	II
1		3,714.00	II
5		3,714.16	II
6	d	3,714.27	II
8		3,714.29	II
7	d	3,714.40	II
10		3,714.41	II
1		3,714.51	II
0		3,714.58	II
3		3,718.7	I
6		3,794.72	I
20		3,915.30	I
20		3,915.35	I
10		3,985.48	I
10		3,985.54	I
40		4,132.56	I
40		4,132.62	I
		4.196.	I
20		4,273.07	I
20		4,273.13	I
5		4,325.42	II
5		4,325.47	II
1		4,325.54	II
		4,516.45	II
		4,590.	
13		4,602.83	I
13		4,602.89	I
		4,607.34	II
0		4,671.51	II
6		4,671.65	II
2		4,671.70	II
3		4,678.06	II
1		4,678.29	II
		4,760.	I
		4,763.	II
		4,788.36	II
		4,843.0	II
4		4,881.32	II
4		4,881.39	II
1		4,881.49	II
8		4,971.66	I
8		4,971.75	I
		5,037.92	II
		5,095.	
		5,114.	
		5,190.	
		5,271.	I
		5,315.	I
		5,395.	I
		5,440.	I
600	c	5,483.55	II
600	c	5,485.65	II
320		6,103.54	I
320		6,103.65	I
3,600		6,707.76	I
3,600		6,707.91	I
48		8,126.23	I
48		8,126.45	I
		8,517.37	II
		9,581.42	II
		10,120.	II
		12,232.	I
		12,782.	I
		13,566.	I
		17,552.	I
		18,697.	I
		19,290.	I
		24,467.	I
		40,475.	I

Li III

Ref. 335 — J.O.S.

Intensity	Wavelength	
	Air	
	102.9	III
	103.4	III
	104.1	III
	105.5	III
	108.0	III
	113.9	III
	135.0	III
	540.0	III
	729.1	III

LUTETIUM (LU)

Z = 71

Lu I and II

Ref. 1 — C.H.C.

Intensity		Wavelength	
		Air	
1,700	h	2,195.54	II
95		2,276.94	II
190		2,297.41	II
1,300		2,392.19	II
120		2,399.14	II
80		2,419.21	II
55		2,430.26	II
130		2,459.64	II
80		2,469.27	II
21	h	2,481.72	II
370		2,536.95	II
40		2,546.87	II
20		2,549.44	I
20		2,549.72	I
35		2,561.80	II
930		2,571.23	II
1,700		2,578.79	II
80	h	2,582.13	II
1,800		2,613.40	II
18,000		2,615.42	II
1,800		2,619.26	II
90		2,657.05	
2,700		2,657.80	II
90	h	2,677.25	I
570	h	2,685.08	I
90	h	2,685.54	I
4,200		2,701.71	II
90	h	2,715.91	I
180	d	2,719.09	I
480	h	2,728.95	I
75	c	2,738.17	II
3,600		2,754.17	II

750	h	2,765.74	I
2,700		2,796.63	II
35		2,821.23	II
270	c	2,834.35	II
330	h	2,845.13	I
3,000		2,847.51	II
570	h	2,885.14	I
6,300		2,894.84	II
4,500		2,900.30	II
300		2,903.05	I
9,000		2,911.39	II
270	h	2,949.73	I
1,200		2,951.69	II
60		2,955.78	II
4,200		2,963.32	II
2,400		2,969.82	II
1,800		2,989.27	I
3,000		3,020.54	II
120		3,027.29	II
2,100		3,056.72	II
7,500		3,077.60	II
390		3,080.11	I
5,100	h	3,081.47	I
3,000		3,118.43	I
2,400		3,171.36	I
100		3,183.73	II
260		3,191.80	II
1,400		3,198.12	II
4,800		3,254.31	II
3,800		3,278.97	I
7,600		3,281.74	I
6,200		3,312.11	I
7,600		3,359.56	I
6,200		3,376.50	I
950		3,385.50	I
160	h	3,391.55	I
1,400		3,396.82	I
4,100		3,397.07	II
4,800		3,472.48	II
8,300	c	3,507.39	II
1,600		3,508.42	I
4,800		3,554.43	II
4,800		3,567.84	I
340		3,596.34	I
800		3,623.99	II
680		3,636.25	I
2,600		3,647.77	I
60		3,684.32	I
60		3,710.95	I
110		3,756.70	I
110		3,756.79	I
30	h	3,786.18	I
150		3,800.67	I
75	h	3,829.07	I
2,700		3,841.18	I
75		3,843.61	I
95		3,853.29	I
40		3,874.61	I
530		3,876.65	II
29		3,911.77	I
50		3,918.86	I
35	h	3,926.62	I
480		3,968.46	I
50		3,991.38	I
670		4,054.45	I
75	b	4,096.13	L
35	h	4,107.44	I
95	h	4,112.67	I
310		4,122.49	I
3,100		4,124.73	I
150	c	4,131.79	I
460		4,154.08	I
24		4,158.98	I
1,600		4,184.25	II
150		4,277.50	I
250		4,281.03	I
330	d	4,295.97	I
		4,296.09	I
150		4,309.57	I
75		4,332.72	I
29		4,341.98	II
65	h	4,420.96	I
190	c	4,430.48	I
35		4,438.79	I
190		4,450.81	I
50	h	4,471.55	I
60	h	4,498.85	I
3,300		4,518.57	I

24	b	4,560.95	LUO
24	b	4,575.31	LUO
85	c	4,605.39	I
95	h	4,645.47	I
100	h	4,648.21	I
95	h	4,648.85	I
65	b	4,654.03	LUO
1,000		4,658.02	I
85	h	4,659.03	I
630	b	4,661.75	LUO
310	b	4,672.31	LUO
420	b	4,684.16	LUO
270	b	4,695.46	LUO
190	b	4,708.00	LUO
30		4,716.70	I
65	b	4,720.86	LUO
65	h	4,726.20	I
100	b	4,735.00	LUO
75	b	4,749.11	LUO
40	b	4,764.22	LUO
150		4,785.42	II
85		4,815.05	I
50	c	4,839.62	II
18		4,865.36	II
460		4,904.88	I
180		4,942.34	I
800		4,994.13	II
800		5,001.14	I
55	h	5,057.60	I
140		5,134.05	I
2,700		5,135.09	I
130	b	5,170.11	LUO
170		5,196.61	I
90		5,206.47	I
40		5,304.40	I
80		5,349.12	I
500		5,402.57	I
140	c	5,421.90	I
100		5,437.88	I
35		5,453.57	I
2,100		5,476.69	II
9		5,664.89	II
14		5,713.49	I
550		5,736.55	I
55		5,775.40	I
80		5,800.59	I
40	h	5,860.79	I
9		5,866.30	I
690	c	5,983.9	II
140		5,997.13	I
1,400		6,004.52	I
35	h	6,041.66	I
440		6,055.03	I
11		6,140.71	I
150		6,159.94	II
160		6,199.66	II
2,100		6,221.87	II
35		6,228.14	II
80		6,235.36	II
160		6,242.34	II
16	h	6,248.80	I
70	h	6,345.35	I
18	h	6,354.85	I
9		6,365.79	I
16		6,366.00	I
22		6,441.14	I
11		6,444.89	II
1,100		6,463.12	II
29		6,477.67	I
55	c	6,523.18	I
35	c	6,611.28	II
		6,611.58	II
		6,611.80	II
		6,611.95	II
		6,612.04	II
11		6,619.15	II
23	c	6,677.14	I
9	h	6,735.76	I
30	c	6,793.77	I
11		6,826.59	II
45		6,917.31	I
8		6,943.96	II
23		7,031.24	I
14	c	7,096.34	I
45		7,125.84	II
9		7,142.79	I
7		7,143.10	I
8		7,165.94	II

14	h	7,237.98	I
5		7,409.70	II
35	c	8,610.98	I
11	c	7,441.52	I
8		7,456.96	II
7	h	7,640.08	I
7	w	7,758.30	I
7	h	7,815.9	I
9	c	8,178.16	I
17		8,382.08	I
35		8,459.19	II
10	d	8,478.50	I
29	c	8,508.08	I
35	c	8610.98	I

Lu III

Ref. 148 — J.R.

Intensity		Wavelength	
		Vacuum	
1		677.34	III
7		691.05	III
30		700.25	III
50		714.89	III
100		738.76	III
200		755.03	III
3		755.16	III
500		810.73	III
100		830.53	III
2,000		832.28	III
10		972.66	III
2		991.26	III
100		996.44	III
400		1,001.18	III
1		1,022.40	III
100		1,029.83	III
200		1,030.33	III
100		1,031.54	III
50		1,056.53	III
20		1,061.99	III
3		1,092.84	III
200		1,187.34	III
50		1,228.7	III
200		1,277.53	III
30		1,283.41	III
100		1,331.93	III
1,000		1,854.57	III
		Air	
40		2,050.72	III
1,500		2,065.35	III
1,500	c	2,070.56	III
100		2,083.34	III
200		2,099.44	III
1,000		2,236.14	III
2,000		2,236.22	III
500	c	2,381.59	III
300		2,563.51	III
4,500	c	2,603.35	III
200		2,721.65	III
2,000		2,772.55	III
20		2,781.16	III
10		2,788.37	III
500		2,800.90	III
20	p	2,993.21	III
1,000		3,057.86	III
200		4,251.44	III
300		4,271.91	III
200		4,490.00	III
400		4,956.43	III
10		5,046.12	III
150	c	5,145.86	III
70		5,419.42	III
60		5,519.88	III
5		5,526.80	III
5		5,748.71	III
70		5,786.46	III
80		5,869.71	III
60		5,889.76	III
300		6,197.96	III
600		6,198.13	III
100		7,309.95	III
200		7,310.25	III
50		7,534.27	III
70	l	7,936.45	III
3		8,008.59	III

Lu IV

Ref. 310 — J.R.

Intensity		Wavelength	
		Vacuum	
400		876.80	IV
100		902.06	IV
300		1,015.18	IV
20		1,136.17	IV
50		1,189.27	IV
20	p	1,194.59	IV
60		1,213.08	IV
15		1,220.74	IV
20		1,223.75	IV
20		1,240.07	IV
20		1,248.10	IV
40		1,266.27	IV
100		1,272.42	IV
20		1,273.02	IV
40		1,274.77	IV
20		1,276.54	IV
50		1,289.38	IV
60		1,310.08	IV
50		1,323.02	IV
15		1,331.04	IV
800		1,333.79	IV
300		1,334.94	IV
50		1,338.20	IV
100		1,339.49	IV
300		1,342.58	IV
200		1,351.68	IV
300		1,353.74	IV
200		1,355.85	IV
20		1,359.67	IV
100		1,363.24	IV
50		1,363.37	IV
20		1,367.34	IV
20		1,373.54	IV
15		1,375.36	IV
100		1,376.02	IV
30		1,379.56	IV
15		1,383.18	IV
15		1,389.85	IV
60		1,390.07	IV
200		1,390.30	IV
50		1,390.69	IV
40		1,392.38	IV
40		1,397.18	IV
30		1,401.32	IV
100		1,401.46	IV
200		1,406.64	IV
100		1,407.00	IV
250		1,407.04	IV
20		1,420.32	IV
100		1,421.59	IV
200		1,429.08	IV
400		1,429.38	IV
40		1,430.80	IV
100		1,440.62	IV
100		1,448.14	IV
200		1,452.33	IV
30		1,462.65	IV
100		1,483.79	IV
200		1,493.24	IV
400		1,511.26	IV
200		1,521.06	IV
100		1,522.21	IV
100		1,537.77	IV
30		1,549.35	IV
20		1,551.59	IV
20		1,562.06	IV
30		1,592.55	IV
15		1,594.92	IV
100	c	1,607.72	IV
50	c	1,631.65	IV
20		1,684.50	IV
60	c	1,693.67	IV
400	c	1,721.42	IV
100		1,725.14	IV
100	c	1,735.79	IV
100		1,736.78	IV
100		1,741.74	IV
50	c	1,743.84	IV
40	c	1,752.60	IV
200		1,759.61	IV
300		1,772.08	IV

600		1,772.57	IV
200		1,782.45	IV
100		1,797.52	IV
20	c	1,901.63	IV
100		1,983.92	IV
20		1,990.52	IV
40		1,996.18	IV
		Air	
100		2,003.18	IV
100		2,020.94	IV
20		2,071.10	IV
100		2,081.09	IV
400	c	2,085.70	IV
600	c	2,086.47	IV
400	c	2,092.16	IV
100		2,103.63	IV
1,000	c	2,104.41	IV
200	c	2,107.85	IV
1,000	c	2,108.31	IV
100	c	2,127.43	IV

MAGNESIUM (MG)

Z = 12

Mg I and II

Ref. 49, 83, 103, 21

Intensity		Wavelength	
		Vacuum	
		184.05	II
		184.31	II
		184.68	II
		184.81	II
		185.26	II
		185.59	II
		185.98	II
		186.47	II
		186.84	II
		187.19	II
		187.38	II
		188.54	II
		188.91	II
		189.01	II
		189.23	II
		189.37	II
		191.30	II
		191.56	II
		191.65	II
		192.40	II
		192.55	II
		192.84	II
		193.09	II
		193.31	II
		193.40	II
		193.64	II
		197.76	II
		199.31	II
		200.29	I
		202.00	II
		202.27	II
		202.51	II
		202.94	II
		203.15	I
		203.42	II
		203.53	II
		204.22	I
		209.09	II
		209.43	II
		209.84	I
		213.53	I
		215.12	I
		215.31	I
		215.45	I
		215.66	I
		215.79	I
		216.22	I
		216.36	I
		216.68	I
		217.21	I
		217.37	I
		218.19	I
		218.34	I
		218.42	I
		218.74	I
		219.04	I
		219.28	I
		220.03	I
		220.33	I
		222.03	I
		222.67	I
		223.45	I
		223.74	I
		225.18	I
		225.54	I
		226.26	I
		247.14	II
		248.47	II
		884.70	II
		884.72	II
		907.38	II
		907.41	II
8		946.70	II
9		946.77	II
14		1,025.96	II
12		1,026.11	II
25		1,239.94	II
20		1,240.40	II
6		1,248.51	II
8		1,249.93	II
8		1,271.24	II
9		1,271.94	II
8		1,272.72	II
11		1,273.43	II
11		1,306.71	II
12		1,307.88	II
12		1,308.28	II
14		1,309.44	II
14		1,365.45	II
		1,365.54	II
15		1,367.26	II
15		1,367.70	II
18		1,369.42	II
20		1,476.00	II
25		1,478.01	II
20		1,480.89	II
30		1,482.90	II
		1,625.22	I
		1,625.50	I
		1,625.81	I
		1,626.16	I
		1,626.36	I
		1,626.56	I
		1,626.79	I
		1,627.02	I
		1,627.27	I
		1,627.53	I
		1,627.82	I
		1,628.12	I
		1,628.46	I
		1,628.80	I
		1,629.21	I
		1,629.59	I
		1,630.52	I
		1,631.62	I
		1,632.93	I
		1,634.52	I
		1,636.48	I
		1,638.90	I
		1,641.97	I
1		1,645.93	I
1		1,651.16	I
2		1,658.31	I
5		1,668.43	I
10		1,683.41	I
15		1,707.06	I
40		1,734.84	II
50		1,737.62	II
20		1,747.80	I
40		1,750.65	II
50		1,753.46	II
30		1,827.93	I
		Air	
9		2,025.82	I
3		2,329.58	II
6		2,449.57	II
1		2,557.23	I
1		2,560.94	I
1		2,562.26	I
1		2,564.94	I
1		2,570.91	I
1		2,572.25	I
2		2,574.94	I
1		2,577.89	I
1		2,580.59	I
1		2,584.22	I
2		2,585.56	I
3		2,588.28	I
1		2,591.89	I
1		2,593.23	I
2		2,595.97	I
2		2,602.50	I
4		2,603.85	I
5		2,606.62	I
1		2,613.36	I
2		2,614.73	I
3		2,617.51	I
3		2,628.66	I
6		2,630.05	I
8		2,632.87	I
2		2,644.80	I
3		2,646.21	I
4		2,649.06	I
8		2,660.76	II
8		2,660.82	II
6		2,668.12	I
8		2,669.55	I
10		2,672.46	I
3		2,693.72	I
5		2,695.18	I
6		2,698.14	I
8		2,731.99	I
10		2,733.49	I
12		2,736.53	I
5		2,765.22	I
7		2,768.34	I
38		2,776.69	I
32		2,778.27	I
90		2,779.83	I
8		2,781.29	I
32		2,781.42	I
36		2,782.97	I
13		2,790.79	II
1,000		2,795.53	II
16		2,798.06	II
600		2,802.70	II
12		2,846.75	I
14		2,848.42	I
16		2,851.65	I
6,000		2,852.13	I
3		2,915.45	I
10		2,936.74	I
12		2,938.47	I
2		2,942.00	I
3		2,809.76	I
2		2,811.11	I
1		2,811.78	I
12		2,846.72	I
14		2,848.34	I
16		2,851.66	I
2		2,902.92	I
4		2,906.36	I
3		2,915.45	I
2		2,928.75	II
3		2,936.54	II
10		2,936.74	I
12		2,938.47	I
13		2,942.00	I
1		2,967.87	II
1		2,971.70	II
20		3,091.08	I
22		3,092.99	I
14		3,096.90	I
9		3,104.71	II
8		3,104.81	II
6		3,168.98	II
6		3,172.71	II
7		3,175.78	II
2		3,197.62	I
17		3,329.93	I
6		3,332.15	I
9		3,336.68	I
7		3,535.04	II
8		3,538.86	II
7		3,549.52	II
8		3,553.37	II
140		3,829.30	I
300		3,832.30	I
500		3,838.29	I
8		3,848.24	II
1		3,848.91	I
7		3,850.40	II
2		3,853.96	I
1		3,854.96	I
2		3,858.86	I
3		3,878.31	I
2		3,891.91	I
2		3,893.30	I
3		3,895.57	I
4		3,903.86	I
6		3,938.40	I
1		3,984.21	I
8		3,986.75	I
2		4,054.69	I
10		4,057.50	I
3		4,075.06	I
2		4,081.83	I
4		4,165.10	I
15		4,167.27	I
20		4,351.91	I
6		4,354.53	I
6		4,380.38	I
9		4,384.64	II
10		4,390.59	II
8		4,428.00	II
9		4,433.99	II
5		4,436.49	II
4		4,436.60	II
14		4,481.16	II
13		4,481.33	II
6		4,534.29	II
28		4,571.10	I
3		4,621.30	I
7		4,702.99	I
10		4,730.03	I
6		4,739.59	II
5		4,739.71	II
7		4,851.10	II
75		5,167.33	I
220		5,172.68	I
400		5,183.61	I
8		5,264.21	II
7		5,264.37	II
1		5,345.98	I
9		5,401.54	II
2		5,509.60	I
6		5,528.41	I
30		5,711.09	I
5		5,785.31	I
4		5,785.56	I
7		5,916.43	II
6		5,918.16	II
10		6,318.72	I
9		6,319.24	I
7		6,319.49	I
10		6,346.74	II
9		6,346.96	II
11		6,545.97	II
5		6,620.44	II
6		6,620.57	II
2		6,630.83	I
7		6,781.45	II
8		6,787.85	II
7		6,812.86	II
8		6,819.27	II
4		6,894.90	I
6		6,965.40	I
8		7,060.41	I
10		7,193.17	I
10		7,291.06	I
5		7,387.00	I
12		7,387.69	I
4		7,580.76	II
20		7,657.60	I
19		7,659.15	I
17		7,659.90	I
8		7,690.16	I
15		7,691.55	I
1		7,722.61	I
1		7,746.34	I
1		7,759.30	I
5		7,786.50	II
4		7,790.98	II
3		7,811.14	I
12		7,877.05	II
2		7,881.67	I
13		7,896.37	II
7		7,930.81	I
3		8,047.73	I
5		8,049.85	I
7		8,054.23	I

10		8,098.72	I
9		8,115.22	II
8		8,120.43	II
1		8,154.64	I
2		8,159.13	I
10		8,209.84	I
20		8,213.03	I
10		8,213.99	II
7		8,222.92	II
7		8,233.19	II
11		8,234.64	II
7		8,303.31	I
9		8,305.60	I
10		8,310.26	I
15		8,346.12	I
2		8,466.48	I
5		8,468.84	I
7		8,473.69	I
10		8,710.18	I
12		8,712.69	I
13		8,717.83	I
10		8,734.99	II
17		8,736.02	I
11		8,745.66	II
14		8,806.76	I
10		8,824.32	II
11		8,835.08	II
20		8,923.57	I
7		8,989.03	I
9		8,991.69	I
10		8,997.16	I
14		9,218.25	II
13		9,244.27	II
12		9,246.50	I
30		9,255.78	I
10		9,327.54	II
10		9,340.54	II
25		9,414.96	I
17		9,429.81	I
19		9,432.76	I
20		9,438.78	I
8		9,502.45	I
7		9,503.11	I
5		9,503.43	I
12		9,631.89	II
11		9,632.43	II
15		9,953.20	I
15		9,983.20	I
17		9,986.47	I
18		9,993.21	I
14		10,092.16	II
5		10,391.76	II
6		10,392.23	II
35		10,811.08	I
11		10,914.23	II
7		10,915.27	II
10		10,951.78	II
25		10,953.32	I
27		10,957.30	I
28		10,965.45	I
15		11,032.10	I
14		11,033.66	I
5		11,255.93	II
4		11,256.35	II
45		11,828.18	I
30		12,083.66	I
28		14,877.62	I
35		15,024.99	I
30		15,040.24	I
25		15,047.70	I
6		15,740.71	I
8		15,748.99	I
10		15,765.84	I
30		17,108.66	I
5		26,392.90	I

Mg III

Ref. 4, 83, 177 — J.O.S.

Intensity		Wavelength	
		Vacuum	
		106.30	III
		106.92	III
		108.08	III
		110.16	III
		114.32	III
		126.50	III
15		170.80	III
15		171.39	III
15		182.24	III
12		182.97	III
20		186.51	III
20		187.20	III
10		188.53	III
100		231.73	III
80		234.26	III
10		1,274.83	III
11	h	1,280.70	III
12		1,391.27	III
15		1,393.39	III
10		1,431.14	III
16		1,572.71	III
12		1,586.24	III
13		1,687.09	III
13		1,697.28	III
10		1,722.04	III
22		1,738.84	III
12		1,747.56	III
18		1,748.93	III
15		1,772.98	III
20		1,783.25	III
14		1,794.58	III
15		1,800.66	III
13		1,858.19	III
12		1,879.49	III
10		1,908.50	III
12		1,923.90	III
13		1,930.67	III
11		1,937.84	III
		Air	
15		2,039.55	III
15		2,055.49	III
25		2,064.90	III
15		2,085.90	III
20		2,091.96	III
13		2,097.93	III
15		2,112.77	III
16		2,134.06	III
20		2,177.70	III
20		2,395.15	III
15		2,467.75	III
10		2,490.54	III
10		2,529.19	III
12		3,299.05	III
13		3,306.39	III
12		3,335.90	III
11		3,342.58	III
12		3,361.41	III
10		3,381.24	III
11		3,382.90	III
11		3,387.37	III
10		3,706.74	III
10		4,916.00	III
10		5,839.82	III
15		6,256.75	III

Mg IV

Ref. 7, 128, 129 — J.O.S.

Intensity		Wavelength	
		Vacuum	
40		118.16	IV
80	p	118.81	IV
70		123.59	IV
240		124.65	IV
300		129.86	IV
300		132.81	IV
400		146.95	IV
300		147.41	IV
300		147.54	IV
350		180.07	IV
400		180.62	IV
400		180.80	IV
350		181.34	IV
4,000		320.99	IV
3,000		323.31	IV
40		800.41	IV
150		857.29	IV
30		866.74	IV
50		919.03	IV
30		929.78	IV
40		1,008.76	IV
30		1,026.41	IV
250		1,037.41	IV
80		1,044.37	IV
60		1,055.76	IV
300		1,210.99	IV
300		1,342.19	IV
800		1,346.57	IV
300		1,346.68	IV
600		1,352.05	IV
900		1,384.46	IV
500		1,385.77	IV
800		1,387.53	IV
300		1,404.68	IV
1,000		1,409.36	IV
500		1,437.53	IV
1,000		1,437.64	IV
300		1,447.42	IV
300		1,459.54	IV
400		1,459.62	IV
400		1,481.51	IV
350		1,490.45	IV
300		1,495.50	IV
300		1,607.11	IV
500		1,683.02	IV
400		1,698.81	IV
300		1,844.17	IV
		Air	
12	p	2,518.40	IV
4		2,534.79	IV

Mg V

Ref. 128 — J.O.S.

Intensity		Wavelength	
		Vacuum	
5		251.58	V
35		276.58	V
10		312.30	V
20		351.09	V
18		352.20	V
30		353.09	V
15		353.30	V
18		354.22	V
20		355.33	V

MANGANESE (MN)

Z = 25

Mn I and II

Ref. 1, 126 — C.H.C.

Intensity		Wavelength	
		Vacuum	
20		1,726.47	II
30		1,732.70	II
50		1,733.55	II
40		1,734.49	II
30		1,737.93	II
20		1,740.16	II
20		1,742.00	II
30		1,853.27	II
20		1,857.92	II
50		1,902.95	II
20		1,907.84	II
30		1,911.41	II
20	d	1,914.68	II
100		1,915.10	II
20		1,918.64	II
30		1,919.64	II
80		1,921.25	II
20		1,923.07	II
20		1,923.34	II
30		1,925.52	II
50		1,926.59	II
30		1,931.40	II
20		1,945.15	II
20		1,947.93	II
20		1,950.14	II
30		1,953.23	II
20	d	1,954.81	II
30		1,959.25	II
20		1,969.24	II
30		1,994.23	II
9,700		1,996.06	I
14,000		1,999.81	I
		Air	
18,000		2,003.85	I
50		2,037.31	II
40		2,037.64	II
40		2,039.97	II
30		2,076.21	II
1,500		2,092.16	I
20		2,097.46	II
20		2,102.50	II
1,700		2,109.58	I
30		2,113.96	II
290		2,208.81	I
540		2,213.85	I
770		2,221.84	I
20		2,373.36	II
20		2,427.38	II
50		2,427.72	II
30		2,427.94	II
30		2,437.37	II
20		2,437.84	II
30		2,452.49	II
50		2,499.00	II
30		2,507.60	II
20		2,516.60	II
30		2,516.74	II
20		2,521.66	II
20		2,530.72	II
20		2,531.80	II
50		2,532.78	II
75		2,533.06	I
50		2,533.33	II
30		2,534.10	II
80		2,534.22	II
100		2,535.66	II
30		2,535.98	II
100		2,537.92	II
50		2,541.11	II
80		2,542.92	II
50		2,543.45	II
100		2,548.75	II
50		2,551.85	II
30		2,553.27	II
75		2,556.57	II
30		2,556.89	II
50		2,557.54	II
95		2,558.59	II
30		2,559.41	II
150		2,563.65	II
30		2,565.22	II
580		2,572.76	I
480		2,575.51	I
12,000		2,576.10	II
550		2,584.31	I
30		2,588.97	II
45		2,589.71	II
250		2,592.94	I
6,200		2,593.73	II
250		2,595.76	I
95		2,598.90	II
40		2,602.14	I
30		2,602.72	II
45		2,603.72	II
4,300		2,605.69	II
190		2,610.20	II
500		2,618.14	II
140		2,622.90	I
150		2,624.04	I
40		2,624.80	II
200		2,625.58	II
95		2,626.64	I
30		2,630.26	I
60		2,630.57	I
190		2,632.35	II
130		2,638.17	II
80		2,639.84	II
27		2,650.99	II
60		2,655.91	II
30		2,666.77	II
45		2,667.00	I
30		2,667.03	II
110		2,672.59	II
55		2,673.37	II
55		2,674.43	II
30		2,676.33	I
45		2,680.34	II
30		2,680.68	II
30		2,681.25	II

40		2,681.72	I
45		2,683.02	I
23		2,683.75	I
55		2,684.55	II
55		2,685.94	II
110		2,688.25	II
85		2,692.66	I
27		2,693.19	II
55		2,695.36	II
27		2,698.97	II
85		2,701.00	II
50		2,701.17	II
160		2,701.70	II
100		2,703.98	II
130		2,705.74	II
80		2,707.53	II
110		2,708.45	II
45		2,709.96	II
80		2,710.33	II
110		2,711.58	II
30		2,716.80	II
30		2,717.53	II
30		2,719.01	II
50		2,719.74	II
30		2,722.10	II
30		2,724.46	II
55		2,728.61	II
30		2,738.86	I
45	h	2,760.93	I
30		2,771.44	I
30	h	2,776.23	I
30		2,780.00	I
55		2,789.20	I
60		2,790.36	I
60		2,791.08	I
6,200		2,794.82	I
5,100		2,798.27	I
220		2,799.84	I
3,700		2,801.06	I
70		2,804.10	I
60		2,806.14	I
55		2,808.02	I
110		2,809.11	I
60		2,812.84	I
70		2,813.47	I
60		2,815.02	II
30		2,816.33	II
85		2,817.97	I
40		2,818.77	I
55		2,821.45	I
55		2,822.55	I
80		2,830.79	I
27		2,836.31	I
60		2,870.08	II
30		2,872.94	II
80		2,879.49	II
40		2,882.90	I
70		2,886.68	II
160		2,889.58	II
55		2,892.39	II
50		2,898.70	II
80		2,900.16	II
40		2,907.22	I
140	h	2,914.60	I
190	h	2,925.57	I
27		2,928.68	I
1,100		2,933.06	II
27		2,934.02	I
1,500		2,939.30	II
250	h	2,940.39	I
		2,940.48	I
60		2,941.04	I
1,900		2,949.20	II
40		3,007.66	I
40		3,011.16	I
40		3,011.38	I
40		3,014.67	I
60		3,016.45	I
30		3,019.92	II
70		3,022.75	I
55		3,031.06	II
30		3,035.35	II
95		3,040.60	I
27		3,042.73	I
85		3,043.36	I
330		3,044.57	I
120		3,045.59	I
200		3,047.04	I
40		3,048.86	I
30		3,050.65	II
250		3,054.36	I
140		3,062.12	I
170		3,066.02	I
170		3,070.27	I
160		3,073.13	I
90		3,079.63	I
50		3,081.33	I
23		3,082.05	I
40		3,097.06	I
40		3,110.68	I
60	h	3,148.18	I
90	h	3,161.04	I
140	h	3,178.50	I
220		3,212.88	I
65		3,216.95	I
1,000		3,228.09	I
300		3,230.72	I
850		3,236.78	I
330		3,243.78	I
650		3,248.52	I
100		3,251.14	I
310		3,252.95	I
65		3,254.04	I
310		3,256.14	I
220		3,258.41	I
180		3,260.23	I
180		3,264.71	I
65		3,296.88	I
65		3,298.22	I
65		3,320.69	I
70		3,330.67	I
200		3,330.78	II
100		3,336.39	II
30		3,365.02	II
30		3,400.12	II
50		3,438.97	II
720		3,441.99	II
50		3,460.03	II
360		3,460.33	II
360	h	3,474.04	II
		3,474.13	II
290		3,482.91	II
180		3,488.68	II
140		3,495.84	II
50		3,496.81	II
100		3,497.54	II
360		3,531.85	I
		3,532.00	I
1,100		3,532.12	I
1,300		3,547.80	I
1,100		3,548.03	I
390		3,548.20	I
2,200		3,569.49	I
720		3,569.80	I
		3,570.04	I
1,400		3,577.88	I
720		3,586.54	I
290		3,595.12	I
420		3,607.54	I
420		3,608.49	I
360		3,610.30	I
290		3,619.28	I
220		3,623.79	I
140		3,629.74	I
100		3,660.40	I
70		3,670.52	I
70		3,676.96	I
50		3,682.09	I
280		3,693.67	I
180		3,696.57	I
70		3,701.73	I
210		3,706.08	I
130		3,718.93	I
55		3,728.89	I
130		3,731.93	I
260		3,790.22	I
55		3,799.26	I
110		3,800.55	I
55		3,801.91	I
3,200		3,806.72	I
700		3,809.59	I
55		3,810.69	I
90		3,816.75	I
2,100		3,823.51	I
390		3,823.89	I
200		3,829.68	I
480		3,833.86	I
1,300		3,834.36	I
350		3,839.78	I
670		3,841.08	I
350		3,843.98	I
65		3,918.32	I
120		3,926.47	I
65		3,952.84	I
55		3,975.89	I
65		3,977.08	I
130		3,982.58	I
150		3,985.24	I
190		3,986.83	I
150		3,987.10	I
1,500		4,018.10	I
150		4,026.44	I
27,000		4,030.76	I
19,000		4,033.07	I
11,000		4,034.49	I
1,500		4,035.73	I
55		4,038.73	I
5,600		4,041.36	I
210	d	4,045.13	I
		4,045.21	I
1,100		4,048.76	I
80		4,049.00	I
55		4,051.73	I
65		4,052.47	I
150		4,055.21	I
1,900		4,055.54	I
210		4,057.95	I
1,100		4,058.93	I
150		4,059.39	I
730		4,061.74	I
730		4,063.53	I
80		4,065.08	I
80		4,068.00	I
290		4,070.28	I
730		4,079.24	I
730		4,079.42	I
1,100		4,082.94	I
1,100		4,083.63	I
65		4,089.94	I
55		4,105.36	I
200		4,110.90	I
150		4,131.12	I
120		4,135.04	I
80		4,141.06	I
55		4,147.53	I
80		4,148.80	I
150		4,176.60	I
120		4,189.99	I
65		4,201.76	I
65		4,211.75	I
370		4,235.14	I
510		4,235.29	I
190		4,239.72	I
290		4,257.66	I
290		4,265.92	I
270		4,281.10	I
65		4,284.08	I
65		4,312.55	I
50		4,323.63	II
45		4,374.95	I
45		4,381.70	I
55		4,411.88	I
350		4,414.88	I
55		4,419.78	I
210		4,436.35	I
800		4,451.59	I
160		4,453.00	I
130		4,455.01	I
160		4,455.32	I
110		4,455.82	I
55		4,457.04	I
210		4,457.55	I
270		4,458.26	I
55		4,460.38	I
150		4,461.08	I
510		4,462.02	I
290		4,464.68	I
200		4,470.14	I
130		4,472.79	I
40		4,479.40	I
170		4,490.08	I
240		4,498.90	I
240		4,502.22	I
80		4,605.36	I
80		4,626.54	I
35		4,671.69	I
50		4,701.16	I
160		4,709.72	I
180		4,727.48	I
130		4,739.11	I
1,000		4,754.04	I
180		4,761.53	I
750		4,762.38	I
300		4,765.86	I
500		4,766.43	I
940		4,783.42	I
1,000		4,823.52	I
25		4,844.32	I
35		4,965.88	I
19		5,004.91	I
30		5,074.79	I
60		5,117.94	I
50		5,150.89	I
50		5,196.59	I
85		5,255.32	I
160		5,341.06	I
19		5,349.88	I
95		5,377.63	I
95		5,394.67	I
50		5,399.49	I
95		5,407.42	I
35		5,413.69	I
85		5,420.36	I
35		5,432.55	I
12		5,457.47	I
60		5,470.64	I
40		5,481.40	I
30		5,505.87	I
50		5,516.77	I
40		5,537.76	I
21		5,551.98	I
8		5,567.76	I
7		5,573.01	I
8		5,573.68	I
7		5,738.29	I
7		5,780.19	I
7		5,816.84	I
140		6,013.50	I
200		6,016.64	I
290		6,021.80	I
7		6,384.67	I
17		6,440.97	I
24		6,491.71	I
14	h	6,942.52	I
12		6,989.96	I
14		7,069.84	I
12		7,184.25	I
10		7,247.82	I
24	h	7,283.82	I
35	h	7,302.89	I
50		7,326.51	I
12		7,680.20	I
10		7,712.42	I
10	h	7,764.72	I
10	h	8,670.92	I
12	h	8,672.06	I
10	h	8,673.97	I
12	h	8,701.05	I
17	h	8,703.76	I
30	h	8,740.93	I

MERCURY (HG)

Z = 80

Hg I and II

Ref. 10, 34, 43, 45, 50, 69, 90, 117, 133, 14, 229, 235, 242, 304, 327, 328, 339 — R

Intensity	Wavelength
	Vacuum

Note: Wavelengths are for natural mercury, except where no figures are given to the left of the decimal point; in these cases the decimal places are those that apply to the 198 isotopes.

400	893.08	II
300	915.83	II
150	923.39	II
200	940.80	II
100	962.74	II
50	969.13	II
800	1,099.26	II
80	1,250.58	I
	0.564	I
8	1,259.24	I
	0.242	I
100	1,268.82	I
	0.825	I
5	1,307.75	I
	0.751	I
300	1,307.93	II
400	1,321.71	II
400	1,331.74	II
80	1,350.07	II
200	1,361.27	II
20	1,402.62	I
	0.619	I
200	1,414.43	II
10	1,435.51	II
	0.503	I
15	1,619.46	II
120	1,623.95	II
20	1,628.25	II
150	1,649.94	II
50	1,653.64	II
200	1,672.41	II
100	1,702.73	II
100	1,707.40	II
120	1,727.18	II
250	1,732.14	II
20	1,775.68	I
40	1,783.70	II
30	1,796.22	II
200	1,796.90	II
60	1,798.74	II
30	1,803.89	II
40	1,808.29	II
400	1,820.34	II
5	1,832.74	I
1,000	1,849.50	I
	0.492	I
160	1,869.23	II
300	1,870.55	II
200	1,875.54	II
20	1,900.28	II
30	1,927.60	II
300	1,942.27	II
100	1,972.94	II
200	1,973.89	II
150	1,987.98	II
	Air	
90	2,026.97	II
90	2,052.93	II
70	2,148.00	II
5	2,247.55	I
60	2,262.23	II
	0.210	II
20	2,302.06	I
	0.065	I
15	2,323.20	I
5	2,340.57	I
20	2,345.43	I
	0.440	I
20	2,352.48	I
100	2,378.32	I
	0.325	I
20	2,380.00	I
40	2,399.45	I
	0.349	I
20	2,399.73	I
	0.729	I
10	2,400.49	I
60	2,407.35	II
50	2,414.13	II
5	2,441.06	I
20	2,446.90	I
	0.900	I
15	2,464.06	I
	0.064	I
40	2,482.00	I
	0.713	I
30	2,482.72	I
	1.999	I
40	2,483.82	I
	0.821	I
90	2,534.77	I
	0.769	I
15,000	2,536.52	I
	0.506	I
25	2,563.86	I
	0.861	I
25	2,576.29	I
	0.290	I
5	2,578.91	I
15	2,625.19	I
5	2,639.78	I
250	2,652.04	I
	0.043	I
400	2,653.69	I
	0.683	I
100	2,655.13	I
	0.130	I
5	2,674.91	I
50	2,698.83	I
	0.831	I
50	2,699.38	I
80	2,705.36	II
80	2,752.78	I
	0.783	I
20	2,759.71	I
	0.710	I
40	2,803.46	I
	0.471	I
30	2,804.43	I
	0.438	I
2	2,805.34	I
2	2,806.77	I
150	2,814.93	II
750	2,847.68	II
	0.675	I
50	2,856.94	I
	0.939	I
150	2,893.60	II
	0.598	I
150	2,916.27	II
	0.227	II
60	2,925.41	I
	0.413	I
150	2,935.94	II
400	2,947.08	II
1,200	2,967.28	I
	0.283	I
300	3,021.50	I
	0.500	I
120	3,023.47	I
	0.476	I
30	3,025.61	I
	0.608	I
50	3,027.49	I
	0.490	I
400	3,125.67	I
	0.670	I
320	3,131.55	I
	0.551	I
320	3,131.84	I
	0.842	I
400	3,208.20	II
400	3,264.06	II
80	3,341.48	I
	0.481	I
100	3,385.25	II
400	3,451.69	II
200	3,549.42	II
2,800	3,650.15	I
	0.157	I
300	3,654.84	I
	0.839	I
80	3,662.88	I
240	3,663.28	I
	0.883	I
30	3,701.44	I
	0.281	I
35	3,704.17	I
	0.432	I
30	3,801.66	I
	0.170	I
100	3,806.38	II
20	3,901.87	I
	0.660	I
60	3,906.37	I
	0.867	I
100	3,918.92	II
200	3,983.96	II
	0.372	I
1,800	4,046.56	I
	0.839	II
150	4,077.83	I
	0.572	I
40	4,108.05	I
	0.838	I
250	4,339.22	I
	0.057	I
400	4,347.49	I
	0.224	I
4,000	4,358.33	I
	0.496	I
100	4,398.62	II
90	4,660.28	II
80	4,855.72	II
5	4,883.00	I
5	4,889.91	I
80	4,916.07	I
	0.337	I
5	4,970.37	I
5	4,980.64	I
20	5,102.70	I
40	5,120.64	I
100	5,128.45	II
20	5,137.94	I
20	5,290.74	I
5	5,316.78	I
60	5,354.05	I
30	5,384.63	I
1,100	5,460.74	I
	0.068	I
30	5,549.63	I
160	5,675.86	I
	0.753	I
240	5,769.60	I
	0.922	I
100	5,789.66	I
280	5,790.66	I
	0.598	I
140	5,803.78	I
60	5,859.25	I
60	5,871.73	II
20	5,871.98	I
20	6,072.72	I
	0.663	I
1,000	6,149.50	II
30	6,234.40	I
	0.713	I
80	6,521.13	II
160	6,716.43	I
	0.402	I
250	6,907.52	I
	0.429	I
250	7,081.90	I
200	7,091.86	I
40	7,346.37	II
100	7,485.87	II
20	7,728.82	I
100	7,944.66	II
2,000	10,139.75	I
240	11,287.40	I
	0.407	I
120	13,209.95	I
140	13,426.57	I
60	13,468.38	I
80	13,505.58	I
500	13,570.21	I
450	13,673.51	I
200	13,950.55	I
500	15,295.82	I
100	16,881.48	I
400	16,920.16	I
300	16,942.00	I
500	17,072.79	I
400	17,109.93	I
20	17,116.75	I
20	17,198.67	I
20	17,213.20	I
70	17,329.41	I
30	17,436.18	I
50	18,130.38	I
40	19,700.17	I
	22,493.28	I
250	23,253.07	I
	32,148.06	I
	36,303.03	I

Hg III
Ref. 343 — C.H.C

Intensity	Wavelength	
	Vacuum	
3	621.44	III
2	679.68	III
2	878.59	III
1	886.48	III
1	988.89	III
2	1,009.29	III
5	1,068.03	III
2	1,161.95	III
9	1,681.40	III
15	1,759.75	III
1	1,894.77	III
	Air	
7	2,314.15	III
4	2,380.55	III
8	2,431.65	III
5	2,480.56	III
7	2,484.50	III
2	2,612.92	III
4	2,617.97	III
3	2,670.49	III
70	2,724.43	III
6	2,769.22	III
3	2,844.76	III
15	3,090.05	III
5	3,283.02	III
12	3,312.28	III
8	3,389.01	III
5	3,450.77	III
3	3,500.35	III
4	3,538.88	III
5	3,557.24	III
15	3,803.51	III
70	4,122.07	III
10	4,140.34	III
100	4,216.74	III
15	4,470.58	III
12	4,552.84	III
50	4,797.01	III
10	4,869.85	III
80	4,973.57	III
30	5,210.82	III
6	5,695.71	III
25	6,220.35	III
35	6,418.98	III
40	6,501.38	III
10	6,584.26	III
6	6,610.12	III
30	6,709.29	III
12	7,517.46	III
7	7,808.10	III
25	7,946.75	III
50	7,984.51	III
5	8,151.64	III

MOLYBDENUM (MO)

Z = 42

Mo I and II
Ref. 1 — C.H.C.

Intensity	Wavelength	
	Vacuum	
19,000	2,015.11	II
40,000	2,020.30	II
21,000	2,038.44	II
17,000	2,045.98	II
4,800	2,081.68	II
2,400	2,089.52	II
2,200	2,092.50	II
4,000	2,093.11	II
2,700	2,100.84	II
1,500	2,104.29	II
1,400	2,108.02	II
400	2,269.69	II
160	2,304.25	II
160	2,306.97	II
130	2,325.94	I

240		2,330.46	I	410		2,640.99	I	40		2,868.32	II	110		3,138.72	II
110		2,332.12	II	600		2,644.35	II	1,700		2,871.51	II	220		3,147.35	I
190		2,340.47	I	370		2,646.49	II	85		2,872.88	II	220		3,152.82	II
190		2,341.59	II	640		2,649.46	I	220		2,879.05	II	55		3,155.64	II
80		2,352.61	I	480		2,653.35	II	65		2,888.15	II	6,000		3,158.16	I
80		2,355.22	I	560	h	2,655.03	I	95		2,891.28	II	120		3,164.53	I
80		2,355.42	II	290		2,658.11	I	1,300		2,890.99	II	8,700		3,170.35	I
70		2,364.37	I	640		2,660.58	II	190		2,892.81	II	95		3,172.03	II
50		2,366.09	II	110		2,665.10	I	950		2,894.45	II	160		3,172.74	II
140		2,372.27	I	55		2,671.83	II	140		2,897.63	II	370		3,183.03	I
100		2,380.41	II	720		2,672.84	II	70		2,900.80	II	120		3,184.57	I
150		2,383.52	I	250		2,673.27	II	290		2,903.07	II	370		3,185.10	I
110		2,389.20	II	1,000		2,679.85	I	160		2,905.27	I	180		3,185.71	I
140		2,403.61	II	95		2,681.36	II	80		2,907.12	II	120	d	3,187.59	II
80		2,404.66	II	640		2,683.23	II	600		2,909.12	II	7,600		3,193.97	I
140		2,405.86	I	880		2,684.14	II	1,100		2,911.92	II	290		3,195.96	I
40		2,408.39	I	560		2,687.99	II	55		2,913.81	II	120		3,198.85	I
40		2,412.84	II	30		2,692.61	II	120		2,918.83	II	40		3,201.50	II
120		2,413.01	II	55		2,695.22	II	1,300		2,923.39	II	330		3,205.22	I
70		2,415.33	I	30		2,696.83	II	140		2,924.32	II	880		3,205.88	I
80		2,417.96	II	55		2,699.41	II	65		2,927.54	II	3,000		3,208.83	I
65		2,419.01	II	140		2,701.03	I	50		2,930.06	II	240		3,210.97	I
80		2,420.18	II	480		2,701.42	II	1,100		2,930.50	II	560		3,215.07	I
70		2,424.00	II	30		2,701.87	II	55		2,930.77	II	350		3,221.74	I
65		2,430.43	I	30		2,704.93	II	800		2,934.30	II	880		3,228.22	I
65		2,435.96	II	40		2,710.19	II	65		2,935.20	II	600		3,229.79	I
65		2,440.28	II	30		2,711.49	II	120		2,937.66	I	1,100		3,233.14	I
40		2,461.81	II	50		2,712.35	II	40		2,938.30	II	950		3,237.08	I
50		2,466.68	II	190		2,713.51	II	95		2,940.10	II	65		3,240.71	II
50		2,466.97	II	290		2,717.35	II	110		2,941.22	II	950		3,256.21	I
50		2,468.78	II	110		2,724.41	I	140		2,944.21	I	300		3,262.63	I
30		2,470.04	II	180		2,725.15	I	150		2,944.82	II	480		3,264.40	I
150	h	2,471.97	I	85		2,726.97	II	140		2,945.66	II	800		3,270.90	I
70		2,477.57	II	140		2,729.68	II			2,945.95	II	240		3,285.02	I
70	h	2,481.81	I	80		2,730.20	II	190		2,946.01	I	320		3,285.36	I
65		2,482.57	II	330		2,732.88	II	140		2,946.42	I	1,100		3,289.02	I
40		2,484.75	II	250		2,733.39	I	140		2,946.69	II	950		3,290.82	I
40	h	2,485.31	I	160		2,736.96	II	95		2,947.28	II	190		3,292.31	II
24		2,496.24	II	80	h	2,737.88	II	95		2,955.84	II	320		3,305.56	I
85		2,498.28	II	50		2,738.60	II	240		2,956.06	II	320		3,307.12	I
40		2,500.44	II	40		2,741.32	II	70		2,956.90	II	100		3,313.62	II
65		2,502.84	II	55		2,741.62	II	95		2,960.24	II	190		3,320.90	II
50		2,511.80	II	240		2,743.07	I	140		2,962.89	II	640		3,323.95	I
65		2,515.08	II	290		2,746.30	II	250		2,963.79	II	360		3,325.67	I
70		2,527.14	II	320		2,751.47	I	50		2,964.96	II	360		3,327.30	I
50		2,530.34	II	110		2,756.07	II	210		2,965.27	II	240		3,340.17	I
70		2,532.31	II	65	d	2,758.63	II	70		2,971.91	II	1,300		3,344.75	I
440		2,538.46	II	20		2,760.53	II	250		2,972.61	II	95		3,346.40	II
50		2,539.44	II	190		2,761.53	I	80		2,975.40	II	320		3,347.02	I
110	h	2,540.45	I	220		2,763.62	II	180		2,978.28	I	1,600		3,358.12	I
330		2,542.67	II	110		2,766.26	I	120		2,981.52	I	250		3,361.37	I
40		2,543.61	II	240		2,769.76	II	110		2,987.92	I	950		3,363.78	I
330		2,548.22	I	160		2,773.78	II	160		2,988.68	I	950		3,379.97	I
110	h	2,550.85	I	190		2,774.39	II	190		2,989.80	I	320		3,382.48	I
65		2,555.42	II	1,700		2,775.40	II	95		2,992.84	II	1,900		3,384.62	I
40		2,556.75	II	130		2,777.74	I	50		2,993.52	II	130		3,395.36	II
80		2,558.88	II	65		2,777.86	II	190		3,002.21	I	640		3,404.34	I
65		2,562.08	II	880		2,780.04	II	40		3,004.46	II	1,300		3,405.94	I
85		2,564.34	II	400		2,784.99	II	130		3,013.39	I	240		3,418.52	I
40		2,566.26	II	180		2,787.83	I	140	h	3,013.76	I	250		3,420.04	I
250		2,567.05	I	40		2,791.54	II	250		3,025.00	I	250		3,422.31	I
20		2,571.45	II	240	d	2,797.93	I	95		3,027.77	II	380		3,434.79	I
320		2,572.34	I	220		2,801.47	I	100		3,036.31	I	320		3,435.45	I
50		2,574.42	II	400		2,807.76	II	300		3,041.70	I	640		3,437.22	I
40		2,576.56	II	28		2,812.58	II	150		3,046.80	I	250		3,438.87	I
40		2,578.36	II	24		2,814.67	II	210		3,047.31	I	250		3,441.44	I
250		2,582.16	I	1,700		2,816.15	II	210		3,055.32	I	250		3,443.26	I
30		2,585.95	II	220		2,817.44	II	100		3,060.78	II	130		3,446.08	II
65		2,588.78	II	50		2,822.03	II	160		3,061.59	I	3,200		3,447.12	I
40		2,591.77	II	240		2,826.54	I	800		3,064.28	I	640		3,449.07	I
250		2,593.70	II	80		2,827.74	II	250		3,065.04	I	300		3,451.75	I
100		2,595.40	I	40		2,831.44	II	100		3,068.00	I	250		3,452.60	I
40		2,597.38	II	30		2,832.07	II	250		3,070.90	I	950		3,456.39	I
250		2,602.80	II	80		2,834.39	II	800		3,074.37	I	640		3,460.78	I
40		2,605.08	II	80		2,835.33	II	85		3,077.66	II	320		3,466.83	I
40		2,605.93	II	160		2,842.15	II	150		3,079.88	I	250		3,467.85	I
250		2,607.37	I	24		2,843.73	II	210		3,080.41	I	320		3,469.22	I
190		2,611.20	I	220		2,844.39	I	800		3,085.62	I	240		3,485.93	I
290		2,613.08	I	1,700		2,848.23	II	270		3,087.62	II	800		3,504.41	I
130		2,615.39	I	160		2,849.38	I	100		3,089.12	I	240		3,505.32	I
400		2,616.78	I	370		2,853.23	II	100		3,089.71	I	560		3,508.12	I
70		2,619.34	II	50		2,856.00	II	190		3,092.07	II	480		3,521.41	I
140		2,621.07	I	24		2,863.20	II	560		3,094.66	I	240		3,524.98	I
320		2,627.55	I	370		2,863.81	II	110		3,099.93	I	640		3,537.28	I
160		2,628.74	I	160		2,864.31	I	110		3,100.88	I	320		3,542.17	I
440		2,629.85	I	140		2,864.66	I	560		3,101.34	I	520		3,558.10	I
330		2,636.67	II	40		2,865.62	II	1,400		3,112.12	I	400		3,563.14	I
250		2,638.30	I	220		2,866.69	II	290		3,122.00	II	300		3,566.05	I
720		2,638.76	II	40		2,868.11	II	14,000		3,132.59	I	240		3,570.65	I

320		3,573.88	I
1,400		3,581.89	I
200		3,590.74	I
210		3,598.88	I
270		3,602.94	I
210		3,608.37	I
200		3,623.23	I
1,400		3,624.46	I
330		3,626.18	I
28		3,635.14	II
1,000		3,635.43	I
400		3,657.35	I
540		3,664.81	I
290		3,666.72	I
590		3,672.82	I
1,300		3,680.60	I
45		3,684.22	II
65		3,688.31	II
240		3,690.59	I
180		3,692.64	II
1,400		3,694.94	I
220		3,702.03	I
220		3,715.65	I
500		3,727.69	I
330	d	3,732.71	I
240		3,742.28	I
80		3,744.37	II
360		3,770.45	I
220		3,779.77	I
360		3,781.59	I
250		3,797.30	I
29,000		3,798.25	I
290		3,801.84	I
520		3,826.70	I
940		3,828.87	I
1,700		3,833.75	I
380		3,847.25	I
29,000		3,864.11	I
580		3,869.08	I
580		3,886.82	I
380		3,901.77	I
19,000		3,902.96	I
65		3,941.48	II
230		3,943.04	I
270		4,056.01	I
1,400		4,062.08	I
2,300		4,069.88	I
1,300		4,081.44	I
940		4,084.38	I
250		4,102.15	I
730		4,107.47	I
630		4,120.10	I
2,900		4,143.55	I
230		4,148.94	I
250		4,155.28	I
200		4,157.40	I
200		4,178.27	I
480		4,185.82	I
2,500		4,188.32	I
250		4,194.56	I
1,500		4,232.59	I
270		4,269.28	I
890		4,276.91	I
1,200		4,277.24	I
1,400		4,288.64	I
680		4,292.13	I
890		4,293.21	I
360		4,293.88	I
840		4,326.14	I
250		4,326.74	I
230		4,350.34	I
230		4,369.04	I
1,900		4,381.64	I
2,500		4,411.57	I
210		4,423.62	I
990		4,434.95	I
200		4,442.20	I
340		4,449.74	I
480		4,457.36	I
630		4,474.56	I
230		4,491.28	I
120		4,504.90	I
140		4,512.15	I
230		4,517.13	I
230		4,524.34	I
120		4,529.40	I
400		4,536.80	I
110		4,558.11	I
210		4,576.50	I
170		4,595.16	I
360		4,609.88	I
100		4,621.38	I
460		4,626.47	I
100		4,627.48	I
220		4,662.76	I
130		4,671.90	I
130		4,688.22	I
640		4,707.26	I
150		4,708.22	I
220		4,717.92	I
100		4,729.14	I
700		4,731.44	I
100		4,750.39	I
770		4,760.19	I
150		4,776.34	I
100		4,796.52	I
410		4,819.25	I
410		4,830.51	I
360		4,868.00	I
110		4,950.62	I
150		4,957.54	I
210		4,979.12	I
110		4,999.91	I
20		5,010.81	I
180		5,014.60	I
26		5,016.78	I
20		5,019.85	I
80		5,029.00	I
65		5,030.78	I
23		5,038.91	I
26		5,046.52	I
100		5,047.71	I
50		5,055.00	I
35		5,058.07	I
200		5,059.88	I
35		5,062.52	I
29		5,064.64	I
35		5,079.87	I
100		5,080.02	I
35		5,081.26	I
40		5,090.97	I
35		5,091.34	I
35		5,092.16	I
40		5,095.89	I
100		5,096.65	I
130		5,097.52	I
35		5,098.03	I
130		5,109.71	I
80		5,114.97	I
35		5,116.97	I
29		5,123.83	I
150		5,145.38	I
110		5,147.39	I
80		5,163.19	I
100		5,167.76	I
160	d	5,171.08	I
		5,171.25	I
230	h	5,172.94	I
160	h	5,174.18	I
40		5,191.44	I
110		5,200.17	I
50		5,200.74	I
26		5,210.44	I
50		5,211.86	I
80		5,219.40	I
65		5,231.06	I
26		5,232.36	I
100		5,234.26	I
460	h	5,238.20	I
230	h	5,240.88	I
110	h	5,242.81	I
100		5,245.51	I
150		5,259.04	I
16		5,260.17	I
65		5,261.14	I
20		5,268.95	I
35		5,271.80	I
35		5,276.28	I
65		5,279.65	I
210		5,280.86	I
20		5,283.84	I
55		5,292.08	I
35		5,293.46	I
55		5,295.47	I
20		5,306.26	I
55		5,313.89	I
35		5,315.04	I
20		5,319.89	I
20		5,324.47	I
35		5,327.06	I
20		5,352.35	I
80		5,354.88	I
35		5,355.51	I
65		5,356.48	I
560	h	5,360.56	I
110	h	5,364.28	I
35	h	5,367.11	I
35		5,372.40	I
26		5,388.69	I
65		5,394.52	I
35		5,397.38	I
50		5,400.47	I
35		5,405.79	I
35		5,406.39	I
40		5,417.38	I
23		5,426.89	I
55		5,435.68	I
65		5,437.75	I
40		5,450.51	I
35		5,456.46	I
26		5,460.53	I
23		5,465.57	I
35		5,475.90	I
35		5,490.28	I
20		5,492.17	I
26	h	5,493.80	I
26		5,498.49	I
50		5,501.54	I
23		5,501.87	I
26	h	5,503.54	I
7,800		5,506.49	I
23		5,520.04	I
26		5,520.64	I
40		5,526.52	I
40		5,526.97	I
5,200		5,533.05	I
40		5,539.41	I
50		5,543.12	I
40		5,544.49	I
55		5,556.28	I
26		5,556.72	I
20		5,564.05	I
40		5,568.62	I
26		5,569.48	I
2,500		5,570.45	I
35		5,575.19	I
20		5,591.58	I
40		5,602.76	I
23		5,608.62	I
23		5,609.23	I
100		5,610.93	I
23		5,613.07	I
20		5,618.45	I
23		5,619.38	I
330		5,632.47	I
50		5,634.86	I
230		5,650.13	I
23		5,673.63	I
55		5,674.47	I
40		5,677.89	I
35		5,682.89	I
460		5,689.14	I
23		5,699.28	I
80		5,705.72	I
23		5,711.80	I
210		5,722.74	I
23		5,728.77	I
26	d	5,729.45	I
		5,729.59	I
620		5,751.40	I
23		5,774.55	I
40		5,779.36	I
23	h	5,783.33	I
520		5,791.85	I
23	h	5,795.77	I
26		5,800.46	I
35		5,802.67	I
23		5,825.20	I
23		5,835.59	I
20		5,839.99	I
20	h	5,848.86	I
55	h	5,849.73	I
50	h	5,851.52	I
520		5,858.27	I
20		5,861.38	I
50		5,869.33	I
26		5,876.59	I
820		5,888.33	I
23		5,892.29	I
50	h	5,893.38	I
20		5,898.78	I
		5,898.82	I
40		5,901.47	I
40	h	5,926.36	I
160	h	5,928.88	I
40		5,988.17	I
35		6,025.49	I
16		6,027.27	I
1,300		6,030.66	I
20		6,047.83	I
20		6,054.81	I
20		6,079.58	I
10		6,081.27	I
40		6,101.87	I
10		6,130.63	I
10		6,197.66	I
20		6,217.89	I
10		6,264.27	I
16		6,265.88	I
15		6,290.74	I
13		6,301.75	I
11		6,323.54	I
40		6,357.22	I
16		6,389.11	I
11		6,391.12	I
35		6,401.07	I
26		6,409.11	I
10		6,412.39	I
100		6,424.37	I
20		6,446.34	I
20		6,471.20	I
20		6,473.99	I
10		6,493.13	I
23		6,519.84	I
15	h	6,611.20	I
230		6,619.13	I
10		6,624.57	I
50		6,650.38	I
13		6,659.68	I
18		6,690.47	I
110		6,733.98	I
21		6,746.08	I
50		6,746.27	I
35		6,753.97	I
13		6,763.50	I
10	h	6,799.88	I
10		6,802.62	I
10		6,812.03	I
13		6,825.63	I
18	d	6,828.87	I
		6,829.05	I
40		6,838.88	I
16		6,848.92	I
21		6,886.28	I
16		6,892.36	I
10		6,898.01	I
10		6,898.98	I
13		6,908.20	I
35		6,914.01	I
13		6,934.10	I
10		6,947.39	I
10		6,960.64	I
16		6,978.71	I
26		6,988.94	I
12		6,999.13	I
16		7,001.60	I
22		7,037.98	I
22		7,060.21	I
13		7,063.34	I
13		7,081.22	I
110		7,109.87	I
27		7,134.08	I
150		7,242.50	I
40		7,245.85	I
22		7,267.62	I
17		7,300.19	I
13		7,348.49	I
13		7,361.65	I
10	h	7,364.41	I
40		7,391.36	I
10		7,434.10	I
13		7,447.34	I
13	h	7,452.85	I
140		7,485.74	I
13		7,504.47	I
11		7,572.64	I

11	h	7,595.16	I
11		7,601.84	I
17	h	7,656.76	I
13		7,679.49	I
27		7,720.77	I
17		7,829.65	I
15		7,854.45	I
11		7,923.15	I
15		7,986.60	I
22	h	8,245.06	I
40	h	8,328.44	I
45	h	8,389.32	I
45	h	8,483.39	I

NEODYMIUM (ND)

Z = 60

Nd I and II

Ref. 1 — C.H.C.

Intensity		Wavelength	
		Vacuum	
		Air	
75		2,702.46	
75		2,704.54	
75		2,764.98	I
60		2,785.79	I
50		2,863.95	
50		2,921.26	
55		2,962.88	II
65		2,963.58	II
80		2,993.20	II
40		2,994.73	
95		3,007.97	II
95		3,014.19	II
95		3,018.35	II
80		3,026.47	II
50		3,038.98	II
50		3,043.29	
50		3,051.11	II
80		3,052.15	II
140		3,056.71	II
130		3,069.73	II
65	d	3,071.43	II
		3,071.50	II
160		3,075.38	II
95		3,079.38	II
95		3,080.94	II
95		3,092.73	
240		3,092.92	II
140		3,098.48	II
55		3,099.52	II
130		3,105.43	II
95		3,106.18	II
65		3,108.01	II
260		3,115.18	II
190		3,116.15	II
50		3,119.75	II
160		3,123.06	II
190		3,124.58	II
290		3,133.60	II
220		3,134.90	II
100		3,137.24	II
170		3,141.46	II
170		3,142.44	II
100		3,144.55	II
100		3,144.82	II
100		3,148.51	II
100		3,149.29	II
100		3,149.51	II
100		3,162.62	II
100		3,175.99	II
50		3,181.54	II
50		3,188.73	II
50		3,200.62	II
150		3,203.47	II
85		3,211.00	II
100		3,217.12	II
50		3,222.62	I
50		3,228.04	II
60		3,234.62	
40		3,237.91	II
100		3,254.08	II
50		3,256.91	II
220		3,259.24	II
100		3,260.66	II
220		3,265.12	II
50		3,265.38	II
170		3,267.25	II
100		3,273.18	II
320		3,275.22	II
50		3,281.49	II
50		3,282.78	II
290		3,285.10	II
100		3,286.62	II
50		3,289.52	
100		3,290.65	II
70		3,293.84	II
70		3,294.68	II
70		3,298.61	
300		3,300.16	II
200		3,312.75	II
200		3,325.90	II
410		3,328.28	II
250		3,331.57	II
290		3,334.48	II
290		3,339.07	II
320		3,353.59	II
200		3,355.93	II
270		3,364.96	II
290		3,393.63	II
120	h	3,484.88	I
200		3,527.53	II
290		3,543.35	II
200		3,555.77	II
410		3,560.75	II
340		3,568.87	II
470		3,587.51	II
300		3,592.59	II
340		3,598.02	II
300		3,600.91	II
320		3,609.79	II
370		3,615.82	II
300		3,618.96	II
300		3,631.02	II
340		3,634.30	II
240		3,637.00	II
240		3,637.23	II
240		3,640.24	II
240		3,645.78	II
340		3,648.20	II
240		3,649.46	
240		3,650.42	II
410		3,653.15	II
240		3,654.16	
470		3,662.26	II
540		3,665.18	II
540		3,672.36	II
580		3,673.54	II
240		3,678.18	II
1,200		3,685.80	II
440		3,687.30	II
410		3,689.69	II
300		3,694.81	II
410		3,697.56	II
240		3,702.84	
240		3,704.95	II
200		3,712.81	
470		3,713.70	II
370		3,714.20	II
640	d	3,714.73	II
250		3,715.04	II
200		3,715.39	II
470		3,715.68	II
410		3,718.54	II
410		3,721.35	II
220		3,722.42	II
780		3,723.50	II
410		3,724.87	II
250		3,726.90	II
710		3,728.13	II
470		3,730.58	II
270		3,732.78	II
1,000	d	3,735.54	II
		3,735.60	II
440		3,737.10	II
1,000		3,738.06	II
270		3,741.42	II
200		3,749.85	II
320		3,750.31	II
580		3,752.49	II
370		3,752.67	II
250		3,754.83	II
370		3,755.60	II
510		3,757.82	II
930		3,758.95	II
300		3,759.79	II
930		3,763.47	II
300		3,766.59	II
510		3,769.65	II
1,400		3,775.50	II
250		3,776.34	II
710		3,779.47	II
580		3,780.40	II
510		3,781.32	II
300		3,783.78	II
2,400		3,784.25	II
270		3,784.73	II
340		3,791.50	II
340		3,795.45	II
240		3,799.55	II
370		3,801.12	II
200		3,801.38	II
340		3,802.30	II
1,200		3,803.47	II
200		3,804.10	II
2,500		3,805.36	II
340		3,805.55	II
470		3,807.23	II
540		3,808.77	II
440		3,809.06	II
580		3,810.49	II
240		3,811.06	II
270		3,811.77	II
200		3,812.53	II
710		3,814.73	II
240		3,819.70	II
410		3,822.47	II
1,200		3,826.42	II
240		3,828.00	II
540		3,828.85	II
440		3,829.16	II
510		3,830.47	II
740		3,836.54	II
340		3,837.91	II
1,700		3,838.98	II
340		3,839.51	II
410	d	3,841.82	II
		3,841.88	II
1,700	d	3,848.24	II
		3,848.31	II
1,500		3,848.52	II
470		3,850.22	II
2,400	d	3,851.66	II
		3,851.74	II
340		3,858.55	II
270		3,860.94	II
300		3,862.52	II
3,700	d	3,863.33	II
		3,863.40	II
240		3,866.52	II
220		3,866.81	II
850		3,869.07	II
240		3,875.74	II
470		3,875.87	II
1,100		3,878.58	II
1,000		3,879.55	II
780		3,880.38	II
1,200		3,880.78	II
200		3,881.59	
540		3,887.87	II
370	h	3,889.66	II
1,300		3,889.93	II
1,300		3,890.58	II
1,300		3,890.94	II
580		3,891.51	II
470		3,892.06	II
810		3,894.63	II
270		3,896.13	II
440		3,897.63	II
2,000		3,900.21	II
1,300		3,901.84	II
1,700		3,905.89	II
200		3,907.70	II
510		3,907.84	II
2,000		3,911.16	II
850		3,912.23	II
340		3,913.69	II
440		3,915.13	II
610		3,915.95	II
340		3,917.65	II
220		3,919.92	II
1,100		3,920.96	II
510		3,927.10	II
200		3,929.26	II
610		3,934.82	II
410		3,936.11	II
510		3,938.86	II
2,000		3,941.51	II
2,000		3,951.16	II
810		3,952.20	II
320		3,952.87	II
320		3,953.52	II
240		3,957.45	II
590		3,958.00	II
510		3,962.21	II
1,400		3,963.12	II
270		3,963.90	II
1,100		3,973.30	II
740		3,973.69	II
740		3,976.85	II
740		3,979.49	II
320		3,982.36	II
470		3,986.25	II
1,400		3,990.10	II
1,000		3,991.74	II
1,100		3,994.68	II
410		4,000.50	II
540		4,004.02	II
410		4,007.43	II
3,700		4,012.25	II
540		4,012.70	II
370		4,018.81	II
1,000		4,020.87	II
1,000		4,021.34	II
1,000		4,021.78	II
1,200		4,023.00	II
340		4,024.78	II
410		4,030.47	II
1,200		4,031.82	II
270		4,038.12	II
3,000		4,040.80	II
200		4,041.06	II
410		4,043.59	II
410		4,048.81	II
850		4,051.15	II
850		4,059.96	II
4,700		4,061.09	II
1,100		4,069.28	II
710		4,075.12	II
470		4,075.28	II
240		4,077.62	II
470		4,080.23	II
240		4,085.82	II
270		4,096.13	II
220		4,098.18	II
200		4,106.59	II
1,400		4,109.08	II
2,500		4,109.46	II
510		4,110.48	II
300	h	4,113.83	II
410		4,123.88	II
470		4,133.36	II
510		4,135.33	II
3,000		4,156.08	II
510		4,156.26	II
340		4,160.57	II
410		4,168.00	II
810		4,175.61	II
2,400		4,177.32	II
200		4,178.64	II
640		4,179.59	II
250		4,184.98	II
470		4,205.60	II
470		4,211.29	II
290		4,220.25	II
440		4,227.73	II
1,300		4,232.38	II
250		4,234.19	II
290	h	4,235.24	II
290		4,239.84	II
2,000		4,247.38	II
850		4,252.44	II
290		4,254.29	
410		4,261.84	II
340		4,266.71	II
240		4,270.56	II
340		4,272.79	II
340		4,275.09	II
470		4,282.44	II
240		4,282.57	II
710		4,284.52	II
270		4,297.80	II
5,400		4,303.58	II
340		4,304.45	II

200		4,307.78	II
470		4,314.52	II
1,100		4,325.76	II
510		4,327.93	II
540		4,338.70	II
680		4,351.29	II
850		4,358.17	II
240		4,366.38	II
340		4,368.64	II
470	d	4,374.93	II
		4,375.04	II
710		4,385.66	II
250		4,390.66	II
540		4,400.83	II
510		4,411.06	II
580		4,446.39	II
1,400		4,451.57	II
200		4,451.99	II
300		4,456.40	II
740		4,462.99	II
410		4,501.82	II
200		4,506.59	II
170		4,513.34	II
250		4,516.36	II
120		4,527.25	I
340		4,541.27	II
340		4,542.61	II
100		4,556.14	II
170		4,559.67	I
340		4,563.22	II
200		4,578.89	II
200		4,579.32	II
100		4,586.62	I
200		4,597.02	II
100		4,603.82	I
100		4,609.87	I
300		4,621.94	I
100		4,627.98	I
510		4,634.24	I
340		4,641.10	I
250		4,645.77	II
200		4,646.40	I
300		4,649.67	I
200		4,654.73	I
130		4,670.56	II
170		4,680.74	II
310		4,683.45	I
110		4,684.04	I
110		4,690.35	I
190		4,696.44	I
130		4,703.57	II
470		4,706.54	II
140		4,706.96	I
190		4,709.71	II
190		4,715.59	II
240		4,719.02	I
190		4,724.35	II
140		4,731.77	I
120		4,779.46	I
170		4,789.41	II
120		4,797.15	II
240		4,811.34	II
140		4,820.34	II
350		4,825.48	II
130		4,832.28	II
110		4,849.06	II
280		4,859.02	II
190		4,866.74	I
350		4,883.81	I
140		4,889.10	II
220		4,890.70	II
240		4,891.07	II
280		4,896.93	I
120		4,901.53	I
210		4,901.84	I
110		4,902.03	II
190		4,913.41	I
170		4,914.37	I
330		4,920.68	II
470		4,924.53	I
260		4,944.83	I
290		4,954.78	I
290		4,959.13	I
150		4,961.39	II
250		4,989.94	II
150		5,033.52	II
110		5,063.73	II
360		5,076.59	II
150	h	5,089.84	II
360		5,092.80	II
180		5,102.39	II
150	d	5,105.21	II
		5,105.35	I
360		5,107.59	II
340		5,123.79	II
680		5,130.60	II
170		5,132.33	II
170		5,165.14	II
130		5,181.17	II
120		5,182.60	II
500		5,191.45	II
630		5,192.62	II
330		5,200.12	II
310		5,212.37	II
150		5,213.23	I
130		5,225.05	II
130		5,228.43	II
450		5,234.20	II
250		5,239.79	II
720		5,249.59	II
200		5,250.82	II
360		5,255.51	II
120		5,269.48	II
590		5,273.43	II
150		5,276.88	II
110		5,291.67	I
680		5,293.17	II
160		5,302.28	II
110		5,306.47	II
220		5,311.46	II
500		5,319.82	II
180		5,356.98	II
290		5,361.47	II
150		5,371.94	II
110		5,385.90	II
160		5,431.53	II
110		5,451.12	II
170		5,485.70	II
35		5,501.47	I
45		5,525.72	I
90		5,533.82	I
55		5,535.27	II
55		5,543.24	I
55		5,548.47	II
55		5,561.17	I
27		5,575.50	I
27		5,576.70	I
27		5,577.70	I
27		5,587.61	I
240		5,594.43	II
55		5,601.43	I
45		5,601.92	I
220		5,620.54	I
65		5,635.76	I
45		5,639.54	I
35		5,653.57	I
70		5,668.87	II
65		5,669.77	I
140	d	5,675.97	I
55		5,676.33	I
220		5,688.53	II
23		5,689.51	I
30		5,701.57	I
130		5,702.24	II
80		5,706.21	II
160		5,708.28	II
80		5,718.12	II
65		5,726.83	II
100		5,729.29	I
23		5,734.55	I
70		5,740.86	II
55		5,749.19	I
27		5,749.66	I
23		5,767.33	I
45		5,776.12	I
45		5,784.96	I
45		5,788.22	I
45		5,800.09	I
160		5,804.02	I
80		5,811.57	II
45		5,813.89	I
27		5,820.37	I
70		5,825.87	II
30		5,826.74	I
80		5,842.39	II
30		5,844.66	I
23		5,845.95	I
55		5,858.91	I
35		5,867.08	I
30		5,868.90	I
27		5,871.04	I
30		5,883.29	I
23		5,886.24	I
30		5,887.91	I
27		5,921.22	I
27		5,955.87	I
30		5,994.76	I
27		5,996.47	I
45		6,007.67	I
35		6,031.27	II
27		6,033.29	I
45		6,034.24	II
55		6,066.03	I
27		6,071.70	I
30		6,073.97	I
23	d	6,133.47	II
27		6,149.28	I
27		6,155.06	I
35		6,157.83	II
23		6,166.67	II
35		6,170.49	II
45		6,178.59	I
27		6,183.91	I
27		6,208.24	I
45		6,223.39	I
27		6,226.50	I
23		6,238.50	II
35		6,244.08	I
23		6,257.49	I
27		6,258.73	II
23		6,277.29	II
27		6,285.79	I
23		6,292.84	I
23		6,297.07	I
55		6,310.49	I
27		6,341.51	II
23		6,382.07	II
65		6,385.20	II
35		6,485.69	I
45		6,630.14	I
35		6,637.96	II
45		6,650.57	II
30		6,655.67	I
25		6,737.79	II
40		6,740.11	II
25		6,742.54	I
30		6,790.37	II
30		6,804.00	II
25		6,846.72	II
40		6,900.43	II
24		6,941.39	II
17	h	7,010.80	II
8		7,018.85	II
17		7,020.92	II
17		7,024.58	II
10		7,033.21	
35		7,037.30	II
7		7,052.14	II
7		7,054.74	II
7		7,061.47	II
40		7,066.89	II
8		7,082.93	II
12	h	7,089.71	II
12	h	7,092.09	
12	h	7,092.74	II
12	h	7,092.94	
17	h	7,093.98	I
20	h	7,095.42	I
29		7,129.35	II
12	h	7,142.04	II
10		7,143.72	
8		7,151.03	II
6		7,153.09	I
6	h	7,185.01	
10		7,189.09	II
24		7,189.42	II
20		7,192.01	II
10		7,199.00	II
8	h	7,227.01	I
15		7,236.54	II
7	h	7,261.64	II
9		7,285.29	II
9		7,288.56	II
6		7,291.38	II
7		7,298.72	II
12		7,316.81	II
7	h	7,321.43	I
7		7,323.12	II
6		7,334.54	I
6		7,357.10	I
6		7,374.04	II
7		7,381.79	II
9		7,401.31	I
10		7,406.62	II
6		7,411.20	II
10		7,418.18	II
9		7,427.41	II
9		7,448.71	II
5		7,481.28	II
12		7,511.16	II
17		7,513.73	II
7	h	7,514.44	II
7		7,516.02	II
9		7,526.45	II
12		7,528.99	II
10		7,538.26	II
5		7,540.97	II
7		7,547.00	II
5		7,577.54	II
7		7,587.65	II
6		7,590.75	II
6		7,603.73	II
5		7,605.92	II
5		7,614.72	I
9		7,639.79	II
8		7,646.00	II
6		7,663.52	II
12		7,696.56	II
6		7,718.20	II
4		7,743.90	II
4		7,748.92	II
10		7,750.95	II
6		7,773.06	II
7		7,792.22	II
6		7,796.40	II
8		7,797.32	II
5		7,798.32	II
10		7,808.47	II
7		7,818.83	II
5		7,825.20	II
12		7,863.04	II
5	h	7,872.03	I
7		7,886.60	II
4	h	7,896.50	II
9		7,900.40	II
5	h	7,906.03	I
12		7,917.01	II
10		7,925.03	II
5		7,947.93	II
10		7,949.68	II
5	h	7,955.38	II
12		7,958.95	I
12		7,965.73	II
15		7,982.09	II
12		7,982.68	II
12		8,000.76	II
9		8,007.70	I
4	h	8,020.07	
8		8,026.35	
10		8,043.24	I
8		8,051.33	II
5		8,064.00	II
10		8,099.17	I
10		8,120.93	II
12		8,122.07	II
12		8,141.75	II
12		8,143.27	II
7	h	8,164.97	I
8		8,172.56	II
9		8,179.83	II
9		8,182.41	II
4		8,185.58	II
7	h	8,205.38	II
10		8,231.52	II
4		8,248.76	II
5	h	8,249.68	II
4	h	8,262.80	II
7	h	8,266.72	II
4		8,272.79	II
4	h	8,302.74	II
10		8,307.72	II
6		8,324.50	II
4		8,332.01	II
12		8,346.36	II
4		8,375.16	II
4		8,375.33	II

4		8,394.71	II
7		8,400.85	II
5	h	8,456.87	II
4		8,530.53	II
5		8,582.03	II
5		8,591.53	II
7		8,594.87	II
8	c	8,643.43	II
5		8,667.07	II
5		8,677.48	I
6		8,691.29	II
6		8,695.07	II
6		8,712.82	II
6		8,715.03	I
17		8,839.10	II

NEON (NE)

Z = 10

Ne I and II

Ref. 56, 58, 118, 150, 230 — S.P.

Intensity		Wavelength	
		Vacuum	
90		352.956	I
60		354.962	I
90		361.433	II
60		362.455	II
150		405.854	II
120		407.138	II
200		445.040	II
300		446.256	II
250		446.590	II
180		447.815	II
150		454.654	II
200		455.274	II
10		456.275	II
120		456.348	II
90		456.896	II
1,000		460.728	II
500		462.391	II
35		587.213	I
35		589.179	I
35		589.911	I
70		591.830	I
100		595.920	I
75		598.706	I
35		598.891	I
70		600.036	I
170		602.726	I
170		615.628	I
170		618.672	I
120		629.102	I
200		626.823	I
200		629.739	I
1,000		735.896	I
400		743.720	I
60		993.88	I
70		1,068.65	I
90		1,131.72	I
100		1,131.85	II
90		1,229.83	I
90		1,418.38	I
90		1,428.58	I
90		1,436.09	I
120		1,681.68	II
180		1,688.36	II
100		1,888.11	II
100		1,889.71	II
200		1,907.49	II
500		1,916.08	II
300		1,930.03	II
200		1,938.83	II
100	c	1,945.46	II
		Air	
80		2,007.01	II
80		2,025.56	II
150		2,085.47	II
180		2,096.11	II
120		2,096.25	II
80	p	2,562.12	II
90	w	2,567.12	II
80		2,623.11	II
80		2,629.89	II
90	w	2,636.07	II
80		2,638.29	II
80		2,644.10	II
80		2,762.92	II
90		2,792.02	II
80		2,794.22	II
100		2,809.48	II
80		2,906.59	II
80		2,906.82	II
90		2,910.06	II
90		2,910.41	II
80		2,911.14	II
80		2,915.12	II
80		2,925.62	II
80	w	2,932.10	II
80		2,940.65	II
90		2,946.04	II
150		2,955.72	II
150		2,963.24	II
150		2,967.18	II
100		2,973.10	II
15		2,974.72	I
100		2,979.46	II
12		2,982.67	I
150		3,001.67	II
120	p	3,017.31	II
300		3,027.02	II
300		3,028.86	II
100		3,030.79	II
120		3,034.46	II
100		3,035.92	II
100		3,037.72	II
100		3,039.59	II
100		3,044.09	II
100		3,045.56	II
120		3,047.56	II
100		3,054.34	II
100		3,054.68	II
100		3,059.11	II
100		3,062.49	II
100		3,063.30	II
100		3,070.89	II
100		3,071.53	II
100		3,075.73	II
120		3,088.17	II
100		3,092.09	II
120		3,092.90	II
100		3,094.01	II
100		3,095.10	II
100		3,097.13	II
100		3,117.98	II
120		3,118.16	II
10		3,126.199	I
300		3,141.33	II
100		3,143.72	II
100	p	3,148.68	II
100		3,164.43	II
100		3,165.65	II
100		3,188.74	II
120		3,194.58	II
500		3,198.59	II
60		3,208.96	II
120		3,209.36	II
120		3,213.74	II
150		3,214.33	II
150		3,218.19	II
120		3,224.82	II
120		3,229.57	II
200		3,230.07	II
120		3,230.42	II
120		3,232.02	II
150		3,232.37	II
100		3,243.40	II
100		3,244.10	II
100		3,248.34	II
100		3,250.36	II
150		3,297.73	II
150		3,309.74	II
300		3,319.72	II
1,000		3,323.74	II
150		3,327.15	II
100		3,329.16	II
200		3,334.84	II
150		3,344.40	II
300		3,345.45	II
150		3,345.83	II
200		3,355.02	II
120		3,357.82	II
200		3,360.60	II
120		3,362.16	II
100		3,362.71	II
120		3,367.22	II
12		3,369.808	I
40		3,369.908	I
100		3,371.80	II
500		3,378.22	II
150		3,388.42	II
120		3,388.94	II
300		3,392.80	II
100		3,404.82	II
120		3,406.95	II
100		3,413.15	II
120		3,416.91	II
120		3,417.69	II
50		3,417.904	I
15		3,418.006	I
120		3,428.69	II
60		3,447.703	I
50		3,454.195	I
100		3,456.61	II
100		3,459.32	II
25		3,460.524	I
30		3,464.339	I
30		3,466.579	I
60		3,472.571	I
150		3,479.52	II
200		3,480.72	II
200		3,481.93	II
25		3,498.064	I
30		3,501.216	I
25		3,515.191	I
150		3,520.472	I
120		3,542.85	II
120		3,557.80	II
100		3,561.20	II
250		3,568.50	II
100		3,574.18	II
200		3,574.61	II
50		3,593.526	I
30		3,593.640	I
15		3,600.169	I
20		3,633.665	I
150		3,643.93	II
200		3,664.07	II
20		3,682.243	I
12		3,685.736	I
200		3,694.21	II
10		3,701.225	I
150		3,709.62	II
250		3,713.08	II
250		3,727.11	II
800		3,766.26	II
1,000		3,777.13	II
100		3,818.43	II
120		3,829.75	II
150		4,219.74	II
100		4,233.85	II
120		4,250.65	II
120		4,369.86	II
70		4,379.40	II
150		4,379.55	II
100		4,385.06	II
200		4,391.99	II
150		4,397.99	II
150		4,409.30	II
100		4,413.22	II
100		4,421.39	II
100		4,428.52	II
100		4,428.63	II
150	p	4,430.90	II
150	p	4,430.94	II
120		4,457.05	II
100		4,522.72	II
10		4,537.754	I
10		4,540.380	I
100		4,569.06	II
15		4,704.395	I
12		4,708.862	I
10		4,710.067	I
10		4,712.066	I
15		4,715.347	I
10		4,752.732	I
12		4,788.927	I
10		4,790.22	I
10		4,827.344	I
10		4,884.917	I
4		5,005.159	I
10		5,037.751	I
10		5,144.938	I
25		5,330.778	I
20		5,341.094	I
8		5,343.283	I
60		5,400.562	I
5		5,562.766	I
10		5,656.659	I
5		5,719.225	I
12		5,748.298	I
80		5,764.419	I
12		5,804.450	I
40		5,820.156	I
500		5,852.488	I
100		5,872.828	I
100		5,881.895	I
60		5,902.462	I
60		5,906.429	I
100		5,944.834	I
100		5,965.471	I
100		5,974.627	I
120		5,975.534	I
80		5,987.907	I
100		6,029.997	I
100		6,074.338	I
80		6,096.163	I
60		6,128.450	I
100		6,143.063	I
120		6,163.594	I
250		6,182.146	I
150		6,217.281	I
150		6,266.495	I
60		6,304.789	I
100		6,334.428	I
120		6,382.992	I
200		6,402.246	I
150		6,506.528	I
60		6,532.882	I
150		6,598.953	I
70		6,652.093	I
90		6,678.276	I
20		6,717.043	I
100		6,929.467	I
90		7,024.050	I
100		7,032.413	I
50		7,051.292	I
80		7,059.107	I
100		7,173.938	I
150		7,213.20	II
150		7,235.19	II
100		7,245.167	I
150		7,343.94	II
40		7,472.439	I
90		7,488.871	I
100		7,492.10	II
150		7,522.82	II
80		7,535.774	I
60		7,544.044	I
100		7,724.628	I
120		7,740.74	II
300		7,839.055	I
120		7,926.20	II
400		7,927.118	I
700		7,936.996	I
2,000		7,943.181	I
2,000		8,082.458	I
100		8,084.34	II
1,000		8,118.549	I
600		8,128.911	I
3,000		8,136.406	I
2,500		8,259.379	I
100		8,264.81	II
2,500		8,266.077	I
800		8,267.117	I
6,000		8,300.326	I
100		8,315.00	II
1,500		8,365.749	I
100		8,372.11	II
8,000		8,377.606	I
1,000		8,417.159	I
4,000		8,418.427	I
1,500		8,463.358	I
800		8,484.444	I
5,000		8,495.360	I
600		8,544.696	I
1,000		8,571.352	I
4,000		8,591.259	I
6,000		8,634.647	I
3,000		8,647.041	I

15,000	8,654.383	I
4,000	8,655.522	I
100	8,668.26	II
5,000	8,679.492	I
5,000	8,681.921	I
2,000	8,704.112	I
4,000	8,771.656	I
12,000	8,780.621	I
10,000	8,783.753	I
500	8,830.907	I
7,000	8,853.867	I
1,000	8,865.306	I
1,000	8,865.755	I
3,000	8,919.501	I
2,000	8,988.57	I
100	9,079.46	II
6,000	9,148.67	I
6,000	9,201.76	I
4,000	9,220.06	I
2,000	9,221.58	I
2,000	9,226.69	I
1,000	9,275.52	I
200	9,287.56	II
6,000	9,300.85	I
1,500	9,310.58	I
3,000	9,313.97	I
6,000	9,326.51	I
2,000	9,373.31	I
5,000	9,425.38	I
3,000	9,459.21	I
5,000	9,486.68	I
5,000	9,534.16	I
3,000	9,547.40	I
120	9,577.01	II
1,000	9,665.42	I
100	9,808.86	II
800	10,295.42	I
2,000	10,562.41	I
1,500	10,798.07	I
2,000	10,844.48	I
3,000	11,143.020	I
3,500	11,177.528	I
1,600	11,390.434	I
1,100	11,409.134	I
3,000	11,522.746	I
1,500	11,525.020	I
950	11,536.344	I
500	11,601.537	I
1,200	11,614.081	I
300	11,688.002	I
2,000	11,766.792	I
1,500	11,789.044	I
500	11,789.889	I
1,000	11,984.912	I
3,000	12,066.334	I
800	12,459.389	I
1,000	12,689.201	I
1,100	12,912.014	I
700	13,219.241	I
800	15,230.714	I
400	17,161.930	I
400	18,035.80	I
1,000	18,083.21	I
350	18,221.11	I
250	18,227.02	I
2,500	18,276.68	I
2,000	18,282.62	I
1,200	18,303.97	I
250	18,359.12	I
1,200	18,384.85	I
2,000	18,389.95	I
1,000	18,402.84	I
1,200	18,422.39	I
300	18,458.65	I
400	18,475.79	I
900	18,591.55	I
1,600	18,597.70	I
350	18,618.96	I
550	18,625.16	I
1,200	21,041.295	I
750	21,708.145	I
300	22,247.35	I
350	22,428.13	I
2,250	22,530.40	I
400	22,661.81	I
600	23,100.51	I
1,000	23,260.30	I
1,050	23,373.00	I
850	23,565.36	I
3,500	23,636.52	I
300	23,701.64	I
1,100	23,709.2	I
1,800	23,951.42	I
600	23,956.46	I
1,000	23,978.12	I
200	24,098.54	I
500	24,161.42	I
600	24,249.64	I
1,500	24,365.05	I
800	24,371.60	I
400	24,447.85	I
700	24,459.4	I
300	24,776.46	I
550	24,928.88	I
250	25,161.69	I
650	25,524.37	I
125	28,386.21	I
150	30,200.	I
250	33,173.09	I
450	33,352.35	I
1,300	33,901.	I
2,200	33,912.10	I
600	34,131.31	I
100	34,471.44	I
120	35,834.78	I

NEPTUNIUM (NP)

Z = 93

Np I and II

Ref. 93 — J.G.C.

Intensity		Wavelength	
		Air	
300		3,481.93	I
300	h	3,501.50	I
300	l	3,986.89	I
300	s	5,044.66	I
300	l	5,601.70	I
300	l	5,652.75	I
300	l	5,784.39	I
300	l	5,878.04	I
300	s	6,011.22	I
300		6,056.09	I
300	s	6,073.90	I
300	s	6,080.05	I
300	l	6,120.49	I
300		6,188.59	I
300	l	6,200.00	I
300	s	6,215.90	I
300	s	6,317.84	I
300	l	6,341.38	I
300	l	6,566.11	I
300	l	6,720.68	I
300	s	6,751.32	I
300	s	6,795.21	I
300	l	6,802.62	I
300	l	6,805.81	I
300	s	6,816.44	I
300	l	6,865.45	I
300	s	6,907.13	I
300	h	6,912.91	I
1,000	s	6,930.31	I
300	l	6,963.63	I
3,000	s	6,972.09	I
300		7,014.02	I
300	l	7,018.91	I
300	s	7,039.14	I
300	s	7,080.01	I
300	l	7,174.83	I
300	l	7,184.93	I
300	l	7,284.28	I
300	l	7,292.29	I
300	l	7,332.52	I
300	s	7,370.60	I
300	l	7,381.03	I
300	l	7,381.65	I
300	l	7,402.70	I
300	s	7,512.22	I
300	l	7,515.15	I
300	l	7,546.05	I
300	l	7,624.83	I
300		7,626.85	I
300	s	7,681.01	I
300	s	7,685.25	I
1,000	l	7,735.14	I
300	l	7,761.61	I
1,000	l	7,765.75	I
300	s	7,776.07	I
300		7,787.46	I
1,000	l	7,791.38	I
300	l	7,851.44	I
300	l	7,887.88	I
300	l	7,901.71	I
300	l	7,975.98	I
300	h	8,080.32	I
300	s	8,124.59	I
300		8,155.11	I
300	l	8,167.42	I
300	l	8,183.06	I
300	l	8,188.61	I
300	l	8,247.82	I
300	l	8,287.11	I
300	s	8,287.75	I
300	l	8,306.22	I
300	s	8,313.66	I
1,000	l	8,339.12	I
300		8,356.79	I
300	l	8,367.11	I
3,000		8,372.88	I
3,000		8,529.96	I
1,000	s	8,696.23	I
1,000	s	8,906.02	I
1,000		8,942.70	I
1,000	s	9,004.75	I
1,000	l	9,006.31	I
10,000	l	9,016.18	I
3,000	l	9,141.30	I
3,000	s	9,379.33	I
3,000	l	9,468.66	I
3,000	s	9,679.13	I
3,000	l	9,930.55	I
10,000	l	10,091.99	I
10,000	s	10,817.45	I
10,000	l	11,695.15	I
10,000	l	11,776.64	I
10,000	s	12,148.18	I
10,000	s	12,377.42	I
10,000	l	12,407.99	I
10,000	l	13,834.33	I

NICKEL (NI)

Z = 28

Ni I and II

Ref. 1, 294 — C.H.C.

Intensity	Wavelength	
	Vacuum	
500	1,317.22	II
400	1,335.20	II
500	1,370.14	II
1,000	1,741.55	II
500	1,748.28	II
	Air	
1,000	2,165.55	II
2,000	2,169.10	II
2,000	2,174.67	II
1,500	2,175.15	II
500	2,177.09	II
400	2,177.36	II
400	2,179.35	II
800	2,180.47	II
800	2,184.60	II
2,500	2,185.50	II
3,000	2,192.09	II
600	2,201.41	II
5,000	2,205.55	II
4,000	2,206.72	II
6,000	2,216.48	II
800	2,220.40	II
500	2,221.06	II
900	2,222.96	II
500	2,242.68	II
500	2,253.85	II
1,000	2,264.46	II
2,000	2,270.21	II
800	2,277.28	II
400	2,278.32	II
800	2,278.77	II
500	2,287.65	II
1,600	2,289.98	I
400	2,296.55	II
400	2,297.14	II
630	2,300.78	I
1,000	2,303.00	II
2,000	2,310.96	I
1,700	2,312.34	I
1,400	2,313.66	I
1,400	2,313.98	I
1,000	2,316.04	II
1,400	2,317.16	I
500	2,319.75	II
2,600	2,320.03	I
1,900	2,321.38	I
240	2,322.68	I
1,400	2,325.79	I
940	2,329.96	I
500	2,334.58	II
460	2,337.49	I
160	2,337.82	I
500	2,341.20	II
1,200	2,345.54	I
190	2,346.63	I
400	2,347.52	I
160	2,360.63	I
200	2,362.06	I
1,000	2,375.42	II
240	2,386.58	I
1,000	2,394.52	II
2,000	2,416.13	II
240	2,419.31	I
85	2,421.23	I
70	2,423.33	I
70	2,423.66	I
70	2,424.03	I
500	2,437.89	II
85	2,453.99	I
160	2,472.06	I
85	2,476.87	I
500	2,510.87	II
500	2,565.92	II
500	2,606.26	II
500	2,609.94	II
500	2,615.06	II
45	2,696.49	I
150	2,798.65	I
250	2,821.29	I
500	2,864.02	II
50	2,865.50	I
60	2,907.46	I
25	2,914.01	I
500	2,943.91	I
570	2,981.65	I
250	2,984.13	I
500	2,992.60	I
1,000	2,994.46	I
4,000	3,002.49	I
2,200	3,003.63	I
3,700	3,012.00	I
350	3,019.14	I
120	3,031.87	I
1,700	3,037.94	I
150	3,045.01	I
3,500	3,050.82	I
1,500	3,054.32	I
1,900	3,057.64	I
500	3,064.62	I
420	3,080.76	I
260	3,097.12	I
210	3,099.12	I
2,600	3,101.55	I
1,300	3,101.88	I
220	3,105.47	I
270	3,114.12	I
2,900	3,134.11	I
55	3,145.72	I
55	3,181.74	I
100	3,184.37	I
55	3,195.57	I
150	3,197.11	I
55	3,202.14	I
180	3,214.06	I
180	3,217.83	I
100	3,221.27	I
150	3,221.65	I
210	3,225.02	I
1,100	3,232.96	I
290	3,234.65	I
600	3,243.06	I

100		3,248.46	I
120		3,250.74	I
100		3,271.12	I
120		3,282.70	I
400		3,292.87	II
500		3,297.60	II
400		3,305.71	II
660		3,315.66	I
330		3,320.26	I
310		3,322.31	I
2,000		3,331.88	II
400		3,335.64	II
500		3,338.09	II
500		3,348.84	II
500		3,349.24	II
600		3,358.68	II
330		3,361.56	I
500		3,363.45	II
330		3,365.77	I
330		3,366.17	I
65		3,366.81	I
65		3,367.89	I
2,900		3,369.57	I
400		3,371.99	I
260		3,374.22	I
130		3,374.64	I
500		3,378.97	II
3,300		3,380.57	I
240		3,380.85	I
1,300		3,391.05	I
3,300		3,392.99	I
500		3,401.05	II
130		3,409.58	I
330		3,413.48	I
330		3,413.94	I
8,200		3,414.76	I
1,600		3,423.71	I
2,600		3,433.56	I
990		3,437.28	I
4,800		3,446.26	I
1,300		3,452.89	I
5,000		3,458.47	I
5,000		3,461.65	I
200		3,467.50	I
240		3,469.49	I
1,600		3,472.54	I
550		3,483.77	I
130		3,485.89	I
5,500		3,492.96	I
660		3,500.85	I
65		3,502.60	I
55		3,507.69	I
2,600		3,510.34	I
260		3,513.93	I
6,600		3,515.05	I
660		3,519.77	I
8,200		3,524.54	I
110		3,527.98	I
330		3,548.18	I
55		3,551.53	I
65		3,561.75	I
5,000		3,566.37	I
990		3,571.87	I
130		3,587.93	I
1,300		3,597.70	I
1,300		3,610.46	I
530		3,612.74	I
6,600		3,619.39	I
130		3,624.73	I
200		3,664.10	I
130		3,669.24	I
180		3,670.43	I
260		3,674.15	I
160		3,688.42	I
80		3,693.93	I
120		3,722.48	I
150		3,736.81	I
60		3,739.23	I
600		3,775.57	I
700		3,783.53	I
700		3,807.14	I
110		3,831.69	I
1,200		3,858.30	I
30		3,889.67	I
35		3,972.17	I
110		3,973.56	I
110		4,401.55	I
85		4,459.04	II
18		4,462.46	I
55		4,470.48	I
35		4,592.53	I
18		4,600.37	I
65		4,605.00	I
18		4,606.23	I
75		4,648.66	I
23		4,686.22	I
110		4,714.42	I
22		4,715.78	I
30		4,756.52	I
15		4,763.95	I
45		4,786.54	I
22		4,807.00	I
22	h	4,829.03	I
19		4,831.18	I
45		4,855.41	I
30		4,866.27	I
17		4,873.44	I
40		4,904.41	I
22		4,918.36	I
16		4,935.83	I
45		4,980.16	I
45		4,984.13	I
500		4,992.02	II
16	h	5,000.34	I
18		5,012.46	I
50		5,017.59	I
100		5,035.37	I
16		5,048.85	I
100		5,080.52	I
65		5,081.11	I
26	h	5,084.08	I
18		5,099.32	I
26	h	5,099.95	I
21		5,115.40	I
18	h	5,129.38	I
23		5,137.08	I
23	h	5,142.77	I
40	h	5,146.48	I
40	h	5,155.76	I
16		5,168.66	I
13		5,176.56	I
8		5,435.87	I
180		5,476.91	I
6		5,510.00	I
6		5,578.73	I
9		5,587.86	I
13		5,592.28	I
9		5,614.79	I
5	h	5,625.33	I
4		5,649.70	I
5		5,664.02	I
12		5,682.20	I
8		5,695.00	I
23		5,709.56	I
10		5,711.90	I
10		5,715.09	I
16		5,754.68	I
8		5,760.85	I
10		5,857.76	I
10		5,892.88	I
10		6,108.12	I
10		6,176.81	I
10		6,191.18	I
13		6,256.36	I
10		6,314.66	I
16		6,643.64	I
22		6,767.77	I
9		6,772.32	I
10		6,914.56	I
5		7,110.90	I
26		7,122.20	I
6		7,182.00	I
5		7,197.02	I
5		7,261.93	I
5		7,291.45	I
4		7,385.24	I
16		7,393.60	I
16		7,409.35	I
5		7,414.51	I
23		7,422.28	I
13		7,522.76	I
9		7,525.12	I
19		7,555.60	I
8		7,574.05	I
23		7,617.00	I
9		7,619.21	I
16		7,714.32	I
5	h	7,715.58	I
19		7,727.61	I
19		7,748.89	I
10		7,788.94	I
13		7,797.59	I
2		7,917.44	I
1,000		8,096.75	II
500		8,114.21	II
700		8,121.48	II
2		8,809.42	I
9		8,862.55	I
500	w	9,900.92	II

NIOBIUM (NB)

Z = 41

Nb I and II

Ref. 1 — C.H.C.

Intensity		Wavelength Air	
3,300		2,029.32	II
3,000		2,032.99	II
2,000		2,109.42	II
1,700		2,125.21	II
1,100		2,126.54	II
1,500		2,131.18	II
370		2,295.68	II
280		2,302.08	II
170		2,376.40	II
110		2,387.09	II
140		2,387.52	II
45		2,388.27	II
160		2,398.48	II
55		2,405.34	II
55		2,405.85	II
140		2,412.46	II
160		2,416.99	II
140		2,418.69	II
75		2,433.80	II
40		2,435.95	II
35		2,436.33	I
45		2,437.42	II
40		2,442.14	II
28		2,442.68	II
65		2,451.87	II
65		2,453.95	II
55		2,458.09	II
65		2,462.89	I
35		2,466.73	I
55		2,469.08	I
110		2,477.38	II
65		2,478.29	II
65		2,479.94	II
35		2,483.88	II
110		2,504.65	I
110		2,511.00	II
110		2,521.40	II
390		2,544.80	II
110		2,551.38	II
130		2,556.94	II
130		2,562.41	II
130		2,565.41	I
100		2,569.03	I
110		2,571.33	II
200		2,578.74	I
390		2,583.99	II
390		2,590.94	II
270		2,592.20	I
130		2,616.48	I
130		2,623.51	I
130		2,627.44	I
130		2,628.49	I
200		2,642.24	II
320		2,646.26	II
330		2,647.50	II
240		2,649.52	I
330		2,654.45	I
310		2,656.08	II
160		2,657.62	I
110		2,665.25	II
110		2,666.59	II
110		2,667.30	II
130		2,668.29	I
400		2,671.93	II
200		2,673.57	II
200		2,675.94	II
130		2,687.15	I
160		2,691.77	II
1,000		2,697.06	II
320		2,698.86	II
320		2,702.20	II
150		2,702.52	II
470		2,716.62	II
470		2,721.98	II
310		2,733.26	II
110		2,737.09	II
200		2,746.91	I
200		2,748.85	I
190		2,753.01	I
280		2,758.61	I
240		2,768.13	II
310		2,773.20	I
270		2,780.24	II
130		2,782.36	I
110		2,793.05	II
190		2,827.08	II
150		2,836.24	I
110		2,840.94	I
250		2,841.15	II
280		2,842.65	II
160		2,846.28	II
110		2,851.45	I
240		2,861.09	II
100		2,864.32	I
100		2,865.61	II
500		2,868.52	II
800		2,875.39	II
270		2,876.95	II
530		2,877.03	II
100		2,880.72	II
570		2,883.18	II
280		2,888.83	II
470		2,897.81	II
400		2,899.24	II
470		2,908.24	II
670		2,910.59	II
470		2,911.74	II
1,100		2,927.81	II
110		2,931.47	II
870		2,941.54	II
110	h	2,945.88	II
110		2,946.12	II
110		2,946.90	II
1,100		2,950.88	II
400		2,972.57	II
320		2,974.10	II
210		2,977.68	II
200		2,982.11	II
330		2,990.26	II
470		2,994.73	II
140		3,024.74	II
350		3,028.44	II
300		3,032.77	II
100		3,044.76	II
150		3,048.10	I
110		3,053.09	I
100		3,055.52	II
220		3,064.53	II
110		3,069.68	II
100		3,070.90	II
110		3,071.56	II
100		3,073.24	II
400		3,076.87	II
110		3,080.35	II
1,800		3,094.18	II
140		3,099.19	II
150		3,111.45	I
270		3,127.53	II
1,500		3,130.79	II
390		3,145.40	II
140		3,151.87	I
1,200		3,163.40	II
150		3,175.78	II
390		3,180.29	II
200		3,187.49	I
300		3,191.10	II
150		3,191.43	II
1,000		3,194.98	II
120		3,203.35	II
300		3,206.34	II
390		3,215.60	II
800		3,225.48	II
140		3,229.56	II
400		3,236.40	II
200		3,247.47	II
120		3,248.94	II

160		3,249.52	I
320		3,254.07	II
230		3,260.56	II
160		3,263.37	II
160		3,264.59	I
120		3,270.47	I
100		3,270.76	I
200		3,272.07	I
160		3,277.67	I
200		3,283.46	II
230		3,285.66	I
200		3,287.59	I
160		3,287.92	I
160		3,292.02	II
320		3,296.01	I
160		3,299.61	I
120		3,304.83	I
120		3,308.05	I
120		3,310.47	I
400		3,312.60	I
200		3,315.22	I
200		3,318.98	I
120		3,319.26	I
120		3,319.58	II
240		3,326.62	I
170		3,329.36	I
110		3,332.16	I
130		3,341.60	II
1,300		3,341.97	I
1,300		3,343.71	I
130		3,346.93	I
1,700		3,349.06	I
420		3,349.52	I
340		3,354.74	I
130		3,357.04	I
1,700		3,358.42	II
130		3,365.58	II
340		3,366.96	I
130		3,369.16	II
170		3,371.33	I
350		3,374.92	I
270		3,380.41	I
130		3,380.86	I
170		3,386.24	II
350		3,392.34	I
170		3,395.93	I
120		3,399.40	I
230		3,405.41	I
130		3,406.13	I
270		3,408.38	I
230		3,408.68	II
180		3,409.19	II
230		3,412.94	II
180		3,415.97	I
180		3,423.76	I
230		3,425.42	II
130		3,425.85	I
230		3,426.57	II
230		3,427.45	I
130		3,429.04	I
180		3,432.70	II
180		3,440.59	II
170		3,463.81	I
180		3,465.86	I
130		3,469.44	I
100		3,471.19	I
140		3,473.02	I
290		3,478.69	I
200		3,479.56	II
100		3,484.05	II
230		3,491.03	I
200		3,497.81	I
500		3,498.63	I
460		3,507.96	I
200		3,510.26	II
200		3,515.42	II
200		3,517.67	II
200		3,520.06	I
2,000		3,535.30	I
1,300		3,537.48	I
250		3,540.96	II
500		3,544.02	I
250		3,544.65	I
300		3,550.45	I
250		3,554.52	I
1,000		3,554.66	I
630		3,563.50	I
630		3,563.62	I
1,500		3,575.85	I

200		3,577.72	I
5,000		3,580.27	I
500		3,584.97	I
750		3,589.11	I
500		3,589.36	I
500		3,593.97	I
500		3,602.56	I
300		3,619.51	II
200		3,621.03	I
200		3,639.33	I
420		3,649.85	I
250		3,650.81	I
400		3,651.19	II
200		3,659.61	II
630		3,660.37	I
900		3,664.70	I
220		3,669.01	I
270		3,674.78	I
1,500		3,697.85	I
330		3,711.34	I
3,300		3,713.01	I
480		3,716.99	I
2,700		3,726.24	I
270		3,738.42	I
2,700		3,739.80	I
670		3,740.73	II
270		3,741.78	I
1,700		3,742.39	I
250		3,753.18	I
210		3,755.77	I
530		3,763.49	I
350		3,765.08	I
250		3,766.13	I
530		3,771.85	I
870		3,781.01	I
1,700		3,787.06	I
1,300		3,790.15	I
3,500		3,791.21	I
2,700		3,798.12	I
270		3,801.30	I
2,700		3,802.92	I
670		3,803.88	I
530		3,804.74	I
670		3,810.49	I
530		3,811.03	I
530		3,815.51	I
210		3,818.86	II
210		3,819.15	I
670		3,824.88	I
350		3,835.18	I
250		3,836.45	I
210		3,845.90	I
290		3,858.95	I
350		3,863.38	I
270		3,867.92	I
530		3,877.56	I
870		3,878.82	I
670		3,883.14	I
1,100		3,885.44	I
670		3,885.68	I
210		3,886.07	I
580		3,891.30	I
210		3,908.97	I
670		3,914.70	I
530		3,920.20	I
670		3,937.44	I
520		3,943.67	I
250		3,965.69	I
910	d	3,966.09	I
210		3,971.85	I
1,100		4,032.52	I
250		4,039.53	I
16,000	c	4,058.94	I
210		4,059.51	I
350		4,060.79	I
210		4,068.26	I
12,000		4,079.73	I
270		4,084.86	I
440		4,100.40	I
6,700		4,100.92	I
310		4,116.90	I
5,300		4,123.81	I
670		4,129.43	I
770		4,129.93	I
2,300		4,137.10	I
440		4,139.44	I
2,700		4,139.71	I
350		4,143.21	I
870		4,150.12	I

4,400		4,152.58	I
870		4,163.47	I
4,400		4,163.66	I
4,000		4,164.66	I
3,500		4,168.13	I
310		4,184.44	I
1,200		4,190.88	I
870		4,192.07	I
870		4,195.09	I
1,300		4,195.66	I
310		4,198.51	I
350		4,201.52	I
870		4,205.31	I
350		4,214.73	I
420		4,217.94	I
420		4,229.15	I
250		4,255.44	I
770		4,262.05	I
420		4,266.02	I
290		4,270.69	I
400		4,286.99	I
580		4,299.60	I
580		4,300.99	I
120		4,309.56	I
390		4,311.27	I
120		4,312.45	I
350		4,326.33	I
120		4,327.38	I
390		4,331.37	I
140		4,342.82	I
420		4,348.65	I
110		4,349.03	I
290		4,351.57	I
210		4,368.43	I
140		4,377.96	I
130		4,388.36	I
160		4,392.69	I
330		4,410.21	I
190		4,419.44	I
230	c	4,437.22	I
290		4,447.18	I
140		4,456.80	I
140		4,457.42	I
140		4,469.71	I
140		4,471.29	I
140		4,472.53	I
150		4,503.04	I
530		4,523.41	I
480		4,546.82	I
370		4,564.53	I
720		4,573.08	I
480		4,581.62	I
1,200		4,606.77	I
170		4,616.17	I
450		4,630.11	I
450		4,648.95	I
110		4,649.27	I
450		4,663.83	I
340		4,666.24	I
240		4,667.22	I
580		4,672.09	I
530		4,675.37	I
110		4,678.48	I
320		4,685.14	I
130	c	4,706.14	I
260		4,708.29	I
150		4,713.50	I
110	c	4,733.89	I
220	c	4,749.70	I
110		4,816.38	I
110	c	4,848.37	I
130	c	4,967.78	I
110		4,973.14	I
190		4,988.97	I
85		5,000.95	I
65		5,002.25	I
40		5,013.27	I
230		5,017.75	I
40		5,019.51	I
150		5,026.36	I
40		5,030.13	I
210		5,039.04	I
40		5,047.96	I
170		5,058.01	I
65		5,059.35	I
130		5,065.25	I
40		5,077.40	I
750		5,078.96	I
40	c	5,094.41	I

420		5,095.30	I
170		5,100.16	I
170		5,120.30	I
85		5,121.80	I
85		5,127.66	I
40		5,133.34	I
210		5,134.75	I
75		5,140.58	I
75		5,147.54	I
40		5,150.64	I
75		5,152.63	I
250		5,160.33	I
250		5,164.38	I
230		5,180.31	I
110		5,186.98	I
190		5,189.20	I
170		5,193.08	I
150		5,195.84	I
65		5,203.22	I
35		5,205.13	I
85	c	5,219.10	I
65		5,225.16	I
150		5,232.81	I
85	c	5,237.43	I
29		5,240.39	I
150	d	5,251.62	I
		5,251.81	I
75		5,253.03	I
85		5,253.93	I
50		5,269.92	I
270		5,271.53	I
25		5,272.48	I
130	c	5,276.20	I
29	c	5,279.43	I
50		5,285.26	I
35		5,296.34	I
50		5,315.55	I
17		5,317.01	I
250		5,318.60	I
50		5,319.49	I
75		5,334.87	I
25		5,336.81	I
50		5,340.80	I
25		5,343.58	I
460		5,344.17	I
340		5,350.74	I
40		5,353.28	I
25		5,355.31	I
40		5,355.70	I
29		5,359.19	I
17		5,362.01	I
40		5,375.27	I
40		5,381.34	I
17		5,388.30	I
21		5,395.86	I
29		5,396.33	I
29		5,411.24	I
21		5,416.30	I
65		5,422.44	I
21		5,431.26	I
110		5,437.27	I
19		5,448.31	I
19		5,456.19	I
40		5,458.04	I
19	h	5,468.10	I
40		5,481.00	I
13		5,483.09	I
19		5,483.49	I
13		5,491.06	I
17		5,499.53	I
40		5,504.58	I
17		5,509.12	I
35	c	5,512.82	I
17		5,517.39	I
50		5,523.57	I
25		5,541.47	I
85		5,551.35	I
29		5,563.00	I
17	c	5,571.44	I
35	c	5,576.16	I
35		5,578.29	I
50		5,586.97	I
17	c	5,590.95	I
13		5,594.89	I
17	c	5,599.59	I
40		5,603.52	I
13		5,603.93	I
25		5,628.26	I
65		5,629.17	I

35	c	5,635.42	I
170		5,642.11	I
35		5,645.30	I
17		5,654.14	I
130		5,664.71	I
170		5,665.63	I
17		5,666.86	I
65	c	5,671.02	I
85		5,671.91	I
25		5,677.47	I
25		5,693.09	I
35	d	5,697.90	I
		5,698.03	I
40		5,706.16	I
85		5,706.48	I
29		5,709.33	I
17		5,715.59	I
65		5,716.35	I
25		5,725.66	I
130		5,729.19	I
21		5,737.36	I
13		5,738.20	I
85		5,751.44	I
110		5,760.34	I
65		5,764.99	I
29		5,771.08	I
50	c	5,776.07	I
17		5,780.34	I
85		5,787.54	I
17		5,789.79	I
50		5,794.24	I
50		5,804.03	I
29	h	5,815.33	I
110		5,819.43	I
35		5,820.62	I
75		5,834.90	I
25		5,838.15	I
130	d	5,838.64	I
50		5,842.47	I
17		5,846.09	I
65		5,866.47	I
35		5,874.70	I
17		5,877.79	I
40		5,893.44	I
190	c	5,900.62	I
40	c	5,903.80	I
29		5,927.41	I
40	c	5,934.16	I
40		5,957.70	I
150		5,983.22	I
65		5,986.08	I
85	c	5,997.93	I
50		6,029.75	I
50		6,031.84	I
50		6,045.50	I
25		6,048.72	I
29		6,056.65	I
29		6,107.71	I
40		6,142.51	I
50		6,148.13	I
50		6,164.32	I
29		6,213.06	I
75		6,221.96	I
40	c	6,251.76	I
21		6,260.77	I
85	c	6,430.46	I
50	c	6,433.22	I
17		6,497.84	I
65		6,544.61	I
15		6,574.73	I
19	c	6,591.00	I
19		6,606.16	I
19		6,607.28	I
35		6,614.15	I
19		6,626.98	I
210	c	6,660.84	I
150	c	6,677.33	I
65		6,701.20	I
130	c	6,723.62	I
75		6,739.88	I
25		6,795.31	I
85		6,828.11	I
25	c	6,849.35	I
19		6,870.92	I
40		6,876.36	I
25	c	6,902.89	I
35		6,908.07	I
40		6,918.32	I
17		6,946.07	I
17		6,972.49	I
25		6,986.09	I
85		6,990.32	I
17	c	6,996.11	I
21		7,023.48	I
17		7,038.04	I
190	c	7,046.81	I
8		7,066.41	I
8		7,075.23	I
40	c	7,098.94	I
17	c	7,102.01	I
19		7,119.31	I
15		7,122.95	I
35		7,126.17	I
17		7,130.06	I
130		7,159.43	I
17		7,191.37	I
19	c	7,208.94	I
50		7,252.35	I
15		7,274.81	I
13		7,317.03	I
17	c	7,323.92	I
29	c	7,328.38	I
65	c	7,353.16	I
190	c	7,372.50	I
13		7,419.83	I
15		7,436.02	I
19		7,478.20	I
65		7,515.93	I
29	c	7,519.77	I
170	c	7,574.58	I
17	c	7,583.21	I
13		7,639.81	I
13		7,647.71	I
25		7,703.33	I
75	c	7,726.68	I
25		7,757.31	I
6		7,787.11	I
13	c	7,873.41	I
35		7,885.31	I
25		7,938.89	I
8		7,954.76	I
40		8,135.20	I
13	c	8,240.00	I
29	c	8,320.93	I
29		8,346.08	I
10		8,350.04	I
17		8,439.77	I
17	c	8,475.98	I
25		8,526.99	I
13	c	8,547.25	I
17	c	8,560.54	I
17		8,575.87	I
21	c	8,697.55	I
21		8,740.96	I
21		8,767.97	I
29	c	8,815.56	I
35		8,905.78	I

NITROGEN (N)

Z = 7

N I and II
Ref. 213 — R.L.K.

Intensity	Wavelength	
	Vacuum	
285	644.634	II
360	644.837	II
450	645.178	II
140	647.50	I
360	660.286	II
170	671.016	II
285	671.386	II
150	671.630	II
160	671.773	II
170	672.001	II
350	692.70	I
285	746.984	II
650	775.965	II
90	885.67	I
90	909.697	I
80	910.278	I
40	910.645	I
450	915.612	II
450	915.962	II
550	916.012	II
650	916.701	II
90	953.415	I
100	953.655	I
130	953.970	I
130	963.990	I
115	964.626	I
70	965.041	I
90	1,067.614	I
60	1,068.612	I
450	1,083.990	II
600	1,084.580	II
430	1,085.546	II
650	1,085.701	II
175	1,097.237	I
115	1,098.095	I
115	1,098.260	I
105	1,100.360	I
40	1,100.465	I
90	1,101.291	I
360	1,134.165	I
385	1,134.415	I
410	1,134.980	I
105	1,143.65	I
130	1,163.884	I
60	1,164.206	I
105	1,164.325	I
270	1,167.448	I
105	1,168.334	I
60	1,168.417	I
195	1,168.536	I
230	1,176.510	I
105	1,176.630	I
195	1,177.695	I
410	1,199.550	I
385	1,200.223	I
360	1,200.710	I
175	1,225.026	I
160	1,225.37	I
130	1,228.41	I
160	1,228.79	I
360	1,243.179	I
315	1,243.306	I
290	1,310.540	I
250	1,310.95	I
230	1,319.00	I
315	1,319.68	I
115	1,326.57	I
115	1,327.92	I
360	1,411.94	I
700	1,492.625	I
490	1,492.820	I
640	1,494.675	I
775	1,742.729	I
700	1,745.252	I
	Air	
160	2,095.53	II
70	2,096.20	II
110	2,096.86	II
110	2,130.18	II
160	2,142.78	II
160	2,206.09	II
160	2,286.69	II
110	2,288.44	II
220	2,316.49	II
160	2,316.69	II
285	2,317.05	II
160	2,461.27	II
110	2,496.83	II
70	2,496.97	II
110	2,520.22	II
160	2,520.79	II
220	2,522.23	II
110	2,590.94	II
160	2,709.84	II
110	2,799.22	II
110	2,823.64	II
160	2,885.27	II
220	3,006.83	II
360	3,437.15	II
285	3,838.37	II
360	3,919.00	II
450	3,955.85	II
1,000	3,995.00	II
360	4,035.08	II
550	4,041.31	II
360	4,043.53	II
140	4,099.94	I
185	4,109.95	I
285	4,176.16	II
285	4,227.74	II
285	4,236.91	II
220	4,237.05	II
450	4,241.78	II
285	4,432.74	II
650	4,447.03	II
360	4,530.41	II
550	4,601.48	II
450	4,607.16	II
360	4,613.87	II
450	4,621.39	II
870	4,630.54	II
550	4,643.08	II
285	4,788.13	II
450	4,803.29	II
180	4,847.38	I
285	4,895.11	II
160	4,914.94	I
210	4,935.12	I
160	4,950.23	I
350	4,963.98	I
285	4,987.37	II
450	4,994.36	II
650	5,001.48	II
360	5,002.70	II
870	5,005.15	II
550	5,007.32	II
450	5,010.62	II
360	5,016.39	II
360	5,025.66	II
550	5,045.10	II
185	5,281.20	I
140	5,292.68	I
450	5,495.67	II
285	5,535.36	II
650	5,666.63	II
550	5,676.02	II
870	5,679.56	II
450	5,686.21	II
450	5,710.77	II
285	5,747.30	II
700	5,752.50	I
240	5,764.75	I
265	5,829.54	I
235	5,854.04	I
360	5,927.81	II
550	5,931.78	II
285	5,940.24	II
650	5,941.65	II
285	5,952.39	II
160	5,999.43	I
210	6,008.47	I
285	6,167.76	II
360	6,379.62	II
185	6,411.65	I
210	6,420.64	I
210	6,423.02	I
210	6,428.32	I
185	6,437.68	I
235	6,440.94	I
185	6,457.90	I
300	6,468.44	I
750	6,482.05	II
360	6,482.70	I
300	6,483.75	I
265	6,481.71	I
325	6,484.80	I
160	6,491.22	I
210	6,499.54	I
185	6,506.31	I
750	6,610.56	II
185	6,622.54	I
185	6,636.94	I
235	6,644.96	I
185	6,646.50	I
235	6,653.46	I
210	6,656.51	I
185	6,722.62	I
210	7,398.64	I
160	7,406.12	I
265	7,406.24	I
685	7,423.64	I
785	7,442.29	I
900	7,468.31	I
185	7,608.80	I
450	7,762.24	II
400	8,184.87	I

400		8,188.02	I
250		8,200.36	I
300		8,210.72	I
570		8,216.34	I
400		8,223.14	I
400		8,242.39	I
550		8,438.74	II
500		8,567.74	I
570		8,594.00	I
650		8,629.24	I
500		8,655.89	I
220		8,676.08	II
700		8,680.28	I
650		8,683.40	I
500		8,686.15	I
110		8,687.43	II
110	h	8,699.00	II
500		8,703.25	I
160	h	8,710.54	II
570		8,711.70	I
500		8,718.83	I
250		8,728.89	I
200		8,747.36	I
500		9,386.80	I
570		9,392.79	I
250		9,460.68	I
200		9,863.33	I
160	h	9,865.41	II
110	h	9,868.21	II
160	h	9,887.39	II
220	h	9,891.09	II
160	h	9,961.86	II
220	h	9,969.34	II
285	h	10,023.27	II
220	h	10,035.45	II
220	h	10,065.15	II
160	h	10,070.12	II
250		10,105.13	I
300		10,108.89	I
350		10,112.48	I
400		10,114.64	I
110	h	10,126.27	II
250		10,539.57	I
200		12,074.51	I
380		12,186.82	I
225		12,288.97	I
290		12,328.76	I
310		12,381.65	I
180		12,438.40	I
510		12,461.25	I
920		12,469.62	I
500		13,429.61	I
840		13,581.33	I
180		13,587.73	I
180		13,602.27	I
290		13,624.18	I
250		14,757.07	I
250		14,757.07	I
100		14,868.87	I
160		14,966.60	I
160		14,966.60	I
180		15,582.27	I
180		15,582.27	I
120	s	17,516.58	I
100	l	17,584.86	I
100		17,878.26	I

N III

Ref. 66, 213 — R.L.K.

Intensity		Wavelength	
		Vacuum	
500		257.95	III
650		258.50	III
700		259.19	III
800		260.09	III
800		261.28	III
500		262.91	III
500		265.23	III
500		265.27	III
500		268.70	III
150		314.715	III
200		314.850	III
90		314.877	III
600		323.26	III
500		338.35	III
500		340.20	III
500		351.98	III
120		362.833	III
150		362.881	III
150		362.946	III
90		362.985	III
300		374.204	III
350		374.441	III
500		387.48	III
250		451.869	III
300		452.226	III
500		684.996	III
570		685.513	III
650		685.816	III
500		686.335	III
500		763.336	III
570		764.359	III
250		771.544	III
300		771.901	III
350		772.385	III
200		772.891	III
150		772.975	III
650		979.842	III
700		979.919	III
900		989.790	III
700		991.514	III
1,000		991.579	III
500		1,183.031	III
570		1,184.550	III
150		1,387.371	III
250		1,729.945	III
570		1,747.848	III
350		1,751.218	III
650		1,751.657	III
150		1,804.486	III
200		1,805.669	III
150		1,846.42	III
350		1,885.06	III
400		1,885.22	III
200		1,907.99	III
150		1,919.55	III
150		1,919.77	III
300		1,920.65	III
150		1,920.84	III
200		1,921.30	III
		Air	
200		2,064.01	III
250		2,064.42	III
120		2,068.68	III
90		2,071.09	III
90		2,117.59	III
90		2,121.50	III
90		2,147.31	III
200		2,188.20	III
150		2,188.38	III
250	w	2,682.18	III
90		2,689.20	III
120		3,367.34	III
90		3,754.67	III
120		3,771.05	III
90		3,938.52	III
150		3,998.63	III
200		4,003.58	III
250		4,097.33	III
200		4,103.43	III
120		4,195.76	III
150		4,200.10	III
90		4,332.91	III
120		4,345.68	III
300		4,379.11	III
90		4,510.91	III
120		4,514.86	III
90		4,634.14	III
120		4,640.64	III
90		4,858.82	III
150		4,867.15	III
90		5,314.35	III
200		5,320.82	III
150		5,327.18	III
90		6,454.11	III
120		6,467.02	III

N IV

Ref. 108, 212 — R.L.K.

Intensity		Wavelength	
		Vacuum	
400		181.75	IV
400		191.7	IV
400		192.9	IV
500		196.87	IV
500		197.23	IV
500		202.60	IV
500		205.94	IV
500		205.97	IV
500		206.03	IV
500		217.20	IV
500	d	217.90	IV
500	d	223.4	IV
800	w	225.12	IV
800		225.21	IV
600	w	234.12	IV
600	w	234.20	IV
600	w	234.25	IV
550		236.07	IV
500		237.99	IV
500	w	238.7	IV
600		238.80	IV
500	w	239.62	IV
900		247.20	IV
500	w	248.43	IV
500	w	248.46	IV
500	w	248.48	IV
600		260.45	IV
650		270.99	IV
250		283.42	IV
300		283.48	IV
350		283.58	IV
600		285.56	IV
600	w	297.7	IV
700		297.82	IV
650		300.32	IV
90		303.123	IV
500		303.28	IV
150		315.053	IV
120		322.503	IV
150		322.570	IV
200		322.724	IV
120		323.175	IV
300		335.050	IV
500	w	351.93	IV
700		353.06	IV
500		420.77	IV
650		463.74	IV
570		765.148	IV
500		922.519	IV
520		921.992	IV
480		923.057	IV
650		923.220	IV
500		923.675	IV
520		924.283	IV
1,000		955.335	IV
150	w	1,036.16	IV
90		1,078.71	IV
90		1,188.01	IV
1,000		1,718.55	IV
		Air	
90		2,080.34	IV
90	w	2,318.09	IV
150		2,477.69	IV
250		2,645.65	IV
300		2,646.18	IV
350		2,646.96	IV
90		3,078.25	IV
90		3,463.37	IV
570		3,478.71	IV
500		3,482.99	IV
400		3,484.96	IV
90		3,747.54	IV
150		4,057.76	IV
90		4,606.33	IV
150		6,380.77	IV

N V

Ref. 66, 107, 318 — R.L.K.

Intensity		Wavelength	
		Vacuum	
52		166.947	V
52		186.069	V
62		186.153	V
90		209.303	V
90		247.561	V
120		247.706	V
150		266.196	V
200		266.379	V
90		713.518	V
150		713.860	V
150		748.195	V
200		748.291	V
1,000		1,238.821	V
900		1,242.804	V
90		1,549.336	V
200	l	1,616.33	V
350	l	1,619.69	V
90	w	1,860.37	V
		Air	
60	l	2,859.16	V
90	l	2,974.52	V
150	w	2,980.78	V
250	w	2,981.31	V
60	w	2,998.43	V
350		4,603.73	V
250		4,619.98	V
200	w	4,944.56	V
60	w	7,618.46	V

OSMIUM (OS)

Z = 76

Os I and II

Ref. 1 — C.H.C.

Intensity	Wavelength	
	Air	
9,600	2,001.45	I
13,000	2,003.73	I
9,000	2,004.78	
17,000	2,010.15	I
29,000	2,018.14	I
29,000	2,020.26	
14,000	2,022.76	I
14,000	2,028.23	I
18,000	2,034.44	I
26,000	2,045.36	I
7,800	2,048.28	I
7,800	2,049.42	I
8,600	2,058.69	I
13,000	2,061.69	I
7,800	2,067.21	II
4,200	2,070.67	II
7,200	2,076.95	I
7,200	2,078.09	
14,000	2,079.97	I
2,900	2,082.54	I
2,900	2,089.03	I
2,900	2,089.21	I
6,000	2,097.60	I
5,300	2,100.63	I
2,100	2,117.66	I
4,800	2,117.96	I
6,600	2,119.79	
1,900	2,123.84	I
5,300	2,137.11	I
2,400	2,149.97	
2,600	2,154.59	I
1,300	2,157.84	I
1,200	2,158.53	I
2,400	2,161.00	
3,100	2,166.90	I
1,100	2,167.75	I
2,100	2,171.65	I
960	2,184.68	I
840	2,194.39	II
760	2,202.49	I
600	2,227.98	I
1,100	2,234.61	I
1,300	2,252.15	I
2,000	2,255.85	II
1,400	2,264.60	I
360	2,268.28	I
960	2,270.17	I
1,400	2,282.26	II
840	2,283.67	I
570	2,289.32	I
380	2,297.31	I
660	2,308.31	I
190	2,313.75	II
550	2,320.18	I

310		2,323.98	I
660		2,324.24	I
330		2,326.99	I
310		2,334.56	I
720		2,336.80	II
430		2,338.63	I
290		2,340.69	I
430		2,343.74	I
260		2,345.75	I
430		2,347.38	I
230		2,350.23	II
360		2,352.99	I
120		2,355.28	II
240		2,356.92	I
240		2,357.25	I
310		2,362.41	I
900		2,362.77	I
500		2,367.35	II
290		2,369.24	I
500		2,370.70	I
480		2,371.18	I
95		2,375.06	II
2,600		2,377.03	I
260		2,377.61	I
900		2,379.39	I
240		2,384.62	I
1,700		2,387.29	I
330		2,394.29	I
290		2,395.39	I
1,100		2,395.88	I
220		2,396.78	I
960		2,401.13	I
260		2,402.23	I
200		2,403.54	I
330		2,403.85	I
95		2,405.08	II
290		2,405.45	I
200		2,405.96	I
360		2,408.67	I
240		2,410.98	I
290		2,414.52	I
530		2,417.99	I
530		2,418.53	I
95		2,420.02	II
200		2,423.07	II
70		2,424.02	II
500		2,424.56	I
1,400		2,424.97	I
240		2,426.81	I
70		2,427.90	II
380		2,431.19	I
380		2,431.61	I
360		2,446.02	I
900		2,450.74	I
530		2,451.73	I
530		2,453.90	I
110		2,454.91	II
530		2,456.46	I
1,800		2,461.42	I
110		2,468.90	II
290		2,472.28	I
290		2,474.78	I
900		2,476.84	I
360		2,482.43	I
530		2,486.24	II
4,500		2,488.55	I
290		2,491.02	I
290		2,491.69	I
360		2,492.42	I
2,600		2,498.41	I
330		2,499.92	I
330		2,502.29	I
500		2,504.39	I
260		2,504.51	I
35		2,507.18	II
70		2,509.71	II
660		2,512.87	I
2,400		2,513.25	I
660		2,515.04	I
500		2,517.92	I
660		2,518.44	I
200		2,519.29	I
330		2,519.79	I
200		2,532.44	I
780		2,538.00	II
240		2,538.10	I
1,000		2,542.51	I
30		2,548.83	II
310		2,554.46	I
190		2,563.16	II
600		2,566.49	I
290		2,566.88	I
480		2,568.83	I
340		2,571.78	I
150		2,578.32	II
130		2,580.03	II
360		2,581.05	I
740		2,581.96	I
1,000		2,590.76	I
200		2,591.98	I
170		2,596.00	II
210		2,609.20	I
380		2,609.56	I
400		2,610.78	I
470		2,612.63	I
1,800		2,613.06	I
800		2,619.94	I
230		2,620.62	I
530		2,621.82	I
380		2,628.48	I
27		2,631.22	II
3,800		2,637.13	I
1,900		2,644.11	I
340		2,646.89	I
380		2,647.73	I
380		2,649.34	I
490		2,656.68	I
1,900		2,658.60	I
640		2,659.83	I
380		2,661.18	I
40		2,664.29	II
580		2,674.57	I
400		2,674.88	I
2,100		2,689.82	I
510		2,699.59	I
580		2,706.70	I
3,000		2,714.64	I
580		2,715.36	I
1,300		2,720.04	I
850		2,721.86	I
580		2,730.61	I
40		2,731.36	II
580		2,732.80	I
690		2,761.42	I
470		2,763.27	I
340		2,765.04	I
960		2,770.71	I
300		2,776.91	I
740		2,782.55	I
40		2,783.88	II
640		2,786.31	I
230		2,793.99	I
230		2,794.19	I
530		2,796.73	I
320		2,804.07	I
2,800		2,806.91	I
470		2,808.94	I
420		2,813.84	I
740		2,814.20	I
300		2,815.78	I
420		2,829.27	I
230		2,837.42	I
470		2,838.17	I
5,100		2,838.63	I
740		2,841.60	I
2,300		2,844.40	I
420		2,846.39	I
420		2,848.25	I
1,500		2,850.76	I
1,500		2,860.96	I
35		2,863.37	II
360		2,874.96	I
300		2,878.40	I
35		2,879.39	II
30		2,880.20	II
260		2,896.06	I
9,600		2,909.06	I
2,100		2,912.33	I
530		2,917.26	I
2,100		2,919.79	I
300		2,925.57	I
360		2,929.51	I
510		2,931.28	I
260		2,934.64	I
200		2,942.85	I
1,100	h	2,948.23	I
1,400		2,949.53	I
210	d	2,949.81	I
300		2,961.01	I
530		2,962.15	I
450		2,964.06	I
740		2,970.97	I
450		2,977.64	I
510		2,982.90	I
340		2,983.49	I
260		2,997.65	I
330		3,013.07	I
570		3,017.25	I
4,400		3,018.04	I
480		3,019.38	I
1,100		3,030.70	I
2,900		3,040.90	I
210		3,043.50	I
120		3,042.74	II
230		3,049.46	I
210		3,050.39	I
8,600		3,058.66	I
290		3,060.30	I
570		3,062.19	I
210		3,069.94	I
360		3,074.08	I
290		3,074.96	I
290		3,077.44	I
1,100		3,077.72	I
360		3,078.11	I
230		3,078.38	I
230		3,090.08	I
270		3,093.59	I
310		3,101.53	I
360		3,105.99	I
310		3,108.98	I
620		3,109.38	I
250		3,111.09	I
310		3,118.33	I
480		3,131.12	I
250		3,152.67	I
290		3,153.61	I
3,100		3,156.25	I
250		3,156.78	I
310		3,166.51	I
180		3,173.93	II
420		3,178.06	I
230		3,181.88	I
230		3,185.33	I
310		3,186.98	I
310		3,189.46	I
310		3,194.23	I
150		3,213.31	II
1,900		3,232.06	I
290		3,238.63	I
190		3,241.04	I
120		3,248.00	I
190		3,254.91	I
190		3,256.92	I
190		3,260.30	I
3,100		3,262.29	I
380		3,262.75	I
3,100		3,267.94	I
620		3,269.21	I
190		3,272.16	I
530		3,275.20	I
330		3,277.97	I
190		3,288.84	I
1,200		3,290.26	I
7,600		3,301.56	I
250		3,306.23	I
620		3,310.91	I
120		3,315.42	I
250		3,324.33	I
310		3,327.42	I
960		3,336.15	I
110		3,351.74	I
120		3,353.91	I
230		3,357.97	I
250		3,361.15	I
190		3,364.12	I
120		3,370.20	I
960		3,370.59	I
160		3,372.08	I
120		3,378.68	I
310		3,384.00	I
190		3,385.94	I
620		3,387.84	I
120		3,401.17	I
620		3,401.86	I
250		3,402.51	I
120		3,408.76	I
120		3,412.74	I
120		3,421.69	I
150		3,427.67	I
250		3,440.60	I
120		3,444.46	I
160		3,445.55	I
310		3,449.20	I
120		3,458.38	I
120		3,465.44	I
120		3,478.53	I
120		3,482.11	I
120		3,487.46	I
120		3,490.33	
160		3,498.54	I
250		3,501.16	I
620		3,504.66	I
440		3,512.99	I
310		3,518.72	I
120		3,520.00	I
480		3,523.64	I
120		3,526.04	I
1,200		3,528.60	I
230		3,530.06	I
230		3,532.80	I
120		3,533.41	I
230		3,542.71	I
960		3,559.79	I
1,200		3,560.86	I
120		3,562.34	I
310		3,569.78	I
120		3,574.08	I
120		3,587.32	I
620		3,598.11	I
190		3,601.83	I
95		3,604.48	II
250		3,616.57	I
120		3,619.43	I
450		3,640.33	I
230		3,654.49	I
330		3,656.90	I
120		3,666.31	I
480		3,670.89	I
120		3,675.45	I
250		3,689.06	I
190		3,703.25	I
120		3,706.56	I
120		3,709.14	I
230		3,713.73	I
210		3,719.52	I
230		3,720.13	I
180		3,746.47	I
3,700		3,752.52	I
100		3,757.12	I
130		3,766.30	I
120		3,768.14	I
120		3,774.40	I
110		3,774.62	I
120		3,776.25	I
290		3,776.99	I
2,100		3,782.20	I
620		3,790.14	I
180		3,790.73	I
370		3,793.91	I
250		3,836.06	I
150		3,840.30	I
150		3,841.29	I
190		3,849.94	I
230		3,857.09	I
230		3,865.47	I
730		3,876.77	I
250		3,881.86	I
140		3,900.39	I
190		3,901.71	I
100		3,930.00	I
250		3,938.59	I
100		3,949.78	I
200		3,961.02	I
1,000		3,963.63	I
100		3,964.96	I
150		3,969.67	I
110	h	3,975.44	I
730		3,977.23	I
100		3,988.18	I
150		4,003.48	I
100		4,004.02	I
150		4,005.16	I
160		4,018.26	I
100		4,037.84	I
280		4,041.92	I

160		4,048.05	I
960		4,066.69	I
250		4,070.86	I
190		4,071.56	I
230		4,074.68	I
490		4,091.82	I
120		4,100.30	I
1,200		4,112.02	I
180		4,124.60	I
180		4,128.96	I
2,500		4,135.78	I
150		4,137.84	I
180		4,172.57	I
1,200		4,173.23	I
620		4,175.63	I
120		4,184.13	I
320		4,189.91	I
180		4,201.45	I
250		4,202.06	I
1,200		4,211.86	I
120		4,213.86	I
100		4,215.16	I
170		4,233.46	I
4,900		4,260.85	I
100		4,264.75	I
120		4,269.61	I
100		4,285.90	I
560		4,293.95	I
560		4,311.40	I
110		4,326.25	I
340		4,328.68	I
100		4,338.75	I
100		4,351.53	I
210		4,365.67	I
110		4,370.66	I
520		4,394.86	I
160		4,397.26	I
160		4,402.74	I
4,900		4,420.47	I
100		4,432.41	I
290		4,436.32	I
100		4,439.64	I
230		4,447.35	I
120		4,484.76	I
110		4,548.66	I
540		4,550.41	I
140		4,551.30	I
170		4,616.78	I
170		4,631.83	I
140		4,663.82	I
670		4,793.99	I
110		4,865.60	I
55		5,031.83	I
45		5,039.12	I
35		5,072.88	I
35		5,074.77	I
35		5,079.09	I
90		5,103.50	I
55		5,110.81	I
22		5,122.23	I
22		5,145.54	I
140		5,149.74	I
28		5,152.01	I
28		5,168.98	I
40		5,193.52	I
270		5,202.63	I
35		5,203.23	I
20		5,250.46	I
45		5,255.82	I
55		5,265.15	I
20		5,283.89	I
20		5,295.65	I
40		5,298.78	I
13		5,302.58	I
18		5,336.23	I
11		5,346.03	I
13		5,352.25	I
110		5,376.79	I
16		5,403.43	I
13		5,412.14	I
120		5,416.34	I
45		5,416.69	I
28		5,417.51	I
16		5,441.82	I
55		5,443.31	I
22		5,446.93	I
11		5,447.76	I
20		5,449.37	I
20		5,453.40	I
22		5,457.30	I
28		5,470.00	I
13		5,474.58	I
13		5,475.13	I
9		5,477.27	I
16		5,481.85	I
22		5,509.33	I
9		5,516.01	I
270		5,523.53	I
22		5,546.82	I
9		5,549.79	I
13		5,552.88	I
11		5,560.62	I
16		5,580.66	I
80		5,584.44	I
8		5,600.50	I
35		5,620.08	I
9		5,637.41	I
22		5,642.56	I
28		5,645.25	I
7		5,648.98	I
9		5,660.21	I
7		5,674.38	I
28		5,680.88	I
11		5,709.37	I
170		5,721.93	I
8		5,737.89	I
8		5,739.72	I
22		5,765.05	I
170		5,780.82	I
40		5,800.60	I
8		5,842.49	I
110		5,857.76	I
28		5,860.64	I
11		5,882.92	I
11		5,903.98	I
11		5,906.84	I
7		5,908.95	I
7		5,981.36	I
11		5,983.22	I
65		5,996.00	I
20		6,015.79	I
7		6,054.63	I
20		6,144.53	I
11		6,158.03	I
35		6,227.70	I
7		6,241.70	I
22		6,269.41	I
11		6,274.94	I
11		6,286.83	I
9		6,398.86	I
22		6,403.15	I
9		6,448.13	I
6		6,520.85	I
7		6,528.87	I
7		6,533.14	I
11		6,538.30	I
11		6,576.83	I
8		6,614.56	I
4		6,615.43	I
7		6,661.81	I
27		6,729.56	I
18		6,791.53	I
14		6,806.61	I
5		6,878.70	I
4		6,901.58	I
11		6,956.02	I
6		6,984.95	I
15		7,060.67	I
22		7,145.54	I
10		7,149.89	I
4		7,184.10	I
10		7,206.33	I
5		7,209.96	I
9		7,251.16	I
6		7,253.49	I
6		7,375.07	I
9		7,407.95	I
26		7,602.95	I
4		7,701.46	I
7		7,789.96	I
7		7,852.17	I
6		7,981.20	I
7		8,041.29	I

OXYGEN (O)

Z = 8

O I and II

Ref. 66, 69, 209, 210, 215 — R.L.K.

Intensity		Wavelength	
		Vacuum	
250		537.83	II
300		538.26	II
220		539.09	II
200		539.55	II
150		539.85	II
150		644.148	II
200		672.95	II
150		673.77	II
70		685.544	I
900		718.484	II
800		718.56	II
600		718.562	II
70		744.794	I
70		770.793	I
90		771.056	I
70		775.321	I
70		791.973	I
300		796.66	II
90		804.267	I
70		804.848	I
70		805.295	I
80		805.810	I
240		832.762	II
450		833.332	II
600		834.467	II
40		877.879	I
80		922.008	I
90		935.193	I
40		948.686	I
90		971.738	I
40		976.448	I
160		988.773	I
40		990.204	I
250		1,025.762	I
90		1,027.431	I
160		1,039.230	I
60		1,040.942	I
40		1,152.152	I
900		1,302.168	I
600		1,304.858	I
300		1,306.029	I
		Air	
30	d	2,283.42	II
30	d	2,284.89	II
110		2,293.32	II
200		2,300.35	II
30	d	2,313.05	II
30	d	2,316.12	II
30	d	2,316.79	II
50	d	2,319.68	II
30	d	2,322.15	II
30	d	2,339.31	II
110		2,411.60	II
80		2,425.55	II
250		2,433.56	II
80	d	2,436.06	II
80		2,444.26	II
300		2,445.55	II
300		2,733.34	II
110		2,747.46	II
265		2,972.29	I
160		3,122.62	II
220		3,129.44	II
450		3,134.82	II
285		3,138.44	II
220		3,270.98	II
220		3,273.52	II
220		3,277.69	II
360		3,287.59	II
160		3,305.15	II
160		3,306.60	II
220		3,377.20	II
285		3,390.25	II
220		3,407.38	II
160		3,409.84	II
285		3,470.81	II
220		3,712.75	II
285		3,727.33	II
160		3,739.92	II
360		3,749.49	II
160		3,803.14	II
120		3,823.41	I
450		3,911.96	II
160		3,919.29	II
185		3,947.29	I
160		3,947.48	I
140		3,947.59	I
220		3,954.37	II
100		3,954.61	I
450		3,973.26	II
220		3,982.20	II
160		4,069.90	II
285		4,072.16	II
450		4,075.87	II
80	d	4,083.91	II
50	d	4,087.14	II
150	d	4,089.27	II
110		4,097.24	II
220		4,105.00	II
285		4,119.22	II
160		4,132.81	II
50		4,146.06	II
220		4,153.30	II
285		4,185.46	II
450		4,189.79	II
80		4,233.27	I
50	d	4,253.74	II
50	d	4,253.98	II
50	d	4,275.47	II
50	d	4,303.78	II
285		4,317.14	II
160		4,336.86	II
220		4,345.56	II
285		4,349.43	II
220		4,366.90	II
100		4,368.25	I
220		4,395.95	II
450		4,414.91	II
285		4,416.98	II
160		4,448.21	II
160		4,452.38	II
50		4,465.45	II
50	d	4,466.28	II
50		4,467.83	II
50		4,469.41	II
360		4,590.97	II
285		4,596.17	II
80	d	4,609.39	II
160		4,638.85	II
360		4,641.81	II
450		4,649.14	II
160		4,650.84	II
360		4,661.64	II
285		4,676.23	II
220		4,699.21	II
285		4,705.36	II
160		4,924.60	II
220		4,943.06	II
135		5,329.10	I
160		5,329.68	I
190		5,330.74	I
90		5,435.18	I
110		5,435.78	I
135		5,436.86	I
120		5,577.34	I
160		5,958.39	I
190		5,958.58	I
80		5,995.28	I
160		6,046.23	I
190		6,046.44	I
110		6,046.49	I
100		6,106.27	I
400		6,155.98	I
450		6,156.77	I
490		6,158.18	I
80		6,256.83	I
100		6,261.55	I
100		6,366.34	I
100		6,374.32	I
320		6,453.60	I
360		6,454.44	I
400		6,455.98	I
80		6,604.91	I
100		6,653.83	I
360		7,001.92	I
450		7,002.23	I
210		7,156.70	I
400		7,254.15	I
450		7,254.45	I
320		7,254.53	I
210		7,476.44	I
100		7,477.24	I
120		7,479.08	I
120		7,480.67	I
100		7,706.75	I
870		7,771.94	I
810		7,774.17	I
750		7,775.39	I
80		7,886.27	I
100		7,943.15	I
100		7,947.17	I
235		7,947.55	I
210		7,950.80	I
185		7,952.16	I

Intensity		Wavelength	Spectrum
110		7,981.94	I
135		7,982.40	I
190		7,986.98	I
135		7,987.33	I
250		7,995.07	I
400		8,221.82	I
265		8,227.65	I
265		8,230.02	I
325		8,233.00	I
120		8,235.35	I
120		8,426.16	I
810		8,446.25	I
1,000		8,446.36	I
935		8,446.76	I
325		8,820.43	I
160	d	9,057.01	I
120		9,118.29	I
80		9,134.71	I
80		9,150.14	I
80		9,151.48	I
235		9,156.01	I
450		9,260.81	I
490		9,260.84	I
450		9,260.94	I
400		9,262.58	I
540		9,262.67	I
590		9,262.77	I
490		9,265.94	I
640		9,266.01	I
185		9,399.19	I
120		9,481.16	I
120	d	9,482.88	I
235		9,487.43	I
140		9,492.71	I
265		9,497.97	I
160		9,499.30	I
235		9,505.59	I
210		9,521.96	I
120		9,523.36	I
120		9,523.96	I
100		9,528.28	I
100		9,622.13	I
120		9,625.29	I
160		9,677.38	I
80		9,694.66	I
65		9,694.91	I
235		9,741.50	I
235		9,760.65	I
120		9,909.05	I
140		9,936.98	I
120		9,940.41	I
160		9,995.31	I
120	d	10,421.18	I
590		11,286.34	I
640		11,286.91	I
490		11,287.02	I
490		11,287.32	I
490		11,295.10	I
540		11,297.68	I
590		11,302.38	I
265		11,358.69	I
490		12,464.02	I
450		12,570.04	I
120		12,990.77	I
160		13,076.91	I
700		13,163.89	I
750		13,164.85	I
640		13,165.11	I
160		16,212.06	I
120		17,966.70	I
590		18,021.21	I
120		18,041.48	I
120		18,042.19	I
120		18,046.23	I
140		18,229.23	I
540		18,243.63	I
140		26,173.56	I

O III

Ref. 23, 66 — R.L.K.

Intensity		Wavelength	
		Vacuum	
80	d	264.34	III
110		264.48	III
110		266.97	III
150		266.98	III
150		267.03	III
150		277.38	III
80		295.62	III
110		295.66	III
120		295.72	III
150		303.41	III
150		303.46	III
140		303.52	III
160		303.62	III
160		303.69	III
250		303.80	III
200		305.60	III
250		305.66	III
190		305.70	III
300		305.77	III
190		305.84	III
450		320.979	III
300		328.45	III
250		328.74	III
300		345.31	III
110		355.14	III
90		355.33	III
80		355.47	III
200		359.02	III
190		359.22	III
150		359.38	III
210		373.80	III
200		374.00	III
300		374.08	III
190		374.16	III
200		374.33	III
210		374.44	III
450		395.558	III
300		434.98	III
800		507.391	III
900		507.683	III
1,000		508.182	III
1,000		525.795	III
700		597.818	III
1,000		599.598	III
110		609.70	III
160		610.04	III
200		610.75	III
100		610.85	III
800		702.332	III
800		702.822	III
900		702.899	III
1,000		703.850	III
600		832.927	III
780		833.742	III
600		835.096	III
800		835.292	III
160		1,476.89	III
285		1,590.01	III
160		1,591.33	III
220		1,760.12	III
110		1,760.42	III
220		1,763.22	III
220		1,764.48	III
750		1,767.78	III
550		1,768.24	III
360		1,771.67	III
110		1,773.00	III
110		1,773.85	III
220		1,779.16	III
160		1,781.03	III
160		1,784.85	III
220		1,789.66	III
110		1,848.26	III
110		1,856.62	III
285		1,872.78	III
285		1,872.87	III
285		1,874.94	III
160		1,920.04	III
110		1,920.75	III
110		1,921.52	III
220		1,923.49	III
110		1,923.82	III
110		1,926.94	III
		Air	
360		2,013.27	III
160		2,026.96	III
220		2,045.67	III
160		2,052.74	III
200	d	2,390.44	III
80		2,394.33	III
80		2,422.84	III
80	d	2,438.83	III
200		2,454.99	III
200		2,558.06	III
80		2,687.53	III
110		2,695.49	III
80		2,959.68	III
250		2,983.78	III
80		3,017.63	III
80		3,023.45	III
80		3,043.02	III
200		3,047.13	III
110		3,059.30	III
80		3,121.71	III
110		3,132.86	III
80		3,238.57	III
200		3,260.98	III
300		3,265.46	III
80		3,267.31	III
80		3,312.30	III
110		3,340.74	III
80		3,444.10	III
80		3,455.12	III
80		3,698.70	III
80		3,702.75	III
80		3,703.37	III
110		3,707.24	III
110		3,715.08	III
110		3,744.00	III
150		3,754.67	III
80		3,757.21	III
250		3,759.87	III
110		3,791.26	III
200		3,961.59	III
110		5,592.37	III

O IV

Ref. 36, 66 — R.L.K.

Intensity		Wavelength	
		Vacuum	
150		195.86	IV
200		196.01	IV
110		207.18	IV
150		207.24	IV
140		233.46	IV
150		233.50	IV
110		233.52	IV
200		233.56	IV
110		233.60	IV
90		238.36	IV
180		238.57	IV
110		252.56	IV
110		252.95	IV
150		253.08	IV
300		260.39	IV
250		260.56	IV
300		279.63	IV
375		279.94	IV
110		285.71	IV
150		285.84	IV
200		306.62	IV
150		306.88	IV
700		553.330	IV
775		554.075	IV
850		554.514	IV
700		555.261	IV
580		608.398	IV
640		609.829	IV
270		616.952	IV
150		617.005	IV
200		617.036	IV
520		624.617	IV
580		625.130	IV
640		625.852	IV
200		779.734	IV
315		779.821	IV
360		779.912	IV
200		779.997	IV
640		787.711	IV
520		790.109	IV
700		790.199	IV
200		802.200	IV
160		802.255	IV
130		921.296	IV
160		921.366	IV
200		923.367	IV
130		923.433	IV
200		1,338.612	IV
130		1,342.922	IV
230		1,343.512	IV
		Air	
200		2,449.372	IV
200		2,450.040	IV
200		2,493.44	IV
200		2,493.77	IV
200		2,507.73	IV
230		2,509.19	IV
200		2,517.2	IV
160		2,836.26	IV
160		2,921.45	IV
460		3,063.42	IV
410		3,071.61	IV
160		3,209.66	IV
230		3,348.08	IV
270		3,349.11	IV
160		3,354.27	IV
200		3,375.40	IV
130		3,378.06	IV
360		3,381.20	IV
360		3,385.52	IV
270		3,396.79	IV
360		3,403.52	IV
230		3,409.66	IV
410		3,411.69	IV
230		3,413.64	IV
200		3,489.83	IV
160		3,492.24	IV
230		3,560.39	IV
270		3,563.33	IV
315	w	3,725.93	IV
360		3,729.03	IV
410		3,736.85	IV
230		3,744.89	IV

O V

Ref. 24, 66 — R.L.K.

Intensity		Wavelength	
		Vacuum	
80		124.616	V
110		135.523	V
80		138.109	V
110		139.029	V
80		151.447	V
110		151.477	V
150		151.546	V
80		164.574	V
110		164.657	V
80		164.709	V
80		166.235	V
150		167.99	V
110		170.219	V
450		172.169	V
250		185.745	V
375		192.751	V
450		192.799	V
520		192.906	V
80		193.003	V
200		194.593	V
80		202.161	V
80		202.224	V
80		202.283	V
80		202.334	V
150		202.393	V
110		203.78	V
150		203.82	V
100		203.85	V
200		203.89	V
100		203.94	V
300		207.794	V
150		215.040	V
200		215.103	V
250		215.245	V
250		216.018	V
520		220.352	V
80		227.372	V
80		227.469	V
150		227.511	V
80		227.549	V
80		227.634	V
80		227.689	V
150		231.823	V
110		248.459	V
110		286.448	V
1,000		629.730	V

230		681.272	V
700		758.678	V
640		759.441	V
580		760.228	V
775		760.445	V
640		761.128	V
700		762.003	V
520		774.518	V
640		1,371.292	V
160	w	1,506.72	V
315	w	1,643.68	V
160		1,707.996	V

Air

1,000		2,781.01	V
920		2,786.99	V
775		2,789.85	V
200		2,941.33	V
210		2,941.65	V
160		3,144.66	V
100		4,123.99	V
230	w	4,930.27	V
130		5,597.91	V
130		6,500.24	V

PALLADIUM (PD)

Z = 46

Po I and II

Ref. 1, 287 — C.H.C.

Vacuum

Intensity		Wavelength	
50		2,162.27	II
50		2,182.35	II
50		2,212.15	II
100	r	2,231.59	II
200	r	2,296.53	II
50		2,351.32	II
50		2,362.31	II
75		2,367.92	II
60		2,372.16	II
50		2,388.29	II
60		2,414.73	II
75		2,418.72	II
75		2,424.49	II
100		2,426.87	II
100		2,430.94	II
100		2,433.11	II
100		2,435.32	II
150		2,446.17	II
75		2,446.72	II
1,100		2,447.91	I
80		2,448.15	II
100		2,457.29	II
60		2,457.76	II
150		2,469.29	II
80		2,470.06	II
100		2,471.18	II
50		2,472.55	II
1,700		2,476.42	I
250		2,486.52	II
300		2,488.92	II
75		2,489.61	II
200		2,498.81	II
150		2,505.73	II
50		2,514.47	II
80		2,534.57	II
50	h	2,539.44	II
150		2,551.84	II
150		2,565.51	II
100		2,569.56	II
60		2,593.24	II
50		2,628.24	II
70		2,635.92	II
150		2,658.75	II
1,900		2,763.09	I
150	h	2,776.85	II
100	h	2,787.92	II
50	h	2,800.64	II
50	h	2,807.59	II
200		2,854.59	II
100	h	2,871.37	II
100	h	2,878.01	II
520		2,922.49	I
50		2,980.63	II
650		3,002.65	I
45		3,009.78	I
1,500		3,027.91	I
1,100		3,065.31	I
2,600		3,114.04	I
270		3,142.81	I
11,000		3,242.70	I
2,700		3,251.64	I
3,500		3,258.78	I
460		3,287.25	I
3,600		3,302.13	I
5,000		3,373.00	I
24,000		3,404.58	I
13,000		3,421.24	I
5,000		3,433.45	I
6,400		3,441.40	I
7,700		3,460.77	I
10,000		3,481.15	I
2,000		3,489.77	I
12,000		3,516.94	I
12,000		3,553.08	I
4,500		3,571.16	I
20,000		3,609.55	I
20,000		3,634.70	I
5,500		3,690.34	I
1,400		3,718.91	I
1,500		3,799.19	I
1,500		3,832.29	I
2,200		3,894.20	I
1,500		3,958.64	I
290		4,087.34	I
90		4,169.84	I
2,500		4,212.95	I
180		4,473.59	I
55	h	4,788.18	I
45	h	4,817.51	I
35		4,875.43	I
55		5,110.81	I
75		5,117.02	I
160		5,163.84	I
55		5,234.86	I
120		5,295.63	I
18		5,312.57	I
15		5,345.10	I
35		5,395.24	I
55		5,542.80	I
35		5,547.02	I
27		5,619.44	I
15		5,642.69	I
14		5,655.42	I
75		5,670.07	I
11		5,690.14	I
55	h	5,695.09	I
18		5,736.61	I
23		6,774.54	I
65		6,784.52	I
4	h	6,833.42	I
11		7,016.44	I
13	h	7,310.06	I
75		7,368.12	I
27		7,391.92	I
16		7,486.90	I
120		7,764.03	I
27		7,786.67	I
45		7,915.80	I
18		7,961.08	I
55		8,132.82	I
45		8,300.83	I
9	h	8,353.58	I
18	h	8,532.74	I
16	h	8,599.10	I
65		8,761.35	I

PHOSPHORUS (P)

Z = 15

P I and II

Ref. 182 — R.L.K.

Vacuum

Intensity		Wavelength	
10		810.24	II
10		865.44	II
20		1,249.82	II
20		1,301.87	II
20		1,304.47	II
15		1,304.68	II
35		1,305.48	II
25		1,309.87	II
60		1,310.70	II
20		1,373.49	I
20		1,377.06	I
20		1,377.93	I
25		1,379.40	I
25		1,381.47	I
50	d	1,430.13	I
30		1,452.89	II
50		1,491.36	I
40		1,492.99	I
80		1,532.51	II
120		1,535.90	II
80		1,536.39	II
120		1,542.29	II
60		1,548.43	I
70		1,671.07	I
180		1,671.68	I
90		1,672.48	I
230		1,674.61	I
300		1,679.71	I
120		1,685.99	I
100		1,694.06	I
60		1,706.41	I
250		1,774.99	I
200		1,782.87	I
180		1,787.68	I
100		1,847.19	I
80		1,851.22	I
150		1,858.91	I
150		1,859.43	I

Air

100		2,023.48	I
150		2,033.47	I
100		2,135.47	I
200		2,136.18	I
200		2,149.14	I
100		2,152.94	I
150		2,154.08	I
100		2,484.19	II
500		2,533.99	I
700		2,535.61	I
600		2,553.25	I
500		2,554.90	I
150		2,606.06	II
100		2,626.18	II
90		2,636.76	II
150		3,308.92	II
125		3,419.34	II
100		3,425.00	II
100		4,178.48	II
200		4,288.60	II
200		4,385.35	II
400		4,420.71	II
100		4,452.46	II
150		4,463.00	II
120		4,467.98	II
200		4,475.26	II
200		4,499.24	II
120		4,530.81	II
120		4,554.83	II
120		4,558.07	II
120		4,581.71	II
500		4,588.04	II
500		4,589.86	II
600		4,602.08	II
300		4,626.70	II
300		4,658.31	II
200		4,864.42	II
150		4,927.20	II
500		4,943.53	II
300		4,954.39	II
300		4,969.71	II
150		5,191.41	II
300		5,253.52	II
400		5,296.13	II
250		5,316.07	II
300		5,344.75	II
250		5,378.20	II
300		5,386.88	II
200		5,409.72	II
400		5,425.91	II
400		5,450.74	II
100		5,458.31	I
125		5,461.20	II
200		5,477.75	I
200		5,483.55	II
200		5,499.73	II
200		5,507.19	II
200		5,541.14	II
200		5,583.27	II
250		5,588.34	II
100		5,727.71	II
500		6,024.18	II
400		6,034.04	II
500		6,043.12	II
250		6,055.50	II
100		6,057.86	II
350		6,087.82	II
150		6,097.68	I
350		6,165.59	II
180		6,199.01	I
100		6,232.29	II
200		6,367.27	II
250		6,435.32	II
130		6,436.31	II
600		6,459.99	II
600		6,503.46	II
600		6,507.97	II
150		6,713.28	II
120		6,717.42	I
100		7,165.45	I
150		7,175.12	I
120		7,176.66	I
100		7,505.76	I
250		7,845.63	II
120		8,278.07	I
150		8,367.84	I
100		8,531.46	I
100		8,613.85	I
150		8,637.62	I
250		8,741.54	I
50		8,872.17	I
30		9,525.78	I
25		9,593.54	I
20		9,734.74	I
25		9,750.73	I
50		9,796.79	I
25		10,084.22	I
6		10,529.45	I
8		10,581.52	I

P III

Ref. 180 — R.L.K.

Vacuum

Intensity	Wavelength	
90	471.146	III
90	484.278	III
120	498.180	III
200	569.853	III
200	581.831	III
200	844.646	III
150	845.038	III
250	845.664	III
300	847.669	III
200	848.016	III
120	848.465	III
150	848.639	III
250	852.686	III
350	855.624	III
200	859.406	III
500	859.652	III
250	859.729	III
300	913.971	III
300	917.120	III
350	918.665	III
250	921.849	III
200	997.999	III
250	1,003.598	III
500	1,334.808	III
650	1,344.327	III
300	1,344.845	III
250	1,380.463	III
150	1,381.089	III
350	1,502.228	III
250	1,504.663	III
150	1,618.632	III
200	1,618.907	III

Air

200	2,611.147	III
300	2,632.713	III
200	2,680.133	III

Intensity	Wavelength	
250	2,895.241	III
250	3,186.186	III
300	3,219.307	III
150	3,233.536	III
400	3,233.602	III
200	3,556.546	III
200	3,577.526	III
200	3,904.812	III
250	3,914.314	III
300	3,957.641	III
350	3,978.307	III
200	4,057.440	III
400	4,059.312	III
300	4,080.084	III
500	4,222.195	III
350	4,246.720	III
200	4,428.171	III
200	4,463.668	III
250	4,479.776	III
150	6,083.409	III
150	6,409.204	III
150	6,484.440	III
150	6,486.381	III
150	6,992.690	III
150	8,113.528	III

P IV

Ref. 336 — R.L.K.

Intensity	Wavelength	
	Vacuum	
90	282.301	IV
90	304.996	IV
120	359.293	IV
150	359.899	IV
120	361.514	IV
150	361.629	IV
120	371.299	IV
150	371.504	IV
200	372.001	IV
500	388.318	IV
120	414.604	IV
200	414.999	IV
250	415.805	IV
250	444.245	IV
300	445.158	IV
250	568.038	IV
350	629.008	IV
400	629.914	IV
500	631.779	IV
350	648.482	IV
300	756.510	IV
300	776.353	IV
650	823.179	IV
700	824.730	IV
800	827.932	IV
250	847.019	IV
350	849.799	IV
200	850.392	IV
700	877.476	IV
1,000	950.655	IV
570	1,025.563	IV
500	1,028.096	IV
570	1,030.517	IV
500	1,033.111	IV
500	1,035.517	IV
570	1,118.551	IV
200	1,206.422	IV
200	1,335.705	IV
500	1,366.695	IV
400	1,372.674	IV
350	1,377.282	IV
500	1,484.507	IV
400	1,487.788	IV
300	1,489.098	IV
250	1,862.762	IV
120	1,862.893	IV
200	1,863.580	IV
650	1,888.523	IV
200	1,910.183	IV
120	1,985.682	IV
150	1,985.851	IV
200	1,986.114	IV
150	1,987.022	IV
	Air	
200	2,477.823	IV
150	2,478.070	IV
250	2,478.256	IV
250	2,605.506	IV
400	2,644.295	IV
300	2,724.764	IV
400	2,728.770	IV
200	2,729.120	IV
500	2,739.309	IV
250	2,739.872	IV
200	2,740.223	IV
200	2,961.242	IV
650	3,347.736	IV
570	3,364.467	IV
400	3,371.122	IV
200	3,413.543	IV
200	3,733.393	IV
300	4,249.656	IV
250	4,540.288	IV
250	4,541.112	IV
150	4,548.056	IV
200	4,548.449	IV
150	5,235.499	IV
150	5,989.774	IV
150	6,142.605	IV
150	6,713.939	IV
120	6,715.906	IV
200	7,443.657	IV

P V

Ref. 179 — R.L.K.

Intensity	Wavelength	
	Vacuum	
80	255.59	V
50	255.67	V
110	310.58	V
150	311.34	V
300	328.47	V
250	328.78	V
150	347.23	V
200	348.20	V
110	378.56	V
250	389.50	V
300	390.70	V
150	410.03	V
375	475.60	V
110	534.63	V
80	534.99	V
520	542.57	V
600	544.92	V
450	673.90	V
450	865.45	V
600	871.39	V
250	997.62	V
150	1,000.38	V
900	1,117.98	V
700	1,128.01	V
150	1,379.62	V
250	1,385.05	V
375	1,447.83	V
450	1,610.50	V
	Air	
200	2,180.29	V
150	2,186.42	V
375	2,424.40	V
450	2,440.93	V
200	2,441.24	V
300	2,961.00	V
450	2,978.55	V
700	3,175.09	V
520	3,204.04	V
150	4,083.18	V
110	4,094.95	V
110	5,156.72	V

PLATINUM (PT)

Z = 78

Pt I and II

Ref. 1, 298 — C.H.C.

Intensity		Wavelength	
		Vacuum	
30		1,621.66	II
30		1,723.13	II
30		1,751.70	II
50	r	1,777.09	II
30		1,781.86	II
30		1,879.09	II
40		1,883.05	II
50		1,889.52	II
50		1,911.70	II
30		1,929.25	II
30		1,929.68	II
30		1,939.80	II
30		1,949.90	II
30		1,983.74	II
		Air	
40		2,014.93	II
3,200		2,030.63	I
4,400		2,032.41	I
100		2,036.46	II
40		2,041.57	II
5,500		2,049.37	I
1,500		2,067.50	I
3,000		2,084.59	I
1,000		2,103.33	I
30		2,115.57	II
950		2,128.61	I
30		2,130.69	II
1,900		2,144.23	I
100		2,144.24	II
600		2,165.17	I
1,500		2,174.67	I
30		2,190.32	II
400		2,202.22	I
50	h	2,202.58	II
320		2,222.61	I
50	h	2,233.11	II
150		2,249.30	I
30	h	2,240.99	II
100		2,245.52	II
30		2,251.52	II
30	h	2,251.92	II
190		2,268.84	I
30	h	2,271.72	II
280		2,274.38	I
50	h	2,287.50	II
30		2,288.20	II
150		2,289.27	I
150		2,292.40	I
240		2,308.04	I
50		2,310.96	II
90		2,315.50	I
220		2,318.29	I
100		2,326.10	I
170		2,340.18	I
280		2,357.10	I
180		2,368.28	I
50		2,377.28	II
130		2,383.64	I
40		2,386.81	I
120		2,389.53	I
35		2,396.17	I
70		2,401.87	I
200		2,403.09	I
100		2,418.06	I
50		2,424.87	II
80		2,428.04	I
50		2,428.20	I
25		2,429.10	I
180		2,436.69	I
650		2,440.06	I
60		2,450.97	I
440		2,467.44	I
35		2,471.01	I
1,000		2,487.17	I
25		2,488.74	II
200		2,490.12	I
160		2,495.82	I
240		2,498.50	I
50		2,505.93	I
120		2,508.50	I
50		2,514.07	I
60		2,515.03	I
240		2,515.58	I
140		2,524.30	I
40		2,529.41	I
50		2,536.49	I
160		2,539.20	I
18		2,549.46	I
50		2,552.25	I
50		2,596.00	I
70		2,603.14	I
30		2,616.76	II
50		2,619.57	I
30		2,625.34	II
1,100		2,628.03	I
130		2,639.35	I
1,000		2,646.89	I
500		2,650.86	I
20		2,658.17	I
2,800		2,659.45	I
40		2,674.57	I
440		2,677.15	I
200		2,698.43	I
2,000		2,702.40	I
1,600		2,705.89	I
60		2,713.13	I
1,300		2,719.04	I
130		2,729.92	I
1,800		2,733.96	I
70		2,738.48	I
70		2,747.61	I
80		2,753.86	I
200		2,754.92	I
30		2,769.84	I
500		2,771.67	I
40		2,773.24	I
20		2,774.00	I
50		2,774.77	II
50		2,793.27	I
100		2,794.21	II
40	h	2,799.98	II
140		2,803.24	I
10		2,808.51	I
50		2,818.25	I
30	h	2,822.27	II
1,400		2,830.30	I
70		2,834.71	I
16		2,853.11	I
80	h	2,860.68	II
40	h	2,865.05	II
40	h	2,875.85	II
100	h	2,877.52	II
25		2,888.20	I
25		2,893.22	I
600		2,893.86	I
300		2,897.87	I
60		2,905.90	I
120		2,912.26	I
120		2,913.54	I
70		2,919.34	I
30		2,921.38	I
1,700		2,929.79	I
30		2,942.76	I
30		2,944.75	I
25		2,959.10	I
60		2,960.75	I
1,800		2,997.97	I
35		3,001.17	II
220		3,002.27	I
30		3,017.88	I
30	h	3,031.22	II
130		3,036.45	I
800		3,042.64	I
3,200		3,064.71	I
30		3,071.94	I
130		3,100.04	I
320		3,139.39	I
140		3,156.56	I
120		3,200.71	I
320		3,204.04	I
30		3,230.29	I
20		3,233.42	I
20		3,250.36	I
40		3,251.98	I
160		3,255.92	I
25		3,268.42	I
25		3,281.97	I
120		3,290.22	I
500		3,301.86	I
60		3,315.05	I
35		3,323.80	I
340		3,408.13	I
35		3,427.93	I
60		3,483.43	I
160		3,485.27	I
120		3,628.11	I
70		3,638.79	I
70		3,643.17	I
50		3,663.10	I
80		3,671.99	I
80		3,674.04	I

Intensity	Wavelength	
35	3,699.91	I
18	3,706.53	I
20	3,801.05	
80	3,818.69	I
40	3,900.73	I
110	3,922.96	I
35	3,948.40	I
100	3,966.36	I
20	3,996.57	I
110	4,118.69	I
80	4,164.56	I
40	4,192.43	I
18	4,327.06	I
18	4,391.83	I
80	4,442.55	I
14	4,445.55	I
25	4,498.76	I
12	4,520.90	I
35	4,552.42	I
12	4,879.53	I
14	5,044.04	I
30	5,059.48	I
35	5,227.66	I
40	5,301.02	I
12	5,368.99	I
12	5,390.79	I
14	5,475.77	I
14	5,478.50	I
6	5,763.57	I
20	5,840.12	I
8	5,844.84	I
6	6,026.04	I
7	6,318.37	I
8	6,326.58	I
9	6,523.45	I
10	6,710.42	I
20	6,760.02	I
60	6,842.60	I
20	7,113.73	I
10	8,224.74	I

PLUTONIUM (PU)

Z = 94

Pu I and II

Ref. 91 — J.G.C.

Intensity	Wavelength	
	Air	
	2,781.40	II
	2,784.48	II
	2,806.11	II
	2,815.77	II
	2,897.97	II
	2,898.94	II
	2,904.25	II
	2,904.94	II
	2,910.40	II
	2,918.00	II
	2,918.80	II
	2,926.08	II
	2,928.25	II
	2,929.71	II
	2,930.98	II
	2,932.32	II
	2,933.30	II
	2,938.54	II
	2,938.95	II
	2,941.39	II
	2,945.26	II
	2,946.00	II
	2,950.06	II
	2,951.62	
	2,954.46	II
	2,963.47	II
	2,966.84	II
	2,967.54	II
	2,972.50	II
	2,977.81	II
	2,978.37	II
	2,980.23	II
	2,981.23	II
	2,986.95	II
	2,988.21	II
	2,991.31	II
	2,996.40	II
	3,000.31	II
	3,009.57	II
	3,028.85	II
	3,042.61	II
	3,043.12	II
	3,060.32	II
	3,069.32	II
	3,091.33	II
	3,091.94	II
	3,092.59	II
	3,104.12	II
	3,105.04	II
	3,106.03	
	3,123.87	II
	3,159.21	II
	3,161.73	II
	3,163.18	II
	3,174.49	II
	3,179.41	II
	3,185.12	II
	3,187.60	II
	3,189.23	II
	3,193.54	
	3,193.55	II
	3,194.56	II
	3,198.47	II
	3,200.23	II
	3,201.00	II
	3,201.66	II
	3,204.48	II
	3,206.80	II
	3,207.97	II
	3,215.08	I
	3,216.15	II
	3,220.94	II
	3,224.87	II
	3,231.86	II
	3,232.24	
	3,232.63	II
	3,241.39	II
	3,242.96	II
	3,243.40	II
	3,244.16	I
	3,245.25	II
	3,245.71	
	3,246.35	II
	3,247.50	
	3,247.56	II
	3,252.08	I
	3,260.54	II
	3,265.17	
	3,273.11	II
	3,274.71	II
	3,275.24	I
	3,292.56	I
	3,293.61	I
	3,296.91	I
	3,297.87	I
	3,298.47	II
	3,301.76	I
	3,306.59	I
	3,306.66	I
	3,307.66	II
	3,308.75	I
	3,312.65	II
	3,315.34	II
	3,316.96	II
	3,320.61	I
	3,320.84	I
	3,323.48	
	3,327.19	I
	3,330.11	
	3,331.52	II
	3,332.34	I
	3,333.03	
	3,337.71	II
	3,338.40	II
	3,338.94	I
	3,347.87	II
	3,349.63	I
	3,351.82	II
	3,356.61	II
	3,358.41	II
	3,358.84	II
	3,362.26	II
	3,365.20	I
	3,365.66	
	3,368.86	I
	3,370.64	II
	3,371.19	I
	3,375.80	I
	3,376.76	II
	3,376.94	II
	3,377.37	II
	3,379.51	I
	3,381.82	I
	3,381.97	II
	3,382.70	I
	3,390.33	II
	3,391.41	II
	3,393.67	I
	3,394.32	I
	3,418.88	II
	3,465.10	II
	3,473.64	II
	3,483.20	I
	3,585.87	II
	3,632.21	II
	3,699.19	II
	3,720.59	I
	3,725.98	I
	3,726.11	II
	3,726.79	II
	3,732.03	II
	3,744.78	I
	3,753.63	I
	3,755.94	I
	3,757.82	I
	3,758.34	I
	3,774.38	I
	3,776.71	I
	3,792.22	I
	3,799.37	I
	3,805.93	I
	3,811.40	I
	3,812.30	II
	3,827.57	I
	3,835.52	I
	3,836.96	I
	3,838.92	I
	3,842.10	I
	3,851.01	I
	3,851.85	I
	3,878.54	I
	3,895.89	I
	3,928.53	I
	3,975.43	II
	4,097.12	I
	4,101.96	I
	4,105.95	II
	4,111.07	I
	4,114.91	I
	4,128.12	I
	4,129.93	II
	4,133.01	I
	4,135.97	I
	4,140.04	I
	4,141.20	II
	4,151.09	I
	4,151.45	I
	4,155.46	I
	4,159.39	II
	4,167.77	I
	4,170.95	I
	4,178.28	II
	4,189.90	II
	4,190.06	II
	4,196.20	II
	4,206.48	I
	4,208.23	I
	4,221.87	I
	4,224.20	I
	4,229.77	II
	4,254.76	II
	4,261.88	I
	4,269.77	I
	4,273.34	II
	4,281.17	I
	4,289.08	II
	4,337.18	II
	4,352.71	II
	4,367.41	I
	4,379.91	II
	4,385.35	II
	4,393.93	II
	4,404.90	I
	4,441.65	II
	4,468.54	II
	4,472.79	II
	4,493.78	II
	4,504.91	II
	4,536.15	II
	4,735.40	I
	4,989.34	I
	5,269.86	I
	5,381.02	I
	5,498.50	I
	5,510.72	I
	5,537.59	I
	5,549.62	I
	5,590.54	I
	5,592.33	I
	5,712.39	I
	5,770.26	I
	5,839.05	I
	5,983.35	I
	6,012.78	I
	6,192.80	I
	6,304.66	I
	6,449.75	I
	6,486.71	I
	6,488.86	I
	6,488.89	
	6,535.27	I
	6,544.21	I
	6,608.95	I
	6,672.72	I
	6,784.66	I
	6,880.16	I
	6,891.38	I
	7,059.23	I
	7,068.90	I
	7,092.46	I
	7,116.88	I
	7,141.66	I
	7,177.14	I
	7,231.09	I
	7,258.06	I
	7,322.23	I
	7,325.97	I
	7,331.81	I
	7,431.18	I
	7,447.99	I
	7,507.80	I
	7,526.93	I
	7,547.45	I
	7,564.50	I
	7,571.87	I
	7,572.93	I
	7,609.77	I
	7,689.40	I
	7,758.20	I
	7,798.54	I
	7,953.17	I
	8,102.54	I
	8,130.86	I
	8,309.61	I
	8,435.47	I
	8,476.13	I
	8,495.75	
	8,597.26	I
	8,665.02	I
	8,691.94	I
	8,729.82	I
	8,836.16	I
	9,533.07	I
	10,046.75	I
	11,114.82	I
	12,144.46	I
	12,231.22	I
	15,377.31	I
	16,897.38	I

POLONIUM (PO)

Z = 84

Po I and II

Ref. 47, 48 — E.F.W.

Intensity		Wavelength	
		Air	
250	w	2,139.02	I
300	h	2,203.80	
300		2,220.67	I
200		2,222.13	I
200		2,284.22	
250		2,344.61	I

Intensity		Wavelength	Spectrum
250		2,421.72	I
300		2,426.09	I
1,500	w	2,450.08	I
700		2,483.94	I
700		2,490.53	I
200	h	2,502.18	
300		2,534.95	I
300		2,557.33	I
1,500	w	2,558.01	I
400		2,562.31	
300		2,578.80	
400		2,587.64	I
200		2,637.01	
300		2,645.36	I
700	h	2,663.33	
200		2,671.67	I
600		2,761.92	I
400		2,800.26	I
250		2,824.11	I
300		2,866.01	I
400		2,919.31	I
600		2,958.92	I
2,500	w	3,003.21	I
450		3,069.31	I
200		3,115.95	
400		3,189.02	I
600		3,240.24	I
250		3,286.38	I
600		3,328.60	I
300		3,489.79	
200		3,493.65	
400		3,588.33	
200		3,671.36	
500		3,861.93	I
200		4,051.98	
1,200		4,170.52	I
250		4,236.13	
200	h	4,415.58	
800		4,493.21	I
350		4,611.44	I
200		4,867.12	
400		4,876.24	I
450		4,946.81	
350		5,323.23	I
300		5,744.85	I
600		7,962.62	I
300		8,433.87	I
500		8,618.26	I
250		9,227.87	

POTASSIUM (K)

Z = 19

K I and II

Ref. 59, 60, 76, 172, 218

Intensity	Wavelength	
Vacuum		
5	261.20	II
25	441.81	II
5	465.08	II
	469.50	II
10	476.03	II
30	495.14	II
30	600.77	II
25	607.93	II
30	612.62	II
3	1,725.0	II
Air		
6	2,190.00	II
4	2,210.53	II
5	2,265.04	II
6	2,550.02	II
4	2,743.55	II
	2,992.12	I
	2,992.22	I
	3,034.76	I
	3,034.92	I
5	3,062.18	II
4	3,101.79	I
3	3,102.04	I
6	3,105.00	II
5	3,190.07	II
7	3,217.16	I
6	3,217.62	I
4	3,220.60	II
5	3,290.65	II
6	3,345.32	II
6	3,373.60	II
6	3,380.62	II
6	3,384.86	II
6	3,404.24	II
7	3,440.05	II
11	3,446.37	I
10	3,447.38	I
6	3,481.11	II
7	3,530.75	II
5	3,608.88	II
6	3,618.49	II
4	3,626.42	II
3	3,648.84	I
4	3,648.98	I
6	3,681.54	II
5	3,716.60	II
5	3,721.34	II
5	3,739.13	II
5	3,744.42	II
6	3,767.36	II
6	3,783.19	II
6	3,800.14	II
6	3,816.56	II
7	3,817.50	II
5	3,873.74	II
4	3,878.62	II
8	3,897.92	II
5	3,923.00	II
5	3,926.36	II
6	3,942.53	II
6	3,955.21	II
6	3,966.72	II
6	3,972.58	II
6	3,995.10	II
7	4,001.24	II
5	4,012.10	II
6	4,042.59	II
18	4,044.14	I
17	4,047.21	I
5	4,093.69	II
6	4,114.99	II
7	4,134.72	II
7	4,149.19	II
8	4,186.24	II
7	4,222.97	II
7	4,225.67	II
7	4,263.40	II
7	4,305.00	II
7	4,309.10	II
5	4,340.03	II
7	4,388.16	II
5	4,466.65	II
6	4,505.33	II
5	4,595.65	II
8	4,608.45	II
10	4,641.88	I
11	4,642.37	I
5	4,659.38	II
4	4,740.91	I
6	4,744.35	I
5	4,753.93	I
7	4,757.39	I
5	4,786.49	I
7	4,791.05	I
6	4,799.75	I
8	4,804.35	I
9	4,829.23	II
7	4,849.86	I
8	4,856.09	I
8	4,863.48	I
9	4,869.76	I
8	4,942.02	I
6	4,943.29	II
9	4,950.82	I
9	4,956.15	I
10	4,965.03	I
8	5,005.60	II
7	5,056.27	II
10	5,084.23	I
11	5,097.17	I
11	5,099.20	I
12	5,112.25	I
5	5,310.24	II
12	5,323.28	I
13	5,339.69	I
12	5,342.97	I
14	5,359.57	I
6	5,470.13	II
5	5,642.73	II
4	5,772.32	II
16	5,782.38	I
17	5,801.75	I
15	5,812.15	I
17	5,831.89	I
2	5,969.64	II
8	6,120.27	II
6	6,246.59	II
7	6,307.29	II
5	6,427.96	II
2	6,595.00	II
19	6,911.08	I
12	6,936.28	I
20	6,938.77	I
7	6,964.18	I
12	6,964.67	I
25	7,664.90	I
24	7,698.96	I
5	7,955.37	I
4	7,956.83	I
7	8,078.11	I
6	8,079.62	I
9	8,250.18	I
8	8,251.74	I
3	8,390.22	I
2	8,417.54	I
1	8,420.00	I
11	8,503.45	I
10	8,505.11	I
4	8,763.96	I
3	8,767.05	I
13	8,902.19	I
12	8,904.02	I
5	8,923.31	I
4	8,925.44	I
7	9,347.24	I
3	9,349.25	I
6	9,351.59	I
15	9,595.70	I
14	9,597.83	I
6	9,949.67	I
5	9,954.14	I
9	10,479.63	I
5	10,482.15	I
8	10,487.11	I
17	11,019.87	I
16	11,022.67	I
17	11,690.21	I
16	11,769.62	I
17	11,772.83	I
	12,432.24	I
	12,522.11	I
	13,377.86	I
	13,397.09	I
	15,163.08	I
	15,168.40	I
	40,158.37	I

K III

Ref. 60, 76 — L.J.R.

Intensity	Wavelength	
Vacuum		
2	325.28	III
5	327.60	III
25	330.68	III
30	341.92	III
15	348.00	III
30	379.12	III
25	380.48	III
30	382.23	III
15	398.63	III
20	402.10	III
30	406.48	III
40	408.96	III
50	413.79	III
30	414.87	III
30	416.00	III
30	417.54	III
30	418.62	III
75	434.72	III
50	435.68	III
75	444.34	III
75	448.60	III
75	466.79	III
100	470.09	III
75	471.57	III
45	474.92	III
40	479.18	III
10	482.11	III
10	482.41	III
75	497.10	III
10	514.94	III
50	520.61	III
25	523.79	III
40	529.80	III
15	539.71	III
15	546.12	III
20	708.84	III
20	765.31	III
30	765.64	III
35	778.53	III
20	872.31	III
10	873.86	III
15	874.04	III
Air		
6	2,550.02	III
5	2,635.11	III
1	2,736.96	III
5	2,689.90	III
1	2,898.90	III
5	2,938.45	III
1	2,948.94	III
5	2,986.20	III
6	2,992.24	III
6	3,052.07	III
5	3,056.84	III
6	3,201.95	III
6	3,209.34	III
6	3,278.79	III
6	3,289.06	III
6	3,322.40	III
6	3,364.22	III
6	3,420.82	III
4	3,421.83	III
6	3,468.32	III
6	3,481.11	III
5	3,513.88	III
1	3,885.50	III

K IV

Ref. 32, 82, 250, 160, 314

Intensity	Wavelength	
Vacuum		
150	271.82	IV
100	273.06	IV
300	340.46	IV
150	340.74	IV
300	354.93	IV
150	356.26	IV
300	359.73	IV
200	359.91	IV
250	362.08	IV
150	362.15	IV
150	363.02	IV
300	375.96	IV
300	379.88	IV
250	380.48	IV
200	381.70	IV
300	382.23	IV
150	382.49	IV
200	382.65	IV
300	382.91	IV
250	384.10	IV
200	386.61	IV
250	388.92	IV
250	389.07	IV
250	390.42	IV
300	390.57	IV
200	391.46	IV
200	392.47	IV
500	393.14	IV
400	400.21	IV
300	402.91	IV
250	403.97	IV
150	404.41	IV
250	408.08	IV
150	417.28	IV
200	442.30	IV
300	443.57	IV
200	445.61	IV
250	446.83	IV

750		448.60	IV
400		456.33	IV
250		523.00	IV
200		526.45	IV
150		527.62	IV
750		646.19	IV
500		737.14	IV
500		741.95	IV
500		745.26	IV
400		746.35	IV
300		749.99	IV
150		754.19	IV
400		754.67	IV

K V

Ref. 32, 75, 76, 150, 322

Intensity		Wavelength	
		Vacuum	
100		214.35	V
150		282.35	V
150		293.33	V
300		294.84	V
200		296.17	V
200		297.06	V
200		300.25	V
200		300.50	V
200		311.24	V
250		312.77	V
200		315.18	V
250		327.38	V
250		389.07	V
200		349.50	V
500		372.15	V
200		372.46	V
200		372.77	V
300		375.96	V
250		377.76	V
300		379.12	V
300		387.80	V
250		390.11	V
250		395.40	V
200		398.36	V
200		398.88	V
200		399.75	V
250		415.05	V
200		415.79	V
400		422.18	V
300		425.16	V
500		425.59	V
250		438.02	V
200		449.71	V
200		452.90	V
250		455.67	V
400		456.33	V
200		482.71	V
200		483.75	V
750		580.32	V
250		585.51	V
500		586.32	V
250		602.27	V
400		603.43	V
250		638.67	V
300		687.50	V
300		720.43	V
400		724.42	V
600		731.86	V
150		770.29	V
150		771.46	V

PRASEODYMIUM (PR)

$Z = 59$

Pr I and II

Ref. 1 — C.H.C.

Intensity		Wavelength	
		Vacuum	
25		2,558.58	II
25		2,578.27	I
30		2,579.31	I
40	h	2,598.04	II
25		2,608.92	II
25		2,615.75	II
25		2,648.48	II
30		2,654.75	II
25		2,666.70	II
20		2,672.52	II
30		2,685.19	II
45		2,685.70	II
50		2,698.92	II
60		2,700.38	II
30		2,702.25	II
100	h	2,707.37	II
20		2,714.16	II
60		2,720.17	II
30		2,721.90	II
50		2,726.50	II
12		2,731.78	II
25		2,733.12	II
50		2,734.30	II
25		2,737.90	II
40		2,742.12	II
25		2,744.66	II
20		2,746.28	II
60		2,760.35	II
50		2,769.60	II
50	d	2,775.94	II
		2,776.03	II
40		2,778.80	II
50		2,783.31	II
30		2,789.05	II
35		2,792.51	II
50		2,802.05	II
20		2,823.17	II
20		2,824.14	II
20		2,828.29	II
20		2,842.98	I
20		2,844.01	II
20		2,850.62	I
25		2,853.99	II
30		2,865.64	II
50		2,881.60	I
30		2,882.31	II
30		2,884.89	II
30		2,943.97	II
30		2,967.58	II
30		2,971.13	II
40	d	2,971.40	II
		2,971.46	II
50		2,984.98	II
30		2,986.18	II
30		2,990.22	II
110		3,082.11	II
100		3,111.34	II
140		3,121.58	II
140		3,163.73	II
270		3,168.24	II
160		3,172.31	II
110		3,191.42	II
200	d	3,195.99	II
110		3,199.04	II
100		3,207.89	II
190		3,219.48	II
100		3,234.27	II
100		3,245.48	II
140		3,355.67	II
110		3,394.62	II
110		3,465.74	II
200		3,584.21	II
130		3,611.94	II
170		3,630.96	II
100		3,645.55	II
250		3,645.66	II
250		3,646.30	II
100		3,648.30	II
150	c	3,660.36	II
100		3,661.62	II
370		3,668.83	II
250		3,687.03	II
150		3,687.19	II
100		3,689.71	II
150		3,698.06	II
230		3,706.75	II
170	c	3,711.10	II
290		3,714.05	II
120	c	3,733.03	II
210	c	3,734.41	II
250		3,735.76	II
190		3,736.49	II
410		3,739.18	II
150		3,740.99	II
120		3,743.98	II
190		3,750.98	II
140		3,759.60	II
120		3,760.08	II
680		3,761.87	II
230		3,764.77	II
230		3,768.94	II
170	c	3,772.82	II
170		3,774.06	II
140		3,777.62	II
170		3,780.66	II
150		3,785.46	II
150		3,786.86	II
210		3,792.51	II
190		3,794.93	II
680		3,800.30	II
290		3,804.84	II
140		3,809.18	II
390		3,811.84	II
1,300	h	3,816.02	II
120		3,817.66	II
680		3,818.28	II
120		3,819.14	II
310		3,821.80	II
150	c	3,823.18	II
120		3,826.67	II
960		3,830.72	II
140		3,834.93	II
480		3,840.99	II
270		3,842.34	II
150	c	3,844.54	II
580		3,846.59	II
1,200		3,850.79	II
720	c	3,851.55	II
960		3,852.80	II
120		3,858.25	II
110		3,859.14	II
480	c	3,865.45	II
210		3,867.52	II
210		3,870.72	II
480		3,876.19	II
1,700	c	3,877.18	II
270		3,879.20	II
680		3,880.47	II
440	c	3,885.19	II
440	c	3,889.34	II
120	c	3,891.71	II
190		3,897.25	II
210		3,898.84	II
250		3,902.45	II
770	c	3,908.05	II
630		3,912.90	II
310		3,913.55	II
210		3,914.76	II
1,300	c	3,918.85	II
420		3,919.63	II
250		3,920.53	II
960		3,925.47	II
480		3,927.46	II
370		3,929.29	II
370		3,935.82	II
250		3,938.30	II
730	c	3,947.63	II
900	c	3,949.43	II
900	c	3,953.51	II
380		3,956.75	II
190		3,959.44	I
470		3,962.45	II
560		3,964.26	II
1,600	c	3,964.81	II
560	c	3,966.57	II
500		3,971.16	II
320		3,971.67	II
620	c	3,972.14	II
320		3,974.85	II
1,300	c	3,989.68	II
230		3,991.91	II
340		3,992.16	II
1,600		3,994.79	II
270		3,995.83	II
560	c	3,997.04	II
230		3,997.96	II
320		3,999.12	II
620	c	4,000.17	II
730		4,004.70	II
1,900		4,008.69	II
620		4,010.60	II
730		4,015.39	II
620		4,020.96	II
470		4,022.71	II
360		4,025.54	II
230		4,026.83	II
230		4,029.00	II
360	c	4,029.72	II
730	c	4,031.75	II
230		4,032.47	II
960		4,033.83	II
230		4,034.33	II
230		4,038.22	II
730		4,038.45	II
470		4,039.34	II
1,300		4,044.81	II
230		4,045.70	II
230		4,046.63	II
340		4,047.08	II
450		4,051.13	II
2,200		4,054.88	II
2,200		4,056.54	II
450		4,058.80	II
230		4,062.22	II
3,400		4,062.81	II
210		4,068.80	II
500	c	4,079.77	II
500	c	4,080.98	II
790		4,081.85	II
500		4,083.34	II
200	c	4,087.21	II
560		4,096.82	II
380		4,098.40	II
2,900	c	4,100.72	II
270	c	4,113.89	II
1,700	c	4,118.46	II
250		4,129.15	II
340		4,130.77	II
200		4,133.61	II
1,500	c	4,141.22	II
2,700		4,143.11	II
270	c	4,146.50	II
270		4,148.44	II
200		4,156.50	II
1,700	c	4,164.16	II
270		4,168.04	II
230		4,169.45	II
620		4,171.82	II
730		4,172.25	II
250		4,175.32	II
250		4,175.62	II
200		4,178.63	II
5,200		4,179.39	II
2,500		4,189.48	II
560	c	4,191.60	II
290		4,201.17	II
2,500	c	4,206.72	II
500		4,208.32	II
320		4,211.86	II
320		4,217.81	II
3,800		4,222.93	II
3,800		4,225.35	II
320		4,233.11	II
320	c	4,236.15	II
270		4,240.02	II
960		4,241.01	II
340		4,243.51	II
840	c	4,247.63	II
500		4,254.40	II
270	c	4,263.78	II
320		4,269.09	II
790	c	4,272.27	II
470	c	4,280.07	II
790	c	4,282.42	II
450	c	4,298.98	II
290		4,303.61	II
1,500		4,305.76	II
210		4,323.55	II
270		4,329.41	II
1,300		4,333.97	II
200		4,335.74	II
360		4,338.70	II
620	c	4,344.30	II
470	c	4,347.49	II
340		4,350.40	II
450		4,354.91	II
410	c	4,359.79	II
1,200		4,368.33	II
320		4,371.62	II
270		4,396.08	II
170		4,403.60	II

100		4,405.12	II
430		4,405.83	II
1,700		4,408.82	II
410		4,413.77	II
160		4,419.04	II
190		4,419.65	II
160	c	4,421.22	II
160		4,424.58	II
1,200	c	4,429.13	II
110		4,432.28	II
730		4,449.83	II
140		4,451.90	II
140		4,454.68	II
100		4,465.97	II
960		4,468.66	II
140	c	4,477.26	II
1,100		4,496.46	II
790		4,510.15	II
200	c	4,517.58	II
340	c	4,534.15	II
340		4,535.92	II
200		4,563.12	II
140		4,612.08	II
270	c	4,628.74	II
140		4,632.28	I
140		4,635.68	I
200		4,639.55	I
110	c	4,643.49	II
140		4,646.05	II
200	c	4,651.50	II
140		4,664.65	II
270	c	4,672.09	II
180		4,687.80	I
290		4,695.77	I
140	c	4,708.07	II
140		4,709.52	I
180		4,730.67	I
250		4,736.69	I
100		4,744.16	I
150		4,746.92	II
100		4,762.72	II
110		4,783.35	II
110		4,906.99	I
140		4,914.02	I
200		4,924.60	I
140		4,936.00	I
320		4,939.74	I
160		4,940.30	I
380		4,951.37	I
110		4,975.75	I
120		5,018.59	I
200		5,019.76	I
200		5,026.96	I
100		5,033.38	I
270		5,034.41	II
110		5,043.83	I
320		5,045.52	I
160		5,053.40	I
180		5,087.12	I
360		5,110.38	II
560		5,110.76	II
410		5,129.52	II
270		5,133.44	I
270		5,135.14	II
100		5,139.81	I
100	c	5,152.30	II
200		5,161.74	II
620		5,173.90	II
200		5,191.32	II
120		5,194.43	I
150		5,195.11	II
200		5,195.31	II
360		5,206.55	II
150		5,207.90	II
360		5,219.05	II
560		5,220.11	II
110		5,227.97	I
680		5,259.73	II
180		5,263.88	II
340	c	5,292.02	II
340		5,292.62	II
230		5,298.09	II
430		5,322.76	II
200		5,352.40	II
16		5,501.50	I
40		5,508.79	I
65		5,509.15	II
16	c	5,511.63	II
55		5,513.58	II
28		5,515.12	II
13		5,519.38	II
20	c	5,520.31	II
45	c	5,522.79	II
28	c	5,524.15	I
28	c	5,525.91	II
16	c	5,527.93	I
13		5,530.21	I
45		5,531.16	I
150		5,535.17	II
28		5,538.37	I
20		5,538.78	II
55		5,545.01	II
20		5,548.33	II
11		5,553.42	II
22		5,561.46	II
45	c	5,562.06	I
13		5,565.52	I
13		5,566.91	II
45		5,571.83	II
11		5,574.61	II
11		5,578.81	I
13		5,582.35	II
11		5,584.02	I
22		5,594.92	I
22		5,597.29	II
13		5,601.30	II
90		5,605.65	II
13		5,606.68	I
28		5,608.93	II
55		5,610.22	II
11		5,620.06	II
20		5,620.26	I
45	c	5,621.89	II
110		5,623.05	II
90		5,624.45	II
11	h	5,633.03	I
22		5,636.46	II
55	c	5,638.79	II
16		5,640.37	II
16	c	5,643.16	I
22		5,645.41	II
35		5,654.23	II
55		5,659.84	II
35	h	5,661.57	I
16		5,662.19	II
65	c	5,668.46	I
45		5,669.55	II
35		5,669.99	II
16		5,674.14	II
16		5,677.03	II
55		5,681.89	II
13		5,685.60	II
16		5,686.52	I
22	h	5,687.17	II
65		5,688.44	II
22		5,689.21	II
55	h	5,690.97	II
22		5,695.90	II
22		5,704.38	I
65		5,707.61	I
40		5,711.63	II
22		5,713.83	II
16		5,716.08	II
45		5,719.08	II
45	d	5,719.63	II
		5,719.80	II
11		5,728.38	I
40		5,731.88	II
20		5,747.13	II
11		5,747.74	I
11		5,747.95	II
22		5,753.02	II
90		5,756.17	II
16		5,759.40	II
22		5,760.20	I
22		5,769.16	II
16		5,769.79	II
45		5,773.16	II
11		5,775.91	II
16		5,777.29	II
90		5,779.28	I
65	c	5,785.28	II
65		5,786.17	II
16	h	5,788.29	II
16		5,788.92	II
16		5,790.86	II
45		5,791.36	II
22		5,792.95	I
40		5,810.58	II
16		5,813.55	II
160	d	5,815.17	II
		5,815.33	II
55		5,818.57	II
40		5,820.62	II
16	h	5,821.36	I
55		5,822.59	II
90		5,823.72	II
45		5,830.94	II
40		5,835.13	I
35	c	5,844.65	II
40		5,844.98	II
65		5,847.13	II
65	c	5,850.64	II
45		5,852.63	II
11	c	5,854.44	I
45		5,856.07	II
55		5,856.90	II
90		5,859.68	II
80		5,868.83	II
22		5,873.83	II
35		5,874.72	I
35		5,878.10	I
35		5,879.04	I
80		5,879.25	II
35	c	5,884.72	I
55		5,892.23	II
22		5,894.22	II
40		5,903.11	II
45		5,904.45	II
40		5,908.67	II
11		5,915.31	I
11		5,915.97	I
40		5,920.76	I
40		5,930.66	II
16		5,936.33	II
160		5,939.90	II
65		5,940.72	II
22		5,941.65	I
35		5,947.16	II
22	c	5,949.76	I
55		5,951.27	II
20		5,951.76	II
90		5,956.60	II
		5,956.70	I
13		5,959.25	I
20		5,962.18	I
28		5,963.00	I
110		5,967.82	II
13	c	5,976.95	I
13		5,978.88	I
65		5,981.19	II
40		5,986.14	I
45	c	5,987.14	I
		5,987.29	II
13		5,991.27	I
13	c	5,994.89	I
11		5,996.06	I
29		6,002.44	II
90		6,006.33	II
13		6,008.54	I
55		6,016.48	II
150		6,017.80	II
28	c	6,019.85	I
150		6,025.72	II
35		6,042.87	II
55		6,046.66	II
35		6,049.26	I
28		6,050.04	II
11		6,050.88	I
140		6,055.13	I
13		6,067.27	II
13		6,085.81	I
28		6,086.16	II
65		6,087.52	II
20		6,090.38	II
28		6,093.09	II
18		6,096.28	I
22		6,106.72	II
18		6,109.08	I
65		6,114.38	II
22	c	6,118.02	I
22	c	6,122.15	I
35		6,141.51	II
65		6,148.23	I
		6,148.24	II
22		6,157.82	II
13		6,159.10	II
190		6,161.18	II
18		6,165.38	I
270		6,165.94	II
55		6,182.34	II
13		6,187.96	I
35		6,197.45	II
35		6,200.81	II
13		6,205.63	II
13		6,210.59	I
22		6,212.73	I
18		6,218.06	I
20	h	6,236.80	I
20	h	6,241.05	I
45		6,244.35	II
35		6,255.10	II
40		6,262.55	II
18		6,264.54	II
22	c	6,274.66	II
		6,274.81	II
40		6,278.68	II
110		6,281.28	II
18	c	6,289.02	I
11	c	6,298.01	I
11		6,302.05	I
35		6,302.35	II
16		6,304.05	I
35		6,305.23	II
11	h	6,318.13	II
45	c	6,322.36	I
22	h	6,343.88	I
28		6,347.11	II
18	c	6,350.98	I
22	c	6,357.20	I
55	c	6,359.03	I
11	h	6,363.62	II
16		6,377.61	I
16		6,378.59	I
11		6,389.57	I
18	c	6,391.99	I
40		6,393.18	I
45		6,397.96	II
10	h	6,410.69	I
55		6,411.23	I
40		6,413.68	II
10		6,415.43	I
45		6,429.63	II
45		6,431.84	II
7	h	6,442.78	II
9	h	6,443.91	II
16	c	6,453.44	I
9		6,454.84	II
9		6,456.18	I
9	h	6,460.19	I
18		6,467.72	II
9	h	6,475.26	II
35	c	6,478.02	II
45		6,486.55	I
9	h	6,486.97	II
40	h	6,491.75	I
9		6,493.49	I
11		6,494.89	I
22	c	6,497.11	I
18		6,498.94	II
22		6,500.72	I
9		6,504.09	I
8		6,517.14	I
16		6,518.79	II
8		6,534.52	I
16		6,540.47	I
7	h	6,553.30	I
22		6,564.62	II
45		6,566.77	II
7		6,571.03	I
6		6,578.00	I
6		6,584.56	II
9	h	6,593.74	II
11		6,595.48	I
15		6,609.86	I
55		6,616.67	I
11		6,618.34	II
7	h	6,631.00	I
13	h	6,632.06	I
14		6,647.12	I
75		6,656.83	II
55		6,673.41	II
75		6,673.78	II

5	h	6,687.51	II
4	h	6,699.25	I
13		6,736.79	I
35	c	6,747.09	I
19	c	6,749.19	I
7	c	6,784.99	II
55	c	6,798.60	I
11		6,811.76	II
17	c	6,812.87	II
13		6,814.04	II
9		6,817.61	I
35	c	6,827.60	II
19		6,830.50	II
9		6,844.39	I
9	h	6,845.47	II
9		6,846.59	II
17	c	6,850.46	II
11		6,852.77	I
11	c	6,870.44	I
7		6,884.66	I
8		6,892.71	I
8	h	6,970.38	I
8	c	6,980.12	I
40		7,021.51	II
10		7,024.53	I
13		7,042.40	I
8		7,044.45	II
7		7,051.07	I
10		7,079.99	I
11	c	7,095.18	I
20		7,114.55	I
10	h	7,116.90	I
11		7,118.24	II
7		7,137.33	II
10	h	7,159.88	I
7		7,167.77	II
7	h	7,189.95	I
10	c	7,208.85	II
24		7,227.70	II
13		7,231.53	I
7	c	7,243.26	I
7	c	7,259.21	I
7	c	7,287.61	I
7	h	7,289.19	I
7	c	7,324.42	I
7		7,328.47	I
7		7,344.86	
16		7,407.56	II
20	c	7,451.74	II
11	h	7,495.59	I
6	h	7,499.42	I
14		7,541.02	II
6		7,574.86	I
20		7,645.66	II
7		7,704.98	II
16		7,721.84	I
6	h	7,786.16	II
6	c	7,841.27	I
14		7,871.67	I
6		7,881.09	I
6	c	7,888.56	II
6		7,915.19	II
6		8,031.92	I
6		8,055.43	I
14		8,067.44	I
10	c	8,122.78	II
11		8,141.10	I
5		8,181.34	II
5	c	8,211.93	I
6		8,289.93	I
6		8,379.84	I
6	h	8,427.82	I
6	h	8,605.27	II
10		8,714.59	II

Pr III

Ref. 306, 308 — J.R.

Intensity		Wavelength	
		Vacuum	
25		1,008.61	III
50		1,021.35	III
25		1,026.18	III
100		1,029.03	III
50		1,038.29	III
50		1,042.96	III
25		1,043.80	III
25		1,044.03	III
25		1,046.20	III
150		1,047.24	III
100		1,049.09	III
50		1,052.63	III
25		1,061.60	III
25		1,066.03	III
25	p	1,068.85	III
25		1,069.88	III
25		1,084.42	III
25		1,088.66	III
100		1,104.84	III
25		1,108.82	III
30		1,352.70	III
25		1,881.22	III
		Air	
50		2,031.46	III
100		2,033.30	III
25		2,043.12	III
100		2,052.30	III
50		2,052.87	III
50		2,053.85	III
200		2,058.59	III
50		2,064.08	III
25		2,075.08	III
200		2,090.75	III
200		2,093.49	III
25		2,096.85	III
50		2,096.94	III
10	w	2,148.14	III
10	w	2,194.24	III
10	w	2,197.25	III
10	w	2,205.48	III
10	w	2,206.26	III
10	w	2,214.45	III
10	w	2,215.25	III
10	w	2,217.12	III
10	w	2,223.23	III
10	w	2,230.35	III
10	w	2,237.26	III
10		2,239.06	III
10		2,239.42	III
10	w	2,242.15	III
10	w	2,284.62	III
10		2,307.59	III
10		2,307.77	III
10		2,308.41	III
10		2,311.29	III
10		2,311.44	III
10		2,314.18	III
10	w	2,315.46	III
10		2,318.15	III
10		2,318.36	III
10		2,318.64	III
10		2,318.82	III
10		2,318.97	III
10	w	2,319.40	III
10		2,320.41	III
10	w	2,328.56	III
10	w	2,336.13	III
10		2,365.52	III
10		2,368.78	III
10		2,369.08	III
10	w	2,378.06	III
10	w	2,378.97	III
10		2,395.44	III
10		2,399.70	III
10	w	2,405.56	III
10		2,408.19	III
10		2,409.80	III
10		2,412.40	III
10		2,417.69	III
10		2,418.95	III
10		2,426.14	III
10		2,426.85	III
10	w	2,430.32	III
10	w	2,434.18	III
10		2,434.39	III
10		2,435.91	III
10		2,436.89	III
10		2,438.63	III
10	w	2,444.93	III
10		2,445.49	III
10		2,446.77	III
10		2,448.16	III
10		2,452.02	III
10	w	2,452.81	III
10	w	2,452.85	III
10		2,454.60	III
10		2,454.82	III
10	w	2,459.77	III
10	w	2,460.72	III
10		2,462.18	III
10		2,462.90	III
10		2,468.20	III
10		2,468.97	III
10	w	2,473.42	III
10	w	2,478.32	III
10		2,479.98	III
10		2,481.02	III
10		2,483.30	III
10		2,483.99	III
10	w	2,484.60	III
10	w	2,485.16	III
10	w	2,488.72	III
10		2,491.97	III
10	w	2,494.20	III
10	w	2,495.37	III
10	w	2,495.51	III
10		2,499.97	III
20	w	2,587.71	III
40	w	2,624.91	III
20	w	2,644.62	III
20	w	2,656.88	III
20	w	2,667.51	III
70	w	2,679.47	III
40	w	2,710.30	III
20	w	2,718.65	III
100	w	2,724.03	III
20	l	2,841.94	III
70	s	2,910.61	III
50	s	2,911.77	III
100	l	2,914.49	III
50	s	2,930.19	III
70	s	2,942.43	III
70	s	2,953.58	III
90	w	2,954.40	III
90	s	2,964.85	III
150	s	2,968.83	III
80	s	2,969.41	III
150	l	2,976.86	III
150	s	2,977.06	III
500	s	2,980.54	III
100	l	2,981.65	III
150	c	2,982.42	III
500	s	2,985.82	III
150	s	2,997.12	III
70	l	2,998.79	III
150	l	3,000.46	III
150	s	3,003.20	III
60	l	3,006.47	III
150	s	3,008.04	III
150	l	3,010.61	III
90	s	3,014.60	III
100	l	3,015.13	III
90	s	3,016.26	III
70	s	3,021.77	III
70	s	3,025.26	III
100	l	3,029.38	III
100	l	3,033.31	III
90	l	3,034.25	III
70		3,040.02	III
60	c	3,040.94	III
60	l	3,041.78	III
100	s	3,042.35	III
100	l	3,045.81	III
70	l	3,046.98	III
120	l	3,050.30	III
70	s	3,055.30	III
150	l	3,058.90	III
150	l	3,066.71	III
70	s	3,078.68	III
150	l	3,080.20	III
50	w	3,248.39	III
90	s	3,280.92	III
90	l	3,292.58	III
90	s	3,296.10	III
100	l	3,306.14	III
60	s	3,333.26	III
90	s	3,340.58	III
500	s	3,341.43	III
70	s	3,341.68	III
70	s	3,345.38	III
70	l	3,345.44	III
50	w	3,351.07	III
50	s	3,353.87	III
50	s	3,365.80	III
200	s	3,367.35	III
500	s	3,367.58	III
100	l	3,371.92	III
100	s	3,377.14	III
50	s	3,379.13	III
150	s	3,380.21	III
150	s	3,381.26	III
300	s	3,381.84	III
100	s	3,391.08	III
1,000	w	3,394.22	III
600	d	3,396.07	III
300	s	3,396.62	III
300	l	3,397.46	III
50	l	3,402.97	III
500	c	3,413.21	III
300	s	3,415.15	III
150	s	3,420.07	III
300	l	3,422.22	III
50	c	3,426.27	III
500	l	3,427.02	III
300	l	3,436.36	III
150	l	3,440.62	III
50	l	3,445.29	III
70	s	3,454.05	III
180		3,653.58	III
60		3,817.25	III
60	l	3,861.80	III
150		3,980.51	III
200		4,000.20	III
90		4,018.36	III
180		4,029.60	III
90		4,142.46	III
120		4,144.48	III
90		4,147.85	III
90		4,172.15	III
150		4,179.77	III
180		4,184.18	III
240		4,197.01	III
120		4,219.45	III
180		4,231.45	III
180		4,275.07	III
120		4,286.32	III
90		4,298.27	III
90		4,301.73	III
90		4,316.34	III
60		4,354.28	III
90		4,379.82	III
90		4,381.47	III
120		4,404.71	III
120		4,421.10	III
120		4,431.85	III
150	w	4,447.93	III
180	w	4,450.14	III
120	w	4,451.00	III
120		4,461.02	III
200		4,461.81	III
300	w	4,500.31	III
450	w	4,612.02	III
600	w	4,625.18	III
120		4,654.16	III
600	w	4,713.70	III
300	w	4,725.55	III
270		4,728.21	III
300		4,747.11	III
300	w	4,771.83	III
450	w	4,775.30	III
600		4,857.39	III
150		5,208.51	III
150		5,261.68	III
1,000		5,264.44	III
1,500		5,284.70	III
1,500		5,299.99	III
1,500		5,340.02	III
100		5,427.70	III
100		5,581.74	III
150		5,646.80	III
600		5,765.27	III
1,500		5,844.41	III
200		5,947.98	III
7,000	w	5,956.05	III
900	w	5,998.94	III
1,500	w	6,053.01	III
900		6,071.09	III

Intensity		Wavelength	
9,000	w	6,090.02	III
5,000		6,160.24	III
1,500	w	6,161.22	III
100		6,195.05	III
2,000		6,195.63	III
200		6,310.36	III
100		6,361.65	III
300		6,429.26	III
300		6,444.74	III
600		6,500.04	III
300		6,501.49	III
200		6,578.90	III
100		6,616.46	III
600		6,706.70	III
100		6,727.63	III
200		6,827.96	III
100		6,854.63	III
200		6,857.30	III
1,000		6,866.80	III
1,000		6,899.06	III
500		6,903.52	III
7,000		6,910.14	III
150	w	6,934.55	III
500		6,970.96	III
100		6,979.83	III
5,000		7,030.39	III
100		7,075.21	III
4,500		7,076.62	III
100		7,083.99	III
500		7,112.53	III
100	w	7,165.64	III
250		7,231.62	III
100	w	7,238.26	III
250		7,240.21	III
100	w	7,262.32	III
150	w	7,340.69	III
350		7,343.70	III
200		7,349.75	III
300	w	7,350.61	III
100	w	7,355.52	III
2,000		7,426.48	III
4,000		7,429.05	III
100		7,463.96	III
250		7,487.40	III
200		7,493.20	III
100		7,511.17	III
500		7,529.11	III
100	w	7,549.20	III
150	w	7,588.64	III
300		7,596.41	III
100	w	7,625.63	III
100	w	7,648.34	III
100	w	7,670.65	III
200	w	7,674.65	III
500		7,742.34	III
250	w	7,745.59	III
500		7,754.31	III
100	w	7,755.48	III
3,000		7,781.98	III
200	w	7,814.74	III
1,500	w	7,866.14	III
1,000		7,888.12	III
400		7,897.09	III
1,000		7,914.00	III
100		7,923.16	III
1,000	w	7,972.75	III
250		8,001.14	III
3,000		8,102.90	III
250		8,119.54	III
400		8,132.23	III
100		8,138.34	III
150	w	8,235.33	III
250	w	8,244.89	III
100		8,409.10	III
100	w	8,494.99	III
200	w	8,567.63	III
5,000	w	8,602.74	III
500		8,691.58	III
500		8,771.38	III
1,000		8,854.05	III
125		8,886.17	III
100		8,908.70	III
250		9,099.98	III
250		9,131.90	III
200		9,222.32	III
250	w	9,265.56	III
125		9,320.54	III
250		9,334.33	III
175		9,377.44	III
175		9,388.56	III
175		9,549.77	III
100		9,579.74	III
150		9,802.98	III
175		9,806.37	III
500	w	9,991.16	III
500		10,031.10	III
500		10,160.33	III
500		10,238.63	III
500		10,301.58	III
500		10,324.59	III
500		10,716.58	III

Pr IV

Ref. 337, 338 — J.R.

Intensity		Wavelength	
		Vacuum	
20		718.23	IV
30		721.34	IV
60		722.41	IV
30		722.58	IV
20		726.04	IV
50		730.37	IV
30		731.77	IV
30		734.86	IV
20		735.04	IV
20		736.19	IV
50		736.32	IV
100		737.17	IV
100		741.45	IV
20		743.15	IV
20		743.89	IV
20		746.14	IV
40		763.16	IV
20		764.00	IV
300		1,226.40	IV
2,000		1,228.59	IV
500		1,230.69	IV
400		1,238.19	IV
200		1,249.35	IV
200		1,255.64	IV
200		1,261.27	IV
400		1,268.32	IV
300		1,270.58	IV
1,000		1,275.10	IV
200		1,275.40	IV
1,000		1,278.65	IV
200		1,279.34	IV
1,000		1,287.44	IV
300		1,290.93	IV
1,000		1,292.30	IV
5,000		1,293.22	IV
5,000		1,295.28	IV
400		1,296.50	IV
200		1,298.26	IV
300		1,298.54	IV
200		1,304.71	IV
300		1,306.86	IV
200		1,308.08	IV
200		1,310.71	IV
500		1,314.96	IV
300		1,315.28	IV
300		1,316.96	IV
500		1,320.10	IV
1,000		1,320.70	IV
5,000		1,321.36	IV
500		1,322.51	IV
500		1,326.38	IV
5,000		1,333.57	IV
300		1,335.96	IV
500		1,339.29	IV
1,000		1,340.74	IV
200		1,341.32	IV
300		1,344.23	IV
1,000		1,347.07	IV
1,000		1,352.81	IV
500		1,354.35	IV
5,000		1,354.66	IV
2,000		1,360.64	IV
1,000		1,364.81	IV
2,000		1,365.77	IV
400		1,368.90	IV
5,000		1,374.41	IV
1,000		1,382.62	IV
200		1,384.23	IV
200		1,385.91	IV
300		1,394.11	IV
500		1,397.11	IV
1,000		1,399.31	IV
1,000		1,400.96	IV
400		1,410.90	IV
1,000		1,424.36	IV
500		1,426.59	IV
5,000		1,435.56	IV
200		1,459.95	IV
200		1,461.76	IV
1,000		1,474.91	IV
200		1,477.32	IV
500		1,485.88	IV
500		1,503.35	IV
400		1,516.86	IV
200		1,520.71	IV
2,000		1,520.98	IV
400		1,523.46	IV
500		1,553.62	IV
500		1,559.49	IV
500		1,570.13	IV
200		1,572.80	IV
5,000		1,574.55	IV
5,000		1,575.10	IV
3,000		1,578.38	IV
500		1,585.10	IV
300		1,613.00	IV
400		1,613.65	IV
1,000		1,618.03	IV
2,000		1,622.30	IV
300		1,634.77	IV
400		1,676.08	IV
200		1,688.49	IV
200		1,713.53	IV
500		1,732.86	IV
300		1,762.86	IV
1,000		1,766.88	IV
1,000		1,771.14	IV
500		1,841.08	IV
10,000		1,884.87	IV
400		1,951.23	IV
200		1,954.61	IV
		Air	
200		2,025.06	IV
1,000		2,039.15	IV
200		2,047.05	IV
200		2,050.73	IV
200		2,058.48	IV
2,000		2,083.23	IV
500		2,100.42	IV
1,000		2,154.31	IV
300		2,193.37	IV
1,000		2,205.13	IV
300		2,265.70	IV
200	c	2,334.46	IV
200	c	2,339.08	IV
500	c	2,376.09	IV
2,000	c	2,378.98	IV
1,000	c	2,379.66	IV
500	c	2,427.07	IV
500	c	2,428.13	IV
500	c	2,438.57	IV
500	c	2,455.64	IV
500	c	2,705.19	IV
200	c	2,708.01	IV
200	c	2,753.47	IV
200	c	2,767.60	IV

Pr V

Ref. 149 — J.R.

Intensity		Wavelength	
		Vacuum	
200		843.78	V
7,000		865.90	V
5,000		869.17	V
80		869.66	V
1,000		896.65	V
750		922.29	V
250		1,234.07	V
250		1,342.78	V
200		1,958.09	V
400		1,958.20	V
		Air	
300		2,246.06	V
300		2,246.20	V

PROMETHIUM (PM)

Z = 61

Pm I and II

Ref. 196, 260 — C.H.C.

Intensity		Wavelength	
		Air	
40	w	2,502.12	II
40		2,608.24	II
150		2,632.00	II
70		2,638.46	II
100		2,671.05	II
50	w	2,787.72	II
40		2,808.05	II
100	h	2,820.10	II
100	w	2,840.82	II
150	w	2,841.86	II
200	c	2,857.46	II
100		3,004.59	II
100		3,008.85	II
300		3,072.41	II
150		3,086.02	II
120		3,090.19	II
150		3,091.86	II
150		3,108.11	II
100		3,115.36	II
100		3,117.22	II
100		3,118.76	II
35		3,162.23	I
35		3,168.82	I
100		3,172.77	II
35		3,222.04	I
35		3,238.55	I
60		3,239.62	II
75		3,296.63	I
60		3,311.76	II
50		3,313.38	I
50		3,329.22	I
75		3,331.57	I
100		3,354.45	I
100		3,358.14	II
80		3,360.21	II
80		3,364.44	II
300		3,366.03	I
90		3,377.68	II
100		3,391.28	II
100		3,408.06	II
500		3,427.40	II
120		3,441.15	II
400		3,449.80	II
250		3,460.25	II
200		3,462.91	II
200		3,480.61	II
150		3,497.13	II
100		3,514.85	II
200		3,546.81	I
100		3,559.43	II
200		3,565.31	II
150		3,580.10	II
200		3,610.76	II
200		3,629.84	II
300		3,634.20	II
300		3,659.39	II
300	r	3,669.22	I
200		3,674.85	I
200		3,678.51	II
300	r	3,679.85	I
200		3,687.65	II
400		3,689.79	II
300		3,692.50	II
300		3,697.50	II
300	r	3,697.63	I
400		3,702.63	II
800		3,711.72	II
200		3,715.75	II
200		3,721.72	II
500		3,726.01	I
200		3,738.43	I
300		3,740.68	I
300		3,742.52	II
300	r	3,742.97	I
500		3,745.86	II
300		3,747.09	II

500		3,750.09	II
200		3,761.68	I
300		3,765.75	I
300		3,775.42	I
300		3,780.77	I
200		3,781.43	I
400		3,795.66	II
250		3,806.06	II
300	r	3,809.20	I
400		3,810.93	I
200		3,819.26	II
300		3,820.53	II
300		3,839.52	I
200		3,842.88	II
300		3,842.98	II
250		3,845.38	II
300		3,874.03	I
800		3,877.62	II
300	r	3,885.79	I
250		3,890.97	I
1,000		3,892.15	II
300		3,898.73	I
400		3,899.78	II
250		3,909.50	II
1,000		3,910.26	II
1,000		3,919.10	II
800		3,936.48	II
300		3,944.21	II
300	r	3,954.76	I
1,000		3,957.74	II
500		3,980.74	II
300		3,995.05	II
1,000	r	3,998.96	II
500		4,009.96	II
200		4,012.72	II
250		4,014.20	II
200		4,019.34	II
250		4,028.20	II
200		4,045.36	II
300		4,051.54	II
600	r	4,055.20	II
200	r	4,056.56	I
600		4,075.84	II
200	r	4,085.31	I
500		4,086.10	II
250		4,140.46	II
200		4,185.74	II
300		4,192.92	II
200		4,194.70	II
200		4,222.15	II
300	r	4,264.32	I
300	r	4,284.37	I
600		4,297.78	II
200		4,303.89	II
200	r	4,305.64	I
400		4,318.80	I
250		4,325.92	II
200		4,332.05	II
300		4,336.54	II
200		4,337.48	II
300		4,342.12	II
200		4,347.72	I
350	r	4,363.92	I
300	r	4,369.64	I
200		4,381.88	II
400	r	4,388.49	I
200		4,388.76	II
400	r	4,409.42	I
500	r	4,412.47	I
1,000		4,417.96	II
400		4,432.51	II
250	r	4,435.86	I
300	r	4,436.55	I
300	r	4,438.68	I
500		4,445.41	II
600		4,446.90	II
800		4,453.95	II
200		4,459.97	II
250	r	4,468.16	I
200		4,471.48	II
300		4,473.23	II
200		4,477.46	II
350	r	4,478.58	I
300	r	4,481.60	I
300	r	4,485.05	I
300	r	4,490.50	I
250		4,492.05	II
600		4,500.15	II
350	r	4,500.33	I
250		4,506.84	I
100		4,509.38	II
100		4,513.56	II
200		4,517.31	I
200		4,523.32	I
600		4,525.20	II
250	r	4,526.12	I
250	r	4,526.76	I
400	r	4,527.70	I
800		4,529.21	II
300		4,540.06	I
300	r	4,541.42	I
450	r	4,541.75	I
500	r	4,544.08	I
200		4,545.17	I
400	r	4,549.78	I
300	r	4,554.03	I
200		4,554.63	I
500	r	4,555.34	I
200		4,556.06	I
300		4,557.03	I
300		4,559.21	I
100		4,564.83	II
300	r	4,568.14	I
200		4,570.37	I
300	r	4,572.15	I
400	r	4,575.27	I
300		4,578.28	I
200	r	4,578.41	I
300	r	4,579.48	I
300		4,581.14	I
300	r	4,585.49	I
200		4,593.82	I
400	r	4,595.82	I
800		4,597.55	I
500	r	4,600.25	I
400	r	4,602.96	I
400		4,604.59	I
600	r	4,605.66	I
500	r	4,609.85	I
100		4,615.87	II
600	r	4,617.02	I
200		4,618.40	I
400	r	4,618.49	I
500	r	4,619.75	I
500		4,621.57	I
500		4,623.31	I
700	r	4,623.68	I
900		4,624.41	I
500	r	4,625.29	I
400	r	4,627.60	I
200		4,630.93	I
600	r	4,633.45	I
400	r	4,640.96	I
700	r	4,643.36	I
700	r	4,643.76	I
400		4,645.94	I
600	r	4,647.03	I
600	r	4,650.42	I
500		4,650.52	I
400	r	4,653.41	I
400		4,654.50	I
500	r	4,655.05	I
300		4,659.38	I
500	r	4,660.79	I
300	r	4,663.26	I
600	r	4,663.46	I
400	r	4,665.19	I
500	r	4,671.23	I
400		4,671.76	I
500	r	4,674.42	I
200		4,677.46	I
500	r	4,677.92	I
400	r	4,678.09	I
700	r	4,682.92	I
500	r	4,696.80	I
200		4,699.51	I
250		4,722.06	II
300		4,727.06	I
900	r	4,728.36	I
400	r	4,728.68	I
800	r	4,734.27	I
200	r	4,737.99	I
100		4,739.08	II
200		4,739.78	I
350	r	4,745.13	I
500	r	4,757.73	I
800	r	4,759.00	I
700	r	4,762.57	I
700	r	4,773.46	I
900	r	4,781.29	I
250		4,794.59	I
200		4,795.43	I
700	r	4,798.98	I
900	r	4,801.36	I
700	r	4,809.54	I
900	r	4,811.96	I
400	r	4,817.12	I
400	r	4,827.72	I
800	r	4,837.66	I
400	r	4,838.92	I
300		4,839.62	I
200		4,844.01	I
350	r	4,852.73	I
400	r	4,860.62	I
700	r	4,860.74	I
300	r	4,865.30	I
500	r	4,865.72	I
400	r	4,869.80	I
700	r	4,872.42	I
500	r	4,887.02	I
700	r	4,892.52	I
400	r	4,900.30	I
300	r	4,904.28	I
400	r	4,918.28	I
600	r	4,932.99	I
700	r	4,959.46	I
100		4,971.40	II
500	r	4,997.10	I
200	r	5,030.80	I
300	r	5,058.31	I
100		5,067.35	II
150		5,080.52	II
150		5,089.35	II
200		5,092.42	I
400	r	5,094.83	I
200		5,096.18	I
150		5,097.30	II
400	r	5,100.77	I
250		5,121.47	II
400	r	5,127.34	I
200		5,129.75	I
400	r	5,145.13	I
500	r	5,146.30	I
400		5,153.86	II
300		5,169.71	II
500		5,171.58	II
300		5,194.05	II
500		5,208.09	II
150		5,215.96	II
250		5,225.12	II
500		5,236.26	II
300		5,236.66	II
400		5,246.33	II
150		5,262.42	II
500		5,270.64	II
200		5,293.92	II
100		5,308.86	II
150		5,318.58	II
200		5,410.45	II
200		5,424.54	II
180		5,424.79	II
150		5,429.04	II
100		5,467.64	II
150		5,495.45	II
100		5,516.42	II
180		5,534.96	II
200		5,537.38	I
800		5,546.08	II
120		5,556.88	II
150		5,558.39	II
200		5,561.73	II
800		5,576.02	II
200		5,641.29	II
200		5,730.81	I
200		5,768.16	II
200		5,776.99	I
500		5,823.93	II
300	c	5,868.79	II
200	c	5,875.31	II
100	c	5,878.76	II
150		5,899.76	II
250		5,904.71	I
100		5,905.90	I
125		5,914.96	I
250	c	5,927.17	II
150		5,939.66	I
400	c	5,946.49	II
800		5,956.42	I
200		5,956.69	I
100	c	5,960.08	II
150		5,963.00	II
400		5,967.89	I
200		5,979.73	I
200		5,984.82	I
100	c	5,987.13	II
400		5,997.12	I
200		6,027.11	II
300		6,030.06	I
400		6,031.32	I
500		6,043.39	I
150	c	6,052.57	II
100	c	6,067.00	II
500		6,069.06	I
100		6,076.40	II
200		6,085.41	II
900		6,100.21	I
400		6,106.40	I
100		6,114.90	II
400	h	6,151.76	I
100		6,159.53	II
400		6,163.16	I
100		6,184.52	II
200		6,208.91	II
500		6,229.64	I
400		6,237.79	I
100		6,263.25	II
400		6,272.69	I
400		6,286.06	I
500		6,308.29	I
100		6,314.20	II
700		6,323.84	I
500		6,390.31	I
100		6,429.64	II
500	h	6,431.93	I
100		6,436.57	II
400		6,487.61	I
400		6,510.34	I
500		6,517.25	I
200		6,519.43	II
1,000	d	6,520.45	I
500		6,542.20	I
100	h	6,558.48	II
100		6,586.39	II
100		6,592.29	II
900		6,598.15	I
800		6,598.66	I
700		6,606.37	I
800	w	6,625.23	I
100	h	6,625.54	II
700		6,649.81	I
400		6,659.05	II
100		6,661.25	II
500		6,661.68	I
400		6,663.76	I
800	c	6,667.51	I
700	h	6,677.47	I
200		6,680.89	II
500		6,685.55	I
500		6,685.68	I
150		6,690.09	II
600		6,700.33	I
100		6,706.27	II
700		6,714.67	I
500		6,717.26	I
500		6,720.71	I
700		6,727.50	I
600		6,743.71	I
900		6,749.91	I
900		6,750.48	I
200		6,756.45	II
300		6,772.29	II
400		6,778.78	I
100		6,783.09	II
100		6,796.87	II
200		6,811.68	II
800		6,833.30	I
400		6,848.37	I
50		6,858.58	II

PROTACTINIUM (PA)

Z = 91

Pa I and II

Ref. 96 — J.G.C.

Intensity		Wavelength	
		Air	
3,000	h	2,466.85	
3,000	h	2,492.85	
3,000		2,599.16	II
3,000		2,699.22	II
3,000		2,822.79	II
3,000	l	2,832.14	
3,000		2,870.01	
3,000	h	2,871.42	II
3,000	h	2,891.14	II
3,000	h	2,906.93	
3,000	l	3,011.10	II
3,000	s	3,033.59	II
3,000	l	3,071.24	II
3,000	h	3,083.19	
3,000	l	3,093.23	II
3,000	l	3,126.23	II
3,000	l	3,146.28	II
3,000	l	3,170.89	II
3,000	l	3,171.54	II
3,000		3,204.16	
3,000	l	3,240.58	II
3,000		3,274.46	II
3,000	l	3,332.69	II
3,000	s	3,346.66	II
3,000	l	3,394.49	
3,000	l	3,452.82	II
3,000		3,504.97	I
3,000	s	3,530.65	II
3,000		3,570.56	I
3,000		3,571.82	I
3,000		3,618.07	I
10,000		3,636.52	I
3,000		3,702.74	I
3,000		3,752.67	I
3,000		3,873.35	I
3,000		3,931.83	I
3,000	s	3,952.62	II
10,000	l	3,957.85	II
3,000	s	3,970.07	II
3,000		3,981.82	I
10,000		3,982.23	I
3,000	l	4,012.96	II
3,000	s	4,018.21	II
3,000		4,030.16	II
3,000	s	4,046.93	II
10,000	s	4,056.20	II
10,000	s	4,070.40	II
3,000		4,117.62	
3,000	l	4,176.18	II
10,000	l	4,217.23	II
10,000	s	4,248.08	II
3,000	s	4,291.34	II
3,000		4,400.77	
3,000		4,436.13	
3,000	s	4,601.43	II
3,000		4,628.19	
3,000		4,820.34	
3,000	s	4,861.49	
3,000	l	6,035.78	I
3,000		6,162.56	I
3,000		6,216.35	
3,000	l	6,358.61	I
3,000		6,379.25	I
3,000	l	6,438.97	I
3,000	h	6,792.75	I
10,000		6,945.72	I
3,000		6,960.09	I
3,000	h	6,961.78	I
3,000	s	6,992.73	I
3,000		7,076.27	I
3,000	h	7,100.94	I
10,000	s	7,114.89	I
3,000	h	7,171.55	I
3,000		7,227.13	I
3,000		7,318.79	I
10,000	l	7,368.25	I
3,000	h	7,471.89	I
10,000	h	7,493.15	I
3,000	h	7,558.26	I
10,000	h	7,608.20	I
10,000		7,626.79	I
10,000	s	7,635.18	I
10,000		7,669.34	I
3,000		7,679.20	I
10,000	h	7,749.19	I
3,000		7,872.95	I
3,000	l	7,945.56	I
10,000		8,039.34	I
10,000	h	8,099.84	I
10,000		8,199.04	I
10,000		8,271.87	I
3,000	s	8,358.98	I
3,000	s	8,369.60	I
3,000	h	8,441.04	I
10,000	h	8,532.66	I
10,000	s	8,572.96	I
3,000	h	8,639.91	I
3,000	h	8,653.51	I
10,000		8,735.27	I
3,000		10,594.38	
3,000		10,923.32	I
3,000		11,646.78	
10,000		11,791.73	I
3,000		12,279.01	
3,000		13,234.09	
10,000		13,522.40	
10,000		14,344.76	I
3,000		18,478.61	I

RADIUM (RA)

Z = 88

Ra I and II

Ref. 253, 254 — E.F.W.

Intensity		Wavelength	
		Air	
8		2,369.73	II
8		2,460.55	II
10		2,475.50	II
8		2,586.61	II
10		2,643.73	II
20		2,708.96	II
10		2,795.21	II
30		2,813.76	II
10		3,033.44	II
5		3,101.80	I
100		3,649.55	II
200		3,814.42	II
8		4,194.09	II
8		4,244.72	II
100		4,340.64	II
20		4,436.27	II
30		4,533.11	II
8		4,641.29	I
100		4,682.28	II
8		4,699.28	I
100		4,825.91	I
10		4,856.07	I
10		4,859.41	II
10		4,927.53	II
10		5,097.56	I
10		5,205.93	I
10		5,283.28	I
10		5,320.29	I
10		5,399.80	I
20		5,400.23	I
20		5,406.81	I
8		5,482.13	I
10		5,501.98	I
10		5,553.57	I
20		5,555.85	I
10		5,616.66	I
50		5,660.81	I
20		5,813.63	II
30		6,200.30	I
10	p	6,336.90	I
20		6,446.20	I
20		6,487.32	I
10		6,593.34	II
10		6,719.32	II
20		6,980.22	I
20		7,118.50	I
50		7,141.21	I
20		7,225.16	I
10		7,310.27	I
20		7,838.12	I
50		8,019.70	II
6		8,177.31	I
5		8,335.07	I
5		9,932.21	I

RADON (RN)

Z = 86

Rn I

Ref. 251 — E.F.W.

Intensity	Wavelength	
	Air	
5	3,514.60	I
10	3,739.89	I
20	3,753.65	I
10	3,917.20	I
10	3,941.72	I
10	3,952.36	I
10	4,226.06	I
80	4,307.76	I
7	4,335.78	I
100	4,349.60	I
40	4,435.05	I
50	4,459.25	I
50	4,508.48	I
50	4,577.72	I
50	4,609.38	I
30	4,721.76	I
6	5,722.58	I
10	6,061.92	I
6	6,200.75	I
6	6,380.45	I
10	6,557.49	I
10	6,606.43	I
15	6,627.23	I
6	6,669.60	I
8	6,704.28	I
20	6,751.81	I
6	6,806.79	I
8	6,836.95	I
8	6,837.57	I
10	6,891.16	I
10	6,998.90	I
200	7,055.42	I
100	7,268.11	I
20	7,291.00	I
6	7,320.98	I
10	7,419.04	I
300	7,450.00	I
8	7,470.89	I
8	7,483.13	I
8	7,514.13	I
8	7,516.92	I
6	7,523.93	I
6	7,597.55	I
8	7,601.28	I
10	7,657.48	I
10	7,738.43	I
20	7,746.64	I
100	7,809.82	I
20	8,049.00	I
100	8,099.51	I
6	8,173.84	I
100	8,270.96	I
8	8,314.51	I
6	8,349.74	I
10	8,381.05	I
10	8,487.48	I
10	8,494.89	I
20	8,520.95	I
100	8,600.07	I
10	8,639.76	I
15	8,675.83	I
10	8,807.75	I
50	9,327.02	I
6	9,948.57	I
5	10,106.13	I

RHENIUM (RE)

Z = 75

Re I and II

Ref. 1 — C.H.C.

Intensity		Wavelength	
		Air	
25,000		2,003.53	I
16,000		2,017.87	I
27,000		2,049.08	I
4,200		2,074.70	I
3,700		2,083.92	I
10,000		2,085.59	I
4,700		2,092.41	II
9,800		2,097.12	I
2,700		2,109.22	I
3,400		2,139.04	II
1,600		2,142.74	II
		2,142.97	I
3,700		2,156.67	I
4,900		2,167.94	I
3,400		2,176.21	I
4,200	c	2,214.26	II
2,200		2,214.58	I
1,700		2,226.42	I
920		2,235.44	I
440		2,255.73	I
860		2,256.19	I
2,000		2,264.39	I
2,100		2,274.62	I
5,200	c	2,275.25	II
1,600		2,281.62	I
2,900		2,287.51	I
2,700		2,294.49	I
390		2,298.09	II
390		2,299.77	I
610		2,302.99	I
680		2,306.54	I
230		2,312.97	I
220		2,313.34	I
220		2,319.19	I
370		2,320.16	I
800		2,322.49	I
300		2,328.66	I
270		2,334.33	I
270		2,335.73	I
220		2,336.10	I
270		2,337.95	I
860		2,344.78	I
230		2,349.39	I
220	d	2,350.46	I
680		2,352.07	I
210	d	2,353.95	I
250		2,356.50	I
200		2,365.32	I
1,200		2,365.90	I
570		2,367.68	I
180		2,368.53	II
520		2,369.27	I
220		2,370.76	II
210		2,371.52	I
150		2,373.48	II
320		2,375.07	I
75		2,378.53	II
370		2,379.77	I
180		2,386.90	II
340		2,388.57	I
230		2,393.65	I
320		2,394.37	I
320		2,396.79	I
200		2,397.31	I
210	d	2,400.72	I
		2,400.89	I
210		2,401.68	I
75		2,403.04	II
1,500		2,405.06	I
740		2,405.60	I
320		2,406.70	I
270		2,410.37	I
60		2,418.20	II
1,200		2,419.81	I
300		2,421.73	I
300		2,421.88	I
60		2,423.84	II
2,500		2,428.58	I
490		2,431.54	I
420		2,432.18	I
340	c	2,441.47	I
230		2,442.51	I
250		2,444.94	I
610		2,446.98	I
85		2,449.03	II
85		2,449.52	II
610		2,449.71	I
200		2,455.83	II
390		2,461.20	I
800	c	2,461.84	II

200		2,467.57	II
120		2,467.85	II
150	c	2,469.36	II
120		2,470.61	II
75		2,471.05	II
150		2,473.72	II
160		2,475.17	II
75		2,477.43	II
200		2,479.02	I
1,200		2,483.92	I
390		2,485.81	I
980		2,487.33	I
75		2,490.16	II
200		2,492.84	I
370		2,496.04	I
200		2,498.22	I
370		2,501.72	I
570		2,502.35	II
230		2,504.60	II
270		2,505.94	I
1,800	c	2,508.99	I
570		2,520.01	I
540		2,521.50	I
150		2,534.10	II
370		2,534.80	I
570		2,540.51	I
740	d	2,544.74	I
370		2,545.48	I
160		2,550.09	II
300		2,552.02	I
150	c	2,553.59	II
370		2,554.63	I
1,000		2,556.51	I
250		2,559.08	I
340		2,564.19	I
540		2,568.64	II
370		2,571.81	I
380		2,586.79	I
290		2,599.86	I
290		2,603.89	I
660		2,608.50	II
610	d	2,611.54	I
160	c	2,616.72	II
200		2,622.76	I
310		2,635.83	II
550		2,636.64	I
190		2,637.01	II
90		2,641.02	II
270		2,642.75	I
65		2,648.46	II
270		2,649.05	I
660		2,651.90	I
400		2,654.12	I
220		2,663.63	I
940		2,674.34	I
220		2,688.53	I
1,300		2,715.47	I
200		2,731.56	II
220		2,732.21	I
610		2,733.04	II
110	h	2,753.64	II
220		2,758.00	I
210		2,763.79	I
200		2,766.39	I
310		2,767.74	I
220		2,768.85	I
220		2,769.32	I
350		2,770.42	I
550		2,783.57	I
220		2,791.29	I
120		2,803.28	II
220		2,814.68	I
75		2,819.78	II
880		2,819.95	I
310		2,834.08	I
200		2,837.55	I
200		2,840.35	I
220		2,843.00	I
270		2,850.98	I
240		2,867.19	I
200		2,875.28	I
200		2,883.44	I
2,900		2,887.68	I
130	c	2,888.06	II
490		2,896.01	I
830	c	2,902.48	I
210		2,905.58	I
550		2,909.82	I
65	h	2,916.73	II

830	c	2,927.42	I
270		2,930.61	I
440		2,943.14	I
130	h	2,957.91	II
270		2,962.27	I
720		2,965.11	I
1,500		2,965.76	I
90		2,968.98	II
310		2,976.29	I
210		2,978.15	I
220		2,980.82	I
220		2,982.19	I
220		2,988.47	I
1,800		2,992.36	I
5,500		2,999.60	I
350		3,001.14	I
220		3,004.14	I
200		3,006.42	I
500		3,016.02	I
300		3,016.49	I
380		3,030.45	I
240		3,047.25	I
200		3,058.78	I
1,600		3,067.40	I
320		3,069.94	I
260		3,071.16	I
200		3,072.96	I
550		3,082.43	I
340		3,088.76	I
200		3,093.64	I
200		3,095.06	I
700		3,100.67	I
140		3,103.06	II
700		3,108.81	I
340		3,110.86	I
340	c	3,118.19	I
340		3,121.36	I
420		3,128.94	I
260		3,134.02	I
250		3,141.38	I
440		3,151.64	I
330		3,153.79	I
360	c	3,158.31	I
220		3,164.52	I
700		3,168.37	I
220		3,174.61	I
440		3,177.71	I
260		3,178.61	I
600		3,182.87	I
1,100		3,184.76	I
1,100		3,185.57	I
260		3,190.78	I
260		3,192.36	I
200		3,194.50	I
220		3,198.58	I
1,100	c	3,204.25	I
380		3,235.94	I
600		3,258.85	I
600		3,259.55	I
200		3,261.56	I
300		3,268.89	I
200		3,294.83	I
280		3,296.70	I
280		3,296.99	I
280		3,301.60	I
240		3,302.23	I
320		3,303.21	II
280		3,303.75	I
240		3,313.95	I
600		3,322.48	I
200		3,331.52	I
2,000		3,338.18	I
1,600		3,342.24	I
810		3,344.32	I
320		3,346.20	I
240	d	3,356.33	I
200		3,358.02	I
200		3,362.74	I
240		3,377.74	I
320		3,379.06	II
320		3,379.70	I
200		3,385.76	I
240		3,389.43	I
200		3,390.25	I
4,000		3,399.30	I
650		3,404.72	I
650		3,405.89	I
240		3,408.67	I
320		3,409.83	I

320		3,417.77	I
810		3,419.41	I
8,000		3,424.62	I
400		3,426.19	I
300		3,427.61	I
320		3,437.71	I
400		3,449.37	I
16,000	c	3,451.88	I
240		3,453.50	I
55,000	c	3,460.46	I
40,000	c	3,464.73	I
400		3,467.96	I
240		3,476.44	I
400		3,480.38	I
320		3,480.85	I
240		3,482.23	I
560		3,503.06	I
100	c	3,512.28	I
320		3,516.65	I
320		3,517.33	I
120		3,534.82	I
320		3,537.46	I
160		3,539.33	I
240		3,549.89	I
160		3,551.29	I
160		3,553.65	I
160		3,558.94	I
160		3,568.23	I
240		3,570.26	I
360		3,579.12	I
810	c	3,580.15	II
650		3,580.97	I
810		3,583.02	I
160		3,596.39	I
160		3,610.49	I
320		3,617.08	I
160		3,621.46	I
160		3,625.91	I
140		3,637.06	I
810		3,637.84	I
440		3,651.97	I
120		3,669.78	I
320		3,670.53	I
860	c	3,689.50	I
1,500	c	3,691.48	I
100		3,697.71	I
520		3,703.24	I
100		3,705.02	I
240		3,709.93	I
360	c	3,717.28	I
4,000		3,725.76	I
140		3,731.87	I
140		3,732.28	I
240	c	3,735.01	I
810		3,735.31	I
910		3,740.10	I
140		3,740.41	I
130		3,742.26	II
300	c	3,745.44	I
140		3,766.48	I
120		3,768.26	I
140		3,777.66	I
700		3,787.52	I
160		3,796.59	I
160		3,797.59	I
190		3,807.74	I
120		3,815.66	I
120		3,836.30	I
240		3,869.94	I
240		3,875.26	I
240		3,876.86	I
100		3,908.21	I
130		3,913.92	I
380	c	3,917.27	I
550		3,929.85	I
140		3,936.90	I
110		3,944.72	I
180		3,945.91	I
280		3,961.04	I
350	c	3,962.48	I
100		4,004.93	I
140		4,022.96	I
100		4,023.31	I
110	c	4,029.63	I
220		4,033.31	I
110		4,037.49	I
200		4,048.99	I
240		4,081.43	I
140		4,104.42	I
240	c	4,110.89	I

190		4,121.64	I
240	c	4,133.42	I
1,800		4,136.45	I
700		4,144.36	I
140		4,149.96	I
160		4,170.40	I
220		4,182.90	I
220		4,183.06	I
650		4,221.08	I
3,600	c	4,227.46	I
150		4,241.39	I
260	c	4,257.60	I
120	c	4,291.17	I
200		4,304.40	I
200		4,332.25	I
40		4,357.98	II
380		4,358.69	I
190		4,367.58	I
140		4,391.34	I
360	c	4,394.38	I
110	c	4,406.40	I
180		4,415.82	I
150		4,475.08	I
120		4,478.39	I
120	c	4,507.04	I
2,600		4,513.31	I
260		4,516.64	I
500		4,522.73	I
120		4,523.88	I
120		4,529.95	I
100		4,545.17	I
120		4,580.68	I
120		4,605.73	I
100		4,621.38	I
190	c	4,791.42	I
2,200	c	4,889.14	I
220		4,923.90	I
40		5,058.56	I
70		5,096.50	I
20		5,120.32	I
25		5,161.65	I
40	c	5,178.89	I
20		5,181.74	I
35		5,234.31	I
50		5,248.86	I
1,300		5,270.95	I
1,600	c	5,275.56	I
100		5,278.24	I
30		5,305.56	I
20		5,317.28	I
35		5,321.28	I
50		5,327.46	I
20		5,331.90	I
20		5,332.76	I
20		5,333.85	I
35		5,369.48	I
50	c	5,369.80	I
100	c	5,377.10	I
25		5,431.90	I
14		5,437.03	I
14		5,447.92	I
25		5,460.64	I
14	h	5,520.05	I
25		5,521.10	I
50	c	5,532.68	I
50	c	5,563.24	I
25		5,573.47	I
25		5,584.72	I
10	h	5,607.21	I
12	h	5,612.27	I
100		5,667.88	I
25		5,711.43	I
18		5,716.95	I
110	c	5,752.93	I
110	c	5,776.83	I
18		5,791.60	I
10	c	5,815.92	I
550		5,834.31	I
10		5,919.86	I
60		5,943.24	I
10		5,950.21	I
18	h	5,969.77	I
10		5,989.99	I
18	h	5,995.73	I
30		6,114.22	I
35	c	6,145.81	I
50		6,146.82	I
18		6,203.24	I
25		6,217.97	I
30	c	6,229.42	I

35	c	6,243.24	I
35	d	6,260.02	I
		6,260.24	I
18	c	6,271.37	I
18		6,278.76	I
10		6,286.41	I
10		6,303.42	I
200		6,307.70	I
200		6,321.90	I
80	d	6,350.75	I
16	h	6,382.94	I
14		6,411.47	I
50		6,511.47	I
14		6,515.25	I
12		6,544.91	I
35	c	6,577.11	I
40	c	6,592.52	I
100	c	6,605.19	I
30	c	6,623.91	I
10	c	6,637.25	I
27	c	6,652.39	I
15	h	6,683.28	I
9	c	6,711.30	I
30		6,751.22	I
5	c	6,761.19	I
180	c	6,813.41	I
260		6,829.90	I
85		6,971.53	I
35	c	7,006.63	I
65	c	7,024.15	I
65	c	7,246.67	I
13		7,292.72	I
40	c	7,578.73	I
13	c	7,611.89	I
7	c	7,620.25	I
50	c	7,640.94	I
65	c	7,912.94	I
35	c	7,980.77	I
40		8,417.13	I
29	c	8,527.73	I

RHODIUM (RH)

Z = 45

Rh I and II

Ref. 1 — C.H.C.

Intensity		Wavelength	
		Air	
150		2,276.21	II
140		2,288.57	I
110		2,309.82	I
55		2,318.36	I
95		2,319.10	I
95		2,321.73	I
350		2,322.58	I
140		2,326.47	I
80		2,328.64	I
190		2,334.77	II
55		2,345.41	I
55		2,352.47	I
55		2,359.18	I
300		2,361.92	I
110		2,368.34	I
270		2,382.89	I
230		2,383.40	I
40		2,384.65	I
270		2,386.14	II
80		2,407.88	I
27		2,408.19	I
27		2,410.25	I
80		2,415.84	II
55		2,418.64	
45		2,419.75	I
45		2,420.18	II
65		2,420.98	II
75		2,423.94	
65		2,427.11	II
130		2,427.68	I
230		2,429.52	I
40		2,431.85	II
40		2,432.66	I
18		2,437.08	I
110		2,437.90	I
330		2,440.34	I
50	h	2,444.27	I
65		2,448.84	I
50		2,449.04	I
75		2,450.56	I
30		2,455.70	II
65		2,458.90	II
90		2,461.04	II
30		2,463.61	I
75		2,470.39	I
90		2,471.47	I
30		2,472.51	I
130		2,473.09	I
15		2,475.64	II
15		2,477.54	II
25		2,482.04	I
50		2,483.33	I
150		2,487.47	I
100		2,490.77	II
30		2,492.30	I
75	h	2,494.51	
15		2,499.02	I
40		2,500.58	
130		2,502.46	I
15		2,503.84	II
300		2,504.29	II
40		2,505.10	II
150		2,505.67	I
350		2,509.70	I
50		2,510.66	II
300		2,511.03	II
75		2,513.36	II
200		2,515.75	I
130		2,520.53	II
13		2,525.99	I
13		2,531.74	I
50		2,532.66	
13		2,533.59	
50		2,534.07	
110		2,536.71	
110		2,537.04	II
30		2,539.72	
40		2,544.22	
350		2,545.70	I
13		2,548.60	
550		2,555.36	I
25		2,558.62	I
50		2,565.79	I
45		2,566.04	I
25		2,566.92	II
50		2,567.28	
25		2,574.66	
25		2,575.75	I
13		2,576.23	
40		2,587.29	II
30		2,598.07	
30		2,603.32	II
75		2,606.44	II
75		2,613.60	
150		2,622.58	I
230		2,625.88	I
100		2,630.42	I
40		2,634.99	I
30		2,638.74	II
75		2,643.00	I
110		2,647.28	I
400		2,652.66	I
30		2,659.01	I
30		2,671.06	I
65		2,676.11	I
25		2,680.28	I
100		2,680.63	I
30		2,681.78	I
30	h	2,686.50	
30	h	2,686.91	
50		2,694.31	I
400		2,703.73	I
40		2,705.63	II
40		2,707.23	I
75		2,714.41	I
100		2,715.31	II
75		2,717.51	I
180		2,718.54	I
65		2,720.14	I
30		2,720.52	I
160		2,728.94	I
40		2,736.76	I
75		2,741.75	I
50		2,767.73	I
100		2,771.51	I
50		2,778.06	I
75		2,779.54	I
130		2,783.03	I
25		2,791.16	I
75		2,796.63	I
150		2,826.43	I
180		2,826.68	I
30		2,827.31	I
75		2,834.12	I
45		2,835.44	I
75		2,836.69	I
50		2,856.16	I
50	d	2,860.68	I
		2,860.76	I
280		2,862.94	I
65		2,864.40	I
50		2,871.35	I
30		2,873.62	I
110		2,878.66	I
75		2,880.76	I
140		2,882.37	I
75		2,885.97	I
75		2,889.11	I
75		2,889.84	I
65		2,899.96	I
25		2,904.81	I
160		2,907.21	I
65		2,910.17	II
75		2,912.62	I
90		2,915.42	I
30		2,923.10	I
180		2,924.02	I
130		2,929.11	I
130		2,931.94	I
30		2,955.41	I
230		2,968.66	I
25		2,974.03	I
160		2,977.68	I
450		2,986.20	I
90		2,986.99	I
50		2,987.45	I
110		3,004.46	I
50		3,019.54	I
130		3,023.91	I
50		3,028.43	I
30		3,045.77	I
30		3,046.76	I
25		3,057.89	I
65		3,067.30	I
180		3,083.96	I
29		3,087.42	I
70		3,114.91	I
140		3,121.76	I
240		3,123.70	I
35		3,130.79	I
95		3,137.71	I
45		3,151.36	I
45		3,152.60	I
130		3,155.78	I
70		3,179.73	I
80		3,185.59	I
140		3,189.05	I
470		3,191.19	I
190		3,197.13	I
70		3,214.32	I
80		3,237.66	I
520		3,263.14	I
520		3,271.61	I
2,300		3,280.55	I
110		3,281.70	I
2,300		3,283.57	I
280		3,289.14	I
45		3,289.64	I
210		3,294.28	I
45		3,296.72	I
260		3,300.46	I
4,200		3,323.09	I
60		3,331.09	I
45		3,331.24	I
330		3,338.54	I
70		3,342.90	I
80		3,344.20	I
60		3,359.90	I
280		3,360.80	I
60		3,362.18	I
420		3,368.38	I
45		3,369.68	I
1,100		3,372.25	I
110		3,377.14	I
80		3,377.71	I
110		3,385.78	I
5,600		3,396.82	I
820		3,399.70	I
160		3,406.55	I
820		3,412.27	I
60		3,420.16	I
330		3,421.22	I
120	d	3,424.38	I
8,200		3,434.89	I
1,400		3,440.53	I
35		3,442.63	I
120		3,447.74	I
60		3,448.58	I
120		3,450.29	I
60		3,451.15	I
400		3,455.22	I
60		3,455.42	I
180		3,457.07	I
220		3,457.93	I
5,900		3,462.04	I
180		3,469.62	I
4,700		3,470.66	I
120		3,472.25	I
4,700		3,474.78	I
2,100		3,478.91	I
95		3,484.04	I
80		3,491.07	I
110		3,494.44	I
1,200		3,498.73	I
5,900		3,502.52	I
60		3,505.41	I
2,800		3,507.32	I
60		3,511.78	I
60		3,513.10	I
60		3,519.54	I
8,800		3,528.02	I
880	d	3,538.14	I
		3,538.26	I
280		3,541.91	I
1,200		3,543.95	I
1,800		3,549.54	I
240		3,564.13	I
1,200		3,570.18	I
4,700		3,583.10	I
120		3,583.53	I
4,700		3,596.19	I
5,900		3,597.15	I
310		3,605.86	I
3,100		3,612.47	I
240		3,614.78	I
200		3,620.46	I
1,800		3,626.59	I
95		3,627.80	I
310		3,639.51	I
350		3,654.87	I
8,200		3,657.99	I
280		3,661.86	I
1,300		3,666.22	I
180		3,666.91	I
140		3,674.76	I
560		3,681.04	I
1,900		3,690.70	I
9,400		3,692.36	I
60		3,694.95	I
940		3,695.52	I
280		3,698.26	I
380		3,698.60	I
7,600		3,700.91	I
940		3,713.02	I
60		3,713.43	I
45		3,714.83	I
16		3,724.94	I
650		3,735.28	I
420		3,737.27	I
420		3,744.17	I
1,200		3,748.22	I
240		3,754.12	I
380		3,754.27	I
490		3,755.58	I
1,000		3,760.40	I
2,300		3,765.08	I
490		3,769.97	I
70		3,775.72	I
380		3,778.13	I
1,000		3,788.47	I
1,300		3,792.18	I
3,800		3,793.22	I
4,900		3,799.31	I
760		3,805.92	I
1,300		3,806.76	I

45		3,809.50	I
95		3,812.45	I
470		3,815.01	I
760		3,816.47	I
1,300		3,818.19	I
3,800		3,822.26	I
2,300		3,828.48	I
2,000		3,833.89	I
45		3,834.75	I
5,900		3,856.52	I
490		3,870.01	I
70		3,872.39	I
380		3,877.34	I
70		3,888.34	I
29		3,904.22	I
23		3,912.83	I
120		3,913.51	I
240		3,922.19	I
2,000		3,934.23	I
45		3,934.98	I
50		3,935.84	I
590		3,942.72	I
95		3,958.24	I
3,800		3,958.86	I
45		3,964.54	II
380		3,975.31	I
240		3,984.40	I
240		3,995.61	I
380		3,996.15	I
120		4,023.14	I
60		4,048.41	I
23		4,049.04	I
40		4,053.44	I
23		4,056.34	I
70		4,077.57	I
560		4,082.78	I
19		4,084.28	I
45		4,087.79	I
60		4,088.50	I
140		4,097.52	I
45		4,107.49	I
70		4,116.33	I
120		4,119.68	I
1,100		4,121.68	I
1,500		4,128.87	I
2,100		4,135.27	I
240		4,154.37	I
330		4,196.50	I
70		4,206.62	I
3,300		4,211.14	I
29		4,230.20	I
40		4,244.44	I
60		4,273.43	I
60		4,278.60	I
820		4,288.71	I
70		4,296.77	I
23		4,342.44	I
45		4,373.04	I
4,200		4,374.80	I
95		4,379.92	I
23		4,433.32	I
35		4,492.47	I
29		4,503.78	I
23		4,528.72	I
16		4,544.27	I
35		4,548.73	I
40		4,551.64	I
19		4,560.89	I
16		4,565.19	I
130		4,569.00	I
14		4,571.31	I
29		4,608.12	I
14		4,619.91	I
23		4,643.18	I
150		4,675.03	I
19		4,721.00	I
70		4,745.11	I
12		4,755.58	I
23		4,810.49	I
21		4,842.43	I
45		4,843.99	I
60		4,851.63	I
60		4,963.71	I
60		4,977.75	I
40		4,979.18	I
14		5,085.52	I
70		5,090.63	I
23		5,120.69	I
19		5,130.76	I
60		5,155.54	I
14		5,157.09	I
40		5,158.69	I
60		5,175.97	I
12		5,177.27	I
35		5,184.19	I
95		5,193.14	I
16		5,206.95	I
16		5,211.52	I
19		5,212.73	I
16		5,214.79	I
19		5,222.66	I
19		5,230.62	I
45		5,237.16	I
9		5,237.80	I
14		5,269.27	I
11	h	5,280.12	I
14		5,292.14	I
14		5,314.79	I
40	h	5,329.74	I
14	h	5,331.08	I
9		5,349.31	I
130		5,354.40	I
23		5,356.47	I
45		5,379.10	I
95		5,390.44	I
23	h	5,404.73	I
60	h	5,424.07	I
19		5,424.72	I
19	h	5,425.45	I
12		5,439.58	I
12	h	5,441.36	I
9	h	5,444.32	I
35	h	5,445.23	I
23	h	5,468.11	I
35	h	5,470.85	I
12		5,476.12	I
12		5,481.42	I
16		5,484.23	I
9		5,504.65	I
29		5,535.04	I
21	l	5,544.58	I
160		5,599.42	I
7		5,607.71	I
16		5,608.35	I
5		5,632.77	I
9		5,659.62	I
40		5,686.38	I
9	h	5,702.47	I
6		5,727.30	I
29		5,792.66	I
9		5,795.79	I
9		5,803.34	I
40		5,806.91	I
6		5,821.84	I
35		5,831.58	I
7		5,907.31	I
9		5,918.54	I
7		5,941.46	I
130		5,983.60	I
9		5,991.19	I
35		6,102.72	I
6		6,116.15	I
8		6,128.06	I
8		6,186.89	I
14		6,199.99	I
16		6,253.72	I
5		6,276.66	I
8		6,277.46	I
6		6,293.38	I
29		6,319.53	I
12		6,414.72	I
16		6,510.41	I
19		6,519.70	I
9		6,627.80	I
19		6,630.16	I
40		6,752.35	I
9		6,796.65	I
13		6,827.33	I
11		6,857.68	I
20		6,879.94	I
65		6,965.67	I
8		6,972.91	I
16		6,979.15	I
16		7,001.58	I
11		7,038.76	I
18		7,101.64	I
15		7,104.45	I
6		7,142.55	I
9		7,219.06	I
18		7,268.18	I
35		7,270.82	I
12		7,271.94	I
5		7,273.03	I
9		7,375.57	I
5	h	7,386.64	I
9		7,430.80	I
18	h	7,442.39	I
7		7,446.77	I
12		7,475.74	I
12		7,495.24	I
8		7,542.02	I
11		7,557.67	I
8		7,577.22	I
11		7,690.05	I
18		7,772.90	I
29		7,791.61	I
55		7,824.91	I
15		7,830.05	I
15		7,846.50	I
21		8,029.91	I
11	h	8,036.09	I
29		8,045.36	I
7		8,063.50	I
15		8,136.20	I
7	h	8,193.67	I
5	h	8,369.67	I
8		8,425.59	I

RUBIDIUM (RB)

Z = 37

Rb I and II

Ref. 12, 130, 241, 257, 264

Intensity		Wavelength	
		Vacuum	
10		474.88	II
40		481.118	II
90		497.430	II
20		508.434	II
150		513.266	II
300		530.173	II
75		533.801	II
40		542.887	II
200		555.036	II
2,500		589.419	II
1,500		643.878	II
3,000		697.049	II
6,000		711.187	II
10,000		741.456	II
1,000		1,604.12	II
200		1,644.96	II
200		1,707.52	II
600		1,716.85	II
5,000		1,760.50	II
200		1,803.47	II
500		1,809.68	II
500		1,865.33	II
500		1,889.42	II
500		1,954.24	II
300		1,956.54	II
200		1,971.42	II
500		1,983.19	II
		Air	
300		2,042.23	II
300		2,052.21	II
500		2,052.80	II
2,000		2,068.92	II
1,000		2,071.50	II
10,000		2,075.95	II
1,000		2,090.29	II
200		2,108.06	II
300		2,116.50	II
1,000		2,125.25	II
400		2,129.82	II
200		2,143.10	II
30,000		2,143.83	II
200		2,190.36	II
600		2,197.99	II
600		2,198.26	II
300		2,207.86	II
10,000		2,217.08	II
200		2,223.79	II
400		2,237.72	II
500		2,250.65	II
200		2,251.43	II
800		2,254.19	II
200		2,254.55	II
200		2,263.54	II
500		2,263.94	II
500		2,286.82	II
5,000		2,291.71	II
300		2,298.80	II
250		2,333.01	II
2,000		2,333.39	II
350		2,353.11	II
300		2,353.96	II
400		2,356.97	II
300		2,358.04	II
300		2,364.27	II
200		2,364.32	II
200		2,365.15	II
300		2,367.51	II
200		2,373.21	II
2,000		2,385.34	II
250		2,405.94	II
400		2,434.17	II
800		2,459.14	II
50,000		2,472.20	II
300		2,484.56	II
700		2,484.70	II
2,000		2,496.38	II
200		2,502.67	II
250		2,514.18	II
1,000		2,524.24	II
200		2,594.56	II
400		2,623.76	II
400		2,645.58	II
1,000		2,684.10	II
1,000		2,711.76	II
250		2,741.01	II
500		2,812.15	II
350		2,838.51	II
750		2,873.88	II
1,000		3,051.36	II
2		3,082.02	I
250		3,088.58	II
10		3,112.57	I
3		3,113.06	I
5,000	c	3,148.90	II
25		3,157.54	I
5		3,158.26	I
1,200		3,161.00	II
50		3,227.98	I
6		3,229.16	I
2,000		3,270.99	II
1,500		3,321.49	II
1,200		3,340.55	II
60		3,348.72	I
75		3,350.82	I
750		3,353.89	II
1,200		3,393.03	II
750		3,415.58	II
1,000		3,434.18	II
1,500		3,461.50	II
3,000		3,521.39	II
3,000	l	3,531.55	II
1,000		3,541.15	II
100		3,587.05	I
40		3,591.57	I
5,000		3,600.60	II
10,000		3,600.64	II
600	c	3,639.80	II
400	c	3,646.26	II
350	c	3,647.56	II
1,000	c	3,662.74	II
900	c	3,663.81	II
350		3,666.72	II
300		3,675.66	II
2,500	c	3,699.58	II
350		3,746.33	II
3,500		3,796.81	II
2,500		3,801.90	II
1,000		3,826.66	II
450		3,860.74	II
250		3,907.29	II
500		3,922.20	II
2,500	l	3,926.44	II
25,000		3,940.51	II
1,000	c	3,978.15	II
1,700		4,029.49	II
2,500	c	4,083.88	II

Intensity		Wavelength	Spectrum
2,000	c	4,104.28	II
1,700	c	4,136.11	II
3,500		4,193.08	II
1,000		4,201.80	I
500		4,215.53	I
90,000		4,244.40	II
500		4,266.58	II
250	c	4,270.25	II
15,000		4,273.14	II
2,500	c	4,287.97	II
1,500		4,293.97	II
500	c	4,306.26	II
1,000		4,346.96	II
2,500		4,377.12	II
300		4,440.10	II
1,000		4,469.47	II
400	c	4,493.92	II
700		4,519.04	II
3,000		4,530.34	II
500	l	4,533.79	II
400		4,540.74	II
20,000		4,571.77	II
3,000	c	4,622.42	II
350	c	4,631.89	II
10,000		4,648.57	II
500		4,659.28	II
1,000		4,730.45	II
1,000		4,755.30	II
400	c	4,757.82	II
30,000		4,775.95	II
5,000	c	4,782.83	II
300	c	4,855.34	II
1,500	c	4,885.59	II
2		5,087.987	I
2		5,132.471	I
10		5,150.134	I
10,000		5,152.08	II
300		5,164.58	II
1		5,165.023	I
2		5,165.142	I
1		5,169.65	I
15		5,195.278	I
2		5,233.968	I
20		5,260.034	I
1		5,260.228	I
200		5,270.51	II
3		5,322.380	I
40		5,362.601	I
4		5,390.568	I
75		5,431.532	I
3		5,431.830	I
500		5,512.55	II
5,000		5,522.78	II
6		5,578.788	I
5,000	c	5,635.99	II
40		5,647.774	I
20		5,653.750	I
3,000	d	5,699.15	II
60		5,724.121	I
3		5,724.614	I
200		5,739.64	II
75		6,070.755	I
200		6,135.27	II
30	c	6,159.626	I
1,000	c	6,199.08	II
75	c	6,206.309	I
300		6,269.40	II
120	c	6,298.325	I
5		6,299.224	I
10,000		6,458.33	II
1,000		6,555.62	II
5,000		6,560.81	II
3,000	l	6,775.07	II
100	l	7,279.997	I
300	c	7,316.52	II
150		7,408.173	I
200	l	7,618.933	I
300		7,757.651	I
60		7,759.436	I
90,000	c	7,800.27	I
5	l	7,925.26	I
4		7,925.54	I
45,000	c	7,947.60	I
40	l	8,271.41	I
30		8,271.71	I
2,000		8,603.96	II
40	l	8,868.512	I
30		8,868.852	I
300		8,978.88	II
300		9,021.77	II
3		9,224.64	I
2		9,234.25	I
500	c	9,246.41	II
300		9,338.87	II
200	w	9,373.50	II
300		9,391.36	II
1,000		9,479.32	II
700	l	9,493.72	II
30	l	9,522.65	I
5		9,523.05	I
20	l	9,540.18	I
300		9,612.99	II
300		9,671.54	II
2,000	c	9,689.05	II
200		9,776.06	II
200		9,934.76	II
35	l	10,075.282	I
30	l	10,075.708	I
100		13,235.17	I
20		13,442.81	I
30		13,443.57	I
75		13,665.01	I
1,000		14,752.41	I
800		15,288.43	I
150		15,289.48	I
20		22,529.65	I
10		22,932.47	I
4		27,314.31	I
2		27,905.37	I

Rb III

Ref. 258, 262 — J.R.

Intensity		Wavelength	
		Vacuum	
30		465.85	III
35	p	482.43	III
30	p	482.47	III
500		482.83	III
300		484.84	III
500		489.66	III
100		489.96	III
600		493.48	III
50		497.82	III
100		500.28	III
30		508.33	III
400		516.79	III
800		533.64	III
1,200		535.86	III
1,200		556.19	III
500		558.36	III
700		564.77	III
1,500		566.71	III
1,000		572.82	III
1,500		576.65	III
2,500		579.63	III
1,500		581.26	III
500		582.34	III
800		586.77	III
100		591.42	III
900		593.65	III
1,000		594.94	III
1,300		595.88	III
1,200		598.49	III
450		602.09	III
50		605.51	III
500		607.28	III
400		613.31	III
500		619.67	III
20		620.83	III
100		622.24	III
250		630.06	III
500		645.67	III
20		674.81	III
5,000		769.04	III
2,500		815.28	III
		Air	
100		2,153.21	III
250		2,164.59	III
100		2,268.00	III
150		2,300.12	III
500		2,304.14	III
150		2,304.45	III
250		2,312.46	III
200		2,337.07	III
100		2,341.90	III
200		2,345.37	III
100		2,349.81	III
150		2,380.44	III
100		2,381.29	III
150		2,418.46	III
300		2,561.86	III
100		2,573.71	III
100		2,577.07	III
200		2,586.83	III
1,000		2,631.75	III
350		2,636.83	III
100		2,656.68	III
100		2,713.86	III
500		2,798.86	III
150		2,800.27	III
500		2,807.58	III
100		2,845.44	III
150		2,869.77	III
500		2,903.69	III
150		2,949.62	III
100		2,951.01	III
2,000		2,956.07	III
500	l	2,967.45	III
150		2,968.13	III
500		2,970.74	III
250		2,987.40	III
350		3,023.61	III
200		3,039.62	III
200		3,041.48	III
250		3,070.70	III
500		3,086.84	III
100		3,098.49	III
500		3,111.36	III
250	s	3,114.82	III
120		3,118.92	III
100		3,169.34	III
200		3,222.60	III
500		3,286.41	III
100		3,330.16	III
200		3,346.92	III
250		3,439.26	III
100		3,492.68	III

Rb IV

Ref. 109 — J.R.

Intensity	Wavelength	
	Vacuum	
10	595.18	IV
25	663.76	IV
25	716.24	IV
20	733.41	IV
50	740.85	IV
20	749.86	IV
20	753.75	IV
10	771.54	IV
25	776.89	IV
9	817.92	IV
15	850.18	IV
10	988.00	IV

RUTHENIUM (RU)

Z = 44

Ru I and II

Ref. 1 — C.H.C.

Intensity		Wavelength	
		Air	
2,400		2,076.43	I
2,600		2,083.77	I
2,400		2,090.89	I
690		2,255.52	I
290		2,259.53	I
780		2,272.09	I
240		2,278.19	I
780		2,279.57	I
170		2,285.38	I
290		2,302.54	I
480		2,317.80	I
150		2,322.01	I
120		2,334.96	II
240		2,340.69	I
190	h	2,342.85	II
190		2,349.34	I
310		2,351.33	I
170		2,357.91	II
140		2,360.56	I
170		2,370.17	I
240		2,375.27	I
80		2,375.63	II
160		2,392.42	I
95		2,396.71	II
780		2,402.72	II
150		2,407.92	II
55		2,410.89	I
55		2,414.82	II
130		2,420.82	I
55		2,422.92	I
45		2,429.60	I
65		2,432.93	I
30		2,447.45	I
30		2,450.58	I
65		2,454.92	I
180		2,455.53	II
150		2,456.44	II
370		2,456.57	II
65	h	2,458.62	I
55		2,462.94	I
85		2,464.70	I
30		2,474.04	I
110		2,475.41	I
100		2,476.88	I
280		2,478.93	II
28		2,481.11	I
30		2,489.91	I
18		2,491.78	I
65		2,493.69	II
85		2,494.02	I
45		2,494.48	II
85		2,495.69	II
65		2,496.56	I
140		2,498.42	II
140		2,498.57	II
85		2,499.78	I
260		2,507.01	II
130		2,508.27	I
110		2,509.07	I
110		2,512.81	I
110		2,513.32	II
110		2,517.32	II
150		2,535.59	II
65		2,543.25	II
280		2,544.22	I
120		2,546.67	I
280		2,549.48	I
550		2,549.58	I
130		2,560.26	I
120		2,560.83	I
110		2,563.15	I
160		2,568.77	I
100		2,570.97	I
100		2,578.57	I
100		2,579.53	I
100		2,589.57	I
170		2,591.12	I
120		2,592.02	I
100		2,593.70	I
110		2,594.85	I
370		2,609.06	I
830		2,612.07	I
100		2,615.09	I
220		2,631.30	I
220		2,635.86	I
170		2,636.67	I
110		2,640.33	I
460		2,642.96	I
110		2,647.32	I
110		2,651.29	I
330		2,651.84	I
28		2,656.25	II
400		2,659.62	I
23		2,661.17	II
330		2,661.61	II
200		2,664.76	I
30		2,667.40	II
690		2,678.76	II
220		2,686.29	I
28		2,687.50	II
		2,688.16	I
330		2,692.06	II
110		2,701.34	I
110		2,702.83	I
170		2,709.20	I

200	2,712.41	II	120	3,241.24	I	600	3,782.74	I	1,600	4,410.03	I
690	2,719.52	I	120	3,243.50	I	3,900	3,786.06	I	160	4,421.46	I
130	2,722.65	I	280	3,260.35	I	6,000	3,790.51	I	330	4,428.46	I
140	2,725.47	II	120 d	3,264.55	I	240	3,794.92	I	460	4,439.76	I
310	2,734.35	II	120	3,266.44	I	760	3,798.05	I	440	4,449.34	I
1,800	2,735.72	I	200	3,268.21	I	7,600	3,798.90	I	1,100	4,460.04	I
170	2,739.22	I	200	3,273.08	I	7,600	3,799.35	I	190	4,473.93	I
130	2,744.45	I	200	3,274.71	I	310	3,800.26	I	150	4,480.45	I
35	2,747.97	II	100	3,277.57	I	310	3,808.68	I	350	4,498.14	I
75	2,752.45	II	490	3,294.11	I	600	3,812.72	I	120	4,510.10	I
75	2,752.77	II	370	3,301.59	I	760	3,817.27	I	220	4,516.89	I
260	2,763.42	I	220	3,306.17	I	760	3,819.03	I	220	4,517.82	I
35	2,765.44	II	290	3,315.23	I	650	3,822.09	I	110	4,520.95	I
90	2,768.93	II	290	3,316.39	I	550	3,824.93	I	170	4,547.33	I
100	2,778.38	II	100	3,325.00	I	760	3,831.80	I	110	4,547.85	I
110	2,787.83	II	120	3,335.69	I	220	3,835.05	I	5,400	4,554.51	I
140	2,802.81	I	930	3,339.55	I	310	3,838.07	I	110	4,559.98	I
35	2,806.74	II	240	3,341.66	I	930	3,839.70	I	1,700	4,584.44	I
350	2,810.03	I	200	3,361.15	I	480	3,846.68	I	110	4,591.10	I
1,700	2,810.55	I	370	3,368.45	I	760	3,850.43	I	150	4,592.52	I
350	2,818.36	I	100	3,371.86	I	480	3,856.46	I	330	4,599.08	I
110	2,822.03	I	130	3,374.65	I	1,300	3,857.55	I	170	4,635.69	I
200	2,827.87	I	120	3,378.02	I	220	3,860.72	I	200	4,645.09	I
400	2,829.16	I	100	3,379.60	I	650	3,862.69	I	720	4,647.61	I
130	2,834.00	I	130	3,380.18	I	1,300	3,867.84	I	290	4,654.32	I
150	2,840.54	I	130	3,385.14	I	260	3,873.52	I	290	4,681.79	I
35	2,841.68	II	130	3,388.71	I	650	3,892.21	I	190	4,684.02	I
640	2,854.07	I	100	3,389.50	I	760	3,909.08	I	290	4,690.11	I
180	2,860.02	I	370	3,392.54	I	260	3,920.92	I	1,400	4,709.48	I
420	2,861.41	I	310	3,401.74	I	1,500	3,923.47	I	140	4,731.33	I
550	2,866.64	I	310	3,409.28	I	3,300	3,925.92	I	120	4,733.52	I
110	2,868.31	I	3,100	3,417.35	I	600	3,931.76	I	500	4,757.84	I
1,800	2,874.98	I	4,900	3,428.31	I	310	3,933.55	I	260	4,815.52	I
220	2,879.76	I	490	3,430.77	I	760	3,945.57	I	120	4,844.56	I
55	2,882.12	II	310	3,432.74	I	460	3,950.21	I	550	4,869.15	I
130	2,883.60	I	6,400	3,436.74	I	310	3,952.68	I	160	4,895.60	I
740	2,886.54	I	260	3,438.37	I	460	3,964.90	I	470	4,903.05	I
180	2,892.56	I	220	3,440.20	I	600	3,978.44	I	120	4,907.89	I
110	2,901.94	I	260	3,473.75	I	600	3,979.42	I	260	4,921.07	I
140	2,905.65	I	240	3,481.30	I	870	3,984.86	I	180	4,938.43	I
370	2,908.88	I	8,300	3,498.94	I	280	3,995.98	I	160	4,968.90	I
1,100	2,916.26	I	640	3,514.49	I	1,500	4,022.16	I	160	4,980.35	I
150	2,919.61	I	330	3,519.64	I	600	4,023.83	I	120	4,992.74	I
35	2,927.54	II	200	3,528.68	I	310	4,039.21	I	160	5,011.23	I
180	2,945.67	II	240	3,532.81	I	1,400	4,051.40	I	90	5,014.95	I
180	2,946.99	I	390	3,537.95	I	710	4,054.05	I	90	5,026.18	I
370	2,949.50	I	790	3,539.37	I	370	4,064.46	I	65	5,028.16	I
150	2,954.49	I	200	3,541.63	I	200	4,067.61	I	35	5,040.35	I
18	2,963.40	II	690	3,570.59	I	760	4,068.37	I	35	5,040.74	I
550	2,965.16	I	200	3,574.58	I	200	4,073.00	I	65	5,047.31	I
170	2,965.55	II	390	3,587.20	I	980	4,076.73	I	450	5,057.33	I
140	2,976.59	II	6,400	3,589.22	I	6,000	4,080.60	I	21	5,062.64	I
550	2,976.92	I	6,900	3,593.02	I	310	4,085.43	I	90	5,072.97	I
45	2,977.23	II	6,400	3,596.18	I	930	4,097.79	I	120	5,076.32	I
75	2,979.96	II	1,300	3,599.76	I	350	4,101.74	I	200	5,093.83	I
1,400	2,988.95	I	350	3,625.20	I	1,900	4,112.74	I	80	5,107.07	I
35	2,991.62	II	370	3,626.74	I	2,000	4,144.16	I	24	5,123.73	I
110	2,993.27	I	3,100	3,634.93	I	650	4,145.74	I	55	5,127.26	I
460	2,994.96	I	210	3,637.47	I	260	4,146.77	I	65	5,133.89	I
440	3,006.59	I	200	3,640.64	I	870	4,167.51	I	530	5,136.55	I
330	3,017.24	I	290	3,650.32	I	550	4,197.58	I	170	5,142.76	I
310	3,020.88	I	310	3,654.40	I	550	4,198.88	I	250	5,147.24	I
240	3,033.45	I	6,200	3,661.35	I	7,600	4,199.90	I	110	5,151.07	I
200	3,040.31	I	830	3,663.37	I	1,500	4,206.02	I	55	5,153.20	I
220	3,042.48	I	650	3,669.49	I	5,400	4,212.06	I	500	5,155.14	I
110	3,045.71	I	240	3,678.32	I	760	4,214.44	I	55	5,160.00	I
110	3,048.78	I	260	3,696.59	I	930	4,217.27	I	920	5,171.03	I
150	3,054.94	I	410	3,717.00	I	370	4,220.68	I	180	5,195.02	I
390	3,064.84	I	260	3,719.33	I	550	4,230.31	I	80	5,199.87	I
170	3,089.14	I	550	3,726.10	I	760	4,241.05	I	45	5,202.12	I
120	3,089.80	I	8,700	3,726.93	I	760	4,243.06	I	45	5,213.43	I
330	3,096.57	I	11,000	3,728.03	I	370	4,246.73	I	65	5,223.55	I
120	3,097.60	I	7,100	3,730.43	I	310	4,258.99	I	40	5,242.38	I
830	3,099.28	I	280	3,737.40	I	760	4,284.33	I	55	5,251.67	I
740	3,100.84	I	410	3,739.46	I	270	4,293.28	I	40	5,257.07	I
120	3,125.96	I	3,500	3,742.28	I	260	4,294.79	I	40	5,266.47	I
120	3,153.82	I	870	3,742.78	I	550	4,295.93	I	40	5,266.83	I
290	3,159.92	I	280	3,744.22	I	3,700	4,297.71	I	40	5,280.82	I
200	3,168.52	I	410	3,744.40	I	930	4,307.60	I	130	5,284.08	I
60	3,177.05	II	2,800	3,745.59	I	370	4,318.43	I	40	5,291.16	I
180	3,186.04	I	760	3,753.54	I	550	4,319.87	I	80	5,304.86	I
240	3,188.34	I	310	3,755.09	I	550	4,342.07	I	260	5,309.27	I
240	3,189.98	I	870	3,755.93	I	350	4,349.70	I	13	5,315.33	I
180	3,196.59	I	1,200	3,759.84	I	710	4,354.13	I	40	5,332.93	I
180	3,223.27	I	370	3,760.03	I	870	4,361.21	I	45 h	5,334.70	I
110	3,226.37	I	600	3,761.51	I	2,400	4,372.21	I	110	5,335.93	I
100	3,227.88	I	600	3,767.35	I	870	4,385.39	I	130	5,361.77	I
220	3,228.53	I	1,500	3,777.59	I	1,300	4,385.65	I	65	5,377.84	I
220	3,238.53	I	460	3,781.18	I	1,700	4,390.44	I	65	5,385.88	I

110	h	5,401.04	I
40		5,401.39	I
40		5,418.86	I
55		5,427.59	I
26	l	5,439.21	I
13		5,452.71	I
80	h	5,454.82	I
90		5,456.13	I
13	h	5,475.18	I
55		5,479.40	I
26		5,480.30	I
80		5,484.32	I
18		5,484.64	I
26		5,496.69	I
13		5,501.02	I
130		5,510.71	I
20		5,512.37	I
8		5,517.86	I
12		5,521.78	I
12		5,530.99	I
24		5,540.66	I
12		5,556.52	I
90		5,559.75	I
11		5,569.03	I
21		5,578.40	I
21		5,603.14	I
8		5,603.55	I
13		5,606.73	I
11		5,629.79	I
290		5,636.24	I
11		5,641.66	I
7		5,649.56	I
7		5,653.30	I
11		5,665.20	I
16		5,679.63	I
180		5,699.05	I
13		5,724.82	I
13		5,725.73	I
16		5,745.99	I
16		5,747.47	I
11		5,752.02	I
11		5,756.83	I
11		5,767.92	I
16		5,804.39	I
65		5,814.98	I
8		5,828.06	I
16	h	5,833.21	I
55		5,919.34	I
80		5,921.45	I
21		5,926.87	I
26		5,932.38	I
8		5,936.65	I
8		5,951.15	I
21	h	5,973.38	I
8		5,974.17	I
16		5,988.67	I
35		5,993.65	I
18		6,116.77	I
26		6,199.42	I
26		6,225.20	I
9		6,284.49	I
18		6,295.22	I
13		6,330.62	I
9		6,336.12	I
9	h	6,363.41	I
9		6,376.45	I
16		6,390.23	I
8		6,417.57	I
26	h	6,444.84	I
8		6,496.44	I
11		6,528.74	I
4		6,560.45	I
4		6,593.74	I
9		6,618.20	I
21		6,663.14	I
55		6,690.00	I
11		6,707.52	I
15		6,718.30	I
15		6,730.45	I
7		6,756.54	I
21		6,766.95	I
30		6,775.02	I
13		6,787.23	I
8		6,813.51	I
15		6,823.88	I
21		6,824.17	I
7		6,831.52	I
26		6,911.48	I
110		6,923.23	I
26		6,982.01	I
26		7,027.98	I
9		7,086.06	I
12		7,087.35	I
4		7,141.72	I
6		7,219.26	I
35		7,238.92	I
7		7,266.96	I
8		7,323.56	I
16		7,393.93	I
18		7,468.91	I
12		7,475.40	I
26		7,485.79	I
70		7,499.75	I
7		7,532.07	I
26		7,559.61	I
5		7,612.94	I
18		7,621.50	I
18		7,722.87	I
5		7,729.91	I
22		7,791.86	I
4		7,797.89	I
4		7,806.82	I
3		7,813.43	I
4		7,829.81	I
5	h	7,833.39	I
6	h	7,841.90	I
30		7,847.80	I
80		7,881.49	I
16		7,890.37	I
16		7,924.43	I
5		7,948.15	I
9		7,967.84	I
9		8,112.47	I
18		8,264.96	I
11		8,348.98	I
6		8,352.94	I
4		8,435.77	I
11		8,473.64	I
11		8,483.56	I
22		8,710.84	I
14		8,724.98	I
9		8,777.36	I

SAMARIUM (SM)
Z = 62

Sm I and II

Ref. 1 — C.H.C.

Intensity		Wavelength	
		Vacuum	
45		2,610.07	
90		2,640.27	
35		2,649.17	
45		2,657.68	
70		2,662.42	
120		2,675.15	
100		2,688.60	
45		2,690.90	II
130		2,693.34	
45		2,693.74	
60		2,696.08	
85		2,707.96	
50		2,732.42	
35		2,739.87	
29		2,762.28	
35		2,764.18	
85		2,767.85	II
60		2,774.77	
85		2,776.11	
85		2,779.23	II
85		2,786.64	
150		2,789.38	II
130	h	2,796.70	
85		2,807.36	
150		2,809.50	
120		2,810.86	II
85		2,817.20	II
29		2,820.96	II
220		2,830.94	
60		2,840.30	
60		2,847.49	II
60		2,851.35	
120		2,866.09	II
70		2,868.40	II
70		2,881.34	
85		2,881.68	
60		2,883.09	
45		2,889.06	
60		2,891.34	
85	d	2,907.88	
		2,907.99	II
130		2,910.28	II
85		2,937.48	II
70		2,943.49	II
150		2,953.19	II
85		2,962.74	II
160		2,969.02	II
100		2,983.43	II
60		2,991.57	II
100		3,021.01	
150		3,034.84	II
100		3,039.13	II
120		3,046.93	II
150		3,067.54	
120		3,071.29	II
100		3,086.45	II
120		3,096.88	II
100	h	3,102.30	II
250		3,106.52	II
220		3,110.20	II
200		3,117.72	II
270		3,136.30	II
150		3,139.97	II
150		3,147.19	II
180		3,152.10	II
410		3,152.52	II
150		3,162.30	II
360		3,169.88	II
180		3,178.12	II
720		3,183.92	II
310		3,187.01	II
430		3,187.22	II
360		3,187.79	II
360		3,193.01	II
360		3,196.18	II
150		3,201.80	II
150		3,204.90	II
360		3,207.18	II
180		3,208.17	II
600		3,211.73	II
150		3,214.12	II
270		3,215.26	II
530		3,216.85	II
600		3,218.61	II
150		3,219.43	II
270		3,226.84	II
180		3,228.50	II
270		3,228.78	II
720		3,230.56	II
360		3,231.53	II
150		3,231.95	II
430		3,233.68	II
720		3,236.64	II
150		3,237.89	II
720		3,239.66	II
530		3,241.16	II
180		3,241.59	II
180		3,242.04	II
150		3,244.69	II
240		3,249.75	II
720		3,250.37	II
360		3,253.40	II
270		3,253.94	II
850		3,254.38	II
110		3,255.63	II
360		3,262.28	II
430		3,264.94	II
180		3,270.49	II
180		3,270.68	II
430	d	3,272.48	II
		3,272.60	II
430		3,272.81	II
430		3,273.48	II
430		3,276.75	II
270		3,280.84	II
180		3,285.66	II
430		3,286.23	II
720	d	3,290.28	II
		3,290.39	II
180		3,290.65	II
240		3,293.37	II
360		3,295.44	II
430		3,295.81	II
720		3,298.10	II
170		3,300.98	II
340		3,301.68	II
340		3,304.52	II
340		3,305.18	II
1,700		3,306.39	II
170		3,306.61	II
850		3,307.02	II
340		3,309.52	II
850		3,310.66	II
600		3,312.42	II
410		3,316.58	II
430		3,320.16	II
110		3,320.59	II
1,200		3,321.18	II
340		3,323.77	II
340		3,325.26	II
170		3,325.48	II
340		3,327.88	II
170		3,333.64	II
170		3,336.12	II
850		3,340.58	II
240		3,343.49	II
110		3,343.64	II
240		3,344.35	II
170		3,347.30	II
240		3,348.68	II
220		3,350.88	II
410	d	3,354.18	II
		3,354.30	II
170		3,354.72	II
1,200		3,365.86	II
150		3,367.27	II
340		3,368.57	II
340		3,369.46	II
170		3,370.59	II
340		3,371.21	II
150		3,376.48	II
1,200		3,382.40	II
510		3,384.66	II
150		3,384.86	
150		3,387.66	II
410		3,389.32	II
150		3,391.11	II
410		3,396.19	II
150		3,397.76	II
150		3,399.84	II
600		3,402.46	II
210		3,403.09	II
850		3,408.68	II
270		3,418.15	II
430		3,418.51	II
170		3,419.77	II
120		3,424.78	II
170		3,426.20	II
170		3,433.68	II
150		3,437.10	II
170		3,438.06	II
240		3,440.50	II
170		3,453.56	II
170		3,459.20	II
120		3,459.42	II
240		3,461.13	II
120		3,464.07	II
170		3,467.87	II
130		3,473.96	II
130		3,479.53	II
130		3,480.26	II
170		3,480.56	II
170		3,487.41	II
170		3,493.61	II
220		3,499.84	II
340		3,511.23	II
310		3,530.60	II
220		3,532.57	II
270		3,535.65	II
240		3,554.15	II
510		3,559.10	II
220		3,566.84	II
4,200		3,568.27	II
270		3,577.79	II
390		3,580.94	II
310		3,583.39	II
4,200		3,592.60	II
340		3,601.69	II
1,700		3,604.28	II
3,400		3,609.49	II
240		3,620.58	II
1,700		3,621.23	II
240		3,623.32	II
850		3,627.01	II
850		3,631.13	II

3,400		3,634.29	II
240		3,634.93	II
410		3,638.77	II
360		3,645.29	II
300		3,645.39	II
660		3,649.53	II
340		3,650.19	II
340		3,656.22	II
2,200		3,661.36	II
220		3,662.69	II
340		3,667.93	II
340		3,670.66	II
2,200		3,670.84	II
340		3,677.79	II
270		3,681.73	II
270		3,688.42	II
270		3,692.22	II
1,100		3,693.99	II
480		3,706.75	II
480		3,706.98	II
480		3,708.41	II
930		3,708.65	II
480		3,711.54	II
350		3,712.76	II
930		3,718.88	II
930		3,721.85	II
420		3,724.90	II
1,600		3,728.47	II
2,100		3,731.26	II
1,600		3,735.98	II
800		3,737.14	II
320		3,737.48	II
2,900		3,739.12	II
		3,739.20	II
800		3,741.29	II
1,200		3,743.87	II
930		3,745.46	I
		3,745.60	II
480		3,747.62	II
800		3,755.28	II
800		3,756.41	I
1,200		3,757.53	II
450		3,758.45	II
660		3,758.97	II
350		3,760.04	II
1,900		3,760.69	II
660		3,762.59	II
1,100		3,764.37	II
480		3,767.36	II
480		3,767.76	II
370	d	3,773.33	I
		3,773.42	II
1,100		3,778.14	II
660		3,780.76	II
420		3,780.93	II
320		3,787.20	II
1,500		3,788.12	II
1,600		3,793.97	II
420		3,797.28	II
1,600		3,797.73	II
500		3,799.54	II
800		3,800.89	II
320		3,805.63	II
420		3,808.46	II
320		3,809.75	II
320		3,809.88	II
420		3,810.43	II
500		3,812.07	II
480		3,813.63	II
420		3,814.63	II
930	d	3,820.82	
530		3,824.18	II
1,600		3,826.20	II
530		3,830.29	II
1,100		3,831.50	II
530		3,833.83	II
560		3,834.48	I
560		3,834.60	II
370		3,835.72	II
500		3,838.94	II
400		3,840.45	II
1,600		3,843.50	II
530		3,847.51	II
640		3,848.78	II
420		3,851.88	II
530		3,853.30	I
2,700		3,854.21	II
480		3,854.56	I
800		3,855.90	II
480		3,857.91	II
400		3,858.74	I
660		3,862.05	II
350		3,862.23	II
320		3,865.24	II
800		3,871.78	II
400		3,875.19	II
560		3,875.54	II
800		3,880.77	II
450		3,881.38	II
450		3,881.79	II
320		3,882.50	II
3,700		3,885.29	II
660		3,889.16	II
		3,889.22	II
610		3,890.08	II
320		3,891.21	II
400		3,894.05	II
1,600		3,896.98	II
1,300		3,903.42	II
620		3,917.44	II
2,500		3,922.40	II
1,900		3,928.28	II
470		3,935.76	II
1,300		3,941.87	II
620		3,943.24	II
500		3,946.51	II
740		3,948.11	II
470		3,951.89	I
370		3,959.53	II
1,500		3,963.00	II
620		3,966.04	II
470		3,967.68	II
740		3,970.53	II
1,500		3,971.40	II
620		3,974.66	I
960		3,976.27	II
1,000		3,976.43	II
960		3,979.20	II
740		3,983.14	II
740		3,986.68	II
370		3,987.43	II
1,500		3,990.00	II
		3,990.02	I
740		3,993.31	II
280		4,003.46	II
470		4,007.48	II
280		4,019.98	II
880		4,023.23	II
740		4,035.11	II
590		4,041.68	II
740		4,042.72	II
880		4,042.90	II
240		4,044.11	II
560		4,045.05	II
440		4,046.16	II
740		4,047.16	II
210		4,048.62	II
590		4,049.81	II
440		4,058.87	II
560		4,063.54	II
280		4,064.32	II
1,400		4,064.58	II
810		4,066.74	II
710		4,068.33	II
810		4,075.84	II
280		4,076.65	II
240		4,080.56	II
410		4,082.60	II
280		4,083.58	II
220		4,084.40	II
1,000		4,092.27	II
290		4,094.05	II
240		4,104.13	II
810		4,107.28	II
		4,107.39	II
410		4,109.40	II
280		4,110.19	II
410		4,113.90	II
1,900		4,118.55	II
410		4,121.36	II
280		4,122.51	II
710		4,123.96	II
280		4,129.23	II
250		4,135.14	II
320		4,147.71	II
810		4,149.83	II
1,200		4,152.21	II
530		4,153.33	II
560		4,155.22	II
810		4,169.48	II
410		4,171.57	II
440		4,178.02	II
530		4,181.10	II
210		4,183.33	I
530		4,183.76	II
1,000		4,188.13	II
410		4,191.93	II
270		4,199.45	II
650		4,202.92	II
1,100		4,203.05	II
660		4,206.13	II
270		4,206.62	II
660		4,210.35	II
740		4,220.66	II
1,000		4,225.33	II
740		4,229.70	II
620		4,234.57	II
1,200		4,236.74	II
500		4,237.66	II
620		4,244.70	II
210		4,249.55	II
250		4,251.78	II
2,100		4,256.39	II
210		4,258.58	II
1,300		4,262.68	II
500		4,265.08	II
1,200		4,279.68	II
		4,279.75	II
240		4,279.94	II
2,200		4,280.79	II
710		4,282.21	I
470		4,282.83	I
240		4,283.50	I
350		4,286.64	II
350		4,292.18	II
1,600		4,296.74	I
320		4,304.94	II
880		4,309.01	II
240		4,312.85	I
1,900		4,318.94	II
470		4,319.53	I
590		4,323.28	II
240		4,324.46	I
1,800		4,329.02	II
440		4,330.02	I
1,300		4,334.15	II
880		4,336.14	I
560		4,345.86	II
1,100		4,347.80	II
560		4,350.46	II
560		4,352.10	II
560		4,360.72	II
220		4,361.07	II
810		4,362.04	II
440		4,362.91	I
220		4,363.45	II
500		4,368.03	II
210		4,369.92	II
440		4,373.46	II
320		4,374.98	II
880		4,378.24	II
530		4,380.42	I
290		4,384.29	II
1,600		4,390.86	II
210		4,393.35	I
290		4,397.34	I
410		4,401.17	I
810	d	4,403.06	II
		4,403.13	I
410		4,403.36	II
520		4,409.33	II
290		4,411.58	I
380		4,417.58	II
470		4,419.33	I
1,500		4,420.53	II
960		4,421.14	II
2,900		4,424.34	II
470		4,429.66	I
1,600		4,433.88	II
1,800		4,434.32	II
530		4,441.81	I
440		4,442.28	I
710		4,444.26	II
710		4,445.15	I
1,300		4,452.73	II
250		4,452.95	I
1,200		4,454.63	II
1,000		4,458.52	II
250		4,459.29	I
2,200		4,467.34	II
810		4,470.89	I
470		4,472.43	II
620		4,473.02	II
740		4,478.66	II
370		4,499.11	I
370		4,499.48	II
240		4,503.38	I
180		4,505.05	II
120		4,511.33	I
560		4,511.83	II
440		4,515.09	II
880		4,519.63	II
440		4,523.04	II
		4,523.18	I
650		4,523.91	II
290		4,533.80	I
270		4,536.51	II
710		4,537.95	II
150		4,538.53	II
290		4,540.19	II
380		4,542.06	II
810		4,543.95	II
100		4,544.83	II
410		4,552.66	II
270		4,554.45	II
240		4,560.43	II
470		4,566.21	II
590		4,577.69	II
290		4,581.58	I
440		4,581.73	I
560		4,584.83	II
290		4,591.82	II
380		4,593.54	II
560		4,595.29	II
240		4,596.74	I
220		4,604.18	II
290		4,606.51	II
290		4,615.44	II
470		4,615.69	II
150		4,630.21	II
880		4,642.24	II
290		4,645.40	I
290		4,646.68	II
240		4,648.16	II
380		4,649.49	I
150		4,655.13	II
290		4,663.56	I
740		4,669.40	II
620		4,669.65	II
470	d	4,670.75	I
		4,670.83	I
1,100		4,674.60	II
680		4,676.91	II
210		4,681.55	I
370		4,687.18	II
370		4,688.73	I
130		4,693.63	II
120		4,699.34	II
530		4,704.40	II
270		4,713.06	II
130		4,715.26	II
730		4,716.10	I
270		4,717.07	I
210		4,717.72	II
190		4,718.33	II
270		4,719.84	II
130		4,726.02	II
770		4,728.42	I
470		4,745.68	II
150		4,750.72	I
730		4,760.27	I
110		4,770.20	I
110		4,774.15	II
190		4,777.85	II
580		4,783.10	I
350		4,785.86	I
160		4,789.96	I
230		4,791.58	II
430		4,815.81	II
130		4,829.57	II
970		4,841.70	I
310		4,844.21	II
140		4,847.76	II
270		4,848.32	I
120		4,854.36	II
210		4,883.77	I

730		4,883.97	I
170		4,904.97	I
630		4,910.40	I
350		4,913.25	II
430		4,918.99	I
110		4,924.04	I
120		4,938.10	II
170		4,948.63	II
120		4,952.37	II
170		4,961.94	II
170		4,975.98	I
140		5,028.44	II
400		5,044.28	I
200		5,052.76	II
170		5,069.46	II
540		5,071.20	I
170		5,100.22	II
		5,100.39	I
260		5,103.09	II
140		5,104.48	II
140		5,116.70	II
510		5,117.16	I
350		5,122.14	I
360		5,155.03	II
250		5,172.74	I
470		5,175.42	I
250		5,200.59	I
260		5,251.92	I
400		5,271.40	I
250		5,282.91	I
190		5,320.60	I
110		5,341.29	I
140		5,368.36	I
130		5,405.23	I
220		5,453.00	I
140		5,466.72	I
230		5,493.72	I
80		5,512.10	I
230		5,516.09	I
50		5,548.95	I
140		5,550.40	I
45		5,573.42	I
35		5,588.20	I
50		5,600.86	II
50		5,621.79	I
70		5,626.01	I
85		5,644.10	I
140		5,659.86	I
120		5,696.73	I
85		5,706.20	I
35		5,710.93	I
50		5,732.95	I
50		5,743.35	II
45		5,759.52	II
70		5,773.77	I
60		5,778.33	I
45		5,779.24	I
45		5,781.93	II
70	d	5,786.98	II
60		5,788.38	I
60		5,800.52	I
65		5,802.84	I
45		5,814.89	I
45		5,831.02	II
45		5,836.37	II
35		5,860.78	I
65		5,867.79	I
45		5,868.61	I
35		5,871.06	I
50		5,874.21	I
45		5,897.39	II
50		5,898.96	I
35		5,938.90	II
65		5,965.71	II
35	h	5,968.82	II
35		5,984.29	I
50		6,045.00	I
45		6,045.39	I
50		6,070.06	I
45		6,084.12	I
35	h	6,091.40	I
45		6,110.66	II
45	h	6,159.56	I
45		6,246.76	II
45		6,256.54	I
45		6,256.66	II
100		6,267.28	II
50		6,291.82	II
35		6,307.06	II
70		6,327.47	II
45		6,426.64	II
45		6,472.34	II
35		6,484.52	II
35		6,498.67	II
50		6,542.76	II
140		6,569.31	II
35	h	6,570.67	II
40	h	6,585.21	II
110		6,589.72	II
40		6,601.83	II
95		6,604.56	II
40	h	6,632.28	II
50		6,671.51	I
70		6,679.21	II
70		6,693.55	II
40	d	6,723.07	I
120	d	6,731.84	II
70	d	6,734.06	II
40	d	6,734.81	II
55		6,741.47	II
40	h	6,778.61	II
60		6,790.00	II
95		6,794.20	II
55		6,844.71	II
75		6,856.03	II
120		6,860.93	I
40		6,862.82	II
30		6,950.51	II
120		6,955.29	II
90		7,020.44	II
13		7,036.73	II
90		7,039.22	II
90		7,042.24	II
13		7,049.15	II
90		7,051.52	II
16		7,054.97	II
19		7,074.67	I
90		7,082.37	II
40	d	7,085.52	II
26		7,088.30	I
16		7,091.16	I
30		7,095.50	I
16		7,096.33	I
30		7,104.54	I
19		7,106.23	I
26		7,115.96	I
23		7,117.51	II
26	h	7,119.81	II
12		7,122.40	II
23	h	7,125.11	II
13		7,131.80	I
10		7,136.01	I
12		7,139.39	II
40	d	7,143.98	II
85	d	7,149.60	II
10		7,172.67	I
10		7,189.57	II
9		7,210.95	I
23		7,213.82	I
26	d	7,218.09	II
13		7,220.07	I
13		7,237.02	II
60		7,240.90	II
9		7,257.11	II
9	d	7,261.52	II
13		7,279.25	I
26		7,281.47	II
8		7,282.21	I
19		7,283.33	II
16		7,288.92	II
13		7,290.23	I
26	h	7,300.72	II
13		7,327.08	II
13		7,332.65	I
8		7,338.04	I
26		7,347.30	I
26		7,376.69	II
13		7,393.98	II
30		7,444.56	I
26		7,445.41	I
26	d	7,453.03	II
13		7,470.76	I
26		7,481.99	II
23	h	7,502.39	II
10	h	7,517.00	II
23	h	7,541.42	II
9		7,544.74	I
10		7,546.57	I
12	h	7,560.03	II
19		7,562.94	II
23		7,570.95	II
23		7,572.29	II
19		7,578.09	II
30		7,585.85	II
23		7,588.31	II
10		7,598.01	I
23	d	7,607.48	II
		7,607.74	I
12		7,613.94	II
10	h	7,631.77	II
23		7,637.94	II
45		7,645.09	II
12		7,645.82	I
19		7,648.02	II
10		7,655.78	II
19		7,667.20	II
8		7,672.49	II
10	h	7,678.79	II
10	h	7,695.78	I
23		7,712.04	II
30		7,728.56	II
30		7,736.26	II
30		7,749.30	II
23		7,755.20	II
10		7,794.50	I
10		7,801.54	I
8	h	7,812.75	II
16		7,820.15	II
10		7,831.40	II
40	w	7,835.08	II
26		7,837.27	II
10		7,844.82	II
6		7,859.53	I
19		7,863.65	II
10	h	7,880.01	II
16		7,895.96	I
26		7,914.96	II
90		7,928.14	II
9		7,931.92	I
19		7,937.09	II
16		7,948.12	II
19	w	8,001.61	II
19	w	8,014.92	II
23		8,025.12	II
23	w	8,026.32	II
16		8,032.03	II
40		8,048.70	II
16		8,065.16	I
45		8,068.46	II
9	w	8,117.16	II
9		8,125.12	II
26		8,161.82	II
19	w	8,195.50	II
6		8,206.30	II
26	w	8,218.76	II
9		8,230.33	I
16		8,240.98	II
19	w	8,289.26	II
10		8,300.88	II
40	w	8,305.79	II
10		8,315.45	I
19	w	8,348.68	II
19		8,383.71	I
19		8,387.77	II
30	w	8,432.64	II
19	w	8,473.54	II
45	w	8,485.99	II
30	w	8,510.90	II
23		8,543.22	II
23	w	8,617.03	II
23	w	8,632.82	II
12	w	8,677.81	II
13		8,706.32	II
45	w	8,708.43	II
30	w	8,717.89	II
30	w	8,758.28	II
16	w	8,780.59	II
23	w	8,788.83	II
26		8,859.76	II
95		8,913.66	II

SCANDIUM (SC)
Z = 21

Sc I and II
Ref. 1, 88 — C.H.C.

Intensity		Wavelength	
		Vacuum	
65		2,429.16	I
110		2,438.62	I
560		2,545.22	II
2,900		2,552.37	II
560		2,555.82	II
2,300		2,560.25	II
1,100		2,563.21	II
40		2,611.22	II
19		2,684.23	II
120		2,692.78	I
360		2,706.77	I
210		2,707.95	I
580		2,711.35	I
30		2,819.54	II
35		2,822.15	II
60		2,826.68	II
340		2,965.86	I
1,200		2,974.01	I
1,400		2,980.75	I
340		2,988.95	I
2,200		3,015.36	I
2,700		3,019.34	I
360		3,030.76	I
30		3,039.93	II
70		3,045.72	II
85		3,052.93	II
120	h	3,056.31	I
130		3,065.11	II
45		3,139.75	II
990		3,251.32	II
1,500		3,255.69	I
4,400		3,269.91	I
5,500		3,273.63	I
110	d	3,343.28	II
270		3,352.05	II
9,900		3,353.73	II
65	d	3,357.30	II
2,000		3,359.68	II
1,700		3,361.27	II
1,700		3,361.94	II
4,000		3,368.95	II
6,600		3,372.15	II
90		3,416.68	I
130		3,418.51	I
65		3,419.36	I
200		3,429.21	I
200		3,429.48	I
270		3,431.36	I
530		3,435.56	I
90		3,439.41	I
65		3,440.18	I
65		3,448.49	I
270		3,457.45	I
180		3,462.19	I
130	d	3,469.65	I
110		3,471.13	I
200		3,498.91	I
2,700		3,535.73	II
6,600		3,558.55	II
6,100		3,567.70	II
13,000		3,572.53	II
9,900		3,576.35	II
7,700		3,580.94	II
4,000		3,589.64	II
4,000		3,590.48	II
28,000		3,613.84	II
110		3,617.43	I
20,000		3,630.75	II
13,000		3,642.79	II
6,600		3,645.31	II
110		3,646.90	I
5,300		3,651.80	II
110		3,664.25	II
290		3,666.54	II
55		3,675.26	II
40		3,678.35	II
75	h	3,717.10	I
270		3,833.07	II
610		3,843.03	II
90		3,894.97	I
20,000		3,907.49	I
23,000		3,911.81	I
45		3,923.51	II
4,400		3,933.38	I
45		3,952.27	I
45		3,989.06	II
5,500		3,996.61	I
530		4,014.49	II
20,000		4,020.40	I
20,000		4,023.69	I
220		4,030.67	I
140		4,031.39	I
100		4,034.23	I

220		4,043.80	I
200		4,046.48	I
2,700		4,047.79	I
120		4,049.95	I
5,500		4,054.55	I
220		4,056.59	I
160	h	4,074.97	I
160		4,078.57	I
6,100		4,082.40	I
200		4,086.67	I
400		4,087.16	I
40	h	4,093.13	I
65		4,094.85	I
55	h	4,098.35	I
65		4,100.33	I
440	h	4,133.00	I
530	h	4,140.30	I
65	h	4,147.40	I
720		4,152.36	I
55	h	4,154.72	I
90	d	4,161.88	I
1,100	h	4,165.19	I
65	h	4,171.56	I
45	h	4,186.45	I
65	h	4,187.62	I
75		4,205.20	I
65		4,212.34	I
45		4,212.49	I
75	h	4,216.10	I
110	h	4,218.26	I
110	h	4,219.73	I
40		4,221.88	I
90	d	4,225.59	I
180		4,231.93	I
200		4,233.61	I
100		4,237.82	I
400		4,238.05	I
90		4,239.57	I
100		4,246.12	I
15,000		4,246.83	II
55		4,283.56	I
290		4,294.77	II
350		4,305.71	II
4,200		4,314.09	II
3,300		4,320.74	II
2,400		4,325.01	II
28		4,348.53	I
180		4,354.61	II
110		4,358.64	I
55		4,359.08	I
28		4,364.92	I
2,000		4,374.46	II
130		4,384.81	II
45	h	4,389.60	I
1,100		4,400.37	II
880		4,415.56	II
28		4,420.66	II
45		4,431.36	II
65		4,542.55	I
90		4,544.68	I
120	h	4,557.24	I
160	h	4,573.99	I
65	h	4,592.94	I
65	h	4,598.45	I
55	h	4,604.72	I
45		4,609.53	I
45		4,609.95	I
350		4,670.40	II
40	h	4,680.49	I
50		4,698.29	II
120		4,706.97	I
120		4,709.34	I
200		4,728.77	I
490		4,729.23	I
40	h	4,732.30	I
590		4,734.10	I
60		4,735.08	
690		4,737.65	I
790		4,741.02	I
1,200		4,743.81	I
200		4,753.16	I
220		4,779.35	I
90		4,791.50	I
100		4,827.28	I
100		4,833.67	I
170		4,839.44	I
40		4,840.47	I
80		4,847.68	I
80		4,852.68	I
140	b	4,857.79	S
80		4,906.67	I
90		4,909.76	I
90		4,922.84	I
90		4,934.25	I
45		4,935.74	I
70		4,941.33	I
170		4,954.06	I
120		4,973.66	I
150		4,980.37	I
80		4,983.45	I
140		4,991.92	I
80		5,018.39	I
70		5,020.14	I
80		5,021.51	I
530		5,031.02	II
55		5,032.74	I
250		5,064.32	I
80		5,068.86	I
530		5,070.23	I
250		5,075.81	I
2,100		5,081.56	I
1,200		5,083.72	I
1,100		5,085.55	I
750		5,086.95	I
390		5,087.14	I
270		5,089.89	I
45		5,092.46	I
390		5,096.73	I
620		5,099.23	I
370		5,101.12	I
180		5,109.06	I
150		5,112.86	I
320		5,116.69	I
70	b	5,133.68	S
45	b	5,171.06	S
390		5,210.52	I
45		5,211.28	I
280		5,219.67	I
350		5,239.82	II
280		5,258.33	I
35		5,284.97	I
210		5,285.76	I
35		5,301.94	I
22		5,318.35	II
70		5,331.77	I
14		5,334.23	II
95		5,339.41	I
120		5,341.05	I
95		5,342.96	I
350		5,349.30	I
120		5,349.71	I
60		5,350.30	I
210		5,355.75	I
530		5,356.10	I
14		5,357.19	II
270		5,375.35	I
370		5,392.08	I
45		5,416.12	I
45		5,425.57	I
45		5,429.41	I
35		5,432.94	I
55		5,433.23	I
45		5,438.22	I
55		5,439.03	I
55	h	5,442.60	I
270		5,446.20	I
18		5,447.39	I
120		5,451.34	I
30		5,455.21	I
18		5,465.20	I
55		5,468.40	I
60		5,472.19	I
18		5,474.64	I
750		5,481.99	I
530		5,484.62	I
570		5,514.22	I
16		5,515.39	I
660		5,520.50	I
18		5,526.06	I
660		5,526.82	II
55		5,541.04	I
30		5,546.40	I
18		5,550.40	I
5		5,552.25	II
35		5,553.59	I
16		5,561.10	I
70		5,564.86	I
18		5,571.24	I
14		5,579.76	I
110		5,591.33	I
35	h	5,593.38	I
22		5,604.19	I
22		5,631.02	I
80		5,640.98	II
45		5,646.36	I
16		5,647.60	I
55		5,649.56	I
250		5,657.88	II
60		5,658.34	II
55		5,667.16	II
70		5,669.04	II
1,500		5,671.81	I
95		5,684.20	II
1,200		5,686.84	I
1,100		5,700.21	I
190		5,708.61	I
880		5,711.75	I
230		5,717.28	I
180		5,724.08	I
55	b	5,736.85	S
55	b	5,764.45	S
95	b	5,772.74	S
55	b	5,775.32	S
70	b	5,809.84	S
70	b	5,811.60	S
95	b	5,847.73	S
70	b	5,849.07	S
70	b	5,887.38	S
35	b	5,918.04	S
30		5,919.11	I
60	b	5,928.10	S
35		5,961.49	I
60	b	5,968.25	S
35		5,969.19	I
90		5,988.42	I
160	b	6,017.07	S
60		6,026.18	I
620	b	6,036.17	S
490	b	6,064.31	S
440	b	6,072.65	S
620	b	6,079.30	S
320	b	6,101.87	S
370	b	6,109.93	S
370	b	6,115.97	S
180	b	6,148.70	S
150	b	6,153.93	S
150	b	6,188.09	S
150	b	6,192.90	S
620		6,210.68	I
90		6,239.41	I
320		6,239.78	I
120		6,245.63	II
110		6,249.96	I
250		6,258.96	I
60		6,262.25	I
55		6,276.31	I
45		6,279.76	II
18		6,300.70	II
750		6,305.67	I
26		6,309.90	II
16		6,320.85	II
26		6,344.83	I
60		6,378.82	I
55	b	6,408.41	S
90		6,413.35	I
26	b	6,437.08	S
55	b	6,446.24	S
26	b	6,457.78	S
35	b	6,485.40	S
26	b	6,495.90	S
55	b	6,525.62	S
22	b	6,535.30	S
45	b	6,557.84	S
35	b	6,566.88	S
18	b	6,575.85	S
60		6,604.60	II
26	b	6,609.99	S
18	b	6,617.94	S
18	b	6,645.08	S
22	b	6,654.42	S
26	b	6,661.01	S
18	b	6,700.48	S
18	b	6,705.93	S
65		6,737.87	I
35		6,739.40	I
35		6,817.08	I
50		6,819.52	I
29		6,829.54	I
50		6,835.03	I
5	b	6,963.12	S
5	b	6,990.68	S
5	b	7,025.72	S
8	b	7,035.77	S
5	b	7,072.37	S
5	b	7,094.38	S
12	h	7,138.14	I
14		7,169.13	I
12		7,257.57	I
8		7,275.57	I
3	h	7,300.62	I
12	h	7,524.13	I
14	h	7,553.96	I
15	h	7,574.44	I
11		7,617.45	I
14	h	7,665.72	I
30		7,697.73	I
18		7,729.72	I
55	h	7,741.17	I
5	h	7,750.37	I
5		7,752.72	I
6	h	7,771.06	I
15		7,785.17	I
8		7,794.68	I
30		7,800.44	I
11		7,821.64	I
11	h	8,196.98	I
15		8,241.13	I
19	h	8,761.40	I
11	h	8,774.8	
15	h	8,794.72	I
15	h	8,823.8	I
30	h	8,834.35	I
70		0,616.32	
30		0,985.81	
400		2,051.86	I
150		2,065.05	I

Sc III

Ref. 323 — C.H.C.

Intensity	Wavelength	
	Vacuum	
10	730.60	III
15	731.65	III
15	1,148.24	III
20	1,154.52	III
20	1,162.44	III
25	1,168.61	III
10	1,168.88	III
80	1,598.00	III
180	1,603.06	III
150	1,610.19	III
40	1,895.44	III
60	1,912.62	III
90	1,993.89	III
	Air	
160	2,010.42	III
50	2,012.26	III
350	2,699.07	III
230	2,734.05	III
10	2,831.75	III
80	4,061.21	III
100	4,068.66	III
40	4,309.47	III
10	4,740.95	III
15	4,780.87	III
50	4,992.89	III
60	5,032.09	III
80	6,256.01	III
60	6,307.60	III
90	7,449.16	III
70	7,548.15	III
70	7,868.65	III
35	8,814.29	III
50	8,829.78	III
30	8,865.89	III
15	8,881.58	III

Sc IV

Ref. 298 — C.H.C.

Intensity	Wavelength	
	Vacuum	
8	220.28	IV
15	289.85	IV
15	296.31	IV
15	299.04	IV
10	371.16	IV
9	438.80	IV
8	557.50	IV

8		584.83	IV
9		617.08	IV
8		761.43	IV
8		769.70	IV
10		785.12	IV
8		789.00	IV
8		791.71	IV
8		861.24	IV
8		861.30	IV
8		890.87	IV
8		1,219.40	IV
9		1,228.20	IV
9		1,424.66	IV
9		1,444.10	IV
8		1,489.64	IV
8		1,514.96	IV
8		1,535.76	IV
9		1,543.86	IV
9		1,549.55	IV
15		1,550.80	IV
8		1,555.72	IV
9		1,563.81	IV
10		1,574.92	IV
9		1,583.41	IV
8		1,584.64	IV
8		1,592.23	IV
8		1,660.71	IV
10		1,665.92	IV
8		1,746.23	IV
		Air	
10		2,056.06	IV
8		2,078.93	IV
12		2,118.97	IV
9		2,164.43	IV
11		2,185.43	IV
11		2,205.46	IV
14		2,222.22	IV
11		2,271.33	IV
9		2,464.45	IV
8		2,520.93	IV
11		2,586.93	IV
9		2,595.17	IV
8		2,678.01	IV
8		2,723.52	IV
8		2,773.04	IV
8	d	4,594.42	IV
8	d	4,639.96	IV
8		5,501.74	IV
9		5,620.72	IV
10		5,706.82	IV
14		5,771.63	IV
9		6,548.03	IV

Sc V

Ref. 150 — C.H.C.

Intensity		Wavelength	
		Vacuum	
150		179.42	V
350		180.14	V
200		180.82	V
200		180.96	V
50		181.55	V
200		182.39	V
300		228.56	V
100		230.85	V
40		243.82	V
500		243.87	V
400		246.42	V
400		250.98	V
500		252.85	V
500		253.73	V
50		255.38	V
300		255.64	V
200		257.16	V
150		258.24	V
40		258.81	V
50		260.05	V
400		281.00	V
900		283.91	V
800		284.45	V
600		288.29	V
900		289.59	V
1,000	d	291.93	V
800		293.25	V
400		296.17	V
700		300.00	V
400		375.05	V
100		378.68	V
200		388.68	V
400		395.32	V
200		399.50	V
1,000		573.36	V
600		587.94	V

SELENIUM (SE)

Z = 34

Se I and II

Ref. 80, 181, 216, 275 —R.L.K.

Intensity	Wavelength	
	Vacuum	
285	828.5	II
360	832.7	II
285	906.6	II
360	912.9	II
360	1,013.4	II
360	1,014.0	II
450	1,033.6	II
450	1,049.6	II
360	1,057.4	II
285	1,097.8	II
360	1,141.9	II
220	1,156.0	II
285	1,156.9	II
285	1,168.5	II
450	1,192.3	II
220	1,205.7	II
220	1,234.9	II
285	1,291.0	II
285	1,308.9	II
100	1,405.4	I
100	1,406.4	I
100	1,406.6	I
120	1,435.3	I
120	1,435.8	I
100	1,444.8	I
100	1,446.8	I
100	1,447.0	I
150	1,449.2	I
120	1,456.3	I
150	1,500.9	I
250	1,530.4	I
150	1,531.3	I
200	1,531.8	I
120	1,547.1	I
120	1,560.3	I
150	1,575.3	I
150	1,577.6	I
150	1,577.9	I
150	1,579.5	I
200	1,580.0	I
150	1,587.5	I
150	1,593.2	I
250	1,606.5	I
100	1,610.7	I
100	1,611.3	I
200	1,617.4	I
150	1,621.2	I
100	1,622.7	I
120	1,626.2	I
150	1,643.4	I
250	1,671.2	I
250	1,675.3	I
250	1,690.7	I
250	1,793.3	I
300	1,795.3	I
300	1,855.2	I
250	1,858.8	I
400	1,898.6	I
350	1,913.8	I
300	1,919.2	I
500	1,960.9	I
150	1,995.1	I
	Air	
500	2,039.8	I
500	2,074.8	I
500	2,164.2	I
150	2,332.8	I
600	2,413.5	I
300	2,548.0	I
220	3,038.7	II
220	3,041.3	II
285	4,070.2	II
360	4,175.3	II
450	4,180.9	II
120	4,328.7	I
100	4,330.3	I
285	4,382.9	II
285	4,446.0	II
220	4,449.2	II
285	4,467.6	II
500	4,730.8	I
400	4,739.0	I
300	4,742.2	I
285	4,840.6	II
360	4,845.0	II
450	5,227.5	II
360	5,305.4	II
100	5,365.5	I
120	5,369.9	I
110	5,374.1	I
285	5,522.4	II
285	5,566.9	II
285	5,866.3	II
450	6,056.0	II
200	6,325.6	I
360	6,444.2	II
285	6,490.5	II
285	6,535.0	II
150	6,831.3	I
120	6,990.690	I
100	6,991.792	I
200	7,010.809	I
150	7,013.875	I
300	7,062.065	I
200	7,575.1	I
250	7,583.4	I
150	7,592.2	I
120	7,606.8	I
300	8,001.0	I
200	8,036.4	I
120	8,060.9	I
120	8,065.3	I
120	8,081.1	I
150	8,093.2	I
150	8,094.7	I
180	8,149.3	I
150	8,152.0	I
200	8,157.7	I
180	8,163.1	I
150	8,182.9	I
100	8,185.0	I
120	8,194.6	I
150	8,440.47	I
150	8,450.38	I
150	8,742.33	I
300	8,918.86	I
100	8,969.69	I
200	9,001.97	I
200	9,038.61	I
80	9,083.14	I
120	9,088.79	I
80	9,140.83	I
60	9,181.88	I
60	9,271.12	I
100	9,432.50	I
60	9,825.58	I
200	10,217.25	I
377	10,307.45	I
900	10,327.26	I
640	10,386.36	I
124	10,650.30	I
125	11,934.56	I
275	11,946.87	I
100	11,047.92	I
105	11,952.27	I
170	11,952.64	I
100	11,966.04	I
205	11,972.93	I
115	11,973.07	I
315	14,817.93	I
410	14,917.47	I
500	15,151.44	I
115	15,469.06	I
320	15,471.00	I
265	15,520.97	I
395	15,618.40	I
115	15,620.38	I
360	16,650.44	I
505	16,813.78	I
165	16,817.76	I
205	16,866.54	I
115	16,972.71	I
235	21,374.24	I
680	21,442.56	I
415	21,473.48	I
270	21,716.36	I
240	21,730.60	I
105	23,133.66	I
150	23,388.85	I
110	23,628.17	I
265	24,148.18	I
170	24,159.23	I
185	24,204.44	I
375	24,385.99	I
160	24,413.67	I
225	24,471.17	I
255	25,017.51	I
510	25.127.43	I

Se III

Ref. 9, 247 — R.L.K.

Intensity	Wavelength	
	Vacuum	
220	709.2	III
220	709.4	III
220	720.6	III
360	724.3	III
285	726.4	III
220	737.2	III
220	741.9	III
285	777.3	III
220	790.8	III
360	843.0	III
220	879.2	III
285	953.7	III
220	954.4	III
220	954.7	III
160	974.1	III
360	974.8	III
285	1,079.8	III
360	1,099.1	III
450	1,119.2	III
	Air	
285	2,057.5	III
285	2,767.2	III
220	2,773.8	III
285	3,379.8	III
450	3,387.2	III
450	3,413.9	III
285	3,428.4	III
450	3,457.8	III
360	3,543.6	III
285	3,570.2	III
450	3,637.6	III
360	3,711.7	III
450	3,738.7	III
285	3,743.0	III
450	3,800.9	III
360	4,046.7	III
220	4,083.2	III
450	4,169.1	III
220	4,637.9	III
285	6,303.8	III

Se IV

Ref. 245 — R.L.K.

Intensity	Wavelength	
	Vacuum	
285	636.0	IV
285	654.2	IV
360	652.7	IV
450	670.1	IV
285	671.9	IV
220	722.8	IV
285	734.6	IV
450	746.4	IV
285	759.0	IV
285	776.5	IV
285	803.8	IV
360	959.6	IV

Intensity	Wavelength	
450	996.7	IV
220	1,307.2	IV
285	1,314.4	IV

Air

Intensity	Wavelength	
220	2,090.0	IV
285	2,136.6	IV
160	2,165.2	IV
160	2,166.6	IV
360	2,665.5	IV
285	2,724.3	IV
160	2,951.6	IV

Se V

Ref. 245 — R.L.K.

Intensity	Wavelength	
	Vacuum	
285	596.0	V
285	601.0	V
220	608.7	V
360	613.0	V
285	614.3	V
450	759.1	V
285	785.8	V
285	804.3	V
360	808.7	V
220	814.8	V
220	820.7	V
360	830.3	V
450	839.5	V
360	845.8	V
360	1,094.7	V
220	1,151.0	V
450	1,227.6	V

SILICON (SI)

Z = 14

Si I and II

Ref. 170, 237, 292 — L.J.R.

Intensity		Wavelength	
		Vacuum	
10	h	805.10	II
20	h	820.52	II
20	h	843.72	II
40	h	845.77	II
10		850.14	II
100		889.72	II
200		892.00	II
10		899.41	II
20		901.74	II
10		913.01	II
20		913.85	II
20		929.81	II
100		989.87	II
200		992.68	II
25		1,020.70	II
50		1,023.69	II
30		1,057.05	II
15		1,057.50	II
20	h	1,127.44	II
40	h	1,127.91	II
100		1,190.42	II
200		1,193.28	II
250		1,194.50	II
100		1,197.39	II
10	h	1,216.12	II
20		1,223.91	II
20		1,224.25	II
10		1,224.97	II
50		1,226.81	II
20		1,226.89	II
40		1,226.99	II
100		1,227.60	II
10		1,228.44	II
25		1,228.62	II
150		1,228.75	II
200		1,229.39	II
10		1,235.92	II
100		1,246.74	II
150		1,248.43	II
100		1,250.09	II
150		1,250.43	II
200		1,251.16	II
10		1,255.28	I
40		1,256.49	I
50		1,258.80	I
1,000		1,260.42	II
2,000		1,264.73	II
200		1,265.02	II
100		1,304.37	II
50	h	1,305.59	II
200		1,309.27	II
20	h	1,309.46	II
100		1,346.87	II
100		1,348.54	II
150		1,350.06	II
20		1,350.52	II
20		1,350.66	II
100		1,352.64	II
100		1,353.72	II
10	h	1,409.07	II
20	h	1,410.22	II
10	h	1,416.97	II
15	h	1,474.65	II
15		1,484.87	II
90	h	1,485.02	II
30		1,485.22	II
100	h	1,485.51	II
100	h	1,509.10	II
50	h	1,512.07	II
30	p	1,513.57	II
60	p	1,516.91	II
500		1,526.72	II
1,000		1,533.45	II
10		1,562.45	II
15		1,562.85	II
10		1,563.77	II
50		1,573.87	I
50		1,574.82	I
50		1,592.41	I
150		1,594.55	I
50		1,594.93	I
30		1,597.95	I
100		1,622.87	I
30		1,625.71	I
300		1,629.43	I
200		1,629.92	I
75		1,631.13	I
50		1,633.98	I
30	h	1,653.35	I
30		1,664.52	I
50		1,666.37	I
100		1,667.62	I
100		1,668.52	I
100		1,672.59	I
200		1,675.20	I
30		1,682.68	I
30		1,686.82	I
50	h	1,689.29	I
30	h	1,690.79	I
50		1,693.29	I
50		1,695.51	I
200		1,696.20	I
200		1,697.94	I
50		1,700.42	I
30		1,700.63	I
30		1,702.86	I
50		1,704.43	I
10	h	1,710.83	II
20	h	1,711.30	II
30	h	1,743.88	I
50		1,747.40	I
30	h	1,753.11	I
50		1,763.66	I
40		1,765.03	I
30	h	1,765.60	I
30		1,766.06	I
30		1,770.63	I
100	h	1,770.92	I
100	h	1,776.83	I
50	h	1,783.23	I
100	h	1,799.12	I
150		1,808.00	II
50	h	1,809.09	I
500	h	1,814.07	I
200		1,816.92	II
10		1,817.45	II
50		1,822.45	I
200		1,836.51	I
30	h	1,838.01	I
100	h	1,841.15	I
200		1,841.44	I
200		1,843.77	I
300		1,845.51	I
100		1,846.10	I
400		1,847.47	I
200		1,848.14	I
100		1,848.74	I
500		1,850.67	I
30	h	1,851.79	I
200		1,852.46	I
50		1,853.15	I
20		1,869.32	II
15		1,870.23	II
100		1,873.10	I
500	h	1,874.84	I
100		1,875.81	I
200		1,881.85	I
200		1,887.70	I
200	h	1,893.25	I
1,000	h	1,901.33	I
100	h	1,902.46	II
50	h	1,904.66	I
50	h	1,910.62	II
50		1,941.67	II
15		1,944.59	II
10		1,949.33	II
100		1,949.56	II
100		1,954.97	I
30		1,984.43	I
50		1,991.85	I
		Air	
30		2,010.97	I
50		2,054.83	I
50		2,058.65	II
50		2,059.01	II
40		2,061.19	I
30		2,065.52	I
200		2,072.02	II
200		2,072.70	II
30	h	2,103.21	I
30		2,114.63	I
100		2,124.12	I
10	h	2,133.99	II
30	h	2,136.40	II
50	h	2,136.56	II
50	h	2,147.91	I
110		2,207.98	I
115		2,210.89	I
110		2,211.74	I
120		2,216.67	I
120		2,218.06	I
50		2,218.91	I
35		2,291.03	I
55		2,303.06	I
30		2,334.40	II
30		2,334.61	II
10		2,344.20	II
10	h	2,349.54	II
20		2,350.17	II
20	h	2,353.09	II
100	h	2,356.30	II
30	h	2,357.18	II
50	h	2,357.97	II
10	h	2,360.20	II
30		2,366.97	II
20		2,374.26	II
10	h	2,428.45	II
300		2,435.15	I
65		2,438.77	I
65		2,443.36	I
70		2,452.12	I
425		2,506.90	I
375		2,514.32	I
500		2,516.113	I
350		2,519.202	I
425		2,524.108	I
450		2,528.509	I
110		2,532.381	I
30		2,563.679	I
85		2,568.641	I
45		2,577.151	I
190		2,631.282	I
10	h	2,682.21	II
1,000		2,881.579	I
10	h	2,887.51	II
300		2,904.28	II
500		2,905.69	II
55		2,970.355	I
150		2,987.645	I
50		3,006.739	I
100	h	3,030.00	II
75		3,020.004	I
20	h	3,021.55	II
20	h	3,041.57	II
30	h	3,042.19	II
100	h	3,043.69	II
10	h	3,043.85	II
10	h	3,045.77	II
50	h	3,048.30	II
150	h	3,053.18	II
150		3,188.97	II
50		3,192.25	II
150		3,193.09	II
50		3,194.21	II
50		3,194.69	II
100		3,195.41	II
200		3,199.51	II
20		3,202.49	II
100	h	3,203.87	II
200	h	3,210.03	II
75		3,214.66	II
15	h	3,217.99	II
10		3,220.44	II
20		3,223.01	II
300		3,333.14	II
500		3,339.82	II
100	h	3,853.66	II
500	h	3,856.02	II
200	h	3,862.60	II
300		3,905.523	I
10	h	3,955.74	II
10	h	3,977.46	II
15	h	3,991.77	II
10	h	3,998.01	II
20	h	4,075.45	II
15	h	4,076.78	II
70		4,102.936	I
300	h	4,128.07	II
500	h	4,130.89	II
10	h	4,183.35	II
100	h	4,190.72	II
50		4,198.13	II
100		4,621.42	II
150		4,621.72	II
50		4,782.991	I
35		4,792.212	I
80		4,792.324	I
15	h	4,883.20	II
20	h	4,906.99	II
20	h	4,932.80	II
30		4,947.607	I
40		5,006.061	I
1,000		5,041.03	II
1,000		5,055.98	II
100		5,181.90	II
100	h	5,185.25	II
200	h	5,192.86	II
500	h	5,202.41	II
30	h	5,295.19	II
100	h	5,405.34	II
15	h	5,417.24	II
15	h	5,428.92	II
15	h	5,432.89	II
100	h	5,438.62	II
20	h	5,447.26	II
15	h	5,454.49	II
100	h	5,456.45	II
500	h	5,466.43	II
500	h	5,466.87	II
100	h	5,469.21	II
40		5,493.23	I
200	h	5,496.45	II
35		5,517.535	I
100	h	5,540.74	II
150	h	5,576.66	II
30		5,622.221	I
100	h	5,632.97	II
200	h	5,639.48	II
90		5,645.611	I
150	h	5,660.66	II
80		5,665.554	I
1,000	h	5,669.56	II
30	h	5,681.44	II
120		5,684.484	I
300	h	5,688.81	II
100		5,690.425	I

90		5,701.105	I
200	h	5,701.37	II
100	h	5,706.37	II
160		5,708.397	I
45		5,747.667	I
45		5,753.625	I
45		5,754.220	I
45		5,762.977	I
70		5,772.145	I
70		5,780.384	I
30	h	5,785.73	II
90		5,793.071	I
30	h	5,794.90	II
100		5,797.859	I
150	h	5,800.47	II
200		5,806.74	II
30		5,827.80	II
50		5,846.13	II
10		5,867.48	II
300	h	5,868.40	II
40		5,873.764	I
150		5,915.22	II
200		5,948.545	I
500		5,957.56	II
500		5,978.93	II
10	h	6,067.45	II
20	h	6,080.06	II
10	h	6,086.67	II
90		6,125.021	I
85		6,131.574	I
90		6,131.850	I
100		6,142.487	I
100		6,145.015	I
160		6,155.134	I
160		6,237.320	I
40		6,238.287	I
125		6,243.813	I
125		6,244.468	I
180		6,254.188	I
45		6,331.954	I
1,000		6,347.10	II
1,000		6,371.36	II
45		6,526.609	I
45		6,527.199	I
45		6,555.462	I
50	h	6,660.52	II
15		6,665.00	II
100		6,671.88	II
20		6,699.38	II
50	h	6,717.04	II
100		6,721.853	I
30		6,741.64	I
20	h	6,750.28	II
30		6,818.45	II
50		6,829.82	II
30		6,848.568	I
80		6,976.523	I
180		7,003.567	I
180		7,005.883	I
30		7,017.28	I
90		7,017.646	I
250		7,034.903	I
70		7,164.69	I
200		7,165.545	I
70		7,184.89	I
65		7,193.58	I
30		7,193.90	I
100		7,226.206	I
100		7,235.326	I
60		7,235.82	I
180		7,250.625	I
160		7,275.294	I
40		7,282.81	I
400		7,289.173	I
55		7,290.26	I
35		7,373.00	I
375		7,405.774	I
200		7,409.082	I
40		7,415.35	I
275		7,415.946	I
425		7,423.497	I
85		7,424.60	I
100		7,680.267	I
40		7,742.71	I
30		7,800.008	I
400		7,848.80	II
500		7,849.72	II
30		7,849.967	I
90		7,918.386	I
120		7,932.349	I
140		7,944.001	I
35		7,970.306	I
35		8,035.619	I
70		8,093.241	I
35		8,230.642	I
40		8,443.982	I
40		8,501.547	I
60		8,502.221	I
40		8,536.165	I
120		8,556.780	I
50		8,648.462	I
40		8,728.011	I
75		8,742.451	I
100		8,752.009	I
35		8,790.389	I
100		9,412.72	II
100		9,413.506	I
30		10,371.269	I
120		10,585.141	I
120		10,603.431	I
120		10,660.975	I
30		10,694.251	I
30		10,727.408	I
60		10,749.384	I
30		10,784.550	I
80		10,786.856	I
140		10,827.091	I
60		10,843.854	I
30		10,868.79	I
130		10,869.541	I
30		10,882.802	I
30		10,885.336	I
80		10,979.308	I
30		10,982.061	I
80		11,017.965	I
13		11,187.60	I
12		11,289.84	I
12		11,611.09	I
370		11,984.19	I
220		11,991.57	I
440		12,031.51	I
150		12,103.53	I
120		12,270.68	I
11		13,176.90	I
190		15,888.39	I
40		15,960.04	I
95		16,060.03	I
20		16,094.80	I
60		16,163.71	I
11		16,215.68	I
16		16,381.55	I
29		16,680.77	I
28		17,327.29	I
26		18,722.90	I
15		19,385.94	I
48		19,432.97	I
13		19,493.38	I
110		19,722.50	I
31		19,928.88	I
12		20,917.13	I
21		21,354.24	I
12		22,062.71	I

Si III

Ref. 320 — L.J.R.

Intensity		Wavelength	
		Vacuum	
8		566.61	III
6		652.22	III
8		653.33	III
5		673.48	III
5		800.07	III
9		823.41	III
5		883.40	III
7		939.09	III
9		967.95	III
10		993.52	III
13		994.79	III
16		997.39	III
7		1,005.37	III
7		1,031.16	III
8		1,033.92	III
7		1,037.05	III
6		1,083.22	III
14		1,108.37	III
16		1,109.97	III
18		1,113.23	III
6		1,140.55	III
7		1,141.58	III
6		1,142.28	III
8		1,144.31	III
6		1,144.96	III
8		1,145.11	III
7		1,145.18	III
6		1,155.00	III
6		1,155.96	III
7		1,158.10	III
6		1,160.26	III
8		1,161.58	III
5		1,174.37	III
6		1,174.43	III
8		1,178.00	III
30		1,206.51	III
30		1,206.53	III
9		1,207.52	III
10		1,210.46	III
7		1,235.43	III
6		1,280.35	III
17		1,294.54	III
14		1,296.73	III
15		1,298.89	III
18		1,298.96	III
14		1,301.15	III
16		1,303.32	III
13		1,312.59	III
8		1,341.47	III
7		1,342.39	III
6		1,343.39	III
8		1,361.60	III
5		1,362.37	III
7		1,363.47	III
8		1,365.26	III
7		1,367.05	III
5		1,369.44	III
5		1,373.03	III
5		1,387.99	III
13		1,417.24	III
6		1,433.69	III
8		1,435.77	III
7		1,436.17	III
5		1,441.73	III
6		1,447.20	III
5		1,457.25	III
12		1,500.24	III
10		1,501.19	III
9		1,501.87	III
6		1,506.06	III
7		1,673.32	III
9		1,842.55	III
		Air	
5		2,176.89	III
6		2,295.48	III
10		2,296.87	III
8		2,300.93	III
10		2,308.19	III
11		2,449.48	III
6		2,483.20	III
25		2,541.82	III
10		2,546.09	III
14		2,559.21	III
11		2,640.79	III
14		2,655.51	III
9		2,817.11	III
7		2,831.49	III
5		2,839.62	III
5		2,959.15	III
5		2,980.52	III
5		3,013.09	III
6		3,034.73	III
8		3,037.29	III
9		3,040.93	III
7		3,043.93	III
5		3,045.08	III
7		3,068.24	III
25		3,086.24	III
6		3,086.46	III
20		3,093.42	III
5		3,093.65	III
16		3,096.83	III
6		3,126.27	III
7		3,147.37	III
8		3,161.61	III
16		3,185.13	III
13		3,186.02	III
14		3,196.50	III
15		3,210.55	III
7		3,216.25	III
12		3,230.50	III
14		3,233.95	III
15		3,241.62	III
7		3,253.40	III
5		3,253.74	III
7		3,254.80	III
12		3,258.66	III
6		3,270.46	III
10		3,276.26	III
7		3,279.26	III
15		3,486.91	III
9		3,525.94	III
8		3,569.67	III
20		3,590.47	III
8	h	3,622.54	III
5	h	3,639.45	III
6	h	3,645.12	III
7	h	3,681.40	III
5	h	3,682.15	III
20	c	3,791.41	III
25		3,796.11	III
30		3,806.54	III
7		3,842.46	III
20		3,924.47	III
6	h	3,947.49	III
6		3,963.84	III
5		3,981.24	III
5	h	4,101.86	III
8		4,102.42	III
5	h	4,115.50	III
9		4,338.50	III
8		4,341.40	III
8	h	4,377.63	III
6	h	4,405.90	III
8	h	4,406.72	III
6		4,494.05	III
30		4,552.62	III
8		4,554.00	III
25		4,567.82	III
20		4,574.76	III
7		4,619.66	III
7		4,638.28	III
8		4,665.87	III
9		4,683.02	III
7		4,683.80	III
16		4,716.65	III
7		4,730.52	III
8		4,800.43	III
15		4,813.33	III
16		4,819.72	III
18		4,828.97	III
10	h	5,091.42	III
7	h	5,113.76	III
8	h	5,114.12	III
5		5,197.26	III
6		5,451.46	III
7		5,473.05	III
7		5,704.60	III
8		5,716.29	III
20		5,739.73	III
10	h	5,898.79	III
7		6,314.46	III
6	h	6,524.36	III
6	h	6,831.56	III
6	h	6,851.65	III
5	h	7,461.89	III
8	h	7,462.62	III
9	h	7,466.32	III
12	h	7,612.36	III
9	h	8,102.86	III
11	h	8,103.45	III
7	h	8,190.43	III
6	h	8,191.16	III
8	h	8,191.68	III
9	h	8,262.57	III
5	h	8,265.64	III
8	h	8,269.32	III
5	h	8,271.38	III
6	h	8,271.94	III

Si IV

Ref. 319 — L.J.R.

Intensity		Wavelength	
		Vacuum	
4		457.82	IV
3		458.16	IV
2		515.12	IV
3		516.35	IV
2		645.76	IV
5		749.94	IV
7		815.05	IV

8		818.13	IV
8		1,066.63	IV
8		1,122.49	IV
10		1,128.34	IV
15		1,393.76	IV
12		1,402.77	IV
1		1,634.61	IV
6		1,722.53	IV
5		1,727.38	IV
3		2,120.18	IV
4		2,127.47	IV
5	h	2,287.04	IV
2	h	2,328.56	IV
2		2,366.76	IV
3		2,370.99	IV
2		2,482.82	IV
1		2,485.38	IV
7		2,517.51	IV
1		2,672.19	IV
4		2,675.12	IV
4		2,675.25	IV
1		2,677.57	IV
3	h	2,723.81	IV
3	h	2,895.13	IV
2	h	2,904.47	IV
1	h	2,971.52	IV
7		3,149.56	IV
9		3,165.71	IV
1	h	3,244.19	IV
8		3,762.44	IV
6		3,773.15	IV
1	h	4,031.39	IV
2	h	4,038.06	IV
10		4,088.85	IV
9		4,116.10	IV
7	h	4,212.41	IV
3		4,314.10	IV
5		4,328.18	IV
2	h	4,403.73	IV
		4,411.65	IV
1	h	4,611.27	IV
3	h	4,628.62	IV
9	h	4,631.24	IV
10	h	4,654.32	IV
3	h	4,656.92	IV
1	h	4,667.14	IV
2	h	4,673.30	IV
1	h	4,947.45	IV
3		4,950.11	IV
2	h	5,304.97	IV
1	h	5,309.49	IV
5		6,667.56	IV
7		6,701.21	IV
3	h	6,998.36	IV
6	h	7,047.94	IV
4	h	7,068.41	IV
2	h	7,630.50	IV
4	h	7,654.56	IV
4	h	7,678.75	IV
5	h	7,718.79	IV
6	h	7,723.82	IV
2	h	7,725.64	IV
1	h	7,730.47	IV
1	h	7,752.91	IV
1	h	8,240.61	IV
2	h	8,957.25	IV
1	h	9,018.16	IV

Si V

Ref. 87 — L.J.R.

Intensity		Wavelength	
		Vacuum	
1		78.61	V
1		78.90	V
2		80.81	V
2		81.11	V
10		85.18	V
6		85.58	V
4		90.45	V
4		90.85	V
15		96.44	V
10		97.14	V
2		98.21	V
20		117.86	V
20		118.97	V

SILVER (AG)

Z = 47

Ag I and II

Ref. 13, 99, 255, 286, 289 — C.H.C.

Intensity		Wavelength	
		Vacuum	
25		730.83	II
30		752.80	II
15		1,005.32	II
10		1,065.49	II
12		1,072.23	II
250		1,074.22	II
150		1,107.03	II
150		1,112.46	II
60		1,195.83	II
50		1,223.33	II
50		1,240.80	II
50		1,246.87	II
55		1,256.81	II
55		1,257.55	II
50		1,266.63	II
70		1,273.67	II
65		1,297.51	II
85		1,311.20	II
55		1,313.81	II
50		1,314.61	II
60		1,323.84	II
60		1,342.09	II
50		1,342.57	II
70		1,346.62	II
50		1,353.54	II
150		1,364.50	II
100		1,396.00	II
100		1,410.93	II
90		1,419.72	II
95		1,432.60	II
100		1,464.72	II
50		1,466.23	II
50	r	1,507.37	I
100	r	1,515.63	I
50	r	1,548.58	I
100		1,555.16	II
100		1,644.50	II
60		1,651.52	I
50		1,652.10	I
120		1,682.82	II
10		1,708.11	I
50		1,709.27	I
125		1,736.44	II
10	h	1,766.14	I
75		1,790.37	II
20		1,847.71	I
100		1,967.38	II
150		2,015.96	II
150		2,033.98	II
200		2,061.17	I
100		2,069.85	I
80	r	2,113.82	II
60		2,145.60	II
15		2,170.00	I
50		2,186.76	II
60		2,229.53	II
100	r	2,246.43	II
75	r	2,248.74	II
75		2,280.03	II
30	h	2,309.56	I
10	h	2,312.60	I
70	r	2,317.05	II
80	r	2,320.29	II
70	r	2,324.68	II
80	r	2,331.40	II
70		2,357.92	II
50	h	2,375.02	I
75		2,411.41	II
90	r	2,413.23	II
100	r	2,437.81	II
100		2,447.93	II
80		2,473.84	II
60		2,506.63	II
50	h	2,575.63	I
60		2,660.49	II
60		2,721.77	II
75		2,767.54	II
100	h	2,824.39	I
10	h	2,926.77	I
20	h	2,938.42	I
20		3,099.10	I
30	h	3,130.02	I
10	h	3,170.58	I
90		3,180.70	II
15	h	3,215.67	I
10		3,225.15	I
15		3,233.18	I
100		3,267.35	II
55,000	r	3,280.68	I
10	h	3,305.67	I
28,000	r	3,382.89	I
10	h	3,403.78	I
30		3,469.16	I
70		3,475.82	II
80		3,495.28	II
20	h	3,501.92	I
20		3,508.03	I
15	h	3,513.38	I
10		3,521.12	I
50		3,542.61	I
10	h	3,547.16	I
10	h	3,557.01	I
20	h	3,586.67	I
10	h	3,623.49	I
50	h	3,624.68	I
75		3,682.46	II
30		3,682.50	I
80		3,683.34	II
50	h	3,709.20	I
10	h	3,727.42	I
20	h	3,753.14	I
200		3,810.94	I
50		3,811.78	I
100	h	3,840.74	I
15		3,847.85	I
50	h	3,907.41	I
50		3,909.31	II
50	h	3,914.40	I
70		3,920.10	II
10	h	3,928.01	I
10	h	3,940.43	I
10	h	3,942.97	I
60		3,949.43	II
100	h	3,981.58	I
70		3,985.19	II
10	h	3,992.15	I
100	h	4,055.48	I
10	h	4,083.43	I
80		4,085.91	II
100		4,185.48	II
90	h	4,210.96	I
100		4,212.82	I
50		4,311.07	I
20		4,396.23	I
50	h	4,476.04	I
20	h	4,556.0	I
30	h	4,615.69	I
80		4,620.04	II
50		4,620.46	II
60	h	4,668.48	I
30	h	4,677.60	I
100		4,788.40	II
20	h	4,796.2	I
30	h	4,847.82	I
100		4,874.10	I
20		4,888.21	I
10	h	4,917.5	I
10		4,935.75	I
20	h	4,992.89	I
80		5,027.35	II
15	h	5,123.50	I
1,000		5,209.08	I
10	h	5,333.62	I
1,000		5,465.50	I
100		5,471.55	I
20		5,475.38	I
20	h	5,545.67	I
10	h	5,559.58	I
100		5,667.34	I
10	h	6,083.78	I
10	h	6,268.50	I
20		6,621.08	I
20		7,359.96	I
320		7,687.78	I
25		8,005.4	II
15		8,254.7	II
500		8,273.52	I
20		8,324.4	II
15		8,379.5	II
25		8,403.8	II
15		8,492.5	II
30	h	8,645.70	I
10	h	8,704.85	I
12		8,747.6	II
15		9,000.9	II
10		12,551.0	I
60		16,819.5	I
20		17,416.7	I
15		18,307.9	I
15		18,382.3	I

SODIUM (NA)

Z = 11

Na I and II

Ref. 268, 334 — T.K.M.

Intensity		Wavelength	
		Vacuum	
160		300.15	II
160		300.20	II
90		301.32	II
100		301.44	II
60		302.45	II
300		372.08	II
350		376.38	II
60		1,293.97	II
50		1,327.74	II
45		1,347.54	II
90		1,374.69	II
90		1,404.68	II
45		1,495.21	II
40		1,496.01	II
45		1,497.73	II
80		1,506.41	II
60		1,506.91	II
70		1,513.10	II
60		1,519.63	II
60		1,657.92	II
90		1,776.57	II
40		1,778.24	II
60		1,783.04	II
80		1,787.19	II
45		1,788.85	II
80		1,798.41	II
45		1,801.26	II
90		1,807.09	II
60		1,808.38	II
50		1,821.70	II
45		1,833.87	II
80		1,835.22	II
45		1,837.89	II
60		1,841.82	II
70		1,845.02	II
45		1,850.15	II
70		1,851.19	II
80		1,853.17	II
45		1,866.45	II
45		1,873.37	II
60		1,875.08	II
160		1,881.91	II
50		1,885.09	II
45		1,885.74	II
		Air	
80		2,228.53	II
80		2,303.58	II
300		2,315.65	II
130		2,393.28	II
100		2,401.01	II
300		2,420.99	II
300		2,424.73	II
200		2,439.14	II
250		2,441.50	II
200		2,448.72	II
200		2,452.18	II
1,000		2,493.15	II

300	2,502.84	II
450	2,506.30	II
600	2,515.46	II
600	2,531.54	II
550	2,586.31	II
600	2,594.96	II
850	2,611.81	II
300	2,627.41	II
300	2,631.81	II
160	2,648.53	II
200	2,651.31	II
200	2,659.81	II
850	2,661.00	II
200	2,663.46	II
350	2,666.46	II
1,000	2,671.83	II
200	2,674.04	II
850	2,678.09	II
650	2,808.71	II
850	2,809.52	II
350	2,818.29	II
600	2,829.87	II
800	2,839.56	II
1,000	2,841.72	II
16	2,852.81	I
15	2,853.01	I
650	2,856.51	II
800	2,859.49	II
350	2,861.02	II
750	2,871.28	II
650	2,872.95	II
900	2,881.15	II
850	2,886.26	II
700	2,893.95	II
900	2,901.14	II
800	2,904.72	II
1,100	2,904.92	II
1,100	2,917.52	II
1,100	2,919.05	II
1,200	2,919.85	II
1,300	2,920.95	II
1,000	2,923.49	II
750	2,930.88	II
850	2,934.08	II
950	2,937.74	II
450	2,942.66	II
800	2,945.70	II
950	2,947.50	II
1,200	2,951.24	II
1,100	2,952.40	II
850	2,960.12	II
450	2,965.74	II
500	2,970.73	II
600	2,974.24	II
750	2,974.99	II
1,000	2,977.13	II
1,100	2,979.66	II
1,100	2,980.63	II
1,300	2,984.19	II
550	3,004.15	II
750	3,007.44	II
750	3,009.14	II
450	3,009.48	II
600	3,015.40	II
450	3,017.34	II
400	3,029.07	II
400	3,037.08	II
400	3,045.60	II
550	3,053.67	II
550	3,055.35	II
550	3,056.16	II
550	3,057.38	II
550	3,057.95	II
550	3,058.72	II
700	3,060.25	II
800	3,061.35	II
500	3,064.38	II
500	3,066.22	II
500	3,066.54	II
550	3,074.33	II
550	3,078.32	II
550	3,080.25	II
550	3,087.06	II
450	3,088.26	II
550	3,092.04	II
550	3,092.73	II
650	3,094.45	II
650	3,095.55	II
500	3,103.58	II

500	3,104.40	II
450	3,111.45	II
500	3,113.69	II
400	3,122.94	II
1,700	3,124.42	II
600	3,125.21	II
600	3,129.38	II
2,500	3,135.48	II
1,700	3,137.86	II
950	3,145.71	II
2,000	3,149.28	II
2,000	3,163.74	II
700	3,175.09	II
1,000	3,179.06	II
1,700	3,189.79	II
1,600	3,212.19	II
700	3,234.93	II
1,500	3,257.96	II
650	3,260.21	II
950	3,274.22	II
1,700	3,285.60	II
1,700	3,301.35	II
19	3,302.37	I
18	3,302.98	I
1,500	3,304.96	II
1,000	3,318.04	II
950	3,327.69	II
6	3,426.86	I
1,500	3,533.05	II
1,200	3,631.27	II
850	3,711.07	II
200	4,081.37	II
300	4,113.70	II
250	4,123.08	II
250	4,233.26	II
2	4,238.99	I
200	4,240.37	II
250	4,240.90	II
3	4,242.08	I
4	4,273.64	I
5	4,276.79	I
2	4,287.84	I
3	4,291.01	I
250	4,292.48	II
250	4,292.86	II
250	4,308.81	II
250	4,309.04	II
250	4,320.91	II
6	4,321.40	I
7	4,324.62	I
250	4,337.29	II
4	4,341.49	I
250	4,344.11	II
5	4,344.74	I
200	4,368.60	II
200	4,375.22	II
200	4,387.49	II
8	4,390.03	I
250	4,392.81	II
9	4,393.34	I
200	4,405.12	II
6	4,419.89	I
7	4,423.25	I
200	4,446.70	II
200	4,447.41	II
200	4,454.74	II
200	4,455.23	II
200	4,457.21	II
200	4,474.63	II
200	4,478.80	II
200	4,481.67	II
200	4,490.15	II
200	4,490.87	II
10	4,494.18	I
11	4,497.66	I
200	4,499.62	II
200	4,506.97	II
200	4,519.21	II
200	4,524.98	II
200	4,533.32	II
7	4,541.63	I
8	4,545.19	I
200	4,551.53	II
160	4,590.92	II
160	4,722.23	II
160	4,731.10	II
100	4,732.50	II
160	4,741.67	II
160	4,768.79	II

100	4,788.79	II
50	4,814.75	II
50	4,835.26	II
60	5,143.11	II
100	5,191.65	II
50	5,203.33	II
80	5,208.55	II
60	5,390.63	II
70	5,400.46	II
90	5,414.55	II
5	5,682.633	I
9	5,688.204	I
32	5,889.950	I
16	5,895.924	I
60	6,175.25	II
70	6,199.26	II
70	6,234.68	II
80	6,260.01	II
80	6,274.74	II
60	6,310.80	II
60	6,352.83	II
60	6,358.05	II
70	6,361.15	II
70	6,366.41	II
60	6,378.91	II
50	6,475.29	II
90	6,514.21	II
80	6,524.68	II
130	6,530.70	II
130	6,544.04	II
130	6,545.75	II
80	6,552.43	II
4	7,809.78	I
3	7,810.24	I
5	8,183.256	I
9	8,194.824	I
7	8,649.92	I
6	8,650.89	I
4	9,153.88	I
6	9,465.94	I
7	9,961.28	I
3	10,572.28	I
10	10,746.44	I
9	10,749.29	I
8	10,834.87	I
11	11,381.45	I
12	11,403.78	I
115	14,767.48	I
30	22,056.44	I
27	22,083.67	I
24	23,348.41	I
24	23,379.13	I

Na III

Ref. 178, 205, 287 — T.K.M.

Intensity		Wavelength	
		Vacuum	
5		183.95	III
5	h	189.35	III
5		193.80	III
5	h	194.04	III
5	h	194.17	III
5		194.29	III
6		194.68	III
6		195.53	III
6		202.15	III
6		202.19	III
8		202.49	III
5	d	202.71	III
7	d	202.72	III
8		202.76	III
8	p	203.06	III
8		203.28	III
8		203.33	III
10		207.30	III
10	c	215.34	III
12		215.86	III
12		216.12	III
15		229.87	III
12		230.59	III
50	c	250.52	III
30		251.37	III
25		266.90	III
70		267.65	III
50		267.87	III
50		268.63	III
20	p	272.08	III
20		272.45	III
100		378.14	III
70		380.10	III
7		1,336.76	III
7		1,337.36	III
8		1,340.67	III
9	d	1,342.39	III
10		1,342.73	III
11		1,355.28	III
12		1,361.90	III
11		1,372.34	III
10		1,420.89	III
10		1,444.19	III
12		1,449.31	III
11		1,562.87	III
10		1,565.29	III
15		1,598.18	III
11		1,688.94	III
10		1,699.29	III
10		1,711.12	III
11		1,728.27	III
10		1,731.11	III
10		1,755.48	III
15		1,807.07	III
10		1,810.77	III
11		1,811.67	III
10		1,816.81	III
10	d	1,835.22	III
10		1,838.94	III
11		1,844.36	III
12	d	1,847.53	III
10	d	1,847.59	III
15		1,849.56	III
12		1,850.38	III
10		1,855.92	III
10		1,856.71	III
10		1,861.21	III
10		1,880.66	III
10	d	1,887.39	III
20	d	1,887.47	III
15	d	1,890.75	III
15		1,900.16	III
10		1,918.45	III
11		1,923.96	III
14		1,926.26	III
11		1,927.24	III
12		1,932.74	III
13		1,933.89	III
10		1,943.52	III
12		1,946.43	III
12		1,950.91	III
14		1,951.24	III
10		1,977.16	III
13		1,985.57	III
10		1,995.68	III
		Air	
10		2,004.21	III
11		2,005.22	III
11		2,008.47	III
15		2,011.87	III
11		2,014.17	III
12		2,017.03	III
12		2,028.56	III
12		2,031.13	III
11		2,035.90	III
12		2,041.66	III
12		2,043.29	III
12		2,044.82	III
10		2,045.44	III
11		2,051.48	III
10		2,060.36	III
15		2,066.60	III
13		2,082.91	III
15		2,140.72	III
14		2,144.54	III
15		2,202.83	III
15		2,225.93	III
30		2,230.33	III
16		2,232.19	III
20	h	2,246.70	III
14		2,251.47	III
15		2,278.42	III
13		2,285.66	III
15		2,309.99	III
18		2,386.99	III
17		2,394.03	III
15		2,406.59	III

Intensity		Wavelength	
25		2,459.31	III
18		2,468.85	III
20		2,474.73	III
25		2,497.03	III
17		2,510.26	III
15		2,530.25	III
14		2,542.80	III

Na IV

Ref. 206 — T.K.M.

Intensity		Wavelength	
		Vacuum	
4		136.551	IV
4		136.854	IV
4		139.961	IV
7		142.232	IV
6		142.359	IV
8		146.064	IV
7		146.302	IV
9		150.298	IV
7		150.543	IV
7	c	150.64	IV
8		150.687	IV
7		151.299	IV
7		155.083	IV
7		155.240	IV
7		155.448	IV
8		155.510	IV
8		156.537	IV
12		162.448	IV
10		163.190	IV
12		168.411	IV
10		168.546	IV
10		190.445	IV
10		199.772	IV
10	c	205.49	IV
10		319.644	IV
10		360.76	IV
12		408.684	IV
10		409.614	IV
15		410.372	IV
10		411.334	IV
13		412.242	IV
10		1,580.50	IV
11		1,582.18	IV
10		1,582.33	IV
11	d	1,583.98	IV
12		1,584.14	IV
10	d	1,586.99	IV
12	d	1,587.05	IV
10		1,613.95	IV
11		1,615.92	IV
12		1,618.57	IV
11		1,655.47	IV
15	c	1,701.97	IV
10		1,702.41	IV
12		1,960.76	IV
11		1,965.08	IV
10		1,967.60	IV
		Air	
10		2,018.39	IV
12	d	2,106.33	IV
10		2,114.53	IV
10		2,155.76	IV

Na V

Ref. 299 — T.K.M.

Intensity		Wavelength	
		Vacuum	
100		106.28	V
100		106.30	V
100		106.40	V
100		106.49	V
200	c	107.93	V
200		108.02	V
200	c	110.82	V
200		110.88	V
100		111.51	V
300	c	112.01	V
100	h	114.70	V
100		114.74	V
400		117.99	V
100		120.04	V
400		125.18	V
400		125.22	V
500		125.29	V
300		125.43	V
300		125.46	V
200		125.90	V
100		126.21	V
200		126.56	V
100		126.61	V
400		127.44	V
400		127.47	V
400		128.03	V
400		128.05	V
200		130.68	V
300		131.35	V
200		131.41	V
300	h	131.64	V
500		133.16	V
400		133.39	V
200		134.27	V
300		135.79	V
300		135.85	V
200		138.81	V
300		138.92	V
400		148.64	V
300		148.86	V
400		151.13	V
300		157.21	V
300		163.62	V
800		307.15	V
1,000		308.26	V
800		332.55	V
900		333.91	V
800		360.32	V
800		360.37	V
1,000		400.72	V
500		445.05	V
600		445.19	V
600		459.90	V
850		461.05	V
1,000		463.26	V

STRONTIUM (SR)

Z = 38

Sr I and II

Ref. 1, 218, 279, 313 — J.J.W.

Intensity		Wavelength	
		Air	
1,400		2,152.84	II
1,400		2,165.96	II
160		2,428.10	I
120		2,569.47	I
200		2,931.83	I
300		3,301.73	I
300		3,329.99	I
400		3,351.25	I
300		3,366.33	I
650		3,380.71	II
950		3,464.46	II
120		3,474.89	II
300	h	3,940.80	I
600		3,969.26	I
300		3,970.04	I
1,300		4,030.38	I
300		4,032.38	I
46,000		4,077.71	II
200		4,161.80	II
32,000		4,215.52	II
340		4,305.45	II
350	h	4,438.04	I
		4,526.10	II
		4,585.91	II
65,000		4,607.33	I
3,200		4,722.28	I
2,200		4,741.92	I
1,400		4,784.32	I
4,800		4,811.88	I
3,600		4,832.08	I
500		4,855.04	I
600		4,868.70	I
3,000		4,872.49	I
600		4,876.06	I
2,000		4,876.32	I
1,000		4,891.98	I
8,000		4,962.26	I
1,300		4,967.94	I
800	h	5,156.07	I
1,400		5,222.20	I
2,000		5,225.11	I
2,000		5,229.27	I
2,800		5,238.55	I
4,800		5,256.90	I
		5,303.13	II
350	h	5,329.82	I
		5,379.13	II
		5,385.45	II
1,500		5,450.84	I
7,000		5,480.84	I
1,100		5,486.12	I
3,500		5,504.17	I
2,600		5,521.83	I
2,000		5,534.81	I
2,000		5,540.05	I
250	h	5,543.36	I
		5,622.94	II
		5,650.54	II
		5,723.70	II
		5,819.00	II
200	h	5,970.10	I
250	h	6,345.75	I
250	h	6,363.94	I
350	h	6,369.96	I
1,000		6,380.75	I
900	h	6,386.50	I
600	h	6,388.24	I
9,000		6,408.47	I
250		6,446.68	I
250	h	6,465.79	I
		6,483.17	II
5,500		6,504.00	I
		6,509.20	II
1,000		6,546.79	I
1,700		6,550.26	I
3,000		6,617.26	I
800		6,643.54	I
1,800		6,791.05	I
4,800		6,878.38	I
1,200		6,892.59	I
5,500		7,070.10	I
60		7,153.09	I
250	h	7,167.24	I
200		7,232.27	I
2,500		7,309.41	I
500		7,621.50	I
400	h	7,673.06	I
50	h	7,850.00	I
30	h	7,866.90	I
20	h	7,874.00	I
200	h	8,422.80	I
120		8,505.69	II
200		8,688.91	II
30		8,719.56	II
40		9,170.00	I
30		9,204.50	I
20		9,283.90	I
100		9,294.10	I
15		9,306.60	I
30		9,319.20	I
60		9,380.45	I
40	h	9,411.25	I
400	h	9,448.95	I
600		9,596.00	I
300		9,624.70	I
100		9,638.10	I
100	h	9,647.70	II
300		10,036.66	II
1,000		10,327.31	II
7		10,872.70	II
200		10,914.88	II
10		10,984.00	I
13		11,224.57	II
700		11,241.25	I
100		12,014.76	II
20		12,236.20	I
60		12,445.90	II
20		12,479.60	I
40		12,495.00	I
15		12,652.20	I
75		12,974.70	II
100		13,123.80	II
15		13,522.80	I
15		17,140.90	I
30		17,170.50	I
50		17,447.40	I
4		17,626.00	II
30		17,743.00	I
15		19,759.60	I
230		20,261.40	I
120		20,700.70	I
40		20,764.50	I
15		20,778.70	I
30		26,023.60	I

Sr III

Ref. 231, 265 — J.J.W.

Intensity	Wavelength	
	Vacuum	
25	307.18	III
50	316.11	III
50	321.61	III
125	330.67	III
500	351.62	III
75	358.80	III
250	363.49	III
150	371.21	III
1,000	437.24	III
1,875	491.79	III
1,250	507.04	III
3,750	514.38	III
2,500	562.75	III
20	968.37	III
20	975.78	III
25	992.98	III
50	1,025.23	III
35	1,044.91	III
20	1,057.74	III
25	1,060.20	III
20	1,098.77	III
35	1,125.49	III
20	1,140.24	III
20	1,168.27	III
20	1,182.09	III
50	1,236.23	III
20	1,940.58	III
30	1,958.44	III
30	1,966.92	III
	Air	
25	2,068.63	III
50	2,099.59	III
25	2,114.31	III
30	2,118.48	III
50	2,119.52	III
50	2,133.12	III
30	2,142.80	III
20	2,145.74	III
30	2,178.91	III
30	2,180.14	III
50	2,190.88	III
50	2,203.86	III
50	2,219.50	III
50	2,220.05	III
50	2,267.03	III
100	2,273.71	III
50	2,277.87	III
30	2,310.33	III
50	2,314.95	III
50	2,334.79	III
100	2,340.13	III
50	2,404.17	III
30	2,410.52	III
50	2,454.03	III
100	2,486.52	III
50	2,503.59	III
30	2,599.10	III
35	2,622.69	III
30	2,642.96	III
30	2,648.51	III
35	2,654.66	III
40	2,722.47	III
30	2,786.00	III
50	2,821.42	III
30	2,874.86	III
30	2,929.34	III
30	2,983.00	III
100	3,002.61	III
200	3,012.32	III
100	3,021.73	III
30	3,059.83	III
50	3,061.43	III
30	3,104.25	III
50	3,182.61	III
100	3,235.39	III

30	3,302.72	III
50	3,430.76	III
30	3,874.26	III
30	3,936.40	III
30	3,936.72	III
30	3,958.75	III
30	4,094.03	III
30	4,097.02	III
30	4,105.63	III
35	4,335.80	III
30	5,071.09	III
30	5,130.34	III
35	5,158.26	III
40	5,257.71	III
30	5,262.21	III
30	5,288.32	III
30	5,391.03	III
40	5,443.48	III
30	5,463.90	III
30	5,664.66	III
30	5,689.72	III

Sr IV

Ref. 110 — J.J.W.

Intensity	Wavelength	
	Vacuum	
12	284.31	IV
12	291.09	IV
12	291.19	IV
12	293.22	IV
15	298.12	IV
15	300.12	IV
12	300.27	IV
12	301.67	IV
20	378.53	IV
75	392.44	IV
50	393.00	IV
45	394.90	IV
50	396.22	IV
40	399.92	IV
35	403.85	IV
35	406.94	IV
30	412.93	IV
40	413.07	IV
40	415.32	IV
30	419.78	IV
25	430.21	IV
30	430.65	IV
25	442.73	IV
25	471.76	IV
25	484.20	IV
25	508.14	IV
25	534.19	IV
200	664.43	IV
100	710.35	IV
20	1,189.21	IV
30	1,244.14	IV
20	1,244.75	IV
20	1,244.87	IV
20	1,257.78	IV
20	1,268.62	IV
20	1,331.13	IV
30	1,347.90	IV
20	1,361.15	IV
25	1,408.67	IV
20	1,592.74	IV
25	1,677.03	IV
20	1,705.16	IV
20	1,724.23	IV
25	1,729.53	IV
20	1,732.12	IV
20	1,777.25	IV
20	1,994.61	IV
	Air	
20	2,104.38	IV
20	2,117.90	IV
20	2,217.99	IV
20	2,230.41	IV
20	2,240.49	IV
20	2,253.38	IV
50	2,346.97	IV
20	2,357.34	IV
20	2,438.93	IV
25	2,441.41	IV
30	2,482.79	IV
25	2,483.57	IV
18	2,500.57	IV
20	2,508.02	IV
20	2,534.03	IV
18	2,548.02	IV
40	2,555.60	IV
40	2,571.04	IV
25	2,571.58	IV
15	2,589.34	IV
25	2,620.35	IV
20	2,621.16	IV
20	2,642.16	IV
15	2,830.53	IV
9	2,934.60	IV
10	3,019.29	IV
9	3,266.52	IV
9	3,566.43	IV
9	3,741.05	IV
9	4,298.57	IV
9	4,685.08	IV

Sr V

Ref. 8 — J.J.W.

Intensity	Wavelength	
	Vacuum	
10	517.28	V
6	540.51	V
25	578.01	V
30	624.93	V
25	642.23	V
50	649.21	V
20	659.15	V
25	660.94	V
9	669.93	V
35	686.23	V
6	715.79	V
12	747.82	V
9	862.32	V

SULFUR (S)

Z = 16

S I and II

Ref. 144, 209, 210 — R.L.K.

Intensity	Wavelength	
	Vacuum	
40	906.9	II
40	910.5	II
40	912.7	II
40	937.4	II
40	937.7	II
20	996.0	II
20	1,000.5	II
20	1,014.4	II
20	1,019.5	II
20	1,096.6	II
40	1,102.3	II
20	1,131.0	II
20	1,131.6	II
40	1,234.1	II
40	1,250.5	II
110	1,253.8	II
110	1,259.5	II
275	1,270.782	I
250	1,277.216	I
280	1,295.653	I
275	1,302.337	I
235	1,302.863	I
235	1,303.110	I
245	1,303.430	I
260	1,305.883	I
265	1,310.194	I
355	1,316.542	I
290	1,316.618	I
375	1,323.515	I
355	1,326.643	I
775	1,381.552	I
710	1,385.510	I
960	1,388.435	I
640	1,389.154	I
775	1,392.588	I
1,000	1,396.112	I
300	1,409.337	I
510	1,425.030	I
425	1,433.280	I
300	1,436.968	I
300	1,448.229	I
425	1,472.972	I
550	1,473.995	I
300	1,474.380	I
355	1,481.665	I
485	1,483.039	I
300	1,483.233	I
330	1,485.622	I
390	1,487.150	I
680	1,666.688	I
640	1,687.530	I
710	1,807.311	I
680	1,820.343	I
640	1,826.245	I
710	1,900.286	I
550	1,914.698	I
	Air	
20	2,629.1	II
40	2,670.0	II
40	2,847.7	II
285	3,867.6	I
285	3,902.0	I
360	3,933.3	II
450	4,120.8	I
280	4,142.3	II
360	4,145.1	II
450	4,153.1	II
450	4,162.7	II
450	4,694.1	I
285	4,695.4	I
160	4,696.2	I
280	4,716.2	II
450	4,815.5	II
360	4,924.1	II
450	4,925.3	II
285	4,993.5	I
360	5,428.6	II
650	5,432.8	II
1,000	5,453.8	II
1,000	5,473.6	II
1,000	5,509.7	II
280	5,564.9	II
1,000	5,606.1	II
450	5,640.0	II
450	5,640.3	II
280	5,647.0	II
650	5,659.9	II
450	5,664.7	II
160	5,706.1	I
450	5,819.2	II
450	6,052.7	I
280	6,286.4	II
450	6,287.1	II
450	6,305.5	II
450	6,312.7	II
280	6,384.9	II
280	6,397.3	II
280	6,398.0	II
360	6,413.7	II
160	6,743.6	I
285	6,748.8	I
450	6,757.2	I
450	7,579.0	I
450	7,629.8	I
285	7,686.1	I
450	7,696.7	I
1,000	7,924.0	I
160	7,928.8	I
285	7,930.3	I
450	7,931.7	I
450	7,967.4	I
450	7,967.4	II
450	8,314.7	I
450	8,314.7	II
450	8,585.6	I
285	8,680.5	I
450	8,694.7	I
360	8,874.5	I
110	8,882.5	I
220	8,884.2	I
160	9,035.9	I
450	9,212.9	I
450	9,228.1	I
450	9,237.5	I
285	9,413.5	I
285	9,421.9	I
285	9,437.1	I
650	9,649.9	I
450	9,672.3	I
450	9,680.8	I
450	9,693.7	I
285	9,697.3	I
285	9,739.7	I
110	9,741.9	I
285	9,932.3	I
285	9,949.8	I
285	9,958.9	I
285	10,455.5	I
70	10,456.8	I
285	10,459.5	I

S III

Ref. 209, 210 — R.L.K.

Intensity	Wavelength	
	Vacuum	
70	729.5	III
110	732.42	III
70	735.2	III
70	738.5	III
70	789.0	III
70	796.7	III
70	824.9	III
70	836.3	III
285	1,077.1	III
70	1,194.0	III
70	1,201.0	III
	Air	
110	2,460.5	III
110	2,489.6	III
160	2,496.2	III
160	2,499.1	III
220	2,508.2	III
70	2,636.9	III
220	2,665.4	III
70	2,680.5	III
110	2,691.8	III
110	2,702.8	III
220	2,718.9	III
110	2,721.4	III
220	2,726.8	III
220	2,731.1	III
110	2,741.0	III
285	2,756.9	III
110	2,775.2	III
160	2,785.5	III
70	2,797.4	III
70	2,856.0	III
110	2,863.5	III
160	2,904.3	III
70	2,964.8	III
160	2,986.0	III
70	3,234.2	III
70	3,324.9	III
110	3,497.3	III
160	3,632.0	III
70	3,662.0	III
110	3,709.4	III
160	3,717.8	III
160	3,838.3	III
160	3,928.6	III
360	4,253.6	III
110	4,285.0	III
70	4,332.7	III

S IV

Ref. 29, 202, 209 — R.L.K.

Intensity	Wavelength	
	Vacuum	
20	519.3	IV
20	520.1	IV
40	520.8	IV
20	522.0	IV
20	522.5	IV
20	551.2	IV
40	652.5	IV
40	653.0	IV
70	653.6	IV
40	654.0	IV

70	655.6	IV
20	655.9	IV
110	657.3	IV
40	660.9	IV
160	661.4	IV
40	663.7	IV
40	664.8	IV
70	666.1	IV
110	744.9	IV
110	748.4	IV
110	750.2	IV
110	753.8	IV
40	798.3	IV
70	800.5	IV
70	804.0	IV
70	809.7	IV
110	816.0	IV
160	1,062.7	IV
160	1,073.0	IV
70	1,073.5	IV
20	1,108.4	IV
20	1,110.9	IV
20	1,624.0	IV
20	1,629.2	IV

Air

20	2,387.0	IV
40	2,398.9	IV
110	3,097.5	IV
40	3,117.7	IV

S V

Ref. 29 — R.L.K.

Vacuum

Intensity	Wavelength	
5	437.4	V
5	438.2	V
5	439.6	V
40	658.3	V
70	659.8	V
110	663.2	V
5	676.2	V
5	677.3	V
20	678.1	V
40	680.3	V
110	680.9	V
40	681.6	V
5	686.2	V
5	686.9	V
5	689.8	V
5	691.7	V
20	693.5	V
285	786.5	V
160	849.2	V
110	852.2	V
220	854.8	V
110	857.9	V
110	860.5	V
20	883.6	V
20	884.5	V
5	885.8	V
20	900.9	V
5	902.8	V
20	905.9	V

TANTALUM (TA)

Z = 73

Ta I and II

Ref. 1 — C.H.C.

Air

Intensity		Wavelength	
1,100		2,140.13	II
1,500		2,146.87	II
740		2,150.62	II
600		2,165.01	II
740		2,178.03	II
1,200		2,182.71	II
540		2,193.20	II
1,100		2,193.88	II
1,500		2,196.03	II
1,500		2,199.67	II
500		2,207.14	II
1,400	d	2,210.03	II
		2,210.19	II
420		2,215.60	II
1,400		2,239.48	II
240		2,248.48	II
480		2,249.79	II
1,200		2,250.76	II
260		2,254.86	II
440		2,255.77	II
360		2,256.51	II
500		2,258.71	II
840		2,261.42	II
260		2,261.62	II
990		2,262.30	II
220		2,269.56	II
740		2,271.85	II
990		2,272.59	II
200		2,279.85	I
320		2,282.19	II
130		2,285.02	II
790		2,285.25	II
600		2,286.59	II
240		2,287.27	II
990		2,289.16	II
180		2,292.54	II
160		2,295.18	
160		2,301.47	II
440		2,302.24	II
440		2,302.93	II
300		2,303.49	II
100		2,308.46	II
440		2,312.60	II
420		2,315.46	II
260		2,319.16	II
100		2,331.29	II
690		2,331.98	II
550		2,332.19	II
110		2,334.13	II
180		2,334.88	II
140		2,335.75	II
300		2,338.28	II
200		2,340.94	II
200		2,341.61	II
130		2,343.64	II
100		2,346.42	II
90		2,351.99	II
170		2,353.86	II
120		2,355.22	II
170		2,356.05	II
140		2,356.90	II
250		2,357.30	I
170		2,359.16	II
260		2,361.09	I
160		2,362.78	II
130		2,363.32	II
600		2,364.24	II
50		2,367.24	II
150		2,369.32	II
300		2,370.76	II
320		2,371.58	I
			II
70		2,372.80	II
100		2,373.94	II
70		2,375.91	I
150		2,378.31	II
440		2,381.13	II
240		2,381.52	II
170		2,383.72	II
240		2,384.28	II
130		2,385.73	I
1,400		2,387.06	II
80		2,388.37	II
160		2,389.11	II
70		2,396.30	I
110		2,399.15	I
50		2,399.92	II
2,400		2,400.63	II
140		2,402.13	II
100		2,403.68	II
130		2,406.55	I
130		2,408.26	II
120		2,414.32	I
240		2,415.21	II
320		2,416.89	II
220		2,417.86	II
150		2,418.77	II
140		2,421.03	I
150		2,421.85	II
170		2,423.48	II
130		2,425.91	II
360		2,427.64	I
360		2,429.71	II
170		2,431.06	II
480		2,432.70	II
130		2,433.59	II
130		2,436.51	II
110		2,437.07	I
110		2,438.64	II
200		2,439.91	I
130		2,442.39	I
100		2,444.13	II
100		2,447.17	I
100		2,454.48	I
100		2,458.68	I
100		2,460.55	I
160		2,463.82	II
130		2,466.99	II
130		2,467.37	II
380		2,470.90	II
120		2,471.38	I
120		2,472.13	I
150		2,473.13	I
120		2,473.31	II
600		2,474.62	I
120		2,475.33	I
200		2,476.67	II
150		2,478.22	I
120		2,481.86	II
100		2,482.10	I
100		2,482.58	II
100		2,484.04	II
500		2,484.95	I
120		2,486.70	I
600		2,488.70	II
500		2,490.46	I
600		2,504.45	I
600		2,507.45	I
240		2,512.65	I
1,200	d	2,526.35	I
600		2,532.12	II
240		2,545.49	II
240		2,546.80	I
460	d	2,551.07	I
460		2,554.62	II
240		2,555.05	I
1,200		2,559.43	I
460		2,562.10	I
340		2,571.51	II
430		2,573.54	I
390		2,573.79	I
600		2,577.37	II
340		2,577.78	I
210		2,580.16	I
340		2,584.03	II
430		2,593.08	I
410		2,593.66	II
310		2,594.25	II
560		2,595.26	I
310	l	2,596.45	II
220		2,600.14	I
600		2,603.49	II
1,400		2,608.63	I
210		2,609.00	I
310	d	2,611.34	I
340		2,615.46	I
310		2,615.66	I
1,200		2,635.58	II
470		2,636.67	I
860		2,636.90	I
510		2,646.22	I
600		2,646.37	I
2,400		2,647.47	I
270		2,651.22	II
2,600		2,653.27	I
1,900		2,656.61	I
1,500		2,661.34	I
220		2,665.60	II
220		2,668.07	I
600		2,668.62	I
770		2,675.90	II
270		2,680.06	II
220		2,680.66	II
600		2,684.28	I
1,500		2,685.17	II
340		2,691.31	I
260		2,692.40	I
470		2,694.52	II
240		2,696.81	I
1,000		2,698.30	I
470		2,706.69	I
310		2,709.27	II
1,200		2,710.13	I
2,600		2,714.67	I
240		2,717.18	I
470		2,720.76	I
470		2,727.44	II
410		2,727.78	I
310		2,736.25	II
210		2,739.26	II
210		2,743.59	I
510		2,746.68	I
1,200		2,748.78	I
860		2,749.83	I
410		2,752.49	II
1,000		2,758.31	I
430		2,761.68	II
770		2,775.88	I
390		2,787.69	I
680		2,796.34	I
680		2,797.76	II
380		2,802.07	I
430		2,806.30	I
510		2,806.58	I
260		2,817.10	II
260		2,842.82	I
640		2,844.25	I
290		2,844.46	II
290	c	2,845.35	I
560		2,848.52	I
1,500		2,850.49	I
1,900		2,850.98	I
220		2,858.44	II
360		2,861.98	I
310		2,868.65	I
470		2,871.42	I
270		2,873.36	I
260		2,873.56	I
210		2,874.17	I
380		2,880.02	I
770		2,891.84	I
260		2,899.04	I
560		2,902.05	I
210		2,914.12	I
310		2,915.49	I
410		2,925.19	I
310		2,932.70	I
1,700		2,933.55	I
470		2,940.06	I
1,200		2,940.22	I
240		2,942.14	I
510		2,951.92	I
340		2,953.56	I
1,500		2,963.32	I
770		2,965.13	II
770		2,965.54	I
340		2,969.47	I
430		2,975.56	I
210		3,011.88	I
1,800		3,012.54	II
290	d	3,027.48	I
290		3,042.06	II
530		3,049.56	I
530		3,069.24	I
360		3,077.24	I
560		3,103.25	I
380		3,124.97	I
380		3,130.58	I
270		3,132.64	I
320		3,170.29	I
270		3,173.59	I
200		3,176.29	I
600		3,180.95	I
240		3,184.55	I
200		3,198.67	I
200		3,213.91	II
300		3,223.83	I
230		3,229.24	I
200		3,242.05	I
200		3,242.83	I
210		3,274.95	II
1,100		3,311.16	I
210		3,317.93	I
680		3,318.84	I
330	d	3,330.99	II
230		3,358.47	I
640		3,371.54	I
360		3,385.05	I

230		3,398.33	I
450		3,406.94	I
230		3,463.77	I
490		3,480.52	I
380		3,497.85	I
240		3,503.87	I
490		3,511.04	I
200		3,513.61	I
750		3,607.41	I
980		3,626.62	I
500		3,642.06	I
100		3,686.18	I
100		3,689.73	I
130		3,731.02	I
140		3,736.76	I
130		3,746.36	I
110		3,754.52	I
110		3,777.10	I
110		3,792.02	I
210		3,833.74	II
100		3,848.05	I
100		3,885.20	I
210		3,918.51	I
140		3,922.78	I
140		3,922.92	I
210		3,970.10	I
210		3,996.17	I
100		3,999.28	I
190		4,006.84	I
190		4,026.94	I
140		4,029.94	I
120		4,040.87	I
410		4,061.40	I
210		4,064.63	I
100		4,067.24	I
310		4,067.91	I
120		4,105.02	I
210		4,129.38	I
230		4,136.20	I
230		4,147.89	I
210		4,175.21	I
100		4,177.92	I
130		4,181.15	I
300		4,205.88	I
120		4,206.40	I
130		4,245.35	I
130		4,268.26	I
160	c	4,302.98	I
110		4,355.14	I
100		4,378.82	I
150		4,386.07	I
110		4,398.45	I
180		4,402.50	I
130		4,415.74	I
360	c	4,510.98	I
190		4,530.85	I
130		4,551.95	I
170		4,565.85	I
340		4,574.31	I
260		4,619.51	I
130		4,669.14	I
450		4,681.88	I
130		4,691.90	I
150		4,740.16	I
220		4,756.51	I
120		4,768.98	I
220		4,812.75	I
110		4,920.11	I
100		4,921.27	I
110		4,926.00	I
150		4,936.42	I
200		5,037.37	I
100		5,067.87	I
110		5,115.84	I
100		5,141.62	I
100		5,143.69	I
330		5,156.56	I
110		5,212.74	I
110	d	5,218.45	I
		5,218.66	I
140		5,341.05	I
200		5,402.51	I
130		5,419.19	I
18		5,500.68	I
20		5,505.66	I
15	c	5,516.27	I
90		5,518.91	I
9		5,521.15	I
10		5,523.98	I
13		5,528.36	I
10		5,545.20	I
20		5,548.32	I
30		5,584.02	I
15		5,598.75	I
30		5,599.52	I
9	c	5,605.50	I
9	c	5,617.71	I
40		5,620.68	I
13		5,628.20	I
20		5,635.71	I
40		5,640.18	I
150		5,645.91	I
130		5,664.90	I
30		5,688.25	I
40		5,699.24	I
15		5,704.31	I
25		5,706.28	I
30		5,715.24	I
8		5,716.53	I
23		5,746.71	I
30		5,755.81	I
15	h	5,761.61	I
25		5,766.56	I
30	c	5,767.91	I
10		5,771.93	I
130		5,776.77	I
25		5,780.02	I
90		5,780.71	I
130		5,811.10	I
25		5,816.51	I
45	c	5,843.94	I
13		5,849.68	I
15		5,866.61	I
240		5,877.36	I
130		5,882.30	I
90		5,901.91	I
30		5,916.51	I
90		5,918.95	I
15		5,925.90	I
15		5,930.62	I
23		5,931.05	I
20		5,931.68	I
18		5,935.54	I
130		5,939.76	I
240		5,944.02	I
25		5,951.78	I
18		5,960.13	I
190	c	5,997.23	I
25	h	6,009.89	I
25		6,015.90	I
100		6,020.72	I
250		6,045.39	I
100		6,047.25	I
25		6,053.70	I
30		6,090.82	I
18		6,092.06	I
100		6,101.58	I
25		6,140.07	I
65		6,144.56	I
30		6,152.54	I
130		6,154.50	I
40		6,158.84	I
40		6,170.46	I
15		6,189.66	I
15		6,193.11	I
25		6,208.37	I
40		6,249.79	I
150		6,256.68	I
150		6,268.70	I
50		6,278.34	I
65		6,281.33	I
15		6,287.36	I
40		6,287.91	I
40		6,289.34	I
50		6,309.06	I
150		6,309.58	I
25	c	6,312.22	I
75		6,325.08	I
50		6,332.91	I
65		6,341.17	I
30		6,346.02	I
75		6,356.16	I
65		6,360.84	I
40		6,373.06	I
15		6,379.07	I
90		6,389.45	I
23	h	6,392.21	I
65		6,428.60	I
250		6,430.79	I
13		6,437.36	I
40		6,444.61	I
30		6,445.87	I
200		6,450.36	I
20		6,455.83	I
30		6,459.92	I
380		6,485.37	I
18	h	6,502.43	I
65		6,505.52	I
100		6,514.39	I
100		6,516.10	I
25	h	6,561.60	I
25	c	6,564.26	I
100		6,574.84	I
10		6,585.13	I
15	c	6,587.16	I
110		6,611.95	I
75		6,621.30	I
15	c	6,662.24	I
100		6,673.73	I
180		6,675.53	I
30		6,684.00	I
15		6,693.61	I
15		6,706.46	I
25		6,709.39	I
10		6,714.44	I
15	h	6,723.61	I
75	c	6,740.73	I
40		6,754.91	I
13	c	6,755.85	I
13		6,770.37	I
75		6,771.74	I
40	c	6,774.25	I
40	c	6,788.99	I
13	c	6,790.06	I
13		6,799.27	I
40	c	6,810.46	I
160	c	6,813.25	I
20		6,819.36	I
18	c	6,824.96	I
13		6,832.00	I
15		6,850.83	I
15		6,865.13	I
210		6,866.23	I
180		6,875.27	I
40		6,877.49	I
15		6,896.77	I
40		6,900.55	I
150		6,902.10	I
140		6,927.38	I
140		6,928.54	I
8	h	6,939.33	I
20	c	6,946.87	I
65		6,951.26	I
45		6,953.88	I
180		6,966.13	I
8		6,969.49	I
8	c	6,971.31	I
9		6,971.53	I
23		6,983.52	I
110	d	6,995.22	I
		6,995.49	I
20		7,000.21	I
40		7,005.07	I
75		7,006.96	I
50		7,025.03	I
13		7,031.51	I
40		7,039.07	I
15	h	7,081.30	I
20		7,085.40	I
23		7,093.02	I
8		7,108.05	I
15	c	7,117.52	I
20		7,121.27	I
40		7,125.72	I
150		7,148.63	I
110		7,172.90	I
13	c	7,174.91	I
13		7,191.35	I
8		7,233.45	I
30	h	7,250.27	I
11	h	7,264.82	I
6		7,272.29	I
30		7,276.96	I
5		7,277.54	I
9		7,286.36	I
13	c	7,296.32	I
140		7,301.74	I
20		7,319.84	I
11	c	7,322.72	I
13		7,325.95	I
11	c	7,340.19	I
160		7,346.41	I
140	c	7,352.86	I
100		7,356.96	I
90	c	7,369.09	I
160		7,407.89	I
11	c	7,435.19	I
23		7,440.17	I
30		7,467.75	I
23		7,486.01	I
30		7,520.56	I
6		7,569.23	I
9		7,590.22	I
6		7,649.62	I
11	h	7,722.02	I
11		7,763.11	I
9	c	7,779.67	I
20	c	7,842.76	I
100		7,882.37	I
30		7,950.19	I
5		7,952.07	I
6		7,998.75	I
6		8,022.09	I
75		8,026.50	I
5		8,029.04	I
15		8,039.08	I
8		8,053.93	I
15		8,068.98	I
5		8,100.11	I
13		8,128.76	I
9		8,158.54	I
5	c	8,180.74	I
13	c	8,248.95	I
20	d	8,264.85	I
75		8,281.62	I
11		8,389.06	I
5		8,415.73	I
25	c	8,447.62	I
11	h	8,550.49	I
15		8,575.92	I
10	c	8,595.84	I

TECHNETIUM (TC)
Z = 43

Tc I and II
Ref. 35 — C.H.C.

Intensity		Wavelength Air	
15		2,106.23	II
20		2,116.44	II
15		2,119.41	I
30		2,156.27	I
30		2,185.39	I
30		2,189.06	I
40		2,193.35	I
10		2,266.22	II
10		2,282.12	I
10		2,282.71	I
50		2,285.45	I
100		2,298.08	I
30		2,416.22	I
50		2,423.23	I
20		2,424.54	I
20		2,435.83	I
10		2,436.99	I
80	w	2,463.69	I
20		2,465.09	I
30		2,466.87	I
20		2,475.11	I
50		2,480.70	I
50		2,483.22	I
20		2,486.50	I
25		2,492.72	I
20		2,493.43	I
100		2,496.77	II
30		2,510.17	I
80		2,529.34	II
500		2,543.23	II
60		2,544.81	II
50		2,547.92	II
50		2,558.61	II
50		2,567.01	II

30		2,575.06	II	1,000		2,928.20	I	200		3,451.05	I	3,000		3,771.03	I
80		2,576.28	II	80		2,933.89	I	200		3,456.85	I	500		3,777.27	I
40		2,577.86	II	200		2,955.93	I	400		3,457.24	I	2,000		3,779.37	I
300	h	2,578.79	I	200		2,973.65	I	40		3,457.60	I	3,000	c	3,780.68	I
200		2,589.86	I	100		2,979.34	I	5,000	c	3,466.28	I	500		3,784.06	I
20	w	2,590.19	I	150		2,985.36	I	150		3,470.51	I	200		3,786.06	I
100		2,592.82	I	100		3,010.83	I	80		3,475.18	I	500		3,791.28	I
20		2,597.19	II	300		3,017.23	I	1,000		3,475.59	I	300		3,791.73	I
500		2,608.86	I	150		3,021.56	I	60		3,484.62	I	200		3,797.44	I
1,000	c	2,609.99	II	100		3,022.66	I	1,000	c	3,486.23	I	1,000		3,797.77	I
1,500		2,614.23	I	200		3,023.68	I	100		3,490.30	I	200		3,814.67	I
1,000		2,615.87	I	80		3,025.26	I	400		3,493.39	I	300		3,816.89	I
30		2,618.28	I	300	w	3,026.89	I	500		3,494.62	I	300		3,824.47	I
200		2,634.91	II	50		3,033.16	I	40		3,499.14	I	500		3,828.54	I
80		2,636.36	I	80		3,034.57	I	1,000		3,500.70	I	200		3,830.35	I
30		2,641.26	I	40		3,036.88	I	200		3,501.24	I	200		3,832.45	I
100		2,642.37	I	20		3,037.90	II	800	c	3,502.70	I	600		3,832.82	I
40		2,644.50	II	100		3,038.23	I	100	c	3,507.19	I	1,500		3,837.56	I
1,000	c	2,647.01	II	40		3,042.64	I	100		3,508.27	I	800		3,841.31	I
300	c	2,649.21	I	100	h	3,051.55	I	100		3,510.91	I	800		3,845.97	I
100		2,652.35	II	40		3,052.47	I	800		3,525.83	I	500		3,847.60	I
30		2,653.57	I	80		3,062.11	I	300		3,526.18	I	300		3,851.22	I
100		2,654.31	I	200		3,062.36	I	100		3,529.83	I	500	c	3,856.73	I
120		2,660.88	I	300		3,064.67	I	150		3,534.88	I	200		3,863.07	I
100		2,662.30	I	100	c	3,066.60	I	500		3,535.51	I	400		3,864.11	I
80		2,681.19	II	120	c	3,068.34	I	300		3,538.12	I	1,000		3,868.24	I
60		2,683.14	I	80		3,076.24	I	800		3,538.68	I	200		3,875.66	I
80		2,683.89	I	150		3,089.34	I	2,000	c	3,541.77	I	500	c	3,879.16	I
80		2,693.11	I	1,000		3,099.10	I	6,000	c	3,549.72	I	600	c	3,880.72	I
50		2,696.64	I	200		3,099.52	I	4,000	c	3,550.64	I	300	w	3,892.12	I
70		2,702.27	I	60		3,108.25	I	300		3,559.75	I	200		3,893.22	I
40		2,702.96	II	40		3,109.15	I	800		3,560.32	I	600		3,899.83	I
100		2,707.90	II	60		3,115.98	I	100		3,565.22	I	300		3,919.38	I
1,000		2,708.78	I	80		3,119.17	I	800		3,568.85	I	300	c	3,923.66	I
30		2,715.20	I	40		3,119.66	I	100		3,570.65	I	200		3,927.57	I
30		2,723.55	I	700		3,122.64	I	100		3,575.42	I	200		3,933.70	I
1,000		2,726.69	I	1,500		3,131.23	I	1,000		3,580.06	I	4,000	c	3,946.57	I
30	c	2,728.47	I	40	c	3,150.26	I	600		3,581.26	I	2,000		3,947.09	I
500		2,730.53	I	300		3,161.67	I	800		3,582.08	I	200		3,955.73	I
300		2,732.87	I	3,000		3,173.30	I	2,000		3,582.63	I	300		3,979.64	I
150		2,736.23	I	200		3,180.30	I	4,000		3,587.94	I	500		3,980.35	I
60	c	2,736.83	II	2,000		3,182.37	I	200		3,593.47	I	10,000	c	3,984.97	I
100		2,737.97	II	2,000		3,183.11	I	300		3,594.57	I	400		3,987.78	I
20	c	2,738.83	II	800	c	3,195.20	II	1,000	c	3,595.66	I	300		3,994.04	I
100		2,755.76	I	40	w	3,197.53	I	1,000	c	3,607.32	I	2,000		3,994.51	I
100		2,762.13	I	300	c	3,202.83	I	200		3,607.62	I	200		3,996.97	I
200		2,762.34	I	1,000		3,212.02	II	2,000	c	3,608.27	I	300		4,004.69	I
60		2,765.95	I	40		3,220.74	I	200		3,618.94	I	500		4,007.14	I
500		2,766.89	I	60		3,230.02	I	1,000	c	3,627.36	I	1,000		4,012.00	I
20		2,777.31	II	1,000		3,237.02	II	200		3,630.39	I	400		4,016.68	I
150		2,778.91	I	100		3,241.84	I	3,000	c	3,635.15	I	600		4,017.22	I
25		2,781.22	I	500		3,244.19	I	10,000	c	3,636.07	I	2,000		4,020.76	I
1,000		2,782.05	I	300		3,252.05	I	1,000	c	3,638.22	I	20,000	c	4,031.63	I
500		2,785.59	I	40		3,256.10	I	200		3,638.85	I	1,000		4,039.25	I
40		2,788.89	I	40		3,261.94	I	900		3,639.38	I	200		4,041.78	I
500		2,789.25	I	100		3,287.14	I	400		3,640.23	I	10,000	c	4,049.11	I
100		2,794.53	I	30		3,298.84	II	1,000	c	3,648.04	I	500		4,051.95	I
80		2,795.65	I	100		3,300.77	I	600		3,651.47	I	200		4,053.18	I
200		2,795.78	II	80		3,305.89	I	1,000	c	3,658.59	I	200	c	4,056.08	I
1,000		2,802.81	I	200		3,310.65	I	400	c	3,661.45	I	400		4,083.54	I
150		2,803.02	I	150		3,313.65	I	200		3,664.92	I	10,000		4,088.71	I
500		2,808.36	I	200		3,325.55	I	1,000		3,679.15	I	200		4,093.69	I
50	c	2,809.65	II	150	c	3,327.10	I	300		3,680.32	I	15,000		4,095.67	I
500		2,811.61	II	100		3,330.77	I	5,000		3,684.74	I	1,000		4,110.22	I
30		2,814.86	I	50		3,332.47	I	300		3,692.76	I	10,000		4,115.08	I
40		2,819.46	I	60		3,350.56	I	800		3,703.83	I	600		4,119.27	I
100		2,821.35	II	50	c	3,350.83	I	300		3,704.80	I	8,000		4,124.22	I
200		2,828.04	I	400		3,366.75	I	200		3,706.70	I	1,000		4,128.27	I
60		2,831.18	II	40		3,386.67	I	200		3,707.63	I	300		4,134.81	I
50		2,840.38	II	50		3,392.23	I	200		3,708.26	I	300		4,139.12	I
60		2,845.04	I	300		3,394.18	I	1,000		3,712.26	I	800		4,139.85	I
10		2,846.39	II	60		3,396.90	I	300		3,712.82	I	400		4,141.27	I
60		2,849.20	I	40		3,397.83	I	500		3,715.94	I	6,000		4,144.95	I
150		2,850.96	I	300		3,398.33	I	10,000		3,718.86	I	3,000		4,145.08	I
500	h	2,857.13	I	200		3,402.10	I	1,500		3,723.67	I	200		4,147.62	I
2,000	c	2,859.11	I	200	c	3,403.93	I	2,000		3,724.40	I	10,000		4,165.61	I
500		2,864.49	I	80		3,405.33	I	5,000		3,726.35	I	500		4,167.42	I
100		2,868.09	I	80		3,407.28	I	200		3,727.36	I	1,000		4,169.68	I
1,000		2,887.73	I	50		3,408.33	I	400	c	3,729.18	I	4,000		4,170.27	I
100		2,888.46	I	40		3,411.80	I	500		3,731.74	I	5,000		4,172.53	I
30		2,889.20	II	40	c	3,418.20	I	300		3,737.42	I	1,000		4,176.28	I
200		2,893.16	I	100		3,419.10	I	400		3,745.01	I	800		4,186.51	I
150		2,893.45	I	60		3,427.85	I	1,000		3,746.15	I	300		4,218.61	I
200		2,894.32	I	40		3,431.75	I	5,000		3,746.84	I	10,000	c	4,238.19	I
1,000		2,896.34	I	200		3,434.70	I	1,000		3,752.13	I	20,000		4,262.27	I
40		2,903.81	I	40		3,435.68	I	4,000		3,754.37	I	1,000		4,262.69	I
1,000		2,913.15	I	150		3,437.44	I	1,000		3,758.54	I	800		4,274.97	II
500		2,921.91	I	80		3,438.73	I	2,000		3,761.81	I	800		4,278.90	I
20	c	2,923.34	II	200	c	3,443.47	I	5,000		3,768.77	I	30,000		4,297.06	I

400	c	4,336.86	I
400		4,358.49	I
200		4,359.26	I
1,000		4,429.59	I
1,000		4,481.53	I
3,000		4,487.06	I
400		4,495.03	I
1,000		4,515.98	I
10,000		4,522.84	I
2,000		4,539.53	I
400		4,542.09	I
400		4,552.20	I
800		4,552.85	I
1,000		4,557.05	I
2,000		4,564.54	I
1,000		4,578.45	I
1,000		4,593.35	I
300	c	4,609.16	I
1,000		4,616.86	I
200	c	4,622.69	I
300		4,624.96	I
1,000		4,630.57	I
200		4,633.15	I
3,000		4,637.50	I
500		4,643.28	I
2,000		4,648.33	I
2,000	c	4,660.21	I
2,000		4,669.30	I
400		4,672.17	I
200		4,678.90	I
400		4,689.36	I
300		4,694.28	I
1,000		4,706.92	I
200		4,714.22	I
2,000		4,717.77	I
500	c	4,719.02	I
4,000	c	4,719.28	I
200	c	4,736.51	I
10,000		4,740.61	I
500		4,749.61	I
1,000		4,752.72	I
200		4,762.36	I
4,000		4,771.54	I
200		4,773.89	I
200		4,783.92	I
500		4,785.60	I
200	c	4,790.48	I
250		4,791.62	I
300		4,799.98	I
100		4,805.69	I
100		4,809.42	I
500		4,816.79	I
10,000		4,820.74	I
300		4,831.35	I
1,000		4,834.37	I
1,000		4,835.39	I
100		4,841.36	I
20,000		4,853.59	I
100		4,857.21	I
100	c	4,862.19	I
10,000		4,866.73	I
200		4,870.77	I
100		4,888.70	I
150	c	4,890.88	I
8,000		4,891.92	I
150		4,892.49	I
1,000		4,908.51	I
2,000		4,909.57	I
500		4,913.02	I
150	c	4,914.70	I
200	c	4,920.67	I
300		4,923.60	I
400		4,948.06	I
5,000		4,976.34	I
400		4,995.00	I
200		5,002.67	I
100		5,005.74	I
200	c	5,014.52	I
500		5,026.24	I
300		5,026.79	I
150		5,027.89	I
80		5,032.45	I
300		5,055.27	I
60		5,058.33	I
500		5,060.69	I
80		5,090.74	I
5,000		5,096.28	I
200	c	5,103.24	I
500		5,104.32	I
200		5,109.81	I
100		5,120.60	I
500		5,139.26	I
500		5,150.63	I
2,000		5,161.81	I
2,000		5,174.81	I
100		5,206.56	I
200		5,225.55	I
200	c	5,260.22	I
200	c	5,261.44	I
1,000		5,275.51	I
800		5,285.07	I
100		5,305.31	I
400		5,314.96	I
600		5,320.20	I
200		5,334.79	I
500	c	5,353.48	I
200		5,356.63	I
300		5,358.65	I
200		5,360.14	I
500	h	5,375.20	I
150	c	5,423.05	I
200		5,447.40	I
500	c	5,451.90	I
100		5,455.95	I
300		5,471.96	I
70		5,483.01	I
60		5,485.37	I
80	c	5,506.89	I
150		5,524.11	I
100		5,528.23	I
200		5,541.94	I
80		5,543.63	I
100		5,550.53	I
3,000	c	5,589.02	I
200		5,602.23	I
2,000	c	5,620.45	I
300		5,629.94	I
1,500		5,642.13	I
800		5,644.94	I
100		5,656.00	I
60		5,672.15	I
200		5,687.30	I
200		5,689.05	I
700		5,725.31	I
500	c	5,771.47	I
100		5,794.65	I
80	c	5,799.85	I
100	c	5,814.24	I
200		5,831.48	I
150	c	5,836.33	I
150		5,923.36	I
1,000	c	5,924.47	I
200		5,926.29	I
600	c	5,931.93	I
60		6,032.36	I
60		6,047.99	I
200		6,065.09	I
800		6,085.23	I
300		6,099.39	I
500	c	6,102.96	I
1,000		6,120.68	I
1,000		6,130.80	I
150	c	6,132.23	I
100		6,184.70	I
800		6,192.66	I
600	c	6,244.18	I
100		6,312.18	I
100		6,354.86	I
100		6,356.73	I
80		6,389.87	I
100		6,408.83	I
1,000		6,455.90	I
600	c	6,461.93	I
100		6,470.27	I
200	c	6,491.68	I
200		6,526.82	I
150		6,579.24	I
500	c	6,625.57	I
300	c	6,673.66	I
100		6,687.10	I
80		6,786.00	I
70		6,798.63	I
60		6,856.90	I
150		7,002.37	I
100		7,016.57	I
500	c	7,086.18	I
60		7,093.12	I
200		7,141.28	I
200	c	7,157.62	I
70		7,256.08	I
100		7,322.38	I
80		7,329.14	I
100		7,396.80	I
100	c	7,402.61	I
200		7,405.36	I
60		7,427.15	I
150		7,434.12	I
600		7,452.49	I
60		7,461.59	I
80		7,534.95	I
800		7,540.26	I
80		7,543.39	I
200		7,574.02	I
500		7,579.26	I
90		7,624.53	I
100		7,684.45	I
500		7,697.37	I
80	c	7,698.19	I
800	c	7,793.04	I
60		7,798.28	I
60		7,816.74	I
800		7,817.72	I
100		7,856.38	I
200		7,861.44	I
400	d	7,871.25	I
60	c	7,874.76	I
70		7,965.45	I
500		7,999.73	I
200		8,126.55	I
200		8,170.55	I
150		8,205.27	I
100		8,206.49	I
150		8,211.31	I
500	c	8,237.08	I
200		8,308.15	I
200		8,309.16	I
60		8,315.50	I
100		8,531.06	I
100		8,543.61	I
100	c	8,707.21	I
100	c	8,737.93	I
200	c	8,829.82	I

TELLURIUM (TE)

Z = 52

Te I and II

Ref. 1, 344—347 — C.H.C.

Intensity		Wavelength	
		Vacuum	
6		799.60	II
8		802.28	II
6		942.62	II
6		1,003.73	II
6		1,007.80	II
5		1,014.27	II
5		1,022.79	II
6		1,057.00	II
8		1,059.51	II
6		1,068.86	II
8		1,077.66	II
6		1,090.11	II
6		1,144.04	II
5		1,153.10	II
10		1,161.42	II
10		1,174.34	II
12		1,175.79	II
9		1,208.54	II
5		1,213.00	II
9		1,220.98	II
9		1,253.62	II
9		1,270.52	II
7		1,274.76	II
8		1,306.53	II
10		1,324.92	II
7		1,336.42	II
7		1,345.20	II
9		1,363.24	II
8		1,366.73	II
10		1,374.80	II
6		1,395.22	II
6		1,439.52	II
6		1,465.25	II
7		1,489.56	II
8		1,607.99	II
10		1,608.41	II
10		1,613.15	II
6		1,638.91	II
5		1,655.4	I
5		1,688.5	I
6		1,700.0	I
6		1,701.58	II
5		1,708.0	I
6		1,751.0	I
5		1,759.4	I
5		1,775.0	I
6		1,795.7	I
6		1,796.3	I
10		1,822.4	I
6		1,825.5	I
6		1,850.6	I
6		1,852.1	I
6		1,853.8	I
8		1,857.2	I
6		1,860.4	I
3		1,962.88	II
7		1,994.83	I
		Air	
6		2,000.2	I
26,000		2,002.02	I
8		2,070.9	I
6,500		2,081.16	I
18,000		2,142.81	I
3,200		2,147.25	I
360		2,159.85	I
9		2,208.74	I
10		2,255.49	I
500		2,259.02	I
10		2,265.52	I
20		2,373.06	II
1,200		2,383.26	I
1,500		2,385.78	I
20		2,387.82	II
10		2,401.63	II
10		2,436.47	II
50		2,438.69	II
120		2,530.72	I
20		2,567.82	II
10		2,574.96	II
5		2,576.10	II
7		2,579.24	II
10		2,591.12	II
10		2,592.85	II
10		2,605.72	II
5		2,621.92	II
10		2,624.86	II
20		2,627.96	II
20		2,641.89	II
20		2,648.48	II
100		2,649.66	II
40		2,657.70	II
80		2,661.10	II
110		2,677.13	I
20		2,711.58	II
6		2,769.65	I
10		2,841.17	II
10		2,846.15	II
100		2,858.29	II
20		2,861.00	II
40		2,868.82	II
150		2,895.41	II
30		2,919.89	II
50		2,942.11	II
50		2,946.68	II
70		2,967.29	II
20		2,973.67	II
50		2,975.90	II
15		2,997.04	II
10		3,012.02	II
50		3,017.58	II
20		3,023.31	II
70		3,047.00	II
20		3,052.46	II
10		3,063.16	II
15		3,073.56	II
8		3,104.44	II
10		3,132.58	II
20		3,160.66	II
100		3,175.14	I
10		3,189.83	II
5		3,211.21	II

60		3,256.80	II
30		3,268.77	II
30		3,282.63	II
40		3,321.92	II
40		3,323.11	II
60		3,329.22	II
60		3,352.10	II
60		3,362.79	II
25		3,374.10	II
150		3,406.79	II
20		3,419.63	II
50		3,442.25	II
40		3,455.12	II
20		3,456.88	II
20		3,480.32	II
40		3,483.67	II
20		3,486.11	II
50		3,521.11	II
50		3,552.19	II
100		3,611.78	II
50		3,617.57	II
40		3,644.46	II
20		3,679.26	II
30		3,725.66	II
40		3,797.22	II
20		3,800.92	II
20		3,905.67	II
20		3,918.54	II
30		3,931.49	II
20		3,947.98	II
40		3,969.22	II
25		3,975.94	II
20		3,981.77	II
50		4,006.52	II
20		4,011.69	II
30		4,029.73	II
40		4,047.17	II
30		4,048.88	II
15		4,073.48	II
30		4,101.04	II
70		4,127.32	II
30		4,163.55	II
100		4,169.77	II
30		4,179.29	II
25		4,211.31	II
80		4,225.73	II
30		4,246.47	II
20		4,251.15	II
100		4,261.11	II
30		4,264.36	II
60		4,273.43	II
80		4,285.85	II
40		4,320.90	II
30		4,361.28	II
150		4,364.00	II
30		4,377.12	II
75		4,385.10	II
60		4,396.00	II
170		4,478.63	II
80		4,537.07	II
100		4,557.78	II
70		4,630.62	II
100		4,641.12	II
180		4,654.37	II
200		4,686.91	II
100		4,696.38	II
100		4,706.53	II
100		4,766.05	II
70		4,771.56	II
100		4,784.87	II
100		4,827.14	II
150		4,831.28	II
150		4,842.90	II
130		4,865.12	II
200		4,866.24	II
80		4,885.22	II
80		4,904.44	II
60		4,961.88	II
60		5,000.82	II
8		5,083.0	I
7		5,148.7	I
50		5,449.84	II
50		5,487.95	II
150		5,576.35	II
150		5,649.26	II
100		5,666.20	II
200		5,708.12	II
7		5,733.5	I
150		5,755.85	II
8		5,789.1	I
50		5,936.15	II
100		5,974.68	II
8	d	6,273.5	I
8	h	6,349.7	I
50		6,367.13	II
8		6,405.9	I
7	h	6,456.7	I
8	h	6,613.4	I
10		6,648.58	II
8		6,660.2	I
8	h	6,690.0	I
10	h	6,790.0	I
20	h	6,837.6	I
20	h	6,854.7	I
10		7,016.06	II
10		7,039.13	II
15	h	7,191.1	I
10		7,236.62	II
20	h	7,263.5	I
8		7,280.9	I
10		7,289.26	II
10		7,445.39	II
12		7,460.98	II
15		7,468.75	II
10		7,481.26	II
10		7,556.8	I
6		7,688.61	II
15		7,759.1	I
8		7,818.79	II
8		7,861.61	II
15		7,921.69	II
15		7,943.14	II
10		7,950.34	II
20		7,972.9	I
6		8,056.15	II
30	h	8,061.4	I
10	h	8,082.5	I
10		8,122.44	II
8		8,130.39	II
8		8,154.47	II
20		8,186.44	II
6		8,190.94	II
10		8,251.5	I
15		8,273.53	II
10		8,276.6	I
10		8,291.1	I
15		8,355.8	I
10		8,372.12	II
7		8,469.8	I
8		8,492.2	I
8		8,500.8	I
12		8,521.4	I
10		8,535.68	II
12		8,575.78	II
10		8,604.63	II
8		8,621.68	II
7	h	8,632.1	I
15		8,672.95	II
12		8,701.09	I
10		8,733.81	II
205		8,758.18	I
12		8,831.52	I
18		8,851.15	I
6		8,897.92	II
81		9,004.37	I
18		9,043.39	I
12		9,196.80	I
15		9,206.78	I
17		9,207.64	I
12		9,469.00	I
5,660		9,722.74	I
185		9,785.54	I
109		9,842.30	I
532		9,868.92	I
118		9,902.61	I
689		9,956.30	I
37		9,959.93	I
325		9,977.13	I
136		9,979.31	I
45		9,985.85	I
5,950		10,051.41	I
4,097		10,091.01	I
104		10,099.57	I
279		10,106.05	I
381		10,118.08	I
296		10,151.06	I
397		10,300.56	I
205		10,323.05	I
745		10,493.57	I
197		10,509.86	I
1,880		10,918.34	I
298		11,007.80	I
10,200		11,089.56	I
508		11,163.74	I
6,620		11,487.23	I
280		11,978.96	I
188		12,566.24	I
389		12,589.19	I
161		12,805.50	I
400		13,104.18	I
1,580		13,247.75	I
483		13,316.63	I
217		14,037.09	I
144		14,072.53	I
434		14,335.74	I
220		14,417.46	I
1,050		14,513.51	I
129		14,554.68	I
1,480		15,452.45	I
2,430		15,546.23	I
3,760		16,403.90	I
1,960		17,303.54	I
2,780		18,291.59	I
394		18,777.30	I
269		19,623.52	I
239		20,147.54	I
1,020		21,043.73	I
464		21,602.50	I
37		21,799.64	I
74		22,555.29	I
48		22,755.66	I
27		23,294.94	I
17		23,978.70	I
25		24,059.04	I
13		26,428.62	I
38		26,539.17	I
15		26,553.74	I
7		27,179.26	I

TERBIUM (TB)

Z = 65

Tb I and II
Ref. 1 — C.H.C.

Intensity		Wavelength	
		Air	
29		2,577.73	II
110		2,584.61	II
29		2,590.31	II
29		2,591.42	II
24		2,592.64	II
55		2,597.71	II
40		2,602.93	II
110		2,608.57	II
40		2,616.90	II
130		2,628.69	II
55		2,655.96	II
50		2,661.40	II
24		2,661.64	II
55		2,667.64	II
50		2,668.86	II
140		2,669.29	II
40		2,674.13	II
40		2,674.69	II
29		2,678.15	II
40		2,683.97	II
35		2,687.82	II
50		2,691.90	II
35		2,693.05	II
55		2,693.41	II
35		2,695.46	II
50		2,696.83	II
190		2,704.07	II
130		2,736.24	II
160		2,759.47	II
270		2,769.53	II
130		2,784.49	II
180		2,800.51	II
250		2,802.75	II
250		2,809.30	II
180		2,812.64	II
190		2,852.14	II
110		2,857.68	II
230		2,886.29	II
160		2,894.45	II
320		2,897.44	II
160		2,898.86	II
110		2,901.54	II
110		2,910.30	II
160		2,914.75	II
160		2,915.30	II
190		2,915.60	II
120		2,916.24	II
120		2,918.89	II
120		2,924.16	II
120		2,924.53	II
160		2,932.89	II
150		2,940.05	II
250		2,956.21	II
170		2,968.87	II
170		2,977.78	II
110		2,987.03	II
110		2,988.57	II
130		2,996.00	II
110		2,999.03	II
130		3,005.52	II
170		3,009.30	II
230		3,010.59	II
230		3,016.18	II
130		3,019.17	II
170		3,020.29	II
110		3,023.43	II
170		3,027.33	II
230		3,031.60	II
230		3,044.96	II
190		3,051.13	II
130		3,053.24	II
460		3,053.55	II
130		3,062.78	II
230		3,064.09	II
110		3,065.69	II
230		3,067.20	II
270		3,069.03	II
460		3,070.05	II
270		3,072.60	II
670		3,078.86	II
480		3,082.36	II
120		3,086.78	II
250		3,088.43	II
480		3,089.58	II
230		3,102.54	II
480		3,102.96	II
290		3,117.89	II
290		3,119.62	II
230		3,121.94	II
230		3,123.05	II
160		3,124.54	II
110		3,131.35	II
250		3,134.26	II
440		3,139.64	II
190		3,140.06	II
230		3,145.22	II
150		3,146.67	II
310		3,147.04	II
310		3,147.15	II
310		3,148.71	II
120		3,155.62	II
130		3,162.42	II
290		3,162.93	II
190		3,165.74	II
380		3,167.52	II
140		3,168.32	II
230		3,169.84	II
190		3,173.76	II
380		3,174.66	II
380		3,180.54	II
140		3,183.88	II
480		3,187.26	II
290		3,188.03	II
190		3,194.69	II
380		3,195.60	II
480		3,199.56	II
1,100		3,218.93	II
1,200		3,219.98	II
250		3,230.03	II
250		3,231.06	II
210	d	3,239.60	II
250		3,240.00	II
480		3,252.32	II
250		3,262.97	II
230	d	3,263.87	II
230		3,264.90	II

400		3,266.40	II
250		3,274.14	II
250		3,274.33	II
210		3,277.32	II
760		3,280.31	II
760		3,281.40	II
520		3,283.10	II
1,000		3,285.04	II
310		3,287.55	II
310		3,291.56	II
1,500		3,293.07	II
210		3,295.33	II
310		3,298.66	II
210		3,304.95	II
420	d	3,307.44	II
210		3,308.51	II
210		3,314.38	II
340	d	3,321.15	II
420		3,322.28	II
210		3,323.38	II
210		3,323.89	II
3,800		3,324.40	II
520		3,329.08	II
210		3,334.48	II
250		3,336.70	II
310		3,338.03	II
250		3,339.00	II
210		3,347.27	II
210		3,348.07	II
760		3,349.42	II
320		3,362.25	II
760		3,364.93	II
230	d	3,370.61	II
320		3,371.50	II
520		3,372.36	II
460	d	3,372.72	II
520		3,375.03	II
320		3,378.73	II
520		3,378.86	II
320		3,382.80	II
210		3,390.60	II
380		3,391.28	II
270		3,398.35	II
320		3,399.10	II
270		3,400.53	II
210	d	3,400.86	II
420		3,402.33	II
210		3,410.40	II
210		3,410.68	II
520		3,413.76	II
270		3,416.24	II
400	d	3,420.34	II
210		3,430.61	II
320		3,433.26	II
270		3,439.72	II
520		3,440.37	II
320		3,444.58	II
210		3,446.40	II
270		3,449.46	II
810		3,454.06	II
380		3,460.38	II
230	d	3,462.97	II
620		3,468.03	II
270		3,471.73	II
270		3,472.37	II
810	d	3,472.79	II
210		3,473.00	II
380		3,480.17	II
230		3,483.04	II
230		3,483.69	II
290	d	3,489.51	II
210	d	3,492.00	II
270		3,494.21	II
270		3,495.36	II
810		3,500.84	II
570		3,507.45	II
5,700		3,509.17	II
380		3,510.10	II
320		3,513.10	II
570		3,519.76	II
1,300		3,523.66	II
380		3,525.14	II
440		3,525.61	II
440		3,536.32	II
570		3,537.94	II
1,100		3,540.24	II
810		3,543.89	II
310	d	3,551.03	II
320		3,551.96	II
460	d	3,558.77	
3,200		3,561.74	II
480		3,562.90	II
570		3,565.74	
810		3,567.35	II
4,200		3,568.52	II
1,600		3,568.98	II
320		3,572.07	II
1,100		3,579.20	II
710		3,585.03	II
570		3,587.44	II
810		3,596.38	II
440		3,598.06	II
1,600		3,600.44	II
320		3,604.90	II
320		3,611.33	II
320		3,614.63	II
320	d	3,615.66	II
320		3,616.58	II
380		3,617.86	II
380		3,619.73	II
810		3,625.54	II
570		3,626.50	II
380		3,629.44	II
670		3,633.29	II
670		3,638.46	II
670		3,641.66	II
440		3,647.06	II
570		3,647.75	II
2,300		3,650.40	II
810		3,654.88	II
2,000		3,658.88	II
450		3,663.12	II
3,800		3,676.35	II
300		3,677.89	II
810		3,682.26	II
320		3,688.15	II
610		3,691.15	II
300		3,692.95	II
450		3,693.58	I
320		3,696.85	II
450		3,700.12	I
4,700		3,702.86	II
300		3,703.12	I
2,400		3,703.92	II
370		3,709.30	II
1,000	d	3,711.76	II
300		3,719.45	II
650		3,729.91	II
430		3,732.39	II
430		3,743.09	
650		3,745.04	I
870		3,747.17	
870		3,747.34	II
1,100		3,755.24	II
430		3,757.44	II
430	d	3,757.90	II
650		3,759.35	I
350		3,761.14	I
1,700		3,765.14	I
2,100		3,776.49	II
330		3,779.22	I
600		3,783.53	I
410	d	3,787.22	II
410		3,789.92	I
390		3,792.20	I
600		3,793.55	I
330		3,801.80	II
760	d	3,806.85	II
1,500		3,830.26	I
540		3,833.42	I
920	d	3,842.50	II
370	d	3,845.61	II
3,700		3,848.73	II
450	d	3,869.75	II
3,500	w	3,874.17	II
330		3,883.34	I
480		3,888.22	I
490		3,894.64	I
330		3,895.99	I
330		3,896.58	II
330		3,897.89	I
2,400		3,899.20	II
1,600		3,901.33	I
480		3,908.06	I
380		3,909.14	I
330		3,909.55	I
650		3,915.43	I
480		3,919.52	II
300		3,922.10	II
480		3,922.74	II
760		3,925.45	II
650		3,935.24	II
810	d	3,939.52	II
650		3,946.89	II
350	d	3,958.36	II
2,200	d	3,976.84	II
1,800		3,981.87	II
300		3,983.85	II
350		3,999.40	II
350	d	4,002.19	II
970		4,002.59	II
1,900		4,005.47	II
300		4,010.04	I
760		4,012.75	II
330		4,013.26	I
370		4,019.14	II
540		4,020.47	II
220		4,022.88	I
370		4,024.77	I
520		4,031.66	II
870		4,032.28	I
2,100		4,033.03	II
350		4,036.22	I
210		4,038.86	I
300		4,051.86	II
300		4,052.87	II
430		4,054.12	I
410		4,060.37	I
220		4,060.87	II
1,300		4,061.58	I
220		4,063.89	II
390		4,066.22	II
260		4,075.22	I
390		4,081.24	I
210		4,086.60	I
210		4,092.19	I
260		4,094.37	II
260		4,094.49	I
260		4,103.90	II
650		4,105.37	I
300		4,112.50	I
260		4,119.92	I
280		4,143.51	I
1,100		4,144.41	II
350		4,158.53	I
240		4,169.09	I
240		4,169.32	I
240		4,171.05	I
240		4,172.60	I
240		4,172.82	I
260		4,173.47	I
240		4,186.21	I
300		4,187.16	I
390		4,196.74	I
450		4,201.00	II
650		4,203.74	I
600		4,206.49	I
300	c	4,213.50	I
300		4,214.42	II
480		4,215.09	I
300		4,217.56	I
260		4,219.16	I
260		4,224.28	I
480	c	4,226.45	II
260		4,231.89	I
480		4,232.82	I
300		4,235.35	I
370		4,255.24	I
480		4,258.23	II
260		4,263.66	I
650		4,266.34	I
330		4,269.69	I
220		4,275.21	I
760	c	4,278.52	II
300		4,285.13	II
300		4,289.70	I
370		4,298.36	I
300		4,299.90	I
240		4,302.95	I
240		4,307.18	I
450		4,310.42	I
300		4,311.56	I
370		4,313.25	I
2,200		4,318.83	I
600		4,322.23	I
600		4,325.83	II
3,000		4,326.43	I
240		4,328.90	I
600		4,332.12	I
870		4,336.43	I
600		4,337.64	I
1,700		4,338.41	I
700		4,340.62	I
430	c	4,342.53	I
430	d	4,353.20	II
280		4,356.09	I
870		4,356.81	I
280		4,360.16	I
220		4,367.30	II
220		4,372.02	I
330		4,382.45	I
300		4,388.23	I
260		4,390.91	I
200		4,416.27	II
140		4,420.19	I
350		4,423.10	I
110		4,432.72	I
240		4,436.12	I
110		4,439.38	I
240		4,448.04	I
110		4,467.69	I
430		4,493.07	I
45	d	4,509.04	II
150	h	4,511.52	I
45		4,512.96	II
75		4,514.31	II
45		4,519.72	II
45		4,525.01	II
45		4,529.76	
45	h	4,531.83	II
45		4,534.13	I
45		4,537.14	I
45		4,537.23	I
110		4,549.07	I
45		4,549.72	II
110		4,550.45	I
110		4,556.46	I
55		4,562.24	II
110		4,563.69	II
30		4,564.85	II
55		4,573.19	II
210		4,578.69	II
65		4,584.84	II
65		4,591.56	II
45		4,592.38	I
45	h	4,604.10	II
30		4,611.96	I
45	h	4,615.92	II
27		4,617.49	I
30		4,619.36	II
75	d	4,626.32	II
95		4,626.94	II
65		4,632.07	I
65	h	4,636.59	I
30		4,636.99	II
85		4,641.00	II
210		4,641.98	II
260	c	4,645.31	II
80		4,647.23	I
60		4,658.38	I
20		4,658.73	
80		4,662.79	I
50	c	4,665.45	I
40		4,669.40	I
80		4,676.90	I
70	c	4,681.87	I
50		4,682.52	I
25	c	4,682.79	II
80		4,688.63	II
80		4,693.11	II
30	h	4,693.39	II
200		4,702.41	II
110		4,707.94	II
40	w	4,716.07	II
40		4,728.16	II
60	c	4,734.20	I
80		4,739.93	I
70		4,747.80	I
410	c	4,752.53	II
40		4,758.44	II
40		4,760.19	II
30		4,762.37	II
25		4,764.47	II
35		4,778.36	II
35		4,778.80	II
180		4,786.78	I

40	c	4,789.91	II
30		4,801.87	II
100		4,813.77	I
60		4,837.59	II
25		4,840.39	I
30	c	4,842.69	II
30		4,844.89	II
30		4,854.81	I
20		4,856.54	II
30		4,858.87	II
80		4,875.57	II
25		4,876.12	II
80		4,881.15	II
29		4,894.33	
95		4,915.90	I
35		4,924.09	I
35		4,926.83	I
50		4,928.93	I
65		4,931.79	I
29		4,970.99	II
29		4,971.42	I
29		4,973.04	I
29		4,980.16	II
29		4,980.56	I
85		4,993.82	II
50		4,995.84	II
55		4,997.95	I
29		5,006.10	II
50		5,022.16	I
29		5,024.24	II
29		5,024.65	I
50		5,033.12	I
50	w	5,042.06	II
55		5,054.30	I
55		5,065.79	I
110		5,078.25	I
24		5,080.05	II
24		5,081.11	I
75		5,089.12	II
24		5,089.66	I
24		5,101.09	I
24		5,108.56	I
35		5,118.39	I
24		5,120.18	I
50	w	5,131.69	I
50	w	5,141.08	II
50		5,147.58	I
24		5,164.27	I
29		5,170.13	I
24		5,170.61	I
50		5,176.51	I
50		5,179.97	I
50		5,184.59	I
85		5,186.13	I
50		5,188.48	I
50		5,198.86	I
35	w	5,202.77	I
40		5,204.55	I
40		5,207.97	I
40		5,214.28	I
40		5,221.99	I
120		5,228.12	I
40		5,235.11	I
75		5,248.71	I
75	w	5,262.11	II
24		5,275.03	I
75		5,281.05	I
65		5,304.72	I
29		5,308.19	I
29		5,309.46	I
110		5,319.23	I
35		5,331.04	I
65	w	5,337.90	I
35	d	5,338.59	I
24		5,347.83	II
160		5,354.88	I
75		5,369.72	I
75		5,375.98	I
29	d	5,402.06	II
29		5,413.65	I
29		5,416.20	I
50		5,424.10	II
29	c	5,426.43	I
35		5,443.38	I
29		5,457.00	I
55		5,459.81	I
29	w	5,470.34	II
24		5,481.45	I
55		5,509.61	I
50		5,514.54	I
65		5,524.12	I
24	c	5,525.62	I
35		5,565.93	I
29	c	5,638.80	I
29	c	5,685.74	II
40	c	5,686.48	I
85	c	5,747.58	I
24		5,762.66	I
24		5,785.18	II
75		5,795.64	I
75		5,803.13	II
65		5,815.36	I
29		5,842.97	I
65		5,851.07	I
65		5,870.62	I
35		5,898.84	I
24		5,902.40	I
35		5,904.71	I
65	c	5,920.78	I
50	c	5,939.38	I
35		5,940.17	I
24		5,951.17	I
75		5,967.34	II
29		6,038.97	I
29		6,039.38	I
24		6,104.29	II
24	c	6,292.43	I
35		6,331.68	II
24		6,334.91	II
24		6,446.87	II
35	c	6,518.68	I
24	c	6,574.04	II
35		6,581.82	I
30		6,607.17	II
90		6,677.94	II
40	c	6,702.61	I
20	c	6,706.79	II
30		6,785.12	II
130		6,794.58	II
40		6,874.18	II
55		6,896.37	II
45	h	6,899.95	I
40		6,901.98	I
9		7,005.99	II
17		7,082.85	II
11		7,089.22	II
11		7,112.69	II
10		7,187.48	I
10	h	7,195.89	II
65		7,204.28	I
19	h	7,234.98	I
40		7,257.73	I
17		7,311.57	I
45		7,348.88	II
10		7,398.27	II
15	h	7,424.24	II
10	h	7,429.62	II
9		7,472.15	I
22		7,484.54	I
9		7,495.45	I
45		7,496.12	I
17		7,499.69	II
27		7,511.40	II
9		7,519.77	II
6		7,557.59	I
27	h	7,582.03	II
27		7,587.49	I
45		7,590.24	I
65		7,596.44	I
17	h	7,601.18	II
17		7,616.01	II
22	h	7,624.05	I
30		7,627.81	I
9	h	7,639.05	II
8		7,672.72	II
8	h	7,694.74	II
22	h	7,706.16	II
22	h	7,726.97	II
30		7,737.63	I
22		7,793.20	I
8		7,807.33	II
16		7,832.91	II
30		7,855.79	II
15		7,864.99	I
6	h	7,885.70	I
6		7,913.11	I
27		7,927.90	II
13		7,955.31	I
11		7,998.03	I
17		8,001.04	I
13	h	8,010.16	II
30		8,025.42	II
6		8,053.80	I
19		8,067.35	II
30		8,085.06	II
27		8,164.17	I
13		8,171.70	I
65		8,194.82	II
95		8,212.57	I
11		8,214.33	I
8		8,259.08	I
40		8,450.06	II
8	h	8,465.80	II
13		8,502.70	I
30	h	8,511.80	I
45		8,583.45	II
30		8,603.40	I
9		8,678.25	I
65		8,765.74	II

Tb IV

Ref. 302 — J.R.

Intensity		Wavelength	
		Vacuum	
30		1,176.58	IV
30		1,192.01	IV
70		1,200.58	IV
50		1,213.94	IV
80		1,221.22	IV
500		1,235.04	IV
1,000		1,259.40	IV
300		1,301.48	IV
300		1,308.30	IV
600		1,311.70	IV
700		1,315.12	IV
500		1,325.56	IV
1,000		1,327.67	IV
100		1,367.56	IV
400		1,367.71	IV
700		1,369.64	IV
1,000		1,373.86	IV
400		1,376.46	IV
200		1,378.23	IV
300		1,381.00	IV
100		1,382.83	IV
20		1,389.92	IV
200		1,516.17	IV
50		1,530.10	IV
5,000		1,595.39	IV
2,000		1,633.19	IV
300		1,649.38	IV
400		1,654.75	IV
400		1,667.58	IV
200		1,672.55	IV
400		1,681.98	IV
5		1,684.46	IV
100		1,685.37	IV
400		1,691.95	IV
10		1,695.23	IV
300		1,698.36	IV
30		1,701.60	IV
50		1,704.79	IV
20		1,705.05	IV
3		1,943.94	IV
50		1,970.90	IV
		Air	
2,000		2,027.79	IV
200		2,029.22	IV
400		2,048.88	IV
200		2,078.83	IV
1,000		2,089.98	IV
1,000		2,332.54	IV
100		2,436.01	IV

THALLIUM (TL)

Z = 81

Tl I and II

Ref. 1, 195, 348, 354 — C.H.C.

Intensity		Wavelength	
		Vacuum	
3		650.90	II
5	r	670.87	II
4		674.10	II
15	r	696.30	II
5	r	709.23	II
10	r	817.18	II
5	r	836.34	II
8	r	1,018.85	II
10	r	1,049.73	II
8	r	1,050.30	II
5	r	1,074.97	II
10	r	1,130.17	II
15	r	1,162.55	II
10	r	1,167.43	II
10	r	1,183.41	II
12	r	1,194.84	II
8		1,231.81	II
5	r	1,246.00	II
15	r	1,307.50	II
8	r	1,310.20	II
25	r	1,321.71	II
8	r	1,330.40	II
10	r	1,373.52	II
1		1,423.2	I
8	r	1,489.65	I
5		1,490.50	II
10	r	1,499.30	II
10	r	1,507.82	II
15	r	1,561.58	II
10	r	1,568.57	II
7	r	1,593.26	II
5	h	1,616.	I
1		1,650.2	I
5		1,685.40	I
1		1,728.	I
10	r	1,792.76	II
12	r	1,814.85	II
3	h	1,847.	I
8		1,892.72	II
25	r	1,908.64	II
		Air	
100	r	2,007.56	I
2		2,209.75	I
100	r	2,210.71	I
3		2,287.6	I
30		2,298.04	II
140		2,315.98	I
900	h	2,379.69	I
8		2,451.83	II
6		2,469.03	II
1		2,508.2	I
20		2,530.86	II
700		2,580.14	I
60		2,608.99	I
80		2,665.57	I
420		2,709.23	I
50	h	2,710.67	I
4,400	d	2,767.87	I
280		2,826.16	I
10		2,849.80	II
2,800		2,918.32	I
440		2,921.52	I
5		3,029.01	II
20		3,091.56	II
15		3,185.51	II
15		3,186.56	II
15		3,187.74	II
1,200		3,229.75	I
15		3,291.01	II
12		3,319.91	II
12		3,321.04	II
8		3,322.25	II
15		3,369.15	II
8		3,381.00	II
6		3,381.80	II
6	d	3,460.48	II
20,000		3,519.24	I
5,000		3,529.43	I
8		3,540.08	II
9		3,560.68	II
5		3,567.67	II
12,000	c	3,775.72	I
8		3,793.95	II
10		3,832.30	II
6		3,869.15	II
10		3,887.15	II
8		4,223.05	II

20		4,274.98	II
40		4,306.80	II
2		4,359.9	I
8		4,490.77	II
20		4,737.05	II
15		4,981.35	II
25		5,078.54	II
25		5,152.14	II
6		5,181.95	II
6		5,183.10	II
18,000		5,350.46	I
15	d	5,384.85	II
7		5,409.92	II
10		5,410.97	II
25		5,949.48	II
10		6,179.98	II
8	d	6,239.03	II
10		6,378.32	II
16	h	6,549.84	I
6	h	6,713.80	I
10		6,966.5	II
3		7,493.6	I
2		7,678.93	I
10		7,815.80	I
8		8,130.0	I
20		8,373.6	I
8		8,445.8	II
10		8,474.27	I
8		8,632.9	II
10		8,664.1	II
4		8,850.4	I
5		8,976.75	I
3		9,038.4	I
20		9,130.	II
20		9,130.5	I
2	h	9,183.1	I
4		9,225.	II
2	h	9,252.6	I
3		9,254.	II
40		9,509.4	I
10		9,863.4	I
20		9,930.4	I
2		9,937.4	I
30		10,011.9	I
40		10,488.80	I
5		11,101.61	I
4		11,483.7	I
1,000		11,512.82	I
5		11,592.9	I
15		12,491.8	I
150		12,736.4	I
700		13,013.2	I

Tl III

Ref. 355 — C.H.C.

Intensity	Wavelength	
	Vacuum	
7	1,231.57	III
10	1,266.33	III
4	1,332.36	III
10	1,477.14	III
4	1,506.37	III
8	1,558.67	III
8	1,660.05	III
	Air	
6	3,163.53	III
3	3,300.80	III
9	3,456.34	III
4	3,507.41	III
6	3,933.05	III
2	3,946.02	III
7	4,109.85	III
4	4,155.75	III
6	4,269.81	III
2	4,380.57	III
4	5,086.99	III
4	5,362.40	III
2	5,499.4	III
5	5,927.8	III
4	8,001.	III

Tl IV

Ref. 356 — C.H.C.

Intensity	Wavelength	
	Vacuum	
7	531.26	IV
10	570.49	IV
4	597.01	IV
1	868.99	IV
3	912.74	IV
8	917.31	IV
30	1,028.69	IV
20	1,034.73	IV
20	1,036.61	IV
10	1,049.48	IV
10	1,057.56	IV
20	1,068.04	IV
20	1,070.47	IV
30	1,079.68	IV
5	1,079.70	IV
2	1,092.90	IV
4	1,094.41	IV
6	1,099.60	IV
4	1,125.52	IV
5	1,139.30	IV
3	1,144.07	IV
3	1,225.45	IV
6	1,273.03	IV
3	1,304.55	IV
6	1,323.66	IV
7	1,337.10	IV
7	1,358.56	IV
5	1,374.62	IV
7	1,377.75	IV
8	1,404.60	IV
5	1,412.93	IV
6	1,434.72	IV
5	1,449.37	IV
5	1,883.2	IV
3	1,974.6	IV

THORIUM (TH)

Z = 90

Th I and II

Ref. 1, 98 — J.G.C.

Intensity	Wavelength	
	Air	
190	2,377.84	II
500	2,565.593	II
270	2,566.588	II
200	2,576.688	II
230	2,589.059	II
230	2,597.047	II
230	2,600.882	II
230	2,618.91	II
270	2,623.448	II
270	2,625.737	II
270	2,641.488	II
170	2,650.583	II
360	2,684.288	II
480	2,692.416	II
100	2,695.553	II
270	2,703.957	II
170	2,708.176	II
230	2,721.691	II
170	2,722.380	II
250	2,729.327	II
250	2,732.808	II
520	2,747.155	II
100	2,749.531	II
410	2,752.166	II
130	2,760.391	II
100	2,765.124	II
270	2,768.841	II
200	2,770.815	II
200	2,771.51	II
70	2,773.951	II
70	2,797.737	II
70	2,814.319	II
45	2,816.071	II
100	2,820.336	II
100	2,822.025	II
170	2,826.856	II
70	2,830.442	II
800	2,832.315	II
,200	2,837.295	II
320	2,842.812	II
270	2,851.260	II
220	2,861.42	II
550	2,870.406	II
320	2,884.289	II
360	2,885.048	II
360	2,887.818	II
250	2,899.720	II
200	2,910.60	II
140	2,919.840	II
250	2,925.051	II
250	2,928.254	II
340	2,942.860	II
150	2,949.068	II
270	2,968.686	II
220	2,974.011	II
360	2,988.232	II
220	3,002.400	II
180	3,008.497	II
150	3,026.575	II
370	3,034.065	II
170	3,035.110	II
420	3,049.092	II
220	3,061.699	II
450	3,067.729	II
670	3,078.828	II
480	3,080.217	II
240	3,088.470	II
200	3,102.664	II
510	3,108.296	II
510	3,119.526	II
510	3,122.963	II
370	3,124.387	II
480	3,125.507	II
150	3,131.070	II
420	3,139.306	II
420	3,142.835	II
310	3,146.044	II
150	3,150.455	II
310	3,154.301	II
310	3,154.775	II
140	3,166.099	II
110	3,169.328	II
420	3,175.726	II
270	3,179.048	II
1,100	3,180.193	II
310	3,184.948	II
770	3,188.233	II
170	3,210.308	II
560	3,221.292	II
560	3,229.009	II
480	3,235.84	II
590	3,238.116	II
280	3,251.915	II
910	3,256.274	II
910	3,262.668	II
130	3,285.752	I
620	3,287.789	II
910	3,291.739	II
620	3,292.520	II
480	3,304.238	I
510	3,321.45	II
390	3,324.752	II
840	3,325.120	II
620	3,334.604	II
620	3,337.870	II
980	3,351.228	II
620	3,358.602	II
390	3,367.819	II
390	3,378.573	II
310	3,385.531	II
1,300	3,392.035	II
980	3,402.70	II
980	3,433.998	II
770	3,435.976	II
530	3,439.71	II
340	3,462.851	II
450	3,465.76	II
390	3,468.219	II
1,300	3,469.92	II
250	3,479.173	II
270	3,493.518	II
450	3,511.67	II
670	3,539.587	II
180	3,544.018	I
140	3,551.401	I
530	3,559.451	II
530	3,575.32	II
270	3,592.780	I
170	3,608.377	I
980	3,609.445	II
480	3,615.133	II
400	3,617.12	II
140	3,632.831	I
270	3,635.943	I
420	3,658.06	II
560	3,659.51	II
140	3,668.140	I
700	3,675.567	II
280	3,711.304	II
590	3,719.435	I
450	3,719.98	II
770	3,721.82	II
50	3,733.672	I
1,300	3,741.183	II
310	3,747.539	I
650	3,752.569	II
140	3,757.694	I
510	3,762.88	II
50	3,780.966	I
340	3,785.600	II
590	3,803.075	I
340	3,813.068	II
450	3,828.384	I
840	3,839.74	II
280	3,841.960	II
390	3,854.511	II
140	3,859.840	II
450	3,863.405	II
210	3,875.374	I
170	3,905.186	II
590	3,929.669	II
200	3,948.964	II
200	3,972.155	I
150	3,980.089	I
530	3,994.549	II
240	4,003.307	II
250	4,007.02	II
220	4,008.210	I
220	4,009.056	I
280	4,012.495	I
4,200	4,019.129	II
210	4,022.07	I
210	4,025.65	II
250	4,030.842	I
250	4,036.047	I
240	4,036.565	II
240	4,041.20	II
110	4,050.887	I
250	4,063.407	I
910	4,069.202	II
700	4,085.04	II
700	4,086.520	II
700	4,094.747	II
270	4,105.34	II
840	4,108.420	II
240	4,112.754	I
280	4,115.758	I
1,100	4,116.713	II
200	4,127.411	I
110	4,131.002	I
340	4,132.753	II
200	4,134.067	I
220	4,140.235	II
250	4,142.701	II
200	4,148.18	II
450	4,149.986	II
340	4,156.51	II
140	4,165.766	I
220	4,170.47	II
620	4,178.060	II
250	4,179.71	II
620	4,208.890	II
280	4,273.357	II
480	4,277.313	II
700	4,282.042	II
55	4,297.306	I
85	4,299.839	I
100	4,307.176	I
200	4,309.99	II
110	4,318.416	I
85	4,342.255	II
130	4,344.327	II
85	4,365.930	I
1,300	4,381.860	II
1,100	4,391.110	II
55	4,392.974	I
210	4,412.74	II
55	4,414.486	I
250	4,432.963	II
220	4,465.341	II
280	4,510.526	II
22	4,530.319	I
40	4,535.255	I
30	4,545.915	II
65	4,570.972	I
140	4,631.761	II
190	4,740.529	II

Intensity		Wavelength	
140		4,752.414	II
40		4,826.700	I
40		4,858.333	II
280		4,863.16	II
240		4,919.816	II
50		5,002.097	I
50		5,015.889	II
260		5,017.255	II
130		5,028.61	II
50		5,044.719	I
240		5,049.796	II
85		5,055.347	II
70		5,058.56	II
110		5,067.974	I
50		5,098.043	II
50		5,110.86	II
95		5,143.267	II
120		5,148.21	II
50		5,151.612	I
50		5,154.243	II
85		5,158.604	I
70		5,160.730	I
50		5,164.98	I
50		5,176.961	I
50		5,190.87	II
50	h	5,195.814	I
50		5,198.800	I
95		5,199.164	I
50		5,211.230	I
95		5,216.59	II
50		5,218.53	II
110		5,231.160	I
85		5,233.225	II
95		5,247.654	II
70		5,277.500	II
30		5,307.466	II
60		5,325.145	II
50		5,326.976	I
60		5,343.581	I
14		5,351.126	I
70		5,390.466	II
50		5,392.57	II
70		5,415.46	II
60		5,425.678	II
50		5,435.893	II
40		5,449.479	II
50		5,539.262	I
70		5,539.911	II
50		5,558.342	I
60		5,564.20	II
60		5,587.026	I
50		5,604.51	II
70		5,639.746	II
50		5,645.89	II
65		5,700.918	II
95		5,707.103	II
50		5,720.183	I
70		5,760.551	I
19		5,815.422	II
17		5,914.671	I
85		5,989.045	II
24		5,994.129	I
21		6,007.072	I
30		6,015.422	II
24		6,044.433	II
30		6,073.104	II
30		6,087.262	II
24		6,099.084	II
30		6,104.57	I
40		6,112.838	II
30		6,120.56	II
60		6,169.822	I
50		6,182.622	I
24		6,193.86	II
24	h	6,234.856	I
21		6,261.064	II
21		6,261.418	I
50		6,274.117	II
30		6,279.16	II
21		6,327.278	I
35		6,342.860	I
40		6,376.931	I
30		6,411.899	I
24		6,413.615	I
60		6,457.283	I
50		6,462.614	I
50	h	6,531.342	I
24		6,583.907	I
24		6,588.540	I
24		6,593.940	I
24		6,605.417	II

Intensity	Wavelength	
24	6,619.947	II
21	6,644.66	II
30	6,662.269	I
20	6,756.453	I
20	6,889.30	II
24	6,911.227	I
35	6,943.611	I
55	6,989.656	I
24	6,993.03	II
18	7,000.806	I
10	7,018.569	I
30	7,045.80	II
15	7,053.61	II
24	7,075.33	II
30	7,084.171	I
24	7,089.33	II
10	7,100.51	II
11	7,124.562	I
10	7,154.954	I
30	7,168.896	I
40	7,191.13	II
35	7,208.006	I
11	7,212.69	I
10	7,217.76	II
11	7,218.054	I
11	7,305.40	II
18	7,385.501	I
21	7,428.940	I
10	7,430.254	I
10	7,481.35	I
50	7,525.51	II
18	7,567.74	I
12	7,585.69	I
12	7,585.78	I
30	7,647.380	I
21	7,685.30	II
10	7,731.72	II
15	7,787.79	II
15	7,788.937	I
21	7,817.771	I
15	7,847.540	I
12	7,865.95	I
11	7,900.31	I
11	7,941.72	I
24	7,978.974	I
11	7,987.97	I
11	8,032.433	I
11	8,062.64	I
11	8,138.477	I
18	8,143.139	I
12	8,159.729	I
10	8,163.12	II
15	8,186.914	I
12	8,203.19	II
18	8,275.629	I
15	8,320.857	I
30	8,330.451	I
18	8,403.79	II
15	8,416.74	I
12	8,421.227	I
21	8,446.52	I
18	8,478.360	I
11	8,573.122	I
12	8,591.83	II
10	8,665.487	I
18	8,748.033	I
15	8,758.244	I
18	8,842.07	II
15	8,868.834	I
15	8,957.97	II
40	8,967.641	I
15	9,399.09	I
15	9,474.882	I
15	9,495.50	I
15	9,497.19	I
15	9,700.56	I
15	9,746.46	I
10	9,812.70	I
10	9,826.45	I
20	9,833.42	I
15	10,039.364	I
15	10,089.138	I
15	10,133.56	II
15	10,419.57	II
15	10,556.45	I
15	10,723.92	II
20	10,726.93	I
20	10,942.24	II
15	11,051.90	I
30	11,230.259	I
20	11,354.719	I

Intensity	Wavelength	
15	11,703.46	I
15	11,864.25	I
15	11,940.64	I
20	11,984.67	II
15	12,018.72	I
20	12,127.30	I
20	12,194.16	II
15	12,206.89	I
20	12,231.94	I
15	12,338.00	I
15	12,477.30	I
20	12,646.54	I
15	12,866.64	I
15	12,940.65	II
15	12,959.82	I
15	13,145.90	II
15	13,565.67	I
15	14,090.25	I
15	14,168.67	I
15	14,424.54	I
15	14,618.98	I
15	14,654.91	I
15	14,940.49	II
15	15,240.24	II
15	15,429.78	I
15	15,831.75	I
20	17,208.22	II
15	17,307.66	I
15	17,381.91	I
15	17,481.04	I
15	17,584.52	I
15	17,936.43	II
15	18,811.88	I
15	19,145.60	II
15	19,338.98	II
10	19,774.30	II
10	20,634.36	I
10	20,692.06	II
10	22,264.35	II

Th III

Ref. 157 — J.G.C.

Intensity	Wavelength	
	Vacuum	
100	1,888.12	III
	Air	
50	2,149.18	III
50	2,162.82	III
50	2,199.74	III
50	2,206.62	III
50	2,291.59	III
100	2,301.18	III
100	2,319.52	III
80	2,324.68	III
150	2,335.50	III
100	2,340.58	III
100	2,363.06	III
50	2,368.91	III
100	2,371.42	III
80	2,381.47	III
100	2,391.48	III
200	2,413.50	III
50	2,424.54	III
200	2,427.94	III
200	2,431.68	III
200	2,441.24	III
100	2,463.66	III
50	2,473.93	III
100	2,501.08	III
60	2,512.69	III
50	2,514.31	III
100	4,555.73	III
50	4,589.28	III
100	5,376.13	III
50	5,447.18	III
50	6,242.95	III
50	6,599.39	III
50	7,461.59	III
50	8,105.14	III

Th IV

Ref. 156, 165 — J.G.C.

Intensity	Wavelength	
	Vacuum	
4	797.53	IV
1	835.55	IV
30	846.91	IV
1	854.02	IV
30	882.39	IV
12	886.66	IV
100	1,565.85	IV
70	1,682.22	IV
30	1,684.01	IV
150	1,707.37	IV
200	1,959.02	IV
	Air	
200	2,002.34	IV
100	2,066.70	IV
20	2,143.91	IV
30	2,146.81	IV
1	2,242.11	IV
2	2,261.26	IV
5	2,296.81	IV
100	2,693.99	IV
2	4,937.09	IV
4	4,952.52	IV
3	5,420.38	IV
2	6,711.87	IV
3	6,740.37	IV
50	6,901.16	IV

THULIUM (TM)

Z = 69

Th I and II

Ref. 1 — C.H.C.

Intensity		Wavelength	
		Air	
360		2,284.79	II
120		2,329.77	II
70		2,340.92	II
120		2,363.91	II
45		2,365.96	II
160		2,367.11	II
150		2,383.68	II
110		2,388.95	II
450		2,409.02	II
110		2,412.44	II
120		2,421.65	II
450		2,426.17	II
140		2,445.47	II
770		2,480.13	II
150		2,481.15	II
130		2,487.52	II
250		2,491.60	II
100		2,499.54	II
130		2,507.15	II
1,300		2,509.08	II
200		2,520.87	II
250		2,522.17	II
180		2,524.11	II
130		2,527.02	I
110		2,527.42	II
120		2,542.66	II
360		2,552.76	I
540		2,561.65	II
150		2,563.86	II
430		2,588.27	II
170	h	2,596.49	I
110		2,601.09	I
220		2,606.02	II
810		2,607.06	II
730		2,624.33	II
210		2,640.76	II
130		2,646.45	II
160		2,650.27	II
190		2,658.48	II
250		2,660.09	II
140		2,668.20	II
310		2,679.57	II
170		2,697.50	II
540		2,721.19	II
200		2,744.08	II
270		2,779.55	II
350		2,785.07	II
680		2,794.60	II
730		2,797.27	II
250		2,818.47	II
250		2,827.02	II

580		2,827.92	II	340		3,480.98	I	140		4,396.50	I	35		5,085.09	I
200		2,831.55	II	340		3,481.75	II	55		4,437.40	II	40		5,107.53	I
310		2,844.67	II	420		3,487.38	I	80		4,442.74	I	95		5,113.97	I
200		2,854.17	I	210		3,492.58	II	50		4,447.58	I	50		5,114.55	II
200		2,860.12	II	340		3,499.95	I	120		4,454.03	I	22		5,120.67	I
200		2,861.74	II	250		3,513.02	II	80		4,459.99	I	22		5,140.28	II
1,600		2,869.23	II	250		3,517.60	I	50		4,467.98	I	40		5,149.40	II
630		2,890.94	II	250		3,534.85	II	540		4,481.26	II	19		5,182.68	I
210		2,918.27	II	1,700		3,535.52	II	80		4,489.70	II	40		5,185.25	I
270		2,925.65	II	490		3,536.21	II	150		4,519.60	I	14		5,204.51	II
680		2,926.74	II	850		3,536.58	II	260		4,522.57	II	80		5,213.38	I
630		2,935.99	II	420		3,537.91	II	180		4,529.38	I	22		5,228.23	II
350		2,951.26	II	210		3,555.82	I	80		4,532.15	I	14		5,260.93	II
430		2,965.86	II	420		3,557.79	II	110		4,548.60	I	24		5,267.34	II
490		2,973.22	I	340		3,560.92	I	40		4,556.68	II	40		5,291.14	I
540		2,981.48	II	420		3,563.88	I	40		4,561.86	II	40		5,294.32	I
350		2,986.52	II	490		3,565.91	II	80		4,564.68	I	35		5,300.21	I
630		2,990.54	II	1,300		3,566.47	II	40		4,567.11	II	35		5,302.69	I
200		2,993.26	II	420		3,567.36	I	95		4,596.63	I	55		5,305.87	II
230		3,013.71	II	280		3,574.06	II	270		4,599.02	I	650		5,307.12	I
430		3,014.65	II	280		3,586.07	I	35		4,601.29	I	16		5,322.99	II
1,500		3,015.30	II	2,100		3,608.77	II	55		4,603.43	II	35		5,338.90	I
270		3,017.09	II	250		3,609.53	II	40		4,604.85	I			5,339.03	I
330		3,026.07	II	380		3,638.41	I	50		4,613.97	I	80		5,346.49	II
280		3,042.35	II	950		3,643.65	II	40		4,614.47	II	27		5,372.98	II
340	d	3,046.76	II	240		3,647.72	II	300		4,615.94	I	14		5,391.96	II
320		3,050.73	II	600		3,653.61	II	35		4,619.06	II	27		5,400.46	II
340		3,056.07	II	500		3,665.81	II	40		4,621.72	I	27		5,402.23	I
580		3,073.08	II	1,100		3,668.09	II	80		4,626.33	II	14		5,405.98	II
360		3,081.12	I	410		3,677.98	II	95		4,626.56	II	14		5,461.95	II
740		3,098.60	II	450	d	3,678.85	II	40		4,626.97	I	14		5,464.14	I
7,400		3,131.26	II	410		3,694.74	II	110		4,634.26	II	14		5,465.54	II
2,300		3,133.89	II	4,800		3,700.26	II	40		4,642.96	II	16		5,500.30	II
230		3,144.90	II	3,800		3,701.36	II	95		4,643.12	I	14		5,526.82	II
230		3,146.16	II	330		3,704.85	II	35		4,644.58	I	24		5,528.34	I
1,900		3,151.04	II	7,700		3,717.91	I	120		4,655.09	I	14		5,539.03	II
1,500		3,157.34	II	890		3,725.06	II	35		4,666.70	II	27		5,566.00	I
450		3,172.65	I	2,400		3,734.12	II	35		4,671.99	II	22		5,581.37	I
2,300		3,172.83	II	5,000		3,744.06	II	35		4,675.10	I	14		5,586.65	II
380		3,173.58	II	1,700		3,751.81	II	80		4,675.31	I	14		5,589.94	II
230		3,195.33	II	310		3,756.86	II	40		4,677.86	II	14		5,606.64	I
320		3,210.56	II	6,000		3,761.33	II	160		4,681.92	I	270		5,631.41	I
320		3,210.82	II	4,800		3,761.91	II	70		4,685.11	I	40		5,642.60	I
320		3,212.01	II	260		3,783.55	II	120		4,691.11	I	27		5,645.40	I
230		3,231.51	II	380		3,795.16	II	110		4,724.26	I	70		5,658.30	I
470		3,235.44	II	7,100		3,795.75	II	680		4,733.34	I	520		5,675.84	I
1,200		3,236.81	II	770		3,798.54	I	35		4,750.75	II	14		5,683.59	I
1,600		3,240.23	II	240		3,798.75	II	70		4,759.90	I	40		5,684.76	II
2,300		3,241.54	II	600		3,807.72	I	27		4,789.92	II	14		5,696.42	II
320		3,246.96	I	380		3,810.72	II	27		4,807.48	I	35		5,709.97	II
420		3,247.46	II	550		3,817.39	II	35		4,808.68	I	22		5,715.79	I
1,900		3,258.05	II	290		3,826.39	I	35		4,813.50	I	14		5,733.81	II
400		3,261.65	II	1,300		3,838.20	II	27		4,826.99	II	11	d	5,737.20	II
320		3,264.10	II	290		3,840.87	I	27		4,828.97	I			5,737.25	II
1,600		3,266.64	II	8,900		3,848.02	II	80		4,831.20	II	14	h	5,738.92	II
1,200		3,267.40	II	140		3,857.84	II	35		4,835.75	I	27		5,758.02	I
790		3,268.99	II	6,800		3,883.13	I	27	d	4,851.76	I	55		5,760.20	I
1,100		3,276.81	II	1,800		3,883.44	II			4,851.90	II	190		5,764.29	I
1,200		3,283.40	II	5,400		3,887.35	I	19		4,872.28	II	5		5,778.82	II
1,200		3,285.61	II	440		3,890.53	II	27		4,879.19	I	19		5,782.36	II
2,300		3,291.00	II	440		3,896.62	II	27		4,891.64	I	22		5,784.46	II
2,000		3,302.46	II	680		3,900.79	II	24		4,909.74	I	11		5,799.97	II
210		3,306.01	II	3,500		3,916.48	I	55		4,923.83	I	14		5,811.19	II
210		3,306.91	II	120		3,928.66	II	140		4,957.18	I	14	h	5,816.46	I
210		3,308.01	II	570		3,929.58	II	40		4,970.87	II	35		5,838.76	II
1,200		3,309.80	II	1,500		3,949.27	I	27		4,971.26	I	240		5,895.63	I
640		3,310.59	II	1,500		3,958.10	II	40		4,975.12	II	35		5,899.47	I
400		3,316.88	II	440		3,995.58	II	50		4,978.90	I	24		5,901.57	I
210		3,318.65	II	1,800		3,996.52	II	40		4,980.68	II	8		5,912.58	I
230		3,349.99	I	220		4,024.23	I	55		4,989.32	II	11		5,931.70	I
230		3,354.86	II	380		4,044.47	I	27		4,993.79	II	27		5,935.90	I
4,000		3,362.61	II	10,000		4,094.19	I	19		4,994.72	II	140		5,971.26	I
490		3,374.50	II	9,500		4,105.84	I	35		5,001.02	I	27		5,975.02	I
420	d	3,384.99	II	120		4,132.69	II	27		5,001.59	I	11		5,984.87	I
1,700		3,397.50	II	1,100		4,138.33	I	160		5,009.77	II	19		6,025.44	I
420		3,399.95	II	120		4,149.14	I	35		5,014.56	II	11		6,067.78	II
850		3,410.05	I	120		4,158.60	I	27		5,017.87	II	16		6,131.53	I
340		3,412.59	I	8,800		4,187.62	I	160		5,034.22	II	14		6,175.29	I
340		3,416.59	I	520		4,199.92	II	27	h	5,041.00	II	14		6,181.41	II
6,400		3,425.08	II	6,000		4,203.73	I	22		5,043.50	I	14		6,299.46	II
950		3,425.63	II	220		4,206.00	I	35		5,045.41	I	27		6,352.66	I
340		3,429.33	I	380		4,222.67	I	27		5,060.42	II	22		6,401.44	I
850		3,429.96	II	3,000		4,242.15	II	150		5,060.90	I	8		6,430.94	II
420		3,431.19	II	270		4,271.71	I	27		5,062.25	I	14		6,440.54	I
4,900		3,441.50	II	150		4,298.36	I	27		5,065.88	I	200		6,460.26	I
4,900		3,453.66	II	2,700		4,359.93	I	80		5,066.67	I	14		6,490.70	I
8,500		3,462.20	II	1,400		4,386.43	I	27		5,072.42	I	14		6,519.78	I
210		3,467.51	I	200		4,394.42	I	27		5,076.36	I	8		6,575.54	I
340		3,476.69	I	120		4,395.96	I	27		5,077.18	I	95		6,604.96	I

8		6,627.25	I
35		6,657.72	I
11		6,658.64	I
11		6,692.93	I
30		6,721.36	I
9		6,726.34	I
9		6,727.94	II
18		6,739.22	I
9	h	6,767.48	I
9		6,777.93	I
110		6,779.77	I
14	h	6,782.00	I
18		6,788.52	I
13	h	6,820.27	I
14		6,826.95	I
14		6,829.12	II
23		6,831.09	I
120		6,844.26	I
80		6,845.76	I
18		6,854.12	I
6		6,898.56	I
6		6,915.86	I
10		6,937.37	I
5		6,949.54	I
5	h	6,976.69	II
5		7,010.79	I
6	h	7,014.31	II
10		7,017.90	I
6	h	7,029.40	I
12		7,034.34	I
10		7,056.43	II
5		7,060.97	I
6		7,079.78	II
10		7,106.14	I
5	h	7,231.33	I
5		7,233.74	II
4		7,257.72	I
17		7,272.62	I
8		7,284.30	I
11	h	7,286.16	I
14		7,310.51	I
11		7,336.63	II
14		7,432.18	I
5		7,434.51	II
5		7,439.95	II
75		7,481.08	I
75		7,490.20	I
10	h	7,507.28	I
14		7,545.78	I
140		7,558.33	I
17	h	7,580.61	I
20	h	7,593.74	I
17		7,595.07	II
5		7,629.85	I
5	h	7,648.76	II
17		7,655.00	I
4		7,660.32	I
7	h	7,666.24	I
8		7,676.04	II
8	h	7,701.46	I
80		7,731.53	I
4	h	7,778.27	I
12	h	7,782.35	I
8	h	7,785.51	I
		7,785.90	I
17		7,803.93	I
4		7,829.22	I
40		7,856.08	I
3		7,861.67	I
5		7,918.10	I
55		7,927.51	I
110		7,930.84	I
6		7,971.56	I
11	h	7,985.93	I
14	h	8,014.77	I
95		8,017.90	I
3	h	8,021.33	I
14		8,194.19	I
5		8,294.52	I
7		8,365.75	I
7		8,460.79	II
27		8,472.01	II
7	h	8,546.07	II
11		8,565.73	II

Th III

Ref. 307 — J.R.

Intensity	Wavelength	
	Air	
500	2,099.11	III
500	2,107.10	III
200	2,136.67	III
200	2,156.29	III
800	2,182.98	III
300	2,183.91	III
5,000	2,185.94	III
100	2,212.25	III
300	2,230.86	III
400	2,231.25	III
200	2,243.34	III
400	2,243.98	III
200	2,246.68	III
200	2,269.39	III
1,000	2,276.91	III
100	2,280.08	III
100	2,281.27	III
100	2,282.86	III
200	2,282.98	III
200	2,286.57	III
400	2,287.21	III
500	2,294.73	III
20,000	2,296.21	III
200	2,297.43	III
100	2,304.64	III
400	2,304.82	III
5,000	2,305.03	III
20,000	2,311.16	III
5,000	2,312.72	III
200	2,314.88	III
400	2,317.35	III
500	2,320.96	III
200	2,322.83	III
100	2,323.71	III
100	2,323.77	III
100	2,324.43	III
500	2,324.62	III
5,000	2,326.19	III
100	2,327.02	III
300	2,327.25	III
6,000	2,328.50	III
6,000	2,329.29	III
200	2,330.87	III
3,000	2,331.80	III
400	2,335.01	III
1,000	2,338.36	III
500	2,341.74	III
300	2,342.04	III
100	2,344.59	III
500	2,345.61	III
300	2,347.43	III
400	2,353.10	III
100	2,355.65	III
3,000	2,357.05	III
1,000	2,361.23	III
500	2,363.97	III
1,000	2,375.32	III
700	2,375.83	III
400	2,389.52	III
4,000	2,406.63	III
500	2,435.31	III
500	2,457.86	III
500	2,471.23	III
30,000	2,489.44	III
200	2,496.25	III
2,000	2,504.71	III
3,000	2,519.78	III
10,000	2,552.46	III
500	2,557.90	III
1,000	2,574.52	III
500	2,574.98	III
100	2,581.84	III
100	2,585.48	III
300	2,589.20	III
500	2,608.96	III
300	2,609.66	III
500	2,617.22	III
500	2,618.78	III
1,000	2,621.12	III
400	2,621.35	III
400	2,622.31	III
100	2,627.09	III
100	2,628.83	III
300	2,634.66	III
200	2,636.68	III
200	2,637.30	III
100	2,640.32	III
500	2,643.58	III
100	2,645.05	III
200	2,649.27	III
100	2,650.82	III
100	2,654.05	III
100	2,656.30	III
500	2,661.51	III
1,000	2,663.00	III
500	2,664.76	III
500	2,664.88	III
200	2,665.05	III
1,000	2,666.93	III
200	2,668.59	III
100	2,668.66	III
200	2,669.18	III
100	2,671.42	III
100	2,675.30	III
1,000	2,676.64	III
500	2,676.91	III
100	2,678.28	III
100	2,680.49	III
5,000	2,682.32	III
300	2,682.64	III
300	2,687.14	III
300	2,695.69	III
400	2,698.21	III
1,000	2,699.49	III
1,000	2,699.80	III
100	2,703.63	III
200	2,703.68	III
100	2,704.93	III
2,000	2,707.03	III
300	2,707.19	III
200	2,707.44	III
500	2,707.60	III
1,000	2,709.74	III
200	2,710.79	III
1,000	2,713.38	III
200	2,715.81	III
300	2,717.56	III
100	2,718.02	III
3,000	2,719.47	III
3,000	2,724.44	III
4,000	2,727.56	III
200	2,728.13	III
1,000	2,731.38	III
300	2,732.11	III
400	2,737.98	III
800	2,744.74	III
400	2,745.99	III
500	2,752.46	III
400	2,753.20	III
400	2,756.15	III
800	2,765.98	III
700	2,769.92	III
200	2,772.64	III
100	2,777.43	III
400	2,781.12	III
2,000	2,806.77	III
300	2,821.12	III
200	2,849.52	III
700	2,882.02	III
100	2,899.29	III
100	2,912.33	III
400	2,921.08	III
200	2,947.02	III
1,000	2,947.72	III
500	2,953.18	III
100	2,966.15	III
500	2,966.85	III
400	2,972.61	III
100	2,974.85	III
1,000	2,998.28	III
100	3,048.11	III
700	3,078.87	III
200	3,120.15	III
200	3,277.26	III
200	3,407.73	III
100	3,415.40	III
100	3,415.96	III
100	3,436.93	III
400	3,467.93	III
200	3,529.29	III
200	3,533.28	III
100	3,537.47	III
300	3,562.41	III
100	3,563.42	III
200	3,587.74	III
600	3,617.96	III
1,000	3,629.09	III
100	3,706.11	III
100	3,799.41	III
300	3,998.84	III
200	4,021.92	III
200	4,026.03	III
700	4,032.13	III
100	4,076.15	III
200	4,335.47	III
500	4,385.41	III

TIN (SN)

Z = 50

Sn I and II

Ref. 187, 191 — C.H.C.

Intensity		Wavelength	
		Vacuum	
1		899.92	II
2		917.40	II
1		935.63	II
3		945.83	II
4		954.50	II
7		985.13	II
4		997.21	II
2		1,016.26	II
4		1,040.78	II
1		1,041.32	II
3		1,062.10	II
8		1,108.19	II
4		1,159.05	II
10		1,161.43	II
3		1,162.94	II
4		1,180.51	II
9		1,219.07	II
13		1,223.70	II
11		1,243.00	II
20		1,290.86	II
20		1,316.59	II
25		1,400.52	II
20		1,475.15	II
9		1,489.22	II
7		1,699.47	II
10	r	1,737.21	I
15	r	1,751.46	I
10	h	1,753.3	I
7		1,758.00	II
20	r	1,764.98	I
20		1,773.40	I
30	r	1,790.75	I
80	r	1,804.60	I
15		1,811.34	II
30		1,813.04	I
40	r	1,815.74	I
25		1,819.31	I
120	r	1,823.00	I
9		1,831.89	II
50	r	1,848.75	I
30		1,852.00	I
200	r	1,860.32	I
20		1,861.42	I
20		1,865.52	I
30		1,865.96	I
15		1,873.29	I
30		1,882.64	I
80		1,886.05	I
100		1,891.40	I
20		1,897.29	I
12		1,899.91	II
50		1,909.30	I
40		1,911.61	I
20		1,913.52	I
80		1,925.31	I
20		1,926.77	I
15		1,927.95	I
40	h	1,928.9	I
25		1,933.17	I
20		1,942.69	I
150		1,952.15	I
15		1,960.21	I
30		1,971.46	I
50	h	1,977.6	I
80		1,984.20	I
15		1,991.88	I
20		1,994.98	I
		Air	
25		2,008.05	I
30		2,015.76	I
30		2,026.98	I
50		2,040.66	I

20		2,040.90	I
50		2,054.03	I
70		2,058.31	I
20		2,064.00	I
80		2,068.58	I
100		2,072.89	I
100		2,073.08	I
25		2,080.62	I
30		2,091.58	I
40		2,094.35	I
200		2,096.39	I
100		2,100.93	I
100	r	2,113.93	I
50		2,121.26	I
25		2,140.73	I
20		2,141.43	I
15		2,148.46	I
1		2,148.63	II
40	r	2,148.73	I
20	r	2,151.43	I
30		2,151.54	II
80		2,171.32	I
150	r	2,194.49	I
300	r	2,199.34	I
400	r	2,209.65	I
4		2,209.67	II
40		2,211.05	I
80	r	2,231.72	I
400	r	2,246.05	I
6		2,246.07	II
60		2,251.17	I
30		2,267.19	I
400	r	2,268.91	I
20		2,282.26	I
200	r	2,286.68	I
600	r	2,317.23	I
300	r	2,334.80	I
1,000	r	2,354.84	I
20		2,357.90	I
3		2,360.34	II
22		2,368.33	II
60		2,380.72	I
4		2,384.54	II
100		2,408.15	I
800	r	2,421.70	I
1,000	r	2,429.49	I
1		2,433.52	II
15		2,448.98	II
60		2,455.24	I
20		2,476.40	I
300		2,483.39	I
13		2,483.48	II
10		2,486.99	II
200		2,495.70	I
5		2,522.61	II
90		2,523.92	I
80	h	2,531.17	I
400		2,546.55	I
40	h	2,558.01	I
500	r	2,571.58	I
200		2,594.42	I
50	h	2,636.94	I
200	r	2,661.24	I
2		2,664.93	II
700	r	2,706.51	I
2		2,727.82	II
20		2,761.78	I
150		2,779.81	I
80		2,785.03	I
60		2,787.96	I
60		2,812.59	I
80		2,813.58	I
2		2,825.52	II
1,400	r	2,839.99	I
1		2,846.42	II
200		2,850.62	I
1,000	r	2,863.32	I
1		2,912.80	II
200		2,913.54	I
6		2,919.82	II
3		2,991.00	II
7		2,994.44	II
700	r	3,009.14	I
1		3,012.18	II
8		3,023.94	II
200		3,032.80	I
850	r	3,034.12	I
12		3,047.50	II
6		3,094.69	II
60		3,141.84	I
550	r	3,175.05	I
40		3,218.71	I
550	r	3,262.34	I
50		3,283.21	II
110		3,330.62	I
60		3,351.97	II
2		3,407.48	II
10		3,472.46	II
7		3,537.57	II
11		3,575.45	II
3		3,582.39	II
2		3,620.08	II
6		3,620.54	II
40		3,655.78	I
6		3,715.23	II
280	r	3,801.02	I
4		3,841.44	II
1		4,294.65	II
40		4,524.74	I
1		4,579.13	II
1		4,580.29	II
2		4,877.22	II
3		4,944.31	II
20		4,979.73	I
2		5,071.14	II
2		5,072.67	II
20		5,174.54	I
10		5,332.36	II
20		5,561.95	II
25		5,588.92	II
2		5,596.20	II
500		5,631.71	I
15		5,753.59	I
1		5,797.20	II
15		5,799.18	II
50		5,925.44	I
100		5,970.30	I
150		6,037.70	I
200		6,054.86	I
250		6,069.00	I
100		6,073.46	I
6		6,077.48	II
5		6,079.70	II
400		6,149.71	I
200		6,154.60	I
150		6,171.50	I
100		6,310.78	I
40		6,354.35	I
70		6,453.50	II
8		6,761.45	II
25		6,844.05	II
20		7,191.40	II
10		7,387.79	II
20	h	7,398.6	I
1		7,408.62	II
30		7,685.30	I
13		7,741.80	II
100		7,754.97	I
3		7,904.00	II
100	h	8,030.5	I
30	h	8,039.3	I
200		8,114.09	I
30	h	8,121.0	I
30		8,349.35	I
80		8,357.04	I
300		8,422.72	I
400		8,552.60	I
50	h	8,681.7	I
30	h	9,018.95	I
50	h	9,410.86	I
80	h	9,415.37	I
150		9,616.40	I
50		9,741.1	I
100	h	9,742.8	I
300	h	9,805.38	I
500		9,850.52	I
25		10,456.47	I
11		10,807.58	I
54		10,894.00	I
70		11,191.85	I
56		11,277.66	I
17		11,336.97	I
200		11,454.59	I
200		11,616.26	I
76		11,670.77	I
25		11,694.45	I
258		11,739.78	I
96		11,825.18	I
106		11,835.82	I
254		11,932.99	I
48		12,009.50	I
111		12,313.24	I
33		12,335.6	I
42		12,530.87	I
42		12,536.5	I
37		12,788.2	I
89		12,888.5	I
187		12,981.7	I
20		13,000.3	I
187		13,018.5	I
68		13,081.5	I
378		13,460.2	I
144		13,608.2	I
13		15,018.2	I
30		15,464.2	I
20		17,000.5	I
10		17,807.5	I
20		20,622.2	I
40		20,861.7	I
8		21,686.2	I
4		22,131.7	I
3		22,997.2	I
4		24,327.2	I
4		24,738.2	I

TITANIUM (Ti)

Z = 22

Ti I and II

Ref. 1 — C.H.C.

Intensity		Wavelength	
		Air	
140		2,272.61	I
180		2,273.28	I
130		2,276.70	I
190		2,279.96	I
150		2,299.85	I
140		2,302.73	I
190		2,305.67	I
65		2,380.81	I
35		2,384.52	I
55		2,418.36	I
75		2,421.30	I
95		2,424.24	I
40		2,428.23	I
35		2,433.22	I
19		2,434.10	I
35		2,440.21	II
65		2,440.98	I
24		2,450.44	II
24		2,504.54	I
75		2,517.43	II
40		2,519.04	I
140		2,520.54	I
75		2,524.64	II
360		2,525.60	II
29		2,527.98	I
210		2,529.85	I
190		2,531.25	II
190		2,534.62	II
130		2,535.87	II
190		2,541.92	I
65		2,555.99	II
110		2,571.03	II
50		2,572.65	II
50		2,580.82	I
35		2,590.26	I
190		2,593.64	I
65		2,596.58	I
270		2,599.92	I
340		2,605.15	I
510		2,611.28	I
75		2,611.48	I
300		2,619.94	I
170		2,631.54	I
170		2,632.42	I
640		2,641.10	I
800		2,644.26	I
950		2,646.64	I
30		2,649.30	I
15		2,654.93	I
35		2,657.19	I
85		2,661.97	I
95		2,669.60	I
130		2,679.93	I
26		2,684.80	I
30		2,685.14	I
65		2,688.82	I
26		2,716.25	II
85		2,725.07	I
75		2,727.42	I
21		2,731.13	I
40		2,731.58	I
170		2,733.26	I
55		2,735.29	I
40		2,735.61	I
85		2,739.81	I
250		2,742.32	I
40		2,749.06	I
65		2,757.40	I
95		2,758.08	I
15		2,761.29	II
250		2,802.50	I
55		2,805.70	I
30		2,806.50	II
40		2,809.17	I
75		2,810.30	II
30		2,812.98	I
30		2,817.40	I
65		2,817.84	I
		2,817.87	II
65		2,828.07	I
		2,828.15	II
130		2,832.16	II
190		2,841.94	II
110		2,851.10	II
40		2,853.93	II
95		2,862.32	II
55		2,868.74	II
180		2,877.44	II
280		2,884.11	II
65		2,888.93	II
55		2,891.07	II
55		2,905.66	I
30		2,909.92	II
450		2,912.08	I
340		2,928.34	I
15		2,931.03	
180		2,933.55	I
26		2,935.96	
150		2,937.32	I
1,100		2,942.00	I
1,300		2,948.26	I
30		2,954.58	
1,600		2,956.13	I
170		2,956.80	I
30		2,958.77	
26		2,959.71	I
35		2,959.99	I
170		2,965.71	I
190		2,967.22	I
26		2,968.23	I
75		2,970.38	I
30		2,974.93	I
170		2,983.31	I
35		3,000.87	I
120		3,017.19	II
140		3,029.73	II
110		3,046.68	II
130		3,056.74	II
130		3,057.40	II
170		3,058.09	II
85		3,059.74	II
1,300	d	3,066.22	II
		3,066.35	II
70		3,071.24	II
600		3,072.11	II
1,100		3,072.97	II
1,600		3,075.22	II
2,300		3,078.64	II
3,600		3,088.02	II
180		3,089.40	II
180		3,097.19	II
180		3,100.67	I
230		3,103.80	II
230		3,105.08	II
260		3,106.23	II
70		3,106.81	I
50		3,110.67	II
50		3,112.48	I
140		3,117.67	II
720		3,119.72	I
		3,119.80	II

190		3,123.07	I
240		3,130.80	II
140		3,141.54	I
95		3,141.67	I
220		3,143.76	II
240		3,148.04	II
240		3,152.25	II
240		3,154.20	II
240		3,155.67	II
500		3,161.20	II
780		3,161.77	II
1,000		3,162.57	II
1,600		3,168.52	II
2,400		3,186.45	I
1,000		3,190.87	II
3,100		3,191.99	I
50		3,197.52	II
3,800		3,199.92	I
780		3,202.54	II
50		3,203.44	II
240		3,203.83	I
50		3,204.87	I
110		3,213.14	II
260		3,214.24	I
190		3,214.75	II
1,100		3,217.06	II
110		3,217.94	I
260		3,218.27	II
110		3,219.21	I
110		3,221.38	I
1,300		3,222.84	II
220		3,223.52	I
240		3,224.24	II
140		3,226.13	I
530		3,228.60	II
780		3,229.19	II
530		3,229.42	II
110		3,231.32	II
240		3,232.28	II
6,600		3,234.52	II
220		3,236.12	II
5,200		3,236.57	II
4,100		3,239.04	II
220		3,239.66	II
2,600		3,241.99	II
1,200		3,248.60	II
950		3,251.91	II
1,200		3,252.91	II
1,200		3,254.25	II
1,200		3,261.60	II
310		3,271.65	II
310		3,272.08	II
200		3,278.29	II
260		3,278.92	II
220		3,282.33	II
530		3,287.66	II
290		3,292.08	I
170		3,299.41	I
170		3,306.88	I
220		3,308.39	I
220		3,308.81	II
260		3,309.50	I
60		3,309.73	I
110		3,312.69	I
840		3,314.42	I
		3,314.52	I
290		3,315.32	II
330		3,318.02	II
550		3,321.70	II
2,900		3,322.94	II
380		3,326.76	II
2,100		3,329.46	II
550		3,332.11	II
1,800		3,335.20	II
1,100		3,340.34	II
5,700		3,341.88	I
120		3,342.15	I
260		3,343.77	II
330		3,346.73	II
4,300		3,349.04	II
12,000		3,349.41	II
120		3,352.94	I
4,100		3,354.64	I
290		3,358.28	I
290		3,360.99	I
7,200		3,361.21	II
		3.361.26	I
120		3,361.84	I
1,100		3,370.44	I
4,300		3,371.45	I
140		3,372.21	II
5,700		3,372.80	II
60		3,374.35	II
2,900	d	3,377.48	I
		3,377.58	I
290		3,379.22	I
1,400		3,380.28	II
170		3,382.31	I
5,700		3,383.76	II
170		3,385.66	I
1,400		3,385.95	I
1,400		3,387.84	II
60		3,388.76	II
140		3,390.68	I
140		3,392.71	I
1,100		3,394.58	II
60		3,398.63	I
60		3,402.42	II
60		3,407.20	II
95		3,409.81	II
60		3,439.30	I
890		3,444.31	II
60		3,452.47	II
180		3,456.39	II
600		3,461.50	II
95		3,467.26	I
600		3,477.18	II
60		3,478.92	I
240		3,480.53	I
60		3,485.69	I
60		3,489.74	II
480		3,491.05	II
60		3,495.75	I
95		3,499.10	I
890		3,504.89	II
120		3,506.64	I
600		3,510.84	II
60		3,520.25	II
310		3,535.41	II
190		3,547.03	I
120		3,573.74	II
60		3,574.24	I
60		3,587.13	II
240		3,596.05	II
190		3,598.72	I
600		3,610.16	I
190		3,624.82	II
95		3,635.20	I
4,800		3,635.46	I
120		3,637.97	I
190		3,641.33	II
6,600		3,642.68	I
180		3,646.20	I
7,200		3,653.50	I
290		3,654.59	I
660		3,658.10	I
120		3,659.76	II
380		3,660.63	I
190		3,662.24	II
380		3,668.97	I
600		3,671.67	I
3,100		3,685.20	II
120		3,685.96	I
95		3,687.35	I
600		3,689.91	I
140		3,694.45	I
30		3,698.18	I
60		3,698.43	I
60		3,700.08	I
120		3,702.29	I
190		3,704.30	I
140		3,706.23	II
50		3,707.53	I
290		3,709.96	I
30		3,715.40	I
450		3,717.40	I
140		3,721.64	II
330		3,722.57	I
600		3,724.57	I
380		3,725.16	I
2,900		3,729.82	I
50		3,735.67	I
60		3,738.90	I
3,300		3,741.06	I
330		3,741.64	II
160		3,748.10	I
5,200		3,752.86	I
600		3,753.64	I
140		3,757.69	II
3,300		3,759.30	II
2,900		3,761.32	II
50		3,761.89	II
60		3,766.45	I
600		3,771.66	I
30		3,776.06	II
840		3,786.04	I
120		3,789.30	I
70		3,795.90	I
60		3,798.31	I
70		3,818.22	I
60		3,822.03	I
240		3,828.19	I
95		3,833.68	I
95		3,836.78	I
60		3,846.45	I
130		3,853.05	I
130		3,853.73	I
170		3,858.14	I
240		3,866.44	I
170		3,868.40	I
120		3,873.21	I
260		3,875.26	I
170		3,882.15	I
170		3,882.33	I
500		3,882.89	I
60	h	3,888.02	I
70		3,889.95	I
200	h	3,895.25	I
85		3,898.49	I
530		3,900.54	II
180		3,900.96	I
2,600		3,904.78	I
110	h	3,911.19	I
500		3,913.46	II
500		3,914.34	I
24		3,914.74	I
35		3,919.82	I
290		3,921.42	I
1,100		3,924.53	I
110		3,926.32	I
890		3,929.88	I
35		3,932.02	II
70		3,934.24	I
1,100		3,947.78	I
4,500		3,948.67	I
4,500		3,956.34	I
5,200		3,958.21	I
950		3,962.85	I
950		3,964.27	I
4,800		3,981.76	I
570		3,982.48	I
60		3,984.33	I
35		3,985.25	I
60		3,985.59	I
5,700		3,989.76	I
35		3,994.70	I
7,800		3,998.64	I
70		3,999.36	I
70		4,002.49	I
70		4,003.81	I
35		4,005.97	I
70		4,008.06	I
950		4,008.93	I
190		4,009.66	I
70		4,012.39	II
180		4,013.58	I
70		4,015.38	I
35		4,016.28	I
120	h	4,017.77	I
140		4,021.83	I
1,200		4,024.57	I
40		4,025.14	II
190	h	4,026.54	I
40		4,027.48	I
40		4,028.34	II
190	h	4,030.51	I
40		4,033.91	I
30		4,034.91	I
110		4,035.83	I
35	h	4,040.32	I
290		4,055.02	I
85		4,057.62	I
85		4,058.14	I
410		4,060.26	I
200		4,064.22	I
200		4,065.10	I
840		4,078.47	I
40		4,079.72	I
290		4,082.46	I
85		4,099.17	I
220		4,112.71	I
85		4,122.17	I
40		4,123.31	I
85		4,123.57	I
130		4,127.54	I
40		4,129.17	I
40		4,131.25	I
140		4,137.29	I
85		4,143.05	I
170		4,150.96	I
85		4,159.64	I
70		4,163.65	II
35		4,164.14	I
40		4,166.32	I
85		4,169.35	I
120		4,171.03	I
40		4,171.90	II
35		4,183.30	I
360		4,186.12	I
40		4,188.69	I
70		4,200.75	I
85		4,203.46	I
35		4,211.73	I
40		4,224.79	I
40		4,227.65	I
130		4,237.89	I
85		4,249.12	I
130		4,256.04	I
70		4,258.54	I
70		4,261.60	I
330		4,263.13	I
35		4,265.71	I
40		4,266.22	I
70		4,270.14	I
85		4,272.43	I
240		4,274.58	I
120		4,276.43	I
120		4,278.23	I
30		4,278.81	I
110		4,281.38	I
220		4,282.71	I
160		4,284.99	I
890		4,286.01	I
840		4,287.40	I
30		4,288.16	I
950		4,289.07	I
120		4,290.23	II
840		4,290.94	I
120		4,291.14	I
140		4,294.12	II
840		4,295.76	I
2,000		4,298.66	I
200		4,299.23	I
200		4,299.64	I
200		4,300.05	II
2,900		4,300.56	I
4,100		4,301.09	I
85		4,301.93	II
6,000		4,305.92	I
180		4,307.90	II
35		4,308.50	I
40		4,311.65	I
85		4,312.87	II
85		4,314.35	I
1,200		4,314.80	I
360		4,318.64	I
180		4,321.66	I
190		4,325.13	I
160		4,326.36	I
30		4,334.84	I
160		4,337.92	II
24		4,344.29	II
70		4,346.11	I
35		4,354.06	I
95		4,360.49	I
24		4,368.94	I
95		4,369.68	I
60		4,372.38	I
30		4,388.08	I
170		4,393.92	I
330		4,395.04	II
60		4,399.77	II
240		4,404.28	I
60		4,404.90	I
30		4,405.68	I
60		4,416.54	I

220		4,417.28	I
60		4,417.72	II
120		4,421.76	I
120		4,422.82	I
24		4,424.39	I
30		4,425.83	I
120		4,426.06	I
890		4,427.10	I
21		4,430.02	I
85		4,430.37	I
50		4,431.28	I
30		4,432.60	I
24		4,433.58	I
170		4,434.00	I
70		4,436.59	I
30		4,438.23	I
130		4,440.35	I
50		4,441.27	I
230		4,443.80	II
24		4,444.27	I
840		4,449.15	I
30		4,450.49	II
550		4,450.90	I
840		4,453.32	I
290		4,453.71	I
950		4,455.33	I
1,100		4,457.43	I
21		4,462.09	I
70		4,463.38	I
95		4,463.54	I
290		4,465.81	I
240		4,468.50	II
240		4,471.24	I
95		4,474.85	I
95		4,479.70	I
50		4,480.59	I
530		4,481.26	I
95		4,482.69	I
19		4,488.32	II
260		4,489.09	I
24		4,492.55	I
40		4,495.01	I
240		4,496.15	I
24		4,497.73	I
200		4,501.27	II
40		4,503.78	I
21		4,506.36	I
50		4,511.17	I
780		4,512.74	I
19		4,515.62	I
1,000		4,518.03	I
95		4,518.70	I
1,000		4,522.80	I
780		4,527.31	I
6,000		4,533.24	I
240		4,533.97	II
3,600		4,534.78	I
2,400		4,535.58	I
1,200		4,535.92	I
1,200		4,536.05	I
24		4,537.23	I
24		4,539.10	I
720		4,544.69	I
950		4,548.77	I
240		4,549.63	II
950		4,552.46	I
24		4,555.08	I
720		4,555.49	I
19		4,557.86	I
19		4,558.11	I
60		4,559.92	I
50		4,562.63	I
35		4,563.43	I
110		4,563.77	II
35		4,570.91	I
240		4,571.98	II
19		4,585.84	I
24		4,589.95	II
60		4,599.23	I
21		4,609.37	I
950		4,617.27	I
24		4,619.52	I
480		4,623.09	I
190		4,629.34	I
50	d	4,634.87	
60		4,637.88	I
240		4,639.37	I
220		4,639.67	I
190		4,639.95	I

140		4,645.19	I
120		4,650.02	I
24		4,656.04	I
720		4,656.47	I
840		4,667.59	I
70		4,675.12	I
950		4,681.92	I
21		4,686.92	I
24		4,690.80	I
190		4,691.34	I
40		4,693.68	I
24		4,696.94	I
190		4,698.76	I
120		4,710.19	I
24		4,715.30	I
65		4,722.62	I
65		4,723.17	I
55		4,731.17	I
45		4,733.43	I
18		4,734.68	I
22		4,742.11	I
170		4,742.79	I
22		4,747.68	I
310		4,758.12	I
310		4,759.28	I
45		4,766.33	I
28		4,769.77	I
65		4,778.26	I
45		4,781.72	I
110		4,792.49	I
45		4,796.22	I
35		4,797.98	I
110		4,799.80	I
28		4,805.10	II
110		4,805.43	I
45		4,808.53	I
22		4,811.08	I
40		4,812.25	I
200		4,820.42	I
22		4,825.46	I
40		4,836.13	I
470		4,840.87	I
65		4,848.47	I
290		4,856.01	I
35		4,864.18	I
200		4,868.26	I
250		4,870.14	I
28		4,880.91	I
45		4,882.35	I
400		4,885.08	I
380		4,899.91	I
320		4,913.62	I
55		4,915.24	I
130		4,919.87	I
180		4,921.77	I
55		4,925.41	I
30		4,926.16	I
150		4,928.34	I
30		4,937.74	I
95		4,938.29	I
30		4,941.58	I
21		4,948.19	I
21		4,958.25	I
55		4,964.75	I
21		4,966.04	I
65		4,968.58	I
75		4,973.05	I
120		4,975.35	I
65		4,977.74	I
120		4,978.20	I
5,800		4,981.73	I
150		4,989.15	I
4,600		4,991.07	I
30		4,995.08	I
140		4,997.10	I
4,000		4,999.51	I
230		5,001.01	I
3,600		5,007.21	I
120		5,009.65	I
230		5,013.30	I
3,200	d	5,014.19	I
		5,014.24	I
580		5,016.17	I
840		5,020.03	I
840		5,022.87	I
580		5,024.84	I
300		5,025.58	I
1,200		5,035.91	I
840		5,036.47	I

740		5,038.40	I
1,200		5,039.95	I
75		5,040.62	I
85		5,043.59	I
35		5,044.27	I
55		5,045.41	I
26		5,048.21	I
110		5,052.87	I
21		5,054.08	I
110		5,062.11	I
35		5,064.07	I
1,400		5,064.66	I
95		5,065.99	I
35	h	5,068.33	I
65		5,069.35	I
130		5,071.48	I
40		5,085.34	I
130		5,087.07	I
21		5,103.15	I
55		5,109.44	I
190		5,113.44	I
270		5,120.42	I
30		5,129.15	II
270		5,145.47	I
230		5,147.48	I
210		5,152.20	I
21	b	5,166.86	T
1,100		5,173.75	I
40		5,186.34	I
85		5,188.70	II
30		5,189.58	I
1,300		5,192.98	I
85	h	5,194.04	I
65		5,201.10	I
120		5,206.08	I
75		5,207.87	I
65		5,208.42	
1,400		5,210.39	I
65		5,212.29	I
150		5,219.71	I
95		5,222.69	I
85		5,223.64	I
250		5,224.32	I
95		5,224.56	I
190		5,224.95	I
65		5,226.56	II
120		5,238.58	I
21		5,246.15	I
55		5,246.57	I
75		5,247.31	I
21		5,250.95	I
110		5,252.11	I
75		5,255.83	I
55		5,259.99	I
55		5,263.50	I
150		5,265.98	I
40		5,282.39	I
140		5,283.45	I
35		5,284.39	I
26		5,288.81	
65		5,295.79	I
120		5,297.26	I
65		5,298.44	I
26		5,336.81	II
17		5,341.50	I
75		5,351.08	I
26		5,366.65	I
55		5,369.64	I
40		5,389.18	I
55		5,389.99	I
17		5,396.60	I
85		5,397.09	I
35		5,404.02	I
110		5,409.61	I
40		5,426.26	I
75		5,429.15	I
26		5,436.73	I
17		5,438.32	I
40		5,446.64	I
11	b	5,448.34	T
30		5,448.90	I
21		5,449.16	I
35		5,453.65	I
55		5,460.51	I
75		5,471.21	I
35		5,472.70	I
40	h	5,473.55	I
85		5,474.23	I
30		5,474.46	I

120	h	5,477.71	I
110		5,481.43	I
75		5,481.87	I
85	h	5,488.20	I
150		5,490.15	I
26		5,490.84	I
110		5,503.90	I
40		5,511.78	I
340		5,512.53	I
270		5,514.35	I
320		5,514.54	I
26		5,530.49	
110		5,565.49	I
13		5,579.16	
21	h	5,582.98	
30	h	5,585.68	
65	b	5,597.85	T
55	b	5,629.28	T
17		5,635.84	
250		5,644.14	I
75		5,648.58	I
26	b	5,661.55	T
190		5,662.16	I
75		5,662.91	I
21		5,673.42	I
130		5,675.44	I
30	h	5,679.94	I
95		5,689.47	I
75		5,702.68	I
35		5,708.23	I
65		5,711.88	I
40	h	5,713.92	I
95		5,715.13	I
55		5,716.48	I
35		5,720.48	I
85		5,739.51	I
40		5,740.02	I
19		5,741.22	I
21		5,752.84	I
19		5,756.86	I
40	h	5,762.27	I
55	h	5,766.35	I
75	h	5,774.05	I
30		5,780.78	I
75	h	5,785.98	I
65	h	5,804.26	I
21	b	5,814.96	T
40		5,823.71	I
21	h	5,841.18	
21		5,852.34	
400		5,866.46	I
65		5,880.31	I
21	h	5,888.68	
230		5,899.32	I
55		5,903.33	I
120		5,918.55	I
150		5,922.12	I
75		5,937.82	I
120		5,941.76	I
300		5,953.17	I
200		5,965.84	I
270		5,978.56	I
340		5,999.04	I
65		5,999.68	I
21		6,012.73	
110		6,064.63	I
120		6,085.23	I
120		6,091.17	I
40		6,092.81	I
40	h	6,098.67	I
35	h	6,121.01	I
120		6,126.22	I
19		6,138.38	I
30		6,146.22	I
21		6,149.74	I
30	b	6,162.23	T
35		6,186.15	I
95	h	6,215.28	I
75	h	6,220.49	I
65	h	6,221.41	I
380		6,258.10	I
380		6,258.70	I
300		6,261.10	I
65		6,303.75	I
55		6,312.24	I
26		6,318.03	I
30		6,336.10	I
35		6,366.35	I
11		6,419.10	I

17		6,497.69	I
19		6,508.14	I
55		6,546.28	I
65		6,554.23	I
11	h	6,554.83	
75		6,556.07	I
19	h	6,565.62	I
14	h	6,575.18	I
35		6,599.11	I
18	b	6,651.46	TIO
18	h	6,666.55	I
22	h	6,667.74	
9		6,668.39	
18		6,677.18	I
22	b	6,691.21	TIO
26		6,716.68	I
16	b	6,723.95	TIO
80		6,743.12	I
22		6,745.52	I
18		6,844.64	
18		6,860.39	
35		6,861.47	I
9		6,873.92	I
12		6,913.19	I
14	h	6,933.15	I
14	h	6,943.70	I
23		6,996.63	I
15		7,004.66	I
14		7,008.35	I
14		7,010.94	I
14	h	7,035.86	I
40		7,038.80	I
14		7,050.65	I
40	b	7,054.51	TIO
23		7,069.11	I
23		7,072.05	
45	b	7,087.89	TIO
30	b	7,124.9	TIO
40	b	7,125.61	TIO
26		7,138.91	I
26		7,167.13	
23		7,171.53	
55		7,189.89	I
26	b	7,203.64	TIO
260		7,209.44	I
60		7,216.20	I
130		7,244.86	I
130		7,251.72	I
19		7,263.40	
19		7,266.29	I
19	b	7,269.05	TIO
15		7,315.56	I
26		7,318.39	I
120		7,344.72	I
11		7,352.16	I
90		7,357.74	I
60		7,364.11	I
26		7,440.60	I
9		7,474.94	I
26		7,489.61	I
19		7,496.12	I
12		7,580.55	I
9	b	7,589.62	TIO
15		7,614.50	I
23		7,654.44	I
11	b	7,705.21	TIO
30		7,949.17	I
26	h	7,961.58	I
60		7,978.88	I
9		7,979.07	I
30		7,996.53	I
7	h	8,003.55	
55		8,024.84	I
30		8,068.24	I
8		8,267.62	
14	h	8,306.31	I
9	h	8,307.41	I
9	h	8,311.76	I
8	h	8,312.85	I
12		8,334.37	I
14		8,353.15	I
75		8,364.24	I
100		8,377.85	I
100		8,382.54	I
55		8,382.82	I
75		8,396.87	I
120		8,412.36	I
19		8,416.98	I
15		8,424.41	I

170		8,426.52	I
490		8,434.94	I
240		8,435.70	I
40		8,438.93	I
40		8,450.89	I
9	h	8,457.10	I
19	h	8,467.15	I
45		8,468.50	I
15		8,496.04	I
19	h	8,518.05	I
40		8,518.32	I
14		8,539.38	I
40		8,548.12	I
9		8,569.77	I
9	h	8,598.18	I
90		8,675.39	I
45		8,682.99	I
23		8,692.33	I
19		8,734.69	I
23		8,766.64	I
15	h	8,778.71	I

TUNGSTEN (W)

Z = 74

W I and II

Ref. 1 — C.H.C.

Intensity		Wavelength	
		Air	
5,800		2,001.71	II
13,000		2,008.07	II
5,100		2,009.98	II
4,100		2,010.23	II
4,100		2,014.23	II
7,300		2,026.08	II
15,000		2,029.98	II
2,700		2,035.03	II
5,300		2,049.63	II
2,300		2,065.57	II
3,400		2,071.21	II
2,200		2,075.59	II
9,700		2,079.11	II
3,600		2,088.19	II
2,200		2,089.14	II
1,700		2,090.48	I
6,100		2,094.75	II
2,400		2,098.60	II
2,200		2,100.67	II
1,500		2,101.54	I
1,500		2,106.18	II
1,300		2,110.34	II
2,100		2,118.87	II
2,400		2,121.59	II
850		2,153.56	II
850		2,157.80	II
1,500		2,166.32	II
480		2,182.90	I
440		2,194.52	II
1,300		2,204.48	II
460		2,248.75	II
460		2,249.80	I
180		2,270.24	II
95		2,271.37	I
510		2,277.58	I
160		2,284.91	I
320		2,285.17	I
530	d	2,294.49	I
		2,294.54	II
270		2,298.33	I
240		2,303.83	II
240		2,306.59	I
340		2,309.02	I
440		2,313.17	I
220		2,314.17	I
190		2,315.02	II
460		2,321.63	I
290		2,326.09	II
390	d	2,326.56	I
		2,326.70	I
75		2,328.31	II
130		2,333.77	II
210		2,341.37	I
75		2,349.26	II
120	d	2,350.37	II
320		2,354.61	I
60		2,358.81	II
580		2,360.44	I

850		2,363.07	I
60		2,364.22	II
510		2,374.47	I
210		2,382.99	I
670		2,384.82	I
240		2,389.08	I
120		2,390.37	II
120		2,392.93	II
730		2,397.09	II
560		2,397.73	I
560		2,397.98	I
75		2,404.24	II
1,700	d	2,405.58	I
		2,405.69	I
75		2,411.54	II
320		2,414.04	I
610		2,415.68	I
50		2,419.34	II
50		2,421.01	II
870		2,424.21	I
190		2,427.49	II
170		2,429.39	II
580		2,431.08	I
630		2,433.98	I
60		2,435.01	II
1,800		2,435.96	I
250		2,436.62	I
580		2,444.06	I
160		2,446.39	II
270		2,448.39	I
270		2,451.35	I
780		2,451.48	II
870		2,452.00	I
430		2,454.72	I
630		2,454.98	I
780		2,455.51	I
780		2,456.53	I
1,100		2,459.30	I
270		2,460.16	I
480		2,462.79	I
270		2,464.30	I
230		2,466.52	II
1,400		2,466.85	I
75		2,470.80	II
480		2,472.51	I
1,200		2,474.15	I
290		2,477.80	II
870		2,480.13	I
390		2,480.96	I
1,500		2,481.44	I
480	d	2,482.10	I
		2,482.21	I
29		2,484.40	II
580		2,484.74	I
390		2,487.50	I
270		2,488.77	II
390		2,489.23	II
75		2,492.93	II
630		2,495.26	I
230		2,496.64	II
95		2,497.48	II
140		2,499.69	II
40		2,500.11	II
		2,501.90	II
680		2,504.70	I
270		2,506.02	I
24		2,508.00	II
250		2,510.17	I
75		2,510.47	II
60		2,518.14	II
310		2,520.46	I
780		2,521.32	I
270		2,522.04	II
780		2,523.41	I
430		2,527.76	I
780		2,533.64	I
50		2,534.82	II
580		2,545.34	I
1,200		2,547.14	I
50		2,549.09	II
40		2,550.10	II
780		2,550.38	I
2,700		2,551.35	I
450		2,553.82	I
410		2,554.86	II
580		2,555.09	II
		2,555.21	I
310		2,556.75	I
290		2,560.12	I

730		2,561.97	I
230		2,563.16	II
110		2,563.91	II
530		2,571.44	II
170	d	2,572.24	II
		2,572.35	II
75		2,573.95	II
190		2,579.26	II
290		2,580.34	I
870		2,580.49	I
40		2,581.20	II
390		2,584.39	I
390		2,589.17	II
170		2,591.49	II
110		2,598.74	II
370		2,601.96	I
75		2,602.51	II
75		2,603.02	II
270		2,603.54	I
680		2,606.39	I
320		2,607.38	I
370		2,608.32	I
970		2,613.08	I
480		2,613.82	I
230		2,615.12	I
70		2,615.44	II
210		2,619.18	I
400		2,620.25	I
400		2,622.21	I
400		2,625.22	I
210		2,628.26	I
400		2,632.48	I
400		2,632.70	I
810		2,633.13	I
290		2,636.54	I
400	d	2,638.62	I
		2,638.75	I
210		2,645.69	I
650		2,646.18	I
400		2,646.73	I
75		2,647.74	II
40		2,653.42	II
80		2,653.57	II
1,600		2,656.54	I
400		2,657.38	I
400	d	2,658.04	II
		2,658.18	I
810		2,662.84	I
260		2,664.97	I
75		2,666.49	II
210		2,669.30	I
810		2,671.47	I
80		2,673.59	II
650		2,677.28	I
160	d	2,677.79	II
		2,677.91	I
400		2,678.88	I
2,100		2,681.42	I
290		2,683.35	I
210		2,691.09	I
650		2,695.67	I
210		2,697.71	II
650		2,699.59	I
400		2,700.01	I
40		2,701.48	II
160		2,702.11	II
210		2,702.52	I
400		2,706.58	I
400		2,708.59	I
400	d	2,708.80	I
		2,708.93	I
80		2,709.58	II
40		2,710.78	II
400		2,715.50	I
80		2,716.32	II
80		2,718.04	II
2,100		2,718.91	I
320		2,719.33	I
210		2,719.86	I
		2,722.81	II
2,600		2,724.35	I
210		2,724.62	I
400		2,725.03	I
		2,725.06	I
80		2,729.62	II
75		2,740.79	II
650		2,748.84	I
40	d	2,760.74	II
80		2,761.59	II

400		2,762.34	I
400		2,764.27	II
210		2,768.98	I
400		2,769.74	I
810		2,770.88	I
210		2,773.70	I
810		2,774.00	I
810		2,774.48	I
160		2,776.50	II
40		2,778.69	II
210		2,787.98	I
340		2,791.96	I
810		2,792.70	I
80		2,799.03	II
400		2,799.93	I
160	d	2,801.05	II
		2,801.17	I
130		2,805.92	II
40		2,812.25	II
810		2,818.06	I
160		2,822.57	II
260		2,829.82	I
1,600		2,831.38	I
810		2,833.63	I
210		2,835.64	I
400		2,841.57	I
810		2,848.02	I
650		2,856.03	I
650		2,866.06	I
230		2,878.72	I
610		2,879.11	I
610		2,879.40	I
440		2,896.01	I
1,500		2,896.44	I
230		2,910.48	I
270		2,911.00	I
360		2,918.25	I
50		2,918.63	II
360		2,923.10	I
230		2,923.54	I
230		2,925.13	I
690		2,935.00	I
2,400		2,944.40	I
2,400		2,946.99	I
480		2,947.39	I
210		2,952.29	II
440		2,964.52	I
480		2,977.11	I
		2,977.21	I
730	d	2,979.71	I
		2,979.86	I
400		2,993.61	I
240		2,995.26	I
190		3,009.09	I
360		3,013.79	I
520		3,016.47	I
770		3,017.44	I
110		3,024.50	II
210		3,024.93	I
310	d	3,026.67	I
		3,026.79	I
160		3,033.56	I
160		3,034.19	I
160		3,039.31	I
440	d	3,041.73	I
		3,041.86	I
270		3,043.80	I
440		3,046.44	I
110		3,048.66	I
810		3,049.69	I
110		3,064.93	I
180		3,073.28	I
110		3,077.52	II
180	d	3,084.83	I
		3,084.91	I
370		3,093.50	I
240		3,107.23	I
240		3,108.02	I
230		3,117.57	I
260		3,120.18	I
160		3,133.88	I
130		3,141.42	I
65		3,149.85	II
290		3,163.42	I
130		3,164.44	I
130		3,165.38	I
320		3,176.60	I
130		3,179.06	I
190		3,181.82	I
130		3,184.05	I
130		3,184.42	I
65		3,189.24	II
390		3,191.57	I
390		3,198.84	I
520		3,207.25	I
140		3,208.28	I
1,000		3,215.56	I
140		3,221.21	I
140		3,221.91	I
190		3,232.49	I
140		3,237.09	I
140		3,242.03	I
140		3,252.29	I
210		3,254.36	I
140		3,259.43	I
210		3,259.66	I
210	d	3,266.62	I
		3,266.77	I
150		3,281.94	I
150		3,293.71	I
730		3,300.82	I
440		3,311.38	I
440		3,326.20	I
440		3,331.69	I
150		3,354.45	I
150		3,371.04	I
390		3,373.75	I
150		3,412.96	I
150		3,413.53	I
150		3,422.42	I
150		3,427.71	I
230		3,429.59	I
240		3,443.00	I
160		3,477.94	I
400		3,495.24	I
160		3,508.73	I
160		3,510.02	I
160		3,526.85	I
160		3,535.54	I
160		3,537.45	I
650		3,545.22	I
160		3,568.04	I
240		3,570.65	I
80		3,572.48	II
160		3,575.22	I
80		3,592.42	II
240		3,606.06	I
80		3,613.79	II
1,900		3,617.52	I
160		3,622.34	I
130		3,627.24	I
320		3,631.94	I
240		3,641.41	II
80		3,646.52	II
80		3,657.59	II
160		3,675.55	I
650		3,682.08	I
400		3,683.30	I
		3,683.39	I
160		3,683.93	I
570		3,688.06	I
810		3,707.92	I
60		3,716.08	II
100		3,719.39	I
50		3,736.22	II
120		3,741.71	I
510		3,757.92	I
680		3,760.13	I
1,000		3,768.45	I
120		3,769.21	I
120		3,769.86	I
340		3,773.71	I
1,000		3,780.77	I
170		3,792.76	I
290		3,809.22	I
190		3,810.38	I
260		3,810.79	I
1,400		3,817.48	I
110		3,829.13	I
1,100		3,835.06	I
290		3,838.51	I
730		3,846.22	I
250		3,847.49	I
27		3,851.57	II
150		3,855.55	I
150		3,859.30	I
180		3,864.34	I
1,800		3,867.99	I
250		3,872.84	I
110		3,874.41	I
730		3,881.41	I
110		3,892.72	I
140	h	3,897.91	I
150		3,935.03	I
120		3,936.97	I
120		3,947.98	I
120		3,952.52	I
120		3,952.90	I
160		3,953.15	I
200		3,955.30	I
160		3,965.14	I
130		3,968.59	I
150	h	3,970.80	I
130		3,979.29	I
130		3,980.64	I
250		3,983.29	I
8,600		4,008.75	I
540		4,015.22	I
170	h	4,016.52	I
220		4,019.23	I
130		4,022.12	I
180		4,028.79	I
180		4,036.86	I
140		4,039.85	I
140	h	4,044.28	I
910		4,045.59	I
180		4,064.79	I
150		4,069.79	I
730		4,069.95	I
340		4,070.61	I
100		4,071.93	I
5,000		4,074.36	I
150		4,082.96	I
130		4,088.33	I
100		4,095.69	I
1,000		4,102.70	I
150		4,109.75	I
100		4,111.82	I
150		4,118.05	I
100		4,118.19	I
100		4,120.85	I
100		4,125.16	I
150		4,126.80	I
100		4,133.48	I
540		4,137.46	I
150		4,138.02	I
110		4,142.25	I
140		4,145.16	I
110		4,145.95	I
160		4,154.66	I
160		4,170.53	I
450		4,171.17	I
160		4,204.40	I
220		4,207.05	I
110		4,215.38	I
250		4,219.37	I
110		4,222.04	I
150		4,234.34	I
290		4,241.44	I
540		4,244.36	I
290		4,259.35	I
200		4,260.29	I
200		4,263.30	I
1,400		4,269.38	I
110		4,269.77	I
220		4,274.55	I
160		4,275.49	I
160		4,276.74	I
110		4,282.34	I
110		4,286.01	I
110		4,294.10	I
4,100		4,294.61	I
2,200		4,302.11	I
160		4,306.87	I
110		4,307.64	I
110		4,332.13	I
100		4,347.00	I
150		4,355.17	I
100		4,361.81	I
150		4,364.78	I
100	d	4,365.95	I
		4,366.07	I
150		4,372.52	I
200		4,378.48	I
180		4,384.85	I
100		4,403.95	I
200		4,408.28	I
130		4,412.19	I
160		4,436.90	I
140		4,460.49	I
140		4,466.34	I
140		4,466.74	I
640		4,484.19	I
160		4,504.84	I
130		4,512.88	I
120		4,513.25	I
150		4,543.54	I
150		4,546.47	I
150		4,551.82	I
140		4,570.64	I
170		4,588.73	I
140		4,599.94	I
140		4,609.89	I
160		4,613.30	I
100		4,642.53	I
130		4,657.42	I
640		4,659.87	I
640		4,680.51	I
100		4,693.72	I
140		4,757.54	I
790		4,843.81	I
380		4,886.90	I
220		4,982.59	I
330		5,006.15	I
220		5,015.30	I
820		5,053.28	I
210		5,054.60	I
210		5,069.12	I
120		5,071.74	I
770		5,224.66	I
27		5,500.49	I
27		5,503.44	I
10		5,508.61	I
220		5,514.68	I
15		5,531.38	I
15		5,537.72	I
13		5,568.09	I
13		5,604.31	I
11		5,631.27	I
27		5,631.94	I
65		5,648.37	I
35		5,660.72	I
27		5,674.39	I
13		5,676.60	I
15		5,676.90	I
15		5,697.79	I
55		5,735.09	I
13		5,749.24	I
11		5,756.10	I
13		5,793.06	I
13		5,796.49	I
45		5,804.85	I
13	d	5,806.05	I
		5,806.24	I
13		5,833.61	I
13		5,838.97	I
17		5,845.27	I
28		5,851.58	I
11		5,856.61	I
22		5,864.63	I
11		5,874.22	I
13		5,880.21	I
13		5,891.61	I
13		5,901.20	I
40		5,902.64	I
13		5,928.58	I
55		5,947.57	I
13		5,953.96	I
13		5,956.19	I
27		5,960.83	I
55		5,965.86	I
27		5,972.51	I
20		5,978.86	I
20		5,983.82	I
13		6,009.01	I
55		6,012.78	I
40		6,021.52	I
20		6,028.32	I
20		6,043.31	I
13		6,049.92	I
13		6,065.08	I
22		6,081.44	I
13		6,111.66	I
13		6,115.52	I
22		6,128.25	I
13		6,143.94	I

20		6,153.72	I
20		6,154.87	I
20		6,203.51	I
20		6,254.28	I
27		6,285.88	I
45		6,292.02	I
20		6,303.21	I
13		6,386.47	I
35		6,404.21	I
40		6,445.12	I
11		6,508.05	I
15		6,532.39	I
13		6,538.11	I
13		6,563.20	I
20		6,573.93	I
11		6,607.13	I
11		6,609.05	I
17		6,611.62	I
11		6,621.74	I
13		6,678.42	I
15		6,693.08	I
5		6,746.56	I
5		6,764.45	I
7		6,805.31	I
9		6,814.92	I
9		6,820.27	I
8		6,828.43	I
4		6,853.74	I
4		6,876.01	I
5		6,908.29	I
9		6,934.23	I
8		6,964.12	I
13		6,984.27	I
8		6,993.27	I
4		6,994.06	I
8		7,017.88	I
3		7,028.68	I
3		7,098.22	I
4	h	7,111.18	I
15		7,140.52	I
9		7,162.64	I
5	h	7,191.33	I
5		7,198.62	I
11		7,200.16	I
5		7,216.35	I
4		7,226.06	I
8		7,237.12	I
5		7,274.47	I
10		7,278.24	I
15		7,285.81	I
15		7,296.55	I
3		7,298.25	I
7		7,385.08	I
4		7,451.39	I
3		7,456.37	I
8		7,483.35	I
7		7,504.13	I
10		7,509.00	I
3		7,520.66	I
9		7,537.45	I
9		7,550.48	I
17		7,569.92	I
5		7,582.88	I
3		7,612.18	I
17		7,614.15	I
3		7,631.29	I
3		7,654.81	I
13		7,688.97	I
4		7,701.01	I
5		7,761.16	I
3		7,776.73	I
11		7,784.15	I
7		7,808.96	I
2		7,823.82	I
4		7,863.47	I
2		7,867.04	I
4		7,880.40	I
5		7,886.48	I
9		7,940.92	I
3		7,957.06	I
22		8,017.19	I
7		8,054.89	I
22		8,055.64	I
5		8,060.38	I
13		8,123.82	I
5		8,143.19	I
3		8,165.72	I
5		8,210.22	I
4		8,322.05	I
10		8,338.08	I
4		8,348.81	I
7		8,358.72	I
3		8,382.94	I
4		8,402.60	I
4		8,475.14	I
27		8,585.11	I
10		8,594.42	I
8		8,613.27	I
3		8,614.50	I
13		8,865.53	I

URANIUM (U)

Z = 92

U I and II

Ref. 1, 303 — J.G.C.

Intensity		Wavelength	
		Air	
440		2,565.41	II
340		2,569.71	II
340		2,591.25	II
610		2,635.53	II
470		2,645.47	II
340		2,669.17	II
470		2,683.28	II
320		2,691.04	II
370		2,706.95	II
370		2,733.97	II
470		2,754.16	II
340		2,762.85	II
390		2,770.04	II
410		2,784.45	II
830		2,793.94	II
870		2,802.56	II
630		2,807.05	II
440		2,808.98	II
630		2,817.96	II
870		2,821.12	II
390		2,824.37	II
680		2,828.90	II
920		2,832.06	II
360		2,837.19	II
460		2,839.89	II
360		2,842.09	II
360		2,849.48	II
390		2,860.47	II
970		2,865.68	II
340		2,870.97	II
490		2,882.74	II
460		2,887.25	II
410		2,888.26	II
1,200		2,889.62	II
320		2,894.14	II
410		2,894.51	II
780		2,906.80	II
780		2,908.28	II
320		2,914.25	II
360		2,914.63	II
440	p	2,921.68	II
320		2,927.38	II
490		2,928.60	II
580		2,931.41	II
440		2,932.61	II
340		2,933.86	II
530	p	2,940.37	II
1,300		2,941.92	II
830		2,943.90	II
340		2,948.09	II
390		2,954.77	II
580		2,956.06	II
460		2,965.03	II
580		2,967.94	II
580		2,971.06	II
410		2,976.35	II
320		2,982.74	II
530		2,984.61	II
410		2,992.72	II
360		3,007.91	II
320		3,021.22	II
630		3,022.21	II
320		3,024.51	II
320		3,028.19	II
630		3,031.99	II
490		3,033.18	II
490		3,044.16	II
580		3,050.20	II
630		3,057.91	II
460		3,061.62	II
630		3,062.54	II
580		3,072.78	II
580		3,093.01	II
320		3,095.75	II
320		3,098.01	II
580		3,102.39	II
460		3,104.15	II
970		3,111.62	II
530		3,119.35	II
680		3,124.95	II
530		3,139.61	II
410		3,144.97	II
490		3,145.56	II
680		3,149.24	II
530		3,153.11	II
340		3,176.21	II
340		3,177.33	II
340		3,206.05	II
730		3,229.50	II
680		3,232.16	II
440		3,244.22	II
340		3,265.79	II
440		3,270.12	II
440		3,288.21	II
730		3,291.33	II
1,100		3,305.89	II
390		3,337.79	II
440		3,341.66	II
390		3,357.84	I
730		3,390.38	I
340		3,394.77	II
580		3,424.56	II
580		3,435.49	I
360		3,453.55	II
320		3,454.23	II
320		3,457.05	II
320		3,457.71	II
360		3,459.92	I
320		3,462.22	I
460		3,463.55	I
630		3,466.30	I
390		3,472.52	II
320		3,473.43	I
360		3,480.36	I
680		3,482.49	II
1,600		3,489.37	I
390		3,493.33	II
340		3,494.00	I
320		3,494.84	II
530		3,496.41	II
630		3,500.08	I
320		3,504.01	I
320		3,505.07	II
780		3,507.34	I
320		3,508.84	II
390		3,509.66	II
320		3,513.67	I
1,600		3,514.61	I
390		3,519.96	II
390		3,531.11	II
630		3,533.57	II
320		3,534.33	I
530		3,540.47	II
320		3,542.57	I
390		3,547.19	II
320		3,549.20	I
1,200		3,550.82	II
320		3,552.17	II
680		3,555.32	I
320		3,561.41	I
1,200		3,561.80	I
390		3,563.66	I
2,300		3,566.59	I
530		3,569.08	I
320		3,574.76	I
360		3,577.92	I
630		3,578.72	II
360		3,581.84	II
3,200		3,584.88	I
320		3,590.50	II
390		3,591.74	I
460		3,593.52	II
460		3,605.27	I
360		3,606.32	II
320		3,616.33	I
320		3,616.76	II
320		3,620.08	I
320		3,622.70	I
390		3,623.06	II
460		3,630.73	II
840		3,638.20	I
310		3,640.76	II
420		3,644.24	I
310		3,645.03	II
660		3,651.54	I
490		3,652.06	I
960		3,659.15	I
2,800		3,670.07	II
380		3,678.75	II
540		3,691.92	II
330		3,693.70	II
540		3,700.57	II
1,100		3,701.52	II
350		3,713.55	I
300		3,717.42	II
350		3,718.11	II
350		3,729.82	II
350		3,732.62	II
350		3,733.07	II
600		3,738.04	II
300		3,744.25	II
680		3,746.42	II
350		3,747.14	II
950		3,748.68	II
600		3,751.17	I
350		3,752.66	II
350		3,755.48	II
490		3,758.35	I
350		3,759.24	II
330		3,763.26	I
490		3,764.57	II
430		3,766.89	I
330		3,769.53	II
540		3,773.43	I
300		3,776.48	I
380		3,780.71	II
1,900		3,782.84	II
430		3,783.84	II
570		3,793.10	II
380		3,793.26	I
380		3,793.57	II
380		3,808.92	I
380		3,809.22	II
1,900		3,811.99	I
380		3,813.79	II
380		3,814.06	II
850		3,826.51	II
2,000		3,831.46	II
1,200		3,839.63	I
490	p	3,848.60	II
620		3,854.22	I
2,400		3,854.64	II
4,900		3,859.57	II
490		3,861.17	II
1,900		3,865.92	II
380		3,866.80	II
1,500		3,871.03	I
620		3,874.04	II
620		3,878.08	II
1,000		3,881.45	II
490		3,882.36	II
380		3,883.28	II
2,200		3,890.36	II
620		3,892.68	II
490		3,894.12	I
490		3,896.77	II
620		3,899.78	II
410		3,902.55	II
460		3,904.30	II
380		3,906.45	I
330		3,911.67	II
380		3,915.88	II
330		3,926.21	I
330		3,926.72	I
430		3,930.98	II
2,000		3,932.02	II
490		3,935.38	II
330		3,940.48	II
1,200		3,943.82	I
300		3,948.44	I
300		3,953.58	II
360		3,954.67	II
350		3,964.21	I
600		3,966.52	II
1,200		3,985.79	II

460		3,990.42	II
380		3,992.53	II
350		3,998.24	II
350		4,004.06	II
430		4,005.21	I
570		4,017.72	II
300		4,018.99	II
1,000		4,042.75	I
520		4,044.41	II
410		4,047.61	I
1,600		4,050.04	II
540		4,051.91	II
300		4,054.30	II
430		4,058.19	II
880		4,062.54	II
520		4,067.75	II
410		4,071.12	II
300		4,074.48	II
330		4,076.69	II
330		4,080.60	II
2,200		4,090.13	II
460		4,093.03	II
380		4,106.38	II
810		4,116.10	II
410		4,124.73	II
410		4,128.34	II
460		4,141.22	II
880		4,153.97	II
380		4,156.65	I
350		4,163.68	II
1,400		4,171.59	II
300		4,189.27	II
350		4,222.37	I
1,000		4,241.67	II
520		4,244.37	II
680		4,341.69	II
430	h	4,355.74	I
430		4,362.05	I
330		4,393.59	I
600		4,472.33	II
240		4,515.28	II
620		4,543.63	II
300		4,620.21	I
240		4,627.07	II
210		4,631.62	I
220		4,646.60	II
140		4,666.85	II
100		4,671.40	II
170		4,689.07	II
100		4,702.51	II
160		4,722.72	II
120		4,731.59	II
100		4,755.74	II
150		4,756.81	I
100		4,772.70	II
100		4,860.99	II
110		5,008.21	II
170		5,027.38	I
70		5,117.24	II
80		5,160.32	II
55		5,164.14	I
55		5,184.57	II
45		5,204.31	II
45		5,247.75	II
45		5,257.04	II
70		5,280.38	I
55		5,386.19	II
80		5,475.70	II
70		5,480.26	II
70		5,481.20	II
45		5,482.53	II
160		5,492.95	II
70		5,527.82	II
70		5,564.17	I
45		5,581.59	II
55		5,620.78	I
70		5,780.59	I
70		5,798.53	II
45		5,836.02	I
55		5,837.68	II
230		5,915.39	I
55		5,971.50	I
100		5,976.32	I
45		5,997.31	I
28		6,017.38	II
55		6,051.74	II
45		6,067.22	II
90		6,077.29	I
28		6,087.34	II
40		6,171.86	I
35		6,175.39	I
28		6,280.18	II
28		6,359.29	I
55		6,372.46	I
28		6,378.52	II
28		6,392.77	I
90		6,395.42	I
110		6,449.16	I
35		6,464.98	I
90		6,826.92	I
35		6,876.74	II
23		7,074.79	I
27		7,101.61	I
30		7,128.90	I
16		7,147.89	I
16		7,254.45	I
23		7,425.50	I
45		7,533.93	I
16		7,619.35	I
50		7,881.94	I
18		7,970.46	I
16		8,174.66	I
18		8,262.06	I
16		8,318.35	I
16		8,337.50	II
18		8,381.87	I
16		8,441.21	I
35		8,445.39	I
18		8,450.03	I
16		8,570.52	I
75		8,607.95	I
23		8,691.28	I
18		8,710.76	I
18		8,753.69	I
30		8,757.76	I
16		8,951.96	I
16		8,989.92	I
10		9,093.67	I
10		9,139.56	I
10		9,201.51	I
10		9,265.34	I
10		9,276.44	I
10		9,385.90	I
10		9,653.26	I
10		9,819.00	I
10		9,819.05	I
10		9,868.36	I
10		9,932.76	I
10		9,964.11	I
50		10,157.91	I
50		10,259.55	I
100		10,554.93	I
250		10,799.78	I
25		10,823.93	I
25		11,095.77	I
75		11,167.84	I
50		11,294.13	I
100		11,384.13	I
25		11,410.43	I
50		11,503.38	I
25		11,568.81	I
20		11,784.72	II
100		11,859.42	I
100		11,908.83	I
100		12,250.46	I
25		13,088.28	I
100		13,185.16	I
75		13,306.23	I
100		13,961.58	I
50		16,906.00	I
50		17,451.11	I
50		18,136.65	I
50		18,366.96	I
75		18,634.43	I
25		19,029.39	I
10		20,201.13	I
10		20,271.41	I
10		20,374.13	I
10		20,517.29	I
10		20,690.64	I
10		20,772.19	I
10		21,008.38	I
10		21,099.98	I
10		21,112.14	I
20		21,144.90	II
10		21,674.51	I
10		21,693.38	I
75		21,910.22	I
10		22,110.73	I
10		23,156.76	I
10		23,948.19	I
10		29,557.07	I

VANADIUM (V)

Z = 23

V I and II
Ref. 1 — C.H.C.

Intensity		Wavelength	
		Air	
2,100		2,092.44	I
40		2,384.00	II
40		2,384.28	I
60		2,386.96	I
60		2,388.92	I
75		2,390.87	I
75		2,391.26	I
85		2,392.90	I
70		2,397.78	I
70		2,398.27	I
70		2,399.96	I
120		2,406.75	I
110		2,407.90	I
120		2,415.33	I
120		2,416.75	I
100		2,420.12	I
100		2,421.06	I
100		2,421.98	I
110		2,428.28	I
110		2,435.52	I
140		2,501.61	I
150		2,506.90	I
240		2,507.78	I
180		2,511.65	I
180		2,511.95	I
180		2,517.14	I
240		2,519.62	I
410		2,526.22	I
210		2,527.90	I
120		2,528.47	II
150		2,528.84	II
240		2,530.18	I
110		2,549.28	II
120		2,552.65	I
210		2,562.13	I
110		2,564.82	I
230		2,574.02	I
140		2,630.67	II
130		2,642.21	II
150		2,645.26	II
140		2,651.90	I
150		2,656.22	I
180		2,661.42	I
290		2,672.00	II
380		2,677.80	II
270		2,678.57	II
380		2,679.32	II
180		2,682.87	II
180		2,683.09	II
1,100		2,687.96	II
170		2,688.72	II
150		2,689.88	II
230		2,690.24	II
240		2,690.79	II
120		2,696.99	I
120		2,697.74	I
680		2,700.94	II
380		2,702.19	II
530		2,706.17	II
150		2,706.70	II
110		2,707.86	II
170		2,711.74	II
120		2,714.20	II
640		2,715.69	II
150		2,722.56	I
240		2,728.64	II
180		2,731.35	II
100		2,739.71	II
140		2,753.40	II
140		2,765.67	II
140		2,777.73	II
120		2,803.47	II
120		2,846.57	II
110		2,847.57	II
140		2,852.87	I
140		2,854.34	II
200		2,855.22	I
180		2,859.97	I
240		2,864.36	I
170		2,866.59	I
210		2,868.10	I
140		2,869.13	II
210		2,870.55	I
110		2,877.69	II
110		2,879.16	II
350		2,880.03	II
380		2,882.50	II
380		2,884.78	II
140		2,888.25	II
380		2,889.62	II
900		2,891.64	II
530		2,892.44	II
900		2,892.66	II
1,400		2,893.32	II
360		2,896.21	II
110		2,899.60	II
360		2,903.08	II
150		2,906.13	I
900		2,906.46	II
490		2,907.47	II
2,400		2,908.82	II
710		2,910.02	II
530		2,910.39	II
560		2,911.06	II
380		2,914.93	I
120		2,917.37	II
210		2,919.99	II
380		2,920.38	II
710		2,923.62	I
2,400		2,924.02	II
1,700		2,924.64	II
710		2,930.81	II
210		2,934.40	II
110		2,935.87	II
900		2,941.37	II
450		2,941.49	II
230	d	2,942.33	I
230		2,943.20	II
1,100		2,944.57	II
110		2,946.53	II
230		2,949.63	II
300		2,950.35	II
640		2,952.08	II
120		2,954.33	I
260		2,957.52	II
410		2,962.77	I
600		2,968.38	II
120		2,972.25	II
120		2,976.20	II
380		2,976.52	II
240		2,977.54	I
260		3,001.20	II
140		3,014.82	II
180		3,016.78	II
270		3,033.45	II
290		3,033.82	II
230		3,043.12	I
230		3,043.56	I
230		3,044.94	I
230		3,048.22	II
170		3,050.89	I
180		3,053.39	II
450		3,053.65	I
1,200		3,056.33	I
1,400		3,060.46	I
140		3,063.25	II
2,400		3,066.38	II
200		3,067.12	I
140		3,069.64	I
170		3,073.82	I
100		3,075.27	I
150		3,082.11	I
3,800		3,093.11	II
200		3,094.20	II
180		3,100.94	II
3,000		3,102.30	II
2,600		3,110.71	II
2,000		3,118.38	II
380		3,121.14	II
150		3,122.90	II
1,500		3,125.28	II
260		3,126.22	II
530		3,130.27	II
410		3,133.33	II

210		3,134.93	II
150		3,136.51	II
150		3,139.74	II
200		3,142.48	II
150		3,145.34	II
3,200		3,183.41	I
5,300		3,183.98	I
3,800		3,185.40	I
410		3,187.71	II
530		3,188.51	II
750		3,190.68	II
530		3,198.01	I
750		3,202.38	I
450		3,205.58	I
450		3,207.41	I
410		3,212.43	I
210		3,217.11	II
150		3,237.87	II
140		3,254.77	II
140		3,263.24	I
1,100		3,267.70	II
900		3,271.12	II
750		3,276.12	II
110		3,279.84	II
140		3,298.14	I
110		3,329.86	I
110		3,365.55	I
110		3,377.62	I
170		3,400.40	I
110		3,425.07	I
110		3,485.92	II
210		3,504.44	II
560		3,517.30	II
150		3,520.02	II
110		3,524.72	II
230		3,529.74	I
230		3,530.77	II
560		3,533.68	I
110		3,543.50	I
560		3,545.20	II
110		3,553.27	I
560		3,556.80	I
110		3,566.18	I
560		3,589.76	II
490		3,592.02	II
560		3,592.53	I
270		3,593.33	II
110		3,606.69	I
110		3,639.02	I
110		3,644.71	I
250		3,663.59	I
250		3,667.74	I
110		3,669.41	II
170		3,671.20	I
280		3,673.40	I
280		3,675.70	I
170		3,676.68	I
300		3,680.11	I
570		3,683.13	I
190		3,686.26	I
470		3,687.47	I
1,300		3,688.07	I
1,000		3,690.28	I
1,500		3,692.22	I
450		3,695.34	I
1,000		3,695.86	I
3,800		3,703.58	I
1,800		3,704.70	I
570		3,705.04	I
130		3,708.72	I
320		3,715.47	II
250		3,727.34	II
280		3,732.76	II
150		3,734.43	II
230		3,745.80	II
210		3,750.87	II
210		3,770.97	II
270		3,778.68	I
520		3,790.32	I
1,100		3,794.96	I
570		3,799.91	I
570		3,803.47	I
190		3,806.80	I
300		3,807.50	I
520		3,808.52	I
230		3,809.60	I
1,000		3,813.49	I
140	d	3,817.84	I
1,300		3,818.24	I

230		3,819.96	I
230		3,821.49	I
570		3,822.01	I
450		3,822.89	I
300		3,823.21	I
1,700		3,828.56	I
280		3,834.22	I
160		3,839.00	I
110		3,839.38	I
570		3,840.44	I
2,600		3,840.75	I
110		3,841.89	I
380		3,844.44	I
320		3,847.33	I
110		3,849.32	I
1,200		3,855.37	I
3,000		3,855.84	I
150		3,862.22	I
130		3,863.87	I
1,300		3,864.86	I
230		3,867.60	I
170		3,871.08	I
1,500		3,875.08	I
420		3,875.90	I
570		3,876.09	I
130		3,878.71	II
700		3,890.18	I
460		3,892.86	I
280	h	3,898.02	I
140		3,899.13	II
140	h	3,900.18	I
140	h	3,901.15	I
2,400		3,902.25	I
100		3,906.75	I
700		3,909.89	I
100		3,910.79	I
220		3,912.21	I
140		3,914.33	I
100		3,916.41	II
100		3,920.49	I
100		3,921.90	I
230		3,922.43	I
240		3,924.66	I
150		3,925.24	I
200		3,927.93	I
260		3,930.02	I
150		3,931.34	I
260		3,934.01	I
150		3,935.14	I
100		3,936.28	I
150		3,943.66	I
100		3,950.23	I
140		3,951.97	II
100		3,973.64	II
540		3,990.57	I
260		3,992.80	I
430		3,998.73	I
170		4,005.71	II
120		4,023.39	II
120		4,031.83	I
150		4,035.63	II
120		4,042.64	I
360		4,050.96	I
360		4,051.35	I
280		4,057.07	I
130		4,057.82	I
230		4,063.93	I
230		4,071.54	I
1,100		4,090.58	I
180		4,092.41	I
1,800		4,092.69	I
120		4,093.50	I
890		4,095.49	I
2,800		4,099.80	I
590		4,102.16	I
230		4,104.40	I
260		4,104.78	I
2,800		4,105.17	I
120		4,108.22	I
2,300		4,109.79	I
8,900		4,111.78	I
120		4,112.33	I
230		4,113.52	I
4,300		4,115.18	I
1,800		4,116.47	I
180		4,118.18	I
180		4,118.64	I
230		4,119.46	I
180		4,120.54	I

180		4,123.19	I
2,000		4,123.57	I
120		4,124.07	I
3,100		4,128.07	I
120		4,128.86	I
3,100		4,132.02	I
2,300		4,134.49	I
150		4,159.69	I
100		4,174.01	I
230		4,179.42	I
150		4,182.59	I
180		4,189.84	I
180		4,191.56	I
230		4,209.86	I
120		4,226.62	I
360		4,232.46	I
180		4,232.95	I
180		4,234.00	I
120		4,235.76	I
100		4,257.37	I
120		4,259.31	I
120		4,262.16	I
560		4,268.64	I
460		4,271.55	I
460		4,276.96	I
430		4,284.06	I
330		4,291.82	I
220		4,296.11	I
170		4,297.68	I
170		4,298.03	I
170		4,306.21	I
140		4,307.18	I
170		4,309.80	I
460		4,330.02	I
510		4,332.82	I
760		4,341.01	I
1,000		4,352.87	I
130		4,354.98	I
150		4,355.94	I
150		4,368.04	I
140	d	4,373.23	I
100		4,375.30	I
12,000		4,379.24	I
100		4,380.55	I
7,000		4,384.72	I
4,800		4,389.97	I
3,600		4,395.23	I
1,400		4,400.58	I
2,300		4,406.64	I
2,800		4,407.64	I
3,600		4,408.20	I
4,600		4,408.51	I
140		4,412.14	I
640		4,416.47	I
120		4,419.94	I
640		4,421.57	I
460		4,426.00	I
120		4,427.31	I
310		4,428.52	I
230		4,429.80	I
430		4,436.14	I
640		4,437.84	I
830		4,441.68	I
640		4,444.21	I
610		4,452.01	I
410		4,457.48	I
120		4,457.76	I
1,000		4,459.76	I
2,000		4,460.29	I
610		4,462.36	I
120		4,468.01	I
380		4,469.71	I
120		4,474.04	I
200		4,474.71	I
380		4,488.89	I
100		4,496.06	I
120		4,501.95	I
140		4,524.22	I
360		4,545.39	I
100		4,549.65	I
280		4,560.71	I
200		4,571.78	I
510		4,577.17	I
140		4,578.73	I
640		4,580.40	I
830		4,586.36	I
170		4,591.22	I
1,300		4,594.11	I
100		4,606.15	I

30		4,609.65	I
25		4,611.74	I
230		4,619.77	I
65		4,624.41	I
50		4,626.48	I
100		4,635.18	I
65		4,640.07	I
65		4,640.74	I
130		4,646.40	I
30		4,648.89	I
30		4,666.14	I
160		4,670.49	I
24		4,684.45	I
35		4,686.92	I
55		4,706.16	I
80		4,706.57	I
80		4,710.56	I
65		4,714.12	I
35		4,715.89	I
55		4,717.69	I
40		4,721.51	I
40		4,722.86	I
40		4,729.53	I
27		4,730.38	I
27		4,742.63	I
24		4,746.63	I
40		4,748.52	I
45		4,750.98	I
35		4,751.56	I
40		4,753.93	I
65		4,757.48	I
55		4,766.63	I
130		4,776.36	I
		4,776.52	I
110		4,786.51	I
130		4,796.92	I
19		4,799.77	I
130		4,807.53	I
130		4,827.45	I
150		4,831.64	I
120		4,832.43	I
		4,833.02	I
19		4,848.81	I
320		4,851.48	I
35		4,862.61	I
480		4,864.74	I
21		4,871.26	I
620		4,875.48	I
55		4,880.56	I
740		4,881.56	I
27		4,891.60	I
21		4,894.21	I
55		4,900.62	I
95	d	4,904.29	I
		4,904.34	I
85		4,925.65	I
35		4,932.03	I
23		4,966.12	I
70		5,002.33	I
85		5,014.62	I
28		5,051.63	I
35		5,064.12	I
35		5,105.14	I
110		5,128.53	I
110		5,138.42	I
25		5,139.53	I
70		5,148.72	I
40		5,159.35	I
23		5,169.94	I
70		5,176.77	I
20		5,192.01	I
110		5,192.99	I
23		5,193.62	I
110		5,194.83	I
55		5,195.36	I
20		5,206.61	I
40		5,216.59	I
35		5,225.77	I
35		5,233.75	I
110		5,234.07	I
20		5,240.20	I
110		5,240.87	I
17		5,260.98	I
40		5,353.41	I
35		5,383.43	I
40		5,385.14	I
14		5,388.30	I
11		5,397.87	I
100		5,401.93	I

140		5,415.26	I
28		5,418.09	I
50		5,424.08	I
40		5,434.18	I
11		5,437.66	I
17		5,458.12	I
13		5,471.33	I
25		5,487.22	I
85		5,487.92	I
25		5,489.94	I
28		5,504.87	I
70		5,507.75	I
14		5,511.18	I
23		5,545.93	I
70		5,547.07	I
35		5,558.75	I
28		5,561.66	I
140		5,584.50	I
23		5,586.00	I
100		5,592.42	I
28		5,601.38	I
70		5,604.94	I
13		5,624.20	I
200		5,624.60	I
70		5,624.89	I
55		5,626.01	I
400		5,627.64	I
13		5,632.46	I
10		5,633.90	I
13		5,635.51	I
85		5,646.11	I
110		5,657.44	I
110		5,668.36	I
310		5,670.85	I
20		5,683.22	I
1,200		5,698.52	I
920		5,703.56	I
570		5,706.98	I
11		5,708.95	I
11	h	5,716.21	I
70		5,725.64	I
850		5,727.03	I
170		5,727.66	I
230		5,731.25	I
40		5,734.01	I
230		5,737.06	I
110		5,743.45	I
17		5,747.70	I
40		5,748.87	I
17		5,752.74	I
17		5,761.41	I
70		5,772.42	I
35		5,776.64	I
11		5,782.61	I
11		5,783.50	I
40	h	5,784.38	I
55	h	5,786.16	I
23		5,788.56	I
35	h	5,807.14	I
23		5,817.06	I
35	h	5,817.53	I
55	h	5,830.72	I
85	h	5,846.30	I
11		5,850.32	I
40		5,924.57	I
28		5,978.91	I
20		5,980.78	I
28		6,002.31	I
55		6,002.63	I
28		6,016.12	I
20		6,025.41	I
450		6,039.73	I
100		6,058.14	I
20		6,067.26	I
480		6,081.44	I
1,300		6,090.22	I
28		6,106.98	I
280		6,111.67	I
600		6,119.52	I
20		6,128.34	I
280		6,135.38	I
180		6,150.15	I
85		6,170.36	I
23		6,189.35	I
450		6,199.19	I
130		6,213.87	I
450		6,216.37	I
28	h	6,218.31	I
130		6,224.50	I
430		6,230.74	I
100		6,233.20	I
55		6,240.13	I
170		6,242.81	I
710		6,243.10	I
280		6,251.82	I
85		6,256.90	I
85		6,258.57	I
55		6,261.22	I
85		6,266.32	I
130		6,268.82	I
170		6,274.65	I
17	h	6,282.33	I
200		6,285.16	I
200		6,292.83	I
170		6,296.49	I
28	h	6,311.50	I
14		6,324.66	I
70		6,326.84	I
55		6,339.09	I
50		6,349.48	I
14		6,355.58	I
50		6,357.30	I
25		6,358.82	I
35		6,361.27	I
23		6,379.36	I
14		6,393.28	I
35		6,430.47	I
23		6,431.63	I
14		6,433.18	I
11		6,435.16	I
70		6,452.34	I
11		6,488.05	I
55		6,504.17	I
110		6,531.43	I
28		6,543.51	I
17		6,558.02	I
11		6,565.88	I
50		6,605.97	I
15		6,607.83	I
10		6,623.54	I
50		6,624.85	I
13		6,633.26	I
13		6,643.79	I
8		6,693.66	I
8		6,708.07	I
65	c	6,753.00	I
10		6,760.12	I
50	c	6,766.49	I
40		6,784.98	I
15		6,786.32	I
26		6,812.40	I
9	c	6,829.94	I
15		6,832.44	I
12		6,839.58	I
12		6,841.90	I
10	c	6,870.88	I
8		6,871.56	I
7		6,894.00	I
12		6,974.50	I
21		7,026.07	I
7		7,063.69	I
11	h	7,092.08	I
6		7,102.58	I
24		7,148.15	I
7		7,151.36	I
7		7,182.08	I
14		7,264.29	I
8		7,321.44	I
40		7,338.92	I
35		7,356.54	I
11		7,358.66	I
24		7,361.39	I
12		7,362.49	I
24		7,363.16	I
9		7,385.95	I
6	h	7,393.49	
12	h	7,485.90	I
12	h	7,488.08	I
12	h	7,492.44	
12	h	7,578.75	I
9	h	7,591.24	I
14	h	7,596.92	I
12	h	7,598.28	I
24		7,624.81	I
5		7,701.37	I
8		7,704.81	I
8	h	7,851.18	
14	b	7,865.51	VO
12		7,896.40	
14	h	7,898.81	
24		7,937.92	I
29	c	8,027.39	I
14		8,028.13	I
14	h	8,035.38	
14	h	8,045.71	
12		8,051.89	
14		8,093.48	I
8		8,102.44	I
12		8,108.59	I
9	h	8,109.07	I
120	c	8,116.80	I
11	h	8,136.79	I
29		8,144.59	I
9		8,154.55	I
70	c	8,161.07	I
14		8,171.35	I
7		8,180.21	I
35		8,180.21	I
24		8,187.33	I
29		8,198.87	I
35		8,203.07	I
24		8,241.61	I
29		8,253.51	I
29		8,255.88	I
5		8,280.39	I
19		8,282.37	I
8		8,324.42	I
14		8,331.23	I
14		8,342.03	I
7		8,402.81	I
12		8,499.52	I
6		8,534.49	I
6		8,624.86	VO
60		8,919.85	I
29		8,932.93	I
12		8,971.62	I

XENON (XE)

Z = 54

Xe I and II

Ref. 33, 116, 120, 232 — S.P.D.

Intensity		Wavelength	
Vacuum			
350		740.41	II
350		803.07	II
600		880.80	II
350		885.54	II
600		925.87	II
250		935.40	II
800		972.77	II
700		976.68	II
500		1,032.44	II
700		1,037.68	II
1,100		1,041.31	II
1,000		1,048.27	II
1,200		1,051.92	II
2,000		1,074.48	II
600		1,083.86	II
1,200		1,100.43	II
600		1,158.47	II
250		1,169.63	II
800	p	1,183.05	II
250		1,192.04	I
600		1,244.76	II
250		1,250.20	I
1,000		1,295.59	I
600		1,469.61	I
Air			
200		2,864.73	II
150	h	2,895.22	II
400		2,979.32	II
100	h	3,017.43	II
300		3,128.87	II
200	h	3,366.72	II
2		3,400.07	I
2		3,418.37	I
2		3,420.00	I
3		3,442.66	I
100	h	3,461.26	II
4		3,469.81	I
4		3,472.36	I
5		3,506.74	I
10		3,549.86	I
10		3,554.04	I
15		3,610.32	I
8		3,613.06	I
6		3,633.06	I
10		3,669.91	I
40		3,685.90	I
40		3,693.49	I
100	l	3,907.91	II
100		4,037.59	II
200	l	4,057.46	II
100	h	4,098.89	II
200	l	4,158.04	II
1,000	h	4,180.10	II
500	h	4,193.15	II
300	h	4,208.48	II
100	h	4,209.47	II
300	h	4,213.72	II
100		4,215.60	II
300	h	4,223.00	II
400	h	4,238.25	II
500	h	4,245.38	II
100	l	4,251.57	II
500	h	4,296.40	II
500	h	4,310.51	II
1,000	l	4,330.52	II
200	h	4,369.20	II
100	l	4,373.78	II
500	h	4,393.20	II
500	l	4,395.77	II
200	l	4,406.88	II
150	l	4,416.07	II
500	h	4,448.13	II
1,000	h	4,462.19	II
500	l	4,480.86	II
100	l	4,521.86	II
600		4,734.152	I
150		4,792.619	I
500		4,807.02	I
400		4,829.71	I
300		4,843.29	I
500		4,916.51	I
500		4,923.152	I
200	l	4,971.71	II
400		4,972.71	II
300		4,988.77	II
100	l	4,991.17	II
200		5,028.280	I
200		5,044.92	II
1,000		5,080.62	II
300		5,122.42	II
100		5,125.70	II
100		5,178.82	II
300		5,188.04	II
400		5,191.37	II
100		5,192.10	II
500		5,260.44	II
500		5,261.95	II
2,000		5,292.22	II
300		5,309.27	II
1,000		5,313.87	II
2,000		5,339.33	II
200		5,363.20	II
200		5,368.07	II
500		5,372.39	II
100		5,392.80	I
3,000		5,419.15	II
800		5,438.96	II
300		5,445.45	II
200		5,450.45	II
400		5,460.39	II
1,000		5,472.61	II
100	l	5,494.86	II
200		5,525.53	II
600		5,531.07	II
100		5,566.62	I
300		5,616.67	II
300		5,659.38	II
600		5,667.56	II
150		5,670.91	II
100		5,695.75	I
200		5,699.61	II
200		5,716.10	II
500		5,726.91	II
500		5,751.03	II
300		5,758.65	II
300		5,776.39	II
100		5,815.96	II
300		5,823.89	I
150		5,824.80	I
100		5,875.02	I
300		5,893.29	II
100		5,894.99	I
200		5,905.13	II
100		5,934.17	I
500		5,945.53	II
300		5,971.13	II
2,000		5,976.46	II

Note: Wavelengths above 12,000 A are for the 136 isotope.

200		6,008.92	II
1,000		6,036.20	II
2,000		6,051.15	II
600		6,093.50	II
1,500		6,097.59	II
400		6,101.43	II
100		6,115.08	II
100		6,146.45	II
150		6,178.30	I
120		6,179.66	I
300		6,182.42	I
500		6,194.07	II
100		6,198.26	I
100		6,220.02	II
500		6,270.82	II
400		6,277.54	II
100		6,284.41	II
100		6,286.01	I
250		6,300.86	II
500		6,318.06	I
400		6,343.96	II
600		6,356.35	II
200		6,375.28	II
100		6,397.99	II
300		6,469.70	I
150		6,472.84	I
120		6,487.76	I
100		6,498.72	I
200	h	6,504.18	I
300		6,512.83	II
200		6,528.65	II
100		6,533.16	I
1,000		6,595.01	II
100		6,595.56	I
400		6,597.25	II
100		6,598.84	II
150		6,668.92	I
300		6,694.32	II
200		6,728.01	I
150		6,788.71	II
100		6,790.37	II
1,000		6,805.74	II
200		6,827.32	I
100		6,872.11	I
300		6,882.16	I
80		6,910.22	II
100		6,925.53	I
800	h	6,942.11	II
100		6,976.18	I
2,000		6,990.88	II
150		7,082.15	II
500		7,119.60	I
50	s	7,147.50	II
200		7,149.03	II
500		7,164.83	II
100		7,284.34	II
200		7,301.80	II
200		7,339.30	II
100		7,386.00	I
150		7,393.79	I
300		7,548.45	II
200		7,584.68	I
80		7,618.57	II
500		7,642.02	I
100		7,643.91	I
200		7,670.66	II
60		7,787.04	II
100		7,802.65	I
100		7,881.32	I
300		7,887.40	I
500		7,967.34	I
100		8,029.67	I
200		8,057.26	I
150		8,061.34	I
100		8,101.98	I
150	h	8,151.80	II
100		8,171.02	I
700		8,206.34	I
10,000		8,231.635	I
500		8,266.52	I
7,000		8,280.116	I
2,000		8,346.82	I
100		8,347.24	II
2,000		8,409.19	I
50	h	8,515.19	II
200		8,576.01	I
50	h	8,604.23	II
250		8,648.54	I
100		8,692.20	I
200		8,696.86	I
50	h	8,716.19	II
300		8,739.39	I
100		8,758.20	I
5,000		8,819.41	I
300		8,862.32	I
200		8,908.73	I
200		8,930.83	I
1,000		8,952.25	I
100		8,981.05	I
200		8,987.57	I
400		9,045.45	I
500		9,162.65	I
100		9,167.52	I
100		9,374.76	I
200		9,513.38	I
50	h	9,591.35	II
150		9,685.32	I
50	l	9,698.68	II
100		9,718.16	I
2,000		9,799.70	I
3,000		9,923.19	I
100		10,838.37	I
90		11,742.01	I
375		12,235.24	I
100		12,257.76	I
300		12,590.20	I
2,500		12,623.391	I
250		13,544.15	I
2,000		13,657.055	I
1,250		14,142.444	I
800		14,240.96	I
375		14,364.99	I
140		14,660.81	I
3,000		14,732.806	I
100		15,099.72	I
2,500		15,418.394	I
150		15,557.13	I
250		15,979.54	I
100		16,039.90	I
1,000		16,053.28	I
125		16,554.49	I
1,500		16,728.15	I
1,500		17,325.77	I
350		18,788.13	I
150		20,187.19	I
3,000		20,262.242	I
250		21,470.09	I
1,250		23,193.33	I
110		23,279.54	I
1,800		24,824.71	I
175		25,254.84	I
2,000		26,269.08	I
2,500		26,510.86	I
250		28,381.54	I
750		28,582.25	I
300		29,384.41	I
150		29,448.06	I
100		29,649.58	I
100		29,813.62	I
600		30,253.14	I
1,500		30,475.46	I
100		30,504.12	I
500		30,794.18	I
6,000		31,069.23	I
125		31,336.01	I
550		31,607.91	I
100		32,293.08	I
1,800		32,739.26	I
3,500		33,666.69	I
150		34,014.67	I
450		34,335.27	I
170		34,744.00	I
5,000		35,070.25	I
110		35,246.92	I
250		36,209.21	I
150		36,231.74	I
450		36,508.36	I
850		36,788.83	I
140		38,685.98	I
175		38,737.82	I
270		38,939.60	I
120		39,955.14	I

YTTERBIUM (YB)
Z = 70

Yb I and II
Ref. 1 — C.H.C.

Intensity		Wavelength Air	
2,500		2,116.67	II
3,000		2,126.74	II
370		2,161.60	II
850		2,185.71	II
640		2,224.46	II
140		2,320.81	II
50		2,362.89	II
170		2,390.74	II
18		2,398.02	II
28		2,421.35	II
25		2,447.26	II
28		2,460.25	II
460		2,464.50	I
14		2,484.89	II
70		2,502.02	II
28		2,505.48	II
11		2,508.07	II
140		2,512.06	II
18		2,516.35	II
50		2,522.44	II
65		2,537.65	II
270		2,538.67	II
14		2,550.06	II
70		2,552.15	II
55		2,552.70	II
21		2,565.57	II
28		2,571.36	II
13		2,573.15	II
18		2,596.16	II
28		2,596.32	II
21		2,615.26	II
100		2,617.01	II
55		2,634.31	II
45		2,639.45	II
85		2,641.89	II
110		2,644.31	II
28		2,646.44	II
28		2,647.46	II
28		2,648.80	II
50		2,649.79	II
28		2,650.73	II
990		2,653.75	II
35		2,656.12	II
21		2,659.27	II
200		2,665.04	II
55		2,668.75	II
390		2,671.96	I
390		2,672.66	II
21		2,680.40	II
14		2,683.42	II
70		2,684.75	II
25		2,687.98	II
28		2,695.43	II
14		2,696.62	II
18		2,700.80	II
21		2,708.84	II
65		2,710.54	II
25		2,711.78	II
55		2,712.66	II
170		2,718.35	II
21		2,722.20	II
110		2,732.74	II
21		2,734.09	II
55		2,741.71	II
55		2,747.58	II
18		2,748.04	II
230		2,748.66	II
1,300		2,750.48	II
85		2,751.45	II
21		2,759.00	II
65		2,760.78	II
65		2,761.37	II
35		2,764.41	II
85		2,771.32	II
170		2,776.28	II
100		2,784.66	II
18		2,787.96	II
45		2,793.28	II
25		2,794.44	II
21		2,795.07	II
18		2,795.29	II
35		2,797.80	II
100		2,798.21	II
45		2,799.38	II
50		2,800.00	II
35		2,800.06	II
14		2,810.72	II
65		2,814.53	II
28		2,816.32	II
140		2,821.15	II
100		2,824.97	II
190		2,830.99	II
18		2,832.20	II
28		2,834.97	II
14		2,842.59	II
230	h	2,847.18	II
100		2,848.44	II
21		2,849.34	II
360		2,851.13	II
55		2,851.86	II
21		2,853.41	II
18		2,853.68	II
55		2,854.14	II
45		2,854.49	II
45		2,858.33	II
45		2,858.46	II
100		2,859.39	II
430		2,859.80	II
55		2,860.39	II
140		2,861.21	II
100		2,861.34	II
200		2,867.06	II
25		2,870.06	II
45		2,873.49	I
28		2,885.97	II
70		2,886.26	II
200		2,888.04	II
3,600		2,891.38	II
45		2,893.62	II
28		2,896.90	II
85		2,899.70	II
18		2,902.41	II
21		2,902.92	II
21		2,906.88	II
28		2,908.33	II
35		2,909.19	II
55		2,909.48	II
85		2,911.52	II
18		2,912.86	II
170		2,914.21	II
140		2,915.28	II
18		2,916.43	II
280		2,919.35	II
55		2,921.12	II
45		2,924.24	II
25		2,927.85	II
35		2,934.36	I
55		2,935.11	II
21		2,937.19	II
45		2,939.53	II
45		2,940.52	II
28		2,942.04	II
140		2,945.91	II
45		2,946.30	II
18		2,946.76	II
28		2,950.33	II
45		2,955.32	II
18		2,957.63	II
65		2,962.52	II
21		2,963.26	II
45		2,963.46	II
130		2,964.76	II
2,000		2,970.56	II
45		2,982.49	II
21		2,982.66	II
28		2,983.70	II
200		2,983.99	II
90		2,985.08	II
35		2,985.88	II
45		2,990.37	II
65		2,991.87	II
28		2,993.94	II
170		2,994.80	II
28		2,995.86	II
70		3,000.46	II
25		3,002.61	II
310		3,005.77	II
100		3,009.39	II
65		3,010.62	II
55		3,014.43	II
160		3,017.56	II
160		3,026.67	II
920		3,031.11	II
55		3,034.64	II
25		3,037.99	II
55		3,039.67	II
80		3,042.65	II
21		3,044.00	II
45		3,046.48	II
35		3,047.05	II
45		3,063.12	II
21		3,063.67	II

110		3,065.04	II
18		3,076.01	II
100		3,089.10	II
70		3,093.87	II
28		3,100.74	I
45		3,101.36	II
28		3,102.07	II
55		3,107.76	II
170		3,107.90	II
85		3,115.34	II
55		3,116.70	II
190		3,117.81	II
50		3,136.76	II
230		3,140.94	II
80		3,141.73	II
80		3,145.06	II
28		3,145.54	II
28		3,153.18	II
90		3,153.88	II
50		3,155.18	II
28		3,162.29	I
70		3,163.80	II
50		3,165.21	II
120		3,169.06	II
120		3,180.92	II
390		3,192.88	II
70		3,198.65	II
240		3,201.16	II
80		3,217.18	II
50		3,218.32	II
50		3,225.88	II
45		3,239.20	II
35		3,239.58	I
35		3,246.06	II
130		3,261.51	II
18,000		3,289.37	II
130		3,305.25	I
140		3,305.73	II
50		3,315.10	II
80		3,319.41	I
50		3,333.06	II
240		3,337.17	II
280	d	3,342.93	II
		3,343.07	II
80		3,346.50	
50		3,347.54	II
35		3,351.09	II
50		3,351.26	
100		3,352.49	II
100		3,362.44	II
50		3,363.64	II
240		3,375.48	II
50		3,376.62	II
28		3,382.54	
140		3,387.50	I
50		3,390.25	II
28		3,390.42	II
50		3,391.10	II
50	h	3,394.44	II
50		3,401.01	II
35		3,404.10	II
50		3,412.45	I
140		3,418.39	I
360		3,426.04	I
80		3,428.46	II
240		3,431.11	I
45		3,434.61	II
50		3,438.71	II
100		3,438.85	II
35		3,443.59	
35		3,446.89	II
85		3,452.40	I
500		3,454.08	II
190	d	3,458.29	II
		3,458.39	I
360		3,460.27	I
35		3,462.34	II
2,400		3,464.37	I
500		3,476.30	II
500		3,478.84	II
50		3,482.56	II
85		3,485.76	II
85		3,488.43	II
100	w	3,495.90	II
85		3,507.83	II
50		3,517.00	I
230		3,520.29	II
50		3,545.72	II
100		3,549.82	II
35		3,559.03	I
200		3,560.33	II
170		3,560.70	II
50	h	3,563.94	II
85		3,570.57	II
50		3,572.50	II
50		3,574.58	II
360		3,585.47	II
130		3,606.48	II
50		3,610.23	II
70		3,611.30	II
200		3,619.80	II
110		3,634.52	
240		3,637.76	II
70		3,648.15	I
90		3,655.73	I
240		3,669.69	II
50		3,670.69	II
140		3,675.08	II
50		3,690.56	II
32,000		3,694.19	II
70		3,698.60	II
70		3,700.58	I
50		3,710.34	II
60		3,724.21	II
180		3,734.69	I
550		3,770.10	I
80		3,774.32	I
60	h	3,791.74	I
170		3,839.91	I
340		3,872.85	I
340		3,900.85	I
50		3,904.81	II
140		3,911.27	I
		3,987.99	I
930		3,990.88	I
50		4,007.36	I
70		4,052.28	I
85		4,077.28	II
440		4,089.68	I
120	h	4,119.25	II
70		4,135.09	II
470		4,149.07	I
120		4,174.56	I
340		4,180.81	II
150	d	4,218.56	II
		4,218.69	I
120		4,231.97	I
70		4,277.74	I
120		4,305.97	I
70		4,316.95	II
60	h	4,393.69	I
60	h	4,430.21	I
440		4,439.19	I
85	h	4,482.42	I
85		4,515.16	II
35		4,553.58	II
85	h	4,563.95	I
640		4,576.21	I
200		4,582.36	I
70		4,589.21	I
140		4,590.83	I
40		4,598.36	II
35		4,683.81	II
40		4,684.27	I
190		4,726.08	II
170	h	4,781.87	I
170		4,786.61	II
35		4,816.43	I
40		4,820.24	II
35		4,836.96	II
40		4,837.46	I
17		4,851.15	II
40	h	4,894.60	I
27		4,912.36	I
710		4,935.50	I
24		4,937.22	II
140		4,966.90	I
24		5,009.52	II
17		5,067.30	II
30		5,067.80	I
70		5,069.14	I
220		5,074.34	I
50		5,076.74	I
20		5,135.98	II
14		5,147.02	II
20		5,184.15	II
60		5,196.08	I
85		5,211.60	I
35		5,240.51	II
100		5,244.11	I
40		5,257.49	II
150	h	5,277.04	I
35		5,279.53	II
17		5,300.94	II
170		5,335.15	II
30	d	5,345.66	II
		5,345.83	II
60		5,347.22	II
30	h	5,351.29	I
150		5,352.95	II
30		5,358.64	II
30		5,363.66	I
17		5,389.84	II
14		5,432.71	II
40		5,449.27	II
14		5,478.50	II
60		5,481.92	I
40		5,505.49	I
17		5,524.54	I
85	h	5,539.05	I
2,400		5,556.47	I
35		5,562.09	I
20		5,568.11	I
20		5,586.36	I
40		5,588.45	II
60		5,651.98	II
7		5,686.53	II
220		5,719.99	I
10		5,749.91	II
10	h	5,755.89	I
27		5,771.66	II
10		5,803.44	I
10		5,819.41	II
35		5,833.99	II
35		5,837.14	II
27		5,854.51	I
8		5,897.21	II
20		5,908.36	II
17		5,989.33	I
40		5,991.51	II
10		6,052.88	II
10		6,054.57	I
60		6,152.57	II
30		6,246.97	II
60		6,274.78	II
14		6,308.15	II
35	h	6,400.35	I
35	h	6,417.91	I
20		6,432.73	II
17	h	6,463.15	II
340		6,489.06	I
20		6,643.55	I
180		6,667.82	I
15		6,678.17	I
25		6,727.61	II
25		6,768.70	I
690		6,799.60	I
18		6,934.05	II
20		6,999.88	II
10		7,043.78	II
9	h	7,244.41	I
8	h	7,305.22	I
10	h	7,313.05	I
16	h	7,350.04	I
25		7,448.28	I
30	h	7,527.46	I
750		7,699.48	I
7		7,895.08	I
70	h	8,922.56	II

Yb III

Ref. 40, 192 — J.R.

Intensity		Wavelength	
		Vacuum	
5		968.46	III
20		973.16	III
10		994.56	III
10		1,560.66	III
80		1,561.42	III
30		1,669.60	III
50		1,670.78	III
50		1,719.82	III
60		1,739.18	III
70		1,762.80	III
80	h	1,765.21	III
65		1,775.29	III
70		1,779.74	III
70		1,781.31	III
20		1,793.70	III
60		1,798.85	III
65		1,810.88	III
60		1,826.41	III
20		1,826.77	III
30		1,838.01	III
30		1,847.30	III
30		1,849.24	III
10		1,849.42	III
75		1,852.36	III
75		1,852.94	III
90		1,854.80	III
80		1,857.16	III
100		1,863.32	III
5		1,864.85	III
10		1,867.23	III
10		1,867.63	III
10		1,868.19	III
5		1,868.92	III
10		1,870.07	III
15		1,870.83	III
10		1,871.15	III
200		1,872.03	III
800		1,873.91	III
100		1,875.41	III
75		1,875.92	III
70		1,880.30	III
80		1,884.22	III
70		1,885.07	III
70		1,887.22	III
10		1,890.34	III
10		1,890.87	III
15		1,892.42	III
5		1,895.50	III
100		1,896.18	III
10		1,897.57	III
500		1,898.25	III
7		1,906.74	III
10		1,908.50	III
100		1,909.66	III
70		1,910.86	III
20		1,920.53	III
10		1,926.76	III
70		1,928.09	III
15		1,930.63	III
55		1,942.59	III
40		1,950.34	III
15		1,962.80	III
80		1,967.13	III
20		1,969.47	III
30		1,969.73	III
10		1,973.96	III
10		1,974.18	III
25		1,976.46	III
2		1,981.74	III
5		1,983.88	III
25		1,984.62	III
7		1,985.74	III
80		1,986.43	III
80	h	1,989.82	III
50		1,991.14	III
45		1,995.05	III
55		1,997.28	III
55		1,997.66	III
500		1,998.82	III
		Air	
20		2,054.80	III
10		2,066.49	III
10		2,073.64	III
30		2,078.05	III
10		2,087.37	III
50		2,087.98	III
20		2,091.23	III
20		2,092.26	III
10		2,094.77	III
80		2,095.31	III
15		2,096.79	III
30		2,098.36	III
10		2,106.71	III
50		2,109.54	III
20		2,119.18	III
20		2,198.14	III
80		2,202.27	III
300		2,240.11	III

100		2,244.28	III
200		2,257.03	III
100		2,262.26	III
200		2,265.67	III
150		2,282.99	III
100		2,283.99	III
300		2,305.32	III
100		2,309.27	III
200		2,314.49	III
200		2,337.97	III
40		2,361.08	III
200		2,365.43	III
50		2,367.46	III
30		2,369.99	III
20		2,377.22	III
50		2,403.95	III
20		2,410.04	III
60		2,412.33	III
10		2,429.18	III
20		2,433.43	III
100		2,438.27	III
20		2,439.31	III
20		2,440.43	III
10		2,458.64	III
10		2,464.59	III
200		2,490.42	III
20		2,491.69	III
40		2,506.25	III
300		2,516.82	III
15		2,522.07	III
20		2,529.14	III
40		2,550.39	III
300		2,555.29	III
100		2,560.56	III
10		2,561.66	III
100		2,566.78	III
2,000		2,567.61	III
1,000		2,579.57	III
100		2,588.62	III
20		2,592.69	III
500		2,597.23	III
800		2,599.14	III
30		2,609.14	III
600		2,621.11	III
300		2,627.07	III
30		2,635.37	III
500		2,638.06	III
300		2,640.48	III
1,000		2,642.56	III
100		2,643.62	III
1,000		2,651.74	III
700		2,652.25	III
100		2,659.98	III
70		2,664.89	III
2,000		2,666.13	III
2,000		2,666.99	III
30		2,673.33	III
500		2,677.39	III
500		2,691.01	III
30		2,708.04	III
400		2,712.32	III
500		2,749.91	III
200		2,755.94	III
200		2,756.76	III
100		2,765.50	III
300		2,788.24	III
600		2,795.60	III
400		2,803.32	III
1,000		2,803.43	III
10		2,807.22	III
50		2,808.51	III
600		2,816.92	III
1,000		2,818.72	III
15		2,826.01	III
300		2,842.96	III
400		2,875.86	III
600		2,898.30	III
1,000		2,906.31	III
300		2,928.97	III
50		2,977.84	III
800		2,998.00	III
2,000		3,029.49	III
100		3,031.62	III
30		3,040.65	III
3,000		3,092.50	III
20		3,102.18	III
4,000		3,126.01	III
1,000		3,138.58	III
100		3,151.44	III
70		3,179.34	III
800		3,191.35	III
50		3,216.27	III
2,000		3,228.58	III
2,000		3,325.51	III
50		3,358.25	III
20		3,364.30	III
2,000		3,384.01	III
150		3,392.56	III
100		3,397.66	III
80		3,432.94	III
40		3,456.18	III
150		3,463.51	III
20		3,469.98	III
300		3,550.87	III
200		3,613.89	III
30		3,659.84	III
30		3,663.74	III
200		3,664.74	III
20		3,675.78	III
400		3,711.91	III
20		3,879.98	III
10		3,882.58	III
20		3,887.17	III
150		3,896.55	III
15		3,912.75	III
20		3,913.23	III
500		3,931.23	III
100		3,985.56	III
10		3,991.74	III
10		3,997.67	III
2,000		4,028.14	III
10		4,033.03	III
20		4,074.53	III
20		4,090.67	III
20		4,098.23	III
15		4,121.06	III
10		4,150.04	III
15		4,153.11	III
100		4,162.72	III
60		4,172.95	III
30		4,194.34	III
100		4,194.95	III
10		4,198.74	III
300		4,213.64	III
10		4,220.83	III
15		4,231.07	III
20		4,289.64	III
40		4,301.14	III
20		4,304.01	III
15		4,350.80	III
10		4,380.07	III
100		4,517.58	III
40		4,639.14	III
10		4,834.93	III
15		5,054.94	III
20		5,256.85	III
20		5,331.54	III
15		5,740.83	III
10	d	5,949.02	III
20		5,973.05	III
40		6,055.85	III
100		6,214.22	III
200		6,328.52	III
10		6,365.88	III
150		6,378.33	III
25		6,466.33	III
20		6,985.15	III
10		7,037.04	III
15		7,157.72	III
10		7,311.02	III
10		7,399.98	III
80		7,410.01	III
15		7,456.86	III
70		7,664.41	III
80		7,892.39	III
20		7,893.10	III
100		7,971.46	III
20		8,056.02	III
10		8,117.44	III
10		8,326.86	III
30		8,327.88	III
20		8,400.01	III
30		8,489.90	III
200		10,110.60	III
100		10,830.36	III

Yb IV

Ref. 40, 311 — J.R.

Intensity		Wavelength	
		Vacuum	
200		828.96	IV
200		870.35	IV
300		902.46	IV
300		927.01	IV
300		936.22	IV
400		943.04	IV
400		946.20	IV
200		975.21	IV
1,000		1,050.24	IV
1,000		1,054.46	IV
400		1,092.51	IV
200		1,110.55	IV
5,000		1,134.43	IV
300		1,136.24	IV
500		1,166.01	IV
600		1,185.58	IV
200		1,290.24	IV
600		1,305.58	IV
900		1,316.04	IV
200		1,326.32	IV
800		1,326.36	IV
200		1,340.06	IV
300		1,345.36	IV
900		1,350.26	IV
200		1,353.43	IV
400		1,356.15	IV
200		1,361.75	IV
300		1,365.88	IV
300		1,369.72	IV
400		1,375.42	IV
300		1,376.66	IV
200		1,384.41	IV
250		1,393.93	IV
350		1,398.77	IV
400		1,407.05	IV
400		1,413.14	IV
300		1,416.15	IV
400		1,417.72	IV
200		1,423.99	IV
300		1,430.29	IV
200		1,440.61	IV
400		1,477.92	IV
300		1,491.57	IV
400		1,765.03	IV
200		1,776.18	IV
200		1,778.20	IV
200		1,779.34	IV
300		1,789.71	IV
800		1,791.06	IV
200		1,801.67	IV
250		1,809.63	IV
600		1,813.84	IV
400		1,816.07	IV
250		1,817.58	IV
300		1,819.02	IV
200		1,824.22	IV
		Air	
300		2,106.48	IV
900		2,116.65	IV
500		2,121.29	IV
250		2,122.84	IV
800		2,123.32	IV
600		2,125.72	IV
200		2,129.65	IV
500		2,135.21	IV
300		2,137.58	IV
500		2,138.35	IV
200		2,138.53	IV
800		2,139.99	IV
400		2,141.04	IV
200		2,142.20	IV
300		2,143.42	IV
300		2,143.89	IV
2,000		2,144.77	IV
400		2,148.10	IV
1,500		2,148.52	IV
15,000		2,154.18	IV
250		2,165.55	IV
300		2,169.12	IV
200		2,172.16	IV
300		2,177.53	IV
400		2,183.32	IV
270	h	2,186.13	IV
270		2,187.17	IV
90	h	2,189.90	IV
120		2,193.34	IV
150		2,198.27	IV
90		2,224.64	IV
150		2,231.28	IV
90		2,233.30	IV
90		2,244.20	IV
140		2,331.36	IV

YTTRIUM (Y)

Z = 39

Y I and II

Ref. 1 — C.H.C.

Intensity		Wavelength	
		Air	
350		2,243.06	II
50		2,354.20	I
30		2,373.83	
50		2,385.24	
25		2,413.93	II
560		2,422.20	II
60		2,460.61	II
25		2,490.42	I
12		2,540.28	
14		2,547.57	
10		2,550.17	
20		2,681.65	I
60		2,694.21	I
26		2,695.39	I
95		2,723.00	I
22		2,730.08	I
22		2,734.85	II
70		2,742.53	I
140		2,760.10	I
30		2,785.21	II
12		2,785.59	II
12		2,791.20	I
30		2,800.11	II
26		2,813.64	I
18		2,818.86	I
45		2,822.56	I
22		2,825.37	II
45		2,826.38	II
70		2,854.43	II
26		2,856.30	II
11		2,857.87	II
95		2,886.48	I
18		2,897.69	II
14		2,898.82	II
160		2,919.05	I
18	h	2,930.03	II
390		2,948.40	I
350		2,964.96	I
18		2,973.91	II
480		2,974.59	I
30		2,980.55	II
750		2,984.26	I
70		2,995.26	I
140		2,996.94	I
70		3,005.26	I
55		3,018.95	I
130		3,021.73	I
90		3,022.28	I
26		3,026.49	II
30		3,036.59	II
45		3,044.84	I
190		3,045.37	I
22		3,047.11	I
60		3,055.22	II
60		3,086.85	II
55	h	3,091.70	I
22		3,093.76	II
95		3,095.88	II
45		3,111.81	I
55		3,112.04	II
22		3,114.28	I

60		3,128.77	II	280	h	4,220.63	I	330		4,859.84	I	24	b	5,931.10	YO
80		3,129.93	II	80		4,224.25	I	50		4,879.65	I	90	b	5,939.08	YO
95		3,135.17	II	600		4,235.73	II	1,900		4,883.69	II	45		5,945.72	I
110		3,173.06	II	2,200		4,235.94	I	50		4,886.28	I	24		5,950.02	I
220		3,179.41	II	300		4,251.20	I	40		4,886.65	I	75	b	5,956.41	YO
0		3,191.31	I	360	h	4,302.30	I	95		4,893.44	I	1,300	b	5,972.04	YO
2,300		3,195.62	II	2,800		4,309.63	II	1,100		4,900.12	II	50		5,981.86	I
2,200		3,200.27	II	50		4,316.30	I	100		4,906.11	I	1,000	b	5,987.64	YO
2,200		3,203.32	II	110		4,330.78	I	45		4,909.00	I	740	b	6,003.60	YO
3,900		3,216.69	II	30		4,337.29	I	150		4,921.87	I	120		6,004.65	
6,200		3,242.28	II	60		4,344.65	I	35		4,930.93	I	120		6,009.19	I
310		3,280.91	II	440	h	4,348.79	I	45		4,950.66	I	620	b	6,019.87	YO
19		3,308.47	II	60		4,352.33	I	120		4,974.30	I	120		6,023.41	I
4,700		3,327.89	II	60		4,352.70	I	120		4,982.13	II	500	b	6,036.60	YO
55		3,340.38	I	120		4,357.73	I	100		5,006.97	I	420	b	6,053.81	YO
160		3,362.00	II	800		4,358.73	II	75		5,070.21	I	130	b	6,072.78	YO
85		3,388.59	I	120		4,366.03	I	75		5,072.19	I	50		6,088.00	I
45		3,397.04	I	12,000		4,374.94	II	1,100		5,087.42	II	210	b	6,089.35	YO
85		3,412.47	I	150	h	4,375.61	I	30		5,088.18	I	160	b	6,096.78	YO
200		3,448.82	II	80		4,379.33	I	210		5,119.11	II	130	b	6,107.82	YO
70		3,450.95	I	30		4,385.48	I	450		5,123.21	II	130	b	6,114.73	YO
110		3,467.88	II	100		4,387.74	I	180		5,135.20	I	75	b	6,127.38	YO
170		3,485.73	I	30		4,394.01	I	120		5,196.43	II	1,400	b	6,132.06	YO
1,700		3,496.09	II	30		4,394.67	I	960		5,200.41	II	120		6,135.04	I
80		3,521.53	I	1,800		4,398.02	II	1,500		5,205.72	II	150		6,138.43	I
45		3,546.01	II	890		4,422.59	II	180		5,240.81	I	1,100	b	6,148.36	YO
3,900		3,549.01	II	80		4,437.34	I	60		5,289.82	I	120		6,151.72	YO
130		3,551.80	I	100		4,443.66	I	45		5,320.78	II	820	b	6,165.08	YO
540		3,552.69	I	130		4,446.63	I	75		5,380.62	I	560	b	6,182.23	YO
170		3,558.76	I	20		4,465.27	I	220		5,402.78	II	1,200		6,191.73	I
190		3,571.43	I	40		4,473.89	I	24		5,417.03	I	590	b	6,199.82	YO
260		3,576.05	I	170		4,475.72	I	90		5,424.37	I	450	b	6,217.96	YO
3,300		3,584.52	II	180		4,476.96	I	190		5,438.24	I	300		6,222.59	I
300		3,587.75	I	160		4,477.45	I	710		5,466.46	I	270	b	6,236.72	YO
100		3,589.69	I	110		4,487.28	I	100		5,468.47	I	45		6,251.05	I
2,800		3,592.92	I	300		4,487.47	I	90		5,473.39	II	120	b	6,275.01	YO
10,000		3,600.73	II	30		4,491.75	I	90		5,480.74	II	60	b	6,295.46	YO
6,200		3,601.92	II	25		4,492.42	I	60		5,493.17	I	24	b	6,316.20	YO
7,800		3,611.05	II	500		4,505.95	I	35		5,495.59	I	24	b	6,338.10	YO
4,300		3,620.94	I	50		4,513.58	I	240		5,497.41	II	15	b	6,359.48	YO
1,900		3,628.71	II	80		4,514.01	I	300		5,503.45	I	15		6,369.87	YO
7,800		3,633.12	II	40	h	4,522.05	I	250		5,509.90	II	75		6,402.01	I
3,000		3,664.61	II	890		4,527.25	I	60		5,513.64	I	1,000		6,435.00	I
45		3,668.49	II	440		4,527.80	I	120		5,521.63	I	24		6,437.18	I
170		3,692.53	I	100		4,544.32	I	24		5,526.76	I	18	h	6,501.23	YO
13,000		3,710.30	II	100		4,559.37	I	740		5,527.54	I	18	h	6,518.33	YO
60		3,718.12	I	30		4,564.39	I	35		5,541.63	I	18	h	6,535.84	YO
60		3,738.61	I	60	h	4,573.56	I	120		5,544.50	I	90		6,538.60	I
1,200		3,747.55	II	35		4,581.32	I	90		5,546.02	II	12	h	6,553.84	YO
50		3,749.89	I	30		4,581.77	I	75		5,556.43	I	70		6,557.39	I
10,000		3,774.33	II	130		4,596.55	I	60		5,567.75	I	12	h	6,572.58	
1,400		3,776.56	II	95		4,604.80	I	180		5,577.42	I	35		6,576.85	I
50		3,782.30	II	40		4,613.00	I	24		5,581.08	I	23		6,584.87	I
7,400		3,788.70	II	2,000		4,643.70	I	620		5,581.87	I	95		6,613.75	II
1,300		3,818.35	II	200	h	4,658.32	I	21		5,590.96	I	14		6,622.49	I
4,000		3,832.88	II	70		4,658.89	I	21		5,594.12	I	19	h	6,636.49	I
70		3,847.87	II	85		4,667.47	I	120		5,606.33	I	40		6,650.61	I
80		3,876.82	I	60		4,670.82	I	15		5,623.91	I	21		6,664.40	I
480		3,878.28	II	2,000		4,674.84	I	560		5,630.13	I	150		6,687.58	I
30		3,887.77	I	60		4,678.35	I	24		5,632.25	I	14	h	6,691.83	I
60	h	3,904.59	I	260		4,682.32	II	21		5,632.89	I	7		6,694.75	I
50	h	3,918.25	I	85		4,689.77	I	120		5,644.69	I	16	h	6,699.26	I
60	h	3,930.11	I	180		4,696.81	I	120		5,648.47	I	70		6,700.71	I
240		3,930.66	II	35		4,708.85	I	740		5,662.94	II	35		6,713.20	I
4,400		3,950.36	II	60		4,725.85	I	90		5,675.27	I	40		6,735.99	I
150		3,951.60	II	170		4,728.53	I	18		5,693.63	I	190		6,793.71	I
60	h	3,955.09	I	60	h	4,732.37	I	160		5,706.73	I	70	h	6,795.41	II
3,600		3,982.60	II	85		4,741.40	I	24		5,720.61	I	12	h	6,803.15	I
40		3,987.50	I	160		4,752.79	I	75		5,728.89	II	21		6,815.16	I
940		4,039.83	I	410		4,760.98	I	150	b	5,730.12	YO	14		6,832.49	II
2,400		4,047.64	I	17		4,780.18	I	21		5,732.09	I	45		6,845.24	I
9,400		4,077.38	I	120		4,781.04	I	90		5,743.85	I	14		6,858.24	II
90	h	4,081.22	I	160		4,786.58	II	18	b	5,746.93	YO	29		6,887.22	I
2,000		4,083.71	I	170		4,786.89	I	24	b	5,764.22	YO	21		6,896.00	II
9,900		4,102.38	I	180		4,799.30	I	75		5,765.64	I	9		6,908.26	I
60	h	4,106.39	I	50		4,804.31	I	35		5,773.95	I	14		6,933.52	I
80		4,110.81	I	70		4,804.81	I	100		5,781.69	II	24	h	6,950.31	I
320		4,124.92	II	85	b	4,817.38	YO	15	b	5,800.00	YO	10		6,951.68	II
8,900		4,128.31	I	140	b	4,818.20	YO	15	b	5,818.58	YO	10		6,958.04	I
7,500		4,142.85	I	140		4,819.64	I	30		5,821.87	I	24		6,979.88	I
100	h	4,157.63	I	120		4,822.13	I	21		5,832.27	I	13	h	7,008.97	I
2,400		4,167.52	I	190		4,823.31	II	9	b	5,838.07	YO	10		7,009.93	I
2,000		4,174.14	I	60		4,839.15	I	15		5,858.83	YO	19	h	7,035.18	I
8,000		4,177.54	II	770		4,839.87	I	15		5,871.83	I	29		7,052.94	I
120		4,199.28	II	550		4,845.68	I	24		5,876.14	YO	13	h	7,054.28	I
380		4,204.70	II	410		4,852.69	I	24		5,879.96	I	9		7,075.13	I
80		4,213.02	I	120		4,854.25	I	24	b	5,893.94	YO	11		7,127.92	I
40		4,213.54	I	890		4,854.87	II	35		5,902.96	I	35		7,191.66	I
160		4,217.80	I	50		4,856.70	I	24	b	5,912.19	YO	10	h	7,195.93	I

Intensity		Wavelength	
35		7,264.17	II
9	h	7,293.08	I
9	h	7,330.62	I
5		7,332.96	II
50		7,346.46	I
11	h	7,398.77	I
29		7,450.30	II
17		7,494.88	I
7	h	7,536.71	I
35		7,563.13	I
8	h	7,617.72	I
19	h	7,622.94	I
7		7,652.89	I
5		7,689.49	I
8	h	7,698.00	I
19		7,719.89	I
19		7,724.08	I
13		7,788.42	I
13		7,796.32	I
6		7,802.52	I
17		7,812.16	I
29		7,855.52	I
110		7,881.90	II
10	h	7,999.33	I
9		8,329.61	I
24		8,344.43	I
8	h	8,365.64	I
17		8,450.36	I
8	h	8,528.94	I
95		8,800.62	I
19	h	8,835.85	II

Y III

Ref. 77 — J.R.

Intensity		Wavelength	
		Vacuum	
1		643.68	III
4		646.69	III
6		653.87	III
10		656.98	III
25		668.74	III
40		671.98	III
100		691.72	III
4		693.85	III
200		695.20	III
9		727.91	III
4		728.47	III
2		728.83	III
20		729.73	III
600		730.49	III
15		732.70	III
800		734.36	III
15		770.78	III
10		771.79	III
20		804.26	III
5,000		805.20	III
75		806.18	III
150		808.97	III
7,000		809.92	III
100		855.64	III
60		857.82	III
25		984.23	III
15		987.96	III
15,000		989.21	III
25,000		996.37	III
20		999.19	III
150		1,000.56	III
25		1,003.35	III
1,000		1,006.58	III
1,200		1,007.86	III
120		1,077.52	III
500		1,081.35	III
75	p	1,084.63	III
350	p	1,088.39	III
250		1,095.25	III
25		1,095.87	III
150		1,103.21	III
3,000		1,289.74	III
2,500		1,306.96	III
5,000		1,314.51	III
1,500		1,316.10	III
4,000		1,334.04	III
8		1,549.08	III
15	p	1,553.81	III
30		1,635.14	III
75		1,640.43	III
200		1,779.80	III
600		1,786.05	III
		Air	
10		2,041.93	III
5		2,042.07	III
1,500		2,060.58	III
4,000		2,068.98	III
10,000		2,127.98	III
16,000		2,191.16	III
8,000		2,200.76	III
8,000		2,206.03	III
150		2,261.41	III
80		2,261.57	III
10,000		2,284.34	III
3		2,319.92	III
10,000		2,327.31	III
50,000		2,367.23	III
40,000		2,414.64	III
100	p	2,710.30	III
90	h	2,710.54	III
5		2,780.11	III
70		2,791.44	III
20		2,803.27	III
100		2,807.00	III
90,000		2,817.04	III
6,000		2,867.67	III
6,000		2,913.41	III
1,500		2,917.74	III
1,600		2,918.56	III
15		2,940.53	III
99,000		2,946.01	III
20	p	2,948.48	III
6,000		2,970.42	III
1,400		3,013.93	III
1,500		3,018.85	III
3		3,267.10	III
25	l	3,276.80	III
500		3,866.96	III
3,000		3,900.74	III
4,000		3,914.58	III
3,800		4,039.60	III
3,000		4,040.11	III
120	h	4,121.61	III
2,000	c	4,737.62	III
7,500		5,102.88	III
1,300		5,120.40	III
10,000		5,238.10	III
3,000		5,263.58	III
4,000		5,383.64	III
6,000		5,562.81	III
600		5,567.27	III
4,000		5,572.24	III
400		5,595.48	III
3,000		5,602.08	III
2,000	h	7,254.58	III
9,000		7,558.71	III
6,000		7,864.53	III
8,000		7,916.71	III
400		7,989.41	III
10,000		7,991.43	III
8,000		8,171.41	III
4,000		8,645.09	III
10,000		8,796.21	III
8,000		9,116.59	III

ZINC (ZN)

Z = 30

Zn I and II

Ref. 39, 55, 113, 131, 185, 186

Intensity		Wavelength	
		Vacuum	
60		1,193.23	II
60		1,277.31	II
60	d	1,366.68	II
		1,404.12	I
60		1,410.44	II
60		1,439.09	II
60		1,445.04	II
50		1,456.91	II
		1,457.57	I
60		1,477.02	II
50		1,514.76	II
90		1,572.99	II
		1,589.57	I
60		1,617.68	II
60		1,658.25	II
50		1,713.25	II
60		1,715.76	II
80	d	1,735.61	II
60		1,736.89	II
50		1,737.90	II
75		1,747.12	II
80	c	1,762.19	II
75		1,774.04	II
80		1,790.76	II
100		1,797.64	II
100	d	1,811.05	II
80		1,816.48	II
80		1,831.38	II
100	d	1,833.57	II
70		1,836.01	II
75		1,836.65	II
75		1,847.56	II
100		1,864.12	II
100		1,866.08	II
100		1,872.13	II
75		1,894.26	II
60		1,901.52	II
60		1,914.81	II
100	d	1,918.96	II
70		1,920.27	II
100	d	1,929.67	II
60		1,945.58	II
60		1,951.91	II
80		1,953.00	II
75		1,954.87	II
80		1,964.54	II
100		1,969.40	II
100		1,982.11	II
70		1,985.61	II
100		1,986.99	II
50		1,993.37	II
50		1,996.92	II
		Air	
100		2,011.94	II
500		2,025.48	II
60		2,039.31	II
500		2,062.00	II
200		2,064.23	II
120		2,079.08	I
50		2,079.93	II
60		2,087.33	I
80		2,096.93	I
300		2,099.94	I
200		2,102.18	II
150		2,104.42	I
75		2,122.74	II
800	r	2,138.56	I
75		2,147.42	II
60		2,210.18	II
50		2,273.15	II
1,000		2,501.99	II
150		2,515.81	I
50		2,527.96	II
1,000		2,557.95	II
50		2,567.80	II
50		2,567.98	II
100	h	2,569.87	I
100		2,582.44	I
300		2,582.49	I
200		2,608.56	I
300		2,608.64	I
200		2,670.53	I
300		2,684.16	I
300		2,712.49	I
200		2,756.45	I
300		2,770.86	I
300		2,770.98	I
400		2,800.87	I
100		2,801.06	I
5		2,801.17	I
100		2,801.96	I
100		2,902.30	II
125		3,018.36	I
200		3,035.78	I
200		3,072.06	I
150		3,075.90	I
100		3,171.45	II
100		3,172.23	II
300		3,196.31	II
100		3,197.10	II
500	r	3,282.33	I
50		3,299.42	II
800		3,302.58	I
700	r	3,302.94	I
75		3,306.01	II
800		3,345.02	I
500		3,345.57	I
150		3,345.94	I
5		3,799.00	I
50		3,806.34	II
100		3,840.29	II
50		3,883.34	I
15		3,965.43	I
10		4,113.21	I
25		4,292.88	I
25		4,298.33	I
35		4,629.81	I
300		4,680.14	I
400		4,722.15	I
400		4,810.53	I
800		4,911.62	II
500		4,924.03	II
7		5,068.66	I
15		5,069.58	I
200		5,181.98	I
8		5,308.65	I
7		5,310.24	I
7		5,311.02	I
4		5,772.10	I
4		5,775.50	I
10		5,777.11	I
500		5,894.33	II
500		6,021.18	II
500		6,102.49	II
100		6,111.53	II
500		6,214.61	II
8		6,237.90	I
8		6,239.17	I
1,000	h	6,362.34	I
10		6,479.18	I
15		6,928.32	I
8		6,938.47	I
3		6,943.20	I
200		7,478.8	II
300		7,588.5	II
100		7,612.9	II
300		7,732.5	II
200		7,757.9	II
10		7,799.36	I
100		11,054.25	I
100		13,053.63	I
100		13,150.59	I
20		13,196.61	I
100		14,038.70	I
20		15,680.29	I
20		16,483.45	I
20		16,491.98	I
20		16,505.23	I
5		24,044.16	I
10		24,375.02	I

ZIRCONIUM (ZR)

Z = 40

Zr I and II

Ref. 1 — C.H.C.

Intensity		Wavelength	
		Air	
60		2,374.42	I
60		2,384.17	I
50		2,388.01	I
50		2,389.21	I
45		2,405.52	I
60		2,419.41	II
150		2,449.85	II
21		2,457.44	II
75		2,487.29	II
45		2,496.48	II
180		2,532.46	II
90		2,539.65	II
220		2,542.10	II
45		2,550.51	II
220		2,550.74	II
45		2,556.43	I
60		2,567.45	I
570		2,567.64	II
1,600		2,568.87	II
2,100		2,571.39	II
75		2,583.40	II

130		2,589.07	II	180		3,030.92	II	380		3,399.35	II	200		4,007.60	I
22		2,589.65	I	350	d	3,036.39	II	570		3,404.83	II	200		4,012.25	I
45		2,609.43	I	100		3,045.83	I	760		3,410.25	II	400		4,023.98	I
150		2,630.91	II	690		3,054.84	II	380		3,414.66	I	770		4,024.92	I
80		2,635.42	I	100		3,060.11	II	1,000		3,430.53	II	990		4,027.20	I
210		2,639.09	II	100		3,064.63	II	380		3,437.14	II	240		4,028.95	I
70		2,643.40	II	110		3,085.34	I	4,700		3,438.23	II	400		4,029.68	II
55		2,647.78	I	110		3,094.80	I	600		3,447.36	I	490		4,030.04	I
110		2,650.38	II	250		3,095.07	II	200		3,455.91	I	400		4,035.89	I
70		2,658.69	I	110		3,095.82	I	410		3,457.56	II	240		4,042.22	I
180		2,667.80	II	280		3,099.23	II	200		3,458.93	II	610		4,043.58	I
55		2,669.49	II	690		3,106.58	II	820		3,463.02	II	490		4,044.56	I
120		2,670.96	II	110		3,108.37	I	600		3,471.19	I	400		4,045.61	II
1,800		2,678.63	II	210		3,110.88	II	200		3,478.79	I	610		4,048.67	II
35		2,681.76	II	350		3,120.74	I	1,200		3,479.39	II	200		4,050.33	II
90		2,687.75	I	320		3,125.92	II	1,300		3,481.15	II	200		4,050.48	I
90		2,692.60	II	500		3,129.18	II	760		3,483.54	II	770		4,055.03	I
22		2,692.92	I	500		3,129.76	II	4,100		3,496.21	II	600		4,055.71	I
160		2,693.53	II	140		3,131.11	I	350		3,505.48	II	330		4,061.53	I
180		2,694.06	II	350		3,132.07	I	820		3,505.67	II	1,500		4,064.16	I
70		2,695.43	II	110		3,133.23	I	1,000		3,509.32	I	2,000		4,072.70	I
95		2,699.60	II	350		3,133.48	II	200		3,510.46	II	310		4,074.93	I
750		2,700.13	II	180		3,136.96	I	2,000		3,519.60	I	200		4,076.53	I
280		2,711.51	II	690		3,138.68	II	440		3,525.81	II	240		4,078.31	I
140		2,712.42	II	140		3,139.80	I	440		3,533.22	I	2,000		4,081.22	I
140		2,714.26	II	180		3,148.82	I	210		3,535.16	I	200		4,108.40	I
1,300		2,722.61	II	290		3,155.67	II	630		3,542.62	II	400		4,121.46	I
140		2,725.47	I	150		3,157.00	II	1,800		3,547.68	I	1,200		4,149.20	II
800		2,726.49	II	320		3,157.82	I	210		3,549.74	I	200		4,152.64	I
490		2,732.72	II	540		3,164.31	II	630		3,550.46	I	290		4,156.24	II
1,400		2,734.86	II	150		3,165.45	II	1,800		3,551.95	II	400		4,161.21	II
110		2,740.51	II	880		3,165.97	II	2,100		3,556.60	II	400		4,166.36	I
140		2,741.55	II	150		3,166.26	II	1,100		3,566.10	I	200		4,183.32	I
1,100		2,742.56	II	190		3,178.09	II	210		3,568.88	I	660		4,187.56	I
660		2,745.86	II	190		3,181.58	II	2,100		3,572.47	II	400		4,194.76	I
660		2,752.21	II	150		3,181.92	II	210		3,573.08	II	610		4,199.09	I
530		2,758.81	II	880		3,182.86	II	1,100		3,575.79	I	610		4,201.46	I
200	d	2,768.73	II	540		3,191.21	I	1,300		3,576.85	II	610		4,208.98	II
		2,768.85	II	210		3,191.90	II	880		3,586.29	I	200		4,211.88	II
170	d	2,774.04	I	540		3,212.01	I	440		3,587.98	II	400		4,213.86	I
		2,774.16	II	760		3,214.19	II	3,500		3,601.19	I	2,000		4,227.76	I
200		2,790.14	I	110		3,222.47	II	690		3,611.89	II	200		4,236.06	I
120		2,792.04	I	200		3,228.81	II	1,100		3,613.10	II	2,000		4,239.31	I
160		2,796.90	II	630		3,231.69	II	1,100		3,614.77	II	770		4,240.34	I
110		2,799.15	II	630		3,234.12	I	1,100		3,623.86	I	770		4,241.20	I
180		2,810.91	II	110		3,236.58	II	320		3,634.15	I	1,200		4,241.69	I
620		2,814.90	I	760		3,241.05	II	260		3,661.20	I	310		4,268.02	I
390		2,818.74	II	320		3,250.39	I	1,100		3,663.65	I	550		4,282.20	I
530		2,825.56	II	200		3,254.28	I	390		3,671.27	II	550		4,294.79	I
110		2,833.91	II	200		3,260.11	I	800		3,674.72	II	310		4,302.89	I
710		2,837.23	I	190		3,269.66	I	390		3,697.46	II	550		4,341.13	I
120		2,839.34	II	150		3,271.13	II	960		3,698.17	II	1,000		4,347.89	I
130		2,843.52	II	540		3,272.22	II	720		3,709.26	II	290		4,359.74	II
660		2,844.58	II	1,000		3,273.05	II	270		3,731.26	II	310		4,360.81	I
210		2,848.19	II	1,300		3,279.26	II	560		3,745.98	II	350		4,366.45	I
350		2,848.52	I	320	d	3,282.73	I	880		3,751.60	II	240		4,379.78	II
350		2,851.97	II	880		3,284.71	II	480		3,764.39	I	190		4,413.04	I
340		2,869.81	II	140		3,285.88	II	480		3,766.72	I	240		4,420.46	I
490		2,875.98	I	150		3,288.80	II	340		3,766.82	II	120		4,427.24	I
120		2,892.26	I	540		3,305.15	II	720		3,780.54	I	160		4,431.49	I
160		2,905.23	II	880		3,306.28	II	560		3,791.40	I	140		4,443.00	II
300		2,915.99	II	150		3,313.70	II	210		3,817.58	II	110		4,457.43	I
110		2,916.64	II	210		3,314.50	II	560		3,822.41	I	110		4,466.91	I
270		2,918.24	II	150		3,319.02	II	2,200		3,835.96	I	110		4,470.31	I
320		2,926.99	II	380		3,322.99	II	1,300		3,836.76	II	190		4,470.56	I
160		2,934.61	II	380		3,326.80	II	550		3,843.02	II	200		4,496.97	II
160		2,936.31	II	380		3,334.25	II	550		3,847.01	I	550		4,507.12	I
320		2,948.94	II	210		3,334.62	II	550		3,849.25	I	610		4,535.75	I
210		2,951.48	II	190		3,338.41	II	2,900		3,863.87	I	490		4,542.22	I
320		2,955.78	II	760		3,340.56	II	770		3,864.34	I	200		4,553.01	I
320		2,960.87	I	380		3,344.79	II	990		3,877.60	I	200		4,555.13	I
320		2,962.68	II	130		3,353.66	I	200		3,879.05	I	140		4,555.52	I
320		2,968.96	II	180		3,354.39	II	1,500		3,885.42	I	490		4,575.52	I
120		2,969.19	I	760		3,356.09	II	2,900		3,890.32	I	100		4,582.29	I
230		2,969.63	II	540		3,357.26	II	2,000		3,891.38	I	140		4,590.55	I
130		2,976.61	II	180		3,359.96	II	400		3,900.52	I	350		4,602.57	I
320		2,978.05	II	150		3,360.46	I	310		3,915.94	II	140		4,604.42	I
230		2,979.18	II	150		3,363.82	II	610		3,921.79	I	210		4,626.41	I
160		2,981.02	II	150		3,367.82	II	1,200		3,929.53	I	700		4,633.98	I
820		2,985.39	I	150		3,370.59	I	200		3,934.12	II	210		4,644.83	I
320		3,003.74	II	180		3,373.42	II	200		3,934.79	II	260		4,683.42	I
100		3,005.37	I	380		3,374.73	II	940		3,958.22	II	2,300		4,687.80	I
160		3,005.50	I	110		3,376.27	II	490		3,966.66	I	510		4,688.45	I
820		3,011.75	I	150		3,377.46	II	990		3,968.26	I	110		4,707.79	I
100		3,013.32	II	570		3,387.87	II	660		3,973.50	I	1,900		4,710.08	I
160		3,019.84	II	760		3,388.30	II	200		3,975.29	I	160		4,711.92	I
350		3,020.47	II	5,700		3,391.98	II	200	h	3,981.60	I	120		4,717.62	I
500		3,028.04	II	570		3,393.12	II	770		3,991.13	II	210		4,719.12	I
880		3,029.52	I	160		3,396.33	II	770		3,998.97	II	300		4,732.33	I

1,400		4,739.48	I
190		4,762.78	I
870		4,772.31	I
210		4,784.92	I
160		4,788.67	I
260		4,805.87	I
140		4,809.47	I
190		4,815.04	I
700		4,815.63	I
280		4,824.29	I
190		4,828.04	I
110		4,838.78	I
210		4,851.36	I
160		4,866.06	I
110		4,881.24	I
110		4,883.60	I
100		4,994.76	I
30		5,011.46	I
250		5,046.58	I
85		5,060.39	I
360		5,064.91	I
110		5,065.22	I
100		5,070.26	I
75		5,073.98	I
470		5,078.25	I
85		5,085.26	I
50		5,112.27	II
140		5,115.24	I
50		5,120.42	I
85		5,133.40	I
300		5,155.45	I
200		5,158.00	I
35		5,158.67	I
75		5,160.99	I
85		5,165.96	I
17		5,178.99	I
100		5,183.70	I
30		5,187.03	I
100		5,191.60	II
100		5,201.15	I
85		5,209.30	I
85		5,224.93	I
30		5,243.47	I
120		5,277.41	I
75		5,280.05	I
60		5,294.82	I
120		5,296.79	I
60		5,301.97	I
110		5,311.40	I
25		5,321.26	I
22		5,330.84	I
12		5,338.43	I
30		5,350.09	II
30		5,350.35	II
25		5,350.90	I
25		5,351.92	I
75		5,362.56	I
12		5,363.35	I
17		5,369.39	I
20		5,382.37	I
270		5,385.14	I
30		5,386.65	I
17		5,391.18	I
17		5,395.88	I
25		5,405.13	I
85		5,407.62	I
17		5,413.93	I
20		5,421.86	I
15		5,426.36	I
25		5,428.42	I
25		5,437.76	I
15		5,440.41	I
35		5,448.57	I
10		5,474.92	I
10		5,477.40	I
35		5,478.33	I
35		5,480.83	I
10		5,481.16	I
30		5,486.09	I
140		5,502.12	I
25		5,507.87	I
30		5,517.11	I
10		5,518.05	I
75		5,528.41	I
20		5,532.30	I
45		5,537.46	I
50		5,545.32	I
22	b	5,551.75	ZRO
25	b	5,553.17	ZRO
12		5,612.11	I
120		5,620.14	I
35		5,623.53	I
25	b	5,629.02	ZRO
25	b	5,629.58	ZRO
160		5,664.51	I
20		5,666.28	I
120		5,680.90	I
15		5,685.42	I
30		5,708.89	I
75	b	5,718.21	ZRO
120		5,735.70	I
35	b	5,748.17	ZRO
17	b	5,778.57	ZRO
160		5,797.74	I
30		5,847.32	I
50		5,868.27	I
110		5,869.50	I
340		5,879.80	I
85		5,885.62	I
50		5,901.09	I
30	b	5,908.61	ZRO
140		5,925.13	I
100		5,935.20	I
110		5,955.35	I
30	b	5,977.80	ZRO
100		5,984.23	I
17		5,995.37	I
50		6,001.05	I
30		6,025.36	I
85		6,032.61	I
170		6,045.85	I
100		6,049.24	I
140		6,062.84	I
50		6,120.83	I
170		6,121.91	I
85		6,124.84	I
680		6,127.44	I
340		6,134.55	I
100		6,140.46	I
440		6,143.20	I
30		6,155.61	I
75		6,157.71	I
25		6,160.20	I
35		6,189.40	I
60		6,192.96	I
85		6,213.05	I
100		6,214.69	I
170	b	6,226.51	ZRO
100		6,257.26	I
50	b	6,261.05	ZRO
35		6,267.06	I
45	b	6,292.84	ZRO
120		6,299.66	I
15		6,304.34	I
300		6,313.02	I
30		6,314.71	I
50		6,321.35	I
22		6,340.36	I
50	b	6,345.10	ZRO
75		6,345.22	I
75	b	6,378.56	ZRO
35		6,407.00	I
50	b	6,412.39	ZRO
12		6,426.17	I
35		6,434.33	I
60		6,445.74	I
20		6,451.62	I
20		6,457.63	I
110		6,470.21	I
60	b	6,473.79	ZRO
11		6,484.35	I
110		6,489.64	I
22		6,493.10	I
50		6,503.26	I
50		6,506.36	I
50	b	6,508.15	ZRO
30	b	6,542.90	ZRO
35		6,550.54	I
30		6,569.43	I
20		6,576.56	I
30	b	6,578.06	ZRO
50		6,591.99	I
10		6,596.71	I
10		6,598.84	I
50		6,603.27	I
15		6,620.56	I
11		6,678.01	II
22		6,688.18	I
11		6,702.12	I
17		6,709.61	I
27		6,717.88	I
40		6,752.73	I
75		6,762.38	I
85		6,769.16	I
27		6,772.89	I
15		6,787.15	II
35		6,790.85	I
45		6,828.78	I
45		6,832.89	I
13		6,845.33	I
17		6,846.34	I
100		6,846.97	I
27		6,849.26	I
13		6,852.56	I
120		6,888.29	I
29		6,900.59	I
20		6,904.36	I
29		6,907.37	I
20		6,916.87	I
16		6,932.38	I
29		6,948.46	I
150		6,953.84	I
60		6,966.44	I
10		6,975.91	I
150		6,990.84	I
80		6,994.32	I
10		7,005.46	I
100		7,027.40	I
25		7,057.36	I
14		7,057.96	I
140		7,087.30	I
25		7,089.43	I
35		7,094.46	I
50		7,095.59	I
540		7,097.70	I
280		7,102.91	I
170		7,103.72	I
140		7,111.68	I
40		7,112.82	I
18		7,113.52	I
12		7,132.95	I
16		7,140.74	I
12		7,144.47	I
590		7,169.09	I
50		7,201.62	I
12		7,258.17	I
35		7,264.76	I
20		7,306.21	I
25		7,311.62	I
35		7,313.72	I
90		7,318.08	I
10		7,327.82	I
50		7,335.97	I
50		7,343.96	I
20		7,373.50	I
25		7,383.63	I
14		7,400.90	I
10		7,411.39	I
10		7,422.75	I
10		7,433.10	I
110		7,439.86	I
18		7,467.57	I
16		7,479.58	I
14	h	7,515.70	I
12		7,517.95	I
20	h	7,540.62	I
20		7,544.59	I
29		7,551.46	I
40		7,554.70	I
25		7,558.45	I
12		7,560.09	I
12		7,562.12	I
80		7,607.15	I
14		7,612.08	I
20		7,621.17	I
29		7,658.60	I
18		7,690.83	I
14		7,704.27	I
10		7,708.42	I
10		7,766.55	I
12		7,816.32	I
110		7,819.35	I
35		7,822.94	I
40		7,826.72	I
90		7,849.35	I
35		7,869.99	I
14		7,876.25	I
16		7,882.18	I
10		7,897.98	I
16		7,908.46	I
20		7,940.47	I
160		7,944.61	I
80		7,956.66	I
80		7,959.98	I
20		7,963.63	I
160		8,005.27	I
25		8,046.05	I
16		8,053.06	I
20		8,055.29	I
20		8,055.76	I
60		8,058.08	I
150		8,063.09	I
790		8,070.08	I
10		8,114.28	I
20		8,120.17	I
390		8,132.99	I
20		8,152.58	I
12		8,188.77	I
40		8,194.73	I
60		8,201.73	I
280		8,212.53	I
20		8,240.37	I
40		8,283.81	I
140		8,305.90	I
14		8,332.44	I
50		8,370.23	I
120		8,389.41	I
70		8,414.00	I
50		8,453.17	I
50		8,464.65	I
40		8,498.44	I
18		8,584.21	I
10		8,734.86	I
12		8,749.48	I
10		8,786.23	I
16		8,804.98	I
70		8,836.09	I
60		8,899.52	I

REFERENCES

1. **Meggers, W. F., Corliss, C. H., and Scribner, B. F.,** *Natl. Bur. Stand. (U.S.) Monogr.*, 145, Washington, D.C., 1975.
2. **Aksenov, V. P. and Ryabtsev, A. N.,** *Opt. Spectrosc.*, 37, 860, 1970.
3. **Andersen, N., Bickel, W. S., Carriveau, G. W., Jensen, K., and Veje, E.,** *Phys. Scr.*, 4, 113, 1971.
4. **Andersson, E. and Johannesson, G. A.,** *Phys. Scr.*, 3, 203, 1971.
5. **Andrew, K. L. and Meissner, K. W.,** *J. Opt. Soc. Am.*, 49, 146, 1959.
6. **Artru, M. C. and Brillet, W. U. L.,** *J. Opt. Soc. Am.*, 64, 1063, 1974.
7. **Artru, M. C. and Kaufman, V.,** *J. Opt. Soc. Am.*, 62, 949, 1972.
8. **Artru, M. C. and Kaufman, V.,** *J. Opt. Soc. Am.*, 65, 594, 1975.
9. **Badami, J. S. and Rao, K. R.,** *Proc. R. Soc. London*, 140(A), 387, 1933.
10. **Baird, K. M. and Smith, D. S.,** *J. Opt. Soc. Am.*, 48, 300, 1958.
11. **Bashkin, S. and Martinson, I.,** *J. Opt. Soc. Am.*, 61, 1686, 1971.
12. **Beacham, J. R.,** Ph.D. thesis, Purdue University, 1970.
13. **Benschop, H., Joshi, Y. N., and van Kleef, T. A. M.,** *Can. J. Phys.*, 53, 700, 1975.
14. **Berry, H. G., Bromander, J., and Buchta, R.,** *Phys. Scr.*, 1, 181, 1970.
15. **Berry, H. G., Bromander, J., Martinson, I., and Buchta, R.,** *Phys. Scr.*, 3, 63, 1971.
16. **Berry, H. G., Desesquelles, J., and Dufay, M.,** *Phys. Rev. Sect. A*.6, 600, 1972.
17. **Berry, H. G., Desesquelles, J., and Dufay, M.,** *Nucl. Instrum. Methods*, 110, 43, 1973.
18. **Berry, H. G., Pinnington, E. H., and Subtil, J. L.,** *J. Opt. Soc. Am.*, 62, 767, 1972.
19. **Bidelman, W. P. and Corliss, C. H.,** *Astrophys. J.*, 135, 968, 1962.
20. **Bloch, L. and Bloch, E.,** *Ann. Phys.* (Paris), 10(11), 141, 1929.
21. **Bloch, L., Bloch, E., and Felici, N.,** *J. Phys. Radium*, 8, 355, 1937.
22. **Bockasten, K.,** *Ark. Fys.*, 9, 457, 1955.
23. **Bockasten, K., Hallin, R., Johansson, K. B., and Tsui, P.,** *Phys. Lett.* (Netherlands), 8, 181, 1964.
24. **Bockasten, K. and Johansson, K. B.,** *Ark. Fys.*, 38, 563, 1969.
25. **Borgstrom, A.,** *Ark. Fys.*, 38, 243, 1968.
26. **Borgstrom, A.,** *Phys. Scr.*, 3, 157, 1971.
27. **Bowen, I. S.,** *Phys. Rev.*, 29, 231, 1927.
28. **Bowen, I. S.,** *Phys. Rev.*, 31, 34, 1928.
29. **Bowen, I. S.,** *Phys. Rev.*, 39, 8, 1932.
30. **Bowen, I. S.,** *Phys. Rev.*, 45, 401, 1934.
31. **Bowen, I. S.,** *Phys. Rev.*, 46, 377, 1934.
32. **Bowen, I. S.,** *Phys. Rev.*, 46, 791, 1934.
33. **Boyce, J. C.,** *Phys. Rev.*, 49, 730, 1936.
34. **Boyce, J. C. and Robinson, H. A.,** *J. Opt. Soc. Am.*, 26, 133, 1936.
35. **Bozman, W. R., Meggers, W. F., and Corliss, C. H.,** *J. Res. Natl. Bur. Stand. Sect. A*, 71, 547, 1967.
36. **Bromander, J.,** *Ark. Fys.*, 40, 257, 1969.
37. **Bromander, J. and Buchta, R.,** *Phys. Scr.*, 1, 184, 1970.
38. **Brown, M. and Ginter, M. L.,** *J. Opt. Soc. Am.*, to be published.
39. **Brown, C. M., Tilford, S. G., and Ginter, M. L.,** *J. Opt. Soc. Am.*, 65, 1404, 1975.
40. **Bryant, B. W.,** *Johns Hopkins Spectroscopic Report* No. 21, 1961.
41. **Buchet, J. P., Buchet-Poulizac, M. C., Berry, H. G., and Drake, G. W. F.,** *Phys. Rev. Sect. A*, 7, 922, 1973.
42. **Budhiraja, C. J. and Joshi, Y. N.,** *Can. J. Phys.*, 49, 391, 1971.
43. **Burns, K. and Adams, K. B.,** *J. Opt. Soc. Am.*, 42, 56, 1952.
44. **Burns, K. and Adams, K. B.,** *J. Opt. Soc. Am.*, 46, 94, 1956.
45. **Burns, K., Adams, K. B., and Longwell, J.,** *J. Opt. Soc. Am.*, 40, 339, 1950.
46. **Callahan, W. R.,** Ph.D. thesis, Johns Hopkins University, 1962.
47. **Charles, G. W.,** *J. Opt. Soc. Am.*, 56, 1292, 1966.
48. **Charles, G. W., Hunt, D. J., Pish, G., and Timma, D. L.,** *J. Opt. Soc. Am.*, 45, 869, 1955.
49. **Codling, K.,** *Proc. Phys. Soc.*, 77, 797, 1961.
50. **Comite Consulatif Pour La Definition du Metre,** *J. Phys. Chem. Ref. Data*, 3, 852, 1974.
51. **Conway, J. G., Blaise, J., and Verges, J.,** *Spectrochim. Acta Part B*, 31, 31, 1976.
52. **Conway, J. G., Worden, E. F., Blaise, J., and Verges, J.,** *Spectrochim. Acta Part B*, 32, 97, 1977.
53. **Conway, J. G., Worden, E. F., Blaise, J., Camus, P., and Verges, J.,** *Spectrochim. Acta Part B*, 32, 101, 1977.
54. **Crooker, A. M.,** *Can. J. Res. Sect. A*, 14, 115, 1936.
55. **Crooker, A. M. and Dick, K. A.,** *Can. J. Phys.*, 46, 1241, 1968.
56. **Crosswhite, H. M.,** *J. Res. Natl. Bur. Stand. Sect. A*, 79, 17, 1975.
58. **Crosswhite, H. M. and Dieke, G. H.,** *American Institute of Physics Handbook*, Section 7, 1972.

59. de Bruin, T. L., *Z. Phys.*, 38, 94, 1926.
60. de Bruin, T. L., *Z. Phys.*, 53, 658, 1929.
61. de Bruin, T. L., Humphreys, C. J., and Meggers, W. F., *J. Res. Natl. Bur. Stand.*, 11, 409, 1933.
62. Dick, K. A., *J. Opt. Soc. Am.*, 64, 702, 1973.
63. Dobbie, J. C., *Ann. Solar Phys. Observ.* (Cambridge), 5, 1, 1938.
64. Earls, L. T. and Sawyer, R. A., *Phys. Rev.*, 47, 115, 1935.
65. Edlen, B., *Z. Phys.*, 85, 85, 1933.
66. Edlen, B., *Nova Acta Reglae Soc. Sci. Ups.*, (IV) 9, No. 6, 1934.
67. Edlen, B., *Z. Phys.*, 93, 726, 1935.
68. Edlen, B., *Z. Phys.*, 94, 47, 1935.
69. Edlen, B., *Rep. Prog. Phys.*, 26, 181, 1963.
70. Edlen, B. and Risberg, P., *Ark. Fys.*, 10, 553, 1956.
71. Edlen, B. and Swings, P., *Astrophys. J.*, 95, 532, 1942.
72. Ehrhardt, J. C. and Davis, S. P., *J. Opt. Soc. Am.*, 61, 1342, 1971.
73. Eidelsberg, M., *J. Phys. B*, 5, 1031, 1972.
74. Eidelsberg, M., *J. Phys. B*, 7, 1476, 1974.
75. Ekberg, J. O. and Svensson, L. A., *Phys. Scr.*, 2, 283, 1970.
76. Ekefors, E., *Z. Phys.*, 71, 53, 1931.
77. Epstein, G. L. and Reader, J., *J. Opt. Soc. Am.*, 65, 310, 1975.
78. Epstein, G. L. and Reader, J., *J. Opt. Soc. Am.*, 66, 590, 1976.
79. Epstein, G. L. and Reader, J., unpublished.
80. Eriksson, K. B. S., *Phys. Lett. A.*, 41, 97, 1972.
81. Eriksson, K. B. S. and Isberg, H. B. S., *Ark. Fys.*, 23, 527, 1963.
82. Eriksson, K. B. S. and Wenaker, I., *Phys. Scr.*, 1, 21, 1970.
83. Esteva, J. M. and Mehlman, G., *Astrophys. J.*, 193, 747, 1974.
84. Even-Zohar, M. and Fraenkel, B. S., *J. Phys. B*, 5, 1596, 1972.
85. Fawcett, B. C., *J. Phys. B*, 3, 1732, 1970.
86. Fawcett, B. C., Culham Laboratory Report ARU-R4, 1971.
87. Ferner, E., *Ark. Mat. Astron. Fys.*, 28(A), 4, 1941.
88. Fischer, R. A., Knopf, W. C., and Kinney, F. E., *Astrophys. J.*, 130, 683, 1959.
89. Fowler, A., *Report on Series in Line Spectra*, Fleetway Press, London, 1922.
90. Fowles, G. R., *J. Opt. Soc. Am.*, 44, 760, 1954.
91. Fred, M., unpublished, 1977.
92. Fred, M. and Tomkins, F. S., *J. Opt. Soc. Am.*, 47, 1076, 1957.
93. Fred, M., Tomkins, F. S., Blaise, J. E., Camus, P., and Verges, J., Argonne National Laboratory Report No. 76-68, 1976.
94. Garcia, J. D. and Mack, J. E., *J. Opt. Soc. Am.*, 55, 654, 1965.
96. Giacchetti, A., unpublished, 1975.
97. Giacchetti, A., Blaise, J., Corliss, C. H., and Zalubas, R., *J. Res. Natl. Bur. Stand. Sect. A*, 78, 247, 1974.
98. Giacchetti, A., Stanley, R. W., and Zalubas, R., *J. Opt. Soc. Am.*, 69, 474, 1970.
99. Gilbert, W. P., *Phys. Rev.*, 47, 847, 1935.
100. Gilroy, H. T., *Phys. Rev.*, 38, 2217, 1931.
101. Glad, S., *Ark. Fys.*, 10, 291, 1956.
102. Goldsmith, S., *J. Phys. B*, 2, 1075, 1969.
103. Goorvitch, D., Mehlman-Balloffet, G., and Valero, F. P. J., *J. Opt. Soc. Am.*, 60, 1458, 1970.
104. Goorvitch, D. and Valero, F. P. J., *Astrophys. J.*, 171, 643, 1972.
105. Green, L. C., *Phys. Rev.*, 55, 1209, 1939.
106. Gutman, F., *Diss. Abstr. Int. B*, 31, 363, 1970.
107. Hallin, R., *Ark. Fys.*, 31, 511, 1966.
108. Hallin, R., *Ark. Fys.*, 32, 201, 1966.
109. Hansen, J. E. and Persson, W., *J. Opt. Soc. Am.*, 64, 696, 1974.
110. Hansen, J. E. and Persson, W., *Phys. Scr.*, 13, 166, 1976.
111. Hellintin, P., *Phys. Scr.*, 13, 155, 1976.
112. Herzberg, G. and Moore, H. R., *Can. J. Phys.*, 37, 1293, 1959.
113. Hetzler, C. W., Boreman, R, W., and Burns, K., *Phys. Rev.*, 48, 656, 1935.
114. Holmstrom, J. E. and Johansson, L., *Ark. Fys.*, 40, 133, 1969.
115. Hontzeas, S., Martinson, I., Erman, P., and Buchta, R., *Nucl. Instrum. Methods*, 110, 51, 1973.
116. Humphreys, C. J., *J. Res. Natl. Bur. Stand.*, 22, 19, 1939.
117. Humphreys, C. J., *J. Opt. Soc. Am.*, 43, 1027, 1953.
118. Humphreys, C. J., *J. Phys. Chem. Ref. Data*, 2, 519, 1973.
119. Humphreys, C. J. and Andrew, K. L., *J. Opt. Soc. Am.*, 54, 1134, 1964.
120. Humphreys, C. J. and Meggers, W. F., *J. Res. Natl. Bur. Stand.*, 10, 139, 1933.
121. Humphreys, C. J. and Paul, E., *J. Opt. Soc. Am.*, 60, 200, 1970.
122. Humphreys, C. J. and Paul, E., *J. Opt. Soc. Am.*, 62, 432, 1972.

123. **Humphreys, C. J., Paul, E., Cowan, R. D., and Andrew, K. L.**, *J. Opt. Soc. Am.*, 57, 855, 1967.
124. **Humphreys, C. J., Paul, E., and Minnhagen, L.**, *J. Opt. Soc. Am.*, 61, 110, 1971.
125. **Iglesias, L.**, unpublished, 1977.
126. **Iglesias, L. and Velasco, R.**, *Publ. Inst. Opt. Madrid,* No. 23, 1964.
127. **Isberg, B.**, *Ark. Fys.*, 35, 551, 1967.
128. **Johannesson, G. A., Lundstrom, T., and Minnhagen, L.**, *Phys. Scr.*, 6, 129, 1972.
129. **Johannesson, G. A. and Lundstrom, T.**, *Phys. Scr.*, 8, 53, 1973.
130. **Johansson, I.**, *Ark. Fys.*, 20, 135, 1961.
131. **Johansson, I. and Contreras, R.**, *Ark. Fys.*, 37, 513, 1968.
132. **Johansson, I. and Litzen, U.**, *Ark. Fys.*, 34, 573, 1967.
133. **Johansson, I. and Svensson, K. F.**, *Ark. Fys.*, 16, 353, 1960.
134. **Johansson, L.**, *Ark. Fys.*, 20, 489, 1961.
135. **Johansson, L.**, *Ark. Fys.*, 23, 119, 1963.
136. **Johansson, S. and Litzen, U.**, *Phys. Scr.*, 6, 139, 1972.
137. **Johansson, S. and Litzen, U.**, *Phys. Scr.*, 8, 43, 1973.
138. **Johansson, S. and Litzen, U.**, *Phys. Scr.*, 10, 121, 1974.
139. **Joshi, Y. N.**, unpublished.
140. **Joshi, Y. N., Bhatia, K. S., and Jones, W. E.**, *Sci. Light Tokyo,* 21, 113, 1972.
141. **Joshi, Y. N., Bhatia, K. S., and Jones, W. E.**, *Spectrochim. Acta Part B,* 28, 149, 1973.
142. **Joshi, Y. N. and Budhiraja, C. J.**, *Can. J. Phys.*, 49, 670, 1971.
143. **Joshi, Y. N. and van Kleef, T. A. M.**, *Can. J. Phys.*, 52, 1891, 1974.
144. **Kaufman, V.**, unpublished.
145. **Kaufman, V.**, *J. Opt. Soc. Am.*, 52, 866, 1962.
146. **Kaufman, V., Artru, M. C., and Brillet, W. U. L.**, *J. Opt. Soc. Am.*, 64, 197, 1974.
147. **Kaufman, V. and Humphreys, C. J.**, *J. Opt. Soc. Am.*, 59, 1614, 1969.
148. **Kaufman, V. and Sugar, J.**, *J. Opt. Soc. Am.*, 61, 1693, 1971.
149. **Kaufman, V. and Sugar, J.**, *J. Res. Natl. Bur. Stand. Sect. A,* 71, 583, 1967.
150. **Kelly, R. L. and Palumbo, L. J.**, *National Research Laboratory Report 7599, Washington, D.C.*, 1973.
151. **Kielkopf, J. F.**, unpublished, 1975.
152. **Kielkopf, J. F.**, unpublished, 1976.
153. **Kiess, C. C. and Corliss, C. H.**, *J. Res. Natl. Bur. Stand. Sect. A,* 63, 1, 1959.
154. **Kleiman, H.**, *J. Opt. Soc. Am.*, 52, 441, 1962.
155. **Kleiman, H. and Meissner, K. W.**, *Office of Naval Research, O.N.R. Technical Report I NP-8045,* 1959.
156. **Klinkenberg, P. F. A.**, *Physica,* 15, 774, 1949.
157. **Klinkenberg, P. F. A.**, *Physica,* 16, 618, 1950.
158. **Krishnamurty, S. G.**, *Proc. Phys. Soc. London,* 48, 277, 1936.
159. **Kruger, P. G. and Gilroy, H. T.**, *Phys. Rev.*, 48, 720, 1935.
160. **Kruger, P. G. and Pattin, H. S.**, *Phys. Rev.*, 52, 621, 1937.
161. **Lacroute, P.**, *Ann. Phys.* (Paris), 3, 5, 1935.
162. **Lang, R. J.**, *Phys. Rev.*, 30, 762, 1927.
163. **Lang, R. J.**, *Phys. Rev.*, 32, 737, 1928.
164. **Lang, R. J.**, *Phys. Rev.*, 35, 445, 1930.
165. **Lang, R. J.**, *Can. J. Res. Sect. A,* 14, 43, 1936.
166. **Lang, R. J.**, *Can. J. Res. Sect. A,* 14, 127, 1936.
167. **Lang, R. J. and Vestine, E. H.**, *Phys. Rev.*, 42, 233, 1932.
168. **Li, H. and Andrew, K. L.**, *J. Opt. Soc. Am.*, 61, 96, 1971.
169. **Liden, K.**, *Ark. Fys.*, 1, 229, 1949.
170. **Litzen, U.**, *Ark. Fys.*, 28, 239, 1965.
171. **Litzen, U.**, *Phys. Scr.*, 1, 251, 1970.
172. **Litzen, U.**, *Phys. Scr.*, 1, 253, 1970.
173. **Litzen, U.**, *Phys. Scr.*, 2, 103, 1970.
174. **Litzen, U. and Verges, J.**, *Phys. Scr.*, 13, 240, 1976.
175. **Lofstrand, B.**, *Phys. Scr.*, 8, 57, 1973.
176. **Luc-Koenig, E., Morillon, C., and Verges, J.**, *Phys. Scr.*, 12, 199, 1975.
177. **Lundstrom, T.**, *Phys. Scr.*, 7, 62, 1973.
178. **Lundstrom, T. and Minnhagen, L.**, *Phys. Scr.*, 5, 243, 1972.
179. **Magnusson, C. E. and Zetterberg, P. O.**, *Phys. Scr.*, 10, 177, 1974.
180. **Magnusson, C. E. and Zetterberg, P. O.**, *Phys. Scr.*, 15, 237, 1977.
181. **Martin, D. C.**, *Phys. Rev.*, 48, 938, 1935.
182. **Martin, W. C.**, *J. Opt. Soc. Am.*, 49, 1071, 1959.
183. **Martin, W. C.**, *J. Res. Natl. Bur. Stand. Sect. A,* 64, 19, 1960.

184. Martin, W. C. and Corliss, C. H., *J. Res. Natl. Bur. Stand. Sect. A*, 64, 443, 1960.
185. Martin, W. C. and Kaufman, V., *J. Res. Natl. Bur. Stand. Sect. A*, 74, 11, 1970.
186. Martin, W. C. and Kaufman, V., *J. Opt. Soc. Am.*, 60, 1096, 1970.
187. McCormick, W. W. and Sawyer, R. A., *Phys. Rev.*, 54, 71, 1938.
188. McLaughlin, R., *J. Opt. Soc. Am.*, 54, 965, 1964.
189. McLennan, J. C., McLay, A. B., and Crawford, M. F., *Proc. R. Soc. London Ser. A*, 134, 41, 1931.
190. Meissner, K. W., *Z. Phys.*, 39, 172, 1926.
191. Meggers, W. F., *J. Res. Natl. Bur. Stand.*, 24, 153, 1940.
192. Meggers, W. F. and Corliss, C. H., *J. Res. Natl. Bur. Stand. Sect. A*, 70, 63, 1966.
193. Meggers, W. F., Fred, M., and Tomkins, F. S., *J. Res. Natl. Bur. Stand.*, 58, 295, 1957.
194. Meggers, W. F. and Humphreys, C. J., *J. Res. Natl. Bur. Stand.*, 28, 463, 1942.
195. Meggers, W. F. and Murphy, R. J., *J. Res. Natl. Bur. Stand.*, 48, 334, 1952.
196. Meggers, W. F., Scribner, B. F., and Bozman, W. R., *J. Res. Natl. Bur. Stand.*, 46, 85, 1951.
197. Meggers, W. F., Shenstone, A. G., and Moore, C. E., *J. Res. Natl. Bur. Stand.*, 45, 346, 1950.
198. Mehlman, G. and Esteva, J. M., *Astrophys. J.*, 188, 191, 1974.
199. Meinders, E., *Physica*, 84(C), 117, 1976.
200. Meissner, K. W., *Ann. Phys.*, 50, 713, 1916.
201. Meissner, K. W., *Ann. Phys.*, 65, 378, 1921.
202. Millikan, R. A. and Bowen, I. S. *Phys. Rev.*, 25, 600, 1925.
203. Minnhagen, L., *J. Opt. Soc. Am.*, 61, 1257, 1925.
204. Minnhagen, L., *J. Opt. Soc. Am.*, 63, 1185, 1973.
205. Minnhagen, L., *Phys. Scr.*, 11, 38, 1975.
206. Minnhagen, L., *J. Opt. Soc. Am.*, 66, 659, 1976.
207. Minnhagen, L. and Nietsche, H., *Phys. Scr.*, 5, 237, 1972.
208. Minnhagen, L., Strihed, H., and Petersson, B., *Ark. Fys.*, 39, 471, 1969.
209. Moore, C. E., *Natl. Bur. Standards U.S. Circ.*, I, 488, 1950.
210. Moore, C. E., *Revised Multiplet Table*, Princeton University Observatory No. 20, 1945.
211. Moore, C. E., National Standard Reference Data Series — National Bureau of Standards 3, Sect. 3, 1970.
212. Moore, C. E., National Standard Reference Data Series — National Bureau of Standards 3, Sect. 4, 1971.
213. Moore, C. E., National Standard Reference Data Series — National Bureau of Standards 3, Sect. 5, 1975.
214. Moore, C. E., National Standard Reference Data Series — National Bureau of Standards 3, Sect. 6, 1972.
215. Moore, C. E., *National Standard Reference Data Series — National Bureau of Standards 3, Sect. 7*, 1975.
216. Morillon, C. and Verges, J., *Phys. Scr.*, 10, 227, 1974.
217. Newsom, G. H., *Astrophys. J.*, 166, 243, 1971.
218. Newsom, G. H., O'Connor, S., and Learner, R. C. M., *J. Phys. B*, 6, 2162, 1973.
219. Norlen, G., *Phys. Scr.*, 8, 249, 1973.
220. Odabasi, H., *J. Opt. Soc. Am.*, 57, 1459, 1967.
221. Olme, A., *Ark. Fys.*, 40, 35, 1969.
222. Olme, A., *Phys. Scr.*, 1, 256, 1970.
223. Olthoff, J. and Sawyer, R. A., *Phys. Rev.*, 42, 766, 1932.
224. Palenius, H. P., *Ark. Fys.*, 39, 15, 1969.
225. Palenius, H. P., *Phys. Scr.*, 1, 113, 1970.
226.
227. Paschen, F., *Ann. Phys.*, Series 5, 12, 509, 1932.
228. Paschen, F. and Ritschl, R., *Ann. Phys.*, Series 5, 18, 867, 1933.
229. Peck, E. R., Khanna, B. N., and Anderholm, N. C. *J. Opt. Soc. Am.*, 52, 53, 1962.
230. Persson, W., *Phys. Scr.*, 3, 133, 1971.
231. Persson, W. and Valind S., *Phys. Scr.*, 5, 187, 1972.
232. Petersson, B., *Ark. Fys.*, 27, 317, 1964.
233. Phillips, L. W. and Parker, W. L., *Phys. Rev.*, 60, 301, 1941.
234. Platt, J. R. and Sawyer, R. A., *Phys. Rev.*, 60, 866, 1941.
235. Plyer, E. K., Blaine, L. R., and Tidwell, E., *J. Res. Natl. Bur. Stand.*, 55, 279, 1955.
236. Poppe, R., van Kleef, T. A. M., and Raassen, A. J. J., *Physica*, 77, 165, 1974.
237. Radziemski, L. J., Jr. and Andrew, K. L., *J. Opt. Soc. Am.*, 55, 474, 1965.
238. Radziemski, L. J., Jr. and Kaufman, V., *J. Opt. Soc. Am.*, 59, 424, 1969.
239. Radziemski, L. J., Jr. and Kaufman, V., *J. Opt. Soc. Am.*, 64, 366, 1964.
240. Ramanadham, R. and Rao, K. R., *Indian J. Phys.*, 18, 317, 1944.

241. Ramb, R., *Ann. Phys.*, 10, 311, 1931.
242. Rank, D. H., Bennet, J. M., and Bennet, H. E., *J. Opt. Soc. Am.*, 40, 477, 1956.
243. Rao, A. S. and Krishnamurty, S. G., *Proc. Phys. Soc. London*, 46, 531, 1943.
244. Rao, K. R., *Proc. R. Soc. London, Ser. A*, 134, 604, 1932.
245. Rao, K. R. and Badami, J. S., *Proc. R. Soc. London Ser. A*, 131, 154, 1931.
246. Rao, K. R. and Krishnamurty, S. G., *Proc. R. Soc. London Ser. A*, 161, 38, 1937.
247. Rao, K. R. and Murti, S. G. K., *Proc. R. Soc. London Ser. A*, 145, 681, 1934.
248. Rao, Y. B., *Indian J. Phys.*, 32, 497, 1958.
249. Rao, Y. B., *Indian J. Phys.*, 33, 546, 1959.
250. Rao, Y. B., *Indian J. Phys.*, 35, 386, 1961.
251. Rasmussen, E., *Z. Phys.*, 80, 726, 1933.
252. Rasmussen, E., *Z. Phys.*, 83, 404, 1933.
253. Rasmussen, E., *Z. Phys.*, 86, 24, 1934.
254. Rasmussen, E., *Z. Phys.*, 87, 607, 1934.
255. Rasmussen, E., *Phys. Rev.*, 57, 840, 1940.
256. Rau, A. S. and Narayan, A. L., *Z. Phys.*, 59, 687, 1930.
257. Reader, J., *J. Opt. Soc. Am.*, 65, 286, 1975.
258. Reader, J., *J. Opt. Soc. Am.*, 65, 988, 1975.
259. Reader, J., unpublished.
260. Reader, J. and Davis, S., *J. Res. Natl. Bur. Stand. Sect. A*, 71, 587, 1967, and unpublished.
261. Reader, J. and Ekberg, J. O., *J. Opt. Soc. Am.*, 62, 464, 1972.
262. Reader, J. and Epstein, G. L., *J. Opt. Soc. Am.*, 62, 467, 1972.
263. Reader, J. and Epstein, G. L., *J. Opt. Soc. Am.*, 65, 638, 1975.
264. Reader, J. and Epstein, G. L., unpublished.
265. Reader, J., Epstein, G. L., and Ekberg, J. O., *J. Opt. Soc. Am.*, 62, 273, 1972.
266. Ricard, R., *C. R. Acad. Sci. Paris*, 206, 905, 1938.
267. Ricard, R., Givord, M., and George, F., *C. R. Acad. Sci. Paris*, 205, 1929, 1937.
268. Risberg, P., *Ark. Fys.*, 10, 583, 1956.
269. Risberg, G., *Ark. Fys.*, 28, 381, 1965.
270. Risberg, G., *Ark. Fys.* , 37, 231, 1968.
271. Robinson, H. A., *Phys. Rev.*, 49, 297, 1936.
272. Robinson, H. A., *Phys. Rev.*, 50, 99, 1936.
273. Ross, C. B., Jr., Ph.D. dissertation, Purdue University, 1969.
274. Ross, C. B., Wood, D. R., and Scholl, P. S., *J. Opt. Soc. Am.*, 66, 36, 1976.
275. Ruedy, J. E. and Gibbs, R. C., *Phys. Rev.*, 46, 880, 1934.
276. Russell, H. N., King, R. B., and Moore, C. E., *Phys. Rev.*, 58, 407, 1940.
277. Russell, H. N. and Moore, C. E., *J. Res. Natl. Bur. Stand.*, 55, 299, 1955.
278. Russell, H. N., Moore, C. E., and Weeks, D. W., *Trans. Am. Philos. Soc.*, 34(2), 111, 1944.
279. Saunders, F., Schneider, E., and Buckingham, E., *Proc. Natl. Acad. Sci.*, 20, 291, 1934.
280. Sawyer, R. A. and Humphreys, C. J., *Phys. Rev.*, 32, 583, 1928.
281. Sawyer,R. A. and Lang, R. J., *Phys. Rev.*, 34, 712, 1929.
282. Sawyer, R. A. and Paschen, F. *Ann. Phys.*, 84(4), 1, 1927.
283. Scholl, P. S., unpublished, 1975.
284. Schurmann, D., unpublished, 1975.
285. Seguier, J., *C. R. Acad. Sci. Paris*, 256, 1703, 1963.
286. Shenstone, A. G., *Phys. Rev.*, 31, 317, 1928.
287. Shenstone, A. G., *Phys. Rev.*, 32, 30, 1928.
288. Shenstone, A. G., *Trans. R. Soc. London*, 237(A), 57, 1938.
289. Shenstone, A. G., *Phys. Rev.*, 57, 894, 1940.
290. Shenstone, A. G. *Philos. Trans. R. Soc. London Ser. A*, 241, 297, 1948.
291. Shenstone, A. G., *Can. J. Phys.*, 38, 677, 1960.
292. Shenstone, A. G., *Proc. R. Soc. London*, 261(A), 153, 1961.
293. Shenstone, A. G., *Proc. R. Soc. London*, 276(A), 293, 1963.
294. Shenstone, A. G., *J. Res. Natl. Bur. Stand. Sect. A*, 74, 801, 1970.
295. Shenstone, A. G., *J. Res. Natl. Bur. Stand. Sect. A*, 79, 497, 1975.
296. Shenstone, A. G. and Pittenger, J. T., *J. Opt. Soc. Am.*, 39, 219, 1949.
297. Smith, S., *Phys. Rev.*, 36, 1, 1930.
298. Smitt, R., *Phys. Scr.*, 8, 292, 1973.
299. Soderqvist, J., *Ark. Mat. Astronom. Fys.*, 32(A), 1, 1946.
300. Sommer, L. A., *Ann. Phys.*, 75, 163, 1924.
301. Spector, N., *J. Opt. Soc. Am.*, 63, 358, 1973.
302. Spector, N. and Sugar, J., *J. Opt. Soc. Am.*, 66, 436, 1976.
303. Steinhaus, D. W., Radziemski, L. J., Jr., and Blaise, J., upublished, 1975.

304. Subbaraya, T. S., *Z. Phys.*, 78, 541, 1932.
305. Sugar, J., *J. Opt. Soc. Am.*, 55, 33, 1965.
306. Sugar, J., *J. Res. Natl. Bur. Stand. Sect. A*, 73, 333, 1969.
307. Sugar, J., *J. Opt. Soc. Am.*, 60, 454, 1970.
308. Sugar, J., *J. Res. Natl. Bur. Stand. Sect. A*, 78, 555, 1974.
309. Sugar, J. and Kaufman, V., *J. Opt. Soc. Am.*, 55, 1283, 1965.
310. Sugar, J. and Kaufman, V., *J. Opt. Soc. Am.*, 62, 562, 1972.
311. Sugar, J., Kaufman, V., and Spector, N., *J. Res. Natl. Bur. Stand.*, in press.
312. Sugar, J. and Spector, N., *J. Opt. Soc. Am.*, 64, 1484, 1974.
313. Sullivan, F. J. *Univ. Pittsburgh Bull.*, 35, 1, 1938.
314. Svensson, L. A. and Ekberg, J. O., *Ark. Fys.*, 37, 65, 1968.
315. Swensson, J. W. and Risberg, G., *Ark. Fys.*, 31, 237, 1966.
316. Tech. J. L., *J. Res. Natl. Bur. Stand. Sect. A*, 67, 505, 1963.
317. Tech, J. L. and Ward, J. F., *Phys. Rev. Lett.*, 27, 367, 1971.
318. Tilford, S. G., *J. Opt. Soc. Am.*, 53, 1051, 1963.
319. Toresson, Y. G., *Ark. Fys.*, 17, 179, 1960.
320. Toresson, Y. G., *Ark. Fys.*, 18, 389, 1960.
321. Toresson, Y. G. and Edlen, B., *Ark. Fys.*, 23, 117, 1963.
322. Tsien, W. Z., *Chin. J. Phys.*, Peiping, 3, 117, 1939.
323. van Deurzen, C. H. H., Conway, J., and Davis, S. P., *J. Opt. Soc. Am.*, 63, 158, 1973.
324. van Kleef, T. A. M., Raassen, A. J. J., and Joshi, Y. N., *Physica*, 84(C), 401, 1976.
325. Verges, J., Sansonetti, C. J., and Andrew, K. L., unpublished.
326. Wheatley, M. A. and Sawyer, R. A., *Phys. Rev.*, 61, 591, 1942.
327. Wilkinson, P. G., *J. Opt. Soc. Am.*, 45, 862, 1955.
328. Wilkinson, P. G. and Andrew, K. L., *J. Opt. Soc. Am.*, 53, 710, 1963.
329. Wood, D. and Andrew, K. L., *J. Opt. Soc. Am.*, 58, 818, 1968.
330. Wood, D. R., Ross, C. B., Scholl, P. S., and Hoke, M., *J. Opt. Soc. Am.*, 64, 1159, 1974.
331. Worden, E. F. and Conway, J. G., unpublished, 1977.
332. Worden, E. F., Hulet, E. K., Gutmacher, R. G., Conway, J. G., *At. Data Nucl. Data Tables*, 18, 459, 1976.
333. Worden, E. F., Lougheed, R. W., Gutmacher, R. G., and Conway, J. C., *J. Opt. Soc. Am.*, 64, 77, 1974.
334. Wu, C. M., Ph.D. thesis, University of British Columbia, 1971.
335. Zaidel, A. N., Prokofev, V. K., Raiskii, S. M., Slavnyi, V. A., and Schreider, E. Y., *Tables of Spectral Lines*, 3rd ed., Plenum, New York, 1970.
336. Zetterberg, P. O. and Magnusson, C. E., *Phys. Scr.*, 15, 189, 1977.
337. Sugar, J., *J. Opt. Soc. Am.*, 55, 1058, 1965.
338. Sugar, J., *J. Opt. Soc. Am.*, 61, 727, 1971.
339. Bloch, L. and Bloch, E., *Ann. Phys. Paris*, 6(11), 561, 1936.
340. Kaufman, V. and Edlen, B., *J. Phys. Chem. Ref. Data*, 3, 825, 1974.
341. Lang, R. J., *Phys. Rev.*, 34, 697, 1929.
342. Ryabtsev, A. N., *Opt. Spectros.*, 39, 455, 1975.
343. Foster, E. W., *Proc. R. Soc. London*, 200(A), 429, 1950.
344. Morillon, C. and Verges, J., *Phys. Scr.*, 12, 129, 1975.
345. Ruedy, J. E., *Phys. Rev.*, 41, 588, 1932.
346. McLennan, J. C., McLay, A. B., and McLeod, J. H., *Philos. Mag.*, 4, 486, 1927.
347. Handrup, M. B. and Mack, J. E., *Physica*, 30, 1245, 1964.
348. Clearman, H. E., *J. Opt. Soc. Am.*, 42, 373, 1952.
349. Paschen, F., *Ann. Physik*, 424, 148, 1938.
350. Paschen, F. and Campbell, J. S., *Ann. Phys.*, 31(5), 29, 1938.
351. Nodwell, R., unpublished, 1955.
352. Gibbs, R. C. and White, H. E., *Phys. Rev.*, 31, 776, 1928.
353. Green, M., *Phys. Rev.*, 60, 117, 1941.
354. Ellis, C. B. and Sawyer, R. A., *Phys. Rev.*, 49, 145, 1936.
355. McLennan, J. C., McLay, A. B., and Crawford, M. F., *Proc. R. Soc. London Ser. A*, 125, 50, 1929.
356. Mack, J. E. and Fromer, M., *Phys. Rev.*, 48, 346, 1935.
357. Humphreys, C. J. and Paul, E., U.S. Nav. Ord. Lab., Navord Rep. 4589, 25, 1956.
358. Walters, F. M., *Sci. Pap. Bur. Stand.*, 17, 161, 1921.
359. Crawford, M. F. and McLay, A. B., *Proc. R. Soc. London Ser. A*, 143, 540, 1934.
360. McLay, A. B. and Crawford, M. F., *Phys. Rev.*, 44, 986, 1933.
361. Schoepfle, G. K., *Phys. Rev.*, 47, 232, 1935.

SECONDARY ELECTRON EMISSION

N. R. Whetten
General Electric Research Laboratory, Schenectady, New York

By permission from "Methods of Experimental Physics" Vol. IV (1962)
Academic Press

The secondary emission yield, or secondary emission ratio, δ, is the average number of secondary electrons emitted from a bombarded material for every incident primary electron. The secondary emission yield is a function of the primary electron energy. δ_{max} is the maximum yield corresponding to a primary electron energy $E_{p\,max}$ (see figure). The two primary electron energies corresponding to a yield of unity are denoted the first and second crossovers (E_I and E_{II}). An insulating target, or a

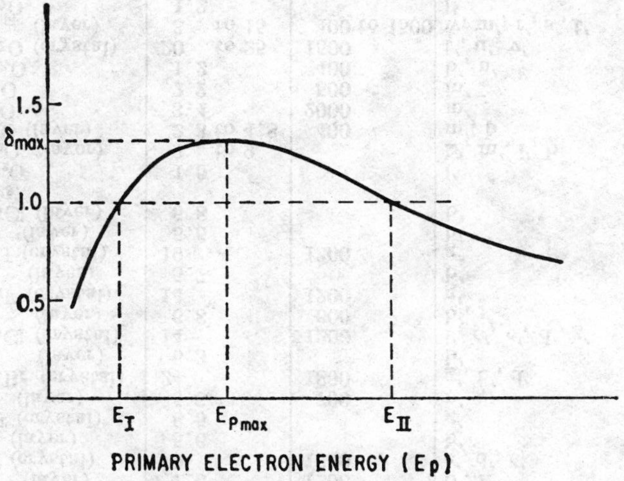

conducting target that is electrically floating, will charge positively or negatively depending on the primary electron energy. For $E_I < E_p < E_{II}$, $\delta > 1$ and the surface charges positively provided there is a collector present that is positive with respect to the target. For $E_p < E_I$, $\delta < 1$, and the surface charges negatively towards the potential of the source of primary electrons. For $E_p > E_{II}$, $\delta < 1$, and the surface charges negatively to the second crossover.

The secondary emission yield is very sensitive to surface contamination, such as oxide films and carbon deposits. Whenever possible, yields believed to be most typical of clean surfaces have been selected. The yields are for measurements at room temperature and normal incidence of the primary electrons.

Table I

Secondary Electron Emission Properties of Elements and Compounds. δ_{max} is the maximum secondary emission yield, $E_{p\,max}$ the primary electron energy for maximum yield, and E_I and E_{II} are the first and second crossovers.

Elements	δ_{max}	$E_{p\,max}$ (ev)	E_I (ev)	E_{II} (ev)	Ref.
Ag	1.5	800	200	>2000	a, b, c
Al	1.0	300	300	300	b
Au	1.4	800	150	>2000	a, c, d
B	1.2	150	50	600	f
Ba	0.8	400	None	None	b
Bi	1.2	550			g, s
Be	0.5	200	None	None	b, h, i, d
C (diamond)	2.8	750		>5000	j
(graphite)	1.0	300	300	300	k
(soot)	0.45	500	None	None	k
Cd	1.1	450	300	700	l, d
Co	1.2	600	200		m, n
Cs	0.7	400	None	None	b, o
Cu	1.3	600	200	1500	a, l, b
Fe	1.3	400	120	1400	n, c, p
Ga	1.55	500	75		q
Ge	1.15	500	150	900	f, r, s
Hg	1.3	600	350	>1200	q
K	0.7	200	None	None	t, u
Li	0.5	85	None	None	b
Mg	0.95	300	None	None	o, b
Mo	1.25	375	150	1200	a, w, c, e, p
Na	0.82	300	None	None	x
Nb	1.2	375	150	1050	a, c
Ni	1.3	550	150	>1500	a, n, m, w, p
Pb	1.1	500	250	1000	g, q
Pd	>1.3	>250	120		v
Pt	1.8	700	350	3000	c
Rb	0.9	350	None	None	t
Sb	1.3	600	250	2000	y
Si	1.1	250	125	500	f
Sn	1.35	500			g, x
Ta	1.3	600	250	>2000	a
Th	1.1	800			b
Ti	0.9	280	None	None	k
Tl	1.7	650	70	>1500	s
W	1.4	650	250	>1500	z, a, a′, c
Zr	1.1	350			k

Table I (Continued)

Compounds	δ_{max}	E_p max (ev)	Ref.
Alkali Halides			
CsCl	6.5		b′
KBr (crystal)	14	1800	c′, e′
KCl (crystal)	12	1600	p′, d′
(layer)	7.5	1200	b′, f′
KI (crystal)	10	1600	c′, d′, e′
(layer)	5.6		b′
LiF (crystal)	8.5		g′
(layer)	5.6	700	b′
NaBr (crystal)	24	1800	g′, h′, d′
(layer)	6.3		b′
NaCl (crystal)	14	1200	i′, e′, c′, d′, g′
(layer)	6.8	600	b′, j′
NaF (crystal)	14	1200	g′
(layer)	5.7		b′
NaI (crystal)	19	1300	g′
(layer)	5.5		b′
RbCl (layer)	5.8		b′
Oxides			
Ag_2O	1.0		l′
Al_2O_3 (layer)	2 to 9		k′, m′, i′, b
BaO (layer)	2.3 to 4.8	400	m′, b
BeO	3.4	2000	m′
CaO	2.2	500	m′
Cu_2O	1.2	400	b′, n′
MgO (crystal)	20 to 25	1500	t′, u′, v′
(layer)	3 to 15	400 to 1500	w, m′, r′, s′, t′
MoO_2	1.2		l′
SiO_2 (quartz)	2.1 to 4	400	k′, g′
SnO_2	3.2	640	o′
Sulfides			
MoS_2	1.1		b′
PbS	1.2	500	p′
WS_2	1.0		b′
ZnS	1.8	350	q′
Others			
BaF_2 (layer)	4.5		b′
CaF_2 (layer)	3.2		b′
$BiCs_3$	6	1000	p′
BiCs	1.9	1000	p′
GeCs	7	700	p′
Rb_3Sb	7.1	450	p′
$SbCs_3$	6	700	p′, x′
Mica	2.4	350	k′
Glasses	2 to 3	300 to 450	k′, y′

References

a R. Warnecke, J. phys. radium, **7**, 270 (1936).
b H. Bruining and J. H. deBoer, Physica, **5**, 17 (1938).
c R. Kollath, Physik. Z., **38**, 202 (1937).
d R. Suhrmann and W. Kundt, Z. Physik, **121**, 118 (1943).
e P. L. Copeland, J. Franklin Inst., **215**, 593 (1933).
f L. R. Koller and J. S. Burgess, Phys. Rev., **70**, 571 (1946).
g P. M. Morozov, J. Exptl. Theoret. Phys. USSR, **11**, 410 (1941).
h R. Kollath, Ann. Physik, **33**, 285 (1938).
i E. G. Schneider, Phys. Rev., **54**, 185 (1938).
j J. B. Johnson, Phys. Rev., **92**, 843 (1953).
k H. Bruining, Philips Tech. Rev., **3**, 80 (1938).
l R. Suhrmann and W. Kundt, Z. Physik, **120**, 363 (1943).
m D. E. Wooldridge, Phys. Rev., **56**, 1062 (1939).
n L. R. G. Treloar and D. H. Landon, Proc. Soc. (London), **B50**, 625 (1938).
o N. S. Klebnikov, Tech. Phys. USSR, **5**, 593 (1938).
p R. L. Petry, Phys. Rev., **26**, 346 (1925).
q J. J. Brophy, Phys. Rev., **83**, 534 (1951).
r J. B. Johnson and K. G. McKay, Phys. Rev., **93**, 668 (1954).
s H. Gobrecht and F. Spear, Z. Physik, **135**, 602 (1953).
t A. Afanasjewa and P. W. Timofeew, Tech. Phys. USSR, **4**, 953 (1937).
u M. S. Joffe and I. V. Nechlaev, J. Exptl. Theoret. Phys. USSR, **11**, 93 (1941).
v H. E. Farnsworth, Phys. Rev., **25**, 41 (1925).
w G. Blankenfeld, Ann. Physik, **9**, 48 (1951).
x J. Woods, Proc. Phys. Soc. (London), **B67**, 843 (1954).
y R. Kollath, *Handbuch der Physik*, vol. **21** (1956), p. 232.
z R. L. Petry, Phys. Rev., **28**, 362 (1926).
a′ E. A. Coomes, Phys. Rev., **55**, 519 (1939).
b′ H. Bruining and J. H. deBoer, Physica, **6**, 834 (1939).
c′ D. N. Dobretzov and A. S. Titkow, Doklady Acad. Nauk USSR, **100**, 33 (1955).
d′ N. R. Whetten, Bull. Am. Phys. Soc. Ser. II, **5**, 347 (1960).
e′ A. R. Shulman and B. P. Dementyev, J. Tech. Phys. USSR, **25**, 2256 (1955).
f′ M. Knoll, O. Hachenberg, and J. Randmer, Z. Physik, **122**, 137 (1944).
g′ D. N. Dobretzov and T. L. Matskevich, J. Tech. Phys. USSR, **27**, 734 (1957).
h′ T. L. Matskevich, J. Tech. Phys. USSR, **26**, 2399 (1956).
i′ A. R. Shulman, W. L. Makedonsky, and J. D. Yaroshetsky, J. Tech. Phys. USSR, **23**, 1152 (1953).
j′ M. M. Vudinsky, J. Tech. Phys. USSR, **9**, 271 (1939).
k′ H. Salów, Z. Tech. Phys., **21**, 8 (1940).
l′ A. Afanasjewa, P. Timofeew, and A. Ignaton, Phys. Z. Sowjet, **10**, 831 (1936).
m′ K. H. Geyer, Ann Phys., **42**, 241 (1942).
n′ N. B. Gornij, J. Exptl. Theoret. Phys. USSR, **26**, 79 (1954).
o′ H. E. Mendenhall, Phys. Rev., **72**, 532 (1947).
p′ O. Hachenberg and W. Brauer in *Advances in Electronics and Electron Physics*, Vol. XI, Academic Press, New York (1959), p. 438.
q′ N. B. Gornij, J. Exptl. Theoret. Phys. USSR, **26**, 88 (1954).
r′ P. Wargo, B. V. Haxby, and W. G. Shepherd, J. Appl. Phys., **27**, 1311 (1956).
s′ P. Rappaport, J. Appl. Phys., **25**, 288 (1954).
t′ N. R. Whetten and A. B. Laponsky, J Appl. Phys., **30**, 432 (1959).
u′ J. B. Johnson and K. G. McKay, Phys. Rev., **91**, 582 (1953).
v′ R. G. Lye, Phys. Rev., **99**, 1647 (1955).
w′ N. R. Whetten and A. B. Laponsky, J. Appl. Phys., **28**, 515 (1957).
x′ N. D. Morgulis and B. I. Djatlowitskaja, J. Tech. Phys. USSR, **10**, 657 (1940).
y′ C. W. Mueller, J. Appl. Phys., **16**, 453 (1945).

SECONDARY ELECTRON EMISSION (Continued)

EMISSIVITY OF TUNGSTEN

Wavelengths in μ

Temperature °K	0.25	0.30	0.35	0.40	0.50	0.60	0.70
1600	.448*	.482	.478	.481	.469	.455	.444
1800	.442*	.478*	.476	.477	.465	.452	.44
2000	.436*	.474	.473	.474	.462	.448	.436
2200	.429*	.470	.470	.471	.458	.445	.431
2400	.422	.465	.466	.468	.455	.441	.427
2600	.418	.461	.464	.464	.451	.437	.423
2800	.411	.456	.461	.461	.448	.434	.419

Temperature °K	0.80	0.90	1.0	1.1	1.2	1.3	1.4
1600	.431	.413	.39	.366	.345	.322*	.300*
1800	.425	.407	.385	.364	.344	.323*	.302*
2000	.419	.401	.381	.361	.343	.323	.305
2200	.415	.396	.378	.359	.342	.324	.306
2400	.408	.391	.372	.355	.340	.324	.309
2600	.404	.386	.369	.352	.338	.325	.310
2800	.400	.383	.367	.352	.337	.325	.313

Temperature °K	1.5	1.6	1.8	2.0	2.2	2.4	2.6
1600	.279*	.263*	.234*	.210*	.19*	.175*	.164*
1800	.282	.267*	.241*	.218*	.20*	.182*	.174*
2000	.288	.273	.247	.227	.209	.197	.175
2200	.291	.278	.254	.235	.218	.205	.194
2400	.296	.283	.262	.244	.228	.215	.205
2600	.299	.288	.269	.251	.236	.224	.214
2800	.302	.292	.274	.259	.245	.233	.224

*** Values by extrapolation.**

LIQUIDS FOR INDEX BY IMMERSION METHOD

Liquid	N_D 24° C
Trimethylene chloride	1.446
Cineole	1.456
Hexahydrophenol	1.466
Decahydronaphthalene	1.477
Isoamylphthalate	1.486
Tetrachloroethane	1.492
Pentachloroethane	1.501
Trimethylene bromide	1.513
Chlorobenzene	1.523
Ethylene bromide + Chlorobenzene	1.533
o-Nitrotoluene	1.544
Xylidine	1.557
o-Toluidine	1.570
Aniline	1.584
Bromoform	1.595
Iodobenzene + Bromobenzene	1.603
Iodobenzene + Bromobenzene	1.613
Quinoline	1.622
α-Chloronaphthalene	1.633
α-Bromonaphthalene + α-Chloronaphthalene	1.640–1.650
α-Bromonaphthalene + α-Iodonaphthalene	1.660–1.690
Methylene iodide + Iodobenzene	1.700–1.730
Methylene iodide	1.738
Methylene iodide saturated with sulfur	1.78
Yellow phosphorus, sulfur and methylene iodide (8:1:1 by weight)	2.06
Can be diluted with methylene iodide to cover range 1.74–2.06. For precautions in use, cf. West, Am. Mineral, **21**, p. 245–9 (1936).	

HEAVY LIQUIDS FOR MINERAL SEPARATION

Liquid	Density
Tetrabromoethane (sym.)	2.964, 20°/4°
Can be diluted with carbon tetrachloride (1.595) or benzene (0.894).	
Methylene iodide	3.325, 20°/4°
Can be diluted with carbon tetrachloride or benzene.	
Thallium formate, aq	3.5
Can be diluted with water.	
Thallium malonate-thallium formate, aq	4.9
Can be diluted with water.	

For preparation and recovery of these liquids, cf. U. S. Bureau Mines, Rept. Inv. #2897 (1928).

INDEX OF REFRACTION

Indices of refraction for elements, inorganic, metal-organic and organic compounds and minerals will be found in the tables of physical constants for the various classes of substances in the section Properties and Physical Constants.

Values for compounds not there listed and data subsequently collected are given below.

Indices not otherwise indicated are for sodium light, λ = 589.3 mμ. Other wave lengths are indicated by the value in millimicrons or symbol in parentheses which follows the index. Wave lengths are indicated as follows: He, λ = 587.6 mμ; Li, λ = 670.8 mμ; Hg, λ = 579.1 mμ; A, λ = 759.4 mμ; C, λ = 656.3 mμ; D, λ = 589.3 mμ; F, λ = 486.1 mμ.

Temperatures are understood to be 20°C for liquids, or ordinary room temperatures in the case of solids. Other temperatures appear as superior figures with the index.

Indices for the elements and inorganic compounds will be understood to be for the solid form except as indicated by the abbreviation liq.

See also under Physical Constants of Inorganic Compounds and index of Refraction of Gases.

Elements

Name	Formula	Index
Bromine (liq.)	Br_2	1.661^{15}
Cadmium (liq.)	Cd	0.82 (579 mμ)
(sol.)		1.13
Chlorine (liq.)	Cl_2	1.385
" (gas)		1.000768
Hydrogen (liq.)	H_2	$1.10974^{-252.83}$(579 mμ)
Iodine (sol.)	I_2	3.34
(gas)		1.001920
Lead	Pb	2.6 (579 mμ)
Mercury (liq.)	Hg	1.6–1.9
Nitrogen (liq.)	N_2	1.2053^{-190}

INDEX OF REFRACTION (Continued)

Elements

Name	Formula	Index
Oxygen (liq.)	O_2	1.221^{-181}
Phosphorus (yel.) (sol.)		2.1442^{25}
Selenium	Se_8	3.00, 4.04
" (amor.) (sol.)		2.92
Sodium (liq.)	Na	0.0045
(sol.)		4.22
Sulfur (liq.)	S_8	1.929^{110}
" (amor.) (sol.)		1.998
" (rhombic, α)		1.957, 2.0377, 2.2454
Tin (liq.)	Sn	2.1

Inorganic Compounds

See also under Physical Constants of Inorganic Compounds.

Name	Formula	Index
Aluminum carbide	Al_4C_3	2.7, 2.75 (700 mμ)
chloride	$AlCl_3.6H_2O$	1.560, 1.507
oxide	Al_2O_3	1.665–1.680, 1.63–1.65
Alums. See under appropriate	element.	
Ammonium antimonyl tartrate	$2(NH_4.SbO.C_4H_4O_6).H_2O$	β1.6229 (C)
*ortho*arsenate, di-H	$NH_4H_2AsO_4$	1.5766, 1.5217
bromide	NH_4Br	1.7108
*per*chlorate	NH_4ClO_4	1.4818, 1.4833, 1.4881
chloroplatinate	$(NH_4)_2PtCl_6$	1.8
fluoride	NH_4F	ω<1.328
" acid	NH_4HF_2	1.385, 1.390, 1.394
hydrogen malate (*d*)	$NH_4C_4H_5O_5$	β1.503
nitrate	NH_4NO_3	1.413, 1.611(He), 1.63
Ammonium sulfate, acid	NH_4HSO_4	1.463, 1.473, 1.510
tartrate (*dl*)	$(NH_4)_2C_4H_4O_6.2H_2O$	β1.564
thiocyanate	NH_4CNS	1.546, 1.685, 1.692
uranyl acetate	$NH_4C_2H_3O_2.UO_2(C_2H_3O_2)_2$	1.4808, 1.4933
Antimony bromide	$SbBr_3$	>1.74+
iodide, tri-	SbI_3	2.78 (Li), 2.36
Barium cadmium bromide	$BaCdBr_4.4H_2O$	β1.702
cadmium chloride	$BaCdCl_4.4H_2O$	β1.651
calcium propionate	$BaCa_2(C_3H_5O_2)_6$	1.4442
fluochloride	$BaCl_2.BaF_2$	1.640, 1.633
fluoride	BaF_2	1.475 also 1.4741
Barium oxide	BaO	1.980
*ortho*phosphate, di-	$BaHPO_4$	1.617, 1.63±, 1.635
propionate	$Ba(C_2H_5CO_2)_2.H_2O$	β1.5175
sulfide, mono-	BaS	2.155
Cadmium ammonium chloride	$CdCl_2.4NH_4Cl$	1.6038, 1.6042
cesium sulfate	$CdSO_4.Cs_2SO_4.6H_2O$	1.498, 1.500, 1.506
fluoride	CdF_2	1.56
magnesium chloride	$(CdCl_2)_2.MgCl_2.12H_2O$	1.49, 1.5331, 1.5769
oxide	CdO	2.49 (Li)
potassium chloride	$CdCl_2.4KCl$	1.5906, 1.5907
" cyanide	$Cd(CN)_2.2KCN$	1.4213
rubidium sulfate	$CdSO_4.Rb_2SO_4.6H_2O$	1.4798, 1.4848, 1.4948
Calcium aluminate	$Ca_3Al_2O_6$	1.710
borate	$CaO.B_2O_3$	1.540, 1.656, 1.682
carbide	CaC_2	>1.75
copper acetate	$CaCu(C_2H_3O_2)_4.6H_2O$	1.436, 1.478
cyanamide	$CaCN_2$	1.60, >1.95
dithionate	$CaS_2O_6.4H_2O$	1.5516, 1.5414
*pyro*phosphate	$Ca_2P_2O_7$	1.585, 1.60±, 1.605
platinocyanide	$CaPt(CN)_4.5H_2O$	1.623, 1.644, 1.767
strontium propionate	$Ca_2Sr(C_3H_5O_2)_6$	1.4871, 1.4956
sulfide (oldhamite)	CaS	2.137
sulfite	$CaSO_3.2H_2O$	1.590, 1.595, 1.628
thiosulfate	$CaS_2O_3.6H_2O$	1.545, 1.560, 1.605
Carbon dioxide (liq.)	CO_2	1.195^{15}
Cerium dithionate	$Ce_2(S_2O_6)_3.15H_2O$	β1.507
Cesium *per*chlorate	$CsClO_4$	1.4752, 1.4788, 1.4804
nitrate	$CsNO_3$	1.55, 1.56
selenate	Cs_2SeO_4	1.5989, 1.5999, 1.6003
thallium chloride	$Cs_3Tl_2Cl_9$	1.784, 1.774
Chromium cesium sulfate	$CrCs(SO_4)_2.12H_2O$	1.4810
oxide (ic)	Cr_2O_3	2.5
potassium cyanide (ic)	$CrK_3(CN)_6$	1.5221, 1.5244, 1.5373
sulfate (ic)	$Cr_2(SO_4)_3.18H_2O$	1.564
thallium sulfate	$CrTl(SO_4)_2.12H_2O$	1.5228
Cobalt acetate	$Co(C_2H_3O_2)_2.4H_2O$	β1.542
aluminate (Thenard's Blue)	$Co(AlO_2)_2$	>1.78 (red), 1.74 (blue)
ammonium selenate	$CoSeO_4.(NH_4)_2SeO_4.6H_2O$	1.5246, 1.5311, 1.5396
cesium sulfate	$CoCs_2(SO_4)_2.6H_2O$	1.5057, 1.5085, 1.5132
chloride (ous)	$CoCl_2.2H_2O$	<1.625, <1.671, >1.67
potassium selenate	$CoSeO_4.K_2SeO_4.6H_2O$	1.5135, 1.5195, 1.5358
rubidium sulfate	$CoSO_4.Rb_2SO_4.6H_2O$	1.4859, 1.4916, 1.5014
selenate	$CoSeO_4.6H_2O$	β1.5225, γ1.5227
Copper ammonium selenate	$CuSeO_4.(NH_4)_2SeO_4.6H_2O$	1.5213, 1.5355, 1.5395
ammonium sulfate	$CuSO_4.(NH_4)_2SO_4.6H_2O$	1.4910, 1.5007, 1.5054
cesium sulfate	$CuSO_4.Cs_2SO_4.6H_2O$	1.5048, 1.5061, 1.5153
chloride (ic)	$CuCl_2.2H_2O$	1.644, 1.684, 1.742
formate	$Cu(CHO_2)_2.4H_2O$	1.4133, 1.5423, 1.5571

INDEX OF REFRACTION (Continued)

Inorganic Compounds (Continued)

Name	Formula	Index
Copper oxide (ous) (cuprite)	Cu_2O	2.705
potassium chloride	$CuCl_2.2KCl.2H_2O$	1.6365, 1.6148
" cyanide (ous)	$CuK_3(CN)_4$	1.5215
" selenate	$CuSeO_4.K_2SeO_4.6H_2O$	1.5096, 1.5235, 1.5387
" sulfate	$CuSO_4.K_2SO_4.6H_2O$	1.4836, 1.4864, 1.5020
strontium formate	$Cu(HCO_2)_2.2[Sr(HCO_2)_2]8H_2O$	1.4995, 1.5199, 1.5801
sulfate (ic)	$CuSO_4$	1.724, 1.733, 1.739
Cyanogen	C_2N_2	1.327^{18} (liq.)
Germanium bromide, tetra-	$GeBr_4$	1.6269
Gold sodium chloride	$AuNaCl_4.2H_2O$	α1.545, γ1.75+
Hafnium oxychloride	$HfOCl_2.8H_2O$	1.557, 1.543
Ice	H_2O	1.3049, 1.3062 (A), 1.3091, 1.3104 (D), 1.3133, 1.3147 (F)
Iron ammonium chloride	$Fe(NH_4)_2Cl_4$	1.6439
ammonium selenate	$FeSeO_4.(NH_4)_2SeO_4.6H_2O$	1.5201, 1.5260, 1.5356
cesium sulfate (ic)	$FeCs(SO_4)_2.12H_2O$	1.4839
" " (ous)	$FeSO_4.Cs_2SO_4.6H_2O$	1.5003, 1.5035, 1.5094
rubidium sulfate	$FeRb(SO_4)_2.12H_2O$	1.48234
sulfate (ic)	$Fe_2(SO_4)_3$	1.802, 1.814, 1.818
thallium sulfate	$FeTl(SO_4)_2.12H_2O$	1.52365
Lanthanum sulfate	$La_2(SO_4)_3.9H_2O$	1.564, 1.569
Lead *ortho*arsenate, di-	$PbHAsO_4$	1.8903, 1.9097, 1.9765
nitrate	$Pb(NO_3)_2$	1.782
Lithium ammonium sulfate	$LiNH_4SO_4$	β1.437 (Li)
ammonium tartrate (*d*)	$LiNH_4(C_4H_4O_6).H_2O$	β1.567, γ1.5673
" " (*dl*)	$LiNH_4(C_4H_4O_6).H_2O$	β1.5287
bromide	$LiBr$	1.784
chloride	$LiCl$	1.662
dithionate	$Li_2S_2O_6.2H_2O$	1.5487, 1.5602, 1.5788
oxide	Li_2O	1.644
potassium sulfate	$LiKSO_4$	1.4723, 1.4717
" tartrate	$LiK(C_4H_4O_6).H_2O$	β1.5226 (red)
rubidium tartrate (*a*)	$LiRb(C_4H_4O_6).H_2O$	β1.552
sodium tartrate (*dl*)	$LiNa(C_4H_4O_6).2H_2O$	β1.4904
Magnesium ammonium selenate	$MgSeO_4.(NH_4)_2SeO_4.6H_2O$	1.5070, 1.5093, 1.5169
ammonium sulfate	$Mg(NH_4)_2.(SO_4)_2.6H_2O$	1.4716, 1.4730, 1.4786
*ortho*borate	$3MgO.B_2O_3$	1.6527, 1.6537, 1.6748
cesium sulfate	$MgCs_2(SO_4)_2.6H_2O$	1.4857, 1.4858, 1.4916
chlorostannate	$MgSnCl_6.6H_2O$	1.5885, 1.5970
fluosilicate	$MgSiF_6.6H_2O$	1.3439, 1.3602
platinocyanide	$MgPt(CN)_4.7H_2O$	1.5608, 1.91
Magnesium potassium selenate	$MgK_2(SeO_4)_2.6H_2O$	1.4969, 1.4991, 1.5139
potassium sulfate	$MgK_2(SO_4)_2.6H_2O$	1.4607, 1.4629, 1.4755
rubidium sulfate	$MgRb_2(SO_4)_2.6H_2O$	1.4672, 1.4689, 1.4779
silicate	$MgSiO_3$	1.651, 1.654 (calc.), 1.660
sulfide	MgS	2.271 also 2.268
Manganese borate	$Mn_3B_4O_9$	1.617, 1.738, 1.776
cesium sulfate	$MnCs_2(SO_4)_2.6H_2O$	1.4946, 1.4966, 1.5025
chloride	$MnCl_2.4H_2O$	1.555, 1.575, 1.607
rubidium sulfate	$MnRb_2(SO_4)_2.6H_2O$	1.4767, 1.4807, 1.4907
sulfate (ous)	$MnSO_4.4H_2O$	1.508, 1.518, 1.522
" "	$MnSO_4.5H_2O$	1.495, 1.508, 1.514
Mercury chloride (ic)	$HgCl_2$	1.725, 1.859, 1.965
cyanide (ic)	$Hg(CN)_2$	1.645, 1.492
iodide (ic) (red)	HgI_2	2.748, 2.455
Nickel ammonium selenate	$Ni(NH_4)_2.(SeO_4)_2.6H_2O$	1.5291, 1.5372, 1.5466
cesium sulfate	$NiCs_2(SO_4)_2.6H_2O$	1.5087, 1.5129, 1.5162
Nickel chloride	$NiCl_2.6H_2O$	α1.535, γ1.61
fluoride, acid	$NiF_2.5HF.6H_2O$	1.392, 1.408
potassium selenate	$NiK_2(SeO_4)_2.6H_2O$	1.5199, 1.5248, 1.5339
rubidium sulfate	$NiRb_2(SO_4)_2.6H_2O$	1.4895, 1.4961, 1.505[illegible]
selenate	$NiSeO_4.6H_2O$	1.5393, 1.5125
Platinum potassium dibromonitrite	$PtK_2(NO_2)_2Br_2.H_2O$	1.626, 1.6684, 1.757
Potassium carbonate	K_2CO_3	1.426, 1.531, 1.541
carbonate, acid	$KHCO_3$	1.380, 1.482, 1.578
*per*chlorate	$KClO_4$	1.4731, 1.4737, 1.4769
chloroplatinate	K_2PtCl_6	1.827 (577 mμ)
chloroplatinite	K_2PtCl_4	1.64, 1.67
*di*chromate	$K_2Cr_2O_7$	1.7202, 1.7380, 1.8197
cyanide	KCN	1.410
fluoborate	KBF_4	1.3239, 1.3245, 1.3247
fluoride	KF	1.352 (1.361)
"	$KF.2H_2O$	1.345, 1.352, 1.363
fluosilicate	K_2SiF_6	1.3391
*per*iodate	KIO_4	1.6205, 1.6479
lithium ferrocyanide	$K_2Li_2Fe(CN)_6.3H_2O$	1.5883, 1.6007, 1.6316
*hypo*phosphate	$K_2H_2P_2O_6.2H_2O$	1.4893, 1.5314, 1.5363
"	$K_2H_2P_2O_6.3H_2O$	1.4768, 1.4843, 1.4870
ruthenium cyanide	$K_4Ru(CN)_6.3H_2O$	β1.5837
silicate	K_2SiO_3	1.520, 1.521, 1.528
thiocyanate	$KCNS$	1.532, 1.660, 1.730
thionate, tetra-	$K_2S_4O_6$	1.5896, 1.6057, 1.6435
" penta-	$2K_2S_5O_6.3H_2O$	1.565, 1.63, 1.655
Rhodium cesium sulfate	$RhCs(SO_4)_2.12H_2O$	1.5077
Rubidium *per*chlorate	$RbClO_4$	1.4692, 1.4701, 1.4731
chromate	Rb_2CrO_4	β1.71, γ1.72
dithionate	$Rb_2S_2O_6$	1.4574, 1.5078
fluoride	RbF	1.396
selenate	Rb_2SeO_4	1.5515, 1.5537, 1.5582
Ruthenium sodium nitrate	$RuNa_2(NO_2)_5.2H_2O$	1.5889, 1.5943, 1.7163
Selenium oxide	SeO_2	>1.76
Silver cyanide	$AgCN$	1.685, 1.94
nitrate	$AgNO_3$	1.729, 1.744, 1.788
phosphate	Ag_2HPO_4	1.8036, 1.7983
potassium cyanide	$AgK(CN)_2$	1.625, 1.63
Sodium ammonium tartrate (*d*)	$NaNH_4(C_4H_4O_6).4H_2O$	1.495, 1.498, 1.499
ammonium tartrate (*dl*)	$NaNH_4(C_4H_4O_6).H_2O$	β1.473 (red)
*ortho*arsenate	$NaH_2AsO_4.H_2O$	1.5382, 1.5535, 1.5607
"	$NaH_2AsO_4.2H_2O$	1.4794, 1.5021, 1.5265
bromide	$NaBr$	1.6412
carbonate	Na_2CO_3	1.415, 1.535, 1.546
Sodium carbonate, acid	$NaHCO_3$	1.376, 1.500, 1.582
cyanide	$NaCN$	1.452
iodide	NaI	1.7745
molybdate	$3Na_2O.7MoO_3.22H_2O$	β1.627
nitrate	$NaNO_3$	1.5874, 1.3361
phosphate	$NaH_2PO_4.2H_2O$	1.4401, 1.4629, 1.4815
"	$Na_2HPO_4.7H_2O$	1.4412, 1.4424, 1.4526
*hypo*phosphate	$Na_3HP_2O_6.9H_2O$	1.4653, 1.4738, 1.4804
silicate	Na_2SiO_3	1.513, 1.520, 1.528
sulfate, acid	$NaHSO_4.H_2O$	1.43, 1.46, 1.47
sulfite	Na_2SO_3	1.565, 1.515
" acid	$NaHSO_3$	1.474, 1.526, 1.685
tartrate, acid (*d*)	$NaH(C_4H_4O_6).H_2O$	β1.533
thiocyanate	$NaCNS$	1.545, 1.625, 1.695
Sodium tungstate	$Na_2WO_4.2H_2O$	1.5526, 1.5533, 1.5695
vanadate	$Na_3VO_4.10H_2O$	1.5305; ω1.5398, ε1.5475
"	$Na_3VO_4.12H_2O$	1.5095, 1.5232
Strontium dichromate	$SrCr_2O_7.3H_2O$	1.7146, 1.7174, 1.812
fluoride	SrF_2	1.442 (1.438)
oxide	SrO	1.870
*ortho*phosphate, acid	$SrHPO_4$	1.608, 1.62±, 1.625
sulfide, mono-	SrS	2.107
Sulfur nitride	S_4N_4	α1.908, β2.046
Thallium chloride, mono-	$TlCl$	2.247
iodide, mono-	TlI	2.78
Tin iodide (ic)	SnI_4	2.106
Uranyl potassium sulfate	$UO_2.SO_4.K_2SO_4.2H_2O$	1.5144, 1.5266, 1.5705 (580 mμ)
Vanadium ammonium sulfate	$VNH_4(SO_4)_2.12H_2O$	1.475
Zinc ammonium selenate	$Zn(SeO_4).(NH_4)_2SeO_4.6H_2O$	1.5240, 1.5300, 1.5385
bromate	$Zn(BrO_3)_2.6H_2O$	1.5452
cesium sulfate	$ZnCs_2(SO_4)_2.6H_2O$	1.5022, 1.5048, 1.5093
chloride	$ZnCl_2$	1.687, 1.713
fluosilicate	$ZnSiF_6.6H_2O$	1.3824, 1.3956
potassium cyanide	$ZnK_2(CN)_4$	1.4115
" selenate	$ZnK_2(SeO_4)_2.6H_2O$	1.5121, 1.5181, 1.5335
" sulfate	$ZnK_2(SO_4)_2.6H_2O$	1.4775, 1.4833, 1.4969
rubidium sulfate	$ZnRb_2(SO_4)_2.6H_2O$	1.4833, 1.4884, 1.4975
silicate	$ZnSiO_3$	1.616, 1.62±, 1.623
Zirconium ammonium fluoride	$Zr(NH_4)_3F_7$	1.433

Organic Compounds

See also under Physical Constants of Organic Compounds.

Name	Index
Allontoin, solid	α1.579, γ1.660
Dimethyl thiophene (α, α′), liq	$1.51693^{13.4}$ (He)
" " (β, β′), liq	1.52217^{15} (He)
Ethyl carbylamine, liq	1.3659^{24}
Ethylidene cyanhydrin, liq	$1.40582^{18.4}$
Hexyl acetylene (n), liq	$1.4208^{12.5}$

Miscellaneous

Albite glass	1.4890	Magdala red	1.90
Amber	1.546	Obsidian	1.482–1.496
Anorthite glass	1.5755	Paraffin	$1.43295^{38.3}$ (C)
Asphalt	1.635	Quartz, fused	1.45640 (656 mμ)
Bell metal	1.0052		1.45843 (589 mμ)
Borax, amorphous, fused	1.4630		1.46190 (509 mμ)
Canada balsam	1.530		1.47503 (361 mμ)
Ebonite	1.66 (red)		1.49634 (275 mμ)
Fuchsin	2.70		1.53386 (214 mμ)
Gelatin, Nelson's No. 1	1.530		1.57464 (185 mμ)
Gelatin, various	1.516–1.534	Resin, aloes	1.619 (red)
Gum Arabic	1.480 (1.514) (red)	colophony	1.548 (red)
		copal	1.528 (red)
Hoffman's violet	2.20	mastic	1.535 (red)
Ivory	1.539, 1.541	Peru balsam	1.593

INDEX OF REFRACTION OF ORGANIC COMPOUNDS

(See also that section of this book which contains data of Physical Constants of Organic Compounds)

The following table contains a list of organic compounds arranged in order of increasing refractive index. Measurements were made at 25°C.

Compound	nD
Trifluoroacetic acid	1.283
2,2,2-Trifluoroethanol	1.290
Octofluoropentanol-1	1.316
Dodecafluoroheptanol-1	1.316
Methanol	1.326
Acetonitrile	1.342
Ethyl ether	1.352
Acetone	1.357
Ethyl formate	1.358
Ethanol	1.359
Methyl acetate	1.360
Propionitrile	1.363
2,2-Dimethylbutane	1.366
Isopropyl ether	1.367
2-methylpentane	1.369
Ethyl acetate	1.370
Acetic acid	1.370
Propionaldehyde	1.371
n-Hexane	1.372
2,3-Dimethylbutane	1.372
3-Methylpentane	1.374
2-Propanol	1.375
Isopropyl acetate	1.375
Propyl formate	1.375
2-Chloropropane	1.376
2-Butanone	1.377
2-Chloropropane	1.377
Methylethyl ketone	1.377
Butyraldehyde	1.378
2,4-Dimethylpentane	1.379
Propyl ether	1.379
Acetaldehyde-diethylacetal	1.379
Butylethyl ether	1.380
Nitromethane	1.380
Trifluoropropanol	1.381
2-Methylhexane	1.382
Butyronitrile	1.382
Propyl acetate	1.382
Ethyl propionate	1.382
2-Methyl-2-propanol	1.383
1-Propanol	1.383
Isobutyl formate	1.383
Diethyl carbonate	1.383
Heptane	1.385
2-Methyl-2-propanol	1.385
Propionic acid	1.385
3-Methylhexane	1.386
n-Propyl amine	1.386
1,1-Dimethyl-2-propanone	1.386
1-Chloropropane	1.386
2,2,3-Trimethylbutane	1.387
Methylpropyl ketone	1.387
sec-Butyl acetate	1.387
Butyl formate	1.387
β-Methylpropyl ethanoate	1.388
2,2,4-Trimethyl pentane	1.389
2,3-Dimethyl pentane	1.389
Acetic anhydride	1.389
Diisopropyl amine	1.390
2-Aminobutane	1.390
2-Pentanone	1.390
3-Pentanone	1.390
Nitroethane	1.390
Methyl-b-butyrate	1.391
Butyl acetate	1.392
2-Nitropropane	1.392
4-Methyl-2-pentanone	1.394
2-Methyl-1-propanol	1.394
Octane	1.395
1-Amino-2-methylpropane	1.395
Valeronitrile	1.395
2-Butanol	1.395
2-Hexanone	1.395
5-Methyl-3-hexanone	1.395
2-Chlorobutane	1.395
Butyric acid	1.396
2,2,2-Trimethylhexane	1.397
n-Dibutyl ether	1.397
1-Butanol	1.397
Acrolein	1.397
1-Chloro-2-methylpropane	1.397
Methacrylonitrile	1.398
3-Methyl-2-pentanone	1.398
Triethyl amine	1.399
n-Butyl amine	1.399
1,1,3,3-Tetramethyl-2-propanone	1.399
Isobutyl-n-butyrate	1.399
1-Nitropropane	1.399
n-Dodecane	1.400
Amyl acetate	1.400
1-Chlorobutane	1.400
2-Methoxy ethanol	1.400
Propionic acid anhydride	1.400
2,2,3-Trimethylpentane	1.401
1-Chlorobutane	1.401
β-Methoxypropionitrile	1.401
3-Methyl butanoic acid	1.402
n-Nonane	1.403
Dipropylamine	1.403
Isoamylacetate	1.403
Cyclopentane	1.404
2-Methyl-2-butanol	1.404
3-Methyl-1-butanol	1.404
Tetrahydrofuran	1.404
Capronitrile	1.405
2-Pentanone	1.405
2-Ethoxyethanol	1.405
2-Heptanone	1.406
Valeric acid	1.406
Diisobutylene	1.407
Methylcyclopentane	1.407
Isoamyl ether	1.407
Methylpropyl carbinol	1.407
Tributyl borate	1.407
1-Pentanol	1.408
3-Methyl-2-butanol	1.408
Diethyl oxalate	1.408
n-Decane	1.409
4-Methyl-2-pentanol	1.409
3-Isopropyl-2-pentanone	1.409
2-Methyl-1-butanol	1.409
Butyric acid anhydride	1.409
Amyl ether	1.410
Isoamyl isovalerate	1.410
1-Chloropentane	1.410
2-Propene-1-ol	1.411
2,4-Dimethyl dioxane	1.412
Ethyl lactate	1.412
Diethyl malonate	1.412
3-Chloropropene	1.413
Ethyleneglycol diacetate	1.413
2-Octanone	1.414
3-Octanone	1.414
3-Methyl-2-heptanone	1.415
Caproic acid	1.415
4-Methyldioxane	1.415
1,2-Propyleneglycol-1-monobutyl ether	1.415
Ethylcyanoacetate	1.415
Dibutylamine	1.416
2-Pentanol	1.416
1,1-Dichloroethane	1.416
Heptachlorodiethyl ether	1.416
1-Hexanol	1.416
1-Amino-3-methoxy propane	1.417
Octyl nitrile	1.418
2-Heptanol	1.418
2-Propenyl amine	1.419
1,2-Propyleneglycol carbonate	1.419
Methylpentyl carbinol	1.420
2-Ethyl-1-butanol	1.420
1-Chloro-2-methyl-1-propene	1.420
p-Dioxane	1.420
Methylcyclohexane	1.421
4-Hydroxy-4-methyl-2-pentanone	1.421
1-Heptanol	1.422
3-Isopropyl-2-heptanone	1.423
Cyclohexane	1.424
2-Bromopropane	1.424
3-Chloro-2-methylprop-1-ene	1.425
Caproic acid	1.426
Glycol carbonate	1.426
1-Octanol	1.427
1,1-Dimethylhexanol	1.427
N,N-Dimethylformamide	1.427
Sulfuric acid	1.427
1-Chlorooctane	1.428
Triisobutylene	1.429
N-Methylaniline nitrile	1.429
Ethylene glycol	1.429
1-Chloro-2-ethylhexane	1.430
Ethylcyclohexane	1.431
1,2-Propanediol	1.431
1-Bromopropane	1.431
2-Methyl-7-ethyl-4-nonanone	1.433
Ethyleneglycol-mono-allyl ether	1.434
Butyral lactone	1.434
2-Methyl-7-ethyl-4-undecanone	1.435
4-n-Propyl-5-ethyldioxane	1.435
1,2-Dichloro-2-methylpropane	1.435
1,2-Propyleneglycol sulfite	1.435
N-Methylmorpholine	1.436
1-Chloro-2-methyl-2-propanol	1.436
Epichlorohydrin	1.436
Triethyleneglycol-mono-butyl ether	1.437
4-Ethyl-7,7,7-trimethyl-1-heptanol	1.438
1-Methyl-3-ethyloctan-1-ol	1.438

Compound	nD
1-Ethyl-3-ethylhexan-1-ol	1.438
Diethyl maleate	1.438
1-Butanethiol	1.440
2-Chloroethanal	1.440
Dibutyl sebacate	1.440
1-Ethyl-3-ethyloctan-1-ol	1.441
Dimethylmaleate	1.441
3-Methylpentane-2,4-diol	1.441
Ethyl sulfide	1.442
Mesityl oxide	1.442
Butyl stearate	1.442
1,2-Dichloroethane	1.444
Chloroform	1.444
trans-1,2-Dichloroethylene	1.444
Diethyleneglycol	1.445
cis-1,2-Dichloroethylene	1.445
3-(α-Butyloctyl)-oxypropyl-1-amine	1.446
2-Methylmorpholine	1.446
Dipropyleneglycol-monoethyl ether	1.446
Formamide	1.446
3-Lauryloxypropyl-1-amine	1.447
Cyclohexanone	1.448
1-Aminopropan-1-ol	1.448
Diethyleneglycol-mono-β-oxypropyl ether	1.448
1-Amino-2-methylpentan-1-ol	1.449
Tetrahydrofurfural alcohol	1.450
2-Propylcyclohexa-1-one	1.452
2-Aminoethanol	1.452
2-Butylcyclohexan-1-one	1.453
Ethylenediamine	1.454
2-(β-Methyl)-propylcyclohexan-1-one	1.454
4-Methylcyclohexanol	1.454
3-Methylcyclohexanol	1.455
bis-2-Chloroethyl ether	1.455
Cyclohexylamine	1.456
1,8-Cineol	1.456
2,2'-Dimethyl-2,2'-dipropyldieththanol amine	1.456
1,1',2,2'-Tetramethyldiethanol amine	1.459
1-Aminopropan-3-ol	1.459
Carbon tetrachloride	1.459
3-Methyl-5-ethylheptan-2,4-diol	1.459
2-(β-Ethyl)-butylcyclohexan-1-one	1.461
2-Methylcyclohexanol	1.461
N-(n-Butyl)-diethanol amine	1.461
4,5-Chloro-1,3-dioxolane-2	1.461
2-Butylcyclohexan-1-ol	1.462
N-β-Oxypropyl morpholine	1.462
2-(β-Ethyl)-hexylcyclohexanone	1.463
2-Ethylcyclohexan-1-ol	1.463
Fluorobenzene	1.463
d-α-Pinene	1.464
1-α-Pinene	1.465
Cyclohexanol	1.465
m-Fluorotoluene	1.465
p-Fluorotoluene	1.467
trans-Decahydronaphthalene	1.468
o-Fluorotoluene	1.468
3-Alloxy-2-oxypropylamine-1	1.469
Ethanol-1-methylisopropanol amine	1.470
d-Limonene	1.471
l,2,3-Trichloroisobutane	1.473
Decahydronaphthalene	1.474
1,2,3-Propanetriol	1.474
Trichloroethylene	1.475
N-β-Oxyethylmorpholine	1.476
Dimethylsulfoxide	1.476
cis-Decahydronaphthalene	1.479
N-β-Chlorallylmorpholine	1.481
n-Dodecyl-4-tertiarybutylphenyl ether	1.482
n-Dodecylphenyl ether	1.482
n-Dodecyl-4-methylphenyl ether	1.483
2-Ethylidene cyclohexanone	1.486
n-Butylbenzene	1.487
p-Cymene	1.488
iso-Propylbenzene	1.489
Furfuralcohol	1.489
tert-Butylcumene	1.490
n-Propylbenzene	1.490
sec-Butylbenzene	1.490
tert-Butylbenzene	1.490
Dibutylphthalate	1.490
tert-Butyltoluene	1.491
1-Penyl-1-oxyphenylethane	1.491
n-Hexylcumene	1.492
n-Octyltoluene	1.492
n-Octylcumene	1.492
p-Xylene	1.493
1,31-Diethylbenzene	1.493
Ethylbenzene	1.493
1,3-Dimorpholylpropan-2-ol	1.493
1,12,2-Tetrachloroethane	1.493
Toluene	1.494
Benzylethyl ether	1.494
m-Xylene	1.495
1,4-Diethylbenzene	1.496
2,3-Dichlorodioxane	1.496
Mesitylene	1.497
2-Iodopropane	1.497
Benzene	1.498
Propyl benzoate	1.498
α-Picoline	1.499
1,2-Diethylbenzene	1.501
Pentachloroethane	1.501
1-Iodopropane	1.502
1,2-Dimethylbenzene	1.503
Ethyl benzoate	1.503
β-Picoline	1.504
Tetrachloroethylene	1.504
Phenetole	1.505
Pyridine	1.507
Iodoethane	1.512
Phenylmethallyl ether	1.514
Anisole	1.515
Methyl benzoate	1.515
Diallylphthalate	1.517
Benzylacetate	1.518
2-Methyl-4-tertiarybutylphenetol	1.521
Phenylacetonitrile	1.521
Methyl salicylate	1.522
Chlorobenzene	1.523
Fufural	1.524
Benzonitrile	1.526
Thiophene	1.526
Nonachlorodiethyl ether	1.529
Iodomethane	1.530
4-Phenyldioxane	1.530
3-Phenylpropan-1-ol	1.532
Acetophenone	1.532
Benzyl alcohol	1.538
1,2-Dibromoethane	1.538
1,2,3,4-Tetrahydronaphthalene	1.539
m-Cresol	1.542
1,3-Dichlorobenzene	1.543
Benzaldehyde	1.544
Styrene	1.545
Nitrobenzene	1.550
o-Dichlorobenzene	1.551
Bromobenzene	1.557
o-Nitroanisole	1.560
m-Toluidine	1.566
Benzyl benzoate	1.568
o-Toluidine	1.570
1-Methoxyphenyl-1-phenyl-ethane	1.571
Aniline	1.583
o-Chloroaniline	1.586
Bromoform	1.587
Benzenethiol	1.588
2,4-Bis(β-phenylethyl)-phenylmethyl ether	1.590
Carbondisulfide	1.628
1,12,2-Tetrabromomethane	1.633
Diiodomethane	1.749

MOLAR REFRACTION OF ORGANIC COMPOUNDS

The molar refraction, R, is defined as:

$$R = \left(\frac{n^2 - 1}{n^2 + 2}\right)\left(\frac{M}{d}\right)$$

where n = refractive index; M = molecular weight; d = density in grams per cm³; and (M/d) is the volume occupied by 1 gram molecular weight of the compound. The units of R will then be cm^{-3}.

R is, to a first approximation, independent of temperature or physical state, and it provides an approximate measure of the actual total volume (without free space) of the molecules in one gram mole.

For a very large number of compounds R is approximately additive for the bonds present in the molecule. Using R_D based on n_D (sodium light), the following atomic, group and structural contributions to R_D are based on Vogel's extensive modern measurements published in the Journal of the Chemical Society, 1948.

C	2.591	Cl	5.844	=S	7.921
H	1.028	Br	8.741	C≡N	5.459
=O	2.122	I	13.954	N (primary aliphatic)	2.376
–O–	1.643	C_6H_5	25.463	N (secondary aliphatic)	2.582
OH	2.553	$C_{10}H_7$	43.00		
F	0.81	–S–	7.729		

Ethylenic bond	1.575	Four membered ring	0.317
Acetylenic bond	1.977	Three membered ring	0.614
N (aromatic)	3.550		

Example:—For C_2H_5COOH: $R_{calc.} = 3(2.591) + 5(1.028) + 2.122 + 2.553 = 17.588$

$R_{obs.}$ for this compound is 17.51

For $C_6H_5NHCH_3$: $R_{calc.} = 25.463 + 3.550 + 2.591 + 4(1.028) = 35.716$

$R_{obs.}$ for this compound is 35.67

INDEX OF REFRACTION OF WATER

Alcohol and Carbon Bisulfide

For sodium light, $\lambda = .5893$

Temp. °C	Water, pure relative to air	Ethyl Alcohol 99.8 relative to air	Carbon Bisulfide relative to air
14	1.33348	....	
15	1.33341	....	1.62935
16	1.33333	1.36210	1.62858
18	1.33317	1.36129	1.62704
20	1.33299	1.36048	1.62546
22	1.33281	1.35967	1.62387
24	1.33262	1.35885	1.62226
26	1.33241	1.35803	1.62064
28	1.33219	1.35721	1.61902
30	1.33192	1.35639	1.61740
32	1.33164	1.35557	1.61577
34	1.33136	1.35474	1.61413
36	1.33107	1.35390	1.61247
38	1.33079	1.35306	1.61080
40	1.33051	1.35222	1.60914
42	1.33023	1.35138	1.60748
44	1.32992	1.35054	1.60582
46	1.32959	1.34969	......
48	1.32927	1.34885	......
50	1.32894	1.34800	......
52	1.32860	1.34715	......
54	1.32827	1.34629	......
56	1.32792	1.34543	......
58	1.32755	1.34456	......
60	1.32718	1.34368	......
62	1.32678	1.34279	......
64	1.32636	1.34189	......
66	1.32596	1.34096	......
68	1.32555	1.34004	......
70	1.32511	1.33912	......
72	1.32466	1.33820	......
74	1.32421	1.33728	......
76	1.32376	1.33626	......
78	1.32332	......	......
80	1.32287	......	......
82	1.32241	......	......
84	1.32195	......	......
86	1.32148	......	......
88	1.32100	......	......
90	1.32050	......	......
92	1.32000	......	......
94	1.31949	......	......
96	1.31897	......	......
98	1.31842	......	......
100	1.31783		

ABSOLUTE INDEX FOR PURE WATER FOR SODIUM LIGHT

Temperature	Index	Temperature	Index
15° C.	1.33377	60° C.	1.32754
20	1.33335	65	1.32652
25	1.33287	70	1.32547
30	1.33228	75	1.32434
35	1.33157	80	1.32323
40	1.33087	85	1.32208
45	1.33011	90	1.32086
50	1.32930	95	1.31959
55	1.32846	100	1.31819

INDEX OF REFRACTION OF GLASS

RELATIVE TO AIR

Variety	Wave length in microns.							
	.361	.434	.486	.589 (Na)	.656	.768	1.20	2.00
Zinc crown...........	1.539	1.528	1.523	1.517	1.514	1.511	1.505	1.497
Higher dispersion crown	1.546	1.533	1.527	1.520	1.517	1.514	1.507	1.497
Light flint...........	1.614	1.594	1.585	1.575	1.571	1.567	1.559	1.549
Heavy flint...........	1.705	1.675	1.664	1.650	1.644	1.638	1.628	1.617
Heaviest flint.........	...	1.945	1.919	1.890	1.879	1.867	1.848	1.832

INDEX OF REFRACTION OF ROCK SALT, SYLVINE, CALCITE, FLUORITE AND QUARTZ

(Compiled from data of Martens, Paschen, and others.)

Wave length.	Rock salt.	Silvine, KCl.	Fluorite.	Calcspar, ordinary ray.	Calcspar, extraordinary ray.	Quartz, ordinary ray.	Quartz, extraordinary ray.
0.185	1.893	1.827	......	......	......	1.676	1.690
0.198	.....	......	1.496	......	1.578	1.651	1.664
0.340	.....	......	......	1.701	1.506	1.567	1.577
0.589	1.544	1.490	1.434	1.658	1.486	1.544	1.553
0.760	.....	......	1.431	1.650	1.483	1.539	1.548
0.884	1.534	1.481	1.430				
1.179	1.530	1.478	1.428				
1.229	.....	......	......	1.639	1.479		
2.324	.....	......	......	......	1.474	1.516	
2.357	1.526	1.475	1.421				
3.536	1.523	1.473	1.414				
5.893	1.516	1.469	1.387				
8.840	1.502	1.461	1.331				

INDEX OF REFRACTION, AQUEOUS SOLUTIONS

Substance	Density	Temp. °C	Index for λ = .5893 (Na)	Observer
Ammonium chloride	1.067	27.05	1.379	Willigen
Ammonium chloride	1.025	29.75	1.351	Willigen
Calcium chloride	1.398	25.65	1.443	Willigen
Calcium chloride	1.215	22.9	1.397	Willigen
Calcium chloride	1.143	25.8	1.374	Willigen
Hydrochloric acid	1.166	20.75	1.411	Willigen
Nitric acid	1.359	18.75	1.402	Willigen
Potash (caustic)	1.416	11.0	1.403	Frauenhofer
Potassium chloride	Normal	solution	1.343	Bender
Potassium chloride	Double	normal	1.352	Bender
Potassium chloride	Triple	normal	1.360	Bender
Soda (caustic)	1.376	21.6	1.413	Willigen
Sodium chloride	1.189	18.07	1.378	Schutt
Sodium chloride	1.109	18.07	1.360	Schutt
Sodium chloride	1.035	18.07	1.342	Schutt
Sodium nitrate	1.358	22.8	1.385	Willigen
Sulfuric acid	1.811	18.3	1.437	Willigen
Sulfuric acid	1.632	18.3	1.425	Willigen
Sulfuric acid	1.221	18.3	1.370	Willigen
Sulfuric acid	1.028	18.3	1.339	Willigen
Zinc chloride	1.359	26.6	1.402	Willigen
Zinc chloride	1.209	26.4	1.375	Willigen

INDEX OF REFRACTION OF FUSED QUARTZ

λ mμ, 15° C	n, 18° C	λ mμ, 15° C	n, 18° C
185.467	1.57436	434.047	1.46690
193.583	1.55999	435.834	1.46675
202.55	1.54727	467.815	1.46435
214.439	1.53386	479.991	1.46355
219.462	1.52907	486.133	1.46318
226.503	1.52308	508.582	1.46191
231.288	1.51941	533.85	1.46067
250.329	1.50745	546.072	1.46013
257.304	1.50379	589.29	1.45845
274.867	1.49617	643.847	1.45674
303.412	1.48594	656.278	1.45640
340.365	1.47867	706.520	1.45517
396.848	1.47061	794.763	1.45340
404.656	1.46968		

INDEX OF REFRACTION OF AIR (15°C, 76 cm Hg)

Corrections for reducing wavelengths and frequencies in air (15°C, 76 cm Hg) to vacuo

The indices were computed from the Cauchy formula $(n-1)10^7 = 2726.43 + 12.288/(\lambda^2 \times 10^{-8}) + 0.3555/(\lambda^4 \times 10^{-16})$. For 0°C and 76 cm Hg the constants of the equation become 2875.66, 13.412 and 0.3777 respectively, and for 30°C and 76 cm Hg 2589.72, 12.259 and 0.2576. Sellmeier's formula for but one absorption band closely fits the observations: $n^2 = 1 + 0.00057378\lambda^2/(\lambda^2 - 595260)$. If $n-1$ were strictly proportional to the density, then $(n-1)_0/(n-1)t$ would equal $1 + \alpha t$ where α should be 0.00367. The following values of α were found to hold:

λ	0.85μ	0.75μ	0.65μ	0.55μ	0.45μ	0.35μ	0.25μ
α	0.003672	0.003674	0.003678	0.003685	0.003700	0.003738	0.003872

The indices are for dry air (0.05 ± % CO_2). Corrections to reduce to dry air the indices for moist air may be made for any wavelength by Lorenz's formula, $+0.000041(m/760)$, where m is the vapor pressure in mm. The corresponding frequencies in waves per cm and the corrections to reduce wavelengths and frequencies in air at 15°C and 76 cm Hg pressure to vacuo are given. E.g., a light wave of 5000 angstroms in dry air at 15°C, 76 cm Hg becomes 5001.391 A in vacuo; a frequency of 20.000 waves per cm correspondingly becomes 19994.44.

Wave-length, λ angstroms	Dry air $(n-1) \times 10^7$ 15°C 76 cm Hg	Vacuo correction for λ in air $(n\lambda - \lambda)$ add	Frequency waves per cm $\frac{1}{\lambda}$ in air	Vacuo correction for $\frac{1}{\lambda}$ in air $\left(\frac{1}{n\lambda} - \frac{1}{\lambda}\right)$ subtract	Wave-length, λ angstroms	Dry air $(n-1) \times 10^7$ 15°C 76 cm Hg	Vacuo correction for λ in air $(n\lambda - \lambda)$ add	Frequency waves per cm $\frac{1}{\lambda}$ in air	Vacuo correction for $\frac{1}{\lambda}$ in air $\left(\frac{1}{n\lambda} - \frac{1}{\lambda}\right)$ subtract
2000	3256	.651	50,000	16.27	3500	2850	.998	28,571	8.14
2100	3188	.670	47,619	15.18	3600	2842	1.023	27,777	7.89
2200	3132	.689	45,454	14.23	3700	2835	1.049	27,027	7.66
2300	3086	.710	43,478	13.41	3800	2829	1.075	26,315	7.44
2400	3047	.731	41,666	12.69	3900	2823	1.101	25,641	7.24
2500	3014	.754	40,000	12.05	4000	2817	1.127	25,000	7.04
2600	2986	.776	38,461	11.48	4100	2812	1.153	24,390	6.86
2700	2962	.800	37,037	10.97	4200	2808	1.179	23,809	6.68
2800	2941	.824	35,714	10.50	4300	2803	1.205	23,255	6.52
2900	2923	.848	34,482	10.08	4400	2799	1.232	22,727	6.36
3000	2907	.872	33,333	9.69	4500	2796	1.258	22,222	6.21
3100	2893	.897	32,258	9.33	4600	2792	1.284	21,739	6.07
3200	2880	.922	31,250	9.00	4700	2789	1.311	21,276	5.93
3300	2869	.947	30,303	8.69	4800	2786	1.338	20,833	5.80
3400	2859	.972	29,411	8.41	4900	2784	1.364	20,406	5.68

INDEX OF REFRACTION OF AIR (15°C, 76 cm Hg) (Continued)

Wave-length, λ angstroms	Dry air $(n-1) \times 10^7$ 15°C 76 cm Hg	Vacuo correction for λ in air $(n\lambda - \lambda)$ add	Frequency waves per cm $\frac{1}{\lambda}$ in air	Vacuo correction for $\frac{1}{\lambda}$ in air $\left(\frac{1}{n\lambda} - \frac{1}{\lambda}\right)$ subtract	Wave-length, λ angstroms	Dry air $(n-1) \times 10^7$ 15°C 76 cm Hg	Vacuo correction for λ in air $(n\lambda - \lambda)$ add	Frequency waves per cm $\frac{1}{\lambda}$ in air	Vacuo correction for $\frac{1}{\lambda}$ in air $\left(\frac{1}{n\lambda} - \frac{1}{\lambda}\right)$ subtract
5000	2781	1.391	20,000	5.56	7000	2753	1.927	14,285	3.93
5100	2779	1.417	19,607	5.45	7100	2752	1.954	14,084	3.88
5200	2777	1.444	19,230	5.34	7200	2751	1.981	13,888	3.82
5300	2775	1.471	18,867	5.23	7300	2751	2.008	13,698	3.77
5400	2773	1.497	18,518	5.13	7400	2750	2.035	13,513	3.72
					7500	2749	2.062	13,333	3.66
5500	2771	1.524	18,181	5.04	7600	2749	2.089	13,157	3.62
5600	2769	1.551	17,857	4.94	7700	2748	2.116	12,987	3.57
5700	2768	1.578	17,543	4.85	7800	2748	2.143	12,820	3.52
5800	2766	1.604	17,241	4.77	7900	2747	2.170	12,658	3.48
5900	2765	1.631	16,949	4.68					
6000	2763	1.658	16,666	4.60	8000	2746	2.197	12,500	3.43
6100	2762	1.685	16,393	4.53	8100	2746	2.224	12,345	3.39
6200	2761	1.712	16,129	4.45	8250	2745	2.265	12,121	3.33
6300	2760	1.739	15,873	4.38	8500	2744	2.332	11,764	3.23
6400	2759	1.766	15,625	4.31	8750	2743	2.400	11,428	3.13
6500	2758	1.792	15,384	4.24	9000	2742	2.468	11,111	3.05
6600	2757	1.819	15,151	4.18	9250	2741	2.536	10,810	2.96
6700	2756	1.846	14,925	4.11	9500	2740	2.604	10,526	2.88
6800	2755	1.873	14,705	4.05	9750	2740	2.671	10,256	2.81
6900	2754	1.900	14,492	3.99	10000	2739	2.739	10,000	2.74

INDEX OF REFRACTION, GASES

Values are relative to a vacuum and for a Temp. of 0° C. and 760 mm. pressure.

(From Smithsonian Tables)

Substance	Kind of light	Indices of refraction	Observer
Acetone	D	1.001079–1.001100	
Air	D	1.0002926	Perreau
Ammonia	white	1.000381–1.000385	
Ammonia	D	1.000373–1.000379	
Argon	D	1.000281	Rayleigh
Benzene	D	1.001700–1.001823	
Bromine	D	1.001132	Mascart
Carbon dioxide	white	1.000449–1.000450	
dioxide	D	1.000448–1.000454	
disulfide	white	1.001500	Dulong
disulfide	D	1.001478–1.001485	
monoxide	white	1.000340	Dulong
monoxide	white	1.000335	Mascart
Chlorine	white	1.000772	Dulong
Chlorine	D	1.000773	Mascart
Chloroform	D	1.001436–1.001464	
Cyanogen	white	1.000834	Dulong
Cyanogen	D	1.000784–1.000825	
Ethyl alcohol	D	1.000871–1.000885	
ether	D	1.001521–1.001544	
Helium	D	1.000036	Ramsay
Hydrochloric acid	white	1.000449	Mascart
Hydrochloric acid	D	1.000447	Mascart
Hydrogen	white	1.000138–1.000143	
Hydrogen	D	1.000132	Burton
sulfide	D	1.000644	Dulong
sulfide	D	1.000623	Mascart
Methane	white	1.000443	Dulong
Methane	D	1.000444	Mascart
Methyl alcohol	D	1.000549–1.000623	
Methyl ether	D	1.000891	Marcast
Nitric oxide	white	1.000303	Dulong
Nitric oxide	D	1.000297	Mascart
Nitrogen	white	1.000295–1.000300	
Nitrogen	D	1.000296–1.000298	
Nitrous oxide	white	1.000503–1.000507	
Nitrous oxide	D	1.000516	Mascart
Oxygen	white	1.000272–1.000280	
Oxygen	D	1.000271–1.000272	
Pentane	D	1.001711	Mascart
Sulfur dioxide	white	1.000665	Dulong
Sulfur dioxide	D	1.000686	Ketteler
Water	white	1.000261	Jamin
Water	D	1.000249–1.000259	

COEFFICIENT OF TRANSPARENCY OF UVIOL GLASS FOR THE ULTRA-VIOLET

For a thickness of 1 mm.

Wave length, microns	0.280	0.309	0.325	0.346	0.361	0.383	0.397
Uviol crown	0.56	0.95	0.990	0.996	0.999	1.000	1.000

INDEX OF REFRACTION OF AQUEOUS SOLUTIONS OF SUCROSE (CANE SUGAR)

The table gives the index of refraction for λ = 0.5893 of aqueous sugar solutions at 20°C from 0–85 % sugar. Corrections for temperatures other than 20° are given at the end of the table.

Per cent sugar		.0	.1	.2	.3	.4	.5	.6	.7	.8	.9
00	1.3	330	331	333	334	336	337	338	340	341	342
1		344	345	347	348	350	351	353	355	356	357
2		359	361	362	363	365	367	368	369	371	373
3		374	375	377	378	380	381	382	384	385	387
4		388	389	391	393	394	395	397	399	400	401
5		403	405	406	407	409	411	412	413	415	417
6		418	419	421	423	424	425	427	429	430	431
7		433	435	436	437	439	441	442	443	445	447
8		448	450	451	453	454	456	458	459	461	462
9		464	465	467	469	470	471	473	475	476	477
10		479	481	482	483	485	487	488	489	491	493
11		494	496	497	499	500	502	504	505	507	508
12		510	512	513	515	516	518	520	521	523	524
13		526	527	529	531	532	533	535	537	538	539
14		541	543	544	546	547	549	551	552	554	555
15		557	559	560	562	563	565	567	568	570	571
16		573	575	576	578	580	582	583	585	587	588
17		590	592	593	595	596	598	600	601	603	604
18		606	608	609	611	612	614	616	617	619	620
19		622	624	625	627	629	631	632	634	636	637
20		639	641	642	644	645	647	649	650	652	653
21		655	657	658	660	662	663	665	667	669	670
22		672	674	675	677	679	681	682	684	686	687
23		689	691	692	694	696	698	699	701	703	704
24		706	708	709	711	713	715	716	718	720	721
25		723	725	726	728	730	731	733	735	737	738
26		740	742	744	745	747	749	751	753	754	756
27		758	760	761	763	765	767	768	770	772	773
28		775	777	779	780	782	784	786	788	789	791
29		793	795	797	798	800	802	804	806	807	809
30		811	813	815	816	818	820	822	824	825	827
31		829	831	833	834	836	838	840	842	843	845
32		847	849	851	852	854	856	858	860	861	863
33		865	867	869	870	872	874	876	878	879	881
34		883	885	887	889	891	893	894	896	898	900
35		902	904	906	907	909	911	913	915	916	918
36		920	922	924	926	928	929	931	933	935	937
37		939	941	943	945	947	949	950	952	954	956
38		958	960	962	964	966	968	970	972	974	976
39		978	980	982	984	986	987	989	991	993	995
40		997	999	*001	*003	*005	*007	*008	*010	*012	*014
41	1.4	016	018	020	022	024	026	028	030	032	034
42		036	038	040	042	044	046	048	050	052	054
43		056	058	060	062	064	066	068	070	072	074
44		076	078	080	082	084	086	088	090	092	094
45		096	098	100	102	104	107	109	111	113	115
46		117	119	121	123	125	127	129	131	133	135
47		137	139	141	143	145	147	150	152	154	156
48		158	160	162	164	166	169	171	173	175	177
49		179	181	183	185	187	189	192	194	196	198
50		200	202	204	206	208	211	213	215	217	219
51		221	223	225	227	229	231	234	236	238	240
52		242	244	246	249	251	253	255	257	260	262
53		264	266	268	270	272	275	277	279	281	283
54		285	287	289	292	294	296	298	300	303	305
55		307	309	311	313	316	318	320	322	325	327
56		329	331	333	336	338	340	342	344	347	349
57		351	353	355	358	360	362	364	366	369	371
58		373	375	378	380	382	385	387	389	391	394
59		396	398	400	403	405	407	409	411	414	416
60		418	420	423	425	427	429	432	434	436	439
61		441	443	446	448	450	453	455	457	459	462
62		464	466	468	471	473	475	477	479	482	484
63		486	488	491	493	495	497	500	502	504	507
64		509	511	514	516	518	521	523	525	527	530
65		532	534	537	539	541	544	546	548	550	553
66		558	561	563	565	567	570	572	574	577	579
67		581	584	586	588	591	593	595	598	600	602
68		605	607	609	612	614	616	619	621	623	625
69		628	630	632	635	637	639	642	644	646	649
70		651	653	656	658	661	663	666	668	671	673
71		676	678	681	683	685	688	690	693	695	698
72		700	703	705	708	710	713	715	717	720	722
73		725	727	730	732	735	737	740	742	744	747
74		749	752	754	757	759	762	764	767	769	772
75		774	777	779	782	784	787	789	792	794	797
76		799	802	804	807	810	812	815	817	820	822
77		825	827	830	832	835	838	840	843	845	848
78		850	853	855	858	860	863	865	868	871	873
79		876	878	881	883	886	888	891	893	896	898
80		901	904	906	909	912	914	917	919	922	925
81		927	930	933	935	938	941	943	946	949	951
82		954	956	959	962	964	967	970	972	975	978
83		980	983	985	988	991	993	996	999	*001	*004
84	1.5	007	009	012	015	017	020	022	025	028	030
85		033									

Correction table for determining the percentage of sucrose by means of the refractometer when the readings are made at temperatures other than 20° C[1]

Temperature	Percentage of sucrose— 0	5	10	15	20	25	30	35	40	45	50	55	60	65	70
	Subtract from the percentage of sucrose														
°C.															
10	0.50	0.54	0.58	0.61	0.64	0.66	0.68	0.70	0.72	0.73	0.74	0.75	0.76	0.78	0.79
11	.46	.49	.53	.55	.58	.60	.62	.64	.65	.66	.67	.68	.69	.70	.71
12	.42	.45	.48	.50	.52	.54	.56	.57	.58	.59	.60	.61	.61	.63	.63
13	.37	.40	.42	.44	.46	.48	.49	.50	.51	.52	.53	.54	.54	.55	.55
14	.33	.35	.37	.39	.40	.41	.42	.43	.44	.45	.45	.46	.46	.47	.48
15	.27	.29	.31	.33	.34	.34	.35	.36	.37	.37	.38	.39	.39	.40	.40
16	.22	.24	.25	.26	.27	.28	.28	.29	.30	.30	.30	.31	.31	.32	.32
17	.17	.18	.19	.20	.21	.21	.21	.22	.22	.23	.23	.23	.23	.24	.24
18	.12	.13	.13	.14	.14	.14	.14	.15	.15	.15	.15	.16	.16	.16	.16
19	.06	.06	.06	.07	.07	.07	.07	.08	.08	.08	.08	.08	.08	.08	.08
	Add to the percentage of sucrose														
21	0.06	0.07	0.07	0.07	0.07	0.08	0.08	0.08	0.08	0.08	0.08	0.08	0.08	0.08	0.08
22	.13	.13	.14	.14	.15	.15	.15	.15	.15	.16	.16	.16	.16	.16	.16
23	.19	.20	.21	.22	.22	.23	.23	.23	.23	.24	.24	.24	.24	.24	.24
24	.26	.27	.28	.29	.30	.30	.31	.31	.31	.31	.31	.32	.32	.32	.32
25	.33	.35	.36	.37	.38	.38	.39	.40	.40	.40	.40	.40	.40	.40	.40
26	.40	.42	.43	.44	.45	.46	.47	.48	.48	.48	.48	.48	.48	.48	.48
27	.48	.50	.52	.53	.54	.55	.55	.56	.56	.56	.56	.56	.56	.56	.56
28	.56	.57	.60	.61	.62	.63	.63	.64	.64	.64	.64	.64	.64	.64	.64
29	.64	.66	.68	.69	.71	.72	.72	.73	.73	.73	.73	.73	.73	.73	.73
30	.72	.74	.77	.78	.79	.80	.80	.81	.81	.81	.81	.81	.81	.81	.81

[1] International Temperature Correction Table, 1936, adopted by the International Commission for Uniform Methods of Sugar Analysis (Int. Sugar J. **39**, 24s, 1937).

INDEX OF REFRACTION OF MALTOSE HYDRATE SOLUTIONS

McDonald: Journal of Research of the National Bureau of Standards, Vol. **46**, 165, 1951.
The following table gives the refractive index of maltose solutions for λ = 5893 A at 20° and 25°C.

Percent	n_D^{20}	n_D^{25}	Δn	$\Delta n/\Delta t$
1	1.33438	1.33389	0.00049	0.00010
2	1.33579	1.33528	.00051	.00010
3	1.33720	1.33668	.00052	.00010
4	1.33862	1.33810	.00052	.00010
5	1.34006	1.33952	.00054	.00011
6	1.34150	1.34095	.00055	.00011
7	1.34295	1.34239	.00056	.00011
8	1.34442	1.34384	.00058	.00012
9	1.34589	1.34530	.00059	.00012
10	1.34738	1.34677	.00061	.00012
11	1.34887	1.34825	.00062	.00012
12	1.35039	1.34974	.00065	.00013
13	1.35190	1.35124	.00066	.00013
14	1.35343	1.35276	.00067	.00013
15	1.35497	1.35428	.00069	.00014
16	1.35652	1.35582	.00070	.00014
17	1.35808	1.35737	.00071	.00014
18	1.35965	1.35893	.00072	.00014
19	1.36124	1.36051	.00073	.00015
20	1.36283	1.36209	.00074	.00015
21	1.36444	1.36369	.00075	.00015
22	1.36606	1.36530	.00076	.00015
23	1.36770	1.36692	.00078	.00016
24	1.36934	1.36856	.00078	.00016
25	1.37100	1.37021	.00079	.00016
26	1.37267	1.37187	.00080	.00016
27	1.37435	1.37355	.00080	.00016
28	1.37604	1.37524	.00080	.00016
29	1.37775	1.37694	.00081	.00016
30	1.37946	1.37865	.00081	.00016
31	1.38120	1.38038	.00082	.00016
32	1.38294	1.38213	.00082	.00016
33	1.38470	1.38388	.00082	.00016
34	1.38647	1.38565	.00081	.00016
35	1.38825	1.38744	.00081	.00016
36	1.39004	1.38924	0.00080	0.00016
37	1.39185	1.39105	.00080	.00016
38	1.39367	1.39287	.00080	.00016
39	1.39551	1.39471	.00080	.00016
40	1.39735	1.39656	.00079	.00016
41	1.39922	1.39843	.00079	.00016
42	1.40109	1.40031	.00078	.00016
43	1.40298	1.40221	.00077	.00016
44	1.40488	1.40411	.00077	.00015
45	1.40680	1.40603	.00077	.00015
46	1.40873	1.40797	.00076	.00015
47	1.41067	1.40992	.00075	.00015
48	1.41263	1.41188	.00075	.00015
49	1.41460	1.41385	.00075	.00015
50	1.41658	1.41584	.00074	.00015
51	1.41858	1.41784	.00074	.00015
52	1.42059	1.41986	.00073	.00015
53	1.42262	1.42189	.00073	.00015
54	1.42466	1.42392	.00074	.00015
55	1.42672	1.42598	.00074	.00015
56	1.42878	1.42804	.00074	.00015
57	1.43087	1.43012	.00075	.00015
58	1.43296	1.43221	.00075	.00015
59	1.43508	1.43431	.00077	.00015
60	1.43720	1.43643	.00077	.00015
61	1.43934	1.43855	.00079	.00016
62	1.44150	1.44069	.00081	.00016
63	1.44367	1.44283	.00084	.00017
64	1.44585	1.44499	.00086	.00017
65	1.44805	1.44716	.00089	.00018

RADIATION FROM AN IDEAL BLACK BODY

From NASA TT-F-783

Temperature dependence of the specific power radiated, Q_T, and of λ_{max} for an ideal black body according to Kirchhoff's law ($\sigma_o = 5.68 \cdot 10^{-8}$ W/m²·deg⁴ = $4.88 \cdot 10^{-8}$ kcal/m²·deg⁴)

T, °K	t, °C	Q_T, W/cm²	Q_T kcal/m²·hr	λmax μ	T, °K	t, °C	Q_T W/cm²	Q_T kcal/m²·hr	λmax μ
100	−173	$5.680 \cdot 10^{-4}$	$4.880 \cdot 10^{0}$	28.96	730	457	$1.613 \cdot 10^{0}$	$1.386 \cdot 10^{1}$	3.967
200	−73	$9.088 \cdot 10^{-3}$	$7.808 \cdot 10^{1}$	14.48	740	467	1.703	1.463	3.914
273	0	$3.155 \cdot 10^{-2}$	$2.711 \cdot 10^{2}$	10.608	750	477	1.797	1.544	3.861
300	27	4.601	3.953	9.655	760	487	1.895	1.628	3.811
310	37	5.246	4.507	9.342	770	497	1.997	1.715	3.761
320	47	5.956	5.117	9.050	780	507	2.102	1.806	3.713
330	57	6.736	5.787	8.766	790	517	2.212	1.901	3.666
340	67	7.590	6.521	8.518	800	527	2.327	1.999	3.620
350	77	8.524	7.323	8.274	810	537	2.445	2.101	3.565
360	87	9.540	8.196	8.044	820	547	2.568	2.206	3.532
370	97	1.065	9.146	7.827	830	557	2.696	2.316	3.489
380	107	$1.184 \cdot 10^{-1}$	$1.018 \cdot 10^{3}$	7.621	840	567	2.828	2.430	3.448
390	117	1.314	1.128	7.426	850	577	2.965	2.547	3.407
400	127	1.454	1.249	7.270	860	587	3.107	2.670	3.367
410	137	1.605	1.379	7.053	870	597	3.254	2.796	3.329
420	147	1.76	1.519	6.865	880	607	3.406	2.927	3.291
430	157	1.942	1.668	6.735	890	617	3.564	3.062	3.254
440	167	2.129	1.829	6.562	900	627	3.727	3.202	3.218
450	177	2.329	2.001	6.436	910	637	3.895	3.346	3.162
460	187	2.543	2.185	6.266	920	647	4.069	3.496	3.148
470	197	2.772	2.381	6.162	930	657	4.249	3.650	3.114
480	207	3.015	2.591	6.033	940	667	4.435	3.810	3.081
490	217	3.274	2.813	5.910	950	677	4.626	3.975	3.048
500	227	3.550	3.05	5.792	960	687	4.824	4.145	3.017
510	237	3.843	3.301	5.668	970	697	5.028	4.320	2.986
520	247	4.163	3.568	5.559	980	707	5.239	4.501	2.955
530	257	4.482	3.851	5.454	990	717	5.456	4.688	2.925
540	267	4.830	4.150	5.363	1000	727	5.680	4.880	2.896
550	277	5.198		5.255	1010	737	5.909	5.07	2.866
560	287	5.586	4.799	5.161	1020	747	6.143	5.278	2.836
570	297	5.996	5.151	5.061	1030	757	6.394	5.494	2.812
580	307	6.428	5.522	4.963	1040	767	6.645	5.709	2.785
590	317	6.883	5.913	4.908	1050	777	6.904	5.932	2.758
600	327	7.361	6.324	4.827	1060	787	7.171	6.161	2.732
610	337	7.864	6.757	4.748	1070	797	7.445	6.397	2.707
620	347	8.393	7.211	4.671	1080	807	7.728	6.640	2.681
630	357	8.948	7.687	4.597	1090	817	8.018	6.888	2.657
640	367	9.529	8.187	4.525	1100	827	8.316	7.145	2.633
650	377	$1.014 \cdot 10^{0}$	8.711	4.455	1110	837	8.623	7.408	2.609
660	387	1.078	9.260	4.388	1120	847	8.937	7.679	2.586
670	397	1.145	9.831	4.322	1130	857	9.261	7.956	2.563
680	407	1.214	$1.013 \cdot 10^{4}$	4.259	1140	867	9.593	8.242	2.540
690	417	1.287	1.106	4.197	1150	877	9.934	8.535	2.516
700	427	1.364	1.172	4.137	1160	887	$1.028 \cdot 10^{1}$	8.836	2.497
710	437	1.443	1.240	4.069	1170	897	1.064	9.145	2.475
720	447	1.526	1.311	4.022	1180	907	1.101	9.461	2.454
1190	917	$1.139 \cdot 10^{1}$	$9.786 \cdot 10^{4}$	2.434	1640	1367	$4.109 \cdot 10^{1}$	$3.530 \cdot 10^{5}$	1.766
1200	927	1.178	$1.042 \cdot 10^{5}$	2.413	1650	1377	4.210	3.617	1.755
1210	937	1.218	1.046	2.393	1660	1387	4.313	3.706	1.745
1220	947	1.258	1.081	2.374	1670	1397	4.418	3.796	1.734
1230	957	1.300	1.117	2.354	1680	1407	4.525	3.887	1.724

RADIATION FROM AN IDEAL BLACK BODY *(Continued)*

From NASA TT-F-783

T, °K	t, °C	Q_T, W/cm²	Q_T kcal/m²·hr	λmax μ	T, °K	t, °C	Q_T W/cm²	Q_T kcal/m²·hr	λmax μ
1240	967	1.343	1.154	2.335	1690	1417	4.633	3.981	1.714
1250	977	1.387	1.191	2.317	1700	1427	4.744	4.076	1.704
1260	987	1.431	1.230	2.298	1710	1437	4.858	4.183	1.694
1270	997	1.478	1.270	2.280	1720	1447	4.971	4.271	1.684
1280	1007	1.525	1.310	2.263	1730	1457	5.088	4.371	1.674
1290	1017	1.573	1.351	2.245	1740	1467	5.206	4.473	1.664
1300	1027	1.622	1.394	2.227	1750	1477	5.327	4.577	1.655
1310	1037	1.673	1.437	2.211	1760	1487	5.450	4.682	1.645
1320	1047	1.724	1.482	2.194	1770	1497	5.575	4.790	1.636
1330	1057	1.777	1.527	2.177	1780	1507	5.703	4.900	1.627
1340	1067	1.831	1.573	2.161	1790	1517	5.831	5.010	1.617
1350	1077	1.887	1.621	2.145	1800	1527	5.963	5.123	1.607
1360	1087	1.943	1.669	2.129	1810	1537	6.096	5.238	1.600
1370	1097	2.001	1.719	2.114	1820	1547	6.232	5.354	1.591
1380	1107	2.058	1.768	2.099	1830	1557	6.370	5.473	1.583
1390	1117	2.120	1.822	2.083	1840	1567	6.511	5.594	1.574
1400	1127	2.182	1.875	2.067	1850	1577	6.653	5.717	1.565
1410	1137	2.245	1.929	2.054	1860	1587	6.798	5.841	1.557
1420	1147	2.309	1.984	2.036	1870	1597	6.946	5.967	1.549
1430	1157	2.375	2.041	2.025	1880	1607	7.095	6.096	1.540
1440	1167	2.442	2.098	2.011	1890	1617	7.248	6.227	1.532
1450	1177	2.511	2.157	1.997	1900	1627	7.402	6.360	1.524
1460	1187	2.581	2.217	1.984	1910	1637	7.559	6.495	1.516
1470	1197	2.652	2.279	1.970	1920	1647	7.720	6.632	1.509
1480	1207	2.725	2.341	1.957	1930	1657	7.881	6.771	1.501
1490	1217	2.800	2.405	1.944	1940	1667	8.046	6.912	1.493
1500	1227	2.876	2.471	1.931	1950	1677	8.213	7.056	1.485
1510	1237	2.953	2.537	1.917	1960	1687	8.382	7.202	1.478
1520	1247	3.032	2.605	1.905	1970	1697	8.555	7.350	1.470
1530	1257	3.113	2.674	1.893	1980	1707	8.730	7.500	1.463
1540	1267	3.195	2.745	1.881	1990	1717	8.907	7.653	1.455
1550	1277	3.278	2.817	1.866	2000	1727	9.088	7.808	1.448
1560	1287	3.363	2.890	1.856	2500	2227	2.219·10²	1.906·10⁶	1.156
1570	1297	3.451	2.965	1.845	3000	2727	4.601	3.953	0.965
1580	1307	3.540	3.041	1.833	3500	3227	8.524	7.323	0.826
1590	1317	3.630	3.198	1.821	4000	3727	1.454·10³	1.249	0.722
1600	1327	3.722		1.810	4500	4227	2.329	2.001	0.644
1610	1337	3.816	3.279	1.797	5000	4727	3.550	3.050	0.579
1620	1347	3.912	3.361	1.787	5500	5227	5.198	4.465	0.527
1630	1357	4.010	3.445	1.777	6000	5727	7.361	6.324	0.483

REFLECTION COEFFICIENTS

Coefficients of Reflection of Miscellaneous Surfaces for Monochromatic Radiation in the Visible Spectrum
(J. L. Michaelson)

Material	Wave lengths (μ)			
	0.400	0.500	0.600	0.700
Carbon Black in Oil	0.003	0.003	0.003	0.003
Clay,				
Kaolin (treated)	0.82	0.81	0.82	0.82
Kaolin (untreated)	0.75	0.79	0.85	0.86
White Georgia	0.94	0.92	0.93	0.94
Magnesium oxide	0.97	0.98	0.98	0.98
Paint,				
Lithopone	0.95	0.98	0.98	0.98
$MgCO_3$-Vynal Acetate Lacquer	0.90	0.88	0.88	0.88
ZnO-Milk	0.74	0.84	0.85	0.86
Paper,				
Blotting	0.64	0.72	0.79	0.79
Calendered	0.64	0.69	0.73	0.76
Crepe, green	0.23	0.49	0.19	0.48
Crepe, red	0.03	0.02	0.21	0.69
Crepe, yellow	0.17	0.44	0.75	0.79
News Print Stock	0.38	0.61	0.63	0.78
Peach,				
Green	0.18	0.17	0.62	0.63
Ripe	0.10	0.10	0.41	0.42
Pear,				
Green	0.04	0.12	0.29	0.41
Ripe	0.08	0.19	0.46	0.53
Pigment,				
Chrome Yellow	0.05	0.13	0.70	0.77
French Ochre	0.06	0.14	0.50	0.56
Porcelain Enamel,				
Blue	0.44	0.10	0.05	0.23
Orange	0.09	0.09	0.59	0.69
Red	0.05	0.03	0.08	0.62
White	0.77	0.73	0.72	0.70
Yellow	0.11	0.46	0.62	0.62
Talcum, Italian	0.94	0.89	0.88	0.88
Wheat Flour	0.75	0.87	0.94	0.97

REFLECTION COEFFICIENTS OF SURFACES FOR "INCANDESCENT" LIGHT

Material	Nature of Surface	Coefficient	Authority
Aluminum, "Alzak"	Diffusing	0.77–0.81	3
"Alzak"	Specular	0.79–0.83	3
on Glass	First Surface	0.82–0.86	4
Polished	Specular	0.69	3
Black Paper	Diffusing	0.05–0.06	4
Chromium	Specular	0.62	4
Copper	Specular	0.63	4
Gold	Specular	0.75	1
Magnesium oxide	Diffusing	0.98	5
Nickel	Specular	0.62–0.64	1, 3
Platinum	Specular	0.62	1
Porcelain Enamel	Glossy	0.76–0.79	3
Porcelain Enamel	Ground	0.81	3
Porcelain Enamel	Matt.	0.72–0.76	3
Silver	Polished	0.93	1
Silvered Glass	Second Surface	0.88–0.93	3
Snow	Diffusing	0.93	2
Steel	Specular	0.55	1
Stellite	Specular	0.58–0.65	4

(1) Hagen and Rubens. (2) Nutting, Jones, and Elliot. (3) J. E. Bock. (4) Frank Benford. (5) J. L. Michaelson.

EMISSIVITY AND ABSORPTION

These data are the result of investigations made by the Bureau of Standards, the British National Physical Laboratory, General Electric Research Laboratories, and several eastern universities, and were collected by W. J. King of the General Electric Company.

Low Temperature Total Emissivities

Silver, highly polished	0.02	Brass, polished	0.60
Platinum " "	0.05	Oxidized copper	0.60
Zinc " "	0.05	Oxidized steel	0.70
Aluminum, " "	0.08	Bronze paint	0.80
*Monel metal, polished	0.09	Black gloss paint	0.90
Nickel "	0.12	White lacquer	0.95
Copper "	0.15	White vitreous enamel	0.95
Stellite "	0.18	Asbestos paper	0.95
Cast iron "	0.25	Green paint	0.95
Monel metal, oxidized	0.43	Gray paint	0.95
Aluminum paint	0.55	Lamp black	0.95

Coefficient of Absorption of Solar Radiation

Silver, highly polished	0.07	Stellite, polished	0.3
Platinum " "	0.10	Light cream paint	0.3
Nickel " "	0.15	Monel metal, polished	0.4
*Aluminum	0.15	Light yellow paint	0.4
Magnesium carbonate	0.15	Light green paint	0.5
Zinc oxide	0.15	Aluminum paint	0.5
*Steel	0.20	Zinc, polished metal	0.5
Copper	0.25	Gray paint	0.7
White lead paint	0.25	Black matte	0.9
Zinc oxide paint	0.30		

* Questionable because of scant or inconsistent data.

EMISSIVITY OF TOTAL RADIATION, ε_{tot}, FOR VARIOUS MATERIALS

Material	Temperature (° C)	ε_{tot}
Alloys		
Nickel-Chromium		
20 Ni—25 Cr—55 Fe, oxidized	200	0.90
	500	0.97
60 Ni—12 Cr—28 Fe, oxidized	270	0.89
	560	0.82
80 Ni—20 Cr	100	0.87
	600	0.87
	1300	0.89
Aluminum		
Polished	50—500	0.04—0.06
Rough surface	20—50	0.06—0.07
Strongly oxidized	55—500	0.2—0.3
	25	0.022
	100	0.028
	500	0.060
Oxidized	200	0.11
	600	0.19
Asbestos board	20	0.96
Bismuth		
Unoxidized	25	0.048
Brass		
Dull tarnished	200	0.61
Oxidized at 600° C	200	0.61
	600	0.59
Unoxidized	25	0.035
	100	0.035
Polished	200	0.03
Rolled sheet	20	0.06
Bronze		
Polished	50	0.1
Carbon		
Filament	1000—1400	0.53
Graphite	0—3600	0.7—0.8
Lamp black	20—400	0.96
Soot applied to solid	50—1000	0.96
Soot with water glass	20—200	0.96
Unoxidized	100	0.81
Chromium		
Polished	50	0.1
	500—1000	0.28—0.38
Colbalt		
Unoxidized	500	0.13
	1000	0.23
Columbium		
Unoxidized	1500	0.19
Copper		
Calorized	100	0.26
Calorized, oxidized	200	0.18
	600	0.19
Commercial, scoured to a shine	20	0.07
Oxidized	50	0.6—0.7
	500	0.88
Polished	50—100	0.02
Unoxidized	100	0.02
Unoxidized, liquid	—	0.15
Fire brick	1000	0.75

EMISSIVITY OF TOTAL RADIATION, ε_{tot}, FOR VARIOUS MATERIALS (continued)

Material	Temperature (° C)	ε_{tot}
Glass	20—100	0.94—0.91
	250—1000	0.87—0.72
	1100—1500	0.7—0.67
Gold		
Carefully polished	200—600	0.02—0.03
Unoxidized	100	0.02
Enamel	100	0.37
Graphite	0—3600	0.7—0.8
Gypsum	20	0.93
Iron		
Cast		
Oxidized	200	0.64
	600	0.78
Strongly oxidized	40	0.95
	250	0.95
Unoxidized	100	0.21
Unoxidized, liquid	—	0.29
Oxidized	100	0.74
	500	0.84
	1200	0.89
Rusted	25	0.65
Wrought, dull	100	0.05
Lamp black	25	0.94
Lead	20—400	0.96
Oxidized		
Unoxidized	200	0.05
Mercury	200	0.63
Unoxidized		
	25	0.10
Molybdenum	100	0.12
	600—1000	0.08—0.13
Monel metal	1500—2200	0.19—0.26
Oxidized		
	200	0.43
	600	0.43
Nichrome		
Wire		
Clean		
	50	0.65
Oxidized	500—1000	0.71—0.79
Nickel	50—500	0.95—0.98
Industrial, polished		
Oxidized	200—400	0.07—0.09
Oxidized at 600°C	200	0.37
Unoxidized	200—600	0.37—0.48
	25	0.045
	100	0.06
	500	0.12
Platinum	1000	0.19
Clean, polished		
Unoxidized	200—600	0.05—0.1
	25	0.037
	100	0.047

Material	Temperature (° C)	ε_{tot}
	500	0.096
	1000	0.152
	1500	0.191
Wire	50—200	0.06—0.07
	500—1000	0.1—0.16
	1400	0.18
Porcelain		
Glazed	20	0.92
Rubber		
Hard	20	0.95
Soft, gray, rough	20	0.86
Silica brick	1000	0.80
	1100	0.85
Silver		
Clean, polished	200—600	0.02—0.03
Unoxidized	100	0.02
	500	0.035
Soot applied to a solid surface	50—1000	0.94—0.91
Soot with water glass	20—200	0.96
Steel		
Alloyed (8% Ni, 18% Cr)	500	0.35
Aluminized	50—500	0.79
Dull nickel plated	20	0.11
Flat, rough surface	50	0.95—0.98
Cast, polished	750—1050	0.52—0.56
Sheet, ground	50	0.56
	950—1100	0.55—0.61
Oxidized	200—600	0.8
Calorized, oxidized	200	0.52
	600	0.57
Sheet with shiny layer of oxide	20	0.82
Strongly oxidized	50	0.88
	500	0.98
Unoxidized	100	0.08
Unoxidized, liquid	—	0.28
Tantalum		
Unoxidized	1500	0.21
	2000	0.26
Tungsten		
Unoxidized	25	0.024
	100	0.032
	500	0.071
	1000	0.15
	1500	0.23
	2000	0.28
Varnish	40—100	0.8—0.95
Dull black	40—100	0.96—0.98
Glossy black sprayed on iron	20	0.87
Zinc		
Polished	200—300	0.04—0.05
Unoxidized	300	0.05

SPECTRAL EMISSIVITY

Prepared by Roeser and Wensel, National Bureau of Standards

Spectral Emissivity of Materials, Surface Unoxidized for 0.65μ

Element	Solid	Liquid	Element	Solid	Liquid
Beryllium	0.61	0.61	Thorium	0.36	0.40
Carbon	0.80–0.93	....	Titanium	0.63	0.65
Chromium	0.34	0.39	Tungsten	0.43	
Cobalt	0.36	0 37	Uranium	0.54	0.34
Columbium	0.37	0.40	Vanadium	0.35	0.32
Copper	0.10	0.15	Yttrium	0.35	0.35
Erbium	0.55	0.38	Zirconium	0.32	0.30
Gold	0.14	0.22	Steel	0.35	0.37
Iridium	0.30	....	Cast Iron	0.37	0.40
Iron	0.35	0.37	Constantan	0.35	
Manganese	0.59	0.59	Monel	0.37	
Molybdenum	0.37	0.40	Chromel P (90Ni-10Cr)	0.35	
Nickel	0.36	0.37	80Ni-20Cr	0.35	
Palladium	0.33	0.37	60Ni-24Fe-16Cr	0.36	
Platinum	0.30	0.38	Alumel (95Ni; Bal. Al, Mn, Si)	0.37	
Rhodium	0.24	0.30	90Pt-10Rh	0.27	
Silver	0.07	0.07			
Tantalum	0.49	....			

Spectral Emissivity of Oxides

The emissivity of oxides and oxidized metals depends to a large extent upon the roughness of the surface. In general, higher values of emissivity are obtained on the rougher surfaces.

Material	Range of observed values	Probable value for oxide formed on smooth metal	Material	Range of observed values	Probable value for oxide formed on smooth metal
Aluminum oxide	0.22–0.40	0.30	Alumel (oxidized)	....	0.87
Beryllium oxide	0.07–0.37	0.35	Cast Iron (oxidized)	....	0.70
Cerium oxide	0.58–0.80	....	Chromel P (90Ni-10Cr) (oxidized)	....	0.87
Chromium oxide	0.60–0.80	0.70	80Ni-20Cr (oxidized)	....	0.90
Cobalt oxide	....	0.75	60Ni-24Fe-16Cr (oxidized)	....	0.83
Columbium oxide	0.55–0.71	0.70	55Fe-37.5Cr-7.5 Al (oxidized)	....	0.78
Copper oxide	0.60–0.80	0.70	70Fe-23Cr-5Al-2Co (oxidized)	....	0.75
Iron oxide	0.63–0.98	0.70	Constantan (55Cu-45Ni) (oxidized)	....	0.84
Magnesium oxide	0.10–0.43	0.20	Carbon Steel (oxidized)	....	0.80
Nickel oxide	0.85–0.96	0.90	Stainless Steel (18-8) (oxidized)	....	0.85
Thorium oxide	0 20–0.57	0.50	Porcelain	0.25–0.50	
Tin oxide	0.32–0.60	....			
Titanium oxide	....	0.50			
Uranium oxide	....	0 30			
Vanadium oxide	....	0.70			
Yttrium oxide	....	0.60			
Zirconium oxide	0.18–0.43	0.40			

PROPERTIES OF TUNGSTEN

JONES AND LANGMUIR, GENERAL ELECTRIC REVIEW

Temp. °K	Resistivity microhm cm	Electron emission amp./cm²	Evaporation g/cm² sec	Vapor pressure dynes/cm²	Thermal expansion per cent l_0 at 293°	Atomic heat cal./g. atom./°C.
300	5.65	....	....	....	.003	6.0
400	8.06	....	....	....	.044	6.0
500	10.56	....	....	....	.086	6.1
600	13.23	....	....	....	.130	6.1
700	16.09	....	....	....	.175	6.2
800	19.00	....	....	....	.222	6.2
900	21.94	....	....	....	.270	6.3
1000	24.93	1.07×10^{-15}	5.32×10^{-34}	1.98×10^{-29}	.320	6.4
1100	27.94	1.52×10^{-13}	2.17×10^{-30}	1.22×10^{-25}	.371	6.4
1200	30.98	9.73×10^{-12}	3.21×10^{-27}	1.87×10^{-22}	.424	6.5
1300	34.08	3.21×10^{-10}	1.35×10^{-24}	$8.18 \times .0^{-20}$	.479	6.7
1400	37.19	6.62×10^{-9}	2.51×10^{-22}	1.62×10^{-17}	.535	6.8
1500	40.36	9.14×10^{-8}	2.37×10^{-20}	1.54×10^{-15}	.593	7.0
1600	43.55	9.27×10^{-7}	1.25×10^{-18}	8.43×10^{-14}	.652	7.1
1700	46.78	7.08×10^{-6}	4.17×10^{-17}	2.82×10^{-12}	.713	7.2
1800	50.05	4.47×10^{-5}	8.81×10^{-16}	6.31×10^{-11}	.775	7.4
1900	53.35	2.28×10^{-4}	1.41×10^{-14}	1.01×10^{-9}	.839	7.6
2000	56.67	1.00×10^{-3}	1.76×10^{-13}	1.33×10^{-8}	.904	7.7
2100	60.06	3.93×10^{-3}	1.66×10^{-12}	1.28×10^{-7}	.971	7.8
2200	63.48	1.33×10^{-2}	1.25×10^{-11}	9.88×10^{-7}	1.039	8.0
2300	66.91	4.07×10^{-2}	8.00×10^{-11}	6.47×10^{-6}	1.109	8.2
2400	70.39	1.16×10^{-1}	4.26×10^{-10}	3.52×10^{-5}	1.180	8.3
2500	73.91	2.98×10^{-1}	2.03×10^{-9}	1.71×10^{-4}	1.253	8.4
2600	77.49	7.16×10^{-1}	8.41×10^{-9}	7.24×10^{-4}	1.328	8.6
2700	81.04	1.63	3.19×10^{-8}	2.86×10^{-3}	1.404	8.7
2800	84.70	3.54	1.10×10^{-7}	9.84×10^{-3}	1.479	8.9
2900	88.33	7.31	3.30×10^{-7}	3.00×10^{-2}	1.561	9.0
3000	92.04	1.42×10	9.95×10^{-7}	9.20×10^{-2}	1.642	9.2
3100	95.76	2.64×10	2.60×10^{-6}	2.50×10^{-1}	1.724	9.4
3200	99.54	4.78×10	6.38×10^{-6}	6.13×10^{-1}	1.808	9.5
3300	103.3	8.44×10	1.56×10^{-5}	1.51	1.893	9.6
3400	107.2	1.42×10^{2}	3.47×10^{-5}	3.41	1.980	9.8
3500	111.1	2.33×10^{2}	7.54×10^{-5}	7.52	2.068	9.9
3600	115.0	3.73×10^{2}	1.51×10^{-4}	1.53×10	2.158	10 1
3655	117.1	4.79×10^{2}	2.28×10^{-4}	2.33×10	2.209	10.2

PROPERTIES OF TUNGSTEN

JONES AND LANGMUIR GENERAL ELECTRIC REVIEW

ROESER AND WENSEL, NATIONAL BUREAU OF STANDARDS

Temp. °K	Normal brightness new candles per cm²	Spectral emissivity 0.65μ	Spectral emissivity 0.467μ	Color emissivity	Total emissivity	Brightness temp. 0.65μ	Color temp.
300	....	0.472	0.505	.....	0.032		
400	....	.....	....	.....	.042		
500	....	.....	....	.....	.053		
600	....	.....	....	.....	.064		
700	....	.....	....	.....	.076		
800	....	.....	....	.....	.088		
900	....	.....	....	.....	.101		
1000	0.0001	.458	.486	.395	.114	966	1007
1100	0.001	.456	.484	.392	.128	1059	1108
1200	0.006	.454	.482	.390	.143	1151	1210
1300	0.029	.452	.480	.387	.158	1242	1312
1400	0.11	.450	.478	.385	.175	1332	1414
1500	0.33	.448	.476	.382	.192	1422	1516
1600	0.92	.446	.475	.380	.207	1511	1619
1700	2.3	.444	.473	.377	.222	1599	1722
1800	5.1	.442	.472	.374	.236	1687	1825
1900	10.4	.440	.470	.371	.249	1774	1928
2000	20.0	.438	.469	.368	.260	1861	2032
2100	36	.436	.467	.365	.270	1946	2136
2200	61	.434	.466	.362	.279	2031	2241
2300	101	.432	.464	.359	.288	2115	2345
2400	157	430	.463	.356	.296	2198	2451
2500	240	.428	.462	.353	.303	2280	2556
2600	350	.426	.460	.349	.311	2362	2662
2700	500	.424	.459	.346	.318	2443	2769
2800	690	.422	.458	.343	.323	2523	2876
2900	950	.420	.456	.340	.329	2602	2984
3000	1260	.418	.455	.336	.334	2681	3092
3100	1650	.416	.454	.333	.337	2759	3200
3200	2100	.414	.452	.330	.341	2837	3310
3300	2700	.412	.451	.326	.344	2913	3420
3400	3400	.410	.450	.323	.348	2989	3530
3500	4200	.408	.449	.320	.351	3063	3642
3600	5200	.406	.447	.317	.354	3137	3754

TRANSMISSION OF CORNING COLORED FILTERS

Supplied by R. G. Saxton

If I_o is the intensity of radiation entering a layer of some medium and I the intensity reaching the opposite surface, the ratio I/I_o is called the transmittance. In practice the ratio of intensity of radiation passing through a glass sample to that incident on its surface is often measured and plotted as transmission. The transmission is the result of two factors, the transmittance of the glass and the losses by reflection. These losses amount to about 4 % for each glass-air surface; the transmission of a sample is about 92 % of its transmittance. Since the reflection losses differ slightly with different samples, the correction is often determined and applied when the transmission is measured. Values in this table have been corrected for reflection losses.

The identifying glass number, CS number, color and properties, and nominal thickness for the Corning glasses in this table are:

Glass No.	CS	Color and properties	Nominal thickness
0160	0-54	Clear; Ultraviolet transmitting	2.0
2030	2-64	Red; Sharp cut	3.0
2403	2-58	Red; Sharp cut	3.0
2404	2-59	Red; Sharp cut	3.0
2408	2-60	Red; Sharp cut	3.0
2412	2-61	Red; Sharp cut	3.0
2418	2-62	Red; Sharp cut	3.0
2424	2-63	Red; Sharp cut	3.0
2434	2-73	Red; Sharp cut	3.0
2540	7-56	Black; IR transmitting; Visible absorbing	2.5
2550	7-57	Black; IR transmitting; Visible absorbing	2.0
2600	7-69	Black; IR transmitting; Visible absorbing	3.0
3060	3-75	Straw	2.0
3304	3-76	Dark amber	3.0
3307	3-77	Dark amber	3.0
3384	3-70	Yellow	3.0
3385	3-71	Yellow	3.0
3387	3-72	Straw	3.0
3389	3-73	Straw	3.0
3391	3-74	Straw	3.0
3480	3-66	Yellow; Sharp cut	3.0
3482	3-67	Yellow; Sharp cut	3.0
3484	3-68	Yellow; Sharp cut	3.0
3486	3-69	Yellow; Sharp cut	3.0
3718	3-94	Yellow	3.0
3750	3-79	Yellow; Yellow green fluorescing	5.0
3780	3-80	Yellow	2.0
3850	0-51	Clear; UV transmitting	4.0
3961	1-56	Bluish; IR absorbing; Visible transmitting	2.5
3962	1-57	Bluish; IR absorbing; Visible transmitting	2.5
3965	1-58	Bluish; IR absorbing; Visible transmitting	2.5
3966	1-59	Bluish; IR absorbing; Visible transmitting	2.5
4010	4-64	Green	4.0
4015	4-65	Yellow green	3.0
4060	4-67	Green	2.0
4084	4-68	Green	4.5
4303	4-72	Blue green	4.0
4305	4-71	Blue green	4.0
4308	4-70	Blue green	4.0
4309	4-69	Blue green	4.0
4445	4-74	Green	2.5
4602	1-75	Bluish; IR absorbing; Visible transmitting	3.0
4784	4-94	Blue green	5.0
5030	5-57	Blue	5.0
5031	5-56	Blue	4.5
5070	7-62	Amethyst	3.9
5071	7-63	Amethyst	3.9
5073	7-64	Amethyst	3.9
5113	5-58	Blue	4.0
5120	1-60	Smoky violet; Absorbs yellow	5.2
5300	4-106	Green	3.9
5330	1-64	Blue	4.5
5433	5-59	Blue	5.0
5543	5-60	Blue	5.0
5562	5-61	Blue	5.0
5572	1-61	Blue	5.0
5840	7-60	Black; UV transmitting; Visible absorbing	4.5
5850	7-59	Purple; UV transmitting; Visible absorbing	4.0
5860	7-37	Black; UV transmitting; Visible absorbing	5.0
5874	7-39	Black; UV transmitting; Visible absorbing	5.0
5900	1.62	Blue	5.5
5970	7-51	Black; UV transmitting; Visible absorbing	5.0
7380	0-52	Clear; UV transmitting	2.0
7740	0-53	Clear; UV transmitting	2.0
7905	9-30	Clear; UV transmitting; Long Range IR transmitting	2.0
7910	9-54	Clear; UV transmitting	2.0
8364	7-98	Gray	2.0
9780	4-76	Blue green	5.0
9782	4-96	Blue green	5.0
9788	4-97	Blue green	5.0
9830	4-77	Green	3.4
9863	7-54	Black; UV transmitting; Visible absorbing	3.0

TRANSMISSION OF CORNING COLORED FILTERS (Continued)

	Transmittance											
	Corning Glass Number											
λ(μ)	0160	2030	2403	2404	2408	2412	2418	2424	2434	2540	2550	2600
.22	.000	.000	.000	.000	.000	.000	.000	.000	.000	.000	.000	.000
.24	.000	.000	.000	.000	.000	.000	.000	.000	.000	.000	.000	.000
.26	.000	.000	.000	.000	.000	.000	.000	.000	.000	.000	.000	.000
.28	.000	.000	.000	.000	.000	.000	.000	.000	.000	.000	.000	.000
.30	.005	.000	.000	.000	.000	.000	.000	.000	.000	.000	.000	.000
.32	.642	.000	.000	.000	.000	.000	.000	.000	.000	.000	.000	.000
.34	.850	.000	.000	.000	.000	.000	.000	.000	.000	.000	.000	.000
.36	.882	.000	.000	.000	.000	.000	.000	.000	.000	.000	.000	.000
.38	.890	.000	.000	.000	.000	.000	.000	.000	.000	.000	.000	.000
.40	.892	.000	.000	.000	.000	.000	.000	.000	.000	.000	.000	.000
.41	.893	.000	.000	.000	.000	.000	.000	.000	.000	.000	.000	.000
.42	.896	.000	.000	.000	.000	.000	.000	.000	.000	.000	.000	.000
.43	.896	.000	.000	.000	.000	.000	.000	.000	.000	.000	.000	.000
.44	.898	.000	.000	.000	.000	.000	.000	.000	.000	.000	.000	.000
.45	.899	.000	.000	.000	.000	.000	.000	.000	.000	.000	.000	.000
.46	.900	.000	.000	.000	.000	.000	.000	.000	.000	.000	.000	.000
.47	.900	.000	.000	.000	.000	.000	.000	.000	.000	.000	.000	.000
.48	.900	.000	.000	.000	.000	.000	.000	.000	.000	.000	.000	.000
.49	.900	.000	.000	.000	.000	.000	.000	.000	.000	.000	.000	.000
.50	.900	.000	.000	.000	.000	.000	.000	.000	.000	.000	.000	.000
.51	.900	.000	.000	.000	.000	.000	.000	.000	.000	.000	.000	.000
.52	.900	.000	.000	.000	.000	.000	.000	.000	.000	.000	.000	.000
.53	.900	.000	.000	.000	.000	.000	.000	.000	.000	.000	.000	.000
.54	.900	.000	.000	.000	.000	.000	.000	.000	.000	.000	.000	.000
.55	.900	.000	.000	.000	.000	.000	.000	.000	.000	.000	.000	.000
.56	.901	.000	.000	.000	.000	.000	.000	.000	.000	.000	.000	.000
.57	.904	.000	.000	.000	.000	.000	.000	.000	.005	.000	.000	.000
.58	.904	.000	.000	.000	.000	.000	.000	.005	.200	.000	.000	.000
.59	.908	.000	.000	.000	.000	.000	.008	.170	.615	.000	.000	.000
.60	.910	.000	.000	.000	.000	.006	.250	.575	.808	.000	.000	.000
.61	.910	.000	.000	.000	.018	.190	.660	.790	.856	.000	.000	.000
.62	.910	.000	.000	.015	.265	.625	.822	.848	.872	.000	.000	.000
.63	.910	.000	.018	.295	.670	.828	.862	.870	.881	.000	.000	.000
.64	.910	.006	.260	.660	.828	.868	.874	.880	.887	.000	.001	.000
.65	.910	.028	.675	.796	.866	.881	.881	.887	.892	.000	.003	.000
.66	.910	.110	.838	.828	.877	.885	.885	.893	.895	.000	.005	.000
.67	.910	.305	.871	.842	.883	.887	.887	.897	.897	.000	.006	.000
.68	.910	.550	.880	.847	.886	.889	.889	.900	.899	.000	.009	.000
.69	.910	.735	.885	.851	.888	.900	.900	.901	.900	.000	.012	.000
.70	.910	.820	.886	.852	.888	.900	.900	.903	.900	.000	.017	.000
.71	.910	.853	.888	.854	.888	.889	.889	.903	.900	.000	.023	.000
.72	.910	.864	.889	.853	.888	.888	.888	.903	.899	.000	.031	.040
.73	.910	.867	.900	.851	.887	.887	.887	.903	.897	.000	.041	.175
.74	.910	.867	.900	.850	.885	.886	.886	.903	.896	.000	.055	.372
.75	.910	.866	.900	.849	.884	.885	.885	.902	.895	.000	.069	.547
.80	.910	.839	.875	.827	.870	.858	.857	.881	.866	.005	.225	.770
1.00	.912	.801	.840	.772	.840	.828	.822	.857	.842	.562	.780	.350
1.20	.908	.799	.845	.786	.845	.837	.827	.859	.849	.790	.870	.000
1.40	.909	.811	.854	.809	.854	.848	.840	.862	.857	.850	.895	.000
1.60	.913	.839	.873	.837	.873	.869	.858	.880	.879	.872	.904	.000
1.80	.909	.844	.870	.829	.870	.864	.854	.877	.872	.880	.900	.000
2.00	.904	.841	.868	.827	.868	.861	.851	.874	.871	.880	.897	.000
2.20	.888	.833	.820	.773	.825	.820	.810	.837	.835	.860	.875	.005
2.40	.875	.832	.803	.757	.809	.803	.792	.818	.812	.868	.870	.049
2.60	.868	.822	.750	.695	.754	.750	.723	.772	.767	.858	.850	.058
2.80	.690	.600	.100	.100	.100	.100	.050	.050	.260	.450	.480	.030
3.00	.630	.470	.070	.020	.070	.070	.072	.122	.400	.465	.383	.022
3.20	.500	.340	.140	.074	.150	.140	.142	.180	.470	.390	.330	.020
3.40	.379	.260	.140	.078	.140	.140	.120	.150	.350	.310	.245	.018
3.60	.320	.247	.000	.000	.000	.000	.000	.006	.015	.280	.220	.021
3.80	.310	.257	.000	.000	.000	.000	.000	.000	.020	.285	.250	.035
4.00	.311	.274	.000	.000	.000	.000	.000	.000	.015	.285	.275	.068
4.20	.251	.200	.000	.000	.000	.000	.000	.000	.017	.190	.190	.065
4.40	.110	.060	.000	.000	.000	.000	.000	.000	.002	.050	.100	.020
4.60	.012	.000	.000	.000	.000	.000	.000	.000	.000	.000	.000	.000
4.80	.004	.000	.000	.000	.000	.000	.000	.000	.000	.000	.000	.000
5.00	.000	.000	.000	.000	.000	.000	.000	.000	.000	.000	.000	.000

TRANSMISSION OF CORNING COLORED FILTERS (Continued)

	Transmittance											
	Corning Glass Number											
λ(μ)	3060	3304	3307	3384	3385	3387	3389	3391	3480	3482	3484	3486
.22	.000	.000	.000	.000	.000	.000	.000	.000	.000	.000	.000	.000
.24	.000	.000	.000	.000	.000	.000	.000	.000	.000	.000	.000	.000
.26	.000	.000	.000	.000	.000	.000	.000	.000	.000	.000	.000	.000
.28	.000	.000	.000	.000	.000	.000	.000	.000	.000	.000	.000	.000
.30	.000	.000	.000	.000	.000	.000	.000	.000	.000	.000	.000	.000
.32	.000	.000	.000	.000	.000	.000	.005	.000	.000	.000	.000	.000
.34	.000	.000	.038	.000	.000	.000	.010	.000	.000	.000	.000	.000
.36	.000	.000	.050	.000	.005	.000	.015	.000	.000	.000	.000	.000
.38	.060	.000	.027	.005	.010	.010	.020	.000	.000	.000	.000	.000
.40	.410	.000	.016	.011	.016	.020	.026	.075	.000	.000	.000	.005
.41	.517	.000	.014	.011	.016	.020	.025	.425	.000	.000	.000	.005
.42	.604	.000	.014	.010	.015	.019	.105	.655	.000	.000	.000	.005
.43	.665	.000	.016	.009	.013	.017	.437	.747	.000	.000	.000	.005
.44	.710	.000	.022	.005	.011	.050	.620	.801	.000	.000	.000	.005
.45	.748	.000	.033	.003	.010	.325	.714	.838	.000	.000	.000	.005
.46	.778	.000	.049	.002	.008	.565	.780	.860	.000	.000	.000	.004
.47	.800	.000	.070	.001	.060	.690	.820	.874	.000	.000	.000	.003
.48	.819	.000	.101	.005	.410	.763	.848	.884	.000	.000	.000	.003
.49	.836	.003	.143	.088	.640	.803	.866	.890	.000	.000	.000	.002
.50	.850	.009	.193	.350	.727	.834	.878	.895	.000	.000	.000	.001
.51	.860	.019	.250	.595	.780	.854	.886	.898	.000	.000	.000	.045
.52	.870	.037	.315	.725	.817	.868	.890	.900	.000	.000	.003	.425
.53	.875	.063	.379	.789	.840	.876	.892	.901	.000	.000	.175	.710
.54	.881	.102	.447	.825	.856	.883	.894	.902	.000	.015	.600	.792
.55	.884	.146	.504	.846	.866	.887	.895	.903	.000	.230	.774	.823
.56	.886	.200	.560	.860	.873	.889	.894	.902	.020	.675	.818	.844
.57	.885	.255	.607	.869	.876	.890	.893	.901	.325	.850	.839	.859
.58	.883	.310	.648	.873	.878	.889	.892	.900	.710	.885	.854	.868
.59	.882	.360	.680	.876	.877	.887	.890	.898	.829	.894	.865	.876
.60	.882	.404	.705	.877	.877	.884	.886	.896	.858	.900	.873	.882
.61	.882	.438	.722	.877	.877	.881	.884	.893	.869	.903	.880	.886
.62	.882	.466	.735	.875	.876	.875	.880	.890	.876	.905	.885	.888
.63	.882	.488	.744	.871	.874	.871	.876	.886	.881	.906	.888	.889
.64	.883	.505	.748	.865	.872	.866	.872	.884	.884	.907	.890	.890
.65	.885	.519	.750	.860	.867	.860	.868	.881	.885	.908	.892	.890
.66	.886	.531	.750	.856	.863	.856	.865	.876	.886	.908	.893	.890
.67	.888	.543	.749	.850	.858	.851	.860	.873	.886	.908	.894	.890
.68	.890	.552	.745	.844	.853	.846	.856	.869	.886	.908	.893	.889
.69	.891	.561	.740	.837	.847	.839	.852	.865	.885	.907	.892	.887
.70	.892	.569	.734	.831	.842	.834	.847	.860	.884	.907	.891	.885
.71	.893	.574	.727	.825	.837	.827	.842	.856	.882	.906	.890	.883
.72	.893	.575	.720	.819	.831	.822	.837	.852	.880	.905	.888	.880
.73	.892	.576	.712	.813	.825	.816	.831	.848	.877	.905	.886	.877
.74	.891	.574	.702	.807	.820	.810	.826	.844	.874	.904	.885	.875
.75	.890	.570	.694	.800	.814	.805	.820	.840	.870	.903	.882	.873
.80	.871	.526	.642	.770	.865	.780	.775	.815	.837	.878	.846	.829
1.00	.830	.435	.516	.715	.830	.725	.716	.772	.801	.857	.811	.781
1.20	.860	.429	.500	.718	.858	.735	.730	.782	.807	.859	.819	.793
1.40	.901	.475	.540	.750	.900	.768	.768	.810	.828	.870	.837	.817
1.60	.917	.580	.635	.795	.918	.812	.812	.843	.852	.884	.856	.841
1.80	.916	.627	.675	.808	.915	.818	.820	.849	.847	.882	.852	.834
2.00	.908	.620	.668	.800	.909	.822	.817	.846	.847	.884	.852	.835
2.20	.900	.630	.675	.802	.900	.823	.810	.840	.805	.865	.829	.798
2.40	.885	.651	.690	.800	.885	.825	.811	.842	.787	.853	.817	.777
2.60	.860	.650	.690	.785	.858	.815	.800	.840	.725	.818	.757	.718
2.80	.550	.345	.390	.325	.670	.360	.340	.440	.050	.060	.110	.060
3.00	.379	.320	.360	.318	.348	.348	.322	.423	.080	.088	.190	.088
3.20	.315	.240	.290	.268	.332	.324	.290	.395	.150	.158	.270	.145
3.40	.250	.151	.190	.218	.289	.288	.255	.353	.120	.132	.140	.090
3.60	.231	.130	.150	.217	.266	.280	.249	.351	.000	.000	.000	.000
3.80	.258	.140	.160	.228	.270	.298	.267	.376	.000	.000	.000	.000
4.00	.283	.140	.175	.220	.280	.290	.260	.365	.000	.000	.000	.000
4.20	.200	.090	.125	.143	.210	.210	.178	.288	.000	.000	.000	.000
4.40	.100	.020	.030	.025	.080	.070	.040	.115	.000	.000	.000	.000
4.60	.008	.008	.010	.007	.002	.003	.000	.009	.000	.000	.000	.000
4.80	.000	.005	.009	.000	.000	.000	.000	.000	.000	.000	.000	.000
5.00	.000	.001	.008	.000	.000	.000	.000	.000	.000	.000	.000	.000

TRANSMISSION OF CORNING COLORED FILTERS (Continued)

λ(μ)	Transmittance											
	Corning Glass Number											
	3718	3750	3780	3850	3961	3962	3965	3966	4010	4015	4060	4084
.22	.000	.000	.000	.000	.000	.000	.000	.000	.000	.000	.000	.000
.24	.000	.000	.000	.000	.000	.000	.000	.000	.000	.000	.000	.000
.26	.000	.000	.000	.000	.000	.000	.000	.000	.000	.000	.000	.000
.28	.000	.000	.000	.000	.000	.000	.000	.000	.000	.000	.000	.000
.30	.000	.000	.000	.000	.000	.000	.000	.000	.000	.000	.000	.000
.32	.004	.000	.000	.000	.000	.000	.018	.055	.000	.000	.000	.000
.34	.030	.000	.000	.000	.000	.018	.192	.375	.000	.000	.001	.018
.36	.550	.215	.000	.005	.020	.125	.430	.630	.000	.000	.021	.128
.38	.665	.327	.000	.350	.085	.270	.558	.710	.000	.000	.080	.248
.40	.480	.113	.000	.675	.185	.395	.636	.781	.000	.000	.178	.216
.41	.443	.088	.000	.749	.218	.426	.651	.788	.000	.000	.228	.180
.42	.465	.088	.000	.788	.248	.453	.666	.795	.000	.000	.281	.151
.43	.560	.135	.000	.812	.269	.474	.678	.800	.000	.000	.335	.136
.44	.675	.255	.006	.828	.290	.494	.693	.806	.000	.000	.388	.140
.45	.748	.410	.028	.841	.313	.519	.709	.816	.000	.005	.434	.163
.46	.780	.472	.058	.850	.331	.538	.724	.824	.006	.025	.473	.200
.47	.803	.570	.092	.858	.346	.556	.737	.831	.021	.073	.506	.247
.48	.800	.555	.088	.865	.361	.570	.748	.836	.050	.145	.527	.303
.49	.802	.550	.095	.870	.370	.582	.756	.840	.100	.245	.535	.370
.50	.824	.597	.152	.874	.376	.590	.762	.842	.160	.350	.528	.430
.51	.862	.720	.325	.878	.377	.594	.765	.843	.220	.455	.503	.465
.52	.894	.825	.595	.881	.373	.593	.765	.841	.252	.537	.460	.467
.53	.904	.853	.717	.883	.364	.588	.761	.838	.247	.582	.400	.438
.54	.905	.860	.763	.884	.354	.579	.764	.833	.207	.594	.325	.376
.55	.906	.864	.783	.884	.342	.569	.746	.827	.153	.572	.252	.303
.56	.907	.867	.795	.883	.331	.559	.736	.821	.096	.525	.183	.225
.57	.907	.869	.799	.882	.317	.547	.724	.813	.054	.457	.125	.157
.58	.907	.870	.804	.880	.298	.529	.706	.802	.026	.385	.083	.107
.59	.907	.870	.811	.877	.276	.508	.685	.790	.011	.311	.053	.073
.60	.908	.876	.826	.876	.251	.481	.662	.775	.004	.245	.033	.048
.61	.908	.880	.835	.875	.225	.452	.636	.757	.000	.190	.020	.034
.62	.909	.881	.842	.874	.299	.423	.610	.739	.000	.145	.012	.024
.63	.909	.884	.848	.875	.217	.392	.577	.719	.000	.115	.007	.018
.64	.910	.885	.854	.876	.147	.359	.546	.698	.000	.098	.004	.015
.65	.910	.887	.859	.877	.125	.326	.515	.675	.000	.084	.001	.012
.66	.911	.891	.864	.880	.104	.297	.482	.652	.000	.075	.000	.010
.67	.913	.896	.869	.883	.086	.265	.450	.630	.000	.075	.000	.009
.68	.914	.900	.873	.885	.063	.235	.418	.603	.000	.071	.000	.008
.69	.915	.901	.877	.887	.055	.206	.385	.576	.000	.065	.000	.007
.70	.915	.904	.880	.888	.042	.279	.352	.550	.000	.067	.000	.007
.71	.915	.905	.883	.888	.032	.155	.322	.524	.000	.070	.000	.007
.72	.915	.906	.885	.889	.025	.133	.294	.496	.000	.075	.000	.007
.73	.915	.907	.885	.888	.018	.114	.266	.472	.000	.080	.000	.007
.74	.915	.907	.882	.887	.014	.097	.242	.448	.000	.084	.000	.008
.75	.914	.906	.882	.886	.010	.084	.220	.424	.000	.086	.000	.008
.80	.902	.875	.855	.863	.000	.033	.120	.310	.000	.109	.000	.013
1.00	.899	.860	.882	.820	.000	.002	.038	.158	.000	.215	.018	.100
1.20	.898	.882	.898	.850	.000	.002	.040	.161	.007	.393	.158	.303
1.40	.880	.810	.855	.894	.002	.018	.100	.270	.058	.549	.404	.548
1.60	.882	.805	.855	.905	.007	.050	.190	.390	.162	.663	.612	.710
1.80	.900	.844	.907	.904	.008	.057	.201	.408	.299	.740	.740	.791
2.00	.897	.819	.909	.895	.011	.070	.228	.435	.422	.783	.817	.830
2.20	.888	.720	.900	.870	.021	.105	.277	.475	.518	.803	.840	.792
2.40	.865	.668	.890	.850	.037	.140	.320	.512	.597	.817	.862	.808
2.60	.840	.570	.860	.800	.050	.005	.339	.515	.634	.813	.870	.740
2.80	.460	.075	.620	.200	.022	.092	.135	.085	.270	.460	.520	.010
3.00	.282	.028	.465	.165	.033	.133	.200	.230	.260	.418	.620	.033
3.20	.252	.016	.420	.120	.054	.150	.247	.294	.204	.345	.642	.128
3.40	.175	.007	.370	.070	.048	.112	.165	.200	.131	.235	.631	.130
3.60	.150	.003	.351	.045	.003	.008	.016	.018	.121	.192	.634	.006
3.80	.168	.000	.368	.067	.007	.015	.035	.048	.125	.200	.600	.016
4.00	.182	.000	.370	.080	.006	.010	.020	.022	.123	.200	.557	.006
4.20	.120	.000	.300	.040	.005	.013	.022	.021	.070	.130	.422	.001
4.40	.020	.000	.140	.005	.001	.001	.000	.001	.002	.015	.135	.000
4.60	.002	.000	.010	.000	.000	.000	.000	.000	.000	.000	.012	.000
4.80	.000	.000	.000	.000	.000	.000	.000	.000	.000	.000	.002	.000
5.00	.000	.000	.000	.000	.000	.000	.000	.000	.000	.000	.000	.000

TRANSMISSION OF CORNING COLORED FILTERS (Continued)

	Transmittance											
	Corning Glass Number											
λ(μ)	4303	4305	4308	4309	4445	4602	4784	5030	5031	5070	5071	5073
.22	.000	.000	.000	.000	.000	.000	.000	.000	.000	.000	.000	.000
.24	.000	.000	.000	.000	.000	.000	.000	.000	.000	.000	.000	.000
.26	.000	.000	.000	.000	.000	.000	.000	.000	.000	.000	.000	.000
.28	.000	.000	.000	.000	.000	.000	.000	.000	.000	.000	.000	.000
.30	.000	.000	.000	.000	.000	.001	.000	.000	.016	.000	.000	.000
.32	.000	.000	.000	.022	.001	.106	.000	.000	.145	.012	.045	.090
.34	.011	.060	.190	.394	.018	.505	.009	.038	.420	.310	.330	.540
.36	.188	.380	.580	.740	.114	.755	.200	.285	.685	.628	.340	.757
.38	.390	.590	.740	.831	.248	.827	.450	.595	.820	.745	.420	.810
.40	.545	.723	.826	.884	.400	.835	.596	.770	.884	.712	.665	.830
.41	.588	.750	.840	.887	.454	.845	.627	.799	.894	.600	.712	.786
.42	.624	.770	.850	.890	.505	.846	.648	.808	.895	.430	.716	.705
.43	.654	.786	.857	.892	.550	.851	.666	.797	.890	.290	.694	.620
.44	.680	.798	.862	.894	.593	.856	.680	.767	.872	.170	.655	.525
.45	.698	.809	.867	.897	.631	.857	.697	.738	.865	.097	.612	.436
.46	.712	.815	.869	.898	.659	.856	.717	.702	.864	.055	.568	.365
.47	.715	.814	.866	.897	.678	.861	.735	.628	.845	.035	.531	.313
.48	.705	.802	.856	.894	.689	.866	.750	.522	.805	.023	.501	.275
.49	.678	.780	.838	.885	.687	.869	.763	.406	.750	.017	.482	.252
.50	.636	.740	.810	.872	.673	.870	.768	.288	.684	.015	.469	.237
.51	.570	.685	.770	.850	.641	.869	.767	.186	.601	.013	.463	.231
.52	.480	.610	.714	.817	.586	.866	.753	.105	.495	.013	.461	.230
.53	.387	.525	.650	.781	.520	.863	.725	.053	.388	.014	.464	.235
.54	.288	.430	.576	.736	.437	.865	.676	.022	.295	.016	.473	.245
.55	.205	.340	.502	.683	.355	.869	.615	.007	.198	.018	.486	.260
.56	.132	.255	.422	.627	.275	.868	.525	.000	.113	.023	.502	.278
.57	.082	.184	.345	.565	.202	.863	.427	.000	.057	.030	.522	.300
.58	.047	.127	.277	.505	.144	.856	.328	.000	.025	.038	.540	.324
.59	.026	.087	.218	.447	.102	.848	.235	.000	.008	.048	.555	.348
.60	.013	.057	.170	.393	.068	.838	.157	.000	.000	.058	.571	.373
.61	.006	.036	.131	.341	.046	.824	.102	.000	.000	.070	.587	.395
.62	.001	.022	.100	.296	.031	.806	.058	.000	.000	.083	.600	.415
.63	.000	.013	.074	.256	.020	.787	.032	.000	.000	.096	.612	.435
.64	.000	.007	.056	.221	.013	.767	.017	.000	.000	.108	.622	.450
.65	.000	.004	.042	.191	.008	.745	.007	.000	.000	.120	.633	.466
.66	.000	.002	.033	.167	.006	.722	.002	.000	.000	.135	.644	.482
.67	.000	.001	.025	.146	.003	.695	.000	.000	.000	.148	.658	.498
.68	.000	.000	.020	.128	.001	.665	.000	.000	.000	.165	.674	.515
.69	.000	.000	.016	.116	.000	.634	.000	.000	.000	.182	.686	.531
.70	.000	.000	.013	.104	.000	.600	.000	.000	.000	.200	.700	.548
.71	.000	.000	.010	.096	.000	.565	.000	.000	.000	.220	.712	.566
.72	.000	.000	.009	.088	.000	.531	.000	.004	.024	.245	.725	.586
.73	.000	.000	.007	.083	.000	.496	.000	.047	.119	.268	.736	.606
.74	.000	.000	.006	.079	.000	.463	.000	.190	.330	.295	.749	.675
.75	.000	.000	.005	.075	.000	.430	.000	.440	.580	.323	.759	.642
.80	.000	.000	.008	.080	.003	.258	.000	.890	.917	.505	.815	.750
1.00	.000	.005	.045	.188	.056	.019	.000	.753	.868	.860	.885	.872
1.20	.013	.060	.166	.375	.269	.009	.000	.455	.720	.890	.897	.890
1.40	.080	.180	.342	.542	.527	.016	.003	.100	.285	.892	.902	.892
1.60	.210	.342	.500	.658	.701	.035	.038	.056	.162	.890	.902	.892
1.80	.350	.482	.617	.732	.792	.059	.142	.052	.140	.878	.890	.877
2.00	.475	.590	.690	.772	.839	.048	.275	.075	.175	.860	.865	.860
2.20	.560	.653	.720	.778	.850	.038	.345	.172	.295	.840	.830	.840
2.40	.635	.704	.752	.791	.870	.045	.404	.330	.483	.812	.795	.805
2.60	.663	.710	.748	.770	.861	.066	.340	.382	.530	.804	.752	.775
2.80	.260	.370	.370	.250	.280	.018	.045	.030	.030	.550	.500	.500
3.00	.249	.250	.203	.212	.395	.000	.000	.010	.001	.390	.308	.340
3.20	.202	.212	.145	.168	.451	.000	.000	.065	.026	.222	.145	.180
3.40	.135	.136	.084	.107	.449	.000	.000	.002	.006	.120	.070	.090
3.60	.124	.120	.078	.077	.470	.000	.000	.000	.000	.078	.032	.063
3.80	.132	.125	.082	.083	.470	.000	.000	.000	.000	.078	.029	.060
4.00	.132	.131	.094	.090	.448	.000	.000	.000	.000	.093	.037	.075
4.20	.078	.073	.050	.042	.320	.000	.000	.000	.000	.079	.020	.048
4.40	.008	.009	.008	.002	.100	.000	.000	.000	.000	.020	.004	.014
4.60	.000	.000	.000	.000	.007	.000	.000	.000	.000	.002	.000	.000
4.80	.000	.000	.000	.000	.000	.000	.000	.000	.000	.000	.000	.000
5.00	.000	.000	.000	.000	.000	.000	.000	.000	.000	.000	.000	.000

	Transmittance											
	Corning Glass Number											
λ(μ)	5113	5120	5300	5330	5433	5543	5562	5572	5840	5850	5860	5874
.22	.000	.000	.000	.000	.000	.000	.000	.000	.000	.000	.000	.000
.24	.000	.000	.000	.000	.000	.000	.000	.000	.000	.000	.000	.000
.26	.000	.000	.000	.000	.000	.000	.000	.000	.000	.000	.000	.000
.28	.000	.000	.000	.000	.000	.000	.000	.000	.000	.000	.000	.000
.30	.000	.000	.000	.002	.000	.000	.000	.000	.001	.039	.000	.000
.32	.000	.000	.000	.250	.000	.000	.000	.045	.242	.490	.008	.031
.34	.000	.000	.000	.622	.000	.000	.012	.325	.600	.790	.179	.228
.36	.035	.018	.000	.796	.100	.120	.205	.660	.682	.858	.340	.447
.38	.200	.540	.000	.835	.350	.380	.495	.805	.392	.850	.085	.378
.40	.371	.670	.000	.865	.585	.600	.717	.874	.000	.788	.000	.032
.41	.371	.790	.000	.850	.636	.635	.748	.873	.000	.720	.000	.004
.42	.337	.805	.000	.823	.665	.646	.761	.865	.000	.630	.000	.000
.43	.272	.560	.000	.783	.674	.635	.759	.857	.000	.522	.000	.000
.44	.198	.386	.000	.725	.665	.602	.742	.845	.000	.410	.000	.000
.45	.118	.485	.000	.650	.635	.550	.713	.832	.000	.290	.000	.000
.46	.055	.475	.000	.555	.577	.465	.662	.808	.000	.175	.000	.000
.47	.013	.370	.008	.455	.467	.335	.565	.765	.000	.125	.000	.000
.48	.000	.385	.026	.355	.327	.190	.435	.693	.000	.022	.000	.000
.49	.000	.685	.085	.270	.205	.090	.300	.605	.000	.005	.000	.000
.50	.000	.660	.125	.197	.120	.040	.197	.523	.000	.000	.000	.000
.51	.000	.390	.106	.145	.060	.013	.110	.430	.000	.000	.000	.000
.52	.000	.305	.094	.110	.024	.002	.051	.328	.000	.000	.000	.000
.53	.000	.175	.064	.085	.008	.000	.022	.247	.000	.000	.000	.000
.54	.000	.610	.149	.068	.005	.000	.012	.216	.000	.000	.000	.000
.55	.000	.817	.147	.055	.006	.000	.015	.246	.000	.000	.000	.000
.56	.000	.230	.087	.043	.007	.000	.018	.285	.000	.000	.000	.000
.57	.000	.125	.013	.032	.000	.000	.011	.258	.000	.000	.000	.000
.58	.000	.000	.000	.022	.000	.000	002	.175	.000	.000	.000	.000
.59	.000	.006	.000	.016	.000	.000	.000	.116	.000	.000	.000	.000
.60	.000	.180	.000	.012	.000	.000	.000	.113	.000	.000	.000	.000
.61	.000	.545	.000	.009	.000	.000	.000	.123	.000	.000	.000	.000
.62	.000	.825	.000	.010	.000	.000	.000	.126	.000	.000	.000	.000
.63	.000	.838	.000	.013	.000	.000	.000	.120	.000	.000	.000	.000
.64	.000	.878	.000	.015	.000	.000	.000	.111	.000	.000	.000	.000
.65	.000	.893	.000	.015	.000	.000	.000	.120	.000	.000	.000	.000
.66	.000	.883	.000	.014	.000	.000	.000	.165	.000	.000	.000	.000
.67	.000	.820	.000	.013	.000	.000	.000	.265	.000	.000	.000	.000
.68	.000	.705	.000	.015	.000	.000	.000	.425	.000	.000	.000	.000
.69	.000	.743	.000	.022	.000	.000	.001	.615	.000	.029	.000	.000
.70	.000	.860	.000	.041	.000	.000	.004	.756	.007	.160	.000	.017
.71	.000	.876	.000	.085	.000	.001	.004	.837	.020	.385	.000	.075
.72	.000	.815	.000	.165	.000	.002	.004	.874	.037	.615	.000	.168
.73	.000	.435	.000	.285	.000	.002	.002	.889	.060	.760	.000	.257
.74	.000	.045	.000	.460	.000	.002	.002	.895	.086	.843	.000	.320
.75	.000	.055	.000	.645	.000	.001	.001	.898	.080	.878	.000	.335
.80	.000	.030	.000	.900	.005	.002	.003	.895	.009	.890	.000	.218
1.00	.000	.860	.003	.800	.010	.015	.020	.880	.000	.716	.000	.050
1.20	.000	.770	.026	.570	.030	.020	.047	.690	.000	.169	.000	.011
1.40	.000	.550	.072	.410	.060	.040	.095	.640	.004	.042	.004	.012
1.60	.000	.620	.200	.405	.116	.060	.154	.625	.002	.036	.002	.010
1.80	.000	.580	.265	.425	.172	.090	.223	.635	.000	.048	.000	.010
2.00	.010	.580	.374	.476	.343	.247	.410	.750	.000	.168	.000	.018
2,20	.071	.747	.533	.535	.475	.400	.528	.780	.000	.338	.000	.036
2.40	.190	.400	.370	.552	.575	.520	.600	.780	.000	.492	.000	.074
2.60	.203	.530	.523	.540	.580	.532	.603	.745	.000	.510	.000	.088
2.80	.100	.250	.265	.007	.162	.132	.360	.470	.000	.170	.000	.003
3.00	.080	.080	.202	.012	.195	.151	.280	.315	.000	.128	.000	.020
3.20	.068	.072	.180	.082	.131	.105	.172	.200	.000	.084	.000	.053
3.40	.030	.029	.131	.003	.079	.065	.100	.100	.000	.065	.000	.050
3.60	.021	.017	.103	.000	.061	.030	.053	.049	.002	.070	.000	.000
3.80	.023	.021	.093	.000	.068	.032	.040	.037	.001	.082	.000	.000
4.00	.040	.025	.112	.000	.071	.039	.050	.042	.005	.090	.000	.000
4.20	.019	.010	.069	.000	.020	.015	.017	.020	.002	.055	.000	.000
4.40	.001	.001	.007	.000	.000	.000	.000	.002	.000	.002	.000	.000
4.60	.000	.000	.000	.000	.000	.000	.000	.000	.000	.000	.000	.000
4.80	.000	.000	.000	.000	.000	.000	.000	.000	.000	.000	.000	.000
5.00	.000	.000	.000	.000	.000	.000	.000	.000	.000	.000	.000	.000

TRANSMISSION OF CORNING COLORED FILTERS (Continued)

	Transmittance											
	Corning Glass Number											
λ (μ)	5900	5970	7380	7740	7905	7910	8364	9780	9782	9788	9830	9863
.22	.000	.000	.000	.000	.000	.012	.000	.000	.000	.000	.000	.000
.24	.000	.000	.000	.000	.360	.505	.000	.000	.000	.000	.000	.054
.26	.000	.000	.000	.000	.495	.780	.000	.000	.000	.000	.000	.482
.28	.000	.000	.000	.004	.590	.855	.000	.000	.000	.000	.000	.731
.30	.000	.000	.000	.321	.720	.877	.000	.000	.000	.000	.000	.831
.32	.000	.138	.000	.722	.825	.900	.000	.000	.000	.000	.000	.862
.34	.008	.600	.000	.851	.880	.903	.002	.015	.000	.060	.001	.854
.36	.150	.799	.440	.889	.910	.905	.083	.290	.060	.470	.059	.816
.38	.445	.742	.795	.900	.915	.906	.136	.590	.445	.770	.160	.620
.40	.678	.190	.892	.916	.920	.920	.296	.725	.747	.885	.130	.090
.41	.688	.029	.904	.915	.920	.920	.273	.705	.790	.895	.044	.018
.42	.635	.000	.910	.915	.920	.921	.232	.770	.818	.902	.004	.003
.43	.586	.000	.913	.914	.920	.923	.191	.778	.836	.905	.000	.000
.44	.522	.000	.915	.913	.920	.924	.157	.801	.847	.906	.000	.000
.45	.458	.000	.916	.913	.922	.925	.144	.814	.855	.906	.000	.000
.46	.400	.000	.917	.914	.922	.925	.140	.823	.860	.906	.000	.000
.47	.350	.000	.917	.915	.922	.925	.141	.832	.863	.906	.000	.000
.48	.306	.000	.917	.915	.923	.925	.146	.839	.863	.905	.000	.000
.49	.275	.000	.918	.915	.930	.926	.152	.843	.859	.904	.000	.000
.50	.246	.000	.918	.915	.925	.926	.166	.843	.848	.900	.014	.000
.51	.223	.000	.919	.915	.925	.927	.178	.838	.825	.893	.180	.000
.52	.196	.000	.919	.916	.925	.928	.190	.824	.784	.880	.175	.000
.53	.172	.000	.919	.916	.923	.928	.196	.798	.720	.862	.018	.000
.54	.154	.000	.919	.916	.926	.929	.198	.756	.627	.831	.000	.000
.55	.148	.000	.920	.917	.923	.929	.197	.697	.515	.787	.050	.000
.56	.151	.000	.920	.917	.925	.930	.199	.615	.380	.728	.265	.000
.57	.146	.000	.919	.918	.925	.930	.206	.518	.255	.655	.165	.000
.58	.125	.000	.918	.919	.925	.930	.217	.414	.150	.570	.035	.000
.59	.102	.000	.918	.920	.925	.930	.222	.302	.075	.475	.004	.000
.60	.093	.000	.920	.920	.925	.930	.215	.215	.032	.380	.000	.000
.61	.087	.000	.920	.920	.925	.930	.196	.135	.010	.290	.000	.000
.62	.081	.000	.920	.920	.925	.930	.175	.080	.002	.210	.000	.000
.63	.070	.000	.920	.919	.926	.930	.161	.042	.000	.145	.000	.000
.64	.061	.000	.920	.919	.927	.931	.156	.021	.000	.094	.000	.000
.65	.055	.000	.920	.918	.927	.931	.162	.008	.000	.059	.000	.000
.66	.055	.000	.920	.917	.927	.932	.176	.003	.000	.035	.000	.000
.67	.059	.000	.920	.916	.928	.932	.200	.000	.000	.020	.000	.000
.68	.065	.000	.920	.916	.927	.932	.228	.000	.000	.010	.075	.022
.69	.068	.007	.921	.915	.927	.932	.248	.000	.000	.005	.380	.106
.70	.068	.036	.921	.915	.928	.932	.251	.000	.000	.001	.642	.234
.71	.066	.085	.922	.914	.926	.933	.237	.000	.000	.000	.694	.332
.72	.064	.145	.922	.912	.926	.933	.223	.000	.000	.000	.666	.383
.73	.060	.222	.922	.910	.926	.933	.210	.000	.000	.000	.607	.384
.74	.057	.323	.921	.909	.927	.934	.197	.000	.000	.000	.531	.358
.75	.055	.385	.921	.907	.928	.934	.180	.000	.000	.000	.445	.322
.80	.050	.287	.918	.890	.930	.932	.110	.000	.000	.000	.045	.175
1.00	.085	.032	.910	.860	.930	.928	.032	.000	.000	.000	.000	.119
1.20	.180	.021	.910	.860	.925	.928	.032	.000	.000	.005	.007	.016
1.40	.295	.109	.906	.870	.925	.930	.062	.018	.000	.080	.000	.005
1.60	.405	.088	.910	.892	.931	.930	.137	.131	.011	.266	.000	.007
1.80	.495	.040	.903	.896	.931	.930	.182	.317	.085	.430	.157	.011
2.00	.590	.008	.900	.897	.934	.929	.171	.440	.216	.512	.041	.029
2.20	.628	.002	.898	.875	.934	.835	.184	.440	.278	.455	.000	.048
2.40	.640	.009	.890	.850	.930	.890	.225	.440	.325	.433	.000	.060
2.60	.630	.018	.860	.820	.920	.780	.258	.280	.212	.252	.057	.051
2.80	.470	.030	.375	.140	.908	.180	.145	.060	.040	.060	.020	.000
3.00	.248	.030	.425	.360	.880	.695	.130	.000	.000	.000	.000	.000
3.20	.121	.030	.380	.490	.861	.760	.125	.000	.000	.000	.000	.000
3.40	.032	.020	.310	.270	.670	.620	.099	.000	.000	.000	.000	.000
3.60	.010	.015	.270	.010	.111	.080	.115	.000	.000	.000	.000	.000
3.80	.010	.020	.275	.040	.270	.240	.142	.000	.000	.000	.000	.000
4.00	.010	.030	.260	.013	.170	.150	.172	.000	.000	.000	.000	.000
4.20	.006	.017	.180	.026	.250	.230	.158	.000	.000	.000	.000	.000
4.40	.001	.002	.040	.004	.085	.080	.078	.000	.000	.000	.000	.000
4.60	.000	.000	.000	.000	.050	.020	.010	.000	.000	.000	.000	.000
4.80	.000	.000	.000	.000	.000	.000	.003	.000	.000	.000	.000	.000
5.00	.000	.000	.000	.000	.000	.000	.000	.000	.000	.000	.000	.000

TRANSMISSION OF WRATTEN FILTERS

Compiled by Allie C. Peed, Jr. for The Eastman Kodak Company

Data condensed from Kodak Wratten Filters for Scientific and Technical Use published by the Eastman Kodak Company, manufacturers of the filters.

The following pages give (1) percentage luminous transmittance at wave lengths from 400 to 700μ at intervals of 10μ for the standard illuminant "C" adopted by the International Commission of Illumination, (2) dominant wave length in millimicrons, and (3) percentage of excitation purity. Values of wave length followed by "c" indicate the complementary wave lengths of purple filters which do not have a dominant wave length.

All colorimetric specifications are based on the 1931 standard ICI colorimetric and luminosity data.

The transmittance data are given as representing standard samples of the filters. They are intended only for the information of users in choosing filters which will meet their requirements. Values taken from the tables of data should not be used by research workers as representing precisely the absorption characteristics of a particular filter. If such precise data are needed, they should be determined for the particular filter being used.

Where the spectra extend into the ultraviolet this fact is indicated by an asterisk (*) in the transmission tables immediately beneath the filter number, and quantitative data are not given. The manufacturer should be consulted for this information. Transmission in the ultraviolet of wave lengths less than 330μ will be eliminated in the case of cemented filters, as glass absorbs ultraviolet radiation of wave lengths shorter than about 330μ.

Stability ratings are given as three letter combinations following the filter description in the table below. In establishing the stability classifications each filter is exposed to a selected light source for a specific time interval. The following grading system is used to describe the result:

Class A—stable
Class B—relatively stable
Class C—somewhat unstable
Class D—unstable

The classification letters, for example, AAA, describe the stability to the following three exposure tests in this order:

1. Two weeks' exposure to daylight in a south window
2. Twenty-four hours' exposure to a "Fade-Ometer"
3. Two weeks' exposure at two feet from a 1000-watt tungsten lamp.

Filters are supplied in two forms; as lacquered gelatin film, or as a gelatin film cemented between pieces of optical glass. Filters in glass are cemented between sheets of plane-parallel glass, which is surfaced in quantities and is of sufficient accuracy for general photographic work, and for most scientific purposes.

Most Wratten Gelatin Filters are stocked in 2- or 3-inch squares. Stocks of 2- or 3-inch square filters cemented in glass are maintained only in filters usually used for general photographic work.

The booklet "Kodak Filters and Lens Attachments" gives more valuable information on this subject.

FILTER DATA

No.	Description, use, and stability
	Colorless
0	For compensating thickness of other gelatin filters in optical systems, AAA.
1	Absorbs ultraviolet below 360 mμ, DDD.
1A	Kodak Skylight Filter—Reduces excess bluishness in outdoor color photographs in open shade under a clear, blue sky, ACA.
	Yellows
2B	Absorbs ultraviolet below 410 mμ, ACA.
3	Light yellow, CCD.
3N5	No. 3 plus 0.5 neutral density, AAA.
4	Light yellow—Approximate correction on panchromatic materials for outdoor scenes, including sky, CCC.
6	K1—Light yellow—Partial correction outdoors, BBA.
8	K2—Yellow—Full correction outdoors on Type B panchromatic materials. Widely used for proper sky, cloud, and foliage rendering. Green separation for Fluorescence Process, AAA.
8N5	No. 8 plus 0.5 neutral density, AAA.
9	K3—Deep yellow. Moderate contrast in outdoor photography (with black-and-white films), AAA.
11	X1—Greenish yellow. Correction for tungsten light on Type B panchromatic materials: also for daylight correction with Type C panchromatic materials in making outdoor portraits, darkening skies, or lightening foliage, AAA.
12	Minus blue. Haze cutting in aerial photography, AAA.
13	X2—Yellow green. Correction for Type C panchromatic materials in tungsten light, ABA.
15	G—Deep yellow. Overcorrection in landscape photography. Contrast control in copying and in aerial infrared photography, AAA.
16	Blue absorption, AAB.
18A	Transmits ultraviolet and infrared only (glass), AAA.
	Oranges and Reds
21	Blue and blue-green absorption, CBB.
22	Yellow-orange. For increasing contrast in blue preparations in microscopy. Mercury yellow, BAC.
23A	Light red. Two-color projection—contrast effects, BAB.
24	Red for two-color photography (daylight or tungsten). White-flame-arc tricolor projection, AAB.
25	A—Tricolor red for direct color separation. Contrast effects in commercial photography and in outdoor scenes. Two-color general viewing. Aerial infrared photography and haze cutting. AAA.
26	Stereo red, AAA.
29	Red color separation from transparencies and for the Kodak Fluorescence Process. Strong contrast effects. Copying blue-prints. Tungsten tricolor projection, AAA.
	Magentas and Violets
30	Green absorption, BBC.
31	Green absorption, CCA.
32	Minus green, CCD.
33	Strong green absorption, CCB.
34	Violet, CDD.
34A	Blue separation—Kodak Fluorescence Process, DCC.
35	Contrast in microscopy, CDD.
36	Dark violet, CCC.
	Blues and Blue-greens
38	Red absorption, BCA.
38A	Red absorption. Increasing contrast in visual microscopy, BBB
39	Contrast control in printing motion-picture duplicates (glass) AAA.
40	Green for two-color photography (tungsten), CBC.
44	Minus red—Two-color general viewing, DDD.
44A	Minus red, DDD.
45	Contrast in microscopy, DDD.
45A	Highest resolving power in visual microscopy, CDC.
46	Blue projection (experimental), DDD.
47	Tricolor blue for direct color separation and from Kodak Ektacolor Film for Dye Transfer. Contrast effects in commercial photography. Tungsten and white-flame-arc tricolor projection BBC.
47B	Tricolor blue for color separation from transparencies and from Kodak Ektacolor Film for Graphic Arts, BBB.
48	Green and red absorption, CBC.
48A	Green and red absorption, AAB.
49	Dark blue, BCB.
49B	Very dark blue, BBB.
50	Very dark blue. Mercury violet, CCC.
	Greens
52	Light green, AAB.
53	Medium green, CCB.
54	Very dark green, AAA.
55	Stereo green, BBC.
56	Very light green, CBC.

FILTER DATA

No.	Description, use, and stability
	Greens (Continued)
57	Green for two-color photography (daylight), CBC.
57A	Light green, BBC.
58	Tricolor green for direct color separation. Contrast effects in commercial photography and microscopy, BBC.
59	Green for tricolor projection (white-flame-arc), BBB.
59A	Very light green, BBB.
60	Green for two-color photography (tungsten), BDC.
61	Green color separation from transparencies and Kodak Ekta-color Film. Tricolor projection (tungsten), ABC.
64	Red absorption (light), CDB.
65	Red absorption, ADB.
65A	Red absorption, CCD.
66	Contrast effects in microscopy and medical photography, DDC.
67A	Red absorption (light). Two-color projection, CDC.
	Narrow-band
70	Dark red. Infrared photography. Color separation for Kodak Ektacolor Film (with tungsten), ABC.
72B	Dark orange-yellow, CCC.
73	Dark yellow-green, ABB.
74	Dark green. Mercury green, BBC.
75	Dark blue-green, ACC.
76	Dark violet (compound filter), DDD.
77	Transmits 546 mμ mercury line (glass plus gelatin), AAA.
77A	Transmits 546 mμ mercury line (glass plus gelatin), AAA.
	Photometrics
78	Bluish. Photometric filter (visual), BAB.
78AA	Bluish. Photometric filter (visual), BAA.
78A	Bluish. Photometric filter (visual), AAA.
78B	Bluish. Photometric filter (visual), AAA.
78C	Bluish. Photometric filter (visual), BAA.
86	Yellowish. Photometric filter (visual), BBA.
86A	Yellowish. Photometric filter (visual), AAA.
†86B	Yellowish. Photometric filter (visual), BCA.
†86C	Yellowish. Photometric filter (visual), AAA.

No.	Description, use, and stability
	Light Balancing
80A	For Kodachrome Film, Daylight Type, and photographic flood lamps, ABA.
81	Yellowish. For warmer color rendering.
81A	Yellowish. For Kodak Ektachrome Film, Type B, with photographic flood lamps.
81B	Yellowish. For warmer color rendering.
81C	Yellowish. For Kodachrome Film, Type A, with flash lamps.
81D	Yellowish. For Kodachrome Film, Type A, with flash lamps.
81EF	Yellowish. For Kodak Ektachrome Film, Type B, with flash lamps.
82	Bluish. For cooler color rendering.
82A	Bluish. For Kodachrome Film, Type A, with 3200 K lamps.
82B	Bluish. For cooler color rendering.
82C	Bluish. For cooler color rendering.
83	Yellowish. For 16 mm Commercial Kodachrome Film and daylight exposure, BBB.
85	Orange. For Type A Kodak color films and daylight exposure, BAA.
85B	Orange. For Kodak Ektachrome Film, Type B, and daylight exposure, BAB.
	Miscellaneous
79	Photographic sensitometry. Corrects 2360 K to 5500 K, AAA.
87	For infrared photography. Absorbs visual.
87C	Absorbs visual, transmits infrared.
88A	For infrared photography. Absorbs visual.
89B	For infrared photography, AAA.
90	Narrow-band viewing filter for judging brightness scale of scenes CCD.
96	Neutral filters for controlling luminance, AAB.
97	Dichroic absorption, AAA.
102	Correction filter for Barrier-layer photocell, ABA.
106	Correction filter for S-4 type photocell, AAA.

Wave length	Percent transmittance												
	No. 0 *	No. 1 *	No. 1A *	No. 2B *	No. 3	No. 3N5 *	No. 4	No. 6 *	No. 8	No. 8N5	No. 9	No. 11 *	No. 12 *
400	88.0	85.0	59.0	19.0	.....	.....	.....	7.40	.....	.....	.....	.....	.....
10	88.5	85.5	76.0	48.0	.....	.....	.....	8.32	.....	.....	.....	.....	.....
20	88.9	86.0	82.0	67.0	.....	.....	.....	10.4	.....	.....	.....	0.16	.....
30	89.3	86.5	84.6	75.3	0.36	.....	.....	13.5	.....	.....	.....	0.29	.....
40	89.6	87.0	86.0	80.0	1.78	.....	.....	18.9	.....	.....	.....	0.56	.....
50	89.8	87.4	86.8	83.0	11.5	1.59	.....	27.6	.....	.....	.....	1.32	.....
60	89.9	87.8	87.2	85.2	38.0	9.40	6.9	39.0	0.25	0.16	.....	4.00	.....
70	90.1	88.2	87.5	86.7	68.0	18.5	42.0	52.3	5.50	2.0	1.78	12.0	.....
80	90.3	88.5	87.3	88.1	80.8	23.5	74.0	65.8	19.0	6.3	8.31	26.0	.....
90	90.4	88.7	86.8	88.8	85.2	25.5	84.7	76.8	41.0	13.2	20.7	43.7	.....
500	90.5	88.9	86.3	89.5	86.9	26.3	87.5	83.5	63.5	20.0	34.5	55.0	1.50
10	90.6	89.1	85.5	89.9	87.8	26.7	88.5	87.0	78.0	24.3	48.8	60.0	17.3
20	90.7	89.3	84.8	90.3	88.4	27.0	89.1	88.4	84.1	26.7	62.0	60.2	55.0
30	90.7	89.5	84.3	90.5	89.0	27.2	89.4	89.0	86.5	28.0	76.0	57.8	77.8
40	90.8	89.7	84.0	90.6	89.5	27.5	89.6	89.4	87.7	28.6	83.8	54.2	86.0
50	90.8	89.9	83.9	90.7	89.8	27.8	89.8	89.7	88.4	29.0	87.0	50.0	88.4
60	90.9	90.1	84.1	90.8	90.1	27.9	90.0	89.9	88.8	29.3	88.3	44.8	89.4
70	90.9	90.2	84.8	90.9	90.4	28.0	90.2	90.1	89.2	29.5	88.8	38.9	89.7
80	90.9	90.3	86.0	90.9	90.6	28.4	90.4	90.3	89.5	29.6	89.1	33.1	90.1
90	91.0	90.4	87.4	91.0	90.7	29.0	90.6	90.5	89.8	29.8	89.3	27.6	90.3
600	91.0	90.5	88.5	91.1	90.8	29.5	90.8	90.6	90.1	29.9	89.5	22.7	90.4
10	91.0	90.5	89.5	91.2	90.9	29.5	90.9	90.7	90.3	29.6	89.7	19.0	90.5
20	91.0	90.6	90.2	91.3	91.0	29.3	91.0	90.8	90.5	29.4	89.8	14.9	90.7
30	91.0	90.6	90.6	91.3	91.0	29.1	91.1	90.9	90.7	29.1	89.9	11.4	90.8
40	91.1	90.7	90.8	91.4	91.1	29.0	91.2	91.0	90.9	28.8	90.0	9.10	90.9
50	91.1	90.7	91.0	91.4	91.2	29.4	91.3	91.1	91.0	28.9	90.1	8.05	91.0
60	91.1	90.8	91.1	91.5	91.3	29.6	91.4	91.2	91.1	29.2	90.1	7.50	91.1
70	91.1	90.8	91.1	91.5	91.4	29.8	91.5	91.2	91.2	29.4	90.2	7.05	91.2
80	91.1	90.9	91.1	91.6	91.5	30.0	91.5	91.3	91.3	29.5	90.2	6.50	91.2
90	91.1	90.9	91.1	91.7	91.6	30.2	91.6	91.4	91.4	29.7	90.3	6.10	91.2
700	91.1	91.0	91.1	91.8	91.7	31.0	91.6	91.5	91.5	30.2	90.3	6.20	91.3
Luminous transmit.	90.8	89.9	85.9	90.5	88.3	27.4	87.8	87.5	82.7	27.0	76.6	40.2	73.8
Dominant wave lgth.	571.0	575.0	498.0	570.0	569.5	570.5	569.5	570.3	571.8	572.0	574.4	550.3	576.1
Excitation purity.	0.8	1.5	1.2	5.7	50.0	56.3	64.0	44.7	85.2	84.0	91.4	60.7	97.8

* Some transmission below 400 mμ. Consult the manufacturer.

TRANSMISSION OF WRATTEN FILTERS (Continued)

Percent transmittance

Wave length	No. 13 *	No. 15 *	No. 16	No. 18A *	No. 21	No. 22	No. 23A	No. 24	No. 25	No. 26	No. 29	No. 30 *	No. 31 *
400	.....	.....	.....	.....	.....	.....	.....	.....	.....	.....	.....	48.6	13.8
10	.....	.....	.....	.....	.....	.....	.....	.....	.....	.....	.....	47.4	14.5
20	.....	.....	.....	.....	.....	.....	.....	.....	.....	.....	.....	48.5	16.4
30	0.18	.....	.....	.....	.....	.....	.....	.....	.....	.....	.....	50.1	25.5
40	0.50	.....	.....	.....	.....	.....	.....	.....	.....	.....	.....	49.4	42.7
50	1.35	.....	.....	.....	.....	.....	.....	.....	.....	.....	.....	43.0	50.2
60	4.08	.....	.....	.....	.....	.....	.....	.....	.....	.....	.....	26.5	40.4
70	11.0	.....	.....	.....	.....	.....	.....	.....	.....	.....	.....	13.8	22.6
80	23.5	.....	.....	.....	.....	.....	.....	.....	.....	.....	.....	5.00	8.20
90	39.0	.....	.....	.....	.....	.....	.....	.....	.....	.....	.....	0.63	1.85
500	50.8	.....	.....	.....	.....	.....	.....	.....	.....	.....	.....	.....	0.12
10	55.2	1.00	.....	.....	.....	.....	.....	.....	.....	.....	.....	.....	.....
20	56.5	16.0	3.00	.....	.....	.....	.....	.....	.....	.....	.....	.....	.....
30	55.0	52.1	22.0	.....	.....	.....	.....	.....	.....	.....	.....	.....	.....
40	51.0	70.7	48.0	.....	2.50	.....	.....	.....	.....	.....	.....	.....	.....
50	46.0	84.3	69.5	.....	29.0	0.25	.....	.....	.....	.....	.....	.....	.....
60	39.2	87.5	79.5	.....	65.0	19.0	.....	.....	.....	.....	.....	0.10	.....
70	32.0	88.7	84.0	.....	80.6	60.0	11.0	.....	.....	.....	.....	10.0	.....
80	25.1	89.3	86.3	.....	85.4	81.0	47.0	4.55	.....	.....	.....	45.0	.....
90	18.2	89.7	87.8	.....	87.3	87.0	69.6	37.3	12.6	2.90	.....	76.0	0.63
600	13.5	90.0	89.0	.....	88.1	88.5	82.7	72.3	50.0	30.0	.....	87.4	26.0
10	9.60	90.1	89.6	.....	88.7	89.0	85.8	82.9	75.0	63.2	10.0	89.5	67.2
20	6.40	90.2	90.0	.....	89.0	89.5	87.2	86.4	82.6	78.9	45.3	90.2	84.0
30	3.66	90.3	90.2	.....	89.5	89.8	87.9	87.8	85.5	84.0	71.4	90.5	88.1
40	2.20	90.4	90.3	.....	89.9	90.0	88.5	88.5	86.7	86.1	82.7	90.7	89.8
50	1.58	90.5	90.4	.....	90.2	90.1	89.0	89.0	87.6	87.2	86.6	90.8	90.2
60	1.74	90.6	90.5	.....	90.4	90.2	89.4	89.3	88.2	88.1	88.4	90.9	90.4
70	2.62	90.6	90.6	.....	90.5	90.3	89.6	89.7	88.5	88.5	89.4	91.0	90.5
80	3.55	90.7	90.7	.....	90.5	90.4	89.8	89.9	89.0	88.9	90.0	91.1	90.7
90	4.48	90.7	90.8	0.25	90.6	90.5	90.0	90.2	89.3	89.2	90.3	91.1	90.8
700	5.25	90.8	90.8	1.20	90.6	90.6	90.2	90.3	89.5	89.5	90.4	91.1	91.0
Luminous transmit.	34.5	66.2	57.7	0.0014	45.6	35.8	25.0	17.8	14.0	11.7	6.3	26.6	12.9
Dominant wave lgth.	542.0	579.3	582.7	700.0	588.9	595.1	602.7	610.6	615.1	619.0	631.6	498.6_c	513.1_c
Excitation purity.	57.5	99.0	99.3	100.0	99.9	99.9	100.0	100.0	100.0	100.0	100.0	62.4	81.9

Percent transmittance

Wave length	No. 32 *	No. 33 *	No. 34 *	No. 34A *	No. 35 *	No. 36 *	No. 38 *	No. 38A *	No. 39 *	No. 40	No. 44 *	No. 44A *	No. 45
400	38.0	0.85	64.0	.....	48.0	36.5	60.5	33.4	85.2	.....	0.44	2.52	.....
10	37.9	0.71	70.1	0.1	57.0	45.5	66.5	41.2	78.2	.....	0.36	3.39	.....
20	40.0	1.17	72.0	40.0	57.6	45.5	72.5	53.0	70.5	.....	0.63	6.30	.....
30	43.0	1.69	68.4	69.7	47.5	32.7	75.3	58.0	63.3	.....	3.63	17.4	.....
40	55.5	5.36	58.2	68.7	29.5	15.2	76.2	58.8	53.6	.....	13.1	32.7	5.00
50	66.0	14.3	42.3	56.2	12.3	3.7	75.9	57.6	42.5	.....	25.4	41.8	19.0
60	66.0	12.4	25.2	40.5	3.5	0.35	74.8	55.2	28.5	3.16	36.5	48.1	29.5
70	57.0	5.00	12.1	23.8	0.25	.....	73.4	51.9	17.3	21.6	46.5	51.7	34.4
80	40.0	0.50	2.7	9.2	.....	.....	71.6	48.5	10.2	44.7	53.6	52.9	35.7
90	21.0	.....	0.2	2.3	.....	.....	69.5	44.6	4.00	61.4	56.8	52.2	34.5
500	9.56	.....	.....	0.33	.....	.....	66.7	40.2	1.33	70.2	55.8	49.8	29.7
10	2.51	.....	.....	.....	.....	.....	63.9	35.8	0.35	72.4	50.9	44.8	21.5
20	0.13	.....	.....	.....	.....	.....	60.8	31.7	.....	70.5	42.1	36.8	11.5
30	.....	.....	.....	.....	.....	.....	57.0	27.2	.....	64.8	30.5	26.8	3.80
40	.....	.....	.....	.....	.....	.....	52.6	22.3	.....	55.5	18.6	16.8	0.85
50	.....	.....	.....	.....	.....	.....	48.0	17.6	.....	44.2	8.99	8.20	.....
60	.....	.....	.....	.....	.....	.....	42.8	12.9	.....	32.5	3.59	2.95	.....
70	.....	.....	.....	.....	.....	.....	37.0	8.78	.....	20.3	0.80	0.91	.....
80	.....	.....	.....	.....	.....	.....	30.6	5.65	.....	9.56	.....	0.10	.....
90	.....	.....	.....	.....	.....	.....	25.5	3.48	.....	3.20	.....	.....	.....
600	6.04	.....	.....	.....	.....	.....	20.9	2.09	.....	1.10	.....	.....	.....
10	41.0	0.80	.....	0.13	.....	.....	16.8	1.15	.....	0.32	.....	.....	.....
20	75.0	24.9	.....	1.0	.....	.....	12.9	0.59	.....	.....	.....	.....	.....
30	86.1	60.8	.....	6.3	.....	.....	10.0	0.28	.....	.....	.....	.....	.....
40	89.0	78.0	0.4	22.0	.....	.....	7.79	0.13	.....	.....	.....	.....	.....
50	90.0	85.0	4.0	45.0	0.1	.....	6.68	.....	.....	.....	.....	.....	.....
60	90.6	87.5	20.7	65.0	3.0	0.21	6.20	.....	.....	.....	.....	.....	.....
70	90.7	88.7	45.2	77.3	19.0	7.5	5.91	.....	.....	.....	.....	.....	.....
80	90.8	89.4	66.5	85.0	43.5	29.0	5.41	.....	0.50	0.80	.....	.....	.....
90	90.9	89.8	78.8	88.2	66.0	55.0	4.90	.....	4.06	6.99	0.18	.....	.....
700	91.0	90.0	85.0	89.8	77.7	71.3	5.00	.....	17.8	23.5	1.60	.....	1.00
Luminous transmit.	12.5	5.2	1.3	2.9	0.45	0.25	42.5	17.3	1.2	33.6	15.6	14.4	5.2
Dominant wave lgth.	551.7_c	498.0_c	424.0	564.8_c	566.8_c	566.4_c	483.5	478.9	450.6	516.2	589.1	483.4	481.5
Excitation purity.	79.6	88.3	94.4	91.4	96.3	97.8	41.8	69.8	98.9	48.5	72.9	77.2	88.4

* Some transmission below 400 mμ. Consult the manufacturer.

Percent transmittance

Wave length	No. 45A	No. 46 *	No. 47 *	No. 47B *	No. 48 *	No. 48A *	No. 49 *	No. 49B *	No. 50 *	No. 52 *	No. 53 *	No. 54	No. 55
400	......	1.20	7.80	16.0	0.96	5.65	3.30	1.70	0.45	2.18	.....	.....	......
10	......	0.60	17.4	29.5	3.16	10.0	4.28	2.00	0.39	1.51	.....	.....	......
20	......	0.80	34.0	43.6	8.25	16.0	6.93	3.55	0.59	0.80	.....	.....	......
30	1.00	5.98	47.0	50.0	15.0	21.0	11.2	7.00	2.63	0.44	.....	.....	......
40	8.81	19.0	50.3	47.2	22.6	25.0	18.9	13.0	8.90	0.41	.....	.....	......
50	17.4	30.1	48.3	36.0	30.3	26.2	25.6	17.4	14.0	0.69	.....	.....	......
60	20.9	33.8	43.4	25.0	33.2	22.9	24.0	14.8	12.3	1.45	.....	.....	0.20
70	21.6	32.1	36.2	13.2	29.6	16.5	15.7	7.60	5.36	2.70	0.10	.....	2.90
80	20.5	27.0	28.5	4.5	22.4	9.55	6.93	2.76	1.55	4.90	0.71	.....	13.1
90	18.0	20.2	19.6	1.3	14.1	4.27	2.14	0.40	0.10	8.50	2.14	.....	34.2
500	14.4	11.1	11.3	0.17	7.30	1.58	0.46	.....	.....	13.3	4.47	.....	53.4
10	10.1	4.39	5.64	.....	2.64	0.48	.....	.....	.....	18.2	7.24	0.10	67.0
20	5.60	1.66	1.91	.....	0.50	.....	.....	.....	.....	23.7	10.7	0.31	69.3
30	2.52	0.35	0.36	.....	.....	.....	.....	.....	.....	28.5	14.0	0.64	65.1
40	0.64	.....	.....	.....	.....	.....	.....	.....	.....	32.1	16.6	0.89	56.7
50	0.10	.....	.....	.....	.....	.....	.....	.....	.....	33.1	17.3	0.93	45.0
60	......	.....	.....	.....	.....	.....	.....	.....	.....	31.0	15.4	0.62	33.1
70	......	.....	.....	.....	.....	.....	.....	.....	.....	25.6	11.4	0.21	20.7
80	......	.....	.....	.....	.....	.....	.....	.....	.....	19.1	6.90	.....	9.00
90	......	.....	.....	.....	.....	.....	.....	.....	.....	12.6	3.60	.....	2.70
600	......	.....	.....	.....	.....	.....	.....	.....	.....	7.78	1.41	.....	0.40
10	......	.....	.....	.....	.....	.....	.....	.....	.....	4.17	0.40	.....	......
20	......	.....	.....	.....	.....	.....	.....	.....	.....	2.34	0.15	.....	......
30	......	.....	.....	.....	.....	.....	.....	.....	.....	1.38	.....	.....	......
40	......	.....	.....	.....	.....	.....	.....	.....	.....	0.80	.....	.....	......
50	......	.....	.....	.....	.....	.....	.....	.....	.....	0.54	.....	.....	......
60	......	.....	.....	.....	.....	.....	.....	.....	.....	0.36	.....	.....	......
70	......	.....	.....	.....	.....	.....	.....	.....	.....	0.27	.....	.....	......
80	......	.....	.....	.....	.....	.....	.....	.....	.....	0.23	.....	.....	0.66
90	0.20	0.25	.....	.....	.....	.....	.....	.....	.....	0.19	.....	.....	6.90
700	2.24	0.85	.....	.....	.....	.....	.....	.....	.....	0.17	.....	.....	27.8
Luminous transmit.	2.8	2.4	2.8	0.78	1.86	0.88	0.69	0.36	0.26	20.1	9.0	0.032	31.4
Dominant wave lgth.	477.6	470.4	463.7	479.8	466.5	458.0	457.9	455.5	455.9	553.3	551.1	546.1	530.2
Excitation purity.	89.7	94.9	95.8	69.1	96.1	98.3	98.9	99.3	99.4	77.3	89.7	97.0	68.4

Percent transmittance

Wave length	No. 56	No. 57	No. 57A	No. 58	No. 59 *	No. 59A *	No. 60	No. 61	No. 64 *	No. 65	No. 65A *	No. 66 *	No. 67A *
400	.....	.....	......	.....	.....	......	.....	.....	9.00	......	......	12.3	1.10
10	.....	.....	......	.....	.....	......	.....	.....	9.20	......	......	13.0	0.93
20	.....	.....	......	.....	.....	......	.....	.....	8.75	0.23	......	15.0	1.28
30	.....	.....	......	.....	.....	0.16	.....	.....	9.20	0.61	0.16	18.4	3.16
40	.....	.....	0.19	.....	.....	0.37	.....	.....	11.3	1.58	1.32	23.2	6.40
50	.....	.....	0.87	.....	0.40	1.26	0.19	.....	15.5	4.10	5.50	31.2	10.5
60	0.16	0.44	2.56	.....	1.90	4.57	1.38	.....	23.3	9.00	13.0	42.2	17.7
70	3.12	3.10	7.80	0.23	7.70	13.2	5.38	.....	34.4	16.8	24.9	55.5	28.5
80	13.0	13.1	21.6	1.38	21.0	30.0	15.0	0.33	46.8	24.9	36.6	68.4	41.4
90	34.5	31.9	41.7	4.90	41.5	50.8	32.0	4.00	56.6	31.3	45.1	77.6	52.1
500	59.0	50.5	58.8	17.7	59.0	66.0	48.4	16.6	62.1	33.7	45.8	82.7	57.9
10	73.0	60.6	67.9	38.8	67.7	73.0	57.2	32.3	62.9	32.4	39.7	84.6	58.8
20	79.0	63.3	70.1	52.2	69.8	75.1	59.2	40.0	59.1	27.5	29.7	84.0	55.4
30	79.9	61.0	67.6	53.6	67.2	73.2	55.5	39.6	51.6	20.7	17.8	82.6	47.5
40	77.5	55.0	61.8	47.6	61.5	68.5	47.5	34.5	41.3	13.7	7.90	79.1	36.6
50	72.6	47.1	53.5	38.4	54.0	62.0	36.8	26.3	28.0	6.50	2.40	73.7	25.0
60	66.1	37.3	43.3	27.8	45.0	54.4	25.2	17.3	16.2	1.66	0.32	67.1	14.2
70	58.0	26.5	31.6	17.4	35.0	44.5	14.4	9.70	7.95	0.40	......	58.8	5.50
80	46.1	16.6	19.4	9.0	24.0	33.0	6.3	4.40	3.10	......	......	47.2	1.40
90	33.8	8.69	9.70	3.50	14.0	22.0	1.82	1.66	0.80	......	......	34.5	0.28
600	24.0	3.70	4.50	1.50	7.95	14.6	0.48	0.38	.....	......	......	24.4	......
10	18.7	1.60	2.00	0.41	4.90	10.5	0.10	.....	.....	......	......	18.5	......
20	13.2	0.49	0.87	.....	2.70	6.92	.....	.....	.....	......	......	13.7	......
30	7.22	.....	0.22	.....	1.00	3.16	.....	.....	.....	......	......	7.70	......
40	3.02	.....	......	.....	0.17	1.07	.....	.....	.....	......	......	3.00	......
50	1.48	.....	......	.....	.....	0.50	.....	.....	.....	......	......	1.46	......
60	1.91	.....	......	.....	.....	0.91	.....	.....	.....	......	......	1.91	......
70	7.95	.....	......	.....	0.63	3.00	.....	.....	.....	......	......	6.17	......
80	23.0	.....	0.16	.....	4.00	10.0	.....	.....	.....	......	......	19.9	......
90	44.1	.....	1.15	.....	12.0	20.0	2.10	.....	0.10	......	0.20	42.6	......
700	64.8	.....	3.17	0.53	22.6	30.0	8.70	.....	4.50	......	2.18	63.1	0.40
Luminous transmit.	52.8	32.5	37.2	23.7	38.7	45.8	26.1	16.8	25.0	9.6	9.8	58.3	22.4
Dominant wave lgth.	552.3	536.4	534.0	540.2	538.3	541.4	525.7	536.8	497.3	496.6	492.7	512.3	499.8
Excitation purity.	78.2	69.2	62.1	88.1	66.0	59.3	62.2	85.4	55.0	67.8	77.4	21.5	55.8

* Some transmission below 400 mμ. Consult the manufacturer.

Percent transmittance

Wave length	No. 70	No. 72B	No. 73	No. 74	No. 75	No. 76 *	No. 77	No. 77A	No. 78 *	No. 78AA *	No. 78A *	No. 78B *
400	......	......	......	......	......	0.22	......	......	37.2	43.0	56.0	64.1
10	......	......	......	......	......	0.18	......	......	41.7	46.0	58.6	66.5
20	......	......	......	......	......	0.29	......	......	44.2	48.7	61.0	68.4
30	......	......	......	......	......	1.38	......	......	44.6	49.8	61.8	69.5
40	......	......	......	......	......	3.50	......	......	44.2	49.7	61.8	70.0
50	......	......	......	......	......	3.50	......	......	41.7	48.0	61.0	69.4
60	......	......	......	......	1.97	1.92	......	......	38.0	44.9	58.7	67.5
70	......	......	......	......	10.0	0.51	......	......	33.8	40.3	55.0	65.4
80	......	......	......	......	17.4	......	......	......	27.5	35.6	51.0	62.9
90	......	......	......	......	18.0	......	......	......	23.5	30.9	47.1	59.8
500	......	......	......	......	13.0	......	......	......	19.5	26.5	43.5	57.0
10	......	......	......	0.96	7.35	......	0.30	0.10	15.8	23.4	40.0	54.2
20	......	......	......	7.95	3.20	......	9.10	5.35	13.8	20.3	36.9	51.4
30	......	......	......	14.6	0.83	......	13.5	1.90	11.8	17.8	34.4	49.3
40	......	......	......	12.9	0.14	......	46.1	35.0	10.5	16.6	32.7	48.1
50	......	......	......	7.60	......	......	78.0	71.8	9.56	14.9	31.2	46.7
60	......	......	2.24	3.06	......	......	75.8	63.1	8.53	13.2	29.4	45.0
70	......	......	5.97	0.83	......	......	8.00	......	7.77	12.1	28.0	43.6
80	......	......	4.56	0.12	......	......	1.00	......	7.41	11.6	27.5	43.1
90	......	1.26	2.00	......	......	......	0.32	......	6.93	11.1	27.0	42.9
600	......	5.89	0.56	......	......	......	16.2	1.60	6.45	10.40	26.0	41.8
10	......	5.25	0.10	......	......	......	52.1	32.1	5.50	9.20	24.1	40.0
20	......	2.88	......	......	......	......	83.0	78.0	4.80	7.70	21.8	37.6
30	......	1.26	......	......	......	......	84.9	79.5	3.94	6.50	19.7	35.5
40	......	0.48	......	......	......	......	88.1	86.5	3.46	5.60	18.6	34.2
50	0.63	0.14	......	......	......	......	89.8	89.2	3.24	5.50	18.4	33.6
60	10.5	......	......	......	......	......	89.8	89.0	3.16	5.60	18.5	34.0
70	35.0	......	......	......	......	......	85.5	79.5	3.39	5.80	18.7	34.1
80	55.2	......	......	......	......	......	76.1	62.5	3.45	6.10	19.0	34.5
90	70.0	......	......	......	......	0.13	75.0	62.4	3.51	6.10	19.3	34.8
700	79.0	......	......	......	0.14	1.24	86.5	83.0	3.90	6.50	20.2	36.0
Luminous transmit.	0.31	0.74	1.3	4.0	1.9	0.046	32.3	25.5	10.7	15.8	31.6	46.7
Dominant wave lgth.	675.6	604.9	574.9	538.6	487.7	449.2	579.9	581.5	471.1	473.4	475.7	477.2
Excitation purity.	100.0	100.0	100.0	96.7	90.4	99.7	99.0	99.1	63.0	54.5	33.7	20.7

Percent transmittance

Wave length	No. 78C *	No. 79 *	No. 80A *	No. 81 *	No. 81A *	No. 81B *	No. 81C *	No. 81D *	No. 81EF *	No. 82 *	No. 82A *	No. 82B *
400	74.9	24.0	67.6	77.7	65.1	55.1	46.1	38.2	30.7	83.0	80.1	76.7
10	76.6	26.0	73.1	78.1	65.9	55.8	46.6	38.4	31.5	83.7	80.8	78.0
20	77.9	29.0	76.8	79.0	67.6	57.7	49.0	41.0	34.3	84.6	81.6	79.2
30	78.9	31.0	7.77	80.5	70.2	61.0	52.5	45.5	38.6	85.1	82.2	79.7
40	79.4	32.2	76.5	81.9	72.8	64.5	57.2	50.0	43.2	85.4	82.4	79.7
50	79.5	32.7	73.0	83.0	74.8	67.2	60.5	53.9	47.4	85.4	82.4	79.2
60	79.3	31.4	69.0	83.7	76.0	69.1	63.0	56.5	50.2	85.0	81.7	78.0
70	78.6	28.8	63.6	84.3	77.1	70.6	64.2	58.1	52.0	84.6	80.7	76.3
80	77.8	25.6	57.6	84.6	77.8	71.3	65.0	59.0	53.0	84.0	79.3	74.4
90	76.7	22.2	51.3	84.9	78.3	71.8	65.7	60.0	54.0	83.3	78.0	72.1
500	75.5	19.3	45.2	85.3	78.6	72.6	66.4	60.8	55.4	82.6	76.6	70.2
10	74.2	16.8	39.4	85.4	79.0	72.9	66.5	61.1	56.2	82.0	75.3	68.3
20	73.0	14.2	34.2	85.5	79.5	73.2	67.0	61.6	57.0	81.4	74.0	66.5
30	72.1	12.7	30.0	86.0	80.4	74.5	68.8	62.5	59.5	81.0	73.1	65.5
40	71.5	11.8	27.1	86.5	81.5	76.0	71.0	66.1	62.7	80.8	72.7	65.0
50	70.7	11.0	24.8	86.8	82.3	77.0	72.0	67.3	64.5	80.6	72.4	64.5
60	69.8	9.76	23.5	87.0	82.6	77.6	72.5	68.0	65.3	80.4	71.8	63.8
70	69.0	8.81	22.6	87.1	82.7	77.8	72.7	68.3	65.8	80.2	71.5	63.2
80	68.8	8.50	22.6	87.1	82.8	78.0	73.0	68.5	66.0	80.2	71.5	63.2
90	68.6	8.29	23.2	87.4	83.1	78.2	74.0	69.5	66.5	80.3	71.7	63.4
600	68.0	7.56	23.7	87.6	84.0	79.1	75.6	72.0	68.1	80.2	71.5	63.0
10	66.7	6.45	23.2	88.1	85.0	81.0	78.5	75.0	71.6	79.3	70.3	61.5
20	65.0	5.13	21.0	88.8	86.1	83.1	80.8	78.0	74.7	78.4	68.5	59.0
30	63.8	4.17	18.2	89.2	87.0	84.2	82.1	79.8	77.0	77.5	66.9	56.9
40	63.0	3.47	15.8	89.4	87.4	85.1	83.0	80.8	78.4	76.8	65.5	55.0
50	62.7	3.16	14.5	89.5	87.7	85.6	83.5	81.5	79.2	76.5	64.8	54.1
60	63.0	3.09	13.8	89.8	88.0	86.0	84.1	82.1	80.1	76.2	64.6	53.7
70	63.3	3.16	13.4	90.0	88.2	86.5	84.8	83.0	80.9	76.1	64.5	53.7
80	63.4	3.16	12.7	90.1	88.5	87.0	85.5	83.7	81.8	76.1	64.4	53.5
90	63.6	3.16	11.7	90.3	89.0	87.5	86.1	84.6	82.9	76.2	64.2	53.4
700	65.0	3.31	11.5	90.5	89.2	88.0	86.8	85.5	84.0	77.1	64.6	54.1
Luminous transmit.	70.4	11.3	28.4	86.8	82.0	76.9	72.0	67.4	64.0	80.7	72.5	64.6
Dominant wave lgth.	479.8	474.8	471.7	576.7	577.5	577.8	577.4	579.5	579.0	477.5	476.6	475.6
Excitation purity.	6.8	52.8	45.9	2.9	6.0	8.7	10.7	14.7	19.0	3.0	6.3	10.2

* Some transmission below 400 mμ. Consult the manufacturer.

Percent transmittance

Wave length	No. 82C *	No. 83 *	No. 85 *	No. 85B *	No. 86 *	No. 86A *	No. 86B *	No. 86C *	No. 89B	No. 90	No. 96 *	No. 97	No. 102 *
400	73.4	13.5	6.0	1.59	0.50	8.00	20.0	44.0	.....	.....	4.28	...	1.12
10	75.0	13.1	18.0	9.32	0.81	12.2	26.1	55.0	.....	.....	4.91	...	0.96
20	76.4	13.5	28.4	15.5	1.55	16.7	31.6	62.0	.....	.....	5.50	...	0.89
30	77.2	14.1	33.4	19.0	2.88	21.5	37.5	66.6	.....	.....	6.17	...	0.96
40	77.2	15.6	36.2	20.8	5.50	27.8	44.0	70.8	.....	.....	6.92	...	1.23
50	76.6	17.8	38.1	22.1	9.10	34.2	50.1	74.3	.....	.....	7.50	...	1.86
60	75.2	21.0	40.4	24.3	13.5	40.4	55.4	76.8	.....	.....	7.81	...	3.23
70	73.2	25.5	43.0	27.5	17.8	45.0	59.5	78.7	.....	.....	8.15	0.22	6.45
80	70.7	30.2	45.3	30.9	21.3	48.7	62.5	80.2	.....	.....	8.47	0.43	14.0
90	68.1	35.8	47.2	34.3	24.5	51.2	64.6	81.2	.....	.....	8.60	0.39	21.6
500	65.7	43.5	48.9	38.3	26.8	52.8	66.0	81.7	.....	.....	8.73	0.15	30.7
10	63.5	46.3	49.2	40.7	27.9	53.4	66.4	81.9	.....	.....	8.85	...	41.4
20	61.5	47.2	48.2	40.6	28.6	53.7	66.6	82.0	.....	.....	8.90	...	51.3
30	59.9	48.3	48.3	40.7	30.4	55.0	67.6	82.4	.....	.....	9.01	...	59.4
40	59.1	49.6	49.2	41.6	32.5	56.5	69.0	83.0	.....	.....	9.07	...	64.2
50	58.3	51.8	51.0	43.2	35.0	58.5	70.2	83.5	.....	.....	9.20	...	66.7
60	57.2	56.5	55.8	47.1	41.2	63.0	73.0	84.6	.....	9.00	9.30	...	66.3
70	56.2	65.0	64.5	56.0	53.0	70.9	78.1	86.8	.....	30.5	9.20	...	63.0
80	56.1	75.5	75.0	68.1	67.5	79.0	84.0	88.9	.....	34.3	9.19	...	58.0
90	56.0	83.0	83.0	78.1	76.5	85.2	87.5	89.9	.....	25.2	9.54	...	51.9
600	55.0	87.3	87.2	85.0	85.0	88.1	89.3	90.6	.....	16 1	9.64	...	45.2
10	53.0	89.3	88.9	88.0	88.1	89.8	90.3	91.0	.....	11.3	9.73	...	37.8
20	50.2	90.4	90.0	89.6	89.6	90.5	90.7	91.1	.....	7.40	9.56	...	30.5
30	47.4	90.8	90.5	90.3	90.4	90.8	90.9	91.2	.....	2.94	9.27	...	25.0
40	45.2	91.0	90.7	90.7	90.7	91.1	91.1	91.3	.....	0.76	9.10	...	20.6
50	44.1	91.1	90.9	90.9	91.0	91.2	91.2	91.4	.....	0.29	9.07	...	17.5
60	43.6	91.3	91.0	91.0	91.1	91.3	91.3	91.5	.....	0.41	9.00	...	15.2
70	43.5	91.5	91.0	91.2	91.2	91.4	91.4	91.6	.....	2.30	9.13	...	13.7
80	43.1	91.5	91.0	91.3	91.3	91.4	91.5	91.6	0.10	9.52	9.08	0.44	12.8
90	42.8	91.5	91.0	91.3	91.3	91.5	91.6	91.6	1.58	28.5	9.21	5.02	12.1
700	43.5	91.5	91.0	91.3	91.3	91.5	91.6	91.6	11.2	51.9	9.52	18.7	12.0
Luminous transmit.	58.1	61.4	62.5	55.5	49.7	67.1	75.5	85.4	0.017	9.8	9.1	0.041	50.8
Dominant wave lgth.	477.2	581.5	587.7	585.7	585.7	581.7	579.6	577.6	700	583.1	572.4	555.0c	564.9
Excitation purity.	14.5	55.4	30.3	48.0	69.7	37.1	24.1	9.0	100	100.0	12.1	48.0	80.0

Percent transmittance

Wave length	No. 106	CC-05R *	CC-10R *	CC-20R *	CC-30R *	CC-40R *	CC-50R *	CC-05B *	CC-10B *	CC-20B *	CC-30B *	CC-40B *	CC-50B *
400	......	81.0	73.0	61.5	51.6	42.5	36.4	87.0	85.5	82.2	80.2	77.0	74.1
10	......	81.0	72.4	60.0	50.0	40.0	33.9	87.5	86.4	84.0	82.5	80.3	78.4
20	0.10	81.0	72.0	58.6	48.2	38.2	31.9	87.7	87.2	85.0	84.0	82.2	80.7
30	0.20	81.1	71.6	57.7	47.0	36.8	30.5	88.0	87.5	85.3	84.3	82.5	81.1
40	0.35	81.2	71.5	57.2	46.4	36.0	29.7	88.1	87.5	85.0	83.5	81.3	79.8
50	0.58	81.4	71.6	57.2	46.4	36.1	29.6	88.1	87.2	83.9	81.9	78.7	76.6
60	0.98	81.7	72.4	58.5	47.5	37.5	31.0	87.9	86.4	82.5	79.5	75.9	72.9
70	1.5	82.3	73.7	60.6	49.9	40.0	33.6	87.5	85.3	80.3	76.2	72.6	67.9
80	2.3	82.8	74.9	62.0	52.0	42.5	35.9	87.0	84.0	77.8	72.5	67.5	62.7
90	3.5	83.2	75.8	63.5	53.9	44.8	37.9	86.2	82.4	74.2	68.3	62.3	56.6
500	5.2	83.3	76.6	64.6	55.2	46.1	39.4	85.2	80.5	71.2	63.8	56.7	50.1
10	7.7	83.0	76.1	64.0	54.5	46.0	38.5	84.4	78.6	67.7	58.7	51.0	44.5
20	10.7	82.4	74.9	61.5	51.6	42.5	35.0	83.5	77.0	64.0	54.4	46.0	38.6
30	15.1	81.6	73.5	59.4	48.5	38.5	31.5	82.6	75.2	61.5	50.7	41.6	34.1
40	20.2	81.2	72.5	57.8	46.5	36.6	29.4	82.1	73.9	59.5	48.3	39.0	31.3
50	25.7	81.1	72.4	57.1	45.6	35.6	28.7	81.5	73.0	58.0	46.6	36.9	29.5
60	31.0	81.4	72.8	58.0	46.9	36.4	29.7	81.4	72.7	57.5	45.9	35.9	28.6
70	35.6	82.5	74.5	60.6	49.2	39.2	32.7	81.4	73.0	57.9	46.3	36.1	28.7
80	43.2	83.9	77.3	65.0	54.6	45.0	38.6	81.9	73.9	59.3	47.9	37.8	30.5
90	53.8	85.7	80.7	71.0	61.8	53.5	47.6	82.7	75.1	61.6	50.3	40.8	33.4
600	65.6	87.6	84.0	77.0	70.5	64.0	58.7	83.4	76.3	63.5	53.0	43.5	36.0
10	77.0	89.0	86.5	82.0	77.8	73.0	69.8	83.6	76.7	64.5	54.6	44.7	37.8
20	82.8	90.0	88.9	86.2	83.5	80.8	78.3	83.5	76.5	64.3	54.4	44.3	37.5
30	86.0	90.6	89.9	88.1	87.2	85.1	84.2	83.2	76.6	63.1	53.2	42.5	35.6
40	87.6	91.1	90.5	89.8	89.2	88.0	87.5	82.8	74.5	61.6	51.5	40.2	33.7
50	88.7	91.2	90.8	90.5	90.3	89.5	89.2	82.5	74.0	60.6	50.3	39.0	32.4
60	89.5	91.3	91.1	90.8	90.6	90.4	90.1	82.5	73.8	60.1	49.5	38.4	31.8
70	90.0	91.4	91.3	91.0	90.9	90.8	90.7	82.3	73.3	59.6	49.0	37.7	31.0
80	90.5	91.6	91.5	91.2	91.1	91.0	91.1	82.0	72.8	58.6	48.2	36.5	30.0
90	90.8	91.7	91.7	91.4	91.4	91.3	91.1	81.9	72.5	58.1	47.2	35.4	29.0
700	91.0	91.9	91.9	91.5	91.5	91.4	91.2	82.2	73.0	58.5	47.5	35.6	29.0
Luminous transmit.	34.6	83.7	77.0	65.3	55.9	47.3	41.3	82.8	75.5	62.3	52.0	42.8	35.7
Dominant wave lgth.	589.4	605.0	597.8	604.2	605.8	605.5	608.5	459.0	462.0	460.0	461.0	463.2	462.5
Excitation purity.	95.2	2.0	4.7	8.5	12.3	17.3	21.4	2.8	6.3	13.2	20.2	27.7	34.2

* Some transmission below 400 mμ. Consult the manufacturer.

Percent transmittance

Wave length	CC-05G *	CC-10G *	CC-20G *	CC-30G *	CC-40G *	CC-50G *	CC-05Y *	CC-10Y *	CC-20Y *	CC-30Y *	CC-40Y *	CC-50Y *
400	80.0	73.1	58.8	48.0	39.7	32.0	81.0	74.5	61.3	50.5	43.0	34.5
10	80.7	72.9	57.8	46.5	38.1	30.3	80.6	73.2	59.0	47.4	39.5	30.5
20	81.0	72.8	57.3	45.8	37.3	29.5	80.4	72.6	57.8	46.0	37.5	29.0
30	81.4	72.7	57.0	45.5	36.5	29.0	80.4	72.5	57.5	45.6	36.5	28.7
40	81.6	73.0	57.3	45.8	36.6	29.1	80.6	72.8	57.8	46.5	36.8	29.5
50	82.1	73.9	58.4	46.9	38.1	30.6	81.2	74.0	59.5	48.5	38.5	31.5
60	83.0	75.5	61.4	50.3	41.5	34.3	82.5	76.0	63.0	52.5	42.5	36.2
70	84.4	78.0	65.4	55.8	47.0	40.5	83.9	78.5	67.5	48.2	48.8	43.5
80	85.6	80.4	70.0	61.8	53.5	47.8	85.3	81.2	72.3	64.9	56.2	54.0
90	86.8	83.0	75.2	68.9	61.3	57.0	87.0	84.4	78.0	72.4	66.0	64.0
500	87.9	85.9	80.3	76.4	70.7	68.0	88.4	87.2	84.0	81.0	77.0	75.5
10	88.7	87.5	83.8	80.9	77.8	75.3	89.5	89.0	88.0	86.6	85.5	84.2
20	89.0	88.1	84.9	82.3	79.5	77.5	90.0	90.0	89.6	89.1	89.0	88.5
30	89.0	88.0	84.6	81.7	79.4	77.0	90.4	90.4	90.0	89.7	89.9	89.6
40	89.0	87.6	83.7	80.5	77.8	74.8	90.7	90.7	90.6	90.4	90.2	90.0
50	88.6	87.1	82.4	78.6	75.8	72.2	90.9	90.9	90.8	90.6	90.4	90.3
60	88.1	86.3	80.9	76.2	72.9	68.8	91.0	91.0	90.9	90.8	90.7	90.6
70	87.5	85.3	79.0	73.5	69.3	64.8	91.3	91.3	91.0	90.9	90.8	90.7
80	87.0	84.1	77.0	70.4	65.3	60.3	91.4	91.4	91.1	91.0	90.8	90.7
90	86.4	82.8	74.5	67.2	61.9	55.9	91.4	91.4	91.2	91.1	90.9	90.8
600	85.7	81.5	72.0	64.1	57.7	51.7	91.4	91.4	91.3	91.2	90.9	90.8
10	85.0	80.0	69.5	60.7	53.7	47.3	91.4	91.4	91.3	91.2	90.9	90.9
20	84.0	78.5	66.5	57.2	49.8	42.5	91.4	91.4	91.3	91.2	91.0	90.9
30	83.0	76.9	63.8	53.7	45.5	38.0	91.5	91.5	91.4	91.3	91.0	91.0
40	82.2	75.6	61.5	50.8	42.0	34.6	91.5	91.5	91.4	91.3	91.0	91.0
50	81.9	74.9	60.1	49.1	40.2	32.5	91.5	91.5	91.4	91.3	91.1	91.1
60	81.5	74.4	59.4	48.1	39.5	31.5	91.5	91.5	91.4	91.3	91.1	91.1
70	81.4	74.0	58.8	47.5	38.5	31.0	91.5	91.5	91.4	91.4	91.2	91.2
80	81.1	73.5	58.1	46.6	37.6	29.9	91.5	91.5	91.4	91.4	91.2	91.2
90	81.1	73.2	57.6	46.0	36.6	28.9	91.5	91.5	91.4	91.4	91.3	91.3
700	81.5	73.5	58.0	46.4	36.5	28.7	91.5	91.5	91.4	91.4	91.3	91.3
Luminous transmit.	87.2	84.5	77.8	72.2	67.7	63.3	90.4	90.1	89.1	88.2	87.4	86.9
Dominant wave lgth.	553.0	555.5	555.0	554.0	554.3	553.4	572.0	571.3	571.4	571.3	571.3	571.2
Excitation purity.	2.3	5.2	10.9	15.8	21.1	25.9	5.3	9.6	18.8	28.3	35.7	42.0

Percent transmittance

Wave length	CC-05M *	CC-10M *	CC-20M *	CC-30M *	CC-40M *	CC-50M *	CC-05C *	CC-10C *	CC-20C *	CC-30C *	CC-40C *	CC-50C *
400	87.6	86.6	85.6	84.2	82.3	80.9	87.3	86.0	83.9	82.3	80.4	78.8
10	88.2	87.7	86.6	85.7	84.6	83.6	88.2	87.5	85.2	84.5	83.4	82.7
20	88.6	88.0	87.0	85.9	85.2	84.4	88.7	88.1	86.5	86.0	85.3	84.8
30	88.7	88.0	86.9	85.6	84.4	83.6	89.0	88.6	87.5	87.0	86.3	85.9
40	88.7	87.9	86.0	84.7	82.5	81.4	89.3	89.0	87.7	87.3	86.6	86.1
50	88.6	87.5	84.9	82.8	80.0	78.1	89.5	89.1	87.8	87.5	86.6	86.0
60	88.4	86.5	83.1	80.0	76.1	73.7	89.6	89.1	87.7	87.3	86.4	85.7
70	87.8	85.2	80.8	76.4	71.3	68.0	89.7	89.0	87.5	87.0	85.8	85.2
80	87.0	83.6	77.9	72.1	65.8	61.7	89.7	89.0	87.2	86.5	85.3	84.3
90	86.0	81.8	74.4	67.0	60.0	55.0	89.7	89.0	87.0	86.0	84.4	83.4
500	85.0	79.7	70.5	61.7	53.7	48.1	89.6	89.0	86.5	85.2	83.5	82.3
10	83.8	77.5	66.7	56.5	47.7	41.6	89.6	88.7	86.0	84.4	82.4	80.8
20	82.7	75.3	63.4	52.0	42.8	36.3	89.5	88.5	85.2	83.5	81.1	79.2
30	81.8	73.7	60.5	48.6	39.0	31.9	89.4	88.0	84.3	82.4	79.6	77.3
40	81.3	72.5	58.6	46.6	36.7	29.8	89.2	87.5	83.4	81.0	77.7	75.0
50	81.2	72.2	58.0	46.0	36.0	29.1	88.9	87.0	82.3	79.0	75.3	72.2
60	81.5	72.8	58.3	46.5	36.7	29.7	88.5	86.1	80.5	76.7	72.7	69.0
70	82.5	74.6	60.5	49.8	40.2	32.3	88.0	85.0	78.5	74.0	69.3	65.0
80	84.0	77.3	64.9	55.6	46.2	39.0	87.5	83.8	76.1	70.9	65.4	60.5
90	85.8	80.8	70.6	63.3	54.9	48.7	87.0	82.5	73.9	67.5	61.6	55.8
600	88.0	84.5	77.1	71.6	64.9	59.9	86.4	81.0	71.2	64.1	57.6	51.3
10	89.3	87.0	82.2	79.2	74.9	70.7	85.5	79.5	68.5	60.4	53.4	36.2
20	90.2	88.9	86.1	84.1	81.4	79.2	84.5	77.9	65.5	56.7	49.2	42.3
30	90.6	90.0	88.7	87.4	86.0	84.5	83.8	76.3	62.7	53.1	45.0	38.0
40	90.8	90.5	90.0	89.3	88.7	87.6	83.3	75.1	60.8	50.4	42.0	34.9
50	91.0	90.7	90.5	90.2	90.0	89.7	82.8	74.4	59.5	48.8	40.2	32.9
60	91.1	91.0	90.8	90.8	90.4	90.4	82.5	74.0	58.5	48.0	39.4	32.0
70	91.2	91.2	91.0	91.0	90.7	90.7	82.4	73.6	57.9	47.2	38.6	31.0
80	91.3	91.3	91.2	91.1	91.0	91.0	82.0	73.0	57.5	46.0	37.5	29.9
90	91.4	91.4	91.4	91.3	91.3	91.3	82.0	72.8	57.4	45.5	36.7	29.1
700	91.5	91.5	91.5	91.5	91.5	91.5	82.5	74.0	58.5	46.4	37.3	29.8
Luminous transmit.	84.2	77.9	67.1	58.1	50.0	44.0	88.0	85.1	78.9	74.8	70.5	66.7
Dominant wave lgth.	541.0$_c$	547.5$_c$	551.2$_c$	550.0$_c$	550.3$_c$	551.2$_c$	489.2	487.5	486.5	486.2	486.1	485.5
Excitation purity.	3.5	7.4	14.4	21.5	28.3	34.0	1.6	4.1	8.9	12.8	17.5	20.2

* Some transmission below 400 mμ. Consult the manufacturer.

TRANSMISSION OF WRATTEN FILTERS (Continued)

Wave length	Percent transmittance			
	No. 87	No. 87C	No. 88A	No. 89B
700	.....	.....	....	11.2
10	.....	.....	....	32.4
20	.....	.....	....	57.6
30	.....	.....	7.4	69.1
40	0.10	.....	32.8	77.6
50	2.19	.....	56.3	83.1
60	7.95	.....	69.2	85.0
70	17.4	.....	74.2	86.1
80	31.6	.....	77.6	87.0
90	43.7	.....	79.7	87.7
800	53.8	0.32	81.4	88.1
10	61.7	3.20	82.6	88.4
20	69.2	8.90	83.7	88.6
30	74.1	17.8	84.7	88.8
40	77.7	28.2	85.5	89.0
50	81.4	41.0	86.1	89.2
60	84.0	53.8	86.6	89.4
70	85.4	61.6	87.2	89.6
80	86.8	69.2	87.5	89.8
90	87.8	74.1	87.8	89.9
900	88.4	78.5	88.0	90.0
10	88.8	81.5	88.2	90.1
20	89.1	83.6	88.4	90.2
30	89.1	85.1	88.6	90.3
40	89.1	86.0	88.8	90.4
50	89.1	87.0	89.0	90.5

* Some transmission below 400 mμ. Consult the manufacturer.

Ratio of the transmitted light to the incident light for a definite thickness of the substance, usually 1 cm.

GLASS.

Glass in general is opaque to the ultra-violet and infra-red. Uviol glass is transparent to the longer radiations of the ultra-violet.

Coefficient of transparency of glass for visible and ultra-violet radiations.

Normal incidence, thickness 1 cm.

Wave length microns......	0.309	0.330	0.347	0.357	0.361	0.375	0.384	0.388	0.396
Crown, ordinary..	...	...	...	...	...	.947			
Crown, borosilicate..........	0.08	0.65	0.88	...	0.95	...	0.972	0.975	0.986
Flint, ordinary...	...	...	...	0.72	...	...	...	0.904	
Flint, heavy.....	...	...	0.01	...	0.16	...	0.58		

Normal incidence, thickness 1 cm.

Wave length, microns......	0.400	0.415	0.419	0.425	0.434	0.455	0.500	0.580	0.677
Crown, ordinary.	0.964	...	0.952	...	0.960	0.981	...	0.986	0.990
Crown, borosilicate.........	...	0.985	...	0.993	...	...	0.993		
Flint, ordinary..	...	0.959	...	...	...	...	1.00		
Flint, heavy....	...	...	...	0.905					

QUARTZ

Quartz is very transparent to the ultra-violet and to the visible spectrum, but opaque for the infra-red beyond 7.0μ.

(Pflüger.)

Wave length, microns.........	0.19	0.20	0.21	0.22
Transmission for 1 mm..........	.67	.84	.92	.94

FLUORITE

Fluorite is very transparent to the ultra-violet, nearly to 0.10μ. Coefficient of transparency at $\lambda = 186$ is found by Pflüger to be 0.80.

For the infra-red the values are given in a table below.

ROCK SALT AND SYLVINE AND FLUORITE
TRANSPARENCY FOR THE INFRA-RED.

Thickness 1 cm.

Wave length, microns.	Rock salt	Sylvine KCl	Fluorite
8.	....	....	.844
9.	0.995	1.000	.543
10.	.995	.988	.164
12.	.993	.995	.010
14.	.931	.975	.000
16.	.661	.936	
18.	.275	.862	
19.	.096	.758	
20.7	.006	.585	
23.7	.000	.155	

COLORIMETRY

Selected from Judd, Jour. Opt. Soc. Amer. 23, 359 (1933)

Recommendations of the International Commission on Illumination

Standard Illuminants

A. Gas-filled tungsten incandescent lamp of color temperature 2848° K.

B. Noon Sunlight. Lamp as above in combination with the Davis-Gibson filter for converting color temperature 2848° to 4800° K.

The filter is to be composed of a layer one centimeter thick of each of two separate solutions B_1 and B_2, contained in a double cell of colorless optical glass.

Solution B_1

Copper sulfate ($CuSO_4 \cdot 5H_2O$)........	2.452	g
Mannite ($C_6H_8(OH)_6$)........	2.452	g
Pyridine (C_5H_5N)........	30.0	cc
Distilled water to make........	1000	cc

Solution B_2

Cobalt ammonium sulfate ($CoSO_4 \cdot (NH_4)_2SO_4 \cdot 6H_2O$)........	21.71	g
Copper sulfate ($CuSO_4 \cdot 5H_2O$)........	16.11	g
Sulfuric acid (density 1.835)........	10.0	cc
Distilled water to make........	1000	cc

C. Average Daylight. Lamp as in A in combination with Davis-Gibson filter for converting color temperature 2848° to 6500° K.

The filter is composed of a layer one centimeter thick of each of two separate solutions C_1 and C_2, contained in a double cell made of colorless optical glass.

Solution C_1

Copper sulfate ($CuSO_4 \cdot 5H_2O$)........	3.412	g
Mannite ($C_6H_8(OH)_6$)........	3.412	g
Pyridine (C_5H_5N)........	30.0	cc
Distilled water to make........	1000	cc

Solution C_2

Cobalt ammonium sulfate ($CoSO_4 \cdot (NH_4)_2SO_4 \cdot 6H_2O$)........	30.580	g
Copper sulfate ($CuSO_4 \cdot 5H_2O$)........	22.520	g
Sulfuric acid (density 1.835)........	10.0	cc
Distilled water to make........	1000	cc

See R. Davis and K. S. Gibson Bur. Stds. Misc. Pub. No. 114, Jan. 1931 or Bur. Stds. Jour. Research 7, 796 (1931).

Standard Coordinate System

The tristimulus system of color specification is based on four chosen stimuli consisting of homogeneous radiant energy of wave lengths

700.0 **546.1** **435.8**

mμ and of standard illuminant B (see above).

To establish the system of specification coordinates are assigned as follows:

Stimulus	x	y	z
700.0 mμ	0.73467	0.26533	0.00000
546.1 mμ	0.27376	0.71741	0.00883
435.8 mμ	0.16658	0.00886	0.82456
Standard illuminant B:	0.34842	0.35161	0.29997

The Standard Observer

The "standard observer" is determined below by the specification for the equal energy spectrum both in fractions, x, y, z of the total amount for each wave length interval of 5 mμ and directly $\bar{x}$, $\bar{y}$, $\bar{z}$. The fractional values are known as the **trilinear coordinates** or **trichromatic coefficients** of the spectrum; the direct values as the **distribution functions** or coefficients. The sum of the trichromatic coefficients is unity, that is $x + y + z = 1$. Therefore the value of z may be and often is omitted from a specification.

Relative Visibility

The value of $\bar{y}$ given in the table is the standard visibility function or relative visibility.

Wave length mμ	Trichromatic coefficients x	y	z	Distribution coefficients for equal energy $\bar{x}$	$\bar{y}$ (Rel. Vis.)	$\bar{z}$	Wave length mμ
380	0.1741	0.0050	0.8209	0.0014	0.0000	0.0065	380
385	0.1740	0.0050	0.8210	0.0022	0.0001	0.0105	385
390	0.1738	0.0049	0.8213	0.0042	0.0001	0.0201	390
395	0.1736	0.0049	0.8215	0.0076	0.0002	0.0362	395
400	0.1733	0.0048	0.8219	0.0143	0.0004	0.0679	400
405	0.1730	0.0048	0.8222	0.0232	0.0006	0.1102	405
410	0.1726	0.0048	0.8226	0.0435	0.0012	0.2074	410
415	0.1721	0.0048	0.8231	0.0776	0.0022	0.3713	415
420	0.1714	0.0051	0.8235	0.1344	0.0040	0.6456	420

The Standard Observer (Continued)

Wave length mμ	Trichromatic coefficients x	y	z	Distribution coefficients for equal energy $\bar{x}$	$\bar{y}$ (Rel. Vis.)	$\bar{z}$	Wave length mμ
425	0.1703	0.0058	0.8239	0.2148	0.0073	1.0391	425
430	0.1689	0.0069	0.8242	0.2839	0.0116	1.3856	430
435	0.1669	0.0086	0.8245	0.3285	0.0168	1.6230	435
440	0.1644	0.0109	0.8247	0.3483	0.0230	1.7471	440
445	0.1611	0.0138	0.8251	0.3481	0.0298	1.7826	445
450	0.1566	0.0177	0.8257	0.3362	0.0380	1.7721	450
455	0.1510	0.0227	0.8263	0.3187	0.0480	1.7441	455
460	0.1440	0.0297	0.8263	0.2908	0.0600	1.6692	460
465	0.1355	0.0399	0.8246	0.2511	0.0739	1.5281	465
470	0.1241	0.0578	0.8181	0.1954	0.0910	1.2876	470
475	0.1096	0.0868	0.8036	0.1421	0.1126	1.0419	475
480	0.0913	0.1327	0.7760	0.0956	0.1390	0.8130	480
485	0.0687	0.2007	0.7306	0.0580	0.1693	0.6162	485
490	0.0454	0.2950	0.6596	0.0320	0.2080	0.4652	490
495	0.0235	0.4127	0.5638	0.0147	0.2586	0.3533	495
500	0.0082	0.5384	0.4534	0.0049	0.3230	0.2720	500
505	0.0039	0.6548	0.3413	0.0024	0.4073	0.2123	505
510	0.0139	0.7502	0.2359	0.0093	0.5030	0.1582	510
515	0.0389	0.8120	0.1491	0.0291	0.6082	0.1117	515
520	0.0743	0.8338	0.0919	0.0633	0.7100	0.0782	520
525	0.1142	0.8262	0.0596	0.1096	0.7932	0.0573	525
530	0.1547	0.8059	0.0394	0.1655	0.8620	0.0422	530
535	0.1929	0.7816	0.0255	0.2257	0.9149	0.0298	535
540	0.2296	0.7543	0.0161	0.2904	0.9540	0.0203	540
545	0.2658	0.7243	0.0099	0.3597	0.9803	0.0134	545
550	0.3016	0.6923	0.0061	0.4334	0.9950	0.0087	550
555	0.3373	0.6589	0.0038	0.5121	1.0002	0.0057	555
560	0.3731	0.6245	0.0024	0.5945	0.9950	0.0039	560
565	0.4087	0.5896	0.0017	0.6784	0.9786	0.0027	565
570	0.4441	0.5547	0.0012	0.7621	0.9520	0.0021	570
575	0.4788	0.5202	0.0010	0.8425	0.9154	0.0018	575
580	0.5125	0.4866	0.0009	0.9163	0.8700	0.0017	580
585	0.5448	0.4544	0.0008	0.9786	0.8163	0.0014	585
590	0.5752	0.4242	0.0006	1.0263	0.7570	0.0011	590
595	0.6029	0.3965	0.0006	1.0567	0.6949	0.0010	595
600	0.6270	0.3725	0.0005	1.0622	0.6310	0.0008	600
605	0.6482	0.3514	0.0004	1.0456	0.5668	0.0006	605
610	0.6658	0.3340	0.0002	1.0026	0.5030	0.0003	610
615	0.6801	0.3197	0.0002	0.9384	0.4412	0.0002	615
620	0.6915	0.3083	0.0002	0.8544	0.3810	0.0002	620
625	0.7006	0.2993	0.0001	0.7514	0.3210	0.0001	625
630	0.7079	0.2920	0.0001	0.6424	0.2650	0.0000	630
635	0.7140	0.2859	0.0001	0.5419	0.2170	0.0000	635
640	0.7190	0.2809	0.0001	0.4479	0.1750	0.0000	640
645	0.7230	0.2770	0.0000	0.3608	0.1382	0.0000	645
650	0.7260	0.2740	0.0000	0.2835	0.1070	0.0000	650
655	0.7283	0.2717	0.0000	0.2187	0.0816	0.0000	655
660	0.7300	0.2700	0.0000	0.1649	0.0610	0.0000	660
665	0.7311	0.2689	0.0000	0.1212	0.0446	0.0000	665
670	0.7320	0.2680	0.0000	0.0874	0.0320	0.0000	670
675	0.7327	0.2673	0.0000	0.0636	0.0232	0.0000	675
680	0.7334	0.2666	0.0000	0.0468	0.0170	0.0000	680
685	0.7340	0.2660	0.0000	0.0329	0.0119	0.0000	685
690	0.7344	0.2656	0.0000	0.0227	0.0082	0.0000	690
695	0.7346	0.2654	0.0000	0.0158	0.0057	0.0000	695
700	0.7347	0.2653	0.0000	0.0114	0.0041	0.0000	700
705	0.7347	0.2653	0.0000	0.0081	0.0029	0.0000	705
710	0.7347	0.2653	0.0000	0.0058	0.0021	0.0000	710
715	0.7347	0.2653	0.0000	0.0041	0.0015	0.0000	715
720	0.7347	0.2653	0.0000	0.0029	0.0010	0.0000	720
725	0.7347	0.2653	0.0000	0.0020	0.0007	0.0000	725
730	0.7347	0.2653	0.0000	0.0014	0.0005	0.0000	730
735	0.7347	0.2653	0.0000	0.0010	0.0004	0.0000	735
740	0.7347	0.2653	0.0000	0.0007	0.0003	0.0000	740
745	0.7347	0.2653	0.0000	0.0005	0.0002	0.0000	745
750	0.7347	0.2653	0.0000	0.0003	0.0001	0.0000	750
755	0.7347	0.2653	0.0000	0.0002	0.0001	0.0000	755
760	0.7347	0.2653	0.0000	0.0002	0.0001	0.0000	760
765	0.7347	0.2653	0.0000	0.0001	0.0000	0.0000	765
770	0.7347	0.2653	0.0000	0.0001	0.0000	0.0000	770
775	0.7347	0.2653	0.0000	0.0000	0.0000	0.0000	775
780	0.7347	0.2653	0.0000	0.0000	0.0000	0.0000	780
			Totals	21.3713	21.3714	21.3715	

SPECIFIC ROTATION

Specific rotation or rotatory power is given in degrees per decimeter for liquids and solutions and in degrees per millimeter for solids; + signifies right handed rotation, − left. Specific rotation varies with the wave length of light used, with temperature and, in the case of solutions, with the concentration. When sodium light is used, indicated by D in the wave length column, a value of λ = 0.5893 may be assumed.

Optical rotatory power for a large number of organic compounds will be found in the International Critical Tables, Vol. VII; for sugars, Vol. II.

SOLIDS

Substance	Wave length μ	Rotation deg./mm	Substance	Wave length μ	Rotation deg./mm
Cinnabar (HgS)...	D	+32.5	Quartz (continued)........	0.3609	+63.628
Lead hyposulfate..	D	5.5		0.3582	64.459
Potassium hyposulphate........	D	8.4		0.3466	69.454
Quartz............	0.7604	12.668		0.3441	70.587
	0.7184	14.304		0.3402	72.448
	0.6867	15.746		0.3360	74.571
	0.6562	17.318		0.3286	78.579
	0.5895	21.684		0.3247	80.459
	0.5889	21.727		0.3180	84.972
	0.5269	27.543		0.2747	121.052
	0.4861	32.773		0.2571	143.266
	0.4307	42.604		0.2313	190.426
	0.4101	47.481		0.2265	201.824
	0.3968	51.193		0.2194	220.731
	0.3933	52.155		0.2143	235.972
	0.3820	55.625	Sodium bromate	D	2.8
	0.3726	58.894	Sodium chlorate	D	3.13

LIQUID

Liquid	Temp. °C	Wave length μ	Specific rotation deg./dm
Amyl alcohol........................	......	D	− 5.7
Camphor............................	204	D	+ 70.33
Cedar oil............................	15	D	− 30 to −40
Citron oil............................	15	D	+ 62
Ethyl malate $(C_2H_5)_2C_4H_4O_5$......	11	D	− 10.3 to −12.4
Menthol............................	35.2	D	− 49.7
Nicotine $C_{10}H_{14}N_2$..................	10–30	D	−162
	20	0.6563	−126
	20	0.5351	−207.5
	20	0.4861	−253.5
Turpentine $C_{10}H_{16}$................	20	D	− 37
	20	0.6563	− 29.5
	20	0.5351	− 45
	20	0.4861	− 54.5

SOLUTIONS

Corrections for values of the specific rotation for concentration are given in the last column. *c* indicates concentration in grams per 100 milliliters of solution; *d* indicates the concentration in grams per 100 grams of solution.

Substance	Solvent	Temp. °C	Wave length μ	Specific rotation deg./dm	Correction for concentration or temperature
Albumen............	water	..	D	− 25 to −38	
Arabinose............	water	20	D	− 105.0	
Camphor............	alcohol	20	D	+ 54.4	− .135*d* for *d* = 45–91
	benzene	20	D	+ 56	− .166*d* for *d* = 47–90
	ether	..	D	+ 57	
Dextrose *d*-glucose $C_6H_{12}O_6$	water	20	D	+ 52.5	+ .025*d* for *d* = 1–18
			.5461	+ 62.03	+ .04257*c* for *c* = 6–32
Galactose............	water	..	D	+ 83.9	+ .078*d* − .21*t* for *d* = 4–36 and *t* = 10–30°C
l-Glucose (*β*).........	water	20	D	− 51.4	
Invert sugar $C_6H_{12}O_6$	water	20	D	− 19.7	− .036*c* for *c* = 9–35; $\alpha_t = \alpha_{20} + .304(t - 20) + .00165(t - 20)^2$ for *t* = 3–30°C
		25	5461	− 21.5	
Lactose............	water	20	D	+ 52.4	+ .072 (20° − *t*) for *c* = 5
			.5461	+ 61.9	+ .085(20° − *t*) for *c* = 5
Levulose fruit sugar...	water	25	D	− 88.5	− .145*d* for *d* = 2.6–18.6
		25	.5461	− 105.30	

SPECIFIC ROTATION (Continued)

SOLUTIONS (Continued)

Substance	Solvent	Temp °C	Wave length μ	Specific rotation deg./dm	Correction for concentration or temperature
Maltose............	water	20	D	+ 138.48	− .01837*d* for *d* = 5–35
		25	5461	+ 153.75	
Mannose............	water	20	D	+ 14.1	*c* = 10.2
Nicotine............	water	20	D	− 77	for *d* = 1–16
	benzene	20	D	− 164	for *d* = 8–100
Potassium tartrate...	water	20	D	+ 27.14	+ .0992*c* − .00094c^2 for *c* = 8–50
Quinine sulfate......	water	17	D	− 214	
Santonin............	alcohol	20	D	− 161.0	*c* = 1.78
		20	D	+ 693	*c* = 4.05
	chloroform	20	D	− 202.7	+ .309*d* for *d* = 75–96.5
	alcohol	20	.6867	+ 442	*c* = 4.05
			.5269	+ 991	*c* = 4.05
			.4861	+1323	*c* = 4.05
Sodium potassium tartrate (Rochelle salt)	water	20	D	+ 29.75	− .0078*c*
Sucrose (cane sugar) $C_{12}H_{22}O_{11}$	water	20	D	+ 66.412	+ .01267*d* − .000376d^2 for *d* = 0–50; $\alpha_t = \alpha_{20}[1 - .00037(t - 20)]$ for *t* = 14–30°C

Sucrose dissolved in water, 20°C.

μ	Spec. rot.	μ	Spec. rot.	μ	Spec. rot.
670.8 (Li)	+50.51	510.6 (Cu)	+90.46	435.3 (Fe)	+128.5
643.8 (Cd)	55.04	508.6 (Cd)	91.16	433.7 (Fe)	129.8
636.2 (Zn)	56.51	481.1 (Zn)	103.07	431.5 (Fe)	130.7
589.3 (Na)	66.45	480.0 (Cd)	103.62	428.2 (Fe)	133.6
578.2 (Cu)	69.10	472.2 (Zn)	107.38	427.2 (Fe)	134.2
578.0 (Hg)	69.22	468.0 (Zn)	109.49	426.1 (Fe)	134.9
570.0 (Cu)	71.24	467.8 (Cd)	109.69	419.1 (Fe)	140.0
546.1 (Hg)	78.16	438.4 (Fe)	126.5	414.4 (Fe)	144.2
521.8 (Cu)	86.21	437.6 (Fe)	127.2	388.9 (Fe)	166.7
515.3 (Cu)	88.68	435.8 (Hg)	128.49	383.3 (Fe)	171.8
				382.6 (Fe)	173.1

Substance	Solvent	°C	μ	Spec. rot.	Correct.
Tartaric acid (ord.)...	water	20	D	+15.06	− .131*c*
		20	.6563	7.75	for *d* = 41
		20	D	8.86	for *d* = 41
		20	.5351	9.65	for *d* = 41
		20	.4861	9.37	for *d* = 41
Turpentine...........	alcohol	20	D	−37	− .00482*d* − .00013d^2 for *d* = 0–90
	benzene	20	D	−37	− .0265*d* for *d* = 0–91
Xylose...............	water	20	D	+19.13	*d* = 2.7

OPTICAL ROTATION OF ACIDS AND BASES

Optical rotation of acids and bases commonly used in the resolution of racemic substances. Compiled by F. E. Ray.

Name	Formula	Solvent	Conc. %	a_D
Bromocamphor-sulfonic acid. K salt.........	$C_{10}H_{15}O_4BrS$	H_2O	....	72.1
Camphorsulfonic acid...	$C_{10}H_{16}O_4S$	H_2O	....	23.9
Chlorocamphor-sulfonic acid.............	$C_{10}H_{15}ClO_4S$	H_2O	....	49.6
Codeinesulfonic acid.....	$C_{18}H_{21}NO_6S$	H_2O	3	−190.1
Hydroxybutyric acid....	$C_4H_8O_3$	H_2O	3.3	− 24.8
Lactic acid............	$C_3H_6O_3$	H_2O	10.5	3.8
Malic acid............	$C_4H_6O_5$	H_2O	....	2.4
Mandelic acid..........	$C_8H_8O_3$	H_2O	2.01	155.5
Methylene-camphor.....	$C_{11}H_{16}O$	C_2H_5OH	....	127
Phenylsuccinic acid.....	$C_{10}H_{10}O_4$	C_2H_5OH	1.5	148
Tartaric acid..........	$C_4H_6O_6$	C_2H_5OH and H_2O	....	3 to 25*
Brucine...............	$C_{23}H_{26}N_2O_4$	C_2H_5OH	5.4	− 85
Cinchonidine..........	$C_{19}H_{22}N_2O$	C_2H_5OH	1.0	−111.0
Cinchonine............	$C_{19}H_{22}N_2O$	$CHCl_3$	0.6	+209.6
Cocaine...............	$C_{17}H_{21}NO_4$	50 % C_2H_5OH	1.1	− 35.4
Coniine...............	$C_8H_{17}N$	$CHCl_3$	4	8.0
Codeine...............	$C_{18}H_{21}NO_3$	C_2H_5OH	5	−135.8
Hydrastine............	$C_{21}H_{21}NO_6$	50 % C_2H_5OH	0.2	115
Menthol...............	$C_{10}H_{20}O$	C_2H_5OH	9.6	− 50.6
Menthylamine..........	$C_{10}H_{21}N$	C_2H_5OH	11.3	− 31.9
Narcotine.............	$C_{22}H_{23}NO_7$	$CHCl_3$	2.6	±200.0
Quinidine.............	$C_{20}H_{24}N_2O_4$	C_2H_5OH	1.0	+233.6
Quinine...............	$C_{20}H_{24}N_2O_2$	C_6H_6	0.6	−136
Thebaine..............	$C_{19}H_{21}NO_3$	$CHCl_3$	5	−229.5
Strychnine............	$C_{21}H_{22}N_2O_2$	C_2H_5OH	0.9	−128

* Varies greatly with temperature, solvent, and conc.

The rotation of the plane of polarization of light by transparent substances subjected to a magnetic field may be applied to problems of molecular structure. This rotatory effect is known as the Faraday effect. Investigations of this effect by E. Verdet showed the angle of rotation (α) to depend on the nature of the substance and to be proportional to the length (l) of the column of the substance which the light traverses and to the strength (H) of the magnetic field. Thus

$$\alpha = \Lambda l H$$

where Λ is the Verdet constant for the experimental material.

Other symbols in the table have the following significance:

λ = Wavelength in μ
t = Degrees centigrade
ρ_M = Molecular magnetic rotation of the substance under consideration compared to that of water determined in the same apparatus in the same magnetic field.

Thus

$$\rho_M = M'\alpha'\rho'/M'\alpha\rho$$

where M is the molecular weight, α the angle of rotation and ρ the density of the given substance, and M′, α', and ρ' are the same quantities of water.

$[\Lambda]_M^{\lambda,t}$ = Molecular rotatory value of the substance at the given temperature and at the wavelength λ. Values are in (radians)(gauss^{-1})(cm^{-1}).

$\Lambda^{\lambda,t}$ = Verdet constant at the temperature t and at the wavelength λ. Values are in (minutes)(gauss^{-1})(cm^{-1}).

Values in this table are reproduced by permission from Volume 3, Tables de Constantes et Donnees Nuimeriques, "Pouvoir Rotatorie Magnetique, Effet Magneto-Optique de Kerr."

A much greater listing of values for both organic and inorganic compounds is in the above publication.

Formula	Name	λ	t	ρ_M	$10^5[\Lambda]_M^{\lambda,t}$	$10^5\Lambda^{\lambda,t}$ Verdet
CH_4	Methane	578	0	......	....	*17.4
CCl_4	Tetrachloromethane	589	25.1	6.58	45.3	1.60
$CHCl_3$	Trichloromethane	589	20.0	5.535	37.98	1.60
$CHBr_3$	Tribromomethane	589	17.9	11.63	80.00	3.13
CH_2O_2	Formic acid	589	20.8	1.671	11.50	1.046
CH_2Cl_2	Dichloromethane	589	11.9	4.31	29.7	1.60
CH_2Br_2	Dibromomethane	589	15.9	8.11	55.8	2.74
CH_2I_2	Diiodomethane	589	15.0	10.83	129.5	1.51
CH_3Cl	Monochloromethane	589	23.0	2.99	20.5	1.37
CH_3Br	Monobromomethane	589	1.5	4.64	31.9	2.04
CH_3I	Monoiodomethane	589	19.5	9.01	63.2	3.35
CH_4O	Methylalcohol	589	18.7	1.640	11.28	0.958
CH_5N	Methylamine	578	.....	......	....	*22.7
CF_2Cl_2	Difluorodichloromethane (Freon)	...	.....	......	....	*32.7
CH_3O_2N	Mononitromethane	589	9.9	1.86	12.8	0.826
CH_4ON_2	Urea (carbamide) 40 % aqueous solution	578	20.0	2.38	22.7	....
C_2H_2	Acetylene (ethyne)	578	.....	......	....	*33.0
C_2H_4	Ethylene (ethene)	578	.....	......	....	*34.5
C_2H_6	Ethane	578	.....	......	....	*23.5
C_2N_2	Cyanogen	...	.....	......	....	....
$C_2H_2O_4$	Oxalic acid (aqueous sol. 8.3 %)	578	20	2.88	20.6	....
$C_2H_2O_4 \cdot 2H_2O$	Oxalic acid (alcohol sol. 16.5 %)	578	24	2.82	20.2	....
C_2H_3Br	Vinylbromide	589	7.8	6.22	42.8	2.10
C_2H_3N	Acetonitrile (ethanonitrile)	589	25.0	2.32	16.0	*21.0
C_2H_4O	Ethyleneoxide (1,2-epoxyethane)	589	8.0	1.935	13.3	0.92
C_2H_4O	Acetaldehyde (ethanal)	589	16.3	2.38	16.4	1.00
$C_2H_4O_2$	Acetic acid	589	21.0	2.525	17.37	1.044
$C_2H_4O_2$	Formic acid methylester (methylformate)	589	16.5	2.49	17.1	0.96
$C_2H_4Cl_2$	Ethylidenechloride (1,1-dichloroethane)	589	14.4	5.33	36.7	1.51
$C_2H_4Cl_2$	Ethylenechloride (1,2-dichloroethane)	589	14.4	5.49	37.7	1.65
$C_2H_4Br_2$	Ethylenebromide (1,2-dibromoethane)	589	15.2	9.70	66.7	2.66
C_2H_5Cl	Monochloroethane	589	5.0	4.04	27.8	1.36
C_2H_5Br	Monobromoethane	589	19.7	5.85	40.2	1.82
C_2H_5I	Monoiodoethane	589	18.1	10.07	69.3	2.95
C_2H_6O	Ethylalcohol (ethanol)	589	16.8	2.780	19.13	1.131
$C_2H_6O_2$	Glycol (1,2-ethanediol)	589	15.1	2.94	20.2	1.25
C_2H_6S	Ethylmercaptan	578	16.0	5.52	39.5	1.85
C_2H_7N	Ethylamine (aminoethane)	589	5.8	3.61	24.8	*34.5
$C_2H_2O_2Cl_2$	Dichloroacetic acid (dichloroethanoic acid)	589	13.5	5.30	3.65	1.52
$C_2H_3O_2Cl$	Chloroacetic acid (chloroethanoic acid)	589	64.5	3.89	26.7	1.33
$C_2H_3O_2Cl_3$	Chloralhydrate (2-2-2-trichloro-1,1-ethanediol)	589	54.6	7.10	48.8	1.65
$C_2H_5O_2N$	Mononitroethane	589	10.2	2.84	19.5	0.946
C_3H_8	Propane	578	.....	......	....	*34.0
C_3H_4O	Acrolein (propenal)	578	20.0	4.74	34.0	1.76
$C_3H_4O_3$	Pyruvic acid (2-oxopropanoic acid)	589	14.5	3.56	24.2	1.21
$C_3H_4O_4$	Malonic acid (propandioic acid) 2*n*-aqueous sol.	589	23.0	3.47	....	....
C_3H_6O	Allyl alcohol	589	18.3	4.68	32.2	1.60
C_3H_8O	Propyl alcohol (1-propanol)	589	13.6	3.33	22.9	1.09
C_3H_6O	Acetone (2-propanone)	589	20.0	3.472	23.89	1.1136
$C_3H_6O_2$	Propionic acid (propanoic acid)	589	20.3	3.462	23.82	1.10
$C_3H_6O_2$	Formic acid ethylester (ethylmethanoate)	589	18.8	3.56	24.5	1.05
$C_3H_6O_2$	Acetic acid methylester (methylacetate)	589	20.0	3.42	23.5	1.03
C_3H_7Cl	Propylchloride (1-chloropropane)	589	16.1	5.04	34.7	1.34
C_3H_7Cl	Isopropylchloride (2-chloropropane)	589	17.2	5.16	35.5	1.34
C_3H_7Br	Propylbromide (1-bromopropane)	589	19.2	6.88	47.8	1.79
C_3H_7Br	Isopropylbromide (2-bromopropane)	589	17.1	7.00	48.2	1.77
C_3H_7I	Propyliodide (1-iodopropane)	589	18.1	11.08	76.2	2.69
C_3H_7I	Isopropyliodide (2-iodopropane)	589	26.3	11.18	76.9	2.63
C_3H_8O	*n*-Propyl alcohol (1-propanol)	589	15.6	3.77	25.9	1.20
C_3H_8O	Isopropyl alcohol (2-propanol)	589	20.0	3.90	26.8	1.23
$C_3H_8O_3$	Glycerine (1,2,3-propanetriol)	589	16.0	4.11	28.3	1.33
C_3H_9N	*n*-Propylamine	589	9.6	4.56	31.4	1.33
$C_3H_5O_9N_3$	Nitroglycerine	589	13.5	5.405	37.2	0.900

* Verdet constant factor = 10^6.

Formula	Name	λ	t	ρ_M	$10^5[\Lambda]_M^{\lambda,t}$	$10^5\Lambda^{\lambda,t}$ Verdet
$C_3H_7O_2N$	1-Nitropropane	589	18.9	3.82	26.3	1.018
C_4H_6	1,3-Butadiene (erythrene)	589	15.0	7.94	54.6	2.16
C_4H_8	1-Butene (α-butylene)	589	15.0	5.53	38.0	1.39
C_4H_8	cis-2-Butene (β-butylene)	589	15.0	5.27	36.3	1.38
C_4H_8	trans-2-Butene	589	15.0	5.07	34.9	1.29
C_4H_{10}	Butane	589	15.0	4.59	31.6	1.09
C_4H_{10}	Isobutane (2-methylpropane)	589	15.0	4.87	33.5	1.11
$C_4H_2O_3$	Maleic anhydride (cis-butenedioic anhydride)	589	25.0	4.5	31.0	....
C_4H_4O	Furan (furfuran)	589	20.0	5.48	37.7	1.78
$C_4H_4O_4$	Maleic acid (cis-butenedioic acid) 2*n* aqueous solution	589	25.0	5.63	38.7	....
C_4H_4S	Thiophene (thiofuran)	589	20.0	9.40	64.7	2.83
$C_4H_6O_3$	Acetic anhydride (ethanoic anhydride)	589	20.0	4.28	29.5	1.115
$C_4H_6O_4$	Succinic acid (butanedioic acid) 5.9 % aqueous sol.	578	20.0	4.68	33.5	....
$C_4H_6O_6$	Tartaric acid (47.8 % aqueous sol.)	578	20.0	4.79	34.3	....
$C_4H_8O_2$	*n*-Butyric acid (butanoic acid)	589	18.8	4.47	30.8	1.15
$C_4H_8O_2$	Ethylacetate (ethyl ethanoate)	589	20.0	4.47	30.8	1.08
$C_4H_8O_2$	Propionic acid methylester (methylpropanoate)	589	20.0	4.37	30.1	1.07
$C_4H_8O_3$	Lactic acid methylester (methyl 2-hydroxypropanoate)	589	.....	4.66	32.1	....
$C_4H_{10}O$	Ethyl ether (ethoxyethane)	589	20.0	4.78	32.9	1.09
$C_4H_{10}O$	*n*-Butyl alcohol (1-butanol)	589	20.0	4.60	31.6	1.23
$C_4H_{10}O$	Isobutyl alcohol (2-methyl-1-propanol)	589	17.7	4.94	34.0	1.27
$C_4H_{10}O$	sec-Butyl alcohol (methylethylcarbinol)	589	20.0	4.91	33.8	1.27
C_5H_6	Cyclopentadiene	589	15.0	7.03	48.8	2.02
C_5H_8	1,3-Pentadiene	589	15.0	8.80	60.5	2.08
C_5H_8	Isoprene (2-methyl-1,3-butadiene)	589	15.0	8.80	60.5	2.08
C_5H_8	Cyclopentene	589	15.0	5.69	39.1	1.52
C_5H_{10}	1-Pentene	589	15.0	6.45	44.4	1.39
C_5H_{10}	Isopentane (2-methyl-1-butane)	589	15.0	6.36	43.8	1.39
C_5H_{10}	Cyclopentane	589	20.0	4.89	33.6	1.23
C_5H_{12}	Pentane	589	15.0	5.60	38.5	1.15
C_5H_{12}	Isopentane (2-methylbutane)	589	15.0	5.75	39.6	1.17
$C_5H_4O_2$	Furfural (2-furancarbonal)	578	20.0	7.01	50.2	2.06
C_5H_5N	Pyridine	589	11.9	8.76	60.3	2.58
$C_5H_8O_4$	Glutaric acid (pentanedioic acid) 2 N aqueous sol.	589	16.0	5.48	37.7	....
$C_5H_{10}O_2$	Propionic acid ethylester	589	20.0	5.46	37.6	1.13
$C_5H_{10}O_2$	Acetic acid propylester	589	15.7	5.45	37.5	1.13
$C_5H_{14}N_2$	Cadaverine (1,5-pentanediamine)	589	14.7	7.49	51.6	1.53
C_6H_6	Benzene	589	15.0	11.27	77.5	3.00
C_6H_{12}	Cyclohexane	589	20.0	5.66	39.0	1.24
C_6H_{14}	Hexane	589	15.0	6.62	45.5	1.20
$C_6H_4Cl_2$	1,4-Dichlorobenzene (*p*-dichlorobenzene)	589	64.5	13.55	93.2	2.69
C_6H_5F	Fluorobenzene (phenylfluoride)	589	19.0	9.96	68.5	2.51
C_6H_5Cl	Chlorobenzene (phenylchloride)	589	15.0	12.51	86.1	2.92
C_6H_5Br	Bromobenzene (phenylbromide)	589	15.0	14.51	99.8	3.26
C_6H_5I	Iodobenzene (phenyliodide)	589	15.0	19.11	131.4	4.06
C_6H_6O	Phenol (hydroxybenzene)	589	39.0	12.07	83.5	3.21
C_6H_7N	Aniline (aminobenzene)	589	15.0	16.08	110.6	4.18
$C_6H_{11}Cl$	Chlorocyclohexane (cyclohexylchloride)	589	13.0	7.50	51.6	1.46
$C_6H_{12}O_3$	Paraldehyde (paraacetaldehyde)	589	17.3	6.66	45.8	1.19
$C_6H_{12}O_6$	Glucose $11H_2O$ (dextrose) (1 M aqueous sol.)	589	15.0	6.72	46.2	....
$C_6H_{12}O_6$	Galactose $10H_2O$ (1 M aqueous sol.)	589	15.0	6.89	47.4	....
$C_6H_{12}O_6$	Fructose $10H_2O$ (levulose) (1 M aqueous sol.)	589	15.0	6.73	46.3	....
$C_6H_{14}O$	2-Hexanol (butylmethylcarbinol)	589	20.0	6.89	47.4	1.31
$C_6H_{14}O$	3-Hexanol (ethylpropylcarbinol)	589	20.0	6.85	47.1	1.30
$C_6H_{14}O$	2-Methyl-3-Pentanol (ethylisopropylcarbinol)	589	20.0	6.90	47.5	1.32
$C_6H_4O_4N_2$	1,3-Dinitrobenzene (*m*-Dinitrobenzene)	589	17.1	9.65	66.4	2.17
$C_6H_5O_2N$	Nitrobenzene	589	15.0	9.36	64.4	2.17
C_7H_8	Toluene (methylbenzene)	589	15.0	12.16	83.7	2.71
C_7H_{14}	1-Heptene (α-heptylene)	589	18.0	8.48	58.3	1.43
C_7H_{16}	Heptane	589	15.0	7.61	52.7	1.23
C_7H_5N	Benzonitrile (benzenecarbonitrile)	589	15.7	11.85	81.5	2.74
$C_7H_6O_2$	Benzoic acid (20 % alcohol sol.)	578	20.0	11.8	84.7	....
C_7H_7Cl	*o*-Chlorotoluene (2-chloro-1-methylbenzene)	589	15.4	13.72	94.3	2.95
C_7H_7Cl	*p*-Chlorotoluene (4-chloro-1-methylbenzene)	589	15.2	13.25	90.8	2.65
C_7H_7Br	*o*-Bromotoluene (2-bromo-1-methylbenzene)	589	16.7	15.67	107.3	3.08
C_7H_7Br	*p*-Bromotoluene (4-bromo-1-methylbenzene)	589	39.0	15.09	103.6	2.88
C_7H_8O	*o*-Cresol (*o*-methylphenol)	589	16.0	13.38	92.1	3.07
C_7H_8O	*m*-Cresol (*m*-methylphenol)	589	17.9	12.77	87.6	2.89
C_7H_8O	*p*-Cresol (*p*-methylphenol)	589	17.0	12.86	88.5	2.91
C_7H_9N	*o*-Toluidine (*o*-methylaniline)	589	17.3	17.18	118.2	3.79
C_7H_9N	*m*-Toluidine (*m*-methylaniline)	589	15.0	16.21	111.5	3.56
C_7H_9N	*p*-Toluidine (*p*-methylaniline)	589	50.0	15.92	109.5	3.37
$C_7H_{14}O$	Enanthaldehyde (heptanal)	589	16.2	7.42	51.1	1.26
$C_7H_{16}O$	1-Heptanol (*n*-heptylalcohol)	589	12.6	7.85	54.0	1.33
$C_7H_{16}O$	2-Heptanol (amylmethylcarbinol)	589	20.0	7.94	54.6	1.32
$C_7H_{16}O$	3-Heptanol (butylethylcarbinol)	589	20.0	7.86	54.1	1.37
$C_7H_7O_2N$	*o*-Nitrotoluene	589	18.0	10.80	74.3	2.16
$C_7H_7O_2N$	*p*-Nitrotoluene	589	54.3	10.20	70.2	1.97
C_8H_{10}	Ethylbenzene (phenylethane)	589	15.0	13.41	92.3	2.80
C_8H_{10}	*o*-Xylene (1,2-dimethylbenzene)	589	15.0	13.36	91.9	2.62
C_8H_{10}	*m*-Xylene (1,3-dimethylbenzene)	589	15.0	12.82	88.2	2.47
C_8H_{10}	*p*-Xylene (1,4-dimethylbenzene)	589	15.0	12.80	88.1	2.46
C_8H_{16}	1-Octene (α-octylene)	589	15.0	9.00	65.5	1.44
C_8H_{16}	2-Octene (β-octylene)	589	15.0	9.33	64.2	1.43
C_8H_{18}	Octane	589	15.0	8.65	59.5	1.26
$C_8H_{18}O$	1-Octanol (*n*-octylalcohol)	589	20.0	8.88	61.1	1.33
$C_8H_{18}O$	2-Octanol (methylhexylcarbinol)	589	20.0	9.00	61.8	1.34
$C_8H_{18}O$	3-Octanol (ethylamylcarbinol)	589	20.0	8.90	61.2	1.33
C_9H_{12}	*o*-Ethyltoluene (1-ethyl-2-ethylbenzene)	589	15.0	14.56	100.2	2.32
C_9H_{12}	*m*-Ethyltoluene (1-ethyl-3-ethylbenzene)	589	15.0	14.18	97.6	2.91
C_9H_{12}	*p*-Ethylbenzene (1-ethyl-4-ethylbenzene)	589	15.0	13.98	96.2	2.37
C_9H_{12}	Mesitylene (1-3-5-trimethylbenzene)	589	15.0	13.36	91.9	2.28
C_9H_{20}	Nonane	589	15.0	9.70	66.7	1.28
$C_{10}H_8$	Naphthalene	589	89.5	24.98	171.8	4.47
$C_{10}H_{20}$	1-Decene (*n*-decylene)	589	21.0	11.65	80.1	1.45
$C_{10}H_{22}$	Decane	589	15.0	10.70	73.6	1.30

Formula	Name	λ	t	ρ_M	$10^5[A]_M^{\lambda,t}$	$10^5\Lambda^{\lambda,t}$ Verdet
$C_{10}H_7Cl$	1-Chloronaphthalene (α-chloronaphthalene)	578	18.0	28.15	201.5	4.91
$C_{10}H_7Br$	1-Bromonaphthalene (α-bromonaphthalene)	578	20.0	31.05	222.0	5.19
$C_{10}H_8O$	β-Naphthol (2-hydroxynaphthalene)	578	136.0	27.1	194.0	4.80
$C_{10}H_9N$	1-Naphthylamine (α-naphthylamine)	589	32.6	37.23	256.1	6.84
$C_{10}H_{12}O_2$	Isoeugenol (4-propenylguaiacol)	589	19.3	21.44	147.5	3.55
$C_{10}H_{12}O_2$	Eugenol (4-allylguaiacol)	589	15.4	18.72	128.8	2.88
$C_{10}H_{12}O_2$	Benzoic acid propylester (*n*-propylbenzoate)	589	15.4	14.87	102.3	2.20
$C_{10}H_{12}O_2$	*o*-Toluic acid ethylester	589	15.2	15.06	103.6	2.25
$C_{10}H_{12}O_2$	*p*-Toluic acid ethylester	589	15.0	14.74	101.4	2.18
$C_{10}H_{12}O_2$	α-Toluic acid ethylester (ethylphenylacetate)	589	14.0	14.99	103.1	2.25
$C_{10}H_{12}O_3$	Methylsalicylic acid ethylester	589	18.6	17.14	117.9	2.50
$C_{10}H_{15}N$	N,N-Diethylaniline (N-phenyldiethylamine)	589	15.3	25.16	173.1	3.74
$C_{10}H_{16}O$	Camphor	589	14.0	9.26	63.7	....
$C_{10}H_{18}O$	α-Terpineol	589	16.0	10.84	74.6	1.56
$C_{10}H_{18}O$	Citronellal	589	14.5	11.48	79.0	1.51
$C_{10}H_{18}O_4$	Dipropylsuccinate	589	11.4	10.36	71.3	1.22
$C_{10}H_{18}O_6$	Tartaric acid dipropylester (propyltartrate)	589	15.4	10.83	74.5	1.24
$C_{10}H_{20}O$	Menthol	589	45.2	10.51	72.3	1.40
$C_{11}H_{24}$	Undecane	589	20.5	11.65	80.1	1.31
$C_{12}H_{26}$	Dodecane	589	21.5	12.71	87.4	1.32
$C_{12}H_{22}O_{11}$	Saccharose 19H_2O (1 M aqueous sol.)	589	15.0	12.59	86.6	....
$C_{12}H_{22}O_{11}$	Maltose 20H_2O (1 M aqueous sol.)	589	15.0	12.69	87.3	....
$C_{12}H_{22}O_{11}$	Lactose 41H_2O (1 M aqueous sol.)	589	18.4	12.71	87.5	....
$C_{14}H_{10}$	Phenanthrene	578	100.0	39.7	284.0	5.84
$C_{16}H_{34}$	Hexadecane	589	15.0	16.8	115.6	1.35
$C_{18}H_{14}$	1,2-Diphenylbenzene	589	15.0	40.2	276.3	4.70
$C_{18}H_{14}$	1,3-Diphenylbenzene (*m*-phenyldiphenyl)	589	15.0	41.0	282.4	....
$C_{18}H_{22}$	1,6-Diphenylhexane	589	20.0	......	....	2.75

TRANSPARENCY TO OPTICAL DENSITY CONVERSION TABLE

Transparency of a layer of material is defined as the ratio of the intensity of the transmitted light to that of the incident light. Opacity is the reciprocal of the transparency. Optical density is the common logarithm of the opacity.

Thus,

$$\text{Transparency} = \frac{I_t}{I_i}$$

$$\text{Opacity} = \frac{1}{\text{Transparency}} = \frac{I_i}{I_t}$$

$$\text{Optical density} = \log_{10}\left(\frac{I_i}{I_t}\right)$$

where I_i = Intensity of incident light
I_t = Intensity of transmitted light.

Trans.	Density	Trans.	Density	Trans.	Density	Trans.	Density	Trans.	Density	Trans.	Density	Trans.	Density	Trans.	Density
0.000	—	**.030**	1.523	**.060**	1.222	**.090**	1.046	**.120**	.9208	**.150**	.8239	**.180**	.7447	**.210**	.6778
.001	3.000	.031	1.509	.061	1.215	.091	1.041	.121	.9172	.151	.8210	.181	.7423	.211	.6757
.002	2.699	.032	1.495	.062	1.208	.092	1.036	.122	.9137	.152	.8182	.182	.7399	.212	.6737
.003	2.523	.033	1.482	.063	1.201	.093	1.032	.123	.9101	.153	.8153	.183	.7375	.213	.6716
.004	2.398	.034	1.469	.064	1.194	.094	1.027	.124	.9066	.154	.8125	.184	.7352	.214	.6696
.005	2.301	.035	1.456	.065	1.187	.095	1.022	.125	.9031	.155	.8097	.185	.7328	.215	.6676
.006	2.222	.036	1.444	.066	1.180	.096	1.018	.126	.8996	.156	.8069	.186	.7305	.216	.6655
.007	2.155	.037	1.432	.067	1.174	.097	1.013	.127	.8962	.157	.8041	.187	.7282	.217	.6635
.008	2.097	.038	1.420	.068	1.168	.098	1.009	.128	.8928	.158	.8013	.188	.7258	.218	.6615
.009	2.046	.039	1.409	.069	1.161	.099	1.004	.129	.8894	.159	.7986	.189	.7235	.219	.6596
.010	2.000	**.040**	1.398	**.070**	1.155	**.100**	1.000	**.130**	.8861	**.160**	.7959	**.190**	.7212	**.220**	.6576
.011	1.959	.041	1.387	.071	1.149	.101	.9957	.131	.8827	.161	.7932	.191	.7190	.221	.6556
.012	1.921	.042	1.377	.072	1.143	.102	.9914	.132	.8794	.162	.7905	.192	.7167	.222	.6536
.013	1.886	.043	1.367	.073	1.137	.103	.9872	.133	.8761	.163	.7878	.193	.7144	.223	.6517
.014	1.854	.044	1.357	.074	1.131	.104	.9830	.134	.8729	.164	.7852	.194	.7122	.224	.6498
.015	1.824	.045	1.347	.075	1.125	.105	.9788	.135	.8697	.165	.7825	.195	.7100	.225	.6478
.016	1.796	.046	1.337	.076	1.119	.106	.9747	.136	.8665	.166	.7799	.196	.7077	.226	.6459
.017	1.770	.047	1.328	.077	1.114	.107	.9706	.137	.8633	.167	.7773	.197	.7055	.227	.6440
.018	1.745	.048	1.319	.078	1.108	.108	.9666	.138	.8601	.168	.7747	.198	.7033	.228	.6421
.019	1.721	.049	1.310	.079	1.102	.109	.9626	.139	.8570	.169	.7721	.199	.7011	.229	.6402
.020	1.699	**050**	1.301	**.080**	1.097	**.110**	.9586	**.140**	.8539	**.170**	.7696	**.200**	.6990	**.230**	.6383
.021	1.678	.051	1.292	.081	1.092	.111	.9547	.141	.8508	.171	.7670	.201	.6968	.231	.6364
.022	1.658	.052	1.284	.082	1.086	.112	.9508	.142	.8477	.172	.7645	.202	.6946	.232	.6345
.023	1.638	.053	1.276	.083	1.081	.113	.9469	.143	.8447	.173	.7620	.203	.6925	.233	.6326
.024	1.620	.054	1.268	.084	1.076	.114	.9431	.144	.8416	.174	.7594	.204	.6904	.234	.6308
.025	1.602	.055	1.260	.085	1.071	.115	.9393	.145	.8386	.175	.7570	.205	.6882	.235	.6289
.026	1.585	.056	1.252	.086	1.066	.116	.9356	.146	.8356	.176	.7545	.206	.6861	.236	.6271
.027	1.569	.057	1.244	.087	1.060	.117	.9318	.147	.8327	.177	.7520	.207	.6840	.237	.6253
.028	1.553	.058	1.237	.088	1.055	.118	.9281	.148	.8297	.178	.7496	.208	.6819	.238	.6234
.029	1,538	.059	1.229	.089	1.051	.119	.9244	.149	.8268	.179	.7471	.209	.6799	.239	.6216

TRANSPARENCY TO OPTICAL DENSITY CONVERSION TABLE (Continued)

Trans.	Density	Trans.	Density	Trans.	Density	Trans.	Density	Trans.	Density
.240	.6198	.315	.5017	**.390**	.4089	.465	.3325	**.540**	.2676
.241	.6180	.316	.5003	.391	.4078	.466	.3316	.541	.2668
.242	.6162	.317	.4989	.392	.4067	.467	.3307	.542	.2660
.243	.6144	.318	.4976	.393	.4056	.468	.3298	.543	.2652
.244	.6126	.319	.4962	.394	.4045	.469	.3288	.544	.2644
.245	.6108	**.320**	.4949	.395	.4034	**.470**	.3279	.545	.2636
.246	.6091	.321	.4935	.396	.4023	.471	.3270	.546	.2628
.247	.6073	.322	.4921	.397	.4012	.472	.3260	.547	.2620
.248	.6056	.323	.4908	.398	.4001	.473	.3251	.548	.2612
.249	.6038	.324	.4895	.399	.3990	.474	.3242	.549	.2604
.250	.6021	.325	.4881	**.400**	.3979	.475	.3233	**.550**	.2596
.251	.6003	.326	.4868	.401	.3969	.476	.3224	.551	.2589
.252	.5986	.327	.4855	.402	.3958	.477	.3215	.552	.2581
.253	.5969	.328	.4841	.403	.3947	.478	.3206	.553	.2573
.254	.5952	.329	.4828	.404	.3936	.479	.3197	.554	.2565
.255	.5935	**.330**	.4815	.405	.3925	**.480**	.3188	.555	.2557
.256	.5918	.331	.4802	.406	.3915	.481	.3179	.556	.2549
.257	.5901	.332	.4789	.407	.3904	.482	.3170	.557	.2541
.258	.5884	.333	.4776	.408	.3893	.483	.3161	.558	.2534
.259	.5867	.334	.4763	.409	.3883	.484	.3152	.559	.2526
.260	.5850	.335	.4750	**.410**	.3872	.485	.3143	**.560**	.2518
.261	.5834	.336	.4737	.411	.3862	.486	.3134	.561	.2510
.262	.5817	.337	.4724	.412	.3851	.487	.3125	.562	.2503
.263	.5800	.338	.4711	.413	.3840	.488	.3116	.563	.2495
.264	.5784	.339	.4698	.414	.3830	.489	.3107	.564	.2487
.265	.5768	**.340**	.4685	.415	.3819	**.490**	.3098	.565	.2479
.266	.5751	.341	.4673	.416	.3809	.491	.3089	.566	.2472
.267	.5735	.342	.4660	.417	.3799	.492	.3080	.567	.2464
.268	.5719	.343	.4647	.418	.3788	.493	.3072	.568	.2457
.269	.5702	.344	.4634	.419	.3778	.494	.3063	.569	.2449
.270	.5686	.345	.4622	**.420**	.3768	.495	.3054	**.570**	.2441
.271	.5670	.346	.4609	.421	.3757	.496	.3045	.571	.2434
.272	.5654	.347	.4597	.422	.3747	.497	.3036	.572	.2426
.273	.5638	.348	.4584	.423	.3737	.498	.3028	.573	.2418
.274	.5622	.349	.4572	.424	.3726	.499	.3019	.574	.2411
275	.5607	**.350**	.4559	.425	.3716	**.500**	.3010	.575	.2403
.276	.5591	.351	.4547	.426	.3706	.501	.3002	.576	.2396
.277	.5575	.352	.4535	.427	.3696	.502	.2993	.577	.2388
.278	.5560	.353	.4522	.428	.3685	.503	.2984	.578	.2381
.279	.5544	.354	.4510	.429	.3675	.504	.2975	.579	.2373
.280	.5528	.355	.4498	**.430**	.3665	.505	.2967	**.580**	.2366
.281	.5513	.356	.4486	.431	.3655	.506	.2959	.581	.2358
282	.5498	.357	.4473	.432	.3645	.507	.2950	.582	.2351
283	.5482	.358	.4461	.433	.3635	.508	.2941	.583	.2343
284	.5467	.359	.4449	.434	.3625	.509	.2933	.584	.2336
.285	.5452	**.360**	.4437	.435	.3615	**.510**	.2924	.585	.2328
.286	.5436	.361	.4425	.436	.3605	.511	.2916	.586	.2321
.287	.5421	.362	.4413	.437	.3595	.512	.2907	.587	.2314
.288	.5406	.363	.4401	.438	.3585	.513	.2899	.588	.2306
.289	.5391	.364	.4389	.439	.3575	.514	.2890	.589	.2299
.290	.5376	.365	.4377	**.440**	.3565	.515	.2882	**.590**	.2291
.291	.5361	.366	.4365	.441	.3556	.516	.2873	.591	.2284
.292	.5346	.367	.4353	.442	.3546	.517	.2865	.592	.2277
.293	.5331	.368	.4342	.443	.3536	.518	.2857	.593	.2269
.294	.5317	.369	.4330	.444	.3526	.519	.2848	.594	.2262
295	.5302	**.370**	.4318	.445	.3516	**.520**	.2840	.595	.2255
.296	.5287	.371	.4306	.446	.3507	.521	.2831	.596	.2248
.297	.5272	.372	.4295	.447	.3497	.522	.2823	.597	.2240
.298	.5258	.373	.4283	.448	.3487	.523	.2815	.598	.2233
.299	.5243	.374	.4271	.449	.3478	.524	.2807	.599	.2226
.300	.5229	.375	.4260	**.450**	.3468	.525	.2798	**.600**	.2219
.301	.5215	.376	.4248	.451	.3458	.526	.2790	.601	.2211
.302	.5200	.377	.4237	.452	.3449	.527	.2782	.602	.2204
.303	.5186	.378	.4225	.453	.3439	.528	.2774	.603	.2197
.304	.5171	.379	.4214	.454	.3429	.529	.2766	.604	.2190
.305	.5157	**.380**	.4202	.455	.3420	**.530**	.2757	.605	.2182
.306	.5143	.381	.4191	.456	.3410	.531	.2749	.606	.2175
.307	.5128	.382	.4179	.457	.3401	.532	.2741	.607	.2168
.308	.5114	.383	.4168	.458	.3391	.533	.2733	.608	.2161
.309	.5100	.384	.4157	.459	.3382	.534	.2725	.609	.2154
.310	.5086	.385	.4145	**.460**	.3372	.535	.2717	**.610**	.2147
.311	.5072	.386	.4134	.461	.3363	.536	.2708	.611	.2140
.312	.5058	.387	.4123	.462	.3354	.537	.2700	.612	.2132
.313	.5045	.388	.4112	.463	.3344	.538	.2692	.613	.2125
.314	.5031	.389	.4101	.464	.3335	.539	.2684	.614	.2118

TRANSPARENCY TO OPTICAL DENSITY CONVERSION TABLE (Continued)

Trans.	Density	Trans.	Density	Trans.	Density	Trans.	Density	Trans.	Density
.615	.2111	.695	.1580	.775	.1107	.855	.0680	.935	.0292
.616	.2104	.696	.1574	.776	.1102	.856	.0675	.936	.0287
.617	.2097	.697	.1568	.777	.1096	.857	.0670	.937	.0282
.618	.2090	.698	.1562	.778	.1090	.858	.0665	.938	.0278
.619	.2083	.699	.1555	.779	.1085	.859	.0660	.939	.0273
.620	.2076	**.700**	.1549	**.780**	.1079	**.860**	.0655	**.940**	.0269
.621	.2069	.701	.1543	.781	.1073	.861	.0650	.941	.0264
.622	.2062	.702	.1537	.782	.1068	.862	.0645	.942	.0260
.623	.2055	.703	.1531	.783	.1062	.863	.0640	.943	.0255
.624	.2048	.704	.1524	.784	.1057	.864	.0635	.944	.0250
.625	.2041	.705	.1518	.785	.1051	.865	.0630	.945	.0246
.626	.2034	.706	.1512	.786	.1046	.866	.0625	.946	.0241
.627	.2027	.707	.1506	.787	.1040	.867	.0620	.947	.0237
.628	.2020	.708	.1500	.788	.1035	.868	.0615	.948	.0232
.629	.2013	.709	.1493	.789	.1029	.869	.0610	.949	.0227
.630	.2007	**.710**	.1487	**.790**	.1024	**.870**	.0605	**.950**	.0223
.631	.2000	.711	.1481	.791	.1018	.871	.0600	.951	.0218
.632	.1993	.712	.1475	.792	.1013	.872	.0595	.952	.0214
.633	.1986	.713	.1469	.793	.1007	.873	.0590	.953	.0209
.634	.1979	.714	.1463	.794	.1002	.874	.0585	.954	.0204
.635	.1972	.715	.1457	.795	.0996	.875	.0580	.955	.0200
.636	.1965	.716	.1451	.796	.0991	.876	.0575	.956	.0195
.637	.1959	.717	.1445	.797	.0985	.877	.0570	.957	.0191
.638	.1952	.718	.1439	.798	.0980	.878	.0565	.958	.0186
.639	.1945	.719	.1433	.799	0975	.879	.0560	.959	.0182
.640	.1938	**.720**	.1427	**.800**	.0969	**.880**	.0555	**.960**	.0177
.641	.1932	.721	.1421	.801	.0964	.881	.0550	.961	.0173
.642	.1925	.722	.1415	.802	.0958	.882	.0545	.962	.0168
.643	.1918	.723	.1409	.803	.0953	.883	.0540	.963	.0164
.644	.1911	.724	.1403	.804	.0948	.884	.0535	.964	.0159
.645	.1904	.725	.1397	.805	.0942	.885	.0530	.965	.0155
.646	.1898	.726	.1391	.806	.0937	.886	.0526	.966	.0150
.647	.1891	.727	.1385	.807	.0931	.887	.0521	.967	.0146
.648	.1884	.728	.1379	.808	.0926	.888	.0516	.968	.0141
.649	.1877	.729	.1373	.809	.0921	.889	.0511	.969	.0137
.650	.1871	**.730**	.1367	**.810**	.0915	**.890**	.0506	**.970**	.0132
.651	.1864	.731	.1361	.811	.0910	.891	.0501	.971	.0128
.652	.1857	.732	.1355	.812	.0904	.892	.0496	.972	.0123
653	.1851	.733	.1349	.813	.0899	.893	.0491	.973	.0119
.654	.1844	.734	.1343	.814	.0894	.894	.0487	.974	.0114
.655	.1838	.735	.1337	.815	.0888	.895	.0482	.975	.0110
.656	.1831	.736	.1331	.816	.0883	.896	.0477	.976	.0106
.657	.1824	.737	.1325	.817	.0878	.897	.0472	.977	.0101
.658	.1818	.738	.1319	.818	.0872	.898	.0467	.978	.0097
.659	.1811	.739	.1314	.819	.0867	.899	.0462	.979	.0092
.660	.1805	**.740**	.1308	**.820**	.0862	**.900**	.0458	**.980**	.0088
.661	.1798	.741	.1302	.821	.0856	.901	.0453	.981	.0083
.662	.1791	.742	.1296	.822	.0851	.902	.0448	.982	.0079
.663	.1785	.743	.1290	.823	.0846	.903	.0443	.983	.0074
.664	.1778	.744	.1284	.824	.0841	.904	.0438	.984	.0070
.665	.1772	.745	.1278	.825	.0835	.905	.0434	.985	.0066
.666	.1765	.746	.1273	.826	.0830	.906	.0429	.986	.0061
.667	.1759	.747	.1267	.827	.0825	.907	.0424	.987	.0057
.668	.1752	.748	.1261	.828	.0820	.908	.0419	.988	.0052
.669	.1746	.749	.1255	.829	.0815	.909	.0414	.989	.0048
.670	.1739	**.750**	.1249	**.830**	.0809	**.910**	.0410	**.990**	.0044
.671	.1733	.751	.1244	.831	.0804	.911	.0405	.991	.0039
.672	.1726	.752	.1238	.832	.0799	.912	.0400	.992	.0035
.673	.1720	.753	.1232	.833	.0794	.913	.0395	.993	.0030
.674	.1713	.754	.1226	.834	.0788	.914	.0391	.994	.0026
.675	.1707	.755	.1221	.835	.0783	.915	.0386	.995	.0022
.676	.1701	.756	.1215	.836	.0778	.916	.0381	.996	.0017
.677	.1694	.757	.1209	.837	.0773	.917	.0376	.997	.0013
.678	.1688	.758	.1203	.838	.0767	.918	.0371	.998	.0009
.679	.1681	.759	.1198	.839	.0762	.919	.0367	.999	.0004
.680	.1675	**.760**	.1192	**.840**	.0757	**.920**	.0362	**1.000**	.0000
.681	.1668	.761	.1186	.841	.0752	.921	.0357	.....	.....
.682	.1662	.762	.1180	.842	.0747	.922	.0353	.....	.....
.683	.1655	.763	.1175	.843	.0742	.923	.0348	.....	.....
.684	.1649	.764	.1169	.844	.0736	.924	.0343	.....	.....
.685	.1643	.765	.1163	.845	.0731	.925	.0339	.....	.....
.686	.1637	.766	.1158	.846	.0726	.926	.0334	.....	.....
.687	.1630	.767	.1152	.847	.0721	.927	.0329	.....	.....
.688	.1624	.768	.1146	.848	.0716	.928	.0325	.....	.....
.689	.1618	.769	.1141	.849	.0711	.929	.0320	.....	.....
.690	.1612	**.770**	.1135	**.850**	.0706	**.930**	.0315	.....	.....
.691	.1605	.771	.1129	.851	.0701	.931	.0310	.....	.....
.692	.1599	.772	.1124	.852	.0696	.932	.0306	.....	.....
.693	.1593	.773	.1118	.853	.0690	.933	.0301	.....	.....
.694	.1586	.774	.1113	.854	.0685	.934	.0296	.....	.....

DENSITY OF VARIOUS SOLIDS

The approximate density of various solids at ordinary atmospheric temperature.
In the case of substances with voids such as paper or leather the bulk density is indicated rather than the density of the solid portion.

(Selected principally from the Smithsonian Tables.)

Substance	Grams per cu. cm	Pounds per cu. ft.
Agate	2.5–2.7	156–168
Alabaster, carbonate	2.69–2.78	168–173
sulfate	2.26–2.32	141–145
Albite	2.62–2.65	163–165
Amber	1.06–1.11	66–69
Amphiboles	2.9–3.2	180–200
Anorthite	2.74–2.76	171–172
Asbestos	2.0–2.8	125–175
Asbestos slate	1.8	112
Asphalt	1.1–1.5	69–94
Basalt	2.4–3.1	150–190
Beeswax	0.96–0.97	60–61
Beryl	2.69–2.7	168–169
Biotite	2.7–3.1	170–190
Bone	1.7–2.0	106–125
Brick	1.4–2.2	87–137
Butter	0.86–0.87	53–54
Calamine	4.1–4.5	255–280
Calcspar	2.6–2.8	162–175
Camphor	0.99	62
Caoutchouc	0.92–0.99	57–62
Cardboard	0.69	43
Celluloid	1.4	87
Cement, set	2.7–3.0	170.190
Chalk	1.9–2.8	118–175
Charcoal, oak	0.57	35
pine	0.28–0.44	18–28
Cinnabar	8.12	507
Clay	1.8–2.6	112–162
Coal, anthracite	1.4–1.8	87–112
bituminous	1.2–1.5	75–94
Cocoa butter	0.89–0.91	56–57
Coke	1.0–1.7	62–105
Copal	1.04–1.14	65–71
Cork	0.22–0.26	14–16
Cork linoleum	0.54	34
Corundum	3.9–4.0	245–250
Diamond	3.01–3.52	188–220
Dolomite	2.84	177
Ebonite	1.15	72
Emery	4.0	250
Epidote	3.25–3.50	203–218
Feldspar	2.55–2.75	159–172
Flint	2.63	164
Fluorite	3.18	198
Galena	7.3–7.6	460–470
Gamboge	1.2	75
Garnet	3.15–4.3	197–268
Gas carbon	1.88	117
Gelatin	1.27	79

Substance	Grams per cu. cm	Pound per cu. ft.
Glass, common	2.4–2.8	150–175
flint	2.9–5.9	180–370
Glue	1.27	79
Granite	2.64–2.76	165–172
Graphite*	2.30–2.72	144–170
Gum arabic	1.3–1.4	81–87
Gypsum	2.31–2.33	144–145
Hematite	4.9–5.3	306–330
Hornblende	3.0	187
Ice	0.917	57.2
Ivory	1.83–1.92	114–120
Leather, dry	0.86	54
Lime, slaked	1.3–1.4	81–87
Limestone	2.68–2.76	167–171
Linoleum	1.18	74
Magnetite	4.9–5.2	306–324
Malachite	3.7–4.1	231–256
Marble	2.6–2.84	160–177
Meerschaum	0.99–1.28	62–80
Mica	2.6–3.2	165–200
Muscovite	2.76–3.00	172–187
Ochre	3.5	218
Opal	2.2	137
Paper	0.7–1.15	44–72
Paraffin	0.87–0.91	54–57
Peat blocks	0.84	52
Pitch	1.07	67
Porcelain	2.3–2.5	143–156
Porphyry	2.6–2.9	162–181
Pressed wood pulp board	0.19	12
Pyrite	4.95–5.1	309–318
Quartz	2.65	165
Resin	1.07	67
Rock salt	2.18	136
Rubber, hard	1.19	74
Rubber, soft commercial	1.1	69
pure gum	0.91–0.93	57–58
Sandstone	2.14–2.36	134–147
Serpentine	2.50–2.65	156–165
Silica, fused transparent	2.21	138
translucent	2.07	129
Slag	2.0–3.9	125–240
Slate	2.6–3.3	162–205
Soapstone	2.6–2.8	162–175
Spermacéti	0.95	59
Starch	1.53	95
Sugar	1.59	99
Talc	2.7–2.8	168–174

Substance	Grams per cu. cm	Pounds per cu. ft.
Tallow, beef	0.94	59
mutton	0.94	59
Tar	1.02	66
Topaz	3.5–3.6	219–223
Tourmaline	3.0–3.2	190–200
Wax, sealing	1.8	112
Wood (seasoned)		
alder	0.42–0.68	26–42
apple	0.66–0.84	41–52
ash	0.65–0.85	40–53
balsa	0.11–0.14	7–9
bamboo	0.31–0.40	19–25
basswood	0.32–0.59	20–37
beech	0.70–0.90	43–56
birch	0.51–0.77	32–48
blue gum	1.00	62
box	0.95–1.16	59–72
butternut	0.38	24
cedar	0.49–0.57	30–35
cherry	0.70–0.90	43–56
dogwood	0.76	47
ebony	1.11–1.33	69–83
elm	0.54–0.60	34–37
hickory	0.60–0.93	37–58
holly	0.76	47
juniper	0.56	35
larch	0.50–0.56	31–35
lignum vitae	1.17–1.33	73–83
locust	0.67–0.71	42–44
logwood	0.91	57
mahogany		
Honduras	0.66	41
Spanish	0.85	53
maple	0.62–0.75	39–47
oak	0.60–0.90	37–56
pear	0.61–0.73	38–45
pine, pitch	0.83–0.85	52–53
white	0.35–0.50	22–31
yellow	0.37–0.60	23–37
plum	0.66–0.78	41–49
poplar	0.35–0.5	22–31
satinwood	0.95	59
spruce	0.48–0.70	30–44
sycamore	0.40–0.60	24–37
teak, Indian	0.66–0.88	41–55
African	0.98	61
walnut	0.64–0.70	40–43
water gum	1.00	62
willow	0.40–0.60	24–37

* Some values reported as low as 1.6.

WEIGHT OF ONE GALLON OF WATER (U.S. GALLONS)

The weights are for dry air at the same temperature as the water up to 40°C and at a barometric pressure corrected to 760 mm and against brass weights of 8.4 density at 0°C. Above 40°C the temperature of the air is assumed to be 20°C, i.e., the water is allowed to cool to 20°C prior to the weighings being made. The volumetric computations are based upon the relations that one liter = 1 dm^3 and that 1 dm^3 = 61.023744 $in.^3$.

Temperature (°C)	Weight in vacuo (g)	Weight in vacuo (lb)	Weight in air (g)	Weight in air (lb)
0	3784.856	8.34417	3780.543	8.33467
1	3785.078	8.34466	3780.781	8.33518
2	3785.233	8.34500	3780.953	8.33556
3	3785.326	8.34520	3781.060	8.33580
4	3785.355	8.34527	3781.105	8.33590
5	3785.325	8.34520	3781.090	8.33587
6	3785.235	8.34500	3781.015	8.33570
7	3785.089	8.34468	3780.884	8.33541
8	3784.887	8.34424	3780.698	8.33500
9	3784.633	8.34368	3780.458	8.33447
10	3784.326	8.34300	3780.167	8.33383
11	3783.966	8.34221	3779.821	8.33307
12	3783.557	8.34130	3779.426	8.33220
13	3783.099	8.34030	3778.983	8.33122
14	3782.597	8.33919	3778.495	8.33014
15	3782.049	8.33798	3777.962	8.32897
16	3781.458	8.33668	3777.415	8.32770
17	3780.824	8.33528	3776.764	8.32633
18	3780.148	8.33379	3776.103	8.32487
19	3779.430	8.33221	3775.398	8.32332
20	3778.672	8.33054	3774.653	8.32167
21	3777.873	8.32877	3773.868	8.31994
22	3777.035	8.32693	3773.044	8.31813
23	3776.158	8.32499	3772.180	8.31622
24	3775.243	8.32298	3771.279	8.31424
25	3774.291	8.32088	3770.340	8.31217
26	3773.320	8.31870	3769.364	8.31001
27	3772.277	8.31644	3769.352	8.30778
28	3771.218	8.31410	3767.306	8.30548
29	3770.123	8.31169	3766.224	8.30309
30	3768.995	8.30920	3765.109	8.30063
31	3768.995	8.30664	3763.961	8.29810
32	3766.641	8.30401	3762.780	8.29550
33	3765.416	8.30131	3761.568	8.29283
34	3764.160	8.29854	3760.324	8.29008
35	3762.874	8.29571	3759.050	8.28728
40	3756.018	8.28059	3752.255	8.27230
45	3748.41	8.2638	3744.42	8.2550
50	3740.19	8.2457	3736.22	8.2369
55	3731.34	8.2261	3727.37	8.2174
60	3721.91	8.2054	3717.95	8.1966
65	3711.88	8.1832	3707.93	8.1745
70	3701.35	8.1600	3697.42	8.1514
75	3690.30	8.1357	3686.38	8.1270
80	3678.72	8.1101	3674.81	8.1015
85	3666.68	8.0836	3662.78	8.0750
90	3654.15	8.0560	3650.27	8.0474
95	3641.21	8.0274	3637.34	8.0189
100	3627.81	7.9979	3623.95	7.9894

TEMPERATURE CORRECTION FOR VOLUMETRIC SOLUTIONS

This table gives the correction to various observed volumes of water, measured at the designated temperatures to give the volume at the standard temperature, 20°C. Conversely, by subtracting the corrections from the volume desired at 20°C., the volume that must be measured out at the designated temperatures in order to give the desired volume at 20°C., will be obtained. It is assumed that the volumes are measured in glass apparatus having a coefficient of cubical expansion of 0.000025 per degree centigrade. The table is applicable to dilute aqueous solutions having the same coefficient of expansion as water.

Temperature of measurement, °C.	Capacity of apparatus in milliliters at 20°C.						
	2,000	1,000	500	400	300	250	150
	Correction in milliliters to give volume of water at 20°C.						
15	+1.54	+0.77	+0.38	+0.31	+0.23	+0.19	+0.12
16	+1.28	+ .64	+ .32	+ .26	+ .19	+ .16	+ .10
17	+ .99	+ .50	+ .25	+ .20	+ .15	+ .12	+ .07
18	+ .68	+ .34	+ .17	+ .14	+ .10	+ .08	+ .05
19	+ .35	+ .18	+ .09	+ .07	+ .05	+ .04	+ .03
21	− .37	− .18	− .09	− .07	− .06	− .05	− .03
22	− .77	− .38	− .19	− .15	− .12	− .10	− .06
23	−1.18	− .59	− .30	− .24	− .18	− .15	−.09
24	−1.61	− .81	− .40	− .32	− .24	− .20	−.12
25	−2.07	−1.03	− .52	− .41	− .31	− .26	−.15
26	−2.54	−1.27	− .64	− .51	− .38	− .32	−.19
27	−3.03	−1.52	− .76	− .61	− .46	− .38	−.23
28	−3.55	−1.77	− .89	− .71	− .53	− .44	−.27
29	−4.08	−2.04	−1.02	− .82	− .61	− .51	−.31
30	−4.62	−2.31	−1.16	− .92	− .69	− .58	−.35

In using the above table to correct the volume of certain standard solutions to 20°C. more accurate results will be obtained if the numerical values of the corrections are increased by the percentages given below:

Solution	Normality N	N/2	N/10
HNO_3	50	25	6
H_2SO_4	45	25	5
NaOH	40	25	5
KOH	40	20	4

TEMPERATURE CORRECTION FOR GLASS VOLUMETRIC APPARATUS

This table gives the correction to be added to actual capacity (determined at certain temperatures) to give the capacity at the standard temperature, 20°C. Conversely, by subtracting the corrections from the indicated capacity of an instrument standard at 20°C. the corresponding capacity at other temperatures is obtained. The table assumes for the cubical coefficient of expansion of glass 0.000025 per degree centigrade. The coefficients of expansion of glasses used for volumetric instruments vary from 0.000023 to 0.000028.

Temperature in degrees C.	2,000 ml	1,000 ml	500 ml	400 ml	300 ml	250 ml
15	+0.25	+0.12	+0.06	+0.05	+0.04	+0.031
16	+ .20	+ .10	+ .05	+0.04	+ .03	+ .025
17	+ .15	+ .08	+ .04	+ .03	+ .02	+ .019
18	+ .10	+ .05	+ .02	+ .02	+ .02	+ .012
19	+ .05	+ .02	+ .01	+ .01	+ .01	+ .006
21	− .05	− .02	− .01	− .01	− .01	− .006
22	− .10	− .05	− .02	− .02	− .02	− .012
23	− .15	− .08	− .04	− .03	− .02	− .019
24	− .20	− .10	− .05	− .04	− .03	− .025
25	− .25	− .12	− .06	− .05	− .04	− .031
26	− .30	− .15	− .08	− .06	− .04	− .038
27	− .35	− .18	− .09	− .07	− .05	− .044
28	− .40	− .20	− .10	− .08	− .06	− .050
29	− .45	− .22	− .11	− .09	− .07	− .056
30	− .50	− .25	− .12	− .10	− .08	− .062

DENSITY OF VARIOUS LIQUIDS

(Selected from Smithsonian Tables.)

Liquid	Grams per cu. cm	Pounds per cu. ft.	Temp ° C
Acetone	0.792	49.4	20°
Alcohol, ethyl	0.791	49.4	20
methyl	0.810	50.5	0
Benzene	0.899	56.1	0
Carbolic acid	0.950–0.965	59.2–60.2	15
Carbon disulfide	1.293	80.7	0
tetrachloride	1.595	99.6	20
Chloroform	1.489	93.0	20
Ether	0.736	45.9	0
Gasoline	0.66–0.69	41.0–43.0	
Glycerin	1.260	78.6	0
Kerosene	0.82	51.2	
Mercury	13.6	849.0	
Milk	1.028–1.035	64.2–64.6	
Naphtha, petroleum ether	0.665	41.5	15
wood	0.848–0.810	52.9–50.5	0
Oils:			
castor	0.969	60.5	15
cocoanut	0.925	57.7	15
cotton seed	0.926	57.8	16
creosote	1.040–1.100	64.9–68.6	15
linseed, boiled	0.942	58.8	15
olive	0.918	57.3	15
Sea water	1.025	63.99	15
Turpentine (spirits)	0.87	54.3	
Water	1.00	62.43	4

DENSITY OF ALCOHOL

DENSITY OF ETHYL ALCOHOL IN GRAMS PER CUBIC CENTIMETER, COMPUTED FROM MENDELEEFF'S FORMULA

(Selected from Smithsonian Tables.)

Temp. ° C	0	1	2	3	4	5	6	7	8	9
0	.80625	.80541	.80457	.80374	.80290	.80207	.80123	.80039	.79956	.79872
10	.79788	.79704	.79620	.79535	.79451	.79367	.79283	.79198	.79114	.79029
20	.78945	.78860	.78775	.78691	.78606	.78522	.78437	.78352	.78267	.78182
30	.78097	.78012	.77927	.77841	.77756	.77671	.77585	.77500	.77414	.77329

HYDROMETERS AND DENSITY UNITS

Alcoholometer. — For testing alcoholic solutions; the scale shows the per cent of alcohol by volume; 0°–100° is the per cent.

Ammoniameter. — For testing ammonia solutions; scale 0°–40°; to convert to sp. gr. multiply by 3 and deduct from 1000.

Barktrometer or *Barkometer.* — For testing tanning liquor; scale 0°–80° Bk; the number to the right of the decimal point of the sp. gr. is the degree Bk; thus, 1.025 sp. gr. is 25° Bk.

Baumé. — There are two kinds in use; heavy Bé, for liquids heavier than water and light Bé for liquids lighter than water. In the former, 0° corresponds to a sp. gr. 1.000 (water at 4°C.) and 66° corresponds to a sp. gr. 1.842; in the lighter than water scale, 0° Bé is equivalent to the gravity of a 10% solution of sodium chloride and 60° Bé corresponds to a sp. gr. of 0.745. For Baumé degrees on the scale of densities greater than unity, the following equation gives the means of conversion:

$$\text{Sp. gr.} = \frac{m}{m - d}$$

where m = 145 (in the United States)
m = 144 (old scale used in Holland)
m = 146.78 (New scale or Gerlach scale)
d = Baumé reading

Beck's Hydrometer has 0° corresponding to sp. gr. 1.000 and 30° to sp. gr. 0.850; equal divisions on the scale are continued as far as required in both directions.

Brix Saccharometer or *Balling Saccharometer* shows directly the per cent of sugar (sucrose) by weight at the temperature indicated on the instrument, usually 17.5°C.; i.e., degrees Brix is the per cent sugar.

Cartier's Hydrometer floats in water at the 10° scale division and at 30° corresponds to 32° Bé.

Oleometer.—For vegetable and sperm oils; scale 50°–0° corresponds to sp. gr. 0.870–0.970.

Sozhlet's Lactometer, for determining the density of milk, has a scale from 25° (sp. gr. 1.025) to 35° (sp. gr. 1.035) divided into suitable scale divisions.

Twaddell Hydrometers have the scale so arranged that the reading multiplied by 5 and added to 1000 gives the sp. gr. with reference to water as 1000; it is always used for densities greater than water.

HYDROMETER CONVERSION TABLES

SHOWING THE RELATION BETWEEN DENSITY (C. G. S.) AND DEGREES BAUMÉ FOR DENSITIES LESS THAN UNITY.

Density.	Degrees Baumé. .00	.01	.02	.03	.04
0.60	103.33	99.51	95.81	92.22	88.75
.70	70.00	67.18	64.44	61.78	59.19
.80	45.00	42.84	40.73	38.68	36.67
.90	25.56	23.85	22.17	20.54	18.94
1.00	10.00	.....	.....	.....	.....

Density.	Degrees Baumé. .05	.06	.07	.08	.09
0.60	85.38	82.12	78.95	75.88	72.90
.70	56.67	54.21	51.82	49.49	47.22
.80	34.71	32.79	30.92	29.09	27.30
.90	17.37	15.83	14.33	12.86	11.41
1.00	......	.....	.....	.....	.....

HYDROMETER CONVERSION TABLES
(Continued)

SHOWING THE RELATION BETWEEN DENSITY (C. G. S.) AND BAUMÉ AND TWADDELL SCALES FOR DENSITIES ABOVE UNITY.

Density	Degrees Baume.	Degrees Twaddell.	Density.	Degrees Baume.	Degrees Twaddell	Density	Degrees Baume.	Degrees Twaddell.	Density.	Degrees Baume.	Degrees Twaddell.
1.00	0.00	0	1.20	24.17	40	1.41	42.16	82	1.61	54.94	122
1.01	1.44	2	1.21	25.16	42	1.42	42.89	84	1.62	55.49	124
1.02	2.84	4	1.22	26.15	44	1.43	43.60	86	1.63	56.04	126
1.03	4.22	6	1.23	27.11	46	1.44	44.31	88	1.64	56.58	128
1.04	5.58	8	1.24	28.06	48	1.45	45.00	90	1.65	57.12	130
1.05	6.91	10	1.25	29.00	50	1.46	45.68	92	1.66	57.65	132
1.06	8.21	12	1.26	29.92	52	1.47	46.36	94	1.67	58.17	134
1.07	9.49	14	1.27	30.83	54	1.48	47.03	96	1.68	58.69	136
1.08	10.74	16	1.28	31.72	56	1.49	47.68	98	1.69	59.20	138
1.09	11.97	18	1.29	32.60	58	1.50	48.33	100	1.70	59.71	140
1.10	13.18	20	1.30	33.46	60	1.51	48.97	102	1.71	60.20	142
1.11	14.37	22	1.31	34.31	62	1.52	49.60	104	1.72	60.70	144
1.12	15.54	24	1.32	35.15	64	1.53	50.23	106	1.73	61.18	146
1.13	16.68	26	1.33	35.98	66	1.54	50.84	108	1.74	61.67	148
1.14	17.81	28	1.34	36.79	68	1.55	51.45	110	1.75	62.14	150
1.15	18.91	30	1.35	37.59	70	1.56	52.05	112	1.76	62.61	152
1.16	20.00	32	1.36	38.38	72	1.57	52.64	114	1.77	63.08	154
1.17	21.07	34	1.37	39.16	74	1.58	53.23	116	1.78	63.54	156
1.18	22.12	36	1.38	39.93	76	1.59	53.80	118	1.79	63.99	158
1.19	23.15	38	1.39	40.68	78	1.60	54.38	120	1.80	64.44	160
			1.40	41.43	80				....	.....	...

DENSITY OF D_2O

G. S. Kell

t, ° C.	ρ, G./Cc.	t, ° C.	ρ, G./Cc.	t, ° C.	ρ, G./Cc.	t, ° C.	ρ, G./Cc.
0	1.10469	20	1.10534	50	1.09570	80	1.07824
3.813	1.10546	25	1.10445	55	1.09325	85	1.07475
5	1.10562	30	1.10323	60	1.09060	90	1.07112
10	1.10599	35	1.10173	65	1.08777	95	1.06736
11.185	1.10600	40	1.09996	70	1.08475	100	1.06346
15	1.10587	45	1.09794	75	1.08158	101.431	1.06232

VOLUME PROPERTIES OF WATER AT 1 atm*

t	ρ, kg m^{-3}, Equation 1	$10^6\ \alpha$, K^{-1}, Equation 1	$10^6\ \kappa T$/bar^{-1} Equation 2	t	ρ, kg m^{-3}, Equation 1	$10^6\ \alpha$, K^{-1}, Equation 1	$10^6\ \kappa T$/bar^{-1} Equation 2
−30	983.857	−1400.0	80.79	38	992.9683	369.79	44.3051
−25	989.588	−955.9	70.94	39	992.5973	377.59	44.2697
−20	993.550	−660.6	64.25	40	992.2187	385.30	44.2391
−15	996.286	−450.3	59.44	41	991.8327	392.91	44.2131
−10	998.120	−292.4	55.83	42	991.4394	400.43	44.1917
−9	998.398	−265.3	55.22	43	991.0388	407.85	44.1747
−8	998.650	−239.5	54.64	44	990.6310	415.19	44.1620
−7	998.877	−214.8	54.08	45	990.2162	422.45	44.1536
−6	999.080	−191.2	53.56	46	989.7944	429.63	44.1494
−5	999.259	−168.6	53.06	47	989.3657	436.73	44.1494
−4	999.417	−146.9	52.58	48	988.9303	443.75	44.1533
−3	999.553	−126.0	52.12	49	988.4881	450.71	44.1613
−2	999.669	−106.0	51.69	50	988.0393	457.59	44.1732
−1	999.765	−86.7	51.28	51	987.5839	464.40	44.189
0	999.8425	−68.05	50.8850	52	987.1220	471.15	44.209
1	999.9015	−50.09	50.5091	53	986.6537	477.84	44.232
2	999.9429	−32.74	50.1505	54	986.1791	484.47	44.259
3	999.9672	−15.97	49.8081	55	985.6982	491.04	44.290
4	999.9750	0.27	49.4812	56	985.2111	497.55	44.324
5	999.9668	16.00	49.1692	57	984.7178	504.01	44.362
6	999.9432	31.24	48.8712	58	984.2185	510.41	44.403
7	999.9045	46.04	48.5868	59	983.7132	516.76	44.448
8	999.8512	60.41	48.3152	60	983.2018	523.07	44.496
9	999.7838	74.38	48.0560	61	982.6846	529.32	44.548
10	999.7026	87.97	47.8086	62	982.1615	535.53	44.603
11	999.6081	101.20	47.5726	63	981.6327	541.70	44.662
12	999.5004	114.08	47.3474	64	981.0981	547.82	44.723
13	999.3801	126.65	47.1327	65	980.5578	553.90	44.788
14	999.2474	138.90	46.9280	66	980.0118	559.94	44.857
15	999.1026	150.87	46.7331	67	979.4603	565.95	44.928
16	998.9460	162.55	46.5475	68	978.9032	571.91	45.003
17	998.7779	173.98	46.3708	69	978.3406	577.84	45.081
18	998.5986	185.15	46.2029	70	977.7726	583.74	45.162
19	998.4082	196.08	46.0433	71	977.1991	589.60	45.246
20	998.2071	206.78	45.8918	72	976.6203	595.43	45.333
21	997.9955	217.26	45.7482	73	976.0361	601.23	45.424
22	997.7735	227.54	45.6122	74	975.4466	607.00	45.517
23	997.5415	237.62	45.4835	75	974.8519	612.75	45.614
24	997.2995	247.50	45.3619	76	974.2520	618.46	45.714
25	997.0479	257.21	45.2472	77	973.6468	624.15	45.817
26	996.7867	266.73	45.1392	78	973.0366	629.82	45.922
27	996.5162	276.10	45.0378	79	972.4212	635.46	46.031
28	996.2365	285.30	44.9427	80	971.8007	641.08	46.143
29	995.9478	294.34	44.8537	81	971.1752	646.67	46.258
30	995.6502	303.24	44.7707	82	970.5446	652.25	46.376
31	995.3440	312.00	44.6935	83	969.9091	657.81	46.497
32	995.0292	320.63	44.6221	84	969.2686	663.34	46.621
33	994.7060	329.12	44.5561	85	968.6232	668.86	46.748
34	994.3745	337.48	44.4956	86	967.9729	674.37	46.878
35	994.0349	345.73	44.4404	87	967.3177	679.85	47.011
36	993.6872	353.86	44.3903	88	966.6576	685.33	47.148
37	993.3316	361.88	44.3452	89	965.9927	690.78	47.287

VOLUME PROPERTIES OF WATER AT 1 atm* (continued)

	ρ, kg m⁻³,	10^6 α, K⁻¹,	10^6 κT/bar⁻¹	
	Equation 1	Equation 1	Equation 2	Equation 3
90	965.3230	696.23	47.429	47.428
91	964.6486	701.66	47.574	47.574
92	963.9693	707.08	47.722	47.722
93	963.2854	712.49	47.874	47.873
94	962.5967	717.89	48.028	48.028
95	961.9033	723.28	48.185	48.185
96	961.2052	728.67	48.346	48.346
97	960.5025	734.04	48.509	48.510
98	959.7951	739.41	48.676	48.677
99	959.0831	744.78	48.846	48.847
100	958.3665	750.14	49.019	49.020
101	957.645	755.5		49.20
102	956.920	760.8		49.38
103	956.189	766.2		49.56
104	955.454	771.5		49.74
105	954.715	776.9		49.93
106	953.971	782.2		50.13
107	953.222	787.6		50.32
108	952.469	792.9		50.52
109	951.712	798.3		50.72
110	950.950	803.6		50.93
115	947.073	830.4		52.01
120	943.085	857.4		53.17
125	938.987	884.7		54.43
130	934.778	912.3		55.79
135	930.459	940.3		57.24
140	926.029	968.9		58.80
145	921.487	998.0		60.47
150	916.832	1027.8		62.25

DENSITY AND VOLUME OF MERCURY

BASED ON THE DENSITY OF MERCURY AT 0° C. BY THIESEN AND SCHEEL

(Selected from Smithsonian Tables.)

Temp. °C.	Mass in gr. per ml.	Vol. of 1 gr. in ml.	Temp. °C.	Mass in gr. per ml.	Vol. of 1 gr. in ml.	Temp. °C.	Mass in gr. per ml.	Vol. of 1 gr. in ml.	Temp. °C.	Mass in gr. per ml.	Vol. of 1 gr. in ml.
−10	13.6202	0.0734205	11	5684	7011	30°	13.5217	0.0739552	150	2330	5688
−9	6177	4338	12	5659	7145	31	5193	9686	160	2093	7044
−8	6152	4472	13	5634	7278	32	5168	9820	170	1856	8402
−7	6128	4606	14	5610	7412	33	5144	9953	180	1620	9764
−6	6103	4739	15	13.5585	0.0737546	34	5119	40087	190	13.1384	0.0761128
−5	13.6078	0.0734873	16	5561	7680	35	13.5095	0.0740221	200	1148	2495
−4	6053	5006	17	5536	7813	36	5070	0354	210	0913	3865
−3	6029	5140	18	5512	7947	37	5046	0488	220	0678	5239
−2	6004	5273	19	5487	8081	38	5021	0622	230	0443	6616
−1	5979	5407	20	13.5462	0.0738215	39	4997	0756	240	13.0209	0.0767996
0	13.5955	0.0735540	21	5438	8348	40	13.4973	0.0740891	250	12.9975	9381
1	5930	5674	22	5413	8482	50	4729	2229	260	9741	70769
2	5906	5808	23	5389	8616	60	4486	3569	270	9507	2161
3	5881	5941	24	5364	8750	70	4244	4910	280	9273	3558
4	5856	6075	25	13.5340	0.0738883	80	4003	6252	290	12.9039	0.0774958
5	13.5832	0.0736209	26	5315	9017	90	13.3762	0.0747594	300	8806	6364
6	5807	6342	27	5291	9151	100	3522	8939	310	8572	7774
7	5782	6476	28	5266	9285	110	3283	50285	320	8339	9189
8	5758	6610	29	5242	9419	120	3044	1633	330	8105	80609
9	5733	6744	30	13.5217	0.0739552	130	2805	2982	340	12.7872	0.0782033
10	13.5708	0.0736877				140	13.2567	0.0754334	350	7638	3464
									360	7405	4900

SULFURIC ACID

SPECIFIC GRAVITY OF AQUEOUS SULFURIC ACID SOLUTIONS

AT $\frac{20°}{4°}$ C.

Be.	Sp. gr.	Per cent H_2SO_4	G. per liter	Lbs. per cu. ft.	Lbs. per gal.
0.7	1.0051	1	10.05	0.6275	0.0839
1.7	1.0118	2	20.24	1.263	0.1689
2.6	1.0184	3	30.55	1.907	0.2550
3.5	1.0250	4	41.00	2.560	0.3422
4.5	1.0317	5	51.59	3.220	0.4305
5.4	1.0385	6	62.31	3.890	0.5200
6.3	1.0453	7	73.17	4.568	0.6106
7.2	1.0522	8	84.18	5.255	0.7025
8.1	1.0591	9	95.32	5.950	0.7955
9.0	1.0661	10	106.6	6.655	0.8897
9.9	1.0731	11	118.0	7.369	0.9851
10.8	1.0802	12	129.6	8.092	1.082
11.7	1.0874	13	141.4	8.825	1.180
12.5	1.0947	14	153.3	9.567	1.279
13.4	1.1020	15	165.3	10.32	1.379
14.3	1.1094	16	177.5	11.08	1.481
15.2	1.1168	17	189.9	11.85	1.584
16.0	1.1243	18	202.4	12.63	1.689
16.9	1.1318	19	215.0	13.42	1.795
17.7	1.1394	20	227.9	14.23	1.902
18.6	1.1471	21	240.9	15.04	2.010
19.4	1.1548	22	254.1	15.86	2.120
20.3	1.1626	23	267.4	16.69	2.231
21.1	1.1704	24	280.9	17.54	2.344
21.9	1.1783	25	294.6	18.39	2.458
22.8	1.1862	26	308.4	19.25	2.574
23.6	1.1942	27	322.4	20.13	2.691
24.4	1.2023	28	336.6	21.02	2.809
25.2	1.2104	29	351.0	21.91	2.929
26.0	1.2185	30	365.6	22.82	3.051
26.8	1.2267	31	380.3	23.74	3.173
27.6	1.2349	32	395.2	24.67	3.298
28.4	1.2432	33	410.3	25.61	3.424
29.1	1.2515	34	425.5	26.56	3.551
29.9	1.2599	35	441.0	27.53	3.680
30.7	1.2684	36	456.6	28.51	3.811
31.4	1.2769	37	472.5	29.49	3.943
32.2	1.2855	38	488.5	30.49	4.077
33.0	1.2941	39	504.7	31.51	4.212
33.7	1.3028	40	521.1	32.53	4.349
34.5	1.3116	41	537.8	33.57	4.488
35.2	1.3205	42	554.6	34.62	4.628
35.9	1.3294	43	571.6	35.69	4.770
36.7	1.3384	44	588.9	36.76	4.914
37.4	1.3476	45	606.4	37.86	5.061
38.1	1.3569	46	624.2	38.97	5.209
38.9	1.3663	47	642.2	40.09	5.359
39.6	1.3758	48	660.4	41.23	5.511
40.3	1.3854	49	678.8	42.38	5.665
41.1	1.3951	50	697.6	43.55	5.821
41.8	1.4049	51	716.5	44.73	5.979
42.5	1.4148	52	735.7	45.93	6.140
43.2	1.4248	53	755.1	47.14	6.302
44.0	1.4350	54	774.9	48.37	6.467
44.7	1.4453	55	794.9	49.62	6.634
45.4	1.4557	56	815.2	50.89	6.803
46.1	1.4662	57	835.7	52.17	6.974
46.8	1.4768	58	856.5	53.47	7.148
47.5	1.4875	59	877.6	54.79	7.324
48.2	1.4983	60	899.0	56.12	7.502
48.9	1.5091	61	920.6	57.47	7.682
49.6	1.5200	62	942.4	58.83	7.865
50.3	1.5310	63	964.5	60.21	8.049
51.0	1.5421	64	986.9	61.61	8.236
51.7	1.5533	65	1010	63.03	8.426
52.3	1.5646	66	1033	64.46	8.618
53.0	1.5760	67	1056	65.92	8.812
53.7	1.5874	68	1079	67.39	9.008
54.3	1.5989	69	1103	68.87	9.207
55.0	1.6105	70	1127	70.38	9.408
55.6	1.6221	71	1152	71.90	9.611
56.3	1.6338	72	1176	73.44	9.817
56.9	1.6456	73	1201	74.99	10.02
57.5	1.6574	74	1226	76.57	10.24
58.1	1.6692	75	1252	78.15	10.45
58.7	1.6810	76	1278	79.75	10.66
59.3	1.6927	77	1303	81.37	10.88
59.9	1.7043	78	1329	82.99	11.09
60.5	1.7158	79	1355	84.62	11.31
61.1	1.7272	80	1382	86.26	11.53
61.6	1.7383	81	1408	87.90	11.75
62.1	1.7491	82	1434	89.54	11.97
62.6	1.7594	83	1460	91.16	12.19
63.0	1.7693	84	1486	92.78	12.40
63.5	1.7786	85	1512	94.38	12.62
63.9	1.7872	86	1537	95.95	12.83
64.2	1.7951	87	1562	97.49	13.03
64.5	1.8022	88	1586	99.01	13.23
64.8	1.8087	89	1610	100.5	13.42
65.1	1.8144	90	1633	101.9	13.63
65.3	1.8195	91	1656	103.4	13.82
65.5	1.8240	92	1678	104.8	14.00
65.7	1.8279	93	1700	106.1	14.19
65.8	1.8312	94	1721	107.5	14.36
65.9	1.8337	95	1742	108.7	14.54
66.0	1.8355	96	1762	110.0	14.70
66.0	1.8364	97	1781	111.2	14.87
66.0	1.8361	98	1799	112.3	15.02
65.9	1.8342	99	1816	113.4	15.15
65.8	1.8305	100	1831	114.3	15.28

DENSITY AND COMPOSITION OF FUMING SULFURIC ACID

Actual H_2SO_4, %	Specific gravity	Equiv. H_2SO_4, %	Weight, lb./cu. ft.	Weight, lb. per U.S. gal.	Comb. H_2O, %	Free SO_3, %	Total SO_3, %	SO_3, lb./cu. ft.
100	1.839	100.00	114.70	15.33	18.37	0	81.63	93.63
99	1.845	100.22	115.07	15.38	18.19	1	81.81	94.14
98	1.851	100.45	115.33	15.41	18.00	2	82.00	94.57
97	1.855	100.67	115.70	15.46	17.82	3	82.18	95.08
96	1.858	100.89	115.88	15.49	17.64	4	82.36	95.44
95	1.862	101.13	116.13	15.52	17.45	5	82.55	95.87
94	1.865	101.35	116.32	15.55	17.27	6	82.73	96.23
93	1.869	101.58	116.57	15.58	17.08	7	82.92	96.66
92	1.873	101.80	116.82	15.61	16.90	8	83.10	97.12
91	1.877	102.02	117.07	15.64	16.72	9	83.28	97.50
90	1.880	102.25	117.26	15.67	16.57	10	83.47	97.88
89	1.884	102.47	117.51	15.70	16.35	11	83.65	98.30
88	1.887	102.71	117.69	15.73	16.17	12	83.83	98.66
87	1.891	102.92	117.94	15.76	15.98	13	84.02	99.09
86	1.895	103.15	118.19	15.79	15.80	14	84.20	99.52
85	1.899	103.38	118.44	15.82	15.61	15	84.39	99.95
84	1.902	103.60	118.63	15.86	15.43	16	84.57	100.33
83	1.905	103.82	118.81	15.89	15.25	17	84.75	100.69
82	1.909	104.05	119.06	15.92	15.06	18	84.94	101.13
81	1.911	104.28	119.28	15.95	14.88	19	85.12	101.45
80	1.915	104.50	119.50	15.98	14.70	20	85.30	101.93
79	1.920	104.73	119.75	16.01	14.51	21	85.49	102.37
78	1.923	104.95	119.94	16.04	14.33	22	85.67	102.75
77	1.927	105.18	120.19	16.07	14.14	23	85.86	103.20
76	1.931	105.40	120.44	16.10	13.96	24	86.04	103.63
75	1.934	105.62	120.62	16.12	13.78	25	86.22	104.00
74	1.939	105.85	120.94	16.16	13.59	26	86.41	104.50
73	1.943	106.08	121.18	16.19	13.41	27	86.59	104.93
72	1.946	106.29	121.37	16.22	13.28	28	86.72	105.31
71	1.949	106.53	121.56	16.25	13.04	29	86.96	105.71
70	1.952	106.75	121.75	16.28	12.86	30	87.14	106.09
69	1.955	106.97	121.93	16.30	12.68	31	87.32	106.47
68	1.958	107.20	122.12	16.33	12.49	32	87.51	106.87
67	1.961	107.42	122.31	16.35	12.31	33	87.69	107.25
66	1.965	107.65	122.56	16.38	12.12	34	87.88	107.71
65	1.968	107.87	122.74	16.40	11.94	35	88.06	108.08
64	1.972	108.10	122.99	16.43	11.76	36	88.24	108.53
63	1.976	108.33	123.24	16.46	11.57	37	88.43	108.98
62	1.979	108.55	123.43	16.50	11.39	38	88.61	109.37
61	1.981	108.77	123.55	16.52	11.21	39	88.79	109.70
60	1.983	109.00	123.74	16.54	11.02	40	88.98	110.10
59	1.985	109.22	123.80	16.55	10.84	41	89.16	110.38
58	1.987	109.45	123.93	16.56	10.65	42	89.35	110.83
57	1.989	109.68	124.05	16.58	10.47	43	89.53	111.06
56	1.991	109.90	124.18	16.60	10.29	44	89.71	111.40
55	1.993	110.13	124.30	16.62	10.10	45	89.90	111.75
50	2.001	111.25	124.80	16.68	9.18	50	90.72	113.34
40	2.102	113.50	131.10	17.53	7.35	60	92.65	121.46
30	1.982	115.75	123.62	16.50	5.51	70	94.49	116.81
20	1.949	118.00	121.56	16.25	3.67	80	96.33	117.10
10	1.911	120.25	119.19	15.92	1.84	90	98.16	117.00
0	1.857	122.50	115.83	15.50	0.00	100	100.00	115.83

*** By permission from the 7th edition of Chemical Plant Control Data, Chemical Construction Corporation (1957).**

DENSITY OF MOIST AIR

The density of dry air may be determined by computation from the general relation $D = D_0(T_0/T)(P/P_0)$ where D_0 represents a known density at absolute temperature T_0 and pressure P_0 and D, the density at absolute temperature T and pressure P.

The density of **moist** air may be determined by a similar relation: $D = 1.2929\ (273.13/T)\ [(B - 0.3783e)/760]$ where T is the absolute temperature; B, the barometric pressure in mm, and e the vapor pressure of the moisture in the air in mm. The density will then be the product of two terms, each of which may be found by use of the tables which follow.

The **first factor,** 1.2929 $(273.13/T)$, may be found directly in Table I for various temperatures. For convenience, temperatures are given in the table in °C although the values of the factor have been computed with absolute temperatures. The tabular values actually represent the density of dry air at various temperatures and 760 mm pressure.

The **second factor,** $[(B - 0.3783e)/760]$, must be obtained in two steps: **First**—the numerator of the expression is obtained by subtracting $0.3783e$ from the barometric pressure. The quantity $0.3783e$ may be found directly from the dew point in Table II. If the wet and dry bulb thermometer readings are known e may be found in the table Reduction of Psychrometric Observations given in the section Hygrometric and Barometric Tables. $0.3783e$ may then be found by calculation or read from the table. **Second**—the value of the whole factor for any value of $B - 0.3783e$ may be obtained from Table III.

The product of the above two factors will give the required density in g/l.

To facilitate obtaining **approximate values** of the density for ordinary pressures and temperatures, a table of products is given which may be entered with the temperature in °C and the corrected (for moisture) value of the barometric pressure in mm to obtain density.

As an illustration of the use of the tables, let it be desired to find the density of air for a barometric pressure of 750 mm, a dew point of 10° C, and air temperature of 20° C.

From the dew point, the value of $0.3783e$ is found in Table II to be 3.48 mm. 750 − 3.48 = 746.52, the corrected pressure. The pressure factor for this value found in Table III by interpolation is 0.98226.

The temperature factor from Table I is 1.2047.

$$1.2047 \times 0.98224 = 1.1833 \text{ g/l.}$$

To obtain the value directly from Table IV, enter it for 20° C and 746.5 mm which gives by interpolation 1.183 g/l.

TABLE I

(1.2929 × 273.13/*T*)

(Besides being a necessary part of the determination of the density of moist air, the values in this table are actually the density of dry air in g/l at 760 mm pressure for various temperatures.)

Temp. °C	0	1	2	3	4	5	6	7	8	9
−50	1.5 826	897	969	*042	*115	*189	*264	*339	*415	*491
−40	1.5 147	213	278	345	412	479	547	616	686	756
−30	1.4 524	584	645	706	767	829	892	955	*019	*083
−20	1.3 951	*006	*062	*118	*175	*232	*289	*347	*406	*465
−10	1.3 420	472	523	575	628	680	734	787	841	896
− 0	1.2 929	977	*024	*073	*121	*170	*219	*269	*319	*370
+ 0	1.2 929	882	835	789	742	697	651	606	561	517
10	1.2 472	428	385	342	299	256	214	171	130	088
20	1.2 047	006	*965	*925	*885	*845	*805	*766	*727	*688
30	1.1 649	611	573	535	498	460	423	387	350	314
40	1.1 277	242	206	170	135	100	065	031	*996	*962
50	1.0 928	895	861	828	795	762	729	697	664	632
60	1.0 600	569	537	506	475	444	413	382	352	322

TABLE II

Vapor Pressure—Value of $0.3783e$

Dew point °C	Vap. press. *e* mm (ice)	0.3783*e*	Dew point °C	Vap. press. *e* mm (water)	0.3783*e*	Dew point °C	Vap. press. *e* mm (water)	0.3783*e*	Dew point °C	Vap. press. *e* mm (ice)	0.3783*e*	Dew point °C	Vap. press. *e* mm (water)	0.3783*e*	Dew point °C	Vap. press. *e* mm (water)	0.3783*e*
−50	0.029	0.01	**−15**	1.252	0.47	**0**	4.58	1.73	**15**	12.79	4.84	**30**	31.86	12.05	**45**	71.97	27.23
−45	.054	.02	−14	1.373	.52	1	4.92	1.86	16	13.64	5.16	31	33.74	12.76	46	75.75	28.66
−40	.096	.04	−13	1.503	.57	2	5.29	2.00	17	14.54	5.50	32	35.70	13.51	47	79.70	30.15
−35	.169	.06	−12	1.644	.62	3	5.68	2.15	18	15.49	5.86	33	37.78	14.29	48	83.83	31.71
−30	.288	.11	−11	1.798	.68	4	6.10	2.31	19	16.49	6.24	34	39.95	15.11	49	88.14	33.34
−25	0.480	0.18	**−10**	1.964	0.74	**5**	6.54	2.47	**20**	17.55	6.64	**35**	42.23	15.98	**50**	92.6	35.03
−24	.530	.20	− 9	2.144	.81	6	7.01	2.65	21	18.66	7.06	36	44.62	16.88	51	97.3	36.81
−23	.585	.22	− 8	2.340	.89	7	7.51	2.84	22	19.84	7.51	37	47.13	17.83	52	102.2	38.66
−22	.646	.24	− 7	2.550	.96	8	8.04	3.04	23	21.09	7.98	38	49.76	18.82	53	107.3	40.59
−21	.712	.27	− 6	2.778	1.05	9	8.61	3.26	24	22.40	8.47	39	52.51	19.86	54	112.7	42.63
−20	0.783	0.30	**− 5**	3.025	1.14	**10**	9.21	3.48	**25**	23.78	9.00	**40**	55.40	20.96	**55**	118.2	44.72
−19	.862	.33	− 4	3.291	1.24	11	9.85	3.73	26	25.24	9.55	41	58.42	22.10	56	124.0	46.91
−18	.947	.36	− 3	3.578	1.35	12	10.52	3.98	27	26.77	10.13	42	61.58	23.30	57	130.0	49.18
−17	1.041	.39	− 2	3.887	1.47	13	11.24	4.25	28	28.38	10.74	43	64.89	24.55	58	136.3	51.56
−16	1.142	.43	− 1	4.220	1.60	14	11.99	4.54	29	30.08	11.38	44	68.35	25.86	59	142.8	54.02
															60	149.6	56.59

TABLE III

Pressure Factor.—$[(B - 0.3783e)/760]$

The figures in the body of the table give values of the whole term $(B - 0.3783e)/760]$ for various values of the numerator $(B - 0.3783e)$ expressed at the left and top.

Press. mm corr.	0	1	2	3	4	5	6	7	8	9	Press. mm corr.	0	1	2	3	4	5	6	7	8	9
80	.10526	.10658	.10789	.10921	.11053	.11184	.11316	.11447	.11579	.11711	. . .	. . . . .	. . . . .	. . . . .	. . . . .	. . . . .	. . . . .	. . . . .	. . . . .	. . . . .	. . . . .
90	.11842	.11974	.12105	.12237	.12368	.12500	.12632	.12763	.12895	.13026	. . .	. . . . .	. . . . .	. . . . .	. . . . .	. . . . .	. . . . .	. . . . .	. . . . .	. . . . .	. . . . .
100	.13158	.13289	.13421	.13553	.13684	.13816	.13947	.14079	.14211	.14342	**200**	.26316	.26447	.26579	.26711	.26842	.26974	.27105	.27237	.27368	.27500
110	.14474	.14605	.14737	.14868	.15000	.15132	.15263	.15395	.15526	.15658	210	.27632	.27763	.27895	.28026	.28158	.28289	.28421	.28553	.28684	.28816
120	.15789	.15921	.16053	.16184	.16316	.16447	.16579	.16711	.16842	.16974	220	.28947	.29079	.29211	.29342	.29474	.29605	.29737	.29868	.30000	.30132
130	.17105	.17237	.17368	.17500	.17632	.17763	.17895	.18026	.18158	.18289	230	.30263	.30395	.30526	.30658	.30789	.30921	.31053	.31184	.31316	.31447
140	.18421	.18553	.18684	.18816	.18947	.19079	.19211	.19342	.19474	.19605	240	.31579	.31711	.31842	.31974	.32105	.32237	.32368	.32500	.32632	.32763
150	.19737	.19868	.20000	.20132	.20263	.20395	.20526	.20658	.20789	.20921	**250**	.32895	.33026	.33158	.33289	.33421	.33553	.33684	.33816	.33947	.34079
160	.21053	.21184	.21316	.21447	.21579	.21711	.21842	.21974	.22105	.22237	260	.34211	.34342	.34474	.34605	.34737	.34868	.35000	.35132	.35263	.35395
170	.22368	.22500	.22632	.22763	.22895	.23026	.23158	.23289	.23421	.23553	270	.35526	.35658	.35789	.35921	.36053	.36184	.36316	.36447	.36579	.36711
180	.23684	.23816	.23947	.24079	.24211	.24342	.24474	.24605	.24737	.24868	280	.36842	.36974	.37105	.37237	.37368	.37500	.37632	.37763	.37895	.38026
190	.25000	.25132	.25263	.25395	.25526	.25658	.25789	.25921	.26053	.26184	290	.38158	.38289	.38421	.38553	.38684	.38816	.38947	.39079	.39211	.39342

DENSITY OF MOIST AIR (Continued)

TABLE III (Continued)

Press. mm corr.	0	1	2	3	4	5	6	7	8	9
300	.39474	.39605	.39737	.39868	.40000	.40132	.40263	.40395	.40526	.40658
310	.40789	.40921	.41053	.41184	.41316	.41447	.41579	.41711	.41842	.41974
320	.42105	.42237	.42368	.42500	.42632	.42763	.42895	.43026	.43158	.43289
330	.43421	.43553	.43684	.43816	.43947	.44079	.44211	.44342	.44474	.44605
340	.44737	.44868	.45000	.45132	.45263	.45395	.45526	.45658	.45789	.45921
350	.46053	.46184	.46316	.46447	.46579	.46711	.46842	.46974	.47105	.47237
360	.47368	.47500	.47632	.47763	.47895	.48026	.48158	.48289	.48421	.48553
370	.48684	.48816	.48947	.49079	.49211	.49342	.49474	.49605	.49737	.49868
380	.50000	.50132	.50263	.50395	.50526	.50658	.50789	.50921	.51053	.51184
390	.51316	.51447	.51579	.51711	.51842	.51974	.52105	.52237	.52368	.52500
400	.52632	.52763	.52895	.53026	.53158	.53289	.53421	.53553	.53684	.53816
410	.53947	.54079	.54211	.54342	.54474	.54605	.54737	.54868	.55000	.55132
420	.55263	.55395	.55526	.55658	.55789	.55921	.56053	.56184	.56316	.56447
430	.56579	.56711	.56842	.56974	.57105	.57237	.57368	.57500	.57632	.57763
440	.57895	.58026	.58158	.58289	.58421	.58553	.58684	.58816	.58947	.59079
450	.59211	.59342	.59474	.59605	.59737	.59868	.60000	.60132	.60263	.60395
460	.60526	.60658	.60789	.60921	.61053	.61184	.61316	.61447	.61579	.61711
470	.61842	.61974	.62105	.62237	.62368	.62500	.62632	.62763	.62895	.63026
480	.63158	.63289	.63421	.63553	.63684	.63816	.63947	.64079	.64211	.64342
490	.64474	.64605	.64737	.64868	.65000	.65132	.65263	.65395	.65526	.65658
500	.65790	.65921	.66053	.66184	.66316	.66447	.66579	.66711	.66842	.66974
510	.67105	.67237	.67368	.67500	.67632	.67763	.67895	.68026	.68158	.68290
520	.68421	.68553	.68684	.68816	.68947	.69079	.69211	.69342	.69474	.69605
530	.69737	.69868	.70000	.70132	.70263	.70395	.70526	.70658	.70790	.70921
540	.71053	.71184	.71316	.71447	.71579	.71711	.71842	.71974	.72105	.72237

Press. mm corr.	0	1	2	3	4	5	6	7	8	9
550	.72368	.72500	.72632	.72763	.72895	.73026	.73158	.73290	.73421	.73553
560	.73684	.73816	.73947	.74079	.74211	.74342	.74474	.74605	.74737	.74868
570	.75000	.75132	.75263	.75395	.75526	.75658	.75790	.75921	.76053	.76184
580	.76316	.76447	.76579	.76711	.76842	.76974	.77105	.77237	.77368	.77500
590	.77632	.77763	.77895	.78026	.78158	.78290	.78421	.78553	.78684	.78816
600	.78947	.79079	.79211	.79342	.79474	.79605	.79737	.79868	.80000	.80132
610	.80263	.80395	.80526	.80658	.80790	.80921	.81053	.81184	.81316	.81447
620	.81579	.81711	.81842	.81974	.82105	.82237	.82368	.82500	.82632	.82763
630	.82895	.83026	.83158	.83290	.83421	.83553	.83684	.83816	.83947	.84079
640	.84211	.84342	.84474	.84605	.84737	.84868	.85000	.85132	.85263	.85395
650	.85526	.85658	.85790	.85921	.86053	.86184	.86316	.86447	.86579	.86711
660	.86842	.86974	.87105	.87237	.87368	.87500	.87632	.87763	.87895	.88026
670	.88158	.88290	.88421	.88553	.88684	.88816	.88947	.89079	.89211	.89342
680	.89474	.89605	.89737	.89868	.90000	.90132	.90263	.90395	.90526	.90658
690	.90790	.90921	.91053	.91184	.91316	.91447	.91579	.91711	.91842	.91974
700	.92105	.92237	.92368	.92500	.92632	.92763	.92895	.93026	.93158	.93290
710	.93421	.93553	.93684	.93816	.93947	.94079	.94211	.94342	.94474	.94605
720	.94737	.94868	.95000	.95132	.95263	.95395	.95526	.95658	.95790	.95921
730	.96053	.96184	.96316	.96447	.96579	.96711	.96842	.96974	.97105	.97237
740	.97368	.97500	.97632	.97763	.97895	.98026	.98158	.98290	.98421	.98553
750	.98684	.98816	.98947	.99079	.99211	.99342	.99474	.99605	.99737	.99868
760	1.0000	1.0013	1.0026	1.0039	1.0053	1.0066	1.0079	1.0092	1.0105	1.0118
770	1.0132	1.0145	1.0158	1.0171	1.0184	1.0197	1.0211	1.0224	1.0237	1.0250
780	1.0263	1.0276	1.0289	1.0303	1.0316	1.0329	1.0342	1.0355	1.0368	1.0382
790	1.0395	1.0408	1.0421	1.0434	1.0447	1.0461	1.0474	1.0487	1.0500	1.0513

TABLE IV

Density of Moist Air

Values in the body of the table give the density of moist air in g/l for a limited range of temperatures and corrected pressure values ($B - 0.3783e$). The latter may be obtained by use of Table II.

°C	600	610	620	630	640	650	660	670	680	690	700	710	720	730	740	750	760	770	780	790
5	1.0024	1.0191	1.0358	1.0525	1.0692	1.0859	1.1026	1.1193	1.1361	1.1528	1.1695	1.1862	1.2029	1.2196	1.2363	1.2530	1.2697	1.2864	1.3031	1.3198
6	.99876	1.0154	1.0321	1.0487	1.0654	1.0820	1.0986	1.1153	1.1319	1.1486	1.1652	1.1819	1.1985	1.2152	1.2318	1.2485	1.2651	1.2817	1.2984	1.3150
7	.99521	1.0118	1.0284	1.0450	1.0616	1.0781	1.0947	1.1113	1.1279	1.1445	1.1611	1.1777	1.1943	1.2108	1.2274	1.2440	1.2606	1.2772	1.2938	1.3104
8	.99165	1.0082	1.0247	1.0412	1.0578	1.0743	1.0908	1.1074	1.1239	1.1404	1.1569	1.1735	1.1900	1.2065	1.2230	1.2396	1.2561	1.2726	1.2892	1.3057
9	.98818	1.0047	1.0211	1.0376	1.0541	1.0705	1.0870	1 1035	1.1199	1.1364	1.1529	1.1694	1.1858	1.2023	1.2188	1.2352	1.2517	1.2682	1.2846	1.3011
10	.98463	1.0010	1.0175	1.0339	1.0503	1.0667	1.0831	1.0995	1.1159	1.1323	1.1487	1.1651	1.1816	1.1980	1.2144	1.2308	1.2472	1.2636	1.2800	1.2964
11	.98115	.99751	1.0139	1.0302	1.0466	1.0629	1.0793	1.0956	1.1120	1.1283	1.1447	1.1610	1.1774	1.1937	1.2101	1.2264	1.2428	1.2592	1.2755	1.2919
12	.97776	.99406	1.0104	1.0267	1.0430	1.0592	1.0755	1.0918	1.1081	1.1244	1.1407	1.1570	1.1733	1.1896	1.2059	1.2222	1.2385	1.2548	1.2711	1.2874
13	.97436	.99061	1.0068	1.0231	1.0393	1.0556	1.0718	1.0880	1.1043	1.1205	1.1368	1.1530	1.1692	1.1855	1.2017	1.2180	1.2342	1.2504	1.2667	1.2829
14	.97097	.98715	1.0033	1.0195	1.0357	1.0519	1.0681	1.0843	1.1004	1.1166	1.1328	1.1490	1.1652	1.1814	1.1975	1.2137	1.2299	1.2461	1.2623	1.2784
15	.96757	.98370	.99983	1.0160	1.0321	1.0482	1.0643	1.0805	1.0966	1.1127	1.1288	1.1450	1.1611	1.1772	1.1933	1.2095	1.2256	1.2417	1.2579	1.2740
16	.96426	.98033	.99641	1.0125	1.0286	1.0446	1.0607	1.0768	1.0928	1.1089	1.1250	1.1410	1.1571	1.1732	1.1893	1.2053	1.2214	1.2375	1.2535	1.2696
17	.96086	.97688	.99290	1.0089	1.0249	1.0409	1.0570	1.0730	1.0890	1.1050	1.1210	1.1370	1.1530	1.1691	1.1851	1.2011	1.2171	1.2331	1.2491	1.2651
18	.95763	.97359	.98955	1.0055	1.0215	1.0374	1.0534	1.0694	1.0853	1.1013	1.1172	1.1332	1.1492	1.1651	1.1811	1.1970	1.2130	1.2290	1.2449	1.2609
19	.95431	.97022	.98613	1.0020	1.0179	1.0338	1.0497	1.0656	1.0816	1.0975	1.1134	1.1293	1.1452	1.1611	1.1770	1.1929	1.2088	1.2247	1.2406	1.2565
20	.95107	.96693	.98278	.99864	1.0145	1.0303	1.0462	1.0620	1.0779	1.0937	1.1096	1.1254	1.1413	1.1572	1.1730	1.1888	1.2047	1.2206	1.2364	1.2522
21	.94784	.96364	.97944	.99524	1.0110	1.0294	1.0426	1.0584	1.0742	1.0900	1.1058	1.1216	1.1374	1.1532	1.1690	1.1848	1.2006	1.2164	1.2322	1.2480
22	.94460	.96035	.97609	.99184	1.0076	1.0233	1.0391	1.0548	1.0706	1.0863	1.1020	1.1178	1.1335	1.1493	1.1650	1.1808	1.1965	1.2122	1.2280	1.2437
23	.94144	.95714	.97283	.98852	1.0042	1.0199	1.0356	1.0513	1.0670	1.0827	1.0984	1.1140	1.1297	1.1454	1.1611	1.1768	1.1925	1.2082	1.2239	1.2396
24	.93829	.95393	.96957	.98521	1.0008	1.0165	1.0321	1.0478	1.0634	1.0790	1.0947	1.1103	1.1259	1.1416	1.1572	1.1729	1.1885	1.2041	1.2198	1.2354
25	.93513	.95072	.96630	.98189	.99748	1.0131	1.0286	1.0442	1.0598	1.0754	1.0910	1.1066	1.1222	1.1377	1.1533	1.1689	1.1845	1.2001	1.2157	1.2313
26	.93197	.94750	.96304	.97858	.99411	1.0096	1.0252	1.0407	1.0562	1.0718	1.0873	1.1028	1.1184	1.1339	1.1494	1.1650	1.1805	1.1960	1.2116	1.2271
27	.92889	.94437	.95986	.97534	.99083	1.0063	1.0218	1.0373	1.0528	1.0682	1.0837	1.0992	1.1147	1.1302	1.1456	1.1611	1.1766	1.1921	1.2076	1.2230
28	.92581	.94124	.95668	.97211	.98754	1.0030	1.0184	1.0338	1.0493	1.0647	1.0801	1.0955	1.1110	1.1264	1.1418	1.1573	1.1727	1.1881	1.2036	1.2190
29	.92273	.93811	.95350	.96888	.98426	.99963	1.0150	1.0304	1.0458	1.0612	1.0765	1.0919	1.1073	1.1227	1.1380	1.1534	1.1688	1.1842	1.1996	1.2149
30	.91965	.93498	.95031	.96564	.98097	.99629	1.0116	1.0270	1.0423	1.0576	1.0729	1.0883	1.1036	1.1189	1.1342	1.1496	1.1649	1.1802	1.1956	1.2109
31	.91665	.93193	.94721	.96249	.97777	.99304	1.0083	1.0236	1.0389	1.0542	1.0694	1.0847	1.1000	1.1153	1.1305	1.1458	1.1611	1.1764	1.1917	1.2069
32	.91365	.92888	.94411	.95934	.97457	.98979	1.0050	1.0203	1.0355	1.0507	1.0659	1.0812	1.0964	1.1116	1.1268	1.1421	1.1573	1.1725	1.1878	1.2030
33	.91065	.92583	.94101	.95619	.97137	.98654	1.0017	1.0169	1.0321	1.0473	1.0624	1.0776	1.0928	1.1080	1.1231	1.1383	1.1535	1.1687	1.1839	1.1990
34	.90773	.92286	.93800	.95313	.96826	.98338	.99851	1.0136	1.0288	1.0439	1.0590	1.0742	1.0893	1.1044	1.1195	1.1347	1.1498	1.1649	1.1801	1.1952
35	.90473	.91981	.93490	.94998	.96506	.98013	.99521	1.0103	1.0254	1.0405	1.0555	1.0706	1.0857	1.1008	1.1158	1.1309	1.1460	1.1611	1.1762	1.1912

DENSITY OF DRY AIR

AT THE TEMPERATURE t, AND UNDER THE PRESSURE H CM OF MERCURY THE DENSITY OF AIR

$$= \frac{0.001293}{1+0.00367\,t} \frac{H}{76}.$$

Units of this table are grams per milliliter

(From Miller's Laboratory Physics, Ginn & Co., publishers, by permission.)

t	Pressure H in Centimeters						Proportional Parts	
	72.0	73.0	74.0	75.0	76.0	77.0		
°								17
10	0.001182	0.001198	0.001215	0.001231	0.001247	0.001264	cm	
11	178	193	210	227	243	259	0.1	2
12	173	190	206	222	239	255	0.2	3
13	169	186	202	218	234	251	0.3	5
14	165	181	198	214	230	246	0.4	7
							0.5	8
							0.6	10
							0.7	12
15	0.001161	0.001177	0.001193	0.001210	0.001226	0.001242	0.8	14
16	157	173	189	205	221	238	0.9	15
17	153	169	185	201	217	233		16
18	149	165	181	197	213	229	cm	
19	145	161	177	193	209	225	0.1	2
							0.2	3
							0.3	5
							0.4	6
20	0.001141	0.001157	0.001173	0.001189	0.001205	0.001221	0.5	8
21	137	153	169	185	201	216	0.6	10
22	134	149	165	181	197	212	0.7	11
23	130	145	161	177	193	208	0.8	13
24	126	142	157	173	189	204	0.9	14
								15
							cm	
25	0.001122	0.001138	0.001153	0.001169	0.001185	0.001200	0.1	1
26	118	134	149	165	181	196	0.2	3
27	115	130	146	161	177	192	0.3	4
28	111	126	142	157	173	188	0.4	6
29	107	123	138	153	169	184	0.5	7
							0.6	9
							0.7	10
							0.8	12
30	0.001104	0.001119	0.001134	0.001150	0.001165	0.001180	0.9	13

DENSITY OF WATER

The temperature of maximum density for pure water, free from air = 3.98C (277.13K)

t, °C	d, gm/ml	t, °C	d, gm/ml
0	0.99987	40	0.99224
3.98	1.00000	45	0.99025
5	0.99999	50	0.98807
10	0.99973	55	0.98573
15	0.99913	60	0.98324
18	0.99862	65	0.98059
20	0.99823	70	0.97781
25	0.99707	75	0.97489
30	0.99567	80	0.97183
35	0.99406	85	0.96865
38	0.99299	90	0.96534
		95	0.96192
		100	0.95838

THERMODYNAMIC AND TRANSPORT PROPERTIES OF AIR

From NASA Technical Note D-7488 by David J. Poferl and Roger Svehla (1973). The following three tables list the thermodynamic and transport properties of air over the temperature range of 300-2800K at pressures of 20, 30, and 40 atm. Factors for converting viscosity, specific heat at constant pressure, thermal conductivity, and enthalpy from cgs units to SI and English units are

Viscosity:

$$1\ \frac{g}{(cm)(sec)} = 0.1\ \frac{(N)(sec)}{m^2}$$

$$= 6.72\text{x}10^{-2}\ \frac{lbm}{(ft)(sec)}$$

$$= 241.9\ \frac{lbm}{(ft)(hr)}$$

$$= 2.089\text{x}10^{-3}\ \frac{(lbf)(sec)}{ft^2}$$

Thermal conductivity:

$$1\ \frac{cal}{(cm)(sec)(K)} = 418.4\ \frac{W}{(m)(K)}$$

$$= 0.8064\ \frac{Btu}{(ft)^2\,(sec)(^\circ F/in.)}$$

$$= 6.72\text{x}10^{-2}\ \frac{Btu}{(ft^2)(sec)(^\circ F/ft)}$$

$$= 241.9\ \frac{Btu}{(ft)^2\,(hr)(^\circ F/ft)}$$

Specific heat at constant pressure:

$$1\ \frac{cal}{(g)(K)} = 4.184\ \frac{J}{(g)(K)}$$

$$= 1\ \frac{Btu}{(lbm)(^\circ F)}$$

Enthalpy:

$$1\ \frac{cal}{g} = 4.184\ \frac{J}{g}$$

$$= 1.8\ \frac{Btu}{lbm}$$

PROPERTIES AT 20 ATM

Temperature, T, K	Isentropic exponent, γ	Molecular weight, m	Viscosity, μ g/(cm)(sec)	Specific heat at constant pressure, c_p, cal/(g)(K)	Thermal conductivity, k, cal/(cm)(sec)(K)	Prandtl number, Pr	Enthalpy, h, cal/g
2800	1.2309	28.890	821×10^{-6}	0.3850	473×10^{-6}	0.669	747.0
2700	1.2374	28.915	800	.3707	439	.676	709.2
2600	1.2437	28.933	779	.3588	409	.683	672.7
2500	1.2498	28.945	758	.3488	384	.688	637.4
2400	1.2556	28.953	736	.3404	362	.692	602.9
2300	1.2612	28.958	715	.3332	343	.696	569.3
2200	1.2666	28.961	694	.3270	325	.698	536.2
2100	1.2719	28.963	672	.3214	309	.699	503.8
2000	1.2772	28.964	651	.3163	294	.700	471.9
1900	1.2825	28.965	629	.3115	280	.701	440.6
1800	1.2879	↓	607	.3070	266	.702	409.6
1700	1.2933	↓	585	.3025	252	.702	379.2
1600	1.2989	↓	563	.2981	239	.703	349.1
1500	1.3045	↓	540	.2939	226	.703	319.5
1400	1.3103	↓	517	.2897	213	.704	290.3
1300	1.3162	↓	494	.2855	200	.704	261.6
1200	1.3224	↓	470	.2814	188	.705	233.2
1100	1.3288	↓	445	.2773	175	.705	205.3
1000	1.3356	↓	419	.2730	162	.705	177.8
900	1.3439	28.964	391	.2681	148	.706	150.7
800	1.3537	↓	362	.2626	135	↓	124.2
700	1.3646	↓	331	.2568	121	↓	98.2
600	1.3759	↓	299	.2511	106	↓	72.8
500	1.3865	↓	265	.2461	92	↓	48.0
400	1.3951	↓	227	.2422	78	↓	23.6
300	1.4000	↓	184	.2401	63	↓	–.5

THERMODYNAMIC AND TRANSPORT PROPERTIES OF AIR *(Continued)*

PROPERTIES AT 30 ATM

Temperature, T, K	Isentropic exponent, γ	Molecular weight, m	Viscosity, μ, g/(cm)(sec)	Specific heat at constant pressure, c_p, cal/(g)(K)	Thermal conductivity, k, cal/(cm)(sec)(K)	Prandtl number, Pr	Enthalpy, h, cal/g
2800	1.2340	28.904	821×10^{-6}	0.3773	460×10^{-6}	0.674	745.0
2700	1.2399	28.924	800	.3652	430	.680	707.9
2600	1.2456	28.939	779	.3548	403	.686	671.9
2500	1.2512	28.949	758	.3462	380	.690	636.8
2400	1.2566	28.956	736	.3387	359	.694	602.6
2300	1.2619	28.960	715	.3321	341	.696	569.1
2200	1.2671	28.962	694	.3263	324	.698	536.1
2100	1.2722	28.964	672	.3211	309	.699	503.8
2000	1.2773	28.965	651	.3161	294	.700	471.9
1900	1.2826		629	.3115	280	.701	440.5
1800	1.2879		607	.3069	266	.702	409.6
1700	1.2933		585	.3025	252	.702	379.2
1600	1.2988		563	.2981	239	.703	349.1
1500	1.3045		540	.2939	226	.703	319.5
1400	1.3103		517	.2897	213	.704	290.3
1300	1.3162		494	.2855	200	.704	261.6
1200	1.3223		470	.2814	188	.705	233.2
1100	1.3288		445	.2773	175	.705	205.3
1000	1.3356		419	.2730	162	.705	177.8
900	1.3439	28.964	391	.2681	148	.706	150.7
800	1.3537		362	.2626	135		124.2
700	1.3646		331	.2568	121		98.2
600	1.3759		299	.2511	106		72.8
500	1.3865		265	.2461	92		48.0
400	1.3951		227	.2422	78		23.6
300	1.4000		184	.2401	63		−.5

THERMODYNAMIC AND TRANSPORT PROPERTIES OF AIR *(Continued)*

PROPERTIES AT 40 ATM

Temperature, T, K	Isentropic exponent, γ	Molecular weight, m	Viscosity, μ, g/(cm)(sec)	Specific heat at constant pressure, c_p, cal/(g)(K)	Thermal conductivity, k, cal/(cm)(sec)(K)	Prandtl number, Pr	Enthalpy, h, cal/g
2800	1.2360	28.913	821×10^{-6}	0.3727	452×10^{-6}	0.677	743.8
2700	1.2414	28.930	800	.3619	424	.683	707.1
2600	1.2468	28.943	779	.3526	399	.688	671.4
2500	1.2521	28.951	758	.3446	377	.692	636.5
2400	1.2573	28.957	736	.3376	358	.695	602.4
2300	1.2623	28.961	715	.3315	340	.697	569.0
2200	1.2673	28.963	694	.3260	324	.699	536.1
2100	1.2724	28.964	672	.3209	308	.700	503.8
2000	1.2774	28.965	651	.3160	294	.700	471.9
1900	1.2826		629	.3114	279	.701	440.5
1800	1.2879		607	.3069	266	.702	409.6
1700	1.2933		585	.3025	252	.702	379.2
1600	1.2988		563	.2981	239	.703	349.1
1500	1.3045		540	.2939	226	.703	319.5
1400	1.3103		517	.2897	213	.704	290.4
1300	1.3162		494	.2855	200	.704	261.6
1200	1.3223		470	.2814	188	.705	233.2
1100	1.3288		445	.2773	175	.705	205.3
1000	1.3356		419	.2730	162	.705	177.8
900	1.3439		391	.2681	148	.706	150.7
800	1.3537	28.964	362	.2626	135		124.2
700	1.3646		331	.2568	121		98.2
600	1.3759		299	.2511	106		72.8
500	1.3865		265	.2461	92		48.0
400	1.3951		227	.2422	78		23.6
300	1.4000		184	.2401	63		–.5

ISOTHERMAL COMPRESSIBILITY OF LIQUIDS

J. C. McGowan

The figures in this table are for isothermal compressibilities in cgs units. The compiler suggests that, provided the pressure is not too high, the reciprocal of the isothermal compressibility varies linearly with the pressure. This suggestion also appears in papers by J. R. Macdonald, *Rev. Mod. Phys.*, 38, 669, 1966, and O. L. Anderson, *J. Phys. Chem. Solids*, 27, 547, 1966. The papers vary somewhat as to how far this linearity persists.

Liquid	Temp., °C[a]	Isothermal compressibility × 10^{10} $m^2 N^{-1}$ At 1 atm (or 1.013 × 10^5 Nm^2)	At 1000 atm (or 1.013 × 10^8 Nm^3)	Ref.
Acetic acid	15	8.75	–	1
	20	9.08	–	1
	30	9.72	–	1
	40	10.37	–	1
	50	11.11	–	1
	60	11.91	–	1
	70	12.77	–	1
	80	13.68	–	1
Acetic acid, ethyl ester	0	9.78	–	1
	10	10.36	–	1
	20	11.32	–	1
	30	12.37	–	1
	40	13.52	–	1
	50	14.78	–	1
	60	16.21	–	1
	70	17.90	–	1
Acetone	20	12.75	–	3
	20	12.29	–	10
	25	12.39	6.02	4
	30	13.34	–	10
	40	15.61	–	3
	40	14.64	–	10
	50	16.03	–	10
Aniline	0	4.08	–	1
	10	4.30	–	1
	20	4.53	–	1
	25	4.67	3.23	5
	40	5.04	–	1
	45	5.22	3.48	5
	50	5.33	–	1
	60	5.64	–	1
	65	5.84	3.76	5
	70	5.97	–	1
	80	6.32	–	1
	85	6.56	4.04	5
	90	6.70	–	1
Anisole	21	6.67	–	2
	30	7.04	–	2
	45	7.72	–	2
	60	8.50	–	2
	81	9.79	–	2
	100	11.25	–	2
	120	13.07	–	2
	140	15.45	–	2
Benzene	0	8.09	–	1
	10	8.73	–	1
	10	8.64	–	3
	20	9.44	–	1
	20	9.37	–	3
	20	9.54	–	10
	25	9.67	5.07	6
	25	9.7	–	15
	30	10.27	–	10
	30	10.18	–	1
	30	10.12	–	3
	35	10.43	5.28	6
	39.5	10.91	–	15
	40	11.05	–	10
	40	11.00	–	1
	40	10.96	–	3
	45	11.32	5.50	6
	50	11.90	–	10
	50	11.89	–	1
	50	11.83	–	3
	50.1	11.91	–	15
	55	12.29	5.73	6
	60	12.78	–	10
	60	12.83	–	1
	60	12.96	–	15
	65	13.39	5.98	6
	70	13.72	–	10
	70	14.13	–	1
	75.9	14.95	–	15
	80	15.44	–	1

ISOTHERMAL COMPRESSIBILITY OF LIQUIDS (continued)

Liquid	Temp., °C[a]	Isothermal compressibility × 10^{10} $m^2 N^{-1}$ At 1 atm (or 1.013 × 10^5 Nm^2)	At 1000 atm (or 1.013 × 10^8 Nm^3)	Ref.
Benzene, bromo-	25	6.68	4.09	5
	45	7.52	4.39	5
	65	8.50	4.72	5
	85	9.65	5.06	5
Benezene, chloro-	0	6.61	–	1
	10	7.02	–	1
	20	7.45	–	1
	20	7.38	–	3
	25	7.51	4.39	5
	30	7.89	–	1
	30	7.84	–	3
	40	8.39	–	1
	40	8.32	–	3
	45	8.55	4.73	5
	50	8.92	–	1
	50	8.83	–	3
	60	9.50	–	1
	65	9.76	5.10	5
	70	10.13	–	1
	80	10.79	–	1
	85	11.23	5.49	5
Benezene, nitro-	0	4.41	–	1
	10	4.67	–	1
	20	4.93	–	1
	25	5.03	3.39	5
	30	5.23	–	1
	40	5.49	–	1
	45	5.59	3.64	5
	65	6.24	3.91	5
	85	6.99	4.20	5
n-Butyl alcohol	0	8.10	–	11
Carbon disulphide	0	8.04	–	1
	0	7.95	–	3
	10	8.64	–	1
	10	8.54	–	3
	20	9.13	–	10
	20	9.26	–	1
	20	9.19	–	3
	30	9.72	–	10
	30	9.92	–	1
	30	9.96	–	3
	40	10.65	–	1
	40	10.57	–	10
	40	10.89	–	3
	50	11.48	–	1
	50	11.95	–	3
Carbon tetrachloride	–22.9	7.84	–	12
	–13.1	8.31	–	12
	–3.1	8.87	–	12
	0	8.98	–	1
	0	8.85	–	3
	6.9	9.59	–	12
	10	9.70	–	1
	10	9.57	–	3
	10	9.45	–	16
	16.9	10.35	–	12
	20	10.46	–	1
	20	10.34	–	3
	20	10.40	–	10
	25	10.67	5.30	7
	25	10.77	–	15
	25	10.58	–	16
	26.9	11.15	–	12
	30	11.28	–	10
	30	11.29	–	1
	30	11.18	–	3
	35	11.95	5.52	7
	37.5	11.99	–	15
	40	12.20	–	10
	40	12.23	–	1
	40	12.16	–	3
	40	11.96	–	16
	45	12.54	5.75	7
	50	13.20	–	10
	50	13.32	–	1
	50	13.26	–	3
	50.3	13.28	–	15
	55	13.63	5.97	7
	55	13.51	–	15
	60	14.26	–	10
	60	14.52	–	1
	62.6	14.84	–	15
	65	14.87	6.22	7
	70	15.43	–	10
	70	15.77	–	1
	75	16.70	–	15
Chloroform	–33.1	7.13	–	12
	–23.1	7.61	–	12
	–13.1	8.12	–	12
	–3.1	8.66	–	12
	0	8.48	–	1
	0	8.55	–	3
	6.9	9.30	–	12
	10	9.17	–	1
	10	9.19	–	3

Liquid	Temp., °C[a]	Isothermal compressibility × 10^{10} m^2N^{-1} At 1 atm (or 1.013 × 10^5 Nm^2)	At 1000 atm (or 1.013 × 10^8 Nm^3)	Ref.
Chloroform (continued)	16.9	10.35	–	12
	20	9.98	–	1
	20	9.94	–	3
	20	10.15	–	10
	25	9.74	5.34	4
	26.9	11.15	–	12
	30	10.86	–	1
	30	10.81	–	3
	30	10.99	–	10
	40	11.84	–	1
	40	11.79	–	3
	40	11.94	–	10
	50	12.90	–	1
	50	12.99	–	10
	60	14.06	–	1
Cyclohexane	10	9.88	–	16
	25	11.10	–	16
	25	11.40	–	15
	35	12.02	–	16
	37.6	12.67	–	15
	40	12.56	–	16
	45	13.14	–	16
	50.1	14.15	–	15
	55	14.35	–	16
	60	14.88	–	16
	62.4	15.76	–	15
	75	17.84	–	15
n-Decane	25	11.63	–	17
	35	12.34	–	17
	50	13.26	–	17
Dodecane	37.8	9.9	5.3	13
	60.0	11.3	5.7	13
	79.4	12.8	6.1	13
	98.9	14.4	6.4	13
	115.0	16.1	6.8	13
	135.0	18.3	7.3	13
Ethane 1,1,2,2-tetrachloro-	25	6.17	3.88	4
Ethyl alcohol	0	9.87	–	1
	0	9.87	–	11
	0	9.63	–	3
	10	10.49	–	1
	10	10.30	–	3
	20	11.19	–	1
	20	10.98	–	3
	30	11.91	–	1
	30	11.80	–	3

Liquid	Temp. °C[a]	Isothermal compressibility × 10^{10} m^2N^{-1} At 1 atm (or 1.013 × 10^5 Nm^2)	At 1,000 atm (or 1.013 × 10^8 Nm^3)	Ref.
Ethyl alcohol (continued)	40	12.74	–	1
	40	12.61	–	3
	50	13.70	–	1
	50	13.60	–	3
	60	14.74	–	1
	70	15.93	–	1
	75	16.67	–	1
Ethyl bromide	0	10.76	–	1
	10	11.78	–	1
	20	12.94	–	1
	30	14.23	–	1
	40	15.52	–	1
Ethylene, 1,2-dichloro-(trans)	25	11.19	5.62	4
Ethylene, tetrachloro-	25	7.56	4.45	4
Ethylene, trichloro-	25	8.57	4.99	4
Ethylene, chloride	0	6.91	–	1
	10	7.42	–	1
	20	7.97	–	1
	20	7.82	–	10
	25	7.78	4.54	4
	30	8.41	–	10
	30	8.58	–	1
	40	9.09	–	10
	40	9.25	–	1
	50	9.86	–	10
	50	9.99	–	1
	60	10.66	–	10
	60	10.83	–	1
	70	11.54	–	10
	70	11.76	–	1
	80	12.79	–	1
Ethyl ether	0	15.10	–	1
	0	15.07	–	3
	10	16.81	–	1
	10	16.52	–	3
	20	18.65	–	1
	20	18.44	–	3
	30	20.90	–	1
	30	20.80	–	3
	35	24.15	–	1
Ethyl iodide	0	8.45	–	1
	10	9.12	–	1
	20	9.82	–	1

ISOTHERMAL COMPRESSIBILITY OF LIQUIDS (continued)

Liquid	Temp. °C[a]	Isothermal compressibility × 10^{10} $m^2 N^{-1}$ At 1 atm (or 1.013 × 10^5 Nm^2)	At 1000 atm (or 1.013 × 10^8 Nm^3)	Ref.
Ethyl iodide (continued)	30	10.59	–	1
	40	11.44	–	1
	50	12.38	–	1
	60	13.40	–	1
	70	14.49	–	1
Glycol	25	3.72	2.73	7
	45	4.00	2.89	7
	65	4.32	3.05	7
	85	4.70	3.24	7
	105	5.14	3.44	7
n-Hendecane	25	10.81	–	17
	35	11.50	–	17
	50	12.47	–	17
n-Heptane	0	11.80	5.64	8
	25	14.24	6.18	8
	25	14.95	–	17
	35	16.30	–	17
	40	15.96	6.50	8
	50	18.59	–	17
	60	18.67	6.94	8
1-Heptanol	0	7.05	–	11
n-Hexadecane	25	8.67	–	17
	35	9.43	–	17
	50	10.29	–	17
n-Hexane	0	13.04	5.92	8
	25	16.06	6.51	8
	25	17.09	–	17
	35	18.96	–	17
	40	18.31	6.89	8
	50	22.32	–	17
	60	21.93	8.87	8
1-Hexanol	0	7.47	–	11
Mercury	0	0.40	–	14
	20	0.40	0.39	14
	40	0.41	–	14
	80	0.42	–	14
	120	0.44	–	14
	160	0.46	–	14
Methanol	0	10.62	–	1
	0	10.68	–	11
	0	10.78	–	3
	10	11.34	–	1

Liquid	Temp., °C[a]	Isothermal compressibility × 10^{10} $m^2 N^{-1}$ At 1 atm (or 1.013 × 10^5 Nm^2)	At 1000 atm (or 1.013 × 10^8 Nm^3)	Ref.
Methanol (continued)	10	11.45	–	3
	20	12.11	–	1
	20	12.18	–	3
	30	12.93	–	1
	30	12.98	–	3
	40	13.85	–	1
	40	13.82	–	3
	50	14.76	–	3
Methylene bromide	–33.1	4.95	–	12
	–23.1	5.11	–	12
	–13.1	5.50	–	12
	–3.1	5.81	–	12
	6.9	6.13	–	12
	16.9	6.47	–	12
	26.9	6.85	–	12
Methylene chloride	25	9.74	5.31	4
Methyl iodide	–33.1	6.86	–	12
	–23.1	7.41	–	12
	–13.1	7.97	–	12
	–3.1	8.55	–	12
	6.9	9.13	–	12
	16.9	9.71	–	12
	26.9	10.33	–	12
n-Nonane	25	12.31	–	17
	35	13.20	–	17
	50	14.51	–	17
*n*Octadecane	60	9.4	5.1	13
	79.4	10.4	5.5	13
	98.9	11.6	5.8	13
	115.0	12.8	6.1	13
	135	14.4	6.4	13
n-Octane	0	9.99	5.11	8
	25	11.98	5.69	8
	40	13.36	6.00	8
	60	15.53	6.40	8
1-Octanol	0	6.82	–	11
n-Pentadecane	37.8	9.1	–	13
	60	10.2	5.2	13
	79.4	11.7	5.5	13
	98.9	13.2	5.8	13
	115	14.7	6.1	13
	135	16.8	6.4	13
Pentanol	0	7.71	–	11

ISOTHERMAL COMPRESSIBILITY OF LIQUIDS (continued)

Liquid	Temp., °C[a]	Isothermal compressibility × 10^{10} $m^2 N^{-1}$: At 1 atm (or 1.013 × 10^5 Nm^2)	At 1000 atm (or 1.013 × 10^8 Nm^3)	Ref.
Phenol	46	5.61	–	2
	60	6.05	–	2
	80	6.78	–	2
	110	8.12	–	2
	125	8.88	–	2
	150	10.30	–	2
	175	12.35	–	2
n-Propyl alcohol	0	8.43	–	11
Toluene	–59.3	5.27	–	9
	–41.1	5.93	–	9
	–19.2	6.87	–	9
	0	7.83	–	1
	0	7.97	–	3
	0	7.84	–	9
	10	8.38	–	1
	10	8.44	–	3
	20	8.96	–	1
	20	8.94	–	3
	30	9.60	–	1
	30	9.49	–	3
	40	10.33	–	1
	40	10.14	–	3
	50	11.13	–	1
	50	10.90	–	3
	60	11.99	–	1
	70	12.95	–	1

Liquid	Temp. °C[a]	Isothermal compressibility × 10^{10} $m^2 N^{-1}$: At 1 atm (or 1.013 × $10^5 Nm^2$)	At 1,000 atm (or 1.013 × $10^8 Nm^3$)	Ref.
Water	0	5.01	–	1
	0	5.04	–	9
	10	4.78	–	1
	20	4.58	–	1
	25	4.57	3.48	7
	30	4.46	–	1
	34.8	4.44	–	9
	35	4.48	3.42	7
	40	4.41	–	1
	45	4.41	3.40	7
	50	4.40	–	1
	55	4.44	3.40	7
	60	4.43	–	1
	65	4.48	3.42	7
	70	4.49	–	1
	75	4.55	3.47	7
	80	4.57	–	1
	85	4.65	3.53	7
	90	4.68	–	1
	100	4.80	–	1
m-Xylene	0	7.44	–	1
	10	7.94	–	1
	20	8.46	–	1
	30	9.03	–	1
	40	9.63	–	1
	50	10.25	–	1
	60	11.01	–	1
	70	11.77	–	1
	80	12.56	–	1

REFERENCES

1. Tyrer, D., *J. Chem. Soc.*, 105, 2534, 1914.
2. Lutskii, A. E. and Solonko, V. N., *Russ. J. Phys. Chem.*, 38, 602, 1964.
3. Fryer, E. B., Hubbard, J. C., and Andrews, D. H., *J. Am. Chem. Soc.*, 51, 759, 1929.
4. Newitt, D. M. and Weale, K. E., *J. Chem. Soc.*, p. 3092, 1951.
5. Gibson, R. E. and Loeffler, O. H., *J. Am. Chem. Soc.*, 61, 2515, 1939.
6. Gibson, R. E. and Kincaid, J. F., *J. Am. Chem. Soc.*, 60, 511, 1938.
7. Gibson, R. E. and Loeffler, O. H., *J. Am. Chem. Soc.*, 63, 898, 1941.
8. Eduljee, H. E., Newitt, D. M., and Weale, K. E., *J. Chem. Soc.*, p. 3086, 1951.
9. Marshall, J. G., Staveley, L. A. K., and Hart, K. R., *Trans. Faraday Soc.*, 52, 23, 1956.
10. Staveley, L. A. K., Tupman, W. I., and Hart, K. R., *Trans. Faraday Soc.*, 51, 323, 1955.
11. McKinney, W. P., Skinner, G. F., and Staveley, L. A. K., *J. Chem. Soc.*, p. 2415, 1959.
12. Harrison, D. and Moelwyn-Hughes, E. A., *Proc. R. Soc. London Ser. A*, 239, 230, 1957.
13. Cutler, W. G., McMickle, R. H., Webb, W., and Schiessler, R. W., *J. Chem. Phys.*, 29, 727, 1958.
14. Moelwyn-Hughes, E. A., *J. Phys. Colloid Chem.*, 55, 1246, 1951.
15. Holder, G. A. and Walley, E., *Trans. Faraday Soc.*, 58, 2095, 1962.
16. Diaz Pena, M. and McGlashan, M. L., *Trans. Faraday Soc.*, 57, 1511, 1961.
17. Blinowska, A. and Brostow, W., *J. Chem. Thermodyn.*, 7, 787, 1975.

COEFFICIENT OF FRICTION

Compiled by Harold Minshall

The coefficient of friction between two surfaces is the ratio of the force required to move one over the other to the total force pressing the two together. If F is the force required to move one surface over another and W, the force pressing the surfaces together, the coefficient of friction,

$$\mu = \frac{F}{W}$$

Materials	Condition	Temperature °C	μ (Static)
A. STATIC FRICTION			
Non Metals			
Glass on glass	clean	—	0.9–1.0
,, ,, ,,	lubricated with paraffin oil	—	0.5–0.6
,, ,, ,,	,, ,, liquid fatty acids	—	0.3–0.6
,, ,, ,,	,, ,, solid hydrocarbons, alcohols or fatty acids	—	0.1
,, ,, metal	clean	—	0.5–0.7
,, ,, ,,	lubricated	—	0.2–0.3
Diamond on diamond	clean	—	0.1
,, ,, ,,	lubricated	—	0.05–0.1
,, ,, metal	clean	—	0.1–0.15
,, ,, ,,	lubricated	—	0.1
Sapphire on sapphire	clean or lubricated	—	0.2
,, ,, steel	,, ,, ,,	—	0.15
Hard carbon on carbon	clean	—	0.16
,, ,, ,, ,,	lubricated	—	0.12–0.14
Graphite on graphite	clean or lubricated	—	0.1
,, ,, ,,	outgassed	—	0.5–0.8
,, ,, steel	clean or lubricated	—	0.1
Mica on mica	freshly cleaved	—	1.0
,, ,, ,,	contaminated	—	0.2–0.4
Crystals of $NaNO_3$, KNO_3, NH_4Cl on self	clean	—	0.5
,, ,, ,, ,, ,, ,, ,,	lubricated with long chain polar compounds	—	0.12
Tungsten carbide on tungsten carbide	clean	room	0.17
Tungsten carbide on tungsten carbide	outgassed	room	0.58
,, ,, ,, ,, ,,	clean	820	0.35
,, ,, ,, ,, ,,	,,	970	0.40
,, ,, ,, ,, ,,	,,	1010	0.45
,, ,, ,, ,, ,,	,,	1160	0.5
,, ,, ,, ,, ,,	,,	1220	0.7
,, ,, ,, ,, ,,	,,	1440	1.2
,, ,, ,, ,, ,,	,,	1600	1.8
,, ,, ,, graphite	outgassed	room	0.62
,, ,, ,, ,,	clean	,,	0.15
,, ,, ,, ,,	,,	800	0.32
,, ,, ,, ,,	,,	910	0.30
,, ,, ,, ,,	,,	1000	0.25
,, ,, ,, ,,	,,	1120	0.29
,, ,, ,, ,,	,,	1220	0.26
,, ,, ,, ,,	,,	1300	0.25
,, ,, ,, ,,	,,	1410	0.25
,, ,, ,, ,,	,,	1800	0.24
,, ,, ,, ,,	,,	2030	0.25
,, ,, ,, steel	,,	—	0.4–0.6
,, ,, ,, ,,	lubricated	—	0.1–0.2
Polymethyl methacrylate on self	clean	—	0.8
,, ,, ,, steel	,,	—	0.4–0.5
Polystyrene on self	,,	—	0.5
,, ,, steel	,,	—	0.3–0.35
Polyethylene on self	,,	—	0.2
,, ,, steel	,,	—	0.2
Polytetrafluoroethylene on self	,,	—	0.04
,, ,, steel	,,	—	0.04
Nylon on nylon[1]	,,	—	0.15–0.25
Silk on silk	commercially clean	—	0.2–0.3
Cotton on cotton (thread)	,, ,,	—	0.3
,, ,, ,, (from cotton wool)	,, ,,	—	0.6
Rubber on solids	,, ,,	—	1–4
Wood on wood	,, ,, and dry	—	0.25–0.5
,, ,, ,,	,, ,, ,, wet	—	0.2
,, ,, metals	,, ,, ,, dry	—	0.2–0.6
,, ,, ,,	,, ,, ,, wet	—	0.2
,, ,, brick	,, ,,	—	0.6
,, ,, leather	,, ,,	—	0.3–0.4
Leather on metal	,, ,,	—	0.6
,, ,, ,,	,, ,, and wet	—	0.4
,, ,, ,,	greasy	—	0.2
Brake material on cast iron	commercially clean	—	0.4
,, ,, ,, ,, ,,	,, ,, and wet	—	0.2
,, ,, ,, ,, ,,	lubricated with mineral oil	—	0.1
Wool fiber on horn	clean (against scales)	—	0.8–1.0
,, ,, ,, ,,	,, (with scales)	—	0.4–0.6
,, ,, ,, ,,	greasy (against scales)	—	0.5–0.8
,, ,, ,, ,,	,, (with scales)	—	0.3–0.4
Metals			
Steel on steel	clean	20	0.58
,, ,, ,,	vegetable oil lubricant (a) castor oil	20	0.095
		100	0.105
,, ,, ,,	(b) rape	20	0.105
		100	0.105
,, ,, ,,	(c) olive	20	0.105
		100	0.105
,, ,, ,,	(d) coconut	20	0.08
		100	0.08

[1] Registered trade name.

Materials	Condition	Temperature °C	μ (Static)
A. STATIC FRICTION (Cont.)			
Metals (Cont.)			
Steel on steel	Animal oil lubricant		
" " "	(a) sperm	20	0.10
		100	0.10
" " "	(b) pale whale	20	0.095
		100	0.095
" " "	(c) neatsfoot	20	0.095
		100	0.095
" " "	(d) lard	20	0.085
		100	0.085
" " "	Mineral oil lubricant		
" " "	(a) light machine	20	0.16
		100	0.19
" " "	(b) thick gear	20	0.125
		100	0.15
" " "	(c) solvent refined	20	0.15
		100	0.20
" " "	(d) heavy motor	20	0.195
		100	0.205
" " "	(e) extreme pressure	20	0.09–0.1
		100	0.09–0.1
" " "	(f) graphited oil	20	0.13
		100	0.15
" " "	(g) B.P. Paraffin	20	0.18
		100	0.22
" " "	lubricated with trichloroethylene	20	0.33
" " "	" " benzene	20	0.48
" " "	" " glycerol	20	0.2
" " "	" " ethyl alcohol	20	0.43
" " "	" " butyl alcohol	room	0.3
" " "	" " octyl	"	0.23
" " "	" " decyl	"	0.16
" " "	" " cetyl	"	0.10
" " "	lubricated with nonane	room	0.26
" " "	" " decane	"	0.23
" " "	" " acetic acid	"	0.5
" " "	" " proprionic acid	"	0.4
" " "	" " valeric acid	"	0.17
" " "	" " caproic acid	"	0.12
" " "	" " pelargonic acid	"	0.11
" " "	" " capric acid	"	0.11
" " "	" " lauric acid	"	0.11
" " "	" " myristic acid	"	0.11
" " "	" " oleic acid	20–100	0.08
" " "	" " palmitic acid	room	0.11
" " "	" " stearic acid	"	0.10
" " hard steel	" " rape oil	—	0.14
" " " "	" " castor oil	—	0,12
" " " "	" " mineral oil	—	0.16
" " " "	" " long chain fatty acid	—	0.09
" " cast iron	" " rape oil	—	0.11
" " " "	" " castor oil	—	0.15
" " " "	" " mineral oil	—	0.21
" " " "	clean	—	0.4
" " gun metal	lubricated with rape oil	—	0.15
" " " "	" " castor oil	—	0.16
" " " "	" " mineral oil	—	0.21
" " bronze	" " rape oil	—	0.12
" " "	" " caster oil	—	0.12
" " "	" " mineral oil	—	0.16
" " lead	" "	—	0.5
" " "	" " long chain fatty acid	—	0.22
" " base white metal	" " mineral oil	—	0.1
" " " " "	" " long chain fatty acid	—	0.08
" " "	clean	—	0.55
" " tin	lubricated with mineral oil	—	0.6
" " "	" " long chain fatty acid	—	0.21
" " white metal, tin base	" " mineral oil	—	0.1
" " " " " "	" " long chain fatty acid	—	0.07
" " " " " "	clean	—	0.8
" " sintered bronze	lubricated with mineral oil	—	0.13
" " brass	" "	—	0.19
" " "	" " castor oil	—	0.11
" " "	" " long chain fatty acid	—	0.13
" " "	clean	—	0.35
" " copper–lead alloy	"	—	0.22
" " Wood's alloy	"	—	0.7
" " phosphor bronze	"	—	0.35
" " aluminum bronze	"	—	0.45
" " constantan	"	—	0.4
[2] " " indium film deposited on steel	4 kg load, clean	—	0.08
[2] " " " " " " "	8 kg " , "	—	0.04
[2] " " " " " " silver	4 kg " "	—	0.1
[2] " " " " " " "	8 kg " , "	—	0.07
[2] " " lead film deposited on copper	4 kg " , "	—	0.18
[2] " " " " " " "	8 kg " , "	—	0.12
[2] " " copper film deposited on steel	4 kg " , "	—	0.3
[2] " " " " " " "	8 kg " , "	—	0.2
[2]Al on Al	in air or O_2	—	1.9
" " "	" H_2O vapor	—	1.1
[3]Cu on Cu	" H_2 or N_2	—	4.0
" " "	" air or O_2	—	1.6

[2] Hemispherical steel slider having 0.6 cm. diameter. The thin, 10^{-3} to 10^{-4} cm., thin metallic films were deposited on various substrates as indicated. Amonton's Law is not obeyed in this case.

[3] The metals which were spectroscopically pure were outgassed in a vacuum prior to other gases being admitted. When clean and in vacuum there is gross seizure.

Materials	Condition	Temperature °C	μ (Static)
A. STATIC FRICTION (Cont.)			
Metals (Cont.)			
[3]Au on Au	in H_2 or N_2	—	4.0
,, ,, ,,	,, air or O_2	—	2.8
,, ,, ,,	,, H_2O vapor	—	2.5
[3]Fe on Fe	,, in air or O_2	—	1.2
,, ,, ,,	,, H_2O vapor	—	1.2
[3]Mo on Mo	,, air or O_2	—	0.8
,, ,, ,,	,, H_2O vapor	—	0.8
[3]Ni on Ni	,, H_2 or N_2	—	5.0
,, ,, ,,	,, air or O_2	—	3.0
,, ,, ,,	,, H_2O vapor	—	1.6
[3]Pt on Pt	,, air or O_2	—	3.0
,, ,, ,,	,, H_2O vapor	—	3.0
[3]Ag on Ag	,, air or O_2	—	1.5
,, ,, ,,	,, H_2O vapor	—	1.5
Various Materials on Snow and Ice			
Ice on ice	clean	0	0.05–0.15
,, ,, ,,	,,	−12	0.3
,, ,, ,,	,,	−71	0.5
,, ,, ,,	,,	−82	0.5
,, ,, ,,	,,	−110	0.5
Polymethylmethylacrylate	on wet snow	0	0.5
,,	,, dry ,,	0	0.3
,,	,, ,, ,,	−10	0.34
	,, ,, ,,	−32	0.4
Polyester of teraphthalic acid and ethylene glycol	,, wet ,,	0	0.5
Polyester of teraphthalic acid and ethylene glycol	,, dry ,,	0	0.35
Polyester of teraphthalic acid and ethylene glycol	,, ,, ,,	−10	0.38
[1]Nylon	on wet snow	0	0.4
,,	,, dry ,,	0	0.3
,,	,, ,, ,,	−10	0.3
Polytetrafluoroethylene	,, wet ,,	0	0.05
,,	,, dry ,,	0	0.02
,,	,, ,, ,,	−10	0.08
,,	,, ,, ,,	−32	0.1
Paraffin wax	,, wet ,,	0	0.06
,, ,,	,, dry ,,	0	0.06
,, ,,	,, ,, ,,	−10	0.35
,, ,,	,, ,, ,,	−32	0.4
Swiss wax	,, wet ,,	0	0.05
,, ,,	,, dry ,,	0	0.03
,, ,,	,, ,, ,,	−10	0.2
,, ,,	,, ,, ,,	−32	0.2
Ski wax	,, wet ,,	0	0.1
,, ,,	,, dry ,,	0	0.04
,, ,,	,, ,, ,,	−10	0.2
,, ,,	,, ,, ,,	−32	0.2
,, laquer	,, wet ,,	0	0.2
,, ,,	,, dry ,,	0	0.1
,, ,,	,, ,, ,,	−10	0.4
,, ,,	,, ,, ,,	−32	0.4
Aluminum	,, wet ,,	0	0.4
,,	,, dry ,,	0	0.35
,,	,, ,, ,,	−10	0.38

Materials	Condition	Temperature °C	μ (Kinetic)
B. KINETIC FRICTION Various Materials			
Unwaxed hickory	4 m/sec on dry snow	−3	0.08
Waxed ,,	0.1 m/sec ,, wet ,,	0	0.14
,, ,,	0.1 m/sec ,, dry ,,	0	0.04
,, ,,	0.1 m/sec ,, ,, ,,	−3	0.09
,, ,,	4 m/sec ,, ,, ,,	−3	0.03
Waxed hickory	0.1 m/sec on dry snow	−10	0.18
,, ,,	0.1 m/sec ,, ,, ,,	−40	0.4
Ice on ice	4m/sec, clean	0	0.02
,, ,, ,,	,, ,,	−10	0.035
,, ,, ,,	,, ,,	−20	0.050
,, ,, ,,	,, ,,	−40	0.075
,, ,, ,,	,, ,,	−60	0.085
,, ,, ,,	,, ,,	−80	0.09
Ebonite	4m/sec on ice	0	0.02
,,	,, ,, ,,	−10	0.05
,,	,, ,, ,,	−20	0.065
,,	,, ,, ,,	−40	0.085
,,	,, ,, ,,	−60	0.10
,,	,, ,, ,,	−80	0.11
Brass	4m/sec on ice	0	0.02
,,	,, ,, ,,	−10	0.075
,,	,, ,, ,,	−20	0.085
,,	,, ,, ,,	−40	0.115
,,	,, ,, ,,	−60	0.14
,,	,, ,, ,,	−80	0.15
Natural rubber, vulcanized	100m/min on ground glass, clean	—	1.07
,, ,, , ,,	100m/min ,, ,, ,, , wetted with water	—	0.94
,, ,, , ,,	100m/min on concrete, clean	—	1.02
,, ,, , ,,	100m/min ,, ,, , wetted with water	—	0.97
,, ,, , ,,	100m/min on bitumen, clean	—	1.07
,, ,, , ,,	100m/min ,, ,, , wetted with water	—	0.95
,, ,, , ,,	100m/min on rubber flooring or rubber tread vulcanisate, clean	—	1.16
,, ,, , ,,	100m/min on bitumen containing rubber powder, clean	—	1.15 (Varies with quantity of powder)
,, ,, , ,,	100m/min on bitumen containing rubber powder, wetted with water	—	1.03

[1]Registered trade name. [3] The metals which were spectroscopically pure were outgassed in a vacuum prior to other gases being admitted. When clean and in vacuum there is gross seizure.

LOW MELTING POINT ALLOYS

Melting point °C	Name	Composition, wt %				
−48	Binary Eutectic	Cs 77.0	K 23.0			
−40	Binary Eutectic	Cs 87.0	Rb 13.0			
−30	Binary Eutectic	Cs 95.0	Na 5.0			
−11	Binary Eutectic	K 78.0	Na 22.0			
−8	Binary Eutectic	Rb 92.0	Na 8.0			
10.7	Ternary Eutectic	Ga 62.5	In 21.5	Sn 16.0		
10.8	Ternary Eutectic	Ga 69.8	In 17.6	Sn 12.5		
17	Ternary Eutectic	Ga 82.0	Sn 12.0	Zn 6.0		
33	Binary Eutectic	Rb 68.0	K 32.0			
46.5	Quinternary Eutectic	Sn 10.65	Bi 40.63	Pb 22.11	In 18.1	Cd 8.2
47	Quinternary Eutectic	Bi 44.7	Pb 22.6	Sn 8.3	Cd 5.3	In 19.1
58.2	Quaternary Eutectic	Bi 49.5	Pb 17.6	Sn 11.6	In 21.3	
60.5	Ternary Eutectic	In 51.0	Bi 32.5	Sn 16.5		
70	Wood's Metal	Bi 50.0	Pb 25.0	Sn 12.5	Cd 12.5	
70	Lipowitz's Metal	Bi 50.0	Pb 26.7	Sn 13.3	Cd 10.0	
70	Binary Eutectic	In 67.0	Bi 33.0			
91.5	Ternary Eutectic	Bi 51.6	Pb 40.2	Cd 8.2		
95	Ternary Eutectic	Bi 52.5	Pb 32.0	Sn 15.5		
97	Newton's Metal	Bi 50.0	Sn 18.8	Pb 31.2		
98	D'Arcet's Metal	Bi 50.0	Sn 25.0	Pb 25.0		
100	Onion's or Lichtenberg's Metal	Bi 50.0	Sn 20.0	Pb 30.0		
102.5	Ternary Eutectic	Bi 54.0	Sn 26.0	Cd 20.0		
109	Rose's Metal	Bi 50.0	Pb 28.0	Sn 22.0		
117	Binary Eutectic	In 52.0	Sn 48.0			
120	Binary Eutectic	In 75.0	Cd 25.0			
123	Malotte's Metal	Bi 46.1	Sn 34.2	Pb 19.7		
124	Binary Eutectic	Bi 55.5	Pb 44.5			
130	Ternary Eutectic	Bi 56.0	Sn 40.0	Zn 4.0		
140	Binary Eutectic	Bi 58.0	Sn 42.0			
140	Binary Eutectic	Bi 60.0	Cd 40.0			
183	Eutectic solder	Sn 63.0	Pb 37.0			
185	Binary Eutectic	Tl 52.0	Bi 48.0			
192	Soft solder	Sn 70.0	Pb 30.0			
198	Binary Eutectic	Sn 91.0	Zn 9.0			
199	Tin foil	Sn 92.0	Zn 8.0			
199	White Metal	Sn 92.0	Sb 8.0			
221	Binary Eutectic	Sn 96.5	Ag 3.5			
226	Matrix	Bi 48.0	Pb 28.5	Sn 14.5	Sb 9.0	
227	Binary Eutectic	Sn 99.25	Cu 0.75			
240	Antimonial Tin solder	Sn 95.0	Sb 5.0			
245	Tin-silver solder	Sn 95.0	Ag 5.0			

MOHS HARDNESS SCALE

Hardness number	Original scale	Modified scale
1	Talc	Talc
2	Gypsum	Gypsum
3	Calcite	Calcite
4	Fluorite	Fluorite
5	Apatite	Apatite
6	Orthoclase	Orthoclase
7	Quartz	Vitreous silica
8	Topaz	Quartz or Stellite
9	Corundum	Topaz
10	Diamond	Garnet
11	....	Fused Zirconia
12	....	Fused Alumina
13	....	Silicon Carbide
14	....	Boron Carbide
15	....	Diamond

HARDNESS OF MATERIALS

Agate	6–7	Indium	1.2
Alabaster	1.7	Iridium	6–6.5
Alum	2–2.5	Iridosmium	7
Aluminum	2–2.9	Iron	4–5
Alundum	9+	Kaolinite	2.0–2.5
Amber	2–2.5	Lead	1.5
Andalusite	7.5	Lithium	0.6
Anthracite	2.2	Loess (0°)	0.3
Antimony	3.0–3.3	Magnesium	2.0
Apatite	5	Magnetite	6
Aragonite	3.5	Manganese	5.0
Arsenic	3.5	Marble	3–4
Asbestos	5	Meerschaum	2–3
Asphalt	1–2	Mica	2.8
Augite	6	Opal	4–6
Barite	3.3	Orthoclase	6
Bell-metal	4	Osmium	7.0
Beryl	7.8	Palladium	4.8
Bismuth	2.5	Phosphorus	0.5
Boric acid	3	Phosphorbronze	4
Boron	9.5	Platinum	4.3
Brass	3–4	Plat-iridium	6.5
Cadmium	2.0	Potassium	0.5
Calamine	5	Pumice	6
Calcite	3	Pyrite	6.3
Calcium	1.5	Quartz	7
Carbon	10.0	Rock salt (halite)	2
Carborundum	9–10	Ross' metal	2.5–3.0
Cesium	0.2	Rubidium	0.3
Chromium	9.0	Ruthenium	6.5
Copper	2.5–3	Selenium	2.0
Corundum	9	Serpentine	3–4
Diamond	10	Silicon	7.0
Diatomaceous earth	1–1.5	Silver	2.5–4
Dolomite	3.5–4	Silver chloride	1.3
Emery	7–9	Sodium	0.4
Feldspar	6	Steel	5–8.5
Flint	7	Stibnite	2
Fluorite	4	Strontium	1.8
Galena	2.5	Sulfur	1.5–2.5
Gallium	1.5	Talc	1
Garnet	6.5–7	Tellurium	2.3
Glass	4.5–6.5	Tin	1.5–1.8
Gold	2.5–3	Topaz	8
Graphite	0.5–1	Tourmaline	7.3
Gypsum	1.6–2	Wax (0°)	0.2
Hematite	6	Wood's metal	3
Hornblende	5.5	Zinc	2.5

COMPARISON OF HARDNESS VALUES OF VARIOUS MATERIALS ON MOHS AND KNOOP SCALES*

Compiled by Laurence S. Foster

Substance	Formula	Mohs value	Knoop value
Talc	$3MgO·4SiO_2·H_2O$	1	
Gypsum	$CaSO_4·2H_2O$	2	32
Cadmium	Cd	..	37
Silver	Ag	..	60
Zinc	Zn	..	119
Calcite	$CaCO_3$	3	135
Fluorite	CaF_2	4	163
Copper	Cu	..	163
Magnesia	MgO	..	370
Apatite	$CaF_2·3Ca_3(PO_4)_2$	5	430
Nickel	Ni	..	557
Glass (soda lime)	....	..	530
Feldspar (orthoclase)	$K_2O·Al_2O_3·6SiO_2$	6	560
Quartz	SiO_2	7	820
Chromium	Cr	..	935
Zirconia	ZrO_2	..	1160
Beryllia	BeO	..	1250
Topaz	$(AlF)_2SiO_4$	8	1340
Garnet	$Al_2O_3·3FeO·3SiO_2$	..	1360
Tungsten carbide alloy	WC, Co	..	1400–1800
Zirconium boride	ZrB_2	..	1550
Titanium nitride	TiN	9	1800
Tungsten carbide	WC	..	1880
Tantalum carbide	TaC	..	2000
Zirconium carbide	ZrC	..	2100
Alumina	Al_2O_3	..	2100
Beryllium carbide	Be_2C	..	2410
Titanium carbide	TiC	..	2470
Silicon carbide	SiC	..	2480
Aluminum boride	AlB	..	2500
Boron carbide	B_4C	..	2750
Diamond	C	10	7000

* Acknowledgment is made to N. W. Thibault, Norton Company, Worcester, Massachusetts, for many of Knoop hardness values. *Cf.* R. F. Geller, "A Study of Ceramics for Nuclear Reactors," Nucleonics, Vol. 7, No. 4, Table 1, pp. 8–9 (Oct. 1950). V. E. Lysaght, *Indentation Hardness Testing*, Reinhold 1949.

SURFACE TENSION OF LIQUID ELEMENTS

Gernot Lang

The following data were collected from many sources. As a result their accuracy varies. Users of data in this table are advised that:

1. As a rule, results from the "sessile drop" and "maximum bubble pressure" as well as from the "pendant drop" methods are preferable to results obtained from other methods for metals with very high melting points.
2. Values of single measurements are usually not as well supported by experiments as those of serial measurements at various temperatures.
3. Values in parentheses can be considered improbable.

Element	Purity (wt.%)	σ_{mp} (dyn/cm)	Atm.	σ_{t1} t_1 °C	σ_{t1} σ (dyn/cm)	σ_{t2} t_2 °C	σ_{t2} σ (dyn/cm)	σ_{t3} t_3 °C	σ_{t3} σ (dyn/cm)	Method	Ref.
Ag	99.99		H_2	1000	916					Bubble pressure	134
	–	(785)	vac.							Pendant drop	182
	99.96		H_2	1000	893	1150	862	1250	849	Bubble pressure	127
	–		vac.	1000	908					Sessile drop	51
	99.995		H_2	1000	907	1100	894	1200	876	Bubble pressure	128
	99.999	(828)	vac.							Pendant drop	65
	99.99		Ar, H_2	1000	890					Bubble pressure	169
	spect. pure	921			$\sigma = 1136 - 0.174\ T$ (°K) (1300–2200°K)					Bubble pressure	26
		918	–		$\sigma = 918 - 0.149\ (t - t_{mp})$ (°C)					Bubble pressure	204
	99.999		Ar	980	905±10	1108	890±10			Sessile drop	Z3
Al	99.99	860±20	Ar							Bubble pressure	102
	99.72		vac.	950	840					Bubble pressure	46
	99.7	863±25	Ar		$\sigma_t = (863 \pm 25) - 0.33\ (t - t_{mp})$ (°C)					Bubble pressure	113
	99.99	865	vac.		$\sigma_t = 865 - 0.14\ (t - t_{mp})$ (°C)					Sessile drop	137, 56
	99.99	(825)	Ar		$(\sigma_t = 825 - 0.05\ (T - 993))$ (T°K)					Bubble pressure	38
	99.99	866	He		$\sigma_t = 866 - 0.15\ (t - t_{mp})$ (°C)					Sessile drop	10
	99.999	873	He	1600	725 $\sigma_t = 873 - 0.15\ (t - t_{mp})$ (°C)					Sessile drop	199
	99.99	(760)	vac.		$(\sigma_t = 948 - 0.202T)$ (T°K.980–1090°K)					Sessile drop	Z4
	99.998	(915)	Ar		$(\sigma_t = 915 - 0.51\ (t - t_{mp}))$ (°C) (660–800°C)					Sessile drop	Z6
	99.996	855±6	Ar		$\sigma_t = 855 - 0.104\ (t - t_{mp})$ (°C) (660–911°C)					Bubble pressure	Z13

SURFACE TENSION OF LIQUID ELEMENTS (*Continued*)

Element	Purity (wt.-%)	σ_{mp} (dyn/cm)	Atm.	σ_{t1} t_1 °C	σ_{t1} σ (dyn/cm)	σ_{t2} t_2 °C	σ_{t2} σ (dyn/cm)	σ_{t3} t_3 °C	σ_{t3} σ (dyn/cm)	Method	Ref.
Au		(754)	vac.							Pendant drop	182
	99.999	1130	He	1200	1070	1300	1020			Sessile drop	86
	99.999	(731)	vac.							Pendant drop	65
	99.999		Ar	1108	1130±10					Sessile drop	Z3
B	99.8	1060±50	vac.							Sessile & pendant drop	187
Ba	–		Ar	720	224					Bubble pressure	3
	99.5	276			σ = 351–0.075 T (°K) (1410–1880°K)					Bubble pressure	25
Be	99.98		vac.	1500	1100					Sessile drop	58
Bi		376	vac.							Drop pressure	155
	99.99	376	vac.							Drop pressure	156
	99.90		H_2	800	343	1000	328			Bubble pressure	134
			vac.	450	(382)					Electro-capillarity	118
	99.9	380±10	Ar							Bubble pressure	103
			–	350	362					Drop pressure	76
			vac.	700	350					Sessile drop	29
	99.98	380±10	Ar							Bubble pressure	105
			–	450	380					Drop pressure	120
			vac.	300	379					Sessile drop	64
		378	vac., Ar, H_2							Drop weight	4
	99.99995	375			σ = 423–0.088 T (T°K) (1352–1555°K)					Bubble pressure	26
	99.999	380±3	Ar		$\sigma_t = 380-0.142\,(t-t_{mp})$ (MP–555°C) (t°C)					Bubble pressure	Z13a
Ca	–		Ar	850	337					Bubble pressure	3
	p.a.	360			σ = 472–0.100 T (T°K) (1445–1655°K)					Bubble pressure	25
Cd			–	450	600					Drop pressure	119
			–	400	600					Bubble pressure	13
			–	350	586					Drop pressure	69
	99.9	(550±10)	Ar							Bubble pressure	103

SURFACE TENSION OF LIQUID ELEMENTS (*Continued*)

Element	Purity (wt.-%)	σ_{mp} (dyn/cm)	Atm.	σ_{t1} t_1 °C	σ (dyn/cm)	σ_{t2} t_2 °C	σ (dyn/cm)	σ_{t3} t_3 °C	σ (dyn/cm)	Method	Ref.
				390	604					Bubble pressure	1
		(525±30)	H_2							Solid state curvature	Z8
	99.9999	590±5	–				(non linear)			Sessile drop	Z12
Co			Ar	1550	1836					Sessile drop	108
	99.99		vac., Al_2O_3	1520	1800					Sessile drop	46
	99.99		He, Al_2O_3	1520	(1630)					Sessile drop	46
	99.99		He, BeO	1520	(1640)					Sessile drop	46
	99.99		He, MgO	1520	(1560)					Sessile drop	46
	99.99		H_2, Al_2O_3	1520	1780					Sessile drop	46
	99.99		He	1520	(1620)					Bubble pressure	46
	99.99		H_2	1520	(1590)					Bubble pressure	46
			vac.	1500	1870					Sessile drop	7
				1600	(1640)					Sessile drop	135
				1600	(1600)					Sessile drop	135
			vac.	1600	1815					Sessile drop	53
	99.99		vac., Al_2O_3	1600	1812					Sessile drop	54
		(1520)	H_2, He							Bubble pressure	61
	99.99		H_2, He	1550	1845					Bubble pressure	63
	99.9983	1880	vac.							Pendant drop	5
	99.99			1550	1780					Bubble pressure	59
Cr	–		vac.	1950	1590±50					Sessile drop	47
	99.9997	1700±50	Ar							Dynam. drop weight	6
Cs			Ar	62	68.4					Pendant drop	195
			Ar	62	67.5	146	62.9			Bubble pressure	193
	99.95		Ar	39	69.5	494	42.8	642	34.6	Bubble pressure	24
	99.99	73.74	Ar		$\sigma = 73.74 - 1.791 \cdot 10^{-2} (t-t_{mp}) - 9.610 \cdot 10^{-5} (t-t_{mp})^2 + 6.629 \cdot 10^{-8} (t-t_{mp})^3$ (t°C) (71–1011°C)					Bubble pressure	26
	99.995	68.6	He		$\sigma = 68.6 - 0.047 (t-t_{mp})$ (t°C) (52–1100°C)					Bubble pressure	121
Cu			Ar	1120	1269±20					Sessile drop	11
	–	(1150)	vac.							Pendant drop	35
			Ar	1120	1285±10					Sessile drop	108

SURFACE TENSION OF LIQUID ELEMENTS (*Continued*)

Element	Purity (wt.–%)	σ_{mp} (dyn/cm)	Atm.	σ_{t1} t_1 °C	σ_{t1} σ (dyn/cm)	σ_{t2} t_2 °C	σ_{t2} σ (dyn/cm)	σ_{t3} t_3 °C	σ_{t3} σ (dyn/cm)	Method	Ref.
	99.98		H_2	1100	1301	1165	1295	1255	1287	Bubble pressure	134
		1270	vac.							Sessile drop	7
			vac.	1120	1285					Sessile drop	198
				1440	1298					Bubble pressure	129
		(1085)	vac.							Pendant drop	182
	99.99		He	1250	1290					Bubble pressure	62
	99.99		H_2	1250	1300					Bubble pressure	62
	99.9	(1180±40)	Ar							Bubble pressure	103
	—			1100	1220					Sessile drop	136
	—			1183	(1130)					Sessile drop	200
	—		vac.	1150	1370					Sessile drop	52
	99.997	1355	He, H_2							Sessile drop + Bubble pressure	138
	99.997	1352	vac.	$\sigma_t = 1352 - 0.17\ (t - t_{mp})$ (t°C)						Sessile drop	137
	99.997	1358	Ar	$\sigma_t = 1358 - 0.20\ (t - t_{mp})$ (t°C)						Bubble pressure	137
	—		Ar, He	1120	1285±10					Sessile drop	110
	99.99		H_2, He	1550	1265					Bubble pressure	63
	99.99999	1300	vac.							Pendant drop	5
			vac.	1130	1268±60					Sessile drop	15
	99.98		Ar	1600	1230					Bubble pressure	205
	99.999		N_2	1100	1341	1150	1338	1200	1335	Bubble pressure	128
	99.9	(1127)	vac.							Pendant drop	65
	99.99		Ar, H_2	1100	1320					Bubble pressure	169
Fe				1570	(1731)					Sessile drop	95
	—			1550	1860					Sessile drop	74
		1720	He							Sessile drop	74
				1580–1760	(880)					Bubble pressure	100
				1570	(1632)					Sessile drop	93
	99.99		He	1650	(1610)					Sessile drop	46
	99.99		He	1650	(1430)					Sessile drop	46
	99.99		H_2	1650	(1400)					Sessile drop	46
	—	(1384)	vac.							Pendant drop	182
		(1700)	vac.							Sessile drop	7
			vac., He	1550	1865					Sessile drop	209

SURFACE TENSION OF LIQUID ELEMENTS (*Continued*)

Element	Purity (wt.–%)	σ_{mp} (dyn/cm)	Atm.	σ_{t1} t_1 °C	σ_{t1} σ (dyn/cm)	σ_{t2} t_2 °C	σ_{t2} σ (dyn/cm)	σ_{t3} t_3 °C	σ_{t3} σ (dyn/cm)	Method	Ref.
	99.99		He	1650	(1430)					Bubble pressure	62
	99.99		H_2	1650	(1400)					Bubble pressure	62
				1650	(1640)					Sessile drop	136
		(1650)	He, H_2							Bubble pressure	61
	99.985		Ar	1550	1788					Sessile drop	110
	–	(1560)								Sessile drop	68
	99.94		vac., Al_2O_3	1560	(1710)					Sessile drop	207
	99.9998	1880	vac.							Pendant drop	5
	99.93	1860±40	He							Sessile drop	116
		(1510)	vac.							Drop weight	139
	99.97		vac., BeO	1550	1830±6					Sessile drop	45
	Armco		Ar, N_2	1550	1795					Bubble pressure	164
			vac.	1550	1754					Sessile drop	42
	99.987		vac.	1550	(1730)					Sessile drop	159
	99.85	(1619)	vac.							Pendant drop	65
	99.69		He, Al_2O_3	1550	(1727)					Sessile drop	158
	99.69		H_2, Al_2O_3	1550	(1734)					Sessile drop	158
	–	1760±20	He, H_2		$\sigma = 1760-0.35\ (t-t_{mp})$ (t°C)					Sessile drop	157
	99.9992	1773	He, H_2		$\sigma = 773+0.65\ t$ (t°C) (1550–1780°C)					Oscillating drop	Z2
	–		–	1550	1780					Sessile drop	Z11
Fr	–		–	100	58.4					calculated	145
Ga		704	Ar, He							Bubble pressure	197
	–	725±10	Ar							Bubble pressure	104
			vac.	350	718					Sessile drop	64
	–		He, Al_2O_3	1500	559					Sessile drop	57
	99.9998	718	vac., Al_2O_3		$\sigma = 718-0.101\ (t-t_{mp})$ (t°C)					Sessile drop	85
Ge		621.4								Drop weight	126
		650	vac.	1200	530					Sessile drop	99
			vac.	1000	650					Sessile drop	Z5
		632±5	N_2, He							Solid state curvature	Z8

SURFACE TENSION OF LIQUID ELEMENTS (*Continued*)

Element	Purity (wt.–%)	σ_{mp} (dyn/cm)	Atm.	σ_{t1} t_1 °C	σ_{t1} σ (dyn/cm)	σ_{t2} t_2 °C	σ_{t2} σ (dyn/cm)	σ_{t3} t_3 °C	σ_{t3} σ (dyn/cm)	Method	Ref.
Hf		(1460)	vac.							Pendant drop	151
	97.5+2.5 Zr	1630	vac.							Pendant drop	5
Hg			H_2	20	(542)					Pendant drop	165
			air	16	(410)					Sessile drop	181
			air	20	(435.5)					Oscillating jet	175
			vac.	20	472					Oscillating jet	73
			vac.	20	(402)					Drop pressure	146
				25	476					Drop weight	75
			vac.	20	(436)					Sessile drop	160
			vac.	20	(432)					Sessile drop	170
			H_2	25	476					Drop pressure	78
				25	472					Drop weight	81
				25	(464)					Sessile drop	82
			H_2	19	473					Bubble pressure	173
				20	(437)					Capillary depression	144
			vac.	20	480					Drop weight	21
				25	(516)					Sessile drop	37
				25	(435)					Sessile drop	91
			vac.	25	473					Drop weight	31
				25	488					Sessile drop	32
				25	(498)					Sessile drop	30
			vac.	20	(420)					Sessile drop	174
				25	476					Sessile drop	90
			vac.	20	(410)					Drop pressure	178
			vac.	20	(455)					Drop pressure	39
				25	484±1.5					Sessile drop	89
			vac.	22	(468)					Drop pressure	162
			vac.	20	(465.2)	103	449.7	350	387.1	Drop pressure	162
				25	484.9±1.8					Sessile drop	Z7
			vac.	20	485.5±1.0	$\sigma_t = 489.5–0.20\ (t°C)$				Drop pressure	17
			Ar	20	(454.7)					Bubble pressure	188
				21	(350.5)					Bubble pressure	23

SURFACE TENSION OF LIQUID ELEMENTS (*Continued*)

Element	Purity (wt.-%)	σ_{mp} (dyn/cm)	Atm.	σ_{t1}: t_1 °C	σ_{t1}: σ (dyn/cm)	σ_{t2}: t_2 °C	σ_{t2}: σ (dyn/cm)	σ_{t3}: t_3 °C	σ_{t3}: σ (dyn/cm)	Method	Ref.
	99.99		He, H_2	20	475					Bubble pressure	62
	99.9		Ar	20	(500±15)					Bubble pressure	103
			vac.	–10	487					Drop pressure	69
			vac.	25	483.5±1.0					Sessile drop	140
				22	(465)					Bubble pressure	194
				25	485.1					Sessile drop	142
				16.5	487.3	25	485.4±1.2			Sessile drop	171
				23–25	482.8±9.7					Contact angle	18
					$\sigma_t = 468.7 - 1.61 \cdot 10^{-1} t - 1.815 \cdot 10^{-4} t^2$ (t°C)					–	206
			vac.	20	484.6±1.3					Pendant drop	132
			Ar	25	480					Bubble pressure	172
			vac.	20	482.5±3.0						
					$\sigma_t = 485.5 - 0.149t - 2.84 \cdot 10^{-4} t^2$ (t°C)					Bubble pressure	177
			Ar	21.5	484.9±0.3					Bubble pressure	Z13a
In	99.95	559	H_2	600	515					Capillary method	133
				623	540					Capillary method	77
	99.995	556.0	Ar, He							Bubble pressure	196
			vac.	185	592					Drop pressure	69
			H_2	600	514					–	16
				300	541					Sessile drop	83
	99.999		Ar	200	556	400	535	550	527.8	Bubble pressure	123
	99.9994		vac.	350	539					Drop pressure	106
	99.9999	560±5	–		$\sigma = 568.0 - 0.04t - 7.08 \cdot 10^{-5} t^2$ (t°C)					Sessile drop	Z12
Ir	99.9980	2250	vac.							Pendant drop	5
K	99.895	101	Ar							Bubble pressure	189
		110.3±1	–							–	84
		117	vac.		$\sigma = 117 - 0.66\,(t - t_{mp})$					Drop weight	185
	–		Ar	87	112	457	80	677	64.8	Bubble pressure	24
	99.986	116.95	Ar		$\sigma = 116.95 - 6.742 \cdot 10^{-2} (t - t_{mp}) - 3.836 \cdot 10^{-5} (t - t_{mp})^2 + 3.707 \cdot 10^{-8} (t - t_{mp})^3$ (t°C) (77–983°C)					Bubble pressure	172

SURFACE TENSION OF LIQUID ELEMENTS (*Continued*)

Element	Purity (wt.-%)	σ_{mp} (dyn/cm)	Atm.	σ_{t1}: t_1 °C	σ_{t1}: σ (dyn/cm)	σ_{t2}: t_2 °C	σ_{t2}: σ (dyn/cm)	σ_{t3}: t_3 °C	σ_{t3}: σ (dyn/cm)	Method	Ref.
	99.936	(79.2)	He		$(\sigma = 76.8 - 70.3 \cdot 10^{-4}(t-400))$ (t°C) (600–1126°C)					Bubble pressure	121
		95±9.5	–							Drop weight	Z9
	99.97	111.35 ±0.64	He		$\sigma = 115.51 - 0.0653 \cdot$ (t°C) (70–713°C)					Bubble pressure	Z10
Li	99.95		Ar	180	397.5	300	380	500	351.5	Bubble pressure	189
	99.98		Ar	287	386	922	275	1077	253	Bubble pressure	24
Mg			Ar	681	563	789	532	894	502	Bubble pressure	208
	99.8		N_2	670	552	700	542	740	528	Bubble pressure	149
	99.9		Ar	700	550±15					Bubble pressure	114
	99.91	(525±10)	Ar							Bubble pressure	105
	99.5	583			$\sigma = 721 - 0.149$ T (T°K) (1125–1326°K)					Bubble pressure	25
Mn	99.9985	1100 ± 50	Ar							Dynam. drop weight	6
	99.94		vac.	1550	1030					Sessile drop	159
	–		–	1550	1010					Sessile drop	Z11
Mo	99.7	(1915)	vac.							Pendant drop	148
		2080	vac.							Drop weight	139
	99.9996	2250	vac.							Pendant drop	5
	99.98	2049	vac.							Pendant drop	65
	–	2130	vac.							Pendant drop	Z1
Na			Ar	110	205.7	263	198.2			Bubble pressure	208
	99.995	191	Ar							Bubble pressure	188
			vac.	123	198	129	198.5			Drop volume	2
				140	190					Sessile drop	28
		200.2±0.6								–	84
		202	vac.		$\sigma = 202 - 0.092\,(t - t_{mp})$; 100–1000C					Drop weight	185
	p.a.		Ar	617	144	764	130	855	120.4	Bubble pressure	24
	99.96	210.12	Ar		$\sigma = 210.12 - 8.105 \cdot 10^{-2}\,(t-t_{mp}) - 8.064 \cdot 10^{-5}\,(t-t_{mp})^2 + 3.380 \cdot 10^{-8}\,(t-t_{mp})^3$ (t°C) (141–992°C)					Bubble pressure	172
	99.982	187.4	He		$\sigma = 144 - 0.108\,(t-500)$ (t°C) (400–1125°C)					Bubble pressure	121
Nb, Cb	99.9986	1900	vac.							Pendant drop	5
	99.99	(1827)	vac.							Pendant drop	65
	–	2020	vac.							Pendant drop	Z1

SURFACE TENSION OF LIQUID ELEMENTS (*Continued*)

Element	Purity (wt.-%)	σ_{mp} (dyn/cm)	Atm.	σ_{t1}		σ_{t2}		σ_{t3}		Method	Ref.
				t_1 °C	σ (dyn/cm)	t_2 °C	σ (dyn/cm)	t_3 °C	σ (dyn/cm)		
Nd		688	Ar	1186	674					Bubble pressure	124
Ni	99.7		He	1470	(1615)					Sessile drop	94
	99.7		H_2	1470	(1570)					Sessile drop	94
	99.7		vac.	1470	1735					Sessile drop	94
		1725	vac.							Sessile drop	141
			vac.	1475	1725					Sessile drop	115
	–		Ar	1550	(1934)					Sessile drop	108
	99.99		vac., Al_2O_3	1520	1740					Sessile drop	46
	99.99		He, Ar, Al_2O_3	1520	1770					Sessile drop	46
	99.99		H_2, Al_2O_3	1520	(1600)					Sessile drop	46
	99.99		He, MgO	1470	(1530)					Sessile drop	46
	99.99		He, BeO	1470	(1500)					Sessile drop	46
	99.99		H_2	1530	(1650)					Bubble pressure	46
	99.99		H_2	1470	(1530)					Bubble pressure	46
	99.99		He	1470	(1490)					Bubble pressure	46
		1725	vac.							Sessile drop	7
	99.99			1600	(1600)					Sessile drop	135
			vac.	1500	1720					Sessile drop	50
	99.99		vac., Al_2O_3	1550	1780					Sessile drop	54
				1550	1735					Bubble pressure	60
			H_2, He	1470	1700					Bubble pressure	61
	99.99		vac.	1640	1705					Sessile drop	56
	–		vac., Al_2O_3	1560	1810					Sessile drop	207
	99.9991	1770±13	vac.							Drop weight	5
	99.9991	1728±10	vac.							Drop weight	5
	99.9991	1822±8	vac.							Pendant drop	5
		(1670)	vac.							Drop weight	139
		1760	vac.							Sessile drop	55
		(1687)	vac.							Pendant drop	65
	–		He	1500	1745					Sessile drop	57
	–	1809±20	H_2, He Al_2O_3		σ = 1770–0.39 (t–1550) (t°C)					Sessile drop	157
	99.99975	(1977)	He		σ = 1665+0.215t (t°C) (1475–1650°C)					Oscillating drop	Z2

SURFACE TENSION OF LIQUID ELEMENTS (*Continued*)

Element	Purity (wt.-%)	σ_{mp} (dyn/cm)	Atm.	σ_{t1} t_1 °C	σ_{t1} σ (dyn/cm)	σ_{t2} t_2 °C	σ_{t2} σ (dyn/cm)	σ_{t3} t_3 °C	σ_{t3} σ (dyn/cm)	Method	Ref.
Os	99.9998	2500	vac.							Pendant drop	5
P(white)				50	69.7	68.7	64.95			Bubble pressure	80
Pb	99.98		H_2, N_2	340	448	390	442	440	439	Bubble pressure	71
			air	360	452					Ring removal	97
		451	vac.	425	440					Drop pressure	101
		450	He							Pendant drop	87
				350–450	450					Ring removal	13
	99.998	480	H_2							Capillary method	133
				623	474					Capillary method	77
			vac.	362	455					Drop pressure	69
				700	428					Sessile drop	27
	99.9		H_2	1000	388					Bubble pressure	134
	99.9	(410±5)	Ar							Bubble pressure	104
				350	445					Drop pressure	153
	99.98		vac.	340	442	400	435			Drop pressure	70
	99.9995	470			$\sigma = 538-0.114\ T$ (T°K) (1440–1970°K)					Bubble pressure	26
	99.9994		vac.	450	438					Drop pressure	107
	99.999		He	1600	310					Sessile drop	199
				390	456					Bubble pressure	1
		424±10	Air							Solid state curvature	Z8
	99.999	470	Ar		$\sigma_t = 470-0.164\ (t-t_{mp})$ (MP–535°C) (t°C)					Bubble pressure	Z13a
Pd		1470	vac.							Sessile drop	50
	99.998	1500	vac.							Pendant drop	5
	99.998	1460	He							Sessile drop	199
Pt		1869	CO_2							Drop weight	167
	99.84	(1740±20)	vac.							Drop volume	48
	99.999		Ar	1800	(1699±20)					Sessile drop	109
	99.9980	1865	vac.							Pendant drop	5
Pu		550±55								–	186
Rb		(77±5)	vac.							Drop diffusion in quartz tube	201

SURFACE TENSION OF LIQUID ELEMENTS (*Continued*)

Element	Purity (wt.-%)	σ_{mp} (dyn/cm)	Atm.	σ_{t1} t_1 °C	σ_{t1} σ (dyn/cm)	σ_{t2} t_2 °C	σ_{t2} σ (dyn/cm)	σ_{t3} t_3 °C	σ_{t3} σ (dyn/cm)	Method	Ref.
	99.8		Ar	52	84	477	55	632	46.8	Bubble pressure	24
	99.92	91.17	Ar		$\sigma = 91.17 - 9.189 \cdot 10^{-2}\,(t-t_{mp}) + 7.228 \cdot 10^{-5}\,(t-t_{mp})^2 - 3.830 \cdot 10^{-8}\,(t-t_{mp})^3$ (t°C) (104–1006°C)					Bubble pressure	172
	99.997	85.7	He		$\sigma = 85.7 - 0.054\,(t-t_{mp})$ (t°C) (53–1115°C)					Bubble pressure	121
Re	99.4	2610	vac.							Pendant drop	148
	99.9999	2700	vac.							Pendant drop	5
Ru	99.9980	2250	vac.							Pendant drop	5
Rh		1940	vac.							Sessile drop	50
	99.9975	2000	vac.							Pendant drop	5
S	–	60.9	vac.	250	51.1					Pendant drop	143
Sb			H_2	640	349	700	349	974	342	Drop weight	20
			H_2	750	368	900	361	1100	348	Bubble pressure	40
			vac.	640	367.9	762	364.9			Drop weight	131
	99.5	383	H_2, N_2	675	384	800	380			Bubble pressure	71
	99.99	395±20	Ar							Bubble pressure	104
	99.15	395±20	Ar							Bubble pressure	105
	99.999		N_2	800	359	1000	351	1100	345	Bubble pressure	128
	99.995		Ar	650	350.2	700	347.6	800	345.0	Bubble pressure	123
	99.999		He	1600	320					Sessile drop	199
Se	–		Ar	230–250	88.0±5					Bubble pressure	105
Si			He	1450	725					Pendant drop	87
			vac.	1550	720					Sessile drop	68
	99.99		vac.	1550	750					Sessile drop	41
	99.9999		Ar	1500	825					Pendant drop	43
Sn	–		N_2	275	612	500	572	800	520	Bubble pressure	149
			air	280	523	340	520			Ring removal	97

SURFACE TENSION OF LIQUID ELEMENTS (*Continued*)

Element	Purity (wt.-%)	σ_{mp} (dyn/cm)	Atm.	σ_{t1}: t_1 °C	σ_{t1}: σ (dyn/cm)	σ_{t2}: t_2 °C	σ_{t2}: σ (dyn/cm)	σ_{t3}: t_3 °C	σ_{t3}: σ (dyn/cm)	Method	Ref.
	99.99	537	vac.	500	524	600	508			Drop pressure	154
		530	He							Pendant drop	87
			H_2	489	543	572	528	692	503	Conical capillaries	9
			–	250	536					Drop pressure	179
			–	450	530					Drop pressure	119
			–	250	545					Drop pressure	112
	99.93		vac.	250	549	400	539	600	526	Drop pressure	156
	99.998	566	H_2							Capillary method	133
			–	623	559					Capillary method	77
		610	vac.							Pendant drop	182
			–	800	500					Sessile drop	7
			–	300	538					Drop pressure	190
			–	300	(527)					Drop pressure	76
			–	290	546					Sessile drop	190
	99.99		H_2, He	600	530					Bubble pressure	62
	99.9	(526±10)	Ar							Bubble pressure	104
			vac.	290	600					Sessile drop	147
	99.965		H_2	740	508	950	489.5	1115	479.5	Bubble pressure	127
	99.89	543.7								Bubble pressure	33
		562	vac.							Sessile drop	55
			vac.	300	554					Sessile drop	64
	99.999	590	vac.							Sessile drop	86
			H_2	290	(520)	290	(524)	(vac.)		Sessile drop	67
	99.9999		H_2	246	552.7					Sessile drop	203
	99.9994		vac.	350	537					Drop pressure	106
	99.999	555.8±1.9		$\sigma = 566.84 - 4.76 \cdot 10^{-2}\, t$ (t°C)						Bubble pressure	176
	99.96	552	vac.	1000	470					Sessile drop	Z5
	99.96	552	Ar	$\sigma_t = 552 - 0.167\,(t - t_{mp})$ (MP–500°C) (t°C)						Bubble pressure	Z13a
Sr			Ar	775	288	830	282	893	282	Bubble pressure	125
	99.5	303		$\sigma = 392 - 0.085T$ (T°K) (1152–1602°K)						Bubble pressure	25
Ta		2360	vac.							Pendant drop	88
		2030	vac.							Pendant drop	88
		1910	vac.							Drop weight	139
	99.9983	2150	vac.							Pendant drop	5
	99.9	(1884)	vac.							Pendant drop	65

SURFACE TENSION OF LIQUID ELEMENTS (*Continued*)

Element	Purity (wt.-%)	σ_{mp} (dyn/cm)	Atm.	σ_{t1} t_1 °C	σ_{t1} σ (dyn/cm)	σ_{t2} t_2 °C	σ_{t2} σ (dyn/cm)	σ_{t3} t_3 °C	σ_{t3} σ (dyn/cm)	Method	Ref.
Te	99.4	186±2	Ar							Bubble pressure	105
			vac.	460	178±1.5					Capillary method	184
	–		vac.	475	(162)					Electro-capillarity	117
		178			$\sigma = 178-0.024\ (t-t_{mp})$ (t°C)					Bubble pressure	204
Ti	98.7	1510	vac.							Capillary method	44
	99.92	1390	Ar							Pendant drop	151
		1460	vac.							Drop weight	139
	99.9991	1650	vac.							Pendant drop	5
	99.0		vac.	1680	1576					Drop weight	191
	99.99999		vac.	1680	1588					Drop weight	191
	99.85	(1880)	vac.							Pendant drop	65
	99.69	1402	vac.							Pendant drop	65
Tl		464.5	Ar							Bubble pressure	192
			–	450	452					Drop pressure	119
	–		vac.	450	450					Electro-capillarity	117
	99.999	467			$\sigma = 536-0.119T$ (T°K) (1270–1695°K)					Bubble pressure	26
	99.999		vac.	450	450					Drop pressure	107
U		1500±75								–	186
		1550	Ar							Bubble pressure	34
	99.94	(1294)	vac.							Pendant drop	65
V	99.9977	1950	vac.							Pendant drop	5
	–	(1760)	vac.							Pendant drop	Z1
W	–	2310	vac.							Pendant drop	35
	99.9999	2500	vac.							Pendant drop	5
	99.8	2220	vac.							Pendant drop	148
	99.9	(2000)	vac.							Pendant drop	65

SURFACE TENSION OF LIQUID ELEMENTS (*Continued*)

Element	Purity (wt.–%)	σ_{mp} (dyn/cm)	Atm.	σ_{t1} t_1 °C	σ_{t1} σ (dyn/cm)	σ_{t2} t_2 °C	σ_{t2} σ (dyn/cm)	σ_{t3} t_3 °C	σ_{t3} σ (dyn/cm)	Method	Ref.
	99.99	537	vac.	500	524	600	508			Drop pressure	154
		530	He							Pendant drop	87
			H_2	489	543	572	528	692	503	Conical capillaries	9
			–	250	536					Drop pressure	179
			–	450	530					Drop pressure	119
			–	250	545					Drop pressure	112
	99.93		vac.	250	549	400	539	600	526	Drop pressure	156
	99.998	566	H_2							Capillary method	133
			–	623	559					Capillary method	77
		610	vac.							Pendant drop	182
			–	800	500					Sessile drop	7
			–	300	538					Drop pressure	190
			–	300	(527)					Drop pressure	76
			–	290	546					Sessile drop	190
	99.99		H_2, He	600	530					Bubble pressure	62
	99.9	(526±10)	Ar							Bubble pressure	104
			vac.	290	600					Sessile drop	147
	99.965		H_2	740	508	950	489.5	1115	479.5	Bubble pressure	127
	99.89	543.7								Bubble pressure	33
		562	vac.							Sessile drop	55
			vac.	300	554					Sessile drop	64
	99.999	590	vac.							Sessile drop	86
			H_2	290	(520)	290	(524)	(vac.)		Sessile drop	67
	99.9999		H_2	246	552.7					Sessile drop	203
	99.9994		vac.	350	537					Drop pressure	106
	99.999	555.8±1.9			$\sigma = 566.84 – 4.76 \cdot 10^{-2} t$ (t°C)					Bubble pressure	176
	99.96	552	vac.	1000	470					Sessile drop	Z5
	99.96	552	Ar		$\sigma_t = 552 – 0.167 (t – t_{mp})$ (MP–500°C) (t°C)					Bubble pressure	Z13a
Sr			Ar	775	288	830	282	893	282	Bubble pressure	125
	99.5	303			$\sigma = 392 – 0.085T$ (T°K) (1152–1602°K)					Bubble pressure	25
Ta		2360	vac.							Pendant drop	88
		2030	vac.							Pendant drop	88
		1910	vac.							Drop weight	139
	99.9983	2150	vac.							Pendant drop	5
	99.9	(1884)	vac.							Pendant drop	65

References

1. Abdel-Aziz Abol Hassan, Neue Hütte, **15,** 304 (1970).
2. Addison, Addison, Kerridge and Lewis, J. Chem. Soc., 2262 (1955).
3. Addison, Coldrey and Pulham, J. Chem. Soc., 1227 (1963).
4. Addison and Raynor, J. Chem. Soc., 965 (1966).
5. Allen, Trans. AIME, **227,** 1175 (1963).
6. Allen, Trans. AIME, **230,** 1357 (1964).
7. Allen and Kingery, Trans. AIME, **215,** 30 (1959).
8. Astakhov, Penin and Dobkina, Zh. Fiz. Khim., **20,** 403 (1946).
9. Atterton and Hoar, J. Inst. Met., **81,** 541 (1953).
10. Ayushina, Levin and Geld, Zh. Fiz. Khim., **42,** 2799 (1968).
11. Baes and Kellogg, J. Metals, **15,** 643 (1953).
12. Baker and Gilbert, J. Amer. Chem. Soc., **62,** 2479 (1940).
13. Bakradse and Pines, Zh. Tekhn. Fiz., **23,** 1548 (1953).
14. Becker, Harders and Kornfeld, Arch. f. Eisenh., **20,** 363 (1949).
15. Belforti and Lepie, Trans. AIME, **227,** 80 (1963).
16. Bergh, J. Electrochem. S., **109,** 1199 (1962).
17. Bering and Ioileva, Doklady A.N., **93,** 85 (1953).
18. Biery and Oblak, Ind. Eng. Chem., Fund., **5,** 121 (1966).
19. Bircumshaw, Phil. Mag., **2,** 341 (1926).
20. Bircumshaw, Phil. Mag., **3,** 1286 (1927).
21. Bircumshaw, Phil. Mag., **6,** 510 (1928).
22. Bircumshaw, Phil. Mag., **12,** 596 (1931).
23. Bobyk, Przemysl Chem., **39,** 423 (1960).
24. Bohdanski and Schins, J. Inorg. Nucl. Chem., **29,** 2173 (1967).
25. Bohdanski and Schins, J. Inorg. Nucl. Chem., **30,** 2331 (1968).
26. Bohdanski and Schins, J. Inorg. Nucl. Chem., **30,** 3362 (1968).
27. Bradhurst and Buchanan, J. Phys. Chem., **63,** 1486 (1959).
28. Bradhurst and Buchanan, Austral. J. Chem., **14,** 397 (1961).
29. Bradhurst and Buchanan, Austral. J. Chem., **14,** 409 (1961).
30. Bradley, J. Phys. Chem., **38,** 234 (1934).
31. Brown, Phil. Mag., **13,** 578 (1932).
32. Burdon, Trans. Farad. Soc., **28,** 866 (1932).
33. Cahill and Kirshenbaum, J. Inorg. Nucl. Chem., **26,** 206 (1964).
34. Cahill and Kirshenbaum, J. Inorg. Nucl. Chem., **27,** 73 (1965).
35. Calverley, Proc. Phys. Soc., **70,** 1040 (1957).
36. Coffman and Parr, Ind. Eng. Chem., **19,** 1308 (1927).
37. Cook, Phys. Review, **34,** 513 (1929).
38. de L. Davies and West, J. Inst. Met., **92,** 208 (1964).
39. Didenko and Pokrovski, Doklady A.N., **31,** 233 (1941).
40. Drath and Sauerwald, Z. allg. anorg. Chem., **162,** 301 (1927).
41. Dshemilev, Popel and Zarevski, Fiz. Met.i Met., **18,** 83 (1964).
42. Dyson, Trans. AIME, **227,** 1098 (1963).
43. Eljutin, Kostikov and Levin, Izv. Vys. Uch. Sav., Tsvetn. Met., **2,** 131 (1970).
44. Eljutin and Maurakh, Izv. A.N., OTN, **4,** 129 (1956).
45. Eremenko, Ivashchenko and Bogatyrenko, The Role of Surface Phenomena in Metallurgy, 37 (1963).
46. Eremenko, Ivashchenko, Fessenko and Nishenko, Izv. A.N., OTN, **7,** 144 (1958).
47. Eremenko and Naidich, Izv. A.N., OTN, **2,** 111 (1959).
48. Eremenko and Naidich, Izv. A.N., OTN, **6,** 129 (1959).
49. Eremenko and Naidich, Izv. A.N., OTN, **6,** 100 (1961).
50. Eremenko and Naidich, Izv. A.N., OTN, **2,** 53 (1960).
51. Eremenko and Naidich, The Role of Surface Phenomena in Metallurgy, 65 (1963).
52. Eremenko, Naidich and Nossonovich, Zh. Fiz. Khim., **34,** 1018 (1960).
53. Eremenko and Nishenko, Ukr. Khim. Zh., **26,** 423 (1960).
54. Eremenko and Nishenko, Zh. Fiz. Khim., **35,** 1301 (1961).
55. Eremenko and Nishenko, Ukr. Khim. Zh., **30,** 125 (1964).
56. Eremenko, Nishenko and Naidich, Izv. A.N., OTN, **3,** 150 (1961).
57. Eremenko, Nishenko and Skljarenko, Izv. A. N., OTN, **2,** 188 (1966).

References (Continued)

58. Eremenko, Nishenko and Taj-Shou-Vej., Izv. A.N., OTN, **3**, 116 (1960).
59. Eremenko and Vassiliu, Ukr. Khim. Zh., **31**, 557 (1965).
60. Eremenko, Vassiliu and Fessenko, Zh. Fiz. Khim., **35**, 1750 (1961).
61. Fessenko, Zh. Fiz. Khim., **35**, 707 (1961).
62. Fessenko and Eremenko, Ukr. Khim. Zh., **26**, 198 (1960).
63. Fessenko, Vassiliu and Eremenko, Zh. Fiz. Khim., **36**, 518 (1962).
64. Flechsig, Thesis, Techn. Univ. Berlin, (1964).
65. Flint, J. Nucl. Mat., **16**, 260 (1965).
66. Gans, Pawlek and Roepenack, Z. Metallkunde, **54**, 147 (1963).
67. Gans and Parthey, Z. Metallkunde, **57**, 19 (1966).
68. Geld and Petrushevski, Izv. A.N., OTN, **3**, 160 (1961).
69. Gratzianski and Rjabov, Zh. Fiz. Khim., **33**, 487, 1253 (1959).
70. Gratzianski, Rjabov and Tobolich, Ukr. Khim. Zh., **29**, 1219 (1963).
71. Greenaway, J. Inst. Met., **74**, 133 (1947).
72. Grunmach, Ann.d. Physik, **3**, 660 (1900).
73. Hagemann, Thesis, Univ. of Freiburg/Br., (1914).
74. Halden and Kingery, J. Phys. Chem., **59**, 557 (1955).
75. Harkins and Ewing, J. Amer. Chem., Soc., **42**, 2539 (1920).
76. Herczynska, Z. Phys. Chem., **214**, 355 (1960).
77. Hoar and Melford, Trans. Farad. Soc., **53**, 315 (1957).
78. Hogness, J. Amer. Chem. Soc., **43**, 1621 (1921).
79. Humenik and Kingery, J. Amer. Ceram. Soc., **37**, 18 (1954).
80. Hutchinson, Trans. Farad. Soc., **39**, 229 (1943).
81. Iredale, Phil. Mag., **45**, 1088 (1923).
82. Iredale, Phil. Mag., **48**, 177 (1924).
83. Jacobj, Thesis, Univ. of Braunschweig, (1962).
84. Jordan and Lane, Austral. J. Chem., **18**, 1711 (1965).
85. Karasaev, Sadumkin and Kukhno, Zh. Fiz. Khim., **41**, 654 (1967).
86. Kaufman and Whalen, Acta Metallurg., **13**, 797 (1965).
87. Keck and van Horn, Phys. Review, **91**, 512 (1953).
88. Kelly and Calverley, SERL-Report, **80**, 53 (1959).
89. Kemball, Trans. Farad. Soc., **42**, 526 (1946).
90. Kernaghan, Phys. Review, **37**, 990 (1931).
91. Kernaghan, Phys. Review, **49**, 414 (1936).
92. Kingery, J. Amer. Ceram. Soc., **37**, 42 (1954).
93. Kingery, Kolloid-Z., **161**, 95 (1958).
94. Kingery and Humenik, J. Phys. Chem., **57**, 359 (1953).
95. Kingery and Norton, AEC-Progr. Rep., **NYO-6296**, (1954).
96. Klyachko, Zavodskaja Labor., **6**, 1376 (1937).
97. Klyachko and Kunin, Zavodskaja Labor., **14**, 66 (1948).
98. Klyachko and Kunin, Doklady A.N., **64**, 64 (1949).
99. Kolesnikova, Izv. Vys. Uch. Sav., Tsvetn. Met., **9** (1960).
100. Kolesnikova and Samarin, Izv. A.N., OTN, **5**, 63 (1956).
101. Konstantinov, Thesis, State Univ. Moscow, (1950).
102. Korolkov, Izv. A.N., OTN, **2**, 35 (1956).
103. Korolkov: Litein'e Svojstva Metallov i Splavov, Isdatelstvo, A.N., SSSR, 37 (1960).
104. Korolkov and Bychkova, Issled. Splav. Tsvetn. Met., **2**, 122 (1960).
105. Korolkov and Igumnova, Izv. A.N., OTN, **6**, 95 (1961).
106. Kovalchuk, Kusnezov and Kotlovanova, Zh. Fiz. Khim., **42**, 1754 (1968).
107. Kovalchuk, Kusnezov and Butuzova, Zh. Fiz. Khim., **42**, 2265 (1968).
108. Kozakevitch and Urbain, J. Iron-& Steel-Inst., **186**, 167 (1957).
109. Kozakevitch and Urbain, C.R., Paris, **253**, 2229 (1961).
110. Kozakevitch and Urbain, Mém. Sci. Rev. Met., **58**, 401, 517, 931 (1961).
111. Krause, Sauerwald and Michalke, Z. anorg. allg. Chem., **181**, 353 (1929).
112. Kristian, Thesis, State Univ. Moscow, (1954).
113. Kubichek, Izv. A.N., OTN, **2**, 96 (1959).
114. Kubichek and Malzev, Izv. A.N., OTN, **3**, 144 (1959).
115. Kurkjian and Kingery, J. Phys. Chem., **60**, 961 (1956).
116. Kurochkin, Baum and Borodulin, Fiz. Met.i Met., **15**, 461 (1963).

References (Continued)

117. Kusnezov, The Role of Surface Phenomena in Metallurgy, 72 (1963).
118. Kusnezov, Djakova and Malzeva, Zh. Fiz. Khim., **33**, 1551 (1959).
119. Kusnezov, Kochergin, Tishchenko and Posdynsheva, Doklady A.N., **92**, 1197 (1953).
120. Kusnezov, Popova and Duplina, Zh. Fiz. Khim., **36**, 880 (1962).
121. Kyrianenko and Solovev, Teplofiz. Vysok. Temp., **8**, 537 (1970).
122. Lasarev, Zh. Fiz. Khim., **36**, 405 (1962).
123. Lasarev, Zh. Fiz. Khim., **38**, 325 (1964).
124. Lasarev and Pershikov, Doklady A.N., **146**, 143 (1962).
125. Lasarev and Pershikov, Zh. Fiz. Khim., **37**, 907 (1963).
126. Lasarev and Pugachevich, Doklady A.N., **134**, 132 (1960).
127. Lauermann, Metzger and Sauerwald, Z. Phys. Chem., **216**, 42 (1961).
128. Lauermann and Sauerwald, Z. Metallkunde, **55**, 605 (1964).
129. Lucas, C.R., Paris, **248**, 2336 (1959).
130. Mack, Davis and Bartell, J. Phys. Chem., **45**, 846 (1941).
131. Matuyama, Sci. Rep. RITU, **16**, 555 (1927).
132. Melik-Gajkazan, Woronchikhina and Sakharova, Elektrokhim., **4**, 1420 (1968).
133. Melford and Hoar, J. Inst. Met., **85**, 197 (1957).
134. Metzger, Z. Phys. Chem., **211**, 1 (1959).
135. Monma and Suto, J. Jap. Inst. Met., **24**, 167 (1960).
136. Monma and Suto, J. Inst. Met., **1**, 69 (1960).
137. Naidich and Eremenko, Fiz. Met.i Met., **11**, 883 (1961).
138. Naidich, Eremenko, Fessenko, Vassiliu and Kirichenko, Zh. Fiz. Khim., **35**, 694 (1961).
139. Namba and Isobe, Sci. Pap. Inst. Phys. Chem. Res., Tokyo, **57**, 5154 (1963).
140. Nicholas, Joyner, Tessem and Olson, J. Phys. Chem., **65**, 1375 (1961).
141. Norton and Kingery, AEC-Progr. Rep., NYO 4632, (1955).
142. Olson and Johnson, J. Phys. Chem., **67**, 2529 (1963).
143. Ono and Matsushima, Sci. Rep. RITU, **9**, 309 (1957).
144. Oppenheimer, Z. anorg. allg. Chem., **171**, 98 (1928).
145. Osminin, Zh. Fiz. Khim., **43**, 2610 (1969).
146. Palacios, An. Soc. Esp. Fis., **18**, 294 (1920).
147. Parthey, Thesis, Techn. Univ. Berlin, (1961).
148. Pekarev, Izv. Vys. Uch. Sav., Tsvetn. Met., **6**, 111 (1963).
149. Pelzel, Berg-u. Hütt. Mon. Hefte, Leoben, **93**, 248 (1948).
150. Pelzel, Berg-u. Hütt. Mon. Hefte, Leoben, **94**, 10 (1949).
151. Peterson, Kedesdy, Keck and Schwarz, J. Appl. Phys., **29**, 213 (1958).
152. Poindexter, Phys. Review, **27**, 820 (1926).
153. Pokrovski, Ukr. Khim. Zh., **7**, 845 (1962).
154. Pokrovski and Galanina, Zh. Fiz. Khim., **23**, 324 (1949).
155. Pokrovski and Kristian, Zh. Fiz. Khim., **28**, (1954).
156. Pokrovski and Saidov, Fiz. Met.i Met., **2**, 546 (1956).
157. Popel, Shergin and Zarevski, Zh. Fiz. Khim., **43**, 2365 (1969).
158. Popel, Smirnov, Zarevski, Dshemilev and Pastukhov, Izv. A.N., **1**, 62 (1965).
159. Popel, Zarevski and Dshemilev, Fiz. Met.i Met., **18**, 468 (1964).
160. Popesco, C.R., Paris, **172**, 1474 (1921).
161. Portevin and Bastien, C.R., Paris, **202**, 1072 (1936).
162. Pugachevich, Zh. Fiz. Khim., **25**, 1365 (1951).
163. Pugachevich and Altynov, Doklady A.N., **86**, 117 (1952).
164. Pugachevich and Yashkichev, The Role of Surface Phenomena in Metallurgy, 46 (1963).
165. Quincke, Ann. d. Physik, **134**, 356 (1868).
166. Quincke, Ann. d. Physik, **135**, 621 (1868).
167. Quincke, Ann. d. Physik, **138**, 141 (1869).
168. Quincke, Ann. d. Physik, **61**, 267 (1897).
169. Raue, Metzger and Sauerwald, Metall, **20**, 1040 (1966).
170. Richards and Boyer, J. Amer. Chem. Soc., **43**, 290 (1921).
171. Roberts, J. Chem. Soc., 1907 (1964).
172. Roehlich jun., Tepper and Rankin, J. Chem. Eng. Data, **13**, 518 (1968).
173. Sauerwald and Drath, Z. anorg. allg. Chem., **154**, 79 (1926).
174. Sauerwald, Schmidt and Pelka, Z. anorg. allg. Chem., **223**, 84 (1935).

References (Continued)

175. Schmidt, Ann. d. Physik, **39,** 1108 (1912).
176. Schwaneke and Falke, US-Bur. Min., Inv. Rep. No. **7372,** (1970).
177. Schwaneke, Falke and Miller, US-Bur. Min., Inv. Rep. No. **7340,** (1970).
178. Semenchenko and Pokrovski, Uspekhii Khim., **6,** 945 (1937).
179. Semenchenko, Pokrovski and Lasarev, Doklady A.N., **89,** 1021 (1953).
180. Shunk and Burr, Trans. ASM, **55,** 786 (1962).
181. Siedentopf, Ann. d. Physik, **61,** 235 (1897).
182. Smirnova and Ormont, Zh. Fiz. Khim., **33,** 771 (1959).
183. Smith, J. Inst. Met., **12,** 168, 20 (1914).
184. Smith and Spitzer, J. Phys. Chem., **66,** 946 (1962).
185. Solovev and Makarova, Teplofiz. Vysok. Temp., **4,** 189 (1966).
186. Spriet, Mém. Sci. Rev. Mét., **60,** 531 (1963).
187. Tavadse, Bairamashvili, Khantadse and Zagareishvili, Doklady A.N., **150,** 544 (1963).
188. Taylor, J. Inst. Met., **83,** 143 (1954).
189. Taylor, Phil. Mag., **46,** 867 (1955).
190. Thyssen, Thesis, Humboldt-Univ. Berlin, (1960).
191. Tille and Kelly, Brit. J. Appl. Phys., **146,** 717 (1963).
192. Timofeyevicheva and Lasarev, Doklady A.N., **138,** 412 (1961).
193. Timofeyevicheva and Lasarev, Doklady A.N., 358 (1962).
194. Timofeyevicheva and Lasarev, Kolloidnyi-Zh., **24,** 227 (1962).
195. Timofeyevicheva, Lasarev and Pershikov, Doklady A.N., **143,** 618 (1962).
196. Timofeyevicheva and Pugachevich, Doklady A.N., **124,** 1093 (1959).
197. Timofeyevicheva and Pugachevich, Doklady A.N., **134,** 840 (1960).
198. Urbain and Lucas, Proc. N.P.L., Teddington, **9,** Pap. 4E (1959).
199. Watolin, Esin, Ukhov and Dubinin, Trudy Inst. Met., Sverdlovsk, **18,** 73 (1969).
200. Whalen and Humenik, Trans. AIME, **218,** 952 (1960).
201. Wegener, Z. Physik, **143,** 548 (1956).
202. White, Trans. AIME, **236,** 796 (1966).
203. White, Met. Review, **124,** 73 (1968).
204. Wobst and Rentzsch, Z. Phys. Chem., **240,** 36 (1969).
205. Yashkichevich and Lasarev, Izv. A.N., OTN, **1,** 170 (1964).
206. Yung Lee, Ind. Eng. Chem., Prod. Res. Dev., **7,** 66 (1968).
207. Zarevski and Popel, Fiz. Met.i Met., **13,** 451 (1962).
208. Zhivov, Trudy WAMI, SSSR, **14,** 99 (1937).
209. Zsin-Tan Wan, Karassev and Samarin, Izv. A.N., OTN, **1,** 30, 49 (1960).
210. Elverum and Doescher, J. Chem. Phys., **20,** 1834 (1952).
211. Fredrickson, J. Colloid Interfac. Sci., **48,** 506 (1974).
212. Johnson and McIntosh, J. Am. Chem. Soc., **31,** 1138 (1909).
213. Chao and Stenger, Talanta, **11,** 271 (1963).

Addendum

Z1. Eljutin, Kostikow and Penkow, Poroshk. Met., **9,** 46 (1970).
Z2. Fraser, Lu, Hamielec and Murarka, Met. Trans., **2,** 817 (1971).
Z3. Bernard and Lupis, Met. Trans., **2,** 555 (1971).
Z4. Rhee, J. Am. Ceram. Soc., **53,** 386 (1970).
Z5. Naidich, Perevertailo and Shuravlev, Zh. Fiz. Chim., **45,** 991 (1971).
Z6. Körber and Löhberg, Gieβereiforschung, **23,** 173 (1971).
Z7. Ziesing, Austral. J. Phys., **6,** 86 (1953).
Z8. Sangster and Carman, J. Chem. Phys., **23,** 1142 (1955).
Z9. Primak and Quarterman, J. Phys. Chem., **58,** 1051 (1954).
Z10. Cooke, HTLMHTTM, Oak Ridge, Vol. I, p. 66, (Nov. 1964).
Z11. Ofizerow, Izv. A.N., Metally, **4,** 91 (1971).
Z12. White, Met. Trans., **3,** 1933 (1972).
Z13. Lang, Aluminum (Germany), **49,** 231 (1972).
Z13a. Lang, J. Inst. Met. (to be published).

SURFACE TENSION

SURFACE TENSION OF INORGANIC SOLUTES IN WATER

% = Weight % of solute
γ = Surface tension in dynes/cm.

Solute	T°C	%/γ							
HCl	20	%	1.78	3.52	6.78	12.81	16.97	23.74	35.29
		γ	72.55	72.45	72.25	71.85	71.75	70.55	65.75
HNO_3	20	%	4.21	8.64	14.99	34.87			
		γ	72.15	71.65	70.95	68.75			
H_2SO_4	25	%	4.11	8.26	12.18	17.66	21.88	29.07	33.63
		γ	72.21	72.55	72.80	73.36	73.91	74.80	75.29
$HClO_4$	25	%	4.86	10.01	20.38	30.36	53.74	63.47	72.25
		γ	71.18	70.34	69.21	68.57	69.02	69.73	69.01
KOH	18	%	2.73	5.31	10.08	17.57			
		γ	73.95	74.85	76.55	79.75			
NaOH	18	%	2.72	5.66	16.66	30.56	35.90		
		γ	74.35	75.85	83.05	96.05	101.05		
NH_4OH	18	%	1.72	3.39	4.99	9.51	17.37	34.47	54.37
		γ	71.65	70.65	69.95	67.85	65.25	61.05	57.05
KCl	20	%	0.74	3.60	6.93	13.88	18.77	22.97	24.70
		γ	72.99	73.45	74.15	75.55	76.95	78.25	78.75
LiCl	25	%	5.46	7.37	10.17	13.95			
		γ	74.23	75.10	76.30	78.10			

Solute	T°C	%/γ							
NaCl	20	%	0.58	2.84	5.43	10.46	14.92	22.62	25.92
		γ	72.92	73.75	74.39	76.05	77.65	80.95	82.55
PbCl	25	%	21.57	28.52	37.74				
		γ	75.20	76.80	79.20				
$BaCl_2$	25	%	9.26	16.73	25.58				
		γ	73.50	74.93	76.38				
$MgCl_2$	20	%	0.94	4.55	8.69	16.00	22.30	25.44	
		γ	73.07	74.00	75.75	79.15	82.95	85.75	
NaBr	20	%	4.89	9.33	13.37	23.00			
		γ	73.45	74.05	74.75	76.55			
$Al_2(SO_4)_3$	25	%	2.54	4.06	9.40	14.60	19.32	23.54	25.50
		γ	72.32	72.92	73.51	74.71	76.06	78.30	79.73
$MgSO_4$	20	%	1.19	5.68	10.75	19.41	24.53		
		γ	73.01	73.78	74.85	77.35	79.25		
Na_2SO_4	20	%	2.76	6.63	12.44				
		γ	73.25	74.15	75.45				
Na_2CO_3	20	%	2.58	5.03	9.59	13.72			
		γ	73.45	74.05	75.45	76.75			
$NaNO_3$	20	%	0.85	4.08	7.84	14.53	29.82	37.30	47.06
		γ	72.87	73.75	73.95	75.15	78.35	80.25	87.05

SURFACE TENSION OF INORGANIC SOLUTES IN ORGANIC SOLVENTS

% = Mol %
γ = Surface tension in dynes/cm.

Solute	Solvent	T°C	%/γ						
LiCl	Ethyl alcohol	14	%	0.72	2.30	4.62			
			γ	22.90	23.17	23.26			
LiBr	Ethyl alcohol	14	%	0.95	2.60				
			γ	23.08	23.35				
LiI	Ethyl alcohol	14	%	1.43	2.87	5.08	10.21	19.47	26.92
			γ	23.11	23.56	24.39	26.03	28.87	31.95
KI	Methyl alcohol	14	%	0.81	1.52	2.68			
			γ	23.76	24.11	24.71			
$CaCl_2$	Ethyl alcohol	24	%	0.94	1.98	3.79	7.47		
			γ	22.62	22.58	23.23	23.97		
NaI	Methyl alcohol	14	%	0.76	1.48	4.33	8.55	12.53	
			γ	22.83	23.29	24.85	27.41	29.75	

Solute	Solvent	T°C	%/γ						
NaI	Ethyl alcohol	24	%	0.45	1.80	3.63	4.54	6.02	10.46
			γ	22.47	22.82	23.41	23.52	24.00	25.07
NaI	Acetone	14	%	0.93	2.08	5.07	6.53		
			γ	24.22	24.40	25.04	25.12		
ZnI_2	Methyl alcohol	22	%	0.90	2.79	5.07			
			γ	22.97	24.23	25.84			
ZnI_2	Ethyl alcohol	24	%	0.41	1.72	3.42	6.90		
			γ	22.70	22.90	23.71	25.49		
H_2SO_4	Ethyl ether	17	%	3	10	40	75	90	
			γ	17.30	19.55	32.50	46.30	46.83	
H_2SO_4	Nitrobenzene	17	%	3	10	40	75	90	
			γ	42.73	43.96	46.05	47.52	48.25	

SURFACE TENSION OF ORGANIC COMPOUNDS IN WATER

% = Weight % of solute
γ = Surface tension in dynes/cm.

Solute	T° C	%/γ							
Acetic acid...	30	%	1.00	2.475	5.001	10.01	30.09	49.96	69.91
		γ	68.00	64.40	60.10	54.60	43.60	38.40	34.30
Acetone......	25	%	5.00	10.00	20.00	50.00	75.00	95.00	100.00
		γ	55.50	48.90	41.10	30.40	26.80	24.20	23.00
Acetonitrile..	20	%	1.13	3.35	11.77	20.20	37.58	61.33	81.22
		γ	69.02	63.03	47.61	39.06	31.84	30.02	29.02
o-Aminobenzoic acid	25	%	12.35	22.36	30.45	37.44			
		γ	71.96	73.23	74.54	75.79			
m-Aminobenzoic acid	25	%	12.35	22.36	30.45	37.44			
		γ	73.30	74.59	76.16	77.89			

Solute	T° C	%/γ							
p-Aminobenzoic acid	25	%	12.35	22.36	30.45	37.44			
		γ	73.38	74.79	76.32	78.20			
Aminobutyric acid	25	%	4.96	9.34	13.43				
		γ	71.91	71.67	71.40				
Ammonium lactate	29	%	30.00	50.00	60.00	70.00	80.00	90.00	
		γ	35.40	34.40	35.40	35.60	38.20	44.50	
n-Butanol....	30	%	0.04	0.41	9.53	80.44	86.05	94.20	97.40
		γ	69.33	60.38	26.97	23.69	23.47	23.29	22.25
n-Butyric acid	25	%	0.14	0.31	1.05	8.60	25.00	79.00	100.00
		γ	69.00	65.00	56.00	33.00	28.00	27.00	26.00

SURFACE TENSION OF ORGANIC COMPOUNDS IN WATER (Continued)

Solute	T°C	% γ							
Dioxan.......	26	%	0.44	2.20	4.70	11.14	20.17	35.20	55.00
		γ	69.83	65.64	62.45	56.90	51.57	45.30	39.27
Dioxan......	26	%	67.68	76.45	83.02	91.90	95.60	97.77	
		γ	36.95	35.80	35.00	33.95	33.60	33.10	
Formic acid..	30	%	1.00	5.00	10.00	25.00	50.00	75.00	100.00
		γ	70.07	66.20	62.78	56.29	49.50	43.40	36.51
Glycerol.....	18	%	5.00	10.00	20.00	30.00	50.00	85.00	100.00
		γ	72.90	72.90	72.40	72.00	70.00	66.00	63.00
Glycine......	25	%	3.62	6.98	10.12	13.10			
		γ	72.54	73.11	73.74	74.18			
Hydrocinnamic acid	21.5	%	7.02	12.62	18.39	26.09	31.06	38.25	47.93
		γ	69.08	66.49	63.63	59.25	56.14	52.96	47.24
Methyl acetate	25	%	0.66	1.29	2.29	3.56			
		γ	66.33	62.92	58.22	55.08			

Solute	T°C	% γ							
Morpholine...	20	%	8.56	19.39	30.41	50.45	69.93	80.14	92.00
		γ	67.80	62.62	59.15	52.85	47.05	43.62	41.60
Potassium lactate	29	%	40.00	50.00	60.00	70.00			
		γ	66.40	66.40	65.40	63.40			
Phenol.......	20	%	0.024	0.047	0.118	0.417	0.941	3.76	5.62
		γ	72.60	72.20	71.30	66.50	61.10	46.00	42.30
n-Propanol...	25	%	0.1	0.5	1.0	50.0	60.0	80.0	90.0
		γ	67.10	56.18	49.30	24.34	24.15	23.66	23.41
Propionic acid	25	%	1.91	5.84	9.80	21.70	49.80	73.90	100.00
		γ	60.00	49.00	44.00	36.00	32.00	30.00	26.00
Sodium lactate	29	%	1.00	10.00	30.00	40.00	50.00	60.00	70.00
		γ	70.40	69.60	68.50	64.80	45.40	56.70	60.70
Sucrose......	25	%	10.00	20.00	30.00	40.00	55.00		
		γ	72.50	73.00	73.40	74.10	75.70		

SURFACE TENSION OF ORGANIC COMPOUNDS IN ORGANIC SOLVENTS

% = Weight % of solvent
γ = Surface tension dynes/cm.

Solute	Solvent	T°C	% γ					
Acetic acid...	Benzene	35	%	10.45	25.53	34.28	43.93	68.77
			γ	25.40	25.21	25.32	25.43	25.99
Acetic acid...	Acetone	25	%	25.63	50.83	75.62		
			γ	27.50	26.61	24.90		
Acetone......	Ethyl ether	30	%	21.83	40.98	61.29	75.69	89.03
			γ	16.75	17.49	19.15	19.80	21.00
Acetonitrile...	Ethyl alcohol	20	%	10.83	32.42	49.90	69.78	80.19
			γ	22.92	23.92	24.36	25.08	26.51
Aniline.......	Cyclohexane	32	%	14.35	37.65	50.67	72.28	96.46
			γ	24.21	24.51	24.50	25.61	37.45
Carbontetrachloride	Benzene	50	%	30.40	51.69	62.38	78.29	88.62
			γ	24.39	24.09	23.78	23.47	23.21
Cyclohexane..	Nitrobenzene	15	%	5.00	10.00			
			γ	37.59	33.03			
Naphthalene..	Benzene	79.5	%	20.00	40.00	50.00	60.00	80.00
			γ	23.42	26.70	27.80	29.20	31.70
Naphthalene..	*p*-Nitrophenol	121	%	10.21	29.74	45.94	58.01	67.57
			γ	29.90	31.80	33.70	34.80	41.80

SURFACE TENSION OF METHYL ALCOHOL IN WATER

% = Volume % of alcohol
γ = Surface tension dynes/cm.

T°C	%	7.5	10.00	25.0	50.0	60.0	80.0	90.0	100.0
20	γ	60.90	59.04	46.38	35.31	32.95	27.26	25.36	22.65
30	γ	59.33	57.27	45.30	34.52	32.26	26.48	24.42	21.58
50	γ	56.19	55.01	43.24	32.95	30.79	25.01	22.55	19.52

SURFACE TENSION OF ETHYL ALCOHOL IN WATER

% = Volume % of alcohol
γ = Surface tension dynes/cm.

T°C	%	5.00	10.00	24.00	34.00	48.00	60.00	72.00	80.00	96.00
20	γ	...	...	...	33.24	30.10	27.56	26.28	24.91	23.04
40	γ	54.92	48.25	35.50	31.58	28.93	26.18	24.91	23.43	21.38
50	γ	53.35	46.77	34.32	30.70	28.24	25.50	24.12	22.56	20.40

WATER AGAINST AIR

Temperature °C	Surface tension dynes/cm.	Temperature °C	Surface tension dynes/cm.	Temperature °C	Surface tension dynes/cm.
−8	77.0	15	73.49	40	69.56
−5	76.4	18	73.05	50	67.91
0	75.6	20	72.75	60	66.18
5	74.9	25	71.97	70	64.4
10	74.22	30	71.18	80	62.6
				100	58.9

INTERFACIAL TENSION

Surface Tension at the Interface Between Two Liquids
(Each liquid saturated with the other)

Liquids	Temperature °C	γ
Benzene- -Mercury. .	20	357
Ethyl ether- -Mercury	20	379
Water- -Benzene.	20	35.00
Water- -Carbon tetrachloride	20	45.
Water- -Ethyl ether. .	20	10.7

Liquids	Temperature °C	γ
Water- -Heptylic acid	20	7.0
Water- -n-Hexane. . . .	20	51.1
Water- -Mercury.	20	375.
Water-n-Octane.	20	50.8
Water-n-Octyl alcohol	20	8.5

SURFACE TENSION OF VARIOUS LIQUIDS

Substance Name	Formula	In contact with	Temperature °C	Surface tension dynes/cm.	References
Acetaldehyde	C_2H_4O	--vapor	20	21.2	AC(12)
Acetaldoxime	C_2H_5NO	--vapor	35	30.1	JP(1); JSG(1)
Acetamide	C_2H_5NO	--vapor	85	39.3	JS(3)
Acetanilide	C_2H_5NO	--vapor	120	35.6	JS(3)
Acetic acid	$C_2H_4O_2$	--vapor	10	28.8	AC(22, 23, 25); GC(1); JS(14); tPRS(1); ZC(1, 6)
" "	C_2H_4O	--vapor	20	27.8	
" "	C_2H_4O	--vapor	50	24.8	
Acetic anhydride	$C_4H_6O_3$	--vapor	20	32.7	GC(1); JS(4)
Acetone	C_3H_6O	--air or vapor	0	26.21	AC(20, 24, 25); AdC(1); BF(1); JP(5); JS(4, 14); ZC(6)
"	C_3H_6O	--air or vapor	20	23.70	
"	C_3H_6O	--air or vapor	40	21.16	
Acetonitrile	C_2H_3N	--vapor	20	29.30	AC(16); BF(1); GC(1); JP(5)
Acetophenone	C_8H_8O	--vapor	20	39.8	AC(27); AS(1); JP(2)
Acetyl chloride	C_2H_3ClO	--vapor	14.8	26.7	JS(4)
Acetylene	C_2H_2	--vapor	−70.5	16.4	AC(14)
Acetylsalicylic acid (in aq. sol.)	$C_9H_8O_4$	--vapor	25.9	60.06	GC(2)
Allyl alcohol	C_3H_6O	--air or vapor	20	25.8	AC(27); AdC(1); JS(4)
Allyl isothiocyanate	C_4H_5NS	--air or vapor	20	34.5	AC(16); GC(1)
Ammonia	NH_3	--vapor	11.1	23.4	JP(7)
"	NH_3	--vapor	34.1	18.1	JP(7)
Aniline	C_6H_7N	--air	10	44.10	AC(6, 17, 28); GC(1); JP(5); JS(12)
"	C_6H_7N	--vapor	20	42.9	
"	C_6H_7N	--air	50	39.4	
Argon	A	--vapor	−188	13.2	JS(15)
Azoxybenzene	$C_{12}H_{10}N_2O$	--vapor	51	43.34	JS(5)
Benzaldehyde	C_7H_6O	--air	20	40.04	AC(6, 24)
Benzene	C_6H_6	--air	10	30.22	AC(3, 5, 31, 32, 34); BF(2); JP(5); JS(4, 9, 10, 11, 14); PRS(2); tRIA(1); tPRS(1); ZC(4, 5)
"	C_6H_6	--air	20	28.85	
"	C_6H_6	--saturated with vapor	20	28.89	
"	C_6H_6	--air	30	27.56	
Benzonitrile	C_7H_5N	--air	20	39.05	AC(28); AS(2); GC(1); JP(5); JS(3,
Benzophenone	$C_{13}H_{10}O$	--air or vapor	20	45.1	AC(27); AS(1); ZC(5)
Benzylamine	C_7H_9N	--vapor	20	39.5	JS(3)
Benzyl alcohol	C_7H_8O	--air or vapor	20	39.0	AC(33); JS(6); PRS(1)
Bromine	Br_2	--air or vapor	20	41.5	AC(17); AdP(3); GC(1)
Bromobenzene	C_6H_5Br	--air	20	36.5	AC(8, 17); GC(1); JS(12)
Bromoform	$CHBr_3$	--vapor	20	41.53	AC(8); JP(2)
p-Bromophenol	C_6H_5BrO	--vapor	74.4	42.36	JS(6)
d-*sec*-Butyl alcohol	$C_4H_{10}O$	--vapor	10	23.5	JS(7)
n-Butyl alcohol	$C_4H_{10}O$	--air or vapor	0	26.2	AC(27, 33); JS(4)
" " "	$C_4H_{10}O$	--air or vapor	20	24.6	AC(27, 33); JS(4)
" " "	$C_4H_{10}O$	--air or vapor	50	22.1	AC(27, 33); JS(4)
tert-Butyl alcohol	$C_4H_{10}O$	--air or vapor	20	20.7	AC(33); JS(8)
n-Butylamine	$C_4H_{11}N$	--nitrogen	41	19.7	ZA(1)
n-Butyric acid	$C_4H_8O_2$	--air	20	26.8	AC(27); GC(1); JS(4)
Carbon bisulfide	CS_2	--vapor	20	32.33	AC(17, 28); GC(1); BF(2); JS(14); PRS(2); ZC(6)
Carbon dioxide	CO_2	--vapor	20	1.16	VK(1, 2)
" "	CO_2	--vapor	−25	9.13	VK(1, 2)
Carbon tetrachloride	CCl_4	--vapor	20	26.95	AC(3, 5, 6, 28, 31); PRS(1, 2); ZC(5)
" "	CCl_4	--vapor	100	17.26	
" "	CCl_4	--vapor	200	6.53	
Carbon monoxide	CO	--vapor	−193	9.8	JS(15)
" "	CO	--vapor	−203	12.1	JS(15)
Chloral	C_2HCl_3O	--vapor	19.4	25.34	JS(4)
Chlorine	Cl_2	--vapor	20	18.4	AC(11); JP(3)
"	Cl_2	--vapor	−30	25.4	AC(11); JP(3)
"	Cl_2	--vapor	−40	27.3	AC(11); JP(3)
"	Cl_2	--vapor	−50	29.2	AC(11); JP(3)
"	Cl_2	--vapor	−60	31.2	AC(11); JP(3)
Chloroacetic acid	$C_2H_2Cl_2O_2$	--nitrogen	25.7	35.4	ZA(1)
Chlorobenzene	C_4H_5Cl	--vapor	20	33.56	AC(6, 20, 28); JP(5); JS(11); PRS(2); tRIA(1); tPRS(1); ZC(5)
Chloroform	$CHCl_3$	--air	20	27.14	AC(6, 28, 31); AdC(1); PRS(2); tRIA(1); ZC(6)
o-Chlorophenol	C_6H_5ClO	--vapor	12.7	42.25	JS(6)
Cyclohexane	C_6H_{12}	--air	20	25.5	PRS(1); ZA(1)
Dichloroacetic acid	$C_2H_2Cl_2O_2$	--nitrogen	25.7	35.4	ZA(1)
Dichloroethane	$C_2H_4Cl_2$	--air	35.0	23.4	AC(17); AdC(1)
Diethylamine	$C_4H_{11}N$	--air	56	16.4	GC(1)
Diethylaniline	$C_{10}H_{15}N$	--vapor	20	34.2	AC(31); AS(1); MfC(1)
Diethyl carbonate	$C_5H_{10}O$	--air	20	26.31	AC(4); JS(5)
Diethyl oxalate	$C_6H_{10}O_4$	--vapor	20	32.0	AC(16); GC(1)
Diethyl phthalate	$C_{12}H_{14}O_4$	--vapor	20	37.5	AC(21); PRS(1)
Diethyl sulfate	$C_4H_{12}O_4S$	--air	13	34.61	JS(5)
Dimethylamine	C_2H_7N	--nitrogen	0	18.1	ZA(1)
"	C_2H_7N	--nitrogen	5	17.7	ZA(1)
Dimethylaniline	C_8H_{11}	--air or vapor	20	36.6	AC(4, 17, 31); JP(5)
1,5-Dimethyl-2-phenyl-3-pyrazolone	$C_{11}H_{12}N_2O$	--vapor	25.9	63.63	GC(2)
Dimethyl sulfate	$C_2H_6O_4S$	--air	18	40.12	JS(5)
Diphenylamine	$C_{12}H_{11}N$	--air or vapor	80	37.7	JP(2, 9); JS(3); ZA(1)
Ethyl acetate	$C_4H_8O_2$	--air	0	26.5	AC(26, 33); AdC(1); AS(2); JP(5); tPRS(1); ZC(6)
" "	$C_4H_8O_2$	--air	20	23.9	
" "	$C_4H_8O_2$	--air	50	20.2	
Ethyl acetoacetate	$C_6H_{10}O_3$	--air or vapor	20	32.51	AC(16); BD(2); GC(1); JS(4)
Ethyl alcohol	C_2H_6O	--air	0	24.05	AC(22, 23, 25, 32); BF(2); JP(5); tRIA(1); tPRS(1)
" "	C_2H_6O	--vapor	10	23.61	
" "	C_2H_6O	--vapor	20	22.75	
" "	C_2H_6O	--vapor	30	21.89	
Ethylamine	C_2H_7N	--nitrogen	0	21.3	PRA(4); ZA(1)
"	C_2H_7N	--nitrogen	9.9	20.4	PRA(4); ZA(1)
Ethylaniline	$C_8H_{11}N$	--air or vapor	20	36.6	AC(17, 33); AS(1); MfC(1)

Substance		In contact with	Temperature °C	Surface tension dynes/cm.	References
Name	Formula				
Ethylbenzene	C_8H_{10}	vapor	20	29.20	AC(**17, 33, 34**); AdC(**1**); PRS(**1**)
Ethylbenzoate	$C_9H_{10}O_2$	vapor	20	35.5	GC(**1**); ZA(**1**); MfC(**1**)
Ethyl bromide	C_2H_5Br	vapor	20	24.15	AC(**6, 33**); GC(**1**)
Ethyl chloroformate	$C_3H_5ClO_2$	vapor	15.1	27.5	JS(**4**)
Ethyl Cinnamate	$C_{11}H_{12}O_2$	air	20	38.37	AC(**16**); PRS(**1**); ZC(**5**)
Ethylene bromide	$C_2H_4Br_2$	vapor	20	38.37	AC(**5, 28, 33**); GC(**1**); JP(**2**); PRS(**2**); tRIA(**1**)
Ethylene chloride	$C_2H_4Cl_2$	air	20	24.15	AC(**17, 33**); AdC(**1**); JS(**14**); ZC(**6**)
Ethylene oxide	C_2H_4O	vapor	−20	30.8	
" "	C_2H_4O	vapor	0.0	27.6	
" "	C_2H_4O	vapor	20	24.3	
Ethyl ether	$C_4H_{10}O$	vapor	20	17.01	AC(**4, 15, 28, 31**); AdC(**1**); tPRS(**1**)
" "	$C_4H_{10}O$	vapor	50	13.47	
Ethyl formate	$C_3H_6O_2$	air or vapor	20	23.6	AC(**20, 26**); AdC(**1**); PRS(**2**)
Ethyl iodide	C_2H_5I	vapor	20	29.4	AC(**33**); GC(**1**); JS(**4**); PRS(**1**); ZC(**6**)
Ethyl nitrate	$C_2H_5NO_3$	air or vapor	20	28.7	AC(**27**); GC(**1**)
dl-Ethyl lactate	$C_5H_{10}O_3$	air	20	29.9	AC(**20, 21**); JP(**8**)
Ethyl mercaptan	C_2H_6S	air or vapor	20	22.5	AC(**4, 16**); JS(**4**)
Ethyl salicylate	$C_9H_{10}O_3$	vapor	20.5	38.33	JS(**6**)
Formamide	CH_3NO	vapor	20	58.2	AC(**27**); JS(**3**); ZC(**4**)
Formic acid	CH_2O_2	air	20	37.6	AC(**23, 27**); GC(**1**); JS(**4**); MfC(**1**); ZA(**1**)
Furfural	$C_5H_4O_2$	air or vapor	20	43.5	AC(**27**); GC(**1**)
Gelatin solution (1%)		water	2.85	8.3	ZK(**1**)
Glycerol	$C_3H_8O_3$	air	20	63.4	JR(**1**); MB(**1**); ZA(**1**); ZC(**3**)
"	$C_3H_8O_3$	air	90	58.6	
"	$C_3H_8O_3$	air	150	51.9	
Glycol	$C_2H_6O_2$	air or vapor	20	47.7	AC(**29**); JS(**4**)
Helium	He	vapor	−269	.12	cUL(**2**); PRA(**2**)
"	He	vapor	−270	.239	cUL(**2**); PRA(**2**)
"	He	vapor	−271.5	.353	cUL(**2**); PRA(**2**)
n-Hexane	C_6H_{14}	air	20	18.43	AC(**5, 6, 16**); AdC(**1**); AS(**1**)
Hydrazine	N_2H_4	vapor	25	91.5	AC(**1**)
Hydrogen	H_2	vapor	−255	2.31	cUL(**1**); PRA(**1**)
Hydrogen cyanide	HCN	vapor	17	18.2	ZE(**1**)
Hydrogen peroxide	H_2O_2	vapor	18.2	76.1	AC(**13**)
Isobutyl alcohol	$C_4H_{10}O$	vapor	20	23.0	AC(**4, 27, 32**); AdC(**1**); JP(**5**); JS(**4**)
Isobutylamine	$C_4H_{11}N$	air	68	17.6	GC(**1**)
Isobutyl chloride	C_4H_9Cl	air	20	21.94	AC(**6**); GC(**1**)
Isobutyric acid	$C_4H_8O_2$	air or vapor	20	25.2	AC(**27**); GC(**1**); JS(**4**)
Isopentane	C_5H_{12}	air	20	13.72	AC(**6**)
Isopropyl alcohol	C_3H_8O	air or vapor	20	21.7	JS(**4**); AdC(**1**)
Methyl acetate	$C_3H_6O_2$	air or vapor	20	24.6	AC(**26, 33**); AdC(**1**); PRS(**2**)
Methyl alcohol	CH_4O	air	0	24.49	AC(**22, 23, 25, 32**); tPRS(**1**)
" "	CH_4O	air	20	22.61	
" "	CH_4O	vapor	50	20.14	
Methylamine	CH_3NH_2	nitrogen	−12	22.2	PRA(**1**); ZA(**1**)
"	CH_3NH_2	vapor	−20	23.0	PRA(**1**); ZA(**1**)
"	CH_3NH_2	nitrogen	−70	29.2	PRA(**1**); ZA(**1**)
N-Methylaniline	C_7H_9N	air or vapor	20	39.6	AC(**17, 32**); AS(**1**)
Methyl benzoate	$C_8H_8O_2$	air or vapor	20	37.6	GC(**1**); JP(**5**)
Methyl chloride	CH_3Cl	air	20	16.2	WN(**1**)
Methyl ether	C_2H_6O	vapor	−10	16.4	AC(**12**)
" "	C_2H_6O	vapor	−40	21	AC(**12**)
Methylene chloride	CH_2Cl_2	air	20	26.52	AC(**6**)
Methylene iodide	CH_2I_2	air	20	50.76	AC(**8**)
Methyl ethyl ketone	C_4H_8O	air or vapor	20	24.6	AC(**24**); AS(**1**)
Methyl formate	$C_2H_4O_2$	vapor	20	25.08	AC(**26**); tPRS(**1**); GC(**1**)
Methyl iodide	CH_3I	air	43.5	25.8	GC(**1**)
Methyl propionate	$C_4H_8O_2$	air or vapor	20	24.9	AC(**20, 26**); AdC(**1**); PRS(**2**)
Methyl salicylate	$C_8H_8O_3$	nitrogen	94	31.9	ZA(**1**)
Methyl sulfide	C_2H_6S	vapor	11.1	26.50	JP(**6**)
Naphthalene	$C_{10}H_8$	air or vapor	127	28.8	AS(**1**); JP(**2**)
Neon	Ne	vapor	−248	5.50	cUL(**3**); PRA(**3**)
Nitric acid (98.8%)	HNO_3	air	11.6	42.7	JS(**2**)
Nitrobenzene	$C_6H_5NO_2$	air or vapor	20	43.9	AC(**6, 23**); AS(**2**); BF(**2**); GC(**1**); JS(**4, 12**)
Nitroethane	$C_2H_5NO_2$	air or vapor	20	32.2	GC(**1**); JS(**4**)
Nitrogen	N_2	vapor	−183	6.6	JS(**15**)
"	N_2	vapor	−193	8.27	JS(**15**)
"	N_2	vapor	−203	10.53	JS(**15**)
Nitrogen tetra oxide	N_2O_4	vapor	19.8	27.5	JS(**4**)
Nitromethane	CH_3NO_2	vapor	20	36.82	AC(**6, 27**); GC(**1**)
Nitrous oxide	N_2O	vapor	20	1.75	VK(**1**)
n-Octane	C_8H_{18}	vapor	20	21.80	AC(**4, 5, 34**); JS(**4**)
n-Octyl alcohol	$C_8H_{18}O$	air	20	27.53	AC(**4, 5**)
Oleic acid	$C_{18}H_{34}O_2$	air	20	32.50	PRS(**1**); AC(**4**)
Oxygen	O_2	vapor	−183	13.2	JS(**15**)
Oxygen (65%)	O_2	air	−190.5	12.2	AD(**1**)
"	O_2	vapor	−193	15.7	JS(**15**)
"	O_2	vapor	−203	18.3	JS(**15**)
Paraldehyde	$C_6H_{12}O_3$	air	20	25.9	AC(**27**); AdC(**1**)
Phenetole	$C_8H_{10}O$	vapor	20	32.74	AC(**6, 28**); GC(**1**); JP(**5**)
Phenol	C_6H_6O	air or vapor	20	40.9	AC(**18, 19, 25**); JS(**2, 6, 13**); JP(**4**)
	C_6H_6O	air or vapor	30	39.88	
Phenylhydrazine	$C_6H_8N_2$	vapor	20	46.1	JS(**3**)
Phosphorus tribromide	PBr_3	air	24	45.8	JS(**5**)
Phosphorus trichloride	PCl_3	vapor	20	29.1	AC(**17**); GC(**1**); JP(**2**); JS(**4**)
Phosphorus triiodide	PI_3	vapor	75.3	56.5	ZA(**1**)
Propionic acid	$C_3H_6O_2$	vapor	20	26.7	AC(**27**); GC(**1**); JS(**4**)
n-Propyl acetate	$C_5H_{10}O_2$	air or vapor	20	24.3	AC(**20, 26**); AdC(**1**); PRS(**2**)
n-Propyl alcohol	C_3H_8O	vapor	20	23.78	AC(**24, 33**); AdC(**1**); JP(**5**); JS(**4**)
n-Propylamine	C_3H_9N	air	20	22.4	GC(**1**); JS(**3**)

Substance — Name	Substance — Formula	In contact with	Temperature °C	Surface tension dynes/cm.	References
n-Propyl bromide	C_3H_7Br	vapor	71	19.65	GC(**1**)
n-Propyl chloride	C_3H_7Cl	air	47	18.2	AdC(**1**)
n-Propyl formate	$C_4H_8O_2$	vapor	20	24.5	AC(**26, 33**); AdC(**1**); PRS(**2**)
Pyridine	C_5H_5N	air	20	38.0	AC(**17, 26, 28**); GC(**1**); JP(**5**); JS(**4, 14**)
Quinoline	C_9H_7N	air	20	45.0	AC(**17, 22**); GC(**1**); JP(**4, 5**)
Ricinoleic acid	$C_{18}H_{34}O_3$	air	16	35.81	PRS(**1**)
Selenium	Se	air	217	92.4	AdP(**3**)
Styrene	C_8H_8	air	19	32.14	PRS(**1**)
Sulfuric acid (98.5%)	H_2SO_4	air or vapor	20	55.1	AC(**17a**); AdP(**7**); JS(**2**)
Tetrabromoethane 1,1,2,2-	$C_2H_2Br_4$	air	20	49.67	AC(**6**); ZC(**2**)
Tetrachloroethane 1,1,2,2-	$C_2H_2Cl_4$	air	22.5	36.03	ZC(**2**)
Tetrachloroethylene	C_2Cl_4	vapor	20	31.74	AC(**8**); GC(**1**)
Toluene	C_7H_8	vapor	10	27.7	AC(**4, 17, 20, 31**
"	C_7H_8	vapor	20	28.5	AdC(**1**); BF(**2**)
"	C_7H_8	vapor	30	27.4	JP(**5**); PRS(**2**); ZC(**5, 6**)
m-Toluidine	C_7H_9N	vapor	20	36.9	AC(**33**)
o-Toluidine	C_7H_9N	air or vapor	20	40.0	AC(**27, 33**); AS(**1, 3**); ZA(**1**)
p-Toluidine	C_7H_9N	air	50	34.6	AC(**27**); AS(**1**); JS(**12**)
Trichloroacetic acid	$C_2HCl_3O_2$	nitrogen	80.2	27.8	ZA(**1**)
Trichloroethane 1,1,2-	$C_2H_3Cl_3$	air	114	22.0	GC(**1**)
Triethyl phosphate	$C_6H_{15}O_4P$	air	15.5	30.61	JS(**5**)
Trimethylamine	C_3H_9N	nitrogen	−4	17.3	ZA(**1**)
Triphenylcarbinol	$C_{19}H_{16}O$	vapor	165.8	30.38	JS(**6**)
Vinyl acetate	$C_4H_6O_2$	vapor	20	23.95	
" "	$C_4H_6O_2$	vapor	25	23.16	
" "	$C_4H_6O_2$	vapor	30	22.54	
Water	H_2O	air	18	73.05	
m-Xylene	C_8H_{10}	vapor	20	28.9	AC(**4, 33, 34**); AdC(**1**); JP(**5**); ZC(**5**)
o-Xylene	C_8H_{10}	air	20	30.10	AC(**6, 17, 34**)
p-Xylene	C_8H_{10}	vapor	20	28.37	AC(**6, 17, 33, 34**); AdC(**1**)

REFERENCE KEY

AC—

Journal of the American Chemical Society. (**1**) Baker and Gilbert, 62: 2479–2480; 40. (**2**) H. Brown, 56: 2564–2568; 38. (**3**) Harkins and Brown, 41: 449; 19. (**4**) Harkins, Brown and Davies, 39: 354; 17. (**5**) Harkins and Cheng, 43: 35; 21. (**6**) Harkins, Clark and Roberts, 42: 700; 20. (**7**) Harkins and Ewing, 42: 2539; 20. (**8**) Harkins and Feldman, 44: 2665; 22. (**9**) Harkins and Grafton, 42: 2534; 20. (**10**) Hogness, 43: 1621; 21. (**11**) Johnson and McIntosh, 31: 1139; 09. (**12**) Maass and Boomer, 44: 1709; 22. (**13**) Maass and Hatcher, 42: 2548; 20. (**14**) Maass and McIntosh, *36*, 741 (1914). (**15**) Maass and Wright, 43: 1098; 21. (**16**) Morgan and Chazel, 35: 1821, 13. (**17**) Morgan and Daghlian, 33: 672; 11. (**17a**) Morgan and Davis, 38: 555; 16. (**18**) Morgan and Egloff, 38; 844: 16. (**19**) Morgan and Evans, 39: 2151; 17. (**20**) Morgan and Griggs, 39: 2261; 17. (**21**) Morgan and Kramer, 35: 1834; 13. (**22**) Morgan and McAfee, 33: 1275; 11. (**23**) Morgan and Neidle, 35: 1856; 13. (**24**) Morgan and Owen, 33: 1713; 11. (**25**) Morgan and Scarlett, 39, 2275; 17. (**26**) Morgan and Schartz, 33: 1041; 11. (**27**) Morgan and Stone, 35: 1505; 13. (**28**) Morgan and Thomssen, 33: 657; 11. (**29**) Morgan and Woodward, 35: 1249; 13. (**30**) Richards and Boyer, 43: 274; 21. (**31**) Richards and Carver 43: 827; 21. (**32**) Richards and Coombs, 37: 1656; 15. (**33**) Richards and Matthews, 30: 8; 08. (**34**) Richards, Speyers and Carver, 46: 1196: 24.

AD—

Atti della reale accademia nazionale dei Lincei. (**1**) Magini, 19 II: 184; 10.

AdC—Annalen der Chemie, Justus Liebig's. (**1**) Schiff. 223: 47; 84

AdP—

Annalen der Physik. (**1**) Heydweiller, 62: 694; 97. (**2**) Quincke, 134: 356; 68. (**3**) Quincke, 135: 621; 68. (**4**) Gradenwitz, 67: 467; 99. (**5**) Meyer, 66: 523; 98. (**6**) Stöckle 66: 499; 98. (**7**) Rontgen and Schneider, 29: 165; 86.

AS—

Archives des sciences physicques at naturelles. (**1**) Dutoit and Friederich, 9: 105; 00. (**2**) Guye and Baud, 11: 449; 01. (**3**) Herzen, 14: 232; 02.

BD—

Berichte der deutschen Chemischen Gesellschaft. (**1**) Lorenz and Kauffler, 41: 3727; 08. (**2**) Schenck and Ellenberger, 37: 3443; 04.

BF—

Bulletin de la société Chimique de France. (**1**) Dutoit and Frederich, 19: 321; 98. (**2**) Sentis Ann. Univ. Grenoble, 27: 593; 04.

CR—

Comptes Rendus. (**1**) A Portevin and P. Bastien, 202; 1072–1074; 36. (**2**) Popesco, 172: 1474; 21.

cUL—

Communications from the Physical Laboratory at the University of Leiden. (**1**) No. 142. (**2**) No. 179a. (**3**) No. 182b.

GC—

Gazzetta chimica italiana. (**1**) Schitt, 14: 368; 84. (**2**) A. Giacolone and D. DiMaggio, 3: 198–206: 39

JI—

Journal of the Institute of Metals, London. (**1**) Smith, 12: 168; 14.

JP—

Journal de Chimie physique. (**1**) Dutoit and Fath, 1: 358; 03. (**2**) Dutoit and Mojoiu, 7: 169; 09. (**3**) Marchand, 11: 573; 13. (**4**) Bolle and Guye, 3: 38; 05. (**5**) Renard and Guye, 5: 81; 07. (**6**) Berthoud, 21: 143; 24. (**7**) Berthoud, 16: 429; 18. (**8**) Homfray and Guye, 1: 505; 03. (**9**) Przyluska, 7: 511: 09.

JR—

Journal of the Russian Physico Chem. Soc. (Chem. part). (**1**) Elisseev and Kurbatov, 41: 1426; 09.

JS—

Journal of the Chemical Society of London. (**1**) Kellas, 113: 903; 18. (**2**) Aston and Ramsay, 65: 167; 94. (**3**) Turner and Merry, 97: 2069; 10. (**4**) Ramsay and Shields, 63: 1089; 93. (**5**) Sugden, Reed and Wilkins, 127: 1525; 25. (**6**) Hewitt and Winmill, 91: 441; 07. (**7**) Smith, 105: 1703; 14. (**8**) Atkins, 99: 10; 11. (**9**) Sugden, 119: 1483; 21. (**10**) Sugden, 121: 858, 22. (**11**) Sugden, 125: 32; 24. (**12**) Sugden, 125: 1167; 24. (**13**) Worley, 105; 260; 14. (**14**) Worley, 105: 273; 14. (**15**) Baly and Donnan, 81: 907; 02.

JSG—Jahresb. Schles. Ges. Vaterl. Kultur. (**1**) Wilborn, 1912: 56.

MB—Metron. Beit. (**1**) Weinstein, No. 6: 89.

MfC—

Monatshefte für Chemie und verwandte Teile anderer Wissenschaften. (**1**) Kremann and Meingast, 35: 1323; 14.

PM—

Philosophical Magazine and Journal of Science, Lond, Edinburg and Dublin. (**1**) Bircumshaw, 2: 341; 26. 3: 1286; 27. (**2**) R. C. Brown, 13: 578–584; 32. (**3**) A. E. Bate, 28: 252–255; 39.

PR—The Physical Review. (**1**) Poindexter, 27: 820; 26.

PRA—

Proceedings of the Royal Academy of Sciences of Amsterdam. (**1**) Kamerlingh Onnes and Kuypers, 17: 528; 14. (**2**) Van Urk, Keesom and Onnes, 28: 958; 25. (**3**) Van Urk, Keesom and Nijhoff 29: 914; 26. (**4**) Jaeger and Kahn, 18: 75; 15.

PRS—

Proceedings of the Royal Society, London. (**1**) Hardy 88: 303; 13. (**2**) Ramsay and Aston, 56: 162; 182; 94.

tPRS—

Philosophical Transactions Royal Society of London, Series A. (**1**) Ramsay and Shields, 184: 647; 93.

tRIA—

Royal Irish Academy Transactions. (**1**) Ramsay and Aston, 32A: 93; 02.

VK—

Verslag koninklijke Academie van Wetenschappen te Amsterdam. (**1**) Verschaffelt, No. 18: 1895; 74. (**2**) Verschaffelt, No. 28, 1896: 94.

WN—

Wissenschaftliche Natuurk. Tydschr. (**1**) Verschaffelt, 2: 231, 25.

ZA—

Zeitschrift fur anorganische and allgemeine Chemie. (**1**) Jaeger, 101: 1; 17. (**2**) Motylewski, 38: 410, 04. (**3**) Lorenz, Liebmann and Hochberg, 94: 301; 16. (**4**) Sauerwald and Drath 154: 79; 26.

ZC—

Zeitschrift für physikalische Chemie. (**1**) Bennett and Mitchell, 84: 475; 13. (**2**) Walden and Swinne, 82: 271; 13. (**3**) Drucker, 52: 641; 05. (**4**) Walden, 75: 555; 10. (**5**) Walden and Swinne, 79: 700; 12. (**6**) Whatmough, 39: 129: 02.

ZE—

Zeitschrift fur Elektrochemie und angewandte physikalische Chemie. (**1**) Bredig and Tiechmann. 31: 449: 25.

ZK—

Zeitschrift Kolloid. (**1**) N. Jermolanko, 48: 14–146; 29.

VISCOSITY

Viscosity.—All fluids possess a definite resistance to change of form and many solids show a gradual yielding to forces tending to change their form. This property, a sort of internal friction, is called viscosity; it is expressed in dyne-seconds per cm^2 or poises. Dimensions,—$[m\,l^{-1}\,t^{-1}]$. If the tangential force per unit area, exerted by a layer of fluid upon one adjacent is one dyne for a space rate of variation of the tangential velocity of unity, the viscosity is one poise.

Kinematic viscosity is the ratio of viscosity to density. The c. g. s. unit of kinematic viscosity is the **stoke.**

Flow of liquids through a tube; where l is the length of the tube, r its radius, p the difference of pressure at the ends, η the coefficient of viscosity, the volume escaping per second,

$$v = \frac{\pi p r^4}{8 l \eta} \text{ (Poiseuille).}$$

The volume will be given in cm^3 per second if l and r are in cm, p in dynes per cm^2 and η in poises or dyne-seconds per cm^2.

VISCOSITY OF WATER BELOW 0°C

White-Twining 1914

Temperature	Viscosity centipoises	Temperature	Viscosity centipoises
0°C	1.798	−7.23	2.341
−2.10	1.930	−8.48	2.458
−4.70	2.121	−9.30	2.549
−6.20	2.250		

ABSOLUTE VISCOSITY OF WATER AT 20°C

Swindells, J. R. Coe, Jr., and T. B. Godfrey, Journal of Research, National Bureau of Standards **48**, 1, 1952.

The value found for the viscosity of water at 20°C was 0.010019 ± 0.000003 poise.

The value **0.01002** poise is to be used as the absolute value of the viscosity of water for calibration purposes.

VISCOSITY CONVERSION TABLE

Poise = c.g.s. unit of absolute viscosity = $\frac{gm}{sec \times cm}$

Stoke = c.g.s. unit of kinematic viscosity = $\frac{gm}{sec \times cm \times density_{(t°F)}}$

Centipoise = 0.01 poise
Centistoke = 0.01 stoke
Centipoises = Centistokes × density (at given temperature)

To convert poises to $\frac{lb}{sec \times ft}$ or $\frac{lb}{hr \times ft}$ multiply by **0.0672** or **242** respectively.

Centistokes	Saybolt Seconds at 100°F	Saybolt Seconds at 130°F	Saybolt Seconds at 210°F	Redwood Seconds at 70°F	Redwood Seconds at 140°F	Redwood Seconds at 200°F	Engler Degrees at all temps.
2.0	32.6	32.7	32.8	30.2	31.0	31.2	1.14
3.0	36.0	36.1	36.3	32.7	33.5	33.7	1.22
4.0	39.1	39.2	39.4	35.3	36.0	36.3	1.31
5.0	42.3	42.4	42.6	37.9	38.5	38.9	1.40
6.0	45.5	45.6	45.8	40.5	41.0	41.5	1.48
7.0	48.7	48.8	49.0	43.2	43.7	44.2	1.56
8.0	52.0	52.1	52.4	46.0	46.4	46.9	1.65
9.0	55.4	55.5	55.8	48.9	49.1	49.7	1.75
10.0	58.8	58.9	59.2	51.7	52.0	52.6	1.84
11.0	62.3	62.4	62.7	54.8	55.0	55.6	1.93
12.0	65.9	66.0	66.4	57.9	58.1	58.8	2.02
14.0	73.4	73.5	73.9	64.4	64.6	65.3	2.22
16.0	81.1	81.3	81.7	71.0	71.4	72.2	2.43
18.0	89.2	89.4	89.8	77.9	78.5	79.4	2.64
20.0	97.5	97.7	98.2	85.0	85.8	86.9	2.87
22.0	106.0	106.2	106.7	92.4	93.3	94.5	3.10
24.0	114.6	114.8	115.4	99.9	100.9	102.2	3.34
26.0	123.3	123.5	124.2	107.5	108.6	110.0	3.58
28.0	132.1	132.4	133.0	115.3	116.5	118.0	3.82
30.0	140.9	141.2	141.9	123.1	124.4	126.0	4.07
32.0	149.7	150.0	150.8	131.0	132.3	134.1	4.32
34.0	158.7	159.0	159.8	138.9	140.2	142.2	4.57
36.0	167.7	168.0	168.9	146.9	148.2	150.3	4.83
38.0	176.7	177.0	177.9	155.0	156.2	158.3	5.08
40.0	185.7	186.0	187.0	163.0	164.3	166.7	5.34
42.0	194.7	195.1	196.1	171.0	172.3	175.0	5.59
44.0	203.8	204.2	205.2	179.1	180.4	183.3	5.85
46.0	213.0	213.4	214.5	187.1	188.5	191.7	6.11
48.0	222.2	222.6	223.8	195.2	196.6	200.0	6.37
50.0	231.4	231.8	233.0	203.3	204.7	208.3	6.63
60.0	277.4	277.9	279.3	243.5	245.3	250.0	7.90
70.0	323.4	324.0	325.7	283.9	286.0	291.7	9.21
80.0	369.6	370.3	372.2	323.9	326.6	333.4	10.53
90.0	415.8	416.6	418.7	364.4	367.4	375.0	11.84
*100.0	462.0	462.9	465.2	404.9	408.2	416.7	13.16

*** At higher values use the same ratio as above for 100 centistokes; e.g., 110 centistokes = 110 × 4.620 Saybolt seconds at 100°F.**

To obtain the Saybolt Universal viscosity equivalent to a kinematic viscosity determined at t°F, multiply the equivalent Saybolt Universal viscosity at 100°F by $1 + (t - 100)0.000064$; e.g., 10 centistokes at 210°F are equivalent to 58.8 × 1.0070, or 59.2 Saybolt Universal seconds at 210°F.

VISCOSITY CONVERSION

Kinematic

To convert from	To	Multiply by
cm^2/sec (Stokes)	Centistokes	10^2
	ft^2/hr.	3.875
	ft^2/sec.	1.076×10^{-3}
	in.^{2}/sec.	1.550×10^{-1}
	m^2/hr.	3.600×10^{-1}
cm^2/sec $\times 10^2$ (Centistokes)	cm^2/sec. (Stokes)	1×10^{-2}
	ft^2/hr.	3.875×10^{-2}
	ft^2/sec.	1.076×10^{-5}
	in.^{2}/sec.	1.550×10^{-3}
	m^2/hr.	3.600×10^{-3}
ft^2/hr	cm^2/sec (Stokes)	2.581×10^{-1}
	cm/sec $\times 10^2$ (centistokes)	2.581×10
	ft^2/sec.	2.778×10^{-4}
	in.^{2}/sec.	4.00×10^{-2}
	m^2/hr.	9.290×10^{-2}
ft^2/sec	cm^2/sec. (Stokes)	9.29×10^2
	cm^2/sec. $\times 10^2$ (centistokes)	9.29×10^4
	ft^2/hr	3.60×10^3
	in.^{2}/sec.	1.44×10^2
	m^2/hr.	3.345×10^2
in.^{2}/sec	cm^2/sec. (Stokes)	6.452
	cm^2/sec. $\times 10^2$ (centistokes)	6.452×10^2
	ft^2/hr.	2.50×10
	ft^2/sec.	6.944×10^{-3}
	m^2/hr.	2.323
m^2/hr	cm^2/sec. (Stokes)	2.778
	cm^2/sec. $\times 10^2$ (centistokes)	2.778×10^2
	ft^2/hr.	1.076×10
	ft^2/sec.	2.990×10^{-3}
	in.^{2}/sec.	4.306×10^{-1}

VISCOSITY CONVERSION

Absolute

Absolute viscosity = kinematic viscosity × density; lb = mass pounds; lb$_F$ = force pounds

To convert from	To	Multiply by
gm/(cm)(sec) [Poise]	gm/(cm)(sec)(10^2) [Centipoise]	10^2
	kg/(m)(hr)	3.6×10^2
	lb/(ft)(sec)	6.72×10^{-2}
	lb/(ft)(hr)	2.419×10^2
	lb/(in.)(sec)	5.6×10^{-3}
	(gm$_F$)(sec)/cm^2	1.02×10^{-3}
	(lb$_F$)(sec)/in.^{2} [Reyn]	1.45×10^{-5}
	(lb$_F$)(sec)/ft^2	2.089×10^{-3}
gm/(cm)(sec)(10^2) [Centipoise]	gm/(cm)(sec) [Poise]	10^{-2}
	kg/(m)(hr)	3.6
	lb/(ft)(sec)	6.72×10^{-4}
	lb/(ft)(hr)	2.419
	lb/(in.)(sec)	5.60×10^{-5}
	(gm$_F$)(sec)/cm^2	1.02×10^{-5}
	(lb$_F$)(sec)/in.^{2} [Reyn]	1.45×10^{-7}
	(lb$_F$)(sec)/ft^2	2.089×10^{-5}
kg/(m)(hr)	gm/(cm)(sec)	2.778×10^{-3}
	gm/(cm)(sec)(10^2) [Centipoise]	2.778×10^{-1}
	lb/(ft)(sec)	1.867×10^{-4}
	lb/(ft)(hr)	6.720×10^{-1}
	lb/(in.)(sec)	1.555×10^{-5}
	(gm$_F$)(sec)/cm^2	2.833×10^{-4}

VISCOSITY CONVERSION

Absolute (Continued)

To convert from	To	Multiply by
kg/(m)(hr) (Cont.)	(lb$_F$)(sec)/in.^{2} [Reyn]	4.029×10^{-8}
	(lb$_F$)(sec)/ft^2	5.801×10^{-6}
lb/(ft)(sec)	gm/(cm)(sec) [Poise]	1.488×10^1
	gm/(cm)(sec)(10^2) [Centipoise]	1.488×10^3
	kg/(m)(hr)	5.357×10^3
	lb/(ft)(hr)	3.60×10^3
	lb/(in.)(sec)	8.333×10^{-2}
	(gm$_F$)(sec)/cm^2	1.518×10^{-2}
	(lb$_F$)(sec)/in.^{2} [Reyn]	2.158×10^{-4}
	(lb$_F$)(sec)/ft^2	3.108×10^{-2}
lb/(ft)(hr)	gm/(cm)(sec) [Poise]	4.134×10^{-3}
	gm/(cm)(sec)(10^2) [Centipoise]	4.134×10^{-1}
	kg/(m)(hr)	1.488
	lb/(ft)(sec)	2.778×10^{-4}
	lb/(in.)(sec)	2.315×10^{-5}
	(gm$_F$)(sec)/cm^2	4.215×10^{-6}
	(lb$_F$)(sec)/in.^{2} [Reyn]	5.996×10^{-8}
	(lb$_F$)(sec)/ft^2	8.634×10^{-6}
lb/(in.)(sec)	gm/(cm)(sec) [Poise]	1.786×10^2
	gm/(cm)(sec)(10^2) [Centipoise]	1.786×10^4
	kg/(m)(hr)	6.429×10^4
	lb/(ft)(sec)	1.20×10
	lb/(ft)(hr)	4.32×10^4
	(gm$_F$)(sec)/cm^2	1.821×10^{-1}
	(lb$_F$)(sec)/in.^{2} [Reyn]	2.590×10^{-3}
	(lb$_F$)(sec)/ft^2	3.73×10^{-1}
(gm$_F$)(sec)/cm^2	gm/(cm)(sec)	9.807×10^2
	gm/(cm)(sec)(10^2) [Centipoise]	9.807×10^4
	kg/(m)(hr)	3.530×10^5
	lb/(ft)(sec)	6.590×10
	lb/(ft)(hr)	2.372×10^5
	lb/(in.)(sec)	5.492
	(lb$_F$)(sec)/in.^{2} [Reyn]	1.422×10^{-2}
	(lb$_F$)(sec)/ft^2	2.048
(lb$_F$)(sec)/in.^{2} [Reyn]	gm/(cm)(sec) [Poise]	6.895×10^4
	gm/(cm)(sec)(10^2) [Centipoise]	6.895×10^6
	kg/(m)(hr)	2.482×10^7
	lb/(ft)(sec)	4.633×10^3
	lb/(ft)(hr)	1.668×10^7
	lb/(in.)(sec)	3.861×10^2
	(gm$_F$)(sec)/cm^2	7.031×10
	(lb$_F$)(sec)/ft^2	1.440×10^2
(lb$_F$)(sec)/ft^2	gm/(cm)(sec) [Poise]	4.788×10^2
	gm/(cm)(sec)(10^2) [Centipoise]	4.788×10^4
	kg/(m)(hr)	1.724×10^5
	lb/(ft)(sec)	3.217×10
	lb/(ft)(hr)	1.158×10^5
	lb/(in.)(sec)	2.681
	(gm$_F$)(sec)/cm^2	4.882×10^{-1}
	(lb$_F$)(sec)/in.^{2} [Reyn]	6.944×10^{-3}

THE VISCOSITY OF WATER 0°C TO 100°C

Contribution from the National Bureau of Standards, not subject to copyright.

°C	η(cp)	°C	η(cp)	°C	η(cp)	°C	η(cp)
0	1.787	26	0.8705	52	0.5290	78	0.3638
1	1.728	27	.8513	53	.5204	79	.3592
2	1.671	28	.8327	54	.5121	80	.3547
3	1.618	29	.8148	55	.5040	81	.3503
4	1.567	30	.7975	56	.4961	82	.3460
5	1.519	31	.7808	57	.4884	83	.3418
6	1.472	32	.7647	58	.4809	84	.3377
7	1.428	33	.7491	59	.4736	85	.3337
8	1.386	34	.7340	60	.4665	86	.3297
9	1.346	35	.7194	61	.4596	87	.3259
10	1.307	36	.7052	62	.4528	88	.3221
11	1.271	37	.6915	63	.4462	89	.3184
12	1.235	38	.6783	64	.4398	90	.3147
13	1.202	39	.6654	65	.4335	91	.3111
14	1.169	40	.6529	66	.4273	92	.3076
15	1.139	41	.6408	67	.4213	93	.3042
16	1.109	42	.6291	68	.4155	94	.3008
17	1.081	43	.6178	69	.4098	95	.2975
18	1.053	44	.6067	70	.4042	96	.2942
19	1.027	45	.5960	71	.3987	97	.2911
20	1.002	46	.5856	72	.3934	98	.2879
21	0.9779	47	.5755	73	.3882	99	.2848
22	.9548	48	.5656	74	.3831	100	.2818
23	.9325	49	.5561	75	.3781		
24	.9111	50	.5468	76	.3732		
25	.8904	51	.5378	77	.3684		

The above table was calculated from the following empirical relationships derived from measurements in viscometers calibrated with water at 20°C (and one atmosphere), modified to agree with the currently accepted value for the viscosity at 20° of 1.002 cp:

$$0^\circ \text{ to } 20^\circ\text{C: } \log_{10} \eta_T = \frac{1301}{998.333 + 8.1855(T-20) + 0.00585(T-20)^2} - 3.30233$$

(R. C. Hardy and R. L. Cottington, J.Res.NBS *42*, 573 (1949).)

$$20^\circ \text{ to } 100^\circ\text{C: } \log_{10} \frac{\eta_T}{\eta_{20}} = \frac{1.3272(20-T) - 0.001053(T-20)^2}{T + 105}$$

(J. F. Swindells, NBS, unpublished results.)

VISCOSITY OF LIQUIDS

Viscosity of liquids in centipoises (cp) including elements, inorganic and organic compounds and mixtures.

Liquid	Temp. °C	Viscosity cp
Acetaldehyde	0	.2797
	10	.2557
	20	.22
Acetanilide	120	2.22
	130	1.90
Acetic acid	15	1.31
	18	1.30
	25.2	1.155
	30	1.04
	41	1.00
	59	.70
	70	.60
	100	.43
anhydride	0	1.24
	15	.971
	18	.90
	30	.783
	100	.49
Acetone	−92.5	2.148
	−80.0	1.487
	−59.6	.932
	−42.5	.695
	−30.0	.575
	−20.9	.510
	−13.0	.470
	−10.0	.450
	0	.399
	15	.337
	25	.316
	30	.295
	41	.280
Acetonitrile	0	.442
	15	.375
	25	.345
Acetophenone	11.9	2.28
	23.5	1.59
	25.0	1.617
	50.0	1.246
	80.0	.734
Air, liq	−192.3	.172
Alcohol. See *Ethyl, Methyl,* etc.		
Allyl alcohol	0	2.145
	15	1.49
	20	1.363
	30	1.07
	40	.914
	70	.553
Allylamine	130	.506
Allyl chloride	15	.347
	30	.300
Ammonia	−69	.475
	−50	.317
	−40	.276
	−33.5	.255
n-Amyl acetate	11	1.58
	45	.805
alcohol	15	4.65
	30	2.99
ether	15	1.188
Aniline	−6	13.8
	0	10.2
	5	8.06
	10	6.50
	15	5.31
	20	4.40
	25	3.71
	30	3.16
	35	2.71
	40	2.37
	50	1.85
	60	1.51
	70	1.27
	80	1.09
	90	.935
	100	.825
Anisol	0	1.78
	20	1.32
	40	1.12
Antimony, liq	645	1.55
	700	1.26
	800	1.08
	850	1.05
Benzaldehyde	25	1.39
Benzene	0	.912
	10	.758
	20	.652
	30	.564
	40	.503
	50	.442
	60	.392
	70	.358
	80	.329
Benzonitrile	25	1.24
Benzophenone	55	4.79
	120	1.38
Benzyl alcohol	20	5.8
Benzylamine	25	1.59
Benzylaniline	33	2.18
	130	1.20
Benzyl ether	0	10.5
	20	5.33
	40	3.21
Bismuth	285	1.61
	304	1.662
	365	1.46
	451	1.280
	600	.998
Bromine, liq	−4.3	1.31
	0	1.241

VICOSITY OF LIQUIDS (Continued)

Liquid	Temp. °C	Viscosity cp
Bromine, liq	12.6	1.07
	16	1.0
	19.5	.995
	28.9	.911
o-Bromoaniline	40	3.19
m-Bromoaniline	20	6.81
	40	3.70
	80	1.70
p-Bromoaniline	80	1.81
Bromobenzene	15	1.196
	30	.985
Bromoform	15	2.152
	25	1.89
	30	1.741
Butyl acetate	0	1.004
	20	.732
	40	.563
n-Butyl alcohol	−50.9	36.1
	−30.1	14.7
	−22.4	11.1
	−14.1	8.38
	0	5.186
	15	3.379
	20	2.948
	30	2.30
	40	1.782
	50	1.411
	70	.930
	100	.540
sec-Butyl alcohol	15	4.21
n-Butyl bromide	15	.626
n-Butyl chloride	15	.469
Butyl chloride, tertiary	15	.543
n-Butyl formate	0	.940
	20	.689
Butyric acid	0	2.286
	15	1.81
	20	1.540
	40	1.120
	50	.975
	70	.760
	100	.551
Cadmium, liq	349	1.44
	506	1.18
	603	1.10
Carbolic acid. See *Phenol.*		
Carbon dioxide, liq., pressure that of saturated vapor	0	.099
	10	.085
	20	.071
	30	.053
disulfide	−13	.514
	−10	.495
	0	.436
	5	.380
	20	.363
	40	.330
Carbon tetrachloride	0	1.329
	15	1.038
Carbon tetrachloride	20	.969
	30	.843
	40	.739
	50	.651
	60	.585
	70	.524
	80	.468
	90	.426
	100	.384
Cetyl alcohol	50	13.4
Chlorine, liq	−76.5	.729
	−70.5	.680
	−60.2	.616
	−52.4	.566
	−35.4	.494
	0	.385
Chlorobenzene	15	.900
	20	.799
	40	.631
	80	.431
	100	.367
Chloroform	−13	.855
	0	.700
	8.1	.643
	15	.596
	20	.58
	25	.542
	30	.514
	39	.500
o-Chlorophenol	25	4.11
	50	2.015
m-Chlorophenol	25	11.55
p-Chlorophenol	50	4.99
Copper, liq	1,085	3.36
	1,100	3.33
	1,150	3.22
	1,200	3.12
o-Cresol	40	4.49
m-Cresol	10	43.9
	20	20.8
	40	6.18
p-Cresol	40	7.00
Creosote	20	12.0
Cycloheptane	13.5	1.64
Cyclohexane	17	1.02
Cyclohexanol	20	68
Cyclohexene	13.5	.696
	20	.66
Cyclooctane	13.5	2.35
Cyclopentane	13.5	.493
*n*Decane	20	.92
Diethylamine	25	.346
	25	.367
Diethylaniline	.5	3.84
	20.0	2.18
	25.0	1.95
Diethylcarbinol	15.0	7.34
Diethylketone	15	.493

Liquid	Temp. °C	Viscosity cp	Liquid	Temp. °C	Viscosity cp
Dimethylaniline	10	1.69	Ethyl bromide	−80	1.81
	20	1.41		0	.487
	25	1.285		10	.441
	30	1.17		15	.418
	40	1.04		20	.402
	50	.91		30	.348
Dimethyl-α-naphthylamine	130	.868	*n*-Ethyl butyrate	15	.711
Dimethyl-β-naphthylamine	130	.952	Ethyl carbonate	15	.868
Diphenyl	70	1.49	Ethylene bromide	0	2.438
	100	.97		17	1.95
Diphenylamine	130	1.04		20	1.721
Dodecane	25	1.35		40	1.286
Ether (diethyl-)	−100	1.69		67.3	.922
	−80	.958		70	.903
	−60	.637		82.2	.750
	−40	.461		99.0	.648
	−20	.362	chloride	0	1.077
	0	.2842		15	.887
	17	.240		19.4	.800
	20	.2332		40	.652
	25	.222		50	.565
	40	.197		70	.479
	60	.166	glycol	20	19.9
	80	.140		40	9.13
	100	.118		60	4.95
Ethyl acetate	0	.582		80	3.02
	8.96	.516		100	1.99
	10	.512	oxide	−49.8	.577
	15	.473		−38.2	.488
	20	.455		−21.0	.394
	25	.441		0	.320
	30	.400	Ethyl formate	20	.402
	50	.345	iodide	0	.727
	75	.283		15	.617
Ethyl alcohol	−98.11	44.0		20	.592
	−89.8	28.4		40	.495
	−71.5	13.2		70	.391
	−59.42	8.41	malate	24.7	3.016
	−52.58	6.87	oxalate	15	2.31
	−32.01	3.84	propionate	15	.564
	−17.59	2.68	Eugenol	0	29.9
	−.30	1.80		20	9.22
	0	1.773		40	4.22
	10	1.466	Fluorobenzene	20	.598
	20	1.200		40	.478
	30	1.003		60	.389
	40	.834		80	.329
	50	.702		100	.275
	60	.592	Formamide	0	7.55
	70	.504		25	3.30
Ethyl alcohol, anh.	−148	8,470	Formic acid	7.59	2.3868
	−146	5,990		10	2.262
	−130	467		20	1.804
Ethyl aniline	25	2.04		30	1.465
Ethylbenzene	17	.691		40	1.219
Ethyl benzoate	20	2.24		70	.780
Ethyl bromide	−120	5.6		100	.549
	−100	2.89			

Liquid	Temp. °C	Viscosity cp
Furfural	0	2.48
	25	1.49
Glucose	22	9.1×10^{15}
	30	6.6×10^{13}
	40	2.8×10^{11}
	60	9.3×10^{7}
	80	6.6×10^{5}
	100	2.5×10^{4}
Glycerin	−42	6.71×10^{6}
	−36	2.05×10^{6}
	−25	2.62×10^{5}
	−20	1.34×10^{5}
	−15.4	6.65×10^{4}
	−10.8	3.55×10^{4}
	−4.2	1.49×10^{4}
	0	12,110
	6	6,260
	15	2,330
	20	1,490
	25	954
	30	629
Glycerin trinitrate	10	69.2
	20	36.0
	30	21.0
	40	13.6
	60	6.8
Heptane	0	.524
	17	.461
	20	.409
	25	.386
	40	.341
	70	.262
n-Heptyl alcohol	15	8.53
Hexadecane	20	3.34
Hexane	0	.401
	17	.374
	20	.326
	25	.294
	40	.271
	50	.248
Hydrazine	1	1.29
	10	1.12
	20	.97
Hydrogen, liq		.011
Iodine, liq	116	2.27
Iodobenzene	15	1.74
Iron, 2.5% carbon, liq	1,400	2.25
Isoamyl acetate	8.97	1.030
	19.91	.872
alcohol	10	6.20
amine	25	.724
Isobutyl alcohol	15	4.703
amine	25	.553
Isobutyric acid	15	1.44
	30	1.13
Isoeugenol	25	26.72
Isoheptane	0	.481
	20	.384
Isoheptane	40	.315
Isohexane	0	.376
	20	.306
	40	.254
Isopentane	0	.273
	20	.223
Isopropyl alcohol	15	2.86
	30	1.77
Isoquinoline	25	3.57
Isosafrol	25	3.981
Lead, liq	350	2.58
	400	2.33
	441	2.116
	500	1.84
	551	1.70
	600	1.38
	703	1.349
	844	1.185
Menthol, liq	55.6	6.29
	74.6	2.47
	99.0	1.04
Mercury	−20	1.855
	−10	1.764
	0	1.685
	10	1.615
	19.02	1.56
	20	1.554
	20.2	1.55
	30	1.499
	40	1.450
	40.8	1.45
	41.86	1.44
	50	1.407
	60	1.367
	70	1.331
	80	1.298
	90	1.268
	100	1.240
	150	1.130
	200	1.052
	250	.995
	300	.950
	340	.921
Methyl acetate	0	.484
	20	.381
	40	.320
Methyl alcohol	−98.30	13.9
(Methanol)	−84.23	6.8
	−72.55	4.36
	−44.53	1.98
	−22.29	1.22
	0	.82
	15	.623
	20	.597
	25	.547
	30	.510
	40	.456
	50	.403

Liquid	Temp. °C	Viscosity cp
Methyl amine	0	.236
aniline	25	2.02
	30	1.55
chloride	20	.1834
Methylene bromide	15	1.09
	30	0.92
chloride	15	.449
	30	.393
Methyl iodide	0	.606
	15	.518
	20	.500
	30	.460
	40	.424
Naphthalene	80	.967
	100	.776
Nitric acid	0	2.275
	10	1.770
Nitrobenzene	2.95	2.91
	5.69	2.71
	5.94	2.71
	9.92	2.48
	14.94	2.24
	20.00	2.03
Nitromethane	0	.853
	25	.620
o-Nitrotoluene	0	3.83
	20	2.37
	40	1.63
	60	1.21
m-Nitrotoluene	20	2.33
	40	1.60
	60	1.18
p-Nitrotoluene	60	1.20
n-Nonane	20	.711
n-Octane	0	.706
	16	.574
	20	.542
	40	.433
Octodecane	40	2.86
n-Octylalcohol	15	10.6
Oil, castor	10	2,420
	20	986
	30	451
	40	231
	100	16.9
cottonseed	20	70.4
cylinder, filtered	37.8	240.6
	100	18.7
cylinder, dark	37.8	422.4
	100	24.0
linseed	30	33.1
	50	17.6
	90	7.1
machine, light	15.6	113.8
	37.8	34.2
	100	4.9
machine, heavy	15.6	660.6
	37.8	127.4

Liquid	Temp. °C	Viscosity cp
Oil, olive	10	138.0
	20	84.0
	40	36.3
	70	12.4
rape	0	2,530
	10	385
	20	163
	30	96
soya bean	20	69.3
	30	40.6
	50	20.6
	90	7.8
sperm	15.6	42.0
	37.8	18.5
	100.0	4.6
Oleic acid	30	25.6
Pentadecane	22	2.81
Pentane	0	.289
	20	.240
o-Phenetidine	0	16.5
	20	6.08
	30	4.22
m-Phenetidine	30	12.9
p-Phenetidine	20	12.9
	30	8.3
Phenol	18.3	12.7
	50	3.49
	60	2.61
	70	2.03
	90	1.26
Phenylcyanide	.28	1.96
	20.0	1.33
Phosphorus, liq	21.5	2.34
	31.2	2.01
	43.2	1.73
	50.5	1.60
	60.2	1.45
	69.7	1.32
	79.9	1.21
Potassium bromide, liq	745	1.48
	775	1.34
	805	1.19
nitrate, liq	334	2.1
	358	1.7
	333	2.97
	418	2.00
Propionic acid	10	1.289
	15	1.18
	20	1.102
	40	.845
Propyl acetate	10	.66
	20	.59
	40	.44
n-Propyl alcohol	0	3.883
	15	2.52
	20	2.256
	30	1.72
	40	1.405

Liquid	Temp. °C	Viscosity cp	Liquid	Temp. °C	Viscosity cp
n-Propyl alcohol	50	1.130	Sulfuric acid	50	8.82
	70	.760		60	7.22
Propyl aldehyde	10	.47		70	6.09
	20	.41		80	5.19
	40	.33	Tetrachloroethane	15	1.844
bromide	0	.651	Tetradecane	20	2.18
	20	.524	Tin, liq	240	2.12
	40	.433		280	1.678
chloride	0	.436		300	1.73
	20	.352		301	1.680
	40	.291		400	1.43
n-Propyl ether	15	.448		450	1.270
Pyridine	20	.974		500	1.20
Salicylic acid	10	3.20		600	1.08
	20	2.71		604	1.045
	40	1.81		750	.905
Salol	45	.746	Toluene	0	.772
Sodium bromide	762	1.42		17	.61
	780	1.28		20	.590
chloride, liq	841	1.30		30	.526
	896	1.01		40	.471
	924	.97		70	.354
nitrate, liq	308	2.919	*o*-Toluidine	20	4.39
	348	2.439	*m*-Toluidine	20	3.81
	398	1.977	*p*-Toluidine	50	1.80
	418	1.828	Triacetin	17	28.0
Stearic acid	70	11.6	Tributyrin	20	11.6
Sucrose (cane sugar)	109	2.8×10^6	Trichlorethane	20	1.2
	124.6	1.9×10^5	Tridecane	23.3	1.55
Sulfur (gas free)	123.0	10.94	Triethylcarbinol	20	6.75
	135.5	8.66	Tripalmitin	70	16.8
	149.5	7.09	Tristearin	75	18.5
	156.3	7.19	Turpentine	0	2.248
	158.2	7.59		10	1.783
	159.2	9.48		20	1.487
	159.5	14.45		30	1.272
	160.0	22.83		40	1.071
	160.3	77.32		70	.728
	165.0	500.0	Turpentine, Venice	17.3	1.3×10^5
	171.0	4,500.0	*n*-Undecane	20	1.17
	184.0	16,000.00	*o*-Xylene (xylol)	0	1.105
	190.5	19,700.0		16	.876
	197.5	21,300.0		20	.810
	200.0	21,500.0		40	.627
	210.0	20,500.0	*m*-Xylene (xylol)	0	.806
	217.0	19,100.0		15	.650
	220.0	18,600.0		20	.620
Sulfur dioxide, liq	−33.5	.5508		40	.497
	−10.5	.4285	*p*-Xylene (xylol)	16	.696
	0.1	.3936		20	.648
Sulfuric acid	0	48.4		40	.513
	15	32.8	Zinc, liq	280	1.68
	20	25.4		357	1.42
	30	15.7		389	1.31
	40	11.5			

VISCOSITY OF GASES

Gas or vapor	Temp. °C	Viscosity micro-poises
Acetic acid, vap	119.1	107.0
Acetone, vap	100	93.1
	119.0	99.1
	190.4	118.6
	247.7	133.4
	306.4	148.1
Acetylene	0	93.5
Air	−194.2	55.1
	−183.1	62.7
	−104.0	113.0
	−69.4	133.3
	−31.6	153.9
	0	170.8
	18	182.7
	40	190.4
	54	195.8
	74	210.2
	229	263.8
	334	312.3
	357	317.5
	409	341.3
	466	350.1
	481	358.3
	537	368.6
	565	375.0
	620	391.6
	638	401.4
	750	426.3
	810	441.9
	923	464.3
	1034	490.6
	1134	520.6
Alcohol. See *Ethyl*, *Methyl*, etc.		
Ammonia	−78.5	67.2
	0	91.8
	20	98.2
	50	109.2
	100	127.9
	132.9	139.9
	150	146.3
	200	164.6
	250	181.4
	300	198.7
Argon	0	209.6
	20	221.7
	100	269.5
	200	322.3
	302	368.5
	401	411.5
	493	448.4
	584	481.5
	714	525.7
	827	563.2
Arsenic hydride (Arsine)	0	145.8
	15	114.0
	100	198.1
Benzene, vap	14.2	73.8
	131.2	103.1
	194.6	119.8
	252.5	134.3
	312.8	148.4
Bromine, vap	12.8	151
	65.7	170
	99.7	188
	139.7	208
	179.7	227
	220.3	248
Bromoform, vap	151.2	253.0
Butyl alcohol, *n*, vap	116.9	143
tert, vap	82.9	160
chloride, *n*, vap	78	149.5
iodide, vap	130	202
β-Butylene	18.8	74.4
	100.4	94.5
	200	119.2
Butyric acid, vap	161.7	130.0
Carbon dioxide	−97.8	89.6
	−78.2	97.2
	−60.0	106.1
	−40.2	115.5
	−21	129.4
	−19.4	126.0
	0	139.0
	15	145.7
	19	149.9
	20	148.0
	30	153
	32	155
	35	156
	40	157
	99.1	186.1
	104	188.9
	182.4	222.1
	235	241.5
	302.0	268.2
	490	330.0
	685	380.0
	850	435.8
	1052	478.6
disulfide, vap	0	91.1
	14.2	96.4
	114.3	130.3
	190.2	156.1
	309.8	196.6
monoxide	−191.5	56.1
	−78.5	127
	0	166
	15	172
	21.7	175.3
	126.7	218.3
	227.0	254.8
	276.9	271.4
tetrachloride, vap	76.7	195.0

Gas or vapor	Temp. °C	Viscosity micro-poises
Carbon tetrachloride, vap	127.9	133.4
	200.2	156.2
	314.9	190.2
Chlorine	12.7	129.7
	20	132.7
	50	146.9
	100	167.9
	150	187.5
	200	208.5
	250	227.6
Chloroform, vap	0	93.6
	14.2	98.9
	100	129
	121.3	135.7
	189.1	157.9
	250.0	177.6
	307.5	194.7
Cyanogen	0	92.8
	17	98.7
	100	127.1
Ethane	−78.5	63.4
	0	84.8
	17.2	90.1
	50.8	100.1
	100.4	114.3
	200.3	140.9
Ether (diethyl), vap	0	67.8
	14.2	71.6
	100	95.5
	121.8	98.3
	159.4	107.9
	189.9	115.2
	251.0	130.0
	277.8	135.8
Ethyl acetate, vap	0	68.4
	100	94.3
	128.1	101.8
	158.6	109.8
	192.9	119.5
	212.5	126
alcohol, vap	100	108
	130.2	117.3
	170.7	129.3
	191.8	135.5
	212.5	140
	251.7	151.9
	308.7	167.0
bromide, vap	38.4	186.5
butyrate, vap	119.8	160.0
chloride, vap	0	93.7
Ethylene	−75.7	69.9
	−44.1	76.9
	−38.6	78.5
	0	90.7
	13.8	95.4
	20	100.8
Ethylene	50	110.3
	100	125.7
	150	140.3
	200	154.1
	250	166.6
bromide, vap	131.6	221.0
chloride, vap	83.5	168.0
Ethyl formate, vap	99.8	92
iodide, vap	72.3	216.0
Helium	−257.4	27.0
	−252.6	35.0
	−191.6	87.1
	0	186.0
	20	194.1
	100	228.1
	200	267.2
	250	285.3
	282	299.2
	407	343.6
	486	370.6
	606	408.7
	676	430.3
	817	471.3
Hydrogen	−257.7	5.7
	−252.5	8.5
	−198.4	33.6
	−183.4	38.8
	−113.5	57.2
	−97.5	61.5
	−31.6	76.7
	0	83.5
	20.7	87.6
	28.1	89.2
	129.4	108.6
	229.1	126.0
	299	138.1
	412	155.4
	490	167.2
	601	182.9
	713	198.2
	825	213.7
bromide	18.7	181.9
	100.2	234.4
chloride	12.5	138.5
	16.5	140.7
	18	142.6
	100.3	182.2
iodide	20	165.5
	50	201.8
	100	231.6
	150	262.7
	200	292.4
	250	318.9
phosphide	0	106.1
	15	112.0
	100	143.8
sulfide	0	116.6

Gas or vapor	Temp. °C	Viscosity micro-poises
Hydrogen sulfide	17	124.1
	100	158.7
Iodine, vap	124.0	184
	170.0	204
	205.4	220
	247.1	240
Isobutyl acetate, vap	16.1	76.4
	116.4	155.0
alcohol, vap	108.4	144.5
bromide, vap	92.3	179.5
butyrate, vap	156.9	167.0
chloride, vap	68.5	150.0
iodide, vap	120	204.7
Isopentane, vap	25	69.5
	100	86.0
Isopropyl alcohol, vap	99.8	109
	120.3	103.1
	198.4	124.8
	293.1	148.8
bromide, vap	60	176.0
chloride, vap	37.0	148.5
iodide, vap	89.3	201.5
Krypton	0	232.7
	15	246
Mercury, vap	273	494
	313	551
	369	641
	380	654
Methane	−181.6	34.8
	−78.5	76.0
	0	102.6
	20	108.7
	100.0	133.1
	200.5	160.5
	284	181.3
	380	202.6
	499	226.4
Methyl acetate, vap	99.8	98
	100	100
	143.3	113.9
	218.5	134.8
alcohol, vap	66.8	135.0
	111.3	125.9
	217.5	162.0
	311.5	192.1
chloride	−15.3	92
	0	96.9
	15.0	104
	99.1	137
	182.4	168
	302.0	211
iodide, vap	44	232
Neon	0	297.3
	20	311.1
	100	364.6
	200	424.8
	250	453.2

Gas or vapor	Temp. °C	Viscosity micro-poises
Neon	285	470.8
	429	545.4
	502	580.2
	594	623.0
	686	662.6
	827	721.0
Nitric oxide (NO)	0	178
	20	187.6
	100	227.2
	200	268.2
Nitrogen	−21.5	156.3
	10.9	170.7
	27.4	178.1
	127.2	219.1
	226.7	255.9
	299	279.7
	490	337.4
	825	419.2
Nitrosyl chloride	15	113.9
	100	150.4
	200	192.0
Nitrous oxide (N_2O)	0	135
	26.9	148.8
	126.9	194.3
n-Nonane, vap	100.3	63.3
	202.1	78.1
n-Octane, vap	100.4	67.5
	202.2	84.8
Oxygen	0	189
	19.1	201.8
	127.7	256.8
	227.0	301.7
	283	323.3
	402	369.3
	496	401.3
	608	437.0
	690	461.2
	829	501.2
n-Pentane, vap	25	67.6
	100	84.1
Propane	17.9	79.5
	100.4	100.9
	199.3	125.1
n-Propyl alcohol, vap	99.9	93
	121.7	102.5
	209.7	126.7
	273.0	143.4
bromide, vap	99.8	119
Propylene	16.7	83.4
	49.9	93.5
	100.1	107.6
	199.4	133.8
Propyl iodide, vap	102	210.0
Sulfur dioxide	−75.0	85.8
	−20.0	107.8
	0	115.8
	0	117

VISCOSITY OF GASES (Continued)

Gas or vapor	Temp. °C	Viscosity micro-poises	Gas or vapor	Temp. °C	Viscosity micro-poises
Sulfur dioxide	18	124.2	Water, vap	150	144.5
	20.5	125.4		200	163.5
	100.4	161.2		250	182.7
	199.4	203.8		300	202.4
	293	244.7		350	221.8
	490	311.5		400	241.2
Trimethylbutane. (2,2,3-),			Xenon	0	210.1
vap	70.3	73.4		16.5	223.5
	132.2	82.7		20	226.0
	262.1	104.8		127	300.9
Trimethylethylene, vap	25	70.1		177	335.1
	100	86.9		227	365.2
Water, vap	100	125.5		277	395.4

DIFFUSIVITIES OF GASES IN LIQUIDS

Solute	Solvent	Temp., °C	Diffusivity x 10^5, cm^2/sec
H_2	*n*-Hexane	25.4	16.36
H_2	Cyclohexane	25.4	7.08
H_2	Ethylene glycol	25.4	0.75
H_2	Carbon tetrachloride	0	6.28
O_2	Cyclohexane	29.6	5.31
O_2	Carbon tetrachloride	25.4	3.71
O_2	Ethanol	29.6	2.64
N_2	Carbon tetrachloride	0	2.44
CH_4	Glycerol	25.4	0.95
C_2H_6	*n*-Hexane	30	6.00
C_2H_6	*n*-Heptane	30	5.60
C_2H_2	Water	0	1.10
H_2S	Water	16	1.77
CO_2	Amyl alcohol	25	1.91
CO_2	Isobutyl alcohol	25	2.20
SO_2	*n*-Heptane	20	2.70
SO_2	*n*-Nonane	20	2.50
SO_2	*n*-Decane	20	2.40

DIFFUSION COEFFICIENTS IN AQUEOUS SOLUTIONS AT 25°

The diffusion coefficient D may be defined by either of the equations

$$J = -D\frac{\partial c}{\partial x}$$

or

$$\frac{\partial c}{\partial t} = D\frac{\partial^2 c}{\partial x^2}$$

when diffusion occurs in the x-direction only. Here J is the diffusion-flux across unit area normal to the x-direction, $\frac{\partial c}{\partial x}$ is the concentration-gradient at a fixed time, $\frac{\partial c}{\partial t}$ is the rate of change of concentration with time at a fixed distance. If J is expressed in mole cm^{-2} sec^{-1} and c in mole cm^{-3}, x in cm, and t in sec, D will be given in units of cm^2 sec^{-1}. In general D varies somewhat with concentration. The values below are a selection from measurements by modern high-precision methods, mainly by H. S. Harned and collaborators, R. H. Stokes and collaborators, L. J. Gosting and collaborators, and L. G. Longsworth.

For strong electrolytes at infinite dilution, limiting diffusion coefficients may be calculated by the Nernst relation:

$$D = \frac{RT}{F^2}\left[\frac{(\nu_1 + \nu_2)(\lambda_1^0\lambda_2^0)}{\nu_1|Z_1|(\lambda_1^0 + \lambda_2^0)}\right]$$

where R = gas constant, F = Faraday, T = absolute temperature, λ_1^0 and λ_2^0 are cation and anion limiting equivalent conductances, ν_1 and ν_2 are the numbers of cations and anions formed from one "molecule" of electrolyte, and Z_1 is the cation valency. Concentrations, unless expressed otherwise are as molarities and the diffusion coefficients are expressed as $10^5 D/cm^2\ sec^{-1}$ at 25°C.

DIFFUSION COEFFICIENTS OF STRONG ELECTROLYTES

Molarity

Solute	0.01	0.1	1.0
HCL	....	3.050	3.436
HBr	....	3.156	3.87
LiCl	1.312	1.269	1.302
LiBr	....	1.279	1.404
$LiNO_3$	1.276	1.240	1.293
NaCl	1.545	1.483	1.484
NaBr	....	1.517	1.596
NaI	....	1.520	1.662
KCl	1.917	1.844	1.892
KBr	....	1.874	1.975
KI	....	1.865	2.065
KNO_3	1.846	....	....
$KClO_4$	1.790	....	....
$CaCl_2$	1.188	1.110	1.203

C = 0.005M

Solute	0.01	0.1	1.0
$BaCl_2$	1.265	1.159	1.179
Na_2SO_4	1.123	....	....
$MgSO_4$	0.710	....	....
$LaCl_3$	1.105	....	....
$K_4Fe(CN)_6$	1.183	....	....

DIFFUSION COEFFICIENTS OF WEAK AND NON-ELECTROLYTES

Concentration

Solute	Concentration	Coefficient
Glucose	0.39%	0.673
Sucrose	0.38%	0.521
Raffinose	0.38%	0.434
Sucrose	Zero	0.5226
Mannitol	Zero	0.682
Penta-erythritol	Zero	0.761
Glycolamide	Zero	1.142
Glycine	Zero	1.064
α-alanine	0.32%	0.910
β-alanine	0.31%	0.933
Amino-benzoic acid ortho	0.24%	0.840
Amino-benzoic acid meta	0.24%	0.774
Amino-benzoic acid para	0.23%	0.843
Citric acid	0.1 M	0.661

DIFFUSION OF GASES INTO AIR

Gas or vapor	Temp. °C	Coefficient of diffusion, sq. cm/sec	Observer
Alcohol, vapor........	40.4	0.137	Winkelmann
Carbon dioxide........	0.0	0.139	Mean of various
Carbon disulfide......	19.9	0.102	Winkelmann
Ether, vapor........	19.9	0.089	Winkelmann
Hydrogen............	0.0	0.634	Obermayer
Oxygen............	0.0	0.178	Obermayer
Water, vapor........	8.0	0.239	Guglielmo

RADIOACTIVE TRACER DIFFUSION DATA FOR PURE METALS

John Askill

The data in these tables are the most reliable set of radioactive tracer diffusion data for pure metals published in the literature from 1938 through December, 1970. For a complete listing of all published data on this subject up to December 1968 see "Tracer Diffusion Data for Metals, Alloys and Simple Oxides" by John Askill, published by Plenum Press, New York, 1970.

The diffusion coefficient D_T at a temperature T(°K) is given by the following relation:

$$D_T = D_0 e^{-Q/RT}$$

Abbreviations used in the tables are:

A.R.G. = Autoradiography
R.A. = Residual Activity
S.D. = Surface Decrease
S.S. = Serial Sectioning
P. = Polycrystalline
S. = Single Crystal
⊥c. = Perpendicular to c Direction
∥c. = Parallel to c Direction
99.95 = 99.95%

Solute (tracer)	Material (metal, crystalline form and purity)		Temperature range, °C	Form of analysis	Activation energy, Q, Kcal/mole	Frequency factor, D_0, cm²/sec	Reference
Aluminum							
Ag^{110}	S	99.999	371–655	S.S.	27.83	0.118	1
Al^{27}	S		450–650	S.S.	34.0	1.71	2
Au^{198}	S	99.999	423–609	S.S.	27.0	0.077	3
Cd^{115}	S	99.999	441–631	S.S.	29.7	1.04	3
Ce^{141}	P	99.995	450–630	R.A.	26.60	1.9×10^{-6}	5
Co^{60}	S	99.999	369–655	S.S.	27.79	0.131	1
Cr^{51}	S	99.999	422–654	S.S.	41.74	464	1
Cu^{64}	S	99.999	433–652	S.S.	32.27	0.647	1
Fe^{59}	S	99.99	550–636	S.S.	46.0	135	3
Ga^{72}	S	99.999	406–652	S.S.	29.24	0.49	1
Ge^{71}	S	99.999	401–653	S.S.	28.98	0.481	1
In^{114}	P	99.99	400–600	S.S., R.A.	27.6	0.123	4
La^{140}	P	99.995	500–630	R.A.	27.0	1.4×10^{-6}	5
Mn^{54}	P	99.99	450–650	S.S.	28.8	0.22	2
Mo^{99}	P	99.995	400–630	R.A.	13.1	1.04×10^{-9}	6
Nb^{95}	P	99.95	350–480	R.A.	19.65	1.66×10^{-7}	7
Nd^{147}	P	99.995	450–630	R.A.	25.0	4.8×10^{-7}	5
Ni^{63}	P	99.99	360–630	R.A.	15.7	2.9×10^{-8}	8
Pd^{103}	P	99.995	400–630	R.A.	20.2	1.92×10^{-7}	9
Pr^{142}	P	99.995	520–630	R.A.	23.87	3.58×10^{-7}	5
Sb^{124}	P		448–620	R.A.	29.1	0.09	10
Sm^{153}	P	99.995	450–630	R.A.	22.88	3.45×10^{-7}	5
Sn^{113}	P		400–600	S.S., R.A.	28.5	0.245	4
V^{48}	P	99.995	400–630	R.A.	19.6	6.05×10^{-8}	11
Zn^{65}	S	99.999	357–653	S.S.	28.86	0.259	1
Beryllium							
Ag^{110}	S⊥c	99.75	650–900	R.A.	43.2	1.76	12
Ag^{110}	S∥c	99.75	650–900	R.A.	39.3	0.43	12
Be^{7}	S⊥c	99.75	565–1065	R.A.	37.6	0.52	13
Be^{7}	S∥c	99.75	565–1065	R.A.	39.4	0.62	13
Fe^{59}	S	99.75	700–1076	R.A.	51.6	0.67	12
Ni^{63}	P		800–1250	R.A.	58.0	0.2	14
Cadmium							
Ag^{110}	S	99.99	180–300	—	25.4	2.21	15
Cd^{115}	S	99.95	110–283	R.A.	19.3	0.14	16
Zn^{65}	S	99.99	180–300	—	19.0	0.0016	15
Calcium							
C^{14}		99.95	550–800	R.A.	29.8	3.2×10^{-5}	17
Ca^{45}		99.95	500–800	R.A.	38.5	8.3	17

Solute (tracer)	Material (metal, crystalline form and purity)		Temperature range, °C	Form of analysis	Activation energy, Q, Kcal/mole	Frequency factor, D_0, cm^2/sec	Reference
Fe^{59}		99.95	500–800	R.A.	23.3	2.7×10^{-3}	17
Ni^{63}		99.95	550–800	—	28.9	1.0×10^{-5}	17
U^{235}		99.95	500–700	R.A.	34.8	1.1×10^{-5}	17
Carbon							
Ag^{110}	⊥c		750–1050	R.A.	64.3	9280	18
C^{14}			2000–2200	—	163	5	19
Ni^{63}	⊥c		540–920	R.A.	47.2	102	18
Ni^{63}	‖c		750–1060	R.A.	53.3	2.2	18
Th^{228}	⊥c		1400–2200	R.A.	145.4	1.33×10^{-5}	18
Th^{228}	‖c		1800–2200	R.A.	114.7	2.48	18
U^{232}	⊥c		1400–2200	R.A.	115.0	6760	18
U^{232}	‖c		1400–1820	R.A.	129.5	385	18
Chromium							
C^{14}	P		1200–1500	R.A.	26.5	9.0×10^{-3}	20
Cr^{51}	P	99.98	1030–1545	S.S.	73.7	0.2	21
Fe^{59}	P	99.8	980–1420	R.A.	79.3	0.47	22
Mo^{99}	P		1100–1420	R.A.	58.0	2.7×10^{-3}	20
Cobalt							
C^{14}	P	99.82	600–1400	R.A.	34.0	0.21	23
Co^{60}	P	99.9	1100–1405	S.S.	67.7	0.83	24
Fe^{59}	P	99.9	1104–1303	S.S.	62.7	0.21	24
Ni^{63}	P		1192–1297	R.A.	60.2	0.10	25
S^{35}	P	99.99	1150–1250	R.A.	5.4	1.3	26
Copper							
Ag^{110}	S, P		580–980	R.A.	46.5	0.61	27
As^{76}	P		810–1075	R.A.	42.13	0.20	28
Au^{198}	S, P		400–1050	S.S.	42.6	0.03	29
Cd^{115}	S	99.98	725–950	S.S.	45.7	0.935	30
Ce^{141}	P	99.999	766–947	R.A.	27.6	2.17×10^{-8}	31
Cr^{51}	S, P		800–1070	R.A.	53.5	1.02	32
Co^{60}	S	99.998	701–1077	S.S.	54.1	1.93	33
Cu^{67}	S	99.999	698–1061	S.S.	50.5	0.78	34
Eu^{152}	P	99.999	750–970	S.S., R.A.	26.85	1.17×10^{-7}	31
Fe^{59}	S, P		460–1070	R.A.	52.0	1.36	32
Ga^{72}			—	—	45.90	0.55	35
Ge^{68}	S	99.998	653–1015	S.S.	44.76	0.397	36
Hg^{203}	P		—	—	44.0	0.35	35
Lu^{177}	P	99.999	857–1010	R.A.	26.15	4.3×10^{-9}	31
Mn^{54}	S	99.99	754–950	S.S.	91.4	10^7	37
Nb^{95}	P	99.999	807–906	R.A.	60.06	2.04	38
Ni^{63}	P		620–1080	R.A.	53.8	1.1	39
Pd^{102}	S	99.999	807–1056	S.S.	54.37	1.71	40
Pm^{147}	P	99.999	720–955	R.A.	27.5	3.62×10^{-8}	31
Pt^{195}	P		843–997	S.S.	37.5	4.8×10^{-4}	41
S^{35}	S	99.999	800–1000	R.A.	49.2	23	42
Sb^{124}	S	99.999	600–1000	S.S.	42.0	0.34	43
Sn^{113}	P		680–910	—	45.0	0.11	44
Tb^{160}	P	99.999	770–980	R.A.	27.45	8.96×10^{-9}	31
Tl^{204}	S	99.999	785–996	S.S.	43.3	0.71	45
Tm^{170}	P	99.999	705–950	R.A.	24.15	7.28×10^{-9}	31
Zn^{65}	P	99.999	890–1000	S.S.	47.50	0.73	46
Germanium							
Cd^{115}	S		750–950	R.A.	102.0	1.75×10^{9}	47
Fe^{59}	S		775–930	R.A.	24.8	0.13	48
Ge^{71}	S		766–928	S.S.	68.5	7.8	49
In^{114}	S		600–920	—	39.9	2.9×10^{-4}	50
Sb^{124}	S		720–900	—	50.2	0.22	51
Te^{125}	S		770–900	S.S.	56.0	2.0	52
Tl^{204}	S		800–930	S.S.	78.4	1700	53

RADIOACTIVE TRACER DIFFUSION DATA FOR PURE METALS (Continued)

Solute (tracer)	Material (metal, crystalline form and purity)		Temperature range, °C	Form of analysis	Activation energy, Q, Kcal/mole	Frequency factor, D_0, cm^2/sec	Reference
Gold							
Ag^{110}	S	99.99	699–1007	S.S.	40.2	0.072	54
Au^{198}	S	99.97	850–1050	S.S.	42.26	0.107	224
Co^{60}	P	99.93	702–948	R.A.	41.6	0.068	55
Fe^{59}	P	99.93	701–948	R.A.	41.6	0.082	55
Hg^{203}	S	99.994	600–1027	—	37.38	0.116	56
Ni^{63}	P	99.96	880–940	S.S.	46.0	0.30	57
Pt^{195}	P, S	99.98	800–1060	S.S.	60.9	7.6	58
β-Hafnium							
Hf^{181}	P	97.9	1795–1995	S.S.	38.7	1.2×10^{-3}	59
Indium							
Ag^{110}	S⊥c	99.99	25–140	S.S.	12.8	0.52	60
Ag^{110}	S∥c	99.99	25–140	S.S.	11.5	0.11	60
Au^{198}	S	99.99	25–140	S.S.	6.7	9×10^{-3}	60
In^{114}	S⊥c	99.99	44–144	S.S.	18.7	3.7	61
In^{114}	S∥c	99.99	44–144	S.S.	18.7	2.7	61
Tl^{204}	S	99.99	49–157	S.S.	15.5	0.049	62
α-Iron							
Ag^{110}	P		748–888	S.S.	69.0	1950	63
Au^{198}	P	99.999	800–900	R.A.	62.4	31	64
C^{14}	P	99.98	616–844	R.A.	29.3	2.2	65
Co^{60}	P	99.995	638–768	R.A.	62.2	7.19	62
Cr^{51}	P	99.95	775–875	R.A.	57.5	2.53	66
Cu^{64}	P	99.9	800–1050	R.A.	57.0	0.57	67
Fe^{55}	P	99.92	809–889	—	60.3	5.4	68
K^{42}	P	99.92	500–800	R.A.	42.3	0.036	69
Mn^{54}	P	99.97	800–900	R.A.	52.5	0.35	70
Mo^{99}	P		750–875	R.A.	73.0	7800	71
Ni^{63}	P	99.97	680–800	R.A.	56.0	1.3	72
P^{32}	P		860–900	R.A.	55.0	2.9	73
Sb^{124}	P		800–900	R.A.	66.6	1100	74
V^{48}	P		755–875	R.A.	55.4	1.43	75
W^{185}	P		755–875	R.A.	55.1	0.29	75
γ-Iron							
Be^{7}	P	99.9	1100–1350	R.A.	57.6	0.1	76
C^{14}	P	99.34	800–1400	—	34.0	0.15	23
Co^{60}	P	99.98	1138–1340	S.S.	72.9	1.25	77
Cr^{51}	P	99.99	950–1400	R.A.	69.7	10.8	78
Fe^{59}	P	99.98	1171–1361	S.S.	67.86	0.49	79
Hf^{181}	P	99.99	1110–1360	R.A.	97.3	3600	78
Mn^{54}	P	99.97	920–1280	R.A.	62.5	0.16	70
Ni^{63}	P	99.97	930–2050	R.A.	67.0	0.77	72
P^{32}	P	99.99	950–1200	R.A.	43.7	0.01	80
S^{35}	P		900–1250	R.A.	53.0	1.7	81
V^{48}	P	99.99	1120–1380	R.A.	69.3	0.28	78
W^{185}	P	99.5	1050–1250	R.A.	90.0	1000	82
δ-Iron							
Co^{60}	P	99.995	1428–1521	R.A.	61.4	6.38	83
Fe^{59}	P	99.95	1428–1492	S.S.	57.5	2.01	83
P^{32}	P	99.99	1370–1460	R.A.	55.0	2.9	73
Lanthanum							
Au^{198}	P	99.97	600–800	S.S.	45.1	1.5	84
La^{140}	P	99.97	690–850	S.S.	18.1	2.2×10^{-2}	84
Lead							
Ag^{110}	P	99.9	200–310	R.A.	14.4	0.064	85
Au^{198}	S	99.999	190–320	S.S.	10.0	8.7×10^{-3}	86

RADIOACTIVE TRACER DIFFUSION DATA FOR PURE METALS (Continued)

Solute (tracer)	Material (metal, crystalline form and purity)	Temperature range, °C	Form of analysis	Activation energy, Q, Kcal/mole	Frequency factor, D_0, cm^2/sec	Reference
Cd^{115}	S 99.999	150–320	S.S.	21.23	0.409	87
Cu^{64}	S	150–320	S.S.	14.44	0.046	88
Pb^{204}	S 99.999	150–320	S.S.	25.52	0.887	87
Tl^{205}	P 99.999	207–322	S.S.	24.33	0.511	89
Lithium						
Ag^{110}	P 92.5	65–161	S.S.	12.83	0.37	90
Au^{195}	P 92.5	47–153	S.S.	10.49	0.21	90
Bi	P 99.95	141–177	S.S.	47.3	5.3×10^{13}	91
Cd^{115}	P 92.5	80–174	S.S.	16.05	2.35	90
Cu^{64}	P 99.98	51–120	S.S.	9.22	0.47	93
Ga^{72}	P 99.98	58–173	S.S.	12.9	0.21	93
Hg^{203}	P 99.98	58–173	S.S.	14.18	1.04	93
In^{114}	P 92.5	80–175	S.S.	15.87	0.39	90
Li^{6}	P 99.98	35–178	S.S.	12.60	0.14	94
Na^{22}	P 92.5	52–176	S.S.	12.61	0.41	90
Pb^{204}	P 99.95	129–169	S.S.	25.2	160	91
Sb^{124}	P 99.95	141–176	S.S.	41.5	1.6×10^{10}	91
Sn^{113}	P 99.95	108–174	S.S.	15.0	0.62	91
Zn^{65}	P 92.5	60–175	S.S.	12.98	0.57	92
Magnesium						
Ag^{110}	P 99.9	476–621	S.S.	28.50	0.34	95
Fe^{59}	P 99.95	400–600	R.A.	21.2	4×10^{-6}	96
In^{114}	P 99.9	472–610	S.S.	28.4	5.2×10^{-2}	95
Mg^{28}	S⊥c	467–635	S.S.	32.5	1.5	97
Mg^{28}	S∥c	467–635	S.S.	32.2	1.0	97
Ni^{63}	P 99.95	400–600	R.A.	22.9	1.2×10^{-5}	96
U^{235}	P 99.95	500–620	R.A.	27.4	1.6×10^{-5}	96
Zn^{65}	P 99.9	467–620	S.S.	28.6	0.41	95
Molybdenum						
C^{14}	P 99.98	1200–1600	R.A.	41.0	2.04×10^{-2}	99
Co^{60}	P 99.98	1850–2350	S.S.	106.7	18	100
Cr^{51}	P	1000–1500	R.A.	54.0	2.5×10^{-4}	20
Cs^{134}	S 99.99	1000–1470	R.A., A.R.G.	28.0	8.7×10^{-11}	101
K^{42}	S	800–1100	R.A.	25.04	5.5×10^{-9}	102
Mo^{99}	P	1850–2350	S.S.	96.9	0.5	103
Na^{24}	S	800–1100	R.A.	21.25	2.95×10^{-9}	102
Nb^{95}	P 99.98	1850–2350	S.S.	108.1	14	100
P^{32}	P 99.97	2000–2200	S.S.	80.5	0.19	104
Re^{186}	P	1700–2100	A.R.G.	94.7	0.097	105
S^{35}	S 99.97	2220–2470	S.S.	101.0	320	106
Ta^{182}	P	1700–2150	R.A.	83.0	3.5×10^{-4}	20
U^{235}	P 99.98	1500–2000	R.A.	76.4	7.6×10^{-3}	107
W^{185}	P 99.98	1700–2260	S.S.	110	1.7	108
Nickel						
Au^{198}	S, P 99.999	700–1075	S.S.	55.0	0.02	109
Be^{7}	P 99.9	1020–1400	R.A.	46.2	0.019	76
C^{14}	P 99.86	600–1400	—	34.0	0.012	23
Co^{60}	P 99.97	1149–1390	R.A.	65.9	1.39	110
Cr^{51}	P 99.95	1100–1270	S.S.	65.1	1.1	111
Cu^{64}	P 99.95	1050–1360	S.S.	61.7	0.57	111
Fe^{59}	P	1020–1263	S.S.	58.6	0.074	112
Mo^{99}	P	900–1200	R.A.	51.0	1.6×10^{-3}	20
Ni^{63}	P 99.95	1042–1404	S.S.	68.0	1.9	111
Pu^{238}	P	1025–1125	A.R.G.	51.0	0.5	113
Sb^{124}	P 99.97	1020–1220	—	27.0	1.8×10^{-5}	114
Sn^{113}	P 99.8	700–1350	A.R.G.	58.0	0.83	115
V^{48}	P 99.99	800–1300	R.A.	66.5	0.87	11
W^{185}	P 99.95	1100–1300	S.S.	71.5	2.0	116

Solute (tracer)	Material (metal, crystalline form and purity)	Temperature range, °C	Form of analysis	Activation energy, Q, Kcal/mole	Frequency factor, D_0, cm^2/sec	Reference
Niobium						
C^{14}	P	800–1250	R.A.	32.0	1.09×10^{-5}	117
Co^{60}	P 99.85	1500–2100	A.R.G.	70.5	0.74	118
Cr^{51}	S	943–1435	S.S.	83.5	0.30	119
Fe^{55}	P 99.85	1400–2100	A.R.G.	77.7	1.5	118
K^{42}	S	900–1100	R.A.	22.10	2.38×10^{-7}	102
Nb^{95}	P, S 99.99	878–2395	S.S.	96.0	1.1	120
P^{32}	P 99.0	1300–1800	S.S.	51.5	5.1×10^{-2}	104
S^{35}	S 99.9	1100–1500	R.A.	73.1	2600	121
Sn^{113}	P 99.85	1850–2400	S.S.	78.9	0.14	122
Ta^{182}	P, S 99.997	878–2395	S.S.	99.3	1.0	120
Ti^{44}	S	994–1492	S.S.	86.9	0.099	123
U^{235}	P 99.55	1500–2000	R.A.	76.8	8.9×10^{-3}	107
V^{48}	S 99.99	1000–1400	R.A.	85.0	2.21	124
W^{185}	P 99.8	1800–2200	R.A.	91.7	5×10^{-4}	125
Palladium						
Pd^{103}	S 99.999	1060–1500	S.S.	63.6	0.205	126
Phosphorus						
P^{32}	P	0–44	S.S.	9.4	1.07×10^{-3}	127
Platinum						
Co^{60}	P 99.99	900–1050	—	74.2	19.6	129
Cu^{64}	P	1098–1375	S.S.	59.5	0.074	41
Pt^{195}	P 99.99	1325–1600	S.S.	68.2	0.33	130
Potassium						
Au^{198}	P 99.95	5.6–52.5	S.S.	3.23	1.29×10^{-3}	131
K^{42}	S 99.7	−52–61	S.S.	9.36	0.16	132
Na^{22}	P 99.7	0–62	S.S.	7.45	0.058	133
Rb^{86}	P 99.95	0.1–59.9	S.S.	8.78	0.090	134
γ-Plutonium						
Pu^{238}	P	190–310	S.S.	16.7	2.1×10^{-5}	135
δ-Plutonium						
Pu^{238}	P	350–440	S.S.	23.8	4.5×10^{-3}	136
ε-Plutonium						
Pu^{238}	P	500–612	R.A.	18.5	2.0×10^{-2}	137
α-Praseodymium						
Ag^{110}	P 99.93	610–730	S.S.	25.4	0.14	138
Au^{195}	P 99.93	650–780	S.S.	19.7	4.3×10^{-2}	138
Co^{60}	P 99.93	660–780	S.S.	16.4	4.7×10^{-2}	138
Zn^{65}	P 99.96	766–603	S.S.	24.8	0.18	139
β-Praseodymium						
Ag^{110}	P 99.93	800–900	S.S.	21.5	3.2×10^{-2}	138
Au^{195}	P 99.93	800–910	S.S.	20.1	3.3×10^{-2}	138
Ho^{166}	P 99.96	800–930	S.S.	26.3	9.5	140
In^{114}	P 99.96	800–930	S.S.	28.9	9.6	140
La^{140}	P 99.96	800–930	S.S.	25.7	1.8	140
Pr^{142}	P 99.93	800–900	S.S.	29.4	8.7	140
Zn^{65}	P 99.96	822–921	S.S.	27.0	0.63	139
Selenium						
Fe^{59}	P	40–100	R.A.	8.88	—	141
Hg^{203}	P 99.996	25–100	R.A.	1.2	—	141
S^{35}	S⊥c	60–90	S.D.	29.9	1700	142

RADIOACTIVE TRACER DIFFUSION DATA FOR PURE METALS (Continued)

Solute (tracer)	Material (metal, crystalline form and purity)	Temperature range, °C	Form of analysis	Activation energy, Q, Kcal/mole	Frequency factor, D_0, cm^2/sec	Reference
S^{35}	S‖c	60–90	S.D.	15.6	1100	142
Se^{75}	P	35–140	—	11.7	1.4×10^{-4}	143
Silicon						
Au^{198}	S	700–1300	S.S.	47.0	2.75×10^{-3}	145
C^{14}	P	1070–1400	R.A.	67.2	0.33	146
Cu^{64}	P	800–1100	R.A.	23.0	4×10^{-2}	147
Fe^{59}	S	1000–1200	R.A.	20.0	6.2×10^{-3}	148
Ni^{63}	P	450–800	—	97.5	1000	149
P^{32}	S	1100–1250	R.A.	41.5	—	150
Sb^{124}	S	1190–1398	R.A.	91.7	12.9	151
Si^{31}	S 99.99999	1225–1400	S.S.	110.0	1800	146
Silver						
Au^{198}	P 99.99	718–942	S.S.	48.28	0.85	54
Ag^{110}	S 99.999	640–955	S.S.	45.2	0.67	152
Cd^{115}	S 99.99	592–937	S.S.	41.69	0.44	153
Co^{60}	S 99.999	700–940	—	48.75	1.9	154
Cu^{64}	P 99.99	717–945	S.S.	46.1	1.23	155
Fe^{59}	S 99.99	720–930	S.S.	49.04	2.42	156
Ge^{77}	P	640–870	S.S.	36.5	0.084	157
Hg^{203}	P 99.99	653–948	S.S.	38.1	0.079	155
In^{114}	S 99.99	592–937	S.S.	40.80	0.41	153
Ni^{63}	S 99.99	749–950	S.S.	54.8	21.9	158
Pb^{210}	P	700–865	S.S.	38.1	0.22	159
Pd^{102}	S 99.999	736–939	S.S.	56.75	9.56	140
Ru^{103}	S 99.99	793–945	S.S.	65.8	180	160
S^{35}	S 99.999	600–900	R.A.	40.0	1.65	161
Sb^{124}	P 99.999	780–950	S.S., R.A.	39.07	0.234	162
Sn^{113}	S 99.99	592–937	S.S.	39.30	0.255	153
Te^{125}	P	770–940	R.A.	38.90	0.47	163
Tl^{204}	P	640–870	S.S.	37.9	0.15	157
Zn^{65}	S 99.99	640–925	S.S.	41.7	0.54	164
Sodium						
Au^{198}	P 99.99	1.0–77	S.S.	2.21	3.34×10^{-4}	165
K^{42}	P 99.99	0–91	S.S.	8.43	0.08	133
Na^{22}	P 99.99	0–98	S.S.	10.09	0.145	166
Rb^{86}	P 99.99	0–85	S.S.	8.49	0.15	133
Tantalum						
C^{14}	P	1450–2200	S.S.	40.3	1.2×10^{-2}	167
Fe^{59}	P	930–1240	—	71.4	0.505	168
Mo^{99}	P	1750–2220	R.A.	81.0	1.8×10^{-3}	20
Nb^{95}	P, S 99.996	921–2484	S.S.	98.7	0.23	169
S^{35}	P 99.0	1970–2110	R.A.	70.0	100	170
Ta^{182}	P, S 99.996	1250–2200	S.S.	98.7	1.24	226
Tellurium						
Hg^{203}	P	270–440	—	18.7	3.14×10^{-5}	171
Se^{75}	P	320–440	—	28.6	2.6×10^{-2}	171
Tl^{204}	P	360–430	—	41.0	320	172
Te^{127}	S⊥c 99.9999	300–400	S.S.	46.7	3.91×10^{4}	173
Te^{127}	S‖c 99.9999	300–400	S.S.	35.5	130	173
α-Thallium						
Ag^{110}	P⊥c 99.999	80–250	S.S.	11.8	3.8×10^{-2}	174
Ag^{110}	P‖c 99.999	80–250	S.S.	11.2	2.7×10^{-2}	174
Au^{198}	P⊥c 99.999	110–260	S.S.	2.8	2.0×10^{-5}	174
Au^{198}	P‖c 99.999	110–260	S.S.	5.2	5.3×10^{-4}	174
Tl^{204}	S⊥c 99.9	135–230	S.S.	22.6	0.4	175
Tl^{204}	S‖c 99.9	135–230	S.S.	22.9	0.4	175

Solute (tracer)	Material (metal, crystalline form and purity)	Temperature range, °C	Form of analysis	Activation energy, Q, Kcal/mole	Frequency factor, D_0, cm^2/sec	Reference
β-Thallium						
Ag^{110}	P 99.999	230–310	S.S.	11.9	4.2×10^{-2}	174
Au^{198}	P 99.999	230–310	S.S.	6.0	5.2×10^{-4}	174
Tl^{204}	S 99.9	230–280	S.S.	20.7	0.7	175
α-Thorium						
Pa^{231}	P 99.85	770–910	—	74.7	126	176
Th^{228}	P 99.85	720–880	—	71.6	395	176
U^{233}	P 99.85	700–880	—	79.3	2210	176
Tin						
Ag^{110}	S⊥c	135–225	S.S.	18.4	0.18	177
Ag^{110}	S∥c	135–225	S.S	12.3	7.1×10^{-3}	177
Au^{198}	S⊥c	135–225	S.S.	17.7	0.16	177
Au^{198}	S∥c	135–225	S.S.	11.0	5.8×10^{-3}	177
Co^{60}	S, P	140–217	R.A.	22.0	5.5	178
In^{114}	S⊥c 99.998	181–221	S.S.	25.8	34.1	179
In^{114}	S∥c 99.998	181–221	S.S.	25.6	12.2	179
Sn^{113}	S⊥c 99.999	160–226	S.S.	25.1	10.7	180
Sn^{113}	S∥c 99.999	160–226	S.S.	25.6	7.7	180
Tl^{204}	P 99.999	137–216	S.S.	14.7	1.2×10^{-3}	181
α-Titanium						
Ti^{44}	P 99.99	700–850	R.A.	35.9	8.6×10^{-6}	182
β-Titanium						
Ag^{110}	P 99.95	940–1570	S.S.	43.2	3×10^{-3}	183
Be^{7}	P 99.96	915–1300	R.A.	40.2	0.8	184
C^{14}	P 99.62	1100–1600	R.A.	20.0	3.02×10^{-3}	185
Cr^{51}	P 99.7	950–1600	A.R.G.	35.1	5×10^{-3}	186
				61.0	4.9	
Co^{60}	P 99.7	900–1600	S.S.	30.6	1.2×10^{-2}	186
				52.5	2.0	
Fe^{59}	P 99.7	900–1600	A.R.G.	31.6	7.8×10^{-3}	186
				55.0	2.7	
Mo^{99}	P 99.7	900–1600	S.S.	43.0	8.0×10^{-3}	186
				73.0	20	
Mn^{54}	P 99.7	900–1600	S.S.	33.7	6.1×10^{-3}	186
				58.0	20	
Nb^{95}	P 99.7	1000–1600	A.R.G.	39.3	5.0×10^{-3}	186
				73.0	20	
Ni^{63}	P 99.7	925–1600	A.R.G.	29.6	9.2×10^{-3}	186
				52.5	2.0	
P^{32}	P 99.7	950–1600	S.S.	24.1	3.62×10^{-3}	187
				56.5	5	
Sc^{46}	P 99.95	940–1590	S.S.	32.4	4.0×10^{-3}	183
Sn^{113}	P 99.7	950–1600	S.S.	31.6	3.8×10^{-4}	187
				69.2	10	
Ti^{44}	P 99.95	900–1540	S.S.	31.2	3.58×10^{-4}	188
				60.0	1.09	
U^{235}	P 99.9	900–1400	R.A.	29.3	5.1×10^{-4}	189
V^{48}	P 99.95	900–1545	S.S.	32.2	3.1×10^{-4}	190
				57.2	1.4	
W^{185}	P 99.94	900–1250	R.A.	43.9	3.6×10^{-3}	191
Zr^{95}	P 98.94	920–1500	R.A.	35.4	4.7×10^{-3}	191
Tungsten						
C^{14}	P 99.51	1200–1600	R.A.	53.5	8.91×10^{-3}	99
Fe^{59}	P	940–1240	—	66.0	1.4×10^{-2}	168
Mo^{99}	P	1700–2100	R.A.	101.0	0.3	20
Nb^{95}	P 99.99	1305–2367	S.S.	137.6	3.01	192
Re^{186}	S	2100–2400	R.A.	141.0	19.5	193

Solute (tracer)	Material (metal, crystalline form and purity)	Temperature range, °C	Form of analysis	Activation energy, Q, Kcal/mole	Frequency factor, D_0, cm²/sec	Reference
Ta^{182}	P 99.99	1305–2375	S.S.	139.9	3.05	192
W^{185}	P 99.99	1800–2403	S.S.	140.3	1.88	192
α-Uranium						
U^{234}	P	580–650	—	40.0	2×10^{-3}	194
β-Uranium						
Co^{60}	P 99.999	692–763	S.S.	27.45	1.5×10^{-2}	195
U^{235}	P	690–750	R.A.	44.2	2.8×10^{-3}	196
γ-Uranium						
Au^{195}	P 99.99	785–1007	S.S.	30.4	4.86×10^{-3}	197
Co^{60}	P 99.99	783–989	S.S.	12.57	3.51×10^{-4}	198
Cr^{51}	P 99.99	797–1037	S.S.	24.46	5.37×10^{-3}	198
Cu^{64}	P 99.99	787–1039	S.S.	24.06	1.96×10^{-3}	198
Fe^{55}	P 99.99	787–990	S.S.	12.0	2.69×10^{-4}	198
Mn^{54}	P 99.99	787–939	S.S.	13.88	1.81×10^{-4}	198
Nb^{95}	P 99.99	791–1102	S.S.	39.65	4.87×10^{-2}	198
Ni^{63}	P 99.99	787–1039	S.S.	15.66	5.36×10^{-4}	198
U^{233}	P 99.99	800–1070	S.S.	28.5	2.33×10^{-3}	227
Zr^{95}	P	800–1000	R.A.	16.5	3.9×10^{-4}	228
Vanadium						
C^{14}	P 99.7	845–1130	S.S.	27.3	4.9×10^{-3}	199
Cr^{51}	P 99.8	960–1200	R.A.	64.6	9.54×10^{-3}	200
Fe^{59}	P	960–1350	S.S.	71.0	0.373	201
P^{32}	P 99.8	1200–1450	R.A.	49.8	2.45×10^{-2}	202
S^{35}	P 99.8	1320–1520	R.A.	34.0	3.1×10^{-2}	184
V^{48}	S, P 99.99	880–1360	S.S.	73.65	0.36	203
V^{48}	S, P 99.99	1360–1830	S.S.	94.14	214.0	203
Yttrium						
Y^{90}	S⊥c	900–1300	R.A.	67.1	5.2	204
Y^{90}	S∥c	900–1300	R.A.	60.3	0.82	204
Zinc						
Ag^{110}	S⊥c 99.999	271–413	S.S.	27.6	0.45	205
Ag^{110}	S∥c 99.999	271–413	S.S.	26.0	0.32	205
Au^{198}	S⊥c 99.999	315–415	S.S.	29.72	0.29	206
Au^{198}	S∥c 99.999	315–415	S.S.	29.73	0.97	206
Cd^{115}	S⊥c 99.999	225–416	S.S.	20.12	0.117	206
Cd^{115}	S∥c 99.999	225–416	S.S.	20.54	0.114	206
Cu^{64}	S⊥c 99.999	338–415	S.S.	29.92	2.0	206
Cu^{64}	S∥c 99.999	338–415	S.S.	29.53	2.22	207
Ga^{72}	S⊥c	240–403	S.S.	18.15	0.018	207
Ga^{72}	S∥c	240–403	S.S.	18.4	0.016	207
Hg^{203}	S⊥c	260–413	S.S.	20.18	0.073	208
Hg^{203}	S∥c	260–413	S.S.	19.70	0 056	208
In^{114}	S⊥c	271–413	S.S.	19.60	0.14	205
In^{114}	S∥c	271–413	S.S.	19.10	0.062	205
Sn^{113}	S⊥c	298–400	S.S.	18.4	0.13	209
Sn^{113}	S∥c	298–400	S.S.	19.4	0.15	209
Zn^{65}	S⊥c 99.999	240–418	S.S.	23.0	0.18	210
Zn^{65}	S∥c 99.999	240–418	S.S.	21.9	0.13	210
α-Zirconium						
Cr^{51}	P 99.9	700–850	R.A.	18.0	1.19×10^{-8}	211
Fe^{55}	P	750–840	—	48.0	2.5×10^{-2}	212
Mo^{99}	P	600–850	R.A.	24.76	6.22×10^{-8}	213
Nb^{95}	P 99.99	740–857	R.A.	31.5	6.6×10^{-6}	182
Sn^{113}	P	300–700	A.R.G.	22.0	1.0×10^{-8}	214
Ta^{182}	P 99.6	700–800	R.A.	70.0	100	215

Solute (tracer)	Material (metal, crystalline form and purity)		Temperature range, °C	Form of analysis	Activation energy, Q, Kcal/mole	Frequency factor, D_0, cm^2/sec	Reference
V^{48}	P	99.99	600–850	R.A.	22.9	1.12×10^{-8}	124
Zr^{95}	P	99.95	750–850	S.S.	45.5	5.6×10^{-4}	216
β-Zirconium							
Be^{7}	P	99.7	915–1300	R.A.	31.1	8.33×10^{-2}	184
C^{14}	P	96.6	1100–1600	R.A.	34.2	3.57×10^{-2}	217
Ce^{141}	P		880–1600	R.A.	41.4	3.16	218
					74.1	42.17	
Co^{60}	P	99.99	920–1600	S.S.	21.82	3.26×10^{-3}	219
Cr^{51}	P	99.9	700–850	R.A.	18.0	1.19×10^{-8}	211
Fe^{55}	P		750–840	—	48.0	2.5×10^{-2}	212
Mo^{99}	P		900–1635	R.A.	35.2	1.99×10^{-6}	218
					68.55	2.63	
Nb^{95}	P		1230–1635	R.A.	36.6	7.8×10^{-4}	220
P^{32}	P	99.94	950–1200	R.A.	33.3	0.33	221
Sn^{113}	P		300–700	A.R.G.	22.0	1×10^{-8}	214
Ta^{182}	P	99.6	900–1200	R.A.	27.0	5.5×10^{-5}	215
U^{235}	P		900–1065	S.S.	30.5	5.7×10^{-4}	222
V^{48}	P	99.99	870–1200	R.A.	45.8	7.59×10^{-3}	223
V^{48}	P	99.99	1200–1400	R.A.	57.7	0.32	223
W^{185}	P	99.7	900–1250	R.A.	55.8	0.41	223
Zr^{95}	P		1100–1500	S.S.	30.1	2.4×10^{-4}	225

REFERENCES

1. N. L. Peterson and S. J. Rothman, *Phys. Rev.* **B1** (8), 3264 (1970).
2. T. S. Lundy and J. F. Murdock, *J. Appl. Phys.* **33** (5), 1671 (1962).
3. W. B. Alexander and L. M. Slifkin, *Phys. Rev.* **B1** (8), 3274 (1970).
4. M. S. Anand and R. P. Agarwala *Phys. Stat. Solidi* **A1** (1), 41K (1970).
5. S. P. Murarka and R. P. Agarwala, *Indian At. Energy Comm.* **Report BARC-368** (1968).
6. A. R. Paul and R. P. Agarwala, *J. Appl. Phys.* **38** (9), 3790 (1967).
7. G. P. Tiwari and B. D. Sharma, *Trans. Indian Inst. Metals* **20,** 83 (1967).
8. K. Hirano, R. P. Agarwala, and M. Cohen, *Acta Met.* **10** (9), 857 (1962).
9. M. S. Anand and R. P. Agarwala, *Trans. AIME* **239** (11), 1848 (1967).
10. S. Badrinarayanan and H. B. Mathers, *Int. J. Appl. Radiat. Isotopes* **19** (4), 353 (1968).
11. S. P. Murarka, M. S. Anand, and R. P. Agarwala, *Acta Met.* **16** (1), 69 (1968).
12. M. C. Naik, J. M. Dupony, and Y. Adda, *Mem. Sci. Rev. Met.* **63,** 488 (1966).
13. J. M. Dupony, J. Mathie, and Y. Adda, *Mem. Sci. Rev. Met.* **63,** 481 (1966).
14. V. M. Ananyn, V. P. Gladkov, V. S. Zotov, and D. M. Skorov, *At. Energ:* **29** (3), 220 (1970).
15. W. Hirschwald and W. Schroedter, *Z. Physik. Chem. N.F.* **53,** 392 (1967).
16. W. Chomka, *Zess. Nauk Politech Gdansk Fizyka* **1,** 39 (1967).
17. L. V. Pavlinov, A. M. Gladyshev, and V. N. Bikov, *Fiz. Metal Metalloved.* **26** (5), 823 (1968).
18. J. R. Wolfe, D. R. McKenzie, and R. J. Borg, *J. Appl. Phys.* **36** (6), 1906 (1965).
19. M. A. Kanter, *Phys. Rev.* **107,** 655 (1957).
20. E. V. Borisov, P. L. Gruzin, and S. V. Zemskii, *Zashch Pokryt Metal* **2,** 104 (1968).
21. J. Askill and D. H. Tomlin, *Phil. Mag.* **11** (111), 467 (1965).
22. H. W. Paxton and R. A. Wolfe, *Trans. AIME* **230,** 1426 (1964).
23. I. I. Kovenski, *Fiz. Metal. i Metalloved.* **16,** 613 (1963).
24. H. W. Mead and C. E. Birchenall, *Trans. Met. Soc. AIME* **203** (9), 994 (1955).
25. K. Hirano, R. P. Agarwala, B. L. Averbach, and M. Cohen, *J. Appl. Phys.* **33** (10), 3049 (1962).
26. M. M. Pavlyuchenko and I. F. Kononyuk, *Dokl. Akad. Nauk Belorusskoi* **SSR 8,** 157 (1964).
27. G. Barreau, G. Brunel, G. Azeron, P. Lacombe, *C. R. Acad. Sci. Paris* **Ser C 270** (6), 514 (1970).
28. S. M. Klotsman, A. Ya Rabovskii, V. K. Talinskii, and A. N. Timofeeu, *Fiz. Met. Metalloved* **29** (4), 803 (1970).
29. A. Chatterjee and D. J. Fabian, *Acta Met.* **17** (9), 1141 (1969).
30. T. Hirone, N. Kunitomi, M. Sakamoto, and H. Yamaki, *J. Phys. Soc. Japan* **13** (8), 838 (1958).
31. S. Badrinarayanan and H. B. Mathur, *Indian J. Pure Appl. Phys.* **8** (6), 324 (1970).
32. G. Barreau, G. Brunel, and G. Cizeron, *C. R. Acad. Sci.* **Ser. C 272** (7), 618 (1971).
33. C. A. Machiet, *Phys. Rev.* **109** (6), 1964 (1958).
34. S. J. Rothman and N. L. Peterson, *Phys. Stat. Solidi* **35,** 305 (1969).
35. C. T. Tomizuka, cited by D. Lazarus, *Solid State Phys.* **Vol. 10** (1960).
36. F. D. Reinke and C. E. Dahlstrom, *Phil. Mag.* **22** (175), 57 (1970).
37. A. Ikushima, *J. Phys. Soc. Japan* **14,** 111 (1959).
38. M. C. Saxena and B. D. Sharma, *Trans. Indian Inst. Metals* **23** (3), 16 (1970).
39. G. Brunel, G. Cizeron, P. Lacombe, *C. R. Acad. Sci. Paris* **Ser. C 270** (4), 393 (1970).
40. N. L. Peterson, *Phys. Rev.* **132** (6), 2471 (1963).
41. R. D. Johnson and B. H. Faulkenberry, *ASD-TDR*-63-625 (July 1963).
42. F. Moya, G. E. Moya-Gontier and F. Cabane-Brouty, *Phys. Stat. Solidi* **35** (2), 893 (1969).
43. M. C. Inman and L. W. Barr, *Acta Met.* **8** (2), 112 (1960).
44. P. P. Kuzmenko, L. F. Ostrovskii, and V. S. Kovalchuk, *Fiz. Tverd. Tela* **4,** 490 (1962).
45. S. Komura and N. Kunitomi, *J. Phys. Soc. Japan* **18 (Suppl. 2),** 208 (1963).
46. S. M. Klotsman, Ya A. Rabovskii, V. K. Talinskii, and A. N. Timofeev, *Fiz. Metal. Metalloved.* **28** (6), 1025 (1969).
47. V. E. Kosenko, *Fiz. Tverd. Tela* **1**:1622 (1959); *Soviet Phys.—Solid State* (*English transl.*) **1,** 1481 (1959).
48. A. A. Bugai, V. E. Kosenko, and E. G. Miselynuk, *Sh. Tekhn. Fiz.* **27** (1), 207 (1957); *NP-tr*-448, **P.** 219 (1960).
49. H. Letaw, W. M. Portnoy, and L. Slifkin, *Phys. Rev.* **102,** 636 (1956).
50. A. V. Sandulova, M. I. Droniuk, and V. M. P'dak, *Fiz. Tverd. Tela* **3,** 2913 (1961).
51. B. I. Boltaks, V. P. Grabtchak, and T. D. Dzafarov, *Fiz. Tverd. Tela* **6,** 3181 (1964).
52. V. D. Ignatkov and V. E. Kosenko, *Fiz. Tverd. Tela* **4** (6), 1627 (1962): *Soviet Phys.—Solid State* (*Eng. transl.*) **4** (6), 1627 (1962).
53. V. I. Tagirov and A. A. Kuliev, *Fiz. Tverd. Tela* **4** (1), 272 (1962) *Soviet Phys.—Solid State* (*Eng. transl.*) **4** (1), 196 (1962).
54. W. C. Mallard, A. B. Gardner, R. F. Bass, and L. M. Slifkin, *Phys. Rev.* **129** (2), 617 (1963).
55. D. Duhl, K. Hirano, and M. Cohen, *Acta Met.* **11** (1), 1 (1963).
56. A. J. Mortlock and A. H. Rowe, *Phil. Mag.* **11** (114), 115 (1965).
57. J. E. Reynolds, B. L. Averbach, and M. Cohen, *Acta Met.* **5,** 29 (1957).
58. A. J. Mortlock, A. H. Rowe, and A. D. LeClaire, *Phil. Mag.* **5,** 803 **(1960).**

59. F. R. Winslow and T. S. Lundy, *Trans. Met. Soc. AIME* **233**, 1790 (1965).
60. T. R. Anthony and D. Turnbull, *Phys. Rev.* **151**, 495 (1966).
61. V. M. Amonenko, A. M. Blinkin, and I. G. Ivantsov, *Fiz. Metal. i Metalloved.* **17** (1), 56 (1964); *Phys. Metals Metallog.* (*USSR*) (*Eng. transl.*) **17** (1), 54 (1964).
62. D. W. James and G. M. Leak, *Phil. Mag.* **14**, 701 (1966).
63. A. Bondy and V. Levy, *C. R. Acad. Sci. Ser C* **272** (1), 19 (1971).
64. R. J. Borg and D. Y. F. Lai, *Acta Met.* **11** (8), 861 (1963).
65. C. G. Homan, *Acta Met.* **12**, 1071 (1964).
66. A. M. Huntz, M. Aucouturier, and P. Lacombe, *C. R. Acad. Sci. Paris Ser. C* **265** (10), 554 (1967).
67. M. S. Anand and R. P. Agarwala, *J. Appl. Phys.* **37**, 4248 (1966).
68. R. Angers and F. Claisse, *Can. Met. Quart.* **7** (2), 73 (1968).
69. A. V. Tomilov and G. V. Shcherbedinskii, *Fiz. Khim. Met. Mater.* **3** (3), 261 (1967).
70. K. Nohara and K. Hirano, *Proc. Intl. Conf. Sci. Technol. Iron Steel Tokyo*, Sept 1970.
71. V. T. Borisov, V. M. Golikov, and G. V. Shcherbedinskii, *Fiz. Metal. i Metalloved.* **22** (1), 159 (1966).
72. K. Hirano, M. Cohen, and B. L. Averbach, *Acta Met.* **9** (5), 440 (1961).
73. G. Seibel, *Compt. Rend.* **256** (22), 4661 (1963).
74. G. Bruggeman and J. Roberts, *J. Met.* **20** (8), 54 (1968).
75. V. D. Lyubimov, *Izv. Akad. Nauk. SSSR. Met.* **3** 201 (1969); V. D. Lyubimov, V. A. Tskhai, and G. B. Bogomolov, *Trudy Inst. Khim. Akad. Nauk. SSSR Ural Filial* **17**, 44, 48 (1970).
76. G. V. Grigorev and L. V. Pavlinov, *Fiz. Metal. i Metalloved.* **25** (5), 836 (1968).
77. T. Suzuoka, *Japan. Inst. Metals* **2**, 176 (1961).
78. A. W. Bowen and G. M. Leak, *Met. Trans.* **1** (6), 1695 (1970).
79. Th. Heumann and R. Imm, *J. Phys. Chem. Solids* **29** (9), 1613 (1968).
80. P. L. Gruzin and V. V. Mural, *Fiz. Metal. i Metalloved.* **16** (4), 551 (1963); *Phys. Metals Metallog.* (*USSR*) (*Eng. transl.*) **16** (4), 50 (1963).
81. A. Hoshino and R. Araki, *Tetsu to Hagane* **56** (2), 252 (1970).
82. K. Sato, *Trans. Japan Inst. Metals* **5**, 91 (1964).
83. D. W. James and G. M. Leak, *Phil. Mag.* **14**, 701 (1966).
84. M. P. Dariel, G. Erez, and G. M. J. Schmidt, *Phil. Mag.* **19** (161), 1053 (1969).
85. P. P. Kuzmenko, G. P. Grinevich and B. A. Danilchenko, *Fiz. Met. Metalloved* **29** (2), 318 (1970).
86. G. V. Kidson, *Phil. Mag.* **13**, 247 (1966).
87. J. W. Miller, *Phys. Rev.* **181** (3), 1095 (1969).
88. B. F. Dyson, T. Anthony, and D. Turnbull, *J. Appl. Phys.* **37**, 2370 (1966).
89. H. A. Resing and N. H. Nachtrieb, *Phys. Chem. Solids* **21** (1/2), 40 (1961).
90. A. Ott and A. Lodding, *Int. Conf. Vac. and Interstitials in metals.* **Julich 1**, 43 (1968); *Z. Naturforsch* **23A**, 1683 (1968), 2126 (1968).
91. A. Ott, A. Lodding, and D. Lazarus, *Phys. Rev.* **188** (3), 1088 (1969).
92. A. Ott, J. Appl. Phys. **40** (6), 2395 (1969); J. N. Mundy, A. Ott, L. Lowenberg, and A. Lodding, *Phys. Stat. Solidi.* **35** (1), 359 (1969).
93. A. Ott, *Z. Naturforschg* **25a** (10), 1477 (1970).
94. A. Lodding, J. N. Mundy, and A. Ott, *Phys. Stat. Solidi* **38** (2), 559 (1970).
95. K. Lal, *Commis. Energ. At. Report.* **CEA-R** 3136 (1967).
96. L. V. Pavlinov, A. M. Gladyshev, and V. N. Bikov, *Fiz. Metal Metalloved* **26** (5), 823 (1968).
97. P. G. Shewmon, *Trans. Met. Soc. AIME* **206**, 918 (1956).
98. Y. V. Borisov, P. L. Gruzin, and L. V. Pavlinov, *Met. i Metalloved. Chistykh Metal.* **1** 213 (1959), *translated in JPRS-5195.*
99. A. Y. Nakonechnikov, L. V. Pavlinov, and V. N. Bikov, *Fiz. Metal i Metalloved.* **22**, 234 (1966).
100. J. Askill, *Phys. Status Solidi* **9** (2), K113 (1965).
101. L. V. Pavlinov and A. A. Kordev, *Fiz. Metal. Metalloved.* **29** (6), 1326 (1970).
102. G. N. Dubinin, G. P. Benediktova, M. G. Karpman, and G. V. Shcherbedinskii, *Khim—Term Obrab Stali Splavov* **6**, 129 (1969).
103. J. Askill and D. H. Tomlin, *Phil. Mag.* **8** (90), 997 (1963).
104. B. A. Vandyshev and A. S. Panov, *Fiz. Met. Metalloved.* **26** (3), 517 (1968).
105. S. Z. Benediktova, G. N. Dubinin, M. G. Kapman, and G. V. Shcherbedinskii, *Metalloved. i Term. Obrabotka Metal.* **5**, 5 (1966).
106. B. A. Vandyshev and A. S. Panov, *Fiz. Metal. i Metalloved.* **25** (2), 321 (1968).
107. L. V. Pavlinov, A. Y. Nakonechnikov, and V. N. Bikov, *At. Energ.* (*USSR*) **19** 521 (1965).
108. J. Askill, *Phys. Status Solidi* **23**, K21 (1967).
109. A. Chatterjee and D. J. Fabian, *J. Inst. Metals* **96** (6), 186 (1968).
110. A. Hassner and W. Lange, *Phys. Status Solidi* **8**, 77 (1965).
111. K. Monma, H. Suto, and H. Oikawa, *J. Japan Inst. Metals* **28**, 188 (1964).
112. A. Ya. Shinyaev, *Izv. Akad. Nauk. S.S.S.R. Metl.* **4**, 182 (1969).
113. J. J. Blechet, A. VanGryeynest, and D. Calais, *J. Nucl. Mater.* **28** (2), 177 (1968).
114. P. P. Kuzmenko and G. P. Grinevich, *Fiz. Tverd. Tela.* **4** (11), 3266 (1962); *Soviet Physics—Solid State* **4** (11), 2390 (1962).
115. S. Z. Bokshtein, S. T. Kishkin, and L. M. Moroz, *Investigation of the Structure of Metals by Radioactive Isotope Methods; State Publishing House of the Ministry of Defense Industry, Moscow* (1959); *AEC-tr-4505* (1961).
116. M. S. Anand, S. P. Murarka, and R. P. Agarwala, *J. Appl. Phys.* **36** (12), 3860 (1965).
117. V. D. Lyubimov, *Izv. Akad. Nauk. S.S.S.R. Metal* **3**, 201 (1969); V. D. Lyubimov, V. A. Tskhai, and G. B. Bogmolov, *Toudy Inst. Khim. Akad. Nauk. S.S.S.R. Ural Filial* **17**, 44, 48 (1970).
118. R. F. Peart, D. Graham, and D. H. Tomlin, *Acta Met.* **10**, 519 (1962).
119. J. Pelleg, *J. Metals* **20** (8), 54 (1968); *J. Less Common Metals* **17**, 319 (1969); *Phil. Mag.* **19** (157), 25 (1969).
120. T. S. Lundy, F. E. Winslow, R. E. Pawel, and C. J. McHargue, *Trans. Met. Soc. AIME* **223**, 1533 (1965).
121. B. A. Vandyshev and A. S. Panov, *Izv. Akad. Nauk. S.S.S.R.* **1**, 206 (1968).
122. J. Askill, *Phys. Status Solidi* **9** (3), K167 (1965).
123. J. Pelleg, *Phil. Mag.* **21** (172), 735 (1970).
124. R. P. Agarwala, S. P. Murarka, and M. S. Anand, *Acta Met.* **16** (1), 61 (1968).
125. G. B. Fedorov, F. I. Zhomov, and E. A. Smirnov, *Met. Metalloved. Chist. Metal* **8**, 145 (1969).
126. N. L. Peterson, *Phys. Rev.* **136** (2A), A568 (1964).
127. N. H. Nachtrieb and G. S. Handler, *J. Chem. Phys.* **23**, 1187 (1955).
129. J. Kucera and T. Zemcik, *Can. Met. Quart.* **7** (2), 83 (1968).
130. G. V. Kidson and R. Ross, *Proc. 1st UNESCO Int. Conf. Radioisotopes in Scientific Res.*, p. 185 (1958).
131. F. A. Smith and L. W. Barr, *Phil. Mag.* **21** (171), 633 (1970).
132. J. N. Mundy, T. E. Miller, and R. J. Porte, *Phys. Rev.* **B3** (8), 2445 (1971).
133. L. W. Barr, J. N. Mundy, and F. A. Smith, *Phil. Mag.* **16**, 1139 (1967).
134. F. A. Smith and L. W. Barr, *Phil. Mag.* **20** (163), 205 (1969).
135. R. E. Tate and G. R. Edwards, *Symposium on Thermodynamics with Emphasis on Nuclear Materials and Atomic Solids*, **pp. 105–113** in **IAEA Symposium Vol. II**, International Atomic Energy Agcy., Vienna (1966).
136. R. E. Tate and E. M. Cramer, *Trans. Met. Soc. AIME* **230**, 639 (1964).
137. M. Dupuy and D. Calais, *Trans. AIME* **242**, 1679 (1968).
138. M. P. Dariel, G. Erez, and G. M. J. Schmidt, *J. Appl. Phys.* **40** (7), 2746 (1969).
139. M. P. Dariel, *Phil. Mag.* **22** (177), 653 (1970).
140. M. P. Dariel, G. Erez, and G. M. J. Schmidt, *Phil. Mag.* **18** (161), 1045 (1969).
141. A. A. Kuliev and D. N. Nasledov, *Zh. Tekhn. Fiz.* **28** (2), 259 (1958); *Soviet Phys.—Tech. Phys.* (*Eng. transl.*) **3** (2), 235 (1958).
142. P. Braetter and H. Gobrecht, *Phys. Stat. Sol* **41** (2), 631 (1970).
143. B. I. Boltaks, and B. T. Plachenov, *Soviet Phys.—Tech. Phys.* (*Eng. transl.*) **27** (10), 2071 (1957).
144. R. F. Peart, *Phys. Status Solidi* **15**, K119 (1966).
145. W. R. Wilcox and T. J. LaChapelle, *J. Appl. Phys.* **35** (1) 240 (1964).
146. R. C. Newman and J. Wakefield, *Phys. Chem. Solids* **19** (3), 230 (1961).
147. B. I. Boltaks and I. I. Sozinov, *Zh. Tekhn. Fiz.* **28** (3), 679 (1958); *Soviet Phys.—Tech. Phys.* (*Eng. transl.*) **3** (3), 636 (1958).
148. J. D. Struthers, *J. Appl. Phys.* **27** (12), 1560 (1958); errata **28** (4), 516 (1957).
149. H. P. Bonzel, *Phys. Status Solidi* **20**, 493 (1967).
150. S. Mackawa, *J. Phys. Soc. Japan* **17** (10), 1592 (1962).
151. J. J. Rahan, M. E. Pickering, and J. Kennedy, *J. Electrochem. Soc.* **106** (8), 705 (1959).
152. S. J. Rothman, N. L. Peterson, and J. T. Robinson, *Phys. Stat. Sol.* **39** (2), 635 (1970).
153. C. T. Tomizuka and L. M. Slifkin, *Phys. Rev.* **96**, 610 (1954).
154. J. Bernardini, A. Combe-Brun, and J. Cabane, *Ser. Met.* **4** (12), 985 (1970).
155. A. Sawatskii and F. E. Jaumot, *Trans. AIME* **209**, 1207 (1957).
156. J. G. Mullen, *Phys. Rev.* **121**, 1649 (1961).
157. R. E. Hoffman, *Acta Met.* **6**, 95 (1958).
158. T. Hirone, S. Miura, and T. Suzuoka, *J. Phys. Soc. Japan* **16** (12), 2456 (1961).
159. R. E. Hoffman, D. Turnbull, and E. W. Hart, *Acta Met.* **3**, 417 (1955).
160. C. B. Pierce and D. Lazarus, *Phys. Rev.* **114**, 686 (1959).
161. N. Barbouth, J. Ouder, and J. Cabane, *C. R. Acad. Sci. Paris Ser. C* **264** (12), 1029 (1967).
162. V. N. Kaigorodov, Ya. A. Rabovskii, and V. K. Talinskii, *Fiz. Metal. i Metalloved.* **24** (4), 661 (1967).
163. V. N. Kaigorodov, S. M. Klotsman, A. N. Timofeev, and I. Sh. Traktenberg, *Fiz. Metal. Metalloved.* **28** (1), 120 (1969).
164. A. Sawatskii and F. E. Jaumot, *Phys. Rev.* **100**, 1627 (1955).
165. L. W. Barr, J. N. Mundy, and F. A. Smith, *Phil. Mag.* **20** (164), 389 (1969).
166. J. N. Mundy, L. W. Barr, and F. A. Smith, *Phil. Mag.* **14**, 785 (1966).
167. P. Son, S. Ihara, M. Miyake, and T. Sano, *J. Japan Inst. Met.* **33** (1), 1 (1969).
168. V. P. Vasilev, I. F. Kamardin, V. I. Skatskill, S. G. Chernomorchenko, and G. N. Shuppe, *Trudy Sred. Gos. Uni. im V.I. Lenina* **65**, 47 (1955); *Translation AEC-tr-4272.*
169. R. E. Pawel and T. S. Lundy, *J. Phys. Chem. Solids* **26**, 937 (1965).

REFERENCES (Continued)

170. B. A. Vandyshev and A. S. Panov, *Izv. Akad. Nauk. SSSR Metal* **1**, 244 (1969).
171. Sh. Movalanov and A. A. Kuliev, *Fiz. Tverd. Tela* **4** (2), 542 (1962).
172. N. I. Ibraginov, M. G. Shachtachtinskii, and A. A. Kuliev, *Fiz. Tverd. Tela* **4**, 3321 (1962).
173. R. N. Ghoshtagore, *Phys. Rev.* **155** (3), 698 (1967).
174. T. R. Anthony, B. F. Dyson, and D. Turnbull, *J. Appl. Phys.* **39** (3), 1391 (1968).
175. G. A. Shirn, *Acta Met.* **3**, 87 (1955).
176. F. Schmitz and M. Fock, *J. Nucl. Materials* **21**, 317 (1967).
177. B. F. Dyson, *J. Appl. Phys.* **37**, 2375 (1966).
178. W. Chomka and J. Andruszkiewicz, *Nukleonika* **5** (10), 611 (1960).
179. A. Sawatskii, *J. Appl. Phys.* **29** (9), 1303 (1958).
180. D. Yolokoff, S. May, and Y. Adda, *Compt. Rend.* **251** (3), 2341 (1960).
181. L. Bartha and T. Szalay, *Int. J. Appl. Radiat. Isotopes* **20** (2), 825 (1969).
182. F. Dyment and C. M. Libanati, *J. Mager. Sci.* **3** (4), 349 (1968).
183. J. Askill, *Phys. Stat Sol.* **B43** (1), K1 (1971).
184. L. V. Pavlinov, G. V. Grigorev, and G. O. Gromyko, *Izv. Akad. Nauk. SSSR Metal* **3**, 207 (1969).
185. A. Y. Nakonechnikov, L. V. Pavlinov, and V. N. Bikov, *Fiz. Metal i Metalloved* **22**, 234 (1966).
186. G. B. Gibbs, D. Graham, and D. H. Tomlin, *Phil. Mag.* **8** (92), 1269 (1963).
187. J. Askill and G. B. Gibbs, *Phys. Status Solidi* **11**, 557 (1965).
188. J. F. Murdock, T. S. Lundy, and E. E. Stansbury, *Acta Met.* **12** (9), 1033 (1964).
189. L. V. Pavlinov, *Fiz. Metal Metalloved* **30** (4), 800 (1970).
190. J. F. Murdock, T. S. Lundy, and E. E. Stansbury, *Acta Met.* **12** (9), 1033 (1964).
191. L. V. Pavlinov, *Fiz. Metal. i Metalloved.* **24** (2), 272 (1967).
192. R. E. Pawel and T. S. Lundy, *Acta Met.* **17** (8), 979 (1969).
193. L. N. Larikov, V. M. Tyshkevich, and L. F. Chorna, *Ukr. Fiz. Zh.* **12** (6), 983 (1967).
194. Y. Adda, A. Kirianenko, and C. Mairy, *Compt. Rend.* **253**, 445 (1961); *J. Nucl. Mater.* **6** (1), 130 (1962).
195. M. P. Dariel, M. Blumenfeld, and G. Kimmel, *J. Appl. Phys.* **41** (4), 1480 (1970).
196. G. B. Federov, E. A. Smirnov, and S. S. Moiseenko, *Met. Metalloved. Chist. Metal.* **7**, 124 (1968).
197. S. J. Rothman, *J. Nucl. Mater.* **3** (1), 77 (1961).
198. N. L. Peterson and S. J. Rothman, *Phys. Rev.* **136** (3A), A842 (1964).
199. P. Son, S. Ihara, M. Miyake, and T. Sano, *J. Japan Inst. Met.* **33** (1), 1 (1969).
200. H. W. Paxton and R. A. Wolfe, *Trans. AIME* **230**, 1426 (1964).
201. M. G. Coleman, *Ph.D. Thesis Univ. of Illinois* (1967); *Univ. Microf.* **68–1725** (1968).
202. B. A. Vandychev and A. S. Panov, *Izv. Akad. Nauk SSSR Metal* **2**, 231 (1970).
203. R. F. Peart, *J. Phys. Chem. Solids* **26**, 1853 (1965).
204. D. S. Gorney and R. M. Altovskii, *Fiz. Metal Metalloved* **30** (1), 85 (1970).
205. J. H. Rosolowski, *Phys. Rev.* **124** (6), 1828 (1961).
206. P. B. Ghate, *Phys. Rev.* **130** (1), 174 (1963).
207. A. P. Batra and H. B. Huntington, *Phys. Rev.* **145**, 542 (1966).
208. A. P. Batra and H. B. Huntington, *Phys. Rev.* **154** (3), 569 (1967).
209. J. S. Warford and M. B. Huntington, *Phys. Rev.* **B1** (4), 1867 (1970).
210. N. L. Peterson and S. J. Rothman, *Phys. Rev.* **163** (3), 645 (1967).
211. R. P. Agarwala, S. P. Murarka, and M. S. Anand, *Trans. Met. Soc. AIME* **233**, 986 (1965).
212. A. M. Blinkin and V. V. Vorobiov, *Ukr. Fiz. Zh.* **9** (1), 91 (1964).
213. R. P. Agarwala and A. R. Paul, *Proc. Nucl. Rad. Chem. Symp. Poona, India* **p. 542** (1967).
214. P. L. Gruzin, V. S. Emelyanov, G. G. Ryabova, and G. B. Federov, *Proc. 2nd Intern. Conf. Peaceful Uses At. Energy*, **19:187** (1958).
215. Y. V. Borisov, Y. G. Godin, P. L. Gruzin, A. I. Evstyukhin, and V. S. Yemelyanov, *Met. i Met. Izdatel. Akad. Nauk. SSSR Moscow* 1958. *Translation NP-TR-448* **p. 196** (1960).
216. P. Flubacher, *E. I. R. Bericht* **49**, May (1963).
217. R. A. Andriyevskii, V. N. Zagraykin, and G. Ya. Meshcheryakov, *Phys. Met. Metalloved.* **19** (3), 146 (1966).
218. R. P. Agarwala and A. R. Paul, Report CONF-670335, *Int. Conf. Vacancies and Interstitials in Metals*, Julich (1968) **1**, 105 (1968), *Report BARC*-377 (1968).
219. G. V. Kidson and G. J. Young, *Phil. Mag.* **20** (167), 1047 (1969).
220. G. B. Federov, E. A. Smirnov, and S. M. Novikov, *Met. Metalloved Chist. Metal* **8**, 41 (1969).
221. B. A. Vandychev and A. S. Panov, *Izv. Akad. Nauk. SSSR Metal* **2**, 231 (1970).
222. G. B. Federov, E. A. Smirnov, and F. I. Zhomov, *Met. Metalloved. Chist. Metal.* **7**, 116 (1968).
223. L. V. Pavlinov, *Fiz. Metal i Metalloved* **24** (2), 272 (1967).
224. H. M. Gilder and D. Lazarus, *J. Phys. Chem. Solids* **26**, 2081 (1965).
225. G. V. Kidson and J. McGurn, *Can. J. Phys.* **39** (8), 1146 (1961).
226. R. E. Pawel and T. S. Lundy, *J. Phys. Chem. Solids* **26**, 937 (1965).
227. S. J. Rothman, L. T. Lloyd, and A. L. Harkness, *Trans. Met. Soc. AIME* **218** (4), 605 (1960).
228. G. B. Federov, E. A. Smirnov, and V. N. Gusev, *At. Energ.* **27** (2), 149 (1969).

VISCOSITY AND THERMAL CONDUCTIVITY OF OXYGEN AS A FUNCTION OF TEMPERATURE

Conversion Factors

T, °K	→ T, °F: multiply by (9/5) then subtract 459.67
T, °K	→ T, °C: subtract 273.15
T, °K	→ T, °R: multiply by (9/5)
P, atm	→ P, psia: multiply by 14.69595
P, atm	→ p, N/m^2: multiply by 1.01325 × 10^5
η, g/cm-s	→ η, N-s/m^2
η, g/cm-s	→ η, lb_m/ft-s
λ, W/m-°K	→ λ, cal/cm-s-°K
λ, W/m-°K	→ λ, Btu ft-hr-°R

Temperature, °K	Viscosity, g/cm-s $10^3\eta_0$	Thermal conductivity, W/m-K $10^3\lambda_0$
80	0.0585	6.94
90	0.0663	7.95
100	0.0740	8.96
110	0.0818	9.96
120	0.0894	10.94
130	0.0970	11.92
140	0.1045	12.89
150	0.1118	13.85
160	0.1191	14.78
170	0.1261	15.70
180	0.1331	16.60
190	0.1399	17.48
200	0.1465	18.35
210	0.1530	19.21
220	0.1595	20.05
230	0.1658	20.88
240	0.1719	21.69
250	0.1780	22.49
260	0.1840	23.29
270	0.1898	24.08
280	0.1955	24.86
290	0.2012	25.62
300	0.2068	26.38
310	0.2122	27.14
320	0.2176	27.89
330	0.2230	28.62
340	0.2282	29.36
350	0.2334	30.10
360	0.2385	30.85
370	0.2435	31.58

[a]Data for all temperatures in excess of 1,000 K are extrapolated.

VISCOSITY AND THERMAL CONDUCTIVITY OF OXYGEN AS A FUNCTION OF TEMPERATURE (Continued)

Temperature, °K	Viscosity, g/cm-s $10^3\ \eta_0$	Thermal conductivity, W/m-K $10^3\ \lambda_0$
380	0.2485	32.32
390	0.2534	33.06
400	0.2583	33.79
410	0.2631	34.54
420	0.2678	35.28
430	0.2725	36.02
450	0.2818	37.49
470	0.2908	38.95
490	0.2997	40.41
510	0.3085	41.86
530	0.3170	43.29
550	0.3255	44.72
570	0.3338	46.14
590	0.3420	47.56
610	0.3501	48.97
630	0.3580	50.37
650	0.3659	51.77
670	0.3736	53.16
690	0.3813	54.54
710	0.3888	55.89
730	0.3963	57.25
750	0.4037	58.59
770	0.4110	59.90
790	0.4183	61.22
820	0.4290	63.16
850	0.4396	65.09
880	0.4500	66.98
910	0.4603	68.89
940	0.4705	70.72
970	0.4805	72.52
1,000	0.4905	74.32
1,030[a]	0.5002	76.07
1,060	0.5100	77.84
1,090	0.5196	79.56
1,120	0.5291	81.27
1,150	0.5384	82.97
1,180	0.5478	84.66
1,210	0.5571	86.32
1,240	0.5662	89.97
1,270	0.5753	89.61
1,300	0.5843	91.24
1,330	0.5932	92.85
1,360	0.6021	94.45
1,390	0.6109	96.06
1,420	0.6196	97.64

VISCOSITY AND THERMAL CONDUCTIVITY OF OXYGEN AS A FUNCTION OF TEMPERATURE (Continued)

Temperature, °K	Viscosity, g/cm-s $10^3\ \eta_0$	Thermal conductivity, W/m-K $10^3\ \lambda_0$
1,450	0.6282	99.22
1,480	0.6368	100.78
1,510	0.6454	102.35
1,540	0.6538	103.90
1,570	0.6622	105.45
1,600	0.6706	106.98
1,630	0.6789	108.52
1,660	0.6871	110.05
1,690	0.6953	111.58
1,720	0.7034	113.09
1,750	0.7116	114.61
1,780	0.7196	116.12
1,810	0.7276	117.63
1,840	0.7355	119.12
1,870	0.7435	120.63
1,900	0.7514	122.13
1,930	0.7592	123.62
1,960	0.7670	125.11
1,980	0.7721	126.11
2,000	0.7773	127.10

From Hanley, H. J. M., McCarty, R. D., and Sengers, J. V., *Viscosity and Thermal Conductivity Coefficients of Gaseous and Liquid Oxygen,* NASA CR-2440, National Aeronautics and Space Administration, Washington, D.C., August 1974 (available from National Technical Information Service, Springfield, Va. 22151).

THERMAL CONDUCTIVITY OF COMPRESSED OXYGEN

In this table, thermal conductivity, λ, is in the unit milliwatt/m-K.

Conversion Factors

T, °K → T, °F: multiply by (9/5) then subtract 459.67
T, °K → T, °C: subtract 273.15
T, °K → T, °R: multiply by (9/5)
P, atm → P, psia: multiply by 14.69595
P, atm → P, N/m²: multiply by 1.01325×10^5
η, g/cm-s → η, Ns/m²: multiply by 10^{-1}
η, g/cm-s → η, lb_m/ft-s: multiply by 0.0671969
λ, W/m-K → λ, cal/cm-s-K: multiply by (1/418.4)
λ, W/m-K → λ, Btu/ft-hr-°R: multiply by 0.578176

T, °K \ P, atm	1	5	10	15	20	25	30	35	40	45	50	55
80	164.7	164.9	165.2	165.4	165.7	165.9	166.2	166.4	166.7	166.9	167.1	167.4
90	151.7	152.0	152.3	152.6	152.9	153.3	153.6	153.9	154.2	154.5	154.8	155.1
100	9.5	138.4	138.8	139.2	139.6	140.0	140.4	140.8	141.1	141.5	141.9	142.3
110	10.4	11.4	124.9	125.3	125.8	126.3	126.8	127.3	127.7	128.2	128.6	129.1
120	11.4	12.2	13.5	110.9	111.5	112.1	112.8	113.3	113.9	114.5	115.1	115.6
130	12.3	13.1	14.2	15.6	96.3	97.1	98.0	98.8	99.6	100.3	101.1	101.8
140	13.2	14.0	15.0	16.1	17.5	19.3	81.6	82.9	84.1	85.2	86.3	87.3
150	14.1	14.8	15.7	16.7	17.9	19.2	20.9	23.3	27.8	68.4	70.4	72.1
160	15.8	15.7	16.5	17.4	18.4	19.5	20.7	22.3	24.2	26.7	30.5	37.3
170	15.9	16.5	17.3	18.1	19.0	19.9	20.9	22.1	23.4	24.9	26.8	29.0
180	16.8	17.4	18.1	18.8	19.6	20.4	21.3	22.3	23.3	24.4	25.7	27.1
190	17.7	18.2	18.9	19.5	20.3	21.0	21.8	22.6	23.5	24.4	25.4	26.4
200	18.5	19.0	19.6	20.3	20.9	21.6	22.3	23.0	23.8	24.6	25.4	26.3
210	19.3	19.8	20.4	21.0	21.6	22.3	22.9	23.6	24.2	24.9	25.7	26.4
220	20.2	20.6	21.2	21.7	22.3	22.9	23.5	24.1	24.7	25.4	26.0	26.7
230	21.0	21.4	21.9	22.5	23.0	23.6	24.1	24.7	25.3	25.9	26.5	27.1
240	21.7	22.2	22.7	23.2	23.7	24.2	24.8	25.3	25.8	26.4	26.9	27.5
250	22.5	22.9	23.4	23.9	24.4	24.9	25.4	25.9	26.4	26.9	27.5	28.0
260	23.3	23.7	24.1	24.6	25.1	25.6	26.0	26.5	27.0	27.5	28.0	28.5
270	24.0	24.4	24.9	25.3	25.8	26.2	26.7	27.1	27.6	28.1	28.5	29.0
280	24.8	25.1	25.6	26.0	26.4	26.9	27.3	27.8	28.2	28.6	29.1	29.5
290	25.5	25.9	26.3	26.7	27.1	27.5	28.0	28.4	28.8	29.2	29.6	30.1
300	26.2	26.6	27.0	27.4	27.8	28.2	28.6	29.0	29.4	29.8	30.2	30.6
310	27.0	27.3	27.7	28.1	28.5	28.8	29.2	29.6	30.0	30.4	30.8	31.2
320	27.7	28.0	28.4	28.7	29.1	29.5	29.9	30.2	30.6	31.0	31.4	31.7
330	28.4	28.7	29.0	29.4	29.8	30.1	30.5	30.9	31.2	31.6	31.9	32.3
340	29.1	29.3	29.7	30.1	30.4	30.8	31.1	31.5	31.8	32.2	32.5	32.9
350	29.7	30.0	30.4	30.7	31.1	31.4	31.7	32.1	32.4	32.8	33.1	33.4
360	30.4	30.7	31.0	31.4	31.7	32.0	32.4	32.7	33.0	33.3	33.7	34.0
370	31.1	31.4	31.7	32.0	32.3	32.7	33.0	33.3	33.6	33.9	34.3	34.6
380	31.8	32.0	32.3	32.7	33.0	33.3	33.6	33.9	34.2	34.5	34.8	35.1
390	32.4	32.7	33.0	33.3	33.6	33.9	34.2	34.5	34.8	35.1	35.4	35.7
400	33.1	33.3	33.6	33.9	34.2	34.5	34.8	35.1	35.4	35.7	36.0	36.3

THERMAL CONDUCTIVITY OF COMPRESSED OXYGEN (Continued)

T, °K \ P, atm	60	65	70	80	90	100	110	120	130	150	175
80	167.6	167.9	168.1	168.6	169.1	169.5	170.0	170.4	170.9	171.8	172.8
90	155.4	155.7	156.0	156.6	157.1	157.7	158.3	158.8	159.4	160.4	161.8
100	142.6	143.0	143.3	144.0	144.7	145.4	146.1	146.8	147.4	148.7	150.3
110	129.5	130.0	130.4	131.2	132.1	132.9	133.7	134.5	135.3	136.8	138.6
120	116.2	115.7	117.2	118.3	119.3	120.3	121.2	122.2	123.1	124.9	127.0
130	102.5	103.2	103.9	105.2	106.4	107.6	108.8	109.9	111.0	113.1	115.6
140	88.3	89.2	90.1	91.8	93.4	95.0	96.4	97.8	99.1	101.6	104.5
150	73.5	74.9	76.2	78.5	80.5	82.5	84.2	85.9	87.5	90.5	93.8
160	50.2	57.8	61.3	65.6	68.6	71.0	73.1	75.1	76.9	80.2	83.9
170	31.7	35.1	39.2	47.9	54.7	59.3	62.7	65.3	67.6	71.3	75.3
180	26.7	38.4	32.4	36.9	42.0	47.0	51.3	55.0	58.1	63.0	67.7
190	27.8	28.8	30.1	33.0	36.2	39.6	43.2	46.5	49.7	55.1	60.5
200	27.2	28.1	29.1	31.3	33.6	36.1	38.7	41.4	44.0	48.9	54.3
210	27.2	28.0	28.8	30.5	32.4	34.4	36.4	38.5	40.6	44.9	49.8
220	27.4	28.1	28.8	30.3	31.9	33.5	35.2	36.9	38.7	42.3	46.6
230	27.7	28.3	29.0	30.3	31.7	33.1	34.6	36.1	37.6	40.7	44.5
240	28.1	28.7	29.2	30.5	31.7	33.0	34.3	35.6	36.9	39.7	43.1
250	28.5	29.1	29.6	30.7	31.8	33.0	34.2	35.4	36.6	39.0	42.1
260	29.0	29.5	30.0	31.0	32.1	33.1	34.2	35.3	36.4	38.7	41.5
270	29.5	29.9	30.4	31.4	32.4	33.3	34.4	35.4	36.4	38.4	41.0
280	30.0	30.4	30.9	31.8	32.7	33.6	34.6	35.5	36.5	38.4	40.8
290	30.5	30.9	31.3	32.2	33.1	33.9	34.8	35.7	36.6	38.4	40.6
300	31.0	31.4	31.8	32.7	33.5	34.3	35.1	36.0	36.8	38.5	40.6
310	31.6	31.9	32.3	33.1	33.9	34.7	35.5	36.3	37.1	38.7	40.6
320	32.1	32.5	32.8	33.6	34.3	35.1	35.8	36.6	37.4	38.9	40.8
330	32.7	33.0	33.4	34.1	34.8	35.5	36.2	37.0	37.7	39.1	40.9
340	33.2	33.5	33.9	34.6	35.3	36.0	36.7	37.3	38.0	39.4	41.1
350	33.8	34.1	34.4	35.1	35.8	36.4	37.1	37.7	38.4	39.7	41.4
360	34.3	34.6	35.0	35.6	36.2	36.9	37.5	38.2	38.6	40.1	41.7
370	34.9	35.2	35.5	36.1	36.7	37.9	38.0	38.6	39.2	40.4	42.0
380	35.4	35.7	36.0	36.6	37.2	37.8	38.4	39.0	39.6	40.8	42.3
390	36.0	36.3	36.6	37.2	37.8	38.3	38.9	39.5	40.1	41.2	42.6
400	36.6	36.9	37.1	37.7	38.3	38.8	39.4	40.0	40.5	41.6	43.0

From Hanley, H. J. M., McCarty, R. D., and Sengers, J. V., *Viscosity and Thermal Conductivity Coefficients of Gaseous and Liquid Oxygen,* NASA CR-2440, National Aeronautics and Space Administration, Washington, D.C., August 1974 (available from National Technical Information Service, Springfield, Va. 22151).

TRANSPORT PROPERTIES OF OXYGEN AT SATURATION

Conversion Factors

T, °K → T, °F: multiply by (9/5) then subtract 459.67
T, °K → T, °C: subtract 273.15
T, °K → T, °R: multiply by (9/5)
P, atm → P, psia: multiply by 14.69595
P, atm → P, N/m²: multiply by 1.01325×10^5
η, g/cm-s → η, Ns/m²: multiply by 10^{-1}
η, g/cm-s → η, lb_m/ft-s: multiply by 0.0671969
λ, W/m-K → λ, cal/cm-s-K: multiply by (1/418.4)
λ, W/m-K → λ, Btu/ft-hr-°R: multiply by 0.578176

Temperature, °K	Pressure, atm	Viscosity, mg/cm-s		Thermal conductivity, mW/m-K	
		Vapor	Liquid	Vapor	Liquid
80	0.30	0.059	2.652	7.4	164.7
90	0.98	0.068	1.971	8.5	151.8
100	2.51	0.079	1.504	9.9	138.2
110	5.36	0.092	1.188	11.5	124.3
120	10.09	0.108	0.971	13.6	110.3
130	17.26	0.129	0.824	16.3	95.9
140	27.52	0.158	0.696	20.1	80.6
142	30.00	0.166	0.668	21.1	77.4
144	32.64	0.175	0.638	22.3	74.0
146	35.45	0.185	0.607	23.6	70.6
148	38.44	0.197	0.574	25.2	67.0
150	41.61	0.211	0.537	27.2	63.2
152	45.00	0.231	0.494	29.9	58.7
154	48.65	0.269	0.424	35.2	52.0

From Hanley, H. J. M., McCarty, R. D., and Sengers, J. V., *Viscosity and Thermal Conductivity Coefficients of Gaseous and Liquid Oxygen,* NASA CR-2440, National Aeronautics and Space Administration, Washington, D.C., August 1974 (available from National Technical Information Service, Springfield, Va. 22151).

PHYSICAL CONSTANTS OF OZONE AND OXYGEN

Physical Constant	Ozone (O_3)	Oxygen (O_2)
Molecular Weight	47.9982 g/g-mol	31.9988 g/g-mol
Boiling Point (760 mm)	−111.9 ± 0.3°C	−182.97°C
Melting Point	−192.7 ± 0.2°C	−218.4°C
Critical Temperature	−12.1 ± 0.1°C	−118.574°C
Critical Pressure	54.6 atm	49.77 atm
Critical Density	0.437 g/cc	0.436 g/cc
Critical Volume	147.1 cc/mol	73.37 cc/mol
Gas Density (0°C) (760 mm pressure)	2.144 g/liter	1.429 g/liter
Liquid Density		
−112°C	1.358 g/cc	
−183°C	1.571 g/cc	1.14 g/cc
−195.4°C	1.614 g/cc	1.201 g/cc
Surface Tension		
−195°C	43.8 ± 0.1 dyne/cm	
−182.7°C	38.1 ± 0.2 dyne/cm	
−183.0°C	38.4 ± 0.7 dyne/cm	13.2 dyne/cm
Heat Capacity of Liquid		
−183 to −145°C	0.45 cal/g°C	
Heat Capacity of Gas		
−173°C	7.95 cal/g mol°C	
0°C	9.10 cal/g mol°C	
25°C	9.37 cal/g mol°C	
100°C		6.979 cal/g mol°C
127°C	10.44 cal/g mol°C	
Viscosity of Liquid		
−195.6°C	4.14 ± 0.05 cp	
−183.0°C	1.57 ± 0.02 cp	0.1958 cp
Heat of Vaporization		
−112°C	75.6 cal/g	
−182.9°C		50.9 cal/g
Heat of Formation		
25°C	−34.4 Kg cal/mol	
Free Energy		
25°C	32.4 Kg cal/mol	
Van der Waals Constant (a)	3.545 atm liter²/mol²	1.36 atm liter²/mol²
Van der Waals Constant (b)	0.04903 liter/mol	0.03803 liter/mol
Magnetic Susceptibility		
gas ($\times 10^{-6}$)	0.002 cgs units	10.6.2 cgs units
liq ($\times 10^{-6}$)	0.150 cgs units	260.0 cgs units
Thermal Conductivity of Liquid		
−195.8°C	5.21 cal/sec cm°C $\times 10^4$	
−183.0°C	5.31 cal/sec cm°C $\times 10^4$	
−165.0°C	5.42 cal/sec cm°C $\times 10^4$	
−128.0°C	5.52 cal/sec cm°C $\times 10^4$	

Phase Boundaries—Ozone-Oxygen System
−183°C 29.8 and 72.4 wt % O_3
−195.4°C 9 and 90.8 wt %O_3
Consolute Temperature—Ozone-Oxygen System
−180 ± 0.5°C
Coefficient of Thermal Expansion for Liquid Ozone

Temp. °C	α
−195.6	1.62
−183.0	1.58
−148.0	1.47
−123.0	1.41
−112.0	1.35
−98.8	1.31

PHYSICAL CONSTANTS OF CLEAR FUSED QUARTZ

Based on information contained in Fused Quartz Catalogue Q-7A General Electric Company.

Property		Clear fused quartz	Property		Clear fused quartz
Density		2.2 g./c.c.	Annealing Point	(approx.)	1140°C
Hardness		4.9 (Mohs')	Strain Point		1070°C
Tensile Strength		7,000 p.s.i.	Electrical Resistance		9.5 $\log_{10}$ R for cm.³ at 350°C
Compressive Strength		>160,000 p.s.i.	Dielectric Constant		3.75 at 20°C. 1 Mc.
Bulk Modulus	(approx.)	5.3×10^6 p.s.i.	Dielectric Loss Factor	less than	.0004 at 20°C. 1 Mc.
Rigidity Modulus		4.5×10^6 p.s.i.	Dissipation Factor	less than	.0001 at 20°C. 1 Mc.
Young's Modulus		10.4×10^6 p.s.i.	Index of Refraction		1.4585
Poisson's Ratio		.16	Velocity of Sound—Shear Wave		3.75×10^5 cm./sec.
Coefficient of Thermal Expansion	(av.)	5.5×10^{-7} cm./cm./°C {20°C–320°C}	Velocity of Sound—Compressional Wave		5.90×10^5 cm./sec.
Thermal Conductivity		.0033 g. cal./cm.²/sec./°C/cm.	Sonic Attenuation	less than	.033 db/ft./mc.
Specific Heat		.18 g. cal./gm.			
Softening Point	(approx.)	1665°C			

FIXED POINT PROPERTIES OF OXYGEN

From NASA SP-3071, "ASRDI Oxygen Technology Survey, Thermophysical Properties", Volume I (1972), edited by Hans M. Roder and Lloyd A. Weber. This NASA publication contains an extensive bibliography and a discussion of basis for selection of these data for oxygen. The publication is available from the National Technical Information Service, Springfield, Virginia 22151.

PROPERTIES ↓ CONDITIONS →	Triple Point Solid	Triple Point Liquid	Triple Point Vapor	Normal Boiling Point Liquid	Normal Boiling Point Vapor	Critical Point ††	Standard Conditions STP (0°C)	Standard Conditions NTP (20°C)
Temperature (K)		54.351		90.180		154.576	273.15	293.15
Pressure (mmHg)		1.138		760		37,823	760	760
Density (mole/cm³) x 10³	42.46	40.83	0.000336	35.65	0.1399	13.63	0.04466	0.04160
Specific Volume (cm³/mole) x 10⁻³	0.02355	0.02449	2975	0.028047	7.1501	0.07337	22.392	24.038
Compressibility Factor, $Z = \frac{PV}{RT}$	–	0.0000082	0.9986	0.00379	0.9662	0.2879	0.9990	0.9992
Heats of Fusion & Vaporization (J/mole)	444.8	7761.4		6812.3		0	–	–
Specific Heat (J/mole-K) C_s, @saturation	46.07	53.313	−108.7	54.14	−53.2	(very large)	–	–
C_p, @ constant pressure	–	53.27	29.13	54.28	30.77	(very large)	29.33	29.40
C_v, @ constant volume	–	35.65	20.81	29.64	21.28	(38.7)	20.96	21.04
Specific Heat Ratio, $\gamma = C_p/C_v$	–	1.494	1.400	1.832	1.446	(large)	1.40	1.40
Enthalpy (J/mole)	−6634.4	−6189.6	1571.8	−4270.3	2542.0	1032.2	7937.8	8525.1
Internal Energy (J/mole)	−6634.4	−6189.6	1120.0	−4273.1	1817.5	662.3	5668.9	6089.5
Entropy (J/mole-K)	58.92	67.11	209.54	94.17	169.68	134.42	202.4	204.5
Velocity of Sound (m/sec)		1159	141	903	178	164	315	326
Viscosity, μ, (N-sec/m³) x 10³	–	0.6194	0.003914	0.1958	0.00685	(0.031)	0.01924	0.02036
(centipoise)‡‡	–	0.6194	0.003914	0.1958	0.00685	(0.031)	0.01924	0.02036
Thermal Conductivity (mW/cm-K), k		1.929	0.04826	1.515	0.08544	(*)	0.2428	0.2575
Prandtl Number, $N_{pr} = \mu C_p/k$	–	5.344	0.7392	2.193	0.7714		0.7259	0.7265
Dielectric Constant, ϵ	(1.614)	1.5687	1.000004	1.4870	1.00166	1.17082	1.00053	1.00049
Index of Refraction, $n = \sqrt{\epsilon}$ †	(1.271)	1.2525	1.000002	1.219	1.00083	1.0820	1.00027	1.00025
Surface Tension (N/m) x 10³	–	22.65	–	13.20	–	0	–	–
Equiv. Vol./Vol. Liquid at NBT	0.8397	0.8732	106,068	1	254.9	2.616	798.4	857.1

† Long Wavelengths
* Anomalously Large

Gas Constant: R = 62, 365.4 cm³-mm Hg/mole-K[1]
†† Values in parenthesis are estimates

Molecular Weight = 31.9988[3]
"mole" = gram mole
‡‡ Units for poise are: g/cm-sec

FIXED POINTS AND PHASE EQUILIBRIUM BOUNDARIES FOR PARAHYDROGEN

a. Triple point

P_t = 0.0704 bar

T_t = 13.803 K

ρ_t (liquid) = 38.21 mol/ℓ

b. Normal boiling point

P_b = 1.01325 bar

T_b = 20.268 K

ρ_b (liquid) = 35.11 mol/ℓ

ρ_b (gas) = 0.6636 mol/ℓ

c. Critical point

P_c = 12.928 bar

T_c = 32.976 K

ρ_c = 15.59 mol/ℓ

Note: Some data indicate that the true critical temperature is probably closer to 32.93 K. However, that value is pending further verfication.

d. Melting pressures: in atmospheres

$$P = P_t + (T - T_t)\,[A_1 e^{-\alpha/T} + A_2 T]$$

A_1 = 30.3312

A_2 = 0.6667

α = 5.693

e. Liquid-vapor coexistence densities

Liquid, density in mol/cm³:

$$\rho \text{ sat } \ell = \rho_c + A_1(\Delta T)^{0.380} + A_2(\Delta T) + A_3(\Delta T)^{4/3} + A_4(\Delta T)^{5/3} + A_5(\Delta T)^2$$

A_1 = 7.323 4603 × 10^{-3}

A_2 = −4.407 4261 × 10^{-4}

A_3 = 6.620 7946 × 10^{-4}

A_4 = −2.922 6363 × 10^{-4}

A_5 = 4.008 4907 × 10^{-5}

$\Delta T = T_c - T$

Vapor $T_b \leqslant T \leqslant T_c$, density in mol/cm³:

$$\rho \text{ sat } G = \rho_c + A_1(\Delta T)^{0.370} + A_2(\Delta T) + A_3(\Delta T)^{0.7} + A_4(\Delta T)^{0.8}$$

A_1 = −7.196 7724 × 10^{-3}

A_2 = 1.449 5527 × 10^{-3}

A_3 = 3.240 3120 × 10^{-3}

A_4 = −4.464 0177 × 10^{-3}

f. Vapor pressure: in atmospheres

For T ⩽ 29 K:

$$\log_{10} P_a = A_1 + \frac{A_2}{T + A_3} + A_4 T$$

A_1 = 2.000 620

A_2 = −50.09 708

A_3 = 1.0044

A_4 = 1.748 495 × 10^{-2}

For T > 29 K:

$$P = P_a + A_5(T - 29)^3 + A_6(T - 29)^5 + A_7(T - 29)^7$$

A_5 = 1.317 × 10^{-3}

A_6 = −5.926 × 10^{-5}

A_7 = 3.913 × 10^{-6}

From Weber, L. A., Thermodynamic and Related Properties of Parahydrogen from the Triple Point to 300 K at Pressures to 1000 Bar, NASA SP-3008, NBSIR 74-374, 1975.

Equations:

$$\rho/\text{kg m}^{-3} = (999.83952 + 16.945176\ t - 7.9870401 \times 10^{-3}\ t^2 - 46.170461 \times 10^{-6}\ t^3 + 105.56302 \times 10^{-9}\ t^4 - 280.54253 \times 10^{-12}\ t^5)/(1 + 16.879850 \times 10^{-3}\ t) \quad (1)$$

$$10^6\ \kappa_T/\text{bar}^{-1} = (50.88496 + 0.6163813\ t + 1.459187 \times 10^{-3}\ t^2 + 20.08438 \times 10^{-6}\ t^3 - 58.47727 \times 10^{-9}\ t^4 + 410.4110 \times 10^{-12}\ t^5)/(1 + 19.67348 \times 10^{-3}\ t) \quad (2)$$

$$10^6\ \kappa_T/\text{bar}^{-1} = (50.884917 + 0.62590623\ t + 1.3848668 \times 10^{-3}\ t^2 + 21.603427 \times 10^{-6}\ t^3 - 72.087667 \times 10^{-9}\ t^4 + 465.45054 \times 10^{-12}\ t^5)/(1 + 19.859983 \times 10^{-3}\ t) \quad (3)$$

$$\kappa_S = (\partial \ln \rho/\partial P)_S = \frac{1}{\rho U^2} \quad (4)$$

* Density ρ, thermal expansivity $\alpha = -(\partial \ln \rho/\partial T)_p$, and isothermal compressibility $\kappa T = (\partial \ln \rho/\partial p)T$. For purposes of this table, ordinary water is that with a maximum density of 999.972 kg m^{-3}. Equation 4 for the compressibility should be used for temperatures $0 \leqslant t \leqslant 100$°C, and Equation 3 for $100 \leqslant + \leqslant 150$°C. The liquid is metastable below 0°C and above 100°C. Values below 0°C were obtained by extrapolation, and no claim is made for their accuracy.

Reprinted with permission from Kell, G. S., *J. Chem. Eng. Data*, 20(1), 97, 1975. Copyright by the American Chemical Society.

PHYSICAL PROPERTIES OF SODIUM, POTASSIUM AND Na-K ALLOYS

Temperature °C	Density (g/cm^3)								Viscosity (centipoise)		
	Na			K	Alloys (wt % K)				Na	Alloys (wt % K)	
					43.4		78.6			43.3	66.0
	(a)	(b)	(c)		Experimental	(d) Calculated	Experimental	(d) Calculated			
100	.927	.927	.9265	.819	.887	.890	.847	.850	.705	.540	.529
200	.904	.904	.9037	.795	.862	.867	.823	.827	.450	.379	.354
300	.882	.880	.8805	.771	.838	.843	.799	.802	.345	.299	.276
400	.859	.856	.8570	.747	.814	.818	.775	.778	.284	.245	.229
500	.834	.831	.8331	.723	.789	.794	.751	.754	.234	.207	.195[e]
600	.809	.808	.8089	.701	.765	.771	.727	.732	.210	.178[e]	.168[e]
700	.783	.784	.....	.676	.740	.745	.703	.705	.186[e]	.257[e]	.146[e]
800	.757	.760	.....	....	....	....	....	....	.165[e]		
900	....	....	.....	....	....	....	....	....	.150[e]	....	.. ..

Temperature, °C	Thermal conductivity ($watts/cm^2$-°C/cm)				Electrical resistivity (microohms)			Heat capacity[g] (cal/°C-g)		
	Na		Alloy (wt % K)		Na[f]	Alloy (wt % K)		K	Alloy (wt % K)	
	Experimental	Calculated	56.5	77.0		56.5	78.0		44.8	78.26
100	......	....	....	.238[h]	8.99	41.61	45.63	.1940	.2690	.2248
200	.815	.808	.249	.247	13.52	47.23	51.33	.1887	.2612	.2169
300	.757	.755	.262	.259	17.52	54.33	58.58	.1894	.2553	.2122
400	.712	.710	.269	.262	21.93	62.21	65.65	.126	.2512	.2097
500	.668	.672	.271	.259	26.96	69.37	73.48	.1818	.2498	.2088
600	.627[e]	.639	....	.255	32.65	78.29	82.61	.1825	.2484	.2092
700	.590[e]	.610	....	......	39.05	88.23	91.76	.1846	.2497	.2108
800	.547[e]	.583	....	......	46.15	99.68	104.51	.1883	.2529	.2133
900	......	....	....	......	....	.....	......	....	......	.....

Vapor Pressure, mm Hg

Temperature °C	Na	K	Alloys (wt % K) 56	78
127	2.23×10^{-6}	....	....	....
227	1.15×10^{-3}	2.88×10^{-2}	1.57×10^{-3}	1.81×10^{-3}
327	5.03×10^{-2}	9.27×10^{-1}	5.73×10^{-2}	6.14×10^{-2}
427	.881	9.26	3.53	5.06
527	7.53	52.22	23.0	31.87
627	39.98	201.25	101.35	136.50
727	148.5	588.62	328.7	431.85
827	453.7	1421.0	864.2	1099.52
927	1127.8	....	....	....
1027	2522.4	....	....	....
1127	4696.8	....	....	....

NOTES: (a) From plotted data.

(b) Epstein equation: $d_t = 0.9514 - 2.392 \times 10^{-4}\, t$°C.

(c) Thomson and Garelis: $d_t = 0.9490 - 22.3 \times 10^{-5} t°\text{C} - 1.75 \times 10^{-8} t^2 °\text{C}$.

(d) Formula to calculate density: $V = M_K \cdot V_K + M_{Na} \cdot V_{Na}$ (where V, V_K and V_{Na} are the specific volumes (reciprocal of density) of the alloy, K and Na respectively, M_K and M_{Na} the mole fraction of the elements).

(e) Extrapolated by calculation, Epstein equation:

$$K = \frac{2.433 \times 10^{-2}(t + 273.16)}{6.8393 + 3.3873 \times 10^{-2} t + 1.7235 \times 10^{-5} t^2}$$

(f) Epstein equation: $r_t = 10.892 + 0.015272t + 3.6746 \times 10^{-5} t^2 - \dfrac{379.26}{t}$

(g) Formula to calculate heat capacity: $C = W_{Na} \cdot C_{Na} + W_K \cdot C_K$ (where C, C_{Na} and C_K are the heat capacity of the alloy, Na and K respectively, W_{Na} and W_{Ka} the weight fractions of Na and K respectively in the alloy).

(h) 150°C.

MECHANICAL AND PHYSICAL PROPERTIES OF WHISKERS

From NASA SP-5055

This table lists some nominal values for physical and mechanical properties of some whiskers. The strength of whiskers is influenced by temperature, time, surface conditions, surface films or corrosion, crystallographic orientation, impurities and testing techniques.

Material	Tensile Strength (σ_f) lb/in.² × 10^{-6}	Young's Modulus (E) lb/in.² × 10^{-6}	Specific Gravity (S) lb/in.² × 10^{-6}	$\frac{(\sigma_f)}{(S)}$ inch	$\frac{(E)}{(S)}$ inch
Graphite	3.0	..	137	21,700	......
Al_2O_3	3.0	76	250	12,000	300,000
Iron	1.8	28	485	3,700	58,000
Si_3N_4	2.0	55	193	10,000	285,000
SiC	3.0	70	187	16,000	380,000
Si	1.1	26	143	7,000	182,000

ABSOLUTE VISCOSITY OF LIQUID SODIUM AND POTASSIUM

Sodium

T (°K)	η (10^{-2} poise)	ν (cm³/g)	$\eta\nu^{1/3}$ (10^3 poise cm/g$^{-1/3}$)	$1/T\nu$ [10^3g/(cm³ · °K)]
Experimental range				
371.00	0.690	1.078,75	7.0766	2.4987
473	.450	1.106,56	4.6544	1.9106
573	.340	1.135,72	3.5482	1.5366
673	.278	1.166,86	2.9268	1.2734
773	.239	1.200,34	2.54057	1.0776
873	.212	1.236,25	2.2754	0.9266
973	.193	1.274,37	2.0925	.8065
1073	.179	1.315,79	1.9615	.7083
1173	.167	1.360,54	1.8505	.6266
1203	.164	1.373,62	1.8230	.6052

Potassium

T (°K)	η (10^{-2} poise)	ν (cm³/g)	$\eta\nu^{1/3}$ (10^3 poise cm/g$^{-1/3}$)	$1/T\nu$ [10^3g/(cm³ · °K)]
Experimental range				
336.9	0.560	1.2062_7	5.961_2	2.4606
400	.384	1.22911	4.113_4	2.0340
500	.276	1.26711	2.986_6	1.5784
600	.221	1.3075_3	2.416_6	1.2747
700	.185	1.3506_2	2.045_0	1.0577
800	.162	1.3966_5	1.810_8	0.8949_7
900	.147	1.4459_2	1.6623	$.7684_6$
1000	.132	1.4988_0	1.5106	$.6672_0$
1100	.121	1.5556_9	1.4020	$.5843_5$
1200	.113	1.6170_7	1.3264	$.5153_3$
1300	.106	1.6835_0	1.2610	$.4569_1$
1400	.100	1.7556_1	1.2064	$.4068_7$

PHYSICAL PROPERTIES OF PIGMENTS

From Rutherford J. Gettens and George L. Stout. *Painting Materials*, Dover Publications, Inc., New York, 1966. Reprinted by permission of the Publisher.

OPAQUE WHITE PIGMENTS

Pigment Name and Chemical Composition[1]	Specific Gravity[2]	Particle Characteristics[3]	Refractive Index[4]
Titanium calcium white, TiO_2 (25%) + $CaSO_4$ (75%)	3.10	prism. or ragged gr.	mostly 1.8–2.0 (irr.) (bi.) [M*]
Titanium dioxide (rutile) + $CaSO_4$	3.25	. . .	Av. 1.98 [TPC]
Titanium barium white, TiO_2 (25%) + $BaSO_4$ (75%)	4.30	min. round. gr.	$n_{\Sigma c}$ 1.7–2.5 [M*]
White lead (basic sulfate) $PbSO_4 \cdot PbO$	6.46	. . .	Av. 1.93 [TPC]
Lithopone	4.3	. . .	Av. 1.84 [TPC]
Lithopone (regular), ZnS (28–30%), $BaSO_4$ (72–70%)	4.30	fine comp. gr.	2.3 (ZnS)–1.64 ($BaSO_4$) [M]
Zinc white (ordinary), ZnO	5.65	v. fine cryst. gr.	ϵ2.02, ω2.00 [M]
(acicular), ZnO	. . .	spicules, fourlets	ϵ2.02, ω2.00 [M*]
White lead (basic carbonate), $2PbCO_3 \cdot Pb(OH)_2$	6.70	v. fine cryst.	ϵ1.94, ω2.09 [M]
Antimony oxide, Sb_2O_3	5.75	v. fine cryst.	valentinite, α2.18, γ and β2.35 [LB, M*] senarmonite, 2.09 (isot.)
Zirconium oxide (baddeleyite), ZrO_2	5.69	. . .	χ2.13, γ2.20, β2.19 [LB], Av. 2.40 [TPC]
Titanium dioxide (anatase), TiO_2	3.9	min. round. gr.	ϵ and ω 2.5 (w. bi.) [M*]
(rutile), TiO_2	4.2	round. or prism. gr.	ϵ2.9, ω2.6 [M*]

TRANSPARENT WHITE PIGMENTS†

Pigment Name and Chemical Composition	Specific Gravity	Particle Characteristics	Refractive Index
Diatomaceous earth, SiO_2	2.31	min. fossil forms	n mostly 1.435, some 1.40 [M*]
Aluminum stearate, $Al(C_{18}H_{35}O_2)_3$	0.99	agg. of spher. gr.	1.49 (w. bi.) [W]
Pumice (volcanic glass), Na, K, Al, silicate	. . .	vesicular vitr. frag.	c 1.50 (isot.) [M*]
Aluminum hydrate, $Al(OH)_3$	2.45	v. fine amorph. part.	$n_{\Sigma c}$ 1.50–1.56 [M*]
Gypsum, $CaSO_4 \cdot 2H_2O$	2.36	fine cryst. gr.	α1.520, γ1.530, β1.523 [LB]
Silica quartz), SiO_2	2.66	cryst. frag.	ϵ1.553, ω1.544 [LB]
(chalcedony), SiO_2	2.6	crypt. agg.	ϵ, ω1.54 [LB, M*]
China clay (kaolinite), $Al_2O_3 \cdot 2SiO_2 \cdot 2H_2O$	2.60	fine, vermicular cryst.	α1.558, γ1.565, β1.564 (all ± .005) [LB, M*]
Talc, $3MgO \cdot 4SiO_2 \cdot H_2O$	2.77	platy frag.	α1.539, γ1.589, β1.589 [LB]
Mica (muscovite), $H_2KAl_3(SiO_4)_3$	2.89	platy frag.	α1.563, γ1.604, β1.599 [LB]
Anhydrite, $CaSO_4$	2.93	cryst. frag.	α1.570, γ1.614, β1.575 [LB]
Chalk (whiting), $CaCO_3$	2.70	hollow spherulites	$\epsilon_{\Sigma c}$ 1.510, $\omega_{\Sigma c}$ 1.645 [M*]
Barytes (barite, nat.), $BaSO_4$	4.45	cryst. frag.	α1.636, γ1.648, β1.637 [LB]
(blanc fixe, art.), $BaSO_4$	4.36	v. fine cryst. agg.	1.62–1.64 [M*]
Barium carbonate, $BaCO_3$ (witherite)	4.3	. . .	α1.529, γ1.677, β1.676 [LB]

†Also called "Extender" or "Inert" White Pigments.

IRON OXIDE PIGMENTS

Pigment Name and Chemical Composition	Specific Gravity	Particle Characteristics	Refractive Index
Ochre, yellow (goethite), $Fe_2O_3 \cdot H_2O$, clay, etc.	2.9–4.0	irr. spherulites	n_Σ2.0 (isot. part); (α, β)$_\Sigma$ 2.05–2.31; γ_Σ 2.08–2.40 (bi. part) [M*]
Sienna, raw (goethite), $Fe_2O_3 \cdot H_2O$, clay, etc.	3.14	uneven spherulites	1.87–2.17 (mostly 2.06) (isot.) [M*]
Sienna, burnt, Fe_2O_3, clay, etc.	3.56	uneven, round. part.	c1.85 (var.) (isot.) [M]
Umber, raw, Fe_2O_3 + MnO_2+ H_2O, clay, etc.	3.20	uneven, round. gr.	mostly 1.87–2.17 [M*]
Umber, burnt, Fe_2O_3 + MnO_2, clay, etc.	3.64	uneven, round. gr.	mostly 2.2–2.3 [M*]
Iron oxide red (haematite), Fe_2O_3	5.2	min. cryst.	ϵ_{Li} 2.78, ω_{Li} 3.01 [M]

RED AND ORANGE PIGMENTS

Pigment Name and Chemical Composition[1]	Specific Gravity[2]	Particle Characteristics[3]	Refractive Index[4]
Red lead, Pb_3O_4 (c95%)	8.73	crypt. agg.	2.42_{Li} (w. bi.; pleo.) [M]
**Realgar, As_2S_2	3.56	cryst. frag.	α_{Li} 2.46, γ_{Li} 2.61, β_{Li} 2.59 [LB]
Molybdate orange, $Pb(Mo,S,Cr,P)O_4$	. . .	min. round. gr.	β_{Li} 2.55 (s. bi.) [M*]
Chrome orange, $PbCrO_4 \cdot Pb(OH)_2$	6.7	tabular cryst.	α2.42, γ2.7 +, β2.7 [M*]
Cadmium red lithopone, $CdS(Se) + BaSO_4$	4.30	min. round. gr.	2.50–2.76 (for CdS(Se)part) (isot.) [M*]
Cadmium red, CdS(Se)	4.5	min. round. gr.	2.64 (bright red)–2.77 (deep red) (isot.) [M*]
Antimony vermilion, Sb_2S_3	. . .	v. fine red glob.	n_{Li} 2.65 (isot.) [M*]
Vermilion (art.), HgS	8.09	hexagonal gr. and prisms	ϵ_{Li} 3.14, ω_{Li} 2.81 [M]
**(nat., cinnabar), HgS	8.1	cryst. frag.	ϵ_{Li} 3.146, ω_{Li} 2.819 [LB]
Quinacridone red, $C_{20}H_{12}O_2N_2$ (gamma)	1.5	thin plates	n_{Na} Av. 2.04 [Du P]

YELLOW PIGMENTS

Pigment Name and Chemical Composition[1]	Specific Gravity[2]	Particle Characteristics[3]	Refractive Index[4]
**Gamboge, organic resin	. . .	irr. amorph. part.	1.582–1.586 [W]
**Indian yellow, $C_{19}H_{18}O_{11}Mg \cdot 5H_2O$	. . .	prisms, plates	1.67 (w. bi.) [M*]
Cobalt yellow, $CoK_3(NO_2)_6 \cdot H_2O$	. . .	fine dendritic cryst.	1.72–1.76 (isot.) [W]
Zinc yellow, $4ZnO \cdot 4CrO_3 \cdot K_2O \cdot 3H_2O$	3.46	min. spher. gr.	1.84–1.9 (irr.; bi.) [M*]
Strontium yellow, $SrCrO_4$	. . .	small needles	α, β (or ω) 1.92, γ (or ϵ) 2.01 ($\parallel$ext.) [M*]
Barium yellow, $BaCrO_4$	4.49	v. fine cryst. gr.	1.94–1.98 (bi.) [M]
**Naples yellow, $Pb_3(SbO_4)_2$	. . .	round. gr.	2.01–2.28 (isot.) [M*]
Chrome yellow (med.), $PbCrO_4$	5.96	fine prism. gr.	$\alpha_{620m\mu} < 2.31$, $\gamma_{650m\mu}$ 2.49 [M]
Cadmium yellow lithopone, $CdS + BaSO_4$	4.25	fine comp. gr.	2.39–2.40 (for CdS part) [M*]
Cadmium yellow, CdS	4.35	min. round. gr.	2.35–2.48 (isot.) [M*]
Massicot (litharge), PbO	9.40	min. flakes	α_{Li} 2.51, γ_{Li} 2.71, β_{Li} 2.61 [M]
**Orpiment, As_2S_3	3.4	min. flakes	α_{Li} 2.4 ±, γ_{Li} 3.02, β_{Li} 2.81 [LB]

GREEN PIGMENTS

Pigment Name and Chemical Composition[1]	Specific Gravity[2]	Particle Characteristics[3]	Refractive Index[4]
Phthalocyanine green, chloro-copper phthalocyanine	2.1	laths	$n_{580m\mu}$ 1.40 [ACC]
**Verdigris (copper basic acetate), $Cu(C_2H_3O_2)_2 \cdot 2Cu(OH)_2$	. . .	cryst. frag.	α1.53, γ1.56 [M]
**Chrysocolla, $CuSiO_3 \cdot \eta H_2O$	2.4	crypt. agg.	α1.575, γ1.598, β1.597 [LB]
Green earth (celadonite and glauconite), Fe, Mg, Al, K, hydrosilicate	2.5–2.7	round. irr. gr.	n var. c 1.62, (porous) [M*]
Emerald green (Paris green), $Cu(C_2H_3O_2)_2 \cdot 3Cu(AsO_2)_2$	3.27	spherulites and disks	α_Σ1.71, γ_Σ1.78 (w. pleo.) [M*]
**Malachite, $CuCO_3 \cdot Cu(OH)_2$	4.0	cryst. frag.	α1.655, γ1.909, β1.875 [LB]
Cobalt green, $CoO \cdot \eta ZnO$	. . .	spher. gr.	1.94–2.0 (w. bi.) [M*]
Viridian (chromium oxide, transparent), $Cr_2O_3 \cdot 2H_2O$	3.32	spherul. gr.	α, β_Σ 1.82, γ_Σ 2.12 [M*]
Chrome green (med.), $Fe_4[Fe(CN)_6]_3 + PbCrO_4$	4.06	fine green agg.	c 2.4 (cf. Prussian blue and chrome yellow)
Chromium oxide green, opaque, Cr_2O_3	5.10	fine cryst. agg.	n_{Li} 2.5 [M]

BLUE PIGMENTS

Pigment Name and Chemical Composition[1]	Specific Gravity[2]	Particle Characteristics[3]	Refractive Index[4]
Phthalocyanine blue, copper phthalocyanine	1.6	laths	Av. 1.38 [DuP]
Ultramarine blue (art.), $Na_{8-10}Al_6Si_6O_{24}S_{2-4}$	2.34	uniform small round. gr.	n 1.51 $_{green}$, 1.63 $_{red}$ (isot.) [M]
**(nat., lazurite), $3Na_2O \cdot 3Al_2O_3 \cdot 6SiO_2 \cdot 2Na_2S$	2.4	angular, broken frag.	1.50± (isot.) [LB]
**Maya blue, Fe, Mg, Ca, Al, silicate (?)	...	porous irr. agg.	β_Σ1.54 (irr.; bi. and pleo.) [M*]
Smalt, K, Co(Al), silicate (glass)	...	splintery, vitr. frag.	1.49–1.52 [M*]
Prussian blue, $Fe_4[Fe(CN)_6]_3$	1.83	colloidal agg.	$1.56_{460m\mu}$ [M*]
**Egyptian blue, $CaO \cdot CuO \cdot 4SiO_2$	...	cryst. frag.	ϵ1.605, ω1.635 [APL]
Manganese blue, $BaMnO_4 + BaSO_4$	...	gr. and stubby prisms	c 1.65 [W]
**Blue verditer, $2CuCO_3 \cdot Cu(OH)_2$	...	fibrous agg.	α_Σ1.72, $\gamma_\Sigma$slightly > 1.74 [M*]
Cobalt blue, $CoO \cdot Al_2O_3$	3.83	round. gr.	n var.; max. c 1.74_{blue} (isot.) [M]
**Azurite, $2CuCO_3 \cdot Cu(OH)_2$	3.80	cryst. frag.	α1.730, γ1.838, β1.758 [LB]
Cerulean blue, $CoO \cdot \eta SnO_2$	...	round gr.	1.84 (isot.) [M*]

VIOLET PIGMENTS

Ultramarine violet	...	round. gr. (blue, rose and violet)	c 1.56 (isot.) [M*]
Cobalt violet, $Co_3(PO_4)_2$	...	round. gr.	ϵ1.65–1.79 (dull violet), ω1.68–1.81 (salmon) (s. bi.) [M*]
Manganese violet, $(NH_4)_2Mn_2(P_2O_7)_2$	...	fine cryst. gr.	α1.67, γ1.75, β1.72 (for violet) [M]
Quinacridone violet, $C_{20}H_{12}O_2N_2$ (beta)	1.5	thin plates	Av. 2.02 [DuP]

BROWN PIGMENTS

Sepia (organic)	...	angular frag.	(opaque) [M*]
Asphaltum (bitumen)	...	irr. amorph. part.	1.64–1.66 [M*]
Van Dyke brown (bituminous earth)	1.66	irr. amorph. part.	1.62–1.69 [M*]

BLACK PIGMENTS

Bone black, $C + Ca_3(PO_4)_2$	2.29	irr. coarse grains	1.65–1.70 (for larger translucent gr.) [M]
Lamp black, C	1.77	min. round. part.	(opaque)
Charcoal black, C	...	irr. splintery part.	(opaque)
Graphite, C	2.36	irr. plates	(opaque) [M]

[1]Abbreviations: art. = artificial; med. = medium; nat. = natural. The chemical formulas are those commonly accepted in chemical and mineralogical literature, but they may not compare exactly with structural formulas based on x-ray diffraction data or even on critical chemical analysis.

[2]The figures for specific gravity of the artificial pigments are mainly from H. A. Gardner, pp. 710–712, and those on the mineral pigments are chiefly from E. S. Larsen and H. Berman.

[3]Symmetry terms (monoclinic, orthorhombic, etc.) are omitted because pigments are so finely divided that it is rare when observations on crystal symmetry can be made. The term "spherulitic," as used here means aggregates that tend toward radial structure and spherical shape. "Amorphous" describes materials that are microscopically formless but may be truly crystalline on the basis of x-ray diffraction data. Abbreviations: agg. = aggregate(s); amorph. = amorphous; comp. = composite; crypt. = cryptocrystalline; cryst. = crystal(s); frag. = fragment(s); glob. = globule(s); gr. = grain(s); irr. = irregular; min. = minute; part. = particle(s); prism. = prismatic; round. = rounded; spher. = spheroidal; spherul. = spherulitic; var. = variable; v. = very; vitr. = vitreous.

[4]Unless otherwise indicated, all refractive index measurements are by sodium light. Σ is the symbol used by H. E. Merwin to indicate greater or less indefiniteness or irregularity in the case of aggregates, especially in respect to refractive index. Abbreviations: bi. = birefringent; c = *circa*; ext. = extinction; isot. = isotropic; ‖ = parallel; pleo. = pleochroic; s. = strongly; w. = weakly. The letters in brackets refer to the authorities for the refractive index data: M = H. E. Merwin; M* = H. E. Merwin, data by private communication, hitherto unpublished; W = C. D. West data by private communication, hitherto unpublished; LB = E. S. Larsen and H. Berman; APL = A. P. Laurie and co-authors; ACC = A. C. Cooper, "The refractive index of organic pigments. Its determination and significance," *Journal Oil & Colour Chemists Association*, Vol. 31 (1948), pp. 343–357; TPC = Titanium Pigment Corporation; Du P = E. I. Du Pont de Nemours & Co.

**Chiefly of historical interest.

CRITICAL TEMPERATURES AND PRESSURES

Compiled by Rudolf Loebel

Table I —Organic compounds
Table II—Inorganic compounds

Table I

Formula	Name	Critical temp. T_c °C	Critical press. P_c atm.
$CHClF_2$	Methane, monochlorodifluoro-	96	48.5
$CHCl_2F$	Methane, dichloromonofluoro-	178.5	51
$CHCl_3$	Methane, trichloro- (Chloroform)	263	54
CHF_3	Methane, trifluoro- (Fluoroform)	25.9	46.9
CH_2Cl_2	Methylene chloride	237	60
CH_3NO_2	Methane, nitro-	314.8	62.3
CH_3Br	Methane, monobromide-	194	83.4
CH_3Cl	Methane, monochloro-	143.8	65.9
CH_3F	Methane, monofluoro-	44.6	58
CH_3I	Methane, monoiodo-	254.8	72.7
CH_4	Methane	−82.1	45.8
CH_4O	Methanol (Methyl alcohol)	240	78.5
CH_4S	Methylmercaptan	196.8	71.4
CH_5N	Methylamine	156.9	40.2
$CBrF_3$	Methane, monobromotrifluoro-	67	50.3
$CClF_3$	Methane, monochlorotrifluoro-	28.85	38.2
CCl_2F_2	Methane, dichlorodifluoro-	111.5	39.6
CCl_3F	Methane, trichloromonofluoro-	198	43.2
CCl_4	Methane, tetrachloro- (Carbon tetrachloride)	283.1	45
CF_4	Methane, tetrafluoro-	−45.7	41.4
C_2H_2	Acetylene	35.5	61.6
C_2H_2	Ethyne (see Acetylene)	35.5	61.6
$C_2H_2F_2$	Ethylene, 1,1-difluoro-	30.1	—
C_2H_3N	Acetonitrile	274.7	47.7
$C_2H_3F_3$	Ethane, 1,1,1-trifluoro-	73.1	—
C_2H_4	Ethene	9.9	50.5
C_2H_4	Ethylene (see Ethene)	9.9	50.5
C_2H_4O	Acetaldehyde	187.8	54.7
C_2H_4O	Ethylene oxide	195.8	71
$C_2H_4O_2$	Acetic acid	321.6	57.1
$C_2H_4O_2$	Formic acid, methyl- (Methyl formate)	214	59.2
$C_2H_4Cl_2$	Ethane, 1,1-dichloro-	249.8	50
$C_2H_4F_2$	Ethane, 1,1-difluoro-	386.7	—
C_2H_5Br	Ethane, monobromo-	230.8	61.5
C_2H_5Cl	Ethane, monochloro-	187.2	52
C_2H_5F	Ethane, monofluoro-	102.16	49.6
C_2H_6	Ethane	32.2	48.2
C_2H_6O	Ether, dimethyl-	127	52.6
C_2H_6O	Ethanol (Ethyl alcohol)	243	63
C_2H_6O	Glycol, ethylene-	(374)*	—
C_2H_6S	Dimethylsulfide	229.9	54.5
C_2H_6S	Ethylmercaptan	225.5	54.2
C_2H_7N	Dimethylamine	164.6	52.4
C_2H_7N	Ethylamine	183.2	55.5
$C_2Br_2F_4$	Ethane, dibromotetrafluoro-	214.5	—
C_2ClF_5	Ethane, chloropentafluoro-	80	—
$C_2Cl_2F_4$	Ethane, 1,2-dichlorotetrafluoro-	145.7	—
$C_2Cl_3F_3$	Ethane, trichlorotrifluoro-	214.2	33.7
$C_2Cl_4F_2$	Ethane, tetrachlorodifluoro-	278	32.9
C_2F_6	Ethane, hexafluoro-	24.3	—
C_3H_4	Propadiene	120	43.6
C_3H_4	Allene (see Propadiene)	120	43.6
C_3H_4	Acetylene, methyl-	127.8	52.8
C_3H_4	Propyne (see Acetylene, methyl-)	127.8	52.8
C_3H_5N	Ethyl cyanide	290.8	41.3
C_3H_5N	Propionitrile (see Ethyl cyanide)	290.8	41.3
C_3H_5OCl	Epichlorohydrin	(323)	—
C_3H_5Cl	Propene, 3-chloro- (allyl-chloride)	240.3	46.5
C_3H_6	Propylene	91.9	45.4
C_3H_6	Cyclopropane	124.7	—
C_3H_6O	Acetone	235.5	47
C_3H_6O	Allylalcohol	272	55.5
C_3H_6O	Propylene oxide	209	48.6
$C_3H_6O_2$	Formic acid, ethyl- (Ethyl formate)	235.3	46.3
$C_3H_6O_2$	Acetic acid, methyl- (Methyl acetate)	233.7	46.3
$C_3H_6O_2$	Propanoic acid	337.6	53
C_3H_7Cl	n-Propane, monochloro-	230	45.2
C_3H_8	Propane	96.8	42
C_3H_8O	Ether, ethyl methyl- (Methoxyethane)	164.7	43.4
C_3H_8O	Glycol, 1,2-propylene-	351	—
C_3H_8O	Isopropyl alcohol	235	47
C_3H_8O	n-Propyl alcohol	263.6	51
$C_3H_8O_3$	Glycerol	(452)	—
C_3H_9N	Isopropylamine	209.7	—
C_3H_9N	n-Propylamine	223.8	46.8
	*( ) uncertain		
C_4H_4O	Furan	213.8	52.5
C_4H_4S	Thiophene	307	56.2
C_4H_6	1,2-Butadiene	171	44.4
C_4H_6	1,3-Butadiene	152	42.7
C_4H_8	n-Butene	146	39.7
C_4H_8	2-Butene, cis-	160	40.5
C_4H_8	2-Butene, trans-	155	41.5
C_4H_8O	1,2-Butylene oxide	243	—
C_4H_8O	Ketone, ethyl methyl- (2-Butanone)	262	41
$C_4H_8O_2$	Butanoic acid	355	52
$C_4H_8O_2$	p-Dioxane	314.8	51.4
$C_4H_8O_2$	Acetic acid, ethyl- (Ethyl acetate)	250.4	37.8
$C_4H_8O_2$	Propanoic acid, methyl- (Methyl propionate)	257.4	39.3
$C_4H_8O_2$	Formic acid, propyl-(Propyl formate)	264.9	40.1
C_4H_9Cl	n-Butane, monochloride-	269	—
C_4H_{10}	n-Butane	152	37.5
C_4H_{10}	Isobutane	135	36
$C_4H_{10}O$	Butanol (n-Butyl alcohol)	289.8	43.6
$C_4H_{10}O$	sec-Butyl alcohol	263	41.4
$C_4H_{10}O$	tert-Butyl alcohol	235	39.2
$C_4H_{10}O$	Ether, diethyl-	192.6	35.6
$C_4H_{10}O$	Isobutyl alcohol	277	42.4
$C_4H_{10}O_3$	Glycol, diethylene-	407	—
$C_4H_{10}S$	Diethyl sulfide	283.8	39.1
$C_4H_{11}N$	Butyl amine	287.9	—
$C_4H_{11}N$	Diethyl amine	223.3	36.6
$C_4H_{11}N$	Isobutyl amine	266.7	—
C_4F_{10}	Butane, perfluoro-	113.2	23
C_5H_5N	Pyridine	346.8	—
C_5H_8	Cyclopentene	232.94	47.2
C_5H_{10}	Cyclopentane	238.6	44.6
C_5H_{10}	1-Pentene	191	39.9
$C_5H_{10}O$	Ketone, diethyl-	287.8	36.9
$C_5H_{10}O$	Propanoic acid, ethyl- (Ethyl propionate)	272.9	33
$C_5H_{10}O_2$	n-Butanoic acid, methyl- (n-Methyl butyrate)	281.3	34.3
$C_5H_{10}O_2$	Acetic acid, n-propyl (n-Propyl acetate)	276	32.9
$C_5H_{10}O_2$	n-Valeric acid	378	37.6
$C_5H_{11}N$	Piperidine	320.8	44.1
C_5H_{12}	Butane, 2-methyl- (See Isopentane)	187.8	32.9
C_5H_{12}	Isopentane	187.8	32.9
C_5H_{12}	Neopentane	160.6	31.6
C_5H_{12}	n-Pentane	196.6	33.3
C_5H_{12}	Propane, 2,2-dimethyl (see Neopentane)	160.6	31.6
$C_5H_{12}O$	Isoamyl alcohol	309.77	—
C_6H_5Br	Benzene, bromo-	397	44.6
C_6H_5Cl	Benzene, chloro-	359.2	44.6
C_6H_5F	Benzene, fluoro-	286.95	44.6
C_6H_6	Benzene	288.9	48.6
C_6H_6O	Phenol	421.1	60.5
C_6H_7N	Aniline	425.6	52.3
C_6H_7N	α-Picoline	348	—
C_6H_7N	β-Picoline	371.7	—
C_6H_7N	γ-Picoline	372.5	—
C_6H_{10}	Cyclohexene	287.3	—
C_6H_{10}	1,5-Hexadiene	234.4	32.6
C_6H_{12}	Cyclohexane	280.4	40
C_6H_{12}	Cyclopentane, methyl-	259.5	37.4
C_6H_{12}	1-Hexene	231	31.1
$C_6H_{12}O_2$	Formic acid, n-amyl (n-Amyl formate)	302.6	34.1
$C_6H_{12}O_2$	Acetic acid, n-butyl- (n-Butyl acetate)	305.9	30.7
$C_6H_{12}O_2$	Butanoic acid, ethyl- (Ethyl butyrate)	293	30.2
$C_6H_{12}O_2$	Propanoic acid, propyl- (Propyl propionate)	304.8	30.7
$C_6H_{12}O_3$	Paraldehyde	290	—
C_6H_{14}	Butane, 2,3-dimethyl-	226.8	30.9
C_6H_{14}	n-Hexane	234.2	29.9
C_6H_{14}	Pentane, 2-methyl-	224.3	30
$C_6H_{14}O$	Ether, isopropyl-	226.9	28.4
$C_6H_{14}O$	1-Hexyl alcohol	313.5	—
$C_6H_{14}O_4$	Glycol, triethylene-	(437)	—
$C_6H_{15}N$	Dipropyl amine	277	31
$C_6H_{15}N$	Hexyl amine	318.8	—
$C_6H_{15}N$	Triethyl amine	258.9	30
C_7H_5N	Benzonitrile	426.2	41.6
C_7H_8	Benzene, methyl- (see Toluene)	320.8	41.6

CRITICAL TEMPERATURES AND PRESSURES (Continued)

Table I (Continued)

Formula	Name	Critical temp. T_c °C	Critical press. P_c atm.
C_7H_8	Toluene	320.8	41.6
C_7H_8O	Anisole, (see Benzene, methoxy-)	368.5	41.3
C_7H_8O	Benzene, methoxy-	368.5	41.3
C_7H_8O	o-Cresol	424.4	49.4
C_7H_8O	m-Cresol	432	45
C_7H_8O	p-Cresol	431.4	50.8
C_7H_9N	Aniline, methyl-	428.4	51.3
C_7H_9N	2,3-Lutidine	382.3	—
C_7H_9N	2,4-Lutidine	374	—
C_7H_9N	2,6-Lutidine	350.6	—
C_7H_9N	3,4-Lutidine	410.6	—
C_7H_9N	3,5-Lutidine	394.1	—
C_7H_{14}	Cyclohexane, methyl-	299.1	34.3
C_7H_{14}	Cyclopentane, ethyl-	296.3	33.5
C_7H_{14}	1-Heptene	264.1	—
$C_7H_{14}O_2$	Acetic acid, isoamyl- (Isoamyl acetate)	326.1	28
C_7H_{16}	Butane, 2,2,3-trimethyl-	258.3	29.8
C_7H_{16}	n-Heptane	267.1	27
C_7H_{16}	Hexane, 2-methyl-	257.9	27.2
C_7H_{16}	Pentane, 3-ethyl-	267.6	28.6
C_7H_{16}	Pentane, 2,4-dimethyl-	247.1	27.4
$C_7H_{16}O$	1-Heptyl alcohol	365.3	29.4
C_7F_{16}	n-Heptane, perfluoro-	201.7	16
C_8H_8	Styrene	374.4	39.4
C_8H_{10}	Benzene, ethyl-	343.9	36.9
C_8H_{10}	o-Xylene	359	35.7
C_8H_{10}	m-Xylene	346	34.7
C_8H_{10}	p-Xylene	345	33.9
$C_8H_{10}O$	Xylenol	449.7	56.4
$C_8H_{10}O$	Phenetol (see Benzene, ethoxy-)	374	33.8
$C_8H_{10}O$	Benzene, ethoxy-	374	33.8
$C_8H_{11}N$	Aniline, N,N-dimethyl-	414.4	35.8
$C_8H_{11}N$	Aniline, N-ethyl-	425.4	—
C_8H_{16}	n-Octene	305	25.5
C_8H_{18}	Butane, 2,2,3,3-tetramethyl-	270.8	24.5
C_8H_{18}	Heptane, 4-methyl-	290	25.6
C_8H_{18}	Hexane, 2,4-dimethyl-	282	25.8
C_8H_{18}	n-Octane	296	24.8
C_8H_{18}	Pentane, 2,2,3-trimethyl-	294	28.2
$C_8H_{18}O$	1-Octyl alcohol	385.5	26.5
$C_8H_{19}N$	Dibutyl amine	322.6	—
C_9H_7N	Quinoline	508.8	—
C_9H_{12}	Benzene, n-propyl-	365	31.2
C_9H_{12}	Benzene, 1,2,3-trimethyl-	391.3	31
C_9H_{12}	Cumene (see Benzene, isopropyl-)	362.7	31.2
C_9H_{12}	Benzene, isopropyl-	362.7	31.2
C_9H_{20}	n-Nonane	321	22.5
C_9H_{21}	Tripropyl amine	304.3	—
$C_{10}H_8$	Naphthalene	474.8	40.6
$C_{10}H_{12}O$	Ether, diphenyl-	494	30.9
$C_{10}H_{14}$	Benzene, 1-isopropyl-4-methyl-	385.5	27.7
$C_{10}H_{14}$	Benzene, 1,2,3,5-tetramethyl-	402.8	28.6
$C_{10}H_{14}$	p-Cymene, (see Benzene, 1-isopropyl-4-methyl-)	—	—
$C_{10}H_{14}$	Isodurene, (see Benzene, 1,2,3,5-tetramethyl-)	—	—
$C_{10}H_{14}O$	3-p-Cymenol	425.1	33
$C_{10}H_{14}O$	Thymol, (see 3-p-Cymenol)	425.1	33
$C_{10}H_{18}$	Decalin, cis-	418	28.7
$C_{10}H_{18}$	Decalin, trans-	408	28.7
$C_{10}H_{22}$	n-Decane	344.4	20.8
$C_{11}H_{10}$	Naphthalene, 1-methyl-	498.8	32.1
$C_{12}H_{10}$	Biphenyl	495	31.8
$C_{12}H_{11}N$	Diphenylamine	615.5	—
$C_{12}H_{18}$	Benzene, hexamethyl-	494	23.5
$C_{12}H_{18}$	Mellitine, (see Benzene, hexamethyl-)	—	—
$C_{12}H_{26}$	n-Dodecane	386	17.9
$C_{12}H_{27}N$	Tributyl amine	365.2	—

Table II

Name	Formula	Critical temp. T_c °C	Critical press. P_c atm.
Ammonia	NH_3	132.5	112.5
Argon	Ar	−122.3	48
Boron tribromide	BBr_3	300	—
Boron trichloride	BCl_3	178.8	38.2
Boron trifluoride	BF_3	−12.26	49.2
Carbon dioxide	CO_2	31	72.9
Carbon disulfide	CS_2	279	78
Carbon monoxide	CO	−140	34.5
Carbonyl sulfide	COS	104.8	65
Chlorine	Cl_2	144	76.1
Cyanogen	C_2N_2	126.6	—
Deuterium	D_2	−234.8	16.4
Fluorine	F_2	−129	55
Germanium tetrachloride	$GeCl_4$	276.9	38
Helium	He	−267.9	2.26
Hydrogen	H_2	−239.9	12.8
Hydrogen bromide	HBr	90	84.5
Hydrogen chloride	HCl	51.4	82.1
Hydrogen deuteride	HD	237.3	14.6
Hydrogen cyanide	HCN	183.5	48.9
Hydrogen fluoride	HF	188	64
Hydrogen iodide	HI	150	81.9
Hydrogen sulfide	H_2S	100.4	88.9
Hydrazine	N_2H_2	380	—
Iodine	I_2	512	116
Krypton	Kr	−63.8	54.3
Neon	Ne	−228.7	26.9
Nitric oxide	NO	−93	64
Nitrogen dioxide	NO_2	157.8	100
Nitrogen	N_2	−147	33.5
Nitrous oxide	N_2O	36.5	71.7
Oxygen	O_2	−118.4	50.1
Ozone	O_3	−5.16	67
Phosphine	PH_3	51.3	64.5
Radon	Rn	104.04	62
Silane, chlorotrifluoro-	$SiClF_3$	34.5	34.2
Silane	SiH_4	−3.46	47.8
Silicon tetrachloride	$SiCl_4$	232.8	—
Silicon tetrafluoride	SiF_4	−14.06	36.7
Sulfur dioxide	SO_2	157.8	77.7
Stannic chloride	$SnCl_4$	318.7	37
Water	H_2O	374.1	218.3
Xenon	Xe	16.6	58

DISSOCIATION PRESSURE OF CALCIUM CARBONATE

Temp. °C	mm/Hg	Temp. °C	mm/Hg	Temp. °C	mm/Hg	Temp. °C	mm/Hg
550	0.41	727	44	819	235	894	716
587	1.0	736	54	830	255	898	760 atm.
605	2.3	743	60	840	311	906.5	1.151
671	13.5	748	70	852	381	937	1.770
680	15.8	749	72	857	420	1082.5	8.892
691	19.0	777	105	871	537	1157.7	18.687
701	23.0	786	134	881	603	1226.3	34.333
703	25.5	795	150	891	684	1241	39.094
711	32.7	800	183				

DEFINITIONS AND FORMULAS

The chemical terms have been compiled with the collaboration of B. Clifford Hendricks

AB-.—A prefix attached to the names of the practical electric units to indicate the corresponding unit in the cgs electromagnetic system (emu), e.g. abampere, abvolt.

Abcoulomb.—The abcoulomb, the emu of charge, is defined as the charge which passes a given surface in one second if a steady current of one abampere flows across the surface. Its dimensions are, therefore, $cm^{\frac{1}{2}}\ gm^{\frac{1}{2}}$, which differ from the dimensions of the statcoulomb by a factor which has the dimensions of a speed. This relationship is connected with the fact that the ratio $2K_e/K_m$ must have the value of the square of the speed of light in any consistent system of units. It follows further that

$$1 \text{ abcoulomb} = 2.99793 \times 10^{10} \text{ statcoulomb},$$

the speed of light in vacuo being $(2.99793 \pm 0.000003) \times 10^{10}$ cm/sec.

Abegg's rule.—For use in regard to a helical periodic system. If the maximum positive valence exhibited by an element be numerically added to its maximum negative valence, there is evidently a tendency for the sum to equal 8. This tendency is exhibited especially by the elements of the 4th, 5th, 6th and 7th groups and is known as Abegg's rule.

Absolute humidity.—See *Humidity*

Absolute pressure.—See *Pressure*

Absolute temperature.—Temperature reckoned from the absolute zero. See *Temperature*

Absolute units.—A system of units based on the smallest possible number of independent units. Specifically, units of force, work, energy and power not derived from or dependent on gravitation.

Absolute zero.—The temperature at which a gas would show no pressure if the general law for gases would hold for all temperatures. It is equal to −273.15°C or −459.67°F.

Absorption.—1. Penetration of a substance into the body of another. 2. Transformation into other forms suffered by radiant energy passing through a material substance.

Absorption coefficient.—See *Absorption factor*

Absorption factor.—The ratio of the intensity loss by absorption to the total original intensity of radiation. If I_o represents the original intensity, I_r, the intensity of reflected radiation, I_t, the intensity of the transmitted radiation, the absorption factor is given by the expression

$$\frac{I_o - (I_r + I_t)}{I_o}$$

Also called coefficient of absorption.

Permission was granted by D. Van Nostrand Company, Inc. publishers of The Dictionary of Physics and Electronics for the inclusion of 18 definitions in this section.

Absorption, Lambert's law.—If I_o is the original intensity, I the intensity after passing through a thickness x of a material whose absorption coefficient is k,

$$I = I_o e^{-kx}$$

The **index of absorption** k' is given by the relation $k = (4\pi k'n)/\lambda$ where n is the index of refraction and λ the wave length in vacuo. The **mass absorption** is given by k/d when d is the density. The transmission factor is given by I/I_o.

Absorption spectrum.—The spectrum obtained by the examination of light from a source, itself giving a continuous spectrum, after this light has passed through an absorbing medium in the gaseous state. The absorption spectrum will consist of dark lines or bands, being the reverse of the emission spectrum of the absorbing substance.

When the absorbing medium is in the solid or liquid state the spectrum of the transmitted light shows broad dark regions which are not resolvable into lines and have no sharp or distinct edges.

Absorptive power or absorptivity for any body is measured by the fraction of the radiant energy falling upon the body which is absorbed or transformed into heat. This ratio varies with the character of the surface and the wave length of the incident energy. It is the ratio of the radiation absorbed by any substance to that absorbed under the same conditions by a black body.

Abvolt.—The cgs electromagnetic unit of potential difference and electromotive force. It is the potential difference that must exist between two points in order that one erg of work be done when one abcoulomb of charge is moved from one point to the other. One abvolt is 10^{-8} volt.

Acceleration.—The time rate of change of velocity in either speed or direction. Cgs unit,—one centimeter per second per second. Dimensions,—$[l\ t^{-2}]$.—See *Angular acceleration*

Acceleration due to gravity.—The acceleration of a body freely falling in a vacuum. The International Committee on Weights and Measures has adopted as a standard or accepted value, 980.665 cm/sec² or 32.174 ft/sec².

Acceleration due to gravity at any latitude and elevation.—If ϕ is the latitude and H the elevation in centimeters the acceleration in cgs units is, $g = 980.616 - 2.5928 \cos 2\phi + 0.0069 \cos^2 2\phi - 3.086 \times 10^{-6} H$. (Helmert's equation)

Accelerators.—Machines for speeding up subatomic particles to energies running into millions of electron volts.—See *Betatron, cyclotron,* etc.

Achromatic.—A term applied to lenses signifying their more or less complete correction for chromatic aberration.

Acid.—For many purposes it is sufficient to say that an acid is a hydrogen-containing substance which dissociates on solu-

DEFINITIONS AND FORMULAS
The chemical terms have been compiled with the collaboration of B. Clifford Hendricks

tion in water to produce one or more hydrogen ions. More generally, however, acids are defined according to other concepts. The Brönsted concept states that an acid is any compound which can furnish a proton. Thus NH_4^+ is an acid since it can give up a proton:

$$NH_4^+ \rightleftarrows NH_3 + H^+$$

and NH_3 is a base since it accepts a proton.

A still more general concept is that of G. N. Lewis which defines an acid as anything which can attach itself to something with an unshared pair of electrons. Thus in the reaction

$$H^+ + :\overset{\diagup H}{\underset{\diagdown H}{N}}—H \rightleftarrows NH_4^+$$

the NH_3 is a base because it possesses an unshared pair of electrons. This latter concept explains many phenomena, such as the effect of certain substances other than hydrogen ions in the changing of the color of indicators. It also explains acids and bases in non-aqueous sytems as liquid NH_3 and SO_2.

Actinide Series.—Elements of atomic numbers 89 to 103 analogous to the lanthanide series of the so-called rare earths.

Action is measured by the product of work by time. Cgs units of action are the erg-second and the joule-second. Dimensions,—$[m\ l^2\ t^{-1}]$. Planck's quantum or constant of action is $(6.62517 - 0.00023) \times 10^{-27}$ erg-sec.

Active mass of a substance is the number of gram molecular weights per liter in solution, or in gaseous form.

Activity coefficient.—A factor which, when multiplied by the molecular concentration yields the active mass. The activity coefficient is evaluated by thermodynamic calculations, usually from data on the emf of certain cells, or the lowering of the freezing point of certain solutions. It is a correction factor which makes the thermodynamic calculations correct.

Adiabatic.—A body is said to undergo an adiabatic change when its condition is altered without gain or loss of heat. The line on the pressure volume diagram representing the above change is called an adiabatic line.

Adsorption.—The condensation of gases, liquids, or dissolved substances on the surfaces of solids is called adsorption.

Air columns, frequency of vibration in—See *Organ pipes*

Albedo. — The fraction of electromagnetic radiation reflected by a surface.

Allobar.—A form of an element differing in isotopic composition from the naturally occurring form.

Allotropy.—The property shown by certain elements or being capable of existence in more than one form, due to differences in the arrangement of atoms or molecules.—See *Monotropic* and *Enantiotropic*

Alpha (α)-particle, or alpha-ray.—One of the particles emitted in radioactive decay. It is identical with the nucleus of the helium atom and consists, therefore, of two protons plus two neutrons bound together. A moving alpha particle is strongly ionizing and so loses energy rapidly in traversing through matter. Natural alpha particles will traverse only a few centimeters of air before coming to rest.

Alternating current, (A-C).—Current in which the charge-flow periodically reverses, as opposed to direct current, and whose average value is zero. Alternating current usually implies a sinusoidal variation of current and voltage. This behavior is represented mathematically in various ways:

$$I = I_0 \cos (2\pi ft + \phi)$$
$$I = I_0 \angle\phi$$
$$I = I_1 e^{j\omega t}$$

where f is the frequency; $\omega \equiv 2\pi f$, the pulsatance, or radian frequency; ϕ the phase angle; I_0 the amplitude; and I_1 the complex amplitude. In the complex rotation, it is understood that the actual current is the real part of I. For circuits involving also a capacitance C in farads and L in henrys, the impedance becomes,

$$\sqrt{R^2 + \left(2\pi fL - \frac{1}{2\pi fC}\right)^2}$$

Altitudes with the barometer.—If b_1 and b_2 denote the corrected barometer readings at two stations, t the mean of the temperatures, t_1 and t_2 of the air at the two stations, e_1 and e_2 the tension of water vapor at the two stations, h the mean height above sea level, ϕ the latitude; then the difference in elevation in centimeters is $H = 1{,}843{,}000\ (\log b_1 - \log b_2)\ (1 + 0.00367t)\ (1 + 0.0026 \cos 2\phi + 0.00002h + \frac{3}{8}k)$, where

$$k = \tfrac{1}{2}\left(\frac{e_1}{b_1} + \frac{e_2}{b_2}\right)$$

An approximate formula, sufficient for differences not over 1000 meters is

$$H = 1{,}600{,}000\,\frac{b_1 - b_2}{b_1 + b_2}\,(1 + 0.004t).$$

Amorphous.—Without definite form, not crystallized.

Ampere's rule.—A positive charge moving horizontally is deflected by a force to the right if it is moving in a region where the magnetic field is vertically upward. This may be generalized to currents in wires by recalling that a current in a certain direction is equivalent to the motion of positive charges in that direction. The force felt by a negative charge is opposite to that felt by a positive charge.

Amplitude.—The maximum value of the displacement in an oscillatory motion.

DEFINITIONS AND FORMULAS (Continued)

AMU.—The atomic mass unit (amu), a unit of mass equal to 1/12 the mass of the carbon atom of mass number 12. On the atomic mass scale $^{12}C \equiv 12$.

$$1 \text{ amu} = 931.4812(52) \text{ MeV}$$
$$= 1.660531(11)\, 10^{-27} \text{ kg (SI units)}$$
$$= 1.660531(11)\, 10^{-24} \text{ g (cgs units)}$$

The numbers in parentheses are the standard deviation uncertainties in the last digits of the quoted value, computed on the basis of internal consistency.

Angle.—The ratio between the arc and the radius of the arc. Units of angle,—the radian, the angle subtended by an arc equal to the radius; the degree, $\frac{1}{360}$ part of the total angle about a point. Dimensions,—a numeric.

Angstrom.—A unit of length, used especially in expressing the length of light waves, equal to one ten-thousandth of a micron, or one hundred-millionth of a centimeter (1×10^{-8} cm).

Angular acceleration.—The time rate of change of angular velocity either in angular speed or in direction of the axis of rotation (precession). Cgs unit,—one radian per second per second. Dimensions,—$[t^{-2}]$.

If the initial angular velocity is ω_o, and the velocity after time t is ω_t, the angular acceleration,

$$\alpha = \frac{\omega_t - \omega_o}{t}$$

The angular velocity after time t,

$$\omega_t = \omega_o + \alpha t$$

The angle swept out in time t,

$$\theta = \omega_o t + \tfrac{1}{2}\alpha t^2$$

The angular velocity after movement through the arc θ,

$$\omega = \sqrt{\omega_o^2 + 2\alpha\theta}$$

In the above equations, for angular displacement in radians, angular velocity will be in radians per second and angular acceleration in radians per second per second.

Angular aperture of an objective is the largest angular extent of wave surface which it can transmit.

Angular harmonic motion or harmonic motion of rotation.—Periodic, oscillatory angular motion in which the restoring torque is proportional to the angular displacement. Torsional vibration.

Angular momentum or moment of momentum.—Quantity of angular motion measured by the product of the angular velocity and the moment of inertia. Cgs unit,—unnamed, its nature is expressed by g-cm²/sec. Dimensions,—$[m\, l^2\, t^{-1}]$.

The angular momentum of a mass whose moment of inertia is I, rotating with angular velocity ω, is $I\omega$.

Angular velocity.—Time rate of angular motion about an axis. Cgs unit,—one radian per second. Dimensions,—$[t^{-1}]$.

If the angle described in time t is θ, the angular velocity,

$$\omega = \frac{\theta}{t}$$

θ in radians and t in seconds gives ω in radians per second.

Anhydride (of acid or base).—An oxide which when combined with water gives an acid or base.

Anion.—A negatively charged ion.

Anode.—The electrode at which oxidation occurs in a cell. It is also the electrode toward which anions travel due to the electrical potential. In spontaneous cells the anode is considered negative. In non-spontaneous or electrolytic cells the anode is considered positive.

Apochromat.—A term applied to photographic and microscope objectives indicating the highest degree of color correction.

Archimedes principle.—A body wholly or partly immersed in a fluid is buoyed up by a force equal to the weight of the fluid displaced. A body of volume V cm³ immersed in a fluid of density ρ grams per cm³ is buoyed up by a force in dynes,

$$F = \rho g V$$

where g is the accleration due to gravity.

A floating body displaces its own weight of liquid.

Area, unit of.—The square centimeter. The area of a square whose sides are one centimeter in length. Other units of area are similarly derived. Dimensions,—$[l^2]$.

Arrhenius theory of electrolytic dissociation states that the molecule of an electrolyte can give rise to two or more electrically charged atoms or ions.

Astigmatism is an error of spherical lenses peculiar to the formation of images by oblique pencils. The image of a point when astigmatism is present will consist of two focal lines at right angles to each other and separated by a measurable distance along the axis of the pencil. The error is not eliminated by reduction of aperture as is spherical aberration.

Atom.—The smallest particle of an element which can enter into a chemical *combination*. All chemical compounds are formed of atoms, the difference between compounds being attributable to the nature, number, and arrangement of their constituent atoms.—See *isotopes, nuclear atom*

Atomic bomb.—An explosive that derives its energy from the fission or fusion of atomic nuclei.

Atomic energy.—1. The constitutive internal energy of the atom which was absorbed when it was formed. 2. Energy derived from the mass converted into energy in nuclear transformations.—See *Einstein's formula*

Atomic mass (atomic weight).—The mass of a neutral atom of a nuclide. It is usually expressed in terms of the physical scale of atomic masses, that is, in atomic mass units (amu). See AMU

Atomic number.—The number (Z) of protons within the atomic nucleus. The electrical charge of these protons determines the number and arrangement of the outer electrons of the

atom, and thereby the chemical and physical properties of the element.

Atomic structure.—According to the currently accepted view, the atom consists of a central part, called nucleus, and a number of *electrons* (called orbital or planetary electrons) circling about the latter, like planets about the sun. The nucleus is of a high specific weight; it contains most of the mass of the entire atom (its mass is considered equal to the atomic mass) and is composed of positively charged particles, called *protons* (the number of which always equals the atomic number, (Z), and particles of 0 charge, called neutrons (the number of which equals the difference between the atomic weight and the atomic number, $A - Z$). The diameter of the nucleus is between 10^{-13} and 10^{-12} cm, and the relatively vast distance in which the orbital electrons circle about it is illustrated by the fact that this nuclear diameter is only 10^{-4} to 10^{-5} of the entire atomic diameter. While the nucleus carries an integral number of positive charges (an integral number of protons) each of $1{\cdot}6 \times 10^{-19}$ coulomb, each electron carries one negative charge of $1{\cdot}6 \times 10^{-19}$ coulomb, and the number of orbital electrons is equal to the number of protons in the nucleus (i.e. to the atomic number, Z), so that the atom as a whole has a net charge of 0. The electrons are arranged in successive shells (q.v.) around the nucleus; the maximum number of electrons in each shell is determined by natural laws, and the extranuclear electronic structure of the atom is characteristic of the element. The electrons in the inner shells are tightly bound to the nucleus; this inner structure can be altered by high-energy particles, γ-rays of radium, or x-rays. The electrons in the outer shells are responsible for the chemical properties of the element.—See *Bohr's atomic theory, Heisenberg's theory, shell and subshell*

Atomic theory.—All elementary forms of matter are composed of very small unit quantities called atoms. The atoms of a given element all have the same size and weight. The atoms of different elements have different sizes and weights. Atoms of the same or different elements unite with each other to form very small unit quantities of compound substances called molecules.

Atomic weight.—Atomic weight is the relative weight of the atom on the basis of $^{12}C \equiv 12$. For a pure isotope, the atomic weight rounded off to the nearest integer gives the total number of nucleons (neutrons and protons) making up the atomic nucleus. If these weights are expressed in grams they are called gram atomic weights.—See *Isotopes and Atomic Mass*

Avogadro's law.—Equal volumes of different gases at the same pressure and temperature contain the same number of molecules.

Avogadro's number.—The number of molecules in one mole or gram-molecular weight of a substance. A number of values of the Avogadro number, which is usually denoted by N, have been found by various methods, generally lying within a range of 1% about the value 6.02252×10^{23} per gram mole.

Avogadro's principle (or theory).—The numbers of molecules present in equal volumes of gases at the same temperature and pressure are equal.

Babo's law.—The addition of a non-volatile solid to a liquid in which it is soluble lowers the vapor pressure of the solvent in proportion to the amount of substance dissolved.

Balmer series of spectral lines. The wave lengths of a series of lines in the spectrum of hydrogen are given in angstroms by the equation

$$\lambda = 3646 \frac{N^2}{N^2 - 4}$$

where N is an integer having values greater than 2.

Bar.—International unit of pressure 10^6 dyne/cm². Unfortunately some writers have used this term for 1 dyne/cm². 1 bar = 0.987 atmosphere.

Barn.—Unit for measuring capture cross sections (q.v.) of elements. One barn = 10^{-24} cm² per nucleus.

Barye.—Cgs pressure unit = one dyne/cm².

Bases.—For many purposes it is sufficient to say that a base is a substance which dissociates on solution in water to produce one or more hydroxyl ions. More generally, however, bases are defined according to other concepts. The Brönsted concept states that a base is any compound which can accept a proton. Thus NH_3 is a base since it can accept a proton to form ammonium ions.

$$NH_3 + H^+ \rightleftarrows NH_4^+$$

A still more general concept is that of G. N. Lewis which defines a base as anything which has an unshared pair of electrons. Thus in the reaction

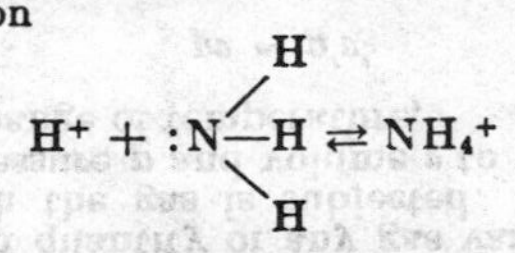

the NH_3 is a base because it possesses an unshared pair of electrons. This latter concept explains many phenomena, such as the effect of certain substances other than hydrogen ions in the changing of the color of indicators. It also explains acids and bases in non-aqueous systems as liquid NH_3 and SO_2.

Beat(s).—Two vibrations of slightly different frequencies f_1 and f_2 when added together, produce in a detector sensitive to both these frequencies, a regularly varying response which rises and falls at the "beat" frequency $f_b = |f_1 - f_2|$. It is important to note that a resonator which is sharply tuned to f_b alone will not resound at all in the presence of these two beating frequencies.—See *Combination Frequencies*

DEFINITIONS AND FORMULAS (Continued)

Beat frequencies.—The beat of two different frequencies of signals on a non-linear circuit when they combine or beat together. It has a frequency equal to the difference of the two applied frequencies.

Beer's law.—If two solutions of the same colored compound be made in the same solvent, one of which is, say, twice the concentration of the other, the absorption due to a given thickness of the first solution should be equal to that of twice the thickness of the second.

Mathematically this may be expressed $l_1c_1 = l_2c_2$ when the intensity of light passing through the two solutions is a constant and if the intensity and wave length of light incident upon each solution are the same.

Bernoulli's theorem.—At any point in a tube through which a liquid is flowing the sum of the pressure energy, potential energy, and kinetic energy is constant. If p is pressure; h, height above a reference plane; d, density of the liquid, and v, velocity of flow,

$$p + hdg + \tfrac{1}{2}dv^2 = \text{a constant.}$$

Berthelot principle of maximum work.—Of all possible chemical processes which can proceed without the aid of external energy, that process always takes place which is accompanied by the greatest evolution of heat. This law holds good for low temperatures only and does not account for endothermic reactions.

Beta (β)-particle, (Beta ray).—One of the particles which can be emitted by a radioactive atomic nucleus. It has a mass about $\frac{1}{1837}$ that of the proton. The negatively charged beta particle is identical with the ordinary electron, while the positively charged type (positron) differs from the electron in having equal but opposite electrical properties. The emission of an electron entails the change of a neutron into a proton inside the nucleus. The emission of a positron is similarly associated with the change of a proton into a neutron. Beta particles have no independent existance inside the nucleus, but are created at the instant of emission.—See *Neutrino*

Betatron.—An accelerator used to impart high velocities to electrons (beta particles). Propellant is an electromagnetic field. A five to six Mev betatron can produce X-rays equivalent to the gamma readiation of 10 to 20 grams of radium.

Bevatron.—A six or more billion electron volt accelerator of protons and other atomic particles. Makes use of a Cockcroft-Walton transformer cascade accelerator and a linear (q.v.) as well as an electromagnetic field in the build-up.

Black body.—If, for all values of the wave length of the incident radiant energy, all of the energy is absorbed the body is called a black body.

Bohr's atomic theory.—The theory that atoms can exist for a duration solely in certain states, characterized by definite electronic orbits, i.e., by definite energy levels of their extranuclear electrons, and in these stationary states they do not emit radiation; the jump of an electron from an orbit to another of a smaller radius is accompanied by monochromatic radiation.

Boyle's law for gases.—At a constant temperature the volume of a given quantity of any gas varies inversely as the pressure to which the gas is subjected. For a perfect gas, changing from pressure p and volume v to pressure p' and volume v' without change of temperature,

$$pv = p'v'$$

Breeder, Reactor (Breeder pile).—A nuclear chain reactor in which transmutation produces a greater number of fissionable atoms than the number of parent atoms consumed.

Brewster's law.—The tangent of the polarizing angle for a substance is equal to the index of refraction. The polarizing angle is that angle of incidence for which the reflected polarized ray is at right angles to the refracted ray. If n is the index of refraction and θ the polarizing angle, $n = \tan\theta$.

Brightness is measured by the flux emitted per unit emissive area as projected on a plane normal to the line of sight. The unit of brightness is that of a perfectly diffusing surface giving out one lumen per square centimeter of projected surface and is called the lambert. The millilambert (0.001 lambert) is a more convenient unit. **Candle per square centimeter** is the brightness of a surface which has, in the direction considered, a luminous intensity of one candle per cm².

British thermal unit.— (Btu) is equal to 1054.350 J. A Btu (mean) is the quantity of energy required to raise the temperature of 1 pound mass of water 1°F, averaged from 32 to 212°F. 1 Btu (mean) = 1055.87 J = 0.252 Kcal (mean). 1 Btu = 0.999233 Btu (mean) = 1.0003601 Btu (60°F).

Brownian movement.—A continuous agitation of particles in a colloidal solution caused by unbalanced impacts with molecules of the surrounding medium. The motion may be observed with a microscope when a strong beam of light is caused to traverse the solution across the line of sight.

Bulk modulus.—The modulus of volume elasticity,

$$M_B = \frac{p_2 - p_1}{\dfrac{v_1 - v_2}{v_1}}$$

where p_1, p_2; v_1, v_2 are the initial and final pressure and volume respectively.

Calorie.— By definition, 1 cal = 4.184 J (exactly). The joule has been adopted internationally as the unit of mechanical, electrical, and thermal energy. 1 cal = 860.42075 Wh. A calorie (mean) is the energy required to raise 1 gram mass of water 1°C, averaged from 0 to 100°C. 1 cal × 1.001439 = 1 cal (mean). 1 cal × 1.00043 = 1 c al (15°C). 1 cal × 0.999498 = 1 cal (20°C).

Calutron.—An apparatus operating on the principle of the mass spectrograph and used for separating U^{235} from U^{238}.

Candela.–The candela is the luminous intensity, in the direction of the normal, of a black body surface 1/600,000 square meter in area, at the temperature of the solidification of platinum (2042K) under a pressure of 101,325 newtons per square meter.

Candle (new unit).—$\frac{1}{60}$ of the intensity of one square centimeter of a blackbody radiator at the temperature of solidification of platinum (2042K).

Capacitance is measured by the charge which must be communicated to a body to raise its potential one unit. Electrostatic unit capacitance is that which requires one electrostatic unit of charge to raise the potential one electrostatic unit. The farad = 9×10^{11} electrostatic units. A capacitance of one farad requires one coulomb of electricity to raise its potential one volt. Dimensions,—$[\epsilon\, l]$; $[\mu^{-1} l^{-1} t^2]$.

A conductor charged with a quantity Q to a potential V has a capacitance,

$$C = \frac{Q}{V}$$

Capacitance of a spherical conductor of radius r,

$$C = Kr$$

Capacitance of two concentric spheres of radii r and r'

$$C = K\frac{rr'}{r - r'}$$

Capacitance of a parallel plate condenser, the area of whose plates is A and the distance between them d,

$$C = \frac{KA}{4\pi d}$$

Capacitances will be given in electrostatic units if the dimensions of condensers are substituted in cm. K is the dielectric constant of the medium.

Capillary constant or specific cohesion,

$$a^2 = \frac{2T}{(d_1 - d_2)g} = hr$$

where T is surface tension, d_1 and d_2, the densities of the two fluids, g the acceleration due to gravity, h the height of rise in a capillary tube of radius r.—See *Surface tension*

Carnot cycle.—A sequence of operations forming the working cycle of an ideal heat engine of maximum thermal efficiency. It consists of isothermal expansion, adiabatic expansion, isothermal compression, and adiabatic compression to the initial state.

Catalytic agent.—A substance which by its mere presence alters the velocity of a reaction, and may be recovered unaltered in nature or amount at the end of the reaction.

Cathode.—The electrode at which reduction occurs. It is the negative electrode in a cell through which current is being forced, but it is the positive pole of a battery. In a vacuum tube, the cathode is the electrode from which electrons are liberated.—See *Anode*

Cation.—A positively charged ion.

Cauchy's dispersion formula.

$$n = A + \frac{B}{\lambda^2} + \frac{C}{\lambda^4} + \cdots$$

An empirical expression giving an approximate relation between the refractive index n of a medium and the wavelength λ of the light; A, B, and C being constants for a given medium.

Celsius.—See temperature, Celsius, in this section.

Centipoise.—A standard unit of *viscosity*, equal to 0.01 poise, the c.g.s. unit of viscosity. Water at 20°C has a viscosity of 1.002.

Centripetal force.—The force required to keep a moving mass in a circular path. Centrifugal force is the name given to the reaction against centripetal force.

Chain reaction.—In general, any self-sustaining process, whether molecular or nuclear, the products of which are instrumental in, and directly contribute to the propagation of the process. Specifically, a *fission chain reaction*, where the energy liberated or particles produced (fission products) by the fission of an atom cause the fission of other atomic nuclei, which in turn propagate the fission recation in the same manner.

Charles' law or Gay-Lussac's law.—The volumes assumed by a given mass of a gas at different temperatures, the pressure remaining constant, are, within moderate ranges of temperature, directly proportional to the corresponding absolute temperatures.

Chemiluminescence.—Emission of light during a chemical reaction.

Christiansen effect.—When finely powdered substances, such as glass or quartz, are immersed in a liquid of the same index of refraction complete transparency can only be obtained for monochromatic light. If white light is employed the transmitted color corresponds to the particular wave-length for which the two substances, solid and liquid have exactly the same index of refraction. Due to differences in dispersion the indices of refraction will match for only a narrow band of the spectrum.

DEFINITIONS AND FORMULAS (Continued)

DEFINITIONS AND FORMULAS (Continued)

Chromatic aberration.—Due to the difference in the index of refraction for different wave lengths, light of various wave lengths from the same source cannot be focused at a point by a simple lens. This is called chromatic aberration.

Cloud chamber.—An apparatus containing moist air or other gas which on sudden expansion condenses moisture to droplets on dust particles or other nuclei. Thus charged particles or ions in the space become nuclei and their numbers and behavior, when properly illuminated, may be studied.

Colligative property.—A property numerically the same for a group of substances, independent of their chemical nature.

Colloid.—A phase dispersed to such a degree that the surface forces become an important factor in determining its properties.

In general particles of colloidal dimensions are approximately 10 angstroms to 1 micron in size. Colloidal particles are often best distinguished from ordinary molecules due to the fact that colloidal particles cannot diffuse through membranes which do allow ordinary molecules and ions to pass freely.

Coma.—An aberration of spherical lenses, occurring in the case of oblique incidence, when the bundle of rays forming the image is unsymmetrical. The image of a point is comet shaped, hence the name.

Combination frequencies.—Two vibrations of arbitrary frequencies f_1 and f_2 when applied simultaneously to a nonlinear (distorting) device will excite it to a motion containing not only the original frequencies, but also members of a set of "combination" frequencies given by $f_c = mf_1 + nf_2$ where m and n are integers. A resonator sharply tuned to any one of these frequencies which may be produced in the nonlinear device will resound to it with an amplitude depending on the type of nonlinearity. The superheterodyne radio receiver depends on this phenomenon.

Combining volumes.—Under comparable conditions of pressure and temperature the volume ratios of gases involved in chemical reactions are simple whole numbers.

Combining weight of an element or radical is its atomic weight divided by its valence.

Combining weights, law of.—If the weights of elements which combine with each other be called their "combining weights," then elements always combine either in the ratio of their combining weights or of simple multiples of these weights.

Component substances, law of.—Every material consists of one substance, or is a mixture of two or more substances, each of which exhibits a specific set of properties, independent of the other substances.

Compounds are substances containing more than one constituent element and having properties, on the whole, different from those which their constituents had as elementary substances. The composition of a given pure compound is perfectly definite, and is always the same no matter how that compound may have been formed.

Compressibility.—Reciprocal of the bulk modulus.

Compton effect, (Compton recoil effect).—Elastic scattering of photons by electrons results in decrease in frequency and increase of wave length of x-rays and gamma-rays when scattered by free electrons.

Concentration.—The amount of a substance in weight, moles, or equivalents contained in unit volume.

Condensers in parallel and series.—If c_1, c_2, c_3, etc. represent the capacitances of a series of condensers and C their combined capacitance,—

when in parallel, $C = c_1 + c_2 + c_3 \cdots$

when in series, $\frac{1}{C} = \frac{1}{c_1} + \frac{1}{c_2} + \frac{1}{c_3} \cdots$

Conductance, the reciprocal of resistance, is measured by the ratio of the current flowing through a conductor to the difference of potential between its ends. The practical unit of conductance, the mho, the conductance of a body through which one ampere of current flows when the potential difference is one volt. The conductance of a body in mho is the reciprocal of the value of its resistance in ohms. Dimensions,—$[\epsilon\ l\ t^{-1}]$. $[\mu^{-1}\ l^{-1}\ t]$.

Conductivity, electrical, is measured by the quantity of electricity transferred across unit area, per unit potential gradient per unit time. Reciprocal of resistivity. **Volume conductivity** or specific conductance, $k = 1/\rho$ where ρ is the volume resistivity. **Mass conductivity** $= k/d$ where d is density. **Equivalent conductivity** $\Lambda = k/c$ where c is the number of equivalents per unit volume of solution. **Molecular conductivity** $\mu = k/m$ where m is the number of moles per unit volume of solution. Dimensions: volume conductivity,—$[\epsilon\ t^{-1}]$; $[\mu^{-1}\ l^{-2}\ t]$,—mass conductivity,—$[\epsilon\ m^{-1}\ l^3\ t^{-1}]$; $[\mu^{-1}\ m^{-1}\ lt]$.

Conductivity, thermal.—Time rate of transfer of heat by conduction, through unit thickness, across unit area for unit difference of temperature. It is measured as calories per second per square centimeter for a thickness of one centimeter and a difference of temperature of 1°C. Dimensions,—W/m K.

If the two opposite faces of a rectangular solid are maintained at temperatures t_1 and t_2 the heat conducted across the solid of section a and thickness d in a time T will be,

$$Q = \frac{K(t_2 - t_1)aT}{d}$$

K is a constant depending on the nature of the substance, designated as the specific heat conductivity. K is usually given for Q in calories, t_1 and t_2 in °C, a in cm², T in sec, and d in cm.—See *Heat conductivity*

DEFINITIONS AND FORMULAS (Continued

Conductors.—A class of bodies which are incapable of supporting electric strain. A charge given to a conductor spread to all parts of the body.

Conjugate foci.—Under proper conditions light divergent from a point on or near the axis of a lens or spherical mirror is focused at another point. The point of convergence and the position of the source are interchangeable and are called conjugate foci.

Conservation of energy, (Chem).—In a chemical change there is no loss or gain but merely a transformation of energy from one form to another.

Conservation of energy, law of.—Energy can neither be created nor destroyed and therefore the total amount of energy in the universe remains constant.

Conservation of mass.—In all ordinary chemical changes, the total of the reactants is always equal to the total mass of the products.

Conservation of momentum, law of.—For any collision, the vector sum of the momenta of the colliding bodies after collison equals the vector sum of their momenta before collision. If two bodies of masses m_1 and m_2 have, before impact velocities v_1 and v_2 and after impact velocities u_1 and u_2

$$m_1u_1 + m_2u_2 = m_1v_1 + m_2v_2$$

Constitutive property.—A property which depends on the constitution or structure of the molecule.

Cooling.—Processing highly radioactive materials to attain lesser radioactivity for subsequent use or handling.

Cosmic rays.—Highly penetrating radiations which strike the earth, assumed to originate in interstellar space. They are classed as: primary, coming from the assumed source, and secondary, those induced in upper atmospheric nuclei by collisions with primary cosmic rays.

Cosmotron.—A particle accelerator capable of giving them energies to billions of electron volts.

Coulomb.—A unit quantity of electricity. It is the quantity of electricity which must pass through a circuit to deposit 0.0011180 grams of silver from a solution of silver nitrate. An ampere is one coulomb per second. A coulomb is also the quantity of electricity on the positive plate of a condenser of one-farad capacity when the electromotive force is one volt.

Couple.—Two equal and oppositely directed parallel but not colinear forces acting upon a body form a couple. The moment of the couple or torque is given by the product of one of the forces by the perpendicular distance between them. Dimension, —$[m\, l^2\, t^{-2}]$.

Couple acting on a magnet of magnetic moment ml in a field of strength H. If the magnet is perpendicular to the direction of the field

$$C = Hml = HM$$

If the angle between the magnet and the field is θ

$$C = Hml \sin \theta$$

The couple will be in dyne-cm for cgs electromagnetic units of H, m and l.

Critical mass.—The minimum mass the fissile material must have in order to maintain a spontaneous fission chain reaction. For pure U^{235} it is computed to be about 20 pounds.

Critical temperature is that temperature above which a gas cannot be liquefied by pressure alone. The pressure under which a substance may exist as a gas in equilibrium with the liquid at the critical temperature is the **critical pressure.**

Cross section, (Nuclear cross section).—A measure of the probability of a particular process. The nuclear cross section is expressed by a/bc, where a is the number of processes occurring, b the number of incident particles, and c the number of target nuclei per cm². There are nuclear cross sections for fission, for slow neutron capture, for Compton collision, and for ionization by electron impact.

Cryohydrate.—The solid which separates when a saturated solution freezes. It contains the solvent and the solute in the same proportions as they were in the saturated solution.

Crystal.—The "ideal crystal" is a homogeneous portion of crystalline matter, (q.v.) whether bounded by faces or not.

Crystalline matter is matter that possesses a triperiodic structure on the atomic scale. It is characterized by discontinuous vectorial properties that give rise to "crystal planes" [(1) crystal growth (faces); (2) cohesion (cleavage planes); (3) twinning (twin planes); (4) gliding (gliding planes); (5) x-ray, electron, or neutron diffraction ("reflecting" planes); all of which are parallel to lattice planes.]

Curie.—Unit for measuring radioactivity. One Curie = that quantity of any radioactive isotope undergoing 3.7×10^{10} disintegrations per second.

Curie's law.—The intensity of magnetization,

$$I = \frac{AH}{T}$$

where H, is the magnetic field strength, T the absolute temperature and A Curie's constant. Used for paramagnetic substances.

Curie point.—All ferro-magnetic substances have a definite temperature of transition at which the phenomena of ferromagnetism disappear and the substances become merely paramagnetic. This temperature is called the "Curie Point" and is usually lower than the melting point.

Curie-Weiss law.—The Curie law was modified by Weiss to state that the susceptibility of a paramagnetic substance above the Curie point varies inversely as the excess of the temperature above that point.

This law is not valid at or below the Curie point.

Current (electric).—The rate of transfer of electricity. The transfer at the rate of one electrostatic unit of electricity in one second is the electrostatic unit of current. The electromagnetic unit of current is a current of such strength that one centimeter of the wire in which it flows is pushed sideways with a force of one dyne when the wire is at right angles to a magnetic field of unit intensity. The practical unit of current is the **ampere,** a transfer of one coulomb per second, which is one tenth the electromagnetic unit. The **international ampere** is the unvarying electric current which, when passed through a solution of silver nitrate in accordance with certain specifications, deposits silver at the rate of 0.00111800 gram per second. The international ampere is equivalent to 0.999835 absolute ampere. The **ampere-turn** is the magnetic potential produced between the two faces of a coil of one turn carrying one ampere. Dimensions,—$[\epsilon^{\frac{1}{2}}\, m^{\frac{1}{2}}\, l^{\frac{3}{2}}\, t^{-2}]$; $[\mu^{-\frac{1}{2}}\, m^{\frac{1}{2}}\, l^{\frac{1}{2}}\, t^{-1}]$.

Current in a simple circuit.—The current in a circuit including an external resistance R and a cell of electromotive force E and internal resistance r,

$$I = \frac{E}{R + r}$$

If E is in volts and r and R in ohms the current will be in amperes.

For two cells in parallel,

$$I = \frac{E}{R + \frac{r}{2}}$$

For two cells in series,

$$I = \frac{2E}{R + 2r}$$

Cyclotron.—The magnetic resonance accelerator for imparting very great velocities to heavier nuclear particles without the use of excessive voltages.

Dalton's law of partial pressures.—The pressure exerted by a mixture of gases is equal to the sum of the separate pressures which each gas would exert if it alone occupied the whole volume. This fact is expressed in the following formula:

$$PV = V(p_1 + p_2 + p_3, \text{etc.})$$

Decay.—Diminution of a radioactive substance due to nuclear emission of alpha or beta particles, gamma rays or positrons.

Declination.—The angle between the vertical plane containing the direction of the earth's field at any point and a plane containing the geographic north and south meridian.

Decomposition is the chemical separation of a substance into two or more substances, which may differ from each other and from the original substances.

Definite proportions, law of.—In every sample of each compound substance the proportions by weight of the constituent elements are always the same.

Degree of freedom.—The number of the variables determining the state of a system (usually pressure, temperature, and concentrations of the components) to which arbitrary values can be assigned.

Density.—Concentration of matter, measured by the mass per unit volume. Dimensions,—$[m\, l^{-3}]$.

Deuteron.—Nucleus of the deuterium atom or the ion of deuterium. Its structure—one neutron and one proton.

Dew point.—The temperature at which condensation of water vapor in the air takes place.

Diamagnetic materials.—Are those within which an externally applied magnetic field is slightly reduced because of an alteration of the atomic electron orbits produced by the field. Diamagnetism is an atomic-scale consequence of the Lenz law of induction. The permeability of diagmagnetic materials is slightly less than that of empty space.

Dielectric constant of a medium is defined by ϵ in the equation

$$F = \frac{QQ'}{\epsilon r^2}$$

where F is the force of attraction between two charges Q and Q' separated by a distance r in a uniform medium.

Dielectrics or insulators or non-conductors.—A class of bodies supporting an electric strain. A charge on one part of a non-conductor is not communicated to any other part.

Diffraction.—That phenomena produced by the spreading of waves around and past obstacles which are comparable in size to their wavelength.

Diffraction grating.—If s is the distance between the rulings, d the angle of diffraction, then the wave length where the angle of incidence is 90° is (for the nth order spectrum),

$$\lambda = \frac{s \sin d}{n}$$

If i is the angle of incidence, d the angle of diffraction, s the distance between the rulings, n the order of the spectrum, the wave length is, $\lambda = \frac{s}{n}(\sin i + \sin d)$.

Diffusion.—If the concentration (mass of solid per unit volume of solution) at one surface of a layer of liquid is d_1 and at the other surface d_2, the thickness of the layer h and the area under consideration A, then the mass of the substance which diffuses through the cross-section A in time t is,

$$m = \Delta A\, \frac{(d_2 - d_1)t}{h}$$

where Δ is the coefficient of diffusion.

Diffusivity or coefficient of diffusion is also given by Δ in the equation

$$\frac{dQ}{dt} = -\Delta\left(\frac{dc}{dx}\right) dy\, dz$$

where dQ is the amount passing through an area $dy\, dz$ in the direction of x in a time dt where dc/dx is the rate of increase of volume concentration in the direction of x. Dimensions,—$[l^2\, t^{-1}]$.

Diffusivity of heat is given by Δ in the equation

$$\frac{dH}{dt} = -\Delta sd \frac{dT}{dx} dy\, dz$$

where dH is the quantity of heat passing through the area $dy\, dz$ in the direction of x in a time dt. The rate of variation of temperature along x is given by dT/dx, s is specific heat and d, density. Dimensions,—$[l^2\, t^{-1}]$.

Dimensional formulae.—If mass, length, and time are considered fundamental quantities, the relation of other physical quantities and their units to these three may be expressed by a formula involving the symbols l, m and t respectively, with appropriate exponents. For example; the dimensional formula for volume would be expressed,—$[l^3]$; velocity,—$[lt^{-1}]$; force—$[m\, lt^{-2}]$. Other fundamental quantities used in dimensional formulae may be indicated as follows: θ, temperature, ϵ the dielectric constant of a vacuum; μ, the magnetic permeability of a vacuum.

Diminution of pressure at the side of a moving stream. If a fluid of density d moves with a velocity v, the diminution of pressure due to the motion is (neglecting viscosity),

$$p = \tfrac{1}{2}\, dv^2$$

Dip.—The angle measured in a vertical plane between the direction of the earth's magnetic field and the horizontal.

Dipole.—(1) A combination of two electrically or magnetically charged particles of opposite sign which are separated by a very small distance. (2) Any system of charges, such as a circulating current, which has the properties that: (a) no forces act on it in a uniform field; (b) a torque proportional to sin θ, where θ is the angle between the dipole axis and a uniform field, does act on it; (c) it produces a potential which is proportional to the inverse square of the distance from it.

Dipole moment.—A mathematical entity; the product of one of the charges of a dipole unit by the distance separating the two dipolar charges. In terms of the definition of a dipole (2), the dipole moment $\mathbf{p}$ is related to the torque $\mathbf{T}$, and the field strength $\mathbf{E}$ (or $\mathbf{B}$) through the equation:

$$\mathbf{T} = \mathbf{p} \times \mathbf{E}.$$

Dipole moment, molecular.—It is found from measurements of dielectric constant (i.e. by its temperature dependence, as in the **Debye equation for total polarization**) that certain molecules have permanent dipole moments. These moments are associated with transfer of charge within the molecule, and provide valuable information as to the molecular structure.

Dispersion.—The difference between the index of refraction of any substance for any two wave lengths is a measure of the dispersion for these wave lengths, called the coefficient of dispersion.

Dispersion forces.—The force of attraction between molecules possessing no permanent **dipole.** The interaction energy is given by

$$U_D = -\frac{3}{4} h \frac{V_0 \alpha^2}{r^6}$$

where h is Planck's constant, V_0 a characteristic frequency of the molecule, r the distance between the molecules, and α the polarizability.

Dispersive power.—If n_1 and n_2 are the indices of refraction for wave lengths λ_1 and λ_2 and n the mean index or that for sodium light, the dispersive power for the specified wave length is,

$$\omega = \frac{n_2 - n_1}{n - 1}$$

Displacement is a reaction in which an elementary substance displaces and sets free a constituent element from a compound.

Displacement or elongation at any instant. The distance of a vibrating or oscillating particle from its position of equilibrium.

Distribution law.—A substance distributes itself between two immiscible solvents so that the ratio of its concentrations in the two solvents is approximately a constant (and equal to the ratio of the solubilities of the substance in each solvent). The above statement requires modification if more than one molecular species is formed.

Doppler effect, (Light).—The apparent change in the wave-length of light produced by the motion in the line of sight of either the observer or the source of light.

Doppler effects.—Effects on the apparent frequency of a wave train produced (1) by motion of the source toward or away from the stationary observer, and (2) by motion of the observer toward or from the stationary source; the motion in each case being with reference to the (supposedly stationary) medium.

For sound waves, the observed frequency f_0, in cycles/sec., is given by

$$f_0 = \frac{v + w - v_0}{v + w - v_s} f_s$$

where v is the velocity of sound in the medium, v_0 is the velocity

DEFINITIONS AND FORMULAS (Continued)

of the observer, v_s is the velocity of the source, w is the velocity of the wind in the direction of sound propagation, and f_s is the frequency of source.

For optical waves

$$f_0 = f_s \sqrt{\frac{c + v_r}{c - v_r}}$$

where v_r is the velocity of the source relative to the observer and c is the speed of light.

Dosimeter.—A gadget worn by those who work about radioactive sources to record the amount of exposure the wearer has had to radioactivity.

Double decomposition consists of a simple exchange of the parts of two substances to form two new substances.

Dulong and Petit, law of.—The specific heats of the several elements are inversely proportional to their atomic weights. The atomic heats of solid elements are constant and approximately equal to 6.3. Certain elements of low atomic weight and high melting point have, however, much lower atomic heats at ordinary temperatures.

Dyne.—The force necessary to give acceleration of one centimeter per second per second to one gram of mass.

Eddy current.—A current induced in a mass of conducting material by a varying magnetic field. Also called *Foucault current.*

Einstein theory for mass-energy equivalence.—The equivalence of a quantity of mass m and a quantity of energy E by the formula $E = mc^2$. The conversion factor c^2 is the square of the velocity of light.

Elasticity.—The property by virtue of which a body resists and recovers from deformation produced by force.

Elastic limit.—The smallest value of the stress producing permanent alteration.

Elastic moduli.

Young's modulus by stretching.—If an elongation s is produced by the weight of the mass m, in a wire of length l, and radius r, the modulus,

$$M = \frac{mgl}{\pi r^2 s}$$

Young's modulus by bending, bar supported at both ends. If a flexure s is produced by the weight of mass m, added midway between the supports separated by a distance l, for a rectangular bar with vertical dimensions of cross-section a and horizontal dimension b, the modulus is,

$$M = \frac{mgl^3}{4sa^3b}$$

For a cylindrical bar of radius r,

$$M = \frac{mgl^3}{12\pi r^4 s}$$

For a bar supported at one end. In the case of a rectangular bar as described above,

$$M = \frac{4mgl^3}{sa^3b}$$

For a round bar supported at one end,

$$M = \frac{4mgl^3}{3\pi r^4 s}$$

Modulus of rigidity.—If a couple C ($= mgx$) produces a twist of θ radians in a bar of length l and radius r, the modulus is

$$M = \frac{2Cl}{\pi r^4 \theta}$$

The substitution in the above formulae for the elastic coefficients of m in grams, g in cm per sec², l, a, b, and r in cm, s in cm, and C in dyne-cm will give moduli in dynes per cm². The dimensions of elastic moduli are the same as of stress,—$[m\, l^{-1}\, t^{-2}]$.

Coefficient of restitution.—Two bodies moving in the same straight line, with velocities v_1 and v_2 respectively, collide and after impact move with velocities v_3 and v_4. The coefficient of restitution is

$$C = \frac{v_4 - v_3}{v_2 - v_1}$$

Electric field intensity is measured by the force exerted on unit charge. Unit field intensity is the field which exerts the force of one dyne on unit positive charge. Dimensions,—$[\epsilon^{-\frac{1}{2}}\, m^{\frac{1}{2}}\, l^{-\frac{1}{2}}\, t^{-1}]$; $[\mu^{\frac{1}{2}}\, m^{\frac{1}{2}}\, l^{\frac{1}{2}}\, t^{-2}]$.

The field intensity or force exerted on unit charge at a point distant r from a charge q in a vacuum

$$H = \frac{q}{r^2}$$

If the dielectric in the above cases is not a vacuum the dielectric constant ϵ must be introduced. The formula becomes

$$H = \frac{q}{\epsilon r^2}$$

The value of ϵ is frequently considered unity for air. If the dielectric constant of a vacuum is considered unity the value for air at 0°C and 760 mm pressure is 1.000576.

Electrochemical equivalent of an ion is the mass liberated by the passage of unit quantity of electricity.

DEFINITIONS AND FORMULAS (Continued)

Electrolysis.—If a current i flows for a time t and deposits a metal whose electrochemical equivalent is e, the mass deposited is

$$m = eit$$

The value of e is usually given for mass in grams, i in amperes and t in seconds.

Electrolytic dissociation or ionization theory.—When an acid, base or salt is dissolved in water or any other dissociating solvent, a part or all of the molecules of the dissolved substance are broken up into parts called ions, some of which are charged with positive electricity and are called cations, and an equivalent number of which are charged with negative electricity and are called anions.

Electrolytic solution tension theory (or the Helmholtz double layer theory).—When a metal, or any other substance capable of existing in solution as ions, is placed in water or any other dissociating solvent, a part of the metal or other substances passes into solution in the form of ions, thus leaving the remainder of the metal or substances charged with an equivalent amount of electricity of opposite sign from that carried by the ions. This establishes a difference in potential between the metal and the solvent in which it is immersed.

Electromotive force is defined as that which causes a flow of current. The electromotive force of a cell is measured by the maximum difference of potential between its plates. The electromagnetic unit of potential difference is that against which one erg of work is done in the transfer of electromagnetic unit quantity. The **volt** is that potential difference against which one joule of work is done in the transfer of one coulomb. One volt is equivalent to 10^8 electromagnetic units of potential. The **international** volt is the electrical potential which when steadily applied to a conductor whose resistance is one international ohm will cause a current of one international ampere to flow. The international volt = 1.00033 absolute volts. The electromotive force of a Weston standard cell is 1.0183 int. volts at 20°C. Dimensions,—$[\epsilon^{-\frac{1}{2}} m^{\frac{1}{2}} l^{\frac{1}{2}} t^{-1}]$, $[\mu^{\frac{1}{2}} m^{\frac{1}{2}} l^{\frac{3}{2}} t^{-2}]$.

Electromotive series is a list of the metals arranged in the decreasing order of their tendencies to pass into ionic form by losing electrons.

Electron.—The electron is a small particle having a unit negative electrical charge, a small mass, and a small diameter. Its charge is $(4.80294 \pm .00008) \times 10^{-10}$ absolute electrostatic units, its mass $\frac{1}{1837}$ that of the hydrogen nucleus, and its diameter about 10^{-12} cm. Every atom consists of one nucleus and one or more electrons. Cathode rays and Beta rays are electrons.

Electron-volt (ev).—Energy acquired by any charged particle carrying unit electronic charge when it falls through a potential difference of one volt. 1 electron-volt = $(1.60207 \pm .00007 \times 10^{-12}$ erg. Multiples of this unit are also in common use: the kilo-, million-, and billion electron-volt. 1 kev = 10^3 ev; 1 mev = 10^6 ev; and 1 bev = 10^9 ev.

An ev is associated through the Planck constant with a photon of wave length 1.2395 microns.

Elements are substances which cannot be decomposed by the ordinary types of chemical change, or made by chemical union.

Emissive power or emissivity is measured by the energy radiated from unit area of a surface in unit time for unit difference of temperature between the surface in question and surrounding bodies. For the cgs system the emissive power is given in ergs per second per square centimeter with the radiating surface at 1° absolute and the surroundings at absolute zero.—See *Radiation formula*

Enantiotropic.—Crystal forms capable of existing in reversible equilibrium with each other.

Energy.—The capability of doing work. **Potential energy** is energy due to position of one body with respect to another or to the relative parts of the same body. **Kinetic energy** is energy due to motion. Cgs units,—the erg, the energy expended when a force of one dyne acts through a distance of one centimeter; the joule is 1×10^7 ergs. Dimensions,—$[m\, l^2\, t^{-2}]$.

The potential energy of a mass m, raised through a distance h, where g is the acceleration due to gravity is

$$E = mgh.$$

The kinetic energy of mass m, moving with a velocity v, is

$$E = \frac{1}{2}mv^2$$

Energy will be given in ergs if m is in grams, g in cm per sec^2, h in cm and v in cm per sec.

Energy of a charge in ergs where Q is the charge and V the potential in electrostatic units.

$$E = \frac{1}{2}QV.$$

Energy of the electric field.—If H is the electric field intensity in electrostatic units and K the specific inductive capacity, the energy of the field in ergs per cm^3 is

$$E = \frac{KH^2}{8\pi}$$

Energy of rotation.—If a mass whose moment of inertia about an axis is I, rotates with angular velocity ω about this axis, the kinetic energy of rotation will be,

$$E = \frac{1}{2}I\omega^2$$

Energy will be given in ergs if I is in g-cm^2 and ω in radians per sec.

Enthalpy, or heat content, is a thermodynamic quantity. It is equal to the sum of the internal energy of a system plus

DEFINITIONS AND FORMULAS (Continued)

the product of the pressure-volume work done on the system. Thus

$$H = E + pv$$

where H = enthalpy or heat content
E = internal energy of the system
p = pressure
v = volume.

Entropy.—Entropy is the capacity factor for isothermally unavailable energy. The increase in the entropy of a body during an infinitesimal stage of a reversible process is equal to the infinitesimal amount of heat absorbed divided by the absolute temperature of the body. Thus for a reversible process

$$dS = \frac{Q}{T}$$

Every spontaneous process in nature is characterized by an increase in the total entropy of the bodies concerned in the process.

Equilibrium, chemical.—A state of affairs in which a chemical reaction and its reverse reaction are taking place at equal velocities, so that the concentrations of reacting substances remain constant.

Equilibrium constant.—The product of the concentrations (or activities) of the substances produced at equilibrium in a chemical reaction divided by the product of concentrations of the reacting substances, each concentration raised to that power which is the coefficient of the substance in the chemical equation.

Equivalent conductance of an electrolyte is defined as the conductance of a volume of solution containing one equivalent weight of dissolved substance when placed between two parallel electrodes 1 cm apart, and large enough to contain between them all of the solution. Λ is never determined directly, but is calculated from the specific conductance. If C is the concentration of a solution in gram equivalents per liter, then the concentration per cubic centimeter is C/1000, and the volume containing one equivalent of the solute is, therefore, 1000/C. Since L_s is the conductance of a centimeter cube of the solution, the conductance of 1000/C cc, and hence Λ will be

$$\Lambda = \frac{1000L_s}{C}$$

Equivalent weight or combining weight of an element or ion is its atomic or formula weight divided by its valence. Elements entering into combination always do so in quantities proportional to their equivalent weights.

In oxidation-reduction reactions the equivalent weight of the reacting substances is dependent upon the change in oxidation number of the particular substance.

Ettinghausen's effect (Von Ettinghausen's).—When an electric current flows across the lines of force of a magnetic field an electromotive force is observed which is at right angles to both the primary current and the magnetic field: a temperature gradient is observed which has the opposite direction to the Hall electromotive force.

Eutectic.—A term applied to the mixture of two or more substances which has the lowest melting point.

Expansion of gases.

Charles' law or Gay-Lussac's law.—The volume of a gas at constant pressure increases proportionately to the absolute temperature. If V_1 and V_2 are volumes of the same mass of gas at absolute temperatures, T_1 and T_2,

$$\frac{V_1}{V_2} = \frac{T_1}{T_2}$$

For an original volume V_o at 0°C the volume at t°C (at constant pressure) is

$$V_t = V_o(1 + 0.00367t).$$

General law for gases.

$$p_t v_t = p_o v_o \left(1 + \frac{t}{273}\right)$$

where p_o, v_o, p_t, v_t represent the pressure and volume at 0° and t°C or

$$\frac{p_1 v_1}{T_1} = \frac{p_2 v_2}{T_2}$$

where p_1, v_1 and T_1 represent pressure, volume and absolute temperature in one case and p_2, v_2 and T_2 the same quantities for the same mass of gas in another.

The law may also be expressed:

$$pv = RmT$$

where m is the mass of gas at absolute temperature T. R is the **gas constant** which depends on the units used. **Boltzmann's molecular gas constant** is obtained by expressing m in terms of the number of molecules.

For volume in cm³, pressure in dynes per cm² and temperature in Centigrade degrees on the absolute scale $R = 8.3136 \times 10^7$.

Reduction of a gas volume to 0°C, 760 mm pressure.—If V is the original volume of a gas at temperature t and pressure H, the volume at 0°C and 760 mm pressure will be,

$$V_o = \frac{V}{(1 + \alpha t)} \frac{H}{760}$$

If d is the original density the density at 0°C and 760 mm pressure will be

$$d_o = d(1 + \alpha t) \frac{760}{H}$$

$$\alpha = 0.00367 \text{ approximately.}$$

Falling bodies.—For bodies falling from rest conditions are as for uniformly accelerated motion except that $v_o = O$ and g is the acceleration due to gravity. The formulae become,—air resistance neglected,

$$v_t = gt,\ s = \tfrac{1}{2}gt^2,\ v_s = \sqrt{2gs}.$$

For bodies projected vertically upward,—if v is the velocity of projection, the time to reach greatest height, neglecting the resistance of the air,

$$t = \frac{v}{g}$$

Greatest height,

$$h = \frac{v^2}{2g}$$

—See *Projectiles*

Faraday's laws.—In the process of electrolytic changes equal quantities of electricity charge or discharge equivalent quantities of ions at each electrode.

One gram equivalent weight of matter is chemically altered at each electrode for 96,487 int. coulombs, or one faraday, of electricity passed through the electrolyte.

Faraday effect.—The rotation of the plane of polarization produced when plane-polarized light is passed through a substance in a magnetic field, the light traveling in a direction parallel to the lines of force. For a given substance, the rotation is proportional to the thickness traversed by the light and to the magnetic field strength.

Faraday, new determination of (1960).—By the National Bureau of Standards which uses an electrochemical method that dissolves, rather than deposits, silver from a solution. The new value, 96,516 ± 2 coulombs (physical scale) or 96,489 ± 2 coulombs (chemical scale). NBS used its mass, time, and electrical standards in measuring the faraday, and have found that its value agreed within 22 p.p.m. with the one obtained by an independent physical method using the omegatron.

Fermat's principle.—The path followed by light (or other waves) passing through any collection of media from one specified point to another, is that path for which the time of travel is least.

Fission.—A nuclear reaction from which the atoms produced are each approximately half the mass of the parent nucleus. In other words, the atom is split into two approximately equal masses. There is also the emission of extremely great quantities of energy since the sum of the masses of the two new atoms is less than the mass of the parent heavy atom. The energy released is expressed by Einstein's equation.

Fleming's rule.—A simple rule for relating the directions of the flux, motion, and e.m.f. in an electric machine. The forefinger, second finger and thumb, placed at right-angles to each other, represent respectively the directions of flux, e.m.f., and motion or torque. If the right hand is used the conditions are those obtaining in a generator and if the left hand is used the conditions are those obtaining in a motor.

Fluidity.—The reciprocal of viscosity. The cgs unit is the rhe, the reciprocal of the poise. Dimensions,—$[m^{-1}\ lt]$.

Fluorescence.—The property of emitting radiation as the result of absorption of radiation from some other source. The emitted radiation persists only as long as the exposure is subjected to radiation which may be either electrified particles or waves. The fluorescent radiation generally has a longer wave length than that of the absorbed radiation. If the fluorescent radiation includes waves of the same length as that of the absorbed radiation it is termed resonance radiation.

Foot-candle.—Foot-candle = 1 lumen incident per foot^2 = 1.076 milliphots = 10.76 lux.

Foot-lambert is the unit of photometric brightness (luminance) equal to $1/\pi$ candle per square foot.

Force.—That which changes the state of rest or motion in matter, measured by the rate of change of momentum. Absolute unit,—the **dyne,** the force which will produce an acceleration of one centimeter per second per second in a gram mass. The gram weight or weight of a gram mass is the cgs gravitational unit. The poundal is that force which will give an acceleration of one foot per second per second to a pound mass. Dimensions,—$[m\ lt^{-2}]$.

The force F required to produce an acceleration a in a mass m is given by

$$F = ma.$$

If m is substituted in grams and a in cm per sec^2, F will be given in dynes.

Force between two charges, Coulomb's law.—If two charges q and q' are at a distance r in a vacuum, the force between them is,

$$F = \frac{qq'}{r^2}$$

The force will be given in dynes if q and q' are in electrostatic units and r in cm.

Force between two magnetic poles.—If two poles of strength m and m' are separated by a distance r in a medium whose permeability is μ (unity for a vacuum), the force between them is,

$$F = \frac{mm'}{\mu r^2}$$

Force will be given in dynes if r is in cm and m and m' are in cgs units of pole strength.

The strength of a magnetic field at a point distance r from an isolated pole of strength m is

$$H = \frac{m}{\mu r^2}$$

The field will be given in gauss if m and r are in cgs units.

Formula, chemical.—A combination of symbols with their subscripts representing the constituents of a substance and their proportions by weight.

Fraunhofer's lines.—When sunlight is examined through a spectroscope it is found that the spectrum is traversed by an enormous number of dark lines parallel to the length of the slit. These dark lines are known as Fraunhofer's lines. Kirchoff conceived the idea that the sun is surrounded by layers of vapors which act as filters of the white light arising from incandescent solids within and which abstract those rays which correspond in their periods of vibration to those of the components of the vapors. Thus reversed or dark lines are obtained due to the absorption by the vapor envelop, in place of the bright lines found in the emission spectrum.

Frequency in uniform circular motion or in any periodic motion is the number of revolutions or cycles completed in unit time. Cgs units,—cycles per second. Dimension,—$[t^{-1}]$.

Frequency of vibrating strings.—The fundamental frequency of a stretched string is given by

$$n = \frac{1}{2l}\sqrt{\frac{T}{m}}$$

where l is the length, T, the tension and m the mass per unit length.

For a string or wire of circular section of length l, tension T, density d, and radius r, the frequency of the fundamental is

$$n = \frac{1}{2rl}\sqrt{\frac{T}{\pi d}}$$

The frequency in vibrations per second will be given if T is in dynes, r and l in cm and d in g per cm^3.

Friction, coefficient of.—The coefficient of friction between two surfaces is the ratio of the force required to move one over the other to the total force pressing the two together.

If F is the force required to move one surface over another and W, the force pressing the surfaces together, the coefficient of friction,

$$k = \frac{F}{W}$$

Fundamental units.—See *Mass, Length and Time*

Fusion (atomic).—A nuclear reaction involving the combination of smaller atomic nuclei or particles into larger ones with the release of energy from mass transformation. This is also called a thermo-nuclear reaction by reason of the extremely high temperature required to initiate it.

Gal.—1 gal = cm/sec./sec. Therefore, where the value of gravity is 980 this is the same as 980 gals. The milligal is now quite commonly used since it is approximately one part in a thousand of the normal gravity of the earth.

Gamma (γ) rays (nuclear x-rays).—May be emitted from radioactive substances. They are quanta of electromagnetic wave energy similar to but of much higher energy than ordinary x-rays. The energy of a quantum is equal to $h\nu$ ergs, where h is Planck's constant (6.6254×10^{-27} erg sec) and ν is the frequency of the radiation. Gamma rays are highly penetrating, an appreciable fraction being able to traverse several centimeters of lead.

Gas.—A state of matter in which the molecules are practically unrestricted by cohesive forces. A gas has neither definite shape nor volume.

Gas thermometer.—Where P_o, P_s and P_x represent the total pressure with the bulb at 0°C, at the boiling-point of water and at the unknown temperature respectively, t_s the temperature of steam and t_x the unknown temperature,

$$t_x = t_s \frac{P_x - P_o}{P_s - P_o}$$

(approximately). The total pressure on the gas in the bulb is the algebraic sum of barometric pressure at the time and that measured by the manometer.

Gauss.—The cgs emu of magnetic induction (flux density). It is equal to 1 maxwell per cm^2. It has such a value that if a conductor 1 cm long moves through a magnetic field at a velocity of 1 cm, in an induction mutually perpendicular, the induced emf is one abvolt.

Gay-Lussac's law.—See *Charles' law*

Gay-Lussac's law of combining volumes.—If gases interact and form a gaseous product, the volumes of the reacting gases and the volumes of the gaseous products are to each other in simple proportions, which can be expressed by small whole numbers.

Geiger counter.—Detector for radioactivity depending upon ionized particles that affect its mechanism. As its name indicates, it both detects and makes a count of them possible.

Geopotential. – The geopotential, Φ, of a point at a height z above mean sea level is the work which must be accomplished against gravity in elevating a unit mass from sea level to the height z.

$$\Phi = \int_0^z g\, dz$$

where g is the local acceleration of gravity at height z. For most metrological work geopotential is given in the units geopotential meter (gpm). By definition, 1 gpm = 9.8 ×

DEFINITIONS AND FORMULAS (Continued)

10^4 cm^2 sec^{-2}. For most purposes the geopotential can be assumed to equal the geometric height.

Gibbs' phase rule.—$F = C + 2 - P$, **F, the number of degrees of freedom of a system, is the number of variable factors (temperature, pressure and concentration) of the components, which must be arbitrarily fixed in order that the condition of the system may be perfectly defined. C, the number of the components of the system, is chosen equal to the smallest number of independently variable constituents by means of which the composition of each phase participating** in the state of equilibrium can be expressed in the form of a chemical equation; the components must be chosen from among the constituents which are present when the system is in a state of true equilibrium and which take part in that equilibrium; as components are chosen the smallest number of such constituents necessary to express the composition of each phase participating in the equilibrium, zero and negative quantities of components being permissible; in any system the number of components is definite, but may alter with changes in conditions of experiment; a qualitative but not quantitative freedom of selection of components is allowed, the choice being influenced by suitability and simplicity of application. *P*, the number of phases of the system, are the homogeneous, mechanically separable and physically distinct portions of a heterogeneous system; the number of phases capable of existence varies greatly in different systems; there can never be more than one gas or vapor phase since all gases are miscible in all proportions, a heterogeneous mixture of solid substances forms as many phases as there are substances present.

Gilbert.—The cgs emu of magnetomotive force. 1 gilbert $= 10/4\pi$ ampere-turns.

Graham's law.—The relative rates of diffusion of gases under the same conditions are inversely proportional to the square roots of the densities of those gases.

Gram atom or gram atomic weight.—The mass in grams numerically equal to the atomic weight.

Gram equivalent of a substance is the weight of a substance displacing or otherwise reacting with 1.008 grams of hydrogen or combining with one-half of a gram atomic weight (8.00 grams) of oxygen.

Gram mole, gram formula weight, gram equivalent.—Mass in grams numerically equal to the molecular weight, formula weight or chemical equivalent, respectively.

Gram molecular weight or gram molecule.—A mass in grams of a substance numerically equal to its molecular weight. Gram mole.

Gravitation.—The universal attraction existing between all material bodies. The force of attraction between two masses *m* and *m'*, separated by a distance *r*, *k* being the constant of gravitation,

$$F = k\frac{mm'}{r^2}$$

(If *m* and *m'* are given in grams, and *r* in centimeters, *F* will be in dynes if $k = 6.670 \times 10^{-8}$.)

Half-life.—Used to measure the rate of radioactive decay of disintegration. The time lapse during which a radioactive mass loses one half of its radioactivity.

Hall effect.—When a steady current is flowing in a steady magnetic field, electromotive forces are developed which are at right angles both to the magnetic force and to the current and are proportional to the product of the intensity of the current, the magnetic force and the sine of the angle between the directions of these quantities.

Hardness.—Property of substances determined by their ability to abrade or indent one another. An arbitrary scale of hardness is based upon ten selected minerals. For metals the diameter of the indentation made by a hardened steel sphere (Brinnell) or the height of rebound of a small drop hammer (Shore Scleroscope) serve to measure hardness.

Harmonic motion.—See *Simple harmonic motion and Angular harmonic motion*

Heat capacity.—That quantity of heat required to increase the temperature of a system or substance one degree of temperature. It is usually expressed in calories per degree centigrade. Molar heat capacity is the quantity of heat necessary to raise the temperature of one molecular weight of the substance one degree.

Heat effect.—The heat in calories developed in a circuit by an electric current of *I* amperes flowing through a resistance of *R* ohms, with a difference of potential *E* volts for a time *t* seconds.

$$H = \frac{RI^2t}{4.18} = \frac{EIT}{4.18}$$

Heat equivalent, or latent heat, of fusion.—The quantity of heat necessary to change one gram of solid to a liquid with no temperature change. Dimensions,—$[l^2\, t^{-2}]$.

Heat of combustion of a substance is the amount of heat evolved by the combustion of 1 gram molecular weight of the substance.

Heat quantity.—The cgs unit of heat is the **calorie,** the quantity of heat necessary to change the temperature of one gram of water from 3.5°C to 4.5°C (called a small calorie). If the temperature change involved is from 14.5 to 15.5°C, the unit is the normal calorie. The mean calorie is $\frac{1}{100}$ the quantity of heat necessary to raise one gram of water from 0°C to 100°C. The large calorie is equal to 1000 small calories. The British thermal unit is the heat required to raise the temperature of one pound of water at its maximum density, 1°F. It is equal to about 252 calories. Dimensions of energy,—$[m\, l^2\, t^{-2}]$.

Hehner number (value).—A number expressing the percentage (i.e. grams per hundred grams) of water-insoluble fatty acids in an oil or fat.

Heisenberg's theory of atomic structure.—The currently accepted view of the structure of atom, formulated by

DEFINITIONS AND FORMULAS (Continued)

Heisenberg in 1934, according to which the atomic nuclei are built of nucleons, which may be protons or neutrons, while the extranuclear shells consist of electrons only. The nucleons are held together by nuclear forces of attraction, with exchange forces operating between them. The number of protons is equal to the atomic number (Z) of the element, the number of neutrons is equal to the difference between the mass number and the atomic number ($A - Z$). The number of excess neutrons, i.e. the excess of neutrons over protons, is of paramount importance for the radioactive properties or stability of the element.

Henry's law.—The mass of a slightly soluble gas that dissolves in a definite mass of a liquid at a given temperature is very nearly directly proportional to the partial pressure of that gas. This holds for gases which do not unite chemically with the solvent.

Hess' law of constant heat summation.—The amount of heat generated by a chemical reaction is the same whether reaction takes place in one step or in several steps, or all chemical reactions which start with the same original substances and end with the same final substances liberate the same amounts of heat, irrespective of the process by which the final state is reached.

Hooke's law.—Within the elastic limit of any body the ratio of the stress to the strain produced is constant.

Humidity, absolute.—Mass of water vapor present in unit volume of the atmosphere, usually measured as grams per cubic meter. It may also be expressed in terms of the actual pressure of the water vapor present.

Huygens' theory of light.—This theory states that light is a disturbance traveling through some medium, such as the ether. Thus light is due to wave motion in ether.

Every vibrating point on the wave-front is regarded as the center of a new disturbance. These secondary disturbances traveling with equal velocity, are enveloped by a surface identical in its properties with the surface from which the secondary disturbances start and this surface forms the new wave-front.

Hydrogen equivalent of a substance is the number of replaceable hydrogen atoms in 1 molecule or the number of atoms of hydrogen with which 1 molecule could react.

Hydrogen ion concentration.—The concentration of hydrogen ions in solution when the concentration is expressed as gram-ionic weights per liter. A convenient form of expressing hydrogen ion concentration is in terms of the negative logarithm of this concentration. The negative logarithm of the hydrogen ion concentration is called pH. The significance of pH is still in dispute (ref. J. Am. Chem. Soc. 60, 1094 (1938)).

Water at 25°C has a concentration of H ion of 10^{-7} and of OH ion of 10^{-7} moles per liter. Thus the pH of water is 7 at 24°C. A greater accuracy is obtained if one substitutes the thermodynamic activity of the ion for its concentration.

Hydrolysis is a double decomposition reaction involving the splitting of water into its ions and the formation of a weak acid or base or both.

Hydrostatic pressure at a distance h from the surface of a liquid of density d,

$$P = hdg.$$

The total force on an area A due to hydrostatic pressure,

$$F = PA = Ahdg.$$

Force in dynes and pressure in dynes per cm² will be given if h is in cm, d in g per cm³ and g in cm per sec².

Hyperon.—Any article with mass intermediate between that of the neutron and the deuteron.—See *Meson*

Hysteresis.—The magnetization of a sample of iron or steel due to a magnetic field which is made to vary through a cycle of values, lags behind the field. This phenomenon is called hysteresis.

Steinmetz' equation for hysteresis gives the loss of energy in ergs per cycle per cm³,

$$W = \eta B^{1.6}$$

where B is the maximum induction in maxwells per cm² and η the coefficient of hysteresis.

Illumination on any surface is measured by the luminous flux incident on unit area. The units in use are: the **lux,** (abbreviation lx) one lumen per square meter; the **phot,** (abbreviation ph) one lumen per square centimeter; the **footcandle,** (abbreviation fc) one lumen per square foot.

Indeterminancy principle (Uncertainty principle).—The postulate that it is impossible to determine simultaneously both the exact position and the exact momentum of an electron. So this aspect of electronics can only be expressed as a probability.

Index of refraction The refractive index n of an isotropic medium for reversible transport of a given wave or particle is the reciprocal of the phase velocity times a constant which is often arbitrarily chosen for convenience, as in the case of electrostatic electron lenses. For electromagnetic waves or photons this constant is conventionally the vacuum speed of light. (See Velocity and Snell's law.)

Indicators are substances which change from one color to another when the hydrogen ion concentration reaches a certain value, different for each indicator.

Induced electromotive force in a circuit is proportional to the rate of change of magnetic flux through the circuit.

$$E = -\frac{d\phi}{dt}$$

where $d\phi$ is the change of magnetic flux in a time dt. The induced current will be given by

$$I = \frac{d\phi}{Rdt}$$

where R is the resistance of the circuit.

Inductance.—The change in magnetic field due to the variation of a current in a conducting circuit causes an induced counter electromotive force in the circuit itself. This phenomenon is known as **self-induction.** If an electromotive force is induced in a neighboring circuit the term mutual induction is used. Inductance may thus be distinguished as self- or mutual and is measured by the electromotive force produced in a conductor by unit rate of variation of the current. Units of inductance are the centimeter (absolute electromagnetic) and the henry, which is equal to 10^9 centimeters of inductance. The **henry** is that inductance in which an induced electromotive force of one volt is produced when the inducing current is changed at the rate of one ampere per second. Dimensions,—$[\epsilon^{-1} l^{-1} t^2]$; $[\mu l]$.

Induction.—Any change in the intensity or direction of a magnetic field causes an electromotive force in any conductor in the field. The induced electromotive force generates an induced current if the conductor forms a closed circuit.

Inertia.—The resistance offered by a body to a change of its state of rest or motion, a fundamental property of matter. Dimension,—$[m]$.

Intensity of Illumination (Properly called **Illumination**). Illumination in **lux** of a screen by a source of illuminating power P at a distance r meters, for normal incidence,

$$I = \frac{P}{r^2}$$

If two sources of illuminating power P_1 and P_2 produce equal illumination on a screen when at distances r_1 and r_2 respectively,

$$\frac{P_1}{r_1^2} = \frac{P_2}{r_2^2} \quad \text{or} \quad \frac{P_1}{P_2} = \frac{r_1^2}{r_2^2}$$

If I_o is the illumination when the screen is normal to the incident light, then I is the illumination when the screen is at an angle θ. Thus,

$$I = I_o \cos \theta$$

Intensity of magnetization is given by the quotient of the magnetic moment of a magnet by its volume. Unit intensity of magnetization is the intensity of a magnet which has unit magnetic moment per cubic centimeter. Dimensions,—$[\epsilon^{-\frac{1}{2}} m^{\frac{1}{2}} l^{-\frac{3}{2}}]$; $[\mu^{\frac{1}{2}} m^{\frac{1}{2}} l^{-\frac{1}{2}} t^{-1}]$.

Intensity of radiation is the rate of transfer of energy across unit areas by the radiation. In all forms of energy transfer by waves (radiation) the intensity I is given by $I = Uv$ where U is the energy density of the wave in the medium, and v is the velocity of propagation of the wave. The energy density U is always proportional to the square of the wave amplitude.

Intensity of sound depends upon the energy of the wave motion. The intensity is measured by the energy in ergs transmitted per second through one square centimeter of surface. The energy in ergs per cm^3 in a sound wave is given by

$$E = 2\pi^2 dn^2 a^2$$

where d is density in g per cm^3, n is frequency in vib. per sec and a is amplitude in cm. The energy reaching the ear in unit time will also be proportional to the velocity of propagation.

Iodine number (value).—A number expressing the percentage (i.e. grams per 100 grams) of iodine absorbed by a substance. It is a measure of the proportion of unsaturated linkages present and is usually determined in the analysis of oils and fats.

Ion.—An ion is an atom or group of atoms that is not electrically neutral but instead carries a positive or negative electric charge. Positive ions are formed when neutral atoms or molecules lose valence electrons; negative ions are those which have gained electrons.

Ionization potential.—The work (expressed in electron volts) required to remove a given electron from its atomic orbit and place it at rest at an infinite distance. It is customary to list values in electron volts (ev.) 1 ev. = 23,053 calories per mole.

Isobars.—For chemistry, elements of the same atomic mass but of different atomic numbers. The sum of their nucleons is the same but there are more protons in one than in the other.

Isomerism.—Existence of molecules having the same number and kinds of atoms but in different configurations.

Isothermal.—When a gas passes through a series of pressure and volume variations without change of temperature the changes are called isothermal. A line on a pressure-volume diagram representing these changes is called an isothermal line.

Isotopes.—Two or more nuclides having the same atomic number, hence constituting the same element, but differing in mass number. Isotopes of a given element have the same number of nuclear protons but differing numbers of neutrons. Naturally occurring chemical elements are usually mixtures of isotopes so that observed (non-integer) atomic weights are average values for the mixture.—See *Atomic number*

Joule-Thomson effect.—The cooling which occurs when a highly compressed gas is allowed to expand in such a way that no external work is done is known as the Joule-Thomson effect. This cooling is inversely proportional to the square of the absolute temperature.

DEFINITIONS AND FORMULAS (Continued)

Kepler's laws.
I. The planets move about the sun in ellipses, at one focus of which the sun is situated.

II. The radius vector joining each planet with the sun describes equal areas in equal times.

III. The cubes of the mean distances of the planets from the sun are proportional to the squares of their times of revolution about the sun.

Kerr effect.—When plane polarized light is incident on the pole of an electromagnet, polished so as to act like a mirror, the plane of polarization of the reflected light is not the same when the magnet is "on" as when it is "off." It was found that the direction of rotation was opposite to that of the currents exciting the pole from which the light was reflected.

Kinetic theory, expression for pressure.

$$P = \tfrac{1}{3}Nmv^2$$

where N is the number of molecules in unit volume, m the mass of each molecule and v^2 the mean square of the velocity of the molecules.

Kinetic theory of gases.—Gases are considered to be made up of minute, perfectly elastic particles which are ceaselessly moving about with high velocities, colliding with each other and with the walls of the containing vessel. The pressure exerted by a gas is due to the combined effect of the impacts of the moving molecules upon the walls of the containing vessel, the magnitude of the pressure being dependent upon the kinetic energy of the molecules and their number.

Kirchhoff's laws.
I. The algebraic sum of the currents which meet at any point is zero.

II. In any closed circuit the algebraic sum of the products of the current and the resistance in each conductor in the circuit is equal to the electromotive force in the circuit.

Kirchhoff's laws of radiation.—The relation between the powers of emission and the powers of absorption for rays of the same wave-length is constant for all bodies at the same temperature. First, a substance when excited by some means or other possess a certain power of emission; it tends to emit definite rays, whose wave-lengths depend upon the nature of the substance and upon the temperature. Second, the substance exerts a definite absorptive power, which is a maximum for the rays it tends to emit. Third, at a given temperature the ratio between the emissive and the absorptive power for a given wave-length is the same for all bodies, and is equal to the emissive power of a perfectly black body.

Kohlrausch's law.—When ionization is complete, the conductivity of an electrolyte is equal to the sum of the conductivities of the ions into which the substance dissociates.

Kundt's law.—On approaching an absorption band from the red side of the spectrum the refractive index is abnormally increased by the presence of the band, while the approach is from the blue side and the index is abnormally decreased.

Lambert is the unit of brightness equal to $1/\pi$ candle per square centimeter.

Lambert's law of absorption.—Each layer of equal thickness absorbs an equal fraction of the light which traverses it.

Lambert's law of illumination.—The illumination of a surface on which the light falls normally from a point source is inversely proportional to the square of the distance of the surface from the source. If the normal to the surface makes an angle with the direction of the rays, the illumination is proportional to the cosine of that angle.

Lanthanide series (Lanthanides).—Rare earth elements of atomic numbers 57 through 71, which have chemical properties similar to lanthanum (#57).

Latent heat of vaporization.—The quantity of heat necessary to change one gram of liquid to vapor without change of temperature, measured as calories per gram. Dimensions,—$[l^2 t^{-2}]$.

Lattice energy.—The energy required to separate the ions of a crystal to an infinite distance from each other.

LeChatelier's principle.—If some stress is brought to bear upon a system in equilibrium, a change occurs, such that the equilibrium is displaced in a direction which tends to undo the effect of the stress.

Length, units of.—Meter.1. (Abbr m) The basic unit of length of the metric system, defined as 1,650,763.73 wavelengths in vacuo of the unperturbed transition $2p^{10}$-$5d^5$ in krypton^{86}.

Effective 1 July 1959 in the U. S. system of measures, 1 yard = 0.9144 meter, exactly, or 1 meter = 1.094 yards = 39.37 inches. The standard inch is exactly 25.4 millimeters.

Lenses.—For a single thin lens whose surfaces have radii of curvature r_1 and r_2 whose principal focus is F, the index of refraction n, and conjugate focal distances f_1 and f_2,

$$\frac{1}{F} = \frac{1}{f_1} + \frac{1}{f_2} = (n-1)\left(\frac{1}{r_1} + \frac{1}{r_2}\right)$$

For a thick lens, of thickness t,

$$F = \frac{nr_1r_2}{(n-1)[n(r_1+r_2) - t(n-1)]}$$

Combinations of lenses.—If f_1 and f_2 are the focal lengths of two thin lenses separated by a distance d the focal length of the system,

$$F = \frac{f_1 f_2}{f_1 + f - d}$$

Lenz's law.—When an electromotive force is induced in a conductor by any change in the relation between the conductor and the magnetic field, the direction of the electromotive force is such as to produce a current whose magnetic field will oppose the change.

Line of force.—A term employed in the description of an electric or magnetic field. A line such that its direction at every point is the same as the direction of the force which would act on a small positive charge (or pole) placed at that point. A line of force is defined as starting from a positive charge (or pole) and ending on a negative charge (or pole).

The line (of force) is also used as a unit of magnetic flux, equivalent to the maxwell.

Lissajous figures.—The path described by a particle which is simultaneously displaced by two simple harmonic motions at right angles, when the periods of the two motions are in the ratio of two small whole numbers, shows a variety of characteristic curves called Lissajous figures.

Liquid.—A state of matter in which the molecules are relatively free to change their positions with respect to each other but restricted by cohesive forces so as to maintain a relatively fixed volume.

Loschmidt's number.—The number of molecules per unit volume of an ideal gas at 0°C and normal atmospheric pressure.

$$n_0 = (2.68719 \pm 0.0001) \times 10^{19}\ \text{cm}^{-3}$$

Loudness.—The psychological response of the ear which is related to the physical quantity *intensity*. The loudness of a sound depends on frequency also since the ear responds more strongly to some frequency bands than to others. The loudness is roughly related to the cube root of the intensity, and for many purposes it is convenient to represent loudness as proportional to the logarithm of the intensity.

Lumen.—The lumen is the unit of luminous flux. It is equal to the luminous flux through a unit-solid angle (steradian) from a uniform point source of one candle, or to the flux on a unit surface all points of which are at unit distance from a uniform point source of one candle.

Luminous flux.—The total visible energy emitted by a source per unit time is called the total luminous flux from the source. The unit of flux, the **lumen,** is the flux emitted in unit solid angle (steradian) by a point source of one candle luminous intensity. A uniform point source of one candle intensity thus emits 4π lumens.

Luminous intensity or candle-power is the property of a source of emitting luminous flux and may be measured by the luminous flux emitted per unit solid angle. The accepted unit of luminous intensity is the **international candle.** The **Hefner unit,** which is equivalent to 0.9 international candles, is the intensity of a lamp of specified design burning amyl acetate, called the Hefner lamp.

The mean horizontal candle-power is the average intensity measured in a horizontal plane passing through the source. The mean spherical candle-power is the average candle-power measured in all directions and is equal to the total luminous flux in lumens divided by 4π.

Magnetic field due to a current.—The intensity of the magnetic field in oersted at the center of a circular conductor of radius r in which a current I in absolute electromagnetic units is flowing,

$$H = \frac{2\pi I}{r}$$

If the circular coil has n turns the magnetic intensity at the center is,

$$H = \frac{2\pi n I}{r}$$

The magnetic field in a long solenoid of n turns per centimeter carrying a current I in absolute electromagnetic units

$$H = 4\pi n I$$

If I is given in amperes the above formulae become,—

$$H = \frac{2\pi I}{10r}, \quad H = \frac{2\pi n I}{10r}, \quad H = \frac{4\pi n I}{10}$$

Magnetic field due to a magnet.—At a point on the magnetic axis prolonged, at a distance r cm from the center of the magnet of length $2l$ whose poles are $+m$ and $-m$ and magnetic moment M, the field strength in oersted is,

$$H = \frac{4mlr}{(r^2 - l^2)^2}$$

If r is large compared with l,

$$H = \frac{2M}{r^3}$$

At a point on a line bisecting the magnet at right angles, with corresponding symbols,

$$H = \frac{2ml}{(r^2 + l^2)^{\frac{3}{2}}}$$

For large value of r,

$$H = \frac{M}{r^3}$$

Magnetic field intensity or magnetizing force.—Is measured by the force acting on unit pole. Unit field intensity, the oersted, is that field which exerts a force of one dyne on unit magnetic pole. The field intensity is also specified

DEFINITIONS AND FORMULAS (Continued)

by the number of lines of force intersecting unit area normal to the field, equal numerically to the field strength in oersted. Magnetizing force is measured by the space rate of variation of magnetic potential and as such its unit may be the **gilbert per centimeter.** The gamma (γ) is equivalent to 0.00001 oersted. Dimensions,—$[\epsilon^{\frac{1}{2}} m^{\frac{1}{2}} l^{\frac{1}{2}} t^{-2}]$; $[\mu^{-\frac{1}{2}} m^{\frac{1}{2}} l^{-\frac{1}{2}} t^{-1}]$.

Magnetic flux through any area perpendicular to a magnetic field is measured as the product of the area by the field strength. The units of magnetic flux, the **maxwell** is the flux through a square centimeter normal to a field of one gauss. The line is also a unit of flux. It is equivalent to the maxwell. Dimensions,—$[\epsilon^{-\frac{1}{2}} m^{\frac{1}{2}} l^{\frac{1}{2}}]$; $[\mu^{\frac{1}{2}} m^{\frac{1}{2}} l^{\frac{3}{2}} t^{-1}]$.

Magnetic induction resulting when any substance is subjected to a magnetic field is measured as the magnetic flux per unit area taken perpendicular to the direction of the flux. The unit is the maxwell per square centimeter or its equivalent, the gauss. Dimensions,—$[\epsilon^{-\frac{1}{2}} m^{\frac{1}{2}} l^{-\frac{3}{2}}]$; $[\mu^{\frac{1}{2}} m^{\frac{1}{2}} l^{-\frac{1}{2}} t^{-1}]$.

If a substance of permeability μ is placed in a magnetic field H the magnetic induction in the substance,

$$\mathbf{M} = \mu H.$$

If I is the magnetic moment for unit volume, or intensity of magnetization,

$$\mathbf{M} = H + 4\pi I.$$

The susceptibility,

$$\kappa = \frac{I}{H}, \qquad \mu = 1 + 4\pi\kappa.$$

Magnetic moment of a magnet is measured by the torque experienced when it is at right angles to a uniform field of unit intensity. The value of the magnetic moment is given by the product of the magnetic pole strength by the distance between the poles. Unit magnetic moment is that possessed by a magnet formed by two poles of opposite sign and of unit strength, one centimeter apart. Dimensions,—$[\mu^{\frac{1}{2}} m^{\frac{1}{2}} l^{\frac{5}{2}} t^{-1}]$; $[\epsilon^{-\frac{1}{2}} m^{\frac{1}{2}} l^{\frac{3}{2}}]$.

If the poles are separated by a distance which is great compared with the dimensions of the magnet, the magnetic moment of a magnet of length 1 whose poles have values of $+m$ and $-m$ is,

$$\mathrm{m} = ml.$$

Magnetic permeability is a property of materials modifying the action of magnetic poles placed therein and modifying the magnetic induction resulting when the material is subjected to a magnetic field or magnetizing force. The permeability of a substance may be defined as the ratio of the magnetic induction in the substance to the magnetizing field to which it is subjected. The permeability of a vacuum is unity. Dimensions,—$[\epsilon^{-1} l^{-2} t^2]$; $[\mu]$.

Magnetic pole or quantity of magnetism.—Two unit quantities of magnetism concentrated at points unit distance apart in a vacuum repel each other with unit force. If the distance involved is one centimeter and the force one dyne, the quantity of magnetism at each point is one cgs unit of magnetism. Dimensions,—$[\epsilon^{-\frac{1}{2}} m^{\frac{1}{2}} l^{\frac{1}{2}}]$; $[\mu^{\frac{1}{2}} m^{\frac{1}{2}} l^{\frac{3}{2}} t^{-1}]$.

Magnetic potential or magnetomotive force at a point is measured by the work required to bring unit positive pole from an infinite distance (zero potential) to the point. The unit is the *gilbert*, that magnetic potential against which an erg of work is done when unit magnetic pole is transferred. Dimensions,—$[\epsilon^{\frac{1}{2}} m^{\frac{1}{2}} l^{\frac{3}{2}} t^{-2}]$; $[\mu^{-\frac{1}{2}} m^{\frac{1}{2}} l^{\frac{1}{2}} t^{-1}]$.

Magnifying power of an optical instrument is the ratio of the angle subtended by the image of the object seen through the instrument to the angle subtended by the object when seen by the unaided eye. In the case of the microscope or simple magnifier the object as viewed by the unaided eye is supposed to be a distance of 25 cm (10 in.).

Mass.—Quantity of matter. **Units of mass**—the gram is $\frac{1}{1000}$ the quantity of matter in the International Prototype Kilogram; one of the three fundamental units of the cgs system. The British standard of mass is the pound, of which a standard is preserved by the government. The United States standard mass is the avoirdupois pound defined as 1/2.20462 kilogram.

Mass action, law of.—At a constant temperature the product of the active masses on one side of a chemical equation when divided by the product of the active masses on the other side of the chemical equation is a constant, regardless of the amounts of each substance present at the beginning of the action.

At constant temperature the rate of the reaction is proportional to the concentration of each kind of substance taking part in the reaction.

Mass by weighing on a balance with unequal arms.—If W_1 is the value for one side, W_2 the value for the other, the true mass,

$$W = \sqrt{W_1 W_2}.$$

Mass defect.—Difference between atomic mass and mass number of a nuclide.—See *Packing fraction*

Mass-energy equivalence.—The equivalence of a quantity of mass and a quantity of energy when the two quantities are related by the equation $E = mc^2$. The conversion factor c^2 is the square of the velocity of light. The relationship was developed from **relativity theory,** but has been experimentally confirmed.

Maxwell.—The cgs emu magnetic flux is the flux through a cm^2 normal to a field at 1 cm from a unit magnetic pole.

Maxwell's rule.—A law stating that every part of an electric circuit is acted upon by a force tending to move it in such a direction as to enclose the maximum amount of magnetic flux.

DEFINITIONS AND FORMULAS (Continued)

Mechanical equivalent of heat is the quantity of energy which, when transformed into heat, is equivalent to unit quantity of heat; 4.18×10^7 ergs = 1 calorie (20°C).

Meson.—Two types of particles of mass intermediate between that of the electron and proton have been discovered in cosmic radiation and in the laboratory. The one particle with mass about $215m_e$ is called μ-meson, the other with about $280m_e$ π-meson. Mesons of both positive and negative charge have been found and there is now reasonably good evidence for neutral mesons. Both types of mesons decay spontaneously. Some evidence exists for a meson of mass about $1000m_e$.

Metallic elements in general are distinguished from the non-metallic elements by their lustre, malleability, conductivity and usual ability to form positive ions. **Non-metallic** elements are not malleable, have low conductivity and never form positive ions.

Mev.—Mev = million electron volts.

Minimum deviation.—The deviation or change of direction of light passing through a prism is a minimum when the angle of incidence is equal to the angle of emergence. If D is the angle of minimum deviation and A the angle of the prism, the index of refraction of the prism for the wave length used is,

$$n = \frac{\sin \frac{1}{2}(A + D)}{\sin \frac{1}{2}A}$$

Mixtures consist of two or more substances intermingled with no constant percentage composition, and with each component retaining its essential original properties.

Moderator.—A material used for slowing down neutrons in an atomic pile or reactor. Usually graphite or "heavy water" (deuterium oxide).

Modulus of elasticity.—The stress required to produce unit strain, which may be a change of length (Young's modulus): a twist or shear (modulus of rigidity or modulus of torsion) or a change of volume (bulk modulus), expressed in dynes per square centimeter. Dimensions,—the same as of stress, $[m\, l^{-1} t^{-2}]$.

Mol volume.—The volume occupied by a mol or a gram molecular weight of any gas measured at standard conditions is 22.414 liters.

A **molal solution** contains one mole per 1000 grams of solvent.

A **molar solution** contains one mole or gram molecular weight of the solute in one liter of solution.

Mole.—Mass numerically equal to the molecular weight. It is most frequently expressed as the gram molecular weight, i. e. as the weight of one mole expressed in grams.

Molecular volume.—Volume occupied by one mole. Numerically equal to the molecular weight divided by the density.

Molecular weight.—The sum of the atomic weights of all the atoms in a molecule.

Molecule.—The smallest unit quantity of matter which can exist by itself and retain all the properties of the original substance.

Moment of force or torque.—The effectiveness of a force to produce rotation about an axis, measured by the product of the force and the perpendicular distance from the line of action of the force to the axis. Cgs unit—the dyne-centimeter. Dimensions,—$[m\, l^2 t^{-2}]$. If a force F acts to produce rotation about a center at a distance d from the line in which the force acts, the force has a torque,

$$L = Fd.$$

Moment of inertia.—A measure of the effectiveness of mass in rotation. In the rotation of a rigid body not only the body's mass, but the distribution of the mass about the axis of rotation determines the change in the angular velocity resulting from the action of a given torque for a given time. Moment of inertia in rotation is analogous to mass (inertia) in simple translation. The cgs unit is g-cm². Dimensions,—$[m\, l^2]$.

If m_1, m_2, m_3, etc. represent the masses of infinitely small particles of a body; r_1, r_2, r_3, etc. their respective distances from an axis of rotation, the moment of inertia about this axis will be

$$I = (m_1r_1^2 + m_2r_2^2 + m_3r_3^2 + \cdots)$$

or

$$I = \Sigma(mr^2)$$

Momentum.—Quantity of motion measured by the product of mass and velocity. Cgs unit—, one gram-centimeter per second. Dimensions,—$[m\, l\, t^{-1}]$.

A mass m moving with velocity v has a momentum,

$$M = mv.$$

If a mass m has its velocity changed from v_1 to v_2 by the action of a force F for a time t,

$$mv_2 - mv_1 = Ft.$$

Monochromatic emissive power is the ratio of the energy of certain defined wave lengths radiated at definite temperatures to the energy of the same wave lengths radiated by a black body at the same temperature and under the same conditions.

Monotropic.—Crystal forms one of which is always metastable with respect to the other.

Mosley's law.—The frequencies of the characteristic X-rays of the elements show a strict linear relationship with the square of the atomic number.

DEFINITIONS AND FORMULAS (Continued)

Motion, laws of.—See *Newton's law of motion*

Multiple proportions, law of.—If two elements form more than one compound, the weights of the first element which combine with a fixed weight of the second element are in the ratio of integers to each other.

Negatron.—1. A term used for electron when it is necessary to distinguish between (negative) electrons and positrons. 2. A four element vacuum tube which displays a negative resistance characteristic.

Nernst effect.—When heat flows across the lines of magnetic force, there is observed an electromotive force in the mutually perpendicular direction.

Neutralization is a reaction in which the hydrogen ion of an acid and the hydroxyl ion of a base unite to form water, the other product being a salt.

Neutrino.—An electrically neutral particle of very small (probably zero) rest mass and of spin quantum number $\frac{1}{2}$. When the spin is oriented parallel to the linear momentum the particle is the antineutrino. When the spin is oriented anti-parallel to the linear momentum the particle is the neutrino. Postulated by Pauli in explaining the beta decay process.

Whenever a beta (positron) particle is created in a radioactive decay so is an antineutrino (neutrino). The two particles and the parent nucleus share between them the available energy and momentum. Neutrinos and antineutrinos can penetrate amounts of matter measured in light years without appreciable attenuation. Detected by Reines and Cowan using antineutrinos from fission reactors and large scintillation detectors.

Neutron.—A neutral elementary particle of mass number 1. It is believed to be a constituent particle of all nuclei of mass number greater than 1. It is unstable with respect to beta-decay, with a half life of about 12 minutes. It produces no detectable primary ionization in its passage through matter, but interacts with matter predominantly by collisions and, to a lesser extent, magnetically. Some properties of the neutron are: rest mass, 1.00894 atomic mass unit; charge, 0; spin quantum number, $\frac{1}{2}$; magnetic moment, −1.9125 nuclear Bohr *magnetrons*.

Neutron cross section.—See *cross section*

Newton.—The force necessary to give acceleration of one meter per second per second to one kilogram of mass.

Newton's law of cooling.—The rate of cooling of a body under given conditions is proportional to the temperature difference between the body and its surroundings.

Newton's law of motion.

I. Every body continues in its state of rest or of uniform motion in a straight line except in so far as it may be compelled to change that state by the action of some outside force.

II. Change of motion is proportional to force applied and takes place in the direction of the line of action of the force.

III. To every action there is always an equal and opposite reaction.

Nodal points.—Two points on the axis of a lens such that a ray entering the lens in the direction of one, leaves as if from the other and parallel to the original direction.

A **normal salt** is an ionic compound containing neither replaceable hydrogen nor hydroxyl ions.

A **normal solution** contains one gram molecular weight of the dissolved substance divided by the hydrogen equivalent of the substance (that is, one gram equivalent) per liter of solution.

Nuclear atom.—The atom of each element consists of a small dense nucleus which includes most of the mass of the atom. The nucleus is made up of roughly equal numbers of neutrons and protons. The positive charges of the protons enables the nucleus to surround itself with a set of negatively charged electrons which move around the nucleus in complicated orbits with well defined energies. The outermost electrons which are least tightly bound to the nucleus play the dominant part in determining the physical and chemical properties of the atom. There are as many electrons in orbits as there are protons in the nucleus.

Nuclear fusion.—See *fusion*

Nuclear isomers.—Isotopes of elements having the same mass number and atomic number but differing in radioactive properties such as half-life period.

Nucleon.—Any particle found in the structure of an atom's nucleus. The most plentiful ones are neutrons and protons.

Nucleus.—The dense central core of the atom, in which most of the mass and all of the positive charge is concentrated. The charge on the nucleus, an integral multiple of Z of the electronic charge, is the essential factor which distinguishes one element from another. Z is called the atomic number and gives the number of protons in the nucleus, which includes a roughly equal number of neutrons. The mass number A gives the total number of neutrons plus protons.—See *Isotopes and Nuclear Atom*

Nuclide.—A species of atom distinguished by the constitution of its nucleus. The nuclear constitution is specified by the number of protons, Z; number of neutrons, N; and energy content. (Or, by the atomic number, Z; mass number A $(= N + Z)$ and atomic mass.)

Numerical aperture is the sine of half the angular aperture, used as a measure of the optical power of an objective.

Oersted.—The cgs emu of magnetic intensity exists at a point where a force of 1 dyne acts upon a unit magnetic pole at that point, i.e., the intensity 1 cm from a unit magnetic pole.

Ohm's law.—Current in terms of electromotive force E and resistance R.

$$I = \frac{E}{R}.$$

The current is given in amperes when E is in volts and R in ohms.

Organ pipes.—The frequency of vibration of a closed pipe or other air column of length l, where V is the velocity of sound in air, for the fundamental and first three overtones respectively is,

$$n_o = \frac{V}{4l}, \quad n_1 = \frac{3V}{4l}, \quad n_2 = \frac{5V}{4l}, \quad n_3 = \frac{7V}{4l}$$

For an open pipe,

$$n_o = \frac{V}{2l}, \quad n_1 = \frac{2V}{2l}, \quad n_2 = \frac{3V}{2l}, \quad n_3 = \frac{4V}{2l}$$

Oxidation is any process which increases the proportion of oxygen or acid-forming element or radical in a compound.

Packing fraction.—Packing fraction is the difference between the actual mass of the isotope and the nearest whole number divided by the mass number. Thus the packing fraction is equal to $(M - A)/A$ where M is the actual mass and A is the mass number. For example, one of the chlorine isotopes has a mass of 32.9860. The packing fraction for this isotope is

$$\frac{32.9860 - 33.0000}{33.0000} = -0.00042$$

Packing fractions are usually expressed as parts per 10,000 and so the packing fraction for this isotope of chlorine is written as -4.2. Since oxygen is taken as the standard, elements with positive packing fractions are less stable than oxygen, those with negative packing fractions are more stable.

It is positive for nuclides with mass number less than 16 or greater than 180, and negative for most others.

Paramagnetic materials.—Those within which an applied magnetic field is slightly increased by the alignment of electron orbits. The slight diamagnetic effect in materials having magnetic dipole moments is overshadowed by this paramagnetic alignment. As the temperature increases this paramagnetism disappears leaving only diamagnetism. The permeability of paramagnetic materials is slightly greater than that of empty space.

Pascal's law.—Pressure exerted at any point upon a confined liquid is transmitted undiminished in all directions.

Peltier effect.—When a current flows across the junction of two unlike metals it gives rise to an absorption or liberation of heat. If the current flows in the same direction as the current at the hot junction in a thermoelectric circuit of the two metals, heat is absorbed; if it flows in the same direction as the current at the cold junction of the thermoelectric circuit heat is liberated.

Pendulum.—For a simple pendulum of length l, for a small amplitude, the complete period,

$$T = 2\pi\sqrt{\frac{l}{g}} \quad \text{or} \quad g = 4\pi^2 \frac{l}{T^2}$$

T will be given in seconds if l is in cm and g in cm per sec². For a sphere suspended by a wire of negligible mass where d is the distance from the knife edge to the center of the sphere whose radius is r, the length of the equivalent simple pendulum,

$$l = d + \frac{2r^2}{5d}$$

If the period is P for an arc θ, the time of vibration in an infinitely small arc is approximately

$$T = \frac{P}{1 + \frac{1}{4}\sin^2\frac{\theta}{4}}$$

For a compound pendulum, if a body of mass m be suspended from a point about which its moment of inertia is I with its center of gravity a distance h below the point of suspension, the period

$$T = 2\pi\sqrt{\frac{I}{mgh}}$$

Period in uniform circular motion is the time of one complete revolution. In any oscillatory motion it is the time of a complete oscillation. Dimension,—$[t]$.

Periodic law.—Elements when arranged in the order of their atomic weights or atomic numbers show regular variations in most of their physical and chemical properties.

Permeance, the reciprocal of reluctance. Unit permeance is the permeance of a cylinder one square centimeter cross-section and one centimeter length taken in a vacuum. Dimensions—$[\epsilon^{-1} l^{-1} t^2]$; $[\mu l]$.

Phase of oscillatory motion.—The fraction of a whole period which has elapsed since the moving particle last passed through its middle position in a positive direction.

Photographic density.—The density D of silver deposit on a photographic plate or film is defined by the relation

$$D = \log O$$

where O is the opacity. If I_o and I are the incident and transmitted intensities respectively the opacity is given by I_o/I. The transparency is the reciprocal of the opacity or I/I_o.

Photon.—A photon (or γ-ray) is a quantum of electromagnetic radiation which has zero rest mass and an energy of h

(Planck's constant) times the frequency of the radiation. Photons are generated in collisions between nuclei or electrons and in any other process in which an electrically charged particle changes its momentum. Conversely photons can be absorbed (i.e., annihilated) by any charged particle.

Piezo-electric effect.—The phenomenon exhibited by certain crystals of expansion along one axis and contraction along another when subjected to an electric field. Conversely compression of certain crystals generate an electrostatic voltage across the crystal. Piezoelectricity is only possible in crystal classes which do not possess a center of symmetry.

Pinch effect.—When an electric current, either direct or alternating, passes through a liquid conductor, that conductor tends to contract in cross-section, due to *electromagnetic* forces.

Pitch.—Psychological response of the ear, primarily dependent upon the frequency of vibration of the air. The intensity of the sound also has a certain effect on the pitch. Pitch of a screw is the axial distance between adjacent turns of a single thread on the screw.

Planck's constant (h).—A universal constant of nature which relates the energy of a quantum of radiation to the frequency of the oscillator which emitted it. It has the dimensions of action (energy × time). Expressed by $E = h\nu$ where E is the energy of the quantum and ν is its frequency. Its numerical value is $6.626196\ (.50) \times 10^{-27}$ erg sec.

Plutonium.—A fissile element, artificially produced in the pile by neutron bombardment of U^{238}.

Poise.—A unit of coefficient of viscosity, defined as the tangential force per unit area (dynes/cm²) required to maintain unit difference in velocity (1 cm/sec) between two parallel planes separated by 1 cm of fluid;

$$1 \text{ poise} = 1 \text{ dyne sec/cm}^2 = 1 \text{ gm/cm sec.}$$

Poisson's ratio is the ratio of the transverse contraction per unit dimension of a bar of uniform cross-section to its elongation per unit length, when subjected to a tensile stress.

Polarized light.—Light which exhibits different properties in different directions at right angles to the line of propagation is said to be polarized. Specific rotation is the power of liquids to rotate the plane of polarization. It is stated in terms of specific rotation or the rotation in degrees per decimeter per unit density.

Polymorphism.—The ability to exist in two or more crystalline forms.

Positron.—A particle of the same mass M_e, as an ordinary electron. It has a positive electrical charge of exactly the same amount as that of an ordinary electron (which is sometimes called negatron). Positrons are created either by the radioactive decay of certain unstable nuclei or, together with a negatron, in a collision between an energetic (more than one Mev) photon and an electrically charged particle (or another photon). A positron does not decay spontaneously but on passing through matter it sooner or later collides with an ordinary electron and in this collision the positron-negatron pair is annihilated. The rest energy of the two particles, which is given by Einstein's relation $E = mc^2$ and amounts to 1.0216 mev altogether, is converted into electromagnetic radiation in the form of one or more photons.

Potential (electric) at any point is measured by the work necessary to bring unit positive charge from an infinite distance. Difference of potential between two points is measured by the work necessary to carry unit positive charge from one to the other. If the work involved is one erg we have the electrostatic unit of potential. Dimensions,—$[\epsilon^{-\frac{1}{2}}\, m^{\frac{1}{2}}\, l^{\frac{1}{2}}\, t^{-1}]$, $[\mu^{\frac{1}{2}}\, m^{\frac{1}{2}}\, l^{\frac{3}{2}}\, t^{-2}]$.

The potential at a point due to a charge q at a distance r in a medium whose dielectric constant is ϵ is,

$$V = \frac{q}{\epsilon r}$$

Power.—The time rate at which work is done. Units of power,—the watt, one joule (ten million ergs) per second; the kilowatt is equal to 1000 watts; the horse-power, 33,000 foot-pounds per minute, is equal to 746 watts. Dimensions,—$[m\, l^2\, t^{-3}]$.

If an amount of work W is done in time t the power or rate of doing work is

$$P = \frac{W}{t}$$

Power will be obtained in watts if W is expressed in joules (10^7 ergs) and t in sec.

Power in watts for alternating current,

$$P = EI \cos \phi$$

where E and I are the effective values of the electromotive force and current in volts and amperes respectively and ϕ the phase angle between the current and the impressed electromotive force.

The ratio,

$$\frac{P}{EI} = \cos \phi$$

is called the power factor.

Power developed by a direct current.—The power in watts developed by an electric current flowing in a conductor, where E is the difference of potential at its terminals in volts, R its resistance in ohms, and I the current in amperes,

$$P = EI = RI^2$$

The work done in joules in a time t sec is,

$$W = EIt{:}RI^2t.$$

DEFINITIONS AND FORMULAS (Continued)

Power ratios in telephone engineering are measured in **decibels.** The gain or loss of power expressed in decibels is ten times the logarithm of the power ratio. By reference to an arbitrarily chosen "power level" the actual power may be expressed in decibels. The numerical values thus used will not be proportional to the actual power level but roughly to the sensation on the ear produced when the electrical power is converted into sound. A difference of 1 decibel in the power supply to a telephone receiver produces approximately the smallest change in volume of sound which a normal ear can detect.

Pressure.—Force applied to, or distributed, over a surface; measured as force per unit area. Cgs unit,—the barye, one dyne per square centimeter. The megabarye is equal to 10^6 dynes per square centimeter. Pressure is also measured by the height of the column of mercury or water which it supports. Dimensions,—$[m\, l^{-1}\, t^{-2}]$.

The pressure due to a force F distributed over an area A,

$$P = \frac{F}{A}$$

Absolute pressure.—Pressure measured with respect to zero pressure. *Gauge pressure*—pressure measured with respect to that of the atmosphere.

Principal focus of a lens or spherical mirror is the point of convergence of light coming from a source at an infinite distance.

Projectiles.—For bodies projected with velocity v at an angle a above the horizontal, the time to highest point of flight.

$$t = \frac{v \sin a}{g}$$

Total time of flight to reach the original horizontal plane,

$$T = \frac{2v \sin a}{g}$$

Maximum height,

$$h = \frac{v^2 \sin^2 a}{2g}$$

Horizontal range,

$$R = \frac{v^2 \sin 2a}{g}$$

In the above equations the resistance of the air is neglected. g is the acceleration due to gravity.

Proton.—An elementary particle having a positive charge equivalent to the negative charge of the electron but possessing a mass approximately 1837 times as great. The proton is in effect the positive nucleus of the hydrogen atom.

Purkinje effect.—A phenomenon associated with the human eye, making it more sensitive to blue light when the illumination is poor (less than about 0.1 lumen per sq. ft.) and to yellow light when the illumination is good.

Quality or timbre of a sound depends on the coexistence with the fundamental of other vibrations of various frequencies and amplitudes.

Quantity of electricity or charge.—The electrostatic unit of charge, the quantity which when concentrated at a point and placed at unit distance from an equal and similarly concentrated quantity, is repelled with unit force. If the distance is one centimeter and force of repulsion one dyne and the surrounding medium a vacuum, we have the electrostatic unity of quantity. The electrostatic unit of quantity may be defined as that transferred by electrostatic unit current in unit time. The quantity transferred by one ampere in one second is the coulomb, the practical unit. The faraday is the electrical charge carried by one gram equivalent. The coulomb $= 3 \times 10^9$ electrostatic units. Dimensions,—

$$[\epsilon^{\frac{1}{2}}\, m^{\frac{1}{2}}\, l^{\frac{3}{2}}\, t^{-1}];\ [\mu^{-\frac{1}{2}}\, m^{\frac{1}{2}}\, l^{\frac{1}{2}}].$$

Quantum.—Unit quantity of energy postulated in the quantum theory. The *photon* is a quantum of the electromagnetic field, and in nuclear field theories, the *meson* is considered to be the quantum of the nuclear field.

Radiation.—The emission and propagation of energy through space or through a material medium in the form of waves.

The term may be extended to include streams of sub-atomic particles as alpha-rays, or beta-rays, and cosmic rays as well as electromagnetic radiation. Often used to designate the energy alone without reference to its character. In the case of light this energy is transmitted in bundles (photons).

Radiation formula, Planck's.—The emissive power of a black body at wave length λ may be written

$$E_\lambda = \frac{c_1 \lambda^{-5}}{e^{c_2/\lambda T} - 1}$$

where c_1 and c_2 are constants with c_1 being 3.7403×10^{10} microwatts microns^4 per cm^2 or 3.7403×10^{-12} watt cm^2, c_2 being 14384 micron degrees and T the absolute temperature.

Radioactive nuclides.—Atoms that disintegrate by emission of corpuscular or electromagnetic radiations. The rays most commonly emitted are alpha or beta or gamma rays. The three classes are:

1. Primary, which have half-life times exceeding 10^8 years. These may be alpha-emitters or beta-emitters.

2. Secondary, which are formed in radioactive transformations starting with U^{238}, U^{235}, or Th^{232}.

3. Induced, having geologically short lifetimes and formed by induced nuclear reactions occurring in nature. All these reactions result in transmutation.

DEFINITIONS AND FORMULAS (Continued)

Radius of gyration may be defined as the distance from the axis of rotation at which the total mass of a body might be concentrated without changing its moment of inertia. The product of total mass and the square of the radius of gyration will give (the) moment of inertia.

Rankine scale of temperature.—The absolute Fahrenheit scale.

$$°F + 459.69 = °R; \text{ thus } 0°F \text{ Rankine} = -459.69°F.$$

Raoult's law.—Molar weights of non-volatile non-electrolytes when dissolved in a definite weight of a given solvent under the same conditions lower the solvent's freezing point, elevate its boiling point and reduce its vapor pressure equally for all such solutes.

Reduction is any process which increases the proportion of hydrogen or base-forming elements or radicals in a compound. Reduction is also the gaining of electrons by an atom, an ion, or an element thereby reducing the positive valence of that which gained the electron.

Reflection coefficient or reflectivity is the ratio of the light reflected from a surface to the total incident light. The coefficient may refer to diffuse or to specular reflection. In general it varies with the angle of incidence and with the wave-length of the light.

Reflection of light by a transparent medium in air (Fresnel's formulae).—If i is the angle of incidence, r the angle of refraction, n_1 the index of refraction for air (nearly equal to unity), n_2 index of refraction for a medium, then the ratio of the reflected light to the incident light is,

$$R = \tfrac{1}{2}\left(\frac{\sin^2(i-r)}{\sin^2(i+r)} + \frac{\tan^2(i-r)}{\tan^2(i+r)}\right)$$

If $i = 0$ (normal incidence), and $n_1 = 1$ (approximate for air),

$$R = \left(\frac{n_2 - 1}{n_2 + 1}\right)^2$$

Refraction at a spherical surface.—If u be the distance of a point source, v the distance of the point image or the intersection of the refracted ray with the axis, n_1 and n_2 the indices of refraction of the first and second medium, and r the radius of curvature of the separating surface,

$$\frac{n_2}{v} + \frac{n_1}{u} = \frac{n_2 - n_1}{r}$$

If the first medium is air the equation becomes,

$$\frac{n}{v} + \frac{1}{u} = \frac{n - 1}{r}$$

Refractive index. – See Index of refraction; Snell's law.

Refractivity is given by $(n - 1)$ when n is the index of refraction; the **specific refractivity** is given by $\frac{n-1}{d}$ where d is the density. **Molecular refractivity** is the product of specific refractivity by the molecular weight.

Relative humidity.—The ratio of the quantity of water vapor present in the atmosphere to the quantity which would saturate at the existing temperature. It is also the ratio of the pressure of water vapor present to the pressure of saturated water vapor at the same temperature.

Reluctance is that property of a magnetic circuit which determines the total magnetic flux in the circuit when a given magnetomotive force is applied. Unit, the reluctance of one centimeter length and one square centimeter cross-section of space taken in a vacuum. Dimensions,—$[\epsilon l\, t^{-2}]$; $[\mu^{-1}\, l^{-1}]$.

Reluctivity or specific reluctance is the reciprocal of magnetic permeability. The reluctivity of empty space is taken as unity. Dimensions,—$[\epsilon l^2\, t^{-2}]$; $[\mu^{-1}]$.

Resistance is a property of conductors depending on their dimensions, material and temperature which determines the current produced by a given difference of potential. The practical unit of resistance, the **ohm** is that resistance through which a difference of potential of one volt will produce a current of one ampere. The **international ohm** is the resistance offered to an unvarying current by a column of mercury at 0°C, 14.4521 grams in mass, of constant cross-sectional area and 106.300 centimeters in length, sometimes called the legal ohm. Dimensions,—$[\epsilon^{-1}\, l^{-1}\, t]$; $[\mu l\, t^{-1}]$.

Resistance of a conductor at 0°C, of length l, cross-section s and specific resistance ρ

$$R_o = \rho \frac{l}{s}$$

The resistivity may be expressed as ohm-cm when R is in ohms, l in cm and s in cm².

Resistance of a conductor at a temperature t whose resistance at 0°C is R_o and whose temperature resistance coefficient is α

$$R_t = R_o(1 + \alpha t)$$

Resistance of conductors in series and parallel.—The total resistance of any number of resistances joined in series is the sum of the separate resistances. The total resistance of conductors in parallel whose separate resistances are r_1, r_2, r_3 . . r_n is given by the formula

$$\frac{1}{R} = \frac{1}{r_1} + \frac{1}{r_2} + \frac{1}{r_3} \cdots + \frac{1}{r_n}$$

Where R is the total resistance. For two terms this becomes,

$$R = \frac{r_1 r_2}{r_1 + r_2}$$

Resistance, specific (Resistivity).—A proportionality factor characteristic of different substances equal to the **resistance** that a centimeter cube of the substance offers to the passage of electricity, the current being perpendicular to two parallel faces. It is defined by the expression:

$$R = \rho \frac{l}{A}$$

where R is the resistance of a uniform conductor, l is its length, A is its cross sectional area, and ρ is its resistivity. Resistivity is usually expressed in ohm-centimeters.

Resolving power of a telescope or microscope is indicated by the minimum separation of two objects for which they appear distinct and separate when viewed through the instrument.

Resonance (chemical).—The moving of electrons from one atom of a molecule or ion to another atom of that molecule or ion. It is simply the oriented movement of the bonds between atoms.

Restitution, coefficient of, for two bodies on impact.—The ratio of the difference in velocity, after impact to the difference before impact.

Reversible reaction.—One which can be caused to proceed in either direction by suitable variation in the conditions of temperature, volume, pressure or of the quantities of reacting substances.

Roentgen.—A radiation unit, that amount of radiation that will produce one electrostatic unit of ions per cubic centimeter volume.

Rotatory power is the power of rotating the plane of polarized light, given in general by θ/l where θ is the total rotation which occurs in a distance l.

The **molecular** or **atomic rotatory power** is the product of the specific rotatory power by the molecular or atomic weight. Magnetic rotatory power is given by

$$\theta/e\ H \cos \alpha$$

where H the intensity of the magnetic field, and α the angle between the field and the direction of the light.

Rydberg formula.—A formula, similar to that of Balmer, for expressing the wave-numbers (ν) of the lines in a spectral series:

$$\nu = R\left[\frac{1}{(n+a)^2} - \frac{1}{(m+b)^2}\right]$$

where n and m are integers and $m > n$, a and b are constants for a particular series, and R is the *Rydberg constant*, 109737.3 cm^{-1} for hydrogen.

Salt.—Any substance which yields ions, other than hydrogen or hydroxyl ions. A salt is obtained by displacing the hydrogen of an acid by a metal.

Scintillometer.—An instrument which detects radiation by emitting flashes of light.

Seebeck effect.—If a circuit consists of two metals, one junction hotter than the other, a current flows in the circuit. The direction of the flow depends on the metals and the temperature of the junctions.

Sensitiveness of a balance.—Assuming the three knife edges of a balance to lie on a straight line,—if M is the weight of the beam, h the distance of the center of gravity below the knife edge, L the length of the balance arms and m a small mass added to one pan, the deflection θ produced is given by

$$\tan \theta = \frac{mL}{Mh}.$$

Shell.—According to *Pauli's exclusion principle* (q.v.), the extranuclear electrons do not circle around the nucleus all in orbits of the same radius, but are arranged in orbits at various distances from the nucleus. The extranuclear orbital electrons are thus assumed to be arranged in a series of concentric spheres, called *shells*, which are designated, in the order of increasing distance from the nucleus, as K, L, M, N, O, P, and Q shells. The number of the electrons which each of these shells can contain is limited. All electrons arranged in the same shell have the same principal quantum number. The electrons in the same shell are grouped into various subshells (q.v.), and all the electrons in the same subshell have the same orbital angular momentum.—See *Subshell*

Simple harmonic motion.—Periodic oscillatory motion in a straight line in which the restoring force is proportional to the displacement. If a point moves uniformly in a circle, the motion of its projection on the diameter (or any straight line in the same plane) is simple harmonic motion.

If r is the radius of the reference circle, ω the angular velocity of the point in the circle, θ the angular displacement at the time t after the particle passes the mid-point of its path, the linear displacement,

$$x = r \sin \theta = r \sin \omega t.$$

The velocity at the same instant,

$$v = r\omega \cos \theta = \omega \sqrt{r^2 - x^2}$$

The acceleration,

$$a = -\omega^2 x.$$

The force for a mass m,

$$F = m\omega^2 x = -\frac{4\pi^2 mx}{T^2}$$

The period,

$$T = 2\pi \sqrt{\frac{x}{a}}$$

In the above equations the cgs system calls for x and r in cm, v in cm per sec, a in cm per sec², T in sec, m in grams, F in dynes, θ in radians and ω in radians per sec.

Simple machine.—A contrivance for the transfer of energy and for increased convenience in the performance of work.

Mechanical advantage is the ratio of the resistance overcome to the force applied. Velocity ratio is the ratio of the distance through which force is applied to the distance through which resistance is overcome.

Efficiency is the ratio of the work done by a machine to the work done upon it.

If a force f applied to a machine through a distance S results in a force F exerted by the machine through a distance s, neglecting friction,

$$fS = Fs.$$

The theoretical mechanical advantage or velocity ratio in the above case is, $\frac{S}{s}$.

Actually the force obtained from the machine will have a smaller value than will satisfy the equation above. If F' be the actual force obtained, the practical mechanical advantage will be, $\frac{F'}{f}$.

The efficiency of the machine,

$$E = \frac{F's}{fS}.$$

Snell's law of refraction.—If i is the angle of incidence and r the angle of refraction, then for a wave or particle undergoing reversible refraction between isotropic media

$$\frac{\sin i}{\sin r} = \frac{n'}{n} = \frac{u}{u'} = \frac{p'}{p},$$

where, in the medium of incidence, n is the refractive index, u is the phase velocity, and p is the generalized momentum. The prime implies the medium of refraction. (See Refractive index and Velocity.)

Solid.—A state of matter in which the relative motion of the molecules is restricted and they tend to retain a definite fixed position relative to each other, giving rise to crystal structure. A solid may be said to have a definite shape and volume.

Solid angle.—Measured by the ratio of the surface of the portion of a sphere enclosed by the conical surface forming the angle, to the square of the radius of the sphere. Unit of solid angle,—the steradian, the solid angle which encloses a surface on the sphere equivalent to the square of the radius. Dimensions,—unity.

Solubility of one liquid or solid in another is the mass of a substance contained in a solution which is in equilibrium with an excess of the substance. Under these conditions the solution is said to be saturated. Solubility of a gas is the ratio of concentration of gas in the solution to the concentration of gas above the solution.

Solubility product or precipitation value is the product of the concentrations of the ions of a substance in a saturated solution of the substance. These concentrations are frequently expressed as moles of solute per liter of solution.

Solute.—That constituent of a solution which is considered to be dissolved in the other, the solvent. The solvent is usually present in larger amount than the solute.

A **solution** is **saturated** if it contains at given temperature as much of a solute as it can retain in the presence of an excess of that solute.

A **true solution** is a mixture, liquid, solid or gaseous, in which the components are uniformly distributed throughout the mixture. The proportion of the constituents may be varied within certain limits.

Solvent.—That constituent of a solution which is present in larger amount; or, the constituent which is liquid in the pure state, in the case of solutions of solids or gases in liquids.

Specific gravity.—The ratio of the mass of a body to the mass of an equal volume of water at 4°C or other specified temperature. Dimensions,—unity.

Specific heat of a substance is the ratio of its thermal capacity to that of water at 15°C. Dimensions,—unity.

If a quantity of heat H calories is necessary to raise the temperature of m grams of a substance from t_1 to t_2°C, the specific heat, or more properly, thermal capacity of the substance,

$$s = \frac{H}{m(t_2 - t_1)}.$$

Specific heat by the method of mixtures.—Where a mass m_1 of the substance is heated to a temperature t_1, then placed in a mass of water m_2 at a temperature t_2 contained in a calorimeter with stirrer (of same material) of mass m_3, specific heat of the calorimeter c, t_3 the final temperature

$$m_1 s(t_1 - t_3) = (m_3 c + m_2)(t_3 - t_2).$$

Black's ice calorimeter.—If a body of mass m and temperature t melts a mass m' of ice, its temperature being reduced to 0°C, the specific heat of the substance is,

$$s = \frac{80\ 1m'}{mt}$$

Bunsen's ice calorimeter.—A body of mass m at temperature t causes a motion of the mercury column of l centimeters in a tube whose volume per unit length is v. The specific heat is

$$s = \frac{884lv}{mt}$$

Specific inductive capacity.—The ratio of the capacitance of a condenser with a given substance as dielectric to the

DEFINITIONS AND FORMULAS (Continued)

capacitance of the same condenser with **air or a** vacuum as dielectric is called the specific inductive capacity. The ratio of the dielectric constant of a substance to that of a vacuum.

Specific rotation.—If there are n grams of active substance in v cubic centimeters of solution and the light passes through l decimeters, r being the observed rotation in degrees, the specific rotation (for 1 centimeter),

$$[\alpha] = \frac{rv}{nl}$$

Specific volume is the reciprocal of density. Dimensions,—$[m^{-1}\,l^3]$.

Spectral series are spectral lines or groups of lines which occur in an orderly sequence.

Speed.—Time rate of motion measured by the distance moved over in unit time. Cgs unit,—one centimeter per second. Dimension,—$[l\,t^{-1}]$.

Spherical aberration.—When large surfaces of spherical mirrors or lenses are used the light divergent from a point source is not exactly focused at a point. The phenomenon is known as spherical aberration. For axial pencils the error is known as axial spherical aberration; for oblique pencils, coma.

Spherical mirrors.—If R is the radius of curvature, F principal focus, and f_1 and f_2 any two conjugate focal distances,

$$\frac{1}{f_1} + \frac{1}{f_2} = \frac{1}{F} = \frac{2}{R}$$

If the linear dimensions of the object and image be O and I respectively and u and v their distances from the mirror,

$$\frac{O}{I} = \frac{u}{v}$$

Spin.—In nuclear physics, used to describe the angular momentum of elementary particles or of nuclei.

Standard conditions for gases.—Measured volumes of gases are quite generally recalculated to 0 degrees C temperature and 760 mm pressure, which have been arbitrarily chosen as standard conditions.

Stark effect.—The splitting of a single spectrum line into multiple lines which occurs when the emitting material is placed in a strong electric field. The observed effect depends on the angle between the direction of the field and the direction of observation. The effect is due to the shifting of the energy states of certain orbits which all have the same energy in zero field.

Statcoulomb.—The unit of electric charge in the metric system. 3×10^9 statcoulombs = 1 coulomb.

Stationary or standing waves are produced in a medium by the simultaneous transmission, in opposite directions of two similar wave motions. Fixed points of minimum amplitude are called **nodes.** A **segment** extends from one node to the next. An **antinode** or **loop** is the point of maximum amplitude between two nodes.

Stefan-Boltzmann law of radiation.—The energy radiated in unit time by a black body is given by, $E = K(T^4 - T_o^4)$, where T is the absolute temperature of the body, T_o the absolute temperature of the surroundings, and K a constant.

Stoichiometric.—Pertaining to weight relations in chemical reactions.

Stoke.—See under viscosity.

Stokes' law.—1. Gives the rate of fall of a small sphere in a viscous fluid. When a small sphere falls under the action of gravity through a viscous medium it ultimately acquires a constant velocity,

$$V = \frac{2ga^2(d_1 - d_2)}{9\eta}$$

where a is the radius of the sphere, d_1 and d_2 the densities of the sphere and the medium respectively, and η the coefficient of viscosity. V will be in cm per sec if g is in cm per sec^2, a in cm, d_1 and d_2 in g per cm^3 and η in dyne-sec per cm^2 or poises.

2. The empirical law stating that the wavelength of light emitted by a fluorescent material is longer than that of the radiation used to excite the fluorescence. In modern language the emitted photons carry off less energy than is brought in by the exciting photons; the details accord with the energy conservation principle.

Strain.—The deformation resulting from a stress measured by the ratio of the change to the total value of the dimension in which the change occurred. Dimensions,—unity.

Stress.—The force producing or tending to produce deformation in a body measured by the force applied per unit area Cgs units,—one dyne per square centimeter. Dimensions,—$[m\,l^{-1}\,t^{-2}]$.

Subshell.—The electrons within the same shell (energy level) of the atom are characterized by the same principal quantum number (n), and are further divided into groups according to the value of their azimuthal quantum numbers (l); the electrons which possess the same azimuthal quantum number for the same principal quantum number are considered to occupy the same subshell (or sublevel). The individual subshells are designated with the letters s, p, d, f, g, and h, as follows:

l value	designation of subshell
0	s
1	p
2	d
3	f
4	g
5	h

An electron assigned to the *s*-subshell is called an *s*-electron, one assigned to the *p*-subshell is referred to as a *p*-electron, etc. In formulae of electron structure, the value of the principal quantum number (n) is prefixed to the letter indicating the azimuthal quantum number (l) of the electron; thus, e.g., a $4f$-electron is an electron which has the principal quantum number 4 (i.e. assigned to the *N*-shell) and the orbital angular momentum 3 (*f*-subshell).

Surface density of electricity.—Quantity of electricity per unit area. Dimensions,—$[\epsilon^{\frac{1}{2}}\, m^{\frac{1}{2}}\, l^{-\frac{3}{2}}\, t^{-1}]$; $[\mu^{-\frac{1}{2}}\, m^{\frac{1}{2}}\, l^{-\frac{5}{2}}]$.

Surface density of magnetism.—Quantity of magnetism per unit area. Dimensions,—$[\epsilon^{-1}\, m^{\frac{1}{2}}\, l^{-\frac{1}{2}}]$, $[\mu^{\frac{1}{2}}\, m^{\frac{1}{2}}\, l^{-\frac{3}{2}}\, t^{-1}]$.

Surface tension.—Two fluids in contact exhibit phenomena, due to molecular attractions which appear to arise from a tension in the surface of separation. It may be expressed as dynes per cm or as ergs per square centimeter. Dimensions, —$[m\, t^{-2}]$.

The total force along a line of length l on the surface of a liquid whose surface tension is T,

$$F = lT.$$

Capillary tubes.—If a liquid of density d rises a height h in a tube of internal radius r the surface tension is,

$$T = \frac{rhdg}{2}$$

The tension will be in dynes per cm if r and h are in cm, d in g per cm³ and g in cm per sec².

Drops and bubbles.—Pressure in dynes per cm² due to surface tension on a drop of radius r cm for a liquid whose surface tension is T dynes per cm, $P = \frac{2T}{r}$

For a bubble of mean radius r cm,

$$P = \frac{4T}{r}$$

Susceptibility (magnetic) is measured by the ratio of the intensity of magnetization produced in a substance to the magnetizing force or intensity of field to which it is subjected. The susceptibility of a substance will be unity when unit intensity of magnetization is produced by a field of one gauss. Dimensions,—$[\epsilon^{-1}\, l^{-2}\, t^{2}]$, $[\mu]$.

Tangent galvanometer.—A tangent galvanometer with n turns, of radius r, in the earth's field H, has a deflection θ. The current flowing is, $i = \frac{Hr}{2\pi n} \tan \theta$

If $\frac{2\pi n}{r} = G$ (the galvanometer constant).

$$i = \frac{H}{G} \tan \theta$$

Temperature may be defined as the condition of a body which determines the transfer of heat to or from other bodies. Particularly it is a manifestation of the average translational kinetic energy of the molecules of a substance due to heat agitation.

The scientific unit of temperature is the **Kelvin** (K), a more customary unit is the degree **Celsius** (°C). The magnitude of 1K is equal to that of 1°C, both being 1/100 of the difference between the temperature of boiling water and that of melting ice (both under standard atmospheric pressure). The latter temperature is also the zero point of the Celsius scale; while the zero point of the Kelvin scale is at the "absolute zero point" at about −273.15°C. Therefore, the Kelvin scale is also called the absolute scale. The word "centigrade" instead of "degree Celsius" is confusing (because the prefix "centi" is applied in it in a meaning different from that in centimeter, centiliter, etc.) and should not be used. In the Kelvin scale, the temperature measure is based on the average kinetic energy per molecule of a perfect (ideal) gas.

The size of the degree **Fahrenheit** (°F) and that of the degree Rankine are both 1/180 of the ice-boiling temperature difference, with the zero point of the Fahrenheit scale being at −32°C and that of the Rankine scale at about −273.15°C = −459.67°F. The boiling point is 212°F = 100°C. The size of the (nearly outmoded) degree Reaumur (°R) is that of 1/80 of the ice-boiling temperature difference and 0°R = 0°C.

The fundamental temperature scale is now defined by means of the equation

$$\theta(X) = 273.15°K \frac{X}{X_3}$$

where θ denotes the temperature; X the thermometric property (P, V, ...); the subscript 3 refers to the triple point of water; and 273.16°K is the arbitrary fixed point for the temperature associated with the triple point of water.

The ideal gas temperature θ, (numerically equal to the Kelvin temperature), in particular, is defined by either of the two equations:

$$\theta = \begin{cases} 273.15° \lim\limits_{P_3 \to 0} \dfrac{P}{P_3}, \text{ const. V} \\[2ex] 273.15° \lim\limits_{P \to 0} \dfrac{V}{V_3}, \text{ const. P} \end{cases}$$

Temperature resistance coefficient.—The ratio of the change of resistance in a wire due to a change of temperature of 1°C to its resistance at 0°C. Dimension,—$[\theta^{-1}]$.

Thermal capacity of a substance is the quantity of heat necessary to produce unit change of temperature in unit mass.

DEFINITIONS AND FORMULAS (Continued)

It is ordinarily expressed as calories per gram per degree Centigrade. Numerically equivalent to specific heat.

Thermal capacity or water equivalent.—The total quantity of heat necessary to raise any body or system unit temperature, measured as calories per degree centigrade in the cgs system. Dimension,—$[m]$.

Thermal expansion.—The coefficient of linear expansion or expansivity is the ratio of the change in length per degree C to the length at 0°C. The coefficient of volume expansion (for solids) is approximately three times the linear coefficient. The coefficient of volume expansion for liquids is the ratio of the change in volume per degree to the volume at 0°C. The value of the coefficient varies with temperature. The coefficient of volume expansion for a gas under constant pressure is nearly the same for all gases and temperatures and is equal to 0.00367 for 1°C. Dimension,—$[\theta^{-1}]$.

If l_o is the length at 0°C, α the coefficient of linear expansion, the length at t°C is,

$$l_t = l_o(1 + \alpha t).$$

General formula for thermal expansion.—The rate of thermal expansion varies with the temperature. The general equation giving the magnitude m_t (length or volume) at a temperature t, where m_o is the magnitude at 0°C, is

$$m_t = m_o(1 + \alpha t + \beta t^2 + \gamma t^3 \cdot \cdot \cdot)$$

where α, β, γ, etc. are empirically determined coefficients.

Volume expansion.—If V represents volume and β the coefficient of expansion,

$$V_t = V_o(1 + \beta t).$$

For solids,

$$\beta = 3\alpha \text{ (approximately).}$$

Thermal neutrons.—Neutrons slowed down by a moderator to an energy of a fraction of an electron volt—about 0.025 ev. at 15°.

Thermionic emission.—Electron or ion emission due to the temperature of the emitter. The rate of emission increases rapidly with the increase of temperature. It is also very sensitive to the state of the surface.

Thermodynamics, law of.

I. When mechanical work is transformed into heat or heat into work, the amount of work is always equivalent to the quantity of heat.

II. It is impossible by any continuous self-sustaining process for heat to be transferred from a colder to a hotter body.

Thermoelectric power is measured by the electromotive force produced by a thermocouple for unit difference of temperature between the two junctions. It varies with the average temperature and is usually expressed in microvolts per degree C. It is customary to list the thermoelectric power of the various metals with respect to lead. Dimensions,—$[\epsilon^{-\frac{1}{2}} m^{\frac{1}{2}} l^{\frac{1}{2}} t^{-1} \theta^{-1}]$; $[\mu^{\frac{1}{2}} m^{\frac{1}{2}} t^{-1} \theta^{-1}]$.

Thomson thermoelectric effect is the designation of the potential gradient along a conductor which accompanies a temperature gradient. The magnitude and direction of the potential varies with the substance.

The coefficient of the Thomson effect or specific heat of electricity is expressed in joules per coulomb per degree Centigrade. Dimensions,—$[\epsilon^{-\frac{1}{2}} m^{\frac{1}{2}} l^{\frac{1}{2}} t^{-1} \theta^{-1}]$, $[\mu^{\frac{1}{2}} m^{\frac{1}{2}} l^{\frac{3}{2}} t^{-2} \theta^{-1}]$.

Time, unit of.—The fundamental invariable unit of time is the ephemeris second, which is defined as 1/31,556,925.9747 of the tropical year for 1900 January 0^d12^h ephemeris time. The ephemeris day is 86,400 ephemeris seconds.

The former unit of time was the mean solar second, defined as 1/86,400 of the mean solar day. (See supplementary list).

Torque produced by the action of one magnet on another.—The turning moment experienced by a magnet of pole strength m' and length $2l'$ placed at a distance r from another magnet of length $2l$ and pole strength m, where the center of the first magnet is on the axis (extended) of the second and the axis of the first is perpendicular to the axis of the second,

$$C = 8\frac{mm'll'}{r^3} = \frac{2MM'}{r^3}$$

If the first magnet is deflected through an angle θ, the expression becomes,

$$C = \frac{2MM'}{r^3} \cos \theta$$

Torsional vibration.—See *Angular harmonic motion*

Total reflection.—When light passes from any medium to one in which the velocity is greater, refraction ceases and total reflection begins at a certain critical angle of incidence θ such that

$$\sin \theta = \frac{1}{n}$$

where n is the index of the first medium with respect to the second. If the second medium is air n has the ordinary value for the first medium. For any other second medium,

$$n = \frac{n_1}{n_2}$$

where n_1 and n_2 are the ordinary indices of refraction for the first and second medium respectively.

Tractive force of a magnet.—If a magnet with induction B has a pole face of area A the force, is,

$$F = \frac{B^2A}{8\pi}$$

If B and A are in cgs units, F will be in dynes.

DEFINITIONS AND FORMULAS (Continued)

Transmutation.—A nuclear change producing a new element from an old one.

Transuranic elements.—Elements of atomic numbers above 92. All of them are radioactive and are products of artificial nuclear changes. All are members of the actinide group.

Triangle or polygon of forces.—If three or more forces acting on the same point are in equilibrium, the vectors representing them form, when added, a closed figure.

Tritium.—An isotope of hydrogen with a mass of three, structure, two neutrons, and one proton in its nucleus.

Uncertainty principle.—See *Indeterminancy*

Uniform circular motion.—If r is the radius of a circle, v the linear speed in the arc, ω the angular velocity and T the period or time of one revolution,

$$\omega = \frac{v}{r} = \frac{2\pi}{T}$$

The acceleration toward the center is

$$a = \frac{v^2}{r} = \omega^2 r = \frac{4\pi^2 r}{T^2}$$

The centrifugal force for a mass m,

$$F = \frac{mv^2}{r} = m\omega^2 r = \frac{4\pi^2 mr}{T^2}$$

In the above equations ω will be in radians per second and a in cm per sec² if r is in cm, v in cm per sec and T in sec. F will be in dynes if mass is in grams and other units as above.

Application to the solar system.—If M is the mass of the sun, G the constant of gravitation, P the period of the planet and r the distance of the planet from the sun, then the mass of the sun

$$M = \frac{4\pi^2 r^3}{GP^2}, \ (G = 6.670 \times 10^{-8} \text{ for cgs units}).$$

If P is the period and r the distance of a satellite revolving around the planet, the above expression for M gives the mass of the planet. The formula is written on the assumption that the orbit of the planet or satellite is circular, which is only approximately true.

Uniformly accelerated rectilinear motion.—If v_o is the initial velocity, v_t the velocity after time t, the acceleration,

$$a = \frac{v_t - v_o}{t}$$

The velocity after time t,

$$v_t = v_o + at$$

Space passed over in time t,

$$s = v_o t + \tfrac{1}{2}at^2$$

Velocity after passing over space s,

$$v = \sqrt{v_o^2 + 2as}$$

Space passed over in the nth second

$$s = v_o + \tfrac{1}{2}a(2n - 1).$$

In the above and following similar equations the values of the space, velocity, and acceleration must be substituted in the same system. For space in *cm*, velocity will be in *cm* per sec and acceleration in *cm* per sec per sec.

Unit.—Specific magnitude of a quantity, set apart by appropriate definition, which is to serve as a basis of comparison or measurement for other quantities of the same nature.

Valence of an atom of an element is that property which is measured by the number of atoms of hydrogen (or its equivalent) one atom of that element can hold in combination if negative, or can displace in a reaction if it is positive.

Valence electrons of the atom are electrons which are gained, lost or shared in chemical reactions.

Van der Waals' equation of state.—This equation is expressed by:

$$\left(p + \frac{a}{v^2}\right)(v - b) = RT$$

It makes allowance both for the volume occupied by the molecules and for the attractive force between the molecules. b is the effective volume of molecules in one mole of gas. a is a measure of the attractive force between the molecules. For values of R, a, and b see index for table of Van der Waal's constants for gasses.

Van't Hoff's principle.—If the temperature of interacting substances in equilibrium is raised, the equilibrium concentrations of the reaction are changed so that the products of that reaction which absorb heat are increased in quantity, or if the temperature for such an equilibrium is lowered, the products which evolve heat in their formation are increased in amounts.

Vapor.—The words **vapor** and **gas** are often used interchangeably. **Vapor** is more frequently used for a substance which, though present in the gaseous phase, generally exists as a liquid or solid at room temperature. **Gas** is more frequently used for a substance that generally exists in the gaseous phase at room temperature. Thus one would speak of iodine or carbon tetrachloride vapors and of oxygen gas.

Vapor pressure.—The pressure exerted when a solid or liquid is in equilibrium with its own vapor. The vapor pressure is a function of the substance and of the temperature.

DEFINITIONS AND FORMULAS (Continued)

Vectors, composition of.—If the angle between two vectors is A, and their magnitude a and b, their resultant,

$$C = \sqrt{a^2 + b^2 - 2ab\cos A}.$$

Velocity.—Time rate of motion in a fixed direction. Cgs units,—one centimeter per second. Dimensions,—$[l\,t^{-1}]$.

If s is space passed over in time t, the velocity,

$$\bar{v} = \frac{s}{t}$$

Velocity of a compressional wave.—The velocity of a compressional wave in an elastic medium, in terms of elasticity E (bulk modulus) and density d,

$$V = \sqrt{\frac{E}{d}}$$

For the velocity of sound in air, where p is the pressure and d the density,

$$V = \sqrt{\frac{1.4p}{d}}$$

Velocity of efflux of a liquid.—If h is the distance from the opening to the free surface of the liquid, the velocity of efflux is

$$V = \sqrt{2gh}$$

The above is the theoretical discharge velocity disregarding friction and the shape of orifice. For water issuing through a circular opening with sharp edges of area, A, the volume discharged per second is given approximately by,

$$Q = 0.62A\,\sqrt{2gh}$$

Velocity of sound, variation with temperature.—The velocity in meters per sec at any temperature t in °C is given approximately by

$$V = V_o\sqrt{1 + \frac{t}{273}}$$

$$V = 331.5 + .607t$$

The **variation with humidity** is given by the equation

$$V_d = V_h\sqrt{1 - \frac{e}{p}\left(\frac{\gamma_\omega}{\gamma_a} - \frac{5}{8}\right)}$$

where V_d is the velocity in dry air, V_h that in air at barometric pressure p in which the pressure of water vapor is e. γ_ω and γ_a are the specific heat ratios for water vapor and for air respectively.

Velocity, phase and group. – A particle of energy E and generalized momentum p or a wave of frequency ν, angular frequency $\omega = 2\pi\nu$, wavelength λ, and wave or propagation vector k = $2\pi/\lambda$ has, when undergoing reversible motion in an isotropic medium, a phase velocity $u = \omega/k = E/p$ and a group or particle velocity $v = \partial\omega/\partial k = \partial E/\partial p$. Energy and signals are transported at the group velocity in the absence of dissipation, whereas the phase velocity is not the propagation velocity of any physical quantity. The de Broglie relations state that $p = h/\lambda$ and $E = h\nu$ where h is Planck's constant; thus any particle may be regarded as a wave, and conversely.

Velocity of a transverse wave in a stretched cord. If T is the tension of the cord and m the mass per unit length,

$$V = \sqrt{\frac{T}{m}}$$

Velocity of water waves.—If the depth h is small compared with the wave length, the velocity,

$$V = \sqrt{gh}$$

In deep water for a wave length λ,

$$V = \sqrt{\frac{g\lambda}{2\pi}}$$

If the wavelength is very small, less than about 1.6 cm, the velocity increases as the wave length decreases and is expressed by the following,

$$V = \sqrt{\frac{2\pi T}{\lambda d} + \frac{g\lambda}{2\pi}}$$

where T is the surface tension and d the density of the liquid V will be given in cm per sec if h and λ are in cm, g in cm per sec², T in dynes per cm and d in g per cm³.

Velocity of a wave.—The velocity of propagation in terms of wavelength λ and the period T or frequency n is,

$$V = \frac{\lambda}{T} = n\lambda$$

Viscosity.—All fluids possess a definite resistance to change of form and many solids show a gradual yielding to forces tending to change their form. This property, a sort of internal friction, is called viscosity; it is expressed in dyne-seconds per cm² or poises. Dimensions,—$[m\,l^{-1}\,t^{-1}]$. If the tangential force per unit area, exerted by a layer of fluid upon one adjacent is one dyne for a space rate of variation of the tangential velocity of unity, the viscosity is one poise.

Kinematic viscosity is the ratio of viscosity to density. The c. g. s. unit of kinematic viscosity is the **stoke.**

Flow of liquids through a tube; where l is the length of the tube, r its radius, p the difference of pressure at the ends, η the coefficient of viscosity, the volume escaping per second,

$$v = \frac{\pi p r^4}{8 l \eta} \text{ (Poiseuille).}$$

The volume will be given in cm^3 per second if l and r are in cm, p in dynes per cm^2 and η in poises or dyne-seconds per cm^2.

Visibility is measured by the ratio of the luminous flux in lumens to the total radiant energy in ergs per second or in watts.

Volt.—The unit of electromotive force. It is the difference in potential required to make a current of one ampere flow through a resistance of one ohm.

Volume, unit of.—The cubic centimeter, the volume of a cube whose edges are one centimeter in length. Other units of volume are derived in a similar manner. Dimension,—$[l^3]$.

Wave motion.—A progressive disturbance propagated in a medium by the periodic vibration of the particles of the medium. Transverse wave motion is that in which the vibration of the particles is perpendicular to the direction of propagation. Longitudinal wave motion is that in which the vibration of the particles is parallel to the direction of propagation.

Weight.—The force with which a body is attracted toward the earth. Cgs unit,—the dyne. Dimensions,—$[m\, l\, t^{-2}]$.

Although the weight of a body varies with its location, the weights of various standards of mass are often used as units of force as,—pound weight, or pound force, gram weight, etc. The weight of mass m, where g is the acceleration due to gravity,

$$W = mg.$$

The weight will be given in dynes when m is in grams and g in cm per sec^2.

Wheatstone's bridge.—If the resistances r_1, r_2, r_3, and r_4 form the arms of a Wheatstone's bridge in order as the circuit (omitting cell and galvanometer connections) is traced, when the bridge is balanced,

$$\frac{r_1}{r_2} = \frac{r_4}{r_3} \quad \text{or} \quad \frac{r_1}{r_4} = \frac{r_2}{r_3}$$

Wien's displacement law.—When the temperature of a radiating black body increases, the wave length corresponding to maximum energy decreases in such a way that the product of the absolute temperature and wave length is constant.

$$\lambda_{max} T = w$$

w is known as **Wien's displacement constant.**

Work.—When a force acts against resistance to produce motion in a body the force is said to do work. Work is measured by the product of the force acting and the distance moved through against the resistance. Cgs units of work,—the erg, a force of one dyne acting through a distance of one centimeter. The joule is 1×10^7 ergs. Dimensions,—$[m\, l^2\, t^{-2}]$. The foot-pound is the work required to raise a mass of one pound a vertical distance of one foot where $g = 32.174$ ft./sec^2. The foot-poundal is the work done by a force of one poundal acting through a distance of one foot. The International joule, a unit of electrical energy, is the work expended per second by a current of one International ampere flowing through one International ohm. The kilowatt-hour is the total amount of energy developed in one hour by a power of one kilowatt.

If a force F act through a space s, the work done is

$$W = Fs$$

Work will be given in ergs if F is in dynes and s in cm.

Work done in rotation. If a torque L dyne-cm acts through an angle θ radians, the work done in ergs is

$$W = L\theta$$

X-rays.—A type of radiation of higher frequency than visible light but lower than gamma rays. Usually produced by high energy electrons impinging upon a metal target.

X units.—X-ray wavelengths have been measured in two kinds of units. The older measurements are given in X units (XU) which are based on the effective lattice constant of rock salt being 2,814.00 XU. More recently X-ray wavelengths have been directly connected, through measurements with ruled gratings, to the wavelengths in the optical region and through them to the standard meter. It turned out that the XU which was originally intended as 10^{-11} cm was 0.202 per cent larger than this value. It has become customary to give X-ray wavelengths in Angstrom units (Å) when the absolute scale is used (1 Å = 10^{-8} cm). The two are related by

$$1{,}000 \text{ XU} = 1.00202 \pm 0.00003) \text{ Å}$$

and wavelengths given in XU must be multiplied by 1.00202 and then divided by 1,000 in order to convert them into Angstrom units.

Zeeman effect.—The splitting of a spectrum line into several symmetrically disposed components, which occurs when the source of light is placed in a strong magnetic field. The components are polarized, the directions of polarization and the appearance of the effect depending on the direction from which the source is viewed relative to the lines of force.

Supplementary Tables

Units for a System of Measures for International Relations

Length	meter	m
Mass	kilogram	kg
Time	second	s
Electric current	ampere	A
Temperature	Kelvin	K
Luminous intensity	candela	cd

Prefix Names of Multiples and Submultiples of Units

Power of 10	Prefix	Symbol
−18	atto	a
−15	femto	f
−12	pico	p
−9	nano	n
−6	micro	μ
−3	milli	m
−2	centi	c
−1	deci	d
+1	deca	da
+2	hecto	h
+3	kilo	k
+6	mega	M
+9	giga	G
+12	tera	T
+15	exa	E
+18	peta	P

Derived Units

Acceleration	meter per second squared	m/s^2	
Activity (of radioactive source)	1 per second	s^{-1}	
Angular acceleration	radian per second squared	rad/s^2	
Angular velocity	radian per second	rad/s	
Area	square meter	m^2	
Density	kilogram per cubic meter	kg/m^3	
Dynamic viscosity	newton-second per sq meter	$N{\cdot}s/m^2$	
Electric capacitance	farad	F	(A·s/V)
Electric charge	coulomb	C	(A·s)
Electric field strength	volt per meter	V/m	
Electric resistance	ohm	Ω	(V/A)
Entropy	joule per kelvin	J/K	
Force	newton	N	$(kg{\cdot}m/s^2)$
Frequency	hertz	Hz	(s^{-1})
Illumination	lux	lx	(lm/m^2)
Inductance	henry	H	(V·s/A)
Kinematic viscosity	sq meter per second	m^2/s	
Luminance	candela per sq meter	cd/m^2	
Luminous flux	lumen	lm	(cd·sr)
Magnetomotive force	ampere	A	
Magnetic field strength	ampere per meter	A/m	
Magnetic flux	weber	Wb	(V·s)
Magnetic flux density	tesla	T	(Wb/m^2)
Power	watt	W	(J/s)
Pressure	newton per square meter (Pascal)	N/m^2	
Radiant intensity	watt per steradian	W/sr	
Specific heat	joule per kilogram kelvin	J/kg K	
Thermal conductivity	watt per meter kelvin	W/m K	
Velocity	meter per second	m/s	
Volume	cubic meter	m^3	
Voltage, Potential difference, Electromotive force	volt	V	(W/A)
Wave number	1 per meter	m^{-1}	
Work, energy, quantity of heat	joule	J	(N·m)

Supplementary Tables

Definition of Most Important International System (SI) Units

The *ampere* (unit of electric current) is the constant current which, if maintained in two straight parallel conductors of infinite length, of negligible circular sections, and placed 1 meter apart in a vacuum, will produce between these conductors a force equal to 2×10^{-7} newton per meter of length.

Antiferromagnetic Materials: Those in which the magnetic moments of atoms or ions tend to assume an ordered arrangement in zero applied field, such that the vector sum of the moments is zero, below a characteristic temperature called the Néel Point. The permeability of antiferromagnetic materials is comparable to that of paramagnetic materials. Above the Néel Point, these materials become paramagnetic.

Candela. The candela is the luminous intensity, in the direction of the normal, of a black body surface 1/600,000 square meter in area, at the temperature of solidification of platinum under a pressure of 101,325 newtons per square meter.

Compensation Point: The temperature (below the Néel Point) at which, in some ferrimagnetic compounds, the saturation magnetization becomes zero.

The *coulomb* (unit of quantity of electricity) is the quantity of electricity transported in 1 second by a current of 1 ampere.

Diamagnetic Materials: Those within which the magnetic induction is slightly less than the applied magnetic field. Diamagnetism is an atomic scale consequence of the Lenz law of induction. For diamagnetic materials, the permeability is slightly less than that of empty space, and the magnetic susceptibility is negative and small.

The ephemeris *second* (unit of time) is exactly 1/31 556 925.974 7 of the tropical year of 1900, January, 0 days, and 12 hours ephemeris time.

The *farad* (unit of electric capacitance) is the capacitance of a capacitor between the plates of which there appears a difference of potential of 1 volt when it is charged by a quantity of electricity equal to 1 coulomb.

Ferrimagnetic Materials: Those in which the magnetic moments of atoms or ions tend to assume an ordered but nonparallel arrangement in zero applied field, below a characteristic temperature called the Néel Point. In the usual case, within a magnetic domain, a substantial net magnetization results from the antiparallel alignment of neighboring *nonequivalent* sublattices. The macroscopic behavior is similar to that in ferromagnetism. Above the Néel Point, these materials become paramagnetic.

Ferromagnetic Materials: Those in which the magnetic moments of atoms or ions in a magnetic domain tend to be aligned parallel to one another in zero applied field, below a characteristic temperature called the Curie Point. Complete ordering is achieved only at the absolute zero of temperature. Within a magnetic domain, at absolute zero, the magnetization is equal to the sum of the magnetic moments of the atoms or ions per unit volume. Bulk matter, consisting of many small magnetic domains, has a net magnetization which depends upon the magnetic history of the specimen (hysteresis effect). The permeability depends on the magnetic field, and can reach values of the order of 10^6 times that of free space. Above the Curie Point, these materials become paramagnetic.

The *henry* (unit of electric inductance) is the inductance of a closed circuit in which an electromotive force of 1 volt is produced when the electric current in the circuit varies uniformly at a rate of 1 ampere per second.

International Practical Kelvin Temperature Scale of 1960 and the *International Practical Celsius Temperature Scale* of 1960 are defined by a set of interpolation equations based on the following reference temperatures:

	°K	°C
Oxygen, liquid-gas equilibrium	90.18	−182.97
Water, solid-liquid equilibrium	273.15	0.00
Water, solid-liquid-gas equilibrium	273.16	0.01
Water, liquid-gas equilibrium	373.15	100.00
Zinc, solid-liquid equilibrium	692.655	419.505
Sulphur, liquid-gas equilibrium	717.75	444.6
Silver, solid-liquid equilibrium	1233.95	960.8
Gold, solid-liquid equilibrium	1336.15	1063.0

The *joule* (unit of energy) is the work done when the point of application of 1 newton is displaced a distance of 1 meter in the direction of the force.

Kelvin: The kelvin, the unit of thermodynamic temperature, is the fraction 1/273.16 of the thermodynamic temperature of the triple point of water. The decision was made at the 13th General Conference on Weights and Measures on October 13, 1967 that the name of the unit of thermodynamic temperature would be changed from *degree Kelvin* (symbol: °*K*) to *kelvin* (symbol: *K*). The name (*kelvin*) and symbol (*K*) are to be used for expressing temperature intervals. The former convention which expressed a temperature interval in *degrees Kelvin* or, abbreviated, *deg. K* is dropped. However, the old designations are acceptable temporarily as alternatives to the new ones. One may also express temperature intervals in *degrees Celsius.*

The *kilogram* (unit of mass) is the mass of a particular cylinder of platinum- iridium alloy, called the International Prototype Kilogram, which is preserved in a vault at Sèvres, France, by the International Bureau of Weights and Measures.

Length: The name "micron", for a unit of length equal to 10^{-6} meter, and the symbol "μ" which has been used for it were dropped by action of the 13th General Conference on Weights and Measures on October 13, 1967. The symbol "μ" is to be used solely as an abbreviation for the prefix "micro-", standing for the multiplication by 10^{-6}. Thus the length previously designated as 1 micron, should be designated 1 μm.

The *lumin* (unit of luminous flux) is the luminous flux emitted in a solid angle of 1 steradian by a uniform point source having an intensity of 1 candela.

Magnetic Anisotropy: In ferro- or ferrimagnetic crystals, it is found that the magnetization prefers to lie along certain crystal directions. These are termed *easy directions* of magnetization. Work must be expended to turn the magnetization away from these easy directions. That work as a function of crystal direction defines the *anisotropy energy surface.* Directions associated with a maximum of the anisotropy energy are termed *hard directions* of magnetization. In general, the energy difference between easy and hard directions decreases as the temperature is increased, and vanishes at the Curie or Néel point.

Magnetic Domains: The magnetization of a ferromagnetic or a ferrimagnetic material tends to break up into regions called *domains* separated by thin transition regions called *domain walls.* Within the volume of a domain, the magnetization has its saturation value, and is directed along a single direction. The magnetizations of other domains are directed along different directions in such a way that the net magnetization of the whole sample may be zero. The application of an external magnetic field first causes some domains to grow by the motion of their walls. At higher fields, the magnetizations of the resulting domains rotate toward parallelism with the field.

Magnetostriction: Change in sample dimensions as the magnitude or the direction of the magnetization in a crystal is changed.

Metamagnetic Materials: Those which are antiferromagnetic in weak fields, but which become ferromagnetically ordered in strong applied fields.

The *meter* (unit of length) is the length of exactly 1 650 763.73 wavelengths of the radiation in vacuum corresponding to the unperturbed transition between the levels $2p_{10}$ and $5d_5$ of the atom of Krypton 86, the orange-red line.

Néel Point: The temperature at which ferrimagnetic and antiferromagnetic materials become paramagnetic.

The *newton* (unit of force) is that force which gives to a mass of 1 kilogram an acceleration of 1 meter per second per second (m/sec^2). 1 newton/m^2 = 1 pascal.

The *ohm* (unit of electric resistance) is the electric resistance between two points of a conductor when a constant difference of potential of 1 volt, applied between these two points, produces in this conductor a current of 1 ampere, this conductor not being the source of any electromotive force.

Paramagnetic Materials: Those within which the magnetic induction is slightly greater than the applied magnetic field. Paramagnetism arises from the partial alignment of the permanent magnetic dipole moments of atoms or ions. The permeability is slightly greater than that of empty space, and the magnetic susceptibility is positive and small.

Second: The second is the unit of time of the International System of Units. The definition adopted at the October 13, 1967 meeting of the 13th General Conference on Weights and Measures is: "The second is the duration of 9,192,631,770 periods of the radiation corresponding to the transition between the two hyperfine levels of the fundamental state of the atom of cesium 133." The frequency (9,192,631,770 Hz) which the definition assigns to the cesium radiation was carefully chosen to make it impossible, by any existing experimental evidence, to distinguish the new second from the "ephemeris second" based on the earth's motion. Therefore, no changes need to be made in data stated in terms of the old standard in order to convert them to the new one. The atomic definition has two important advantages over the previous definition: (1) it can be realized (i.e., generated by a suitable clock) with sufficient precision, ±1 part per hundred billion (10^{11}) or better, to meet the most exacting demands of modern metrology; and (2) it is available to anyone who has access to or who can build an atomic clock controlled by the specified cesium radiation. (A description of such clocks is given in "Atomic Frequency Standards," *NBS Tech. News Bull.* 45, 8–11 (Jan., 1961). For more recent developments and technical details, see R. E. Beehler, R. C. Mockler, and J. M. Richardson, "Cesium Beam Atomic Time and Frequency Standards," *Metrologia* 1, 114–131 (July, 1965)). In addition one can compare other high-precision clocks directly with such a standard in a relatively short time—an hour or so compared against years with the astronomical standard. Laboratory-type atomic clocks are complex and expensive, so that most clocks and frequency generators will continue to be calibrated against a standard such as the NBS Frequency Standard, controlled by a cesium atomic beam, at the Radio Standards Laboratory in Boulder, Colorado. In most cases the comparison will be by way of the standard-frequency and time-interval signals broadcast by NBS radio stations WWV, WWVH, WWVB, and WWVL.

The *volt* (unit of electric potential difference and electromotive force) is the difference of electric potential between two points of a conducting wire carrying a constant current of 1 ampere, when the power dissipated between these points is equal to 1 watt.

The *watt* (unit of power) is the power which gives rise to the production of energy at the rate of 1 joule per second.

The *weber* (unit of magnetic flux) is the magnetic flux which, linking a circuit of one turn, produces in it an electromotive force of 1 volt as it is reduced to zero at a uniform rate in 1 second.

Definitions of the most important SI Units are given in the above Supplementary Tables. These definitions have been extracted from the records of the International Committee and the General Conferences.

DEFINITIONS FOR ELECTROLYTIC SOLUTIONS

Conductance, σ.—The conductance of a conductor of electricity is the reciprocal of its electrical resistance (R) and its unit is the reciprocal "absolute" ohm, ohm^{-1}, or mho.

Debye-Falkenhagen effect.—The increase in the conductance of an electrolytic solution produced by alternating currents of sufficiently high frequencies over that observed with low frequencies or with direct current.

Degree of association, $(1 - \alpha)$.—The degree of association of an electrolytic solution is the percentage of ions associated into nonconducting species, such as ion-pairs. (See ionophores).

Degree of dissociation (or ionization) in general, α.—The degree of dissociation (or ionization) of an electrolytic solution is the percentage of solute (or electrolyte) in the dissociated (or ionized) state in solution. Classically this degree is obtained from conductance measurements from the ratio, Λ/Λ_i where Λ_i is the equivalent conductance an electrolytic solution would have at some *finite* concentration if it were completely dissociated into ions at that concentration. (See ionogens). This symbol is also used to denote the fraction of free ions in a solution when simple ions, ion pairs, and clusters higher than ion pairs are present. (See ionophores).

Dissociation-field effect.—The increased dissociation (or ionization) of the molecules of weak electrolytes under the influence of high electrical fields (potential gradients).

Electrolytic cell constant, J_c.—The cell constant of an electrolytic cell is the resistance in ohms of that cell when filled with a liquid of unit resistance.

Electrophoretic effect.—The slowing down, owing to interionic attraction and repulsion, of the movement of an ion with its solvent molecules in the forward direction by ions of opposite charge with their solvent molecules moving in the reverse direction under an applied electrical field (potential gradient).

Equivalent conductance, Λ.—The equivalent conductance of an electrolytic solution is the conductance of the amount of solution that contains one gram-equivalent of a solute (or electrolyte) when measured between parallel electrodes which are one centimeter apart and large enough in area to include the necessary volume of solution. Equivalent conductance is numerically equal to the conductivity multiplied by the volume in cubic centimeters containing one gram-equivalent of the electrolyte. The unit of equivalent conductance is $ohm^{-1}\ cm^2\ equiv^{-1}$ (frequently, in the literature the unit is given simply, although incorrectly, as ohm^{-1}, so that it may be comparable to the unit for conductance, in general).

Interionic attraction.—The electrostatic attraction between ions of unlike charge (sign).

Interionic repulsion.—The electrostatic repulsion between ions of like charge (sign).

Ion atmosphere (or continuous charge distribution).—In the electrostatic effects between ions the term ion atmosphere denotes a continuous charge distribution, or charge density, $\rho\ (r)$, which is a continuous function of r, the distance from the reference ion, rather than a discrete or discontinuous charge distribution. The ion atmosphere extends from $r = a$ to $r = 0(V^{1/3}) \approx \infty$, where V is the volume of the system, and acts electrostatically somewhat like a sphere of charge $-e$ at a distance, κ^{-1}, from the reference ion of charge $+e$ (see below for definition of κ^{-1}).

Ion size or "ion-size" parameter, a (or a_i).—The ion size is formally considered to be the sum of the ionic radii of the oppositely charged ions in contact. The ion size is also called the "distance of closest approach" of the ions, or the "ion-size" parameter. Generally the ion size is greater than the sum of the crystal radii, and the "ion-size" parameter may include several factors which contribute to its numerical value.

Ionic equivalent conductance, λ.—The ionic equivalent conductance is the equivalent conductance of an individual ion constituent of the solute (or electrolyte) of an electrolytic solution. This symbol is also used to designate the equivalent conductance of complex ions, ion pairs, ion clusters, etc., in combination with simple ions.

Ionic mobility, u.—The mobility of an ion at any finite equivalent concentration is the velocity with which the ion moves under unit potential gradient. Its unit is $cm^2\ sec^{-1}\ volt^{-1}\ equiv^{-1}$ or $cm^2\ ohm^{-1}\ F^{-1}$ where F is the Faraday expressed in coulombs (or ampere seconds) $equiv^{-1}$.

Ionogens.—Substances, like acetic acid (HAc), which, although in the pure state are nonelectrolytic neutral molecules, can react with certain solvents to form products which rearrange to ion pairs which then dissociate to give conducting solutions. As an example:

$$HAc + H_2O \rightleftarrows HAc \cdot H_2O \rightleftarrows H_3O^+ \cdot Ac^- \rightleftarrows H_3O^+ + Ac^-$$

Ionophores.—Substances, like sodium chloride, which exist only as ionic lattices in the pure crystalline form, and which when dissolved in an appropriate solvent give conductances which change according to some fractional power of the concentration. Such solutions possess no neutral molecules which can dissociate, but may contain associated ions.

Kohlrausch law of independent migration of ions.—The value of the equivalent conductance, as the concentration approaches zero, is equal to the sum of the limiting ionic equivalent conductances of the ions constituting the solute of the electrolytic solution.

Limiting equivalent conductance, Λ_0.—The limiting equivalent conductance of an electrolytic solution, Λ_0, is expressed by $\Lambda_0 \equiv \lim_{c \to 0} (\sigma_{corr}/c)$ where σ_{corr} is solution conductance corrected for solvent conductance and c is the equivalent concentration. Λ_0 is the value which Λ approaches as the solution is diluted so far that the effects of interionic

forces become negligible (and dissociation, in the case of ionogens, is essentially complete).

Limiting ionic equivalent conductance, λ_0.—The limiting ionic equivalent conductance of an individual ion constituent of the solute (or electrolyte) of an electrolytic solution is given by $\lambda_0 \equiv \lim_{c \to 0} (\lambda/c)$. This symbol is also used to designate the limiting equivalent conductances of complex ions, ion pairs, ion clusters, etc., in combination with simple ions.

Limiting ionic mobility, u^0.—The limiting mobility of an individual ion of a solute (or electrolyte) is given by $u^\circ \equiv \lim_{c \to 0} u$.

Limiting molar conductance, Λ_0^m.—The limiting molar conductance of an electrolytic solution, Λ_0^m, is expressed by $\Lambda_0^m \equiv \lim_{m \to 0} (\sigma_{corr}/m)$ where σ_{corr} is solution conductance corrected for solvent conductance and m is the molar concentration. Λ_0^m is the value which Λ^m approaches as the solution is diluted so far that the effects of interionic forces become negligible. Seldomly used.

Molar conductance, Λ^m.—The molar conductance of an electrolytic solution is the conductance of a solution containing one gram mole of the solute (or electrolyte) when measured in a like manner to equivalent conductance. Seldomly used.

Osmotic-pressure effect.—An enhancement in the velocity of the central ion, in the direction of the applied external field, as a result of more collisions on the central ion from ions behind the central ion than from ions in front of it.

Relaxation-field effect.—The delay in the ion atmosphere in maintaining its symmetry around a central ion as the central ion moves in the forward direction under an applied electrical field (potential gradient).

Specific conductance, σ_{sp}.—The specific conductance, or conductivity, of a conductor of electricity is the conductance of the material between opposite sides of a cube, one centimeter in each direction. The unit of specific conductance is $ohm^{-1}\ cm^{-1}$ or mho cm^{-1}.

Thickness or average radius of ion atmosphere, κ^{-1}.—The average distance of the ion atmosphere from the reference ion in angstrom units. This average distance decreases in magnitude with the square root of the ionic concentration. Mathematically, κ^{-1} is the distance at which the average charge, $d\bar{q}$, in a spherical shell of volume $4\pi r^2 dr$ reaches a maximum using the continuous density, $\rho\ (r)$, approximation.

Transference (or transport) number, t.—The transference number of each ion of a solute (or electrolyte) in an electrolytic solution is the fraction of the total current carried by that ion, and is given by the ratio of the mobility of the ion to the sum of the mobilities of the ions of the solute constituting the electrolytic solution.

Viscosity effect.—An alteration in the velocity of a given ion as a result of the contribution to the bulk viscosity owing to the ions of opposite charge. This effect applies to ions of large size.

Walden's rule, $\Lambda_0 \eta_0$.—Walden's rule states that the product of the limiting equivalent conductance of an electrolytic solution, Λ_0, and the viscosity of the solvent, η_0, in which the solute (or electrolyte) is dissolved is a constant at a particular temperature. Walden's rule is an approximation which would be valid only for ions which behave hydrodynamically like Stokes spheres in a continuum.

Wien effect.—The increase in the conductance of an electrolytic solution produced by high electrical fields (potential gradients).

DEFINITIONS FOR ELECTROLYTIC SOLUTIONS (Continued)

PRIMARY FIXED POINTS OF INTERNATIONAL PRACTICAL TEMPERATURE SCALE-68 (IPTS-68)

Primary fixed points of IPTS-68 are given in degrees Celsius and kelvin. Also shown are the comparative values from previous international temperature scales.

Fixed point	ITS-27, °C	ITS-48, °C	IPTS-48, °C	IPTS-68		
				°C	K	Uncertainty (K)
Equilibrium between the liquid and vapor phases of water (boiling point of water)	100.000	100	100	100	373.15	0.005
Equilibrium between the solid and liquid phases of zinc (freezing point of zinc)				419.58	692.73	0.03
Equilibrium between the liquid and vapor phases of sulfur (boiling point of sulfur)	444.60	444.600	444.6			
Equilibrium between the solid and liquid phases of silver (freezing point of silver)	960.5	960.8	960.8	961.93	1,235.08	0.2
Equilibrium between the solid and liquid phases of gold (freezing point of gold)	1,063	1,063.0	1,063	1,064.43	1,337.58	0.2

From Sparks, L. L., *ASRDI Oxygen Technology Survey,* Vol. 4, Scientific and Technical Information Office, National Aeronautics and Space Administration, Washington, D. C., 1974.

TEMPERATURE CONVERSION TABLE

This table permits one to convert from degrees Celsius to degrees Fahrenheit or from degrees Fahrenheit to degrees Celsius. The conversion is accomplished by first locating in a column printed in bold face type the number that is to be converted. If the number to be converted is in degrees Fahrenheit, one may find its equivalent in degrees Celsius by reading to the left. If the number to be converted is in degrees Celsius, one may find its equivalent in degrees Fahrenheit by reading to the right. Degrees Celsius are identical to degrees Centigrade. However, the word Celsius is preferred for international use.

The approved international symbolic abbreviation for degrees Celsius is °C, whereas for degrees Fahrenheit it is °F. Absolute zero on the Celsius scale is -273.15°C; on the Fahrenheit scale it is -459.67°F. The relation between degrees Fahrenheit and degrees Celsius may be expressed by

$$°C = 5/9(°F - 32) \text{ or}$$
$$°F = 9/5(°C) + 32.$$

To Convert			To Convert			To Convert		
To °C	←°F or °C→	To °F	To °C	←°F or °C→	To °F	To °C	←°F or °C→	To °F
-273.15	**-459.67**	—	-245.56	**-410**	—	-217.78	**-360**	—
-272.78	**-459**	—	-245	**-409**	—	-217.22	**-359**	—
-272.22	**-458**	—	-244.44	**-408**	—	-216.67	**-358**	—
-271.67	**-457**	—	-243.89	**-407**	—	-216.11	**-357**	—
-271.11	**-456**	—	-243.33	**-406**	—	-215.56	**-356**	—
-270.56	**-455**	—	-242.78	**-405**	—	-215	**-355**	—
-270	**-454**	—	-242.22	**-404**	—	-214.44	**-354**	—
-269.44	**-453**	—	-241.67	**-403**	—	-213.89	**-353**	—
-268.89	**-452**	—	-241.11	**-402**	—	-213.33	**-352**	—
-268.33	**-451**	—	-240.56	**-401**	—	-212.78	**-351**	—
-267.78	**-450**	—	-240	**-400**	—	-212.22	**-350**	—
-267.22	**-449**	—	-239.44	**-399**	—	-211.67	**-349**	—
-266.67	**-448**	—	-238.89	**-398**	—	-211.11	**-348**	—
-266.11	**-447**	—	-238.33	**-397**	—	-210.56	**-347**	—
-265.56	**-446**	—	-237.78	**-396**	—	-210	**-346**	—
-265	**-445**	—	-237.22	**-395**	—	-209.44	**-345**	—
-264.44	**-444**	—	-236.67	**-394**	—	-208.89	**-344**	—
-263.89	**-443**	—	-236.11	**-393**	—	-208.33	**-343**	—
-263.33	**-442**	—	-235.56	**-392**	—	-207.78	**-342**	—
-262.78	**-441**	—	-235	**-391**	—	-207.22	**-341**	—
-262.22	**-440**	—	-234.44	**-390**	—	-206.67	**-340**	—
-261.67	**-439**	—	-233.89	**-389**	—	-206.11	**-339**	—
-261.11	**-438**	—	-233.33	**-388**	—	-205.56	**-338**	—
-260.56	**-437**	—	-232.78	**-387**	—	-205	**-337**	—
-260	**-436**	—	-232.22	**-386**	—	-204.44	**-336**	—
-259.44	**-435**	—	-231.67	**-385**	—	-203.89	**-335**	—
-258.89	**-434**	—	-231.11	**-384**	—	-203.33	**-334**	—
-258.33	**-433**	—	-230.56	**-383**	—	-202.78	**-333**	—
-257.78	**-432**	—	-230	**-382**	—	-202.22	**-332**	—
-257.22	**-431**	—	-229.44	**-381**	—	-201.67	**-331**	—
-256.67	**-430**	—	-228.89	**-380**	—	-201.11	**-330**	—
-256.11	**-429**	—	-228.33	**-379**	—	-200.56	**-329**	—
-255.56	**-428**	—	-227.78	**-378**	—	-200	**-328**	—
-255	**-427**	—	-227.22	**-377**	—	-199.44	**-327**	—
-254.44	**-426**	—	-226.67	**-376**	—	-198.89	**-326**	—
-253.89	**-425**	—	-226.11	**-375**	—	-198.33	**-325**	—
-253.33	**-424**	—	-225.56	**-374**	—	-197.78	**-324**	—
-252.78	**-423**	—	-225	**-373**	—	-197.22	**-323**	—
-252.22	**-422**	—	-224.44	**-372**	—	-196.67	**-322**	—
-251.67	**-421**	—	-223.89	**-371**	—	-196.11	**-321**	—
-251.11	**-420**	—	-223.33	**-370**	—	-195.56	**-320**	—
-250.56	**-419**	—	-222.78	**-369**	—	-195	**-319**	—
-250	**-418**	—	-222.22	**-368**	—	-194.44	**-318**	—
-249.44	**-417**	—	-221.67	**-367**	—	-193.89	**-317**	—
-248.89	**-416**	—	-221.11	**-366**	—	-193.33	**-316**	—
-248.33	**-415**	—	-220.56	**-365**	—	-192.78	**-315**	—
-247.78	**-414**	—	-220	**-364**	—	-192.22	**-314**	—
-247.22	**-413**	—	-219.44	**-363**	—	-191.67	**-313**	—
-246.67	**-412**	—	-218.89	**-362**	—	-191.11	**-312**	—
-246.11	**-411**	—	-218.33	**-361**	—	-190.56	**-311**	—

TEMPERATURE CONVERSION TABLE (Continued)

To °C	←°F or °C→	To °F
−190	**−310**	—
−189.44	**−309**	—
−188.89	**−308**	—
−188.33	**−307**	—
−187.78	**−306**	—
−187.22	**−305**	—
−186.67	**−304**	—
−186.11	**−303**	—
−185.56	**−302**	—
−185	**−301**	—
−184.44	**−300**	—
−183.89	**−299**	—
−183.33	**−298**	—
−182.78	**−297**	—
−182.22	**−296**	—
−181.67	**−295**	—
−181.11	**−294**	—
−180.56	**−293**	—
−180	**−292**	—
−179.44	**−291**	—
−178.89	**−290**	—
−178.33	**−289**	—
−177.78	**−288**	—
−177.22	**−287**	—
−176.67	**−286**	—
−176.11	**−285**	—
−175.56	**−284**	—
−175	**−283**	—
−174.44	**−282**	—
−173.89	**−281**	—
−173.33	**−280**	—
−172.78	**−279**	—
−172.22	**−278**	—
−171.67	**−277**	—
−171.11	**−276**	—
−170.56	**−275**	—
−170	**−274**	—
—	**−273.15**	−459.67
−169.44	**−273**	−459.4
−168.89	**−272**	−457.6
−168.33	**−271**	−455.8
−167.78	**−270**	−454
−167.22	**−269**	−452.2
−166.67	**−268**	−450.4
−166.11	**−267**	−448.6
−165.56	**−266**	−446.8
−165	**−265**	−445
−164.44	**−264**	−443.2
−163.89	**−263**	−441.4
−163.33	**−262**	−439.6
−162.78	**−261**	−437.8
−162.22	**−260**	−436
−161.67	**−259**	−434.2
−161.11	**−258**	−432.4
−160.56	**−257**	−430.6
−160	**−256**	−428.8
−159.44	**−255**	−427
−158.89	**−254**	−425.2
−158.33	**−253**	−423.4
−157.78	**−252**	−421.6
−157.22	**−251**	−419.8

To °C	←°F or °C→	To °F
−156.67	**−250**	−418
−156.11	**−249**	−416.2
−155.56	**−248**	−414.4
−155	**−247**	−412.6
−154.44	**−246**	−410.8
−153.89	**−245**	−409
−153.33	**−244**	−407.2
−152.78	**−243**	−405.4
−152.22	**−242**	−403.6
−151.67	**−241**	−401.8
−151.11	**−240**	−400
−150.56	**−239**	−398.2
−150	**−238**	−396.4
−149.44	**−237**	−394.6
−148.89	**−236**	−392.8
−148.33	**−235**	−391
−147.78	**−234**	−389.2
−147.22	**−233**	−387.4
−146.67	**−232**	−385.6
−146.11	**−231**	−383.8
−145.56	**−230**	−382
−145	**−229**	−380.2
−144.44	**−228**	−378.4
−143.89	**−227**	−376.6
−143.33	**−226**	−374.8
−142.78	**−225**	−373
−142.22	**−224**	−371.2
−141.67	**−223**	−369.4
−141.11	**−222**	−367.6
−140.56	**−221**	−365.8
−140	**−220**	−364
−139.44	**−219**	−362.2
−138.89	**−218**	−360.4
−138.33	**−217**	−358.6
−137.78	**−216**	−356.8
−137.22	**−215**	−355
−136.67	**−214**	−353.2
−136.11	**−213**	−351.4
−135.56	**−212**	−349.6
−135	**−211**	−347.8
−134.44	**−210**	−346
−133.89	**−209**	−344.2
−133.33	**−208**	−342.4
−132.78	**−207**	−340.6
−132.22	**−206**	−338.8
−131.67	**−205**	−337
−131.11	**−204**	−335.2
−130.56	**−203**	−333.4
−130	**−202**	−331.6
−129.44	**−201**	−329.8
−128.89	**−200**	−328
−128.33	**−199**	−326.2
−127.78	**−198**	−324.4
−127.22	**−197**	−322.6
−126.67	**−196**	−320.8
−126.11	**−195**	−319
−125.56	**−194**	−317.2
−125	**−193**	−315.4
−124.44	**−192**	−313.6
−123.89	**−191**	−311.8

To °C	←°F or °C→	To °F
−123.33	**−190**	−310
−122.78	**−189**	−308.2
−122.22	**−188**	−306.4
−121.67	**−187**	−304.6
−121.11	**−186**	−302.8
−120.56	**−185**	−301
−120	**−184**	−299.2
−119.44	**−183**	−297.4
−118.89	**−182**	−295.6
−118.33	**−181**	−293.8
−117.78	**−180**	−292
−117.22	**−179**	−290.2
−116.67	**−178**	−288.4
−116.11	**−177**	−286.6
−115.56	**−176**	−284.8
−115	**−175**	−283
−114.44	**−174**	−281.2
−113.89	**−173**	−279.4
−113.33	**−172**	−277.6
−112.78	**−171**	−275.8
−112.22	**−170**	−274
−111.67	**−169**	−272.2
−111.11	**−168**	−270.4
−110.56	**−167**	−268.6
−110	**−166**	−266.8
−109.44	**−165**	−265
−108.89	**−164**	−263.2
−108.33	**−163**	−261.4
−107.78	**−162**	−259.6
−107.22	**−161**	−257.8
−106.67	**−160**	−256
−106.11	**−159**	−254.2
−105.56	**−158**	−252.4
−105	**−157**	−250.6
−104.44	**−156**	−248.8
−103.89	**−155**	−247
−103.33	**−154**	−245.2
−102.78	**−153**	−243.4
−102.22	**−152**	−241.6
−101.67	**−151**	−239.8
−101.11	**−150**	−238
−100.56	**−149**	−236.2
−100	**−148**	−234.4
−99.44	**−147**	−232.6
−98.89	**−146**	−230.8
−98.33	**−145**	−229
−97.78	**−144**	−227.2
−97.22	**−143**	−225.4
−96.67	**−142**	−223.6
−96.11	**−141**	−221.8
−95.56	**−140**	−220
−95	**−139**	−218.2
−94.44	**−138**	−216.4
−93.89	**−137**	−214.6
−93.33	**−136**	−212.8
−92.78	**−135**	−211
−92.22	**−134**	−209.2
−91.67	**−133**	−207.4
−91.11	**−132**	−205.6
−90.56	**−131**	−203.8

TEMPERATURE CONVERSION TABLE (Continued)

To Convert			To Convert			To Convert		
To °C	←°F or °C→	To °F	To °C	←°F or °C→	To °F	To °C	←°F or °C→	To °F
−90	**−130**	−202	−56.67	**−70**	−94	−23.33	**−10**	14
−89.44	**−129**	−200.2	−56.11	**−69**	−92.2	−22.78	**−9**	15.8
−88.89	**−128**	−198.4	−55.56	**−68**	−90.4	−22.22	**−8**	17.6
−88.33	**−127**	−196.6	−55	**−67**	−88.6	−21.67	**−7**	19.4
−87.78	**−126**	−194.8	−54.44	**−66**	−86.8	−21.11	**−6**	21.2
−87.22	**−125**	−193	−53.89	**−65**	−85	−20.56	**−5**	23
−86.67	**−124**	−191.2	−53.33	**−64**	−83.2	−20	**−4**	24.8
−86.11	**−123**	−189.4	−52.78	**−63**	−81.4	−19.44	**−3**	26.6
−85.56	**−122**	−187.6	−52.22	**−62**	−79.6	−18.89	**−2**	28.4
−85	**−121**	−185.8	−51.67	**−61**	−77.8	−18.33	**−1**	30.2
−84.44	**−120**	−184	−51.11	**−60**	−76	−17.78	**0**	32
−83.89	**−119**	−182.2	−50.56	**−59**	−74.2	−17.22	**1**	33.8
−83.33	**−118**	−180.4	−50	**−58**	−72.4	−16.67	**2**	35.6
−82.78	**−117**	−178.6	−49.44	**−57**	−70.6	−16.11	**3**	37.4
−82.22	**−116**	−176.8	−48.89	**−56**	−68.8	−15.56	**4**	39.2
−81.67	**−115**	−175	−48.33	**−55**	−67	−15	**5**	41
−81.11	**−114**	−173.2	−47.78	**−54**	−65.2	−14.44	**6**	42.8
−80.56	**−113**	−171.4	−47.22	**−53**	−63.4	−13.89	**7**	44.6
−80	**−112**	−169.6	−46.67	**−52**	−61.6	−13.33	**8**	46.4
−79.44	**−111**	−167.8	−46.11	**−51**	−59.8	−12.78	**9**	48.2
−78.89	**−110**	−166	−45.56	**−50**	−58	−12.22	**10**	50
−78.33	**−109**	−164.2	−45	**−49**	−56.2	−11.67	**11**	51.8
−77.78	**−108**	−162.4	−44.44	**−48**	−54.4	−11.11	**12**	53.6
−77.22	**−107**	−160.6	−43.89	**−47**	−52.6	−10.56	**13**	55.4
−76.67	**−106**	−158.8	−43.33	**−46**	−50.8	−10	**14**	57.2
−76.11	**−105**	−157	−42.78	**−45**	−49	−9.44	**15**	59
−75.56	**−104**	−155.2	−42.22	**−44**	−47.2	−8.89	**16**	60.8
−75	**−103**	−153.4	−41.67	**−43**	−45.4	−8.33	**17**	62.6
−74.44	**−102**	−151.6	−41.11	**−42**	−43.6	−7.78	**18**	64.4
−73.89	**−101**	−149.8	−40.56	**−41**	−41.8	−7.22	**19**	66.2
−73.33	**−100**	−148	−40	**−40**	−40	−6.67	**20**	68
−72.78	**−99**	−146.2	−39.44	**−39**	−38.2	−6.11	**21**	69.8
−72.22	**−98**	−144.4	−38.89	**−38**	−36.4	−5.56	**22**	71.6
−71.67	**−97**	−142.6	−38.33	**−37**	−34.6	−5	**23**	73.4
−71.11	**−96**	−140.8	−37.78	**−36**	−32.8	−4.44	**24**	75.2
−70.56	**−95**	−139	−37.22	**−35**	−31	−3.89	**25**	77
−70	**−94**	−137.2	−36.67	**−34**	−29.2	−3.33	**26**	78.8
−69.44	**−93**	−135.4	−36.11	**−33**	−27.4	−2.78	**27**	80.6
−68.89	**−92**	−133.6	−35.56	**−32**	−25.6	−2.22	**28**	82.4
−68.33	**−91**	−131.8	−35	**−31**	−23.8	−1.67	**29**	84.2
−67.78	**−90**	−130	−34.44	**−30**	−22	−1.11	**30**	86
−67.22	**−89**	−128.2	−33.89	**−29**	−20.2	−0.56	**31**	87.8
−66.67	**−88**	−126.4	−33.33	**−28**	−18.4	0	**32**	89.6
−66.11	**−87**	−124.6	−32.78	**−27**	−16.6	.56	**33**	91.4
−65.56	**−86**	−122.8	−32.22	**−26**	−14.8	1.11	**34**	93.2
−65	**−85**	−121	−31.67	**−25**	−13	1.67	**35**	95
−64.44	**−84**	−119.2	−31.11	**−24**	−11.2	2.22	**36**	96.8
−63.89	**−83**	−117.4	−30.56	**−23**	−9.4	2.78	**37**	98.6
−63.33	**−82**	−115.6	−30	**−22**	−7.6	3.33	**38**	100.4
−62.78	**−81**	−113.8	−29.44	**−21**	−5.8	3.89	**39**	102.2
−62.22	**−80**	−112	−28.89	**−20**	−4	4.44	**40**	104
−61.67	**−79**	−110.2	−28.33	**−19**	−2.2	5	**41**	105.8
−61.11	**−78**	−108.4	−27.78	**−18**	−0.4	5.56	**42**	107.6
−60.56	**−77**	−106.6	−27.22	**−17**	1.4	6.11	**43**	109.4
−60	**−76**	−104.8	−26.67	**−16**	3.2	6.67	**44**	111.2
−59.44	**−75**	−103	−26.11	**−15**	5	7.22	**45**	113
−58.89	**−74**	−101.2	−25.56	**−14**	6.8	7.78	**46**	114.8
−58.33	**−73**	−99.4	−25	**−13**	8.6	8.33	**47**	116.6
−57.78	**−72**	−97.6	−24.44	**−12**	10.4	8.89	**48**	118.4
−57.22	**−71**	−95.8	−23.89	**−11**	12.2	9.44	**49**	120.2

TEMPERATURE CONVERSION TABLE (Continued)

To Convert			To Convert			To Convert		
To °C	←°F or °C→	To °F	To °C	←°F or °C→	To °F	To °C	←°F or °C→	To °F
10	**50**	122	43.33	**110**	230	76.67	**170**	338
10.56	**51**	123.8	43.89	**111**	231.8	77.22	**171**	339.8
11.11	**52**	125.6	44.44	**112**	233.6	77.78	**172**	341.6
11.67	**53**	127.4	45	**113**	235.4	78.33	**173**	343.4
12.22	**54**	129.2	45.56	**114**	237.2	78.89	**174**	345.2
12.78	**55**	131	46.11	**115**	239	79.44	**175**	347
13.33	**56**	132.8	46.67	**116**	240.8	80	**176**	348.8
13.89	**57**	134.6	47.22	**117**	242.6	80.56	**177**	350.6
14.44	**58**	136.4	47.78	**118**	244.4	81.11	**178**	352.4
15	**59**	138.2	48.33	**119**	246.2	81.67	**179**	354.2
15.56	**60**	140	48.89	**120**	248	82.22	**180**	356
16.11	**61**	141.8	49.44	**121**	249.8	82.78	**181**	357.8
16.67	**62**	143.6	50	**122**	251.6	83.33	**182**	359.6
17.22	**63**	145.4	50.56	**123**	253.4	83.89	**183**	361.4
17.78	**64**	147.2	51.11	**124**	255.2	84.44	**184**	363.2
18.33	**65**	149	51.67	**125**	257	85	**185**	365
18.89	**66**	150.8	52.22	**126**	258.8	85.56	**186**	366.8
19.44	**67**	152.6	52.78	**127**	260.6	86.11	**187**	368.6
20	**68**	154.4	53.33	**128**	262.4	86.67	**188**	370.4
20.56	**69**	156.2	53.89	**129**	264.2	87.22	**189**	372.2
21.11	**70**	158	54.44	**130**	266	87.78	**190**	374
21.67	**71**	159.8	55	**131**	267.8	88.33	**191**	375.8
22.22	**72**	161.6	55.56	**132**	269.6	88.89	**192**	377.6
22.78	**73**	163.4	56.11	**133**	271.4	89.44	**193**	379.4
23.33	**74**	165.2	56.67	**134**	273.2	90	**194**	381.2
23.89	**75**	167	57.22	**135**	275	90.56	**195**	383
24.44	**76**	168.8	57.78	**136**	276.8	91.11	**196**	384.8
25	**77**	170.6	58.33	**137**	278.6	91.67	**197**	386.6
25.56	**78**	172.4	58.89	**138**	280.4	92.22	**198**	388.4
26.11	**79**	174.2	59.44	**139**	282.2	92.78	**199**	390.2
26.67	**80**	176	60	**140**	284	93.33	**200**	392
27.22	**81**	177.8	60.56	**141**	285.8	93.89	**201**	393.8
27.78	**82**	179.6	61.11	**142**	287.6	94.44	**202**	395.6
28.33	**83**	181.4	61.67	**143**	289.4	95	**203**	397.4
28.89	**84**	183.2	62.22	**144**	291.2	95.56	**204**	399.2
29.44	**85**	185	62.78	**145**	293	96.11	**205**	401
30	**86**	186.8	63.33	**146**	294.8	96.67	**206**	402.8
30.56	**87**	188.6	63.89	**147**	296.6	97.22	**207**	404.6
31.11	**88**	190.4	64.44	**148**	298.4	97.78	**208**	406.4
31.67	**89**	192.2	65	**149**	300.2	98.33	**209**	408.2
32.22	**90**	194	65.56	**150**	302	98.89	**210**	410
32.78	**91**	195.8	66.11	**151**	303.8	99.44	**211**	411.8
33.33	**92**	197.6	66.67	**152**	305.6	100	**212**	413.6
33.89	**93**	199.4	67.22	**153**	307.4	100.56	**213**	415.4
34.44	**94**	201.2	67.78	**154**	309.2	101.11	**214**	417.2
35	**95**	203	68.33	**155**	311	101.67	**215**	419
35.56	**96**	204.8	68.89	**156**	312.8	102.22	**216**	420.8
36.11	**97**	206.6	69.44	**157**	314.6	102.78	**217**	422.6
36.67	**98**	208.4	70	**158**	316.4	103.33	**218**	424.4
37.22	**99**	210.2	70.56	**159**	318.2	103.89	**219**	426.2
37.78	**100**	212	71.11	**160**	320	104.44	**220**	428
38.33	**101**	213.8	71.67	**161**	321.8	105	**221**	429.8
38.89	**102**	215.6	72.22	**162**	323.6	105.56	**222**	431.6
39.44	**103**	217.4	72.78	**163**	325.4	106.11	**223**	433.4
40	**104**	219.2	73.33	**164**	327.2	106.67	**224**	435.2
40.56	**105**	221	73.89	**165**	329	107.22	**225**	437
41.11	**106**	222.8	74.44	**166**	330.8	107.78	**226**	438.8
41.67	**107**	224.6	75	**167**	332.6	108.33	**227**	440.6
42.22	**108**	226.4	75.56	**168**	334.4	108.89	**228**	442.4
42.78	**109**	228.2	76.11	**169**	336.2	109.44	**229**	444.2

TEMPERATURE CONVERSION TABLE (Continued)

To Convert			To Convert			To Convert		
To °C	←°F or °C→	To °F	To °C	←°F or °C→	To °F	To °C	←°F or °C→	To °F
110	**230**	446	143.33	**290**	554	176.67	**350**	662
110.56	**231**	447.8	143.89	**291**	555.8	177.22	**351**	663.8
111.11	**232**	449.6	144.44	**292**	557.6	177.78	**352**	665.6
111.67	**233**	451.4	145	**293**	559.4	178.33	**353**	667.4
112.22	**234**	453.2	145.56	**294**	561.2	178.89	**354**	669.2
112.78	**235**	455	146.11	**295**	563	179.44	**355**	671
113.33	**236**	456.8	146.67	**296**	564.8	180	**356**	672.8
113.89	**237**	458.6	147.22	**297**	566.6	180.56	**357**	674.6
114.44	**238**	460.4	147.78	**298**	568.4	181.11	**358**	676.4
115	**239**	462.2	148.33	**299**	570.2	181.67	**359**	678.2
115.56	**240**	464	148.89	**300**	572	182.22	**360**	680
116.11	**241**	465.8	149.44	**301**	573.8	182.78	**361**	681.8
116.67	**242**	467.6	150	**302**	575.6	183.33	**362**	683.6
117.22	**243**	469.4	150.56	**303**	577.4	183.89	**363**	685.4
117.78	**244**	471.2	151.11	**304**	579.2	184.44	**364**	687.2
118.33	**245**	473	151.67	**305**	581	185	**365**	689
118.89	**246**	474.8	152.22	**306**	582.8	185.56	**366**	690.8
119.44	**247**	476.6	152.78	**307**	584.6	186.11	**367**	692.6
120	**248**	478.4	153.33	**308**	586.4	186.67	**368**	694.4
120.56	**249**	480.2	153.89	**309**	588.2	187.22	**369**	696.2
121.11	**250**	482	154.44	**310**	590	187.78	**370**	698
121.67	**251**	483.8	155	**311**	591.8	188.33	**371**	699.8
122.22	**252**	485.6	155.56	**312**	593.6	188.89	**372**	701.6
122.78	**253**	487.4	156.11	**313**	595.4	189.44	**373**	703.4
123.33	**254**	489.2	156.67	**314**	597.2	190	**374**	705.2
123.89	**255**	491	157.22	**315**	599	190.56	**375**	707
124.44	**256**	492.8	157.78	**316**	600.8	191.11	**376**	708.8
125	**257**	494.6	158.33	**317**	602.6	191.67	**377**	710.6
125.56	**258**	496.4	158.89	**318**	604.4	192.22	**378**	712.4
126.11	**259**	498.2	159.44	**319**	606.2	192.78	**379**	714.2
126.67	**260**	500	160	**320**	608	193.33	**380**	716
127.22	**261**	501.8	160.56	**321**	609.8	193.89	**381**	717.8
127.78	**262**	503.6	161.11	**322**	611.6	194.44	**382**	719.6
128.33	**263**	505.4	161.67	**323**	613.4	195	**383**	721.4
128.89	**264**	507.2	162.22	**324**	615.2	195.56	**384**	723.2
129.44	**265**	509	162.78	**325**	617	196.11	**385**	725
130	**266**	510.8	163.33	**326**	618.8	196.67	**386**	726.8
130.56	**267**	512.6	163.89	**327**	620.6	197.22	**387**	728.6
131.11	**268**	514.4	164.44	**328**	622.4	197.78	**388**	730.4
131.67	**269**	516.2	165	**329**	624.2	198.33	**389**	732.2
132.22	**270**	518	165.56	**330**	626	198.89	**390**	734
132.78	**271**	519.8	166.11	**331**	627.8	199.44	**391**	735.8
133.33	**272**	521.6	166.67	**332**	629.6	200	**392**	737.6
133.89	**273**	523.4	167.22	**333**	631.4	200.56	**393**	739.4
134.44	**274**	525.2	167.78	**334**	633.2	201.11	**394**	741.2
135	**275**	527	168.33	**335**	635	201.67	**395**	743
135.56	**276**	528.8	168.89	**336**	636.8	202.22	**396**	744.8
136.11	**277**	530.6	169.44	**337**	638.6	202.78	**397**	746.6
136.67	**278**	532.4	170	**338**	640.4	203.33	**398**	748.4
137.22	**279**	534.2	170.56	**339**	642.2	203.89	**399**	750.2
137.78	**280**	536	171.11	**340**	644	204.44	**400**	752
138.33	**281**	537.8	171.67	**341**	645.8	205	**401**	753.8
138.89	**282**	539.6	172.22	**342**	647.6	205.56	**402**	755.6
139.44	**283**	541.4	172.78	**343**	649.4	206.11	**403**	757.4
140	**284**	543.2	173.33	**344**	651.2	206.67	**404**	759.2
140.56	**285**	545	173.89	**345**	653	207.22	**405**	761
141.11	**286**	546.8	174.44	**346**	654.8	207.78	**406**	762.8
141.67	**287**	548.6	175	**347**	656.6	208.33	**407**	764.6
142.22	**288**	550.4	175.56	**348**	658.4	208.89	**408**	766.4
142.78	**289**	552.2	176.11	**349**	660.2	209.44	**409**	768.2

TEMPERATURE CONVERSION TABLE (Continued)

To Convert			To Convert			To Convert		
To °C	←°F or °C→	To °F	To °C	←°F or °C→	To °F	To °C	←°F or °C→	To °F
210	**410**	770	243.33	**470**	878	276.67	**530**	986
210.56	**411**	771.8	243.89	**471**	879.8	277.22	**531**	987.8
211.11	**412**	773.6	244.44	**472**	881.6	277.78	**532**	989.6
211.67	**413**	775.4	245	**473**	883.4	278.33	**533**	991.4
212.22	**414**	777.2	245.56	**474**	885.2	278.89	**534**	993.2
212.78	**415**	779	246.11	**475**	887	279.44	**535**	995
213.33	**416**	780.8	246.67	**476**	888.8	280	**536**	996.8
213.89	**417**	782.6	247.22	**477**	890.6	280.56	**537**	998.6
214.44	**418**	784.4	247.78	**478**	892.4	281.11	**538**	1000.4
215	**419**	786.2	248.33	**479**	894.2	281.67	**539**	1002.2
215.56	**420**	788	248.89	**480**	896	282.22	**540**	1004
216.11	**421**	789.8	249.44	**481**	897.8	282.78	**541**	1005.8
216.67	**422**	791.6	250	**482**	899.6	283.33	**542**	1007.6
217.22	**423**	793.4	250.56	**483**	901.4	283.89	**543**	1009.4
217.78	**424**	795.2	251.11	**484**	903.2	284.44	**544**	1011.2
218.33	**425**	797	251.67	**485**	905	285	**545**	1013
218.89	**426**	798.8	252.22	**486**	906.8	285.56	**546**	1014.8
219.44	**427**	800.6	252.78	**487**	908.6	286.11	**547**	1016.6
220	**428**	802.4	253.33	**488**	910.4	286.67	**548**	1018.4
220.56	**429**	804.2	253.89	**489**	912.2	287.22	**549**	1020.2
221.11	**430**	806	254.44	**490**	914	287.78	**550**	1022
221.67	**431**	807.8	255	**491**	915.8	288.33	**551**	1023.8
222.22	**432**	809.6	255.56	**492**	917.6	288.89	**552**	1025.6
222.78	**433**	811.4	256.11	**493**	919.4	289.44	**553**	1027.4
223.33	**434**	813.2	256.67	**494**	921.2	290	**554**	1029.2
223.89	**435**	815	257.22	**495**	923	290.56	**555**	1031
224.44	**436**	816.8	257.78	**496**	924.8	291.11	**556**	1032.8
225	**437**	818.6	258.33	**497**	926.6	291.67	**557**	1034.6
225.56	**438**	820.4	258.89	**498**	928.4	292.22	**558**	1036.4
226.11	**439**	822.2	259.44	**499**	930.2	292.78	**559**	1038.2
226.67	**440**	824	260	**500**	932	293.33	**560**	1040
227.22	**441**	825.8	260.56	**501**	933.8	293.89	**561**	1041.8
227.78	**442**	827.6	261.11	**502**	935.6	294.44	**562**	1043.6
228.33	**443**	829.4	261.67	**503**	937.4	295	**563**	1045.4
228.89	**444**	831.2	262.22	**504**	939.2	295.56	**564**	1047.2
229.44	**445**	833	262.78	**505**	941	296.11	**565**	1049
230	**446**	834.8	263.33	**506**	942.8	296.67	**566**	1050.8
230.56	**447**	836.6	263.89	**507**	944.6	297.22	**567**	1052.6
231.11	**448**	838.4	264.44	**508**	946.4	297.78	**568**	1054.4
231.67	**449**	840.2	265	**509**	948.2	298.33	**569**	1056.2
232.22	**450**	842	265.56	**510**	950	298.89	**570**	1058
232.78	**451**	843.8	266.11	**511**	951.8	299.44	**571**	1059.8
233.33	**452**	845.6	266.67	**512**	953.6	300	**572**	1061.6
233.89	**453**	847.4	267.22	**513**	955.4	300.56	**573**	1063.4
234.44	**454**	849.2	267.78	**514**	957.2	301.11	**574**	1065.2
235	**455**	851	268.33	**515**	959	301.67	**575**	1067
235.56	**456**	852.8	268.89	**516**	960.8	302.22	**576**	1068.8
236.11	**457**	854.6	269.44	**517**	962.6	302.78	**577**	1070.6
236.67	**458**	856.4	270	**518**	964.4	303.33	**578**	1072.4
237.22	**459**	858.2	270.56	**519**	966.2	303.89	**579**	1074.2
237.78	**460**	860	271.11	**520**	968	304.44	**580**	1076
238.33	**461**	861.8	271.67	**521**	969.8	305	**581**	1077.8
238.89	**462**	863.6	272.22	**522**	971.6	305.56	**582**	1079.6
239.44	**463**	865.4	272.78	**523**	973.4	306.11	**583**	1081.4
240	**464**	867.2	273.33	**524**	975.2	306.67	**584**	1083.2
240.56	**465**	869	273.89	**525**	977	307.22	**585**	1085
241.11	**466**	870.8	274.44	**526**	978.8	307.78	**586**	1086.8
241.67	**467**	872.6	275	**527**	980.6	308.33	**587**	1088.6
242.22	**468**	874.4	275.56	**528**	982.4	308.89	**588**	1090.4
242.78	**469**	876.2	276.11	**529**	984.2	309.44	**589**	1092.2

To Convert			To Convert			To Convert		
To °C	←°F or °C→	To °F	To °C	←°F or °C→	To °F	To °C	←°F or °C→	To °F
310	**590**	1094	343.33	**650**	1202	376.67	**710**	1310
310.56	**591**	1095.8	343.89	**651**	1203.8	377.22	**711**	1311.8
311.11	**592**	1097.6	344.44	**652**	1205.6	377.78	**712**	1313.6
311.67	**593**	1099.4	345	**653**	1207.4	378.33	**713**	1315.4
312.22	**594**	1101.2	345.56	**654**	1209.2	378.89	**714**	1317.2
312.78	**595**	1103	346.11	**655**	1211	379.44	**715**	1319
313.33	**596**	1104.8	346.67	**656**	1212.8	380	**716**	1320.8
313.89	**597**	1106.6	347.22	**657**	1214.6	380.56	**717**	1322.6
314.44	**598**	1108.4	347.78	**658**	1216.4	381.11	**718**	1324.4
315	**599**	1110.2	348.33	**659**	1218.2	381.67	**719**	1326.2
315.56	**600**	1112	348.89	**660**	1220	382.22	**720**	1328
316.11	**601**	1113.8	349.44	**661**	1221.8	382.78	**721**	1329.8
316.67	**602**	1115.6	350	**662**	1223.6	383.33	**722**	1331.6
317.22	**603**	1117.4	350.56	**663**	1225.4	383.89	**723**	1333.4
317.78	**604**	1119.2	351.11	**664**	1227.2	384.44	**724**	1335.2
318.33	**605**	1121	351.67	**665**	1229	385	**725**	1337
318.89	**606**	1122.8	352.22	**666**	1230.8	385.56	**726**	1338.8
319.44	**607**	1124.6	352.78	**667**	1232.6	386.11	**727**	1340.6
320	**608**	1126.4	353.33	**668**	1234.4	386.67	**728**	1342.4
320.56	**609**	1128.2	353.89	**669**	1236.2	387.22	**729**	1344.2
321.11	**610**	1130	354.44	**670**	1238	387.78	**730**	1346
321.67	**611**	1131.8	355	**671**	1239.8	388.33	**731**	1347.8
322.22	**612**	1133.6	355.56	**672**	1241.6	388.89	**732**	1349.6
322.78	**613**	1135.4	356.11	**673**	1243.4	389.44	**733**	1351.4
323.33	**614**	1137.2	356.67	**674**	1245.2	390	**734**	1353.2
323.89	**615**	1139	357.22	**675**	1247	390.56	**735**	1355
324.44	**616**	1140.8	357.78	**676**	1248.8	391.11	**736**	1356.8
325	**617**	1142.6	358.33	**677**	1250.6	391.67	**737**	1358.6
325.56	**618**	1144.4	358.89	**678**	1252.4	392.22	**738**	1360.4
326.11	**619**	1146.2	359.44	**679**	1254.2	392.78	**739**	1362.2
326.67	**620**	1148	360	**680**	1256	393.33	**740**	1364
327.22	**621**	1149.8	360.56	**681**	1257.8	393.89	**741**	1365.8
327.78	**622**	1151.6	361.11	**682**	1259.6	394.44	**742**	1367.6
328.33	**623**	1153.4	361.67	**683**	1261.4	395	**743**	1369.4
328.89	**624**	1155.2	362.22	**684**	1263.2	395.56	**744**	1371.2
329.44	**625**	1157	362.78	**685**	1265	396.11	**745**	1373
330	**626**	1158.8	363.33	**686**	1266.8	396.67	**746**	1374.8
330.56	**627**	1160.6	363.89	**687**	1268.6	397.22	**747**	1376.6
331.11	**628**	1162.4	364.44	**688**	1270.4	397.78	**748**	1378.4
331.67	**629**	1164.2	365	**689**	1272.2	398.33	**749**	1380.2
332.22	**630**	1166	365.56	**690**	1274	398.89	**750**	1382
332.78	**631**	1167.8	366.11	**691**	1275.8	399.44	**751**	1383.8
333.33	**632**	1169.6	366.67	**692**	1277.6	400	**752**	1385.6
333.89	**633**	1171.4	367.22	**693**	1279.4	400.56	**753**	1387.4
334.44	**634**	1173.2	367.78	**694**	1281.2	401.11	**754**	1389.2
335	**635**	1175	368.33	**695**	1283	401.67	**755**	1391
335.56	**636**	1176.8	368.89	**696**	1284.8	402.22	**756**	1392.8
336.11	**637**	1178.6	369.44	**697**	1286.6	402.78	**757**	1394.6
336.67	**638**	1180.4	370	**698**	1288.4	403.33	**758**	1396.4
337.22	**639**	1182.2	370.56	**699**	1290.2	403.89	**759**	1398.2
337.78	**640**	1184	371.11	**700**	1292	404.44	**760**	1400
338.33	**641**	1185.8	371.67	**701**	1293.8	405	**761**	1401.8
338.89	**642**	1187.6	372.22	**702**	1295.6	405.56	**762**	1403.6
339.44	**643**	1189.4	372.78	**703**	1297.4	406.11	**763**	1405.4
340	**644**	1191.2	373.33	**704**	1299.2	406.67	**764**	1407.2
340.56	**645**	1193	373.89	**705**	1301	407.22	**765**	1409
341.11	**646**	1194.8	374.44	**706**	1302.8	407.78	**766**	1410.8
341.67	**647**	1196.6	375	**707**	1304.6	408.33	**767**	1412.6
342.22	**648**	1198.4	375.56	**708**	1306.4	408.89	**768**	1414.4
342.78	**649**	1200.2	376.11	**709**	1308.2	409.44	**769**	1416.2

To Convert			To Convert			To Convert		
To °C	←°F or °C→	To °F	To °C	←°F or °C→	To °F	To °C	←°F or °C→	To °F
410	**770**	1418	443.33	**830**	1526	476.67	**890**	1634
410.56	**771**	1419.8	443.89	**831**	1527.8	477.22	**891**	1635.8
411.11	**772**	1421.6	444.44	**832**	1529.6	477.78	**892**	1637.6
411.67	**773**	1423.4	445	**833**	1531.4	478.33	**893**	1639.4
412.22	**774**	1425.2	445.56	**834**	1533.2	478.89	**894**	1641.2
412.78	**775**	1427	446.11	**835**	1535	479.44	**895**	1643
413.33	**776**	1428.8	446.67	**836**	1536.8	480	**896**	1644.8
413.89	**777**	1430.6	447.22	**837**	1538.6	480.56	**897**	1646.6
414.44	**778**	1432.4	447.78	**838**	1540.4	481.11	**898**	1648.4
415	**779**	1434.2	448.33	**839**	1542.2	481.67	**899**	1650.2
415.56	**780**	1436	448.89	**840**	1544	482.22	**900**	1652
416.11	**781**	1437.8	449.44	**841**	1545.8	482.78	**901**	1653.8
416.67	**782**	1439.6	450	**842**	1547.6	483.33	**902**	1655.6
417.22	**783**	1441.4	450.56	**843**	1549.4	483.89	**903**	1657.4
417.78	**784**	1443.2	451.11	**844**	1551.2	484.44	**904**	1659.2
418.33	**785**	1445	451.67	**845**	1553	485	**905**	1661
418.89	**786**	1446.8	452.22	**846**	1554.8	485.56	**906**	1662.8
419.44	**787**	1448.6	452.78	**847**	1556.6	486.11	**907**	1664.6
420	**788**	1450.4	453.33	**848**	1558.4	486.67	**908**	1666.4
420.56	**789**	1452.2	453.89	**849**	1560.2	487.22	**909**	1668.2
421.11	**790**	1454	454.44	**850**	1562	487.78	**910**	1670
421.67	**791**	1455.8	455	**851**	1563.8	488.33	**911**	1671.8
422.22	**792**	1457.6	455.56	**852**	1565.6	488.89	**912**	1673.6
422.78	**793**	1459.4	456.11	**853**	1567.4	489.44	**913**	1675.4
423.33	**794**	1461.2	456.67	**854**	1569.2	490	**914**	1677.2
423.89	**795**	1463	457.22	**855**	1571	490.56	**915**	1679
424.44	**796**	1464.8	457.78	**856**	1572.8	491.11	**916**	1680.8
425	**797**	1466.6	458.33	**857**	1574.6	491.67	**917**	1682.6
425.56	**798**	1468.4	458.89	**858**	1576.4	492.22	**918**	1684.4
426.11	**799**	1470.2	459.44	**859**	1578.2	492.78	**919**	1686.2
426.67	**800**	1472	460	**860**	1580	493.33	**920**	1688
427.22	**801**	1473.8	460.56	**861**	1581.8	493.89	**921**	1689.8
427.78	**802**	1475.6	461.11	**862**	1583.6	494.44	**922**	1691.6
428.33	**803**	1477.4	461.67	**863**	1585.4	495	**923**	1693.4
428.89	**804**	1479.2	462.22	**864**	1587.2	495.56	**924**	1695.2
429.44	**805**	1481	462.78	**865**	1589	496.11	**925**	1697
430	**806**	1482.8	463.33	**866**	1590.8	496.67	**926**	1698.8
430.56	**807**	1484.6	463.89	**867**	1592.6	497.22	**927**	1700.6
431.11	**808**	1486.4	464.44	**868**	1594.4	497.78	**928**	1702.4
431.67	**809**	1488.2	465	**869**	1596.2	498.33	**929**	1704.2
432.22	**810**	1490	465.56	**870**	1598	498.89	**930**	1706
432.78	**811**	1491.8	466.11	**871**	1599.8	499.44	**931**	1707.8
433.33	**812**	1493.6	466.67	**872**	1601.6	500	**932**	1709.6
433.89	**813**	1495.4	467.22	**873**	1603.4	500.56	**933**	1711.4
434.44	**814**	1497.2	467.78	**874**	1605.2	501.11	**934**	1713.2
435	**815**	1499	468.33	**875**	1607	501.67	**935**	1715
435.56	**816**	1500.8	468.89	**876**	1608.8	502.22	**936**	1716.8
436.11	**817**	1502.6	469.44	**877**	1610.6	502.78	**937**	1718.6
436.67	**818**	1504.4	470	**878**	1612.4	503.33	**938**	1720.4
437.22	**819**	1506.2	470.56	**879**	1614.2	503.89	**939**	1722.2
437.78	**820**	1508	471.11	**880**	1616	504.44	**940**	1724
438.33	**821**	1509.8	471.67	**881**	1617.8	505	**941**	1725.8
438.89	**822**	1511.6	472.22	**882**	1619.6	505.56	**942**	1727.6
439.44	**823**	1513.4	472.78	**883**	1621.4	506.11	**943**	1729.4
440	**824**	1515.2	473.33	**884**	1623.2	506.67	**944**	1731.2
440.56	**825**	1517	473.89	**885**	1625	507.22	**945**	1733
441.11	**826**	1518.8	474.44	**886**	1626.8	507.78	**946**	1734.8
441.67	**827**	1520.6	475	**887**	1628.6	508.33	**947**	1736.6
442.22	**828**	1522.4	475.56	**888**	1630.4	508.89	**948**	1738.4
442.78	**829**	1524.2	476.11	**889**	1632.2	509.44	**949**	1740.2

TEMPERATURE CONVERSION TABLE (Continued)

To Convert			To Convert			To Convert		
To °C	←°F or °C→	To °F	To °C	←°F or °C→	To °F	To °C	←°F or °C→	To °F
510	**950**	1742	543.33	**1010**	1850	576.67	**1070**	1958
510.56	**951**	1743.8	543.89	**1011**	1851.8	577.22	**1071**	1959.8
511.11	**952**	1745.6	544.44	**1012**	1853.6	577.78	**1072**	1961.6
511.67	**953**	1747.4	545	**1013**	1855.4	578.33	**1073**	1963.4
512.22	**954**	1749.2	545.56	**1014**	1857.2	578.89	**1074**	1965.2
512.78	**955**	1751	546.11	**1015**	1859	579.44	**1075**	1967
513.33	**956**	1752.8	546.67	**1016**	1860.8	580	**1076**	1968.8
513.89	**957**	1754.6	547.22	**1017**	1862.6	580.56	**1077**	1970.6
514.44	**958**	1756.4	547.78	**1018**	1864.4	581.11	**1078**	1972.4
515	**959**	1758.2	548.33	**1019**	1866.2	581.67	**1079**	1974.2
515.56	**960**	1760	548.89	**1020**	1868	582.22	**1080**	1976
516.11	**961**	1761.8	549.44	**1021**	1869.8	582.78	**1081**	1977.8
516.67	**962**	1763.6	550	**1022**	1871.6	583.33	**1082**	1979.6
517.22	**963**	1765.4	550.56	**1023**	1873.4	583.89	**1083**	1981.4
517.78	**964**	1767.2	551.11	**1024**	1875.2	584.44	**1084**	1983.2
518.33	**965**	1769	551.67	**1025**	1877	585	**1085**	1985
518.89	**966**	1770.8	552.22	**1026**	1878.8	585.56	**1086**	1986.8
519.44	**967**	1772.6	552.78	**1027**	1880.6	586.11	**1087**	1988.6
520	**968**	1774.4	553.33	**1028**	1882.4	586.67	**1088**	1990.4
520.56	**969**	1776.2	553.89	**1029**	1884.2	587.22	**1089**	1992.2
521.11	**970**	1778	554.44	**1030**	1886	587.78	**1090**	1994
521.67	**971**	1779.8	555	**1031**	1887.8	588.33	**1091**	1995.8
522.22	**972**	1781.6	555.56	**1032**	1889.6	588.89	**1092**	1997.6
522.78	**973**	1783.4	556.11	**1033**	1891.4	589.44	**1093**	1999.4
523.33	**974**	1785.2	556.67	**1034**	1893.2	590	**1094**	2001.2
523.89	**975**	1787	557.22	**1035**	1895	590.56	**1095**	2003
524.44	**976**	1788.8	557.78	**1036**	1896.8	591.11	**1096**	2004.8
525	**977**	1790.6	558.33	**1037**	1898.6	591.67	**1097**	2006.6
525.56	**978**	1792.4	558.89	**1038**	1900.4	592.22	**1098**	2008.4
526.11	**979**	1794.2	559.44	**1039**	1902.2	592.78	**1099**	2010.2
526.67	**980**	1796	560	**1040**	1904	593.33	**1100**	2012
527.22	**981**	1797.8	560.56	**1041**	1905.8	593.89	**1101**	2013.8
527.78	**982**	1799.6	561.11	**1042**	1907.6	594.44	**1102**	2015.6
528.33	**983**	1801.4	561.67	**1043**	1909.4	595	**1103**	2017.4
528.89	**984**	1803.2	562.22	**1044**	1911.2	595.56	**1104**	2019.2
529.44	**985**	1805	562.78	**1045**	1913	596.11	**1105**	2021
530	**986**	1806.8	563.33	**1046**	1914.8	596.67	**1106**	2022.8
530.56	**987**	1808.6	563.89	**1047**	1916.6	597.22	**1107**	2024.6
531.11	**988**	1810.4	564.44	**1048**	1918.4	597.78	**1108**	2026.4
531.67	**989**	1812.2	565	**1049**	1920.2	598.33	**1109**	2028.2
532.22	**990**	1814	565.56	**1050**	1922	598.89	**1110**	2030
532.78	**991**	1815.8	566.11	**1051**	1923.8	599.44	**1111**	2031.8
533.33	**992**	1817.6	566.67	**1052**	1925.6	600	**1112**	2033.6
533.89	**993**	1819.4	567.22	**1053**	1927.4	600.56	**1113**	2035.4
534.44	**994**	1821.2	567.78	**1054**	1929.2	601.11	**1114**	2037.2
535	**995**	1823	568.33	**1055**	1931	601.67	**1115**	2039
535.56	**996**	1824.8	568.89	**1056**	1932.8	602.22	**1116**	2040.8
536.11	**997**	1826.6	569.44	**1057**	1934.6	602.78	**1117**	2042.6
536.67	**998**	1828.4	570	**1058**	1936.4	603.33	**1118**	2044.4
537.22	**999**	1830.2	570.56	**1059**	1938.2	603.89	**1119**	2046.2
537.78	**1000**	1832	571.11	**1060**	1940	604.44	**1120**	2048
538.33	**1001**	1833.8	571.67	**1061**	1941.8	605	**1121**	2049.8
538.89	**1002**	1835.6	572.22	**1062**	1943.6	605.56	**1122**	2051.6
539.44	**1003**	1837.4	572.78	**1063**	1945.4	606.11	**1123**	2053.4
540	**1004**	1839.2	573.33	**1064**	1947.2	606.67	**1124**	2055.2
540.56	**1005**	1841	573.89	**1065**	1949	607.22	**1125**	2057
541.11	**1006**	1842.8	574.44	**1066**	1950.8	607.78	**1126**	2058.8
541.67	**1007**	1844.6	575	**1067**	1952.6	608.33	**1127**	2060.6
542.22	**1008**	1846.4	575.56	**1068**	1954.4	608.89	**1128**	2062.4
542.78	**1009**	1848.2	576.11	**1069**	1956.2	609.44	**1129**	2064.2

TEMPERATURE CONVERSION TABLE (Continued)

To Convert			To Convert			To Convert		
To °C	←°F or °C→	To °F	To °C	←°F or °C→	To °F	To °C	←°F or °C→	To °F
610	**1130**	2066	643.33	**1190**	2174	676.67	**1250**	2282
610.56	**1131**	2067.8	643.89	**1191**	2175.8	677.22	**1251**	2283.8
611.11	**1132**	2069.6	644.44	**1192**	2177.6	677.78	**1252**	2285.6
611.67	**1133**	2071.4	645	**1193**	2179.4	678.33	**1253**	2287.4
612.22	**1134**	2073.2	645.56	**1194**	2181.2	678.89	**1254**	2289.2
612.78	**1135**	2075	646.11	**1195**	2183	679.44	**1255**	2291
613.33	**1136**	2076.8	646.67	**1196**	2184.8	680	**1256**	2292.8
613.89	**1137**	2078.6	647.22	**1197**	2186.6	680.56	**1257**	2294.6
614.44	**1138**	2080.4	647.78	**1198**	2188.4	681.11	**1258**	2296.4
615	**1139**	2082.2	648.33	**1199**	2190.2	681.67	**1259**	2298.2
615.56	**1140**	2084	648.89	**1200**	2192	682.22	**1260**	2300
616.11	**1141**	2085.8	649.44	**1201**	2193.8	682.78	**1261**	2301.8
616.67	**1142**	2087.6	650	**1202**	2195.6	683.33	**1262**	2303.6
617.22	**1143**	2089.4	650.56	**1203**	2197.4	683.89	**1263**	2305.4
617.78	**1144**	2091.2	651.11	**1204**	2199.2	684.44	**1264**	2307.2
618.33	**1145**	2093	651.67	**1205**	2201	685	**1265**	2309
618.89	**1146**	2094.8	652.22	**1206**	2202.8	685.56	**1266**	2310.8
619.44	**1147**	2096.6	652.78	**1207**	2204.6	686.11	**1267**	2312.6
620	**1148**	2098.4	653.33	**1208**	2206.4	686.67	**1268**	2314.4
620.56	**1149**	2100.2	653.89	**1209**	2208.2	687.22	**1269**	2316.2
621.11	**1150**	2102	654.44	**1210**	2210	687.78	**1270**	2318
621.67	**1151**	2103.8	655	**1211**	2211.8	688.33	**1271**	2319.8
622.22	**1152**	2105.6	655.56	**1212**	2213.6	688.89	**1272**	2321.6
622.78	**1153**	2107.4	656.11	**1213**	2215.4	689.44	**1273**	2323.4
623.33	**1154**	2109.2	656.67	**1214**	2217.2	690	**1274**	2325.2
623.89	**1155**	2111	657.22	**1215**	2219	690.56	**1275**	2327
624.44	**1156**	2112.8	657.78	**1216**	2220.8	691.11	**1276**	2328.8
625	**1157**	2114.6	658.33	**1217**	2222.6	691.67	**1277**	2330.6
625.56	**1158**	2116.4	658.89	**1218**	2224.4	692.22	**1278**	2332.4
626.11	**1159**	2118.2	659.44	**1219**	2226.2	692.78	**1279**	2334.2
626.67	**1160**	2120	660	**1220**	2228	693.33	**1280**	2336
627.22	**1161**	2121.8	660.56	**1221**	2229.8	693.89	**1281**	2337.8
627.78	**1162**	2123.6	661.11	**1222**	2231.6	694.44	**1282**	2339.6
628.33	**1163**	2125.4	661.67	**1223**	2233.4	695	**1283**	2341.4
628.89	**1164**	2127.2	662.22	**1224**	2235.2	695.56	**1284**	2343.2
629.44	**1165**	2129	662.78	**1225**	2237	696.11	**1285**	2345
630	**1166**	2130.8	663.33	**1226**	2238.8	696.67	**1286**	2346.8
630.56	**1167**	2132.6	663.89	**1227**	2240.6	697.22	**1287**	2348.6
631.11	**1168**	2134.4	664.44	**1228**	2242.4	697.78	**1288**	2350.4
631.67	**1169**	2136.2	665	**1229**	2244.2	698.33	**1289**	2352.2
632.22	**1170**	2138	665.56	**1230**	2246	698.89	**1290**	2354
632.78	**1171**	2139.8	666.11	**1231**	2247.8	699.44	**1291**	2355.8
633.33	**1172**	2141.6	666.67	**1232**	2249.6	700	**1292**	2357.6
633.89	**1173**	2143.4	667.22	**1233**	2251.4	700.56	**1293**	2359.4
634.44	**1174**	2145.2	667.78	**1234**	2253.2	701.11	**1294**	2361.2
635	**1175**	2147	668.33	**1235**	2255	701.67	**1295**	2363
635.56	**1176**	2148.8	668.89	**1236**	2256.8	702.22	**1296**	2364.8
636.11	**1177**	2150.6	669.44	**1237**	2258.6	702.78	**1297**	2366.6
636.67	**1178**	2152.4	670	**1238**	2260.4	703.33	**1298**	2368.4
637.22	**1179**	2154.2	670.56	**1239**	2262.2	703.89	**1299**	2370.2
637.78	**1180**	2156	671.11	**1240**	2264	704.44	**1300**	2372
638.33	**1181**	2157.8	671.67	**1241**	2265.8	705	**1301**	2373.8
638.89	**1182**	2159.6	672.22	**1242**	2267.6	705.56	**1302**	2375.6
639.44	**1183**	2161.4	672.78	**1243**	2269.4	706.11	**1303**	2377.4
640	**1184**	2163.2	673.33	**1244**	2271.2	706.67	**1304**	2379.2
640.56	**1185**	2165	673.89	**1245**	2273	707.22	**1305**	2381
641.11	**1186**	2166.8	674.44	**1246**	2274.8	707.78	**1306**	2382.8
641.67	**1187**	2168.6	675	**1247**	2276.6	708.33	**1307**	2384.6
642.22	**1188**	2170.4	675.56	**1248**	2278.4	708.89	**1308**	2386.4
642.78	**1189**	2172.2	676.11	**1249**	2280.2	709.44	**1309**	2388.2

To Convert			To Convert			To Convert		
To °C	←°F or °C→	To °F	To °C	←°F or °C→	To °F	To °C	←°F or °C→	To °F
710	**1310**	2390	743.33	**1370**	2498	776.67	**1430**	2606
710.56	**1311**	2391.8	743.89	**1371**	2499.8	777.22	**1431**	2607.8
711.11	**1312**	2393.6	744.44	**1372**	2501.6	777.78	**1432**	2609.6
711.67	**1313**	2395.4	745	**1373**	2503.4	778.33	**1433**	2611.4
712.22	**1314**	2397.2	745.56	**1374**	2505.2	778.89	**1434**	2613.2
712.78	**1315**	2399	746.11	**1375**	2507	779.44	**1435**	2615
713.33	**1316**	2400.8	746.67	**1376**	2508.8	780	**1436**	2616.8
713.89	**1317**	2402.6	747.22	**1377**	2510.6	780.56	**1437**	2618.6
714.44	**1318**	2404.4	747.78	**1378**	2512.4	781.11	**1438**	2620.4
715	**1319**	2406.2	748.33	**1379**	2514.2	781.67	**1439**	2622.2
715.56	**1320**	2408	748.89	**1380**	2516	782.22	**1440**	2624
716.11	**1321**	2409.8	749.44	**1381**	2517.8	782.78	**1441**	2625.8
716.67	**1322**	2411.6	750	**1382**	2519.6	783.33	**1442**	2627.6
717.22	**1323**	2413.4	750.56	**1383**	2521.4	783.89	**1443**	2629.4
717.78	**1324**	2415.2	751.11	**1384**	2523.2	784.44	**1444**	2631.2
718.33	**1325**	2417	751.67	**1385**	2525	785	**1445**	2633
718.89	**1326**	2418.8	752.22	**1386**	2526.8	785.56	**1446**	2634.8
719.44	**1327**	2420.6	752.78	**1387**	2528.6	786.11	**1447**	2636.6
720	**1328**	2422.4	753.33	**1388**	2530.4	786.67	**1448**	2638.4
720.56	**1329**	2424.2	753.89	**1389**	2532.2	787.22	**1449**	2640.2
721.11	**1330**	2426	754.44	**1390**	2534	787.78	**1450**	2642
721.67	**1331**	2427.8	755	**1391**	2535.8	788.33	**1451**	2443.8
722.22	**1332**	2429.6	755.56	**1392**	2537.6	788.89	**1452**	2645.6
722.78	**1333**	2431.4	756.11	**1393**	2539.4	789.44	**1453**	2647.4
723.33	**1334**	2433.2	756.67	**1394**	2541.2	790	**1454**	2649.2
723.89	**1335**	2435	757.22	**1395**	2543	790.56	**1455**	2651
724.44	**1336**	2436.8	757.78	**1396**	2544.8	791.11	**1456**	2652.8
725	**1337**	2438.6	758.33	**1397**	2546.6	791.67	**1457**	2654.6
725.56	**1338**	2440.4	758.89	**1398**	2548.4	792.22	**1458**	2656.4
726.11	**1339**	2442.2	759.44	**1399**	2550.2	792.78	**1459**	2658.2
726.67	**1340**	2444	760	**1400**	2552	793.33	**1460**	2660
727.22	**1341**	2445.8	760.56	**1401**	2553.8	793.89	**1461**	2661.8
727.78	**1342**	2447.6	761.11	**1402**	2555.6	794.44	**1462**	2663.6
728.33	**1343**	2449.4	761.67	**1403**	2557.4	795	**1463**	2665.4
728.89	**1344**	2451.2	762.22	**1404**	2559.2	795.56	**1464**	2667.2
729.44	**1345**	2453	762.78	**1405**	2561	796.11	**1465**	2669
730	**1346**	2454.8	763.33	**1406**	2562.8	796.67	**1466**	2670.8
730.56	**1347**	2456.6	763.89	**1407**	2564.6	797.22	**1467**	2672.6
731.11	**1348**	2458.4	764.44	**1408**	2566.4	797.78	**1468**	2674.4
731.67	**1349**	2460.2	765	**1409**	2568.2	798.33	**1469**	2676.2
732.22	**1350**	2462	765.56	**1410**	2570	798.89	**1470**	2678
732.78	**1351**	2463.8	766.11	**1411**	2571.8	799.44	**1471**	2679.8
733.33	**1352**	2465.6	766.67	**1412**	2573.6	800	**1472**	2681.6
733.89	**1353**	2467.4	767.22	**1413**	2575.4	800.56	**1473**	2683.4
734.44	**1354**	2469.2	767.78	**1414**	2577.2	801.11	**1474**	2685.2
735	**1355**	2471	768.33	**1415**	2579	801.67	**1475**	2687
735.56	**1356**	2472.8	768.89	**1416**	2580.8	802.22	**1476**	2688.8
736.11	**1357**	2474.6	769.44	**1417**	2582.6	802.78	**1477**	2690.6
736.67	**1358**	2476.4	770	**1418**	2584.4	803.33	**1478**	2692.4
737.22	**1359**	2478.2	770.56	**1419**	2586.2	803.89	**1479**	2694.2
737.78	**1360**	2480	771.11	**1420**	2588	804.44	**1480**	2696
738.33	**1361**	2481.8	771.67	**1421**	2589.8	805	**1481**	2697.8
738.89	**1362**	2483.6	772.22	**1422**	2591.6	805.56	**1482**	2699.6
739.44	**1363**	2485.4	772.78	**1423**	2593.4	806.11	**1483**	2701.4
740	**1364**	2487.2	773.33	**1424**	2595.2	806.67	**1484**	2703.2
740.56	**1365**	2489	773.89	**1425**	2597	807.22	**1485**	2705
741.11	**1366**	2490.8	774.44	**1426**	2598.8	807.78	**1486**	2706.8
741.67	**1367**	2492.6	775	**1427**	2600.6	808.33	**1487**	2708.6
742.22	**1368**	2494.4	775.56	**1428**	2602.4	808.89	**1488**	2710.4
742.78	**1369**	2496.2	776.11	**1429**	2604.2	809.44	**1489**	2712.2

TEMPERATURE CONVERSION TABLE (Continued)

To Convert			To Convert			To Convert		
To °C	←°F or °C→	To °F	To °C	←°F or °C→	To °F	To °C	←°F or °C→	To °F
810	**1490**	2714	843.33	**1550**	2822	876.67	**1610**	2930
810.56	**1491**	2715.8	843.89	**1551**	2823.8	877.22	**1611**	2931.8
811.11	**1492**	2717.6	844.44	**1552**	2825.6	877.78	**1612**	2933.6
811.67	**1493**	2719.4	845	**1553**	2827.4	878.33	**1613**	2935.4
812.22	**1494**	2721.2	845.56	**1554**	2829.2	878.89	**1614**	2937.2
812.78	**1495**	2723	846.11	**1555**	2831	879.44	**1615**	2939
813.33	**1496**	2724.8	846.67	**1556**	2832.8	880	**1616**	2940.8
813.89	**1497**	2726.6	847.22	**1557**	2834.6	880.56	**1617**	2942.6
814.44	**1498**	2728.4	847.78	**1558**	2836.4	881.11	**1618**	2944.4
815	**1499**	2730.2	848.33	**1559**	2838.2	881.67	**1619**	2946.2
815.56	**1500**	2732	848.89	**1560**	2840	882.22	**1620**	2948
816.11	**1501**	2733.8	849.44	**1561**	2841.8	882.78	**1621**	2949.8
816.67	**1502**	2735.6	850	**1562**	2843.6	883.33	**1622**	2951.6
817.22	**1503**	2737.4	850.56	**1563**	2845.4	883.89	**1623**	2953.4
817.78	**1504**	2739.2	851.11	**1564**	2847.2	884.44	**1624**	2955.2
818.33	**1505**	2741	851.67	**1565**	2849	885	**1625**	2957
818.89	**1506**	2742.8	852.22	**1566**	2850.8	885.56	**1626**	2958.8
819.44	**1507**	2744.6	852.78	**1567**	2852.6	886.11	**1627**	2960.6
820	**1508**	2746.4	853.33	**1568**	2854.4	886.67	**1628**	2962.4
820.56	**1509**	2748.2	853.89	**1569**	2856.2	887.22	**1629**	2964.2
821.11	**1510**	2750	854.44	**1570**	2858	887.78	**1630**	2966
821.67	**1511**	2751.8	855	**1571**	2859.8	888.33	**1631**	2967.8
822.22	**1512**	2753.6	855.56	**1572**	2861.6	888.89	**1632**	2969.6
822.78	**1513**	2755.4	856.11	**1573**	2863.4	889.44	**1633**	2971.4
823.33	**1514**	2757.2	856.67	**1574**	2865.2	890	**1634**	2973.2
823.89	**1515**	2759	857.22	**1575**	2867	890.56	**1635**	2975
824.44	**1516**	2760.8	857.78	**1576**	2868.8	891.11	**1636**	2976.8
825	**1517**	2762.6	858.33	**1577**	2870.6	891.67	**1637**	2978.6
825.56	**1518**	2764.4	858.89	**1578**	2872.4	892.22	**1638**	2980.4
826.11	**1519**	2766.2	859.44	**1579**	2874.2	892.78	**1639**	2982.2
826.67	**1520**	2768	860	**1580**	2876	893.33	**1640**	2984
827.22	**1521**	2769.8	860.56	**1581**	2877.8	893.89	**1641**	2985.8
827.78	**1522**	2771.6	861.11	**1582**	2879.6	894.44	**1642**	2987.6
828.33	**1523**	2773.4	861.67	**1583**	2881.4	895	**1643**	2989.4
828.89	**1524**	2775.2	862.22	**1584**	2883.2	895.56	**1644**	2991.2
829.44	**1525**	2777	862.78	**1585**	2885	896.11	**1645**	2993
830	**1526**	2778.8	863.33	**1586**	2886.8	896.67	**1646**	2994.8
830.56	**1527**	2780.6	863.89	**1587**	2888.6	897.22	**1647**	2996.6
831.11	**1528**	2782.4	864.44	**1588**	2890.4	897.78	**1648**	2998.4
831.67	**1529**	2784.2	865	**1589**	2892.2	898.33	**1649**	3000.2
832.22	**1530**	2786	865.56	**1590**	2894	898.89	**1650**	3002
832.78	**1531**	2787.8	866.11	**1591**	2895.8	899.44	**1651**	3003.8
833.33	**1532**	2789.6	866.67	**1592**	2897.6	900	**1652**	3005.6
833.89	**1533**	2791.4	867.22	**1593**	2899.4	900.56	**1653**	3007.4
834.44	**1534**	2793.2	867.78	**1594**	2901.2	901.11	**1654**	3009.2
835	**1535**	2795	868.33	**1595**	2903	901.67	**1655**	3011
835.56	**1536**	2796.8	868.89	**1596**	2904.8	902.22	**1656**	3012.8
836.11	**1537**	2798.6	869.44	**1597**	2906.6	902.78	**1657**	3014.6
836.67	**1538**	2800.4	870	**1598**	2908.4	903.33	**1658**	3016.4
837.22	**1539**	2802.2	870.56	**1599**	2910.2	903.89	**1659**	3018.2
837.78	**1540**	2804	871.11	**1600**	2912	904.44	**1660**	3020
838.33	**1541**	2805.8	871.67	**1601**	2913.8	905	**1661**	3021.8
838.89	**1542**	2807.6	872.22	**1602**	2915.6	905.56	**1662**	3023.6
839.44	**1543**	2809.4	872.78	**1603**	2917.4	906.11	**1663**	3025.4
840	**1544**	2811.2	873.33	**1604**	2919.2	906.67	**1664**	3027.2
840.56	**1545**	2813	873.89	**1605**	2921	907.22	**1665**	3029
841.11	**1546**	2814.8	874.44	**1606**	2922.8	907.78	**1666**	3030.8
841.67	**1547**	2816.6	875	**1607**	2924.6	908.33	**1667**	3032.6
842.22	**1548**	2818.4	875.56	**1608**	2926.4	908.89	**1668**	3034.4
842.78	**1549**	2820.2	876.11	**1609**	2928.2	909.44	**1669**	3036.2

To Convert			To Convert			To Convert		
To °C	←°F or °C→	To °F	To °C	←°F or °C→	To °F	To °C	←°F or °C→	To °F
910	**1670**	3038	943.33	**1730**	3146	976.67	**1790**	3254
910.56	**1671**	3039.8	943.89	**1731**	3147.8	977.22	**1791**	3255.8
911.11	**1672**	3041.6	944.44	**1732**	3149.6	977.78	**1792**	3257.6
911.67	**1673**	3043.4	945	**1733**	3151.4	978.33	**1793**	3259.4
912.22	**1674**	3045.2	945.56	**1734**	3153.2	978.89	**1794**	3261.2
912.78	**1675**	3047	946.11	**1735**	3155	979.44	**1795**	3263
913.33	**1676**	3048.8	946.67	**1736**	3156.8	980	**1796**	3264.8
913.89	**1677**	3050.6	947.22	**1737**	3158.6	980.56	**1797**	3266.6
914.44	**1678**	3052.4	947.78	**1738**	3160.4	981.11	**1798**	3268.4
915	**1679**	3054.2	948.33	**1739**	3162.2	981.67	**1799**	3270.2
915.56	**1680**	3056	948.89	**1740**	3164	982.22	**1800**	3272
916.11	**1681**	3057.8	949.44	**1741**	3165.8	982.78	**1801**	3273.8
916.67	**1682**	3059.6	950	**1742**	3167.6	983.33	**1802**	3275.6
917.22	**1683**	3061.4	950.56	**1743**	3169.4	983.89	**1803**	3277.4
917.78	**1684**	3063.2	951.11	**1744**	3171.2	984.44	**1804**	3279.2
918.33	**1685**	3065	951.67	**1745**	3173	985	**1805**	3281
918.89	**1686**	3066.8	952.22	**1746**	3174.8	985.56	**1806**	3282.8
919.44	**1687**	3068.6	952.78	**1747**	3176.6	986.11	**1807**	3284.6
920	**1688**	3070.4	953.33	**1748**	3178.4	986.67	**1808**	3286.4
920.56	**1689**	3072.2	953.89	**1749**	3180.2	987.22	**1809**	3288.2
921.11	**1690**	3074	954.44	**1750**	3182	987.78	**1810**	3290
921.67	**1691**	3075.8	955	**1751**	3183.8	988.33	**1811**	3291.8
922.22	**1692**	3077.6	955.56	**1752**	3185.6	988.89	**1812**	3293.6
922.78	**1693**	3079.4	956.11	**1753**	3187.4	989.44	**1813**	3295.4
923.33	**1694**	3081.2	956.67	**1754**	3189.2	990	**1814**	3297.2
923.89	**1695**	3083	957.22	**1755**	3191	990.56	**1815**	3299
924.44	**1696**	3084.8	957.78	**1756**	3192.8	991.11	**1816**	3300.8
925	**1697**	3086.6	958.33	**1757**	3194.6	991.67	**1817**	3302.6
925.56	**1698**	3088.4	958.89	**1758**	3196.4	992.22	**1818**	3304.4
926.11	**1699**	3090.2	959.44	**1759**	3198.2	992.78	**1819**	3306.2
926.67	**1700**	3092	960	**1760**	3200	993.33	**1820**	3308
927.22	**1701**	3093.8	960.56	**1761**	3201.8	993.89	**1821**	3309.8
927.78	**1702**	3095.6	961.11	**1762**	3203.6	994.44	**1822**	3311.6
928.33	**1703**	3097.4	961.67	**1763**	3205.4	995	**1823**	3313.4
928.89	**1704**	3099.2	962.22	**1764**	3207.2	995.56	**1824**	3315.2
929.44	**1705**	3101	962.78	**1765**	3209	996.11	**1825**	3317
930	**1706**	3102.8	963.33	**1766**	3210.8	996.67	**1826**	3318.8
930.56	**1707**	3104.6	963.89	**1767**	3212.6	997.22	**1827**	3320.6
931.11	**1708**	3106.4	964.44	**1768**	3214.4	997.78	**1828**	3322.4
931.67	**1709**	3108.2	965	**1769**	3216.2	998.33	**1829**	3324.2
932.22	**1710**	3110	965.56	**1770**	3218	998.89	**1830**	3326
932.78	**1711**	3111.8	966.11	**1771**	3219.8	999.44	**1831**	3327.8
933.33	**1712**	3113.6	966.67	**1772**	3221.6	1000	**1832**	3329.6
933.89	**1713**	3115.4	967.22	**1773**	3223.4	1000.56	**1833**	3331.4
934.44	**1714**	3117.2	967.78	**1774**	3225.2	1001.11	**1834**	3333.2
935	**1715**	3119	968.33	**1775**	3227	1001.67	**1835**	3335
935.56	**1716**	3120.8	968.89	**1776**	3228.8	1002.22	**1836**	3336.8
936.11	**1717**	3122.6	969.44	**1777**	3230.6	1002.78	**1837**	3338.6
936.67	**1718**	3124.4	970	**1778**	3232.4	1003.33	**1838**	3340.4
937.22	**1719**	3126.2	970.56	**1779**	3234.2	1003.89	**1839**	3342.2
937.78	**1720**	3128	971.11	**1780**	3236	1004.44	**1840**	3344
938.33	**1721**	3129.8	971.67	**1781**	3237.8	1005	**1841**	3345.8
938.89	**1722**	3131.6	972.22	**1782**	3239.6	1005.56	**1842**	3347.6
939.44	**1723**	3133.4	972.78	**1783**	3241.4	1006.11	**1843**	3349.4
940	**1724**	3135.2	973.33	**1784**	3243.2	1006.67	**1844**	3351.2
940.56	**1725**	3137	973.89	**1785**	3245	1007.22	**1845**	3353
941.11	**1726**	3138.8	974.44	**1786**	3246.8	1007.78	**1846**	3354.8
941.67	**1727**	3140.6	975	**1787**	3248.6	1008.33	**1847**	3356.6
942.22	**1728**	3142.4	975.56	**1788**	3250.4	1008.89	**1848**	3358.4
942.78	**1729**	3144.2	976.11	**1789**	3252.2	1009.44	**1849**	3360.2

TEMPERATURE CONVERSION TABLE (Continued)

To Convert			To Convert			To Convert		
To °C	←°F or °C→	To °F	To °C	←°F or °C→	To °F	To °C	←°F or °C→	To °F
1010	**1850**	3362	1043.33	**1910**	3470	1076.67	**1970**	3578
1010.56	**1851**	3363.8	1043.89	**1911**	3471.8	1077.22	**1971**	3579.8
1011.11	**1852**	3365.6	1044.44	**1912**	3473.6	1077.78	**1972**	3581.6
1011.67	**1853**	3367.4	1045	**1913**	3475.4	1078.33	**1973**	3583.4
1012.22	**1854**	3369.2	1045.56	**1914**	3477.2	1078.89	**1974**	3585.2
1012.78	**1855**	3371	1046.11	**1915**	3479	1079.44	**1975**	3587
1013.33	**1856**	3372.8	1046.67	**1916**	3480.8	1080	**1976**	3588.8
1013.89	**1857**	3374.6	1047.22	**1917**	3482.6	1080.56	**1977**	3590.6
1014.44	**1858**	3376.4	1047.78	**1918**	3484.4	1081.11	**1978**	3592.4
1015	**1859**	3378.2	1048.33	**1919**	3486.2	1081.67	**1979**	3594.2
1015.56	**1860**	3380	1048.89	**1920**	3488	1082.22	**1980**	3596
1016.11	**1861**	3381.8	1049.44	**1921**	3489.8	1082.78	**1981**	3597.8
1016.67	**1862**	3383.6	1050	**1922**	3491.6	1083.33	**1982**	3599.6
1017.22	**1863**	3385.4	1050.56	**1923**	3493.4	1083.89	**1983**	3601.4
1017.78	**1864**	3387.2	1051.11	**1924**	3495.2	1084.44	**1984**	3603.2
1018.33	**1865**	3389	1051.67	**1925**	3497	1085	**1985**	3605
1018.89	**1866**	3390.8	1052.22	**1926**	3498.8	1085.56	**1986**	3606.8
1019.44	**1867**	3392.6	1052.78	**1927**	3500.6	1086.11	**1987**	3608.6
1020	**1868**	3394.4	1053.33	**1928**	3502.4	1086.67	**1988**	3610.4
1020.56	**1869**	3396.2	1053.89	**1929**	3504.2	1087.22	**1989**	3612.2
1021.11	**1870**	3398	1054.44	**1930**	3506	1087.78	**1990**	3614
1021.67	**1871**	3399.8	1055	**1931**	3507.8	1088.33	**1991**	3615.8
1022.22	**1872**	3401.6	1055.56	**1932**	3509.6	1088.89	**1992**	3617.6
1022.78	**1873**	3403.4	1056.11	**1933**	3511.4	1089.44	**1993**	3619.4
1023.33	**1874**	3405.2	1056.67	**1934**	3513.2	1090	**1994**	3621.2
1023.89	**1875**	3407	1057.22	**1935**	3515	1090.56	**1995**	3623
1024.44	**1876**	3408.8	1057.78	**1936**	3516.8	1091.11	**1996**	3624.8
1025	**1877**	3410.6	1058.33	**1937**	3518.6	1091.67	**1997**	3626.6
1025.56	**1878**	3412.4	1058.89	**1938**	3520.4	1092.22	**1998**	3628.4
1026.11	**1879**	3414.2	1059.44	**1939**	3522.2	1092.78	**1999**	3630.2
1026.67	**1880**	3416	1060	**1940**	3524	1093.33	**2000**	3632
1027.22	**1881**	3417.8	1060.56	**1941**	3525.8	1093.89	**2001**	3633.8
1027.78	**1882**	3419.6	1061.11	**1942**	3527.6	1094.44	**2002**	3635.6
1028.33	**1883**	3421.4	1061.67	**1943**	3529.4	1095	**2003**	3637.4
1028.89	**1884**	3423.2	1062.22	**1944**	3531.2	1095.56	**2004**	3639.2
1029.44	**1885**	3425	1062.78	**1945**	3533	1096.11	**2005**	3641
1030	**1886**	3426.8	1063.33	**1946**	3534.8	1096.67	**2006**	3642.8
1030.56	**1887**	3428.6	1063.89	**1947**	3536.6	1097.22	**2007**	3644.6
1031.11	**1888**	3430.4	1064.44	**1948**	3538.4	1097.78	**2008**	3646.4
1031.67	**1889**	3432.2	1065	**1949**	3540.2	1098.33	**2009**	3648.2
1032.22	**1890**	3434	1065.56	**1950**	3542	1098.89	**2010**	3650
1032.78	**1891**	3435.8	1066.11	**1951**	3543.8	1099.44	**2011**	3651.8
1033.33	**1892**	3437.6	1066.67	**1952**	3545.6	1100	**2012**	3653.6
1033.89	**1893**	3439.4	1067.22	**1953**	3547.4	1100.56	**2013**	3655.4
1034.44	**1894**	3441.2	1067.78	**1954**	3549.2	1101.11	**2014**	3657.2
1035	**1895**	3443	1068.33	**1955**	3551	1101.67	**2015**	3659
1035.56	**1896**	3444.8	1068.89	**1956**	3552.8	1102.22	**2016**	3660.8
1036.11	**1897**	3446.6	1069.44	**1957**	3554.6	1102.78	**2017**	3662.6
1036.67	**1898**	3448.4	1070	**1958**	3556.4	1103.33	**2018**	3664.4
1037.22	**1899**	3450.2	1070.56	**1959**	3558.2	1103.89	**2019**	3666.2
1037.78	**1900**	3452	1071.11	**1960**	3560	1104.44	**2020**	3668
1038.33	**1901**	3453.8	1071.67	**1961**	3561.8	1105	**2021**	3669.8
1038.89	**1902**	3455.6	1072.22	**1962**	3563.6	1105.56	**2022**	3671.6
1039.44	**1903**	3457.4	1072.78	**1963**	3565.4	1106.11	**2023**	3673.4
1040	**1904**	3459.2	1073.33	**1964**	3567.2	1106.67	**2024**	3675.2
1040.56	**1905**	3461	1073.89	**1965**	3569	1107.22	**2025**	3677
1041.11	**1906**	3462.8	1074.44	**1966**	3570.8	1107.78	**2026**	3678.8
1041.67	**1907**	3464.6	1075	**1967**	3572.6	1108.33	**2027**	3680.6
1042.22	**1908**	3466.4	1075.56	**1968**	3574.4	1108.89	**2028**	3682.4
1042.78	**1909**	3468.2	1076.11	**1969**	3576.2	1109.44	**2029**	3684.2

To Convert			To Convert			To Convert		
To °C	←°F or °C→	To °F	To °C	←°F or °C→	To °F	To °C	←°F or °C→	To °F
1110	**2030**	3686	1143.33	**2090**	3794	1176.67	**2150**	3902
1110.56	**2031**	3687.8	1143.89	**2091**	3795.8	1177.22	**2151**	3903.8
1111.11	**2032**	3689.6	1144.44	**2092**	3797.6	1177.78	**2152**	3905.6
1111.67	**2033**	3691.4	1145	**2093**	3799.4	1178.33	**2153**	3907.4
1112.22	**2034**	3693.2	1145.56	**2094**	3801.2	1178.89	**2154**	3909.2
1112.78	**2035**	3695	1146.11	**2095**	3803	1179.44	**2155**	3911
1113.33	**2036**	3696.8	1146.67	**2096**	3804.8	1180	**2156**	3912.8
1113.89	**2037**	3698.6	1147.22	**2097**	3806.6	1180.56	**2157**	3914.6
1114.44	**2038**	3700.4	1147.78	**2098**	3808.4	1181.11	**2158**	3916.4
1115	**2039**	3702.2	1148.33	**2099**	3810.2	1181.67	**2159**	3918.2
1115.56	**2040**	3704	1148.89	**2100**	3812	1182.22	**2160**	3920
1116.11	**2041**	3705.8	1149.44	**2101**	3813.8	1182.78	**2161**	3921.8
1116.67	**2042**	3707.6	1150	**2102**	3815.6	1183.33	**2162**	3923.6
1117.22	**2043**	3709.4	1150.56	**2103**	3817.4	1183.89	**2163**	3925.4
1117.78	**2044**	3711.2	1151.11	**2104**	3819.2	1184.44	**2164**	3927.2
1118.33	**2045**	3713	1151.67	**2105**	3821	1185	**2165**	3929
1118.89	**2046**	3714.8	1152.22	**2106**	3822.8	1185.56	**2166**	3930.8
1119.44	**2047**	3716.6	1152.78	**2107**	3824.6	1186.11	**2167**	3932.6
1120	**2048**	3718.4	1153.33	**2108**	3826.4	1186.67	**2168**	3934.4
1120.56	**2049**	3720.2	1153.89	**2109**	3828.2	1187.22	**2169**	3936.2
1121.11	**2050**	3722	1154.44	**2110**	3830	1187.78	**2170**	3938
1121.67	**2051**	3723.8	1155	**2111**	3831.8	1188.33	**2171**	3939.8
1122.22	**2052**	3725.6	1155.56	**2112**	3833.6	1188.89	**2172**	3941.6
1122.78	**2053**	3727.4	1156.11	**2113**	3835.4	1189.44	**2173**	3943.4
1123.33	**2054**	3729.2	1156.67	**2114**	3837.2	1190	**2174**	3945.2
1123.89	**2055**	3731	1157.22	**2115**	3839	1190.56	**2175**	3947
1124.44	**2056**	3732.8	1157.78	**2116**	3840.8	1191.11	**2176**	3948.8
1125	**2057**	3734.6	1158.33	**2117**	3842.6	1191.67	**2177**	3950.6
1125.56	**2058**	3736.4	1158.89	**2118**	3844.4	1192.22	**2178**	3952.4
1126.11	**2059**	3738.2	1159.44	**2119**	3846.2	1192.78	**2179**	3954.2
1126.67	**2060**	3740	1160	**2120**	3848	1193.33	**2180**	3956
1127.22	**2061**	3741.8	1160.56	**2121**	3849.8	1193.89	**2181**	3957.8
1127.78	**2062**	3743.6	1161.11	**2122**	3851.6	1194.44	**2182**	3959.6
1128.33	**2063**	3745.4	1161.67	**2123**	3853.4	1195	**2183**	3961.4
1128.89	**2064**	3747.2	1162.22	**2124**	3855.2	1195.56	**2184**	3963.2
1129.44	**2065**	3749	1162.78	**2125**	3857	1196.11	**2185**	3965
1130	**2066**	3750.8	1163.33	**2126**	3858.8	1196.67	**2186**	3966.8
1130.56	**2067**	3752.6	1163.89	**2127**	3860.6	1197.22	**2187**	3968.6
1131.11	**2068**	3754.4	1164.44	**2128**	3862.4	1197.78	**2188**	3970.4
1131.67	**2069**	3756.2	1165	**2129**	3864.2	1198.33	**2189**	3972.2
1132.22	**2070**	3758	1165.56	**2130**	3866	1198.89	**2190**	3974
1132.78	**2071**	3759.8	1166.11	**2131**	3867.8	1199.44	**2191**	3975.8
1133.33	**2072**	3761.6	1166.67	**2132**	3869.6	1200	**2192**	3977.6
1133.89	**2073**	3763.4	1167.22	**2133**	3871.4	1200.56	**2193**	3979.4
1134.44	**2074**	3765.2	1167.78	**2134**	3873.2	1201.11	**2194**	3981.2
1135	**2075**	3767	1168.33	**2135**	3875	1201.67	**2195**	3983
1135.56	**2076**	3768.8	1168.89	**2136**	3876.8	1202.22	**2196**	3984.8
1136.11	**2077**	3770.6	1169.44	**2137**	3878.6	1202.78	**2197**	3986.6
1136.67	**2078**	3772.4	1170	**2138**	3880.4	1203.33	**2198**	3988.4
1137.22	**2079**	3774.2	1170.56	**2139**	3882.2	1203.89	**2199**	3990.2
1137.78	**2080**	3776	1171.11	**2140**	3884	1204.44	**2200**	3992
1138.33	**2081**	3777.8	1171.67	**2141**	3885.8	1205	**2201**	3993.8
1138.89	**2082**	3779.6	1172.22	**2142**	3887.6	1205.56	**2202**	3995.6
1139.44	**2083**	3781.4	1172.78	**2143**	3889.4	1206.11	**2203**	3997.4
1140	**2084**	3783.2	1173.33	**2144**	3891.2	1206.67	**2204**	3999.2
1140.56	**2085**	3785	1173.89	**2145**	3893	1207.22	**2205**	4001
1141.11	**2086**	3786.8	1174.44	**2146**	3894.8	1207.78	**2206**	4002.8
1141.67	**2087**	3788.6	1175	**2147**	3896.6	1208.33	**2207**	4004.6
1142.22	**2088**	3790.4	1175.56	**2148**	3898.4	1208.89	**2208**	4006.4
1142.78	**2089**	3792.2	1176.11	**2149**	3900.2	1209.44	**2209**	4008.2

TEMPERATURE CONVERSION TABLE (Continued)

To Convert			To Convert			To Convert		
To °C	←°F or °C→	To °F	To °C	←°F or °C→	To °F	To °C	←°F or °C→	To °F
1210	**2210**	4010	1243.33	**2270**	4118	1276.67	**2330**	4226
1210.56	**2211**	4011.8	1243.89	**2271**	4119.8	1277.22	**2331**	4227.8
1211.11	**2212**	4013.6	1244.44	**2272**	4121.6	1277.78	**2332**	4229.6
1211.67	**2213**	4015.4	1245	**2273**	4123.4	1278.33	**2333**	4231.4
1212.22	**2214**	4017.2	1245.56	**2274**	4125.2	1278.89	**2334**	4233.2
1212.78	**2215**	4019	1246.11	**2275**	4127	1279.44	**2335**	4235
1213.33	**2216**	4020.8	1246.67	**2276**	4128.8	1280	**2336**	4236.8
1213.89	**2217**	4022.6	1247.22	**2277**	4130.6	1280.56	**2337**	4238.6
1214.44	**2218**	4024.4	1247.78	**2278**	4132.4	1281.11	**2338**	4240.4
1215	**2219**	4026.2	1248.33	**2279**	4134.2	1281.67	**2339**	4242.2
1215.56	**2220**	4028	1248.89	**2280**	4136	1282.22	**2340**	4244
1216.11	**2221**	4029.8	1249.44	**2281**	4137.8	1282.78	**2341**	4245.8
1216.67	**2222**	4031.6	1250	**2282**	4139.6	1283.33	**2342**	4247.6
1217.22	**2223**	4033.4	1250.56	**2283**	4141.4	1283.89	**2343**	4249.4
1217.78	**2224**	4035.2	1251.11	**2284**	4143.2	1284.44	**2344**	4251.2
1218.33	**2225**	4037	1251.67	**2285**	4145	1285	**2345**	4253
1218.89	**2226**	4038.8	1252.22	**2286**	4146.8	1285.56	**2346**	4254.8
1219.44	**2227**	4040.6	1252.78	**2287**	4148.6	1286.11	**2347**	4256.6
1220	**2228**	4042.4	1253.33	**2288**	4150.4	1286.67	**2348**	4258.4
1220.56	**2229**	4044.2	1253.89	**2289**	4152.2	1287.22	**2349**	4260.2
1221.11	**2230**	4046	1254.44	**2290**	4154	1287.78	**2350**	4262
1221.67	**2231**	4047.8	1255	**2291**	4155.8	1288.33	**2351**	4263.8
1222.22	**2232**	4049.6	1255.56	**2292**	4157.6	1288.89	**2352**	4265.6
1222.78	**2233**	4051.4	1256.11	**2293**	4159.4	1289.44	**2353**	4267.4
1223.33	**2234**	4053.2	1256.67	**2294**	4161.2	1290	**2354**	4269.2
1223.89	**2235**	4055	1257.22	**2295**	4163	1290.56	**2355**	4271
1224.44	**2236**	4056.8	1257.78	**2296**	4164.8	1291.11	**2356**	4272.8
1225	**2237**	4058.6	1258.33	**2297**	4166.6	1291.67	**2357**	4274.6
1225.56	**2238**	4060.4	1258.89	**2298**	4168.4	1292.22	**2358**	4276.4
1226.11	**2239**	4062.2	1259.44	**2299**	4170.2	1292.78	**2359**	4278.2
1226.67	**2240**	4064	1260	**2300**	4172	1293.33	**2360**	4280
1227.22	**2241**	4065.8	1260.56	**2301**	4173.8	1293.89	**2361**	4281.8
1227.78	**2242**	4067.6	1261.11	**2302**	4175.6	1294.44	**2362**	4283.6
1228.33	**2243**	4069.4	1261.67	**2303**	4177.4	1295	**2363**	4285.4
1228.89	**2244**	4071.2	1262.22	**2304**	4179.2	1295.56	**2364**	4287.2
1229.44	**2245**	4073	1262.78	**2305**	4181	1296.11	**2365**	4289
1230	**2246**	4074.8	1263.33	**2306**	4182.8	1296.67	**2366**	4290.8
1230.56	**2247**	4076.6	1263.89	**2307**	4184.6	1297.22	**2367**	4292.6
1231.11	**2248**	4078.4	1264.44	**2308**	4186.4	1297.78	**2368**	4294.4
1231.67	**2249**	4080.2	1265	**2309**	4188.2	1298.33	**2369**	4296.2
1232.22	**2250**	4082	1265.56	**2310**	4190	1298.89	**2370**	4298
1232.78	**2251**	4083.8	1266.11	**2311**	4191.8	1299.44	**2371**	4299.8
1233.33	**2252**	4085.6	1266.67	**2312**	4193.6	1300	**2372**	4301.6
1233.89	**2253**	4087.4	1267.22	**2313**	4195.4	1300.56	**2373**	4303.4
1234.44	**2254**	4089.2	1267.78	**2314**	4197.2	1301.11	**2374**	4305.2
1235	**2255**	4091	1268.33	**2315**	4199	1301.67	**2375**	4307
1235.56	**2256**	4092.8	1268.89	**2316**	4200.8	1302.22	**2376**	4308.8
1236.11	**2257**	4094.6	1269.44	**2317**	4202.6	1302.78	**2377**	4310.6
1236.67	**2258**	4096.4	1270	**2318**	4204.4	1303.33	**2378**	4312.4
1237.22	**2259**	4098.2	1270.56	**2319**	4206.2	1303.89	**2379**	4314.2
1237.78	**2260**	4100	1271.11	**2320**	4208	1304.44	**2380**	4316
1238.33	**2261**	4101.8	1271.67	**2321**	4209.8	1305	**2381**	4317.8
1238.89	**2262**	4103.6	1272.22	**2322**	4211.6	1305.56	**2382**	4319.6
1239.44	**2263**	4105.4	1272.78	**2323**	4213.4	1306.11	**2383**	4321.4
1240	**2264**	4107.2	1273.33	**2324**	4215.2	1306.67	**2384**	4323.2
1240.56	**2265**	4109	1273.89	**2325**	4217	1307.22	**2385**	4325
1241.11	**2267**	4110.8	1274.44	**2326**	4218.8	1307.78	**2386**	4326.8
1241.67	**2267**	4112.6	1275	**2327**	4220.6	1308.33	**2387**	4328.6
1242.22	**2268**	4114.4	1275.56	**2328**	4222.4	1308.89	**2388**	4330.4
1242.78	**2269**	4116.2	1276.11	**2329**	4224.2	1309.44	**2389**	4332.2

To Convert			To Convert			To Convert		
To °C	←°F or °C→	To °F	To °C	←°F or °C→	To °F	To °C	←°F or °C→	To °F
1310	**2390**	4334	1343.33	**2450**	4442	1376.67	**2510**	4550
1310.56	**2391**	4335.8	1343.89	**2451**	4443.8	1377.22	**2511**	4551.8
1311.11	**2392**	4337.6	1344.44	**2452**	4445.6	1377.78	**2512**	4553.6
1311.67	**2393**	4339.4	1345	**2453**	4447.4	1378.33	**2513**	4555.4
1312.22	**2394**	4341.2	1345.56	**2454**	4449.2	1378.89	**2514**	4557.2
1312.78	**2395**	4343	1346.11	**2455**	4451	1379.44	**2515**	4559
1313.33	**2396**	4344.8	1346.67	**2456**	4452.8	1380	**2516**	4560.8
1313.89	**2397**	4346.6	1347.22	**2457**	4454.6	1380.56	**2517**	4562.6
1314.44	**2398**	4348.4	1347.78	**2458**	4456.4	1381.11	**2518**	4564.4
1315	**2399**	4350.2	1348.33	**2459**	4458.2	1381.67	**2519**	4566.2
1315.56	**2400**	4352	1348.89	**2460**	4460	1382.22	**2520**	4568
1316.11	**2401**	4353.8	1349.44	**2461**	4461.8	1382.78	**2521**	4569.8
1316.67	**2402**	4355.6	1350	**2462**	4463.6	1383.33	**2522**	4571.6
1317.22	**2403**	4357.4	1350.56	**2463**	4465.4	1383.89	**2523**	4573.4
1317.78	**2404**	4359.2	1351.11	**2464**	4467.2	1384.44	**2524**	4575.2
1318.33	**2405**	4361	1351.67	**2465**	4469	1385	**2525**	4577
1318.89	**2406**	4362.8	1352.22	**2466**	4470.8	1385.56	**2526**	4578.8
1319.44	**2407**	4364.6	1352.78	**2467**	4472.6	1386.11	**2527**	4580.6
1320	**2408**	4366.4	1353.33	**2468**	4474.4	1386.67	**2528**	4582.4
1320.56	**2409**	4368.2	1353.89	**2469**	4476.2	1387.22	**2529**	4584.2
1321.11	**2410**	4370	1354.44	**2470**	4478	1387.78	**2530**	4586
1321.67	**2411**	4371.8	1355	**2471**	4479.8	1388.33	**2531**	4587.8
1322.22	**2412**	4373.6	1355.56	**2472**	4481.6	1388.89	**2532**	4589.6
1322.78	**2413**	4375.4	1356.11	**2473**	4483.4	1389.44	**2533**	4591.4
1323.33	**2414**	4377.2	1356.67	**2474**	4485.2	1390	**2534**	4593.2
1323.89	**2415**	4379	1357.22	**2475**	4487	1390.56	**2535**	4595
1324.44	**2416**	4380.8	1357.78	**2476**	4488.8	1391.11	**2536**	4596.8
1325	**2417**	4382.6	1358.33	**2477**	4490.6	1391.67	**2537**	4598.6
1325.56	**2418**	4384.4	1358.89	**2478**	4492.4	1392.22	**2538**	4600.4
1326.11	**2419**	4386.2	1359.44	**2479**	4494.2	1392.78	**2539**	4602.2
1326.67	**2420**	4388	1360	**2480**	4496	1393.33	**2540**	4604
1327.22	**2421**	4389.8	1360.56	**2481**	4497.8	1393.89	**2541**	4605.8
1327.78	**2422**	4391.6	1361.11	**2482**	4499.6	1394.44	**2542**	4607.6
1328.33	**2423**	4393.4	1361.67	**2483**	4501.4	1395	**2543**	4609.4
1328.89	**2424**	4395.2	1362.22	**2484**	4503.2	1395.56	**2544**	4611.2
1329.44	**2425**	4397	1362.78	**2485**	4505	1396.11	**2545**	4613
1330	**2426**	4398.8	1363.33	**2486**	4506.8	1396.67	**2546**	4614.8
1330.56	**2427**	4400.6	1363.89	**2487**	4508.6	1397.22	**2547**	4616.6
1331.11	**2428**	4402.4	1364.44	**2488**	4510.4	1397.78	**2548**	4618.4
1331.67	**2429**	4404.2	1365	**2489**	4512.2	1398.33	**2549**	4620.2
1332.22	**2430**	4406	1365.56	**2490**	4514	1398.89	**2550**	4622
1332.78	**2431**	4407.8	1366.11	**2491**	4515.8	1399.44	**2551**	4623.8
1333.33	**2432**	4409.6	1366.67	**2492**	4517.6	1400	**2552**	4625.6
1333.89	**2433**	4411.4	1367.22	**2493**	4519.4	1400.56	**2553**	4627.4
1334.44	**2434**	4413.2	1367.78	**2494**	4521.2	1401.11	**2554**	4629.2
1335	**2435**	4415	1368.33	**2495**	4523	1401.67	**2555**	4631
1335.56	**2436**	4416.8	1368.89	**2496**	4524.8	1402.22	**2556**	4632.8
1336.11	**2437**	4418.6	1369.44	**2497**	4526.6	1402.78	**2557**	4634.6
1336.67	**2438**	4420.4	1370	**2498**	4528.4	1403.33	**2558**	4636.4
1337.22	**2439**	4422.2	1370.56	**2499**	4530.2	1403.89	**2559**	4638.2
1337.78	**2440**	4424	1371.11	**2500**	4532	1404.44	**2560**	4640
1338.33	**2441**	4425.8	1371.67	**2501**	4533.8	1405	**2561**	4641.8
1338.89	**2442**	4427.6	1372.22	**2502**	4535.6	1405.56	**2562**	4643.6
1339.44	**2443**	4429.4	1372.78	**2503**	4537.4	1406.11	**2563**	4645.4
1340	**2444**	4431.2	1373.33	**2504**	4539.2	1406.67	**2564**	4647.2
1340.56	**2445**	4433	1373.89	**2505**	4541	1407.22	**2565**	4649
1341.11	**2446**	4434.8	1374.44	**2506**	4542.8	1407.78	**2566**	4650.8
1341.67	**2447**	4436.6	1375	**2507**	4544.6	1408.33	**2567**	4652.6
1342.22	**2448**	4438.4	1375.56	**2508**	4546.4	1408.89	**2568**	4654.4
1342.78	**2449**	4440.2	1376.11	**2509**	4548.2	1409.44	**2569**	4656.2

To Convert			To Convert			To Convert		
To °C	←°F or °C→	To °F	To °C	←°F or °C→	To °F	To °C	←°F or °C→	To °F
1410	**2570**	4658	1443.33	**2630**	4766	1476.67	**2690**	4874
1410.56	**2571**	4659.8	1443.89	**2631**	4767.8	1477.22	**2691**	4875.8
1411.11	**2572**	4661.6	1444.44	**2632**	4769.6	1477.78	**2692**	4877.6
1411.67	**2573**	4663.4	1445	**2633**	4771.4	1478.33	**2693**	4879.4
1412.22	**2574**	4665.2	1445.56	**2634**	4773.2	1478.89	**2694**	4881.2
1412.78	**2575**	4667	1446.11	**2635**	4775	1479.44	**2695**	4883
1413.33	**2576**	4668.8	1446.67	**2636**	4776.8	1480	**2696**	4884.8
1413.89	**2577**	4670.6	1447.22	**2637**	4778.6	1480.56	**2697**	4886.6
1414.44	**2578**	4672.4	1447.78	**2638**	4780.4	1481.11	**2698**	4888.4
1415	**2579**	4674.2	1448.33	**2639**	4782.2	1481.67	**2699**	4890.2
1415.56	**2580**	4676	1448.89	**2640**	4784	1482.22	**2700**	4892
1416.11	**2581**	4677.8	1449.44	**2641**	4785.8	1482.78	**2701**	4893.8
1416.67	**2582**	4679.6	1450	**2642**	4787.6	1483.33	**2702**	4895.6
1417.22	**2583**	4681.4	1450.56	**2643**	4789.4	1483.89	**2703**	4897.4
1417.78	**2584**	4683.2	1451.11	**2644**	4791.2	1484.44	**2704**	4899.2
1418.33	**2585**	4685	1451.67	**2645**	4793	1485	**2705**	4901
1418.89	**2586**	4686.8	1452.22	**2646**	4794.8	1485.56	**2706**	4902.8
1419.44	**2587**	4688.6	1452.78	**2647**	4796.6	1486.11	**2707**	4904.6
1420	**2588**	4690.4	1453.33	**2648**	4798.4	1486.67	**2708**	4906.4
1420.56	**2589**	4692.2	1453.89	**2649**	4800.2	1487.22	**2709**	4908.2
1421.11	**2590**	4694	1454.44	**2650**	4802	1487.78	**2710**	4910
1421.67	**2591**	4695.8	1455	**2651**	4803.8	1488.33	**2711**	4911.8
1422.22	**2592**	4697.6	1455.56	**2652**	4805.6	1488.89	**2712**	4913.6
1422.78	**2593**	4699.4	1456.11	**2653**	4807.4	1489.44	**2713**	4915.4
1423.33	**2594**	4701.2	1456.67	**2654**	4809.2	1490	**2714**	4917.2
1423.89	**2595**	4703	1457.22	**2655**	4811	1490.56	**2715**	4919
1424.44	**2596**	4704.8	1457.78	**2656**	4812.8	1491.11	**2716**	4920.8
1425	**2597**	4706.6	1458.33	**2657**	4814.6	1491.67	**2717**	4922.6
1425.56	**2598**	4708.4	1458.89	**2658**	4816.4	1492.22	**2718**	4924.4
1426.11	**2599**	4710.2	1459.44	**2659**	4818.2	1492.78	**2719**	4926.2
1426.67	**2600**	4712	1460	**2660**	4820	1493.33	**2720**	4928
1427.22	**2601**	4713.8	1460.56	**2661**	4821.8	1493.89	**2721**	4929.8
1427.78	**2602**	4715.6	1461.11	**2662**	4823.6	1494.44	**2722**	4931.6
1428.33	**2603**	4717.4	1461.67	**2663**	4825.4	1495	**2723**	4933.4
1428.89	**2604**	4719.2	1462.22	**2664**	4827.2	1495.56	**2724**	4935.2
1429.44	**2605**	4721	1462.78	**2665**	4829	1496.11	**2725**	4937
1430	**2606**	4722.8	1463.33	**2666**	4830.8	1496.67	**2726**	4938.8
1430.56	**2607**	4724.6	1463.89	**2667**	4832.6	1497.22	**2727**	4940.6
1431.11	**2608**	4726.4	1464.44	**2668**	4834.4	1497.78	**2728**	4942.4
1431.67	**2609**	4728.2	1465	**2669**	4836.2	1498.33	**2729**	4944.2
1432.22	**2610**	4730	1465.56	**2670**	4838	1498.89	**2730**	4946
1432.78	**2611**	4731.8	1466.11	**2671**	4839.8	1499.44	**2731**	4947.8
1433.33	**2612**	4733.6	1466.67	**2672**	4841.6	1500	**2732**	4949.6
1433.89	**2613**	4735.4	1467.22	**2673**	4843.4	1500.56	**2733**	4951.4
1434.44	**2614**	4737.2	1467.78	**2674**	4845.2	1501.11	**2734**	4953.2
1435	**2615**	4739	1468.33	**2675**	4847	1501.67	**2735**	4955
1435.56	**2616**	4740.8	1468.89	**2676**	4848.8	1502.22	**2736**	4956.8
1436.11	**2617**	4742.6	1469.44	**2677**	4850.6	1502.78	**2737**	4958.6
1436.67	**2618**	4744.4	1470	**2678**	4852.4	1503.33	**2738**	4960.4
1437.22	**2619**	4746.2	1470.56	**2679**	4854.2	1503.89	**2739**	4962.2
1437.78	**2620**	4748	1471.11	**2680**	4856	1504.44	**2740**	4964
1438.33	**2621**	4749.8	1471.67	**2681**	4857.8	1505	**2741**	4965.8
1438.89	**2622**	4751.6	1472.22	**2682**	4859.6	1505.56	**2742**	4967.6
1439.44	**2623**	4753.4	1472.78	**2683**	4861.4	1506.11	**2743**	4969.4
1440	**2624**	4755.2	1473.33	**2684**	4863.2	1506.67	**2744**	4971.2
1440.56	**2625**	4757	1473.89	**2685**	4865	1507.22	**2745**	4973
1441.11	**2626**	4758.8	1474.44	**2686**	4866.8	1507.78	**2746**	4974.8
1441.67	**2627**	4760.6	1475	**2687**	4868.6	1508.33	**2747**	4976.6
1442.22	**2628**	4762.4	1475.56	**2688**	4870.4	1508.89	**2748**	4978.4
1442.78	**2629**	4764.2	1476.11	**2689**	4872.2	1509.44	**2749**	4980.2

To Convert			To Convert			To Convert		
To °C	←°F or °C→	To °F	To °C	←°F or °C→	To °F	To °C	←°F or °C→	To °F
1510	**2750**	4982	1543.33	**2810**	5090	1576.67	**2870**	5198
1510.56	**2751**	4983.8	1543.89	**2811**	5091.8	1577.22	**2871**	5199.8
1511.11	**2752**	4985.6	1544.44	**2812**	5093.6	1577.78	**2872**	5201.6
1511.67	**2753**	4987.4	1545	**2813**	5095.4	1578.33	**2873**	5203.4
1512.22	**2754**	4989.2	1545.56	**2814**	5097.2	1578.89	**2874**	5205.2
1512.78	**2755**	4991	1546.11	**2815**	5099	1579.44	**2875**	5207
1513.33	**2756**	4992.8	1546.67	**2816**	5100.8	1580	**2876**	5208.8
1513.89	**2757**	4994.6	1547.22	**2817**	5102.6	1580.56	**2877**	5210.6
1514.44	**2758**	4996.4	1547.78	**2818**	5104.4	1581.11	**2878**	5212.4
1515	**2759**	4998.2	1548.33	**2819**	5106.2	1581.67	**2879**	5214.2
1515.56	**2760**	5000	1548.89	**2820**	5108	1582.22	**2880**	5216
1516.11	**2761**	5001.8	1549.44	**2821**	5109.8	1582.78	**2881**	5217.8
1516.67	**2762**	5003.6	1550	**2822**	5111.6	1583.33	**2882**	5219.6
1517.22	**2763**	5005.4	1550.56	**2823**	5113.4	1583.89	**2883**	5221.4
1517.78	**2764**	5007.2	1551.11	**2824**	5115.2	1584.44	**2884**	5223.2
1518.33	**2765**	5009	1551.67	**2825**	5117	1585	**2885**	5225
1518.89	**2766**	5010.8	1552.22	**2826**	5118.8	1585.56	**2886**	5226.8
1519.44	**2767**	5012.6	1552.78	**2827**	5120.6	1586.11	**2887**	5228.6
1520	**2768**	5014.4	1553.33	**2828**	5122.4	1586.67	**2888**	5230.4
1520.56	**2769**	5016.2	1553.89	**2829**	5124.2	1587.22	**2889**	5232.2
1521.11	**2770**	5018	1554.44	**2830**	5126	1587.78	**2890**	5234
1521.67	**2771**	5019.8	1555	**2831**	5127.8	1588.33	**2891**	5235.8
1522.22	**2772**	5021.6	1555.56	**2832**	5129.6	1588.89	**2892**	5237.6
1522.78	**2773**	5023.4	1556.11	**2833**	5131.4	1589.44	**2893**	5239.4
1523.33	**2774**	5025.2	1556.67	**2834**	5133.2	1590	**2894**	5241.2
1523.89	**2775**	5027	1557.22	**2835**	5135	1590.56	**2895**	5243
1524.44	**2776**	5028.8	1557.78	**2836**	5136.8	1591.11	**2896**	5244.8
1525	**2777**	5030.6	1558.33	**2837**	5138.6	1591.67	**2897**	5246.6
1525.56	**2778**	5032.4	1558.89	**2838**	5140.4	1592.22	**2898**	5248.4
1526.11	**2779**	5034.2	1559.44	**2839**	5142.2	1592.78	**2899**	5250.2
1526.67	**2780**	5036	1560	**2840**	5144	1593.33	**2900**	5252
1527.22	**2781**	5037.8	1560.56	**2841**	5145.8	1593.89	**2901**	5253.8
1527.78	**2782**	5039.6	1561.11	**2842**	5147.6	1594.44	**2902**	5255.6
1528.33	**2783**	5041.4	1561.67	**2843**	5149.4	1595	**2903**	5257.4
1528.89	**2784**	5043.2	1562.22	**2844**	5151.2	1595.56	**2904**	5259.2
1529.44	**2785**	5045	1562.78	**2845**	5153	1596.11	**2905**	5261
1530	**2786**	5046.8	1563.33	**2846**	5154.8	1596.67	**2906**	5262.8
1530.56	**2787**	5048.6	1563.89	**2847**	5156.6	1597.22	**2907**	5264.6
1531.11	**2788**	5050.4	1564.44	**2848**	5158.4	1597.78	**2908**	5266.4
1531.67	**2789**	5052.2	1565	**2849**	5160.2	1598.33	**2909**	5268.2
1532.22	**2790**	5054	1565.56	**2850**	5162	1598.89	**2910**	5270
1532.78	**2791**	5055.8	1566.11	**2851**	5163.8	1599.44	**2911**	5271.8
1533.33	**2792**	5057.6	1566.67	**2852**	5165.6	1600	**2912**	5273.6
1533.89	**2793**	5059.4	1567.22	**2853**	5167.4	1600.56	**2913**	5275.4
1534.44	**2794**	5061.2	1567.78	**2854**	5169.2	1601.11	**2914**	5277.2
1535	**2795**	5063	1568.33	**2855**	5171	1601.67	**2915**	5279
1535.56	**2796**	5064.8	1568.89	**2856**	5172.8	1602.22	**2916**	5280.8
1536.11	**2797**	5066.6	1569.44	**2857**	5174.6	1602.78	**2917**	5282.6
1536.67	**2798**	5068.4	1570	**2858**	5776.4	1603.33	**2918**	5284.4
1537.22	**2799**	5070.2	1570.56	**2859**	5178.2	1603.89	**2919**	5286.2
1537.78	**2800**	5072	1571.11	**2860**	5180	1604.44	**2920**	5288
1538.33	**2801**	5073.8	1571.67	**2861**	5181.8	1605	**2921**	5289.8
1538.89	**2802**	5075.6	1572.22	**2862**	5183.6	1605.56	**2922**	5291.6
1539.44	**2803**	5077.4	1572.78	**2863**	5185.4	1606.11	**2923**	5293.4
1540	**2804**	5079.2	1573.33	**2864**	5187.2	1606.67	**2924**	5295.2
1540.56	**2805**	5081	1573.89	**2865**	5189	1607.22	**2925**	5297
1541.11	**2806**	5082.8	1574.44	**2866**	5190.8	1607.78	**2926**	5298.8
1541.67	**2807**	5084.6	1575	**2867**	5192.6	1608.33	**2927**	5300.6
1542.22	**2808**	5086.4	1575.56	**2868**	5194.4	1608.89	**2928**	5302.4
1542.78	**2809**	5088.2	1576.11	**2869**	5196.2	1609.44	**2929**	5304.2

To Convert			To Convert			To Convert		
To °C	←°F or °C→	To °F	To °C	←°F or °C→	To °F	To °C	←°F or °C→	To °F
1610	**2930**	5306	1643.33	**2990**	5414	1676.67	**3050**	5522
1610.56	**2931**	5307.8	1643.89	**2991**	5415.8	1677.22	**3051**	5523.8
1611.11	**2932**	5309.6	1644.44	**2992**	5417.6	1677.78	**3052**	5525.6
1611.67	**2933**	5311.4	1645	**2993**	5419.4	1678.33	**3053**	5527.4
1612.22	**2934**	5313.2	1645.56	**2994**	5421.2	1678.89	**3054**	5529.2
1612.78	**2935**	5315	1646.11	**2995**	5423	1679.44	**3055**	5531
1613.33	**2936**	5316.8	1646.67	**2996**	5424.8	1680	**3056**	5532.8
1613.89	**2937**	5318.6	1647.22	**2997**	5426.6	1680.56	**3057**	5534.6
1614.44	**2938**	5320.4	1647.78	**2998**	5428.4	1681.11	**3058**	5536.4
1615	**2939**	5322.2	1648.33	**2999**	5430.2	1681.67	**3059**	5538.2
1615.56	**2940**	5324	1648.89	**3000**	5432	1682.22	**3060**	5540
1616.11	**2941**	5325.8	1649.44	**3001**	5433.8	1682.78	**3061**	5541.8
1616.67	**2942**	5327.6	1650	**3002**	5435.6	1683.33	**3062**	5543.6
1617.22	**2943**	5329.4	1650.56	**3003**	5437.4	1683.89	**3063**	5545.4
1617.78	**2944**	5331.2	1651.11	**3004**	5439.2	1684.44	**3064**	5547.2
1618.33	**2945**	5333	1651.67	**3005**	5441	1685	**3065**	5549
1618.89	**2946**	5334.8	1652.22	**3006**	5442.8	1685.56	**3066**	5550.8
1619.44	**2947**	5336.6	1652.78	**3007**	5444.6	1686.11	**3067**	5552.6
1620	**2948**	5338.4	1653.33	**3008**	5446.4	1686.67	**3068**	5554.4
1620.56	**2949**	5340.2	1653.89	**3009**	5448.2	1687.22	**3069**	5556.2
1621.11	**2950**	5342	1654.44	**3010**	5450	1687.78	**3070**	5558
1621.67	**2951**	5343.8	1655	**3011**	5451.8	1688.33	**3071**	5559.8
1622.22	**2952**	5345.6	1655.56	**3012**	5453.6	1688.89	**3072**	5561.6
1622.78	**2953**	5347.4	1656.11	**3013**	5455.4	1689.44	**3073**	5563.4
1623.33	**2954**	5349.2	1656.67	**3014**	5457.2	1690	**3074**	5565.2
1623.89	**2955**	5351	1657.22	**3015**	5459	1690.56	**3075**	5567
1624.44	**2956**	5352.8	1657.78	**3016**	5460.8	1691.11	**3076**	5568.8
1625	**2957**	5354.6	1658.33	**3017**	5462.6	1691.67	**3077**	5570.6
1625.56	**2958**	5356.4	1658.89	**3018**	5464.4	1692.22	**3078**	5572.4
1626.11	**2959**	5358.2	1659.44	**3019**	5466.2	1692.78	**3079**	5574.2
1626.67	**2960**	5360	1660	**3020**	5468	1693.33	**3080**	5576
1627.22	**2961**	5361.8	1660.56	**3021**	5469.8	1693.89	**3081**	5577.8
1627.78	**2962**	5363.6	1661.11	**3022**	5471.6	1694.44	**3082**	5579.6
1628.33	**2963**	5365.4	1661.67	**3023**	5473.4	1695	**3083**	5581.4
1628.89	**2964**	5367.2	1662.22	**3024**	5475.2	1695.56	**3084**	5583.2
1629.44	**2965**	5369	1662.78	**3025**	5477	1696.11	**3085**	5585
1630	**2966**	5370.8	1663.33	**3026**	5478.8	1696.67	**3086**	5586.8
1630.56	**2967**	5372.6	1663.89	**3027**	5480.6	1697.22	**3087**	5588.6
1631.11	**2968**	5374.4	1664.44	**3028**	5482.4	1697.78	**3088**	5590.4
1631.67	**2969**	5376.2	1665	**3029**	5484.2	1698.33	**3089**	5592.2
1632.22	**2970**	5378	1665.56	**3030**	5486	1698.89	**3090**	5594
1632.78	**2971**	5379.8	1666.11	**3031**	5487.8	1699.44	**3091**	5595.8
1633.33	**2972**	5381.6	1666.67	**3032**	5489.6	1700	**3092**	5597.6
1633.89	**2973**	5383.4	1667.22	**3033**	5491.4	1700.56	**3093**	5599.4
1634.44	**2974**	5385.2	1667.78	**3034**	5493.2	1701.11	**3094**	5601.2
1635	**2975**	5387	1668.33	**3035**	5495	1701.67	**3095**	5603
1635.56	**2976**	5388.8	1668.89	**3036**	5496.8	1702.22	**3096**	5604.8
1636.11	**2977**	5390.6	1669.44	**3037**	5498.6	1702.78	**3097**	5606.6
1636.67	**2978**	5392.4	1670	**3038**	5500.4	1703.33	**3098**	5608.4
1637.22	**2979**	5394.2	1670.56	**3039**	5502.2	1703.89	**3099**	5610.2
1637.78	**2980**	5396	1671.11	**3040**	5504	1704.44	**3100**	5612
1638.33	**2981**	5397.8	1671.67	**3041**	5505.8	1705	**3101**	5613.8
1638.89	**2982**	5399.6	1672.22	**3042**	5507.6	1705.56	**3102**	5615.6
1639.44	**2983**	5401.4	1672.78	**3043**	5509.4	1706.11	**3103**	5617.4
1640	**2984**	5403.2	1673.33	**3044**	5511.2	1706.67	**3104**	5619.2
1640.56	**2985**	5405	1673.89	**3045**	5513	1707.22	**3105**	5621
1641.11	**2986**	5406.8	1674.44	**3046**	5514.8	1707.78	**3106**	5622.8
1641.67	**2987**	5408.6	1675	**3047**	5516.6	1708.33	**3107**	5624.6
1642.22	**2988**	5410.4	1675.56	**3048**	5518.4	1708.89	**3108**	5626.4
1642.78	**2989**	5412.2	1676.11	**3049**	5520.2	1709.44	**3109**	5628.2

To Convert			To Convert			To Convert		
To °C	←°F or °C→	To °F	To °C	←°F or °C→	To °F	To °C	←°F or °C→	To °F
1710	**3110**	5630	1743.33	**3170**	5738	1776.67	**3230**	5846
1710.56	**3111**	5631.8	1743.89	**3171**	5739.8	1777.22	**3231**	5847.8
1711.11	**3112**	5633.6	1744.44	**3172**	5741.6	1777.78	**3232**	5849.6
1711.67	**3113**	5635.4	1745	**3173**	5743.4	1778.33	**3233**	5851.4
1712.22	**3114**	5637.2	1745.56	**3174**	5745.2	1778.89	**3234**	5853.2
1712.78	**3115**	5639	1746.11	**3175**	5747	1779.44	**3235**	5855
1713.33	**3116**	5640.8	1746.67	**3176**	5748.8	1780	**3236**	5856.8
1713.89	**3117**	5642.6	1747.22	**3177**	5750.6	1780.56	**3237**	5858.6
1714.44	**3118**	5644.4	1747.78	**3178**	5752.4	1781.11	**3238**	5860.4
1715	**3119**	5646.2	1748.33	**3179**	5754.2	1781.67	**3239**	5862.2
1715.56	**3120**	5648	1748.89	**3180**	5756	1782.22	**3240**	5864
1716.11	**3121**	5649.8	1749.44	**3181**	5757.8	1782.78	**3241**	5865.8
1716.67	**3122**	5651.6	1750	**3182**	5759.6	1783.33	**3242**	5867.6
1717.22	**3123**	5653.4	1750.56	**3183**	5761.4	1783.89	**3243**	5869.4
1717.78	**3124**	5655.2	1751.11	**3184**	5763.2	1784.44	**3244**	5871.2
1718.33	**3125**	5657	1751.67	**3185**	5765	1785	**3245**	5873
1718.89	**3126**	5658.8	1752.22	**3186**	5766.8	1785.56	**3246**	5874.8
1719.44	**3127**	5660.6	1752.78	**3187**	5768.6	1786.11	**3247**	5876.6
1720	**3128**	5662.4	1753.33	**3188**	5770.4	1786.67	**3248**	5878.4
1720.56	**3129**	5664.2	1753.89	**3189**	5772.2	1787.22	**3249**	5880.2
1721.11	**3130**	5666	1754.44	**3190**	5774	1787.78	**3250**	5882
1721.67	**3131**	5667.8	1755	**3191**	5775.8	1788.33	**3251**	5883.8
1722.22	**3132**	5669.6	1755.56	**3192**	5777.6	1788.89	**3252**	5885.6
1722.78	**3133**	5671.4	1756.11	**3193**	5779.4	1789.44	**3253**	5887.4
1723.33	**3134**	5673.2	1756.67	**3194**	5781.2	1790	**3254**	5889.2
1723.89	**3135**	5675	1757.22	**3195**	5783	1790.56	**3255**	5891
1724.44	**3136**	5676.8	1757.78	**3196**	5784.8	1791.11	**3256**	5892.8
1725	**3137**	5678.6	1758.33	**3197**	5786.6	1791.67	**3257**	5894.6
1725.56	**3138**	5680.4	1758.89	**3198**	5788.4	1792.22	**3258**	5896.4
1726.11	**3139**	5682.2	1759.44	**3199**	5790.2	1792.78	**3259**	5898.2
1726.67	**3140**	5684	1760	**3200**	5792	1793.33	**3260**	5900
1727.22	**3141**	5685.8	1760.56	**3201**	5793.8	1793.89	**3261**	5901.8
1727.78	**3142**	5687.6	1761.11	**3202**	5795.6	1794.44	**3262**	5903.6
1728.33	**3143**	5689.4	1761.67	**3203**	5797.4	1795	**3263**	5905.4
1728.89	**3144**	5691.2	1762.22	**3204**	5799.2	1795.56	**3264**	5907.2
1729.44	**3145**	5693	1762.78	**3205**	5801	1796.11	**3265**	5909
1730	**3146**	5694.8	1763.33	**3206**	5802.8	1796.67	**3266**	5910.8
1730.56	**3147**	5696.6	1763.89	**3207**	5804.6	1797.22	**3267**	5912.6
1731.11	**3148**	5698.4	1764.44	**3208**	5806.4	1797.78	**3268**	5914.4
1731.67	**3149**	5700.2	1765	**3209**	5808.2	1798.33	**3269**	5916.2
1732.22	**3150**	5702	1765.56	**3210**	5810	1798.89	**3270**	5918
1732.78	**3151**	5703.8	1766.11	**3211**	5811.8	1799.44	**3271**	5919.8
1733.33	**3152**	5705.6	1766.67	**3212**	5813.6	1800	**3272**	5921.6
1733.89	**3153**	5707.4	1767.22	**3213**	5815.4	1800.56	**3273**	5923.4
1734.44	**3154**	5709.2	1767.78	**3214**	5817.2	1801.11	**3274**	5925.2
1735	**3155**	5711	1768.33	**3215**	5819	1801.67	**3275**	5927
1735.56	**3156**	5712.8	1768.89	**3216**	5820.8	1802.22	**3276**	5928.8
1736.11	**3157**	5714.6	1769.44	**3217**	5822.6	1802.78	**3277**	5930.6
1736.67	**3158**	5716.4	1770	**3218**	5824.4	1803.33	**3278**	5932.4
1737.22	**3159**	5718.2	1770.56	**3219**	5826.2	1803.89	**3279**	5934.2
1737.78	**3160**	5720	1771.11	**3220**	5828	1804.44	**3280**	5936
1738.33	**3161**	5721.8	1771.67	**3221**	5829.8	1805	**3281**	5937.8
1738.89	**3162**	5723.6	1772.22	**3222**	5831.6	1805.56	**3282**	5939.6
1739.44	**3163**	5725.4	1772.78	**3223**	5833.4	1806.11	**3283**	5941.4
1740	**3164**	5727.2	1773.33	**3224**	5835.2	1806.67	**3284**	5943.2
1740.56	**3165**	5729	1773.89	**3225**	5837	1807.22	**3285**	5945
1741.11	**3166**	5730.8	1774.44	**3226**	5838.8	1807.78	**3286**	5946.8
1741.67	**3167**	5732.6	1775	**3227**	5840.6	1808.33	**3287**	5948.6
1742.22	**3168**	5734.4	1775.56	**3228**	5842.4	1808.89	**3288**	5950.4
1742.78	**3169**	5736.2	1776.11	**3229**	5844.2	1809.44	**3289**	5952.2

To Convert			To Convert			To Convert		
To °C	←°F or °C→	To °F	To °C	←°F or °C→	To °F	To °C	←°F or °C→	To °F
1810	**3290**	5954	1954.44	**3550**	6422	2121.11	**3850**	6962
1810.56	**3291**	5955.8	1957.22	**3555**	6431	2123.89	**3855**	6971
1811.11	**3292**	5957.6	1960.00	**3560**	6440	2126.67	**3860**	6980
1811.67	**3293**	5959.4	1962.78	**3565**	6449	2129.44	**3865**	6989
1812.22	**3294**	5961.2	1965.56	**3570**	6458	2132.22	**3870**	6998
1812.78	**3295**	5963	1968.33	**3575**	6467	2135.00	**3875**	7007
1813.33	**3296**	5964.8	1971.11	**3580**	6476	2137.78	**3880**	7016
1813.89	**3297**	5966.6	1973.89	**3585**	6485	2140.56	**3885**	7025
1814.44	**3298**	5968.4	1976.67	**3590**	6494	2143.33	**3890**	7034
1815	**3299**	5970.2	1979.44	**3595**	6503	2145.11	**3895**	7043
1815.56	**3300**	5972	1982.22	**3600**	6512	2148.89	**3900**	7052
1818.33	**3305**	5981	1985.00	**3605**	6521	2151.67	**3905**	7061
1821.11	**3310**	5990	1987.78	**3610**	6530	2154.44	**3910**	7070
1823.89	**3315**	5999	1990.56	**3615**	6539	2157.22	**3915**	7079
1826.67	**3320**	6008	1993.33	**3620**	6548	2160.00	**3920**	7088
1829.44	**3325**	6017	1996.11	**3625**	6557	2162.78	**3925**	7097
1832.22	**3330**	6026	1998.89	**3630**	6566	2165.56	**3930**	7106
1835.00	**3335**	6035	2001.67	**3635**	6575	2168.33	**3935**	7115
1837.78	**3340**	6044	2004.44	**3640**	6584	2171.11	**3940**	7124
1840.56	**3345**	6053	2007.22	**3645**	6593	2173.89	**3945**	7133
1843.33	**3350**	6062	2010.00	**3650**	6602	2176.67	**3950**	7142
1846.11	**3355**	6071	2012.78	**3655**	6611	2179.44	**3955**	7151
1848.89	**3360**	6080	2015.56	**3660**	6620	2182.22	**3960**	7160
1851.67	**3365**	6089	2018.33	**3665**	6629	2185.00	**3965**	7169
1854.44	**3370**	6098	2021.11	**3670**	6638	2187.78	**3970**	7178
1857.22	**3375**	6107	2023.89	**3675**	6647	2190.56	**3975**	7187
1860.00	**3380**	6116	2026.67	**3680**	6656	2193.33	**3980**	7196
1862.78	**3385**	6125	2029.44	**3685**	6665	2196.11	**3985**	7205
1865.56	**3390**	6134	2032.22	**3690**	6674	2198.89	**3990**	7214
1868.33	**3395**	6143	2035.00	**3695**	6683	2201.67	**3995**	7223
1871.11	**3400**	6152	2037.78	**3700**	6692	2204.44	**4000**	7232
1873.89	**3405**	6161	2040.56	**3705**	6701	2207.22	**4005**	7241
1876.67	**3410**	6170	2043.33	**3710**	6710	2210.00	**4010**	7250
1879.44	**3415**	6179	2046.11	**3715**	6719	2212.78	**4015**	7259
1882.22	**3420**	6188	2048.89	**3720**	6728	2215.56	**4020**	7268
1885.00	**3425**	6197	2051.67	**3725**	6737	2218.33	**4025**	7277
1887.78	**3430**	6206	2054.44	**3730**	6746	2221.11	**4030**	7286
1890.56	**3435**	6215	2057.22	**3735**	6755	2223.89	**4035**	7295
1893.33	**3440**	6224	2060.00	**3740**	6764	2226.67	**4040**	7304
1896.11	**3445**	6233	2062.78	**3745**	6773	2229.44	**4045**	7313
1898.89	**3450**	6242	2065.56	**3750**	6782	2232.22	**4050**	7322
1901.67	**3455**	6251	2068.33	**3755**	6791	2235.00	**4055**	7331
1904.44	**3460**	6260	2071.11	**3760**	6800	2237.78	**4060**	7340
1907.22	**3465**	6269	2073.89	**3765**	6809	2240.56	**4065**	7349
1910.00	**3470**	6278	2076.67	**3770**	6818	2243.33	**4070**	7358
1912.78	**3475**	6287	2079.44	**3775**	6827	2246.11	**4075**	7367
1915.56	**3480**	6296	2082.22	**3780**	6836	2248.89	**4080**	7376
1918.33	**3485**	6305	2085.00	**3785**	6845	2251.67	**4085**	7385
1921.11	**3490**	6314	2087.78	**3790**	6854	2254.44	**4090**	7394
1923.89	**3495**	6323	2090.56	**3795**	6863	2257.22	**4095**	7403
1926.67	**3500**	6332	2093.33	**3800**	6872	2260.00	**4100**	7412
1929.44	**3505**	6341	2096.11	**3805**	6881	2262.78	**4105**	7421
1932.22	**3510**	6350	2098.89	**3810**	6890	2265.56	**4110**	7430
1935.00	**3515**	6359	2101.67	**3815**	6899	2268.33	**4115**	7439
1937.78	**3520**	6368	2104.44	**3820**	6908	2271.11	**4120**	7448
1940.56	**3525**	6377	2107.22	**3825**	6917	2273.89	**4125**	7457
1943.33	**3530**	6386	2110.00	**3830**	6926	2276.67	**4130**	7466
1946.11	**3535**	6395	2112.78	**3835**	6935	2279.44	**4135**	7475
1948.89	**3540**	6404	2115.56	**3840**	6944	2282.22	**4140**	7484
1951.67	**3545**	6413	2118.33	**3845**	6953	2285.00	**4145**	7493

To Convert			To Convert			To Convert		
To °C	←°F or °C→	To °F	To °C	←°F or °C→	To °F	To °C	←°F or °C→	To °F
2287.78	**4150**	7502	2354.44	**4270**	7718	2421.11	**4390**	7934
2290.56	**4155**	7511	2357.22	**4275**	7727	2423.89	**4395**	7943
2293.33	**4160**	7520	2360.00	**4280**	7736	2426.67	**4400**	7952
2296.11	**4165**	7529	2362.78	**4285**	7745	2429.44	**4405**	7961
2298.89	**4170**	7538	2365.56	**4290**	7754	2432.22	**4410**	7970
2301.67	**4175**	7547	2368.33	**4295**	7763	2435.00	**4415**	7979
2304.44	**4180**	7556	2371.11	**4300**	7772	2437.78	**4420**	7988
2307.22	**4185**	7565	2373.89	**4305**	7781	2440.56	**4425**	7997
2310.00	**4190**	7574	2376.67	**4310**	7790	2443.33	**4430**	8006
2312.78	**4195**	7583	2379.44	**4315**	7799	2446.11	**4435**	8015
2315.56	**4200**	7592	2382.22	**4320**	7808	2448.89	**4440**	8024
2318.33	**4205**	7601	2385.00	**4325**	7817	2451.67	**4445**	8033
2321.11	**4210**	7610	2387.78	**4330**	7826	2454.44	**4450**	8042
2323.89	**4215**	7619	2390.56	**4335**	7835	2457.22	**4455**	8051
2326.67	**4220**	7628	2393.33	**4340**	7844	2460.00	**4460**	8060
2329.44	**4225**	7637	2396.11	**4345**	7853	2462.78	**4465**	8069
2332.22	**4230**	7646	2398.89	**4350**	7862	2465.56	**4470**	8078
2335.00	**4235**	7655	2401.67	**4355**	7871	2468.33	**4475**	8087
2337.78	**4240**	7664	2404.44	**4360**	7880	2471.11	**4480**	8096
2340.56	**4245**	7673	2407.22	**4365**	7889	2473.89	**4485**	8105
2343.33	**4250**	7682	2410.00	**4370**	7898	2476.67	**4490**	8114
2346.11	**4255**	7691	2412.78	**4375**	7907	2479.44	**4495**	8123
2348.89	**4260**	7700	2415.56	**4380**	7916	2482.22	**4500**	8132
2351.67	**4265**	7709	2418.33	**4385**	7925			

STANDARD TYPES OF STAINLESS AND HEAT RESISTING STEELS

Chemical Ranges and Limits

Subject to Tolerances for Check Analyses

By permission of American Iron and Steel Institute

Type Number	Chemical Composition, per cent C	Mn Max.	P Max.	S Max.	Si Max.	Cr	Ni	Mo	Zr	Se	Cb-Ta	Ta	Al	N
-201	0.15 Max.	5.50/ 7.50	0.060	0.030	1.00	16.00/ 18.00	3.50/ 5.50							0.25 Max.
-202	0.15 Max.	7.50/ 10.00	0.060	0.030	1.00	17.00/ 19.00	4.00/ 6.00							0.25 Max.
-301	0.15 Max.	2.00	0.045	0.030	1.00	16.00/ 18.00	6.00/ 8.00							
-302	0.15 Max.	2.00	0.045	0.030	1.00	17.00/ 19.00	8.00/ 10.00							
-302B	0.15 Max.	2.00	0.045	0.030	2.00/ 3.00	17.00/ 19.00	8.00/ 10.00							
-303	0.15 Max.	2.00	0.20	0.15 Min.	1.00	17.00/ 19.00	8.00/ 10.00	0.60* Max.	0.60* Max.					
-303 Se	0.15 Max.	2.00	0.20	0.06	1.00	17.00/ 19.00	8.00/ 10.00			0.15 Min.				
-304	0.08 Max	2.00	0.045	0.030	1.00	18.00/ 20.00	8.00/ 12.00							
-304L	0.03 Max.	2.00	0.045	0.030	1.00	18.00/ 20.00	8.00/ 12.00							
-305	0.12 Max.	2.00	0.045	0.030	1.00	17.00/ 19.00	10.00/ 13.00							
-308	0.08 Max.	2.00	0.045	0.030	1.00	19.00/ 21.00	10.00/ 12.00							
-309	0.20 Max.	2.00	0.045	0.030	1.00	22.00/ 24.00	12.00/ 15.00							
-309S	0.08 Max.	2.00	0.045	0.030	1.00	22.00/ 24.00	12.00/ 15.00							
-310	0.25 Max.	2.00	0.045	0.030	1.50	24.00/ 26.00	19.00/ 22.00							
-310S	0.08 Max.	2.00	0.045	0.030	1.50	24.00/ 26.00	19.00/ 22.00							
-314	0.25 Max.	2.00	0.045	0.030	1.50/ 3.00	23.00/ 26.00	19.00/ 22.00							
-316	0.08 Max.	2.00	0.045	0.030	1.00	16.00/ 18.00	10.00/ 14.00	2.00/ 3.00						
-316L	0.03 Max.	2.00	0.045	0.030	1.00	16.00/ 18.00	10.00/ 14.00	2.00 3.00						
-317	0.08 Max.	2.00	0.045	0.030	1.00	18.00/ 20.00	11.00/ 15.00	3.00/ 4.00						
-321	0.08 Max.	2.00	0.045	0.030	1.00	17.00/ 19.00	9.00/ 12.00				5 × C Min.			
-347	0.08 Max.	2.00	0.045	0.030	1.00	17.00/ 19.00	9.00/ 13.00					10 × C Min.		
-348	0.08 Max.	2.00	0.045	0.030	1.00	17.00/ 19.00	9.00/ 13.00				10 × C Min.	0.10 Max.		
**403	0.15 Max.	1.00	0.040	0.030	0.50	11.50/ 13.00								
°405	0.08 Max.	1.00	0.040	0.030	1.00	11.50/ 14.50							0.10/ 0.30	
**410	0.15 Max.	1.00	0.040	0.030	1.00	11.50/ 13.50								
**414	0.15 Max.	1.00	0.040	0.030	1.00	11.50/ 13.50	1.25/ 2.50							
**416	0.15 Max.	1.25	0.06	0.15 Min.	1.00	12.00/ 14.00		0.60* Max.	0.60* Max.					
**416 Se	0.15 Max.	1.25	0.06	0.06	1.00	12.00/ 14.00				0.15 Min.				
**420	Over 0.15	1.00	0.040	0.030	1.00	12.00/ 14.00								

* At producer's option; reported only when intentionally added.
** Heat treatable.
- Not heat treatable.
° Essentially not heat treatable.

Chemical Composition, per cent

Type Number	C	Mn Max.	P Max.	S Max.	Si Max.	Cr	Ni	Mo	Zr	Se	Cb-Ta	Ta	Al	N
°430	0.12 Max.	1.00	0.040	0.030	1.00	14.00/ 18.00								
°430F	0.12 Max.	1.25	0.06	0.15 Min.	1.00	14.00/ 18.00		0.60* Max.	0.60* Max.					
°430F Se	0.12 Max.	1.25	0.06	0.06	1.00	14.00/ 18.00				0.15 Min.				
**431	0.20 Max.	1.00	0.040	0.030	1.00	15.00/ 17.00	1.25/ 2.50							
**440A	0.60/ 0.75	1.00	0.040	0.030	1.00	16.00/ 18.00		0.75 Max.						
**400B	0.75/ 0.95	1.00	0.040	0.030	1.00	16.00/ 18.00		0.75 Max.						
**440C	0.95/ 1.20	1.00	0.040	0.030	1.00	16.00/ 18.00		0.75 Max.						
°446	0.20 Max.	1.50	0.040	0.030	1.00	23.00/ 27.00								0.25 Max.
**501	Over 0.10	1.00	0.040	0.030	1.00	4.00/ 6.00		0.40/ 0.65						
**502	0.10 Max.	1.00	0.040	0.030	1.00	4.00/ 6.00		0.40/ 0.65						

* At producer's option; reported only when intentionally added.
** Heat treatable.
⁻ Not heat treatable.
° Essentially not heat treatable.

STANDARD TEST SIEVES (WIRE CLOTH)

Sieve Designation		Nominal Sieve Opening in	Permissible Variation of Average Opening from the Standard Sieve Designation	Maximum Opening Size for Not More than 5 percent of Openings	Maximum Individual Opening	Nominal Wire Diameter, mm[a]
Standard	Alternative					
(1)	(2)	(3)	(4)	(5)	(6)	(7)
125 mm	5 in.	5	±3.7 mm	130.0 mm	130.9 mm	8.0
106 mm	4.24 in.	4.24	±3.2 mm	110.2 mm	111.1 mm	6.40
100 mm	4 in.	4	±3.0 mm	104.0 mm	104.8 mm	6.30
90 mm	$3\frac{1}{2}$ in.	3.5	±2.7 mm	93.6 mm	94.4 mm	6.08
75 mm	3 in.	3	±2.2 mm	78.1 mm	78.7 mm	5.80
63 mm	$2\frac{1}{2}$ in.	2.5	±1.9 mm	65.6 mm	66.2 mm	5.50
53 mm	2.12 in.	2.12	±1.6 mm	55.2 mm	55.7 mm	5.15
50 mm	2 in.	2	±1.5 mm	52.1 mm	52.6 mm	5.05
45 mm	$1\frac{3}{4}$ in.	1.75	±1.4 mm	46.9 mm	47.4 mm	4.85
37.5 mm	$1\frac{1}{2}$ in.	1.5	±1.1 mm	39.1 mm	39.5 mm	4.59
31.5 mm	$1\frac{1}{4}$ in.	1.25	±1.0 mm	32.9 mm	33.2 mm	4.23
26.5 mm	1.06 in.	1.06	±0.8 mm	27.7 mm	28.0 mm	3.90
25.0 mm	1 in.	1	±0.8 mm	26.1 mm	26.4 mm	3.80
22.4 mm	$\frac{7}{8}$ in.	0.875	±0.7 mm	23.4 mm	23.7 mm	3.50
19.0 mm	$\frac{3}{4}$ in.	0.750	±0.6 mm	19.9 mm	20.1 mm	3.30
16.0 mm	$\frac{5}{8}$ in.	0.625	±0.5 mm	16.7 mm	17.0 mm	3.00
13.2 mm	0.530 in.	0.530	±0.41 mm	13.83 mm	14.05 mm	2.75
12.5 mm	$\frac{1}{2}$ in.	0.500	±0.39 mm	13.10 mm	13.31 mm	2.67
11.2 mm	$\frac{7}{16}$ in.	0.438	±0.35 mm	11.75 mm	11.94 mm	2.45
9.5 mm	$\frac{3}{8}$ in.	0.375	±0.30 mm	9.97 mm	10.16 mm	2.27
8.0 mm	$\frac{5}{16}$ in.	0.312	±0.25 mm	8.41 mm	8.58 mm	2.07
6.7 mm	0.265 in.	0.265	±0.21 mm	7.05 mm	7.20 mm	1.87
6.3 mm	$\frac{1}{4}$ in.	0.250	±0.20 mm	6.64 mm	6.78 mm	1.82
5.6 mm	No. $3\frac{1}{2}$	0.223	±0.18 mm	5.90 mm	6.04 mm	1.68
4.75 mm	No. 4	0.187	±0.15 mm	5.02 mm	5.14 mm	1.54
4.00 mm	No. 5	0.157	±0.13 mm	4.23 mm	4.35 mm	1.37
3.35 mm	No. 6	0.132	±0.11 mm	3.55 mm	3.66 mm	1.23
2.80 mm	No. 7	0.111	±0.095 mm	2.975 mm	3.070 mm	1.10
2.36 mm	No. 8	0.0937	±0.080 mm	2.515 mm	2.600 mm	1.00
2.00 mm	No. 10	0.0787	±0.070 mm	2.135 mm	2.215 mm	0.900
1.70 mm	No. 12	0.0661	±0.060 mm	1.820 mm	1.890 mm	0.810
1.40 mm	No. 14	0.0555	±0.050 mm	1.505 mm	1.565 mm	0.725
1.18 mm	No. 16	0.0469	±0.045 mm	1.270 mm	1.330 mm	0.650
1.00 mm	No. 18	0.0394	±0.040 mm	1.080 mm	1.135 mm	0.580
850 μm	No. 20	0.0331	±35 μm	925 μm	970 μm	0.510
710 μm	No. 25	0.0278	±30 μm	775 μm	815 μm	0.450
600 μm	No. 30	0.0234	±25 μm	660 μm	695 μm	0.390
500 μm	No. 35	0.0197	±20 μm	550 μm	585 μm	0.340
425 μm	No. 40	0.0165	±19 μm	471 μm	502 μm	0.290
355 μm	No. 45	0.0139	±16 μm	396 μm	425 μm	0.247
300 μm	No. 50	0.0117	±14 μm	337 μm	363 μm	0.215
250 μm	No. 60	0.0098	±12 μm	283 μm	306 μm	0.180
212 μm	No. 70	0.0083	±10 μm	242 μm	263 μm	0.152
180 μm	No. 80	0.0070	±9 μm	207 μm	227 μm	0.131
150 μm	No. 100	0.0059	±8 μm	174 μm	192 μm	0.110
125 μm	No. 120	0.0049	±7 μm	147 μm	163 μm	0.091
106 μm	No. 140	0.0041	±6 μm	126 μm	141 μm	0.076
90 μm	No. 170	0.0035	±5 μm	108 μm	122 μm	0.064
75 μm	No. 200	0.0029	±5 μm	91 μm	103 μm	0.053
63 μm	No. 230	0.0025	±4 μm	77 μm	89 μm	0.044
53 μm	No. 270	0.0021	±4 μm	66 μm	76 μm	0.037
45 μm	No. 325	0.0017	±3 μm	57 μm	66 μm	0.030
38 μm	No. 400	0.0015	±3 μm	48 μm	57 μm	0.025

[a] The average diameter of the warp and of the shoot wires, taken separately, of the cloth of any sieve shall not deviate from the nominal values by more than the following:

Sieves coarser than 600 μm	5 percent
Sieves 600 to 125 μm	$7\frac{1}{2}$ percent
Sieves finer than 125 μm	10 percent

PHYSICAL PROPERTIES OF GLASS SEALING AND LEAD WIRE MATERIALS

From General Electric Company

	Glass Sealing Materials												
	Copper		Dumet	42 Ni	Gas Free 42% Ni	42-6 Ni Cr.	46 Ni	52 Ni.	27 Cr.	® Kovar	W	Mo	Nickel 200
	OFC	0.02P											
Analysis: Carbon				0.10	0.05	0.10	0.10	0.10	0.15	0.02			0.06
Manganese				0.50	0.50	0.50	0.50	0.50	0.60	0.30			0.25
Silicon			*Copper Clad*	0.25	0.25	0.25	0.25	0.25	0.40	0.20			0.05
Chromium			*42% Nickel*	—	—	5.75	—	—	28.00	—			
Nickel			*Iron*	42	42	42.5	46	51	0.50	29			99.5
Copper	99.95	99.90	*See GE Spec.*	—	—	—	—	—	—	—			0.05
		P	*DS 8311-01*		Ti0.4		Ti0.4			Co 17			
Other		0.02		Bal.Fe	Bal.Fe	Bal.Fe	Bal.Fe	Bal.Fe	Bal.Fe	Bal.Fe			
Density: grams/cc	8.94	8.94	8.26 to 8.32	8.12	8.12	8.12	8.17	8.30	7.60	8.36	19.3	10.2	8.89
lbs. per cu. in.	0.323	0.323	0.298—0.301	0.293	0.293	0.293	0.295	0.300	0.274	0.302	0.697	0.369	0.321
Thermal Conductivity 20-100°/C													
Cal/cm/sec/cm^2/°C	0.948	0.8	0.2—0.3	0.025	0.025	0.029	0.028	0.032	0.054	0.04	0.31	0.34	0.15
Btu/in/hr/sq. ft/°F	2750	2320	580—870	74	74	84	81	93	158	116	900	1000	435
Electrical Resistivity (20°C) Microhm—cm	1.71	2.03	7.3 to 12.0	72	72	95	46	43	63	49	5.5	5.2	9.5
Ohm per cir. mil ft.	10.3	12.2	44 to 72	430	430	570	275	258	380	294	33	31	57
Elec. Cond. % IACS	101	85	23 to 14	2.3	2.3	1.8	3.6	3.9	2.8	3.4	31	33	18
Curie Temperature °C	—	—	380	380	380	295	460	530	610	435	—	—	360
Melting Temperature °C	1083	1083	— —	1425	1425	1425	1425	1425	1425	1450	3410	2610	1455
°F	1981	1981	— —	2597	2597	2597	2597	2597	2597	2642	6170	4730	2651
Specific Heat cal/gr.	0.092	0.092	0.11	0.12	0.12	0.12	0.12	0.12	0.14	0.11	0.033	0.066	0.13
Thermal Expansion in/in/°Cx10^7			Radial										
25-100°C	168	168	60 to 80	50.1	43.4	65.5	71.0	99.5	94.6	58.6	45	51	133
25-200°C	172	172		47.1	44.1	70.8	73.7	101.0	100.5	52.0	46		139
25-300°C	177	177	60 to 80	47.6	46.1	82.6	75.0	101.0	105.3	51.3	46		144
25-350°C	178	178	65 to 85	50.5		90.4	74.4	100.0	107.0	48.9	46		146
25-400°C	181	181	80 to 100	62.5	64.1	100.0	74.3	100.0	107.8	50.6	46		148
25-500°C	183	183	100 to 140	83.2	85.6	115.0	86.8	102.1	111.2	61.5	46	57	172
25-600°C	188	188		99.0	100.1	125.8	100.2	110.0	112.6	78.0			
Mechanical Properties (Annealed)													
Ultimate Str. (1000 psi)	35	35	74	82	80	80	82	80	85	75	490	120	70
Yield Str. (1000 psi)			50	34	34	40	34	40	55	50	360	110	25
% Elong. (2″)			30	30	30	30	27	35	25	30	8	30	45
Rockwell Hardness				B76	B76	B80	B76	B83	B85	B68	C25	B88	B62
Elastic Modulus (10^6 psi)	16	16		21.5	21.0	23	23	24	30	20	50	47	30

PHYSICAL PROPERTIES OF GLASS SEALING AND LEAD WIRE MATERIALS (Continued)

Metal Alloy Lead Wires									Clad Materials (Also see Dumet)			Pure Metals (Also see W and Mo glass sealing materials)						
Nickel 211	AISI Type 302	AISI Type 316	AISI Type 430	400 Monel	600 Inco-nel	Ad-vance ®	CDA 752	70-30 Brass	Kul-grid ®	40% CCFe	30% CCFe	Au	Ag	Al	Pt	Ta	Fe	Ti
0.10	0.15Max	0.08Max	0.12Max	0.12	0.04													
4.75	2.00Max	2.00Max	1.00Max	0.90	0.20		0.50Max											
0.05	1.00Max	1.00Max	1.00Max	0.15														
	17–19	16–18	14–18		16													
95.0	8–10	10–14	0.50Max	66	76	43	18		27									
		—		31.5		Bal.	65	70		37.5	26							
		Mo 2–3			Fe 7.20		Bal. Zn	Bal. Zn	Cu Core	Fe Core	Fe Core							
8.72	7.9	7.9	7.7	8.84	8.41	8.9	8.73	8.53	8.89	8.15	8.15	19.3	10.5	2.69	21.45	16.6	7.87	4.51
0.315	0.285	0.285	0.278	0.319	0.304	0.322	0.316	0.308	0.321	0.294	0.294	0.698	0.379	0.097	0.775	0.600	0.284	0.163
0.12	0.04	0.04	0.05	0.062	0.036	0.051	0.08	0.29		0.46	0.38	0.71	1.00	0.57	0.165	0.13	0.18	0.43
350	116	116	145	180	104	148	232	845		1330	1100	2050	2900	1650	480	380	520	1250
18.3	72	74	60	48.2	98.1	49	28.7	6.16	2.3	4.4	5.9	2.19	1.629	2.65	9.83	12.45	9.71	7.0
110	433	445	361	290	590	294	173	37	14	26.4	35.3	13	9.8	16	59	75	58	42
9	2.3	2.3	2.9	3.6	1.7	3.5	6	28	70	40	30	78	105	65	16	14	18	25
352			610	43/60	−125	—	—	—		770	770	—	—	—	—	—	770	—
1427	1421	1399	1510	1349	1427	1210	1110	954				1063	960	660	1769	3000	1536	1668
2600	2590	2550	2750	2460	2600	2210	2030	1750				1945	1760	1220	3217	5425	2797	3035
0.13	0.12	0.12	0.11	0.102	0.109	0.094	0.09	0.09		0.10	0.10	0.031	0.056	0.225	0.031	0.034	0.11	0.124
133	166	166	101	140	115	149						142	196	239	91	65	122	88
139				145										243			129	91
144	171	175	110	150			160	199						253				
146																		
148				155													138	94
153	180	180	120	160		163						152	206		96	66		
				165										287			145	97
	90	85	75	85	100	60	60	50				20	22	13	40	55	40	
	37	35	45	40	45		30	—					8	5	12		20	
	55	55	30	40	40		40	60				45	48	40	30	30	45	
	B82	B80	B82	B68	B74		—	—										
30	29	29	29	26	31	18	18	16		24	24	11.6	11	9	21.3	27	30	16.8

WIRE TABLES

COMPARISON OF WIRE GAUGES

Diameter of Wire in Inches

Gauge No.	Brown & Sharpe	Birmingham or Stubs'.	Washburn & Moen	Imperial or Brit. Std.	Stubs' Steel	U. S. Std. plate
00000000	......	......	......	......	......	......
0000000	......	......	......	.500	......	......
000000	......	......	......	.464	......	.46875
00000	......	......	......	.432	......	.4375
0000	.4600	.454	.3938	.400	......	.40625
000	.4096	.425	.3625	.372	......	.375
00	.3648	.380	.3310	.348	......	.34375
0	.3249	.340	.3065	.324	......	.3125
1	.2893	.300	.2830	.300	.227	.28125
2	.2576	.284	.2625	.276	.219	.265625
3	.2294	.259	.2437	.252	.212	.25
4	.2043	.238	.2253	.232	.207	.234375
5	.1819	.220	.2070	.212	.204	.21875
6	.1620	.203	.1920	.192	.201	.203125
7	.1443	.180	.1770	.176	.199	.1875
8	.1285	.165	.1620	.160	.197	.171875
9	.1144	.148	.1483	.144	.194	.15625
10	.1019	.134	.1350	.128	.191	.140625
11	.09074	.120	.1205	.116	.188	.125
12	.08081	.109	.1055	.104	.185	.109375
13	.07196	.095	.0915	.092	.182	.09375
14	.06408	.083	.0800	.080	.180	.078125
15	.05707	.072	.0720	.072	.178	.0703125
16	.05082	.065	.0625	.064	.175	.0625
17	.04526	.058	.0540	.056	.172	.05625
18	.04030	.049	.0475	.048	.168	.05
19	.03589	.042	.0410	.040	.164	.04375
20	.03196	.035	.0348	.036	.161	.0375
21	.02846	.032	.0318	.032	.157	.034375
22	.02535	.028	.0286	.028	.155	.03125
23	.02257	.025	.0258	.024	.153	.028125
24	.02010	.022	.0230	.022	.151	.025

Gauge No.	Brown & Sharpe	Birmingham or Stubs'.	Washburn & Moen	Imperial or Brit. Std.	Stubs' steel	U. S. Std. plate
25	0.01790	0.020	0.0204	0.020	0.148	0.021875
26	0.01594	0.018	0.0181	0.018	0.146	0.01875
27	0.01419	0.016	0.0173	0.0164	0.143	0.0171875
28	0.01264	0.014	0.0162	0.0149	0.139	0.015625
29	0.01126	0.013	0.0150	0.0136	0.134	0.0140625
30	0.01003	0.012	0.0140	0.0124	0.127	0.0125
31	0.008928	0.010	0.0132	0.0116	0.120	0.0109375
32	0.007950	0.009	0.0128	0.0108	0.115	0.01015625
33	0.007080	0.008	0.0118	0.0100	0.112	0.009375
34	0.006304	0.007	0.0104	0.0092	0.110	0.00859375
35	0.005614	0.005	0.0095	0.0084	0.108	0.0078125
36	0.005000	0.004	0.0090	0.0076	0.106	0.00703125
37	0.004453	......	0.0085	0.0068	0.103	0.006640625
38	0.003965	......	0.0080	0.0060	0.101	0.00625
39	0.003531	......	0.0075	0.0052	0.099	......
40	0.003145	......	0.0070	0.0048	0.097	......
41	......	......	0.0066	0.0044	0.095	......
42	......	......	0.0062	0.0040	0.092	......
43	......	......	0.0060	0.0036	0.088	......
44	......	......	0.0058	0.0032	0.085	......
45	......	......	0.0055	0.0028	0.081	......
46	......	......	0.0052	0.0024	0.079	......
47	......	......	0.0050	0.0020	0.077	......
48	......	......	0.0048	0.0016	0.075	......
49	......	......	0.0046	0.0012	0.072	......
50	......	......	0.0044	0.0010	0.069	......

Diameter of Wire in Centimeters

Gauge No.	Brown & Sharpe	Birmingham or Stubs.	Washburn & Moen	Imperial or Brit. Std.	Stubs' steel	U. S. Std. plate
00000000	......	......	......	......	......	......
0000000	......	......	1.245	1.27	......	1.27
000000	......	......	1.172	1.18	......	1.191
00000	......	......	1.093	1.10	......	1.111
0000	1.168	1.15	1.000	1.02	......	1.032
000	1.040	1.08	0.9208	0.945	......	0.9525
00	0.9266	0.965	0.8407	0.884	......	0.8731
0	0.8252	0.864	0.7785	0.823	......	0.7938
1	0.7348	0.762	0.7188	0.762	0.577	0.7144
2	0.6543	0.721	0.6668	0.701	0.556	0.6747
3	0.5827	0.658	0.6190	0.640	0.538	0.6350
4	0.5189	0.605	0.5723	0.589	0.526	0.5953
5	0.4620	0.559	0.5258	0.538	0.518	0.5556
6	0.4115	0.516	0.4877	0.488	0.511	0.5159
7	0.3665	0.457	0.4496	0.447	0.505	0.4763
8	0.3264	0.419	0.4115	0.406	0.500	0.4366
9	0.2906	0.376	0.3767	0.366	0.493	0.3969
10	0.2588	0.340	0.3429	0.325	0.485	0.3572
11	0.2305	0.305	0.3061	0.295	0.478	0.3175
12	0.2053	0.277	0.2680	0.264	0.470	0.2778
13	0.1828	0.241	0.232	0.234	0.462	0.2381
14	0.1628	0.211	0.203	0.203	0.457	0.1984
15	0.1450	0.183	0.183	0.183	0.452	0.1786
16	0.1291	0.165	0.159	0.163	0.445	0.1588
17	0.1150	0.147	0.137	0.142	0.437	0.1429
18	0.1024	0.124	0.121	0.122	0.427	0.1270
19	0.09116	0.107	0.104	0.102	0.417	0.1111
20	0.08118	0.089	0.0884	0.0914	0.409	0.09525
21	0.07229	0.081	0.0808	0.0813	0.399	0.08731
22	0.06439	0.071	0.0726	0.0711	0.394	0.07938
23	0.05733	0.064	0.0655	0.0610	0.389	0.07144
24	0.05105	0.056	0.0584	0.0559	0.384	0.06350

Gauge No.	Brown & Sharpe	Birmingham or Stubs'.	Washburn & Moen	Imperial or Brit. Std.	Stubs' steel	U. S. Std. plate
25	0.04547	0.051	0.0518	0.0508	0.376	0.05556
26	0.04049	0.046	0.0460	0.0457	0.371	0.04763
27	0.03604	0.041	0.0439	0.0417	0.363	0.04366
28	0.03211	0.036	0.0411	0.0378	0.353	0.03969
29	0.02860	0.033	0.0381	0.0345	0.340	0.03572
30	0.02548	0.030	0.0356	0.0315	0.323	0.03175
31	0.02268	0.025	0.0335	0.0295	0.305	0.02778
32	0.02019	0.023	0.0325	0.0274	0.292	0.02580
33	0.01798	0.020	0.0300	0.0254	0.284	0.02381
34	0.01601	0.018	0.0264	0.0234	0.279	0.02183
35	0.01426	0.013	0.024	0.0213	0.274	0.01984
36	0.01270	0.010	0.023	0.0193	0.269	0.01786
37	0.01131	......	0.022	0.0173	0.262	0.01687
38	0.01007	......	0.020	0.0152	0.257	0.01588
39	0.008969	......	0.019	0.0132	0.251	......
40	0.007988	......	0.018	0.0122	0.246	......
41	......	......	0.017	0.0112	0.241	......
42	......	......	0.016	0.0102	0.234	......
43	......	......	0.015	0.0091	0.224	......
44	......	......	0.015	0.0081	0.216	......
45	......	......	0.014	0.0071	0.206	......
46	......	......	0.013	0.0061	0.201	......
47	......	......	0.013	0.0051	0.196	......
48	......	......	0.012	0.0041	0.191	......
49	......	......	0.012	0.0030	0.183	......
50	......	......	0.011	0.0025	0.175	......

TWIST DRILL AND STEEL WIRE GAUGE

Inches

No.	Size	No.	Size	No.	Size	No.	Size	No.	Size
1	0.2280	17	0.1730	33	0.1130	49	0.0730	65	0.0350
2	0.2210	18	0.1695	34	0.1110	50	0.0700	66	0.0330
3	0.2130	19	0.1660	35	0.1100	51	0.0670	67	0.0320
4	0.2090	20	0.1610	36	0.1065	52	0.0635	68	0.0310
5	0.2055	21	0.1590	37	0.1040	53	0.0595	69	0.02925
6	0.2040	22	0.1570	38	0.1015	54	0.0550	70	0.0280
7	0.2010	23	0.1540	39	0.0995	55	0.0520	71	0.0260
8	0.1990	24	0.1520	40	0.0980	56	0.0465	72	0.0250
9	0.1960	25	0.1495	41	0.0960	57	0.0430	73	0.0240
10	0.1935	26	0.1470	42	0.0935	58	0.0420	74	0.0225
11	0.1910	27	0.1440	43	0.0890	59	0.0410	75	0.0210
12	0.1890	28	0.1405	44	0.0860	60	0.0400	76	0.0200
13	0.1850	29	0.1360	45	0.0820	61	0.0390	77	0.0180
14	0.1820	30	0.1285	46	0.0810	62	0.0380	78	0.0160
15	0.1800	31	0.1200	47	0.0785	63	0.0370	79	0.0145
16	0.1770	32	0.1160	48	0.0760	64	0.0360	80	0.0135

Centimeters

No.	Size	No.	Size	No.	Size	No.	Size	No.	Size
1	0.5791	17	0.4394	33	0.2870	49	0.1854	65	0.0889
2	0.5613	18	0.4305	34	0.2819	50	0.1778	66	0.0838
3	0.5410	19	0.4216	35	0.2794	51	0.1702	67	0.0813
4	0.5309	20	0.4089	36	0.2705	52	0.1613	68	0.0787
5	0.5220	21	0.4039	37	0.2642	53	0.1511	69	0.0743
6	0.5182	22	0.3988	38	0.2578	54	0.1397	70	0.0711
7	0.5105	23	0.3912	39	0.2527	55	0.1321	71	0.0660
8	0.5055	24	0.3861	40	0.2489	56	0.1181	72	0.0635
9	0.4978	25	0.3797	41	0.2438	57	0.1092	73	0.0610
10	0.4915	26	0.3734	42	0.2375	58	0.1067	74	0.0572
11	0.4851	27	0.3658	43	0.2261	59	0.1041	75	0.0533
12	0.4801	28	0.3569	44	0.2184	60	0.1016	76	0.0508
13	0.4699	29	0.3454	45	0.2083	61	0.0991	77	0.0457
14	0.4623	30	0.3264	46	0.2057	62	0.0965	78	0.0406
15	0.4572	31	0.3048	47	0.1994	63	0.0940	79	0.0368
16	0.4496	32	0.2946	48	0.1930	64	0.0914	80	0.0343

DIMENSIONS OF WIRE

Stubs' Gauge

Giving the diameter and cross-section in English and metric system for the Birmingham or Stubs' gauge.

Gauge No.	Diameter in in.	Section in sq. in.	Diameter in cm	Section in sq. cm
0000	0.454	0.16188	1.1532	1.0444
000	.425	.14186	1.0795	0.9152
00	.380	.11341	0.9652	.7317
0	0.340	0.09079	0.8636	0.5858
1	.300	.07069	.7620	.4560
2	.284	.06335	.7214	.4087
3	.259	.05269	.6579	.3399
4	.238	.04449	.6045	.2870
5	0.220	0.03801	0.5588	0.2452
6	.203	.03237	.5156	.20881
7	.180	.02545	.4572	.16147
8	.165	.02138	.4191	.13795
9	.148	.01720	.3759	.11099
10	0.134	0.01410	0.3404	0.09098
11	.120	.011310	.3048	.07297
12	.109	.009331	.2769	.06160
13	.095	.007088	.2413	.04573
14	.083	.005411	.2108	.03491
15	0.072	0.004072	0.1829	0.02627
16	.065	.0033183	.16510	.021409
17	.058	.0026421	.14732	.017046
18	.049	.0018857	.12446	.012166
19	.042	.0013854	.10668	.008938
20	0.035	0.0009621	0.08890	0.006207
21	.032	.0008042	.08128	.005189
22	.028	.0006158	.07112	.003973
23	.025	.0004909	.06350	.003167
24	.022	.0003801	.05588	.002452
25	0.020	0.0003142	0.05080	0.002027
26	.018	.0002545	.04572	.0016417
27	.016	.0002011	.04064	.0012972
28	.014	.0001539	.03556	.0009932
29	.013	.0001327	.03302	.0008563
30	0.012	0.0001181	0.03048	0.0007297
31	.010	.00007854	.02540	.0005067
32	.009	.00006362	.02286	.0004104
33	.008	.00005027	.02032	.0003243
34	.007	.00003848	.01778	.0002483
35	0.005	0.00001963	0.01270	0.0001267
36	.004	.00001257	.01016	.0000811

British Standard Gauge

Giving the diameter and cross-section in English and metric system for the British Standard Gauge.

Gauge No.	Diameter in in.	Section in sq. in.	Diameter in cm	Section in sq. cm
0000000	0.500	0.1963	1.2700	1.267
000000	.464	.1691	1.1786	1.091
00000	0.432	0.1466	1.0973	0.9456
0000	.400	.1257	1.0160	.8107
000	.372	.1087	0.9449	.7012
00	.348	.0951	.8839	.6136
0	0.324	0.0825	0.8230	0.5319
1	.300	.07069	.7620	.4560
2	.276	.05983	.7010	.3858
3	.252	.04988	.6401	.3218
4	.232	.04227	.5893	.2727
5	0.212	0.03530	0.5385	0.2277
6	.192	.02895	.4877	.18679
7	.176	.02433	.4470	.15696
8	.160	.02010	.4064	.12973
9	.144	.01629	.3658	.10507
10	0.128	0.01287	0.3251	0.08302
11	.116	.010568	.2946	.06818
12	.104	.008495	.2642	.05480
13	.092	.006648	.2337	.04289
14	.080	.005027	.2032	.03243
15	0.072	0.004071	0.1829	0.02627
16	.064	.003217	.16256	.020755
17	.056	.002463	.14224	.015890
18	.048	.001810	.12192	.011675
19	.040	.001257	.10160	.008107
20	0.036	0.001018	0.09144	0.006567
21	.032	.0008042	.08128	.005189
22	.028	.0006158	.07112	.003973
23	.024	.0004524	.06096	.002922
24	.022	.0003801	.05588	.002452
25	0.020	0.0003142	0.05080	0.002027
26	.0180	.0002545	.04572	.0016417
27	.0164	.0002112	.04166	.0013628
28	.0148	.0001728	.03759	.0011099
29	.0136	0001453	.03454	.0009363
30	0.0124	0.0001208	0.03150	0.0007791
31	.0116	.00010568	.02946	.0006818
32	.0108	.00009161	.02743	.0005910
33	.0100	.00007854	.02540	.0005067
34	.0092	.00006648	.02337	.0004289
35	0.0084	0.00005542	0.02134	0.0003575
36	.0076	.00004536	.01930	.0002927
37	.0068	.00003632	.01727	.0002343
38	.0060	.00002827	.01524	.0001824
39	.0052	.00002124	.01321	.0001370
40	0.0048	0.00001810	0.01219	0.0001167
41	.0044	.00001521	.01118	.0000982
42	.0040	.00001257	.01016	.0000811
43	.0036	.00001018	.00914	.0000656
44	.0032	.00000804	.00813	.0000519
45	0.0028	0.00000616	0.00711	0.0000397
46	.0024	.00000452	.00610	.0000212
47	.0020	.00000314	.00508	.0000203
48	.0016	.00000201	.00406	.0000129
49	.0012	.00000113	.00305	.0000073
50	0.0010	0.00000079	0.00254	0.0000051

PLATINUM WIRE

Mass in Grams per Foot

B. & S. Gauge	Diameter, inches	Mass, g per ft.	B. & S. Gauge	Diameter, inches	Mass, g per ft.
10	.1019	37.5	23	.02257	1.8
11	.09074	28.0	24	.02010	1.4
12	.08081	22.0	25	.01790	1.1
13	.07196	17.5	26	.01594	0.9
14	.06408	14.0	27	.01420	0.7
15	.05707	11.0	28	.01264	0.6
16	.05082	9.0	29	.01126	0.45
17	.04526	7.0	30	.01003	0.35
18	.04030	5.7	31	.008928	0.28
19	.03589	4.4	32	.007950	0.22
20	.03196	3.4	33	.007080	0.17
21	.02846	2.9	34	.006305	0.15
22	.02535	2.3	35	.005615	0.11

ALLOWABLE CARRYING CAPACITIES OF CONDUCTORS
(National Electrical Code)

The ratings in the following tabulation are those permitted by the National Electrical Code for flexible cords and for interior wiring of houses, hotels, office buildings, industrial plants, and other buildings.

The values are for copper wire. For aluminum wire the allowable carrying capacities shall be taken as 84% of those given in the table for the respective sizes of copper wire with the same kind of covering.

Size A.W.G.	Area Circular Mils	Diameter of Solid Wires Mils	Rubber Insulation Amperes	Varnished Cambric Insulation Amperes	Other Insulations and Bare Conductors Amperes
18	1,624.	40.3	3*	...	6†
16	2,583.	50.8	6*	...	10†
14	4,107.	64.1	15	18	20
12	6,530.	80.8	20	25	30
10	10,380.	101.9	25	30	35
8	16,510.	128.5	35	40	50
6	26,250.	162.0	50	60	70
5	33,100.	181.9	55	65	80
4	41,740.	204.3	70	85	90
3	52,630.	229.4	80	95	100
2	66,370.	257.6	90	110	125
1	83,690.	289.3	100	120	150
0	105,500.	325.0	125	150	200
00	135,100.	364.8	150	180	225
000	167,800.	409.6	175	210	275
0000	211,600.	460	225	270	325

1 Mil = 0.001 inch.

* The allowable carrying capacities of No. 18 and 16 are 5 and 7 amperes respectively, when in flexible cords.

† The allowable carrying capacities of No. 18 and 16 are 10 and 15 amperes respectively, when in cords for portable heaters. Types AFS, AFSJ, HC, HPD, and HSJ.

WIRE TABLE, STANDARD ANNEALED COPPER
American Wire Gauge (B. & S.) English Units

Gauge No.	Diameter in mils at 20°C	Cross section at 20°C Circular mils	Cross section at 20°C Sq. inches	Ohms per 1000 feet* 0°C (32°F)	Ohms per 1000 feet* 20°C (68°F)	Ohms per 1000 feet* 50°C (122°F)	Ohms per 1000 feet* 75°C (167°F)
0000	460.0	211600	0.1662	0.04516	0.04901	0.05479	0.05961
000	409.6	167800	.1318	.05695	.06180	.06909	.07516
00	364.8	133100	.1045	.07181	.07793	.08712	.09478
0	324.9	105500	.08289	.09055	.09827	.1099	.1195
1	289.3	83690	.06573	.1142	.1239	.1385	.1507
2	257.6	66370	.05213	.1440	.1563	.1747	.1900
3	229.4	52640	.04134	.1816	.1970	.2203	.2396
4	204.3	41740	.03278	.2289	.2485	.2778	.3022
5	181.9	33100	.02600	.2887	.3133	.3502	.3810
6	162.0	26250	.02062	.3640	.3951	.4416	.4805
7	144.3	20820	.01635	.4590	.4982	.5569	.6059
8	128.5	16510	.01297	.5788	.6282	.7023	.7640
9	114.4	13090	.01028	.7299	.7921	.8855	.9633
10	101.9	10380	.008155	.9203	.9989	1.117	1.215
11	90.74	8234	.006467	1.161	1.260	1.408	1.532
12	80.81	6530	.005129	1.463	1.588	1.775	1.931
13	71.96	5178	.004067	1.845	2.003	2.239	2.436
14	64.08	4107	.003225	2.327	2.525	2.823	3.071
15	57.07	3257	.002558	2.934	3.184	3.560	3.873
16	50.82	2583	.002028	3.700	4.016	4.489	4.884
17	45.26	2048	.001609	4.666	5.064	5.660	6.158
18	40.30	1624	.001276	5.883	6.385	7.138	7.765
19	35.89	1288	.001012	7.418	8.051	9.001	9.792
20	31.96	1022	.0008023	9.355	10.15	11.35	12.35
21	28.45	810.1	.0006363	11.80	12.80	14.31	15.57
22	25.35	642.4	.0005046	14.87	16.14	18.05	19.63
23	22.57	509.5	.0004002	18.76	20.36	22.76	24.76
24	20.10	404.0	.0003173	23.65	25.67	28.70	31.22
25	17.90	320.4	.0002517	29.82	32.37	36.18	39.36
26	15.94	254.1	.0001996	37.61	40.81	45.63	49.64
27	14.20	201.5	.0001583	47.42	51.47	57.53	62.59
28	12.64	159.8	.0001255	59.80	64.90	72.55	78.93
29	11.26	126.7	.00009953	75.40	81.83	91.48	99.52
30	10.03	100.5	.00007894	95.08	103.2	115.4	125.5
31	8.928	79.70	.00006260	119.9	130.1	145.5	158.2
32	7.950	63.21	.00004964	151.2	164.1	183.4	199.5
33	7.080	50.13	.00003937	190.6	206.9	231.3	251.6
34	6.305	39.75	.00003122	240.4	260.9	291.7	317.3
35	5.615	31.52	.00002476	303.1	329.0	367.8	400.1
36	5.000	25.00	.00001964	382.2	414.8	463.7	504.5
37	4.453	19.83	.00001557	482.0	523.1	584.8	636.2
38	3.965	15.72	.00001235	607.8	659.6	737.4	802.2
39	3.531	12.47	.000009793	766.4	831.8	929.8	1012
40	3.145	9.888	.000007766	966.5	1049	1173	1276

* Resistance at the stated temperatures of a wire whose length is 1000 feet at 20°C.

WIRE TABLE, STANDARD ANNEALED COPPER
American Wire Gauge (B. & S.) English Units (Continued)

Gauge No.	Pounds per 1000 feet	Feet per pound	Feet per ohm* 0°C (32°F)	Feet per ohm* 20°C (68°F)	Feet per ohm* 50°C (122°F)	Feet per ohm* 75°C (167°F)
0000	640.5	1.561	22140	20400	18250	16780
000	507.9	1.968	17560	16180	14470	13300
00	402.8	2.482	13930	12830	11480	10550
0	319.5	3.130	11040	10180	9103	8367
1	253.3	3.947	8758	8070	7219	6636
2	200.9	4.977	6946	6400	5725	5262
3	159.3	6.276	5508	5075	4540	4173
4	126.4	7.914	4368	4025	3600	3309
5	100.2	9.980	3464	3192	2855	2625
6	79.46	12.58	2747	2531	2264	2081
7	63.02	15.87	2179	2007	1796	1651
8	49.98	20.01	1728	1592	1424	1309
9	39.63	25.23	1370	1262	1129	1038
10	31.43	31.82	1087	1001	895.6	823.2
11	24.92	40.12	861.7	794.0	710.2	652.8
12	19.77	50.59	683.3	629.6	563.2	517.7
13	15.68	63.80	541.9	499.3	446.7	410.6
14	12.43	80.44	429.8	396.0	354.2	325.6
15	9.858	101.4	340.8	314.0	280.9	258.2
16	7.818	127.9	270.3	249.0	222.8	204.8
17	6.200	161.3	214.3	197.5	176.7	162.4
18	4.917	203.4	170.0	156.6	140.1	128.8
19	3.899	256.5	134.8	124.2	111.1	102.1
20	3.092	323.4	106.9	98.50	88.11	80.99
21	2.452	407.8	84.78	78.11	69.87	64.23
22	1.945	514.2	67.23	61.95	55.41	50.94
23	1.542	648.4	53.32	49.13	43.94	40.39
24	1.223	817.7	42.28	38.96	34.85	32.03
25	0.9699	1031	33.53	30.90	27.64	25.40
26	.7692	1300	26.59	24.50	21.92	20.15
27	.6100	1639	21.09	19.43	17.38	15.98
28	.4837	2067	16.72	15.41	13.78	12.67
29	.3836	2607	13.26	12.22	10.93	10.05
30	.3042	3287	10.52	9.691	8.669	7.968
31	.2413	4145	8.341	7.685	6.875	6.319
32	.1913	5227	6.614	6.095	5.452	5.011
33	.1517	6591	5.245	4.833	4.323	3.974
34	.1203	8310	4.160	3.833	3.429	3.152
35	.09542	10480	3.299	3.040	2.719	2.499
36	.07568	13210	2.616	2.411	2.156	1.982
37	.06001	16660	2.075	1.912	1.710	1.572
38	.04759	21010	1.645	1.516	1.356	1.247
39	.03774	26500	1.305	1.202	1.075	0.9886
40	.02993	33410	1.035	0.9534	0.8529	.7840

* Length at 20°C of a wire whose resistance is 1 ohm at the stated temperatures.

WIRE TABLE, STANDARD ANNEALED COPPER (Continued)

American Wire Gauge (B. & S.) English Units

Gauge No.	Diameter in mils at 20°C	Ohms per pound 0°C (32°F)	Ohms per pound 20°C (68°F)	Ohms per pound 50°C (122°F)	Lbs. per ohm 20°C (68°F)
0000	460.0	0.00007051	0.00007652	0.00008554	13070
000	409.6	.0001121	.0001217	.0001360	8219
00	364.8	.0001783	.0001935	.0002163	5169
0	324.9	.0002835	.0003076	.0003439	3251
1	289.3	.0004507	.0004891	.0005468	2044
2	257.6	.0007166	.0007778	.0008695	1286
3	229.4	.001140	.001237	.001383	808.6
4	204.3	.001812	.001966	.002198	508.5
5	181.9	.002881	.003127	.003495	319.8
6	162.0	.004581	.004972	.005558	201.1
7	144.3	.007284	.007905	.008838	126.5
8	128.5	.01158	.01257	.01405	79.55
9	114.4	.01842	.01999	.02234	50.03
10	101.9	.02928	.03178	.03553	31.47
11	90.74	.04656	.05053	.05649	19.79
12	80.81	.07404	.08035	.08983	12.45
13	71.96	.1177	.1278	.1428	7.827
14	64.08	.1872	.2032	.2271	4.922
15	57.07	.2976	.3230	.3611	3.096
16	50.82	.4733	.5136	.5742	1.947
17	45.26	.7525	.8167	.9130	1.224
18	40.30	1.197	1.299	1.452	0.7700
19	35.89	1.903	2.065	2.308	.4843
20	31.96	3.025	3.283	3.670	.3046
21	28.46	4.810	5.221	5.836	.1915
22	25.35	7.649	8.301	9.280	.1205
23	22.57	12.16	13.20	14.76	.07576
24	20.10	19.34	20.99	23.46	.04765
25	17.90	30.75	33.37	37.31	.02997
26	15.94	48.89	53.06	59.32	.01885
27	14.20	77.74	84.37	94.32	.01185
28	12.64	123.6	134.2	150.0	.007454
29	11.26	196.6	213.3	238.5	.004688
30	10.03	312.5	339.2	379.2	.002948
31	8.928	497.0	539.3	602.9	.001854
32	7.950	790.2	857.6	958.7	.001166
33	7.080	1256	1364	1524	.0007333
34	6.305	1998	2168	2424	.0004612
35	5.615	3177	3448	3854	.0002901
36	5.000	5051	5482	6128	.0001824
37	4.453	8032	8717	9744	.0001147
38	3.965	12770	13860	15490	.00007215
39	3.531	20310	22040	24640	.00004538
40	3.145	32290	35040	39170	.00002854

Gauge No.	Diameter in mm at 20°C	Cross section in mm² at 20°C	Ohms per kilometer* 0°C	Ohms per kilometer* 20°C	Ohms per kilometer* 50°C	Ohms per kilometer* 75°C
0000	11.68	107.2	0.1482	0.1608	0.1798	0.1956
000	10.40	85.03	.1868	.2028	.2267	.2466
00	9.266	67.43	.2356	.2557	.2858	.3110
0	8.252	53.48	.2971	.3224	.3604	.3921
1	7.348	42.41	.3746	.4066	.4545	.4944
2	6.544	33.63	.4724	.5127	.5731	.6235
3	5.827	26.67	.5956	.6465	.7227	.7862
4	5.189	21.15	.7511	.8152	.9113	.9914
5	4.621	16.77	.9471	1.028	1.149	1.250
6	4.115	13.30	1.194	1.296	1.449	1.576
7	3.665	10.55	1.506	1.634	1.827	1.988
8	3.264	8.366	1.899	2.061	2.304	2.506
9	2.906	6.634	2.395	2.599	2.905	3.161
10	2.588	5.261	3.020	3.277	3.663	3.985
11	2.305	4.172	3.807	4.132	4.619	5.025
12	2.053	3.309	4.801	5.211	5.825	6.337
13	1.828	2.624	6.054	6.571	7.345	7.991
14	1.628	2.081	7.634	8.285	9.262	10.08
15	1.450	1.650	9.627	10.45	11.68	12.71
16	1.291	1.309	12.14	13.17	14.73	16.02
17	1.150	1.038	15.31	16.61	18.57	20.20
18	1.024	0.8231	19.30	20.95	23.42	25.48
19	0.9116	.6527	24.34	26.42	29.53	32.12
20	.8118	.5176	30.69	33.31	37.24	40.51
21	.7230	.4105	38.70	42.00	46.95	51.08
22	.6438	.3255	48.80	52.96	59.21	64.41
23	.5733	.2582	61.54	66.79	74.66	81.22
24	.5106	.2047	77.60	84.21	94.14	102.4
25	.4547	.1624	97.85	106.2	118.7	129.1
26	.4049	.1288	123.4	133.9	149.7	162.9
27	.3606	.1021	155.6	168.9	188.8	205.4
28	.3211	.08098	196.2	212.9	238.0	258.9
29	.2859	.06422	247.4	268.5	300.1	326.5
30	.2546	.05093	311.9	338.6	378.5	411.7
31	.2268	.04039	393.4	426.9	477.2	519.2
32	.2019	.03203	496.0	538.3	601.8	654.7
33	.1798	.02540	625.5	678.8	758.8	825.5
34	.1601	.02014	788.7	856.0	956.9	1041
35	.1426	.01597	994.5	1079	1207	1313
36	.1270	.01267	1254	1361	1522	1655
37	.1131	.01005	1581	1716	1919	2087
38	.1007	.007967	1994	2164	2419	2632
39	.08969	.006318	2514	2729	3051	3319
40	.07987	.005010	3171	3441	3847	4185

* Resistance at the stated temperatures of a wire whose length is 1 kilometer at 20°C.

Gauge No.	Diameter in mm at 20°C	Kilograms per kilometer	Meters per gram	Meters per ohm* 0°C	Meters per ohm* 20°C	Meters per ohm* 50°C	Meters per ohm* 75°C
0000	11.68	953.2	0.001049	6749	6219	5563	5113
000	10.40	755.9	.001323	5352	4932	4412	4055
00	9.266	599.5	.001668	4245	3911	3499	3216
0	8.252	475.4	.002103	3366	3102	2774	2550
1	7.348	377.0	.002652	2669	2460	2200	2022
2	6.544	299.0	.003345	2117	1951	1745	1604
3	5.827	237.1	.004217	1679	1547	1384	1272
4	5.189	188.0	.005318	1331	1227	1097	1009
5	4.621	149.1	.006706	1056	972.9	870.2	799.9
6	4.115	118.2	.008457	837.3	771.5	690.1	634.4
7	3.665	93.78	.01066	664.0	611.8	547.3	503.1
8	3.264	74.37	.01345	526.6	485.2	434.0	399.0
9	2.906	58.98	.01696	417.6	384.8	344.2	316.4
10	2.588	46.77	.02138	331.2	305.1	273.0	250.9
11	2.305	37.09	.02696	262.6	242.0	216.5	199.0
12	2.053	29.42	.03400	208.3	191.9	171.7	157.8
13	1.828	23.33	.04287	165.2	152.2	136.1	125.1
14	1.628	18.50	.05406	131.0	120.7	108.0	99.24
15	1.450	14.67	.06816	103.9	95.71	85.62	78.70
16	1.291	11.63	.08595	82.38	75.90	67.90	62.41
17	1.150	9.226	.1084	65.33	60.20	53.85	49.50
18	1.024	7.317	.1367	51.81	47.74	42.70	39.25
19	0.9116	5.803	.1723	41.09	37.86	33.86	31.13
20	.8118	4.602	.2173	32.58	30.02	26.86	24.69
21	.7230	3.649	.2740	25.84	23.81	21.30	19.58
22	.6438	2.894	.3455	20.49	18.88	16.89	15.53
23	.5733	2.295	.4357	16.25	14.97	13.39	12.31
24	.5106	1.820	.5494	12.89	11.87	10.62	9.764
25	.4547	1.443	.6928	10.22	9.417	8.424	7.743
26	.4049	1.145	.8736	8.105	7.468	6.680	6.141
27	.3606	0.9078	1.102	6.428	5.922	5.298	4.870
28	.3211	.7199	1.389	5.097	4.697	4.201	3.862
29	.2859	.5709	1.752	4.042	3.725	3.332	3.063
30	.2546	.4527	2.209	3.206	2.954	2.642	2.429
31	.2268	.3590	2.785	2.542	2.342	2.095	1.926
32	.2019	.2847	3.512	2.016	1.858	1.662	1.527
33	.1798	.2258	4.429	1.599	1.473	1.318	1.211
34	.1601	.1791	5.584	1.268	1.168	1.045	0.9606
35	.1426	.1420	7.042	1.006	0.9265	0.8288	.7618
36	.1270	.1126	8.879	0.7974	.7347	.6572	.6041
37	.1131	.08931	11.20	.6324	.5827	.5212	.4791
38	.1007	.07083	14.12	.5015	.4621	.4133	.3799
39	.08969	.05617	17.80	.3977	.3664	.3278	.3013
40	.07987	.04454	22.45	.3154	.2906	.2600	.2390

* Length at 20°C of a wire whose resistance is 1 ohm at the stated temperatures.

American Wire Gauge (B. & S.) Metric Units

Gauge No.	Ohms per kilogram			Grams per ohm
	0°C	20°C	50°C	20°C
0000	0.0001554	0.0001687	0.0001886	5928000
000	.0002472	.0002682	.0002999	3728000
00	.0003930	.0004265	.0004768	2344000
0	.0006249	.0006782	.0007582	1474000
1	.0009936	.001078	.001206	927300
2	.001580	.001715	.001917	583200
3	.002512	.002726	.003048	366800
4	.003995	.004335	.004846	230700
5	.006352	.006893	.007706	145100
6	.01010	.01096	.01225	91230
7	.01606	.01743	.01948	57380
8	.02553	.02771	.03098	36080
9	.04060	.04406	.04926	22690
10	.06456	.07007	.07833	14270
11	.1026	.1114	.1245	8976
12	.1632	.1771	.1980	5645
13	.2595	.2817	.3149	3550
14	.4127	.4479	.5007	2233
15	.6562	.7122	.7961	1404
16	1.043	1.132	1.266	883.1
17	1.659	1.801	2.013	555.4
18	2.638	2.863	3.201	349.3
19	4.194	4.552	5.089	219.7
20	6.670	7.238	8.092	138.2
21	10.60	11.51	12.87	86.88
22	16.86	18.30	20.46	54.64
23	26.81	29.10	32.53	34.36
24	42.63	46.27	51.73	21.61
25	67.79	73.57	82.25	13.59
26	107.8	117.0	130.8	8.548
27	171.4	186.0	207.9	5.376
28	272.5	295.8	330.6	3.381
29	433.3	470.3	525.7	2.126
30	689.0	747.8	836.0	1.337
31	1096	1189	1329	0.8410
32	1742	1891	2114	.5289
33	2770	3006	3361	.3326
34	4404	4780	5344	.2092
35	7003	7601	8497	.1316
36	11140	12090	13510	.08274
37	17710	19220	21480	.05204
38	28150	30560	34160	.03273
39	44770	48590	54310	.02058
40	71180	77260	86360	.01294

ALUMINUM WIRE TABLE

Hard-Drawn Aluminum Wire at 20°C (or, 68°F)

American Wire Gauge (B. & S.) English Units

Gauge No.	Diameter in mils	Cross section		Ohms per 1000 ft.	Pounds per 1000 ft.	Pounds per ohm	Feet per ohm
		Circular mils	Square inches				
0000	460	212000	0.166	0.0804	195	2420	12400
000	410	168000	.132	.101	154	1520	9860
00	365	133000	.105	.128	122	957	7820
0	325	106000	.0829	.161	97.0	602	6200
1	289	83700	.0657	.203	76.9	379	4920
2	258	66400	.0521	.256	61.0	238	3900
3	229	52600	.0413	.323	48.4	150	3090
4	204	41700	.0328	.408	38.4	94.2	2450
5	182	33100	.0260	.514	30.4	59.2	1950
6	162	26300	.0206	.648	24.1	37.2	1540
7	144	20800	.0164	.817	19.1	23.4	1220
8	128	16500	.0130	1.03	15.2	14.7	970
9	114	13100	.0103	1.30	12.0	9.26	770

ALUMINUM WIRE TABLE (Continued)

Gauge No.	Diameter in mils	Cross section		Ohms per 1000 ft.	Pounds per 1000 ft.	Pounds per ohm	Feet per ohm
		Circular mils	Square inches				
10	102	10400	.00815	1.64	9.55	5.83	610
11	91	8230	.00647	2.07	7.57	3.66	484
12	81	6530	.00513	2.61	6.00	2.30	384
13	72	5180	.00407	3.29	4.76	1.45	304
14	64	4110	.00323	4.14	3.78	0.911	241
15	57	3260	.00256	5.22	2.99	.573	191
16	51	2580	.00203	6.59	2.37	.360	152
17	45	2050	.00161	8.31	1.88	.227	120
18	40	1620	.00128	10.5	1.49	.143	95.5
19	36	1290	.00101	13.2	1.18	.0897	75.7
20	32	1020	.000802	16.7	0.939	.0564	60.0
21	28.5	810	.000636	21.0	.745	.0355	47.6
22	25.3	642	.000505	26.5	.591	.0223	37.8
23	22.6	509	.000400	33.4	.468	.0140	29.9
24	20.1	404	.000317	42.1	.371	.00882	23.7
25	17.9	320	.000252	53.1	.295	.00555	18.8
26	15.9	254	.000200	67.0	.234	.00349	14.9
27	14.2	202	.000158	84.4	.185	.00219	11.8
28	12.6	160	.000126	106.	.147	.00138	9.39
29	11.3	127	.0000995	134.	.117	.000868	7.45
30	10.0	101	.0000789	169.	.0924	.000546	5.91
31	8.9	79.7	.0000626	213.	.0733	.000343	4.68
32	8.0	63.2	.0000496	269.	.0581	.000216	3.72
33	7.1	50.1	.0000394	339.	.0461	.000136	2.95
34	6.3	39.8	.0000312	428.	.0365	.0000854	2.34
35	5.6	31.5	.0000248	540.	.0290	.0000537	1.85
36	5.0	25.0	.0000196	681.	.0230	.0000338	1.47
37	4.5	19.8	.0000156	858.	.0182	.0000212	1.17
38	4.0	15.7	.0000123	1080.	.0145	.0000134	0.924
39	3.5	12.5	.00000979	1360.	.0115	.0000840	.733
40	3.1	9.9	.00000777	1720.	.0091	.0000528	.581

Gauge No.	Diameter in mm	Cross section in mm²	Ohms per kilometer	Kilograms per kilometer	Grams per ohm	Meters per ohm
0000	11.7	107	0.264	289	1100000	3790
000	10.4	85.0	.333	230	690000	3010
00	9.3	67.4	.419	182	434000	2380
0	8.3	53.5	.529	144	273000	1890
1	7.3	42.4	.667	114.	172000	1500
2	6.5	33.6	.841	90.8	108000	1190
3	5.8	26.7	1.06	72.0	67900	943
4	5.2	21.2	1.34	57.1	42700	748
5	4.6	16.8	1.69	45.3	26900	593
6	4.1	13.3	2.13	35.9	16900	470
7	3.7	10.5	2.68	28.5	10600	373
8	3.3	8.37	3.38	22.6	6680	296
9	2.91	6.63	4.26	17.9	4200	235
10	2.59	5.26	5.38	14.2	2640	186
11	2.30	4.17	6.78	11.3	1660	148
12	2.05	3.31	8.55	8.93	1050	117
13	1.83	2.62	10.8	7.08	657	92.8
14	1.63	2.08	13.6	5.62	413	73.6
15	1.45	1.65	17.1	4.46	260	58.4
16	1.29	1.31	21.6	3.53	164	46.3
17	1.15	1.04	27.3	2.80	103	36.7
18	1.02	0.823	34.4	2.22	64.7	29.1
19	0.91	.653	43.3	1.76	40.7	23.1
20	.81	.518	54.6	1.40	25.6	18.3
21	.72	.411	68.9	1.11	16.1	14.5
22	.64	.326	86.9	0.879	10.1	11.5
23	.57	.258	110	.697	6.36	9.13
24	.51	.205	138	.553	4.00	7.24
25	.45	.162	174	.438	2.52	5.74
26	.40	.129	220	.348	1.58	4.55
27	.36	.102	277	.276	0.995	3.61
28	.32	.0810	349	.219	.626	2.86
29	.29	.0642	440	.173	.394	2.27
30	.25	.0509	555	.138	.248	1.80
31	.227	.0404	700	.109	.156	1.43
32	.202	.0320	883	.0865	.0979	1.13
33	.180	.0254	1110	.0686	.0616	0.899
34	.160	.0201	1400	.0544	.0387	.712
35	.143	.0160	1770	.0431	.0244	.565
36	.127	.0127	2230	.0342	.0153	.448
37	.113	.0100	2820	.0271	.00963	.355
38	.101	.0080	3550	.0215	.00606	.282
39	.090	.0063	4480	.0171	.00381	.223
40	.080	.0050	5640	.0135	.00240	.177

CROSS-SECTION AND MASS OF WIRES

U. S. Measure

Diameters are given in mils (1 mil = .001 in.), and area in square mils (1 sq. mil = .000001 sq. in.). For sections and masses for one-tenth the diameters given, divide by 100 and for sections and masses for ten times the diameter multiply by 100.

Diam. in mils	Cross-sec. in sq. mils	Pounds per foot			
		Copper, density 8.90	Iron, density 7.80	Brass, density 8.56	Aluminum, density 2.67
10	78.54	0.000303	0.0002656	0.0002915	0.0000909
11	95.03	0367	03214	03527	01100
12	113.10	0436	03825	04197	01309
13	132.73	0512	04488	04926	01536
14	153.94	0594	05206	05713	01782
15	176.71	0.000682	0.0005976	0.0006558	0.0002045
16	201.06	0776	06799	07461	02327
17	226.98	0876	07675	08423	02627
18	254.47	0982	08605	09443	02946
19	283.53	1094	09588	10522	03282
20	314.16	0.001212	0.001062	0.001166	0.0003636
21	346.36	1336	1171	1285	04009
22	380.13	1467	1286	1411	04400
23	415.48	1603	1405	1542	04809
24	452.39	1746	1530	1679	05237
25	490.87	0.001894	0.001660	0.001822	0.0005682
26	530.93	2046	1795	1970	06147
27	572.56	2209	1936	2125	06628
28	615.75	2376	2082	2285	07127
29	660.52	2549	2234	2451	07646
30	706.86	0.002727	0.002390	0.002623	0.0008182
31	754.77	2912	2552	2801	08737
32	804.25	3103	2720	2985	09309
33	855.30	3300	2892	3174	09900
34	907.92	3503	3070	3369	10509
35	962.11	0.003712	0.003253	0.003570	0.001114
36	1017.88	3927	3442	3777	1178
37	1075.21	4149	3636	3990	1245
38	1134.11	4376	3844	4218	1316
39	1194.59	4609	4040	4433	1383
40	1256.64	0.004849	0.004249	0.004664	0.001455
41	1320.25	5094	4465	4900	1528
42	1385.44	5346	4685	5141	1604
43	1452.20	5603	4911	5389	1681
44	1520.53	5867	5142	5643	1760
45	1590.43	0.006137	0.005378	0.005902	0.001841
46	1661.90	6412	5620	6167	1924
47	1734.94	6694	5867	6438	2008
48	1809.56	6982	6119	6715	2095
49	1885.74	7276	6377	6998	2183
50	1963.50	0.007576	0.006640	0.007287	0.002273
51	2042.82	7882	6908	7581	2365
52	2123.72	8194	7181	7881	2458
53	2206.18	8512	7460	8187	2554
54	2290.22	8837	7744	8499	2651
55	2375.83	0.009167	0.008034	0.008817	0.002750
56	2463.01	09504	08329	09140	2851
57	2551.76	09846	08629	09470	2954
58	2642.08	10195	08934	09805	3058
59	2733.97	10549	09245	10146	3165
60	2827.43	0.01091	0.00956	0.01049	0.003273
61	2922.47	1128	0988	1085	3383
62	3019.07	1165	1021	1120	3495
63	3117.25	1203	1054	1157	3608
64	3216.99	1241	1088	1194	3724
65	3318.31	0.01280	0.01122	0.01231	0.003841
66	3421.19	1320	1157	1270	3960
67	3525.65	1360	1192	1308	4081
68	3631.68	1401	1228	1348	4204
69	3739.28	1443	1264	1388	4328
70	3848.45	0.01485	0.01302	0.01429	0.004456
71	3959.19	1528	1339	1469	4583
72	4071.50	1571	1377	1511	4713
73	4185.39	1615	1415	1553	4845
74	4300.84	1660	1454	1596	4978
75	4417.86	0.01705	0.01494	0.01639	0.005114
76	4536.46	1751	1534	1684	5251
77	4656.63	1797	1575	1728	5390
78	4778.36	1844	1616	1773	5531
79	4901.67	1892	1658	1819	5674
80	5026.55	0.01939	0.01700	0.01865	0.005818
81	5153.00	1988	1743	1912	5965
82	5281.02	2038	1786	1960	6113
83	5410.61	2088	1830	2008	6263
84	5541.77	2138	1874	2057	6415
85	5674.50	0.02189	0.01919	0.02106	0.006568
86	5808.80	2241	1964	2156	6724
87	5944.68	2294	2010	2206	6881
88	6082.12	2347	2057	2257	7040
89	6221.14	2400	2104	2309	7201
90	6361.73	0.02455	0.02151	0.02360	0.007364
91	6503.88	2509	2199	2414	7528
92	6647.61	2565	2248	2467	7695
93	6792.91	2621	2297	2521	7863
94	6939.78	2678	2347	2575	8033
95	7088.22	0.02735	0.02397	0.02630	0.008205
96	7238.23	2793	2448	2686	8378
97	7389.81	2851	2499	2742	8554
98	7542.96	2910	2551	2799	8731
99	7697.69	2970	2603	2857	8910
100	7853.98	0.03030	0.02656	0.02915	0.009091

CROSS-SECTION AND MASS OF WIRES (Continued)

Metric Measure

Diameters are given in thousandths of a centimeter and area of section in square thousandths of a centimeter. 1 $(cm/1000)^2$ = .000001 sq. cm For sections and masses for diameters 1/10 or 10 times those of the table, divide or multiply by 100.

Diam. in thousandths of a cm	Cross-section in square thousandths of a cm	Grams per meter			
		Copper, density 8.90	Iron, density 7.80	Brass, density 8.56	Aluminum, density 2.67
10	78.54	0.06990	0.06126	0.06723	0.02097
11	95.03	.08458	.07412	.08135	.02537
12	113.10	.10065	.08822	.09681	.03020
13	132.73	.11813	.10353	.11362	.03544
14	153.94	.13701	.12008	.13177	.04110
15	176.71	0.1573	0.1378	0.1513	0.04718
16	201.06	.1789	.1568	.1721	.05368
17	226.98	.2020	.1770	.1943	.06060
18	254.47	.2265	.1985	.2178	.06794
19	283.53	.2523	.2212	.2427	.07570
20	314.16	0.2796	0.2450	0.2689	0.08388
21	346.36	.3083	.2702	.2965	.09248
22	380.13	.3383	.2965	.3254	.10149
23	415.48	.3698	.3241	.3557	.11093
24	452.39	.4026	.3529	.3872	.12079
25	490.87	0.4369	0.3829	0.4202	0.1311
26	530.93	.4725	.4141	.4545	.1418
27	572.56	.5096	.4466	.4901	.1529
28	615.75	.5480	.4803	.5271	.1644
29	660.52	.5879	.5152	.5654	.1764
30	706.86	0.6291	0.5514	0.6051	0.1887
31	754.77	.6717	.5887	.6461	.2015
32	804.25	.7158	.6273	.6884	.2147
33	855.30	.7612	.6671	.7321	.2284
34	907.92	.8081	.7082	.7772	.2424
35	962.11	0.856	0.7504	0.8236	0.2569
36	1017.88	.906	.7939	.8713	.2718
37	1075.21	.957	.8387	.9204	.2871
38	1134.11	1.012	.8866	.9730	.3035
39	1194.59	.063	.9318	1.0230	.3190
40	1256.64	1.118	0.980	1.076	0.3355
41	1320.25	.175	1.030	.130	.3525
42	1385.44	.233	.081	.186	.3699
43	1452.20	.292	.133	.243	.3877
44	1520.53	.353	.186	.302	.4060
45	1590.43	1.415	1.241	1.361	0.4246
46	1661.90	.479	.296	.423	.4437
47	1734.94	.544	.353	.485	.4632
48	1809.56	.611	.411	.549	.4832
49	1885.74	.678	.471	.614	.5035
50	1963.50	1.748	1.532	1.681	.5243
51	2042.82	.818	.593	.753	.5454
52	2123.72	.890	.657	.818	.5670
53	2206.18	.964	.721	.888	.5891
54	2290.22	2.038	.786	.960	.6115

CROSS-SECTION AND MASS OF WIRES (Continued)

Metric Measure (Continued)

Diameters are given in thousandths of a centimeter and area of section in square thousandths of a centimeter. 1 $(cm/1000)^2$ = .000001 sq. cm. For sections and masses for diameters 1/10 or 10 times those of the table, divide or multiply by 100.

Diam. in thousandths of a cm	Cross-section in square thousandths of a cm	Grams per meter: Copper, density 8.90	Iron, density 7.80	Brass, density 8.56	Aluminum, density 2.67
55	2375.83	2.114	1.853	2.034	0.6343
56	2463.01	.192	.921	.108	.6576
57	2551.76	.271	.990	.184	.6813
58	2642.08	.351	2.061	.262	.7054
59	2733.97	.433	.132	.340	.7300
60	2827.43	2.516	2.205	2.420	0.7549
61	2922.47	.601	.280	.502	.7803
62	3019.07	.687	.355	.584	.8061
63	3117.25	.774	.431	.668	.8323
64	3216.99	.863	.509	.760	.8589
65	3318.31	2.953	2.588	2.840	0.8860
66	3421.19	3.045	.669	.929	.9135
67	3525.65	.138	.750	3.018	.9413
68	3631.68	.232	.833	.109	.9697
69	3739.28	.328	.917	.201	.9984
70	3848.45	3.426	3.003	3.295	1.028
71	3959.19	.524	.088	.389	.057
72	4071.50	.624	.176	.485	.087
73	4185.39	.725	.265	.583	.117
74	4300.84	.828	.355	.682	.148
75	4417.86	3.932	3.446	3.782	1.180
76	4536.46	4.037	.538	.883	.211
77	4656.63	.144	.632	.986	.243
78	4778.36	.253	.727	4.090	.276
79	4901.67	.362	.823	.177	.309
80	5026.55	4.474	3.921	4.303	1.342
81	5153.00	.586	4.019	.411	.376
82	5281.02	.700	.119	.521	.410
83	5410.61	.815	.220	.631	.445
84	5541.77	.932	.323	.744	.480
85	5674.50	5.050	4.426	4.857	1.515
86	5808.80	.170	.531	.972	.551
87	5944.68	.291	.637	5.089	.587
88	6082.12	.413	.744	.206	.624
89	6221.14	.537	.852	.325	.661
90	6361.73	5.662	4.962	5.446	1.699
91	6503.88	.788	5.073	.567	.737
92	6647.61	.916	.185	.690	.775
93	6792.91	6.046	.298	.815	.814
94	6939.78	.176	.413	.940	.853
95	7088.22	6.309	5.529	6.068	1.893
96	7238.23	.442	.646	.196	.933
97	7389.81	.577	.764	.326	.973
98	7542.96	.713	.884	.457	2.014
99	7697.69	.851	6.004	.589	.055
100	7853.98	6.990	6.126	6.723	2.097

RESISTANCE OF WIRES

The following table gives the approximate resistance of various metallic conductors. The values have been computed from the resistivities at 20°C, except as otherwise stated, and for the dimensions of wire indicated. Owing to differences in purity in the case of elements and of composition in alloys, the values can be considered only as approximations.

The following dimensions have been adopted in the computations.

B. & S. gauge	Diameter: mm	Diameter: mils 1 mil = .001 in.	B. & S. gauge	Diameter: mm	Diameter: mils 1 mil = .001 in.
10	2.588	101.9	26	0.4049	15.94
12	2.053	80.81	27	0.3606	14.20
14	1.628	64.08	28	0.3211	12.64
16	1.291	50.82	30	0.2546	10.03
18	1.024	40.30	32	0.2019	7.950
20	0.8118	31.96	34	0.1601	6.305
22	0.6438	25.35	36	0.1270	5.000
24	0.5106	20.10	40	0.07987	3.145

B. & S. No.	Ohms per cm	Ohms per ft.	B. & S. No.	Ohms per cm	Ohms per ft.
*Advance (0°C) $\rho = 48. \times 10^{-6}$ ohm cm			Aluminum $\rho = 2.828 \times 10^{-6}$ ohm cm		
10	.000912	.0278	10	.0000538	.00164
12	.00145	.0442	12	.0000855	.00260
14	.00231	.0703	14	.000136	.00414
16	.00367	.112	16	.000216	.00658
18	.00583	.178	18	.000344	.0105
20	.00927	.283	20	.000546	.0167
22	.0147	.449	22	.000869	.0265
24	.0234	.715	24	.00138	.0421
26	.0373	1.14	26	.00220	.0669
27	.0470	1.43	27	.00277	.0844
28	.0593	1.81	28	.00349	.106
30	.0942	2.87	30	.00555	.169
32	.150	4.57	32	.00883	.269
34	.238	7.26	34	.0140	.428
36	.379	11.5	36	.0223	.680
40	.958	29.2	40	.0564	1.72

* Trade mark.

RESISTANCE OF WIRES

B. & S. No.	Ohms per cm	Ohms per ft.	B. & S. No.	Ohms per cm	Ohms per ft.
Brass $\rho = 7.00 \times 10^{-6}$ ohm cm			Climax $\rho = 87. \times 10^{-6}$ ohm cm		
10	.000133	.00406	10	.00165	.0504
12	.000212	.00645	12	.00263	.0801
14	.000336	.0103	14	.00418	.127
16	.000535	.0163	16	.00665	.203
18	.000850	.0259	18	.0106	.322
20	.00135	.0412	20	.0168	.512
22	.00215	.0655	22	.0267	.815
24	.00342	.104	24	.0425	1.30
26	.00543	.166	26	.0675	2.06
27	.00686	.209	27	.0852	2.60
28	.00864	.263	28	.107	3.27
30	.0137	.419	30	.171	5.21
32	.0219	.666	32	.272	8.28
34	.0348	1.06	34	.432	13.2
36	.0552	1.68	36	.687	20.9
40	.140	4.26	40	1.74	52.9
Constantan (0°C) $\rho = 44.1 \times 10^{-6}$ ohm cm			Copper, annealed $\rho = 1.724 \times 10^{-6}$ ohm cm		
10	.000838	.0255	10	.0000328	.000999
12	.00133	.0406	12	.0000521	.00159
14	.00212	.0646	14	.0000828	.00253
16	.00337	.103	16	.000132	.00401
18	.00536	.163	18	.000209	.00638
20	.00852	.260	20	.000333	.0102
22	.0135	.413	22	.000530	.0161
24	.0215	.657	24	.000842	.0257
26	.0342	1.04	26	.00134	.0408
27	.0432	1.32	27	.00169	.0515
28	.0545	1.66	28	.00213	.0649
30	.0866	2.64	30	.00339	.103
32	.138	4.20	32	.00538	.164
34	.219	6.67	34	.00856	.261
36	.348	10.6	36	.0136	.415
40	.880	26.8	40	.0344	1.05

Eureka (0°C) $\rho = 47. \times 10^{-6}$ ohm cm

B. & S. No.	Ohms per cm	Ohms per ft.
10	.000893	.0272
12	.00142	.0433
14	.00226	.0688
16	.00359	.109
18	.00571	.174
20	.00908	.277
22	.0144	.440
24	.0230	.700
26	.0365	1.11
27	.0460	1.40
28	.0580	1.77
30	.0923	2.81
32	.147	4.47
34	.233	7.11
36	.371	11.3
40	.938	28.6

German silver $\rho = 33. \times 10^{-6}$ ohm cm

B. & S. No.	Ohms per cm	Ohms per ft.
10	.000627	.0191
12	.000997	.0304
14	.00159	.0483
16	.00252	.0768
18	.00401	.122
20	.00638	.194
22	.0101	.309
24	.0161	.491
26	.0256	.781
27	.0323	.985
28	.0408	1.24
30	.0648	1.97
32	.103	3.14
34	.164	4.99
36	.260	7.94
40	.659	20.1

Iron $\rho = 10. \times 10^{-6}$ ohm cm

B. & S. No.	Ohms per cm	Ohms per ft.
10	.000190	.00579
12	.000302	.00921
14	.000481	.0146
16	.000764	.0233
18	.00121	.0370
20	.00193	.0589
22	.00307	.0936
24	.00489	.149
26	.00776	.237
27	.00979	.299
28	.0123	.376
30	.0196	.598
32	.0312	.952
34	.0497	1.51
36	.0789	2.41
40	.200	6.08

Magnesium $\rho = 4.6 \times 10^{-6}$ ohm cm

B. & S. No.	Ohms per cm	Ohms per ft.
10	.0000874	.00267
12	.000139	.00424
14	.000221	.00674
16	.000351	.0107
18	.000559	.0170
20	.000889	.0271
22	.00141	.0431
24	.00225	.0685
26	.00357	.109
27	.00451	.137
28	.00568	.173
30	.00903	.275
32	.0144	.438
34	.0228	.696
36	.0363	1.11
40	.0918	2.80

Molybdenum $\rho = 5.7 \times 10^{-6}$ ohm cm

B. & S. No.	Ohms per cm	Ohms per ft.
10	.000108	.00330
12	.000172	.00525
14	.000274	.00835
16	.000435	.0133
18	.000693	.0211
20	.00110	.0336
22	.00175	.0534
24	.00278	.0849
26	.00443	.135
27	.00558	.170
28	.00704	.215
30	.0112	.341
32	.0178	.542
34	.0283	.863
36	.0450	1.37
40	.114	3.47

Excello $\rho = 92. \times 10^{-6}$ ohm cm

B. & S. No.	Ohms per cm	Ohms per ft.
10	.00175	.0533
12	.00278	.0847
14	.00442	.135
16	.00703	.214
18	.0112	.341
20	.0178	.542
22	.0283	.861
24	.0449	1.37
26	.0714	2.18
27	.0901	2.75
28	.114	3.46
30	.181	5.51
32	.287	8.75
34	.457	13.9
36	.726	22.1
40	1.84	56.0

Gold $\rho = 2.44 \times 10^{-6}$ ohm cm

B. & S. No.	Ohms per cm	Ohms per ft.
10	.0000464	.00141
12	.0000737	.00225
14	.000117	.00357
16	.000186	.00568
18	.000296	.00904
20	.000471	.0144
22	.000750	.0228
24	.00119	.0363
26	.00189	.0577
27	.00239	.0728
28	.00301	.0918
30	.00479	.146
32	.00762	.232
34	.0121	.369
36	.0193	.587
40	.0487	1.48

Lead $\rho = 22. \times 10^{-6}$ ohm cm

B. & S. No.	Ohms per cm	Ohms per ft.
10	.000418	.0127
12	.000665	.0203
14	.00106	.0322
16	.00168	.0512
18	.00267	.0815
20	.00425	.130
22	.00676	.206
24	.0107	.328
26	.0171	.521
27	.0215	.657
28	.0272	.828
30	.0432	1.32
32	.0687	2.09
34	.109	3.33
36	.174	5.29
40	.439	13.4

Manganin $\rho = 44. \times 10^{-6}$ ohm cm

B. & S. No.	Ohms per cm	Ohms per ft.
10	.000836	.0255
12	.00133	.0405
14	.00211	.0644
16	.00336	.102
18	.00535	.163
20	.00850	.259
22	.0135	.412
24	.0215	.655
26	.0342	1.04
27	.0431	1.31
28	.0543	1.66
30	.0864	2.63
32	.137	4.19
34	.218	6.66
36	.347	10.6
40	.878	26.8

Monel Metal $\rho = 42. \times 10^{-6}$ ohm cm

B. & S. No.	Ohms per cm	Ohms per ft.
10	.000798	.0243
12	.00127	.0387
14	.00202	.0615
16	.00321	.0978
18	.00510	.156
20	.00811	.247
22	.0129	.393
24	.0205	.625
26	.0326	.994
27	.0411	1.25
28	.0519	1.58
30	.0825	2.51
32	.131	4.00
34	.209	6.36
36	.331	10.1
40	.838	25.6

***Nichrome** $\rho = 115. \times 10^{-6}$ ohm cm

B. & S. No.	Ohms per cm	Ohms per ft.
10	.0021281	.06488
12	.0033751	.1029
14	.0054054	.1648
16	.0085116	.2595
18	.0138383	.4219
20	.0216218	.6592
22	.0346040	1.055
24	.0548088	1.671
26	.0875760	2.670
28	.1394328	4.251
30	.2214000	6.750
32	.346040	10.55
34	.557600	17.00
36	.885600	27.00
38	1.383832	42.19
40	2.303872	70.24

Platinum $\rho = 10. \times 10^{-6}$ ohm cm

B. & S. No.	Ohms per cm	Ohms per ft.
10	.000190	.00579
12	.000302	.00921
14	.000481	.0146
16	.000764	.0233
18	.00121	.0370
20	.00193	.0589
22	.00307	.0936
24	.00489	.149
26	.00776	.237
27	.00979	.299
28	.0123	.376
30	.0196	.598
32	.0312	.952
34	.0497	1.51
36	.0789	2.41
40	.200	6.08

Steel, piano wire (0°C) $\rho = 11.8 \times 10^{-6}$ ohm cm

B. & S. No.	Ohms per cm	Ohms per ft.
10	.000224	.00684
12	.000357	.0109
14	.000567	.0173
16	.000901	.0275
18	.00143	.0437
20	.00228	.0695
22	.00363	.110
24	.00576	.176
26	.00916	.279
27	.0116	.352
28	.0146	.444
30	.0232	.706
32	.0368	1.12
34	.0586	1.79
36	.0931	2.84
40	.236	7.18

Tantalum $\rho = 15.5 \times 10^{-6}$ ohm cm

B. & S. No.	Ohms per cm	Ohms per ft.
10	.000295	.00898
12	.000468	.0143
14	.000745	.0227
16	.00118	.0361
18	.00188	.0574
20	.00299	.0913
22	.00476	.145
24	.00757	.231
26	.0120	.367
27	.0152	.463
28	.0191	.583
30	.0304	.928
32	.0484	1.47
34	.0770	2.35
36	.122	3.73
40	.309	9.43

Nickel $\rho = 7.8 \times 10^{-6}$ ohm cm

B. & S. No.	Ohms per cm	Ohms per ft.
10	.000148	.00452
12	.000236	.00718
14	.000375	.0114
16	.000596	.0182
18	.000948	.0289
20	.00151	.0459
22	.00240	.0730
24	.00381	.116
26	.00606	.185
27	.00764	.233
28	.00963	.294
30	.0153	.467
32	.0244	.742
34	.0387	1.18
36	.0616	1.88
40	.156	4.75

Silver (18°C) $\rho = 1.629 \times 10^{-6}$ ohm cm

B. & S. No.	Ohms per cm	Ohms per ft.
10	.0000310	.000944
12	.0000492	.00150
14	.0000783	.00239
16	.000124	.00379
18	.000198	.00603
20	.000315	.00959
22	.000500	.0153
24	.000796	.0243
26	.00126	.0386
27	.00160	.0486
28	.00201	.0613
30	.00320	.0975
32	.00509	.155
34	.00809	.247
36	.0129	.392
40	.0325	.991

Steel, invar (35 % Ni) $\rho = 81. \times 10^{-6}$ ohm cm

B. & S. No.	Ohms per cm	Ohms per ft.
10	.00154	.0469
12	.00245	.0746
14	.00389	.119
16	.00619	.189
18	.00984	.300
20	.0156	.477
22	.0249	.758
24	.0396	1.21
26	.0629	1.92
27	.0793	2.42
28	.100	3.05
30	.159	4.85
32	.253	7.71
34	.402	12.3
36	.639	19.5
40	1.62	49.3

Tin $\rho = 11.5 \times 10^{-6}$ ohm cm

B. & S. No.	Ohms per cm	Ohms per ft.
10	.000219	.00666
12	.000348	.0106
14	.000553	.0168
16	.000879	.0268
18	.00140	.0426
20	.00222	.0677
22	.00353	.108
24	.00562	.171
26	.00893	.272
27	.0113	.343
28	.0142	.433
30	.0226	.688
32	.0359	1.09
34	.0571	1.74
36	.0908	2.77
40	.230	7.00

B. & S. No.	Ohms per cm	Ohms per ft.	B. & S. No.	Ohms per cm	Ohms per ft.
Tungsten $\rho = 5.51 \times 10^{-6}$ ohm cm			Zinc (0°C) $\rho = 5.75 \times 10^{-6}$ ohm cm		
10	.000105	.00319	10	.000109	.00333
12	.000167	.00508	12	.000174	.00530
14	.000265	.00807	14	.000276	.00842
16	.000421	.0128	16	.000439	.0134
18	.000669	.0204	18	.000699	.0213
20	.00106	.0324	20	.00111	.0339
22	.00169	.0516	22	.00177	.0538
24	.00269	.0820	24	.00281	.0856
26	.00428	.130	26	.00446	.136
27	.00540	.164	27	.00563	.172
28	.00680	.207	28	.00710	.216
30	.0108	.330	30	.0113	.344
32	.0172	.524	32	.0180	.547
34	.0274	.834	34	.0286	.870
36	.0435	1.33	36	.0454	1.38
40	.110	3.35	40	.115	3.50

COLOR CODE FOR RESISTORS

GENERAL REQUIREMENTS

Code Colors. Colors used for color coding shall conform to Standard MIL–STD–174, and shall be permanent and nonfading. The color-code marking shall remain legible after the resistor has been subjected to all the tests specified in the individual resistor specification.

Body Colors (Background). The exterior body color of resistors shall be any color other than black; for composition-type resistors, tan is preferred, for wire-wound-type resistors, brown is preferred.

Conflict of Colors. When the body color is the same as any of the band colors, then either the body color or the band color shall be differentiated by shade or gloss.

COLOR CODE FOR RESISTORS (Not wire wound)

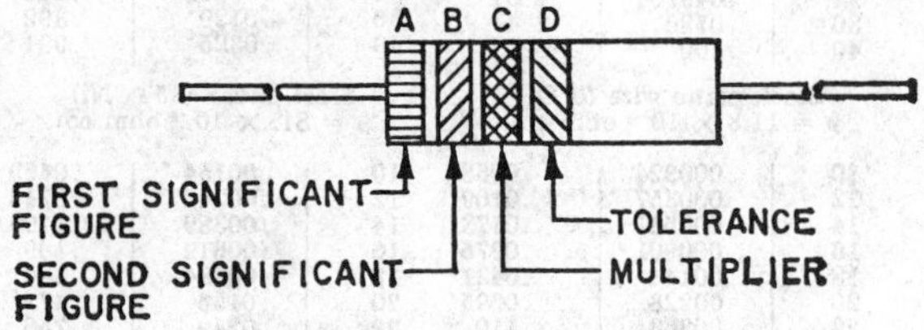

COLOR CODE FOR RESISTORS (Wire wound)

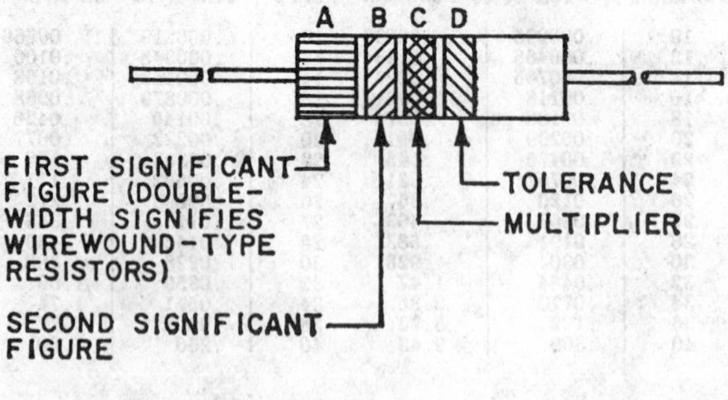

COLOR CODE FOR RESISTORS

Band A		Band B		Band C		Band D	
Color	First significant figure	Color	Second significant figure	Color	Multiplier	Color	Resistance tolerance (per cent)
Black	0	Black	0	Black	1	Silver	±10
Brown	1	Brown	1	Brown	10	Gold	±5
Red	2	Red	2	Red	100		
Orange	3	Orange	3	Orange	1,000		
Yellow	4	Yellow	4	Yellow	10,000		
Green	5	Green	5	Green	100,000		
Blue	6	Blue	6	Blue	1,000,000		
Purple (violet)	7	Purple (violet)	7	...	...		
Gray	8	Gray	8	Silver	0.01		
White	9	White	9	Gold	0.1		

Band A—The first significant figure of the resistance value. (For composition-type resistors, all bands shall be of equal width. For wire-wound-type resistors, band A shall be double width.)

Band B—The second significant figure of the resistance value.

Band C—The multiplier. (The multiplier is the factor by which the two significant figures are multiplied to yield the nominal resistance value.)

Band D—The resistance tolerance.

Examples of color coding

3,900 ohms ±10 percent—Band A. orange; band B. white; band C, red; band D, silver.

43,000 ohms ±5 percent—Band A. yellow; band B. orange; band C orange; band D, gold.

ELECTRICAL RESISTIVITY AND TEMPERATURE COEFFICIENTS OF ELEMENTS

Element	Temperature °C	Microhm-Cm	Temperature Coefficient per °C
Aluminum, 99.996%	20	2.6548	$0.00429^{20\,z}$
Antimony	0	39.0	
Arsenic	20	33.3	
Beryllium[a]	20	4.0	$0.025^{20\,z}$
Bismuth	0	106.8	
Boron	0	1.8×10^{12}	
Cadmium	0	6.83	$0.0042^{0\,z}$
Calcium	0	3.91	$0.00416^{0\,z}$
Carbon[b]	0	1375.0	
Cerium	25	75.0	0.00087^{0-25}
Cesium	20	20	
Chromium	0	12.9	$0.003^{0\,z}$
Cobalt	20	6.24	0.00604^{0-100}
Copper	20	1.6730	$0.0068^{20\,z}$
Dysprosium[c]	25	57.0	0.00119^{0-25}
Erbium	25	107.0	0.00201^{0-25}
Europium	25	90.0	
Gadolinium	25	140.5	0.00176^{0-25}
Gallium[d]	20	17.4	
Germanium[e]	22	46×10^{6}	
Gold	20	2.35	0.004^{0-100}
Hafnium	25	35.1	$0.0038^{25\,z}$
Holmium	25	87.0	0.00171^{0-25}
Indium	20	8.37	
Iodine	20	1.3×10^{15}	
Iridium	20	5.3	0.003925^{0-100}
Iron, 99.99%	20	9.71	$0.00651^{20\,z}$
Lanthanum	25	5.70	0.00218^{0-25}
Lead	20	20.648	0.00336^{20-40}
Lithium	0	8.55	
Lutetium	25	79.0	0.00240^{0-25}
Magnesium[f]	20	4.45	$0.0165^{20\,z}$
Manganese α	23–100	185.0	
Mercury	50	98.4	
Molybdenum	0	5.2	
Neodymium	25	64.0	0.00164^{0-25}
Nickel	20	6.84	0.0069^{0-100}
Niobium (Columbium)[g]	0	12.5	
Osmium	20	9.5	0.0042^{0-100}
Palladium	20	10.8	0.00377^{0-100}
Phosphorus, white	11	1×10^{17}	
Platinum, 99.85%	20	10.6	0.003927^{0-100}
Plutonium	107	141.4	
Potassium	0	6.15	
Praseodymium	25	68	0.00171^{0-25}
Rhenium	20	19.3	0.00395^{0-100}
Rhodium	20	4.51	0.0042^{0-100}
Rubidium	20	12.5	
Ruthenium	0	7.6	
Samarium	25	88.0	0.00184^{0-25}
Scandium[h]	22	61.0	0.00282^{0-25}
Selenium[k]	0	10^{6}	
Silicon	0	3–4[j]	
Silver	20	1.59	0.0041^{0-100}
Sodium	0	4.2	
Strontium	20	23.0	
Sulfur, yellow	20	2×10^{23}	
Tantalum	25	12.45	0.00383^{0-100}

ELECTRICAL RESISTIVITY AND TEMPERATURE COEFFICIENTS OF ELEMENTS (Continued)

Element	Temperature °C	Microhm-Cm	Temperature Coefficient per °C
Tellurium	25	4.36×10^5	
Thallium	0	18.0	
Thorium	0	13.0	0.0038^{0-100}
Thulium	25	79.0	0.00195^{0-25}
Tin	0	11.0	0.0047^{0-100}
Titanium	20	42.0	
Tungsten	27	5.65	
Uranium		30.0	
Vanadium	20	24.8–26.0	
Ytterbium	25	29.0	0.0013^{0-25}
Yttrium	25	57.0	0.0027^{0-25}
Zinc	20	5.916	0.00419^{0-100}
Zirconium	20	40.0	0.0044^{20} [s]

[a] Annealed, comm. pure.
[b] Graphite.
[c] Polycrystalline.
[d] Hard Wire.
[e] Intrinsic Ge.
[f] Polycrystalline.
[g] High Purity.
[h] Zone refined bar.
[s] Data not available to indicate range over which coefficient is valid.
[j] Very sensitive to purity.
[k] Crystalline.

ELECTRICAL RESISTIVITY OF THE ALKALI METALS

Metal	At. No.	At. Wt.	Sp. Gr.	M.P., °C	Resistivity	
					μ-ohm cm	at °C
Li	3	6.939	0.534^{20}	179	12.17	86.6
					13.36	120.5
					14.73	153.3
					15.54	178.2
Na	11	22.9898	0.971^{20}	97.81	5.23	29.4
					5.72	50.6
					6.53	83.3
					6.71	91.6
					6.83	97.1
K	19	39.102	0.86^{20}	63.65	7.01	22.8
					7.32	35.5
					7.54	41.1
					8.05	56.1
Rb	37	85.47	$1.475^{38.89}$	38.89	11.28	0
					12.52	53
Cs	55	132.905	1.8785^{15}	28.5	20.29	18.2
					37.39	28.1

LOW MELTING POINT SOLDERS
(From N.B.S. Circular 492)

Nominal Composition, Weight Percent			Liquidus Temperature, °F
Pb	Sn	Bi	
25	25	50	266
50	37.5	12.5	374
25	50	25	336

BRAZING FILLER METALS (SOLDERS)

AWS-ASTM Classification	Solidus, °F	Liquidus, °F	Brazing Temperature Range, °F	Ag	Al	As	Au	B	Be	Bi	C	Cd	Cr	Cu	Fe	Li	Mg	Mn	Ni	P	Pb	Sb	Si	Sn	Ti	Zn	Other Each	Other Total	AWS-ASTM Classification
ALUMINUM-SILICON																													
BAlSi-2	1070	1135	1110-1150		Bal.									0.25	0.8			0.10					6.8-8.2			0.20	0.05	0.15	BAlSi-2
BAlSi-3	970	1085	1060-1120		Bal.								0.15	3.3-4.7	0.8		0.15	0.15					9.3-10.7			0.20	0.05	0.15	BAlSi-3
BAlSi-4	1070	1080	1080-1120		Bal.									0.30	0.8		0.10	0.15					11.0-13.0			0.20	0.05	0.15	BAlSi-4
BAlSi-5	1070	1095	1090-1120		Bal.									0.30	0.8		0.05	0.05					9.0-11.0		0.20	0.10	0.05	0.15	BAlSi-5
COPPER-PHOSPHORUS																													
BCuP-1	1310	1650	1450-1700											Bal.						4.75-5.25								0.15	BCuP-1
BCuP-2	1310	1460	1350-1550											Bal.						7.00-7.25								0.15	BCuP-2
BCuP-3	1190	1485	1300-1500	4.75-5.25										Bal.						5.75-6.25								0.15	BCuP-3
BCuP-4	1190	1335	1300-1450	5.75-6.25										Bal.						7.00-7.50								0.15	BCuP-4
BCuP-5	1190	1475	1300-1500	14.50-15.50										Bal.						4.75-5.25								0.15	BCuP-5
SILVER																													
BAg-1	1125	1145	1145-1400	44-46								23-25		14-16												14-18		0.15	BAg-1
BAg-1a	1160	1175	1175-1400	49-51								17-19		14.5-16.5												14.5-18.5		0.15	BAg-1a
BAg-2	1125	1295	1295-1550	34-36								17-19		25-27												19-23		0.15	BAg-2
BAg-3	1170	1270	1270-1500	49-51								15-17		14.5-16.5					2.5-3.5							13.5-17.5		0.15	BAg-3
BAg-4	1240	1435	1435-1650	39-41										29-31					1.5-2.5							26-30		0.15	BAg-4
BAg-5	1250	1370	1370-1550	44-46										29-31												23-27		0.15	BAg-5
BAg-6	1270	1425	1425-1600	49-51										33-35												14-18		0.15	BAg-6
BAg-7	1145	1205	1205-1400	55-57										21-23										4.5-5.5		15-19		0.15	BAg-7
BAg-8	1435	1435	1435-1650	71-73										Bal.														0.15	BAg-8
BAg-8a	1410	1410	1410-1600	71-73										Bal.		0.15-0.3												0.15	BAg-8a
BAg-13	1325	1575	1575-1775	53-55										Bal.												4.0-6.0		0.15	BAg-13
BAg-18	1115	1125	1325-1550	59-61										Bal.						0.025				9.5-10.5				0.15	BAg-18
BAg-19	1435	1635	1610-1800	92-93										Bal.		0.15-0.3												0.15	BAg-19
PRECIOUS METALS																													
BAu-1	1815	1860	1860-2000				37.0+1 −0							Bal.														0.15	BAu-1
BAu-2	1635	1635	1635-1850				79.5+1 −0							Bal.														0.15	BAu-2
BAu-3	1785	1885	1885-1995				34.5+1 −0							Bal.					2.5-3.5									0.15	BAu-3
BAu-4	1740	1740	1740-1840				81.5+1 −0												Bal.									0.15	BAu-4
COPPER AND COPPER-ZINC																													
BCu-1	1980	1980	2000-2100		0.01									99.90 min.						0.075	0.02							0.10	BCu-1
BCu-1a	1980	1980	2000-2100											99.0 min.														0.30[a]	BCu-1a
BCu-2	1980	1980	2000-2100											86.5 min.														0.50[c]	BCu-2[b]
RBCuZn-A	1630	1650	1670-1750		0.01*									57-61	*			*			0.05*		*	0.25-1.00		Bal.		0.50[d]	RBCuZn-A[e]
RBCuZn-D	1690	1715	1720-1800		0.01*									46-50					9.0-11.0	0.25	0.05*		0.04-0.25			Bal.		0.50[d]	RBCuZn-D[e]
MAGNESIUM																													
BMg-1	830	1110	1120-1160		8.3-9.7				0.0002-0.0008					0.05	0.005		Bal.	0.15 min.	0.005				0.05			1.7-2.3		0.30	BMg-1
BMg-2	770	1050	1080-1130		11.0-13.0												Bal.									4.5-5.5		0.30	BMg-2
BMg-2a	770	1050	1080-1130		11.0-13.0				0.0002-0.0008								Bal.									4.5-5.5		0.30	BMg-2a

[a] Total other elements requirement pertains only to the metallic elements for this filler metal.

[b] These chemical requirements pertain only to the copper oxide and do not include requirements for the organic vehicle in which the copper oxide is suspended.

[c] Total other elements requirement pertains only to metallic elements for this filler metal. The following limitations are placed on the nonmetallic elements:

Constituent	per cent (max)
Chlorides	0.4
Sulfates	0.1
Oxygen	remainder
Nitric acid insoluble	0.3
Acetone soluble matter	0.5

[d] Total other elements, including the elements marked with an asterisk (*), shall not exceed the value specified.

[e] This AWS-ASTM classification is intended to be identical with the same classification that appears in the Specification for Copper and Copper-Alloy Welding Rods (AWS Designation A5.7; ASTM Designation B 259).[3]

AWS-ASTM Classification	Solidus, °F	Liquidus, °F	Brazing Temperature Range, °F	Ag	Al	As	Au	B	Be	Bi	C	Cd	Cr	Cu	Fe	Li	Mg	Mn	Ni	P	Pb	Sb	Si	Sn	Ti	Zn	Other Each	Other Total	AWS-ASTM Classification
NICKEL																													
BNi-1	1790	1900	1950-2200					2.75-4.00			0.6-0.9		13.0-15.0		4.0-5.0				Bal.				3.0-5.0					0.50	BNi-1
BNi-2	1780	1830	1850-2150					2.75-3.5			0.15		6.0-8.0		2.0-4.0				Bal.				4.0-5.0					0.50	BNi-2
BNi-3	1800	1900	1850-2150					2.75-3.5			0.06				1.5				Bal.				4.0-5.0					0.50	BNi-3
BNi-4	1800	1950	1850-2150					1.0-2.2			0.06				1.5				Bal.				3.0-4.0					0.50	BNi-4
BNi-5	1975	2075	2100-2200								0.15		18.0-20.0						Bal.				9.75-10.5					0.50	BNi-5
BNi-6	1610	1610	1700-1875								0.15								Bal.	10.0-12.0								0.50	BNi-6
BNi-7	1630	1630	1700-1900										11.0-15.0						Bal.	9.0-11.0								0.50	BNi-7
SOLDERS—ASTM DESIGNATION B-32-60T, REVISED 1966																													
70A	361	378			0.005 max	0.03 max				0.25 max				0.08 max	0.02 max						30			70		0.005 max			70A
70B					0.005 max	0.03 max				0.25 max				0.08 max	0.02 max						30	0.2-0.5		70		0.005 max			70B
63A	361	361			0.005 max	0.03 max				0.25 max				0.08 max	0.02 max						37	0.12 max		63		0.005 max			63A
63B					0.005 max	0.03 max				0.25 max				0.08 max	0.02 max						37	0.2-0.5		63		0.005 max			63B
60A	361	374			0.005 max	0.03 max				0.25 max				0.08 max	0.02 max						40	0.12 max		60		0.005 max			60A
60B					0.005 max	0.03 max				0.25 max				0.08 max	0.02 max						40	0.2-0.5		60		0.005 max			60B
50A	361	421			0.005 max	0.03 max				0.25 max				0.08 max	0.02 max						50	0.12 max		50		0.005 max			50A
50B	360	420			0.005 max	0.03 max				0.25 max				0.08 max	0.02 max						50	0.2-0.5		50		0.005 max			50B
45A	361	441			0.005 max	0.03 max				0.25 max				0.08 max	0.02 max						55	0.12 max		45		0.005 max			45A
45B					0.005 max	0.03 max				0.25 max				0.08 max	0.02 max						55	0.2-0.5		45		0.005 max			45B
40A	361	460			0.005 max	0.03 max				0.25 max				0.08 max	0.02 max						60	0.12 max		40		0.005 max			40A
40B	360	460			0.005 max	0.03 max				0.25 max				0.08 max	0.02 max						60	0.2-0.5		40		0.005 max			40B
40C	365	448			0.005 max	0.03 max				0.25 max				0.08 max	0.02 max						58	1.8-2.4		40		0.005 max			40C
35A	361	477			0.005 max	0.03 max				0.25 max				0.08 max	0.02 max						65	0.25 max		35		0.005 max			35A
35B					0.005 max	0.03 max				0.25 max				0.08 max	0.02 max						65	0.2-0.5		35		0.005 max			35B
35C	365	470			0.005 max	0.03 max				0.25 max				0.08 max	0.02 max						63.2	1.6-2.0		35		0.005 max			35C
30A	361	491			0.005 max	0.03 max				0.25 max				0.08 max	0.02 max						70	0.25 max		30		0.005 max			30A
30B					0.005 max	0.03 max				0.25 max				0.08 max	0.02 max						70	0.2-0.5		30		0.005 max			30B
30C	364	482			0.005 max	0.03 max				0.25 max				0.08 max	0.02 max						68.4	1.4-1.8		30		0.005 max			30C
25A	361	511			0.005 max	0.03 max				0.25 max				0.08 max	0.02 max						75	0.25 max		25		0.005 max			25A
25B	360	510			0.005 max	0.03 max				0.25 max				0.08 max	0.02 max						75	0.2-0.5		25		0.005 max			25B
25C					0.005 max	0.03 max				0.25 max				0.08 max	0.02 max						73.7	1.1-1.5		25		0.005 max			25C
20B	361	531			0.005 max	0.03 max				0.25 max				0.08 max	0.02 max						80	0.2-0.5		20		0.005 max			20B
20C	363	517			0.005 max	0.03 max				0.25 max				0.08 max	0.02 max						79	0.8-1.2		20		0.005 max			20C
15B	440	550			0.005 max	0.03 max				0.25 max				0.08 max	0.02 max						85	0.2-0.5		15		0.005 max			15B
10B					0.005 max	0.03 max				0.25 max				0.08 max	0.02 max						90	0.2-0.5		10		0.005 max			10B
5A	518	594			0.005 max	0.03 max				0.25 max				0.08 max	0.02 max						95	0.12 max		5*		0.005 max			5A
5B					0.005 max	0.03 max				0.25 max				0.08 max	0.02 max						95	0.2-0.5		5*		0.005 max			5B
2A					0.005 max	0.03 max				0.25 max				0.08 max	0.02 max						98	0.12 max		2**		0.005 max			2A
2B					0.005 max	0.03 max				0.25 max				0.08 max	0.02 max						98	0.2-0.5		2**		0.005 max			2B
2.5S	579	579			0.005 max	0.03 max				0.25 max				0.08 max	0.02 max						97.5	0.4 max		0***		0.005 max			2.5S
1.5S	588	588		2.3-2.7	0.005 max	0.03 max				0.25 max				0.08 max	0.02 max						97.5	0.4 max		1****		0.005 max			1.5S
95TA	452	464		1.3-1.7	0.005 max	0.03 max				0.25 max				0.08 max	0.02 max						0.2	4.5-5.5		95		0.005 max			95TA
96.5TS				3.3-3.7	0.005 max	0.03 max				0.25 max				0.08 mas	0.02 max						0.2	0.2-0.5		96.5		0.005 max			96.5TS

* Permissible tin range, 4.5–5.5%.
** Permissible tin range, 1.5–2.5%.
*** Tin maximum, 0.25%.
**** Permissible tin range, 0.75–1.25%.

PHYSICAL DATA FOR THE PLANETS, THEIR SATELLITES AND SOME ASTEROIDS

Compiled by W. Joseph Armento, 1970

Data was gathered for inclusion in this table from some of the latest available references. Calculations were performed to verify the internal consistency of the interrelated quantities. The deviations are based on the spread of the data as gathered from the selected references and their respective uncertainties.

Certain data which were calculated include the length of the solar day, rotational velocity, centrifical force, isosynchronous satellite data, escape velocity, orbital velocity, synodic period, equilibrium temperature, solar constant, gravitational constant, and all conversions to "earth" units. In the last calculations, a standard deviation is carried in the data to indicate the error arising from the uncertainties in the Earth measurements.

Sidereal time is the time required for the body to return to the equivalent position as defined by the stars. An isosynchronous satellite placed over a point on an equator with no perturbations in the orbit, no eccentricity, and no deflection from the equator will appear to remain stationary in the sky (or remain over the same position on the body); the angular velocity of the body equals the angular velocity of the satellite. The synodic period is the time from conjunction to conjunction. The letter "R" indicates a retrograde or opposite motion to that of the Sun/Earth system.

The exponents for large numbers are expressed in the following form:

$$N \times 10^{m} = N\,EM$$

i.e. $1.991 \pm 0.002\ E^{30}$ is equivalent
to $1.991 \times 10^{30} \pm 0.002 \times 10^{30}$

PHYSICAL DATA FOR THE PLANETS, THEIR SATELLITES AND SOME ASTEROIDS (Continued)

Body	Radius of Body (km) Average	Equatorial	Polar	Ellipticity (Oblateness)	Mass (kg)	Average Density (g/cc.)
Sun (Sol)	695950.0 ± 810.0	695950.0 ± 810.0	695950.0 ± 810.0	⩽ 0.0017	1.991 ± 0.002 E30	1.410 ± 0.002
Mercury	2433.0 ± 35.0	2433.0 ± 35.0	Unknown	⩽ 0.0292	3.181 ± 0.110 E23	5.431 ± 0.300
Venus	6053.0 ± 7.8	6053.0 ± 7.8	6053.0 ± 7.8	0000 ± 0.0018	4.883 ± 0.034 E24	5.256 ± 0.043
Earth (Terra)	6371.315 ± 0.437	6378.533 ± 0.437	6356.912 ± 0.437	0.003393 ± 0.000097	5.979 ± 0.004 E24	5.519 ± 0.004
Moon (Luna)	1738.3 ± 1.1	1738.7 ± 1.1	1737.6 ± 1.1	0.006327 ± 0.008947	7.354 ± 0.066 E22	3.342 ± 0.031
Mars	3380.0 ± 20.0	3386.0 ± 20.0	3369.0 ± 20.0	0.0050 ± 0.0084	6.418 ± 0.024 E23	3.907 ± 0.079
Phobos	10.4 ± 5.0	11.3 ± 5.0	8.8 ± 5.0	0.22 ± 0.63	2.72 E16†	3.5 est.
Deimos	5.0 ± 4.0	5.0 ± 4.0	Unknown	—	1.83 E15†	3.5 est.
Jupiter	69758.0 ± 139.0	71370.0 ± 254	66644.0 ± 42.0	0.0662 ± 0.0036	1.901 ± 0.003 E27	1.337 ± 0.015
Io	1726.0 ± 109.0	1726.0 ± 109.0	Unknown	—	7.87 ± 0.15 E22	3.654 ± 0.696
Europa	1488.0 ± 70.0	1488.0 ± 70.0	Unknown	—	4.78 ± 0.15 E22	3.464 ± 0.501
Ganymede	2529.0 ± 54.0	2529.0 ± 54.0	Unknown	—	1.54 ± 0.03 E23	2.273 ± 0.152
Callisto	2416.0 ± 137.0	2416.0 ± 137.0	Unknown	—	7.35 ± 2.65 E22	1.244 ± 0.496
V (Amalthea)	87.0 ± 23.0	87.0 ± 23.0	Unknown	—	8.3 E18†	3.0 est.
VI (Hestia)	67.0 ± 16.0	67.0 ± 16.0	Unknown	—	3.8 E18†	3.0 est.
VII (Hera)	20.0 ± 9.0	20.0 ± 9.0	Unknown	—	1.0 E17†	3.0 est.
VIII (Poseidon)	18.0 ± 8.0	18.0 ± 8.0	Unknown	—	7.3 E16†	3.0 est.
IX (Hades)	12.0 ± 3.0	12.0 ± 3.0	Unknown	—	2.2 E16†	3.0 est.
X (Demeter)	9.0 ± 3.0	9.0 ± 3.0	Unknown	—	9.2 E15†	3.0 est.
XI (Pan)	12.0 ± 4.0	12.0 ± 4.0	Unknown	—	2.2 E16†	3.0 est.
XII (Adrastea)	10.0 ± 2.0	10.0 ± 2.0	Unknown	—	1.3 E16†	3.0 est.
Saturn	58219.0 ± 262.0	60369.0 ± 218.0	54148.0 ± 379.0	0.1030 ± 0.0073	5.684 ± 0.022 E26	0.688 ± 0.010
Mimas	272.0 ± 23.0	272.0 ± 23.0	Unknown	—	3.7 ± 2.0 E19	0.439 ± 0.262
Enceladus	299.0 ± 21.0	299.0 ± 21.0	Unknown	—	7.4 ± 3.0 E19	0.661 ± 0.302
Tethys	581.0 ± 74.0	581.0 ± 74.0	Unknown	—	4.9 E20†	0.6 est.
Dione	598.0 ± 86.0	598.0 ± 86.0	Unknown	—	5.4 E20†	0.6 est.
Rhea	890.0 ± 37.0	890.0 ± 37.0	Unknown	—	1.8 E21†	0.6 est.
Titan	2379.0 ± 194.0	2379.0 ± 194.0	Unknown	—	1.19 ± 0.55 E23	2.110 ± 1.103
Hyperion	201.0 ± 17.0	201.0 ± 17.0	Unknown	—	6.8 E19†	2.0 est.
Iapetus	647.0 ± 166.0	647.0 ± 166.0	Unknown	—	2.3 E21†	2.0 est.
Phoebe	132.0 ± 23.0	132.0 ± 23.0	Unknown	—	1.9 E19†	2.0 est.
X (Themis/Janus)	241.0 ± 16.0	241.0 ± 16.0	Unknown	—	1.2 E20†	2.0 est.
Uranus	23470.0 ± 732.0	24045.0 ± 559.0	22362.0 ± 1254.0	0.0700 ± 0.0572	8.682 ± 0.031 E25	1.603 ± 0.149
Ariel	311.0 ± 25.0	311.0 ± 25.0	Unknown	—	5.0 E20†	4.0 est.
Umbriel	201.0 ± 20.0	201.0 ± 20.0	Unknown	—	1.4 E20†	4.0 est.
Titania	500.0 ± 32.0	500.0 ± 32.0	Unknown	—	2.1 E21†	4.0 est.
Oberon	401.0 ± 28.0	401.0 ± 28.0	Unknown	—	1.1 E21†	4.0 est.
Miranda	121.0 ± 11.0	121.0 ± 11.0	Unknown	—	3.0 E19†	4.0 est.
Neptune	22716.0 ± 869.0	22716.0 ± 869.0	Unknown	⩽ 0.0796	1.027 ± 0.014 E26	2.272 ± 0.263
Triton	2008.0 ± 55.0	2008.0 ± 55.0	Unknown	—	1.46 ± 0.05 E23	4.305 ± 0.383
Nereid	136.0 ± 34.0	136.0 ± 34.0	Unknown	—	5.0 ± 3.0 E19	4.7 ± 4.6
Pluto	5700.0 ± 413.0	5700.0 ± 413.0	Unknown	⩽ 0.1562	1.08 ± 1.00 E24	1.65 ± 1.57
Typical Asteroid	20.0 ± 10.0	20.0 ± 10.0	Unknown	—	1.17 E17†	3.5 est.
1—Ceres	372.0 ± 22.0	372.0 ± 22.0	Unknown	—	6.0 ± 3.0 E20	2.8 ± 1.5
2—Pallas	238.0 ± 22.0	238.0 ± 22.0	Unknown	—	1.8 ± ? E20	3.2 ± ?
3—Juno	103.0 ± 14.0	103.0 ± 14.0	Unknown	—	2.0 ± ? E19	4.4 ± ?
4—Vesta	195.0 ± 100.0	195.0 ± 100.0	Unknown	—	1.0 ± ? E20	3.2 ± ?
5—Astraea	Unknown	Unknown	Unknown	—	Unknown	3.5 est.
6—Hebe	110.0 ± 10.0	110.0 ± 10.0	Unknown	—	2.0 ± ? E19	3.6 ± ?
7—Iris	100.0 ± 10.0	100.0 ± 10.0	Unknown	—	1.5 ± ? E19	3.6 ± ?
10—Hygiea	160.0 ± 13.0	160.0 ± 13.0	Unknown	—	6.0 ± ? E19	3.5 ± ?
15—Eunomia	140.0 ± 12.0	140.0 ± 12.0	Unknown	—	4.0 ± ? E19	3.5 ± ?
16—Psyche	140.0 ± 12.0	140.0 ± 12.0	Unknown	—	4.0 ± ? E19	3.5 ± ?
51—Nemansa	40.0 ± 6.0	40.0 ± 6.0	Unknown	—	9.0 ± ? E17	3.4 ± ?
323—Brucia	Unknown	Unknown	Unknown	—	Unknown	3.5 est.
433—Eros	12.0 ± 5.0	12.0 ± 5.0	Unknown	—	5.0 ± ? E15	0.7 ± ?
511—Davida	130.0 ± 11.0	130.0 ± 11.0	Unknown	—	3.0 ± ? E19	3.3 ± ?
588—Achilles	28.0 ± 5.0	28.0 ± 5.0	Unknown	—	3.22 E14†	3.5 est
617—Patrochus	Unknown	Unknown	Unknown	—	Unknown	3.5 est.
744—Hidalgo	18.0 ± 8.0	18.0 ± 8.0	Unknown	—	3.96 E14†	3.5 est.
1048—Feodosia	Unknown	Unknown	Unknown	—	Unknown	3.5 est.
1221—Amor	3.0 ± 1.5	3.0 ± 1.5	Unknown	—	3.96 E14†	3.5 est.
1566—Icarus	0.53 ± 0.13	0.7 ± 0.1	0.3 ± 0.2	0.57 ± 0.36	5.0 ± ? E12	8.1 ± ?
1620—Geographos	1.5 ± 0.8	1.5 ± 0.8	Unknown	—	5.0 ± ? E13	3.5 ± ?
—Adonis	1.0 ± 0.5	1.0 ± 0.5	Unknown	—	1.47 E13†	3.5 est.
—Apollo	2.0 ± 1.0	2.0 ± 1.0	Unknown	—	1.17 E14†	3.5 est.
—Hermes	1.0 ± 0.5	1.0 ± 0.5	Unknown	—	1.47 E13†	3.5 est.

† Value of mass calculated from an estimated density.

PHYSICAL DATA FOR THE PLANETS, THEIR SATELLITES AND SOME ASTEROIDS (Continued)

Body	Semimajor axis of Revolution (km) a	Eccentricity (x/a)	Aphelion (km) (a + x)	Perhelion (km) (a − x)	Longitude of Perhelion (Deg.)	Longitude of Ascending Node (Deg.)
Sun (Sol)	—	—	—	—	—	—
Mercury	5.795 ± 0.013 E7	0.2056 ± 0.0002	6.986 ± 0.016 E7	4.604 ± 0.010 E7	76.583	47.667
Venus	1.0811 ± 0.0010 E8	0.0068 ± 0.0001	1.0885 ± 0.0010 E8	1.0737 ± 0.0010 E8	130.783	76.183
Earth (Terra)	1.4957 ± 0.0007 E8	0.0167 ± 0.0001	1.5207 ± 0.0007 E8	1.4707 ± 0.0007 E8	101.983	—
Moon (Luna)	3.84403 ± 0.00006 E5	0.0549 ± 0.0000	4.05506 ± 0.00006 E5	3.63299 ± 0.00006 E5	Not tabulated	Unknown
Mars	2.2784 ± 0.0022 E8	0.0934 ± 0.0002	2.4912 ± 0.0024 E8	2.0656 ± 0.0020 E8	335.033	49.133
Phobos	9.408 ± 0.034 E3	Unknown	—	—	Not tabulated	None*
Deimos	2.3457 ± 0.0265 E4	Unknown	—	—	Not tabulated	None*
Jupiter	7.7814 ± 0.0017 E8	0.0484 ± 0.0002	8.1580 ± 0.0024 E8	7.4048 ± 0.0022 E8	13.417	99.883
Io	4.219 ± 0.003 E5	Unknown	—	—	Not tabulated	None*
Europa	6.712 ± 0.002 E5	Unknown	—	—	Not tabulated	None*
Ganymede	1.071 ± 0.001 E6	Unknown	—	—	Not tabulated	None*
Callisto	1.8830 ± 0.005 E6	Unknown	—	—	Not tabulated	None*
V (Amalthea)	1.814 ± 0.005 E5	Unknown	—	—	Not tabulated	None*
VI (Hestia)	1.14 ± 0.02 E7	Unknown	—	—	Not tabulated	Unknown
VII (Hera)	1.16 ± 0.04 E7	Unknown	—	—	Not tabulated	Unknown
VIII (Poseidon)	2.35 ± 0.01 E7	Unknown	—	—	Not tabulated	Unknown
IX (Hades)	2.37 ± 0.02 E7	Unknown	—	—	Not tabulated	Unknown
X (Demeter)	1.178 ± 0.005 E7	Unknown	—	—	Not tabulated	Unknown
XI (Pan)	2.26 ± 0.02 E7	Unknown	—	—	Not tabulated	Unknown
XII (Adrastea)	2.11 ± 0.01 E7	Unknown	—	—	Not tabulated	Unknown
Saturn	1.4270 ± 0.0009 E9	0.0543 ± 0.0021	1.5045 ± 0.0031 E9	1.3495 ± 0.0031 E9	91.950	113.167
Mimas	1.816 ± 0.007 E5	Unknown	—	—	Not tabulated	Unknown
Enceladus	2.394 ± 0.016 E5	Unknown	—	—	Not tabulated	Unknown
Tethys	2.952 ± 0.004 E5	Unknown	—	—	Not tabulated	Unknown
Dione	3.781 ± 0.015 E5	Unknown	—	—	Not tabulated	Unknown
Rhea	5.284 ± 0.019 E5	Unknown	—	—	Not tabulated	Unknown
Titan	1.227 ± 0.007 E6	Unknown	—	—	Not tabulated	Unknown
Hyperion	1.4873 ± 0.0066 E6	Unknown	—	—	Not tabulated	Unknown
Iapetus	3.571 ± 0.019 E6	Unknown	—	—	Not tabulated	Unknown
Phoebe	1.2983 ± 0.0072 E7	0.17 ± 0.01	1.5190 ± 0.0155 E7	1.0776 ± 0.0143 E9	Not tabulated	Unknown
X (Themis/Janus)	1.58 ± 0.15 E5	Unknown	—	—	Not tabulated	Unknown
Uranus	2.8703 ± 0.0015 E9	0.0460 ± 0.0015	3.0023 ± 0.0046 E9	2.7383 ± 0.0045 E9	169.75	73.717
Ariel	1.918 ± 0.003 E5	Unknown	—	—	Not tabulated	None*
Umbriel	2.672 ± 0.002 E5	Unknown	—	—	Not tabulated	None*
Titania	4.386 ± 0.009 E5	Unknown	—	—	Not tabulated	None*
Oberon	5.861 ± 0.006 E5	Unknown	—	—	Not tabulated	None*
Miranda	1.294 ± 0.014 E5	Unknown	—	—	Not tabulated	None*
Neptune	4.4999 ± 0.0038 E9	0.0082 ± 0.0007	4.5368 ± 0.0050 E9	4.4630 ± 0.0049 E9	44.117	131.167
Triton	3.531 ± 0.004 E5	Unknown	—	—	Not tabulated	Unknown
Nereid	5.9 ± 0.2 E6	Unknown	—	—	Not tabulated	Unknown
Pluto	5.909 ± 0.019 E9	0.2481 ± 0.0006	7.375 ± 0.024 E9	4.443 ± 0.015 E9	270.059	131.278
Typical Asteroid	2.9 ± 1.5 E8	0.10 ± 0.05	3.2 ± 1.7 E8	2.6 ± 1.4 E8	Unknown	Unknown
1—Ceres	4.141 ± 0.003 E8	0.079 ± 0.002	4.468 ± 0.009 E8	3.814 ± 0.009 E8	Unknown	Unknown
2—Pallas	4.141 ± 0.003 E8	0.236 ± 0.003	5.118 ± 0.013 E8	3.164 ± 0.013 E8	Unknown	Unknown
3—Juno	3.994 ± 0.001 E8	0.259 ± 0.001	5.028 ± 0.004 E8	2.920 ± 0.004 E8	Unknown	Unknown
4—Vesta	3.531 ± 0.001 E8	0.089 ± 0.001	3.845 ± 0.004 E8	3.217 ± 0.004 E8	Unknown	Unknown
5—Astraea	3.848 ± ? E8	0.190 ± ?	4.580 ± ? E8	3.117 ± ? E8	Unknown	Unknown
6—Hebe	3.629 ± ? E8	0.20 ± ?	4.35 ± ? E8	2.90 ± ? E8	Unknown	Unknown
7—Iris	3.567 ± ? E8	0.23 ± ?	4.39 ± ? E8	2.75 ± ? E8	Unknown	Unknown
10—Hygiea	4.173 ± ? E8	0.10 ± ?	4.59 ± ? E8	3.76 ± ? E8	Unknown	Unknown
15—Eunomia	3.956 ± ? E8	0.18 ± ?	4.67 ± ? E8	3.24 ± ? E8	Unknown	Unknown
16—Psyche	4.392 ± ? E8	0.14 ± ?	4.98 ± ? E8	3.76 ± ? E8	Unknown	Unknown
51—Nemansa	3.539 ± ? E8	0.06 ± ?	3.75 ± ? E8	3.33 ± ? E8	Unknown	Unknown
323—Brucia	3.564 ± ? E8	0.301 ± ?	4.637 ± ? E8	2.491 ± ? E8	Unknown	Unknown
433—Eros	2.179 ± 0.004 E8	0.222 ± 0.002	2.663 ± 0.007 E8	1.695 ± 0.005 E8	Unknown	Unknown
511—Davida	4.759 ± ? E8	0.18 ± ?	5.616 ± ? E8	3.902 ± ? E8	Unknown	Unknown
588—Achilles	7.793 ± 0.005 E8	0.148 ± 0.001	8.946 ± 0.010 E8	6.640 ± 0.009 E8	Unknown	Unknown
617—Patrochus	7.785 ± 0.005 E8	0.141 ± 0.002	8.883 ± 0.017 E8	6.687 ± 0.016 E8	Unknown	Unknown
744—Hidalgo	8.660 ± 0.004 E8	0.656 ± 0.001	1.4341 ± 0.0011 E9	2.979 ± 0.009 E8	Unknown	Unknown
1048—Feodosia	4.088 ± ? E8	0.180 ± ?	4.824 ± ? E8	3.352 ± ? E8	Unknown	Unknown
1221—Amor	2.878 ± ? E8	0.437 ± ?	4.135 ± ? E8	1.620 ± ? E8	Unknown	Unknown
1566—Icarus	1.613 ± 0.003 E8	0.828 ± 0.002	2.949 ± 0.006 E8	2.77 ± 0.03 E7	Unknown	Unknown
1620—Geographos	1.861 ± ? E8	0.34 ± ?	2.494 ± ? E8	1.228 ± ? E8	Unknown	Unknown
—Adonis	2.942 ± ? E8	0.779 ± ?	5.234 ± ? E8	6.50 ± ? E7	Unknown	Unknown
—Apollo	2.221 ± 0.005 E8	0.566 ± 0.001	3.473 ± 0.008 E8	9.64 ± 0.03 E7	Unknown	Unknown
—Hermes	1.934 ± ? E8	0.630 ± ?	3.152 ± ? E8	7.15 ± ? E7	Unknown	Unknown

* To equator of primary.

PHYSICAL DATA FOR THE PLANETS, THEIR SATELLITES AND SOME ASTEROIDS (Continued)

Body	Sidereal Period of Revolution (Solar Seconds)	Equatorial Rotation Period (Solar Seconds)	Average Solar Day (Solar Seconds)	Rotational Velocity at Equator (km/sec)	Centrifugal Force at Equator (cm/sec²)	Altitude of an Isosynchronous Satellite (km)
Sun (Sol)	—	2.125 ± ? E6†	—	2.0578 ± 0.0024	0.6085 ± 0.0016	24054600.0 ± 1100.0
Mercury	7.60234 ± 0.00173 E6	5.06775 ± 0.00020 E6	1.52004 ± 0.00110 E7	0.0030 ± 0.0000	0.00037 ± 0.00001	237298.0 ± 49.0
Venus	1.94141 ± 0.00017 E7	R 2.10747 ± 0.09243 E7	R 1.01052 ± 0.04996 E7	0.0018 ± 0.0000	0.0000538 ± 0.0000001	1534693.0 ± 11.0
Earth (Terra)	3.1558150 ± 0.0000000 E7	8.616406 ± 0.000001 E4	8.639988 ± 0.000010 E4	0.4651 ± 0.0000	3.39177 ± 0.00052	35767.0 ± 0.6
Moon (Luna)	2.36055 ± 0.00009 E6	2.36055 ± 0.00009 E6	2.55139 ± 0.00010 E6	0.0046 ± 0.0000	0.0012319 ± 0.0000017	86673.1 ± 1.6
Mars	5.93553 ± 0.00016 E7	8.86427 ± 0.00001 E4	8.87753 ± 0.00034 E4	0.2400 ± 0.0014	1.70126 ± 0.02247	17049.4 ± 6.4
Phobos	2.75468 ± 0.00097 E4	*	2.75596 ± 0.00098 E4	0.0026 ± 0.0011	0.05879 ± 0.05817	232.0 ± ?
Deimos	1.09077 ± 0.00009 E5	*	1.09278 ± 0.00010 E5	0.00029 ± 0.00023	0.00166 ± 0.00297	283.0 ± ?
Jupiter	3.74320 ± 0.00037 E8	3.54300 ± 0.00000 E4‡	3.54334 ± 0.00050 E4	12.6568 ± 0.0450	224.46 ± 1.79	87688.0 ± 359.0
Io	1.52859 ± 0.00009 E5	*	1.52922 ± 0.00023 E5	0.0709 ± 0.0045	0.2916 ± 0.0412	12860.0 ± 1718.0
Europa	3.06824 ± 0.00017 E5	*	3.07076 ± 0.00046 E5	0.0305 ± 0.0014	0.0624 ± 0.0066	18175.0 ± 2226.0
Ganymede	6.18175 ± 0.00017 E5	*	6.19198 ± 0.00088 E5	0.02570 ± 0.00055	0.0262 ± 0.0012	43740.0 ± 4155.0
Callisto	1.44193 ± 0.00000 E6	*	1.44751 ± 0.00020 E6	0.01053 ± 0.00060	0.00459 ± 0.00058	61263.0 ± 19485.0
V (Amalthea)	4.30358 ± 0.00000 E4	*	4.30408 ± 0.00060 E4	0.01270 ± 0.00336	0.1854 ± 0.1096	2087.0 ± ?
VI (Hestia)	2.16639 ± 0.00199 E7	Unknown	—	Unknown	Unknown	Unknown
VII (Hera)	2.24467 ± 0.00173 E7	Unknown	—	Unknown	Unknown	Unknown
VIII (Poseidon)	R 6.38496 ± 0.00086 E7	Unknown	—	Unknown	Unknown	Unknown
IX (Hades)	R 6.54048 ± 0.01469 E7	Unknown	—	Unknown	Unknown	Unknown
X (Demeter)	2.23085 ± 0.03110 E7	Unknown	—	Unknown	Unknown	Unknown
XI (Pan)	R 6.03677 ± 0.01987 E7	Unknown	—	Unknown	Unknown	Unknown
XII (Adrastea)	R 5.4000 ± 0.01296 E7	Unknown	—	Unknown	Unknown	Unknown
Saturn	9.29604 ± 0.00035 E8	3.68400 ± 0.00000 E4§	3.6414 ± 0.00020 E4	10.2961 ± 0.0372	175.60 ± 1.42	48797.0 ± 308.0
Mimas	8.17603 ± 0.06912 E4	*	8.17674 ± 0.06913 E4	0.0209 ± 0.0018	0.1606 ± 0.0305	475.0 ± 297.0
Enceladus	1.18515 ± 0.00242 E5	*	1.18530 ± 0.00242 E5	0.0159 ± 0.0011	0.0840 ± 0.0132	907.0 ± 397.0
Tethys	1.62743 ± 0.00639 E5	*	1.62772 ± 0.00639 E5	0.0224 ± 0.0029	0.0866 ± 0.0241	2222.0 ± ?
Dione	2.36848 ± 0.00648 E5	*	2.36905 ± 0.00648 E5	0.0158 ± 0.0023	0.0421 ± 0.0135	3107.0 ± ?
Rhea	3.89820 ± 0.00890 E5	*	3.89983 ± 0.00891 E5	0.0143 ± 0.0006	0.0231 ± 0.0022	6796.0 ± ?
Titan	1.37912 ± 0.00225 E6	*	1.38117 ± 0.00225 E6	0.0108 ± 0.0009	0.00494 ± 0.00090	70161.0 ± 22891.0
Hyperion	1.83876 ± 0.00078 E6	*	1.84240 ± 0.00079 E6	0.000687 ± 0.000058	0.000235 ± 0.000044	7092.0 ± ?
Iapetus	6.84720 ± 0.01469 E6	Unknown	—	Unknown	Unknown	Unknown
Phoebe	R 4.74509 ± 0.1814 E7	Unknown	—	Unknown	Unknown	Unknown
X (Themis/Janus)	6.4145 ± ? E4	Unknown	—	Unknown	Unknown	Unknown
Uranus	2.65114 ± 0.00026 E9	R 3.88080 ± 0.01920	R 3.88074 ± 0.01921 E4	3.8930 ± 0.0901	63.03 ± 3.26	36370.0 ± 791.0
Ariel	R 2.17745 ± 0.00017 E5	*	R 2.17727 ± 0.00035 E5	0.00897 ± 0.00072	0.0259 ± 0.0047	3117.0 ± ?
Umbriel	R 3.58050 ± 0.00000 E5	*	R 3.58002 ± 0.00050 E5	0.00353 ± 0.00035	0.0062 ± 0.0014	2886.0 ± ?
Titania	R 7.52189 ± 0.00000 E5	*	R 7.51976 ± 0.00104 E5	0.00418 ± 0.00027	0.00349 ± 0.00050	?
Oberon	R 1.16321 ± 0.00001 E6	*	R 1.16270 ± 0.00016 E6	0.00217 ± 0.00015	0.00117 ± 0.00018	?
Miranda	R 1.22169 ± 0.00120 E5	*	R 1.22163 ± 0.00121 E5	0.00622 ± 0.00057	0.03201 ± 0.00651	?
Neptune	5.20027 ± 0.00037 E9	5.66520 ± 0.02410 E4	5.66526 ± 0.00241 E4	2.5194 ± 0.0968	27.94 ± 2.40	59506.0 ± 1229.0
Triton	R 5.07712 ± 0.00095 E5	Unknown	R 5.07663 ± 0.00108 E5	Unknown	Unknown	Unknown
Nereid	3.91390 ± 0.69980 E7	Unknown	3.94358 ± 0.70513 E7	Unknown	Unknown	Unknown
Pluto	7.83735 ± 0.01058 E9	5.52320 ± 0.00430 E5	5.52349 ± 0.01139 E5	0.0648 ± 0.0047	0.0738 ± 0.0120	76510.0 ± 584.0
Typical Asteroid	8.5224 ± ? E7	3.96 ± 3.60 E4	4.0 ± 3.8 E4	0.0032 ± 0.0030	0.0503 ± 0.0994	11259.0 ± 11019.0
1—Ceres	1.4515 ± 0.0009 E8	Unknown	—	Unknown	Unknown	Unknown
2—Pallas	1.4558 ± 0.0017 E8	Unknown	—	Unknown	Unknown	Unknown
3—Juno	1.3772 ± 0.0017 E8	Unknown	—	Unknown	Unknown	Unknown
4—Vesta	1.1457 ± 0.0009 E8	1.92 ± 0.06 E4	1.92 ± 0.06 E4	0.0635 ± 0.0337	2.088 ± 2.455	130109.0 ± 119463.0
5—Astraea	1.3029 ± 0.0035 E8	Unknown	—	Unknown	Unknown	Unknown
6—Hebe	1.1923 ± 0.0009 E8	Unknown	—	Unknown	Unknown	Unknown
7—Iris	1.1612 ± 0.0009 E8	Unknown	—	Unknown	Unknown	Unknown
10—Hygiea	1.7643 ± 0.0009 E8	Unknown	—	Unknown	Unknown	Unknown
15—Eunomia	1.3556 ± 0.0009 E8	Unknown	—	Unknown	Unknown	Unknown
16—Psyche	1.5777 ± 0.0009 E8	Unknown	—	Unknown	Unknown	Unknown

PHYSICAL DATA FOR THE PLANETS, THEIR SATELLITES AND SOME ASTEROIDS (Continued)

Body	Sidereal Period of Revolution (Solar Seconds)	Equatorial Rotation Period (Solar Seconds)	Average Solar Day (Solar Seconds)	Rotational Velocity at Equator (km/sec)	Centrifugal Force at Equator (cm/sec²)	Altitude of an Isosynchronous Satellite (km)
51—Nemansa	1.1491 ± 0.0009 E8	Unknown	—	Unknown	Unknown	Unknown
323—Brucia	1.1612 ± 0.0035 E8	Unknown	—	Unknown	Unknown	Unknown
433—Eros	5.5556 ± 0.0086 E7	Unknown	—	Unknown	Unknown	Unknown
511—Davida	1.7902 ± 0.0009 E8	Unknown	—	Unknown	Unknown	Unknown
588—Achilles	3.7688 ± 0.0190 E8	Unknown	—	Unknown	Unknown	Unknown
617—Patrochus	3.7489 ± 0.0035 E8	Unknown	—	Unknown	Unknown	Unknown
744—Hidalgo	4.3943 ± 0.0112 E8	Unknown	—	Unknown	Unknown	Unknown
1048—Feodosia	1.4265 ± 0.0035 E8	Unknown	—	Unknown	Unknown	Unknown
1221—Amor	8.4240 ± 0.0346 E7	Unknown	—	Unknown	Unknown	Unknown
1566—Icarus	3.5338 ± 0.0086 E7	9.72 ± 3.51 E4	9.72 ± 3.51 E4	0.0000453 ± 0.0000088	0.000287 ± 0.000144	264.0 ± 186.0
1620—Geographos	4.3805 ± 0.0086 E7	Unknown	—	Unknown	Unknown	Unknown
—Adonis	8.7091 ± 0.0346 E7	Unknown	—	Unknown	Unknown	Unknown
—Apollo	5.7110 ± 0.0346 E7	Unknown	—	Unknown	Unknown	Unknown
—Hermes	4.6397 ± 0.0346 E7	Unknown	—	Unknown	Unknown	Unknown

* Estimated to be equal to the period of revolution (shows same face to primary).
† At latitude 30°—2.290 E6; 60°—2.678 ± 6; 90°—2.851 E6.
‡ Polar—3.5741 E4.
§ Polar—3.8280 E4.

Body	Gravity (cm/sec²)				Escape Velocity (km/sec)	Orbital Velocity (km/sec)		
	Mean	Polar	Equatorial	Equatorial—Uncorrected		Semimajor	Perhelion	Aphelion
Sun (Sol)	27372.0 ± 7.0	27370.0 ± 11.0	27369.0 ± 116.0	27369.0 ± 116.0	617.23 ± 0.44	—	—	—
Mercury	357.8 ± 16.1	357.8 ± 19.1	357.8 ± 21.7	357.8 ± 21.7	4.1725 ± 0.0493	47.828 ± 0.566	58.921 ± 0.697	38.824 ± 0.460
Venus	887.4 ± 6.6	887.4 ± 7.4	887.4 ± 7.4	887.4 ± 7.4	10.365 ± 0.020	35.017 ± 0.066	35.256 ± 0.067	34.780 ± 0.066
Earth (Terra)	980.7 ± 0.9	985.1 ± 1.0	975.0 ± 1.0	978.4 ± 1.0	11.179 ± 0.003	29.771 ± 0.007	30.272 ± 0.008	29.278 ± 0.007
Moon (Luna)	162.0 ± 1.5	162.2 ± 1.5	162.0 ± 1.5	162.0 ± 1.5	2.3735 ± 0.0054	1.0176 ± 0.0023	1.0751 ± 0.0024	0.9632 ± 0.0022
Mars	374.0 ± 4.6	376.5 ± 7.8	355.7 ± 7.8	372.7 ± 7.8	5.0282 ± 0.0173	24.121 ± 0.083	26.490 ± 0.091	21.964 ± 0.076
Phobos	1.67 ± ?	2.33 ± ?	Unknown	1.41 ± ?	0.01865 ± 0.00501	0.6739 ± 0.1818	Unknown	Unknown
Deimos	0.49 ± ?	Unknown	Unknown	Unknown	0.00699 ± 0.00312	0.4268 ± 0.2156	Unknown	Unknown
Jupiter	2601.0 ± 11.0	2850.0 ± 17.0	2260.0 ± 23.0	2485.0 ± 23.0	60.238 ± 0.007	13.052 ± 0.016	13.700 ± 0.017	12.435 ± 0.015
Io	176.0 ± 22.0	Unknown	Unknown	Unknown	2.464 ± 0.088	17.320 ± 0.617	Unknown	Unknown
Europa	144.0 ± 14.0	Unknown	Unknown	Unknown	2.068 ± 0.057	13.732 ± 0.377	Unknown	Unknown
Ganymede	160.0 ± 8.0	Unknown	Unknown	Unknown	0.900 ± 0.012	10.871 ± 0.140	Unknown	Unknown
Callisto	84.0 ± 32.0	Unknown	Unknown	Unknown	2.013 ± 0.192	8.194 ± 0.783	Unknown	Unknown
V (Amalthea)	7.28 ± ?	Unknown	Unknown	Unknown	0.1125 ± ?	26.414 ± 3.914	Unknown	Unknown
VI (Hestia)	5.61 ± ?	Unknown	Unknown	Unknown	0.08666 ± ?	3.332 ± 0.453	Unknown	Unknown
VII (Hera)	1.67 ± ?	Unknown	Unknown	Unknown	0.02587 ± ?	3.303 ± 0.860	Unknown	Unknown
VIII (Poseidon)	1.51 ± ?	Unknown	Unknown	Unknown	0.02329 ± ?	2.321 ± 0.579	Unknown	Unknown
IX (Hades)	1.00 ± ?	Unknown	Unknown	Unknown	0.01552 ± ?	2.311 ± 0.326	Unknown	Unknown
X (Demeter)	0.73 ± ?	Unknown	Unknown	Unknown	0.01164 ± ?	3.278 ± 0.613	Unknown	Unknown
XI (Pan)	1.00 ± ?	Unknown	Unknown	Unknown	0.01552 ± ?	2.366 ± 0.445	Unknown	Unknown
XII (Adrastea)	0.84 ± ?	Unknown	Unknown	Unknown	0.01294 ± ?	2.449 ± 0.275	Unknown	Unknown
Saturn	1117.0 ± 11.0	1291.0 ± 25.0	863.0 ± 16.0	1038.0 ± 16.0	36.056 ± 0.097	9.6383 ± 0.0261	10.177 ± 0.031	9.1284 ± 0.0282
Mimas	3.33 ± 1.89	Unknown	Unknown	Unknown	0.1346 ± 0.0193	14.436 ± 2.075	Unknown	Unknown
Enceladus	5.51 ± 2.36	Unknown	Unknown	Unknown	0.1815 ± 0.0197	12.573 ± 1.376	Unknown	Unknown
Tethys	9.72 ± ?	Unknown	Unknown	Unknown	0.3361 ± ?	11.322 ± 0.807	Unknown	Unknown
Dione	10.01 ± ?	Unknown	Unknown	Unknown	0.3460 ± ?	10.004 ± 0.807	Unknown	Unknown
Rhea	14.89 ± ?	Unknown	Unknown	Unknown	0.5149 ± ?	8.4628 ± 0.1974	Unknown	Unknown
Titan	140.0 ± 69.0	Unknown	Unknown	Unknown	0.8161 ± 0.1014	5.5536 ± 0.6938	Unknown	Unknown
Hyperion	11.21 ± ?	Unknown	Unknown	Unknown	0.2123 ± ?	5.0442 ± 0.2396	Unknown	Unknown
Iapetus	36.09 ± ?	Unknown	Unknown	Unknown	0.6834 ± ?	3.2554 ± 0.4694	Unknown	Unknown
Phoebe	7.36 ± ?	Unknown	Unknown	Unknown	0.1394 ± ?	1.7073 ± 0.1672	2.0270 ± 0.1991	1.4380 ± 0.1412
X (Themis/Janus)	13.45 ± ?	Unknown	Unknown	Unknown	0.2546 ± ?	1.5476 ± 0.0575	Unknown	Unknown
Uranus	1049.0 ± 66.0	1156.0 ± 164.0	937.0 ± 101.0	1000.0 ± 101.0	22.194 ± 0.387	6.7951 ± 0.1187	7.1161 ± 0.1245	6.4902 ± 0.1135
Ariel	34.69 ± ?	Unknown	Unknown	Unknown	0.4645 ± ?	5.4898 ± 0.2471	Unknown	Unknown
Umbriel	22.43 ± ?	Unknown	Unknown	Unknown	0.3003 ± ?	4.6512 ± 0.2589	Unknown	Unknown
Titania	55.77 ± ?	Unknown	Unknown	Unknown	0.7468 ± ?	3.6304 ± 0.1302	Unknown	Unknown
Oberon	44.72 ± ?	Unknown	Unknown	Unknown	0.5989 ± ?	3.1405 ± 0.1227	Unknown	Unknown
Miranda	13.50 ± ?	Unknown	Unknown	Unknown	0.1807 ± ?	6.6837 ± 0.3433	Unknown	Unknown
Neptune	1325.0 ± 103.0	1325.0 ± 145.0	1297.0 ± 177.0	1325.0 ± 177.0	24.536 ± 0.531	5.4276 ± 0.1176	5.4723 ± 0.1186	5.3833 ± 0.1167
Triton	241.0 ± 16.0	Unknown	Unknown	Unknown	3.1116 ± 0.0546	4.4006 ± 0.0773	Unknown	Unknown
Nereid	18.0 ± 14.0	Unknown	Unknown	Unknown	0.2213 ± 0.0454	1.0765 ± 0.2282	Unknown	Unknown
Pluto	221.0 ± 207.0	221.0 ± 210.0	221.0 ± 212.0	221.0 ± 212.0	5.0230 ± 1.1804	4.7365 ± 1.1166	6.1024 ± 1.4387	3.6763 ± 0.86670
Typical Asteroid	1.95 ± 1.95	1.95 ± 2.76	1.95 ± 3.38	1.95 ± 3.38	0.02795 ± 0.00781	21.380 ± 9.067	23.637 ± 10.060	19.339 ± 8.231
1—Ceres	28.9 ± 14.8	Unknown	Unknown	Unknown	0.4634 ± 0.0599	17.892 ± 2.315	19.366 ± 2.506	16.530 ± 2.139
2—Pallas	21.2 ± ?	Unknown	Unknown	Unknown	0.3174 ± ?	17.892 ± 0.925	22.757 ± 1.178	14.067 ± 0.728
3—Juno	12.6 ± ?	Unknown	Unknown	Unknown	0.1608 ± ?	18.218 ± 1.385	23.747 ± 1.805	13.977 ± 1.062
4—Vesta	17.5 ± ?	17.5 ± ?	15.4 ± ?	17.5 ± ?	0.2613 ± ?	19.376 ± 5.556	21.184 ± 6.075	17.722 ± 5.082
5—Astraea	Unknown	Unknown	Unknown	Unknown	Unknown	Unknown	Unknown	Unknown
6—Hebe	11.0 ± ?	Unknown	Unknown	Unknown	0.1556 ± ?	19.113 ± 0.971	23.408 ± 1.190	15.605 ± 0.793
7—Iris	10.0 ± ?	Unknown	Unknown	Unknown	0.1413 ± ?	19.278 ± 1.078	24.365 ± 1.362	15.233 ± 0.853
10—Hygiea	15.6 ± ?	Unknown	Unknown	Unknown	0.2235 ± ?	17.823 ± 0.810	19.704 ± 0.895	16.122 ± 0.732
15—Eunomia	13.6 ± ?	Unknown	Unknown	Unknown	0.1951 ± ?	18.306 ± 0.877	21.959 ± 1.052	15.260 ± 0.731
16—Psyche	13.6 ± ?	Unknown	Unknown	Unknown	0.1951 ± ?	17.413 ± 0.834	20.048 ± 0.961	15.124 ± 0.725
51—Nemansa	37.5 ± ?	Unknown	Unknown	Unknown	0.1731 ± ?	19.354 ± 1.623	20.552 ± 1.723	18.226 ± 1.528
323—Brucia	Unknown	Unknown	Unknown	Unknown	Unknown	Unknown	Unknown	Unknown
433—Eros	0.0172 ± ?	Unknown	Unknown	Unknown	0.00203 ± ?	24.665 ± 5.756	30.912 ± 7.214	19.681 ± 4.593
511—Davida	0.880 ± ?	Unknown	Unknown	Unknown	0.0478 ± ?	16.690 ± 0.789	20.021 ± 0.947	13.913 ± 0.658
588—Achilles	0.204 ± ?	Unknown	Unknown	Unknown	0.0107 ± ?	13.042 ± 1.303	15.139 ± 1.512	11.236 ± 1.122
617—Patrochus	Unknown	Unknown	Unknown	Unknown	Unknown	Unknown	Unknown	Unknown
744—Hidalgo	0.606 ± ?	Unknown	Unknown	Unknown	0.0148 ± ?	12.372 ± 3.075	27.146 ± 6.748	5.639 ± 1.402
1048—Feodosia	Unknown	Unknown	Unknown	Unknown	Unknown	Unknown	Unknown	Unknown
1221—Amor	0.29 ± ?	Unknown	Unknown	Unknown	0.1325 ± ?	21.46 ± ?	34.29 ± ?	13.43 ± ?
1566—Icarus	0.120 ± ?	0.357 ± ?	0.069 ± ?	0.069 ± ?	0.00112 ± ?	28.668 ± 3.833	93.458 ± 12.507	8.794 ± 1.177
1620—Geographos	0.015 ± ?	Unknown	Unknown	Unknown	0.00067 ± ?	26.689 ± 7.957	38.029 ± 11.337	18.731 ± 5.584
—Adonis	0.098 ± ?	Unknown	Unknown	Unknown	0.0441 ± ?	21.23 ± ?	60.23 ± ?	7.483 ± ?
—Apollo	0.29 ± ?	Unknown	Unknown	Unknown	0.1073 ± ?	24.431 ± 6.844	46.408 ± 13.001	12.861 ± 3.603
—Hermes	0.098 ± ?	Unknown	Unknown	Unknown	0.0441 ± ?	26.18 ± ?	54.95 ± ?	12.47 ± ?

PHYSICAL DATA FOR THE PLANETS, THEIR SATELLITES AND SOME ASTEROIDS (Continued)

Body	Inclination of Equator to Orbit (Deg.)	Inclination of Orbit to Ecliptic (Deg.)	Inclination of Equator to Ecliptic (Deg.)	Inclination of Orbit to Equator of Primary (Deg.)	Inclination of Orbit to Orbit of Primary (Deg.)	Average Synodic Period (Solar Seconds)	Albedo	Average Temperatures (°K) Night	Day	Equilibrium	Solar Constant (cal/cm²/min)	Surface Pressure (atmospheres)	Atmospheric Composition (in cm-atm) (1 cm-atm = 2.69 E19 molecules per cm²)
Sun (Sol)	—	—	7.0	—	—	—	—	—	—	5776	88682.0 (est.)	Unknown	H_2/He major
Mercury	0.0	7.0028	Unknown	—	—	1.00149 E7	0.076 ± 0.022	13	683	633	12.79	1 E-4 (est.)	150–3550 CO_2 (surface outgassings) SO_2?, Ar?
Venus	R 179.0 ± 1.0	3.40	Unknown	—	—	5.04505 E7	0.76	233	720	463	3.675	10–25‡	90% CO_2, 0.4% H_2O, 0.4% O_2, balance N_2 + inerts
Earth (Terra)	23.45	—	23.45	—	—	—	0.36 ± 0.06	275	295	394	1.920	1.0	78% N_2, 21% O_2, 1% Ar (see supplementary tables)
Moon (Luna)	6.683	5.12	1.533	Unknown	Unknown	—	0.086 ± 0.025	100	365	394	1.920	~0	small amounts of outgassing, SO_2?, Ar?, Xe?
Mars	25.20	1,850	Unknown	—	—	6.73861 E7	0.152 ± 0.003	170	300	319	0.827	0.006–0.009	8300 ± 800 CO_2, 3900 CO, 4.35 ± 1.87 H_2O, ≤**20.** O_2, bal N_2
Phobos	0.0*	Unknown	Unknown	2.0	Unknown	Not tabulated	0.065	Unknown	Unknown	319	0.827	None	—
Deimos	0.0*	Unknown	Unknown	2.0	Unknown	Not tabulated	Unknown	Unknown	Unknown	319	0.827	None	—
Jupiter	3.117	1.309	Unknown	—	—	3.44637 E7	0.54 ± 0.04	123	313	173	0.0709	10^4–10^5	1.20 E7 H_2, 2.35 E6 He, 1.5 E4 CH_4, 7 E2 NH_3, 0.198 H_2O, no O_2/N_2 ? (CH_4 likely)
Io	0.0*	Unknown	Unknown	0.0	3.1	Not tabulated	0.37	Unknown	Unknown	173	0.0709	Trace	None detected
Europa	0.0*	Unknown	Unknown	0.0	3.1	Not tabulated	0.39	Unknown	Unknown	173	0.0709	None?	None detected
Ganymede	0.0*	Unknown	Unknown	0.0	3.1	Not tabulated	0.20	Unknown	Unknown	173	0.0709	None?	None detected
Callisto	0.0*	Unknown	Unknown	0.0	3.1	Not tabulated	0.03	Unknown	Unknown	173	0.0709	None?	—
V (Amalthea)	0.0*	Unknown	Unknown	0.0	3.1	Not tabulated	Unknown	Unknown	Unknown	173	0.0709	None	—
VI (Hestia)	Unknown	Unknown	Unknown	28.5	Unknown	Not tabulated	Unknown	Unknown	Unknown	173	0.0709	None	—
VII (Hera)	Unknown	Unknown	Unknown	28.0	Unknown	Not tabulated	Unknown	Unknown	Unknown	173	0.0709	None	—
VIII (Poseidon)	Unknown	Unknown	Unknown	R 147.5 ± 1.0	Unknown	Not tabulated	Unknown	Unknown	Unknown	173	0.0709	None	—
IX (Hades)	Unknown	Unknown	Unknown	R 156.5 ± 1.0	Unknown	Not tabulated	Unknown	Unknown	Unknown	173	0.0709	None	—
X (Demeter)	Unknown	Unknown	Unknown	28.3	Unknown	Not tabulated	Unknown	Unknown	Unknown	173	0.0709	None	—
XI (Pan)	Unknown	Unknown	Unknown	R 163.4	Unknown	Not tabulated	Unknown	Unknown	Unknown	173	0.0709	None	—
XII (Adrastea)	Unknown	Unknown	Unknown	R ?	Unknown	Not tabulated	Unknown	Unknown	Unknown	173	0.0709	None	—
Saturn	26.75	2.493	Unknown	—	—	3.26671 E7	0.57 ± 0.09	103	223	128	0.0211	10^2–10^4	H_2/He major, 2.5 E4 CH_4, 2.5 E2 NH_3
Mimas	0.0*	Unknown	Unknown	1.5	27 ± 2	Not tabulated	Unknown	Unknown	Unknown	128	0.0211	None	—
Enceladus	0.0*	Unknown	Unknown	0.0	26.8	Not tabulated	Unknown	Unknown	Unknown	128	0.0211	None	—
Tethys	0.0*	Unknown	Unknown	1.1	27 ± 1	Not tabulated	Unknown	Unknown	Unknown	128	0.0211	None	—
Dione	0.0*	Unknown	Unknown	0.0	26.8	Not tabulated	Unknown	Unknown	Unknown	128	0.0211	None	—
Rhea	0.0*	Unknown	Unknown	0.3	26.8 ± 0.3	Not tabulated	Unknown	Unknown	Unknown	128	0.0211	None	—
Titan	0.0*	Unknown	Unknown	0.3	26.8 ± 0.6	Not tabulated	Unknown	Unknown	Unknown	128	0.0211	?	traces NH_3, CH_4 (est. at 2 E4)
Hyperion	0.0*	Unknown	Unknown	0.6	27.0 ± 1.0	Not tabulated	Unknown	Unknown	Unknown	128	0.0211	None	—
Iapetus	Unknown	Unknown	Unknown	14.7	Unknown	Not tabulated	Unknown	Unknown	Unknown	128	0.0211	None	—
Phoebe	0.0*	Unknown	Unknown	R 174.7	R 150.0	Not tabulated	Unknown	Unknown	Unknown	128	0.0211	None	—
X (Themis/Janus)	Unknown	Unknown	Unknown	Unknown	Unknown	Not tabulated	Unknown	Unknown	Unknown	128	0.0211	None	—
Uranus	R 97.983	0.773	R ?	—	—	3.19383 E7	0.65 ± 0.03	103	123	90	0.0052	2–3	H_2/He major, 1.5 E5 CH_4, trace NH_3 (?)
Ariel	0.0*	R ?	R ?	0.0	R 98.0	Not tabulated	Unknown	Unknown	Unknown	90	0.0052	None	—
Umbriel	0.0*	R ?	R ?	0.0	R 98.0	Not tabulated	Unknown	Unknown	Unknown	90	0.0052	None	—
Titania	0.0*	R ?	R ?	0.0	R 98.0	Not tabulated	Unknown	Unknown	Unknown	90	0.0052	None	—
Oberon	0.0*	R ?	R ?	0.0	R 98.0	Not tabulated	Unknown	Unknown	Unknown	90	0.0052	None	—
Miranda	0.0*	R ?	R ?	0.0	R 98.0	Not tabulated	Unknown	Unknown	Unknown	90	0.0052	None	—
Neptune	29.0	1.779	Unknown	—	—	3.17508 E7	0.68 ± 0.08	103	123	72	0.0021	5–10 (est.)	H_2/He major, 2.5 E5 CH_4, trace NH_3 (?)
Triton	Unknown	Unknown	Unknown	R 150.0 ± 10.0	Unknown	Not tabulated	Unknown	Unknown	Unknown	72	0.0021	Trace?	trace CH_4 (at <8 E2)
Nereid	Unknown	Unknown	Unknown	Unknown	Unknown	Not tabulated	Unknown	Unknown	Unknown	72	0.0021	None	—
Pluto	90.0 (?)	17.146	Unknown	—	—	3.16857 E7	0.13 ± 0.05	43	63	63	0.0012	Unknown	None detected

PHYSICAL DATA FOR THE PLANETS, THEIR SATELLITES AND SOME ASTEROIDS (Continued)

Body	Inclination of Equator to Orbit (Deg.)	Inclination of Orbit to Ecliptic (Deg.)	Inclination of Equator to Ecliptic (Deg.)	Inclination of Orbit to Equator of Primary (Deg.)	Inclination of Orbit to Orbit of Primary (Deg.)	Average Synodic Period (Solar Seconds)	Albedo	Average Temperatures (°K) Night	Day	Equilibrium	Solar Constant (cal/cm²/min)	Surface Pressure (atmospheres)	Atmospheric Composition (in cm-atm) (1 cm-atm = 2.69 E19 molecules per cm²)
Typical Asteroid	Unknown	10.0	Unknown	—	—	5.01157 E7	Unknown	Unknown	Unknown	283	0.275	None	—
1—Ceres	Unknown	10.4 ± 0.3	Unknown	—	—	4.03256 E7	Unknown	Unknown	Unknown	237	0.250	None	—
2—Pallas	Unknown	34.8 ± 0.1	Unknown	—	—	4.02926 E7	Unknown	Unknown	Unknown	237	0.250	None	—
3—Juno	Unknown	13.0 ± 0.1	Unknown	—	—	4.09392 E7	Unknown	Unknown	Unknown	241	0.269	None	—
4—Vesta	Unknown	7.11 ± 0.03	Unknown	—	—	4.35554 E7	0.20 ± 0.11	Unknown	Unknown	256	0.345	None	—
5—Astraea	Unknown	5.33 ± ?	Unknown	—	—	4.16452 E7	Unknown	Unknown	Unknown	246	0.290	None	—
6—Hebe	Unknown	14.8 ± ?	Unknown	—	—	4.29177 E7	Unknown	Unknown	Unknown	253	0.326	None	—
7—Iris	Unknown	5.5 ± ?	Unknown	—	—	4.33355 E7	Unknown	Unknown	Unknown	255	0.338	None	—
10—Hygiea	Unknown	3.8 ± ?	Unknown	—	—	3.84326 E7	Unknown	Unknown	Unknown	236	0.247	None	—
15—Eunomia	Unknown	11.8 ± ?	Unknown	—	—	4.11341 E7	Unknown	Unknown	Unknown	242	0.274	None	—
16—Psyche	Unknown	3.1 ± ?	Unknown	—	—	3.94490 E7	Unknown	Unknown	Unknown	230	0.225	None	—
51—Nemansa	Unknown	9.9 ± ?	Unknown	—	—	4.35065 E7	Unknown	Unknown	Unknown	256	0.343	None	—
323—Brucia	Unknown	24.2 ± ?	Unknown	—	—	4.33355 E7	Unknown	Unknown	Unknown	255	0.338	None	—
433—Eros	Unknown	10.8 ± 0.1	Unknown	—	—	7.30584 E7	Unknown	Unknown	Unknown	326	0.905	None	—
511—Davida	Unknown	15.7 ± ?	Unknown	—	—	3.83119 E7	Unknown	Unknown	Unknown	221	0.190	None	—
588—Achilles	Unknown	10.3 ± 0.1	Unknown	—	—	3.44422 E7	Unknown	Unknown	Unknown	173	0.071	None	—
617—Patrochus	Unknown	22.1 ± ?	Unknown	—	—	3.44589 E7	Unknown	Unknown	Unknown	173	0.071	None	—
744—Hidalgo	Unknown	42.5 ± 0.1	Unknown	—	—	3.39999 E7	Unknown	Unknown	Unknown	164	0.058	None	—
1048—Feodosia	Unknown	53.8 ± ?	Unknown	—	—	4.05229 E7	Unknown	Unknown	Unknown	238	0.257	None	—
1221—Amor	Unknown	11.9 ± ?	Unknown	—	—	5.04625 E7	Unknown	Unknown	Unknown	284	0.519	None	—
1566—Icarus	Unknown	23.0 ± 0.1	Unknown	—	—	2.95038 E8	Unknown	Unknown	Unknown	377	1.651	None	—
1620—Geographos	Unknown	13.3 ± ?	Unknown	—	—	1.12878 E8	Unknown	Unknown	Unknown	353	1.240	None	—
—Adonis	Unknown	1.5 ± ?	Unknown	—	—	4.94920 E7	Unknown	Unknown	Unknown	281	0.496	None	—
—Apollo	Unknown	6.4 ± 0.1	Unknown	—	—	7.05344 E7	Unknown	Unknown	Unknown	323	0.871	None	—
—Hermes	Unknown	6.2 ± ?	Unknown	—	—	9.86736 E7	Unknown	Unknown	Unknown	346	1.140	None	—

*Assumed value.
‡Conflicting satellite data.

PHYSICAL DATA FOR THE PLANETS, THEIR SATELLITES AND SOME ASTEROIDS (Continued)

Body	Radius of Body (Earth Radii)			Period of Rotation (Sidereal Days)	Mass (in Earth Masses)	Density (in Earth Units)	Gravitational Constant (gR²)	
	Mean	Equatorial	Polar				km³/sec²	in Earth Units
Sun (Sol)	109.227 ± 0.128	109.232 ± 0.127	109.232 ± 0.129	24.66225 ± 0.00003	332999.0 ± 402.0	0.2555 ± 0.0009	1.3256 ± 0.0047 E + 11	3.330 ± 0.012 E + 5
Mercury	0.3819 ± 0.0055	0.3819 ± 0.0055	Unknown	58.8151 ± 0.0023	0.0532 ± 0.0018	0.9554 ± 0.0528	2.118 ± 0.113 E + 4	5.32 ± 0.28 E − 2
Venus	0.9500 ± 0.0012	0.9500 ± 0.0012	0.9500 ± 0.0012	R 224.588 ± 10.727	0.8167 ± 0.0057	0.9524 ± 0.0076	3.251 ± 0.026 E + 5	0.817 ± 0.006
Earth (Terra)	1.00000	1.00114 ± 0.00010	0.99774 ± 0.00010	1.000000 ± 0.000002	1.0000	1.0000	3.981 ± 0.004 E + 5	1.000 ± 0.001
Moon (Luna)	0.27283 ± 0.00017	0.27290 ± 0.00017	0.27272 ± 0.00017	27.3960 ± 0.0010	0.01230 ± 0.00011	0.6056 ± 0.0056	4.896 ± 0.045 E + 3	0.0123 ± 0.0001
Mars	0.5306 ± 0.0031	0.5314 ± 0.031	0.5288 ± 0.0031	1.028766 ± 0.000002	0.10734 ± 0.00041	0.7188 ± 0.0130	4.273 ± 0.073 E + 4	0.1073 ± 0.0018
Phobos	0.0016 ± 0.0008	0.0018 ± 0.0008	0.0014 ± 0.0008	0.319702 ± 0.000113	4.5426 ± ? E − 9	0.63 ± ?	1.81 ± ? E − 3	4.54 ± ? E − 9
Deimos	0.0008 ± 0.0006	0.0008 ± 0.0006	Unknown	1.26592 ± 0.00005	3.0657 ± ? E − 10	0.63 ± ?	0.12 ± ? E − 4	3.07 ± ? E − 10
Jupiter	10.949 ± 0.022	11.202 ± 0.040	10.460 ± 0.007	0.411192 ± 0.000000	317.929 ± 0.545	0.2422 ± 0.0015	1.266 ± 0.007 E + 8	317.9 ± 1.9
Io	0.2709 ± 0.0171	0.2709 ± 0.0171	Unknown	1.77405 ± 0.00010	0.01316 ± 0.00025	0.662 ± 0.126	5.240 ± 0.941 E + 3	0.0132 ± 0.0024
Europa	0.2335 ± 0.0110	0.2335 ± 0.0110	Unknown	3.56093 ± 0.00020	0.00799 ± 0.00025	0.628 ± 0.091	3.183 ± 0.435 E + 3	0.0080 ± 0.0011
Ganymede	0.3969 ± 0.0085	0.3969 ± 0.0085	Unknown	7.17440 ± 0.00020	0.02580 ± 0.00052	0.412 ± 0.003	1.025 ± 0.065 E + 3	0.0026 ± 0.0002
Callisto	0.3792 ± 0.0215	0.3792 ± 0.0215	Unknown	16.7347 ± 0.0000	0.01239 ± 0.00443	0.225 ± 0.090	4.894 ± 1.931 E + 3	0.0123 ± 0.0049
V (Amalthea)	0.0137 ± 0.0036	0.0137 ± 0.0036	Unknown	0.499463 ± 0.000000	1.3840 ± ? E − 6	0.5 ± ?	0.551 ± ?	1.38 ± ? E − 6
VI (Hestia)	0.0105 ± 0.0025	0.0105 ± 0.0025	Unknown	Unknown	6.3205 ± ? E − 7	0.5 ± ?	0.252 ± ?	6.32 ± ? E − 7
VII (Hera)	0.0031 ± 0.0014	0.0031 ± 0.0014	Unknown	Unknown	1.6809 ± ? E − 8	0.5 ± ?	6.69 ± ? E − 3	1.68 ± ? E − 8
VIII (Poseidon)	0.0028 ± 0.0013	0.0028 ± 0.0013	Unknown	R ?	1.2258 ± ? E − 8	0.5 ± ?	4.88 ± ? E − 3	1.23 ± ? E − 8
IX (Hades)	0.0019 ± 0.0005	0.0019 ± 0.0005	Unknown	R ?	3.6310 ± ? E − 9	0.5 ± ?	1.45 ± ? E − 3	3.63 ± ? E − 9
X (Demeter)	0.0014 ± 0.0005	0.0014 ± 0.0005	Unknown	Unknown	1.5322 ± ? E − 9	0.5 ± ?	6.10 ± ? E − 4	1.53 ± ? E − 9
XI (Pan)	0.0019 ± 0.0006	0.0019 ± 0.0006	Unknown	R ?	3.6310 ± ? E − 9	0.5 ± ?	1.45 ± ? E − 3	3.63 ± ? E − 9
XII (Adrastea)	0.0016 ± 0.0003	0.0016 ± 0.0003	Unknown	R ?	2.1024 ± ? E − 9	0.5 ± ?	8.37 ± ? E − 4	2.10 ± ? E − 9
Saturn	9.1377 ± 0.0411	9.4751 ± 0.0342	8.4987 ± 0.0595	0.427556 ± 0.000000	95.0661 ± 0.3734	0.1246 ± 0.0018	3.784 ± 0.050 E + 7	95.06 ± 1.27
Mimas	0.0427 ± 0.0036	0.0427 ± 0.0036	Unknown	0.948890 ± 0.008022	6.188 ± 3.345 E − 6	0.080 ± 0.047	2.464 ± 1.456	6.188 ± 3.658 E − 6
Enceladus	0.0469 ± 0.0033	0.0469 ± 0.0033	Unknown	1.37546 ± 0.00281	1.238 ± 0.502 E − 5	0.120 ± 0.055	4.927 ± 2.224	1.238 ± 0.559 E − 5
Tethys	0.0912 ± 0.0116	0.0912 ± 0.0116	Unknown	1.88876 ± 0.00742	8.244 ± ? E − 5	0.11 ± ?	3.282 ± ? E + 1	8.24 ± ? E − 5
Dione	0.0939 ± 0.0135	0.0939 ± 0.0135	Unknown	2.74880 ± 0.00752	8.990 ± ? E − 5	0.11 ± ?	3.579 ± ? E + 1	8.99 ± ? E − 5
Rhea	0.1397 ± 0.0058	0.1397 ± 0.0058	Unknown	4.52416 ± 0.00453	2.964 ± ? E − 4	0.11 ± ?	1.180 ± ? E + 2	2.96 ± ? E − 4
Titan	0.3734 ± 0.0304	0.3734 ± 0.0304	Unknown	16.0057 ± 0.0261	0.01991 ± 0.00092	0.382 ± 0.020	7.923 ± 4.093 E + 3	1.990 ± 1.028 E − 2
Hyperion	0.0315 ± 0.0027	0.0315 ± 0.0027	Unknown	21.3402 ± 0.0091	1.138 ± ? E − 5	0.36 ± ?	4.53 ± ?	1.14 ± ? E − 5
Iapetus	0.1015 ± 0.0261	0.1015 ± 0.0261	Unknown	Unknown	3.795 ± ? E − 4	0.36 ± ?	1.51 ± ? E + 2	3.79 ± ? E − 4
Phoebe	0.0207 ± 0.0036	0.0207 ± 0.0036	Unknown	R ?	3.223 ± ? E − 6	0.36 ± ?	1.78 ± ?	3.22 ± ? E − 6
X (Themis/Janus)	0.0378 ± 0.0025	0.0378 ± 0.0025	Unknown	Unknown	1.962 ± ? E − 5	0.36 ± ?	7.81 ± ?	1.96 ± ? E − 5
Uranus	3.6837 ± 0.1149	3.7740 ± 0.0877	3.5098 ± 0.1968	R 0.450397 ± 0.002228	14.521 ± 0.053	0.2905 ± 0.0272	5.781 ± 0.510 E + 6	14.52 ± 1.28
Ariel	0.0488 ± 0.0039	0.0488 ± 0.0039	Unknown	R 25.2710 ± 0.0002	8.430 ± ? E − 5	0.72 ± ?	3.36 ± ? E + 1	8.43 ± ? E − 5
Umbriel	0.0315 ± 0.0031	0.0315 ± 0.0031	Unknown	R 4.15544 ± 0.00000	2.276 ± ? E − 5	0.72 ± ?	9.06 ± ?	2.28 ± ? E − 5
Titania	0.0785 ± 0.0050	0.0785 ± 0.0050	Unknown	R 8.72973 ± 0.00000	3.502 ± ? E − 4	0.72 ± ?	1.39 ± ? E + 2	3.50 ± ? E − 4
Oberon	0.0629 ± 0.0044	0.0629 ± 0.0044	Unknown	R 13.4999 ± 0.0001	1.806 ± ? E − 4	0.72 ± ?	7.19 ± ? E + 1	1.81 ± ? E − 4
Miranda	0.0190 ± 0.0017	0.0190 ± 0.0017	Unknown	R 1.41787 ± 0.00139	4.964 ± ? E − 6	0.72 ± ?	1.98 ± ?	4.96 ± ? E − 6
Neptune	3.5654 ± 0.1364	3.5654 ± 0.1364	Unknown	0.657490 ± 0.002797	17.177 ± 0.234	0.3790 ± 0.0438	6.838 ± 0.746 E + 6	17.18 ± 1.87
Triton	0.3152 ± 0.0086	0.3152 ± 0.0086	Unknown	Unknown	0.02442 ± 0.00084	0.780 ± 0.069	9.721 ± 0.823 E + 3	2.442 ± 0.207 E − 2
Nereid	0.0213 ± 0.0053	0.0213 ± 0.0053	Unknown	Unknown	8.363 ± 5.018 E − 6	0.860 ± 0.826	3.329 ± 3.087	8.363 ± 7.755 E − 6
Pluto	0.8946 ± 0.0648	0.8946 ± 0.0648	Unknown	6.4101 ± 0.0050	0.1806 ± 0.1673	0.2523 ± 0.2399	7.191 ± 6.819 E + 4	0.1806 ± 0.1713
Typical Asteroid	0.0031 ± 0.0016	0.0031 ± 0.0016	Unknown	0.4596 ± 0.4178	1.962 ± ? E − 8	0.63 ± ?	7.81 ± ? E − 3	1.96 ± ? E − 8
1—Ceres	0.0584 ± 0.0035	0.0584 ± 0.0035	Unknown	Unknown	1.004 ± 0.502 E − 5	0.504 ± 0.267	3.995 ± 2.106	1.004 ± 0.529 E − 5
2—Pallas	0.0374 ± 0.0035	0.0374 ± 0.0035	Unknown	Unknown	3.011 ± ? E − 6	0.58 ± ?	1.20 ± ?	3.01 ± ? E − 6
3—Juno	0.0162 ± 0.0022	0.0162 ± 0.0022	Unknown	Unknown	3.345 ± ? E − 6	0.79 ± ?	1.33 ± ?	3.35 ± ? E − 6
4—Vesta	0.0306 ± 0.0157	0.0306 ± 0.0157	Unknown	0.2228 ± 0.0070	1.673 ± ? E − 5	0.58 ± ?	6.66 ± ?	1.67 ± ? E − 5
5—Astraea	Unknown	Unknown	Unknown	Unknown	Unknown	Unknown	Unknown	Unknown
6—Hebe	0.0173 ± ?	0.0173 ± ?	Unknown	Unknown	3.345 ± ? E − 6	0.65 ± ?	1.33 ± ?	3.35 ± ? E − 6
7—Iris	0.0153 ± ?	0.0153 ± ?	Unknown	Unknown	2.509 ± ? E − 6	0.65 ± ?	1.00 ± ?	2.51 ± ? E − 6
10—Hygiea	0.0251 ± ?	0.0251 ± ?	Unknown	Unknown	1.004 ± ? E − 5	0.63 ± ?	3.99 ± ?	1.00 ± ? E − 5
15—Eunomia	0.0220 ± ?	0.0220 ± ?	Unknown	Unknown	6.690 ± ? E − 6	0.63 ± ?	2.66 ± ?	6.69 ± ? E − 6

PHYSICAL DATA FOR THE PLANETS, THEIR SATELLITES AND SOME ASTEROIDS (Continued)

Body	Radius of Body (Earth Radii)			Period of Rotation (Sidereal Days)	Mass (in Earth Masses)	Density (in Earth Units)	Gravitational Constant (gR^2)	
	Mean	Equatorial	Polar				km^3/sec^2	in Earth Units
16—Psyche	0.0220 ± ?	0.0220 ± ?	Unknown	Unknown	6.690 ± ? E − 6	0.63 ± ?	2.66 ±	6.69 ± ? E − 6
51—Nemansa	0.0063 ± ?	0.0063 ± ?	Unknown	Unknown	1.505 ± ? E − 7	0.61 ± ?	0.060 ± ?	1.51 ± ? E − 7
323—Brucia	Unknown	Unknown	Unknown	Unknown	Unknown	Unknown	Unknown	Unknown
433—Eros	0.0019 ± 0.0008	0.0019 ± 0.0008	Unknown	Unknown	8.363 ± ? E − 10	0.13 ± ?	3.33 ± ? E − 4	8.36 ± ? E − 10
511—Davida	0.0204 ± ?	0.0204 ± ?	Unknown	Unknown	5.018 ± ? E − 6	0.59 ± ?	2.00 ± ?	5.02 ± ? E − 6
588—Achilles	0.0044 ± 0.0008	0.0044 ± 0.0008	Unknown	Unknown	5.382 ± ? E − 11	0.63 ± ?	2.14 ± ? E − 5	5.38 ± ? E − 11
617—Patrochus	Unknown	Unknown	Unknown	Unknown	Unknown	Unknown	Unknown	Unknown
744—Hidalgo	0.0028 ± 0.0013	0.0028 ± 0.0013	Unknown	Unknown	6.620 ± ? E − 11	0.63 ± ?	2.64 ± ? E − 5	6.62 ± ? E − 11
1048—Feodosia	Unknown	Unknown	Unknown	Unknown	Unknown	Unknown	Unknown	Unknown
1221—Amor	0.00047 ± ?	0.00047 ± ?	Unknown	Unknown	6.620 ± ? E − 11	0.63 ± ?	2.64 ± ? E − 5	6.67 ± ? E − 11
1566—Icarus	0.00008 ± 0.00002	0.00011 ± 0.00002	0.00005 ± 0.00003	1.128 ± 0.407	8.363 ± ? E − 13	1.47 ± ?	3.33 ± ? E − 7	8.36 ± ? E − 13
1620—Geographos	0.00024 ± ?	0.00024 ± ?	Unknown	Unknown	8.363 ± ? E − 12	0.64 ± ?	3.33 ± ? E − 6	8.36 ± ? E − 12
—Adonis	0.00016 ± ?	0.00016 ± ?	Unknown	Unknown	2.452 − ? E − 12	0.63 ± ?	9.76 ± ? E − 7	2.45 ± ? E − 12
—Apollo	0.00031 ± 0.00016	0.00031 ± 0.00016	Unknown	Unknown	2.892 ± ? E − 11	0.63 ± ?	1.15 ± ? E − 5	2.89 ± ? E − 11
—Hermes	0.00016 ± ?	0.00016 ± ?	Unknown	Unknown	2.452 ± ? E − 12	0.63 ± ?	9.76 ± ? E − 7	2.45 ± ? E − 12

PHYSICAL DATA FOR THE PLANETS, THEIR SATELLITES AND SOME ASTEROIDS (Continued)

Body	Radius of Orbit (Au)			Period of Revolution (Sidereal Years)	Velocity (in Earth Units)			Gravity (in g-s)		Volume (in Earth Units)
	Mean	Aphelion	Perhelion		Rotational	Orbital	Escape	Polar	Equatorial	
Sun (Sol)	—	—	—	—	4.4241	—	55.089	27.915	27.908	1303150.0 ± 4580.0
Mercury	0.38744 ± 0.00049	0.46710 ± 0.00107	0.30779 ± 0.00071	0.240899 ± 0.000055	0.00649	1.6040	0.3723	Unknown	0.3648	0.0557 ± 0.0024
Venus	0.72281 ± 0.00075	0.72772 ± 0.00076	0.71789 ± 0.00075	0.615185 ± 0.000005	R 0.00388	1.1745	0.9286	0.9049	0.9049	0.8575 ± 0.0033
Earth (Terra)	1.00000	1.01670 ± 0.00068	0.98330 ± 0.00066	1.000000 ± 0.000000	1.0000	1.0000	1.0000	1.005	0.9942	1.0000
Moon (Luna)	0.0025701 ± 0.0000012	0.0027111 ± 0.0000013	0.0024290 ± 0.0000011	0.074800 ± 0.000003	0.00995	0.03423	0.2116	0.1654	0.1652	0.02031 ± 0.00004
Mars	1.5233 ± 0.0016	1.6656 ± 0.0018	1.3810 ± 0.0015	1.88082 ± 0.00005	0.5160	0.8087	0.4489	0.3839	0.3627	0.1493 ± 0.0027
Phobos	0.0000629 ± 0.0000002	Unknown	Unknown	0.0008729 ± 0.0000003	0.00554	0.02262	0.001666	Unknown	Unknown	4.344 ± 6.268 E – 9
Deimos	0.000157 ± 0.000001	Unknown	Unknown	0.0034564 ± 0.0000001	0.00062	0.01433	0.000624	Unknown	Unknown	4.833 ± 11.599 E – 10
Jupiter	5.2025 ± 0.0027	5.4543 ± 0.0030	4.9507 ± 0.0028	11.8613 ± 0.0117	27.212	0.4396	5.375	2.906	2.305	1312.5 ± 7.9
Io	0.002821 ± 0.000002	Unknown	Unknown	0.0048437 ± 0.0000003	0.1525	0.5805	0.2196	Unknown	Unknown	0.01988 ± 0.00377
Europa	0.004488 ± 0.000002	Unknown	Unknown	0.0097225 ± 0.0000005	0.0655	0.4597	0.1848	Unknown	Unknown	0.0127 ± 0.0018
Ganymede	0.007161 ± 0.000007	Unknown	Unknown	0.0195884 ± 0.0000005	0.0553	0.3658	0.0804	Unknown	Unknown	0.0625 ± 0.0040
Callisto	0.012589 ± 0.000007	Unknown	Unknown	0.0456912 ± 0.0000007	0.0226	0.2752	0.1795	Unknown	Unknown	0.0545 ± 0.0093
V (Amalthea)	0.001213 ± 0.000003	Unknown	Unknown	0.0013637 ± 0.0000005	0.0273	0.8860	0.01005	Unknown	Unknown	2.546 ± 2.019 E – 6
VI (Hestia)	0.07622 ± 0.00134	Unknown	Unknown	0.686475 ± 0.000631	Unknown	0.1117	0.00774	Unknown	Unknown	1.163 ± 0.833 E – 6
VII (Hera)	0.07756 ± 0.00267	Unknown	Unknown	0.711280 ± 0.000548	Unknown	0.1107	0.00231	Unknown	Unknown	3.093 ± 4.176 E – 8
VIII (Poseidon)	0.15712 ± 0.00067	Unknown	Unknown	R 2.02324 ± 0.00027	R ?	R 0.0779	0.00208	Unknown	Unknown	2.255 ± 3.007 E – 8
IX (Hades)	0.15845 ± 0.00134	Unknown	Unknown	R 2.07252 ± 0.00465	R ?	R 0.0775	0.00139	Unknown	Unknown	6.681 ± 5.011 E – 9
X (Demeter)	0.07876 ± 0.00034	Unknown	Unknown	0.706901 ± 0.009855	Unknown	0.1101	0.00104	Unknown	Unknown	2.819 ± 2.819 E – 9
XI (Pan)	0.1151 ± 0.0013	Unknown	Unknown	R 1.91290 ± 0.00630	R ?	R 0.0795	0.00139	Unknown	Unknown	6.681 ± 6.681 E – 9
XII (Adrastea)	1.14107 ± 0.00067	Unknown	Unknown	R 1.71113 ± 0.00411	R ?	R 0.0822	0.00115	Unknown	Unknown	3.866 ± 2.320 E – 9
Saturn	9.5407 ± 0.0075	10.0587 ± 0.0215	9.0226 ± 0.0213	29.4568 ± 0.0111	22.136	0.3235	3.2232	1.316	0.8800	763.0 ± 10.3
Mimas	0.001214 ± 0.000005	Unknown	Unknown	0.0025908 ± 0.0000219	0.0449	0.4832	0.0121	Unknown	Unknown	7.781 ± 1.974 E – 5
Enceladus	0.001601 ± 0.000011	Unknown	Unknown	0.0037554 ± 0.0000077	0.0341	0.4228	0.0162	Unknown	Unknown	1.034 ± 0.218 E – 4
Tethys	0.001874 ± 0.000003	Unknown	Unknown	0.0051569 ± 0.0000202	0.0482	0.3792	0.02873	Unknown	Unknown	7.584 ± 2.897 E – 4
Dione	0.002528 ± 0.000010	Unknown	Unknown	0.0075051 ± 0.0000205	0.0341	0.3365	0.03078	Unknown	Unknown	8.268 ± 3.567 E – 4
Rhea	0.003533 ± 0.000013	Unknown	Unknown	0.0123524 ± 0.0000124	0.0308	0.2839	0.04602	Unknown	Unknown	2.726 ± 0.340 E – 3
Titan	0.008204 ± 0.000047	Unknown	Unknown	0.0437009 ± 0.0000713	0.0233	0.1862	0.0729	Unknown	Unknown	5.206 ± 1.274 E – 2
Hyperion	0.009944 ± 0.000044	Unknown	Unknown	0.0582658 ± 0.0000247	0.00148	0.1691	0.01895	Unknown	Unknown	3.140 ± 0.797 E – 5
Iapetus	0.02388 ± 0.00013	Unknown	Unknown	0.216971 ± 0.000466	Unknown	0.1094	0.06099	Unknown	Unknown	1.047 ± 0.806 E – 3
Phoebe	0.08680 ± 0.00048	0.10156 ± 0.00104	0.07205 ± 0.00096	R 1.50360 ± 0.00575	R ?	R 0.0574	0.01245	Unknown	Unknown	8.893 ± 4.648 E – 6
X (Themis/Janus)	0.001056 ± 0.000001	Unknown	Unknown	0.002033 ± ?	Unknown	0.5201	0.02273	Unknown	Unknown	5.412 ± 1.078 E – 5
Uranus	19.190 ± 0.013	20.073 ± 0.032	18.308 ± 0.032	84.0081 ± 0.0824	R 8.3697	0.2282	1.9821	1.179 ± 0.168	0.9554 ± 0.168	49.987 ± 4.676
Ariel	0.001282 ± 0.000002	Unknown	Unknown	R 0.0068998 ± 0.0000005	R 0.01929	R 0.1842	0.04150	Unknown	Unknown	1.163 ± 0.280 E – 4
Umbriel	0.001786 ± 0.000002	Unknown	Unknown	R 0.0113457 ± 0.0000005	R 0.0075	R 0.1560	0.02682	Unknown	Unknown	3.140 ± 0.937 E – 5
Titania	0.002932 ± 0.000006	Unknown	Unknown	R 0.0238350 ± 0.0000000	R 0.00898	R 0.1218	0.06663	Unknown	Unknown	4.833 ± 0.928 E – 4
Oberon	0.003919 ± 0.000004	Unknown	Unknown	R 0.0368593 ± 0.0000003	R 0.00466	R 0.1054	0.05336	Unknown	Unknown	2.493 ± 0.522 E – 4
Miranda	0.0008651 ± 0.0000094	Unknown	Unknown	R 0.0038712 ± 0.0000038	R 0.01338	R 0.2242	0.01615	Unknown	Unknown	6.850 ± 1.868 E – 6
Neptune	30.086 ± 0.029	30.332 ± 0.036	29.839 ± 0.036	164.784 ± 0.012	5.4166	0.1822	2.1875	Unknown	1.323 ± 0.210	45.322 ± 5.201
Triton	0.002361 ± 0.000003	Unknown	Unknown	0.0160881 ± 0.0000030	Unknown	R 0.1477	0.2777	Unknown	Unknown	3.130 ± 0.257 E – 2
Nereid	0.03945 ± 0.00134	Unknown	Unknown	1.24022 ± 0.22175	Unknown	0.0362	0.0197	Unknown	Unknown	9.726 ± 7.294 E – 6
Pluto	39.507 ± 0.128	49.308 ± 0.162	29.705 ± 0.099	248.35 ± 0.34	0.1394	0.1591	0.4482	Unknown	0.225 ± 0.217	0.7160 ± 0.1556
Typical Asteroid	1.94 ± 1.00	2.13 ± 1.11	1.75 ± 0.91	2.70 ± ?	0.00682	0.7181	0.00250	Unknown	Unknown	3.093 ± 4.640 E – 8
1—Ceres	2.7686 ± 0.0024	2.9873 ± 0.0061	2.5499 ± 0.0060	4.5994 ± 0.0029	Unknown	0.6007	0.0413	Unknown	Unknown	1.990 ± 0.353 E – 4
2—Pallas	2.7686 ± 0.0024	3.4220 ± 0.0088	2.1152 ± 0.0085	4.6131 ± 0.0054	Unknown	0.6007	0.0283	Unknown	Unknown	5.213 ± 1.445 E – 5
3—Juno	2.6703 ± 0.0014	3.3619 ± 0.0032	1.9787 ± 0.0029	4.3640 ± 0.0054	Unknown	0.6107	0.0144	Unknown	Unknown	4.225 ± 1.723 E – 6
4—Vesta	2.3608 ± 0.0013	2.5709 ± 0.0027	2.1507 ± 0.0026	3.6304 ± 0.0029	0.1372	0.6510	0.0233	Unknown	Unknown	2.867 ± 4.411 E – 5
5 Astraea	2.57 ± ?	3.06 ± ?	2.08 ± ?	3.5816 ± 0.0111	Unknown	0.6229	Unknown	Unknown	Unknown	Unknown
6—Hebe	2.43 ± ?	2.91 ± ?	1.94 ± ?	3.7781 ± 0.0029	Unknown	0.6409	0.0139	Unknown	Unknown	5.146 ± 1.404 E – 6
7—Iris	2.38 ± ?	2.93 ± ?	1.84 ± ?	3.6796 ± 0.0029	Unknown	0.6477	0.0126	Unknown	Unknown	3.866 ± 1.160 E – 6
10—Hygiea	2.79 ± ?	3.07 ± ?	2.51 ± ?	5.5906 ± 0.0029	Unknown	0.5973	0.0199	Unknown	Unknown	1.584 ± 0.386 E – 5
15—Eunomia	2.64 ± ?	3.12 ± ?	2.17 ± ?	4.2956 ± 0.0029	Unknown	0.6141	0.0174	Unknown	Unknown	1.061 ± 0.273 E – 5

PHYSICAL DATA FOR THE PLANETS, THEIR SATELLITES AND SOME ASTEROIDS (Continued)

Body	Radius of Orbit (Au)			Period of Revolution	Velocity (in Earth Units)			Gravity (in g-s)		Volume
	Mean	Aphelion	Perhelion	(Sidereal Years)	Rotational	Orbital	Escape	Polar	Equatorial	(in Earth Units)
16—Psyche	2.92 ± ?	3.33 ± ?	2.51 ± ?	4.9993 ± 0.0029	Unknown	0.5839	0.0174	Unknown	Unknown	1.061 ± 0.273 E − 5
51—Nemansa	2.37 ± ?	2.51 ± ?	2.22 ± ?	3.6412 ± 0.0029	Unknown	0.6510	0.0154	Unknown	Unknown	2.475 ± 1.114 E − 7
323—Brucia	2.38 ± ?	3.10 ± ?	1.67 ± ?	3.6796 ± 0.0111	Unknown	0.6473	Unknown	Unknown	Unknown	Unknown
433—Eros	1.4568 ± 0.0028	1.7803 ± 0.0045	1.1334 ± 0.0036	1.7604 ± 0.0027	Unknown	0.8289	0.000210	Unknown	Unknown	6.681 ± 8.351 E − 9
511—Davida	3.18 ± ?	3.75 ± ?	2.61 ± ?	5.6727 ± 0.0029	Unknown	0.5604	0.00495	Unknown	Unknown	8.495 ± 2.156 E − 6
588—Achilles	5.2103 ± 0.0041	5.9814 ± 0.0071	4.4391 ± 0.0063	11.942 ± 0.060	Unknown	0.4362	0.00111	Unknown	Unknown	8.488 ± 4.547 E − 8
617—Patrochus	5.2049 ± 0.0041	5.9388 ± 0.0114	4.4710 ± 0.0110	11.879 ± 0.011	Unknown	0.4362	Unknown	Unknown	Unknown	Unknown
744—Hidalgo	5.7899 ± 0.0038	9.5881 ± 0.0086	1.9917 ± 0.0059	13.924 ± 0.035	Unknown	0.4161	0.0015	Unknown	Unknown	2.255 ± 3.007 E − 8
1048—Feodosia	2.73 ± ?	3.23 ± ?	2.24 ± ?	4.5202 ± 0.0111	Unknown	0.6038	Unknown	Unknown	Unknown	Unknown
1221—Amor	1.92 ± ?	2.77 ± ?	1.08 ± ?	2.6694 ± 0.0110	Unknown	0.7205	0.0119	Unknown	Unknown	1.044 ± 1.566 E − 10
1566—Icarus	1.0784 ± 0.0021	1.9714 ± 0.0044	0.1855 ± 0.0022	1.1198 ± 0.0027	0.000098	0.9631	0.0001	Unknown	3.772 ± 5.637 E − 4	5.684 ± 4.070 E − 13
1620—Geographos	1.24 ± ?	1.67 ± ?	0.82 ± ?	1.3881 ± 0.0027	Unknown	0.8960	0.000059	Unknown	Unknown	1.305 ± 2.088 E − 11
—Adonis	1.97 ± ?	3.50 ± ?	0.43 ± ?	2.7597 ± 0.0110	Unknown	0.7120	0.0039	Unknown	Unknown	3.866 ± 5.800 E − 12
—Apollo	1.4849 ± 0.0034	2.3254 ± 0.0055	0.6445 ± 0.0071	1.3097 ± 0.0110	Unknown	0.8188	0.0096	Unknown	Unknown	3.093 ± 4.640 E − 11
—Hermes	1.29 ± ?	2.11 ± ?	0.48 ± ?	1.4702 ± 0.0110	Unknown	0.7205	0.0039	Unknown	Unknown	3.866 ± 5.800 E − 12

PHYSICAL DATA FOR THE PLANETS, THEIR SATELLITES AND SOME ASTEROIDS (Continued)

		Escape Velocity							Isosynchronous Satellite Data			
		Solar Escape from Position of		Solar Escape from Surface of		Mean Escape Velocity	Mean Orbital Velocity	Mean Rotational Velocity				
Body	Symbol	km/sec	in Earth Units	km/sec	in Earth Units	km/sec	km/sec	km/sec	Altitude (km)	Orbital Velocity (km/sec)	Escape Velocity from (km/sec)	Solar Escape from (km/sec)
Sun	⊙	—	—	617.2 ± 0.9	14.17 ± 0.02	617.2 ± 0.9	—	2.058 ± 0.002	24055600.0 ± 29000.0	73.19 ± 0.43	103.5 ± 0.6	—
Mercury	☿	67.64 ± 0.10	1.607 ± 0.002	67.77 ± 0.14	1.556 ± 0.003	4.173 ± 0.070	47.83 ± 0.10	0.0030 ± 0.0000	237301.0 ± 4268.0	0.297 ± 0.003	0.420 ± 0.004	67.64 ± 0.14
Venus	♀	49.52 ± 0.06	1.176 ± 0.001	50.60 ± 0.06	1.161 ± 0.001	10.36 ± 0.03	35.02 ± 0.04	0.0018 ± 0.0000	1534720.0 ± 45230.0	0.459 ± 0.017	0.650 ± 0.024	49.53 ± 0.31
Earth	⊕ or ♁	42.10 ± 0.05	1.000 ± 0.001	43.56 ± 0.04	1.000 ± 0.001	11.18 ± 0.00	29.77 ± 0.01	0.4651 ± 0.0000	35775.0 ± 13.0	3.073 ± 0.002	4.346 ± 0.003	42.33 ± 0.05
Moon*	☾	1.439 ± 0.000	0.0342 ± 0.0000	2.775 ± 0.001	0.0637 ± 0.0000	2.373 ± 0.011	1.018 ± 0.002	0.0046 ± 0.0000	86667.0 ± 275.0	0.235 ± 0.002	0.333 ± 0.003	1.477 ± 0.002*
Mars	♂	34.11 ± 0.04	0.8102 ± 0.0009	34.48 ± 0.05	0.7915 ± 0.0011	5.028 ± 0.024	24.12 ± 0.044	0.2400 ± 0.0014	17032.0 ± 118.0	1.447 ± 0.012	2.046 ± 0.017	34.18 ± 0.08
Jupiter	♃	18.46 ± 0.02	0.4384 ± 0.0004	63.00 ± 0.17	1.446 ± 0.004	60.24 ± 0.10	13.05 ± 0.03	12.66 ± 0.05	89304.0 ± 338.0	28.21 ± 0.26	39.89 ± 0.37	43.96 ± 2.92
Saturn	♄	13.63 ± 0.02	0.3238 ± 0.0004	38.55 ± 0.14	0.8850 ± 0.0032	36.06 ± 0.14	9.639 ± 0.017	10.30 ± 0.04	50964.0 ± 552.0	18.62 ± 0.34	26.33 ± 0.49	29.65 ± 2.56
Uranus	⛢	9.611 ± 0.012	0.2283 ± 0.0002	24.18 ± 0.35	0.5551 ± 0.0080	22.19 ± 0.55	6.796 ± 0.012	3.893 ± 0.091	36938.0 ± 1939.0	9.780 ± 0.825	13.83 ± 1.17	16.84 ± 3.23
Neptune	♆	7.676 ± 0.010	0.1823 ± 0.0002	25.71 ± 0.52	0.5901 ± 0.0119	24.54 ± 0.75	5.428 ± 0.010	2.519 ± 0.096	59504.0 ± 3122.0	9.119 ± 1.031	12.82 ± 1.46	15.01 ± 3.76
Pluto	♇	6.699 ± 0.011	0.1591 ± 0.0002	8.371 ± 0.237	0.1921 ± 0.0054	5.019 ± 1.67	4.737 ± 0.011	0.0648 ± 0.0047	76470.0 ± 25960.0	0.935 ± 0.847	1.322 ± 1.198	6.828 ± 0.317

*Escape from Earth

*Earth Escape from

HOHMANN ELLIPSE TRANSFER DATA (MINIMUM ENERGY OF TRANSFER)

W. Joseph Armento

The "from" body is the body of departure and the "to" body is the body of arrival. The three lines of data for each planetary entry are:

top—from perhelion of inner planet to aphelion of outer planet
middle—from mean of inner planet to mean of outer planet
bottom—from aphelion of inner planet to perhelion of outer planet

The orbital parameters and velocities are given for the transfer orbit. The orbital velocity, change in velocity necessary to reach the new Hohmann orbital velocity, and the maximum change in velocity to reach the Hohmann orbital velocity are given for departure. For arrival data, look in the table listing the arrival planet as "from" and the departure planet as "to". The maximum Hohmann velocity change is assumed to be a single impulse to overcome the necessary energy for change in velocity and escape velocity from the surface of the departure planet.

Body	Mean orbital Semimajor axis (km × 10^{-6})	Eccentricity ε	Aphelion (km × 10^{-6})	Perhelion (km × 10^{-6})	Orbital Velocity (km/sec) Mean	Aphelion	Perhelion	Orbital Velocity of Departure Body (km/sec)	Time of one way trip (½ orbit) (Solar Seconds)	Departure Δ_V orbit (km/sec)	Maximum Δ_V (km/sec)
From Mercury to:											
	77.440	0.4055	103.85	46.035	41.38	26.91	63.62	58.92	5.880 E6	4.699	6.284
Venus	83.030	0.3021	108.11	57.950	39.96	29.26	54.58	47.83	6.528 E6	6.748	7.934
	88.620	0.2116	107.37	69.864	38.68	31.20	47.95	38.83	7.198 E6	9.124	10.03
	99.052	0.5352	152.07	46.035	36.58	20.13	66.49	58.92	8.506 E6	7.569	8.643
Earth	103.76	0.4415	149.57	57.950	35.75	22.25	57.43	47.83	9.119 E6	9.596	10.46
	108.47	0.3559	147.07	69.864	34.96	24.10	50.72	38.83	9.747 E6	11.90	12.61
	147.58	0.6881	249.12	46.035	29.97	12.88	69.72	58.92	1.547 E7	10.80	11.58
Mars	142.89	0.5945	227.84	57.950	30.46	15.36	60.40	47.83	1.474 E7	12.57	13.24
	138.21	0.4945	206.56	69.864	30.97	18.01	53.25	38.83	1.402 E7	14.43	15.02
	430.92	0.8932	815.80	46.035	17.54	4.167	73.84	58.92	7.718 E7	14.91	15.49
Jupiter	418.04	0.8614	778.14	57.950	17.81	4.860	65.26	47.83	7.375 E7	17.43	17.92
	405.17	0.8276	740.48	69.864	18.09	5.556	58.89	38.83	7.037 E7	20.06	20.49
	775.26	0.9406	1504.5	46.035	13.08	2.287	74.76	58.92	1.862 E8	15.83	16.37
Saturn	742.47	0.9220	1427.0	57.950	13.36	2.693	66.31	47.83	1.746 E8	18.48	18.94
	709.69	0.9016	1349.5	69.864	13.67	3.110	60.07	38.83	1.631 E8	21.24	21.65
	1524.2	0.9698	3002.3	46.035	9.326	1.155	75.32	58.92	5.134 E8	16.39	16.92
Uranus	1464.1	0.9604	2870.3	57.950	9.515	1.352	66.97	47.83	4.834 E8	19.14	19.59
	1404.1	0.9502	2738.3	69.864	9.717	1.552	60.83	38.83	4.539 E8	22.01	22.40
	2291.4	0.9799	4536.8	46.035	7.606	0.7662	75.51	58.92	9.464 E8	16.59	17.10
Neptune	2278.9	0.9746	4499.9	57.950	7.627	0.8655	67.21	47.83	9.337 E8	19.38	19.83
	2266.4	0.9692	4463.0	69.864	7.648	0.9569	61.13	38.83	9.310 E8	22.30	22.69
	3710.5	0.9876	7375.0	46.035	5.977	0.4723	75.66	58.92	1.950 E9	16.73	17.24
Pluto	2983.5	0.9806	5909.0	57.950	6.666	0.6601	67.31	47.83	1.406 E9	19.48	19.93
	2256.4	0.9690	4443.0	69.864	7.665	0.9612	61.13	38.83	9.248 E8	22.30	22.69
From Venus to:											
	77.440	0.4055	108.85	46.035	41.38	26.91	63.62	34.78	5.880 E6	7.873	13.02
Mercury	83.030	0.3021	108.11	57.950	39.96	29.26	54.59	35.02	6.528 E6	5.763	11.86
	88.620	0.2116	107.37	69.864	38.68	31.20	47.95	35.26	7.198 E6	4.058	11.13
	129.72	0.1723	152.07	107.37	31.97	26.86	38.04	35.26	1.275 E7	2.787	10.73
Earth	128.84	0.1609	149.57	108.11	32.08	27.27	37.73	35.02	1.262 E7	2.712	10.71
	127.96	0.1494	147.07	108.85	32.19	27.69	37.42	34.78	1.249 E7	2.635	10.69
	178.25	0.3976	249.12	107.37	27.27	17.90	41.54	35.26	2.053 E7	6.283	12.12
Mars	167.97	0.3564	227.84	108.11	28.09	19.35	40.78	35.02	1.878 E7	5.766	11.86
	157.70	0.3098	206.56	108.85	28.99	21.05	39.94	34.78	1.709 E7	5.161	11.58
	461.59	0.7674	815.80	107.37	16.95	6.148	46.71	35.26	8.557 E7	11.46	15.45
Jupiter	443.12	0.7560	778.14	108.11	17.30	6.447	46.40	35.02	8.048 E7	11.39	15.40
	424.66	0.7437	740.48	108.85	17.67	6.774	46.09	34.78	7.551 E7	11.30	15.34
	805.93	0.8668	1504.5	107.37	12.83	3.426	48.01	35.26	1.974 E8	12.75	16.43
Saturn	767.55	0.8592	1427.0	108.11	13.14	3.617	47.75	35.02	1.835 E8	12.73	16.42
	729.18	0.8507	1349.5	108.85	13.48	3.829	47.48	34.78	1.699 E8	12.70	16.39
	1554.9	0.9309	3002.3	107.37	9.234	1.746	48.83	35.26	5.290 E8	13.57	17.08
Uranus	1489.2	0.9274	2870.3	108.11	9.435	1.831	48.62	35.02	4.958 E8	13.60	17.10
	1423.6	0.9235	2738.3	108.85	9.650	1.924	48.40	34.78	4.634 E8	13.62	17.12
	2322.6	0.9538	4536.8	107.37	7.556	1.162	49.12	35.26	9.655 E8	13.86	17.31
Neptune	2304.0	0.9531	4499.9	108.11	7.586	1.176	48.94	35.02	9.542 E8	13.92	17.36
	2285.9	0.9524	4463.0	108.85	7.616	1.189	48.77	34.78	9.430 E8	13.98	17.41
	3741.2	0.9713	7375.0	107.37	5.953	0.7183	49.34	35.26	1.974 E9	14.08	17.48
Pluto	3008.6	0.9641	5909.0	108.11	6.638	0.8979	49.08	35.02	1.424 E9	14.06	17.47
	2275.9	0.9522	4443.0	108.85	7.632	1.195	48.76	34.78	9.368 E8	13.98	17.40

HOHMANN ELLIPSE TRANSFER DATA (Continued)

Body	Mean orbital Semimajor axis (km × 10^{-6})	Eccentricity ε	Aphelion (km × 10^{-6})	Perihelion (km × 10^{-6})	Orbital Velocity (km/sec) Mean	Orbital Velocity (km/sec) Aphelion	Orbital Velocity (km/sec) Perihelion	Orbital Velocity of Departure Body (km/sec)	Time of one way trip (½ orbit) (Solar Seconds)	Departure Δ_V orbit (km/sec)	Departure Maximum Δ_V (km/sec)
From Earth to:											
	99.052	0.5352	152.07	46.035	36.58	20.13	66.49	29.28	8.506 E6	9.150	14.45
Mercury	103.76	0.4415	149.57	57.950	35.75	22.25	57.43	29.77	9.119 E6	7.523	13.47
	108.47	0.3559	147.07	69.864	34.96	24.10	50.72	30.27	9.747 E6	6.178	12.77
	129.72	0.1723	152.07	107.37	31.97	26.86	38.04	29.28	1.275 E7	2.416	11.44
Venus	128.84	0.1609	149.57	108.11	32.08	27.27	37.73	29.77	1.262 E7	2.500	11.46
	127.96	0.1494	147.07	108.85	32.19	27.69	37.42	30.27	1.249 E7	2.583	11.47
	198.10	0.2576	249.12	147.07	25.87	19.88	33.67	30.27	2.406 E7	3.396	11.68
Mars	188.70	0.2074	227.84	149.57	26.51	21.48	32.71	29.77	2.237 E7	2.942	11.56
	179.31	0.1519	206.56	152.07	27.19	23.33	31.69	29.28	2.072 E7	2.411	11.44
	481.44	0.6945	815.80	147.07	16.59	7.046	39.08	30.27	9.114 E7	8.810	14.23
Jupiter	463.85	0.6776	778.14	149.57	16.91	7.412	38.56	29.77	8.620 E7	8.789	14.22
	446.27	0.6592	740.48	152.07	17.24	7.811	38.03	29.28	8.134 E7	8.755	14.20
	825.78	0.8219	1504.5	147.07	12.67	3.962	40.53	30.27	2.047 E8	10.25	15.17
Saturn	788.28	0.8103	1427.0	149.57	12.97	4.199	40.06	29.77	1.910 E8	10.29	15.19
	750.79	0.7975	1349.5	152.07	13.29	4.461	39.59	29.28	1.775 E8	10.31	15.21
	1574.7	0.9066	3002.3	147.07	9.176	2.031	41.46	30.27	5.392 E8	11.18	15.81
Uranus	1509.9	0.9009	2870.3	149.57	9.370	2.139	41.05	29.77	5.062 E8	11.28	15.88
	1445.2	0.8948	2738.3	152.07	9.578	2.257	40.64	29.28	4.740 E8	11.36	15.94
	2341.9	0.9372	4536.8	147.07	7.524	1.355	41.79	30.27	9.779 E8	11.51	16.05
Neptune	2324.7	0.9357	4499.9	149.57	7.552	1.377	41.42	29.77	9.671 E8	11.65	16.15
	2307.5	0.9341	4463.0	152.07	7.580	1.399	41.06	29.28	9.564 E8	11.78	16.24
	3761.0	0.9609	7375.0	147.07	5.937	0.8384	42.04	30.27	1.990 E9	11.77	16.23
Pluto	3029.3	0.9506	5909.0	149.57	6.615	1.053	41.58	29.77	1.439 E9	11.81	16.26
	2297.5	0.9338	4443.0	152.07	7.596	1.405	41.06	29.28	9.502 E8	11.78	16.24
From Mars to:											
	147.58	0.6881	249.12	46.035	29.97	12.88	69.72	21.97	1.547 E7	9.081	10.38
Mercury	142.89	0.5945	227.84	57.950	30.46	15.36	60.40	24.12	1.474 E7	8.761	10.10
	138.21	0.4945	206.56	69.864	30.97	18.01	53.25	26.49	1.402 E7	8.479	9.858
	178.25	0.3976	249.12	107.37	27.27	17.90	41.54	21.97	2.053 E7	4.060	6.463
Venus	167.97	0.3564	227.84	108.11	28.09	19.35	40.78	24.12	1.878 E7	4.770	6.931
	157.70	0.3098	206.56	108.85	28.99	21.05	39.94	26.49	1.709 E7	5.444	7.411
	198.10	0.2576	249.12	147.07	25.87	19.88	33.67	21.97	2.406 E7	2.088	5.444
Earth	188.70	0.2074	227.84	149.57	26.51	21.48	32.71	24.12	2.237 E7	2.646	5.682
	179.31	0.1519	206.56	152.07	27.19	23.33	31.69	26.49	2.072 E7	3.161	5.939
	511.18	0.5959	815.80	206.56	16.10	8.104	32.00	26.49	9.972 E7	5.514	7.462
Jupiter	502.99	0.5470	778.14	227.84	16.23	8.785	30.00	24.12	9.733 E7	5.881	7.737
	494.80	0.4965	740.48	249.12	16.37	9.494	28.22	21.97	9.496 E7	6.256	8.026
	855.52	0.7586	1504.5	206.56	12.45	4.613	33.60	26.49	2.159 E8	7.105	8.704
Saturn	827.42	0.7246	1427.0	227.84	12.66	5.058	31.68	24.12	2.054 E8	7.556	9.076
	799.32	0.6883	1349.5	249.12	12.38	5.533	29.97	21.97	1.950 E8	8.010	9.457
	1604.4	0.8713	3002.3	206.56	9.090	2.384	34.66	26.49	5.545 E8	8.165	9.589
Uranus	1549.1	0.8529	2870.3	227.84	9.251	2.606	32.84	24.12	5.260 E8	8.713	10.06
	1493.7	0.8332	2738.3	249.12	9.421	2.842	31.23	21.97	4.981 E8	9.269	10.55
	2371.7	0.9129	4536.8	206.56	7.477	1.595	35.04	26.49	9.966 E8	8.548	9.917
Neptune	2363.9	0.9036	4499.9	227.84	7.489	1.685	33.28	24.12	9.916 E8	9.160	10.45
	2356.1	0.8943	4463.0	249.12	7.501	1.772	31.75	21.97	9.867 E8	9.785	11.00
	3790.8	0.9455	7375.0	206.56	5.914	0.9897	35.34	26.49	2.014 E9	8.846	10.17
Pluto	3068.4	0.9257	5909.0	227.84	6.573	1.291	33.47	24.12	1.467 E9	9.352	10.62
	2346.0	0.8938	4443.0	249.12	7.517	1.780	31.75	21.97	9.804 E8	9.781	11.00
From Jupiter to:											
	430.92	0.8932	815.80	46.036	17.54	4.167	73.84	12.44	7.718 E7	8.269	60.80
Mercury	418.04	0.8614	778.14	57.950	17.81	4.860	65.26	13.05	7.375 E7	8.193	60.79
	405.17	0.8276	740.48	69.865	18.09	5.556	58.89	13.70	7.037 E7	8.144	60.79
	461.59	0.7674	815.80	107.37	16.95	6.148	46.71	12.44	8.557 E7	6.287	60.57
Venus	443.12	0.7560	778.14	108.11	17.30	6.447	46.40	13.05	8.048 E7	6.606	60.60
	424.66	0.7437	740.48	108.85	17.67	6.774	46.09	13.70	7.551 E7	6.926	60.64
	481.44	0.6945	815.80	147.07	16.59	7.046	39.08	12.44	9.114 E7	5.390	60.48
Earth	463.85	0.6776	778.14	149.57	16.91	7.412	38.56	13.05	8.620 E7	5.641	60.50
	446.27	0.6592	740.48	152.07	17.24	7.811	38.03	13.70	8.134 E7	5.890	60.53
	511.18	0.5959	815.80	206.56	16.10	8.104	32.00	12.44	9.972 E7	4.332	60.40
Mars	502.99	0.5470	778.14	227.84	16.23	8.785	30.00	13.05	9.733 E7	4.268	60.39
	494.80	0.4965	740.48	249.12	16.37	9.494	28.22	13.70	9.496 E7	4.206	60.39
	1122.5	0.3403	1504.5	740.48	10.87	7.624	15.49	13.70	3.245 E8	1.790	60.27
Saturn	1102.6	0.2942	1427.0	778.14	10.97	8.097	14.85	13.05	3.159 E8	1.797	60.27
	1082.7	0.2465	1349.5	815.80	11.07	8.604	14.23	12.44	3.074 E8	1.797	60.27
	1871.4	0.6043	3007.3	740.48	8.417	4.180	16.95	13.70	6.985 E8	3.248	60.33
Uranus	1824.2	0.5743	2870.3	778.14	8.525	4.439	16.37	13.05	6.723 E8	3.320	60.33
	1777.0	0.5409	2738.3	815.80	8.637	4.715	15.82	12.44	6.463 E8	3.389	60.33
	2638.6	0.7194	4536.8	740.48	7.088	2.864	17.55	13.70	1.169 E9	3.845	60.36
Neptune	2639.0	0.7051	4499.9	778.14	7.088	2.947	17.04	13.05	1.170 E9	3.992	60.37
	2639.4	0.6909	4463.0	815.80	7.087	3.030	16.58	12.44	1.170 E9	4.141	60.38

HOHMANN ELLIPSE TRANSFER DATA (Continued)

Body	Mean orbital Semimajor axis (km × 10^{-6})	Eccentricity ε	Aphelion (km × 10^{-6})	Perhelion (km × 10^{-6})	Orbital Velocity (km/sec) Mean	Aphelion	Perhelion	Orbital Velocity of Departure Body (km/sec)	Time of one way trip (½ orbit) (Solar Seconds)	Departure Δ_V orbit (km/sec)	Departure Maximum Δ_V (km/sec)
	4057.7	0.8175	7375.0	740.48	5.716	1.811	18.04	13.70	2.230 E9	4.339	60.40
Pluto	3343.6	0.7673	5909.0	778.14	6.297	2.285	17.35	13.05	1.668 E9	4.299	60.39
	2629.4	0.6897	4443.0	815.80	7.101	3.043	16.57	12.44	1.163 E9	4.135	60.38
From Saturn to:											
	775.26	0.9406	1504.5	46.036	13.08	2.287	74.76	9.129	1.862 E8	6.841	36.71
Mercury	742.47	0.9220	1427.0	57.950	13.36	2.693	66.31	9.639	1.746 E8	6.946	36.73
	709.69	0.9016	1349.5	69.860	13.67	3.110	60.07	10.18	1.631 E8	7.067	36.75
	805.93	0.8668	1504.5	107.37	12.83	3.426	45.01	9.129	1.974 E8	5.702	36.51
Venus	767.55	0.8592	1427.0	108.11	13.14	3.617	47.75	9.639	1.835 E8	6.021	36.56
	729.18	0.8507	1349.5	108.85	13.48	3.829	47.48	10.18	1.699 E8	6.348	36.62
	825.78	0.8219	1504.5	147.07	12.67	3.962	40.53	9.129	2.047 E8	5.167	36.43
Earth	788.28	0.8103	1427.0	149.57	12.97	4.199	40.06	9.639	1.910 E8	5.440	36.47
	750.79	0.7975	1349.5	152.07	13.29	4.461	39.59	10.18	1.775 E8	5.716	36.51
	855.52	0.7586	1504.5	206.56	12.45	4.613	33.60	9.129	2.159 E8	4.516	36.35
Mars	827.42	0.7246	1427.0	227.84	12.66	5.058	31.68	9.639	2.054 E8	4.581	36.35
	799.32	0.6883	1349.5	249.12	12.88	5.533	29.97	10.18	1.950 E8	4.644	36.36
	1122.5	0.3403	1504.5	740.48	10.87	7.624	15.49	9.129	3.245 E8	1.504	36.10
Jupiter	1102.6	0.2942	1427.0	778.14	10.97	8.097	14.85	9.639	3.159 E8	1.541	36.10
	1082.7	0.2465	1349.5	815.80	11.07	8.604	14.23	10.18	3.074 E8	1.573	36.10
	2175.9	0.3798	3002.3	1349.5	7.806	5.233	11.64	10.18	8.758 E8	1.465	36.09
Uranus	2148.6	0.3359	2870.3	1427.0	7.855	5.539	11.14	9.639	8.593 E8	1.502	36.10
	2121.4	0.2908	2738.3	1504.5	7.905	5.860	10.67	9.129	8.430 E8	1.536	36.10
	2943.2	0.5415	4536.8	1349.5	6.712	3.660	12.31	10.18	1.378 E9	2.129	36.13
Neptune	2963.4	0.5185	4499.9	1427.0	6.689	3.767	11.88	9.639	1.392 E9	2.239	36.13
	2983.7	0.4958	4463.0	1504.5	6.666	3.870	11.48	9.129	1.406 E9	2.352	36.14
	4362.3	0.6906	7375.0	1349.5	5.513	2.358	12.89	10.18	2.486 E9	2.710	36.17
Pluto	3668.0	0.6110	5909.0	1427.0	6.012	2.954	12.23	9.639	1.917 E9	2.595	36.16
	2973.7	0.4941	4443.0	1504.5	6.677	3.885	11.47	9.129	1.399 E9	2.345	36.14
From Uranus to:											
	1524.2	0.9698	3002.3	46.036	9.326	1.155	75.32	6.490	5.134 E8	5.336	22.82
Mercury	1464.1	0.9604	2870.3	57.950	9.516	1.352	66.97	6.796	4.834 E8	5.444	22.85
	1404.1	0.9502	2738.3	69.865	9.717	1.552	60.83	7.116	4.539 E8	5.564	22.88
	1554.9	0.9309	3002.3	107.37	9.234	1.746	48.83	6.490	5.290 E8	4.744	22.69
Venus	1489.2	0.9274	2870.3	108.11	9.435	1.831	48.62	6.796	4.958 E8	4.965	22.74
	1423.6	0.9235	2738.3	108.85	9.650	1.924	48.40	7.116	4.634 E8	5.192	22.69
	1574.7	0.9066	3002.3	147.07	9.176	2.031	41.46	6.490	5.392 E8	4.460	22.63
Earth	1509.9	0.9009	2870.3	149.57	9.370	2.139	41.05	6.796	5.062 E8	4.657	22.67
	1445.2	0.8948	2738.3	152.07	9.578	2.257	40.64	7.116	4.740 E8	4.859	22.72
	1604.4	0.8813	3002.3	206.56	9.090	2.384	34.66	6.490	5.545 E8	4.106	22.57
Mars	1549.1	0.8529	2870.3	227.84	9.251	2.606	32.84	6.796	5.260 E8	4.190	22.58
	1493.7	0.8332	2738.3	249.12	9.421	2.842	31.23	7.116	4.981 E8	4.275	22.60
	1817.4	0.6043	3002.3	740.48	8.417	4.180	16.95	6.490	6.985 E8	2.310	22.31
Jupiter	1824.2	0.5743	2870.3	778.14	8.525	4.439	16.37	6.796	6.723 E8	2.358	22.31
	1777.0	0.5409	2738.3	815.80	8.637	4.715	15.82	7.116	6.463 E8	2.402	22.32
	2175.9	0.3798	3002.3	1349.5	7.806	5.233	11.64	6.490	8.758 E8	1.257	22.23
Saturn	2148.6	0.3359	2870.3	1427.0	7.855	5.539	11.14	6.796	8.593 E8	1.258	22.23
	2121.4	0.2908	2738.3	1504.5	7.905	5.860	10.67	7.116	8.430 E8	1.257	22.23
	3637.5	0.2472	4536.8	2738.3	6.037	4.690	7.771	7.116	1.893 E9	0.6544	22.20
Neptune	3685.1	0.2211	4499.9	2870.3	5.998	4.790	7.510	6.796	1.930 E9	0.7139	22.20
	3732.7	0.1957	4463.0	3002.3	5.960	4.888	7.266	6.490	1.968 E9	0.7757	22.20
	5056.6	0.4585	7375.0	2738.3	5.120	3.120	8.403	7.116	3.102 E9	1.287	22.23
Pluto	4389.6	0.3461	5909.0	2870.3	5.496	3.830	7.885	6.796	2.509 E9	1.089	22.22
	3722.7	0.1935	4443.0	3002.3	5.968	4.906	7.260	6.490	1.960 E9	0.7691	22.20
From Neptune to:											
	2291.4	0.9799	4536.8	46.037	7.606	0.7662	75.51	5.384	9.464 E8	4.617	24.97
Mercury	2278.9	0.9746	4499.9	57.948	7.627	0.8655	67.21	5.428	9.387 E8	4.562	24.96
	2266.4	0.9692	4463.0	69.864	7.648	0.9569	61.13	5.473	9.310 E8	4.516	24.95
	2322.1	0.9538	4536.8	107.37	7.556	1.162	49.12	5.384	9.655 E8	4.221	24.90
Venus	2304.0	0.9531	4499.9	108.11	7.586	1.176	48.94	5.428	9.542 E8	4.252	24.90
	2285.9	0.9524	4463.0	108.85	7.616	1.189	48.77	5.473	9.430 E8	4.283	24.91
	2341.9	0.9372	4536.8	147.07	7.524	1.355	41.79	5.384	9.779 E8	4.029	24.86
Earth	2324.7	0.9357	4499.9	149.57	7.552	1.377	41.42	5.428	9.667 E8	4.051	24.87
	2307.5	0.9341	4463.0	152.07	7.580	1.399	41.06	5.473	9.564 E8	4.073	24.87
	2371.7	0.9129	4536.8	206.56	7.477	1.595	35.04	5.384	9.966 E8	3.788	24.83
Mars	2363.9	0.9036	4499.9	227.84	7.489	1.685	33.28	5.428	9.916 E8	3.743	24.82
	2356.1	0.8943	4463.0	249.12	7.501	1.772	31.75	5.473	9.867 E8	3.700	24.81
	2638.6	0.7194	4536.8	740.48	7.088	2.864	17.55	5.384	1.169 E9	2.520	24.66
Jupiter	2639.0	0.7051	4499.9	778.14	7.088	2.947	17.04	5.428	1.170 E9	2.480	24.66
	2639.4	0.6909	4463.0	815.80	7.088	3.030	16.58	5.473	1.170 E9	2.442	24.66

HOHMANN ELLIPSE TRANSFER DATA (Continued)

Body	Mean orbital Semimajor axis (km × 10^{-6})	Eccentricity ε	Aphelion (km × 10^{-6})	Perhelion (km × 10^{-6})	Orbital Velocity (km/sec)			Orbital Velocity of Departure Body (km/sec)	Time of one way trip (½ orbit) (Solar Seconds)	Departure	
					Mean	Aphelion	Perhelion			Δ_V orbit (km/sec)	Maximum Δ_V (km/sec)
	2943.2	0.5415	4536.8	1349.5	6.712	3.660	12.31	5.384	1.378 E9	1.723	24.60
Saturn	2963.4	0.5185	4499.9	1427.0	6.689	3.767	11.88	5.428	1.392 E9	1.661	24.59
	2983.7	0.4958	4463.0	1504.5	6.666	3.870	11.48	5.473	1.406 E9	1.602	24.59
	3637.5	0.2472	4536.8	2738.3	6.037	4.690	7.771	5.384	1.893 E9	0.6933	24.54
Uranus	3685.1	0.2211	4499.9	2870.3	5.998	4.790	7.510	5.428	1.930 E9	0.6375	24.54
	3732.7	0.1957	4463.0	3002.3	5.960	4.888	7.266	5.473	1.968 E9	0.5845	24.54
	5919.0	0.2460	7375.0	4463.0	4.733	3.682	6.084	5.473	3.929 E9	0.6112	24.54
Pluto	5204.4	0.1354	5909.0	4499.9	5.047	4.404	5.784	5.428	3.240 E9	0.3557	24.54
	4489.9	−0.1045	4443.0	4536.8	5.434	5.491	5.377	5.384	2.596 E9	−0.0061	24.54
From Pluto to:											
	3710.5	0.9876	7375.0	46.037	5.977	0.4723	75.66	3.676	1.950 E9	3.204	5.955
Mercury	2983.5	0.9806	5909.0	57.948	6.666	0.6601	67.31	4.737	1.406 E9	4.077	6.466
	2256.4	0.9690	4443.0	69.864	7.665	0.9612	67.13	6.103	9.248 E8	5.141	7.185
	3741.2	0.9713	7375.0	107.37	5.953	0.7183	49.34	3.676	1.974 E9	2.958	5.826
Venus	3008.6	0.9641	5909.0	108.11	6.638	0.8979	49.08	4.737	1.424 E9	3.839	6.319
	2275.9	0.9522	4443.0	108.85	7.632	1.195	48.76	6.103	9.368 E8	4.908	7.020
	3761.0	0.9609	7375.0	147.07	5.937	0.8384	42.04	3.676	1.990 E9	2.838	5.766
Earth	3029.3	0.9506	5909.0	149.57	6.615	1.053	41.58	4.737	1.439 E9	3.684	6.226
	2297.5	0.9338	4443.0	152.07	7.596	1.405	41.06	6.103	9.502 E8	4.697	6.874
	3790.8	0.9455	7375.0	206.56	5.914	0.9897	35.34	3.676	2.014 E9	2.687	5.693
Mars	3068.4	0.9257	5909.0	227.84	6.573	1.291	33.47	4.737	1.467 E9	3.446	6.088
	2346.0	0.8938	4443.0	249.12	7.517	1.780	31.75	6.103	9.804 E8	4.323	6.624
	4057.7	0.8175	7375.0	740.48	5.716	1.811	18.04	3.676	2.230 E9	1.865	5.355
Jupiter	3343.6	0.7673	5909.0	778.14	6.297	2.285	17.35	4.737	1.668 E9	2.452	5.586
	2629.4	0.6897	4443.0	815.80	7.101	3.043	16.57	6.103	1.163 E9	3.060	5.879
	4362.3	0.6906	7375.0	1349.5	5.513	2.358	12.89	3.676	2.486 E9	1.318	5.190
Saturn	3668.0	0.6110	5909.0	1427.0	6.012	2.954	12.23	4.737	1.917 E9	1.782	5.326
	2973.7	0.4941	4443.0	1504.5	6.677	3.885	11.47	6.103	1.399 E9	2.217	5.487
	5056.6	0.4585	7375.0	2738.3	5.120	3.120	8.403	3.676	3.102 E9	0.5564	5.050
Uranus	4389.6	0.3461	5909.0	2870.3	5.496	3.830	7.885	4.737	2.509 E9	0.9065	5.101
	3722.7	0.1935	4443.0	3002.3	5.968	4.906	7.260	6.103	1.960 E9	1.197	5.160
	5919.0	0.2460	7375.0	4463.0	4.733	3.682	6.084	3.676	3.929 E9	−5.164	5.019
Neptune	5204.4	0.1354	5909.0	4499.9	5.047	4.404	5.784	4.737	3.240 E9	0.3323	5.030
	4489.9	−0.1045	4443.0	4536.8	5.434	5.491	5.377	6.103	2.596 E9	0.6117	5.056

CONSTANTS FOR SATELLITE GEODESY

Defining Constants

1. Number of ephemeris seconds in 1 tropical year (1900) .... s = 31 556 925.9747
2. Gaussian gravitational constant, defining the a. u. .... k = 0.017 202 09895

Primary Constants

3. Velocity of light in meters per second .... c = 299 792.5 × 10³
4. Dynamical form-factor for Earth .... J_2 = 0.001 082 7
5. Sidereal mean motion of Moon in radians per second (1900) .... $n_☾{}^*$ = 2.661 699 489 × 10⁻⁶
6. General precession in longitude per tropical century (1900) .... p = 5025″.64
7. Constant of nutation (1900) .... N = 9″.210

Auxiliary Constants and Factors

$k/86400$, for use when the unit of time is 1 second .... k' = 1.990 983 675 × 10⁻⁷
Number of seconds of arc in 1 radian .... = 206 264.806
Factor for constant of aberration (note 10) .... F_1 = 1.000 142
Factor for mean distance of Moon (note 12) .... F_2 = 0.999 093 142
Factor for parallactic inequality (note 15) .... F_3 = 49853″.2

Derived Constants

8. Solar parallax .... $\arcsin(a_e/A) = \pi_\odot$ = 8″.79405 (8″.794)
9. Light-time for unit distance .... $A/c' = \tau_A$ = 499ˢ.012
   = 1ˢ/0.002 003 96
10. Constant of aberration .... $F_1 k' \tau_A = \kappa$ = 20″.4958 (20″.496)
11. Ratio of masses of Sun and Earth + Moon .... $S/E(1 + \mu)$ = 328 912
12. Perturbed mean distance of Moon, in meters .... $F_2(GE(1 + \mu)/n_☾{}^{*2})^{1/3} = a_☾$ = 384 400 × 10³
13. Constant of sine parallax for Moon .... $a_e/a_☾ = \sin \pi_☾$ = 3422″.451
14. Constant of lunar inequality .... $\frac{\mu}{1 + \mu}\frac{a_☾}{A} = L$ = 6″.43987 (6″.440)
15. Constant of parallactic inequality .... $F_3 \frac{1 - \mu}{1 + \mu}\frac{a_☾}{A} = P_☾$ = 124″.986

THE EARTH: ITS MASS, DIMENSIONS AND OTHER RELATED QUANTITIES

From NASA Technical Translation NASA TT-F-533

TABLE A

Quantity	Unit of Measurement	Symbol	Numerical Value	Sources; Remarks[1]
Mass	Proportion of the mass of the sun	M	1/331950	I. D. Zhongolovich
	gram		$5.9763 \cdot 10^{27}$	
Major Orbital semi-axis	Astronomical unit	a_{orb}	1.000000	1961 data of Soviet radar determinations
	km		149,457,000	
Distance from sun at perihelion	a.u.	r_{π}	0.983298	for 1962
Distance from sun at aphelion	a.u.	r_{α}	1.016744	for 1962
Moment of perihelion passage		T_{π}	Jan. 2, $4^{hr}52^{m}$	for 1962 USSR Astron. Yearbook for 1962
Moment of aphelion passage		T_{α}	Jul 4 $5^{hr}05^{m}$	for 1962
Siderial rotation period around sun	sec	P_{orb}	$31.558 \cdot 10^{6}$	
Mean rotational velocity	km/sec	U_{orb}	29.8	
Mean equatorial radius	km	$\bar{a}$	6,378.245 6,378.077	A. A. Izotov, 1950 I. D. Zhongolovich, 1956
Mean polar compression		α	1/298.3 1/296.6 1/298.2	A. A. Izotov, 1950 I. D. Zhongolovich, 1952 D. G. King-Hele, R. Merson, 1959. On observations on the movements of artificial earth satellites.
Difference in equatorial and polar semi-axes	km	$a - c$	21.382 21.500	A. A. Izotov, 1950 I. D. Zhongolovich, 1956
Compression of meridian of major equatorial axis		α_a	1/295.2	I. D. Zhongolovich, 1952
Compression of meridian of minor equatorial axis		α_b	1/298.0	I. D. Zhongolovich, 1952
Equatorial compression		ϵ	1/30 000 1/32 000	A. A. Izotov, 1950 I. D. Zhongolovich, 1952

[1] The source does not indicate whether the value given is generally accepted or has merely been calculated by the author, G. N. Katterfel'd.

TABLE A Continued

Quantity	Unit of Measurement	Symbol	Numerical Value	Sources; Remarks[1]
Difference in equatorial semi-axes	m	$a - b$	213 199	A. A. Izotov, 1950 I. D. Zhongolovich, 1952
Meridian of longitude of minor equatorial semi-axis		λ_a	15°E - 6°W	A. A. Izotov, 1950 I. D. Zhongolovich, 1952
Meridian of longitude of minor equatorial axis		λ_b	105°E - 75° W; 84°E - 96°W	A. A. Izotov, 1950 I. D. Zhongolovich, 1952
Difference in polar semi-axes	m	$C_N - C_S$	~ 70 <100	I. D. Zhongolovich, 1952 Based on observations of artificial satellites[1]
Polar asymmetry		η	$\sim 1.10^{-5}$	
Mean acceleration of gravity at equator		g_e	978,057.3	I. D. Zhongolovich, 1952
Mean acceleration of gravity at poles	milligals (mgl)	g_ρ	983,225.1	
Difference in acceleration of gravity at pole and at equator		$g_p - g_e$	+5,167.8	
Difference in acceleration of gravity at equator	mgl	$g_a - g_b$	+30.2	I. D. Zhongolovich, 1952
Difference in acceleration of gravity at poles		$g_N - g_S$	+30	I. D. Zhongolovich, 1952
Mean acceleration of gravity for entire surface of terrestrial ellipsoid	mgl	g	979,783.0	I. D. Zhongolovich, 1952
Mean radius	km	R	6,370.949	
Area of surface	km²	S	$510.0501 \cdot 10^6$	
Volume	km³	V	$1{,}083.1579 \cdot 10^9$	
Mean density	gr/cm³	δ	5.5170	I. D. Zhongolovich, 1952
Siderial rotational period	sec	P	86,164.09	
Angular rotational velocity	rad/sec	ω	$7.292116 \cdot 10^{-5}$	
Mean equatorial rotational velocity	km/sec	v	0.465	

TABLE A Continued

Quantity	Unit of Measurement	Symbol	Numerical Value	Sources; Remarks[1]
Ratio of centrifugal force to attractive force at equator		q	$\frac{1}{289}$	
Ratio of centrifugal force to force of gravity at equator		q_c	$0.0034677 = \frac{1}{288}$	I. D. Zhongolovich, 1952
Coefficients characterizing the radial distribution of densities within the earth		κ_1 κ	0.966 0.331	Based on observations of artificial satellites; several larger values were given earlier[1]
Radius of inertia	km; proportion of mean radius	R_i	3,674.735 0.5768	
Geocentric latitude of inertial parallel		ϕ_i	54°47′	
Moment of inertia	$gr \cdot cm^2$	I	$8.070 \cdot 10^{44}$	
Moment of rotation	$gr \cdot cm^2/sec$	L	$5.885 \cdot 10^{40}$	
Relative true secular braking of earth's rotation due to tidal friction		$\frac{\Delta\omega_e}{\omega}$	$-4.2 \cdot 10^{-8}$ per century	
Relative proper secular acceleration of earth's rotation		$\frac{\Delta\omega_i}{\omega}$	$+1.4 \cdot 10^{-8}$ per century	N. N. Pariyskiy, 1955
Relative observed secular braking of earth's rotation		$\frac{\Delta\omega}{\omega}$	$-2.8 \cdot 10^{-8}$ per century	
Mean rotational velocity of terrestrial radius due to abyssal compression	cm/century	$\frac{\Delta R}{\Delta t}$	~5 Assumed invariability of mass (M) 4.5 and distribution of masses (κ)	B. Meyermann, 1928, 1928a N. N. Pariyskiy, 1955
Secular variation in potential gravitational energy of earth accompanying reduction of terrestrial radius by 5 cm and corresponding increase in earth's kinetic energy	erg/century	ΔE	$\sim 17 \cdot 10^{30}$	Assumed uniformity of compression of entire planet. P. N. Kropotkin, 1948 and A. T. Aslanyan, 1955

[1] On the basis of materials published in the **Astron. J., Vol. 64, 1272, 1959.**

TABLE A Continued

Quantity	Unit of Measurement	Symbol	Numerical Value	Sources; Remarks[1]
Probable value of total energy of tectonic deformation of earth	erg/century	E_t	$\sim 1 \cdot 10^{30}$	With allowance for earthquakes, volcanic eruptions and other forms of tectonic activity P. N. Kropotkin, 1948
Secular loss of heat of earth through radiation into space	erg/century cal/century	$\Delta' E_k$	$1 \cdot 10^{30}$ $2.4 \cdot 10^{22}$	P. N. Kropotkin, 1948
Portion of earth's kinetic energy transformed into heat as a result of lunar and solar tides in the hydrosphere	erg/century cal/century	$\Delta'' E_k$	$0.11 \cdot 10^{30}$ $0.26 \cdot 10^{22}$	Heiskanen, 1922 and de Sitter, 1927
Difference in duration of days in March and August	sec	ΔP	0.0025 (March-Aug.)	N. N. Pariyskiy, 1955
Corresponding relative annual variation in earth's rotational velocity		$\frac{\Delta^* \omega}{\omega}$	$2.9 \cdot 10^{-8}$ (Aug.-March)	N. N. Pariyskiy, 1955
Presumed variation in earth's radius between August and March	cm	$\Delta^* R$	−9.2 (Aug.-March)	
Annual variation in level of world ocean	cm	Δh_0	~10 (Sept.-March)	N. N. Pariyskiy, 1955

TABLE B

Quantity	Unit of Measurement	Symbol	Numerical Value	Source
Area of continents	km²; in % of area of surface of earth	S_C	$149 \cdot 10^6$ 29.2	E. Kossina, 1933
Area of world ocean	km²; in % of area of surface of earth	S_o	$361 \cdot 10^6$ 70.8	E. Kossina, 1933
Mean height of continents above sea level	m	h_C	875	E. Kossina, 1933
Mean depth of world ocean	m	h_o	=3794	Morskoy Atlas, Vol. II, 1953
Mean position of earth's surface with respect to sea level	m	h_m	=2430	E. Kossina, 1933
Mean thickness of lithosphere within the limits of the continents	km	$h_{c.l.}$	35	M. Yuing and F. Press, 1955
Mean thickness of lithosphere within the limits of the ocean	km	$h_{o.l.}$	4.7	Kh. Khess, 1955
Mean rate of thickening of continental lithosphere	m/10⁶ yr	$\frac{\Delta h}{\Delta t}$	10 – 40	V. I. Popov, 1955
Mean rate of horizontal extension of continental lithosphere	km/10⁶ yr	$\frac{\Delta l}{\Delta t}$	0.75 – 20	V. I. Popov, 1955
Mass of lithosphere	gr	m_l	$2.367 \cdot 10^{25}$	A. Poldervart, 1955
Amount of water released from the mantle and core in the course of geological time	gr		$3.400 \cdot 10^{24}$	Kalp, 1951
Total reserve of water in the mantle	gr		$2 \cdot 10^{26}$	A. P. Vinogradov, 1959
Present day content of free and bound water in the earth's lithosphere	gr		$2.2 - 2.6 \cdot 10^{24}$ $1.8 - 2.7 \cdot 10^{24}$	Kalp, 1951 A. Poldervart, 1955
Mass of hydrosphere	gr	m_h	$1.664 \cdot 10^{24}$	A. Poldervart, 1955
Amount of oxygen bound in the earth's crust	gr		$1.300 \cdot 10^{24}$	A. Poldervart, 1955
Amount of free oxygen	gr		$1.5 \cdot 10^{21}$	A. Poldervart, 1955

TABLE B Continued

Quantity	Unit of Measurement	Symbol	Numerical Value	Source
Mass of atmosphere	gr	m_a	$5.136 \cdot 10^{21}$	A. Poldervart, 1955
Mass of biosphere	gr	m_b	$1.148 \cdot 10^{19}$	A. Poldervart, 1955
Mass of living matter in the biosphere	gr		$3.6 \cdot 10^{17}$	A. Poldervart, 1955
Density of living matter on dry land	gr/cm^2		0.1	A. Poldervart, 1955
Density of living matter in ocean	gr/cm^3		$15 \cdot 10^{-8}$	A. Poldervart, 1955

ESTIMATED AGE OF EARTH

TABLE C

Object of Study	Age in billions of years	Source
Fossils of the most ancient organisms	2.7	L. Arens, 1955
Most ancient known terrestrial rocks:		
Mica (biotite) found within migmatites on the Kola Peninsula and at the Great Rapids of the Voron'ya River in 1958;	3.6	A. A. Polkanov and E. K. Gerling, 1961
Rock found in South Africa	4	A. L. Heils, 1960[1]
Lithosphere	~ 4	V. I. Baranov, 1958
Earth	~ 5	A P. Vinogradov, 1959 and others

[1] See *Priroda*, No. 11, 1960, p. 113.

Approximate Scale of Geologic Time,[1]
Based on the Data of Soviet Research, 1960.

TABLE D

Era	Period or Epoch	Beginning and end, in millions of years	Approximate Duration, in millions of years		
Cenozoic	Quaternian				
	Contemporary	0 – 10,000 yrs ± 2,000 yrs	8 – 12,000 years		
	Pleistocene	10,000 – 1,000,000 yrs ± 50,000 yrs	1		
	Tertiary				
	Pliocene	1 – 10	9	69	570
	Miocene	10 – 25	15		
	Oligocene	25 – 40	15		
	Eocene	40 – 60	20		
	Paleocene	60 – 70	10		
Mesozoic	Cretaceous	70 – 140	70		
	Jurassic	140 – 185	45		
	Triassic	185 – 225	40		
Paleozoic	Permian	225 – 270	45		
	Carboniferous	270 – 320	50		
	Devonian	320 – 400	80		
	Silurian	400 – 420	20		
	Ordovician	420 – 480	60		
	Cambrian	480 – 570	90		
	Pre-Cambrian IV (Riphean)[2]	570 – 1,200	630		
	Pre-Cambrian III (Proterozoic)[3]	1,200 – 1,900	700		
	Pre-Cambrian II (Archean)	1,900 – 2,700	800		
	Pre-Cambrian I (Catarchean)	2,700 – 3,500	800		
	Pregeological era	3,500 – 5,000	1,500		

[1] Based on the geochronological scale of the Commission for Determining the Absolute Age of Geological Formations, published in *Izv. AN SSSR*, Geological series, No. 10, 1960.

[2] Proterozoic II.

[3] Proterozoic I.

ACCELERATION DUE TO GRAVITY AND LENGTH OF THE SECONDS PENDULUM

For Sea Level at Various Latitudes

Based on the formula of the U. S. Coast and Geodetic Survey. The length of the simple pendulum whose period is two seconds, that is which beats seconds, is computed in each case from the corresponding value of the acceleration.

Latitude	Acceleration due to gravity		Length of seconds pendulum	
°	cm/sec.^{2}	ft./sec.^{2}	cm	in.
0	978.039	32.0878	99.0961	39.0141
5	978.078	32.0891	99.1000	39.0157
10	978.195	32.0929	99.1119	39.0204
15	978.384	32.0991	99.1310	39.0279
20	978.641	32.1076	99.1571	39.0382
25	978.960	32.1180	99.1894	39.0509
30	979.329	32.1302	99.2268	39.0656
31	979.407	32.1327		
32	979.487	32.1353		
33	979.569	32.1380		
34	979.652	32.1407		
35	979.737	32.1435	99.2681	39.0819
36	979.822	32.1463		
37	979.908	32.1491		
38	979.995	32.1520		
39	980.083	32.1549		
40	980.171	32.1578	99.3121	39.0992
41	980.261	32.1607		
42	980.350	32.1636		
43	980.440	32.1666		
44	980.531	32.1696		
45	980.621	32.1725	99.3577	39.1171
46	980.711	32.1755		
47	980.802	32.1785		
48	980.892	32.1814		
49	980.981	32.1844		
50	981.071	32.1873	99.4033	39.1351
51	981.159	32.1902		
52	981.247	32.1931		
53	981.336	32.1960		
54	981.422	32.1988		
55	981.507	32.2016	99.4475	39.1525
56	981.592	32.2044		
57	981.675	32.2071		
58	981.757	32.2098		
59	981.839	32.2125		
60	981.918	32.2151	99.4891	39.1689
65	982.288	32.2272	99.5266	39.1836
70	982.608	32.2377	99.5590	39.1964
75	982.868	32.2463	99.5854	39.2068
80	983.059	32.2525	99.6047	39.2144
85	983.178	32.2564	99.6168	39.2191
90	983.217	32.2577	99.6207	39.2207

Free Air Correction for Altitude

−0.0003086 cm/sec.^{2}/m for altitude in meters.
−0.000003086 ft./sec.^{2}/ft. for altitude in feet.

Altitude meters	Correction cm/sec.^{2}	Altitude feet	Correction ft./sec.^{2}
200	−0.0617	200	−0.000617
300	0.0926	300	0.000926
400	0.1234	400	0.001234
500	0.1543	500	0.001543
600	0.1852	600	0.001852
700	0.2160	700	0.002160
800	0.2469	800	0.002469
900	0.2777	900	0.002777

DATA IN REGARD TO THE EARTH

Quadrant of the equator, 10,019,150 meters, 6,225.60 miles.
Quadrant of the meridian, 10,002,290 meters, 6,215.12 miles.
1° latitude at the equator = 69.41 miles.
1° latitude at the pole = 68.70 miles.
Mean surface density of the continents, 2.67 g/cm^3, 166.7 lb./ft.^{3}
Land area, 148.847 × 10^6 km^2, 57.470 × 10^6 sq. mi.
Ocean area, 361.254 × 10^6 km^2, 139.480 × 10^6 sq. mi.

DATA IN REGARD TO THE EARTH (Continued)

Highest mountain, Everest, 8,840 meters, 29,003 ft.
Greatest sea depth, 10,430 meters, 34,219 ft.
Thermal gradient of the earth, higher at increasing depths, 30°C per km, 48°C per mi. (uncertain).

THE COMMONER CHEMICAL ELEMENTS IN THE EARTH'S CRUST

Reprinted from "Principles of Geochemistry" (1952) with the permission of Brian Mason, author, and John Wiley and Sons, publishers.
The "Earth's Crust" refers to the rocks only and does not include atmosphere or the oceans. The atom percent column is obtained by dividing the weight percent by the atomic weights and reducing to 100%. The radius is the ionic radius. The volume percent is the atomic percent multiplied by $\frac{4}{3}\pi r^3$ and reducing to 100%.

Element	Weight %	Atom %	Ion Radius (Å)	Volume %
O	46.60	62.55	1.32	91.97
Si	27.72	21.22	0.39	0.80
Al	8.13	6.47	0.57	0.77
Fe	5.00	1.92	0.82	0.68
Mg	2.09	1.84	0.78	0.56
Ca	3.63	1.94	1.06	1.48
Na	2.83	2.64	0.98	1.60
K	2.59	1.42	1.33	2.14

THE AVERAGE AMOUNTS OF THE ELEMENTS IN EARTH'S CRUST IN GRAMS PER METRIC TON OR PARTS PER MILLION

Reprinted from "Principles of Geochemistry" (1952) with the permission of Brian Mason, author, and John Wiley and Sons, publishers.

O	466,000	N	46	Br	1.6
Si	277,200	Ce	46	Ho	1.2
Al	81,300	Sn	40	Eu	1.1
Fe	50,000	Y	28	Sb	1?
Ca	36,300	Nd	24	Tb	0.9
Na	28,300	Nb	24	Lu	0.8
K	25,900	Co	23	Tl	0.6
Mg	20,900	La	18	Hg	0.5
Ti	4,400	Pb	16	I	0.3
H	1,400	Ga	15	Bi	0.2
P	1,180	Mo	15	Tm	0.2
Mn	1,000	Th	12	Cd	0.15
S	520	Cs	7	Ag	0.1
C	320	Ge	7	In	0.1
Cl	314	Sm	6.5	Se	0.09
Rb	310	Gd	6.4	A	0.04
F	300	Be	6	Pd	0.01
Sr	300	Pr	5.5	Pt	0.005
Ba	250	Sc	5	Au	0.005
Zr	220	As	5	He	0.003
Cr	200	Hf	4.5	Te	0.002?
V	150	Dy	4.5	Rh	0.001
Zn	132	U	4	Re	0.001
Ni	80	B	3	Ir	0.001
Cu	70	Yb	2.7	Os	0.001?
W	69	Er	2.5	Ru	0.001?
Li	65	Ta	2.1		

CHEMICAL COMPOSITION OF ROCKS

Reprinted from "Sedimentary Rocks" (1948) with the permission of F. J. Pettijohn, author, and Harper Brothers, publishers.

Element	Average igneous rock	Average shale	Average sandstone	Average limestone	Average sediment
SiO_2	59.14	58.10	78.33	5.19	57.95
TiO_2	1.05	0.65	0.25	0.06	0.57
Al_2O_3	15.34	15.40	4.77	0.81	13.39
Fe_2O_3	3.08	4.02	1.07	0.54	3.47
FeO	3.80	2.45	0.30		2.08
MgO	3.49	2.44	1.16	7.89	2.65
CaO	5.08	3.11	5.50	42.57	5.89
Na_2O	3.84	1.30	0.45	0.05	1.13
K_2O	3.13	3.24	1.31	0.33	2.86
H_2O	1.15	5.00	1.63	0.77	3.23
P_2O_5	0.30	0.17	0.08	0.04	0.13
CO_2	0.10	2.63	5.03	41.54	5.38
SO_3		0.64	0.07	0.05	0.54
BaO	0.06	0.05	0.05		
C		0.80			0.66
	99.56	100.00	100.00	99.84	99.93

SOLAR SPECTRAL IRRADIANCE

From NASA Technical Report R-351, "The Solar Constant and the Solar Spectrum Measured from a Research Aircraft". Report edited by Matthew P. Thekaekara, Goddard Space Flight Center, Greenbelt, Maryland 20771. Discussion of previously reported values and of measurements and calculations leading to the following table are contained in the NASA Technical Report R-351.

SOLAR SPECTRAL IRRADIANCE—PROPOSED STANDARD CURVE

λ —Wavelength in microns
P_λ —Solar spectral irradiance averaged over small bandwidth centered at λ, in Watts $cm^{-2}\mu^{-1}$
A_λ—Area under the solar spectral irradiance curve in the wavelength range 0 to λ, mW cm^{-2}
D_λ—Percentage of the solar constant associated with wavelengths shorter than λ
Solar Constant—135.30 mW cm^{-2}, or 1.940 cal min^{-1} cm^{-2}

λ	P_λ	A_λ	D_λ	λ	P_λ	A_λ	D_λ	λ	P_λ	A_λ	D_λ
0.120	0.000010	0.00059992	0.00044	0.475	0.2044	25.6001	18.921	2.4	0.0064	129.695	95.858
0.140	0.000003	0.00072999	0.00053	0.480	0.2074	26.6296	19.681	2.5	0.0054	130.285	96.294
0.150	0.000007	0.00077999	0.00057	0.485	0.1976	27.6421	20.430	2.6	0.0048	130.795	96.671
0.160	0.000023	0.00092999	0.00068	0.490	0.1950	28.6236	21.155	2.7	0.0043	131.250	97.007
0.170	0.000063	0.00135999	0.00100	0.495	0.1960	29.6011	21.878	2.8	0.00390	131.660	97.3103
0.180	0.000125	0.00229999	0.00169	0.500	0.1942	30.5766	22.599	2.9	0.00350	132.030	97.5838
0.190	0.000271	0.00427999	0.00316	0.505	0.1920	31.5421	23.312	3.0	0.00310	132.360	97.8277
0.200	0.00107	0.010984	0.0081	0.510	0.1882	32.4926	24.015	3.1	0.00260	132.645	98.0383
0.210	0.00229	0.027784	0.0205	0.515	0.1833	33.4214	24.701	3.2	0.00226	132.888	98.2179
0.220	0.00575	0.067984	0.0502	0.520	0.1833	34.3379	25.379	3.3	0.00192	133.097	98.3724
0.225	0.00649	0.098584	0.0728	0.525	0.1852	35.2591	26.059	3.4	0.00166	133.276	98.5047
0.230	0.00667	0.131484	0.0971	0.530	0.1842	36.1826	26.742	3.5	0.00146	133.432	98.6200
0.235	0.00593	0.162984	0.1204	0.535	0.1818	37.0976	27.418	3.6	0.00135	133.573	98.7238
0.240	0.00630	0.193559	0.1430	0.540	0.1783	37.9979	28.084	3.7	0.00123	133.702	98.8192
0.245	0.00723	0.227384	0.1680	0.545	0.1754	38.8821	28.737	3.8	0.00111	133.819	98.9056
0.250	0.00704	0.263059	0.1944	0.550	0.1725	39.7519	29.380	3.9	0.00103	133.926	98.9847
0.255	0.0104	0.306659	0.226	0.555	0.1720	40.6131	30.017	4.0	0.00095	134.025	99.0579
0.260	0.0130	0.365159	0.269	0.560	0.1695	41.4669	30.648	4.1	0.00087	134.116	99.1252
0.265	0.0185	0.443909	0.328	0.565	0.1705	42.3169	31.276	4.2	0.00078	134.198	99.1861
0.270	0.0232	0.548159	0.405	0.570	0.1712	43.1711	31.907	4.3	0.00071	134.273	99.2412
0.275	0.0204	0.657159	0.485	0.575	0.1719	44.0289	32.541	4.4	0.00065	134.341	99.2915
0.280	0.0222	0.763659	0.564	0.580	0.1715	44.8874	33.176	4.5	0.00059	134.403	99.3373
0.285	0.0315	0.897909	0.663	0.585	0.1712	45.7441	33.809	4.6	0.00053	134.459	99.3787
0.290	0.0482	1.09715	0.810	0.590	0.1700	46.5971	34.439	4.7	0.00048	134.509	99.4160
0.295	0.0584	1.36365	1.007	0.595	0.1682	47.4426	35.064	4.8	0.00045	134.556	99.4504
0.300	0.0514	1.63815	1.210	0.600	0.1666	48.2796	35.683	4.9	0.00041	134.599	99.482195
0.305	0.0603	1.91740	1.417	0.605	0.1647	49.1079	36.295	5.0	0.0003830	134.63905	99.511500
0.310	0.0689	2.24040	1.655	0.610	0.1635	49.9284	36.902	6.0	0.0001750	134.91805	99.717708
0.315	0.0764	2.60365	1.924	0.620	0.1602	51.5469	38.098	7.0	0.0000990	135.05505	99.818965
0.320	0.0830	3.00215	2.218	0.630	0.1570	53.1329	39.270	8.0	0.0000600	135.13455	99.877723
0.325	0.0975	3.45340	2.552	0.640	0.1544	54.6899	40.421	9.0	0.0000380	135.18355	99.913939
0.330	0.1059	3.96190	2.928	0.650	0.1511	56.2174	41.550	10.0	0.0000250	135.21505	99.937220
0.335	0.1081	4.49690	3.323	0.660	0.1486	57.7159	42.657	11.0	0.0000170	135.23605	99.952742
0.340	0.1074	5.03565	3.721	0.670	0.1456	59.1869	43.744	12.0	0.0000120	135.25055	99.963458
0.345	0.1069	5.57140	4.117	0.680	0.1427	60.6284	44.810	13.0	0.0000087	135.26090	99.971108
0.350	0.1093	6.11190	4.517	0.690	0.1402	62.0429	45.855	14.0	0.0000055	135.26800	99.976356
0.355	0.1083	6.65590	4.919	0.700	0.1369	63.4284	46.879	15.0	0.0000049	135.27320	99.980199
0.360	0.1068	7.19365	5.316	0.710	0.1344	64.7849	47.882	16.0	0.0000038	135.27755	99.983414
0.365	0.1132	7.74365	5.723	0.720	0.1314	66.1139	48.864	17.0	0.0000031	135.28100	99.985964
0.370	0.1181	8.32190	6.150	0.730	0.1290	67.4159	49.826	18.0	0.0000024	135.28375	99.987997
0.375	0.1157	8.90640	6.582	0.740	0.1260	68.6909	50.769	19.0	0.0000020	135.28595	99.989623
0.380	0.1120	9.47565	7.003	0.750	0.1235	69.9384	51.691	20.0	0.0000016	135.28775	99.990953
0.385	0.1098	10.0301	7.413	0.800	0.1107	75.7934	56.018	25.0	0.000000610	135.29328	99.995036
0.390	0.1098	10.5791	7.819	0.850	0.0988	81.0309	59.889	30.0	0.000000300	135.29555	99.996718
0.395	0.1189	11.1509	8.241	0.900	0.0889	85.7234	63.358	35.0	0.000000160	135.29670	99.997568
0.400	0.1429	11.8054	8.725	0.950	0.0835	90.0334	66.543	40.0	0.000000094	135.29734	99.998037
0.405	0.1644	12.5736	9.293	1.000	0.0746	93.9859	69.464	50.0	0.000000038	135.29800	99.998525
0.410	0.1751	13.4224	9.920	1.100	0.0592	100.675	74.409	60.0	0.000000019	135.29828	99.998736
0.415	0.1774	14.3036	10.571	1.200	0.0484	106.055	78.385	80.0	0.000000007	135.29854	99.998928
0.420	0.1747	15.1839	11.222	1.300	0.0396	110.455	81.637	100.0	0,000000003	135.29864	99.999002
0.425	0.1693	16.0439	11.858	1.400	0.0336	114.115	84.342	1000.0	0	135.30000	100.000000
0.430	0.1639	16.8769	12.473	1.500	0.0287	117.230	86.645				
0.435	0.1663	17.7024	13.083	1.600	0.0244	119.885	88.607				
0.440	0.1810	18.5706	13.725	1.700	0.0202	122.115	90.255				
0.445	0.1922	19.5036	14.415	1.800	0.0159	123.920	91.589				
0.450	0.2006	20.4856	15.140	1.900	0.0126	125.345	92.642				
0.455	0.2057	21.5014	15.891	2.000	0.0103	126.490	93.489				
0.460	0.2066	22.5321	16.653	2.100	0.0090	127.455	94.202				
0.465	0.2048	23.5606	17.413	2.200	0.0079	128.300	94.826				
0.470	0.2033	24.5809	18.167	2.300	0.0068	129.035	95.370				

TOTAL MONTHLY SOLAR RADIATION IN A CLOUDLESS SKY

Total radiation is the sum of the direct and scattered radiation that strikes the earth's surface. It is influenced by the degree of cloudiness, atmospheric transparency, duration of sunshine, and height of elevation at which the measurements are taken. Deviations from values in this table are bound to occur; however, it is believed that the deviations do not exceed ±10% for the summer months and ±15% for the winter months. The radiation units in this table are kcal/cm².

$\varphi°$	January	February	March	April	May	June	July	August	September	October	November	December
90 N	0	0	0.1	10.0	21.9	26.0	23.8	12.9	2.4	0	0	0
85	0	0	0.7	10.2	21.8	25.8	23.4	13.1	3.0	0	0	0
80	0	0	2.4	10.8	21.4	25.2	23.0	13.4	4.3	0.5	0	0
75	0	0.5	4.0	11.7	21.0	24.5	22.2	13.8	5.8	1.3	0	0
70	0	1.6	6.0	13.1	20.5	23.6	21.2	14.6	7.5	2.7	0.5	0
65	0.7	2.8	8.0	14.5	20.1	22.8	21.0	15.6	9.5	4.3	1.4	0.2
60	1.8	4.3	9.9	16.0	20.8	22.9	21.4	16.7	11.3	6.1	2.6	1.1
55	3.1	6.2	11.7	17.3	21.4	23.4	21.9	17.9	12.9	7.8	4.0	2.3
50	4.8	8.2	13.3	18.5	22.2	23.7	22.6	19.1	14.4	9.7	5.8	3.9
45	6.7	10.3	14.8	19.5	22.6	23.9	23.2	20.1	15.8	11.5	7.8	5.9
40	8.8	12.2	16.4	20.3	23.0	24.0	23.4	20.9	17.0	13.2	9.7	7.7
35	10.7	14.0	17.6	21.0	23.0	24.0	23.6	21.6	18.1	14.7	11.4	9.7
30	12.5	15.5	18.6	21.4	23.0	23.8	23.4	21.8	19.1	16.1	13.1	11.5
25	14.1	16.8	19.5	21.6	23.0	23.4	23.1	21.8	19.8	17.4	14.6	13.1
20	15.5	17.9	20.2	21.6	22.5	22.8	22.6	21.6	20.4	18.5	16.1	14.7
15	16.9	19.0	20.8	21.4	21.9	22.0	21.9	21.2	20.9	19.3	17.4	16.1
10	18.1	19.8	21.1	21.2	21.2	21.0	21.1	20.6	21.2	20.1	18.5	17.5
5	19.3	20.4	21.4	21.0	20.2	19.9	20.0	20.0	21.3	20.6	19.5	18.8
0	20.2	20.9	21.5	20.4	19.3	18.8	19.1	19.3	21.2	21.2	20.4	19.2
5S	20.1	21.4	21.4	19.9	18.3	17.6	17.9	18.3	20.8	21.4	21.8	21.0
10	22.0	21.8	21.1	19.2	17.7	16.3	16.3	17.3	20.4	21.4	21.8	22.0
15	22.6	22.0	20.6	18.3	16.0	14.9	15.4	16.1	19.7	21.3	22.4	22.9
20	23.2	22.0	20.0	17.2	14.7	13.4	13.8	14.9	18.9	21.0	22.6	23.6
25	23.6	22.0	19.4	16.0	13.0	12.0	12.6	13.6	18.0	20.6	23.0	24.1
30	23.9	21.8	18.6	14.9	11.9	10.6	11.1	12.1	17.0	20.1	23.0	24.6
35	24.0	21.3	17.6	13.6	10.4	9.0	9.6	10.6	15.9	19.5	23.0	25.0
40	24.0	20.6	16.4	12.2	8.7	7.3	8.1	9.0	14.3	18.7	22.8	25.2
45	24.0	19.9	15.2	10.7	7.1	5.5	6.3	7.3	13.4	17.7	22.4	25.2
50	23.6	18.9	13.8	9.2	5.4	3.8	4.6	5.5	12.0	16.6	21.8	25.0
55	23.2	17.8	12.3	7.5	3.8	2.3	3.0	3.8	10.3	15.4	21.2	24.6

TOTAL MONTHLY SOLAR RADIATION IN A CLOUDLESS SKY (Continued)

φ°	January	February	March	April	May	June	July	August	September	October	November	December
60	22.6	16.6	10.8	5.6	2.4	1.0	1.6	2.3	8.5	14.1	21.0	24.4
65	22.4	15.3	9.1	3.9	1.1	0.1	0.4	1.0	6.7	12.6	20.8	24.5
70	22.6	14.2	7.3	2.2	0.1	0	0	0	5.0	11.4	21.0	24.9
75	23.2	13.4	5.7	0.9	0	0	0	0	3.5	10.4	21.2	25.4
80	24.0	12.8	4.3	0	0	0	0	0	2.1	9.7	21.9	26.0
85	24.6	12.4	2.9	0	0	0	0	0	0.9	9.2	22.4	26.6
90	24.9	12.3	1.7	0	0	0	0	0	0	9.0	22.6	27.0

ELEMENTS PRESENT IN SOLUTION IN SEA WATER EXCLUDING DISSOLVED GASES

Reprinted by permission of the publishers from "The Oceans" by Sverdrup, Johnson, and Fleming. Copyright, 1942, by Prentice-Hall, Inc.

Element	Concentration (grams/metric ton) or parts per million
Cl	18,980
Na	10,561
Mg	1,272
S	884
Ca	400
K	380
Br	65
C (inorganic)	28
Sr	13
(SiO_2)	0.01–7.0
B	4.6
Si	0.02–4.0
C (organic)	1.2–3.0
Al	0.16–1.9
F	1.4
N (as nitrate)	0.001–0.7
N (as organic nitrogen)	0.03–0.2
Rb	0.2
Li	0.1
P (as phosphate)	>0.001–0.10
Ba	0.05
I	0.05
N (as nitrite)	0.0001–0.05
N (as ammonia)	>0.005–0.05
As (as arsenite)	0.003–0.024
Fe	0.002–0.02
P (as organic phosphorus)	0–0.016
Zn	0.005–0.014
Cu	0.001–0.09
Mn	0.001–0.01
Pb	0.004–0.005
Se	0.004
Sn	0.003
Cs	0.002 (approximate)
U	0.00015–0.0016
Mo	0.0003–0.002
Ga	0.0005
Ni	0.0001–0.0005
Th	<0.0005
Ce	0.0004
V	0.0003
La	0.0003
Y	0.0003
Hg	0.00003
Ag	0.00015–0.0003
Bi	0.0002
Co	0.0001
Sc	0.00004
Au	0.000004–0.000008
Fe (in true solution)	$<10^{-9}$
Ra	2.10^{-11}–3.10^{-10}
Ge	Present
Ti	Present
W	Present
Cd	Present in marine organisms
Cr	Present in marine organisms
Tl	Present in marine organisms
Sb	Present in marine organisms
Zr	Present in marine organisms
Pt	Present in marine organisms

THE pH OF NATURAL MEDIA AND ITS RELATION TO THE PRECIPITATION OF HYDROXIDES

Reprinted from "Principles of Geochemistry" (1952) with the permission of Brian Mason, author, and John Wiley and Sons, publishers.

pH	Precipitation of hydroxides	Natural media	pH
11 }	Magnesium		11
10 }		Alkali soils	10
9 }	Bivalent manganese		{ 9
8 }		Seawater	{ 8
7	Bivalent iron	River water	7
6	Zinc copper	Rain water	6
5	Aluminum		5
4		Peat water	4
3 }	Trivalent iron	Mine waters	3
2 }		Acid thermal springs	{ 2
1			{ 1

PROPERTIES OF THE EARTH'S ATMOSPHERE AT ELEVATIONS UP TO 160 KILOMETERS

The average atmosphere up to 160 km based on pressure and density data obtained on rocket flights above White Sands, New Mexico.
Havens, Koll, and LaGow, Journal of Geophysical Research, March, 1952.

Altitude km above sea level	Pressure mm Hg	Density gm/meters^3	Temperatures °K (N_2, O_2) M = 29	Temperatures °K (N_2, O) M = 24	Velocity of Sound m. sec	Mean Free Path cm (N_2)
0	760	1220	290		345	6.5×10^{-6}
10	210	425	230		310	1.9×10^{-5}
20	42	92	210		295	8.6×10^{-5}
30	9.5	19	235		315	4.2×10^{-4}
40	2.4	4.3	260		325	1.8×10^{-3}
50	7.5×10^{-1}	1.3	270		330	6.1×10^{-3}
60	2.1×10^{-1}	3.8×10^{-1}	260		325	2.1×10^{-2}
70	5.4×10^{-2}	1.2×10^{-1}	210		295	6.6×10^{-1}
80	1.0×10^{-2}	2.5×10^{-2}	190		280	3.2×10^{-1}
90	1.9×10^{-3}	4.0×10^{-3}	210		295	2.0
100	4.2×10^{-4}	8.0×10^{-4}	240		315	10.0
110	1.2×10^{-4}	2.0×10^{-4}	270	220	330	40.0
120	3.5×10^{-5}	5.0×10^{-5}	330	270	370	1.5×10^{2}
130	1.5×10^{-5}	2.0×10^{-5}	390	320	400	4.0×10^{2}
140	7×10^{-6}	7.0×10^{-6}	450	370	430	1.0×10^{3}
150	3×10^{-6}	3.0×10^{-6}	510	420	460	2.5×10^{3}
160	2×10^{-6}	1.5×10^{-6}	570	470	480	5.0×10^{3}

VELOCITY OF SEISMIC WAVES

Depth km	Longitudinal or condensational km/sec.	Transverse or distortional km/sec.
0–20	5.4 –5.6	3.2
20–45	6.25–6.75	3.5
1300	12.5	6.9
2400	13.5	7.5

ATMOSPHERIC AND METEOROLOGICAL DATA

Total mass of the atmosphere, estimated by Ekholm, 5.2×10^{21} g, 11.4×10^{18} pounds, 5.70×10^{16} tons.
Evidence of extent: twilight, 63 km, 39 mi.: meteors, 200 km, 124 mi.: aurora 44–360 km, 27–224 mi.

*Distance to Earth.

STANDARD ATMOSPHERE

U. S. Extension to International Civil Aviation Organization Standard Atmosphere, 1958

The atmosphere was classified in 1958 into three altitude regions designated as

a. Standard 0 to 32 standard geopotential kilometers
b. Tentative 32 to 75 " " "
c. Speculative 75 to 300 " " "

Properties of the atmosphere were calculated as functions of geometric altitude as well as geopotential, the potential being established for the latitude where the acceleration of gravity has a sea-level value of 9.80665 meters per second per second. Symbols and abbreviations used in these tables are as follows:

H – Altitude in geopotential measure
L_M = Molecular-scale-temperature gradient
M = Mean molecular weight of air
m = Meter
m′ = Standard geopotential meter
P = Pressure
ρ = Mass density
T = Temperature in absolute thermodynamic scales
T_M = Molecular-scale temperature in absolute thermodynamic scales
Z = Altitude in geometric measure

Relationship between Geopotential and Geometric Altitude

The concept originally introduced by Bjerknes expresses vertical displacement in units of geopotential. Geopotential at an altitude Z is the potential energy of a unit mass at that altitude relative to the potential energy of that same unit mass at sea level. *Geopotential* H of a point at altitude Z may be rigorously defined as the increase in potential energy of a unit mass lifted from mean sea level to Z against the local force of gravity. Mathematically this definition becomes,

$$GH = \frac{\Delta E}{m} = \int_o^Z g(Z)dZ$$

where

ΔE = increase in potential energy in joules,
m = mass of the body in kilograms,
$g(Z)$ = acceleration of gravity in m sec^{-2} expressed as a function of Z,
H = geopotential of a point at altitude Z,
G = proportionality factor depending on the units of H.

Solving this equation for H in terms of Z yields:

$$H = \frac{1}{G}\int_o^Z g(Z)dZ.$$

The differential form of this relationship to be used later is

$$GdH = g(Z)dZ.$$

The basic unit of geopotential in these tables is one *standard geopotential* meter, m′, which is defined as 9.80665 $m^2 sec^{-2}$ and which is equal to the vertical distance through which a one-kilogram mass must be lifted against the local force of gravity to increase its potential energy by 9.80665 joules. If the acceleration of gravity were constant at 9.80665 m sec^{-2} over an altitude interval of one geometric meter, one *standard geopotential* meter would be exactly equal to one geometric meter; this condition is true within two parts in a million at 45° 32′ 33″ at sea level, where g is equal to 9.80665 m sec^{-2}.

Relationship between Temperature and Molecular-scale Temperature

The molecular-scale temperature introduced by Minzner and Ripley is the defining atmospheric property of this Extension. This property is a composite of temperature and molecular weight, and is defined by the equation:

$$T_M = \left(\frac{T}{M}\right) M_o$$

where

T = temperature in the absolute thermodynamic scales,
T_M = molecular-scale temperature in the absolute thermodynamic scales,
M = molecular weight (nondimensional),
M_o = sea-level value of molecular weight.

No direct measurements of temperature have been made at altitudes above those which are reached by balloons; instead, the temperature is derived from values of the velocity of sound, or by substitution of measured pressures or densities into the barometric equation. The molecular-scale temperature can be derived in this way without specifying the molecular weight, whereas the temperature can be derived only if the molecular weight is known. Since the molecular weight is not well known at altitudes above 90 km, the molecular-scale temperature is more precisely known than temperature. Thus, the introduction of molecular-scale temperature increases the validity of some of the tabulated properties while simultaneously decreasing the complexity of the mathematics relating the basic atmospheric properties. The use of molecular-scale temperature also avoids the necessity for changing the defined atmosphere each time new values for the inadequately known molecular-weight distribution may be adopted.

SEA LEVEL ATMOSPHERIC COMPOSITION FOR A DRY ATMOSPHERE[a]

Constituent gas	Molecular fraction, %	Molecular weight ($O = 15.999_4$)
Nitrogen (N_2)	78.09	14.0067
Oxygen (O_2)	20.95	31.998_8
Argon (Ar)	0.93	39.94_8
Carbon dioxide (CO_2)	0.03	44.01
Neon (Ne)	1.8×10^{-3}	20.117_9
Helium (He)	5.24×10^{-4}	4.00260
Krypton (Kr)	1.0×10^{-4}	83.80
Hydrogen (H_2)	5.0×10^{-5}	2.016
Xenon (Xe)	8.0×10^{-6}	131.30
Ozone (O_3)	1.0×10^{-6}	47.999_4
Radon (Rn)	6.0×10^{-18}	222

[a]These values are taken as standard and do not necessarily indicate the exact condition of the atmosphere. Ozone and radon particularly are known to vary at sea level and above, but these variations would not appreciably affect the value of M_O.

SEA LEVEL VALUES OF ICAO ATMOSPHERE

Property	Metric units
Collision frequency	6.9204049×10^{9} sec^{-1}
Conductivity, thermal	2.5339053×10^{-2} J m^{-1} sec^{-1} (°K)$^{-1}$
Conductivity, thermal	6.0532182×10^{-6} kcal m^{-1} sec^{-1} (°K)$^{-1}$
Conductivity, thermal	2.5838643×10^{-3} kg*f*[a] sec^{-1} (°K)$^{-1}$
Density, mass	1.2250140 kg m^{-3}
Density, mass	0.12491666 kg*f* sec^{2} m^{-4}
Gravitational acceleration	9.80665 m sec^{-2}
Kinematic viscosity	1.4607413×10^{-5} m^{2} sec^{-1}
Mean free path	6.6317223×10^{-8} m
Molar volume	23.645444 m^{3} (kg-mol)$^{-1}$
Molar volume	231.88259 m^{3} [(kg*f* sec^{2} m^{-1})-mol]
Molecular weight	28.966 (dimensionless)
Number density	2.5475521×10^{25} m^{-3}
Particle speed	458.94204 m sec^{-1}
Pressure	0.760 m Hg
Pressure	1,013.2500 mbar
Pressure	101,325.00 nt m^{-2}
Pressure	10,332.275 kg*f* m^{-2}
Scale height	8,434.4134 m
Sound speed	340,29205 m sec^{-1}
Specific weight	12.013284 kg m^{-2} sec^{-2}
Specific weight	1.2250140 kg*f* m^{-3}
Temperature	15.0°C
Temperature, absolute	288.16 K
Temperature, molecular scale	288.16 K
Viscosity, coefficient of	1.7894285×10^{-5} kg m^{-1} sec^{-1}
Viscosity, coefficient of	1.8247093×10^{-6} kg*f* sec m^{-2}

[a]kg*f* = kilogram (force)

STANDARD ATMOSPHERE

Nature of Ions in Unpolluted Air

Initially, all molecules are ionized with equal probability according to their abundance. After a sequence of charge transfer, molecular switching reactions, etc., these ions transform within less than 1 μs into the "fast ions" with the following "core composition:"

Dominant positive ions:	$(NH_4)^+ \cdot (NH_3)_1 \cdot (H_2O)_n$	$0 < H \lesssim 25$ km
	$(H_3O)^+ \cdot (H_2O)_n$	$25 \lesssim H < 90$ km

the n, 1 depending on atmospheric temperature, water vapor concentration, and ammonia concentration;

Dominant negative ions: $O_2^- \cdot (H_2O)_m$ and $CO_4^- \cdot (H_2O)_m$ $H < 25$ km

the m depending on temperature and the ratio of water vapor to carbon dioxide concentration. Collisional interaction of these ions with atmospheric trace gases (with permanent dipoles or high polarizability) will lead to larger cluster ions through attachment and detachment reactions around the "core ion."

Compiled by V. A. Mohnen.

Abbreviated Metric Table of the U. S. Extension to the ICAO Standard Atmosphere

H m′	Z m	L_M °K m′$^{-1}$	T_M °K	T °K	M	P mb	ρ kg m^{-3}
−5,000	−4,996.070		320.66	320.66	28.966	1.7776×10^{3}	1.9312
		−0.0065					
0	0		288.16	288.16	28.966	1.01325×10^{3}	1.2250
		−0.0065					
11,000	11,019.067		216.66	216.66	28.966	2.2632×10^{2}	3.6391×10^{-1}
		zero					
20,000	20,063.124		216.66	216.66	28.966	5.4748×10^{1}	8.8034×10^{-2}
		zero					
25,000	25,098.710		216.66	216.66	28.966	2.4886×10^{1}	4.0016×10^{-2}
		+0.0030					
32,000	32,161.906		237.66	237.66	28.966	8.6776×10^{0}	1.2721×10^{-2}
		+0.0030					
47,000	47,350.101		282.66	282.66	28.966	1.2044×10^{0}	1.4845×10^{-3}
		zero					
53,000	53,445.620		282.66	282.66	28.966	5.8320×10^{-1}	7.1881×10^{-4}
		−0.0039					
75,000	75,895.488		196.86	196.86	28.966	2.452×10^{-2}	4.339×10^{-5}
		zero					
90,000	91,292.601		196.86	196.86	28.966	1.815×10^{-3}	3.213×10^{-6}
		+0.0035					
126,000	128,548.193		322.86	273.6	24.54	1.451×10^{-5}	1.566×10^{-8}
		+0.0100					
175,000	179,954.614		812.86	669.0	23.84	6.190×10^{-7}	2.655×10^{-10}
		+0.0058					
300,000	314,862.257		1,537.86	973.5	18.34	1.447×10^{-8}	3.279×10^{-12}

Abbreviated English Table of the U. S. Extension to the ICAO Standard Atmosphere

H ft′	Z ft	L_M °R ft′$^{-1}$	T_M °R	T °R	M	P lb ft^{-2}	ρ slugs ft^{-3}
−16,404.199	−16,391.307		577.188	577.188	28.966	3.7110×10^{3}	3.7457×10^{-3}
		−0.003566160					
0	0		518.688	518.688	28.966	2.1162×10^{3}	2.3769×10^{-3}
		−0.003566160					
36,089.239	36,151.798		389.988	389.988	28.966	4.7268×10^{2}	7.0611×10^{-4}
		zero					
65,616.798	65,823.897		389.988	389.988	28.966	1.1548×10^{2}	1.7251×10^{-4}
		zero					
82,020.997	82,344.849		389.988	389.988	28.966	3.1975×10^{1}	7.7644×10^{-5}
		+0.001645920					
104,986.877	105,518.055		427.788	427.788	28.966	1.8124×10^{1}	2.4682×10^{-5}
		+0.001645920					
154,199.475	155,348.103		508.788	508.788	28.966	2.5155×10^{0}	2.8803×10^{-6}
		zero					
173,884.514	175,346.523		508.788	508.788	28.966	1.2180×10^{0}	1.3947×10^{-6}
		−0.002139696					
246,062.992	249,000.945		354.348	354.348	28.966	5.121×10^{-2}	8.420×10^{-8}
		zero					
295,275.591	299,516.408		354.348	354.348	28.966	3.792×10^{-3}	6.234×10^{-9}
		+0.001920240					
413,385.827	421,746.041		581.148	492.4	24.54	3.031×10^{-5}	3.038×10^{-11}
		+0.005486400					
574,146.982	590,402.278		1,463.148	1,204.000	23.84	1.293×10^{-6}	5.147×10^{-13}
		+0.003182112					
984,251.969	1,033,012.654		2,768.148	1,752.000	18.3	3.023×10^{-8}	6.362×10^{-15}

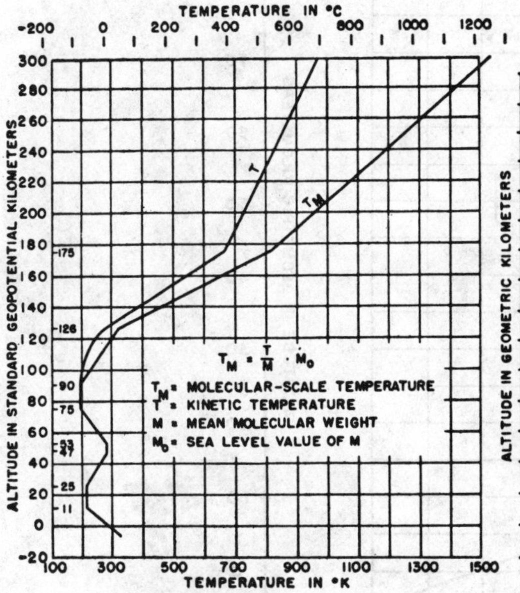

TEMPERATURE VS ALTITUDE

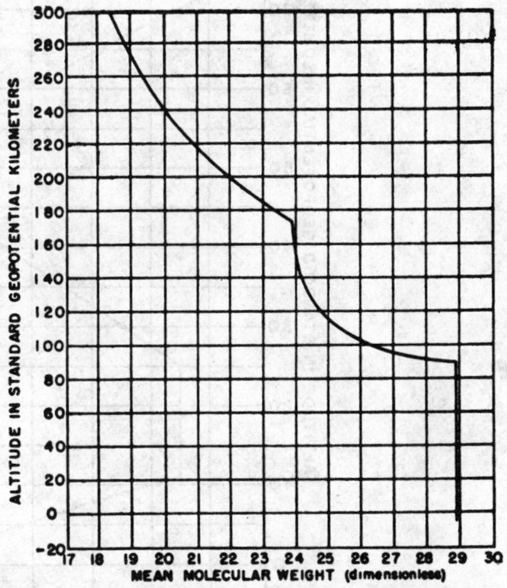

MEAN MOLECULAR WEIGHT VS ALTITUDE

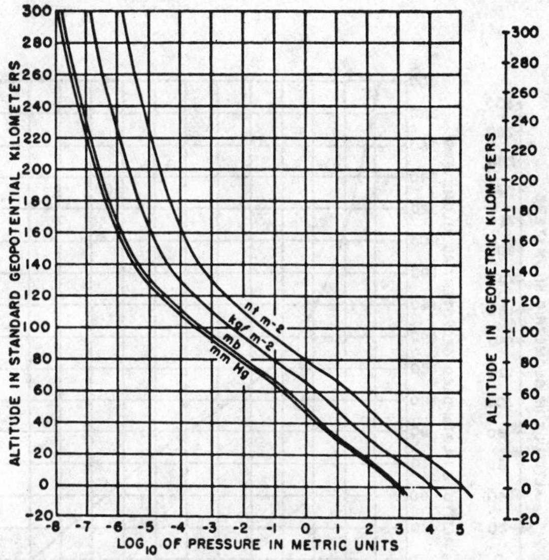

PRESSURE VS. ALTITUDE

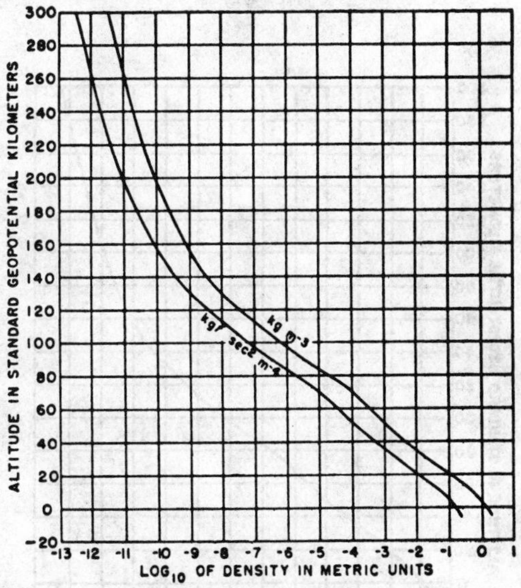

DENSITY VS. ALTITUDE

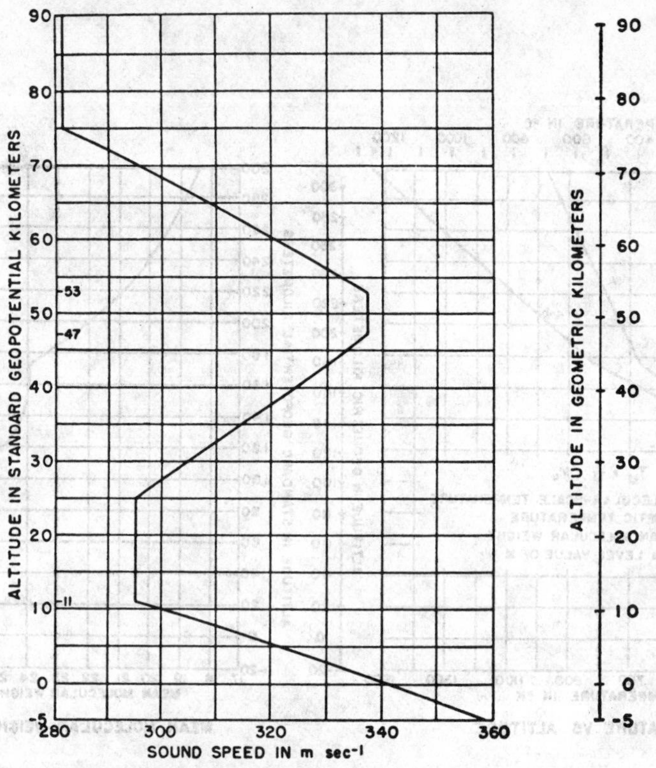

SPEED OF SOUND VS. ALTITUDE

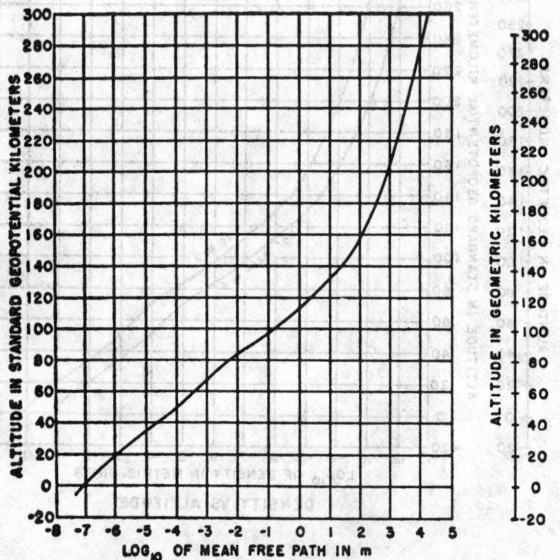

MEAN FREE PATH VS. ALTITUDE

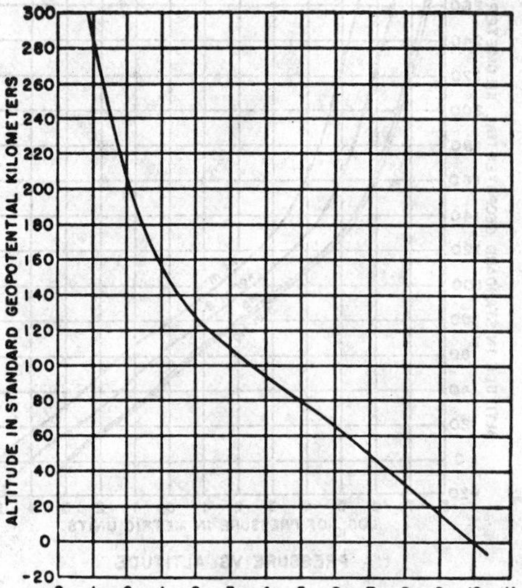

COLLISION FREQUENCY VS ALTITUDE

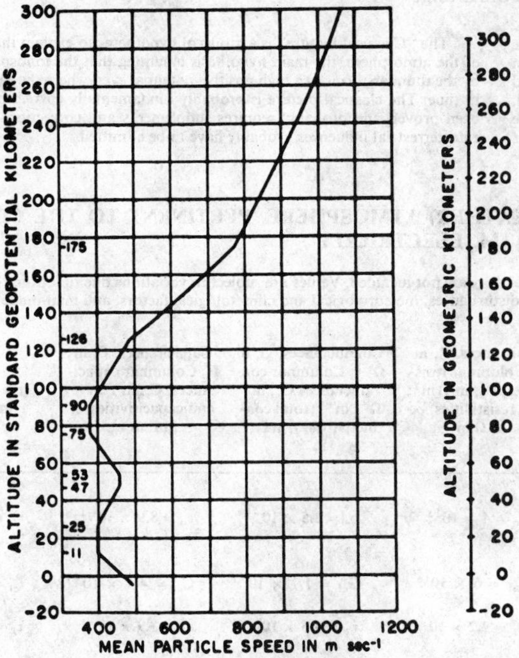

MEAN PARTICLE SPEED VS. ALTITUDE

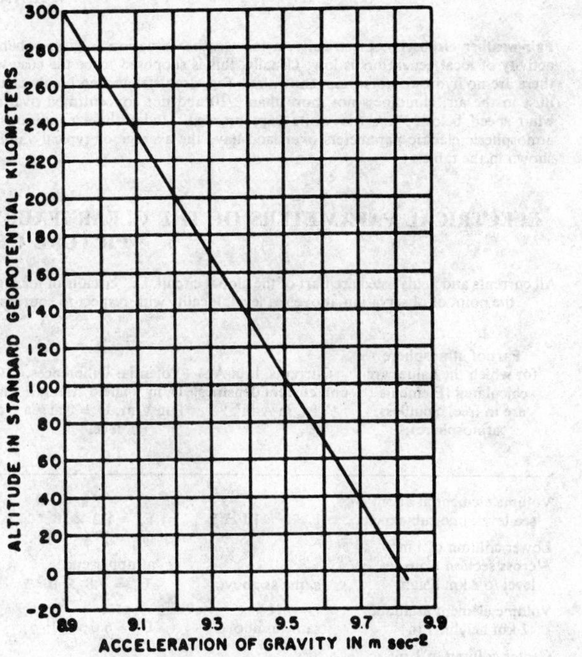

ACCELERATION OF GRAVITY VS. ALTITUDE

Angular Radius of Halos and Rainbows

Coronae due to small water drops	1° to 10°
Small halo, due to 60° angles of ice crystals	22°
Large halo, due to 90° angles of ice crystals	46°
Rainbow, primary	41° 20′
Rainbow, secondary	52° 15′

Solar Constant

The energy falling on one sq. cm. area at normal incidence, outside the earth's atmosphere, at the mean distance of the earth from the sun, equals 2.00 small calories per minute. This value varies ± 2 %.

COMPONENTS OF ATMOSPHERIC AIR

(Exclusive of water vapor)

Constituent	Content (per cent) by volume	Content (ppm) by volume
N_2	78.084 ± 0.004	
O_2	20.946 ± 0.002	
CO_2	0.033 ± 0.001	
Ar	0.934 ± 0.001	
Ne		18.18 ± 0.04
He		5.24 ± 0.004
Kr		1.14 ± 0.01
Xe		0.087 ± 0.001
H_2		0.5
CH_4		2
N_2O		0.5 ± 0.1

MOLECULAR CONSTANTS

The following gives the arithmetical average velocity, the mean free path, molecular diameter and collision frequency for the temperatures indicated and at standard pressure, 760 mm Hg. except as otherwise stated.

Gas	Average velocity in cm/sec.		Mean free path in cm at 75 cm Hg.		
			Boltzmann		Meyer
	0° C	20° C	0° C	20° C	20° C
Air	447×10^2	463×10^2	......	......	......
Ammonia	583	604	5.92×10^{-6}	6.60×10^{-6}	5.83×10^{-6}
Argon	381	395	8.98	9.88	8.73
Carbon monoxide	454	471	8.46	9.23	8.16
Carbon dioxide	362	376	5.56	6.15	5.44
Helium	1208	1252	25.25	27.45	33.10
Hydrogen	1696	1755	16.00	17.44	15.40
Krypton	263	272	9.5	......	......
Mercury	170	176	......	(14.70)	(13.0)
Neon	538	557	......	......	......
Nitrogen	454	471	8.50	9.29	8.21
Oxygen	425	440	9.05	9.93	8.78
Water vapor	566	587	......	......	......
Xenon	210	218	5.6	......	......

Gas	Collision frequency 20° C	Molecular diameter, cm		
		From viscosity	From van der Waal's equation	From heat conductivity
Ammonia	9150×10^6	2.97×10^{-8}	3.08×10^{-8}	......
Argon	4000	2.88	2.94	2.86×10^{-8}
Carbon monoxide	5100	3.19	3.12	......
Carbon dioxide	6120	3.34	3.23	3.40
Helium	4540	1.90	2.65	2.30
Hydrogen	10060	2.40	2.34	2.32
Krypton	......	......	(3.69)	3.14
Mercury	......	......	3.01	......
Nitrogen	5070	3.15	3.15	3.53
Oxygen	4430	2.98	2.92	......
Xenon	......	......	4.02	3.42

ATMOSPHERIC ELECTRICITY

HANS DOLEZALEK

Fair-weather electricity. Electrically, a fair-weather situation is given when the activity of local generators is low. Usually, this is supposed to be the case when there are no hydrometeors (rain, snow, hail, fog, droplets) and no highly charged dust in the air, cloudiness not more than 3/10 and not concentrated overhead, wind speed below 5 m/s (no white caps on sea). Under these conditions the atmospheric electric parameters over land have the average or typical values as shown in the table.

The "Classical Picture" is a group of hypotheses to explain the electrification of the atmosphere, the main hypothesis assuming that the ionosphere is charged by the thunderstorms to a high positive potential everywhere the same but varying in time. The classical picture is probably fundamentally correct but it has never been proven and probably requires supplements and corrections. In particular, extraterrestrial influences probably have to be admitted.

ELECTRICAL PARAMETERS OF THE CLEAR (FAIR-WEATHER) ATMOSPHERE, PERTINENT TO THE CLASSICAL PICTURE OF ATM. ELECTRICITY

All currents and fields listed are part of the global circuit, i.e., circuits of local generators are not included. Values are subject to variations due to latitude and altitude of the point of observation above sea level, locality with respect to sources of disturbances, meteorological and climatological factors, and man-made changes.

Part of atmosphere for which the values are calculated (Elements are in free, cloudless atmosphere):	Currents, I, in A; and current densities, i, in A/m²	Potential Differences, U, in V; field strength E in V/m; U = 0 at sea level	Resistances, R, in Ω; Columnar resistances, R_c, in Ω m²; and resistivities, ρ, in Ω m	Conductances, G, in Ω^{-1}; Columnar conductances G_c, in $\Omega^{-1} m^{-2}$; total conductivities, γ, in $\Omega^{-1} m^{-1}$	Capacitances, C, in F; Columnar capacitances, C_c, in F m^{-2} and capacitivities, ε, in F m^{-1}	Time constants τ, in seconds
Volume element at about sea level, one cubic meter	$i = 3 \times 10^{-12}$*	$E_0 = 1.2 \times 10^2$*	$\rho_0 = 4 \times 10^{13}$	$\gamma_0 = 2.5 \times 10^{-14}$	$\varepsilon_0 = 8.9 \times 10^{-12}$*	$\tau_0 = 3.6 \times 10^2$
Lower column of 1 m² cross section from sea level to 2 km height	same as above	at upper end: $U_l = 1.8 \times 10^5$	$R_{cl} = 6 \times 10^{16}$	$G_{cl} = 1.7 \times 10^{-17}$	$C_{cl} = 4.4 \times 10^{-15}$	$\tau_{cl} = 2.6 \times 10^2$
Volume element at about 2 km height, 1 m³	same as above	$E_2 = 6.6 \times 10^1$	$\rho_2 = 2.2 \times 10^{13}$(*)	$\gamma_2 = 4.5 \times 10^{-14}$	$\varepsilon_2 = 8.9 \times 10^{-12}$	$\tau_2 = 2 \times 10^2$
Center column of 1 m² cross section from 2 to 12 km	same as above	at upper end: $U_m = 3.15 \times 10^5$	$R_{cm} = 4.5 \times 10^{16}$	$G_{cm} = 5 \times 10^{-17}$	$C_{cm} = 8.8 \times 10^{-16}$	$\tau_{cm} = 1.8 \times 10^1$
Volume element at about 12 km height, 1 m³	same as above	$E_{12} = 4.2 \times 10^0$	$\rho_{12} = 1.3 \times 10^{12}$(*)	$\gamma_{12} = 4.0 \times 10^{-13}$	$\varepsilon_{12} = 8.9 \times 10^{-12}$	$\tau_{12} = 1.2 \times 10^1$
Upper column of 1 m² cross section from 12 to 65 km height	same as above	at upper end: $U_u = 3.5 \times 10^5$	$R_{cu} = 1.5 \times 10^{16}$	$G_{cu} = 2.5 \times 10^{-17}$	$C_{cu} = 1.67 \times 10^{-16}$	$\tau_{cu} = 6.7 \times 10^0$
Whole column of 1 m² cross section from 0 to 65 km height	same as above	at upper end: $U = 3.5 \times 10^5$	$R_c = 1.2 \times 10^{17}$	$G_c = 8.3 \times 10^{-18}$	$C_c = 1.36 \times 10^{-16}$	$\tau_c = 1.64 \times 10^1$
Total spherical capacitor area: 5×10^{14} m²	$1 = 1.5 \times 10^3$	$U = 3.5 \times 10^5$*	$R = 2.4 \times 10^2$	$G = 4.2 \times 10^{-3}$	$C = 6.8 \times 10^{-2}$	$\tau = 1.64 \times 10^1$

Values with a star, *, are rough average values from measurement. A star in parentheses, (*), points to a typical value from one or a few measurements. All other values have been calculated from starred values, under the assumption, that at 2 km 50% and at 12 km 90% of the columnar resistance is reached. Voltage drop along one of the partial columns can be calculated by subtracting the value for the lower column from that of the upper one. Columnar resistances, conductances, and capacitances are valid for that particular part of the column which is indicated at left. Capacitances are calculated with the formula for plate capacitors, and this fact must be considered also for the time constants for columns.

According to measurements, U, the potential difference between 0 m and 65 km may vary by a factor of approx. 2 (two). The total columnar resistance, R_c, is estimated to vary up to a factor of 3 (three), the variation being due to either reduction of conductivity in the exchange layer (about lowest 2 km of this table) or to the presence of high mountains, in both cases the variation is caused in the troposphere. Smaller variations in the stratosphere and mesosphere discussed because of aerosol there. The air-earth current density in fair-weather varies by a factor of 3 to 6 accordingly. Conductivity near the ground varies by a factor of about 3 (three) but only decreasing, increase of conductivity due to extraordinary radioactivity is a singular event. The field strength near the ground varies as a consequence of variations of air-earth current density and conductivity from about 1/3 to about 10 of the value quoted in the table. Conductivity near the ground shows a diurnal and an annual variation which depends strongly on the locality; air-earth current density shows a diurnal and annual variation because the earth-ionosphere potential difference undergoes such variations, and because also the columnar resistance is supposed to have a diurnal and probably an annual variation.

Conductivities and air-earth current densities on high mountains are by factors of up to 10 greater than at sea level. Conductivity decreases when atmospheric humidity increases. Values for space charges not quoted because measurements are too few to allow calculation of average values. Values of parameters over the oceans still rather uncertain.

Ions. Ionization rate at ground level over land: from radioactive substances in the ground: 4.0; from radioactive substances in the air: 4.6; from cosmic radiation 1.5–1.8, total (10.1–10.4) × 10^6 ion pairs per m³ and sec. Over ocean far from shore: cosmic radiation only, but in spite of smaller ionization over the ocean conductivity is nearly equal to that over land because of longer ion life time (a consequence of smaller aerosol density).

Contribution from radioactive material over land decreases with altitude in the free air, from cosmic radiation increases (up to a certain level) and depends on latitude. Number densities of fast ions over land and ocean average at (4–5) × 10^8 m^{-3}, varying with height by only about one order of magnitude up to ionospheric C-layer. Their life time varies from about 10 s in aerosol-rich air to 300 s in pure air. All values quoted are approximate averages.

Ion classes in lower atmosphere	Designation	Mobilities in m²/(Vs)	Size ranges, radii in m
Primary atmospheric ions:	Fast ions (also called small or light ions)	$k \geq 10^{-4}$	$r < 6.6 \times 10^{-10}$
Secondary atmospheric ions: slow ions (also called large or heavy ions	Fast intermediate ions	$10^{-4} > k \geq 10^{-2}$	$6.6 \times 10^{-10} < r < 7.8 \times 10^{-9}$
	Slow intermediate ions	$10^{-2} > k \geq 10^{-1}$	$7.8 \times 10^{-9} < r < 2.5 \times 10^{-8}$
	Langevin ions	$10^{-1} > k \geq 2.5$	$2.5 \times 10^{-8} < r < 5.7 \times 10^{-8}$
	Ultralarge ions	$k < 2.5$	$5.7 \times 10^{-8} < r < 10^{-7}$

Electric conduction current in atmosphere carried nearly exclusively by the fast ions because number densities of the intermediate ions generally one order of magnitude smaller.

Nature of ions in pure air ("Standard Atmosphere"): initially all molecules are ionized with equal probability according to their abundance, but within nanoseconds these ions transform into the "fast ions" with the following composition:

positive ions: $(H_3O)^+(H_2O)_n$

negative ions: $O_2^-(H_2O)_n$, or $CO_4^-(H_2O)_n$;

the n depending on atmospheric temperature and water vapor pressure, varies very often in the life time of an ion. Most probable values of n for positive ions: troposphere 6 or 7, stratosphere 4, mesosphere 3 or 2.

Thunderstorm electricity. Clouds are generally more or less electrified. In thunderstorms, mechanical, thermodynamical and maybe chemical energy is transformed into electrical energy (generator) with separate poles. Most thunderstorms have during most of their life time (phases: development 10–20, mature 15–30, decay 30 min) a positive charge center above and a negative one below, probably another positive one near the base. Electrification occurs (according to processes not yet fully understood) in periods of tens of minutes and spaces of about 1 km^3, and there may be several such centers in a cloud. Thunderclouds mostly reach beyond freezing level, in moderate latitudes from about 3 to 10 km, in the tropics often up to 16–20 km, seldom more. 1000 to 2000 thunderstorms active at any time. Updrafts in storms up to 30 m/s. Water content 10^8–10^9 kg. Total energy about 10^{15} joule, electric energy 10^{12}–10^{13} J. Electric potential vs. ground at negative and positive charge centers: minus and plus 10–100 MV, resp.; field strength not yet measured reliably. Electric resistances of an atmospheric vertical column of 50 km^2 cross section: several 100 MΩ each between ground and negative center, negative and positive center, and positive center and ionosphere. Total current flowing upwards towards ionosphere derived from (still unsufficient) measurements of current densities: 0.5–2 A per thunderstorm cell. Current below cloud carried by conduction by fair-weather ions, conduction by corona ions, convection by vertical winds, precipitation, and lightning. Field strength at ground under storm: varying in strength and polarity, often 10 kV/m and more.

Atmospheric Electricity by H. Israël. Tabular material reproduced from this publication. Permission received from Israel Program for Scientific Translations.

Lightning, during its life time of up to a second, undergoes numerous variations and all its parameters change by orders of magnitude in these. Also, there are many different forms of lightning. Even disregarding particular forms such as ball lightning, we find at least four classification criteria which, taken together, give a great number of differences: intercloud, intracloud, and cloud-to-ground discharges and combinations of these; short high-current and long low-current flashes; lightnings beginning in the cloud and moving towards ground, and lightnings moving upwards; lightnings lowering positive and lightnings lowering negative charges, and those which do first the one and then the other. The usual cloud-to-ground lightning begins with a stepped leader of low luminosity, probably a meter or so in diameter, followed by a return stroke with a diameter in the order of centimeters, temperatures of about 30,000 K, and pressures of up to a MN/m^2. After this, a dart leader may again move downwards causing a second return stroke. This may repeat several times (multiple stroke flash).

Typical voltage drop in ground or other conductors after lightning impact in the neighborhood: 10 kV/m (dangerous!).

Intracloud lightnings observed with up to 100 km length.

There are probably about 100 lightnings occurring on earth at any time. Lightning frequency over oceans only about 1/10 from that over continents. Diurnal variations on continents show maximum in late afternoon and early evening, over oceans late evening until after midnight.

Energy delivered to a stroke about 100 kJ/m.

Upward lightnings initiated mostly from high masts and towers and mountains. Over flat terrain mostly downward flashes, probably generally met by short (decameters) upwards moving darts.

Height from which a lightning points at a target probably several decameters, depending on conductivity distribution in ground.

Long-lasting low-current (hundreds of amperes) flashes more dangerous to man and more damaging to objects (forest fires) than short high-current flashes.

Sferics, etc. Lightning generates a sudden variation of electric field (decreasing with r^{-3}), a quickly varying magnetic field (r^{-2}) and an electromagnetic radiation (r^{-1}) from a few hertz (in case of excitation of ionosphere-earth resonance range around the globe) through VLF (maximum intensity, range up to 10 Mm and more depending on direction and time of day) into VHF. Main frequency of thunder about 200 Hz.

Data for a normal cloud-to-ground lightning discharge bringing negative charge to earth. The values listed are intended to convey a rough feeling for the various physical parameters of lightning. No great accuracy is claimed since the results of different investigators are often not in good agreement. These values may in fact, depend on the particular environment in which the lightning discharge is generated. The choice of some of the entries in the table is arbitrary.

	Minimum*	Representative	Maximum*
Stepped leader			
Length of step, *m*	3	50	200
Time interval between steps, μsec	30	50	125
Average velocity of propagation of stepped leader, m/sec†	1.0×10^5	1.5×10^5	2.6×10^6
Charge deposited on stepped-leader channel, coulombs	3	5	20
Dart leader			
Velocity of propagation, m/sec†	1.0×10^6	2.0×10^6	2.1×10^7
Charge deposited on dart-leader channel, coulombs	0.2	1	6
Return stroke‡			
Velocity of propagation, m/sec†	2.0×10^7	5.0×10^7	1.4×10^8
Current rate of increase, kA/μsec§	<1	10	>80
Time to peak current, μsec§	<1	2	30
Peak current, kA §		10–20	110
Time to half of peak current μsec	10	40	250
Charge transferred excluding continuing current, coulombs	0.2	2.5	20
Channel length, km	2	5	14
Lightning flash			
Number of strokes per flash	1	3–4	26
Time interval between strokes in absence of continuing current, msec	3	40	100
Time duration of flash, sec	10^{-2}	0.2	2¶
Charge transferred including continuing current, coulombs	3	25	90

* The words maximum and minimum are used in the sense that most measured values fall between these limits.

† Velocities of propagation are generally determined from photographic data and thus represent "two-dimensional" velocities. Since many lightning flashes are not vertical, values stated are probably slight underestimates of actual values.

‡ First return strokes have slower average velocities of propagation, slower current rates of increase, longer times to current peak, and generally larger charge transfer than subsequent return strokes in a flash.

§ Current measurements are made at the ground.

¶ A lightning flash lasting 15 to 20 sec has been reported by Godlonton (1896).

Lightning by M. A. Uman, Table 1.1, page 4. Tabular material reproduced from this publication. Permission received from McGraw-Hill Book Company.

CRYSTAL IONIC RADII OF THE ELEMENTS

Numerical values of the radii of the ions may vary depending on how they were measured. They may have been calculated from wavefunctions and determined from the lattice spacings or crystal structure of various salts. Different values are obtained depending on the kind of salt used or the method of calculating. Data for many of the rare-earth ions were furnished by F. H. Spedding and K. Gschneidner.

Element	Charge	Atomic number	Radius in A	Element	Charge	Atomic number	Radius in A	Element	Charge	Atomic number	Radius in A
Ac	+3	89	1.18	Fe	+2	26	0.74	Os	+4	76	0.6
Ag	+1	47	1.26		+3		0.64		+6		0.8
	+2		0.89	Fr	+1	87	1.80	P	−3	15	2.12
Al	+3	13	0.51	Ga	+1	31	0.81		+3		0.44
Am	+3	95	1.07		+3		0.62		+5		0.35
	+4		0.92	Gd	+3	64	0.938	Pa	+3	91	1.13
Ar	+1	18	1.54	Ge	−4	32	2.72		+4		0.98
As	−3	33	2.22		+2		0.73		+5		0.89
	+3		0.58		+4		0.53	Pb	+2	82	1.20
	+5		0.46	H	−1	1	1.54		+4		0.84
At	+7	85	0.62	Hf	+4	72	0.78	Pd	+2	46	0.80
Au	+1	79	1.37	Hg	+1	80	1.27		+4		0.65
	+3		0.85		+2		1.10	Pm	+3	61	0.979
B	+1	5	0.35	Ho	+3	67	0.894	Po	+6	84	0.67
	+3		0.23	I	−1	53	2.20	Pr	+3	59	1.013
Ba	+1	56	1.53		+5		0.62		+4		0.90
	+2		1.34		+7		0.50	Pt	+2	78	0.80
Be	+1	4	0.44	In	+3	49	0.81		+4		0.65
	+2		0.35	Ir	+4	77	0.68	Pu	+3	94	1.08
Bi	+1	83	0.98	K	+1	19	1.33		+4		0.93
	+3		0.96	La	+1	57	1.39	Ra	+2	88	1.43
	+5		0.74		+3		1.016	Rb	+1	37	1.47
Br	−1	35	1.96	Li	+1	3	0.68	Re	+4	75	0.72
	+5		0.47	Lu	+3	71	0.85		+7		0.56
	+7		0.39	Mg	+1	12	0.82	Rh	+3	45	0.68
C	−4	6	2.60		+2		0.66	Ru	+4	44	0.67
	+4		0.16	Mn	+2	25	0.80	S	−2	16	1.84
Ca	+1	20	1.18		+3		0.66		+2		2.19
	+2		0.99		+4		0.60		+4		0.37
Cd	+1	48	1.14		+7		0.46		+6		0.30
	+2		0.97	Mo	+1	42	0.93	Sb	−3	51	2.45
Ce	+1	58	1.27		+4		0.70		+3		0.76
	+3		1.034		+6		0.62		+5		0.62
	+4		0.92	N	−3	7	1.71	Sc	+3	21	0.732
Cl	−1	17	1.81		+1		0.25	Se	−2	34	1.91
	+5		0.34		+3		0.16		−1		2.32
	+7		0.27		+5		0.13		+1		0.66
Co	+2	27	0.72	NH_4	+1		1.43		+4		0.50
	+3		0.63	Na	+1	11	0.97		+6		0.42
Cr	+1	24	0.81	Nb	+1	41	1.00	Si	−4	14	2.71
	+2		0.89		+4		0.74		−1		3.84
	+3		0.63		+5		0.69		+1		0.65
	+6		0.52	Nd	+3	60	0.995		+4		0.42
Cs	+1	55	1.67	Ne	+1	10	1.12	Sm	+3	62	0.964
Cu	+1	29	0.96	Ni	+2	28	0.69	Sn	−4	50	2.94
	+2		0.72	Np	+3	93	1.10		−1		3.70
Dy	+3	66	0.908		+4		0.95		+2		0.93
Er	+3	68	0.881		+7		0.71		+4		0.71
Eu	+3	63	0.950	O	−2	8	1.32	Sr	+2	38	1.12
	+2		1.09		−1		1.76	Ta	+5	73	0.68
F	−1	9	1.33		+1		0.22	Tb	+3	65	0.923
	+7		0.08		+6		0.09		+4		0.84

CRYSTAL IONIC RADII OF THE ELEMENTS (Continued)

Element	Charge	Atomic number	Radius in A	Element	Charge	Atomic number	Radius in A	Element	Charge	Atomic number	Radius in A
Tc	+7	43	0.979	Ti	+4		0.68	W	+4	74	0.70
Te	−2	52	2.11	Tl	+1	81	1.47		+6		0.62
	−1		2.50		+3		0.95	Y	+3	39	0.893
	+1		0.82	Tm	+3	69	0.87	Yb	+2	70	0.93
	+4		0.70	U	+4	92	0.97		+3		0.858
	+6		0.56		+6		0.80	Zn	+1	30	0.88
Th	+4	90	1.02	V	+2	23	0.88		+2		0.74
Ti	+1	22	0.96		+3		0.74	Zr	+1	40	1.09
	+2		0.94		+4		0.63		+4		0.79
	+3		0.76		+5		0.59				

BOND LENGTHS BETWEEN CARBON AND OTHER ELEMENTS

Prepared by Olga Kennard.

The tables are based on bond distance determinations, by experimental methods, mainly X-ray and electron diffraction, and include values published up to January 1, 1956. In the present tables, for the sake of completeness individual values of bond distances of lower accuracy are quoted with limits of error indicated where possible. Values for tungsten and bismuth should be treated with particular caution.

According to the statistical theory of errors if an average quantity $\bar{\mu}$ and a standard deviation σ can be evaluated there is a 95% probability that the true value lies within the interval $\bar{\mu} \pm 2\sigma$. Too much reliance should, however, not be placed on σ values in bond distance determinations since the derivation of these certain sources of error may have been neglected.

Values of the bond lengths and the limits of error are each given in Ångstrom units.

Reproduced by permission from International Tables for X-ray Crystallography.

BOND LENGTHS BETWEEN CARBON AND OTHER ELEMENTS

Reference: HCP and "Tables of interatomic distances" Chem. Soc. of London, 1958

Group	Bond type	Element						
I		H**						
	All types	1.056 − 1.115						
II		Be	Hg					
		1.93	2.07 ± 0.01					
III		B	Al	In				
		1.56 ± 0.01	2.24 ± 0.04	2.16 ± 0.04				
IV		C**	Ge	Si	Sn	Pb		
	All types	1.54 − 1.20						
	Alkyls (CH_3XH_3)		1.98 ± 0.03	1.865 ± 0.008	2.143 ± 0.008	2.29 ± 0.05		
	Aryl ($C_6H_5XH_3$)			1.84 ± 0.01				
	Neg. Subst. (CH_3XCl_3)			1.88 ± 0.01	2.18 ± 0.02			
V		N**	P	As	Sb	Bi		
	All types	1.47 − 1.1						
	Paraffinic $(CH_3)_3X$		1.87 ± 0.02	1.98 ± 0.02	2.202 ± 0.016	2.30*		
VI		O**	S**	Cr	Se	Te	Mo	W
		1.43 − 1.15	1.81 − 1.55	1.92 ± 0.04	1.98 − 1.71	2.05 ± 0.14	2.08 ± 0.04	2.06 ± 0.01*
VII		F	Cl	Br	I			
	Paraffinic (monosubstituted) (CH_3X)	1.381 ± 0.005	1.767 ± 0.002	1.937 ± 0.003	2.13_5 ± 0.01			
	Paraffinic (disubstituted) (CH_2X_2)	1.334 ± 0.004	1.767 ± 0.002	1.937 ± 0.003	2.13_5 ± 0.1			
	Olefinic (CH_2:CHX)	1.32_5 ± 0.1	1.72 ± 0.01	1.89 ± 0.01	2.092 ± 0.005			
	Aromatic (C_6H_5X)	1.30 ± 0.01	1.70 ± 0.01	1.85 ± 0.01	2.05 ± 0.01			
	Acetylenic (HC:CX)		1.635 ± 0.004	1.79_5 ± 0.01	1.99 ± 0.02			
VIII		Fe	Co	Ni	Pd			
		1.84 ± 0.02	1.83 ± 0.02	1.82 ± 0.03	2.27 ± 0.04			

* Error uncertain.
** See following individual tables.

CARBON-CARBON

Single Bond

Paraffinic	1.541	± 0.003
In diamond (18°C)	1.54452	± 0.00014

Partial Double Bond

(1) Shortening of single bond in presence of carbon carbon double bond, e.g. $(CH_3)_2$-C:CH_2; or of aromatic ring e.g. C_6H_5.-CH_3	1.53	± 0.01
(2) Shortening in presence of a carbon oxygen double bond e.g. CH_3CHO	1.516	± 0.005
(3) Shortening in presence of two carbon-oxygen double bonds, e.g. $(CO_2H)_2$	1.49	± 0.01
(4) Shortening in presence of one carbon-carbon triple bond, e.g. CH_3.C:CH	1.460	± 0.003
(5) In compounds with tendency to dipole formation, e.g. C:C.C:N	1.44	± 0.01
(6) In graphite (at 15°C)	1.4210	± 0.0001
(7) In aromatic compounds	1.395	± 0.003
(8) In presence of two carbon carbon triple bonds, e.g. HC:C.C:CH	1.373	± 0.004

Double Bond

(1) Simple	1.337 ± 0.006
(2) Partial triple bond, e.g. CH_2:C:CH_2	1.309 ± 0.005

Triple Bond

(1) Simple, e.g. C_2H_2	1.204 ± 0.002
(2) Conjugated, e.g. CH_3.(C:C)$_2$.H	1.206 ± 0.004

CARBON-HYDROGEN

(1) Paraffinic (a) in methane	1.091
(b) in monosubstituted carbon	1.101 ± 0.003
(c) in disubstituted carbon	1.073 ± 0.004
(d) in trisubstituted carbon	1.070 ± 0.007
(2) Olefinic, e.g. CH_2:CH_2	1.07 ± 0.01
(3) Aromatic in C_6H_6	1.084 ± 0.006
(4) Acetylenic, e.g. CH:C.X	1.056 ± 0.003
(5) Shortening in presence of a carbon triple bond, e.g. CH_3CN	1.115 ± 0.004
(6) In small rings, e.g. $(CH_2)_2S$	1.081 ± 0.007

BOND LENGTHS BETWEEN CARBON AND OTHER ELEMENTS (Continued)

CARBON-NITROGEN

Single Bond

(1) Paraffinic (a) 4 co-valent nitrogen	1.479	± 0.005
(b) 3 co-valent nitrogen	1.472	± 0.005
(2) In C—N= e.g. CH_3NO_2	1.475	± 0.010
(3) Aromatic in $C_6H_5NHCOCH_3$	1.426	± 0.012
(4) Snortened (partial double bond) in heterocyclic systems, e.g. C_5H_5N	1.352	± 0.005
(5) Shortened (partial double bond) in N—C=O e.g. $HCONH_2$	1.322	± 0.003

Triple Bond

(1) In R.C⋮N	1.158	± 0.002

CARBON-OXYGEN

Single Bond

(1) Paraffinic	1.43	± 0.01
(2) Strained e.g. epoxides	1.47	± 0.01
(3) Shortened (partial double bond) as in carboxylic acids or through influence of aromatic ring, e.g. salicylic acid	1.36	± 0.01

Double Bond

(1) In aldehydes, ketones, carboxylic acids, esters	1.23	± 0.01
(2) In zwitterion forms, e.g. DL serine	1.26	± 0.01
(3) Shortened (partial triple bond) as in conjugated systems	1.207	± 0.006
(4) Partial triple bond as in acyl halides or isocyanates	1.17	± 0.01

CARBON-SULPHUR

Single Bond

(1) Paraffinic, e.g. CH_3SH	1.81(5)	± 0.01
(2) Lengthened in presence of fluorine, e.g. $(CF_3)_2S$	1.83(5)	± 0.01
(3) Shortened (partial double bond) as in heterocyclic systems, e.g. C_4H_4S	1.73	± 0.01

Double Bond

(1) In ethylene thiourea	1.71	± 0.02
(2) Shortened (partial triple bond) in presence of second carbon double bond, e.g. COS	1.558	± 0.003

BOND LENGTHS OF ELEMENTS

Element	Bond	Å
Ac	Ac—Ac	3.756
Ag (25°C)	Ag—Ag	2.8894
Al (25°C)	Al—Al	2.863
As	As—As	2.49
As_4	As—As	2.44 ± 0.03
Au (25°C)	Au—Au	2.8841
B_2	B—B	1.589
Ba (room temp.)	Ba—Ba	4.347
Be (α-form, 20°C)	Be—Be	2.2260
Bi (25°C)	Bi—Bi	3.09
Br_2	Br—Br	2.290
Ca (α-form, 18°C)	Ca—Ca	3.947 (f.c.c.)
(β-form, 500°C		3.877 (b.c.c.)
Cd (21°C)	Cd—Cd	2.9788
Cl_2	Cl—Cl	1.988
Ce	Ce—Ce	3.650
Co (18°C)	Co—Co	2.5061
Cr (α-form, 20°C)	Cr—Cr	2.4980
(β-form, >1850°C		2.61
Cs (−10°C)	Cs—Cs	5.309
Cu (20°C)	Cu—Cu	2.5560
Dy	Dy—Dy	3.503

BOND LENGTHS OF ELEMENTS (Continued)

Element	Bond	Å
Er	Er—Er	3.468
Eu	Eu—Eu	3.989
F_2	F—F	1.417 ± 0.001
Fe (α-form, 20°C)	Fe—Fe	2.4823 (b.c.c.)
(γ-form, 916°C)		2.578 (f.c.c.)
(δ-form, 1394°C)		2.539 (b.c.c.)
Ga (20°C)	Ga—Ga	2.442
Gd (20°C)	Gd—Gd	3.573
Ge (20°C)	Ge—Ge	2.4498
H_2	H—H in H_2	0.74611
	H—D in HD	0.74136
	D—D in D_2	0.74164
He	He—He in $[He_2]^+$	1.08_0
Hf (α-form, 24°C)	Hf—Hf	3.1273 (h.c.p.)
Hg (−46°C)	Hg—Hg	3.005
Ho	Ho—Ho	3.486
I_2	I—I	2.662
In (20°C)	In—In	3.2511
Ir (room temp.)	Ir—Ir	2.714
K (78°K)	K—K	4.544
La (α-form)	La—La	3.739 (h.c.p.)
(β-form)		3.745 (f.c.c.)
Li (20°C)	Li—Li	3.0390
Lu	Lu—Lu	3.435
Mg (25°C)	Mg—Mg	3.1971
Mn (γ-form, 1095°C)	Mn—Mn	2.7311 (f.c.c.)
(δ-form, 1134°C)		2.6679 (b.c.c.)
Mo (20°C)	Mo—Mo	2.7251
N_2	N—N	1.0975_8 ± 0.0001
Na (20°C)	Na—Na	3.7157
Nb (20°C)	Nb—Nb	2.8584
Nd	Nd—Nd	3.628
Ni (18°C)	Ni—Ni	2.4916
Np (α-form, 20°C)	Np—Np	2.60 (orthorhombic)
(β-form, 313°C)		2.76 (tetragon.)
(γ-form, 600°C)		3.05 (b.c.c.)
O_2	O—O	1.208
O_3 angle 116.8 ± 0.5°		1.278 ± 0.003
Os (20°C)	Os—Os	2.6754
P black	P—P	2.18
P_4	P—P	2.21 ± 0.02
Pa	Pa—Pa	3.212
Pb (25°C)	Pb—Pb	3.5003
Pd (25°C)	Pd—Pd	2.7511
Po (α-form, 10°C)	Po—Po	3.345 (cubic)
(β-form, 75°C)		3.359 (rh. hedr.)
Pr (α-form)	Pr—Pr	3.640 (tetrag.)
(β-form)		3.649 (f.c.c.)
Pt (20°C)	Pt—Pt	2.746
Pu (γ-form, 235°C)	Pu—Pu	3.026 (f.c.c.)
(δ-form, 313°C)		3.279 (f.c.c.)
(ε-form, 500°C)		3.150 (b.c.c.)
Rb (20°C)	Rb—Rb	4.95
Re (room temp.)	Re—Re	2.741
Rh (20°C)	Rh—Rh	2.6901
Ru (25°C)	Ru—Ru	2.6502
S_2	S—S	1.887
S_8	S—S	2.07 ± 0.02
Sb (25°C)	Sb—Sb	2.90
Sc (room temp.)	Sc—Sc	3.212
Se (20°C)	Se—Se	2.321
Se_2	Se—Se	2.152 ± 0.003
Se_8	Se—Se	2.32 ± 0.003
Si (20°C)	Si—Si	2.3517
Sn (α-form, 20°C)	Sn—Sn diamond type lattice	2.8099
(β-form, 25°C)		3.022 (tetrag.)
Sr (α-form, 25°C)	Sr—Sr	4.302 (f.c.c.)
(β-form, 248°C)		4.32 (h.c.p.)
(γ-form, 614°C)		4.20 (b.c.c.)
Ta (20°C)	Ta—Ta	2.86
Tb	Tb—Tb	3.525
Tc (room temp.)	Tc—Tc	2.703
Te (25°C)	Te—Te	2.864
Th (α-form, 25°C)	Th—Th	3.595 (f.c.c.)
(β-form, 1450°C)		3.56 (b.c.c.)
Ti (α-form, 25°C)	Ti—Ti	2.8956 (h.c.p.)
(β-form, 900°C)		2.8636 (b.c.c.)
Tl (α-form, 18°C)	Tl—Tl	3.4076 (h.c.p.)
(β-form, 262°C)		3.362 (b.c.c.)
Tm	Tm—Tm	3.447
U (α-form)	U—U	2.77
(β-form, 805°C)		3.058 (b.c.c.)
V (30°C)	V—V	2.6224
W (25°C)	W—W	2.7409
Y	Y—Y	3.551
Yb	Yb—Yb	3.880
Zn (25°C)	Zn—Zn	2.6694
Zr	Zr—Zr	3.179

BOND LENGTH AND ANGLE VALUES BETWEEN ELEMENTS

Elements	In	Bond length (Å)	Bond angle (°)
Boron			
B—B	B_2H_6	1.770 ± 0.013	H—B—H 121.5 ± 7.5
B—Br	BBr	1.88_7	—
	BBr_3	1.87 ± 0.02	Br—B—Br 120 ± 6
B—Cl	BCl	1.715_7	—
	BCl_3	1.72 ± 0.01	Cl—B—Cl 120 ± 3
B—F	BF	1.262	—
	BF_3	1.29_5 ± 0.01	F—B—F 120
B—H	Hydrides	1.21 ± 0.02	—
B—H Bridge	Hydrides	1.39 ± 0.02	—
B—N	$(BClNH)_3$	1.42 ± 0.01	B—N—B 121
B—O	BO	1.2049	—
	$B(OH)_3$	1.362 ± 0.005 (av.)	O—B—O 119.7
Nitrogen			
N—Cl	NO_2Cl	1.79 ± 0.02	—
N—F	NF_3	1.36 ± 0.02	F—N—F 102.5 ± 1.5
N—H	$[NH_4]^+$	1.034 ± 0.003	—
	NH	1.038	—
	ND	1.041	—
	HNCS	1.013 ± 0.005	H—N—C 130.25 ± 0.25
N—N	N_3H	1.02 ± 0.01	H—N—N′ 112.65 ± 0.5
	N_2O	1.126 ± 0.002	—
	$[N_2]^+$	1.116_2	—
N—O	NO_2Cl	1.24 ± 0.01	O—N—O 126 ± 2
	NO_2	1.188 ± 0.005	O—N—O 134.1 ± 0.25
N=O	N_2O	1.186 ± 0.002	—
	$[NO]^+$	1.0619	—
N—Si	SiN	1.572	—
Oxygen			
O—H	$[OH]^+$	1.0289	—
	OD	0.9699	—
	H_2O_2	0.960 ± 0.005	O—O—H 100 ± 2
O—O	H_2O_2	1.48 ± 0.01	—
	$[O_2]^+$	1.227	—
	$[O_2]^-$	1.26 ± 0.02	—
	$[O_2]^{--}$	1.49 ± 0.02	—
Phosphorus			
P—D	PD	1.429	—
P—H	$[PH_4]^+$	1.42 ± 0.02	—
P—N	PN	1.4910	—
P—S	$PSBr_3(Cl_3, F_3)$	1.86 ± 0.02	—
Sulfur			
S—Br	$SOBr_2$	2.27 ± 0.02	Br—S—Br 96 ± 2
S—F	SOF_2	1.585 ± 0.005	F—S—F 92.8 ± 1
S—D	SD	1.3473	—
	SD_2	1.345	—
S—O	SO_2	1.4321	O—S—O 119.54
	$SOCl_2$	1.45 ± 0.02	—
S—S	S_2Cl_2	2.04 ± 0.01	—
Silicon			
Si—Br	$SiBr_4$	2.17 ± 1.01	—
Si—Cl	$SiCl_4$	2.03 ± 1.01 (av.)	—
Si—F	SiF_4	1.561 ± 0.003 (av.)	—
SiH	SiH_4	1.480 ± 0.005	—
Si—O	$[SiO]^+$	1.504	—
Si—Si	Si_2Cl_2	2.30 ± 0.02	—

BOND LENGTHS AND ANGLES OF CHEMICAL COMPOUNDS

A. Inorganic Compounds

Compound	Formula	Bond lengths in Å		Bond angles (°)	
Ammonia	NH_3	N—H	1.008 ± 0.004	H—N—H	107.3 ± 0.2
Antimony tribromide	$SbBr_3$	Sb—Br	2.51 ± 0.02	Br—Sb—Br	97 ± 2
Antimony trichloride	$SbCl_3$	Sb—Cl	2.352 ± 0.005	Cl—Sb—Cl	99.5 ± 1.5
Antimony triiodide	SbI_3	Sb—I	2.67 ± 0.03	I—Sb—I	99.0 ± 1
Arsenic tribromide	$AsBr_3$	As—Br	2.33 ± 0.02	Br—As—Br	100.5 ± 1.5
Arsenic trichloride	$AsCl_3$	As—Cl	2.161 ± 0.004	Cl—As—Cl	98.4 ± 0.5
Arsenic trifluoride	AsF_3	As—F	1.712 ± 0.005	F—As—F	102.0 ± 2
Arsenic triiodide	AsI_3	As—I	2.55 ± 0.03	I—As—I	101.0 ± 1.5
Arsenic trioxide	As_4O_6	As—O	1.78 ± 0.02	O—As—O	99.0 ± 2
				As—O—As	128.0 ± 2
Arsine	AsH_3	As—H	1.5192 ± 0.002	H—As—H	91.83 ± 0.33
Bismuthum tribromide	$BiBr_3$	Bi—Br	2.63 ± 0.02	Br—Bi—Br	100.0 ± 4
Bismuthum trichloride	$BiCl_3$	Bi—Cl	2.48 ± 0.02	Cl—Bi—Cl	100.0 ± 6
Bromosilane	SiH_3Br	Si—H	1.57 ± 0.03	H—Si—H	111.3 ± 1
Chlorine dioxide	ClO_2	Cl—O	1.49	O—Cl—O	118.5
Chlorogermane	GeH_3Cl	Ge—H	1.52 ± 0.03	H—Ge—H	110.9 ± 1.5
Chlorosilane	SiH_3Cl	Si—H	1.483 ± 0.001	H—Si—H	110 ± 0.03
		Si—Cl	2.0479 ± 0.0007		
Chromium oxychloride	$Cr(OCl)_2$	Cr—O	1.57 ± 0.03	O—Cr—O	105 ± 4
		Cr—Cl	2.12 ± 0.02	Cl—Cr—Cl	113 ± 3
				Cl—Cr—O	109 ± 3
Cyanuric triazide	C_3N_{12}	C—N	1.38	C—N=C	113.0
Dichlorosilane	SiH_2Cl_2	Si—H	1.46		
		SiCl	2.02 ± 0.03	Cl—Si—Cl	110 ± 1
Difluorodiazine	N_2F_2	N—F	1.44 ± 0.04		
		N—N	1.25 ± 0.02	F—N=N	115 ± 5
Difluoromethylsilane	Ch_3SiHf_2			F—Si—F	106 ± 0.5
				H—Si—C	116.2 ± 1
				C—Si—F	109.8 ± 0.5
Disilicon hexachloride (hexachlorosilane)	Si_2Cl_6	Si—Cl	2.02 ± 0.02	Cl—Si—Cl	109.5 ± 1
		Si—Si	2.34 ± 0.06		
Fluorosilane	SiH_3F	Si—H	1.460 ± 0.01	H—Si—H	109.3 ± 0.3
		Si—F	1.595 ± 0.002		
Hydrogen phosphide	PH_3	P—H	1.415 ± 0.003	H—P—H	93.3 ± 0.2
Hydrogen selenide	SeH_2	Se—H	1.47	H—Se—H	91.0
Hydrogen sulfide	SH_2	S—H	1.3455	H—S—H	93.3
Hydrogen telluride	$Te—H_2$			H—Te—H	89.5 ± 1
Iodo silane	$Si—H_3I$	Si—H	1.48 ± 0.01	H—Si—H	109.9 ± 0.4
		Si—I	2.45 ± 0.09		
Methylgermane	Ch_3GeH_3			H—C—H	108.2 ± 0.5
				H—Ge—H	108.6 ± 0.5
Nitrosyl bromide	NOBr	O—N	1.15 ± 0.04	Br—N=O	117 ± 3
		N—Br	2.14 ± 0.02		
Nitrosyl chloride	NOCl	N—O	1.14 ± 0.02	Cl—N=O	113.0 ± 2
		N—Cl	1.97 ± 0.01		
Nitrosyl fluoride	NOF	N—O	1.13	F—N=O	110.0
		N—F	1.52		
cis-Nitrous acid	NO(OH)	H—O	0.98	O—N=O	114 ± 2
		O—N	1.46		
		N—O′	1.20		
trans-Nitrous acid	NO(OH)	H—O	0.98	O—N=O	118 ± 2
		O—N	1.46		
		NO′	1.20		
Oxygen chloride	OCl_2	O—Cl	1.70_1 ± 0.02	Cl—O—Cl	110.8 ± 1
Oxygen fluoride	OF_2	O—F	1.418	F—O—F	103.2
Phosphorus oxychloride	$POCl_3$	P—Cl	1.99_8 ± 0.02	Cl—P—Cl	103.5 ± 1
Phosphorus oxysulfide	$P_4O_6S_4$	P—O	1.61 ± 0.02	P—O—P	128.5 ± 1.5
				O—P—O	101.5 ± 1
		P—S	1.85 ± 0.02	O—P—S	116.5 ± 1
Phosphorus pentoxide	P_4O_{10}	P—O	1.62 ± 0.02	O—P—O	101.5 ± 1
		P—O′	1.38 ± 0.02	O—P—O′	116.5 ± 1
				P—O—P	123.5 ± 1

A. Inorganic Compounds (Continued)

Compound	Formula	Bond lengths in Å	Bond angles (°)
Phosphorus tribromide	PBr_3	P—Br 2.18 ± 0.03	Br—P—Br 101.5 ± 1.5
Phosphorus trichloride	PCl_3	P—Cl 2.043 ± 0.003	Cl—P—Cl 100.1 ± 0.3
Phosphorus trifluoride	PF_3	P—F 1.535	F—P—F 100.0
Phosphorus trioxide	P_4O_6	P—O 1.65 ± 0.02	{ O—P—O 99.0 ± 1
			{ P—O—P 127.5 ± 1
Stibine	SbH_3	Sb—H 1.7073 ± 0.0025	H—Sb—H 91.3 ± 0.33
Sulfur dichloride	SCl_2	S—Cl 1.99 ± 0.03	Cl—S—Cl 101.0 ± 4
Sulfur dioxide	SO_2	S—O 1.4321	O—S—O 119.536
Sulfur monochloride	S_2Cl_2	S—Cl 2.01 ± 0.07	Cl—S—S 104.5 ± 0.25
Sulfurylchloride	SO_2Cl_2	S—O 1.43 ± 0.02	O—S—O 119.75 ± 5
		S—Cl 1.99 ± 0.02	Cl—S—Cl 111.20 ± 2
			Cl—O—O 106.5 ± 2
Tellurium bromide	$TeBr_2$	Te—Br 2.51 ± 0.02	Br—Te—Br 98.0 ± 3
Tribromo silane	$SiHBr_3$	Si—Br 2.16 ± 0.03	Br—Si—Br 110.5 ± 1.5
Trichloro germane	$GeHCl_3$	Ge—H 1.55 ± 0.04	Cl—Ge—Cl 108.3 ± 0.2
Trichloro silane	$Si—HCl_3$	Si—H 1.47	Cl—Si—Cl 109.4 ± 0.3
Trifluorochlorosilane	$SiClF_3$	Si—F 1.560 ± 0.005	F—Si—F 108.5 ± 1
		Si—Cl 1.989 ± 0.018	
Trifluorochlorogermane	$GeClF_3$	Ge—F 1.688 ± 0.0017	F—Ge—F 107.7 ± 1.5
		Gr—Cl 2.067 ± 0.005	
Trifluorosilane	$SiHF_3$	Si—H 1.455 ± 0.01	
		Si—F 1.565 ± 0.005	F—Si—F 108.3 ± 0.5
Vanadium oxytrichloride	$VOCl_3$	V—O 1.56 ± 0.04	Cl—V—Cl 111.2 ± 2
		V—Cl 2.12 ± 0.03	Cl—V—O 108.2 ± 2
Water	H_2O	O—H 0.958_4	H—O—H 104.45

B. Organic Compounds

Compound	Formula	Bond lengths in Å	Bond angles (°)
Acetaldehyde	CH_3COH	C—H 1.09	C—C═O 121 ± 2
		C—C 1.50 ± 0.02	
		C—O 1.22 ± 0.02	
Bromomethane	Ch_3Br	C—H 1.11 ± 0.01	H—C—H 111.2 ± 0.5
		C—Br 1.929	
Carbon tetrachloride	CCl_4	C—Cl 1.766 ± 0.003	Cl—C—Cl 109.5
		Cl—Cl 2.887 ± 0.004	
Chloromethane	CH_3Cl	C—H 1.11 ± 0.01	H—C—H 110 ± 2
		C—Cl 1.784 ± 0.003	
Dichloromethane	CH_2Cl_2	C—H 1.068 ± 0.005	H—C—H 112 ± 0.3
		C—Cl 1.7724 ± 0.0005	Cl—C—Cl 111.8
Difluorochloromethane	$CHCLF_2$	C—H 1.06	F—C—F 110.5 ± 1
		C—Cl 1.73 ± 0.03	Cl—C—Cl 110.5 ± 1
		C—F 1.36 ± 0.03	
1,1-Difluoroethylene	$C_2H_2F_2$	C—H 1.07 ± 0.02	F—C═C 125.2 ± 0.2
		C—F 1.321 ± 0.015	H—C—H 117 ± 7
		C—C 1.311 ± 0.035	
Difluoromethane	CH_2F_2	C—F 1.360 ± 0.005	F—C—F 108.2 ± 0.8
		C—H 1.09 ± 0.03	H—C—H 112.5 ± 6
p-Dinitrobenzene	$C_6H_4(NO_2)_2$	C—C 1.38	O—N═O 124.0
		C—N 1.48	
		N—O 1.21	
Dithio oxamide	$NH_2CSCSNH_2$	C—C 1.54	N—C═S 124.8
		C—N 1.30	S—C—C 124.87
		C—S 1.66	N—C—C 115.25
Ethane	C_2H_6	C—H 1.107	H—C—H 109.3
		C—C 1.536	
Ethylidene fluoride	CH_3CHF_2	C—F 1.345 ± 0.001	F—C—F 109.15 ± 0.001
		C—C 1.540	C—C—F 109.4
		C—H 1.100	H—C—C 110.2

B. Organic Compounds (Continued)

Compound	Formula	Bond lengths in Å	Bond angles (°)
Fluorochloromethane	CH_2ClF	C—H 1.078 ± 0.005	Cl—C—F 100.0 ± 0.1
		C—Cl 1.759 ± 0.003	
		C—F 1.378 ± 0.006	
Fluorotrichloromethane	$CFCl_3$	C—Cl 1.76 ± 0.02	Cl—C—Cl 113 ± 3
		C—F 1.44 ± 0.04	
Formaldehyde	CH_2O	C—H 1.060 ± 0.038	H—C—H 125.8 ± 7
		C—O 1.230 ± 0.017	
Formamide	$HCONH_2$	C—O 1.25_5	N—C═O 121.5
		C—N 1.300	
Formic acid	$HCOOH$	C—O′ 1.245	O—C═O′ 124.3
		C—O 1.312	H—C—O′ 117.8
		C—H 1.085	C—O—H 107.8
		O—H 0.95	
Glycine	NH_2CH_2COOH	C—C 1.52	C—C—O 119.0
		C—O 1.27	O—C—O 122.0
		C—N 1.39	C—C—N 112.0
Hexachloroethane	C_2Cl_6	C—Cl 1.74 ± 0.01	Cl—C—Cl 109.3 ± 0.01
		C—C 1.57 ± 0.06	
Iodomethane	CH_3I	C—H 1.11 ± 0.01	H—C—H 111.4 ± 0.1
		C—I 2.139	
Methane	CH_4	C—H 1.091	
Methanethiol	CH_3SH	C—H 1.1039 ± 0.002	110.3 ± 0.2
		C—S 1.8177 ± 0.0002	H—S—C 100.3 ± 0.2
		S—H 1.329 ± 0.004	
Methanol	CH_3OH	C—H 1.096 ± 0.01	H—C—H 109.3 ± 0.75
		C—O 1.427 ± 0.007	C—O—H 108.9 ± 2
		O—H 0.956 ± 0.015	
Methylamine	CH_3NH_2	C—H 1.093	H—C—H 109.5 ± 1
		C—N 1.474 ± 0.005	H—N—H 105.8 ± 1
		N—H 1.014	
Methylether	$(CH_3)_2O$	C—O 1.43 ± 0.03	C—O—C 110.0 ± 3
Methylnitrite	CH_3NO_2	C—H 1.00	O—N═O 127 ± 4
		C—N 1.49 ± 0.02	
		N—O 1.22 ± 0.1	
Methylsulfide	$(CH_3)_2S$	C—S 1.82 ± 0.01	C—S—C 105 ± 3
		C—H 1.06	H—C—H 109.5
Oxamide	$NH_2COCONH_2$	C—C 1.54	N—C═O 125.7 ± 0.3
		C—O 1.24	
		C—N 1.32	
Phosgene	CCl_2O	C—Cl 1.746 ± 0.004	Cl—C═O 124.3 ± 0.1
		C—O 1.166 ± 0.002	Cl—C—Cl 111.3 ± 0.1
Propylene	C_3H_6		C—C═C 124.75 ± 0.3
Propynal	$CHC.COH$	C_1—C_2 1.204	C—C═O 123.0
		C_2—C_3 1.46	C—C—H 120.0
		C—O 1.21	
		C_1—H 1.06	
		C_3—H 1.08	
Tribromomethane	$CHBr_3$	C—H 1.068 ± 0.01	Br—C—Br 110.8 ± 0.3
		C—Br 1.930 ± 0.003	
Trichlorobromomethane	$CBrCl_3$	C—Br 1.936	Cl—C—Cl 111.2 ± 1
		C—Cl 1.764	
Trifluorochloromethane	$CClF_3$	C—Cl 1.751 ± 0.004	F—C—F 108.6 ± 0.4
		C—F 1.328 ± 0.02	
Trifluoromethane	CHF_3	C—H 1.098	F—C—F 108 ± 0.75
		C—F 1.332 ± 0.008	
Triiodomethane	CHI_3	C—I 2.12 ± 0.04	I—C—I 113.0 ± 1
Trimethylamine	$(CH_3)_3N$	C—N 1.47 ± 0.01	C—N—C 108.0 ± 4
		C—H 1.06	H—C—H 109.5
Trimethylarsine	$(CH_3)_3As$	C—H 1.09	C—As—C 96 ± 5
		C—As 1.98 ± 0.02	
Trimethylphosphine	$(CH_3)_3P$	C—H 1.09	C—P—C 100.0 ± 4
		C—P 1.87 ± 0.02	

BOND LENGTHS AND ANGLES OF CHEMICAL COMPOUNDS (Continued)

STRENGTHS OF CHEMICAL BONDS*

J. A. Kerr and A. F. Trotman-Dickenson

The strength of a chemical bond, D(R − X), often known as the bond dissociation energy, is defined as the heat of the reaction: RX → R + X. It is given by D(R − X) = ΔHf°(R) + ΔHf°(X) − ΔHf°(RX). Some authors list bond strengths for O K but here the values for 298 K are given because more thermodynamic data are available for this temperature. Bond strengths, or bond dissociation energies, are not equal to, and may differ considerably from, mean bond energies derived solely from thermochemical data on molecules and atoms.

Bond Strengths in Diatomic Molecules

These have usually been measured spectroscopically or by mass spectrometric analysis of hot gases effusing from a Knudsen cell. Excellent accounts of these and other methods are given in Reference 61. The errors quoted are those given in the original paper or review article. The references have been chosen primarily as a key to the literature. It should not be assumed that the author referred to was responsible for the determination quoted, as the reference may be only to a review article.

D°_o values have been converted to D°_{298} by use of the simple relation

$$D^\circ_{298} = D^\circ_O + (3/2)RT$$

The table has been arranged in alphabetical order with no regard to electronegativity or any other property of the atoms.

Table 1
BOND STRENGTHS IN DIATOMIC MOLECULES

Molecule	D°_{298}/kcal mol^{-1}	D°_{298}/kJ mol^{-1}	Ref.
Ag—Ag	39 ± 2	163 ± 8	37
Ag—Al	43.9 ± 2.2	183.7 ± 9.2	36
Ag—Au	48.5 ± 2.2	202.9 ± 9.2	4
Ag—Bi	46 ± 10	193 ± 42	117
Ag—Br	70 ± 7	293 ± 29	61
Ag—Cl	81.6	341.4	82
Ag—Cu	41.6 ± 2.2	174.1 ± 9.2	65
Ag—Eu	30.3 ± 3	126.8 ± 12.6	28a
Ag—F	84.7 ± 3.9	354.4 ± 16.3	61
Ag—Ga	43 ± 4	180 ± 15	21
Ag—Ge	41.7 ± 5.0	174.9 ± 20.9	131
Ag—H	54 ± 2	226 ± 8	61
Ag—Ho	29.5 ± 4.0	123.4 ± 16.7	25
Ag—I	56 ± 7	234 ± 29	61
Ag—In	42 ± 4	176 ± 17	15
Ag—Li	42.4 ± 1.5	177.4 ± 6.3	131b
Ag—Mn	24 ± 5	100 ± 21	117
Ag—Na	32.7 ± 3.0	136.8 ± 12.6	142
Ag—Nd	<50	<209	106
Ag—O	51 ± 20	213 ± 84	17
Ag—Sn	32.5 ± 5.0	136.0 ± 20.9	1
Ag—Te	70 ± 23	293 ± 96	61
Al—Al	44.5 ± 2.2	186.2 ± 9.2	82a, 22b

* Revised to 30 June 1977.

Table 1
BOND STRENGTHS IN DIATOMIC MOLECULES

Molecule	$D°_{298}$/kcal mol^{-1}	$D°_{298}$/kJ mol^{-1}	Ref.
Al—As	42.9	179.5	139b
Al—Au	77.9 ± 1.5	325.9 ± 6.3	77
Al—Br	106 ± 2	444 ± 8	37
Al—Cl	118 ± 3	494 ± 13	37
Al—Cu	51.8 ± 2.5	216 ± 10.5	137
Al—D	69.5	290.8	117
Al—F	158.6 ± 1.5	663.6 ± 6.3	37
Al—H	68.1 ± 1.5	284.9 ± 6.3	37
Al—I	88 ± 1	368± 4	37
Al—Li	42.0 ± 3.5	175.7 ± 14.6	82a
Al—N	71 ± 23	297 ± 96	61
Al—O	122.4 ± 1	512.1 ± 4.2	36a, 93
Al—P	51.8 ± 3.0	216.7 ± 12.6	37
Al—Pd	60.8 ± 2.9	254.4 ± 12.0	27a
Al—S	89.3 ± 1.9	373.6 ± 8.0	169
Al—Se	80.7 ± 2.4	337.7 ± 10.1	169
Al—Si	60 ± 3	251 ± 3	22b
Al—Te	64.0 ± 2.4	267.8 ± 10.1	169
Al—U	78 ± 7	326 ± 29	76
Ar—Ar	1.13 ± 0.01	4.73 ± 0.04	122
Ar—He	0.93	3.89	117
Ar—Hg	1.47	6.15	117
Ar—I	2.4	10.0	20
As—As	91.3 ± 2.5	382.0 ± 10.5	118
As—Cl	107	448	37
As—Ga	50.1 ± 0.3	209.6 ± 1.2	42
As—H	65 ± 3	272 ± 12	44
As—N	139 ± 30	582 ± 126	37
As—O	115 ± 2	481 ± 8	37
As—P	103.6 ± 3.0	433.5 ± 12.6	78
As—S	~114	~478	57, 134a
As—Se	23	96	136
As—Tl	47.4 ± 3.5	198.3 ± 14.6	144
At—At	~19	~80	49
Au—Au	53.8 ± 0.5	221.3 ± 2.1	120
Au—B	87.9 ± 2.5	367.8 ± 10.5	71
Au—Ba	38 ± 14	159 ± 59	61
Au—Be	68 ± 2	285 ± 8	61
Au—Bi	70 ± 20	293 ± 84	117
Au—Ca	46 ± 23	193 ± 96	61
Au—Ce	77.9 ± 3.5	325.9 ± 14.6	80
Au—Cl	82 ± 2.3	343 ± 9.6	61
Au—Co	51.3 ± 3.0	214.6 ± 12.6	104
Au—Cr	51.4 ± 1.5	215.1 ± 6.3	37
Au—Cu	55.4 ± 2.2	231.8 ± 9.2	4
Au—Eu	57.6 ± 2.5	241.0 ± 10.5	28a
Au—Fe	44.7 ± 4.0	187.0 ± 16.7	104
Au—Ga	70.2 ± 3.6	293.7 ± 15.1	21
Au—Ge	66.2 ± 3.5	277.0 ± 14.6	131
Au—H	75 ± 2.3	314 ± 9.6	61
Au—Ho	63.9 ± 8.0	267.4 ± 33.5	25, 120
Au—La	80.4 ± 5.0	336.4 ± 20.9	80
Au—Li	68.0 ± 1.6	284.5 ± 6.5	131b
Au—Lu	79.4 ± 4.0	332.2 ± 16.7	63
Au—Mg	58 ± 10	243 ± 42	117
Au—Mn	44.3 ± 3.0	185.4 ± 12.6	157
Au—Nd	71.5 ± 5.0	299.2 ± 20.9	80
Au—Ni	59 ± 5	247 ± 21	104
Au—Pb	31 ± 10	130 ± 42	117

Table 1 (continued)
BOND STRENGTHS IN DIATOMIC MOLECULES

Molecule	$D°_{298}$/kcal mol^{-1}	$D°_{298}$/kJ mol^{-1}	Ref.
Au—Pd	34.2 ± 5.0	143.1 ± 20.9	5
Au—Pr	72.9 ± 5.0	305.0 ± 20.9	80
Au—Rh	55.2 ± 7	230.9 ± 29	28a
Au—S	100 ± 6	418 ± 25	62
Au—Sc	67.0 ± 4.0	280.3 ± 16.7	79
Au—Si	74.5 ± 2.9	311.7 ± 12.1	173
Au—Sn	58.4 ± 4.0	244.3 ± 16.7	1
Au—Sr	63 ± 10	264 ± 42	117
Au—Tb	70.1	293.3	65, 120
Au—Te	59 ± 16	247 ± 67	61
Au—U	76 ± 7	318 ± 29	76
Au—Y	72.0 ± 4.0	301.3 ± 16.7	79
B—B	71 ± 5	297 ± 21	37
B—Br	101 ± 5	423 ± 21	61
B—C	107 ± 7	448 ± 29	117
B—Ce	73 ± 5	305 ± 21	117
B—Cl	131 ± 5	548 ± 21	117
B—D	81.5 ± 1.5	341.0 ± 6.3	117
B—F	183 ± 3	766 ± 13	37
B—H	80.7 ± 1.3	337.7 ± 5.4	117
B—I	91 ± 5	381 ± 21	117
B—Ir	122.9 ± 4.1	514.2 ± 17.2	173
B—La	81 ± 15	339 ± 63	117
B—N	93 ± 5	389 ± 21	37
B—O	192.7 ± 1.2	806.3 ± 5.0	167
B—P	82.9 ± 4.0	346.9 ± 16.7	73
B—Pd	78.7 ± 5.0	329.3 ± 20.9	173
B—Pt	114.2 ± 4.0	477.8 ± 16.7	127
B—Rh	113.7 ± 5.0	475.7 ± 20.9	173
B—Ru	106.8 ± 5.0	446.9 ± 20.9	173
B—S	138.8 ± 2.2	580.7 ± 9.2	167
B—Sc	66 ± 15	276 ± 63	117
B—Se	110.4 ± 3.5	461.9 ± 14.6	167
B—Si	69 ± 7	289 ± 29	117
B—Te	84.7 ± 4.8	354.4 ± 20.1	167
B—Th	71	297	66
B—Ti	66 ± 15	276 ± 63	117
B—U	77 ± 8	322 ± 34	117
B—Y	70 ± 15	293 ± 63	117
Ba—Br	88.4 ± 2.0	369.9 ± 8.4	114
Ba—Cl	106 ± 3	444 ± 13	96
Ba—F	140.3 ± 1.6	587.0 ± 6.7	95
Ba—H	42 ± 3.5	176 ± 15	61
Ba—I	⩾103 ± 1	⩾431 ± 4	43a
Ba—O	⩾134.5 ± 3.5	⩾562.8 ± 14.6	54a, 55a
Ba—S	95.6 ± 4.5	400.0 ± 18.8	31
Be—Be	14	59	48
Be—Br	91 ± 20	381 ± 84	117
Be—Cl	92.8 ± 2.2	388.3 ± 9.2	55c, 101
Be—F	138 ± 10	577 ± 42	37, 55c
Be—H	50.7 ± 0.3	212.1 ± 1.3	28b
Be—O	107 ± 5	448 ± 21	37
Be—S	89 ± 14	372 ± 59	61
Bi—Bi	47.9 ± 1.8	200.4 ± 7.5	148
Bi—Br	63.9 ± 1.0	267.4 ± 4.2	34
Be—Cl	73 ± 2	305 ± 8	117
Bi—D	67.8	283.7	125a
Bi—F	62 ± 7	259 ± 29	61
Bi—Ga	38 ± 4	159 ± 17	141

Table 1 (continued)
BOND STRENGTHS IN DIATOMIC MOLECULES

Molecule	$D°_{298}$/kcal mol^{-1}	$D°_{298}$/kJ mol^{-1}	Ref.
Bi—H	66.7	279.1	125a
Bi—I	52.1 ± 1.1	218.0 ± 4.6	35
Bi—O	81.9 ± 1.4	342.7 ± 5.9	168
Bi—P	67 ± 3	280 ± 13	78
Bi—Pb	33.9 ± 3.5	141.8 ± 14.6	148
Bi—S	75.4 ± 1.1	315.5 ± 4.6	168
Bi—Sb	60 ± 1	251 ± 4	115
Bi—Se	67.0 ± 1.4	280.3 ± 5.9	168
Bi—Te	55.5 ± 2.7	232.2 ± 11.3	168
Bi—Tl	29 ± 3	121 ± 13	43
Br—Br	46.336 ± 0.001	193.870 ± 0.004	9a, 123
Br—C	67 ± 5	280 ± 21	61
Br—Ca	76.6 ± 5.5	320.5 ± 23.0	114
Br—Cd	38 ± 23	159 ± 96	61
Br—Cl	52.3 ± 0.2	218.8 ± 0.8	61
Br—Co	79 ± 10	331 ± 42	117
Br—Cr	78.4 ± 5.8	328.0 ± 24.3	61
Br—Cs	95.0 ± 1.0	397.5 ± 4.2	117
Br—Cu	79 ± 6	331 ± 25	61
Br—F	55.89 ± 0.05	233.84 ± 0.21	23a, 61
Br—Fe	59 ± 23	247 ± 96	61
Br—Ga	106 ± 4	444 ± 17	37
Br—Ge	61 ± 7	255 ± 29	61
Br—H	87.4 ± 0.5	365.7 ± 2.1	61
Br—Hg	17.4 ± 1	72.8 ± 4.2	37
Br—I	42.8 ± 0.1	179.1 ± 0.4	61
Br—In	100 ± 5	418 ± 21	37
Br—K	91.5 ± 2.0	382.8 ± 8.4	37
Br—Li	101 ± 5	423 ± 21	37
Br—Mg	71 ± 15	297 ± 63	117
Br—Mn	75.1 ± 2.3	314.2 ± 9.6	61
Br—N	66 ± 5	276 ± 21	61
Br—Na	88.5 ± 3.0	370.3 ± 12.6	37
Br—Ni	86 ± 3	360 ± 13	61
Br—O	56.2 ± 0.1	235.1 ± 0.4	37
Br—Pb	59 ± 9	247 ± 38	61
Br—Rb	93 ± 3	389 ± 13	61
Br—Sb	75 ± 14	314 ± 59	61
Br—Sc	106 ± 15	444 ± 63	117
Br—Se	71 ± 20	297 ± 84	117
Br—Si	82 ± 12	343 ± 50	61
Br—Sn	81 ± 1	339 ± 4	117
Br—Sr	79.3 ± 4.5	331.8 ± 18.8	114
Br—Ti	105	439	117
Br—Tl	79.8 ± 0.4	333.9 ± 1.7	13
Br—V	105 ± 10	439 ± 42	117
Br—Y	116 ± 20	485 ± 84	117
Br—Zn	34 ± 7	142 ± 29	117
C—C	145 ± 5	607 ± 21	37
C—Ce	109 ± 7	456 ± 29	70
C—Cl	95 ± 7	397 ± 29	133
C—F	128 ± 5	536 ± 21	61
C—Ge	110 ± 5	460 ± 21	61
C—H	80.6 ± 0.2	337.2 ± 0.8	91
C—Hf	129 ± 6	540 ± 25	162
C—I	50 ± 5	209 ± 21	61
C—Ir	149.3 ± 3.0	624.7 ± 12.6	127
C—La	121 ± 15	506 ± 63	117
C—N	184 ± 1	770 ± 4	38

Table 1 (continued)
BOND STRENGTHS IN DIATOMIC MOLECULES

Molecule	$D°_{298}$/kcal mol^{-1}	$D°_{298}$/kJ mol^{-1}	Ref.
C—O	257.3 ± 0.1	1076.5 ± 0.4	37
C—P	122.7 ± 2	513.4 ± 8	159
C—Pt	146.3 ± 1.5	612.1 ± 6.3	171
C—Rh	139.5 ± 1.5	583.7 ± 6.3	171
C—Ru	154.9 ± 3	648.1 ± 13	64
C—S	167.0 ± 2.0	698.7 ± 8.4	92
C—Sc	94 ± 15	393 ± 63	117
C—Se	139 ± 23	582 ± 96	61
C—Si	104 ± 5	435 ± 21	46
C—Th	116 ± 6	485 ± 25	162
C—Ti	104 ± 6	435 ± 25	162
C—U	111 ± 7	464 ± 29	70
C—V	112 ± 15	469 ± 63	70
C—Y	100 ± 15	418 ± 63	70
C—Zr	134 ± 6	561 ± 25	162
Ca—Ca	3.58 ± 0.11	14.98 ± 0.46	7
Ca—Cl	95 ± 3	398 ± 13	96
Ca—F	126 ± 5	527 ± 21	16
Ca—H	40.1	167.8	61
Ca—I	68 ± 15	285 ± 63	117
Ca—O	⩾110.6 ± 3.5	⩾462.8 ± 14.6	54a, 55b
Ca—S	75 ± 4.5	314 ± 18.8	31
Cd—Cd	2.7 ± 0.2	11.3 ± 0.8	61
Cd—Cl	49.4 ± 0.8	206.7 ± 3.4	117
Cd—F	73 ± 5	305 ± 21	14
Cd—H	16.5 ± 0.1	69.0 ± 0.4	61
Cd—I	33 ± 5	138 ± 21	61
Cd—In	33	138	117
Cd—O	<88	<368	61
Cd—S	48	201	128
Cd—Se	<75	<314	61
Ce—Ce	58 ± 5	243 ± 21	117
Ce—F	139 ± 10	582 ± 42	117
Ce—N	124 ± 5	519 ± 21	72
Ce—O	190 ± 3	795 ± 13	117
Ce—Pd	77.0	322.2	26
Ce—S	137 ± 3	573 ± 13	32
Ce—Se	118.2 ± 3.5	494.6 ± 14.6	130b
Ce—Te	93 ± 10	389 ± 42	117
Cl—Cl	57.978 ± 0.004	242.580 ± 0.016	123, 187
Cl—Co	95 ± 2	398 ± 8	117
Cl—Cr	87.5 ± 5.8	366.1 ± 24.3	61
Cl—Cs	105 ± 5	439 ± 21	37
Cl—Cu	91.5 ± 1.1	382.8 ± 21	84
Cl—Eu	~78	~326	59
Cl—F	59.88 ± 0.02	250.54 ± 0.08	61, 131c
Cl—Fe	~84	~352	61
Cl—Ga	115 ± 3	481 ± 13	37
Cl—Ge	82 ± 5	343 ± 21	61
Cl—H	103.2 ± 0.1	431.8 ± 0.4	37
Cl—Hg	24 ± 2	100 ± 8	61
Cl—I	50.5 ± 0.1	211.3 ± 0.4	61
Cl—In	105 ± 2	439 ± 8	37
Cl—K	102 ± 2	427 ± 8	37
Cl—Li	112 ± 3	469 ± 13	37
Cl—Mg	76.1 ± 3.0	318.4 ± 12.6	96
Cl—Mn	86.2 ± 2.3	360.7 ± 9.6	61
Cl—N	93 ± 12	389 ± 50	61
Cl—Na	98 ± 2	410 ± 8	37

Table 1 (continued)
BOND STRENGTHS IN DIATOMIC MOLECULES

Molecule	$D°_{298}$/kcal mol^{-1}	$D°_{298}$/kJ mol^{-1}	Ref.
Cl—Ni	89 ± 5	372 ± 21	61
Cl—O	65 ± 1	272 ± 4	37
Cl—P	69 ± 10	289 ± 42	117
Cl—Pb	72 ± 7	301 ± 29	61
Cl—Ra	82 ± 18	343 ± 75	61
Cl—Rb	107 ± 5	448 ± 21	37
Cl—Sb	86 ± 12	360 ± 50	61
Cl—Sc	76	318	117
Cl—Se	77	322	117
Cl—Si	109 ± 10	456 ± 42	117
Cl—Sm	⩾101 ± 3	423 ± 13	181a
Cl—Sn	99 ± 4	414 ± 17	117
Cl—Sr	97 ± 3	406 ± 13	96
Cl—Ti	118	494	117
Cl—Tl	89.1 ± 0.5	372.8 ± 2.1	13
Cl—V	114 ± 15	477 ± 63	117
Cl—W	101 ± 10	423 ± 42	117
Cl—Y	126 ± 20	527 ± 42	117
Cl—Yb	~77	~322	59
Cl—Zn	54.7 ± 4.7	228.9 ± 19.7	33
Cm—O	176	736	155
Co—Co	40 ± 6	167 ± 25	107
Co—Cu	38.7 ± 4.0	161.9 ± 16.7	110
Co—F	104 ± 15	435 ± 63	117
Co—Ge	57 ± 6	239 ± 25	109
Co—I	68 ± 5	285 ± 21	117
Co—O	88 ± 5	368 ± 21	17
Co—S	82 ± 5	343 ± 21	117
Co—Si	66 ± 4	276 ± 17	172
Cr—Cr	37 ± 5	155 ± 21	108
Cr—Cu	37 ± 5	155 ± 21	110
Cr—F	104.5 ± 4.7	437.2 ± 19.7	113
Cr—Ge	40.6 ± 7	169.9 ± 29	109
Cr—H	67 ± 12	280 ± 50	61
Cr—I	68.6 ± 5.8	287.0 ± 24.3	61
Cr—N	90.3 ± 4.5	377.8 ± 18.8	160
Cr—O	102 ± 7	427 ± 29	37
Cr—S	81 ± 5	339 ± 21	117
Cs—Cs	9.97 ± 0.22	41.75 ± 0.93	117
Cs—F	122.9 ± 2	514.2 ± 8	117
Cs—H	42.6 ± 0.9	178.2 ± 3.8	147, 165a
Cs—I	81 ± 1	339 ± 4	117
Cs—O	71 ± 6	297 ± 25	117
Cu—Cu	48.2 ± 2	201.7 ± 4	137
Cu—F	103 ± 3	431.0 ± 13	94
Cu—Ga	51.6 ± 3.6	215.9 ± 15.1	21
Cu—Ge	49.9 ± 5	208.8 ± 21	131
Cu—H	67 ± 2	280 ± 8	19
Cu—I	47 ± 5	197 ± 21	61
Cu—Li	46.1 ± 2.1	192.9 ± 8.8	131b
Cu—Na	42.1 ± 4.0	176.2 ± 16.7	143
Cu—Ni	49.2 ± 4	205.9 ± 17	110
Cu—O	82 ± 15	343 ± 63	17
Cu—S	68 ± 4	285 ± 17	117
Cu—Se	70 ± 9	293 ± 38	61
Cu—Sn	42.3 ± 4	177.0 ± 17	1
Cu—Te	42 ± 9	176 ± 38	61
D—D	106.010 ± 0.001	443.546 ± 0.004	90
D—Ga	<65.2	<272.8	121

Table 1 (continued)
BOND STRENGTHS IN DIATOMIC MOLECULES

Molecule	D°_{298}/kcal mol^{-1}	D°_{298}/kJ mol^{-1}	Ref.
D—H	105.030 ± 0.001	439.446 ± 0.004	90
D—Li	57.4066 ± 0.0011	240.1892 ± 0.0046	164, 101a
Dy—F	126 ± 5	527 ± 21	117
Dy—O	146 ± 10	611 ± 42	17
Dy—S	99 ± 10	414 ± 42	117
Dy—Se	77 ± 10	322 ± 42	117
Dy—Te	56 ± 10	234 ± 42	117
Er—F	135 ± 4	565 ± 17	117
Er—O	146 ± 3	611 ± 13	156
Er—S	100 ± 10	418 ± 42	117
Er—Se	78 ± 10	326 ± 42	117
Er—Te	57 ± 10	239 ± 42	117
Eu—Eu	8.0 ± 4.0	33.5 ± 16.7	28a
Eu—F	126.1 ± 4.4	527.6 ± 18.4	184, 43b
Eu—O	133 ± 3	557 ± 13	156, 43b
Eu—Rh	55.9 ± 8.0	233.9 ± 33.5	28a
Eu—S	86.9 ± 3.5	363.6 ± 14.6	156, 130b
Eu—Se	72 ± 3.5	301 ± 14.6	11, 130b
Eu—Te	58 ± 3.5	243 ± 14.6	11, 130b
F—F	37.5 ± 2.3	156.9 ± 9.6	39
F—Ga	138 ± 3.5	577 ± 14.6	130
F—Gd	141.1 ± 6.5	590.4 ± 27.2	184
F—Ge	116 ± 5	485 ± 21	54
F—H	135.9 ± 0.3	568.6 ± 1.3	61
F—Hg	31 ± 9	130 ± 38	61
F—Ho	131 ± 4	548 ± 17	117
F—I	67 ± 1	280 ± 4	61, 5a, 14a, 33a
F—In	121 ± 3.5	506 ± 14.6	130
F—K	118.9 ± 0.6	497.5 ± 2.5	9
F—La	143 ± 10	598 ± 42	117
F—Li	138 ± 5	577 ± 21	37
F—Lu	136 ± 10	569 ± 42	117
F—Mg	110.4 ± 1.2	461.9 ± 5.0	95
F—Mn	101.2 ± 3.5	423.4 ± 14.6	112
F—N	72 ± 10	301 ± 42	37
F—Na	115 ± 2	481 ± 8	117
F—Nd	130.3 ± 3.0	545.2 ± 12.6	183
F—Ni	104	435	117
F—O	53 ± 4	222 ± 17	24
F—P	105 ± 23	439 ± 96	61
F—Pb	85 ± 2	356 ± 8	182
F—Pm	129 ± 10	540 ± 42	117
F—Pr	139 ± 11	582 ± 46	117
F—Pu	128.7 ± 7	538.5 ± 29	111
F—Rb	118 ± 5	493.7 ± 21	37
F—S	81.9 ± 1.2	342.7 ± 5.0	98
F—Sb	105 ± 23	439 ± 96	61
F—Sc	140.8 ± 3	589.1 ± 13	185
F—Se	81 ± 10	339 ± 42	117
F—Si	129 ± 3	540 ± 13	87
F—Sm	126.9 ± 4.4	531.0 ± 18.4	184,43b,181a
F—Sn	111.5 ± 3	466.5 ± 13	182
F—Sr	129.5 ± 1.6	541.8 ± 6.7	95
F—Tb	134 ± 10	561 ± 42	117
F—Ti	136 ± 8	569 ± 34	186
F—Tl	106.4 ± 4.6	445.2 ± 19.3	13
F—Tm	136 ± 10	569 ± 42	117

Table 1 (continued)
BOND STRENGTHS IN DIATOMIC MOLECULES

Molecule	$D°_{298}$/kcal mol^{-1}	$D°_{298}$/kJ mol^{-1}	Ref.
F—V	141 ± 15	590 ± 63	117
F—W	131 ± 15	548 ± 63	117
F—Xe	3.1 ± 0.1	13.0 ± 0.4	166a
F—Y	144.6 ± 5.0	605.0 ± 20.9	185
F—Yb	⩾124.6 ± 2.3	⩾521.3 ± 9.6	59, 181a, 9a
F—Zn	88 ± 15	368 ± 63	117
F—Zr	149 ± 15	623 ± 63	117
Fe—Fe	24 ± 5	100 ± 21	125
Fe—Ge	50.4 ± 7	210.9 ± 29	109
Fe—O	97.7 ± 3	408.8 ± 13	93a
Fe—S	81 ± 5	339 ± 21	117
Fe—Si	71 ± 6	297 ± 25	172
Ga—Ga	33 ± 5	138 ± 21	117
Ga—H	<65.5	<274.1	121
Ga—I	81 ± 2.3	339 ± 9.6	61
Ga—Li	31.8 ± 3.5	133.1 ± 14.6	82a
Ga—O	68 ± 15	285 ± 63	17
Ga—P	54.9 ± 3.0	229.7 ± 12.6	81
Ga—Sb	45.9 ± 3.0	208.8 ± 12.6	139a
Ga—Te	60 ± 6	251 ± 25	170
Gd—O	171 ± 4	716 ± 17	156
Gd—S	125.5 ± 3.5	525.1 ± 14.6	156
Gd—Se	103 ± 3.5	431 ± 14.6	11
Gd—Te	82 ± 3.5	343 ± 14.6	11
Ge—Ge	65.4 ± 5	273.6 ± 21	131
Ge—H	76.8 ± 0.2	321.3 ± 0.8	10
Ge—Ni	67 ± 3	280 ± 13	110
Ge—O	158.2 ± 3.0	661.9 ± 12.6	17
Ge—Pd	62.3 ± 3.0	260.7 ± 12.6	135
Ge—S	131.7 ± 0.6	551.0 ± 2.5	47, 32
Ge—Se	117 ± 5	490 ± 21	117
Ge—Si	72 ± 5	301 ± 21	61
Ge—Te	96 ± 2	402 ± 8	117
H—H	104.207 ± 0.001	436.002 ± 0.004	90, 187
H—Hg	9.5	39.8	61
H—I	71.4 ± 0.2	298.7 ± 0.8	61
H—In	59 ± 2.3	247 ± 9.6	61
H—K	43.8 ± 3.5	183.3 ± 14.6	61
H—Li	56.895 ± 0.001	238.049 ± 0.004	177
H—Mg	47 ± 12	197 ± 50	61, 53
H—Mn	56 ± 7	234 ± 29	61
H—N	75 ± 4	314 ± 17	61
H—Na	48 ± 5	201 ± 21	61
H—Ni	69 ± 3	289 ± 13	117
H—O	102.3 ± 0.5	428.0 ± 2.1	37
H—P	82 ± 7	343 ± 29	61
H—Pb	42 ± 5	176 ± 21	61
H—Pt	84 ± 9	352 ± 38	61
H—Rb	40 ± 5	167 ± 21	61
H—S	82.3 ± 2.9	344.3 ± 12.1	103
H—Sc	~43	~180	151
H—Se	73 ± 0.5	305 ± 2.1	85
H—Si	71.34 ± 0.11	298.49 ± 0.46	117
H—Sn	63 ± 4	264 ± 17	61
H—Sr	39 ± 2	163 ± 8	61
H—Te	64 ± 0.5	268 ± 2.1	85
H—Ti	~38	~159	150
H—Tl	45 ± 2	188 ± 8	61

Table 1 (continued)
BOND STRENGTHS IN DIATOMIC MOLECULES

Molecule	$D°_{298}$/kcal mol^{-1}	$D°_{298}$/kJ mol^{-1}	Ref.
H—Yb	38 ± 9	159 ± 38	61
H—Zn	20.5 ± 0.5	85.8 ± 2.1	61
He—Hg	1.58	6.61	117
Hf—C	131 ± 15	548 ± 63	117
Hf—N	128 ± 7	534 ± 29	116
Hf—O	189.8 ± 2.0	794.1 ± 8.4	3
Hg—Hg	4.1 ± 0.5	17.2 ± 2.1	61
Hg—I	9	38	180
Hg—K	1.97 ± 0.05	8.24 ± 0.21	117
Hg—Na	>1.6	>6.7	186a
Hg—S	51	213	128
Hg—Se	≤40	≤167	117
Hg—Te	≤34	≤142	117
Hg—Tl	1	4	89
Ho—Ho	20 ± 4	84 ± 17	25
Ho—O	149 ± 10	623 ± 42	17
Ho—S	102.4 ± 3.5	428.4 ± 14.6	156
Ho—Se	80 ± 4	335 ± 17	11
Ho—Te	62 ± 4	259 ± 17	11
I—I	36.460 ± 0.002	152.549 ± 0.008	123
I—In	80	335	8
I—K	79 ± 3	331 ± 13	37
I—Li	84 ± 3	352 ± 13	37
I—Mg	~68	~285	12
I—Mn	67.6 ± 2.3	282.8 ± 9.6	61
I—N	38 ± 4	159 ± 17	117
I—Na	72 ± 2	301 ± 8	37
I—Ni	70 ± 5	293 ± 21	61
I—O	44 ± 5	184 ± 21	37
I—Pb	47 ± 9	197 ± 38	61
I—Rb	80 ± 3	335 ± 13	37
I—Si	81 ± 20	339 ± 84	117
I—Sn	56 ± 10	234 ± 42	117
I—Sr	63 ± 10	263 ± 42	117
I—Te	46 ± 10	193 ± 42	117
I—Ti	74 ± 10	310 ± 42	117
I—Tl	65 ± 2	272 ± 8	12
I—Zn	33 ± 7	138 ± 29	61
In—In	24 ± 2	100 ± 8	117
In—Li	22.1 ± 3.5	92.5 ± 14.6	82a
In—O	86 ± 5	360 ± 21	37
In—P	47.3 ± 2.0	197.9 ± 8.5	139b
In—S	69 ± 4	289 ± 17	30
In—Sb	36.3 ± 2.5	151.9 ± 10.5	40
In—Se	59 ± 4	247 ± 17	30
In—Te	52 ± 4	218 ± 17	30
Ir—O	84 ± 5	352 ± 21	117
Ir—Si	110.6 ± 5.0	462.8 ± 20.9	173
K—K	13.7 ± 1.0	57.3 ± 4.2	117
K—Na	15.2 ± 0.7	63.6 ± 2.9	61
K—O	57 ± 8	239 ± 34	17
Kr—Kr	1.3 ± 0.2	5.4 ± 0.8	22
Kr—O	<2	<8	117
La—La	59 ± 5	247 ± 21	176
La—N	124 ± 10	519 ± 42	117
La—O	191 ± 3	799 ± 13	156
La—Rh	126 ± 4	527 ± 17	28
La—S	136.9 ± 3.0	572.8 ± 12.6	163a, 32

Table 1 (continued)
BOND STRENGTHS IN DIATOMIC MOLECULES

Molecule	$D°_{298}$/kcal mol^{-1}	$D°_{298}$/kJ mol^{-1}	Ref.
La—Se	114 ± 4	477 ± 17	11, 130b
La—Te	91 ± 4	381 ± 17	11
La—Y	48.3	202.1	176
Li—Li	26.34 ± 1	106.48 ± 4.2	181, 163b, 174
Li—Na	21	88	117
Li—O	81.4 ± 1.5	340.6 ± 6.3	97
Lu—Lu	34 ± 8	142 ± 34	117
Lu—O	166 ± 3	695 ± 13	156
Lu—Pt	96 ± 8	402 ± 34	67
Lu—S	121.2 ± 3.5	507.1 ± 14.6	156
Lu—Se	100 ± 4	418 ± 17	11
Lu—Te	78 ± 4	326 ± 17	11
Mg—Mg	2.044 ± 0.001	8.552 ± 0.004	124
Mg—O	94.1 ± 8.4	393.7 ± 35	54b
Mg—S	74 ± 18	310 ± 75	117
Mn—Mn	10 ± 7	42 ± 29	117
Mn—Cu	37.9 ± 4	158.6 ± 17	110
Mn—O	96 ± 8	402 ± 34	17
Mn—S	72 ± 4	301 ± 17	178
Mn—Se	48 ± 3	201 ± 13	179
Mo—O	145.1 ± 8	607.1 ± 34	22c
N—N	225.94 ± 0.14	945.33 ± 0.59	117, 187
N—O	150.71 ± 0.03	630.57 ± 0.13	117
N—P	147.5 ± 5.0	617.1 ± 20.9	75
N—Pu	113 ± 15	473 ± 63	117
N—S	111 ± 5	464 ± 21	117
N—Sb	72 ± 12	301 ± 50	61
N—Sc	112 ± 20	469 ± 84	117
N—Se	91 ± 15	381 ± 63	117
N—Si	105 ± 9	439 ± 38	61
N—Ta	146 ± 20	611 ± 84	117
N—Th	138.0 ± 0.5	577.4 ± 2.1	69
N—Ti	111	464	161
N—U	127.0 ± 0.5	531.4 ± 2.1	68
N—V	114.1 ± 2	477.4 ± 8	55
N—Xe	5.5	23.0	88
N—Y	115 ± 15	481 ± 63	117
N—Zr	135 ± 6	565 ± 25	117
Na—Na	18.4	77.0	154
Na—O	61.3 ± 4	256.5 ± 17	99
Na—Rb	14 ± 0.9	59 ± 3.8	61
Nb—O	180 ± 3	753 ± 13	117
Nd—Nd	<39	<163	117
Nd—O	168 ± 8	703 ± 34	17
Nd—S	113.2 ± 3.5	473.6 ± 14.6	156
Nd—Se	92 ± 4	385 ± 17	11, 130b
Nd—Te	73 ± 4	305 ± 17	11
Ne—Ne	0.94	3.93	166
Ni—Ni	62.6 ± 0.6	261.9 ± 2.5	149
Ni—O	93.6 ± 0.9	391.6 ± 3.8	149
Ni—S	86 ± 5	360 ± 21	117
Ni—Si	76 ± 4	318 ± 17	172
Np—O	172 ± 7	720 ± 29	2
O—O	119.106 ± 0.048	498.340 ± 0.200	18, 187
O—Os	<142	<594	17
O—P	142.6	596.6	50
O—Pb	90.3 ± 1.0	377.8 ± 4.2	168, 132a

Table 1 (continued)
BOND STRENGTHS IN DIATOMIC MOLECULES

Molecule	D°_{298}/kcal mol^{-1}	D°_{298}/kJ mol^{-1}	Ref.
O—Pd	56 ± 7	234 ± 29	17
O—Pm	161 ± 15	674 ± 63	117
O—Pr	180 ± 4	753 ± 17	156
O—Pt	83 ± 8	347 ± 34	17
O—Pu	163 ± 15	682 ± 63	17
O—Rb	61 ± 20	255 ± 84	17
O—Rh	90 ± 15	377 ± 63	17
O—Ru	115 ± 15	481 ± 63	17
O—S	124.69 ± 0.03	521.70 ± 0.13	17
O—Sb	89 ± 20	372 ± 84	17
O—Sc	161 ± 3	674 ± 13	117
O—Se	101 ± 3	423 ± 13	117
O—Si	190.9 ± 2.0	798.7 ± 8.4	117
O—Sm	148 ± 3	619 ± 13	156, 43b
O—Sn	131 ± 5	548 ± 21	37
O—Sr	≥108.6 ± 3.5	≥454.4 ± 14.6	54a, 55b
O—Ta	192.4 ± 3	805.0 ± 13	157a
O—Tb	169 ± 3	707 ± 13	156
O—Te	93.4 ± 2.0	390.8 ± 8.4	129
O—Th	204 ± 3	854 ± 13	100, 131a
O—Ti	158.2 ± 3.7	661.9 ± 15.5	126a
O—Tm	133 ± 3	557 ± 13	156
O—U	181.9 ± 4.0	761.1 ± 16.7	163a
O—V	154 ± 5	644 ± 21	17
O—W	156 ± 6	653 ± 25	17
O—Xe	8.7	36.4	117
O—Y	170.9 ± 3.0	715.1 ± 3.0	163a
O—Yb	95.1 ± 1.5	397.9 ± 6.3	181a
O—Zn	67.9	284.1	81a
O—Zr	181.6 ± 2	759.8 ± 8.4	3, 130a
P—P	117.0 ± 2.5	489.5 ± 10.5	75
P—Pt	≤99.6 ± 4	≤416.7 ± 17	158
P—Rh	84.4 ± 4	353.1 ± 17	158
P—S	106 ± 2	444 ± 8	51
P—Sb	85.3	356.9	119
P—Se	86.9 ± 2.4	363.6 ± 10	51
P—Te	71.2 ± 2.4	297.9 ± 10.0	51
P—Th	90	377	66
P—U	71 ± 5	297 ± 21	117
P—W	73 ± 1	305 ± 4	74
Pb—Pb	81 ± 6	339 ± 25	78a
Pb—S	82.7 ± 0.4	346.0 ± 1.7	168
Pb—Se	72.4 ± 1	302.9 ± 4	168
Pb—Te	60 ± 3	251 ± 13	168
Pd—Pd	16 ± 5	67 ± 21	117
Pd—Si	74.9 ± 3.3	313.4 ± 13.8	173
Pm—S	101 ± 15	423 ± 63	117
Pm—Se	81 ± 15	339 ± 63	117
Pm—Te	61 ± 15	255 ± 63	117
Po—Po	44.4 ± 2.3	185.8 ± 9.6	61
Pr—S	117.7 ± 1.1	492.5 ± 4.6	58
Pr—Se	106.7 ± 5.5	446.4 ± 23.0	130b
Pr—Te	78 ± 10	326 ± 42	117
Pt—Si	119.8 ± 4.3	501.2 ± 18.0	173
Rb—Rb	10.9 ± 0.5	45.6 ± 2.1	140
Rh—Rh	68.2 ± 5.0	285.4 ± 20.9	27, 139
Rh—Si	94.4 ± 4.3	395.0 ± 18.0	173

Table 1 (continued)
BOND STRENGTHS IN DIATOMIC MOLECULES

Molecule	$D°_{298}$/kcal mol^{-1}	$D°_{298}$/kJ mol^{-1}	Ref.
Rh—Ti	93.4 ± 3.5	390.8 ± 14.6	27
Ru—Si	94.9 ± 5.0	397.1 ± 20.9	173
Ru—Th	141.4 ± 10	591.6 ± 42	64
S—S	101.58 ± 0.01	425.01 ± 0.04	117
S—Sb	90.5	378.7	56
S—Sc	114.3 ± 3.0	478.2 ± 12.6	163a, 32,
S—Se	91 ± 5	381 ± 21	47
S—Si	148 ± 3	619 ± 13	61
S—Sm	93	389	58
S—Sn	111 ± 0.8	464 ± 3.2	47
S—Sr	75 ± 5	314 ± 21	31
S—Tb	123 ± 10	515 ± 42	117
S—Te	81 ± 5	339 ± 21	47
S—Ti	101.8 ± 1.8	425.9 ± 7.5	52
S—Tm	88 ± 10	368 ± 42	117
S—U	124.9 ± 2.3	522.6 ± 9.6	163a
S—V	117 ± 4	490 ± 16	134
S—Y	126.3 ± 2.5	528.4 ± 10.5	163, 32
S—Yb	40	167	117
S—Zn	49 ± 3	205 ± 13	41
S—Zr	137.5 ± 4.0	575.3 ± 16.7	163a
Sb—Sb	71.5 ± 1.5	299.2 ± 6.3	40
Sb—Te	66.3 ± 0.9	277.4 ± 3.8	145, 165
Sc—Sc	38.9 ± 5	162.8 ± 21	65
Sc—Se	92 ± 4	385 ± 17	117
Sc—Te	69 ± 4	289 ± 17	117
Se—Se	79.5 ± 0.1	332.6 ± 0.4	168
Se—Si	127 ± 6	531 ± 25	61
Se—Sm	79.1 ± 3.5	331.0 ± 14.6	130b
Se—Sn	95.9 ± 1.4	401.3 ± 5.9	29
Se—Tb	101 ± 10	423 ± 42	117
Se—Te	64 ± 2	268 ± 8	47
Se—Ti	91 ± 10	381 ± 42	117
Se—Tm	66 ± 10	276 ± 42	117
Se—V	83 ± 5	347 ± 21	117
Se—Y	104 ± 3	435 ± 13	117
Se—Zn	32.6 ± 3.0	136.4 ± 12.6	41
Si—Si	78.1 ± 2.4	326.8 ± 10.0	22b
Si—Te	121 ± 9	506 ± 38	61
Sm—Te	65.1 ± 3.5	272.4 ± 14.6	130b
Sn—Sn	46.7 ± 4	195.4 ± 17	1
Sn—Te	76.3 ± 0.2	319.2 ± 0.8	61
Tb—Tb	31.4 ± 6.0	131.4 ± 25.1	120
Tb—Te	81 ± 10	339 ± 42	117
Te—Te	63.2 ± 0.2	264.4 ± 0.8	168
Te—Ti	69 ± 4	289 ± 17	117
Te—Tm	66 ± 10	276 ± 42	117
Te—Y	81 ± 3	339 ± 13	117
Te—Zn	~49	~205	61
Th—Th	≤69	≤289	66
Ti—Ti	33.8 ± 5	141.4 ± 21	105
Tl—Tl	15	63	117
U—U	53 ± 5	222 ± 21	117
V—V	57.9 ± 5	242.3 ± 21	105
Xe—Xe	1.56 ± 0.07	6.53 ± 0.30	22a, 152
Y—Y	38 ± 5	159 ± 21	117
Yb—Yb	4.9 ± 4	20.5 ± 17	83
Zn—Zn	7	29	154

REFERENCES

1. Ackeman, M., Drowart, J., Stafford, F. E., and Verhaegen, G., *J. Chem. Phys.*,36, 1557, 1962.
2. Ackermann, R. J., Faircloth, R. F., Rauh, E. G., and Thorn, R. J., *J. Inorg. Nucl. Chem.*, 28, 111, 1966.
3. Ackermann, R. J. and Rauh, E. G., *J. Chem. Phys.*, 60, 2266, 1974.
4. Ackerman, M., Stafford, F. E., and Drowart, J., *J. Chem. Phys.*, 33, 1784, 1960.
5. Ackerman, M., Stafford, F. E., and Verhaegen, G., *J. Chem. Phys.*, 36, 1560, 1962.
6. Appelman, E. H. and Clyne, M. A. A., *J. Chem. Soc. Faraday Trans. 1*, 71, 2072, 1975.
7. Balfour, W. J. and Whitlock, R. E., *J. Chem. Soc. D*, p. 1231, 1971.
8. Barrow, R. F., *Trans. Faraday Soc.*, 56, 952, 1960.
9. Barrow, R. F. and Caunt, A. D., *Proc. R. Soc. London Ser. A*, 219, 120, 1953.

9a. Barrow, R. F., Clark, T. C., Coxon, J., and Yee, K. K., *J. Mol. Spectrosc.*, 51, 428, 1974.

9b. Barrow, R. F. and Chojnicki, A. H., *J. Chem. Soc. Faraday Trans. 2*, 71, 728, 1975.

10. Barrow, R. F. and Deutsch, E. W., *Proc. Chem. Soc.*, p. 122, 1960.
11. Bergman, C., Coppens, P., Drowart, J., and Smoes, S., *Trans. Faraday Soc.*, 66, 800, 1970.
12. Berkowitz, J. and Chupka, W. A., *J. Chem. Phys.*, 45, 1287, 1966.
13. Berkowitz, J. and Walter, T., *J. Chem. Phys.*, 49, 1184, 1968.
14. Besenbruch, G., Kana'an, A. S., and Margrave, J. L., *J. Phys. Chem.*, 69, 3174, 1965.

14a. Birks, J. W., Gabelnick, S. D., and Johnston, H. S., *J. Mol. Spectrosc.*, 57, 23, 1975.

15. Biron, M., *C. R. Acad. Ser. B*, 265, 1026, 1427, 1967.
16. Blue, G. D., Green, J. W., Bautista, R. G., and Margrave, J. L., *J. Phys. Chem.*,67, 877, 1963.
17. Brewer, L., and Rosenblatt, G. M., *Adv. High Temp. Sci.*, 2, 1, 1969.
18. Brix, P. and Herzberg, G., *J. Chem. Phys.*, 21, 2240, 1953.
19. Bulewicz, E. M. and Sugden, T. M., *Trans. Faraday Soc.*, 52, 1475, 1956.
20. Burns, G., Le Roy, L. J., Morris, D. J., and Blake, J. A., *Proc. R. Soc. London Ser. A*, 316, 81, 1970.
21. Carbonel, M., Bergman, C., and Laffite, M., *Colloq. Int. Cent. Nat. Rech. Sci.*, 210, 311, 1972.
22. Chashchina, G. I. and Shreider, E. Ya., *Zh. Prikl. Spektrosk.*, 21, 696, 1974.

22a. Chashchina, G. I. and Shreider, E. Ya., *Zh. Prikl. Spektrosk.*, 25, 163, 1976.

22b. Chatillon, C., Allibert, M., and Pattoret, A., *C. R. Acad. Sci. Ser. C.*, 280, 1505, 1975.

22c. Choudary, U. V., Gingerich, K. A., and Kingcade, J. E., *J. Less Common Met.*, 42, 111, 1975.

23. Clements, R. M. and Barrow, R. F., *Trans. Faraday Soc.*, 64, 2893, 1968.

23a. Clyne, M. A. A., Curran, A. H., and Coxon, J. A., *J. Mol. Spectrosc.*, 63, 43, 1976.

24. Clyne, M. A. A. and Watson, R. T., *Chem. Phys. Lett.*, 12, 344, 1971.
25. Cocke, D. L. and Gingerich, K. A., *J. Phys. Chem.*, 75, 3264, 1971.
26. Cocke, D. L. and Gingerich, K. A., *J. Phys. Chem.*, 76, 2332, 1972.
27. Cocke, D. L. and Gingerich, K. A., *J. Chem. Phys.*, 60, 1958, 1974.

27a. Cocke, D. L., Gingerich, K. A., and Chang, C.-A., *J. Chem. Soc. Faraday Trans. 1*, 72, 268, 1976.

28. Cocke, D. L., Gingerich, K. A., and Kordis, J., *High Temp. Sci.*, 5, 474, 1973.

28a. Cocke, D. L., Gingerich, K. A., and Kordis, J., *High Temp. Sci.*, 7, 61, 1975.

28b. Colin, R. and De Greef, D., *Can. J. Phys.*, 53, 2142, 1975.

29. Colin, R. and Drowart, J., *Trans. Faraday Soc.*, 60, 673, 1964.
30. Colin, R. and Drowart, J., *Trans. Faraday Soc.*, 64, 2611, 1968.

31. Colin, R., Goldfinger, P., and Jeunehomme, M., *Trans. Faraday Soc.*, 60, 306, 1964.
32. Coppens, P., Smoes, S., and Drowart, J., *Trans. Faraday Soc.*, 63, 2140, 1967.
33. Corbett, J. D. and Lynde, R. A., *Inorg. Chem.*, 6, 2199, 1967.
33a. Coxon, J. A., *Chem. Phys. Lett.*, 33, 136, 1975.
34. Cubicciotti, D., *Inorg. Chem.*, 7, 208, 1968.
35. Cubicciotti, D., *Inorg. Chem.*, 7, 211, 1968.
36. Cuthill, A. M., Fabian, D. J., and Shu-Shou-Shen, S., *J. Phys. Chem.*, 77, 2008, 1973.
36a. Dagdigian, P. J., Cruze, H. W., and Zare, R. N., *J. Chem. Phys.*, 62, 1824, 1975.
37. Darwent, B. de B., *Bond Dissociation Energies in Simple Molecules*, NSRDS-NBS 31, National Bureau of Standards, Washington, D. C., 1970.
38. Davis, D. D. and Okabe, H., *J. Chem. Phys.*, 49, 5526, 1968.
39. De Corpo, J. J., Steiger, R. P., Franklin, J. L., and Margrave, J. L., *J. Chem. Phys.*, 53, 936, 1970.
40. De Maria, G., Drowart, J., and Inghram, M. G., *J. Chem. Phys.*, 31, 1076, 1959.
41. De Maria, G., Goldfinger, P., Malaspina, L., and Piacente, V., *Trans. Faraday Soc.*, 61, 2146, 1965.
42. De Maria, G., Malaspina, L., and Piacente, V., *J. Chem. Phys.*, 52, 1019, 1970.
43. De Maria, G., Malaspina, L., and Piacente, V., *J. Chem. Phys.*, 56, 1978, 1972.
43a. Dickson, C. R., Kinney, J. B., and Zare, R. N., *Chem. Phys.*, 15, 243, 1976.
43b. Dickson, C. R. and Zare, R. N., *Chem. Phys.*, 7, 361, 1975.
44. Dixon, R. N. and Lambertson, H. M., *J. Mol. Spectrosc.*, 25, 12, 1968.
45. Drowart, J., in *Phase Stability in Metals and Alloys*, Rudman, P. S., Ed., McGraw-Hill, New York, 1967, 305.
46. Drowart, J., De Maria, G., and Inghram, M. G., *J. Chem. Phys.*, 29, 1015, 1958.
47. Drowart, J. and Goldfinger, P., *Q. Rev.* (London), 20, 545, 1966.
48. Drowart, J. and Goldfinger, P., *Angew. Chem.*, 6, 581, 1967.
49. Drowart, J. and Honig, R. E., *J. Phys. Chem.*, 61, 980, 1957.
50. Drowart, J., Myers, C. E., Szwarc, R., Vander Auwera-Mahieu, A., and Uy, O. M., *J. Chem. Soc. Faraday Trans. 2*, 68, 1749, 1972.
51. Drowart, J., Myers, C. E., Szwarc, R., Vander Auwera-Mahieu, A., and Uy, O. M., *High Temp. Sci.*, 5, 482, 1973.
52. Edwards, J. G., Franklin, H. F., and Gilles, P. W., *J. Chem. Phys.*, 54, 545, 1971.
53. Ehlert, T. C. Hilmer, R. M., and Beauchamp, E. A., *J. Inorg. Nucl. Chem.*, 30, 3112, 1968.
54. Ehlert, T. C. and Margrave, J. L., *J. Chem. Phys.*, 41, 1066, 1964.
54a. Engelke, F., Sander, R. K., and Zare, R. N., *J. Chem. Phys.*, 65, 1146, 1976.
54b. Evans, P. J. and Mackie, J. C., *Chem. Phys.*, 5, 277, 1974.
55. Farber, M. and Srivastava, R. D., *J. Chem. Soc. Faraday Trans. 1*, 69, 390, 1973.
55a. Farber, M. and Srivastava, *High Temp. Sci.*, 7, 74, 1975.
55b. Farber, M. and Srivastava, *High Temp. Sci.*, 8, 73, 1976.
55c. Farber, M. and Srivastava, R.D.,J. *Chem. Soc. Faraday Trans. 1*,70,1581,1974
56. Faure, F. M., Mitchell, M. J., and Bartlett, R. W., *High Temp. Sci.*, 4, 181, 1972.
57. Faure, F. M., Mitchell, M. J., and Bartlett, R. W., *High Temp. Sci.*, 5, 128, 1973.
58. Fenochka, B. V. and Gorkienko, S. P., *Zh. Fiz, Khim.*, 47, 2445, 1973.
59. Filippenko, N. V., Motozov, E. V., Giricheva, N. I., and Krasnev, K. S., *Izv. Vyssh. Ucheb. Zaved Khim. Technol.*, 15, 1416, 1972.
60. Fujishiro, S., *Trans. Jap. Inst. Metals*, 1, 125, 1960.
61. Gaydon, A. G., *Dissociation Energies and Spectra of Diatomic Molecules*, 3rd ed., Chapman and Hall, London, 1968.

62. Gingerich, K. A., *Chem. Commun.*, 580, 1970.
63. Gingerich, K. A., *Chem. Phys. Lett.*, 13, 262, 1972.
64. Gingerich, K. A., *Chem. Phys. Lett.*, 25, 523, 1974.
65. Gingerich, K. A., *Chimia*, 26, 619, 1972.
66. Gingerich, K. A., *High Temp. Sci.*, 1, 258, 1969.
67. Gingerich, K. A., *High Temp. Sci.*, 3, 415, 1971.
68. Gingerich, K. A., *J. Chem. Phys.*, 47, 2192, 1967.
69. Gingerich, K. A., *J. Chem. Phys.*, 49, 19, 1968.
70. Gingerich, K. A., *J. Chem. Phys.*, 50, 2255, 1969.
71. Gingerich, K. A., *J. Chem. Phys.*, 54, 2646, 1971.
72. Gingerich, K. A., *J. Chem. Phys.*, 54, 3720, 1971.
73. Gingerich, K. A., *J. Chem. Phys.*, 56, 4239, 1972.
74. Gingerich, K. A., *J. Phys. Chem.*, 68, 768, 1964.
75. Gingerich, K. A., *J. Phys. Chem.*, 73, 2734, 1969.
76. Gingerich, K. A. and Blue, G. D., *J. Chem. Phys.*, 47, 5447, 1967.
77. Gingerich, K. A. and Blue, G. D., *J. Chem. Phys.*, 59, 186, 1973.
78. Gingerich, K. A., Cocke, D. L., and Kordis, J., *J. Phys. Chem.*, 78, 603, 1974.
78a. Gingerich, K. A., Cocke, D. L., and Miller, F., *J. Chem. Phys.*, 64, 4027, 1976.
79. Gingerich, K. A. and Finkbeiner, H. C., Proc 9th Rare Earth Res. Conf. 2, 795, 1971.
80. Gingerich, K. A. and Finkbeiner, H. C., *J. Chem. Phys.*, 54, 2621, 1971.
81. Gingerich, K. A. and Piacente, V., *J. Chem. Phys.*, 54, 2498, 1971.
81a. Grade, M., Hirschwald, W., and Stolze, F., *Z. Phys. Chem. Frankfurt am Main*, 100, 165, 1976.
82a. Guggi, D.J., Neubert, A., and Zmbov, K.F., Conf. Int. Thermodynamique Chimie
82a. Guggi, D. J., Neubert, A., and Zmbov, K. F., Conf. Int. Thermodyn. Chim. [C.R.] 4th, 3, 124, 1975.
83. Guido, M. and Balducci, G., *J. Chem. Phys.*, 57, 5611, 1972.
84. Guido, M., Gigli, G., and Balducci, G., *J. Chem. Phys.*, 57, 3731, 1972.
85. Gunn, S. R., *J. Phys. Chem.*, 68, 949, 1964.
86. Hariharan, A. V. and Eick, H. A., *J. Chem. Thermodyn.*, 6, 373, 1974.
87. Hastie, J. W., *J. Chem. Phys.*, 57, 4556, 1972.
88. Herman, R. and Herman, L., *J. Phys. Radium*, 24, 73, 1963.
89. Herzberg, G., *Molecular Spectra and Molecular Structure. I. Spectra of Diatomic Molecules*, 2nd ed., Van Nostrand, New York, 1950.
90. Herzberg, G., *J. Mol. Spectrosc.*,33, 147, 1970.
91. Herzberg, G. J. and Johns, J. W. G., *Astrophys. J.*, 158, 399, 1969.
92. Hildenbrand, D. L., *Chem. Phys. Lett.*, 15, 379, 1972.
93. Hildenbrand, D. L., *Chem. Phys. Lett.*, 20, 127, 1973.
93a. Hildenbrand, D. L., *Chem. Phys. Lett.*, 34, 352, 1975.
94. Hildenbrand, D. L., *J. Chem. Phys.*, 48, 2457, 1968.
95. Hildenbrand, D. L., *J. Chem. Phys.*, 48, 3657, 1968.
96. Hildenbrand, D. L., *J. Chem. Phys.*, 52, 5751, 1970.
97. Hildenbrand, D. L., *J. Chem. Phys.*, 57, 4556, 1972.
98. Hildenbrand, D. L., *J. Phys. Chem.*, 77, 897, 1973.
99. Hildenbrand, D. L. and Murad, E., *J. Chem. Phys.*, 53, 3403, 1970.
100. Hildenbrand, D. L. and Murad, E., *J. Chem. Phys.*, 61, 1232, 1974.
101. Hildenbrand, D. L. and Theard, L. P., *J. Chem. Phys.*, 50, 5350, 1969.
102. Ihle, H. R. and Wu, C. H., *J. Chem. Phys.*, 63, 1605, 1975.
103. Johns, J. W. C. and Ramsey, D. A., *Can J. Phys.*, 39, 210, 1961.
104. Kant, A., *J. Chem. Phys.*, 49, 5144, 1968.
105. Kant, A. and Lin, S-S., *J. Chem. Phys.*, 51, 1644, 1969.
106. Kant, A., Lin, S-S., and Strauss, B., *J. Chem. Phys.*, 49, 1983, 1968.
107. Kant, A. and Strauss, B. H., *J. Chem. Phys.* 41, 3806, 1964.
108. Kant, A. and Strauss, B., *J. Chem. Phys.*, 45, 3161, 1966.
109. Kant, A. and Strauss, B., *J. Chem. Phys.*, 49, 3579, 1968.
110. Kant, A., Strauss, B., and Lin, S-S., *J. Chem. Phys.*, 52, 2384, 1970.
111. Kent, R. A., *J. Am. Chem. Soc.*, 90, 5657, 1968.
112. Kent, R. A., Ehlert, T. C., and Margrave, J. L., *J. Am. Chem. Soc.*, 86, 5090, 1964.
113. Kent, R. A. and Margrave, J. L., *J. Am. Chem. Soc.*, 87, 5382, 1965.

114. Khitrov, A. N., Ryabova, V. G., and Gurvich, L. V., *Teplofiz Vys. Tempo.*, 11, 1126, 1973.
115. Kohl, F. J. and Carlson, K. D., *J. Am. Chem. Soc.*, 90, 4814, 1968.
116 Kohl, F. J. and Stearns, C. A., *J. Phys. Chem.*, 78, 273, 1974.
117. Kondratiev, V. N., *Bond Dissociation Energies, Ionization Potentials and Electron Affinities,* Mauka Publsihing House, Moscow, 1974.
118. Kordis, J. and Gingerich, K. A., *J. Chem. Eng. Data*, 18, 135, 1973.
119. Kordis, J. and Gingerich, K. A., *J. Phys. Chem.*, 76, 2336, 1972.
120. Kordis, J., Gingerich, K. A., and Seyse, R. J., *J. Chem. Phys.*, 61, 5114, 1974.
121. Kronekvist, M., Lagerqvist, A., and Neuhaus, H., *J. Mol. Spectrosc.*, 39, 516, 1971.
122. LeRoy, R. J., *J. Chem. Phys.*, 57, 573, 1972.
123. LeRoy, R. J. and Bernstein, R. B., *Chem. Phys. Lett.*, 5, 42, 1970.
124. Li, K. C. and Stwalley, W. C., *J. Chem. Phys.*, 59, 4423, 1973.
125. Lin, S-S, and Kant, A., *J. Phys. Chem.*, 73, 2450, 1969.
125a. Lindgren, B. and Nilsson, Ch., *J. Mol. Spectrosc.*, 55, 407, 1975.
126. Liu, M. B. and Wahlbeck, P. G., *J. Chem. Phys.*, 63, 1694, 1975.
127. McIntyre, N. S., Vander Auwera-Mahieu, A., and Drowart, J., *Trans. Faraday Soc.*, 64, 3006, 1968.
128. Marquart, J. R. and Berkowitz, J., *J. Chem. Phys.*, 39, 283, 1963.
129. Muenow, D. W., Hastie, J. W., Hauge, R., Bautista, R., and Margrave, J. L., *Trans Faraday Soc.*, 65, 3210, 1969.
130. Murad, E., Hildenbrand, D. L., and Main, R. P., *J. Chem. Phys.*, 45, 263, 1966.
130a. Murad, E. and Hildenbrand, D. L., *J. Chem. Phys.*, 63, 1139, 1975.
130b. Nagai, S. , Shinmei, M., and Yokokawa, T., *J. Inorg. Nucl. Chem.*, 36, 1904, 1974.
131. Neckel, A. and Sodeck, G., *Monatsh. Chem.*, 103, 367, 1972.
131a. Neubert, A. and Zmbov, K. F., *High Temp. Sci.*, 6, 303, 1974.
131b. Neubert, A. and Zmbov, K. F., *J. Chem. Soc. Faraday Trans. I*, 70, 2219, 1974.
131c. Nordine, P. C., *J. Chem. Phys.*, 61, 224, 1974.
132. O'Hare, P. A. G., *J. Chem. Phys.*,52, 2992, 1970.
132a. Oldenberg, R. C., Dickson C. R., and Zare, R. N., *J. Mol. Spectrosc.*, 58, 283, 1975.
133. Ovcharenko, I. E., Ya. Kuzyankov, Y., and Tatevaskii, V. M., *Opt. Spektrosk.*, 19, 528, 1965.
134. Owzarski, T. P. and Franzen, H. F., *J. Chem. Phys.*, 60, 1113, 1974.
134a. Pashinkin, A. S., Molodyk, A. D., Belousov, V. I., Strel'chenko, S. S., and Fedorova, V. A., *Izv. Akad. Nauk. USSR Neorg. Mater.*, 10, 1600, 1974.
135. Peeters, R., Vander Auwera-Mahieu, A., and Drowart, J., Z. *Naturforsch. Teil A*, 26, 327, 1971.
136. Pelevin, O. V., Mil'vidskii, M. G., Belyaev, A. I., and Khotin, B. A., *Izv. Akad. Nauk SSSR Neorg. Mater.*, 2, 942, 1966.
137. Perakis, J., Chatilion, C., and Pattoret, A., *C. R. Acad. Sci. Ser. C*, 276, 1357, 1973.
138. Petzel, T., *High Temp. Sci.*, 6, 246, 1974.
139. Piacente, V., Balducci, G., and Bardi, G., *J. Less-Common Met.*, 37, 123, 1974.
139a. Piacente, V. and Balducci, G., *High Temp. Sci.*, 6, 254, 1974.
139b. Piacente, V. and Balducci, G., *Dyn. Mass Spectrom.*, 4, 295, 1976.
140. Piacente, V., Bardi, G., and Malaspina, L., *J. Chem. Thermodyn.*, 5, 219, 1973.
141. Piacente, V. and Desideri, A., *J. Chem. Phys.*, 57, 2213, 1972.
142. Piacente, V. and Gingerich, K. A., *High Temp. Sci.*, 4, 312, 1972.
143. Piacente, V. and Gingerich, K. A., *Z. Naturforsch. Teil A*, 28, 316, 1973.
144. Piacente, V. and Malaspina, L., *J. Chem. Phys.*, 56, 1780, 1972.
145. Porter, R. F. and Spencer, C. W. J., *J. Chem. Phys.*, 32, 943, 1960.
146. Ringstrom, U., *Ark, Fys.*, 27, 227, 1964.
147. Ringstrom, U., *J. Mol. Spectrosc.*, 36, 232, 1970.
148. Rovner, L., Drowart, A., and Drowart, J., *Trans. Faraday Soc.*, 63, 2910, 1967.

149. Rutner, E. and Haurey, G. L., *J. Chem. Eng. Data*, 19, 19, 1974.
150. Scott, P. R. and Richards, W. G., *J. Phys. B*, 7, 500, 1974.
151. Scott, P. R. and Richards, W. G., *J. Phys. B*, 7, 1679, 1974.
152. Shardanand, A., *Phys. Rev.*, 160, 67, 1967.
153. Shenyavskaya, E. A., Mal'tsev, A. A., Kataev, D. I., and Gurvich, L. V., *Opt. Spektrosk.*, 26, 937, 1969,
154. Siegel, B., *Q. Rev.* (London), 19, 77, 1965.
155. Smith, P. K. and Peterson, D. E., *J. Chem. Phys.*, 52, 4963, 1970.
156. Smoes, S., Coppens, P., Bergman, C., and Drowart, J., *Trans. Faraday Soc.*, 65, 682, 1969.
157. Smoes, S. and Drowart, J., *Chem. Commun.*, p. 534, 1968.
157a. Smoes, S., Drowart, J., and Myers, C. E., *J. Chem. Thermodyn.*, 8, 225, 1976.
158. Smoes, S., Huguet, R., and Drowart, J., *Z. Naturforsch. Teil A*, 26, 1934, 1971.
159. Smoes, S., Myers, C. E., and Drowart, J., *Chem. Phys. Lett.*, 8, 10, 1971.
160. Srivastava, R. D.and Farber, M., *High Temp. Sci.*, 5, 489, 1973.
161. Stearns, C. A. and Kohl, F. J., *High Temp. Sci.* 2, 146, 1970; NASA Tech. Note 1969, NASA-TN-D-5027.
162. Stearns, C. A. and Kohl, F., *High Temp. Sci.*, 6, 284, 1974.
163. Steiger, R. A. and Cater, E. D., *High Temp. Sci.*, 7, 204, 1975.
163a. Steiger, R. A. and Cater, E. D., *High Temp. Sci.*, 7, 288, 1975.
163b. Stwalley, W. C., *J. Chem. Phys.*, 65, 2038, 1970.
164. Stwalley, W. C., Way, K. R., and Velasco, R., *J. Chem. Phys.*, 60, 3611, 1974.
165. Sullivan, C. L., Zehe, M. J., and Carlson, K. D., *High Temp. Sci.*, 6, 80, 1974.
165a. Tam, A. C. and Happer, W., *J. Chem. Phys.*, 64, 2456, 1976.
166. Tanaka, Y., Yushina, K., and Freeman, D. E., *J. Chem. Phys.*, 59, 564, 1973.
166a. Tellinghuisen, J., Tisone, G. C., Hoffman, J. M., and Hays, A. K., *J. Chem. Phys.*, 64, 4796, 1976.
166b. Tuenge, R. T., Laabs, F., and Franzen, H. F., *J. Chem. Phys.*, 65, 2400, 1976.
167. Uy, O. M. and Drowart, J., *High Temp. Sci.*, 2, 293, 1970.
168. Uy, O. M. and Drowart, J., *Trans. Faraday Soc.*, 65, 3221, 1969.
169. Uy, O. M. and Drowart, J., *Trans. Faraday Soc.*, 67, 1293, 1971.
170. Uy, O. M., Muenow, D. W., Ficalora, P. J., and Margrave, J. L., *Trans. Faraday Soc.*, 64, 2998, 1968.
171. Vander Auwera-Mahieu, A. and Drowart, J., *Chem. Phys. Lett.*, 1, 311, 1967.
172. Vander Auwera-Mahieu, A., McIntyre, N. S., and Drowart, J., *Chem. Phys. Lett.*, 4, 198, 1969.
173. Vander Auwera-Mahieu, A., Peeters, R., McIntyre, N. S., and Drowart, J., *Trans. Faraday Soc.*, 66, 809, 1970.
174. Velasco, R., Ottinger, C., and Zare, R. N., *J. Chem. Phys.*, 51, 5522, 1969.
175.
176. Verhaegen, G., Smoes, S., and Drowart, J., *J. Chem. Phys.*, 40, 239, 1964.
177. Way, K. R. and Stwalley, W. C., *J. Chem. Phys.*, 59, 5298, 1973.
178. Wiedemeier, H. and Gilles, P. W., *J. Chem. Phys.*, 42, 2765, 1965.
179. Wiedemeier, H. and Goyette, W. J., *J. Chem. Phys.*, 48, 2936, 1968.
180. Wieland, Von K., *Z. Elektrochem.*, 64, 761, 1960.
181. Wu, C. H., *J. Chem. Phys.*, 65, 3181, 1976; 65, 2040, 1976.
181a. Yokozeki, A. and Menzinger, M., *Chem. Phys.*, 14, 427, 1976.
182. Zmbov, K. F., Hastie, J. W., and Margrave, J. L., *Trans. Faraday Soc.*, 64, 861, 1968.
183. Zmbov, K. F. and Margrave, J. L., *J. Chem. Phys.*, 45, 3167, 1966.
184. Zmbov, K. F. and Margrave, J. L., *J. Inorg. Nucl. Chem.*, 29, 59, 1976.
185. Zmbov, K. F. and Margrave, J. L. *J. Chem. Phys.*, 47, 3122, 1967.
186. Zmbov, K. F. and Margrave, J. L., *J. Phys. Chem.*, 71, 2893, 1967.
186a. Zollweg, R.J., *Contrib. Pap. Int. Conf. Phenomena of Ionization Gases, 11th*, 402,
187. CODATA recommended key values for thermodynamics, 1973, *J. Chem. Thermodyn.*, 7, 1, 1975.

Heats of Formation of Gaseous Atoms from Elements in Their Standard States

For elements that are diatomic gases in their standard states these are readily obtained from the bond strength. For elements that are crystalline in their standard states they are derived from vapor pressure data.

Table 2
HEATS OF FORMATION OF GASEOUS ATOMS FROM ELEMENTS IN THEIR STANDARD STATES

Atom	$\Delta H^\circ_{f(298)}$/kcal mol^{-1}	$\Delta H^\circ_{f(298)}$/kJ mol^{-1}	Ref.	Atom	$\Delta H^\circ_{f(298)}$/kcal mol^{-1}	$\Delta H^\circ_{f(298)}$/kJ mol^{-1}	Ref.
Ag	68.1 ± 0.2	284.9 ± 0.8	1	Na	108.16 ± 0.63	25.85 ± 0.15	2
Al	78.8 ± 1.0	329.7 ± 4.0	1	Nb	721.3 ± 4	172.4 ± 1	2
As	72.3 ± 3	302.5 ± 13	2	Ni	430.1 ± 2.1	102.8 ± 0.5	2
Au	88.0 ± 0.5	368.2 ± 2.1	2	O	249.17 ± 0.10	59.553 ± 0.024	1
B	139 ± 3	560 ± 12	1	Os	787 ± 6.3	188 ± 1.5	2
Ba	42.5 ± 1	177.8 ± 4	2	P	79.4 ± 1.0	332.2 ± 4.2	2
Be	77.5 ± 1.5	324.3 ± 6.3	2	Pb	46.62 ± 0.3	195.06 ± 1.3	2
Bi	50.1 ± 0.5	209.6 ± 2.1	2	Pd	90.0 ± 0.5	376.6 ± 2.1	2
Br	26.74 ± 0.03	111.86 ± 0.12	1	Pt	135.2 ± 0.3	565.7 ± 1.3	2
C	171.29 ± 0.11	716.67 ± 0.44	1	Pu	87.1 ± 4	364.4 ± 17	2
Ca	42.6 ± 0.4	178.2 ± 1.7	2	Rb	19.6 ± 0.1	82.0 ± 0.4	2
Cd	26.72 ± 0.15	111.80 ± 0.63	2	Re	185 ± 1.5	774 ± 6.3	2
Ce	101 ± 3	423 ± 13	2	Rh	133 ± 1	557 ± 4	2
Cl	28.992 ± 0.002	121.302 ± 0.008	1	Ru	155.5 ± 1.5	648.5 ± 6.3	2
Co	102.4 ± 1	428.4 ± 4	2	S	66.20 ± 0.06	276.98 ± 0.25	1
Cr	95 ± 1	398 ± 4	2	Sb	63.2 ± 0.6	264.4 ± 2.5	2
Cs	18.7 ± 0.1	78.2 ± 0.4	2	Sc	90.3 ± 1	377.8 ± 4	2
Cu	80.7 ± 0.3	337.6 ± 1.2	1	Se	54.3 ± 1	227.2 ± 4	2
Er	75.8 ± 1	317.1 ± 4	2	Si	108 ± 2	450 ± 8	1
F	18.9	79.1	2	Sm	49.4 ± 0.5	206.7 ± 2.1	2
Fe	99.3 ± 0.3	415.5 ± 1.3	2	Sn	72.2 ± 0.5	302.1 ± 2.1	2
Ga	65.4 ± 0.5	273.6 ± 2.1	2	Sr	39.1 ± 0.5	163.6 ± 2.1	2
Ge	89.5 ± 0.5	374.5 ± 2.1	2	Ta	186.9 ± 0.6	782.0 ± 2.5	2
H	52.103 ± 0.001	217.997 ± 0.006	1	Te	47.0 ± 0.5	196.7 ± 2.1	2
Hf	148 ± 1	619 ± 4	2	Th	137.5 ± 0.5	575.3 ± 2.1	2
Hg	14.69 ± 0.03	61.46 ± 0.13	2	Ti	112.3 ± 0.5	469.9 ± 2.1	2
I	25.517 ± 0.010	106.762 ± 0.040	1	Tl	43.55 ± 0.1	182.21 ± 0.4	2
In	58 ± 1	243 ± 4	2	U	126 ± 3	527 ± 13	2
Ir	160 ± 1	669 ± 4	2	V	122.9 ± 0.3	514.2 ± 1.3	2
K	21.42 ± 0.05	89.62 ± 0.21	2	W	203.1 ± 1	849.8 ± 4	2
Li	38.6 ± 0.4	161.5 ± 1.7	2	Y	101.5 ± 0.5	424.7 ± 2.1	2
Mg	35.0 ± 0.3	146.4 ± 1.3	2	Yb	36.35 ± 0.2	152.09 ± 0.8	2
Mn	67.7 ± 1	283.3 ± 4	2	Zn	31.17 ± 0.05	130.42 ± 0.20	1
Mo	157.3 ± 0.5	658.1 ± 2.1	2	Zr	145.5 ± 1	608.8 ± 4	2
N	472.68 ± 0.40	112.97 ± 0.10	1				

REFERENCES

1. CODATA recommended Key values for thermodynamics, 1975, *J. Chem. Thermodyn.*, 8, 603, 1976.
2. Brewer, L. and Rosenblatt, G. M., *Adv. High Temp. Chem.*, 2, 1, 1969.

Bond Strengths in Polyatomic Molecules

The values below refer to a temperature of 298 K and have mostly been determined by kinetic methods (see References 9 and 48 following Table 3 for a full discussion of the methods).

Some have been calculated from the heats of formation of the species involved according to the equations:

$$D(R - X) = \Delta H_f^\circ(\dot{R}) + \Delta H_f^\circ(\dot{X}) - \Delta H_f^\circ(RX)$$

$$D(R - R) = 2\Delta H_f^\circ(\dot{R}) - \Delta H_f^\circ(RR)$$

The sources of the data on the heats of formation are given in the references following Table 3.

An attempt has been made to list all the important values obtained by methods that are considered to be valid. The references are intended to serve as a guide to the literature.

Table 3
BOND STRENGTHS IN POLYATOMIC MOLECULES

Bond	D°_{298}/kcal mol^{-1}	D°_{298}/kJ mol^{-1}	Ref.
H—CH	102 ± 2	427 ± 8	1, 19
H—CH_2	110 ± 2	460 ± 8	1, 19
H—CH_3	104 ± 1	435 ± 4	43
H—ethynyl	125 ± 1	523 ± 4	62
H—vinyl	⩾108 ± 2	⩾452 ± 8	43
H—C_2H_5	98 ± 1	410 ± 4	43
H—propargyl	93.9 ± 1.2	392.9 ± 5.0	89
H—allyl	89 ± 1	372 ± 4	43
H—cyclopropyl	100.7 ± 1	421.3 ± 4	31
H—*n*-C_3H_7	98 ± 1	410 ± 4	43
H—*i*-C_3H_7	95 ± 1	398 ± 4	43
H—cyclobutyl	96.5 ± 1	403.8 ± 4	31, 60
H—cyclopropylcarbinyl	97.4 ± 1.6	407.5 ± 6.7	59
H—$CH(CH_3)CH_2$	83 ± 1	347 ± 4	43
H—$CH_2C(CH_3)_2$	86.0	359.8	84
H—*s*-C_4H_9	95 ± 1	398 ± 4	43
H—*t*C_4H_9	92 ± 1	385 ± 4	43
H—cyclopentadien-1,3-yl-5	81.2 ± 1.2	339.7 ± 5.0	41
H—pentadien-1,4-yl-3	80 ± 1	335 ± 4	43
H—cyclopentenyl-3	82.3 ± 1	344.3 ± 4	40
H—spiropentyl	98.8 ± 1	413.4 ± 4	31
H—$C(CH_3)_2C{:}CH$	79	331	76
H—cyclopentyl	94.5 ± 1	395.4 ± 4	31, 39
H—$C(CH_3)_2CH{:}CH_2$	77.2 ± 1.5	323.0 ± 6.3	83
H—neo-C_5H_{11}	100.3 ± 1	419.7 ± 4	56
H—C_6H_5	110.2 ± 2.0	461.1 ± 8.4	16
H—cyclohexadien-1,3-yl-5	70 ± 5	293 ± 21	47
H—cyclohexyl	95.5 ± 1	400 ± 4	31
H—$CH_2C(CH_3){:}C(CH_3)_2$	78.0 ± 1.1	326.4 ± 4.6	75
H—$C(CH_3)_2C(CH_3)_2$	76.3 ± 1.1	319.2 ± 4.6	75
H—$CH_2C_6H_5$	85 ± 1	356 ± 4	43
H—cycloheptatrien-1,3-yl-7	73.4	307.1	86
H—norbornyl	96.7 ± 2.5	404.6 ± 10.5	65
H—cycloheptyl	92.5 ± 1	387.0 ± 4	31
H—CN	120 ± 1	502 ± 4	27
H—Ch_2CN	~93	~389	49b

Table 3 (continued)
BOND STRENGTHS IN POLYATOMIC MOLECULES

Bond	$D°_{298}$/kcal mol^{-1}	$D°_{298}$/kJmol^{-1}	Ref.
H—CH(CH$_3$)CN	89.9 ± 2.3	376.1 ± 9.6	49a
H—C(CH$_3$)$_2$CN	86.5 ± 2.0	361.9 ± 8.4	49c
H—CH$_2$NH$_2$	94.6 ± 2.0	395.8 ± 8.4	20c, 76
H—CHO	87 ± 1	364 ± 4	43
H—CH$_2$OH	94 ± 2	393 ± 4	43
H—COCH$_3$	86.0 ± 0.8	359.8 ± 3.4	28
H—CH$_2$OCH$_3$	93 ± 1	389 ± 4	43
H—CH(CH$_3$)OH	93.0 ± 1.0	389.1 ± 4.2	2
H—COCH$_2$	87.1 ± 1.0	364.4 ± 4.2	3
H—CH$_2$COCH$_3$	98.3 ± 1.8	411.3 ± 7.5	50, 81
H—COC$_2$H$_5$	87.4 ± 1.0	365.7 ± 4.2	91
H—CH(OH)CH$_2$	81.6 ± 1.8	341.4 ± 7.5	4
H—C(CH$_3$)$_2$OH	91 ± 1	381 ± 4	43
H—tetrahydrofuran-2-yl	92 ± 1	385 ± 4	43
H—CH(CH$_3$)COCH$_3$	92.3 ± 1.4	386.2 ± 5.9	82
H—COC$_6$H$_5$	86.9 ± 1	363.6 ± 4	80
H—COOCH$_3$	92.7 ± 1	387.9 ± 4	78
H—CH$_2$OCOC$_6$H$_5$	100.2 ± 1.3	419.2 ± 5.4	79
H—COCF$_3$	91.0 ± 2	380.7 ± 8	6
H—CH$_2$F	101 ± 2	423 ± 8	49
H—CHF$_2$	101 ± 2	423 ± 8	49
H—CF$_3$	106 ± 1	444 ± 4	43
H—CF$_2$Cl	104 ± 1	435 ± 4	58
H—CH$_2$Cl	100.9	422.2	38
H—CHCl$_2$	99.0	414.2	38
H—CCl$_3$	95.8 ± 1	400.8 ± 4	61
H—CH$_2$Br	102.0	426.8	38
H—CHBr$_2$	103.7	433.9	38
H—CBr$_3$	96.0 ± 1.6	401.7 ± 6.7	51
H—CH$_2$I	103 ± 2	431 ± 8	43
H—CHI$_2$	103 ± 2	431 ± 8	43
H—CH$_2$CF$_3$	106.7 ± 1.1	446.4 ± 4.6	93
H—CF$_2$CH$_3$	99.5 ± 1	416.3 ± 4.2	69b
H—C$_2$F$_5$	103.1 ± 1.5	431.4 ± 6.3	7
H—CCl$_2$CHCl$_2$	94 ± 2	393 ± 8	23,37
H—C$_2$Cl$_5$	95 ± 2	398 ± 8	23,36
H—*n*-C$_3$F$_7$	104 ± 2	435 ± 8	48
H—CHClCH$_2$	88.6 ± 1.4	370.7 ± 5.9	4
H—CH$_2$Si(CH$_3$)$_3$	99.2 ± 1	415.1 ± 4.2	29a
H—NH$_2$	110 ± 2	460 ± 8	44
H—NHCH$_3$	103 ± 2	431 ± 8	44
H—N(CH$_3$)$_2$	95 ± 2	398 ± 8	44
H—NHC$_6$H$_5$	80 ± 3	335 ± 13	11, 48
H—N(CH$_3$)C$_6$H$_5$	74 ± 3	310 ± 13	11, 48
H—N0	⩽49.5	⩽207.1	20
H—NF$_2$	75.7 ± 2.5	316.7 ± 10.5	67
H—N$_3$	85	356	46
H—OH	119 ± 1	498 ± 4	48
H—OCH$_3$	104.4 ± 1	436.8 ± 4	8, 23
H—OC$_2$H$_5$	104.2 ± 1	436.0 ± 4	8, 23
H—OC(CH$_3$)$_3$	105.1 ± 1	439.7 ± 4	8, 23, 72
H—OCH$_2$C(CH$_3$)$_3$	102.3 ± 1.5	428.0 ± 6.3	72
H—OC$_6$H$_5$	88 ± 5	368 ± 21	15
H—O$_2$H	90 ± 2	377 ± 8	48
H—O$_2$CCH$_3$	112 ± 4	469 ± 8	48
H—O$_2$CC$_2$H$_5$	110 ± 4	460 ± 8	48
H—O$_2$C*n*-C$_3$H$_7$	103 ± 4	431 ± 8	48
H—ONO	78.3 ± 0.5	327.6 ± 2.1	1, 10

Table 3 (continued)
BOND STRENGTHS IN POLYATOMIC MOLECULES

Bond	D°_{298}/kcal mol^{-1}	D°_{298}/kJ mol^{-1}	Ref.
H—ONO_2	101.2 ± 0.5	423.4 ± 2.1	1, 10
H—SH	90 ± 2	377 ± 8	48
H—SCH_3	≥88	≥368	48
H—SiH_3	94 ± 3	393 ± 13	73
H—$Si(CH_3)_3$	90 ± 2.6	376 ± 11	90
H—$SiCl_3$	91.3 ± 1.4	382 ± 6	90a
CH≡CH	230 ± 2	962 ± 8	1, 19
CH_2=CH_2	172 ± 2	720 ± 8	1, 19
CH_3—CH_3	88 ± 2	368 ± 8	48
CH_3—$C(CH_3)_2CH_2$	69.4	290.4	11
$C_6H_5CH_2$—C_2H_5	69 ± 2	289 ± 8	48
$C_6H_5CH(CH_3)$—CH_3	71	297	11
$C_6H_5CH_2$—n-C_3H_7	67 ± 2	280 ± 8	48
CH_3—CN	123.9 ± 0.7	518.4 ± 2.9	62
C_2H—CN	144 ± 1	603 ± 4	62
CH_3—CH_2CN	72.7 ± 2	304.2 ± 8	11, 48
CH_3—$CH(CH_3)CN$	78.8 ± 2	329.7 ± 8	49a
C_2H_5—CH_2CN	76.9 ± 1.7	321.8 ± 7.1	49b
CH_3—$C(CH_3)_2CN$	74.7 ± 1.6	312.6 ± 6.7	49c, 11, 48
$C_6H_5C(CH_3)(CN)$—CH_3	59.9	250.6	11
CN—CN	128 ± 1	536 ± 4	27
$C_6H_5CH_2$—CH_2NH_2	65.7	274.9	20c
$C_6H_5CH_2CO$—$CH_2C_6H_5$	65.4	273.6	11
CH_3CO—CF_3	73.8	308.8	11
CH_3CO—$COCH_3$	67.4 ± 2.3	282.0 ± 9.6	52
$C_6H_5CH_2$—COOH	68.1	284.9	11
$C_6H_5CH_2$—O_2CCH_3	67	280	11
C_6H_5CO—COC_6H_5	66.4	277.8	11
$C_6H_5CH_2$—$O_2CC_6H_5$	69	289	11
$(C_6H_5CH_2)_2CH$—COOH	59.4	248.5	11
CH_2F—CH_2F	88 ± 2	368 ± 8	49
CH_3—CF_3	101.2 ± 1.1	423.4 ± 4.6	74
CF_2=CF_2	76.3 ± 3	319.2 ± 13	94
CF_3—CF_3	96.9 ± 2	405.4 ± 8	21
$C_6H_5CH_2$—NH_2	71.9 ± 1	300.8 ± 4	44
C_6H_5NH—CH_3	67.7	283.3	11
$C_6H_5CH_2$—$NHCH_3$	68.7 ± 1	287.4 ± 4	44
$C_6H_5N(CH_3)$—CH_3	65.2	272.8	11
$C_6H_5CH_2$—$N(CH_3)_2$	60.9 ± 1	254.8 ± 4	44
CF_3—NF_2	65 ± 2.5	272 ± 10.5	87
CH_2=N_2	≤41.7 ± 1	≤174.5 ± 4	57
CH_3N:N—CH_3	52.5	219.7	11
C_2H_5N:N—C_2H_5	50.0	209.2	11
i-C_3H_7N:N—i-C_3H_7	47.5	198.7	11
n-C_4H_9N:N—n-C_4H_9	50.0	209.2	11
i-C_4H_9N:N—i-C_4H_9	49.0	205.0	11
s-C_4H_9N:N—s-C_4H_9	46.7	195.4	11
t-C_4H_9N:N—t-C_4H_9	43.5	182.0	11
$C_6H_5CH_2N$:N—$CH_2C_6H_5$	37.6	157.3	11
CF_3N:N—CF_3	55.2	231.0	11
CH_3—NO	41.8 ± 0.9	174.9 ± 3.8	8
C_2H_5—NO	42.0 ± 1.3	175.7 ± 5.5	8
n-C_3H_7—NO	40.1 ± 1.8	167.8 ± 7.5	8
i-C_3H_7—NO	41.0 ± 1.3	171.5 ± 5.5	8
n-C_4H_9—NO	42.5 ± 1.5	177.8 ± 6.3	8
i-C_4H_9—NO	42.0 ± 1.5	175.7 ± 6.3	8
s-C_4H_9—NO	41.5 ± 0.8	173.6 ± 3.4	8b, 8

Table 3 (continued)
BOND STRENGTHS IN POLYATOMIC MOLECULES

Bond	$D°_{298}$/kcal mol^{-1}	$D°_{298}$/kJ mol^{-1}	Ref.
t-C_4H_9—NO	40.9 ± 0.8	171.1 ± 3.4	8b, 8, 18
C_6H_5—NO	51.5 ± 1	215.5 ± 4	18
CF_3—NO	31	130	14
C_6F_5—NO	50.5 ± 1	211.3 ± 4	14, 18
CCl_3—NO	32	134	14
CN—NO	28.8 ± 2.5	120.5 ± 10.5	45
t-C_4H_9—NO*t*-C_4H_9	29	121	14
C_2H_5—NO_2	62	259	1, 77
O=CO	127.2 ± 0.1	532.2 ± 0.4	25
CH_3—OC_6H_5	57 ± 2	239 ± 8	67a
CH_3—$OCH_2C_6H_5$	67.0	280.3	20d
C_2H_5—OC_6H_5	64.0	267.8	20d
CH_2CH_2—OC_6H_5	50.6	211.7	20d
CH_3—O_2SCH_3	66.8	279.5	11, 48
allyl—O_2SCH_3	49.6	207.5	11, 48
$C_6H_5CH_2$—O_2SCH_3	52.9	221.3	11, 48
C_6H_5S—CH_3	67.5 ± 2.0	282.4 ± 8.4	20b
$C_6H_5CH_2$—SCH_3	59.4 ± 2.0	248.5 ± 8.4	20b
F—CH_3	108 ± 3	452 ± 13	29, 48
F—CF_2Cl	117 ± 6	490 ± 25	35a
F—$CFCl_2$	110 ± 6	460 ± 25	35a
Cl—CN	97 ± 1	406 ± 4	27
Cl—COC_6H_5	74 ± 3	310 ± 13	48
Cl—CF_3	86.1 ± 0.8	360.2 ± 3.3	22
Cl—CF_2Cl	76 ± 2	318 ± 8	35a
Cl—CCl_2F	73 ± 2	305 ± 8	35
Cl—CCl_3	70.4 ± 1	294.6 ± 4	61
Cl—C_2F_5	82.7 ± 1.7	346.0 ± 7.1	22
Cl—CF_2CF_2Cl	78 ± 2	326 ± 8	35a
Cl—NF_2	⩽65	⩽272	20a
Cl—$SiCl_3$	111	466	90a
Br—CH_3	70.0 ± 1.2	292.9 ± 5.0	30
Br—CN	83 ± 1	347 ± 4	27
Br—COC_6H_5	64.2	268.6	11
Br—CHF_2	69 ± 2	289 ± 8	63
Br—CF_3	70.6 ± 1.0	295.4 ± 4.2	32, 34
Br—CCl_3	55.7 ± 1	233.1 ± 4.2	61
Br—CBr_3	56.2 ± 1.8	235.1 ± 7.5	51
Br—C_2F_5	68.7 ± 1.5	287.4 ± 6.3	22, 33, 34
Br—*n*-C_3F_7	66.5 ± 2.5	278.2 ± 10.5	22
Br—NF_2	⩽53	⩽222	20a
I—CH_3	56.3 ± 1	235.6 ± 4	48
I—norbornyl	62.5 ± 2.5	261.5 ± 10.5	65
I—CN	73 ± 1	305 ± 4	27
I—CF_3	53.2 ± 1.0	222.6 ± 4.2	64
I—CF_2CH_3	52.1 ± 1.0	218.0 ± 4.2	69a
I—CH_2CF_3	56.3 ± 1	235.6 ± 4	92
I—C_2F_5	51.2 ± 1.0	214.2 ± 4.2	64
I—*n*-C_3F_7	49.8 ± 1.0	208.4 ± 4.2	64
I—*i*-C_3F_7	49.8 ± 1.0	208.4 ± 4.2	64
I—*n*-C_4F_9	49.0 ± 1.0	205.0 ± 4.2	64
I—C_6H_5	63.7 ± 1.2	266.3 ± 4.8	52a
I—C_6F_5	~66.2	~277	54
C_2H_5—ZnC_2H_5	~47.5	~198.7	53a
CH_3—$Ga(CH_3)_2$	59.5	249.0	11
CH_3—$CdCH_3$	54.4	227.6	11
C_2H_5—$Sn(C_2H_5)_3$	~57	~239	24a

Table 3 (continued)
BOND STRENGTHS IN POLYATOMIC MOLECULES

Bond	$D°_{298}$/kcal mol^{-1}	$D°_{298}$/kJ mol^{-1}	Ref.
CH_3—$HgCH_3$	57.5	240.6	53
C_2H_5—HgC_2H_5	43.7 ± 1	182.8 ± 4	55
n-C_3H_7—Hg*n*-C_3H_7	47.1	197.1	11
i-C_3H_7—Hg*i*-C_3H_7	40.7	170.3	11
C_6H_5—HgC_6H_5	68	285	11
CH_3—$Tl(CH_3)_2$	36.4 ± 0.6	152.3 ± 2.5	71
CH_3—$Pb(CH_3)_3$	49.4 ± 1	206.7 ± 4	42
BH_3—BH_3	35	146	11
NH_2—NH_2	70.8 ± 2	296.2 ± 8	11
NH_2—$NHCH_3$	64.8	271.1	11
NH_2—$N(CH_3)_2$	62.7	262.3	11
NH_2—NHC_6H_5	51.1	213.8	11
NO—NO_2	9.5 ± 0.5	39.8 ± 2.1	48
NO_2—NO_2	12.9 ± 0.5	54.0 ± 2.1	48
NF_2—NF_2	21 ± 1	88 ± 4	25
O—N_2	40	167	1, 10
O—NO	73	305	1, 10
HO—NCH_3	49.7	208.0	11
Cl—NF_2	~32	~134	1, 69
HO—OH	51 ± 1	213 ± 4	48
CH_3O—OCH_3	37.6 ± 0.2	157.3 ± 0.8	8a, 8, 6a
HO—$OC(CH_3)_3$	46.3 ± 1.9	193.7 ± 8.0	1, 8, 23
C_2H_5O—OC_2H_5	37.9	158.6	8
n-C_3H_7O—O*n*-C_3H_7	37.1	155.2	8
i-C_3H_7O—O*i*-C_3H_7	37.7	157.7	8
s-C_4H_9O—O*s*-C_4H_9	36.4 ± 1	152.3 ± 4	88
t-C_4H_9O—O*t*-C_4H_9	38.0	159.0	8
neo-$C_5H_{11}O$—Onec-C_5H_{11}	36.4 ± 1	152.3 ± 4	68
CF_3O—OCF_3	46.2	193.3	27a
O—O_2ClF	58.4	244.3	11
CH_3CO_2—O_2CCH_3	30.4 ± 2	127.2 ± 8	11, 48
$C_2H_5CO_2$—$O_2CC_2H_5$	30.4 ± 2	127.2 ± 8	11, 48
n-$C_3H_7CO_2$—O_2C*n*-C_3H_7	30.4 ± 2	127.2 ± 8	11, 48
O—SO	132 ± 2	552 ± 8	25
F—OCF_3	43.5 ± 0.5	182.0 ± 2.1	24
HO—Cl	60 ± 3	251 ± 13	48
O—ClO	59 ± 3	247 ± 13	25
HO—Br	56 ± 3	234 ± 13	48
HO—I	56 ± 3	234 ± 13	48
ClO_3—ClO_4	58.4	244.3	11
O=PF_3	130 ± 5	543.9 ± 21	48
O=PCl_3	122 ± 5	511 ± 21	48
O=PBr_3	119 ± 5	498 ± 21	48
I—NO	17.3 ± 1	72.4 ± 4	84a
I—NO_2	18.3 ± 1	76.6 ± 4	84a
SiH_3—SiH_3	81 ± 4	339 ± 17	73
$(CH_3)_3Si$—$Si(CH_3)_3$	80.5	336.8	26
$(C_6H_5)_3Si$—$Si(C_6H_5)_3$	88 ± 7	368 ± 29	13

REFERENCES

1. A value calculated from one of the above thermochemical equations taking data from the references quoted.
2. Alfassi, Z. B. and Golden, D. M., *J. Phys. Chem.*, 76, 3314, 1972.
3. Alfassi, Z. B. and Golden, D. M., *J. Am. Chem. Soc.*, 95, 319, 1973.
4. Alfassi, Z. B., Golden, D. M., and Benson, S. W., *Int. J. Chem. Kinet.*, 5, 155, 1973; Trenwith, A. B., *Int. J. Chem. Kinet.*, 5, 67, 1973.

5. **Alfassi, Z. B. and Golden, D. M.,** *Int. J. Chem. Kinet.*, 5, 295, 1973; **Trenwith, A. B.,** *J. Chem. Soc. Faraday Trans. 1,* 69, 1737, 1973.
6. **Amphlett, J. C. and Whittle, E.,** *Trans. Faraday Soc.*, 66, 2016, 1970.
6a. **Barker, J. R., Benson, S. W., and Golden, D. M.,** *Int. J. Chem. Kinet.*, 9, 31, 1977.
7. **Bassett, J. E. and Whittle, E.,** *J. Chem. Soc. Faraday Trans. 1,* 68, 492, 1972.
8. **Batt, L., Christie, K., Milne, R. T., and Summers, A. J.,** *Int. J. Chem. Kinet.*, 6, 877, 1974.
8a. **Batt, L. and McCulloch, R. D.,** *Int. J. Chem. Kinet.*, 8, 491, 1976.
8b. **Batt, L. and McCulloch, R. D.,** *Int. J. Chem. Kinet.*, 8, 911, 1976.
8c. **Batt, L. and Milne, R. T.,** *Int. J. Chem. Kinet.*, 8, 59, 1976.
9. **Benson, S. W.,** *J. Chem. Educ.*, 42, 502, 1965.
10. **Benson, S. W.,** *Thermochemical Kinetics,* John Wiley & Sons, New York, 1968.
11. **Benson, S. W. and O'Neal, H. E.,** *Kinetic Data on Gas Phase Unimolecular Reactions,* National Bureau of Standards, Washington, D.C., NSRDS-NBS 21, 1970.
12. **Bohme, D. K., Hemsworth, R. S., and Rundle, H. W.,** *J. Chem. Phys.*, 59, 77, 1973.
13. **Calle, L. M. and Kana'an, A. S.,** *J. Chem. Thermodyn.*, 6, 935, 1974.
14. **Carmichael, P. J., Gowenlock, B. G., and Johnson, C. A. F.,** *J. Chem. Soc. Perkin Trans.*, 2, 1853, 1973.
15. **Carson, A. S., Fine, D. H., Gray, P., and Laye, P. G.,** *J. Chem. Soc. B*, p. 1611, 1971.
16. **Chamberlain, G. A. and Whittle, E.,** *Trans. Faraday Soc.*, 67, 2077, 1971.
17. **Choo, K. Y., Golden, D. M., and Benson, S. W.,** *Int. J. Chem. Kinet.*, 7, 713, 1975.
18. **Choo, K. Y., Mendenhall, G. D., Golden, D. M., and Benson, S. W.,** *Int. J. Chem. Kinet.*, 6, 813, 1974.
19. **Chupka, W. A. and Lifshitz, C.,** *J. Chem. Phys.*, 48, 1109, 1968.
20. **Clement, M. J. Y. and Ramsay, D. A.,** *Can. J. Phys.*, 39, 205, 1961.
20a. **Clyne, M. A. A. and Connor, J.,** *J. Chem. Soc. Faraday Trans. 2,* 68, 1220, 1972.
20b. **Colussi, A. J. and Benson, S. W.,** *Int. J. Chem. Kinet.*, 9, 295, 1977.
20c. **Colussi, A. J. and Benson, S. W.,** *Int. J. Chem. Kinet.*, 9, 307, 1977.
20d. **Colussi, A. J., Zabel, F., and Benson, S. W.,** *Int. J. Chem. Kinet.*, 9, 161, 1977.
21. **Coomber, J. W. and Whittle, E.,** *Trans. Faraday Soc.*, 63, 1394, 1967.
22. **Coomber, J. W. and Whittle, E.,** *Trans. Faraday Soc.*, 63, 2656, 1967.
23. **Cox, J. D. and Pilcher, G.,** *Thermochemistry of Organic and Organometallic Compounds,* Academic Press, New York, 1970.
24. **Czarnarski, J., Castellano, E., and Shumacher, H. J.,** *Chem. Commun.*, p. 1255, 1968.
24a. **Daly, M. and Price, S. J.,** *Can. J. Chem.*, 54, 1814, 1976.
25. **Darwent, B. de B.,** *Bond Dissociation Energies in Simple Molecules,* National Bureau of Standards, Washington, D.C., 31, 1970.
26. **Davidson, I. M. T. and Howard, A. B.,** *J. Chem. Soc. Faraday Trans.* 1, 71, 69, 1975.
27. **Davis, D. D. and Okabe, H.,** *J. Chem. Phys.*, 49, 5526, 1968.
27a. **Descamps, B. and Forst, W.,** *Can. J. Chem.*, 53, 1442, 1975.
28. **Devore, J. A. and O'Neal, H. E.,** *J. Phys. Chem.*, 73, 2644, 1969.
29. **Dibeler, V. H. and Reese, R. M.,** *J. Res. Nat. Bur. Stand.*, 54, 127, 1955.
29a. **Doncaster, A. M. and Walsh, R.,** *J. Chem. Soc. Faraday Trans.* 1, 72, 2908, 1976.
30. **Ferguson, K. C., Okafo, E. N., and Whittle, E.,** *J. Chem. Soc. Faraday Trans.* 1, 69, 295, 1973.
31. **Ferguson, K. C. and Whittle, E.,** *Trans. Faraday Soc.*, 67, 2618, 1971; **Jones, S. H. and Whittle, E.,** *Int. J. Chem. Kinet.* 2, 479, 1970.
32. **Ferguson, K. C. and Whittle, E.,** *J. Chem. Soc. Faraday Trans.* 1, 68, 295, 1972.
33. **Ferguson, K. C. and Whittle, E.,** *J. Chem. Soc. Faraday Trans.* 1, 68, 306, 1972.
34. **Ferguson, K. C. and Whittle, E.,** *J. Chem. Soc. Faraday Trans.* 1, 68, 641, 1972.
35. **Foon, R. and Tait, K. B.,** *J. Chem. Soc. Faraday Trans.* 1, 68, 104, 1972.
35a. **Foon, R. and Tait, K. B.,** *J. Chem. Soc. Faraday Trans.* 1, 68, 1121, 1972.
36. **Franklin, J. A., Huybrechts, G. H., and Cillien, C.,** *Trans. Faraday Soc.*, 65, 2094, 1969.
37. **Franklin, J. A. and Huybrechts, G. H.,** *Int. J. Chem. Kinet.*, 1, 1, 1969.
38. **Furuyama, S., Golden, O. M., and Benson, S. W.,** *J. Am. Chem. Soc.*, 91, 7564, 1969.
39. **Furuyama, S., Golden, D. M., and Benson, S. W.,** *Int. J. Chem. Kinet.*, 2, 83, 1970.
40. **Furuyama, S., Golden, D. M., and Benson, S. W.,** *Int. J. Chem. Kinet.*, 2, 93, 1970.
41. **Furuyama, S., Golden, D. M., and Benson, S. W.,** *Int. J. Chem. Kinet.*, 3, 237, 1971.
42. **Gilroy, K. M., Price, S. J., and Webster, N. J.,** *Can. J. Chem.*, 50, 2639, 1972.
43. **Golden, D. M. and Benson, S. W.,** *Chem. Rev.*, 69, 125, 1969.
44. **Golden, D. M., Solly, R. K., Gac, N. A., and Benson, S. W.,** *J. Am. Chem. Soc.*, 94, 363, 1972.
45. **Gowenlock, B. G., Johnson, C. A. F., Keary, C. M., and Pfab, J.,** *J. Chem. Soc. Perkin Trans.*, 2, 351, 1975.
46. **Gray, P.,** *Q. Rev.* (London), 17, 441, 1963.
47. **James, D. G. L. and Suart, R. D.,** *Trans. Faraday Soc.*, 64, 2752, 1968.

48. Kerr, J. A., *Chem. Rev.*, 66, 465, 1966.
49. Kerr, J. A. and Timlin, D. M., *Int. J. Chem. Kinet.*, 3, 427, 1971.
49a. King, K. D. and Goddard, R. D., *J. Am. Chem. Soc.*, 97, 4504, 1975.
49b. King, K. D. and Goddard, R. D., *Int. J. Chem. Kinet.*, 7, 837, 1975.
49c. King, K. D. and Goddard, R. D., *J. Phys. Chem.*, 80, 546, 1976.
50. King, K. D., Golden, D. M., and Benson, S. W., *J. Am. Chem. Soc.*, 92, 5541, 1970.
51. King, K. D., Golden, D. M., and Benson, S. W., *J. Phys. Chem.*, 75, 987, 1971.
52. Knoll, H., Scherker, K., and Geiseler, G., *Int. J. Chem. Kinet.*, 5, 271, 1973.
53. Kominar, R. J. and Price, S. J., *Can. J. Chem.*, 47, 991, 1969.
53a. Koski, A. A., Price, S. J. W., and Trudell, B. C., *Can. J. Chem.*, 54, 482, 1976.
54. Krech, M. J., Price, S. J. W., and Yared, W. F., *Int. J. Chem. Kinet.*, 6, 257, 1974.
55. LaLonde, A. C. and Price, S. J. W., *Can. J. Chem.*, 49, 3367, 1971.
56. Larson, C. W., Hardwidge, E. A., and Rabinovitch, B. S., *J. Chem. Phys.*, 50, 2769, 1969.
57. Laufer, A. H. and Okabe, H., *J. Am. Chem. Soc.*, 93, 4137, 1971.
58. Leyland, L. M., Majer, J. R., and Robb, J. C., *Trans. Faraday Soc.*, 66, 898, 1970.
59. McMillen, D. F., Golden, D. M., and Benson, S. W., *Int. J. Chem. Kinet.*, 3, 359, 1971.
60. McMillen, D. F., Golden, D. M., and Benson, S. W., *Int. J. Chem. Kinet.*, 4, 487, 1972.
61. Mendenhall, G. D., Golden, D. M., and Benson, S. W., *J. Phys. Chem.*, 77, 2707, 1973.
62. Okabe, H. and Dibeler, V. H., *J. Chem. Phys.*, 59, 2430, 1973.
63. Okafo, E. N. and Whittle, E., *J. Chem. Soc. Faraday Trans.* 1, 70, 1366, 1974.
64. Okafo, E. N. and Whittle, E., *Int. J. Chem. Kinet.*, 7, 287, 1975.
65. O'Neal, H. E., Bagg, J. W., and Richardson, W. H., *Int. J. Chem. Kinet.*, 2, 493, 1970.
66. O'Neal, H. E. and Benson, S. W., in *Free Radicals*, Kochi, J. K., Ed., John Wiley Sons, New York, 1973, 275.
67. Pankratov, A. V., Zercheninov, A. N., Chesnokov, V. I., and Zhdanova, N. N., *Zh. Fiz. Khim.*, 43, 394, 1969.
67a. Paul, S. and Back, M. H., *Can. J. Chem.*, 53, 3330, 1975.
68. Perona, M. J. and Golden, D. M., *Int. J. Chem. Kinet.*, 5, 55, 1973.
69. Petry, R. C., *J. Am. Chem. Soc.*, 89, 4600, 1967.
69a. Pickard, J. M. and Rodgers, A. S., *Int. J. Chem. Kinet.*, 8, 809, 1976.
69b. Pickard, J. M. and Rodgers, A. S., *J. Am. Chem. Soc.*, 99, 695, 1977.
70. Potzinger, P. and Lampe, F. W., *J. Phys. Chem.*, 74, 719, 1970.
71. Price, S. J., Richard, J. P., Rufeldt, R. C., and Jacko, M. G., *Can. J. Chem.*, 51, 1397, 1973.
72. Reed, K. J. and Brauman, J. L., *J. Am. Chem. Soc.*, 97, 1625, 1975.
73. Ring, M. A., Puentes, M. J., and O'Neal, H. E., *J. Am. Chem. Soc.*, 92, 4845, 1970; Steele, W. C. and Stone, F. G. A., *J. Am. Chem. Soc.*, 84, 3599, 1962; Steele, W. C., Nichols, L. D., and Stone, F. G. A., *J. Am. Chem. Soc.*, 84, 441, 1962.
74. Rodgers, A. S. and Ford, W. G. F., *Int. J. Chem. Kinet.*, 5, 965, 1973.
75. Rodgers, A. S. and Wu, M. C. R., *J. Am. Chem. Soc.*, 95, 6913, 1973.
76. Sen Sharma, D. K. and Franklin, J. L., *J. Am. Chem. Soc.*, 95, 6562, 1973.
77. Shaw, R., *Int. J. Chem. Kinet.*, 5, 261, 1973.
77a. Simmons, J. W., Hase, W. L., Phillips, R. J., Porter, E. J., and Growcock, F. B., *Int. J. Chem. Kinet.*, 7, 879, 1975.
78. Solly, R. K. and Benson, S. W., *Int. J. Chem. Kinet.*, 1, 427, 1969.
79. Solly, R. K. and Benson, S. W., *Int. J. Chem. Kinet.*, 3, 509, 1971.
80. Solly, R. K. and Benson, S. W., *J. Am. Chem. Soc.*, 93, 1592, 1971.
81. Solly, R. K., Golden, D. M., and Benson, S. W., *Int. J. Chem. Kinet.*, 2, 11, 1970.
82. Solly, R. K., Golden, D. M., and Benson, S. W., *Int. J. Chem. Kinet.*, 2, 381, 1970.
83. Trenwith, A. B., *Trans. Faraday Soc.*, 66, 2805, 1970.
84. Tsang, W., *Int. J. Chem. Kinet.*, 5, 929, 1973.
84a. van der Bergh, H. and Troe, J., *J. Chem. Phys.*, 64, 736, 1976.
85. Vanderwielen, A. J., Ring, M. A., and O'Neal, H. E., *J. Am. Chem. Soc.*, 97, 993, 1975.
86. Vincow, G., Dauben, H. J., Hunter, F. R., and Volland, W. V., *J. Am. Chem. Soc.*, 91, 2823, 1969.
87. Walker, L. C., *J. Chem. Thermodyn.*, 4, 219, 1972.
88. Walker, R. F. and Phillips, L., *J. Chem. Soc. A*, 2103, 1968.
89. Walsh, R., *Trans. Faraday Soc.*, 67, 2085, 1971.
90. Walsh, R. and Wells, J. M., *J. Chem. Soc. Chem. Commun.*, p. 513, 1973.
90. Walsh, R. and Wells, J. M., *J. Chem. Soc. Faraday Trans. 1*, 72, 100, 1976.
90a. Walsh, R. and Wells, J. M., *J. Chem. Soc. Faraday Trans. 1*, 72, 1212, 1976.
91. Watkins, K. W. and Thompson, W. W., *Int. J. Chem. Kinet.*, 5, 791, 1973.
92. Wu, E-C. and Rodgers, A. S., *Int. J. Chem. Kinet.*, 5, 1001, 1973.
93. Wu, E-C. and Rodgers, A. S., *J. Phys. Chem.*, 78, 2315, 1974.
93a. Wu, E-C. and Rodgers, A. S., *J. Am. Chem. Soc.*, 98, 6112, 1976.
94. Zmbov, K. F., Uy, O. M., and Margrave, J. L., *J. Am. Chem. Soc.*, 90, 5090, 1968.

Heats of Formation of Free Radicals

The heats of formation of the free radicals are related to the corresponding bond strengths by the equations

$$D(R-X) = \Delta H_f^\circ(\dot{R}) + \Delta H_f^\circ(\dot{X}) - \Delta H_f^\circ(RX)$$

or

$$D(R-R) = 2\Delta H_f^\circ(\dot{R}) - \Delta H_f^\circ(RR)$$

For an excellent review of the methods of determining the heats of formation of free radicals the reader is referred to "Thermochemistry of Free Radicals" by H. E. O'Neal and S. W. Benson in *Free Radicals*, Kochi, J. K., Ed., John Wiley and Sons, New York, 1973, 275.

The references are the same as given for Table 3.

Table 4
HEATS OF FORMATION OF FREE RADICALS

Radical	$\Delta H^\circ_{f(298)}$/kcal mol^{-1}	$\Delta H^\circ_{f(298)}$/kJ mol^{-1}	Ref.
CH	142 ± 1	594 ± 4	1, 19
CH_2	92 ± 1	385 ± 4	1, 19
$^1CH_2$	101 ± 3	423 ± 13	77a
CH_3	34.0 ± 1.2	142.3 ± 5.0	43
Ethynyl	128 ± 1	536 ± 4	62
Vinyl	≥68	≥285	43
C_2H_5	25.9 ± 1.3	108.4 ± 5.4	43
Propargyl	86.5 ± 1.2	361.9 ± 5.0	89
Allyl	41.4 ± 1.1	173.2 ± 4.6	43
Cyclopropyl	61.3 ± 1	256.5 ± 4	1, 23, 31
n-C_3H_7	22.6 ± 1.8	94.6 ± 7.5	43
i-C_3H_7	18.2 ± 1.5	76.2 ± 6.3	43
Methallyl	30.4 ± 1.3	127.2 ± 5.4	43
Isobutenyl	29.6	123.9	84
Cyclopropylcarbinyl	51.5 ± 1.6	215.5 ± 6.7	59
Cyclobutyl	51.2 ± 1.0	214.2 ± 4.2	1, 23, 31, 60
s-C_4H_9	13.0 ± 2	54.4 ± 8	43
t-C_4H_9	7.6 ± 1.2	31.8 ± 5.0	43
Cyclopentadien-1,3-yl-5	60.9 ± 1.2	254.8 ± 5.0	41
Pentadienyl	52.9 ± 1.2	221.3 ± 5.0	43
Cyclopentenyl-3	38.4 ± 1	160.7 ± 4	40
Spiropentyl	91.0 ± 1	380.7 ± 4	1, 23, 31
$(CH_3)_2CC{\equiv}CH$	59	247	76
Dimethylallyl	18.3 ± 1.5	76.6 ± 6.3	83
Cyclopentyl	24.3 ± 1	101.7 ± 4	1, 23, 31, 39
Neo-C_5H_{11}	7.6 ± 1	31.8 ± 4	1, 23, 56
Phenyl	77.7 ± 2.0	325.1 ± 8.4	16
Cyclohexadien-1,3-yl-5	44 ± 5	184 ± 21	47
Cyclohexyl	13.9 ± 1	58.2 ± 4	1, 23, 31
$(CH_3)_2C{=}C(CH_3)CH_2$	9.6 ± 1.1	40.2 ± 4.6	75
Benzyl	45.1 ± 1	188.7 ± 4	43
Cycloheptatrien-1,3-yl-7	64.8	271.1	86
Norbornyl	32.6 ± 2.5	136.4 ± 10.5	65
Cycloheptyl	12.2 ± 1	51.1 ± 4	1, 23, 31
$C_6H_5CHCH_3$	38	159	11, 66
CN	101 ± 1	423 ± 4	27
CH_2NH_2	37.0 ± 2.0	154.8 ± 8.4	20c, 76
CH_3NH	45.4 ± 1	190.0 ± 4	44
CH_2CN	58.5 ± 2.2	209.6 ± 9.6	49a

Table 4 (continued)
HEATS OF FORMATION OF FREE RADICALS

Radical	$\Delta h^\circ_{f(298)}$/kcal mol^{-1}	$\Delta H^\circ_{f(298)}$/kJ mol^{-1}	Ref.
$(CH_3)_2N$	38.5 ± 1	161.1 ± 4	44
CH_3CHCN	50.1 ± 2.3	209.6 ± 9.6	49a
$(CH_3)_2CCN$	39.8 ± 2.0	166.5 ± 8.4	49c
C_6H_5NH	55.0	230.1	11, 66
$C_6H_5NCH_3$	53.5	223.8	11, 66
$C_6H_5C(CH_3)CN$	54.5	227.6	11, 66
CHO	7.7 ± 1.2	32.2 ± 5.0	43, 66
CH_2OH	−6.2 ± 1.5	−25.9 ± 6.3	43, 66
CH_3O	3.8 ± 0.2	15.9 ± 0.8	8b, 8
CH_3CO	−5.8 ± 0.4	−24.3 ± 1.7	43, 66
CH_3CHOH	−15.2 ± 1.0	−63.6 ± 4.2	2
CH_3OCH_2	−2.8 ± 1.2	−11.7 ± 5.0	43, 66
C_2H_5O	−4.1	−17.2	8
CH_2CO	17.3	72.4	3
C_2H_5CO	−10.2 ± 1.0	−42.7 ± 4.2	91
CH_3COCH_2	−5.7 ± 2.6	−23.9 ± 10.9	50
CH_2CHOH	0.0	0.0	5
$(CH_3)_2COH$	−26.6 ± 1.1	−111.3 ± 4.6	43, 66
$CH_3COCHCH_3$	−16.8 ± 1.7	−70.3 ± 7.1	82
n-C_3H_7O	−9.9	−41.4	8
i-C_3H_7O	−12.5	−52.3	8
Tetrahydrofuran-2-yl	−4.3 ± 1.5	−18.0 ± 6.3	11, 66
n-C_4H_9O	−14.7	−61.5	66
s-C_4H_9O	−16.6 ± 0.8	−69.5 ± 3.3	8b
$(CH_3)_3CO$	−21.7	−90.8	8, 8b
C_6H_5O	11.4 ± 2.0	47.7 ± 8.4	67a, 15, 20d
C_6H_5CO	26.1 ± 2	109.2 ± 8	66, 80
COOH	−53.3	−223.0	11, 80
$COOCH_3$	−40.4 ± 1	−169.0 ± 4	66, 78
CH_3COO	−49.6 ± 1	−207.5 ± 4	66
C_2H_5COO	−54.6 ± 1	−228.5 ± 4	66
n-C_3H_7COO	−59.6 ± 1	−249.4 ± 4	66
$C_6H_5COOCH_2$	−16.7 ± 2.0	−69.9 ± 8.4	66, 79
CFO	⩽−44	⩽−184	66
CH_3S	34.2 ± 2.0	143.1 ± 8.4	20b
C_6H_5S	56.8 ± 2.0	237.7 ± 8.4	20b
CH_3SO_2	−57.2	−239.3	66
CH_2F	−7.8 ± 2	−32.6 ± 8	49
CHF_2	−59.2 ± 2	−247.7 ± 8	49
CF_2	−46.4 ± 2.2	−194.1 ± 9.2	63, 94
CF_3	−112.0 ± 3.6	−468.6 ± 15.1	43, 66
CCl_2	57	239	66
$CHCl_2$	24.1	100.8	38
CF_2Cl	−64.3 ± 2	−269.0 ± 8	58, 35a
$CFCl_2$	−23	−96	35
CCl_3	19.0 ± 1	79.5 ± 4	61
CH_2Br	41.5	173.6	38
$CHBr_2$	54.3	227.2	38
CH_2I	55.0 ± 1.6	230.1 ± 6.7	43, 66
CHI_2	79.8 ± 2.2	333.9 ± 9.2	43, 66
CH_3CF_2	−72.3 ± 2	−302.5 ± 8.4	69b
CF_3CH_2	−123.6 ± 1.2	−517.1 ± 5.0	93
C_2F_5	−217.3 ± 1	−909.2 ± 4	66, 93a
$CHCl_2CCl_2$	5.6 ± 2	23.4 ± 8	1, 23, 37
CF_2ClCF_2	−164 ± 4	−686 ± 17	35a
C_2Cl_5	8.4 ± 2	35.2 ± 8	1, 23, 36
C_6F_5	−130.9 ± 2	−547.7 ± 8	17

Table 4 (continued)
HEATS OF FORMATION OF FREE RADICALS

Radical	$\Delta H^\circ_{f(298)}$/kcal mol^{-1}	$\Delta H^\circ_{f(298)}$/kJ mol^{-1}	Ref.
NH	86	360	66
NH_2	47.2 ± 1	197.5 ± 4	12, 44, 66
NF_2	8 ± 1	34 ± 4	1, 10, 25
PH_2	33 ± 2	138 ± 8	66
HO	9.2 ± 1	38.5 ± 4	48
HO_2	5.3 ± 2	22.2 ± 8	48
HS	33.1 ± 2	138.5 ± 8	48
SiH_2	58.6 ± 3.5	245.2 ± 14.6	85
SiH_3	49.1 ± 3	205.4 ± 13	70, 73
CH_3SiH	50.9 ± 3.5	213.0 ± 14.6	85
$(CH_3)_3Si$	−2.6	−10.9	26
$C_6H_5Si(CH_3)_2$	39	163	66
$(C_6H_5)_2SiCH_3$	78	326	66
$(C_6H_5)_3Si$	116.2	486.2	13
SiH_3SiH	64.5 ± 3.5	269.9 ± 14.6	85
GeH_3	57	239	66

Bond Strengths of Some Organic Molecules

Bond strengths at 298 K expressed in kcal/mol for some organic molecules of the general formula R—X are presented below. Some are experimental values taken from the preceding tables; the remainder are calculated from the heats of formation of the radicals, listed above, and the heats of formation of the parent compounds from sources indicated by the references below. The table also includes the bond strengths for ammonia, water, and the hydrogen halides.

Table 5
BOND STRENGTHS OF SOME ORGANIC MOLECULES

X / R	H	F	Cl	Br	I	OH	NH_2	OCH_3	CH_3	CH_3CO	NO	CF_3
H	104.207	135.9	103.2	87.4	71.4	119	110	104	104	86	50	106
CH_3	104	109[b]	84[a]	70[a]	56[a]	91[a]	87[a]	81[a]	88[a]	80[a]	42	101
C_2H_5	98	—	81[a]	68[a]	53[a]	91[a]	84[a]	81[a]	85[a]	77[a]	42	—
$i\text{-}C_3H_7$	95	106[a]	81[a]	68[a]	53[a]	92[a]	85[a]	81[a]	84[a]	75[a]	41	—
$t\text{-}C_4H_9$	92	—	81[a]	67[a]	50[a]	91[a]	84[a]	81[a]	82[a]	71[a]	41	—
C_6H_5	110	125[a]	95[a]	80	64[a]	110[a]	104[a]	98[a]	100[a]	93[a]	52	—
$C_6H_5CH_2$	85	—	69[a]	55[a]	45[a]	78[a]	—	—	72[a]	63[a]	—	—
CCl_3	96	102[a]	70	56	—	—	—	—	88[d]	—	32	—
CF_3	106	130[a]	86	71	53	—	—	—	99[a]	—	31	97
C_2F_5	103[f]	127[f]	85[f]	68[f]	53[f]	—	—	—	—	—	—	—
CH_3CO	86	119[a]	81[a]	67[a]	50[a]	106[a]	99[c]	97	80[a]	67	—	—
CN	120	—		83	73	—	—	—	—	—	29	—
C_6F_5	114	114[e]	92[e]	—	66[e]	107[e]	—	—	105[e]	—	50[e]	60[e]

[a] Cox, J. D. and Pilcher, G., *Thermochemistry of Organic and Organometallic Compounds*, Academic Press, London, 1970.
[b] Lacher, J. R. and Skinner, H. A., *J. Chem. Soc. A*,1034, 1968.
[c] Benson, S. W. et al., *Chem. Rev.*, 69, 279, 1969.
[d] Hu, A. T., Sinke, G. C., and Mintz, M. J., *J. Chem. Thermodyn.*, 4, 239, 1972.
[e] Choo, K. Y., Mendenhall, G. D., Golden, D. M., and Benson, S. W., *Int. J. Chem. Kinet.*, 6, 813, 1974.
[f] Wu, E.-C. and Rodgers, A. S., *J. Am. Chem. Soc.*, 98, 6112, 1976.

THE MADELUNG CONSTANT AND CRYSTAL LATTICE ENERGY

Donald F. Swinehart

If U is the crystal lattice energy and *M* is the Madelung constant, then

$$U = \frac{N M z_i z_j e^2}{r}(1 - 1/n)^a$$

Substance	Ion type	Crystal form[b]	*M*
Sodium chloride, NaCl	M^+, X^-	FCC	1.74756
Cesium chloride, CsCl	M^+, X^-	BCC	1.76267
Calcium chloride, $CaCl_2$	$M^{++}, 2X^-$	Cubic	2.365
Calcium fluoride, fluorite, CaF_2	$M^{++}, 2X^-$	Cubic	2.51939
Cadmium chloride, $CdCl_2$	$M^{++}, 2X^-$	Hexagonal	2.244[c]
Cadmium iodide, $CdI_2(\alpha)$	$M^{++}, 2X^-$	Hexagonal	2.355[c]
Magnesium fluoride, MgF_2	$M^{++}, 2X^-$	Tetragonal	2.381[c]
Cuprous oxide, cuprite, Cu_2O	$2M^+, X^=$	Cubic	2.22124
Zinc oxide, ZnO	$M^{++}, X^=$	Hexagonal	1.4985[c]
Sphalerite, zinc blende, ZnS	$M^{++}, X^=$	FCC	1.63806
Wurtzite, ZnS	$M^{++}, X^=$	Hexagonal	1.64132[c]
Titanium dioxide, anatase, TiO_2	$M^{4+}, 2X^=$	Tetragonal	2.400[c]
Titanium dioxide, rutile, TiO_2	$M^{4+}, 2X^=$	Tetragonal	$(2.408) - \frac{(2.055)}{(0.721 - c/a)^2}$[d]
β-Quartz, SiO_2	$M^{4+}, 2X^=$	Hexagonal	2.2197[c]
Corundum, Al_2O_3	$2M^{3+}, 3X^=$	Rhombohedral	4.1719

[a] N is Avogadro's number, z_i and z_j are the integral charges on the ions (i.e., in units of the electronic charge), and e is the charge on the electron in electrostatic units ($e = 4.803 \times 10^{-10}$ ESU). r is the shortest distance between cation-anion pairs in centimeters. Then U is in ergs.

[b] FCC = face centered cubic, BCC is body centered cubic.

[c] For tetragonal and hexagonal crystals the value of *M* depends on the details of the lattice parameters.

[d] For rutile-type structures, *M* is given to a good approximation by inserting the proper c/a ratio. c/a for rutile itself is 0.721 and *M* = 2.408.

Note: Several variations of the equation for U appear in the literature with corresponding variations in *M*. In general, using literature values for *M* requires caution. The Born exponent, n, may be obtained from the following table:

THE BORN EXPONENT, n

Ion type	n
He, Li^+	5
Ne, Na^+, F^-	7
Ar, K^+, Cu^+, Cl^-	9
Kr, Rb^+, Ag^+, Br^-	10
Xe, Cs^+, Au^+, I^-	12

For a crystal with a mixed-ion type, an average of the values of n in this table is to be used (6 for LiF, for example).

VALUES OF THE GAS CONSTANT, R

Coleman J. Major

The numerical value of the gas constant, R, defined by the equation PV = nRT, depends upon the units of P, V, n, and T. A large number of values of the constant may be calculated. The accompanying table gives 84 values of R in a convenient form using the most common units of pressure and volume. It also incorporates both the pound and gram mole and both Rankine and Kelvin temperature scales. Various combinations of metric and English units may, therefore, be used without the necessity of converting each variable to a common system of units. Conversion factors and constants used for computing the values of R are listed at the bottom of the table.

The following example illustrates the use of the table:

Calculate: The volume in ft^3 occupied by 2 lb. moles of a gas at 15°C at a pressure of 32.2 ft. of water, assuming the ideal gas law.

Solution: 15°C+273.2 = 288.2°K

Enter the top of the table under the column headed "ft H_2O" and proceed downward to the value of 44.6 for R. (Note that this lines up horizontally with the desired units of ft^3, °K, and lb. moles shown on the left side of the table)

$$V = \frac{nRT}{P} = \frac{2 \times 44.6 \times 288.2}{32.2} = 798 \text{ ft}^3$$

$$\text{Values of Gas Constant, } R = \frac{PV}{nT}$$

Volume	Temp.	moles	Absolute Pressure						
			Atm	psia	mm Hg	cm Hg	in Hg	in H_2O	ft H_2O
ft^3	°K	gm	0.00290	0.0426	2.20	0.220	0.0867	1.18	0.0982
		lb	1.31	19.31	999	99.9	39.3	535	44.6
	°R	gm	0.00161	0.02366	1.22	0.122	0.0482	0.655	0.0546
		lb	0.730	10.73	555	55.5	21.8	297	24.8
cm^3	°K	gm	82.05	1206	62,400	6240	2450	33,400	2780
		lb	37,200	547,000	2.83×10^7	2.83×10^6	1.11×10^6	1.51×10^7	1.26×10^6
	°R	gm	45.6	670	34,600	3460	1360	18,500	1550
		lb	20,700	304,000	1.57×10^7	1.57×10^6	619,000	8.41×10^6	701,000
liters	°K	gm	0.08205	1.206	62.4	6.24	2.45	33.4	2.78
		lb	37.2	547	28,300	2830	1113	15,140	1262
	°R	gm	0.0456	0.670	34.6	3.46	1.36	18.5	1.55
		lb	20.7	304	15,700	1570	619	8410	701

Conversion Factors and Constants

1 lb. = 453.59 gm	359.0 ft^3/lb mole
1 atm = 14.696 psia	22,414 cm^3/gm mole
1 atm = 760 mm Hg	1 inch = 2.54 cm
1 atm = 76 cm Hg	Std. temp. = 273.16°K or 491.69°R
1 atm = 29.921 in Hg	28.31605 liters = 1 ft^3
1 atm = 406.79 in H_2O	$R = 8.31432 \pm 0.00034 \times 10^7$ erg $°K^{-1}$ mol^{-1}
1 atm = 33.90 ft H_2O	

RECOMMENDED CONSISTENT VALUES OF THE FUNDAMENTAL PHYSICAL CONSTANTS

The numbers in parentheses are the standard deviation uncertainties in the last digits of the quoted value, computed on the basis of internal consistency.

Quantity	Symbol	Value	Uncertainty, ppm
1. Permeability of Vacuum	μ_0	$4\pi \times 10^{-7}$ H m^{-1}	
		= $12.5663706144 \times 10^{-7}$ H m^{-1}	
2. Speed of Light in Vacuum	c	$2.99792458(1.2) \times 10^{8}$ m s^{-1}	0.004
3. Permittivity of Vacuum	$\epsilon_0 = (\mu_0 c^2)^{-1}$	$8.85418782(7) \times 10^{-12}$ F m^{-1}	0.008
4. Fine Structure Constant, $\mu_o ce^2/2h$	α	0.0072973506(60)	0.82
	α^{-1}	137.03604(11)	0.82
5. Elementary Charge	e	$1.6021892(46) \times 10^{-19}$ C	2.9
6. Planck Constant	h	$6.626176(36) \times 10^{-34}$ J Hz^{-1}	5.4
	$\hbar = h/2\pi$	$1.0545887(57) \times 10^{-34}$ J s	5.4
7. Avogadro Constant	N_A	$6.022045(31) \times 10^{23}$ mol^{-1}	5.1
8. Atomic Mass Unit	$1u = (10^{-3}$ kg $mol^{-1})/N_A$	$1.6605655(86) \times 10^{-27}$ kg	5.1
9. Electron Rest Mass	m_e	$0.9109534(47) \times 10^{-30}$ kg	5.1
		$5.4858026(21) \times 10^{-4}$ u	0.38
10. Muon Rest Mass	m_μ	$1.883566(11) \times 10^{-28}$ kg	5.6
		0.11342920(26) u	2.3
11. Proton Rest Mass	m_p	$1.6726485(86) \times 10^{-27}$ kg	5.1
		1.007276470(11) u	0.011
12. Neutron Rest Mass	m_n	$1.6749543(86) \times 10^{-27}$ kg	5.1
		1.008665012(37) u	0.037
13. Ratio, Proton Mass to Electron Mass	m_p/m_e	1,836.15152(70)	0.38
14. Ratio, Muon Mass to Electron Mass	m_μ/m_e	206.76865(47)	2.3
15. Specific Electron Charge	e/m_e	$1.7588047(49) \times 10^{11}$ C kg^{-1}	2.8
16. Faraday Constant	$\mathcal{F} = N_A e$	$9.648456(27) \times 10^{4}$ C mol^{-1}	2.8
17. Magnetic Flux Quantum	$\Phi_0 = h/2e$	$2.0678506(54) \times 10^{-15}$ Wb	2.6
	h/e	$4.135701(11) \times 10^{-15}$ J Hz^{-1} C^{-1}	2.6
18. Josephson Frequency-Voltage Ratio	$2e/h$	483.5939(13) THz V^{-1}	2.6
19. Quantum of Circulation	$h/2m_e$	$3.6369455(60) \times 10^{-4}$ J Hz^{-1} kg^{-1}	1.6
	h/m_e	$7.273891(12) \times 10^{-4}$ J Hz^{-1} kg^{-1}	1.6
20. Rydberg Constant	R_∞	$1.097373177(83) \times 10^{7}$ m^{-1}	0.075
21. Bohr Radius	$a_0 = \alpha/4\pi R_\infty$	$0.52917706(44) \times 10^{-10}$ m	0.82
22. Electron Compton Wavelength	$\lambda_c = \alpha^2/2R_\infty$	$2.4263089(40) \times 10^{-12}$ m	1.6
	$\bar{\lambda}_c = \lambda_c/2\pi = \alpha a_0$	$3.8615905(64) \times 10^{-13}$ m	1.6
23. Classical Electron Radius	$r_e = \mu_o e^2/4\pi m_e = \alpha\bar{\lambda}_c$	$2.8179380(70) \times 10^{-15}$ m	2.5
24. Electron g-Factor	$\frac{1}{2}g_e = \mu_e/\mu_B$	1.0011596567(35)	0.0035
25. Muon g-Factor	$\frac{1}{2}g_\mu$	1.00116616(31)	0.31

RECOMMENDED CONSISTENT VALUES OF THE FUNDAMENTAL PHYSICAL CONSTANTS (Continued)

Quantity	Symbol	Value	Uncertainty, ppm
26. Proton Moment in Nuclear Magnetons	μ_p/μ_N	2.7928456(11)	0.38
27. Bohr Magneton	$\mu_B = e\hbar/2m_e$	$9.274078(36) \times 10^{-24}$ J T^{-1}	3.9
28. Nuclear Magneton	$\mu_N = e\hbar/2m_p$	$5.050824(20) \times 10^{-27}$ J T^{-1}	3.9
29. Electron Magnetic Moment	μ_e	$9.284832(36) \times 10^{-24}$ J T^{-1}	3.9
30. Proton Magnetic Moment	μ_p	$1.4106171(55) \times 10^{-26}$ J T^{-1}	3.9
31. Proton Magnetic Moment in Bohr Magnetons	μ_p/μ_B	$1.521032209(16) \times 10^{-3}$	0.011
32. Ratio, Electron to Proton Magnetic Moments	μ_e/μ_p	658.2106880(66)	0.010
33. Ratio, Muon Moment to Proton Moment	μ_μ/μ_p	3.1833402(72)	2.3
34. Muon Magnetic Moment	μ_μ	$4.490474(18) \times 10^{-26}$ J T^{-1}	3.9
35. Proton Gyromagnetic Ratio	γ_p	$2.6751987(75) \times 10^{8}$ s^{-1} T^{-1}	2.8
36. Diamagnetic Shielding Factor, Spherical H_2O Sample	$1 + \sigma(H_2O)$	1.000025637(67)	0.067
37. Proton Gyromagnetic Ratio (uncorrected)	γ'_p	$2.6751301(75) \times 10^{8}$ s^{-1} T^{-1}	2.8
	$\gamma'_p/2\pi$	42.57602(12) MHz T^{-1}	2.8
38. Proton Moment in Nuclear Magnetons (uncorrected)	μ'_p/μ_N	2.7927740(11)	0.38
39. Proton Compton Wavelength	$\lambda_{C,p} = h/m_pc$	$1.3214099(22) \times 10^{-15}$ m	1.7
	$\bar{\lambda}_{C,p} = \lambda_{C,p}/2\pi$	$2.1030892(36) \times 10^{-16}$ m	1.7
40. Neutron Compton Wavelength	$\lambda_{C,n} = h/m_nc$	$1.3195909(22) \times 10^{-15}$ m	1.7
	$\bar{\lambda}_{C,n} = \lambda_{C,n}/2\pi$	$2.1001941(35) \times 10^{-16}$ m	1.7
41. Molar Gas Constant	R	8.31441(26) J mol^{-1} K^{-1}	31
42. Molar Volume, Ideal Gas (T_0 = 273.15 K, p_0 = 1 atm)	$V_m = RT_0/p_0$	0.02241383(70) m^3 mol^{-1}	31
43. Boltzmann Constant	$k = R/N_A$	$1.380662(44) \times 10^{-23}$ J K^{-1}	32
44. Stefan-Boltzmann Constant	$\sigma = (\pi^2/60)k^4/\hbar^3c^2$	$5.67032(71) \times 10^{-8}$ W m^{-2} K^{-4}	125
45. First Radiation Constant	$c_1 = 2\pi hc^2$	$3.741832(20) \times 10^{-16}$ W m^2	5.4
46. Second Radiation Constant	$c_2 = hc/k$	0.01438786(45) m K	31
47. Gravitational Constant	G	$6.6720(41) \times 10^{-11}$ N m^2 kg^{-2}	615

Data from CODATA Bulletin No. 11, ICSU CODATA Central Office, 19 Westendstrasse, 6 Frankfurt/Main, German Federal Republic (copies of this bulletin are available at no cost from this office).

ENERGY CONVERSION FACTORS

Quantity	Value	Unit	Error (ppm)
1 kg	5.609538(24)	10^{29} MeV	4.4
1 amu	931.4812(52)	MeV	5.5
Electron mass	0.5110041(16)	MeV	3.1
Proton mass	938.2592(52)	MeV	5.5
Neutron mass	939.5527(52)	MeV	5.5
1 electron volt	1.6021917(70)	10^{-19} J 10^{-12} erg	4.4
	2.4179659(81)	10^{14} Hz	3.3
	8.065465(27)	10^{5} m^{-1} 10^{3} cm^{-1}	3.3
	1.160485(49)	10^{4} K	42
Energy–wavelength conversion	1.2398541(41)	10^{-6} eV·m 10^{-4} eV·cm	3.3
Rydberg constant, R_{∞}	2.179914(17)	10^{-18} J 10^{-11} erg	7.6
	13.605826(45)	eV	3.3
	3.2898423(11)	10^{15} Hz	0.35
	1.578936(67)	10^{5} K	43
Bohr magneton, μ_B	5.788381(18)	10^{-5} eV T^{-1}	3.1
	1.3996108(43)	10^{10} Hz T^{-1}	3.1
	46.68598(14)	m^{-1}·T^{-1} 10^{-2} cm^{-1}·T^{-1}	3.1
	0.671733(29)	K T^{-1}	43
Nuclear magneton, μ_n	3.152526(21)	10^{-8} eV T^{-1}	6.8
	7.622700(42)	10^{6} Hz T^{-1}	5.5
	2.542659(14)	10^{-2} m^{-1}·T^{-1} 10^{-4} cm^{-1}·T^{-1}	5.5
	3.65846(16)	10^{-4} K T^{-1}	44
Gas constant, R_0	8.20562(35)	10^{-2} m^3·atm $kmole^{-1}$·K^{-1}	42
Standard volume of ideal gas, V_0	22.4136	m^3 $kmole^{-1}$	

INFRARED CORRELATION CHART No. 1

Prepared from information supplied by Beckman Instruments

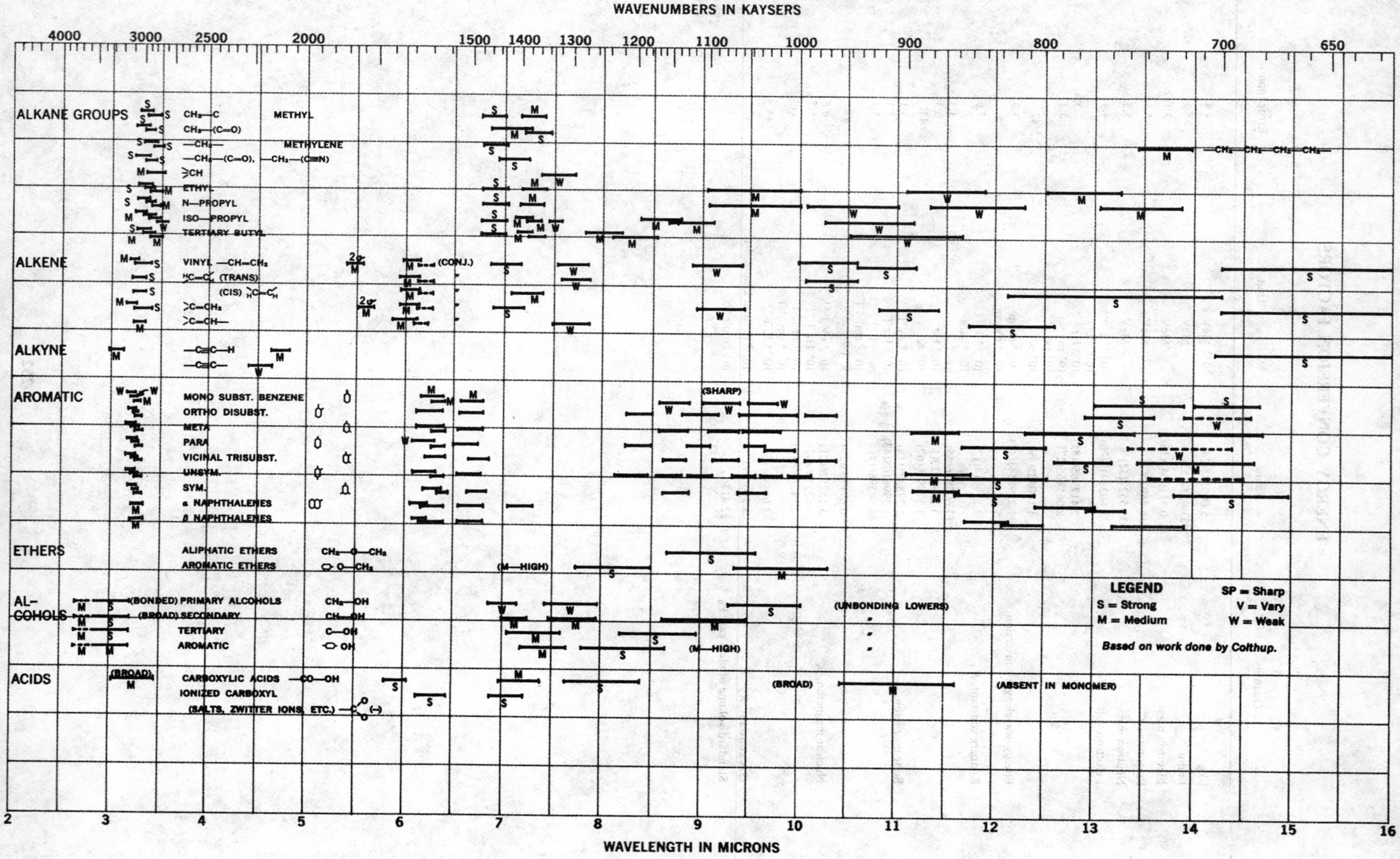

INFRARED CORRELATION CHART No. 1 (Con't.)

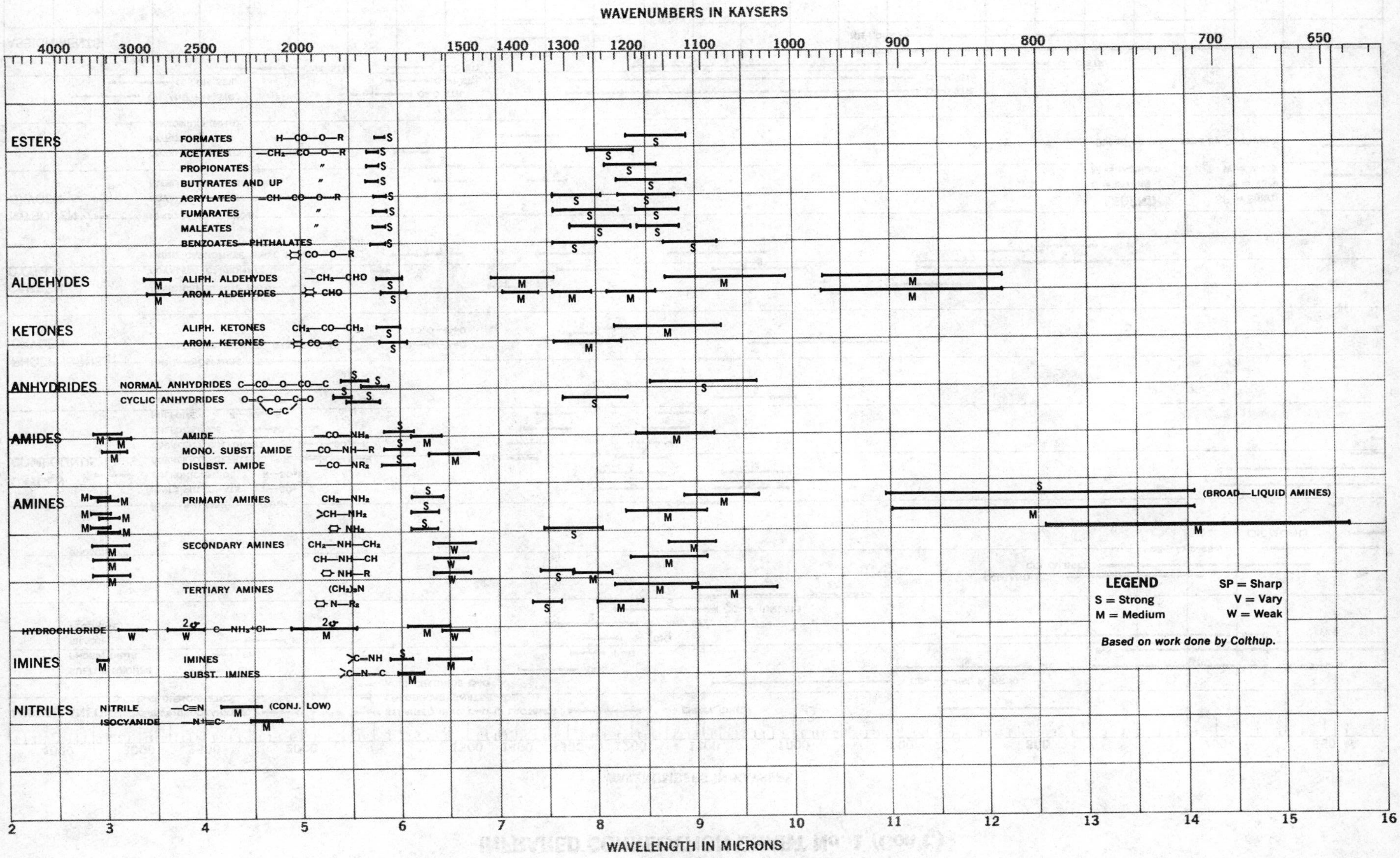

INFRARED CORRELATION CHART No. 1 (Con't.)

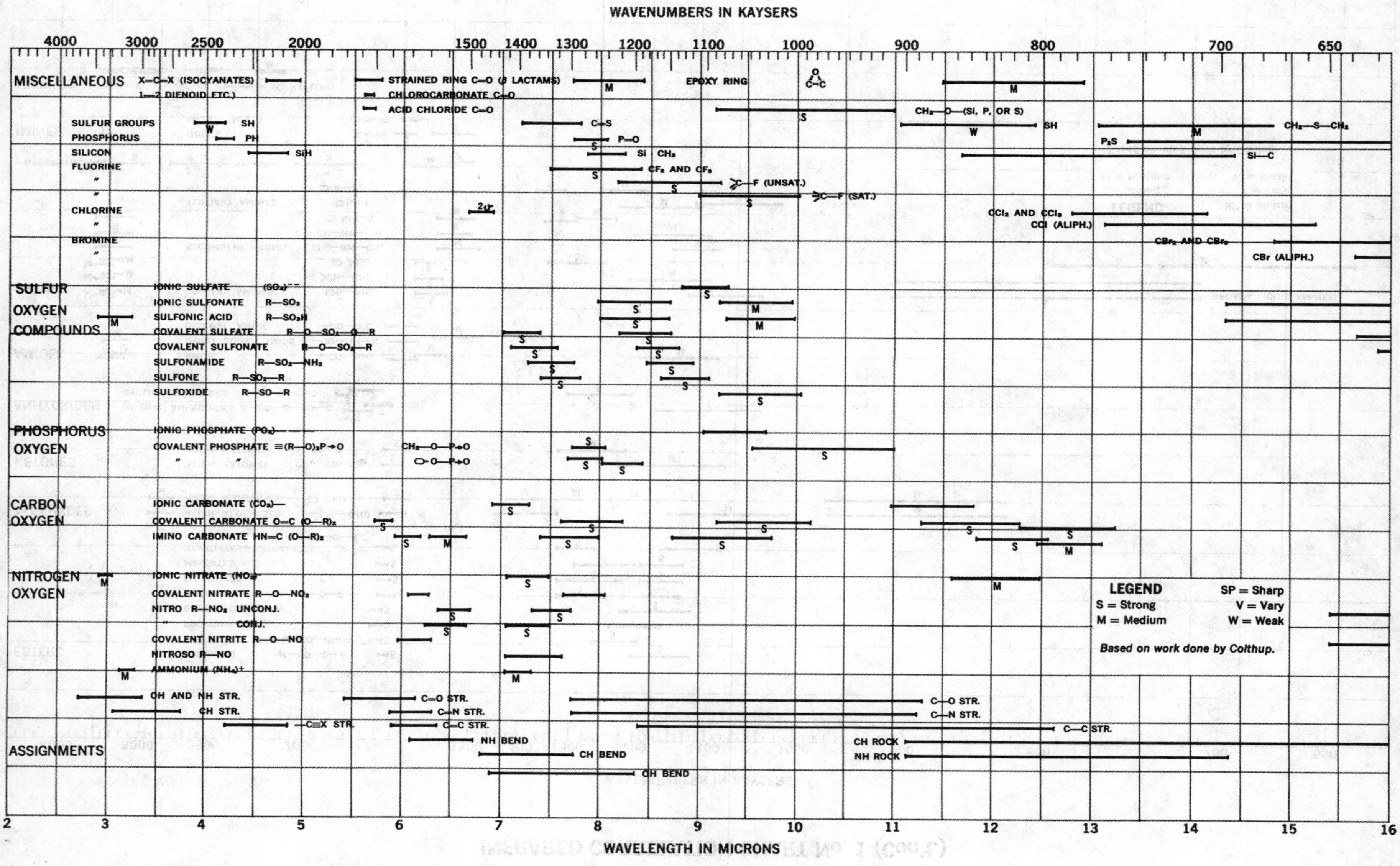

INFRARED CORRELATION CHART No. 2

Prepared from information supplied by Beckman Instruments

This chart presents some information regarding structure, double-bond vibrations, hydrogen stretching and triple-bond vibrations.

HYDROGEN STRETCHING AND TRIPLE-BOND VIBRATIONS, 3750-2000 $CM.^{-1}$ DOUBLE-BOND VIBRATIONS, ETC. 2000-1500 $CM.^{-1}$

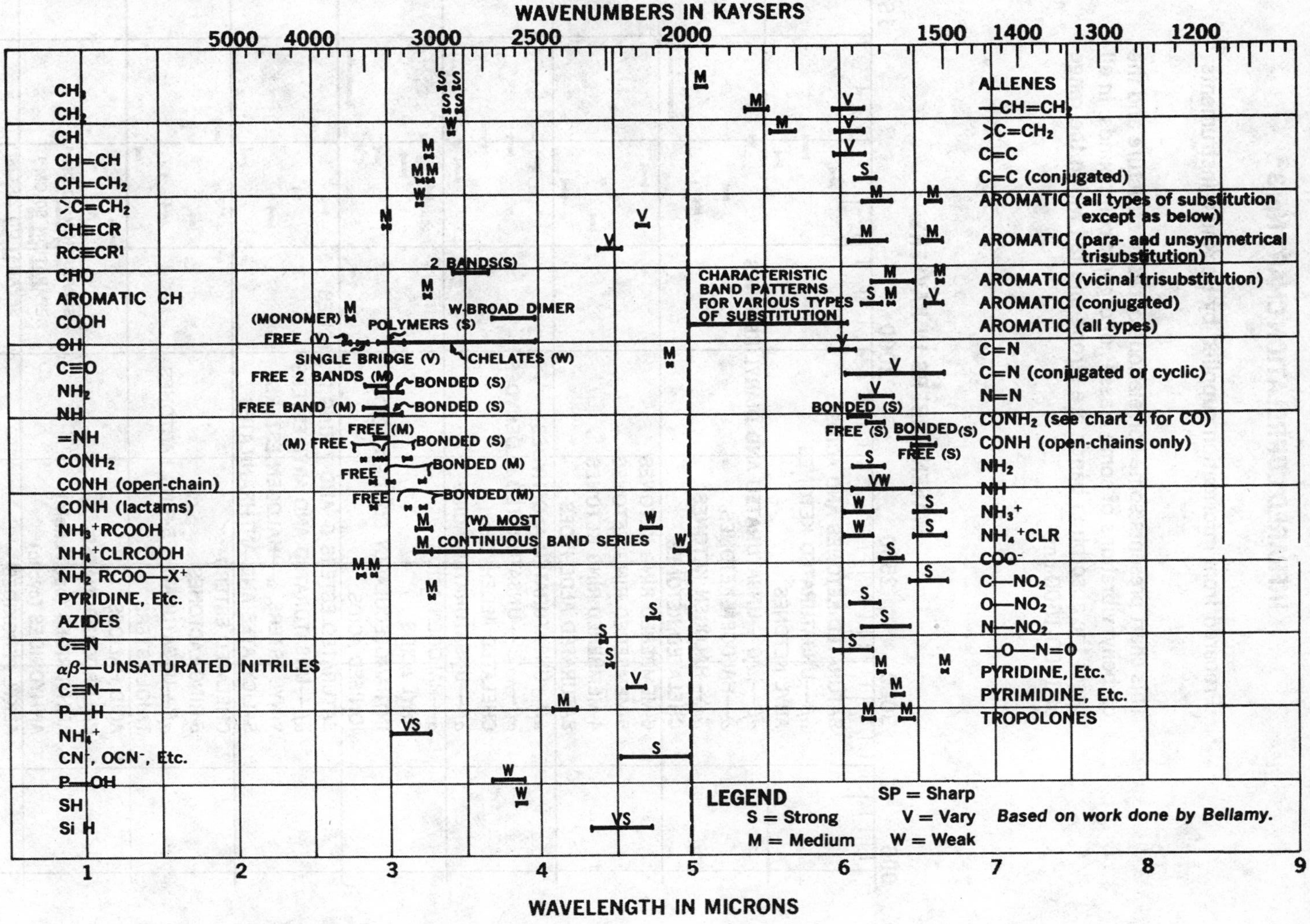

INFRARED CORRELATION CHART No. 3

Prepared from information supplied by Beckman Instruments

This chart presents some correlations between structure and the carbonyl vibrations of some classes of organic compounds. In all cases the absorption bands are strong and fall within the range of 1900-1500 cm^{-1}.

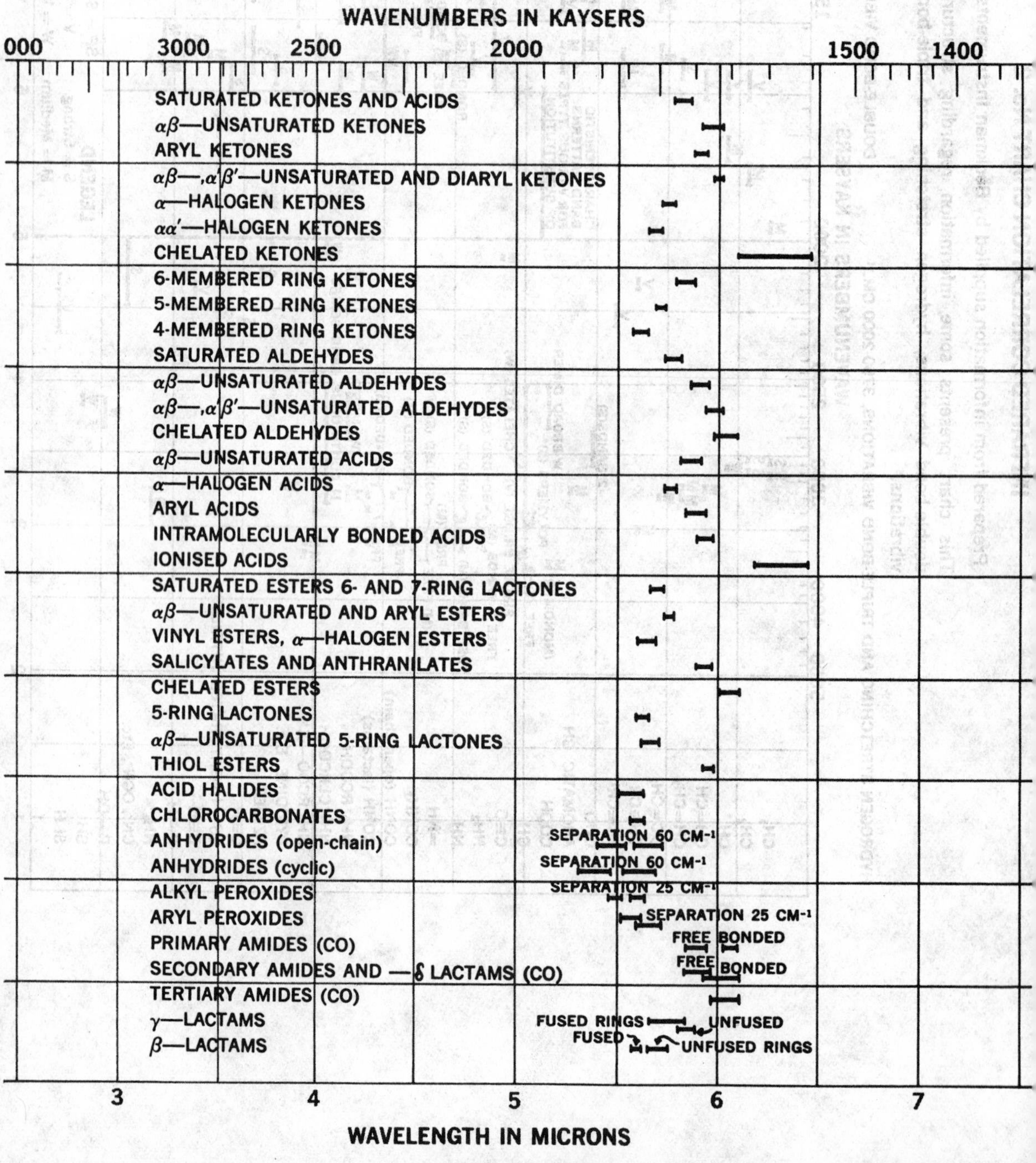

INFRARED CORRELATION CHART No. 4

Prepared from information supplied by Beckman Instruments

This chart presents some correlations between structure and single-bond vibrations for a number of classes of compounds having absorption between 1500-650 cm^{-1}

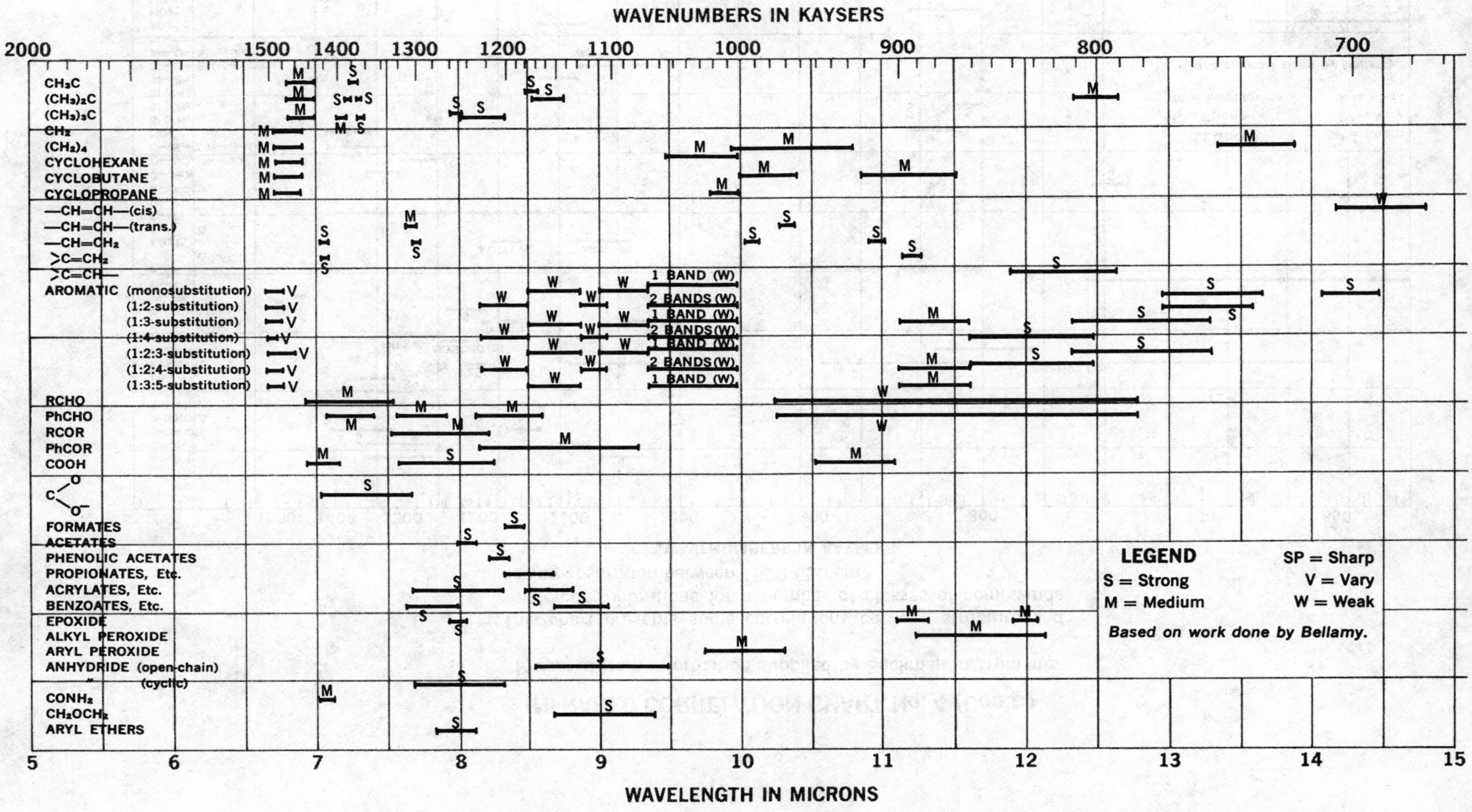

INFRARED CORRELATION CHART No. 4 (Con't.)

Prepared from information supplied by Beckman Instruments

This chart presents some correlations between structure and single-bond vibrations for a number of classes of compounds having absorption between 1500-650 cm^{-1}.

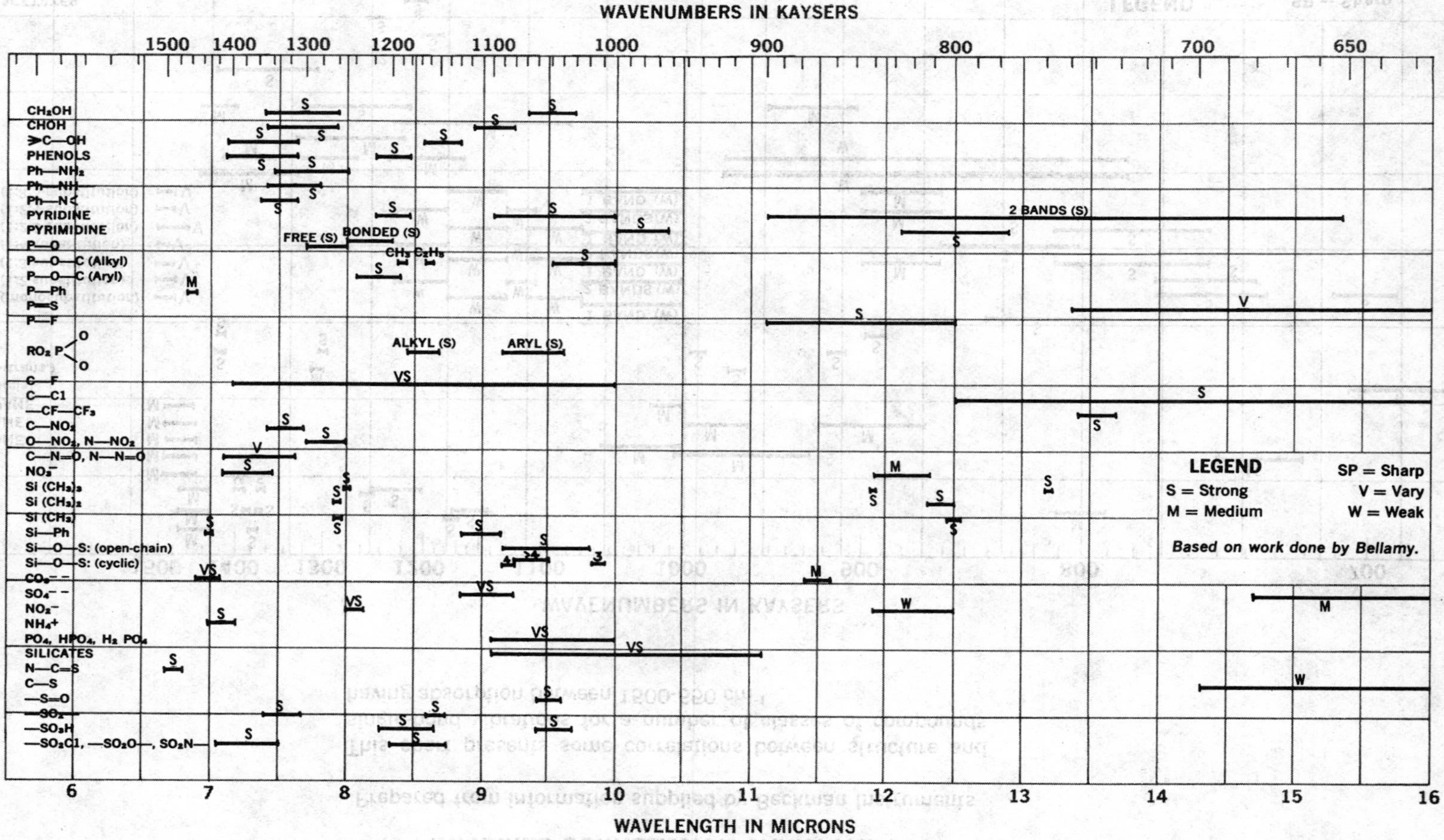

FAR INFRARED VIBRATIONAL FREQUENCY CORRELATION CHART

Based on evidence compiled by James E. Stewart of Beckman Instruments. This chart shows the vibrational frequency correlation in the far infrared region. Because research is continuing in the far infrared region, this chart is not all-inclusive.

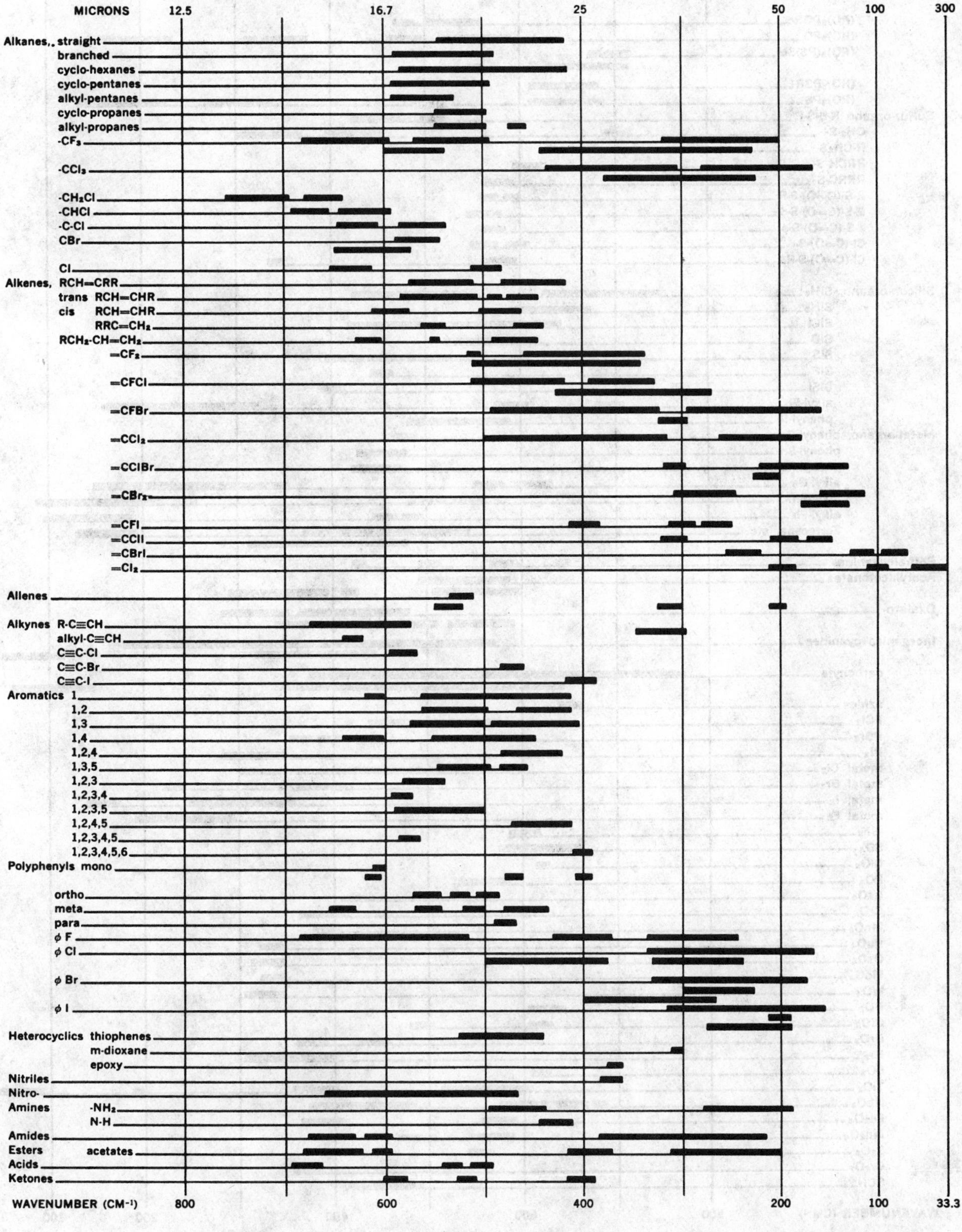

FAR INFRARED VIBRATIONAL FREQUENCY CORRELATION CHART (Con't.)

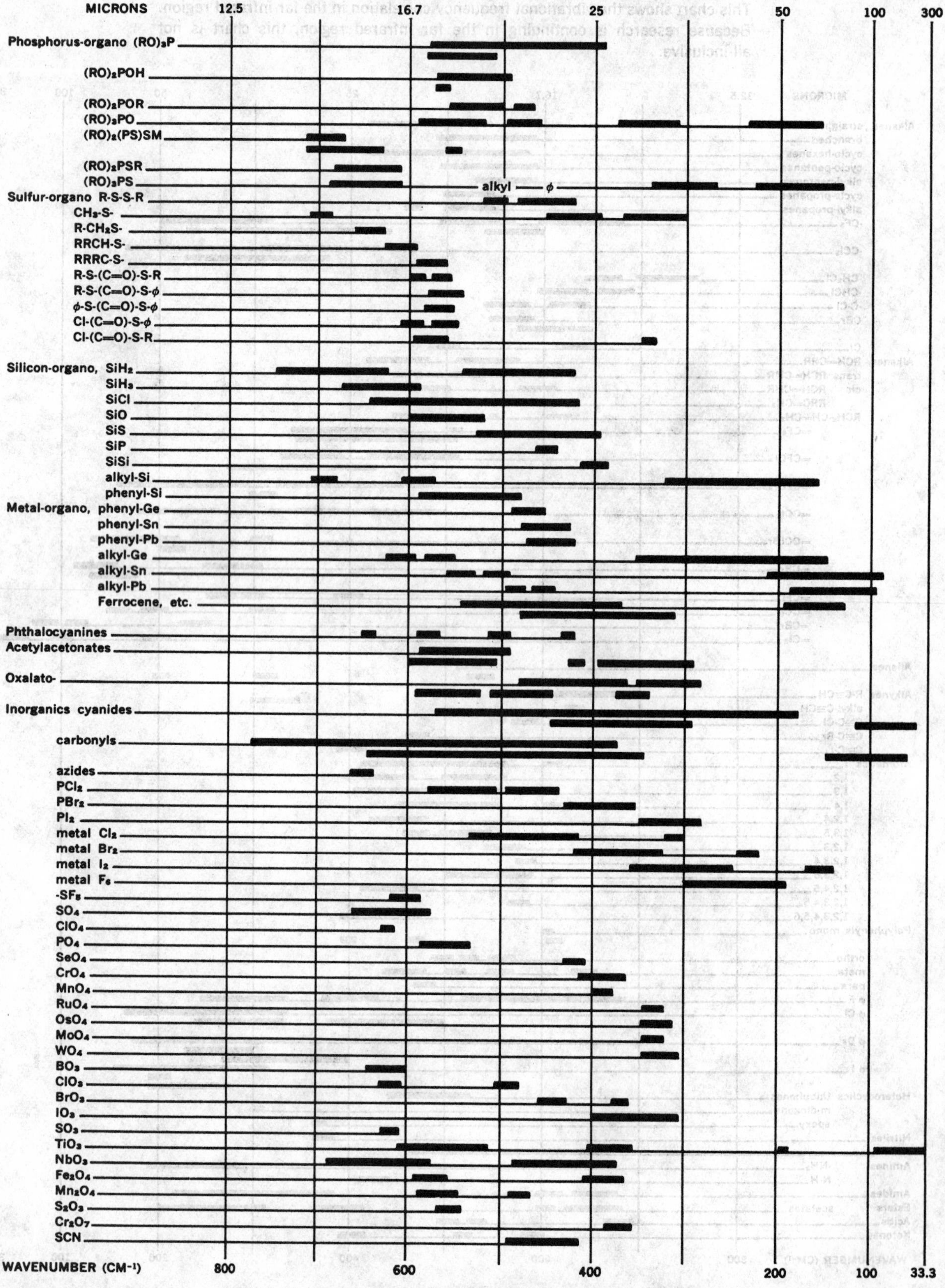

CHART OF CHARACTERISTIC FREQUENCIES BETWEEN ~700–300 cm^{-1}

Freeman F. Bentley, Lee D. Smithson, and Adele L. Rozek

This chart summarizes the characteristic frequencies known to occur between approximately 700–300^{-1}. Those who anticipate using this region of the spectrum should consult "Infrared Spectra and Characteristic Frequencies between ~700–300 cm^{-1}" by Interscience Publishers, a division of John Wiley and Sons, Inc. for a complete discussion of the characteristic frequencies summarized in this chart, a large collection of infrared spectra (700–300 cm^{-1}) of most of the common organic and inorganic compounds, and an extensive bibliography of references to infrared data below ~700 cm^{-1}.

In this chart the black horizontal bars indicate the range of the spectrum in which the characteristic frequencies have been observed to occur in the compounds investigated. The number of compounds investigated is given immediately to the right of the names or structures of the compounds. Obviously those characteristic frequency ranges based upon a limited number of compounds should be used with caution.

The letters above the bars indicate the relative intensities of the absorption bands. These intensities are based upon the strongest band in the spectra (700–300 cm^{-1}) of specific classes of compounds investigated, and they cannot be compared accurately with the intensities given for other classes.

When known, the specific vibration giving rise to the characteristic frequency is printed in abbreviated form immediately to the right of the bar indicating the frequency range except when lack of space prevents this. When there can be no ambiguity, this information may be printed other than to the right. In doubtful cases, arrows are used for clarification.

Naturally, the characteristic frequencies vary in their specificity and analytical value. The user is, therefore, cautioned to use this chart with some reserve. After reviewing this chart, the reader should be aware that there are many characteristic frequencies in the 700–300 cm^{-1} region. Used cautiously, this chart can be of considerable value in the elucidation of structures of unknown compounds.

It is important to emphasize that the region of the infrared between ~700–300 cm^{-1} should be used in conjunction with the more conventional 5000–700 cm^{-1} region. Much of the value of the 700–300 cm^{-1} region can only be realized after interpreting the spectrum between 5000–700 cm^{-1}.

The following symbols and abbreviations are used:

Symbol or Abbreviation	Definition
αCCC	In-plane bending of benzene ring
Antisym.	Antisymmetrical (Asymmetrical)
~	Approximately
β	In-plane bending of ring substituent bond
δ	In-plane bending
γ	Out-of-plane bending
i.p.	In-plane
m	Medium
ν	Stretching
ν_s	Symmetrical stretching
ν_{as}	Antisymmetrical stretching
o.p.	Out-of-plane
$\parallel$	Parallel
$\perp$	Perpendicular
ϕ	Phenyl
ϕCC	Out-of-plane bending of aromatic ring
r	Rocking
s	Strong
sh	Shoulder
Sym.	Symmetrical
v	Variable
w	Weak
"X" Sensitive	An aromatic vibrational mode whose frequency position is greatly dependent on the nature of the substituent.

CHART OF CHARACTERISTIC FREQUENCIES BETWEEN ~ 700–300 cm^{-1}

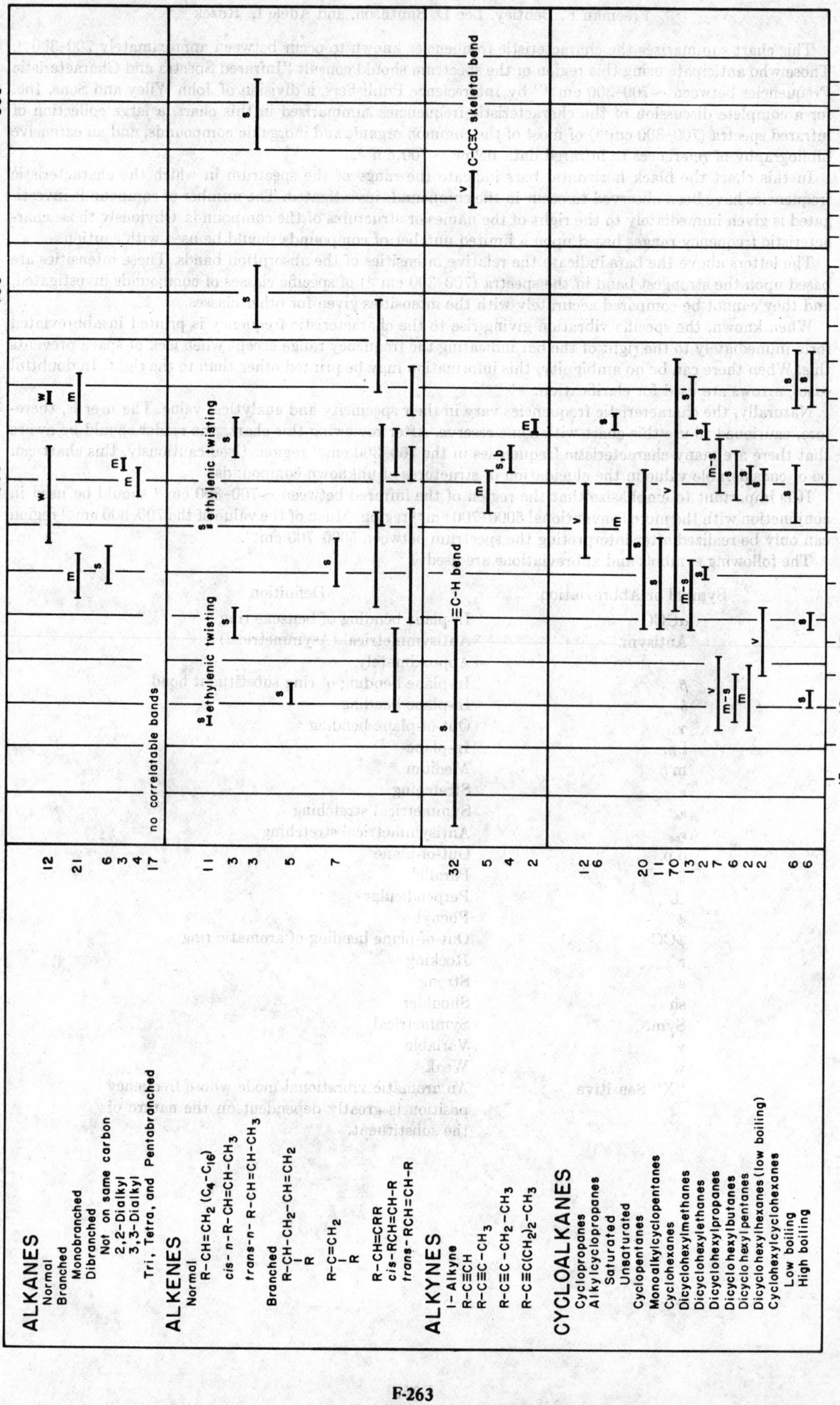

CHART OF CHARACTERISTIC FREQUENCIES BETWEEN ~ 700–300 cm⁻¹

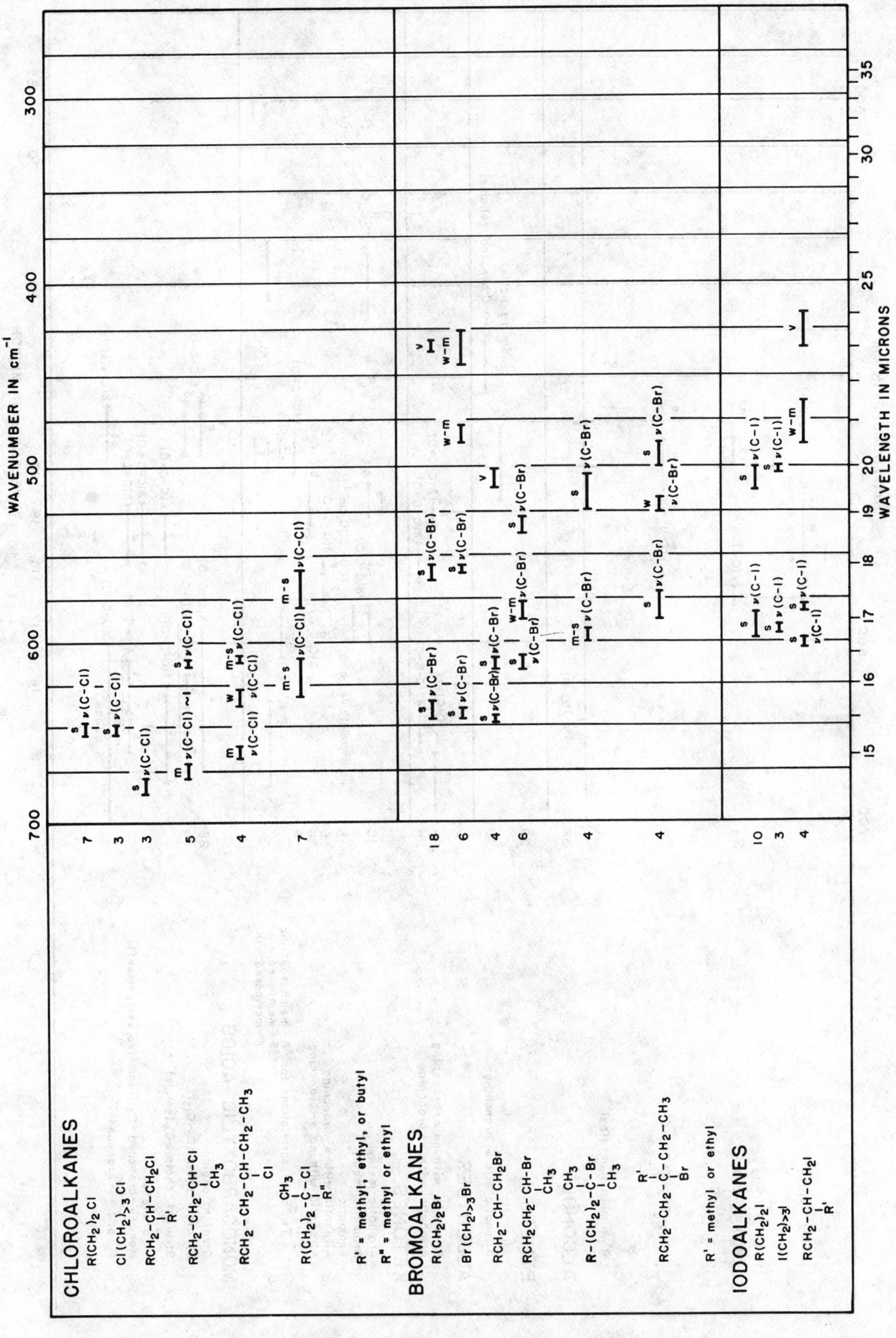

CHART OF CHARACTERISTIC FREQUENCIES BETWEEN ∽ 700–300 cm⁻¹

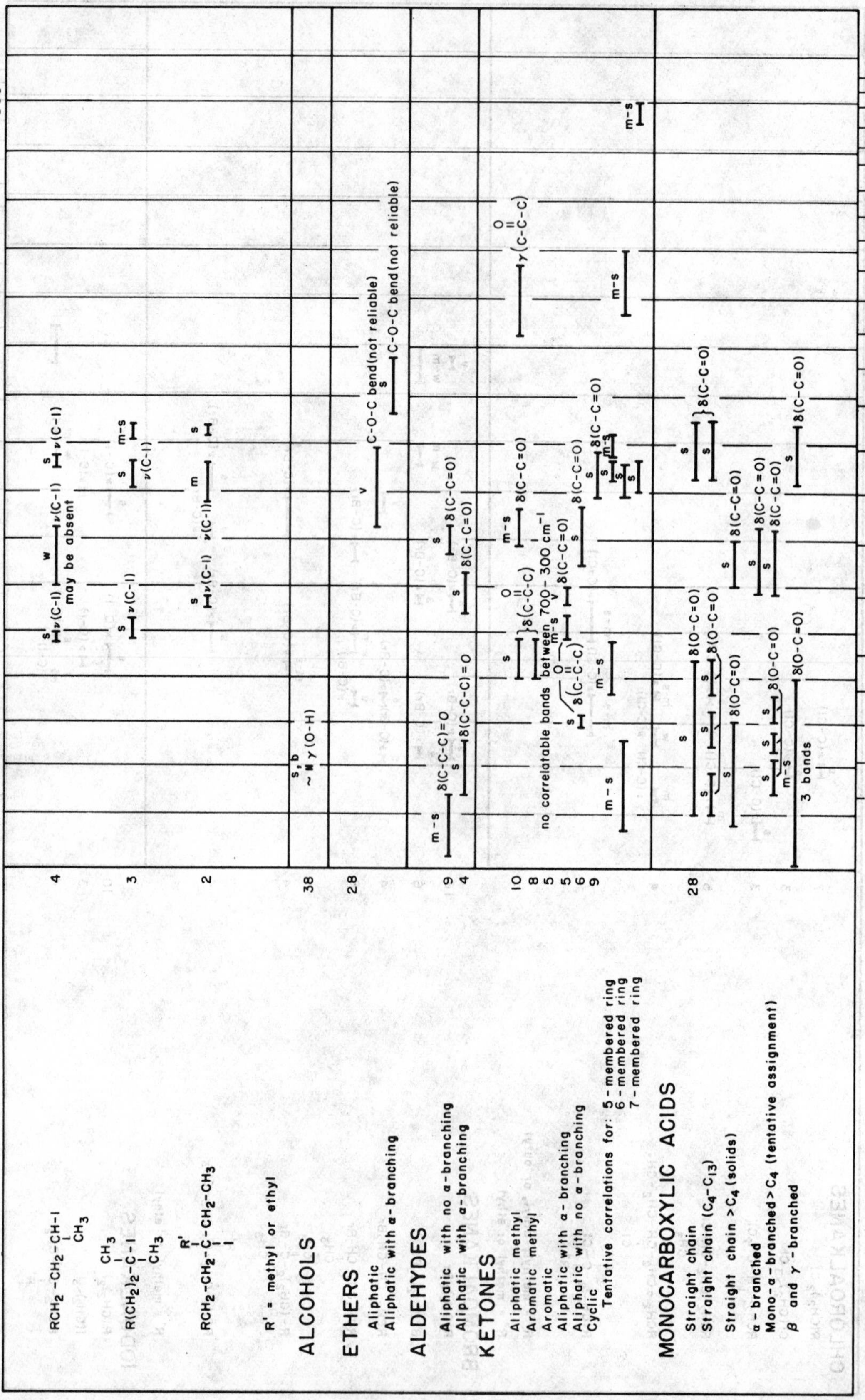

CHART OF CHARACTERISTIC FREQUENCIES BETWEEN ~ 700–300 cm⁻¹

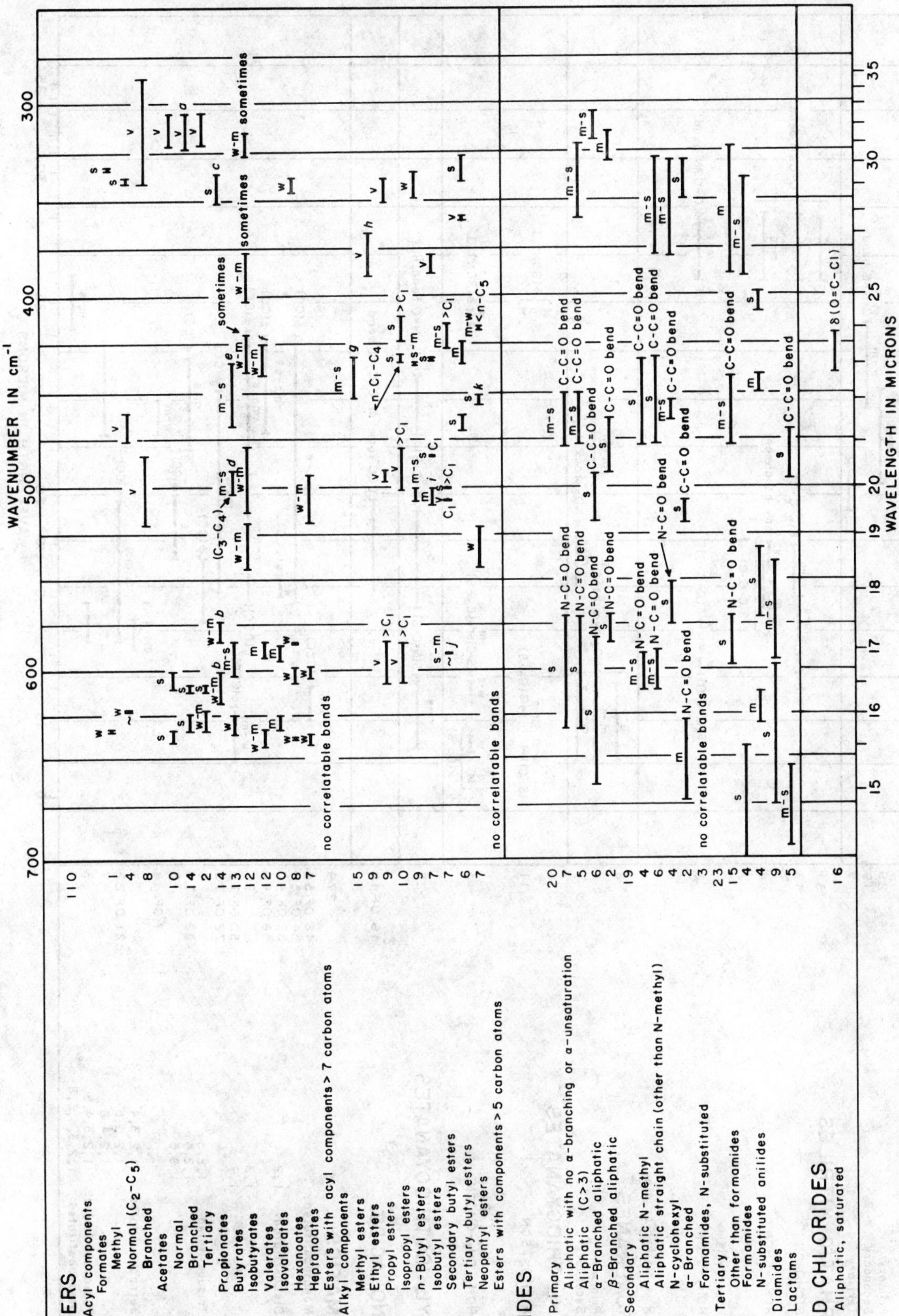

a - Except isopropyl and *sec*-butyl acetate
b - Except isoamyl propionate
c - Except isopropyl and *sec*-butyl propionate
d - Except *t*-butyl butyrate
e - Except propyl butyrate
f - Except isopropyl and *sec*-butyl valerate
g - Except formate and isobutyrate
h - Difficult to characterize as a class
i - Except formate and isobutyrate
j - Except formate and acetate
k - Except formate and isobutyrate

CHART OF CHARACTERISTIC FREQUENCIES BETWEEN ~ 700–300 cm⁻¹

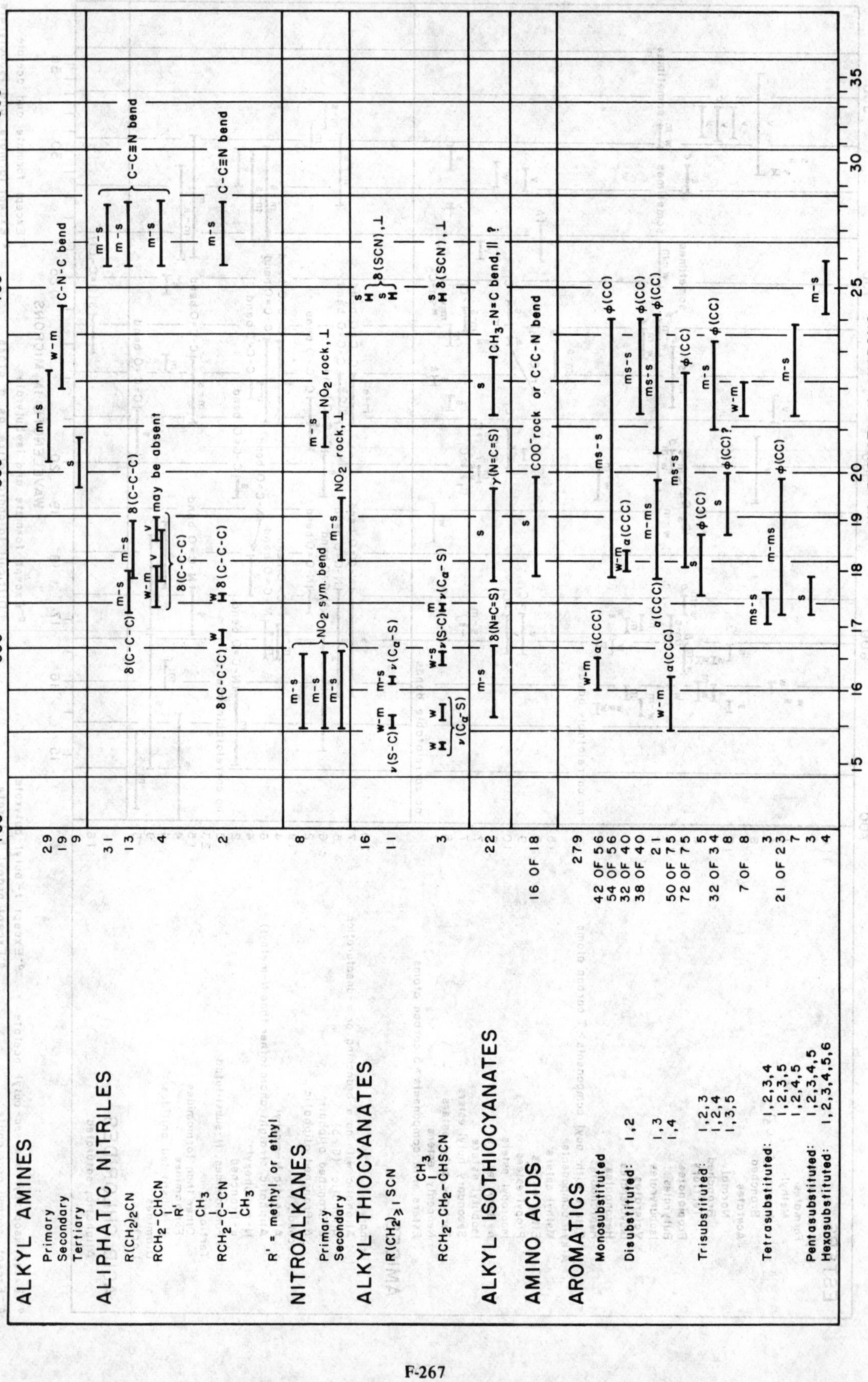

CHART OF CHARACTERISTIC FREQUENCIES BETWEEN ~ 700–300 cm^{-1}

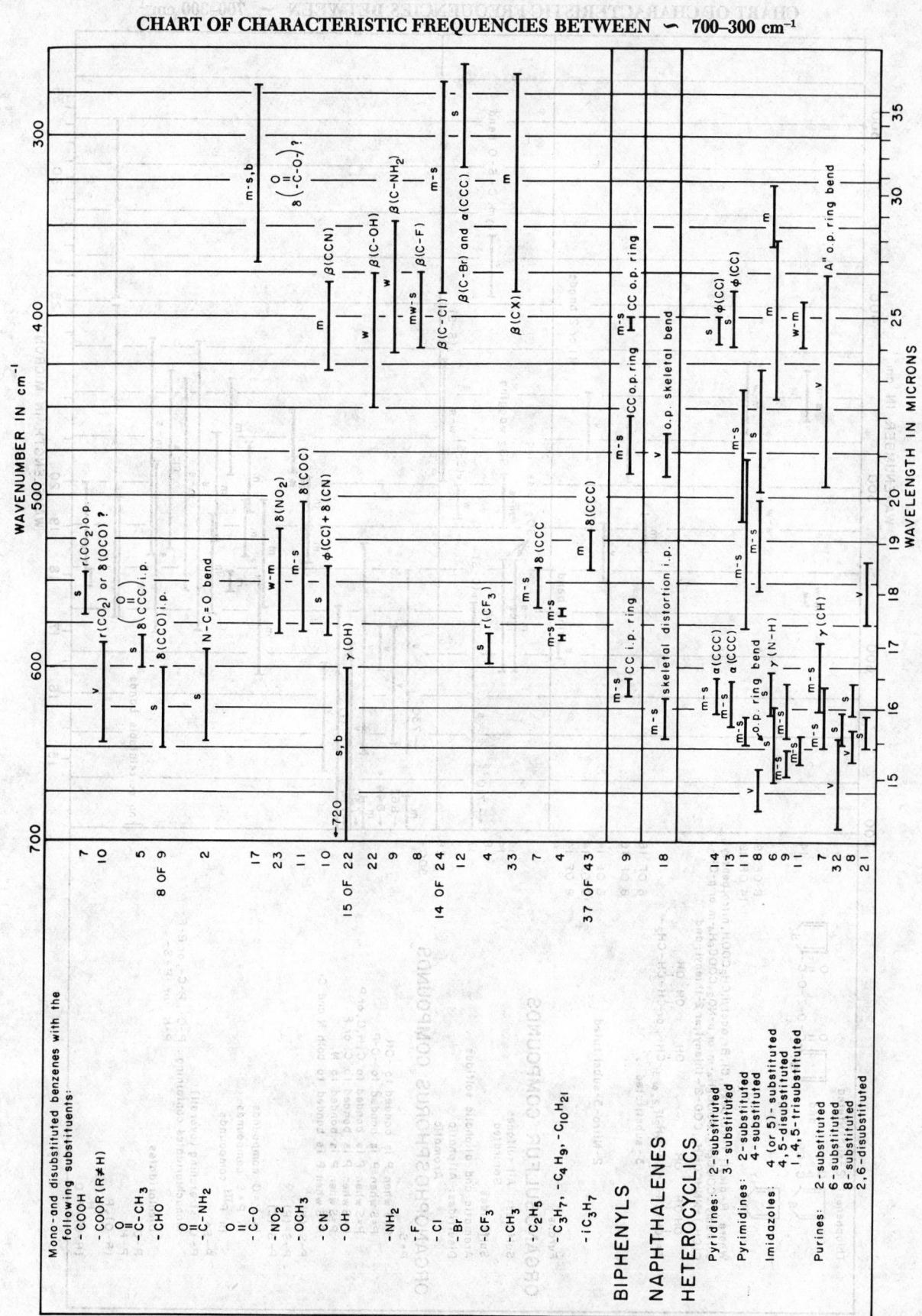

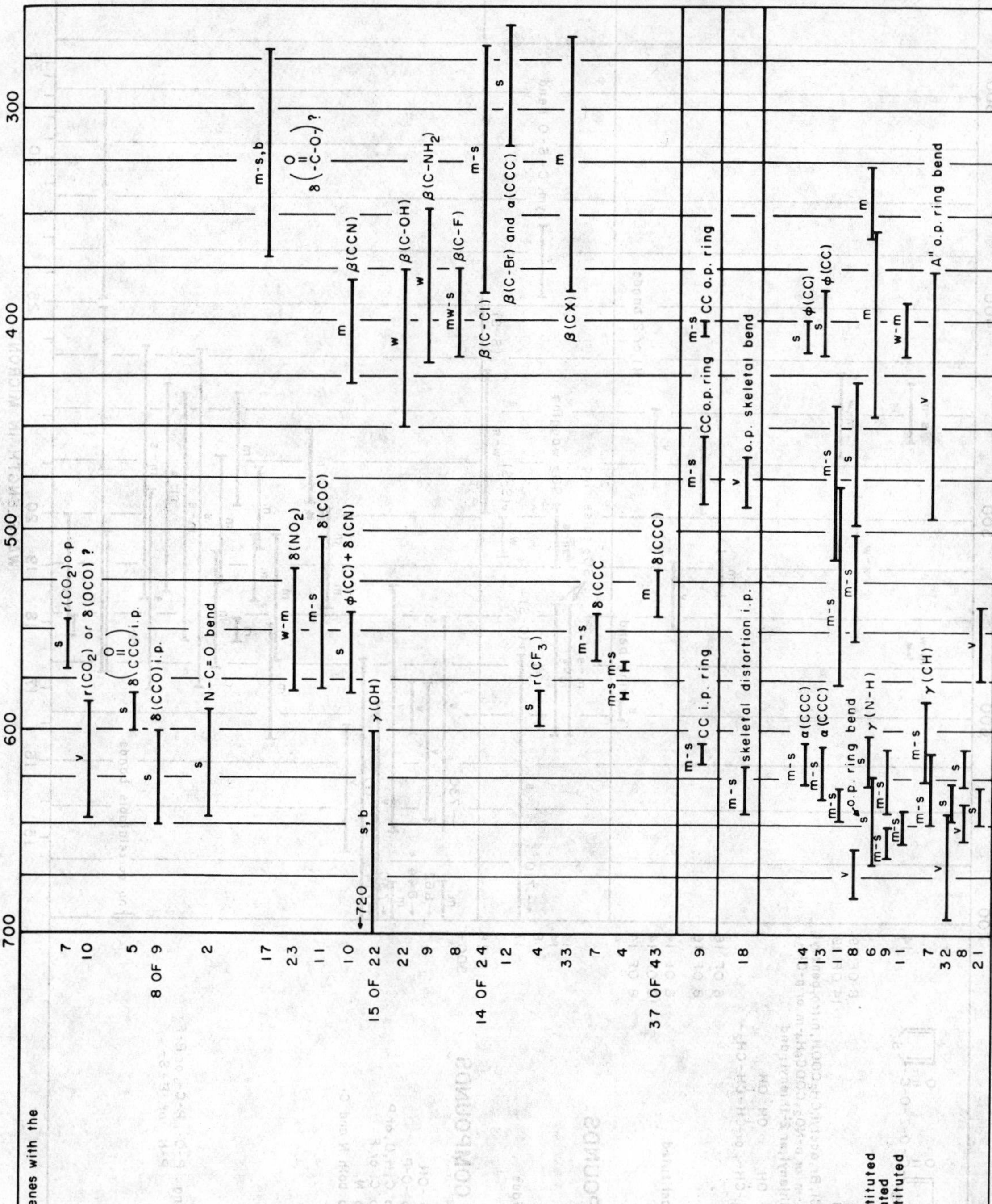

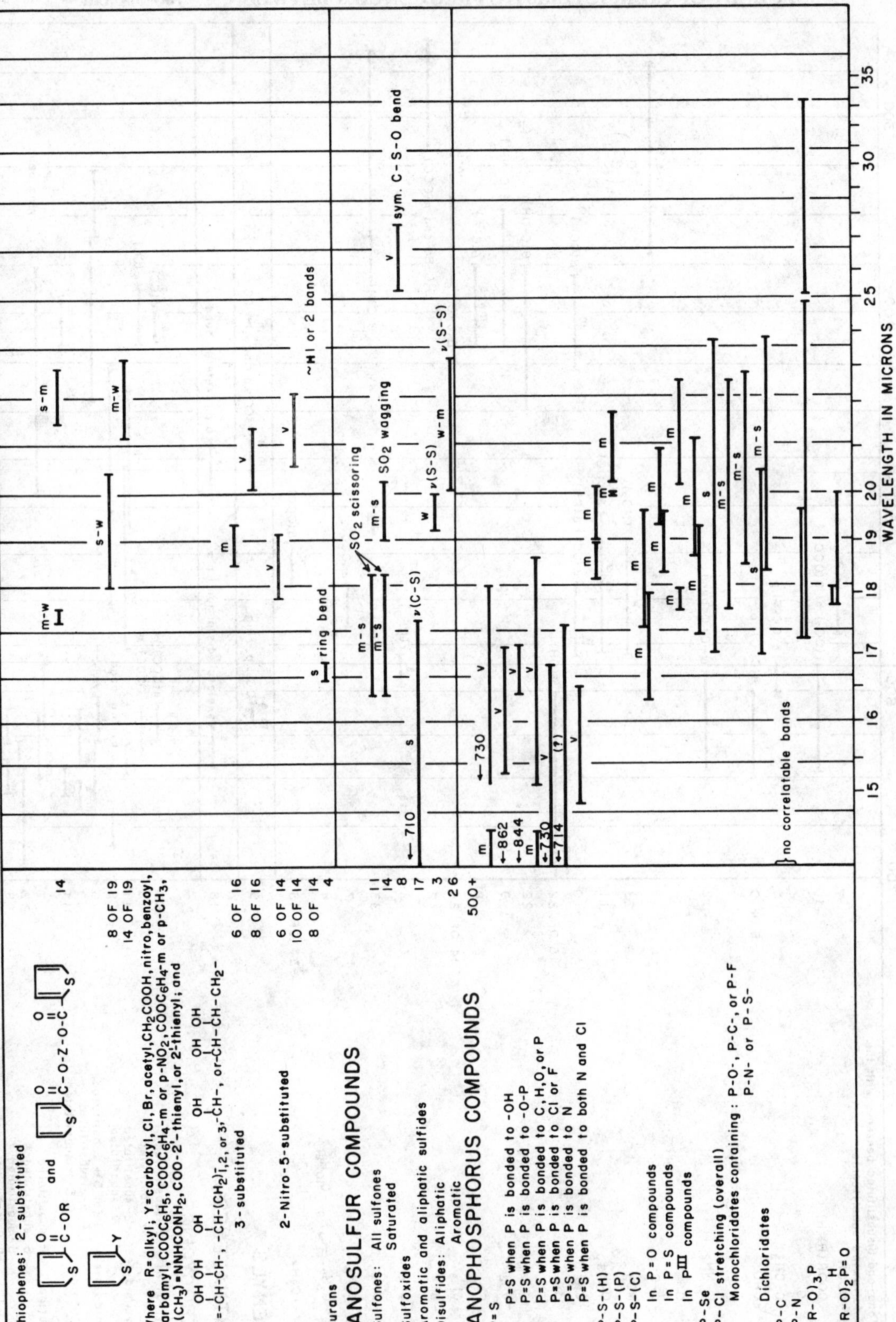
WAVENUMBER IN cm⁻¹
700
600
500
400
300
WAVELENGTH IN MICRONS
15
16
17
18
19
20
25
30
35
Thiophenes: 2-substituted
C-OR and C-O-Z-O-C
14
8 OF 19
14 OF 19
Where R=alkyl; Y=carboxyl, Cl, Br, acetyl, CH2COOH, nitro, benzoyl, carbamyl, COOC6H5, COOC6H4-m or p-NO2, COOC6H4-m or p-CH3, C(CH3)=NNHCONH2, COO-2'-thienyl, or 2'-thienyl; and
Z=-CH-CH-, -CH-(CH2)1,2, or 3-CH-, or-CH-CH-CH2-
6 OF 16
8 OF 16
3-substituted
2-Nitro-5-substituted
6 OF 14
10 OF 14
8 OF 14
Furans
4
ORGANOSULFUR COMPOUNDS
Sulfones: All sulfones
Saturated
Sulfoxides
Aromatic and aliphatic sulfides
Disulfides: Aliphatic
Aromatic
11
14
8
17
3
26
ORGANOPHOSPHORUS COMPOUNDS
P=S
500+
P=S when P is bonded to -OH
P=S when P is bonded to -O-P
P=S when P is bonded to C, H, O, or P
P=S when P is bonded to Cl or F
P=S when P is bonded to N
P=S when P is bonded to both N and Cl
P-S-(H)
P-S-(P)
P-S-(C)
In P=O compounds
In P=S compounds
In PIII compounds
P-Se
P-Cl stretching (overall)
Monochloridates containing: P-O-, P-C-, or P-F
P-N- or P-S-
Dichloridates
P-C
P-N
(R-O)3P
(R-O)2P=O
m-w
s-m
m-w
s-w
m
v
v
v
~1 or 2 bands
s ring bend
m-s
m-s
SO2 scissoring
m-s
SO2 wagging
← 710
s
ν(C-S)
w
ν(S-S)
ν(S-S)
w-m
v sym. C-S-O bend
m
←862
←844
←730
←730
←714
m
←730
v
v
v
v
v
(?)
m
m
m
w
m
m
m
m
m
m
s
m-s
m-s
m-s
s
no correlatable bands

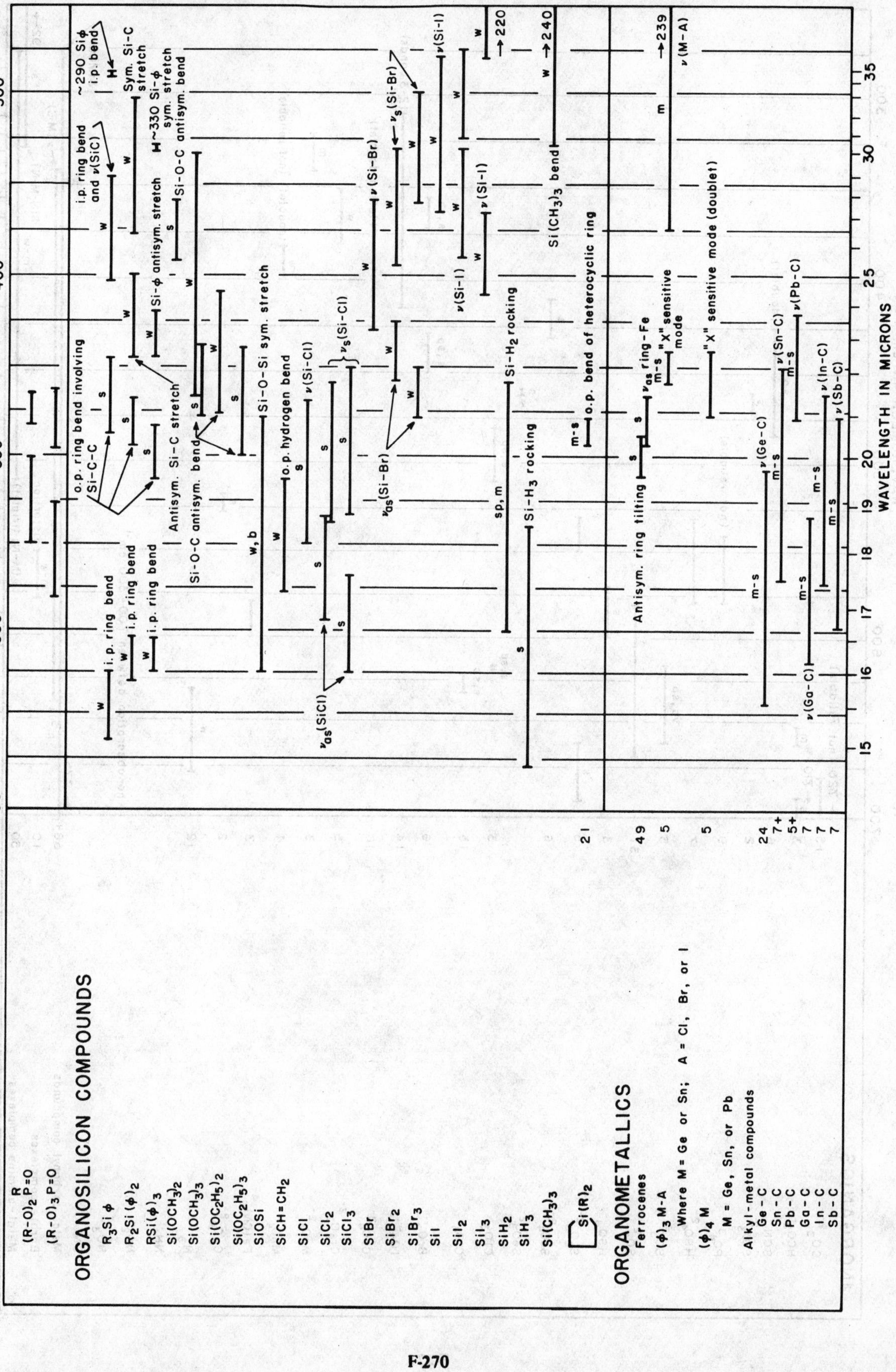
WAVENUMBER IN cm⁻¹
700
600
500
400
300
WAVELENGTH IN MICRONS
15
16
17
18
19
20
25
30
35
(R-O)2 P=O R
(R-O)3 P=O
ORGANOSILICON COMPOUNDS
R3Si φ
R2Si(φ)2
RSi(φ)3
Si(OCH3)2
Si(OCH3)3
Si(OC2H5)2
Si(OC2H5)3
SiOSi
SiCH=CH2
SiCl
SiCl2
SiCl3
SiBr
SiBr2
SiBr3
SiI
SiI2
SiI3
SiH2
SiH3
Si(CH3)3
Si(R)2
21
ORGANOMETALLICS
Ferrocenes
49
(φ)3 M-A
Where M = Ge or Sn; A = Cl, Br, or I
5
(φ)4 M
M = Ge, Sn, or Pb
5
Alkyl-metal compounds
Ge-C
Sn-C
Pb-C
Ga-C
In-C
Sb-C
24
7+
5+
7
7
7
i.p. ring bend
o.p. ring bend involving Si-C-C
i.p. ring bend and ν(SiC)
~290 Si φ i.p. bend
Sym. Si-C stretch
Antisym. Si-C stretch
Si-φ antisym. stretch
~330 Si-φ sym. stretch
Si-O-C antisym. bend
Si-O-C antisym. stretch
Si-O-Si sym. stretch
w, b
o.p. hydrogen bend
ν(Si-Cl)
νs(Si-Cl)
νas(SiCl)
νas(Si-Br)
ν(Si-Br)
νs(Si-Br)
ν(Si-I)
→220
Si-H2 rocking
Si-H3 rocking
sp, m
Si(CH3)3 bend
→240
o.p. bend of heterocyclic ring
Antisym. ring tilting
νas ring-Fe
"X" sensitive mode
"X" sensitive mode (doublet)
→239
ν(M-A)
ν(Ge-C)
ν(Sn-C)
ν(Pb-C)
ν(Ga-C)
ν(In-C)
ν(Sb-C)

CHART OF CHARACTERISTIC FREQUENCIES BETWEEN ~ 700–300 cm⁻¹

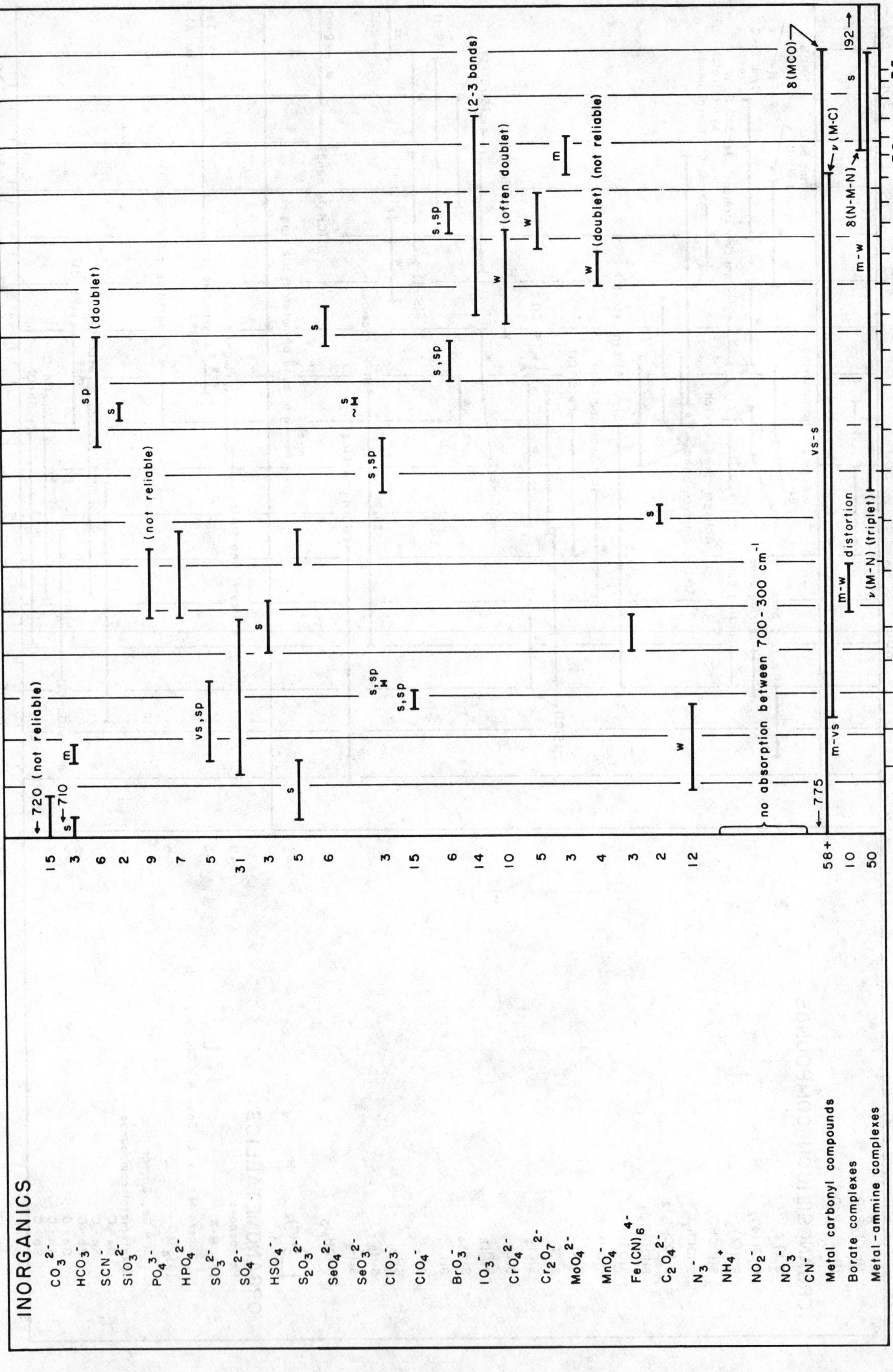

CHARACTERISTIC NMR SPECTRAL POSITIONS FOR HYDROGEN IN ORGANIC STRUCTURES

By permission from Erno Mohacsi, J. of Chemical Education, 41, 38 (1964)

This table is useful for quick qualitative determination of proton spectrum lines by providing a tabulation of line positions obtained using tetramethylsilane as an internal reference. The listing has been kept as simple as possible for this purpose. The proton spectrum lines are arranged according to the chemical shift relative to tetramethylsilane and are given in values of τ and δ. The purpose of this table is to supplement tables available in standard references and to summarize information available in the literature.

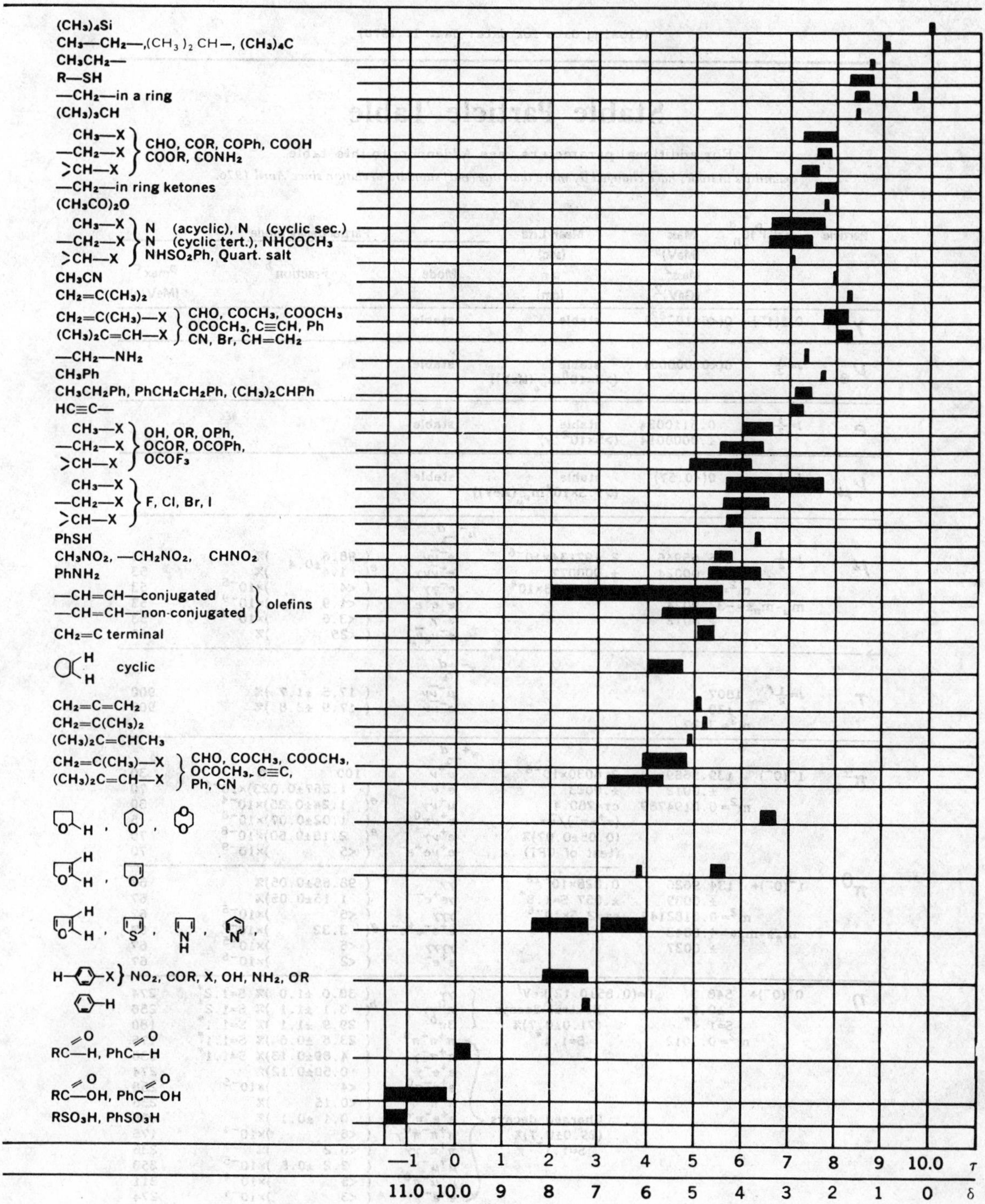

TABLES OF PARTICLE PROPERTIES

April 1978

Reprinted from Physics Letters, Vol. 75B, No. 1, April 1978.

N. Barash-Schmidt, A. Barbaro-Galtieri, C. Bricman, R. L. Crawford,
C. Dionisi, R. J. Hemingway, C. P. Horne, R. L. Kelly,
M. J. Losty, M. Mazzucato, L. Montanet, A. Rittenberg,
M. Roos, T. G. Trippe, G. P. Yost

(Closing date for data: Jan. 1, 1978)

Stable Particle Table

For additional parameters, see Addendum to this table.

Quantities in italics have changed by more than one (old) standard deviation since April 1976.

Particle	$I^G(J^P)C_n$ [a]	Mass (MeV) / Mass^2 (GeV)^2	Mean Life (sec) / $c\tau$ (cm)	Partial decay mode: Mode	Fraction [b]	p or p_{max} [c] (MeV/c)
γ	0,1(1⁻)−	0(<6×10^{-22})	stable	stable		
ν_e	J=½	0(<0.00006)	stable	stable		
			(>3×10^8 m_{ν_e}(MeV))			
e	J=½	0.5110034	stable	stable		
		±.0000014	(>5×10^{21}y)			
ν_μ	J=½	0(<0.57)	stable	stable		
			(>1.3×10^4 m_{ν_μ}(MeV))			
				$\mu^- \xrightarrow{d}$		
μ	J=½	105.65946	2.197134×10^{-6}	$e^-\bar{\nu}\nu$	(98.6)%	53
		±.00024	±.000077	$e^-\bar{\nu}\nu\gamma$	[e](1.4±0.4)%	53
		m^2= 0.01116392	$c\tau$=6.5868×10^4	$e^-\gamma\gamma$	(<4)×10^{-6}	53
		$m_\mu - m_{\pi^\pm}$= −33.9074		$e^-e^+e^-$	(<1.9)×10^{-9}	53
		±.0012		$e^-\gamma$	(<3.6)×10^{-9}	53
				$e^-\nu_e\bar{\nu}_\mu$	(<25)%	53
→				$\tau^- \xrightarrow{d}$		
τ	J=½ [f]	1807		$\mu^-\bar{\nu}\nu$	(17.5 ±1.7)%	900
		±20		$e^-\bar{\nu}\nu$	(17.9 ±2.8)%	903
		m^2= 3.27				
				$\pi^+ \xrightarrow{d}$		
$\pi^\pm$	1⁻(0⁻)	139.5669	2.6030×10^{-8}	$\mu^+\nu$	100 %	30
		±.0012	±.0023	$e^+\nu$	(1.267±0.023)×10^{-4}	70
		m^2= 0.0194789	$c\tau$=780.4	$\mu^+\nu\gamma$	[e](1.24±0.25)×10^{-4}	30
			$(\tau^+ - \tau^-)/\bar{\tau}$=	$e^+\nu\pi^0$	(1.02±0.07)×10^{-8}	5
			(0.05±0.07)%	$e^+\nu\gamma$	[e](2.15±0.50)×10^{-8}	70
			(test of CPT)	$e^+\nu e^+e^-$	(<5)×10^{-9}	70
π^0	1⁻(0⁻)+	134.9626	0.828×10^{-16}	$\gamma\gamma$	(98.85±0.05)%	67
		±.0039	±.057 S=1.8*	γe^+e^-	(1.15±0.05)%	67
		m^2= 0.0182149	$c\tau$=2.5×10^{-6}	$\gamma\gamma\gamma$	(<5)×10^{-6}	67
		$m_{\pi^\pm} - m_{\pi^0}$= 4.6043		$e^+e^-e^+e^-$	[g](3.32)×10^{-5}	67
		±.0037		$\gamma\gamma\gamma\gamma$	(<6)×10^{-5}	67
				e^+e^-	(<2)×10^{-6}	67
η	0⁺(0⁻)+	548.8	Γ=(0.85±0.12)keV [i]	$\gamma\gamma$	(38.0 ±1.0)% S=1.2*	274
		±0.6	Neutral decays	$\pi^0\gamma\gamma$	[h](3.1 ±1.1)% S=1.2*	258
		S=1.4*	(71.0±0.7)%	$3\pi^0$	(29.9 ±1.1)% S=1.1*	180
		m^2= 0.3012	S=1.1*	$\pi^+\pi^-\pi^0$	(23.6 ±0.6)% S=1.1*	175
				$\pi^+\pi^-\gamma$	(4.89±0.13)% S=1.1*	236
				$e^+e^-\gamma$	(0.50±0.12)%	274
				$e^+e^-\pi^0$	(<4)×10^{-5}	258
				$\pi^+\pi^-$	(<0.15)%	236
			Charged decays	$e^+e^-\pi^+\pi^-$	(0.1 ±0.1)%	236
			(29.0±0.7)%	$\pi^+\pi^-\pi^0\gamma$	(<6)×10^{-4}	175
			S=1.1*	$\pi^+\pi^-\gamma\gamma$	(<0.2)%	236
				$\mu^+\mu^-$	(2.2 ±0.8)×10^{-5}	253
				$\mu^+\mu^-\pi^0$	(<5)×10^{-4}	211
				e^+e^-	(<3)×10^{-4}	274

Stable Particle Table *(cont'd)*

Particle	$I^G(J^P)C_n$ [a]	Mass (MeV) Mass² (GeV)²	Mean life (sec) $c\tau$ (cm)	Mode	Fraction [b]	p or p_{max} [c] (MeV/c)
				$K^+ \rightarrow$ [d]		
$K^\pm$	$\frac{1}{2}(0^-)$	*493.668*	1.2371×10^{-8}	$\mu^+\nu$	$(63.50\pm0.16)\%$	236
		±0.018	±.0026 S=1.9*	$\pi^+\pi^0$	$(21.16\pm0.15)\%$	205
		$m^2=0.24371$	$c\tau=370.9$	$\pi^+\pi^+\pi^-$	$(5.59\pm0.03)\%$ S=1.1*	125
			$(\tau^+-\tau^-)/\bar{\tau}=$	$\pi^+\pi^0\pi^0$	$(1.73\pm0.05)\%$ S=1.3*	133
			(.11±.09)%	$\mu^+\nu\pi^0$	$(3.20\pm0.09)\%$ S=1.7*	215
			(test of CPT)	$e^+\nu\pi^0$	$(4.82\pm0.05)\%$ S=1.1*	228
			S=1.2*	$\mu^+\nu\gamma$	[e] $(5.8\pm3.5)\times10^{-3}$	236
				$e^+\nu\pi^0\pi^0$	$(1.8\ ^{+2.4}_{-0.6})\times10^{-5}$	207
		$m_{K^\pm}-m_{K^0}=-4.01$		$e^+\nu\pi^+\pi^-$	$(3.90\pm0.15)\times10^{-5}$	203
		±0.13		$e^-\bar{\nu}\pi^+\pi^+$	$(<5)\times10^{-7}$	203
		S=1.1*		$\mu^+\nu\pi^+\pi^-$	$(0.9\pm0.4)\times10^{-5}$	151
				$\mu^-\bar{\nu}\pi^+\pi^+$	$(<3.0)\times10^{-6}$	151
				$e^+\nu$	$(1.54\pm0.09)\times10^{-5}$	247
				$e^+\nu\gamma$	[e] $(1.62\pm0.47)\times10^{-5}$	247
				$\pi^+\pi^0\gamma$	[i,e] $(2.75\pm0.16)\times10^{-4}$	205
				$\pi^+\pi^+\pi^-\gamma$	[e] $(1.0\pm0.4)\times10^{-4}$	125
				$\mu^+\nu\pi^0\gamma$	[e] $(<6)\times10^{-5}$	215
				$e^+\nu\pi^0\gamma$	[e] $(3.7\pm1.4)\times10^{-4}$	228
				$e^+e^-\pi^+$	$(2.6\pm0.5)\times10^{-7}$	227
				$e^+e^+\pi^-$	$(<1)\times10^{8}$	227
				$\mu^+\mu^-\pi^+$	$(<2.4)\times10^{-6}$	172
				$\pi^+\gamma\gamma$	[e] $(<3.5)\times10^{-5}$	227
				$\pi^+\gamma\gamma\gamma$	[e] $(<3.0)\times10^{-4}$	227
				$\pi^+\nu\bar{\nu}$	$(<0.6)\times10^{-6}$	227
				$\pi^+\gamma$	$(<4)\times10^{-6}$	227
				$e^+\mu^\pm\pi^\mp$	$(<7)\times10^{-9}$	214
				$e^-\mu^+\pi^+$	$(<5)\times10^{-9}$	214
				$\mu^+\nu\nu\bar{\nu}$	$(<6)\times10^{-6}$	236
				$\mu^+\nu e^+e^-$	$(11\pm3)\times10^{-7}$	236
				$\mu^-\nu e^+e^+$	$(<2.0)\times10^{-8}$	236
				$e^+\nu e^+e^-$	$(2\ ^{+2}_{-1})\times10^{-7}$	247
K^0	$\frac{1}{2}(0^-)$	497.67	50% K_{Short}, 50% K_{Long}			
$\bar{K}^0$		±0.13				
		S=1.1*				
		$m^2=0.24768$				
K^0_S	$\frac{1}{2}(0^-)$		0.8923×10^{-10} [j]	$\pi^+\pi^-$	$(68.61\pm0.24)\%$ S=1.1*	206
			±.0022	$\pi^0\pi^0$	$(31.39\pm0.24)\%$ S=1.1*	209
			$c\tau=2.675$	$\mu^+\mu^-$	$(<3.2)\times10^{-7}$	225
				e^+e^-	$(<3.4)\times10^{-4}$	249
				$\pi^+\pi^-\gamma$	[e] $(1.85\pm0.10)\times10^{-3}$	206
				$\gamma\gamma$	$(<0.4)\times10^{-3}$	249
K^0_L	$\frac{1}{2}(0^-)$		5.183×10^{-8}	$\pi^0\pi^0\pi^0$	$(21.5\pm0.7)\%$ S=1.3*	139
			±.040	$\pi^+\pi^-\pi^0$	$(12.39\pm0.18)\%$ S=1.2*	133
			$c\tau=1554$	$\pi^\pm\mu^\mp\nu$	$(27.0\pm0.5)\%$ S=1.1*	216
				$\pi^\pm e^\mp\nu$	[k] $(38.8\pm0.5)\%$ S=1.1*	229
				$\pi e\nu\gamma$	[k,e] $(1.3\pm0.8)\%$	229
		$m_{K_L}-m_{K_S}=0.5349\times10^{10}\hbar\ sec^{-1}$		$\pi^+\pi^-$	[j] $(0.203\pm0.005)\%$	206
		±0.0022		$\pi^0\pi^0$	$(0.094\pm0.018)\%$ S=1.5*	209
				$\pi^+\pi^-\gamma$	[e] $(6.0\pm2.0)\times10^{-5}$	206
				$\pi^0\gamma\gamma$	$(<2.4)\times10^{-4}$	231
				$\gamma\gamma$	$(4.9\pm0.5)\times10^{-4}$	249
				$e\mu$	$(<2.0)\times10^{-9}$	238
				$\mu^+\mu^-$	$(9.1\pm1.8)\times10^{-9}$	225
				$\mu^+\mu^-\gamma$	$(<7.8)\times10^{-6}$	225
				$\mu^+\mu^-\pi^0$	$(<5.7)\times10^{-5}$	177
				e^+e^-	$(<2.0)\times10^{-9}$	249
				$e^+e^-\gamma$	$(<2.8)\times10^{-5}$	249
				$\pi^+\pi^-e^+e^-$	$(<8.8)\times10^{-6}$	206
				$\pi^0\pi^\pm e^\mp\nu$	$(<2.2)\times10^{-3}$	207
				$D^+ \rightarrow$ [d]		
$D^\pm$	$\frac{1}{2}(0^-)$ [f]	1868.3 [l]		$K^-\pi^+\pi^+$	$(3.9\pm1.0)\%$	845
		±0.9		$K^0\pi^+$	$(1.5\pm0.6)\%$	862
		$m^2=3.491$		$e^\pm$anything	[m] $(9.8\pm1.4)\%$	934
		$m_{D^\pm}-m_{D^0}=5.0$		$\pi^+\pi^+\pi^-$	$(<0.31)\%$	908
		±0.8		$\pi^+K^+K^-$	$(<0.6)\%$	743
				$K^+\pi^+\pi^-$	$(<0.20)\%$	845
				$D^0 \rightarrow$ [d]		
D^0	$\frac{1}{2}(0^-)$ [f]	1863.3 [l]		$K^-\pi^+$	$(1.8\pm0.5)\%$	860
$\bar{D}^0$		±0.9		$K^-\pi^+\pi^0$	$(12\pm6)\%$	843
		$m^2=3.472$		$K^-\pi^+\pi^+\pi^-$	$(3.5\pm0.9)\%$	812
				$K^0\pi^0$	$(<6)\%$	860
		$\frac{\Gamma(D^0\rightarrow\bar{D}^0\rightarrow K^+\pi^-)}{\Gamma(D^0\rightarrow K\pi)}<0.16$		$K^0\pi^+\pi^-$	$(4.4\pm1.1)\%$	841
				$e^\pm$anything	[m] $(9.8\pm1.4)\%$	932
				$\pi^+\pi^-$	$(<0.13)\%$	921
				K^+K^-	$(<0.13)\%$	790

→

Stable Particle Table *(cont'd)*

Particle	$I^G(J^P)C_n$ [a]	Mass (MeV) Mass^2 (GeV)^2	Mean life (sec) $c\tau$ (cm)	Partial decay mode: Mode	Fraction [b]	p or p_{max} [c] (MeV/c)
p	$\frac{1}{2}(\frac{1}{2}^+)$	938.2796	stable ($>2\times10^{30}$y)			
		±0.0027				
		m^2= 0.880369				
n	$\frac{1}{2}(\frac{1}{2}^+)$	939.5731	918±14	$pe^-\nu$	100 %	1
		±0.0027	$c\tau=2.75\times10^{13}$			
		m^2= 0.882798				
		m_p-m_n=−1.29343				
		±0.00004				
Λ	$0(\frac{1}{2}^+)$	1115.60	*2.632×10^{-10}*	$p\pi^-$	(64.2 ± 0.5)%	100
		±0.05	*±.020* S=1.6*	$n\pi^0$	(35.8 ± 0.5)%	104
		S=1.2*	$c\tau$=*7.89*	$pe^-\nu$	(8.07±0.28)×10^{-4}	163
		m^2= 1.2446		$p\mu^-\nu$	(1.57±0.35)×10^{-4}	131
		$m_\Lambda-m_{\Sigma^0}$= −76.87		$p\pi^-\gamma$	[e](0.85±0.14)×10^{-3}	100
		±0.08				
Σ^+	$1(\frac{1}{2}^+)$	1189.37	0.802×10^{-10}	$p\pi^0$	(51.6 ±0.7)%	189
		±0.06	±.005	$n\pi^+$	(48.4 ±0.7)%	185
		S=1.8*	$c\tau$=2.40	$p\gamma$	(1.24±0.18)×10^{-3} S=1.4*	225
		m^2= 1.4146		$n\pi^+\gamma$	[e](0.93±0.10)×10^{-3}	185
				$\Lambda e^+\nu$	(2.02±0.47)×10^{-5}	71
		$m_{\Sigma^+}-m_{\Sigma^-}$= −7.98	$\frac{\Gamma(\Sigma^+\to\ell^+n\nu)}{\Gamma(\Sigma^-\to\ell^-n\nu)}<.04$ ←	$n\mu^+\nu$	(<3.0)×10^{-5}	202
		±.08		$ne^+\nu$	(<0.5)×10^{-5}	224
		S=1.2*		pe^+e^-	(<7)×10^{-6}	225
Σ^0	$1(\frac{1}{2}^+)$	1192.47	5.8×10^{-20}	$\Lambda\gamma$	100 %	74
		±0.08	±1.3	Λe^+e^-	[g](5.45)×10^{-3}	74
		m^2= 1.4220	$c\tau$=1.7×10^{-9}	$\Lambda\gamma\gamma$	(<3)%	74
Σ^-	$1(\frac{1}{2}^+)$	1197.35	1.483×10^{-10}	$n\pi^-$	100 %	193
		±0.06	±.015 S=1.4*	$ne^-\nu$	(1.08±0.04)×10^{-3}	230
		m^2= 1.4336	$c\tau$=4.45	$n\mu^-\nu$	(0.45±0.04)×10^{-3}	210
				$\Lambda e^-\nu$	(0.60±0.06)×10^{-4}	79
		$m_{\Sigma^0}-m_{\Sigma^-}$= −4.88		$n\pi^-\gamma$	[e](4.6 ±0.6)×10^{-4}	193
		±.06				
Ξ^0	$\frac{1}{2}(\frac{1}{2}^+)^n$	1314.9	2.90×10^{-10}	$\Lambda\pi^0$	100 %	135
		±0.6	±.10	$\Lambda\gamma$	(0.5 ±0.5)%	184
		m^2= 1.7290	$c\tau$=8.69	$\Sigma^0\gamma$	(<7)%	117
				$p\pi^-$	(<3.6)×10^{-5}	299
				$pe^-\nu$	(<1.3)×10^{-3}	323
				$\Sigma^+e^-\nu$	(<1.1)×10^{-3}	120
				$\Sigma^-e^+\nu$	(<0.9)×10^{-3}	112
		$m_{\Xi^0}-m_{\Xi^-}$=−6.4		$\Sigma^+\mu^-\nu$	(<1.1)×10^{-3}	64
		±.6		$\Sigma^-\mu^+\nu$	(<0.9)×10^{-3}	49
				$p\mu^-\nu$	(<1.3)×10^{-3}	309
Ξ^-	$\frac{1}{2}(\frac{1}{2}^+)^n$	1321.32	1.654×10^{-10}	$\Lambda\pi^-$	100 %	139
		±0.13	±.021	$\Lambda e^-\nu$	[o](0.69±0.18)×10^{-3}	190
		m^2= 1.7459	$c\tau$=4.96	$\Sigma^0e^-\nu$	(<0.5)×10^{-3}	123
				$\Lambda\mu^-\nu$	(3.5 ±3.5)×10^{-4}	163
				$\Sigma^0\mu^-\nu$	(<0.8)×10^{-3}	70
				$n\pi^-$	(<1.1)×10^{-3}	303
				$ne^-\nu$	(<3.2)×10^{-3}	327
				$n\mu^-\nu$	(<1.5)%	313
				$\Sigma^-\gamma$	(<1.2)×10^{-3}	118
				$p\pi^-\pi^-$	(<4)×10^{-4}	223
				$p\pi^-e^-\nu$	(<4)×10^{-4}	304
				$p\pi^-\mu^-\nu$	(<4)×10^{-4}	250
				$\Xi^0e^-\nu$	(<2.3)×10^{-4}	6
Ω^-	$0(\frac{3}{2}^+)^n$	1672.2	$1.1^{+0.4}_{-0.3}\times10^{-10}$ [p]	$\Xi^0\pi^-$	100%	293
		±.4	S=2.5*	$\Xi^-\pi^0$		290
		m^2= 2.7963	$c\tau$=3	ΛK^-		211

→

ADDENDUM TO

Stable Particle Table

e — **Magnetic moment**
1.001 159 652 41 $\frac{e\hbar}{2m_ec}$
±.000 000 000 20

μ — 1.001 165 922 $\frac{e\hbar}{2m_\mu c}$
±.000 000 009

μ Decay parameters [q]

$\rho = 0.752 \pm 0.003$ $\eta = -0.12 \pm 0.21$
$\xi = 0.972 \pm 0.013$ $\delta = 0.755 \pm 0.009$ $h = 1.00 \pm 0.13$
$|g_A/g_V| = 0.86^{+0.33}_{-0.11}$ $\phi = 180° \pm 15°$

η

Mode	Left-right asymmetry	Sextant asymmetry	Quadrant asymmetry
$\pi^+\pi^-\pi^0$	(0.12±.17)%	(0.19±0.16)%	(−0.17±0.17)%
$\pi^+\pi^-\gamma$	(0.88±.40)%		β=0.047±0.062

K$^\pm$

Mode	Partial rate (sec^{-1})	
$\mu\nu$	$(51.33 \pm 0.17) \times 10^6$	S=1.2*
$\pi\pi^0$	$(17.10 \pm 0.13) \times 10^6$	S=1.1*
$\pi\pi^+\pi^-$	$(4.52 \pm 0.02) \times 10^6$	S=1.1*
$\pi\pi^0\pi^0$	$(1.40 \pm 0.04) \times 10^6$	S=1.3*
$\mu\pi^0\nu$	$(2.58 \pm 0.07) \times 10^6$	S=1.7*
$e\pi^0\nu$	$(3.90 \pm 0.04) \times 10^6$	S=1.1*

Slope parameters for K → 3π [r]

$K^+ \to \pi^+\pi^+\pi^-$	g=−0.215±.004	S=1.5*
$K^- \to \pi^-\pi^-\pi^+$	g=−0.214±.007	S=2.7*
$K^\pm \to \pi^0\pi^0\pi^\pm$	g= 0.561±.021	S=1.7*
$K^0_L \to \pi^+\pi^-\pi^0$	g= *0.670±.014*	S=1.6*

See Data Card Listings for quadratic coefficients.

$K^+_{\ell 3}$: $\lambda^e_+ = 0.029 \pm .004$; $\lambda^\mu_+ = 0.026 \pm .008$ S=1.5*; $\lambda^\mu_0 = -0.003 \pm .007$ S=1.5*

$K^0_{\ell 3}$: $\lambda^e_+ = 0.0300 \pm .0018$ S=1.2*; $\lambda^\mu_+ = 0.034 \pm .006$ S=2.5*; $\lambda^\mu_0 = 0.020 \pm .007$ S=2.5*

See Data Card Listings for ξ, f_s, and f_t.

K^0_S

Mode	Partial rate (sec^{-1})	
$\pi^+\pi^-$	[s] $(0.7689 \pm .0033) \times 10^{10}$	
$\pi^0\pi^0$	[s] $(0.3517 \pm .0029) \times 10^{10}$	S=1.1*

K^0_L

Mode	Partial rate (sec^{-1})	
$\pi^0\pi^0\pi^0$	$(4.14 \pm 0.15) \times 10^6$	S=1.3*
$\pi^+\pi^-\pi^0$	$(2.39 \pm 0.04) \times 10^6$	S=1.2*
$\pi\mu\nu$	$(5.21 \pm 0.10) \times 10^6$	S=1.1*
$\pi e\nu$	$(7.49 \pm 0.11) \times 10^6$	S=1.1*
$\pi^+\pi^-$	[j,s] $(3.91 \pm 0.10) \times 10^4$	
$\pi^0\pi^0$	[s] $(1.81 \pm 0.35) \times 10^4$	S=1.5*

CP violation parameters [t,s,j]

$|\eta_{+-}| = (2.274 \pm .022) \times 10^{-3}$ $|\eta_{00}| = (2.32 \pm .09) \times 10^{-3}$ S=1.1*
$\phi_{+-} = (45.0 \pm 1.2)°$ $\phi_{00} = (48 \pm 13)°$
$|\eta_{+-0}|^2 < 0.12$ $|\eta_{000}|^2 < 0.28$ $\delta = (0.330 \pm .012) \times 10^{-2}$

ΔS = −ΔQ
Re x=0.009±.020 S=1.4* Im x = −0.004±.026 S=1.1*

	Magnetic moment ($e\hbar/2m_pc$)	Mode	Decay parameters [u] Measured α	φ(degree)	Derived γ	Δ(degree)	g_A/g_V	g_V/g_A
p	2.7928456 ±.0000011							
n	−1.91304211 ±.00000088	$pe^-\nu$					−1.253±0.007 δ=(180.20±0.19)°	
Λ	−0.606 ±.034	$p\pi^-$	0.642±0.013	(−6.5±3.5)°	0.76	$(7.7^{+4.0}_{-4.1})°$		
		$n\pi^0$	0.646±0.044					
		$pe\nu$					−0.62±0.05 S=1.2*	
Σ$^+$	2.83 ±.25	$p\pi^0$	−0.978±0.016	(36±34)°	0.17	(187±6)°		
		$n\pi^+$	+0.072±0.015	(167±20)°	−0.97	$(-72^{+132}_{-11})°$		
		$p\gamma$	$-1.03^{+0.52}_{-0.42}$	S=1.1*				
Σ$^-$	−1.48 ±.37	$n\pi^-$	−0.069±0.008	(10±15)°	0.98	$(249^{+12}_{-115})°$		
		$ne^-\nu$					±(*0.385±0.070*) S=2.3*	
		$\Lambda e^-\nu$						0.24±0.23 S=1.3*
Ξ^0		$\Lambda\pi^0$	−0.44±0.08 S=1.3*	(21±12)°	0.84	$(216^{+13}_{-19})°$		
Ξ$^-$	−1.85 ±.75	$\Lambda\pi^-$	−0.392±0.021	(2±6)° S=1.1*	0.92	(185±13)°		
Ω$^-$		ΛK^-	$-0.66^{+0.36}_{-0.30}$					

Stable Particle Table *(cont'd)*

→ Indicates an entry in the Stable Particle Data Card Listings not entered in the Stable Particle Table. We do not regard these as established particles.

*S = Scale factor = $\sqrt{\chi^2/(N-1)}$, where N ≈ number of experiments. S should be ≈ 1. If S > 1, we have enlarged the error of the mean, $\delta\bar{x}$; i.e., $\delta\bar{x} \to S\delta\bar{x}$. This convention is still inadequate, since if S >> 1 the experiments are probably inconsistent, and therefore the real uncertainty is probably even greater than $S\delta x$. See text, and ideograms in Stable Particle Data Card Listings.

a. The baryon number B, strangeness S, and charm C of the hadrons which appear in the tables are as follows:

Mesons (B=0)	S	C	Baryons (B=1)	S	C
π,η	0	0	p,n	0	0
K^+,K^0	+1	0	Λ,Σ	−1	0
$K^-,\bar{K}^0$	−1	0	Ξ	−2	0
D^+,D^0	0	+1	Ω^-	−3	0
$D^-,\bar{D}^0$	0	−1			

b. Quoted upper limits correspond to a 90% confidence level.

c. In decays with more than two bodies, p_{max} is the maximum momentum that any particle can have.

d. For simplicity, decay mode charge states are written for the particle shown. For antiparticle modes all particles must be charge conjugated.

e. See Stable Particle Data Card Listings for energy limits used in this measurement.

f. Quantum numbers shown are favored but not yet established. See Data Card Listings.

g. Theoretical value; see also Stable Particle Data Card Listings.

h. See note in Stable Particle Data Card Listings.

i. The direct emission branching fraction is $(1.56\pm.35)\times10^{-5}$.

j. The $\tau(K_S^0)$ and $|\eta_{+-}|$ averages (and the related $K_L^0 \to \pi^+\pi^-$ branching fraction and rate averages) contain only post−1971 results. The pre−1971 averages were $|\eta_{+-}| = (1.95\pm0.03)\times10^{-3}$ and $\tau(K_S^0) = (0.862\pm0.006)\times10^{-10}$ sec. See notes on $|\eta_{+-}|$ and $\tau(K_S^0)$ discrepancies in Stable Particle Data Card Listings.

k. The branching fraction for $K_L^0 \to \pi e \nu$ includes the radiative events $K_L^0 \to \pi e \nu \gamma$.

l. Error does not include 0.13% uncertainty in the absolute SPEAR energy calibration. Assumes m_ψ=3095 MeV.

m. This is a weighted average of $D^\pm$ and D^0 branching fractions with undetermined weighting.

n. P for Ξ and J^P for Ω^- not yet measured. Values reported are SU(3) predictions.

o. Assumes rate for $\Xi^- \to \Sigma^0 e^-\nu$ small compared with $\Xi^- \to \Lambda e^-\nu$.

p. Warning. This is an average of two incompatible results: ABCLV collaboration $(1.41^{+0.15}_{-0.24})\times10^{-10}$ sec and ACNO collaboration $(0.75^{+0.14}_{-0.11})\times10^{-10}$ sec. See note in Data Card Listings.

q. $|g_A/g_V|$ defined by $g_A^2 = |C_A|^2+|C'_A|^2$, $g_V^2 = |C_V|^2+|C'_V|^2$, and $\Sigma\langle\bar{e}|\Gamma_i|\mu\rangle\langle\bar{\nu}|\Gamma_i(C_i+C'_i\gamma_5)|\nu\rangle$;
ϕ defined by $\cos\phi = -\mathrm{Re}(C_A^*C'_V+C'_AC_V^*)/g_Ag_V$ [for more details, see text Section VI A].

r. The definition of the slope parameter of the Dalitz plot is as follows [see also text Section VI B.1]:

$$|M|^2 = 1 + g\left(\frac{s_3-s_0}{m_{\pi^+}^2}\right) .$$

s. The $K_S^0 \to \pi\pi$ and $K_L^0 \to \pi\pi$ rates (and branching fractions) are from independent fits and do not include results of K_L^0–K_S^0 interference experiments. The $|\eta_{+-}|$ and $|\eta_{00}|$ values given in the addendum are these rates combined with the $|\eta_{+-}|$ and $|\eta_{00}|$ results from interference experiments.

t. The definition for the CP violation parameters is as follows [see also text Section VI B.3]:

$$\eta_{+-} = |\eta_{+-}|e^{i\phi_{+-}} = \frac{A(K_L^0 \to \pi^+\pi^-)}{A(K_S^0 \to \pi^+\pi^-)} \qquad \eta_{00} = |\eta_{00}|e^{i\phi_{00}} = \frac{A(K_L^0 \to \pi^0\pi^0)}{A(K_S^0 \to \pi^0\pi^0)}$$

$$\delta = \frac{\Gamma(K_L^0\to\ell^+)-\Gamma(K_L^0\to\ell^-)}{\Gamma(K_L^0\to\ell^+)+\Gamma(K_L^0\to\ell^-)} , \quad |\eta_{+-0}|^2 = \frac{\Gamma(K_S^0\to\pi^+\pi^-\pi^0)^{\text{CP viol.}}}{\Gamma(K_L^0\to\pi^+\pi^-\pi^0)} , \quad |\eta_{000}|^2 = \frac{\Gamma(K_S^0\to\pi^0\pi^0\pi^0)^{\text{CP viol.}}}{\Gamma(K_L^0\to\pi^0\pi^0\pi^0)} .$$

u. The definition of these quantites is as follows [for more details on sign convention, see text Section VI B]:

$\alpha = \dfrac{2\lvert s\rvert\lvert p\rvert\cos\Delta}{\lvert s\rvert^2+\lvert p\rvert^2}$	$\beta = \sqrt{1-\alpha^2}\sin\phi$	g_A/g_V defined by $\langle B_f\lvert\gamma_\lambda(g_V-g_A\gamma_5)\rvert B_i\rangle$
$\beta = \dfrac{-2\lvert s\rvert\lvert p\rvert\sin\Delta}{\lvert s\rvert^2+\lvert p\rvert^2}$	$\gamma = \sqrt{1-\alpha^2}\cos\phi$	δ defined by $g_A/g_V = \lvert g_A/g_V\rvert e^{i\delta}$

Meson Table

April 1978

In addition to the entries in the Meson Table, the Meson Data Card Listings contain all substantial claims for meson resonances. See Contents of Meson Data Card Listings below.

Quantities in italics are new or have changed by more than one (old) standard deviation since April 1976.

Name key: G \ I: 0, 1; −: ω/φ, π; +: η, ρ. $I^G(J^P)C_n$ underlined = estab.

Name	$I^G(J^P)C_n$	Mass M (MeV)	Full Width Γ (MeV)	M^2 $\pm\Gamma M$[a] $(GeV)^2$	Partial decay mode: Mode	Fraction (%) [Upper limits are 1σ (%)]		p or p_{max}[b] (MeV/c)
$\pi^\pm$ / π^0	$1^-(0^-)+$	139.57 / 134.96	0.0 / 7.95 eV ±.55 eV	0.019479 / 0.018215	See Stable Particle Table			
η	$0^+(0^-)+$	548.8 ±0.6	0.85 keV ±.12 keV	0.301 ±.000	Neutral	71.0 See Stable		
					Charged	29.0 Particle Table		
ρ(770)	$1^+(1^-)-$	776¶ ±3§	155¶ ±3§	0.602 ±.120	$\pi\pi$	≈100		362
					$\pi\gamma$	0.024 ±.007		375
					e^+e^-	0.0043±.0005 (d)		388
					$\mu^+\mu^-$	0.0067±.0012 (d)		373
					$\eta\gamma$	*seen*¶		194
M and Γ from neutral mode.					For upper limits, see footnote (e)			
ω(783)	$0^-(1^-)-$	782.6 ±0.3 S=1.3*	10.1 ±.3	0.612 ±.008	$\pi^+\pi^-\pi^0$	89.9±0.6	S=1.2*	327
					$\pi^+\pi^-$	1.3±0.3	S=1.5*	366
					$\pi^0\gamma$	8.8±0.5		380
					e^+e^-	0.0076±.0017	S=1.9*	391
					$\eta\gamma$	*seen*¶		199
					For upper limits, see footnote (f)			
η′(958)	$0^+(0^-)+$¶	957.6 ±0.3	< 1	0.917 <.001	$\eta\pi\pi$	66.2±1.7		231
					$\rho^0\gamma$	29.8±1.7	S=1.1*	165
					$\omega\gamma$	*2.1±0.4*		159
					$\gamma\gamma$	2.0±0.3		479
					For upper limits, see footnote (g)			
δ(980)	$1^-(0^+)+$	980(h) ±5§	50(h) ±10§	0.960 ±.049	$\eta\pi$	seen		318
					$K\bar{K}$	*seen*¶		
S*(980)	$0^+(0^+)+$	∼ *980*(c)§ *±10*§	40(c)§ ±10§	0.960 ±.039	$K\bar{K}$	seen¶		
					$\pi\pi$	seen		470
See note on ππ and K̄K S wave¶.								
Φ(1020)	$0^-(1^-)-$	1019.6 ±0.2 S=1.5*	4.1 ±.2	1.040 ±.004	K^+K^-	48.6±1.2	S=1.3*	128
					K_LK_S	35.1±1.2	S=1.5*	111
					$\pi^+\pi^-\pi^0$ (incl. ρπ)	*14.7±0.7*	S=1.2*	462
					$\eta\gamma$	1.6±0.2		362
					$\pi^0\gamma$	0.14±0.05		501
					e^+e^-	.031±.001	S=1.1*	510
					$\mu^+\mu^-$	.025±.003		499
					For upper limits, see footnote (i)			
A_1(1100)	$1^-(1^+)+$	∼ 1100¶	∼ 300¶	1.21 ±.33	$\rho\pi$	∼ 100		249
B(1235)	$1^+(1^+)-$	1231§ ±10§	128§ ±10§	1.52 ±.16	$\omega\pi$	only mode seen		347
					[D/S amplitude ratio = .29±.05]			
					For upper limits, see footnote (j)			
f(1270)	$0^+(2^+)+$	1271§ ±5§	180§ ±20§	1.62 ±.23	$\pi\pi$	80.3±0.3		620
					$2\pi^+2\pi^-$	2.8±0.3	S = 1.1*	557
					$K\bar{K}$	3.1±0.4	S = 1.3*	395
					$\pi^+\pi^-2\pi^0$	seen		560
					For upper limits, see footnote (ℓ)			
D(1285)	$0^+(1^+)+$	1282§ ±5§	25§ *±10*§	1.64 ±.03	$K\bar{K}\pi$	seen		301
					$\eta\pi\pi$	seen		481
					†[$\delta\pi$	seen]		238
					$2\pi^+2\pi^-$ (prob. $\rho^0\pi^+\pi^-$)	seen		563
ε(1300)	$0^+(0^+)+$	∼ *1300*	*200–400*		$\pi\pi$	seen		
					$K\bar{K}$	*seen*		
See note on ππ and K̄K S wave¶.								

Meson Table *(cont'd)*

Name	$I^G(J^P)C_n$ ⊢ estab.	Mass M (MeV)	Full Width Γ (MeV)	M^2 $\pm\Gamma M$[a] $(GeV)^2$	Mode	Fraction (%) [Upper limits are 1σ (%)]	p or p_{max}[b] (MeV/c)
$A_2(1310)$	$1^-(2^+)+$	1312§ ±5§	102§ ±5§	1.72 ±.13	ρπ	70.3±2.1	411
					ηπ	14.4±0.9	531
					ωππ	10.6±2.5	356
					$\bar{K}K$	4.7±0.5	430
					η′π	<1	281
					πγ	*0.45±0.11*	649
E(1420)	$0^+(A\ )+$	1416 ±10§	60 ±20§	2.01 ±.08	$K\bar{K}\pi$	seen	421
					†[$K^*\bar{K} + \bar{K}^*K$	seen]	130
					ηππ	seen	564
Not a well established resonance.					†[δπ	possibly seen]	349
f′(1515)	$0^+(2^+)+$	1516 ±10§	*65* ±10§	2.30 ±.10	$K\bar{K}$	dominant	572
					ππ	*seen*	745
					For upper limits, see footnote (k)		
ρ′(1600)	$1^+(1^-)-$	∼ 1600¶	∼ *300*¶	2.56 ±.48	4π	*75§±10§*	738
					†[$\rho\pi^+\pi^-$	seen with $\pi^+\pi^-$ in S-wave]	572
					ππ	*25§±10§*	788
$A_3(1640)$	$1^-(2^-)+$	∼ 1640	∼ 300	2.69 ±.49	fπ	dominant	304
Not a well established resonance.¶							
ω(1670)	$0^-(3^-)-$	1668 ±10§	160 ±15§	2.78 ±.27	ρπ	seen	645
					3π	possibly seen	806
					5π	possibly seen	740
					†[ωππ	possibly seen]	615
g(1680)¶	$1^+(3^-)-$	1688 ±20§	180 ±30§	2.85 ±.30	2π	24±5§	832
					4π (incl. ππρ, ρρ, $A_2\pi$, ωπ)	large	786
					$K\bar{K}$	small	682
J^P, M and Γ from the 2π mode.					$K\bar{K}\pi$ (incl. $K^*\bar{K}$)	small	623
S(1935)¶ J < 4		*1935* ±*2*§	*9* ±*4*§	3.74 ±.02	$N\bar{N}$	*dominant*	236
h(2040)	$0^+(4^+)+$	2040 ±20	193 ±50	4.16 ±.39	ππ	seen	1010
					$K\bar{K}$	seen	890
T(2190)¶	$1^+(3^-)-$	*2192* ±*10*§	*150*§ ±*50*§	4.80 ±.33	$N\bar{N}$	*dominant*	564
					ππ	*seen*	1086
U(2350)¶	$0^+(4^+)+$	*2350*§ ±*25*§	∼ *200*§	5.52 ±.47	$N\bar{N}$	*dominant*	707
					ππ	*seen*	1167
ψ(3100) or J	$0^-(1^-)-$	3097±2	0.067±0.012	9.598 ±.000	e^+e^-	7±1	1549
					$\mu^+\mu^-$	7±1	1545
					hadrons	86±2	
					†[$2(\pi^+\pi^-)\pi^0$	*3.7±0.5*	1496
					$3(\pi^+\pi^-)\pi^0$	*2.9±0.7*	1433
					$\pi^+\pi^-\pi^0K^+K^-$	*1.2±0.3*	1369
					ρπ	*1.1±0.2*	1448
					$4(\pi^+\pi^-)\pi^0$	*0.9±0.3*	1345
					$K^{*0}(890)\ \bar{K}^{*0}(1430)$	*0.67±0.26*	1007
					$K\bar{K}^*$	*0.61±0.08*	1373
					$p\bar{p}\pi^+\pi^-$	*0.41±0.08*	1108
					$2(\pi^+\pi^-)$	*0.4±0.1*	1517
					$3(\pi^+\pi^-)$	*0.4±0.2*	1466
					$p\bar{n}\pi^-$	*0.38±0.08*	1174
					$2(\pi^+\pi^-)K^+K^-$	*0.31±0.13*	1320
					$K^0K^{-+}\pi^{-+}$	*0.26±0.07*	1440
					$\phi\pi^+\pi^-$	*0.21±0.09*	1365
					$p\bar{p}$	*0.21±0.02*	1232
					$p\bar{p}\eta$	*0.19±0.04*	948
					$\phi K\bar{K}$	*0.18±0.08*	1176
					$\Lambda\bar{\Lambda}$	*0.16±0.08*	1075
					$p\bar{p}\pi^+\pi^-\pi^0$	*0.11±0.04*	1033
					$p\bar{p}\pi^0$	*0.10±0.02*	1175
					φη	*0.10±0.06*]	1320
					†[γη′	*0.25±0.06*	1401
					γf	*0.20±0.07*	1288
					γX(2830)→3γ	*0.14±0.04*]	256

For smaller branching ratios, upper limits, and resonance subchannels of the above modes, see listing.¶

Meson Table *(cont'd)*

Name	$I^G(J^P)C_n$ ⊢ estab.	Mass M (MeV)	Full Width Γ (MeV)	M^2 $\pm\Gamma M$[a] $(GeV)^2$	Partial decay mode: Mode	Fraction (%) [Upper limits are 1σ (%)]	p or p_{max}[b] (MeV/c)
χ(3415)	$0^+(0^+)+$	3413±5		11.649	$2(\pi^+\pi^-)$ (incl. ππρ)	*4.4±0.8*	1678
					$\pi^+\pi^-K^+K^-$ (incl. $\pi K\bar{K}^*$)	*3.7±1.0*	1579
					γJ/ψ(3100)	*3.3±1.0*	300
					$3(\pi^+\pi^-)$	*1.9±0.7*	1632
					$\pi^+\pi^-$	*1.0±0.3*	1701
					K^+K^-	*1.0±0.3*	1634
					$p\bar{p}\pi^+\pi^-$	*0.5±0.2*	1319
→							
P_C or χ(3510)	$0^+(A)+$	3508±4		12.306	γJ/ψ(3100)	*23.4±0.8* S = *2.4**	388
					$3(\pi^+\pi^-)$	*2.4±0.8*	1682
					$2(\pi^+\pi^-)$ (incl. ππρ)	*1.5±0.6*	1727
					$\pi^+\pi^-K^+K^-$ (incl. $\pi K\bar{K}^*$)	*0.9±0.4*	1632
$J^P = 1^+$ preferred.					$\pi^+\pi^-p\bar{p}$	*0.14±0.11*	1381
χ(3555)	$0^+(N)+$	3554±5		12.631	γJ/ψ(3100)	*16±3* S = *1.3**	427
					$\pi^+\pi^-K^+K^-$ (incl. $\pi K\bar{K}^*$)	*2.0±0.6*	1655
					$3(\pi^+\pi^-)$	*1.1±0.7*	1706
					$\pi^+\pi^-$ and K^+K^-	*0.29±0.15*	
					$\pi^+\pi^-p\bar{p}$	*0.29±0.14*	1408
$J^P = 2^+$ preferred.					$2(\pi^+\pi^-)$ (incl. ππρ)	*0.23±0.06*	1750
ψ(3685)	$0^-(1^-)-$	3686±3	0.228±0.056	13.587 ±.001	e^+e^-	0.9±0.1	1842
					$\mu^+\mu^-$	0.8±0.2	1839
					hadrons	98.1±0.3	
					+[J/ψ $\pi^+\pi^-$	33±3]	474
	$m_{\psi(3685)} - m_{\psi(3100)}$ = 588.6±0.8				+[J/ψ $\pi^0\pi^0$	17±2]	478
					+[J/ψ η	4.2±0.7]	189
					+[$2(\pi^+\pi^-)\pi^0$	*0.4±0.2*]	1798
					+[$\pi^+\pi^-K^+K^-$	*0.14±0.04*]	1725
					+[$2(\pi^+\pi^-)$	*0.08±0.02*]	1816
					+[γ χ(3415)	7±2]	261
					+[γ χ(3510)	7±2]	172
					+[γ χ(3555)	7±2]	128
ψ(3770)	$(1^-)-$	*3772* ±*6*	*28* ±*5*	14.228 ±.106	e^+e^-	*0.0013±0.0002*	1885
					$D\bar{D}$	*dominant*	184
→							
ψ(4415)	$(1^-)-$	4414±7	33±10	19.483 ±.146	e^+e^-	0.0013±0.0003	2207
					hadrons	dominant	
Υ(9500)	$(1^-)-$	∼ *9500*		90.25	$\mu^+\mu^-$	*seen*	4750
					e^+e^-	*seen*	4750
Seen split into two peaks m_1 = 9410±13, m_2=10060±30. Additional structure may be present¶.							
K^+	$1/2(0^-)$	493.67		0.244	See Stable Particle Table		
K^0		497.67		0.248			
$K^*(892)$	$1/2(1^-)$	892.2 ±0.4	49.5 ±1.5	0.796 ±.044	Kπ	≈ 100	288
					Kππ	< 0.2	216
					Kγ	0.15±0.07	309
M and Γ from charged mode; $m^0 - m^\pm$ = 4.1±0.6 MeV.							
$Q_1(1280)$	$1/2(1^+)$	∼ *1280*	∼ *120*	1.64 ±.19	Kππ	*dominant*	501
					+[Kρ	*large*]	62
					+[$K^*\pi$	*possibly seen*]	307
Existence of a second resonance, $Q_2(1400)$, decaying mainly into $K^*\pi$, not well established¶.					Kω	*possibly seen*	
→							
κ(1400)	$1/2(0^+)$	*1400-1450*	*200-300*		Kπ	seen	
See note on Kπ S wave¶.							
$K^*(1430)$	$1/2(2^+)$	*1434*§ ±*5*§	100§ ±10§	2.06 ±.14	Kπ	*49.1±1.6*	623
					$K^*\pi$	*27.0±2.2*	424
					$K^*\pi\pi$	*11.2±2.5*	374
					Kρ	6.6±1.5	327
					Kω	3.7±1.6	320
					Kη	2.5±2.5	492
→							

Meson Table *(cont'd)*

Name ($G\backslash I$: 0, 1; −: ω/φ, π; +: η, ρ)	$I^G(J^P)C_n$ ⊢— estab.	Mass M (MeV)	Full Width Γ (MeV)	$M^2 \pm \Gamma M$[a] $(GeV)^2$	Partial decay mode: Mode	Fraction (%) [Upper limits are 1σ (%)]	p or p_{max}[b] (MeV/c)
L(1770)	1/2(A)	1765 ±10§	140 ±50§	3.11 ±.25	$K\pi\pi$	dominant	788
					$K\pi\pi\pi$	seen	757
					†[K^*(1430)π and other subreactions¶]		
Not a well established resonance¶.							
K^*(1780)¶	1/2(3^-)	*1784* *±10*§	*135*§ *±40*§	3.19 ±.24	$K\pi\pi$	*large*	798
					†[$K\rho$	*large*]	619
					†[$K^*\pi$	*large*]	660
					$K\pi$	*19±5*§	817
→→							
D^+	1/2(0^-)	1868.3		3.491	See Stable Particle Table		
D^0		1863.3		3.472			
D^{*+}(2010)	1/2(1^-)	2008.6 ±1.0	<2.0	4.034	$D^0\pi^+$	*60±15*	39
					$D^+\pi^0$ }	*40±15*	37
	$m_{D^{*+}} - m_{D^0}$ = 145.3 ± 0.5 MeV				$D^+\gamma$ }		135
D^{*0}(2010)	1/2(1^-)	2006 ±1.5	< 5	4.024	$D^0\pi^0$	*55±15*	45
					$D^0\gamma$	*45±15*	138
→→							

Contents of Meson Data Card Listings

Non-strange (S = 0, C = 0) entry	$I^G(J^P)C_n$	entry	$I^G(J^P)C_n$	entry	$I^G(J^P)C_n$	Strange (|S| = 1, C = 0) entry	I (J^P)
π	$1^-(0^-)+$	A_2(1310)	$1^-(2^+)+$	→ e^+e^-(1100-3100)		K	1/2(0^-)
η	$0^+(0^-)+$	E (1420)	0^+(A)+	→ X (2830)		K^*(892)	1/2(1^-)
ρ (770)	$1^+(1^-)-$	→ X (1410-1440)		ψ (3100) or J	$0^-(1^-)-$	Q_1(1280)	1/2(1^+)
ω (783)	$0^-(1^-)-$	f′(1515)	$0^+(2^+)+$	χ (3415)	$0^+(0^+)+$	→ Q_2(1400)	1/2(1^+)
→ M (940-953)		→ F_1(1540)	1 (A)	→ χ (3455)		→ K′(1400)	1/2(0^-)
η′ (958)	$0^+(0^-)+$	ρ′(1600)	$1^+(1^-)-$	P_C or χ(3510)	0^+(A)+	κ (1400)	1/2(0^+)
δ (980)	$1^-(0^+)+$	A_3(1640)	$1^-(2^-)+$	χ (3555)	0^+(N)+	K^*(1430)	1/2(2^+)
S^* (980)	$0^+(0^+)+$	ω (1670)	$0^-(3^-)-$	ψ (3685)	$0^-(1^-)-$	→ K_N(1700)	1/2
→ H (990)		g (1680)	$1^+(3^-)-$	ψ (3770)	$(1^-)-$	L (1770)	1/2(A)
φ (1020)	$0^-(1^-)-$	→ X (1690)		→ ψ (4030)	$(1^-)-$	K^*(1780)	1/2(3^-)
→ M (1033-1040)		→ A_4(1900)	1^-	ψ (4415)	$(1^-)-$	→ K^*(2200)	
→ η_N(1080)	0^+(N)+	→ X (1900)	$1^-(4^+)+$	T (9500)	$(1^-)-$	→ I (2600)	
A_1(1100)	$1^-(1^+)+$	S (1935)	1	→ T (10060)	$(1^-)-$	Charmed (|C| = 1)	
→ M (1150-1170)		h (2040)	$0^+(4^+)+$				
B (1235)	$1^+(1^+)-$	T (2190)	$1^+(3^-)-$			D (1870)	1/2(0^-)
→ ρ′(1250)	$1^+(1^-)-$	U (2350)	$0^+(4^+)+$			D^*(2010)	1/2(1^-)
f (1270)	$0^+(2^+)+$	→ $\bar{N}N$(2360)	1			→ F (2030)	
D (1285)	0^+(A)+	→ $\bar{N}N$(1400-3600)				→ F^*(2140)	
ε (1300)	$0^+(0^+)+$	→ X (1900-3600)				→ Exotics	

Meson Table *(cont'd)*

→ Indicates an entry in Meson Data Card Listings not entered in the Meson Table. We do not regard these as established resonances. All the entries in the Listings can be found in the Table of Contents of Meson Data Card Listings.

¶ See Meson Data Card Listings.

* Quoted error includes scale factor $S = \sqrt{\chi^2/(N-1)}$. See footnote to Stable Particle Table.

† Square brackets indicate a subreaction of the previous (unbracketed) decay mode(s).

§ This is only an educated guess; the error given is larger than the error of the average of the published values. (See Meson Data Card Listings for the latter.)

(a) ΓM is approximately the half-width of the resonance when plotted against M^2.

(b) For decay modes into ≥ 3 particles, p_{max} is the maximum momentum that any of the particles in the final state can have. The momenta have been calculated by using the averaged central mass values, without taking into account the widths of the resonances.

(c) From pole position $(M - i\Gamma/2)$.

(d) The e^+e^- branching ratio is from $e^+e^- \to \pi^+\pi^-$ experiments only. The $\omega\rho$ interference is then due to $\omega\rho$ mixing only, and is expected to be small. See note in Meson Data Card Listings. The $\mu^+\mu^-$ branching ratio is compiled from 3 experiments; each possibly with substantial $\omega\rho$ interference. The error reflects this uncertainty; see notes in Meson Data Card Listings. If $e\mu$ universality holds, $\Gamma(\rho^0 \to \mu^+\mu^-) = \Gamma(\rho^0 \to e^+e^-) \times 0.99785$.

(e) Empirical limits on fractions for other decay modes of $\rho(770)$ are $\pi^\pm\eta < 0.8\%$, $\pi^+\pi^+\pi^-\pi^- < 0.15\%$, $\pi^\pm\pi^+\pi^-\pi^0 < 0.2\%$.

(f) Empirical limits on fractions for other decay modes of $\omega(783)$ are $\pi^+\pi^-\gamma < 5\%$, $\pi^0\pi^0\gamma < 1\%$, η + neutral(s) < 1.5%, $\mu^+\mu^- < 0.02\%$, $\pi^0\mu^+\mu^- < 0.2\%$.

(g) Empirical limits on fractions for other decay modes of $\eta'(958)$: $\pi^+\pi^- < 2\%$, $\pi^+\pi^-\pi^0 < 5\%$, $\pi^+\pi^+\pi^-\pi^- < 1\%$, $\pi^+\pi^+\pi^-\pi^-\pi^0 < 1\%$, $6\pi < 1\%$, $\pi^+\pi^-e^+e^- < 0.6\%$, $\pi^0e^+e^- < 1.3\%$, $\eta e^+e^- < 1.1\%$, $\pi^0\rho^0 < 4\%$.

(h) The mass and width are from the $\eta\pi$ mode only. If the $K\bar{K}$ channel is strongly coupled, the width may be 300 MeV or more.

(i) Empirical limits on fractions for othei decay modes of $\phi(1020)$ are $\pi^+\pi^- < 0.03\%$, $\pi^+\pi^-\gamma < 0.7\%$, $\omega\gamma < 5\%$, $\rho\gamma < 2\%$, $2\pi^+2\pi^-\pi^0 < 1\%$.

(j) Empirical limits on fractions for other decay modes of B(1235): $\pi\pi < 15\%$, $K\bar{K} < 2\%$, $4\pi < 50\%$, $\phi\pi < 1.5\%$, $\eta\pi < 25\%$, $(\bar{K}K)^\pm\pi^0 < 8\%$, $K_SK_S\,\pi^\pm < 2\%$, $K_SK_L\,\pi^\pm < 6\%$.

(k) Empirical limits on fractions for other decay modes of f'(1515) are $\eta\eta < 50\%$, $\eta\pi\pi < 30\%$, $K\bar{K}\pi + K^*\bar{K} < 35\%$, $2\pi^+2\pi^- < 32\%$.

(ℓ) Empirical limits on fractions for other decay modes of f(1270) are $\eta\pi\pi < 1\%$, $K^0K^-\pi^+$ + c.c. < 1%, $\eta\eta < 2\%$.

Established Nonets, and octet-singlet mixing angles from Appendix IIB, Eq. (2'). Of the two isosinglets, the "mainly octet" one is written first, followed by a semicolon.

$(J^P)C_n$	Nonet members	$\theta_{lin.}$	$\theta_{quadr.}$
$(0^-)+$	π, K, η; η′	-24 ± 1°	-11 ± 1°
$(1^-)-$	ρ, K*, φ; ω	38 ± 1°	40 ± 1°
$(2^+)+$	A₂, K*(1430), f′; f	24 ± 2°	26 ± 2°

Baryon Table

April 1978

The following short list gives the status of all the Baryon States in the Data Card Listings. In addition to the status, the name, the nominal mass, and the quantum numbers (where known) are shown. States with three- or four-star status are included in the main Baryon Table; the others have been omitted because the evidence for the existence of the effect and/or for its interpretation as a resonance is open to considerable question.

N(939) P11 ****	Δ(1232) P33 ****	Λ(1115) P01 ****	Σ(1193) P11 ****	Ξ(1317) P11 ****
N(1470) P11 ****	Δ(1550) P31 *	Λ(1330) Dead	Σ(1385) P13 ****	Ξ(1530) P13 ****
N(1520) D13 ****	Δ(1650) S31 ****	Λ(1405) S01 ****	Σ(1480) *	Ξ(1630) **
N(1535) S11 ****	Δ(1670) D33 ***	Λ(1520) D03 ****	Σ(1580) D13 **	Ξ(1820) 13 ***
N(1540) P13 *	Δ(1690) P33 ***	Λ(1600) P01 **	Σ(1620) S11 **	Ξ(1940) **
N(1670) D15 ****	Δ(1890) F35 ****	Λ(1670) S01 ****	Σ(1660) P11 ***	Ξ(2030) 1 ***
N(1688) F15 ****	Δ(1900) S31 *	Λ(1690) D03 ****	Σ(1670) D13 ****	Ξ(2120) *
N(1700) S11 ****	Δ(1910) P31 ****	Λ(1800) P01 **	Σ(1670) **	Ξ(2250) *
N(1700) D13 ***	Δ(1950) F37 ****	Λ(1800) G09 *	Σ(1690) **	Ξ(2500) **
N(1780) P11 ***	Δ(1960) D35 ***	Λ(1815) F05 ****	Σ(1750) S11 ***	
N(1810) P13 ***	Δ(2160) ***	Λ(1830) D05 ****	Σ(1765) D15 ****	Ω(1672) P03 ****
N(1990) F17 **	Δ(2420) H3 11 ***	Λ(1860) P03 ***	Σ(1770) P11 *	
N(2000) F15 **	Δ(2850) ***	Λ(1870) S01 ***	Σ(1840) P13 *	Λ_c(2260) *
N(2040) D13 **	Δ(3230) ***	Λ(2010) **	Σ(1880) P11 **	
N(2100) S11 *		Λ(2020) F07 *	Σ(1915) F15 ****	Σ_c(2430) *
N(2100) D15 **		Λ(2100) G07 ****	Σ(1940) D13 ***	
N(2190) G17 ***	Z0(1780) P01 *	Λ(2110) F05 ***	Σ(2000) S11 *	
N(2200) G19 ***	Z0(1865) D03 *	Λ(2325) D03 *	Σ(2030) F17 ****	Dibaryons
N(2220) H19 ***	Z1(1900) P13 *	Λ(2350) ****	Σ(2070) F15 *	S = 0 *
N(2650) I1 11 ***	Z1(2150) *	Λ(2585) ***	Σ(2080) P13 **	S = -1 *
N(3030) ***	Z1(2500) *		Σ(2100) G17 *	S = -2 *
N(3245) *			Σ(2250) ****	
N(3690) *			Σ(2455) ***	
N(3755) *			Σ(2620) ***	
			Σ(3000) **	

**** Good, clear, and unmistakable.
*** Good, but in need of clarification or not absolutely certain.
** Needs confirmation.
* Weak.

[See notes on N's and Δ's, Z*'s, Λ's and Σ's, Ξ*'s, and dibaryons at the beginning of those sections in the Baryon Data Card Listings; also see notes on individual resonances in the Baryon Data Card Listings.]

Particle[a]	I (J^P)[a] estab.	π or K beam[b] p_{beam} (GeV/c) $\sigma = 4\pi\lambda\!\!\!^{-2}$ (mb)	Mass M[c] (MeV)	Full Width Γ[c] (MeV)	M^2 ±ΓM[b] (GeV^2)	Partial decay mode[f]: Mode	Fraction[c] %	p or p_{max}[d] (MeV/c)
p n	$1/2(1/2^+)$		938.3 939.6		0.880 0.883	See Stable Particle Table		
N(1470)[g]	$1/2(1/2^+)P'_{11}$	p = 0.66 σ = 27.8	1390 to 1470	180 to 240 (200)	2.16 ±0.29	Nπ	~60	420
						Nη	~18	*d*
						Nππ	~25	368
						[Nε	~ 7][e]	*d*
						[Δπ	~23][e]	177
						[Nρ	~ 7][e]	*d*
N(1520)[g]	$1/2(3/2^-)D'_{13}$	p = 0.74 σ = 23.5	1510 to 1530	110 to 150 (125)	2.31 ±0.19	Nπ	~55	456
						Nππ	~45	410
						[Nε	< 5][e]	*d*
						[Nρ	~19][e]	*d*
						[Δπ	~23][e]	228
						Nη	< 1	*d*
N(1535)[g]	$1/2(1/2^-)S'_{11}$	p = 0.76 σ = 22.5	1500 to 1545	50 to 150 (100)	2.36 ±0.15	Nπ	~30	467
						Nη	~65	182
						Nππ	~ 5	422
						[Nρ	~ 3][e]	*d*
						[Nε	~ 2][e]	*d*
→						[Δπ	~ 1][e]	243

Baryon Table *(cont'd)*

Particle[a]	I (J^P)[a] — estab.	π or K beam[b] p_{beam} (GeV/c) $\sigma = 4\pi\lambda\!\!\!^{-2}$ (mb)	Mass M[c] (MeV)	Full Width Γ[c] (MeV)	M^2 $\pm\Gamma M$[b] (GeV^2)	Partial decay mode[f] Mode	Fraction[c] %	p or p_{max}[d] (MeV/c)
N(1670)[g]	$1/2(5/2^-)D'_{15}$	p = 1.00	1650 to	145 to	2.79	Nπ	~45	560
		σ = 15.6	1685	170	±0.26	Nππ	~55	525
				(155)		[Δπ	~47][e]	360
						[Nρ	~ 5][e]	*d*
						ΛK	< 0.3	200
						Nη	< 0.5	368
N(1688)[g]	$1/2(5/2^+)F'_{15}$	p = 1.03	1670 to	120 to	2.85	Nπ	~60	572
		σ = 14.9	1690	145	±0.24	Nππ	~40	538
				(140)		[Nε	~15][e]	340
						[Nρ	~13][e]	*d*
						[Δπ	~12][e]	375
						Nη	< 0.3	388
N(1700)[g]	$1/2(1/2^-)S''_{11}$	p = 1.05	1660 to	100 to	2.89	Nπ	~55	580
		σ = 14.3	1700	200	±0.26	Nππ	~30	547
				(150)		[Nε	<10][e]	355
						[Nρ	7-21][e]	*d*
						[Δπ	4-15][e]	385
						ΛK	~10	250
						ΣK	2-7	109
N(1700)[g]	$1/2(3/2^-)D''_{13}$	p = 1.05	1660 to	80 to	2.89	Nπ	~10	580
		σ = 14.3	1710	120[h]	±0.20	Nππ	~90	547
				(120)		[Nε	<40][e]	355
						[Nρ	< 5][e]	*d*
						[Δπ	15-40][e]	385
						ΛK	~ 1	250
N(1780)	$1/2(1/2^+)P''_{11}$	p = 1.20	1650 to	100 to	3.17	Nπ	~20	633
		σ = 12.2	1750	180	±0.28	Nππ	>50	603
				(160)		[Nε	15-40][e]	440
						[Nρ	40-65][e]	249
						[Δπ	10-20][e]	448
						ΛK	< 5	353
						ΣK	~10	267
						Nη	2-20[i]	476
N(1810)	$1/2(3/2^+)P''_{13}$	p = 1.26	1650 to	100 to	3.28	Nπ	~20	652
		σ = 11.5	1750	300	±0.36	Nππ	~70	624
				(200)		[Nε	~20][e]	468
						[Nρ	45-70][e]	297
						[Δπ	~20][e]	471
						ΛK	1-4	386
						ΣK	~ 2	307
						Nη	< 5	503
N(2190)	$1/2(7/2^-)G_{17}$	p = 2.07	2140 to	150 to	4.80	Nπ	15-35	888
		σ = 6.21	2250	300	±0.55			
				(250)				
N(2200)	$1/2(9/2^-)G_{19}$	p = 2.10	2130 to	200 to	4.84	Nπ	~10	894
		σ = 6.12	2270	300	±0.55			
				(250)				
N(2220)[g]	$1/2(9/2^+)H_{19}$	p = 2.14	2200 to	250 to	4.93	Nπ	~20	905
		σ = 5.97	2250	350	±0.67			
				(300)				
N(2650)	$1/2(11/2^-)I_{1\,11}$[j]	p = 3.26	2580 to	~400	7.02	Nπ	~5	1154
		σ = 3.67	2700	(400)	±1.06			
N(3030)	1/2(?)	p = 4.41	~3030	~400	9.18	Nπ	(J + 1/2) x	1366
		σ = 2.62		(400)	±1.21		<0.1[k]	

Baryon Table *(cont'd)*

Particle[a]	I (J^P)[a] — estab.	π or K beam[b] p_{beam} (GeV/c) $\sigma = 4\pi\lambda\!\!\!^{-2}$ (mb)	Mass M[c] (MeV)	Full Width Γ[c] (MeV)	M^2 $\pm\Gamma M$[b] (GeV^2)	Partial decay mode[f] Mode	Fraction[c] %	p or p_{max}[d] (MeV/c)
Δ(1232)[g]	$3/2(3/2^+)P'_{33}$	p = 0.30 σ = 94.3	1230 to 1234	110 to 120 (115)	1.52 ±0.14	Nπ $N\pi^+\pi^-$	~99.4 ~ 0	227 80
	Δ(++) Pole position:[l] $M - i\Gamma/2 = (1211.0 \pm 0.8) - i(49.9 \pm 0.6)$ Δ(0) Pole position:[l] $M - i\Gamma/2 = (1210.9 \pm 1.0) - i(53.1 \pm 1.0)$							
Δ(1650)[g]	$3/2(1/2^-)S'_{31}$	p = 0.96 σ = 16.4	1600 to 1695	120 to 200 (140)	2.72 ±0.23	Nπ Nππ [Nρ [Δπ	~32 ~65 10-25][e] ~50][e]	547 511 *d* 344
Δ(1670)[g]	$3/2(3/2^-)D_{33}$	p = 1.00 σ = 15.6	1620 to 1720	190 to 240 (200)	2.79 ±0.33	Nπ Nππ [Nρ [Δπ	~15 ~85 <40][e] 45-60][e]	560 525 *d* 361
Δ(1690)[g]	$3/2(3/2^+)P''_{33}$	p = 1.03 σ = 14.9	1650 to 1900[m]	150 to 350 (250)	2.86 ±0.42	Nπ Nππ [Nρ [Δπ	10-20 ~80 10-20][e] >45][e]	573 540 *d* 377
Δ(1890)[g]	$3/2(5/2^+)F_{35}$	p = 1.42 σ = 9.88	1860 to 1910	150 to 300 (250)	3.57 ±0.47	Nπ Nππ [Nρ [Δπ ΣK	~15 ~80 ~60][e] 10-30][e] < 3	704 677 403 531 400
Δ(1910)[g]	$3/2(1/2^+)P''_{31}$	p = 1.46 σ = 9.54	1780 to 1960	200 to 280 (220)	3.65 ±0.42	Nπ Nππ [Nρ [Δπ ΣK	15-25 >40 ~40][e] small][e] 2-20	716 691 429 545 420
Δ(1950)[g]	$3/2(7/2^+)F_{37}$	p = 1.54 σ = 8.90	1910 to 1950	200 to 280 (240)	3.80 ±0.47	Nπ Nππ [Nρ [Δπ ΣK	~40 >25 ~10][e] ~20][e] < 1	741 716 471 574 460
Δ(1960)[g]	$3/2(5/2^-)D_{35}$	p = 1.56 σ = 8.75	1890 to 1950	100 to 300 (200)	3.84 ±0.39	Nπ ΣK	7-15 <10	748 469
Δ(2160)[n]	$3/2(\ ?^-)$	p = 2.00 σ = 6.46	2150 to 2240	160 to 440 (300)	4.67 ±0.65	Nπ	(J + 1/2)x =0.4 - 1.4[k]	870
Δ(2420)[g]	$3/2(11/2^+)H_{3\,11}$	p = 2.64 σ = 4.68	2380 to 2450	300 to 500 (300)	5.86 ±0.73	Nπ	10-15	1023
Δ(2850)	$3/2(\ ?^+)$	p = 3.85 σ = 3.05	2800 to 2900	~400 (400)	8.12 ±1.14	Nπ	(J + 1/2)x ~0.25[k]	1266
Δ(3230)	3/2(?)	p = 5.08 σ = 2.25	3200 to 3350	~440 (440)	10.43 ±1.42	Nπ	(J + 1/2)x ~0.05[k]	1475

Z* Evidence for states with strangeness +1 is inconclusive. See the Baryon Data Card Listings for data and discussion.

Baryon Table *(cont'd)*

Particle[a]	I (J^P)[a] — estab.	π or K beam[b] p_{beam} (GeV/c) $\sigma = 4\pi\lambda\!\!\!^{-2}$ (mb)	Mass M[c] (MeV)	Full Width Γ[c] (MeV)	M^2 $\pm\Gamma M$[b] (GeV^2)	Partial decay mode[f]: Mode	Fraction[c] %	p or p_{max}[d] (MeV/c)
Λ	$0(1/2^+)$		1115.6		1.245	See Stable Particle Table		
Λ(1405)	$0(1/2^-)S'_{01}$	Below K^-p threshold	1405 ±5[o]	40 ± 10[o] (40)	1.97 ±0.06	Σπ	100	142
Λ(1520)	$0(3/2^-)D'_{03}$	p = 0.389 σ = 84.5	1520 ±2[o]	16 ± 2[o] (16)	2.31 ±0.02	$N\bar{K}$	46 ± 1	234
						Σπ	42 ± 1	258
						Λππ	10 ± 1	250
						Σππ	0.9 ± 0.1	140
Λ(1670)	$0(1/2^-)S''_{01}$	p = 0.74 σ = 28.5	1660 to 1680	20 to 60 (40)	2.79 ±0.07	$N\bar{K}$	15-25	410
						Λη	15-35	64
						Σπ	20-60	393
Λ(1690)	$0(3/2^-)D''_{03}$	p = 0.78 σ = 26.1	1690 ±10[o]	40 to 80 (60)	2.86 ±0.10	$N\bar{K}$	20-30	429
						Σπ	20-40	409
						Λππ	~25	415
						Σππ	~20	352
Λ(1815)	$0(5/2^+)F'_{05}$	p = 1.05 σ = 16.7	1820 ±5[o]	70 to 90 (80)	3.29 ±0.15	$N\bar{K}$	55-65	542
						Σπ	5-15	508
						Σ(1385)π	5-10	362
Λ(1830)	$0(5/2^-)D_{05}$	p = 1.09 σ = 15.8	1810 to 1830	60 to 110 (95)	3.35 ±0.17	$N\bar{K}$	<10	554
						Σπ	35-75	519
						Σ(1385)π	>15	375
Λ(1860)	$0(3/2^+)P'_{03}$	p = 1.14 σ = 14.7	1850 to 1920	60 to 200 (100)	3.46 ±0.19	$N\bar{K}$	15-40	576
						Σπ	3-10	534
Λ(1870)	$0(1/2^-)S'''_{01}$	p = 1.16 σ = 14.2	1700 to 1850	200 to 400 (300)	3.50 ±0.56	$N\bar{K}$	20-60	582
						Σπ	seen	542
Λ(2100)	$0(7/2^-)G_{07}$	p = 1.68 σ = 8.68	2080 to 2120	100 to 300 (250)	4.41 ±0.53	$N\bar{K}$	~30	748
						Σπ	~5	699
						Λη	<3	617
						ΞK	<3	483
						Λω	<8	443
Λ(2110)	$0(5/2^+)F''_{05}$	p = 1.70 σ = 8.48	2050 to 2150	150 to 300 (200)	4.45 ±0.42	$N\bar{K}$	5-25	756
						Σπ	<40	709
Λ(2350)	$0(9/2^+)$	p = 2.29 σ = 5.85	2340 to 2420	100 to 250 (120)	5.52 ±0.28	$N\bar{K}$	~12	913
						Σπ	~10	865
Λ(2585)	0(?)	p = 2.91 σ = 4.37	~2585	~300 (300)	6.68 ±0.78	$N\bar{K}$	(J+1/2)x ~1.0[k]	1058
Σ	$1(1/2^+)$		(+)1189.4 (0)1192.5 (−)1197.4		1.415 1.422 1.434	See Stable Particle Table		
Σ(1385)	$1(3/2^+)P'_{13}$	Below K^-p threshold	(+)1382.3±0.4 S = 1.6[p] (−)1387.5±0.6 S = 1.0[p] (0)1382.0±2.5 S = 1.6[p]	(+) 35±2 S = 2.2[p] (−) 40±2 S = 1.9[p] (35)	1.92 ±0.05	Λπ	88 ± 2	208
						Σπ	12 ± 2	117

Baryon Table *(cont'd)*

Particle[a]	I (J^P)[a] — estab.	π or K beam[b]: p_{beam} (GeV/c), $\sigma = 4\pi\lambda\!\!\!^{-2}$ (mb)	Mass M[c] (MeV)	Full Width Γ[c] (MeV)	M^2 $\pm\Gamma M$[b] (GeV^2)	Partial decay mode[f]: Mode	Fraction[c] %	p or p_{max}[d] (MeV/c)
$\Sigma(1660)$[q]	$1(1/2^+)P'_{11}$	p = 0.72 σ = 30.1	1580 to 1690	30 to 200 (100)	2.76 ±0.17	$N\bar{K}$ $\Sigma\pi$ $\Lambda\pi$	<30 seen seen	402 383 440
$\Sigma(1670)$	$1(3/2^-)D''_{13}$	p = 0.74 σ = 28.5	1675[o] ±10	35 to 70 (50)	2.79 ±0.08	$N\bar{K}$ $\Sigma\pi$ $\Lambda\pi$	5-15 20-60 <20	410 387 447
$\Sigma(1750)$	$1(1/2^-)S''_{11}$	p = 0.91 σ = 20.7	1730 to 1820	50 to 160 (75)	3.06 ±0.13	$N\bar{K}$ $\Lambda\pi$ $\Sigma\pi$ $\Sigma\eta$	10-40 5-20 <8 15-55	483 507 450 54
$\Sigma(1765)$	$1(5/2^-)D_{15}$	p = 0.94 σ = 19.6	1774 ±7[o]	105 to 135 (120)	3.12 ±0.21	$N\bar{K}$ $\Lambda\pi$ $\Lambda(1520)\pi$ $\Sigma(1385)\pi$ $\Sigma\pi$	~41 ~14 ~19 ~ 9 ~ 1	496 518 187 315 461
$\Sigma(1915)$[g]	$1(5/2^+)F'_{15}$	p = 1.25 σ = 13.0	1905 to 1930	70 to 160 (100)	3.67 ±0.19	$N\bar{K}$ $\Lambda\pi$ $\Sigma\pi$	5-15 10-20 seen	612 619 568
$\Sigma(1940)$[q]	$1(3/2^-)D'''_{13}$	p = 1.32 σ = 12.0	1890 to 1960	100 to 300 (220)	3.76 ±0.43	$N\bar{K}$ $\Lambda\pi$ $\Sigma\pi$ $\Lambda(1520)\pi$	<20 seen seen seen	678 680 589 370
$\Sigma(2030)$[g]	$1(7/2^+)F_{17}$	p = 1.52 σ = 9.93	2020 to 2040	120 to 200 (180)	4.12 ±0.37	$N\bar{K}$ $\Lambda\pi$ $\Sigma\pi$ ΞK $\Lambda(1520)\pi$ $\Sigma(1385)\pi$	~20 ~20 5-10 <2 5-20 12	700 700 652 412 429 530
$\Sigma(2250)$[q]	1(?)[r]	p = 2.04 σ = 6.76	2200 to 2300	50 to 150 (100)	5.06 ±0.22	$N\bar{K}$ $\Lambda\pi$ $\Sigma\pi$	<10 seen seen	849 841 801
$\Sigma(2455)$	1(?)	p = 2.57 σ = 5.09	~2455	~120 (120)	6.03 ±0.29	$N\bar{K}$	(J + 1/2)x ~0.2[k]	979
$\Sigma(2620)$	1(?)	p = 2.95 σ = 4.30	~2600	~200 (200)	6.86 ±0.52	$N\bar{K}$	(J + 1/2)x ~0.3[k]	1064
Ξ	$1/2(1/2^+)$		(0) 1314.9 (−) 1321.3		1.729 1.746	See Stable Particle Table		
$\Xi(1530)$	$1/2(3/2^+)P_{13}$		(0) 1531.8±0.3 S = 1.3[p] (−) 1535.0±0.6	(0) 9.1±0.5 (−) 10.1±1.9 (10)	2.34 ±0.02	$\Xi\pi$	100	144
$\Xi(1820)$	1/2(3/2)		1823 ±6[o]	20^{+15}_{-10}[o] (20)	3.31 ±0.04	$\Lambda\bar{K}$ $\Xi(1530)\pi$ $\Sigma\bar{K}$ $\Xi\pi$	~45 ~45 ~10 small	396 234 306 413
$\Xi(2030)$[s]	1/2(?)		2024 ±6[o]	16^{+15}_{-5}[o] (16)	4.12 ±0.03	$\Sigma\bar{K}$ $\Lambda\bar{K}$ $\Xi\pi$ $\Xi(1530)\pi$	~80 ~20 small small	524 587 573 418
Ω^-	$0(3/2^+)$		1672.2		2.796	See Stable Particle Table		

Baryon Table *(cont'd)*

→ For convenience all Baryon States for which information exists in the Baryon Data Card Listings are listed at the beginning of the Baryon Table. States with only a one or two star (*) rating in that list have been omitted from the main Baryon Table; each omitted state is indicated by an arrow in the left-hand margin of the Table. In the Listings there is an arrow under the name of each state omitted from the Table.

a. The names of the Baryon States in Col. 1 [such as N(1470)] contain a nominal mass which is primarily for purposes of identification. See Col. 4 for actual mass values. The convention for using primes in the spectroscopic notation for the quantun numbers in Col. 2 (such as P_{11}) is as follows: no prime is attached when the Data Card Listings include only one resonance in the given partial wave; when there is more than one resonance the first has been designated with a prime, the second with a double prime, etc. The name and the quantum numbers for each state are also given in large print at the beginning of the Data Card Listings for that state. See footnote a. of the Stable Particle Table for the strangeness quantum numbers of the baryons; in addition to the names listed there, we also use N and Δ for S = 0 baryons, and Z* for S = +1 baryons.

b. The numbers in Col. 3 and Col. 6 are calculated using the nominal mass (see *a.* above) for M and the nominal width (see *c.* below) for Γ.

c. For masses, widths, and branching fractions of most baryons we report here a range instead of an average. Averages are appropriate if each result is based on independent measurements, but inappropriate where the spread in parameters arises because different models or procedures have been applied to a common set of data. The ranges given in the Table are generally chosen to be *conservatively large*. See the Data Card Listings for the individual values obtained in specific analyses. A single value with an approximation sign (~) indicates that there is not enough data to give a meaningful interval. A nominal width is included in parentheses in Col. 5; this nominal width is used to calculate the value of ΓM given in Col. 6.

d. For two-body decay modes we given the momentum, p, of the decay products in the decaying baryon rest frame. For decay modes into ⩾3 particles we give the maximum momentum, p_{max}, that any of the particles in the final state can have in this frame. The momenta are calculated using the nominal mass (see *a.* above) of the decaying baryon, and of any isobars in the final state. Some decays which would be energetically forbidden for the nominal masses actually occur because of the finite widths of the decaying baryon and/or isobars in the final state. In these cases, the decay momentum is omitted from Col. 9 and replaced with a reference to this footnote.

e. Square brackets around an isobar decay mode indicate that it is a sub-reaction of the previous unbracketed decay mode.

f. Many of the branching fractions in the Table are extracted from significantly more accurate results on $\sqrt{xx'}$ type couplings obtained in partial-wave analyses. The original $\sqrt{xx'}$ values are given in the Baryon Data Card Listings. For information on radiative decays of N's and Δ's, see the mini-review preceding the Baryon Data Card Listings.

g. Only information coming from partial-wave analyses has been used here. For the production experiment results see the Baryon Data Card Listings.

h. The range given here does not include the widths of several hundred MeV reported by LONGACRE 75 and LONGACRE 77.

i. The range given here does not include the branching ratio of approximately 80% reported by FELTESSE 75.

j. The existence of an I_{111} resonance at this mass has been confirmed, but the possibility remains that there are also other nearby I = 1/2 resonances. See the mini-review preceding the Baryon Data Card Listings.

k. This state has been seen only in an energy-dependent fit to total, channel, or fixed angle cross-section data. J is not known; x is Γ_{el}/Γ.

l. See note on determination of resonance parameters in the Baryon Data Card Listings. Values of mass and width are dependent upon resonance shape used to fit the data. The pole position is much less dependent upon the parametrization used. The pole positions given here are taken from results (in the Data Card Listings) of fits to the phase shifts of CARTER 73 without Coulomb corrections.

m. There may be more than one P_{33} resonance in or near this mass range.

n. There is probably more than one Δ resonance near 2160 MeV. The parameters in the Table correspond to the observations of REY 74. See the Baryon Data Card Listings for other possibilities.

o. The error given here is only an educated guess; it is larger than the error of the average of the published values (see the Baryon Data Card Listings for the latter).

p. Quoted error includes a S (scale) factor. See second footnote to Stable Particle Table.

q. Because the elastic branching fraction of this resonance is poorly determined, it is not possible to extract inelastic branching fractions from partial-wave couplings. See the Baryon Data Card Listings for the partial-wave couplings.

r. Recent partial-wave analyses of the College de France-Saclay group find evidence for a $5/2^-$ and a $9/2^-$ Σ resonance at this mass. See the Baryon Data Card Listings.

s. This state is now considered to be firmly established even though the quantum numbers and decay rates are not sufficiently well known.

CHARACTERISTICS OF PARTICLES AND PARTICLE DISPERSOIDS

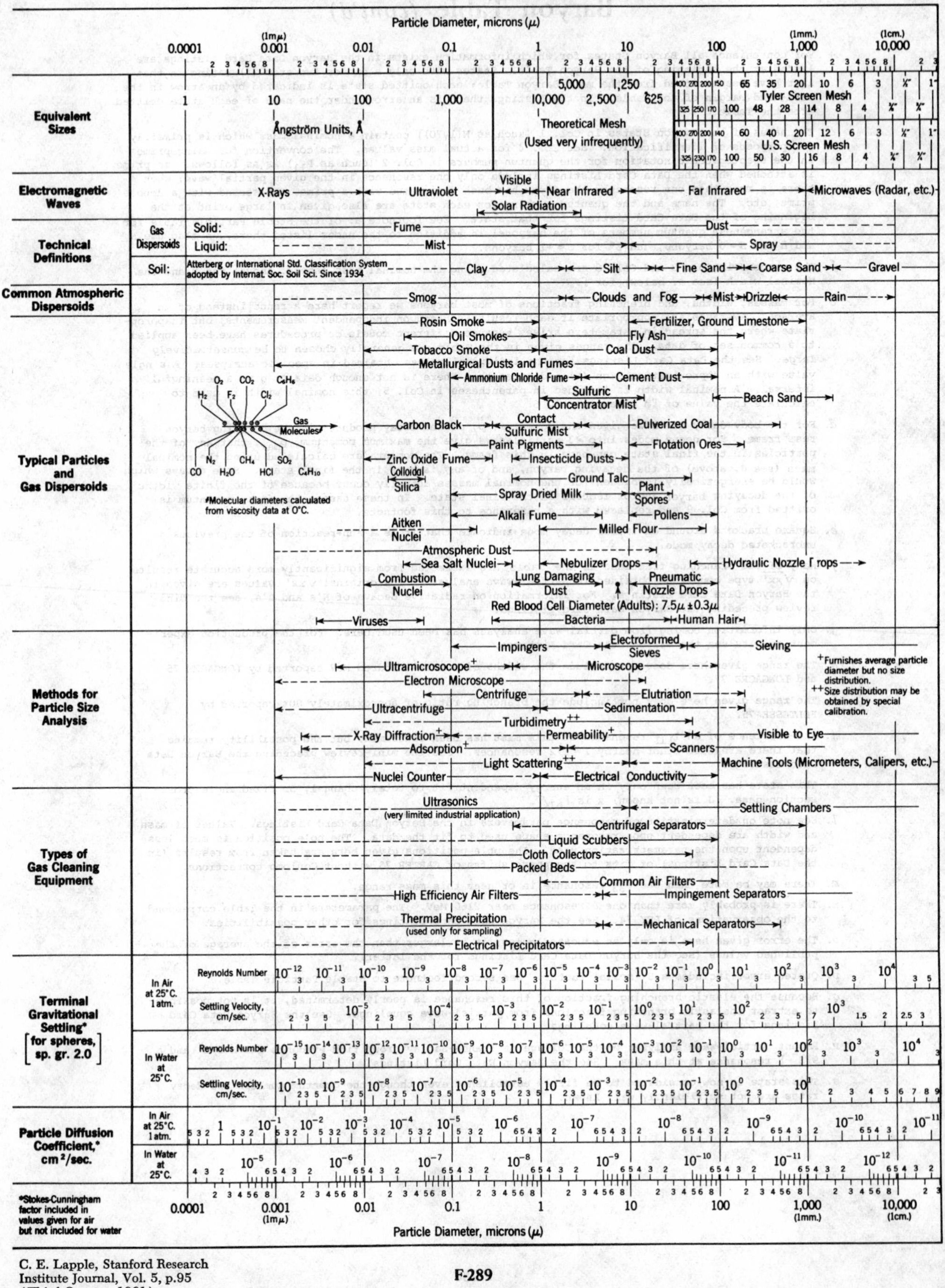

C. E. Lapple, Stanford Research Institute Journal, Vol. 5, p.95 (Third Quarter, 1961)

ABBREVIATIONS AND SYMBOLS

Abbreviations

The following list of abbreviations is intended to cover those in common use in chemistry and physics. Symbols are presented in a separate list following the abbreviations.

Abbreviation	Meaning
A.	Acre
Å	Ångström unit
a	Are
a.	Acid
abs.	Absolute
abt.	About
a.c.	Alternating current
acet.	Acetone
acet. a.	Acetic acid
al.	Alcohol
alk.	Alkali
alt.	Altitude
amal.	Amalgam; amalgamated
amor. *or* amorph.	Amorphous
amp.	Ampere
anh.	Anhydrous
antilog	Antilogarithm
ap.	Apothecaries'
appr.	Approximately
aq.	Aqua; aqueous; water
aq. reg.	Aqua regia
asym.	Asymmetrical
atm. *or* atmos.	Atmosphere (atmospheric)
At. No.	Atomic number
At. Wt.	Atomic weight
aux.	Auxiliary
Av.	Average
av. *or* avoir.	Avoirdupois
bar.	Barometer
bbl.	Barrel
bd.	Board
Bé	Beaumé (degrees)
B.G.	Birmingham gauge (hoop and sheet)
b.h.p.	Brake horse power
bl.	Blue
blk.	Black
B.M.	Board measure
b.p.	Boiling point
br.	Brown
BTU	British thermal unit
bu.	Bushel
B.W.G.	Birmingham wire gauge
bz.	Benzene
C	Centigrade
c	Carat; centi-
c.	Cold
ca	Candle
ca.	*Circa*, about; approximately
cal.	Calorie (gram)
cc. *or* c.c.	Cubic centimeter
cd.	Cord
c. cm	Cubic centimeter
Cent.	Centigrade
centi-	Prefix meaning 1/100
cf.	*Confer*, compare
c.f.m.	Cubic foot per minute
cgs	Centimeter-gram-second system of units
cgse	Cgs electrostatic system
cgsm	Cgs electromagnetic system
ch.	Chain
chl.	Chloroform
cir.	Circular
circum.	Cirumference
cl	Centiliter
cm	Centimeter
cm^2	Square centimeter
cm^3	Cubic centimeter
c.m.	Circular mil
coef.	Coefficient
colog	Cologarithm
colorl.	Colorless
comm'l	Commercial
conc.	Concentrated
cond.	Condensing
const.	Constant
cos	Cosine
cos^{-1}	Arc or angle whose cosine is . . .; anticosine of; inverse cosine of
cosec	Cosecant
cosh	Hyperbolic cosine
cosh^{-1}	Inverse hyperbolic cosine
cot	Cotangent
cot^{-1}	Arc or angle whose cotangent is . . .
coth	Hyperbolic cotangent
coth^{-1}	Inverse hyperbolic cotangent
covers	Coversed sine
c.p.	Candle power; circular pitch; center of pressure
cry. *or* cryst.	Crystalline; crystals
csc	Cosecant
csc^{-1}	Arc or angle whose cosecant is . . .
csch	Hyperbolic cosecant
csch^{-1}	Inverse hyperbolic cosecant
CTU	Centigrade thermal unit
cu.	Cubic
cu. cm	Cubic centimeter
cu. ft.	Cubic foot
cu. in.	Cubic inch
cu. m	Cubic meter
cu. yd.	Cubic yard
cwt.	Hundredweight
cyl.	Cylinder
d	Derivative; deci-
d.	Decomposes; day
d.	Dextrorotary
d.c.	Direct current
dec.	Decomposes
deci-	Prefix meaning 1/10
def.	Definition (s)
deg	Thermometric degree; absolute C unless contrary is indicated
deka-	Prefix meaning 10
deliq.	Deliquescent
den. *or* dens.	Density
dg	Decigram
diam.	Diameter
dil.	Dilute
dissd.	Dissolved
dk	Deka-
dk.	Dark
dkg.	Dekagram
dkl	Dekaliter
dkm	Dekameter
dkm^2	Square dekameter
dkm^3	Cubic dekameter
dks	Dekastere
dl	Deciliter
dm	Decimeter
dm^2	Square decimeter
dm^3	Cubic decimeter
d.p.	Diametral pitch; double pole
dr.	Dram
dr. ap. *or* ʒ ap.	Dram, apothecaries'
dr. av. *or* ʒ av.	Dram, avoirdupois
dr. fl. *or* ʒ fl.	Dram, fluid
dr. t. *or* ʒ t.	Dram, troy
ds	Decistere
dwt.	Pennyweight
efflor.	Efflorescent
e.g.	*Exempli gratia*, for example
e.h.p.	Effective horse power
E.L.	Elastic limit
em	Cgsm unit of quantity of electricity
emf *or* e.m.f.	Electromotive force
es	Electrostatic or cgse unit of quantity of electricity
etc.	*Et. cetera*, and so forth
eth.	Ether
eth. acet.	Ethyl acetate
et. seq.	*Et sequentes*, and the following
evap.	Evaporation
ex.	Excess
exp	Exponential function
exp.	Explodes
exsec	Exterior secant
F	Fahrenheit
f.	From
fahr.	Fahrenheit
fath.	Fathom
feath.	Feathery
f.h.p.	Friction horse power
fir.	Firkin
fl.	Fluid
fl. dr.	Dram, fluid
fl. oz.	Ounce, fluid
fluores.	Fluorescent
fps	Foot-pound-second system of units
fpse	Foot-pound-second electrostatic system
fpsm	Foot-pound-second electromagnetic system
F.S.	Factor of safety
ft.	Foot
ft.^{2}	Square foot
ft.^{3}	Cubic foot
ft.-lb.	Foot-pound
fur.	Furlong
G	Gravitation constant
g	Gram
g-cal. *or* g.-cal.	Gram calorie
gal.	Gallon
gel.	Gelatinous
gi.	Gill
glac.	Glacial
glit.	Glittering
glyc.	Glycerine
gm.	Gram
gr.	Gray; grain
grn.	Green
gyr.	Gyration
h	Hecto-
h.	Hot; hour
ha	Hectare
hecto-	Prefix meaning 100
hex.	Hexagonal
hg	Hectogram
hhd.	Hogshead
hl	Hectoliter
hm	Hectometer
hm^2	Square hectometer
hm^3	Cubic hectometer
hor. *or* horiz.	Horizontal
h.-p.	High-Pressure
HP *or* h.p.	Horse power
h.p.-hr.	Horse power-hour
hr.	Hour
hyg.	Hygroscopic
i.	Insoluble
ibid.	*Ibidem*, in the same place
i.e.	*Id est*, that is
ign.	Ignites
i.h.p.	Indicated horse power
in.	Indigo; inch
in.^{2}	Square inch
in.^{3}	Cubic inch
inc.	Inclusive
in.-lb.	Inch-pound
insol.	Insoluble
Int.	International
iso.	Isotropic
isom.	Isometric
isoth.	Isothermal
k	Kilo-
kg	Kilogram
kg-cal.	Kilogram-calorie
kg-m	Kilogram-meter
kilo-	Prefix meaning 1,000
kl	Kiloliter
km	Kilometer
km^2	Square kilometer
km^3	Cubic kilometer
kva.	Kilovolt-ampere
kw.	Kilowatt
kw.-hr.	Kilowatt-hour
l	Liter
l.	Long
l	Laevorotary
lat.	Latitude
lb.	Pound
lb. ap.	Pound, apothecaries'
lb. av.	Pound, avoirdupois
lb. t.	Pound, troy
leaf.	Leaflets
lgr.	Ligroin
li.	Link
lin.	Linear
liq.	Liquid
lim.	Limit
ln	Natural hyperbolic or Napierian logarithm
log *or* log.	Logarithm
log$_e$	Logarithm to the base *e*; natural, hyperbolic or Napierian logarithm
log$_{10}$	Common logarithm; logarithm to the base 10
long.	Longitude
lng.	Long
l.-p.	Low-pressure
lt.	Light
lust.	Lustrous
♏	Minim or drop
m	Meter; milli-
m^2	Square meter
m^3	Cubic meter
m.	Minute
m.	Meta-
max.	Maximum
med.	Medium
meth.	Methyl
meth. al.	Methyl alcohol
m.e.p.	Mean effective pressure
met.	Metallic
mg	Milligram
m.h.c.p.	Mean horizontal candle power
mi.	Mile
mic.	Microscopic
micro-	Prefix meaning 1/1,000,000 or 10^{-6}
micro-micro	Prefix meaning 10^{-12}
milli-	Prefix meaning 1/1,000
milli-micro-	Prefix meaning 10^{-9}
min *or* min.	Minute
min.	Minim; minimum; mineral
ml	Milliliter
m.l.h.c.p.	Mean lower hemispherical candle power
mm	Millimeter
mm^2	Square millimeter
mm^3	Cubic millimeter
mmf *or* m.m.f.	Magnetomotive force
mol.	Molecule
Mol. Wt.	Molecular weight
monocl.	Monoclinic
m.p.	Melting point
m.s.c.p.	Mean spherical candle power
myria-	Prefix meaning 10,000 or 10^4
mμ	Millimicron; millimicro-
N	Numeric; number (in mathematical tables)
n.	Normal
n	Refractive index
need.	Needles
o	Ortho-
Obs.	Observer
octahdr.	Octahedral
oil turp.	Oil of turpentine
or.	Orange
oz.	Ounce
oz. ap. *or* ℥ ap.	Ounce, apothecaries'
oz. av. *or* ℥ av.	Ounce, avoirdupois
oz. fl. *or* ℥ fl.	Ounce, fluid
oz. t. *or* ℥ t.	Ounce, troy
p	Para-
pa.	Pale
p. ct.	Per cent
perp.	Perpendicular
p.f.	Power factor
pk.	Peck
pl.	Plates
powd.	Powder
pr.	Prisms
precip. *or* p'p't'd	Precipitated
p. sol.	Partly soluble
pt.	Point; pint
purp.	Purple
pyr.	Pyridine
Q	Quantity
q	Quintal
qt.	Quart
q.v.	*Quod vide*, which see
R	Réaumur; radioactive mineral
rac	Racemic
rad	Radian, measure of angle
rad.	Radius
rd.	Rod
reg.	Regular
rev.	Revolution
rhbdr.	Rhombohedral
rhomb.	Rhombic or orthorhombic
R.M.S.	Square root of mean square
r.p.m.	Revolutions per minute
s	Stere
s.	Scruple; soluble; second
s. ap. *or* ℈	Scruple, apothecaries'
sat. *or* sat'd	Saturated
sc.	Scales
S.E.	Siemens unit
sec *or* sec.	Second (mean solar unless contrary is stated)
sec	Secant
sec^{-1}	Arc or angle whose secant is . . .
sech	Hyperbolic secant
sech^{-1}	Inverse hyperbolic secant
segm.	Segment
sh.	Short
sin	Sine
sin^{-1}	Arc or angle whose sine is . . .
sinh	Hyperbolic sine
sinh^{-1}	Inverse hyperbolic sine
sl.	Slightly
sm.	Small
sol.	Solution; soluble
soln.	Solution
sp.	Specific
specif.	Specification
sp. gr.	Specific gravity
sq.	Square
sq. ch.	Square chain
sq. ft.	Square foot
sq. in.	Square inch
sq. mi.	Square mile
sq. rd.	Square rod
sq. yd.	Square yard
std.	Standard
subl.	Sublimes
sym.	Symmetrical
t	Metric ton
t.	Troy
tab. *or* tabl.	Tablets
tan	Tangent
tan^{-1}	Arc or angle whose tangent is . . .
tanh	Hyperbolic tangent
tanh^{-1}	Inverse hyperbolic tangent
temp.	Temperature
tetr. *or* tetrag.	Tetragonal
tn.	Ton
tr.	Transition
tricl.	Triclinic
trig.	Trigonal
trim.	Trimetric
T.S.	Tensile strength
turp.	Turpentine
Tw.°	Degrees Twaddell, hydrometer scale
ult.	Ultimate
uns.	Unsymmetrical
U.S.	United States of America; universal system of lens apertures
v.	Very
v.	*Vide*, see
vel. *or* veloc.	Velocity
vers	Versed sine
vert	Vertical
visc.	Viscous
vol.	Volume
volt.	Volatilizes
w.	Water
wh.	White
wt.	Weight
yd.	Yard
yel.	Yellow
yr.	Year
μ	Micron; micro-
$\mu\mu$	Micromicron; micromicro-

SPELLING AND SYMBOLS FOR UNITS

From "Units of Weight and Measure"
L. B. Chisholm, National Bureau of Standards
Miscellaneous Publication 286 (May, 1967)

The spelling of the names of units as adopted by the National Bureau of Standards is that given in the list below. The spelling of the metric units is in accordance with that given in the law of July 28, 1866, legalizing the Metric System in the United States.

Following the name of each unit in the list below is given the symbol that the Bureau has adopted. Attention is particularly called to the following principles:

1. No period is used with symbols for units. Whenever "in" for inch might be confused with the preposition "in", "inch" should be spelled out.

2. The exponents "2" and "3" are used to signify "square" and "cubic," respectively, instead of the symbols "sq" or "cu," which are, however, frequently used in technical literature for the U. S. Customary units.

3. The same symbol is used for both singular and plural.

Some Units and Their Symbols

Unit	Symbol	Unit	Symbol	Unit	Symbol
acre	acre	fathom	fath	millimeter	mm
are	a	foot	ft	minim	minim
barrel	bbl	furlong	furlong	ounce	oz
board foot	fbm	gallon	gal	ounce, avoirdupois	oz avdp
bushel	bu	grain	grain	ounce, liquid	liq oz
carat	c	gram	g	ounce, troy	oz tr
Celsius, degree	°C	hectare	ha	peck	peck
centare	ca	hectogram	hg	pennyweight	dwt
centigram	cg	hectoliter	hl	pint, liquid	liq pt
centiliter	cl	hectometer	hm	pound	lb
centimeter	cm	hogshead	hhd	pound, avoirdupois	lb avdp
chain	ch	hundredweight	cwt	pound, troy	lb tr
cubic centimeter	cm^3	inch	in	quart, liquid	liq qt
cubic decimeter	dm^3	International Nautical Mile	INM	rod	rod
cubic dekameter	dam^3			second	s
cubic foot	ft^3	Kelvin, degree	°K	square centimeter	cm^2
cubic hectometer	hm^3	kilogram	kg	square decimeter	dm^2
cubic inch	in^3	kiloliter	kl	square dekameter	dam^2
cubic kilometer	km^3	kilometer	km	square foot	ft^2
cubic meter	m^3	link	link	square hectometer	hm^2
cubic mile	mi^3	liquid	liq	square inch	in^2
cubic millimeter	mm^3	liter	liter	square kilometer	km^2
cubic yard	yd^3	meter	m	square meter	m^2
decigram	dg	microgram	μg	square mile	mi^2
deciliter	dl	microinch	μin	square millimeter	mm^2
decimeter	dm	microliter	μl	square yard	yd^2
dekagram	dag	micron	μm	stere	stere
dekaliter	dal	mile	mi	ton, long	long ton
dekameter	dam	milligram	mg	ton, metric	t
dram, avoirdupois	dr avdp	milliliter	ml	ton, short	short ton
				yard	yd

SYMBOLS, UNITS AND NOMENCLATURE IN PHYSICS

From Document U.I.P. 11 (S.U.N. 65–3)
International Union of Pure and Applied Physics
Used by permission of the Secretary

1. PHYSICAL QUANTITIES—GENERAL RECOMMENDATIONS.

1.1 The symbol for a ***physical quantity*** (french: 'grandeur physique', german: 'physikalische Grösse', american sometimes: 'physical magnitude') is equivalent to the product of the *numerical value* (or the measure), a pure number, and a *unit,* i.e.

$$\text{physical quantity} = \text{numerical value} \times \text{unit.}$$

For dimensionless physical quantities the unit often has no name or symbol and is not explicitly indicated.

Examples: $E = 200$ erg, $F = 27$ N, $n = 1.55$ (for quartz), $\nu = 3 \times 10^8\ \mathrm{s}^{-1}$

1.2 Symbols for physical quantities—General rules.

1. *Symbols for physical quantities* should be *single letters* of the latin or greek alphabet with or without modifying signs: subscripts, superscripts, dashes, etc.

   Remark:

   a. An exception to this rule consists of the two letter symbols, which are sometimes used to represent dimensionless combinations of physical quantities. If such a symbol, composed of two letters, appears as a factor in a product, it is recommended to separate this symbol from the other symbols by a dot or by brackets or by a space.

   b. Abbreviations, i.e. shortened forms of names or expressions, such as p.f. for partition function, should not be used in physical equations. These abbreviations in the text should be written in ordinary roman type.

2. *Symbols for physical quantities* should be printed in *italic* (or *sloping*) *type.*

   Remark:

   It is recommended to consider as a guiding principle for the printing of indices the criterion: only indices which are symbols for physical quantities should be printed in italic (sloping) type. *Examples:*

Upright indices:	*Sloping indices:*
C_g (g = gas)	p in C_p
g_n (n = normal)	n in $\Sigma_n a_n \varphi_n$
μ_r (r = relative)	x in $\Sigma_x a_x b_x$
E_k (k = kinetic)	i, k in g_{ik}
χ_e (e = electric)	x in p_x

3. *Symbols for vectors and tensors:* To avoid the usage of subscripts it is often convenient to indicate vectors and tensors of the second rank by letters of a special type. The following choice is recommended:

   a. Vectors should be printed in bold type, by preference bold italic (sloping) type, e.g. ***A, a.***

   b. Tensors of the second rank should be printed in bold face sans serif type, e.g. **S, T.**

   Remark:

   When this is not available, vectors may be indicated by an arrow and tensors by a double arrow on top of the symbol.

1.3 Simple mathematical operations.

1. *Addition and subtraction* of two physical quantities are indicated by:

$$a + b \qquad \text{and} \qquad a - b$$

2. *Multiplication* of two physical quantities may be indicated in one of the following ways:

$$ab \qquad a\ b \qquad a\,.\,b \qquad a \cdot b \qquad a \times b$$

Remark: The various products of vectors and tensors may be written in the following ways:

scalar product of vectors ***A*** and ***B***:	***A . B***	***A · B***
vector product of vectors ***A*** and ***B***:	$\boldsymbol{A} \wedge \boldsymbol{B}$	$\boldsymbol{A} \times \boldsymbol{B}$
dyadic product of vectors ***A*** and ***B***:	***AB***	
scalar product of tensors **S** and **T** ($\Sigma_{i,k} S_{ik} T_{ki}$)	**S : T**	
tensor product of tensors **S** and **T** ($\Sigma_k S_{ik} T_{kl}$)	**S · T**	**S . T**
product of tensor **S** and vector ***A*** ($\Sigma_k S_{ik} A_k$)	**S** · ***A***	**S** . ***A***

3. *Division* of one quantity by another quantity may be indicated in one of the following ways:

$$\frac{a}{b} \qquad a/b \qquad a\ b^{-1}$$

or in any other way of writing the product of a and b^{-1}.

These procedures can be extended to cases where one of the quantities or both are themselves products, quotients, sums or differences of other quantities.
If necessary brackets have to be used in accordance with the rules of mathematics. If the solidus is used to separate the numerator from the denominator and if there is any doubt where the numerator starts or where the denominator ends, brackets should be used.

Examples:

Expressions with a horizontal bar:	*Same expressions with a solidus:*
$\frac{a}{bcd}$	a/bcd
$\frac{2}{9}\sin kx,\ \frac{1}{2}RT$	$(2/9)\sin kx,\ (1/2)RT$ or $RT/2$
$\frac{a}{b} - c$	$a/b - c$
$\frac{a}{b-c}$	$a/(b-c)$
$\frac{a-b}{c-d}$	$(a-b)/(c-d)$
$\frac{a}{c} - \frac{b}{d}$	$a/c - b/d$

Remark: It is recommended that in expressions like:

$$\sin\{2\pi(x - x_0)/\lambda\} \qquad \exp\{(r - r_0)/\sigma\}$$
$$\exp\{-V(r)/kT\} \qquad \surd(\epsilon/c^2)$$

the argument should always be placed between brackets, except when the argument is a simple product of two quantities: e.g. sin kx. When the horizontal bar above the square root is used no brackets are needed.

2. UNITS—GENERAL RECOMMENDATIONS.

2.1 Symbols for units—General rules.

1. Symbols for units of physical quantities should be printed in *roman (upright) type.*

2. Symbols for units should not contain a final full stop and should remain unaltered in the plural, e.g.: 7 cm and *not* 7 cms.

3. Symbols for units should be printed in *lower case* roman (upright) type. However, the symbol for a unit, derived from a proper name, should start with a capital roman letter, e.g.: m (metre); A (ampere); Wb (weber); Hz (hertz).

2.2 Prefixes—General rules.

1. The following *prefixes* should be used to indicate decimal fractions or multiples of a unit

deci; *déci*	$(= 10^{-1})$	d			
centi; *centi*	$(= 10^{-2})$	c			
milli; *milli*	$(= 10^{-3})$	m	kilo; *kilo*	$(= 10^{3})$	k
micro; *micro*	$(= 10^{-6})$	μ	mega; *méga*	$(= 10^{6})$	M
nano; *nano*	$(= 10^{-9})$	n	giga; *giga*	$(= 10^{9})$	G
pico; *pico*	$(= 10^{-12})$	p	tera; *téra*	$(= 10^{12})$	T
femto; *femto*	$(= 10^{-15})$	f			
atto; *atto*	$(= 10^{-18})$	a			

2. The use of *double prefixes* should be avoided when single prefixes are available.

Not: mμs,	*but:* ns	(nanosecond)
Not: kMW	*but:* GW	(gigawatt)
Not: μμF,	*but:* pF	(picofarad)

3. When a prefix is placed before the symbol of a unit, the *combination of prefix and symbol* should be considered as *one new symbol*, which can be raised to a positive or negative power without using brackets.

Examples:

$$\text{cm}^3 \quad , \quad \text{mA}^2 \quad , \quad \mu\text{s}^{-1}$$

Remark:

cm³ *means always* $(0.01\ \text{m})^3$ *but never* $0.01\ \text{m}^3$
μs⁻¹ *means always* $(10^{-6}\text{s})^{-1}$ *but never* $10^{-6}\ \text{s}^{-1}$

2.3 Mathematical operations.

1. *Multiplication* of two units may be indicated in one of the following ways:

Nm   N m   N · m   N . m

2. *Division* of one unit by another unit may be indicated in one of the following ways:

$\frac{\text{m}}{\text{s}}$ m/s   m s^{-1}

or by any other way of writing the product of m and s^{-1}.
Not more than one solidus should be used. *Examples:*

Not: cm/s/s,	*but:* cm/s^2 = cm s^{-2}
Not: 1 poise = 1 g/s/cm,	*but:* 1 poise = 1 g/s cm = 1 gm sec^{-1} cm^{-1}
Not: J/°K/mol,	*but:* J/°K mol = J°K^{-1} mol^{-1}

3. NUMBERS.

1. *Numbers* should be printed in *upright type.*
2. The *decimal sign* between digits in a number should be a comma (,) or (but *only* in *English* texts) a point (.).
3. The *multiplication* sign between numbers should be a cross (×) or (but *only in non-English* texts) a centered dot, e.g. 2.3 × 3.4 or 2,3 · 3,4.
4. *Division* of one number by another number may be indicated in the following ways:

$\frac{136}{273.15}$ 136/273.15

or by writing it as the product of numerator and the inverse first power of the denominator. In such cases the number under the inverse power should always be placed between brackets.

Remark: When the solidus is used and when there is any doubt where the numerator starts or the denominator ends, brackets should be used, as in the case of quantities (see 1.3.3).

5. To facilitate the reading of *long numbers,* the digits may be grouped in *groups of three,* but *no* comma or point should be used except for the decimal sign.

Example: 2 573. 421 736.

4. SYMBOLS FOR CHEMICAL ELEMENTS, NUCLIDES AND PARTICLES.

1. *Symbols for chemical elements* should be written in *roman* (upright) *type.* The symbol is not followed by full stop.

Examples: Ca   C   H   He

2. The attached numerals specifying a *nuclide* are:

mass number $^{14}\text{N}_{2\ \text{atoms/molecule}}$

Remark: The atomic number may be placed as a left subscript, if desired. The right superscript position should be used, if required, for indicating a state of ionisation (e.g. Ca^{2+}, PO_4^{3-}) or an excited state (e.g. $^{110}Ag^m$, He*).

3. *Symbols for particles and quanta.*
It is recommended that the following notation be used:

proton:	p	pion	π
neutron:	n	K-meson	K
Λ-particle	Λ	electron	e
Σ-particle	Σ	muon	μ
Ξ-particle	Ξ	neutrino	ν
deuteron	d		
triton	t	photon	γ
α-particle	α		

A nucleon (proton or neutron) is indicated by N
It is recommended that the charge of particles may be indicated by adding the superscript +, − or 0.
Examples:

$\pi^+\ \pi^-\ \pi^0$, $p^+\ p^-$, $e^+\ e^-$.

If in connection with the symbols p and e no charge is indicated, these symbols should refer to the positive proton and the negative electron respectively.
The tilde ~ above the symbol of a particle is used to indicate the corresponding anti-particle.

Examples:

$$\tilde{n},\ \tilde{\nu},\ \sim p,\ \tilde{N}$$

5. QUANTUM STATES.

5.1 General rules.

A letter-symbol indicating the quantum state of a *system* should be printed in capital upright type.
A letter-symbol indicating the quantum state of a *single particle* should be printed in lower case upright type.

5.2 Atomic spectroscopy.

The letter-symbols indicating atomic quantum states are:

$L, l = 0$: S, s	$L, l = 4$: G, g	$L, l = 8$: L, l
$= 1$: P, p	$= 5$: H, h	$= 9$: M, m
$= 2$: D, d	$= 6$: I, i	$= 10$: N, n
$= 3$: F, f	$= 7$: K, k	$= 11$: O, o

A right hand subscript indicates the total angular momentum quantum number J or j.
A left hand superscript indicates the spin multiplicity $2S + 1$.

Example: ${}^2\mathrm{P}_{3/2}$—state ($J = 3/2$, multiplicity 2)
$\mathrm{p}_{3/2}$—electron ($j = 3/2$)

An atomic electron configuration is indicated symbolically by:

$$(nl)^{\kappa} \quad (n'l')^{\kappa'} \ldots$$

Instead of $l = 0, 1, 2, 3 \ldots$ one uses the quantum state symbol s, p, d, f, . . .

Example: the atomic configuration: $(1s)^2\,(2s)^2\,(2p)^3$.

5.3 Molecular spectroscopy.

The letter symbols, indicating molecular electronic quantum states are in the case of *linear molecules*:

$$\Lambda, \lambda = 0:\ \Sigma, \sigma$$
$$= 1:\ \Pi, \pi$$
$$= 2:\ \Delta, \delta$$

and for *non-linear molecules*

A, a ; B, b ; E, e ; etc.

Remarks: A left hand superscript indicates the spin multiplicity. For molecules having a symmetry center the parity symbol g or u, indicating respectively symmetric or antisymmetric behaviour on inversion, is attached as a right hand subscript. A + or − sign attached as a right hand superscript indicates the symmetry as regards reflection in any plane through the symmetry axis of the molecules.

Examples: Σ_g^+, Π_u, ${}^2\Sigma$, ${}^3\Pi$, etc.

The letter symbols indicating the vibrational angular momentum states in the case of *linear molecules* are

$$l = 0:\ \Sigma$$
$$= 1:\ \Pi$$
$$= 2:\ \Delta$$

5.4 Nuclear spectroscopy.

The spin and parity assignment of a nuclear state is

$$J^{\pi}$$

where the parity symbol π is + for even and − for odd parity.

Examples: 3^+, 2^-, etc.

A shell model configuration is indicated symbolically by:

$$(nlj)^{\kappa} \quad (n'l'j')^{\kappa'}$$

where the first bracket refers to the proton shell and the second to the neutron shell. Negative values of κ or κ' indicate holes in a completed shell. Instead of $l = 0, 1, 2, 3, \ldots$ one uses the quantum state symbols s, p, d, f, . . .

Example: the nuclear configuration: $(1\,\mathrm{d}\,3/2)^3\,(1\,\mathrm{f}\,7/2)^2$

5.5 Spectroscopic transitions.

1. The upper level and the lower level are indicated by ′ and ″ respectively.

Examples: $h\nu = E' - E''$ $\quad \sigma = T' - T''$

2. A spectroscopic transition should be indicated by writing the upper state first and the lower state second, connected by a dash inbetween.

Examples:	$^2P_{1/2} - {}^2S_{1/2}$	for an electronic transition
	$(J', K') - (J'', K'')$	for a rotational transition
	$v' - v''$	for a vibrational transition.

3. Absorption transition and emission transition may be indicated by arrows ← and → respectively.

Examples: $^2P_{1/2} \rightarrow {}^2S_{1/2}$ emission from $^2P_{1/2}$ to $^2S_{1/2}$
$(J', K') \leftarrow (J'', K'')$ absorption from (J'', K'') to (J', K')

4. The difference Δ between two quantum numbers should be that of the upper state minus that of the lower state.

Example: $\Delta J = J' - J''$

5. The indications of the branches of the rotation band should be as follows:

$\Delta J = J' - J'' = -2$:	O – branch
$= -1$:	P – branch
$= 0$:	Q – branch
$= +1$:	R – branch
$= +2$:	S – branch

6. NOMENCLATURE.

1. *Use of the word specific.*

The word 'specific' in English names for physical quantities should be restricted to the meaning 'divided by mass'. *Examples:*

specific volume	volume/mass
specific energy	energy/mass
specific heat capacity	heat capacity/mass

2. *Notation for covariant character of coupling.*

S	Scalar coupling	A	Axial vector coupling
V	Vector coupling	P	Psuedoscalar coupling
T	Tensor coupling		

3. *Abbreviated notation for a nuclear reaction.*

The meaning of the symbolic expression indicating a nuclear reaction should be the following:

initial nuclide (incoming particle(s) or quanta, outgoing particle(s) or quanta) final nuclide

Examples:	$^{14}N(\alpha,p)^{17}O$	$^{59}Co(n,\gamma)^{60}Co$
	$^{23}Na(\gamma,3n)^{20}Na$	$^{31}P(\gamma,pn)^{29}Si$

4. *Character of transitions.*

Multipolarity of transition:

electric or magnetic	monopole	E0 or M0
" " "	dipole	E1 or M1
" " "	quadrupole	E2 or M2
" " "	octupole	E3 or M3
" " "	2^n-pole	E*n* or M*n*

parity change in transition:

transition *with* parity change:	yes
transition *without* parity change:	no

5. *Nuclide:* A species of *atoms*, identical as regards atomic number (proton number) and mass number (nucleon number) should be indicated by the word *nuclide*, not by the word isotope.
Different nuclides having the same atomic number should be indicated as *isotopes* or *isotopic nuclides*.
Different nuclides having the same mass number should be indicated as *isobars* or *isobaric nuclides*.

6. *Sign of polarization vector (Basel Convention).*

In nuclear interactions the positive polarization of particles with spin ½ is taken in the direction of the vector product

$$\boldsymbol{k}_i \times \boldsymbol{k}_0,$$

where $\boldsymbol{k}_i$ and $\boldsymbol{k}_0$ are the circular wave vectors of the incoming and outgoing particles respectively.

7. RECOMMENDED SYMBOLS FOR PHYSICAL QUANTITIES.

Remark:

(1) Where several symbols are given for one quantity, and no special indication is made, they are on equal footing.
(2) In general, no special attention is paid to the name of the quantity.

7.1 Space and time.

space coordinates; *coordonnées d'espace*	(x, y, z)
position vector; *vecteur de position*	$\mathbf{r}$
length; *longueur*	l
breadth; *largeur*	b
height; *hauteur*	h
radius; *rayon*	r
thickness; *épaisseur*	d, δ
diameter; *diamètre:* $d = 2r$	d
path; *parcours:* $L = \int \mathrm{d}s$	L, s
area; *aire; superficie*	A, S
volume; *volume*	$V, (v)$
plane angle, *angle plan*	$\alpha, \beta, \gamma, \theta, \vartheta, \varphi$
solid angle; *angle solide*	ω, Ω
wave length; *longueur d'onde*	λ
wave number; *nombre d'onde* $\sigma = 1/\lambda$	σ *)
wave vector; *vecteur d'onde*	$\boldsymbol{\sigma}$
circular wave number; *nombre d'onde circulaire:* $k = 2\pi/\lambda$	k
circular wave vector; *vecteur d'onde circulaire*	$\mathbf{k}$
attenuation coefficient; *constante d'affaiblissement:* $F(x) = \exp(-\alpha x) \cos \beta x$	α
phase coefficient; *constante de phase*	β
propagation coefficient; *constante de propagation:* $\gamma = \alpha + \mathrm{i}\beta$	γ
time; *temps*	t
period; *période*	T
frequency; *fréquence:* $\nu = 1/T$	ν, f
pulsatance; *pulsation:* $\omega = 2\pi\nu$	ω **)
relaxation time; *temps de relaxation:* $F(t) = \exp(-t/\tau)$	τ
damping coefficient; *coefficient d'amortissement:* $F(t) = \exp(-\delta t) \sin \omega t$	δ
logarithmic decrement; *décrément logarithmique:* $\Lambda = T\delta = T/\tau$	Λ
velocity; *vitesse:* $v = \mathrm{d}s/\mathrm{d}t$	u, v
angular velocity; *vitesse angulaire:* $\omega = \mathrm{d}\varphi/\mathrm{d}t$	ω
acceleration; *accélération:* $a = \mathrm{d}v/\mathrm{d}t$	a
angular acceleration; *accélération angulaire:* $\alpha = \mathrm{d}\omega/\mathrm{d}t$	α
acceleration of free fall; *accélération de la pesanteur*	g
standard — —; — — *normale*	g_n
speed of light in empty space; *vitesse de la lumière dans le vide*	c
v/c	β
relativistic coordinates; *coordonnées relativistes:*	$(x_0 x_1 x_2 x_3)$
$x_0 = ct,\ x_1 = x,\ x_2 = y,\ x_3 = z,\ x_4 = \mathrm{i}ct$	$(x_1 x_2 x_3 x_4)$

7.2 Mechanics.

mass; *masse*	m
(mass) density; *masse volumique:* $\rho = m/V$	ρ
relative density; *densité relative:* $d = \rho/\rho(H_2O)$	d
specific volume; *volume massique:* $v = V/m = 1/\rho$	v
reduced mass; *masse réduite:* $\mu = m_1 m_2/(m_1 + m_2)$	μ
momentum; *quantité de mouvement:* $\mathbf{p} = m\mathbf{v}$	$\mathbf{p}$
angular momentum; *moment cinetique:* $\mathbf{L} = \mathbf{r} \times \mathbf{p}$	$\mathbf{L}$
second moment of plane area; *moment quadratique d'une aire plane:* $I_{a,y} = \int x^2 \mathrm{d}x\mathrm{d}y$	I_a
second polar moment of plane area; *moment quadratique polaire d'une aire plane:* $I_p = \int (x^2 + y^2)\,\mathrm{d}x\mathrm{d}y$	I_p
moment of inertia; *moment d'inertie:* $I_z = \int (x^2 + y^2)\,\mathrm{d}m$	I, J
force; *force*	$\mathbf{F}$
torque, moment of a couple; *torque, moment d'un couple*	$\mathbf{T}$
weight; *poids*	$G, (W, P)$
moment of force; *moment d'une force*	$\mathbf{M}$
pressure; *pression*	p
normal stress; *tension normale*	σ
shear stress; *tension de cisaillement*	τ
gravitational constant; *constante de gravitation:* $F(r) = G\, m_1 m_2/r^2$	G
linear strain, relative elongation; *dilation linéaire relative:* $\epsilon = \Delta l/l_0$	ϵ
modulus of elasticity, Young's modulus; *module d'élasticité:* $\sigma = E\,\epsilon$	E
shear strain; shear angle; *glissement unitaire*	γ
shear modulus; *module de torsion:* $\tau = G \operatorname{tg} \gamma$	G
volume strain, bulk strain; *dilation volumique relative:* $\theta = \Delta V/V_0$	θ
bulk modulus; *module de compression:* $p = -K\theta$	K
Poisson ratio, *rapport de Poisson*	μ, ν
viscosity; *viscosité*	$\eta, (\mu)$
kinematic viscosity; *viscosité cinématique:* $\nu = \eta/\rho$	ν
friction coefficient; *coefficient de frottement*	$\mu, (f)$
surface tension; *tension superficielle*	γ, σ
energy; *énergie*	E, W
potential energy; *énergie potentielle*	E_p, V, Φ
kinetic energy; *énergie cinétique*	E_k, T, K
work; *travail*	W, A

*) In molecular spectroscopy: $\tilde{\nu}$. **) Also called angular frequency.

power; *puissance*	P
efficiency; *rendement*	η
Hamiltonian function; *fonction de Hamilton*	H
Lagrangian function; *fonction de Lagrange*	L
principal function of Hamilton; *fonction principale de Hamilton:* $W = \int L\mathrm{d}t$	W, S_p
characteristic function of Hamilton; *fonction caractéristique de Hamilton:* $S = 2\int T\mathrm{d}t$	S
generalized coordinate; *coordonnée généralisée*	q, q_i
generalized momentum; *moment généralisé*	p, p_i
action integral; *intégrale d'action:* $J = \oint p\mathrm{d}q$	J

7.3 Molecular physics.

number of molecules; *nombre de molécules*	N
number density of molecules; *nombre volumique de molécules* $n = N/V$	n
Avogadro constant; *constante d'Avogadro*	L, N_A
molecular mass; *masse moléculaire*	m
molecular velocity vector with components; *vecteur vitesse moléculaire et ses composantes*	$\boldsymbol{c}$, (c_x, c_y, c_z) $\boldsymbol{u}$, (u_x, u_y, u_z)
molecular position vector with components; *vecteur position moléculaire et ses composantes*	$\mathbf{r}$, (x, y, z)
molecular momentum vector with components; *vecteur quantité de mouvement moléculaire et ses composantes*	$\boldsymbol{p}$, (p_x, p_y, p_z)
average velocity; *vitesse moyenne*	$\boldsymbol{c}_0$, $\boldsymbol{u}_0$, $\langle \boldsymbol{c} \rangle$, $\langle \boldsymbol{u} \rangle$
average speed; *vitesse moyenne*	$\bar{c}$, $\bar{u}$, $\langle c \rangle$, $\langle u \rangle$
most probable speed; *vitesse la plus probable*	$\hat{c}$, $\hat{u}$
mean free path; *libre parcours moyen*	l
molecular attraction energy; *énergie d'attraction moléculaire*	ϵ
interaction energy between molecules i and j; *énergie d'interaction entre les molécules i et j*	φ_{ij}, V_{ij}
velocity distribution function; *fonction de distribution des vitesses:* $n = \int f\, \mathrm{d}c_x \mathrm{d}c_y \mathrm{d}c_z$	$f(c)$
Boltzmann function; *fonction de Boltzman*	H
generalized coordinate; *coordonnée généralisée*	q
generalized momentum; *moment généralisé*	p
volume in γ phase space; *volume dans l'espace* γ	Ω
thermodynamic temperature; *température thermodynamique*	T
Boltzmann constant; *constante de Boltzmann*	k
$1/kT$ (in exponential functions; *dans les fonctions exponentielles*)	β
molar gas constant; *constante molaire des gaz*	R
partition function; *fonction de partitions*	Q, Z
symmetry number; *facteur de symétrie*	s
diffusion coefficient; *coefficient de diffusion*	D
thermal diffusion coefficient; *coefficient de thermodiffusion*	D_T
thermal diffusion ratio; *rapport de thermodiffusion*	K_T
thermal diffusion factor; *facteur de thermodiffusion*	α_T
characteristic temperature; *température caractéristique*	Θ
Debye temperature; *température de Debye:* $\Theta_\mathrm{D} = h\nu_\mathrm{D}/k$	Θ_D
Einstein temperature; *température d'Einstein:* $\Theta_\mathrm{E} = h\nu_\mathrm{E}/k$	Θ_E
rotational temperature; *température de rotation:* $\Theta_\mathrm{r} = h^2/8\pi^2 Ik$	Θ_r
vibrational temperature; *température de vibration:* $\Theta_\mathrm{v} = h\nu/k$	Θ_v

*7.4 Thermodynamics.**)

quantity of heat; *quantité de chaleur*	Q
work; *travail*	W, A
temperature; *température*	t, (ϑ) *)
thermodynamic temperature; *température thermodynamique*	T, (Θ) **)
entropy; *entropie*	S
internal energy; *énergie interne*	U
Helmholtz function; *fonction de Helmholtz, énergie libre:* $F = U - TS$	F
enthalpy; *enthalpie:* $H = U + pV$	H
Gibbs function; *fonction de Gibbs, enthalpie libre:* $G = H - TS$	G
pressure coefficient; *coefficient d'augmentation de pression:* $\beta = (1/p)(\partial p/\partial T)_V$	β
compressibility; *compressibilité:* $\kappa = -(1/V)(\partial V/\partial p)_T$	κ
linear expansion coefficient; *coefficient de dilatation linéique*	α
cubic expansion coefficient; *coefficient de dilatation volumique*	γ
thermal conductivity; *conductivité thermique*	λ
specific heat capacity; *chaleur massique:* $c = C/m$	c_p, c_v
heat capacity; *capacité thermique*	C_p, C_v
Joule-Thomson coefficient; *coefficient de Joule-Thomson*	μ
ratio of specific heat capacities; *rapport des chaleurs massiques*	κ, γ
heat flow rate; *flux thermique*	Φ, (q)
heat current density; *densité de flux thermique*	$\boldsymbol{q}$, (φ)
thermal diffusivity; *diffusivité thermique:* $a = \lambda/\rho c_p$	a

†) The index m is added in the case of molar quantities, if needed, to distinguish them from quantities referring to the whole system. For specific quantities (see 6.1) lower case letters are used.

*) Preferred symbol: t. **) Preferred symbol: T.

7.5 *Electricity and magnetism.*††)

quantity of electricity; *quantité d' électricité*	Q
charge density; *charge volumique*	ρ
surface charge density; *charge surfacique*	σ
electric potential; *potentiel électrique*	V, φ
potential difference, tension; *différence de potentiel, tension*	U, V
electromotive force; *force électromotrice*	E
electric field strength; *champ électrique*	$\boldsymbol{E}$
electric flux; *flux électrique*	Ψ
electric displacement; *déplacement électrique*	$\boldsymbol{D}$
capacitance; *capacité*	C
permittivity; *permittivité:* $\boldsymbol{D} = \epsilon \boldsymbol{E}$	ϵ
permittivity of vacuum; *permittivité du vide*	ϵ_0
relative permittivity; *permittivité relative* $\epsilon_r = \epsilon/\epsilon_0$	ϵ_r
dielectric polarization: *polarisation diélectrique:* $\boldsymbol{D} = \epsilon_0 \boldsymbol{E} + \boldsymbol{P}$	$\boldsymbol{P}$
electric susceptibility; *susceptibilité électrique*	χ_e
polarizability; *polarisabilité*	α, γ
electric dipole moment; *moment dipolaire électrique*	$\boldsymbol{p}$
electric current; *courant électrique*	I
electric current density; *densité de courant électrique*	$\boldsymbol{j}$
magnetic field strength; *champ magnétique*	$\boldsymbol{H}$
magnetic potential difference; *différence de potentiel magnétique*	U_m
magnetomotive force; *force magnétomotrice:* $F_m = \oint H_s \mathrm{d}s$	F_m
magnetic induction, magnetic flux density; *induction magnétique, densité de flux magnétique*	$\boldsymbol{B}$
magnetic flux; *flux magnétique*	Φ
permeability; *perméabilité:* $\boldsymbol{B} = \mu \mathbf{H}$	μ
permeability of vacuum; *perméabilité du vide*	μ_0
relative permeability; *perméabilité relative:* $\mu_r = \mu/\mu_0$	μ_r
magnetization; *aimantation:* $\boldsymbol{B} = \mu_0(\mathbf{H} + \mathbf{M})$	$\mathbf{M}$
magnetic susceptibility; *susceptibilité magnétique*	χ_m
electromagnetic moment; *moment électromagnétique:* $E_p = -\boldsymbol{m} \cdot \boldsymbol{B}$	μ, $\boldsymbol{m}$
magnetic polarization; *polarisation magnétique:* $\boldsymbol{B} = \mu_0 \mathbf{H} + \boldsymbol{J}$	$\boldsymbol{J}$
resistance; *résistance*	R
reactance; *réactance*	X
quality factor: *facteur de qualité:* $Q = \lvert X \rvert / R$	Q
impedance; *impédance:* $Z = R + \mathrm{i}X$	Z
admittance; *admittance:* $Y = 1/Z = G + \mathrm{i}B$	Y
conductance; *conductance*	G
susceptance; *susceptance*	B
resistivity; *résistivité*	ρ
conductivity; *conductivité* $\gamma = 1/\rho$	γ, σ
self inductance; *inductance propre*	L
mutual inductance; *inductance mutuelle*	M, L_{12}
coupling coefficient; *coefficient de couplage:* $k = L_{12}/(L_1 L_2)^{1/2}$	k
phase number; *nombre de phases*	m
loss angle; *angle de pertes*	δ
number of turns; *nombre de tours*	N
power: *puissance*	P
electromagnetic energy density; *énergie électromagnetique volumique*	w
Poynting vector; *vecteur de Poynting*	$\boldsymbol{S}$
magnetic vector potential; *potentiel vecteur magnétique*	$\boldsymbol{A}$

7.6 *Radiation, light.*†)

radiant energy; *énergie rayonnante*	Q, (Q_e), W
radiant flux, radiant power; *flux énergétique, puissance rayonnante*	Φ, (Φ_e), P
radiant intensity; *intensité énergétique*	I, (I_e)
irradiance; *irradiance, éclairement énergétique*	E, (E_e)
radiance; *luminance énergétique, radiance*	L, (L_e)
radiant emittance; *émittance énergétique*	M, (M_e)
quantity of light; *quantité de lumière*	Q, (Q_V)
luminous flux; *flux lumineux*	Φ, (Φ_V)
luminous intensity; *intensité lumineuse:* $\mathrm{d}\Phi/\mathrm{d}\omega$	I, (I_V)
illuminance, illumination; *éclairement lumineux:* $\mathrm{d}\Phi/\mathrm{d}S$	E, (E_V)
luminance; *luminance:* $\mathrm{d}I/\mathrm{d}S \cos\vartheta$	L, (L_V)
luminous emittance; *émittance lumineuse:* $\mathrm{d}\Phi/\mathrm{d}S$	M, (M_V)
absorption factor; *facteur d'absorption:* Φ_a/Φ_0	α
reflection factor; *facteur de réflexion* Φ_r/Φ_0	ρ
transmission factor; *facteur de transmission* Φ_{tr}/Φ_0	τ
(linear) absorption coefficient; *coefficient d'absorption (linéique)*	a
(linear) extinction coefficient; *coefficient d'atténuation (linéique)*	μ
speed of light in empty space; *vitesse de la lumière dans le vide*	c
refractive index; *indice de réfraction:* $n = c/c_n$	n

††) Written in rationalized, 4-dimensional form. See Appendix, section 2.
†) The symbols between brackets are reserve symbols and are to be used whenever it is necessary to distinguish between radiation and light quantities.

7.7 *Acoustics.*

velocity of sound; *vitesse du son*	c
velocity of longitudinal waves; *vitesse d'ondes longitudinale*	c_l
velocity of transversal waves; *vitesse d'ondes transversale*	c_t
group velocity; *vitesse de groupe*	c_g
sound energy flux; *flux énergétique du son*	P
reflexion factor; *facteur de réflexion:* P_r/P_0	ρ
acoustic absorption factor; *facteur d'absorption acoustique* $1 - \rho$	α_a (α)
transmission factor; *facteur de transmission:* P_{tr}/P_c	τ
dissipation factor; *facteur de dissipation:* $\alpha_a - \tau$	δ
loudness level; *niveau d'isosonie*	L_N, (Λ)

7.8 *Quantum mechanics.*

complexe conjugate of Ψ; *complexe conjugué de* Ψ	Ψ^*
probability density; *densité de probabilité:* $P = \Psi^*\Psi$	P
probability current density: *densité de courant de probabilité:* $S = (\hbar/2im)(\Psi^*\Delta\Psi - \Psi\Delta\Psi^*)$	S
charge density of electrons; *charge volumique d'électrons:* $\rho = -eP$	ρ
electric current density of electrons; *densité de courant électrique d'électrons:* $j = -eS$	j
expectation value of A; *valeur moyenne de* A	$\langle A\rangle$, $\bar{A}$
commutator of A and B, *commutateur de A et B:* $[A,B] = AB - BA$	$[A,B]$, $[A,B]$
anticommutator of A and B; *anticommutateur de A et B:* $[A,B]_+ = AB + BA$	$[A,B]_+$
matrix element; *élément de matrice:* $A_{ij} = \int \psi_i^*(A\psi_j)\,d\tau$	A_{ij}
Hermitian conjugate of operator A; *conjugué Hermitien de l'opérateur A*	$A^\dagger$
momentum operator in coordinate representation; *opérateur de quantité de mouvement*	$+(\hbar/i)\Delta$
annihilation operators; *opérateurs d'annihilation*	a, b, α, β
creation operators; *opérateurs de création*	a^+, $b^\dagger$, $\alpha^\dagger$ $\beta^\dagger$

7.9 *Atomic and nuclear physics.*

mass number, nucleon number; *nombre de masse, nombre de nucléons*	A
atomic number, proton number; *nombre atomique, nombre de protons*	Z
neutron number; *nombre de neutrons:* $N = A - Z$	N
elementary charge (of position); *charge élémentaire* (*du position*)	e
electron mass; *masse de l'électron*	m, m_e
proton mass; *masse du proton*	m_p
neutron mass; *masse du neutron*	m_n
meson mass; *masse du méson*	m_π,
nuclear mass; *masse nucléaire* (of nucleus: ^AX)	m_N, m_N (^AX)
atomic; mass; *masse atomique* (of nuclide: ^AX)	m_a, m_a (^AX)
(unified) atomic mass constant; *constante* (*unifiée*) *de masse atomique:* $m_u = m_a(^{12}C)/12$	m_u
relative atomic mass; *masse atomique relative:* m_a/m_u	A_r
Planck constant; *constante de Planck* ($\hbar = h/2\pi$)	h
principal quantum number; *nombre quantique principal*	n, n_i
orbital angular momentum quantum number; *nombre quantique de moment angulaire orbital*	L, l_i
spin quantum number; *nombre quantique de spin*	S, s_i
angular momentum quantum number; *nombre quantique de moment angulaire* (including electron spin)	J, j_i
magnetic quantum number; — — *magnétique*	M, m_i
nuclear spin quantum number; — — *de spin nucléaire*	I, J *)
hyperfine quantum number; — — *hyperfin*	F
rotational quantum number; — — *de rotation*	J, K
vibrational quantum number; — — *de vibration*	v
quadrupole moment; *moment quadripolaire*	Q
Rydberg constant; *constante de Rydberg***)	R_∞
Bohr radius; *rayon de Bohr***)	a_0
fine structure constant; *constante de structure fine***)	α
mass excess; *excès de masse:* $m_a - Am_u$	Δ
packing fraction; *packing fraction:* Δ/Am_u	f
nuclear radius; *rayon nucléaire:* $R = r_0A^{1/3}$	R
magnetic moment of particle: *moment magnétique d'une particule*	μ
magnetic moment of proton; *moment magnétique du proton*	μ_p
magnetic moment of neutron; *moment magnétique du neutron*	μ_n
magnetic moment of electron; *moment magnétique électronique*	μ_e
Bohr magneton; *magnéton de Bohr***)	μ_B
nuclear magneton; *magnéton nucléaire*	μ_N
g-factor; *facteur g:* e.g. $g = \mu/I\mu_N$	g
gyromagnetic ratio; *rapport gyromagnétique***)	γ
Larmor (angular) frequency; *fréquence* (*angulaire*) *de Larmor***)	ω_L
level width; *largeur d'un niveau*	Γ

*) I is used in atomic physics. J in nuclear physics.
**) See for definition: Appendix, section 2.

mean life; *vie moyenne*	τ
reaction energy; *énergie de réaction*	Q
cross section; *section efficace*	σ
macroscopic cross section; *section efficace macroscopique:* $\Sigma = n\sigma$	Σ
impact parameter; *paramètre de collision*	b
scattering angle; *angle de diffusion*	$\vartheta, \theta, \varphi$
internal conversion coefficient; *coefficient de conversion interne*	α
disintegration energy; *énergie de désintégration*	Q
half life; *demi-vie*	$T_{1/2}$
reduced half life; *demi-vie réduite*	$fT_{1/2}$
decay constant, disintegration constant; *constante de désintégration*	λ
activity; *activité*	A
Compton wavelength; *longueur d'onde de Compton:* $\lambda_C = h/mc$	λ_C
electron radius; *rayon de l'électron***)	r_e
linear attenuation coefficient; *coeff. d'atténuation linéaire*	μ, μ_l
atomic attenuation coefficient; *coeff. d'atténuation atomique*	μ_a
mass attenuation coefficient; *coeff. d' atténuation massique*	μ_m
linear stopping power; *pouvoir d' arrêt linéaire*	S, S_l
atomic stopping power; *pouvoir d' arrêt atomique*	S_a
linear range; *distance de pénétration linéaire*	R, R_l
recombination coefficient; *coefficient de recombinaison*	α

7.10 Solid state physics.

fundamental translations for lattice; *translations fondamentales d'un réseau*	$\boldsymbol{a}, \boldsymbol{b}, \boldsymbol{c}$ $\boldsymbol{a}_1, \boldsymbol{a}_2, \boldsymbol{a}_3$
Miller indices; *indices de Miller*	h, k, l h_1, h_2, h_3
plane in lattice; *plan d'un réseau**)	(h, k, l) (h_1, h_2, h_3)
direction in lattice; *direction dans un réseau**)	$[h, k, l]$ $[h_1, h_2, h_3]$
fundamental translations in reciprocal lattice; *translations fondamentales de reseau réciproque*	$\boldsymbol{a}^*, \boldsymbol{b}^*, \boldsymbol{c}^*$ $\boldsymbol{b}_1, \boldsymbol{b}_2, \boldsymbol{b}_3$
vector in crystal lattice; *vecteur dans un réseau*	$\boldsymbol{r}$
distance between successive lattice planes; *distance de plans successifs d'un réseau*	d
Bragg angle; *angle de Bragg*	θ, ϑ
order of reflexion; *ordre du réflexion*	n
short range order parameter; *paramètre d'ordre* (*proche voisin*)	σ
long range order parameter; *paramètre d'ordre* (*longue distance*)	s
Burgers vector; *vecteur de Burgers*	$\boldsymbol{b}$
circular wave vector; propagation vector (of phonons); *vecteur d'onde circulaire, vecteur de propagation* (*de phonons*)	$\boldsymbol{q}$
circular wave vector, propagation vector (of particles); *vecteur d'onde circulaire, vecteur de propagation* (*de particules*)	$\boldsymbol{k}$
effective mass of electron; *masse effective d'une électron*	m^*, m_{eff}
Fermi energy; *énergie de Fermi*	E_F, ϵ_F
Fermi circular wave vector; *vecteur d'onde circulaire de Fermi*	$\boldsymbol{k}_F$
work function; *fonction de travail*	Φ
differential thermoelectric power; *force thermoélectrique différentielle*	$S, (\Sigma)$
Peltier coefficient; *coefficient de Peltier*	Π
Thomson coefficient; *coefficient de Thomson*	μ
piezoelectric coefficient; *coefficient piézoélectrique:* (polarization/stress)	d_{mn}
characteristic (Weiss) temperature; *température caractéristique* (*de Weiss*)	Θ, Θ_W
Curie temperature; *température de Curie*	T_C
Neel temperature; *température de Néel*	T_N
Hall coefficient; *coefficient de Hall*	R_H

7.11 Molecular spectroscopy.

qu.n. of component of electronic orbital angular momentum vector along symmetry axis; *n.qu. de la composante du moment angulaire orbital électronique suivant l'axe de symétrie*	Λ, λ_i
qu.n. of component of electronic spin along symmetry axis; *n.qu. de la composante du spin électronique suivant l'axe de symétrie*	Σ, σ_i
qu.n. of total electronic angular momentum vector along symmetry axis; *n.qu. du moment angulaire total électronique suivant l'axe de symétrie*	Ω, ω_i
qu.n. of electronic spin; *n.qu. du spin électronique*	S
qu.n. of nuclear spin; *n.qu. du spin nucléaire*	I
qu.n. of vibrational mode; *n.qu. d'un mode de vibration*	v
degeneracy of vibrational mode; *degré de dégénérescance d'une mode de vibration*	d
qu.n. of vibrational angular momentum; *n.qu. de moment angulaire vibrationnel* (L.M.)	l
qu.n. of total angular momentum; *n.qu. de moment angulaire total* (excluding nuclear spin)	J
qu.n. of component of $\boldsymbol{J}$ in direction of external field; *n.qu. de la composante de* $\boldsymbol{J}$ *dans la direction du champ extérieur*	M, M_J

*) Instead of the ordinary brackets ( ) and square brackets [ ], curly brackets { } and diamond brackets ⟨ ⟩ are also used respectively.

qu.n. of component of $\boldsymbol{S}$ in direction of external field; *n.qu. de la composante de **S** dans la direction du champ extérieur*	M_S
qu.n. of total angular momentum; *n.qu. du moment angulaire total* (including nuclear spin) $\boldsymbol{F} = \boldsymbol{J} + \boldsymbol{I}$	F
qu.n. of component of $\boldsymbol{F}$ in direction of external field; *n.qu. de la composante de **F** dans la direction du champ extérieur*	M_F
qu.n. of component of $\boldsymbol{I}$ in direction of external field; *n.qu. de la composante de **I** dans la direction du champ extérieur*	M_I
qu.n. of component of angular momentum along axis; *n.qu. de la composante du moment angulaire suivant l'axe* (L.M. and S.T.M.; excluding electron- and nuclear spin; for L.M.: $(K = \lvert\Lambda + l\rvert)$	K
qu.n. of total angular momentum; *n.qu. du moment angulaire total* (L.M. and S.T.M.; excluding electron and nuclear spin: $\boldsymbol{J} = \boldsymbol{N} + \boldsymbol{S}$*)	N
qu.n. of component of angular momentum along symmetry axis; *n.qu. de la composante du moment angulaire suivant l'axe de symétrie* (L.M. and S.T.M., excluding nuclear spin; for L.M.: $P = \lvert K + \Sigma\rvert$**)	P
electronic term; *terme électronique:* $T_e = E_e/hc$†)	T_e
vibrational term; *terme de vibration:* $G = E_{vibr}/hc$	G
coefficients in expression for vibrational term (for D.M.); *coefficients de l'expression d'un terme de vibration:* $G = \sigma_e(v + \frac{1}{2}) - x\sigma_e(v + \frac{1}{2})^2$	σ_e, $x\sigma_e$
coefficients in expression for vibrational term (for P.M.) *coefficients de l'expression d'un terme de vibration:* $G = \Sigma\sigma_i(v_i + \frac{1}{2}d_i) + \frac{1}{2}\Sigma\Sigma x_{ik}(v_i + \frac{1}{2}d_i)(v_k + \frac{1}{2}d_k)$	σ_i, x_{ik}
rotational term; *terme de rotation:* $F = E_{rot}/hc$	F
principal moments of inertia; *moments principaux d'inertie:* $I_A \leqslant I_B \leqslant I_C$††)	I_A, I_B, I_C
rotational constants; *constants de rotation:* $A = h/8\pi^2 cI_A$, etc.††)	A, B, C
total term; *terme total:* $T = T_e + G + F$	T

Remark: L.M. = linear molecules. *S.T.M.* = symmetric top molecules.
D.M. = diatomic molecules. P.M. = polyatomic molecules.
See for further details: Report on Notation for the Spectra of Polyatomic Molecules (Joint Commission for Spectroscopy of I.U.P.A.P. and I.A.U. 1954) J. Chem. Phys. *23* (1955) 1997.

7.12 Chemical physics.

amount of substance; *quantité de substance*	ν, n *)
molar mass of substance B; *masse molaire de la substance* B	M_B
molarity of subst. B; *molarité de la subst.* B	c_B
mole fraction of subst. B; *fraction molaire de la subst.* B	x_B
mass fraction of subst. B; *fraction massique de la subst.* B	w_B
volume fraction of subst. B; *fraction volumique de la subst.* B	φ_B
mole ratio of solution; *rapport molaire d'une solution*	r
molality of solution; *molalité d'une solution*	m
chemical potential of subst. B; *potentiel chimique de la subst.* B	μ_B
absolute activity of subst. B (dimensionless); *activité absolue de la subst.* B (*sans dimension*) $\lambda_B = \exp(\mu_B/kT)$	λ_B
relative activity; *activité relative*	a_B
activity coefficients; *coefficients d'activité*	γ_B, f_B
osmotic pressure; *pression osmotique*	Π
osmotic coefficient; *coefficient osmotique*	g, φ
stoichiometric number of molec. B; *coeff. stœchiométrique de molèc.* B	v_B
affinity; *affinité*	A
extent of reaction; *état d'avancement d'une réaction*	ξ
equilibrium constant; *constante d'équilibre*	K
charge number of ion; *électrovalence d'un ion*	z
Faraday constant; *constante de Faraday*	F
ionic strength; *force ionique*	I
activity of substance B; *activité de la substance* B: $z_B = (2\pi mkT/h^2)^{3/2}\lambda_B$	z_B

8. RECOMMENDED MATHEMATICAL SYMBOLS.

8.1 General symbols.

equal to; *égal à*	$=$
not equal to; *différent de*	$=\!\!\mid\!\!=$ $\neq$
identically equal to; *égal identiquement à*	$\equiv$
corresponds to; *correspond à*	$\hat{=}$

*) case of loosely coupled electron spin. **) case of tightly coupled electron spin
†) All energies are taken here with respect to the ground state as reference level.
††) For diatomic molecules use I and $A = h/8\pi^2 cI$.
*) n is used in chemistry, but v may be used as an alternative to n, when n is used for number density of particles.

approximately equal to; *égal environ à*	$\approx$
asymptotically equal to; *asymptotiquement égal à*	$\simeq$
proportional to; *proportionnel à*	$\propto$
approaches; *tend vers*	$\rightarrow$
larger than; *supérieur á*	$>$
smaller than; *inférieur á*	$<$
much larger than; *très supérieur á*	$\gg$
much smaller than; *très inférieur á*	$\ll$
larger than or equal to; *supérieur ou égal á*	$\geqq$ $\geq$
smaller than or equal to; *inférieur ou égal á*	$\leqq$ $\leq$
plus; *plus*	$+$
minus; *moins*	$-$
plus or minus; *plus ou moins*	$\pm$
a multiplied by *b*; *a multiplié par b*	ab, $a.b$, $a \cdot b$, $a \times b$
a divided by *b*; *a divisé par b*	a/b, $\frac{a}{b}$, ab^{-1}
a raised to the power *n*; *a puissance n*	a^n
magnitude of *a*; *valeur absolue de a*	$\|a\|$
square root of *a*; *racine carrée de a*	$\sqrt{\bar{a}}$, $\sqrt{a}$, $a^{\frac{1}{2}}$
mean value of *a*; *valeur moyenne de a*	$\bar{a}$ $\langle a \rangle$
factorial *p*; *factorielle p*	$p!$
binomial coefficient; *coefficient binomial: n!/p!(n − p)!*	$\binom{n}{p}$
infinity; *infini*	∞

8.2 *Letter symbols* and letter expressions for *mathematical operations* should be written in *roman* (or *upright*) type

exponential of *x*; *exponentielle de x*	$\exp x$, e^x
base of natural logarithms; *base des logarithmes népériens*	e
logarithm to the base *a* of *x*; *logarithme de base a de x*	$\log_a x$
natural logarithm of *x*; *logarithme népérien de x*	$\ln x$
common logarithm of *x*; *logarithme décimal de x*	$\lg x$, $\log x$ *)
binary logarithm of *x*; *logarithme binaire de x*	$\mathrm{lb}\, x$ $\log_2 x$
summation; *somme*	Σ
product; *produit*	Π
finite increase of *x*; *accroissement fini de x*	Δx **)
variation of *x*; *variation de x*	δx
total differential of *x*; *différentielle totale de x*	$\mathrm{d}x$
function of *x*; *fonction de x*	$f(x)$, $\mathrm{f}(x)$
limit of *f(x)*; *limite de f(x)*	$\lim f(x)$
Dirac delta function; *fonction delta de Dirac* $\delta(\boldsymbol{r}) = \delta(\mathrm{x})\delta(\mathrm{y})\delta(\mathrm{z})$	$\delta(x)$, $\delta(\boldsymbol{r})$
Kronecker delta symbol; *symbole delta de Kronecker*	δ_{ij}
Unit step function; *fonction unité*($\epsilon(x) = 1$ for $x > 0$, $\epsilon(x) = 0$ for $x < 0$)	$\epsilon(x)$ †)

8.3 *Trigonometric functions.*

sine of *x*, *sinus x*	$\sin x$
cosine of, *cosinus x*	$\cos x$
tangent of *x*, *tangente x*	$\tan x$, $\mathrm{tg}\, x$
cotangent of *x*, *cotangente x*	$\cot x$, $\mathrm{ctg}\, x$
secant of *x*, *sécante x*	$\sec x$
cosecant of *x*, *cosécante x*	$\mathrm{cosec}\, x$

Remarks:

a. It is recommended to use for the *inverse circular functions* the symbolic expressions for the corresponding circular function preceded by the letters: arc.
Examples: arcsin *x*, arccos *x*, arctan *x* or arctg *x*, etc. Sometimes the notation $\sin^{-1} x$, $\tan^{-1} x$, etc. is used.

b. It is recommended to use for the *hyperbolic functions* the symbolic expressions for the corresponding circular function, followed by the letter: h.
Examples: sinh *x*, cosh *x*, tanh *x* or tgh *x*, etc.

c. It is recommended to use for the *inverse hyperbolic functions* the symbolic expression for the corresponding hyperbolic function preceded by the letters: ar.

Example: arsinh *x*, arcosh *x*, etc.

8.4 *Complex quantities.*

imaginary unit; *unité imaginaire* ($\mathrm{i}^2 = -1$)	i, j
real part of *z*; *partie réelle de z*	Re *z*, z'
imaginary part of *z*; *partie imaginaire de z*	Im *z*, z''
modulus of *z*; *module de z*	$\|z\|$
argument of *z*; *argument de z* : $z = \|z\| \exp \mathrm{i}\varphi$	arg *z*, φ
complex conjugate of *z*, conjugate of *z*; *complexe conjugué de z, conjugué de z*	z^*

Remark: Sometimes the notation $\bar{z}$ is used for the complex conjugate of *z*.

*) In case of ambiguity $\log_{10} x$. **) Greek capital delta, not triangle.
†) $\Theta(t)$ is used for unit step function of time.

8.5 Symbols for special values of periodic quantities.

	Normal case		Exceptional case	
instantaneous value	x	x	x	x
r.m.s. value	x_{eff}	$\tilde{x}$	X	X
maximum value	x_{m}	$\hat{x}$	$x_{\mathrm{m}}, X_{\mathrm{m}}$	$\hat{x}, \hat{X}$
average value	x_{av}	$\bar{x}, \langle x \rangle$	x_{av}	$\bar{x}, \langle x \rangle$

Remarks:

a. The "normal" case is the case in which only a small or only a capital letter may be used for the quantity. The "exceptional" case is the case in which both small and capital letters may be used for the same quantity.

b. The minimum value of x may be indicated by $x_{\min}$ or $\check{x}$.

c. The r.m.s.value is defined as $\tilde{x}^2 = T^{-1} \int_0^T x\,(t)^2 \mathrm{d}t$.

d. The average value is defined as $\bar{x} = T^{-1} \int_0^T x\,(t)\,\mathrm{d}t$.

8.6 Vector calculus. (see also *1.2.3*)

absolute value; *valeur absolue*	$\lvert \boldsymbol{A} \rvert, A$
differential vector operator; *vecteur opérateur différentiel*	$\partial/\partial \mathbf{r}, \nabla$
gradient; *gradient*	grad $\varphi, \nabla\varphi$
divergence; *divergence*	div $\boldsymbol{A}, \nabla\cdot\boldsymbol{A}$
curl; *rotationnel*	curl $\boldsymbol{A}$, rot $\boldsymbol{A}, \nabla \times \boldsymbol{A}$
Laplacian; *Laplacien*	$\triangle\varphi, \nabla^2\varphi$
Dalembertian; *Dalembertien*	$\Box\,\varphi$

8.7 Matrix calculus.

transpose of matrix A; *matrice transposée de A:* $\tilde{A}_{ij} = A_{ji}$	$\tilde{A}$
complex conjugate of A; *matrice complexe conjugueé de A:* $A^*_{ij} = (A_{ij})^*$	A^*
Hermitian conjugate of A; *matrice conjugueé Hermitienne de A:* $A\dagger_{ji} = A_{ji}^*$	$A\dagger$

Pauli matrices; *matrices de Pauli:* — σ, $\sigma_x, \sigma_y, \sigma_z$

$$\sigma_x = \begin{vmatrix} 0 & 1 \\ 1 & 0 \end{vmatrix} \qquad \sigma_y = \begin{vmatrix} 0 & -\mathrm{i} \\ \mathrm{i} & 0 \end{vmatrix} \qquad \sigma_z = \begin{vmatrix} 1 & 0 \\ 0 & -1 \end{vmatrix}$$

$\sigma_1, \sigma_2, \sigma_3$

unit matrix; *matrice unité:* $I = \begin{vmatrix} 1 & 0 \\ 0 & 1 \end{vmatrix}$ — I

Dirac (4 × 4) matrices; (4 × 4) *matrices de Dirac* *)

$$\alpha_x = \begin{vmatrix} 0 & \sigma_x \\ \sigma_x & 0 \end{vmatrix} \qquad x_y = \begin{vmatrix} 0 & \sigma_y \\ \sigma_y & 0 \end{vmatrix} \qquad \alpha_z = \begin{vmatrix} 0 & \sigma_z \\ \sigma_z & 0 \end{vmatrix}$$

$\alpha, \alpha_x, \alpha_y, \alpha_z$

$$\beta = \begin{vmatrix} I & 0 \\ 0 & -I \end{vmatrix}$$

β

9. INTERNATIONAL SYMBOLS FOR UNITS.

9.1 Unit systems.

1. A *coherent system of units* is a system based on a certain set of "basic units" from which all "derived units" are obtained by multiplication or division without introducing numerical factors. In addition there are "dimensionless units," in particular the radian, symbol: rad, for plane angle and the steradian, symbol: sr, for solid angle.

2. The ***CGS system*** or cm-g-s system is a coherent system of units based on *three basic units* for the three basic quantities length, mass and time respectively:

centimetre	cm
gramme	g
second	s

*) Sometimes a different representation is used.

In the field of *mechanics* the following units of this system have special names and symbols, which have been approved by the Conférence Générale des Poids et Mesures:

l, b, h	centimetre; *centimètre*	cm
t	second; *seconde*	s
m	gramme; *gramme*	g
f, v	hertz; *hertz* (= s^{-1})	Hz
F	dyne; *dyne* (= g.cm/s^2)	dyn
E, U, W, A	erg; *erg* (= g.cm^2/s^2)	erg
p	microbar; *microbar* (= dyn/cm^2)	μbar
η	poise; *poise* (= dyn. s/cm^2)	P

In the field of *electricity and magnetism* several variants of the CGS unit system have been developed, in particular the *electrostatic CGS system* and *electromagnetic CGS system*. Special names and symbols for some of the units of the second system are:*)

*H**	oersted; *oersted* (= $cm^{-}.g^{1/2}.s^{-1}$)	Oe
*B**	gauss; *gauss* (= $cm^{-1/2}.g^{1/2}.s^{-1}$)	G
Φ*	maxwell; *maxwell* (= $cm^{3/2}.g^{1/2}s^{-1}$)	Mx

For further information about the units and unit systems in electricity and magnetism see appendix, section 2.

3. The ***MKSA system*** or m-kg-s-A system is a coherent system of units for mechanics, electricity and magnetism, based on *four basic units* for the four basic quantities length, mass, time and electric intensity:

metre	m
kilogramme	kg
second	s
ampere	A

Remark: The system based on these four units has been given the name *Giorgi system* by the International Electrotechenical Committee in 1958. The subsystem for mechanics, which is based on the first three units only, has the name *MKS system.*

The following units of the MKSA system have special names and symbols, which have been approved by the Conférence Générale des Poids et Measures:

l. b. h	metre; *mètre*	m
t	second; *seconde*	s
m	kilogramme; *kilogramme*	kg
v, f	hertz; *hertz* (= s^{-1})	Hz
F	newton; *newton* (= kg.m/s^2)	N
E	joule; *joule* (= kg.m^2/s^2)	J
P	watt; *watt* (= J/s)	W
I	ampere; *ampère*	A
Q	coulomb; *coulomb* (= A.s)	C
V	volt; *volt* (= W/A)	V
C	farad; *farad* (= C/V)	F
R	ohm; *ohm* (= V/A)	Ω
L	henry; *henry* (= V.s/A)	H
Φ	weber; *weber* (= V.s)	Wb
B	tesla; *tesla* (= Wb/m^2)	T

4. The ***degree Kelvin.*** In the field of *thermodynamics* one introduces an additional basic unit, corresponding to the basic quantity:

thermodynamic temperature, the unit being the degree Kelvin, symbol: °K
When the *customary temperature* is used, defined by $t = T - T_0$, where $T_0 = 273.15$°K, this is usually expressed in degree Celsius symbol: °C. For *temperature interval* the name degree, symbol: deg is often used, the indications "Kelvin" or "Celsius," indicating the zeropoint of the temperature scale used, being irrelevant in this case.

5. The ***candela.***

In the field of *photometry* one introduces an additional basic unit, corresponding to the basic quantity: *luminous intensity,* this unit being the candela, symbol: cd. Special names and symbols for units in this field are:

I	candela; *candéla*	cd
Φ	lumen; *lumen*	lm
E	lux; *lux* (= lm/m^2)	lx

*) See also Appendix.

6. The ***International System of Units.*** For the coherent system based on the six basic units:

metre	m	ampere	A
kilogramme	kg	degree Kelvin	°K
second	s	candela	cd

the name *International System of Units* has been recommended by the Conférence Générale des Poids et Mesures in 1960. The units of this system are called: *SI-units*

7. The ***mole.***

In the field of *chemical and molecular physics,* in addition to the basic quantities defined above having units defined by the Conférence Générale des Poids et Mesures, *amount of substance* is also treated as a basic quantity. The recommended basic unit is the mole, symbol: mol. The mole is defined as the amount of substance of a system, which contains the same number of molecules (or ions, or atoms, or electrons, as the case may be), as there are atoms in exactly 12 gramme of the pure carbon nuclide ^{12}C.

9.2 Other units.

l	angström; *angström*	Å
σ	barn; *barn* (= 10^{-24} cm^2)	b
V	litre; *litre*	l
$t, \tau, T_{\frac{1}{2}}$	minute; *minute*	min
$t, \tau, T_{\frac{1}{2}}$	hour; *heure*	h
$t, \tau, T_{\frac{1}{2}}$	day; *jour*	d
$t, \tau, T_{\frac{1}{2}}$	year; *année*	a
p	atmosphere; *atmosphère*	atm
E	kilowatt-hour; *kilowatt-heure*	kWh
Q	calorie; *calorie*	cal
E, Q	electronvolt; *électronvolt*	eV
m	tonne; *tonne* (= 1000 kg)	t
m_a	(unified) atomic mass unit; *unité de masse atomique (unifiée)*	u
p	bar; *bar* (= 10^6 dyn/cm^2)	bar
A	curie, *curie*	Ci

Remark: The (unified) atomic mass unit is defined as $\frac{1}{12}$th of the mass of an atom of the ^{12}C nuclide.

APPENDIX*). SYSTEMS OF QUANTITIES AND UNITS IN ELECTRICITY AND MAGNETISM.

The CGS unit system with three basic units and the MKSA unit system with four basic units correspond respectively to two different sets of equations in the field of electricity and magnetism, which are developed starting from three and from four basic quantities respectively. These systems are denoted as three and four "dimensional" systems of equations respectively.

1. Systems of equations with 3 basic quantities.

Three distinct sets of equations with three basic quantities**) have been developed in the field of electricity and magnetism. These are:

(1.a) *The "electrostatic system" of equations,* defining the electric charge on the basis of Coulomb laws for the force between two electric charges, by taking the permittivity in vacuo equal to a dimensionless quantity, the number unity.

(1.b) *The "electromagnetic system" of equations,* defining the electric current on the basis of the interaction law for the force between two electric current elements, by taking the permeability in vacuo equal to a dimensionless quantity, the number unity.

(1.c) *The "symmetrical system" or Gaussian system of equations,* using the electric quantities from system (1.a) and the magnetic quantities from system (1.b). As a result of combining the two sets of quantities the velocity of light in vacuo appears explicitly in some of the equations interrelating electric and magnetic quantities.

The equations in these three systems are usually written in the *"non-rationalized"* form, which is called "non-rationalized," because in these equations often factors 2π or 4π appear in situations not involving circular or spherical symmetry respectively. These equations are sometimes written in a *"rationalized"* form, in which these factors appear only in those equations, where they could be expected from the geometry of the system. When three dimensional equations are used (e.g. in theoretical physics), one commonly uses those of the *non-rationalized symmetrical system.*

2. Systems of equations with 4 basic quantities.

In the equations with four basic quantities at least one quantity of electric or magnetic nature is included in the basic set. In such a system the permittivity and the permeability in vacuo appear explicitly as physical quantities with dimension in the relevant equations.

* The S.U.N. commission, after reproducing in the previous chapters all the recommendations on symbols, units and nomenclature approved by the I.U.P.A.P., gives in this Appendix some factual information about existing systems of quantities and units in the field of electricity and magnetism.

** Often length, time and mass are chosen as basic quantities, but also other choices, e.g. length, time and energy or length, time and force have been used.

Two different sets of equations are in use:

(2.a) The *"non-rationalized system" of equations*, in which the factors 4π and 2π often appear at unexpected places.

(2.b) The *"rationalized system" of equations*, in which these factors only appear in those equations, where they could be expected from the geometry.

When four-dimensional equations are used, one commonly writes these equations in the rationalized form (2.b).

Some characteristic expressions of the non rationalized three-dimensional "symmetrical system" (1.c) and the corresponding four-dimensional equations in the non-rationalized form (2.a) and the rationalized form (2.b) are given in the table. The quantities of the three-dimensional "symmetrical system" of equations (1.c) have been indicated with an asterisk (*), those of the non-rationalized four-dimensional equations (2.a) with a prime (′), as far as they are different from those of the system (2.b).

Non-rationalized symmetric system with 3 basic quantities	Non-rational. system with 4 basic quantities	Rationalized system with 4 basic quantities
$c \text{ rot } \mathbf{E}^* = -\partial \mathbf{B}^*/\partial t$	$\text{rot } \mathbf{E} = -\partial \mathbf{B}/\partial t$	$\text{rot } \mathbf{E} = -\partial \mathbf{B}/\partial t$
$\text{div } \mathbf{D}^* = 4\pi\rho$	$\text{div } \mathbf{D}' = 4\pi\rho$	$\text{div } \mathbf{D} = \rho$
$\text{div } \mathbf{B}^* = 0$	$\text{div } \mathbf{B} = 0$	$\text{div } \mathbf{B} = 0$
$c \text{ rot } \mathbf{H}^* = 4\pi \mathbf{j}^* + \partial \mathbf{D}^*/\partial t$	$\text{rot } \mathbf{H}' = 4\pi \mathbf{j} + \partial \mathbf{D}'/\partial t$	$\text{rot } \mathbf{H} = \mathbf{j} + \partial \mathbf{D}/\partial t$
$\mathbf{F} = Q^*\mathbf{E}^* + Q^*\mathbf{v} \times \mathbf{B}^*/c$	$\mathbf{F} = Q\mathbf{E} + Q\mathbf{v} \times \mathbf{B}$	$\mathbf{F} = Q\mathbf{E} + Q\mathbf{v} \times \mathbf{B}.$
$w = (\mathbf{E}^*.\mathbf{D}^* + \mathbf{B}^*.\mathbf{H}^*)/8\pi$	$w = (\mathbf{E}.\mathbf{D}' + \mathbf{B}.\mathbf{H}')/8\pi$	$w = (\mathbf{E}.\mathbf{D} + \mathbf{B}.\mathbf{H})/2$
$\mathbf{S} = c(\mathbf{E}^* \times \mathbf{H}^*)/4\pi$	$\mathbf{S} = (\mathbf{E} \times \mathbf{H}')/4\pi$	$\mathbf{S} = \mathbf{E} \times \mathbf{H}$
$\mathbf{E}^* = -\text{grad } V^* - (1/c)\partial \mathbf{A}^*/\partial t$	$\mathbf{E} = -\text{grad } V - \partial \mathbf{A}/\partial t$	$\mathbf{E} = -\text{grad } V - \partial \mathbf{A}/\partial t$
$\mathbf{B}^* = \text{rot } \mathbf{A}^*$	$\mathbf{B} = \text{rot } \mathbf{A}$	$\mathbf{B} = \text{rot } \mathbf{A}$
$\epsilon_r \mathbf{E}^* = \mathbf{D}^*$	$\epsilon_0' \epsilon_r \mathbf{E} = \epsilon' \mathbf{E} = \mathbf{D}'$	$\epsilon_0 \epsilon_r \mathbf{E} = \epsilon \mathbf{E} = \mathbf{D}$
$\mathbf{E}^* = \mathbf{D}^* - 4\pi \mathbf{P}^*$	$\epsilon_0' \mathbf{E} = \mathbf{D}' - 4\pi \mathbf{P}$	$\epsilon_0 \mathbf{E} = \mathbf{D} - \mathbf{P}$
$\mathbf{B}^* = \mu_r \mathbf{H}^*$	$\mathbf{B} = \mu' \mathbf{H}' = \mu_0' \mu_r \mathbf{H}'$	$\mathbf{B} = \mu \mathbf{H} = \mu_0 \mu_r \mathbf{H}$
$\mathbf{B}^* = \mathbf{H}^* + 4\pi \mathbf{M}^*$	$\mathbf{B} = \mu_0'(\mathbf{H}' + 4\pi \mathbf{M})$	$\mathbf{B} = \mu_0(\mathbf{H} + \mathbf{M})$
$a_0 = \hbar^2/me^{*2}$	$a_0 = \hbar^2 \epsilon_0'/me^2$	$a_0 = \hbar^2 4\pi\epsilon_0/me^2$
$hcR_\infty = e^{*2}/2a_0$	$hcR_\infty = e^2/2\epsilon_0' a_0$	$hcR_\infty = e^2/8\pi\epsilon_0' a_0$
$\mu_B^* = e^*\hbar/2mc$	$\mu_B = e\hbar/2m$	$\mu_B = e\hbar/2m$
$\gamma^* = g(e^*/2mc)$	$\gamma = g(e/2m)$	$\gamma = g(\epsilon/2m)$
$\omega_L = (e^*/2mc)B^* = \mu_B^* B^*/\hbar$	$\omega_L = (e/2m)B = \mu_B B/\hbar$	$\omega_L = (e/2m)B = \mu_B B/\hbar$
$\alpha = e^{*2}/\hbar c$	$\alpha = e^2/\epsilon_0' \hbar c$	$\alpha = e^2/4\pi\epsilon_0 \hbar c$
$r_e = e^{*2}/mc^2$	$r_e = e^2/\epsilon_0' mc^2$	$r_e = e^2/4\pi\epsilon_0 mc^2$

Remark on rationalization:

The basis of these considerations in which the rationalization is connected with the writing of the equations between physical quantities is in agreement with the resolution accepted by the I.U.P.A.P. in 1951 in Copenhagen:

"The General Assembly of the Union of Physics considers that, in the case that the equations are rationalized, the rationalization should be effected by the introduction of new quantities."

("L'Assemblée Générale de l'Union de Physique considère que, dans le cas où les équations sont rationalisées, la rationalisation doit étre obtenue par l'introduction de grandeurs nouvelles.")

3. *CGS system of units.*

(3.a) The *electrostatic CGS system of units* forms a coherent system of units in combination with the three-dimensional "electrostatic system" of equations (1.a).

(3.b) The *electromagnetic CGS system of units* forms a coherent system of units in combination with the three-dimensional "electromagnetic system" of equations (1.b).

(3.c) The *mixed CGS system of units*, called *Gaussian units*, consisting of the set of electric units of the electrostatic CGS system on the one hand and the set of magnetic units of the electromagnetic CGS system on the other hand, form together also a coherent system of units, when used in combination with the three-dimensional "symmetrical system" of equations (1.c).

4. *MKSA system of units.*

The MKSA system of units forms a coherent system of units in either of the four-dimensional systems of equations mentioned under 2.

The MKSA system is, however, most commonly used together with the rationalized (four-dimensional) equations (2.b).

5. *centimetre-gramme-second-franklin system and centimetre-gramme-second-biot system*

Several investigators have pointed out over many years the advantages, which result from the use of the four-dimensional equations, i.e. equations with four basic quantities from which at least one is of electrical nature. These advantages are partly of a didactical nature, but many investigators consider the usage of four basic quantities also important for developing a clear representation of the field of electricity and magnetism.

It has often been considered as a disadvantage that the transition from the three- to the four-dimensional system of equations should be tied to the transition from the CGS system of units to the MKSA system of units as explained under 3, and 4, which as a consequence leads to a change in the numerical value of many well known physical quantities.

APPENDIX*). SYSTEMS OF QUANTITIES AND UNITS IN ELECTRICITY AND MAGNETISM (Continued)

For that reason several investigators have advocated the introduction of a four-dimensional system of units, which is a generalization of the CGS system. This unit-system is chosen so that to any unit of the CGS system (with three basic units) there corresponds one particular unit of this "generalized CGS system" (with four basic units), such that the *numerical* values of the physical quantities in the field of electricity are invariant.

The introduction of such a "generalized CGS system" has the advantage that the relation between the units of this "generalized CGS system" and the units of the MKSA system can be expressed by ordinary conversion equations.

The use of the four-dimensional system of equations (with four basic quantities) accompanied by the usage of such a "generalized CGS system" of units (with four basic units) is therefore of advantage as an intermediate representation in this period of coexistence of the CGS system and the MKSA system, where transitions from one system to the other often have to be made.

As a result of all these considerations the General Assembly of the IUPAP, Copenhagen, 1951, approved by its resolution 5 the introduction of the following two "generalized CGS systems" based on four basic units:

(5.a) The system with the *centimetre, gramme, second* and the charge corresponding to one *electrostatic unit*, as basic units, being a 4-dimensional generalization of the electrostatic CGS system (3.a)

(5.b) The system with the *centimetre, gramme second* and the current corresponding to one *electromagnetic unit* or deca-ampere, as basic units, being a 4-dimensional generalization of the electromagnetic CGS system (3.b)

In practical applications of these systems it is of advantage to indicate the *basic electrical unit* in each of these two systems with a name and symbol.

The name franklin was proposed in 1941 for the electrostatic unit of electric charge considered as a basic unit of the system (5.a). The franklin, symbol: Fr, is thus defined as:

The *franklin* is that charge, which exerts on an equal charge at a distance of 1 centimetre in vacuo a force of 1 dyne.

According to this definition 1 franklin = $(10/\zeta)$ coulomb, where $\zeta = 2.997925 \times 10^{10}$ is the numerical value of the velocity of light in vacuum, measured in cm/s.

For the electromagnetic unit of electric current, considered as a basic unit of the system (5.b), the name biot has been used. The biot, symbol: Bi, is thus defined as:

The *biot* is that constant current intensity, which, when maintained in two parallel infinitely long rectilinear conductors of infinite length and of negligible circular section, placed at a mutual distance of 1 centimetre apart in vacuo, would produce between these conductors a force of 2 dyne per centimetre length.

According to this definition 1 biot = 10 ampere).

These unit systems (5.a) and (5.b) are referred to as cm-g-s-Fr system and cm-g-s-Bi system respectively.

The definitions of the franklin in the system (5.a) and that of the biot in the system (5.b) which closely correspond to the electrostatic CGS unit of charge and the electromagnetic CGS unit of current respectively, ensure that the quantities of the *non-rationalized* four-dimensional system of equations (second column, table p. 27), when expressed in these units, have the same numerical value as the corresponding quantities of the non-rationalized three-dimensional system of equations (first column, table p. 27), when expressed in the corresponding CGS units.

Some of the units of the three-dimensional CGS system and the corresponding units of the two four-dimensional systems are given in the table:

	three-dimensional unit of mixed CGS system		corresponding four-dimensional unit of generalized CGS system
$\boldsymbol{Q}^*$	$\mathrm{erg}^{\frac{1}{2}} \cdot \mathrm{cm}^{\frac{1}{2}}$	$\boldsymbol{Q}$	Fr
$\boldsymbol{I}^*$	$\mathrm{erg}^{\frac{1}{2}} \cdot \mathrm{cm}^{\frac{1}{2}}\ \mathrm{s}^{-1\frac{1}{2}}$	$\boldsymbol{I}$	Fr/s
$\boldsymbol{V}^*$	$\mathrm{erg}^{\frac{1}{2}} \cdot \mathrm{cm}^{-\frac{1}{2}}$	$\boldsymbol{V}$	erg/Fr
$\boldsymbol{E}^*$	$\mathrm{erg}^{\frac{1}{2}} \cdot \mathrm{cm}^{-\frac{3}{2}}$	$\boldsymbol{E}$	dyn/Fr
$\boldsymbol{D}^*$	$\mathrm{erg}^{\frac{1}{2}} \cdot \mathrm{cm}^{-\frac{3}{2}}$	$\boldsymbol{D}'$	$\mathrm{Fr/cm^2}$
$\boldsymbol{C}^*$	cm	$\boldsymbol{C}$	$\mathrm{Fr^2/erg}$
ϵ_r	1	ϵ'	$\mathrm{Fr^2/erg \cdot cm}$
$\boldsymbol{B}^*$	G	$\boldsymbol{B}$	dyn/Bi · cm
$\boldsymbol{H}^*$	Oe	$\boldsymbol{H}'$	Bi/cm
μ_r	1	μ'	$\mathrm{dyn/Bi^2}$

ALPHABET TABLE

Greek letter	Greek name	English equivalent
Α α	Alpha	(ä)
Β β	Beta	(b)
Γ γ	Gamma	(g)
Δ δ	Delta	(d)
Ε ϵ	Epsilon	(e)
Ζ ζ	Zeta	(z)
Η η	Eta	(ā)
Θ θ	Theta	(th)
Ι ι	Iota	(ē)
Κ κ	Kappa	(k)
Λ λ	Lambda	(l)
Μ μ	Mu	(m)
Ν ν	Nu	(n)
Ξ ξ	Xi	(ks)
Ο ο	Omicron	(ŏ)
Π π	Pi	(p)
Ρ ρ	Rho	(r)
Σ σ ς	Sigma	(s)
Τ τ	Tau	(t)
Υ υ	Upsilon	(ü, ōō)
Φ ϕ	Phi	(f)
Χ χ	Chi	(H)
Ψ ψ	Psi	(ps)
Ω ω	Omega	(ō)

RUSSIAN letter	English equivalent
А а	(ä)
Б б	(b)
В в	(v)
Г г	(g)
Д д	(d)
Е е	(ye)
Ж ж	(zh)
З з	(z)
И и	(i, ē)
Й й	(ĕ) 7
К к	(k)
Л л	(l)
М м	(m)
Н н	(n)
О о	(ô, o)
П п	(p)
Р р	(r)
С с	(s)
Т т	(t)
У у	(ōō)
Ф ф	(f)
Х х	(kh)
Ц ц	(ts)
Ч ч	(ch)
Ш ш	(sh)
Щ щ	(shch)
Ъ ъ	8
Ы ы	(ē)
Ь ь	9
Э э	(e)
Ю ю	(ū)
Я я	(yä)

AMERICAN STANDARD ABBREVIATIONS FOR SCIENTIFIC AND ENGINEERING TERMS

Reproduced by permission of The American Society of Mechanical Engineers

Introductory Notes

Scope and Purpose

1. The Executive Committee of the Sectional Committee on Scientific and Engineering Symbols and Abbreviations has made the following distinction between symbols and abbreviations: Letter symbols are letters used to represent magnitudes of physical quantities in equations and mathematical formulas. Abbreviations are shortened forms of names or expressions employed in texts and tabulations, and should not be used in equations.

Fundamental Rules

2. Abbreviations should be used sparingly in text and with due regard to the context and to the training of the reader. Terms denoting units of measurement should be abbreviated in the text only when preceded by the amounts indicated in numerals; thus "several inches," "one inch," "12 in." In tabular matter, specifications, maps, drawings, and texts for special purposes, the use of abbreviations should be governed only by the desirability of conserving space.

3. Short words such as ton, day, and mile should be spelled out.

4. Abbreviations should not be used where the meaning will not be clear. In case of doubt, spell out.

5. The same abbreviation is used for both singular and plural, as "bbl" for barrel and barrels.

6. The use of conventional signs for abbreviations in text is not recommended; thus "per," not /; "lb," not #; "in," not ″. Such signs may be used sparingly in tables and similar places for conserving space.

7. The period should be omitted except in cases where the omission would result in confusion.

8. The letters of such abbreviations as ASA should not be spaced (not A S A).

9. The use in text of exponents for the abbreviations of square and cube and of the negative exponents for terms involving "per" is not recommended. The superior figures are usually not available on the keyboards of typesetting and linotype machines and conposition is therefore delayed. There is also the liklihood of confusion with footnote reference numbers. These shorter forms are permissible in tables and are sometimes difficult to avoid in text.

10. A sentence should not begin with a numeral followed by an abbreviation. Abbreviations for names of units are to be used only after numerical values. such as 25 ft or 110 v.

Abbreviations*

absolute .... abs
acre .... spell out
acre-foot .... acre-ft
air horsepower .... air hp
alternating-current (as adjective) .... a-c
ampere .... amp
ampere-hour .... amp-hr
amplitude, an elliptic function .... am.
Angstrom unit .... A
antilogarithm .... antilog
atmosphere .... atm
atomic weight .... at. wt
average .... avg
avoirdupois .... avdp
azimuth .... az or α

barometer .... bar.
barrel .... bbl
Baumé .... Bé
board feet (feet board measure) .... fbm
boiler pressure .... spell out
boiling point .... bp
brake horsepower .... bhp
brake horsepower-hour .... bhp-hr
Brinell hardness number .... Bhn
British thermal unit[1] .... Btu or B
bushel .... bu

calorie .... cal
candle c
candle-hour .... c-hr
candlepower .... cp
cent .... c or ¢
center to center .... c to c
centigram .... cg
centiliter cl
centimeter .... cm
centimeter-gram-second (system) .... cgs
chemical .... chem
chemically pure .... cp
circular .... cir
circular mils .... cir mils
coefficient .... coef
cologarithm .... colog
concentrate .... conc
conductivity .... cond
constant .... const
continental horsepower .... cont hp
cord .... cd
cosecant .... csc
cosine .... cos
cosine of the amplitude, an elliptic function .... cn
cost, insurance, and freight .... cif
cotangent .... cot
coulomb .... spell out
counter electromotive force .... cemf
cubic .... cu
cubic centimeter ... cu cm, cm^3 (liquid, meaning milliliter. ml)
cubic foot .... cu ft
cubic feet per minute .... cfm
cubic feet per second .... cfs
cubic inch .... cu in.
cubic meter .... cu m or m^3
cubic micron .... cu μ or cu mu or μ^3
cubic millimeter .... cu mm or mm^3
cubic yard .... cu yd
current density .... spell out
cycles per second .... spell out or c
cylinder .... cyl

day .... spell out
decibel .... db
degree[1] .... deg or °
degree centigrade C
degree Fahrenheit .... F
degree Kelvin .... K
degree Réaumur .... R
delta amplitude, an elliptic function .... dn
diameter .... diam
direct-current (as adjective) .... d-c
dollar .... $
dozen .... doz
dram .... dr

* These forms are recommended for readers whose familiarity with the terms used makes possible a maximum of abbreviations. For other classes of readers editors may wish to use less contracted combinations made up from this list. For example, the list gives the abbreviation of the term "feet per second" as "fps." To some readers ft per sec will be more easily understood.

[1] Abbreviation recommended by the A.S.M.E. Power Test Codes Committee. B = 1 Btu, kB = 1000 Btu, mB = 1,000,000 Btu. The A.S.H.&V.E. recommends the use of Mb = 1000 Btu and Mbh = 1000 Btu per hr.

AMERICAN STANDARD ABBREVIATIONS (Continued)

efficiency	eff
electric	elec
electromotive force	emf
elevation	el
equation	eq
external	ext
farad	spell out or f
feet board measure (board feet)	fbm
feet per minute	fpm
feet per second	fps
fluid	fl
foot	ft
foot-candle	ft-c
foot-Lambert	ft-L
foot-pound	ft-lb
foot-pound-second (system)	fps
foot-second (see cubic feet per second)	
franc	fr
free aboard ship	spell out
free alongside ship	spell out
free on board	fob
freezing point	fp
frequency	spell out
fusion point	fnp
gallon	gal
gallons per minute	gpm
gallons per second	gps
grain	spell out
gram	g
gram-calorie	g-cal
greatest common divisor	gcd
haversine	hav
hectare	ha
henry	H
high-pressure (adjective)	h-p
hogshead	hhd
horsepower	hp
horsepower-hour	hp-hr
hour	hr
hour (in astronomical tables)	h
hundred	C
hundredweight (112 lb)	cwt
hyperbolic cosine	cosh
hyperbolic sine	sinh
hyperbolic tangent	tanh
inch	in.
inch-pound	in-lb
inches per second	ips
indicated horsepower	ihp
indicated horsepower-hour	ihp-hr
inside diameter	ID
intermediate-pressure (adjective)	i-p
internal	int
joule	J
kilocalorie	kcal
kilocycles per second	kc
kilogram	kg
kilogram-calorie	kg-cal
kilogram-meter	kg-m
kilograms per cubic meter	kg per cu m or kg/m^3
kilograms per second	kgps
kiloliter	kl
kilometer	km
kilometers per second	kmps
kilovolt	kv
kilovolt-ampere	kva
kilowatt	kw
kilowatthour	kwhr
lambert	L
latitude	lat or ϕ
least common multiple	lcm
linear foot	lin ft
liquid	liq
lira	spell out
liter	l
logarithm (common)	log
logarithm (natural)	$\log_e$ or ln
longitude	long. or λ
low-pressure (as adjective)	l-p
lumen	l*
lumen-hour	l-hr*
lumens per watt	lpw
mass	spell out
mathematics (ical)	math
maximum	max
mean effective pressure	mep
mean horizontal candlepower	mhcp
megacycle	spell out
megohm	spell out
melting point	mp
meter	m
meter-kilogram	m-kg
mho	spell out
microampere	μa or mu a
microfarad	μf
microinch	μin.
micromicrofarad	$\mu\mu$f
micromicron	$\mu\mu$ or mu mu
micron	μ or mu
microvolt	μv
microwatt	μw or mu w
mile	spell out
miles per hour	mph
miles per hour per second	mphps
milliampere	ma
milligram	mg
millihenry	mh
millilambert	mL
milliliter	ml
millimeter	mm
millimicron	mμ or m mu
million	spell out
million gallons per day	mgd
millivolt	mv
minimum	min
minute	min
minute (angular measure)	′
minute (time)(in astronomical tables)	m
mole	spell out
molecular weight	mol. wt
month	spell out
National Electrical Code	NEC
ohm	spell out or Ω
ohm-centimeter	ohm-cm
ounce	oz
ounce-foot	oz-ft

[1] There are circumstances under which one or the other of these forms is preferred. In general the sign ° is used where space conditions make it necessary, as in tabular matter, and when abbreviations are cumbersome, as in some angular measurements, i.e., 59° 23′ 42″. In the interest of simplicity and clarity the Committee has recommended that the abbreviation for the temperature scale, F, C, K, etc., always be included in expressions for numerical temperatures, but, wherever feasible, the abbreviation for "degree" be omitted; as 69 F.

* The International Commission on Illumination has changed the symbol for lumen to lm, and the symbol for lumen-hour to lm-hr. This nomenclature is used in American Standard for Illuminating Engineering Nomenclature and Photometric Standards (ASA Z7.1–1942).

AMERICAN STANDARD ABBREVIATIONS (Continued)

ounce-inch	oz-in.
outside diameter	OD
parts per million	ppm
peck	pk
penny (pence – new British)	p.
pennyweight	dwt
per	(See Fundamental Rules)
peso	spell out
pint	pt
potential	spell out
potential difference	spell out
pound	lb
pound-foot	lb-ft
pound-inch	lb-in.
pound sterling	£
pounds per brake horsepower-hour	lb per bhp-hr
pounds per cubic foot	lb per cu ft
pounds per square foot	psf
pounds per square inch	psi
pounds per square inch absolute	psia
power factor	spell out or pf
quart	qt
radian	spell out
reactive kilovolt-ampere	kvar
reactive volt-ampere	var
revolutions per minute	rpm
revolutions per second	rps
rod	spell out
root mean square	rms
secant	sec
second	sec
second (angular measure)	″
second-foot (see cubic feet per second)	
second (time)(in astronomical tables)	s
shaft horsepower	shp
shilling	s
sine	sin
sine of the amplitude, an elliptic function	sn
specific gravity	sp gr
specific heat	sp ht
spherical candle power	scp
square	sq
square centimeter	sq cm or cm^2
square foot	sq ft
square inch	sq in.
square kilometer	sq km or km^2
square meter	sq m or m^2
square micron	sq μ or sq mu or μ^2
square millimeter	sq mm or mm^2
square root of mean square	rms
standard	std
stere	s
tangent	tan
temperature	temp
tensile strength	ts
thousand	M
thousand foot-pounds	kip-ft
thousand pound	kip
ton	spell out
ton-mile	spell out
versed sine	vers
volt	v
volt-ampere	va
volt-coulomb	spell out
watt	w
watthour	whr
watts per candle	wpc
week	spell out
weight	wt
yard	yd
year	yr

ABBREVIATIONS OF COMMON UNITS OF WEIGHT AND MEASURE

From NBS Miscellaneous Publication No. 233.

Spelling and Abbreviations of Units

The spelling of the names of units as adopted by the National Bureau of Standards is that given in the list below. The spelling of the metric units is in accordance with that given in the law of July 28, 1866, legalizing the metric system in the United States.

Following the name of each unit in the list below is given the abbreviation which the Bureau has adopted. Attention is particularly called to the following principles:

1. The period is omitted after all abbreviations of units, except where the abbreviation forms an English word.
2. The exponents "²" and "³" are used to signify "square" and "cubic," respectively, instead of the abbreviations "sq" or "cu," which are, however, frequently used in technical literature for the United States customary units. In conformity with this principle the abbreviation for cubic centimeter is "cm^3" (instead of "cc" or "c cm"). The term "cubic centimeter," as used in chemical work, is, in fact, a misnomer, since the unit actually used is the "milliliter" of which "ml" is the correct abbreviation.
3. The use of the same abbreviation for both singular and plural is recommended. This practice is already established in expressing metric units and is in accordance with the spirit and chief purpose of abbreviations.
4. It is also suggested that, unless all the text is printed in capital letters, only small letters be used for abbreviations, except in such case as, A for angstrom, etc., where the use of capital letters is general.

LIST OF THE MOST COMMON UNITS OF WEIGHT AND MEASURE AND THEIR ABBREVIATIONS

Unit	Abbreviation	Unit	Abbreviation
		kiloliter	kl
acre	acre	kilometer	km
angstrom	A	link	li
are	a	liquid	liq
avoirdupois	avdp	liter	liter
barrel	bbl	meter	m
board foot	fbm	metric ton	t
bushel	bu	microgram*	μg
carat	c	microinch	$\mu in.$
centare	ca	microliter*	μl
centigram	cg	micron	μ
centiliter	cl	mile	mi
centimeter	cm	milligram	mg
chain	ch	milliliter	ml
cubic centimeter	cm^3	millimeter	mm
cubic decimeter	dm^3	millimicron	$m\mu$
cubic dekameter	dkm^3	minim	min *or* ♏
cubic foot	ft^3	ounce	oz
cubic hectometer	hm^3	ounce, apothecaries	oz ap *or* ℥
cubic inch	$in.^3$	ounce, avoirdupois	oz avdp
cubic kilometer	km^3	ounce, fluid	fl oz
cubic meter	m^3	ounce, troy	oz t
cubic mile	mi^3	peck	pk
cubic millimeter	mm^3	pennyweight	dwt
cubic yard	yd^3	pint	pt
decigram	dg	pound	lb
deciliter	dl	pound, apothecaries	lb ap
decimeter	dm	pound, avoirdupois	lb avdp
decistere	ds	pound, troy	lb t
dekagram	dkg	quart	qt
dekaliter	dkl	rod	rd
dekameter	dkm	scruple, apothecaries	s ap *or* ℈
dekastere	dks	square centimeter	cm^2
dram	dr	square chain	ch^2
dram, apothecaries	dr ap *or* ʒ	square decimeter	dm^2
dram, avoirdupois	dr avdp	square dekameter	dkm^2
dram, fluid	fl dr	square foot	ft^2
fathom	fath	square hectometer	hm^2
foot	ft	square inch	$in.^2$
furlong	fur.	square kilometer	km^2
gallon	gal	square link	li^2
grain	grain	square meter	m^2
gram	g	square mile	mi^2
hectare	ha	square millimeter	mm^2
hectogram	hg	square rod	rd^2
hectoliter	hl	square yard	yd^2
hectometer	hm	stere	s
hogshead	hhd	ton	ton
hundredweight	cwt	ton, metric	t
inch	in.	troy	t
kilogram	kg	yard	yd

* The abbreviations γ and λ for microgram and microliter, respectively, have been advocated by some authorities.

CONVERSION FACTORS

(From various U. S. Government and IUPAC publications and from calculations based on values given in these publications)

To convert from	To	Multiply by
Abamperes	Amperes	10
"	E.M. cgs. units of current	1
"	E.S. cgs. units	2.997930×10^{10}
"	Faradays (chem.)/sec	1.036377×10^{-4}
"	Faradays (phys.)/sec	1.036086×10^{-4}
"	Statamperes	2.997930×10^{10}
Abamperes/cm	E.M. cgs. units of surface charge density	1
"	E.S. cgs. units	2.997930×10^{10}
Abamperes/sq. cm	Amperes/circ. mil	5.0670748×10^{-5}
"	Amperes/sq. cm	10
"	Amperes/sq. inch	64.516
Abampere-turns	Ampere-turns	10
Abampere-turns/cm	Ampere-turns/cm	10
Abcoulombs	Ampere-hours	0.0027777
"	Coulombs	10
"	Electronic charges	6.24196×10^{19}
"	E.M. cgs. units of charge	1
"	E.S. cgs. units	2.997930×10^{10}
"	Faradays (chem.)	1.036377×10^{-4}
"	Faradays (phys.)	1.036086×10^{-4}
"	Statcoulombs	2.997930×10^{10}
Abfarads	E.M. cgs. units of capacitance	1
"	E.S. cgs. units	8.987584×10^{20}
"	Farads	1×10^{9}
"	Microfarads	1×10^{15}
"	Statfarads	8.987584×10^{20}
Abhenries	E.M. cgs. units of induction	1
"	E.S. cgs. units	1.112646×10^{-21}
"	Henries	1×10^{-9}
Abmhos	E.M. cgs. units of conductance	1
"	E.S. cgs. units	8.987584×10^{20}
"	Megamhos	1000
"	Mhos	1×10^{9}
"	Statmhos	8.987584×10^{20}
Abohms	E.M. cgs. units of resistance	1
"	Megohms	1×10^{-15}
"	Microhms	0.001
"	Ohms	1×10^{-9}
"	Statohms	1.112646×10^{-21}
Abohm-cm	Circ. mil-ohms/ft	0.0060153049
"	E.M. cgs. units of resistivity	1
"	Microhm-inches	0.00039370079
"	Ohm-cm	1×10^{-9}
Abvolts	Microvolts	0.01
"	Millivolts	1×10^{-5}
"	Volts	1×10^{-8}
"	Volts (Int.)	9.99670×10^{-9}
Abvolts/cm	E.M. cgs. units of electric field intensity	1
"	E.S. cgs. units	3.335635×10^{-11}
"	Volts/cm	1×10^{-8}
"	Volts/inch	2.54×10^{-8}
"	Volts/meter	1×10^{-6}
Acres	Sq. cm	40468564
"	Sq. ft	43560
"	Sq. ft. (U.S. Survey)	43559.826
"	Sq. inches	6272640
"	Sq. kilometers	0.0040468564
"	Sq. links (Gunter's)	1×10^{5}
"	Sq. meters	4046.8564
"	Sq. miles (statute)	0.0015625
"	Sq. perches	160
"	Sq. rods	160
"	Sq. yards	4840
Acre-feet	Cu. feet	43560
"	Cu. meters	1233.4818
"	Cu. yards	1613.333
Acre-inches	Cu. feet	3630
"	Cu. meters	102.79033

To convert from	To	Multiply by
Acre-inches	Gallons (U.S.)	27154.286
Amperes	Abamperes	0.1
"	Amperes (Int.)	1.000165
"	Cgs. units of current	1
"	Mks. units of current	1
"	Coulombs/sec	1
"	Coulombs (Int.)/sec	1.000165
"	Faradays (chem.)/sec	1.036377×10^{-5}
"	Faradays (phys.)/sec	1.036086×10^{-5}
"	Statamperes	2.997930×10^{9}
Amperes (Int.)	Amperes	0.999835
"	Coulombs/sec	0.999835
"	Coulombs (Int.)/sec	1
"	Faradays (chem.)/sec*	1.03623×10^{-5}
"	Faradays (phys.)/sec*	1.03592×10^{-5}
Amperes/meter	Cgs. units of surface current density	0.01
"	E.M. cgs. units	0.001
"	E.S. cgs. units	2.997930×10^{7}
"	Mks. units	1
Amperes/sq. meter	Cgs. units of volume current density	0.0001
"	E.M. cgs. units	1×10^{-5}
"	E.S. cgs. units	299793.0
"	Mks. units	1
Amperes/sq. mil	Abamperes/sq. cm	15500.031
"	Amperes/sq. cm	1.5500031×10^{5}
Ampere-hours	Abcoulombs	360
"	Coulombs	3600
"	Faradays (chem.)*	0.0373096
"	Faradays (phys.)*	0.0372991
Ampere-turns	Cgs. units of magnetomotive force	1.2566371
"	E.M. cgs. units	1.2566371
"	E.S. cgs. units	3.767310×10^{10}
"	Gilberts	1.2566371
Ampere-turns/weber	Cgs. units of reluctance	1.256637×10^{-8}
"	E.M. cgs. units	1.256637×10^{-8}
"	E.S. cgs. units	1.129413×10^{13}
"	Gilberts/maxwell	1.256637×10^{-8}
Ångström units	Centimeters	1×10^{-8}
"	Inches	3.9370079×10^{-9}
"	Microns	0.0001
"	Millimicrons	0.1
"	Wave length of orange-red line of krypton 86	0.000165076373
"	Wave length of red line of cadmium	0.000155316413
Ares	Acres	0.024710538
"	Sq. dekameters	1
"	Sq. feet	1076.3910
"	Sq. ft. (U.S. Survey)	1076.3867
"	Sq. meters	100
"	Sq. miles	3.8610216×10^{-5}
Atmospheres	Bars	1.01325
"	Cm. of Hg (0°C.)	76
"	Cm. of H_2O (4°C.)	1033.26
"	Dynes/sq. cm	1.01325×10^{6}
"	Ft. of H_2O (39.2°F.)	33.8995
"	Grams/sq. cm	1033.23
"	In. of Hg (32°F.)	29.9213
"	Kg./sq. cm	1.03323
"	Mm. of Hg (0°C.)	760
"	Pounds/sq. inch	14.6960
"	Tons (short)/sq. ft	1.05811
"	Torrs	760
Atomic mass units (chem.)*	Electron volts	9.31395×10^{8}
"	Grams*	1.66024×10^{-24}
Atomic mass units (phys.)*	Electron volts	9.31141×10^{8}
"	Grams*	1.65979×10^{-24}

* For factors for C = 12 scale or those which can be derived from same see table on Value for General Physical Constants.

CONVERSION FACTORS (Continued)

To convert from	To	Multiply by
Bags (Brit.)	Bushels (Brit.)	3
Barns	Sq. cm	1×10^{-24}
Barrels (Brit.)*	Bags (Brit.)	1.5
"	Barrels (U.S., dry)	1.415404
"	Barrels (U.S., liq.)	1.372513
"	Bushels (Brit.)	4.5
"	Bushels (U.S.)	4.644253
"	Cu. feet	5.779568
"	Cu. meters	0.1636591
"	Gallons (Brit.)	36
"	Liters	163.6546
Barrels (petroleum, U.S.)	Cu. feet	5.614583
"	Gallons (U.S.)	42
"	Liters	158.98284
Barrels (U.S., dry)	Barrels (U.S. liq.)	0.969696
"	Bushels (U.S.)	3.2812195
"	Cu. feet	4.083333
"	Cu. inches	7056
"	Cu. meters	0.11562712
"	Quarts (U.S., dry)	105
Barrels (U.S., liq.)	Barrels (U.S., dry)	1.03125
"	Barrels (wine)	1
"	Cu. feet	4.2109375
"	Cu. inches	7276.5
"	Cu. meters	0.11924047
"	Gallons (Brit.)	26.22925
"	Gallons (U.S., liq.)	31.5
"	Liters	119.23713
Bars	Atmospheres	0.986923
"	Baryes	1×10^{6}
"	Cm. of Hg (0°C.)	75.0062
"	Dynes/sq. cm	1×10^{6}
"	Ft. of H_2O (60°F.)	33.4883
"	Grams/sq. cm	1019.716
"	In. of Hg (32°F.)	29.5300
"	Kg./sq. cm	1.019716
"	Millibars	1000
"	Pounds/sq. inch	14.5038
Baryes	Atmospheres	9.86923×10^{-7}
"	Bars	1×10^{-6}
"	Dynes/sq. cm	1
"	Grams/sq. cm	0.001019716
"	Millibars	0.001
Bels	Decibels	10
Board feet	Cu. cm	2359.7372
"	Cu. feet	0.083333
"	Cu. inches	144
Bolts of cloth	Linear feet	120
"	Meters	36.576
Bougie decimales	Candles (Int.)	1.00
B.t.u	B.t.u. (IST.)**	0.999346
"	B.t.u. (mean)	0.998563
"	B.t.u. (39°F.)	0.994982
"	B.t.u. (60°F.)	0.999689
"	Cal. *gm.*	251.99576
"	Cal., *gm.* (IST.)	251.831
"	Cal., *gm.* (mean)	251.634
"	Cal., *gm.* (20°C.)	252.122
"	Cu. cm.-atm	10405.6
"	Ergs	1.05435×10^{10}
"	Foot-poundals	25020.1
"	Foot-pounds	777.649
"	Gram-cm	1.07514×10^{7}
"	Hp.-hours	0.000392752
"	Hp.-years	4.48347×10^{-8}
"	Joules	1054.35
"	Joules (Int.)	1054.18
"	Kg.-meters	107.514
"	Kw.-hours	0.000292875
B.t.u	Kw.-hours (Int.)	0.000292827
"	Liter-atm	10.4053
"	Tons of refrig. (U.S. std.)	3.46995×10^{-6}
"	Watt-seconds	1054.35
"	Watt-seconds (Int.)	1054.18
B.t.u. (IST.)	B.t.u	1.00065
B.t.u. (mean)	B.t.u	1.00144
"	B.t.u. (IST.)	1.00078
"	B.t.u. (39°F.)	0.996415
"	B.t.u. (60°F.)	1.00113
"	Hp.-hours	0.000393317
"	Joules	1055.87
"	Kg.-meters	107.669
"	Kw.-hours	0.000293297
"	Kw.-hours (Int.)	0.000293248
"	Liter-atm	10.4203
"	Watt-hours	0.293297
"	Watt-hours (Int.)	0.293248
B.t.u. (39°F.)	B.t.u	1.00504
"	B.t.u. (IST.)	1.00439
"	B.t.u. (mean)	1.00360
"	B.t.u. (60°F)	1.00473
"	Joules	1059.67
B.t.u. (60°F.)	B.t.u	1.00031
"	B.t.u. (IST.)	0.999657
"	B.t.u. (mean)	0.998873
"	B.t.u. (39°F.)	0.995291
B.t.u./hr	Cal., *kg.*/hr	0.251996
"	Ergs/sec	2.928751×10^{6}
"	Foot-pounds/hr	777.649
"	Horsepower	0.000392752
"	Horsepower (boiler)	2.98563×10^{-5}
"	Horsepower (electric)	0.000392594
"	Horsepower (metric)	0.000398199
"	Kilowatts	0.000292875
"	Lb. ice melted/hr	0.0069714
"	Tons of refrig. (U.S. comm.)	8.32789×10^{-5}
"	Watts	0.292875
B.t.u./min	Cal., *kg.*/min	0.251996
"	Ergs/sec	1.75725×10^{8}
"	Foot-pounds/min	777.649
"	Horsepower	0.0235651
"	Horsepower (boiler)	0.00179138
"	Horsepower (electric)	0.0235556
"	Horsepower (metric)	0.0238920
"	Joules/sec	17.5725
"	Kg.-meters/min	107.514
"	Kilowatts	0.0175725
"	Lb. ice melted/hr	0.41828
"	Tons of refrig. (U.S. comm.)	0.00499673
"	Watts	17.5725
B.t.u. (mean)/min	B.t.u. (mean)/hr	60
"	Cal., *kg.* (mean)/hr	15.1197
"	Cal., *kg.* (mean)/min	0.251996
"	Ergs/sec	1.75978×10^{8}
"	Foot-pounds/min	778.768
"	Horsepower	0.0235990
"	Horsepower (boiler)	0.00179396
"	Horsepower (electric)	0.0235895
"	Horsepower (metric)	0.0239264
"	Joules/sec	17.5978
"	Kg.-meters/min	107.669
"	Kilowatts	0.0175978
"	Lb. ice-melted/hr	0.41888
B.t.u./lb	Cal., *gm.*/gram	0.555555
"	Cu. cm.-atm./gram	22.9405
"	Cu. ft.-atm./lb	0.367471
"	Cu. ft.-(lb./sq. in.)/lb	5.40034
"	Foot-pounds/lb	777.649
"	Hp.-hr./lb	0.000392752

* Barrel (Brit., liq.) = Barrel (Brit., dry)

** International Steam Table.

To convert from	To	Multiply by
B.t.u./lb	Joules/gram	2.32444
B.t.u. (mean)/lb	Cal., *gm.* (mean)/gram	0.555555
"	Cu. cm.-atm./gram	22.9735
"	Foot-pounds/lb	778.768
"	Hp.-hr./lb	0.000393317
"	Joules/gram	2.32779
B.t.u./sec	B.t.u./hr	3600
"	B.t.u./min	60
"	Cal., *kg.*/hr	907.185
"	Cal., *kg.*/min	15.1197
"	Cheval-vapeur	1.43352
"	Ergs/sec	1.05435×10^{10}
"	Foot-pounds/sec	777.649
"	Horsepower	1.41391
"	Horsepower (boiler)	0.107483
"	Horsepower (electric)	1.41334
"	Horsepower (metric)	1.43352
"	Kg.-meters/sec	107.514
"	Kilowatts	1.05435
"	Kilowatts (Int.)	1.05418
"	Watts	1054.35
"	Watts (Int.)	1054.18
B.t.u. (mean)/sec	Ergs/sec	1.05587×10^{10}
"	Foot-pounds/sec	778.768
"	Horsepower	1.41594
"	Horsepower (boiler)	0.107637
"	Horsepower (electric)	1.41537
"	Horsepower (metric)	1.43558
"	Watts	1055.87
B.t.u./sq. ft	Cal., *gm.*/sq. cm	0.271246
B.t.u./sq.ft. × min.)	Hp./sq. ft	0.0235651
"	Kw./sq. ft	0.0175725
"	Watts/sq. in	0.122031
Buckets (Brit.)	Cu. cm	18184.35
"	Gallons (Brit.)	4
Bushels (Brit.)	Bags (Brit.)	0.333333
"	Bushels (U.S.)	1.032056
"	Cu. cm	36368.70
"	Cu. feet	1.284348
"	Cu. inches	2219.354
"	Dekaliters	3.636768
"	Gallons (Brit.)	8
"	Hectoliters	0.3636768
"	Liters	36.36768
Bushels (U.S.)*	Barrels (U.S.), dry	0.3047647
"	Bushels (Brit.)	0.9689395
"	Cu. cm	35239.07
"	Cu. feet	1.244456
"	Cu. inches	2150.42
"	Cu. meters	0.03523907
"	Cu. yards	0.04609096
"	Gallons (U.S., dry)	8
"	Gallons (U.S., liq.)	9.309177
"	Liters	35.23907
"	Ounces (U.S., fluid)	1191.575
"	Pecks (U.S.)	4
"	Pints (U.S., dry)	64
"	Quarts (U.S., dry)	32
"	Quarts (U.S., liq.)	37.23671
Butts (Brit.)	Bushels (U.S.)	13.53503
"	Cu. feet	16.84375
"	Cu. meters	0.4769619
"	Gallons (U.S.)	126
Cable lengths	Fathoms	120
"	Feet	720
"	Meters	219.456

To convert from	To	Multiply by
Calories, *gm.***	B.t.u	0.0039683207
"	B.t.u. (IST.)	0.00396573
"	B.t.u. (mean)	0.00396262
"	B.t.u. (39°F.)	0.00394841
"	B.t.u. (60°F.)	0.00396709
"	Cal., *gm.* (IST.)	0.999346
"	Cal., *gm.* (mean)	0.998563
"	Cal., *gm.* (15°C.)	0.999570
"	Cal., *gm.* (20°C.)	1.00050
"	Cal., *kg*	0.001
"	Cal., *kg.* (IST.)	0.000999346
"	Cal., *kg.* (mean)	0.000998563
"	Cal., *kg.* (15°C.)	0.000999570
"	Cal., *kg.* (20°C.)	0.00100050
"	Cu. cm.-atm	41.2929
"	Cu. ft.-atm	0.00145824
"	Ergs	4.184×10^{7}
"	Foot-poundals	99.2878
"	Foot-pounds	3.08596
"	Gram-cm	42664.9
"	Hp.-hours	1.55857×10^{-6}
"	Joules	4.184
"	Joules (Int.)	4.18331
"	Kg.-meters	0.426649
"	Kw.-hours	1.162222×10^{-6}
"	Liter-atm	0.0412917
"	Watt-hours	0.001162222
"	Watt-hours (Int.)	0.00116203
"	Watt-seconds	4.184
Calories, *gm.* (mean)	B.t.u	0.00397403
"	Cal., *gm*	1.00144
"	Cal., *gm.* (IST.)	1.00078
"	Cal., *gm.* (20°C.)	1.00194
"	Cal., *kg.* (mean)	0.001
"	Cu. cm.-atm	41.3523
"	Cu. ft.-atm	0.00146034
"	Ergs	4.19002×10^{7}
"	Foot-poundals	99.4308
"	Foot-pounds	3.09040
"	Hp.-hours	1.56081×10^{-6}
"	Joules	4.19002
"	Joules (Int.)	4.18933
"	Kg.-meters	0.427263
"	Kw.-hours	1.16390×10^{-6}
"	Liter-atm	0.0413511
"	Watt-seconds	4.19002
Calories, *gm.* (15°C.)	B.t.u	0.00397003
"	Cal., *gm*	1.00043
"	Cal., *gm.* (IST.)	0.999776
"	Cal., *gm.* (mean)	0.998992
"	Cal., *gm.* (20°C.)	1.00093
"	Joules	4.18580
"	Joules (Int.)	4.18511
Calories, *gm.* (20°C.)	B.t.u	0.00396633
"	Cal., *gm*	0.999498
"	Cal., *gm.* (IST.)	0.998845
"	Cal., *gm.* (mean)	0.998061
"	Cal., *gm.* (15°C.)	0.999068
"	Joules	4.18190
"	Joules (Int.)	4.18121
Calories, *kg*	B.t.u	3.9683207
"	B.t.u. (IST.)	3.96573
"	B.t.u. (mean)	3.96262
"	B.t.u. (60°F.)	3.96709
"	Cal., *gm*	1000
"	Cal., *kg.* (mean)	0.998563
"	Cal., *kg.* (15°C.)	0.999570
"	Cal., *kg.* (20°C.)	1.00050
"	Cu. cm.-atm	41292.86

* Stricken or struck bushel. A heaped bushel for apples of 2747.715 cu. inches was established by the U.S. Court of Customs Appeals on Feb. 15, 1912. A heaped bushel equal to 1¼ stricken bushels is also known.

** This is the calorie as defined by the U.S. National Bureau of Standards and is equal to 4.18400 joules.

To convert from	To	Multiply by
Calories, *kg.*	Ergs	4.184×10^{10}
"	Foot-poundals	99287.8
"	Foot-pounds	3085.96
"	Gram-cm	4.26649×10^{7}
"	Hp.-hours	0.00155857
"	Joules	4184
"	Kw.-hours	0.001162222
"	Liter-atm	41.2917
"	Watt-hours	1.162222
Calories, *kg.* (mean)	B.t.u.	3.97403
"	B.t.u. (IST.)	3.97144
"	B.t.u. (mean)	3.9683207
"	B.t.u. (60°F.)	3.97280
"	Cal., *gm.*	1001.44
"	Cal., *gm.* (IST.)	1000.78
"	Cal., *gm.* (mean)	1000
"	Cal., *gm.* (15°C.)	1000.10
"	Cal., *gm.* (20°C.)	1001.94
"	Ergs	4.19002×10^{10}
"	Foot-poundals	99430.8
"	Foot-pounds	3090.40
"	Gram-cm	4.27263×10^{7}
"	Hp.-hours	0.00156081
"	Joules	4190.02
"	Kg.-meters	427.263
"	Kw.-hours (Int.)	0.00116370
"	Liter-atm	41.3511
"	Watt-hours	1.16390
Cal., *gm.*/°C	B.t.u./°F	0.00220462
"	Joules/°F	2.324444
"	Joules (Int.)/°F	2.32406
Cal., *gm.*/gram	B.t.u./lb	1.8
"	Foot-pounds/lb	1399.77
"	Joules/gram	4.184
"	Watt-hours/gram	0.001162222
Cal., *gm.*/(gram × °C)	B.t.u./(lb. × °C.)	1.8
"	B.t.u./(lb. × °F.)	1
"	Cal., *kg.*/(kg. × °C.)	1
"	Joules/(gram × °C.)	4.184
"	Joules/(lb. × °F.)	1054.35
Cal., *gm.*/hr	B.t.u./hr	0.0039683207
"	Ergs/sec	11622.222
"	Watts	0.001162222
Cal., *gm.* (mean)/hr	B.t.u. (mean)/hr	0.0039683207
"	Ergs/sec	11639.0
"	Watts	0.00116390
Cal., *kg.*/hr	Watts	1.162222
Cal., *gm.*/min	B.t.u./min	0.0039683207
"	Ergs/sec	697333.3
"	Watts	0.069733
Cal., *gm.* (mean)/min	B.t.u. (mean)/min	0.0039683207
"	Ergs/sec	698337
"	Joules/sec	0.0698337
"	Watts	0.0698337
Cal., *kg.*/min	Kg. ice melted/min	0.012548
"	Lb. ice melted/min	0.027665
"	Watts	69.7333
Cal., *gm.*/sec	B.t.u./sec	0.0039683207
"	Ergs/sec	4.184×10^{7}
"	Foot-pounds/sec	3.08596
"	Horsepower	0.00561084
"	Watts	4.184
Cal., *gm.* (mean)/sec	Ergs/sec	4.19002×10^{7}
"	Watts	4.19002
Cal., *gm.*/(sec. × sq. cm.)	B.t.u./(hr. × sq. ft.)	13272.1
"	Cal., *gm.*/(hr. × sq. cm.)	3600
"	Watts/sq. cm	4.184
Cal., *gm.*/(sec. × sq. cm. × °C.)	B.t.u./(hr. × sq. ft. × °F.)	7373.38
Cal., *gm.*/sq. cm	B.t.u./sq. ft	3.68669

To convert from	To	Multiply by
Cal., *gm.*-cm. / (hr. × sq. cm. × °C.)	B.t.u.-ft. / (hr. × sq. ft. × °F.)	0.0671969
"	B.t.u.-inch / (hr. × sq. ft. × °F.)	0.806363
Cal., *gm.*-cm./sq. cm	B.t.u.-inch/sq. ft	1.4514530
Cal., *gm.*-sec	Planck's constant	6.31531×10^{33}
Cal., *gm.*-sec./Avog. No. (chem.)*	Planck's constant	1.04849×10^{10}
Cal., *gm.*-sec./Avog. No. (phys.)*	Planck's constant	1.04821×10^{10}
Candles (English)	Candles (Int.)	1.04
"	Hefner units	1.16
Candles (German)	Candles (English)	1.01
"	Candles (Int.)	1.05
"	Hefner units	1.17
Candles (Int.)	Candles (English)	0.96
"	Candles (German)	0.95
"	Candles (pentane)	1.00
"	Hefner units	1.11
"	Lumens (Int.)/steradian	1
Candles (pentane)	Candles (Int.)	1.00
Candles/sq. cm	Candles/sq. inch	6.4516
"	Candles/sq. meter	10000
"	Foot-lamberts	2918.6351
"	Lamberts	3.1415927
Candles/sq. ft	Candles/sq. inch	0.0069444
"	Candles/sq. meter	10.763910
"	Foot-lamberts	3.1415927
"	Lamberts	0.0033815822
Candles/sq. inch	Candles/sq. cm	0.15500031
"	Candles/sq. foot	144
"	Foot-lamberts	452.38934
"	Lamberts	0.48694784
Candle power (spher.)	Lumens	12.566370
Carats (parts of gold per 24 of mixture)	Milligrams/gram	41.6666
Carats (1877)	Grains	3.168
"	Milligrams	205.3
Carats (metric)	Grains	3.08647
"	Grams	0.2
"	Milligrams	200
Carcel units	Candles (Int.)	9.61
Centals	Kilograms	45.359237
"	Pounds	100
Centares	Ares	0.01
"	Sq. feet	10.763910
"	Sq. inches	1550.0031
"	Sq. meters	1
"	Sq. yards	1.1959900
Centigrams	Grains	0.15432358
"	Grams	0.01
Centiliters	Cu. cm	10
"	Cu. inches	0.6102545
"	Liters	0.01
"	Ounces (U.S., fluid)	0.3381497
Centimeters	Ångström units	1×10^{8}
"	Feet	0.032808399
"	Feet (U.S. Survey)	0.032808333
"	Hands	0.098425197
"	Inches	0.39370079
"	Links (Gunter's)	0.049709695
"	Links (Ramden's)	0.032808399
"	Meters	0.01
"	Microns	10000
"	Miles (naut., Int.)	5.3995680×10^{-6}
"	Miles (statute)	6.2137119×10^{-6}
"	Millimeters	10
"	Millimicrons	1×10^{7}
"	Mils	393.70079
"	Picas (printer's)	2.3710630
"	Points (printer's)	28.452756

* For factors for C = 12 scale or those which can be derived from same table on Values for General Physical Constants.

To convert from	To	Multiply by
Centimeters	Rods	0.0019883878
"	Wave length of orange-red line of krypton 86	16507.6373
"	Wave length of red line of cadmium	15531.6413
"	Yards	0.010936133
Cm. of Hg (0°C.)	Atmospheres	0.013157895
"	Bars	0.0133322
"	Dynes/sq. cm	13332.2
"	Ft. of H_2O (4°C.)	0.446050
"	Ft. of H_2O (60°F.)	0.446474
"	In. of Hg (0°C.)	0.39370079
"	Kg./sq. meter	135.951
"	Pounds/sq. ft.	27.8450
"	Pounds/sq. inch	0.193368
"	Torrs	10
Cm. of H_2O (4°C.)	Atmospheres	0.000967814
"	Dynes/sq. cm	980.638
"	Pounds/sq. inch	0.0142229
Centimeters/sec.	Feet/min	1.9685039
"	Feet/sec	0.032808399
"	Kilometers/hr	0.036
"	Kilometers/min	0.0006
"	Knots (Int.)	0.019438445
"	Meters/min	0.6
"	Miles/hr	0.022369363
"	Miles/min	0.00037282272
Cm./(sec. × sec.)	Kilometers/(hr. × sec.)	0.036
"	Miles/(hr. × sec.)	0.022369363
Centimeters/year	Inches/year	0.39370079
Centipoises*	Grams/(cm. × sec.)	0.01
"	Poises	0.01
"	Pound/(ft. × hr.)	2.4190883
"	Pounds/(ft. × sec.)	0.00067196898
Centistokes*	Stokes	0.01
Chains (Gunter's)	Centimeters	2011.68
"	Chains (Ramden's)	0.66
"	Feet	66
"	Feet (U.S. Survey)	65.999868
"	Furlongs	0.1
"	Inches	792
"	Links (Gunter's)	100
"	Links (Ramden's)	66
"	Meters	20.1168
"	Miles (statute)	0.0125
"	Rods	4
"	Yards	22
Chains (Ramden's)	Centimeters	3048
"	Chains (Gunter's)	1.515151
"	Feet	100
"	Feet (U.S. Survey)	99.999800
Cheval-vapeur	Horsepower (metric)	1
Cheval-vapeur-heures	Joules	2647795
Circles	Degrees	360
"	Grades	400
"	Minutes	21600
"	Radians	6.2831853
"	Signs	12
Circular inches	Circular mm	645.16
"	Sq. cm	5.0670748
"	Sq. inches	0.78539816
Circular mm	Sq. cm	0.0078539816
"	Sq. inches	0.0012173696
"	Sq. mm	0.78539816
Circular mils	Circular inches	1×10^{-6}
"	Sq. cm	5.0670748×10^{-6}
"	Sq. inches	7.8539816×10^{-7}
"	Sq. mm	0.00050670748
"	Sq. mils	0.78539816
Circumferences	Degrees	360
"	Grades	400

To convert from	To	Multiply by
Circumferences	Minutes	21600
"	Radians	6.2831853
"	Seconds	1296000
Cords	Cord-feet	8
"	Cu. feet	128
"	Cu. meters	3.6245734
Cord-feet	Cords	0.125
"	Cu. feet	16
Coulombs	Abcoulombs	0.1
"	Ampere-hours	0.0002777
"	Ampere-seconds	1
"	Coulombs (Int.)	1.000165
"	Electronic charge	6.24196×10^{18}
"	E.M. cgs. units of electric charge	0.1
"	E.S. cgs. units of electric charge	2.997930×10^{9}
"	Faradays (chem.)	1.036377×10^{-5}
"	Faradays (phys.)	1.036086×10^{-5}
"	Mks. units of electric charge	1
"	Statcoulombs	2.997930×10^{9}
Coulombs/cu. meter	E.M. cgs. units of volume charge density	1×10^{-7}
"	E.S. cgs. units	2997.930
Coulombs/sq. cm	Abcoulombs/sq. cm	0.1
"	Cgs. units of polarization, *and* surface charge density	1
Cubic centimeters	Board feet	0.00042377600
"	Bushels (Brit.)	2.749617×10^{-5}
"	Bushels (U.S.)	2.837759×10^{-5}
"	Cu. feet	3.5314667×10^{-5}
"	Cu. inches	0.061023744
"	Cu. meters	1×10^{-6}
"	Cu. yards	1.3079506×10^{-6}
"	Drachms (Brit., fluid)	0.28156080
"	Drams (U.S., fluid)	0.27051218
"	Gallons (Brit.)	0.0002199694
"	Gallons (U.S., dry)	0.00022702075
"	Gallons (U.S., liq.)	0.00026417205
"	Gills (Brit.)	0.007039020
"	Gills (U.S.)	0.0084535058
"	Liters	0.001
"	Ounces (Brit., fluid)	0.03519510
"	Ounces (U.S., fluid)	0.033814023
"	Pints (U.S., dry)	0.0018161660
"	Pints (U.S., liq.)	0.0021133764
"	Quarts (Brit.)	0.0008798775
"	Quarts (U.S., dry)	0.00090808298
"	Quarts (U.S., liq.)	0.0010566882
Cu. cm./gram	Cu. ft./lb	0.016018463
Cu. cm./sec	Cu. ft./min	0.0021188800
"	Gal. (U.S.)/min	0.015850323
"	Gal. (U.S.)/sec	0.00026417205
Cu. cm.-atm	B.t.u.	9.61019×10^{-5}
"	B.t.u. (mean)	9.59637×10^{-5}
"	Cal., *gm*	0.0242173
"	Cal., *gm.* (mean)	0.0241824
"	Cu. ft.-atm	3.5314667×10^{-5}
"	Joules	0.101325
"	Watt-hours	2.81458×10^{-5}
Cu. cm.-atm./gram	B.t.u./lb	0.0435911
"	Cal., *gm.*/gram	0.0242173
"	Cu. ft.-(lb./sq. in.)/lb	0.235406
"	Ft.-lb./lb	33.8985
"	Joules/gram	0.101325
"	Kg.-meters/gram	0.0103323
"	Kw.-hr./gram	2.81458×10^{-8}
Cubic decimeters	Cu. cm	1000
"	Cu. feet	0.035316667
"	Cu. inches	61.023744
"	Cu. meters	0.001

* See also special table on viscosity.

To convert from	To	Multiply by
Cubic decimeters	Cu. yards	0.0013079506
"	Liters	1
Cubic dekameters	Cu. decimeters	1×10^6
"	Cu. feet	35314.667
"	Cu. inches	6.1023744×10^7
"	Cu. meters	1000
"	Liters	999972
Cubic feet	Acre-feet	2.2956841×10^{-5}
"	Board feet	12
"	Bushels (Brit.)	0.7786049
"	Bushels (U.S.)	0.80356395
"	Cords (wood)	0.0078125
"	Cord-feet	0.0625
"	Cu. centimeters	28316.847
"	Cu. meters	0.028316847
"	Gallons (U.S., dry)	6.4285116
"	Gallons (U.S., liq.)	7.4805195
"	Liters	28.316847
"	Ounces (Brit., fluid)	996.6143
"	Ounces (U.S., fluid)	957.50649
"	Pints (U.S., liq.)	59.844156
"	Quarts (U.S., dry)	25.714047
"	Quarts (U.S., liq.)	29.922078
Cu. ft. of H_2O (39.2°F.)	Pounds of H_2O	62.4262
Cu. ft. of H_2O (60°F.)	Pounds of H_2O	62.3663
Cu. ft./hr	Acre-feet/hr	2.2956841×10^{-5}
"	Cu. cm./sec	7.8657907
"	Cu. ft./day	24
"	Gal. (U.S.)/hr	7.4805195
"	Liters/hr	28.31605
Cu. ft./min	Acre-feet/hr	0.0013774105
"	Acre-feet/min	2.2956841×10^{-5}
"	Cu. cm./sec	471.94744
"	Cu. ft./hr	60
"	Gal. (U.S.)/min	7.4805195
"	Liters/sec	0.4719342
Cu. ft./lb	Cu. cm./gram	62.427961
"	Millimeters/gram	62.42621
Cu. ft./sec.	Acre-inches/hr	0.99173553
"	Cu. cm./sec	28316.847
"	Cu. yards/min	2.222222
"	Gal. (U.S.)/min	448.83117
"	Liters/min	1698.963
"	Liters/sec	28.31605
Cu. ft. of H_2O (60°F.)/sec	Lb. of H_2O/min	3741.98
Cu. ft.-atm	B.t.u.	2.72130
"	Cal., *gm*	685.756
"	Cu. cm.-atm	28316.847
"	Cu. ft.-(lb/sq. in.)	14.6960
"	Foot-pounds	2116.22
"	Hp.-hours	0.00106880
"	Joules	2869.20
"	Kg.-meters	292.577
"	Kw.-hours	0.000797001
Cubic inches	Barrels (Brit.)	0.0001001292
"	Barrels (U.S., dry)	0.00014172336
"	Board feet	0.0069444
"	Bushels (Brit.)	0.0004505815
"	Bushels (U.S.)	0.00046502544
"	Cu. cm	16.387064
"	Cu. feet	0.00057870370
"	Cu. meters	1.6387064×10^{-5}
"	Cu. yards	2.1433470×10^{-5}
"	Drams (U.S., fluid)	4.4329004
"	Gallons (Brit.)	0.003604652
"	Gallons (U.S., dry)	0.0037202035
"	Gallons (U.S., liq.)	0.0043290043
"	Liters	0.016387064
"	Milliliters	16.387064
"	Ounces (Brit., fluid)	0.5767444
Cubic inches	Ounces (U.S., fluid)	0.55411255
"	Pecks (U.S.)	0.0018601017
"	Pints (U.S., dry)	0.029761628
"	Pints (U.S., liq.)	0.034632035
"	Quarts (U.S., dry)	0.014880814
"	Quarts (U.S., liq.)	0.017316017
Cu. in. of H_2O (4°C.)	Pounds of H_2O	0.0361263
Cu. in. of H_2O (60°F.)	Pounds of H_2O	0.0360916
Cubic meters	Acre-feet	0.00081071319
"	Barrels (Brit.)	6.110261
"	Barrels (U.S., dry)	8.648490
"	Barrels (U.S., liq.)	8.3864145
"	Bushels (Brit.)	27.49617
"	Bushels (U.S.)	28.377593
"	Cu. cm	1×10^6
"	Cu. feet	35.314667
"	Cu. inches	61023.74
"	Cu. yards	1.3079506
"	Gallons (Brit.)	219.9694
"	Gallons (U.S., liq.)	264.17205
"	Hogshead	4.1932072
"	Liters	1000
"	Pints (U.S., liq.)	2113.3764
"	Quarts (U.S., liq.)	1056.6882
"	Steres	1
Cu. meters/min	Gal. (Brit.)/min	219.9694
"	Gal. (U.S.)/min	264.1721
"	Liters/min	999.972
Cu. millimeters	Cu. cm	0.001
"	Cu. inches	6.1023744×10^{-5}
"	Cu. meters	1×10^{-9}
"	Minims (Brit.)	0.01689365
"	Minims (U.S.)	0.016230731
Cu. yards	Bushels (Brit.)	21.02233
"	Bushels (U.S.)	21.696227
"	Cu. cm	764554.86
"	Cu. feet	27
"	Cu. inches	46656
"	Cu. meters	0.76455486
"	Gallons (Brit.)	168.1787
"	Gallons (U.S., dry)	173.56981
"	Gallons (U.S., liq.)	201.97403
"	Liters	764.55486
"	Quarts (Brit.)	672.7146
"	Quarts (U.S., dry)	694.27926
"	Quarts (U.S., liq.)	807.89610
Cu. yd./min	Cu. ft./sec	0.45
"	Gal. (U.S.)/sec	3.3662338
"	Liters/sec	12.74222
Cubits	Centimeters	45.72
"	Feet	1.5
"	Inches	18
Daltons (chem.)	Grams	1.66024×10^{-24}
Daltons (phys.)	Grams	1.65979×10^{-24}
Days (mean solar)	Days (sidereal)	1.00273791
"	Hours (mean solar)	24
"	Hours (sidereal)	24.065710
"	Years (calendar)	0.0027397260
"	Years (sidereal)	0.0027378031
"	Years (tropical)	0.0027379093
Days (sidereal)	Days (mean solar)	0.99726957
"	Hours (mean solar)	23.934470
"	Hours (sidereal)	24
"	Minutes (mean solar)	1436.0682
"	Minute (sidereal)	1440
"	Second (sidereal)	86400
"	Years (calendar)	0.0027322454
"	Years (sidereal)	0.0027303277
"	Years (tropical)	0.0027304336
Decibels	Bels	0.1
Decimeters	Centimeters	10

CONVERSION FACTORS (Continued)

To convert from	To	Multiply by
Decimeters	Feet	0.32808399
"	Feet (U.S. Survey)	0.328083333
"	Inches	3.9370079
"	Meters	0.1
Decisteres	Cu. meters	0.1
Degrees	Circles	0.0027777
"	Minutes	60
"	Quadrants	0.0111111
"	Radians	0.017453293
"	Seconds	3600
Degrees/cm	Radians/cm	0.017453293
Degrees/foot	Radians/cm	0.00057261458
Degrees/inch	Radian/cm	0.0068713750
Degrees/min	Degrees/sec	0.0166666
"	Radians/sec	0.00029088821
"	Revolutions/sec	4.629629×10^{-5}
Degrees/sec	Radians/sec	0.017453293
"	Revolutions/min	0.166666
"	Revolutions/sec	0.0027777
Dekaliters	Pecks (U.S.)	1.135136
"	Pints (U.S., dry)	18.16217
Dekameters	Centimeters	1000
"	Feet	32.808399
"	Feet (U.S. Survey)	32.808333
"	Inches	393.70079
"	Kilometers	0.01
"	Meters	10
"	Yards	10.93613
Demals	Gram-equiv./cu. decimeter	1
Drachms (Brit., fluid)	Cu. cm	3.551631
"	Cu. inches	0.2167338
"	Drams (U.S., fluid)	0.9607594
"	Milliliters	3.551531
Drams (apoth. *or* troy)	Drams (avdp.)	2.1942857
"	Grains	60
"	Grams	3887.9346
"	Ounces (apoth. *or* troy)	0.125
"	Ounces (avdp.)	0.13714286
"	Scruples (apoth.)	3
Drams (avdp.)	Drams (apoth. *or* troy)	0.455729166
"	Grains	27.34375
"	Grams	1.7718452
"	Ounces (apoth. *or* troy)	0.056966146
"	Ounces (avdp.)	0.0625
"	Pennyweights	1.1393229
"	Pounds (apoth. *or* troy)	0.0047471788
"	Pounds (avdp.)	0.00390625
"	Scruples (apoth.)	1.3671875
Drams (U.S., fluid)	Cu. cm	3.6967162
"	Cu. inches	0.22558594
"	Drachms (Brit., fluid)	1.040843
"	Gills (U.S.)	0.03125
"	Milliliters	3.696588
"	Minims (U.S.)	60
"	Ounces (U.S., fluid)	0.125
"	Pints (U.S., liq.)	0.0078125
Dynes	Grains	0.01573663
"	Grams	0.001019716
"	Newtons	0.00001
"	Poundals	7.2330138×10^{-5}
"	Pounds	2.248089×10^{-6}
Dynes/cm	Ergs/sq. cm	1
"	Ergs/sq. mm	0.01
"	Grams/cm	0.001019716
"	Poundals/inch	0.00018371855
Dynes/cu. cm	Grams/cu. cm	0.001019716
"	Poundals/cu. inch	0.0011852786
Dynes/sq. cm	Atmospheres	9.86923×10^{-7}
"	Bars	1×10^{-6}
"	Baryes	1
"	Cm. of Hg (0°C.)	7.50062×10^{-5}

To convert from	To	Multiply by
Dynes/sq. cm	Cm. of H_2O (4°C.)	0.001019745
"	Grams/sq. cm	0.001019716
"	In. of Hg (32°F.)	2.95300×10^{-5}
"	In. of H_2O (4°C.)	0.000401474
"	Kg./sq. meter	0.01019716
"	Poundals/sq. in	0.00046664510
"	Pounds/sq. in	1.450377×10^{-5}
Dyne-centimeters	Ergs	1
"	Foot-poundals	2.3730360×10^{-6}
"	Foot-pounds	7.37562×10^{-8}
"	Gram-cm	0.001019716
"	Inch-pounds	8.85075×10^{-7}
"	Kg.-meters	1.019716×10^{-8}
"	Newton-meters	1×10^{-7}
Electron volts	Ergs	1.60219×10^{-12}
"	Grams	1.78253×10^{-33}
Electronic charges	Abcoulombs	1.60209×10^{-20}
"	Coulombs	1.60209×10^{-19}
"	Statcoulombs	4.80296×10^{-10}
Electronic charges/kg.	Statcoulombs/dyne	4.89766×10^{-16}
E.S. cgs. units of induction flux	E.M. cgs. units	2.997930×10^{10}
E.S. cgs. units of magnetic charge	E.M. cgs. units	2.997930×10^{10}
E.S. cgs. units of magnetic field intensity	E.M. cgs. units	3.335635×10^{-11}
Ells	Centimeters	114.3
"	Inches	45
Ergs	B.t.u	9.48451×10^{-11}
"	Cal., *gm*	2.39006×10^{-8}
"	Cal., *kg*	2.39006×10^{-11}
"	Cal., *kg*. (20°C.)	2.39126×10^{-11}
"	Cu. cm.-atm	9.86923×10^{-7}
"	Cu. ft.-atm	3.48529×10^{-11}
"	Cu. ft.-(lb./sq. in.)	5.12196×10^{-10}
"	Dyne-cm	1
"	Electron volts	6.24145×10^{11}
"	Foot-poundals	2.3730360×10^{-6}
"	Foot-pounds	7.37562×10^{-8}
"	Gram-cm	0.001019716
"	Joules	1×10^{-7}
"	Joules (Int.)	9.99835×10^{-8}
"	Kw.-hours	2.777777×10^{-14}
"	Kg.-meters	1.019716×10^{-8}
"	Liter-atm	9.86895×10^{-10}
"	Watt-sec	1×10^{-7}
Ergs/(gram-mol. × °C.)	Foot-pounds/(lb.-mol. × °F.)	1.85863×10^{-5}
Ergs/sec	B.t.u./min	5.69071×10^{-9}
"	Cal., *gm*./min	1.43403×10^{-6}
"	Dyne-cm./sec	1
"	Foot-pounds/min	4.42537×10^{-6}
"	Gram-cm./sec	0.001019716
"	Horsepower	1.34102×10^{-10}
"	Joules/sec	1×10^{-7}
"	Kilowatts	1×10^{-10}
"	Watts	1×10^{-7}
Ergs/sq. cm	Dynes/cm	1
"	Ergs/sq. mm	0.01
Ergs/sq. mm	Dynes/cm	100
"	Ergs/sq. cm	100
Erg-sec	Planck's constant	1.50932×10^{26}
Farads	Abfarads	1×10^{-9}
"	E.M. cgs. units	1×10^{-9}
"	E.S. cgs. units	8.987584×10^{11}
"	Farads (Int.)	1.000495
"	Microfarads	1×10^{6}
"	Statfarads	8.98758×10^{11}
Farads (Int.)	Farads	0.999505
Fathoms	Centimeters	182.88

To convert from	To	Multiply by
Fathoms	Feet	6
"	Inches	72
"	Meters	1.8288
"	Miles (naut., Int.)	0.00098747300
"	Miles (statute)	0.001136363
"	Yards	2
Feet	Centimeters	30.48
"	Chains (Gunter's)	0.01515151
"	Fathoms	0.166666
"	Feet (U.S. Survey)	0.99999800
"	Furlongs	0.00151515
"	Inches	12
"	Meters	0.3048
"	Microns	304800
"	Miles (naut., Int.)	0.00016457883
"	Miles (statute)	0.000189393
"	Rods	0.060606
"	Ropes (Brit.)	0.05
"	Yards	0.333333
Feet (U.S. Survey)	Centimeters	30.480061
"	Chains (Gunter's)	0.015151545
"	Chains (Ramden's)	0.010000020
"	Feet	1.0000020
"	Inches	12.000024
"	Links (Gunter's)	1.5151545
"	Links (Ramden's)	1.0000020
"	Meters	0.30480061
"	Miles (statute)	0.00018939432
"	Rods	0.060606182
"	Yards	0.33333400
Feet of air (1 atm., 60°F.)	Atmospheres	3.6083×10^{-5}
"	Ft. of Hg (32°F.)	0.00089970
"	Ft. of H_2O (60°F.)	0.0012244
"	In. of Hg (32°F.)	0.0010796
"	Pounds/sq. inch	0.00053027
Feet of Hg (32°F.)	Cm. of Hg (0°C.)	30.48
"	Ft. of H_2O (60°F.)	13.6085
"	In. of H_2O (60°F.)	163.302
"	Ounces/sq. inch	94.3016
"	Pounds/sq. inch	5.89385
Feet of H_2O (4°C.)	Atmospheres	0.0294990
"	Cm. of Hg (0°C.)	2.24192
"	Dynes/sq. cm	29889.8
"	Grams/sq. cm	30.4791
"	In. of Hg (32°F.)	0.882646
"	Kg./sq. meter	304.791
"	Pounds/sq. inch	0.433515
Feet/hour	Cm./hr	30.48
"	Cm./min	0.508
"	Cm./sec	0.0084666
"	Feet/min	0.0166666
"	Inches/hr	12
"	Kilometers/hr	0.0003048
"	Kilometers/min	5.08×10^{-6}
"	Knots (Int.)	0.0001645788
"	Miles/hr	0.000189393
"	Miles/min	3.156565×10^{-6}
"	Miles/sec	5.2609428×10^{-8}
Feet/minute	Cm./sec	0.508
"	Feet/sec	0.0166666
"	Kilometers/hr	0.018288
"	Meters/min	0.3048
"	Meters/sec	0.00508
"	Miles/hr	0.01136363
Feet/second	Cm./sec	30.48
"	Kilometers/hr	1.09728
"	Kilometers/min	0.018288
"	Meters/min	18.288
"	Miles/hr	0.68181818
"	Miles/min	0.01136363
Feet/(sec. × sec.)	Kilometers/(hr. × sec.)	1.09728

To convert from	To	Multiply by
Feet/(sec. × sec.)	Meters/(sec. × sec.)	0.3048
"	Miles/(hr. × sec.)	0.68181818
Firkins (Brit.)	Bushels (Brit.)	1.125
"	Cu. cm	40914.79
"	Cu. feet	1.444892
"	Firkins (U.S.)	1.200949
"	Gallons (Brit.)	9
"	Liters	40.91364
"	Pints (Brit.)	72
Firkins (U.S.)	Barrels (U.S., dry)	0.29464286
"	Barrels (U.S., liq.)	0.28571429
"	Bushels (U.S.)	0.96678788
"	Cu. feet	1.203125
"	Firkins (Brit.)	0.8326747
"	Liters	34.06775
"	Pints (U.S., liq.)	72
Foot-candles	Lumens/sq. ft	1
"	Lumens/sq. meter	10.763910
"	Lux	10.763910
"	Milliphots	1.0763910
Foot-lamberts	Candles/sq. cm	0.00034262591
"	Candles/sq. ft	0.31830989
"	Millilamberts	1.0763910
"	Lamberts	0.0010763910
"	Lumens/sq. ft	1
Foot-poundals	B.t.u	3.99678×10^{-5}
"	B.t.u. (IST.)	3.99417×10^{-5}
"	B.t.u. (mean)	3.99104×10^{-5}
"	Cal., *gm*	0.0100717
"	Cal., *gm.* (IST.)	0.0100651
"	Cal., *gm.* (mean)	0.0100573
"	Cu. cm.-atm	0.415890
"	Cu. ft.-atm	1.46870×10^{-5}
"	Dyne-cm	4.2140110×10^{5}
"	Ergs	4.2140110×10^{5}
"	Foot-pounds	0.0310810
"	Hp.-hours	1.56974×10^{-8}
"	Joules	0.042140110
"	Joules (Int.)	0.0421332
"	Kg.-meters	0.00429710
"	Kw.-hours	1.17056×10^{-8}
"	Liter-atm	0.000415879
"	B.t.u	0.00128593
Foot-pounds	B.t.u. (IST.)	0.00128509
"	B.t.u. (mean)	0.00128408
"	Cal., *gm*	0.324048
"	Cal., *gm.* (IST.)	0.323836
"	Cal., *gm.* (mean)	0.323582
"	Cal., *gm.* (20°C.)	0.324211
"	Cal., *kg*	0.000324048
"	Cal., *kg.* (IST.)	0.000323836
"	Cal., *kg.* (mean)	0.000323582
"	Cu. ft.-atm	0.000472541
"	Dyne-cm	1.35582×10^{7}
"	Ergs	1.35582×10^{7}
"	Foot-poundals	32.1740
"	Gram-cm	13825.5
"	Hp.-hours	5.05050×10^{-7}
"	Joules	1.35582
"	Kg.-meters	0.138255
"	Kw.-hours	3.76616×10^{-7}
"	Kw.-hours (Int.)	3.76554×10^{-7}
"	Liter-atm	0.0133805
"	Newton-meters	1.3558180
"	Lb. H_2O evap. from and at 212°F	1.3245×10^{-6}
"	Watt-hours	0.000376616
Foot-pounds/hr	B.t.u./min	2.14321×10^{-5}
"	B.t.u. (mean)/min	2.14013×10^{-5}
"	Cal., gm./min	0.00540080
"	Cal., *gm.* (mean)/min	0.00539304
"	Ergs/min	2.25970×10^{5}

To convert from	To	Multiply by
Foot-pounds/hr	Foot-pounds/min	0.0166666
"	Horsepower	5.050505×10^{-7}
"	Horsepower (metric)	5.12055×10^{-7}
"	Kilowatts	3.76616×10^{-7}
"	Watts	0.000376616
"	Watts (Int.)	0.000376554
Foot-pounds/min	B.t.u./sec	2.14321×10^{-5}
"	B.t.u. (mean)/sec	2.14013×10^{-5}
"	Cal., *gm.*/sec	0.00540080
"	Cal., *gm.* (mean)/sec	0.00539304
"	Ergs/sec	2.25970×10^{5}
"	Foot-pounds/sec	0.0166666
"	Horsepower	3.030303×10^{-5}
"	Horsepower (metric)	3.07233×10^{-5}
"	Joules/sec	0.0225970
"	Joules (Int.)/sec	0.0225932
"	Kilowatts	2.25970×10^{-5}
"	Watts	0.0225970
Foot-pounds/lb	B.t.u./lb	0.00128593
"	B.t.u. (IST.)/lb	0.00128509
"	B.t.u. (mean)/lb	0.00128408
"	Cal., *gm.*/gm	0.000714404
"	Cal., *gm.* (IST.)/gram	0.000713937
"	Cal., *gm.* (mean)/gram	0.000713377
"	Hp.-hr./lb	5.05050×10^{-7}
"	Joules/gram	0.00298907
"	Kg.-meters/gram	0.000304800
"	Kw.-hr./gram	8.30296×10^{-10}
Foot-pounds/sec	B.t.u./min	0.0771556
"	B.t.u. (mean)/min	0.0770447
"	B.t.u./sec	0.00128593
"	B.t.u. (mean)/sec	0.00128408
"	Cal., *gm.*/sec	0.324048
"	Cal., *gm.* (mean)/sec	0.323582
"	Ergs/sec	1.35582×10^{7}
"	Gram-cm./sec	13825.5
"	Horsepower	0.00181818
"	Joules/sec	1.35582
"	Kilowatts	0.00135582
"	Watts	1.35582
"	Watts (Int.)	1.35559
Furlongs	Centimeters	20116.8
"	Chains (Gunter's)	10
"	Chains (Ramden's)	6.6
"	Feet	660
"	Inches	7920
"	Meters	201.168
"	Miles (naut., Int.)	0.10862203
"	Miles (statute)	0.125
"	Rods	40
"	Yards	220
Gallons (Brit.)	Barrels (Brit.)	0.027777
"	Bushels (Brit.)	0.125
"	Cu. centimeters	4546.087
"	Cu. feet	0.1605436
"	Cu. inches	277.4193
"	Drachms (Brit. fluid)	1280
"	Firkins (Brit.)	0.111111
"	Gallons (U.S., liq.)	1.200949
"	Gills (Brit.)	32
"	Liters	4.545960
"	Minims (Brit.)	76800
"	Ounces (Brit., fluid)	160
"	Ounces (U.S., fluid)	153.7215
"	Pecks (Brit.)	0.5
"	Lb. of H_2O (62°F.)	10
Gallons (U.S., dry)	Barrels (U.S., dry)	0.038095592
"	Barrels (U.S., liq.)	0.036941181
"	Bushels (U.S.)	0.125
"	Cu. centimeters	4404.8828
"	Cu. feet	0.15555700
Gallons (U.S., dry)	Cu. inches	268.8025
"	Gallons (U.S., liq.)	1.16364719
"	Liters	4.404760
Gallons (U.S., liq.)	Acre-feet	3.0688833×10^{-6}
"	Barrels (U.S., liq.)	0.031746032
"	Barrels (petroleum, U.S.)	0.023809524
"	Bushels (U.S.)	0.10742088
"	Cu. centimeters	3785.4118
"	Cu. feet	0.133680555
"	Cu. inches	231
"	Cu. meters	0.0037854118
"	Cu. yards	0.0049511317
"	Gallons (Brit.)	0.8326747
"	Gallons (U.S., dry)	0.85936701
"	Gallons (wine)	1
"	Gills (U.S.)	32
"	Liters	3.7854118
"	Minims (U.S.)	61440
"	Ounces (U.S., fluid)	128
"	Pints (U.S., liq.)	8
"	Quarts (U.S., liq.)	4
Gallons (U.S.) of H_2O (4°C.) in air	Lb. of H_2O	8.33585
Gallons (U.S.) of H_2O (60°F.) in air	Lb. of H_2O	8.32823
Gallons (U.S.)/day	Cu. ft./hr	0.0055700231
Gallons (Brit.)/hr	Cu. meters/min	7.576812×10^{-5}
Gallons (U.S.)/hr	Acre-feet/hr	3.0688833×10^{-6}
"	Cu. ft./hr	0.1336805
"	Cu. meters/min	6.3090197×10^{-5}
"	Cu. yd./min	8.2518861×10^{-5}
"	Liters/hr	3.7854118
Gal. (Brit.)/sec	Cu. cm./sec	4546.087
Gal. (U.S.)/sec	Cu. cm./sec	3785.4118
"	Cu. ft./min	8.020833
"	Cu. yd./min	0.29706790
"	Liters/min	227.1183
Gammas	Grams	1×10^{-6}
"	Micrograms	1
Gausses	E.M. cgs. units of magnetic flux density	1
"	E.S. cgs. units	3.335635×10^{-11}
"	Gausses (Int.)	0.999670
"	Maxwells/sq. cm	1
"	Lines/sq. cm	1
"	Lines/sq. inch	6.4516
Gausses (Int.)	Gausses	1.000330
Gausses/oersted	E.M. cgs. units of permeability	1
"	E.S. cgs. units	1.112646×10^{-21}
Geepounds	Slugs	1
"	Kilograms	14.5939
Gigameters	Meters	1×10^{9}
Gilberts	Abampere-turns	0.079577472
"	Ampere-turns	0.79577472
"	E.M. cgs. units of mmf., *or* magnetic potential	1
"	E.S. cgs. units	2.997930×10^{10}
"	Gilberts (Int.)	1.000165
Gilberts (Int.)	Gilberts	0.999835
Gilberts/cm	Ampere-turns/cm	0.79577472
"	Ampere-turns/in	2.0212678
"	Oersteds	1
Gilberts/maxwell	Ampere-turns/weber	7.957747×10^{7}
"	E.M. cgs. units of reluctance	1
"	E.S. cgs. units	8.987584×10^{20}
Gills (Brit.)	Cu. cm	142.0652
"	Gallons (Brit.)	0.03125
"	Gills (U.S.)	1.200949
"	Liters	0.1420613
"	Ounces (Brit., fluid)	5

CONVERSION FACTORS (Continued)

To convert from	To	Multiply by
Gills (Brit.)	Ounces (U.S., fluid)	4.803764
"	Pints (Brit.)	0.25
Gills (U.S.)	Cu. cm	118.29412
"	Cu. inches	7.21875
"	Drams (U.S., fluid)	32
"	Gallons (U.S., liq.)	0.03125
"	Gills (Brit.)	0.8326747
"	Liters	0.1182908
"	Minims (U.S.)	1920
"	Ounces (U.S., fluid)	4
"	Pints (U.S., liq.)	0.25
"	Quarts (U.S., liq.)	0.125
Grades	Circles	0.0025
"	Circumferences	0.0025
"	Degrees	0.9
"	Minutes	54
"	Radians	0.015707963
"	Revolutions	0.0025
"	Seconds	3240
Grains	Carats (metric)	0.32399455
"	Drams (apoth. *or* troy)	0.016666
"	Drams (avdp.)	0.036571429
"	Dynes	63.5460
"	Grams	0.06479891
"	Milligrams	64.79891
"	Ounces (apoth. *or* troy)	0.0020833
"	Ounces (avdp.)	0.0022857143
"	Pennyweights	0.041666
"	Pounds (apoth. *or* troy)	0.000173611
"	Pounds (avdp.)	0.00014285714
"	Scruples (apoth.)	0.05
"	Tons (metric)	6.479891×10^{-8}
Grains/cu. ft	Grams/cu. meter	2.2883519
Grains/gal. (U.S.)	Parts/million*	17.11854
"	Pounds/million gal	142.8571
Grams	Carats (metric)	5
"	Decigrams	10
"	Dekagrams	0.1
"	Drams (apoth. *or* troy)	0.25720597
"	Drams (avdp.)	0.56438339
"	Dynes	980.665
"	Grains	15.432358
"	Kilograms	0.001
"	Micrograms	1×10^{6}
"	Myriagrams	0.0001
"	Ounces (apoth. *or* troy)	0.032150737
"	Ounces (avdp.)	0.035273962
"	Pennyweights	0.64301493
"	Poundals	0.0709316
"	Pounds (apoth. *or* troy)	0.0026792289
"	Pounds (avdp.)	0.0022046226
"	Scruples (apoth.)	0.77161792
"	Tons (metric)	1×10^{-6}
Grams/cm	Dynes/cm	980.665
"	Grams/inch	2.54
"	Kg./km	100
"	Kg./meter	0.1
"	Poundals/inch	0.180166
"	Pounds/ft	0.067196898
"	Pounds/inch	0.0055997415
"	Tons (metric)/km	0.1
Grams/(cm. × sec.)	Poises	1
"	Lb./(ft. × sec.)	0.06719690
Grams/cu. cm	Dynes/cu. cm	980.665
"	Grains/milliliter	15.43279
"	Grams/milliliter	1
"	Poundals/cu. inch	1.16236
"	Pounds/circ. mil-ft	3.4049170×10^{-7}
"	Pounds/cu. ft	62.427961
"	Pounds/cu. inch	0.036127292
"	Pounds/gal. (Brit.)	10.02241

To convert from	To	Multiply by
Grams/cu. cm	Pounds/gal. (U.S., dry)	9.7111064
"	Pounds/gal. (U.S., liq.)	8.3454044
Grams/cu. meter	Grains/cu. ft	0.43699572
Grams/liter	Parts/million*	1000
"	Lb./cu. ft.	0.06242621
"	Lb./gal. (U.S.)	8.345171×10^{-3}
Grams/milliliter	Grams/cu. cm	1
"	Pounds/cu. ft	62.42621
"	Pounds/gallon (U.S.)	8.345171
Grams/sq. cm	Atmospheres	0.000967841
"	Bars	0.000980665
"	Cm. of Hg. (0°C.)	0.0735559
"	Dynes/sq. cm	980.665
"	In. of Hg (32°F.)	0.0289590
"	Kg./sq. meter	10
"	Mm. of Hg (0°C.)	0.735559
"	Poundals/sq. inch	0.457623
"	Pounds/sq. inch	0.014223343
Grams/ton (long)	Milligrams/kg	0.98420653
Grams/ton (short)	Milligrams/kg	1.1023113
Grams-cm	B.t.u.	9.30113×10^{-8}
"	B.t.u. (IST.)	9.29505×10^{-8}
"	B.t.u. (mean)	9.28776×10^{-8}
"	Cal., *gm*	2.34385×10^{-5}
"	Cal., *gm*. (IST.)	2.34231×10^{-5}
"	Cal., *gm*. (mean)	2.34048×10^{-5}
"	Cal., *gm*. (15°C.)	2.34284×10^{-5}
"	Cal., *gm*, (20°C.)	2.34502×10^{-5}
"	Cal., *kg*	2.34385×10^{-8}
"	Cal., *kg*. (IST.)	2.34231×10^{-8}
"	Cal., *kg*. (mean)	2.34048×10^{-8}
"	Dyne-cm	980.665
"	Ergs	980.665
"	Foot-poundals	0.00232715
"	Foot-pounds	7.2330138×10^{-5}
"	Hp.-hours	3.65303×10^{-11}
"	Joules	9.80665×10^{-5}
"	Kw.-hours	2.72407×10^{-11}
"	Kw.-hours (Int.)	2.72362×10^{-11}
"	Newton-meters	9.80665×10^{-5}
"	Watt-hours	2.72407×10^{-8}
Gram-cm./sec	B.t.u./sec	9.30113×10^{-8}
"	Cal., *gm*./sec	2.34385×10^{-5}
"	Ergs-sec	980.665
"	Foot-pounds/sec	7.2330138×10^{-5}
"	Horsepower	1.31509×10^{-7}
"	Joules/sec	9.80665×10^{-5}
"	Kilowatts	9.80665×10^{-8}
"	Kilowatts (Int.)	9.80503×10^{-8}
"	Watts	9.80665×10^{-5}
Gram/sq. cm	Pounds/sq. inch	0.000341717
Gram wt.-sec./sq. cm	Poises	980.665
Gravitational constants	Cm./(sec. × sec.)	980.621
"	Ft./(sec. × sec.)	32.1725
Hands	Centimeters	10.16
"	Inches	4
Hectares	Acres	2.4710538
"	Ares	100
"	Sq. cm	1×10^{8}
"	Sq. feet	107639.10
"	Sq. meters	10000
"	Sq. miles	0.0038610216
"	Sq. rods	395.36861
Hectograms	Grams	100
"	Poundals	7.09316
"	Pounds (apoth *or* troy)	0.26792289
"	Pounds (avdp.)	0.22046226
Hectoliters	Bushels (Brit.)	2.749694
"	Bushels (U.S.)	2.837839

* Based on density of 1 gram/ml.

To convert from	To	Multiply by
Hectoliters	Cu. cm	1.00028×10^5
"	Cu. feet	3.531566
"	Gallons (U.S., liq.)	26.41794
"	Liters	100
"	Ounces (U.S.) fluid	3381.497
"	Pecks (U.S.)	11.35136
Hectometers	Centimeters	10000
"	Decimeters	1000
"	Dekameters	10
"	Feet	328.08399
"	Meters	100
"	Rods	19.883878
"	Yards	109.3613
Hectowatts	Watts	100
Hefner units	Candles (English)	0.86
"	Candles (German)	0.85
"	Candles (Int.)	0.90
"	10-cp. pentane candles	0.090
Henries	Abhenries	1×10^9
"	E.M. cgs. units	1×10^9
"	E.S. cgs. units	1.112646×10^{-12}
"	Henries (Int.)	0.999505
"	Millihenries	1000
"	Mks. (r *or* nr) units	1
"	Stathenries	1.112646×10^{-12}
Henries (Int.)	Henries	1.000495
Henries/meter	Cgs. units of permeability	795774.72
"	E.M. cgs. units	795774.72
"	E.S. cgs. units	8.854156×10^{-16}
"	Gausses/oersted	795774.72
"	Mks. (nr) units	0.079577472
"	Mks. (r) units	1
Hogsheads	Butts (Brit.)	0.5
"	Cu. feet	8.421875
"	Cu. inches	14553
"	Cu. meters	0.23848094
"	Gallons (Brit.)	52.458505
"	Gallons (U.S.)	63
"	Gallons (wine)	63
"	Liters	238.47427
Horsepower*	B.t.u. (mean)/hr	2542.48
"	B.t.u./min	42.4356
"	B.t.u. (mean)/sec	0.706243
"	Cal., *gm.*/hr	6.41616×10^5
"	Cal., *gm.* (IST.)/hr	6.41196×10^5
"	Cal., *gm.* (mean)/hr	6.40693×10^5
"	Cal., *gm.*/min	10693.6
"	Cal., *gm.* (IST.)/min	10686.6
"	Cal., *gm.* (mean)/min	10678.2
"	Ergs/sec	7.45700×10^9
"	Foot-pounds/hr	1980000
"	Foot-pounds/min	33000
"	Foot-pounds/sec	550
"	Horsepower (boiler)	0.0760181
"	Horsepower (electric)	0.999598
"	Horsepower (metric)	1.01387
"	Joules/sec	745.700
"	Kilowatts	0.745700
"	Kilowatts (Int.)	0.745577
"	Tons of refrig. (U.S., comm.)	0.21204
"	Watts	745.700
Horsepower (boiler)	B.t.u. (mean)/hr	33445.7
"	Cal., *gm.*/min	140671.6
"	Cal., *gm.* (mean)/min	140469.4
"	Cal., *gm.* (15°C.)/min	140611.1
"	Cal., *gm.*, (20°C.)/min	140742.2
"	Ergs/sec	9.80950×10^{10}
"	Foot-pounds/min	434107
"	Horsepower	13.1548
"	Horsepower (electric)	13.1495

To convert from	To	Multiply by
Horsepower (boiler)	Horsepower (metric)	13.3372
"	Horsepower (water)	13.1487
"	Joules/sec	9809.50
"	Kilowatts	9.80950
"	Lb. H_2O evap. per hr. from and at 212°F	34.5
Horsepower (electric)	B.t.u./hr	2547.16
"	B.t.u. (IST.)/hr	2545.50
"	B.t.u. (mean)/hr	2543.50
"	Cal., *gm.*/sec	178.298
"	Cal., *kg.*/hr	641.874
"	Ergs/sec	7.46×10^9
"	Foot-pounds/min	33013.3
"	Foot-pounds/sec	550.221
"	Horsepower	1.00040
"	Horsepower (boiler)	0.0760487
"	Horsepower (metric)	1.0142777
"	Horsepower (water)	0.999942
"	Joules/sec	746
"	Kilowatts	0.746
"	Watts	746
Horsepower (metric)	B.t.u./hr	2511.31
"	B.t.u. (IST.)/hr	2509.66
"	B.t.u. (mean)/hr	2507.70
"	Cal., *gm.*/hr	6.32838×10^5
"	Cal., *gm.* (IST.)/hr	6.32425×10^5
"	Cal., *gm.* (mean)/hr	6.31929×10^5
"	Ergs/sec	7.35499×10^9
"	Foot-pounds/min	32548.6
"	Foot-pounds/sec	542.476
"	Horsepower	0.986320
"	Horsepower (boiler)	0.0749782
"	Horsepower (electric)	0.985923
"	Horsepower (water)	0.985866
"	Kg.-meters/sec	75
"	Kilowatts	0.735499
"	Watts	735.499
Horsepower (water)	Foot-pounds/min	33015.2
"	Horsepower	1.00046
"	Horsepower (boiler)	0.0760531
"	Horsepower (electric)	1.00006
"	Horsepower (metric)	1.01434
"	Kilowatts	0.746043
Horsepower-hours	B.t.u	2546.14
"	B.t.u. (IST.)	2544.47
"	B.t.u. (mean)	2542.48
"	Cal., *gm*	641616
"	Cal., *gm.* (IST.)	641196
"	Cal., *gm.* (mean)	640693
"	Foot-pounds	1.98×10^6
"	Joules	2.68452×10^6
"	Kg.-meters	273745
"	Kw.-hours	0.745700
"	Watt-hours	745.700
Hp.-hr./lb	B.t.u./lb	2546.14
"	Cal., *gm.*/gram	1414.52
"	Cu. ft.-(lb./sq. in.)/lb	13750
"	Foot-pounds/lb	1980000
"	Joules/gram	5918.35
Hours (mean solar)	Days (mean solar)	0.0416666
"	Days (sidereal)	0.041780746
"	Hours (sidereal)	1.00273791
"	Minutes (mean solar)	60
"	Minutes (sidereal)	60.164275
"	Seconds (mean solar)	3600
"	Seconds (sidereal)	3609.8565
"	Weeks (mean calendar)	0.0059523809
Hours (sidereal)	Days (mean solar)	0.41552899
"	Days (sidereal)	0.0416666
"	Hours (mean solar)	0.99726957
"	Minutes (mean solar)	59.836174

* Mechanical horsepower, equal to 550 ft.-lb./sec.

To convert from	To	Multiply by
Hours (sidereal)	Minutes (sidereal)	60
Hundredweights (long)	Kilograms	50.802345
"	Pounds	112
"	Quarters (Brit., long)	4
"	Quarters (U.S., long)	0.2
"	Tons (long)	0.05
Hundredweights (short)	Kilograms	45.359237
"	Pounds (advp.)	100
"	Quarters (Brit., short)	4
"	Quarters (U.S., short)	0.2
"	Tons (long)	0.044642857
"	Tons (metric)	0.045359237
"	Tons (short)	0.05
Inches	Ångström units	2.54×10^8
"	Centimeters	2.54
"	Chains (Gunter's)	0.00126262
"	Cubits	0.055555
"	Fathoms	0.013888
"	Feet	0.083333
"	Feet (U.S. Survey)	0.083333167
"	Links (Gunter's)	0.126262
"	Links (Ramden's)	0.083333
"	Meters	0.0254
"	Mils	1000
"	Picas (printer's)	6.0225
"	Points (printer's)	72.27000
"	Wave length of orange-red line of krypton 86	41929.399
"	Wave length of the red line of cadmium	39450.369
"	Yards	0.027777
Inches of Hg (32°F.)	Atmospheres	0.0334211
	Bars	0.0338639
	Dynes/sq. cm	33863.9
	Ft. of air (1 atm., 60°F.)	926.24
"	Ft. of H_2O (39.2°F.)	1.132957
"	Grams/sq. cm	34.5316
"	Kg./sq. meter	345.316
"	Mm. of Hg (60°C.)	25.4
"	Ounces/sq. inch	7.85847
Inches of Hg (32°F.)	Pounds/sq. ft	70.7262
Inches of Hg (60°F.)	Atmospheres	0.0333269
"	Dynes/sq. cm	39768.5
"	Grams/sq. cm.	34.4343
"	Mm. of Hg (60°F.)	25.4
"	Ounces/sq. inch	7.83633
"	Pounds/sq. ft	70.5269
Inches of H_2O(4°C.)	Atmospheres	0.0024582
"	Dynes/sq. cm	2490.82
"	In. of Hg (32°F.)	0.0735539
"	Kg./sq. meter	25.3993
"	Ounces/sq. ft	83.2350
"	Ounces/sq. inch	0.578020
"	Pounds/sq. ft	5.20218
"	Pounds/sq. inch	0.03612628
Inches/hr	Cm./hr	2.54
"	Feet/hr	0.0833333
"	Miles/hr	1.578282×10^{-5}
Inches/min	Cm./hr	152.4
"	Feet/hr	5
"	Miles/hr	0.000946969
Joules (abs.)	B.t.u.	0.000948451
" "	B.t.u. (IST.)	0.000947831
" "	B.t.u. (mean)	0.000947088
" "	Cal., *gm*	0.239006
" "	Cal., *gm.* (IST.)	0.238849
" "	Cal., *gm.* (mean)	0.238662
" "	Cal., *gm.* (15°C.)	0.238903
" "	Cal., *gm.* (20°C.)	0.239126

To convert from	To	Multiply by
Joules (abs.)	Cal., *kg.* (mean)	0.000238662
" "	Cu. ft.-atm	0.000348529
" "	Ergs	1×10^7
" "	Foot-poundals	23.730360
" "	Foot-pounds	0.737562
" "	Gram-cm	10197.16
" "	Hp.-hours	3.72506×10^{-7}
" "	Joules (Int.)	0.999835
" "	Kg.-meters	0.1019716
" "	Kw.-hours	2.7777×10^{-7}
" "	Liter-atm	0.00986895
" "	Volt-coulombs (Int.)	0.999835
" "	Watt-hours (abs.)	0.0002777777
" "	Watt-hours (Int.)	0.000277732
" "	Watt-sec	1
" "	Watt-sec. (Int.)	0.999835
Joules (Int.)	B.t.u	0.000948608
"	B.t.u. (IST.)	0.000947988
"	B.t.u. (mean)	0.000947244
"	Cal. *gm*	0.239045
"	Cal., *gm.* (IST.)	0.238888
"	Cal., *gm.* (mean)	0.238702
"	C.h.u	0.000527004
"	C.h.u. (IST.)	0.000526660
"	C.h.u. (mean)	0.000526247
"	Cu. cm.-atm	9.87086
"	Cu. ft.-atm	0.000348586
"	Dyne-cm	1.000165×10^7
"	Ergs	1.000165×10^7
"	Foot-poundals	23.73428
"	Foot-pounds	0.737684
"	Gram-cm	10198.8
"	Joules (abs.)	1.000165
"	Kw.-hours	2.77824×10^{-7}
"	Liter-atm	0.00987058
"	Volt-coulombs	1.000165
"	Volt-coulombs (Int.)	1
"	Watt-sec	1.000165
"	Watt-sec. (Int.)	1
Joules/(abcoulomb × °F.)	Joules/(coulomb × °C.)	0.18
Joules/amp.-hr	Joules/abcoulomb	0.002777
"	Joules/statcoulomb	9.265653×10^{-14}
Joules/coulomb	Joules/abcoulomb	10
"	Volts	1
Joules/(coulomb × °F.)	Joules/(coulomb × °C.)	1.8
Joules/°C	B.t.u./°F	0.000526917
"	Cal., *gm.*/°C	0.239006
"	Cal., *gm.* (mean)/°C	0.238662
Joules/electronic charge	Joules/abcoulomb	6.24196×10^{19}
Joules/(electronic charge × °C.)	Joules/(coulomb × °C.)	6.24196×10^{18}
Joules/(gram × °C.)	B.t.u./(lb. × °F.)	0.239006
"	Cal., *gm.*/(gram × °C.)	0.239006
Joules (Int.)/(gram °C.)	B.t.u./(lb. × °F.)	0.239045
"	Cal., *gm.* (mean)/(gram × °C.)	0.238702
Joules/sec. (abs.)	B.t.u./min	0.0569071
"	Cal., *gm.*/min	14.3403
"	Cal., *kg.*/min	0.0143403
"	Cal., *kg.* (mean)/min	0.0143197
"	Dyne-cm./sec	1×10^7
"	Ergs/sec	1×10^7
"	Foot-pounds/sec	0.737562
"	Gram-cm./sec	10197.16
"	Horsepower	0.00134102
"	Watts	1
"	Watts (Int.)	0.999835
Joules (Int.)/sec	B.t.u./min	0.0569165

To convert from	To	Multiply by
Joules (Int.)/sec	B.t.u. (mean)/min	0.0568347
"	Cal., *gm.*/min	14.3427
"	Cal., *kg.*/min	0.0143427
"	Dyne-cm./sec	1.000165×10^{7}
"	Ergs/sec	1.000165×10^{7}
"	Foot-pounds/min	44.2610
"	Foot-pounds/sec	0.737684
"	Gram-cm./sec	10198.8
"	Horsepower	0.00134124
"	Watts	1.000165
"	Watts (Int.)	1
Kilderkins (Brit.)	Cu. cm	81829.57
"	Cu. feet	2.889784
"	Cu. inches	4993.55
"	Cu. meters	0.08182957
"	Gallons (Brit.)	18
Kilograms	Drams (apoth. *or* troy)	257.20597
"	Drams (avdp.)	564.38339
"	Dynes	980665
"	Grains	15432.358
"	Hundredweights (long)	0.019684131
"	Hundredweights (short)	0.022046226
"	Ounces (apoth. *or* troy)	32.150737
"	Ounces (avdp.)	35.273962
"	Pennyweights	643.01493
"	Poundals	70.931635
"	Pounds (apoth. *or* troy)	2.6792289
"	Pounds (avdp.)	2.2046226
"	Quarters (Brit., long)	0.078736522
"	Quarters (U.S. long)	0.0039368261
"	Scruples (apoth.)	771.61792
"	Slugs	0.06852177
"	Tons (long)	0.00098420653
"	Tons (metric)	0.001
"	Tons (short)	0.0011023113
Kilograms/cu. meter	Grams/cu. cm	0.001
"	Lb./cu. ft	0.062427961
"	Lb./cu. inch	3.6127292×10^{-5}
Kg. of ice melted/hr	Tons of refrig. (U.S., comm.)	0.026336
Kilograms/sq. cm	Atmospheres	0.967841
"	Bars	0.980665
"	Cm. of Hg (0°C.)	73.5559
"	Dynes/sq. cm	980665
"	Ft. of H_2O (39.2°F.)	32.8093
"	In. of Hg (32°F.)	28.9590
"	Pounds/sq. inch	14.223343
Kilograms/sq. meter	Atmospheres	9.67841×10^{-5}
"	Bars	9.80665×10^{-5}
"	Dynes/sq. cm	98.0665
"	Ft. of H_2O (39.2°F.)	0.00328093
"	Grams/sq. cm	0.1
"	In. of Hg. (32°F.)	0.00289590
"	Mm. of Hg (0°C.)	0.0735559
"	Pounds/sq. ft	0.20481614
"	Pounds/sq. in	0.0014223343
Kilograms/sq. mm	Pounds/sq. ft	204816.14
"	Pounds/sq. in	1422.3343
"	Tons (short)/sq. in	0.71116716
Kilogram sq. cm	Pounds sq. ft	0.0023730360
"	Pounds sq. in	0.34171719
Kilogram-meters	B.t.u. (mean)	0.00928776
"	Cal., *gm.* (mean)	2.34048
"	Cal., *kg.* (mean)	0.00234048
"	Cu. ft.-atm	0.00341790
"	Dynes-cm	9.80665×10^{7}
"	Ergs	9.80665×10^{7}
"	Foot-poundals	232.715
"	Foot-pounds	7.23301
"	Gram-cm	100000
Kilogram-meters	Hp.-hours	3.65304×10^{-6}
"	Joules	9.80665
"	Joules (Int.)	9.80503
"	Kw.-hours	2.72407×10^{-6}
"	Liter-atm	0.0967814
"	Newton-meters	9.80665
"	Watt-hours	0.00272407
"	Watt-hours (Int.)	0.00272362
Kilogram-meters/sec	Watts	9.80665
Kilolines	Maxwells	1000
"	Webers	1×10^{-5}
Kiloliters	Cu. centimeters	1×10^{6}
"	Cu. feet	35.31566
"	Cu. inches	61025.45
"	Cu. meters	1.000028
"	Cu. yards	1.307987
"	Gallons (Brit.)	219.9755
"	Gallons (U.S., dry)	227.0271
"	Gallons (U.S., liq.)	264.1794
"	Liters	1000
Kilometers	Astronomical units	6.68878×10^{-9}
"	Centimeters	100000
"	Feet	3280.8399
"	Feet (U.S. Survey)	3280.833
"	Light years	1.05702×10^{-13}
"	Meters	1000
"	Miles (naut., Int.)	0.53995680
"	Miles (statute)	0.62137119
"	Myriameters	0.1
"	Rods	198.83878
"	Yards	1093.6133
Kilometers/hr	Cm./sec	27.7777
"	Feet/hr	3280.8399
"	Feet/min	54.680665
"	Knots (Int.)	0.53995680
"	Meters/sec	0.277777
"	Miles (statute)/hr	0.62137119
Kilometers/(hr. × sec.)	Cm./(sec. × sec.)	27.7777
"	Ft./(sec. × sec.)	0.91134442
"	Meters/(sec. × sec.)	0.277777
Kilometers/min	Cm./sec	1666.666
"	Feet/min	3280.8399
"	Kilometers/hr	60
"	Knots (Int.)	32.397408
"	Miles/hr	37.282272
"	Miles/min	0.62137119
Kilovolts/cm	Abvolts/cm	1×10^{11}
"	Microvolts/meter	1×10^{11}
"	Millivolts/meter	1×10^{8}
"	Statvolts/cm	3.335635
"	Volts/inch	2540
Kilowatts	B.t.u./hr	3414.43
"	B.t.u. (IST.)/hr	3412.19
"	B.t.u. (mean)/hr	3409.52
"	B.t.u. (mean)/min	56.8253
"	B.t.u. (mean)/sec	0.947088
"	Cal., *gm.* (mean)/hr	859184
"	Cal., *gm.* (mean)/min	14319.7
"	Cal., *gm.* (mean)/sec	238.662
"	Cal., *kg.* (mean)/hr	859.184
"	Cal., *kg.* (mean)/min	14.3197
"	Cal., *kg.* (mean)/sec	0.238662
"	Cu. ft.-atm./hr	1254.70
"	Ergs/sec	1×10^{10}
"	Foot-poundals/min	1.42382×10^{6}
"	Foot-pounds/hr	2.65522×10^{6}
"	Foot-pounds/min	44253.7
"	Foot-pounds/sec	737.562
"	Gram-cm./sec	1.019716×10^{7}
"	Horsepower	1.34102

CONVERSION FACTORS (Continued)

To convert from	To	Multiply by
Kilowatts	Horsepower (boiler)	0.101942
"	Horsepower (electric)	1.34048
"	Horsepower (metric)	1.35962
"	Joules/hr	3.6×10^6
"	Joules (IST.)/hr	3.59941×10^6
"	Joules/sec	1000
"	Kg.-meters/hr	3.67098×10^5
"	Kilowatts (Int.)	0.999835
"	Watts (Int.)	999.835
Kilowatts (Int.)	B.t.u./hr	3414.99
"	B.t.u. (IST.)/hr	3412.76
"	B.t.u. (mean)/hr	3410.08
"	B.t.u. (mean)/min	56.8347
"	B.t.u. (mean)/sec	0.947244
"	Cal., *gm.* (mean)/hr	859326
"	Cal., *gm.* (mean)/min	14322.1
"	Cal., *kg.*/hr	860.563
"	Cal., *kg.* (IST.)/hr	860
"	Cal., *kg.* (mean)/hr	859.326
"	Cu. cm.-atm./hr	3.55351×10^7
"	Cu. ft.-atm./hr	1254.91
"	Ergs/sec	1.000165×10^{10}
"	Foot-poundals/min	1.42406×10^6
"	Foot-pounds/min	44261.0
"	Foot-pounds/sec	737.684
"	Gram-cm./sec	1.01988×10^7
"	Horsepower	1.34124
"	Horsepower (boiler)	0.101959
"	Horsepower (electric)	1.34070
"	Horsepower (metric)	1.35985
"	Joules/hr	3.60059×10^6
"	Joules (Int.)/hr	3.6×10^6
"	Kg.-meters/hr	367158
"	Kilowatts	1.000165
Kilowatt-hours	B.t.u. (mean)	3409.52
"	Cal., *gm.* (mean)	859184
"	Foot-pounds	2.65522×10^6
"	Hp.-hours	1.34102
"	Joules	3.6×10^6
"	Kg.-meters	367098
"	Lb. H_2O evap. from and at 212°F	3.5168
"	Watt-hours	1000
"	Watt-hours (Int.)	999.835
Kilowatt-hours (Int.)	B.t.u. (mean)	3410.08
"	Cal., *gm.* (IST.)	860000
"	Cal., *gm.* (mean)	859326
"	Cu. cm.-atm	3.55351×10^7
"	Cu. ft.-atm	1254.91
"	Foot-pounds	2.65566×10^6
"	Hp.-hours	1.34124
"	Joules	3.60059×10^6
"	Joules (Int.)	3.6×10^6
"	Kg.-meters	367158
Kw.-hr./gram	B.t.u./(lb	1.54876×10^6
"	B.t.u. (IST.)/lb	1.54774×10^6
"	B.t.u. (mean)/lb	1.54653×10^6
"	Cal., *gm.*/gram	860421
"	Cal., *gm.* (mean)/gram	859184
"	Cu. cm.-atm./gram	3.55292×10^7
"	Cu. ft.-atm./lb	569124
"	Hp.-hr./lb	608.277
"	Joules/gram	3.6×10^6
Knots (Int.)	Cm./sec	51.4444
"	Feet/hr	6076.1155
"	Feet/min	101.26859
"	Feet/sec	1.6878099
"	Kilometers/hr	1.852
"	Meters/min	30.8666
"	Meters/sec	0.514444
"	Miles (naut., Int.)/hr	1
"	Miles (statute)/hr	1.1507794

To convert from	To	Multiply by
Lamberts	Candles/sq. cm.	0.31830989
"	Candles/sq. ft.	295.71956
"	Candles/sq. inch	2.0536081
"	Foot-lamberts	929.0304
"	Lumens/sq. cm.	1
Lasts (Brit.)	Liters	2909.414
Leagues (naut., Brit.)	Feet	18240
"	Kilometers	5.559552
"	Leagues (naut., Int.)	1.0006393
"	Leagues (statute)	1.151515
"	Miles (statute)	3.454545
Leagues (naut., Int.)	Fathoms	3038.0577
"	Feet	18228.346
"	Kilometers	5.556
"	Leagues (statute)	1.1507794
"	Miles (statute)	3.4523383
Leagues (statute)	Fathoms	2640
"	Feet	15840
"	Kilometers	4.828032
"	Leagues (naut., Int.)	0.86897625
"	Miles (naut., Int.)	2.6069287
"	Miles (statute)	3
Light years	Astronomical units	63279.5
"	Kilometers	9.46055×10^{12}
"	Miles (statute)	5.87851×10^{12}
Lines	Maxwells	1
Lines (Brit.)	Centimeters	0.211666
"	Inches	0.083333
Lines/sq. cm.	Gausses	1
Lines/sq. inch	Gausses	0.15500031
"	Webers/sq. inch	1×10^{-8}
Links (Gunter's)	Chains (Gunter's)	0.01
"	Feet	0.66
"	Feet (U.S. Survey)	0.65999868
"	Inches	7.92
"	Meters	0.201168
"	Miles (statute)	0.000125
"	Rods	0.04
Links (Ramden's)	Centimeters	30.48
"	Chains (Ramdens)	0.01
"	Feet	1
"	Inches	12
Liters	Bushels (Brit.)	0.02749617
"	Bushels (U.S.)	0.02837759
"	Cu. centimeters	1000
"	Cu. feet	0.035314667
"	Cu. inches	61.023744
"	Cu. meters	0.001
"	Cu. yards	0.0013079506
"	Drams (U.S., fluid)	270.51218
"	Gallons (Brit.)	0.2199694
"	Gallons (U.S., dry)	0.22702075
"	Gallons (U.S., liq.)	0.26417205
"	Gills (Brit.)	7.039020
"	Gills (U.S.)	8.4535058
"	Hogsheads	0.004193212
"	Minims (U.S.)	16320.75
"	Ounces (Brit., fluid)	35.19510
"	Ounces (U.S., fluid)	33.814023
"	Pecks (Brit.)	0.1099848
"	Pecks (U.S.)	0.1135105
"	Pints (Brit.)	1.795756
"	Pints (U.S., dry)	1.8161660
"	Pints (U.S., liq.)	2.1133764
"	Quarts. (Brit.)	0.8798775
"	Quarts (U.S., dry)	0.90808298
"	Quarts (U.S., liq.)	1.0566882
Liters/min	Cu. ft./min	0.0353147
"	Cu. ft./sec	0.000588578
"	Gal. (U.S., liq.)/min	0.26417226
Liters/sec	Cu. ft./min	2.118880
"	Cu. ft./sec	0.0353147

To convert from	To	Multiply by
Liters/sec	Cu. yards/min	0.0784771
"	Gal. (U.S., liq.)/min	15.850342
"	Gal. (U.S., liq.)/sec	0.2641723
Liter-atm	B.t.u.	0.0961045
"	B.t.u. (IST.)	0.0960417
"	B.t.u. (mean)	0.0959664
"	Cal., *gm*	24.2179
"	Cal., *gm.* (IST.)	24.2021
"	Cal., *gm.* (mean)	24.1831
"	Cu. ft.-atm	0.0353157
"	Foot-poundals	2404.55
"	Foot-pounds	74.7356
"	Hp.-hours	3.77452×10^{-5}
"	Joules	101.328
"	Joules (Int.)	101.311
"	Kg.-meters	10.3326
"	Kw.-hours	2.81466×10^{-5}
Liter-atm. (lat. 45°)	Joules	101.323
Lumens	Candle power (spher.)	0.079577472
Lumens (at 5550 Å)	Watts	0.0014705882
Lumens/sq. cm	Lamberts	1
"	Phots	1
Lumens/(sq. cm. × steradian)	Lamberts	3.1415927
Lumens/sq. ft	Foot-candles	1
"	Foot-lamberts	1
"	Lumens/sq. meter	10.763910
Lumens/(sq. ft. × steradian)	Millilamberts	3.3815822
Lumens/sq. meter	Foot-candles	0.09290304
"	Lumens/sq. ft	0.09290304
"	Phots	0.0001
Lux	Foot-candles	0.09290304
"	Lumens/sq. meter	1
"	Phots	0.0001
Maxwells	E.M. cgs. units of induction flux	1
"	E.S. cgs. units	3.335635×10^{-11}
"	Gauss-sq. cm	1
"	Lines	1
"	Maxwells (Int.)	0.999670
"	Volt-seconds	1×10^{-8}
"	Webers	1×10^{-8}
Maxwells (Int.)	Maxwells	1.000330
Maxwells/sq. cm	Maxwells/sq. in	6.4516
"	Maxwells (Int.)/sq. cm	0.999670
Maxwells (Int.)/sq. cm	Maxwells/sq. cm	1.000330
Maxwells/sq. inch	Maxwells/sq. cm	0.15500031
Megalines	Maxwells	1×10^{6}
Megmhos/cm	Abmhos/cm	0.001
"	Megmhos/inch cube	2.54
"	$(\text{Microhm-cm.})^{-1}$	1
Megmhos/inch	Megmhos/cm	0.39370079
"	$(\text{Microhm-inches})^{-1}$	1
Megohms	Microhms	1×10^{12}
"	Ohms	1×10^{6}
"	Statohms	1.112646×10^{-6}
Megohms^{-1}	Micromhos	1
Meters	Ångström units	1×10^{10}
"	Centimeters	100
"	Chains (Gunter's)	0.049709695
"	Chains (Ramden's)	0.032808399
"	Fathoms	0.54680665
"	Feet	3.2808399
"	Feet (U.S. Survey)	3.280833
"	Furlongs	0.0049709695
"	Inches	39.370079
"	Kilometers	0.001
"	Links (Gunter's)	4.9709695
Meters	Links (Ramden's)	3.2808399
"	Megameters	1×10^{-6}
"	Miles (naut., Brit.)	0.00053961182
"	Miles (naut., Int.)	0.00053995680
"	Miles (statute)	0.00062137119
"	Millimeters	1000
"	Millimicrons	1×10^{9}
"	Mils	39370.079
"	Rods	0.19883878
"	Yards	1.0936133
Meters of Hg (0°C.)	Atmospheres	1.3157895
"	Ft. of H_2O (60°F.)	44.6474
"	In. of Hg (32°F.)	39.370079
"	Kg./sq. cm	1.35951
"	Pounds/sq. inch	19.3368
Meters/hr.	Feet/hr	3.2808399
"	Feet/min	0.054680665
"	Knots (Int.)	0.00053995680
"	Miles (statute)/hr	0.00062137119
Meters/min	Cm./sec	1.666666
"	Feet/min	3.2808399
"	Feet/sec	0.054680665
"	Kilometers/hr	0.06
"	Knots (Int.)	0.032397408
"	Miles (statute)/hr	0.037282272
Meters/sec	Feet/min	196.85039
"	Feet/sec	3.2808399
"	Kilometers/hr	3.6
"	Kilometers/min	0.06
"	Miles (statute)/hr	2.2369363
Meters/(sec. × sec.)	Kilometers/(hr. × sec.)	3.6
"	Miles/(hr. × sec.)	2.2369363
Meter-candles	Lumens/sq. meter	1
Mhos	Abmhos	1×10^{-9}
"	Cgs. units of conductance	1
"	E.M. cgs. units	1×10^{-9}
"	E.S. cgs. units	8.987584×10^{11}
"	Mhos (Int.)	1.000495
"	Mks. (r *or* nr) units	1
"	Ohms^{-1}	1
"	Siemen's units	1
"	Statmhos	8.987584×10^{11}
Mhos (Int.)	Abmhos	9.99505×10^{-10}
"	Mhos	0.999505
Mhos/meter	Abmhos/cm	1×10^{-11}
"	Mhos (Int.)/meter	1.000495
Mho-ft./circ. mil	Mhos/cm	6.0153049×10^{6}
Microfarads	Abfarads	1×10^{-15}
"	Farads	1×10^{-6}
"	Statfarads	8.987584×10^{5}
Micrograms	Grams	1×10^{-6}
"	Milligrams	0.001
Microhenries	Henries	1×10^{-6}
"	Stathenries	1.112646×10^{-18}
Microhms	Abohms	1000
"	Megohms	1×10^{-12}
"	Ohms	1×10^{-6}
"	Statohms	1.112646×10^{-18}
Microhm-cm	Abohm-cm	1000
"	Circ. mil-ohms/ft	6.0153049
"	Microhm-inches	0.39370079
"	Ohm-cm	1×10^{-6}
Microhm-inches	Circ. mil-ohms/ft	15.278875
"	Michrom-cm	2.54
Micromicrofarads	Farads	1×10^{-12}
Micromicrons	Ångström units	0.01
"	Centimeters	1×10^{-10}
"	Inches	$3.9370079 \times 10^{-11}$
"	Meters	1×10^{-12}
"	Microns	1×10^{-6}
Microns	Ångström units	10000

To convert from	To	Multiply by
Microns	Centimeters	0.0001
"	Feet	3.2808399×10^{-6}
"	Inches	3.9370079×10^{-5}
"	Meters	1×10^{-6}
"	Millimeters	0.001
"	Millimicrons	1000
Miles (naut., Brit.)	Cable lengths (Brit.)	8.4444
"	Fathoms	1013.333
"	Feet	6080
"	Meters	1853.184
"	Miles (Adm., Brit.)	1
"	Miles (naut., Int.)	1.0006393
"	Miles (statute)	1.151515
Miles (naut., Int.)	Cable lengths	8.4390493
"	Fathoms	1012.6859
"	Feet	6076.1155
"	Feet (U.S. Survey)	6076.1033
"	Kilometers	1.852
"	Leagues (naut., Int.)	0.333333
"	Meters	1852
"	Miles (geographical)	1
"	Miles (naut. Brit.)	0.99936110
"	Miles (statute)	1.1507794
Miles (statute)	Centimeters	160934.4
"	Chains (Gunter's)	80
"	Chains (Ramden's)	52.8
"	Feet	5280
"	Feet (U.S. Survey)	5279.9894
"	Furlongs	8
"	Inches	63360
"	Kilometers	1.609344
"	Light years	1.70111×10^{-13}
"	Links (Gunter's)	8000
"	Meters	1609.344
"	Miles (naut., Brit.)	0.86842105
"	Miles (naut., Int.)	0.86897624
"	Myriameters	0.1609344
"	Rods	320
"	Yards	1760
Miles/hr	Cm./sec	44.704
"	Feet/hr	5280
"	Feet/min	88
"	Feet/sec	1.466666
"	Kilometers/hr	1.609344
"	Knots (Int.)	0.86897624
"	Meters/min	26.8224
"	Miles/min	0.0166666
Miles/(hr. × min.)	Cm./(sec. × sec.)	0.7450666
Miles/(hr. × sec.)	Cm./(sec. × sec.)	44.704
"	Ft./(sec. × sec.)	1.466666
"	Kilometers/(hr. × sec.)	1.609344
"	Meters/(sec. × sec.)	0.44704
Miles/min	Cm./sec	2682.24
"	Feet/hr	316800
"	Feet/sec	88
"	Kilometers/min	1.609344
"	Knots (Int.)	52.138574
"	Meters/min	1609.344
"	Miles/hr	60
Millibars	Atmospheres	0.000986923
"	Bars	0.001
"	Baryes	1000
"	Dynes/sq. cm	1000
"	Grams/sq. cm	1.019716
"	In. of Hg (32°F.)	0.0295300
"	Pounds/sq. ft	2.088543
"	Pounds/sq. inch	0.0145038
Milligrams	Carats (1877)	0.004871
"	Carats (metric)	0.005
"	Drams (apoth. *or* troy)	0.00025720597
"	Drams (advp.)	0.00056438339
Milligrams	Grains	0.015432358
"	Grams	0.001
"	Ounces (apoth. *or* troy)	3.2150737×10^{-5}
"	Ounces (avdp.)	3.5273962×10^{-5}
"	Pennyweights	0.00064301493
"	Pounds (apoth. *or* troy)	2.6792289×10^{-6}
"	Pounds (avdp.)	2.2046226×10^{-6}
"	Scruples (apoth.)	0.00077161792
Milligrams/assay ton	Milligrams/kg	34.285714
"	Ounces (troy)/ton (avdp.)	1
Milligrams/gm	Dynes/cm	0.980665
"	Pounds/inch	5.5997415×10^{-6}
Milligrams/gram	Carats (parts gold per 24 of mixture)	0.024
"	Grams/ton (short)	907.18474
"	Milligrams/assay ton	29.166666
"	Ounces (avdp.)/ton (long)	35.84
"	Ounces (avdp.)/ton (short)	32
"	Ounces (troy)/ton (long)	32.6666
"	Ounces (troy)/ton (short)	29.1666
Milligrams/inch	Dynes/cm	0.386089
"	Dynes/inch	0.980665
"	Grams/cm	0.00039370079
"	Grams/inch	0.0001
Milligrams/kg	Pounds (avdp.)/ton (short)	0.002
Milligrams/liter	Grains/gal. (U.S.)	0.05841620
"	Grams/liter	0.001
"	Parts/million*	1
"	Lb./cu. ft	6.242621×10^{-5}
Milligrams/mm	Dynes/cm	9.80665
Millihenries	Abhenries	1×10^{6}
"	Henries	0.001
"	Stathenries	1.112646×10^{-15}
Millilamberts	Candles/sq. cm	0.00031830989
"	Candles/sq. inch	0.0020536081
"	Foot-lamberts	0.9290304
"	Lamberts	0.001
"	Lumens/sq. cm	0.001
"	Lumens/sq. ft	0.9290304
Milliliters	Cu. cm	1
"	Cu. inches	0.06102545
"	Drams (U.S., fluid)	0.2705198
"	Gills (U.S.)	0.008453742
"	Liters	0.001
"	Minims (U.S.)	16.23119
"	Ounces (Brit., fluid)	0.03519609
"	Ounces (U.S., fluid)	0.03381497
"	Pints (Brit.)	0.001759804
"	Pints (U.S., liq.)	0.002113436
Millimeters	Ångström units	1×10^{7}
"	Centimeters	0.1
"	Decimeters	0.01
"	Dekameters	0.0001
"	Feet	0.0032808399
"	Inches	0.039370079
"	Meters	0.001
"	Microns	1000
"	Mils	39.370079
"	Wave length of orange-red line of krypton 86	1650.76373
"	Wave length of red line of cadmium	1553.16413
Millimeters of Hg (0°C.)	Atmospheres	0.0013157895
"	Bars	0.00133322
"	Dynes/sq. cm	1333.224
"	Grams/sq. cm	1.35951
"	Kg./sq. meter	13,5951
"	Pounds/sq. ft	2.78450
"	Pounds/sq. inch	0.0193368
"	Torrs	1

* Density of 1 gram per milliliter of solvent.

To convert from	To	Multiply by
Millimicrons	Ångström units	10
"	Centimeters	1×10^{-7}
"	Inches	3.9370079×10^{-8}
"	Microns	0.001
"	Millimeters	1×10^{-6}
Milliphots	Foot-candles	0.9290304
"	Lumens/sq. ft	0.9290304
"	Lumens/sq. meter	10
"	Lux	10
"	Phots	0.001
Millivolts	Statvolts	3.335635×10^{-6}
"	Volts	0.001
Minims (Brit.)	Cu. cm	0.05919385
"	Cu. inches	0.003612230
"	Milliliters	0.05919219
"	Ounces (Brit., fluid)	0.0020833333
"	Scruples (Brit., fluid)	0.05
Minims (U.S.)	Cu. cm	0.061611520
"	Cu. inches	0.0037597656
"	Drams (U.S., fluid)	0.0166666
"	Gallons (U.S., liq.)	1.6276042×10^{-5}
"	Gills (U.S.)	0.0005208333
"	Liters	6.160979×10^{-5}
"	Milliliters	0.06160979
"	Ounces (U.S., fluid)	0.002083333
"	Pints (U.S., liq.)	0.0001302083
Minutes (angular)	Degrees	0.0166666
"	Quadrants	0.000185185
"	Radians	0.00029088821
"	Seconds (angular)	60
Minutes (mean solar)	Days (mean solar)	0.0006944444
"	Days (sidereal)	0.00069634577
"	Hours (mean solar)	0.0166666
"	Hours (sidereal)	0.016712298
"	Minutes (sidereal)	1.00273791
Minutes (sidereal)	Days (mean solar)	0.00069254831
"	Minutes (mean solar)	0.99726957
"	Months (mean calendar)	2.2768712×10^{-5}
"	Seconds (sidereal)	60
Minutes/cm	Radians/cm	0.00029088821
Months (lunar)	Days (mean solar)	29.530588
"	Hours (mean solar)	708.73411
"	Minutes (mean solar)	42524.047
"	Second (mean solar)	2.5514428×10^{6}
"	Weeks (mean calendar)	4.2186554
Months (mean calendar)	Days (mean solar)	30.416666
"	Hours (mean solar)	730
"	Months (lunar)	1.0300055
"	Weeks (mean calendar)	4.3452381
"	Years (calendar)	0.08333333
"	Years (sidereal)	0.083274845
"	Years (tropical)	0.083278075
Myriagrams	Grams	10000
"	Kilograms	10
"	Pounds (avdp.)	22.046226
Newtons	Dynes	100000
"	Pounds	0.22480894
Newton-meters	Dyne-cm	1×10^{7}
"	Gram-cm	10197.162
"	Kg.-meters	0.10197162
"	Pound-feet	0.73756215
Noggins (Brit.)	Cu. cm	142.0652
"	Gallons (Brit.)	0.03125
"	Gills (Brit.)	1
Oersteds	Ampere-turns/inch	2.0212678
"	Ampere-turns/meter	79.577472
"	E.M. cgs. units of magnetic field intensity	1
"	E.S. cgs. units	2.997930×10^{10}
Oersteds	Gilberts/cm	1
"	Oersteds (Int.)	1.000165
Oersteds (Int.)	Oersteds	0.999835
Ohms	Abohms	1×10^{9}
"	Cgs. units of resistance	1
"	Megohms	1×10^{-6}
"	Microhms	1×10^{6}
"	Ohms (Int.)	0.999505
"	Statohms	1.112646×10^{-12}
Ohms (Int.)	Ohms	1.000495
Ohms (mil, foot)	Circ. mil-ohms/ft	1
"	Ohm-cm	1.6624261×10^{-7}
Ohm-cm	Circ. mil-ohms/ft	6.0153049×10^{6}
"	Microhm-cm	1×10^{6}
"	Ohm-inches	0.39370079
Ohm-inches	Ohm-cm	2.54
Ohm-meters	Abohm-cm	1×10^{11}
"	E.M. cgs. units	1×10^{11}
"	E.S. cgs. units	1.112646×10^{-10}
"	Mks. units	1
"	Statohm-cm	1.112646×10^{-10}
Ounces (apoth. *or* troy)	Dekagrams	3.1103486
"	Drams (apoth. *or* troy)	8
"	Drams (avdp.)	17.554286
"	Grains	480
"	Grams	31.103486
"	Milligrams	31103.486
"	Ounces (avdp.)	1.0971429
"	Pennyweights	20
"	Pounds (apoth. *or* troy)	0.0833333
"	Pounds (avdp.)	0.068571429
"	Scruples (apoth.)	24
"	Tons (short)	3.4285714×10^{-5}
Ounces (avdp.)	Drams (apoth. *or* troy)	7.291666
"	Drams (avdp.)	16
"	Grains	437.5
"	Grams	28.349523
"	Hundredweights (long)	0.00055803571
"	Hundredweights (short)	0.000625
"	Ounces (apoth. *or* troy)	0.9114583
"	Pennyweights	18.229166
"	Pounds (apoth. *or* troy)	0.075954861
"	Pounds (avdp.)	0.0625
"	Scruples (apoth.)	21.875
"	Tons (long)	2.7901786×10^{-5}
"	Tons (metric)	2.8349527×10^{-5}
"	Tons (short)	3.125×10^{-5}
Ounces (Brit., fluid)	Cu. cm	28.41305
"	Cu. inches	1.733870
"	Drachms (Brit., fluid)	8
"	Drams (U.S., fluid)	7.686075
"	Gallons (Brit.)	0.00625
"	Milliliters	28.41225
"	Minims (Brit.)	480
"	Ounces (U.S., fluid)	0.9607594
Ounces (U.S., fluid)	Cu. cm	29.573730
"	Cu. inches	1.8046875
"	Cu. meters	2.9573730×10^{-5}
"	Drams (U.S., fluid)	8
"	Gallons (U.S., dry)	0.0067138047
"	Gallons (U.S., liq.)	0.0078125
"	Gills (U.S.)	0.25
"	Liters	0.029572702
"	Minims (U.S.)	480
"	Ounces (Brit., fluid)	1.040843
"	Pints (U.S., liq.)	0.0625
"	Quarts (U.S., liq.)	0.03125
Ounces/sq. inch	Dynes/sq. cm	4309.22
"	Grams/sq. cm	4.3941849
"	In. of H_2O (39.2°F.)	1.73004
"	In. of H_2O (60°F.)	1.73166

To convert from	To	Multiply by
Ounces/sq. inch	Pounds/sq. ft.	9
"	Pounds/sq. inch	0.0625
Ounces (avdp.)/ton (long)	Milligrams/kg	27.901786
Ounces (avdp.)/ton (short)	Milligrams/kg	31.25
Paces	Centimeters	76.2
"	Chains (Gunter's)	0.0378788
"	Chains (Ramden's)	0.025
"	Feet	2.5
"	Hands	7.5
"	Inches	30
"	Ropes (Brit.)	0.125
Palms	Centimeters	7.62
"	Chains (Ramden's)	0.0025
"	Cubits	0.1666666
"	Feet	0.25
"	Hands	0.75
"	Inches	3
Parsecs	Kilometers	3.08572×10^{13}
"	Miles (statute)	1.91738×10^{13}
Parts/million*	Grains/gal. (Brit.)	0.07015488
"	Grains/gal. (U.S.)	0.05841620
"	Grams/liter	0.001
"	Milligrams/liter	1
Pecks (Brit.)	Bushels (Brit.)	0.25
"	Coombs (Brit.)	0.0625
"	Cu. cm	9092.175
"	Cu. inches	554.8385
"	Gallons (Brit.)	2
"	Gills (Brit.)	64
"	Hogsheads	0.03812537
"	Kilderkins (Brit.)	0.111111
"	Liters	9.091920
"	Pints (Brit.)	16
"	Quarterns (Brit., dry)	4
"	Quarters (Brit., dry)	0.03125
"	Quarts (Brit.)	8
"	Quarts (U.S., dry)	8.256449
Pecks (U.S.)	Barrels (U.S., dry)	0.076191185
"	Bushels (U.S.)	0.25
"	Cu. cm	8809.7675
"	Cu. feet	0.311114005
"	Cu. inches	537.605
"	Gallons (U.S., dry)	2
"	Gallons (U.S., liq.)	2.3272944
"	Liters	8.809521
"	Pints (U.S., dry)	16
"	Quarts (U.S., dry)	8
Pennyweights	Drams (apoth. *or* troy)	0.4
"	Drams (avdp.)	0.87771429
"	Grains	24
"	Grams	1.55517384
"	Ounces (apoth. *or* troy)	0.05
"	Ounces (avdp.)	0.054857143
"	Pounds (apoth. *or* troy)	0.0041666
"	Pounds (avdp.)	0.0034285714
Perches (masonry)	Cu. feet	24.75
Phots	Foot-candles	929.0304
"	Lumens/sq. cm	1
"	Lumens/sq. meter	10000
"	Lux	10000
Picas (printer's)	Centimeters	0.42175176
"	Inches	0.166044
Pints (Brit.)	Cu. cm	568.26092
"	Gallons (Brit.)	0.125
"	Gills (Brit.)	4
"	Gills (U.S.)	4.803797
"	Liters	0.5682450
Pints (Brit.)	Minims (Brit.)	9600
"	Ounces (Brit., fluid)	20
"	Pints (U.S., dry)	1.032056
"	Pints (U.S., liq.)	1.200949
"	Quarts (Brit.)	0.5
"	Scruples (Brit., fluid)	480
Pints (U.S., dry)	Bushels (U.S.)	0.015625
"	Cu. cm	550.61047
"	Cu. inches	33.6003125
"	Gallons, (U.S., dry)	0.125
"	Gallons (U.S., liq.)	0.14545590
"	Liters	0.5505951
"	Pecks (U.S.)	0.0625
"	Quarts (U.S., dry)	0.5
Pints (U.S., liq.)	Cu. cm	473.17647
"	Cu. feet	0.016710069
"	Cu. inches	28.875
"	Cu. yards	0.00061889146
"	Drams (U.S., fluid)	128
"	Gallons (U.S., liq.)	0.125
"	Gills (U.S.)	4
"	Liters	0.4731632
"	Milliliters	473.1632
"	Minims (U.S.)	7680
"	Ounces (U.S., fluid)	16
"	Pints (Brit.)	0.8326747
"	Quarts (U.S., liq.)	0.5
Planck's constant	Erg-seconds	6.6255×10^{-27}
"	Joule-seconds	6.6255×10^{-34}
"	Joule-sec./Avog. No. (chem.)	3.9905×10^{-10}
Points (printer's)	Centimeters	0.03514598
"	Inches	0.013837
"	Picas	0.0833333
Poises**	Cgs. units of absolute viscosity	1
"	Grams/(cm. × sec.)	1
Poise-cu. cm./gram	Sq. cm./sec	1
Poise-cu. ft./lb.	Sq. cm./sec	62.427960
Poise-cu. in./gram	Sq. cm./sec	16.387064
Poles/sq. cm	E.M. cgs. units of magnetization	1
Pottles (Brit.)	Gallons (Brit.)	0.5
"	Liters	2.272980
Poundals	Dynes	13825.50
"	Grams	14.09808
"	Pounds (avdp.)	0.0310810
Pounds (apoth. *or* troy)	Drams (apoth. *or* troy)	96
"	Drams (avdp.)	210.65143
"	Grains	5760
"	Grams	373.24172
"	Kilograms	0.37324172
"	Ounces (apoth. *or* troy)	12
"	Ounces (avdp.)	13.165714
"	Pennyweights	240
"	Pounds (avdp.)	0.8228571
"	Scruples (apoth.)	288
"	Tons (long)	0.00036734694
"	Tons (metric)	0.00037324172
"	Tons (short)	0.00041142857
Pounds (avdp.)	Drams (apoth. *or* troy)	116.6666
"	Drams (avdp.)	256
"	Grains	7000
"	Grams	453.59237
"	Hundredweights (long)	0.00892857
"	Hundredweights (short)	0.01
"	Kilograms	0.45359237
"	Ounces (apoth. *or* troy)	14.583333
"	Ounces (avdp.)	16

* Based on density of 1 gram/ml. for the solvent.

** See separate table of "viscosity conversion factors".

To convert from	To	Multiply by
Pounds (avdp.).......	Pennyweights..............	291.6666
"	Poundals..................	32.1740
"	Pounds (apoth. *or* troy)....	1.215277
"	Scruples (apoth.).........	350
"	Slugs.....................	0.0310810
"	Tons (long)...............	0.00044642857
"	Tons (metric).............	0.00045359237
"	Tons (short)..............	0.0005
Pounds of H_2O evap. from and at 212°F........	B.t.u.....................	970.9
"	B.t.u. (IST.).............	970.2
"	B.t.u. (mean).............	969.4
"	Joules....................	1.0237×10^6
"	Joules (Int.).............	1.0234×10^6
Pounds/cu. ft........	Grams/cu. cm..............	0.016018463
"	Kg./cu. meter.............	16.018463
Pounds/cu. inch......	Grams/cu. cm..............	27.679905
"	Grams/liter...............	27.68068
"	Kg./cu. meter.............	27679.905
Pounds/gal. (Brit.)...	Pounds/cu. ft.............	6.228839
Pounds/gal. (U.S., liq.)........	Grams/cu. cm..............	0.11982643
"	Pounds/cu. ft.............	7.4805195
Pounds/inch.........	Grams/cm..................	178.57967
"	Grams/ft..................	5443.1084
"	Grams/inch................	453.59237
"	Ounces/cm.................	6.2992
"	Ounces/inch...............	16
"	Pounds/meter..............	39.370079
Pounds/minute.......	Kilograms/hr..............	27.2155422
"	Kilograms/min.............	0.45359237
Pounds of H_2O (39.2°F.)/min........	Cu. ft./min...............	0.01601891
"	Gal. (U.S.)/min...........	0.1198298
"	Liters/min................	0.45359237
Pounds/sq. ft........	Atmospheres...............	0.000472541
"	Bars......................	0.000478803
"	Cm. of Hg (0°C.)..........	0.0359131
"	Dynes/sq. cm..............	478.803
"	Ft. of air (1 atm., 60°F.)...	13.096
"	Grams/sq. cm..............	0.48824276
"	In. of Hg (32°F.).........	0.0141390
"	In. of H_2O (39.2°F.).......	0.192227
"	Kg./sq. meter.............	4.8824276
"	Mm. of Hg (0°C.)..........	0.359131
Pounds/sq. inch......	Atmospheres...............	0.0680460
"	Bars......................	0.0689476
"	Cm. of Hg (0°C.)..........	5.17149
"	Cm. of H_2O (4°C.).........	70.3089
"	Dynes/sq. cm..............	68947.6
"	Grams/sq. cm..............	70.306958
"	In. of Hg (32°F.).........	2.03602
"	In. of H_2O (39.2°F.).......	27.6807
"	Kg./sq. cm................	0.070306958
"	Mm. of Hg (0°C.)..........	51.7149
Pound wt.-sec./sq. ft..	Poises....................	478.803
Pound wt.-sec./sq. in.	Poises....................	68947.6
Puncheons (Brit.)....	Cu. meters................	0.31797510
"	Gallons (Brit.)...........	69.94467
"	Gallons (U.S.)............	84
Quadrants...........	Minutes...................	5400
"	Radians...................	1.5707963
Quarterns (Brit., dry).	Buckets (Brit.)...........	0.125
"	Bushels (Brit.)...........	0.0625
"	Cu. cm....................	2273.044
"	Gallons (Brit.)...........	0.5
"	Liters....................	2.272980
"	Pecks (Brit.).............	0.25
Quarterns (Brit., liq.).	Cu. cm....................	142.0652
Quarterns (Brit., liq.).	Gallons (Brit.)...........	0.03125
"	Liters....................	0.1420613
Quarters (U.S., long)..	Kilograms.................	254.0117272
" ..	Pounds (avdp.)............	560
Quarters (U.S., short).	Kilograms.................	226.796185
"	Pounds....................	500
Quarts (Brit.)........	Cu. cm....................	1136.522
"	Cu. inches................	69.35482
"	Gallons (Brit.)...........	0.25
"	Gallons (U.S., liq.)......	0.3002373
"	Liters....................	1.136490
"	Quarts (U.S., dry)........	1.032056
"	Quarts (U.S., liq.).......	1.200949
Quarts (U.S., dry)....	Bushels (U.S.)............	0.03125
"	Cu. cm....................	1101.2209
"	Cu. feet..................	0.038889251
"	Cu. inches................	67.200625
"	Gallons (U.S., dry).......	0.25
"	Gallons (U.S., liq.)......	0.29091180
"	Liters....................	1.1011901
"	Pecks (U.S.)..............	0.125
"	Pints (U.S., dry).........	2
Quarts (U.S., liq.)....	Cu. cm....................	946.35295
"	Cu. feet..................	0.033420136
"	Cu. inches................	57.75
"	Drams (U.S., fluid).......	256
"	Gallons (U.S., dry).......	0.21484175
"	Gallons (U.S., liq.)......	0.25
"	Gills (U.S.)..............	8
"	Liters....................	0.9463264
"	Ounces (U.S., fluid)......	32
"	Pints (U.S., liq.)........	2
"	Quarts (Brit.)............	0.8326747
"	Quarts (U.S., dry)........	0.8593670
Quintals (metric).....	Grams.....................	100000
"	Hundredweights (long).....	1.9684131
"	Kilograms.................	100
"	Pounds (avdp.)............	220.46226
Radians..............	Circumferences............	0.15915494
"	Degrees...................	57.295779
"	Minutes...................	3437.7468
"	Quadrants.................	0.63661977
"	Revolutions...............	0.15915494
"	Seconds...................	206264.81
Radians/cm..........	Degrees/cm................	57.295779
"	Degrees/ft................	1746.3754
"	Degrees/inch..............	145.53128
"	Minutes/cm................	3437.7468
Radians/sec..........	Degrees/sec...............	57.295779
"	Revolutions/min...........	9.5492966
"	Revolutions/sec...........	0.15915494
Radians/(sec. × sec.).	Revolutions/(min. × min.)	572.95779
" ..	Revolutions/(min. × sec.).	9.5492966
" ..	Revolutions/(sec. × sec.).	0.15915494
Register tons.........	Cu. feet..................	100
"	Cu. meters................	2.8316847
Revolutions..........	Degrees...................	360
"	Grades....................	400
"	Quadrants.................	4
"	Radians...................	6.2831853
Reyns*..............	Centipoises...............	6.89476×10^6
Rhes.................	Poises^{-1}..................	1
Rods.................	Centimeters...............	502.92
"	Chains (Gunter's).........	0.25
"	Chains (Ramden's).........	0.165
"	Feet......................	16.5
"	Feet (U.S. Survey)........	16.499967
"	Furlongs..................	0.025
"	Inches....................	198
"	Links (Gunter's)..........	25

* See also separate table of "viscosity conversion factors".

To convert from	To	Multiply by
Rods	Links (Ramden's)	16.5
"	Meters	5.0292
"	Miles (statute)	0.003125
"	Perches	1
"	Yards	5.5
Rods (Brit., volume)	Cu. feet	1000
"	Cu. meters	28.316847
Roods (Brit.)	Acres	0.25
"	Ares	10.117141
"	Sq. perches	40
"	Sq. yards	1210
Ropes (Brit.)	Feet	20
"	Meters	6.096
"	Yards	6.6666666
Scruples (apoth.)	Drams (apoth. *or* troy)	0.333333
"	Drams (avdp.)	0.73142857
"	Grains	20
"	Grams	1.2959782
"	Ounces (apoth. *or* troy)	0.041666
"	Ounces (avdp.)	0.045714286
"	Pennyweights	0.833333
"	Pounds (apoth. *or* troy)	0.003472222
"	Pounds (avdp.)	0.0028571429
Scruples (Brit., fluid)	Minims (Brit.)	20
Seams (Brit.)	Bushels (Brit.)	8
"	Cu. feet	10.27479
"	Liters	290.9414
Seconds (angular)	Degrees	0.000277777
"	Minutes	0.0166666
"	Radians	4.8481368×10^{-6}
Seconds (mean solar)	Days (mean solar)	1.1574074×10^{-5}
"	Days (sidereal)	1.1605763×10^{-5}
"	Hours (mean solar)	0.0002777777
"	Hours (sidereal)	0.00027853831
"	Minutes (mean solar)	0.0166666
"	Minutes (sidereal)	0.016712298
"	Seconds (sidereal)	1.00273791
Seconds (sidereal)	Days (mean solar)	1.1542472×10^{-5}
"	Days (sidereal)	1.1574074×10^{-5}
"	Hours (mean solar)	0.00027701932
"	Hours (sidereal)	0.000277777
"	Minutes (mean solar)	0.01662159
"	Minutes (sidereal)	0.0166666
"	Seconds (mean solar)	0.99726957
Siemen's units	*Same as* Mhos	
Skeins	Feet	360
"	Meters	109.728
Slugs	Geepounds	1
"	Kilograms	14.5939
"	Pounds (avdp.)	32.1740
Slugs/cu. ft.	Grams/cu. cm	0.515379
Space (entire)	Hemispheres	2
"	Steradians	12.566371
Spans	Centimeters	22.86
"	Fathoms	0.125
"	Feet	0.75
"	Inches	9
"	Quarters (Brit. linear)	1
Spherical right angles	Hemispheres	0.25
"	Spheres	0.125
"	Steradians	1.5707963
Sq. centimeters	Ares	1×10^{-6}
"	Circ. mm	127.32395
"	Circ. mils	197352.52
"	Sq. chains (Gunter's)	2.4710538×10^{-7}
"	Sq. chains (Ramden's)	1.0763910×10^{-7}
"	Sq. decimeters	0.01
"	Sq. feet	0.0010763910
"	Sq. ft. (U.S. Survey)	0.0010763867
"	Sq. inches	0.15500031
"	Sq. meters	0.0001

To convert from	To	Multiply by
Sq. centimeters	Sq. mm	100
"	Sq. mils	155000.31
"	Sq. rods	3.9536861×10^{-6}
"	Sq. yards	0.00011959900
Sq. chains (Gunter's)	Acres	0.1
"	Sq. feet	4356
"	Sq. ft. (U.S. Survey)	4355.9826
"	Sq. inches	627264
"	Sq. links (Gunter's)	10000
"	Sq. meters	404.68564
"	Sq. miles	0.00015625
"	Sq. rods	16
"	Sq. yards	484
Sq. chains (Ramden's)	Acres	0.22956841
"	Sq. feet	10000
"	Sq. ft. (U.S. Survey)	9999.9600
"	Sq. inches	1.44×10^{6}
"	Sq. links (Ramden's)	10000
"	Sq. meters	929.0304
"	Sq. miles	0.00035870064
"	Sq. rods	36.730946
"	Sq. yards	1111.111
Sq. decimeters	Sq. cm	100
"	Sq. inches	15.500031
Square degrees	Steradians	0.00030461742
Sq. dekameters	Acres	0.024710538
"	Ares	1
"	Sq. meters	100
"	Sq. yards	119.59900
Sq. feet	Acres	2.295684×10^{-5}
"	Ares	0.0009290304
"	Sq. cm	929.0304
"	Sq. chains (Gunter's)	0.00022956841
"	Sq. ft. (U.S. Survey)	0.99999600
"	Sq. inches	144
"	Sq. links (Gunter's)	2.2956841
"	Sq. meters	0.09290304
"	Sq. miles	3.5870064×10^{-8}
"	Sq. rods	0.0036730946
"	Sq. yards	0.111111
Sq. feet (U.S. Survey)	Acres	$2.29569330 \times 10^{-5}$
"	Sq. centimeters	929.03412
"	Sq. chains (Ramden's)	0.00010000040
"	Sq. feet	1.0000040
Sq. hectometers	Sq. meters	10000
Sq. inches	Circ. mils	1273239.5
"	Sq. cm	6.4516
"	Sq. chains (Gunter's)	1.5942251×10^{-6}
"	Sq. decimeters	0.064516
"	Sq. feet	0.0069444
"	Sq. ft. (U.S. Survey)	0.0069444167
"	Sq. links (Gunter's)	0.01594225
"	Sq. meters	0.00064516
"	Sq. miles	$2.4909767 \times 10^{-10}$
"	Sq. mm	645.16
"	Sq. mils	1×10^{6}
Sq. inches/sec	Sq. cm./hr	23225.76
"	Sq. cm./sec	6.4516
"	Sq. ft./min	0.416666
Sq. kilometers	Acres	247.10538
"	Sq. feet	1.0763910×10^{7}
"	Sq. ft. (U.S. Survey)	1.0763867×10^{7}
"	Sq. inches	1.5500031×10^{9}
"	Sq. meters	1×10^{6}
"	Sq. miles	0.38610216
"	Sq. yards	1.1959900×10^{6}
Sq. links (Gunter's)	Acres	1×10^{-5}
"	Sq. cm	404.68564
"	Sq. chains (Gunter's)	0.0001
"	Sq. feet	0.4356
"	Sq. ft. (U.S. Survey)	0.43559826
"	Sq. Inches	62.7264

To convert from	To	Multiply by
Sq. links (Ramden's)	Acres	2.2956841×10^{-5}
"	Sq. feet	1
Sq. meters	Acres	0.00024710538
"	Ares	0.01
"	Hectares	0.0001
"	Sq. cm	10000
"	Sq. feet	10.763910
"	Sq. inches	1550.0031
"	Sq. kilometers	1×10^{-6}
"	Sq. links (Gunter's)	24.710538
"	Sq. links (Ramden's)	10.763910
"	Sq. miles	3.8610216×10^{-7}
"	Sq. mm	1×10^{6}
"	Sq. rods	0.039536861
"	Sq. yards	1.1959900
Sq. miles	Acres	640
"	Hectares	258.99881
"	Sq. chains (Gunter's)	6400
"	Sq. feet	2.7878288×10^{7}
"	Sq. ft. (U.S. Survey)	2.78288×10^{7}
"	Sq. kilometers	2.5899881
"	Sq. meters	2589988.1
"	Sq. rods	102400
"	Sq. yards	3.0976×10^{6}
Sq. millimeters	Circ. mm	1.2732395
"	Circ. mils	1973.5252
"	Sq. cm	0.01
"	Sq. inches	0.0015500031
"	Sq. meters	1×10^{-6}
Sq. mils	Circ. mils	1.2732395
"	Sq. cm	6.4516×10^{-6}
"	Sq. inches	1×10^{-6}
"	Sq. mm	0.00064516
Sq. rods	Acres	0.00625
"	Ares	0.2529285264
"	Hectares	0.002529285264
"	Sq. cm	252928.5264
"	Sq. feet	272.25
"	Sq. ft. (U.S. Survey)	272.24891
"	Sq. inches	39204
"	Sq. links (Gunter's)	625
"	Sq. links (Ramden's)	272.25
"	Sq. meters	25.29285264
"	Sq. miles	9.765625×10^{-6}
"	Sq. yards	30.25
Sq. yards	Acres	0.00020661157
"	Ares	0.0083612736
"	Hectares	8.3612736×10^{-5}
"	Sq. cm	8361.2736
"	Sq. chains (Gunter's)	0.0020661157
"	Sq. chains (Ramden's)	0.0009
"	Sq. feet	9
"	Sq. ft. (U.S. Survey)	8.9999640
"	Sq. inches	1296
"	Sq. links (Gunter's)	20.661157
"	Sq. links (Ramden's)	9
"	Sq. meters	0.83612736
"	Sq. miles	$3.228305785 \times 10^{-7}$
"	Sq. perches (Brit.)	0.033057851
"	Sq. rods	0.033057851
Statamperes	Abamperes	3.335635×10^{-11}
"	Amperes	3.335635×10^{-10}
"	E.M. cgs. units of current	3.335635×10^{-11}
"	E.S. cgs. units	1
Statcoulombs	Ampere-hours	9.265653×10^{-14}
"	Coulombs	3.335635×10^{-10}
"	Electronic charges	2.082093×10^{9}
"	E.M. cgs. units of electric charge	3.335635×10^{-11}
Statfarads	E.M. cgs. units of capacitance	1.112646×10^{-21}
"	E.S. cgs. units	1

To convert from	To	Multiply by
Statfarads	Farads	1.112646×10^{-12}
"	Microfarads	1.112646×10^{-6}
Stathenries	Abhenries	8.987584×10^{20}
"	E.M. cgs. units of inductance	8.987584×10^{20}
"	E.S. cgs. units	1
"	Henries	8.987584×10^{11}
"	Millihenries	8.987584×10^{14}
Statohms	Abohms	8.987584×10^{20}
"	E.S. cgs. units	1
"	Ohms	8.987584×10^{11}
Statvolts	Abvolts	2.997930×10^{10}
"	Volts	299.7930
Statvolts/cm	Volts/cm	299.7930
"	Volts/inch	761.4742
Statvolts/inch	Volts/cm	118.0287
Steradians	Hemispheres	0.15915494
"	Solid angles	0.079577472
"	Spheres	0.079577472
"	Spher. right angles	0.63661977
"	Square degrees	3282.8063
Steres	Cubic meters	1
"	Decisteres	10
"	Dekasteres	0.1
"	Liters	999.972
Stilbs	Candles/sq. cm	1
"	Candles/sq. inch	6.4516
"	Lamberts	3.1415927
Stokes*	Cgs. units of kinematic viscosity	1
"	Sq. cm./sec	1
"	Sq. inches/sec	0.15500031
"	Poise cu. cm./gram	1
Stones (Brit., legal)	Centals (Brit.)	0.14
Tons (long)	Dynes	9.96402×10^{8}
"	Hundredweights (long)	20
"	Hundredweights (short)	22.4
"	Kilograms	1016.0469
"	Ounces (avdp.)	35840
"	Pounds (apoth. *or* troy)	2722.22
"	Pounds (avdp.)	2240
"	Tons (metric)	1.0160469
"	Tons (short)	1.12
Tons (metric)	Dynes	9.80665×10^{8}
"	Grams	1×10^{6}
"	Hundredweights (short)	22.046226
"	Kilograms	1000
"	Ounces (avdp.)	35273.962
"	Pounds (apoth. *or* troy)	2679.2289
"	Pounds (avdp.)	2204.6226
"	Tons (long)	0.98420653
"	Tons (short)	1.1023113
Tons (short)	Dynes	8.89644×10^{8}
"	Hundredweights (short)	20
"	Kilograms	907.18474
"	Ounces (avdp.)	32000
"	Pounds (apoth. *or* troy)	2430.555
"	Pounds (avdp.)	2000
"	Tons (long)	0.89285714
"	Tons (metric)	0.90718474
Tons of refrig. (U.S., comm.)	B.t.u. (IST.)/hr	12000
"	B.t.u. (IST.)/min	200
"	Cal., *kg.* (IST.)/hr	3023.949
"	Horsepower	4.71611
"	Kg. of ice melted/hr	37.971
"	Lb. of ice melted/hr	83.711
Tons of refrig. (U.S., std.)	B.t.u. (IST.)	288000
"	B.t.u. (mean)	287774
"	Cal., *kg.* (IST.)	72574.8

* See also separate table of "viscosity conversion factors".

To convert from	To	Multiply by
Tons of refrig. (U.S., std.)	Cal., *kg.* (mean)	72517.9
"	Lb. of ice melted	2009.1
Tons (long)/sq. ft	Atmospheres	1.05849
"	Dynes/sq. cm	1.07252×10^6
"	Grams/sq. cm	1093.6638
"	Pounds/sq. ft	2240
Tons (short)/sq. ft	Atmospheres	0.945082
"	Dynes/sq. cm	957.605
"	Grams/sq. cm	976.486
"	Pounds/sq. inch	13.8888
Tons (long)/sq. in	Atmospheres	152.423
"	Dynes/sq. cm	1.54443×10^8
"	Grams/sq. cm	157487.59
Tons (short)/sq. in	Dynes/sq. cm	1.37895×10^8
"	Kg./sq. mm	1406.139
"	Pounds/sq. inch	2000
Torrs (*or* Tors)	Millimeters of Hg (0°C.)	1
Townships (U.S.)	Acres	23040
"	Sections	36
"	Sq. miles	36
Tuns	Gallons (U.S.)	252
"	Hogsheads	4
Volts	Abvolts	1×10^8
"	Mks. (r *or* nr) units	1
"	Statvolts	0.003335635
"	Volts (Int.)	0.999670
Volts (Int.)	Volts	1.000330
Volts/°C	Joules/(coulomb × °C.)	1
Volt-coulombs	Joules (Int.)	0.999835
Volt-coulombs (Int.)	Joules	1.000165
Volt-electronic charge-seconds	Planck's constant*	2.41814×10^{14}
Volt-faraday (chem.)-seconds	Planck's constant*	1.45650×10^{38}
Volt-faraday (phys.)-seconds	Planck's constant	1.45690×10^{38}
Volt-seconds	Maxwells	1×10^8
Watts	B.t.u./hr	3.41443
"	B.t.u. (mean)/hr	3.40952
"	B.t.u. (mean)/min	0.0568253
"	B.t.u./sec	0.000948451
"	B.t.u. (mean)/sec	0.000947088
"	Cal., *gm.*/hr	860.421
"	Cal., *gm.* (mean)/hr	859.184
"	Cal., *gm.* (20°C.)/hr	860.853
"	Cal., *gm.*/min	14.3403
"	Cal., *gm.* (IST.)/min	14.3310
"	Cal., *gm.* (mean)/min	14.3197
"	Cal., *kh.*/min	0.0143403
"	Cal., *kg.* (IST.)/min	0.0143310
"	Cal., *kg.* (mean)/min	0.0143197
"	Ergs/sec	1×10^7
"	Foot-pounds/min	44.2537
"	Horsepower	0.00134102
"	Horsepower (boiler)	0.000101942
"	Horsepower (elec.)	0.00134048
"	Horsepower (metric)	0.00135962
"	Joules/sec	1
"	Kilowatts	0.001
"	Liter-atm./hr	35.5282
Watts (Int.)	B.t.u./hr	3.41499
"	B.t.u. (mean)/hr	3.41008
"	B.t.u./min	0.569165
"	B.t.u. (mean)/min	0.0568347
"	Cal., *gm.*/hr	860.563
"	Cal., *gm.* (mean)/hr	859.326
"	Cal., *kg.*/min	0.0143427
"	Cal., *kg.* (IST.)/min	0.0143333
"	Cal., *kg.* (mean)/min	0.0143221

To convert from	To	Multiply by
Watts (Int.)	Ergs/sec	1.000165×10^7
"	Joules (Int.)/sec	1
"	Watts	1.000165
Watts/sq. cm	B.t.u./(hr. × sq. ft.)	3172.10
"	Cal., *gm.*/(hr. × sq. cm.)	860.421
"	Ft.-lb./(min. × sq. ft.)	41113.1
Watts/sq. in	B.t.u./(hr. × sq. ft.)	491.677
"	Cal., *gm.*/(hr. × sq. cm.)	133.365
"	Ft.-lb./(min. × sq. ft.)	6372.54
Watt-hours	B.t.u.	3.41443
"	B.t.u. (mean)	3.40952
"	Cal., *gm*	860.421
Watt-hours	Cal., *kg.* (mean)	0.859184
"	Cal., *gm.* (mean)	859.184
"	Foot-pounds	2655.22
"	Hp.-hours	0.00134102
"	Joules	3600
"	Joules (Int.)	3599.41
"	Kg.-meters	367.098
"	Kw.-hours	0.001
"	Watt-hours (Int.)	0.999835
Watt-sec	Foot-pounds	0.737562
"	Gram-cm	10197.16
"	Joules	1
"	Liter-atm	0.00986895
"	Volt-coulombs	1
Wave length of orange-red line of krypton 86	Ångström units	6057.80211
"	Millimeters	0.000605780211
Wave length of red line of cadmium	Ångström units	6438.4696
"	Millimeters	0.00064384696
Webers	Cgs. units of induction flux	1×10^8
"	E.M. cgs. units of induction flux	1×10^8
"	Lines	1×10^8
"	Maxwells	1×10^8
"	Mks. units of induction flux	1
"	Mks. nr units of magnetic charge	0.079577472
"	Mks. r units of magnetic charge	1
"	Volt-seconds	1
Webers/sq. cm	Gausses	1×10^8
"	Lines/sq. cm	1×10^8
"	Lines/sq. inch	6.4516×10^8
Webers/sq. in	Gausses	1.5500031×10^7
Weeks (mean calendar)	Days (mean solar)	7
"	Days (sidereal)	7.0191654
"	Hours (mean solar)	168
"	Hours (sidereal)	168.45997
"	Minutes (mean solar)	10080
"	Minutes (sidereal)	10107.598
"	Months (lunar)	0.23704235
"	Months (mean calendar)	0.23013699
"	Years (calendar)	0.019178082
"	Years (sidereal)	0.019164622
"	Years (tropical)	0.019165365
Weys (Brit., mass.)	Pounds (avdp.)	252
Yards	Centimeters	91.44
"	Chains (Gunter's)	0.4545454
"	Chains (Ramden's)	0.03
"	Cubits	2
"	Fathoms	0.5
"	Feet	3
"	Feet (U.S. Survey)	2.9999940
"	Furlongs	0.00454545
"	Inches	36
"	Meters	0.9144
"	Poles (Brit.)	0.181818

* For factors for C = 12 scale or those which can be derived from same see table on Values for General Physical Constants.

To convert from	To	Multiply by
Yards	Quarters (Brit., linear)	4
"	Rods	0.181818
"	Spans	4
Years (calendar)	Days (mean solar)	365
"	Hours (mean solar)	8760
"	Minutes (mean solar)	525600
"	Months (lunar)	12.360065
"	Months (mean calendar)	12
"	Seconds (mean solar)	3.1536 × 10⁷
"	Weeks (mean calendar)	52.142857
"	Years (sidereal)	0.99929814
"	Years (tropical)	0.99933690
Years (leap)	Days (mean solar)	366
Years (sidereal)	Days (mean solar)	365.25636
Years (sidereal)	Days (sidereal)	366.25640
"	Years (calendar)	1.0007024
"	Years (tropical)	1.0000388
Years (tropical)	Days (mean solar)	365.24219
"	Days (sidereal)	366.24219
"	Hours (mean solar)	8765.8126
"	Hours (sidereal)	8789.8126
"	Months (mean calendar)	12.007963
"	Seconds (mean solar)	3.1556926×10^7
"	Seconds (sidereal)	3.1643326×10^7
"	Weeks (mean calendar)	52.177456
"	Years (calendar)	1.0006635
"	Years (sidereal)	0.99996121

DEFINED VALUES AND EQUIVALENTS

Meter	(m)	1 650 763.73 wave lengths in vacuo of the unperturbed transition $2p_{10} - 5d_5$ in ^{86}Kr
Kilogram	(kg)	mass of the international kilogram at Sèvres, France
Second	(s)	1/31 556 925.974 7 of the tropical year at 12^h ET, 0 January 1900
Degree Kelvin	(°K)	defined in the thermodynamic scale by assigning 273.16 °K to the triple point of water (freezing point, 273.15 °K = 0 °C)
Unified atomic mass unit	(u)	1/12 the mass of an atom of the ^{12}C nuclide
Mole	(mol)	amount of substance containing the same number of atoms as 12 g of pure ^{12}C
Standard acceleration of free fall	(g_n)	9.806 65 m s^{-2}, 980.665 cm s^{-2}
Normal atmospheric pressure	(atm)	101 325 N m^{-2}, 1 013 250 dyn cm^{-2}
Thermochemical calorie	(cal_{th})	4.1840 J, 4.1840×10^7 erg
International Steam Table calorie	(cal_{IT})	4.1868 J, 4.1868×10^7 erg
Liter	(l)	0.001 m^3, 1000 cm^3 (recommended by GCWM, 1964)
Inch	(in)	0.0254 m. 2.54 cm
Pound (avdp)	(lb)	0.453 592 37 kg, 453.592 37 g

FACTORS FOR THE CONVERSION OF ($LOG_{10}X$) TO ($RT\ LOG_eX$)

Units are in Calories

t°C	0	1	2	3	4	5	6	7	8	9	Differences	Tenths	Units	Tens
0	1249.4	1254.0	1258.6	1263.2	1267.7	1272.3	1276.9	1281.45	1286.0	1290.6				
10	1295.2	1299.8	1304.3	1308.9	1313.5	1318.0	1322.6	1327.2	1331.8	1336.3				
20	1340.9	1345.5	1350.1	1354.6	1359.2	1363.8	1368.4	1372.9	1377.5	1382.1				
30	1386.7	1391.2	1395.8	1400.4	1405.0	1409.5	1414.1	1418.7	1423.2	1427.8	1	.5	4.6	45.7
40	1432.4	1437.0	1441.5	1446.1	1450.7	1455.3	1459.8	1464.4	1469.0	1473.6	2	.9	9.1	91.5
50	1478.1	1482.7	1487.3	1491.9	1496.4	1501.0	1505.6	1510.2	1514.7	1519.3	3	1.4	13.7	137.2
60	1523.9	1528.5	1533.0	1537.6	1542.2	1546.7	1551.3	1555.9	1560.5	1565.0	4	1.8	18.3	183.0
70	1569.6	1574.2	1578.8	1583.3	1587.9	1592.5	1597.1	1601.6	1606.2	1610.8	5	2.3	22.9	228.7
80	1615.4	1619.9	1624.5	1629.1	1633.7	1638.2	1642.8	1647.4	1651.9	1656.5	6	2.7	27.4	274.4
90	1661.1	1665.7	1670.2	1674.8	1679.4	1684.0	1688.5	1693.1	1697.7	1702.3				
100	1706.8	1711.4	1716.0	1720.6	1725.1	1729.7	1734.3	1738.9	1743.4	1748.0	7	3.2	32.0	320.2
											8	3.7	36.6	365.9
110	1752.6	1757.2	1761.7	1766.3	1770.9	1775.4	1780.0	1784.6	1789.2	1793.7	9	4.1	41.2	411.7
120	1798.3	1802.9	1807.5	1812.0	1816.6	1821.2	1825.8	1830.3	1834.9	1839.5				
130	1844.1	1848.6	1853.2	1857.8	1862.4	1866.9	1871.5	1876.1	1880.6	1885.2				
140	1889.8	1894.4	1898.9	1903.5	1908.1	1912.7	1917.2	1921.8	1926.4	1931.0				
150	1935.5	1940.1	1944.7	1949.3	1953.8	1958.4	1963.0	1967.6	1972.1	1976.7				
160	1981.3	1985.9	1990.4	1995.0	1999.6	2004.1	2008.7	2013.3	2017.9	2022.4				
170	2027.0	2031.6	2036.2	2040.7	2045.3	2049.9	2054.5	2059.0	2063.6	2068.2				
180	2072.8	2077.3	2081.9	2086.5	2091.1	2095.6	2100.2	2104.8	2109.4	2113.9				
190	2118.5	2123.1	2127.6	2132.2	2136.8	2141.4	2145.9	2150.5	2155.1	2159.7				
200	2164.2	2168.8	2173.4	2178.0	2182.5	2187.1	2191.7	2196.3	2200.8	2205.4				

DIMENSIONLESS GROUPS

John P. Catchpole and George Fulford

Reprinted from Industrial & Engineering Chemistry, Vol. 58, No. 3, pp. 47 to 60. Copyright 1966 by the American Chemical Society and reprinted by permission of the copyright owners and the authors. See also the Supplementary Table which follows this table.

Dimensionless groups are frequently generated in the analysis of a complex engineering problem. The more common groups thus generated are easily recognized, while the less common ones are not. Unless the less common existing groups are recognized, an already named group could unknowingly be renamed. Table A provides a tool that may be used to avoid this occurrence, by listing the groups by the variables of which they consist. These variables—i.e., length, density, diffusivity, viscosity, etc.—are further subdivided into their exponents to which they are raised in the groups in question. Thus, Reynolds number is listed under the exponent +1 for the variables, length, fluid velocity, and density, and the exponent −1 for viscosity.

To illustrate the use of the tables in the analysis of a problem, the group $(kE/\eta\sigma T^3)$ might be generated in the solution of a complex heat transfer problem. From Table A the groups containing the constituent variables are checked and the groups are listed:

Thermal conductivity	(k^{+1})	**F11, L7, R1**
Modulus of elasticity	(E^{+1})	**C1, E13, R1**
Stefan-Boltzmann coefficient	(η^{-1})	**R1, T6**
Surface tension	(σ^{-1})	**B9, C3, E8, L6, R1, W1**
Temperature	(T^{-3})	**R1, T6**

It is immediately apparent that the only group common to all the categories listed is the Radiation number, **R1**, which is equivalent to the previously unidentified group.

The symbol assigned to a dimensionless group is usually the first two letters of its names. Several groups, however, have nonstandard symbols, particularly in the groups which are named after persons. These symbols are listed in the nomenclature.

NOMENCLATURE

a	= annulus or clearance width, L
A	= area, L^2
A^*	= cooling area/unit volume, $1/L$
b	= bearing breadth, L
B	= groups **B6, B11**
c	= specific heat, $L^2/\theta^2 T$
c_A	= concentration, M/L^3
c_b	= specific vapor capacity (mass/unit mass/unit pressure change), $L\theta^2/M$
c_d, c_D	= group **D13**
c_f	= group **R9**
c_H	= group **H4**
c_m	= mass capacity, L^3/M
c_p, c_v	= specific heats at constant pressure and volume, $L^2/\theta^2 T$
c_q	= heat capacity, $L^2/\theta^2 T$
c_Q	= group **F9**
c_S	= group **S11**
C	= group **C10**, dimensional concentration, M/L^3
C_a	= groups **C11, C4**
d	= diameter, L
d_e	= equivalent diameter (of particles, etc.), L
d_h	= hydraulic diameter, L
D	= diffusivity (molecular, unless noted otherwise), L^2/θ
D_{AB}	= binary bulk diffusion coefficient, L^2/θ
D_{KA}	= Knudsen diffusion coefficient, L^2/θ
e	= voidage; porosity (⁻)
e^*	= surface emissivity (⁻)
E	= modulus of elasticity, $M/L\theta^2$
E_a	= activation energy, L^2/θ^2
E_b	= bulk modulus, $M/L\theta^2$
f	= frequency, $1/\theta$, or Group **F1**
$f(M)$	= group **F11**
F	= force, ML/θ^2
F_b	= force per unit length of bearing, M/θ^2
$F(M)$	= group **F6**
F_R	= resistance force in flow, ML/θ^2
g	= acceleration due to gravity, L/θ^2
G	= mass velocity (mass flux density; mass transfer coefficient), $M/\theta L^2$
h	= heat transfer coefficient, $M/T\theta^3$
h_c	= convective heat transfer coefficient, $M/T\theta^3$
H	= energy change per unit mass ($= g \times$ head), L^2/θ^2
H'	= fluid head, L
H_e	= field strength, $Q/L\theta$
H_0	= homochronicity number
$I(M)$	= group **F8**
j	= heat liberated per unit volume per unit time, $M/L\theta^3$
j_H, j_M	= groups **J2, J3**
J	= average free path/average velocity, θ, or group **L6**
k	= thermal conductivity, $ML/T\theta^3$
k_c	= mass transfer coefficient, L/θ
K	= groups **K2, K10, N5**
K_1	= group **A4**
$\bar{K}_E$	= group **E4**
K_F	= group **C2**
K_P	= group **P13**
K_Q	= group **H5**
K_r	= group **E11**
$\bar{K}_g$	= group **A1**
$\bar{K}_E$	= group **E4**
$\bar{K}_r$	= group **E12**
$\bar{K}_{rE}$	= group **E13**
$\bar{K}_s$	= group **R1**
$\bar{K}_\sigma$	= group **C1**
$\bar{K}_{\sigma g}$	= group **C2**
L	= characteristic dimension (except as noted), L
L_m	= distance from midpoint to surface, L
m_T	= group **T3**
M	= group **M10**
M_H	= group **H2**
n	= concentration, wt./wt. (⁻)
n^*	= specific mass content, mass/mass (⁻)
n_m	= moisture content, wt./wt. bone dry gas (⁻)
N	= rate of rotation, $1/\theta$, and groups **M3, N4**
N_{B_0}	= groups **B7, B8**
N_c	= group **C5**
N_{cv}	= group **C10**
N_D	= groups **D7, D14**
N_E	= group **E1**
N_F	= group **F4**
N_H	= groups **H1, H9**
N_K	= group **K1**
N_{KnA}	= Knudsen number for diffusion (see Addendum)
N_1	= group **N2**
N_P	= group **P7**
N_{rf}	= group **R7**
N_{s1}	= group **S8**
N_{s2}	= group **S9**
N_T	= group **N9**
P	= pressure, $M/L\theta^2$
P	= plasticity number (see Addendum)
p_b	= bearing pressure, $M/L\theta^2$
p_s	= static pressure, $M/L\theta^2$
p_v	= vapor pressure, $M/L\theta^2$
p_σ	= capillary pressure, $M/L\theta^2$
Δp_F	= frictional pressure drop, $M/L\theta^2$
q	= heat flux (heat flow/unit time), ML^2/θ^3
q^*	= heat flux density (heat flux/unit area), M/θ^3
Q	= heat liberated/unit mass, L^2/θ^2
r	= latent heat of phase change, L^2/θ^2
r_v	= heat of vaporization, L^2/θ^2
R	= radius, L
R_H	= hydraulic radius, L
R_2'	= group **R5**
R_M	= group **M7**
R_V	= group **V3**
$\mathscr{R}$	= gas constant, $L^2/\theta^2 T$
s	= humid heat, $L^2/\theta^2 T$
S	= particle area/particle volume, L^2/L^3, and group **M6**
St	= group **S14**
t	= temperature, T
T	= absolute temperature, T
Δt, ΔT	= temperature difference, T
U^+	= group **P10**
U	= reaction rate, $M/L^3\theta$
v_s	= velocity of surface (solid), L/θ
V	= fluid velocity, L/θ, and group **V1**
V_A	= velocity of Alfven magnetic waves, L/θ
V_f	= volumetric flow rate, L^3/θ
V_l	= velocity of light, L/θ
V_m	= mass flow rate, M/θ
V_s	= velocity of sound, L/θ
w	= circumferential velocity, L/θ
W	= volume of system, L^3
W^*	= gross volume, L^3
x	= entry length; distance from entrance, L
y^+	= group **P11**
Z	= group **O2**
α	= thermal diffusivity (temperature conductivity), L^2/θ

β = coefficient of bulk expansion, $1/T$, and group **D12**
β^* = Dufour coefficient, T
γ = specific gravity (¯) and group **R3**
$\dot{\gamma}$ = rate of shear, $1/\theta$
Γ = rate of change of temperature of medium, T/θ
δ = Soret or thermogradient coefficient, $1/T$, and group **D11**
Δ = difference in quantity
ε = height of roughness, L and group **A3**
ε_D = eddy mass diffusivity, L^2/θ
ζ = diffusion tortuosity (¯)
η = radiation coefficient (Stefan-Boltzmann coefficient), $M/T^4\theta^3$
θ = time, θ
θ_r = relaxation time, θ
λ = mean free path, L
μ = dynamic viscosity, $M/L\theta$
μ_e = magnetic permeability, ML/Q^2
μ_p = rigidity coefficient, $M/L\theta$
ξ = permeability, L^2
π = 3.1416.
Π = power to agitator or impeller, ML^2/θ^3
ρ = density, M/L^3
ρ_n = group **P3**
σ = surface tension, M/θ^2 and group **S12**
σ_c = group **C6**
σ_e = electrical conductivity, $Q^2\theta/L^3M$
σ_t = group **T4**
τ = group **T8**
τ_w = wall shear stress, $M/L\theta^2$
τ_y = yield stress, $M/L\theta^2$
φ = group **D9**
ψ = groups **N8, P14, R10**
ω = angular velocity (of fluid, unless noted otherwise), $1/\theta$
Ω = mass transfer potential (concn.), M/L^3
— (bar over) = mean value
N.B.: $(F) = \left(\frac{ML}{\theta^2}\right)$; $(H) = \left(\frac{ML^2}{\theta^2}\right)$

TABLES FOR IDENTIFYING DIMENSIONLESS GROUPS

TABLE A

PHYSICAL PROPERTIES
General Physical Properties

Parameter	Symbol	Dimensions	Exponent	Group
Coefficient of bulk expansion	β	$1/T$	-1	E4, 12, G2
			$+1$	G5, R5, K9, 6
Density	ρ	M/L^3	-2	M1
			-1	A1, B1, 10, C1, 2, 6, D3, 13, E3, 9, 10, 13, F1, 11, H5, J1, K4, K10, L7, M3, 6, N2, 3, 4, P7, 9, R10, 13, S4, 13, 15
			$-\frac{2}{3}$	C9, J3
			$-\frac{1}{2}$	E2, L11, O2, P13
			$+\frac{1}{3}$	J3, K5, 53
			$+\frac{1}{2}$	D10, G3, P10, 11, R4, W2
			$+\frac{2}{3}$	F2, N8, S2
			$+1$	A5, B1, B6, 9, C5, 7, D5, 7, 13, E4, 6, 7, 8, H6, 11, J1, 3, 4, K1, L9, P1, R5, 11, 13, S17, T1, T6, V1, W1, W3
			$+2$	C10, G1, 5, K9, R5, 6, T2
Density gradient	$d\rho/dL$	M/L^4	$+1$	R13
Diffusivity (molecular unless noted otherwise)	D, a_m, ε_D	L^2/θ	-1	B5, B7[4], D2, K4[1], K7[2], L2, 7, 10, N7, P2, P9, S4, 13
			$-\frac{2}{3}$	J3
			$-\frac{1}{2}$	T3
			$+1$	D12, F12[1], K7[3], L2[4], L9[1]
Diffusivity (surface)	D_8	ML/θ^2	$+1$	S18
Diffusion tortuosity	ζ	—	-1	K7
Molecular weight	M	—	-1	D14, K9, S6
			$+1$	D14
Permeability (packed bed)	ξ	L^2	$+\frac{1}{2}$	L_6
Porosity (voidage)	e	—	-1	B4
			$-\frac{1}{2}$	L_6
			$+1$	K7
Specific weight	γ	—	$\pm\frac{1}{3}$	F2
Surface tension	σ	M/θ^2	-3	C2
			-2	C1
			-1	B9, C3, 13, E8, L6, R1, W1, 3
			$-\frac{1}{2}$	D10, G3, O2, P13, R4, W2
			$+1$	M9, S17, 18

[1] Coefficient of potential diffusion in mass transfer.
[2] Knudsen diffusion coefficient.
[3] Binary bulk diffusion coefficient.
[4] Effective diffusivity ($D+\varepsilon_D$) (molecular + eddy transfer).

PHYSICAL PROPERTIES
Electrical and Magnetic Properties

Parameter	Symbol	Dimensions	Exponent	Group
Current density	I	$Q/L^2\theta$	$+1$	K10
Electrical conductivity	σ_e	$Q^2\theta/L^3M$	-1	E6
			$+\frac{1}{2}$	H2
			$+1$	L11, M3, 7
Field strength	H_e	$Q/L\theta$	-2	J4
			$+1$	H2, L11
			$+2$	K9, M3, 6, S6
Magnetic permeability	μ_e	ML/Q^2	-1	E5, J4
			$+1$	H2, M6, 7
			$+\frac{3}{2}$	L11
			$+2$	M3
Voltage	E	$ML^2/Q\theta^2$	$+1$	K10

Thermal Properties

Humid heat	S	L^2/θ^2T	-1	P15
Latent heats of phase change	λ, r	L^2/θ^2	-1	A3, E11, 12, 13, J1, K10, M1
			$+1$	B13, C9, K8, 11, N5
Ratio of specific heats	γ	—	$\pm\frac{1}{2}$	L5
Specific heat	C, c	L^2/θ^2T	-1	B2, 13, D3, 15, E1, F11, J2, K8, 11, L7, M10, N5, R3, S13
			$-\frac{1}{3}$	J2
			$+\frac{1}{2}$	F6, 7, 8
			$+\frac{2}{3}$	J2
			$+1$	A3, E4, 12, G4, J1, 4, L9, P1, 8, R3, 5, 6, 7
Surface emissivity	e^*	—	-1	T6
Temperature conductivity (thermal diffusivity)	α	L^2/θ	-1	L9, P1, 12, R5
			$+1$	C13, L7, 10
Thermal conductivity	k or λ	$ML/T\theta^3$	-3	M1
			-2	R6
			-1	B4, 12, C7, 9, 10, D4, G4, K3, L9, N6, P1, 5, 8, R5, S14

Rheological and Elastic Behavior

Modulus of elasticity	E	$M/L\theta^2$	-1	C5, E4, H11
			$+1$	C1, E13, R1
			$+3$	A1
Rate of shear	$\dot{\gamma}$	$1/\theta$	$+1$	T8
Rigidity coefficient	μ_p	$M/L\theta$	-2	H6
			-1	B3
			-1	E5
Shear stress	τ	$M/L\theta^2$	$-\frac{1}{2}$	P10
			$+\frac{1}{2}$	P11
Viscosity (in all cases kinematic viscosity has been written as μ/ρ)	μ	$M/L\theta$	-2	A1, 5, G1, 5, K1, 9, S17, T2
			-1	B6, C10, D5, 7, E6, 7, H8, L1, 3, 9, O1, P4, 11, R5, 6, 11, S9, 19, T1, V1

TABLE A (Continued)

Rheological and Elastic Behavior

Parameter	Symbol	Dimensions	Exponent	Group
			$-\frac{2}{3}$	F2, K5, N8
			$-\frac{1}{2}$	H2
			$-\frac{1}{3}$	S2, 3
			$+\frac{1}{2}$	E2
			$+\frac{2}{3}$	C9, J2, 3
			+1	B12, C3, 13, E3, 5, M1, O2, P8, 9, S4, 8, 15, T8
			+2	C1
			+4	C2
Viscosity (surface)	μ_S	M/θ	+1	S19
Yield stress	τ_y	$M/L\theta^2$	+1	B3, H6

LENGTHS, AREAS AND VOLUMES

Characteristic Linear Dimensions

(In all cases kinematic viscosity has been written as μ/ρ)

Parameter	Symbol	Dimensions	Exponent	Group
General characteristic linear dimension	Various	L	−5	P7
			−3	F9, L1
			−2	D9, 12, E3, F11, 12, H4, 5, N2, 3, 4, O1, R9, S8, 9, 15
			−1	B1, 10, E2, 5, 10, F1, 13, G4, H10, K6, L4, R6, 10, 15, 16, S6, 19, T5, W4
			$-\frac{1}{2}$	B11, D7, F14, O2
			$+\frac{1}{2}$	R4, T1, W2
			+1	B3, 4, 5, 7, 10, C7, D1, 3, 5, 11, 13, E7, 10, F1, 2, 5, G3, H2, K3, 4, 6, 10, L3, 4, 9, 11, M1, 3, 7, N6, 7, 8, O1, P1, 2, 11, R8, 10, 11, 16, S2, 7, 14, 16, 17, 18, T3, W1
			$+\frac{3}{2}$	D7, H7, T1
			+2	B9, D2, 4, E8, H6, K9, O1, P4, 5, 12, S8, 9, V1
			+3	A5, C10, G1, 5, K1, R5
			+4	T2
			+5	R6
Dimension of agitator, impeller, etc.	Various	L	−5	P7
			−3	F9
			−2	D7, H4
			+1	D11, F15
			+3	W3
Film thickness	L_f	L	+1	N8
Furnace half-width	L	L	−1	T7
Larmor radius	L_L	L	+1	L4
Mean free path	λ	L	+1	K6
Particle dimension	d_e	L	−2	S15
			+1	F2
			+3	A5, G1
Pore or nozzle radius	R	L	+1	M12, T7
Reactor length	L	L	+1	B7
			−1	C3, L9
Thickness of liquid layer	L	L	+1	H2, L11
			+2	L9
			+3	R5
			+4	T2

Areas

Parameter	Symbol	Dimensions	Exponent	Group
Area	A	L^2	−1	D9, F6, 7, 8
			+1	B2
Area/unit volume	S, A^*	$1/L$	−1	B6
			+1	M11, 12

Volumes

Parameter	Symbol	Dimensions	Exponent	Group
Volume	—	L^3	+	A5, H9, M11

TIMES AND FREQUENCIES

Parameter	Symbol	Dimensions	Exponent	Group
			−1	D8
Time	θ	θ	+1	D8, 12, E3, F11, 12, H1, 10, S15, T5
Frequency	f'	$1/\theta$	+1	H1, 10, S16, V1

TEMPERATURES AND CONCENTRATIONS (DRIVING FORCES)

Concentrations and Related Quantities

Parameter	Symbol	Dimensions	Exponent	Group
Dimensional concentration	C, c_A	M/L^3	−1	D1, 2, R2
			$-\frac{1}{2}$	T3
			+1	B10
Dimensionless concentration—e.g., wt./wt. inert material, etc.	n	—	−1	P6
			+1	A4, K8
Mass capacity	C_m	L^3/M	−1	R2
Mole fraction	Y, Z	—	±1	D14
Specific mass content, mass/unit mass	n^*	—	−1	K4
Surface concentration	Γ'	M/L^2	±1	S18
Vapor capacity (porous body)	C_b	$L\theta^2/M$	+1	B13, R2

Temperatures, Temperature Differences

Parameter	Symbol	Dimensions	Exponent	Group
Temperature, temperature difference	T, ΔT	T	−3	R1*, T6*
			−1	A4, 6*, B12, 13, C4*, 7, 10, D3*, 4*, 15, E1, G2, 6*, K3, 8, 9, 11, L9, N5, P5*, 12*, R14*, S6
			$-\frac{1}{2}$	L5
			$+\frac{1}{2}$	F6*, 7*, 8*
			+1	C4, G5, 6, J1, 4, L9, M1, P6, R5, 7, 14*
			+3	S14*
Rate of temperature change	—	T/θ	+1	P12

* Absolute temperature; others—temperature differences.

VELOCITIES, RATES, FLUXES, TRANSFER COEFFICIENTS

Velocities

Parameter	Symbol	Dimensions	Exponent	Group
Angular velocity (rate of rotation)	N	$1/\theta$	−3	P7
			−2	H4, L1
			−1	F9, R15
			$-\frac{1}{2}$	E2
			+1	S8, 11, 12, T1
			+2	F15, T2, W3
Fluid velocity	V	L/θ	−3	H5
			−2	C6, 11, D13, E9, 10, F1, M6, N2, 3, 4, R7, 8, 9, 10
			−1	A2, B3, C12, 14, D1, 3, F10, H7, 8, K10, L3, M3, P4, S13, 16
			+1	A2, B6, 7, 11, C3, 12, 14, D5, 7, E5, 7, F10, 14, H10, K2, L5, 8, M2, 4, 7, P1, 2, 3, 10, R4, 11, 15, S1, 2, 3, T5, 6, V2, W2, 4,
			+2	B1, 12, C5, 11, D15, E1, 11, F13, H11, W1
			+3	C7
Impeller or agitator circumference	U_s	L/θ	−2	P14
			−1	D9
Light	V_l	L/θ	−1	L8
Sound	V_s	L/θ	−1	M2, N1, S1
Waves	V_w	L/θ	−1	V1
			−1	P3
Velocity gradient	dV/dL	$1/\theta$	−2	R12
Velocity of Alfven waves	V_A	L/θ	−1	A2, K2, M4
			+1	A2, N1
			+2	C11
Velocity of bearing surface	v_s	L/θ	+1	H8, O1, S9

Flow Rates (Mass Fluxes)

Parameter	Symbol	Dimensions	Exponent	Group
Mass flow rate (mass flux)	V_m	M/θ	−1	B2, M11, T7
			+1	F6, 7, 8, G4, T7
Mass flux density (mass flux/unit area)	G	$M/L^2\theta$	−1	J2, 3, P15, S13
			+1	K4, M11

TABLE A (Continued)

Flow Rates (Mass Fluxes)

Parameter	Symbol	Dimensions	Exponent	Group
Mass flux/unit volume (reaction rate)	U	$M/L^3\theta$	$+\frac{1}{2}$	T3
			+1	D1, 2, 3, 4
Reaction rate constant	K	L/θ	−1	S5
Volumetric flow rate	V_f	L^3/θ	−1	H9
			$-\frac{1}{2}$	D11
			$+\frac{1}{2}$	S11, 12
			+1	D9, F9

Heat Fluxes

Parameter	Symbol	Dimensions	Exponent	Group
Heat flux (heat flow/unit time)	q	ML^2/θ^3	+1	H5
Heat flux/unit area	q^*	M/θ^3	+1	K3, R6
Heat liberated/unit mass	Q	L^2/θ^2 (H/M)	+1	D3, 4
Rate of heat liberation/unit volume (heat source power)	j	$M/L\theta^3$	+1	P5

Transfer Coefficients

Parameter	Symbol	Dimensions	Exponent	Group
Heat transfer coefficient	h	$M/T\theta^3$	+1	B2, 4, C9, J2, N6, P15, S13
		$(H/L^2T\theta)$	+4	M1
Mass transfer coefficient	k_c	L/θ	+1	B5, J3, L9, N7, S5, 7

FORCE, HEAD, POWER, PRESSURE

Forces

Parameter	Symbol	Dimensions	Exponent	Group
Force (resistance)	F, F_R	ML/θ^2	$-\frac{1}{3}$	S2
			$+\frac{1}{3}$	K5
			+1	N2, 3, 4, R9
Force/unit length	F_b	M/θ^2	+1	H8, O1, S9

Heads, Power

Parameter	Symbol	Dimensions	Exponent	Group
Fluid head	H'	L	$-\frac{3}{4}$	S11
			+1	H4

Heads, Power

Parameter	Symbol	Dimensions	Exponent	Group
Head (energy per unit mass of fluid = gH')	H	L^2/θ^2	−1	P3, T4
			$-\frac{3}{4}$	S12
			$+\frac{1}{4}$	D11
			+1	A6, P14, T4
Power	π	ML^2/θ^3 (LF/θ)	+1	L1, P7

Pressures

Parameter	Symbol	Dimensions	Exponent	Group
Pressure	P	$M/L\theta^2$	−1	F6, 7, 8, H9, S8, T8
			+1	B13, C6, L6, P13, R2
Pressure drop	$\Delta P, dP$	$M/L\theta$	+1	E9, 10, F1, H9, L3, R10
Pressure gradient	$\Delta P/L$	$M/L^2\theta^2$	+1	E10, F1, K1, P4, R10

CONSTANTS AND MISCELLANEOUS QUANTITIES

Gravity Acceleration

Parameter	Symbol	Dimensions	Exponent	Group
Gravity acceleration	—	L/θ^2	−2	A1
			−1	B1, F13, 15, M1, S6
			$-\frac{3}{4}$	S11
			$-\frac{1}{2}$	B11, F14, P13
			$-\frac{1}{3}$	C9, S3
			$+\frac{1}{3}$	F2, N8
			+1	A5, B9, C2, 10, D13, E8, G1, 5, H4, R5, 6, 8, 13

Other Quantities

Parameter	Symbol	Dimensions	Exponent	Group
Avogadro's number	N	$1/M$	+1	K9, S6
Boltzmann's constant	k	—	−1	K9, S6
Dufour coefficient	β	T	+1	A3, 4, F3
Energy of activation	E_a	L^2/θ^2	+1	A6
Gas constant	$\mathscr{R}$	L^2/θ^2T	−1	A6
			$-\frac{1}{2}$	L5
Shape factor	ξ^*	—	+1	V2
Soret coefficient	δ	$1/T$	+1	F3, P6
Stefan constant	η	$M/T^4\theta^3$	−1	R1, T6
			+1	S14

TABLE B
ALPHABETICAL LIST OF NAMED GROUPS

Serial No.	Name	Symbol	Definition	Significance	Field of Use	Reference
A1	Acceleration number	$\bar{K}_g$	$E^3/\rho g^2\mu^2 = (N_{Re}N_{Fr1})^2/(H_0)^3$	Group dependent only on physical properties	Accelerated flow	*25*
A2	Alfven number	N_{Al}	V_A/V (or V/V_A) [*cf.* Cowling No. Kármán No. (2), magnetic mach number]	Ratio of Alfven wave velocity/fluid velocity	Magneto-fluid dynamics	*5, 6*
A3	Anonymous group (1)	ε	$\beta^* c/r$ [see also Fedorov No. (2)]		Transfer processes	*39*
A4	Anonymous group (2)	K_1	$\beta^* \Delta n/\Delta t$; Δt = temp. diff. [T]; Δn = conc. diff. [$^-$]		Transfer processes	*67*
A5	Archimedes number	N_{Ar}	$\frac{gL^3\rho_o}{\mu^2}(\rho-\rho_o)$; ρ_o = fluid density; ρ = particle density (*cf.* N_{Ga1})	N_{Re}, gravitational force/viscous force	Fluidization, motion of liquids due to density differences	*6, 13*
A6	Arrhenius group	—	$E_a/\mathscr{R}T$	Activation energy/potential energy of fluid	Reaction rates	*5*
B1	Bagnold number	B	$3c\alpha\rho_g V^2/4d\rho_p g$, ρ_g = gas density· ρ_p = particle density	Drag force/gravitational force	Saltation studies	*12*
B2	Bansen number	N_{Ba}	$h_r A_w/V_{mc}$; h_r = radiant heat transfer coefficient; A_w = wall area of channel; (*cf.* N_{St})	Heat transferred by radiation/thermal capacity of fluid	Radiation	*1*
B3	Bingham number	N_{Bm}	$\tau_y L/\mu_p V$ (L = channel width)	Ratio of yield stress/viscous stress	Flow of Bingham plastics	*5*
B4	Biot number (heat transfer)	N_{Bih}	hL_m/k (in French literature, "Biot No." $= N_{Nu}$)	Midplane thermal internal resistance/surface film resistance	Unsteady state heat transfer	*5, 6, 13*
B5	Biot number (mass transfer)	N_{Bim}	$k_c L/D_{int}$; L = thickness of layer, D_{int} = diffusivity at interface	Mass transfer rate at interface/mass transfer rate in interior of solid wall thickness L	Mass transfer between fluid and solid	*13*
B6	Blake number	B	$V\rho/[\mu(1-e)S]$	Inertial force/viscous force	Beds of particles	*6*
B7	Bodenstein number	N_{Bo}	$VL/D_a = N_{Pe_m}$; L = reactor length, D_a = axial diffusivity (effective) (L^2/θ)		Diffusion in reactors	*6*
B8	Boltzmann number	N_{Bo}	$\equiv$ Thring radiation group			*1*
B9	Bond number	N_{Bo}	$(\rho-\rho')L^2 g/\sigma = N_{We_1}/N_{Fr_1}$ if $\rho-\rho' \cong \rho$ (gas in liq.); ρ = drop or bubble density; ρ' = medium density	Gravitational force/surface tension force	Atomization, motion of bubbles and drops	*5*
B10	Bouguer number	N_{Bu}, B	$3C_D\lambda_r/4\rho DR$; C_D = wt. dust/unit bed volume (M/L^3), λ_r = mean path for radiation (L), ρ_D = dust density, $\bar{R}$ = mean particle radius. Also $N_{Bu} = kL$; L = characteristic dimension, k = absorption coefficient of medium		Radiant heat transfer to dust—gas streams	*66*
B11	Boussinesq number	B	$V/(2gR_H)^{1/2}$ (*cf.* N_{Fr2})	(Inertia force/gravitational force)$^{1/2}$	Wave behavior in open channels	*6*
B12	Brinkman number	N_{Br}	$\mu V^2/k\Delta t$; Δt = temp. diff.	Heat generation/heat transferred	Viscous flow	*6*
B13	Bulygin number	N_{Bu}	$r \cdot \frac{C_b}{C_q} \cdot \frac{P}{t_m - t_o}$ t_m = temp. of medium, t_o = init. temp. of body	Heat for vaporization/heat to bring liquid to boiling point	Heat transfer during evaporation	*6, 13, 14*
C1	Capillary number	$\bar{K}\sigma$	$\mu^2 E/\rho\sigma^2 = (N/_{We_1})^2/H_o \cdot (N_{Re})^2$	Depends only on physical properties	Action of surface tension in flowing media	*25*
C2	Capillarity-buoyancy number (physical properties group) (film No.)[a]	$\bar{K}\sigma_g$, K_F	$g\mu^4/\rho\sigma^3. = \sqrt{\bar{K}\sigma/Kg} = (N_{We_1})^3/(N_{Fr_1})(N_{Re})^4$	Depends only on physical properties and g	Effects of surface tension and acceleration in flowing media (two-phase flow)	*21, 25*
C3	Capillary number	Ca	$\mu V/\sigma = N_{We}/N_{Re}$	Viscous force/surface-tension force	Atomization, two-phase flow	*6*
C4	Carnot number	Ca, N_{Ca}	$(T_2-T_1)/(T_2)$; T_1, T_2 = abs. temp. of two heat sources or sinks	Theoretical efficiency of Carnot cycle operating between T_1 and T_2		*21*
C5	Cauchy number	N_c	$\rho V^2/E_b = (N_{Ma})^2$ = Hooke No.	Inertia force/compressibility force	Compressible flow	*5, 25*
C6	Cavitation number	σ_c	$[(p-P_v)/\rho]/(V^2/2)p$ = local static pressure (abs.); P_v = vapor pressure	Excess of local static head over vapor pressure head/velocity head	Cavitation	*5*
C7	Clausius number	Cl, N_{Cl}	$V^3 L\rho/k\Delta T$; ΔT = temp. diff.		Heat conduction in forced flows	*25*
C8	Colburn number	—	Same as Schmidt number			*5*
C9	Condensation number (1)	N_{Co}	$(h/k)(\mu^2/\rho^2 g)^{1/3}$	$N_{Co} = N_{Nu}\left[\frac{\text{(viscous force)}}{\text{(gravity force)}} \times \frac{1}{Re}\right]^{1/3}$	Condensation	*5*
C10	Condensation number (2)	N_{Cv}	$L^3\rho^2 gr/k\mu\Delta t$; r = latent heat of condensation		Condensation on vertical walls	*5*
C11	Cowling number	C	$(V_A/V)^2 \equiv$ (Alfven number)^2		Magneto-fluid dynamics	*6*
C12	Craya-Curtet number	C_t	$V_k/(V_d^2 - V_k^2/2)^{1/2}$; V_k = kinematic mean velocity, V_d = dynamic mean velocity		Radiant heat transfer	*3, 27*

[a] Very similar to Hu and Kintner's pH factor for drops and bubbles [*A.I.Ch.E. J.* 1, 42 (1955)].

TABLE B (Continued)

Serial No.	Name	Symbol	Definition	Significance	Field of Use	Reference
C13	Crispation group	N_{Cr}	$\mu\alpha/\sigma^*L$; σ^* = undisturbed surface tension; L = layer thickness		Convection currents	55
C14	Crocco number	N_{Cr}	$V/V_{max} = \left[1 + \frac{2}{(\gamma-1)(N_{Ma})^2}\right]^{-1/2}$ V_{max} = maximum velocity of gas expanding adiabatically	Velocity/maximum velocity	Compressible flow	5
D1	Damköhler group I	DaI	UL/V_{cA}	Chemical reaction rate/bulk mass flow rate	Chemical reaction, momentum, and heat transfer	5
D2	Damköhler group II	DaII	UL^2/D_{cA}	Chemical reaction rate/molecular diffusion rate	Chemical reaction, momentum, and heat transfer	5
D3	Damköhler group III	DaIII	$QUL/C_p\rho Vt$	Heat liberated/bulk transport of heat	Chemical reaction, momentum, and heat transfer	5
D4	Damköhler group IV	DaIV	QUL^2/kt	Heat liberated/conductive heat transfer	Chemical reaction, momentum, and heat transfer	5
D5	Damköhler group V	DaV	$= (N_{Re})$			5
D6	Darcy number		$4f$; see Fanning friction factor			5
D7	Dean number	N_D	$(VL\rho/\mu)(L/2R)^{1/2}$; L = pipe diam.; R = radius of curvature of bend	N_{Re} (centrifugal force/inertial force)	Flow in curved channels	5, 6
D8	Deborah number	D	θ_r/θ_o; θ_o = observation time	Relaxation time/observation time	Rheology	49
D9	Delivery number	ϕ	V_f/Aw; A = impeller area = $\pi d^2/4$ ≡ [Diameter No.]$^{-3}$ [Speed No.]$^{-1}$		Flow machines	25
D10	Deryagin number	De	$L(\rho g/2\sigma)^{1/2}$; L = film thickness	Film thickness/capillary length	Coating	58
D11	Diameter group	δ	$(\pi/4)^{1/2}(2H)^{1/4}d/(V_f)^{1/2}$, d = impeller diam. = [pressure No.]$^{1/4}$ × [delivery No.]$^{-1/2}$		Flow machines	25
D12	Diffusion group	β	$D\theta/L_m^2$; D = diffusivity of solute through stationary solution contained in solid; *cf.* N_{Fom}		Mass transfer	37
D13	Drag coefficient	$C_d = C_D$	$(\rho-\rho')L_g/\rho V^2$; ρ = density of object; ρ' = density of medium; *cf.* f, ψ, N_e	Gravity force/inertial force	Free settling velocities, etc.	5
D14	Drew number	N_D	$\frac{Z_A(M_A-M_B)+M_B}{(Z_A-Y_{AW})(M_B-M_A)} \ln \frac{M_V}{M_W}$: M_A, M_B = mol. wt. of components A and B; M_V, M_W = mol. wt. of mixture in vapor and at wall; Y_{AW} = mole fraction of A at wall; Z_A = mole fraction of A in diffusing stream		Boundary layer mass transfer rates; velocity profile distortion; drag coefficients for binary system	22
D15	Dulong number	Du, N_{Du}	$V^2/C_p\Delta T$ = Eckert No.			25
E1	Eckert number	N_E	$V_\infty^2/C_p\Delta T$, V_∞ = velocity of fluid far from body (=2/recovery factor, *q.v.* ≡ Dulong No.)		Compressible flow	5, 24
E2	Ekman number		$(\mu/2\rho\omega L^2)^{1/2} = (N_{Ro}/N_{Re})^{1/2}$	(Viscous force/Coriolis force)$^{1/2}$	Magneto-fluid dynamics	6
E3	Elasticity number (1)	N_{El_1}	$\theta_r\mu/\rho L^2$ = pipe radius	Elastic force/inertial force	Viscoelastic flow	5
E4	Elasticity number (2)	K_E	$\rho C_p/\beta E$ ≡ [Gay Lussac No.] × [Hooke No.] ÷ [Dulong No.]	Depends on physical properties only	Effect of elasticity in flow processes	25
E5	Ellis number	N_{El}	$\mu_o V/2\tau_{1/2}R$; μ_o = zero shear viscosity, $\tau_{1/2}$ = shear stress when $\mu = \mu_o/2[M/L\theta^2]$; R = tube radius		Flow of non-Newtonian liquids	38
E6	Elsasser number	N_{El}	$\rho/\mu\sigma_e\mu_e \equiv N_{Re}$/[magnetic Reynolds No.]		Magneto-fluid dynamics	6
E7	Entry Reynolds Number	K_E	$\chi/d_h \cdot N_{Re} = \frac{\chi V\rho}{\mu}$; χ = entry length	As N_{Re}	Entry or inlet processes	25
E8	Eötvös number	N_{Eo}	$(\rho-\rho')L^2g/\sigma$ ≡ Bond No., *q.v.*			6
E9	Euler number (1)	N_{Eu_1}	$\Delta P_F/\rho V^2$; ΔP_F = pressure drop due to friction	Friction head/2 × velocity head	Fluid friction in conduits	5, 13, 25
E10	Euler number (2)	N_{Eu_2}	$d(-dp/dL)\rho V^2$; d = pipe diam., dp/dL = pressure gradient ≡ 2 × Fanning friction factor		Fluid friction in conduits	
E11	Evaporation number	K_r	V^2/r [r = heat of vaporization (L^2/θ^2)]		Evaporation processes	25
E12	Evaporation number (2)	K_r	$C_p/r\beta$ (r as in E11) ≡ (Gay Lussac No.) × (E11)/(Dulong No.)		Evaporation processes	25
E13	Evaporation-elasticity number	K_{rE}	$E/r\rho = K_r$/Hooke number		Evaporation processes	25
F1	Fanning friction factor	f	$d\Delta p_F/2pV^2L$, d = dimension of cross section; L = length (*cf.* resistance coeff., Ne)	Shear stress at wall expressed as number of velocity heads	Fluid friction in conduits	5
F2	Fedorov number (1)	F_{e_1}, N_{Fe_1}	$d\sqrt[3]{\frac{4g\rho^2}{3\mu^2}\left(\frac{\gamma M}{\nu g}-1\right)}$ d_e = equiv. particle diam.; γM = sp. gr. of particles; γg = sp. gr. of gas (*cf.* N_{Ar})		Fluidized beds	6, 13

TABLE B (Continued)

Serial No.	Name	Symbol	Definition	Significance	Field of Use	Reference
F3	Fedorov number (2)	$F_{e_2}N_{Fe_2}$	$\delta\beta^* = K_1 Pn = \varepsilon \times K_o Pn$	Mass transfer analogy of Posnov number	Transport processes	*39, 67*
F4	Fenske number	N_F		Number of stages in separation process		*6*
F5	Fineness coefficient	ψ	$L/W_D^{1/3}$; W_D = volume displacement $[L^3]$		Ship modeling	*26, 43*
F6, F7, F8	Fliegner numbers	$F(M_a)$, $f(M_a)$	Functions of ratio of specific heats and mach number			
		$I(M_a)$	$V_m(cT)^{1/2}/A(p_s+\rho V^2) = [\gamma M_a/(\gamma-1)^{1/2}]\left[1+\frac{(\gamma-1)Ma^2}{2}\right]^{1/2}$ = impulse Fliegner number; γ = ratio of specific heats, Ma = mach number, A = flow area			*43*
F9	Flow coefficient	C_Q	V_f/Nd, d = impeller diam.		Power required by fans, etc.	*32*
F10	Fluidization number		V/V_{init}, V_{init} = velocity for initial fluidization	Fluid velocity in fluidized bed/that at start of fluidization	Fluidization	*59*
F11	Fourier number (heat transfer)	N_{Foh}	$k\theta/\rho C_p L_m^2$		Unsteady state heat transfer	*5, 13, 14, 25*
F12	Fourier number (mass transfer)	N_{Fom}	$D\theta/L^2 = k_c\theta/L$ (*cf.* **D12**)		Unsteady state mass transfer	*13*
F13	Froude number (1)	N_{Fr_1}	$V^2/gL \equiv (N_{Fr_2})^2$ (*cf.* Reech No., Boussinesq No., Vedernikov No.)	Inertial force/gravitational force	Wave and surface behavior	*5, 13, 25*
F14	Froude number (2)	N_{Fr_2}	$V/\sqrt{gL} \equiv (N_{Fr_1})^{1/2}$ (*cf.* Boussinesq No.)	Velocity of open channel flow/speed of very small gravity wave	Open channel flow; free surfaces	*7, 50*
F15	Froude No. (rotating)	Fr	DN^2/g; D = impeller diam.		Agitation	*64*
G1	Galileo number	N_{Ga_1}	$L^3 g\rho^2/\mu^2$ (*cf.* N_{Ar}, Nusselt thickness group)	$N_{Ga_1} = N_{Re} \times$ gravity force/viscous force	Circulation of viscous liquid, thermal expansion	*5, 13*
G2	Gay Lussac number	Ga, N_{Ga_2}	$1/\beta\Delta T$		Thermal expansion processes	*25*
G3	Goucher number	N_{Go}	$R(\rho g/2\sigma)^{1/2}$; R = wall or wire radius	Gravitational force/surface tension force$^{1/2}$	Coating	*24*
G4	Graetz (Grätz) number	N_{Gz}	$V_m c_p/kL$	Thermal capacity fluid/convective heat transfer	Streamline flow	*5, 25*
G5	Grashof number	N_{Gr}	$L^3\rho^2 g\beta\Delta t/\mu^2 \equiv N_{Ga_1}/N_{Ga_2} \equiv (N_{Re})^2/(N_{Ga_2})(N_{Fr_1})$	$N_{Gr} = N_{Re}$ (buoyancy force/viscous force)	Free convection	*5, 13, 25*
G6	Gukhman number	Gu, N_{Gu}	$(t_o - t_m)/T_o$; t_o, T_o = temp. (°C., °K.) of hot gas stream, t_m = temp. of moist surface (wet bulb temp.)	Thermodynamic criterion of evaporation under isobaric adiabatic conditions	Convective heat transfer in evaporation	*6, 23, 17, 53*
G7	Guldberg-Waage group	N_{Gw}	Given by equation relating volumes of reacting gases and reaction products		Chemical reaction in blast furnaces	*5*
H1	Hall coefficient	N_H	$f_c J$ (f_c = cyclotron frequency, J = av. free path/av. veloc.)		Magneto-fluid dynamics	*7*
H2	Hartmann number	M_H	$(\mu_e^2 H_e^2 \sigma_e L^2/\mu)^{1/2} \equiv (SR_M N_{Re})^{1/2} \equiv (N_{Re}N)^{1/2}$	Magnetically induced stress/hydrodynamic shear stress (magnetic body force/viscous force)$^{1/2}$	Magneto-fluid dynamics	*5, 6*
H3	Hatta number	β	$\gamma/\tanh\ \gamma$; $\gamma = (rCD)^{1/2}/k_c$, r = reaction rate constant $[L^3/M\theta]$ [a modified Hatta number has also been defined *35*)]		Gas absorption with chemical reaction	*41*
H4	Head coefficient	C_H	gH'/N^2d^2 (d = impeller diam.)		Flow in pumps and fans	*32*
H5	Heat transfer number	K_Q	$q/V^3L^2\rho$		Heat transfer in stream	*25*
H6	Hedstrom number	N_{He}	$\tau_y L^2\rho/\mu_p^2 \equiv (N_{Re}) \times (N_{Bm})$		Flow of Bingham plastics	*5*
H7	Helmholtz resonator group		$(d^3/W)^{1/2}/Ma$	Proportional to frequency × residence time	Pulsating combustion	*28*
H8	Hersey number		$F_b/\mu v_s$ (*cf.* truncation number)	Load force/viscous force	Lubrication	*6*
H9	Hodgson number	N_H	$Wf\Delta\rho_F/\overline{V}_{f.\rho s}$	Time constant of system/period of pulsation	Pulsating gas flow	*5*
H10	Homochronous number	H_{o_1}	$V\theta/L$ (θ = time for liquid to move characteristic distance L)	Duration of process/time for liquid to move through L	Choice of time scales	*13*
H11	Hooke number	H_{o_2}	$\rho V^2/E \equiv$ Vauchy No., *q.v.*		Elasticity of flowing media	*25*
J1	Jakob modulus	Ja	$C_P\rho_L\Delta t/r\rho_v$ (ρ_L, ρ_v = densities of liquid and vapor; Δt = liquid superheat temperature diff.)	Maximum bubble radius/thickness of superheated film	Boiling	*6*
J2	*J*-factor (heat transfer)	j_H	$(h/c_pG)(c_p\mu/k)^{2/3} \equiv (N_{Nu})/(N_{Re})(N_{Pr})^{1/3}$		Heat, mass and momentum transfer theory	*5*
J3	*J*-factor (mass transfer)	j_M	$(k_c\rho/G)(\mu/\rho D)^{2/3} \equiv (k_c\rho/G)(N_{Sc})^{2/3}$			*5*
J4	Joule number	J	$2\rho C_p\Delta t/\mu_e H_e^2 \equiv 2(N_{Re})(R_M)/(M_H)^2(N_E)$	Joule heating energy/magnetic field energy	Magneto-fluid dynamics	*6*
K1	Kármán number (1)	N_K	$\rho d^3(-dp/dL)/\mu^2$ (d = pipe diam., dp/dL = pressure gradient) $\equiv 2(N_{Re})^2 f^{1/2}$		Fluid friction in conduits	*5*

TABLE B (Continued)

Serial No.	Name	Symbol	Definition	Significance	Field of Use	Reference
K2	Kármán number (2)	K	V/V_A (see Alfven No.)		Magneto-fluid dynamics	6
K3	Kirpichev number for heat transfer	K_{i_q}, N_{Ki_q}	$q^*L/k\Delta t$ (*cf.* N_{Bih}, N_{Nu})	Intensity external heat transfer/internal heat transfer intensity	Heat transfer	6, 14
K4	Kirpichev number for mass transfer	Ki_m	$GL/D\rho n^*$ (*cf.* N_{Pem}, N_{Bim})	Intensity external mass transfer/internal mass transfer intensity	Mass transfer	13, 14
K5	Kirpitcheff number		$(\rho F_R/\mu^2)^{1/3} = [(N_{Re})^2 cf]^{1/3}$		Flow around obstacles	34
K6	Knudsen number (1)	N_{Kn}	λ/L	Length of mean free path/characteristic dimension	Low pressure gas flow	5
K7	Knudsen number (2)	N_{KnA}	$\rho D/_{AB}D_{KA}\zeta$	Bulk diffusion/Knudsen diffusion	Gaseous diffusion in packed beds	6
K8	Kossovich number	K_o, N_{Ko}	$r_v\Delta n_m/c\Delta t$	Heat used for evaporation/heat used in raising temperature of body	Convective heat transfer during evaporation	6, 14
K9	Kronig number	Kr	$4L^2\beta\rho^2\Delta t E_S^2 N[\alpha + 2/3(p_o^2/kT)]/u^2 M$ E_S = electric field at surface, N = Avogadro's Number, α = polarization coefficient, p_o = molecular dipole moment, k = Boltzmann's constant, M = molecular weight	(N_{Re}) (electrostatic force/viscous force	Convective heat transfer	6, 44
K10	Kutateladze number (1)	Ku	$IEL/\rho Vu'$; I = current density $[Q/L^2\theta]$, E = voltage $[ML/Q\theta^2]$, u' = enthalpy $[L^2/\theta^2]$		Electric arcs in gas streams	65
K11	Kutateladze number (2)	K	$r_v/c_p(t_o - t_w)$, (t_o, t_w = stream, wall temp.)		Combined heat and mass transfer in evaporation	17
L1	Lagrange group (1)	La_1	$\Pi/\mu L^3 N^2$; L = characteristic dimension of agitator = $N_{Re}.N_p$		Agitation	6
L2	Lagrange number (2)	La_2	$(D + \varepsilon_D)/D$	Combined molecular and eddy mass transfer rate/molecular mass transfer rate	Mass transfer in turbulent systems	31
L3	Lagrange number (3)	La_3	$\Delta PR/\mu V$		Magneto-fluid dynamics	6
L4	Larmor number	R_{La}	L_L/L; (L_L = Larmor radius)		Magneto-fluid dynamics	23
L5	Laval number	La	$V\Big/\left(\frac{2\gamma}{\gamma+1}RT\right)^{1/2}$; γ = ratio of specific heats	Linear velocity/critical velocity of sound	Compressible flow	56
L6	Leverett function	J	$(\xi/e)^{1/2}(P\sigma/\sigma$	Characteristic dimension of surface curvature/characteristic dimension of pores	Two-phase flow in porous media	6
L7	Lewis No.	N_{Le}	$k/\rho c_p D = \alpha/D \equiv N_{Sc}/N_{Pr}$ (N.B.: Lewis number is sometimes defined as reciprocal of this quantity)		Combined heat and mass transfer	5, 25
L8	Lorentz number	N_{Lo}	V/V_1; (V_1 = velocity of light)	Fluid velocity/velocity of light	Magneto-fluid dynamics	6
L9	Luikov (Lȳkov) number	Lu	$k_c L/\alpha = k_c L_\rho C_p/k$	Mass diffusivity/thermal diffusivity; rate of extension of mass transfer field/rate of extension of heat transfer field	Combined heat and transfer	6, 13
L10	Lukomskii number	Lu	α/a_m; a_m = potential conductivity of mass transfer $[L^2/\theta]$		Combined heat and mass transfer	36
L11	Lundquist number	N_{Lu}	$\sigma_e H_e \mu_e^{3/2} L/\rho^{1/2} = M_H(R_M/N_{Re})^{1/2}$ (L = thickness of fluid layer)		Magneto-fluid dynamics	6
L12	Lyashchenko number	Ly	$\equiv N_{Re}^3/N_{Ar}$		Fluidization	19
L13	Lykoudis number	N_{Ly}	$(\mu_e H_e)^2 \frac{\sigma_e}{\rho}\left[\frac{L}{g\beta\Delta t}\right]^{1/2} \equiv (M_H)^2/(N_{Gr})^{1/2}$	Magnetic body force/square root of product of the inertia and buoying force.	Magneto-fluid dynamics	6
M1	McAdams group		$h^4 L\mu\Delta t/k^3\rho^2 gr$	Constant for given surface orientation	Condensation	5
M2	Mach number	N_{Ma}, Ma	V/V_s; (V_s = velocity of sound in fluid) $\equiv v/\sqrt{E_b/\rho}$; (E_b = bulk modulus of fluid) (*cf.* Sarrau number)	Linear velocity/sonic velocity	Compressible flow	5–7, 25, 50
M3	Magnetic force parameter	N	$\mu_e^2 H_e^2 \sigma_e L/\rho V$	Magnetic body force/inertia force; resistance time of fluid in field/relaxation time of lines force	Magnetic-fluid dynamics	6
M4	Magnetic mach number	M_{Ma}	V/V_a (see Alfven number)		Magneto-fluid dynamics	6
M5	Magnetic Oseen number	k	$\frac{1}{2}(1 - N_{Al}^2)R_M$	Magnetic force/inertia force	Magneto-fluid dynamics	2
M6	Magnetic pressure number	S	$\mu_e H_e^2/\rho V^2$	Magnetic pressure/2 × dynamic pressure	Magneto-fluid dynamics	5
M7	Magnetic Reynolds number	R_M	$\sigma_e \mu_e LV$ (*cf.* velocity number)	Mass transport diffusivity/magnetic diffusivity	Magneto-fluid dynamics	5
M8	Maievskii number		$\equiv Ma$		Compressible flow	12
M9	Marangoni number	N_{Ma}	$\frac{\Delta\sigma}{\Delta t}\frac{\Delta t}{\Delta L}L^2/\mu\alpha$; L = layer thickness		Cellular convection	46

TABLE B (Continued)

Serial No.	Name	Symbol	Definition	Significance	Field of Use	Reference
M10	Margoulis number	M	$\equiv N_{St}$		Forced convection	*6, 7*
M11	Merkel number	N_{Me}	$GA^* W^*/(V_m)$ gas	Mass of water transferred in cooling per unit humidity difference/mass of dry gas	Cooling towers, liquid-gas contact	*5*
M12	Miniovich number	Mn	SR/e; R = pore radius		Drying	*36*
M13	Mondt number	N_{Mo}		Convective/conductive heat transfer	Heat transfer	*6*
N1	Naze number	N_a	$V_A/V_s \equiv (N_{Ma}.N_{A1})$	Velocity Alfven wave/velocity of sound	Magneto-fluid dynamics	*6*
N2	Newton inertial force group	N_I	$F/\rho V^2 L^2$	Imposed force/inertial group	Agitation	*5*
N3	Newton number	N_e	$F_R/\rho V^2\ L^2$; $(cf.\ f, \psi)$	Resistance force/inertia force	Friction in fluid flow	*25*
N4	Number of velocity heads	N	$(F/\rho L^2)/(V^2/2)$	Imposed head/velocity head	Friction in conduits	*6*
N5	Number for similarity of phys. and chem. changes	K	$r/C_p\Delta t$	Heat flow for phase change/superheat (supercooling) of one of the phases	Changes of phase	*14*
N6	Nusselt number	N_{Nu}	$hL/k \equiv (N_{Re} N_{St})$ $(cf.\ N_{Bi_h}, Ki_g)$	Total heat transfer/conductive heat transfer	Forced convection	*5, 13, 25*
N7	Nusselt number for mass transfer	Nu_m, N_{Nu_m}	$k_c L/D = N_{Sh}$	Intensity of mass flux at interface/specific flux by pure molecular diffusion in layer of thickness, L	Mass transfer	*13, 20*
N8	Nusselt film thickness group	ψ, N_T	$L_f(\rho^2 g/\mu^2)^{1/3} \equiv (N_{Ga})^{1/3}$; ($L_f$ = film thickness)	$= (N_{Re})^{1/3}$ (gravitational force/viscous force)$^{1/3}$	Falling films	*6*
O1	Ocvirk number		$(F_b/\mu V_s)(a/R)^2(D/b)^2$; ($v_s$ = shaft surface velocity; R = shaft radius; D = shaft diam.) $(cf.\ N_S)$	Load force/viscous force	Lubrication	*6*
O2	Ohnesorge number	Z	$\mu/(\rho L\sigma)^{1/2} = (N_{We_1})^{1/2}/(N_{Re})$	Viscous force/(inertia force × surface tension force)$^{1/2}$	Atomization	*5*
P1	Péclet number (heat)	Pe, N_{Pe_h}	$LV\rho C_p/k = LV/\alpha = (N_{Re.}\ N_{Pr})$	Bulk heat transfer/conductive heat transfer	Forced convection	*5, 13, 25*
P2	Péclet number (mass)	N_{Pe_m}	$LV/D = (N_{Re}\ N_{Sc})$	Bulk mass transfer/diffusive mass transfer	Mass transfer	*5*
P3	Pipeline parameter	ρn	$V_W V_o/2H_s'$; (V_W = velocity water-hammer wave, V_o = initial velocity H_s' = static head × $g[L^2/\theta^2]$	Maximum pressure rise in water hammer/2 × static pressure	Water hammer	*5*
P4	Poiseuille number		$D^2(-dp/dL)/\mu V$(D = pipe diam., dp/dL = pressure gradient)	= 32 for laminar flow in round pipe	Laminar fluid friction	*5*
P5	Pomerantsev number	P_o	$jL^2/k(t_m - t_o)$(t_m, t_o = temp. of medium, initial temp. of body) *cf.* Damköhler Group IV)		Heat transfer with heat sources in medium	*6, 14, 60*
P6	Posnov number	Pn	$\delta\Delta t/(\Delta n_m)(cf.\ Fe_2)$		Combined heat and mass transfer	*13, 14, 67*
P7	Power number	N_P	$\Pi/L^5\rho N^3$	Drag on (agitator impeller) or inertial force	Power consumption by agitators, fans, pumps, etc	*5, 32*
P8	Prandtl number	N_{Pr}	$C_p\mu/k$ = Da IV/Da III × Da V	Momentum diffusivity/thermal diffusivity	Forced and free convection	*5, 13, 25*
P9	Prandtl number (mass transfer)	Pr_m	$\mu/\rho D = N_{Sc}$, v (used in Russian, German literature)	See Schmidt number		*13*
P10	Prandtl velocity ratio	u^+	$V/(r_W/\rho)^{1/2}$ (V = local fluid velocity)	Inertial force/wall shear force$^{1/2}$	Turbulence studies	*5*
P11	Prandtl dimensionless	y^+	$L(\rho r_W)^{1/2}/\mu$ (L = distance from wall, etc.)		Turbulence studies	
P12	Predvoditelev number	Pd	$\Gamma L^2/\alpha t_o = \left(\dfrac{dt^*}{\alpha(N_{Fo})}\right)_{max}$ where t_o = init. temp. of body, l^* = temp. of medium relative to its initial temp.	Rate of change of temp. of medium/rate of change of temp. of body	Heat transfer	*13, 14, 60*
P13	Pressure number (1)	K_P	$P/\{g\sigma(\rho' - \rho'')\}^{1/2}$($\rho'$, ρ'' = density of liquid gas)	Absolute pressure in system (pressure jump on interface)		*14*
P14	Pressure number (2)		$H/\frac{1}{2}/_2 U_s^2$(U_s = circumferential velocity) $\equiv$ [diameter No.]$^{-2}$ × [Speed No.]$^{-2}$		Flow machines (turbines, pumps, etc.)	*25*
P15	Psychrometric ratio		h_c/Gs	Heat transfer by convection/heat transfer by mass transfer	Wet and dry bulb thermometry	*5*
R1	Radiation number	$\bar{K}_s$	$kE/\eta\sigma T^3 = (N_{We_1})$/(Hooke No.) × (Stefan No.)		Radiant transfer	*25*
R2	Ramzin number	Ra	$\dfrac{C_b P}{C_m(\Delta\Omega)} = \dfrac{\text{(Bulygin No.)}}{\text{(Kosovich No.)}}$		Molar mass transfer	*40*
R3	Ratio of specific heats	γ	C_p/C_v (specific heats at constant pressure, volume)		Compressible flow	*5*
R4	Rayleigh number (1)	N_{Ra_1}	$V(\rho L/\sigma)^{1/2} = N_{We_2}$ $(q.v.)$	See N_{We}	Breakup of liquid jets	*5*
R5	Rayleigh number (2)	R_2'	$L^3\rho^2 g\beta c_p\Delta t/\mu k = L^3\rho g\beta\Delta t/\mu\alpha = (N_{Gr})\cdot(N_{Pr})$		Free convection	*5, 25*

TABLE B (Continued)

Serial No.	Name	Symbol	Definition	Significance	Field of Use	Reference
R6	Rayleigh number (3)	Ra_3	$q^*L^5\rho^2 g\beta C_p/\mu k^2 x = (N_{Gr})(N_{Pr})(N_{Nu})(L/x)$; ($L$ = pipe diam.)		Combined free and forced convection in vertical tubes	6
R7	Recovery factor	N_{rf}	$C_p(t_{aw}-t_m)/V^2$; t_{aw} = attained adiabatic wall temp. t_m = temp. of moving medium (*cf.* Eckert No.)	Actual temp. recovery/theoretical temp. recovery	Convective heat transfer in compressible flow	5
R8	Reech number		$= 1/(N_{Fr_1})$ *q.v.*		Wave and surface behavior	6
R9	Resistance coefficient (1)	C_f	$F_R/\frac{1}{2}\rho V^2 L^2$ (*cf.* drag coeff., Newton number, Fanning factor)		Flow resistance	25
R10	Resistance coefficient (2)	ψ	$\Delta p.\ D_H/\frac{1}{2}\rho V^2 L$ (Δp = pressure drop over length, L) (*cf.* **R9**)		Fluid friction in conduits	25
R11	Reynolds number	N_{Re}	$LV\rho/\mu$	Inertia force/viscous force	Dynamic similarity	5, 25
R12	Reynolds number (rotating)	Re	$L^2 N\rho/\mu$; L = impeller diam.		Agitation	45
R13	Richardson number	N_{Ri}	$-(g/\rho)(d\rho/dL)/(dV/dL)w^2$ [L = height of liquid layer, $(dV/dL)_w$ = velocity gradient at wall]	Gravity force/inertial force	Stratified flow of multilayer systems	54
R14	Romankov number	R'_o	T_D/T_{PROD}	Dry bulb temperature (abs.)/(product temperature (abs.)	Drying	6, 52
R15	Rossby number	N_{Ro}	$V/2\omega_e L \sin\Lambda$ (ω_e = angular velocity of earth's rotation $[1/\theta]$; Λ = angle between axis of earth's rotation and direction of fluid motion [$^-$])	Inertia force/Coriolis force	Effect of earth's rotation on flow in pipes	5
R16	Roughness factor		ε/L		Fluid friction	5
S1	Sarrau number		≡ mach number, *q.v.*		Compressible flow	6
S2	Schiller number (1)		$LV(\rho^2/\mu F_R)^{1/3}$		Flow around obstacles	34
S3	Schiller number (2)	Sch	$V\left[\frac{3}{4}\cdot\frac{\rho\gamma_m}{g\mu(\gamma_M-\gamma_m)}\right]^{1/3}$; V = velocity in fluidized bed; γ_m, γ_M = specific gravity of medium and material in bed		Fluidization	66
S4	Schmidt number	N_{Sc}	$\mu/\rho D$ (*cf.* N_{Pr_m}) (= Da II/Da I Da V)	Kinetic viscosity/molecular diffusivity	Diffusion in flowing	5, 25
S5	Semenov number	Sm	k_c/K; K = reaction rate constant $[L/\theta]$		Reaction kinetics	11
S6	Senftleben number	S_e	$NE_s^2[\alpha+2/3(\rho_o^2/kT)]\cdot[1/4LM_g]$ Kronig number, *q.v.*		Convective heat transfer	6
S7	Sherwood number	N_{Sh}	$k_c L/D = Nu_m$ (also termed Taylor number)	Mass diffusivity/molecular diffusivity	Mass transfer	5
S8	Sommerfeld number (1)	N_{S_1}	$(\mu N/P_b)(D/a)^2$ (D = shaft diam., (*cf.*) Ocvirk number)	Viscous force/load force	Lubrication	5
S9	Sommerfeld number (2)	N_{S_2}	$(F_b/\mu V_s)(a/R)^2$ (V_s = veloc. of shaft surface; R = shaft radius) ($N_{S_2} = 4/\pi N_{S_1}$)	Viscous force/load force	Lubrication	6
S10	Spalding function	Sp	$-\left(\frac{\delta\theta}{\partial u^+}\right)_{u^+=0}$; $\theta = (T-T_\infty)/(T_w-T_\alpha)$, T_w = wall temperature, T_∞ = free stream temp., u^+ = Prandtl velocity ratio	Dimensionless temp. gradient at wall	Convection	18, 33
S11	Specific speed	C_s	$N(V_f)^{1/2}/(gH')^{3/4}$ (H' = head of liquid produced by one stage) (*cf.* speed number)		Pumps and compressors	8, 32
S12	Speed number	σ	$(4\pi)^{1/2}(V_f)^{1/2}\ N/(2H)^{3/4}$ = (delivery number)$^{1/2}$ × pressure number)$^{-3/4}$ (*cf.* specific speed)		Flow machines	25
S13	Stanton number	N_{St}	$h/C_p\rho V = h/C_p G = (N_{Nu})/(N_{Re})(N_{Pr})$	Heat transferred/thermal capacity of fluid	Forced convection	5, 6, 13, 25
S14	Stefan number	St	$\eta LT^3/k$		Heat radiation	25
S15	Stokes number	St	$\mu\theta_v/\rho L^2$ (θ_v = vibration time) ≡ $(N_{S_1})^{-1}(N_{Re})^{-1}$		Particle dynamics	6
S16	Strouhal number	N_{S_1}, Sr	fL/V (*cf.* N_{Th})		Vortex streets; unsteady-state flow	6, 25
S17	Suratman number	Su	$\rho L\sigma/\mu^2 = (N_{Re})^2/(N_{We_1}) = (Z)^{-2}$		Particle dynamics	6
S18	Surface elasticity number	N_{El}	$-\frac{\Gamma'}{D_S}L\frac{(\partial\sigma)}{(\partial\Gamma')}$; Γ' = surface concentration of surfactant in undisturbed state, D_S = surface diffusivity, L = film thickness		Convection cells	10
S19	Surface viscosity number	N_{Vi}	$\mu_S/\mu L$; μ_S = surface viscosity, $[M/\theta]$, L = film thickness		Convection cells	10
T1	Taylor number (1)	N_{Ta_1}	$\omega_c(R_a)^{1/2}a^{3/2}\rho/\mu$; ($\omega_c$ = angular velocity of cylinder; R_a = mean radius of annulus)		Stability of flow pattern in annulus with rotating cylinder	5
T2	Taylor number (2)	N_{Ta_2}	$(2\omega L^2\rho/\mu)^2$[ω = rate of spin $(1/\theta)$; L = height of fluid layer]	α(Coriolis force/viscous force)^2	Effect of rotation on free convection	6
T3	Thiele modulus	m_T	$Q^{1/2}U^{1/2}L/k^{1/2}t^{1/2}$ = (Da IV)$^{1/2}$		Diffusion in porous catalysts	5
T4	Thoma number	σ_T	$(H_a-H_s-H_v)/H$ (H = total head; H_a = atm. pressure head; H_s = suction head; H_v = vapor pressure head)	Net positive suction head/total head	Cavitation in pumps	5
T5	Thomson number	N_{Th}	$\theta V/L$; θ = characteristic time (*cf.* N_{S_1})		Fluid flow	6
T6	Thring radiation group		$\rho C_p V/e^*\eta T^3$ (*cf.* Boltzmann number)	Bulk heat transport/heat transport by radiation	Radiation	5

TABLE B (Continued)

Serial No.	Name	Symbol	Definition	Significance	Field of Use	Reference
T7	Thring-Newby criterion	θ	$[(V_{m_1}+V_{m0})/V_{m0}](R/L)$; V_{m0}, V_{u_1} = mass flow rates of nozzle fluid and surrounding fluid $[M/\theta]$; R = equivalent nozzle radius; L = furnace half width		Combustion of fuels	*4*
T8	Truncation number	r	$\mu\gamma/P$ (*cf.* Hersey number)	Shear stress/normal stress	Viscous flow	*6*
V1	Valensi number	V	$\omega L^2\rho/\mu$; ω = circular oscillation frequency when $\mu = 0$ $[1/\theta]$		Oscillations of drops and bubble-	*57*
V2	Vedernikov number	V	$\zeta^*\zeta^*\bar{V}/(V_w-\bar{V}) \equiv \zeta^*\zeta^*(N_{Fr_2})$ (ζ^* = exponent of hydraulic radius in formula $[^-]$; ζ^* = shape factor of channel section; V_w = absolute velocity of disturbance wave)	Generalized Froude number	Instability of open-channel flow	*6, 61*
V3	Velocity number	R_v	≡ Magnetic Reynolds number, *q.v.*			*5, 6*
W1	Weber number (1)	N_{We_1}	$V^2\rho L/\sigma = (N_{We_2})^2$	Inertia force/surface tension force	Bubble formation, etc.	*2, 25*
W2	Weber number (2)	N_{We_2}	$V(\rho L/\sigma)^{1/2} = (N_{We_1})^{1/2}$			
W3	Weber number (rotating)	W_e	$L^3N^2\rho/\sigma$; L = impeller diameter		Agitation	*64*
W4	Weissenberg number	N_{W_e}	$\omega_3 V/\omega_1 L$; $\omega_3 = \int_0^\infty sG(s)\,ds$, $\omega_1 = \int_0^\infty G(s)\,ds$, G = relaxation modulus of linear viscoelasticity, s = recoverable elastic strain	Viscoelastic force/viscous force	Viscoelastic flow	*63*

REFERENCES

1. Adrianov, V. N., Shorin, S. N., *AIAA J.* **1**, 1729 (1963).
2. Ahlstrom, H. G., *J. Fluid Mech.* **15**, 205 (1963).
3. Becker, H. A., Hottel, H. C., Williams, G. C., "Ninth Symposium (International) on Combustion," p. 7, Academic Press, New York, 1963.
4. Beer, J. M., Chigier, N. A., Lee, K. B., *Ibid.*, p. 892.
5. Boucher, D. F., Alves, G. E., *Chem. Eng. Progr.* **55** (9), 55 (1959).
6. *Ibid.*, **59** (8), 75 (1963).
7. British Standard 1991, "Recommendations for Letter Symbols, Signs and Abbreviations. Part 2. Chemical Engineering, Nuclear Science, and Applied Chemistry," British Standards Institution, London, 1961.
8. Brown, G. G., *et al.*, "Unit Operations," Wiley, New York, 1950.
9. Buckingham, E., *Phys. Rev.* **4**, 345 (1914).
10. Berg, J. C., Acrivos, A., *Chem. Eng. Sci.* **20**, 737 (1965).
11. Chukhanov, Z. F., *Intern. J. Heat Mass Transfer* **6**, 691 (1963).
12. Dallavalle, J. M., "Micromeritics," 2nd ed., Pitman, New York, 1948.
13. El'perin, I. T., *Inzh. Fiz. Zh. Akad. Nauk Belorussk. SSR* **4** (1), 131 (1963).
14. El'perin, I. T., *Intern. J. Heat Mass Transfer* **5**, 349 (1962).
15. Engel, F. V. A., *Z.V.D.I.* **107**, 671, 793 (1965).
16. Faller, A. J., *J. Fluid Mech.* **15**, 560 (1963).
17. Fedorov, B. I., *Inzh. Fiz. Zh. Akad. Nauk Belorussk. SSR* **7** (1), 21 (1964).
18. Gardner, G. O., Kestin, J., *Intern. J. Heat Mass Transfer* **6**, 289 (1963).
19. Gel'perin, I. T., Aĭnshteĭn, V. G., Goĭkhman, I. D., *Inzh. Fiz. Zh. Akad. Nauk Belorussk. SSR* **7** (7), 15 (1964).
20. Grassmann, P., *Chem. Ing.-Tech.* **31**, 148 (1959).
21. Grassman, P., Lemaire, L. H., *Ibid.*, **30**, 450 (1958).
22. Greene, D. F., Ph.D. Thesis, Columbia Univ., 1961 [*Dissertation Abstr.* **24** (8), 3248 (1964)].
23. Gukhman, A. A., "Introduction to the Theory of Similarity," Academic Press, New York, 1965.
24. Gutfinger, C., Tallmadge, J. A., *A.I.Ch.E. J.* **10**, 774 (1965).
25. Hahnemann, H. W., "Die Umstellung auf das internationale Einheitensystem in Mechanik und Wärmetechnik," VDI-Verlag, Düsseldorf, 1959.
26. Holt, M., "Dimensional Analysis" in "Handbook of Fluid Dynamics," V. L. Streeter, ed., McGraw-Hill, New York, 1961.
27. Hottel, H. C., Sarofim, A. F., *Intern. J. Heat Mass Transfer* **8**, 1153 (1965).
28. Hottel, H. C., Williams, G. C., Jensen, W. P., Tobey, A. C., Burrage, P. M. R., p. 923 in "Ninth Symposium (International) on Combustion," Academic Press, New York, 1963.
29. Huntley, H. E., "Dimensional Analysis," MacDonald & Co., London, 1952.
30. Johnson, S. P., "Survey of Flow Calculation Methods," p. 98, Pre-printed Papers & Program, Aeronautic & Hydraulic Divisions, A.S.M.E. Summer Meeting, June 19–21, Univ. of Calif. and Stanford Univ., 1934.
31. Kafarov, V. V., *Zh. Prikl. Khim.* **29**, 40 (1956).
32. Kay, J. M., "An Introduction to Fluid Mechanics & Heat Transfer," Cambridge Univ. Press, 1957.
33. Kestin, J., Persen, L. N., *Intern. J. Heat Mass Transfer* **5**, 143 (1962).
34. Klinkenberg, A., Mooy, H. H., *Chem. Eng. Progr.* **44**, 17 (1948).
35. Koide, K., Kubota, H., Shindo, M., *Chem. Eng.* (Japan), **28** (8), 657 (1964).
36. Lȳkov, A. V., Mikhaĭlov, Yu. A., "Theory of Energy & Mass Transfer," Prentice-Hall, Englewood Cliffs, N.J., 1961.
37. McCabe, W. L., Smith, J. C., "Unit Operations of Chemical Engineering," McGraw-Hill, New York, 1956.
38. Matsuhisa, S., Bird, R. B., *A.I.Ch.E. J.* **11**, 588 (1965).
39. Mikhaĭlov, Yu. A., Bornikova, R. M., *Inzh. Fiz. Zh. Akad. Nauk Belorussk. SSR* **6** (10), 45 (1963).
40. Mikhaĭlov, Yu. A., Romanina, I. V., *Ibid.*, **7** (1), 49 (1964).
41. Miyauchi, T., Nakano, K., Obata, K., Kimura, S., *Chem. Eng.* (Japan) **26** (9), 999 (1962).
42. Mkhitaryan, A. M., "Hydraulics & Fundamentals of Gas Dynamics," Israel Program for Scientific Translations, Jerusalem, 1964.
43. Mordell, D. L., Wu, J. H. T., *Can. Aeronaut. Space J.* **9** (4), 117 (1963).
44. Motulevich, V. P., Eroshenko, V. M., Petrov, Yu. P., in "Physics of Heat Exchange & Gas Dynamics," A. S. Predvoditelev, ed., Consultants Bureau, New York, 1963.
45. Nagata, S., *Chem. Eng.* (Japan) **27** (8), 592 (1962).
46. Nield, D. A., *J. Fluid Mech.* **19**, 341 (1964).
47. Potter, J. M. F., B.Sc. Thesis, Dept. of Chem. Engrg., Univ. of Birmingham, England, 1959.
48. Rayleigh, Lord, *Phil. Mag.* **48**, 321 (1899).
49. Reiner, M., *Phys. Today* **17** (1), 62 (1964).
50. Rouse, H. (ed.), "Engineering Hydraulics," Wiley, New York, 1950.
51. Rouse, H., Ince, S., "History of Hydraulics," Iowa Institute of Hydraulic Research, State University of Iowa, 1957.
52. Sazhin, B. S., *Inzh. Fiz. Zh. Akad. Nauk Belorussk. SSR* **5** (6), 13 (1962).
53. Sazhin, B. S., Miklin, Yu. A., *Ibid.*, **6** (10), 57 (1963).
54. Schlichting, H., "Boundary Layer Theory," 4th ed., McGraw-Hill, New York, 1960.
55. Scriven, L. E., Sternling, C. V., *J. Fluid Mech.* **19**, 321 (1964).
56. Sillem, H., *Z.V.D.I.* **106**, 398 (1964).
57. Szebehely, V. G., p. 771 in "Proc. 2nd U.S. Nat. Congress of Appl. Mech.," Ann Arbor, Mich., June 1954; A.S.M.E., New York, 1955.
58. Tallmadge, J. A., Labine, R. A., Wood, B. H., *Ind. Eng. Chem. Fundamentals* **4**, 400 (1965).
59. Tamarin, A. I., *Inzh. Fiz. Zh. Akad. Nauk Belorussk. SSR* **6** (7), 19 (1963).
60. Tartakovskii, D. F., *Ibid.*, **7** (1), 71 (1964).
61. Vedernikov, V. V., *Compt. Rend. Acad. Sci. U.R.S.S.* **48**, 239 (1945); 52, 207 (1946).
62. Weber, M., *Jahrb. Schiffbautechn. Ges.* **20**, 355 (1919).
63. White, J. L., *J. Appl. Polymer. Sci.* **8**, 2339 (1964).
64. Yamaguchi, I., Yabuta, S., Nagata, S., *Chem. Eng.* (Japan) **27** (8), 576 (1963).
65. Yas'ko, O. I., *Inzh.-Fiz. Zh. Akad. Nauk Belorussk. SSR* **7** (12), 112 (1964).
66. Zabrodskiĭ, S. S., "Flow & Heat Transfer in Fluidized Beds," to be published shortly by M.I.T. Press.
67. Zhuravleva, V. P., *Inzh.-Fiz. Zh. Akad. Nauk Belorussk. SSR* **6** (9), 73 (1963).

DIMENSIONLESS GROUPS (Supplementary Table)

George D. Fulford and John P. Catchpole

TABLE II. ALPHABETICAL LIST OF NEW GROUPS

Serial No.	Name	Symbol	Definition	Significance	Field of Use	Reference
A0	Absorption No.	*Ab*	$kc_L\sqrt{\frac{xL_f}{DV'_f}}$ kc_L = liquid side mass transfer coefficient; x = length of wetted surface; L_f = film thickness; V'_f = volume flow rate per wetted perimeter $[L^2/\theta]$	Dimensionless mass transfer coefficient	Gas absorption in wetted wall column	*40*
A1a	Advance ratio	*J*	V/ND V = forward speed; D = propeller diameter	Special form of Strouhal No.	Propeller studies	*27*
A1b	Aeroelasticity parameter	—	≡ Cauchy No., *q.v.*	Inertia force/compressibility force	Compressible flow	*27*
A4a	Anonymous group 3	ε	Dx/V'_fL_f (symbols as in Absorption No.). $\varepsilon = (Ab)^2/(N_{Sh})^2$	Dimensionless diffusivity	Gas absorption in wetted wall column	*40*
A4b	Anonymous group 4	$1/\alpha$ $(1/\beta)$	$\tau_w R/V_\infty\mu$; R = cylinder radius; V_∞ = velocity outside boundary layer	Frictional force/viscous force dimensionless skin friction)	Laminar boundary layer flow	*12, 28, 35*
B1a	Bairstow No.	—	$V/V_{s.w}$ $V_{s\,w}$ = velocity of sound at wall (*cf.* Mach. No.)	Previously used for Mach No., now largely obsolete	—	*27*
B2a	Batchelor No.	—	$VL\sigma_e/V_l^2\varepsilon_e$ ε_e = electrical permittivity $[Q^2\theta^2/L^3M]$	—	Magnetofluid dynamics	*27*
B13a	Buoyancy parameter	—	$\frac{\Delta T}{T}\frac{gL}{V^2} = (N_{Gr})/(N_{Re})^2 = \frac{\Delta T}{T}\left(\frac{1}{N_{Fr_1}}\right)$	Buoyancy force/inertia force	Free convection	*27*
D6a	Darcy No. (2)	Da_2	VL/D'; D' = permeability coefficient of porous medium $[L^2/\theta]$	Inertia force/permeation force	Flow in porous media	*19*
D8a	Generalized Deborah No.	N_2	$\sqrt{I_e - I_w}\cdot\theta_n$ I_e = invariant of rate of strain tensor (sec.$^{-2}$); I_w = invariant of vorticity tensor (sec.$^{-2}$); θ_n = natural time (sec.)	Generalization of group **D8**	Rheology	*2*
D14a	Dufour No.	Du_2	$\mathscr{R}\Theta n'_{10}/c_p$ Θ = thermodiffusion constant = $(D_T/D)/n_{10}n_{20}$ [−]; D_T = thermal diffusion coefficient $[L^2/\theta]$; $n'_{10}, n'_{20} = n'_1/n'$, n'_2/n'; n' = total No. of molecules = $n'_1 + n'_2$; n'_1, n'_2 = No. of molecules of components 1, 2, in binary mixture; also $Du_2 = (D_T/D)p/\rho c_p T n_{20}$	Heat of isothermal mass transfer/enthalpy of unit mass of mixture	Thermodiffusion	*22*
E1a	Einstein No.	—	V/V_l (V_l = speed of light) (*cf.* Lorentz No.)	Fluid velocity/velocity of light	Magnetofluid dynamics	*27*
E4a	Electric field parameter	R_E	$E/V\mu_eH_e$		Magnetohydrodynamics	*26*
E4b	Electrical characteristic No.	*El*	$\rho(d\chi/dT)L^2\cdot\Delta T\cdot E_1^2/\mu^2$ E_1 = electrical field strength $[ML/Q\theta^2]$; χ = dielectric susceptibility $[Q^2\theta^2/ML^3]$		Electrical effects on transfer processes	*6*
E4c	Electrical Nusselt No.	N_u	VL/D^*; $D^* = \frac{1}{2}(D^+ + D^-)$ $[L^2/\theta]$ D^+, D^- = diffusion coefficients of ions (*cf.* group *P2*)	Convection current/diffusion current	Electrochemistry	*10*
E4d	Electrical Reynolds No. (1)	—	$\varepsilon_eV/Q'Lb'$ ε_e = electrical permittivity $[Q^2\theta^2/L^3M]$; Q' = space charge density $[Q/L^3]$; b' = carrier mobility $[Q\theta/M]$		Electrical effects in flow	*27*
E4e	Electrical Reynolds No. (2)	—	Alternate name for group **E4c**, *q.v.*			*10*
E10a	Modified Euler No.	*Eu'*	$H_{L'}\rho_Lg/V_G^2\rho_G$ H'_L = head of liquid on tray $[L]$; ρ_L, ρ_G = densities of liquid, vapor; V_G = vapor velocity based on free area, $[L/\theta]$	Friction head/velocity head	Flow of vapor across mass transfer trays	*13*
E13a	Expansion No.	*Ex*	$\left(\frac{gd}{V^2}\right)\left(\frac{\rho_L-\rho_G}{\rho_L}\right)$ d = bubble diam., V = bubble veloc., ρ_L, ρ_G = densities of liquid, gas	$1/N_{Fr_1}$ × density ratio	Rise of bubbles	*13*
F12a	Fourier No. (flow)	*Fo*	$\mu\theta/\rho L^2$		Unsteady state flow problems	*17*
F12b	Frank-Kamenetskii No.	δ	$\frac{Q''}{k}\frac{E_a}{\mathscr{R}T_o^2}L^2k_o\exp(-E_a/\mathscr{R}T)$ Q'' = heat liberated per unit mass of material reacting/unit volume $[1/L\theta^2]$; k = thermal conductivity of reacting mixture $[ML/T\theta^3]$; k_o = preexponential constant in Arrhenius equation $[M/\theta]$	Dimensionless heat effect of reaction	Heat transfer in reacting systems	*24*
F12c	Frequency parameter	—	$\omega'L/V = 2\pi$ × Strouhal No.; $2\pi/\omega'$ = period of motion $[\theta]$	Special form of Strouhal No. (*cf.* also **T5**)	Unsteady state flow, etc.	*27*
F12d	Frequency No. (2)	N_f	ω_rL/V; L = packing element diameter $[L]$; V = interstitial fluid velocity $[L/\theta]$; ω_r = radial frequency (radians/sec.) $[1/\theta]$ (*cf.* groups **H10, S16, T5**)	Special form of group **F12c**	Flow in packed or fluidized beds	*23, 29*

TABLE II. ALPHABETICAL LIST OF NEW GROUPS (Continued)

Serial No.	Name	Symbol	Definition	Significance	Field of Use	Reference
F12e	Frössling No. (heat transfer)	Fs_h	$(N_{Nu} - 2)/(N_{Re}^{1/2} N_{Pr}^{1/3})$	Special dimensionless heat transfer coefficient	Heat transfer to spheres in turbulent streams	*8a*
F12f	Frössling No. (mass transfer)	Fs_m	$(N_{Sh} - 2)/(N_{Re}^{1/2} N_{Se}^{1/3})$	Special dimensionless mass transfer coefficient	Mass transfer to spheres in turbulent streams	*8a*
G2a	Geometric No.	Ge	h^*/H^*; h^* = surface area of packing element/perimeter $[L]$; H = height of packing $[L]$.	Dimensionless packed bed height	Mass transfer in packed beds	*5a*
G2b	Goertler parameter	—	$\frac{VL_b\rho}{\mu}\left(\frac{L_b}{R_c}\right)^{1/2}$ L_b = boundary layer momentum thickness, $[L]$; R_c = radius of curvature $[L]$	Modified Reynolds No.	Boundary layer flow on curved surfaces	*27*
G5a	Diffusional Grashof No.	Gr_{AB}	$L^3\rho^2 g\beta'_A \Delta n'_A/\mu^2$ n'_A = mass fraction of species A, $[-]$; β'_A = coefficient of density change with n'_A, $[-]$	Buoyant forces × inertia forces/(viscous forces)^2	Interphase transfer by free convection (density changes caused by concentration differences)	*37*
H8a	Hess No.	Ge	$(KL^2/a_m)(C_o)^{n-1}$ n = order of reaction $[-]$; C_o = initial concn. $[M/L^3]$; a_m = mass transfer conductivity of reaction products $[L^2/\theta]$; K = reaction rate constant, $\left[\frac{1}{\theta}\frac{L^{3n-3}}{M^{n-1}}\right]$		Heat and mass transfer with chemical and phase changes	*22*
H9a	Homochronicity No.	Ho_3	$N\theta$ (N = mixer r.p.m., θ = mixing time)		Mixing, agitation	*31*
H11a	Hydraulic resistance group	Γc	$\Delta p_p/\rho_L gL = We_4 \times$ Laplace No. $= N_{Eu_1} \times N_{Fr_1}$ Δp_p = pressure drop across liquid on tray $[M/L\theta^2]$; ρ_L = liquid density; L = depth of liquid on tray	Characterizes development of interfacial area per unit area of tray	Pressure drop in distillation columns	*15*
I1	Ilyushin No.	I	$(Vd\rho/\mu) \,.\, 4\tau_D/3V^2\rho = 4\tau_D/3V^2\rho \,.\, N_{Re}$; τ_D = max. dynamic slip resistance $[M/L\theta^2]$		Flow of viscoplastic fluids	*20a*
K7a	Knudsen No. for diffusion	$N_{Kn_{A2}}$	$3eD_{AB}/4\zeta K_{oA}\bar{u}_A$ K_{oA} = Knudsen flow permeability constant; $\bar{u}_A$ = equilibrium mean molecular speed of species A	Differs from **K7** by numerical constant	Gaseous diffusion in packed beds	*34*
K7b	Kondrat'ev No.	Kn	ΨN_{Bih}; N_{Bih} = heat transfer Biot No.; Ψ = temp. field nonuniformity parameter $= (t_s - t_a)/(\bar{t} - t_a)$; t_a = temp. of surrounding medium; t_s = body surface temp.; $\bar{t}$ = body mean temp.		Heat transfer between fluid and body	*21*
L3a	Laplace No.	La_4	$\Delta p_p \,.\, L/\sigma = \Gamma c/We_4$ Δp_p = pressure drop across liquid on tray $[M/L\theta^2]$; L = depth of liquid on tray		Interfacial behavior on distillation trays	*32*
L5a	Lebedev No.	Le_2	$eb_T(t_a - t_o)/c_b p\rho_s$ b_T = intensity of vapor expansion in capillaries of body on heating $[M/L^3T]$; t_a = temp. of surrounding medium; t_o = initial temp.; ρ_s = density of solid $[M/L^3]$	Molar expansion flux/molar vapor transfer flux	Drying of porous materials	*22*
L5b	Leroux No.	—	≡ cavitation No. **C6**			*27*
L7a	Turbulent Lewis No.	Le_T	$c_p\rho\varepsilon_D/k_T = l_D/l_T = \varepsilon_D/\varepsilon_T$ k_T = eddy thermal conductivity $[LM/T\theta^3]$; l_D, l_T = mixing lengths for mass, heat transfer $[L]$; ε_T = eddy thermal diffusivity $[L^2/\theta]$		Combined turbulent heat and mass transfer	*11*
L7b	Lewis-Semenov No.	—	$\equiv 1/N_{Le}$; N_{Le} = group **L7**			*27*
L7c	Lock No.	—	$\rho R^4 ia'/I$ ρ = fluid density; a' = rotor lift curve slope $[L^2/M]$; i = blade chord $[L]$; R = rotor radius $[L]$; I = moment of inertia of blade about hinge $[L^4]$		Rotor blade dynamics	*27*
M3a	Magnetic Grashof No.	—	$4\pi\sigma_e\mu_e(\mu/\rho) \,.\, N_{Gr}$ N_{Gr} = group **G5**		Magnetofluid dynamics	*7*
M4a	Magnetic No.	R_m	$\mu_e H_e(\sigma_e L/\rho V)^{1/2}$ = (magnetic force parameter)$^{1/2}$	See group **M3**	Magnetohydrodynamics	*26*
M5a	Magnetic Prandtl No.	—	$\sigma_e\mu_e\frac{\mu}{\rho}$ *cf.* Magnetic Grashof No., group **M3a**	Magnetic Reynolds No./Reynolds No. (properties of fluid)	Magnetofluid dynamics	*27*
P3a	Plasticity No.	P	≡ Bingham No.			*41*
P7a	Modified Power No.	N'_P	$\frac{\Pi}{L^5\rho N^3}\left(\frac{D'_e}{L'}\right)\frac{(\Delta\omega)^{-1/2}}{(N_b N_s)^{0.67}}$ D'_e = effective agitator diameter $[L]$; L'_e = effective agitator height; $\Delta\omega$ = wall proximity factor; N_b = No. of blades on agitator; N_s = effective No. of blade edges		Agitation	*36*
P8a	Total Prandtl No.	**Pr**	$\frac{\varepsilon_M + (\mu/\rho)}{\varepsilon_T + \alpha}$ $\varepsilon_M, \varepsilon_T$ = eddy transfer coefficient for momentum, heat $[L^2/\theta]$	Total momentum diffusivity/total thermal diffusivity	Heat transfer in combined turbulent and laminar flows	*25*
P8b	Turbulent Prandtl No.	Pr_T	$\varepsilon_M/\varepsilon_T = l/l_T$ $\varepsilon_M, \varepsilon_T$ = eddy viscosity, eddy thermal diffusivity $[L^2/\theta]$; l, l_T = mixing lengths for momentum, heat transfer	Eddy momentum diffusivity/eddy thermal diffusivity	Heat transfer in turbulent flow	*9, 25*

TABLE II. ALPHABETICAL LIST OF NEW GROUPS (Continued)

Serial No.	Name	Symbol	Definition	Significance	Field of Use	Reference
P12a	Predvoditelev No. (mass transfer)	Pd_m	$(\Gamma_m L^2/a_m)(N_{Fo_m})$ Γ_m = rate of change of mass transfer potential of medium, (mass/unit mass)/time $[1/\theta]$; a_m = mass conductivity of material $[L^2/\theta]$; N_{Fo_m} = group **F12**	Rate of change of concn. of medium/ rate of change of concn. of body	Mass transfer	*22*
P15a	Pulsation No.	N_{Pu}	$fd_e\rho/G$ d_e = equiv. diam. of channel		Transfer to pulsed fluid	*20*
R0a	Radial frequency parameter (1)	—	$\omega_r D/V^2$, $\omega_r\alpha/V^2$ D = diffusivity or dispersion coefficient of packed bed $[L^2/\theta]$, ω_r = radial frequency (radians/sec.) $[1/\theta]$		Packed and fluidized beds	*18, 33, 39*
R0b	Radial frequency parameter (2)	—	$\omega_r L^2/\alpha$ ω_r as in group **R0a**; L = tube radius $[L]$		Packed and fluidized beds	*29, 39*
R0c	Radial frequency parameter (3)	—	$\omega_r^2 DL/V^3$ (Quantities as in groups **R0a** and **R0b**)	(Group **R0a**)^2 (Group **P2**)	Packed and fluidized beds	*18, 33, 39*
R0d	Radial frequency parameter (4)	—	$L(\omega_r/2D)^{1/2} \equiv [\frac{1}{2}(\text{group } \mathbf{R0b})]^{1/2}$ (quantities as in groups **R0a** and **R0b**)	Analog of Wave No.	Packed and fluidized beds	*39*
R1a	Radiation parameter	Φ	$e^+\eta T_w^3 d_h/k$ e^+ = function of mean surface emissivity of walls, $[-]$; T_w = wall temp. (abs.) $[T]$		Effect of radiation on convective mass transfer	*14*
R6a	Reaction enthalpy No.	N_H	$(\Delta u)_A(\Delta n_A)/C_p(\Delta T)$ $(\Delta u)_A$ = enthalpy of reaction per unit mass of A produced $[L^2/\theta^2]$, n_A = mass fraction of species A $[-]$	Change in reaction energy/change in sensible energy	Interphase transfer with chemical reaction	*37*
R15a	Rossby No.	—	$V/\omega L$	More general form of group **R15**		*27*
S4a	Schmidt No. (2)	—	$\equiv$ Semenov No. (2)	No longer used		*27*
S4b	Schmidt No. (3)	$(Sc)_3$	$\mu\chi/\rho\sigma_e L^2$ χ as in group **E4b**	Diffusivity of vorticity/mass diffusivity of ions	Electrochemistry	*10*
S4c	Total Schmidt No.	**Sc**	$\dfrac{\varepsilon_M+(\mu/\rho)}{\varepsilon_D+D}$ ε_M = eddy viscosity $[L^2/\theta]$	Total momentum diffusivity/total mass diffusivity	Mass transfer in combined laminar and turbulent flows	*25*
S4d	Turbulent Schmidt No.	Sc_T	$\varepsilon_M/\varepsilon D = l/l_D$ ε_M = eddy viscosity, $[L^2/\theta]$, l, l_D = mixing lengths for momentum, mass transfer $[L]$	Eddy momentum diffusivity/eddy mass diffusivity	Mass transfer in turbulent flow	*25*
S5a	Semenov No. (2)	—	$\equiv 1/N_{Le}$ (see group **L7**)			*27*
S7a	Smoluchowski No.	—	$L/\lambda \equiv 1/N_{Kn}$ N_{Kn} = group **K6**	See group **K6**		*27*
S9a	Soret No.	S_o	$\Theta n'_{20}$ (definitions as group **D14a**)	Dimensionless thermodiffusion coefficient	Coupled heat and mass transfer	*22*
S10a	Spalding No.	B'	$c_p\Delta T/(r_v-q_r/V_m)$; q_r = radiant heat flux $[ML^2/\theta^3]$; V_m = rate of mass transfer $[M/\theta]$	Ratio (sensible heat/ latent heat) for evaporated material	Droplet evaporation	*33a*
S13a	Stark No.	Sk	$\eta T^3L/k$ L = thickness of layer $[L]$; ($\equiv$ Stefan No.)		Radiant heat transfer	*1, 30*
S14a	Stewart No.	—	$\mu_e^2H_e^2\sigma_e L/V\rho$	(Hartmann No.)^2/ Reynolds No.	Magnetofluid dynamics	*38*
S15a	Stokes No. (2)	St_2	$1.042\ m_f g\rho(1-\rho/\rho_f)R^{*3}/\mu^2$; ρ, μ = density, viscosity of fluid; m_f, ρ_f = mass, density of float; R^* = tube radius/float radius $[-]$		Calibration of rotameters	*10a*
S18a	Surface tension No.	T_S	$\mu^2/h^*\sigma\rho$; h^* = surface area of packing element/perimeter $[L]$		Mass transfer in packed columns	*5a*
T5a	Thompson No.	N_{Th_2}	$\dfrac{-\Delta t}{\Delta L}L^2\left(\dfrac{d\sigma}{dT}\right)/\mu\alpha$ (*cf.* Marangoni No.)		Cellular convection	*3*
T7a	Thrust coefficient	T_c	$F_T/\rho V^2d^2$ F_T = thrust force $[ML/\theta^2]$; V = forward speed $[L/\theta]$; d = tip diameter $[L]$		Propeller studies	*27*
T7b	Torque coefficient	Q_c	$F'/\rho V^2d^3$ F' = propeller torque $[ML^2/\theta^2]$		Propeller studies	*27*
T7c	Transiency groups	K_p K_Q	K_p: $\dfrac{1}{\left(\dfrac{\partial p}{\partial L}\right)}\cdot\dfrac{\partial\left(\dfrac{\partial p}{\partial L}\right)}{\partial(Fo_f)}$ K_Q: $\dfrac{1}{(N_{Re})}\cdot\dfrac{\partial(N_{Re})}{\partial(Fo_f)}$ $\partial p/\partial L$ = pressure gradient in flow direction $[M/L^2\theta^2]$; Fo_f = group **F12a**		Transient flow behavior	*17*
W0a	Wave No.	**k**	$L(\omega_r/2\alpha)^{1/2}$ w_r = radial frequency (radians/sec.) $[1/\theta]$	Heat transfer analog group **R0d**	Cyclic heat transfer	*4*
W3a	Weber No. (3)	We_3	$\sigma/\rho_L gL^2$ ρ_L = liquid density $[M/L^3]$; L = depth of liquid on tray $[L]$	Surface tension force/ gravity force	Interfacial area determination in distillation equipment	*32*
W4a	Generalized Weissenberg No.	N_1	$\sqrt{I_e}\theta_n$ (definitions as group **D8a**)	Generalization of group **W4**	Rheology	*2*

TABLE III. TABLES FOR IDENTIFYING DIMENSIONLESS GROUPS

These tables incorporate those from the earlier publication (5)

Basic dimensions are taken to be: Length [L]
Mass [M]
Electrical charge [Q]
Temperature [T]
Time [θ]

PHYSICAL PROPERTIES

General Physical Properties

Parameter	Symbol	Dimensions	Exponent	Groups
Coefficient of thermal bulk expansion	β	$1/T$	−1	E4, E12, G2
			+1	G5, K9, M3a, R5, R6
Coefficient of density change with concn. n_A	$\beta_{A'}$	—	+1	G5a
Density	ρ	M/L^3	−2	M1
			−1	A1, B1, B10, C1, C2, C6, D3, D13, D14a, E3, E9, E10, E10a, E13, E13a, F1, F11, F12a, H5, H11a, I1, J1, K4, K10, L5a, L5b, L7, M3, M3a, M5a, M6, N2, N3, N4, P7, P8a, P9, R0a, R9, R10, R13, S4, S4b, S4c, S13, S14a, S15, S15a, S18a, T7a, T7b, T7c, W3a
			$-\frac{2}{3}$	C9, J3
			$-\frac{1}{2}$	E2, L11, M4a, O2, P13
			$+\frac{1}{3}$	J3, K5, S3
			$+\frac{1}{2}$	D10, G3, P10, P11, R4, W0a, W2
			$+\frac{2}{3}$	F2, N8, S2
			+1	A1b, B1, B6, B9, C5, C7, D5, D7, D13, E4, E4b, E6, E7, E8, E10a, E13a, G2b, H6, H11, I1, J1, J3, J4, K1, L7a, L7b, L7c, L9, M3a, P1, P8a, P15a, R0b, R5, R11, R13, S4a, S5a, S15a, S17, T1, T5a, T6, T7c, V1, W1, W3
			+2	A5, C10, G1, G5, G5a, K9, R5, R6, S15a, T2
Density gradient	$d\rho/dL$	M/L^4	+1	R13
Diffusivity (molecular unless noted otherwise)	D, a_m	L^2/θ	−1	B5, B7[4], D2, D14a, E4c, E4e, H8a[1], K4[1], K7[2], L2, L7, L10, N7, P2, P9, P12a[1], S4, S4c, S9a
			$-\frac{2}{3}$	J3
			$-\frac{1}{2}$	A0, R0d
			+1	A4a, F12[1], K7[3], K7a, L2[4], L7b, L9[1], R0a, R0c, S4a, S5a
Diffusivity (eddy)	E_D, ε_D	L^2/θ	−1	S4c, S4d
			+1	L2, L7a
Diffusivity (surface)	D_S	ML/θ^2	−1	S18
Diffusion tortuosity	ζ	—	−1	K7, K7a
Dispersion (permeability) coefficient	D'	L^2/θ	−1	D6a
			+1	R0a, R0c
Molecular weight	M	M	−1	D14, K9, S6
			+1	D14
Permeability (packed bed)	ξ	L^2	$+\frac{1}{2}$	L6
Porosity (voidage)	e	—	−1	B6, M12
			$-\frac{1}{2}$	L6
			+1	K7, K7a, L5a
Specific gravity	γ	—	$\pm\frac{1}{3}$	F2
Surface tension	σ	M/θ^2	−3	C2
			−2	C1
			−1	B9, C3, C13, E8, L3a, L6, R1, S18a, W1, W3
			$-\frac{1}{2}$	D10, G3, O2, P13, R4, W2
			+1	M9, S17, S18, T5a, W3a

PHYSICAL PROPERTIES

General Physical Properties

Parameter	Symbol	Dimensions	Exponent	Groups
Thermal diffusion coefficient	D_T	L^2/θ	+1	D14a, S9a

[1] Coefficient of potential diffusion in mass transfer.
[2] Knudsen diffusion coefficient.
[3] Binary bulk diffusion coefficient.
[4] Effective diffusivity (molecular and eddy) $= D + \varepsilon_D$.

Electrical and Magnetic Properties

Parameter	Symbol	Dimensions	Exponent	Groups
Carrier suspectibility	b'	$Q\theta/M$	−1	E4d
Current density	I	$Q/L^2\theta$	+1	K10
Dielectric susceptibility	χ	$Q^2\theta^2/L^3M$	+1	E4b, S4b
Electrical conductivity	σ_e	$Q^2\theta/L^3M$	−1	E6, S4b
			$+\frac{1}{2}$	H2, M4a
			+1	B2a, L11, M3, M3a, M5a, M7, S14a
Electrical permittivity	ε_e	$Q^2\theta^2/L^3\theta$	−1	B2a
			+1	E4d
Field strength	H_e	$Q/L\theta$	−2	J4
			−1	E4a
			+1	H2, L11, M4a
			+2	K9, M3, M6, S6, S14a
Magnetic permeability	μ_e	ML/Q^2	−1	E4a, E6, J4
			+1	H2, M3a, M4a, M5a, M6, M7
			$+\frac{1}{2}$	L11
			+2	M3, S14a
Space charge density	Q	Q/L^3	−1	E4d
Voltage	E	ML/θ^2Q	+1	E4a, K10
			+2	E4b

Thermal Properties

Parameter	Symbol	Dimensions	Exponent	Groups
Eddy thermal conductivity	k_T	$ML/T\theta^3$	−1	L7a
Eddy thermal diffusivity	ε_T	L^2/θ	−1	L7a, P8a, P8b
Humid heat	s	L^2/θ^2T	−1	P15
Latent heats of phase change	λ, r	L^2/θ^2	−1	A3, E11, E12, E13, J1, M1, S10a
			+1	B13, C10, K8, K11, N5
Ratio of specific heats	γ	—	$\pm\frac{1}{2}$	L5
Specific heat	C, c_p	L^2/θ^2T	−1	B2, B13, D3, D14a, D15, E1, F11, J2, K8, K11, L7, M10, N5, R0a, R3, R6a, S13
			$-\frac{1}{3}$	J2
			$+\frac{1}{2}$	F6, F7, F8, W0a
			$+\frac{2}{3}$	J2
			+1	A3, E4, E12, G4, J1, J4, L7a, L7b, L9, P1, P8, P8a, R0b, R3, R5, R6, R7, S5a, S10a, T5a
Surface emissivity	e^*	—	−1	T6
			+1	R1a
Temperature conductivity (thermal diffusivity)	$\alpha = \frac{k}{\rho c_p}$	L^2/θ	−1	L7b, L9, P1, P8a, P12, R0b, R5, S5a, T5a
			$-\frac{1}{2}$	W0a
			+1	C13, L7, L10, R0a
Thermal conductivity	k, λ	$ML/T\theta^3$	−3	M1
			−2	R6
			−1	B4, B12, C7, C9, C10, D4, F12b, G4, K3, K7b, L7b, L9, N6, P1, P5,

Thermal Properties

Parameter	Symbol	Dimensions	Exponent	Groups
				P8, R0b, R1a, R5, S4a, S5a, S13a, S14, T5a
			$-\frac{2}{3}$	J2
			$-\frac{1}{2}$	T3, W0a
			+1	F11, L7, R0a, R1

Rheological and Elastic Behavior

Parameter	Symbol	Dimensions	Exponent	Groups
Eddy viscosity	e_M	L^2/θ	+1	P8a, P8b, S4c, S4d
Invariants of rate-of-strain and vorticity tensors	I_e, I_w	θ^{-2}	$+\frac{1}{2}$	D8a, W4a
Maximum dynamic slip resistance	τ_D	$M/L\theta^2$	+1	I1
Modulus of elasticity	E	$M/L\theta^2$	−1	A1b, C5, E4, H11
			+1	C1, E13, R1
			+3	A1
Rate of shear	$\dot{\gamma}$	$1/\theta$	+1	T8
Rigidity coefficient	μ_p	$M/L\theta$	−2	H6
			−1	B3, P3a
Shear stress	τ	$M/L\theta^2$	−1	E5
			$-\frac{1}{2}$	P10
			$+\frac{1}{2}$	P11
			+1	A4b
Viscosity (in all cases, kinematic viscosity has been written as μ/ρ)	μ	$M/L\theta$	−2	A1, A5, E4b, G1, G5, G5a, K1, K9, S15a, S17, T2
			−1	A4b, B6, C10, D5, D7, E6, E7, G2b, H8, I1, L1, L3, M3a, O1, P4, P11, R5, R6, R11, S9, S19, T1, T5a, T7c, V1
			$-\frac{2}{3}$	F2, K5, N8
			$-\frac{1}{2}$	H2
			$-\frac{1}{3}$	S2, S3
			$+\frac{1}{2}$	E2
			$+\frac{2}{3}$	C9, J2, J3
			+1	B12, C3, C13, E3, E5, F12a, M1, M3a, M5a, O2, P8, P8a, P9, S4, S4b, S4c, S8, S15, T7c, T8
			+2	C1, S18a
			+4	C2
Viscosity (surface)	μ_s	M/θ	+1	S19
Yield stress	τ_y	$M/L\theta^2$	+1	B3, H6, P3a

LENGTHS, AREAS, AND VOLUMES

Characteristic Lengths

Parameter	Symbol	Dimensions	Exponent	Groups
General length dimension	Various	L	−5	P7, P7a
			−3	F9, L1, S15a, T7b
			−2	D9, D12, E3, F11, F12, F12a, H4, H5, N2, N3, N4, O1, R9, S4b, S8, S9, S15, T7a, W3a
			−1	A1a, B1, B10, E2, E4d, E5, E10, F1, F13, G2a, G4, H10, H11a, K6, L4, P7a, R6, R10, R15, R15a, R16, S6, S7a, S18a, S19, T5, T5a, T7, T7c, W4
			$-\frac{1}{2}$	B11, D7, F14, G2b, O2
			$+\frac{1}{2}$	A0, G2b, M4a, R4, T1, W2
			+1	A4a, A4b, B2a, B3, B4, B5, B7, B10, B13a, C7, D1, D3, D5, D6a, D11, D13, E4c, E4e, E7, E10, E13a, F1, F2, F5, F12c, F12d, F15, G2a, G2b, G3, H2, I1, K3, K4, K6, K7b, K10, L3, L3a, L4, L7c, L9, L11, M1, M3, M7, N6, N7, N8, P1, P2, P3a, P7a, P11, P15a, R0c, R0d, R1a, R8, R10, R11, R16, S2, S7, S7a, S13a, S14, S14a, S16, S17, S18, T3, T7c, W0a, W1
			$+\frac{3}{2}$	D7, H7, T1
			+2	B9, D2, D4, E4b, E8, F12b, H6, H8a, K9, O1, P4, P5, P12, P12a, R0b, S8, S9, T5a, T7c, V1
			+3	A5, C10, G1, G5, G5a, K1, M3a, R5, S15a, W3
			+4	L7c, T2
			+5	R6
Dimension of agitator, propeller, etc.	Various	L	−5	P7, P7a
			−3	F9, T7b
			−2	H4, T7a
			−1	A1a, P7a
			+1	D11, F15, L7c, P7a
			+3	W3
			+4	L7c
Film thickness	L_f	L	−1	A4a
			+1	N8
Furnace half-width	L	L	−1	T7
Larmor radius	L_L	L	+1	L4
Mean free path	λ	L	−1	S7a
			+1	K6
Mixing length-mass	l_D	L	−1	S4d
			+1	L7a
Mixing length-heat	l_T	L	−1	L7a, P8b
Mixing length-momentum	l	L	+1	P8b, S4d
Particle dimension	d_e	L	−2	S15
			+1	F2, F12d
			+3	A5, G1
Pore or nozzle radius	R	L	+1	B7
Thickness of liquid layer	L	L	−1	C13, M9, T5a
			+1	H2, L11
			+2	M9, T5a
			+3	R5
			+4	T2

Areas

Parameter	Symbol	Dimensions	Exponent	Groups
Area	A	L^2	−1	D9, F6, F7, F8
			+1	B2
Area/unit volume	S, A^*	L^{-1}	−1	B6
			+1	M11, M12

Volumes

Parameter	Symbol	Dimensions	Exponent	Groups
Volume	W, W^*	L^3	+1	A5, H9, M11

TIMES AND FREQUENCIES

Parameter	Symbol	Dimensions	Exponent	Groups
Time	θ	θ	−1	D8, T7c
			+1	D8, D8a, D12, E3, F11, F12, F12a, H1, H9a, H10, P12a, S15, T5, W4a
Frequency	f', ω'	$1/\theta$	+1	F12c, H1, P15a, S16, V1
Radial frequency[1] ($= 2\pi f'$)	ω, (radians/time)	$1/\theta$	$+\frac{1}{2}$	R0d, W0a
			+1	F12d, R0a, R0b
			+2	R0c

[1] Also see angular velocity.

TEMPERATURES AND CONCENTRATIONS (DRIVING FORCES)

Concentrations and Related Quantities

Parameter	Symbol	Dimensions	Exponent	Groups
Dimensional concentration	C, c_A	M/L^3	$(n-1)$	H8a (nth order reaction)
			−1	D1, D2, R2
			+1	B10
Dimensionless concentration (e.g., wt./wt., mass fraction)	n	—	−1	D14a, P6, S9a
			+1	A4, D14a, G5a, K8, R6a, S9a

TABLE III. TABLES FOR IDENTIFYING DIMENSIONLESS GROUPS (Continued)

Concentrations and Related Quantities

Parameter	Symbol	Dimensions	Exponent	Groups
Mole fraction	Y, Z	—	±1	**D14**
Rate of change of mass transfer potential (mass per unit mass/time)	Γ_m	$1/\theta$	+1	**P12a**
Specific mass content (mass/unit mass)	n^*	—	−1	**K4**
Surface concentration	Γ'	M/L^2	±1	**S18**
{Vapor capacity} {Mass capacity} (porous body)	c_b	$L\theta^2/M$	−1	**L5a**
			+1	**B13, R2**
Vapor expansion intensity on heating	b_T	M/L^3T	+1	**L5a**

Temperatures, Temperature Differences

Parameter	Symbol	Dimensions	Exponent	Groups
Dimensional temperature, temperature difference	$T, \Delta T$	T	−3	**R1*, T6***
			−2	**F12b***
			−1	**A4, A6*, B12, B13, B13a*, C4*, C7, C10, D3*, D4*, D14a*, D15, E1, E4b*, F12b*, G2, G6*, K3, K7b, K8, K9, K11, N5, P5*, P12*, R6a, R14*, S6, T5a**
			−½	**L5**
			+½	**F6*, F7*, F8***
			+1	**B13a, C4, E4b, G5, G6, J1, J4, K7b, L5a, M1, M3a, P6, R5, R7, R14*, S10a, T5a**
			+3	**R1a*, S13a*, S14***
Dimensionless temperature difference	Ψ	—	+1	**K7b**
Rate of temperature change	Γ	T/θ	+1	**P12**

* Absolute temperature.

VELOCITIES, RATES, FLUXES, TRANSFER COEFFICIENTS

Velocities

Parameter	Symbol	Dimensions	Exponent	Groups
Angular velocity (rate of rotation) (see also frequency)	N, ω	$1/\theta$	−3	**P7, P7a**
			−2	**H4, L1**
			−1	**A1a, F9, R15, R15a**
			−½	**E2**
			+1	**H9a, S8, S11, S12, T1**
			+2	**F15, T2, W3**
Equilibrium molecular speed of A	u_A	L/θ	−1	**K7a**
General velocity (usually of fluid)	V	L/θ	−3	**H5, R0c**
			−2	**B13a, C6, C11, D13, E9, E10, E13a, F1, I1, L5b, M6, N2, N3, N4, R0a, R7, R8, R9, R10, T7a, T7b**
			−1	**A2, A4b, B3, C12, C14, D1, D3, E4a, F10, F12c, F12d, H7, H8, I1, K10, L3, M3, P3a, P4, S13, S14a, S16, T7c, V2**
			−½	**M4a**
			+1	**A1a, A2, B1a, B2a, B6, B7, B11, C3, C12, C14, D5, D6a, D7, E1a, E4c, E4d, E4e, E5, E7, F10, F14, G2b, H7, H10, I1, K2, L5, L8, M2, M4, M7, P1, P2, P3, P10, R4, R11, R15, R15a, S1, S2, S3, T5, T6, T7c, V2, W2, W4**
			+2	**A1b, B1, B12, C5, C11, D15, E1, E11, F13, H11, P3, W1**
Velocity of impeller or agitator periphery	U_s	L/θ	−2	**P14**
			−1	**D9**
Velocity of light	V_l	L/θ	−2	**B2a**
			−1	**E1a, L8**
Velocity of sound	V_s	L/θ	−1	**B1a, M2, N1, S1**
Velocity of waves	V_w	L/θ	−1	**V2**
			+1	**P3**
Velocity of Alfven waves	V_A	L/θ	−1	**A2, K2, M4**
			+1	**A2, N1**
			+2	**C11**
Velocity of bearing surface	v_s	L/θ	−1	**H8, O1, S9**
Velocity gradient	dV/dL	$1/\theta$	−2	**R13**

Flow Rates, Mass Fluxes, etc.

Parameter	Symbol	Dimensions	Exponent	Groups
Mass flow rate (mass flux)	V_m	M/θ	−1	**B2, M11, T7**
			+1	**F6, F7, F8, G4, S10a, T7**
Mass flow rate/unit area (mass flux density)	G	$M/L^2\theta$	−1	**J2, J3, P15, P15a, S13**
			+1	**K4, M11**
Mass flux/unit volume (reaction rate)	U	$M/L^2\theta$	+½	**T3**
			+1	**D1, D2, D3, D4**
Reaction rate constant (first order)	K	L/θ	−1	**S5**
Reaction rate constant (*n*th order)	K_n	$\frac{1}{\theta}\cdot\frac{L^{3n-3}}{M^{n-1}}$	+1	**H8a**
Volumetric flow rate	V_f	L^3/θ	−1	**H9**
			−½	**D11**
			+½	**S11, S12**
			+1	**D9, F9**
Volumetric flow rate/wetted perimeter	$V_{f'}$	L^2/θ	−1	**A4a**
			−½	**A0**

Heat Fluxes and Related Quantities

Parameter	Symbol	Dimensions	Exponent	Groups
Heat flux (heat flow/unit time)	q	ML^2/θ^3	−1	**S10a**
			+1	**H5**
Heat flux/unit area	q^*	M/θ^2	+1	**K3, R6**
Heat of reaction/unit mass of material reacting	Δu	L^2/θ^2	+1	**R6a**
Heat liberated/unit mass	Q	L^2/θ^2	+1	**D3, D4**
Heat liberated per unit mass of material reacting/unit volume	Q''	$1/L\theta^2$	+1	**F12b**
Heat liberated/unit volume (heat source power)	j	$M/L\theta^3$	+1	**P5**

Transfer Coefficients

Parameter	Symbol	Dimensions	Exponent	Groups
Heat transfer coefficient	h	$M/T\theta^3$	+1	**B2, B4, C9, J2, K7b, N6, P15, S13**
			+4	**M1**
Mass transfer coefficient	k_c	L/θ	+1	**A0, B5, J3, L9, N7, S5, S7**

FORCE, TORQUE, POWER, PRESSURE, HEAD

Force, Torque, Power

Parameter	Symbol	Dimensions	Exponent	Groups
Force	F, F_R	ML/θ^2	−⅓	**S2**
			+⅓	**K5**
			+1	**N2, N3, N4, R9, T7a**
Force/unit length	F_b	M/θ^2	+1	**H8, O1, S9**
Torque	F'	ML^2/θ^2	+1	**T7b**
Power	Π	ML^2/θ^3	+1	**L1, P7, P7a**

TABLE III. TABLES FOR IDENTIFYING DIMENSIONLESS GROUPS (Continued)

Head, Pressure

Parameter	Symbol	Dimensions	Exponent	Groups
Fluid head (ft. fluid)	H'	L	$-\frac{3}{4}$	S12
			+1	E10a, H4
Head (energy/unit mass of fluid = gH')	H	L^2/θ^2	−1	P3, T4
			$-\frac{3}{4}$	S11
			$+\frac{1}{4}$	D11
			+1	P14, T4
Pressure	P, p	$M/L\theta^2$	−1	F6, F7, F8, H9, L5a, S8, T8
			+1	B13, C6, D14a, L5b, L6, P13, R2
Pressure drop	ΔP, dP	$M/L\theta^2$	+1	E9, E10, F1, H9, H11a, K1, L3, L3a, R10
Pressure gradient	$\Delta P/L$	$M/L^2\theta^2$	−1	T7c
			+1	E10, F1, H11a, K1, P4, R10, T7c

CONSTANTS AND MISCELLANEOUS QUANTITIES

Gravity Acceleration

Parameter	Symbol	Dimensions	Exponent	Groups
Gravity acceleration	g	L/θ^2	−2	A1
			−1	B1, F13, F15, H11a, M1, S6, W3a
			$-\frac{3}{4}$	S11
			$-\frac{1}{2}$	B11, F14, P13
			$-\frac{1}{3}$	C9, S3
			$+\frac{1}{3}$	F2, N8
			+1	A5, B9, B13a, C2, C10, D13, E8, E10a, E13a, G1, G5, G5a, H4, M3a, R5, R6, R8, R13, S15a

Other Quantities

Parameter	Symbol	Dimensions	Exponent	Groups
Arrhenius pre-exponential factor	k_0	M/θ	+1	F12b
Avogadro's No.	N	$1/M$	+1	K9, S6
Boltzmann's constant	k	$ML^2/\theta^2 T$	−1	K9, S6
Dufour coefficient	β^*	T	+1	A3, A4, F3
Energy of activation/unit mass	E_a	L^2/θ^2	+1	A6, F12b
Gas constant/unit mass	$\mathscr{R}$	$L^2/\theta^2 T$	−1	A6, F12b
			$-\frac{1}{2}$	L5
			+1	D14a
Knudsen flow permeability	K_{oA}	L	−1	K7a
Mass of float	m_f	M	+1	S15a
Moment of inertia	I	L^4	−1	L7c
Rotor lift curve slope	a'	L^2/M	+1	L7c
Shape factor	ξ^*	—	+1	V2
Soret coefficient	δ	$1/T$	+1	F3, P6
Stefan-Boltzmann coefficient (radiation coefficient)	η	$M/T^4\theta^3$	+1	R1, T6
		–	+1	R1a, S13a, S14
Thermodiffusion constant	Θ	—	+1	D14a, S9a

REFERENCES

1. Adrianov, V. N., "Radiation-Conductive and Radiation-Convective Heat Transfer," in "Teplo- i Massoperenos," A. V. Lykov and B. M. Smol'skii, Eds., Vol. **II**, pp. 92–102, Nauka i Tekhnika, Minsk, 1965.

2. Astarita, G., *Ind. Eng. Chem. Fundamentals* **6**, 257 (1967).

3. Berg, J. C., Acrivos, A., Boudart, M., "Evaporative Convection," in "Advances in Chemical Engineering," T. B. Drew *et al.*, Eds., Vol. **6**, pp. 61–123, Academic Press, New York, 1966.

4. Carslaw, H. S., Jaeger, J. C., "Conduction of Heat in Solids," 2nd ed., Oxford University Press, 1959.

5. Catchpole, J. P., Fulford, G., *Ind. Eng. Chem.* **58** (3), 46 (1966).

5a. Costa Novella, E., Costa Lopez, J., *An. Real. Soc. Espan. Fis. Quim., Ser. B*, **63** (2), 145 (1967).

6. Coulson, J. M., Porter, J. E., *Trans. Inst. Chem. Engrs.* **44**, T388 (1966).

7. Ede, A. J., *Intern. J. Heat Mass Transfer* **9**, 837 (1966).

8. Gall, C. E., Hudgins, R. R., *J. Chem. Educ.* **42**, 611 (1965).

8a. Galloway, T. R., Sage, B. H., *Int. J. Heat Mass Transfer* **10**, 1195 (1967).

9. Gardner, G. O., Kestin, J., *Ibid.*, **6**, 289 (1963).

10. Gibbins, J. C., Hignett, E. T., *Electrochim. Acta* **11**, 815 (1966).

10a. Gilmont, R., Maurer, P. W., *Instr. Control Systems* **34**, 2070 (1961).

11. Ginzburg, I. P., "Heat and Mass Transfer from Bodies Interacting with Gas and Liquid Streams," in "Teplo- i Massoperenos," A. V. Lykov and B. M. Smol'skii, Eds., Vol. **II**, pp. 3–26, Nauka i Tekhnika, Minsk, 1965.

12. Glauert, M. B., Lighthill, M. J., *Proc. Roy. Soc.* **230A**, 188 (1955).

13. Hobler, T., "Mass Transfer and Absorbers," Pergamon Press, New York, 1966.

14. Irvine, T. F., Stein, R. P., Simon, G. A., "Effect of Radiation on Convection in a Flat Channel," in "Teplo- i Massoperenos," A. V. Lykov and B. M. Smol'skii, Eds., Vol. *II*, pp. 78–91, Nauka i Tekhnika, Minsk, 1965.

15. Kasatkin, A. G., Dytnerskii, Yu. N., Kochergin, N. V., "Mass and Heat Transfer in Tray Columns," in "Teplo- i Massoperenos," A. V. Lykov and B. M. Smol'skii, Eds., Vol. **IV**, pp. 12–17, Nauka i Tekhnika, Minsk, 1966.

16. Kerber, R., *Chem.-Ing.-Technik* **38**, 1133 (1966).

17. Kochenov, I. S., Kuznetsov, Yu. N., "Unsteady Flow in Pipes," in "Teplo-Massoperenos," A. V. Lykov and B. M. Smol'skii, Eds., Vol. **II**, pp. 306–314, Nauka i Tekhnika, Minsk, 1965.

18. Kramers, H., Alberda, G., *Chem. Eng. Sci.* **2**, 173 (1953).

19. Laslo, A., *Inzh.-Fiz. Zh.* **10** (1), 60 (1966).

20. Lemlich, R., Armour, J. C., *Chem. Engr. Progr. Symp. Ser.* **61** (57), 83 (1965).

20a. Leshchii, N. P., Mochernyuk, D. Yu, *Izv. Vysshikh Uchebn. Zavedenii, Nefti Gaz* **10** (2), 77 (1967).

21. Luikov (Lykov), A. V., "Heat and Mass Transfer in Capillary-Porous Bodies," Pergamon Press, New York, 1966.

22. Luikov (Lykov), A. V. Mikhailov, Yu. A., "Theory of Energy and Mass Transfer," revised Engl. ed., Pergamon Press, New York, 1965.

23. McHenry, K. W., Wilhelm, R. H., *A.I.Ch.E. J.* **3**, 83 (1957).

24. Merzhanov, A. G., "Problems of Heat Transfer in the Theory of Thermal Detonation," in "Teplo- i Massoperenos," A. V. Lykov and B. M. Smol'skii, Eds., Vol. **IV**, pp. 259–72, Nauka i Tekhnika, Minsk, 1966.

25. Opfell, J. B., Sage, B. H., "Turbulence in Thermal and Material Transport," in "Advances in Chemical Engineering," T. B. Drew *et al.*, Ed., Vol. **I**, pp. 241–88, Academic Press, New York, 1956.

26. Pai, Shih-I., "Magnetohydrodynamics of Channel Flow," in "Advances in Hydroscience," Ven Te Chow, Ed., Vol. **III**, pp. 63–110, Academic Press, New York, 1966.

27. Pankhurst, R. C., "Dimensional Analysis and Scale Factors," Chapman and Hall, London, 1964.

28. Pechoč, V., Research Institute of Chemical Fibers, Svit, Czechoslovakia, private communication, 1966.

29. Philip, J. R., *Austral. J. Phys.* **16** (4), 454 (1963).

30. Postolnik, Yu. S., *Inzh.-Fiz. Zh.* **8** (1), 64 (1965).

31. Prochazka, J., Landau, J., *Collection Czech. Chem. Commun.* **26**, 2961 (1961).

32. Rodionov, A. I., Kashnikov, A. M., Radikovskii, V. M., "Determination of Interfacial Contact Areas and Heat and Mass Transfer Coefficients on Sieve Trays," in "Teplo- i Massoperenos," A. V. Lykov, B. M. Smol'skii, Eds., Vol., **IV**, pp. 28–37, Nauka i Tekhnika, Minsk, 1966.

33. Rosen, J. B., Winsche, W. E., *J. Chem. Phys.* **18**, 1587 (1950).

33a. Ross, L. L., Hoffman, T. W., Proc. 3rd Internat. Heat Transfer Conference, Chicago, August 1966, **5**, 50 (A.I.Ch.E., New York, 1966).

34. Rothfield, L. B., *A.I.Ch.E. J.* **9**, 19 (1963).

35. Sakiadis, B. S., *Ibid.*, **7**, 467 (1961).

36. Skelland, A. H. P., "Non-Newtonian Flow and Heat Transfer," Wiley, New York, 1967.

37. Stewart, W. E., *Chem. Eng. Progr. Symp. Ser.* **61** (58), 16 (1965).

38. Tsinober, A. B., *Appl. Mech. Rev.* **18**, 3751 (1965).

39. Turner, G. A., Dept. of Chemical Engineering, University of Waterloo, Waterloo, Ontario, private communication, 1967.

40. Vyazovov, V. V., *Zh. Tekhn. Fiz.* **10**, 1519 (1940).

41. Wilkinson, W. L., "Non-Newtonian Fluids," Pergamon Press, London, 1960.

NUMERICAL DATA PROJECTS

SOURCES OF CRITICAL DATA

The following is abstracted from *Continuing Numerical Data Projects, A Survey and Analysis*. This publication was prepared under the auspices of the Office of Critical Tables, National Academy of Sciences and National Research Council. Persons requesting further information and details regarding the projects listed may obtain the publication from Printing and Publishing Office, National Academy of Sciences, 2101 Constitution Avenue, Washington, D. C., 20418. Price $5.00.

THERMODYNAMIC PROJECTS

Selected Values of Chemical Thermodynamic Properties, NBS Circular 500, F. D. Rossini, D. D. Wagman, W. H. Evans, S. Levine, and I. Jaffe, Government Printing Office, Washington, D.C. 20402, 1952, iv + 1268 pp, $7.25 (out of print). Paperback reprint issued in two parts: I. Tables; and II. References; Government Printing Office, Washington, D.C. 20402, 1961 (out of print). Revisions of C500 are being issued provisionally as NBS Technical Notes, listed below.

Selected Values of Chemical Thermodynamic Properties

Part 1: Tables for the First Twenty-three Elements in the Standard Order of Arrangement, NBS Technical Note 270-1, D. D. Wagman, W. H. Evans, I. Halow, V. B. Parker, S. M. Bailey, and R. H. Schumm, U.S. Government Printing Office, Washington, D.C. 20402, October 1965, iv + 124 pp, $0.65.

Part 2: Tables for the Elements Twenty-four through Thirty-two in the Standard Order of Arrangement, NBS Technical Note 270–2, D. D. Wagman, W. H. Evans, I. Halow, V. B. Parker, S. M. Bailey, and R. H. Schumm, U.S. Government Printing Office, Washington, D.C. 20402, May 1966, iv + 62 pp, $0.40.

Selected Values of Properties of Hydrocarbons and Related Compounds, American Petroleum Institute Research Project 44, J. B. Zwolinski, dir., Texas A&M Research Foundation, College Station, Texas. As of June 1966 there were 2391 valid data sheets, complete set 6 volumes at $0.30 per sheet.

Selected Values of Properties of Chemical Compounds, Thermodynamics Research Center, B. J. Zwolinski, dir., Texas A&M Research Foundation, College Station, Texas. As of June 1966 there were 754 valid data sheets, complete set 2 volumes at $0.30 per sheet.

JANAF (Joint Army-Navy-Air Force) Thermochemical Tables, PB 168 370, D. R. Stull, dir., Clearinghouse for Federal Scientific and Technical Information, Springfield, Virg., 1965, 945 pp, $10.00.

First Addendum, PB 168 370–1, D. R. Stull, dir., Clearinghouse for Federal Scientific and Technical Information, Springfield, Virg., 1966, vii + 197 pp, $4.00 (Supplements 18, 19, 20, and 21).

Information on further supplements is available from Dr. D. R. Stull, dir., Thermal Research Laboratory, Dow Chemical Company, Midland, Michigan.

Contributions to the Data on Theoretical Metallurgy, I–XV, published as Bureau of Mines bulletins between 1932 and 1962. Many bulletins were superceded; six are revisions and Bulletin 601 is a reprinting of four previous bulletins and out of print for some years, but still in demand as essential in present-day research. The ones currently available are listed below.

Contributions to the Data on Theoretical Metallurgy

XIII High-Temperature Heat-Content, Heat Capacity, and Entropy Data for the Elements and Inorganic Compounds, K. K. Kelley, Bulletin 584 (Bureau of Mines), U.S. Government Printing Office, Washington, D.C. 20402, 1960, 232 pp, $1.25.

XIV Entropies of Elements and Inorganic Compounds, K. K. Kelley and E. G. King, Bulletin 592 (Bureau of Mines), U.S. Government Printing Office, Washington, D.C. 20402, 1961, 149 pp, $0.75.

XV A Reprint of Bulletins 383, 384, 393, and 406 (III, IV, V, and VII), Bulletin 601, Bureau of Mines (Publications Distribution Section, 4800 Forbes Avenue), Pittsburgh, Penna., 1962, 525 pp, single copies free.

383: The Free Energies of Vaporization and Vapor Pressures of Inorganic Substances, K. K. Kelley, 1935.

384: Metal Carbonates, Correlation and Applications of Thermodynamic Data, K. K. Kelley and C. T. Anderson, 1935.

393: Heats of Fusion of Inorganic Substances, K. K. Kelley, 1936.

406: The Thermodynamic Properties of Sulphur and Its Inorganic Compounds, K. K. Kelley, 1937.

Thermodynamic Properties of Chemical Substances, V. P. Glushko et al, eds, Vol. I, 1164 pp, and Vol. II, 916 pp, USSR Academy of Sciences, Moscow, 1962, approx $16.00.

Thermodynamic Constants of Substances, Handbook in 10 parts, V. P. Glushko, ed, USSR Academy of Sciences, All-Union Institute of Scientific and Technological Information, Moscow: Part I, 1965, 146 pp, 72 kopecks ($1.75); Part II, 1966, 96 pp, 52 kopecks ($1.40). Available in the United States from Victor Kamkin, Inc., Bookstore, 1410 Columbia Rd. N.W., Washington, D.C. 20009.

Chemical Thermodynamics in Non-Ferrous Metallurgy, J. I. Gerassimov, A. N. Krestovnikov, and A. S. Shakhov, Metallurgical Publishing House, Moscow.

Vol. I: Theoretical Introduction, Thermodynamic Properties of Important Gases, Thermodynamics of Zinc and Its Important Compounds, 1960, 231 pp.

THERMODYNAMIC PROJECTS (Continued)

Vol. II: Thermodynamics of Copper, Lead, Tin, Silver, and Their Important Compounds, 1960, 231 pp.

Vol. III: Thermodynamics of Tungsten, Molybdenum, Titanium, Zirconium, Niobium, Tantalum, and Their Important Compounds, 1963, 283 pp. English translation available as NASA-TT-F-285 or CFSTI-TT-65-50111, $6.00, from the Clearinghouse for Federal Scientific and Technical Information, Springfield, Virg. 22151.

Vol. IV: Thermodynamics of Aluminum, Antimony, Magnesium, Nickel, Bismuth, Cadmium, and Their Important Compounds, 1966, 428 pp.

Selected Values for the Thermodynamic Properties of Metals and Alloys, Ralph Hultgren, Raymond L. Orr, Philip D. Anderson, and Kenneth K. Kelley, John Wiley & Sons, Inc., New York, N.Y. 10016, 1963, xi + 963 pp, $12.50.

Thermochemistry for Steelmaking, Vol. I, J. F. Elliott and M. Gleiser, Addison-Wesley Publishing Co., Reading, Mass. 01867, 1960, viii + 296 pp, $17.50.

Thermochemistry for Steelmaking, Vol. II: Thermodynamics and Transport Properties, J. F. Elliott, M. Gleiser, and V. Ramakrishna, Addison-Wesley Publishing Co., Reading, Mass. 01867, 1963, xvi + 550 pp (pp 297–846), $25.00.

Thermodynamic Functions of Gases, F. Din, ed, Butterworth & Co. (Publishers) Ltd., 88 Kingsway, London, W.C.2, England, $12.50 per volume.

Vol. 1: Ammonia, Carbon Dioxide, Carbon Monoxide, 1956, viii + 175 pp.

Vol. 2: Air, Acetylene, Ethylene, Propane and Argon, 1956, vi + 201 pp.

Vol. 3: Methane, Nitrogen, Ethane, 1961, vi + 218 pp.

THERMOPHYSICAL PROJECTS

A Compendium of the Properties of Materials at Low Temperatures, V. J. Johnson, ed. Available from Clearinghouse for Federal Scientific and Technical Information, Springfield, Virg. 22151.

Phase I, Part I, Properties of Fluids, July 1960, 489 pp, WADD Technical Report 60-56, Part I (PB-171-618), $6.00.

Phase I, Part II, Properties of Solids, Oct. 1960, 330 pp, WADD Technical Report 60-56, Part II (PB-171-619), $4.00.

Phase I, Part III, Bibliography of References (cross-indexed), Oct. 1960, 161 pp, WADD Technical Report 60-56, Part III (PB-171-620), $3.00.

Phase II, R. B. Stewart and V. J. Johnson, eds, Dec. 1961, 501 pp, WADD Technical Report 60-56, Part IV (AD-272-769), $8.10.

Thermophysical Properties Research Center Data Book. Issued as loose-leaf sheets; sold on subscription basis at $0.10 per sheet, Thermophysical Properties Research Center, Purdue University, Research Park, 2595 Yeager Rd., West Lafayette, Ind. 47906.

Vol. I: Metallic Elements and Their Alloys.

Vol. II: Nonmetallic Elements, Compounds and Mixtures (In Liquid and Gaseous States at Normal Temperature and Pressure).

Vol. III: Nonmetallic Elements, Compounds and Mixtures (In the Solid State at Normal Temperature and Pressure).

Handbook of Thermophysical Properties of Solid Materials, A. Goldsmith, T. E. Waterman, and H. J. Hirschorn, revised edition, 1961 (8½″ × 11¼″), The Macmillan Co., 60 Fifth Ave., New York, N.Y. 10011, $90.00 per set.

Vol. I: Elements, vi + 752 pp.

Vol. II: Alloys, vi + 1,270 pp.

Vol. III: Ceramics, vi + 1,162 pp.

Vol. IV: Cermets, Intermetallics, Polymerics and Composites, vi + 798 pp.

Vol. V: Appendix (includes materials, author indexes, and a list of references), 286 pp.

PHYSICOCHEMICAL PROJECTS

Tables of Chemical Kinetics, Homogeneous Reactions, NBS Circular 510, U.S. Government Printing Office, Washington, D.C. 20402, 1951, xxiv + 732 pp (out of print).

Tables of Chemical Kinetics, Homogeneous Reactions, NBS Circular 510, Supplement 1, U.S. Government Printing Office, Washington, D.C. 20402, 1956, xiv + 422 pp (out of print).

Alphabetical Index to Tables of Chemical Kinetics, Homogeneous Reactions, NBS Circular 510, Supplement 2, U.S. Government Printing Office, Washington, D.C. 20402, 1960, iv + 37 pp, $0.35.

Tables of Chemical Kinetics, Homogeneous Reactions, Supplementary Tables, To Accompany Circular 510 and Supplements 1 and 2, NBS Monograph 34, U.S. Government Printing Office, Washington, D.C. 20402, 1961, $2.75.

PHYSICOCHEMICAL PROJECTS (Continued)

Tables of Chemical Kinetics, Homogeneous Reactions, Supplementary Tables, NBS Monograph 34, Vol. 2, U.S. Government Printing Office, Washington, D.C. 20402, 1964, $2.00.

Phase Diagrams for Ceramists, E. M. Levin, H. F. McMurdie, and F. P. Hall, The American Ceramic Society, 4055 N. High St., Columbus, Ohio 43214, 1956, 286 pp (811 phase diagrams).

Phase Diagrams for Ceramists, Part II, E. M. Levin and H. F. McMurdie, The American Ceramic Society, 4055 N. High St., Columbus, Ohio 43214, 1959, 153 pp (462 phase diagrams).

Phase Diagrams for Ceramists, E. M. Levin, C. R. Robbins, and H. F. McMurdie (7th compilation), The American Ceramic Society, 4055 N. High St., Columbus, Ohio 43214, 1964, 601 pp (2064 phase diagrams), $18.00 (discount to members).

Phase Equilibrium Diagrams of Oxide Systems, revised and redrawn by E. F. Osborn and A. Muan, The American Ceramic Society, 4055 N. High St., Columbus, Ohio 43214, (Ten 19″ × 23″ plates for three-oxide systems containing SiO_2, four of them for oxide phases in equilibrium with metallic iron. Reproductions of these plates appear in the 1964 compilation).

Constitution of Binary Alloys, M. Hansen and K. Anderko, McGraw-Hill Book Company, New York, N.Y. 10036, 1958, xix + 1305 pp, $39.50.

Constitution of Binary Alloys, First Supplement, R. P. Elliott, McGraw-Hill Book Company, New York, N.Y. 10036, 1965, xxxii + 877 pp, $35.00.

Stability Constants of Metal-Ion Complexes, Section I: Inorganic Ligands, compiled by Lars Gunnar Sillen; Section II: Organic Ligands, compiled by Arthur E. Martell; Special Publication No. 17, The Chemical Society, London, 1964, xviii + 754 pp, $23.00.

Solubilities of Inorganic and Metal Organic Compounds, A compilation of Solubility Data from the Periodical Literature, A. Seidell, 4th ed., W. F. Linke, American Chemical Society, 1155 16th St. N.W., Washington, D.C. 20036.

Vol. I, 1958, iv + 1287 pp, $32.50.
Vol. II, 1965, iv + 1914 pp, $32.50.

INDEXES

Consolidated Index of Selected Property Values: Physical Chemistry and Thermodynamics, Prepared by the Office of Critical Tables. A key to the contents of six compilations that present critically evaluated numerical property values. Publication 976, National Academy of Sciences—National Research Council Printing and Publishing Office, 2101 Constitution Ave., N.W., Washington, D.C. 20418, 1962, xxiii + 274 pp, $6.00.

CRYSTALLOGRAPHIC PROJECTS

Crystal Data, Classification of Substances by Space Groups and their Identification from Cell Dimensions, J. D. H. Donnay and Werner Nowacki, Memoir 60 of the Geological Society of America, 1954, ix + 719 pp, Geological Society of America, 419 W. 117 St., New York, N.Y. 10027 (out of print).

Crystal Data, Determinative Tables, second edition, J. D. H. Donnay, General Editor; Gabrielle Donnay, Assistant Editor; E. G. Cox, Inorganic Compounds; Olga Kennard, Organic Compounds; Murray Vernon King, Proteins; Monograph 5 of the American Crystallographic Association, 1963, x + 1,302 pp, $20.00. Polycrystal Book Service, P.O. Box 11567, Pittsburgh, Pa. 15238.

Crystal Structures, first edition, Ralph W. C. Wyckoff, Interscience Publishers, John Wiley & Sons, New York.

Section I: Chapters I to VIII, 1948, 378 pp, $13.50.
Section II: Chapters VIII–X, XIII, 1951, 509 pp, $17.00.
Section III: Chapters XIV, XV, Organic Index, 1953, 465 pp, $25.00.
Section IV: Chapters XI and XII, 1957, 261 pp, $8.00.
Supplement I: Additions to Chapters II to VII, 1951, 143 pp, $5.00.
Supplement II: Additions to Chapter XIII, 1953, 85 pp, $8.50.
Supplement III: Additions to Chapters II to VIII, 1958, 311 pp, $20.00.
Supplement IV: Additions to Chapters IX, X, XIII–XV, 1959, 509 pp, $22.00.
Supplement V: Additions to Chapters II to XV, Indexes, 1960, 513 pp, $26.50.

Crystal Structures, 2nd ed., Ralph W. C. Wyckoff, Interscience Publishers, John Wiley & Sons, New York.

Vol. I: Chapters I–IV, 1963, 467 pp, $17.50.
Vol. II: Chapters V–VII, 1964, 588 pp, $24.00.
Vol. III: Chapters VIII–X, 1965, 981 pp, $27.50.

CRYSTALLOGRAPHIC PROJECTS (Continued)

Powder Diffraction Data File (PD-1), 16 Sets, American Society for Testing and Materials, 1916 Race St., Philadelphia, Pa. 19103.

Sets	Indexes	Cards: Plain ($)	Cards: Keysort ($)	Cards: IBM ($)
1–16	Organic	715	1,125	300
	Inorganic	1,280	1,855	
16 only	Organic	60	95	25
	Inorganic	140	220	
Book Form				
1–5	Organic	25		
	Inorganic	45		
	Both	60		

Structure Reports, Vols. 8–16 and Vol. 18, A. J. C. Wilson, General Editor; Vol. 17 and Vols. 19–23, W. B. Pearson, General Editor; published for the International Union of Crystallography by N. V. A. Oosthoek's Uitgevers, Mij, Utrecht, the Netherlands. Agent in the U.S.—Polycrystal Book Service, Box 11567, Pittsburgh, Pa. 15238.

The years covered, year published, and price of each volume are as follows:

Volume	Years Covered	Year Published	Price: Dutch florin	Price: $
8	1940–1941	1956	80	22.50
9	1942–1944	1955	70	19.50
10	1945–1946	1953	55	15.50
11	1947–1948	1951	100	28.00
12	1949	1952	70	19.50
13	1950	1954	100	28.00
14[a]	1940–1950	1959	35	10.00
15	1951	1957	110	31.00
16	1952	1959	120	33.50
17	1953	1963	125	35.00
18	1954	1961	120	33.50
19	1955	1963	100	28.00
20	1956	1963	100	28.00
21	1957	1964	100	28.00
22	1958	In preparation	....	.....
23	1959	1965	120	33.50

[a]Index.

A Handbook of Lattice Spacings and Structures of Metals and Alloys, No. 4 of Pergamon Press series of monographs on Metal Physics and Physical Metallurgy. W. B. Pearson, Pergamon Press, Ltd., Headington Hill Hall, Oxford, England, also London, Edinburgh, New York, Paris, and Frankfurt, 1958, x + 1044 pp, $38.00.

International Tables for X-Ray Crystallography, Kathleen Lonsdale, general editor, Published for the International Union of Crystallography by the Kynoch Press, Witton, Birmingham 6, England. Also available from Polycrystal Book Service, P.O. Box 11567, Pittsburgh, Pa. 15238.

Vol. I: Symmetry Groups, Norman F. M. Henry and Kathleen Lonsdale, eds, 1952, xii + 558 pp, £5.5.0. ($14.70).

Vol. II: Mathematical Tables, John S. Kasper and Kathleen Lonsdale, eds, 1959, xviii + 444 pp, £5.15.0. ($16.50).

Vol. III: Physical and Chemical Tables, Caroline H. MacGillavry and Gerard D. Rieck, eds, 1962, xvi + 362 pp, £5.15.0. ($16.10).

The Barker Index of Crystals, A Method for the Identification of Crystalline Substances, Published for the Barker Index Committee by W. Heffer & Sons, Ltd., Cambridge, England.

Vol. I, Crystals of the Tetragonal, Hexagonal, Trigonal and Orthorhombic Systems, M. W. Porter and R. C. Spiller, 1951.

Part 1: Introduction and Tables, ix + 350 pp, £1.10.0. net ($4.20).

Part 2: Crystal Descriptions, x + 1,086 pp, £4.10.0. net ($12.60).
The two parts—£6.0.0. net ($16.80).

Vol. II, Crystals of the Monoclinic System, M. W. Porter and R. C. Spiller, 1956.

Part 1: Introduction and Tables, v + 383 pp.

Part 2: Crystal Descriptions, M.1 to M.1800, viii + 760 pp.

Part 3: Crystal Descriptions, M.1801 to M.3572, viii + 688 pp.
The three parts—£10.0.0. net ($28.00).

CRYSTALLOGRAPHIC PROJECTS (Continued)

Vol. III, Crystals of the Anorthic System, M. W. Porter and L. W. Codd, 1964.

Part 1: Introduction and Tables, vi + 50 pp., and unpaginated tables.

Part 2: Crystal Descriptions, A.1 to A.831 and Atlas of Configurations, vii pp + unpaginated descriptions.

The two parts—£12.0.0. net ($33.60).

MINERALOGICAL PROJECTS

The System of Mineralogy of James Dwight Dana and Edward Salibury Dana, Yale University, 1837–1892, 7th edition, John Wiley & Sons, Inc., 605 Third Ave., New York, N.Y. 10016; and Glen House, Stag Place, London, S.W.1., England.

Vol. I: Elements, Sulfides, Sulfosalts, Oxides; Charles Palache, Harry Berman, and Clifford Frondel, 1944, xiii + 834 pp, $14.00.

Vol. II: Halides, Nitrates, Borates, Carbonates, Sulfates, Phosphates, Arsenates, Tungstates, Molybdates, etc.; C. Palache, H. Berman, and C. Frondel, 1951 xi + 1124 pp, $16.00

Vol. III: Silica Minerals, C. Frondel, 1962, xii + 334 pp, $7.95.

Rock-Forming Minerals, W. A. Deer, J. Zussman, and R. A. Howie, Longmans, Green and Co., Ltd., 48 Grosvenor St., London, W.1, England. Rights in USA, Philippines, and Central America—John Wiley & Sons, Inc., 605 Third Ave., New York, N.Y. 10016.

Vol. I: 1962, 333 pp, $16.50.

Vol. II: 1963, 379 pp, $19.50.

Vol. III: 1962, 270 pp, $16.50.

Vol. IV: 1963, 435 pp, $16.50.

Vol. V: 1962, 371 pp, $16.50.

INDEXES

Indexes to Crystallographic Data Compilations. The American Society for Testing and Materials publishes the following indexes to the Powder Diffraction File (see p. 278). All ASTM publications are available from the American Society for Testing and Materials, 1916 Race St., Philadelphia, Pa. 19103.

a. Index to the Powder Diffraction File; Organic (Hanawalt), includes Section 1–16, PDIS–15o, bound, $15.00.

b. Index to the Powder Diffraction File; Inorganic (Hanawalt), includes Section 1–15, PDIS–16i, bound, $20.00.

c. Fink Inorganic Index to the Powder Diffraction File, PDIS–16f, bound, $15.00.

d. Matthews Coordinate Index to the Powder Diffraction File, Inorganic: Termatrex Cards, $675.00; Negative-Line Cards, $185.00.

GENERAL NUCLEAR PROPERTIES PROJECTS

Nuclear Data Sheets, Jan. 1958—. Vols. 5, and 6, 1958, 1959, 1960, 1961, published by the National Academy of Sciences—National Research Council, 2101 Constitution Avenue, Washington, D.C. 20418. Section B of the new journal "Nuclear Data" will be devoted to Nuclear Data Sheets. Vol. 1, No. 1, containing sheets for A-chains from $A=182$ through $A=185$, was issued in Feb. 1966.

1959 Nuclear Data Tables, U.S. Government Printing Office, Washington, D.C. 20402, April 1959, viii + 151 pp, $1.00.

1960 Nuclear Data Tables, U.S. Government Printing Office, Washington, D.C. 20402.

Part 1: Consistent Set of Q Values, $A \leq 66$, 1961, 214 pp, $1.50.

Part 2: Consistent Set of Q Values, $67 \leq A \leq 199$, 1961, 456 pp, $2.75.

Part 3: Nuclear Reaction Graphs, 1960 (out of print).

Part 4: Short Tables, 249 pp, 1961, $1.50.

Energy Levels of Light Nuclei, May 1962, T. Lauritsen and F. Ajzenberg, issued as Sets 5 and 6 of the 1961 Nuclear Data Sheets, essentially an addendum to a review article in **Nucl. Phys., 11,** 1–340 (1959).

Members of the Nuclear Data Project contributed the section **Energy Levels of Nuclei,** $A=21$ to $A=212$ of Group I, Vol. 1 of the New Series of Landolt-Börnstein, **Numerical Data and Functional Relationships in Science and Technology.**

Table of Isotopes, J. J. Livingood and G. T. Seaborg, **Rev. Mod. Phys., 12,** 30–47 (1940).

Table of Isotopes, G. T. Seaborg and I. Perlman, **Rev. Mod. Phys., 16,** 1–32 (1944).

GENERAL NUCLEAR PROPERTIES PROJECTS (Continued)

Table of Isotopes, G. T. Seaborg and I. Perlman, **Rev. Mod. Phys., 20,** 585–666 (1948).

Table of Isotopes, J. M. Hollander, I. Perlman, and G. T. Seaborg, **Rev. Mod. Phys., 25,** 469–650 (1953).

Table of Isotopes, D. Strominger, J. M. Hollander, and G. T. Seaborg, **Rev. Mod. Phys., 30,** No. 2, Part 2, 585–904 (1958). Also available from the American Institute of Physics, 335 E. 45th St., New York, N.Y. 10017.

Tabellen der Atomkerne (Nuclear Tables), W. Kunz and J. Schintlmeister. Teil (Part) I, Eigenschaften der Atomkerne (Nuclear Properties); Band (Volume) 1, Die Elemente Neutron bis Zinn (The Elements Neutron to Tin), 1958, xliv + 465 pp, Akademie-Verlag, G.m.b.H., Berlin W.8, Mohrenstrasse 39, Germany, price, 105 DM ($26.25). Teil I, Band 2, Eigenschaften der Atomkerne, Die Elemente Antimon bis Nobelium (Antimony to Nobelium), 1959, xliv + 641 pp (466–1107), price, 130 DM ($32.50). Akademie-Verlag, Berlin. (Also available from Pergamon Press, Headington Hill Hall, Oxford, England; 4/5 Fitzroy Square, London, W.1, England; 122 E. 55th St., New York, N.Y. 10022, £20.0.0 per set.)

Part II, Nuclear Reactions, Vol. 1, The Elements from Neutron to Magnesium, Text Volume, xliv + 700 pp. Table Volume, 36 tables (five of which require 2 or 3 sheets), 1965. Price £25.0.0. ($75.00) for the two volumes.

CROSS-SECTION PROJECTS

Neutron Cross Sections, BNL (Brookhaven National Laboratory) 325, 2nd ed, U.S. Government Printing Office, Washington, D.C. 20402, July 1958, v + 373 pp, $4.50.

Supplement 1, U.S. Government Printing Office, Washington, D.C. 20402, January 1960, iv + 129 pp, $2.00.

Supplement 2, Vol. I, Z = 1 to 20, U.S. Government Printing Office, Washington, D.C. 20402, May 1964, Sectional pagination, $2.50.

Supplement 2, Vol. IIA, Z = 21 to 40, Clearinghouse for Federal Scientific and Technical Information, Springfield, Virg. 22151, February 1966, sectional pagination, $4.00.

Supplement 2, Vol. III, Z = 88 to 98, Clearinghouse for Federal Scientific and Technical Information, Springfield, Virg. 22151, February 1965, sectional pagination, $3.00.

Neutron Cross Sections—Angular Distributions, BNL 400, Clearinghouse for Federal Scientific and Technical Information, Virg. 22151, June 1956, 102 pp.
2nd ed, Angular Distributions in Neutron-Induced Reactions, Vol. I, Z = 1 to 22; Vol. II, Z = 23 to 94, Clearinghouse for Federal Scientific and Technical Information, Springfield, Virg. 22151, October 1962, 804 pp, sectional pagination, $8.50.

Part I, **Tabulated Neutron Cross Sections,** 0.001–14.5 MeV., R. J. Howerton, Vol. I, $_{1}$H–$_{22}$Ti; Vol. II, $_{23}$V–$_{50}$Sn; Vol. III, $_{51}$Sb–$_{95}$Am; UCRL-5226, first edition, May 1958. UCRL-5226, revised, 1959; Vol. I, $5.00, Vol. II, $5.00, Vol. III, $4.00, Clearinghouse for Federal Scientific and Technical Information, Springfield, Virg. 22151.

Part II, **Semi-Empirical Neutron Cross Sections,** 0.5–15 MeV., $_{1}$H–$_{94}$Pu, R. J. Howerton, Vol. I, Nov. 1958, UCRL-5351. (Photostat copy, $37.80, Microfilm copy, $11.10, available from the Library of Congress.)

Part III, **Tabulated Differential Neutron Cross Sections,** $_{1}$H–$_{94}$Pu, R. J. Howerton, Vol. I, Jan. 1961, UCRL-5573, Clearinghouse for Federal Scientific and Technical Information, Springfield, Virg. 22151, $5.00.

Charged Particle Cross Sections, LA-2014, Nelson Jarmie and John D. Seagrave, eds, Clearinghouse for Federal Scientific and Technical Information, Springfield, Virg. 22151, Feb. 1957, 234 pp, $1.25.

Charged Particle Cross Sections, LA-2424, Neon to Chromium, Darryl B. Smith, Compiler and ed, Nelson Jarmie and John D. Seagrave, associate eds, Clearinghouse for Scientific and Technical Information, Springfield, Virg. 22151, Jan. 1961, iii + 137 pp, $2.50.

Nuclear Cross Sections for Charged-Particle Induced Reactions, ORNL-CPX-1, Mn, Fe, Co, F. K. McGowan, W. T. Milner, and H. J. Kim, compilers, Oak Ridge National Laboratory, P.O. Box X, Oak Ridge, Tenn. 37831, July 1964, 443 pp.

Nuclear Cross Sections for Charged-Particle Induced Reactions, ORNL-CPX-2, Ni, Cu, F. K. McGowan, W. T. Milner, and H. J. Kim, compilers, Oak Ridge National Laboratory, P.O. Box X, Oak Ridge, Tenn. 37831, Sept. 1964, iii + 511 pp.

ENERGY LEVEL PROJECTS

Energy Levels of Light Nuclei

III: W. F. Hornyak, Thomas Lauritsen, P. Morrison, and H. A. Fowler, **Rev. Mod. Phys., 22,** 291–372, (1950).

IV: F. Ajzenberg and T. Lauritsen, **Rev. Mod. Phys., 24,** 321–402 (1952).

ENERGY LEVEL PROJECTS (Continued)

V: F. Ajzenberg and T. Lauritsen, **Rev. Mod. Phys. 27,** 77–166 (1955).

VI: F. Ajzenberg-Selove and T. Lauritsen, **Nucl. Phys., 11,** 1–340 (1959). This report may be obtained for $10.00 from North Holland Publishing Company, P.O. Box 103, Amsterdam, the Netherlands.

Energy Levels of Light Nuclei, Z = 11 to Z = 20

I: P. M. Endt and J. C. Kluyver, **Rev. Mod. Phys., 26,** 95–166, (1954).

II: P. M. Endt and C. M. Braams, **Rev. Mod. Phys., 29,** 683–756 (1957).

III: P. M. Endt and C. van der Leun, **Nucl. Phys., 34,** 1–324 (1962). Reprints of this paper are obtainable for $10.00 from North Holland Publishing Company, P.O. Box 103, Amsterdam, the Netherlands.

Decay Schemes of Radioactive Nuclei, B. S. Dzhelepov and L. K. Peker, 1958, viii + 787 pp, USSR Academy of Sciences Press, Moscow and Leningrad. English translation of above, 1961, vi + 786 pp, Pergamon Press, New York, Oxford, London, Paris.

Decay Schemes of Radioactive Nuclei, A ≥ 100, B. S. Dzhelepov, L. K. Peker, and V. O. Sergejev, 1963, 1060 pp, USSR Academy of Sciences Press, Moscow and Leningrad.

ATOMIC SPECTRA

Atomic Energy Levels As Derived from the Analyses of Optical Spectra, Charlotte E. Moore, NBS Circular 467, U.S. Government Printing Office, Washington, D.C. 20402.

Vol. I: Hydrogen to Vanadium (Z = 1–23), 1949, 309 pp, $5.50.

Vol. II: Chromium to Niobium (Z = 24–41), 1952, 227 pp, $4.00.

Vol. III: Molybdenum to Lanthanum (Z = 42–57) and Hafnium to Actinium (Z = 72–89), 1958, 245 pp, $3.00.

Vol. IV: Lanthanide and Actinide groups; in preparation.

An Ultraviolet Multiplet Table, Charlotte E. Moore, NBS Circular 488, U.S. Government Printing Office, Washington, D.C. 20402.

Section 1, 1950, 85 pp, reprinted 1956.

Section 2, 1952, 120 pp, reprinted 1956.
Combined Sections 1 and 2, reprinted 1963, $1.25;
Section 3, 1962, 98 pp, 60 cents.
The first three sections include spectra of the elements in the corresponding volumes of Circular 467.
Section 4, 1962, 70 pp, 45 cents.
A Finding List for spectra of elements in Sections 1 and 2.

Section 5, 1962, 34 pp, 30 cents.
A Finding List for spectra of elements in Section 3.

Selected Tables of Atomic Spectra, Atomic Energy Levels and Multiplet Tables Si ii, Si iii, Si iv. NSRDS-NBS 3, Section 1, Charlotte E. Moore, U.S. Government Printing Office, Washington, D.C. 20402, June 1965, 40 pp, $0.35.

Atomic Transition Probabilities, A Critical Data Compilation; Vol. I, Hydrogen Through Neon, W. L. Wiese, M. W. Smith, and B. M. Glennon (NSRDS-NBS 4, Vol. I), U.S. Government Printing Office, Washington, D.C. 20402, May 1966, xi + 154 pp, $2.50.

INFRARED AND MICROWAVE SPECTRA

Selected Infrared Spectral Data, American Petroleum Institute Research Project 44, current loose-leaf sheets. As of June 30, 1966, there were 2813 valid sheets; sold as complete set 7 volumes at $0.30 per sheet, Texas A&M Research Foundation, College Station, Texas.

Selected Infrared Spectral Data, Thermodynamics Research Center, current loose-leaf sheets. As of June 30, 1966, there were 342 valid sheets. Sold as complete sets at $0.30 per sheet, Texas A&M Research Foundation, College Station, Texas.

Coblentz Society Spectra, set of 4,000 spectra, 1964, $115 per 1000. Indexes: Alphabetical Index, Molecular Formula Index, Chemical Classes Index, Numerical Index, $5.00 each, set of four, $10.00. Available from Sadtler Research Laboratories, Inc., 1517 Vine St., Philadelphia, Pa. 19102, to whom orders may be sent.

DMS (Documentation of Molecular Spectroscopy) Spectral Cards. The DMS card service is obtained by subscription from Butterworth & Co. (Publishers) Ltd., London, W.C.2, England, from Verlag Chemie, G.m.b.H., Weinheim an der Bergstrasse, West Germany. Volumes 1 through 6 contain 2000 cards, about 80 percent spectral and 20 percent literature. The spectral cards are supplied slotted or unslotted, the unslotted being marked according to the DMS code. The subscription includes a current

INFRARED AND MICROWAVE SPECTRA (Continued)

literature list in a ring binder, the Junior Index, codes for the literature list, instruction and coding manual for the spectral cards, a formula list for each issue and the newsletters. Literature lists for nuclear magnetic resonance were begun in 1965. The prices for the DMS Spectral and Index cards are as follows:

Cards	Price U.K. (£)	Price U.S. ($)	Price W. Germany (DM)
Spectral[a] (volume, each)			
1–4 (unslotted)	67. 0.0.	187.50	750.00
5–8 (unslotted)	75. 0.0.	210.00	840.00
1–4 (slotted)	80. 5.0.	225.00	900.00
5–8 (slotted)	88. 0.0.	247.00	988.00
9– (unslotted)	112.10.0.	315.00	1,260.00
9– (slotted)	130.10.0.	365.00	1,460.00
Index (sets)			
S-1 (1–4999) single	22. 5.0.	62.50	250.00
S-1 (1–4999) multiple	18. 0.0.	50.00	200.00
S-2 (5000–9999) single	26. 5.0.	73.50	295.00
S-2 (5000–9999) multiple	21. 0.0.	59.00	236.00

[a]Each volume consists of four issues, each issue containing about 400 spectral cards (pink) and 100 literature cards (yellow). From Issue 17 (beginning of Vol. 5), spectral cards for inorganic compounds (blue) are included and from Issue 25, the literature cards are replaced by literature lists.

The Infrared Data Committee of Japan Infrared Data Cards, published by Nankodo Co., Ltd., Harukicho, Bunkyo-ku, Tokyo, Japan. They are sold on a subscription basis at $238.00 for 1200 cards (postage included). Asian subscribers receive a special price of $150.00. Export agency is Sanyo Shuppan Bocki Co., Inc., P.O. Box 1705, Tokyo, Central Japan. The agent in the United States is the Preston Technical Abstracts Co., 909 Pitner Ave., Evanston, Ill. 60202. The agent in Europe is Heyden & Son Ltd., Spectrum House, Alderton Crescent, Hendon, London, N.W.4, England.

Standard Infrared Spectra (pure substances), Sadtler Research Laboratories, Inc., 3316 Spring Garden St., Philadelphia, Pa. 19104.

27 volumes of 1000 spectra each (complete through 1965), include alphabetical name, molecular formula, chemical classes, numerical, and commercial alphabetical indexes, $3,500.00 (or each volume $125.00, or indexes, set of five, $50.00).

1966 spectra by subscription, includes 2 volumes of 1000 spectra each, Nos. 27,001—29,000, five indexes and Spec-Finder, $350.00.

Standard spectra on 16-mm microfilm, prices same as above.

Sadtler Standard Grating Spectra, Sadtler Research Laboratories, Inc., 3316 Spring Garden St., Philadelphia, Pa. 19104.

Near Infrared Spectra
Completed set of 2000 spectra in 9 volumes, prices available upon request.

Far Infrared Spectra
Completed set of 500 spectra in 2 volumes, prices available upon request.

Commercial Spectra
Groups priced separately, prices available upon request.
Commercial Spec-Finder, $25.00.

The Red System ($A^2\pi - X^2\Sigma$) of the CN Molecule, Sumner P. Davis and John G. Phillips, University of California Press, Berkeley and Los Angeles, 1963, x + 214 pp, $9.50.

Spectral Data and Physical Constants of Alkaloids, J. Holubek and O. Štrouf, eds. Published by Heyden & Sons, Ltd., London, in cooperation with The Publishing House of the Czechoslovakia Academy of Sciences. Available in Western Europe and the Western Hemisphere from Heyden & Sons, Ltd., Spectrum House, Alderton Crescent, Hendon, London, N.W.4, England.

Vol. I, 1965, spectral cards 1–30 issued in two binders to take Vols. I and II, £23.0.0. ($69.00).

Vol. II, 1966, spectral cards 301–400, punched cards to be inserted in second binder, £7.0.0. ($21.00).

Microwave Spectral Tables, NBS Monograph 70, U.S. Government Printing Office, Washington, D.C. 20402.

Vol. I: Diatomic Molecules, Paul F. Wacker, Masataka Mizushima, Jean D. Peterson, and Joe R. Ballard, Dec. 1, 1964, xviii + 146 pp, 8″ × 10½″, $2.00.

Vol. II: Line Strengths of Asymmetric Rotors, Paul F. Wacker and Marlene R. Pratto, Dec. 15, 1964, xii + 340 pp, 8″ × 10½″, $3.00.

ELECTRONIC SPECTRA (ULTRAVIOLET AND VISIBLE)

Selected Ultraviolet Spectral Data, American Petroleum Institute Research Project 44, current loose-leaf sheets. As of June 30, 1966, there were 1076 valid sheets. Sold in complete sets, 3 volumes, at $0.30 per sheet. Texas A&M Research Foundation, College Station, Texas.

Selected Ultraviolet Spectral Data, Thermodynamics Research Center, current loose-leaf sheets. As of June 30, 1966, there were 126 valid sheets. Sold in complete sets at $0.30 per sheet. Texas A&M Research Foundation, College Station, Texas.

Organic Electronic Spectral Data, Interscience Publishers, a division of John Wiley & Sons, Inc., New York, London, and Sydney.

Vol. I, 1946–1952, M. J. Kamlet, ed, 1961, pp xiii + 1244, $25.00.

Vol. II, 1953–1955, H. E. Ungnade, ed, 1960, pp x + 919, $15.00.

Vol. III, 1956–1957, O. H. Wheeler and L. A. Kaplan, eds, 1966, xii + 1210 pp, $25.00.

Vol. IV, 1958–1959, J. P. Phillips and F. C. Nachod, eds, 1963, viii + 1179 pp, $20.00.

The "Sadtler Standard Ultraviolet Spectra" including alphabetical and numerical indexes are available on a subscription basis—$560.00 for the 1965 subscription (5000 spectra). The 11000 bound spectra previously issued sell for $1,265.00 including indexes. Individual spectra are priced at $0.15 each. The spectra are also available on 16-mm microfilm at the same price.

Orders and inquiries should be sent to Sadtler Research Laboratories, Inc., 3314–3320 Spring Garden St., Philadelphia, Pa. 19104.

Absorption Spectra in the Ultraviolet and Visible Region (a Theoretical and Technical Introduction), elaborated and edited by Dr. L. Láng; published in German and English in 1959, 2nd edition, 1961, 3rd edition, 1963, 80 pp, Co-production of the Academic Press, New York and London, and the Publishing House of the Hungarian Academy of Sciences, Budapest. Printed in English in Hungary. Available in both continents of America from Academic Press, Publishers, 111 Fifth Ave., New York, N.Y. 10003.

Absorption Spectra in the Ultraviolet and Visible Region, L. Láng, ed, Co-production of the Academic Press, New York and London, and the Publishing House of the Hungarian Academy of Sciences, Budapest Printed in English in Hungary. Available in both continents of America from Academic Press, Publishers, 111 Fifth Ave., New York, N.Y. 10003.

Vol. 1: Substances, 170, 1959, 414 pp + Index pamphlet, 24, 2nd edition, 1961, 3rd edition, 1963. Combined introduction and Vol. 1, $18.00.

Vol. 2: Substances, 179, 1961, 408 pp + Index pamphlet, 31, 2nd edition, 1964, $18.00.

Vol. 3: Substances, 172, 1962, 424 pp + Index pamphlet, 24, $20.00.

Vol. 4: Substances, 185, 1963, 414 pp + Index pamphlet, 24, $20.00.

Vol. 5: Substances, 192, 1965, 416 pp + Index pamphlet, 28. Cumulative Index, Vols. 1–5, 1965, 112 pp, $23.00.

Vol. 6: Substances, 197, 1966, 412 pp + Index pamphlet, 30, $23.00.

RAMAN SPECTRA

Selected Raman Spectral Data, American Petroleum Institute Research Project 44, current loose-leaf sheets. As of June 30, 1966, there were 501 valid data sheets. Sold in complete sets, 2 volumes, at $0.30 per sheet, Texas A&M Research Foundation, College Station, Texas.

Selected Raman Spectral Data, Thermodynamics Research Center, current loose-leaf sheets. As of June 30, 1966, there were 40 valid sheets. Sold as complete sets at $0.30 per sheet. Texas A&M Research Foundation, College Station, Texas.

MASS SPECTRA

Selected Mass Spectral Data, American Petroleum Institute Research Project 44, current loose-leaf sheets. As of June 30, 1966, there were 2342 valid sheets in the catalog. Sold as complete sets, 6 volumes, at $0.30 per sheet. Texas A&M Research Foundation, College Station, Texas.

Selected Mass Spectral Data, Thermodynamics Research Center, current loose-leaf sheets. As of June 30, 1966, there were 168 valid sheets. Sold as complete sets at $0.30 per sheet. Texas A&M Research Foundation, College Station, Texas.

Compilation of Mass Spectral Data (Index de Spectres de Masse), A. Cornu and R. Massot, 1966, xv + 617 pp, 8½″ × 11″. Published by Heyden & Sons Ltd., in cooperation with Presses Universitaires de France. Distributed by Heyden & Sons Ltd., Spectrum House, Alderton Crescent, London, N.W.4, England; $42.00.

NUCLEAR MAGNETIC RESONANCE SPECTRA

Selected Nuclear Magnetic Resonance Spectral Data, American Petroleum Institute Research Project 44, current loose-leaf sheets. As of June 30, 1966, there were 649 valid sheets. Sold as complete sets, 2 volumes, at $0.30 per sheet. Texas A&M Research Foundation, College Station, Texas.

Selected Nuclear Magnetic Resonance Spectral Data, Thermodynamics Research Center, current loose-leaf sheets. As of June 30, 1966, there were 551 valid sheets. Sold as complete sets, 2 volumes, at $0.30 per sheet. Texas A&M Research Foundation, College Station, Texas.

High Resolution NMR Spectra Catalog, Vol. I, 1962, N. S. Bhacca, L. F. Johnson, and J. N. Shoolery, compilers, Varian Associates, Instrument Division, Palo Alto, Calif. 94303, 43 pp + 368 spectra, $6.00; 10 or more copies, $4.50 each; student price, $5.00.

High Resolution NMR Spectra Catalog, Vol. II, 1963, N. S. Bhacca, D. P. Hollis, L. F. Johnson, and E. A. Pier, compilers, Varian Associates, Instrument Division, Palo Alto, Calif. 94303, x + 62 pp + 332 spectra; price same as Vol. I; combined edition, Vols. I and II, hard cover, $20.00.

LANDOLT-BÖRNSTEIN

Landolt-Börnstein: Zahlenwerte und Funktionen aus Physik, Chemie, Astronomie, Geophysik und Technik, sixth edition.

Volumes Published

Vol. I: Atom—und Molekularphysik
- Part 1, Atome und Ionen, 1950, xii + 441 pp., DM 126 ($31.50).
- Part 2, Molekeln I: Kerngerüst, 1951, viii + 571 pp., DM 168 ($42.00).
- Part 3, Molekeln II: Elektronenhülle, 1951, xi + 724 pp., DM 218 ($54.50).
- Part 4, Kristalle, 1955, xi + 1,007 pp., DM 318 ($79.50).
- Part 5, Atomkerne und Elementarteilchen, 1952, viii + 470 pp., DM 148 ($37.00).

Vol. II: Eigenschaften der Materie in Ihren Aggregatzuständen
- Part 2, Gleichgewichte ausser Schmelzgleichgewichten.
  - 2a, Gleichgewichte Dampf-Kondensat und Osmotische Phänomene, 1960, xi + 974 pp., DM 448 ($112.00).
  - 2b, Lösungsgleichgewichte I, 1962, x + 983 pp., DM 510 ($127.50).
  - 2c, Lösungsgleichgewichte II, 1964, viii + 731 pp., DM 403.5 ($100.88).
- Part 3, Schmelzgleichgewichte und Grenzflächenerscheinungen, 1956, xi + 535 pp., DM 198 ($49.50).
- Part 4, Kalorische Zustandsgrössen, 1961, xii + 863 pp., DM 438 ($109.50).
- Part 6, Elektrische Eigenschaften I, 1959, xv + 1,018 pp., DM 448 ($112.00).
- Part 7, Elektrische Eigenschaften II, 1960, xii + 959 pp., DM 478 ($119.50).
- Part 8, Optische Konstanten, 1962, xv + 901 pp., DM 476 ($119.00).
- Part 9, Magnetische Eigenschaften I, 1962, xxv + 1,934 pp., DM 496 ($124.00).

Vol. III: Astronomie und Geophysik, 1952, xviii + 795 pp., DM 248 ($62.00).

Vol. IV: Technik
- Part 1, Stoffwerte und mechanisches Verhalten von Nichtmetallen, 1955, xvi + 881 pp., DM 288 ($72.00).
- Part 2, Stoffwerte und Verhalten von metallischen Werkstoffen.
  - 2a, Grundlagen, Prüfverfahren, Eisenwerkstoffe, 1963, xii + 888 pp., DM 468 ($117.00).
  - 2b, Sinterwerkstoffe, Schwermetalle (ohne Sonderwerkstoffe), 1964, xx + 1,000 pp., DM 530 ($132.50).
  - 2c, Leichtmetalle, Sonderwerkstoffe, Halbleiter, Korrosion, 1965, xx + 976 pp., DM 518 ($129.50).
- Part 3, Elektrotechnik, Lichttechnik, Röntgentechnik, 1957, xv + 1,076 pp., DM 396 ($99.00).

Volumes in Preparation

Vol. II: Eigenschaften der Materie in Ihren Aggregatzuständen
- Part 1, Mechanisch-thermische Zustandgrössen.
- Part 5, Physikalische und chemische Kinetik.
- Part 10, Magnetische Eigenschaften II.

Vol. IV: Technik
- Part 4, Wärmetechnik.

SOURCES OF CRITICAL DATA (Continued)

LANDOLT-BÖRNSTEIN

Landolt-Börnstein: Zahlenwerte und Functionen aus Naturwissenschaften und Technik, Neue Serie—Numerical Data and Functional Relationships in Science and Technology, New Series, K. H. Hellwege, editor-in-chief.

Volumes Published

Group I: Kernphysik und Kerntechnik—Nuclear Physics and Technology

Vol. I, Energy Levels of Nuclei: A = 5 to A = 257, A. M. Hellwege and K. H. Hellwege, eds., 1961, xii + 813 pp., DM 212 ($53.00).

Group II: Atom und Molekularphysik—Atomic and Molecular Physics

Vol. I, Magnetische Eigenschaften Freier Radikale—Magnetic Properties of Free Radicals, K. H. Hellwege and A. M. Hellwege, eds., 1965, x + 154 pp., DM 68 ($17.00).

Group VI: Astronomie, Astrophysik und Weltraum-Forschung—Astronomy, Astrophysics and Space Research

Vol. I, Astronomie und Astrophysik—Astronomy and Astrophysics, H. H. Voigt, ed., 1965, xxxix + 711 pp., DM 314 ($78.50).

Other groups planned and volumes expected to be published in 1966 or 1967:

Group I: Nuclear Physics and Technology

Hilfstabellen für α-, β-, γ- Spektroskopie—Auxiliary Tables for α-, β-, γ- Spectroscopy.

Ladungs-und Dichteverteilung im Kern—Charge and Density Distribution in the Nucleus.

Group II: Atomic and Molecular Physics

Vol. I, Lumineszenz organischer Molekeln—Luminescence of Organic Molecules.

Mikrowellenspektroskopie von Molekeln—Microwave Spectroscopy of Molecules.

Molekularakustik—Molecular Acoustics.

Vol. II, Magnetische Eigenschaften der Koordinations und metallorganischen Verbindungen der Übergangselemente—Magnetic Properties of Co-ordination and Organometallic Compounds of Transition Elements.

Group III: Crystal and Solid State Physics

Elastische, elastooptische, piezoelektrische und andere Eigenschaften von Einkristallen—Elastic, Elasto-optic, Piezoelectric and Other Properties of Single Crystals.

Group IV: Macroscopic and Technical Properties of Matter

Phosphoreszenz anorganischer Substanzen—Phosphorescence (Luminescence) of Inorganic Substances.

Group V: Geophysics and Space Research

Aeronomy; Cosmic Radiation.

Published by Springer-Verlag, Berlin, Heidelberg, and New York. Also available from Walter J. Johnson, Inc., 111 Fifth Ave., New York, N.Y. 10003, and Stechert-Hafner, Inc., 31 E. 10th St., New York, N.Y. 10003.

TABLES DE CONSTANTES SELECTIONNEES—TABLES OF SELECTED CONSTANTS

Tables de Constantes Sélectionnées—Tables of Selected Constants

Vol. 1. Longueurs d'Onde d'Emissions X et des Ciscontinuités d'Absorption X-Wavelengths of Emission and Discontinuities in Absorption of X-Rays, Y. Cauchois and H. Hulubei, 1947, 199 pp. (out of print).

Vol. 2. Physique Nucléaire—Nuclear Physics, R. Grégoire, F. Joliot-Curie, and I. Joliot-Curie, 1948, 131 pp. (out of print).

Vol. 3. Pouvoir Rotatoire Magnétique (Effet Faraday)—Magnetic Rotatory Power (Faraday Effect); R. de Mallemann, Effet Magnéto-Optique de Kerr—Magneto-Optic Effect (Kerr), F. Suhner, 1951, 137 pp., 15 F, £1.10.0. ($4.50).

Vol. 4. Données Spectroscopiques Concernant les Molécules Diatomiques—Spectroscopic Data for Diatomic Molecules, B. Rosen, R. F. Barrow, A. D. Caunt, A. R. Downie, R. Herman, E. Huldt, A. MacKellar, E. Miescher, and K. Wieland, 1951, 361 pp., 48 F, £5.0.0. ($15.00).

Vol. 5. Atlas des Longueurs d'Onde Caractéristiques des Bandes d'Emission et d'Absorption des Molécules Diatomiques—Atlas of Characteristic Wavelengths for Emission and Absorption Bands of Diatomic Molecules (a continuation of Vol. 4 by the same authors), 1952, 389 pp., 56 F, £5.15.0 ($17.50).

Vol. 6. Pouvoir Rotatoire Naturel I-Steroïdes—Optical Rotatory Powers I-Steroids, J.-P. Mathieu and A. Petit, 1956, 507 pp. (out of print).

Vol. 7. Diamagnétisme et Paramagnétisme—Diamagnetism and Paramagnetism, G. Foëx; Relaxation Paramagnétique—Paramagnetic Relaxation, C. J. Gorter and L. J. Smits, 1957, 317 pp., 97 F, £9.15.9. ($29.00).

Vol. 8. Potentiels d'Oxydo-Réduction—Oxidation-Reduction Potentials, G. Charlot, D. Bézier, and J. Courtot, 1958, 41 pp., 21.60 F, £1.10.0 ($5.00).

Vol. 9. Pouvoir Rotatoire Naturel II-Triterpénoïdes—Optical Rotatory Power II-Triterpenoids, J.-P. Mathieu and G. Ourisson, 1958, 302 pp., 93.60F, £7.0.0. ($21.00)

Vol. 10. Pouvoir Rotatoire Naturel III-Amino-acides—Optical Rotatory Power III-Amino Acids, J.-P. Mathieu, J. Roche, and P. Desnuelle, 1959, 61 pp., 28 F, £2.0.0. ($6.50).

TABLES DE CONSTANTES SELECTIONNEES—TABLES OF SELECTED CONSTANTS (Continued)

Vol. 11. Pouvoir Rotatoire Naturel IV-Alcaloïdes—Optical Rotatory Power IV-Alkaloids, J.-P. Mathieu and M. M. Janot, 1959, 211 pp., 110.40 F, £8.0.0. ($24.00).

Vol. 12. Semi-Conducteurs—Semi-Conductors, P. Aigrain and J. Balkanski, 1961, 78 pp., 27 F, £2.0.0. ($6.50).

Vol. 13. Rendements Radiolytiques—Radiolytic Yields, M. Haïssinsky and M. Magat, 1963, 230 pp., 114 F, £8.10.0. ($25.50).

Vol. 14. Pouvoir Rotatoire Naturel Ia-Stéroïdes—Optical Rotatory Power Ia-Steroids, J. Jacques, H. Kagan, G. Ourisson, and S. Allard, 1965, 1,046 pp., 258 F.

Vol. 15. Données Relatives aux Sesquiterpenoïdes—Data Relative to Sesquiter-penoids, G. Ourisson, S. Munavalli, and C. Éhret, 1966, 70 pp.

In Preparation

Metals of High Purity, 1967

Refractory Compounds

There have been several publishers for the present series. Volume 8 and later volumes have been published by Pergamon Press from whom the earlier volumes as well as the titles, availability, and prices of the 40 installments of Tables Annuelles, covering the period 1931–1936, may be obtained.

Published by Pergamon Press, 4 and 5 Fitzroy Square, London, W.1, England; 122 E. 55th St., New York, N.Y., U.S.; and 24 rue des Ecoles, Paris V^e, France.

INTERNATIONAL CRITICAL TABLES

International Critical Tables of Numerical Data, Physics, Chemistry and Technology, published for the National Academy of Sciences—National Research Council, by McGraw-Hill Book Co., 330 W. 42nd St., New York, N. Y. 10036.

Vol. I, 1926, xx + 415 pp.
Vol. II, 1927, xviii + 616 pp.
Vol. III, 1928, xiv + 444 pp.
Vol. IV, 1928, viii + 481 pp.
Vol. V, 1929, ix + 465 pp.
Vol. VI, 1929, x + 471 pp.
Vol. VII, 1930, ix + 507 pp.
Index, 1933, vii + 321 pp.

Seven volumes and Index, $250.00; Vols. I-VII and Index, each $35.00.

NATIONAL STANDARD REFERENCE DATA SYSTEM PUBLICATIONS

The National Standard Reference Data System (NSRDS) comprises the set of data centers and other data evaluation projects administered or coordinated by Office of Standard Reference Data in the National Bureau of Standards. The primary aim of this program is to provide critically evaluated physical and chemical data, in a convenient and accessible form, to the scientific and technical community of the United States. The scope of the program is divided into the following areas:

Energy and environmental data
Industrial process data
Materials utilization data
Physical sciences data

The principal output of the program consists of compilations of evaluated data and critical reviews of the status of data in particular technical areas. Evaluated data produced under the NSRDS program are disseminated through the following mechanisms:

Journal of Physical and Chemical Reference Data—A quarterly journal containing data compilations and critical data reviews, published for the National Bureau of Standards by the American Institute of Physics and the American Chemical Society

NSRDS-NBS Series—A publication series distributed by the Superintendent of Documents, U.S. Government Printing Office.

Appropriate publications of technical societies and commercial publishers.

Response by OSRD and individual data centers to inquiries for specific data.

Further information on NSRDS publications, sources of data, and support of data compilation activities can be obtained from:

Office of Standard Reference Data
National Bureau of Standards
Washington, D.C. 20234
Telephone: (301)921-2467

Publications in the *Journal of Physical and Chemical Reference Data*

1. *Gaseous Diffusion Coefficients*, T. R. Marrero and E. A. Mason, Vol. 1, No. 1 (1972). $7.00.
2. *Selected Values of Critical Supersaturation for Nucleation of Liquids from the Vapor*, G. M. Pound, Vol. 1, No. 1 (1972). $3.00.
3. *Selected Values of Evaporation and Condensation Coefficients of Simple Substances*, G. M. Pound, Vol. 1, No. 1 (1972). $3.00.
4. *Atlas of the Observed Absorption Spectrum of Carbon Monoxide between 1060 and 1900 A*, S. G. Tilford and J. D. Simmons, Vol. 1, No. 1 (1972). $4.50.
5. *Tables of Molecular Vibrational Frequencies*, Part 5, T. Shimanouchi, Vol. 1, No. 1 (1972). $4.00.
6. *Selected Values of Heats of Combustion and Heats of Formation of Organic Compounds Containing the Elements C, H, N, O, P, and S*, Eugene S. Domalski, Vol. 1, No. 2 (1972). $5.00.
7. *Thermal Conductivity of the Elements*, C. Y. Ho, R. W. Powell, and P. E. Liley, Vol. 1, No. 2 (1972). $7.50.
8. *The Spectrum of Molecular Oxygen*, Paul H. Krupenie, Vol. 1, No. 2 (1972). $6.50.
9. *A Critical Review of the Gas-Phase Reaction Kinetics of the Hydroxyl Radical*, W. E. Wilson, Jr., Vol. 1, No. 2 (1972). $4.50.
10. *Molten Salts: Volume 3, Nitrates, Nitrites, and Mixtures, Electrical Conductance, Density, Viscosity, and Surface Tension Data*, G. J. Janz, Ursula Krebs, H. F. Siegenthaler and R. P. T. Tomkins, Vol. 1, No. 3 (1972). $8.50.
11. *High Temperature Properties and Decomposition of*

Inorganic Salts — Part 3. Nitrates and Nitrites, Kurt H. Stern, Vol. 1, No. 3 (1972). $4.00.

12. *High-Pressure Calibration: A Critical Review,* D. L. Decker, W. A. Bassett, L. Merrill, H. T. Hall, and J. D. Barnett, Vol. 1, No. 3 (1972). $5.00.

13. *The Surface Tension of Pure Liquid Compounds,* Joseph J. Jasper, Vol. 1, No. 4 (1972). $8.50.

14. *Microwave Spectra of Molecules of Astrophysical Interest. I. Formaldehyde, Formamide, and Thioformaldehyde,* Donald R. Johnson, Frank J. Lovas, and William H. Kirchhoff, Vol. 1, No. 4 (1972). $4.50.

15. *Osmotic Coefficients and Mean Activity Coefficients of Uni-univalent Electrolytes in Water at 25 C,* Walter J. Hamer and Yung Chi Wu, Vol. 1, No. 4 (1972). $5.00.

16. *The Viscosity and Thermal Conductivity Coefficients of Gaseous and Liquid Fluorine,*H. J. M. Hanley and R. Prydz, Vol. 1, No. 4 (1972). $3.00.

17. *Microwave Spectra of Molecules of Astrophysical Interest. II. Methylenimine,* William H. Kirchhoff, Donald R. Johnson, and Frank J. Lovas, Vol. 2, No. 1 (1973). $3.00.

18. *Analysis of Specific Heat Data in the Critical Region of Magnetic Solids,* F. J. Cook, Vol. 2, No. 1 (1973). $3.00.

19. *Evaluated Chemical Kinetic Rate Constants for Various Gas Phase Reactions,* Keith Schofield, Vol. 2, No. 1 (1973). $5.00

20. *Atomic Transition Probabilities for Forbidden Lines of the Iron Group Elements. (A Critical Data Compilation for Selected Lines),* M. W. Smith and W. L. Wiese, Vol. 2, No. 1 (1973). $4.50.

21. *Tables qf Molecular Vibrational Frequencies, Part 6,* T. Shimanouchi, Vol. 2, No. 1 (1973). $4.50.

22. *Compilation of Energy Band Gaps in Elemental and Binary Compound Semiconductors and Insulators,* W. H. Strehlow and E. L. Cook, Vol. 2, No. 1 (1973). $4.50

23. *Microwave Spectra of Molecules of Astrophysical Interest, III. Methanol,* R. M. Lees, F. J. Lovas, W. H. Kirchhoff, and D. R. Johnson, Vol. 2, No. 2 (1973). $3.00.

24. *Microwave Spectra of Molecules of Astrophysical Interest. IV. Hydrogen Sulfide,* Paul Helminger, Frank C. DeLucia and William H. Kirchhoff, Vol. 2, No. 2 (1973). $3.00.

25. *Tables of Molecular Vibrational Frequencies, Part 7,*T. Shimanouchi, Vol. 2, No. 2 (1973). $4.00.

26. *Energy Levels of Neutral Helium (^{4}HE I),* W. C. Martin, Vol. 2, No. 2 (1973). $3.00.

27. *Survey of Photochemical and Rate Data for Twenty-eight Reactions of Interest in Atmospheric Chemistry,* R. F. Hampson, Editor, W. Braun, R. L. Brown, D. Garvin, J. T. Herron, R. E. Huie, M. J. Kurylo, A. H. Laufer, J. D. McKinley, H. Okabe, M. D. Scheer, W. Tsang, and D. H. Stedman, Vol. 2, No. 2 (1973). $4.50.

28. *Compilation of the Static Dielectric Constant of Inorganic Solids,* K. F. Young and H. P. R. Frederikse, Vol. 2, No. 2 (1973). $6.50.

29. *Soft X-Ray Emission Spectra of Metallic Solids: Critical Review of Selected Systems,* A. J. McAlister, R. C. Dobbyn, J. E. Cuthill, and M. L. Williams, Vol. 2, No. 2 (1973). $3.00.

30. *Ideal Gas Thermodynamic Properties of Ethane and Propane,* J. Chao, R. C. Wilhoit, and B. J. Zwolinski, Vol. 2, No. 2 (1973). $3.00.

31. *An Analysis of Coexistence Curve Data for Several Binary Liquid Mixtures Near Their Critical Points,* A. Stein and G. F. Allen, Vol. 2, No. 3 (1973). $4.00.

32. *Rate Constants for the Reactions of Atomic Oxygen (O^3P) with Organic Compounds in the Gas Phase,* John T. Herron and Robert E. Huie, Vol. 2, No. 3 (1973). $5.00.

33. *First Spectra of Neon, Argon, and Xenon 136 in the 1.2—4.0 μm Region,* Curtis J. Humphreys, Vol. 2, No. 3 (1973). $3.00.

34. *Elastic Properties of Metals and Alloys, I. Iron, Nickel, and Iron-Nickel Alloys,* H. M. Ledbetter and R. P. Reed, Vol. 2, No. 3 (1973). $6.00.

35. *The Viscosity and Thermal Conductivity Coefficients of Dilute Argon,* Krypton, and Xenon, H. J. M. Hanley, Vol. 2, No. 3 (1973). $4.00.

36. *Diffusion in Copper and Copper Alloys. Part I. Volume and Surface Self-Diffusion in Copper,* Daniel B. Butrymowicz, John R. Manning, and Michael E. Read, Vol. 2, No. 3 (1973). $3.00.

37. *The 1973 Least-Squares Adjustment of the Fundamental Constants,* E. Richard Cohen and B. N. Taylor, Vol. 2, No. 4 (1973). $5.50.

38. *The Viscosity and Thermal Conductivity Coefficients of Dilute Nitrogen and Oxygen,* H. J. M. Hanley and James F. Ely, Vol. 2, No. 4 (1973). $4.00.

39. *Thermodynamic Properties of Nitrogen Including Liquid and Vapor Phases from 63 K to 2000 K with Pressures to 10,000 Bar,* Richard T. Jacobsen and Richard B. Stewart, Vol. 2, No. 4 (1973). $8.50.

40. *Thermodynamic Properties of Helium 4 from 2 to 1500 K at Pressures to 10^8 Pa,* Robert D. McCarty, Vol. 2, No. 4 (1973) . $7.00.

41. *Molten Salts: Volume 4, Part l, Fluorides and Mixtures, Electrical Conductance, Density, Viscosity, and Surface Tension Data,* G. J. Janz, G. L. Gardner, Ursula Krebs, and R. P. T. Tomkins, Vol. 3, No. 1 (1974). $7.00.

42. *Ideal Gas Thermodynamic Properties of Eight Chloro- and Fluoromethanes,* A. S. Rodgers, J. Chao, R. C. Wilhoit, and B. J. Zwolinski, Vol. 3, No. 1 (1974). $4.00.

43. *Ideal Gas Thermodynamic Properties of Six Chloroethanes,* J. Chao, A. S. Rodgers, R. C. Wilhoit, and B. J. Zwolinski, Vol. 3, No. 1 (1974). $4.00.

44. *Critical Analysis of Heat-Capacity Data and Evaluation of Thermodynamic Properties of Ruthenium, Rhodium, Palladium, Iridium, and Platinum from 0 to 300 K. A Survey of the Literature Data on Osmium,* George T. Furukawa, Martin L. Reilly, and John S. Gallagher, Vol. 3, No. 1 (1974). $4.50.

45. *Microwave Spectra of Molecules of Astrophysical Interest. V. Water Vapor,* Frank C. DeLucia, Paul Helminger, and William H. Kirchhoff, Vol. 3, No. 1 (1974). $3.00.

46. *Microwave Spectra of Molecules of Astrophysical*

Interest. VI. Carbonyl Sulfide and Hydrogen Cyanide, Arthur G. Maki, Vol. 3, No. 1 (1974). $4.00.

47. *Microwave Spectra of Molecules of Astrophysical Interest. VII. Carbon Monoxide, Carbon Monosulfide, and Silicon Monoxide,* Frank J. Lovas and Paul H. Krupenie, Vol. 3, No. 1 (1974). $3.00.

48. *Microwave Spectra of Molecules of Astrophysical Interest. VIII. Sulfur Monoxide,* Eberhard Tiemann, Vol. 3, No. 1 (1974). $3.00.

49. *Tables of Molecular Vibrational Frequencies, Part 8,* T. Shimanouchi, Vol. 3, No. 1 (1974). $4.50.

50. *JANAF Thermochemical Tables, 1974 Supplement,* M. W. Chase, J. L. Curnutt, A. T. Hu, H. Prophet, A. N. Syverud, and L. C. Walker, Vol. 3, No. 2 (1974). $8.50.

51. *High Temperature Properties and Decomposition of Inorganic Salts, Part 4. Oxy-Salts of the Halogens,* Kurt H. Stern, Vol. 3. No. 2 (1974). $4.50.

52. *Diffusion in Copper and Copper Alloys, Part II. Copper-Silver and Copper-Gold Systems,* Daniel B. Butrymowicz, John R. Manning, and Michael E. Read, Vol. 3, No. 2 (1974). $5.50.

53. *Microwave Spectral Tables. I. Diatomic Molecules,* Frank J. Lovas and Eberhard Tiemann, Vol. 3, No. 3(1974). $8.50.

54. *Ground Levels and Ionization Potentials for Lanthanide and Actinide Atoms and Ions,* W. C. Martin, Lucy Hagan, Joseph Reader, and Jack Sugar, Vol. 3, No. 3 (1974). $3.00.

55. *Behavior of the Elements at High Pressure,* John Francis Cannon, Vol. 3, No. 3 (1974). $4.50.

56. *Reference Wavelengths from Atomic Spectra in the Range 15A to 25000A,* Victor Kaufman and Bengt Edlen, Vol. 3, No. 4 (1974). $5.50.

57. *Elastic Properties of Metals and Alloys. II. Copper,* H. M. Ledbetter and E. R. Naimon, Vol. 3, No. 4 (1974). $4.50.

58. *Critical Review of Hydrogen Atom Transfer Reactions in the Liquid Phases,* D. G. Hendry, T. Mill, J. A. Howard, L. Piszkiewicz, and H. K. Eigenmann, Vol. 3, No. 4 (1974). $4.50.

59. *The Viscosity and Thermal Conductivity Coefficients for Dense Gaseous and Liquid Argon, Krypton, Xenon, Nitrogen, and Oxygen,* H. J. M. Hanley, R. D. McCarty, and W. N. Haynes, Vol. 3, No. 4 (1974). $4.50.

60. *JANAF Thermochemical Tables, 1975 Supplement,* M. W. Chase, J. L. Curnutt, H. Prophet, R. A. McDonald, and A. N. Syverud, Vol. 4, No. 1 (1975). $8.50.

61. *Diffusion in Copper and Copper Alloys. Part III. Diffusion in Systems Involving Elements of the Groups IA, IIA, IIIB, IVB, VB, VIB, and VIIB,* Daniel B. Butrymowicz, John R. Manning, and Michael E. Read, Vol. 4, No. 1 (1975). $6.00.

62. *Ideal Gas Thermodynamic Properties of Ethylene and Propylene,* Jing Chao and Bruno J. Zwolinski, Vol. 4, No. 1 (1975). $3.00.

63. *Atomic Transition Probabilities for Scandium and Titanium (A Critical Data Compilation of Allowed Lines),* W. L. Wiese and J. R. Fuhr, Vol. 4, No. 2 (1975). $6.00.

64. *Energy Levels of Iron, Fe I through Fe XXVI,* Joseph Reader and Jack Sugar, Vol. 4, No. 2 (1975). $6.00.

65. *Ideal Gas Thermodynamic Properties of Six Fluoroethanes,* S. S. Chen, A. S. Rodgers, J. Chao, R. C. Wilhoit, and B. J. Zwolinski, Vol. 4, No. 2 (1975). $3.00.

66. *Ideal Gas Thermodynamic Properties of the Eight Bromo— and Iodomethanes,* S. A. Kudchadker and A. P. Kudchadker, Vol. 4, No. 2 (1975). $3.00.

67. *Atomic Form Factors, Incoherent Scattering Functions, and Photon Scattering Cross Sections,* J. H. Hubbell, Wm. J. Veigele, E. A. Briggs, R. T. Brown, D. T. Cromer, and R. J. Howerton, Vol. 4, No. 3 (1975). $5.50.

68. *Binding Energies in Atomic Negative Ions,* H. Hotop and W. C. Lineberger, Vol. 4, No. 3 (1976). $4.50.

69. *A Survey of Electron Swarm Data,* J. Dutton, Vol. 4, No. 3 (1975). $12.00.

70. *Ideal Gas Thermodynamic Properties and Isomerization of n-Butane and Isobutane,* S. S. Chen, R. C. Wilhoit, and B. J. Zwolinski, Vol. 4, No. 4 (1975). $3.00.

71. *Molten Salts: Volume 4, Part 2, Chlorides and Mixtures, Electrical Conductance, Density, Viscosity, and Surface Tension Data,* G. J. Janz, R. P. T. Tomkins, C. B. Allen, J. R. Downey, Jr., G. L. Gardner, U. Krebs, and S. K. Singer, Vol. 4, No. 4 (1975). $13.00.

72. *Property Index to Volumes 1—4 (1972—1975),* Vol. 4, No. 4 (1975). $3.00.

73. *Scaled Equation of State Parameters for Gases in the Critical Region,* J. M. L. Levelt- Sengers, W. L. Greer, and J. V. Sengers, Vol. 5, No. 1 (1975). $5.00.

74. *Microwave Spectra of Molecules of Astrophysical Interest. IX. Acetaldehyde,* A. Bauder, F. J. Lovas, and D. R. Johnson, Vol. 5, No. 1 (1976). $4.00.

75. *Microwave Spectra of Molecules of Astrophysical Interest. X. Isocyanic Acid,* G. Winnewisser, W. H. Hocking, and M. C. L. Gerry, Vol. 5, No. 1 (1976). $4.00.

76. *Diffusion in Copper and Copper Alloys. Part IV. Diffusion in Systems Involving Elements of Group VIII,* Daniel B. Butrymowicz, John R. Manning, and Michael E. Read, Vol. 5, No. 1 (1976). $6.50.

77. *A Critical Review of the Stark Widths and Shifts of Spectral Lines from Non-Hydrogenic Atoms,* N. Konjevic and J. R. Roberts, Vol. 5, No. 2 (1976). $5.00.

78. *Experimental Stark Widths and Shifts for Non-Hydrogenic Spectral Lines of Ionized Atoms,* N. Konjevic and W. L. Wiese, Vol 5, No. 2 (1976). $5.00.

79. *Atlas of the Absorption Spectrum of Nitric Oxide (NO) between 1420 and 1250 A,* E. Miescher and F. Alberti, Vol. 5, No. 2 (1976). $3.00.

80. *Ideal Gas Thermodynamic Properties of Propanone and 2-Butanone,* J. Chao and B. J. Zwolinski, Vol. 5, No. 2 (1976). $3.00.

81. *Refractive Index of Alkali Halides and Its Wavelength and Temperature Derivatives,* H. H. Li, Vol. 5, No. 2 (1976). :9.50.

82. *Tables of Critically Evaluated Oscillator Strengths*

for the Lithium Isoelectronic Sequence, G. A. Martin and W. L. Wiese, Vol. 5, No. 3 (1976). $4.50.

83. *Ideal Gas Thermodynamic Properties of Six Chlorofluoromethanes,* S. S. Chen, R. C. Wilhoit, and B. J. Zwolinski, Vol. 5, No. 3 (1976). $3.00.

84. *Survey of Superconductive Materials and Critical Evaluation of Selected Properties,* B. W. Roberts, Vol. 5, No. 3 (1976). $12.50.

85. *Nuclear Spins and Moments,* Gladys H. Fuller, Vol. 5, No. 4 (1976). $11.50.

86. *Nuclear Moments and Moment Ratios as Determined by Mossbauer Spectroscopy,* J. G. Stevens and B. D. Dunlop, Vol. 5, No. 4 (1976). $4.00.

87. *Rate Coefficients for Ion-Molecule Reactions. I. Ions Containing C and H,* L. Wayne Sieck and Sharon G. Lias, Vol. 5, No. 4 (1976). $4.00.

88. *Microwave Spectra of Molecules of Astrophysical Interest. XI. Silicon Sulfide,* Eberhard Tiemann, Vol. 5, No. 4 (1976). $3.00.

89. *Property Index and Author Index to Volume 1—5 (1972—1976),* Vol. 5, No. 4 (1976). $4.00.

90. *Diffusion in Copper and Copper Alloys. Part V. Diffusion in Systems involving Elements of Group VA,* Daniel B. Butrymowicz, John R. Manning, and Michael E. Read, Vol. 6, No. 1 (1977). $5.00.

91. *The Calculated Thermodynamic Properties of Superfluid Helium-A,* James S. Brooks and Russell J. Donnelly, Vol. 6, No. 1 (1977). $5.00.

92. *Thermodynamic Properties of Normal and Deuterated Methanols,* S. S. Chen, R. C. Wilhoit, and B. J. Zwolinski, Vol. 6, No. 1 (1977). $3.00.

93. *The Spectrum of Molecular Nitrogen,* Alf Lofthus and Paul H. Krupenie, Vol. 6, No. 1 (1977). $9.50.

Supplements to the *Journal of Physical and Chemical Reference Data**

Physical and Thermodynamic Properties of Aliphatic Alcohols, R. C. Wilhoit and B. J. Zwolinski, Vol. 2, Supplement 1, 420 pp. (1973). Hardcover: $33.00; Softcover: members $10.00; nonmembers $30.00.

Thermal Conductivity of the Elements: A Comprehensive Review, C. Y. Ho, R. W. Powell, and P. E. Liley, Vol. 3, Supplement 1, 796 pp. (1974). Hardcover: $60.00; Softcover: members $25.00, nonmembers $50.00.

Energetics of Gaseous Ions, H. M. Rosenstock, K. Draxl, B. Steiner, and J. T. Herron, Vol. 6, Supplement 1, 783 pp. (1977). Hardcover: $70.00; Softcover: member $30.00, nonmember $65.00.

* Available from the Business Operations, Books and Journals Division, American Chemical Society, 1155 16th Street, N. W., Room 604, Washington, D.C. 20036.

Publications in the National Standard Reference Data Series

NSRDS-NBS-1, *National Standard Reference Data System — Plan of Operation,* E. L. Brady and M. B. Wallenstein (1964). SD Catalog No. C13.48:1. $0.55.

NSRDS-NBS-2, *Thermal Properties of Aqueous Uniunivalent Electrolytes,* V. B. Parker (1965). SD Catalog No. C13.48:2. $1.10.

NSRDS-NBS-3, Sec. 1, *Selected Tables of Atomic Spectra, Atomic Energy Levels and Multiplet Tables, Si II, Si III, Si IV,* C. E. Moore (1965).*

NSRDS-NBS-3, Sec. 2, *Selected Tables of Atomic Spectra, Atomic Energy Levels and Multiplet Tables, Si I,* C. E. Moore (1967).**

NSRDS-NBS-3, Sec. 3, *Selected Tables of Atomic Spectra, Atomic Energy Levels and Multiplet Tables, C I , C II, C III, C IV, C V, C VI,* C. F. Moore (1970).*

NSRDS-NBS-3, Sec. 4, *Selected Tables of Atomic Spectra, Atomic Energy Levels and Multiplet Tables, N IV, N V, N VI, N VII,* C. E. Moore (1971). SD Catalog No. C13.48:3/Sec. 4. $1.15.

NSRDS-NBS-3, Sec. 5, *Selected Tables of Atomic Spectra, Atomic Energy Levels and Multiplet Tables, N I, N II, N III,* C. E. Moore (1975). SD Catalog No. C13.48:3/Sec. 5.

NSRDS-NBS-3, Sec. 6, *Selected Tables of Atomic Spectra, Atomic Energy Levels and Multiplet Tables, H I, D, T,* C. E. Moore (1972). SD Catalog No. C13.48:3/Sec. 6. $0.40.

NSRDS-NBS 3, Sec. 7, *Selected Tables of Atomic Spectra, Atomic Energy Levels and Multiplet Tables, O I,* C. E. Moore (1976). SD Catalog No. C13.48:3/Sec. 7. $0.85.

NSRDS-NBS-4, *Atomic Transition Probabilities. Vol. I. Hydrogen Through Neon,* W. L. Wiese, M. W. Smith, and B. M. Glennon (1966). SD Catalog No. C13.48:4/ Vol. I. $2.50.

NSRDS-NBS-5, *The Band Spectrum of Carbon Monoxide,* P. H. Krupenie (1966).**

NSRDS-NBS-6, *Tables of Molecular Vibrational Frequencies, Part 1,* T. Shimanouchi (1967). Superseded by NSRDS-NBS-39.

NSRDS-NBS-7, *High Temperature Properties and Decomposition of Inorganic Salts, Part l. Sulfates,*K. H. Stern and E. L. Weise (1966).*

NSRDS-NBS-8, *Thermal Conductivity of Selected Materials,* R. W. Powell, C. Y. Ho, and P. E. Liley (1966). PB 189 698.**

NSRDS-NBS-9, *Tables of Bimolecular Gas Reactions,* A. F. Trotman-Dickenson and G. S. Milne (1967).**

NSRDS-NBS-10, *Selected Values of Electric Dipole Moments for Molecules in the Gas Phase,* R. D. Nelson, Jr., D. R. Lide, Jr. and A. A. Maryott (1967).**

NSRDS-NBS-11, *Tables of Molecular Vibrational Frequencies, Part 2,* T. Shimanouchi (1967). Superseded by NSRDS-NBS-39.

NSRDS-NBS-12, *Tables for the Rigid Asymmetric Rotor: Transformation Coefficients from Symmetric to Asymmetric Bases and Expectation Values of $P^2{}_2$, $P^4{}_2$, $P^6{}_2$,* R. H. Schwendeman (1968). SD Catalog No. C13.48:12. $1.45.

NSRDS-NBS-13, *Hydrogenation of Ethylene on Metallic Catalysts,* J. Horiuti and K. Miyahara (1968). SD Catalog No. C13.48:13. $3.00.

NSRDS-NBS-14, *X-Ray Wavelengths and X-Ray Atomic Energy Levels,* J. A. Bearden (1967).**

NSRDS-NBS-15, *Molten Salts: Vol. 1. Electrical Conductance, Density, and Viscosity Data,* G. J. Janz, F. W. Dampier, G. R. Lakshminarayana, P. K. Lorenz, and R. P. T. Tomkins (1968).**

NSRDS-NBS-16, *Thermal Conductivity of Selected Materials, Part 2,* C. Y. Ho, R. W. Powell, and P. E. Liley (1968).**

NSRDS-NBS-17, *Tables of Molecular Vibrational Frequencies, Part 3,* T. Shimanouchi (1968). Superseded by NSRDS-NBS-39.

NSRDS-NBS-18, *Critical Analysis of the Heat-Capacity Data of the Literature and Evaluation of Thermodynamic Properties of Copper, Silver, and Gold from 0 to 300 K,* G. T. Furukawa, W. G. Saba, and M. L. Reilly (1968). S. D. Catalog No. C13.48:18. $0.40.

NSRDS-NBS-19, *Thermodynamic Properties of Ammonia as an Ideal Gas,* L. Haar (1968). SD Catalog No. C13.48:19, $0.20.

NSRDS-NBS-20, *Gas Phase Reaction Kinetics of Neutral Oxygen Species,* H. S. Johnston (1968).*

NSRDS-NBS-21, *Kinetic Data on Gas Phase Unimolecular Reactions,* S. W. Benson and H. E. O'Neal (1970). PB191956**

NSRDS-NBS-22, *Atomic Transition Probabilities. Vol. II. Sodium Through Calcium, A. Critical Data Compilation,* W. L. Wiese, M. W. Smith, and B. M. Miles (1969). AD696884**

NSRDS-NBS-23, *Partial Grotrian Diagrams of Astrophysical Interest,* C. E. Moore and P. W. Merrill (1968).**

NSRDS-NBS-24, *Theoretical Mean Activity Coefficients of Strong Electrolytes in Aqueous Solutions from 0 to 100 °C,* Walter J. Hamer (1968).*

NSRDS-NBS-25, *Electron Impact Excitation of Atoms,* B. L. Moiseiwitsch and S. J. Smith (1968).**

NSRDS-NBS-26, *Ionization Potentials, Appearance Potentials and Heats of Formation of Gaseous Positive Ions,* J. L. Franklin, J. G. Dillard, H. M. Rosenstock, J. T. Heron, K. Draxl, and F. H. Field (1969). SD Catalog No. C13.48:26. $6.20.

NSRDS-NBS-27, *Thermodynamic Properties of Argon from the Triple Point to 300 K at Pressures to 1000 Atmospheres,* A. L. Gosman, R. D. McCarty, and J. G. Hust (1969). SD Catalog No. C13.48:27. $1.80.

NSRDS-NBS-28, *Molten Salts: Vol. 2, Section 1. Electrochemistry of Molten Salts: Gibbs Free Energies and Excess Free Energies from Equilibrium-Type Cells,* G. J. Janz and C. G. M. Dijkhuis; *Section 2. Surface Tension Data,* G. J. Janz, G. R. Lakshminarayanan, R. P. T. Tomkins, and J. Wong (1969).**

NSRDS-NBS-29. *Photon Cross Sections, Attenuation Coefficients, and Energy Absorption Coefficients from 10 keV to 100 GeV,* J. H. Hubbell (1969). SD Catalog No. C13.48:29. $1.25.

NSRDS-NBS-30, *High Temperature Properties and Decomposition of Inorganic Salts, Part 2. Carbonates,* H. H. Stern and E. L. Weise (1969).**

NSRDS-NBS-31, *Bond Dissociation Energies in Simple Molecules,* B. deB. Darwent (1970). SD Catalog No. C13.48:31. $0.95.

NSRDS-NBS-32, *Phase Behavior in Binary and Multicomponent Systems at Elevated Pressures: Pentane and Methane-n-Pentane,* V. M. Berry and B. H. Sage (1970). SD Catalog No. C13.48:32. $1.15.

NSRDS-NBS-33, *Electrolytic Conductance and Conductances of the Halogen Acids in Water,* W. J. Hamer and H. J. DeWane (1970). PB192183**

NSRDS-NBS-34, *Ionization Potentials and Ionization Limits Derived from the Analyses of Optical Spectra,* C. E. Moore (1970). SD Catalog No. C13.48:34. $0.75.

NSRDS-NBS-35, *Atomic Energy Levels as Derived from the Analyses of Optical Spectra, Vol. I ^{1}H to ^{23}V; Vol. II. ^{24}Cr to ^{41}Nb; Vol. III. ^{42}Mo to ^{57}La, ^{72}Hf to ^{89}Ac,* C. E. Moore (1971). SD Catalog No. C13.48:35/Vol. I. $9.25; Vol. II. $7.95; Vol. III. $8.30.

NSRDS-NBS-36, *Critical Micelle Concentrations of Aqueous Surfactant Systems,* P. Mukerjee and K. J. Mysels (1971).COM71-50203**

NSRDS-NBS-37, *JANAF Thermochemical Tables, 2nd ed.,* D. R. Stull, II. Prophet, et al. (1971). SD Catalog No. C13.48:37. $15.60.

NSRDS-NBS-38, *Critical Review of Ultraviolet Photoabsorption Cross Sections for Molecules of Astrophysical and Aeronomic Interest,* R. D. Hudson (1971). SD Catalog No. C13.48:38. $1.50.

NSRDS-NBS-39, *Tables of Molecular Vibrational Frequencies, Consolidated Volume I,* T. Shimanouchi (1972). SD Catalog No. C13.48:39. $5.10.

NSRDS-NBS-40, *A Multiplet Table of Astrophysical Interest, Revised Edition, Part I. Table of Multiplets; Part II. Finding List of All Lines in the Table of Multiplets,* C. E. Moore (1972). COM72-50439**

NSRDS-NBS-41, *Crystal Structure Transformations in Binary Halides,* C. N. R. Rao and M. Natarajan (1972). COM72-50849**

NSRDS-NBS-42, *Selected Specific Rates of Reactions of the Solvated Electron in Alcohols,* E. Watson, Jr. and S. Roy (1972). SD Catalog No. C13.48:42. $0.60.

NSRDS-NBS-43, *Selected Specific Rates of Reactions of Transients from Water in Aqueous Solution. I. Hydrated Electron,* M. Anbar, M. Bambenek, and A. B. Ross (1972). SD Catalog No. C13.48:43. 1.05.

NSRDS-NBS-43, Suppl., *Selected Specific Rates of Reactions of Transients from Water in Aqueous Solution. Hydrated Electron, Supplemental Data,* Alberta B. Ross. SD Catalog No. C13.48:43/Suppl. $1.10.

NSRDS-NBS-44, *The Radiation Chemistry of Gaseous Ammonia,* D. B. Peterson (1974). COM74-50175**

NSRDS-NBS-45, *Radiation Chemistry of Nitrous Oxide Gas. Primary Processes, Elementary Reactions, and Yields,* G. R. A. Johnson (1973). SD Catalog No. C13.48:45. $0.60.

NSRDS-NBS-46, *Reactivity of the Hydroxyl Radical in Aqueous Solutions,* Leon M. Dorfman and Gerald E. Adams (1973). COM73-50623**

NSRDS-NBS-47, *Tables of Collision Integrals and Second Virial Coefficients for the (m,6,8) Intermolecular Potential Function,* Max Klein, H. J. M. Hanley, Francis J. Smith, and Paul Holland (1974). SD Catalog No. C13.48:47. $1.90.

NSRDS-NBS-48, *Radiation Chemistry of Ethanol: A Review of Data on Yields, Reaction Rate Parameters, and Spectral Properties of Transients,* Gordon A. Freeman (1974). SD Catalog No. C13.48:48. $0.80.

NSRDS-NBS-49, *Transition Metal Oxides, Crystal Chemistry, Phase Transition, and Related Aspects,* C. N. R. Rao and G. V. Subba Rao (1974). SD Catalog No. C13.48:49. $1.70.

NSRDS-NBS-50, *Resonances in Electron Impact on Atoms and Diatomic Molecules,* George J. Schulz (1973). AD77/200**

NSRDS-NBS-51, *Selected Specific Rates of Reactions of Transients from Water in Aqueous Solution. II. Hydrogen Atom,* Michael Anbar, Farhataziz, and A. B. Ross (1975). SD Catalog No. C13.48:51. $1.20.

NSRDS-NBS-52, *Electronic Absorption and Internal and External Vibrational Data of Atomic and Molecular Ions Doped in Alkali Halide Crystals,* S. C. Jain, A. V. R. Warrier, and S. K. Agarwal (1974). SD Catalog No. C13.48:52. $0.95.

NSRDS-NBS-53, *Crystal Structure Transformations in Inorganic Nitrites, Nitrates and Carbonates,* C. N. R. Rao, B. Prakash, and M. Natarajan. SD Catalog No. C13.48:53. $1.15.

NSRDS-NBS-54, *Radiolysis of Methanol: Product Yields, Rate Constants and Spectroscopic Parameteres of Intermediates,* J. H. Baxendale and Peter Wardman (1975). SD Catalog No. C13.48:54. $0.85.

NSRDS-NBS-55, *Property Index to NSRDS Data Compilations, 1964—1972,* David R. Lide, Jr., Gertrude B. Sherwood, Charles H. Douglass, Jr., and Herman M. Weisman. SD Catalog No. C13.48:55. $0.85.

NSRDS-NBS-56, *Crystal Structure Transformations in Inorganic Sulfates, Phosphates, Perchlorates, and Chromates,* B. Prakash and C. N. R. Rao. SD Catalog No. C13.48:56. $0.85.

NSRDS-NBS-57, *Yields of Free Ions Formed in Liquids by Radiation,* A. O. Allen (1976). SD Catalog No. C13.48:57. $0.55.

NSRDS-NBS-58, *Drift Mobilities and Conduction Band Energies of Excess Electrons in Dielectric Liquids,* A. O. Allen (1976). SD Catalog No. C13.48:58. $0.70.

NSRDS-NBS-59, *Selected Specific Rates of Reactions of Transients from Water in Aqueous Solution, III. Hydroxyl Radical and Perhydroxyl Radical and Their Radical Ions,* Farhataziz and Alberta B. Ross (1977). SD Catalog No. C13.48:59. $1.90.

NSRDS-NBS-60, *Atomic Energy Levels — The Rare Earth Elements, The Spectra of Lanthanum, Cerium, Praseodymium, Neodymium, Promethium, Samarium, Europium, Gadolinium, Terbium, Dysprosium, Holmium, Erbium, Thulium, Ytterbium, and Lutetium,* W. C. Martin, Romuald Zalubas, and Lucy Hagan (1977, in press). SD Catalog No. C13.48:60.

Publications giving SD Catalog Number may be ordered from the Superintendent of Documents, U.S. Government Printing Office, Washington, D.C. 20402, using the SD Catalog Numbers given.

* Available from the Office of Standard Reference Data, National Bureau of Standards, Washington, D.C. 20234.

** Available from the National Technical Information Service, U.S. Department of Commerce, Springfield, VA 22151.

Publications of the Berkeley Particle Data Group

UCRL-20000 YN, *A compilation of YN Reactions,* Particle Data Group (1970).

UCRL-20000 NN, *NN and ND Interactions (above 0.5 GeV/c) — A Compilation,* Particle Data Group (1970).

UCRL-20030 *$\pi N, \pi N$ Partial-Wave Amplitudes,* Particle Data Group (1970).

LBL-53, *$\pi + p$, $\pi + n$, and $\pi + d$ Interactions — A Compilation; Parts I and II,* Particle Data Group (1973).

LBL-55, *$K^\circ{}_l N$ Interactions — A Compilation,* Particle Data Group (1972).

LBL-58, *NN and ND Interactions — A Compilation,* Particle Data Group (1972).

LBL-63, *πN Two-Body Scattering Data I. A. User's Guide to the Lovelace-Almehed Data Tape,* Particle Data Group (1973).

LBL-100, *Review of Particle Properties,* Particle Data Group (1976).

Other NSRDS Data Publications

Crystal Data, Determinative Tables, 3rd ed., Vol. 1. Organic Compounds, J. D. H. Donnay and Helen M. Ondik (1972). Published jointly by the U.S. Department Commerce — National Bureau of Standards and the Joint Committee on Powder Diffraction Standards. $30.00.

Crystal Data, Determinative Tables, 3rd ed., Vol. 2. Inorganic Compounds, J. D. H. Donnay and Helen M. Ondick (1973). Published jointly by the U.S. Department of Commerce — National Bureau of Standards and the Joint Committee on Powder Diffraction Standards. $50.00.

Available from the Joint Committee on Powder Diffraction Standards, 1610 Park Lane, Swarthmore, PA 19081.

Selected Thermodynamic Values and Phase Diagrams for Copper and Some of Its Binary Alloys, R. Hultgren and P. D. Desai (1973). Published by the International Copper Research Association, Inc., New York (Monograph I). $10.00.

Selected Values of the Thermodynamic Properties of Binary Alloys, R. Hultgren, P. D. Desai, D. T. Hawkins, M. Gleiser, and K. K. Kelley (1973). Published by the American Society for Metals, Metals Park, Ohio. $30.00.

Selected Values of the Thermodynamic Properties of the Elements, R. Hultgren, P. D. Desai, D. T. Hawkins, M. Gleiser, K. K. Kelley, and D. D. Wagman (1973). Published by the American Society for Metals, Metals Park, Ohio. $20.00.

Thermodynamic Properties of Copper and Its Inorganic Compounds, E. G. King, Alla D. Mah, and L. B. Pankratz (1973). Published by the International Copper Research Association, Inc. , New York. (Monograph II). $10.00.

Ion-Molecule Reactions, E. W. McDaniel, V. Cermak, A. Dalgarno, E. E. Ferguson, and L. Friedman (1970). Published by Wiley-Interscience Series in Atomic and Molecular Collisional Processes. $21.50.

Theory of Charge Exchange, R. A. Mapleton (1971). Published by Wiley-Interscience Series in Atomic and Molecular Collisional Processes. $19.95.

Dissociation in Heavy Particle Collisions, G. W. McClure and J. M. Peek (1972). Published by Wiley-Interscience Series in Atomic and Molecular Collisional Processes. $13.95.

Excitation in Heavy Particle Collisions, E. W. Thomas (1972). Published by Wiley-Interscience Series in Atomic and Molecular Collisional Processes. $22.50.

Total electron-atom collision cross sections at low energies, — a critical review, Benjamin Bederson, L. J. Kieffer, *Rev. Mod. Phys.,* 43(4), 601, 1971.

Electron impact ionization cross-section data for atoms, atomic ions, and diatomic molecules: I. Experimental data, L. J. Kieffer, G. H. Dunn, *Rev. Mod. Phys.,* 38(1), 1966.

Theory of the ionization of atoms by electron impact, M. R. H. Rudge, *Rev. Mod. Phys.,* 40(3), 564, 1968.

Coblentz Society Evaluated Infrared Reference Spectra, Edited and published by the Coblentz Society, sponsored by the Joint Committee on Atomic and Molecular Physical Data (1969—1974). Available from Sadtler Research Laboratories, Philadelphia, PA 19104. (Microfilm available from the Coblentz Society.)

Vol. 6, 1000 Spectra $295.00
Vol. 7, 1000 Spectra $295.00
Vol. 8, 1000 Spectra $295.00
Vol. 9, 1000 Spectra $295.00
Vol. 10, 1000 Spectra $295.00
Cumulative Coblentz Indexes $50.00

Contributions to the Data on Theoretical Metallurgy: XVI. Thermodynamic Properties of Nickel and Its Inorganic Compounds, Alla D. Mah and L. B. Pankratz (1976) U.S. Bureau of Mines Bulletin 668. SD Catalog No. I28.3.668. $2.70.

Thermodynamic Properties of Aqueous Copper Systems, Paul Duby (1977, in press). To be published by the International Copper Research Association, Inc., New York (Monograph III).

Translations from the Russian

A number of products of the Russian standard reference data system have been translated by the Israel Translation Service with the support of the National Science Foundation at the request of the National Bureau of Standards. These Russian translations are available from NTIS. Send inquiries on prices and orders to National Technical Information Service, Springfield, VA 22151.

Thermodynamic and Thermophysical Properties of Combustion Products, Volume I, Computation Methods, V. E. Alemasov, A. F. Dregalin, A. P. Tishin, and V. A. Khudyakov; V. P. Glushko, Ed. (1971). Original published by the Academy of Sciences of the USSR. TT74-50019, 1974, xx + 433 pp.

Thermodynamic and Thermophysical Properties of Combustion Products, Volume II. Oxygen Based Propellants, V. E. Alemasov, A. F. Dregalin, V. A. Khudyakov and V. M. Kostin; V. P. Glushko, Ed. (1972). Original published by the Academy of Sciences of the USSR. TT74-50032, x + 495 pp.

Thermodynamic and Thermophysical Properties of Combustion Products, Volume III. Oxygen- and Air-Based Propellants, V. E. Alemasov, A. F. Dregalin, A. P. Tishin, V. A. Khudyakov and V. M. Kostin; V. P. Glushko, Ed. (1973). Original published by the Academy of Sciences of the USSR. TT75-50007, 1975, x + 622 pp.

Thermodynamic and Thermophysical Properties of Combustion Products, Volume IV. Nitrogen Tetroxide-Based Propellants, V. E. Alemasov, A. F. Dregalin, A. P. Tishin, V. A. Khudyakov, and V. N. Kostin; V. P. Glushko, Ed. (1973). Original published by the Academy of Sciences of the USSR. TT76-50007, 1976, x + 530 pp.

Properties of Liquid and Solid Hydrogen, B. N. Esel'son, Yu. P. Blagoi, V. N. Grigor'ev, V. G. Manzhelii, S. A. Mikhailenko, and N. P. Neklyudov (1969). Original published by the Committee of Standards, Measures and Measuring Instruments of the USSR Council of Ministers — Government Standards Service. TT70-50179, 1974, iii + 123 pp.

Electric Conductivity of Ferroelectrics, V. M. Gurevich (1969). Original published by the Committee of Standards, Measures and Measuring Instruments of the USSR Council of Ministers — Government Standards Service. TT70-50180, 1971, vi + 362 pp.

Handbook of Hardness Data, A. A. Ivank' ko; G. V. Samsonov, Ed. (1968). Original published by the Academy of Sciences of the Ukrainian SSR, Institute of Materials Sciences. TT70-50177, 1971, iii + 66 pp.

Heavy Water, Thermophysical Properties, Ya. Z. Kazavchinskii et al.; V. A. Kirillin, Ed. (1963). TT70-50094, 1971, vii + 263 pp., 2 charts.

Thermophysical Properties of Freon-22, A. V. Kletskii (1970). Original published by the Committee of Standards, Measures and Measuring Instruments of the USSR Council of Ministers — Government Standards Service. TT70-50178, 1971, iv + 67 pp., 1 chart.

Rate Constants of Gas Phase Reactions, Reference Book, V. N. Kondratiev (1970). Original published by the Academy of Sciences of the USSR Order-of-Lenin Institute of Chemical Physics. Translated by L. Holtschlad and R. Fristrom. COM-72-10014, 1972, vi + 428 pp.

*Thermophysical Properties of Gases and Liquids, No. 1,*V. A. Rabinovich (1968). Original published by the Committee of Standards, Measures and Measuring Instruments of the USSR Council of Ministers — Government Standards Service, Series: Physical Constants and Properties of Substances. TT69-55091. 1970, viii + 207 pp.

Handbook of Phase Diagrams of Silicate Systems, Vol. I. Binary Systems, Second Revised Edition, N. A. Toropov, V. P. Barzakovskii, V. V. Lapin, and N. N. Kurtseva; N. A. Toropov, Ed. (1969). Original published by the Academy of Sciences of the USSR Grebenschikov Institute of Silicate Chemistry. TT71-50040, 1972, viii + 723 pp.

Handbook of Phase Diagrams of Silicate Systems. Vol. II. Metal-Oxygen Compounds in Silicate Systems, Second Revised Edition, N. A. Toropov, V. P. Barzakovskii, I. A. Bondar', and Yu. P. Udalov; N. A. Toropov, Ed. Original published by the Academy of Sciences of the USSR Grebenschikov Institute of Silicate Chemistry. TT71-50041, 1972, iv + 325 pp.

Thermodynamic and Thermophysical Properties of Helium, N. V. Tsederberg, V. N. Popov, and N. A. Morozova; A. F. Alyab'ev, Ed. (1969). Original published by Atomizdat. TT70-50096, 1971, v + 255 pp.

Thermophysical Properties of Air and Air Components, A. A. Vasserman, Ya. A. Kazavchinskii, and V. A. Rabinovich; A. M. Zhuravlev, Ed. (1966). Original published by the Academy of Sciences of the USSR. TT70-50095, 1971, x + 383 pp., 8 charts.

Thermophysical Properties of Liquid Air and Its Components, A. A. Vasserman and V. A. Rabinovitch (1968). Original published by the Committee of Standards, Measures and Measuring Instruments of the USSR Council of Ministers — Government Standards Service. TT69-55092, 1970, viii + 235 pp., 2 charts.

Thermophysical Properties of Gaseous and Liquid Methane, V. A. Zagoruchenko and A. M. Zhuravlev (1969). Original published by the Committee of Standards, Measures and Measuring Instruments of the USSR Council of Ministers — Government Standards Service. TT70-50097, 1970, viii + 243 pp.

Surface Ionization, E. Ya. Zandberg and N. L. Ionov (1969). Original published by the Science Publishing House. Translated by E. Harnik. TT70-50148, 1971, xii + 355 pp.

Thermodynamic and Transport Properties of Ethylene and Propylene, I. A. Neduzhii et al. (1971). Original published by the State Committee of Standards of the Soviet Ministry, USSR, State Office of Standards and Reference Data Series: Monograph No. 8. Translation published June 1972. NBSIR 75-763.

Other NBS Compilations of Data

NBS Tech. Note 270-3, *Selected Values of Chemical Thermodynamic Properties, Tables for the First Thirty-Four Elements in the Standard Order of Arrangement,* D. D. Wagman, W. H. Evans, V. B. Parker, I. Halow, S. M. Bailey, and R. H. Schumm (1968). SD Catalog No. C13.46:270-3. $2.75.

NBS Tech. Note 270-4, *Selected Values of Chemical Thermodynamic Properties, Tables for Elements 35 through 53 in the Standard Order of Arrangement,* D. D. Wagman, W. H. Evans, V. B. Parker, I. Halow, S. M. Bailey, and R. H. Schumm (1969). SD Catalog No. C13.46:270-4. $2.10.

NBS Tech. Note 270-5, *Selected Values of Chemical Thermodynamic Properties, Tables for Elements 54 through 61 in the Standard Order of Arrangement,* D. D. Wagman, W. H. Evans, et al. (1971). SD Catalog No. C13.46:270-5. $0.95.

NBS Tech. Note 270-6, Selected Values of Chemical Thermodynamic Properties, Tables for the Alkaline Earth Elements (Elements 92 through 97 in the Standard Order of Arrangement), *V. B. Parker, D. D. Wagman, and W. H. Evans (1971). SD Catalog No. C13.46:270-6. $1.55.*NBS Tech. Note 270-7, *Selected Values of Chemical Thermodynamic Properties, Tables for the Lanthanide (Rare Earth) Elements (Elements 62 through 76 in the Standard Order Of Arrangement),* R. H. Schumm, D. D. Wagman, S. Bailey, W. H. Evans, and V. B. Parker (1973). SD Catalog No. C13.46:270-7. $1.25.

NBS Tech. Note 438, *Compendium of ab-initio Calculations of Molecular Energies and Properties,* M. Krauss (1967). AD665245.*

NBS Tech. Note 474, *Critically Evaluated Transition Probabilities for Ba I and II,* B. M. Miles and W. L. Wiese (1969). AD681-351*

NBS Tech Note 484, *A Review of Rate Constants of Selected Reactions of Interest in Re-Entry Flow Fields in the Atmosphere,* M. H. Bortner (1969). AD692-231*

NBS Tech. Note 866, *Chemical Kinetic and Photochemical Data for Modelling Atmospheric Chemistry,* Robert F. Hampson, Jr. and David Garvin, Eds. (1975). SD Catalog No. C13.46:866. $1.85.

NBS Tech. Note 928, *Computer Programs for the Evaluation of Activity and Osmotic Coefficients,* Bert R. Staples and Ralph L. Nuttall (1976). SD Catalog No. C13.46:928. $1.10.

NBS Monograph 70. Vol. I. *Microwave Spectral Tables, Diatomic Molecules,* P. F. Wacker, M. Mizushima, J. D. Petersen, and J. R. Ballard (1964). PB168072.*

NBS Monograph 70. Vol. II. *Microwave Spectral Tables, Line Strengths of Asymmetric Rotors,* P. F. Wacker and M. R. Pratto (1964). PB189714.*

NBS Monograph 70. Vol. III. *Microwave Spectral Tables, Polyatomic Molecules with Internal Rotation,* P. F. Wacker, M. S. Cord, D. G. Burkhard, J. D. Petersen, and R. F. Kukol (1969). COM-74-10794.*.

NBS Monograph 70. Vol. IV. *Microwave Spectral Tables Polyatomic Molecules without Internal Rotation,* M. S. Cord, J. D. Petersen, M. S. Lojko, and R. H. Haas (1968). COM-74-10795.*

NBS Monograph 70. Vol. V. *Microwave Spectral Tables, Spectral Line Listing,* M. S. Cord, M. S. Lojko, and J. D. Petersen (1968). COM-74-10796.*

NBS Monograph 94. *Thermodynamic and Related Properties of Parahydrogen from the Triple Point to 100 K at Pressures to 340 Atmospheres,* H. M. Roder, L. A. Weber, and R. D. Goodwin (1965). N65-32001.*

NBS Monograph 115. *The Calculation of Rotational Energy Levels and Rotational Line Intensities in Diatomic Molecules,* Jon T. Hougen (1970). RB192-874.*

NBS Monograph 134, *Space Groups and Lattice Complexes,* Werner Fisher, Hans Burzlaff, Erwin Hellner, and J. D. H. Donnay (1973). COM-73-50582*

NBS Monograph 153, *The First Spectrum of Hafnium (HF I).* William F. Meggers and Charlotte E. Moore (1976). SD Catalog No. C13.44:153. $1.35.

Compilation of Low Energy Electron Collision Cross Section Data. Part I. L. J. Kieffer, JILA Information Center Report No. 6 (1969). PB189127.*

Compilation of Low Energy Electron Collision Cross Section Data. Part II. L. J. Kieffer, JILA Information Center Report No. 7 (1969). AD696467.*

Low Energy Electron-Collision Cross Section Data. Part III: Total Scattering; Differential Elastic Scattering, Atomic Data, L. J. Kieffer, Vol. 2, No. 4, 293—391 (1971).

Compilation of Electron Collision Cross Section Data for Modeling Gas Discharge Lasers, L. J. Kieffer, JILA Information Center Report No. 13 (1973). COM-74-11661.*

Derived Thermodynamic Properties of Ethylene, Roland H. Harrison and Donald R. Douslin, *J. Chem. Eng. Data,* 22(1), 24—30 (Jan. 1977).

Pressure, Volume, Temperature Relations of Ethylene, D.

R. Douslin and R. H. Harrison, *J. Chem. Thermodyn.*, 76(8), 301—330 (Aug. 1976).

* Available from the National Technical Information Service, Springfield, VA 22151.

Nondata Publications from NSRDS Related Projects

NBS Misc. Publ. 281, *Bibliography on Flames Spectroscopy, Analytical Applications, 1800 to 1966*, R. Mavrodineanu (1967).*

NBS Spec. Publ. 306-1, *Bibliography on the Analyses of Optical Atomic Spectra, Section 1, $_1{}^1H$-^{23}V*, C. E. Moore (1968).*

NBS Spec. Publ. 306-2, *Bibliography on the Analyses of Optical Atomic Spectra, Section 2, ^{24}Cr-^{41}Nb*, C. E. Moore (1969).*

NBS Spec. Publ. 306-3, *Bibliography on the Analyses of Optical Atomic Spectra, Section 3, ^{42}Mo-^{57}La, ^{72}Hf-^{89}Ac*, C. E. Moore (1969).*C13.10:306-3. $0.85.

NBS Spec. Publ. 306-4, *Bibliography on the Analyses of Optical Atomic Spectra, Section 4, ^{57}La-^{71}Lu, ^{89}Ac-^{99}Es*, C. E. Moore (1969). COM-73-10870.*

NBS Spec. Publ. 320, *Bibliography on Atomic Transition Probabilities, January 1916 through June 1969*, B. M. Miles and W. L. Wiese (1970). AD 701 614.*

NBS Spec. Publ. 320, Supplement I, *Bibliography on Atomic Transition Probabilities, July 1969 through June 1971*, J. R. Fuhr and W. L. Wiese (1971).**

NBS Spec. Publ. 320, Supplement II, *Bibliography on Atomic Transition Probabilities, July 1971 through June 1973*, J. R. Fuhr and W. L. Wiese (1973). COM-4-50034.*

NBS Spec. Publ. 324, *The NBS Alloy Data Center: Permuted Materials Index*, G. C. Carter, D. J. Kahan, L. H. Bennett, J. R. Cuthill, and R. C. Dobbyn (1971). COM-71-50070.*

NBS Spec. Publ. 349, *Heavy-Atom Kinetic Isotope Effects, An Indexed Bibliography*, M. J. Stern and M. Wolfsberg (1972). COM-72-50807.*

NBS Spec. Publ. 362, *Chemical, Kinetics in the C-O-S and H-N-O-S Systems: A Bibliography — 1899 through June 1971*, F. Westley (1972). COM-72-50466.*

NBS Spec. Publ. 363, *Bibliography on Atomic Energy Levels and Spectra, July 1968 through June 1971*, L. Hagan and W. C. Martin (1972). SD Catalog No. C13.10:363. $1.45

NBS Spec. Publ. 363, Supplement 1, *Bibliography on Atomic Energy Levels and Spectra, July 1971 through June 1975*, Lucy Hagan (1977). SD Catalog No. C13.10:363-1. $2.50.

NBS Spec. Publ. 366, *Bibliography on Atomic Line Shapes and Shifts (1899 through March 1972)*, J. R. Fuhr, W. L. Wiese, and L. J. Roszman (1972). SD Catalog No. C13.10:366. $1.75.

NBS Spec. Publ. 366, Supplement 1, *Bibliography on Atomic Line Shapes and Shifts (April 1972 through June 1973)*. J. R. Fuhr, L. J. Roszman, and W. L. Wiese (1974). SD Catalog No. C13.10:366/Suppl. I. $1.05.

NBS Spec. Publ. 366, Supplement 2, *Bibliography on Atomic Line Shapes and Shifts (July 1973 through May 1975)*, J. R. Fuhr, G. A. Martin, and B. Specht (1975). SD Catalog No. C13.10:366-2. $1.35.

NBS Spec. Publ. 369, *Soft X-Ray Emission Spectra of Metallic Solids: Critical Review of Selected Systems and Annotated Spectral Index*, A. J. McAlister, R. C. Dobbin, J. R. Cuthill, and M. L. Williams (1974). SD Catalog No. C13.10:369. $1.85.

NBS Spec. Publ. 371, *A Supplementary Bibliography of Kinetic Data on Gas Phase Reactions of Nitrogen, Oxygen and Nitrogen Oxides*, F. Westley (1973). COM-73-50245*

NBS Spec. Publ. 371-1, *Supplementary Bibliography of Kinetic Data on Gas Phase Reactions of Nitrogen, Oxygen, and Nitrogen Oxides (1972 — 1973)*, Francis Westley (1975). SD Catalog No. C13.371-1. $1.45.

NBS Spec. Publ. 380, *Photonuclear Reaction Data, 1973*, E. G. Fuller, H. M. Gerstenberg, H. Vander Molen, and T. C. Dunn (1973). SD Catalog No. C13.10:380. $2.10.

NBS Spec. Publ. 381, *Bibliography of Ion-Molecule Reaction Rate Data (January 1950 through October 1971)*, George A. Sinnott (1973). SD Catalog No. C13.10:381. :1.15.

NBS Spec. Publ. 392, *Vibrationally Excited Hydrogen Halides: A Bibliography on Chemical Kinetics of Chemiexcitation and Energy Transfer Processes (1958 through 1973)*, Francis Westley (1974). SD Catalog No. C13.10:392. $1.30.

NBS Spec. Publ. 396-1, *Critical Surveys of Data Sources: Mechanical Properties of Metals*, R. B. Gavert, R. L. Moore, and J. H. Westbrook (1974). SD Catalog No. C13.10:396-1. $1.25.

NBS Spec. Publ. 396-2, *Critical Surveys of Data Sources: Ceramics*, Dorothea M. Johnson and James F. Lynch (1975). SD Catalog No. C13.10:396-2. $1.10.

NBS Spec. Publ. 396-3, *Critical Surveys of Data Sources: Corrosion of Metals*, Ronald B. Diegle and Walter K. Boyd (1976). SD Catalog No. C13.10:396-3. $1.30.

NBS Spec. Publ. 396-4, *Critical Surveys of Data Sources: Electrical and Magnetic Properties of Metals*, M. J. Carr, R. B. Gavert, R. L. Moore, H. W. Wawrousek, and J. H. Westbrook (1976). SD Catalog No. C13.10:396-4. $1.55.

NBS Spec. Publ. 426, *Bibliography of Low Energy Electron and Photon Cross Section Data (through December 1974)*, Lee J. Kieffer (1976). SD Catalog No. C13.10:426. $3.30.

NBS Spec. Publ. 428, *Bibliography of Infrared Spectroscopy through 1960, Parts 1, 2, and 3*, C. N. R. Rao, S. K. Dikshit, S. A. Kudchadker, D. S. Gupta, V. A. Narayan, and J. J. Comeford (1976). SD Catalog No. C13.10:428 parts 1, 2, and 3. Sold in 3-part set only, $28.50.

NBS Spec. Publ. 449, *Chemical Kinetics of the Gas Phase Combustion of Fuels (A Bibliography on the Rates and Mechanisms of the Oxidation of Aliphatic C_1-C_{10} Hydrocarbons and of Their Oxygenated Derivatives)*, Francis Westley (1976). SD Catalog No. C13.10:449. $2.00.

NBS Spec. Publ. 454, *An Annotated Bibliography of Sources of Compiled Thermodynamic Data Relevant to Biochemical and Aqueous Systems (1930 — 1975) Equilibrium, Enthalpy, Heat Capacity, and Entropy Data*, George T.Armstrong and Robert N. Goldberg, (1976). SD Catalog No. C13.10:454. $1.65.

NBS Spec. Publ. 463, *Materials Information Programs, An Interagency Review of Federal Agency Activities on Technical Information about Materials, Proceedings of*

a Conference held at the National Bureau of Standards, Gaithersburg, Maryland, April 16 and 17, 1974, S. A. Rossmassler, Ed. (1977). SD Catalog No. C13.10:463. $3.35.

NBS Tech. Note 291, *A Bibliography on Ion-Molecule Reactions, January 1900 to March 1966,* F. N. Harllee, H. M. Rosenstock, and J. T. Herron (1966). SD Catalog No. C13.46:291. $0.75.

NBS Tech. Note 464, *The NBS Alloy Data Center: Function, Bibliographic System, Related Data Centers, and Reference Books,* G. C. Carter, L. H. Bennett, J. R. Cuthill, and D. J. Kahan (1968).*

NBS Tech. Note 554, *Annotated Accession List of Data Compilations of the NBS Office of Standard Reference Data,* H. M. Weisman and G. B. Sherwood (1970). SD Catalog No. C13.46:554. $2.20.

NBS Tech. Note 848, *A Bibliography of the Russian Reference Data Holdings of the Library of the Office of Standard Reference Data,* Gertrude B. Sherwood and Howard J. White, Jr. (1974). SD Catalog No. C13.46:848. $0.55.

NBS Tech. Note 881, *Critical Evaluation of Data in the Physical Sciences — A Status Report on the National Standard Reference Data System, April 1975,* Stephen A. Rossmassler, Ed. (1975). SD Catalog No. C13.46:881. $1.10.

NBS Tech. Note_, *Critical Evaluation of Data in the Physical Sciences — A Status Report on the National Standard Reference Data System, January 1977,* Stephen A. Rossmassler, Ed. (1977, in press). SD Catalog No. C13.46:_, in press.

A Compilation of Volumes I—IV of Bibliography on High Pressure Research with Author and Subject Indexes, Volumes I and II, J. F. Cannon and L. Merrill (1972). Published by and available from the High Pressure Data Center, Brigham Young University, Provo, UT 84601.

Binary Fluorides, Free Molecular Structures and Force Fields, A Bibliography (1957—1975), D. T. Hawkins, L. S. Bernstein, W. E. Falconer, and W. Klemperer, IFI/Plenum, New York (1976).

Bulletin of Chemical Thermodynamics, IUPAC Commission on Thermodynamics and Thermochemistry, Robert D. Freeman, Ed., former title: Bulletin of Thermodynamics and Thermochemistry. Volume 20, 1977, $20.00.

Subscriptions and orders for single copies and back issues should be addressed to:
Bulletin of Chemical Thermodynamics
Department of Chemistry
Oklahoma State University
Stillwater, OK 74074 USA
Checks should be made payable, in U.S. funds, to:
Bulletin of Chemical Thermodynamics

* Available from National Technical Information Service, Springfield, VA 22151; use PB, AD, or COM number when given or use the series number.

** Available from J. Fuhr, National Bureau of Standards, Phys. Bldg. Room A267, Washington, D.C. 20234.

The Coblentz Society Specifications for Evaluation of Research Quality Analytical Infrared Spectra (Class II), Coblentz Society Board of Managers, *Anal. Chem.,* 47(11) 945—952, 1975.

The Compilation and Evaluation of Chemical Kinetics Data; A Descriptive Survey of Current Efforts, Lewis H. Gevantman and David Garvin, *Int. J. Chem. Kinet.,* 213—230, 1973.

A Compilation of Volumes I—IV of Bibliography on High Pressure Research with Author and Subject Indexes, Volumes I and II, J. F. Cannon and L. Merrill (1972). Published by and available from the High Pressure Data Center, Brigham Young University, Provo, UT 84601.

Cooperation Between Professional Societies and A Government Agency: The Journal of Physical and Chemical Reference Data, David R. Lide, Jr. *IEEE Transactions on Professional Communications PC-18* (3) 127—129 (Sept. 1975).

Equilibrium Properties of Fluid Mixtures, A Bibliography on Fluids of Cryogenic Interest, M. J. Hiza, A. J. Kidnay, and R. C. Miller, IFI/Plenum, New York (1975).

Guide for the Presentation in the Primary Literature of Numerical Data Derived from Experiment, Report of the CODATA Task Group on Publication of Data in the Primary Literature, September 1973, Unesco-UNISIST Guide, Verbatim publication also in *CODATA Bulletin,* 9, December 1973, *NSRDS News,* February 1974.

Liquid Vapor Equilibria on Systems of Interest in Cryogenics, A Survey, A. J. Kidnay, M. J. Hiza, and R. C. Miller, *Cryogenics,*13(10), 575—599, 1973.

The Presentation of Chemical Kinetics Data in the Primary Literature, CODATA Task Group on Data for Chemical Kinetics, S. W. Benson, Chairman, *CODATA Bulletin,*13 (December 1974).

Recommendations for Data Compilations and for the Reporting of Measurements of the Thermal Conductivity of Gases, H. M. Hanley, et al., J. Heat Transfer, 93(4), 479—480, 1971.

Recommendations for the Presentation of NMR Data for Publication in Chemical Journals, IUPAC Commission I.5 on Molecular Structure and Spectroscopy, R. N. Jones, Chairman, *Pure Appl. Chem.,* 29, (4), 627—628, 1972.

Survey of Analytical Spectral Data Sources and Related Data Compilation Activities, Lewis H. Gevantman, *Anal. Chem.,*44(7), 30—48, 1972.

NBS-OSRDB-70-1-Vol. 1, *High Pressure Bibliography 1900—1968. Volume I. Section I — Bibliography, Section II. Author Index,* Leo Merrill (1970). PB 191 174.*

NBS-OSRDB-70-1-Vol. 2, *High Pressure Bibliography 1900—1968. Volume II. Subject Index,* Leo Merrill (1970). PB 191 175.*

NBS-OSRDS-70-2, *The NBS Alloy Data Center: Author Index,* G. C. Carter, D. J. Kahn, L. H. Bennett, J. R. Cuthill, and R. C. Dobbyn (1970). COM-71-00722.*

NBS-OSRDB-70-3, *Semiempirical and Approximate Methods for Molecular Calculations — Bibliography and KWIC Index,* George A. Henderson and Sandra Frattali (1969). AD 705 110.*

NBS-OSRDB-70-4, *Bibliography of Photoabsorption Cross Section Data,* Robert D. Hudson and Lee J. Kieffer (1970). COM-71-00025.*

NBS-OSRDB-71-1, *Bibliography on Properties of Defect Centers in Alkali Halides,* S. C. Jain, S. A. Khan, H. K. Sehgal, V. K. Garg, and R. K. Jain (1971). COM-71-00248.*

NBS-OSRDB-71-2, *A Bibliography of Kinetic Data on Gas Phase Reactions of Nitrogen, Oxygen, and Nitrous Oxides,* Francis Westley (1970). COM-71-00841.*

* Available from National Technical Information Service, Springfield, VA 22151; use PB, AD, or COM number when given or use the series number.

Nondata Publications from NSRDS Related Projects; Computer Programs for Handling Technical Data

NBS Tech. Note 444, *Reform: A General-Purpose Program for Manipulating Formatted Data Files,* R. C. McClenon and J. Hilsenrath (1968).*

NBS Tech. Note 470, *Edpac: Utility Programs for Computer-Assisted Editing, Copy-Production, and Data Retrieval,* C. G. Messina and J. Hilsenrath (1969).*

NBS Tech. Note 500, *Edit-Insertion Programs for Automatic Typesetting of Computer Printout,* C. G. Messina and J. Hilsenrath (1970). PB 191 352.*

NBS Tech. Note 700, *COMBO: A General-Purpose Program for Searching, Annotating, Encoding-Decoding, and Reformatting Data Files,* Robert McClenon and Joseph Hilsenrath (1972). COM-73-50015.*

NBS Tech. 740, *SETAB: An Edit/Insert Program for Automatic Typesetting of Spectroscopic and Other Computerized Tables,* Robert C. Thompson and Joseph Hilsenrath (1973). SD Catalog No. C13.46:740. $0.55.

NBS Tech. Note 760, *Description of the Magnetic Tape Version of the Bulletin of Thermodynamics and Thermochemistry, No. 14,* R. McClenon, W. H. Evans, D. Garvin, and B. C. Duncan (1973). SD Catalog No. C13.46:760. $0.90.

NBS Tech. Note 820, *Complete Clear Text Representation of Scientific Documents in Machine-Readable Form,* Blanton C. Duncan and David Garvin (1974). SD Catalog No. C13.46:820. $0.70.

NBS Handbook 101, *OMNITAB, A Computer Program for Statistical and Numerical Analysis,* J. Hilsenrath, G. G. Ziegler, C. G. Messina, P. J. Walsh, and R. J. Herbold (Revised January 1968).*

NBS Spec. Publ. 424, *A Contribution to Computer Typesetting Techniques: Table of Coordinates for Hershey's Repertory of Occidental Type Fonts and Graphic Symbols,* Norman M. Wolcott and Joseph Hilsenrath (1976). SD Catalog No. C13.10:424. $2.90.

* Available from National Technical Information Service, Springfield, VA 22151; use PB, AD, or COM number when given or use the series number.

NBS Magnetic Tape Series

NBS Magnetic Tape 1, *OMNITAB II Magnetic Tape and Documentation Parcel,* David Hogben, Sally T. Peavy, and Ruth Varner (1970).*

NBS Magnetic Tape 2, *Fortran Programs for Text Editing, File Manipulation and Automatic Typesetting,* C. G. Messina, R. McClenon, and J. Hilsenrath (1973).*

NBS Magnetic Tape 3, *Bibliography and Index to the Literature in the NBS Alloy Data Center,* G. C. Carter and D. J. Kahan (1973).*

NBS Magnetic Tape 4, *Magnetic Tape Version of the Bulletin of Thermodynamics and Thermochemistry No. 14 (1971),* Robert McClenon and Blanton Duncan (1973).*

NBS Magnetic Tape 8, *EPA/NIH Mass Spectral Data Base, 1975 Edition,* Stephen R. Heller, Henry M. Fales, and G. W. A. Milne (1975).*

NBS Magnetic Tape 9, *Crystal Data Tape, Derived from the 3rd Edition of Crystal Data Determinative Tables,* H. M. Ondik and A. D. Mighell (1975).*

NBS Magnetic Tape 10, *Atomic Spectral-Line Intensities,* W. F. Meggers, C. H. Corliss and B. F. Scribner (1975).*

NBS Magnetic Tape 12, *Tables of Coordinates for Hershey's Repertory of Type Fonts and Graphic Symbols,* Herman W. Wolcott and Joseph Hilsenrath (1977).*

* Available from National Technical Information Service, Springfield, VA 22151; use PB, AD, or COM number when given or use the series number.

Bibliographic Publications of the Radiation Chemistry Data Center, University of Notre Dame, Notre Dame, Indiana

Weekly List of Papers on Radiation Chemistry, Index and Cumulation, Volume IV, Numbers 1—26, January through June 1971 (August 1971). COM-71-01103.*

Index and Cumulative list of Papers on Radiation Chemistry, Volume IV, Numbers 27—52, July through December 1971 (January 1972). COM-72-10266.*

Index and Cumulative List of Papers on Radiation Chemistry, Volume V, Numbers 1—16, January through June 1972 (July 1972). COM-72-10621.*

Index and Cumulative List of Papers on Radiation Chemistry, Volume V, Numbers 27—52, July through December 1972 (January 1973). COM-73-10281.*

Index and Cumulative List of Papers on Radiation Chemistry, Volume VI, Numbers 1—26, January through June 1973 (July 1973). COM-73-11878/8AS.*

Index and Cumulative List of Papers on Radiation Chemistry, Volume VI, Numbers 27—52, July through December 1973 (January 1974). COM-74-10899.*

Biweekly List of Papers on Radiation Chemistry, Annual Cumulation with Keyword and Author Indexes, Vol. VII, 1974 (1975). COM-75-11475.*

Biweekly List of Papers on Radiation Chemistry, Annual Cumulation with Keyword and Author Indexes, Vol. VIII, 1975 (1976). PB 257 025.*

Information on the annual cumulations and subscription information for the *Biweekly List* for 1976 * is available from the Radiation Chemistry Data Center, Radiation Laboratory, University of Notre Dame, Notre Dame, IN 46556.

* Available from National Technical Information Service, Springfield, VA 22151; use PB, AD, or COM number when given or use the series number.

INDEX

A

B

C

D

E

F

G

I

J

K

L

N

O

P

Q

R

S

T

U

V

W

XYZ

MELTING AND BOILING POINTS, AND ATOMIC WEIGHTS OF THE ELEMENTS

Based on the Assigned Relative Atomic Mass of $^{12}C = 12$

The following values apply to elements as they exist in materials of terrestrial origin and to certain artificial elements. When used with the footnotes, they are reliable to ±1 in the last digit, or ±3 if that digit is in small type.

Name	Symbol	Atomic number	Atomic weight	Melting point, °C	Boiling point, °C
Actinium[k]	Ac	89	227.028	1,050	3,200 ± 300
Aluminum	Al	13	26.98154[b]	660.37	2,467
Americium	Am	95	(243)	994 ± 4	2,607
Antimony	Sb	51	121.75*	630.74	1,750
Argon[h,i]	Ar	18	39.948*[b,c,d,g]	-189.2	-185.7
Arsenic (gray)	As	33	74.9216[a]	817(28 atm)	613(sub.)
Astatine	At	85	(210)	302	337
Barium[i]	Ba	56	137.33	725	1,640
Berkelium	Bk	97	(247)	–	–
Beryllium	Be	4	9.01218[a]	1,278 ± 5	2,970(5 mm)
Bismuth	Bi	83	208.9804[a]	271.3	1,560 ± 5
Boron[h,j]	B	5	10.81[c,d,e]	2,300	2,550(sub.)
Bromine	Br	35	79.904[c]	-7.2	58.78
Cadmium[i]	Cd	48	112.41	320.9	765
Calcium[i]	Ca	20	40.08	839 ± 2	1,484
Californium	Cf	98	(251)	–	–
Carbon[h]	C	6	12.011[b,d]	~3,550	4,827
Cerium[i]	Ce	58	140.12	798 ± 3	3,257
Cesium	Cs	55	132.9054[c]	28.40 ± 0.01	678.4
Chlorine	Cl	17	35.453[c]	-100.98	-34.6
Chromium	Cr	24	51.996[c]	1,857 ± 20	2,672
Cobalt	Co	27	58.9332[a]	1,495	2,870
Copper[h]	Cu	29	63.546*[c,d]	1,083.4 ± 0.2	2,567
Curium	Cm	96	(247)	1,340 ± 40	
Dysprosium	Dy	66	162.50*	1,409	2,335
Einsteinium	Es	99	(254)	–	–
Erbium	Er	68	167.26*	1,522	2,510
Europium[i]	Eu	63	151.96	822 ± 5	1,597
Fermium	Fm	100	(257)	–	–
Fluorine	F	9	18.998403[a]	-219.62	-188.14
Francium	Fr	87	(223)	(27)	(677)
Gadolinium[i]	Gd	64	157.25*	1,311 ± 1	3,233
Gallium	Ga	31	69.72	29.78	2,403
Germanium	Ge	32	72.59*	937.4	2,830
Gold	Au	79	196.9665[a]	1,064.43	2,807
Hafnium	Hf	72	178.49*	2,227 ± 20	4,602
Helium[i]	He	2	4.00260[b]	-272.2$^{26\ \text{atm}}$	-268.934
Holmium	Ho	67	164.9304[a]	1,470	2,720
Hydrogen[h]	H	1	1.0079[b,d]	-259.14	-252.87
Indium[i]	In	49	114.82	156.61	2,080
Iodine	I	53	126.9045[a]	113.5	184.35
Iridium	Ir	77	192.22*	2,410	4,130
Iron	Fe	26	55.847*	1,535	2,750
Krypton[i,j]	Kr	36	83.80	-156.6	-152.30 ± 0.10
Lanthanum[i]	La	57	138.9055*[b]	920 ± 5	3,454
Lawrencium	Lr	103	(260)	–	–
Lead[h,j]	Pb	82	207.2[d,g]	327.502	1,740
Lithium[h,i,j]	Li	3	6.941*[c,d,e]	180.54	1,347
Lutetium	Lu	71	174.967 ± 0.003	1,656 ± 5	3,315
Magnesium[i]	Mg	12	24.305[c]	648.8 ± 0.5	1,090
Manganese	Mn	25	54.9380[a]	1,244 ± 3	1,962
Mendelevium	Md	101	(258)	–	–
Mercury	Hg	80	200.59*	-38.87	356.58
Molybdenum	Mo	42	95.94	2,617	4,612
Neodymium[i]	Nd	60	144.24*	1,010	3,127
Neon[j]	Ne	10	20.179*[c]	-248.67	-246.048
Neptunium[k]	Np	93	237.0482[b]	640 ± 1	3,902
Nickel	Ni	28	58.70	1,453	2,732
Niobium (Columbium)	Nb	41	92.9064[a]	2,468 ± 10	4,742